Encyclopedia of Physical Sciences and Engineering Information Sources

Encyclopedia of Physical Sciences and Engineering Information Sources

Includes: Abstract Services and Indexes, Annual Reviews and Yearbooks; Associations and Professional Societies; Bibliographies, Directories and Biographical Sources; Encyclopedias and Dictionaries; General Works; Handbooks and Manuals; Online Databases; Periodicals; Research Centers and Institutes; Specifications and Standards; Statistics Sources; and Other Sources of Information on Each Topic

SECOND EDITION

**Martin A Smith,
David E. Wilt,
Judith B. Erickson,**
Editors

GALE

DETROIT • NEW YORK • TORONTO • LONDON

Encyclopedia of Physical Sciences and Engineering Information Sources

Edited by Judith B. Erickson, David E. Wilt, Martin A. Smith

Steven Wasserman, *Consulting Editor*

Gale Research Staff

Jacqueline L. Longe, *Coordinating Editor*
Nicole Beatty, Erika Berry, Kimberley A. McGrath, Bridget Travers, Robyn V. Young, *Assisting Editors*

Mary Beth Trimper, *Production Director*
Shanna Heilveil, *Production Assistant*

Cynthia Baldwin, *Art Director*
C.J. Jonik, *Cover and Page Designer*

ISBN 0-8103-6911-7
Printed in the United States of America
10 9 8 7 6 5 4 3 2 1

Library of Congress Cataloging-in-Publication Data

Encyclopedia of Physical Sciences and Engineering Information Sources / Martin A. Smith, David E. Wilt, and Judith B. Erickson, editors.—2nd ed.
    p.   cm.
    ISBN 0-8103-6911-7
        1. Physical sciences--Bibliography. 2. Engineering--Bibliography.
    3. Physical sciences—Information services. 4. Technology—Information services.
    I. Smith, Martin A. II. Wilt, David E., 1955- . III. Erickson, Judith B.
    Z7401.E56   1996
    [Q158.5]
    016.5002--dc20   96-18663
                                                                            CIP

CONTENTS

INTRODUCTION

Physical scientists and engineers need an easy-to-use tool that leads to the books, periodicals, data bases, and organizations that can answer their technical questions. Those outside the physical science and engineering professions may also require a tool that directs them to more basic scientific books and to the organizations of, and related to, these particular areas.

To meet the specific and unique needs of both groups, this revised second edition of the *Encyclopedia of Physical Science and Engineering Information Sources (EPSEIS II)*, is intended to serve as a starting point for any research project involving the physical sciences or engineering. It provides comprehensive coverage of sources of information in these areas for the non-specialist as well as the specialist working in another field.

Method of Compilation

Extensive research was required in order to gather the sources of information on the hundreds of subjects in *EPSEIS II*. A wide range of sources was consulted, including online data bases, bibliographies, indexes of science and engineering literature, and publishers' promotional catalogs. In each instance, the information have been carefully verified and corroborated.

The editors have placed emphasis on recently published works in a field, but have included important historic sources where it was thought relevant. All printed and electronic sources were available at the time the guide was compiled. Where possible, telephone numbers for the publishers have been included, and toll-free telephone numbers, if available. Prices quoted are intended only as a guide to the relative cost of an item or service.

Subject Headings Narrowly Defined

So that information can be located easily, subject headings are narrowly defined and specific. This second edition of *EPSEIS* includes more than 500 subjects and is further supplemented by numerous cross-references to related subjects.

Subjects selected for inclusion in this directory were compiled from several widely-used science and technology dictionaries, encyclopedias, and indexing services commonly found in college, research, and public libraries. All the major science and engineering subjects, such as astronomy, chemistry, geology, physics, civil engineering, electrical engineering, and mechanical engineering,are included, along with many newer areas of scientific inquiry and applied research.

"Outline of Contents" Speeds Access to Topics

For the most efficient use of this publication, first find the desired topic in the "Outline of Contents," where it is possible to scan the list of headings quickly to determine the exact form of the subject term that has been used. Under each of the major subjects in the directory, there are cross-references to related subjects, and when a topic is not used, a "see reference" directs the reader to the related subjects which have information sources.

Bibliography Covers 14 Kinds of Sources

Under each topic, citations are found for a wide variety of sources, both print and non-print. A complete list of the kinds of sources cited, represented as subheadings in *EPSEIS II*, is outlined below, in the sequence in which they appear (although not all types of sources are listed for each subject).

Abstract Services and Indexes
Annual Reviews and Yearbooks
Associations and Professional Societies
Bibliographies
Directories and Biographical Sources
Encyclopedias and Dictionaries
General Works

Handbooks and Manuals
Online Databases and CD-ROMS
Other Sources
Periodicals
Research Centers
Specifications and Standards
Statistics Sources

Arrangement and Form of Entries

Entries are arranged alphabetically by: (1) subject, and further subdivided by (2) type of source, and (3) publication title or organization name. Complete citations are provided for each information source, and include: title, author or editor, publisher's name and address, publication date or frequency, telephone number, and price. For databases, citations include: database name, producer, producer's address, and telephone number. Citations for organizations include name, address, and telephone number.

Suggestions are Welcome

Considerable care has been taken to keep errors and inconsistencies to a minimum, but they no doubt will occur. The editors would appreciate any suggestions from users that may improve future editions. Please send comments to:

Editor
Encyclopedia of Physical Sciences
and Engineering Information Sources (EPSEIS II)
Gale Research
Penobscot Building
Detroit, MI 48226

Outline of Contents

OUTLINE OF CONTENTS

Ency. of Physical Sciences and Engineering Info. Sources

OUTLINE OF CONTENTS

Ency. of Physical Sciences and Engineering Info. Sources

Encyclopedia of Physical Sciences and Engineering Information Sources

A

ABRASIVES

See: MACHINING

ABSORBERS

See also: CHEMICAL ENGINEERING, CHEMISTRY, DRYING

ABSTRACT SERVICES AND INDEXES

Applied Science and Technology Index; A Cumulative Subject Index To English Language Periodicals in the Fields of Aeronautics and Space Science, Computer Technology, Chemistry, Construction Industry, Energy and Related Areas. H. W. Wilson Co., 950 University Avenue, Bronx, NY 10452-9978. (212) 588-8400. (800) 367-6770. FAX (718) 590-1617. From 1958 to present. Monthly. Inquire about cost and availability. Also available online (BRS and WILSONLINE) and on CD-ROM. ISSN: 0003-6986.

Chemical Abstracts. Chemical Abstracts Service. 2540 Olentangy River Road, Box 3012, Columbus, OH 43210-0012. (614) 447-3600. FAX (614) 447-3713. 1907 to present. Weekly. $16,800.00 per year. Available online and on CD-ROM. CA is also available in five section groupings. Inquire regarding cost and availability.

Chemical Abstracts - Applied Chemistry and Chemical Engineering Sections. Chemical Abstracts Service, 2540 Olentangy River Road, Box 3012, Columbus, OH 43210. (800) 848-6538 or (614) 447-3600. FAX (614) 447-3713. 1963 to present. Biweekly. $2080.00 per year. ISSN: 0090-8363.

Current Contents: Engineering, Technology and Applied Sciences. Institute for Scientific Information, 3501 Market Street, Philadelphia, PA 19104. (215) 386-0100. FAX (215) 386-2991. From 1970 to present. Weekly. $442.00 per year. Also available online. Inquire regarding cost and availability. ISSN: 0095-7917.

Engineering Index Monthly. Engineering Information, Inc., Castle Point on the Hudson, Hoboken, NJ 07030. (201) 216-8500. (800) 221-1044. FAX (201) 216-8532. Monthly. $2300.00 per year for monthly issues. $1980 per year for annual cumulation. Available on CD-ROM and online as COMPENDEX. ISSN: 0742-1974; 0360-8557.

General Science Index. H.W. Wilson Company, 950 University Avenue, Bronx, NY 10452. (212) 588-8400. (800) 367-6770. FAX (718) 590-1617. From 1978 to present. Ten issues per year; quarterly and annual cumulations. Service basis. Available on CD-ROM and online. Inquire regarding cost and availability. ISSN: 0162-1963.

Physics Abstracts. INSPEC. Section A, Science Abstracts. Institution of Electrical Engineers (IEE), London, United Kingdom. Available from: INSPEC/IEEE - Institute of Electrical and Electronic Engineers, Box 1331, 445 Hoes Lane, Piscataway, NJ 08855-1331. (908) 562-5549. 1898 to present. 24 issues per year. $2835.00 per year. ISSN: 0036-8091. Also available online and on CD-ROM.

Process and Chemical Engineering/chemical Engineering Abstracts. Royal Society of Chemistry, Information Services, Thomas Graham House, Science Park, Milton Road, Cambridge, CB4 4WF, England. Contains citations, mostly with abstracts, of theworldwide literature on chemical engineering; from 1982 to present. Monthly. $610.00 per year. Also available online. ISSN: 0960-5045.

Science Citation Index. SCI. Institute for Scientific Information, 3501 Market Street, Philadelphia, PA 19104. (215) 386-0100. (800) 523-1850. FAX (215) 386-2991. 1961 to present. Six issues per year, plus annual cumulation. $11650.00 per year. Also available online and on CD-ROM. Inquire about price and availability. ISSN: 0036-827X.

Theoretical Chemical Engineering. Royal Society of Chemistry, Information Services, Thomas Graham House, Science Park, Milton Road, Cambridge, CB4 4WF, England. Covers theoretical chemical engineering, including theory and laboratory experimentation. 1964 to present. Monthly. $235.00. ISSN: 0960-5053.

ASSOCIATIONS AND PROFESSIONAL SOCIETIES

American Chemical Society, 1155 16th Street, NW, Washington, DC 20036. (202) 872-4600.

American Institute of Chemical Engineers. 345 East 47th Street, New York, NY 10017. (212) 705-7338. FAX (212) 752-3294.

Association of Consulting Chemists and Chemical Engineers, 50 East 41st Street, Suite 92, New York, NY 10017. (212) 684-6255.

Association of official Analytical Chemists. 2200 Wilson Blvd, Suite 400, Arlington, VA 22001. (703) 522-3032.

Institute For Polyacrylate Absorbents. 1100 New York Avenue, no. 1090, Washington, DC 20005. (202) 414-4190. FAX (202) 289-8584.

BIBLIOGRAPHIES

Chemical Information Sources. Gary Wiggins. McGraw-Hill Publishing Company, 1221 Avenue of the Americas, New York, NY 10020. (800) 262-4729 or (212) 512-3825. 1991. $42.50.

Information Sources in Chemistry. R.T. Bottle and J.F.B Rowland, editors. R.R. Bowker Inc., 121 Chanlon Road, New Providence, NJ 07974. (800) 521-8110 or (908) 464-6800. Fourth edition. 1993. $75.00.

DIRECTORIES AND BIOGRAPHICAL SOURCES

American Men and Women of Science. Physical and Biological Sciences. R.R. Bowker Company, Inc., 121 Chanlon Road, New Providence, NJ 07974. (908) 464-6800. (800) 521-8110. 20th edition. 8 volumes. 1996. $850.00

Research Centers Directory. Gale Research, Inc., 835 Penobscot Building, Detroit, MI 48226-4094. (313) 961-2242. (800) 877-4253. 20th edition. 1995. $485.00. ISSN: 0080-1518.

ENCYCLOPEDIAS AND DICTIONARIES

Dictionary of Analytical Reagents. R. Lobinski, editor. Routledge, Chapman and Hall, Inc., 29 West 35th Street, New York, NY 10001-2291. (212) 244-3336. 1993. $1000.00.

Encyclopedia of Chemical Processing and Design. McKetta Marcel Dekker, Inc., 270 Madison Avenue, New York, NY 10016. (212) 696-9000. (800) 228-1160. 1976 - . $175.00 per volume. ISSN:

Industrial Chemical Thesaurus. Michael Ash and Irene Ash. VCH Publishers, Inc., 220 East 23rd Street, New York, NY 10010-4606. (800) 367-8249. 1992. $295.00.

Kirk-Othmer Encyclopedia of Chemical Technology. John Wiley and Sons, Inc., 605 Third Avenue, New York, NY 10158. (800) 526-5368 or (212) 850-6000. Fourth edition. 1991 - . Twenty-seven volumes. $5400.00.

McGraw-Hill Encyclopedia of Science and Technology. McGraw-Hill Book, Incorporated, 1221 Avenue of the Americas, New York, NY 10020. (212) 997-3675. (800) 262-4729. Seventh edition. Twenty volumes. 1992. $1900.00.

GENERAL WORKS

Absorbency: Textile Science and Technology, Volume 7. P. K. Chatterjee, editor. Elsevier Science Publishing Company, Inc., 655 Avenue of the Americas, New York, NY 10010. (212) 989-5800. FAX (914) 333-2444. 1985. $125.75.

Absorbent Polymer Technology. Lisa Brannon-Peppas and Ronald S. Harland. Elsevier Science Publishing Company, Inc.,

655 Avenue of the Americas, New York, NY 10010. (212) 989-5800. FAX (914) 333-2444. 1990.

Absorption: Fundamentals and Applications. Roman Zarzvcki. Pergamon Press, Inc., Maxwell House, Fairview Park, Elmsford, NY 10523. (914) 592-7700. Fax (914) 592-3625. 1993. $60.00.

Absorption of Light By Surface Water. J. H. Hakvoort. Coronet Books, 311 Bainbridge Street, Philadelphia PA 19147. (215) 925-5083. 1994. $52.50.

Analytical Chemistry. Gary D. Christian. John Wiley & Sons, Inc., 605 Third Avenue, New York, NY 10158-0012. (212) 850-6000. (800) 225-5945. Fifth edition. 1993. $53.50.

Distallation and Absorption '92. J. D. Jenkins and D. W. Reay, editors. Hemisphere Publishing Corp., 1900 Frost Road, Suite 101, Bristol PA 19007. (215) 785-5800. (800) 821-8312. 2 volumes. 1992. $225.00.

Molecular Interpretations of Sorption in Polymers, Part 1. L. A. Errede, editor. Springer-Verlag,New York, Inc., 175 Fifth Avenue, New York, NY 10010. (212) 460-1500. (800) 777-4643. FAX (212) 473-6272. 1991. $79.00.

New Directions in Sorption Technology. Butterworth-Heinemann, 313 Washington Street, Newton, MA 02158. (618-928-2500. (800) 366-2665. 1989. $54.95.

Statistics For Analytical Chemistry. Jane C. Miller and James N. Miller. Prentice Hall, 15 Columbus Circle, New York, NY 10023. (212) 373-8500. (800) 223-2348. 3rd edition. 1993. $59.00.

HANDBOOKS AND MANUALS

CRC Handbook of Chemistry and Physics. David R. Kide, editor. CRC Press Inc., 2000 Corporate Blvd., NW, Boca Raton, FL 33431. (407) 994-0555. (800) 333-8300. 77th edition. 1996. $99.95.

Guide To Basic Chemical Compounds. D.R. Lide, Jr. CRC Press, Inc., 2000 Corporate Boulevard, NW, Boca Raton, FL 33431.(407) 994-0555. (800) 272-7737. 1993. $120.00.

Laboratory Handbook of Materials, Equipment, and Techniques. Gary S. Coyne. Prentice Hall, 15 Columbus Circle, New York, NY 10023. (212) 373-8500. (800) 223-2348. 1992. $45.00.

Lange's Handbook of Chemistry. John A. Dean, editor. McGraw-Hill Publishing Company, Inc., 1221 Avenue of the Americas, New York, NY 10020. (212) 512-2000. (800) 262-4729. 14th edition. 1996. $99.95.

Perry's Chemical Engineer's Handbook. Robert H. Perry and Donald W. Green, editors. McGraw-Hill Publishing Company, Inc., 1221 Avenue of the Americas, New York, NY 10020. (212) 512-2000. (800) 262-4729. 6th edition. 1996. $129.50.

Process Engineer's Absorption Pocket Handbook. Robert M. Maddox. Gulf Publishing Co., P.O. Box 2608, Houston, TX 77252-2608. (713) 520-4444. (800) 231-6275. 1985. $19.00.

ONLINE DATABASES AND CD-ROMS

Analytical Abstracts Online. Royal Society of Chemistry, Information Services, Thomas Graham House, Science Park, Milton Road, Cambridge, CB4 4WF, England. Contains citations, mostly with abstracts, of the worldwide literature on analytical chemistry, from 1980 to present. Available on DIALOG, (800) 334-2564, and SON International, FEZ Karlsruhe, P.O. Box 2465, W-7500, Karlsruhe 1, Germany, online services. Inquire as to cost and availability.

CA Search. Chemical Abstracts Service, P.O. Box 3012, Columbus, OH 43210-0012. (614) 447-3600. (800) 848-6533. FAX (614) 447-3709. Very comprehensive guide to worldwide chemical literature and related fields. 1972 to present. Available on BRS,(800) 289-4277, DIALOG, (800) 334-2564, ORBIT (800) 456-7248, and SON International, FEZ Karlsruhe, P.O. Box 2465, W-7500, Karlsruhe 1, Germany, online services. Inquire as to cost and availability.

Compendex Plus. Engineering Information, Inc., 345 East 47th Street, New York, NY 10017. (212) 705-7600 or (800) 221-1044. Contains citations with abstracts to worldwide literature in engineering and technology, from 1970 to present. Available on online BRS,(800) 289-4277, DIALOG, (800) 334-2564, ORBIT (800) 456-7248, and SON International, FEZ Karlsruhe, P.O. Box 2465, W-7500, Karlsruhe 1, Germany, online services. Also available on CD-ROM. Inquire as to cost and availability.

Kirk-Othmer Encyclopedia of Chemical Technology. John Wiley and Sons, Inc., 605 Third Avenue, New York, NY 10158. (800) 526-5368 or (212) 850-6000. Contains the complete text of all chapters in the 27 volume fourth edition of the *Kirk-Othmer Encyclopedia of Chemical Technology.* 1991. Available on BRS,(800) 289-4277, DIALOG,(800) 334-2564, online services. Inquire as to cost and availability.

NTIS Bibliographic Database. National Technical Information Service, 5285 Port Royal Road, Springfield, VA 22161. (703) 487-4929 or FAX (703) 321-8199. Broad coverage of government-sponsored science and technology research reports, 1964 to present. Available on BRS,(800) 289-4277, DIALOG, (800)334-2564, ORBIT, (800) 456-7248, and SON International, FEZ Karlsruhe, P.O. Box 2465, W-7500, Karlsruhe 1, Germany, online services. Also available on CD-ROM. Inquire as to cost and availability.

SCISEARCH. Institute for Scientific Information, 3501 Market Street, Philadelphia, PA 19104. (800) 523-1850 or (215) 386-0100. Broad multi disciplinary title and author index to the international literature of science and technology, 1974 to present. Available on DIALOG, (800) 334-2564, and ORBIT, (800) 456-7248, online services. Also available on CD-ROM. Inquire as to cost and availability.

WILSONLINE. H.W. Wilson Company, 950 University Avenue, Bronx, NY 10452. (800) 367-6770 or (212) 588-8400. Makes available online versions of the printed H.W. Wilson indexes including Applied Science and Technology Index, Business Periodicals Index, General Science Index, and Readers' Guide to Periodical Literature. Period covered is generally 1983 to present. Available on BRS, (800) 289-4277, DIALOG, (800) 334-2564, and OCLC EPIC, (800) 848-5878, online services. Also available on CD-ROM. Inquire as to cost and availability.

PERIODICALS

A I Ch E Journal. American Institute of Chemical Engineers. 345 East 47th Street, New York, NY 10017. (212) 705-7663. FAX (212) 752-3294. 1955 to present. Monthly. $395.00 per year. ISSN: 0001-1541.

American Laboratory. American Laboratory, 30 Controls Drive, Box 870, Shelton, CT 06484-6111. (203) 926-9300. FAX (203) 926-9310. Monthly. $150.00 per year.

Chemical Engineering Progress. American Institute of Chemical Engineers, 345 East 47th Street, New York, NY 10017. (212) 705-7663. FAX (212) 752-3294. 1947 to present. Monthly. $75.00 per year. ISSN: 0360-7275.

Chemical Week. Chemical Week Associates, 810 7th Avenue, New York, NY 10019. 1914 to present. Weekly. $49.00 per year. ISSN: 0009-272X.

Chemtech. American Chemical Society, Drawer 1734, Atlanta, GA 30301-1734. (800) 227-5558. 1970 to present. Monthly. $370.00 per year. ISSN: 0009-2703.

RESEARCH CENTERS AND INSTITUTES

Analytical Chemistry Laboratory. U. S. Department of agriculture. Beltsville Agricultural Research Center - East, Building 306, Beltsville, MD 20705. (301) 344-2495.

Barnett Institute of Chemical Analysis and Materials Science. Northeastern University, 360 Huntington Avenue, Boston, MA 02115. (617) 437-2864.

Chemical Laboratories. Harvard University. Oxford Street, Cambridge, MA 02138. (617) 495-4283. FAX (617) 496-5618.

University/Industry Chemical Research Center. Mississippi State University, Department of Chemistry, P.O. Drawer CH, Mississippi State, MS 39762. (601) 325-3584.

ABSTRACT ALGEBRA

See: ALGEBRA

ACCELERATORS

See: PARTICLE ACCELERATORS

ACCIDENTS

See: SAFETY ENGINEERING

ACETYLENE WELDING

See: WELDING

ACID RAIN

See also: AIR POLLUTION, POLLUTION

ABSTRACT SERVICES AND INDEXES

CA Selects: Environmental Pollution. Chemical Abstracts Service, 2540 Olentangy River Road, Box 3012, Columbus, OH 43210. (800) 848-6538 or (614) 447-3600, FAX (614) 447-3713. Semi-weekly. $210.00 per year.

CA Selects: Pollution Monitoring. Chemical Abstracts Service, 2540 Olentangy River Road, Box 3012, Columbus, OH 43210. (800) 848-6538 or (614) 447-3600, FAX (614) 447-3713. Semi-weekly. $210.00 per year.

Current Contents: Physical, Chemical and Earth Sciences. Institute for Scientific Information, 3501 Market Street, Philadelphia, PA 19104. (215) 386-0100. FAX (215) 386-6362. Weekly. $360.00 per year.

Environmental Abstracts. R.R. Bowker Inc., 121 Chanlon Road, New Providence, NJ 07974. (800) 521-8110 or (908) 464-6800. Contains citations, most with abstracts, to literature relating to environmental issues and problems, from 1971 to present. Inquire as to cost and availability.

NTIS Alerts: Environmental Pollution and Control. National Technical Information Service (NTIS), 5285 Port Royal Road, Springfield, VA 22161. (703) 487-4650. Weekly. $150.00 per year.

Pollution Abstracts, Cambridge Scientific Abstracts, 7200 Wisconsin Avenue, Bethesda, MD 20814-4823. (301) 961-6750. FAX (301) 961-6720. Inquire for cost and availability.

Science Citation Index. Institute for Scientific Information, 3501 Market Street, Philadelphia, PA 19104. (215) 386-0100. FAX (215) 386-6362. Inquire about availability and cost. Also available on CD-ROM.

ASSOCIATIONS AND PROFESSIONAL SOCIETIES

Acid Rain Foundation. 1410 Varsity Drive, Raleigh, NC 27606-2010. (919) 828-9443. FAX (919) 515-3593.

American Academy of Environmental Engineering. 130 Holiday Ct., Suite 100, Annapolis, MD 21401. (410) 266-3311. FAX (410) 266-7653.

Environmental Defense Fund. 257 Park Avenue South, New York, NY 10010. (212) 505-2100. FAX (212) 505-2375.

International Center For the Solution of Environmental Problems. 535 Lovett Blvd., Houston, TX 77006. (713) 527-8711. FAX (713) 527-8025.

BIBLIOGRAPHIES

Environmental Periodicals Bibliography. International Academy at Santa Barbara, Environmental Studies Institute, 800 Garden Street, Suite D, Santa Barbara, CA 93101-1552. (805) 965-5010. 1972 to present. Bi-monthly. Price varies.

Reading About the Environment: An Introductory Guide. P.E. Jansma. Libraries Unlimited, PO Box 6633, Englewood, CO 80155-6633. (800) 237-6124. FAX (303) 220-8843. 1993. $27.50.

DIRECTORIES AND BIOGRAPHICAL SOURCES

Government Research Directory. 9th edition. Gale Research, Inc., 835 Penobscot Building, Detroit, MI 48226-4094. (313) 961-2242. (800) 877-4253. 1996. $465.00.

International Research Centers Directory. Gale Research, Inc., 835 Penobscot Building, Detroit, MI 48226-4094. (313) 961-2242. (800) 877-4253. 8th edition. 1995. $410.00

Research Centers Directory. Gale Research, Inc., 835 Penobscot Building, Detroit, MI 48226-4094. (313) 961-2242. (800) 877-4253. $470.00

Scientific and Technical Organizations and Agencies Directory. Gale Research, Inc., 835 Penobscot Building, Detroit, MI 48226-4094. (313) 961-2242. (800) 877-4253. 4th edition. 1996. $195.00.

ENCYCLOPEDIAS AND DICTIONARIES

Encyclopedia of Environmental Control Technology, Volume 2: Air Pollution Technology. Paul N. Cheremisinoff, editor. Gulf Publishing Company, P.O. Box 2608, Houston, TX 77252-2608. (713) 529-4301 or (800) 231-6275. FAX (713) 525-9647. 1989. $155.00

Encyclopedia of Physical Science and Technology. Robert A. Meyers, ed. 18 volumes. Academic Press, Inc., 1250 Sixth Avenue, San Diego, CA 92101-4311. (619) 231-0926. FAX (619) 699-6715. 1992. $2100.00

McGraw-Hill Encyclopedia of Science and Technology. Sybil P. Parker, ed. 7th edition. 20 volumes. McGraw-Hill Publishing Company, 1221 Avenue of the Americas, New York, NY 10020. (800) 262-4729 or (212) 512-3825. 1992. $1900.00

GENERAL WORKS

Acid Rain and Acid Waters. Gwyneth Parry Howells. 2nd edition. E. Horwood Publishing, distributed by Prentice Hall (Division of Simon and Schuster), 15 Columbus Circle, New York, NY 10023. (212) 373-8500 or (800) 922-0579. 1995. $54.00.

Acid Rain: Current Situation and Remedies. John Rose, editor. Gordon and Breach Science Publishers, 820 Town Center Drive, Langhorne, PA 19047. (215) 750-2642. FAX (215) 750-6343. 1994. Inquire for price.

The Acid Rain Debate: Science and Special Interests in Policy Formation. Bruce A. Forster. Iowa State University Press, 2121 South State Avenue, Ames, IA 50014-8300. (800) 862-6657. FAX (515) 292-3348. 1993. $32.95.

Air Pollution and Climate Change: The Biological Impact. Alan Wellburn. 2nd edition. Longman Publishing Group, the LongmanBuilding, 10 Bank Street, White Plains, NY 10606-1951. (914) 993-5000 or (800) 266-8855. FAX (914) 997-8115. 1994. $39.95.

The Forgiving Air: Understanding Environmental Change. Richard Somerville. University of California Press, 2120 Berkeley Way, Berkeley, CA 94720. (800) 777-4726. FAX (510) 642-7127. 1996. $21.95.

Water: The Vital Resource. Norma H. Jones, et al. Information Plus, 2812 Exchange Street, Wylie, TX 75098. (800) 463-6757. FAX (214) 442-1189.

HANDBOOKS AND MANUALS

Environmental Science and Technology Handbook. Porter C. Knowles, et al. Government Institutes Inc., 4 Research Place, Suite 200, Rockville, MD 20850. (301) 921-2355. 1994. $75.00.

ONLINE DATABASES AND CD-ROMS

NTIS Bibliographic Database. National Technical Information Service, 5285 Port Royal Road, Springfield, VA 22161. (703) 487-4929 or FAX (703) 321-8199. Broad coverage of government-sponsored science and technology research reports, 1964 to present. Available on BRS,(800) 289-4277, DIALOG, (800) 334-2564, ORBIT, (800) 456-7248, and SON International, FEZ Karlsruhe, P.O. Box 2465, W-7500, Karlsruhe 1, Germany, online services. Also available on CD-ROM. Inquire as to cost and availability.

SCISEARCH. Institute for Scientific Information, 3501 Market Street, Philadelphia, PA 19104. (800) 523-1850 or (215) 386-0100. Broad multi disciplinary title and author index to the international literature of science and technology, 1974 to present. Available on DIALOG,(800) 334-2564, and ORBIT,(800) 456-7248, online services. Also available on CD-ROM. Inquire as to cost and availability.

PERIODICALS

Air & Water Pollution Control. Bureau of National Affairs, 1231 25th Street NW, Washington, DC 20037. (202) 452-4200. FAX (202) 822-8092. 1986 to present. Bi-weekly. $301.00 per year.

Bulletin of Environmental Contamination and Toxicology. Springer-Verlag, 175 Fifth Avenue, New York, NY 10010. (212) 460-1500 or (800) 777-4643. FAX (212) 473-6272. 1966 to present. Monthly. $399.00 per year.

Environmental Engineering. Mechanical Engineering Publications Ltd., Northgate Avenue, Bury St. Edmunds, Suffolk IP32 6BW, England. Telephone 0284-763277. FAX 0284-704006. 1972 to present. Quarterly. Inquire for price.

Environmental Monitoring and Assessment. Kluwer Academic Publishers, P.O. Box 358, Accord Station, Hingham, MA 02018-0358. (617) 871-6000. FAX (617) 871-6528. 1981 to present. 15 times a year. $765.00 per year.

Environmental Pollution. Elsevier Science [journals], 660 White Plains Rd., Tarrytown, NY 10159-5153. (919) 524-9200. FAX (919) 333-2444. 1967 to present. 1970 to present. 12 times a year. $1250.00 per year.

Environmental Science and Technology. American Chemical Society, 1155 16th Street, N.W., Washington, DC 20036. (202) 872-4381 or (800) 333-9511. FAX (614) 447-3671. 1967 to present. Monthly. $89.00 per year for non-members.

EPA Journal. Superintendent of Documents, U.S. Government Printing office, Box 371954, Pittsburgh, PA 15250-7954. (202) 783-3238. FAX (202) 512-2233. 1975 to present. Bi-monthly. $10.00 per year.

Water, Air and Soil Pollution. Kluwer Academic Publishers, P.O. Box 358, Accord Station, Hingham, MA 02018-0358. (617) 871-6000. FAX (617) 871-6528. 1971 to present. 28 times a year. $1519.00.

RESEARCH CENTERS AND INSTITUTES

Environmental and Energy Study Institute. 122 C Street NW, Suite 700, Washington, DC 20001. (202) 628-1400.

University of Florida, Environmental Engineering Research Center. College of Engineering, 217 Black Hall, Gainesville, FL 32611. (904) 392-0841. FAX (904) 392-3076.

University of New Mexico, Center For Global Environmental Technologies. Albuquerque, NM 87131-1376. (505) 272-7252. FAX (505) 272-7203.

World Resources Institute. 1709 New York Avenue NW, Suite 700, Washington, DC 20006. (202) 638-6300. FAX (202) 638-0036.

ACIDS AND BASES

See also: CATALYSIS, CHEMICAL ENGINEERING, CHEMISTRY, INORGANIC CHEMISTRY, ORGANIC CHEMISTRY

ABSTRACT SERVICES AND INDEXES

Analytical Abstracts. Royal Society of Chemistry, Information Services, Thomas Graham House, Science Park, Milton Road, Cambridge, CB4 4WF, England. Contains citations, mostly with abstracts, of the worldwide literature on analytical chemistry, from 1954 to present. Monthly. $636.00 per year. Also available on CD-ROM and online.

Applied Science and Technology Index; A Cumulative Subject Index To English Language Periodicals in the Fields of Aeronautics and Space Science, Computer Technology, Chemistry, Construction Industry, Energy and Related Areas. H. W. Wilson Co., 950 University Avenue, Bronx, NY 10452-9978. (212) 588-8400. (800) 367-6770. FAX (718) 590-1617. From 1958 to present. Monthly. Inquire about cost and availability. Also available online (BRS and WILSONLINE) and on CD-ROM. ISSN: 0003-6986.

Chemical Abstracts. Chemical Abstracts Service. 2540 Oleantangy River Road, Box 3012, Columbus, OS 43210-0012. (614) 447-3600. FAX (614) 447-3713. 1907 to present. Weekly. $16,800.00 per year. Available online and on CD-ROM. CA is also available in five section groupings. Inquire regarding cost and availability.

Current Contents: Physical, Chemical and Earth Sciences. Institute for Scientific Information, 3501 Market Street, Philadelphia, PA 19104. (215) 386-0100. (800) 523-1850. FAX (215) 386-2291. 1961 to present. Weekly. $442.00 per year. Also availableonline (BRS, DIALOG) and on CD-ROM. Inquire regarding cost and availability. ISSN: 0163-2574.

ACIDS AND BASES

Ency. of Physical Sciences and Engineering Info. Sources

General Science Index. H.W. Wilson Company, 950 University Avenue, Bronx, NY 10452. (212) 588-8400. (800) 367-6770. FAX (718) 590-1617. From 1978 to present. Ten issues per year; quarterly and annual cumulations. Service basis. Available on CD-ROM and online. Inquire regarding cost and availability. ISSN: 0162-1963.

Physics Abstracts. INSPEC. Section A, Science Abstracts. Institution of Electrical Engineers (IEE), London, United Kingdom. Available from: INSPEC/IEEE - Institute of Electrical and Electronic Engineers, Box 1331, 445 Hoes Lane, Piscataway, NJ 08855-1331. (908) 562-5549. 1898 to present. 24 issues per year. $2835.00 per year. ISSN: 0036-8091. Also available online and on CD-ROM.

Science Citation Index. SCI. Institute for Scientific Information, 3501 Market Street, Philadelphia, PA 19104. (215) 386-0100. (800) 523-1850. FAX (215) 386-2991. 1961 to present. Six issues per year, plus annual cumulation. $11650.00 per year. Also available online and on CD-ROM. Inquire about price and availability. ISSN: 0036-827X.

ANNUAL REVIEWS AND YEARBOOKS

Annual Review of Physical Chemistry. Annual Reviews, Inc., 4139 El Camino Way, Palo Alto, CA 94306-0897. (415) 493-4400. Fax (415) 855-9815. Annual. $55.00 per year.

ASSOCIATIONS AND PROFESSIONAL SOCIETIES

American Chemical Society, 1155 16th Street, Nw, Washington, DC 20036. (202) 872-4600.

Association of Consulting Chemists and Chemical Engineers, 50 East 41st Street, Suite 92, New York, NY 10017. (212) 684-6255.

Association of official Analytical Chemists. 2200 Wilson Blvd, Suite 400, Arlington, VA 22001. (703) 522-3032.

Chemical Manufacturers Association. 2501 M Street, N.W., Washington, DC 20037. (202) 887-1182

BIBLIOGRAPHIES

Chemical Information Sources. Gary Wiggins. McGraw-Hill Publishing Company, 1221 Avenue of the Americas, New York, NY 10020. (212) 512-2000. (800) 262-4729. 1991. $42.50.

Information Sources in Chemistry. R.T. Bottle and J.F.B Rowland, editors. R.R. Bowker Inc., 121 Chanlon Road, New Providence, NJ 07974. (800) 521-8110 or (908) 464-6800. Fourth edition. 1993. $75.00.

DIRECTORIES AND BIOGRAPHICAL SOURCES

American Men and Women of Science. Physical and Biological Sciences. R.R. Bowker Inc., 121 Chanlon Road, New Providence, NJ 07974. (908) 464-6800. (800) 521-8110. 20th edition. 8 volumes. 1996. $850.00

Directory of Chemistry Software 1992. Wendy Warr, Peter Willett, Geoff Downs. American Chemical Society, 1155 16th

Street,NW, Washington, DC 20036. (202) 872-4600. (800) 333-9511. 1992. $35.95.

Research Centers Directory. Gale Research, 835 Penobscot Building, Detroit, MI 48226-4094. (313) 961-2242. (800) 877-4253. 20th edition. 1995. $485.00. ISSN: 0080-1518.

ENCYCLOPEDIAS AND DICTIONARIES

Concise Encyclopedia of Chemical Technology. Kirk-Othmer. John Wiley and Sons, Inc., 605 Third Avenue, New York, NY 10158. (212) 850-6000. (212) 683-8333. (800) 422-8824. 3rd edition. 1985. $215.00.

Illustrated Chemistry Laboratory Terminology. Gerbert W. Ockerman. CRC Press, Inc., 2000 Corporate Boulevard, NW, Boca Raton, FL 33431. (407) 994-0555. (800) 272-7737. 1991. $32.95.

Industrial Chemical thesaurus. Michael Ash and Irene Ash. VCH Publishers, Inc., 220 East 23rd Street, New York, NY 10010-4606. (800) 367-8249. 1992. $295.00.

Kirk-Othmer Encyclopedia of Chemical Technology. John Wiley and Sons, Inc., 605 Third Avenue, New York, NY 10158. (212) 850-6000. (212) 683-8333. (800) 422-8824. Fourth edition. 1991-. Twenty-seven volumes. $5400.00.

McGraw-Hill Encyclopedia of Science and Technology. McGraw-Hill Book, Incorporated, 1221 Avenue of the Americas, New York, NY 10020. (212) 512-2000. (800) 262-4729. Seventeenth editon. 1992. $1900.00.

GENERAL WORKS

Acidity and the Properties of Major Industrial Acids. Ian M. Campbell. Chapman & Hall, 1 Penn Plaza, New York, NY 10119. (212) 564-1060. 1992. $65.00.

Catalysis By Acids and Bases. B. Imelik. Elsevier Science Publishing Company, Inc., 655 Avenue of the Americas, New York, NY 10010. (212) 989-5800. 1985. $141.00.

Hard and Soft Acids and Bases. R.G. Pearson, editor. Van Nostrand Reinhold, 115 Fifth Avenue, New York, NY 10003. (212) 254-3232. (800) 842-3636. 1973. Inquire.

Ionization Constants of Inorganic Acids and Bases in Aqueous Solution. D. D. Perrin, editor. Franklin Book Co., Inc., Elkins Park, PA 19117. (215) 635-5252. 1982. $84.00.

Organic Acids in Geological Processes. E. D. Pittman and M. D. Lewan, editors. Springer-Verlag New York, Inc., 175 Fifth Avenue, New York, NY 10010. (212) 460-1500. (800) 777-4643. 1994. $89.00.

pH, Acids and Bases. H. A. Neidig and J. N. Spencer. Chemical Education Resources, Inc., P. O. Box 357, Palmrya, PA 17078. (717) 838-3165. 1992. $1.25 in paper.

Solid Acids and Bases: Their Catalytic Properties. Kozo Tanabe. Academic Press, Inc., 6277 Sea Harbor Drive, Orlando, FL. (800) 321-5068. 1971. $101.00.

Superacids. George A. Olah, and others. Krieger Publishing Company, P.O. Box 9542, Melbourne, FL 32902-9542. (407) 724-9542. 1985. Inquire

Superacids and Acidic Melts As Inorganic Chemical Reaction Media. Thomas A. O'Donnell. VCH Publications, Inc., 220 East 23rd Street, Suite 909, New York, NY 10010-4606. (212) 683-8333. (800) 422-8824. 1993. $110.00 per year.

HANDBOOKS AND MANUALS

CRC Handbook of Chemistry and Physics. David R. Kide, editor. CRC Press Inc., 2000 Corporate Blvd., NW, Boca Raton, FL 33431. (407) 994-0555. (800) 333-8300. 77th edition. 1996. $99.95. <bodylead>*Guide to Basic Chemical Compounds.* D.R. Lide, Jr. CRC Press, Inc., 2000 Corporate Blvd., N.W., Boca Raton, FL 33431. (407) 994-0555. (800) 272-7737. 1993. $120.00.

Handbook of Laboratory Safety. A. Keith Furr. CRC Press, Inc., 2000 Corporate Boulevard, NW, Boca Raton, FL 33431. (407) 994-0555. (800) 272-7737. 1995. 125.00.

Improving Safety in the Chemical Laboratory: A Practical Guide. Jay A. Young, editor. John Wiley and Sons, Inc., 605 Third Avenue, New York, NY 10158-0012. (212) 850-6000. (800) 223-2348. Second edition. 1991. $75.00.

Laboratory Handbook of Materials, Equipment, and Techniques. Gary S. Coyne. Prentice Hall, 15 Columbus Circle, New York, NY 10023. (212) 373-8500. (800) 223-2348. 1992. $45.00.

Lange's Handbook of Chemistry. John A. Dean, editor. McGraw-Hill Publishing Company, 1221 Avenue of the Americas, New York, NY 10020. (212) 512-2000. (800) 262-4729. Fourteenth edition. 1992. $79.50.

Riegel's Handbook of Industrial Chemistry. James A. Kent, editor. Van Nostrand Reinhold, 115 Fifth Avenue, New York, NY 10003. (212) 254-3232 or (800) 926-2665. Ninth edition. 1992. $114.95.

ONLINE DATABASES AND CD-ROMS

Analytical Abstracts Online. Royal Society of Chemistry, Information Services, Thomas Graham House, Science Park, Milton Road, Cambridge, CB4 4WF, England. Contains citations, mostly with abstracts, of the worldwide literature on analytical chemistry, from 1980 to present. Available on DIALOG, (800) 334-2564, and STN International, FIZ Karlsruhe, P.O. Box 2465, W-7500, Karlsruhe 1, Germany, online services. Inquire as to cost and availability.

CA Search. Chemical Abstracts Service, P.O. Box 3012, Columbus, OH 43210-0012. (614) 447-3600. (800) 848-6533. FAX (614) 447-3709. Very comprehensive guide to worldwide chemical literature and related fields, 1972 to present. Available on BRS, (800) 289-4277, DIALOG, (800) 334-2564, ORBIT (800) 456-7248, and STN International, FIZ Karlsruhe, P.O. Box 2465, W-7500, Karlsruhe 1, Germany, online services. Inquire as to cost and availability.

Chemical Journals of the American Chemical Society. American Chemical Society, 1155 16th Street, N.W., Washington, DC 20036. (202) 872-4381 or (800) 424-6747. Contains complete text of approximately 90,000 articles from 22 primary journals published by the American Chemical Society, from mostly 1982 to present. Available on STN International, FIZ Karlsruhe, P.O. Box 2465, W-7500, Karlsruhe 1, Germany, online service. Inquire as to cost and availability.

Dissertation Abstracts Online. University Microfilms International, 300 North Zeeb Road, Ann Arbor, MI 48106. (800) 521-0600 or (313) 761-4700. Scope includes virtually all doctoral dissertations accepted at accredited American institutions from 1861 to present in 252 subject areas. Available on BRS, (800) 289-4277, DIALOG, (800) 334-2564, and OCLC EPIC, (800) 848-5878, online services. Also available on CD-ROM. Inquire as to cost and availability.

Gmelin Database. Gmelin-Institut fur Anorganische Chemie und Grenzgebiete, Varrentrapperstrasse, 40-42, Carl-Bosch-Haus, D-6000, Frankfurt am Main 90, Germany. Contains structural and factual data relating to inorganic and organometallic chemistry. Provides data from the Gmelin Handbook of Inorganic and Organometallic Chemistry. Covers the period 1817 to 1975; 1988-89. Available on STN International, FIZ Karlsruhe, P.O. Box 2465, W-7500, Karlsruhe 1, Germany, online service. Inquire as to cost and availability.

Kirk-Othmer Encyclopedia of Chemical Technology. John Wiley and Sons, Inc., 605 Third Avenue, New York, NY 10158. (800) 526-5368 or (212) 850-6000. Contains the complete text of all chapters in the 27 volume fourth edition of the Kirk-Othmer Encyclopedia of Chemical Technology. 1991. Available on BRS, (800) 289-4277, DIALOG, (800) 334-2564, online services. Inquire as to cost and availability.

NTIS Bibliographic Database. National Technical Information Service, 5285 Port Royal Road, Springfield, VA 22161. (703) 487-4929 or FAX (703) 321-8199. Broad coverage of government-sponsored science and technology research reports, 1964 to present. Available on BRS, (800) 289-4277, DIALOG, (800) 334-2564, ORBIT, (800) 456-7248, and STN International, FIZ Karlsruhe, P.O. Box 2465, W-7500, Karlsruhe 1, Germany, online services. Also available on CD-ROM. Inquire as to cost and availability.

SCISEARCH. Institute for Scientific Information, 3501 Market Street, Philadelphia, PA 19104. (800) 523-1850 or (215) 386-0100. Broad multidisciplinary title and author index to the international literature of science and technology, 1974 to present. Available on DIALOG, (800) 334-2564, and ORBIT, (800) 456-7248, online services. Also available on CD-ROM. Inquire as to cost and availability.

WILSONLINE. H.W. Wilson Company, 950 University Avenue, Bronx, NY 10452. (800) 367-6770 or (212) 588-8400. Makes available online versions of the printed H.W. Wilson indexes including Applied Science and Technology Index, Business Periodicals Index, General Science Index, and Readers' Guide to Periodical Literature. Period covered is generally 1983 to present. Available on BRS, (800) 289-4277, DIALOG, (800) 334-2564, and OCLC EPIC, (800) 848-5878, online services. Also available on CD-ROM. Inquire as to cost and availability.

PERIODICALS

Analytical Chemistry. American Chemical Society, Drawer 1734, Atlanta, GA 30301-1734. (800) 227-5558. Monthly. $373.00 per year.

ACIDS AND BASES

Ency. of Physical Sciences and Engineering Info. Sources

Chemical Week. Chemical Week Associates, 810 7th Avenue, New York, NY 10019. 1914 to present. Weekly. $49.00 per year. ISSN: 0009-272X.

Chemtech. American Chemical Society, Drawer 1734, Atlanta, GA 30301-1734. (800) 227-5558. 1970 to present. Monthly. $370.00 per year. ISSN: 0009-2703.

Inorganic Chemistry. American Chemical Society, Drawer 1734, Atlanta, GA 30301-1734. (800) 227-5558. Bi-Weekly. $500.00 per year.

Journal of Organic Chemistry. American Chemical Society, Drawer 1734, Atlanta, GA 30301-1734. (800) 227-5558. Bi-Weekly. $575.00 per year.

Journal of Physical Chemistry. American Chemical Society, Drawer 1734, Atlanta, GA 30301-1734. (800) 227-5558. Bi-Weekly. $573.00 per year.

Tetrahedron Letters. Pergamon Press, Maxwell House, Fairview Park, Elmsford, NY 10523. (914) 592-7700. Fax (914) 592-3625. Weekly. $180.00 per year.

RESEARCH CENTERS AND INSTITUTES

Chemical Laboratories. Harvard University, Oxford Street, Cambridge, MA 02138. (617) 495-4283. FAX (617) 496-5618.

Chemistry Laboratories. Rensselaer Polytechnic Institute, Cogswell Laboratory, Troy, NY 12180-3590. (518) 276-8981.

Lawrence Berkeley Laboratory, Chemical Sciences Division. One Cyclotron Road, Building 66, Berkeley, CA 94720. (510) 486-6062. FAX (510) 486-4995.

Research Program in Chemistry and Biochemistry. Southern Illinois University at Carbondale, Carbondale, IL 62901. (618) 453-5721.

ACOUSTICS

See also: AMPLIFIERS, AUDIO ENGINEERING, SIGNAL PROCESS-ING

ABSTRACT SERVICES AND INDEXES

Acoustics Abstracts; Parts A & B. Multi-Science Publishing Co. Ltd., Box 176, Avenel, NJ 07001. 1967 to present. Monthly. L265 per year. ISSN: 0001-4974.

Applied Science and Technology Index; A Cumulative Subject Index To English Language Periodicals in the Fields of Aeronautics and Space Science, Computer Technology, Chemistry, Construction Industry, Energy and Related Areas. H.W. Wilson Co., 950 University Avenue, Bronx, NY 10452. (212) 588-8400. (800) 367-6770. FAX (718) 590-1617. From 1958 to present. Monthly. Inquire about cost and availability. Also available on CD-ROM and online. ISSN: 0003-6986.

Current Contents: Engineering, Technology and Applied Sciences. Institute for Scientific Information, 3501 Market Street, Philadelphia, PA 19104. (215) 386-0100. FAX (215) 386-

6362. 1970 to present. Weekly. $442.00 per year. Also available on CD-ROM and online. Inquire regarding cost and availability. ISSN: 0095-7917.

Electrical and Electronics Abstracts. Institution of Electrical Engineers (IEE), London. Available from INSPEC/IEEE - Institute of Electrical and Electronic Engineers, Box 1331, Hoes Lane, Piscataway, NJ 08855-1331. (908) 562-5549. 1898 to present. Monthly. $2200.00 per year. Also available on CD-ROM and online as INSPEC.

Engineering Index Monthly; Indexes and Abstracts the World's Engineering and Technical Literature. Engineering Information, Inc., Castle Point on the Hudson, Hoboken, NJ 07030. (201) 216-8500. (800) 221-1044. FAX (201) 216-8532. Monthly. $2300.00 per year. Available online as COMPENDEX and also on CD-ROM. ISSN: 0742-1974.

General Science Index. H.W. Wilson Company, 950 University Avenue, Bronx, NY 10452. (212) 588-8400. (800) 367-6770. FAX (718) 590-1617. From 1978 to present. Ten issues per year; quarterly and annual cumulations. Service basis. Available on CD-ROM and online. Inquire regarding cost and availability. ISSN: 0162-1963.

Government Reports Announcements and Index. U. S. National Technical Information Service (NTIS), 5285 Port Royal Road, Springfield, VA 22161. (703) 487-4650. FAX (703) 321-8547. From 1968 to present. Annual. $630.00 per year. Also available online as NTIS BIBLIOGRAPHIC DATABASE and on CD-ROM. ISSN:

Index To IEEE Publications. IEEE Service Center, 445 Hoes Lane, Piscataway, NJ 08855-1331. (908) 981-1393. (800) 678-IEEE. FAX (908) 981-9667. 1973 to present. Annual. ISSN: 0099-1368.

Physics Abstracts. INSPEC. Section A, Science Abstracts. Institution of Electrical Engineers (IEE), London. Available from: INSPEC/IEEE - Institute of Electrical and Electronic Engineers, Box 1331, Hoes Lane, Piscataway, NJ 08855-1331. (908) 562-5549. 1898 to present. 24 issues per year. $2835.00 per year. Also available online and on CD-ROM. ISSN: 0036-8091.

Physics Briefs (Physikalische Berichte). Information Center for Energy, Physics, Mathematics; German Physical Society. V C H Publishers, Inc., 220 East 23rd Street, New York, NY 10010-4606. (212) 683-8333. 1845 to present. 24 issues per year. $2390.00 per year. Also available online. ISSN: 0179-7434.

Science Citation Index. SCI. Institute for Scientific Information, 3501 Market Street, Philadelphia, PA 19104. (215) 386-0100. (800) 523-1850. FAX (215) 386-2991. 1961 to present. Six issues per year, plus annual cumulation. $11650.00 per year. Also available online and on CD-ROM. Inquire about price and availability. ISSN: 0036-827X.

ANNUAL REVIEWS AND YEARBOOKS

Advances in Electronics and Electron Physics. Academic Press, Inc., 6277 Sea Harbor Drive Orlando FL 32887. (800) 321-5068. From 1948 to present. Irregular. Price varies, inquire.ISSN: 0065-2539.

Physical Acoustics. Academic Press, Inc., 6277 Sea Harbor Drive, Orlando, FL 32887. (800) 321-5068. 1964 to present. Irregular. ISSN: 0079-1873.

ASSOCIATIONS AND PROFESSIONAL SOCIETIES

Acoustical Society of America. 500 Sunnyside Boulevard, Woodbury, NY 11797. (516)576-2360.

American Electronics Association. 5201 Great America Way, Suite 520, P.O. 52990, Santa Clara, CA 95056. (408) 987-4200. FAX (408) 970-8565.

American Institute of Physics. One Physics Ellipse, College Park, MD 20740-3843. (301) 209-3100.

Audio Engineering Society. 60 East 42nd Street, New York, NY 10017. (212) 661-8528.

IEEE (Institute of Electrical and Electronics Engineers). 345 East 47th Street, New York, NY 10017. (212) 705-7900. FAX (212) 705-4929.

IEEE Signal Processing Society. c/o IEEE, 345 East 47th Street, New York, NY 10017. (212) 705-7900. FAX (212) 705-4929.

IEEE Ultrasonics, Ferroelectrics and Frequency Control Society. c/o Gary K. Montress, Raytheon Research Division, 131 Spring Street, Lexington, MA 02173. (617) 860-3053. FAX (617) 860-3194.

Society of Professional Audio Recording Services. 4300 10th Avenue North, Lake Worth, FL 33461. (407) 641-6648. FAX (407) 642-8263.

BIBLIOGRAPHIES

Science Books & Films. American Association for the Advancement of Science, 1333 H Street NW, Washington, DC 20005. (202) 326-6434. Reviews of print, film and software materials in all sciences. 1965 to present. Nine issues per year. $40.00 per year. ISSN: 0098-342X.

Scientific and Technical Books and Serials in Print; An Index To Literature in Science and Technology. R. R. Bowker. 121 Chanlon Road, New Providence, NJ 07974. (908) 464-6800. FAX (908) 665-3502. (800) 521-8110. 1972to present. Annual. $299.95 per year. Also available on CD-ROM and online. ISSN: 0000-054X.

DIRECTORIES AND BIOGRAPHICAL SOURCES

American Men and Women of Science: Physical and Biological Sciences. R. R. Bowker Inc., 121 Chanlon Road, New Providence, NJ 07974. (908) 464-6800. (800) 521-8110. 20th edition. 8 volumes. 1996. $850.00.

E E M - Electronic Engineer's Master. Hearst Business Communications, Inc., 645 Stewart Avenue, Garden City NY 11530. (516) 227-1300. Annual. $90.00 ISSN: 0732-9016.

Engineering Research Centers: Incorporating Electronics Research Centers. Stockton Press, 345 Park Avenue, New York, NY 10010. (212) 689-9200. (800) 221-2123. 4th edition. 1995. $515.00.

IEEE Membership Directory. Institute of Electrical and Electronics Engineers, IEEE Service Center, 445 Hoes Lane,Piscataway, NY 08854. (908) 981-1393. (800) 678-IEEE. FAX (908) 981-9667. 2 volumes. Annual. $190.00. ISSN:

Research Centers Directory. Gale Research Inc., 835 Penobscot Building, Detroit, MI 48226-4094. (313) 961-2242. (800) 877-4253. 20th edition. 1995. $485.00. ISSN: 0080-1518.

Who's Who in Engineering. Gordon Davis, editor. American Association of Engineering Societies. 1111 19th Street, NY, Suite 608, Washington, DC 20036. (202) 296-2237. (800) 658-8897. 9th edition. 1995. $220.00.

ENCYCLOPEDIAS AND DICTIONARIES

Applied Technical Dictionary: Acoustics. State Mutual Book Service, Ltd. 521 Fifth Avenue, New York, NY 10175. (212) 682-5844. 1986. $75.00.

The Audio Dictionary. Glenn D. White. University of Washington Press, P. O. Box 50096. Seattle, WA 98145-5096. (206) 543-4050. (800) 441-4115. 2nd revised edition. 1991. $19.95 in paper.

Encyclopedia of Applied Physics. George Trigg, editor. VCH Publications, Inc., 220 East 23rd Street, Suite 909, New York, NY 10010-4606. (212) 683-8333. (800) 422-8824. 20 volume set. 1991-. $5990.00.

IEEE Standard Dictionary of Electrical and Electronics Terms. Christopher J. Booth, editor. IEEE Service Center, 445 Hoes Lane, Piscataway, NJ 08855-1331. (908) 981-1393. (800) 678-IEEE. FAX (908) 981-9667. IEEE Standard 100-1992. 5th edition. 1993. $150.00.

Illustrated Dictionary of Electronics. Stan Gibilsco. TAB Books, P.O. Box 40, Blue Summit, PA 17294-0850. (717) 794-2191. (800) 233-1128. 7th edition. 1994. $34.95.

McGraw-Hill Encyclopedia of Science and Technology. McGraw-Hill Book Company, Inc., 1221 Avenue of the Americas, New York, NY 10020. (212) 512-2000. (800) 262-4729. 7th edition. 20 volume set. 1992. $1900.00.

GENERAL WORKS

Acoustical Measurements. Leo L. Beranek. Acoustical Society of America, 500 Sunnyside Boulevard, Woodbury, NY 11797. (516)576-2360. 1988. $35.00.

Active Control of Sound. P. A. Nelson and S. J. Elliott. Academic Press, Inc., 6277 Sea Harbor Drive, Orlando, FL. (800) 321-5068.

Advanced Audio Production Techniques. Tyree S. Ford. Butterworth-Heinemann, 313 Washington Street, Newton, MA 02158. (618-928-2500. (800) 366-2665. 1993. $24.95 in paper.

Audio Electronics. John L. Hood. Butterworth-Heinemann, 313 Washington Street, Newton, MA 02158. (618-928-2500. (800) 366-2665. 1995. $29.95 in paper.

Basic Acoustics. Donald E. Hall. Krieger Publishing Company, P.O. Box 9542, Melbourne, FL 32902-9542. (407) 724-9542. 1993. Reprint edition. $61.95.

ACOUSTICS

Ency. of Physical Sciences and Engineering Info. Sources

Introduction To Digital Audio. John Watkinson. Butterworth—Heinemann, 313 Washington Street, Newton, MA 02158. (618-928-2500. (800) 366-2665. 1993. $34.95 in paper.

Mechanics of Sound Recording. Anthony J. Zaza. Prentice-Hall, 113 Sylvan Avenue, Route 9W, Englewood Cliffs, NJ 07632. (201) 592-2000. (800) 922-0579. 1991. $50.00.

Origins in Acoustics: The Science of Sound From Antiquity To the Age of Newton. Books on Demand, 300 North Zeeb Road, Ann Arbor, MI 48106-1346. (313) 761-4700. (800) 521-0600. Reprint edition. $62.50.

The Physics of Sound. Richard E. Berg and David G. Stork. Prentice-Hall, 113 Sylvan Avenue, Route 9W, Englewood Cliffs, NJ 07632. (201) 592-2000. (800) 922-0579. 2nd edition. 1994. $63.00.

Principles of Digital Audio. Ken C. Pohlman. McGraw-Hill Publishing Company, Inc., 1221 Avenue of the Americas, New York, NY 10020. (212) 512-2000. (800) 262-4729. 1995. .

Principles of Vibration and Sound. T. D. Rossing and N. H. Fletcher. Springer-Verlag New York, Inc., 175 Fifth Avenue, New York, NY 10010. (212) 460-1500. (800) 777-4643. FAX (212) 473-6272. 1994. $49.95.

Sound Recording and Reproduction. Glyn Alkin. Butterworth-Heinemann, 313 Washington Street, Newton, MA 02158. (618-928-2500. (800) 366-2665. 2nd edition. 1991. $19.95 in paper.

Surface Acoustic Waves and Signal Processing. Michel Feldmann and Jeannine Henaff. Artech House, 685 Canton Street, Norwood, MA 02062. (617) 769-9750. (800) 225-9977. 1989. $89.00.

Vibroacoustical Diagnostics For Machines and Structures. M. F. Dimentberg, et al. John Wiley & Sons, Inc., 605 Third Avenue, New York, NY 10158-0012. (212) 850-6000. (800) 225-5945. 1991. $94.95.

HANDBOOKS AND MANUALS

Audio Engineering Handbook. Blair K. Benson, editor. McGraw-Hill Book Company, 1221 Avenue of the Americas, New York, 10020. (212) 512-2000. (800) 628-0004. 1988. $94.50.

Audio Engineer's Reference Book. Michael Talbot-Smith, editor. Butterworth-Heinemann, 313 Washington Street, Newton, MA 02158. (618-928-2500. (800) 366-2665. 1994. $99.95.

CRC Handbook of Chemistry and Physics. David R. Kide, editor. CRC Press, Inc., 2000 Corporate Boulevard, NW, Boca Raton, FL 33431. (407) 994-0555. (800) 272-7737. 77th edition. 1996. $99.95.

Electrical Engineer's Handbook. Donald Christiansen, editor. McGraw Book Company, 1221 Avenue of the Americas, New York, NY 10020. (212) 512-2000. (800) 262-4729. 4th edition. 1996. $110.00.

Electronics Handbook. Jerry C. Whitaker. CRC Press, Inc., 2000 Corporate Boulevard, NW, Boca Raton, FL 33431. (407) 994-0555. (800) 272-7737. 1996. $120.00.

Handbook For Sound Engineers: The New Audio Cyclopedia. Glen M. Ballou. Macmillian Publishing Company, Inc,. 200 Old Tappan Road, Old Tappan, NJ 07675. (800) 233-2336. 2nd edition. 1991. $99.95.

Handbook of Acoustical Measurements and Noise Control. Cyril M. Harris. McGraw-Hill Publishing Company, 1221 Avenue of the Americas, New York, NY 10020. (212) 512-2000. (800) 262-4729. 1991. $98.50.

Master Handbook of Acoustics. F. Alton Everest. TAB Books, P.O. Box 40, Blue Summit, PA 17294-0850. (717) 794-2191. (800) 233-1128. 2nd edition. 1988. $19.95.

Modern Audio Technology: A Handbook For Technicians and Engineers. Martin Clifford. Prentice Hall , 113 Sylvan Avenue, Route 9W, Englewood Cliffs, NJ 07632. (201) 592-2000. (800) 922-0579.

Sound Recording Handbook. John Woram. Sams, 201 West 103rd Street, Indianapolis, in 46290-1097. (317) 581-3500. (800) 545-5914. 1989. $49.95.

Standard Handbook for Electrical Engineers. Donald Fink. McGraw-Hill Publishing Company, 1221 Avenue of the Americas, New York, NY 10020. (212) 512-2000. (800) 262-4729. 13th edition. 1996. $110.00.

ONLINE DATABASES AND CD-ROMS

CA Search. Chemical Abstracts Service, P.O. Box 3012, Columbus, OH 43210-0012. (614) 447-3600. (800) 848-6533. FAX (614) 447-3709. Very comprehensive guide to worldwide chemical literature and related fields, 1972 to present. Available on BRS,(800) 289-4277, DIALOG, (800) 334-2564, ORBIT (800) 456-7248, and STN International, FIZ Karlsruhe, P.O. Box 2465, W-7500, Karlsruhe 1, Germany, online services. Inquire as to cost and availability.

Compendex Plus. Engineering Information, Inc., 345 East 47th Street, New York, NY 10017. (212) 705-7600 or (800) 221-1044. Contains citations with abstracts to worldwide literature in engineering and technology, from 1970 to present. Available on online BRS,(800) 289-4277, DIALOG, (800) 334-2564, ORBIT (800) 456-7248, and STN International, FIZ Karlsruhe, P.O. Box 2465, W-7500, Karlsruhe 1, Germany, online services. Also available on CD-ROM. Inquire as to cost and availability.

Current Contents Search. Institute for Scientific Information, 3501 Market Street, Philadelphia, PA 19104. (215) 386-0100. FAX (215) 386-6362. Contains citations to articles listed in the table of contents of science and technology journals. Also articles in social sciences and life sciences journals. Available on BRS, (800) 289-4277, DIALOG, (800) 334-2564, online services. Inquire as to cost and availability.

Dissertation Abstracts Online. University Microfilms International, 300 North Zeeb Road, Ann Arbor, MI 48106. (800) 521-0600 or (313) 761-4700. Scope includes virtually all doctoral dissertations accepted at accredited American institutions from 1861 to present in 252 subject areas. Available on BRS, (800) 289-4277, DIALOG, (800) 334-2564, and OCLC EPIC, (800) 848-5878, online services. Also available on CD-ROM. Inquire as to cost and availability.

INSPEC. Institution of Electrical Engineers, Michael Faraday House, Six Hills Way, Stevenage, Herts. SG1 2AY, England.

Telephone: 0438 313311 or FAX 0438 742840. Contains citations to the worldwide literature of physics, electronics and electrical engineering, computer technology, and related fields. Available on BRS, (800) 289-4277, DIALOG, (800) 334-2564, ORBIT, (800) 456-7248, and STN International, FIZ Karlsruhe, P.O. Box 2465, W-7500, Karlsruhe 1, Germany, online services. Inquire as to cost and availability.

NTIS Bibliographic Database. National Technical Information Service, 5285 Port Royal Road, Springfield, VA 22161. (703) 487-4929 or FAX (703) 321-8199. Broad coverage of government-sponsored science and technology research reports, 1964 to present. Available on BRS,(800) 289-4277, DIALOG, (800) 334-2564, ORBIT, (800) 456-7248, and STN International, FIZ Karlsruhe, P.O. Box 2465, W-7500, Karlsruhe 1, Germany, online services. Also available on CD-ROM. Inquire as to cost and availability.

Physics Briefs. American Institute of Physics, 335 East 45th Street, New York, NY 10017. (212) 661-9260 or FAX (212) 661-2036. Contains citations with abstracts of the literature of physics and related fields, 1979 to present. Available on the STN International, FIZ Karlsruhe, P.O. Box 2465, W-7500, Karlsruhe 1, Germany, online service. Inquire as to cost and availability.

Scientific and Technical Books and Serials in Print. R.R. Bowker Inc., 121 Chanlon Road, New Providence, NJ 07974. (800) 521-8110 or (908) 464-6800. List of currently published books and serials in the physical and biological sciences, engineering and technology, with subject, author and titles indexes. Available on ORBIT, (800) 456-7248, online service. Inquire as to cost and availability.

SCISEARCH. Institute for Scientific Information, 3501 Market Street, Philadelphia, PA 19104. (800) 523-1850 or (215) 386-0100. Broad multidisciplinary title and author index to the international literature of science and technology, 1974 to present. Available on DIALOG, (800) 334-2564, and ORBIT, (800) 456-7248, online services. Also available on CD-ROM. Inquire as to cost and availability.

WILSONLINE. H.W. Wilson Company, 950 University Avenue, Bronx, NY 10452. (800) 367-6770 or (212) 588-8400. Makes available online versions of the printed H.W. Wilson indexes including Applied Science and Technology Index, Business Periodicals Index, General Science Index, and Readers' Guide to Periodical Literature. Period covered is generally 1983 to present. Available on BRS, (800) 289-4277, DIALOG, (800) 334-2564, and OCLC EPIC, (800) 848-5878, online services. Also available on CD-ROM. Inquire as to cost and availability.

PERIODICALS

Acoustical Society of America. JOURNAL. American Institute of Physics, 500 Sunnyside Boulevard, Woodbury NY11797-2999. (519) 576-2270. 1929 to present. Monthly. $760.00 per year. ISSN: 0001-4966.

Applied Acoustics. Elsevier Science, 660 White Plains Road, Tarrytown, NY 10591-5153. (914) 524-9200. FAX (914) 333-2444. 1968 to present. Monthly. $770.00 per year. ISSN: 0003-682X.

Audio Engineering Society. JOURNAL. Audio Engineering Society, 60 East 42nd Street, New York, NY 10017. (212) 661-8528. 1953 to present. 10 issues per year. $125.00 per year. ISSN: 0004-7554.

DB, The Sound Engineering Magazine. 203 Commack Road, Suite 1010, Commack, NY 1725. (516) 586-6530. 1967 to present. $15.00 per year. ISSN: 0011-7145.

IEEE - Signal Processing Magazine. Institute of Electrical and Electronics Engineers, Inc., Box 1331, 445 Hoes Lane, Piscataway, NJ 08855-1331. (908) 981-0060. 1984 to present. Quarterly. $70.00 per year. ISSN: 1053-5888.

IEEE Transactions On Signal Processing. Institute of Electrical and Electronics Engineers, Inc., Box 1331, 445 Hoes Lane, Piscataway, NJ 08855-1331. (908) 981-0060. 1951 to present. Monthly. $400.00. ISSN: 1053-587X.

IEEE Transactions On Ultrasonics, Ferroelectrics and frequency Control. Institute of Electrical and Electronics Engineers, Inc., Box 1331, 445 Hoes Lane, Piscataway, NJ 08855-1331. (908) 981-0060. 1954 to present. Bi-monthly. $230.00 per year. ISSN: 0885-3010.

Journal of Low Frequency Noise and Vibration. Multi-Science Publishing Co. Ltd., Box 176 Avenel, NJ 070001. 1982 to present. Quarterly. L102 per year. ISSN: 0263-0923.

Journal of Sound and Vibration. Academic Press Ltd., 6277 Sea Harbor Drive, Orlando, FL. (800) 321-5068. 1964 to present. 50 issues per year. L1345 per year. ISSN: 0022-460X.

Ultrasonics. Butterworth-Heinemann, 313 Washington Street, Newton, MA 02158. (618-928-2500. (800) 366-2665. 1963 to present. Bimonthly. L245 per year. ISSN: 0041-624X

RESEARCH CENTERS AND INSTITUTES

Acoustics and Vibration Laboratory. Massachusetts Institute of Technology, Department of Mechanical Engineering, Building 3, Room 366, Cambridge, MA 02139. (617) 253-2214.

Center for Acoustics and Vibration. Florida Atlantic University, Department of Ocean Engineering, 500 NW 20th Street, Building 36, Boca Raton, FL 33431-0991. (407) 367-3430.

Center For Sound and Vibration. North Carolina State University. Campus Box 7914, Raleigh, NC 27695. (919) 737-3024. FAX (919) 737-7968.

National Center For Physical Acoustics. University of Mississippi, Coliseum Drive, University, MS 38677. (601) 232-5889. FAX (601) 232-7494.

Ray W. Herrick Laboratories. Purdue University. School of Mechanical Engineering, West Lafayette, in 47907. (317) 494-2132. FAX (317) 494-0787.

ACRYLIC RESINS

See: PLASTICS

ADA

See: COMPUTER PROGRAMMING LANGUAGES

ADHESIVES

Ency. of Physical Sciences and Engineering Info. Sources

ADHESIVES

See also: CEMENT, CHEMICAL ENGINEERING, CHEMISTRY,
ORGANIC CHEMISTRY, PLASTICS

ABSTRACT SERVICES AND INDEXES

Adhesives Abstracts. RAPRA Technology Ltd., Shawbury,
Shrewsbury, Shropshire SY4 4NR, England. TEL 0939-250383.
FAX 0939-251118. 1988 to present. Monthly. L195 per year.
ISSN: 0891-7760.

Applied Science and Technology Index; A Cumulative Subject
Index To English Language Periodicals in the Fields of Aeronautics and Space Science, Computer Technology, Chemistry,
Construction Industry, Energy and Related Areas. H. W. Wilson Co., 950 University Avenue, Bronx, NY 10452-9978. (212)
588-8400. (800) 367-6770. FAX (718) 590-1617. From 1958
to present. Monthly. Inquire about cost and availability. Also
available online (BRS and WILSONLINE) and on CD-ROM.
ISSN: 0003-6986.

Chemical Abstracts. Chemical Abstracts Service. 2540
Olentangy River Road, Box 3012, Columbus, OH 43210-0012.
(614) 447-3600. FAX (614) 447-3713. 1907 to present. Weekly.
$16,800.00 per year. Available online and on CD-ROM. CA is
also available in five section groupings. Inquire regarding cost
and availability.

*Current Contents: Engineering, Technology and Applied Sci-
ences.* Institute for Scientific Information, 3501 Market Street,
Philadelphia, PA 19104. (215) 386-0100. FAX (215) 386-2991.
From 1970 to present. Weekly. $442.00 per year. Also avail-
able online. Inquire regarding cost and availability. ISSN: 0095-
7917.

Engineering Index Monthly. Engineering Information, Inc.,
Castle Point on the Hudson, Hoboken, NJ 07030. (201) 216-
8500. (800) 221-1044. FAX (201) 216-8532. Monthly. $2300.00
per year for monthly issues. $1980 per year for annual cumula-
tion. Available on CD-ROM and online as COMPENDEX. ISSN:
0742-1974; 0360-8557.

General Science Index. H.W. Wilson Company, 950 University
Avenue, Bronx, NY 10452. (212) 588-8400. (800) 367-6770.
FAX (718) 590-1617. From 1978 to present. Ten issues per year;
quarterly and annual cumulations. Service basis. Available on
CD-ROM and online. Inquire regarding cost and availability.
ISSN: 0162-1963.

Institute of Paper Science and Technology Abstract Bulletin.
Institute of Paper Science and Technology, 500 10th Street, NW,
Atlanta, GA 30318. (404) 853-9500. 1930 to present. Monthly.
Price varies. ISSN: 1047-2088.

*Process and Chemical Engineering/chemical Engineering Ab-
stracts.* Royal Society of Chemistry, Information Services, Thomas Graham House, Science Park, Milton Road, Cambridge,
CB4 4WF, England. Contains citations, mostly with abstracts,
of the worldwide literature on chemical engineering; from 1982
to present.Monthly. $610.00 per year. Also available online.
ISSN: 0960-5045.

Science Citation Index. SCI. Institute for Scientific Information,
3501 Market Street, Philadelphia, PA 19104. (215) 386-0100.
(800) 523-1850. FAX (215) 386-2991. 1961 to present. Six issues per year, plus annual cumulation. $11650.00 per year. Also

available online and on CD-ROM. Inquire about price and availability. ISSN: 0036-827X.

Theoretical Chemical Engineering. Royal Society of Chemistry, Information Services, Thomas Graham House, Science Park,
Milton Road, Cambridge, CB4 4WF, England. Covers
theoreticalchemical engineering, including theory and laboratory experimentation. 1964 to present. Monthly. $235.00. ISSN:
0960-5053.

ASSOCIATIONS AND PROFESSIONAL SOCIETIES

Adhesive and Sealant Council. 1627 K Street NW, Suite 1000,
Washington, CD 20006-1707. (202) 452-1500. FAX (202) 452-
1501.

Adhesives Manufacturers Association. 401 North Michigan
Avenue, Chicago, IL 60611. (312) 644-6610. FAX (312) 321-
6869.

Adhesion Society. c/o F. James Boerio, University of Cincinnati, Department of Materials Science, Cincinnati, OH 45221-
0012. (203) 486-4629.

American Chemical Society, 1155 16th Street, Nw, Washington, DC 20036. (202) 872-4600.

American Institute of Chemical Engineers. 345 East 47th Street,
New York, NY 10017. (212) 705-7338.

Association of Consulting Chemists and Chemical Engineers,
50 East 41st Street, Suite 92, New York, NY 10017. (212) 684-
6255.

Chemical Manufacturers Association. 2501 M Street, N.W.,
Washington, DC 20037. (202) 887-1182.

Pressure Senstitive Tape Council. 401 North Michigan Avenue,
Chicago IL 60611-4267. (312) 644-6610. FAX (312) 527-6640.

BIBLIOGRAPHIES

Chemical Information Sources. Gary Wiggins. McGraw-Hill
Publishing Company, 1221 Avenue of the Americas, New York,
NY 10020. (800) 262-4729 or (212) 512-3825. 1991. $42.50.

Information Sources in Chemistry. R.T. Bottle and J.F.B
Rowland, editors. R.R. Bowker Inc., 121 Chanlon Road, New
Providence, NJ 07974. (800) 521-8110 or (908) 464-6800.
Fourth edition. 1993. $75.00.

*Scientific and Technical Books and Serials in Print; An Index
To Literature in Science and Technology.* R.R. Bowker Inc.,
121 Chanlon Road, New Providence, NJ 07974. (908) 464-6800.
(800) 521-8110. FAX (908) 665-3502. 1972 to present. Annual.
4 volumes. 1994. $299.95. Also available on compact disc and
online. ISSN: 0000-054X.

DIRECTORIES AND BIOGRAPHICAL SOURCES

Adhesives. International Plastics Selector, Inc., 9889 Willow
Creek Road, San Diego, CA 92126. (619) 578-3910. Annual.$100.00

Ency. of Physical Sciences and Engineering Info. Sources

ADHESIVES

Adhesives Age Directory. Argus Inc., 6151 Powers Perry Road, NW, Atlanta, GA 30339-2941. (404) 955-2500. FAX (404) 955-0400, 1957 to present. Annual. $51.95. ISSN: 0001-821X.

American Men and Women of Science. Physical and Biological Sciences. Fifteenth edition. R.R. Bowker Inc., 121 Chanlon Road, New Providence, NJ 07974. (908) 464-6800. (800) 521-8110. 20th edition. 8 volumes. 1996. $850.00 Directory of Chemistry Software 1992. Wendy Warr, Peter Willett, Geoff Downs. American Chemical Society, 1155 16th Street, NW, Washington, DC 20036. (202) 872-4600. 1992. $35.95.

Research Centers Directory. Gale Research, 835 Penobscot Building, Detroit, MI 48226-4094. (313) 961-2242. (800) 877-4253. 20th edition. 1995. $485.00. ISSN: 0080-1518.

Who's Who in Engineering. Gordon Davis, editor. American Association of Engineering Societies. 1111 19th Street, NY, Suite 608, Washington, DC 20036. (202) 296-2237. (800) 658-8897. 9th edition. 1995. $220.00.

ENCYCLOPEDIAS AND DICTIONARIES

Encyclopedia of Chemical Processing and Design. McKetta Marcel Dekker, Inc., 270 Madison Avenue, New York, NY 10016. (212) 696-9000. (800) 228-1160. 1976 - . $175.00 per volume. ISSN:

Industrial Chemical thesaurus. Michael Ash and Irene Ash. VCH Publishers, Inc., 220 East 23rd Street, New York, NY 10010-4606. (800) 367-8249. 1992. $295.00.

Kirk-Othmer Encyclopedia of Chemical Technology. John Wiley and Sons, Inc., 605 Third Avenue, New York, NY 10158. (800) 526-5368 or (212) 850-6000. Fourth edition. 1991 - . Twenty-seven volumes. $5400.00.

McGraw-Hill Encyclopedia of Science and Technology. McGraw-Hill Book, Incorporated, 1221 Avenue of the Americas, New York, NY 10020. (212) 997-3675. Seventeenth editon. 1992. $1900.00.

Ullman's Encyclopedia of Industrial Chemistry. VCH Publications, Inc., 220 East 23rd Street, Suite 909, New York, NY 10010-4606. (212) 683-8333. (800) 422-8824. 5th edition. 1984 - . Price varies per volume. ISSN:

GENERAL WORKS

Adhesive Bonding. L. H. Lee, editor. Plenum Publishing Corp., 233 Spring Street, New York, NY. (212) 620-8000. (800) 2221-9369. FAX (212) 463-0742. 1991. $115.00.

Adhesives and Adhesive Bonding: Theoretical and Practical Aspects. George Epstein, editor. T-C PR CA 1990. $59.00.

Adhesives and Sealants. American Society for Metals, Volume 3 of Engineered Materials Handbook. ASM International 9639 Kinsman Road, Materials Park, OH 44073. (216) 238-5151. 1990. $153.00.

Adhesives From Renewable Resources. Richard W. Hemmingway, et al, editors. American Chemical Society, 1155 16th Street, NW, Washington, DC 20036. (202) 872-4600. (800) 333-9511. FAX (614) 447-3671. 1989. $99.95.

Construction and Structural Adhesives and Sealants. Ernest W. Flick. Noyes Data Corporation, 120 Mill Road, Park Ridge, NJ 07656. (201) 391-8484. 1988. $78.00.

Industrial Applications of Adhesive Bonding. M. M. Sadek, editor. Elsevier Science Publishing Company, Inc., 655 Avenue of the Americas, New York, NY 10010. (212) 989-5800. FAX (914) 333-2444. 1987. $72.00.

Science and Technology of Building Seals, Sealants, Glazing and Waterproofing. Charles J. Parise, editor. American Society for Testing and Materials, 1916 Race Street, Philadelphia, PA 19103. (215) 299-5419. STP 1168. 1992. $61.00.

Structural Adhesives: Chemistry and Technology. S. R. Hartshorn, editor. Plenum Publishing Corp., 233 Spring Street, New York, NY. (212) 620-8000. (800) 2221-9369. FAX (212) 463-0742. 1986. $115.00.

Technology of Wood Bonding: Principles in Practice. Alan A. Marra. Chapman & Hall, 1 Penn Plaza, New York, NY 10119. (212) 564-1060. 1992. $59.95.

HANDBOOKS AND MANUALS

Adhesives Technology Handbook. Arthur H. Landrock, Noyes Data Corporation, 120 Mill Road, Park Ridge, NJ 07656. (201) 391-8484. 1986. $64.00.

Chemical Formulary. H. Bennett, editor. Chem Publishing Company, Inc., 80 Eighth Avenue, New York, NY 10011. (212) 255-1950. Volumes 1-30. $70.00 per volume.

Comprehensive Organic Synthesis: Selectivity, Strategy, and Efficiency in Modern Organic Chemical. Barry M. Trost and others, editors. Pergamon Press, Maxwell House, Fairview Park, Elmsford, NY 10523. (914) 592-7700. Fax (914) 592-3625. 1991. Nine volumes. $3900.00.

CRC Handbook of Chemistry and Physics. David R. Kide, editor. CRC Press Inc., 2000 Corporate Blvd., NW, Boca Raton, FL 33431. (407) 994-0555. (800) 333-8300. 77th edition. 1996. $99.95.

Guide To Basic Chemical Compounds. D.R. Lide, Jr. CRC Publishers, Inc., 2000 Corporate Blvd., N.W., Boca Raton, FL 33431. (407) 994-0555 or (800) 333-8300. 1993. $120.00.

Handbook of Adhesion. D. E. Packham, editor. Halsted Press, 605 Third Avenue, New York, 10158-0012. (212) 850-6400. 1993. $250.00.

Handbook of Adhesives. Irving Skeist. Chapman & Hall, One Penn Plaza, New York, NY 10119. (212) 564-1060. 1989. $105.00

Handbook of Adhesive Bonded Structural Repair. Raymond F. Wegman and Thomas R. Tullos. Noyes Data Corporation, 120 Mill Road, Park Ridge, NJ 07656. (201) 391-8484. 1992. $48.00.

Handbook of Adhesive Raw Materials. Ernest W. Flick. Noyes Data Corporation, 120 Mill Road, Park Ridge, NJ 07656. (201) 391-8484. 2nd edition. 1989. $69.00.

Handbook of Adhesive Technology. A. Pizzi, editor. Marcel Dekker, Inc., 270 Madison Avenue, New York, NY 10016. (212) 696-9000. (800) 228-1160. 1994. $195.00.

Handbook of Pressure-Sensitive Adhesive Technology. Satas. Chapman & Hall, 1 Penn Plaza, New York, NY 10119. (212) 564-1060. 2nd edition. 1989. $105.00.

Lange's Handbook of Chemistry. John A. Dean, editor. McGraw-Hill Publishing Company, 1221 Avenue of the Americas, New York, NY 10020. (212) 512-2000. (800) 262-4729. 14th edition. 1996. $99.95.

Perry's Chemical Engineers' Handbook. Robert H. Perry and Donald W. Green, editors. McGraw-Hill Publishing Company, Inc., 1221 Avenue of the Americas, New York, NY 10020. (212) 512-2000. (800) 262-4729. 6th edition. 1996. $129.50.

ONLINE DATABASES AND CD-ROMS

CA Search. Chemical Abstracts Service, P.O. Box 3012, Columbus, OH 43210-0012. (614) 447-3600. (800) 848-6533. FAX (614) 447-3709. Very comprehensive guide to worldwide chemical literature and related fields, 1972 to present. Available on BRS,(800) 289-4277, DIALOG, (800) 334-2564, ORBIT (800) 456-7248, and STN International, FIZ Karlsruhe, P.O. Box 2465, W-7500, Karlsruhe 1, Germany, online services. Inquire as to cost and availability.

Chemical Journals of the American Chemical Society. American Chemical Society, 1155 16th Street, NW, Washington, DC 20036. (202) 872-4381 or (800) 424-6747. Contains complete text of approximately 90,000 articles from 22 primary journals published by the American Chemical Society, from mostly 1982 to present. Available on STN International, FIZ Karlsruhe, P.O. Box 2465, W-7500, Karlsruhe 1, Germany, online service. Inquire as to cost and availability.

Compendex Plus. Engineering Information, Inc., 345 East 47th Street, New York, NY 10017. (212) 705-7600 or (800) 221-1044. Contains citations with abstracts to worldwide literature in engineering and technology, from 1970 to present. Available on online BRS,(800) 289-4277, DIALOG, (800) 334-2564, ORBIT (800) 456-7248, and STN International, FIZ Karlsruhe, P.O. Box 2465, W-7500, Karlsruhe 1, Germany, online services. Also available on CD-ROM. Inquire as to cost and availability.

Dissertation Abstracts. University Microfilms International, 300 North Zeeb Road, Ann Arbor, MI 48106. (800) 521-0600 or (313) 761-4700. Scope includes virtually all doctoral dissertations accepted at accredited American institutions from 1861 to present in 252 subject areas. Available on BRS, (800) 289-4277, DIALOG, (800) 334-2564, and OCLC EPIC, (800) 848-5878, online services. Also available on CD-ROM. Inquire as to cost and availability.

Kirk-Othmer Encyclopedia of Chemical Technology. John Wiley and Sons, Inc., 605 Third Avenue, New York, NY 10158. (800) 526-5368 or (212) 850-6000. Contains the complete text of all chapters in the 27 volume fourth edition of the *Kirk-othmer Encyclopedia of Chemical Technology.* 1991. Available on BRS,(800) 289-4277, DIALOG,(800) 334-2564, online services. Inquire as to cost and availability.

NTIS Bibliographic Database. National Technical Information Service, 5285 Port Royal Road, Springfield, VA 22161. (703) 487-4929 or FAX (703) 321-8199. Broad coverage of government-sponsored science and technology research reports, 1964

to present. Available on BRS,(800) 289-4277, DIALOG, (800) 334-2564, ORBIT, (800) 456-7248, and STN International, FIZ Karlsruhe, P.O. Box 2465, W-7500, Karlsruhe 1, Germany, online services. Also available on CD-ROM. Inquire as to cost and availability.

SCISEARCH. Institute for Scientific Information, 3501 Market Street, Philadelphia, PA 19104. (800) 523-1850 or (215) 386-0100. Broad multidisciplinary title and author index to the international literature of science and technology, 1974 to present. Available on DIALOG, (800) 334-2564, and ORBIT, (800) 456-7248, online services. Also available on CD-ROM. Inquire as to cost and availability.

WILSONLINE. H.W. Wilson Company, 950 University Avenue, Bronx, NY 10452. (800) 367-6770 or (212) 588-8400. Makes available online versions of the printed H.W. Wilson indexes including Applied Science and Technology Index, Business Periodicals Index, General Science Index, and Readers' Guide to Periodical Literature. Period covered is generally 1983 to present. Available on BRS, (800) 289-4277, DIALOG, (800) 334-2564, and OCLC EPIC, (800) 848-5878, online services. Also available on CD-ROM. Inquire as to cost and availability.

PERIODICALS

Adhesives Age. Argus Inc., 6151 Powers Perry Road, NW, Atlanta, GA 30339-2941. (404) 955-2500. FAX (404) 955-0400, 1957 to present. Monthly, plus annual directory. $52.00 per year. ISSN: 0001-821X.

Chemical Week. Includes Annual Buyers Guide. Chemical Week Associates, 888 Seventh Avenue, New York, NY 10106. (212) 621-4900. FAX (212) 621-4949. 1914 to present. Weekly. $99.00 per year ISSN: 0009-272X.

Chemtech. American Chemical Society. Box 3337, Columbus, OH 43210. (614) 447-3776. 1970 to present. Monthly. $370.00 per year. ISSN: 0009-2703.

International Journal of Adhesion and Adhesives. Butterworth-Heinemann. Subscriptions to: Turpin Transactions Ltd., Distributin Centre, Blackhorse Road, Letchworth, Herts SG6 1HN, England. TEL 0462-672555. 1980 to present. Quarterly. L180 per year. ISSN: 0143-7496.

Journal of Adhesion Science and Technology. VSP, P.O. Box 346, 3700 AH Zeist, Netherlands. TEL 31-3404-25790. FAX 31-3404-32081. 1987 to present. Monthly. DM 1280 per year. ISSN: 0169-4243.

RESEARCH CENTERS AND INSTITUTES

Center For Adhesion and Sealant Sciences. Virginia Institute for Material Systems, Virginia Polytechnic Institute and State University, 121 Patton Hall, Blacksburg, VA 24060. (703)961-4565.

Chemical Laboratories. Harvard University, Oxford Street, Cambridge, MA 02138. (617) 495-4283. FAX (617) 496-5618.

Chemistry Laboratories. Rensselaer Polytechnic Institute, Cogswell Laboratory, Troy, NY 12180-3590. (518) 276-8981.

Laboratory For Surface Science and Technology. University of Maine, Sawyer Research Center, Orono, ME 044469. (207) 581-2254. FAX (207) 581-2255.

University/industry Chemical Research Center. Mississippi State University, Department of Chemistry, P.O. Drawer CH, Mississippi State, MS 39762. (601) 325-3584.

ADIABATIC PROCESSES

See: THERMODYNAMICS

AERIAL PHOTOGRAPHY

See also: CARTOGRAPHY, PHOTOGRAMMETRY, REMOTE SENSING, SURVEYING

ABSTRACTS AND INDEXES

Applied Science and Technology Index; A Cumulative Subject Index To English Language Periodicals in the Fields of Aeronautics and Space Science, Computer Technology, Chemistry, Construction Industry, Energy and Related Areas. H.W. Wilson Co., 950 University Avenue, Bronx, NY 10452. (800) 367-6770 or (212) 588-8400. FAX (718) 590-1617. From 1958 to present. Monthly. Inquire about cost and availability. Also available on CD-ROM.

Bibliography and Index of Geology. American Geological Institute, 4220 King Street, Alexandria, VA 22302. (703) 379-2480. Fax (703) 379-7563. Monthly. $1295.00 per year. Also available online as GeoRef.

Civil and Structural Engineering Abstracts, Cambridge Scientific Abstracts, 7200 Wisconsin Avenue, Bethesda, MD 20814-4823. (301) 961-6750. FAX (301) 961-6720. Monthly. Topics covered include structural design, construction equipment and methods, civil defense and military engineering, surveying, highway engineering, maritime and port structures, materials, land reclamation, and soil mechanics.

Engineering Index Monthly. Engineering Information, Inc., Castle Point on the Hudson, Hoboken, NJ 07030. (800) 221-1044. FAX (212) 832-1857. Monthly. $2200.00 per year. Also available online as COMPENDEX, and also on CD-ROM.

Geological Abstracts. Elsevier - Geo Abstracts, Regency House, 34 Duke Street, Norwich NR3 3AP, England. Monthly. $760.00 per year. Also available online as GEOBASE.

Imaging Abstracts (formerly *Photographic Abstracts*), Pergamon Press Inc., Maxwell House, Fairview Park, Elmsford, NY 10523. (914) 592-7700. Fax (914) 592-3625. Bimonthly.

Meteorological and Geoastrophysical Abstracts. American Meteorological Society, c/o Inforonics Inc., 550 Newtown Road, Littleton, MA 01460. (508) 486-8976. FAX (508) 486-0027. 1950 to present. Monthly. $950.00.

Physics Abstracts. Institute of Electrical Engineers, Michael Faraday House, Six Hill Way, Stevenage, Herts, SG1 2AY, England. Distributed by IEEE, 445 Hoes Lane, Piscataway, NJ 08854.

(908) 562-5549. 1898 to present. Monthly. $2700.00 per year. Also available online as INSPEC.

Remote Sensing of Earth Resources. University of New Mexico Earth Data Analysis Center, 2500 Yale Blvd., Suite 100, Albuquerque, NM 87131-6031. (505) 277-3622. FAX (505) 277-3614. 1974 to present (1991 to present under current title). Quarterly. $175.00 per year.

Science Citation Index. Institute for Scientific Information, 3501 Market Street, Philadelphia, PA 19104. (215) 386-0100 or (800) 523-1850. FAX (215) 386-6362. Inquire about availability and cost. Also available on CD-ROM.

ASSOCIATIONS AND PROFESSIONAL SOCIETIES

American Society For Photogrammetry and Remote Sensing, 5410 Grosvenor Lane, Suite 210, Bethesda, MD 20814. (301) 493-0290. FAX (301) 493-0208.

The International Society For Optical Engineering (spie), Po Box 10, 1000-20th Street, Bellingham, WA 98227-0010.

Optical Society of America, 2010 Massachusetts Avenue Nw, Washington, DC 20036. (202) 223-8130.

The Society For Imaging Science and Technology, 7003 Kilworth Lane, Springfield, VA 22151. (703) 642-9090.

Society of Photographic Scientists and Engineers. 7003 Kilworth Lane, Springfield, VA 22151. (703) 642-9090.

BIBLIOGRAPHIES

A Selective Bibliography On Imagery Reconnaissance and Related Matters. Robert A. McDonald. 4th ed. Defense Intelligence College. Order from Superintendent of Documents, U.S. Government Printing office, Box 371954, Pittsburgh, PA 15250-7954. (202) 783-3238. FAX (202) 512-2233. 1993. Inquire for price and availability.

DIRECTORIES AND BIOGRAPHICAL SOURCES

Directory of Engineering Societies and Related Organizations. Gordon Davis. 13th edition. American Association of Engineering Societies, 1111 19th Street NW, Suite 608, Washington, DC 20036. (202) 296-2237 or (800) 658-8897. 1989. Inquire for price.

International Research Centers Directory. Gale Research, 835 Penobscot Building, Detroit, MI 48226-4094. (313) 961-2242. (800) 877-4253. FAX (313) 961-6083. 8th edition. 1995. $410.00

Research Centers Directory. Gale Research, 835 Penobscot Building, Detroit, MI 48226-4094. (313) 961-2242. (800) 877-4253. FAX (313) 961-6083. 20th edition. $485.00

Scientific and Technical Organizations and Agencies Directory. Gale Research, 835 Penobscot Building, Detroit, MI 48226-4094. (313) 961-2242. (800) 877-4253. FAX (313) 961-6083. 4th edition. 1996. $195.00

Who's Who in Engineering. American Association of Engineering Societies, 1111 19th Street NW, Suite 608, Washington, DC

AERIAL PHOTOGRAPHY

Ency. of Physical Sciences and Engineering Info. Sources

20036. (202) 296-2237 or (800) 658-8897. 8th edition. 1991. Inquire for price.

ENCYCLOPEDIAS AND DICTIONARIES

The Focal Encyclopedia of Photography. Leslie Stroebel & Richard D. Zakia, editors. 3rd edition. Focal Press, 80 Montvale Avenue, Stoneham, MA 02180. 1993. $125.00.

Thesaurus of Scientific, Technical, and Engineering Terms. Hemisphere Publishing Corporation, 1900 Frost Road, Suite 101, Bristol, PA 19007-1598. (215) 785-5800 or (800) 821-8312. FAX (215) 785-5515. 1987. $173.00.

GENERAL WORKS

Aerial Mapping: Methods and Applications. Edgar Falkner. Lewis Publishers/ CRC Press, 2000 Corporate Blvd., N.W., Boca Raton, FL 33431. (407) 994-0555 or (800) 272-7737. FAX (407) 994-0949. 1994. Inquire for cost.

Aerial Photography: Professional Techniques and Commercial Applications. Harvey Lloyd. Watson-Guptill, 1515 Broadway, New York, NY 10036. (212) 536-5121. (800) 451-1741. FAX (212) 536-5359. 1990. $22.50.

Fundamentals of Remote Sensing and Airphoto Interpretation. Thomas Eugene Avery and Graydon Lennis Berlin. 5th ed. Macmillan Publishing, 200 Old Tappan Road, Old Tappan, NJ 07675. (800) 223-2336. FAX (800) 445-6991. 1992. Inquire for price.

Introduction To Environmental Remote Sensing. E.C. Barrett and L.F. Curtis. 3d ed. Routledge, Chapman and Hall, Inc., 29 West 35th Street, New York, NY 10001-2291. (212) 244-3336 or FAX (212) 563-2267. 1992. $39.95.

A Practical Guide To Aerial Photography: With An Introduction To Surveying. John A. Ciciarelli. Van Nostrand Reinhold, 115 Fifth Avenue, New York, NY 10003. (212) 254-3232. FAX (212) 254-9499. 1991. $49.95.

The Sacred and Secular: A Decade of Aerial Photography. Marilyn Bridges. International Center of Photography, 1130 Fifth Ave., New York, NY 10028. (218) 860-1777. 1990. Inquire for price and availability.

HANDBOOKS AND MANUALS

Manual of Aerial Photography. Ron Graham & Roger E. Read. Focal Press, 80 Montvale Avenue, Stoneham, MA 02180. (617) 438-8464 or (800) 366-2665. 1987. $55.95.

Manual of Photogrammetry. Chester C. Slama, ed. 4th edition. American Society for Photogrammetry and Remote Sensing, 5410 Grosvenor Lane, Suite 210, Bethesda, MD 20814. (301) 493-0290. FAX (301) 493-0208. 1980. Inquire for price and availability.

Manual of Remote Sensing. Robert N. Colwell, ed. 2nd edition. American Society for Photogrammetry and Remote Sensing, 5410 Grosvenor Lane, Suite 210, Bethesda, MD 20814. (301) 493-0290. FAX (301) 493-0208. 1983. Inquire for price and availability.

ONLINE DATABASES AND CD-ROMS

Aerospace Database. American Institute of Aeronautics and Astronautics, 370 L'Enfant Promenade, S.W., Washington, DC 20024.(202) 646-7400. Worldwide published literature on research and development in aerospace and related areas, with abstracts. Covers 1962 to present. Available on DIALOG, (800) 334-2564, online service. Also available on CD-ROM. Inquire as to cost and availability.

Compendex Plus. Engineering Information, Inc., 345 East 47th Street, New York, NY 10017. (212) 705-7600 or (800) 221-1044. Contains citations with abstracts to worldwide literature in engineering and technology, from 1970 to present. Available on online BRS,(800) 289-4277, DIALOG, (800) 334-2564, ORBIT (800) 456-7248, and STN International, FIZ Karlsruhe, P.O. Box 2465, W-7500, Karlsruhe 1, Germany, online services. Also available on CD-ROM. Inquire as to cost and availability.

CSA Engineering. Cambridge Scientific Abstracts, 7200 Wisconsin Avenue, Suite 601, Bethesda, MD 20814. (301) 961-6750 or (800) 843-7751. Contains citations and abstracts of international periodicals and research literature covering all fields of engineering and science and technology,including computer and information science, electronics, mechanical engineering, solid state materials, 1981 to present. Available on BRS,(800) 289-4277, online service. Inquire as to cost and availability.

Current Contents Search. Institute for Scientific Information, 3501 Market Street, Philadelphia, PA 19104. (215) 386-0100. FAX (215) 386-6362. Contains citations to articles listed in the table of contents of science and technology journals. Also articles in social sciences and life sciences journals. Available on BRS,(800) 289-4277, DIALOG,(800) 334-2564, online services. Inquire as to cost and availability.

GEOREF: Bibliography and Index of Geology. American Geological Institute, 4220 King Street, Alexandria, VA 22302. (703) 379-2480. Fax (703) 379-7563. Monthly. Inquire for price and availability.

International Aerospace Abstracts. American Institute of Aeronautics and Astronautics, 370 L'Enfant Promenade, S.W., Washington, DC 20024. (202) 646-7400. Contains references and abstracts of journal and monograph literature relating to aerospace science and technology, from 1963 to present. Available through the NASA/RECON system of the National Aeronautics and Space Administration only.

NASA Database. American Institute of Aeronautics and Astronautics, 370 L'Enfant Promenade, S.W., Washington, DC 20024. (202) 646-7400. Citations and abstracts of aeronautics and astronautics literature, 1962 to present.

Aerospace Reports, and International Aerospace Abstracts. Available through NASA/RECON online service. Inquire as to cost and availability.

NTIS Bibliographic Database. National Technical Information Service, 5285 Port Royal Road, Springfield, VA 22161. (703) 487-4929 or FAX (703) 321-8199. Broad coverage of government-sponsored science and technology research reports, 1964to present. Available on BRS,(800) 289-4277, DIALOG, (800) 334-2564, ORBIT, (800) 456-7248, and STN International, FIZ Karlsruhe, P.O. Box 2465, W-7500, Karlsruhe 1, Germany, online services. Also available on CD-ROM. Inquire as to cost and availability.

SAE Global Mobility Database. Society of Automative Engineers (SAE), Electronic Publishing Division, 400 Commonwealth Drive, Warrendale, PA 15098. (412) 776-4841. Contains citations with abstracts to technical papers on automotive and aerospace technology and vehicular-related industries that have been presented at SAE conferences. Covers 1906 to present. Available on ORBIT,(800) 456-7248, online service. Inquire as to cost and availability.

WILSONLINE. H.W. Wilson Company, 950 University Avenue, Bronx, NY 10452. (800) 367-6770 or (212) 588-8400. Makes available online versions of the printed H.W. Wilson indexes including Applied Science and Technology Index, Business Periodicals Index, General Science Index, and Readers' Guide to Periodical Literature. Period covered is generally 1983 to present. Available on BRS,(800) 289-4277, DIALOG,(800) 334-2564, and OCLC EPIC,(800) 848-5878, online services. Also available on CD-ROM. Inquire as to cost and availability.

PERIODICALS

ISPRS Journal of Photogrammetry and Remote Sensing. Elsevier Science [journals], 660 White Plains Rd., Tarrytown, NY 10159-5153. (919) 524-9200. FAX (919) 333-2444. 1949 to present. Bi-monthly. $225.00 per year.

Photogrammetric Engineering and Remote Sensing. American Society for Photogrammetry and Remote Sensing, 5410 Grosvenor Lane, Suite 210, Bethesda, MD 20814. (301) 493-0290. FAX (301) 493-0208. 1934-present. Monthly. $160.00 per year.

Photogrammetric Record. Photogrammetric Society, Dept. of Photogrammetry & Surveying, University College London, Gower St., London WC1E 6BT England. Telephone 071-387-7050. FAX 071-380-0453. 1953-present. Semi-annual. $88.00 per year.

RESEARCH CENTERS AND INSTITUTES

Boston University Center For Remote Sensing and Cartography. 725 Commonwealth Avenue, Boston, MA 02215. (617) 353-5081. FAX (617) 353-3200.

Michigan State University Center For Remote Sensing. 302 Berkey Hall, East Lansing, MI 48824. (517) 353-7195. FAX (517) 353-1821.

University of Delaware Center For Remote Sensing. College of Marine Studies, Newark, DE 19711. (302) 451-2336.

University of Idaho, Idaho Remote Sensing Research Unit. Moscow, ID 83843. (208) 885-7209.

University of Minnesota Remote Sensing Laboratory. 1530 N. Cleveland Avenue, St. Paul, MN 55108. (612) 624-7764. FAX (612) 625-5212.

University of North Dakota, Institute For Remote Sensing.Geography Department, Grand Forks, ND 58202-8274. (701) 777-4246.

AERODYNAMICS

See also: AERONAUTICAL ENGINEERING, AERONAUTICS, AEROSPACE ENGINEERING

ABSTRACT SERVICES AND INDEXES

Aeronautical Engineering. Scientific and Technical Branch, National Aeronautics and Space Administration. National Technical Information Service (NTIS), 5285 Port Royal Road, Springfield, VA 22161. (703) 487-4650. AX (703) 321-8547. Monthly. A selection of annotated references to unclassified reports and journal articles that were introduced into the NASA scientific and technical information system and announced in STAR and IAA.

Alloys Index. American Society for Metals, Metals Park, OH 44073. (216) 338-5151. FAX (216) 338-4634. 1974 to present. Monthly. $295.00.

Applied Mechanics Reviews: An Assessment of World Literature in Engineering Sciences. American Society of Mechanical Engineers, 345 East 47th Street, New York, NY 10017. (212) 705-7703. 1948 to present. Monthly. $360.00 per year.

Applied Science and Technology Index; A Cumulative Subject Index To English Language Periodicals in the Fields of Aeronautics and Space Science, Computer Technology, Chemistry, Construction Industry, Energy and Related Areas. H.W. Wilson Co., 950 University Avenue, Bronx, NY 10452. (800) 367-6770 or (212) 588-8400. FAX (718) 590-1617. From 1958 to present. Monthly. Inquire about cost and availability. Also available on CD-ROM.

Current Contents: Engin, Tech, and Applied Sciences. Institute for Scientific Information, 3501 Market Street, Philadelphia, PA 19104. (215) 386-0100 or (800) 523-1850. FAX (215) 386-6362. Weekly. $360.00 per year.

Engineering Index Monthly. Engineering Information, Inc., Castle Point on the Hudson, Hoboken, NJ 07030. (800) 221-1044. FAX (212) 832-1857. Monthly. $2200.00 per year. Also available online as COMPENDEX, and also on CD-ROM.

International Aerospace Abstracts. Technical Information Service, American Institute of Aeronautics and Astronautics, Inc., 555 West 57th St., New York, NY 10019. (212) 247-6500. FAX (212) 582-4861. Semi-monthly. $1295.00 per year.

Metals Abstracts and Metals Abstracts Index. American Society for Metals, Metals Park, OH 44073. (216) 338-5151. 1968 to present. Monthly. Abstracts are $1500.00 per year and Index is $460.00 per year.

NASA Patent Abstracts Bibliography. National Aeronautics and Space Administration, Center for Aerospace Information, Box 8757, BWI Airport, Baltimore, MD 21240-0757. (410) 621-0153. Semi-annual. Inquire for price information.

Science Citation Index. Institute for Scientific Information, 3501 Market Street, Philadelphia, PA 19104. (215) 386-0100 or (800) 523-1850. FAX (215) 386-6362. Inquire about availability and cost. Also available on CD-ROM.

Scientific and Technical Aerospace Reports (star). National Aeronautics and Space Administration. National Aeronautics and

AERODYNAMICS

Ency. of Physical Sciences and Engineering Info. Sources

Space Administration, Center for Aerospace Information, Box 8757, BWI Airport, Baltimore, MD 21240-0757. (410) 621-0153. Monthly. Inquire about availability and cost. Also available through the NASA online retrieval service (RECON), and through the Aerospace Database through DIALOG.

World Aluminum Abstracts. Aluminum Association, Inc., 818 Connecticut Ave. NW, Washington, DC 20006. (202) 862-5156.

ASSOCIATIONS AND PROFESSIONAL SOCIETIES

Association of Balloon and Airship Constructors. Box 90864, San Diego, CA 92169. (619) 270-4049.

American Helicopter Society, Inc., 217 N. Washington St., Alexandria, VA 22314. (703) 684-6777. FAX (703) 739-9279.

American Institute of Aeronautics and Astronautics, 370 L'enfant Promenade SW, Washington, DC 20024-2518. (202) 646-7471.

Canadian Aeronautics and Space Institute, 222 Somerset Street West, Suite 601, Ottawa, Ontario, Canada K2p 0j1.

Flight Safety Foundation, Inc., 2200 Wilson Blvd., Suite 500, Arlington, VA 22201-3306. (703) 522-8300. FAX (703) 525-6047.

Society of Automotive Engineers. SAE, Inc., 400 Commonwealth Dr., Warrendale, PA 15096. (412) 776-4841.

BIBLIOGRAPHIES

Aeronautical Engineering: A Continuing Bibliography With. INDEXES. National Aeronautics and Space Administration, Center for Aerospace Information, Box 8757, BWI Airport, Baltimore, MD 21240. (301) 621-0153. Inquire for cost and availability.

Scientific and Technical Books and Serials in Print, R.R. Bowker Inc., 121 Chanlon Road, New Providence, NJ 07974. (800) 521-8110 or (908) 464-6800. FAX (908) 464-3553. 3 volumes, annually. $285.00. Also available on CD-ROM.

DIRECTORIES AND BIOGRAPHICAL SOURCES

Research Centers Directory. Gale Research, 835 Penobscot Building, Detroit, MI 48226-4094. (313) 961-2242. (800) 877-4253. FAX (313) 961-6083. 20th edition. $485.00

Who's Who in Engineering. American Association of Engineering Societies, 1111 19th Street NW, Suite 608, Washington, DC 20036. (202) 296-2237 or (800) 658-8897. 8th edition. 1991. Inquire for price.

World Aviation Directory. McGraw-Hill, Aviation Week Group (NY), 1221 Avenue of the Americas, New York, NY 10020. (215) 237-4112. FAX (215) 586-3232. Semiannual. $250.00.

ENCYCLOPEDIAS AND DICTIONARIES

Dictionary of Aeronautical Terms, Dale Crane, Comp. Aviation Supplies and Academics, Inc., 7005 132nd Place SE, Newcastle, WA 98059. (213) 656-7745. FAX (206) 235-0128. 1991. $15.95.

Encyclopedia of Physical Science and Technology. Robert A. Meyers, ed. Academic Press, Inc., 6277 Sea Harbor Drive, Orlando, FL 32887. (800) 321-5068. 1992. $2100.00.

Jane's Aerospace Dictionary, B. Guston. 3rd ed. Jane's Information Group Dept. DSM, 1340 Braddock Pl., Suite 300, Box 1436, Alexandria, VA 22314-1651. (703) 683-3700. FAX (703) 836-0029. 1991. $45.00.

McGraw-Hill Dictionary of Scientific and Technical Terms. Sybil P. Parker, ed. 5th edition. McGraw-Hill Publishing Company, 1221 Avenue of the Americas, New York, NY 10020. (800) 262-4729 or (212) 512-3825. 1993. $110.50.

McGraw-Hill Encyclopedia of Engineering. Sybil P. Parker, ed. 2nd edition. McGraw-Hill Publishing Company, 1221 Avenue of the Americas, New York, NY 10020. (800) 262-4729 or (212) 512-3825. 1993. $95.50.

McGraw-Hill Encyclopedia of Science and Technology. Sybil P. Parker, ed. 7th edition. 20 volumes. McGraw-Hill Publishing Company, 1221 Avenue of the Americas, New York, NY 10020. (800) 262-4729 or (212) 512-3825. 1992. $1900.00

GENERAL WORKS

Aerodynamics For Engineering Students. E.L. Houghton & P.W. Carpenter. Halsted Press (Division of John Wiley and Sons, Inc.), 605 Third Avenue, New York, NY 10158. (212) 850-6400. FAX (212) 850-6816. 1993. $49.95.

Aerodynamics For Engineers. J.J. Bertin & M.L. Smith. 2nd edition. Prentice Hall, Englewood Cliffs, NJ 07632. (201) 592-2000 or (800) 922-0579. 1989. $60.00.

Analysis of Aircraft Structures: An Introduction. Bruce K. Donaldson. McGraw-Hill Publishing Company, 1221 Avenue of the Americas, New York, NY 10020. (800) 262-4729 or (212) 512-3825. 1992. Inquire for cost and availability.

Fundamentals of Aerodynamics. John David anderson. 2nd edition. McGraw-Hill Publishing Company, 1221 Avenue of the Americas, New York, NY 10020. (800) 262-4729 or (212) 512-3825. 1991. Inquire for cost and availability.

Introduction To Flight. John David anderson. 3rd ed. McGraw-Hill Publishing Company, 1221 Avenue of the Americas, New York, NY 10020. (800) 262-4729 or (212) 512-3825. 1989. $46.63.

HANDBOOKS AND MANUALS

Standard Handbook of Engineering Calculations. Tyler G. Hicks, ed. 2d ed. McGraw-Hill Publishing Company, 1221 Avenue of the Americas, New York, NY 10020. (800) 262-4729 or (212) 512-3825. 1985. $89.50.

The Wiley Engineer's Desk Reference. Sanford I. Heisler. John Wiley and Sons, Inc., 605 Third Avenue, New York, NY 10158. (800) 526-5368 or (212) 850-6000. 1984. $64.95.

Jane's All the World's Aircraft. Janes Info Group (order from:) Dept. DSM, 1340 Braddock Pl., Suite 300, Box 1436, Alexandria, VA 22314-1651. (703) 683-3700. FAX (703) 836-0029. Annual. $245.00. Also available on CD-ROM.

Jane's Avionics. Janes Information Group, Dept. DSM, 1340 Braddock Pl., Suite 300, Box 1436, Alexandria, VA 22314-1651. (703) 683-3700. FAX (703) 836-0029. Annual. $245.00. Also available onCD-ROM.

ONLINE DATABASES AND CD-ROMS

Aerospace Database. American Institute of Aeronautics and Astronautics, 370 L'Enfant Promenade, S.W., Washington, DC 20024. (202) 646-7400. Worldwide published literature on research and development in aerospace and related areas, with abstracts. Covers 1962 to present. Available on DIALOG, (800) 334-2564, online service. Also available on CD-ROM. Inquire as to cost and availability.

Compendex Plus. Engineering Information, Inc., 345 East 47th Street, New York, NY 10017. (212) 705-7600 or (800) 221-1044. Contains citations with abstracts to worldwide literature in engineering and technology, from 1970 to present. Available on online BRS,(800) 289-4277, DIALOG, (800) 334-2564, ORBIT (800) 456-7248, and STN International, FIZ Karlsruhe, P.O. Box 2465, W-7500, Karlsruhe 1, Germany, online services. Also available on CD-ROM. Inquire as to cost and availability.

CSA Engineering. Cambridge Scientific Abstracts, 7200 Wisconsin Avenue, Suite 601, Bethesda, MD 20814. (301) 961-6750 or (800) 843-7751. Contains citations and abstracts of international periodicals and research literature covering all fields of engineering and science and technology,including computer and information science, electronics, mechanical engineering, solid state materials, 1981 to present. Available on BRS,(800) 289-4277, online service. Inquire as to cost and availability.

Current Contents Search. Institute for Scientific Information, 3501 Market Street, Philadelphia, PA 19104. (215) 386-0100. FAX (215) 386-6362. Contains citations to articles listed in the table of contents of science and technology journals. Also articles in social sciences and life sciences journals. Available on BRS,(800) 289-4277, DIALOG,(800) 334-2564, online services. Inquire as to cost and availability.

Dissertation Abstracts. University Microfilms International, 300 North Zeeb Road, Ann Arbor, MI 48106. (800) 521-0600 or (313) 761-4700. Scope includes virtually all doctoral dissertations accepted at accredited American institutions from 1861 to present in 252 subject areas. Available on BRS,(800) 289-4277, DIALOG,(800) 334-2564, and OCLC EPIC,(800) 848-5878, online services. Also available on CD-ROM. Inquire as to cost and availability.

International Aerospace Abstracts. American Institute of Aeronautics and Astronautics, 370 L'Enfant Promenade, S.W., Washington, DC 20024. (202) 646-7400. Contains references and abstracts of journal and monograph literature relating to aerospace science and technology, from 1963 to present. Available through the NASA/RECON system of the National Aeronautics and Space Administration only.

*ISMEC: Mechanical Engineering Abstracts.*Cambridge Scientific Abstracts, 7200 Wisconsin Avenue, Suite 601, Bethesda, MD 20814. (301) 961-6750 or (800) 843-7751. Contains citations tothe literature in mechanical engineering, industrial and production engineering, energy, power, mechanics, devices and related areas, from 1973 to present. Available on the DIALOG,(800) 334-2564, online service. Inquire as to cost and availability.

NASA Database. American Institute of Aeronautics and Astronautics, 370 L'Enfant Promenade, S.W., Washington, DC 20024. (202) 646-7400. Citations and abstracts of aeronautics and astronautics literature, 1962 to present. Also contains citations from *STAR, Scientific and Technical Aerospace Reports,* and *International Aerospace Abstracts.* Available through NASA/RECON online service. Inquire as to cost and availability.

NTIS Bibliographic Database. National Technical Information Service, 5285 Port Royal Road, Springfield, VA 22161. (703) 487-4929 or FAX (703) 321-8199. Broad coverage of government-sponsored science and technology research reports, 1964 to present. Available on BRS,(800) 289-4277, DIALOG, (800) 334-2564, ORBIT, (800) 456-7248, and STN International, FIZ Karlsruhe, P.O. Box 2465, W-7500, Karlsruhe 1, Germany, online services. Also available on CD-ROM. Inquire as to cost and availability.

SAE Global Mobility Database. Society of Automative Engineers (SAE), Electronic Publishing Division, 400 Commonwealth Drive, Warrendale, PA 15098. (412) 776-4841. Contains citations with abstracts to technical papers on automotive and aerospace technology and vehicular-related industries that have been presented at SAE conferences. Covers 1906 to present. Available on ORBIT,(800) 456-7248,online service. Inquire as to cost and availability.

Scientific and Technical Books and Serials in Print. R.R. Bowker Inc., 121 Chanlon Road, New Providence, NJ 07974. (800) 521-8110 or (908) 464-6800. List of currently published books and serials in the physical and biological sciences, engineering and technology, with subject, author and titles indexes. Available on ORBIT,(800) 456-7248, online service.Inquire as to cost and availability.

SCISEARCH. Institute for Scientific Information, 3501 Market Street, Philadelphia, PA 19104. (800) 523-1850 or (215) 386-0100. Broad multidisciplinary title and author index to the international literature of science and technology, 1974 to present. Available on DIALOG,(800) 334-2564, and ORBIT,(800) 456-7248, online services. Also available on CD-ROM. Inquire as to cost and availability.

TRIS (Transportation Research Information). National Academy of Science, 2101 Constitution Avenue, N.W., Washington, DC 20418. (202) 334-3313 or (800) 624-6242. Citations with abstracts of literature on transportation, including air, highway, rail, maritime and other modes from 1968 to present. Available on DIALOG,(800) 334-2564, online service. Inquire as to cost and availability.

WILSONLINE. H.W. Wilson Company, 950 University Avenue,Bronx, NY 10452. (800) 367-6770 or (212) 588-8400. Makes available online versions of the printed H.W. Wilson indexes including Applied Science and Technology Index, Business Periodicals Index, General Science Index, and Readers' Guide to Periodical Literature. Period covered is generally 1983 to present. Available on BRS,(800) 289-4277, DIALOG,(800) 334-2564, and OCLC EPIC,(800) 848-5878, online services. Also available on CD-ROM. Inquire as to cost and availability.

PERIODICALS

Aeronautical Journal. The Royal Aeronautical Society, 4 Hamilton Place, London W1V 0BQ, England. 1963 to present. Monthly except June and August. Inquire for price information in U.S.

AERODYNAMICS

Ency. of Physical Sciences and Engineering Info. Sources

Aerospace America. American Institute of Aeronautics and Astronautics, 370 L'Enfant Promenade SW, Washington, DC 20024-2518. (202) 646-7471. 1932 to present. Monthly. $70 annually (non-members).

Aerospace Engineering. Society of Automotive Engineers, 400 Commonwealth Dr., Warrendale, PA 15096. (412) 776-4841. 1981 to present. Monthly. $48.00 per year.

Aerospace Propulsion. McGraw-Hill Inc., Aviation Week Group, 1200 G Street NW, Suite 200, Washington, DC 20005. (202) 383-2350. 1963 to present. 25 times a year. $550.00 a year.

Aerostation. Association of Balloon and Airship Constructors. Box 90864, San Diego, CA 92169. (619) 270-4049. 1974 to present. Quarterly. Inquire for price and availability.

AIAA Journal. American Institute of Aeronautics and Astronautics, 370 L'Enfant Promenade SW, Washington, DC 20024-2518. (202) 646-7471. 1963 to present. Monthly. $52.00 per year (members) and $435.00 per year (non-members).

Aircraft Engineering and Aerospace Technology. BunHill Publications Ltd., 127 Stanstead Rd., London SE23 1JE, England. 1929 to present. Monthly. $102.00 per year (U.S.).

Canadian Aeronautics and Space Journal. Canadian Aeronautics and Space Institute, Suite 818, the National Building, 130 Slater St., Ottawa, Canada K1P 6E2. 1955 to present. Monthly. $75.00 per year (U.S.).

IEEE Transactions On Aerospace and Electronic Systems. Institute of Electrical and Electronics Engineers, 445 Hoes Lane, PO Box 1331, Piscataway, NJ 08855-1331. (908) 981-0060. 1965 to present. Quarterly. $10.00 per issue (members), $20.00 per issue (non-members).

International Journal of Turbo and Jet Engines. Freund Publishing House, Ltd., PO Box 35010, Tel Aviv 63150, Israel. 972-2-5662925. FX 972-3-5605335. 1983 to present. Quarterly. $190.00 per year.

Journal of Aircraft. American Institute of Aeronautics and Astronautics, 370 L'Enfant Promenade SW, Washington, DC 20024-2518. 1963 to present. Bimonthly. $42.00 (members) and $230.00 (non-members).

Journal of Guidance, Control and Dynamics. AmericanInstitute of Aeronautics and Astronautics, 370 L'Enfant Promenade SW, Washington, DC 20024-2518. 1978 to present. Bimonthly. $42.00 (members) and $250.00 (non-members).

Journal of Propulsion and Power. American Institute of Aeronautics and Astronautics, 370 L'Enfant Promenade SW, Washington, DC 20024-2518. 1985 to present. Bimonthly. $38.00 (members) and $260.00 (non-members).

Journal of the American Helicopter Society. American Helicopter Society, Inc., 217 North Washington St., Alexandria, VA 22314. (703) 684-6777. 1956 to present. Quarterly. $35.00 per year.

Progress in Aerospace Sciences. Pergamon Press, 660 White Plains Rd., Tarrytown, NY 10591-5153. 1961 to present. Quarterly. $320.00 per year.

Vertiflite. American Helicopter Society, 217 N. Washington St., Alexandria, VA 22314. (703) 684-6777. FAX (703) 739-9279. 1955 to present. Bimonthly. $25 per year (members) and $35 per year (non-members).

RESEARCH CENTERS AND INSTITUTES

Aerospace Research Applications Center, Indianapolis Center For Advanced Research, 611 N. Capitol Ave., Indianapolis, in 46204. (317) 262-5003. FAX (317) 262-5044.

Aerospace Research Center, 1250 Eye St., NW, Washington, DC 20005. (202) 371-8400. FAX (202) 371-8470.

Aviation Safety Institute, 6797 N. High St., Suite 316, PO Box 304, Worthington, OH 43085. (614) 885-4242. FAX (614) 885-5891.

Center For Aerospace Technology, Weber State College, Ogden, UT 84408-1805. (801) 626-7272.

Center For Control and Systems Research, University of Texas At Austin, Aerospace Engineering and Engineering Mechanics, Wrw 408, Austin, TX 78712-1085. (512) 471-4908. FAX (512) 471-3788.

Flight Research Laboratory, University of Kansas, Raymond Nichols Hall, Lawrence, KS 66045. (913) 864-3043. FAX (913) 864-7789.

Joint Institute For Advancement of Flight Sciences, George Washington University, Langley Research Center, Mail Stop 269, Hampton, VA 23665. (804) 864-1982. FAX (804) 864-5894.

Machinery and Engine Technology Laboratory, Institute For Mechanical Engineering, Bldg. M-7, Montreal Rd., Ottawa, Ontario, Canada K1A 0R6. (613) 993-2425. (613) 957-3281.

Ohio Aerospace Institute, 2001 Aerospace Parkway, Brook Park, OH 44142. (216) 891-2100. FAX (216) 891-2140.

Transportation Research and Education Center, Georgia Institute of Technology, Mason Civil Engineering Bldg., Atlanta, GA 30332. (404) 894-2236. FAX (404) 894-2278.

AERONAUTICAL ENGINEERING

See also: AERODYNAMICS, AERONAUTICS, AEROSPACE ENGINEERING

ABSTRACT SERVICES AND INDEXES

Aeronautical Engineering. Scientific and Technical Branch, National Aeronautics and Space Administration. NationalTechnical Information Service (NTIS), Springfield, VA 22161. (703) 487-4650. FAX (703) 321-8547. Monthly. A selection of annotated references to unclassified reports and journal articles that were introduced into the NASA scientific and technical information system and announced in STAR and IAA.

Applied Mechanics Reviews: An Assessment of World Literature in Engineering Sciences. American Society of Mechanical

Engineers, 345 East 47th Street, New York, NY 10017. (212) 705-7703. 1948 to present. Monthly. $360.00 per year.

Applied Science and Technology Index; A Cumulative Subject Index To English Language Periodicals in the Fields of Aeronautics and Space Science, Computer Technology, Chemistry, Construction Industry, Energy and Related Areas. H.W. Wilson Co., 950 University Avenue, Bronx, NY 10452. (800) 367-6770 or (212) 588-8400. FAX (718) 590-1617. From 1958 to present. Monthly. Inquire about cost and availability. Also available on CD-ROM.

Current Contents: Engineering, Technology, and Applied Sciences. Institute for Scientific Information, 3501 Market Street, Philadelphia, PA 19104. (215) 386-0100 or (800) 523-1850. FAX (215) 386-6362. Weekly. $360.00 per year.

Engineering Index Monthly. Engineering Information, Inc., Castle Point on the Hudson, Hoboken, NJ 07030. (800) 221-1044. FAX (212) 832-1857. Monthly. $2200.00 per year. Also available online as *COMPENDEX*, and also on CD-ROM.

International Aerospace Abstracts. Technical Information Service, American Institute of Aeronautics and Astronautics, Inc., 555 West 57th St., New York, NY 10019. (212) 247-6500. FAX (212) 582-4861. Semi-monthly. $1295.00 per year.

Metals Abstracts and Metals Abstracts Index. American Society for Metals, Metals Park, OH 44073. (216) 338-5151. 1968 to present. Monthly. Abstracts are $1500.00 per year and Index is $460.00 per year.

NASA Patent Abstracts Bibliography. National Aeronautics and Space Administration, Center for Aerospace Information, Box 8757, BWI Airport, Baltimore, MD 21240-0757. (301) 621-0153. Semi-annual. Inquire for price information.

Science Citation Index. Institute for Scientific Information, 3501 Market Street, Philadelphia, PA 19104. (215) 386-0100 or (800) 523-1850. FAX (215) 386-6362. Inquire about availability and cost. Also available on CD-ROM.

Scientific and Technical Aerospace Reports (star). National Aeronautics and Space Administration, Center for Aerospace Information, Box 8757, BWI Airport, Baltimore, MD 21240-0757. (301) 621-0153. Monthly. Inquire about availability and cost. Also available through the NASA online retrieval service (RECON), and through the Aerospace Database through DIALOG.

ASSOCIATIONS AND PROFESSIONAL SOCIETIES

Association of Balloon and Airship Constructors. Box 90864, San Diego, CA 92169. (619) 270-4049.

American Helicopter Society, Inc., 217 N. Washington St.,Alexandria, VA 22314. (703) 684-6777. FAX (703) 739-9279.

American Institute of Aeronautics and Astronautics, 370 L'enfant Promenade SW , Washington, DC 20024-2518. (202) 646-7471.

Canadian Aeronautics and Space Institute, 222 Somerset Street West, Suite 601, Ottawa, Ontario, Canada K2p 0j1.

Flight Safety Foundation, Inc., 2200 Wilson Blvd., Suite 500, Arlington, VA 22201-3306. (703) 522-8300. FAX (703) 525-

6047.

Society of Automotive Engineers. SAE, Inc., 400 Commonwealth Dr., Warrendale, PA 15096. (412) 776-4841.

BIBLIOGRAPHIES

Aeronautical Engineering: A Continuing Bibliography With. INDEXES. National Aeronautics and Space Administration, Center for Aerospace Information, Box 8757, BWI Airport, Baltimore, MD 21240. (301) 621-0153. Inquire for cost and availability.

Scientific and Technical Books and Serials in Print, R.R. Bowker Inc., 121 Chanlon Road, New Providence, NJ 07974. (800) 521-8110 or (908) 464-6800. FAX (908) 464-3553. 3 volumes, annually. $285.00. Also available on CD-ROM.

DIRECTORIES AND BIOGRAPHICAL SOURCES

Jane's Airport and Atc Equipment, Jane's Information Group, Dept. DSM, 1340 Braddock Pl., Suite 300, Box 1436, Alexandria, VA 22314-1651. (703) 683-3700. FAX (703) 836-0029. Annual. $225.00.

Research Centers Directory. Gale Research, 835 Penobscot Building, Detroit, MI 48226-4094. (313) 961-2242. (800) 877-4253. FAX (313) 961-6083. 20th edition. $485.00

Who's Who in Engineering. American Association of Engineering Societies, 1111 19th St. NW, Suite 608, Washington, DC 20036. 7th ed. 1988. $200.00.

World Aviation Directory. McGraw-Hill, Aviation Week Group (NY), 1221 Avenue of the Americas, New York, NY 10020. (215) 237-4112. FAX (215) 586-3232. Semiannual. $250.00.

ENCYCLOPEDIAS AND DICTIONARIES

Concise Encyclopedia of Aeronautics and Space Systems. M. Pelegrin and W.M. Hollister. Pergamon Press Inc., Maxwell House, Fairview Park, Elmsford, NY 10523. (914) 592-7700. Fax (914) 592-3625. 1993. Inquire for cost.

Dictionary of Aeronautical Terms, Dale Crane, Comp. Aviation Supplies and ACADemics, Inc., 7005 132nd Place SE, Newcastle, WA 98059. (206) 235-1500 or (800) 426-8338. FAX (206) 235-0128. 1991. $15.95.

Encyclopedia of Physical Science and Technology. Robert A. Meyers, ed. Academic Press Inc., 6277 Sea Harbor Drive, Orlando, FL 32887. 1992. $2100.00.

Jane's Aerospace Dictionary, B. Guston. 3rd ed. Jane's Information Group, Dept. DSM, 1340 Braddock Pl., Suite 300, Box 1436, Alexandria, VA 22314-1651. (703) 683-3700. FAX (703) 836-0029. 1991. $45.00.

McGraw-Hill Encyclopedia of Science and Technology, Sybil P. Parker, ed. 7th ed. 20 vols. McGraw-Hill Publishing Company,1221 Avenue of the Americas, New York, NY 10020. (800) 262-4729 or (212) 512-3825. $1900.00

AERONAUTICAL ENGINEERING

Ency. of Physical Sciences and Engineering Info. Sources

SAE Dictionary of Aerospace Engineering. William H. Cubberly, ed. Society of Automotive Engineers, 400 Commonwealth Dr., Warrendale, PA 15096. (412) 776-4841. 1992. $85.00.

GENERAL WORKS

Airplane Design. Jan Roskam. Roskam Aviation and Engineering Corporation, Ottawa, KS 66067. 1989. Write for cost and availability.

Analysis of Aircraft Structures: An Introduction. Bruce K. Donaldson. McGraw-Hill Publishing Company, 1221 Avenue of the Americas, New York, NY 10020. (800) 262-4729 or (212) 512-3825. 1993. Inquire for cost and availability.

Composite Aircraft Design. Martin Hollmann. M. Hollman, Monterey, CA 94940. 1993. Write for cost and availability.

HANDBOOKS AND MANUALS

Standard Handbook of Engineering Calculations. Tyler G. Hicks, ed. 2d ed. McGraw-Hill Publishing Company, 1221 Avenue of the Americas, New York, NY 10020. (800) 262-4729 or (212) 512-3825. 1985. $89.50.

The Wiley Engineer's Desk Reference. Sanford I. Heisler. John Wiley and Sons, Inc., 605 Third Avenue, New York, NY 10158. (800) 526-5368 or (212) 850-6000. 1984. $64.95.

Jane's All the World's Aircraft. Jane's Information Group, Dept. DSM, 1340 Braddock Pl., Suite 300, Box 1436, Alexandria, VA 22314-1651. (703) 683-3700. FAX (703) 836-0029. Annual. $245.00. Also available on CD-ROM.

Jane's Avionics. Jane's Information Group, Dept. DSM, 1340 Braddock Pl., Suite 300, Box 1436, Alexandria, VA 22314-1651. (703) 683-3700. FAX (703) 836-0029. Annual. $245.00. Also available on CD-ROM.

ONLINE DATABASES AND CD-ROMS

Aerospace Database. American Institute of Aeronautics and Astronautics, 370 L'Enfant Promenade, S.W., Washington, DC 20024. (202) 646-7400. Worldwide published literature on research and development in aerospace and related areas, with abstracts. Covers 1962 to present. Available on DIALOG, (800) 334-2564, online service. Also available on CD-ROM. Inquire as to cost and availability.

Compendex Plus. Engineering Information, Inc., 345 East 47th Street, New York, NY 10017. (212) 705-7600 or (800) 221-1044. Contains citations with abstracts to worldwide literature in engineering and technology, from 1970 to present. Available on online BRS,(800) 289-4277, DIALOG, (800) 334-2564, ORBIT (800) 456-7248, and STN International, FIZ Karlsruhe, P.O. Box 2465, W-7500, Karlsruhe 1, Germany, online services. Also available on CD-ROM. Inquire as to cost and availability.

CSA Engineering. Cambridge Scientific Abstracts, 7200 Wisconsin Avenue, Suite 601, Bethesda, MD 20814. (301) 961-6750 or (800) 843-7751. Contains citations and abstracts of internationalperiodicals and research literature covering all fields of engineering and science and technology,including computer and information science, electronics, mechanical engineering,

solid state materials, 1981 to present. Available on BRS,(800) 289-4277, online service.Inquire as to cost and availability.

Current Contents Search. Institute for Scientific Information, 3501 Market Street, Philadelphia, PA 19104. (215) 386-0100. FAX (215) 386-6362. Contains citations to articles listed in the table of contents of science and technology journals. Also articles in social sciences and life sciences journals. Available on BRS,(800) 289-4277, DIALOG,(800) 334-2564, online services. Inquire as to cost and availability.

Dissertation Abstracts. University Microfilms International, 300 North Zeeb Road, Ann Arbor, MI 48106. (800) 521-0600 or (313) 761-4700. Scope includes virtually all doctoral dissertations accepted at accredited American institutions from 1861 to present in 252 subject areas. Available on BRS,(800) 289-4277, DIALOG,(800) 334-2564, and OCLC EPIC,(800) 848-5878, online services. Also available on CD-ROM. Inquire as to cost and availability.

International Aerospace Abstracts. American Institute of Aeronautics and Astronautics, 370 L'Enfant Promenade, S.W., Washington, DC 20024. (202) 646-7400. Contains references and abstracts of journal and monograph literature relating to aerospace science and technology, from 1963 to present. Available through the NASA/RECON system of the National Aeronautics and Space Administration only.

ISMEC: Mechanical Engineering Abstracts. Cambridge Scientific Abstracts, 7200 Wisconsin Avenue, Suite 601, Bethesda, MD 20814. (301) 961-6750 or (800) 843-7751. Contains citations to the literature in mechanical engineering, industrial and production engineering, energy, power, mechanics, devices and related areas, from 1973 to present. Available on the DIALOG,(800) 334-2564, online service. Inquire as to cost and availability.

NASA Database. American Institute of Aeronautics and Astronautics, 370 L'Enfant Promenade, S.W., Washington, DC 20024. (202) 646-7400. Citations and abstracts of aeronautics and astronautics literature, 1962 to present. Also contains citations from STAR, SCIENTIFIC and TECHNICAL AEROSPACE REPORTS, and INTERNATIONAL AEROSPACE ABSTRACTS. Available through NASA/RECON online service. Inquire as to cost and availability.

NTIS Bibliographic Database. National Technical Information Service, 5285 Port Royal Road, Springfield, VA 22161. (703) 487-4929 or FAX (703) 321-8199. Broad coverage of government-sponsored science and technology research reports, 1964 to present. Available on BRS,(800) 289-4277, DIALOG, (800) 334-2564, ORBIT, (800) 456-7248, and STN International, FIZ Karlsruhe, P.O. Box 2465, W-7500, Karlsruhe 1, Germany, online services. Also available on CD-ROM. Inquire as to cost and availability.

SAE Global Mobility Database. Society of Automative Engineers (SAE), Electronic Publishing Division, 400 Commonwealth Drive, Warrendale, PA 15098. (412) 776-4841. Contains citations with abstracts to technical papers on automotive and aerospace technology and vehicular-related industries that have been presented at SAE conferences. Covers 1906 to present. Available on ORBIT,(800) 456-7248,online service. Inquire as to cost and availability.

Scientific and Technical Books and Serials in Print. R.R. Bowker Inc., 121 Chanlon Road, New Providence, NJ 07974. (800) 521-8110 or (908) 464-6800. List of currently published books and serials in the physical and biological sciences, engineering and technology, with subject, author and titles indexes. Available on ORBIT,(800) 456-7248, online service.Inquire as to cost and availability.

SCISEARCH. Institute for Scientific Information, 3501 Market Street, Philadelphia, PA 19104. (800) 523-1850 or (215) 386-0100. Broad multidisciplinary title and author index to the international literature of science and technology, 1974 to present. Available on DIALOG,(800) 334-2564, and ORBIT,(800) 456-7248, online services. Also available on CD-ROM. Inquire as to cost and availability.

TRIS (Transportation Research Information). National Academy of Science, 2101 Constitution Avenue, N.W., Washington, DC 20418. (202) 334-3313 or (800) 624-6242. Citations with abstracts of literature on transportation, including air, highway, rail, maritime and other modes from 1968 to present. Available on DIALOG,(800) 334-2564, online service. Inquire as to cost and availability.

WILSONLINE. H.W. Wilson Company, 950 University Avenue, Bronx, NY 10452. (800) 367-6770 or (212) 588-8400. Makes available online versions of the printed H.W. Wilson indexes including Applied Science and Technology Index, Business Periodicals Index, General Science Index, and Readers' Guide to Periodical Literature. Period covered is generally 1983 to present. Available on BRS,(800) 289-4277, DIALOG,(800) 334-2564, and OCLC EPIC,(800) 848-5878, online services. Also available on CD-ROM. Inquire as to cost and availability.

PERIODICALS

Aeronautical Journal. The Royal Aeronautical Society, 4 Hamilton Place, London W1V 0BQ, England. 1963 to present. Monthly except June and August. Inquire for price information in U.S.

Aerospace America. American Institute of Aeronautics and Astronautics, 370 L'Enfant Promenade SW, Washington, DC 20024-2518. (202) 646-7471. 1932 to present. Monthly. $70 annually (non-members).

Aerospace Engineering. Society of Automotive Engineers, 400 Commonwealth Dr., Warrendale, PA 15096. (412) 776-4841. 1981 to present. Monthly. $48.00 per year.

Aerospace Propulsion. McGraw-Hill Inc., Aviation WeekGroup, 1200 G Street NW, Suite 200, Washington, DC 20005. (202) 383-2350. 1963 to present. 25 times a year. $550.00 a year.

Aerostation. Association of Balloon and Airship Constructors. Box 90864, San Diego, CA 92169. (619) 270-4049. 1974 to present. Quarterly. Inquire for price and availability.

AIAA Journal. American Institute of Aeronautics and Astronautics, 370 L'Enfant Promenade SW, Washington, DC 20024-2518. (202) 646-7471. 1963 to present. Monthly. $52.00 per year (members) and $435.00 per year (non-members).

Aircraft Engineering and Aerospace Technology. BunHill Publications Ltd., 127 Stanstead Rd., London SE23 1JE, England. 1929 to present. Monthly. $102.00 per year (U.S.).

Aviation Mechanics Bulletin. Flight Safety Foundation, Inc., 2200 Wilson Blvd., Suite 500, Arlington, VA 22201-3306. (703) 522-8300. FAX (703) 525-6047. 1953 to present. Bimonthly. $15.00 per year (non-members).

Aviation Week and Space Technology. 1221 Avenue of the Americas, New York, NY 10020. (212) 512-2999 or (800) 525-5003. 1916 to present. Weekly. $82.00 per year.

Canadian Aeronautics and Space Journal. Canadian Aeronautics and Space Institute, Suite 818, the National Building, 130 Slater St., Ottawa, Canada K1P 6E2. 1955 to present. Monthly. US$75.00 per year.

IEEE Transactions On Aerospace and Electronic Systems. Institute of Electrical and Electronics Engineers, 445 Hoes Lane, PO Box 1331, Piscataway, NJ 08855-1331. (908) 981-0060. 1965 to present. Quarterly. $10.00 per issue (members), $20.00 per issue (non-members).

Journal of Aircraft. American Institute of Aeronautics and Astronautics, 370 L'Enfant Promenade SW, Washington, DC 20024-2518. 1963 to present. Bimonthly. $42.00 (members) and $230.00 (non-members).

Journal of Guidance, Control and Dynamics. American Institute of Aeronautics and Astronautics, 370 L'Enfant Promenade SW, Washington, DC 20024-2518. 1978 to present. Bimonthly. $42.00 (members) and $250.00 (non-members).

Journal of Propulsion and Power. American Institute of Aeronautics and Astronautics, 370 L'Enfant Promenade SW, Washington, DC 20024-2518. 1985 to present. Bimonthly. $38.00 (members) and $260.00 (non-members).

Journal of the American Helicopter Society. American Helicopter Society, Inc., 217 North Washington St., Alexandria, VA 22314. (703) 684-6777. 1956 to present. Quarterly. $35.00 per year.

Progress in Aerospace Sciences. Pergamon Press, 660 White Plains Rd., Tarrytown, NY 10591-5153. 1961 to present. Quarterly. $320.00 per year.

Vertiflite. American Helicopter Society, 217 N. Washington St., Alexandria, VA 22314. (703) 684-6777. FAX (703) 739-9279. 1955 to present. Bimonthly. $25 per year (members) and $35 per year (non-members).

RESEARCH CENTERS AND INSTITUTES

Aerospace Research Applications Center, Indianapolis Center For Advanced Research, 611 N. Capitol Ave., Indianapolis, in 46204. (317) 262-5003. FAX (317) 262-5044.

Aerospace Research Center, 1250 Eye St., NW, Washington, DC 20005. (202) 371-8400. FAX (202) 371-8470.

Center For Aerospace Technology, Weber State College, Ogden, UT 84408-1805. (801) 626-7272.

Center For Control and Systems Research, University of Texas At Austin, Aerospace Engineering and Engineering Mechanics, Wrw 408, Austin, TX 78712-1085. (512) 471-4908. FAX (512) 471-3788.

Flight Research Laboratory, University of Kansas, Raymond Nichols Hall, Lawrence, KS 66045. (913) 864-3043. FAX (913) 864-7789.

AERONAUTICAL ENGINEERING

Ency. of Physical Sciences and Engineering Info. Sources

Joint Institute For Advancement of Flight Sciences, George Washington University, Langley Research Center, Mail Stop 269, Hampton, VA 23665. (804) 864-1982. FAX (804) 864-5894.

Machinery and Engine Technology Laboratory, Institute For Mechanical Engineering, Bldg. M-7, Montreal Rd., Ottawa, Ontario, Canada K1A 0R6. (613) 993-2425. (613) 957-3281.

Ohio Aerospace Institute, 2001 Aerospace Parkway, Brook Park, OH 44142. (216) 891-2100. FAX (216) 891-2140.

Pennsylvania State University, Propulsion Engineering Research Center, 106 Research Bldg. E, Bigler Rd., University Park, PA 16802. (814) 863-6272. FAX (814) 865-3389.

Transportation Research and Education Center, Georgia Institute of Technology, Mason Civil Engineering Bldg., Atlanta, GA 30332. (404) 894-2236. FAX (404) 894-2278.

AERONAUTICS

See also: AERONAUTICAL ENGINEERING, AEROSPACE ENGINEERING

ABSTRACT SERVICES AND INDEXES

Alloys Index. American Society for Metals, Metals Park, OH 44073. (216) 338-5151. FAX (216) 338-4634. 1974 to present. Monthly. $295.00.

Applied Mechanics Reviews: An Assessment of World Literature in Engineering Sciences. 1948-present. American Society of Mechanical Engineers, 345 East 47th Street, New York, NY 10017. (212) 705-7703. Monthly. $360.00 per year.

Applied Science and Technology Index; A Cumulative Subject Index To English Language Periodicals in the Fields of Aeronautics and Space Science, Computer Technology, Chemistry, Construction Industry, Energy and Related Areas. H.W. Wilson Co., 950 University Avenue, Bronx, NY 10452. (800) 367-6770 or (212) 588-8400. FAX (718) 590-1617. From 1958 to present. Monthly. Inquire about cost and availability. Also available on CD-ROM and online.

Current Contents: Engineering, Technology, and Applied Sciences. Institute for Scientific Information, 3501 Market Street, Philadelphia, PA 19104. (215) 386-0100 or (800) 523-1850. FAX (215) 386-2991. 1970 to present. Weekly. $442.00 per year.

Engineering Index Monthly. Engineering Information, Inc., Castle Point on the Hudson, Hoboken, NJ 07030. (800) 221-1044. FAX (212) 832-1857. Monthly. $2300.00 per year. Also available online as COMPENDEX, and also on CD-ROM. Covers chemical engineering, computers, electrical engineering, civil engineering, metals and mining, industrial management, and mechanical engineering.

International Aerospace Abstracts. Technical Information Service, American Institute of Aeronautics and Astronautics, Inc., 555 West 57th St., New York, NY 10019. (212) 247-6500. FAX (212) 582-4861. Semi-monthly. $1295.00 per year.

Metals Abstracts and Metals Abstracts Index. American Society for Metals, Metals Park, OH 44073. (216) 338-5151. 1968 to present. Monthly. Abstracts are $1500.00 per year and Index is $460.00 per year.

Science Citation Index. Institute for Scientific Information, 3501 Market Street, Philadelphia, PA 19104. (215) 386-0100 or (800) 523-1850. FAX (215) 386-6362. Inquire about availability and cost. Also available on CD-ROM.

Scientific and Technical Aerospace Reports (star). National Aeronautics and Space Administration. NASA Center for Aerospace Information, Box 8757, BWI Airport, Baltimore, MD 21240. (301) 621-0153. Monthly. Inquire about availability and cost. ALso available through the NASA online retrieval service (RECON), and through the Aerospace Database through DIALOG.

World Aluminum Abstracts. Aluminum Association, Inc., 818 Connecticut Ave. NW, Washington, DC 20006. (202) 862-5156. Inquire for price and availability.

ASSOCIATIONS AND PROFESSIONAL SOCIETIES

Aircraft Electronics Association. PO Box 1963, Independence, MO 64055-0963. (816) 373-6565. FAX (816) 478-3100.

American Helicopter Society, Inc., 217 N. Washington St., Alexandria, VA 22314. (703) 684-6777. FAX (703) 739-9279.

American Institute of Aeronautics and Astronautics, 370 L'enfant Promenade SW , Washington, DC 20024-2518. (202) 646-7471.

Canadian Aeronautics and Space Institute, 222 Somerset Street West, Suite 601, Ottawa, Ontario, Canada K2p 0j1.

Flight Safety Foundation, Inc., 2200 Wilson Blvd., Suite 500, Arlington, VA 22201-3306. (703) 522-8300. FAX (703) 525-6047.

Society of Automotive Engineers. SAE, Inc., 400 Commonwealth Dr., Warrendale, PA 15096. (412) 776-4841.

BIBLIOGRAPHIES

Aeronautical Engineering: A Continuing Bibliography With Indexes. National Aeronautics and Space Administration, Center for Aerospace Information, Box 8757, BWI Airport, Baltimore, MD 21240. (301) 621-0153. Inquire for cost and availability.

Scientific and Technical Books and Serials in Print. R.R. Bowker Inc., 121 Chanlon Road, New Providence, NJ 07974. (800) 521-8110 or (908) 464-6800. FAX (908) 464-3553. List of currently published books and serials in the physical and biological sciences, engineering and technology, with subject, author andtitles indexes. Inquire as to cost and availability.

DIRECTORIES AND BIOGRAPHICAL SOURCES

AAS Directory. American Astronautical Society, 6352 Rolling Mill Place, Suite 102, Springfield, VA 22152-2354. (703) 866-0020. FAX (703) 866-3526. Inquire for cost and availability.

Research Centers Directory. Gale Research, 835 Penobscot Building, Detroit, MI 48226-4094. (313) 961-2242. (800) 877-4253. FAX (313) 961-6083. 20th edition. $485.00

Who's Who in Engineering. American Association of Engineering Societies, 1111 19th Street NW, Suite 608, Washington, DC 20036. (202) 296-2237 or (800) 658-8897. 8th edition. 1991. Inquire for price.

World Aviation Directory. McGraw-Hill, Aviation Week Group (NY), 1221 Avenue of the Americas, New York, NY 10020. (215) 237-4112. FAX (215) 586-3232. Semiannual. $250.00

ENCYCLOPEDIAS AND DICTIONARIES

Concise Encyclopedia of Aeronautics and Space Systems. M. Pelegrin and W.M. Hollister. Pergamon Press Inc., Maxwell House, Fairview Park, Elmsford, NY 10523. (914) 592-7700. Fax (914) 592-3625. 1993. Inquire for cost.

Dictionary of Aerospace Engineering. M.G. Kotik. Elsevier Science Publishing Company, Inc., 655 Avenue of the Americas, New York, NY 10010. (212) 989-5800. 1986. $225.75.

Encyclopedia of Physical Science and Technology. Robert A. Meyers, ed. 18 volumes. Academic Press Inc., 6277 Sea Harbor Drive, Orlando, FL 32887. 1992. $2100.00.

Jane's Aerospace Dictionary, B. Guston. 3d ed. Jane's Information Group, Dept. DSM, 1340 Braddock Pl., Suite 300, Box 1436, Alexandria, VA 22314-1651. (703) 683-3700. FAX (703) 836-0029. 1991. $45.00.

McGraw-Hill Dictionary of Scientific and Technical Terms. Sybil P. Parker, ed. 5th edition. McGraw-Hill Publishing Company, 1221 Avenue of the Americas, New York, NY 10020. (800) 262-4729 or (212) 512-3825. 1993. $110.50.

McGraw-Hill Encyclopedia of Science and Technology. Sybil P. Parker, ed. 7th edition. 20 volumes. McGraw-Hill Publishing Company, 1221 Avenue of the Americas, New York, NY 10020. (800) 262-4729 or (212) 512-3825. 1992. $1900.00

SAE Dictionary of Aerospace Engineering. William H. Cubberly, ed. Society of Automotive Engineers, 400 Commonwealth Drive, Warrendale, PA 15096. (412) 776-4841. 1992. $85.00.

GENERAL WORKS

Analysis of Aircraft Structures: An Introduction. Bruce K. Donaldson. McGraw-Hill Publishing Company, 1221 Avenue of the Americas, New York, NY 10020. (800) 262-4729 or (212) 512-3825. 1992. Inquire for cost.

Fundamentals of Aerodynamics. John David anderson. 2nd edition. McGraw-Hill Publishing Company, 1221 Avenue of the Americas, New York, NY 10020. (800) 262-4729 or (212) 512-3825. 1991. Inquire for cost and availability.

Introduction To Flight. John David anderson. 3d ed. McGraw-Hill Publishing Company, 1221 Avenue of the Americas, New York, NY 10020. (800) 262-4729 or (212) 512-3825. 1989. $46.63.

HANDBOOKS AND MANUALS

Standard Handbook of Engineering Calculations. Tyler G. Hicks, ed. 2d ed. McGraw-Hill Publishing Company, 1221 Avenue of the Americas, New York, NY 10020. (800) 262-4729 or (212) 512-3825. 1985. $89.50.

The Wiley Engineer's Desk Reference. Sanford I. Heisler. John Wiley and Sons, Inc., 605 Third Avenue, New York, NY 10158. (800) 526-5368 or (212) 850-6000. 1984. $64.95.

ONLINE DATABASES AND CD-ROMS

Aerospace Database. American Institute of Aeronautics and Astronautics, 370 L'Enfant Promenade, S.W., Washington, DC 20024. (202) 646-7400. Worldwide published literature on research and development in aerospace and related areas, with abstracts. Covers 1962 to present. Available on DIALOG, (800) 334-2564, online service. Also available on CD-ROM. Inquire as to cost and availability.

Current Contents Search. Institute for Scientific Information, 3501 Market Street, Philadelphia, PA 19104. (215) 386-0100. FAX (215) 386-6362. Contains citations to articles listed in the table of contents of science and technology journals. Also articles in social sciences and life sciences journals. Available on BRS,(800) 289-4277, DIALOG,(800) 334-2564, online services. Inquire as to cost and availability.

Dissertation Abstracts. University Microfilms International, 300 North Zeeb Road, Ann Arbor, MI 48106. (800) 521-0600 or (313) 761-4700. Scope includes virtually all doctoral dissertations accepted at accredited American institutions from 1861 to present in 252 subject areas. Available on BRS,(800) 289-4277, DIALOG,(800) 334-2564, and OCLC EPIC,(800) 848-5878, online services. Also available on CD-ROM. Inquire as to cost and availability.

International Aerospace Abstracts. American Institute of Aeronautics and Astronautics, 370 L'Enfant Promenade, S.W., Washington, DC 20024. (202) 646-7400. Contains references and abstracts of journal and monograph literature relating to aerospace science and technology, from 1963 to present. Available through the NASA/RECON system of the National Aeronautics and Space Administration only.

NASA Database. American Institute of Aeronautics and Astronautics, 370 L'Enfant Promenade, S.W., Washington, DC 20024. (202) 646-7400. Citations and abstracts of aeronautics and astronautics literature, 1962 to present. Also contains citations from *STAR, Scientific and Technical Aerospace Reports, and International Aerospace Abstracts.* Available through NASA/RECON online service. Inquire as to cost and availability.

NTIS Bibliographic Database. National Technical Information Service, 5285 Port Royal Road, Springfield, VA 22161.(703) 487-4929 or FAX (703) 321-8199. Broad coverage of government-sponsored science and technology research reports, 1964 to present. Available on BRS,(800) 289-4277, DIALOG, (800) 334-2564, ORBIT, (800) 456-7248, and STN International, FIZ Karlsruhe, P.O. Box 2465, W-7500, Karlsruhe 1, Germany, online services. Also available on CD-ROM. Inquire as to cost and availability.

SAE Global Mobility Database. Society of Automative Engineers (SAE), Electronic Publishing Division, 400 Commonwealth Drive, Warrendale, PA 15098. (412) 776-4841. Contains citations with abstracts to technical papers on automotive and

AERONAUTICS

Ency. of Physical Sciences and Engineering Info. Sources

aerospace technology and vehicular-related industries that have been presented at SAE conferences. Covers 1906 to present. Available on ORBIT,(800) 456-7248,online service. Inquire as to cost and availability.

SCISEARCH. Institute for Scientific Information, 3501 Market Street, Philadelphia, PA 19104. (800) 523-1850 or (215) 386-0100. Broad multidisciplinary title and author index to the international literature of science and technology, 1974 to present. Available on DIALOG,(800) 334-2564, and ORBIT,(800) 456-7248, online services. Also available on CD-ROM. Inquire as to cost and availability.

TRIS (Transportation Research Information). National Academy of Science, 2101 Constitution Avenue, N.W., Washington, DC 20418. (202) 334-3313 or (800) 624-6242. Citations with abstracts of literature on transportation, including air, highway, rail, maritime and other modes from 1968 to present. Available on DIALOG,(800) 334-2564, online service. Inquire as to cost and availability.

WILSONLINE. H.W. Wilson Company, 950 University Avenue, Bronx, NY 10452. (800) 367-6770 or (212) 588-8400. Makes available online versions of the printed H.W. Wilson indexes including Applied Science and Technology Index, Business Periodicals Index, General Science Index, and Readers' Guide to Periodical Literature. Period covered is generally 1983 to present. Available on BRS,(800) 289-4277, DIALOG,(800) 334-2564, and OCLC EPIC,(800) 848-5878, online services. Also available on CD-ROM. Inquire as to cost and availability.

PERIODICALS

Aeronautical Journal. The Royal Aeronautical Society, 4 Hamilton Place, London W1V 0BQ, England. 1897 to present. Monthly except June and August. Inquire for price information in U.S.

Aerospace America. American Institute of Aeronautics and Astronautics, 370 L'Enfant Promenade SW, Washington, DC 20024-2518. (202) 646-7471. 1932 to present. Monthly. $70 annually (non-members).

Aerospace Engineering. Society of Automotive Engineers, 400 Commonwealth Dr., Warrendale, PA 15096. (412) 776-4841. 1981 to present. Monthly. $48.00 per year.

AIAA Journal. American Institute of Aeronautics andAstronautics, 370 L'Enfant Promenade SW, Washington, DC 20024-2518. (202) 646-7471. 1963 to present. Monthly. $52.00 per year (members) and $435.00 per year (non-members).

Aviation Mechanics Bulletin. Flight Safety Foundation, Inc., 2200 Wilson Blvd., Suite 500, Arlington, VA 22201-3306. (703) 522-8300. FAX (703) 525-6047. 1953 to present. Bimonthly. $15.00 per year (non-members).

Aviation Week and Space Technology. 1221 Avenue of the Americas, New York, NY 10020. (212) 512-2999 or (800) 525-5003. 1916 to present. Weekly. $82.00 per year.

Canadian Aeronautics and Space Journal. Canadian Aeronautics and Space Institute, Suite 818, the National Building, 130 Slater St., Ottawa, Canada K1P 6E2. 1955 to present. Monthly. $75.00 per year (U.S.).

Flying. Hachette Filipacchi Magazines, Inc., 1633 Broadway, New York, NY 10019. (203) 622-2700. FAX (203) 622-2725. 1927 to present. Monthly. $24.00 per year.

IEEE Transactions On Aerospace and Electronic Systems. Institute of Electrical and Electronics Engineers, 445 Hoes Lane, PO Box 1331, Piscataway, NJ 08855-1331. (908) 981-0060. 1965 to present. Quarterly. $10.00 per issue (members), $20.00 per issue (non-members).

International Journal of Turbo and Jet Engines. Freund Publishing House, Ltd., PO Box 35010, Tel Aviv 63150, Israel. 972-2-5662925. FX 972-3-5605335. 1983 to present. Quarterly. $190.00 per year.

Journal of Aircraft. American Institute of Aeronautics and Astronautics, 370 L'Enfant Promenade SW, Washington, DC 20024-2518. 1963 to present. Bimonthly. $42.00 (members) and $230.00 (non-members).

Journal of Guidance, Control and Dynamics. American Institute of Aeronautics and Astronautics, 370 L'Enfant Promenade SW, Washington, DC 20024-2518. 1978 to present. Bimonthly. $42.00 (members) and $250.00 (non-members).

Journal of Propulsion and Power. American Institute of Aeronautics and Astronautics, 370 L'Enfant Promenade SW, Washington, DC 20024-2518. 1985 to present. Bimonthly. $38.00 (members) and $260.00 (non-members).

Journal of the American Helicopter Society. American Helicopter Society, Inc., 217 North Washington St., Alexandria, VA 22314. (703) 684-6777. 1956 to present. Quarterly. $35.00 per year.

Rotor and Wing International. PHillips Publishing Inc., 7811 Montrose Rd., Potomac, MD 20854. (301) 340-2100. FAX (301) 309-3847. 1967 to present. Monthly. $49.00 a year.

Vertiflite. American Helicopter Society, 217 N. Washington St., Alexandria, VA 22314. (703) 684-6777. FAX (703) 739-9279. 1955 to present. Bimonthly. $25 per year (members) and $35 per year (non-members).

RESEARCH CENTERS AND INSTITUTES

Flight Research Laboratory, University of Kansas, RaymondNichols Hall, Lawrence, KS 66045. (913) 864-3043. FAX (913) 864-7789.

Joint Institute For Advancement of Flight Sciences, George Washington University, Langley Research Center, Mail Stop 269, Hampton, VA 23665. (804) 864-1982. FAX (804) 864-5894.

Ohio State University Aeronautical and Astronautical Research Laboratory. 2330 W. Case Road, Columbus, OH 43235. (614) 292-5491.

Stanford University Structures and Composites Laboratory. Department of Aeronautics and Astronomy, Stanford, CA 94305. (415) 723-4135.

University of Texas At Austin, Center For Aeromechanics Research. Aerospace Engineering & Engineering Mechanics, WRW Room 220, Austin, TX 78712. (512) 471-5962.

University of Washington Aeronautical Laboratory. FS-10, Seattle, WA 98105. (206) 543-0439. FAX (206) 542-0217.

AEROSOLS

See also: AIR POLLUTION, CHEMISTRY, COLLOIDS, METEOROLOGY, PHYSICAL CHEMISTRY

ABSTRACT SERVICES AND INDEXES

Applied Science and Technology Index; A Cumulative Subject Index To English Language Periodicals in the Fields of Aeronautics and Space Science, Computer Technology, Chemistry, Construction Industry, Energy and Related Areas. H. W. Wilson Co., 950 University Avenue, Bronx, NY 10452-9978. (212) 588-8400. (800) 367-6770. FAX (718) 590-1617. From 1958 to present. Monthly. Inquire about cost and availability. Also available online (BRS and WILSONLINE) and on CD-ROM. ISSN: 0003-6986.

Chemical Abstracts. Chemical Abstracts Service. 2540 Olentangy River Road, Box 3012, Columbus, OH 43210-0012. (614) 447-3600. FAX (614) 447-3713. 1907 to present. Weekly. $16,800.00 per year. Available online and on CD-ROM. CA is also available in five section groupings. Inquire regarding cost and availability.

Current Contents: Physical, Chemical and Earth Sciences. Institute for Scientific Information, 3501 Market Street, Philadelphia, PA 19104. (215) 386-0100. FAX (215) 386-2291. From 1961 to present. Weekly. $442.00 per year. Also available on CD-ROM and online. Inquire regarding cost and availability. ISSN: 0163-2574.

Engineering Index Monthly. Engineering Information, Inc., Castle Point on the Hudson, Hoboken, NJ 07030. (201) 216-8500. (800) 221-1044. FAX (201) 216-8532. Monthly. $2300.00 per year for monthly issues. $1980 per year for annual cumulation. Available on CD-ROM and online as *COMPENDEX.* ISSN: 0742-1974; 0360-8557.

General Science Index. H.W. Wilson Company, 950 University Avenue, Bronx, NY 10452. (212) 588-8400. (800) 367-6770. FAX (718) 590-1617. From 1978 to present. Ten issues per year; quarterly and annual cumulations. Service basis. Available on CD-ROM and online. Inquire regarding cost and availability. ISSN: 0162-1963.

Physics Abstracts. INSPEC. Section A, Science Abstracts. Institution of Electrical Engineers (IEE), London, United Kingdom. Available from: INSPEC/IEEE - Institute of Electrical and Electronic Engineers, Box 1331, 445 Hoes Lane, Piscataway, NJ 08855-1331. (908) 562-5549. 1898 to present. 24 issues per year. $2835.00 per year. ISSN: 0036-8091. Also available online and on CD-ROM.

Science Citation Index. SCI. Institute for Scientific Information, 3501 Market Street, Philadelphia, PA 19104. (215) 386-0100. (800) 523-1850. FAX (215) 386-2991. 1961 to present. Six issues per year, plus annual cumulation. $11650.00 per year. Also available online and on CD-ROM. Inquire about price and availability. ISSN: 0036-827X.

ASSOCIATIONS AND PROFESSIONAL SOCIETIES

American Association For Aerosol Research. 1330 Kemper Meadow Drive, Number 600, Cincinnati, OH 45240-1643. (513) 742-2227. FAX (513) 742-3355.

American Chemical Society, 1155 16th Street, Nw, Washington, DC 20036. (202) 872-4600.

American Institute of Chemical Engineers. 345 East 47th Street, New York, NY 10017. (212) 705-7663. FAX (212) 752-3294.

Association of Consulting Chemists and Chemical Engineers, 40 East 45th Street, New York, NY 10017. (212) 983-3160. FAX (212) 983-3161.

BIBLIOGRAPHIES

Information Sources in Chemistry. R.T. Bottle and J.F.B Rowland, editors. R.R. Bowker Inc., 121 Chanlon Road, New Providence, NJ 07974. (908) 464-6800. (800) 521-8110. Fourth edition. 1993. $75.00.

Scientific and Technical Books and Serials in Print; An Index To Literature in Science and Technology. R.R. Bowker Inc., 121 Chanlon Road, New Providence, NJ 07974. (908) 464-6800. (800) 521-8110. FAX (908) 665-3502. 1972 to present. Annual. 4 volumes. 1994. $299.95. Also available on compact disc and online. ISSN: 0000-054X.

DIRECTORIES AND BIOGRAPHICAL SOURCES

American Institute of Chemical Engineers. Directory. American Institute of Chemical Engineers. 345 East 47th Street, New York, NY 10017-2396. (212) 705-7663. Annual.

American Institute of Chemists - Professional Directory. American Institute of Chemists, 501 Wythe Street, Alexandria, VA 22314-1917. (703) 836-2090. FAX (703) 836-2091. Annual. $65.00.

American Men and Women of Science. Physical and Biological Sciences. R.R. Bowker Inc., 121 Chanlon Road, New Providence, NJ 07974. (908) 464-6800. (800) 521-8110. 20th edition. 8 volumes. 1996. $850.00

Chemical Week — Buyers Guide Issue. Chemical Week Associates, 888 Seventh Avenue, New York, NY 10106. (212) 621-4900. FAX (212) 621-4949. Annual, October. $50.00.

Chemical Sources International. Mike Desing and KurtGandenberger, editors. Chemical Sources International, Inc., P.O. Box 1824, Clemson, SC 29633. (803) 646-7840. 1994. $285.00.

Research Centers Directory. Gale Research,Inc., 835 Penobscot Building, Detroit, MI 48226-4094. (313) 961-2242. (800) 877-4253. 20th edition. 1995. $485.00. ISSN: 0080-1518.

ENCYCLOPEDIAS AND DICTIONARIES

Dictionary of Named Processes in Chemical Technology. Alan E. Comyns. Oxford University Press, Inc., 200 Madison Avenue, New York, NY 10016. (212) 725-6000. (800) 334-4249. 1994. $75.00.

AEROSOLS

Ency. of Physical Sciences and Engineering Info. Sources

Encyclopedia of Applied Physics. VCH Publications, Inc., 220 East 23rd Street, Suite 909, New York, NY 10010-4606. (212) 683-8333. (800) 422-8824. 1991-. Twenty volumes. $6000.00.

Encyclopedia of Chemical Processing and Design. McKetta Marcel Dekker, Inc., 270 Madison Avenue, New York, NY 10016. (212) 696-9000. (800) 228-1160. 1976 - . $175.00 per volume. ISSN:

Industrial Chemical Thesaurus. Michael Ash and Irene Ash. VCH Publishers, Inc., 220 East 23rd Street, New York, NY 10010-4606. (800) 367-8249. 1992. $295.00.

Kirk-Othmer Encyclopedia of Chemical Technology. John Wiley and Sons, Inc., 605 Third Avenue, New York, NY 10158. (800) 526-5368 or (212) 850-6000. Fourth edition. 1991-. Twenty-seven volumes. $5400.00.

McGraw-Hill Encyclopedia of Science and Technology. McGraw-Hill Book, Incorporated, 1221 Avenue of the Americas, New York, NY 10020. (212) 997-3675. Seventeenth editon. 1992. $1900.00.

Ullman's Encyclopedia of Industrial Chemistry. VCH Publications, Inc., 220 East 23rd Street, Suite 909, New York, NY 10010-4606. (212) 683-8333. (800) 422-8824. 5th edition. 1984 - . Price varies per volume.

GENERAL WORKS

Aerosol Effects On Climate. S.G. Jennings, editor. McGraw-Hill Publishing Company, 1221 Avenue of the Americas, New York, NY 10020. (800) 262-4729 or (212) 512-3825. 1993. $40.00,

Aerosol Measurement: Principles, Techniques, and Application. Klaus Willeke and Paul A. Baron, editors. Van Nostrand Reinhold, 115 Fifth Avenue, New York, NY 10003. (212) 254-3232. (800) 842-3636. 1993. $99.95.

Aerosol Sampling: Science and Practice. James Vincent. John Wiley and Sons, Inc., 605 Third Avenue, New York, NY 10158. (800) 526-5368 or (212) 850-6000. 1989. $250.00.

Aerosol Science: Theory and Practice With Special Applications To the Nuclear Industry. M.M. Williams and Sudarshan K. Loyalka. John Wiley and Sons, Inc., 605 Third Avenue, New York, NY 10158. (800) 526-5368 or (212) 850-6000. 1991. $94.00.

Aerosol Technology: Properties, Behavior, and Measurement of Airborne Particles. William C. Hinds. John Wiley and Sons, Inc., 605 Third Avenue, New York, NY 10158. (800) 526-5368 or (212) 850-6000. 1982. $75.00.

Aerosols: An Industrial and Environmental Science Monograph. George M. Hidy. Academic Press, Inc., 1250 SixthAvenue, San Diego, CA 92101-4311. (619) 231-0926. FAX (619) 699-6715. 1984. $127.00.

Atmospheric Aerosols. S. Twomey. Elsevier Science Publishing Company, Inc., 655 Avenue of the Americas, New York, NY 10010. (212) 989-5800. 1977. $115.00.

HANDBOOKS AND MANUALS

Bioaerosols: Handbook of Samplers and Sampling. Cox. Lewis Pubs., CRC Press, Inc., 2000 Corporate Boulevard, NW, Boca Raton, FL 33431. (407) 994-0555. (800) 272-7737. 1995. $79.95.

CRC Handbook of Chemistry and Physics. David R. Kide, editor. CRC Press Inc., 2000 Corporate Blvd., NW, Boca Raton, FL 33431. (407) 994-0555. (800) 333-8300. 77th edition. 1996. $99.95.

Guide To Basic Chemical Compounds. D.R. Lide, Jr. CRC Publishers, Inc., 2000 Corporate Blvd., N.W., Boca Raton, FL 33431. (407) 994-0555 or (800) 333-8300. 1993. $120.00.

Improving Safety in the Chemical Laboratory: A Practical Guide. Jay A. Young, editor. John Wiley and Sons, Inc., 605 Third Avenue, New York, NY 10158. (800) 526-5368 or (212) 850-6000. Second edition. 1991. $75.00.

Lange's Handbook of Chemistry. John A. Dean, editor. McGraw-Hill Publishing Company, 1221 Avenue of the Americas, New York, NY 10020. (800) 262-4729 or (212) 512-3825. Fourteenth editon. 1992. $79.50.

Perry's Chemical Engineers' Handbook. Robert H. Perry and Donald W. Green, editors. McGraw-Hill Publishing Company, Inc., 1221 Avenue of the Americas, New York, NY 10020. (212) 512-2000. (800) 262-4729. 6th edition. 1996.

ONLINE DATABASES AND CD-ROMS

CA Search. Chemical Abstracts Service, P.O. Box 3012, Columbus, OH 43210-0012. (614) 447-3600. (800) 848-6533. FAX (614) 447-3709. Guide to worldwide chemical literature and related fields, 1972 to present. Available on BRS, DIALOG, ORBIT and STN online services. Inquire as to cost and availability.

Dissertation Abstracts. University Microfilms International, 300 North Zeeb Road, Ann Arbor, MI 48106. (800) 521-0600 or (313) 761-4700. Scope includes virtually all doctoral dissertations accepted at accredited American institutions from 1861 to present in 252 subject areas. Available on BRS, DIALOG, and OCLC EPIC online services. Also available on CD-ROM. Inquire as to cost and availability.

INSPEC. Institution of Electrical Engineers, Michael Faraday House, Six Hills Way, Stevenage, Herts. SG1 2AY, England. Telephone: 0438 313311 or FAX 0438 742840. Contains citations to the worldwide literature of physics, electronics and electrical engineering, computer technology, and related fields. Available on BRS, DIALOG, ORBIT and STN online services. Inquire as to cost and availability.

NASA Database. American Institute of Aeronautics and Astronautics, 370 L'Enfant Promenade, S.W., Washington, DC 20024. (202) 646-7400. Citations and abstracts of aeronautics andastronautics literature, 1962 to present. Also contains citations from *STAR, Scientific and Technical Aerospace Reports, and International Aerospace Abstracts.* Available through NASA/RECON online service. Inquire as to cost and availability.

NTIS. National Technical Information Service, 5285 Port Royal Road, Springfield, VA 22161. (703) 487-4929 or FAX (703)

321-8199. Broad coverage of government-sponsored research reports, 1964 to present. Available on BRS, DIALOG, ORBIT and STN online services. Also available on CD-ROM. Inquire as to cost and availability.

SCISEARCH. Institute for Scientific Information, 3501 Market Street, Philadelphia, PA 19104. (800) 523-1850 or (215) 386-0100. Broad multidisciplinary title and author index to the international literature of science and technology, 1974 to present. Available on DIALOG and ORBIT online services. Also available on CD-ROM. Inquire as to cost and availability.

WILSONLINE. H.W. Wilson Company, 950 University Avenue, Bronx, NY 10452. (800) 367-6770 or (212) 588-8400. Makes available online versions of the printed H.W. Wilson indexes including Applied Science and Technology Index, Business Periodicals Index, and Readers' Guide to Periodical Literature. Period covered is generally 1983 to present. Also available on CD-ROM. Inquire as to cost and availability.

PERIODICALS

Aerosol Science and Technology. Elsevier Science Publishing Company, Inc., 655 Avenue of the Americas, New York, NY 10010. (212) 989-5800. 1982 to present. 8 issues per year, $582.00 per year. ISSN: 0278-6826.

Analytical Chemistry. American Chemical Society, Box 3337, Columbus, OH 43210. (614) 447-3776. 1929 to present. Semi-monthly. $415.00 per year. ISSN: 0003-2700.

Angewandte Chemie. VCH Publishers, Inc., 220 East 23rd Street, New York, NY 10010-4606. (212) 683-8333. (800) 367-8249. 1888 to present. 22 issues per year. $840.00 per year. ISSN: 0044-8249.

Chemical Reviews. American Chemical Society, Box 3337, Columbus, OH 43210. (614) 447-3776. 1924 to present. 8 issues per year. $346.00 per year. ISSN: 0009-2665.

Chemical Week. Includes Annual Buyers Guide. Chemical Week Associates, 888 Seventh Avenue, New York, NY 10106. (212) 621-4900. FAX (212) 621-4949. 1914 to present. Weekly. $99.00 per year. ISSN: 0009-272X.

Chemtech. American Chemical Society. Box 3337, Columbus, OH 43210. (614) 447-3776. 1970 to present. Monthly. $370.00 per year. ISSN: 0009-2703.

Inorganic Chemistry. American Chemical Society, Box 3337, Columbus, OH 43210. (614) 447-3776. 1962 to present. Semi-monthly. $500.00 per year. ISSN: 0020-1669.

Journal of Aerosol Science. Elsevier Science, 660 White Plains Road, Tarrytown, NY 10591-5153. (914) 524-9200. 1970 to present. 9 issues per year. $915.00 per year. ISSN: 0021-8502.

Journal of the American Chemical Society. American Chemical Society, Box 3337, Columbus, OH 43210. (614) 447-3776. 1879 to present. Biweekly. $1055.00 per year. ISSN: 0002-7863.

Spray Technology and Marketing. Industry Publications, Inc., 389 Passaic Avenue, Fairfield, NY 07006. (201) 227-5151, FAX (201) 227-9219. 1956 to present. Monthly. $30.00 per year. ISSN: 1055-2340.

RESEARCH CENTERS AND INSTITUTES

Atmospheric Sciences Research Center. State University of New York, 100 Fuller Road, Albany, NY 12205. (518) 442-3820. FAX (518) 442-3867.

Center For Cloud Physics Research. University of Missouri, Rolla, MO 65401. (314) 341-4332.

Cloud Simulation and Aerosol Laboratory. Colorado State University, Fort Collins, CO 80523. (303) 491-8257. FAX (303) 491-8449.

Particle Technology Laboratory. University of Minnesota, Room 271, Mechanical Engineering Building, 1111 Church Street, SE Minneapolis, MN 55455. (612) 625-7307.

AEROSPACE ENGINEERING

See also: AERONAUTICAL ENGINEERING

ABSTRACT SERVICES AND INDEXES

Aeronautical Engineering. Scientific and Technical Branch, National Aeronautics and Space Administration. National Technical Information Service (NTIS), 5285 Port Royal Road, Springfield, VA 22161. (703) 487-4650. FAX (703) 321-8547. Monthly. A selection of annotated references to unclassified reports and journal articles that were introduced into the NASA scientific and technical information system and announced in STAR and IAA.

Alloys Index. American Society for Metals, Metals Park, OH 44073. (216) 338-5151. FAX (216) 338-4634. 1974 to present. Monthly. $295.00.

Applied Mechanics Reviews: An Assessment of World Literature in Engineering Sciences. American Society of Mechanical Engineers, 345 East 47th Street, New York, NY 10017. (212) 705-7703. 1948 to present. Monthly. $360.00 per year.

Applied Science and Technology Index; A Cumulative Subject Index To English Language Periodicals in the Fields of Aeronautics and Space Science, Computer Technology, chemistry, Construction Industry, Energy and Related Areas. H.W. Wilson Co., 950 University Avenue, Bronx, NY 10452. (800) 367-6770 or (212) 588-8400. FAX (718) 590-1617. From 1958 to present. Monthly. Inquire about cost and availability. Also available on CD-ROM.

Current Contents: Engineering, Technology, and Applied Sciences. Institute for Scientific Information, 3501 Market Street, Philadelphia, PA 19104. (215) 386-0100. FAX (215) 386-6362. (800) 523-1850. Weekly. $360.00 per year.

Engineering Index Monthly. Engineering Information, Inc., Castle Point on the Hudson, Hoboken, NJ 07030. (800) 221-1044. FAX (212) 832-1857. Monthly. $2200.00 per year. Also available online as COMPENDEX, and also on CD-ROM.

International Aerospace Abstracts. Technical Information Service, American Institute of Aeronautics and Astronautics, Inc., 555 West 57th St., New York, NY 10019. (212) 247-6500. FAX (212) 582-4861. Semi-monthly. $1295.00 per year.

Metals Abstracts and Metals Abstracts Index. American Society for Metals, Metals Park, OH 44073. (216) 338-5151. 1968 to present. Monthly. Abstracts are $1500.00 per year and Index is $460.00 per year.

NASA Patent Abstracts Bibliography. National Aeronautics and Space Administration, Center for Aerospace Information, Box 8757, BWI Airport, Baltimore, MD 21240. (301) 621-0153. Semi-annual. Inquire for price information.

Science Citation Index. Institute for Scientific Information, 3501 Market Street, Philadelphia, PA 19104. (215) 386-0100. FAX (215) 386-6362. (800) 523-1850. Inquire about availability and cost. Also available on CD-ROM.

STAR. Scientific and Technical Aerospace Reports. National Aeronautics and Space Administration, Center for Aerospace Information, Box 8757, BWI Airport, Baltimore, MD 21240. (301) 621-0153. Monthly. Inquire about availability and cost. Also available through the NASA online retrieval service (RE-CON), and through the Aerospace Database through DIALOG.

World Aluminum Abstracts. Aluminum Association, Inc., 818 Connecticut Ave. NW, Washington, DC 20006. (202) 862-5156. Inquire for price and availability.

ASSOCIATIONS AND PROFESSIONAL SOCIETIES

Aircraft Electronics Association. PO Box 1963, Independence, MO 64055-0963. (816) 373-6565. FAX (816) 478-3100.

American Astronautical Society. 6352 Rolling Mill Place, Suite 102, Springfield, VA 22152-2354. (703) 866-0020. FAX (703) 866-3526.

American Helicopter Society, Inc., 217 N. Washington St., Alexandria, VA 22314. (703) 684-6777. FAX (703) 739-9279.

American Institute of Aeronautics and Astronautics, 370 L'Enfant Promenade SW, Washington, DC 20024-2518. (202) 646-7471.

Flight Safety Foundation, Inc., 2200 Wilson Blvd., Suite 500, Arlington, VA 22201-3306. (703) 522-8300. FAX (703) 525-6047.

Society of Automotive Engineers. SAE, Inc., 400 Commonwealth Dr., Warrendale, PA 15096. (412) 776-4841.

BIBLIOGRAPHIES

Aeronautical Engineering: A Continuing Bibliography With. INDEXES. National Aeronautics and Space Administration, Center for Aerospace Information, Box 8757, BWI Airport, Baltimore, MD 21240. (301) 621-0153. Inquire for cost and availability.

*Scientific and Technical Books and Serials in Print, R.*R. Bowker Inc., 121 Chanlon Road, New Providence, NJ 07974. (800) 521-8110 or (908) 464-6800. FAX (908)464-3553. 3 volumes, annually. $285.00. Also available on CD-ROM.

DIRECTORIES AND BIOGRAPHICAL SOURCES

AAS Directory. American Astronautical Society, 6352 Rolling Mill Place, Suite 102, Springfield, VA 22152-2354. (703) 866-0020. FAX (703) 866-3526. Inquire for cost and availability.

Aerospace Technical Centres. 2nd edition. Longman Publishing Group, the Longman Building, 10 Bank Street, White Plains, NY 10606-1951. (914) 993-5000 or (800)266-8855. FAX (914) 997-8115. 1991. $475.00.

Research Centers Directory. Gale Research, 835 Penobscot Building, Detroit, MI 48226-4094. (313) 961-2242. (800) 877-4253. FAX (313) 961-6083. $485.00.

SAE Aerospace Sources and Suppliers Directory 1993. SAE International, Inc., 400 Commonwealth Dr., Warrendale, PA 15096. (412) 776-4841. 1993. Inquire for cost and availability.

Who's Who in Engineering. American Association of Engineering Societies, 1111 19th Street NW, Suite 608, Washington, DC 20036. (202) 296-2237 or (800) 658-8897. 8th edition. 1991. Inquire for price.

World Aviation Directory. McGraw-Hill, Aviation Week Group (NY), 1221 Avenue of the Americas, New York, NY 10020. (215) 237-4112. FAX (215) 586-3232. Semiannual. $250.00.

ENCYCLOPEDIAS AND DICTIONARIES

Concise Encyclopedia of Aeronautics and Space Systems. M. Pelegrin and W.M. Hollister. Pergamon Press Inc., Maxwell House, Fairview Park, Elmsford, NY 10523. (914) 592-7700. Fax (914) 592-3625. 1993. Inquire for cost.

Dictionary of Aeronautical Terms, Dale Crane, Comp. Aviation Supplies and Academics, Inc., 7005 132nd Place SE, Newcastle, WA 98059. (213) 656-7745. 1991. $15.95.

Encyclopedia of Physical Science and Technology. Robert A. Meyers, ed. 18 volumes. Academic Press Inc., 6277 Sea Harbor Drive, Orlando, FL 32887. 1992. $2100.00.

Jane's Aerospace Dictionary, B. Guston. 3d ed. Janes Info Group (order from:)Dept. DSM, 1340 Braddock Pl., Suite 300, Box 1436, Alexandria, VA 22314-1651. (703)683-3700. FAX (703) 836-0029. 1991. $45.00. McGraw-Hill DICTIONARY of SCIENTIFIC and TECHNICAL TERMS. Sybil P. Parker, ed. 5th edition. McGraw-Hill Publishing Company, 1221 Avenue of the Americas, New York, NY 10020. (800) 262-4729 or (212) 512-3825. 1993. $110.50.

McGraw-Hill Encyclopedia of Science and Technology, Sybil P. Parker, ed. 7th ed. 20 vols. McGraw-Hill Publishing Company, 1221 Avenue of the Americas, New York, NY 10020. (800) 262-4729 or (212) 512-3825. $1900.00.

SAE Dictionary of Aerospace Engineering. Society of Automotive Engineers, 400 Commonwealth Drive, Warrendale, PA 15096. (412) 776-4841. 1992. $85.00.

GENERAL WORKS

Advanced Aerospace Materials. Horst Buhl, editor. Springer-Verlag, 175 Fifth Avenue, New York, NY 10010. (212) 460-1500 or (800) 777-4643. FAX (212) 473-6272. 1992. $125.00.

Aerospace Science: History of Air Power. Superintendent of Documents, U.S.Government Printing office, Box 371954, Pittsburgh, PA 15250-7954. (202) 783-3238. FAX(202) 512-2233. 1987. $14.00.

Aircraft Systems. Ian Moir. Longman Publishing Group, the Longman Building, 10Bank Street, White Plains, NY 10606-1951. (914) 993-5000 or (800) 266-8855. FAX (914)997-8115. 1992. Inquire for cost and availability.

Applied Mathematics in Aerospace Science and Engineering. A. Miele& A. Salvetti, editors. Plenum Publishing Corporation, 233 Spring Street, New York, NY 10013. (212) 620-8000. FAX (212) 463-0742. 1994. $111.00.

The Future of Aerospace. National Academy Press, Division of the NationalAcademy of Science, 2101 Constitution Avenue, N.W., Lockbox 285, Washington, DC 20055.(800) 624-6242. FAX (202) 334-2793. 1993. $23.00.

Perspectives in Aerospace Design. Conrad F. Newberry, compiler. AmericanInstitute of Aeronautics and Astronautics, 370 L'Enfant Promenade SW, Washington, DC20024-2518. (202) 646-7471. 1991. $79.95.

HANDBOOKS AND MANUALS

Jane's All the World's Aircraft. Jane's Information Group, Dept. DSM,1340 Braddock Pl., Suite 300, Box 1436, Alexandria, VA 22314-1651. (703) 683-3700. FAX(703) 836-0029. Annual. $245.00. Also available on CD-ROM.

Jane's Avionics. Jane's Information Group, Dept. DSM, 1340 Braddock Pl., Suite300, Box 1436, Alexandria, VA 22314-1651. (703) 683-3700. FAX (703) 836-0029. Annual. $245.00. Also available on CD-ROM.

Standard Handbook of Engineering Calculations. Tyler G. Hicks, ed. 2nd ed. McGraw-Hill Publishing Company, 1221 Avenue of the Americas, New York, NY 10020.(800) 262-4729 or (212) 512-3825. 1985. $89.50.

The Wiley Engineer's Desk Reference. Sanford I. Heisler. John Wiley and Sons,Inc., 605 Third Avenue, New York, NY 10158. (800) 526-5368 or (212) 850-6000. 1984. $64.95.

ONLINE DATABASES AND CD-ROMS

Aerospace Database. American Institute of Aeronautics and Astronautics, 370L'Enfant Promenade, S.W., Washington, DC 20024. (202) 646-7400. Worldwide publishedliterature on research and development in aerospace and related areas, with abstracts. Covers 1962 to present. Available on DIALOG, (800) 334-2564, online service. Also available onCD-ROM. Inquire as to cost and availability.

Compendex Plus. Engineering Information, Inc., 345 East 47th Street, New York, NY10017. (212) 705-7600 or (800) 221-1044. Contains citations with abstracts to worldwideliterature in engineering and technology, from 1970 to present. Available on online BRS,(800)289-4277, DIALOG, (800) 334-2564, ORBIT (800) 456-7248, and STN International, FIZKarlsruhe, P.O. Box 2465, W-7500, Karlsruhe 1, Germany, online services. Also available onCD-ROM. Inquire as to cost and availability.

CSA Engineering. Cambridge Scientific Abstracts, 7200 Wisconsin Avenue, Suite 601,Bethesda, MD 20814. (301) 961-6750 or (800) 843-7751. Contains citations and abstracts ofinternational periodicals and research literature covering all fields of engineering and science andtechnology,including computer and information science, electronics, mechanical engineering,solid state materials, 1981 to present. Available on BRS,(800) 289-4277, online service.Inquireas to cost and availability.

Current Contents Search. Institute for Scientific Information, 3501 Market Street,Philadelphia, PA 19104. (215) 386-0100. FAX (215) 386-6362. Contains citations to articles listed in the table of contents of science and technology journals. Also articles in social sciencesand life sciences journals. Available on BRS,(800) 289-4277, DIALOG,(800) 334-2564, online services. Inquire as to cost and availability.

Dissertation Abstracts. University Microfilms International, 300 North Zeeb Road, Ann Arbor, MI 48106. (800) 521-0600 or (313) 761-4700. Scope includes virtually all doctoral dissertations accepted at accredited American institutions from 1861 to present in 252 subject areas. Available on BRS,(800) 289-4277, DIALOG,(800) 334-2564, and OCLC EPIC,(800) 848-5878, online services. Also available on CD-ROM. Inquire as to cost and availability.

International Aerospace Abstracts. American Institute of Aeronautics and Astronautics, 370 L'Enfant Promenade, S.W., Washington, DC 20024. (202) 646-7400. Contains references and abstracts of journal and monograph literature relating to aerospace science and technology, from 1963 to present. Available through the NASA/RECON system of the National Aeronautics and Space Administration only.

*ISMEC: Mechanical Engineering Abstracts.*Cambridge Scientific Abstracts, 7200 Wisconsin Avenue, Suite 601, Bethesda, MD 20814. (301) 961-6750 or (800) 843-7751. Contains citations to the literature in mechanical engineering, industrial and production engineering, energy, power, mechanics, devices and related areas, from 1973 to present. Available on the DIALOG,(800) 334-2564, online service. Inquire as to cost and availability.

NASA Database. American Institute of Aeronautics and Astronautics, 370 L'EnfantPromenade, S.W., Washington, DC 20024. (202) 646-7400. Citations and abstracts of aeronautics and astronautics literature, 1962 to present. Also contains citations from STAR,SCIENTIFIC and TECHNICAL AEROSPACE REPORTS, and INTERNATIONAL AEROSPACE ABSTRACTS. Available through NASA/RECON online service. Inquire as to cost and availability.

NTIS Bibliographic Database. National Technical Information Service, 5285 Port Royal Road, Springfield, VA 22161. (703) 487-4929 or FAX (703) 321-8199. Broad coverage of government-sponsored science and technology research reports, 1964 to present. Available on BRS,(800) 289-4277, DIALOG, (800)334-2564, ORBIT, (800) 456-7248, and STN International, FIZ Karlsruhe, P.O. Box 2465, W-7500, Karlsruhe 1, Germany, online services. Also available on CD-ROM. Inquire as to cost and availability.

SAE Global Mobility Database. Society of Automative Engineers (SAE), Electronic Publishing Division, 400 Commonwealth Drive, Warrendale, PA 15098. (412) 776-4841. Contains citations with abstracts to technical papers on automotive and aerospace technology and vehicular-related industries that have been presented at SAE conferences. Covers 1906 to present.

AEROSPACE ENGINEERING

Ency. of Physical Sciences and Engineering Info. Sources

Available on ORBIT,(800) 456-7248,online service. Inquire as to cost and availability.

Scientific and Technical Books and Serials in Print. R.R. Bowker Inc., 121 Chanlon Road, New Providence, NJ 07974. (800) 521-8110 or (908) 464-6800. List of currently published books and serials in the physical and biological sciences, engineering and technology, with subject, author and titles indexes. Available on ORBIT,(800) 456-7248, online service.Inquire as to cost and availability.

SCISEARCH. Institute for Scientific Information, 3501 Market Street, Philadelphia, PA 19104. (800) 523-1850 or (215) 386-0100. Broad multidisciplinary title and author index to the international literature of science and technology, 1974 to present. Available on DIALOG,(800) 334-2564, and ORBIT,(800) 456-7248, online services. Also available on CD-ROM. Inquire as to cost and availability.

TRIS (Transportation Research Information). National Academy of Science, 2101 Constitution Avenue, N.W., Washington, DC 20418. (202) 334-3313 or (800) 624-6242. Citations with abstracts of literature on transportation, including air, highway, rail, maritime and other modes from 1968 to present. Available on DIALOG,(800) 334-2564, online service. Inquire as to cost and availability.

WILSONLINE. H.W. Wilson Company, 950 University Avenue, Bronx, NY 10452. (800) 367-6770 or (212) 588-8400. Makes available online versions of the printed H.W. Wilson indexes including Applied Science and Technology Index, Business Periodicals Index, General Science Index, and Readers' Guide to Periodical Literature. Period covered is generally 1983 to present. Available on BRS,(800) 289-4277, DIALOG,(800) 334-2564, and OCLC EPIC,(800)848-5878, online services. Also available on CD-ROM. Inquire as to cost and availability.

PERIODICALS

Aerospace America. American Institute of Aeronautics and Astronautics, 370 L'Enfant Promenade SW, Washington, DC 20024-2518. (202) 646-7471. 1932 to present. Monthly. $70 annually (non-members).

Aerospace Daily. McGraw-Hill Inc., Aviation Week Group, 1200 G Street NW, Suite 200, Washington, DC 20005. (202) 383-2350. 1973 to present. Five times a week. $1340.00 per year.

Aerospace Engineering. Society of Automotive Engineers, 400 Commonwealth Dr., Warrendale, PA 15096. (412) 776-4841. 1981 to present. Monthly. $48.00 per year.

Aerospace Propulsion. McGraw-Hill Inc., Aviation Week Group, 1200 G Street NW, Suite 200, Washington, DC 20005. (202) 383-2350. 1963 to present. 25 times a year. $550.00 a year.

Aircraft Engineering and Aerospace Technology. BunHill Publications Ltd., 127 Stanstead Rd., London SE23 1JE, England. 1929 to present. Monthly. $102.00 per year (U.S.).

Aviation & Aerospace. Baxter Publishing Company, 310 Dupont Street, Toronto, Ontario M5R 1V9 Canada. (416) 968-7252. FAX (416) 968-2377. 1928 to present. Bi-monthly. $42.00 per year.

Aviation Week and Space Technology. 1221 Avenue of the Americas, New York, NY 10020. (212) 512-2999 or (800) 525-5003. 1916 to present. Weekly. $82.00 per year.

IEEE Transactions On Aerospace and Electronic Systems. Institute of Electrical and Electronics Engineers, 445 Hoes Lane, PO Box 1331, Piscataway, NJ 08855-1331. (908) 981-0060. 1965 to present. Quarterly. $10.00 per issue (members), $20.00 per issue (non-members).

Interavia/aerospace World. Aerospace Media Publishing, Swissair Centre, 31 Route de l'Aeroport, POB Box 437, 1215 Geneva 15, Switzerland. (022) 788 27 88. FAX (022) 788 27 26. 1987 to present. Monthly. $128.00 per year (U.S.).

Journal of Aircraft. American Institute of Aeronautics and Astronautics, 370 L'Enfant Promenade SW, Washington, DC 20024-2518. 1963 to present. Bimonthly. $42.00 (members) and $230.00 (non-members).

Journal of Guidance, Control and Dynamics. American Institute of Aeronautics and Astronautics, 370 L'Enfant Promenade SW, Washington, DC 20024-2518. 1978 to present. Bimonthly. $42.00 (members) and $250.00 (non-members).

Progress in Aerospace Sciences. Pergamon Press, 660 White Plains Rd., Tarrytown, NY 10591-5153. 1961 to present. Quarterly. $320.00 per year.

RESEARCH CENTERS AND INSTITUTES

Aerospace Research Applications Center, Indianapolis Center For Advanced Research, 611 N. Capitol Ave., Indianapolis, in 46204. (317) 262-5003. FAX (317) 262-5044.

Aerospace Research Center, 1250 Eye St., NW, Washington, DC 20005. (202) 371-8400. FAX (202) 371-8470.

Aviation Safety Institute, 6797 N. High St., Suite 316, PO Box 304, Worthington, OH 43085. (614) 885-4242. FAX (614) 885-5891.

Center For Aerospace Technology, Weber State College, Ogden, UT 84408-1805. (801) 626-7272.

Center For Control and Systems Research, University of Texas At Austin, aerospace Engineering and Engineering Mechanics, Wrw 408, Austin, TX 78712-1085. (512) 471-4908. FAX (512) 471-3788.

Flight Research Laboratory, University of Kansas, RaymondNichols Hall, lawrence, KS 66045. (913) 864-3043. FAX (913) 864-7789.

Joint Institute For Advancement of Flight Sciences, George Washington university, Langley Research Center, Mail Stop 269, Hampton, VA 23665. (804) 864-1982. FAX (804) 864-5894.

Ohio Aerospace Institute, 2001 Aerospace Parkway, Brook Park, OH 44142. (216) 891-2100. FAX (216) 891-2140.

AGGREGATES

See also: BUILDING MATERIALS, CEMENT, CONCRETE

ABSTRACT SERVICES AND INDEXES

Applied Science and Technology Index; A Cumulative Subject Index To English language Periodicals in the Fields of Aeronautics and Space Science, Computer Technology, chemistry, Construction Industry, Energy and Related Areas. H.W. Wilson Co., 950 University Avenue, Bronx, NY 10452. (800) 367-6770 or (212) 588-8400. FAX (718) 590-1617. From 1958 to present. Monthly. Inquire about cost and availability. Also available on CD-ROM and online.

Concrete Abstracts. American Concrete Institute, Box 19150 Redford Station, Detroit, MI 48219. (313) 532-2600. FAX (313) 538-0655. 1972 to present. Bi-monthly. $180.00 per year for non-members.

Engineering Index Monthly. Engineering Information, Inc., Castle Point on the Hudson, Hoboken, NJ 07030. (800) 221-1044. FAX (212) 832-1857. Monthly. $2300.00 per year. Also available online as *COMPENDEX*, and also on CD-ROM. Covers chemical engineering, computers, electrical engineering, civil engineering, metals and mining, industrial management, and mechanical engineering.

ANNUAL REVIEWS AND YEARBOOKS

U.S. Department of the Interior, Bureau of Mines, Minerals Yearbook. Superintendent of Documents, U.S. Government Printing office, Box 371954, Pittsburgh, PA 15250-7954. (202) 783-3238. FAX (202) 512-2233. 1932 to present. Annual. Price varies.

ASSOCIATIONS AND PROFESSIONAL SOCIETIES

American Concrete Institute. Box 19150, Redford Station, Detroit, MI 48219. (313) 532-2600. FAX (313) 538-0655.

Bituminous and Aggregate Equipment Bureau. 111 East Wisconsin Avenue, Suite 940, Milwaukee, WI 53202. (414) 272-0943.

European Cement Association/cimeurope. Rue d'Arlon 55, B-1040 Brussels, Belgium. Telephone 32-2-2341011. FAX 32-2-2304720.

Expanded Shale Clay and Slate Institute. 6218 Montrose Road, Rockville, MD 20852. (301) 231-9497.

National Aggregates Association. 900 Spring Street, Silver Spring, MD 20910. (301) 587-1400. FAX (301) 585-4219.

National Stone Association. 1415 Elliot Place NW, Washington, DC 20007. (202) 342-1100. FAX (202) 342-1415.

BIBLIOGRAPHIES

Aggregates (Building Materials): Monographs. Mary A. Vance. Vance Bibliographies, PO Box 229, Monticello, IL 61856. 1985. Out of print.

DIRECTORIES AND BIOGRAPHICAL SOURCES

American Cement Directory. Bradley Pulverizer Company, Box 1318, 123 S.Third Street, Allentown, PA 18105. (215) 434-5191. FAX (215) 770-9400. 1910 to present. Annual. $62.00.

American Concrete Institute Membership Directory. American ConcreteInstitute, Box 19150, Redford Station, Detroit, MI 48219. (313) 532-2600. FAX (313)538-0655. Inquire for price and availablility.

National Stone Association—Buyer's Guide. National Stone Association,1415 Elliot Place NW, Washington, DC 20007. (202) 342-1100. FAX (202) 342-1415. Free.

World Cement Directory. European Cement Association/Cimeurope, Rue d'Arlon 55, B-1040 Brussels, Belgium. Telephone 32-2-2341011. FAX 32-2-2304720. 1958 to present. Quinquennial. $600.00.

ENCYCLOPEDIAS AND DICTIONARIES

Cement and Concrete Terminology. American Concrete Institute Staff. American Concrete Institute, Box 19150, Redford Station, Detroit, MI 48219. (313) 532-2600. FAX (313) 538-0655. 1990. $33.95. 3

Mcgraw-Hill Encyclopedia of Science and Technology. Sybil P. Parker, ed. 7th edition. 20 volumes. McGraw-Hill Publishing Company, 1221 Avenue of the Americas, New York, NY 10020. (800) 262-4729 or (212) 512-3825. 1992. $1900.00

GENERAL WORKS

Aggregates, Sand, Gravel & Crushed Rock Aggregates For construction Purposes. M.R. Smith & L. Collis, editors. American Association of Petroleum Geologists, PO Box 979, Tulsa, OK 74101-0979. (918) 584-2555 or (800) 364-2274. 1993. $65.00.

Innovation in Aggregate Testing. Transportation Research Board, National Research Council. National Academy Press, Division of the National Academy of Science, 2101 Constitution Avenue, N.W., Lockbox 285, Washington, DC 20055. (800) 624-6242. FAX (202)334-2793. 1989. Inquire for cost and availability.

Natural Aggregate, Building America's Future. William H. Langer and V.M. Glanzman. U.S. Department of the Interior, U.S. Geological Survey, USGS Map Distribution, Denver, CO 80225. 1993. Inquire for cost and availability.

Standards For Aggregates. D.C. Pike, editor. E. Horwood Publishing, distributed by Prentice Hall Inc., 113 Sylvan Avenue, Route 9W, Englewood Cliffs, NJ 07632. (201)592-2000 or (800) 922-0579. 1990. Inquire for cost and availability.

Structural Lightweight Aggregate Concrete. John L. Clarke, editor. Blackie Academic & Professional/Chapman & Hall, 29 West 35th Street, New York, NY 10001. (800) 842-3636. FAX (212) 563-2269. 1993. Inquire for cost.

HANDBOOKS AND MANUALS

Aci Manual of Concrete Practice. 5 volumes. American Concrete Institute, Box 19150, Redford Station, Detroit, MI

AGGREGATES

Ency. of Physical Sciences and Engineering Info. Sources

48219.(313) 532-2600. FAX (313) 538-0655. 1967 to present. Annual. $423.00 per set for non-members.

The Aggregate Handbook. National Stone Association, 1415 Elliot Place NW, Washington, DC 20007. (202) 342-1100. FAX (202) 342-1415. $60.00 for non-members.

Durability Testing of Concrete and Aggregates: Users Manual. David Whiting. Strategic Highway Research Program, National Research Council, 2101 Constitution Avenue NW, Washington, DC 20418. 1994. Inquire for cost.

Pit & Quarry Mining Reference Manual and Buyers Guide. Advanstar Communications Inc., 7500 Old Oak Blvd., Cleveland, OH 44130. (216) 826-2839 or (800) 598-6008. FAX (216) 891-2726. 1907 to present. Annual. $25.00.

ONLINE DATABASES AND CD-ROMS

Compendex Plus. Engineering Information, Inc., 345 East 47th Street, New York, NY 10017. (212) 705-7600 or (800) 221-1044. Contains citations with abstracts to worldwide literature in engineering and technology, from 1970 to present. Available on online BRS,(800) 289-4277, DIALOG, (800) 334-2564, ORBIT (800) 456-7248, and STN International, FIZ Karlsruhe, P.O. Box 2465, W-7500, Karlsruhe 1, Germany, online services. Also available on CD-ROM. Inquire as to cost and availability.

CSA Engineering. Cambridge Scientific Abstracts, 7200 Wisconsin Avenue, Suite 601,Bethesda, MD 20814. (301) 961-6750 or (800) 843-7751. Contains citations and abstracts of international periodicals and research literature covering all fields of engineering and science and technology,including computer and information science, electronics, mechanical engineering, solid state materials, 1981 to present. Available on BRS,(800) 289-4277, online service.Inquire as to cost and availability.

Current Contents Search. Institute for Scientific Information, 3501 Market Street, Philadelphia, PA 19104. (215) 386-0100. FAX (215) 386-6362. Contains citations to articles listed in the table of contents of science and technology journals. Also articles in social sciences and life sciences journals. Available on BRS,(800) 289-4277, DIALOG,(800) 334-2564, online services. Inquire as to cost and availability.

GEOREF: Bibliography and Index of Geology. American Geological Institute, 4220 King Street, Alexandria, VA 22302. (703) 379-2480. Fax (703) 379-7563. Monthly. Inquire for price and availability.

SCISEARCH. Institute for Scientific Information, 3501 Market Street, Philadelphia, PA 19104. (800) 523-1850 or (215) 386-0100. Broad multidisciplinary title and author index to the international literature of science and technology, 1974 to present. Available on DIALOG,(800) 334-2564, and ORBIT,(800) 456-7248, online services. Also available on CD-ROM. Inquire as to cost and availability.

TRIS (Transportation Research Information). National Academy of Science, 2101 Constitution Avenue, N.W., Washington, DC20418. (202) 334-3313 or (800) 624-6242.Citations with abstracts of literature on transportation, including air, highway, rail, maritime and other modes from 1968 to present. Available on DIALOG,(800) 334-2564, online service. Inquire as to cost and availability.

WILSONLINE. H.W. Wilson Company, 950 University Avenue, Bronx, NY 10452. (800) 367-6770 or (212) 588-8400. Makes available online versions of the printed H.W. Wilson indexes including Applied Science and Technology Index, Business Periodicals Index, General Science Index, and Readers' Guide to Periodical Literature. Period covered is generally 1983 to present. Available on BRS,(800) 289-4277, DIALOG,(800) 334-2564, and OCLC EPIC,(800) 848-5878, online services. Also available on CD-ROM. Inquire as to cost and availability.

PERIODICALS

American Concrete Institute Materials Journal. American Concrete Institute, Box 19150, Redford Station, Detroit, MI 48219. (313) 532-2600. FAX (313) 538-0655. 1929 to present. Bi-monthly. $101.00 per year for non-members.

Cement and Concrete Research. Elsevier Science, 660 White Plains Rd., Tarrytown, NY 10159-5153. (919) 524-9200. FAX (919) 333-2444. 1971 to present. Eight times a year. $590.00 per year.

Cement, Concrete and Aggregates. American Society for Testing and Materials, 1916 Race Street, Philadelphia, PA 19103. (215) 299-5585. 1979-present. Semi-annual. $48.00 per year for non-members.

Cement, Quarry & Mineral Aggregate Newsletter. National Safety Council, Periodicals Department, 1121 Spring Lake Drive, Itasca, IL 60143. (708) 285-1121. Bi-monthly. $19.00 per year.

Concrete. The Concrete Society, Framewood Road, Wexham, Slough, Berks. SL3 6PJ England. Telephone 0753-662226. FAX 0753-662126. 1966 to present. Bi-monthly. $75.00 per year.

Concrete International: Design and Construction. American Concrete Institute, Box 19150, Redford Station, Detroit, MI 48219. (313) 532-2600. FAX (313) 538-0655. 1979 to present. Monthly. $101.00 per year.

Pit & Quarry. Advanstar Communications Inc., 7500 Old Oak Blvd., Cleveland, OH 44130. (216) 826-2839. FAX (216) 891-2726. 1916 to present. Monthly. $35.00 per year.

Stone Review. National Stone Association, 1415 Elliot Place NW, Washington, DC 20007. (202) 342-1100. FAX (202) 342-1415. Bi-monthly. $36.00 per year for non-members.

RESEARCH CENTERS AND INSTITUTES

American Society For Testing and Materials. 1916 Race Street, Philadelphia, PA 19103. (215) 299-5585.

Concrete Materials Research Council. PO Box 19150, Detroit, MI 48219. (313) 532-2600.

Concrete Research Laboratory. University of Michigan, G.G. Brown Laboratory, Ann Arbor, MI 48109. (313) 764-9660.

Portland Cement Association. 5420 Old Orchard Road,Skokie, IL 60077. (708) 966-6200. FAX (708) 966-8389.

SPECIFICATIONS AND STANDARDS

Annual Book of ASTM Standards, Volume 04.02. American Society for Testing and Materials (ASTM), 1916 Race Street, Philadelphia, PA 19103. Annual. Inquire for cost and availability.

Cement Standards: Evolution and Trends. P.K. Mehta, editor. American Society for Testing and Materials (ASTM), 1916 Race Street, Philadelphia, PA 19103. (215) 299-5585. 1979. $20.00.

Cement Standards of the World. 4th edition. European Cement Association/ Cimeurope, Rue d'Arlon 55, B-1040 Brussels, Belgium. Telephone 32-2-2341011. FAX 32-2-2304720. 1991. Inquire for cost and availability.

AGRICULTRAL ENGINEERING

See also: BIOTECHNOLOGY, DRAINAGE, MECHANICAL ENGINEERING, METEOROLOGY, WATER

ABSTRACT SERVICES AND INDEXES

Agricultural Engineering Abstracts. CAB International. North American office, 845 North Park Avenue, Tuscson, AZ 85719. (800) 528-4841. 1976 to present. Monthly. $453.00 per year. ISSN: 0308-8863.

Agricultural Engineering Index. American Society of Agricultural Engineers. 2950 Niles Road, St. Joseph, MI 49085-9659. (616) 429-0300. (FAX) 616 429-3852. Contains citations of the literature on agricultural engineeringfrom 1907 to present. Irregular. Vol.5, 1986-1990 latest available. $35.00. ISBN: 0-685-72648-7.

Applied Science and Technology Index; A Cumulative Subject Index To English Language Periodicals in the Fields of Aeronautics and Space Science, Computer Technology, Chemistry , Construction Industry, Energy and Related Areas. H. W. Wilson Co., 950 University Avenue, Bronx, NY 10452-9978. (212) 588-8400. (800) 367-6770. FAX (718) 590-1617. From 1958 to present. Monthly. Inquire about cost and availability. Also available online (BRS and WILSONLINE) and on CD-ROM. ISSN: 0003-6986.

Bibliography and Index of Geology. American Geological Institute, 4220 King Street, Alexandria, VA 22302-1507. (703) 379-2480. FAX (703) 379-7563. From 1969 to present. Monthly. $1295.00 per year. Also available as GEOREF online (CISTI, DIALOG, Orbit, STN) and on CD-ROM. Inquire about price and availability. ISSN: 0098-2784.

Current Contents: Engineering, Technology and Applied Sciences. Institute for Scientific Information, 3501 Market Street, Philadelphia, PA 19104. (215) 386-0100. FAX (215) 386-2991. From 1970 to present. Weekly. $442.00 per year. Also available online. Inquire regarding cost and availability. ISSN: 0095-7917.

Engineering Index. Engineering Information, Inc., Castle Point on the Hudson, Hoboken, NJ 07030. (201) 216-8500. (800) 221-1044. FAX (201) 216-8532. Monthly. $2300.00 per year for monthly issues. $1980 per year for annual cumulation. Available onCD-ROM and online as COMPENDEX. ISSN: 0742-1974; 0360-8557.

General Science Index. H.W. Wilson Company, 950 University Avenue, Bronx, NY 10452-997. (212) 588-8400. (800) 367-6770. FAX (718) 590-1617. From 1978 to present. Ten issues per year; quarterly and annual cumulations. Service basis. Available on CD-ROM and online. Inquire regarding cost and availability. ISSN: 0162-1963.

Meteorological and Geoastrophysical Abstracts. American Meteorological Society, c/o Inforonics, Inc., 550 Newtown Road, Box 458, Littleton, MA 01460. (508) 486-8976. FAX (508) 486-0027. Covers literature in environmental sciences, meteorology, astrophysics, hydrology, glaciology, and physical oceanography. 1950 to present. Monthly. $950.00 per year. ISSN: 0026-1130. Also available online (DIALOG) and on CD-ROM.

ANNUAL REVIEWS AND YEARBOOKS

Advances in Applied Mechanics. Academic Press, Inc., 6277 Sea Harbor Drive, Orlando, FL 32887. (619) 231-0926. (800) 321-5068. FAX (619) 699-6715. 1948 to present. Irregular. Price varies, inquire. ISSN: 0065-2156.

Annual Review of Fluid Mechanics. Annual Reviews, Inc., 4139 Camino Way, P.O. Box 10139, Palo Alto, CA 94303-0897. (800) 523-8635 or (415) 493-4400. 1969 to present. Annual. $47.00 per year. ISSN: 0066-4189.

ASSOCIATIONS AND PROFESSIONAL SOCIETIES

American Society of Agricultural Engineers. 2950 Niles Road, St. Joseph MI 49085-9659. (616) 429-0300. FAX (616) 429-3852.

American National Standards Institute. 1430 Broadway, New York, NY 10018-3363. (212) 354-4916.

American Society For Engineering Education, 2070 Neil Avenue, columbus, OH 43210-1226. (614) 292-2893.

American Society For Testing and Materials. 1916 Race Street, Philadelphia, PA 19103-1180. (215) 299-5400.

Equipment Manufacturers Institute. 10 South Riverside Plaza, Suite 1220, Chicago, IL 60606. (312) 321-1470. FAX (312) 321-1480.

Irrigation Association. 8260 Willow Oak Corporation Drive, Fairfax, VA 22031. (703) 573-3551. FAX (703) 573-1913. Alexandria, VA 22314. (703) 549-2207.

National Society of Professional Engineers, 1420 King Street, Alexandria,VA 22314-2750. (703) 684-2835.

BIBLIOGRAPHIES

Information Sources in Energy Technology. L.J. Anthony, editor. Butterworths Publishers, 80 Montvale Avenue, Stoneham, MA 02180. (617) 438-8464 or (800) 366-2665. 1988. $105.00.

The Literature of Agricultural Engineering. Carl W.Hall and Wallace C. Olsen, editors. Cornell University Press, P.O. Box 6525, Ithaca, NY 14851. (607) 277-2211. (800) 666- 2211. 1993. $62.50. ISBN: 0-8014-2812-2.

AGRICULTURAL ENGINEERING

Ency. of Physical Sciences and Engineering Info. Sources

Scientific An D Technical Books and Serials in Print; An Index To literature in Science and Technology. R.R. Bowker Inc.,121 Chanlon Road, New Providence, NJ 07974. (908) 464-6800. (800) 521-8110. FAX (908) 665-3502. 1972 to present. Annual. 4 volumes. 1994. $299.95. Also available on compact disc and online. ISSN: 0000-054X.

DIRECTORIES AND BIOGRAPHICAL SOURCES

Agricultural Engineering - Technology Issue. American Society of Agricultural Engineers. 2950 Niles Road, St. Joseph, MI 49085. (616) 429-0300. FAX (616) 429-3852. Annual.

Agricultural Research Centers: A Worldwide Directory of Organizations and programs. Stockton Press, 345 Park Avenue, New York, NY 10010. (212) 689-9200. (800) 221- 2123. 12th edition. 1995. $595.00

Agriculture and Veterinary Science International Who's Who. Gale Research Company, 835 Penobscot Building, Detroit, MI 48226-4094. (800) 877-4253. (313) 961-2242. Triennial. 2 volumes. 1990. $550.00.

American Men and Women of Science: Physical and Biological Sciences. R. R. Bowker Inc., 121 Chanlon Road, New Providence, NJ 07974. (908) 464-6800. (800) 521-8110. 20th edition. 8 volumes. 1996. $850.00.

Directory of American Agriculture. Agricultural Resources and Communications, PO Box 67212, Topeka, KS 66667. (913) 456-9705. 1991/92. $50.00.

Directory of Engineering College Research and Graduate Study. American Society for Engineering Education, Ohio State University, 2070 Neil Avenue, Columbus, OH 43210-1226. Annual. $20.00.

Directory of Engineering Societies and Related Organizations. American Association of Engineering Societies, 1111 19th Street, Suite 608, Washington, DC 20036-3603. (202) 296-2237. Semi-annual. $150.00.

Engineering Research Centers: Incorporating Electronics research Centers. Stockton Press, 345 Park Avenue, New York, NY 10010. (212) 689-9200. (800) 221-2123. 4th edition. 1995. $515.00.

Farm and Power Equipment-dealers Buyers Directory Issue. National Farm and Power Equipment Dealers Association, 9701 Gravois Avenue, St. Louis, MO 63123. (314) 638-4050. (800) 451-0914. FAX (314) 638-3880. Annual, in January issue. $14.00.

International Directory of Agricultural Engineering Institutions. Food and Agricultural Organization of the United Nations. Available from Bernan-Unipub, 4611-F Assembly Drive, Lanham, MD. 20706-4391. 1983. $35.00

International Engineering Directory. American Consulting Engineers Council, 1015 15th Street, N.W. Suite 802, Washington, DC 20005-2670. (202) 347-7474. Annual. $10.00.

Professional Workers in State Agricultural Experiment stations and Other Cooperative State Institutions. Science and Education Administration - Cooperative State Research Service, Department of Agriculture, JSM Building, Room 25 Washing-

ton, DC 20250. Annual. $ per year. Send orders to Government Printingoffice, Washington, DC 20402.

Research Centers Directory. Gale Research, 835 Penobscot Building, Detroit, MI 48226-4094. (313) 961- 2242. (800) 877-4253. 20th edition. 1995. $485.00. ISSN: 0080-1518.

Who's Who in Engineering. Gordon Davis, editor. American Association of Engineering Societies. 1111 19th Street, NY, Suite 608, Washington, DC 20036. (202) 296-2237. (800) 658-8897. 9th edition. 1995. $220.00.

Who's Who in Technology. Gale Research Inc., 835 Penobscot Building, Detroit, MI 48226-4094. (313 961-2242. (800) 521-4253. 1995. $195.00. ISBN: 0-8103-7467-6.

ENCYCLOPEDIAS AND DICTIONARIES

ASAE Standards Book. American Society of Agricultural Engineers, 2950 Niles Road, St. Joseph, MI 49085. (616) 429-0300. FAX (616) 429-3852. Annual. $125.00. ISBN: 0- 929355-50-4.

Academic Press Dictionary of Science and Technology. Christopher Morris, editor. Academic Press, Inc., 1250 Sixth Avenue, San Diego, CA 92101. (619) 231-0926. FAX (619) 699-6715. 1991. $115.00.

Chambers Science and Technology Dictionary. Peter M.B. Walker, editor. Cambridge University Press, 40 West 20th Street, New York, NY 10011-4211. (212) 924-3900. 1988. $39.95.

Construction Glossary: An Encyclopedic Reference and Manual. J.Stewart Stein. John Wiley and Sons, Inc., 605 Third Avenue, New York, NY 10158. (800) 526-5368 or (212) 850-6000. Second edition. 1993. $95.00.

Dictionary of Agricultural & Food Engineering. Arthur Farrall and James A. Basselman. Interstate Publishers, Inc. P.O. Box 50, Danville, IL 61834-0050. (217) 446-0500. 1979. $26.60. ISBN: 0-8134-2023-7.

EI Thesaurus. Jessica L. Milstead, editor. Engineering Information, Inc., Castle Point on the Hudson, Hoboken, NJ 07030. (800) 221-1044. FAX (201) 216-8532. 1992. $130.00

McGraw-Hill Encyclopedia of Engineering. Sybil P. Parker, editor. McGraw-Hill Book Company, Inc. 1221 Avenue of the Americas, New York, NY 10020. (212) 997-3675. Second edition. 1993. $95.50. ISBN: 0-07-051392-9.

McGraw-Hill Encyclopedia of Science and Technology. McGraw-Hill Book, Incorporated, 1221 Avenue of the Americas, New York, NY 10020. (212) 997-3675. (800) 262-4729. Seventh edition. Twenty volumes. 1992. $1900.00.

GENERAL WORKS

Advances in Agriculture: a User Friendly Guide to the Latest Technology. Dan Burrus and Patti Thomsen. International Management Publications. Distributed by Burrus Research Associates, 12201 West Burleigh, Milwaukee, WI 53222. (414) 774-7790. (800) 827-6770. 1991. $9.95.

Agricultural and Horticultural Engineering. Clifford J. Studman. Butterworth-Heinemann, 313 Washington Street, Newton, MA

02158. (617) 928-2500. (800) 366-2665. 1990. $59.95. ISBN:0-409-60469-0.

Agricultural Machines: Theory of Operation, Computation of controlling Parameters and the Conditions of Operation. N. I. Klenin. Ashgate Publishing Company, Old Post Road, Brookfield, VT 05036. (802) 276-3162. 1985. $190.00. ISBN: 90-6191-448-5.

Agricultural Mechanics: Fundmentals and Applications. Elmer L. Cooper. Delmar Publishers, 3 Columbia Circle, Box 15015, Albany, NY 12212. (518) 464-3500. (800) 347-7707. 3rd edition. 1996. $39.95. ISBN: 0-8273-6854- 2.

Applied Numerical Methods For Food and Agricultural engineers. Prabir K. Chandra and R. Paul Singh. CRC Press Inc., 2000 Corporate Boulevard, NW, Boca Raton, FL 33431. (407) 994-0555. (800) 272-7737. 1994. $101.50.

Beef Housing and Equipment Handbook. Midwest Plan Service Engineers staff. MidWest Plan Service, 122 Davidson Hall, Ames, IA 50011-3080. (515) 294-4337. (800) 562-3618. 4th edition. 1987. $7.00. ISBN: 0-89373-068-8.

Engineering Applications in Agriculture. W. Bowers and others. Stipes Publishing Company, 1-12 Chester Street, Champaign, IL 61820. (217) 356-8391. 1986. $17.80. ISBN: 0-87563-281-5.

The Mechanical Design Process. David G. Ullman. McGraw-Hill Publishing Company, 1221 Avenue of the Americas, New York, NY 10020. (800) 262-4729 or (212) 512-3825. 1992. $45.95.

Mechanics in Agriculture. Lloyd J. Phipps and Carl I. Reynolds. Interstate Publications Inc., Box 50, Danville, IL 61834-0050. (217) 446-0500. (800) 843-4774. 4th edition. 1992. $43.95. ISBN: 0-8134-2915-0.

Principles of Agricultural Machines. Ajit Sravastava, et. al. American Society of Agricultural Engineers. 2950 Niles Road, St. Joseph, MI 49085. (616) 429-0300. FAX (616) 429-3852. 1993. $54.00. ISBN: 0-929355-33-4.

Principles of Farm Irrigation System Design. Larry G. James. Kriger Publishing Company, P.O. Box 9542, Melbourne, FL 32902-9542. (407) 724-9542. 1993. $81.95. ISBN: 0- 89464-802-0.

HANDBOOKS AND MANUALS

CRC Handbook of Chemistry and Physics. David R. Kide, editor. CRC Press Inc., 2000 Corporate Blvd., NW, Boca Raton, FL 33431. (407) 994-0555. (800) 333-8300. 77th edition. 1996. $99.95.

Engineering Tables and Data. A.M. Howatson, P.G. Lund and J.D. Todd. Van Nostrand Reinhold, 115 Fifth Avenue, New York, NY 10003. (212) 254-3232 or (800) 926-2665. Second Edition. 1991. $39.95.

Handbook of Agricultural Meteorology. John F. Griffith, editor. Oxford University Press, Inc., 200 Madison Avenue, New York, NY 10016. (212) 725-6000. (800) 334-4249. 1994. $95.00.

Machinery's Handbook. Robert Green, editor. Industrial Press, Inc. 200 Madison Avenue, New York, NY 10016. Twenty-fourth edition. 1992. $65.00.

Standard Handbook of Engineering Calculations. Tyler G. Hicks. McGraw-Hill Publishing Company, 1221 Avenue of theAmericas, New York, NY 10020. (800) 262-4729 or (212) 512- 3825. 3rd edition. 1994. $99.50. ISBN: 0-07-028812-7.

The Wiley Engineer' S Desk R Eference: A Guide For the professional Engineer. Sanford I. Heisler. Wiley- Interscience; distributed by John Wiley & Sons, Inc., 605 Third Avenue, New York, NY 10158. (800) 526-5368 or (212) 850-6000 1984. $79.95. ISBN: 0-471-8663206.

ONLINE DATABASES AND CD-ROMS

CA Search. Chemical Abstracts Service, P.O. Box 3012, Columbus, OH 43210-0012. (614) 447-3600. (800) 848-6533. FAX (614) 447-3709. Very comprehensive guide to worldwide chemical literature and related fields, 1972 to present. Available on BRS,(800) 289-4277, DIALOG, (800) 334-2564, ORBIT (800) 456-7248, and STN International, FIZ Karlsruhe, P.O. Box 2465, W-7500, Karlsruhe 1, Germany, online services. Inquire as to cost and availability.

Compendex Plus. Engineering Information, Inc., 345 East 47th Street, New York, NY 10017. (212) 705-7600 or (800) 221-1044. Contains citations with abstracts to worldwide literature in engineering and technology, from 1970 to present. Available on online BRS,(800) 289-4277, DIALOG, (800) 334-2564, ORBIT (800) 456-7248, and STN International, FIZ Karlsruhe, P.O. Box 2465, W-7500, Karlsruhe 1, Germany, online services. Also available on CD-ROM. Inquire as to cost and availability.

Current Contents Search. Institute for Scientific Information, 3501 Market Street, Philadelphia, PA 19104. (215) 386-0100. FAX (215) 386-6362. Contains citations to articles listed in the table of contents of science and technology journals. Also articles in social sciences and life sciences journals. Available on BRS,(800) 289-4277, DIALOG,(800) 334-2564, online services. Inquire as to cost and availability.

Dissertation Abstracts Online. University Microfilms International, 300 North Zeeb Road, Ann Arbor, MI 48106. (800) 521-0600 or (313) 761-4700. Scope includes virtually all doctoral dissertations accepted at accredited American institutions from 1861 to present in 252 subject areas. Available on BRS, (800) 289-4277, DIALOG, (800) 334-2564, and OCLC EPIC,(800) 848-5878, online services. Also available on CD-ROM. Inquire as to cost and availability.

Geomechanics Abstracts. University of London, Imperial College of Science and Technology, Rock Mechanics Information Service, Royal School of Mines, Prince Consort Road, London SW7 2BP, England. Telephone: 071-589 5111 or FAX 071-589 6806. Scope includes worldwide literature on engineering geology, hydrogeology, mining, rock mechanics, soil mechanics, and tunnelling, 1977 to present. Available on the ORBIT (800) 456-7248, online service. Inquire as to cost and availability.

GEOREF. American Geological Institute, 4220 King Street, Alexandria, VA 22302. (800) 336-4764 or (703) 379-2480. Geology and geosciences literature, 1785 to present for North America. Available on DIALOG,(800) 334-2564, ORBIT (800) 456-7248, onlineservices. Also available on CD-ROM. Inquire as to cost and availability.

AGRICULTURAL ENGINEERING

Ency. of Physical Sciences and Engineering Info. Sources

*ISMEC: Mechanical Engineering Abstracts.*Cambridge Scientific Abstracts, 7200 Wisconsin Avenue, Suite 601, Bethesda, MD 20814. (301) 961-6750 or (800) 843-7751. Contains citations to the literature in mechanical engineering, industrial and production engineering, energy, power, mechanics, devices and related areas, from 1973 to present. Available on the DIALOG,(800) 334-2564, online service. Inquire as to cost and availability.

NTIS Bibliographic Database. National Technical Information Service, 5285 Port Royal Road, Springfield, VA 22161. (703) 487-4929 or FAX (703) 321-8199. Broad coverage of government-sponsored science and technology research reports, 1964 to present. Available on BRS,(800) 289-4277, DIALOG, (800) 334-2564, ORBIT, (800) 456-7248, and STN International, FIZ Karlsruhe, P.O. Box 2465, W-7500, Karlsruhe 1, Germany, online services. Also available on CD-ROM. Inquire as to cost and availability.

Scientific and Technical Books and Serials in Print. R.R. Bowker Inc., 121 Chanlon Road, New Providence, NJ 07974. (800) 521-8110 or (908) 464-6800. List of currently published books and serials in the physical and biological sciences, engineering and technology, with subject, author and titles indexes. Available on ORBIT,(800) 456-7248, online service.Inquire as to cost and availability.

SCISEARCH. Institute for Scientific Information, 3501 Market Street, Philadelphia, PA 19104. (800) 523-1850 or (215) 386-0100. Broad multidisciplinary title and author index to the international literature of science and technology, 1974 to present. Available on DIALOG,(800) 334-2564, and ORBIT,(800) 456-7248, online services. Also available on CD-ROM. Inquire as to cost and availability.

Wasteinfo. United Kingdom Atomic Energy Authority, Building 7.12, Harwell Laboratory, Didcot, Oxon, OX11 ORA, England. Contains citations to worldwide literature on non-nuclear waste management, from 1973 to present. Available on ORBIT (800) 456-7248, online service. Also available on CD-ROM. Inquire as to cost and availability.

Water Resources Abstracts (wra). U.S. Geological Survey, Water Resources Scientific Information Center, 12201 Sunrise Valley Drive, Reston, VA 22092-9998. (703) 648-4460. Contains citations with abstracts to scientific and technical literature on the water-sources-related aspects of the physical, social and life sciences, from 1968 to present. Available on DIALOG,(800) 334-2564, and ORBIT,(800) 456-7248, online services. Inquire as to cost and availability.

Waternet. American Water Works Association, Technical Library. Available on DIALOG online services. Citations to literature on water quality, water utility management, water analysis, water pollution, and related areas, 1971 to present.Available on DIALOG,(800) 334-2564, online service. Inquire as to cost and availability.

Who's Who in Technology. Gale Research Inc., 835 Penobscot Building, Detroit, MI 48226-4094. (313) 961-2242. (800) 877-4253. Contains biographical information of contemporary American scieNTISts and engineers. Available on DIALOG,(800) 334-2564, online service. Inquire as to cost and availability.

WILSONLINE. H.W. Wilson Company, 950 University Avenue, Bronx, NY 10452. (800) 367-6770 or (212) 588-8400. Makes available online versions of the printed H.W. Wilson indexes including Applied Science and Technology Index, Business

Periodicals Index, General Science Index, and Readers' Guide to Periodical Literature. Period covered is generally 1983 to present. Available on BRS,(800) 289-4277, DIALOG,(800) 334-2564, and OCLC EPIC,(800) 848-5878, online services. Also available on CD-ROM. Inquire as to cost and availability.

PERIODICALS

Agricultural Engineering. American Society of Agricultural Engineers, 2950 Niles Road, St. Joseph MI 49085-9659. (616) 429-0300. FAX (616) 429-3852. 1920 to present. Monthly. $42.50 per year. ISSN: 0002-1458.

Applied Engineering in Agriculture. American Society of Agricultural Engineers, 2950 Niles Road, St. Joseph MI 49085-9659. (616) 429-0300. FAX (616) 429-3852. 1985 to present. Bi-monthly. $68.00 per year. ISSN: 0883-8542.

Canadian Agricultural Engineering. Canadian Society of Agricultural Engineering. Subscriptions to ASAE, 2950 Niles Road, St. Joseph MI 49085-9659. (616) 429-0300. FAX (616) 429-3852. 1959 to present. Quarterly. $45.00 per year. ISSN: 0045-432X.

Computers and Electronics in Agriculture. Elsevier Science Publishing Company, Inc., 655 Avenue of the Americas, New York, NY 10010. (212) 989-5800. 1984 to present. 8 issues in two volumes. $378.00 per year. ISSN: 0168-1699.

Farm Journal. Farm Journal Incorporated, 230 West Washington, Square, Philadelphia, PA 19106. (215) 829-4700. 1877 to present. 13 issues plus supplements. $14.00 per year. ISSN: 0014-8008.

Implement & Tractor. Farm Press Publications. Box 1420, Clarksdale MS 38614. (601) 624-8503. FAX (601) 627-1977. 1886 to present. 5 issues per year. $25.00 per year. ISSN: 0019-2053.

Irrigation and Drainage Systems. Kluwer Academic Publishers. Box 358, Accord Station, Hingham, MA 02018- 0358. (617) 871-6600. FAX 617-871-6528. 1986 to present. Quarterly. $178.50. ISSN: 0168-6291.

Irrigation Engineering. American Society of Agricultural Engineers, 2950 Niles Road, St. Joseph MI 49085-9659. (616) 429-0300. FAX (616) 429-3852. 1985 to present. Semi-annual. $42.50.

Transactions of the American Society of Agricultural engineers. American Society of Agricultural Engineers, 2950 NilesRoad, St. Joseph MI 49085-9659. (616) 429-0300. FAX (616) 429-3852. 1958 to present. Bimonthly. $171.00 per year. ISSN: 0001-2351.

RESEARCH CENTERS AND INSTITUTES

Agricultural Engineering Department. Michigan State University. East Lansing, MI 48824. (517) 353-7268. FAX (517) 353-8982.

Agricultural Engineering Research Center. Texas A&M University. College Station, TX 77843. (409) 845-3931. FAX (409) 845-3932.

Ency. of Physical Sciences and Engineering Info. Sources

AIR CONDITIONING

Agricultural Experiment Station. Purdue University. Lafayette, IN 47907. (317) 494-8362. FAX (317) 494-0808.

Agricultural Experiment Station. Cornell University. 245 Roberts Hall, Ithaca, NY 14853. (607) 255-5420. FAX (607) 255-9499.

Arizona Agricultural Experiment Station. University of Arizona. Tucson, AZ 85721. (602) 621-3859. FAX (602) 621- 7196.

Maryland Agricultural Experiment Station. University of Maryland. College Park, Md. (301) 405-1210. FAX (301) 405 9098.

New Mexico State University. ENGINEERING RESEARCH CENTER. Box 3449, Las Cruces, NM 88003. (505) 646-3422 or FAX (505) 646-3549.

Ohio Agricultural Research and Development Center. Ohio State University. Wooster, OH 44691. (216) 263-3700. FAX (216) 263-3713.

AI

See: ARTIFICIAL INTELLIGENCE

AIR

See: METEOROLOGY

AIR CONDITIONING

See also: COOLING TOWERS, HEATING AND VENTILATION

ABSTRACT SERVICES AND INDEXES

Applied Science and Technology Index; A cumulative subject index to English language periodicals in the fields of aeronautics and space science, computer technology, chemistry, construction industry, energy and related areas. H.W. Wilson Co., 950 University Avenue, Bronx, NY 10452. (800) 367-6770 or (212) 588-8400. FAX (718) 590-1617. From 1958 to present. Monthly. Inquire about cost and availability. Also available on CD-ROM and online.

Computer Abstracts, MCB University Press Ltd., PO Box 10812, Birmingham, AL 35201-0812. (800) 633-4931. FAX (205) 995-1588. Monthly. Covers computer theory, data, hardware, systems, networks, human-computer interaction, artificial intelligence, as well as applications of computers in aerospace, business, CAD/CAM, cartography, civil engineering, electronics and electrical engineering, industrial engineering, mechanical engineering, medicine, structural engineering, etc.

Current Contents: Engineering, Technology, and Applied Sciences. Institute for Scientific Information, 3501 Market Street, Philadelphia, PA 19104. (215) 386-0100 or (800) 523-1850. FAX (215) 386-2991. 1970 to present. Weekly. $442.00 per year.

Engineering Index Monthly. Engineering Information, Inc., Castle Point on the Hudson, Hoboken, NJ 07030. (800) 221-1044. FAX (212) 832-1857. Monthly. $2300.00 per year. Also available online as COMPENDEX, and also on CD-ROM. Covers chemical engineering, computers, electrical engineering, civil engineering, metals and mining, industrial management, and mechanical engineering.

International Building Services Abstracts. Building Service Research and Information Association, Old Bracknell Lane W, Bracknell, Berks. RG12 7AH, England. Telephone 0344-426511. FAX 0344-487575. 1966 to present. Bi-monthly. Inquire for price. Also available online. A survey of world literature on mechanical and technical services of buildings, including heating, sanitation, ventilation, air conditioning, lighting, and communications.

ASSOCIATIONS AND PROFESSIONAL SOCIETIES

Air-Conditioning and Refrigeration Institute. 1501 Wilson Blvd., Arlington, VA 22209. (703) 524-8800.

Air Conditioning Contractors of America. 1712 New Hampshire Avenue NW, Washington, DC 20009. (202) 483-9370. FAX (202) 234-4721.

American Academy of Environmental Engineers. 132 Holiday Ct., Suite 206, Annapolis, MD 21041.

American Society of Heating, Refrigerating & Air-conditioning engineers. 1791 Tullie Circle, N.E., Atlanta, GA 30329. (404) 636-8400. FAX (404) 321-5478.

Institute of Heating & Air Conditioning Industries. 606 N Larchmont Blvd., Suite 4A, Los Angeles, CA 90004. (213) 467-1158. FAX (213) 461-2588.

National Association of Plumbing, Heating, Cooling Contractors. 180 S. Washington Street, Box 6808, Falls Church, VA 22040-1148. (703) 237-8100. FAX (703) 237-7442.

DIRECTORIES AND BIOGRAPHICAL SOURCES

ACCA Quality Contractor's Catalog of Materials, Products and services. Air Conditioning Contractors of America, 1712 New Hampshire Avenue NW, Washington, DC 20009. (202) 483-9370. FAX (202) 234-4721. 1968 to present. Annual. Inquire for price and availability.

Air Conditioning, Heating and Refrigerating News—Directory Issue. Business News Publishing Company, 755 W. Big Beaver Road, Suite 1000, Troy, MI 48084. (800) 837-1037. FAX (810) 362-0317. Included with subscription to *Air Conditioning,Heating and Refrigerating News* (see below).

Heating, Plumbing and Air Conditioning Buyers' Guide. Cowgate Communications Inc., 1370 Don Mills Road, Suite 300, Don Mills, ON M3B 3N7, Canada. (416) 759-2500. FAX (416) 759-6979. 1923 to present. Annual. $30.00

Research Centers Directory. Gale Research, Inc., 835 Penobscot Building, Detroit, MI 48226-4094. (313) 961-2242. (800) 877-4253. FAX (313) 961-2242. $485.00

Scientific and Technical Organizations and Agencies Directory. Gale Research, 835 Penobscot Building, Detroit, MI48226-

4094. (313) 961-2242. (800) 877-4253. FAX (313) 961-6083. 4th edition. 1996. $195.00.

Who's Who in Engineering. American Association of Engineering Societies, 1111 19th Street NW, Suite 608, Washington, DC 20036. (202) 296-2237 or (800) 658-8897. 8th edition. 1991. Inquire for price.

Who's Who in Technology. Gale Research, 835 Penobscot Building, Detroit, MI 48226-4094. (313) 961-2242. (800) 877-4253. FAX (313) 961-6083. 1995. $195.00.

ENCYCLOPEDIAS AND DICTIONARIES

Dictionary of Refrigerating and Air Conditioning. K.M. Booth, editor. Elsevier Science Publishing Company, Inc., 655 Avenue of the Americas, New York, NY 10010. (212) 989-5800. 1971. $64.00.

GENERAL WORKS

Air Conditioning: A Practical Introduction. David V. Chadderton. Routledge, Chapman and Hall, Inc., 29 West 35th Street, New York, NY 10001-2291. (212) 244-3336 or FAX (212) 563-2267. 1993. Inquire for price.

Air Handling System Design. Tseng-Yao Sun. McGraw-Hill Publishing Company, 1221 Avenue of the Americas, New York, NY 10020. (800) 262-4729 or (212) 512-3825. 1994. Inquire for cost.

Modern Heating and Ventilating Systems Design. George Clifford. Prentice Hall, 113 Sylvan Avenue, Route 9W, Englewood Cliffs, NJ 07632. (201) 592-2000 or (800) 922-0579. 1993. $58.67.

HANDBOOKS AND MANUALS

Air Conditioning & Heating Manual 1991-93. Chilton Staff, editors. Chilton Book Co., One Chilton Way, Radnor PA 19089. (215) 964-4028 or (800) 695-1214. FAX (610) 964-4745. 1993. Inquire for price.

Handbook of Air Conditioning and Refrigeration. Shan K. Wang. McGraw-Hill Publishing Company, 1221 Avenue of the Americas, New York, NY 10020. (800) 262-4729 or (212) 512-3825. 1993. $96.50.

HVAC Engineer's Handbook. Fred Porges. 9th ed. Butterworths Publishers, 80 Montvale Avenue, Stoneham, MA 02180. (617) 438-8464 or (800) 366-2665. 1991. $70.00.

ASHRAE Handbook: Fundamentals. 4 volumes. American Society of Heating, Refrigeration and Air Conditioning Engineers, 1791 Tullie Circle NE, Atlanta, GA 30329. (404) 636-8400. FAX (404) 321-5478. 1922 to present. Annual. $119.00.

ONLINE DATABASES AND CD-ROMS

Compendex Plus. Engineering Information, Inc., 345 East 47th Street, New York, NY 10017. (212) 705-7600 or (800) 221-1044. Contains citations with abstracts to worldwide literature in engineering and technology, from 1970 to present. Available on online BRS,(800) 289-4277, DIALOG, (800) 334-2564,

ORBIT (800) 456-7248, and STN International, FIZ Karlsruhe, P.O. Box 2465, W-7500, Karlsruhe 1, Germany, online services. Also available on CD-ROM. Inquire as to cost and availability.

CSA Engineering. Cambridge Scientific Abstracts, 7200 Wisconsin Avenue, Suite 601, Bethesda, MD 20814. (301) 961-6750 or (800) 843-7751. Contains citations and abstracts of international periodicals and research literature covering all fields of engineering and science and technology,including computer and information science, electronics, mechanical engineering, solid state materials, 1981 to present. Available on BRS,(800) 289-4277, online service.Inquire as to cost and availability.

Current Contents Search. Institute for Scientific Information, 3501 Market Street, Philadelphia, PA 19104. (215) 386-0100. FAX (215) 386-6362. Contains citations to articles listed in the table of contents of science and technology journals. Also articles in social sciences and life sciences journals. Available on BRS,(800) 289-4277, DIALOG,(800) 334-2564, online services. Inquire as to cost and availability.

Dissertation Abstracts. University Microfilms International, 300 North Zeeb Road, Ann Arbor, MI 48106. (800) 521-0600 or (313) 761-4700. Scope includes virtually all doctoral dissertations accepted at accredited American institutions from 1861 to present in 252 subject areas. Available on BRS,(800) 289-4277, DIALOG,(800) 334-2564, and OCLC EPIC,(800) 848-5878, online services. Also available on CD-ROM. Inquire as to cost and availability.

ISMEC: Mechanical Engineering Abstracts. Cambridge Scientific Abstracts, 7200 Wisconsin Avenue, Suite 601, Bethesda, MD 20814. (301) 961-6750 or (800) 843-7751. Contains citations to the literature in mechanical engineering, industrial and production engineering, energy, power, mechanics, devices and related areas, from 1973 to present. Available on the DIALOG,(800) 334-2564, online service. Inquire as to cost and availability.

NTIS Bibliographic Database. National Technical Information Service, 5285 Port Royal Road, Springfield, VA 22161. (703) 487-4929 or FAX (703) 321-8199. Broad coverage of government-sponsored science and technology research reports, 1964 to present. Available on BRS,(800) 289-4277, DIALOG, (800) 334-2564, ORBIT, (800) 456-7248, and STN International, FIZ Karlsruhe, P.O. Box 2465, W-7500, Karlsruhe 1, Germany, online services. Also available on CD-ROM. Inquire as to cost and availability.

SCISEARCH. Institute for Scientific Information, 3501 Market Street, Philadelphia, PA 19104. (800) 523-1850 or (215) 386-0100. Broad multidisciplinary title and author index to the international literature of science and technology, 1974 to present. Available on DIALOG,(800) 334-2564, and ORBIT,(800) 456-7248, online services. Also available on CD-ROM. Inquire as to cost and availability.

WILSONLINE. H.W. Wilson Company, 950 University Avenue, Bronx, NY 10452. (800) 367-6770 or (212) 588-8400. Makes available online versions of the printed H.W. Wilson indexes including Applied Science and Technology Index, Business Periodicals Index,General Science Index, and Readers' Guide to Periodical Literature. Period covered is generally 1983 to present. Available on BRS,(800) 289-4277, DIALOG,(800) 334-2564, and OCLC EPIC,(800) 848-5878, online services. Also available on CD-ROM. Inquire as to cost and availability.

PERIODICALS

Air Conditioning, Heating, and Refrigeration News. Business News Publishing Company, 755 W. Big Beaver Road, Suite 1000, Troy, MI 48084. (313) 362-3700. FAX (313) 362-0317. 1926 to present. Weekly. $72.00 per year. Includes the annual directory and statistical summary issues.

ASHRAE Journal: Heating, Refrigerating, Air Conditioning, and ventilating. American Society of Heating, Refrigeration and Air Conditioning Engineers, 1791 Tullie Circle NE, Atlanta, GA 30329. (404) 636-8400. FAX (404) 321-5478. 1914 to present. Monthly. $49.00 per year for non-members.

ASHRAE Transactions. American Society of Heating, Refrigeration and Air Conditioning Engineers, 1791 Tullie Circle NE, Atlanta, GA 30329. (404) 636-8400. FAX (404) 321-5478. 1895 to present. Semi-annual. $170.00 per volume.

Engineered Systems. Business News Publishing Company, 755 W. Big Beaver Road, Suite 1000, Troy, MI 48084. (313) 362-3700. FAX (313) 362-0317. 1985 to present. 12 times a year. $39.00 per year.

Heating and Air Conditioning Journal. Maclean Hunter Publishing Company, 29 North Wacker Drive, Chicago, IL 60606-3298. (312) 726-2802. 1931 to present. Monthly. Inquire for price and availability.

Heating, Piping, Air Conditioning. Penton Publishing, 110 Superior Ave., Cleveland, OH 44114-2543. 1929 to present. Monthly. $50.00 per year.

Heating, Plumbing, and Air Conditioning. Cowgate Communications Inc., 1370 Don Mills Road, Suite 300, Don Mills, ON M3B 3N7, Canada. (416) 759-2500. FAX (416) 759-6979. 1923 to present. 7 times a year. $43.00 per year.

Indoor Comfort News. Institute of Heating and Air Conditioning Industries, 606 N Larchmont Blvd., Suite 4A, Los Angeles, CA 90004. (213) 467-1158. FAX (213) 461-2588. 1955 to present. Monthly. $12.00 per year.

RESEARCH CENTERS AND INSTITUTES

Kansas State University Mechanical Engineering Research laboratory. Durland Hall, Manhattan, KS 66506. (913) 532-5610.

Lehigh University Institute of thermo-fluid Engineering and science. Bldg. A, Mountain Top Campus, Bethlehem, PA 18015. (215) 758-4091. FAX (215) 758-5057.

Refrigeration Research Foundation. 7315 Wisconsin Avenue, Bethesda, MD 20814. (301) 652-5674. FAX (301) 652-7269.

Texas A&M University Energy Systems Laboratory. College Station, TX 77843. (409) 845-6402.

AIRCRAFT

See also: AERODYNAMICS, AERONAUTICAL ENGINEERING, AERONAUTICS, AIRFRAMES

ABSTRACT SERVICES AND INDEXES

Aeronautical Engineering. Scientific and Technical Branch, National Aeronautics and Space Administration. National Technical Information Service (NTIS), 5285 Port Royal Road, Springfield, VA 22161. (703) 487-4650. FAX (703) 321-8547. Monthly. A selection of annotated references to unclassified reports and journal articles that were introduced into the NASA scientific and technical information system and announced in STAR and IAA.

Aerospace Periodical Index, 1973-1982. The National Air & Space Museum and the Smithsonian Institution. G.K. Hall, 866 Third Avenue, New York, NY 10022. (212) 702-6789 or (800) 257-5755. FAX (212) 605-9350. 1983. Inquire for cost and availability.

Alloys Index. American Society for Metals, Metals Park, OH 44073. (216) 338-5151. FAX (216) 338-4634. 1974 to present. Monthly. $295.00.

Applied Mechanics Reviews: An Assessment of World Literature in Engineering sciences. American Society of Mechanical Engineers, 345 East 47th Street, New York, NY 10017. (212) 705-7703. 1948 to present. Monthly. $360.00 per year.

Applied Science and Technology Index; A cumulative subject index to English language periodicals in the fields of aeronautics and space science, computer technology, chemistry, construction industry, energy and related areas. H.W. Wilson Co., 950 University Avenue, Bronx, NY 10452. (800) 367-6770 or (212) 588-8400. FAX (718) 590-1617. From 1958 to present. Monthly. Inquire about cost and availability. Also available on CD-ROM.

Current Contents: Engineering, Technology, and Applied Sciences. Institute for Scientific Information, 3501 Market Street, Philadelphia, PA 19104. (215) 386-0100 or (800) 523-1850. FAX (215) 386-6362. Weekly. $360.00 per year.

Engineering Index Monthly. Engineering Information, Inc., Castle Point on the Hudson, Hoboken, NJ 07030. (800) 221-1044. FAX (212) 832-1857. Monthly. $2200.00 per year. Also available online as COMPENDEX, and also on CD-ROM.

International Aerospace Abstracts. Technical Information Service, American Institute of Aeronautics and Astronautics, Inc., 555 West 57th St., New York, NY 10019. (212) 247-6500. FAX (212) 582-4861. Semi-monthly. $1295.00 per year.

Metals Abstracts and Metals Abstracts Index. American Society for Metals, Metals Park, OH 44073. (216) 338-5151. 1968 to present. Monthly. Abstracts are $1500.00 per year and Index is $460.00 per year.

Science Citation Index. Institute for Scientific Information, 3501 Market Street, Philadelphia, PA 19104. (215) 386-0100 or (800) 523-1850. FAX (215) 386-6362. Inquire about availability and cost. Also available on CD-ROM.

STAR. Scientific and Technical Aerospace Reports . National Aeronautics and Space Administration. NASA Center for Aerospace Information, Box 8757, BWI Airport, Baltimore, MD 21240.(301) 621-0153. Monthly. Inquire about availability and cost. Also available through the NASA online retrieval service (RECON), and through the Aerospace Database through DIALOG.

World Aluminum Abstracts. Aluminum Association, Inc., 818 Connecticut Ave. NW, Washington, DC 20006. (202) 862-5156. Inquire about price and availability.

ASSOCIATIONS AND PROFESSIONAL SOCIETIES

Aircraft Electronics Association. PO Box 1963, Independence, MO 64055-0963. (816) 373-6565. FAX (816) 478-3100.

American Aviation Historical Society. 2333 Otis St., Santa Ana, CA 92704.

American Helicopter Society, Inc., 217 N. Washington St., Alexandria, VA 22314. (703) 684-6777. FAX (703) 739-9279.

American Institute of Aeronautics and Astronautics, 370 L'Enfant Promenade SW, Washington, DC 20024-2518. (202) 646-7471.

Association of Balloon and Airship Constructors. Box 90864, San Diego, CA 92169. (619) 270-4049.

Canadian Aeronautics and Space Institute, 222 Somerset Street West, Suite 601, Ottawa, Ontario, Canada K2p 0j1.

Flight Safety Foundation, Inc., 2200 Wilson Blvd., Suite 500, Arlington, VA 22201-3306. (703) 522-8300. FAX (703) 525-6047.

Society of Automotive Engineers. SAE, Inc., 400 Commonwealth Dr., Warrendale, PA 15096. (412) 776-4841.

BIBLIOGRAPHIES

Aeronautical Engineering: A Continuing Bibliography With Indexes. National Aeronautics and Space Administration, Center for Aerospace Information, Box 8757, BWI Airport, Baltimore, MD 21240. (301) 621-0153. Inquire for cost and availability.

Scientific and Technical Books and Serials in Print, R.R. Bowker Inc., 121 Chanlon Road, New Providence, NJ 07974. (800) 521-8110 or (908) 464-6800. FAX (908) 464-3553. 3 volumes, annually. $285.00. Also available on CD-ROM.

DIRECTORIES AND BIOGRAPHICAL SOURCES

AAS Directory. American Astronautical Society, 6352 Rolling Mill Place, Suite 102, Springfield, VA 22152-2354. (703) 866-0020. FAX (703) 866-3526. Inquire for cost and availability.

Research Centers Directory. Gale Research, 835 Penobscot Building, Detroit, MI 48226-4094. (313) 961-2242. (800) 877-4253. FAX (313) 961-6083. Irregular. Inquire for price and availability.

Who's Who in Engineering. American Association of Engineering Societies, 1111 19th St. NW, Suite 608, Washington, DC 20036. 7th ed. 1988. $200.00.

World Aviation Directory. McGraw-Hill, Aviation Week Group (NY), 1221 Avenue of the Americas, New York, NY 10020. (215) 237-4112. FAX (215) 586-3232. Semiannual. $250.00.

ENCYCLOPEDIAS AND DICTIONARIES

Dictionary of Aerospace Engineering. M.G. Kotik. Elsevier Science Publishing Company, Inc., 655 Avenue of the Americas, New York, NY 10010. (212) 989-5800. 1986. $225.75.

Encyclopedia of Physical Science and Technology. RobertA. Meyers, ed. 18 volumes. Academic Press Inc., 6277 Sea Harbor Drive, Orlando, FL 32887. 1992. $2100.00

Jane's Aerospace Dictionary, B. Guston. 3rd ed. Jane's Information Group,Dept. DSM, 1340 Braddock Pl., Suite 300, Box 1436, Alexandria, VA 22314-1651. (703) 683-3700. FAX (703) 836-0029. 1991. $45.00.

Jane's Encyclopedia of Aviation. Michael J. Taylor, comp. Outlet Book Co., 40 Engelhard Ave., Avenal, NJ 07001. (908) 827-2700 or (800) 223-6804. FAX (908) 827-2641. 1993. $34.99.

McGraw-Hill Encyclopedia of Engineering. Sybil P. Parker, ed. 2nd edition. McGraw-Hill Publishing Company, 1221 Avenue of the Americas, New York, NY 10020. (800) 262-4729 or (212) 512-3825. 1993. $95.50.

McGraw-Hill Encyclopedia of Science and Technology, Sybil P. Parker, ed. 7th ed. 20 vols. McGraw-Hill Publishing Company, 1221 Avenue of the Americas, New York, NY 10020. (800) 262-4729 or (212) 512-3825. $1900.00

SAE Dictionary of Aerospace Engineering. William H. Cubberly, ed. Society of Automotive Engineers, 400 Commonwealth Drive, Warrendale, PA 15096. (412) 776-4841. 1992. $85.00.

GENERAL WORKS

Advanced Aircraft Systems. David A. Lombardo. McGraw-Hill Publishing Company, 1221 Avenue of the Americas, New York, NY 10020. (800) 262-4729 or (212) 512-3825. 1993. $18.95.

Aircraft Design: A Conceptual Approach. David P. Raymer. American Institute of Aeronautics and Astronautics, 370 L'Enfant Promenade, S.W., Washington, DC 20024. (202) 646-7400. 1992. $66.95.

Analysis of Aircraft Structures: An Introduction. Bruce K. Donaldson. McGraw-Hill Publishing Company, 1221 Avenue of the Americas, New York, NY 10020. (800) 262-4729 or (212) 512-3825. 1993. Inquire for cost and availability.

Design For Flying. David B. Thurston. 2nd edition. TAB Books, P.O. Box 40, Blue Ridge Summit, PA 17294-0850. (717)794-2191 or (800) 822-8138. FAX (717) 794-2080. 1994. $29.95 (hardback), $19.95 (paperback).

Ency. of Physical Sciences and Engineering Info. Sources

AIRCRAFT

HANDBOOKS AND MANUALS

Jane's Airport and Atc Equipment, Jane's Information Group, Dept. DSM, 1340 Braddock Pl., Suite 300, Box 1436, Alexandria, VA 22314-1651. (703) 683-3700. FAX (703) 836-0029. Annual. $225.00.

Jane's All the World's Aircraft. Jane's Information Group, Dept. DSM, 1340 Braddock Pl., Suite 300, Box 1436, Alexandria, VA 22314-1651. (703) 683-3700. FAX (703) 836-0029. Annual. $245.00. Also available on CD-ROM.

Jane's Avionics. Jane's Information Group, Dept. DSM, 1340 Braddock Pl., Suite 300, Box 1436, Alexandria, VA 22314-1651. (703) 683-3700. FAX (703) 836-0029. Annual. $245.00. Also available on CD-ROM.

Standard Aircraft Handbook. Larry Reithmaier, editor. 5thedition. McGraw-Hill Publishing Company, 1221 Avenue of the Americas, New York, NY 10020. (800) 262-4729 or (212) 512-3825. 1991. $12.95.

Standard Handbook of Engineering Calculations. Tyler G. Hicks, ed. 2nd ed. McGraw-Hill Publishing Company, 1221 Avenue of the Americas, New York, NY 10020. (800) 262-4729 or (212) 512-3825. 1985. $89.50.

The Wiley Engineer's Desk Reference. Sanford I. Heisler. John Wiley and Sons, Inc., 605 Third Avenue, New York, NY 10158. (800) 526-5368 or (212) 850-6000. 1984. $64.95.

ONLINE DATABASES AND CD-ROMS

Aerospace Database. American Institute of Aeronautics and Astronautics, 370 L'Enfant Promenade, S.W., Washington, DC 20024. (202) 646-7400. Worldwide published literature on research and development in aerospace and related areas, with abstracts. Covers 1962 to present. Available on DIALOG, (800) 334-2564, online service. Also available on CD-ROM. Inquire as to cost and availability.

Compendex Plus. Engineering Information, Inc., 345 East 47th Street, New York, NY 10017. (212) 705-7600 or (800) 221-1044. Contains citations with abstracts to worldwide literature in engineering and technology, from 1970 to present. Available on online BRS,(800) 289-4277, DIALOG, (800) 334-2564, ORBIT (800) 456-7248, and STN International, FIZ Karlsruhe, P.O. Box 2465, W-7500, Karlsruhe 1, Germany, online services. Also available on CD-ROM. Inquire as to cost and availability.

CSA Engineering. Cambridge Scientific Abstracts, 7200 Wisconsin Avenue, Suite 601, Bethesda, MD 20814. (301) 961-6750 or (800) 843-7751. Contains citations and abstracts of international periodicals and research literature covering all fields of engineering and science and technology,including computer and information science, electronics, mechanical engineering, solid state materials, 1981 to present. Available on BRS,(800) 289-4277, online service.Inquire as to cost and availability.

Current Contents Search. Institute for Scientific Information, 3501 Market Street, Philadelphia, PA 19104. (215) 386-0100. FAX (215) 386-6362. Contains citations to articles listed in the table of contents of science and technology journals. Also articles in social sciences and life sciences journals. Available on BRS,(800) 289-4277, DIALOG,(800) 334-2564, online services. Inquire as to cost and availability.

Dissertation Abstracts. University Microfilms International, 300 North Zeeb Road, Ann Arbor, MI 48106. (800) 521-0600 or (313) 761-4700. Scope includes virtually all doctoral dissertations accepted at accredited American institutions from 1861 to present in 252 subject areas. Available on BRS,(800) 289-4277, DIALOG,(800) 334-2564, and OCLC EPIC,(800) 848-5878, online services. Also available on CD-ROM. Inquire as to cost and availability.

International Aerospace Abstracts. American Institute ofAeronautics and Astronautics, 370 L'Enfant Promenade, S.W., Washington, DC 20024. (202) 646-7400. Contains references and abstracts of journal and monograph literature relating to aerospace science and technology, from 1963 to present. Available through the NASA/RECON system of the National Aeronautics and Space Administration only.

*ISMEC: Mechanical Engineering Abstracts.*Cambridge Scientific Abstracts, 7200 Wisconsin Avenue, Suite 601, Bethesda, MD 20814. (301) 961-6750 or (800) 843-7751. Contains citations to the literature in mechanical engineering, industrial and production engineering, energy, power, mechanics, devices and related areas, from 1973 to present. Available on the DIALOG,(800) 334-2564, online service. Inquire as to cost and availability.

NASA Database. American Institute of Aeronautics and Astronautics, 370 L'Enfant Promenade, S.W., Washington, DC 20024. (202) 646-7400. Citations and abstracts of aeronautics and astronautics literature, 1962 to present. Also contains citations from STAR, SCIENTIFIC and TECHNICAL AEROSPACE REPORTS, and INTERNATIONAL AEROSPACE ABSTRACTS. Available through NASA/RECON online service. Inquire as to cost and availability.

NTIS Bibliographic Database. National Technical Information Service, 5285 Port Royal Road, Springfield, VA 22161. (703) 487-4929 or FAX (703) 321-8199. Broad coverage of government-sponsored science and technology research reports, 1964 to present. Available on BRS,(800) 289-4277, DIALOG, (800) 334-2564, ORBIT, (800) 456-7248, and STN International, FIZ Karlsruhe, P.O. Box 2465, W-7500, Karlsruhe 1, Germany, online services. Also available on CD-ROM. Inquire as to cost and availability.

SAE Global Mobility Database. Society of Automative Engineers (SAE), Electronic Publishing Division, 400 Commonwealth Drive, Warrendale, PA 15098. (412) 776-4841. Contains citations with abstracts to technical papers on automotive and aerospace technology and vehicular-related industries that have been presented at SAE conferences. Covers 1906 to present. Available on ORBIT,(800) 456-7248,online service. Inquire as to cost and availability.

Scientific and Technical Books and Serials in Print. R.R. Bowker Inc., 121 Chanlon Road, New Providence, NJ 07974. (800) 521-8110 or (908) 464-6800. List of currently published books and serials in the physical and biological sciences, engineering and technology, with subject, author and titles indexes. Available on ORBIT,(800) 456-7248, online service.Inquire as to cost and availability.

SCISEARCH. Institute for Scientific Information, 3501 Market Street, Philadelphia, PA 19104. (800) 523-1850 or (215) 386-0100. Broad multidisciplinary title and author index to the international literature of science and technology, 1974 to present. Available on DIALOG,(800) 334-2564, and ORBIT,(800) 456-7248, online services. Also available on CD-ROM. Inquire as tocost and availability.

AIRCRAFT

Ency. of Physical Sciences and Engineering Info. Sources

TRIS (Transportation Research Information). National Academy of Science, 2101 Constitution Avenue, N.W., Washington, DC 20418. (202) 334-3313 or (800) 624-6242. Citations with abstracts of literature on transportation, including air, highway, rail, maritime and other modes from 1968 to present. Available on DIALOG,(800) 334-2564, online service. Inquire as to cost and availability.

WILSONLINE. H.W. Wilson Company, 950 University Avenue, Bronx, NY 10452. (800) 367-6770 or (212) 588-8400. Makes available online versions of the printed H.W. Wilson indexes including Applied Science and Technology Index, Business Periodicals Index, General Science Index, and Readers' Guide to Periodical Literature. Period covered is generally 1983 to present. Available on BRS,(800) 289-4277, DIALOG,(800) 334-2564, and OCLC EPIC,(800) 848-5878, online services. Also available on CD-ROM. Inquire as to cost and availability.

PERIODICALS

AAHS Journal. American Aviation Historical Society, 2333 Otis St., Santa Ana, CA 92704. Quarterly. $25.00 (annual dues includes journal subscription).

Aeronautical Journal. The Royal Aeronautical Society, 4 Hamilton Place, London W1V 0BQ, England. 1897 to present. Monthly except June and August. Inquire for price information in U.S.

Aerospace America. American Institute of Aeronautics and Astronautics, 370 L'Enfant Promenade SW, Washington, DC 20024-2518. (202) 646-7471. 1932 to present. Monthly. $70 annually (non-members).

Aerospace Daily. McGraw-Hill Inc., Aviation Week Group, 1200 G Street NW, Suite 200, Washington, DC 20005. (202) 383-2350. 1973 to present. Five times a week. $1340.00 per year.

Aerospace Engineering. Society of Automotive Engineers, 400 Commonwealth Dr., Warrendale, PA 15096. (412) 776-4841. 1981 to present. Monthly. $48.00 per year.

Aerospace Propulsion. McGraw-Hill Inc., Aviation Week Group, 1200 G Street NW, Suite 200, Washington, DC 20005. (202) 383-2350. 1963 to present. 25 times a year. $550.00 a year.

Aerostation. Association of Balloon and Airship Constructors. Box 90864, San Diego, CA 92169. (619) 270-4049. 1974 to present. Quarterly. Inquire for price and availability.

AIAA Journal. American Institute of Aeronautics and Astronautics, 370 L'Enfant Promenade SW, Washington, DC 20024-2518. (202) 646-7471. 1963 to present. Monthly. $52.00 per year (members) and $435.00 per year (non-members).

Air Progress. Challenge Publications, Inc., 7950 Deering Ave., Canoga Park, CA 91304. (818) 887-0550. FAX (818) 884-1343. 1941 to present. Monthly. $26.50 per year.

Aircraft Engineering and Aerospace Technology. BunHill Publications Ltd., 127 Stanstead Rd., London SE23 1JE, England. 1929 to present. Monthly. $102.00 per year (U.S.).

Airpower Journal. Airpower Research Institute, 401 Chennault Circle, Maxwell AFB, AL 36112-8930. Quarterly. Order from

New Orders, Supt. of Documents, PO Box 371954, Pittsburg, PA 15250-7954. $13.00 per year.

Aviation & Aerospace. Baxter Publishing Company, 310 Dupont Street, Toronto, Ontario M5R 1V9 Canada. (416) 968-7252. FAX (416) 968-2377. 1928 to present. Bi-monthly. $42.00 per year.

Aviation Mechanics Bulletin. Flight Safety Foundation, Inc., 2200 Wilson Blvd., Suite 500, Arlington, VA 22201-3306. (703) 522-8300. FAX (703) 525-6047. 1953 to present. Bimonthly. $15.00 per year (non-members).

Aviation Week and Space Technology. 1221 Avenue of the Americas, New York, NY 10020. (212) 512-2999 or (800) 525-5003. 1916 to present. Weekly. $82.00 per year.

Canadian Aeronautics and Space Journal. Canadian Aeronautics and Space Institute, Suite 818, the National Building, 130 Slater St., Ottawa, Canada K1P 6E2. 1955 to present. Monthly. $75.00 per year (U.S.).

Flying. Hachette Filipacchi Magazines, Inc., 1633 Broadway, New York, NY 10019. (203) 622-2700. FAX (203) 622-2725. 1927 to present. Monthly. $24.00 per year.

IEEE Transactions On Aerospace and Electronic Systems. Institute of Electrical and Electronics Engineers, 445 Hoes Lane, PO Box 1331, Piscataway, NJ 08855-1331. (908) 981-0060. 1965 to present. Quarterly. $10.00 per issue (members), $20.00 per issue (non-members).

Interavia/Aerospace World. Aerospace Media Publishing, Swissair Centre, 31 Route de l'Aeroport, POB Box 437, 1215 Geneva 15, Switzerland. (022) 788 27 88. FAX (022) 788 27 26. 1987 to present. Monthly. $128.00 per year (U.S.).

Journal of Aircraft. American Institute of Aeronautics and Astronautics, 370 L'Enfant Promenade SW, Washington, DC 20024-2518. 1963 to present. Bimonthly. $42.00 (members) and $230.00 (non-members).

Journal of the American Helicopter Society. American Helicopter Society, Inc., 217 North Washington St., Alexandria, VA 22314. (703) 684-6777. 1956 to present. Quarterly. $35.00 per year.

Progress in Aerospace Sciences. Pergamon Press, 660 White Plains Rd., Tarrytown, NY 10591-5153. 1961 to present. Quarterly. $320.00 per year.

Rotor and Wing International. Phillips Publishing Inc., 7811 Montrose Rd., Potomac, MD 20854. (301) 340-2100. FAX (301) 309-3847. 1967 to present. Monthly. $49.00 a year.

Vertiflite. American Helicopter Society, 217 N. Washington St., Alexandria, VA 22314. (703) 684-6777. FAX (703) 739-9279. 1955 to present. Bimonthly. $25 per year (members) and $35 per year (non-members).

RESEARCH CENTERS AND INSTITUTES

Aerospace Research Applications Center, Indianapolis Center For Advanced Research, 611 N. Capitol Ave., Indianapolis, in46204. (317) 262-5003. FAX (317) 262-5044.

Aerospace Research Center, 1250 Eye St., NW, Washington, DC 20005. (202) 371-8400. FAX (202) 371-8470.

Aviation Safety Institute, 6797 N. High St., Suite 316, PO Box 304, Worthington, OH 43085. (614) 885-4242. FAX (614) 885-5891.

Center For Aerospace Technology, Weber State College, Ogden, UT 84408-1805. (801) 626-7272.

Center For Control and Systems Research, University of Texas At Austin, aerospace Engineering and Engineering Mechanics, Wrw 408, Austin, TX 78712-1085. (512) 471-4908. FAX (512) 471-3788.

Flight Research Laboratory, University of Kansas, Raymond Nichols Hall, lawrence, KS 66045. (913) 864-3043. FAX (913) 864-7789.

Joint Institute For Advancement of Flight Sciences, George Washington university, Langley Research Center, Mail Stop 269, Hampton, VA 23665. (804) 864-1982. FAX (804) 864-5894.

Ohio Aerospace Institute, 2001 Aerospace Parkway, Brook Park, OH 44142. (216) 891-2100. FAX (216) 891-2140.

Transportation Research and Education Center, Georgia Institute of technology, Mason Civil Engineering Bldg., Atlanta, GA 30332. (404) 894-2236. FAX (404) 894-2278.

AIRCRAFT ENGINES

See also: AERONAUTICAL ENGINEERING, AIRCRAFT, JET PROPULSION, TURBINES

ABSTRACT SERVICES AND INDEXES

Aeronautical Engineering. Scientific and Technical Branch, National Aeronautics and Space Administration. National Technical Information Service (NTIS), 5285 Port Royal Road, Springfield, VA 22161. (703) 487-4650. FAX (703) 321-8547. Monthly. A selection of annotated references to unclassified reports and journal articles that were introduced into the NASA scientific and technical information system and announced in STAR and IAA.

Alloys Index. American Society for Metals, Metals Park, OH 44073. (216) 338-5151. FAX (216) 338-4634. 1974 to present. Monthly. $295.00.

Applied Mechanics Reviews: An Assessment of World Literature in Engineering sciences. American Society of Mechanical Engineers, 345 East 47th Street, New York, NY 10017. (212) 705-7703. 1948 to present. Monthly. $360.00 per year.

Applied Science and Technology Index; A cumulative subject index to English language periodicals in the fields of aeronautics and space science, computer technology, chemistry, construction industry, energy and related areas. H.W. Wilson Co., 950 University Avenue, Bronx, NY 10452. (800) 367-6770 or (212) 588-8400. FAX (718) 590-1617. From 1958 to present. Monthly. Inquire about cost and availability. Also available on CD-ROM.

Current Contents: Engin, Tech, and Applied Sciences. Institute for Scientific Information, 3501 Market Street, Philadelphia, PA 19104. (215) 386-0100 or (800) 523-1850. FAX (215)386-6362. Weekly. $360.00 per year.

Engineering Index Monthly. Engineering Information, Inc., Castle Point on the Hudson, Hoboken, NJ 07030. (800) 221-1044. FAX (212) 832-1857. Monthly. $2200.00 per year. Also available online as COMPENDEX, and also on CD-ROM.

International Aerospace Abstracts. Technical Information Service, American Institute of Aeronautics and Astronautics, Inc., 555 West 57th St., New York, NY 10019. (212) 247-6500. FAX (212) 582-4861. Semi-monthly. $1295.00 per year.

Science Citation Index. Institute for Scientific Information, 3501 Market Street, Philadelphia, PA 19104. (215) 386-0100. FAX (215) 386-6362. Inquire about availability and cost. Also available on CD-ROM.

ASSOCIATIONS AND PROFESSIONAL SOCIETIES

American Helicopter Society, Inc., 217 N. Washington St., Alexandria, VA 22314. (703) 684-6777. FAX (703) 739-9279.

American Institute of Aeronautics and Astronautics, 370 L'Enfant Promenade SW, Washington, DC 20024-2518. (202) 646-7471.

Flight Safety Foundation, Inc., 2200 Wilson Blvd., Suite 500, Arlington, VA 22201-3306. (703) 522-8300. FAX (703) 525-6047.

Society of Automotive Engineers. SAE, Inc., 400 Commonwealth Dr., Warrendale, PA 15096. (412) 776-4841.

BIBLIOGRAPHIES

Aeronautical Engineering: A Continuing Bibliography With Indexes. National Aeronautics and Space Administration, Center for Aerospace Information, Box 8757, BWI Airport, Baltimore, MD 21240. (301) 621-0153. Inquire for cost and availability.

*Scientific and Technical Books and Serials in Print, R.*R. Bowker Inc., 121 Chanlon Road, New Providence, NJ 07974. (800) 521-8110 or (908) 464-6800. FAX (908) 464-3553. 3 volumes, annually. $285.00. Also available on CD-ROM.

DIRECTORIES AND BIOGRAPHICAL SOURCES

Jane's Airport and Atc Equipment, Jane's Information Group, Dept. DSM, 1340 Braddock Pl., Suite 300, Box 1436, Alexandria, VA 22314-1651. (703) 683-3700. FAX (703) 836-0029. Annual. $225.00.

Research Centers Directory. Gale Research, 835 Penobscot Building, Detroit, MI 48226-4094. (313) 961-2242. (800) 877-4253. FAX (313) 961-6083. 20th edition. $485.00.

Who's Who in Engineering. American Association of Engineering Societies, 1111 19th St. NW, Suite 608, Washington, DC 20036. 7th ed. 1988. $200.00.

World Aviation Directory. McGraw-Hill, Aviation Week Group (NY), 1221 Avenue of the Americas, New York, NY 10020. (215) 237-4112. FAX (215) 586-3232. Semiannual. $250.00.

ENCYCLOPEDIAS AND DICTIONARIES

Dictionary of Aeronautical Terms, Dale Crane, Comp. Aviation Supplies and ACADemics, Inc., 7005 132nd Place SE, Newcastle, WA 98059. (213) 656-7745. 1991. $15.95.

Encyclopedia of Physical Science and Technology. Robert A. Meyers, ed. Academic Press Inc., 6277 Sea Harbor Drive, Orlando, FL 32887. 1992. Inquire for price.

Jane's Aerospace Dictionary, B. Guston. 3d ed. Janes Info Group (order from:) Dept. DSM, 1340 Braddock Pl., Suite 300, Box 1436, Alexandria, VA 22314-1651. (703) 683-3700. FAX (703) 836-0029. 1991. $45.00.

McGraw-Hill Encyclopedia of Science and Technology, Sybil P. Parker, ed. 7th ed. 20 vols. McGraw-Hill Publishing Company, 1221 Avenue of the Americas, New York, NY 10020. (800) 262-4729 or (212) 512-3825. $1900.00

GENERAL WORKS

Aircraft Engine Design. Jack D. Mattingly, William H. Heiser, and Daniel H. Daley. American Institute of Aeronautics and Astronautics, 370 L'Enfant Promenade SW, Washington, DC 20024-2518. (202) 646-7471. 1987. Inquire for cost and availability.

Aircraft Engines and Gas Turbines. Jack L. Kerrebrock. 2d ed. MIT Press, 55 Hayward Street, Cambridge, MA 02142. (617) 625-8569 or (800) 356-0343. FAX (617) 253-1709. 1992. $45.00

Aircraft Powerplants. Ralph D. Bent and James L. McKinley. 5th edition. McGraw-Hill Publishing Company, 1221 Avenue of the Americas, New York, NY 10020. (800) 262-4729 or (212) 512-3825. 1985. Inquire for cost and availability.

Aircraft Propulsion Systems Technology and Design. Gordon C. Oates, editor. American Institute of Aeronautics and Astronautics, 370 L'Enfant Promenade SW, Washington, DC 20024-2518. (202) 646-7471. 1989. Inquire for cost and availability.

Hydrogen Aircraft Technology. G. Daniel Brewer. CRC Press, 2000 Corporate Blvd., N.W., Boca Raton, FL 33431. (407) 994-0555 or (800) 272-7737. FAX (407) 994-0949. 1991. Inquire for cost and availability.

The Jet Engine. 4th edition. Rolls-Royce plc, Derby, England. 1986. Inquire for cost and availability.

Jet Engines and Jet Aircraft: A Source Guide. Gordon Press Publishers, PO Box 459, Bowling Green Station, New York, NY 10004. (718) 624-8419. 1991. $75.00.

HANDBOOKS AND MANUALS

Standard Handbook of Engineering Calculations. Tyler G. Hicks, ed. 2d ed. McGraw-Hill Publishing Company, 1221 Avenue of the Americas, New York, NY 10020. (800) 262-4729 or (212) 512-3825. 1985. $89.50.

The Wiley Engineer's Desk Reference. Sanford I. Heisler. John Wiley and Sons, Inc., 605 Third Avenue, New York, NY 10158. (800) 526-5368 or (212) 850-6000. 1984. $64.95.

Jane's All the World's Aircraft. Jane's Information Group, Dept. DSM, 1340 Braddock Pl., Suite 300, Box 1436, Alexandria, VA 22314-1651. (703) 683-3700. FAX (703) 836-0029. Annual. $245.00. Also available on CD-ROM.

Online Databases and CD-ROMs

Aerospace Database. American Institute of Aeronautics and Astronautics, 370 L'Enfant Promenade, S.W., Washington, DC 20024.(202) 646-7400. Worldwide published literature on research and development in aerospace and related areas, with abstracts. Covers 1962 to present. Available on DIALOG, (800) 334-2564, online service. Also available on CD-ROM. Inquire as to cost and availability.

Compendex Plus. Engineering Information, Inc., 345 East 47th Street, New York, NY 10017. (212) 705-7600 or (800) 221-1044. Contains citations with abstracts to worldwide literature in engineering and technology, from 1970 to present. Available on online BRS,(800) 289-4277, DIALOG, (800) 334-2564, ORBIT (800) 456-7248, and STN International, FIZ Karlsruhe, P.O. Box 2465, W-7500, Karlsruhe 1, Germany, online services. Also available on CD-ROM. Inquire as to cost and availability.

CSA Engineering. Cambridge Scientific Abstracts, 7200 Wisconsin Avenue, Suite 601, Bethesda, MD 20814. (301) 961-6750 or (800) 843-7751. Contains citations and abstracts of international periodicals and research literature covering all fields of engineering and science and technology,including computer and information science, electronics, mechanical engineering, solid state materials, 1981 to present. Available on BRS,(800) 289-4277, online service.Inquire as to cost and availability.

Current Contents Search. Institute for Scientific Information, 3501 Market Street, Philadelphia, PA 19104. (215) 386-0100. FAX (215) 386-6362. Contains citations to articles listed in the table of contents of science and technology journals. Also articles in social sciences and life sciences journals. Available on BRS,(800) 289-4277, DIALOG,(800) 334-2564, online services. Inquire as to cost and availability.

Dissertation Abstracts. University Microfilms International, 300 North Zeeb Road, Ann Arbor, MI 48106. (800) 521-0600 or (313) 761-4700. Scope includes virtually all doctoral dissertations accepted at accredited American institutions from 1861 to present in 252 subject areas. Available on BRS,(800) 289-4277, DIALOG,(800) 334-2564, and OCLC EPIC,(800) 848-5878, online services. Also available on CD-ROM. Inquire as to cost and availability.

*ISMEC: Mechanical Engineering Abstracts.*Cambridge Scientific Abstracts, 7200 Wisconsin Avenue, Suite 601, Bethesda, MD 20814. (301) 961-6750 or (800) 843-7751. Contains citations to the literature in mechanical engineering, industrial and production engineering, energy, power, mechanics, devices and related areas, from 1973 to present. Available on the DIALOG,(800) 334-2564, online service. Inquire as to cost and availability.

NTIS Bibliographic Database. National Technical Information Service, 5285 Port Royal Road, Springfield, VA 22161. (703) 487-4929 or FAX (703) 321-8199. Broad coverage of government-sponsored science and technology research reports, 1964 to present. Available on BRS,(800) 289-4277, DIALOG, (800) 334-2564, ORBIT, (800) 456-7248, and STN International, FIZ

Ency. of Physical Sciences and Engineering Info. Sources

AIRCRAFT ENGINES

Karlsruhe, P.O. Box 2465, W-7500, Karlsruhe 1, Germany, onlineservices. Also available on CD-ROM. Inquire as to cost and availability.

SAE Global Mobility Database. Society of Automative Engineers (SAE), Electronic Publishing Division, 400 Commonwealth Drive, Warrendale, PA 15098. (412) 776-4841. Contains citations with abstracts to technical papers on automotive and aerospace technology and vehicular-related industries that have been presented at SAE conferences. Covers 1906 to present. Available on ORBIT,(800) 456-7248,online service. Inquire as to cost and availability.

Scientific and Technical Books and Serials in Print. R.R. Bowker Inc., 121 Chanlon Road, New Providence, NJ 07974. (800) 521-8110 or (908) 464-6800. List of currently published books and serials in the physical and biological sciences, engineering and technology, with subject, author and titles indexes. Available on ORBIT,(800) 456-7248, online service.Inquire as to cost and availability.

SCISEARCH. Institute for Scientific Information, 3501 Market Street, Philadelphia, PA 19104. (800) 523-1850 or (215) 386-0100. Broad multidisciplinary title and author index to the international literature of science and technology, 1974 to present. Available on DIALOG,(800) 334-2564, and ORBIT,(800) 456-7248, online services. Also available on CD-ROM. Inquire as to cost and availability.

TRIS (Transportation Research Information). National Academy of Science, 2101 Constitution Avenue, N.W., Washington, DC 20418. (202) 334-3313 or (800) 624-6242. Citations with abstracts of literature on transportation, including air, highway, rail, maritime and other modes from 1968 to present. Available on DIALOG,(800) 334-2564, online service. Inquire as to cost and availability.

WILSONLINE. H.W. Wilson Company, 950 University Avenue, Bronx, NY 10452. (800) 367-6770 or (212) 588-8400. Makes available online versions of the printed H.W. Wilson indexes including Applied Science and Technology Index, Business Periodicals Index, General Science Index, and Readers' Guide to Periodical Literature. Period covered is generally 1983 to present. Available on BRS,(800) 289-4277, DIALOG,(800) 334-2564, and OCLC EPIC,(800) 848-5878, online services. Also available on CD-ROM. Inquire as to cost and availability.

PERIODICALS

Aeronautical Journal. The Royal Aeronautical Society, 4 Hamilton Place, London W1V 0BQ, England. 1897 to present. Monthly except June and August. Inquire for price information in U.S.

Aerospace America. American Institute of Aeronautics and Astronautics, 370 L'Enfant Promenade SW, Washington, DC 20024-2518. (202) 646-7471. 1932 to present. Monthly. $70 annually (non-members).

Aerospace Daily. McGraw-Hill Inc., Aviation Week Group, 1200 G Street NW, Suite 200, Washington, DC 20005. (202) 383-2350.1973 to present. Five times a week. $1340.00 per year.

Aerospace Engineering. Society of Automotive Engineers, 400 Commonwealth Dr., Warrendale, PA 15096. (412) 776-4841. 1981 to present. Monthly. $48.00 per year.

Aerospace Propulsion. McGraw-Hill Inc., Aviation Week Group, 1200 G Street NW, Suite 200, Washington, DC 20005. (202) 383-2350. 1963 to present. 25 times a year. $550.00 a year.

AIAA Journal. American Institute of Aeronautics and Astronautics, 370 L'Enfant Promenade SW, Washington, DC 20024-2518. (202) 646-7471. 1963 to present. Monthly. $52.00 per year (members) and $435.00 per year (non-members).

Air Progress. Challenge Publications, Inc., 7950 Deering Ave., Canoga Park, CA 91304. (818) 887-0550. FAX (818) 884-1343. 1941 to present. Monthly. $26.50 per year.

Aircraft Engineering and Aerospace Technology. BunHill Publications Ltd., 127 Stanstead Rd., London SE23 1JE, England. 1929 to present. Monthly. $102.00 per year (U.S.).

Aviation & Aerospace. Baxter Publishing Company, 310 Dupont Street, Toronto, Ontario M5R 1V9 Canada. (416) 968-7252. FAX (416) 968-2377. Bi-monthly. $42.00 per year.

Aviation Mechanics Bulletin. Flight Safety Foundation, Inc., 2200 Wilson Blvd., Suite 500, Arlington, VA 22201-3306. (703) 522-8300. FAX (703) 525-6047. 1953 to present. Bimonthly. $15.00 per year (non-members).

Aviation Week and Space Technology. 1221 Avenue of the Americas, New York, NY 10020. (212) 512-2999 or (800) 525-5003. 1916 to present. Weekly. $82.00 per year.

Flying. Hachette Filipacchi Magazines, Inc., 1633 Broadway, New York, NY 10019. (203) 622-2700. FAX (203) 622-2725. 1927 to present. Monthly. $24.00 per year.

Interavia/Aerospace World. Aerospace Media Publishing, Swissair Centre, 31 Route de l'Aeroport, POB Box 437, 1215 Geneva 15, Switzerland. (022) 788 27 88. FAX (022) 788 27 26. 1987 to present. Monthly. $128.00 per year (U.S.).

International Journal of Turbo and Jet Engines. Freund Publishing House, Ltd., PO Box 35010, Tel Aviv 63150, Israel. 972-2-5662925. FX 972-3-5605335. 1983 to present. Quarterly. $190.00 per year.

Journal of Aircraft. American Institute of Aeronautics and Astronautics, 370 L'Enfant Promenade SW, Washington, DC 20024-2518. 1963 to present. Bimonthly. $42.00 (members) and $230.00 (non-members).

Journal of Propulsion and Power. American Institute of Aeronautics and Astronautics, 370 L'Enfant Promenade SW, Washington, DC 20024-2518. 1985 to present. Bimonthly. $38.00 (members) and $260.00 (non-members).

RESEARCH CENTERS AND INSTITUTES

Aerospace Research Applications Center, Indianapolis Center For Advanced research, 611 N. Capitol Ave., Indianapolis, in 46204. (317) 262-5003. FAX (317) 262-5044.

Aerospace Research Center, 1250 Eye St., NW, Washington, DC 20005. (202) 371-8400. FAX (202) 371-8470.

Center For Aerospace Technology, Weber State College, Ogden, Ut 84408-1805. (801) 626-7272.

AIRCRAFT ENGINES

Ency. of Physical Sciences and Engineering Info. Sources

Machinery and Engine Technology Laboratory, Institute For Mechanical engineering, Bldg. M-7, Montreal Rd., Ottawa, Ontario, Canada K1A 0R6. (613) 993-2425. (613) 957-3281.

Ohio Aerospace Institute, 2001 Aerospace Parkway, Brook Park, OH 44142. (216) 891-2100. FAX (216) 891-2140.

Pennsylvania State University, Propulsion Engineering Research center, 106 Research Bldg. E, Bigler Rd., University Park, PA 16802. (814) 863-6272. FAX (814) 865-3389.

Transportation Research and Education Center, Georgia Institute of technology, Mason Civil Engineering Bldg., Atlanta, GA 30332. (404) 894-2236. FAX (404) 894-2278.

AIR CURRENTS

See: METEOROLOGY

AIRFOILS

See: AERONAUTICAL ENGINEERING

AIRFRAME

See also: AERODYNAMICS, AERONAUTICAL ENGINEERING, AIRCRAFT

ABSTRACT SERVICES AND INDEXES

Aeronautical Engineering. Scientific and Technical Branch, National Aeronautics and Space Administration. National Technical Information Service (NTIS), 5825 Port Royal Road, Springfield, VA 22161. (703) 487-4650. FAX (703) 321-8547. Monthly. A selection of annotated references to unclassified reports and journal articles that were introduced into the NASA scientific and technical information system and announced in STAR and IAA.

Alloys Index. American Society for Metals, Metals Park, OH 44073. (216) 338-5151. FAX (216) 338-4634. 1974 to present. Monthly. $295.00.

Applied Mechanics Reviews: An Assessment of World Literature in Engineering sciences. American Society of Mechanical Engineers, 345 East 47th Street, New York, NY 10017. (212) 705-7703. 1948 to present. Monthly. $360.00 per year.

Applied Science and Technology Index; A Cumulative Subject Index To English language Periodicals in the Fields of Aeronautics and Space Science, Computer Technology, chemistry, Construction Industry, Energy and Related Areas. H.W. Wilson Co., 950 University Avenue, Bronx, NY 10452. (800) 367-6770 or (212) 588-8400. FAX (718) 590-1617. From 1958 to present. Monthly. Inquire about cost and availability. Also available on CD-ROM.

Current Contents: Engineering, Technology, and Applied Sciences. Institute for Scientific Information, 3501 Market Street,

Philadelphia, PA 19104. (215) 386-0100 or (800) 523-1850. FAX (215) 386-6362. Weekly. $360.00 per year.

Engineering Index Monthly. Engineering Information, Inc., Castle Point on the Hudson, Hoboken, NJ 07030. (800) 221-1044. FAX (212) 832-1857. Monthly. $2200.00 per year. Also available onlineas *COMPENDEX,* and also on CD-ROM.

International Aerospace Abstracts. Technical Information Service, American Institute of Aeronautics and Astronautics, Inc., 555 West 57th St., New York, NY 10019. (212) 247-6500. FAX (212) 582-4861. Semi-monthly. $1295.00 per year.

Metals Abstracts and Metals Abstracts Index. American Society for Metals, Metals Park, OH 44073. (216) 338-5151. 1968 to present. Monthly. Abstracts are $1500.00 per year and Index is $460.00 per year.

Science Citation Index. Institute for Scientific Information, 3501 Market Street, Philadelphia, PA 19104. (215) 386-0100 or (800) 523-1850.. FAX (215) 386-6362. Inquire about availability and cost. Also available on CD-ROM.

Scientific and Technical Aerospace Reports (star). National Aeronautics and Space Administration. NASA Center for Aerospace Information, Box 8757, BWI Airport, Baltimore, MD 21240. (301) 621-0153. Monthly. Inquire about availability and cost. Also available through the NASA online retrieval service (RECON), and through the Aerospace Database through DIALOG.

World Aluminum Abstracts. Aluminum Association, Inc., 818 Connecticut Ave. NW, Washington, DC 20006. (202) 862-5156. Inquire about price and availability.

ASSOCIATIONS AND PROFESSIONAL SOCIETIES

American Institute of Aeronautics and Astronautics. 370 L'Enfant Promenade SW, Washington, DC 20024-2518. (202) 646-7471.

Flight Safety Foundation, Inc. 2200 Wilson Blvd., Suite 500, Arlington, VA 22201-3306. (703) 522-8300. FAX (703) 525-6047.

Society of Automotive Engineers. SAE, Inc., 400 Commonwealth Dr., Warrendale, PA 15096. (412) 776-4841.

BIBLIOGRAPHIES

Aeronautical Engineering: A Continuing Bibliography With. INDEXES. National Aeronautics and Space Administration, Center for Aerospace Information, Box 8757, BWI Airport, Baltimore, MD 21240. (301) 621-0153. Inquire for cost and availability.

Bibliography On Propulsion Airframe Integration Technologies for High-speed Civil Transport Applications: 1980-1991. National Aeronautics and Space Administration; distributed by National Technical Information Service, 5285 Port Royal Road, Springfield, VA 22161. (703) 487-4650. 1993. Inquire for cost and availability.

DIRECTORIES AND BIOGRAPHICAL SOURCES

Who's Who in Engineering. American Association of Engineering Societies, 1111 19th St. NW, Suite 608, Washington, DC 20036. 7th ed. 1988. $200.00.

World Aviation Directory. McGraw-Hill, Aviation Week Group (NY), 1221 Avenue of the Americas, New York, NY 10020. (215) 237-4112. FAX (215) 586-3232. Semiannual. $250.00.

ENCYCLOPEDIAS AND DICTIONARIES

Jane's Encyclopedia of Aviation. Michael J. Taylor, comp. Outlet Book Co., 40 Engelhard Ave., Avenal, NJ 07001. (908)827-2700 or (800) 223-6804. FAX (908) 827-2641. 1993. $34.99.

McGraw-Hill Encyclopedia of Engineering. Sybil P. Parker, ed. 2nd edition. McGraw-Hill Publishing Company, 1221 Avenue of the Americas, New York, NY 10020. (800) 262-4729 or (212) 512-3825. 1993. $95.50.

McGraw-Hill Encyclopedia of Science and Technology, Sybil P. Parker, ed. 7th ed. 20 vols. McGraw-Hill Publishing Company, 1221 Avenue of the Americas, New York, NY 10020. (800) 262-4729 or (212) 512-3825. $1900.00

GENERAL WORKS

A & P Technician Airframe Instructor Guide. David Jones. IAP, Inc., 913 N. Foster Road, Casper, WY 82601-1640. (800) 443-9250. FAX (307) 472-5106. 1994. Inquire for cost and availability.

Airplane Design. Jan Roskam. Roskam Aviation and Engineering Corporation, Ottawa, KS 66067. 1989. Inquire for cost and availability.

Aviation Mechanics Airframe Written Test Book. U.S. Department of Transportation, Federal Aviation Administration, Flight Standards Service. Order from Superintendent of Documents, U.S. Government Printing office, Box 371954, Pittsburgh, PA 15250-7954. (202) 783-3238. FAX (202) 512-2233. 1994. Inquire for cost and availability.

Aviation Mechanics Airframe Written Test Book, 1992-1994. IAP, Inc., 913 N. Foster Road, Casper, WY 82601-1640. (800) 443-9250. FAX (307) 472-5106. 1992. $6.00.

Composite Aircraft Design. Martin Hollmann. M. Hollman, Monterey, CA 93940. 1993. Write for cost and availability.

Engine Airframe Integration. Royal Aeronautical Society, 4 Hamilton Place, London W1V 0BQ, England. 1992. Inquire for cost and availability.

HANDBOOKS AND MANUALS

Airframe and Powerplant Mechanics: General Handbook. Advisory Circular Series 65-15A. U.S. Department of Transportation, Federal Aviation Administration, Flight Standards Service. Order from Superintendent of Documents, U.S. Government Printing office, Box 371954, Pittsburgh, PA 15250-7954. (202) 783-3238. FAX (202) 512-2233. 1990. $20.00.

ONLINE DATABASES AND CD-ROMS

Aerospace Database. American Institute of Aeronautics and Astronautics, 370 L'Enfant Promenade, S.W., Washington, DC 20024. (202) 646-7400. Worldwide published literature on research and development in aerospace and related areas, with abstracts. Covers 1962 to present. Available on DIALOG, (800) 334-2564, online service. Also available on CD-ROM. Inquire as to cost and availability.

Compendex Plus. Engineering Information, Inc., 345 East 47th Street, New York, NY 10017. (212) 705-7600 or (800) 221-1044. Contains citations with abstracts to worldwide literature in engineering and technology, from 1970 to present. Available ononline BRS,(800) 289-4277, DIALOG, (800) 334-2564, ORBIT (800) 456-7248, and STN International, FIZ Karlsruhe, P.O. Box 2465, W-7500, Karlsruhe 1, Germany, online services. Also available on CD-ROM. Inquire as to cost and availability.

CSA Engineering. Cambridge Scientific Abstracts, 7200 Wisconsin Avenue, Suite 601, Bethesda, MD 20814. (301) 961-6750 or (800) 843-7751. Contains citations and abstracts of international periodicals and research literature covering all fields of engineering and science and technology,including computer and information science, electronics, mechanical engineering, solid state materials, 1981 to present. Available on BRS,(800) 289-4277, online service.Inquire as to cost and availability.

Current Contents Search. Institute for Scientific Information, 3501 Market Street, Philadelphia, PA 19104. (215) 386-0100. FAX (215) 386-6362. Contains citations to articles listed in the table of contents of science and technology journals. Also articles in social sciences and life sciences journals. Available on BRS,(800) 289-4277, DIALOG,(800) 334-2564, online services. Inquire as to cost and availability.

Dissertation Abstracts. University Microfilms International, 300 North Zeeb Road, Ann Arbor, MI 48106. (800) 521-0600 or (313) 761-4700. Scope includes virtually all doctoral dissertations accepted at accredited American institutions from 1861 to present in 252 subject areas. Available on BRS,(800) 289-4277, DIALOG,(800) 334-2564, and OCLC EPIC,(800) 848-5878, online services. Also available on CD-ROM. Inquire as to cost and availability.

International Aerospace Abstracts. American Institute of Aeronautics and Astronautics, 370 L'Enfant Promenade, S.W., Washington, DC 20024. (202) 646-7400. Contains references and abstracts of journal and monograph literature relating to aerospace science and technology, from 1963 to present. Available through the NASA/RECON system of the National Aeronautics and Space Administration only.

*ISMEC: Mechanical Engineering Abstracts.*Cambridge Scientific Abstracts, 7200 Wisconsin Avenue, Suite 601, Bethesda, MD 20814. (301) 961-6750 or (800) 843-7751. Contains citations to the literature in mechanical engineering, industrial and production engineering, energy, power, mechanics, devices and related areas, from 1973 to present. Available on the DIALOG,(800) 334-2564, online service. Inquire as to cost and availability.

NASA Database. American Institute of Aeronautics and Astronautics, 370 L'Enfant Promenade, S.W., Washington, DC 20024. (202) 646-7400. Citations and abstracts of aeronautics and astronautics literature, 1962 to present. Also contains citations from *STAR, Scientific and Technical Aerospace Reports, and International Aerospace Abstracts.* Available through NASA/RECON online service. Inquire as to cost and availability.

AIRFRAME

Ency. of Physical Sciences and Engineering Info. Sources

NTIS Bibliographic Database. National Technical Information Service, 5285 Port Royal Road, Springfield, VA 22161.(703) 487-4929 or FAX (703) 321-8199. Broad coverage of government-sponsored science and technology research reports, 1964 to present. Available on BRS,(800) 289-4277, DIALOG, (800) 334-2564, ORBIT, (800) 456-7248, and STN International, FIZ Karlsruhe, P.O. Box 2465, W-7500, Karlsruhe 1, Germany, online services. Also available on CD-ROM. Inquire as to cost and availability.

SAE Global Mobility Database. Society of Automative Engineers (SAE), Electronic Publishing Division, 400 Commonwealth Drive, Warrendale, PA 15098. (412) 776-4841. Contains citations with abstracts to technical papers on automotive and aerospace technology and vehicular-related industries that have been presented at SAE conferences. Covers 1906 to present. Available on ORBIT,(800) 456-7248,online service. Inquire as to cost and availability.

Scientific and Technical Books and Serials in Print. R.R. Bowker Inc., 121 Chanlon Road, New Providence, NJ 07974. (800) 521-8110 or (908) 464-6800. List of currently published books and serials in the physical and biological sciences, engineering and technology, with subject, author and titles indexes. Available on ORBIT,(800) 456-7248, online service.Inquire as to cost and availability.

SCISEARCH. Institute for Scientific Information, 3501 Market Street, Philadelphia, PA 19104. (800) 523-1850 or (215) 386-0100. Broad multidisciplinary title and author index to the international literature of science and technology, 1974 to present. Available on DIALOG,(800) 334-2564, and ORBIT,(800) 456-7248, online services. Also available on CD-ROM. Inquire as to cost and availability.

TRIS (Transportation Research Information). National Academy of Science, 2101 Constitution Avenue, N.W., Washington, DC 20418. (202) 334-3313 or (800) 624-6242. Citations with abstracts of literature on transportation, including air, highway, rail, maritime and other modes from 1968 to present. Available on DIALOG,(800) 334-2564, online service. Inquire as to cost and availability.

WILSONLINE. H.W. Wilson Company, 950 University Avenue, Bronx, NY 10452. (800) 367-6770 or (212) 588-8400. Makes available online versions of the printed H.W. Wilson indexes including Applied Science and Technology Index, Business Periodicals Index, General Science Index, and Readers' Guide to Periodical Literature. Period covered is generally 1983 to present. Available on BRS,(800) 289-4277, DIALOG,(800) 334-2564, and OCLC EPIC,(800) 848-5878, online services. Also available on CD-ROM. Inquire as to cost and availability.

PERIODICALS

Aerospace America. American Institute of Aeronautics and Astronautics, 370 L'Enfant Promenade SW, Washington, DC 20024-2518. (202) 646-7471. 1932 to present. Monthly. $70 annually (non-members).

Aerospace Engineering. Society of Automotive Engineers, 400 Commonwealth Dr., Warrendale, PA 15096. (412) 776-4841. 1981 to present. Monthly. $48.00 per year.

AIAA Journal. American Institute of Aeronautics and Astronautics, 370 L'Enfant Promenade SW, Washington, DC 20024-2518. (202) 646-7471. 1963 to present. Monthly. $52.00 per year (members) and $435.00 per year (non-members).

Aircraft Engineering and Aerospace Technology. BunHill Publications Ltd., 127 Stanstead Rd., London SE23 1JE, England. 1929 to present. Monthly. $102.00 per year (U.S.).

Aviation Mechanics Bulletin. Flight Safety Foundation, Inc., 2200 Wilson Blvd., Suite 500, Arlington, VA 22201-3306. (703) 522-8300. FAX (703) 525-6047. 1953 to present. Bimonthly. $15.00 per year (non-members).

Aviation Week and Space Technology. 1221 Avenue of the Americas, New York, NY 10020. (212) 512-2999 or (800) 525-5003. 1916 to present. Weekly. $82.00 per year.

Canadian Aeronautics and Space Journal. Canadian Aeronautics and Space Institute, Suite 818, the National Building, 130 Slater St., Ottawa, Canada K1P 6E2. 1955 to present. Monthly. $75.00 per year (U.S.).

Journal of Aircraft. American Institute of Aeronautics and Astronautics, 370 L'Enfant Promenade SW, Washington, DC 20024-2518. 1963 to present. Bimonthly. $42.00 (members) and $230.00 (non-members).

RESEARCH CENTERS AND INSTITUTES

Aerospace Research Applications Center, Indianapolis Center For Advanced research, 611 N. Capitol Ave., Indianapolis, in 46204. (317) 262-5003. FAX (317) 262-5044.

Aerospace Research Center, 1250 Eye St., NW, Washington, DC 20005. (202) 371-8400. FAX (202) 371-8470.

Aviation Safety Institute, 6797 N. High St., Suite 316, PO Box 304, Worthington, OH 43085. (614) 885-4242. FAX (614) 885-5891.

Center For Aerospace Technology, Weber State College, Ogden, UT 84408-1805. (801) 626-7272.

Flight Research Laboratory, University of Kansas, Raymond Nichols Hall, lawrence, KS 66045. (913) 864-3043. FAX (913) 864-7789.

Joint Institute For Advancement of Flight Sciences, George Washington university, Langley Research Center, Mail Stop 269, Hampton, VA 23665. (804) 864-1982. FAX (804) 864-5894.

Ohio Aerospace Institute, 2001 Aerospace Parkway, Brook Park, OH 44142. (216) 891-2100. FAX (216) 891-2140.

Transportation Research and Education Center, Georgia Institute of technology, Mason Civil Engineering Bldg., Atlanta, GA 30332. (404) 894-2236. FAX (404) 894-2278.

AIRPLANES

See: AIRCRAFT

AIR POLLUTION

See also: POLLUTION

ABSTRACT SERVICES AND INDEXES

Air Pollution Titles. Pennsylvania State University, Environmental Resources Research Lab, Center for Air Environmental Studies. 226 Fenske Lab, University Park, PA 16802. (814) 865-1415. FAX (814) 862-1696. 1965 to present. Bi-monthly. $120.00 per year.

Applied Science and Technology Index; A Cumulative Subject Index To English language periodicals in the fields of aeronautics and space science, computer technology, chemistry, construction industry, energy and related areas. H.W. Wilson Co., 950 University Avenue, Bronx, NY 10452. (800) 367-6770 or (212) 588-8400. From 1958 to present. Monthly. Inquire about cost and availability. Also available on CD-ROM and online.

CA Selects: Air Pollution. Chemical Abstracts Service, 2540 Olentangy River Road, Box 3012, Columbus, OH 43210. (800) 848-6538 or (614) 447-3600, FAX (614) 447-3713. 1988 to present. $210.00 per year.

CA Selects: Environmental Pollution. Chemical Abstracts Service, 2540 Olentangy River Road, Box 3012, Columbus, OH 43210. (800) 848-6538 or (614) 447-3600, FAX (614) 447-3713. Semi-weekly. $210.00 per year.

CA Selects: Pollution Monitoring. Chemical Abstracts Service, 2540 Olentangy River Road, Box 3012, Columbus, OH 43210. (800) 848-6538 or (614) 447-3600, FAX (614) 447-3713. Semi-weekly. $210.00 per year.

Environmental Abstracts. R.R. Bowker Inc., 121 Chanlon Road, New Providence, NJ 07974. (800) 521-8110 or (908) 464-6800. Contains citations, most with abstracts, to literature relating to environmental issues and problems, from 1971 to present. Inquire as to cost and availability.

Environmental Engineering Abstracts, Cambridge Scientific Abstracts, 7200 wisconsin Avenue, Bethesda, MD 20814-4823. (301) 961-6750. FAX (301) 961-6720. Monthly. Covers hazardous materials, environmental impact and protection, treatment of sewage and industrial wastes, hydroelectric power, tidal and wind power, arctic and tropical engineering.

Meteorological and Geoastrophysical Abstracts. American Meteorological Society, c/o Inforonics Inc., 550 Newtown Road, Littleton, MA 01460. (508) 486-8976. FAX (508) 486-0027. 1950 to present. Monthly. $950.00 per year. Current abstracts of books, reports, research papers, and miscellaneous literature on environmental sciences, meteorology, astrophysics, hydrology, glaciology, and physical oceanography.

Pollution Abstracts, Cambridge Scientific Abstracts, 7200 Wisconsin Avenue, bethesda, MD 20814-4823. (301) 961-6750. FAX (301) 961-6720. Inquire for cost and availability.

ASSOCIATIONS AND PROFESSIONAL SOCIETIES

Acid Rain Foundation. 1410 Varsity Drive, Raleigh, NC 27606-2010. (919) 828-9443. FAX (919) 515-3593.

Air & Waste Management Association. 1 Gateway Circle, Third floor, Pittsburgh, PA 15222. (800) 270-3444. FAX (412) 232-3450.

American Academy of Enviromental Engineering. 130 Holiday Ct., Suite 100, Annapolis, MD 21401. (410) 266-3311. FAX (410) 266-7653.

Association of Local Air Pollution Control officials. 444 N. Capitol Street NW, Suite 307, Washington, DC 20001. (202) 624-7864. FAX (202) 624-7863.

Center For Clear Air Policy. 444 N. Capitol Street, Suite 602, Washington, DC 20001. (202) 624-7709. FAX (202) 508-3829.

Environmental Defense Fund. 257 Park Avenue South, New York, NY 10010. (212) 505-2100. FAX (212) 505-2375.

International Center For the Solution of Environmental problems. 535 Lovett Blvd., Houston, TX 77006. (713) 527-8711. FAX (713) 527-8025.

Manufacturers of Emission Controls Association. 1707 L Street NW, Suite 570, Washington, DC 20036. (202) 296-4797. FAX (202) 331-1388.

National Association of Environmental Professionals. 5165 MacArthur Blvd. NW, PO Box 9400, Washington, DC 20016-3315. (202) 966-1500. FAX (202) 966-1977.

State and Territorial Air Pollution Program Administrators. 444 N. Capitol Street NW, Suite 307, Washington, DC 20001. (202) 624-7864. FAX (202) 624-7863.

BIBLIOGRAPHIES

Environmental Periodicals Bibliography. International Academy at Santa Barbara, Environmental Studies Institute, 800 Garden Street, Suite D, Santa Barbara, CA 93101-1552. (805) 965-5010. 1972 to present. Bi-monthly. Price varies.

DIRECTORIES AND BIOGRAPHICAL SOURCES

Government Research Directory. 9th edition. Gale Research, 835 Penobscot Building, Detroit, MI 48226-4094. (313) 961-2242. (800) 877-4253. 1996. $465.00.

International Research Centers Directory. Gale Research, 835 Penobscot Building, Detroit, MI 48226-4094. (313) 961-2242. (800) 877-4253. 8th edition. 1995. $410.00

Research Centers Directory. Gale Research, Inc., 835 Penobscot Building, Detroit, MI 48226-4094. (313) 961-2242. (800) 877-4253. $485.00

Scientific and Technical Organizations and Agencies Directory. Gale Research, 835 Penobscot Building, Detroit, MI 48226-4094. (313) 961-2242. (800) 877-4253. 4th edition. 1996. $195.00.

Who's Who in Technology. Gale Research, 835 Penobscot Building, Detroit, MI 48226-4094. (313) 961-2242. (800) 877-4253. 1995. $195.00.

AIR POLLUTION

Ency. of Physical Sciences and Engineering Info. Sources

ENCYCLOPEDIAS AND DICTIONARIES

Encyclopedia of Environmental Control Technology, Volume 2: Air Pollution Technology. Paul N. Cheremisinoff, editor. GulfPublishing Company, P.O. Box 2608, Houston, TX 77252-2608. (713) 529-4301 or (800) 231-6275. FAX (713) 525-9647. 1989. $155.00

Encyclopedia of Physical Science and Technology. Robert A. Meyers, ed. 18 volumes. Academic Press, Inc., 1250 Sixth Avenue, San Diego, CA 92101-4311. (619) 231-0926. FAX (619) 699-6715. 1992. $2100.00

McGraw-Hill Encyclopedia of Science and Technology. Sybil P. Parker, ed. 7th edition. 20 volumes. McGraw-Hill Publishing Company, 1221 Avenue of the Americas, New York, NY 10020. (800) 262-4729 or (212) 512-3825. 1992. $1900.00

GENERAL WORKS

Air Pollution Control: Traditional and Hazardous Pollutants. Howard E. Hesketh. Revised edition. Technomic Publishing Company, Inc., 851 New Holland Avenue, Box 3535, Lancaster, PA 17604. (717) 291-5609 or (800) 233-9936. FAX (717) 295-4538. 1996. Inquire for price and availability.

Air Pollution Control Engineeering. Noel Denevers. McGraw-Hill Publishing Company, 1221 Avenue of the Americas, New York, NY 10020. (800) 262-4729 or (212) 512-3825. 1995. Inquire for price.

Earth Under Siege: Air Pollution & Global Change. Richard P. Turco. Oxford University Press, Inc., 198 Madison Avenue, New York, NY 10016-4314. (212) 726-6000. FAX (212) 726-6446. 1995. $50.00.

Fundamentals of Air Pollution. Richard W. Boubel, et al. 3d edition. Academic Press Inc., 6277 Sea Harbor Drive, Orlando, FL 32887. (800) 321-5068. 1994. $69.95.

Urban Air Pollution. N. Moussiopoulos and C.A. Brebbia, editors. Computational Mechanics Publications, 25 Bridge Street, Billerica, MA 01821. (508) 667-5841. FAX (508) 667-7582. Volume 1 (1994) $145.00, Volume 2 (1996) $145.00.

HANDBOOKS AND MANUALS

The Air Pollution Control Manual. Air & Waste Management Association staff. Van Nostrand Reinhold, 115 Fifth Avenue, New York, NY 10003. (212) 254-3232. FAX (212) 254-9499. 1992. $129.95.

CRC Handbook of Chemistry and Physics. David R. Lide, editor. 75th edition. CRC Press/Times Mirror Company, 2000 Corporate Blvd. NW, Boca Raton, FL 33431. (407) 994-0555 or (800) 272-7737. $99.50.

Environmental Science and Technology Handbook. Porter C. Knowles, et al. Government Institutes Inc., 4 Research Place, Suite 200, Rockville, MD 20850. (301) 921-2355. 1994. $75.00.

Handbook of Air Pollution Control Engineering. John C. Mycock. CRC Press/Times Mirror Company, 2000 Corporate Blvd. NW, Boca Raton, FL 33431. (407) 994-0555 or (800) 272-7737. 1995. $79.95.

Toxic Air Pollution Handbook. David R. Patrick, editor. Van Nostrand Reinhold, 115 Fifth Avenue, New York, NY 10003. (212) 254-3232. FAX (212) 254-9499. 1994. $104.95.

ONLINE DATABASES AND CD-ROMS

CA Search. Chemical Abstracts Service, P.O. Box 3012, Columbus, OH 43210-0012. (614) 447-3600. (800) 848-6533. FAX (614) 447-3709. Very comprehensive guide to worldwide chemical literature and related fields, 1972 to present. Available on BRS,(800) 289-4277, DIALOG, (800) 334-2564, ORBIT (800) 456-7248, and STN International, FIZ Karlsruhe, P.O. Box 2465, W-7500, Karlsruhe 1, Germany, online services. Inquire as to cost and availability.

NTIS Bibliographic Database. National Technical Information Service, 5285 Port Royal Road, Springfield, VA 22161. (703) 487-4929 or FAX (703) 321-8199. Broad coverage of government-sponsored science and technology research reports, 1964 to present. Available on BRS,(800) 289-4277, DIALOG, (800) 334-2564, ORBIT, (800) 456-7248, and STN International, FIZ Karlsruhe, P.O. Box 2465, W-7500, Karlsruhe 1, Germany, online services. Also available on CD-ROM. Inquire as to cost and availability.

SCISEARCH. Institute for Scientific Information, 3501 Market Street, Philadelphia, PA 19104. (800) 523-1850 or (215) 386-0100. Broad multidisciplinary title and author index to the international literature of science and technology, 1974 to present. Available on DIALOG,(800) 334-2564, and ORBIT,(800) 456-7248, online services. Also available on CD-ROM. Inquire as to cost and availability.

PERIODICALS

Aerosol Science and Technology. Elsevier Science [journals], 660 White Plains Rd., Tarrytown, NY 10159-5153. (919) 524-9200. FAX (919) 333-2444. 1991 to present. Bi-monthly. $395.00 per year.

Air & Water Pollution Control. Bureau of National Affairs, 1231 25th Street NW, Washington, DC 20037. (202) 452-4200. FAX (202) 822-8092. 1986 to present. Bi-weekly. $301.00 to year.

Air Pollution Control. Bureau of National Affairs, 1231 25th Street NW, Washington, DC 20037. (202) 452-4200. FAX (202) 822-8092. 1980 to present. Bi-weekly. $707.00 per year.

Atmospheric Environment. Elsevier Science [journals], 660 White Plains Rd., Tarrytown, NY 10159-5153. (919) 524-9200. FAX (919) 333-2444. 1967 to present. 22 times a year. $1995.00 per year.

Atmospheric Pollution & Abatement News. Business Communications Company, Inc., 25 Van Zant St., Norwalk, CT 06855. (203) 853-4266. FAX (203) 853-0348. Monthly. $295.00 per year.

EPA Journal. Superintendent of Documents, U.S. Government Printing office, Box 371954, Pittsburgh, PA 15250-7954. (202) 783-3238. FAX (202) 512-2233. 1975 to present. Bi-monthly. $10.00 per year.

Environmental Monitoring and Assessment. Kluwer Academic Publishers, P.O. Box 358, Accord Station, Hingham, MA 02018-

Ency. of Physical Sciences and Engineering Info. Sources

AIRPORTS

0358. (617) 871-6000. FAX (617) 871-6528. 1981 to present. 15 times a year. $765.00 per year.

Environmental Pollution. Elsevier Science [journals], 660 White Plains Rd., Tarrytown, NY 10159-5153. (919) 524-9200. FAX (919) 333-2444. 1970 to present. 12 times a year. $1250.00 per year.

Environmental Science and Technology. American Chemical Society, 1155 16th Street, N.W., Washington, DC 20036. (202) 872-4381 or (800) 333-9511. FAX (614) 447-3671. 1967 to present. Monthly. $89.00 per year for non-members.

Pollution Engineering. Cahners Publishing Company (Des Plains), 1350 East Touhy Avenue, Des Plaines, IL 60017-5080. (800) 323-4958. FAX (708) 390-2779. 1969 to prsent. 13 times a year. $70.00 per year.

Techlink: Air Pollution. National Technical Information Service (NTIS), 5285 Port Royal Road, Springfield, VA 22161. (703) 487-4650. FAX (703) 321-8547. 1993 to present. Monthly. $200.00 per year.

Water, Air and Soil Pollution. Kluwer Academic Publishers, P.O. Box 358, Accord Station, Hingham, MA 02018-0358. (617) 871-6000. FAX (617) 871-6528. 1971 to present. 28 times a year. $1519.00.

RESEARCH CENTERS AND INSTITUTES

University of Arizona, Engineering Experiment Station. 100 Civil Engineering Building, Tucson, AZ 85721. (602) 621-6594. FAX (602) 621-2232.

University of California—Davis, Air Quality Group. Crocker Nuclear Laboratories, Davis, CA 95616. (916) 752-1124.

University of California—Riverside, Statewide Air Pollution research Center. Riverside, CA 92521. (714) 787-5124. FAX (714) 787-5004.

University of Illinois, Institute For Environmental Studies. 1101 W Peabody Drive, Urbana, IL 61801. (217) 333-4178. FAX (217) 333-8046.

Renew America. 1400 16th Street NW, Suite 710, Washington, DC 20036. (202) 232-2252.

Resources For the Future, Inc., QUALITY of the ENVIRONMENT DIVISION. 1616 P Street NW, Washington, DC 20036. (202) 328-5000. FAX (202) 265-8069.

AIRPORTS

See also: AERODYNAMICS, AERONAUTICAL ENGINEERING, AERONAUTICS, AEROSPACE ENGINEERING

ABSTRACT SERVICES AND INDEXES

Applied Mechanics Reviews: An Assessment of World Literature in Engineering Sciences. American Society of Mechanical Engineers, 345 East 47th Street, New York, NY 10017. (212) 705-7703. 1948 to present. Monthly. $360.00 per year.

Applied Science and Technology Index. A Cumulative subject index To English language periodicals in the fields of aeronautics and space science, computer technology, chemistry, construction industry, energy and related areas. H.W. Wilson Co., 950 University Avenue, Bronx, NY 10452. (800) 367-6770 or (212) 588-8400. From 1958 to present. Monthly. Inquire about cost andavailability. Also available on CD-ROM.

Civil and Structural Engineering Abstracts, Cambridge Scientific abstracts, 7200 Wisconsin Avenue, Bethesda, MD 20814-4823. (301) 961-6750. FAX (301) 961-6720. 1993 to present. Monthly. $385.00 per year. Topics covered include structural design, construction equipment and methods, civil defense and military engineering, surveying, highway engineering, maritime and port structures, materials, land reclamation, and soil mechanics.

Current Literature in Traffic and Transportation. Northwestern University, Transportation Library, Evanston, IL 60208-2300. (708) 491-5275. FAX (708) 491-8601. 1960 to present. Quarterly. $20.00 per year.

Engineering Index Monthly. Engineering Information, Inc., Castle Point on the Hudson, Hoboken, NJ 07030. (800) 221-1044. FAX (201) 216-8532. Monthly. $2200.00 per year. Also available online as COMPENDEX, and also on CD-ROM.

International Aerospace Abstracts. Technical Information Service, American Institute of Aeronautics and Astronautics, Inc., 555 West 57th St., New York, NY 10019. (212) 247-6500. FAX (212) 582-4861. Semi-monthly. $1295.00 per year.

NTIS Alerts: Transportation. U.S. National Technical Information Service, 5825 Port Royal Road, Springfield, VA 22161. (703) 487-4630. FAX (703) 321-8547. Weekly. $135.00 per year.

STAR. Scientific and Technical Aerospace Reports. National Aeronautics and Space Administration. NASA Center for Aerospace Information, 800 Elkridge Landing Road, Linthicum Heights, MD 21090-2934. Monthly. Inquire about availability and cost. Also available through the NASA online retrieval service (RECON), and through the Aerospace Database through DIALOG.

ASSOCIATIONS AND PROFESSIONAL SOCIETIES

Airport Consultants Council. 421 King Street, Suite 200, Alexandria, VA 22314. (703) 683-5900. FAX (703) 549-4749.

Airports Council International—North America. 1220 19th Street NW, Suite 200, Washington, DC 20036. (202) 293-8500. FAX (202) 331-1362.

Air Transport Association of America. 1709 New York Avenue NW, Washington, DC 2006. (202) 626-4000.

Aircraft Owners and Pilots Association. 421 Aviation Way, Frederick, MD 21701. (301) 695-2000.

Airport Security Council. JFK International Airport, PO Box 3070, Jamaica, NY 11430.

Air Line Pilots Association, Afl-cio. 535 Herndon Parkway, Box 1169, Herndon, VA 22070. (703) 689-4176. FAX (703) 689-4370.

American Association of Airport Executives. 4212 King Street, Alexandria, VA 22302. (703) 824-0500. FAX (703) 820-1395.

American Institute of Aeronautics and Astronautics. 370 L'Enfant Promenade, S.W., Washington, DC 20024. (202) 646-7400.

Flight Safety Foundation, Inc., 2200 Wilson Blvd., Suite 500, Arlington, VA 22201-3306. (703) 522-8300. FAX (703) 525-6047.

Society of Automotive Engineers. SAE, Inc., 400 Commonwealth Dr., Warrendale, PA 15096. (412) 776-4841.

BIBLIOGRAPHIES

Aeronautical Engineering: A Continuing Bibliography With Indexes. National Aeronautics and Space Administration, Center for Aerospace Information, Box 8757, BWI Airport, Baltimore, MD 21240. (301) 621-0153. Inquire for cost and availability.

*Scientific and Technical Books and Serials in Print, R.*R. Bowker Inc., 121 Chanlon Road, New Providence, NJ 07974. (800) 521-8110 or (908) 464-6800. 3 volumes, annually. $285.00. Also available on CD-ROM.

DIRECTORIES AND BIOGRAPHICAL SOURCES

AAS Directory. American Astronautical Society, 6352 Rolling Mill Place, Suite 102, Springfield, VA 22152-2354. (703) 866-0020. FAX (703) 866-3526. Inquire for cost and availability.

AOPA's Aviation USA. Aircraft Owners and Pilots Association, 421 Aviation Way, Frederick, MD 21701. (301) 695-2000. 1962 to present. Annual. $24.95.

Research Centers Directory. Gale Research, 835 Penobscot Building, Detroit, MI 48226-4094. (313) 961-2242. (800) 877-4253. Irregular. Inquire for price and availability.

Who's Who in Engineering. American Association of Engineering Societies, 1111 19th St. NW, Suite 608, Washington, DC 20036. 7th ed. 1988. $200.00.

World Aviation Directory. McGraw-Hill, Aviation Week Group (NY), 1221 Avenue of the Americas, New York, NY 10020. (215) 237-4112. FAX (215) 586-3232. Semiannual. $250.00.

ENCYCLOPEDIAS AND DICTIONARIES

Dictionary of Aeronautical Terms, Dale Crane, Comp. Aviation Supplies and ACADemics, Inc., 7005 132nd Place SE, Renton, WA 98059. (213) 656-7745. 1991. $15.95.

Jane's Aerospace Dictionary, B. Guston. 3d ed. Janes Info Group (order from:) Dept. DSM, 1340 Braddock Pl., Suite 300, Box 1436, Alexandria, VA 22314-1651. 1991. $45.00.

GENERAL WORKS

Airport Design [Advisory Circular AC 150/5300-12]. U.S. Department of Transportation, Federal Aviation Administration. Available from Superintendent of Documents, U.S. Government Printing office, Washington, DC 20402-9325. (202) 783-3238. 1994. Inquire for price.

Airport Planning, Operation, and Management [transportation research Record No. 1423]. National Academy Press, Division of the National Academy of Science, 2101 Constitution Avenue, N.W., Washington, DC 20418. (202) 334-3313 or (800) 624-6242. 1993. Inquire for cost and availability.

Airport Terminals. Christopher J. Blow. Butterworths Publishers, 80 Montvale Avenue, Stoneham, MA 02180. (617) 438-8464 or (800) 366-2665. 1991. $95.00

Airport Engineering. Norman Ashford and Paul H. Wright. 3d ed. John Wiley and Sons, Inc., 605 Third Avenue, New York, NY10158. (800) 526-5368 or (212) 850-6000. 1992. $74.95.

Jane's Airport and Atc Equipment, October '93-94. D. Rider. Jane's Information Group, 1340 Braddock Pl., Suite 300, Alexandria, VA 22314-1651. (703) 683-3700. $245.00.

ONLINE DATABASES AND CD-ROMS

Aerospace Database. American Institute of Aeronautics and Astronautics, 370 L'Enfant Promenade, S.W., Washington, DC 20024. (202) 646-7400. Worldwide published literature on research and development in aerospace and related areas, with abstracts. Covers 1962 to present. Available on DIALOG, (800) 334-2564, online service. Also available on CD-ROM. Inquire as to cost and availability.

Compendex Plus. Engineering Information, Inc., 345 East 47th Street, New York, NY 10017. (212) 705-7600 or (800) 221-1044. Contains citations with abstracts to worldwide literature in engineering and technology, from 1970 to present. Available on online BRS,(800) 289-4277, DIALOG, (800) 334-2564, ORBIT (800) 456-7248, and STN International, FIZ Karlsruhe, P.O. Box 2465, W-7500, Karlsruhe 1, Germany, online services. Also available on CD-ROM. Inquire as to cost and availability.

NASA Database. American Institute of Aeronautics and Astronautics, 370 L'Enfant Promenade, S.W., Washington, DC 20024. (202) 646-7400. Citations and abstracts of aeronautics and astronautics literature, 1962 to present. Also contains citations from *STAR, Scientific and Technical Aerospace Reports, and International Aerospace Abstracts.* Available through NASA/RECON online service. Inquire as to cost and availability.

NTIS Bibliographic Database. National Technical Information Service, 5285 Port Royal Road, Springfield, VA 22161. (703) 487-4929 or FAX (703) 321-8199. Broad coverage of government-sponsored science and technology research reports, 1964 to present. Available on BRS,(800) 289-4277, DIALOG, (800) 334-2564, ORBIT, (800) 456-7248, and STN International, FIZ Karlsruhe, P.O. Box 2465, W-7500, Karlsruhe 1, Germany, online services. Also available on CD-ROM. Inquire as to cost and availability.

TRIS (Transportation Research Information). National Academy of Science, 2101 Constitution Avenue, N.W., Washington, DC 20418. (202) 334-3313 or (800) 624-6242. Citations with abstracts of literature on transportation, including air, highway, rail, maritime and other modes from 1968 to present. Available on DIALOG,(800) 334-2564, online service. Inquire as to cost and availability.

PERIODICALS

Aeronautical Journal. The Royal Aeronautical Society, 4 Hamilton Place, London W1V 0BQ, England. 1897 to present. Monthly except June and August. Inquire for price information in U.S.

Aerospace America. American Institute of Aeronautics and Astronautics, 370 L'Enfant Promenade SW, Washington, DC 20024-2518. (202) 646-7471. 1932 to present. Monthly. $70 annually(non-members).

Aerospace Engineering. Society of Automotive Engineers, 400 Commonwealth Dr., Warrendale, PA 15096. (412) 776-4841. 1981 to present. Monthly. $48.00 per year.

AIAA Journal. American Institute of Aeronautics and Astronautics, 370 L'Enfant Promenade SW, Washington, DC 20024-2518. (202) 646-7471. 1963 to present. Monthly. $52.00 per year (members) and $435.00 per year (non-members).

Airport Highlights. Airports Council International—North America, 1220 19th Street NW, Suite 200, Washington, DC 20036. (202) 293-8500. FAX (202) 331-1362. 1964 to present. 26 times a year. $495.00 per year.

Airport Journal. Box 273, Clarendon Hills, IL 60514-0273. (708) 318-6872. Monthly. $17.00 per year.

Airports. McGraw-Hill Aviation Week Group, 1200 G Street NW, Suite 200, Washington, DC 20005. (202) 383-2350. Weekly. $525.00 per year.

Aviation Mechanics Bulletin. Flight Safety Foundation, Inc., 2200 Wilson Blvd., Suite 500, Arlington, VA 22201-3306. (703) 522-8300. FAX (703) 525-6047. Bimonthly. $15.00 per year (non-members).

Aviation Week and Space Technology. 1221 Avenue of the Americas, New York, NY 10020. (212) 512-2999 or (800) 525-5003. 1916 to present. Weekly. $82.00 per year.

Canadian Aeronautics and Space Journal. Canadian Aeronautics and Space Institute, Suite 818, the National Building, 130 Slater St., Ottawa, Canada K1P 6E2. 1955 to present. Monthly. $75.00 per year (U.S.).

Flying. Hachette Filipacchi Magazines, Inc., 1633 Broadway, New York, NY 10019. (203) 622-2700. FAX (203) 622-2725. 1927 to present. Monthly. $24.00 per year.

IEEE Transactions On Aerospace and Electronic Systems. Institute of Electrical and Electronics Engineers, 445 Hoes Lane, PO Box 1331, Piscataway, NJ 08855-1331. (908) 981-0060. 1965 to present. Quarterly. $10.00 per issue (members), $20.00 per issue (non-members).

Journal of Aircraft. American Institute of Aeronautics and Astronautics, 370 L'Enfant Promenade SW, Washington, DC 20024-2518. 1963 to present. Bimonthly. $42.00 (members) and $230.00 (non-members).

Journal of Guidance, Control and Dynamics. American Institute of Aeronautics and Astronautics, 370 L'Enfant Promenade SW, Washington, DC 20024-2518. 1978 to present. Bimonthly. $42.00 (members) and $250.00 (non-members).

Journal of Transportation Engineering. American Society of Civil Engineers, Air Transport, Highway, Pipeline, Urban Transportation Divisions, 345 East 47th Street, New York, NY 10017-2398. (212) 705-7520 or (800) 548-2723. 1969 to present. Bimonthly. $124.00 per year.

Vertiflite. American Helicopter Society, 217 N.Washington St., Alexandria, VA 22314. (703) 684-6777. FAX (703) 739-9279. 1955 to present. Bimonthly. $25 per year (members) and $35 per year (non-members).

RESEARCH CENTERS AND INSTITUTES

Alaska Department of Transportation and Public Facilities, statewide Research Section. 2301 Peger Road, Fairbanks, AK. (907) 474-2470.

Dolphus E. Milligan Science Research Institute. Atlanta University, 440 Westview Drive SW, Atlanta, GA 30310. (404) 523-5148.

Texas A&M University, Texas Transportation Institute, Systems planning Division. College Station, TX 77843. (409) 845-1525. FAX (409) 845-6008.

AIRSHIPS

See also: AERODYNAMICS, AERONAUTICAL ENGINEERING, AERONAUTICS, AIRCRAFT, BALLOONS

ABSTRACT SERVICES AND INDEXES

Aeronautical Engineering. Scientific and Technical Branch, National Aeronautics and Space Administration. National Technical Information Service (NTIS), 5285 Port Royal Road, Springfield, VA 22161. (703) 487-4650. FAX (703) 321-8547. Monthly. A selection of annotated references to unclassified reports and journal articles that were introduced into the NASA scientific and technical information system and announced in STAR and IAA.

Applied Science and Technology Index. A Cumulative subject index to English language periodicals in the fields of aeronautics and space science, computer technology, chemistry, construction industry, energy and related areas. H.W. Wilson Co., 950 University Avenue, Bronx, NY 10452. (800) 367-6770 or (212) 588-8400. FAX (718) 590-1617. From 1958 to present. Monthly. Inquire about cost and availability. Also available on CD-ROM.

Current Contents: Engineering, Technology, and Applied Sciences. Institute for Scientific Information, 3501 Market Street, Philadelphia, PA 19104. (215) 386-0100 or (800) 523-1850. FAX (215) 386-6362. Weekly. $360.00 per year.

Engineering Index Monthly. Engineering Information, Inc., Castle Point on the Hudson, Hoboken, NJ 07030. (800) 221-1044. FAX (212) 832-1857. Monthly. $2300.00 per year. Also available online as *COMPENDEX*, and also on CD-ROM. Covers chemical engineering, computers, electrical engineering, civil engineering, metals and mining, industrial management, and mechanical engineering.

AIRSHIPS

Ency. of Physical Sciences and Engineering Info. Sources

International Aerospace Abstracts. Technical Information Service, American Institute of Aeronautics and Astronautics, Inc., 555 West 57th St., New York, NY 10019. (212) 247-6500. FAX (212) 582-4861. Semi-monthly. $1295.00 per year.

Science Citation Index. Institute for Scientific Information, 3501 Market Street, Philadelphia, PA 19104. (215) 386-0100 or (800) 523-1850. FAX (215) 386-6362. Inquire about availability and cost. Also available on CD-ROM.

*Scientific and Technical Aerospace Reports (star).*National Aeronautics and Space Administration. NASA Center for Aerospace Information, Box 8757, BWI Airport, Baltimore, MD 21240. (301) 621-0153. Monthly. Inquire about availability and cost. Also available through the NASA online retrieval service (RECON), and through the Aerospace Database through DIALOG.

World Aluminum Abstracts. Aluminum Association, Inc., 818 Connecticut Ave. NW, Washington, DC 20006. (202) 862-5156.

ASSOCIATIONS AND PROFESSIONAL SOCIETIES

Airship Association. 8512 Cedar Street, Silver Spring, MD 20910. (301) 588-7916. FAX (301) 588-2085.

American Aviation Historical Society. 2333 Otis St., Santa Ana, CA 92704.

Association of Balloon and Airship Constructors. Box 90864, San Diego, CA 92169. (619) 270-4049.

American Institute of Aeronautics and Astronautics, 370 L'enfant promenade SW, Washington, DC 20024-2518. (202) 646-7471.

Society of Automotive Engineers. SAE, Inc., 400 Commonwealth Dr., Warrendale, PA 15096. (412) 776-4841.

BIBLIOGRAPHIES

Aeronautical Engineering: A Continuing Bibliography With. INDEXES. National Aeronautics and Space Administration, Center for Aerospace Information, Box 8757, BWI Airport, Baltimore, MD 21240. (301) 621-0153. Inquire for cost and availability.

*Scientific and Technical Books and Serials in Print, R.*R. Bowker Inc., 121 Chanlon Road, New Providence, NJ 07974. (800) 521-8110 or (908) 464-6800. FAX (908) 464-3553. 3 volumes, annually. $285.00. Also available on CD-ROM.

DIRECTORIES AND BIOGRAPHICAL SOURCES

AAS Directory. American Astronautical Society, 6352 Rolling Mill Place, Suite 102, Springfield, VA 22152-2354. (703) 866-0020. FAX (703) 866-3526. Inquire for cost and availability.

Research Centers Directory. Gale Research, 835 Penobscot Building, Detroit, MI 48226-4094. (313) 961-2242. (800) 877-4253. FAX (313) 961-6083. 20th edition. $485.00.

Who's Who in Engineering. American Association of Engineering Societies, 1111 19th St. NW, Suite 608, Washington, DC 20036. 7th ed. 1988. $200.00.

World Aviation Directory. McGraw-Hill, Aviation Week Group (NY), 1221 Avenue of the Americas, New York, NY 10020. (215) 237-4112. FAX (215) 586-3232. Semiannual. $250.00.

ENCYCLOPEDIAS AND DICTIONARIES

Dictionary of Aeronautical Terms, Dale Crane, Comp. Aviation Supplies and Academics, Inc., 7005 132nd Place SE, Newcastle, WA 98059. (206) 235-1500 or (800) 426-8338. FAX (206) 235-0128. 1991. $15.95.

Encyclopedia of Physical Science and Technology. Robert A. Meyers, ed. Academic Press, Inc., 6277 Sea Harbor Drive, Orlando, FL 32887. 1992. $2100.00.

Jane's Aerospace Dictionary, B. Guston. 3rd ed. Jane's Information Group, Dept. DSM, 1340 Braddock Pl., Suite 300, Box1436, Alexandria, VA 22314-1651. (703) 683-3700. FAX (703) 836-0029. 1991. $45.00.

McGraw-Hill Encyclopedia of Science and Technology, Sybil P. Parker, ed. 7th ed. 20 vols. McGraw-Hill Publishing Company, 1221 Avenue of the Americas, New York, NY 10020. (800) 262-4729 or (212) 512-3825. $1900.00

GENERAL WORKS

Airshipmen, Businessmen, and Politics, 1890-1940. Henry Cord Meyer. Smithsonian Institution Press, 470 L'Enfant Plaza, Suite 7100, Washington, DC 20560. (202) 287-3738. 1991. $45.00

Sky Ships: A History of the Airship in the United States Navy. William F. Althoff. Orion Books, Crown Publishing Group, 201 E. 50th St., New York, NY 10022. (800) 726-0600 or (800) 733-3000. 1990. Inquire for cost and availability.

Zeppelin: Rigid Airships 1893-1943. Peter W. Brooks. Smithsonian Institution Press, 470 L'Enfant Plaza, Suite 7100, Washington, DC 20560. (202) 287-3738. 1992. $52.00.

ONLINE DATABASES AND CD-ROMS

Aerospace Database. American Institute of Aeronautics and Astronautics, 370 L'Enfant Promenade, S.W., Washington, DC 20024. (202) 646-7400. Worldwide published literature on research and development in aerospace and related areas, with abstracts. Covers 1962 to present. Available on DIALOG, (800) 334-2564, online service. Also available on CD-ROM. Inquire as to cost and availability.

COMPENDEX Plus. Engineering Information, Inc., 345 East 47th Street, New York, NY 10017. (212) 705-7600 or (800) 221-1044. Contains citations with abstracts to worldwide literature in engineering and technology, from 1970 to present. Available on online BRS,(800) 289-4277, DIALOG, (800) 334-2564, ORBIT (800) 456-7248, and STN International, FIZ Karlsruhe, P.O. Box 2465, W-7500, Karlsruhe 1, Germany, online services. Also available on CD-ROM. Inquire as to cost and availability.

CSA Engineering. Cambridge Scientific Abstracts, 7200 Wisconsin Avenue, Suite 601, Bethesda, MD 20814. (301) 961-6750 or (800) 843-7751. Contains citations and abstracts of international periodicals and research literature covering all fields of engineering and science and technology,including computer and information science, electronics, mechanical engineering,

solid state materials, 1981 to present. Available on BRS,(800) 289-4277, online service.Inquire as to cost and availability.

Current Contents Search. Institute for Scientific Information, 3501 Market Street, Philadelphia, PA 19104. (215) 386-0100. FAX (215) 386-6362. Contains citations to articles listed in the table of contents of science and technology journals. Also articles in social sciences and life sciences journals. Available on BRS,(800) 289-4277, DIALOG,(800) 334-2564, online services. Inquire as to cost and availability.

Dissertation Abstracts. University Microfilms International, 300 North Zeeb Road, Ann Arbor, MI 48106. (800)521-0600 or (313) 761-4700. Scope includes virtually all doctoral dissertations accepted at accredited American institutions from 1861 to present in 252 subject areas. Available on BRS,(800) 289-4277, DIALOG,(800) 334-2564, and OCLC EPIC,(800) 848-5878, online services. Also available on CD-ROM. Inquire as to cost and availability.

International Aerospace Abstracts. American Institute of Aeronautics and Astronautics, 370 L'Enfant Promenade, S.W., Washington, DC 20024. (202) 646-7400. Contains references and abstracts of journal and monograph literature relating to aerospace science and technology, from 1963 to present. Available through the NASA/RECON system of the National Aeronautics and Space Administration only.

NASA Database. American Institute of Aeronautics and Astronautics, 370 L'Enfant Promenade, S.W., Washington, DC 20024. (202) 646-7400. Citations and abstracts of aeronautics and astronautics literature, 1962 to present. Also contains citations from *STAR, Scientific and Technical Aerospace Reports,* and *International Aerospace Abstracts.* Available through NASA/RECON online service. Inquire as to cost and availability.

SAE Global Mobility Database. Society of Automative Engineers (SAE), Electronic Publishing Division, 400 Commonwealth Drive, Warrendale, PA 15098. (412) 776-4841. Contains citations with abstracts to technical papers on automotive and aerospace technology and vehicular-related industries that have been presented at SAE conferences. Covers 1906 to present. Available on ORBIT,(800) 456-7248,online service. Inquire as to cost and availability.

Scientific and Technical Books and Serials in Print. R.R. Bowker Inc., 121 Chanlon Road, New Providence, NJ 07974. (800) 521-8110 or (908) 464-6800. List of currently published books and serials in the physical and biological sciences, engineering and technology, with subject, author and titles indexes. Available on ORBIT,(800) 456-7248, online service.Inquire as to cost and availability.

SCISEARCH. Institute for Scientific Information, 3501 Market Street, Philadelphia, PA 19104. (800) 523-1850 or (215) 386-0100. Broad multidisciplinary title and author index to the international literature of science and technology, 1974 to present. Available on DIALOG,(800) 334-2564, and ORBIT,(800) 456-7248, online services. Also available on CD-ROM. Inquire as to cost and availability.

TRIS (Transportation Research Information). National Academy of Science, 2101 Constitution Avenue, N.W., Washington, DC 20418. (202) 334-3313 or (800) 624-6242. Citations with abstracts of literature on transportation, including air, highway, rail, maritime and other modes from 1968 to present. Available on DIALOG,(800) 334-2564, online service. Inquire as to cost and availability.

WILSONLINE. H.W. Wilson Company, 950 University Avenue,Bronx, NY 10452. (800) 367-6770 or (212) 588-8400. Makes available online versions of the printed H.W. Wilson indexes including Applied Science and Technology Index, Business Periodicals Index, General Science Index, and Readers' Guide to Periodical Literature. Period covered is generally 1983 to present. Available on BRS,(800) 289-4277, DIALOG,(800) 334-2564, and OCLC EPIC,(800) 848-5878, online services. Also available on CD-ROM. Inquire as to cost and availability.

PERIODICALS

AAHS Journal. American Aviation Historical Society, 2333 Otis St., Santa Ana, CA 92704. Quarterly. $25.00 (annual dues includes journal subscription).

Advanced Lighter-Than-Air Review. Airship Association, 8512 Cedar Street, Silver Spring, MD 20910. (301) 588-7916. FAX (301) 588-2085. Quarterly. Inquire for price information.

Aerospace America. American Institute of Aeronautics and Astronautics, 370 L'Enfant Promenade SW, Washington, DC 20024-2518. (202) 646-7471. 1932 to present. Monthly. $70 annually (non-members).

Aerospace Engineering. Society of Automotive Engineers, 400 Commonwealth Dr., Warrendale, PA 15096. (412) 776-4841. 1981 to present. Monthly. $48.00 per year.

Aerostation. Association of Balloon and Airship Constructors. Box 90864, San Diego, CA 92169. (619) 270-4049. 1974 to present. Quarterly. Inquire for price and availability.

AIAA Journal. American Institute of Aeronautics and Astronautics, 370 L'Enfant Promenade SW, Washington, DC 20024-2518. (202) 646-7471. 1963 to present. Monthly. $52.00 per year (members) and $435.00 per year (non-members).

Journal of Aircraft. American Institute of Aeronautics and Astronautics, 370 L'Enfant Promenade SW, Washington, DC 20024-2518. 1963 to present. Bimonthly. $42.00 (members) and $230.00 (non-members).

AIR TRAFFIC CONTROL

See also: AERONAUTICAL ENGINEERING, AERONAUTICS, AEROSPACE ENGINEERING, AIRPORTS

ABSTRACT SERVICES AND INDEXES

Applied Mechanics Reviews: An Assessment of World Literature in Engineering Sciences. 1948-present. American Society of Mechanical Engineers, 345 East 47th Street, New York, NY 10017. (212) 705-7703. Monthly. $360.00 per year.

Applied Science and Technology Index; A cumulative subject index to English language periodicals in the fields of aeronautics and space science, computer technology, chemistry, construction industry, energy and related areas. H.W. Wilson Co., 950 University Avenue, Bronx, NY 10452. (800) 367-6770 or (212) 588-8400. FAX (718) 590-1617. From 1958 to present. Monthly. Inquire about cost and availability. Also available on CD-ROM and online.

Computer Abstracts, MCB University Press Ltd., PO Box10812, Birmingham, AL 35201-0812. (800) 633-4931. FAX (205) 995-1588. Monthly. Covers computer theory, data, hardware, systems, networks, human-computer interaction, artificial intelligence, as well as applications of computers in aerospace, business, CAD/CAM, cartography, civil engineering, electronics and electrical engineering, industrial engineering, mechanical engineering, medicine, structural engineering, etc.

Current Literature in Traffic and Transportation. Northwestern University, Transportation Library, Evanston, IL 60208-2300. (708) 491-5275. FAX (708) 491-8601. 1960 to present. Quarterly. $20.00 per year.

Engineering Index Monthly. Engineering Information, Inc., Castle Point on the Hudson, Hoboken, NJ 07030. (800) 221-1044. FAX (212) 832-1857. Monthly. $2300.00 per year. Also available online as COMPENDEX, and also on CD-ROM. Covers chemical engineering, computers, electrical engineering, civil engineering, metals and mining, industrial management, and mechanical engineering.

International Aerospace Abstracts. Technical Information Service, American Institute of Aeronautics and Astronautics, Inc., 555 West 57th St., New York, NY 10019. (212) 247-6500. FAX (212) 582-4861. Semi-monthly. $1295.00 per year.

NTIS Alerts: Transportation. U.S. National Technical Information Service, 5825 Port Royal Road, Springfield, VA 22161. (703) 487-4630. FAX (703) 321-8547. Weekly. $135.00 per year.

Scientific and Technical Aerospace Reports (star). National Aeronautics and Space Administration. NASA Center for Aerospace Information, Box 8757, BWI Airport, Baltimore, MD 21240. (301) 621-0153. Monthly. Inquire about availability and cost. ALso available through the NASA online retrieval service (RECON), and through the Aerospace Database through DIALOG.

ASSOCIATIONS AND PROFESSIONAL SOCIETIES

Aircraft Owners and Pilots Association. 421 Aviation Way, Frederick, MD 21701. (301) 695-2000.

Air Line Pilots Association, AFL-CIO. 535 Herndon Parkway, Box 1169, Herndon, VA 22070. (703) 689-4176. FAX (703) 689-4370.

Airports Council International—North America. 1220 19th Street NW, Suite 200, Washington, DC 20036. (202) 293-8500. FAX (202) 331-1362.

Airport Security Council. JFK International Airport, PO Box 3070, Jamaica, NY 11430.

Air Traffic Control Association. 2300 Clarendon Blvd., Suite 711, Arlington, VA 22201. (703) 522-5717. FAX (703) 527-7251.

Air Transport Association of America. 1709 New York Avenue NW, Washington, DC 20006. (202) 626-4000.

American Association of Airport Executives. 4212 King Street, Alexandria, VA 22302. (703) 824-0500. FAX (703) 820-1395.

American Helicopter Society, Inc., 217 N. Washington St., Alexandria, VA 22314. (703) 684-6777. FAX (703) 739-9279.

American Institute of Aeronautics and Astronautics, 370L'enfant Promenade SW, Washington, DC 20024-2518. (202) 646-7400.

Flight Safety Foundation, Inc., 2200 Wilson Blvd., Suite 500, Arlington, VA 22201-3306. (703) 522-8300. FAX (703) 525-6047.

Society of Automotive Engineers (sae International). 400 Commonwealth Drive, Warrendale, PA 15096. (412) 776-4841.

BIBLIOGRAPHIES

Aeronautical Engineering: A Continuing Bibliography With. INDEXES. National Aeronautics and Space Administration, Center for Aerospace Information, Box 8757, BWI Airport, Baltimore, MD 21240. (301) 621-0153. Inquire for price.

Scientific and Technical Books and Serials in Print, R.R. Bowker Inc., 121 Chanlon Road, New Providence, NJ 07974. (800) 521-8110 or (908) 464-6800. FAX (908) 464-3553. 3 volumes, annually. $285.00. Also available on CD-ROM.

DIRECTORIES AND BIOGRAPHICAL SOURCES

AAS Directory. American Astronautical Society, 6352 Rolling Mill Place, Suite 102, Springfield, VA 22152-2354. (703) 866-0020. FAX (703) 866-3526. Inquire for cost and availability.

AOPA's Aviation USA. Aircraft Owners and Pilots Association, 421 Aviation Way, Frederick, MD 21701. (301) 695-2000. 1962 to present. Annual. $24.95.

Jane's Airport and ATC Equipment, October '93-94. D. Rider. Jane's Information Group, 1340 Braddock Pl., Suite 300, Alexandria, VA 22314-1651. (703) 683-3700. FAX (703) 836-0029. $245.00.

Research Centers Directory. Gale Research, 835 Penobscot Building, Detroit, MI 48226-4094. (313) 961-2242. (800) 877-4253. FAX (313) 961-6083. $485.00

Who's Who in Engineering. American Association of Engineering Societies, 1111 19th Street NW, Suite 608, Washington, DC 20036. (202) 296-2237 or (800) 658-8897. 8th edition. 1991. Inquire for price.

World Aviation Directory. McGraw-Hill, Aviation Week Group (NY), 1221 Avenue of the Americas, New York, NY 10020. (215) 237-4112. FAX (215) 586-3232. Semiannual. $250.00.

ENCYCLOPEDIAS AND DICTIONARIES

Jane's Aerospace Dictionary, B. Guston. 3rd ed. Jane's Information Group, Dept. DSM, 1340 Braddock Pl., Suite 300, Box 1436, Alexandria, VA 22314-1651. (703) 683-3700. FAX (703) 836-0029. 1991. $45.00.

SAE Dictionary of Aerospace Engineering. Society of Automotive Engineers, 400 Commonwealth Drive, Warrendale, PA 15096. (412) 776-4841. 1992. $85.00.

GENERAL WORKS

Air Traffic Control. G. R. Duke. Fifth edition. Ian Allan/distributed by Motorbooks International, 729 Prospect Avenue, Osceola, WI 54020. (800) 826-6600. 1994. Inquire for cost and availability.

Air Traffic Control. U.S. Air Traffic Rules and Procedures Service. U.S. Department of Transportation, FederalAviation Administration. Order from: Superintendent of Documents, U.S. Government Printing office, Box 371954, Pittsburgh, PA 15250-7954. (202) 783-3238. FAX (202) 512-2233. 1992. Inquire for cost and availability.

Automation and Systems Issues in Air Traffic Control. John A. Wise, V. David Hopkin, Marvin L. Smith, eds. Springer-Verlag, 175 Fifth Avenue, New York, NY 10010. (212) 460-1500 or (800) 777-4643. FAX (212) 473-6272. 1991. Inquire for cost and availability.

Careers in Air Traffic Control. U.S. Department of Transportation, Federal Aviation Administration. Order from: Order from: Superintendent of Documents, U.S. Government Printing office, Box 371954, Pittsburgh, PA 15250-7954. (202) 783-3238. FAX (202) 512-2233. 1991. Inquire for cost and availability.

Free Flight: Reinventing Air Traffic Control. Aviation Systems Research Corporation, Golden, CO 80401. 1994. Inquire for cost and availability.

Fundamentals of Air Traffic Control. Michael S. Nolan. Wadsworth Publishing Company, 10 Davis Dr., Belmont, CA 94002. (415) 595-2350. FAX (606) 525-0978. 1990. Inquire for price and availability.

ONLINE DATABASES AND CD-ROMS

Compendex Plus. Engineering Information, Inc., 345 East 47th Street, New York, NY 10017. (212) 705-7600 or (800) 221-1044. Contains citations with abstracts to worldwide literature in engineering and technology, from 1970 to present. Available on online BRS,(800) 289-4277, DIALOG, (800) 334-2564, ORBIT (800) 456-7248, and STN International, FIZ Karlsruhe, P.O. Box 2465, W-7500, Karlsruhe 1, Germany, online services. Also available on CD-ROM. Inquire as to cost and availability.

NASA Database. American Institute of Aeronautics and Astronautics, 370 L'Enfant Promenade, S.W., Washington, DC 20024. (202) 646-7400. Citations and abstracts of aeronautics and astronautics literature, 1962 to present. Also contains citations from *STAR, Scientific and Technical Aerospace Reports,* and *International Aerospace Abstracts.* Available through NASA/RE-CON online service. Inquire as to cost and availability.

NTIS Bibliographic Database. National Technical Information Service, 5285 Port Royal Road, Springfield, VA 22161. (703) 487-4929 or FAX (703) 321-8199. Broad coverage of government-sponsored science and technology research reports, 1964 to present. Available on BRS,(800) 289-4277, DIALOG, (800) 334-2564, ORBIT, (800) 456-7248, and STN International, FIZ Karlsruhe, P.O. Box 2465, W-7500, Karlsruhe 1, Germany, online services. Also available on CD-ROM. Inquire as to cost and availability.

TRIS (Transportation Research Information). National Academy of Science, 2101 Constitution Avenue, N.W., Washington, DC 20418. (202) 334-3313 or (800) 624-6242. Citations with abstracts of literature on transportation, including air, highway, rail,maritime and other modes from 1968 to present. Available on DIALOG,(800) 334-2564, online service. Inquire as to cost and availability.

PERIODICALS

Aerospace America. American Institute of Aeronautics and Astronautics, 370 L'Enfant Promenade SW, Washington, DC 20024-2518. (202) 646-7471. 1932 to present. Monthly. $70 annually (non-members).

Aerospace Engineering. Society of Automotive Engineers, 400 Commonwealth Dr., Warrendale, PA 15096. (412) 776-4841. 1981 to present. Monthly. $48.00 per year.

AIAA Journal. American Institute of Aeronautics and Astronautics, 370 L'Enfant Promenade SW, Washington, DC 20024-2518. (202) 646-7471. 1963 to present. Monthly. $52.00 per year (members) and $435.00 per year (non-members).

Air Line Pilot. Air Line Pilots Association, AFL-CIO, 535 Herndon Parkway, Box 1169, Herndon, VA 22070. (703) 689-4176. FAX (703) 689-4370. 1932 to present. Monthly. $22.00 per year.

Airport Highlights. Airports Council International—North America, 1220 19th Street NW, Suite 200, Washington, DC 20036. (202) 293-8500. FAX (202) 331-1362. 1964 to present. 26 times a year. $495.00 per year.

Airport Journal. Box 273, Clarendon Hills, IL 60514-0273. (708) 318-6872. Monthly. $17.00 per year.

Airports. McGraw-Hill Aviation Week Group, 1200 G Street NW, Suite 200, Washington, DC 20005. (202) 383-2350. Weekly. $525.00 per year.

Air Traffic Control. U.S. Federal Aviation Administration, 800 Independence Avenue SW, Washington, DC 20591. (202) 267-3484. Base volume and quarterly updates. Inquire for cost and availability.

American Helicopter Society Journal. American Helicopter Society, 217 N. Washington Street, Alexandria, VA 22314. (703) 684-6777. FAX (703) 739-9279. 1956 to present. Quarterly. $35.00 per year.

Aviation Mechanics Bulletin. Flight Safety Foundation, Inc., 2200 Wilson Blvd., Suite 500, Arlington, VA 22201-3306. (703) 522-8300. FAX (703) 525-6047. 1953 to present. Bimonthly. $15.00 per year (non-members).

Aviation Week and Space Technology. 1221 Avenue of the Americas, New York, NY 10020. (212) 512-2999 or (800) 525-5003. 1916 to present. Weekly. $82.00 per year.

Canadian Aeronautics and Space Journal. Canadian Aeronautics and Space Institute, Suite 818, the National Building, 130 Slater St., Ottawa, Canada K1P 6E2. 1955 to present. Monthly. $75.00 per year (U.S.).

IEEE Transactions On Aerospace and Electronic Systems. Institute of Electrical and Electronics Engineers, 445 Hoes Lane, PO Box 1331, Piscataway, NJ 08855-1331. (908) 981-0060. 1965 to present. Quarterly. $10.00 per issue (members), $20.00 per issue(non-members).

Journal of Aircraft. American Institute of Aeronautics and Astronautics, 370 L'Enfant Promenade SW, Washington, DC 20024-2518. 1963 to present. Bimonthly. $42.00 (members) and $230.00 (non-members).

Journal of Air Traffic Control. Air Traffic Control Association, 2300 Clarendon Blvd., Suite 711, Arlington, VA 22201. (703) 522-5717. FAX (703) 527-7251. 1958 to present. Quarterly. $35.00 per year.

Vertiflite. American Helicopter Society, 217 N. Washington St., Alexandria, VA 22314. (703) 684-6777. FAX (703) 739-9279. 1955 to present. Bimonthly. $25 per year (members) and $35 per year (non-members).

RESEARCH CENTERS AND INSTITUTES

Dolphus E. Milligan Science Research Institute. Atlanta University, 440 Westview Drive SW, Atlanta, GA 30310. (404) 523-5148.

Texas A&M University, Texas Transportation Institute, Systems Planning Division. College Station, TX 77843. (409) 845-1525. FAX (409) 845-6008.

ALARM SYSTEMS

ABSTRACTS AND INDEXES

Applied Science and Technology Index; A Cumulative Subject Index To English Language Periodicals in the Fields of Aeronautics and Space Science, Computer Technology, Chemistry, Construction Industry, Energy and Related Areas. H.W. Wilson Co., 950 University Avenue, Bronx, NY 10452. (800) 367-6770 or (212) 588-8400. FAX (718) 590-1617. From 1958 to present. Monthly. Inquire about cost and availability. Also available on CD-ROM.

Computer Abstracts, MCB University Press Ltd., PO Box 10812, Birmingham, AL 35201-0812. (800) 633-4931. FAX (205) 995-1588. Monthly. Covers computer theory, data, hardware, systems, networks, human-computer interaction, artificial intelligence, as well as applications of computers in aerospace, business, CAD/CAM, cartography, civil engineering, electronics and electrical engineering, industrial engineering, mechanical engineering, medicine, structural engineering, etc.

ASSOCIATIONS AND PROFESSIONAL SOCIETIES

American Society For Industrial Security. 1655 N Fort Myer Dr., Suite 1200, Arlington, VA 22209-3198. (703) 522-5800. FAX (708) 243-4954.

Automatic Fire Alarm Association. PO Box 1652, Barrington, IL 60011. (708) 381-6422. FAX (708) 381-0727.

Central Station Alarm Association. 7101 Wisconsin Avenue, Suite 1390, Bethesda, MD 20814. (301) 907-0045. FAX (301) 907-7897.

International Association of Home Safety and Security Professionals. PO Box 2044, Erie, PA 16512-2044. (814) 456-2911. FAX (814) 456-2911.

Locksmith Security Association. 4732 Rochester Road, Royal Oak, MI 48073. (810) 589-0318. FAX (810) 589-3780.

National Alarm Association of America. PO Box 3409, Dayton, OH 45401. (800) 283-6285. FAX (513) 461-4759.

National Association of Fire Equipment Distributors. 401 N. Michigan Ave., Chicago, IL 60611. (312) 644-6610. FAX (312) 321-6869.

National Burglar & Fire Alarm Association. 7101 Wisconsin Avenue, Bethesda, MD 20814. (301) 907-3202. FAX (301) 907-7897.

Security Industry Association. 1801 K St. NW, Suite 1203L, Washington, DC 20006. (202) 466-7420. FAX (202) 466-0010.

Society of Fire Protection Engineers. 1 Liberty Square, Boston, MA 02109-4825. (617) 482-0686. FAX (617) 482-8184.

DIRECTORIES AND BIOGRAPHICAL SOURCES

Automotive, Burglary Protection and Mechanical Equipment Directory. Underwriters Laboratories, 333 Pfingsten Road, Northbrook, IL 60062-2096. (708) 272-8800. Annual. Inquire for cost and availability.

National Burglar and Fire Alarm Association—Membership Directory. National Burglar and Fire Alarm Association, 7101 Wisconsin Avenue, Bethesda, MD 20814. (301) 907-3202. FAX (301) 907-7897. Annual. $195.00.

GENERAL WORKS

The Alarm Book: A Guide To Burglar and Fire Alarms. Dan McTague and Doug Smith. Butterworth-Heinemann, 313 Washington Street, Newton, MA 02158. (617) 928-2500 or (800) 366-2665. FAX (617) 928-2620. 1987. Inquire for cost and availability.

Control Systems Engineering. Norman Nise. 2nd edition. Benjamin-Cummings Publishing Company, 390 Bridge Parkway, Redwood City, CA 94065. (800) 950-2665. FAX (415) 594-4409. $70.95. Inquire for availability.

Design and Application of Security/fire-alarm Systems. John E. Trauster. Rev. ed. McGraw-Hill Publishing Company, 1221 Avenue of the Americas, New York, NY 10020. (800) 262-4729 or (212) 512-3825. 1990. $34.95.

Fire Alarm Signaling Systems. Richard Bukowski. 2md ed. National Fire Protection Association. 1 Batterymarch Park, Quincy, MA 02269. (617) 770-3500. 1994. Write for price information.

Human Factors of Alarm Design. Neville Stanton, editor. Taylor & Francis, 1900 Frost Road, Suite 101, Bristol, PA 19007. (215) 785-5800. FAX (215) 785-5515. 1994. $75.00.

Security: A Guide To Security System Design and Equipment Selection and Installation. 2nd ed. Butterworth-Heinemann, 313

Ency. of Physical Sciences and Engineering Info. Sources

ALCOHOL

Washington Street, Newton, MA 02158. (617) 928-2500 or (800) 366-2665. FAX (617) 928-2620. 1992. $75.00.

Standard For Antitheft Alarms and Devices. Underwriters' Laboratories, 333 Pfingsten Rd., Northbrook, IL 60062-2096. (708) 272-8800. 1994. Write or call for price information.

Understanding and Servicing Alarm Systems. H. William Trimmer. 2nd ed. Butterworth-Heinemann, 313 Washington Street, Newton, MA 02158. (617) 928-2500 or (800) 366-2665. FAX (617) 928-2620. 1990. $32.95.

HANDBOOKS AND MANUALS

Electronic Alarm and Security Systems: A Technician's Guide. TAB Books, P.O. Box 40, Blue Ridge Summit, PA 17294-0850. (717) 794-2191 or (800) 233-1128. FAX (717) 794-2080. 1995. Inquire for cost and availability.

ONLINE DATABASES AND CD-ROMS

ISMEC: Mechanical Engineering Abstracts. Cambridge Scientific Abstracts, 7200 Wisconsin Avenue, Suite 601, Bethesda, MD 20814. (301) 961-6750 or (800) 843-7751. Contains citations to the literature in mechanical engineering, industrial and production engineering, energy, power, mechanics, devices and related areas, from 1973 to present. Available on the DIALOG,(800) 334-2564, online service. Inquire as to cost and availability.

NTIS Bibliographic Database. National Technical Information Service, 5285 Port Royal Road, Springfield, VA 22161. (703) 487-4929 or FAX (703) 321-8199. Broad coverage of government-sponsored science and technology research reports, 1964 to present. Available on BRS,(800) 289-4277, DIALOG, (800) 334-2564, ORBIT, (800) 456-7248, and STN International, FIZ Karlsruhe, P.O. Box 2465, W-7500, Karlsruhe 1, Germany, online services. Also available on CD-ROM. Inquire as to cost and availability.

WILSONLINE. H.W. Wilson Company, 950 University Avenue, Bronx, NY 10452. (800) 367-6770 or (212) 588-8400. Makes available online versions of the printed H.W. Wilson indexes including Applied Science and Technology Index, Business Periodicals Index, General Science Index, and Readers' Guide to Periodical Literature. Period covered is generally 1983 to present. Available on BRS,(800) 289-4277, DIALOG,(800) 334-2564, and OCLC EPIC,(800) 848-5878, online services. Also available on CD-ROM. Inquire as to cost and availability.

PERIODICALS

Access Control. Argus Inc., 6151 Powers Ferry Rd., NW, Atlanta, GA 30339-2491. (404) 955-2500. FAX (404) 955-0400. 1957 to present. Monthly. $49.00 per year.

Corporate Security. Business Research Publications Inc., 817 Broadway, New York, NY 10003. (212) 673-4700. FAX (212) 475-1790. Semi-monthly. $225.00 per year.

Corporate Security Digest. Washington Crime News Services, 3918 Prosperity Avenue, Suite 318, Fairfax, VA 22031-3334. (703) 573-1600. 1987 to present. Weekly. $295.00 per year.

International Security Review. FMJ International Publications, Ltd., Queensway House, 2 Queensway, RedHill, Surrey RH1 1QS England. Telephone 0737-768611. FAX 0737-761685. 1978 to present. Quarterly. Write for price information.

Locksmith Ledger International. Locksmith Publishing Corp., 850 Busse Hwy, Park Ridge, IL 60068. (708) 692-5940. FAX (708) 692-4604. 1939 to present. Monthly. $32.00 per year.

Security. Cahners Publishing Company (Des Plains), 1350East Touhy Avenue, Des Plaines, IL 60017-5080. (708) 635-8800. FAX (708) 635-9950. 1964 to present. 13 times a year. $69.95 per year.

Security and Protection Equipment. Batiste Publications Ltd., Pembroke House, Campsbourne Rd., Hornsey, London N8 7PE, England. Telephone 44-81-340-3291. FAX 44-81-341-4840. 1968 to present. Monthly. Write for price information.

Security Management. American Society for Industrial Security. 1655 N Fort Myer Dr., Suite 1200, Arlington, VA 22209-3198. (703) 522-5800. FAX (708) 522-5226. 1957 to present. Monthly. $48.00 per year.

Security Sales. Bobit Publishing Co., 2512 Artesia Blvd., Redondo Beach, CA 90278-3210. (310) 376-8788. FAX (310) 376-9043. 1979 to present. Monthly. $35.00 per year.

Security Technology & Design. Locksmith Publishing Corp., 850 Busse Hwy, Park Ridge, IL 60068. (708) 692-5940. FAX (708) 692-4604. 1991 to present. Bi-monthly. $38.00 per year.

ALCOHOL

See also: CHEMICAL ENGINEERING, CHEMISTRY, ORGANIC CHEMISTRY

ABSTRACT SERVICES AND INDEXES

Applied Science and Technology Index; A cumulative subject index to English language periodicals in the fields of aeron autics and space science, computer technology, chemistry , construction industry, energy and related areas. H. W. Wilson Co., 950 University Avenue, Bronx, NY 10452-9978. (212) 588-8400. (800) 367-6770. FAX (718) 590-1617. From 1958 to present. Monthly. Inquire about cost and availability. Also available online (BRS and WILSONLINE) and on CD-ROM. ISSN: 0003-6986.

Chemical Abstracts. Chemical Abstracts Service. 2540 Olentangy River Road, Box 3012, Columbus, OH 43210-0012. (614) 447-3600. FAX (614) 447-3713. 1907 to present. Weekly. $16,800.00 per year. Available online and on CD- ROM. CA is also available in five section groupings.   Inquire regarding cost and availability.

Chemical Abstracts - Applied Chemistry and Chemical Engineering Sections. Chemical Abstracts Service, 2540 Olentangy River Road, Box 3012, Columbus, OH 43210. (800) 848-6538 or (614) 447-3600. FAX (614) 447-3713. 1963 to present. Bi-weekly. $2080.00 per year. ISSN: 0090-8363.

Chemical Abstracts - Organic Chemistry Sections. Chemical Abstracts Service, 2540 Olentangy River Road, Box 3012, Columbus, OH 43210. (614) 447-3600. (800) 848-6538.   FAX

ALCOHOL

Ency. of Physical Sciences and Engineering Info. Sources

(614) 447-3713. 1963 to present. Bi-weekly. $2080.00 per year. ISSN: 0009-2282. Also available online.

Current Contents: Physical, Chemical and Earth Sciences. Institute for Scientific Information, 3501 Market Street, Philadelphia, PA 19104. (215) 386-0100. (800) 523-1850. FAX (215) 386-2291. 1961 to present. Weekly. $442.00 per year. Also available online (BRS, DIALOG) and on CD-ROM. Inquire regarding cost and availability. ISSN: 0163-2574.

General Science Index. H.W. Wilson Company, 950University Avenue, Bronx, NY 10452. (212) 588-8400. (800) 367-6770. FAX (718) 590-1617. From 1978 to present. Ten issues per year; quarterly and annual cumulations. Service basis. Available on CD-ROM and online. Inquire regarding cost and availability. ISSN: 0162-1963.

Physics Abstracts. INSPEC. Section A, Science Abstracts. Institution of Elect rical Engineers (IEE), London, United Kingdom. Available from: INSPEC/IEEE - Institute of Electrical and Electronic Engineers, Box 1331, 445 Hoes Lane, Piscataway, NJ 08855-1331. (908) 562-5549. 1898 to present. 24 issues per year. $2835.00 per year. ISSN: 0036-8091. Also available online and on CD-ROM.

Process and Chemical Engineering/Chemical Engineering Abstracts. Royal Society of Chemistry, Information Services, Thomas Graham House, Science Park, Milton Road, Cambridge, CB4 4WF, England. Contains citations, mostly with abstracts, of the worldwide literature on chemical engineering; from 1982 to present. Monthly. $610.00 per year. Also available online. ISSN: 0960-5045.

Science Citation Index. SCI. Institute for Scientific Information, 3501 Market Street, Philadelphia, PA 19104. (215) 386-0100. (800) 523-1850. FAX (215) 386-2991. 1961 to present. Six issues per year, plus annual cumulation. $11650.00 per year. Also available online and on CD-ROM. Inquire about price and availability. ISSN: 0036-827X.

Theoretical Chemical Engineering. Royal Society of Chemistry, Information Services, Thomas Graham House, Science Park, Milton Road, Cambridge, CB4 4WF, England. Covers theoretical chemical engineering, including theory and laboratory experimentation. 1964 to present. Monthly. $235.00. ISSN: 0960-5053.

ANNUAL REVIEWS AND YEARBOOKS

Advances in Physical Organic Chemistry. Academic Press, Inc., 1250 Sixth Avenue, San Diego, CA 92101. (619) 231-0926. FAX (619) 699-6715.1963 to present. Irregular. Price varies, inquire.

Annual Review of Physical Chemistry. Annual Reviews, Inc., 4139 El Camino Way, Palo Alto, CA 94306-0897. (415) 493-4400. Fax (415) 855-9815. Annual. $55.00 per year.

Studies in Organic Chemistry. Marcel Dekker, Inc., 270 Madison Avenue, New York, NY 10016. (212) 696-9000. FAX (212) 685-4540. 1973 to present. Irregular. Price varies, inquire.

ASSOCIATIONS AND PROFESSIONAL SOCIETIES

American Chemical Society, 1155 16th Street, Nw, Washington, DC 20036. (202) 872-4600.

American Institute of Chemical Engineers. 345 East 47th Street, New York, NY 10017. (212) 705-7338.

American Oil Chemists Society. 1608 Broadmoor Drive, Box 3489, Champaign, IL 61921-0489. (217) 359-2344. FAX (217) 351-8091.

Association of Consulting Chemists and Chemical Engineers, 50 East 41st Street, Suite 92, New York, NY 10017. (212)684-6255.

Chemical Manufacturers Association. 2501 M Street, N.W., Washington, DC 20037. (202) 887-1182

BIBLIOGRAPHIES

Chemical Information Sources. Gary Wiggins. McGraw-Hill Publishing Company, 1221 Avenue of the Americas, New York, NY 10020. (800) 262-4729 or (212) 512-3825. 1991. $42.50.

International Bibliogrphy of Studies On Alcohol. Sarah S. Jordy, compiler. Rutgers Center of Alcohol Studies Publications, PO Box 969, Piscataway, NJ 08855. (908) 445- 2190. 3 volumes. 1901-1960. 1981. $150.00 set.

Information Sources in Chemistry. R.T. Bottle and J.F.B Rowland, editors. R.R. Bowker Inc., 121 Chanlon Road, New Providence, NJ 07974. (800) 521-8110 or (908) 464-6800. Fourth edition. 1993. $75.00.

DIRECTORIES AND BIOGRAPHICAL SOURCES

American Institute of Chemical Engineers. DIRECTORY. American Institute of Chemical Engineers. 345 East 47th Street, New York, NY 10017-2396. (212) 705-7663. Annual.

American Men and Women of Science. Physical and Biological Sciences. R.R. Bowker Inc., 121 Chanlon Road, New Providence, NJ 07974. (908) 464-6800. (800) 521-8110. 20th edition. 8 volumes. 1996. $850.00

Directory of Chemistry Software 1992. Wendy Warr, Peter Willett, Geoff Downs. American Chemical Society, 1155 16th Street, NW, Washington, DC 20036. (202) 872-4600. 1992. $35.95.

Research Centers Directory. Gale Research Inc., Detroit, MI 48226. 20th edition. (800) 347-4253. $485.00.

ENCYCLOPEDIAS AND DICTIONARIES

Dictionary of Organic Compounds. John B. Buckingham, edior. Chapman and Hall, 1 Penn Plaza, New York, NY 10119. (212) 564-1060. 1993. 10th supplement. $832.50.

Encyclopedia of Chemical Processing and Design. McKetta Marcel Dekker, Inc., 270 Madison Avenue, New York, NY 10016. (212) 696-9000. (800) 228-1160. 1976 - . $175.00 per volume. ISSN:

Illustrated Chemistry Laboratory Terminology. Gerbert W. Ockerman. CRC Press, Inc., 2000 Corporate Boulevard, NW, Boca Raton, FL 33431. (407) 994-0555. (800) 272-7737. 1991. $32.95.

Industrial Chemical Thesaurus. Michael Ash and Irene Ash. VCH Publishers, Inc., 220 East 23rd Street, New York, NY 10010-4606. (800) 367-8249. 1992. $295.00.

Kirk-Othmer Encyclopedia of Chemical Technology. John Wiley and Sons, Inc., 605 Third Avenue, New York, NY 10158. (800) 526-5368 or (212) 850-6000. Fourth edition. 1991 - . Twenty-seven volumes. $5400.00.

McGraw-Hill Encyclopedia of Science and Technology. McGraw-Hill Book, Incorporated, 1221 Avenue of the Americas, New York, NY 10020. (212) 997-3675. Seventeenth editon. 1992. $1900.00.

Ullman's Encyclopedia of Industrial Chemistry. VCH Publications, Inc., 220 East 23rd Street, Suite 909, New York, NY 10010-4606. (212) 683-8333. (800) 422-8824. 5th edition. 1984 - . Price varies per volume. ISSN:

GENERAL WORKS

Alcohol Fuels: Impacts from Increased Use of Ethanol Blended Fuels. Diane Publishing Co., 600 Upland Avenue, Upland, PA 19015. (610) 499-7415. 1993. $45.00.

Alcohols and Derivatives. J. Tremolieres, editor. Franklin Book Co., Inc., Elkins Park, PA 19117. (215) 635-5252. 2 volumes. 1970. $128.00 set.

Alcohols As Motor Fuels. Society of Automotive Engineers (SAE), 400 Commonwealth Drive, Warrendale, PA 15096. (412) 776-4841. 1980. $36.00.

Corn Into Alcohol: Its Chemistry. Gordon Press Publications, P.O. Box 459, Bowling Green Station, New York, NY 10004. (718) 624-8419. 1991. $69.00.

Distillation of Alcohol and De-Naturing. Gordon Press Publications, P.O. Box 459, Bowling Green Station, New York, NY 10004. (718) 624-8419. 1996. $263.99.

Distillation Process Applications and Operation. Henry Z. Kister. McGraw-Hill Publishing Company, 1221 Avenue of the Americas, New York, NY 10020. (800) 262-4729 or (212) 512-3825. 1990. $79.95.

Exploring Chemical Elements and their Compounds. David L.Heiserman. TAB Books, P.O. Box 40, Blue Summit, PA 17294-0850. (717) 794-2191. (800) 233-1128. 1992. $19.95.

Increasing the Efficiency of Ethanol Production Through the Use of A Membrane Technology. Diane Publishing Co., 600 Upland Avenue, Upland, PA 19015. (610) 499-7415. 1994. $40.00 in paper.

HANDBOOKS AND MANUALS

Comprehensive Organic Synthesis: Selectivity, Strategy, and Efficiency in Modern Organic Chemical. Barry M. Trost and others, editors. Pergamon Press, Maxwell House, Fairview Park, Elmsford, NY 10523. (914) 592-7700. Fax (914) 592-3625. 1991. Nine volumes. $3900.00.

CRC Handbook of Chemistry and Physics. David R. Kide, editor. CRC Press Inc., 2000 Corporate Blvd., NW, Boca Raton,

FL 33431. (407) 994-0555. (800) 333-8300. 77th edition. 1996. $99.95.

Guide To Basic Chemical Compounds. D.R. Lide, Jr. CRC Publishers, Inc., 2000 Corporate Blvd., N.W., Boca Raton, FL 33431. (407) 994-0555 or (800) 333-8300. 1993. $120.00.

Lange's Handbook of Chemistry. John A. Dean, editor. McGraw-Hill Publishing Company, 1221 Avenue of the Americas, New York, NY 10020. (212) 512-2000. (800) 262-4729. 14th edition. 1996. $99.95.

Perry's Chemical Engineers' Handbook. Robert H. Perry and Donald W. Green, editors. McGraw-Hill Publishing Company, Inc., 1221 Avenue of the Americas, New York, NY 10020. (212) 512-2000. (800) 262-4729. 6th edition. 1996. $129.50.

ONLINE DATABASES AND CD-ROMS

CA Search. Chemical Abstracts Service, P.O. Box 3012, Columbus, OH 43210-0012. (614) 447-3600. (800) 848-6533. FAX (614) 447-3709. Very comprehensive guide to worldwide chemical literature and related fields. 1972 to present. Available on BRS,(800) 289-4277, DIALOG, (800) 334-2564, ORBIT (800) 456-7248, and STN International, FIZ Karlsruhe, P.O. Box 2465, W-7500, Karlsruhe 1, Germany, online services. Inquire as to cost and availability.

Chemical Journals of the American Chemical Society. American Chemical Society, 1155 16th Street, N.W., Washington, DC 20036. (202) 872-4381 or (800) 424-6747. Contains complete text of approximately 90,000 articles from 22 primary journals published by the American Chemical Society, from mostly 1982 to present. Available on STN International, FIZ Karlsruhe, P.O. Box 2465, W-7500, Karlsruhe 1, Germany, online service. Inquire as to cost and availability.

Compendex Plus. Engineering Information, Inc., 345 East 47th Street, New York, NY 10017. (212) 705-7600 or (800) 221-1044. Contains citations with abstracts to worldwide literature in engineering and technology, from 1970 to present. Available on online BRS,(800) 289-4277, DIALOG, (800) 334-2564, ORBIT (800) 456-7248, and STN International, FIZ Karlsruhe, P.O. Box 2465, W-7500, Karlsruhe 1, Germany, online services. Also available on CD-ROM. Inquire as to cost and availability.

Dissertation Abstracts. University Microfilms International, 300 North Zeeb Road, Ann Arbor, MI 48106. (800) 521-0600 or (313) 761-4700. Scope includes virtually all doctoral dissertations accepted at accredited American institutions from 1861 to present in 252 subject areas. Available on BRS, (800) 289-4277, DIALOG, (800) 334-2564, and OCLC EPIC, (800) 848-5878, online services. Also available on CD-ROM. Inquire as to cost and availability.

Kirk-Othmer Encyclopedia of Chemical Technology. John Wiley and Sons, Inc., 605 Third Avenue, New York, NY 10158. (800) 526-5368 or (212) 850-6000. Contains the complete text of all chapters in the 27 volume fourth edition of the *Kirk-othmer Encyclopedia of Chemical Technology.* 1991. Available on BRS,(800) 289-4277, DIALOG,(800) 334-2564, online services. Inquire as to cost and availability.

NTIS Bibliographic Database. National Technical Information Service, 5285 Port Royal Road, Springfield, VA 22161. (703) 487-4929 or FAX (703) 321-8199. Broad coverage of government-sponsored science and technology research reports, 1964 to present. Available on BRS,(800) 289-4277, DIALOG, (800)

ALCOHOL

Ency. of Physical Sciences and Engineering Info. Sources

334-2564, ORBIT, (800) 456-7248, and STN International, FIZ Karlsruhe, P.O. Box 2465, W-7500, Karlsruhe 1, Germany, online services. Also available on CD-ROM. Inquire as to cost and availability.

SCISEARCH. Institute for Scientific Information, 3501 Market Street, Philadelphia, PA 19104. (800) 523-1850 or (215) 386-0100. Broad multidisciplinary title and author index to the international literature of science and technology, 1974 topresent. Available on DIALOG, (800) 334-2564, and ORBIT, (800) 456-7248, online services. Also available on CD-ROM. Inquire as to cost and availability.

WILSONLINE. H.W. Wilson Company, 950 University Avenue, Bronx, NY 10452. (800) 367-6770 or (212) 588-8400. Makes available online versions of the printed H.W. Wilson indexes including Applied Science and Technology Index, Business Periodicals Index, General Science Index, and Readers' Guide to Periodical Literature. Period covered is generally 1983 to present. Available on BRS, (800) 289-4277, DIALOG, (800) 334-2564, and OCLC EPIC, (800) 848-5878, online services. Also available on CD-ROM. Inquire as to cost and availability.

PERIODICALS

American Chemical Society Journal. American Chemical Society, Box 3337, Columbus, OH 43210. (614) 447-3776. 1970 to present. 1879 to present. Semi-monthly. $1065.00 ISSN: 0002-7863.

American Oil Chemists' Society Journal. American Oil Chemists' Society, 1608 Broadmoor Drive, Box 3489, Champaign, IL 61921-0489. (217) 359-2344. FAX (217) 351-8091. 1917 to present. Monthly. $185.00. ISSN: 0003-021X.

Carbon. Elsevier Science, 660 White Plains Road, Tarrytown NY 10591-5151. (914) 524-9200. FAX (914) 333-2444. 1963 to present. 8 issues per year. $1045.00. ISSN: 0008-6223.

Chemical Engineering Progress. American Institute of Chemical Engineers, 345 East 47th Street, New York, NY 10017. (212) 705-7338. FAX (212) 752-3294. 1947 to present. Monthly. $75.00 per year. ISSN: 0360-7275.

Chemical Week. Includes Annual Buyers Guide. Chemical Week Associates, 888 Seventh Avenue, New York, NY 10106. (212) 621-4900. FAX (212) 621-4949. 1914 to present. Weekly. $99.00 per year. ISSN: 0009-272X.

Chemtech. American Chemical Society, Box 3337, Columbus, OH 43210. (614) 447-3776. 1970 to present. Monthly. $370.00 per year. ISSN: 0009-2703.

Hydrocarbon Processing. Gulf Publishing Company, P.O. Box 2608, Houston, TX 7725-2608. (713) 520-4301. FAX (713) 520-4433. 1922 to present. Monthly. $22.00 per year. ISSN: 0887-0284.

Industrial and Engineering Chemistry Research. American Chemical Society, Box 3337, Columbus, OH 43210. (614) 447-3776. (800) 333-9511. FAX (614) 447-3671. 1962 to present. Monthly. $567.00 per year. ISSN: 0888-5885.

Journal of Heterocyclic Chemistry. Heterocorporation, Box 20285, Tampa, FL 33622-0285. 1964 to present. Bi-monthly. $420.00 per year. ISSN: 0022-152X.

Journal of Organic Chemistry. American Chemical Society, Box 3337, Columbus, OH 43210. (614) 447-3776. (800) 227-5558. 1936 to present. Bi-Weekly. $785.00 per year. ISSN: 0022-3263.

Tetrahedron Letters. Elsevier Science Publishing Company, Inc., 655 Avenue of the Americas, New York, NY 10010. (212)989-5800. 1959 to present. Weekly. $4745.00 per year. ISSN: 0040-4039.

RESEARCH CENTERS AND INSTITUTES

Center for Catalytic Science and Technology. University of Delaware. Departments of Chemical Engineering and Chemistry, Newark, DE 19716. (302) 451-8056. (302) 451-1048.

Chemical Laboratories. Harvard University, Oxford Street, Cambridge, MA 02138. (617) 495-4283. FAX (617) 496-5618.

Chemistry Laboratories. Renasseslaer Polytechnic Institute. Cogswell Laboratory, Troy, NY 12180-3590. (518) 276-8981.

University/industry Chemical Research Center. Mississippi State University, Department of Chemistry, P.O. Drawer CH, Mississippi State, MS 39762. (601) 325-3584.

ALGEBRA

See also: CALCULUS, GEOMETRY, MATHEMATICS, STATISTICS

ABSTRACT SERVICES AND INDEXES

Applied Mechanics Reviews: An Assessment of World Literature in Engineering Sciences. American Society of Mechanical Engineers, 345 East 47th Street, New York, NY 10017. (212) 705-7703. 1948 to present. Monthly. $360.00 per year. ISSN: 0003-6900.

Applied Science and Technology Index; A cumulative subject index to English language periodicals in the fields of aeronautics and space science, computer technology, chemistry, construction industry, energy and related areas. H.W. Wilson Co., 950 University Avenue, Bronx, NY 10452. (212) 588-8400. (800) 367-6770. FAX (718) 590-1617. From 1958 to present. Monthly. Inquire about cost and availability. Also available on CD-ROM and online. ISSN: 0003-6986.

Compactmath - Compact Mathematics Library. Cumulative CD-ROM edition of Zentralblatt fuer Mathematik - Mathematics Abstracts. Springer-Verlag New York, Inc. 44 Hartz Way. Secaucus. NJ. 07096-2491. (201) 348-4033. FAX (201) 348-4505. 1993 to present. Annual. Available only on CD-ROM. Inquire regarding cost and availability. ISSN: 0938-3174.

Compumath Citation Index. Institute for Scientific Information, 3501 Market Street, Philadelphia, PA 19104. (800) 523-1850 or (215)-386-0100. FAX (215) 386-2991. Three times per year. $1955.00. Also available online and on CD-ROM. Inquire regarding cost and availablity. ISSN: 0730-6199.

Current Mathematical Publications. American Mathematical Soci ety. P.O. Box 1571, Annex Station, Providence, RI 02901-9930. (800) 556-7774 or (401) 455-4000. FAX (401) 331-3842. 1969 to present. Seventeen times per year. $377.00 per year. Available online from BRS, DIALOG, European Space Agency.

Also available on CD-ROM. Inquire regarding cost and availability. ISSN: 0361-4794.

General Science Index. H.W. Wilson Company, 950 University Avenue, Bronx, NY 10452. (212) 588-8400. (800) 367-6770. FAX (718) 590-1617. From 1978 to present. Ten issues per year;quarterly and annual cumulations. Service basis. Available on CD-ROM and online. Inquire regarding cost and availability. ISSN: 0162-1963.

Index To Mathematical Tables. Alan Fletcher. Addison- Wesley Publishing Co., Inc., 1 Jacob Way, Reading, MA 01867. (617) 944-3700. (800) 447-2226. 2nd edition. 1962. 2 volumes.

Mathematical Reviews: A Review journal covering the world literature of mathematical research. American Mathematical Society, P.O. Box 1571, Annex Station, Providence, RI 02901-9930. (800) 556-7774 or (401) 455-4000. 19 40 to present. Monthly. $4594.00 per year. Also available via network (MathSciNet) online and on CD-ROM. Inquire regarding cost and availability. ISSN: 0025-5629.

NTIS Alerts: Mathematical Sciences. U.S. National Technical Information Service, 5285 Port Royal Road, Springfield, VA 22161. (703) 487-4650. FAX (703) 321-8547. Weekly. $140.00.

Science Citation Index. SCI. Institute for Scientific Information, 3501 Market Street, Philadelphia, PA 19104. (215) 386-0100. (800) 523-1850. FAX (215) 386-2991. 1961 to present. Six issues per year, plus annual cumulation. $11650.00 per year. Also available online and on CD-ROM. Inquire about price and availability. ISSN: 0036-827X.

Zentralblatt Fuer Mathematik Und Ihre Grenzebiete/Mathematics Abstracts. Heidelberger Akademie der Wissenschaften/ Springer-Verlag New York, Inc., 44 Hartz Way, Seacacus, NJ 07096-2491. (201) 348-4033. FAX (201) 348-4505. 1931 to present. 30 issues per year. DM 8340. ISSN: 0044-4235. Also available online and on CD-ROM. Inquire regarding cost and availability

ANNUAL REVIEWS AND YEARBOOKS

Advances in Applied Mathematics. Academic Press, Inc., 6277 Sea Harbor Drive, Orlando, FL 32821-8340. (800) 5439534. 1980 to pesent. Irregular, Price varies; inquire.

Applied Mathematical Sciences. Springer-Verlag, 175 Fifth Avenue, New York, NY 10010. (212) 460-2500. FAX (212) 473- 6272. 1972 to present. Irregular. ISSN: 0066-5452.

CBMS-NSF Regional Conference Series in Applied Mathematics. Society for Industrial and Applied Mathematics, 3600 University City Science Center, Philadelphia, PA 19104-2688. (215) 382-9800. FAX (215) 386-7999. 1971 to present. Irregular. ISSN: 0163-9439.

Ergebnisse Der Mathematik Und Ihrer Grenzgebiete. Neue folge. Springer-Verlag New York, Inc. 175 Fifth Avenue, New York, NY 10010. (212) 460-1500. FAX (212) 473-6272. 1953 to present. Irregular. Price varies; inquire. ISSN: 0071-1136.

Lecture Notes in Mathematics. Springer-Verlag New York, Inc. 44 Hartz Way. Secaucus. NJ. 07096-2491. (201) 348-4033. FAX (201) 348-4505. Irregular. Price varies; inquire.

ASSOCIATIONS AND PROFESSIONAL SOCIETIES

American Mathematical Society. P.O. Box 6248, Providence, RI 02940. (401) 455-4000. FAX (401) 331-3842.

Association for Symbolic Logic. University of Illinois, Department of Mathematics, 1409 West Green Street, Urbana IL 61801. (217) 244-7902. FAX (217) 333-9576.

Association for Women in Mathematics. 4114 Computer and Space Science Building, University of Maryland, College Park, MD 20742-2461. (301) 405-7892.

Mathematical Association of America. 1529 18th Street, NW, Washington, DC 20036. (202) 387-5200. FAX (202) 265-2384.

National Council of Teachers of Mathematics. 1906 Association Drive, Reston, VA 22091. (703) 620-9840. FAX (703) 475-2970.

Society For Industrial and Applied Mathematics. 3600 University City Science Center, Philadelphia, PA 19104-2688. (215) 382-9800. FAX (215) 386-7999.

Special Interest Group on Numerical Mathematics (SIGNUM). c/o ACM. 1515 Broadway, New York, NY 10036-5701. (212) 869-7440. FAX (212) 944-1318.

Special Interest Group For Symbolic and Algebraic Manipulation (sigsam). c/o ACM. 1515 Broadway, New York, NY 10036-5701. (212) 869-7440. FAX (212) 944-1318.

BIBLIOGRAPHIES

Bibliography of Early Modern Algebra 1500-1800. Robin E. Rider. Uviversity of California. office for History of Science & Technology, 543 Stephens Hall, Berkeley, CA 94720- 2350. (510) 642-4581. 1982. $8.00. ISBN: 0-918102-08-1

Books You Can Count On: Linking Mathematics & Literature. Rachel Griffiths and Margaret Clyne. Heinemann Publishing, 361 Hanover Street, Portsmouth, NH 03801-3912. (603) 431-7894. (800) 541-2086. 1991. $15.00. ISBN: 0-435-08322-8.

Mathematical Book Review Index, 1800 1940. Louise S. Grinstein. Garland Publ. Fifth Avenue, Suite 2500, New York, NY 10022-8101. (212) 751-7447. 1992. $72.00. ISBN: 0-8240-4114-3.

Bibliography of Mathematical Works Printed in America Through 1850. Louis C. Karpinski. Supplements 1 & 2, Bernard I. Cohen, editor. Ayer Company Publ. Inc. Lower Mill Road, North Stratford, NH 03590. (603) 922-5105. (800) 282-5413. Reprint edition. 1980. $66.95. ISBN: 0-405-12553-4.

International Mathematical Olympiads 1978-1985. Murray S. Klamkin. National Council of Teachers of Mathematics. 1906 Association Drive, Reston, VA 22091. (703) 620-9840. (800) 235-7566. 1987. $11.75. ISBN: 0-88385-631-X.

Mathematical Journals: An Annotated Guide. Diana F. Liang. Scarecrow Press. Distributed by University Press of America, 4720 Boston Way, Lanham, MD 20706. (800) 462-6420. 1987. $29.50. ISBN: 0-8108-2585-6.

DIRECTORIES AND BIOGRAPHICAL SOURCES

Assistantships and Graduate Fellowships in the Mathematical Sciences. American Mathematical Society, P.O. Box 6248, Providence, RI 02940. (401) 455-4000. (800) 321-1267. Annual. $19.00. ISSN: 1040-7650.

Combined Membership List. American Mathematical Society, P.O. Box 6248, Providence, RI 02940. (401) 455-4000. (800) 321-4267. Annual. $50.00. ISSN: 0569-6461.

Mathematical Science Professional Directory. American Mathematical Society, P.O. Box 6248, Providence, RI 02940. Annual. $45.00. ISBN: 0-8218-0173-2.

Research Centers Directory. Gale Research, 835 Penobscot Building, Detroit, MI 48226-4094. (313) 961- 2242. (800) 877-4253. 20th edition. 1995. Annual. $485.00. ISSN: 0080-1518.

World Directory of Mathematicians. G.D. Mostow, editor. International Mathematical Union, Helsinki. 1958 to present. 9th revised edition. 1990. $40.00. Available from: American Mathematical Society, P.O. Box 6248, Providence, RI 02940. (401) 455-4000. ISSN: 0512-2740.

ENCYCLOPEDIAS AND DICTIONARIES

Beyond Numeracy: An Uncommon Dictionary of Mathematics. John A. Paulos. Alfred A. Knopf Inc., 201 East 50th Street, New York, NY 10022. (800) 733-3000. 1991. $21.50. ISBN: 0-394-58640-9. <bodylead>Concise Oxford Dictionary of Mathematics.* Christopher Clapham. Oxford University Press, 200 Madison Avenue, New York, NY 10016. (212) 726-6000. 1990. $10.95. ISBN: 0-19- 286103-4.

Encyclopaedia of Mathematics. An updated and annotated translation of the Soviet Mathematical Encyclopedia. Michael Hazewinkel, editor. Kluwer Academic Publishers. 101 Philip Drive, Assinippi Park, Norwell, MA 02061. (617) 871-6600. 1987 - . 10 volumes. $147.00, volume 1. 1-55608-010-7 set.

Encyclopedia of Mathematics and Its Applications. Addison-Wesley Publishing Co., Inc., 1 Jacob Way, Reading, MA 01867. (617) 944-3700. (800) 447-2226. 1976 to present. irregular. Volumes are individually priced.

Encyclopedic Dictionary of Mathematics. Kiyosi Ito, editor. MIT Press, 55 Hayward Street, Cambridge, MA 02142. (617) 625-8569. 2nd edition, reprinted. 1993. 2 volumes. $70.00. ISBN: 0-262-59020-4.

Harpercollins Dictionary of Mathematics. E. J. Borowski and J. J. Borwein. HarperCollins Publ. 10 East 53rd Street, New York, NY 10022-5299. (800) 331-3761. 1990. $10.95. ISBN: 0-19-286103-4.

Mathematics Dictionary. Robert C. James, editor. Van Nostrand Reinhold, 115 Fifth Avenue, New York, NY 10003. (800)-842-3636. 5th edition. 1992. $42.95. ISBN: 0-442- 00741-8.

Penguin Dictionary of Mathematics. J. Daintith and R. D. Nelson. Viking Penguin, 375 Hudson Street, New York, NY 10014-3657. (800) 331-4624. 1989. $12.00. ISBN: 0-14- 051119-9.

The Words of Mathematics: An Etymological Dictionary of Mathematical Terms Used in English. Steven Schwartzman.

Mathematical Association of America, 18th Street NW, Washington, DC 20036. (800) 331-1622. 1994. $29.50. ISBN: 0-88385-551.

GENERAL WORKS

Algebra. Saunders MacLane and Garrett Birkhoff. ChelseaPublishing Inc., 15 East 26th Street, New York, NY 10010. (212) 889-8095. 3rd edition. 1987. ISBN: 0-8284-0330-9.

Computer Algebra: Systems & Algorithms For Algebraic Computation. J. H. Davenport. Academic Press, Inc., 6277 Sea Harbor Drive, Orlando, FL 32887. (800) 321-5068. 2nd edition. 1993. $45.00. ISBN: 0-12-204232-8.

Functions and Graphs. I. M. Gelfand. Birkhauser Boston. 675 Massachusetts Avenue, Cambridge, MA 02139. (617) 876- 2333. (800) 777-4643. 1993. $16.50. ISBN: 0-8176-3532-7.

Fundamentals of Mathematics. H. Behnke et al, editors. MIT Press, MIT Press, 55 Howard Street, Cambridge, MA 02142. (617) 625-8569. Volume 1, Foundations of Mathematics: Real Number Systems and Algebra; Volume 2, Geometry; Volume 3, Analysis. 1984; reprint edition. $75.00 ISBN: 0-685-07931-7.

Greek Mathematical Thought and the Origin of Algebra. Jacob Klein. Dover Publications, Inc., 180 Varick Street, New York, NY 10014. (212) 255-3755. (800) 223-3130. $9.95. ISBN: 0-486-27289-3.

A History of Algebraic and Differential Topology. Jean A. Dieudonne. Birkhauser Boston. 675 Massachusetts Avenue, Cambridge, MA 02139. (617) 876-2333. (800) 777-4643. 1994. $94.50. ISBN: 0-8176-3388X.

Linear Algebra: An Introductory Approach. Charles W. Curtis. Springer-Verlag New York, Inc., 175 Fifth Avenue, New York, NY 10010. (212) 460-1500. (800) 777-4643. 1993. $32.00 ISBN: 0-387-90992-3.

Mathematica: A System For Doing Mathematics By Computer. Stephen Wolfram. Addison-Wesley Publishing Company, Inc., 1 Jacob Way, Reading, MA 01867. (617) 944-3700. (800) 447-2226. 2nd edition. 1991. ISBN: 0-201-51502-4.

The Maple Handbook. Darren Redfern. Springer Verlag New York, Inc., 175 Fifth Avenue, New York, NY 10010. (212) 460-1500. (800) 777-4643. 2nd edition. 1995. $29.00. 0- 387-94331-5.

Algebra Problem Solver. Research & Education Association, 61 Ethel Road, West, Piscataway, NJ 08854. (908) 819-8880. 1994; revised edition. $29.95.

Practical Algebra: A Self-teaching Guide. Steven L. Slavin. John Wiley & Sons, Inc., 605 Third Avenue, New York, NY 10158-0012. (212) 850-6000. (800) 225-5945. 2nd edition. 1991. $16.95. ISBN: 0-471-53012.

Survey of Modern Algebra. Garrett Birkhoff and Saunders MacLane. Macmillan Publishing Company, Inc. 866 Third Avenue, New York, NY 10022. 4th edition. 1977.

Undergraduate Algebra. Serge Lang. Springer-Verlag New York, Inc., 175 Fifth Avenue, New York, NY 10010. (212) 460-

1500. (800) 777-4643. 2nd edition. 1994. $39.80. ISBN: 0-387-97279-X.

Writing Mathematics Well. Leonard Gillman. Mathematical Association of America, 18th Street NW, Washington, DC 20036. (800) 331-1622. 1987. $8.00. ISBN: 0-88385-443-0.

HANDBOOKS AND MANUALS

Handbook of Boolean Algebras. J. D. Monk editor. Elsevier Science, Inc., 655 Avenue of the Americas, New York, NY 10010. (212) 989-5800. 1989. $72.00. 3 volumes. ISBN: 0-444-70261-X.

Handbook of Mathematical Sciences. William H. Beyer, editor. CRC Press, 2000 Corporate Boulevard, Boca Raton, FL 33431. (800) 272-7737. 6th edition. 1987. $91.95. ISBN: 0-8493-0656-6.

Handbook of Tables For Probability and Statistics. William H. Beyer, editor. CRC Press, 2000 Corporate Boulevard, Boca Raton, FL 33431. (800) 272-7737. 2nd edition. 1968. $92.95. ISBN: 0-8493-0692-2.

Handbook of Writing For the Mathematical Sciences. Society for Industrial and Applied Mathematics, 3600 University City Science Center, Philadelphia, PA 19104-2688. (215) 382-9800. FAX (215) 386-7999. 1993. $21.50. ISBN: 0-89871-314-5.

Standard Mathematical Tables and Formulas. William H. Beyer, editor. CRC Press, 2000 Corporate Boulevard, Boca Raton, FL 33431. (800) 272-7737. 29th edition. 1991. $41.95. ISBN: 084-930-6299.

Handbook of Applicable Mathematics. Walter Ledermann, editor. John Wiley & Sons, Inc., 605 Third Avenue, New York, NY 10158-0012. (212) 850-6000. (800) 225-5945. 1985. $1496.00 6 volumes. ISBN: 0-471-90804-5.

Handbook of Applied Mathematics; Selected Results and Methods. Carl E. Pearsen, editor. Chapman & Hall, 1 Penn Plaza, New York NY 10119. (212) 564-1060. 1990. $46.95. ISBN: 0-442-00521-0.

Handbook of Mathematical Functions With Formulas, Graphs, and Mathmetical Tables. Milton Abramowitz and Irene A. Stegun, editors. U.S. National Bureau of Standards. U.S. Government Printing office, Washington, D.C. 10th edition. 1972. $43.00. ISBN: 0-16-000202-8.

Handbook of Mathematics. I. N. Bronshtein and K. A. Semedyayev. Van Nostrand Reinhold, 115 Fifth Avenue, New York, NY 10003. (800)-842-3636. 20th edition. $59.95. ISBN: 0-442-21171-6.

Tables of Integrals, Series and Products: corrected and enlarged edition. I. S. Gradshteyn and I. M. Ryzhik. Academic Press, Inc., 6277 Sea Harbor Drive, Orlando, FL 32821. (800) 321-5068. 5th edition. 1993. $61.00. ISBN: 0-12-294755-X.

ONLINE DATABASES AND CD-ROMS

CA Search. Chemical Abstracts Service, P.O. Box 3012, Columbus, OH 4321 0-0012. (614) 447-3600. (800) 848-6533. F AX (614) 447-3709. Very comprehensive guide to worldwide chemical literature and related fields, 1972 to present. Available on BRS (800) 289-4277, DIALOG (800) 334-2564, ORBIT (800) 456-7248, and STN International, FIZ Karlsruhe, P.O. Box 2465, W-7500, Karlsruhe 1, Germany, online services. Inquire as to cost and availability.

Compendex Plus. Engineering Information, Inc., 345 East 47th Street, New York, NY 10017. (212) 705-7600 or (800) 221-1044. Conta ins citations with abstracts to worl dwide literature in enginee ring and technology, from 1970 to present. Available ononline BRS, (800) 289-4277, DIALOG, (800) 334-2564, ORBIT (800) 456-7248, and STN International, FIZ Karlsruhe, P.O. Box 2465, W-7500, Karlsruhe 1, Germany, online services. Also available on CD-ROM. Inquire as to cost and availability.

Dissertation Abstracts. University Microfilms International, 300 North Zeeb Road, Ann Arbor, MI 48106. (800) 521-0600 or (313) 761-4700. Scope includes virtually all doctoral dissertations accepted at accredited American institutions from 1861 to present in 252 subject areas. Available on BRS,(800) 289-4277, DIALOG,(800) 334-2564, and OCLC EPIC,(800) 848-5878, online services. Also available on CD-ROM. Inquire as to cost and availability.

INSPEC. Institution of Electrical Engineers, Michael Faraday House, Six Hills Way, Stevenage, Herts. SG1 2AY, England. Telephone: 0438 313311 or FAX 0438 742840. Contains citations to the worldwide literature of physics, electronics and electrical engineering, computer technology, and related fields. Available on BRS, (800) 289-4277, DIALOG, (800) 334-2564, ORBIT, (800) 456-7248, and STN International, FIZ Karlsruhe, P.O. Box 2465, W-7500, Karlsruhe 1, Germany, online services. Inquire as to cost and availability.

Mathsci and Mathscinet. American Mathematical Society, P.O. Box 6248, Providence, RI 02940. (800) 321-4667 or (401) 455-4000 or FAX (401) 331-3842. Scope includes pure and applied mathematics and related areas of physics, statistics, engineering, computer science, and operations research literature since 1959. Available on DIALOG,(800) 334-2564, online service. Also available on CD-ROM and *Mathscinet*, on the Internet. Inquire as to cost and availability.

NTIS Bibliographic Database. National Technical Information Service, 5285 Port Royal Road, Springfield, VA 22161. (703) 487-4929 or FAX (703) 321-8199. Broad coverage of government-sponsored science and technology research reports, 1964 to present. Available on BRS, (800) 289-4277, DIALOG, (800) 334-2564, ORBIT, (800) 456-7248, and STN International, FIZ Karlsruhe, P.O. Box 2465, W-7500, Karlsruhe 1, Germany, online services. Also available on CD-ROM. Inquire as to cost and availability.

Scientific and Technical Books and Serials in Print. R.R. Bowker Inc., 121 Chanlon Road, New Providence, NJ 07974. (800) 521-8110 or (908) 464-6800. List of currently published books and serials in the physical and biological sciences, engineering and technology, with subject, author and titles indexes. Available on ORBIT, (800) 456-7248, online service.Inquire as to cost and availability.

SCISEARCH. Institute for Scientific Information, 3501 Market Street, Philadelphia, PA 19104. (800) 523-1850 or (215) 386-0100. Broad multidisciplinary title and author index to the international literature of science and technology, 1974 to present. Available on DIALOG, (800) 334-2564, and ORBIT,(800)456-7248, online services. Also available on CD-ROM. Inquire as to cost and availability.

ALGEBRA

Ency. of Physical Sciences and Engineering Info. Sources

WILSONLINE. H.W. Wilson Company, 950 University Avenue, Bronx, NY 10452. (800) 367-6770 or (212) 588-8400. Makes available online versions of the printed H.W. Wilson indexes including Applied Science and Technology Index, Business Periodicals Index, General Science Index, and Readers' Guide to Periodical Literature. Period covered is generally 1983 to present. Available on BRS, (800) 289-4277, DIALOG,(800) 334-2564, and OCLC EPIC,(800) 848-5878, online services. Also available on CD-ROM. Inquire as to cost and availability.

PERIODICALS

American Journal of Mathematics. Johns Hopkins University Press, 2715 North Charles Street, Baltimore, MD 21218. (410) 516-6987. FAX (410) 516-6968. 1878 to present. Bimonthly. $162.00 per year. ISSN: 0002-9327.

American Mathematical Monthly. Mathematical Association of America. 1529 Eighteenth Street, NW, Washington, DC 20036. (202) 387-5200. 1894 to present. Ten issues per year. $160.00. ISSN: 0002-9890.

Algebra and Logic. English translation of Algebra i Logika. Plenum Publshing Co., Consultants Bureau, 233 Spring Street, New York NY 10013-1578. (212) 620-8468. FAX (212) 463-0742. 1968 to pr esent. Bimonthly. $860.00. ISSN: 0002-5232.

Applicable Algebra in Engineering, Communication and Computing. Springer-Verlag New York, Inc., 175 Fifth Avenue, New York, NY 10010. (212) 460-1500. (800) 777-4643. FAX (212) 473-6272. 1990 to present. Six issues per year. $161.00 ISSN: 0938-1279.

Canadian Applied Mathematical Quarterly. Canadian Mathematical Society. Distributed by Rocky Mountain Mathematics Consortium, Department of Mathematics, Arizona State University, Tempe AZ 85287-1904. 1992 to present. $175.00 per year. ISSN: 1073-1849.

Communications in Algebra. Marcel Dekker Inc., 270 Madison Avenue, New York NY 10016. (212) 696-9000. FAX (212) 685-4540. 1974 to present. 15 issues per year. $1495.00. ISSN: 0092-7872.

Communications On Pure and Applied Mathematics. Courant Institute of Mathematical Sciences. Distributed by John Wiley & Sons, Inc. 605 Third Avenue, New York, NY 10158- 0012. (212) 850-6000. FAX (212) 850-6088. 1939 to present. $730.00 per year. ISSN: 0010-3640.

IMA Journal of Applied Mathematics. Oxford University Press, 2001 Evans Road, Cary, NC 27513. (919) 677-0977. 1981 to present. Bimonthly. $410.00 per year. ISSN: 0272- 4960.

Industrial Mathematics. Industrial Mathematics Society, Box 159 Roseville, MI 48066. (313) 771-0403. 1950 to present. Semiannual. $15.00 per year. ISSN: 0019-8528.

International Journal of Algebra and Computation. World Scientific Publishing Co., 1060 Main Street St 18, River Edge NJ07661. (800) 227-7562. 1991 to present. $258.00 per year. ISSN: 0218-1067.

Journal of Algebra. Academic Press, Inc., 6277 Sea Harbor Drive, Orlando, FL 32821. (800) 543-9534. 1964 to present. Semi-monthly. $1680.00 per year. ISSN: 0021-8693.

Journal of Algebraic Combinatorics. Kluwer Academic Publishers, Boston, Box 358, Accord Station, Hingham MA 02018-0358. (617) 871-6600. FAX (617) 871-6528. 1992 to present. Quarterly. $200.50 per year. ISSN: 0925-9899.

Journal of Algebraic Geometry. American Mathematical Society, Box 1571, Annex Station, Providence RI 02901-9930. (401) 455-4000. 1991 to present. Quarterly. $188.00 per year. ISSN: 1056-3911.

Journal of Pure and Applied Algebra. Elsevier Science Inc., Box 882, Madison Squ are Station, New York, NY 10159. (212) 989-5800. FAX (212) 633-3990. 1971 to present. 21 issues per year. $1298.00 per year. ISSN: 0022-4049.

Siam Journal on Applied Mathematics. Society for Industrial and Applied Mathematics, 3600 University City Science Center, Philadelphia, PA 19104-2688. (215) 382-9800. FAX (215) 386-7999. 1953 to present. Bimonthly. $242.00 per year. ISSN: 0036-1399.

Siam Journal On Mathematical Analysis. Society for Industrial and Applied Mathematics, 3600 University City Science Center, Philadelphia, PA 19104-2688. (215) 382- 9800. FAX (215) 386-7999. 1970 to present. Bimonthly. $352.00 per year. ISSN: 0036-1410.

Siam Journal On Matrix Analysis and Applications. Society for Industrial and Applied Mathematics, 3600 University City Science Center, Philadelphia, PA 19104-2688. (215) 382- 9800. FAX (215) 386-7999. 1980 to present. Quarterly. $210.00 per year. ISSN: 0895-4798.

Siam Journal on Numerical Analysis. Society for Industrial and Applied Mathematics, 3600 University City Science Center, Philadelphia, PA 19104-2688. (215) 382-9800. FAX (215) 386-7999. Bimonthly. $255.00 per year. ISSN: 0036- 1429.

Siam Review. Society for Industrial and Applied Mathematics, 3600 University City Science Center, Philadelphia, PA 19104-2688. (215) 382-9800. 1959 to present. Quarterly. $148.00 per year. ISSN: 0036-1445.

RESEARCH CENTERS AND INSTITUTES

Center For Mathematical Sciences. University of Wisconsin-Madison. 610 Walnut Street, Madison, WI 53705. (608) 263-2696. FAX (608) 263-2841.

Center For Numerical Analysis. University of Texas at Austin. Austin, TX 78712. (512) 471-1242,

Center For Pure and Applied Mathematics. University of California, Berkeley, 977 Evans Hall, Berkeley, CA 04720. (415) 642-0116. FAX (415) 642-6726.

Computer and Information Science Research Center. OhioState University. 2036 Neil Avenue Mall. Columbus, OH 43210. (614-292-6374. (614) 292-9021.

Ency. of Physical Sciences and Engineering Info. Sources

ALLOYS

Courant Institute of Mathematical Sciences. New York University, 251 Mercer Street, New York, NY 10012. (212) 998-3000.

Institute For Computer Applications in Science and Engineering (icase). Mail Stop 132C, NASA, Langley Research Center, Hampton, VA 23665. (804) 865-2513. FAX (804) 864- 6134.

ALGEBRAIC NUMBER THEORY

See: MATHEMATICS

ALGOL

See: COMPUTER PROGRAMMING LANGUAGES

ALLOYS

See also: BRASS AND BRONZE, FERROALLOYS, MATERIALS
SCIENCE, METALLURGY

ABSTRACT SERVICES AND INDEXES

Alloys Index. American Society for Metals, Metals Park, OH 44073. (216) 338-5151. FAX (216) 338-4634. 1974 to present. Monthly. $380.00. Also available on CD-ROM and online via METADEX, STN International, and DIALOG.

Aluminum Industry Abstracts. Aluminum Association, Materials Park, OH 44073. (216) 338-5151. FAX (216) 338-4634. 1968 to present. Monthly. $525.00 per year.

Applied Mechanics Reviews: An Assessment of World Literature in Engineering Sciences. American Society of Mechanical Engineers, 345 East 47th Street, New York, NY 10017. (212) 705-7703. 1948 to present. Monthly. $360.00 per year.

Applied Science and Technology Index; A cumulative subject index to English language periodicals in the fields of aeronautics and space science, computer technology, chemistry, construction industry, energy and related areas. H.W. Wilson Co., 950 University Avenue, Bronx, NY 10452. (800) 367-6770 or (212) 588-8400. FAX (718) 590-1617. From 1958 to present. Monthly. Inquire about cost and availability. Also available on CD-ROM.

Current Contents: Engineering, Technology, and Applied Sciences. Institute for Scientific Information, 3501 Market Street, Philadelphia, PA 19104. (215) 386-0100. FAX (215) 386-6362. Weekly. $360.00 per year.

Engineering Index Monthly. Engineering Information, Inc., Castle Point on the Hudson, Hoboken, NJ 07030. (800) 221-1044. FAX (212) 832-1857. Monthly. $2200.00 per year. Also available online as COMPENDEX, and also on CD-ROM.

International Copper Information Bulletin. Copper Development Association, Orchard House, Mutton Lane, Potters Bar, Herts EN6 3AP, England. Telephone 0707-50711. FAX 0707-42769. 1976 to present. 3 times a year. Inquire for cost and availability.

Metals Abstracts and Metals Abstracts Index. American Society for Metals, Metals Park, OH 44073. (216) 338-5151. 1968 to present. Monthly. Abstracts are $1925.00 per year and Index is $460.00 per year.

Physics Abstracts. INSPEC/IEEE, Box 1331, 445 Hoes Lane, Piscataway, NJ 08855-1331. (908) 562-5549. 1898 to present. $2835.00 per year. Also available online as INSPEC. Covers recently published primary research in all areas of physics.

ANNUAL REVIEWS AND YEARBOOKS

Aluminum Statistical Review. Aluminum Association Inc., 900 19th Street NW, Suite 300, Washington, DC 20006. (202) 862-5100. FAX (202) 862-5164. 1962 to present. Annual. $50.00 per year.

Annual Review of Materials Science. Annual Reviews Inc., 4139 El Camino Way, PO Box 10139, Palo Alto, CA 94303-0139. (415) 493-4400 or (800) 523-8635. Annual. $75.00 (1994 edition).

Non-Ferrous Metal Data. American Bureau of Metal Statistics, Inc., Box 1405 Plaza Station, 400 Plaza Drive, Secaucus, NJ 07094-0405. (201) 863-6900. FAX (201) 863-6050. 1920 to present. Annual. $350.00 per year.

ASSOCIATIONS AND PROFESSIONAL SOCIETIES

Aluminum Association. 900 19th Street NW, Suite 300, Washington, DC 20006. (202) 862-5100. FAX (202) 862-5164.

American Society For Metals (asm). Materials Park, OH 44073. (216) 338-5151 or FAX (216) 338-4634.

Metallurgical Society of the Aime (american Institute of Mining, Metallurgical, and Petroleum Engineers). 345 E.47th Street, 14th Floor, New York, NY 10017. (212) 705-7695.

BIBLIOGRAPHIES

Scientific and Technical Books in Print. R.R. Bowker Inc., 121 Chanlon Road, New Providence, NJ 07974. (800) 521-8110. FAX (908) 464-3553. List of currently published books and serials in the physical and biological sciences, engineering and technology, with subject, author and titles indexes. Also available on ORBIT, (800) 456-7248, online service. Inquire as to cost and availability.

DIRECTORIES AND BIOGRAPHICAL SOURCES

Dun's Industrial Guide: The Metalworking Directory, 1993-94. Dun & Bradstreet Information Services Staff. 3 volumes. Dun & Bradstreet Information Services, 899 Eaton Avenue, Bethlehem, PA 18025. (610) 882-7000 or (800) 526-0651. FAX (610) 882-7269. 1993. $795.00.

International Research Centers Directory. Gale Research, 835 Penobscot Building, Detroit, MI 48226-4094. (313) 961-2242. (800) 877-4253. FAX (313) 961-6083. 8th edition. 1995. $410.00

ALLOYS

Ency. of Physical Sciences and Engineering Info. Sources

Research Centers Directory. Gale Research, 835 Penobscot Building, Detroit, MI 48226-4094. (313) 961-2242 or (800) 347-4253. FAX (313) 961-6083. $485.00

Scientific and Technical Organizations and Agencies Directory. Gale Research, 835 Penobscot Building, Detroit, MI 48226-4094. (313) 961-2242 or (800) 877-4253. FAX (313) 961-6083. 4th edition. 1996. $195.00.

Who's Who in Engineering. American Association of Engineering Societies, 1111 19th Street NW, Suite 608, Washington, DC 20036. (202) 296-2237 or (800) 658-8897. 8th edition. 1991.Inquire for price.

Who's Who in Technology. Gale Research, 835 Penobscot Building, Detroit, MI 48226-4094. (313) 961-2242 or (800) 347-4253. FAX (313) 961-6083. 1995. $195.00.

ENCYLOPEDIAS AND DICTIONARIES

Encyclopedia of Materials Science and Engineering. M.B. Bever and Robert W. Cahn, editors. 8 volumes (1986) and 3 supplementary volumes (1988-1993). Pergamon/Elsevier Science Inc., 660 White Plains Road, Tarrytown, NY 10591-5153. (914) 524-9200. FAX (914) 333-2444. $3900.00 for 11-volume set.

Encyclopedia of Physical Science and Technology. Robert A. Meyers, ed. 18 volumes. Academic Press Inc., 6277 Sea Harbor Drive, Orlando, FL 32887. 1992. $2100.00

McGraw-Hill Dictionary of Scientific and Technical Terms. Sybil P. Parker, ed. 5th edition. McGraw-Hill Publishing Company, 1221 Avenue of the Americas, New York, NY 10020. (800) 262-4729 or (212) 512-3825. 1993. $110.50.

McGraw-Hill Encyclopedia of Engineering. Sybil P. Parker, ed. 2nd edition. McGraw-Hill Publishing Company, 1221 Avenue of the Americas, New York, NY 10020. (800) 262-4729 or (212) 512-3825. 1993. $95.50.

McGraw-Hill Encyclopedia of Science and Technology. Sybil P. Parker, ed. 7th edition. 20 volumes. McGraw-Hill Publishing Company, 1221 Avenue of the Americas, New York, NY 10020. (800) 262-4729 or (212) 512-3825. 1992. $1900.00

GENERAL WORKS

Aluminum and Aluminum Alloys. J.R. Davis. ASM (American Society for Metals) International, Materials Information, Materials Park, OH 44073. (216) 338-5151 or FAX (216) 338-4634. 1993. Inquire for cost and availability.

Structure and Properties of Engineering Alloys. William F. Smith. 2nd edition. McGraw-Hill Publishing Company, 1221 Avenue of the Americas, New York, NY 10020. (800) 262-4729 or (212) 512-3825. 1993. Inquire for cost and availability.

HANDBOOKS AND MANUALS

ASM Handbook Volume 1: Properties and Selection: Irons, Steels, and High-performance Alloys. ASM International Handbook Committee. ASM International, Materials Information,

Materials Park, OH 44073-0002. (216) 338-5151 or (800) 336-5152. FAX (216) 338-4634. 1990. $147.00.

ASM Handbook Volume 2: Properties and Selection: Nonferrous Alloys and Special-purpose Materials. ASM International Handbook Committee. ASM International, Materials Information, Materials Park, OH 44073-0002. (216) 338-5151 or (800) 336-5152. FAX (216) 338-4634. 1991. $147.00.

ASM Handbook Volume 3: Alloy Phase Diagrams. ASM International Handbook Committee. ASM International, Materials Information, Materials Park, OH 44073-0002. (216) 338-5151 or (800) 336-5152. FAX (216) 338-4634. 1992. $147.00.

*ASM Specialty Handbook: Aluminum and Aluminum Alloys.*J.R. Davis, editor. ASM International Handbook Committee. ASM International, Materials Information, Materials Park, OH 44073-0002. (216) 338-5151 or (800) 336-5152. FAX (216) 338-4634. 1993. $159.00.

CRC Handbook of Chemistry and Physics. David R. Lide, editor. 75th edition. CRC Press, 2000 Corporate Blvd., N.W., Boca Raton, FL 33431. (407) 994-0555 or (800) 272-7737. FAX (407) 994-0949. $99.50.

Smithell's Metals Reference Book. E.A. Brandes & G.B. Brook. 7th edition. Butterworth-Heinemann, 313 Washington Street, Newton, MA 02158. (617) 928-2500 or (800) 366-2665. FAX (617) 928-2620. 1992. $250.00.

Woldman's Engineering Alloys. J. Frick, editor. 8th edition. ASM International, Materials Information, Materials Park, OH 44073-0002. (216) 338-5151 or (800) 336-5152. FAX (216) 338-4634. 1994. $175.00.

ONLINE DATABASES AND CD-ROMS

Compendex Plus. Engineering Information, Inc., 345 East 47th Street, New York, NY 10017. (212) 705-7600 or (800) 221-1044. Contains citations with abstracts to worldwide literature in engineering and technology, from 1970 to present. Available on online BRS,(800) 289-4277, DIALOG, (800) 334-2564, ORBIT (800) 456-7248, and STN International, FIZ Karlsruhe, P.O. Box 2465, W-7500, Karlsruhe 1, Germany, online services. Also available on CD-ROM. Inquire as to cost and availability.

Current Contents Search. Institute for Scientific Information, 3501 Market Street, Philadelphia, PA 19104. (215) 386-0100. FAX (215) 386-6362. Contains citations to articles listed in the table of contents of science and technology journals. Also articles in social sciences and life sciences journals. Available on BRS,(800) 289-4277, DIALOG,(800) 334-2564, online services. Inquire as to cost and availability.

NTIS Bibliographic Database. National Technical Information Service, 5285 Port Royal Road, Springfield, VA 22161. (703) 487-4929 or FAX (703) 321-8199. Broad coverage of government-sponsored science and technology research reports, 1964 to present. Available on BRS,(800) 289-4277, DIALOG, (800) 334-2564, ORBIT, (800) 456-7248, and STN International, FIZ Karlsruhe, P.O. Box 2465, W-7500, Karlsruhe 1, Germany, online services. Also available on CD-ROM. Inquire as to cost and availability.

Metadex. Jointly produced by ASM International and the Institute of Materials. Contains more than 925,000 records from the international literature on metals and alloys, concerning properties, processes, materials classes, applications, and metallurgical systems. Updated monthly. Available from ORBIT-QUESTEL (703) 442-0900.

NASA Database. American Institute of Aeronautics and Astronautics, 370 L'Enfant Promenade, S.W., Washington, DC 20024. (202) 646-7400. Citations and abstracts of aeronautics andastronautics literature, 1962 to present. Also contains citations from *STAR, Scientific and Technical Aerospace Reports,* and *International Aerospace Abstracts.* Available through NASA/RECON online service. Inquire as to cost and availability.

WILSONLINE. H.W. Wilson Company, 950 University Avenue, Bronx, NY 10452. (800) 367-6770 or (212) 588-8400. Makes available online versions of the printed H.W. Wilson indexes including Applied Science and Technology Index, Business Periodicals Index, General Science Index, and Readers' Guide to Periodical Literature. Period covered is generally 1983 to present. Available on BRS,(800) 289-4277, DIALOG,(800) 334-2564, and OCLC EPIC,(800) 848-5878, online services. Also available on CD-ROM. Inquire as to cost and availability.

PERIODICALS

Alloy Digest. Alloy Digest Inc., 27 Canfield Street, Orange, NJ 07050. (201) 677-9161. 1952 to present. Monthly. $140.00 per year.

Aluminum Situation. Aluminum Association Inc., 900 19th Street NW, Suite 300, Washington, DC 20006. (202) 862-5100. FAX (202) 862-5164. Monthly. Free.

Aluminum Standards and Data. Aluminum Association Inc., 900 19th Street NW, Suite 300, Washington, DC 20006. (202) 862-5100. FAX (202) 862-5164. 1968 to present. Biennial. $25.00 per year.

Aluminum Standards and Data—Metric. Aluminum Association Inc., 900 19th Street NW, Suite 300, Washington, DC 20006. (202) 862-5100. FAX (202) 862-5164. Biennial. $12.00 per year.

Journal of Phase Equilibria (formerly *Bulletin of Alloy Phase Diagrams).* ASM International, Materials Park, OH 44073-0002. (216) 338-5151. FAX (216) 338-4634. 6 times a year. $658.00 per year for non-members.

JOM (Journal of Metals). Minerals, Metals and Materials Society, 420 Commonwealth Drive, Warrendale, PA 15086. (412) 776-9080. 1949 to present. Monthly. $50.00 per year.

Light Metal Age. Fellom Publishing, 170 S. Spruce Avenue, Suite 120, San Francisco, CA 94080. (415) 588-8832. FAX (415) 588-0901. 1943 to present. Bi-monthly. $35.00 per year.

Metallurgical Transactions A: Physical Metallurgy and Materials Science. ASM International, Materials Park, OH 44073. (216) 338-5151 or FAX (216) 338-4634. 1970 to present. Monthly. $520.00 per year.

Metals Week. McGraw-Hill Publishing Company, 1221 Avenue of the Americas, New York, NY 10020. (800) 262-4729 or (212) 512-3825. 1930 to present. Weekly. $770.00 per year.

RESEARCH CENTERS AND INSTITUTES

Colorado School of Mines, Advanced Steel Processing and Products Research Center. Golden, CO 80401. (303) 273-3774.

Canada Centre For Mineral and Energy Technology, Materials Technology Laboratories. Energy, Mines & Resources Canada, 568 Booth Street, Ottawa ON, Canada K1A 0G1. (613)995-8248. FAX (613) 992-8735.

Texas A&M University Mechanics and Materials Center. Civil Engineering Department, College Station, TX 77843. (409) 845-7512. FAX (409) 845-6156.

University of Connecticut Institute of Materials Science. U-136, 97 N. Eagleville Road, Storrs, CT 06268. (203) 486-4623. FAX (203) 486-4745.

University of Florida Department of Materials Science and Engineering. Gainesville, FL 32611. (904) 392-1454. FAX (904) 392-6359.

University of Minnesota Corrosion Research Center. 221 Church Street SE, Minneapolis, MN 55455. (612) 625-0717.

University of Wisconsin—Madison, Cast Metals Laboratory. 1509 University Avenue, Madison, WI 53706. (608) 262-2562. FAX (608) 262-6707.

University of Utah, Department of Materials Science and Engineering. 304 EMRO Bldg., Salt Lake City, UT 84112. (801) 581-6863. FAX (801) 581-4816.

ALPHA PARTICLE

See: PARTICLE PHYSICS

ALTERNATING CURRENT

See: ELECTRICITY

ALUMINUM

See also: ALLOYS, METALLURGY

ABSTRACT SERVICES AND INDEXES

Alloys Index. American Society for Metals, Metals Park, OH 44073. (216) 338-5151. FAX (216) 338-4634. 1974 to present. Monthly. $380.00. Also available on CD-ROM and online via METADEX, STN International, and DIALOG.

Aluminum Industry Abstracts. Aluminum Association, Materials Park, OH 44073. (216) 338-5151. FAX (216) 338-4634. 1968 to present. Monthly. $525.00 per year.

Applied Mechanics Reviews: An Assessment of World Literature in Engineering Sciences. 1948-present. American Society of Mechanical Engineers, 345 East 47th Street, New York, NY 10017. (212) 705-7703. Monthly. $360.00 per year.

Applied Science and Technology Index; A cumulative subject index to English language periodicals in the fields of aeronautics and space science, computer technology, chemistry, construction industry, energy and related areas. H.W. Wilson Co., 950 University Avenue, Bronx, NY 10452. (800) 367-6770 or (212) 588-8400. FAX (718) 590-1617. From 1958 to present. Monthly. Inquire about cost and availability. Also available on CD-ROM and online.

Current Contents: Engineering, Technology, and Applied Sciences. Institute for Scientific Information, 3501 Market Street, Philadelphia, PA 19104. (215) 386-0100. FAX (215) 386-6362. 1970 to present. Weekly. $442.00 per year.

Engineering Index Monthly. Engineering Information, Inc., Castle Point on the Hudson, Hoboken, NJ 07030. (800) 221-1044. FAX(212) 832-1857. Monthly. $2300.00 per year. Also available online as COMPENDEX, and also on CD-ROM. Covers chemical engineering, computers, electrical engineering, civil engineering, metals and mining, industrial management, and mechanical engineering.

I.M.M. Abstracts and Index. Institution of Mining and Metallurgy, 44 Portland Pl., London W1N 4BR, England. 071-580-3802. FAX 071-436-5388. Bi-monthly. $364 for non-members.

Manufacturing and Process Engineering Abstracts, Cambridge Scientific Abstracts, 7200 Wisconsin Avenue, Bethesda, MD 20814-4823. (301) 961-6750. FAX (301) 961-6720. 1993 to present. Monthly. $385.00 per year. Covers concurrent engineering, quality control, automated manufacturing, petroleum engineering, oil field operations and equipment, energy management, metallurgy and metallography, foundry practice.

Metals Abstracts and Metals Abstracts Index. American Society for Metals, Metals Park, OH 44073. (216) 338-5151. 1968 to present. Monthly. Abstracts are $1500.00 per year and Index is $460.00 per year.

ANNUAL REVIEWS AND YEARBOOKS

Aluminum Statistical Review. Aluminum Association Inc., 900 19th Street NW, Suite 300, Washington, DC 20006. (202) 862-5100. FAX (202) 862-5164. 1962 to present. Annual. $50.00 per year.

Annual Review of Materials Science. Annual Reviews Inc., 4139 El Camino Way, PO Box 10139, Palo Alto, CA 94303-0139. (415) 493-4400 or (800) 523-8635. FAX (415) 855-9815. Annual. $75.00 (1994 edition).

Minerals Yearbook. Bauxite, Alumina and Aluminum. U.S. Bureau of Mines, Production and Distribution, Cochrans Mill Rd., Box 18070, Pittsburgh, PA 15236. (412) 892-4411. Annual. Inquire for price and availability.

Non-Ferrous Metal Data. American Bureau of Metal Statistics, Inc., Box 1405 Plaza Station, 400 Plaza Drive, Secaucus, NJ 07094-0405. (201) 863-6900. FAX (201) 863-6050. 1920 to present. Annual. $350.00 per year.

ASSOCIATIONS AND PROFESSIONAL SOCIETIES

Aluminum Association. 900 19th Street NW, Suite 300, Washington, DC 20006. (202) 862-5100. FAX (202) 862-5164.

Aluminum Extruders Council. 1000 N. Rand Road, Suite 214, Wauconda, IL 60084. (708) 526-2010.

Aluminum Recycling Association. 1000 16ht Street NW, Suite 400, Washington, DC 20036. (202) 785-0951.

American Society For Metals. Materials Park, OH 44073. (216) 338-5151 or FAX (216) 338-4634.

DIRECTORIES AND BIOGRAPHICAL SOURCES

Dun's Industrial Guide: The Metalworking Directory, 1993-94. Dun & Bradstreet Information Services Staff. Dun & Bradstreet Information Services, 899 Eaton Avenue, Bethlehem, PA 18025. (610) 882-7000 or (800) 526-0651. FAX (610) 882-7269. 3 volumes. 1993. $795.00.

Research Centers Directory. Gale Research, 835 Penobscot Building, Detroit, MI 48226-4094. (313) 961-2242. (800) 877-4253. FAX (313) 961-6083. $485.00

Scientific and Technical Organizations and Agencies Directory. Gale Research, 835 Penobscot Building, Detroit, MI 48226-4094. (313) 961-2242. (800) 877-4253. FAX (313) 961-6083. 4th edition. 1996. $195.00.

Who's Who in Engineering. American Association of Engineering Societies, 1111 19th Street NW, Suite 608, Washington, DC 20036. (202) 296-2237 or (800) 658-8897. 8th edition. 1991. Inquire for price.

Who's Who in Technology. Gale Research, 835 Penobscot Building, Detroit, MI 48226-4094. (313) 961-2242. (800) 877-4253. FAX (313) 961-6083. 1995. $195.00.

ENCYCLOPEDIAS AND DICTIONARIES

Encyclopedia of Physical Science and Technology. Robert A. Meyers, ed. 18 volumes. Academic Press, Inc., 6277 Sea Harbor Drive, Orlando, FL 32887. 1992. $2100.00

McGraw-Hill Dictionary of Scientific and Technical Terms. Sybil P. Parker, ed. 5th edition. McGraw-Hill Publishing Company, 1221 Avenue of the Americas, New York, NY 10020. (800) 262-4729 or (212) 512-3825. 1993. $110.50.

McGraw-Hill Encyclopedia of Engineering. Sybil P. Parker, ed. 2nd edition. McGraw-Hill Publishing Company, 1221 Avenue of the Americas, New York, NY 10020. (800) 262-4729 or (212) 512-3825. 1993. $95.50.

McGraw-Hill Encyclopedia of Science and Technology. Sybil P. Parker, ed. 7th edition. 20 volumes. McGraw-Hill Publishing Company, 1221 Avenue of the Americas, New York, NY 10020. (800) 262-4729 or (212) 512-3825. 1992. $1900.00

GENERAL WORKS

Aluminum and Aluminum Alloys. J.R. Davis. ASM (American Society of Metals) International, Materials Information, Materials Park, OH 44073. (216) 338-5151 or FAX (216) 338-4634. 1993. Inquire for price.

Aluminum Applications For Automotive Design. Society of Automotive Engineers, 400 Commonwealth Drive, Warrendale, PA 15096-0001. (412) 776-4970. FAX (412) 776-0790. 1995. $34.00.

Aluminum Recycling. Date Notes Staff. Prosperity & Profits Unlimited, PO Box 416, Denver, CO 80201. (303) 575-5676. 1992. $17.95.

Behavior and Design of Aluminum Structures. Maurice L. Sharp. McGraw-Hill Publishing Company, 1221 Avenue of the Americas, New York, NY 10020. (800) 262-4729 or (212) 512-3825. 1993. $42.50.

HANDBOOKS AND MANUALS

ASM Specialty Handbook: Aluminum and Aluminum Alloys. J.R. Davis, editor. ASM International Handbook Committee. ASM International, Materials Information, Materials Park, OH 44073-0002. (216) 338-5151 or (800) 336-5152. FAX (216) 338-4634. 1993. $159.00.

CRC Handbook of Chemistry and Physics. David R. Lide, editor. 75th edition. CRC Press, 2000 Corporate Blvd., N.W., Boca Raton, FL 33431. (407) 994-0555 or (800) 272-7737. FAX (407) 994-0949. $99.50.

Smithell's Metals Reference Book. E.A. Brandes & G.B. Brook. 7th edition. Butterworth-Heinemann, 313 Washington Street, Newton, MA 02158. (617) 928-2500 or (800) 366-2665. FAX (617) 928-2620. 1992. $250.00.

ONLINE DATABASES AND CD-ROMS

Compendex Plus. Engineering Information, Inc., 345 East 47th Street, New York, NY 10017. (212) 705-7600 or (800) 221-1044. Contains citations with abstracts to worldwide literature in engineering and technology, from 1970 to present. Available on online BRS,(800) 289-4277, DIALOG, (800) 334-2564, ORBIT (800) 456-7248, and STN International, FIZ Karlsruhe, P.O. Box 2465, W-7500, Karlsruhe 1, Germany, online services. Also available on CD-ROM. Inquire as to cost and availability.

Metadex. Jointly produced by ASM International and the Institute of Materials. Contains more than 925,000 records from the international literature on metals and alloys, concerning properties, processes, materials classes, applications, and metallurgical systems. Updated monthly. Available from ORBIT-QUESTEL (703) 442-0900.

NTIS Bibliographic Database. National Technical Information Service, 5285 Port Royal Road, Springfield, VA 22161. (703) 487-4929 or FAX (703) 321-8199. Broad coverage of government-sponsored science and technology research reports, 1964 to present. Available on BRS,(800) 289-4277, DIALOG, (800) 334-2564, ORBIT, (800) 456-7248, and STN International, FIZ Karlsruhe, P.O. Box 2465, W-7500, Karlsruhe 1, Germany, online services. Also available on CD-ROM. Inquire as to cost and availability.

PERIODICALS

Alloy Digest. Alloy Digest Inc., 27 Canfield Street, Orange, NJ 07050. (201) 677-9161. 1952 to present. Monthly. $140.00 per year.

Aluminum. Aluminum-Verlag GmbH, Postfach 101262, 40003 Dusseldorf, Germany. Telephone 0211-13747-0. FAX 0211-132567. 1919 to present. Monthly. $276.00 per year.

Aluminum Industry. MCB University Press, 62 Toller Lane, Bradford, W. Yorks. BD8 9BY, England. Telephone 0274-499821. FAX 0274-547143. 1982 to present. Bi-monthly. Inquire for cost.

Aluminum Situation. Aluminum Association Inc., 900 19th Street NW, Suite 300, Washington, DC 20006. (202) 862-5100. FAX (202) 862-5164. Monthly. Free.

Aluminum Standards and Data. Aluminum Association Inc., 900 19th Street NW, Suite 300, Washington, DC 20006. (202) 862-5100. FAX (202) 862-5164. 1968 to present. Biennial. $25.00 per year.

Aluminum Standards and Data—Metric. Aluminum Association Inc., 900 19th Street NW, Suite 300, Washington, DC 20006. (202) 862-5100. FAX (202) 862-5164. Biennial. $12.00 per year.

Aluminum Today. FMJ International Publications, Ltd., Queensway House, 2 Queensway, RedHill, Surrey RH1 1QS England. Telephone 0737-768611. FAX 0737-761685. 1989 to present. Quarterly. Inquire for cost.

E M & J : Engineering and Mining Journal. Maclean Hunter Publishing Company, 29 N. Wacker Drive, Chicago, IL 60606. (312) 726-2802. FAX (312) 726-2574. Monthly. $60.00 per year.

JOM (Journal of Metals). Minerals, Metals and Materials Society, 420 Commonwealth Drive, Warrendale, PA 15086. (412) 776-9080. 1949 to present. Monthly. $50.00 per year.

Journal of Phase Equilibria (formerly *Bulletin of Alloy Phase Diagrams*). ASM International, Materials Park, OH 44073-0002. (216) 338-5151. FAX (216) 338-4634. 6 times a year. $658.00 per year for non-members.

Light Metal Age. Fellom Publishing, 170 S. Spruce Avenue, Suite 120, San Francisco, CA 94080. (415) 588-8832. FAX (415) 588-0901. 1943 to present. Bi-monthly. $35.00 per year.

Metallurgical Transactions A: Physical Metallurgy and Materials Science. ASM International, Materials Park, OH 44073. (216) 338-5151 or FAX (216) 338-4634. 1970 to present. Monthly. $520.00 per year.

Mineral Industry Surveys, Aluminum. U.S. Bureau of Mines, Production and Distribution, Cochrans Mill Rd., Box 18070, Pittsburgh, PA 15236. (412) 892-4411. Inquire for price and availability.

Mineral Industry Surveys, Bauxite. U.S. Bureau of Mines, Production and Distribution, Cochrans Mill Rd., Box 18070, Pittsburgh, PA 15236. (412) 892-4411. Quarterly. Inquire for price and availability.

ALUMINUM

Ency. of Physical Sciences and Engineering Info. Sources

RESEARCH CENTERS AND INSTITUTES

Colorado School of Mines, Advanced Steel Processing and Products Research Center. Golden, CO 80401. (303) 273-3774.

Canada Centre For Mineral and Energy Technology, Materials Technology Laboratories. Energy, Mines & Resources Canada, 568 Booth Street, Ottawa ON, Canada K1A 0G1. (613) 995-8248. FAX (613) 992-8735.

Texas A&m University Mechanics and Materials Center. Civil Engineering Department, College Station, TX 77843. (409) 845-7512. FAX (409) 845-6156.

University of Connecticut Institute of Materials Science. U-136, 97 N. Eagleville Road, Storrs, CT 06268. (203) 486-4623. FAX (203) 486-4745.

University of Florida Department of Materials Science and Engineering. Gainesville, FL 32611. (904) 392-1454. FAX (904) 392-6359.

University of Minnesota Corrosion Research Center. 221 Church Street SE, Minneapolis, MN 55455. (612) 625-0717.

University of Wisconsin—Madison, Cast Metals Laboratory.1509 University Avenue, Madison, WI 53706. (608) 262-2562. FAX (608) 262-6707.

University of Utah, Department of Materials Science and Engineering. 304 EMRO Bldg., Salt Lake City, UT 84112. (801) 581-6863. FAX (801) 581-4816.

SPECIFICATIONS AND STANDARDS

1994 Annual Book of ASTM Standards. Volume O2.02 (aluminum and magnesium alloys) American Society for Testing and Materials (ASTM), 1916 Race Street, Philadelphia, PA 19103. (215) 299-5585. Inquire for cost and availability.

AM RADIO

See: RADIO

AMATEUR RADIO

See also: ACOUSTICS, AMPLIFIERS, ANTENNAS, ELECTRICAL ENGINEERING, ELECTRONIC CIRCUITS and COMPONENTS, ELECTRONIC ENGINEERING, ELECTRONICS, RADAR, RADIO, TELEVISION

ABSTRACT SERVICES AND INDEXES

Amateur Radio Technical Abstracts. Graham Thornton, editor. Thornton Publishing, 1502 Howard Street, New Iberia, LA 70560. (318) 364-2752. (800) 551-3076. 1992. $25.00.

Applied Science and Technology Index; A cumulative subject index to English language periodicals in the fields of aeronautics and space science, computer technology, chemistry, construction industry, energy and related areas. H.W. Wilson Co., 950 University Avenue, Bronx, NY 10452. (212) 588-8400. (800) 367-6770. FAX (718) 590-1617. From 1958 to present. Monthly. Inquire about cost and availability. Also available on CD-ROM and online. ISSN: 0003-6986.

Current Contents: Engineering, Technology and Applied Sci Ences. Institute for Scientific in formation, 3501 Market Street, Philadelphia, PA 19104. (215) 386-0100. FAX (215) 386-6362. 1970 to present. Weekly. $442.00 per year. Also available on CD-ROM and online. Inquire regarding cost and availability. ISSN: 0095-7917.

Electrical and Electronics Abstracts. Institution of Elec trical Engineers (IEE), London. Available from INSPEC/IEEE - Institute of Electrical and Electronic Engineers, Box 1331, Hoes Lane, Piscataway, NJ 08855-1331. (908) 562-5549. 1898 to present. Monthly. $2200.00 per year. Also available on CD-ROM and online as INSPEC.

Engineering Index Monthly; Indexes and Abstracts the World's Engineering and Technical Literature. Engineering Information, Inc., Castle Point on the Hudson, Hoboken, NJ 07030. (201) 216-8500. (800) 221-1044. FAX (201) 216-8532. Monthly. $2300.00 per year. Available online as COMPENDEX and also on CD-ROM. ISSN: 0742-1974.

General Science Index. H.W. Wilson Company, 950 University Avenue, Bronx, NY 10452. (212) 588-8400. (800) 367-6770. FAX (718) 590-1617. From 1978 to present. Ten issues per year; quarterly and annual cumulations. Service basis. Available onCD-ROM and online. Inquire regarding cost and availability. ISSN: 0162-1963.

Index To IEEE Publications. IEEE Service Center, 445 Hoes Lane, Piscataway, NJ 08855-1331. (908) 981-1393. (800) 678-IEEE. FAX (908) 981-9667. 1973 to present. Annual. ISSN: 0099-1368.

Physics Abstracts . INSPEC. Section A, Science Abstracts. Instituti on of Electrical Engineers (IEE), London. Available from: INSPEC/IEEE - Institute of Electrical and Electronic Engineers, Box 1331, Hoes Lane, Piscataway, NJ 08855-1331. (908) 562-5549. 1898 to present. 24 issues per year. $2835.00 per year. Also available online and on CD-ROM. ISSN: 0036-8091.

Physics Briefs (*Physikalische Berichte*). Information Center for Energy, Physics, Mathematics; German Physical Society. V C H Publishers, Inc., 220 East 23rd Street, New York, NY 10010-4606. (212) 683-8333. 1845 to present. 24 issues per year. $2390.00 per year. Also available online. ISSN: 0179-7434.

ASSOCIATIONS AND PROFESSIONAL SOCIETIES

American Electronics Association. 5201 Great America Parkway, Suite 520, PO Box 54990, Santa Clara, CA 95056. (408) 987-4200. FAX (408) 970-8565.

American Radio Relay League. 225 Main Street, Newington, CT 06111. (203) 666-1541. FAX (203) 665-7531.

Audio Engineering Society. 60 East 42nd Street, New York, NY 10017. (212) 661-8528.

IEEE (Institute of Electrical and Electronics Engineers). 345 East 47th Street, New York, NY 10017. (212) 705-7900. FAX (212 705-4929.

National Amateur Radio Association. PO Box 201407, Arlington, TX 76006. (817) 860-0978. FAX (817) 860-0979.

National Association of Radio and Telecommunications Engineers. PO Box 678, Medway, MA 02053. (508) 533-8333. FAX (508) 533-3815.

Radio Amateur Satellite Corporation. 850 Sligo Avenue, Suite 600, Silver Spring, MD 20910. (301) 589-6062. FAX (301) 608-3410.

DIRECTORIES AND BIOGRAPHICAL SOURCES

Broadcast Engineering - Equipment Reference Manual. Intertec Publishing Corporation, Box 12901, Overland Park, KS 66212. (913) 888-4664. Annual. $20.00.

E E M - Electronic Engineer's Master. Hearst Business Communications, Inc., 645 Stewart Avenue, Garden City NY 11530. (516) 227-1300. Annual. $90.00. ISSN: 0732-9016.　IEEE MEMBERSHIP DIRECTORY. Institute of Electrical and Electronics Engineers, IEEE Service Center, 445 Hoes Lane, Piscataway, NY 08854. (908) 981-1393. (800) 678-IEEE. FAX (908) 981-9667. 2 volumes. Annual. $190.00. ISSN:

Radio Amateur Callbook 1996: North American Listings. Watson-Guptill Publications, Inc., 1515 Broadway, New York, NY 10036. (2121) 5365121. (800) 451-1741. 75th edition. $35.00.

Shortwave Directory. Larry Van Horn, editor. GroveEnterprises, PO Box 98, Brasstown, NC 28902 (800) 438-8155. 8th edition. 1993. $24.95.

Who's Who in Engineering. Gordon Davis, editor. American Association of Engineering Societies. 1111 19th Street, NY, Suite 608, Washington, DC 20036. (202) 296-2237. (800) 658-8897. 9th edition. 1995. $220.00.

World Ham Net Directory. M. Witkowski. Tiare Publications, P.O. Box 493, Lake Geneva, WI 53147. (800) 420-0579. 3rd edition. 1995. $12.95.

ENCYCLOPEDIAS AND DICTIONARIES

Amateur Radio Encyclopedia. Stan Gibilsco. McGraw-Hill Publishing Company, Inc., 1221 Avenue of the Americas, New York, NY 10020. (212) 512-2000. (800) 262-4729. 1993. $49.95.

IEEE Standard Dictionary of Electrical and Electronics Terms. Christopher J. Booth, editor. IEEE Service Center, 445 Hoes Lane, Piscataway, NJ 08855-1331. (908) 981-1393. (800) 678-IEEE. FAX (908) 981-9667. IEEE Standard 100- 1992. 5th edition. 1993. $150.00.

Illustrated Dictionary of Electronics. Stan Gibilsco. TAB Books, P.O. Box 40, Blue Summit, PA 17294-0850. (717) 794- 2191. (800) 233-1128. 7th edition. 1994. $34.95.

McGraw-Hill Electronics Dictionary. John Markus and Neil Sclater. McGraw-Hill-Hill Publishing Company, Inc., 1221 Avenue of the Americas, New York, NY 10020. (212) 512-2000. (800) 262-4729. 5th edition. 1994. $49.95.

GENERAL WORKS

All about Crystal Sets: How to Build Crystal Set Radios. Charles Green., Allabout Books, PO Box 22366, San Diego, CA 92192. (619 452-6589. 1984. $9.95.

Amateur Radio Equipment Fundamentals. Albert D. Helfrick. Prentice-Hall, 113 Sylvan Avenue, Route 9W, Englewood Cliffs, NJ 07632. (201) 592-2000. (800) 922-0579. 1982. $40.00.

Electric and Electronic Engineering For Scientists and Engineers. K. A. Krishnamurthy. Halsted Press, 605 Third Avenue, New York, 10158-0012. (212) 850-6400. 1944. $48.95.

Hidden Ham Antennas. Frank P. Hughes. Tiare Publications, P.O. Box 493, Lake Geneva, WI 53147. (800) 420-0579. 1994. $12.95.

Hiram Percy Maxim, Father of Amateur Radio. Alice C. Schumacher. Communications Technology, Inc., P.O. Box 209, Rindge, NH 03461. (603) 899-6957. Reprint edition. 1977. $4.50.

Introduction to Radio Frequency and Microwave Engineering. Robert K. Feeney and David R. Hertling. TAB Books, P.O. Box 40, Blue Summit, PA 17294-0850. (717) 794-2191. (800) 233-1128. 1995. $44.50.

Low Power Communications. Richard H. Arland. Tiare Publications, P.O. Box 493, Lake Geneva, WI 53147. (800) 420-0579 volume 1: Basic QRP. 1992. $14.95.

Packet Radio Operator's Manual. Glynn B. Rogers. CQ Communications, Inc., 76 North Broadway, Hicksville, NY 11801.(516) 681-2922. 1993. $15.95 in paper.

Practical Antenna Handbook. Joseph J. Carr. McGraw-Hill Publishing Company, Inc., 1221 Avenue of the Americas, New York, NY 10020. (212) 512-2000. (800) 262-4729. 1990. $32.95.

Shortwave Radio Listening for Beginners. Anita L. McGormic. TAB Books, P.O. Box 40, Blue Summit, PA 17294-0850. (717) 794-2191. (800) 233-1128. 1993. $18.95.

Single-Sideband Systems and Circuits. William E. Sabin and Edgar O. Schoenike. McGraw-Hill Publishing Company, Inc., 1221 Avenue of the Americas, New York, NY 10020. (212) 512- 2000. (800) 262-4729. 1995. $75.00.

Solid State Design For the Radio Amateur. American Radio Relay League, 225 Main Street, Newington, CT 06111. (203) 666-1541. FAX (203) 665-7531. 1986. $15.00 in paper.

HANDBOOKS AND MANUALS

AARL Handbook For Radio Amateurs. American Radio Relay League, 225 Main Street, Newington, CT 06111. (203) 666-1541. FAX (203) 665-7531. 78th edition. 1996. $30.00. ISSN: 0079-9440.

American Electricians Handbook. Terrell Croft and Wilford I. Summers. McGraw-Hill Book Company, 1221 Avenue of the Americas, New York, NY 10020. (212) 512-2000. (800) 262-4729. 12th edition. 1992. $77.50.

AMATEUR RADIO

Ency. of Physical Sciences and Engineering Info. Sources

Antenna Engineering Handbook. Richard C. Johnson. McGraw-Hill Publishing Company, Inc., 1221 Avenue of the Americas, New York, NY 10020. (212) 512-2000. (800) 262-4729. 3rd edition. 1993. $119.50.

The Calculations and Measurement of Inductance and Capacity: A Handbook For Experimenting With Tesla Coils and Radio. Gordon Press Publications, P.O. Box 459, Bowling Green Station, New York, NY 10004. (718) 624-8419. 1991. $88.95.

Complete Handbook of Amateur Radio. Clay Laster. McGraw-Hill Publishing Company, Inc., 1221 Avenue of the Americas, New York, NY 10020. (212) 512-2000. (800) 262-4729. 3rd edition. 1993. $21.95.

Complete Handbook of Amateur Radio. Clay Laster. McGraw-Hill Publishing Company, Inc., 1221 Avenue of the Americas, New York, NY 10020. (212) 512-2000. (800) 262-4729. 3rd edition. 1993. $21.95.

Electronics Handbook. Jerry C. Whitaker. CRC Press, Inc., 2000 Corporate Boulevard, NW, Boca Raton, FL 33431. (407) 994-0555. (800) 272-7737. 1996. $120.00.

The HAMSAT Handbook: A Complete History and Frequency Guide To Amateur Radio Satellites. Anthony R. Curtis. Tiare Publications, P.O. Box 493, Lake Geneva, WI 53147. (800) 420-0579. 1993. $19.95.

Newnes Radio Engineer's Pocket Book. John Davies. Butterworth-Heinemann, 313 Washington Street, Newton, MA 02158. (618-928-2500. (800) 366-2665. 1995. $22.95

Practical Antenna Handbook. Joseph J. Carr. McGraw-Hill Publishing Company, Inc., 1221 Avenue of the Americas, New York, NY10020. (212) 512-2000. (800) 262-4729. 1990. $32.95.

Radio Amateur and Listener's Data Handbook. Steve Money. Butterworth-Heinemann, 313 Washington Street, Newton, MA 02158. (618-928-2500. (800) 366-2665. 1995. $39.95 in paper.

Radio Amateur's Digital Communications Handbook. Jonathan L. Mayo. TAB Books, P.O. Box 40, Blue Summit, PA 17294-0850. (717) 794-2191. (800) 233-1128. 1992. $22.95.

Radio and Radar Reference Data. Gordon Press Publications, P.O. Box 459, Bowling Green Station, New York, NY 10004. (718) 624-8419. 1991. $79.00.

Standard Handbook For Electrical Engineers. Donald Fink. McGraw-Hill Publishing Company, 1221 Avenue of the Americas, New York, NY 10020. (212) 512-2000. (800) 262- 4729. 13th edition. 1996. $110.00.

Standard Radio Communications Manual, with Instrumentation and Testing Techniques. R. Harold Kinley. Prentice-Hall, 113 Sylvan Avenue, Route 9W, Englewood Cliffs, NJ 07632. (201) 592-2000. (800) 922-0579. 1988. $16.95.

ONLINE DATABASES AND CD-ROMS

COMPENDEX PLUS . Engineering Information, Inc., 345 East 47th Street, NE w York, NY 10017. (212) 705-7600 or (800) 221- 1044. Cont ains citations with abstracts to worldwide li terature in engineerin g and technology, from 1970 to present. Available on online BRS,(800) 289-4277, DIALOG, (800) 334-2564, ORBIT (800) 456-7248, and STN International, FIZ

Karlsruhe, P.O. Box 2465, W-7500, Karlsruhe 1, Germany, online services. Also available on CD-ROM. Inquire as to cost and availability.

INSPEC. Institution of Electrical Engineers, Michael Faraday Hou se, Six Hills Way, Stevenage, Herts. SG1 2AY, Englan d. Telephone: 0438 313311 or FAX 0438 742840. Contains citations to the worldwide literature of phy sics, electronics and electrical engineering, computer technology, and related fields. Available on BRS, (800) 289-4277, DIALOG, (800) 334-2564, ORBIT, (800) 456-7248, and STN International, FIZ Karlsruhe, P.O. Box 2465, W-7500, Karlsruhe 1, Germany, online services. Inquire as to cost and availability.

NTIS Bibliographic Database. National Technical Information Service, 5285 Port Royal Road, Springfield, VA 22161 . (703) 487-4929 or FAX (703) 321-8199. Broad covera ge of govern-ment-sponsored science and te chnology research reports, 1964 to present. Available on BRS,(800) 289-4277, DIALOG, (800) 334-2564, ORBIT, (800) 456-7248, and STN International, FIZ Karlsruhe, P.O. Box 2465, W-7500, Karlsruhe 1, Germany, online services. Also available on CD-ROM. Inquire as to cost and availability

WILSONLINE. H.W. W ilson Company, 950 University Avenue, Bronx, NY 10452. (800) 367-6770 or (212) 588-8400. Makes available onlin e versions of the printed H.W. Wilson indexes includin g Applied Science and Technology Index, Business Periodicals Index, General Science Index, and Readers' Guide to Periodical Literature.Period covered is generally 1983 to present. Available on BRS, (800) 289-4277, DIALOG, (800) 334-2564, and OCLC EPIC, (800) 848-5878, online services. Also available on CD-ROM. Inquire as to cost and availability.

PERIODICALS

Audio Engineering Society. Journal. Audio Engineering Society, 60 East 42nd Street, New York, NY 10017. (212) 661-8528. 1953 to present. 10 issues per year. $125.00 per year. ISSN: 0004-7554.

Cq: The Radio Amateur's Journal. CQ Communications, Inc., 76 North Broadway, Hicksville, NY 11801-2953. (516) 681-2922. FAX (516) 681-2926. 1945 to present. Monthly. $24.95. ISSN: 0007-893X.

DB, the Sound Engineering Magazine. 203 Commack Road, Suite 1010, Commack, NY 1725. (516) 586-6530. 1967 to present. $15.00 per year. ISSN: 0011-7145. .

Electronic Design. Penton Publishing, San Jose Gateway, Suite 354. 2025 Gateway Place, San Jose, CA 95110. (408) 441-0550. 1952 to present. Fortnightly. $95.00. ISSN: 0013-4872.

Electronics. Penton Publishing, San Jose Gateway, Suite 354. 2025 Gateway Place, San Jose, CA 95110. (408) 441-0550. 1930 to present. Semi-weekly. $98.00 per year. ISSN: 0883-4989.

Electronics Now. Gernsback Publications, Inc., 500-B Bi-County Boulevard, Farmingdale, NY 11735. (516) 293-3000. FAX (516) 293-3115. 1929 to present. Monthly. $19.95. ISSN: 0033-7862.

Electronics World and Wireless World. Reed Business Publishing, Ltd., Quadrant House, the Quadrant, Sutton, Surrey SM2 5AS, England. 1911 to present. Monthly. L35. ISSN: 0266-3244.

IEEE Antennas and Propagation Magazine. Institute of Electrical and Electronics Engineers, Inc., 445 Hoes Lane, Box 1331, Piscataway, NJ 08855-1331. (908) 981-0060. FAX (908) 981-9667. Bi-monthly. $85.00. ISSN: 1045-9243.

Microwaves and Rf. Penton Publishing Co., Circulation Department, 1100 Superior Avenue, Cleveland, OH 44114. (201) 393-6060. FAX (201) 393-6297. From 1962 to present. Monthly. $60.00. ISSN: 0745-2992.   <subhead1>AMMONIA
See also: CHEMICAL ENGINEERING, CHEMISTRY

ABSTRACT SERVICES AND INDEXES

Chemical Abstracts. Chemical Abstracts Service. 2540 Oleantangy River Road, Box 3012, Columbus, OS 43210-0012. (614) 447-3600. FAX (614) 447-3713. 1907 to present. Weekly. $16,800.00 per year. Available online and on CD-ROM. CA is also available in five section groupings.   Inquire regarding cost and availability.

Current Contents: Physical, Chemical and Earth Sciences. Institute for Scientific Information, 3501 Market Street, Philadelphia, PA 19104. (215) 386-0100. (800) 523-1850. FAX (215) 386-2291. 1961 to present. Weekly. $442.00 per year. Also available online (BRS, DIALOG) and on CD-ROM. Inquire regarding cost and availability. ISSN: 0163-2574.

General Science Index. H.W. Wilson and Company, 950 University Avenue, Bronx, NY 10452. (212) 588-8400. (800) 367-6770. FAX (718) 590-1617. From 1978 to present. Ten issues per year; quarterly and annual cumulations. Service basis. Available on CD-ROM and online. Inquire regarding cost and availability. ISSN: 0162-1963.

Process and Chemical Engineering/chemical Engineering Abstracts. Royal Society of Chemistry, Information Services, Thomas Graham House, Science Park, Milton Road, Cambridge, CB4 4WF, England. Contains citations, mostly with abstracts, of the worldwide literature on chemical engineering; from 1982 to present. Monthly. $610.00 per year. Also available online. ISSN: 0960-5045.

Science Citation Index. SCI. Institute for Scientific Information, 3501 Market Street, Philadelphia, PA 19104.   (215) 386-0100. (800) 523-1850. FAX (215) 386-2991.   1961 to present. Six issues per year, plus annual cumulation. $11650.00 per year. Also available online and on CD-ROM.   Inquire about price and availability. ISSN: 0036-827X.

Theoretical Chemical Engineering. Royal Society of Chemistry, Information Services, Thomas Graham House, Science Park, Milton Road, Cambridge, CB4 4WF, England. Covers theoretical chemical engineering, including theory and laboratory experimentation. 1964 to present. Monthly. $235.00. ISSN: 0960-5053.

ASSOCIATIONS AND PROFESSIONAL SOCIETIES

American Chemical Society. 1155 16th Street, N.W., Washington, DC 20036. (202) 872-4600.

American Institute of Chemical Engineers. 345 East 47th Street, New York, NY 10017. (212) 705-7657. FAX (212) 752-3294.

Association of Consulting Chemists and Chemical Engineers, 50 East 41st Street, Suite 92, New York, NY 10017. (212) 684-6255.

Association of official Analytical Chemists. 1111 North 19th Street, Suite 210, Arlington, VA 22209. (703) 522-3032.

Fertilizer Institute. 1015 18th Street, N.W., Washington, DC 20036. (202) 861-4900.

International Institute of Ammonia Refrigeration. 111 East Wacker Drive, Chicago, IL 60601. (312) 644-6610.

DIRECTORIES AND BIOGRAPHICAL SOURCES

American Institute of Chemical Engineers Directory. American Institute of Chemical Engineers. 345 East 47th Street, New York, NY 10017-2396. (212) 705-7663. Annual.

American Men and Women of Science. Physical and Biological Sciences. R.R. Bowker Inc., 121 Chanlon Road, New Providence, NJ 07974. (908) 464-6800. (800) 521-8110. 20th edition. 8 volumes. 1996. $850.00

ARI Directory. Agricultural Research Institute, 9650 Rockville Pike, Bethesda, MD 20841-3998. (301) 530-7122. $5.00 per copy.

Research Centers Directory. Gale Research, 835 PenobscotBuilding, Detroit, MI 48226-4094. (313) 961-2242. (800) 877-4253. 20th edition. 1995. $485.00. ISSN: 0080-1518.

Who's Who in Engineering. Gordon Davis, editor. American Association of Engineering Societies. 1111 19th Street,   NY, Suite 608, Washington, DC 20036. (202) 296-2237.   (800) 658-8897. 9th edition. 1995. $220.00.

ENCYCLOPEDIAS AND DICTIONARIES

Dictionary of Named Processes in Chemical Technology. Alan E. Comyns. Oxford University Press, Inc., 200 Madison Avenue, New York, NY 10016. (212) 725-6000. (800) 334-4249. 1994. $75.00.

Encyclopedia of Chemical Processing and Design. McKetta Marcel Dekker, Inc., 270 Madison Avenue, New York, NY 10016.   (212) 696-9000. (800) 228-1160. 1976 - . $175.00 per volume.   ISSN:

Industrial Chemical thesaurus. Michael Ash and Irene Ash. VCH Publishers, Inc., 220 East 23rd Street, New York, NY 10010-4606.   (800) 367-8249. 1992. $295.00.

Kirk-Othmer Encyclopedia of Chemical Technology. John Wiley and Sons, Inc., 605 Third Avenue, New York, NY 10158. (800) 526-5368 or (212) 850-6000. Fourth edition. 1991 - . Twenty-seven volumes. $5400.00.

Ullman's Encyclopedia of Industrial Chemistry. VCH Publications, Inc., 220 East 23rd Street, Suite 909, New   York, NY 10010-4606. (212) 683-8333. (800) 422-8824.   5th edition. 1984 - . Price varies per volume. ISSN:

GENERAL WORKS

Ammonia: Catalysis and Manufacture. K. Aika. Springer-Verlag New York, Inc., 175 Fifth Avenue, New York, NY 10010. (212) 460-1500. (800) 777-4643. FAX (212) 473-6272. 1995. .

Ammonia Plant Safety. American Institute of Chemical Engineers, 345 East 47th Street, New York, NY 10017. (212) 705-7657. Volume 31. 1991. $150.00.

Catalytic Ammonia Synthesis: Fundmentals and Practice. J. R. Jennings, editor. Plenum Publishing Corp., 233 Spring Street, New York, NY. (212) 620-8000. (800) 2221-9369. FAX (212) 463-0742. 1991. $69.50.

Odour and Ammonia Emissions From Livestock Farming. V. C. Nielson, et al. Elsevier Science Publishing Company, Inc., 655 Avenue of the Americas, New York, NY 10010. (212) 989-5800. FAX (914) 333-2444. 1991. $76.50.

Technology and Manufacture of Ammonia. Samuel Strelzoff. Krieger Publishing Company, P.O. Box 9542, Melbourne, FL 32902-9542. (407) 724-9542. 1987. $99.50.

HANDBOOKS AND MANUALS

CRC Handbook of Chemistry and Physics. David R. Kide, editor. CRC Press Inc., 2000 Corporate Blvd., NW, Boca Raton, FL 33431. (407) 994-0555. (800) 333-8300. 77th edition. 1996. $99.95.

Handbook of Laboratory Safety. A. Keith Furr. CRC Press, Inc., 2000 Corporate Boulevard, NW, Boca Raton, FL 33431. (407) 994-0555. (800) 272-7737. 1995. 125.00.

Lange's Handbook of Chemistry. John A. Dean, editor. McGraw-Hill Publishing Company, 1221 Avenue of the Americas, New York, NY 10020. (212) 512-2000. (800) 262-4729. 14th edition. 1996. $99.95. PERRY'S CHEMICAL ENGINEERS' HandBOOK. Robert H. Perry and Donald W. Green, editors. McGraw-Hill Publishing Company, Inc., 1221 Avenue of the Americas, New York, NY 10020. (212) 512-2000. (800) 262-4729. 6th edition. 1996. $129.50.

ONLINE DATABASES AND CD-ROMS

CA Search. Chemical Abstracts Service, P.O. Box 3012, Columbus, OH 43210-0012. (614) 447-3600. (800) 848-6533. FAX (614) 447-3709. Very comprehensive guide to worldwide chemical literature and related fields, 1972 to present. Available on BRS, (800) 289-4277, DIALOG, (800) 334-2564, ORBIT (800) 456-7248, and STN International, FIZ Karlsruhe, P.O. Box 2465, W-7500, Karlsruhe 1, Germany, online services. Inquire as to cost and availability.

Compendex Plus. Engineering Information, Inc., 345 East 47th Street, New York, NY 10017. (212) 705-7600 or (800) 221-1044. Contains citations with abstracts to worldwide literature in engineering and technology, from 1970 to present. Available on online BRS, (800) 289-4277, DIALOG, (800) 334-2564,

ORBIT (800) 456-7248, and STN International, FIZ Karlsruhe, P.O. Box 2465, W-7500, Karlsruhe 1, Germany, online services. Also available on CD-ROM. Inquire as to cost and availability.

NTIS. National Tec hnical Information Service, 5285 Port Royal Road, Springfield, VA 22161. (703) 487-4929 or FAX (703) 321-8199. Broad coverage of government-sponsored research reports, 1964 to present. Available on BRS, DIALOG, ORBIT and STN online services. Also available on CD-ROM. Inquire as to cost and availability.

SCISEARCH. Institute for Scientific Information, 3501 Market Street, Philadelphia, PA 19104. (800) 523-1850 or (215) 386-0100. Broad multidisciplinary title and author index to the international literature of science and technology, 1974 to present. Available on BRS, (800) 289-4277, DIALOG, (800) 334-2564, ORBIT, (800) 456-7248, and STN International, FIZ Karlsruhe, P.O. Box 2465, W-7500, Karlsruhe 1, Germany online services. Also available on CD-ROM. Inquire as to cost and availability.

WILSONLINE. H.W. Wilson Company, 950 University Avenue, Bronx, NY 10452. (800) 367-6770 or (212) 588-8400. Makes available online versions of the printed H.W. Wilson indexes including Applied Science and Technology Index, Business Periodicals Index, and Readers' Guide to Periodical Literature. Period covered is generally 1983 to present. Also available on CD-ROM. Inquire as to cost and availability.

PERIODICALS

Chemical Week. Chemical Week Associates, 810 7th Avenue, New York, NY 10019. 1914 to present. Weekly. $49.00 per year. ISSN: 0009-272X.

Chemtech. American Chemical Society, Drawer 1734, Atlanta, GA 30301-1734. (800) 227-5558. 1970 to present. Monthly. $370.00 per year. ISSN: 0009-2703.

Comments on Inorganic Chemistry. Gordon and Breach Science Publishers, Inc., 820 Town Center Drive, Langhorne, PA 19047. (215) 750-2642. Six times per year. 132 ECU per year. ISSN: 0260-3594.

Inorganic Chemistry. American Chemical Society, 1155 16th Street, N.W., Washington, DC 20036. (800) 424-6747. FAX (614) 447-3671. 1962 to present. Biweekly. $949.00 per year. ISSN: 0020-1669.

RESEARCH CENTERS AND INSTITUTES

Chemical Laboratories. Harvard University, Oxford Street, Cambridge, MA 02138. (617) 495-4283. FAX (617) 495-8824.

Chemistry Laboratories. Rensselaer Polytechnic Institute, Cogswell Laboratory, Troy, NY 12180-3590. (518) 276-8981.

University/Industry Chemical Research Center. Mississippi State University, Department of Chemistry, P.O. Drawer CH, Mississippi State, MS 39762. (601) 325-3584.

AMPLIFIERS

See also: ACOUSTICS, AMPLIFIERS, ELECTRICAL ENGINEERING, ELECTRICITY, ELECTRONIC CIRCUITS and COMPONENTS, ELECTRONIC ENGINEERING

ABSTRACT SERVICES AND INDEXES

Acoustics Abstracts; Parts A & B. Multi-Science Publishing Co. Ltd., Box 176, Avenel, NJ 07001. 1967 to present. Monthly. L265 per year. ISSN: 0001-4974.

Applied Science and Technology Index; A cumulative subject index to English language periodicals in the fields of aeronautics and space science, computer technology, chemistry, construction industry, energy and related areas. H.W. Wilson Co., 950 University Avenue, Bronx, NY 10452. (212) 588-8400. (800) 367-6770. FAX (718) 590-1617. From 1958 to present. Monthly. Inquire about cost and availability. Also available on CD-ROM and online. ISSN: 0003-6986.

Current Contents: Engineering, Technolog Y and Applied Sci Ences. Institute for Scientific Inf ormation, 3501 Market Street, Philadelphia, PA 19104. (215) 386-0100. FAX (215) 386-6362. 1970 to present. Weekly. $442.00 per year. Also available on CD-ROM and online. Inquire regarding cost and availability. ISSN: 0095-7917.

Electrical and Elect Ronics Abstracts. Institution of Electrical Engineers (IEE), London. Available from INSPEC/IEEE - Institute of Electrical and Electronic Engineers, Box 1331, Hoes Lane, Piscataway, NJ 08855-1331. (908) 562-5549. 1898 to present. Monthly. $2200.00 per year. Also available on CD-ROM and online as INSPEC.

Engineering Index Monthly; Indexes and Abstracts the World's Engineering and Technical Literature. Engineering Information, Inc., Castle Point on the Hudson, Hoboken, NJ 07030. (201) 216-8500. (800) 221-1044. FAX (201) 216-8532. Monthly. $2300.00 per year. Available online as COMPENDEX and also on CD-ROM. ISSN: 0742-1974.

General Science Index. H.W. Wilson Company, 950 University Avenue, Bronx, NY 10452. (212) 588-8400. (800) 367-6770. FAX (718) 590-1617. From 1978 to present. Ten issues per year; quarterly and annual cumulations. Service basis. Available on CD-ROM and online. Inquire regarding cost and availability. ISSN: 0162-1963.

Government Reports Announcements and Index. U. S. National Technical Information Service (NTIS), 5285 Port Royal Road, Springfield, VA 22161. (703) 487-4650. FAX (703) 321-8547. From 1968 to present. Annual. $630.00 per year. Also available online as *NTIS Bibliographic Database* and on CD-ROM. ISSN:

Index To IEEE Publications. IEEE Service Center, 445 Hoes Lane, Piscataway, NJ 08855-1331. (908) 981-1393. (800) 678-IEEE. FAX (908) 981-9667. 1973 to present. Annual. ISSN: 0099-1368.

Physics Abstracts . INSPEC. Section A, Science Abstracts. Institution of Electrical Engineers (IEE), London. Available from: INSPEC/IEEE - Institute of Electrical and Electronic Engineers, Box 1331, Hoes Lane, Piscataway, NJ 08855-1331. (908) 562-5549. 1898 to present. 24 issues per year. $2835.00 per year. Also available online and on CD-ROM. ISSN: 0036-8091.

Physics Briefs (Physikalische Berichte). Information Center for Energy, Physics, Mathematics; German Physical Society. V C H Publishers, Inc., 220 East 23rd Street, New York, NY 10010-4606. (212) 683-8333. 1845 to present. 24 issues per year. $2390.00 per year. Also available online. ISSN: 0179-7434.

Science Citation Index. SCI. Institute for Scientific Information, 3501 Market Street, Philadelphia, PA 19104. (215) 386-0100. (800) 523-1850. FAX (215) 386-2991. 1961 to present. Six issues per year, plus annual cumulation. $11650.00 per year. Also available online and on CD-ROM. Inquire about price and availability. ISSN: 0036-827X.

Solid State and Superconductivity Abstracts: Covers theory, Production and Application of Solid State Materials. Cambridge Scientific Abstracts, 7200 Wisconsin Avenue, Bethesda, MD 20824. (301) 961-6750. FAX (301) 961-6720. 1957 to present. Bimonthly. $1320.00 per year. Also available online. ISSN: 0896-5900.

ANNUAL REVIEWS AND YEARBOOKS

Advances in Electronics and Electron Physics. Academic Press, Inc., 6277 Sea Harbor Drive Orlando FL 32887. (800) 321-5068. Irregular. 1948 to present. ISSN: 0065-2539.

Physical Acoustics. Academic Press, Inc., 6277 Sea Harbor Drive, Orlando, FL 32887. (800) 321-5068. 1964 to present. Irregular. ISSN: 0079-1873.

ASSOCIATIONS AND PROFESSIONAL SOCIETIES

American Electronics Association. 5201 Great America Way, Suite 520, P.O. 52990, Santa Clara, CA 95056. (408) 987-4200. FAX (408) 970-8565.

American Institute of Physics. One Physics Ellipse, College Park, MD 20740-3843. (301) 209-3100.

Audio Engineering Society. 60 East 42nd Street, New York, NY 10017. (212) 661-8528.

Creative Audio and Music Electronics Organization. 7753 State Boulevard East, Meridian, MS 39305. (601) 693-5185. FAX (601) 485-7117.

IEEE (Institute of Electrical and Electronics Engineers). 345 East 47th Street, New York, NY 10017. (212) 705-7900. FAX (212) 705-4929.

Society of Professional Audio Recording Services. 4300 10th Avenue North, Lake Worth, FL 33461. (407) 641-6648. FAX (407) 642-8263.

DIRECTORIES AND BIOGRAPHICAL SOURCES

American Men and Women of Science: Physical and Biological Sciences. R. R. Bowker Inc., 121 Chanlon Road, New Providence, NJ 07974. (908) 464-6800. (800) 521-8110. 20th edition. 8 volumes. 1996. $850.00.

E E M - Electronic Engineer's Master. Hearst Business Communications, Inc., 645 Stewart Avenue, Garden City NY 11530. (516) 227-1300. Annual. $90.00 ISSN: 0732-9016.

AMPLIFIERS

Ency. of Physical Sciences and Engineering Info. Sources

Engineering Research Centers: Incorporating Electronics Research Centers. Stockton Press, 345 Park Avenue, New York, NY 10010. (212) 689-9200. (800) 221-2123. 4th edition. 1995. $515.00.

IEEE Membership Directory. Institute of Electrical and Electronics Engineers, IEEE Service Center, 445 Hoes Lane, Picastaway, NY 08854. (908) 981-1393. (800) 678-IEEE. FAX (908) 981-9667. 2 volumes. Annual. $190.00. ISSN:

Research Centers Directory. Gale Research Inc., 835 Penobscot Building, Detroit, MI 48226-4094. (313) 961-2242. (800) 877-4253. 20th edition. 1995. $485.00. ISSN: 0080- 1518.

Who's Who in Engineering. Gordon Davis, editor. American Association of Engineering Societies. 1111 19th Street, NY, Suite 608, Washington, DC 20036. (202) 296-2237. (800) 658-8897. 9th edition. 1995. $220.00.

ENCYCLOPEDIAS AND DICTIONARIES

The Audio Dictionary. Glenn D. White. University of Washington Press, P. O. Box 50096. Seattle, WA 98145- 5096. (206) 543-4050. (800) 441-4115. 2nd revised edition. 1991. $19.95 in paper.

Encyclopedia of Applied Physics. George Trigg, editor. VCH Publications, Inc., 220 East 23rd Street, Suite 909, New York, NY 10010-4606. (212) 683-8333. (800) 422-8824. 20 volume set. 1991-. $5990.00.

IEEE Standard Dictionary of Electrical and Electronics Terms. Christopher J. Booth, editor. IEEE Service Center, 445 Hoes Lane, Piscataway, NJ 08855-1331. (908) 981-1393. (800) 678-IEEE.FAX (908) 981-9667. IEEE Standard 100- 1992. 5th edition. 1993. $150.00.

Illustrated Dictionary of Electronics. Stan Gibilsco. TAB Books, P.O. Box 40, Blue Summit, PA 17294-0850. (717) 794- 2191. (800) 233-1128. 7th edition. 1994. $34.95.

McGraw-Hill Encyclopedia of Science and Technology. McGraw- Hill Book Company, Inc., 1221 Avenue of the Americas, New York, NY 10020. (212) 512-2000. (800) 262-4729. 7th edition. 20 volume set. 1992. $1900.00.

GENERAL WORKS

Active Network Analysis; Problems and Solutions. W. K. Chen. World Scientific Publishing Company, Inc., 1060 Main Street, River Edge, NJ 07661. (201) 487-9655. (800) 227- 7562. 1993. $28.00.

Audio Electronics. John L. Hood. Butterworth-Heinemann, 313 Washington Street, Newton, MA 02158. (618-928-2500. (800) 366-2665. 1995. $29.95 in paper.

Electronics Circuits: Amplifiers and Gates. D. V. Bugg. IOP Publishing, Public Ledger Building, Suite 1035, Independence Square, Philadelphia, PA 19106. (215) 627- 0880. (800) 358-4677. 1991. $129.00.

Fundamentals of Distributed Amplification. Thomas T. Wong. Artech House, 685 Canton Street, Norwood, MA 02062. (617) 769-9750. (800) 225-9977. 1993. $55.00.

Integrated Electronics: Operational Amplifiers and Linear Ic's. Joseph J. Carr. SCP: Third World Literature Publishing House, P.O. Box 482, Lithonia, GA 30058-0482. (404) 785-7725. 1990. $52.00.

The Physics of Sound. Richard E. Berg and David G. Stork. Prentice-Hall, 113 Sylvan Avenue, Route 9W, Englewood Cliffs, NJ 07632. (201) 592-2000. (800) 922-0579. 2nd edition. 1994. $63.00.

Practical Guide To Electronic Amplifiers. John D. Lenk. Prentice-Hall, 113 Sylvan Avenue, Route 9W, Englewood Cliffs, NJ 07632. (201) 592-2000. (800) 922-0579. 1991. $36.00.

HANDBOOKS AND MANUALS

Active Electronic Component Handbook. Charles A. Harper and Harold C. Jones. McGraw-Hill Book Company, 1221 Avenue of the Americas, New York, NY 10020. (212) 512-2000. (800) 262-4729. 2nd edition. 1996. $79.50.

American Electricians Handbook. Terrell Croft and Wilford I. Summers. McGraw-Hill Book Company, 1221 Avenue of the Americas, New York, NY 10020. (212) 512-2000. (800) 262-4729. 12th edition. 1992. $77.50.

CRC Handbook of Chemistry and Physics. David R. Kide, editor. CRC Press, Inc., 2000 Corporate Boulevard, NW, Boca Raton, FL 33431. (407) 994-0555. (800) 272-7737. 77th edition. 1996. $99.95.

Electronics Handbook. Jerry C. Whitaker. CRC Press, Inc., 2000 Corporate Boulevard, NW, Boca Raton, FL 33431. (407) 994-0555. (800) 272-7737. 1996. $120.00.

Handbook For Sound Engineers: The New Audio Cyclopedia. Glen M. Ballou. Macmillian Publishing Company, Inc,. 200 Old Tappan Road, Old Tappan, NJ 07675. (800) 233-2336. 2nd edition. 1991. $99.95.

Standard Handbook For Electrical Engineers. Donald Fink. McGraw-Hill Publishing Company, 1221 Avenue of the Americas, New York, NY 10020. (212) 512-2000. (800) 262- 4729. 13th edition. 1996. $110.00.

ONLINE DATABASES AND CD-ROMS

Compendex Plus . Engineering Information, Inc., 345 East 47th Street, NE w York, NY 10017. (212) 705-7600 or (800) 221-1044. Cont ains citations with abstracts to worldwide li terature in engineering and technology, from 1970 to present. Available on online BRS,(800) 289-4277, DIALOG, (800) 334-2564, ORBIT (800) 456-7248, and STN International, FIZ Karlsruhe, P.O. Box 2465, W-7500, Karlsruhe 1, Germany, online services. Also available on CD-ROM. Inquire as to cost and availability.

Current Contents Search. Institute for Scientific Information, 3501 Market Street, Philadelphia, PA 19104. (215) 386 -0100. FAX (215) 386-6362. Contains citations to articles listed in the table of contents of science and technology journals. Also articles in social sciences and life sciences journals. Available on BRS, (800) 289-4277, DIALOG, (800) 334-2564, online services. Inquire as to cost and availability.

Dissertation A Bstracts Online. University Microfilms International, 300 North Zeeb Road, Ann Arbor, MI 48106. (800) 521-

0600 or (313) 761-4700. Scope includes virtual ly all doctoral dissertations accepted at accredited American institutions from 1861 to present in 252 subject areas. Available on BRS, (800) 289-4277, DIALOG, (800) 334-2564, and OCLC EPIC, (800) 848-5878, online services. Also available on CD-ROM. Inquire as to cost and availability.

INSPEC. Institution of Electrical Engineers, Michael Faraday House, Six Hills Way, Stevenage, Herts. SG1 2AY, Englan d. Telephone: 0438 313311 or FAX 0438 742840. Contains citations to the worldwide literature of physics, electronics and electrical engineering, computer technology, and related fields. Available on BRS, (800) 289-4277, DIALOG, (800) 334-2564, ORBIT, (800) 456-7248, and STN International, FIZ Karlsruhe , P.O. Box 2465, W-7500, Karlsruhe 1, Germany, online services. Inquire as to cost and availability.

NTIS Bibliographic Database. National Technical Information Service, 5285 Port Royal Road, Springfield, VA 22161 . (703) 487-4929 or FAX (703) 321-8199. Broad covera ge of government-sponsored science and technology research reports, 1964 to present. Available on BRS,(800) 289-4277, DIALOG, (800) 334-2564, ORBIT, (800) 456-7248, and STN International, FIZ Karlsruhe, P.O. Box 2465, W-7500, Karlsruhe 1, Germany, online services. Also available on CD-ROM. Inquire as to cost and availability.

Physics Briefs. American Institute of Physics, 335 East 45th Street , New York, NY 10017. (212) 661-9260 or FAX (212) 661-2036. Contains citations with abstracts of the literature of physics and related fields, 1979 to present. Available on the STN International, FIZ Karlsruhe, P.O. Box 2465, W-7500, Karlsruhe 1, Germany, online service. Inquire as to cost and availability.

SCISEARCH. Institute for Scientific Information, 3501 Market Street, Philadelphia, PA 19104. (800) 523-1850 or (215) 386-0100. Broad multidisciplinary title and author index to the international literature of science and technology, 1974 to present. Available on DIALOG, (800) 334-2564, and ORBIT, (800) 456-7248, online services. Also available on CD-ROM. Inquire as to cost and availability.

WILSONLINE. H.W. Wilson Company, 950 University Avenue, Bronx, NY 10452. (800) 367-6770 or (212) 588-8400. Makes available onlin e versions of the printed H.W. Wilson indexes includin g Applied Science and Technology Index, Business Periodicals Index, General Science Index, and Readers' Guide to Periodical Literature. Period covered is generally 1983 to present. Available on BRS, (800) 289-4277, DIALOG, (800) 334-2564, and OCLC EPIC, (800) 848-5878, online services. Also available on CD-ROM. Inquire as to cost and availability.

PERIODICALS

Acoustical Society of America Journal. American Institute of Physics, 500 Sunnyside Boulevard, Woodbury NY 11797-2999. (519) 576-2270. 1929 to present. Monthly. $760.00 per year. ISSN: 0001-4966.

Applied Acoustics. Elsevier Science, 660 White Plains Road, Tarrytown, NY 10591-5153. (914) 524-9200. FAX (914) 333-2444. 1968 to present. Monthly. $770.00 per year. ISSN: 0003-682X.

Audio Engineering Society. JOURNAL. Audio Engineering Society, 60 East 42nd Street, New York, NY 10165. (212) 661-2355. 1953 to present. 10 issues per year. $125.00 per year. ISSN: 0004-7554.

DB, the Sound Engineering Magazine. 203 Commack Road, Suite 1010, Commack, NY 1725. (516) 586-6530. 1967 to present. $15.00 per year. ISSN: 0011-7145.

IEEE - Signal Processing Magazine. Institute of Electrical and Electronics Engineers, Inc., Box 1331, 445 Hoes Lane, Piscataway, NJ 08855-1331. (908) 981-0060. 1984 to present. Quarterly. $70.00 per year. ISSN: 1053-5888.

IEEE Transactions On Ultrasonics, Ferroelectrics and Frequency Control. Institute of Electrical and Electronics Engineers, Inc., Box 1331, 445 Hoes Lane, Piscataway, NJ 08855-1331. (908) 981-0060. 1954 to present. Bi-monthly. $230.00 per year. ISSN: 0885-3010.

Journal of Low Frequency Noise and Vibration. Multi-Science Publishing Co. Ltd., Box 176 Avenel, NJ 070001. 1982 to present. Quarterly. L102 per year. ISSN: 0263-0923.

Journal of Sound and Vibration. Academic Press Ltd.,6277 Sea Harbor Drive, Orlando, FL. (800) 321-5068. 1964 to present. 50 issues per year. L1345 per year. ISSN: 0022- 460X.

Ultrasonics. Butterworth-Heinemann, 313 Washington Street, Newton, MA 02158. (618-928-2500. (800) 366-2665. 1963 to present. Bimonthly. L245 per year. ISSN: 0041-624X

RESEARCH CENTERS AND INSTITUTES

Center For Communications and Signal Processing. North Carolina State University. 320 Daniels Hall, Box 7914, Raleigh, NC 27695. (919) 737-3015.

Communications and Signal Processing Laboratory. University of Michigan, 4242 EECS Building, North Campus, Ann Arbor Michigan 48109-2122. (313) 764-5210. FAX (313) 763-1503.

Electronics Research Center. University of Texas at Austin, 132 Engineering Science Building, Austin, TX 78712. (512) 471-3954.

Laboratory For Electromagnetic and Electronic Systems. Massachusetts Institute of Technology, 77 Massachusetts Avenue, Cambridge, MA 02139. (617) 253) 4631.

ANALYTIC GEOMETRY

See also: ALGEBRA, CALCULUS, GEOMETRY, STATISTICS

ABSTRACT SERVICES AND INDEXES

Applied Mechanics Reviews: An Assessment of World Literature in Engineering Sciences. American Society of Mechanical Engineers, 345 East 47th Street, New York, NY 10017. (212) 705-7703. 1948 to present. Monthly. $360.00 ISSN: 0003-6900.

ANALYTIC GEOMETRY

Ency. of Physical Sciences and Engineering Info. Sources

Applied Science and Technology Index; A cumulative subject index to English language periodicals in the fields of aeronautics and space science, computer technology, chemistry, construction industry, energy and related areas. H.W. Wilson Co., 950 University Avenue, Bronx, NY 10452. (212) 588-8400. (800) 367-6770. FAX (718) 590-1617. From 1958 to present. Monthly. Inquire about cost and availability. Also available on CD-ROM and online. ISSN: 0003-6986.

Compactmath - Compact Mathematics Library. Cumulative CD-ROM edi tion of Zentralblatt fuer Mathematik - Mathematics Abstracts. Springer-Verlag New York, Inc. 44 Hartz Way. Secaucus. NJ. 07096-2491. (201) 348-4033. FAX (201) 348-4505. 1993 to present. Annual. Available only on CD-ROM. Inquire regarding cost and availability. ISSN: 0938-3174.

Compumath Citation Index. Institute for Scientific Information, 3501 Market Street, Philadelphia, PA 19104. (800) 523-1850 or (215)-386-0100. FAX (215) 386-2991. Three times per year. $1955.00. Also available online and on CD-ROM. Inquire regarding cost and availablity. ISSN: 0730-6199.

Current Mathematical Publications. American Mathematical Soci ety. P.O. Box 1571, Annex Station, Providence, RI 02901-9930. (800) 556-7774 or (401) 455-4000. FAX (401) 331-3842. 1969 to present. Seventeen times per year. $377.00 per year. Available online from BRS, DIALOG, European Space Agency. Also available on CD-ROM. Inquire regarding cost and availability. ISSN: 0361-4794.

General Science Index. H.W. Wilson Company, 950 University Avenue, Bronx, NY 10452. (212) 588-8400. (800) 367-6770. FAX (718) 590-1617. From 1978 to present. Ten issues per year; quarterly and annual cumulations. Service basis. Available on CD-ROM and online. Inquire regarding cost and availability. ISSN: 0162-1963.

Mathematical Reviews: A Review Journal Covering the World Literature of Mathematical Research. American Mathematical Society, P.O. Box 1571, Annex Stat ion, Providence, RI 02901-9930. (800) 556-7774 or (401) 455-4000. 1940 to present. Monthly. $4594.00 per year. Also available via network (MathSciNet), online and on CD-ROM. Inquire regarding cost and availability. ISSN: 0025-5629.

NTIS Alerts: Mathematical Sciences. U.S. National Technical Information Service, 5285 Port Royal Road, Springfield, VA 22161. (703) 487-4650. FAX (703) 321-8547. Weekly. $140.00.

Science Citation Index. SCI. Institute for Scientific Information, 3501 Market Street, Philadelphia, PA 19104. (215) 386-0100. (800) 523-1850. FAX (215) 386-2991. 1961 to present. Six issues per year, plus annual cumulation. $11650.00 per year. Also available o nline and on CD-ROM. Inquire about price and availability. ISSN: 0036-827X.

Zentralblatt Fuer Mathematik Und Ihre Grenzebiete/mathematics Abstracts. Heidelberger Akademie der Wissenschaften/Springer-Verlag New York, Inc., 44 Hartz Way, Seacaucus, NJ 07096-2491. (201) 348-4033. FAX (201) 348-4505. 1931 to present. 30 issues per year. DM 8340. ISSN: 0044-4235. Also available online and on CD-ROM. Inquire regarding cost and availability.

ANNUAL REVIEWS AND YEARBOOKS

Ergebnisse Der Mathematik Und Ihrer Grenzgebiete. Neue folge. Springer-Verlag New York, Inc. 175 Fifth Avenue, New York,

NY 10010. (212) 460-1500. FAX (212) 473-6272. 1953 to present. Irregular. Price varies; inquire. ISSN: 0071-1136.

Lecture Notes in Mathematics. Springer-Verlag New York, Inc. 44 Hartz Way. Secaucus. NJ. 07096-2491. (201) 348-4033. FAX (201) 348-4505. Irregular. Price varies; inquire.

ASSOCIATIONS AND PROFESSIONAL SOCIETIES

American Mathematical Society. P.O. Box 6248, Providence, RI 02940. (401) 455-4000. FAX (401) 331-3842

Association for Women in Mathematics. 4114 Computer and Space Science Building, University of Maryland, College Park, MD 20742-2461. (301) 405-7892.

Mathematical Association of America. 1529 18th Street, NW, Washington, DC 20036. (202) 387-5200. FAX (202) 265-2384.

Society For Industrial and Applied Mathematics. 3600 University City Science Center, Philadelphia, PA 19104-2688. (215) 382-9800. FAX (215) 386-

BIBLIOGRAPHIES

Mathematical Book Review Index, 1800-1940. Louise S. Grinstein. Garland Publ. Fifth Avenue, Suite 2500, New York, NY 10022-8101. (212) 751-7447. 1992. $72.00. ISBN: 0-8240-4114-3.

Bibliography of Mathematical Works Printed in America Through 1850. Louis C. Karpinski. Supplements 1 & 2, Bernard I. Cohen, editor. Ayer Company Publ. Inc. Lower Mill Road, North Stratford, NH 03590. (603) 922-5105. (800) 282-5413. Reprint edition. 1980. $66.95. ISBN: 0-405-12553-4.

International Mathematical Olympiads 1978-1985. Murray S. Klamkin. National Council of Teachers of Mathematics. 1906 Association Drive, Reston, VA 22091. (703) 620-9840. (800) 235-7566. 1987. $11.75. ISBN: 0-88385-631-X.

DIRECTORIES AND BIOGRAPHICAL SOURCES

Assistantships and Graduate Fellowships in the Mathematical Sciences. American Mathematical Society, P.O. Box 6248, Providence, RI 02940. (401) 455-4000. (800) 321-4267. Annual. $19.00. ISSN: 1040-7650.

Combined Membership List. American Mathematical Society, P.O. Box 6248, Providence, RI 02940. (401) 455-4000. = (800) 321-4267. Annual. $50.00. ISSN:

Mathematical Science Professional Directory. American Mathematical Society, P.O. Box 6248, Providence, RI 02940. Annual. $45.00. ISBN: 0-8218-0173-2.

Research Centers Directory. Gale Research, 835 Penobscot Building, Detroit, MI 48226-4094. (313) 961- 2242. (800) 877-4253. 20th edition. 1995. Annual. $485.00. ISSN: 0080-1518.

World Directory of Mathematicians. G.D. Mostow, editor. International Mathematical Union, Helsinki. 1958 to present. 9th revised edition. 1990. $40.00. Available from: American Mathematical Society, P.O. Box 6248, Providence, RI 02940. (401) 455-4000. ISSN: 0512-2740.

ENCYCLOPEDIAS AND DICTIONARIES

Beyond Numeracy: An Uncommon Dictionary of Mathematics. John A. Paulos. Alfred A. Knopf Inc., 201 East 50th Street, New York, NY 10022. (800) 733-3000. 1991. $21.50. ISBN: 0-394-58640-9.

Concise Oxford Dictionary of Mathematics. Christopher Clapham. Oxford University Press, 200 Madison Avenue, New York, NY 10016. (212) 725-6000. (800) 334-4249. 1990. $10.95. ISBN: 0-19-286103-4.

Encclopedia of Mathematics. An updated and annotated translation of the Soviet Mathematical Encyclopedia. Michael Hazewinkel, editor. Kluwer Academic Publishers. 101 Philip Drive, Assinippi Park, Norwell, MA 02061. (617) 871-6600. 1987 - . 10 volumes. $147.00, volume 1. 1-55608-010-7 set.

Encyclopedia of Mathematics and Its Applications. Addison-Wesley Publishing Co., Inc., 1 Jacob Way, Reading, MA 01867. (617) 944-3700. (800) 447-2226. 1976 to present. irregular. Volumes are individually priced.

Encyclopedic Dictionary of Mathematics. Kiyosi Ito, editor. MIT Press, 55 Hayward Street, Cambridge, MA 02142. (617) 625-8569. 2nd edition, reprinted. 1993. 2 volumes. $70.00. ISBN: 0-262-59020-4.

Harpercollins Dictionary of Mathematics. E. J. Borowski and J. J. Borwein. HarperCollins Publ. 10 East 53rd Street, New York, NY 10022-5299. (800) 331-3761. 1990. $10.95. ISBN: 0-19-286103-4.

Mathematics Dictionary. Robert C. James, editor. Van Nostrand Reinhold, 115 Fifth Avenue, New York, NY 10003. (800)-842-3636. 5th edition. 1992. $42.95. ISBN: 0-442-00741-8.

Penguin Dictionary of Mathematics. J. Daintith and R. D. Nelson. Viking Penguin, 375 Hudson Street, New York, NY 10014-3657. (800) 331-4624. 1989. $12.00. ISBN: 0-14-051119-9

GENERAL WORKS

Analytic Geometry. Gordon Fuller. Addison-Wesley Publishing Company, Inc. 1 Jacob Way, Reading Way MA 01867. (617) 944-3700. 7th edition. 1992. $51.75. ISBN: 0-201-13484-5.

Calculus With Analytic Geometry. Robert Ellis and Denny Gulick. H B Coll Pubs. 4th edition. 1990. $66.50. ISBN: 0-15-505587-5. accompanying disc. $5.00. ISBN: 0-15-505688-3.

Conic Sections. George Salmon. Chelsea Publishing Inc., 15 East 26th Street, New York, NY 10010. (212) 889-8095. 6th edition. $19.95. ISBN: 0-8284-0099.

Fundamentals of Mathematics. H. Behnke et al, editors. MIT Press, MIT Press, 55 Howard Street, Cambridge, MA 02142. (617) Volume 1, Foundations of Mathematics: Real Number Systems and Algebra; Volume 2, Geometry; Volume 3, Analysis. 1984; reprint edition. $75.00 ISBN: 0-685-07931-7.

Introduction To Complex Analytic Geometry. Stanislaw Lojasiewicz. HarperCollins Publ. 10 East 53rd Street, New York, NY 10022-5299. (800) 331-3761. 1991. $148.00. ISBN: 0-8176-1935-6.

Mathematical Description of Shape and Form. E. A. Lord and C. B. Wilson. Prentice-Hall, 113 Sylvan Avenue, Route 9W, Englewood Cliffs, NJ 07632. (201) 592-2000. (800) 922-0579. 1986. $63.95. ISBN: 0-470-20043-X.

Mathematica: a System for Doing Mathematics by Computer. Stephen Wolfram. Addison-Wesley Publishing Co., Inc., 1 Jacob Way, Reading, MA 01867. (617) 944-3700. (800) 447-2226. 2nd edition. 1991. ISBN: 0-201-51

The Maple Handbook. Darren Redfern. Springer-Verlag New York, Inc., 175 Fifth Avenue, New York, NY 10010. (212) 460-1500. (800) 777-4643. FAX (212) 473-6272. 2nd edition. 1995. $29.00. 0-387-94331-5.

Geometry Problem Solver. Research & Education Association, 61 Ethel Road, West Piscataway, NJ 08854. (908) 819-8880. 1994; revised edition. $29.95. ISBN: 0-87891-549-4.

Writing Mathematics Well. Leonard Gillman. Mathematical Association of America, 1987. $8.00. ISBN: 0-88385-443-0.

HANDBOOKS AND MANUALS

Handbook of Mathematical Sciences. William H. Beyer, editor. CRC Press, 2000 Corporate Boulevard, Boca Raton, FL 33431. (800) 272-7737. 6th edition. 1987. $91.95. ISBN: 0-8493-0656-6.

Handbook of Writing For the Mathematical Sciences. Society for Industrial and Applied Mathematics, 3600 University City Science Center, Philadelphia, PA 19104-2688. (215) 382-9800. FAX (215) 386-7999. 1993. $21.50. ISBN: 0-89871-314-5.

Standard Mathematical Tables and Formulas. William H. Beyer, editor. CRC Press, 2000 Corporate Boulevard, Boca Raton, FL 33431. (800) 272-7737. 29th edition. 1991. $41.95. ISBN: 084-930-6299.

Handbook of Applicable Mathematics. Walter Ledermann, editor. John Wiley & Sons, Inc., 605 Third Avenue, New York, NY 10158-0012. (212) 850-6000. (800) 225-5945. 1985. $1496.00. 6 volumes. ISBN: 0-471-90804-5.

Handbook of Applied Mathematics; Selected Results and Methods. Carl E. Pearsen, editor. Chapman & Hall, 1 Penn Plaza, New York NY 10119. (212) 564-1060. 1990. $46.95. ISBN: 0-442-00521-0.

Handbook of Mathematical Functions With Formulas, Graphs, and Mathmetical Tables. Milton Abramowitz and Irene A. Stegun, editors. U.S. National Bureau of Standards. U.S. Government Printing office, Washington, D.C. 10th edition. 1972. $43.00. ISBN: 0-16-000202-8.

Handbook of Mathematics. I. N. Bronshtein and K. A. Semedyayev. Van Nostrand Reinhold, 115 Fifth Avenue, New York, NY 10003. (800)-842-3636. 20th edition. $59.95. ISBN: 0-442-21171-6.

ONLINE DATABASES AND CD-ROMS

Compendex Plus. Engineering Information, Inc., 345 East 47th Stre et, New York, NY 10017. (212) 705-7600 or (800) 221-1044 . Contains citations with abstracts to worldwide litera ture in engineering and technology, from 1970 to present. Available

ANALYTIC GEOMETRY

Ency. of Physical Sciences and Engineering Info. Sources

on online BRS,(800) 289-4277, DIALOG, (800) 334-2564, ORBIT (800) 456-7248, and STN International, FIZ Karlsruhe, P.O. Box 2465, W-7500, Karlsruhe 1, Germany, online services. Also available on CD-ROM. Inquire as to cost and availability.

Dissertation Abstracts. University Microfilms International, 300 North Zeeb Road, Ann Arbor, MI 48106. (800) 521-0600 or (313) 761-4700. Scope includes virtually all doctoral dissertations accepted at accredited American institutions from 1861 to present in 252 subject areas. Available on BRS,(800) 289-4277, DIALOG,(800) 334-2564, and OCLC EPIC,(800) 848-5878, online services. Also available on CD-ROM. Inquire as to cost and availability.

INSPEC. Institution of Electrical Engineers, Michael Faraday House, Six Hills Way, Stevenage, Herts. SG1 2AY, England. Telephone: 0438 313311 or FAX 0438 742840. Contains citations to the worldwide literature of physics, electronics and electrical engineering, computer technology, and related fields. Available on BRS,(800) 289-4277, DIALOG, (800) 334-2564, ORBIT, (800) 456-7248,and STN International, FIZ Karlsruhe, P.O. Box 2465, W-7500, Karlsruhe 1, Germany, online services. Inquire as to cost and availability.

Mathsci and Mathscinet. American Mathematical Society, P.O. Box 6248, Providence, RI 02940. (800) 321-4667 or (401) 455-4000 or FAX (401) 331-3842. Scope includes pure and applied mathematics and related areas of physics, statistics, engineering, computer science, and operations research literature since 1959. Available on DIALOG, (800) 334-2564, online service. Also available on CD-ROM and MATHSCINET, on the Internet. Inquire as to cost and availability.

NTIS Bibliographic Database. National Technical Information Service, 5285 Port Royal Road, Springfield, VA 22161. (703) 487-4929 or FAX (703) 321-8199. Broad coverage of government-sponsored science and technology research reports, 1964 to present. Available on BRS,(800) 289-4277, DIALOG, (800) 334-2564, ORBIT, (800) 456-7248, and STN International, FIZ Karlsruhe, P.O. Box 2465, W-7500, Karlsruhe 1, Germany, online services. Also available on CD-ROM. Inquire as to cost and availability.

SCISEARCH. Institute for Scientific Information, 3501 Market Street, Philadelphia, PA 19104. (800) 523-1850 or (215) 386-0100. Broad multidisciplinary title and author index to the international literature of science and technology, 1974 to present. Available on DIALOG,(800) 334-2564, and ORBIT,(800) 456-7248, online services. Also available on CD-ROM. Inquire as to cost and availability.

WILSONLINE. H.W. Wilson Company, 950 University Avenue, Bronx, NY 10452. (800) 367-6770 or (212) 588-8400. Makes available online versions of the printed H.W. Wilson indexes including Applied Science and Technology Index, Business Periodicals Index, General Science Index, and Readers' Guide to Periodical Literature. Period covered is generally 1983 to present. Available on BRS,(800) 289-4277, DIALOG,(800) 334-2564, and OCLC EPIC,(800) 848-5878, online services. Also available on CD-ROM. Inquire as to cost and availability.

PERIODICALS

American Journal of Mathematics. Johns Hopkins University Press, 2715 North Charles Street, Baltimore, MD 21218. (410) 516-6987. FAX (410) 516-6968. 1878 to present. Bimonthly. $162.00 per year. ISSN: 0002-9327.

American Mathematical Monthly. Mathematical Association of America. 1529 Eighteenth Street, NW, Washington, DC 20036. (202) 387-5200. 1894 to present. Ten issues per year. $160.00. ISSN: 0002-9890.

College Mathematics Journal. Mathematical Association of America, 1529 18th Street NW, Washington D.C. 20036. (202) 387-5200. 1970 to present. Five issues per year. $98.00. ISSN: 0746-8342.

Discrete and Computational Geometry. Springer-Verlag NewYork, Inc., 175 Fifth Avenue, New York, NY 10010. (212) 460- 1500. (800) 777-4643. FAX (212) 473-6272. 1986 to present. 8 issues per year. $254.00 per year. ISSN: 0179-5376.

Geometriae Dedicata. Kluwer Academic Publishers, Box 358, Accord Station, Hingham, MA 02018-0358. (617) 871-6600. FAX (617) 871-6528. 1972 to present. 15 issues per year. $872.50 per year. ISSN: 0046-5755.

Geometric and Functional Analysis. Birkhauser- Verlag/ Springer-Verlag New York, Inc., 175 Fifth Avenue, New York, NY 10010. (212) 460-1500. (800) 777-4643. FAX (212) 473-6272. 1992. Bimonthly. $278.00 per year. ISSN: 1016-443X.

I M A Journal of Applied Mathematics. Oxford University Press, 2001 Evans Road, Cary, NC 27513. (919) 677-0977. 1981 to present. Bimonthly. $410.00 per year. ISSN: 0272- 4960.

Journal of Geometry. Birkhauser Verlag/Springer-Verlag New York, 175 Fifth Avenue, New York, NY 10010. (212) 460-1500. FAX (212) 473-6272. (800) 777-4643. 1971 to present. Bimonthly. $237.00. ISSN: 0047-2468.

Mathematical Intelligencer. Springer-Verlag, 175 Fifth Avenue, New York, NY 10010. (212) 460-1500. FAX (212) 473- 6272. (800) 777-4643. 1978 to present. Quarterly. $39.00 per year. ISSN: -343-6993.

Mathematics Magazine. Mathematical Association of America. 1529 Eighteenth Street, NW, Washington, DC 20036. (202) 387-5200. 1926 to present. Bimonthly. $80.00 per year. ISSN: 0025-570X.

Quarterly Journal of Mechanics and Applied Mathematics. Oxford University Press, 2001 Evans Road, Cary, NC 27513. (919) 677-0977. 1948 to present. Quarterly. $214.00 per year. ISSN: 0033-5614

Siam Journal On Applied Mathematics. Society for Industrial and Applied Mathematics, 3600 University City Science Center, Philadelphia, PA 19104-2688. (215) 382- 9800. FAX (215) 386-7999. 1953 to present. Bimonthly. $242.00 per year. ISSN: 0036-1399.

Siam Journal On Discrete Mathematics. Society for Industrial and Applied Mathematics, 3600 University City Science Center, Philadelphia, PA 19104-2688. (215) 382- 9800. FAX (215) 386-7999. 1980 to present. Quarterly. $220.00 per year. ISSN: 0895-4801.

Siam Journal On Mathematical Analysis. Society for Industrial and Applied Mathematics, 3600 University City Science Center, Philadelphia, PA 19104-2688. (215) 382- 9800. FAX (215) 386-7999. 1970 to present. Bimonthly. $352.00 per year. ISSN: 0036-1410.

Siam Journal on Numerical Analysis. Society for Industrial and Applied Mathematics, 3600 University City Science Center, Philadelphia, PA 19104-2688. (215) 382-9800. FAX (215) 386-7999. Bimonthly. $255.00 per year. ISSN: 0036-1429

Siam Review. Society for Industrial and AppliedMathematics, 3600 University City Science Center, Philadelphia, PA 19104-2688. (215) 382-9800. 1959 to present. Quarterly. $148.00 per year. ISSN: 0036-1445.

RESEARCH CENTERS AND INSTITUTES

Center for Applied Mathematics. University of Georgia, Athens, GA 30602. (404) 542-3491.

Center For Mathematical Sciences. University of Wisconsin-Madison. 610 Walnut Street, Madison, WI 53705. (608) 263-2696. FAX (608) 263-2841.

Center For Pure and Applied Mathematics. University of California, Berkeley. 977 Evans Hall, Berkeley, CA 94720. (415) 642-0116. FAX (415) 642-6726.

Design Research Center. Rensselaer Polytechnic Institute. CII 7015, Troy, NY 12180-3590. (518) 266-6751.

Project in Geometry and Physics. University of California, San Diego, Department of Mathematics, C-012, LA Jolla, CA 92093-0112. (619) 534-2218.

ANALYTICAL CHEMISTRY

See also: CHEMICAL ENGINEERING, CHEMISTRY, PHYSICAL
CHEMISTRY, SPECTROSCOPY

ABSTRACT SERVICES AND INDEXES

Analytical Abstracts. Royal Society of Chemistry, Information Services, Thomas Graham House, Science Park, Milton Road, Cambridge, CB4 4WF, England. Contains citations, mostly with abstracts, of the worldwide literature on analytical chemistry, from 1954 to present. Monthly. $636.00 per year. Also available online.

Applied Science and Technology Index; A cumulative subject index to English language periodicals in the fields of aeron autics and space science, computer technology, chemistry , construction industry, energy and related areas. H. W. Wilson Co., 950 University Avenue, Bronx, NY 10452-9978. (212) 588-8400. (800) 367-6770. FAX (718) 590-1617. From 1958 to present. Monthly. Inquire about cost and availability. Also available online (BRS and WILSONLINE) and on CD-ROM. ISSN: 0003-6986.

Chemical Abstracts. Chemical Abstracts Service. 2540 Olentangy River Road, Box 3012, Columbus, OH 43210-0012. (614) 447-3600. FAX (614) 447-3713. 1907 to present. Weekly. $16,800.00 per year. Available online and on CD- ROM. CA is also available in five section groupings. Inquire regarding cost and availability.

Current Contents: Physical, Chemical and Earth Sciences. Institute for Scientific Information, 3501 Market Street, Philadelphia, PA 19104. (215) 386-0100. FAX (215) 386-2291. From

1961 to present. Weekly. $442.00 per year. Also available on CD-ROM and online. Inquire regarding cost and availability. ISSN: 0163-2574.

General Science Index. H.W. Wilson Company, 950 University Avenue, Bronx, NY 10452. (212) 588-8400. (800) 367-6770. FAX (718) 590-1617. From 1978 to present. Ten issues per year;quarterly and annual cumulations. Service basis. Available on CD-ROM and online. Inquire regarding cost and availability. ISSN: 0162-1963.

Physics Abstracts. INSPEC. Section A, Science Abstracts. Institution of Electrical Engineers (IEE), London, United Kingdom. Available from: INSPEC/IEEE - Institute of Electrical and Electronic Engineers, Box 1331, 445 Hoes Lane, Piscataway, NJ 08855-1331. (908) 562-5549. 1898 to present. 24 issues per year. $2835.00 per year. ISSN: 0036-8091. Also available online and on CD-ROM.

Science Citation Index. SCI. Institute for Scientific Information, 3501 Market Street, Philadelphia, PA 19104. (215) 386-0100. (800) 523-1850. FAX (215) 386-2991. 1961 to present. Six issues per year, plus annual cumulation. $11650.00 per year. Also available online and on CD-ROM. Inquire about price and availability. ISSN: 0036-827X.

ASSOCIATIONS AND PROFESSIONAL SOCIETIES

American Chemical Society, 1155 16th Street, NW, Washington, DC 20036. (202) 872-4600.

American Institute of Chemists. 501 Wythe Street, Alexandria, VA 22314-1917. (703) 836-2090. FAX (703) 836- 2091.

Association of Consulting Chemists and Chemical Engineers, 50 East 41st Street, Suite 92, New York, NY 10017. (212) 684-6255.

Association of official Analytical Chemists. 2200 Wilson Blvd, Suite 400, Arlington, VA 22001. (703) 522-3032.

BIBLIOGRAPHIES

Chemical Information Sources. Gary Wiggins. McGraw-Hill Publishing Company, 1221 Avenue of the Americas, New York, NY 10020. (800) 262-4729 or (212) 512-3825. 1991. $42.50.

Information Sources in Chemistry. R.T. Bottle and J.F.B Rowland, editors. R.R. Bowker Inc., 121 Chanlon Road, New Providence, NJ 07974. (800) 521-8110 or (908) 464-6800. Fourth edition. 1993. $75.00.

Scientific and Technical Books and Serials in Print; An Index To Literature in Science and Technology. R.R. Bowker Inc., 121 Chanlon Road, New Providence, NJ 07974. (908) 464-6800. (800) 521-8110. FAX (908) 665-3502. 1972 to present. Annual. 4 volumes. 1994. $299.95. Also available on compact disc and online. ISSN: 0000-054X.

DIRECTORIES AND BIOGRAPHICAL SOURCES

American Institute of Chemists - Professional Directory. American Institute of Chemists, 501 Wythe Street, Alexandria, VA 22314-1917. (703) 836-2090. FAX (703) 836- 2091. Annual. $65.00.

ANALYTICAL CHEMISTRY

Ency. of Physical Sciences and Engineering Info. Sources

American Men and Women of Science. Physical and Biological Sciences. R.R. Bowker Inc., 121 Chanlon Road, New Providence, NJ 07974. (908) 464-6800. (800) 521-8110. 20th edition. 8 volumes. 1996. $850.00

Chemical Sources International. Mike Desing and Kurt Gandenberger, editors. Chemical Sources International, PO Box 1824,Clemson, SC 29633. (803) 646-7840. 1994. $285.00.

Chemical Week — Buyers Guide Issue. Chemical Week Associates, 888 Seventh Avenue, New York, NY 10106. (212) 621-4900. FAX (212) 621-4949. Annual, October. $50.00.

Directory of Chemistry Software 1992. Wendy Warr, Peter Willett, Geoff Downs. American Chemical Society, 1155 16th Street, NW, Washington, DC 20036. (202) 872-4600. 1992. $35.95.

Research Centers Directory. Gale Research, 835 Penobscot Building, Detroit, MI 48226-4094. (313) 961- 2242. (800) 877-4253. 20th edition. 1995. $485.00. ISSN: 0080-1518.

ENCYCLOPEDIAS AND DICTIONARIES

Academic Press Dictionary of Science and Technology. Christopher Morris, editor. Academic Press, Inc., 1250 Sixth Avenue, San Diego, CA 92101. (619) 231-0926. FAX (619) 699-6715. 1991. $115.00.

Concise Encyclopedia Chemistry. Translated and revised by Mary Eagleson. Walter de Gruyter, Inc., 200 Saw Mill River Road, Hawthorne, New York, 10532. (914) 747-0110 or Fax (914) 747-1326. 1994. $69.95.

Dictionary of Analytical Reagents. R. Lobinski, editor. Routledge, Chapman and Hall, Inc., 29 West 35th Street, New York, NY 10001-2291. (212) 244-3336. 1993. $1000.00.

Grant & Hawks Chemical Dictionary. R. A. Grant. McGraw-Hill Publishing Company, Inc., 1221 Avenue of the Americas, New York, NY 10020. (212) 512-2000. (800) 262-4729. 5th edition. 1987. $64.50.

Hawley's Condensed Chemical Dictionary. Richard J. Lewis. Van Nostrand Reinhold, 115 Fifth Avenue, New York, NY 10003. (212) 254-3232. (800) 842-3636. 12th edition. 1993. $69.95.

Kirk-Othmer Encyclopedia of Chemical Technology. John Wiley and Sons, Inc., 605 Third Avenue, New York, NY 10158. (800) 526-5368 or (212) 850-6000. Fourth edition. 1991 - . Twenty-seven volumes. $5400.00.

McGraw-Hill Encyclopedia of Science and Technology. McGraw-Hill Book, Incorporated, 1221 Avenue of the Americas, New York, NY 10020. (212) 997-3675. 17th edition. Twenty volumes. 1992. $1900.00.

GENERAL WORKS

Analytical Chemistry. Gary D. Christian. John Wiley and Sons, Incorporated, 605 Third Avenue, New York, NY 10158. (800) 526-5368 or (212) 850-6000. Fifth edition. 1993. $53.50.

Analytical Chemistry: Principles and Techniques. Larry G.Hargis. Prentice Hall (Division of Simon and Schuster), 15 Columbus Circle, New York, NY 10023. (212) 373-8500 or (800) 922-0579. 1988. $74.00.

Basic Analytical Chemistry. L. Pataki and E. Zapp. Franklin Book Co., Inc., Elkins Park, PA 19117. (215) 635- 5252. 1981. $191.00 in paper.

Chemical Derivatization in Analytical Chemistry. R.W. Frei, editor. Plenum Publishing Company, 233 Spring Street, NewYork, NY 10013. (800) 221-9369. Volume l: Chromatography. 1981. $95.00. Volume 2: Separation and Continuous Flow Techniques. 1982. $95.00.

Nonparametric Methods in Quantitative Analysis. Jean D. Gibbons. American Sciences Press, 20 Cross Road, Syracuse NY 13224-2144. 3rd edition. 1996. wfi.

Statistics For Analytical Chemistry. Jane C. Miller and James N. Miller. Prentice Hall (Division of Simon and Schuster), 15 Columbus Circle, New York, NY 10023. (212) 373-8500. (800) 922-0579. 3rd edition. 1993. $59.00.

HANDBOOKS AND MANUALS

Chemical Formulary. H. Bennett, editor. Chem Publishing Company, Inc., 80 Eighth Avenue, New York, NY 10011. (212) 255-1950. Volumes 1-30. $70.00 per volume.

Chemometrics: Chemical and Sensory Data. David R. Burgard and James T. Kuznicki. CRC Publishers, Inc., 2000 Corporate Blvd., N.W., Boca Raton, FL 33431. (407) 994-0555 or (800) 333-8300. 1990. $120.00.

Chemical Reference Handbook. Gordon Press Publications, PO Box 459, Bowling Green Station, New York, NY 10004. (718) 624-8419. 1995. $260.00.

Chemical Safety International Reference Manual. Mervyn Richardson, editor. VCH Publications, Inc., 220 East 23rd Street, Suite 909, New York, NY 10010-4606. (212) 683-8333. (800) 422-8824. 1994. $145.50.

Comprehensive Guide To To the Hazardous Properties of Chemical Substances. Pradyot Patnaik. Van Nostrand Reinhold, 115 Fifth Avenue, New York, NY 10003. (212) 254- 3232. (800) 842-3636. 1992. $99.95.

CRC Handbook of Chemistry and Physics. David R. Kide, editor. CRC Press Inc., 2000 Corporate Blvd., NW, Boca Raton, FL 33431. (407) 994-0555. (800) 333-8300. 77th edition. 1996. $99.95.

Guide To Basic Chemical Compounds. D.R. Lide, Jr. CRC Publishers, Inc., 2000 Corporate Blvd., N.W., Boca Raton, FL 33431. (407) 994-0555 or (800) 333-8300. 1993. $120.00.

Handbook of Hazardous Laboratory Chemicals: Information and Disposal. Margaret-Ann Armour. CRC Press, Inc., 2000 Corporate Boulevard, NW, Boca Raton, FL 33431. (407) 994-0555. (800) 272-7737. 1991. $104.95.

Laboratory Handbook of Materials, Equipment, and Techniques. Gary S. Coyne. Prentice Hall (Division of Simon and Schuster), 15 Columbus Circle, New York, NY 10023. (212) 373-8500 or (800) 922-0579. 1992. $45.00.

Lange's Handbook of Chemistry. John A. Dean, editor. McGraw-Hill Publishing Company, 1221 Avenue of the Americas, New York, NY 10020. (800) 262-4729 or (212) 512-3825. Fourteenth editon. 1992. $79.50.

ONLINE DATABASES AND CD-ROMS

Analytical Abstracts Online. Royal Society of Chemistry, Information Services, Thomas Graham House, Science Park, MiltonRoad, Cambridge, CB4 4WF, England. Contains citations, mostly with abstracts, of the worldwide literature on analytical chemistry, from 1980 to present. Available on DIALOG and STN online services. Inquire as to cost and availability.

CA Search. Chemical Abstracts Service, P.O. Box 3012, Columbus, OH 43210-0012. (614) 447-3600. (800) 848-6533. FAX (614) 447-3709. Very comprehensive guide to worldwide chemical literature and related fields. 1972 to present. Available on BRS,(800) 289-4277, DIALOG, (800) 334-2564, ORBIT (800) 456-7248, and STN International, FIZ Karlsruhe, P.O. Box 2465, W-7500, Karlsruhe 1, Germany, online services. Inquire as to cost and availability.

Chemical Journals of the American Chemical Society. American Chemical Society, 1155 16th Street, N.W., Washington, DC 20036. (202) 872-4381 or (800) 424-6747. Contains complete text of approximately 90,000 articles from 22 primary journals published by the American Chemical Society, from mostly 1982 to present. Available on STN International, FIZ Karlsruhe, P.O. Box 2465, W-7500, Karlsruhe 1, Germany, online service. Inquire as to cost and availability.

Dissertation Abstracts. University Microfilms International, 300 North Zeeb Road, Ann Arbor, MI 48106. (800) 521-0600 or (313) 761-4700. Scope includes virtually all doctoral dissertations accepted at accredited American institutions from 1861 to present in 252 subject areas. Available on BRS, (800) 289-4277, DIALOG, (800) 334-2564, and OCLC EPIC, (800) 848-5878, online services. Also available on CD-ROM. Inquire as to cost and availability.

INSPEC. Institution of Electrical Engineers, Michael Faraday House, Six Hills Way, Stevenage, Herts. SG1 2AY, England. Telephone: 0438 313311 or FAX 0438 742840. Contains citations to the worldwide literature of physics, electronics and electrical engineering, computer technology, and related fields. Available on BRS,(800) 289-4277, DIALOG, (800) 334-2564, ORBIT, (800) 456-7248, and STN International, FIZ Karlsruhe, P.O. Box 2465, W-7500, Karlsruhe 1, Germany, online services. Inquire as to cost and availability.

NTIS Bibliographic Database. National Technical Information Service, 5285 Port Royal Road, Springfield, VA 22161. (703) 487-4929 or FAX (703) 321-8199. Broad coverage of government-sponsored science and technology research reports, 1964 to present. Available on BRS,(800) 289-4277, DIALOG, (800) 334-2564, ORBIT, (800) 456-7248, and STN International, FIZ Karlsruhe, P.O. Box 2465, W-7500, Karlsruhe 1, Germany, online services. Also available on CD-ROM. Inquire as to cost and availability.

SCISEARCH. Institute for Scientific Information, 3501 Market Street, Philadelphia, PA 19104. (800) 523-1850 or (215) 386-0100. Broad multidisciplinary title and author index to the international literature of science and technology, 1974 to present. Available on DIALOG, (800) 334-2564, and ORBIT, (800)456-7248, online services. Also available on CD-ROM. Inquire as to cost and availability.

WILSONLINE. H.W. Wilson Company, 950 University Avenue, Bronx, NY 10452. (800) 367-6770 or (212) 588-8400. Makes available online versions of the printed H.W. Wilson indexes including Applied Science and Technology Index, Business Periodicals Index, and Readers' Guide to Periodical Literature. Period covered is generally 1983 to present. Also available on CD-ROM. Inquire as to cost and availability. Available on BRS, (800) 289-4277, DIALOG, (800) 334-2564, and OCLC EPIC, (800) 848-5878, online services. Also available on CD-ROM. Inquire as to cost and availability.

PERIODICALS

American Laboratory. American Laboratory, 30 Controls Drive, Box 870, Shelton, CT 06484-6111. (203) 926-9300. FAX (203) 926-9310. 1968 to present. Monthly. $210.00 per year. ISSN: 0044-7749.

Analyst. Royal Society of Chemistry, Thomas Graham House,Science Park, Milton Road, Cambridge, CB4 4W, England. 1876 to present. Monthly. $641.00 per year. ISSN: 0003-2654.

Analytical Chemistry. American Chemical Society, Box 3337, Columbus, OH 43210. (614) 447-3776. 1929 to present. Monthly. $415.00 per year. ISSN: 0003-2700.

Analytical Methods and Instrumentation. John Wiley and Sons, Inc., 605 Third Avenue, New York, NY 10158. (800) 526-5368 or (212) 850-6000. Bimonthly. $250.00 per year. ISSN: 1063-5246.

Association of Official Analytical Chemists Journal. Association of offical Analytical Chemists. 2200 Wilson Blvd, Suite 400, Arlington, VA 22001. (703) 522-3032. 1915 to present. Bimonthly. $208.00 per year. ISSN: 0004-5756.

Critical Reviews in Analytical Chemistry. CRC Press, Inc., 2000 Corporate Boulevard, NW, Boca Raton, FL 33431. (407) 994-0555. 1970 to present. Bi-monthly. ISSN: 1040-8347.

Fresenius' Journal of Analytical Chemistry. Springer-Verlag, 44 Hartz Way, Seaucus, NJ 07096-2491. (201) 348-4033. 1862 to present. 36 issues per year. $1556.00 per year. ISSN: 0937-0633.

International Laboratory. International Scientific Communications, Inc., 30 Controls Drive, Shelton, CT 06484-0870. (203) 926-9300. 1971 to present. 9 issues per year. L120 per year. ISSN: 0010-2164.

Journal of Chromatographic Science. Preston Publications, Inc., 7800 Merriac Avenue, Box 48312, Niles, IL 60714. (708) 965-0566. 1963 to present. Monthly. $190.00. ISSN: 0021-9665.

Mikrochimica ACTA. Springer-Verlag, 44 Hartz Way, Seaucus, NJ 07096-2491. (201) 348-4033. 1923 to present. 16 issues per year. $1175.00 per year. ISSN: 0026-3672.

Separation Science and Technology. Marcel Dekker Journals, 270 Madison Avenue, New York, NY 10016. (212) 696-9000. 1966 to present. 18 issues per year. $1450.00 per year. ISSN: 0149-6395.

Trends in Analytical Chemistry. Elsevier Science Publishing Company, Inc., Box 882, Madison Square Station, New York,

ANALYTICAL CHEMISTRY

Ency. of Physical Sciences and Engineering Info. Sources

NY 10159. (212) 989-5800. 1981 to present. Monthly. $532.00 per year. ISSN: 0165-9936.

RESEARCH CENTERS AND INSTITUTES

Analytical Chemistry Laboratory. U. S. Department of Agriculture, Beltsville Agricultural Research Center - East, Building 306, Beltsville, MD 20705. (301) 344-2495.

Barnett Institute of Chemical Analysis and Materials Science. Northeastern University, 360 Huntington Avenue, Boston, MA 02115. (617) 437-2864.

Center For Chemical Characterization and Analysis. Texas A & M Univeristy, College Station, TX 77843-3144. (409) 845-2341. FAX (409) 845-1655.

Center For Process Analytical Chemistry. University of Washington, BG-10, Seattle, WA 98195. (206) 545-2326. FAX (206) 543-6506

ANTENNAS

See also: ELECTRONIC CIRCUITS AND COMPONENTS, ELECTRONICS, ELECTRONICS ENGINEERING, MICROWAVES, RADIO, SIGNAL PROCESSING

ABSTRACT SERVICES AND INDEXES

Applied Science and Technology Index; A cumulative subject index to English language periodicals in the fields of aeronautics and space science, computer technology, chemistry, construction industry, energy and related areas. H.W. Wilson Co., 950 University Avenue, Bronx, NY 10452. (212) 588-8400. (800) 367-6770. FAX (718) 590-1617. From 1958 to present. Monthly. Inquire about cost and availability. Also available on CD-ROM and online. ISSN: 0003-6986.

Current Contents: Engineering, Technology and Applied Sci Ences. Institute for Scientific Information, 3501 Market Street, Philadelphia, PA 19104. (215) 386-0100. FAX (215) 386-6362. 1970 to present. Weekly. $442.00 per year. Also available on CD-ROM and online. Inquire regarding cost and availability. ISSN: 0095-7917.

Electrical and Electronics Abstracts. Institution of Elec trical Engineers (IEE), London. Available from INSPEC/IEEE - Institute of Electrical and Electronic Engineers, Box 1331, Hoes Lane, Piscataway, NJ 08855-1331. (908) 562-5549. 1898 to present. Monthly. $2200.00 per year. Also available on CD-ROM and online as INSPEC.

Engineering Index Monthly; Indexes and Abstracts the World's Engineering and Technical Literature. Engineering Information, Inc., Castle Point on the Hudson, Hoboken, NJ 07030. (201) 216-8500. (800) 221-1044. FAX (201) 216-8532. Monthly. $2300.00 per year. Available online as COMPENDEX and also on CD-ROM. ISSN: 0742-1974.

General Science Index. H.W. Wilson Company, 950 University Avenue, Bronx, NY 10452. (212) 588-8400. (800) 367-6770. FAX (718) 590-1617. From 1978 to present. Ten issues per year; quarterly and annual cumulations. Service basis. Available

onCD-ROM and online. Inquire regarding cost and availability. ISSN: 0162-1963.

Government Reports Announcements and Index. U. S. National Technical Information Service (NTIS), 5285 Port Royal Road, Springfield, VA 22161. (703) 487-4650. FAX (703) 321-8547. From 1968 to present. Annual. $630.00 per year. Also available online as *NTIS Bibliographic Database* and on CD-ROM. ISSN:

Index To IEEE Publications. IEEE Service Center, 445 Hoes Lane, Piscataway, NJ 08855-1331. (908) 981-1393. (800) 678-IEEE. FAX (908) 981-9667. 1973 to present. Annual. ISSN: 0099-1368.

Physics Abstracts . INSPEC. Section A, Science Abstracts. Instituti on of Electrical Engineers (IEE), London. Available from: INSPEC/IEEE - Institute of Electrical and Electronic Engineers, Box 1331, Hoes Lane, Piscataway, NJ 08855-1331. (908) 562-5549. 1898 to present. 24 issues per year. $2835.00 per year. Also available online and on CD-ROM. ISSN: 0036-8091.

Physics Briefs (Physikalische Berichte). Information Center for Energy, Physics, Mathematics; German Physical Society. V C H Publishers, Inc., 220 East 23rd Street, New York, NY 10010-4606. (212) 683-8333. 1845 to present. 24 issues per year. $2390.00 per year. Also available online. ISSN: 0179-7434.

Science Citation Index. SCI. Institute for Scientific Information, 3501 Market Street, Philadelphia, PA 19104. (215) 386-0100. (800) 523-1850. FAX (215) 386-2991. 1961 to present. Six issues per year, plus annual cumulation. $11650.00 per year. Also available online and on CD-ROM. Inquire about price and availability. ISSN: 0036-827X.

STAR. Scientific and Technical Aerospace Reports. U.S. National Aeronautics and Space Administration, Scientific and Technical Information Facility, Box 8757, Baltimore-Washington International Airport, MD 21240. (301) 621-0153. Distributed by U. S. Superintendent of Documents, Washington, DC 20402. From 1963 to present. Semimonthly, with semiannual and annual indexes. $114.00 per year. Also available online and on CD-ROM. ISSN: 0036-8741.

ANNUAL REVIEWS AND YEARBOOKS

Advances in Electronics and Electron Physics. ACADemicic Press, Inc., 6277 Sea Harbor Drive, Orlando, FL 32887. (800) 321-5068. From 1948 to present. Irregular. Price varies, inquire. ISSN: 0065-2539.

ASSOCIATIONS AND PROFESSIONAL SOCIETIES

American Association of Engineering Societies, 1111 19th Street, Suite 608, Washington, DC 20036-3603. (202) 296-2237. FAX (202) 296-1151.

American Electronics Association. 5201 Great America Way, Suite 520, P.O. 52990, Santa Clara, CA 95056. (408) 987-4200. FAX (408) 970-8565.

American Institute of Physics. One Physics Ellipse, College Park, MD 20740-3843. (301) 209-3100.

Antennas Measurement Techniques Association. 6065 Roswell Road, Suite 2252. Atlanta, GA 30328. (706) 952-2896. FAX (706)952-3773.

Electronics Industries Association. 2500 Wilson Boulevard, Arlington, VA 22201. (703) 907-7500. FAX (202) 457-4985.

IEEE Antennas and Propagation Society. c/o Institute of Electrical and Electronic Engineers, 345 East 47th Street. New York, NY 10017. (212) 705-7900. FAX (212) 705-4929.

IEEE (Institute of Electrical and Electronic Engineers). 345 East 47th Street. New York, NY 10017. (212) 705-7900. FAX (212) 705-4929,

Optical Society of America, 2010 Massachusetts Avenue, N.W., Washington, DC 20036-1023. (202) 223-8130. FAX (202) 223-1096.

DIRECTORIES AND BIOGRAPHICAL SOURCES

American Men and Women of Science: Physical and Biological Sciences. R. R. Bowker Inc., 121 Chanlon Road, New Providence, NJ 07974. (908) 464-6800. (800) 521-8110. 20th edition. 8 volumes. 1996. $850.00.

Directory of Engineering College Research and Graduate Study. American Society for Engineering Education, Ohio State University, 2070 Neil Avenue, Columbus, OH 43210-1226. Annual. $20.00.

Directory of Engineering Societies and Related Organizations. American Association of Engineering Societies, 1111 19th Street, Suite 608, Washington, DC 20036-3603. (202) 296-2237. Semi-annual. $150.00.

Directory of Engineers in Private Practice. National Society of Professional Engineers, 1420 King Street, Alexandria, VA 22314-2750. (703) 684-2835. Biennial. $30.00.

E E M - Electronic Engineer's Master. Hearst Business Communications, Inc., 645 Stewart Avenue, Garden City NY 11530. (516) 227-1300. Annual. $90.00 ISSN: 0732-9016.

Engineering Research Centers: Incorporating Electronics Research Centers. Stockton Press, 345 Park Avenue, New York, NY 10010. (212) 689-9200. (800) 221-2123. 4th edition. 1995. $515.00.

IEEE Membership Directory. Institute of Electrical and Electronics Engineers, IEEE Service Center, 445 Hoes Lane, Piscataway, NY 08854. (908) 981-1393. (800) 678-IEEE. FAX (908) 981-9667. 2 volumes. Annual. $190.00. ISSN:

International Engineering Directory. American Consulting Engineers Council, 1015 15th Street, N.W. Suite 802, Washington, DC 20005-2670. (202) 347-7474. Annual. $10.00.

Research Centers Directory. Gale Research Inc., 835 Penobscot Building, Detroit, MI 48226-4094. (313) 961-2242. (800) 877-4253. 20th edition. 1995. $485.00. ISSN: 0080- 1518.

Who's Who in Engineering. Gordon Davis, editor. American Association of Engineering Societies. 1111 19th Street, NY, Suite 608, Washington, DC 20036. (202) 296-2237. (800) 658-8897. 9th edition. 1995. $220.00.

ENCYCLOPEDIAS AND DICTIONARIES

Chambers Science and Technology Dictionary. Peter M.B. Walker, editor. Cambridge University Press, 40 West 20th Street, New York, NY 10011-4211. (212) 924-3900. 1988. $39.95.

IEEE Standard Dictionary of Electrical and Electronics Terms. Christopher J. Booth, editor. IEEE Service Center, 445 Hoes Lane, Piscataway, NJ 08855-1331. (908) 981-1393. (800) 678-IEEE. FAX (908) 981-9667. IEEE Standard 100- 1992. 5th edition. 1993. $150.00.

Illustrated Dictionary of Electronics. Stan Gibilsco. TAB Books, P.O. Box 40, Blue Summit, PA 17294-0850. (717) 794- 2191. (800) 233-1128. 7th edition. 1994. $34.95.

McGraw-Hill Electronics Dictionary. J. Markus. McGraw-Hill Book Company, Inc., 1221 Avenue of the Americas, New York, NY 10020. (212) 997-3675. 5th edition. 1994. $49.95.

McGraw-Hill Encyclopedia of Science and Technology. McGraw-Hill Book, Incorporated, 1221 Avenue of the Americas, New York, NY 10020. (212) 997-3675. (800) 262-4729. Seventh edition. Twenty volumes. 1992. $1900.00.

GENERAL WORKS

Analysis, Design and Measurement of Small and Low-Profile Antennas. Kazuhiro Hirasawa and Misao Haneishi, editors. Artech House, 685 Canton Street, Norwood, MA 02062. (617) 769-9750. (800) 225-9977. 1991. $75.00.

Antennas For Radar and Communications: A Polarimetric Approach. Harold Mott. John Wiley & Sons, Inc., 605 Third Avenue, New York, NY 10158-0012. (212) 850-6000. (800) 225-5945. 1992. $99.95.

Fixed and Mobile Terminal Antennas. Akhileshwar Kumar. Artech House, 685 Canton Street, Norwood, MA 02062. (617) 769-9750. (800) 225-9977. 1991. $95.00.

Microwave Antenna theory and Design. S. Silver, et al, editors. Institution of Electrical Engineers, Michael Faraday House, Six Hills Way, Stevenage, Herts. SG1 2AY, England. Telephone: 0438 313311 or FAX 0438 742840. 1984. $75.00.

Principles of Antennas: Wire and Aperture. T. S. Caclean. Cambridge University Press, 40 West 20th Street, New York, NY 10011-4211. (212) 924-3900. (800) 872-7423. 1986. $105.00.

VHF and UHF Antennas. R. A. Burberry. Institution of Electrical Engineers, Michael Faraday House, Six Hills Way, Stevenage, Herts. SG1 2AY, England. Telephone: 0438 313311 or FAX 0438 742840. 1992. $95.00

HANDBOOKS AND MANUALS

Antenna Engineering Handbook. Richard C. Johnson. McGraw-Hill Book Company, 1221 Avenue of the Americas, New York, NY 10020. (212) 512-2000. (800) 262-4729. 3rd edition, 1993. $119.50.

Antenna Handbook. Y. T. Lo and S. W. Lee, editors. Van Nostrand Reinhold, 115 Fifth Avenue, New York, NY 10003. (212) 254-3232. (800) 842-3636. Four volumes: Fundamentals

ANTENNAS

Ency. of Physical Sciences and Engineering Info. Sources

and mathematical techniques, Antenna theory, Applications, Related topics. 1993. $150.95.

CRC Handbook of Chemistry and Physics. David R. Kide, editor. CRC Press Inc., 2000 Corporate Blvd., NW, Boca Raton, FL 33431. (407) 994-0555. (800) 333-8300. 77th edition. 1996. $99.95.

Electrical Engineer's Handbook. Donald Christiansen, editor. McGraw Book Company, 1221 Avenue of the Americas, New York, NY 10020. (212) 512-2000. (800) 262-4729. 4th edition. 1996. $110.00

Engineering Tables and Data. A.M. Howatson, P.G. Lund and J.D. Todd. Van Nostrand Reinhold, 115 Fifth Avenue, New York, NY 10003. (212) 254-3232. (800) 926-2665. Second Edition. 1991. $39.95.

Handbook of Antenna Design. A. W. Rudge, et al, editors. Institution of Electrical Engineers, Michael Faraday House, Six Hills Way, Stevenage, Herts. SG1 2AY, England. Telephone: 0438 313311 or FAX 0438 742840. 2 volumes in 1. 1986. $165.00.

Standard Handbook of Engineering Calculations. Tyler G. Hicks. McGraw-Hill Publishing Company, 1221 Avenue of the Americas, New York, NY 10020. (800) 262-4729 or (212) 512-3825. 3rd edition. 1994. $99.50. ISBN: 0-07-028812-7.

ONLINE DATABASES AND CD-ROMS

Aerospace Database. American Institute of Aeronautics and Astronautics, 370 L'Enfant Promenade, S.W., Washington, DC 20024. (202) 646-7400. Worldwide published literature on research and development in aerospace and related areas, with abstracts. Covers 1962 to present. Available on DIALOG, (800) 334-2564, online service. Also available on CD-ROM. Inquire as to cost and availability.

CA Search. Chemical Abstracts Service, P.O. Box 3012, Columbus, OH 43210-0012. (614) 447-3600. (800) 848-6533. FAX (614) 447-3709. Very comprehensive guide to worldwide chemical literature and related fields, 1972 to present. Available on BRS,(800) 289-4277, DIALOG, (800) 334-2564, ORBIT (800) 456-7248, and STN International, FIZ Karlsruhe, P.O. Box 2465, W-7500, Karlsruhe 1, Germany, online services. Inquire as to cost and availability.

Compendex Plus. Engineering Information, Inc., 345 East 47th Street, New York, NY 10017. (212) 705-7600 or (800) 221-1044. Contains citations with abstracts to worldwide literature in engineering and technology, from 1970 to present. Available on online BRS,(800) 289-4277, DIALOG, (800) 334-2564, ORBIT (800) 456-7248, and STN International, FIZ Karlsruhe, P.O. Box 2465, W-7500, Karlsruhe 1, Germany, online services. Also available on CD-ROM. Inquire as to cost and availability.

Current Contents Search. Institute for Scientific Information, 3501 Market Street, Philadelphia, PA 19104. (215) 386-0100. FAX (215) 386-6362. Contains citations to articles listed in the table of contents of science and technology journals. Also articles in social sciences and life sciences journals. Available on BRS,(800) 289-4277, DIALOG,(800) 334-2564, online services. Inquire as to cost and availability.

Dissertation Abstracts. University Microfilms International, 300 North Zeeb Road, Ann Arbor, MI 48106. (800) 521-0600 or (313) 761-4700. Scope includes virtually all doctoral dissertations accepted at accredited American institutions from 1861 to present in 252 subject areas. Available on BRS,(800) 289-4277, DIALOG,(800) 334-2564, and OCLC EPIC,(800) 848-5878, online services. Also available on CD-ROM. Inquire as to cost and availability.

INSPEC. Institution of Electrical Engineers, Michael Faraday House, Six Hills Way, Stevenage, Herts. SG1 2AY, England. Telephone: 0438 313311 or FAX 0438 742840. Contains citations to the worldwide literature of physics, electronics and electrical engineering, computer technology, and related fields. Available on BRS,(800) 289-4277, DIALOG, (800) 334-2564, ORBIT, (800) 456-7248, and STN International, FIZ Karlsruhe, P.O. Box 2465, W-7500, Karlsruhe 1, Germany, online services. Inquire as to cost and availability.

International Aerospace Abstracts. American Institute of Aeronautics and Astronautics, 370 L'Enfant Promenade, S.W., Washington, DC 20024. (202) 646-7400. Contains references and abstracts of journal and monograph literature relating to aerospace science and technology, from 1963 to present. Available through the NASA/RECON system of the National Aeronautics and Space Administration only.

NTIS Bibliographic Database. National Technical Information Service, 5285 Port Royal Road, Springfield, VA 22161. (703) 487-4929 or FAX (703) 321-8199. Broad coverage of government-sponsored science and technology research reports, 1964 to present. Available on BRS,(800) 289-4277, DIALOG, (800) 334-2564, ORBIT, (800) 456-7248, and STN International, FIZ Karlsruhe, P.O. Box 2465, W-7500, Karlsruhe 1, Germany, online services. Also available on CD-ROM. Inquire as to cost and availability.

Scientific and Technical Books and Serials in Print. R.R. Bowker Inc., 121 Chanlon Road, New Providence, NJ 07974. (800) 521-8110 or (908) 464-6800. List of currently published books and serials in the physical and biological sciences, engineering and technology, with subject, author and titles indexes. Available on ORBIT,(800) 456-7248, online service.Inquire as to cost and availability.

SCISEARCH. Institute for Scientific Information, 3501 Market Street, Philadelphia, PA 19104. (800) 523-1850 or (215) 386-0100. Broad multidisciplinary title and author index to the international literature of science and technology, 1974 to present. Available on DIALOG,(800) 334-2564, and ORBIT,(800) 456-7248, online services. Also available on CD-ROM. Inquire as to cost and availability.

WILSONLINE. H.W. Wilson Company, 950 University Avenue, Bronx, NY 10452. (800) 367-6770 or (212) 588-8400. Makes available online versions of the printed H.W. Wilson indexes includingApplied Science and Technology Index, Business Periodicals Index, General Science Index, and Readers' Guide to Periodical Literature. Period covered is generally 1983 to present. Available on BRS,(800) 289-4277, DIALOG,(800) 334-2564, and OCLC EPIC,(800) 848-5878, online services. Also available on CD-ROM. Inquire as to cost and availability.

PERIODICALS

Chaos: An Interdisciplinary Journal. American Institute of Physics, 335 East 45th Street, New York, NY 10017 (212) 661-9404 or FAX (516)349-9704. 1991 to present. Quarterly. $160.00 per year.

CRC Critical Reviews in Solid State and Materials Science. CRC Publishers, Inc., 2000 Corporate Blvd., N.W., Boca Raton, FL 33431. (407) 994-0555 or (800) 333-8300. Quarterly. $295.00 per year.

Electronic Design. Penton Publishing, San Jose Gateway, Suite 354. 2025 Gateway Place, San Jose, CA 95110. (408) 441-0550. 1952 to present. Fortnightly. $95.00. ISSN: 0013-4872.

Electronics. Penton Publishing, San Jose Gateway, Suite 354. 2025 Gateway Place, San Jose, CA 95110. (408) 441-0550. 1930 to present. Semi-weekly. $98.00 per year. ISSN: 0883-4989.

IEE Proceedings. PART H: MICROWAVES, ANTENNAS and PROPAGATION. Institution of Electrical Engineers. Available from: Institute of Electrical and Electronic Engineers (IEEE) Inc., Box 1331, 445 Hoes Lane, Piscataway, NJ 08855-1331. (908) 981-0060. 1980 to present. Bi-monthly. ISSN: 0950-107X.

IEEE Circuits and Devices Magazine. Institute of Electrical and Electronics Engineers, Inc., Box 1331, 445 Hoes Lane, Piscataway, NJ 08855-1331. (908) 981-0060. 1985 to present. Bi-monthly. $120.00 per year. ISSN: 8755-3996.

IEEE Transactions On Antennas and Propagation. Institute of Electrical and Electronics Engineers, Inc., Box 1331, 445 Hoes Lane, Piscataway, NJ 08855-1331. (908) 981-0060. 1952 to present. Monthly. $212.00 per year. ISSN: 0018-926X.

IEEE Transactions On Electron Devices. Institute of Electrical and Electronics Engineers, Inc., Box 1331, 445 Hoes Lane, Piscataway, NJ 08855-1331. (908) 981-0060. 1952 to present. Monthly. $395.00 per year. ISSN: 0018-9393.

Institute of Electrical and Electronics Engineers Proceedings. Institute of Electrical and Electronics Engineers, Inc., Box 1331, 445 Hoes Lane, Piscataway, NJ 08855-1331. (908) 981-0060. 1913 to present. Monthly. $275.00. ISSN: 0018-9219.

International Journal of Microwave and Millimeter-wave Computer Aided Engineering. John Wiley and Sons, Inc., 605 Third Avenue, New York, NY 10158. (800) 526-5368 or (212) 850-6000. 1991 to present. Quarterly. $175.00 per year.

RESEARCH CENTERS AND INSTITUTES

Electroscience Laboratory. Ohio State University, 1320 Kinnear Road, Columbus, Ohio 43212. (614) 292-7981, FAX (614) 292-7297.

Electrical Engineering Research Laboratories. PurdueUniversity. Electrical Engineering Building, West Lafayette, in 47907. (317) 494-3536. FAX (317) 494-6440.

Electronics Research Center. University of Texas at Austin, 132 Engineering Science Building, Austin, TX 78712. (512) 471-3954.

Systems and Techniques Laboratory. Georgia Institute of Technology, Georgia Tech Research Institute, Atlanta, GA 30332. (404) 528-7010.

Weber Research Institute. Polytechnic University. Route 110, Farmingdale, NY 11735. (516) 755-4250, FAX (516) 755-4404.

ANTHRACITE COAL

See: COAL

ANTICYCLONES

See also: ATMOSPHERE, CLIMATE, CLOUDS, CYCLONES, DROUGHT, FOG, GLOBAL WARMING, GREENHOUSE EFFECT, GROUNDWATER, ICE, JET STREAM, RAIN, SNOW, WATER RESOURCES, WEATHER FORECASTING, WEATHER MODIFICATION

ABSTRACT SERVICES AND INDEXES

General Science Index. H.W. Wilson Co., 950 University Avenue, Bronx, NY 10452. (800) 367-6770 or (212) 588-8400. Inquire about price and availability.

Geophysics Abstracts. Pergamon Press, Inc., Maxwell House, Fairview Park, Elmsford, NY 10523. (914) 592-7700. Fax (914) 592-3625. Twelve times per year. $565.00 per year. Also available in microform.

Government Reports Announcements and Index. National Technical Information Service (NTIS), 5285 Port Royal Road, Springfield, VA 22161. (703) 487-4650. 1968 to present. Annual. $525.00 per year. Also available online as NTIS BIBLIOGRAPHIC DATABASE and on CD-ROM.

Meteorological and Geoastrophysical Abstracts. American Meteorological Society, c/o Inforonics, Inc., 550 Newtown Road, Box 458, Littleton, MA 01460. (508) 486-8976. Monthly. $750.00 per year.

Science Citation Index. Institute for Scientific Information, 3501 Market Street, Philadelphia, PA 19104. (800) 523-1850 or (215) 386-0100. Inquire about price and availability.

Selected Water Resources Abstracts. U.S. Geological Survey. Distributed by National Technical Information Service (NTIS), 5285 Port Royal Road, Springfield, VA 22161. (703) 487-4650. Monthly. $115.00 per year.

ANNUAL REVIEWS AND YEARBOOKS

Annual Review and Earth and Planetary Sciences. Annual Reviews, Inc., 4139 El Camino Way, Palo Alto, CA 94306-0897. (415) 493-4400. Fax (415) 855-9815. Annual. $55.00 per year.

Developments in Atmospheric Science Series. Elsevier Science Publishing Company, Inc., 655 Avenue of the Americas, New York, NY 10010. (212) 989-5800. Irregular. Inquire about price and availability.

ASSOCIATIONS AND PROFESSIONAL SOCIETIES

American Geophysical Union, 2000 Florida Avenue, N.W., Washington, DC 20009. (202) 462-6900.

American Meteorological Society, 45 Beacon Street, Boston, MA 02108-3693. (617) 227-2425. FAX (617) 742-8718. Irregular. Price varies.

ANTICYCLONES

Ency. of Physical Sciences and Engineering Info. Sources

Association of American Weather Observers, 401 Whitney Boulevard, Box 455, Belvedere, IL 61008. (815) 544-5665. FAX (815) 544-6334.

Canadian Meteorological and Oceanographic Society, P.O. Box 334, Newmarket, Ontario, L3Y 4X7, Canada. (416) 898-1040. FAX (416) 898-7937.

National Weather Association, 4400 Stamp Road, Room 404, Temple Hills, MD 20748. (301) 899-3784.

BIBLIOGRAPHIES

Information Sources in the Earth Sciences. David N. Wood, Joan E. Hardy, and Anthony P. Harvey, editors. Bowker-Saur/K.G. Saur. Distributed by R.R. Bowker, 121 Chanlon Road, New Providence, NJ 07974. (800) 521-8110 or (908) 464-6800. Second edition. 1989. $85.00.

Scientific and Technical Books and Serials in Print; An index to Literature in Science and Technology. R.R. Bowker Co., 205 E. 42nd Street, New York, NY 10017. (800) 521-8110 or (212) 916-1600. 1992. $250.00.

DIRECTORIES AND BIOGRAPHICAL SOURCES

American Men and Women of Science: Physical and Biological Sciences. Eighteenth edition. R.R. Bowker Company, 245 West 17th Street, New York, NY 10011. (800) 521-8810 or (212) 916-1600. $750.00.

Research Centers Directory. Gale Research, 835 Penobscot Building, Detroit, MI 48226-4094. (800) 877-4253 or (313) 961-2242. 1995. $485.00.

ENCYCLOPEDIAS AND DICTIONARIES

McGraw-Hill Encyclopedia of Ocean and Atmospheric Sciences. Sybil P. Parker, editor. McGraw-Hill Publishing Company, 1221 Avenue of the Americas, New York, NY 10020. (800) 262-4729 or (212) 512-3825. 1979. $79.95.

McGraw-Hill Encyclopedia of Science and Technology. McGraw-Hill Publishing Company, 1221 Avenue of the Americas, New York, NY 10020. (800) 262-4729 or (212) 512-3825. Seventh edition. 1992. $1900.00.

GENERAL WORKS

The Atmosphere: An Introduction to Meteorology. Frederick K. Lutgens and Edward J. Tarbuck. Prentice Hall (Division of Simon and Schuster), 15 Columbus Circle, New York, NY 10023. (212) 373-8500 or (800) 922-0579. Fifth edition. 1991. $49.00.

Coastal Meteorology. Hsu Shih-Ang, editor. Academic Press, Inc., 1250 Sixth Avenue, San Diego, CA 92101. (619) 231-0926. FAX (619) 699-6715. 1988. $54.00.

Fundamentals of Atmospheric Dynamics and thermodynamics. C. Riegel. World Scientific Publishing Company, Inc., 1060 MainStreet, Unit B, River Edge, NJ 07661. (800) 227-7562 or (201) 487-9655. 1992. $30.00 in paper.

Mesoscale Meteorology and Forecasting. Peter S. Ray, editor. American Meteorological Society, 45 Beacon Street, Boston, MA 02108-3693. (617) 227-2425. FAX (617) 742-8718. 1986. $50.00.

Meteorology and Environmental Sciences. R. Guzzi and others. World Scientific Publishing Company, Inc., 1060 Main Street, Unit B, River Edge, NJ 07661. (800) 227-7562 or (201) 487-9655. 1990. $120.00.

Meteorology: The Atmosphere and the Science of Weather. Joseph M. Moran. Macmillan Publishing, 866 3rd Avenue, New York, NY 10024. (212) 689-9140. Third edition. 1991. $49.95.

Physical Meteorology. Henry G. Houghton. MIT Press, 55 Hayward Street, Cambridge, MA 02142. (617) 253-2884 or (800) 356-0343. 1985. $45.00.

Severe and Unusual Weather. Joe R. Eagleman. Trimedia Publishing Company, 12008 West 87th Street, Suite 117, Lenexa, KS 66215. (913) 599-0505. Second edition. 1990. $41.95.

Weather Companion: An Album of Meteorological History, Science, Legend, and Folklore. Gary Lockhart. John Wiley and Sons, Inc., 605 Third Avenue, New York, NY 10158. (800) 526-5368 or (212) 850-6000. 1988. $12.95.

Weather Cycles: Real OR Imaginary? William James Burroughs. Cambridge University Press, 40 West 20th Street, New York, NY 10011-4211. (212) 924-3900. 1992. $39.95.

HANDBOOKS AND MANUALS

Author's Guide To the Journals of the American Meteorological Society. American Meteorological Society, 45 Beacon Street, Boston, MA 02108-3693. (617) 227-2425. FAX (617) 742-8718. 1983. $15.00 in paper.

Basic Meteorology Lab Manual. Thomas A. Leavy. Allegheny Press, P.O. Box 220, Elgin, PA 19413. (814) 664-8504. Second edition. 1969. $7.95.

Handbook in Applied Meteorology. David D. Houghton, editor. John Wiley and Sons, Inc., 605 Third Avenue, New York, NY 10158. (800) 526-5368 or (212) 850-6000. 1985. $114.00.

ONLINE DATABASES AND CD-ROMS

Climate Assessment Database. National Weather Service, National Meteorological Center, 5200 Auth Road, Suite 101, Camp Springs, MD 20233. (301) 763-8016. Contains daily, weekly and monthly summaries of North American and world climatological data. Also provides five to ten weather forecasts, and 30 to 90 day weather outlook. Subscription required. Inquire as to cost and availability.

International Aerospace Abstracts. American Institute of Aeronautics and Astronautics, 370 L'Enfant Promenade, S.W., Washington, DC 20024. (202) 646-7400. Contains references and abstracts of journal and monograph literature relating to aerospace science and technology, from 1963 to present. Available through the NASA/RECON system of the National Aeronautics and SpaceAdministration only.

Meteorological and Geoastrophysical Abstracts. American Meteorological Society, 45 Beacon Street, Boston, MA 02108-

Ency. of Physical Sciences and Engineering Info. Sources

ANTICYCLONES

3693. (617) 227-2425. FAX (617) 742-8718. Contains citations and abstracts to the worldwide literature on significant research in meteorology and geoastrophysics. Related topics include physical oceanography, hydrology, environmental sciences and glaciology. Covers the period 1972 to present. Available on DIALOG,(800) 334-2564, online service. Inquire as to cost and availability.

NTIS Bibliographic Database. National Technical Information Service, 5285 Port Royal Road, Springfield, VA 22161. (703) 487-4929 or FAX (703) 321-8199. Broad coverage of government-sponsored science and technology research reports, 1964 to present. Available on BRS,(800) 289-4277, DIALOG, (800) 334-2564, ORBIT, (800) 456-7248, and STN International, FIZ Karlsruhe, P.O. Box 2465, W-7500, Karlsruhe 1, Germany, online services. Also available on CD-ROM. Inquire as to cost and availability.

World Climate Disc. Chadwyck-Healey Inc., 1101 King Street, Alexandria, VA 22314. (703) 683-4890. FAX (703) 683-7589. Weather and climate data from approximately 5,000 weather stations worldwide, covering the years 1854 to 1990 on CD-ROM. First edition 1992. Approximately $1200.00 per year with annual updates.

World Weatherdisc. WeatherDisc Associates, Inc., 4584 N.E. 89th Street, Seattle, WA 98115. (206) 524-4314. FAX (206) 543-0308. Meteorological data on CD-ROM which describes the climate of the earth today and for the past few hundred years. First edition 1989. Approximately $295.00 per year with biannual updates.

PERIODICALS

Agricultural and Forest Meteorology. Elsevier Science Publishing Company, Inc., 655 Avenue of the Americas, New York, NY 10010. (212) 989-5800. Twenty times per year. $750.00 per year.

American Meteorological Society Bulletin. American Meteorological Society, 45 Beacon Street, Boston, MA 02108-3693. (617) 227-2425. FAX (617) 742-8718. Monthly. $60.00 per year.

American Meteorological Society. Meteorological Monographs. American Meteorological Society, 45 Beacon Street, Boston, MA 02108-3693. (617) 227-2425. FAX (617) 742-8718. Irregular. Price varies.

American Weather Observer. Association of American Weather Observers, 401 Whitney Boulevard, Box 455, Belvedere, IL 61008. (815) 544-5665. FAX (815) 544-6334. Monthly. $21.00 per year.

Atmosphere - Ocean. Canadian Meteorological and Oceanographic Society, P.O. Box 334, Newmarket, Ontario, L3Y 4X7, Canada. (416) 898-1040. FAX (416) 898-7937. Quarterly. $30.00 per year.

Dynamics of Atmospheres and Oceans. Elsevier Science Publishing Company, Inc., 655 Avenue of the Americas, New York, NY 10010. (212) 989-5800. Six times per year. $205.00 per year.

Earth. Kalmbach Publishing Company, P.O. Box 1612, Waukesha, WI 53187. (414) 796-0126 or (800) 558-1544. 1991 to present. Bimonthly. $14.95 per year.

JGR: Journal of Geophysical Research: Atmosphere. American Geophysical Union, 2000 Florida Avenue, N.W., Washington, DC 20009. (202) 462-6903. Monthly. $90.00 per year to members.

JGR: Journal of Geophysical Research: Oceans. American Geophysical Union, 2000 Florida Avenue, N.W., Washington, DC 20009. (202) 462-6903. Monthly. $1545.00 per year.

Journal of Applied Meteorology. American Meteorological Society, 45 Beacon Street, Boston, MA 02108-3693. (617) 227-2425. FAX (617) 742-8718. Monthly. $165.00 per year.

Journal of Atmospheric Sciences. American Meteorological Society, 45 Beacon Street, Boston, MA 02108-3693. (617) 227-2425. FAX (617) 742-8718. Semi-monthly. $320.00 per year.

Monthly Weather Review. American Meteorological Society, 45 Beacon Street, Boston, MA 02108-3693. (617) 227-2425. FAX (617) 742-8718. Monthly. $205.00 per year.

National Weather Digest. National Weather Association, 4400 Stamp Road, Room 404, Temple Hills, MD 20748. (301) 899-3784.

Royal Meteorological Society. QUARTERLY JOURNAL. Royal Meteorological Society, 104 Oxford Road, Reading, Berks RG1 7LJ, England. Six times per year. $250.00 per year.

Weather. Royal Meteorological Society, 104 Oxford Road, Reading, Berks RG1 7LJ, England. Monthly. $44.00 per year.

Weatherwise. Heldref Publications, 1319 Eighteenth Street, N.W., Washington, DC 20036-1802. (202) 296-6267. FAX (202) 296-5149. Bi-monthly. $28.00 per year.

RESEARCH CENTERS AND INSTITUTES

Atlantic Oceanographic and Meteorological Laboratory. 4301 Rickenbacker Causeway, Miami, FL 33149. (305) 361-4300.

Cooperative Institute For Mesoscale Meteorological Studies. University of Oklahoma, 401 East Boyd, Norman, OK 73019. (405) 325-3041.

Geophysical Fluid Dynamics Laboratory. Princeton University, P.O. Box 308, Princeton, NJ 08542. (609) 452-6500.

Goddard Space Flight Center. Laboratory for Atmospheres, Mail Code 610, Greenbelt, MD 20771. (301) 286-5002.

Joint Institute For Marine and Atmospheric Research. University of Hawaii, 1000 Pope Road, Honolulu, HI 96822. (808) 541-2876.

National Center For Atmospheric Research. P.O. Box 3000, Boulder, CO 80307. (303) 496-1000.

National Meteorological Center. World Weather Building, Room 101, 5200 Auth Road, Camp Springs, MD 20746. (301) 763-8016.

National Weather Service. 1325 East-West Highway, Silver Spring, MD 20910. (301) 427-7689.

ARCHITECTURAL ENGINEERING

Ency. of Physical Sciences and Engineering Info. Sources

APPLIED MATHEMATICS

See: MATHEMATICS

AQUIFER

See: GROUNDWATER

ARC WELDING

See: WELDING

ARCH

See: STRUCTURAL ENGINEERING

ARCHITECTURAL ENGINEERING

See also: BRIDGES, BUILDING MATERIALS, CIVIL ENGINEERING, CONCRETE, CONSTRUCTION ENGINEERING, STRUCTURAL ENGINEERING, STRUCTURES

ABSTRACT SERVICES AND INDEXES

Abstract Journal in Earthquake Engineering. University of California, Berkeley, Earthquake Engineering Research Center, 1301 South 46th Street, Richmond, CA 94804-4698. (510) 231-9413 or FAX (510) 231-9471. 1972 to present. Annual. $80.00 per year.

Applied Science and Technology Index; A cumulative subject index to English language periodicals in the fields of aeronautics and space science, computer technology, chemistry, construction industry, energy and related areas. H.W. Wilson Co., 950 University Avenue, Bronx, NY 10452. (800) 367-6770 or (212) 588-8400. FAX (718) 590-1617. From 1958 to present. Monthly. Inquire about cost and availability. Also available on CD-ROM and online.

ASCE Combined Annual Combined Index. American Society of Civil Engineers, 345 East 47th Street, New York, NY 10017-2398. (212) 705-7520 or (800) 548-2723. Annual. $48.00.

ASCE Publications Information. 1966-present. American Society of Civil Engineers, 345 East 47th Street, New York, NY 10017-2398. (212) 705-7520 or (800) 548-2723. Bi-monthly. $160.00 per year for non-members.

Civil and Structural Engineering Abstracts, Cambridge Scientific Abstracts, 7200 Wisconsin Avenue, Bethesda, MD 20814-4823. (301) 961-6750. FAX (301) 961-6720. 1993 to present. Monthly. $385.00 per year. Topics covered include structural design, construction equipment and methods, civil defense and military engineering, surveying, highway engineering, maritime and port structures, materials, land reclamation, and soil mechanics.

Current Contents: Engineering, Technology, and Applied Sciences. Institute for Scientific Information, 3501 Market Street,

Philadelphia, PA 19104. (215) 386-0100. FAX (215) 386-6362. 1970 to present. Weekly. $442.00 per year.

Engineering Index Monthly. Engineering Information, Inc., Castle Point on the Hudson, Hoboken, NJ 07030. (800) 221-1044. FAX (212) 832-1857. Monthly. $2300.00 per year. Also available online as COMPENDEX, and also on CD-ROM. Covers chemical engineering, computers, electrical engineering, civil engineering, metals and mining, industrial management, and mechanical engineering.

Index To Scientific and Technical Proceedings. Institute for Scientific Information, 3501 Market St., Philadelphia, PA 19104. (215) 386-0100. FAX (215) 386-6362. Monthly. $500.00 per year.

International Civil Engineering Abstracts. CITIS Ltd., 2 Rosemount Terrace, Blackrock, Dublin, Ireland. Telephone 353-1-2886227. FAX 353-1-885-971. 1974 to present. Monthly. $660.00 per year. Also available on CD-ROM.

NTIS Alerts: Building Industry Technology. U.S. National Technical Information Service, 5825 Port Royal Road, Springfield, VA 22161. (703) 487-4630. FAX (703) 321-8547. Weekly. $135.00 per year.

Offshore Engineering Abstracts. STI Limited, 4 Kings Meadow, Ferry Hinksey Road, Oxford, OX2 ODU, England. Distributed in the United States by Air Science Company, Box 143, Corning NY 14830. (607) 962-5591. Contains citations on planning, design, and construction of offshore platforms and pipelines, from 1985 to present. Monthly. $400.00 per year.

Science Citation Index. Institue for Scientific Information, 3501 Market Street, Philadelphia, PA 19104. (215) 386-0100. FAX (215) 386-6362. Inquire about availability and cost. Also available on CD-ROM.

ASSOCIATIONS AND PROFESSIONAL SOCIETIES

American Institute of Architects. 1735 New York Avenue NW, Washington, DC 20006. (202) 626-7300.

American Institute of Building Design. 16 Wilton Road, Bldg. ED, Westport, CT 06880. (800) 366-2423.

American Society For Testing and Materials (astm). 1916 Race Street, Philadelphia, PA 19103. (215) 299-5585.

American Society of Civil Engineers. 345 East 47th Street, New York, NY 10017-2398. (212) 705-7520 or (800) 548-2723.

National Society of Architectural Engineers. PO Box 395, Lawrence, KS 66044. (913) 864-3434.

DIRECTORIES AND BIOGRAPHICAL SOURCES

American Society of Civil Engineers Directory: official Register. 345 East 47th Street, New York, NY 10017-2398. (212) 705-7520 or (800) 548-2723. Inquire for cost and availability.

ENCYCLOPEDIAS AND DICTIONARIES

Concise Encyclopedia of Building and Construction Materials. Fred Moavenzadeh. Pergamon Press Inc., Maxwell House,

Fairview Park, Elmsford, NY 10523. (914) 592-7700. Fax (914) 592-3625. 1990. $210.00.

Construction Glossary: An Encyclopedic Reference and Manual. J. Stewart Stern. 2nd edition. John Wiley and Sons, Inc., 605 Third Avenue, New York, NY 10158. (800) 526-5368 or (212) 850-6000. 1993. $99.95.

Dictionary of Civil Engineering. V.N. Vazirani. South Asia Books, PO Box 502, Columbia, MO 65205. (314) 474-0116. 1992. $20.00.

GENERAL WORKS

Architectural Design: Integration of Structural and Environmental Systems. Carl Bovill. Van Nostrand Reinhold, 115 Fifth Avenue, New York, NY 10003. (212) 254-3232. FAX (212) 254-9499. 1991. $34.95.

Building the Future: Innovation in Design, Materials, and Construction. F.K. Garas, editor. E & FN Spon/ Routledge, Chapman and Hall, Inc., 29 West 35th Street, New York, NY 10001-2291. (212) 244-3336 or FAX (212) 563-2269. 1994. Inquire for cost and availability.

New Directions in Architectural & Engineering Practice. H.G. Birnberg, editor. McGraw-Hill Publishing Company, 1221 Avenue of the Americas, New York, NY 10020. (800) 262-4729 or (212) 512-3825. 1992. $60.00.

Tall Building Structures: Analysis and Design. Bryan Stafford Smith, A. Coull. John Wiley and Sons, Inc., 605 Third Avenue, New York, NY 10158. (800) 526-5368 or (212) 850-6000. 1991. $86.95.

HANDBOOKS AND MANUALS

Building Structural Design Handbook. Richard N. White & Charles G. Salmon, editors. Krieger Publishing Company, P.O. Box 9542, Melbourne, FL 32902-9542. (407) 724-9542. FAX (407) 951-3671. 1987. $95.00.

Civil Engineering Calculations Reference Guide. Tyler G. Hicks. McGraw-Hill Publishing Company, 1221 Avenue of the Americas, New York, NY 10020. (800) 262-4729 or (212) 512-3825. 1987. $46.00.

Civil Engineering Practice. Paul N. Cheremisinoff, et al. 5 volumes. Technomic Publishing Company, Inc., 851 New Holland Avenue, Box 3535, Lancaster, PA 17604. (717) 291-5609. or (800) 233-9936. FAX (717) 295-4538. 1987-88. $199.00.

Handbook of Concrete Engineering. Mark Fintel. 2nd edition. Routledge, Chapman and Hall, Inc., 29 West 35th Street, New York, NY 10001-2291. (212) 244-3336 or FAX (212) 563-2269. 1985. $115.00.

Handbook of Mechanics, Materials, and Structures. Alexander Blake, editor. John Wiley and Sons, Inc., 605 Third Avenue, New York, NY 10158. (800) 526-5368 or (212) 850-6000. 1985. $120.00.

Standard Handbook For Civil Engineers. Frederick S. Merritt, editor. 3rd edition. McGraw-Hill Publishing Company, 1221 Avenue of the Americas, New York, NY 10020. (800) 262-4729 or (212) 512-3825. 1983. $124.50.

Structural Engineering Handbook. Edwin H. Gaylord Jr. and Charles N. Gaylord. 3d ed. McGraw-Hill Publishing Company, 1221 Avenue of the Americas, New York, NY 10020. (800) 262-4729 or (212) 512-3825. 1990. $89.50.

ONLINE DATABASES AND CD-ROMS

Architectural Periodicals Index. Royal Institute of British Architects, British Architectural Library, 66 Portland Place, London, W1N 4AD, England. Citations to worldwide literature on architecture, construction methods and standards, urban planning, and interior design, from 1978 to present. Available on DIALOG, (800) 334-2564, online service. Inquire as to cost and availability.

Compendex Plus. Engineering Information, Inc., 345 East47th Street, New York, NY 10017. (212) 705-7600 or (800) 221-1044. Contains citations with abstracts to worldwide literature in engineering and technology, from 1970 to present. Available on online BRS,(800) 289-4277, DIALOG, (800) 334-2564, ORBIT (800) 456-7248, and STN International, FIZ Karlsruhe, P.O. Box 2465, W-7500, Karlsruhe 1, Germany, online services. Also available on CD-ROM. Inquire as to cost and availability.

CSA Engineering. Cambridge Scientific Abstracts, 7200 Wisconsin Avenue, Suite 601, Bethesda, MD 20814. (301) 961-6750 or (800) 843-7751. Contains citations and abstracts of international periodicals and research literature covering all fields of engineering and science and technology,including computer and information science, electronics, mechanical engineering, solid state materials, 1981 to present. Available on BRS,(800) 289-4277, online service.Inquire as to cost and availability.

ICONDA (CIB International Construction Database). Fraunhofer-Gesellschaft, Informationszentrum RAUM und BAU, Nobelstrasse, 12, D-7000, Stuttgart 80, Germany. Contains citations with abstracts to worldwide technical literature on construction and civil engineering, structural engineering, and engineering geology, from 1976 to present. Available on ORBIT, (800) 456-7248, and STN International, FIZ Karlsruhe, P.O. Box 2465, W-7500, Karlsruhe 1, Germany, online services. Inquire as to cost and availability.

NTIS Bibliographic Database. National Technical Information Service, 5285 Port Royal Road, Springfield, VA 22161. (703) 487-4929 or FAX (703) 321-8199. Broad coverage of government-sponsored science and technology research reports, 1964 to present. Available on BRS,(800) 289-4277, DIALOG, (800) 334-2564, ORBIT, (800) 456-7248, and STN International, FIZ Karlsruhe, P.O. Box 2465, W-7500, Karlsruhe 1, Germany, online services. Also available on CD-ROM. Inquire as to cost and availability.

SCISEARCH. Institute for Scientific Information, 3501 Market Street, Philadelphia, PA 19104. (800) 523-1850 or (215) 386-0100. Broad multidisciplinary title and author index to the international literature of science and technology, 1974 to present. Available on DIALOG,(800) 334-2564, and ORBIT,(800) 456-7248, online services. Also available on CD-ROM. Inquire as to cost and availability.

WILSONLINE. H.W. Wilson Company, 950 University Avenue, Bronx, NY 10452. (800) 367-6770 or (212) 588-8400. Makes available online versions of the printed H.W. Wilson indexes including Applied Science and Technology Index, Business Periodicals Index, General Science Index, and Readers' Guide to Periodical Literature. Period covered is generally 1983 to present. Available on BRS,(800) 289-4277, DIALOG,(800) 334-

ARCHITECTURAL ENGINEERING

Ency. of Physical Sciences and Engineering Info. Sources

2564, and OCLC EPIC,(800) 848-5878, online services. Also available on CD-ROM. Inquire as to cost and availability.

PERIODICALS

American Concrete Institute Materials Journal. American Concrete Institute, Box 19150, Redford Station, Detroit, MI 48219. (313) 532-2600. FAX (313) 538-0655. 1929 to present. Bi-monthly. $101.00 per year for non-members.

American Concrete Institute Structural Journal. American Concrete Institute, Box 19150, Redford Station, Detroit, MI 48219. (313) 532-2600. FAX (313) 538-0655. 1929 to present. Bi-monthly. $101.00 per year for non-members.

Architecture. BPI Communications Inc., 1515 Broadway, New York, NY 10036. (212) 764-7300. FAX (212) 944-1719. 1913 to present. Monthly. $421.00 per year.

Building Design and Construction. Cahners Publishing Company (Des Plains), 1350 East Touhy Avenue, Des Plaines, IL 60017-5080. (708) 635-8800. FAX (708) 635-9950. 1950 to present. Monthly. $90.00 per year.

Civil Engineering ASCE. American Society of Civil Engineers, 345 East 47th Street, New York, NY 10017-2398. (212) 705-7520 or (800) 548-2723. 1930 to present. Monthly. $85.00 per year.

Concrete. The Concrete Society, Framewood Road, Wexham, Slough, Berks. SL3 6PJ England. Telephone 0753-662226. FAX 0753-662126. 1966 to present. Bi-monthly. $75.00 per year.

Concrete International: Design and Construction. American Concrete Institute, Box 19150, Redford Station, Detroit, MI 48219. (313) 532-2600. FAX (313) 538-0655. 1979 to present. Monthly. $101.00 per year.

Journal of Structural Engineering. American Society of Civil Engineers, 345 East 47th Street, New York, NY 10017-2398. (212) 705-7520 or (800) 548-2723. 1956 to present. Monthly. $300.00 per year for non-members.

Structural Engineer. Structural Engineers Trading Organization Ltd., 11 Upper Belgrave Street, London SW1X 8BH, England. Telephone 071-235-4535. FAX 071-235-4294. 1922 to present. 24 times a year. Inquire for cost.

Structural Engineering International. International Association for Bridge and Structural Engineering, ETH-Hoenggerberg, CH-8093 Zurich, Switzerland. Telephone 01-3772647. FAX 01-3712131. Quarterly. Inquire for cost and availability.

Structural Engineering Reviews. Elsevier Science [journals], 660 White Plains Rd., Tarrytown, NY 10159-5153. (919) 524-9200. FAX (919) 333-2444. 1988 to present. 4 times a year. $190.00 per year.

RESEARCH CENTERS AND INSTITUTES

Council On Tall Buildings and Urban Habitat. Fritz Engineering Laboratory #13, Lehigh University, Bethlehem, PA 18015. (215) 758-3515. FAX (215) 758-4522.

Phil M. Ferguson Structural Engineering Laboratory, University of Texas at Austin. Balcones Research Center, 10100 Burnet Road, Bldg. 24, Austin, TX 78758. (512) 471-3062.

Structural Engineering Laboratory, University of Michigan. 2340 G. G. Brown Bldg., Ann Arbor, MI 48109. (313) 763-3046. FAX (313) 764-4292.

Structural Engineering Materials Laboratory, University of California At Berkeley. Davis Hall, Berkeley, CA 94720. (415) 642-3464.

Structural Stability Research Council. Fritz Engineering Laboratory, Lehigh University, Bethlehem, PA 18015. (215) 758-3522. FAX (215) 758-4522.

ARGON LASER

See: LASERS

ARITHMETIC

See: MATHEMATICS

AROMATIZATION

See: CHEMICAL ENGINEERING

ARTESIAN AQUIFER

See: GROUNDWATER

ARTIFICIAL INTELLIGENCE

See also: COMPUTER PROGRAMMING, ROBOTICS

ABSTRACT SERVICES AND INDEXES

ACM Guide To Computing Literature. Association for Computing Machinery, 11 West 42nd Street, New York, NY 10036. (212) 869-7440. Fax (212) 869-0481. 1964 to present. Annual. $175.00 per year.

Applied Science and Technology Index; A cumulative subject index to English language periodicals in the fields of aeronautics and space science, computer technology, chemistry, construction industry, energy and related areas. H.W. Wilson Co., 950 University Avenue, Bronx, NY 10452. (800) 367-6770 or (212) 588-8400. FAX (718) 590-1617. From 1958 to present. Monthly. Inquire about cost and availability. Also available on CD-ROM and online.

Artificial Intelligence Abstracts. R.R. Bowker, Bowker A&I Publishing, 121 Chanlon Road, New Providence, NJ 07974. (908) 771-7714. Fax (908) 771-7725. 1985 to present. Monthly. $495.00 per year.

CAD/CAM Abstracts. R.R. Bowker, Bowker A&I Publishing, 121 Chanlon Road, New Providence, NJ 07974. (908) 771-7714. Fax (908) 771-7725. 1984 to present. Monthly. $495.00 per year.

Computer Abstracts, MCB University Press Ltd., PO Box 10812, Birmingham, AL 35201-0812. (800) 633-4931. FAX (205) 995-1588. Monthly. Covers computer theory, data, hardware, systems, networks, human-computer interaction, artificial intelligence, as well as applications of computers in aerospace, business, CAD/CAM, cartography, civil engineering, electronics and electrical engineering, industrial engineering, mechanical engineering, medicine, structural engineering, etc. 1957 to present. Monthly. $1399.95 per year.

Computer and Control Abstracts (aka *INSPEC*). INSPEC/IEEE, Box 1331, 445 Hoes Lane, Piscataway, NJ 08855-1331. (908) 562-5549.Section C: 1966 to present. Monthly. $1455.00 per year. Abstracts of international technical information.

Computer and Information Systems Abstracts Journal. Cambridge Scientific Abstracts, 7200 Wisconsin Avenue, Bethesda, MD 20814. (301) 961-6750. Fax (301) 961-6720. 1962 to present. Monthly. $1035 per year.

Computer Literature Index. Applied Computer Research Inc., Box 82266, Phoenix, AZ 85071-2266. (800) 234-2227. 1971 to present. Quarterly plus annual cumulation. $198.50 per year. Bibliography of books, articles, and reports.

Computing Journal Abstracts. Techgnosis Ltd., Blade House, Battersea Road, Stockport, Cheshire 3AE, England. Telephone 061-442-2639. FAX 061-443-1162. 1969 to present. Monthly. Inquire for price.

Computing Reviews, Association For Computing Machinery, 11 West 42nd Street, New York, NY 10036. (212) 869-7440. Fax (212) 869-0481. Monthly.

Science Citation Index. Institute for Scientific Information, 3501 Market Street, Philadelphia, PA 19104. (215) 386-0100. FAX (215) 386-6362. Inquire about availability and cost. Also available on CD-ROM.

ASSOCIATIONS AND PROFESSIONAL SOCIETIES

American Association For Artificial Intelligence. 445 Burgess Drive, Menlo Park, CA 94025-3496. (415) 328-3123. FAX (415) 321-4457.

Association For Computing Machinery. 1515 Broadway, New York, NY 10036. (212) 869-7440. FAX (212) 869-0481.

Association For Intelligent Systems Technology. 2-212 Center for Science and Technology, Syracuse, NY 13244. (315) 443-2143. FAX (315) 443-1865.

Society For Machine Intelligence. 100 Farnsworth, Detroit, MI 48202. (313) 832-5400. FAX (313) 832-5920.

Special Interest Group On Artificial Intelligence. c/o ACM, 1515 Broadway, New York, NY 10036. (212) 869-7440. FAX (212) 869-0481.

BIBLIOGRAPHIES

ACM Guide To Computing Literature. Association for Computing Machinery, 1515 Broadway, 17th Floor, New York, NY 10036-5701. (212) 869-7440. FAX (212) 944-1318. 1964 to present. Annual. $190.00 for non-members. Also available online via DIALOG.

DIRECTORIES AND BIOGRAPHICAL SOURCES

The AI Directory. American Association for Artificial Intelligence, 445 Burgess Drive, Menlo Park, CA 94025. (415) 328-3123. FAX (415) 321-4457. 1985 to present (1990 to present under current title). Annual. Inquire for cost.

American Men & Women of Science, 1995-96. R.R. Bowker Staff, eds. 19th edition. 8 volumes. R.R. Bowker/Reed International Publishing Company, 121 Chanlon Road, New Providence, NJ 07974. (908) 464-6800 or (800) 521-8110. 1995. $850.00

ASIS Handbook and Directory, American Society ForInformation Science, 8720 Georgia Avenue, Suite 501, Silver Spring, MD 20910-3602. (301) 495-0900. Annual. $100.00.

Computing Information Directory. Hildebrandt Inc., Box 285, Colville, WA 99114. (509) 684-2324. FAX (509) 684-2324. 1981 to present. Annual. $199.95 per year.

Directory of Engineering Societies and Related Organizations. Gordon Davis. 13th edition. American Association of Engineering Societies, 1111 19th Street NW, Suite 608, Washington, DC 20036. (202) 296-2237 or (800) 658-8897. 1989. Inquire for price.

International Research Centers Directory. Gale Research, 835 Penobscot Building, Detroit, MI 48226-4094. (313) 961-2242 or (800) 877-4253. FAX (313) 961-6083. 8th edition. 1995. $410.00.

Research Centers Directory. Gale Research, 835 Penobscot Building, Detroit, MI 48226-4094. (313) 961-2242. (800) 877-4253. FAX (313) 961-6083. $485.00.

Scientific and Technical Organizations and Agencies Directory. Gale Research, 835 Penobscot Building, Detroit, MI 48226-4094. (313) 961-2242. (800) 877-4253. FAX (313) 961-6083. 4th edition. 1996. $195.00.

Who's Who in Computing. Canadian M I S Database Inc., 268 Lakeshore Road E., Suite 510, Mississauga, ON L5G 1H1, Canada. (905) 271-1601. FAX (905) 271-4522. 1987 to present. Annual. Inquire for price in U.S.

Who's Who in Technology. Gale Research, 835 Penobscot Building, Detroit, MI 48226-4094. (313) 961-2242. (800) 877-4253. FAX (313) 961-6083. 1995. $195.00.

ENCYCLOPEDIAS AND DICTIONARIES

Computer Dictionary. Donald D. Spencer. 4th edition. Camelot Publishing Company, 709 SW 80th Blvd., Gainesville, FL 32607-1537. (904) 331-0952. 1993. $24.95.

Dictionary of Computing. S.M.H. Collin. 2nd edition. Peter Collin Publishing, 8 the Causeway, Teddington, TW11 0HE,

England. FAX 0181-943-3386. 1994. Inquire for cost.

Encyclopedia of Artificial Intelligence. Stuart C. Shapiro, editor. Second edition. John Wiley and Sons, Inc., 605 Third Avenue, New York, NY 10158. (800) 526-5368 or (212) 850-6000. 1992. $297.50.

Encyclopedia of Computer Science and Engineering. Anthony Ralston & Edwin D. Reilly Jr., editors. 3rd revised edition. Van Nostrand Reinhold, 115 Fifth Avenue, New York, NY 10003. (212) 254-3232. FAX (212) 254-9499. 1993. $125.00.

The Encyclopedia of Operations Research and Management Science. Saul I. Gass & Carl M. Harris, editors. Kluwer Academic Publishers, PO Box 358, Accord Station, Hingham, MA 02018-0358. (617) 871-6600. FAX (617) 871-6528. 1996. $295.00.

The McGraw-Hill Illustrated Encyclopedia of Robotics and Artificial Intelligence. Stan Gonilisco. McGraw-Hill Publishing Company, 1221 Avenue of the Americas, New York, NY 10020. (800) 262-4729 or (212) 512-3825. 1994. $34.95 (hardbound), $24.95(paperbound).

GENERAL WORKS

AI: The Tumultuous History of the Search For Artificial Intelligence. Daniel Crevier. Basic Books, 10 E. 53rd Street, New York, NY 10022. (800) 242-7737. FAX (212) 207-7145. 1994. $14.00.

Artificial Intelligence: A Philosophical Introduction. Jack Copeland. Blackwell Scientific Publications, 238 Main Street, Cambridge, MA 02142. (617) 876-7000 or (800) 759-6102. FAX (617) 876-7022. 1993. $17.95.

Artificial Intelligence: Concepts and Applications in Engineering. A.R. Mirzai. MIT Press, 55 Hayward Street, Cambridge, MA 02142. (617) 253-2884 or (800) 356-0343. FAX (617) 253-1709. 1990. Out of print.

Artificial Intelligence Methods and Applications. Nikolaos G. Bourbakis, editor. World Scientific Publishing Company, Inc., 1060 Main Street, Unit B, River Edge, NJ 07661. (800) 227-7562. FAX (201) 487-9656. 1992. $116.00.

Artificial Intelligence: Theory & Practice. Thomas Dean, et al. Benjamin-Cummings Publishing Company, 390 Bridge Parkway, Redwood City, CA 94065. (800) 950-2665. FAX (415) 594-4409. 1995. $54.95.

Essentials of Artificial Intelligence. Matthew L. Ginsberg. Morgan Kaufmann Inc., 340 Pine Street, 6th Floor, San Francisco, CA 94104. (800) 745-7323. FAX (415) 982-2665. 1993. $49.95.

Introduction To Artificial Intelligence and Expert Systems. Dan W. Patterson. Prentice Hall, 113 Sylvan Avenue, Route 9W, Englewood Cliffs, NJ 07632. (201) 592-2000 or (800) 922-0579. 1990. Out of print.

HANDBOOKS AND MANUALS

Artificial Intelligence, Knowledge Engineering, Expert Systems, Natural Language, Human Factors Ergonomics, Man-Machine Interface Design: Handbook of Human-Computer Interaction. M. Helander, editor. Elsevier Science Publishing Company, Inc.,

655 Avenue of the Americas, New York, NY 10010. (212) 989-5800. 1993. $251.00 (hardbound), $83.00 (paperbound).

ONLINE DATABASES AND CD-ROMS

Compuscience. FIZ Karlsruhe, D-7514, Eggenstein-Leopoldshafen 2, Germany. Contains citations with abstracts to European and North American literature on computer science, 1972 to present. Available on STN International, FIZ Karlsruhe, P.O. Box 2465, W-7500, Karlsruhe 1, Germany, online service. Inquire as to cost and availability.

Computer and Information Systems Abstracts. Cambridge Scientific Abstracts, 7200 Wisconsin Avenue, Suite 601, Bethesda, MD 20814. (301) 961-6750 or (800) 843-7751. Contains citations to worldwide literature in theoretical and applied computer science and related areas, from 1981 to present.Inquire as to cost and availability.

Computer and Mathematics Search. Institute for ScientificInformation, 3501 Market Street, Philadelphia, PA 19104. (215) 386-0100. FAX (215) 386-6362. Covers worldwide literature in computer science and mathematics, from 1980 to present. Available on BRS,(800) 289-4277, online service. Inquire as to cost and availability.

Computer Database. Information Access Company, 362 Lakeside Drive, Foster City, CA 94404. (415) 378-5000 or (800) 227-8431. Contains citations with abstracts to literature from trade journals covering the computer,telecommunications,and electronics industries. Available on the BRS, (800) 289-4277, Compuserve Information Service,(800) 848-8990, and DIALOG,(800) 334-2564, online services. Inquire as to cost and availability.

Current Contents Search. Institute for Scientific Information, 3501 Market Street, Philadelphia, PA 19104. (215) 386-0100. FAX (215) 386-6362. Contains citations to articles listed in the table of contents of science and technology journals. Also articles in social sciences and life sciences journals. Available on BRS,(800) 289-4277, DIALOG,(800) 334-2564, online services. Inquire as to cost and availability.

Dissertation Abstracts. University Microfilms International, 300 North Zeeb Road, Ann Arbor, MI 48106. (800) 521-0600 or (313) 761-4700. Scope includes virtually all doctoral dissertations accepted at accredited American institutions from 1861 to present in 252 subject areas. Available on BRS,(800) 289-4277, DIALOG,(800) 334-2564, and OCLC EPIC,(800) 848-5878, online services. Also available on CD-ROM. Inquire as to cost and availability.

INSPEC. Institution of Electrical Engineers, Michael Faraday House, Six Hills Way, Stevenage, Herts. SG1 2AY, England. Telephone: 0438 313311 or FAX 0438 742840. Contains citations to the worldwide literature of physics, electronics and electrical engineering, computer technology, and related fields. Available on BRS,(800) 289-4277, DIALOG, (800) 334-2564, ORBIT, (800) 456-7248, and STN International, FIZ Karlsruhe, P.O. Box 2465, W-7500, Karlsruhe 1, Germany, online services. Inquire as to cost and availability.

NTIS Bibliographic Database. National Technical Information Service, 5285 Port Royal Road, Springfield, VA 22161. (703) 487-4929 or FAX (703) 321-8199. Broad coverage of government-sponsored science and technology research reports, 1964 to present. Available on BRS,(800) 289-4277, DIALOG, (800) 334-2564, ORBIT, (800) 456-7248, and STN International, FIZ

Karlsruhe, P.O. Box 2465, W-7500, Karlsruhe 1, Germany, online services. Also available on CD-ROM. Inquire as to cost and availability.

SCISEARCH. Institute for Scientific Information, 3501 Market Street, Philadelphia, PA 19104. (800) 523-1850 or (215) 386-0100. Broad multidisciplinary title and author index to the international literature of science and technology, 1974 to present. Available on DIALOG,(800) 334-2564, and ORBIT,(800)456-7248, online services. Also available on CD-ROM. Inquire as to cost and availability.

Supertech. Bowker A & I Publishing, 121 Chanlon Road, New Providence, NJ 07974. (800) 521-8110 or (908) 464-6800. Contains citations to the world's published and unpublished literature in the fields of artificial intelligence, biotechnology, computer-aided design and manufacturing, robotics, and telecommunications. Covers the period 1973 to present. Available on DIALOG,(800) 334-2564, and ORBIT,(800) 456-7248, online services. Inquire as to cost and availability.

WILSONLINE. H.W. Wilson Company, 950 University Avenue, Bronx, NY 10452. (800) 367-6770 or (212) 588-8400. Makes available online versions of the printed H.W. Wilson indexes including Applied Science and Technology Index, Business Periodicals Index, General Science Index, and Readers' Guide to Periodical Literature. Period covered is generally 1983 to present. Available on BRS,(800) 289-4277, DIALOG,(800) 334-2564, and OCLC EPIC,(800) 848-5878, online services. Also available on CD-ROM. Inquire as to cost and availability.

PERIODICALS

AI Applications. University of Idaho, Box 3066, Moscow, ID 93943. (208) 885-7033. FAX (208) 885-6226. 1987 to present. Three times a year. $37.00.

AI Expert. Miller Freeman Inc., 600 Harrison Street, San Francisco, CA 94107. (415) 905-2200. FAX (415) 905-2232. 1986 to present. Monthly. $34.00.

AI Magazine. American Association for Artificial Intelligence, 445 Burgess Drive, Menlo Park, CA 94025. (415) 328-3123. FAX (415) 321-4457. 1979 to present. Quarterly. $40.00.

Applied Artificial Intelligence: AAI. Taylor & Francis, 1900 Frost Road, Suite 101, Bristol, PA 19007. (215) 785-5800. FAX (215) 785-5515. 1987 to present. Quarterly. $190.00.

Artificial Intelligence. Elsevier Science Inc., Box 882, Madison Square Station, New York, NY 10159. (212) 989-5800. FAX (212) 633-3990. 1970 to present. 14 times a year. $1279.00.

The Artificial Intelligence Review. Kluwer Academic Publishers, P.O. Box 358, Accord Station, Hingham, MA 02018-0358. (617) 871-6000. 1986 to present. Six times a year. $233.50.

International Journal of Pattern Recognition and Artificial Intelligence. World Scientific Publishing Company, Inc., 1060 Main Street, Unit B, River Edge, NJ 07661. (800) 227-7562. FAX (201) 487-9656. 1987 to present. Six times a year. $350.00.

Law, Computers and Artificial Intelligence. Triangle Journals Ltd., PO Box 65, Wallingford, Oxfordshire, OX10 0YG, England. Telephone 0491-838013. FAX 0491-834968. 1992 to present. Three times a year. Inquire for cost.

Robotica. Cambridge University Press, Journals Department, 40 W 20th Street, New York, NY 10011-4211. (212) 924-3900. 1983 to present. Bi-monthly. $119.00.

Sigart Bulletin. Special Interest Group on ArtificialIntelliegence, c/o ACM, 1515 Broadway, New York, NY 10036. (212) 869-7440. FAX (212) 869-0481. 1990 to present. Quarterly. $40.00 for non-members.

RESEARCH CENTERS AND INSTITUTES

George Washington University Institute For Artificial Intelligence. 2021 K Street NW #710, Washington, DC 20006. (202) 676-5110. FAX (202) 994-4606.

Harvard University Center For Research in Computing Technology. Aiken Computation Laboratory, 33 Oxford Street, Cambridge, MA 02138. (617) 495-4117. FAX (617) 495-9837.

Massachusetts Institute of Technology, Artificial Intelligence Laboratory. 545 Technology Square, Cambridge, MA 02139. (617) 253-6218.

Ohio State University, Laboratory For Artificial Intelligence. CIS Department, 228 Bolz Hall, 2036 Neil Avenue, Columbus, OH 43210. (614) 292-0208.

Stanford University Knowledge Systems Laboratory. 701 C Welch Road, Palo Alto, CA 94304. (415) 723-4878. FAX (415) 723-5850.

University of Pennsylvania, Center For Artificial Intelligence. Computer and Information Sciences Department, Moore School of Electrical Engineering/DZ, 200 S. 33rd Street, Philadelphia, PA 19104-6389. (215) 898-3191.

University of Texas At Austin, Artificial Intelligence Laboratory. Taylor Hal 2.124, Austin, TX 78712-1188. (512) 471-9567. FAX (512) 471-8885.

ASBESTOS

ABSTRACT SERVICES AND INDEXES

Applied Science and Technology Index; A cumulative subject index to English language periodicals in the fields of aeronautics and space science, computer technology, chemistry, construction industry, energy and related areas. H.W. Wilson Co., 950 University Avenue, Bronx, NY 10452. (800) 367-6770 or (212) 588-8400. FAX (718) 590-1617. From 1958 to present. Monthly. Inquire about cost and availability. Also available on CD-ROM and online.

Bibliography and Index of Geology. American Geological Institute, 4220 King Street, Alexandria, VA 22302. (703) 379-2480. Fax (703) 379-7563. Monthly. $1295.00 per year. Also available online as GeoRef.

Engineering Index Monthly. Engineering Information, Inc., Castle Point on the Hudson, Hoboken, NJ 07030. (800) 221-1044. FAX (212) 832-1857. Monthly. $2300.00 per year. Also available online as COMPENDEX, and also on CD-ROM. Cov-

ASBESTOS

Ency. of Physical Sciences and Engineering Info. Sources

ers chemical engineering, computers, electrical engineering, civil engineering, metals and mining, industrial management, and mechanical engineering.

Environmental Engineering Abstracts, Cambridge Scientific Abstracts, 7200 Wisconsin Avenue, Bethesda, MD 20814-4823. (301) 961-6750. FAX (301) 961-6720. Monthly. Covers hazardous materials, environmental impact and protection, treatment of sewage andindustrial wastes, hydroelectric power, tidal and wind power, artic and tropical engineering.

Pollution Abstracts, Cambridge Scientific Abstracts, 7200 Wisconsin Avenue, Bethesda, MD 20814-4823. (301) 961-6750. FAX (301) 961-6720. Inquire for cost and availability.

Science Citation Index. Institue for Scientific Information, 3501 Market Street, Philadelphia, PA 19104. (215) 386-0100. FAX (215) 386-6362. Inquire about availability and cost. Also available on CD-ROM.

ANNUAL REVIEWS AND YEARBOOKS

U.S. Department of the Interior, Bureau of Mines, Minerals Yearbook. Superintendent of Documents, U.S. Government Printing office, Box 371954, Pittsburgh, PA 15250-7954. (202) 783-3238. FAX (202) 512-2233. 1932 to present. Annual. Price varies.

ASSOCIATIONS AND PROFESSIONAL SOCIETIES

Asbestos Information Association of North America. 1745 Jefferson Davis Highway, Suite 509, Arlington, VA 22202. (703) 979-1150. FAX (703) 979-1152.

Asbestos Institute. 1130 Sherbrooke Street West, Suite 410, Montreal PQ, Canada H3A 2M8. (514) 844-3956.

Asbestos International Association. 68 Gloucester Place, London W1H 3HL, England.

DIRECTORIES AND BIOGRAPHICAL SOURCES

American Men & Women of Science, 1995-96. R.R. Bowker Staff, eds. 19th edition. 8 volumes. R.R. Bowker/Reed International Publishing Company, 121 Chanlon Road, New Providence, NJ 07974. (908) 464-6800 or (800) 521-8110. 1995. $850.00.

International Research Centers Directory. Gale Research, 835 Penobscot Building, Detroit, MI 48226-4094. (313) 961-2242. (800) 877-4253. FAX (313) 961-6083. 8th edition. 1995. $410.00

Directory of Engineering Societies and Related Organizations. Gordon Davis. 13th edition. American Association of Engineering Societies, 1111 19th Street NW, Suite 608, Washington, DC 20036. (202) 296-2237 or (800) 658-8897. 1989. Inquire for price.

Research Centers Directory. Gale Research, 835 Penobscot Building, Detroit, MI 48226-4094. (313) 961-2242. (800) 877-4253. FAX (313) 961-6083. $485.00.

Scientific and Technical Organizations and Agencies Directory. Gale Research, 835 Penobscot Building, Detroit, MI 48226-

4094. (313) 961-2242. (800) 877-4253. FAX (313) 961-6083. 4th edition. 1996. $195.00.

Who's Who in Engineering. American Association of Engineering Societies, 1111 19th Street NW, Suite 608, Washington, DC 20036. (202) 296-2237 or (800) 658-8897. 8th edition. 1991. Inquire for price.

ENCYCLOPEDIAS AND DICTIONARIES

Encyclopedia of Health Information Sources. 3d edition. 1996. Gale Research, 835 Penobscot Building, Detroit, MI48226-4094. (313) 961-2242. (800) 877-4253. FAX (313) 961-6083. $165.00.

Encyclopedia of Occupational Health and Safety. Lougi Parmeggiani, ed. 3rd revised edition. 2 volumes. International Labour office, 1828 L Street NW, Suite 801, Washington, DC 20036. (202) 653-7652. FAX (202) 376-5083. Inquire for price and availability.

GENERAL WORKS

Asbestos Risks and Medical Advances. Butterworth Legal Publishers, 8 Industrial Way, Bldg. C, Salem, NH 03079. (603) 898-9664 or (800) 548-4001. 1993. Inquire for cost and availability.

Asbestos: The Hazardous Fiber. Malvin A. Benarde. CRC Publishers, Inc., 2000 Corporate Blvd., N.W., Boca Raton, FL 33431. (407) 994-0555 or (800) 333-8300. 1990. $69.95.

Sourcebook On Asbestos Diseases: Current Asbestos Legal, Medical and Technical Research. George A. and Barbara Peters, editors. Butterworth Legal Publishers, 8 Industrial Way, Bldg. C, Salem, NH 03079. (603) 898-9664 or (800) 548-4001. 1994. Inquire for cost and availability.

ONLINE DATABASES AND CD-ROMS

Compendex Plus. Engineering Information, Inc., 345 East 47th Street, New York, NY 10017. (212) 705-7600 or (800) 221-1044. Contains citations with abstracts to worldwide literature in engineering and technology, from 1970 to present. Available on online BRS,(800) 289-4277, DIALOG, (800) 334-2564, ORBIT (800) 456-7248, and STN International, FIZ Karlsruhe, P.O. Box 2465, W-7500, Karlsruhe 1, Germany, online services. Also available on CD-ROM. Inquire as to cost and availability.

CSA Engineering. Cambridge Scientific Abstracts, 7200 Wisconsin Avenue, Suite 601, Bethesda, MD 20814. (301) 961-6750 or (800) 843-7751. Contains citations and abstracts of international periodicals and research literature covering all fields of engineering and science and technology,including computer and information science, electronics, mechanical engineering, solid state materials, 1981 to present. Available on BRS,(800) 289-4277, online service.Inquire as to cost and availability.

Current Contents Search. Institute for Scientific Information, 3501 Market Street, Philadelphia, PA 19104. (215) 386-0100. FAX (215) 386-6362. Contains citations to articles listed in the table of contents of science and technology journals. Also articles in social sciences and life sciences journals. Available on BRS,(800) 289-4277, DIALOG,(800) 334-2564, online services. Inquire as to cost and availability.

Dissertation Abstracts. University Microfilms International, 300 North Zeeb Road, Ann Arbor, MI 48106. (800) 521-0600 or (313) 761-4700. Scope includes virtually all doctoral dissertations accepted at accredited American institutions from 1861 to present in 252 subject areas. Available on BRS,(800) 289-4277, DIALOG,(800) 334-2564, and OCLC EPIC,(800) 848-5878,online services. Also available on CD-ROM. Inquire as to cost and availability.

GEOREF: Bibliography and Index of Geology. American Geological Institute, 4220 King Street, Alexandria, VA 22302. (703) 379-2480. Fax (703) 379-7563. Monthly. Inquire for price and availability.

NTIS Bibliographic Database. National Technical Information Service, 5285 Port Royal Road, Springfield, VA 22161. (703) 487-4929 or FAX (703) 321-8199. Broad coverage of government-sponsored science and technology research reports, 1964 to present. Available on BRS,(800) 289-4277, DIALOG, (800) 334-2564, ORBIT, (800) 456-7248, and STN International, FIZ Karlsruhe, P.O. Box 2465, W-7500, Karlsruhe 1, Germany, online services. Also available on CD-ROM. Inquire as to cost and availability.

SCISEARCH. Institute for Scientific Information, 3501 Market Street, Philadelphia, PA 19104. (800) 523-1850 or (215) 386-0100. Broad multidisciplinary title and author index to the international literature of science and technology, 1974 to present. Available on DIALOG,(800) 334-2564, and ORBIT,(800) 456-7248, online services. Also available on CD-ROM. Inquire as to cost and availability.

WILSONLINE. H.W. Wilson Company, 950 University Avenue, Bronx, NY 10452. (800) 367-6770 or (212) 588-8400. Makes available online versions of the printed H.W. Wilson indexes including Applied Science and Technology Index, Business Periodicals Index, General Science Index, and Readers' Guide to Periodical Literature. Period covered is generally 1983 to present. Available on BRS,(800) 289-4277, DIALOG,(800) 334-2564, and OCLC EPIC,(800) 848-5878, online services. Also available on CD-ROM. Inquire as to cost and availability.

PERIODICALS

Asbestos Information Association North America—news & Notes. Asbestos Information Association North America, 1745 Jefferson Davis Highway, Suite 509, Arlington, VA 22202. (703) 979-1150. FAX (703) 979-1152. Monthly. Inquire for cost and availability information.

Asbestos Producer. Quebec Asbestos Mining Association, 2000 rue Peel, Suite 750, Montreal, Quebec H3H 2W5, Canada. (514) 844-4751. 1954 to present. Irregular. Inquire for cost and availability.

Asbestos Worker. International Association of Asbestos Workers, Machinists Building, 1776 Massachusetts Avenue NW, Suite 301, Washington, DC 20036. (202) 785-2388. 1916 to present. Quarterly. Free.

Engineering & Mining Journal. Maclean Hunter Publishing Company, 29 N. Wacker Dr., Chicago, IL 60606. (312) 726-2802. FAX (312) 726-2574. Monthly. $60.00 per year.

RESEARCH CENTERS AND INSTITUTES

U.S. Environmental Protection Agency, Research &Development. 401 M Street NW, Washington, DC 20460. (202) 260-7676. FAX (202) 260-9761.

U.S. Environmental Protection Agency, Risk Reduction Engineering Laboratory. 26 W. Martin Luther King Drive, Cincinnati, OH 45268. (513) 569-7418. FAX (513) 569-7680.

ASPHALT

See also: HIGHWAY ENGINEERING

ABSTRACT SERVICES AND INDEXES

Applied Science and Technology Index; A cumulative subject index to English language periodicals in the fields of aeronautics and space science, computer technology, chemistry, construction industry, energy and related areas. H.W. Wilson Co., 950 University Avenue, Bronx, NY 10452. (800) 367-6770 or (212) 588-8400. FAX (718) 590-1617. From 1958 to present. Monthly. Inquire about cost and availability. Also available on CD-ROM and online.

Civil and Structural Engineering Abstracts, Cambridge Scientific Abstracts, 7200 Wisconsin Avenue, Bethesda, MD 20814-4823. (301) 961-6750. FAX (301) 961-6720. 1993 to present. Monthly. $385.00 per year. Topics covered include structural design, construction equipment and methods, civil defense and military engineering, surveying, highway engineering, maritime and port structures, materials, land reclamation, and soil mechanics.

Engineering Index Monthly. Engineering Information, Inc., Castle Point on the Hudson, Hoboken, NJ 07030. (800) 221-1044. FAX (212) 832-1857. Monthly. $2300.00 per year. Also available online as COMPENDEX, and also on CD-ROM. Covers chemical engineering, computers, electrical engineering, civil engineering, metals and mining, industrial management, and mechanical engineering.

ASSOCIATIONS AND PROFESSIONAL SOCIETIES

American Association of State Highway & Transportation officials. 444 N. Capitol Street NW, Suite 249, Washington, DC 20001. (202) 624-5800. FAX (202) 624-5806.

Asphalt Emulsion Manufacturers Association. 3 Church Street, Suite 250, Annapolis, MD 21401. (410) 267-0023.

Asphalt Institute. Research Park Drive, Box 14052, Lexington, KY 40512-4052. (606) 288-4960. FAX (606) 288-4999.

Asphalt Recycling and Reclaiming Association. 3 Church Circle, Suite 250, Annapolis, MD 21401. (410) 267-0023.

Association of Asphalt Paving Technologists. 400 Selby Avenue, Suite 1, St. Paul, MN 55102. (612) 293-9188. FAX (612) 293-9193.

Bituminous and Aggregate Equipment Bureau. 111 East Wisconsin Avenue, Suite 940, Milwaukee, WI 53202. (414) 272-0943.

National Asphalt Pavement Association. NAPA Building, 5100 Forbes Blvd., Lanham, MD 20706-4413. (301) 731-4748. FAX (301) 731-4621.

DIRECTORIES AND BIOGRAPHICAL SOURCES

Asphalt Emulsion Manufacturers Association—Membership Directory. Asphalt Emulsion Manufacturers Association, 3 ChurchStreet, Suite 250, Annapolis, MD 21401. (410) 267-0023. Inquire for cost and availability.

Asphalt Paving Technologists Directory. Association of Asphalt Paving Technologists, 400 Selby Avenue, Suite 1, St. Paul, MN 55102. (612) 293-9188. FAX (612) 293-9193. Inquire for cost and availability.

Asphalt Recycling and Reclaiming Association Membership Directory. Asphalt Recycling and Reclaiming Association, 3 Church Circle, Suite 250, Annapolis, MD 21401. (410) 267-0023. Inquire for availability.

Research Centers Directory. Gale Research, 835 Penobscot Building, Detroit, MI 48226-4094. (313) 961-2242. (800) 877-4253. FAX (313) 961-6083. $485.00

Who's Who in Engineering. American Association of Engineering Societies, 1111 19th Street NW, Suite 608, Washington, DC 20036. (202) 296-2237 or (800) 658-8897. 8th edition. 1991. Inquire for price.

ENCYCLOPEDIAS AND DICTIONARIES

Encyclopedia of Physical Science and Technology. Robert A. Meyers, ed. 18 volumes. Academic Press, Inc., 6277 Sea Harbor Drive, Orlando, FL 32887. 1992. $2100.00

McGraw-Hill Dictionary of Scientific and Technical Terms. Sybil P. Parker, ed. 5th edition. McGraw-Hill Publishing Company, 1221 Avenue of the Americas, New York, NY 10020. (800) 262-4729 or (212) 512-3825. 1993. $110.50.

McGraw-Hill Encyclopedia of Engineering. Sybil P. Parker, ed. 2nd edition. McGraw-Hill Publishing Company, 1221 Avenue of the Americas, New York, NY 10020. (800) 262-4729 or (212) 512-3825. 1993. $95.50.

McGraw-Hill Encyclopedia of Science and Technology. Sybil P. Parker, ed. 7th edition. 20 volumes. McGraw-Hill Publishing Company, 1221 Avenue of the Americas, New York, NY 10020. (800) 262-4729 or (212) 512-3825. 1992. $1900.00

GENERAL WORKS

Asphalt and Asphalt Additives. Transportation Research Board. National Academy Press, Division of the National Academy of Science, 2101 Constitution Avenue, N.W., Lockbox 285, Washington, DC 20055. (800) 624-6242. FAX (202) 334-2793. 1992. $22.00.

Asphalt Cold-mix Recycling. Asphalt Institute, Research Park Drive, Box 14052, Lexington, KY 40512-4052. (606) 288-4960. FAX (606) 288-4999. 1983. $8.00.

Asphalt Usage 1993, United States and Canada. Asphalt Institute, Research Park Drive, Box 14052, Lexington, KY 40512-4052. (606) 288-4960. FAX (606) 288-4999. 1994. Inquire for cost and availability.

HANDBOOKS AND MANUALS

Asphalt Cold-mix Manual. Asphalt Institute, Research Park Drive, Box 14052, Lexington, KY 40512-4052. (606) 288-4960. FAX (606) 288-4999. 1991. $16.00.

Asphalt Handbook. Asphalt Institute, Research Park Drive,Box 14052, Lexington, KY 40512-4052. (606) 288-4960. FAX (606) 288-4999. 1989. $33.00.

Asphalt Pavement Repair Manuals of Practice. Kelly L. Smith, et al. Strategic Highway Research Program/National Research Council, 2101 Constitution Avenue, N.W., Lockbox 285, Washington, DC 20055. (800) 624-6242. FAX (202) 334-2793. $15.00.

Asphalt Paving Manual. Asphalt Institute, Research Park Drive, Box 14052, Lexington, KY 40512-4052. (606) 288-4960. FAX (606) 288-4999. 1983. $12.00.

Asphalt Plant Manual. Asphalt Institute, Research Park Drive, Box 14052, Lexington, KY 40512-4052. (606) 288-4960. FAX (606) 288-4999. 1983. $12.00.

Asphalt Pocketbook of Useful Information. Asphalt Institute, Research Park Drive, Box 14052, Lexington, KY 40512-4052. (606) 288-4960. FAX (606) 288-4999. 1982. $9.00.

ONLINE DATABASES AND CD-ROMS

Compendex Plus. Engineering Information, Inc., 345 East 47th Street, New York, NY 10017. (212) 705-7600 or (800) 221-1044. Contains citations with abstracts to worldwide literature in engineering and technology, from 1970 to present. Available on online BRS,(800) 289-4277, DIALOG, (800) 334-2564, ORBIT (800) 456-7248, and STN International, FIZ Karlsruhe, P.O. Box 2465, W-7500, Karlsruhe 1, Germany, online services. Also available on CD-ROM. Inquire as to cost and availability.

CSA Engineering. Cambridge Scientific Abstracts, 7200 Wisconsin Avenue, Suite 601, Bethesda, MD 20814. (301) 961-6750 or (800) 843-7751. Contains citations and abstracts of international periodicals and research literature covering all fields of engineering and science and technology,including computer and information science, electronics, mechanical engineering, solid state materials, 1981 to present. Available on BRS,(800) 289-4277, online service.Inquire as to cost and availability.

NTIS Bibliographic Database. National Technical Information Service, 5285 Port Royal Road, Springfield, VA 22161. (703) 487-4929 or FAX (703) 321-8199. Broad coverage of government-sponsored science and technology research reports, 1964 to present. Available on BRS,(800) 289-4277, DIALOG, (800) 334-2564, ORBIT, (800) 456-7248, and STN International, FIZ Karlsruhe, P.O. Box 2465, W-7500, Karlsruhe 1, Germany, online services. Also available on CD-ROM. Inquire as to cost and availability.

TRIS (Transportation Research Information). National Academy of Science, 2101 Constitution Avenue, N.W., Washington, DC 20418. (202) 334-3313 or (800) 624-6242. Citations with abstracts of literature on transportation, including air, highway, rail, maritime and other modes from 1968 to present. Available on DIALOG,(800) 334-2564, online service. Inquire as to cost and availability.

PERIODICALS

Asphalt. Asphalt Institute, Research Park Drive, Box14052, Lexington, KY 40512-4052. (606) 288-4960. FAX (606) 288-4999. Three times a year. Write for availability information.

Asphalt Emulsion Manufacturers Association Newsletter. Asphalt Emulsion Manufacturers Association, 3 Church Street, Suite 250, Annapolis, MD 21401. (410) 267-0023. 1973 to present. Quarterly. Free with membership, inquire for cost and availability for non-members.

Asphalt Paving Technology. Association of Asphalt Paving Technologists, 400 Selby Avenue, Suite 1, St. Paul, MN 55102. (612) 293-9188. FAX (612) 293-9193. Annual. Inquire for cost and availability.

Construction Products. Gordon Publications/ Cahners-Reed Elsevier, 301 Gilbraltar Drive, Box 650, Morris Plains, NJ 07950. (201) 252-5100. FAX (201) 539-3476. 1892 to present. 6 times a year. $40.00 per year.

Highway Builder. Naylor Publications Inc., 11350 McCormick Road, Executive Plaza Three, Hunt Valley, MD 21031. (410) 785-2445. 1921 to present. Quarterly. $10.00 per year.

HMAT: Hot Mix Asphalt Technology. National Asphalt Pavement Association, NAPA Building, 5100 Forbes Blvd., Lanham, MD 20706-4413. (301) 731-4748. FAX (301) 731-4621. Quarterly. Write for availability information.

Journal of Materials in Civil Engineering: Properties, Applications, Durability. American Society of Civil Engineers, Materials Engineering Division, 345 East 47th Street, New York, NY 10017-2398. (212) 705-7520 or (800) 548-2723. 1989 to present. Quarterly. $128.00 per year.

Public Roads. U.S. Federal Highway Administration, 400 7th Street NW, Washington, DC 20590. (703) 285-2104. 1918 to present. Quarterly. $12.00 per year.

Public Works. Public Works Journal Corporation, 200 S. Broad Street, Ridgewood, NJ 07451. (201) 445-5800. FAX (201) 445-5170. 1896 to present. 13 times a year. $45.00 per year.

Roads & Bridges. Scranton Gillette Communications Inc., 380 E. Northwest Highway, Des Plaines, IL 60016. (708) 298-6622. FAX (708) 390-0408. 1892 to present. Monthly. $20.00 per year.

Research Centers and Institutes

Asphalt Institute. Research Park Drive, Box 14052, Lexington, KY 40512-4052. (606) 288-4960. FAX (606) 288-4999.

Auburn University National Center For Asphalt Technology. 211 Ramsay Hall, Auburn University, AL 36849. (205) 844-6228. FAX (205) 844-6248.

SPECIFICATIONS AND STANDARDS

*Annual Book of ASTM Standards, Volume 04.*03. American Society for Testing and Materials (ASTM), 1916 Race Street, Philadelphia, PA 19103. (215) 299-5585. Annual. Inquire for cost and availability.

ASTEROIDS

See also: ASTROGEOLOGY, ASTRONOMY, COMETE, METEORITES, NAMES of INDIVIDUAL PLANETS, PLANETARY SCIENCES, SATELLITES (NATURAL),SOLAR SYSTEM,

ABSTRACT SERVICES AND INDEXES

Applied Science and Technology Index; A cumulative subject index to English language periodicals in the fields of aeronautics and space science, computer technology, chemistry, construction industry, energy and related areas. H.W. Wilson Co., 950 University Avenue, Bronx, NY 10452. (212) 588-8400. (800) 367-6770. FAX (718) 590-1617. From 1958 to present. Monthly. Inquire about cost and availability. Also available on CD-ROM and online. ISSN: 0003-6986.

Astronomy and Astrophysics Abstracts. Springer-Verlag New York, 175 F ifth Avenue, New York, NY 10010. (212) 460-1500. FAX (212) 473-6272. Published for the Astronomisches Rechen-Institut. Comprehensive coverage of all aspects of astronomy, astrophysics and related fields. 1969 to present. Two parts per year. Annual. ISSN: 0067-0022.

General Science Index. H.W. Wilson Company, 950 University Avenue, Bronx, NY 10452. (212) 588-8400. (800) 367-6770. FAX (718) 590-1617. From 1978 to present. Ten issues per year; quarterly and annual cumulations. Service basis. Available on CD-ROM and online. Inquire regarding cost and availability. ISSN: 0162-1963.

Government Reports Announcements and Index. U. S. National Technical Information Service (NTIS), 5285 Port Royal Road, Springfield, VA 22161. (703) 487-4650. FAX (703) 321-8547. From 1968 to present. Annual. $630.00 per year. Also available online as *NTIS Bibliographic Database* and on CD-ROM. ISSN:

Meteorological and Geoastrophysical Abstracts. Americ an Meteorological Society, c/o Inforonics, Inc., 550 N ewtown Road, Littleton, MA 01460. (508) 486-8976. FAX (508) 486-0027. Covers literature in environmental sciences, meteorology, astrophysics, hydrology, glaciology, and physical oceanography. From 1950 to present. Monthly. $950.00 per year. Also available on CD-ROM and online. Inquire regarding cost and availability. ISSN: 0026-1130.

NTIS Alerts: Astronomy & Astrophysics. U. S. National Technical Information Service. 5285 Port Royal Road, Springfield, VA 22161. (703) 487-4650. FAX (703) 321-8547. Weekly. $140.00 per year.

Pascal 48: Environnement Cosmique Terrestre, Astronomie Et Geologie Extraterrestre. Centre National de la Recherche Scientifique, Institut de Information Scientifique et Technique, 2 aliee du Parc de Brabois, 54514 Vandoeuvre Les Nancy Cedex, France. TEL 83-50-46-00. FAX: 83-50-46-50. 1985 to present. Ten issues per year. 770 F. Also available on CD-ROM and online.

Physics Abstracts. INSPEC. Section A, Science Abstracts. Institute of Electrical Engineers, London, United Kingdom. Available from: INSPEC/IEEE - Institute of Electrical and Electronic Engineers, Box 1331, Hoes Lane, Piscataway, NJ 08855-1331. (908) 562-5549. 1 898 to present. 24 issues per year. $2835.00 per year. ISSN: 0036-8091. Also available online and onCD-ROM.

Science Citation Index. SCI. Institute for Scientific Information, 3501 Market Street, Philadelphia, PA 19104. (215) 386-0100. (800) 523-1850. FAX (215) 386-2991. 1961 to present. Six issues per year, plus annual cumulation. $11650.00 per year. Also available online and on CD-ROM. Inquire about price and availability. ISSN: 0036-827X.

STAR. Scientific and Technical Aerospace Reports. U.S. National Aeronautics and Space Administration. Distributed by U. S. Superintendent of Documents, Washington, DC 20402. 1963 to present. Semi-monthly, with semiannual and annual indexes. $114.00 per year. ISSN: 0036-8741. Also available online and on CD-ROM.

ANNUAL REVIEWS AND YEARBOOKS

The Astronomical Almanac. Superintendent of Documents, U. S. Government Printing office, Washington, DC 10402. (202) 783-3238. 1981 to present. Supersedes Astronomical Ephemeris. Annual. $44.95. ISSN::

Annual Review of Astronomy and Astrophysics. Original reviews of critical literature and current developments in astronomy and astrophysics. Annual Reviews, Inc., 4139 El Camino Way, Palo Alto, CA 94303-0139. (415) 493-4400. (800) 523-8635. Fax (415) 855-9815. 1963 to date. Annual. $60.00. ISSN: 0066-4146.

Annual Review of Earth and Planetary Sciences. Annual Reviews, Inc., 4139 El Camino Way, Palo Alto, CA 94303-0139. (415) 493-4400. (800) 523-8635. Fax (415) 855-9815. 1973 to date. Annual. $62.00. ISSN: 0084-6597.

Proceedings of the Lunar and Planetary Science Conference. National Technical Information Service (NTIS), 5285 Port Royal Road, Springfield, VA 22161. (703) 487-4650. . 1970 to date. Annual.

ASSOCIATIONS AND PROFESSIONAL SOCIETIES

American Astronomical Society. 2000 Florida Avenue NW, Suite 400, Washington, DC 20009. (202) 328-2010. FAX\; (202) 234-2560.

American Geophysical Union. 2000 Florida Avenue NW, Washington, DC 20009. (202) 462-6900. (800) 966-AGU1. FAX (202) 328-0566.

American Institute of Physics. 1 Physics Ellipse, College Park, MD 20740-3843. (301) 209-3100.

Association of Universities For Research in Astronomy, Inc. (AURA). Suite 701, 1625 Massachusetts Avenue, NW, Washington, DC 20036. (202) 483-2101. FAX (202) 483-2106.

Astronomical Society of the Pacific. 390 Ashton Avenue, San Francisco, CA 94112. (415) 337-1100. FAX: (415) 337-5205. GEOLOGICAL SOCIETY of AMERICA. 3300 Penrose Place,

PO Box 9140. Boulder, CO 80301-9140. (303) 447-2020. FAX (303) 447-1133.

IEEE Geoscience and Remote Sensing Society. c/o Institute of Electrical and Electronics Engineers. 345 East 47th Street, New York, NY 10017. (212) 705-7900. FAX (212) 705- 4929.

Meteoritical Society. University of Massachusetts, 125Marston Hall, Amherst, MA 01003. (413) 545-0300. (413) 545-0724.

Planetary Society. 65 North Catalina Avenue, Pasadena, CA 91106. (818) 793-5100. (800) WOW-MARS. FAX (818) 793-5528.

BIBLIOGRAPHIES

A Bibliography of Astronomy, 1970 - 1979. R. A. Sea and S. S. Martin. Libraries Unlimited, Inc., Littleton, CO 80160. 1982. $37.50.

DIRECTORIES AND BIOGRAPHICAL SOURCES

American Astronomical Society Membership Directory. American Astronomical Society. 2000 Florida Avenue NW, Suite 400, Washington, DC 20009. (202) 328-2010. FAX (202) 234-2560 Annual. Included in membership dues. ISSN: 1061- 9038.

American Men and Women of Science: Physical and Biological Sciences. R. R. Bowker Inc., 121 Chanlon Road, New Providence, NJ 07974. (908) 464-6800. (800) 521-8110. 20th edition. 8 volumes. 1996. $850.00.

The Astronomers. Donald Goldsmith. St. Martin's Press, Inc., 175 Fifth Avenue, New York, NY 10010. (212) 674-5151. (800) 221-7945. 1993. $14.95 in paper.

Astronomical Centers of the World. Kevin Krisciunas. Cambridge University Press, 40 West 20th Street, New York, NY 10011-4211. (212) 924-3900. (800) 872-7423. 1988. $34.95 DIRECTORY of PHYSICS and ASTRONOMY STAFF MEMBERS. American Institute of Physics, One Physics Ellipse, College Park, MD 20740-3843. (301) 209-3100. 1975/76 to present. Annual. $60.00. ISSN: 0361-2228.

Earth and Astronomical Research Centers. Stockton Press, 345 Park Avenue South, New York, NY 10010. 4th edition. 1995. $515.00. ISBN: 1-56169-0967.

Graduate Programs in Physics, Astronomy and Related Fields. 1995-1996. American Institute of Physics, One Physics Ellipse, College Park, MD 20740-3843. (301) 209-3100. 1978 to present. Annual. $45.00. ISSN: 0147-1821.

ENCYCLOPEDIAS AND DICTIONARIES

Asteroid Name Encyclopedia. Jacob Schwartz. Llewellyn Publications, PO. Box 64383, Saint Paul, MN 55164-0383. (612) 291-1970. (800) 843-6666. 1995. $19.95 in paper.

Concise Dictionary of Astronomy. Jacqueline Mitton. Oxford University Press, Inc., 200 Madison Avenue, New York, NY 10016. (212) 725-6000. (800) 334-4249. 1992. $24.95.

Dictionary of Minor Planet Names. Lutz D. Schmadel. Springer-Verlag New York, Inc., 175 Fifth Avenue, New York, NY 10010.

(212) 460-1500. (800) 777-4643. 2nd edition, revised and enlarged. 1993. $69.00.

Encyclopedia of Astronomy and Astrophysics. Stephen Maran, editor. Van Nostrand Reinhold, 115 Fifth Avenue, New York, NY 10003. (212) 254-3232. (800) 842-3636. 1992. $129.95.

McGraw-Hill Encyclopedia of Science and Technology. McGraw- Hill Book Company, Inc., 1221 Avenue of the Americas, New York, NY 10020. (212) 512-2000. (800) 262-4729. 7th edition. 20volume set. 1992. $1900.00.

Moons of the Solar System: an Illustrated Encyclopedia. John Stewart. McFarland & Company, Inc., Box 611, Jefferson, MC 28640. (910) 246-4460. (800) 253-2187. 1991. $49.95.

New Guide To the Planets. Patrick Moore. Trans-Atlantic Publications, Inc., 311 Bainbridge Street, Philadelphia, PA 19147. (215) 925-5083. 1993. $37.50.

Stars and Planets: The Sierra Club Guide To Sky Watching and Direction Finding. W. S. Kals. The Sierra Press, 4988 Gold Leaf Drive, Mariposa, CA 95338. (209) 966-5071. (800) 745-2631. $15.00.

GENERAL WORKS

Asteroids. Tom Gehrels, et al, editors. Books on Demand, 300 North Zeeb Road, Ann Arbor, MI 48106-1346. (313) 761- 4700. (800) 521-0600. $180.00.

Asteroids II. Richard P. Binzel, et al, editors. University of Arizona Press, 1230 North Park Avenue, Number 102, Tucson, AZ 85719. (520) 621-1441. (800) 426- 3797. 1990. $65.00.

Comets, Asteroids and Meteorites. Time-Life Inc., 777 Duke Street, Alexandria, VA 22314. (703) 8388-7000. (800) 621-7026. Revised edition. 1992. $18.95.

Dynamics and Evolution of Minor Bodies With Galactic and Geological Implications. S. V. Clube, et al, editors. Kluwer Academic Publishers, 101 Philip Drive, Assinippi Park, Norwell, MA 02061. (617) 871-6600. 1992. $149.00.

Hazards Due to Comets and Asteroids. Tom Gehrels, editor. University of Arizona Press, 1230 North Park Avenue, Number 102, Tucson, AZ 85719. (520) 621-1441. (800) 426-3797. 1994. $75.00.

Observing Comets, Asteroids, Meteors and the Zodiacal Light. Stephen J. Edberg and David H. Levy. Cambridge University Press, 40 West 20th Street, New York, NY 10011-4211. (212) 924-3900. (800) 872-7423. 1994. $29.95.

Rocks From Space: Meteorites and Meteorite Hunters. O. Richard Norton. Mountain Press, PO Box 2399, Missoula, MT 59806. (406) 728-1900. (800) 234-5308. 1994. $20.00 in paper.

Rogue Astroids and Doomsday Comets. Duncan Stasiuk. John Wiley & Sons, Inc., 605 Third Avenue, New York, NY 10158-0012. (212) 850-6000. (800) 225-5945. 1995. $24.95.

HANDBOOKS AND MANUALS

Astrophysical Data: Planets and Stars. Kenneth R. Lanb. Springer-Verlag New York, Inc., 175 Fifth Avenue, New York, NY 10010. (212) 460-1500. (800) 777-4643. 1993. $59.00. FIELD GUIDE TO the STARS and PLANETS. Jay M. Pasachoff and Donald H. Menzel. Houghton Mifflin Co., 222 Berkeley Street, Boston, MA 02116. (617) 351-5000. (800) 225-3362. Revised edition. 1992. $24.95.

New Guide To the Planets. Patrick Moore. Trans-Atlantic Publications, Inc., 311 Bainbridge Street, Philadelphia, PA 19147. (215) 925-5083. 1993. $37.50.

Planet Observer's Handbook. Fred W. Price. Cambridge University Press, 40 West 20th Street, New York, NY 10011- 4211. (212) 924-3900. (800) 872-7423. 1994. $34.95.

Planets and their Moons. Aububon Society Staff. Alfred A. Knopf, Inc. 201 E. 50th Street, New York, NY 10022. (800) 733-3000. 1995. $7.99.

Tables of Planetary Phenomena. Neil F. Michelsen. ACS Publications, PO Box 34487, San Diego, CA 92163-4487. (619) 492-9919. (800) 888-9983. 2nd revised edition. 1993. $24.95.

ONLINE DATABASES AND CD-ROMS

CA Search. Chemical Abstracts Service, P.O. Box 3012, Columbus, OH 43210-0012. (614) 447-3600. (800) 848-6533. FAX (614) 447-3709. Very comprehensive guide to worldwide chemical literature and related fields, 1972 to present. Available on BRS,(800) 289-4277, DIALOG, (800) 334-2564, ORBIT (800) 456-7248, and STN International, FIZ Karlsruhe, P.O. Box 2465, W-7500, Karlsruhe 1, Germany, online services. Inquire as to cost and availability.

Current Contents Search. Institute for Scientific Information, 3501 Market Street, Philadelphia, PA 19104. (215) 386-0100. FAX (215) 386-6362. Contains citations to articles listed in the table of contents of science and technology journals. Also articles in social scie nces and life sciences journals. Available on BRS, (800) 289-4277, DIALOG, (800) 334-2564, online services. Inquire as to cost and availability.

Dissertation Abstracts Online. University Microfilms International, 300 North Zeeb Road, Ann Arbor, MI 48106. (800) 521-0600 or (313) 761-4700. Scope includes virtually all doctoral dissertations accepted at accredited American institutions from 1861 to present in 252 subject areas. Available on BRS, (800) 289-4277, DIALOG, (800) 334-2564, and OCLC EPIC, (800) 848-5878, online services. Also available on CD-ROM. Inquire as to cost and availability.

INSPEC. Institution of Electrical Engineers, Michael F araday House, Six Hills Way, Stevenage, Herts. SG1 2AY, England. Telephone: 0438 313311 or FAX 0438 7428 40. Contains citations to the worldwide literature of phy sics, electronics and electrical engineering, computer technology, and related fields. Available on BRS, (800) 289-4277, DIALOG, (800) 334-2564, ORBIT, (800) 456-7248, and STN International, FIZ Karlsruhe, P.O. Box 2465, W-7500, Karlsruhe 1, Germany, online services. Inquire as to cost and availability.

International Aerospace Abstracts. American Institute of Aeronautic s and Astronautics, 370 L'Enfant Promenade, S.W., Washington, DC 20024. (202) 646-7400. Contains references and

ASTEROIDS

Ency. of Physical Sciences and Engineering Info. Sources

abstracts of journal and monograph literature relating to aerospace science and technology, from 1963 to present. Available through the NASA/RECON system of the National Aeronautics and Space Administration only.

NTIS Bibliographic Database. National Technical Information Service, 5285 Port Royal Road, Springfield, VA 2216 1. (703) 487-4929 or FAX (703) 321-8199. Broad cover age of government-sponsored science and technology research reports, 1964 to present. Available on BRS,(800) 289-4277, DIALOG, (800)334-2564, ORBIT, (800) 456-7248, and STN International, FIZ Karlsruhe, P.O. Box 2465, W-7500, Karlsruhe 1, Germany, online services. Also available on CD-ROM. Inquire as to cost and availability.

Physics Briefs. American Institute of Physics, 335 East 45th Street , New York, NY 10017. (212) 661-9260 or FAX (212) 661-2036. Contains citations with abstracts of the literature of physics and related fields, 1979 to present. Available on the STN International, FIZ Karlsruhe, P.O. Box 2465, W-7500, Karlsruhe 1, Germany, online service. Inquire as to cost and availability.

Scientific and Technical Books and Serials in Print. R.R. B owker Inc., 121 Chanlon Road, New Providence, NJ 07974. (800) 521-8110 or (908) 464-6800. List of currently published books and serials in the physical and biological sciences, engineering and technology, with subject, author and titles indexes. Available on ORBIT, (800) 456-7248, online service. Inquire as to cost and availability.

SCISEARCH. Institute for Scientific Information, 3501 M arket Street, Philadelphia, PA 19104. (800) 523-1850 or (215) 386-0100. Broad multidisciplinary title and author index to the international literature of science and technology, 1974 to present. Available on DIALOG, (800) 334-2564, and ORBIT, (800) 456-7248, online services. Also available on CD-ROM. Inquire as to cost and availability.

WILSONLINE. H.W. Wilson Company, 950 University Avenue, Bronx, NY 10452. (800) 367-6770 or (212) 588-8400. Makes available onlin e versions of the printed H.W. Wilson indexes includin g Applied Science and Technology Index, Business Periodicals Index, General Science Index, and Readers' Guide to Periodical Literature. Period covered is generally 1983 to present. Available on BRS, (800) 289-4277, DIALOG, (800) 334-2564, and OCLC EPIC, (800) 848-5878, online services. Also available on CD-ROM. Inquire as to cost and availability.

OTHER SOURCES

Atlas of the Planets. Vincent DeCallatay and Audouin Dollfus. Books on Demand, 300 North Zeeb Road, Ann Arbor, MI 48106-1346. (313) 761-4700. (800) 521-0600. $45.60.

Atlas of the Solar System. B. Yenne. Simon & Schuster, Inc. 1230 Avenue of the Americas, New York, NY 10020. (212) 698-7000. (800) 223-2348. 1987. $12.98.

Chronology of Eclipses and Comets. AD1 - 1OOO. D. J. Schove. Boydell and Brewer, Inc., P.O. Box 41026, Rochester, NY 14604-4126. (716-275-0419. 1985. $63.00.

Planetary, Lunar and Solar Positions Six Hundred-one B.C. TO A.D. One at Five-day and Ten-day Intervals. Bryant Tuckerman. American Philosophical Society, 104 South 5th Street, Philadelphia, PA 19106-3387. (215) 440-3400. FAX (215) 440-3436. Memoirs Series, volume 56. 1979. $20.00.

Planetary, Lunar and Solar Positions, A.d.2 to A.d. 1649 at Five Day and Ten Day Intervals. Bryant Tuckerman. AmericanPhilosophical Society, 104 South 5th Street, Philadelphia, PA 19106-3387. (215) 440-3400. FAX (215)440-3436. Memoirs Series, volume 59. 1964. $30.00.

Planetary, Lunar and Solar Positions, 1650 - 1805. Owen Gingerich and Barbara L. Welther. American Philosophical Society, 104 South 5th Street, Philadelphia, PA 19106-3387. (215) 440-3400. FAX (215)440-3436. Memoirs Series, volume 59S. 1983. $20.00.

The Sky: a User's Guide. David H. Levy. Cambridge University Press, 40 West 20th Street, New York, NY 10011- 4211. (212) 924-3900. (800) 872-7423. 1993. $14.95. <bodylead>*The View from Space: Photographic Exploration of the Planets.* Merton Davies and Bruce C. Murray. Columbia University Press, 562 West 113th Street, New York, NY 10025. (212) 666-1000. (800) 944-8648. 1973. $17.50.

PERIODICALS

Astronomical Journal. American Institute of Physics, One Physics Ellipse, College Park, MD 20740-3843. (301) 209- 3000. Published for the American Astronomical Society. 1849 to present. Monthly. $280.00 per year. ISSN: 0004- 6256.

Astronomical Society of the Pacific. PUBLICATIONS. Astronomical Society of the Pacific, 390 Ashton Avenue, San Francisco, CA 94112. (415) 337-1100. FAX (415) 337-5205. 1889 to present. Monthly. $175.00 per year. ISSN: 0004- 6280.

Astronomy. Kalmbach Publishing Company, Box 1612, Waukesha, WI 53187-1612. (414) 796-8776. FAX (414) 796-1142. 1973 to present. Monthly. $27.00 per year. ISSN: 0091-6358.

Astronomy and Astrophysics. Springer-Verlag New York, Inc., 44 Hartz Way, Secaucus, NJ 07096-2491. (201) 348-4033. FAX (201) 348-4505. 1969 to date. 36 issues per year in 12 volumes. $1733.00 per year. ISSN: 0004-6361.

Celestial Mechanics and Dynamical Astronomy; An International Journal of Space Dynamics. Kluwer Academic Publishers, Box 358, Accord Station, Hingham, MA 02018-0358. (617) 871-6600. FAX (617) 871-6528. 1969 to present. Monthly. $703.50 per year. ISSN: 0923-2958.

Earth, Moon and Planets; An International Journal of Comparative Planetology. Kluwer Academic Publishers, Box 358, Accord Station, Hingham, MA 02018-0358. (617) 871- 6600. FAX (617) 871-6528. 1969 to present. Monthly. $840.00 per year. ISSN: 0167-9295.

Geochemica Et Cosmochimica Acta. Elsevier Science. 660 White Plains Road, Tarrytown, NY 10591-5153. (914) 524- 9200. FAX (914) 333-2444. 1950 to the present. Biweekly. $895.00 per year. ISSN: 0016-7037.

ICARUS; International Journal of Solar System Studies. Academic Press, Inc., Journal Division, 525 B Street, Suite 1900, San Diego, CA 92101-4495. (619) 230-1840. FAX (619) 699-6800. 1962 to present. Monthly. $1080.00 per year. ISSN: 0019-1035.

J G R: Journal of Geophysical Research: Planets. American Geophysical Union, 2000 Florida Avenue, NW, Washington,

CD 20009.(202) 462-6900. FAX (202) 328-0566. 1991 to present. Monthly. $597.00 per year. ISSN: 0148-0227.

Lunar and Planetary Information Bulletin. Lunar and Planetary Institute. 3600 Bay Area Boulevard, Houston, TX 77058-1113. (713) 486-2175. FAX (713) 486-2125. 1970 to present. Quarterly. Free. Also available online.

Mercury. Astronomical Society of the Pacific. 390 Ashton Avenue, San Francisco, CA 94112. (415) 337-1100. FAX: (415) 337-5205. 1972 to present. Bimonthly. $175.00 per year. ISSN: 0047-6773.

Meteoritics. Meteoritical Society, Department of Chemistry, University of Arkansas, Fayetteville, AR 72701. (501) 575-7625. FAX (501) 575-7778. 1953 to present. Bimonthly. $210.00 per year. ISSN: 0026-1114.

Monthly Notices of the Royal Astronomical Society. Blackwell Scientific Publications, LT., Osney Mead, Oxford OX2 OEL, England. TEL 0865-240201. FAX 0865-721205. 1827 to present. Fortnightly. $1899.00 per year. ISSN: 0035- 8711.

Planetary and Space Science. Elsevier Science Publishing Company, Inc., 660 White Plains Road, Terrytown, NY 10591- 5153. (914) 524-9200. FAX: (914) 333-2444. 1959 to present. Monthly. $1355.00 per year. ISSN: 0032-0633.

Planetary Report. Planetary Society. 65 North Catalina, Pasadena, CA 91106-2301. (818) 793-5100. FAX (818) 793- 5528. 1980 to present. $25.00 per year. ISSN: 0736-3680.

Sky and Telescope. Sky Publishing Corporation, Box 9111, Belmont, MA 02178. (617) 864-7360. FAX (617) 864-6117. 1941 to present. Monthly. $27.00 per year. ISSN: 0037- 6604.

Vistas in Astronomy; An International Review Journal. Albert C. Beer, editor. Elsevier Science Publishing Company, Inc., 660 White Plains Road, Terrytown, NY 10591- 5153. (914) 524-9200. FAX: (914) 333-2444. 1958 to present. Quarterly. $415.00 per year. ISSN: 0083-6656.

RESEARCH CENTERS AND INSTITUTES

Center for Radiophysics and Space Research. Cornell University, Space Sciences Building, Ithaca NY 14853. (607) 255-4341. FAX (607) 255-9888.

Center For Space Science and Astrophysics. Stanford University. 325 Durand Building, Stanford, CA 04305. (415) 723-3582.

Center For Space Sciences. University of Texas at Dallas, P. O. Box 830688, MS_F022, Richardson, TX 75083-0688. (214) 690-2851. FAX (214) 690-2848.

Earth and Planetary Remote Sensing Laboratory. Washington University, Department of Earth and Planetary Sciences. Campus Box 1169, 1 Brookings Drive, St. Louis, MO 63130. (314) 889-5679. FAX (314) 889-5799.

Institute For Astronomy. University of Hawaii at Manoa, 2680 Woodlawn Drive, Honolulu, HI 96822. (808) 956-8312. FAX (808) 988-2790.

Institute For Terrestrial and Planetary Atmospheres.State University of New York at Stony Brook, Stony Brook, NY 11794-2300. (516) 632-6170. FAX (516) 632-6251.

Institute of Geophysics and Planetary Physics. University of California, Riverside, CA 92521. (714) 787-4503. FAX (714) 787-4529.

Laboratory For Atmospheric and Space Physics. University of Colorado - Boulder, Boulder, CO 80309-0392. (303) 492-7677. (303) 492-6946.

Laboratory For Planetary Geology. Arizona State University, Department of Geology, Tempe, AZ 85281. (601) 965-7029.

Laboratory For Planetary Studies. Cornell University, Space Sciences Building, Ithaca, NY 14853. (607) 255-4971.

Lunar and Planetary Institute. 3303 NASA Road One, Houston, TX 77058-4399. (713) 486-2139. FAX 713-496-2162.

Lunar and Planetary Laboratory. University of Arizona, Tucson, AZ 85721. (602) 621-6963. FAX (602) 621-4933.

McDonnell Center for the Space Sciences. Washington University, Box 1105, One Brookings Drive, St. Louis MO 63130-4899. (314) 889-6255. FAX (314) 889-6219.

Smithsonian Astrophysical Observatory. 60 Garden Street, Cambridge, MA 02138. (617) 495-7461. FAX (617) 495-7105.

ASTROGEOLOGY

See also: ASTEROIDS, ASTRONOMY, ASTROPHYSICS, COMETS, METEORITES, NAMES of INDIVIDUAL PLANETS, PLANETARY SCIENCES, SATELLITES (NATURAL), SOLAR SYSTEM,

ABSTRACT SERVICES AND INDEXES

Applied Science and Technology Index; A cumulative subject index to English language periodicals in the fields of aeronautics and space science, computer technology, chemistry, construction industry, energy and related areas. H. W. Wilson Co., 950 University Avenue, Bronx, NY 10452-9978. (212) 588-8400. (800) 367-6770. FAX (718) 590-1617. From 1958 to present. Monthly. Inquire about cost and availability. Also available online (BRS and WILSONLINE) and on CD-ROM. ISSN: 0003-6986.

Astronomy and Astrophysics Abstracts. Springer-Verlag New York, 175 Fifth Avenue, New York, NY 10010. (212) 460-1500. FAX (212) 473-6272. Published for the Astronomisches Rechen-Institut. Comprehensive coverage of all aspects of astronomy, astrophysics and related fields. 1969 to present. Two parts per year. Annual. ISSN: 0067-0022.

Bibliography and Index of Geology. American Geological Institute, 4220 King Street, Alexandria, VA 22302-1507. (703) 379-2480. FAX (703) 379-7563. From 1969 to present. Monthly. $1295.00 per year. Also available as GEOREF onlin e (CISTI, DIALOG, Orbit, STN) and on CD-ROM. Inquire about price and availability. ISSN: 0098-2784.

ASTROGEOLOGY

Ency. of Physical Sciences and Engineering Info. Sources

Chemical Abstracts. Chemical Abstracts Service. 2540 Oleantangy River Road, Box 3012, Columbus, OS 43210-0012. (614) 447-3600. FAX (614) 447-3713. 1907 to present. Weekly. $16,800.00 per year. Available online and on CD- ROM. CA is also available infive section groupings. Inquire regarding cost and availability.

Current Contents: Physical, Chemical, and Earth Sciences. Institute for Scientific Information, 3501 Market Street, Philadelphia, PA 19104. (215) 386-0100. (800) 523-185 0. FAX (215) 386-2291. 1961 to present. Weekly. $442.00 per year. Also available online (BRS, DIALOG) and on CD-ROM. Inquire about price and availability.

Meteorological and Geoastrophysical Abstracts. American Meteorological Society, c/o Inforonics, Inc., 550 Newtown Road, Box 458, Littleton, MA 01460. (508) 486-8976. FAX (508) 486-0027. Covers literature in environmental sciences, meteorology, astrophsyics, hydrology, glaciology and physical oceanography. 1950 to present. Monthly. $950.00 per year. Also available online (DIALOG) and on CD-ROM. ISSN: 0026-1130.

Pascale 49: Meteorologie, Glaciologie, Physique Des Oceans. Centre National de la Recherche Scientifique, Institut de Information Scientifique et Technique, 2 aliee du Parc de Brabois, 54514 Vandoeuvre Les Nancy Cedex, France. TEL 83- 50-46-00. FAX: 83-50-46-50. 1985 to present. Ten issues per year. 770 F. Also available on CD-ROM and online. ISSN: 1164-5997.

Physics Abstracts . INSPEC. Section A, Science Abstracts. Institute of Electrical Engineers, London, United Kingdom. Available from: INSPEC/IEEE - Institute of Electrical and Electronic Engineers, Box 1331, Hoes Lane, Piscataway, NJ 08855-1331. (908) 562-5549. 1898 to present. 24 issues per year. $2835.00 per year. ISSN: 0036-8091. Also available online and on CD-ROM.

Science Citation Index. SCI. Institute for Scientific Information, 3501 Market Street, Philadelphia, PA 19104. (215) 386-0100. (800) 523-1850. FAX (215) 386-2991. 1961 to present. Six issues per year, plus annual cumulation. $11650.00 per year. Also available online and on CD-ROM. Inquire about price and availability. ISSN: 0036-827X.

STAR. Scientific and Technical Aerospace Reports. U.S. National Aeronautics and Space Administration. Distributed by U. S. Superintendent of Documents, Washington, DC 20402. 1963 to present. Semi-monthly, with semiannual and annual indexes. $114.00 per year. ISSN: 0036-8741. Also available online and on CD-ROM.

ANNUAL REVIEWS AND YEARBOOKS

Annual Review of Astronomy and Astrophysics. Original reviews of critical literature and current developments in astronomy and astrophysics. Annual Reviews, Inc., 4139 El Camino Way, Palo Alto, CA 94303-0139. (415) 493-4400. (800) 523-8635. Fax (415) 855-9815. 1963 to date. Annual. $60.00. ISSN: 0066-4146.

Annual Review of Earth and Planetary Sciences. Annual Reviews, Inc., 4139 El Camino Way, Palo Alto, CA 94303-0139. (415) 493-4400. (800) 523-8635. Fax (415) 855-9815. 1973 to date. Annual. $62.00. ISSN: 0084-6597.

ASSOCIATIONS AND PROFESSIONAL SOCIETIES

American Astronomical Society. 2000 Florida Avenue NW, Suite 400, Washington, DC 20009. (202) 328-2010. FAX\; (202) 234-2560.

American Geophysical Union. 2000 Florida Avenue NW, Washington, DC 20009. (202) 462-6900. (800) 966-AGU1. FAX (202) 328-0566.

American Institute of Physics. One Physics Ellipse, College Park, MD 20740-3843. (301) 209-3100.

American Meteor Society. SUNY-Geneseo, 1 College Circle, Department of Physics and Astronomy, Geneseo, NY 14454. (716) 245-5282.

Astronomical Society of the Pacific. 390 Ashton Avenue, San Francisco, CA 94112. (415) 337-1100. FAX: (415) 337-5205.

Geological Society of America. 3300 Penrose Place, PO Box 9140. Boulder, CO 80301-9140. (303) 447-2020. FAX (303) 447-1133.

IEEE Geoscience and Remote Sensing Society. c/o Institute of Electrical and Electronics Engineers. 345 East 47th Street, New York, NY 10017. (212) 705-7900. FAX (212) 705- 4929.

Meteoritical Society. University of Massachusetts, 125 Marston Hall, Amherst, MA 01003. (413) 545-0300. (413) 545-0724.

Lunar and Planetary Institute. 3600 Bay Area Boulevard, Houston, TX 77058. (713) 486-2143.

Planetary Society. 65 North Catalina Avenue, Pasadena, CA 91106. (818) 793-5100. (800) WOW-MARS. FAX (818) 793-5528.

BIBLIOGRAPHIES

A Bibliography of Astronomy, 1970 - 1979. R. A. Sea and S. S. Martin. Libraries Unlimited, Inc., Littleton, CO 80160. 1982. $37.50.

Science Books & Films. American Association for the Advancement of Science, 1333 H Street NW, Washington, DC 20005. (202) 326-6434. Reviews of print, film and software materials in all sciences.. 1965 to present. Nine issues per year. $40.00 per year. ISSN: 0098-342X.

Scientific and Tech Nical Books and Serials in Print; An Index To Literature in Science and Technology. R. R. Bowker. 121 Chanlon Road, New Providence, NJ 07974. (908) 464-6800. FAX (908) 665-3502. (800) 521-8110. 1972 to present. Annual. $299.95 per year. Also available on CD-ROM. ISSN: 0000-054X.

DIRECTORIES AND BIOGRAPHICAL SOURCES

American Astronomical Society Membership Directory. American Astronomical Society. 2000 Florida Avenue NW, Suite 400, Washington, DC 20009. (202) 328-2010. FAX (202) 234-2560 Annual. Included in membership dues. ISSN: 1061- 9038.

American Men and Women of Science. Physical and Biological Sciences. R. R. Bowker Inc., 121 Chanlon Road, New Providence, NJ 07974. (908) 464-6800. (800) 521-8110. 20th edition. 8 volumes. 1996. $850.00.

Directory of Physics and Astronomy Staff. American Institute of Physics, One Physics Ellipse, College Park, MD 20740-3843. (301) 209-3100. 1975/76 to present. Annual. $60.00. ISSN: 0361-2228.

Earth and Astronomical Research Centers. Stockton Press, 345 Park Avenue South, New York, NY 10010. 4th edition. 1995. $515.00. ISBN: 1-56169-0967.

ENCYCLOPEDIAS AND DICTIONARIES

Asteroid Name Encyclopedia. Jacob Schwartz. Llewellyn Publications, P.O. Box 64383, Saint Paul, MN 55164-0383. (612) 201-1970. (800) 843-6666. 1995. $19.95 in paper.

Concise Dictionary of Astronomy. Jacqueline Mitton. Oxford University Press, Inc., 200 Madison Avenue, New York, NY 10016. (212) 725-6000. (800) 334-4249. 1992. $24.95.

Dictionary of Minor Planet Names. Lutz D. Schmadel. Springer-Verlag, 175 Fifth Avenue, New York, NY 10010. (212) 460-1500. (800) 777-4643. 2nd edition, revised and enlarged. 1993. $69.00.

Encyclopedia of Astronomy and Astrophysics. Stephen Maran, editor. Van Nostrand Reinhold, 115 Fifth Avenue, New York, NY 10003. (212) 254-3232. (800) 842-3636. 1992. $129.95.

McGraw-Hill Encyclopedia of Science and Technology. McGraw-Hill Book Company, Inc., 1221 Avenue of the Americas, New York, NY 10020. (212) 512-2000. (800) 262-4729. 7th edition. 20 volume set. 1992. $1900.00.

Moons of the Solar System: an Illustrated Encyclopedia. John Stewart. McFarland & Company, Inc., Box 611, Jefferson, MC 28640. (910) 246-4460. (800) 253-2187. 1991. $49.95.

New Guide To the Planets. Patrick Moore. Trans-Atlantic Publications, Inc., 311 Bainbridge Street, Philadelphia, PA 19147. (215) 925-5083. 1993. $37.50.

Stars and Planets: The Sierra Club Guide To Sky Watching and Direction Finding. W. S. Kals. The Sierra Press, 4988 Gold Leaf Drive, Mariposa, CA 95338. (209) 966-5071. (800) 745-2631. $15.00.

GENERAL WORKS

Birth of the Earth. David E. Fisher. Columbia University Press, 562 West 113th Street, New York, NY 10025. (212) 666-1000. (800) 944-8648. 1988. $16.50 in paper.

Chemistry and Physics of the Terrestrial Planets. Surendra K. Saxwna. Springer-Verlag New York, Inc., 175 Fifth Avenue, New York, NY 10010. (212) 460-1500. (800) 777-4643. 1986. $109.00.

Dynamics and Evolution of Minor Bodies with Galactic and Geological Implications. S. V. Clube, et al, editors. Kluwer Academic Publishers, 101 Philip Drive, Assinippi Park, Norwell, MA 02061. (617) 871-6600. 1992. $149.00.

Evolution of the Earth and Planets. E. Takahashi, et al, editors. American Geophysical Union, 2000 Florida Avenue, N.W., Washington, DC 20009. (202) 462-6903. (800) 966-2481. 1993. $28.00.

Exploration of Venus & Mars Atmospheres. G. M. Keating, editor. Pergamon Press, Inc., Maxwell House, Fairview Park, Elmsford, NY 10523. (914) 592-7700. Fax (914) 592-3625. 1995. 94.00.

Exploring the Planets. W. Kenneth Hamblin and Eric H. Christiansen. Macmillan Publishing Company, Inc,. 200 Old Tappan Road, Old Tappan, NJ 07675. (800) 233-2336. 2nd edition. 1995. $62.00.

The Geology of Multi-ring Impact Basins: the Moon and Other Planets. P. D. Spudis. Cambridge University Press, 40 West 20th Street, New York, NY 10011-4211. (212) 924-3900. (800) 872-7423. Planetary Sciences Series. 1993. $59.95.

Geology of the Terrestrial Planets. Michael H. Carr. U. S. Government Printing office, Superintendent of Documents, Washington, DC 20402-9325. (202) 783-3238. 1985. $16.00

An Introduction to Cosmochemistry. Charles R. Cowley. Cambridge University Press, 40 West 20th Street, New York, NY 10011-4211. (212) 924-3900. (800) 872-7423. 1995. $29.95 in paper.

Mercury: The Elusive Planet. Robert B. Strom. Smithsonian Institution Press, 470 L'Enfant Plaza, Suite 7100, Washington, DC 20560. (202) 287-3738. (800) 782-4612. 1987. $29.95.

Hazards Due To Comets and Asteroids. Tom Gehrels, editor. University of Arizona Press, 1230 North Park Avenue, Number 102, Tucson, AZ 85719. (520) 621-1441. (800) 426-3797. 1995. $75.00.

Meteorites: An Introduction. Fritz Heide. Springer-Verlag New York, Inc., 175 Fifth Avenue, New York, NY 10010. (212) 460-1500. (800) 777-4643. 1995. $24.00 in paper.

Meteorites and the Origin of Planets. John A. Wood. Books on Demand, 300 North Zeeb Road, Ann Arbor, MI 48106-1346. (313) 761-4700. (800) 521-0600. Reprint edition. $35.00 in paper.

Moons and Planets. William K. Hartmann. Wadsworth Publishing Co., 10 Davis Drive, Belmont, CA 94002. (415) 595-2350. (800) 354-9706. 3rd edition. 1993. $49.95.

The Near Planets. Time-Life Books Editors. Time-Life, Inc., 777 Duke Street, Alexandria, VA 22314. (703) 8388-7000. (800) 621-7026. Voyage through the Universe Series. 1992. $ wfi.

Planetary Nebulae. Hynes . Willmann-Bell, Inc., P.O. Box 35025, Richmond, VA 23235. (804)-320-7016. 1991. $24.95. PLANETARY LandSCAPES. Ronald Greeley. Chapman & Hall, 1 Penn Plaza, New York, NY 10119. (212) 564-1060. Div. of Routledge,.. 2nd edition. 1994. $49.95.

The Planetary System. Tobias Owen and David Morrison. Addison-Wesley Publishing Co., Inc., 1 Jacob Way, Reading, MA 01867. (617) 944-3700. (800) 447-2226. 1988. $49.50.

ASTROGEOLOGY

Ency. of Physical Sciences and Engineering Info. Sources

Planets and their Atmospheres. John S. Lewis and Ronald G. Primm. Academic Press Inc., 6277 Sea Harbor Drive, Orlando, FL. (800) 321-5068. 1983. $65.00 in paper.

Resources of Near-earth Space. John S. Lewis, et al. University of Arizona Press, 1230 North Park Avenue, Number 102, Tucson, AZ 85719. (520) 621-1441. (800) 426-3797. 1993. $75.00.

Story of the Earth. Peter Cattermole and Patrick Moore. Cambridge University Press, 40 West 20th Street, New York, NY 10011-4211. (212) 924-3900. (800) 872-7423. 1986. $5.95 in paper.

Terraforming: Engineering Planetary Environments. Martyn J. Fogg. Society of Automotive Engineers (SAE), 400 Commonwealth Drive, Warrendale, PA 15096. (412) 776-4841. 1995. $49.00.

Uranus and Neptune: The Distant Giants. Eric Burgess. Columbia University Press, 562 West 113th Street, New York, NY 10025. (212) 666-1000. (800) 944-8648. 1988. $40.00.

Venus: An Errant Twin. Eric Burgess. Columbia University Press, 562 West 113th Street, New York, NY 10025. (212) 666-1000. (800) 944-8648. 1985. $43.50.

Venus, A New Geology. Peter Cattermole. Johns Hopkins University Press, 2715 North Charles Street, Baltimore, MD 21218-4319. (410) 516-6900. (800) 537-5487. 1994. $49.95.

HANDBOOKS AND MANUALS

Astrophysical Data: Planets and Stars. Kenneth R. Lanb. Springer-Verlag New York, Inc., 175 Fifth Avenue, New York, NY 10010. (212) 460-1500. (800) 777-4643. 1993. $59.00.

Field Guide To the Stars and Planets. Jay M. Pasachoff and Donald H. Menzel. Houghton Mifflin Co., 222 Berkeley Street, Boston, MA 02116. (617) 351-5000. (800) 225-3362. Revised edition. 1992. $24.95.

ONLINE DATABASES AND CD-ROMS

CA Search. Chemical Abstracts Service, P.O. Box 3012, Columbus, OH 43210-0012. (614) 447-3600. (800) 848-6533. FAX (614) 447-3709. Very comprehensive guide to worldwide chemical literature and related fields, 1972 to present. Available on BRS, (800) 289-4277, DIALOG, (800) 334-2564, ORBIT (800) 456-7248, and STN International, FIZ Karlsruhe, P.O. Box 2465, W-7500, Karlsruhe 1, Germany, online services. Inquire as to cost and availability.

Dissertation Abstracts. University Microfilms International, 300 North Zeeb Road, Ann Arbor, MI 48106. (800) 521-0600 or (313) 761-4700. Scope includes virtually all doctoral dissertations accepted at accredited American institutions from 1861 to present in 252 subject areas. Available on BRS, (800) 289-4277, DIALOG, (800) 334-2564, and OCLC EPIC, (800) 848-5878, online services. Also available on CD-ROM. Inquire as to cost and availability.

GEOREF. American Geological Institute, 4220 King Street, Alexandria, VA 22302. (800) 336-4764 or (703) 379-2480. Geology and geosciences literature, 1785 to present for North America. Available on DIALOG, (800) 334-2564, ORBIT (800)

456-7248, online services. Also available on CD-ROM. Inquire as to cost and availability.

INSPEC. Institution of Electrical Engineers, Michael Faraday House, Six Hills Way, Stevenage, Herts. SG1 2AY, England. Telephone: 0438 313311 or FAX 0438 7428 40. Contains citations to the worldwide literature of physics, electronics and electrical engineering, computer technology, and related fields. Available on BRS, (800) 289-4277, DIALOG, (800) 334-2564, ORBIT, (800) 456-7248, and STN International, FIZ Karlsruhe, P.O. Box 2465, W-7500, Karlsruhe 1, Germany, online services. Inquire as to cost and availability.

NTIS Bibliographic Database. National Technical Information Service, 5285 Port Royal Road, Springfield, VA 2216 1. (703) 487-4929 or FAX (703) 321-8199. Broad cover age of government-sponsored science and technology research reports, 1964 to present. Available on BRS, (800) 289-4277, DIALOG, (800) 334-2564, ORBIT, (800) 456-7248, and STN International, FIZ Karlsruhe, P.O. Box 2465, W-7500, Karlsruhe 1, Germany, online services. Also available on CD-ROM. Inquire as to cost and availability.

SCISEARCH. Institute for Scientific Information, 3501 Market Street, Philadelphia, PA 19104. (800) 523-1850 or (215) 386-0100. Broad multidisciplinary title and author index to the international literature of science and technology, 1974 to present. Available on DIALOG, (800) 334-2564, and ORBIT, (800) 456-7248, online services. Also available on CD-ROM. Inquire as to cost and availability.

WILSONLINE. H.W. Wilson Company, 950 University Avenue, Bronx, NY 10452. (800) 367-6770 or (212) 588-8400. Makes available online versions of the printed H.W. Wilson indexes including Applied Science and Technology Index, Business Periodicals Index, General Science Index, and Readers' Guide to Periodical Literature. Period covered is generally 1983 to present. Available on BRS, (800) 289-4277, DIALOG, (800) 334-2564, and OCLC EPIC, (800) 848-5878, online services. Also available on CD-ROM. Inquire as to cost and availability.

OTHER SOURCES

Atlas of the Planets. Vincent DeCallatay and Audouin Dollfus. Books on Demand, 300 North Zeeb Road, Ann Arbor, MI 48106-1346. (313) 761-4700. (800) 521-0600. $45.60.

Atlas of the Solar System. B. Yenne. Simon & Schuster, Inc. 1230 Avenue of the Americas, New York, NY 10020. (212) 698-7000. (800) 223-2348. 1987. $12.98.

Cambridge Atlas of Astronomy. Cambridge University Press, 40 West 20th Street, New York, NY 10011-4211. (212) 924-3900. (800) 872-7423. 3rd edition. 1994. $74.00.

The Sky: A User's Guide. David H. Levy. Cambridge University Press, 40 West 20th Street, New York, NY 10011- 4211. (212) 924-3900. (800) 872-7423. 1993. $14.95.

PERIODICALS

Astronomical Journal. American Institute of Physics, One Physics Ellipse, College Park, MD 20740-3843. (301) 209- 3000. Published for the American Astronomical Society. 1849 to present. Monthly. $280.00 per year. ISSN: 0004- 6256.

Astronomical Society of the Pacific. PUBLICATIONS. Astronomical Society of the Pacific, 390 Ashton Avenue, San Francisco, CA 94112. (415) 337-1100. FAX (415) 337-5205. 1889 to present. Monthly. $175.00 per year. ISSN: 0004- 6280.

Astronomy. Kalmbach Publishing Company, Box 1612, Waukesha, WI 53187-1612. (414) 796-8776. FAX (414) 796-1142. 1973 to present. Monthly. $27.00 per year. ISSN: 0091-6358.

Astronomy and Astrophysics. Springer-Verlag New York, Inc., 44 Hartz Way, Secaucus, NJ 07096-2491. (201) 348-4033. FAX (201) 348-4505. 1969 to date. 36 issues per year in 12 volumes. $1733.00 per year. ISSN: 0004-6361.

Earth, Moon and Planets; An International Journal of Comparative Planetology. Kluwer Academic Publishers, Box 358, Accord Station, Hingham, MA 02018-0358. (617) 871- 6600. FAX (617) 871-6528. 1969 to present. Monthly. $840.00 per year. ISSN: 0167-9295.

Geochemica Et Cosmochimica Acta. Elsevier Science. 660 White Plains Road, Tarrytown, NY 10591-5153. (914) 524- 9200. FAX (914) 333-2444. 1950 to the present. Biweekly. $895.00 per year. ISSN: 0016-7037.

ICARUS; International Journal of Solar System Studies. Academic Press, Inc., Journal Division, 525 B Street, Suite 1900, San Diego, CA 92101-4495. (619) 230-1840. FAX (619) 699-6800. 1962 to the present. Monthly. $1080.00. ISSN: 0019-1035.

J G R: Journal of Geophysical Research: Planets. American Geophysical Union, 2000 Florida Avenue, NW, Washington, CD 20009. (202) 462-6900. FAX (202) 328-0566. 1991 to present. Monthly. $597.00 per year. ISSN: 0148-0227.

Lunar and Planetary Information Bulletin. Lunar and Planetary Institute. 3600 Bay Area Boulevard, Houston, TX 77058-1113. (713) 486-2175. FAX (713) 486-2125. 1970 to present. Quarterly. Free. Also available online.

Mercury. Astronomical Society of the Pacific. 390 Ashton Avenue, San Francisco, CA 94112. (415) 337-1100. FAX: (415) 337-5205. 1972 to present. Bimonthly. $175.00 per year. ISSN: 0047-6773.

Meteoritics. Meteoritical Society, Department of Chemistry, University of Arkansas, Fayetteville, AR 72701. (501) 575-7625. FAX (501) 575-7778. 1953 to present. Bimonthly. $210.00 per year. ISSN: 0026-1114.

Planetary and Space Science. Elsevier Science Publishing Company, Inc., 660 White Plains Road, Terrytown, NY 10591- 5153. (914) 524-9200. FAX: (914) 333-2444. 1959 to present. Monthly. $1355.00 per year. ISSN: 0032-0633.

Planetary Report. Planetary Society. 65 North Catalina, Pasadena, CA 91106-2301. (818) 793-5100. FAX (818) 793- 5528. 1980 to present. $25.00 per year. ISSN: 0736-3680.

Sky and Telescope. Sky Publishing Corporation, Box 9111, Belmont, MA 02178. (617) 864-7360. FAX (617) 864-6117. 1941 to present. Monthly. $27.00 per year. ISSN: 0037- 6604.

RESEARCH CENTERS AND INSTITUTES.

Center for Space Sciences. University of Texas at Austin, WRW 402, Austin, TX 78712. (512) 471-1356. FAX (512) 471- 3570.

Earth and Planetary Remote Sensing Laboratory. Washington University, Department of Earth and Planetary Sciences. Campus Box 1169, 1 Brookings Drive, St. Louis, MO 63130. (314) 889-5679. FAX (314) 889-5799.

Harvard College Observatory. Harvard University. 60 Garden Street, Cambridge, MA 02138. (617) 495-9059.

Institute For Astronomy. University of Hawaii at Manoa, 2680 Woodlawn Drive, Honolulu, HI 96822. (808) 956-8312. FAX (808) 988-2790.

Institute of Geophysics and Planetary Physics. University of California, 405 Hilgard Avenue, Los Angeles, CA 90024. (213) 825-1580. (213) 206-3051.

Laboratory For Atmospheric and Space Physics. University of Colorado - Boulder, Boulder, CO 80309-0392. (303) 492-7677. (303) 492-6946.

Laboratory For Planetary Geology. Arizona State University, Department of Geology, Tempe, AZ 85281. (601) 965-7029.

Lunar and Planetary Institute. 3303 NASA Road One, Houston, TX 77058-4399. (713) 486-2139. FAX 713-496-2162.

Lunar and Planetary Laboratory. University of Arizona, Tucson, AZ 85721. (602) 621-6963. FAX (602) 621-4933.

McDonnell Center For the Space Sciences. Washington University, Box 1105, One Brookings Drive, St. Louis MO 63130-4899. (314) 889-6255. FAX (314) 889-6219.

Smithsonian Astrophysical Observatory. 60 Garden Street, Cambridge, MA 02138. (617) 495-7461. FAX (617) 495-7105.

ASTROMETRY AND ASTRONOMICAL PHOTOMETRY

See: ASTRONOMY

ASTRONAUTICAL ENGINERING

See: AEROSPACE ENGINEERING

ASTRONAUTICS

See also: AERONAUTICS, MANNED SPACE FLIGHT, SPACECRAFT

ABSTRACT SERVICES AND INDEXES

Aeronautical Engineering. Scientific and Technical Branch, National Aeronautics and Space Administration. National Technical Information Service (NTIS), 5285 Port Royal Road, Spring-

ASTRONAUTICS

Ency. of Physical Sciences and Engineering Info. Sources

field, VA 22161. (703) 487-4650. FAX (703) 321-8547. Monthly. A selection of annotated references to unclassified reports and journal articles that were introduced into the NASA scientific and technical information system and announced in STAR and IAA.

Applied Science and Technology Index; A cumulative subject index to English language periodicals in the fields of aeronautics and space science, computer technology, chemistry, construction industry, energy and related areas. H.W. Wilson Co., 950 University Avenue, Bronx, NY 10452. (800) 367-6770 or (212) 588-8400. FAX (718) 590-1617. From 1958 to present. Monthly. Inquire about cost and availability. Also available on CD-ROM.

International Aerospace Abstracts. Technical Information Service, American Institute of Aeronautics and Astronautics, Inc., 555 West 57th St., New York, NY 10019. (212) 247-6500. FAX (212) 582-4861. Semi-monthly. $1295.00 per year.

Science Citation Index. Institute for Scientific Information, 3501 Market Street, Philadelphia, PA 19104. (215) 386-0100 or (800) 523-1850. FAX (215) 386-6362. Inquire about availability and cost. Also available on CD-ROM.

STAR. Scientific and Technical Aerospace Reports. National Aeronautics and Space Administration. NASA Center for Aerospace Information, Box 8757, BWI Airport, Baltimore, MD 21240. (301) 621-0153. Monthly. Inquire about availability and cost. Also available through the NASA online retrieval service (RECON), and through the Aerospace Database through DIALOG.

ASSOCIATIONS AND PROFESSIONAL SOCIETIES

American Astronautical Society. 6352 Rolling Mill Place, Suite 102, Springfield, VA 22152. (703) 866-0020. FAX (703) 866-3526.

American Institute of Aeronautics and Astronautics. The Aerospace Center, 370 L'Enfant Promenade SW, Washington, DC 20024. (202) 646-7400. FAX (202) 646-7508.

Canadian Aeronautics and Space Institute, 222 Somerset Street West, Suite 601, Ottawa, Ontario, Canada K2P 0J1.

DIRECTORIES AND BIOGRAPHICAL SOURCES

American Astronautical Society—Membership Directory. American Astronautical Society, 6352 Rolling Mill Place, Suite 102, Springfield, VA 22152. (703) 866-0020. FAX (703) 866-3526. Biennial. $60.00.

Research Centers Directory. Gale Research, 835 Penobscot Building, Detroit, MI 48226-4094. (313) 961-2242. (800) 877-4253. FAX (313) 961-6083. $485.00.

Who's Who in Engineering. American Association of Engineering Societies, 1111 19th St. NW, Suite 608, Washington, DC 20036. 7th ed. 1988. $200.00.

ENCYCLOPEDIAS AND DICTIONARIES

Dictionary of Aeronautical Terms, Dale Crane, Comp. Aviation Supplies and ACADemics, Inc., 7005 132nd Place SE,

Newcastle, WA 98059. (206) 235-1500 OR (800) 426-8338. fax (206) 235-0128. 1991. $15.95.

Encyclopedia of Physical Science and Technology. Robert A. Meyers, ed. Academic Press, Inc., 6277 Sea Harbor Drive, Orlando, FL 32887. 1992. $2100.00.

Jane's Aerospace Dictionary, B. Gunston. 3rd ed. Jane's Information Group, Dept. DSM, 1340 Braddock Pl., Suite 300, Box 1436, Alexandria, VA 22314-1651. (703) 683-3700. FAX (703) 836-0029. 1991. $45.00.

McGraw-Hill Encyclopedia of Science and Technology, Sybil P. Parker, ed. 7th ed. 20 vols. McGraw-Hill Publishing Company, 1221 Avenue of the Americas, New York, NY 10020. (800) 262-4729 or (212) 512-3825. $1900.00

GENERAL WORKS

Blueprint For Space: Science Fiction To Science Fact. Frederick I. Ordway III and Randy Liebermann, eds. Smithsonian Institution Press, 470 L'Enfant Plaza, Suite 7100, Washington, DC 20560. (202) 287-3738. 1992. $60.00 (hardback) and $24.95 (paperback).

Breakout Into Space: Mission For A Generation. George Henry Elias. William Morrow and Company, Inc., 1350 Avenue of the Americas, New York, NY 10019. (212) 261-6500 or (800) 843-9389. FAX (212) 779-0965. 1990. $16.95.

Humans & Machines in Space: The Vision, the Challenge, the Payoff. Bradley Johnson, et al., eds. Univelt, Inc. PO Box 28130, San Diego, CA 92198. (619) 746-4005. FAX (619) 746-3139. 1992. $50.00 (hardback) and $35.00 (paperback).

Humans in Space: 21st Century Frontiers. Harry L. Shipman. Plenum Publishing Corporation, 233 Spring Street, New York, NY 10013. (212) 620-8000. FAX (212) 463-0742. 1989. $22.95.

Introduction To Space Flight. Francis J. Hale. Prentice Hall Inc., 113 Sylvan Avenue, Route 9W, Englewood Cliffs, NJ 07632. (201) 592-2000 or (800) 922-0579. 1994. Write for information.

Space. Roy Gibson. Clarendon Press; Oxford University Press, Inc., 198 Madison Avenue, New York, NY 10016-4314. (212) 726-6000. FAX (212) 726-6446. 1992. $36.00.

ONLINE DATABASES AND CD-ROMS

Aerospace Database. American Institute of Aeronautics and Astronautics, 370 L'Enfant Promenade, S.W., Washington, DC 20024. (202) 646-7400. Worldwide published literature on research and development in aerospace and related areas, with abstracts. Covers 1962 to present. Available on DIALOG, (800) 334-2564, online service. Also available on CD-ROM. Inquire as to cost and availability.

Compendex Plus. Engineering Information, Inc., 345 East 47th Street, New York, NY 10017. (212) 705-7600 or (800) 221-1044. Contains citations with abstracts to worldwide literature in engineering and technology, from 1970 to present. Available on online BRS,(800) 289-4277, DIALOG, (800) 334-2564, ORBIT (800) 456-7248, and STN International, FIZ Karlsruhe, P.O. Box 2465, W-7500, Karlsruhe 1, Germany, online services. Also available on CD-ROM. Inquire as to cost and availability.

CSA Engineering. Cambridge Scientific Abstracts, 7200 Wisconsin Avenue, Suite 601, Bethesda, MD 20814. (301) 961-6750 or (800) 843-7751. Contains citations and abstracts of international periodicals and research literature covering all fields of engineering and science and technology, including computer and information science, electronics, mechanical engineering, solid state materials, 1981 to present. Available on BRS,(800) 289-4277, online service. Inquire as to cost and availability.

Current Contents Search. Institute for Scientific Information, 3501 Market Street, Philadelphia, PA 19104. (215) 386-0100. FAX (215) 386-6362. Contains citations to articles listed in the table of contents of science and technology journals. Also articles in social sciences and life sciences journals. Available on BRS,(800) 289-4277, DIALOG,(800) 334-2564, online services. Inquire as to cost and availability.

Dissertation Abstracts. University Microfilms International, 300 North Zeeb Road, Ann Arbor, MI 48106. (800) 521-0600 or (313) 761-4700. Scope includes virtually all doctoral dissertations accepted at accredited American institutions from 1861 to present in 252 subject areas. Available on BRS,(800) 289-4277, DIALOG,(800) 334-2564, and OCLC EPIC,(800) 848-5878, online services. Also available on CD-ROM. Inquire as to cost and availability.

International Aerospace Abstracts. American Institute of Aeronautics and Astronautics, 370 L'Enfant Promenade, S.W., Washington, DC 20024. (202) 646-7400. Contains references and abstracts of journal and monograph literature relating to aerospace science and technology, from 1963 to present. Available through the NASA/RECON system of the National Aeronautics and Space Administration only.

NASA Database. American Institute of Aeronautics and Astronautics, 370 L'Enfant Promenade, S.W., Washington, DC 20024. (202) 646-7400. Citations and abstracts of aeronautics and astronautics literature, 1962 to present. Also contains citations from *STAR, Scientific and Technical Aerospace Reports,* and *International Aerospace Abstracts.* Available through NASA/RECON online service. Inquire as to cost and availability.

SCISEARCH. Institute for Scientific Information, 3501 Market Street, Philadelphia, PA 19104. (800) 523-1850 or (215) 386-0100. Broad multidisciplinary title and author index to the international literature of science and technology, 1974 to present. Available on DIALOG,(800) 334-2564, and ORBIT,(800) 456-7248, online services. Also available on CD-ROM. Inquire as to cost and availability.

WILSONLINE. H.W. Wilson Company, 950 University Avenue, Bronx, NY 10452. (800) 367-6770 or (212) 588-8400. Makes available online versions of the printed H.W. Wilson indexes including Applied Science and Technology Index, Business Periodicals Index, General Science Index, and Readers' Guide to Periodical Literature. Period covered is generally 1983 to present. Available on BRS,(800) 289-4277, DIALOG,(800) 334-2564, and OCLC EPIC,(800) 848-5878, online services. Also available on CD-ROM. Inquire as to cost and availability.

PERIODICALS

ACTA Astronautica. Pergamon Press, Pergamon Press, 660 White Plains Rd., Tarrytown, NY 10591-5153. 14 times a year. Inquire for price and availability.

Ad Astra. National Space Society, 922 Pennsylvania Avenue SE, Washington, DC 20003-2140. (202) 543-1900. FAX (202) 546-4189. 1989 to present. 6 times a year. $35.00 a year.

Advances in the Astronautical Sciences. American Astronautical Society, 6352 Rolling Mill Place, Suite 102, Springfield, VA 22152. (703) 866-0020. FAX (703) 866-3526. Quarterly. Write or call for price information.

Advances in Space Research. Pergamon Press, Pergamon Press, 660 White Plains Rd., Tarrytown, NY 10591-5153. 12 times a year. Inquire for price and availability.

Aeronautical Journal. Royal Aeronautical Society, PO Box 139, Tonbridge, Kent, TN9 1EW, England. 0732-770823. FAX 0723-361708. 1897 to present. Ten times a year. Write for price and availability.

Aerospace. Royal Aeronautical Society, PO Box 139, Tonbridge, Kent, TN9 1EW, England. 0732-770823. FAX 0723-361708. Monthly. Write for price and availability.

Aerospace America. American Institute of Aeronautics and Astronautics, 370 L'Enfant Promenade SW, Washington, DC 20024-2518. (202) 646-7471. 1932 to present. Monthly. $70 annually (non-members).

AIAA Journal. American Institute of Aeronautics and Astronautics, 370 L'Enfant Promenade SW, Washington, DC 20024-2518. (202) 646-7471. 1963 to present. Monthly. $52.00 per year (members) and $435.00 per year (non-members).

Canadian Aeronautics and Space Journal. Canadian Aeronautics and Space Institute, Suite 818, the National Building, 130 Slater St., Ottawa, Canada K1P 6E2. 1955 to present. Monthly. $75.00 per year (U.S.).

Journal of the Astronautical Sciences. American Astronautical Society, 6352 Rolling Mill Place, Suite 102, Springfield, VA 22152. (703) 866-0020. FAX (703) 866-3526. Quarterly. $130.00 per year.

Journal of the British Interplanetary Society. British Interplanetary Society, 27/29 South Lambeth Road, London SW8 1SZ, England. (071) 735-3160. FAX (071) 820-1504. Monthly. $370.00 per year.

Journal of Guidance, Control and Dynamics. American Institute of Aeronautics and Astronautics, 370 L'Enfant Promenade SW, Washington, DC 20024-2518. 1978 to present. Bimonthly. $42.00 (members) and $250.00 (non-members).

Journal of Propulsion and Power. American Institute of Aeronautics and Astronautics, 370 L'Enfant Promenade SW, Washington, DC 20024-2518. 1985 to present. Bimonthly. $38.00 (members) and $260.00 (non-members).

Journal of Spacecraft and Rockets. American Institute of Aeronautics and Astronautics, the Aerospace Center, 370 L'Enfant Promenade SW, Washington, DC 20024. (202) 646-7400. FAX (202) 646-7508. 1964 to present. Bi-monthly. $185.00 per year.

Space Times. American Astronautical Society, 6352 Rolling Mill Place, Suite 102, Springfield, VA 22152. (703) 866-0020. FAX (703) 866-3526. 1962 to present. Bi-monthly. $60.00 per year.

ASTRONAUTICS

Ency. of Physical Sciences and Engineering Info. Sources

RESEARCH CENTERS AND INSTITUTES

Center For Space Research. Massachusetts Institute of Technology, 77 Massachusetts Avenue, Cambridge, MA 02139. (617) 253-7501.

Center For Space Research. University of Texas at Austin, WRW 402, Austin, TX 78712. (512) 471-1356. FAX (512) 471-3570.

Charles Stark Draper Laboratory Inc., 555 Technology Square, Cambridge, MA 02139. (617) 258-1000.

Goddard Space Flight Center, Greenbelt, MD 20771. (301) 286-7351.

Graduate Aeronautical Laboratories. California Institute of Technology, 105-50, Pasadena, CA 91125. (818) 356-4551. FAX (818) 449-2677.

Instrumentation and Control Laboratory. Princeton University, Department of MAE, Engineering Quadrangle, Princeton, NJ 08544. (609) 452-5154.

George C. Marshall Space Flight Center, Huntsville, AL 35812. (205) 544-0024.

Space Science and Engineering Center. University of Wisconsin—Madison, 1150 University Avenue, Madison, WI 53706. (608) 263-4680.

ASTRONOMICAL OBSERVATORIES

See: OBSERVATORIES

ASTRONOMY

See also: ASTROPHYSICS, COSMOLOGY, GALAXIES, PLANETS, SOLAR SYSTEM, TELESCOPES

ABSTRACT SERVICES AND INDEXES

Applied Science and Technology Index; A cumulative subject index to English language periodicals in the fields of aeronautics and space science, computer technology, chemistry, construction industry, energy and related areas. H.W. Wilson Co., 950 University Avenue, Bronx, NY 10452. (800) 367-6770 or (212) 588-8400. From 1958 to present. Monthly. Inquire about cost and availability. Also available on CD-ROM.

Astronomy and Astrophysics Abstracts. Springer-Verlag, 175 Fifth Avenue, New York, NY 10010. (212) 460-1500. FAX (212) 473-6272. From 1969 to present. Irregular. Price varies; inquire. Also available on CD-ROM. ISSN: 0067- 0022. ISSN: 0067-0022.

Chemical Abstracts — Physical, Inorganic and Analytical Chemistry. Chemical Abstracts Service, 2540 Olentangy River Road, Box 3012, Columbus, OH 43210. (800) 848-6538 or (614) 447-3600, FAX (614) 447-3713. Bi-weekly. $1410.00 per year. Also available on CD-ROM and online. Inquire regarding cost and availability.

Current Contents: Physical, Chemical and Earth Sciences. Institute for Scientific Information, 3501 Market Street, Philadelphia, PA 19104. (215) 386-0100. FAX (215) 386-6362. Weekly. $360.00 per year

Current Papers in Physics. Institute of Electrical Engineers, Michael Faraday House, Six Hill Way, Stevenage, Herts, SG1 2AY, England. Distributed by IEEE, Box 1331, 445 Hoes Lane, Piscataway, NJ 08855-1331. (908) 562-5549. From 1966 to present. Fortnightly. $410.00 per year.

General Science Index. H.W. Wilson Company, 950 University Avenue, Bronx, NY 10452. (212) 588-8400. (800) 367-6770. FAX (718) 590-1617. From 1978 to present. Ten issues per year; quarterly and annual cumulations. Service basis. Available on CD-ROM and online. Inquire regarding cost and availability. ISSN: 0162-1963.

Government Reports Announcements and Index. U. S. National Technical Information Service (NTIS), 5285 Port Royal Road, Springfield, VA 22161. (703) 487-4650. FAX (703) 321-8547. From 1968 to present. Annual. $630.00 per year. Also available online as *NTIS Bibliographic Database* and on CD-ROM. ISSN:

International Aerospace Abstracts. American Institute of Aeronautics and Astronautics. Technical Information Service, 555 West 57th Street, New York 10019. (212) 247- 6500. FAX (212) 582-4961. 1961 to present. Semi-monthly with annual index. $1295.00 per year. Also available online and on CD-ROM. ISSN: 0020-5842.

Meteorological and Geoastrophysical Abstracts. American Meteorological Society, c/o Inforonics, Inc., 550 N ewtown Road, Littleton, MA 01460. (508) 486-8976. FAX (508) 486-0027. Covers literature in environmental sciences, meteorology, astrophysics, hydrology, glaciology, and physical oceanography. From 1950 to present. Monthly. $950.00 per year. Also available on CD-ROM and online. Inquire regarding cost and availability. ISSN: 0026-1130.

NTIS Alerts: Astronomy & Astrophysics. U. S. National Technical Information Service. 5285 Port Royal Road, Springfield, VA 22161. (703) 487-4650. FAX (703) 321-8547. Weekly. $140.00 per year.

Pascal 48: Environnement Cosmique Terrestre, Astronomie Et Geologie Extraterrestre. Centre National de la Recherche Scientifique, Institut de Information Scientifique et Technique, 2 aliee du Parc de Brabois, 54514 Vandoeuvre Les Nancy Cedex, France. TEL 83-50-46-00. FAX: 83-50-46-50. 1985 to present. Ten issues per year. 770 F. Also available on CD-ROM and online.

Physics Abstracts. INSPEC. Section A, Science Abstracts. Institution of Electrical Engineers, London, United Kingdom. Available from: INSPEC/IEEE - Institute of Electrical and Electronic Engineers, Box 1331, Hoes Lane, Piscataway, NJ 08855-1331. (908) 562-5549. 1898 to present. 24 issues per year. $2835.00 per year. Also available on CD-ROM and online. ISSN: 0036-8091.

Physics Briefs (physikalische Berichte). Information Center for Energy, Physics, Mathematics; German Physical Society. V C H Publishers, Inc., 220 East 23rd Street, New York, NY 10010-4606. (212) 683-8333. 1845 to present. 24 issues per year. $2390.00 per year. Also available online. ISSN: 0179-7434.

Science Citation Index. SCI. Institute for Scientific Information, 3501 Market Street, Philadelphia, PA 19104. (215) 386-0100. (800) 523-1850. FAX (215) 386-2991. 1961 to present. Six issues per year, plus annual cumulation. $11650.00 per year. Also available online and on CD-ROM. Inquire about price and availability. ISSN: 0036-827X.

STAR. Scientific and Technical Aerospace Reports. U.S. National Aeronautics and Space Administration. Distributed by U. S. Superintendent of Documents, Washington, DC 20402. 1963 to present. Semi-monthly, with semiannual and annual indexes. $114.00 per year. ISSN: 0036-8741. Also available online and on CD-ROM.

ANNUAL REVIEWS AND YEARBOOKS

The Astronomical Almanac. Superintendent of Documents, U.S. Government Printing office, Washington, DC 10402. (202) 783-3238. 1981 to present. Superceeds Astronomical Ephemeris. Annual. $44.95.

Annual Review of Astronomy and Astrophysics; Original R Eviews of Critical Literature and Current Developments in Astronomy and Astrophysics. Annual Reviews, Inc., 4139 El Camino Way, Palo Alto, CA 94306-0139. (415) 493-4400. (800) 523-8635. FAX (415) 855-9815. 1963 to date. Annual. $60.00. ISSN: 0066-4146.

Annual Review of Earth and Planetary Sciences. Annual Reviews, Inc., 4139 El Camino Way, Palo Alto CA 94306-0139. (415) 403-4400. (800) 523-8635. Fax (415) 855-9815. 1973 to date. Annual. $62.00. ISSN: 0084-6597.

Highlights of Astronomy. I. Appenzeller, editor. Kluwer Academic Publishers, Box 358, Accord Station, Hingham, MA 02018-0358. (617) 871-6600. International Astronomical Union Highlights series. ISBN: 07923-3553-8. Triennal. 1995. Price varies.

ASSOCIATIONS AND PROFESSIONAL SOCIETIES

Amateur Astronomers Association. 1010 Park Avenue, New York, NY 10028. (212) 535-2922.

American Association of Variable Star Observers. 23 Birch Street, Cambridge, MA 02138. (617) 354-0484. FAX: (617) 354-0665.

American Astronomical Society. 2000 Florida Avenue NW, Suite 400, Washington, DC 20009. (202) 328-2010. FAX: (202) 234-2560.

American Geophysical Union. 2000 Florida Avenue, NW, Washington, DC 20009. (202) 462-6900. (800) 966-AGU1. FAX (202) 328-0566.

American Institute of Physics. One Physics Ellipse, College Park, MD 20740-3843. (301) 209-3100.

Astronomical League. 2112 Kingfisher Lane East, Rolling Meadows, IL 60008. (708) 398-0562.

Astronomical Society of the Pacific. 390 Ashton Avenue, San Francisco, CA 94112. (415) 337-1100. FAX: (415) 337-5205.

International Astronomical Union. 98 bis Boulevard Arago, F-75014 Paris, France. Tel 1-43-25-83-58. FAX 1-43-25-26-16.

International Dark-sky Association. 3545 North Stewart, Tucson, AZ 95716. (602) 795-1381

International Planetarium Society. 15 South State Street, Salt Lake City, UT 84111. (801) 538-2104. FAX (801) 538- 2059.

Joint Institute For Laboratory Astrophysics. University of Colorado, Box 440, Boulder, CO 80309-0440. (303) 492-7780. FAX (303) 492-5235.

Maria Mitchell Association. 2 Vestal Street, Nantucket, MA 02554. (508 228-9198.

Royal Astronomical Society of Canada. 136 Dupont Street, Toronto. ON N5R 1V2 Canada. (416) 924-7973. FAX (416) 968-6687.

Von Braun Astronomical Society. PO Box 1142. Huntsville, AL 35807. (205) 539-0316.

Webb Society. 1440 South Marmora Avenue, Tuscon, AZ95713. (602) 628-1077.

BIBLIOGRAPHIES

Astronomy and Astrophysics: a Source Guide. Gordon Press, P.O. Box 459, Bowling Green Station, New York, NY 10004. (718) 624-8419. 1991. $250.00. .

A Bibliography of Astronomy, 1970-1979. R. A. Sea and S. S. Martin. Libraries Unlimited, Inc., Littleton, CO 80160. 1982. $37.50.

Catalog of the Naval Observatory Library. Naval Observatory Staff. G. K. Hall & Company, 866 Third Avenue, New York, NY 10022. (212)702-6789. (800) 257-5755. 6 volumes. 1977. $655.00

Science Books & Films. American Association for the Advancement of Science, 1333 H Street NW, Washington, DC 20005. (202) 326-6434. Reviews of print, film and software materials in all sciences. 1965 to present. Nine issues per year. $40.00 per year. ISSN: 0098-342X.

Scientific and Techni Cal Books and Serials in Print; An Index To Literature in Science and Technology. R. R. Bowker. 121 Chanlon Road, New Providence, NJ 07974. (908) 464-6800. FAX (908) 665-3502. (800) 521-8110. 1972 to present. Annual. $299.95 per year. Also available on CD-ROM. ISSN: 0000-054X.

A Source Book On Astronomy and Astrophysics, 1900 - 1975. Kenneth Lang and Owen Gingerich, editors. Harvard University Press, 79 Garden Street, Cambridge MA. (617) 495-2600. 1980. $41.95.

DIRECTORIES AND BIOGRAPHICAL SOURCES

American Astronomical Society Membership Directory. American Astronomical Society. 2000 Florida Avenue NW, Suite 400, Washington, DC 20009. (202) 328-2010. FAX (202) 234-2560. Annual. Included in membership dues. ISSN: 1061-9038.

ASTRONOMY

Ency. of Physical Sciences and Engineering Info. Sources

American Men and Women of Science: Physical and Biological Sciences. R. R. Bowker Inc., 121 Chanlon Road, New Providence, NJ 07974. (908) 464-6800. (800) 521-8110. 20th edition. 8 volumes. 1996. $850.00.

The Astronomers. Donald Goldsmith. St. Martin's Press, Inc., 175 Fifth Avenue, New York, NY 10010. (212) 674-5151. (800) 221-7945. 1993. $14.95 in paper.

Astronomical Centers of the World. Kevin Krisciunas. Cambridge University Press, 40 West 20th Street, New York, NY 10011-4211. (212) 924-3900. (800) 872-7423. 1988. $34.95

The Biographical Dictionary of Scientists: Astronomers. D. Abbott, editor. Peter Bedrick Books, Inc., 2112 Broadway, Room 318, New York, NY 10023. (212) 496-0751. 1984.

Directory of Physics and Astronomy Staff. American Institute of Physics, One Physics Ellipse, College Park, MD 20740-3843. (301) 209-3100. 1975/76 to present. Annual. $60.00. ISSN: 0361-2228.

Earth and Astronomical Research Centers. Stockton Press, 345 Park Avenue South, New York, NY 10010. 4th edition. 1995. $515.00. ISBN: 1-56169-0967.

Graduate Programs in Physics, Astronomy and Related Fields. American Institute of Physics, One Physics Ellipse, College Park, MD 20740-3843. (301) 209-3100. 1978 to present. Annual. $45.00. ISSN: 0147-1821. 1995-1996.

Research Centers Directory. Gale Research Inc., 835 Penobscot Building, Detroit, MI 48226-4094. (313) 961-2242. (800) 877-4253. 20th edition. 1995. $485.00.

ENCYCLOPEDIAS AND DICTIONARIES

Astronomy From A To Z: A Dictionary of Celestial Objects and Ideas. Charles A. Schweighauser. Illinois Issues, Sangamon State University, K-10, Springfield, IL 62794-9243. (217) 786-9243. 1991. $14.95 in paper.

Concise Dictionary of Astronomy. Jacqueline Mitton. Oxford University Press, Inc., 200 Madison Avenue, New York, NY 10016. (212) 725-6000. (800) 334-4249. 1992. $24.95.

Dictionary of Astronomical Names. Adrian Room. Routledge, Chapman and Hall, Inc., 29 West 35th Street, New York, NY 10001-3291. (212) 244-3336. 1988. $27.50.

Facts On File Dictionary of Astronomy. Valerie Illingworth. Facts-on-File, Inc., 460 Park Avenue South, New York, NY 10016-7382. (212) 683-2244. (800) 322-8755. Fax (800) 683-3633. 1994. $27.95.

Oxford Illustrated Encyclopedia. VOLUME 8, the UNIVERSE. Archie Roy, editor. Oxford University Press, Inc., 200 Madison Avenue, New York, NY 10016. (212) 725-6000. (800) 334-4249. 1993. $45.00.

McGraw-Hill Encyclopedia of Science and Technology. McGraw-Hill Book Company, Inc., 1221 Avenue of the Americas, New York, NY 10020. (212) 512-2000. (800) 262-4729. 7th edition. 20 volume set. 1992. $1900.00.

Penguin Dictionary of Astronomy. Jacqueline Mitton. Viking Penguin, 375 Hudson Street, New York, NY 10014-3657. (212) 366-2000. (800) 331-4624. 1993. $12.95 in paper.

Quick Reference Guide To Astronomy Magazine 1973 - 1990. Jack Hobart. Geoimages Publishing Co., P.O. Box 45677, 8005 Dunbarton Avenue, Los Angeles, CA 90045. (310) 645-5419. 1990. $8.00.

Stars, Galaxies, Cosmos. William R. Corliss. Sourcebook Project, P.O. Box 107, Glen Arm, MD 21057. (410) 668-6047. 1987. $17.95.

GENERAL WORKS

Applying Fractals in Astronomy. andre Heck, et al, editors. Springer Verlag New York, Inc., 175 Fifth Avenue, New York, NY 10010. (212) 460-1500. (800) 777-4643. 1991. $30.00.

Astronomer's Universe: Stars, Galaxies and Cosmos. Herbert Friedman. W. W. Norton & Company, 500 Fifth Avenue, New York, NY 10110. (212) 354-5500. (800) 223-2584 1990. $24.95.

Astronomy and Cosmology. John North. W. W. Norton & Company, 500 Fifth Avenue, New York, NY 10110. (212) 354-5500. (800) 223-2584 1990. 1994. $18.95.

Compendium of Practical Astronomy. Gunter D. Roth, editor. Springer-Verlag New York, Inc., 175 Fifth Avenue, New York, NY 10010. (212) 460-1500. (800) 777-4643. 3 volumes. Revised edition. 1994. $125.00.

Comets in the Post Halley Era. R. L. Newburn, et al, editors. Kluwer Academic Publishers, 101 Philip Drive, Assinippi Park, Norwell, MA 02061. (617) 871-6600. 2 volumes. 1991. $61.50 in paper.

Cosmic Catastrophies. C. P. Chapman and D. Morrison, editors. Plenum Publishing Corp., 233 Spring Street, New York, NY. (212) 620-8000. (800) 2221-9369. 1989. $22.95.

Cycles of Fire: Stars, Galaxies and the Wonder of Deep Space. William K. Hartmann. Workman Publishing Company, Inc., 708 Broadway, New York, NY 10003. (212) 254-5900. (800) 722-7202. 1988. $15.95 in paper.

Databases and Online Data in Astronomy. Migual A. Albrecht and Daniel Egret, editors. Kluwer Academic Publishers, 101 Philip Drive, Assinippi Park, Norwell, MA 02061. (617) 871-6600. 1991. $77.50.

Farthest Things in the Universe. Jay M. Pasachoff. Cambridge University Press 40 West 20th Street, New York, NY 10011-4211. (212) 924-3900. (800) 872-7423. 1995. $29.95.

Fireside Astronomy: An Anecdotal Tour Through the History and Lore of Astronomy. Patrick Moore. John Wiley & Sons, Inc., 605 Third Avenue, New York, NY 10158-0012. (212) 850-6000. (800) 225-5945. 1993. $14.95 in paper.

An Introduction To Astronomy: Stars, Galaxies and the Universe. Gerald D. Waxman. Kendall-Hunt Publishing Co., 4050 Westmark Drive, P.O. Box 1840 Dubuque, IA 52004-1840. (319) 589-1000. (800) 228-0810. 1992. $28.95.

Discovering the Cosmos. Robert C. Bless. University Science Books, 55D Gate Five Road, Sausalito, CA 94965. (415) 332-5390. 1995. $58.00.

Exploring the Sun: Solar Science Since Galileo. Karl Hufbauer. Johns Hopkins University Press, 2715 North Charles Street, Baltimore, MD 21218-4319. (410) 516-6900. (800) 537- 5487. 1991. $50.00.

Introduction To Astronomical Photometry. Edwin Budding. Cambridge University Press, 40 West 20th Street, New York, NY 10011-4211. (212) 924-3900. (800) 872-7423. 1993. $44.95.

Kepler's Physical Astronomy. Bruce Stephenson. Princeton University Press, 41 William Street, Princeton, NJ 08540. (609) 258-4900. (800) 777-4726. 1994. $14.95.

Millimeter and Submillimeter Astronomy. R. D. Wolstencroft and W. Butler Burton, editors. Kluwer Academic Publishers, 101 Philip Drive, Assinippi Park, Norwell, MA 02061. (617) 871-6600. 1988. $136.50.

Modern Astrometry. Jean Kovalevsky. Springer-Verlag New York, Inc., 175 Fifth Avenue, New York, NY 10010. (212) 460-1500. (800) 777-4643 . 1994. $64.50.

The New Cosmos. A. Unsold and B. Baschek. Springer-Verlag New York, Inc., 175 Fifth Avenue, New York, NY 10010. (212) 460-1500. (800) 777-4643 . 4th revised edition. 1991. $59.00.

Observational Astronomy. D. Scott Birney. Cambridge University Press, 40 West 20th Street, New York, NY 10011- 4211. (212) 924-3900. (800) 872-7423. 1991. $27.95 in paper.

Quasar Astronomy. Daniel W. Weedman. Cambridge University Press, 40 West 20th Street, New York, NY 10011-4211. (212) 924-3900. (800) 872-7423. 1988. $24.95 in paper.

The Red Limit: The Search For the Edge of the Universe. Timothy Ferris. Morrow, William & Company, 1350 Avenue of the Americas, New York, NY 10019. (212) 261-6500. (800) 843-9389. Revised edition. 1993. $12.95.

Space Technology and Planetary Astronomy. Joseph N. Tatarewicz. Indiana University Press, 601 North Morton Street, Bloomington, in 47404-3797. (812) 855-6804. (800) 842-6796. 1990. $29.95.

Stars and Atoms: From the Big Bang To the Solar System. Stuart Clark. Oxford University Press, Inc., 200 Madison Avenue, New York, NY 10016. (212) 725-6000. (800) 334-4249. 1995. $35.00.

Stephen Hawking's Universe. John Boslough. Morrow, William & Company, 1350 Avenue of the Americas, New York, NY 10019. (212) 261-6500. (800) 843-9389. 1985. $7.95.

Visual Astronomy of the Deep Sky. Roger N. Clark. Cambridge University Press, 40 West 20th Street, New York, NY 10011-4211. (212) 924-3900. (800) 872-7423. 1991. $39.95.

Universe. William J. Kaufmann. W. H. Freeman & Co., 41 Madison Avenue, East 26th Avenue, 35th Floor, New York, NY 10010. (212) 576-0400. 4th edition. 1995. $54.95.

HANDBOOKS AND MANUALS

Burnham's Celestial Handbook: an Observer's Guide to the Universe Beyond the Solar System. Robert Burnham, Jr. Dover Publications, Inc., 180 Varick Street, New York, NY 10014. (212) 255-3755. (800) 223-3130. 3 volumes. 1978. $41.85 for the set.

Cambridge Guide To the Constellations. Michael E. Bakich. Cambridge University Press, 40 West 20th Street, New York, NY 10011-4211. (212) 924-3900. (800) 872-7423. 1995. $69.95.

Encyclopedia of Astronomy and Astrophysics. Stephen Maran, editor. Van Nostrand Reinhold, 115 Fifth Avenue, New York, NY 10003. (212) 254-3232. (800) 842-3636. 1992. $129.95.

Greenwich Guide To Stars, Galaxies and Nebulae. Stuart Malin. Cambridge University Press, 40 West 20th Street, New York, NY 10011-4211. (212) 924-3900. (800) 872-7423. 1990. $10.95.

Handbook of Space Astronomy and Astrophsyics. Martin V. Zumbeck. Cambridge University Press, 40 West 20th Street, New York, NY 10011-4211. (212) 924-3900. (800) 872-7423. 2nd edition. 1990. $79.95.

Landolt-borenstein Numerical Data and Functional Relationships in Science and Technology: Astronomy, Astrophsyics and Space Research. GROUP VI. Springer- Verlag New York, Inc., 175Fifth Avenue, New York, NY 10010. (212) 460-1500. (800) 777-4643. volumes priced individually.

McGraw-Hill Encyclopedia of Astronomy. Sybil P. Parker and Jay M. Pasachoff, editors. McGraw-Hill Publishing Company, Inc., 1221 Avenue of the Americas, New York, NY 10020. (212) 512-2000. (800) 262-4729. 2nd edition. 1993. $75.00

Observer's Guide To Astronomy. Patrick Martinez, editor. Cambridge University Press, 40 West 20th Street, New York, NY 10011-4211. (212) 924-3900. (800) 872-7423. 2 volumes. 1994. $159.90.

ONLINE DATABASES AND CD-ROMS

CA Search. Chemical Abstracts Service, P.O. Box 3012, Columbus, OH 43210-0012. (614) 447-3600. (800) 848-6533. FAX (614) 447-3709. Very comprehensive guide to worldwide chemical literature a nd related fields, 1972 to present. Available on BRS,(800) 289-4277, DIALOG, (800) 334-2564, ORBIT (800) 456-7248, and STN International, FIZ Karlsruhe, P.O. Box 2465, W-7500, Karlsruhe 1, Germany, online services. Inquire as to cost and availability.

Compendex Plus. Engineering Information, Inc., 345 East 47th Street, New York, NY 10017. (212) 705-7600 or (800) 221- 1 044. Contains citations with abstracts to worldwide lite rature in engineering and technology, from 1970 to present. Available on online BRS,(800) 289-4277, DIALOG, (800) 334-2564, ORBIT (800) 456-7248, and STN International, FIZ Karlsruhe, P.O. Box 2465, W-7500, Karlsruhe 1, Germany, online services. Also available on CD-ROM. Inquire as to cost and availability.

Current Contents Search. Institute for Scientific Informatio n, 3501 Market Street, Philadelphia, PA 19104. (215) 386-0 100. FAX (215) 386-6362. Contains citations to articles listed in the table of contents of science and technology journals. Also articles in social sciences and life sciences journals. Available on

BRS, (800) 289-4277, DIALOG, (800) 334-2564, online services. Inquire as to cost and availability.

Dissertation Abstracts Online. University Microfilms Internatio nal, 300 North Zeeb Road, Ann Arbor, MI 48106. (800) 521-0600 or (313) 761-4700. Scope includes virtually all doctoral dissertations accepted at accredited American institutions from 1861 to present in 252 subject areas. Available on BRS, (800) 289-4277, DIALOG, (800) 334-2564, and OCLC EPIC, (800) 848-5878, online services. Also available on CD-ROM. Inquire as to cost and availability.

INSPEC. Institution of Electrical Engineers, Michael Faraday House, Six Hills Way, Stevenage, Herts. SG1 2AY, Eng land. Telephone: 0438 313311 or FAX 0438 742840 . Contains citations to the worldwide literature of phys ics, electronics and electrical engineering, computer technology, and related fields. Available on BRS, (800) 289-4277, DIALOG, (800) 334-2564, ORBIT, (800) 456-7248, and STN International, FIZ Karlsruhe, P.O. Box 2465, W-7500, Karlsruhe 1, Germany, online services. Inquire as to cost and availability.

International Aerospace Abstracts. American Institute of Aeronautics and Astronautics, 370 L'Enfant Promenade, S.W., Washington, DC 20024. (202) 646-7400. Contains references and abstracts of journal and monograph literature relating to aerospace science and technology, from 1963 to present. Available through the NASA/RECON system of the National Aeronautics and Space Administration only.

NTIS Bibliographic Database. National Technical Information Service, 5285 Port Royal Road, Springfield, VA 2216 1. (703) 487-4929 or FAX (703) 321-8199. Broad cover age of government-sponsored science and technology research reports, 1964 to present. Available on BRS,(800) 289-4277, DIALOG, (800) 334-2564, ORBIT, (800) 456-7248, and STN International, FIZ Karlsruhe, P.O. Box 2465, W-7500, Karlsruhe 1, Germany, online services. Also available on CD-ROM. Inquire as to cost and availability.

Physics Briefs. American Institute of Physics, 335 East 45th Street , New York, NY 10017. (212) 661-9260 or FAX (212) 661-2036. Contains citations with abstracts of the literature of physics and related fields, 1979 to present. Available on the STN International, FIZ Karlsruhe, P.O. Box 2465, W-7500, Karlsruhe 1, Germany, online service. Inquire as to cost and availability.

Scientific and Technical Books and Serials in Print. R.R. B owker Inc., 121 Chanlon Road, New Providence, NJ 07974. (800) 521-8110 or (908) 464-6800. List of currently published books and serials in the physical and biological sciences, engineering and technology, with subject, author and titles indexes. Available on ORBIT, (800) 456-7248, online service. Inquire as to cost and availability.

SCISEARCH. Institute for Scientific Information, 3501 M arket Street, Philadelphia, PA 19104. (800) 523-1850 or (215) 386-0100. Broad multidisciplinary title and author index to the international literature of science and technology, 1974 to present. Available on DIALOG, (800) 334-2564, and ORBIT, (800) 456-7248, online services. Also available on CD-ROM. Inquire as to cost and availability.

WILSONLINE. H.W. W ilson Company, 950 University Avenue, Bronx, NY 10452. (800) 367-6770 or (212) 588-8400. Makes available onlin e versions of the printed H.W. Wilson indexes includin g Applied Science and Technology Index, Business Periodicals Index, General Science Index, and Readers' Guide to Periodical Literature. Period covered is generally

1983 to present. Available on BRS, (800) 289-4277, DIALOG, (800) 334-2564, and OCLC EPIC, (800) 848-5878, online services. Also available on CD-ROM. Inquire as to cost and availability.

OTHER SOURCES

Atlas of Deep-Sky Spendors. Hans Vehrenberg. Sky Publishing Corp., 49 Bay State Road, Cambridge, MA 02138. (617) 864-7360. (800) 253-0245. 4th edition. 1983. $24.95.

Atlas of the Ultraviolet Sky. Richard C. Henry, Wayne B.Landsman, et al. Johns Hopkins University Press, 2715 North Charles Street, Baltimore, MD 21218-4319. (410) 516-6900. (800) 537-5487. 1988. $70.00.

Atlas of the Universe. Patrick Moore. Rand McNally & Company. 8255 North Central Park, Skokie, IL 60076-2970. (708) 673-9100. 1994. $29.95.

The Bright Star Catalogue. Dorrit Hoffleit and Carlos Jaschek. Yale University Observatory, 260 Whitney Avenue, P.O. Box 6666, New Haven, CT 06511. (203) 432-3000. 4th revised edition. 1982. $35.00 in paper.

Cambridge Atlas of Astronomy. Jean Audouze, et al. Cambridge University Press, 40 West 20th Street, New York, NY 10011-4211. (212) 924-3900. (800) 872-7423. 3rd edition. 1994. $75.00.

Cambridge Guide To the Constellations. Michael E. Bakich. Cambridge University Press, 40 West 20th Street, New York, NY 10011-4211. (212) 924-3900. (800) 872-7423. 1995. $19.95 in paper.

Cambridge Star Atlas 2000.0. Wil Tirion. Cambridge University Press, 40 West 20th Street, New York, NY 10011- 4211. (212) 924-3900. (800) 872-7423. 2nd edition. 1996. $19.95.

Color Atlas of Galaxies. James Wray. Cambridge University Press, 40 West 20th Street, New York, NY 10011-4211. (212) 924-3900. (800) 872-7423. 1988. $94.95.

Exploring the Southern Sky: A Pictorial Atlas From the European Southern Observatory. S. Lausten, et al. Springer-Verlag New York, 175 Fifth Avenue, New York, NY 10010. (212) 460-1500. (800) 777-4643. 1987. $59.00.

Observer's Sky Atlas. E. Karkoschka. Springer-Verlag New York, 175 Fifth Avenue, New York, NY 10010. (212) 460-1500. (800) 777-4643. 1995. $15.95.

Planetary, Lunar and Solar Positions Six Hundred-one B.C. to A.D. One at Five-day and Ten-day Intervals. Bryant Tuckerman. American Philosophical Society, 104 South 5th Street, Philadelphia, PA 19106-3387. (215) 440-3400. FAX (215) 440-3436. Memoirs Series, volume 56. 1979. $20.00.

Planetary, Lunar and Solar Positions, A.D.2 to A.D. 1649 at Five Day and Ten Day Intervals. Bryant Tuckerman. American Philosophical Society, 104 South 5th Street, Philadelphia, PA 19106-3387. (215) 440-3400. FAX (215) 440-3436. Memoirs Series, volume 59. 1964. $30.00.

Planetary, Lunar and Solar Positions, 1650 - 1805. Owen Gingerich and Barbara L. Welther. American Philosophical Society, 104 South 5th Street, Philadelphia, PA 19106-3387. (215)

440-3400. FAX (215) 440-3436. Memoirs Series, volume 59S. 1983. $20.00.

NGC 2000.0: the Complete New General Catalogue and Index Catalogues of Nebulae and Star Clusters. J. L. E. Dreyer. editor. Cambridge University Press, 40 West 20th Street, New York, NY 10011-4211. (212) 924-3900. (800) 872-7423 1989. $21.95.

Norton's 2000.0 Star Atlas and Reference Handbook. Ian Ridpath, editor. Halsted Press, 605 Third Avenue, New York, 10158-0012. (212) 850-6400. 1989. $49.95.

Observing Handbook and Catalog of Deep-sky Objects. Christian B. Luginbuhl and Brian A. Skiff. Cambridge University Press, 40 West 20th Street, New York, NY 10011- 4211. (212) 924-3900. (800) 872-7423. 1990. $49.95.

The Sky: A User's Guide. David H. Levy. Cambridge University Press, 40 West 20th Street, New York, NY 10011- 4211. (212) 924-3900. (800) 872-7423. 1993. $14.95 in paper.

Sky Catalog 2000.0: Volume 1: Stars to Magnitude 8.0. Alan Hirshfeld, et al. Cambridge University Press, 40 West 20th Street, New York, NY 10011-4211. (212) 924-3900. (800) 872-7423. 2nd edition. 1991. $42.95 in paper.

Starlist Two Thousand: A Quick Reference Star Catalog For Astronomers. Richard Dibon-Smith. John Wiley & Sons, Inc., 605 Third Avenue, New York, NY 10158-0012. (212) 850-6000. (800) 225-5945. 1992. $29.95.

PERIODICALS

Astronomical Journal. American Institute of Physics, One Physics Ellipse, College Park, MD 20740-3843. (301) 209- 3000. Published for the American Astronomical Society. 1849 to present. Monthly. $280.00 per year. ISSN: 0004- 6256.

Astronomical Society of the Pacific. PUBLICATIONS. Astronomical Society of the Pacific, 390 Ashton Avenue, San Francisco, CA 94112. (415) 337-1100. FAX (415) 337-5205. 1889 to present. Monthly. $175.00 per year. ISSN: 0004- 6280.

Astronomy. Kalmbach Publishing Company, Box 1612, Waukesha, WI 53187-1612. (414) 796-8776. FAX (414) 796-1142. 1973 to present. Monthly. $27.00 per year. ISSN: 0091-6358.

Astronomy and Astrophysics. Springer-Verlag New York, Inc., 44 Hartz Way, Secaucus, NJ 07096-2491. (201) 348-4033. FAX (201) 348-4505. 1969 to date. 36 issues per year in 12 volumes. $1733.00 per year. ISSN: 0004-6361.

Astronomy Reports. American Institute of Physics, One Physics Ellipse, College Park, MD 20740-3843. (301) 209- 3000. English translation of Astronomicheksii Zhurnal. Formerly Soviet Astronomy AJ. 1957 to present. Bimonthly. $1160.00 per year. ISSN: 1063-7729.

Astrophysical Journal; An International Review of Astronomy and Astronomical Physics. University of Chicago Press, Journals Division, 5720 South Woodlawn Avenue, Chicago, IL 60637. (312) 753-3347. FAX (312) 753-0811. 1895 to date. three issues per month. $740.00 per year. ISSN: 0004-637X.

Astrophysics and Space Science; An International Journal of Cosmic Physics. Kluwer Academic Publishers, Box 358, Ac-

cord Station, Hingham, MA 02018-0358. (617) 871-6600. FAX (617) 871-6528. 1968 to present. 24 issues per year. $2544.00 per year. ISSN: 0004-640X.

Celestial Mechanics; An International Journal of Space Dynamics. Kluwer Academic Publishers, Box 358, Accord Station, Hingham, MA 02018-0358. (617) 871-6600. FAX (617) 871-6528. 1969 topresent. Monthly. $703.50 per year. ISSN: 0923-2958.

Ciel Et Terre. Societe Royale Belge d'Astronomie, de Meteorologie et de Physique du Globe, 3 avenue Circulaire, Uccle, Brusels, Belgium. FAX 32-2-374-9822. 1880 to present. Bimonthly. 1600 BEF per year. ISSN: 0009-6709.

Earth, Moon and Planets; An International Journal of Comparative Planetology. Kluwer Academic Publishers, Box 358, Accord Station, Hingham, MA 02018-0358. (617) 871- 6600. FAX (617) 871-6528. 1969 to present. Monthly. $840.00 per year. ISSN: 0167-9295.

Experimental Astronomy; An International Journal On Astronomical Instrumentation and Data Analysis. Kluwer Academic Publishers, Box 358, Accord Station, Hingham, MA 02018-0358. (617) 871-6600. FAX (617) 871-6528. 1989 to present. Quarterly. $161.00 per year. ISSN: 0922-6435.

Icarus; International Journal of Solar System Studies. Academic Press, Inc., Journal Division, 525 B Street, Suite 1900, San Diego, CA 92101-4495. (619) 230-1840. FAX (619) 699-6800. 1962 to present. Monthly. $1080.00 per year. ISSN: 0019-1035.

J G R: Journal of Geophysical Research: Planets. American Geophysical Union, 2000 Florida Avenue, NW, Washington, DC 20009. (202) 462-6900. FAX (202) 328-0566. 1991 to present. Monthly. $597.00 per year. ISSN: 0148-0227.

Journal For the History of Astronomy; Includes Archeoastronomy. Science History Publications Ltd., 16 Rutherford Road, Cambridge CB2 2HH, England. TEL 0223-565532. 1970 to present. Quarterly. $144.00 per year. ISSN: 0021-8286.

Landolt-Boernstein, Zahlenwerte Und Funktionen Aus Naturwissenschaften Und Technik. GROUP 6: ASTRONOMY. Springer-Verlag New York, Inc., 175 Fifth Avenue, New York, NY 10010. (212) 460-1500. FAX (212) 473-6272. 1965 - . Irregular. Price varies. ISSN: 0075-7896.

Mercury. Astronomical Society of the Pacific. 390 Ashton Avenue, San Francisco, CA 94112. (415) 337-1100. FAX: (415) 337-5205. 1972 to present. Bimonthly. $175.00 per year. ISSN: 0047-6773.

Meteoritics. Meteoritical Society, Department of Chemistry, University of Arkansas, Fayetteville, AR 72701. (501) 575-7625. FAX (501) 575-7778. 1953 to present. Bimonthly. $210.00 per year. ISSN: 0026-1114.

Monthly Notices of the Royal Astronomical Society. Blackwell Scientific Publication LT., Osney Mead, Oxford OX2 OEL, England. TEL 0865-240201. FAX 0865-721205. 1827 to present. Fortnightly. $1899.00 per year. ISSN: 0035-8711.

Observatory. c/o Dr. D. J. Stickland, Space and Astrophysics Division, Rutherford Appleton Laboratory, Chilton, Didcot,

ASTRONOMY

Ency. of Physical Sciences and Engineering Info. Sources

Oxon OS11 OQX England. FAX 0235-445848. 1877 to present. $42.00 per year. ISSN: 0029-7704.

Planetary and Space Science. Elsevier Science Publishing Company, Inc., 660 White Plains Road, Terrytown, NY 10591-5153.(914) 524-9200. FAX: (914) 333-2444. 1959 to present. Monthly. $1355.00 per year. ISSN: 0032-0633.

Sky and Telescope. Sky Publishing Corporation, Box 9111, Belmont, MA 02178. (617) 864-7360. FAX (617) 864-6117. 1941 to present. Monthly. $27.00 per year. ISSN: 0037- 6604.

Space Science Reviews. Kluwer Academic Publishers, Box 358, Accord Station, Hingham, MA 02018- 0358. (617) 871- 6600. FAX (617) 871-6528. 1962 to present. Sixteen issues per year. $940.00 per year. ISSN: 0038-6308.

Vistas in Astronomy; An International Review Journal. Albert C. Beer, editor. Elsevier Science Publishing Company, Inc., 660 White Plains Road, Terrytown, NY 10591- 5153. (914) 524-9200. FAX: (914) 333-2444. 1958 to present. Quarterly. $415.00 per year. ISSN: 0083-6656.

Transactions of the International Astronomical Union. Proceedings of the General Assembly. Kluwer Academic Publishers, Box 358, Accord Station, Hingham, MA 02018- 0358. (617) 871-6600. Issues in two parts: Part A, Reports, and Part B, Proceedings. Triennial. XXI - 1994. Price varies. ISSN: 0080-1372.

Webb Society Quarterly Journal. Webb Society, 194 Foundry Lane, Freemantle, Southampton, Hampshire SO1 3EQ, England. 1969 to present. Quarterly. $19.50 per year. ISSN: 0043- 1680.

RESEARCH CENTERS AND INSTITUTES

Astronomy Program. University of Maryland, 1207 Computer & Space Sciences Building, College Park, MD 20742. (301) 405- 3001. FAX (301) 314-9067.

Astrophysical Research Consortium. University of Chicago, Department of Astrophysics, 5640 South Ellis, Chicago, IL 60637. Includes University of Washington, Washington State University, Princeton University and New Mexico State University.

Center For Astronomy and Space Astrophysics. University of Colorado - Boulder, Campus Box 391, Boulder, CO 80309. (303) 492-4050. (FAX) 492-7178.

Center for Space Research. Massachusetts Institute of Technology, 77 Massachusetts Avenue, Camridge, MA 02139. (617) 253-7501.

Center For Space Science and Astrophysics. Stanford University, 325 Durand Building, Stanford, CA 94305. (415) 723-3592.

National Astronomy and Ionosphere Center. Cornell University, Space Sciences Building, Ithaca, NY 14853. (607) 255-3735. (607) 255-8803.

Smithsonian Center For Astrophysics. Harvard University, 60 Garden Street, Cambridge, MA 02138. (617) 495-7461. FAX (617) 495-7326.

Space Telescope Science Institute. 3700 San Martin Drive, Baltimore, MD 21218. (301) 338-4700.

ASTROPHYSICS

See also: ASTRONOMY, CELESTIAL MECHANICS, COSMOLOGY, GALAXIES, INTERSTELLAR MATTER, PLANETS, SOLAR SYSTEM, TELESCOPES

ABSTRACT SERVICES AND INDEXES

Applied Science and Technology Index; A cumulative subject index to English language periodicals in the fields of aeronautics and space science, computer technology, chemistry, construction industry, energy and related areas. H.W. Wilson Co., 950 University Avenue, Bronx, NY 10452. (212) 588-8400. (800) 367-6770. FAX (718) 590-1617. From 1958 to present. Monthly. Inquire about cost and availability. Also available on CD-ROM and online. ISSN: 0003-6986.

Astronomy and Astrophysics Abstracts. Springer-Verlag New York, 175 F ifth Avenue, New York, NY 10010. (212) 460-1500. FAX (212) 473-6272. Published for the Astronomisches Rechen-Institut. Comprehensive coverage of all aspects of astronomy, astrophysics and related fields. 1969 to present. Two parts per year. Annual. ISSN: 0067-0022.

Bibliography and Index of Geology. American Geological Institute, 4220 King Street, Alexandria, VA 22302-1507. (703) 379-2480. FAX (703) 379-7563. From 1969 to present. Monthly. $1295.00 per year. ISSN: 0098-2784. Also available as GEOREF online (CISTI, DIALOG, Orbit, STN) and on CD-ROM. Inquire about price and availability.

Chemical Abstracts. Chemical abstracts Service. 2540 Oleant angy River Road, Box 3012, Columbus, OS 43210-0012. (614) 447-3600. (800) 848-6538. FAX (614) 447-3713. 1907 to present. Weekly. $16,800.00 per year. Available online and on CD-ROM. CA is also available in five section groupings. Inquire regarding cost and availability. ISSN:

Current Contents: Physical, Chemical, and Earth Sciences. Institute for Scientific Information, 3501 Market Street, Philadelphia, PA 19104. (215) 386-0100. (800) 523-1850. FAX (215) 386-2291. 1961 to present. Weekly. $442.00 per year. Also available online (BRS, DIALOG) and on CD-ROM. Inquire about price and availability.

Current Contents: Physical, Chemical and Earth Sciences. Institute for Scientific Information, 3501 Market Street, Philadelphia, PA 19104. (215) 386-0100. (800) 523-1850. FAX (215) 386-2291. 1961 to present. Weekly. $442.00 per year. Also available online (BRS, DIALOG) and on CD-ROM. Inquire regarding cost and availability. ISSN: 0163-2574.

Current Papers in Physics. Institute of Electrical Engineers, Michael Faraday House, Six Hill Way, Stevenage, Herts, SG1 2AY, England. Distributed by IEEE, Box 1331, 445 Hoes Lane, Piscataway, NJ 08855-1331. (908) 562-5549. From 1966 to present. Fortnightly. $410.00 per year.

General Science Index. H.W. Wilson Company, 950 University Avenue, Bronx, NY 10452. (212) 588-8400. (800) 367-6770. FAX (718) 590-1617. From 1978 to present. Ten issues per year; quarterly and annual cumulations. Service basis. Available on CD-ROM and online. Inquire regarding cost and availability. ISSN: 0162-1963.

G Overnment Reports Announcements and Index. U. S. National Technical Information Service (NTIS), 5285 Port Royal Road, Springfield, VA 22161. (703) 487-4650. FAX (703) 321-8547.

From 1968 to present. Annual. $630.00 per year. Also available online as *NTIS Bibliographic Database* and on CD-ROM.

International Aerospace Abstracts. American Institute of Aeronautics and Astronautics. Technical Information Service, 555 West 57th Street, New York 10019. (212) 247- 6500. FAX (212) 582-4961. 1961 to present. Semi-monthly with annual index. $1295.00 per year. Also available online and on CD-ROM. ISSN: 0020-5842.

Meteorological and Geoastrophysical Abstracts. American Meteorological Society, c/o Inforonics, Inc., 550 Newtown Road, Littleton, MA 01460. (508) 486-8976. FAX (508) 486- 0027. Cove rs literature in environmental sciences, meteorology, astrophysics, hydrology, glaciology, and physical oceanography. From 1950 to present. Monthly. $950.00 per year. Also available on CD-ROM and online. Inquire regarding cost and availability. ISSN: 0026-1130.

NTIS Alerts: Astronomy & Astrophysics. U. S. National Technical Information Service. 5285 Port Royal Road, Springfield, VA 22161. (703) 487-4650. FAX (703) 321-8547. Weekly. $140.00 per year.

Pascal 48: Environnement Cosmique Terrestre, Astronomie Et Geologie Extraterrestre. Centre National de la Recherche Scientifique, Institut de Information Scientifique et Technique, 2 aliee du Parc de Brabois, 54514 Vandoeuvre Les Nancy Cedex, France. TEL 83-50-46-00. FAX: 83-50-46-50. 1985 to present. Ten issues per year. 770 F. Also available on CD-ROM and online.

Physics Abstracts . INSPEC. Section A, Science Abstracts. Institution of Electrical Engineers, London, United Kingdom. Available from: INSPEC/IEEE - Institute of Electrical and Electronic Engineers, Box 1331, Hoes Lane, Piscataway, NJ 08855-1331. (908) 562-5549. 1898 to present. 24 issues per year. $2835.00 per year. Also available on CD-ROM and online. ISSN: 0036-8091.

Physics Briefs (Physikalische Berichte). Information Center for Energy, Physics, Mathematics; German Physical Society. VCH Publishers, Inc., 220 East 23rd Street, New York, NY 10010-4606. (212) 683-8333. 1845 to present. 24 issues per year. $2390.00 per year. Also available online. ISSN: 0179-7434.

Science Citation Index. SCI. Institute for Scientific Information, 3501 Market Street, Philadelphia, PA 19104. (215) 386-0100. (800) 523-1850. FAX (215) 386-2991. 1961 to present. Six issues per year, plus annual cumulation. $11650.00 per year. Also available online and on CD-ROM. Inquire about price and availability. ISSN: 0036-827X.

STAR. Scientific and Technical Aerospace Reports. U .S. National Aeronautics and Space Administration. Distributed by U. S. Superintendent of Documents, Washington, DC 20402. 1963 to present. Semi-monthly, with semiannual and annual indexes. $114.00 per year. ISSN: 0036-8741. Also available online and on CD-ROM.

ANNUAL REVIEWS AND YEARBOOKS

The Astronomical Almanac. Superintendent of Documents, U. S. Government Printing office, Washington, DC 10402. (202) 783-3238. 1981 to present. Superceeds Astronomical Ephemeris and the American Ephermis and Nautical Almanac. Annual. $44.95. ISSN: 0737-6421.

Annual Review of Astronomy and Astrophysics; Original Reviews of Critical Literature and Current Developments in Astronomy and Astrophysics. Annual Reviews, Inc., 4139 El Camino Way, Palo Alto, CA 94306-0139. (415) 493-4400. (800) 523-8635. FAX (415) 855-9815. 1963 to date. Annual. $60.00. ISSN: 0066-4146.

Annual Review of Earth and Planetary Sciences. Annual Reviews, Inc., 4139 El Camino Way, Palo Alto CA 94306-0139. (415) 403-4400. (800) 523-8635. Fax (415) 855-9815. 1973 to date. Annual. $62.00. ISSN: 0084-6597.

Highlights of Astronomy. I. Appenzeller, editor. Kluwer Academic Publishers, Box 358, Accord Station, Hingham, MA 02018-0358. (617) 871-6600. International Astronomical Union Highlights series. ISBN: 07923-3553-8. Triennal. 1995. Price varies.

ASSOCIATIONS AND PROFESSIONAL SOCIETIES

Amateur Astronomers Association. 1010 Park Avenue, New York, NY 10028. (212) 535-2922.

American Association of Variable Star Observers. 23 Birch Street, Cambridge, MA 02138. (617) 354-0484. FAX: (617) 354-0665.

American Astronomical Society. 2000 Florida Avenue NW, Suite 400, Washington, DC 20009. (202) 328-2010. FAX: (202) 234-2560.

American Geophysical Union. 2000 Florida Avenue, NW, Washington, DC 20009. (202) 462-6900. (800) 966-AGU1. FAX (202) 328-0566.

American Institute of Physics. One Physics Ellipse, College Park, MD 20740-3843. (301) 209-3100.

Association of University for Research in Astronomy. Suite 701, 1625 Massachusetts Avenue, NW, Washington, DC 20036. (202) 483-2101. FAX (202) 483-2106.

Astronomical League. 2112 Kingfisher Lane East, Rolling Meadows, IL 60008. (708) 398-0562.

Astronomical Society of the Pacific. 390 Ashton Avenue, San Francisco, CA 94112. (415) 337-1100. FAX: (415) 337-5205.

International Astronomical Union. 98 bis Boulevard Arago, F-75014 Paris, France. Tel 1-43-25-83-58. FAX 1-43-25-26-16.

Joint Institute For Laboratory Astrophysics. University of Colorado, Box 440, Boulder, CO 80309-0440. (303) 492-7780. FAX (303) 492-5235.

Royal Astronomical Society of Canada. 136 Dupont Street, Toronto. ON N5R 1V2 Canada. (416) 924-7973. FAX (416) 968-6687.

Universities Research Association, 1111 19th Street Nw, Suite 400, Washington, DC 20036. (202) 293-1382.

Webb Society. 1440 South Marmora Avenue, Tuscon, AZ 95713. (602) 628-1077.

ASTROPHYSICS

Ency. of Physical Sciences and Engineering Info. Sources

BIBLIOGRAPHIES

Astronomy and Astrophysics: a Source Guide. Gordon Press, P.O. Box 459, Bowling Green Station, New York, NY 10004. (718) 624-8419. 1991. $250.00. .

A Bibliography of Astronomy, 1970-1979. R. A. Sea and S. S. Martin. Libraries Unlimited, Inc., Littleton, CO 80160. 1982.

Catalog of the Naval Observatory Library. Naval Observatory Staff. G. K. Hall & Company, 866 Third Avenue, New York, NY 10022. (212)702-6789. (800) 257-5755. 6 volumes. 1977. $655.00

Science Books and Films. American Association for the Advancement of Science, 1333 H Street NW, Washington, DC 20005. (202) 326-6454. 1965 to present. Nine issues per year. $40.00 per year. ISSN: 0098-342X

Sc Ientific and Technical Books and Serials in Print; An Index To Literature in Science and Technology. R. R. Bowker. 121 Chanlon Road, New Providence, NJ 07974. (908) 464-6800. FAX (908) 665-3502. (800) 521-8110. 1972 to present. Annual. $299.95 per year. Also available on CD-ROM. ISSN: 0000-054X.

A Source Book On Astronomy and Astrophysics, 1900 - 1975. Kenneth Lang and Owen Gingerich, editors. Harvard University Press, 79 Garden Street, Cambridge MA. (617) 495-2600. 1980. $41.95.

DIRECTORIES AND BIOGRAPHICAL SOURCES

American Astronomical Society Membership Directory. American Astronomical Society. 2000 Florida Avenue NW, Suite 400, Washington, DC 20009. (202) 328-2010. FAX (202) 234-2560 Annual. Included in membership dues. ISSN: 1061- 9038.

American Men and Women of Science: Physical and Biological Sciences. R. R. Bowker Inc., 121 Chanlon Road, New Providence, NJ 07974. (908) 464-6800. (800) 521-8110. 20th edition. 8 volumes. 1996. $850.00.

The Astronomers. Donald Goldsmith. St. Martin's Press, Inc., 175 Fifth Avenue, New York, NY 10010. (212) 674-5151. (800) 221-7945. 1993. $14.95 in paper.

Astronomical Centers of the World. Kevin Krisciunas. Cambridge University Press, 40 West 20th Street, New York, NY 10011-4211. (212) 924-3900. (800) 872-7423. 1988. $34.95

The Biographical Dictionary of Scientists: Astronomers. D. Abbott, editor. Peter Bedrick Books, Inc., 2112 Broadway, Room 318, New York, NY 10023. (212) 496-0751. 1984.

Directory of Physics and Astronomy Staff Members. American Institute of Physics, One Physics Ellipse, College Park, MD 20740-3843. (301) 209-3100. 1975/76 to present. Annual. $60.00. ISSN: 0361-2228.

Earth and Astronomical Research Centers. Stockton Press, 345 Park Avenue South, New York, NY 10010. 4th edition. 1995. $515.00. ISBN: 1-56169-0967.

Graduate Programs in Physics, Astronomy and Related Fields. 1995-1996. American Institute of Physics, One PhysicsEllipse,

College Park, MD 20740-3843. (301) 209-3100. 1978 to present. Annual. $45.00. ISSN: 0147-1821.

Research Centers Directory. Gale Research Inc., 835 Penobscot Building, Detroit, MI 48226-4094. (313) 961-2242. (800) 877-4253. 20th edition. 1995. $485.00.

ENCYCLOPEDIAS AND DICTIONARIES

Astronomy From A To Z: A Dictionary of Celestial Objects and Ideas. Charles A. Schweighauser. Illinois Issues, Sangamon State University, K-10, Springfield, IL 62794-9243. (217) 786-9243. 1991. $14.95 in paper.

Concise Dictionary of Astronomy. Jacqueline Mitton. Oxford University Press, Inc., 200 Madison Avenue, New York, NY 10016. (212) 725-6000. (800) 334-4249. 1992. $24.95.

Dictionary of Astronomical Names. Adrian Room. Routledge, Chapman and Hall, Inc., 29 West 35th Street, New York, NY 10001-3291. (212) 244-3336. 1988. $27.50.

Encyclopedia of Astronomy and Astrophysics. Stephen Maran, editor. Van Nostrand Reinhold, 115 Fifth Avenue, New York, NY 10003. (212) 254-3232. (800) 842-3636. 1992. $129.95.

Facts On File Dictionary of Astronomy. Valerie Illingworth. Facts-on-File, Inc., 460 Park Avenue South, New York, NY 10016-7382. (212) 683-2244. (800) 322-8755. Fax (800) 683-3633. 1994. $27.95.

Illustrated Encyclopedia of the Universe.

Landolt-boernstein Numerical Data and Functional Relationships in Science and Technology: Astronomy, Astrophysics and Space Research; A. O. Madelung and H. H. Boigt, editors. Springer Verlag . 1993. $550.00.

McGraw-Hill Encyclopedia of Science and Technology. McGraw- Hill Book Company, Inc., 1221 Avenue of the Americas, New York, NY 10020. (212) 512-2000. (800) 262-4729. 7th edition. 20 volume set. 1992. $1900.00.

Oxford Illustrated Encyclopedia. Volume 8. Archie Roy, editor. Oxford University Press, Inc., 200 Madison Avenue, New York, NY 10016. (212) 725-6000. (800) 334-4249. 1993. $45.00.

Penguin Dictionary of Astronomy. Jacqueline Mitton. Viking Penguin, 375 Hudson Street, New York, NY 10014-3657. (212) 366-2000. (800) 331-4624. 1993. $12.95 in paper.

Stars, Galaxies, Cosmos. William R. Corliss. Sourcebook Project, P.O. Box 107, Glen Arm, MD 21057. (410) 668-6047. 1987. $17.95.

GENERAL WORKS

Astronomer's Universe: Stars, Galaxies and Cosmos. Herbert Friedman. W. W. Norton & Company, 500 Fifth Avenue, New York, NY 10110. (212) 354-5500. (800) 223-2584. 1990. $24.95.

Astronomy and Cosmology. John North. W. W. Norton & Company, 500 Fifth Avenue, New York, NY 10110. (212) 354- 5500. (800) 223-2584 1990. 1994. $18.95.

Astrophysics and Twentieth-Century Astronomy to 1950. Owen Gingerich. Cambridge University Press, 40 West 20th Street,New York, NY 10011-4211. (212) 924-3900. (800) 872-7423 1984. $44.95.

Astrophysics Today. A. G. Cameron, editor. American Institute of Physics, One Physics Ellipse, College Park, MD 20740-3843. (301) 209-3100., 1984. $39.00 in paper.

Astrophysics With A PC: An Introduction To Computational Astrophysics. Willmann-Bell, Inc., P.O. Box 35025, Richmond, VA 23235. (804) 320-7016. 1994.

Atoms, Stars, and Nebulae. Lawrence H. Aller. Cambridge University Press, 40 West 20th Street, New York, NY 10011- 4211. (212) 924-3900. (800) 872-7423. 1991. $34.95 in paper.

Classical General Relativity. Subrahmanyan Chandrasekhar and Morton D. Hull, editors. Oxford University Press, Inc., 200 Madison Avenue, New York, NY 10016. (212) 725-6000. (800) 334-4249. 1993. $45.00.

Compendium of Practical Astronomy. Gunter D. Roth, editor. Springer-Verlag New York, Inc., 175 Fifth Avenue, New York, NY 10010. (212) 460-1500. (800) 777-4643. 3 volumes. Revised edition. 1994. $125.00.

Cosmic Catastrophies. C. P. Chapman and D. Morrison, editors. Plenum Publishing Corp., 233 Spring Street, New York, NY. (212) 620-8000. (800) 2221-9369. 1989. $22.95.

Cycles of Fire: Stars, Galaxies and the Wonder of Deep Space. William K. Hartmann. Workman Publishing Company, Inc., 708 Broadway, New York, NY 10003. (212) 254-5900. (800) 722-7202. 1988. $15.95 in paper.

Farthest Things in the Universe. Jay M. Pasachoff. Cambridge University Press 40 West 20th Street, New York, NY 10011-4211. (212) 924-3900. (800) 872-7423. 1995. $29.95.

Frontiers of Fundamental Physics. M. Barone and F. Selleri, editors. Plenum Publishing Corp., 233 Spring Street, New York, NY. (212) 620-8000. (800) 2221-9369. 1994. $145.00.

Gravitation. Charles W. Misner, et al. W. H. Freeman & Co., 41 Madison Avenue, East 26th Avenue, 35th Floor, New York, NY 10010. (212) 576-0400. 1995. .

Discovering the Cosmos. Robert C. Bless. University Science Books, 55D Gate Five Road, Sausalito, CA 94965. (415) 332-5390. 1995. $58.00.

Universe. William J. Kaufmann. W. H. Freeman & Co., 41 Madison Avenue, East 26th Avenue, 35th Floor, New York, NY 10010. (212) 576-0400. 4th edition. 1995. $54.95.

An Introduction To Astronomy: Stars, Galaxies and the Universe. Gerald D. Waxman. Kendall-Hunt Publishing Co., 4050 Westmark Drive, P.O. Box 1840 Dubuque, IA 52004-1840. (319) 589-1000. (800) 228-0810. 1992. $28.95.

An Introduction To Cosmochemistry. Charles R. Cowley. Cambridge University Press, 40 West 20th Street, New York, NY

10011-4211. (212) 924-3900. (800) 872-7423. 1995. $29.95 in paper.

Millimeter and Submillimeter Astronomy. R. D. Wolstencroft and W. Butler Burton, editors. Kluwer Academic Publishers, 101 Philip Drive, Assinippi Park, Norwell, MA 02061. (617) 871-6600. 1988. $136.50.

The New Cosmos. A. Unsold and B. Baschek. Springer-Verlag York, Inc., 175 Fifth Avenue, New York, NY 10010. (212) 460-1500. (800) 777-4643 . 4th revised edition. 1991. $59.00.

Observational Astrophysics. Robert C. Smith. Cambridge University Press, 40 West 20th Street, New York, NY 10011- 4211. (212) 924-3900. (800) 872-7423. 1995. $69.95.

Origin of the Universe. John D. Barrow. Basic Books, 10 East 53rd Street, New York, NY 10022. (212) 207-7057. (800) 242-7737. 1994. $20.00.

The Physics of Stars. A. C. PHillips. John Wiley & Sons, Inc., 605 Third Avenue, New York, NY 10158-0012. (212) 850- 6000. (800) 225-5945. 1994. $32.95 in paper.

Quasar Astronomy. Daniel W. Weedman. Cambridge University Press, 40 West 20th Street, New York, NY 10011-4211. (212) 924-3900. (800) 872-7423. 1988. $24.95 in paper.

The Red Limit: The Search For the Edge of the Universe. Timothy Ferris. Morrow, William & Company, 1350 Avenue of the Americas, New York, NY 10019. (212) 261-6500. (800) 843-9389. Revised edition. 1993. $12.95.

Stars and Atoms: From the Big Bang To the Solar System. Stuart Clark. Oxford University Press, Inc., 200 Madison Avenue, New York, NY 10016. (212) 725-6000. (800) 334-4249. 1995. $35.00.

Stephen Hawking's Universe. John Boslough. Morrow, William & Company, 1350 Avenue of the Americas, New York, NY 10019. (212) 261-6500. (800) 843-9389. 1985. $7.95.

Visual Astronomy of the Deep Sky. Roger N. Clark. Cambridge University Press, 40 West 20th Street, New York, NY 10011-4211. (212) 924-3900. (800) 872-7423. 1991. $39.95.

Workings of the Universe. Time-Life, Inc., 777 Duke Street, Alexandria, VA 22314. (703) 8388-7000. (800) 621-7026. . 1991. $25.95.

HANDBOOKS AND MANUALS

Astrophysical Data: Planets and Stars. Kenneth R. Lanb. Springer-Verlag New York, Inc., 175 Fifth Avenue, New York, NY 10010. (212) 460-1500. (800) 777-4643. 1993. $59.00.

Burnham's Celestial Handbook: an Observer's Guide to the Universe Beyond the Solar System. Robert Burnham, Jr. Dover Publications, Inc., 180 Varick Street, New York, NY 10014. (212) 255-3755. (800) 223-3130. 3 volumes. 1978. $41.85 for the set.

Cambridge Guide To the Constellations. Michael E. Bakich. Cambridge University Press, 40 West 20th Street, New York, NY 10011-4211. (212) 924-3900. (800) 872-7423. 1995. $69.95.

ASTROPHYSICS

Ency. of Physical Sciences and Engineering Info. Sources

Encyclopedia of Astronomy and Astrophysics. Stephen Maran, editor. Van Nostrand Reinhold, 115 Fifth Avenue, New York, NY 10003. (212) 254-3232. (800) 842-3636. 1992. $129.95.

Greenwich Guide To Stars, Galaxies and Nebulae. Stuart Malin. Cambridge University Press, 40 West 20th Street, New York, NY 10011-4211. (212) 924-3900. (800) 872-7423. 1990. $10.95.

Handbook of Astronomy, Astrophysics and Geophysics; Volume 2: Galaxies and Cosmology. V. M. Canuto and B. G. Elmegreen. Gordon & Breach Science Publishers, Inc., P.O.Box 200029, Riverfront Plaza Station, Newark, NJ 07102- 0301. (201) 643-7500. 1988. $194.00.

Handbook of Space Astronomy and Astrophsyics. Martin V. Zumbeck. Cambridge University Press, 40 West 20th Street, New York, NY 10011-4211. (212) 924-3900. (800) 872-7423. 2nd edition. 1990. $79.95.

Landolt-Borenstein Numerical Data and Functional Relationships in Science and Technology: Astronomy, Astrophysics and Space Research. GROUP VI. Springer-Verlag New York, Inc., 175 Fifth Avenue, New York, NY 10010. (212) 460-1500. (800) 777-4643. volumes priced individually.

McGraw-Hill Encyclopedia of Astronomy. Sybil P. Parker and Jay M. Pasachoff, editors. McGraw-Hill Publishing Company, Inc., 1221 Avenue of the Americas, New York, NY 10020. (212) 512-2000. (800) 262-4729. 2nd edition. 1993. $75.00

Observer's Guide To Astronomy. Patrick Martinez, editor. Cambridge University Press, 40 West 20th Street, New York, NY 10011-4211. (212) 924-3900. (800) 872-7423. 2 volumes. 1994. $159.90.

ONLINE DATABASES AND CD-ROMS

CA Search. Chemical Abstracts Service, P.O. Box 3012, Colum bus, OH 43210-0012. (614) 447-3600. (800) 848-6533. FAX (614) 447-3709. Very comprehensive guide to worldwide chemical literature and related fields, 1972 to present. Available on BRS,(800) 289-4277, DIALOG, (800) 334-2564, ORBIT (800) 456-7248, and STN International, FIZ Karlsruhe, P.O. Box 2465, W-7500, Karlsruhe 1, Germany, online services. Inquire as to cost and availability.

Current Contents Search. Institute for Scientific Informatio n, 3501 Market Street, Philadelphia, PA 19104. (215) 386-0 100. FAX (215) 386-6362. Contains citations to articles listed in the table of contents of science and technology journals. Also articles in social sciences and life sciences journals. Available on BRS, (800) 289-4277, DIALOG, (800) 334-2564, online services. Inquire as to cost and availability.

Dissertation Abstracts Online. University Microfilms Internatio nal, 300 North Zeeb Road, Ann Arbor, MI 48106. (800) 521-0600 or (313) 761-4700. Scope includes virtually all doctoral dissertations accepted at accredited American institutions from 1861 to present in 252 subject areas. Available on BRS, (800) 289-4277, DIALOG, (800) 334-2564, and OCLC EPIC, (800) 848-5878, online services. Also available on CD-ROM. Inquire as to cost and availability.

INSPEC. Institution of Electrical Engineers, Michael Faraday House, Six Hills Way, Stevenage, Herts. SG1 2AY, England.Telephone: 0438 313311 or FAX 0438 742840. Contains citations to the worldwide literature of physics, electron-

ics and electrical engineering, computer technology, and related fields. Available on BRS, (800) 289-4277, DIALOG, (800) 334-2564, ORBIT, (800) 456-7248, and STN International, FIZ Karlsruhe, P.O. Box 2465, W-7500, Karlsruhe 1, Germany, online services. Inquire as to cost and availability.

International Aerospace Abstracts. American Institute of Aeronautic s and Astronautics, 370 L'Enfant Promenade, S.W., Washington, DC 20024. (202) 646-7400. Contains references and abstracts of journal and monograph literature relating to aerospace science and technology, from 1963 to present. Available through the NASA/RECON system of the National Aeronautics and Space Administration only.

NTIS Bibl Iographic Database. National Technical Information Service, 5285 Port Royal Road, Springfield, VA 2216 1. (703) 487-4929 or FAX (703) 321-8199. Broad cove rage of government-sponsored science and technology research reports, 1964 to present. Available on BRS,(800) 289-4277, DIALOG, (800) 334-2564, ORBIT, (800) 456-7248, and STN International, FIZ Karlsruhe, P.O. Box 2465, W-7500, Karlsruhe 1, Germany, online services. Also available onCD-ROM. Inquire as to cost and availability.

Physics Briefs. American Institute of Physics, 335 East 45th Street , New York, NY 10017. (212) 661-9260 or FAX (212) 661-2036. Contains citations with abstracts of the literature of physics and related fields, 1979 to present. Available on the STN International, FIZ Karlsruhe, P.O. Box 2465, W-7500, Karlsruhe 1, Germany, online service. Inquire as to cost and availability.

Scientific and Technical Books and Serials in Print. R.R. B owker Inc., 121 Chanlon Road, New Providence, NJ 07974. (800) 521-8110 or (908) 464-6800. List of currently published books and serials in the physical and biological sciences, engineering and technology, with subject, author and titles indexes. Available on ORBIT, (800) 456-7248, online service. Inquire as to cost and availability.

SCISEARCH. Institute for Scientific Information, 3501 Market Street, Philadelphia, PA 19104. (800) 523-1850 or (215) 386-0100. Broad multidisciplinary title and author index to the international literature of science and technology, 1974 to present. Available on DIALOG, (800) 334-2564, and ORBIT, (800) 456-7248, online services. Also available on CD-ROM. Inquire as to cost and availability.

WILSONLINE. H.W. W ilson Company, 950 University Avenue, Bronx, NY 10452. (800) 367-6770 or (212) 588-8400. Makes available onlin e versions of the printed H.W. Wilson indexes includin g Applied Science and Technology Index, Business Periodicals Index, General Science Index, and Readers' Guide to Periodical Literature. Period covered is generally 1983 to present. Available on BRS, (800) 289-4277, DIALOG, (800) 334-2564, and OCLC EPIC, (800)848-5878, online services. Also available on CD-ROM. Inquire as to cost and availability.

OTHER SOURCES

Atlas of Deep-Sky Spendors. Hans Vehrenberg. Sky Publishing Corp., 49 Bay State Road, Cambridge, MA 02138. (617) 864-7360. (800) 253-0245. 4th edition. 1983. $24.95.

Atlas of the Ultraviolet Sky. Richard C. Henry, Wayne B. Landsman, et al. Johns Hopkins University Press, 2715 North Charles Street, Baltimore, MD 21218-4319. (410) 516-6900. (800) 537-5487. 1988. $70.00.

Atlas of the Universe. Patrick Moore. Rand McNally & Company. 8255 North Central Park, Skokie, IL 60076-2970. (708) 673-9100. 1994. $29.95.

The Bright Star Catalogue. Dorrit Hoffleit and Carlos Jaschek. Yale University Observatory, 260 Whitney Avenue, P.O. Box 6666, New Haven, CT 06511. (203) 432-3000. 4th revised edition. 1982. $35.00 in paper.

Cambridge Atlas of Astronomy. Jean Audouze, et al. Cambridge University Press, 40 West 20th Street, New York, NY 10011-4211. (212) 924-3900. (800) 872-7423. 3rd edition. 1994. $75.00.

Cambridge Guide To the Constellations. Michael E. Bakich. Cambridge University Press, 40 West 20th Street, New York, NY 10011-4211. (212) 924-3900. (800) 872-7423. 1995. $19.95 in paper.

Cambridge Star Atlas 2000.0. Wil Tirion. Cambridge University Press, 40 West 20th Street, New York, NY 10011- 4211. (212) 924-3900. (800) 872-7423. 2nd edition. 1996. $19.95.

Color Atlas of Galaxies. James Wray. Cambridge University Press, 40 West 20th Street, New York, NY 10011-4211. (212) 924-3900. (800) 872-7423. 1988. $94.95.

Exploring the Southern Sky: A Pictorial Atlas From the European Southern Observatory. S. Lausten, et al. Springer-Verlag, 175 Fifth Avenue, New York, NY 10010. (212) 460-1500. (800) 777-4643. 1987. $59.00.

Observer's Sky Atlas. E. Karkoschka. Springer-Verlag New York, Inc., 175 Fifth Avenue, New York, NY 10010. (212) 460-1500. (800) 777-4643. 1995. $15.95.

NGC 2000.0: the Complete New General Catalogue and Index Catalogues of Nebulae and Star Clusters. J. L. E. Dreyer. editor. Cambridge University Press, 40 West 20th Street, New York, NY 10011-4211. (212) 924-3900. (800) 872-7423 1989. $21.95.

Norton's 2000.0 Star Atlas and Reference Handbook. Ian Ridpath, editor. Halsted Press, 605 Third Avenue, New York, 10158-0012. (212) 850-6400. 1989. $49.95.

Observing Handbook and Catalog of Deep-Sky Objects. Christian B. Luginbuhl and Brian A. Skiff. Cambridge University Press, 40 West 20th Street, New York, NY 10011- 4211. (212) 924-3900. (800) 872-7423. 1990. $49.95.

Starlist Two Thousand: A Quick Reference Star Catalog For Astronomers. Richard Dibon-Smith. John Wiley & Sons, Inc., 605Third Avenue, New York, NY 10158-0012. (212) 850-6000. (800) 225-5945. 1992. $29.95.

The Sky: A User's Guide. David H. Levy. Cambridge University Press, 40 West 20th Street, New York, NY 10011- 4211. (212) 924-3900. (800) 872-7423. 1993. $14.95 in paper.

Sky Catalog 2000.0: Volume 1: Stars to Magnitude 8.0. Alan Hirshfeld, et al. Cambridge University Press, 40 West 20th Street, New York, NY 10011-4211. (212) 924-3900. (800) 872-7423. 2nd edition. 1991. $42.95 in paper.

PERIODICALS

ACTA Astronomica. Copernicus Foundation for Polish Astronomy, AL Ujazdowskie 4, 00-478 Warsaw, Poland. TEL 48-22-295346. FAX 48-22-294967. 1925 to present. Quarterly. $80.00 per year. ISSN: 0001-5237.

Astronomical Journal. American Institute of Physics, One Physics Ellipse, College Park, MD 20740-3843. (301) 209- 3000. Published for the American Astronomical Society. 1849 to present. Monthly. $280.00 per year. ISSN: 0004- 6256.

Astronomical Society of the Pacific Publications. Astronomical Society of the Pacific, 390 Ashton Avenue, San Francisco, CA 94112. (415) 337-1100. FAX (415) 337-5205. 1889 to present. Monthly. $175.00 per year. ISSN: 0004- 6280.

Astronomy. Kalmbach Publishing Company, Box 1612, Waukesha, WI 53187-1612. (414) 796-8776. FAX (414) 796-1142. 1973 to present. Monthly. $27.00 per year. ISSN: 0091-6358.

Astronomy and Astrophysics. Springer-Verlag New York, Inc., 44 Hartz Way, Secaucus, NJ 07096-2491. (201) 348-4033. FAX (201) 348-4505. 1969 to date. 36 issues per year in 12 volumes. $1733.00 per year. ISSN: 0004-6361.

Astronomy Reports. American Institute of Physics, One Physics Ellipse, College Park, MD 20740-3843. (301) 209- 3000. English translation of Astronomicheksii Zhurnal. Formerly Soviet Astronomy AJ. 1957 to present. Bimonthly. $1160.00 per year. ISSN: 1063-7729.

Astroparticle Physics. Elsevier Science Publishing Company, Inc., Box 882, Madison Square Station, New York, NY 10159. (212) 989-5800. FAX (212) 633-3990. 1992 to present. Quarterly. $180.00 per year. ISSN: 0927-6505.

Astrophysical Journal; An International Review of Astronomy and Astronomical Physics. University of Chicago Press, Journals Division, 5720 South Woodlawn Avenue, Chicago, IL 60637. (312) 753-3347. FAX (312) 753-0811. 1895 to date. three issues per month. $740.00 per year. ISSN: 0004-637X.

Astrophysics and Space Science; An International Journal of Cosmic Physics. Kluwer Academic Publishers, Box 358, Accord Station, Hingham, MA 02018-0358. (617) 871-6600. FAX (617) 871-6528. 1968 to present. 24 issues per year. $2544.00 per year. ISSN: 0004-640X.

Celestial Mechanics and Dynamical Astronomy; An International Journal of Space Dynamics. Kluwer AcademicPublishers, Box 358, Accord Station, Hingham, MA 02018-0358. (617) 871-6600. FAX (617) 871-6528. 1969 to present. Monthly. $703.50 per year. ISSN: 0923-2958.

Ciel Et Terre. Societe Royale Belge d'Astronomie, 'de Meteorologie et de Physique du Globe, 3 avenue Circulaire, Uccle, Brusels, Belgium. FAX 32-2-374-9822. 1880 to present. Bimonthly. 1600 BEF per year. ISSN: 0009-6709.

Experimental Astronomy; An International Journal On Astronomical Instrumentation and Data Analysis. Kluwer Academic Publishers, Box 358, Accord Station, Hingham, MA 02018-0358. (617) 871-6600. FAX (617) 871-6528. 1989 to present. Quarterly. $161.00 per year. ISSN: 0922-6435.

ICARUS; International Journal of Solar System Studies. Academic Press, Inc., Journal Division, 525 B Street, Suite 1900, San Diego, CA 92101-4495. (619) 230-1840. FAX (619) 699-6800. 1962 to present. Monthly. $1080.00 per year. ISSN: 0019-1035.

Land Olt-boernstein, Zahlenwerte Und Funktionen Aus Naturwissenschaften Und Technik. Group 6: Astronomy. Springer-Verlag New York, Inc., 175 Fifth Avenue, New York, NY 10010. (212) 460-1500. FAX (212) 473-6272. 1965 - . Irregular. Price varies. ISSN: 0075-7896.

Mercury. Astronomical Society of the Pacific. 390 Ashton Avenue, San Francisco, CA 94112. (415) 337-1100. FAX: (415) 337-5205. 1972 to present. Bimonthly. $175.00 per year. ISSN: 0047-6773.

Monthly Notices of the Royal Astronomical Society. Blackwell Scientific Publication LT., Osney Mead, Oxford OX2 OEL, England. TEL 0865-240201. FAX 0865-721205. 1827 to present. Fortnightly. $1899.00 per year. ISSN: 0035-8711.

Observatory. c/o Dr. D. J. Stickland, Space and Astrophysics Division, Rutherford Appleton Laboratory, Chilton, Didcot, Oxon OX11 OQX England. FAX 0235-445848. 1877 to present. Bimonthly. $42.00 per year. ISSN: 0029-7704.

Sky and Telescope. Sky Publishing Corporation, Box 9111, Belmont, MA 02178. (617) 864-7360. FAX (617) 864-6117. 1941 to present. Monthly. $27.00 per year. ISSN: 0037-6604.

Soviet Scientific Reviews, Section E: Astrophysics and Space Physics Reviews. Harwood Academic Publishers, 820 Town Center Drive, Langhorne, PA 19047. (215) 740-2642. FAX (215) 750-6343. 1984 to present. 2 issues per year. 216 ECU per year. ISSN: 0143-0432.

Space Science Reviews. Kluwer Academic Publishers, Box 358, Accord Station, Hingham, MA 02018- 0358. (617) 871- 6600. FAX (617) 871-6528. 1962 to present. Sixteen issues per year. $940.00 per year. ISSN: 0038-6308.

Vistas in Astronomy; An International Review Journal. Albert C. Beer, editor. Elsevier Science Publishing Company, Inc., 660 White Plains Road, Terrytown, NY 10591- 5153. (914) 524-9200. FAX: (914) 333-2444. 1958 to present. Quarterly. $415.00 per year. ISSN: 0083-6656.

Transactions of the International Astronomical Union. Proceedings of the General Assembly. Kluwer Academic Publishers, Box 358, Accord Station, Hingham, MA 02018- 0358. (617) 871-6600. Issues in two parts: Part A, Reports, and Part B, Proceedings. Triennial. XXI - 1994. Price varies. ISSN: 0080-1372.

Webb Society Quarterly Journal. Webb Society, 194 Foundry Lane, Freemantle, Southampton, Hampshire SO1 3EQ, England. 1969 to present. Quarterly. $19.50 per year. ISSN: 0043- 1680.

RESEARCH CENTERS AND INSTITUTES

Astronomy Program. University of Maryland, 1207 Computer & Space Sciences Building, College Park, MD 20742. (301) 405- 3001. FAX (301) 314-9067.

Astrophysical Research Consortium. University of Chicago, Department of Astrophysics, 5640 South Ellis, Chicago, IL 60637. Includes University of Washington, Washington State

University, Princeton University and New Mexico State University.

Carnegie Institution of Washington Observatories, 813 Santa Barbara Street, Pasadena, CA 91101. (818) 577-1122. FAX (818) 759-8136.

Center For Astronomy and Space Astrophysics. University of Colorado - Boulder, Campus Box 391, Boulder, CO 80309. (303) 492-4050. (FAX) 492-7178.

Center For Astrophysics and Space Sciences. University of California, San Diego, C-0111, LA Jolla, CA 92093-0111. (619) 534-3460. FAX (619) 534-2294.

Center For Space Research. Massachusetts Institute of Technology, 77 Massachusetts Avenue, Camridge MA 02139. (617) 253-7501.

Center For Space Science and Astrophysics. Stanford University, 325 Durand Building, Stanford, CA 94305. (415) 723-3592.

Institute For Space and Terrestrial Science. 4850 Keele Street, North York, ON, Canada M3J 3K1. (416) 665-3311. FAX (416) 665-2032.

Joint Institute For Laboratory Astrophysics. University of Colorado - Boulder, Boulder CO 80309-0440. (303) 492-7789. FAX (303) 492-5235.

National Astronomy and Ionosphere Center. Cornell University, Space Sciences Building, Ithaca, NY 14853. (607) 255-3735. (607) 255-8803.

Smithsonian Center For Astrophysics. Harvard University, 60 Garden Street, Cambridge, MA 02138. (617) 495-7461. FAX (617) 495-7326.

Space Sciences Laboratory. University of California, Berkeley, Berkeley, CA 94720. (415) 642-1361. FAX (415) 643-7629.

Space Telescope Science Institute. 3700 San Martin Drive, Baltimore, MD 21218. (301) 338-4700.

Theoretical Astrophysics Program. University of Arizona, Planetary Sciences Department, Tuscon, AZ 85721. (602) 621- 6963.

ATMOSPHERE

See also: ACID RAIN, AIR POLLLUTION, GLOBAL WARMING, GREENHOUSE EFFECT, METEOROLOGY, OZONE, WEATHER, WEATHER FORECASTING

ABSTRACT SERVICES AND INDEXES

Applied Science and Technology Index; A cumulative subject index to English language periodicals in the fields of aeronautics and space science, computer technology, chemistry, construction industry, energy and related areas. H.W. Wilson Co., 950 U niversity Avenue, Bronx, NY 10452-9978. (212) 588-8400. (800) 367-6770. FAX (718) 590-1617. From 1958 to present. Monthly. Inquire about cost and availability. Also available online (BRS and WILSONLINE) and on CD-ROM. ISSN: 0003-6986.

Bibliography and Index of Geology. American Geological Institute, 4220 King Street, Alexandria, VA 22302-1507. (703) 379-2480. FAX (703) 379- 7563. From 1969 to present. Monthly. $1295.00 per year. ISSN: 0098-2784. Also available as GEOREF online (CISTI, DIALOG, Orbit, STN) and on CD-ROM. Inquire about price and availability.

Chemical Abstracts. Chemical Abstracts Service. 2540 Oleantangy River Road, Box 3012, Columbus, OS 43210-0012. (614) 447-3600. FAX (614) 447-3713. 1907 to present. Weekly. $16,800.00 per year. Available online and on CD-ROM. CA is also available in five section groupings. Inquire regarding cost and availability.

Current Contents: Physical and Chemical Science. Institute for Scientific Information, 3501 Market Street, Philadelphia, PA 19104. (215) 386-0100. (800) 523-1850. FAX (215) 386-2291. 1961 to present. Weekly. $442.00 per year. Also available online (BRS, DIALOG) and on CD-ROM. Inquire regarding cost and availability. ISSN: 0163-2574.

Deep-Sea Research. Part A: Oceanographic Research Papers.
Deep-Sea Research. Part B: Oceanographic Literature Review. Pergamon Press, Inc., Maxwell House, Fairview Park, Elmsford, NY 10523. (914) 592-7700. Fax (914) 592-3625. Twelve times per year. $2000.00 per year for Parts A and B. Oceanographic Literature Review also available on CD-ROM. Inquire about price and availability

General Science Index. H.W. Wilson Co., 950 University Avenue, Bronx, NY 10452. (212) 588-8400. (800) 367-6770. FAX (718) 590-1617. From 1978 to present. Ten issues per year; quarterly and annual cumulations. Service basis. Available on CD-ROM and online. Inquire regarding cost and availability. ISSN: 0162-1963.

Government Reports Announcements and Index. National Technical Information Service (NTIS), 5285 Port Royal Road, Springfield, VA 22161. (703) 487-4650. 1968 to present. Annual. $525.00 per year. Also available online as NTIS BIBLIO-GRAPHIC DATABASE and on CD-ROM.

Meteorological and Geoastrophysical Abstracts. American Meteorological Society, c/o Inforonics, Inc., 550 Newtown Road, Box 458, Littleton, MA 01460. (508) 486-8976. FAX (508) 486-0027.Covers literature in environmental sciences, meteorology, astrophysics, hydrology, glaciology, and physical oceanography. 1950 to present. Monthly. $950.00 per year. ISSN: 0026-1130. Also available online (DIALOG) and on CD-ROM.

Oceanic Abstracts. Cambridge Scientific Abstracts, 7200 Wisconsin Avenue, Bethesda, MD 20814. (301) 961-6750. Fax (301) 961-6720. Bimonthly. $995.00 per year.

Physics Abstracts. Institute of Electrical Engineers, London, United Kingdom. Available from: Institute of Electrical and Electronic Engineers (IEEE), Box 1331, 445 Hoes Lane, Piscataway, NJ 08855-1331. (908) 562-5549. 1898 to present. 24 issues per year. $2835.00 per year. ISSN: 0036-8091. Also available online and on CD-ROM.

Science Citation Index. Institute for Scientific Information, 3501 Market Street, Philadelphia, PA 19104. (215) 386-0100. (800) 523-1850. FAX (215) 386-2991. 1961 to present. Six issues per year, plus annual cumulation. $11650.00 per year. Also available online and on CD-ROM. Inquire about price and availability. ISSN: 0036-827X.

Selected Water Resources Abstracts. U.S. Geological Survey. Distributed by National Technical Information Service (NTIS), 5285 Port Royal Road, Springfield, VA 22161. (703) 487-4650. Monthly. $115.00 per year.

ANNUAL REVIEWS AND YEARBOOKS

American Meteorological Society Meteorological Monographs. American Meteorological Society, 45 Beacon Street, Boston, MA 02108-3693. (617) 227-2425. FAX (617) 742-8718. Irregular. Price varies.

Annual Review and Earth and Planetary Sciences. Annual Reviews, Inc., 4139 El Camino Way, Palo Alto, CA 94306-0897. (415) 493-4400. Fax (415) 855-9815. Annual. $55.00 per year.

Developments in Atmospheric Science Series. Elsevier Science Publishing Company, Inc., 655 Avenue of the Americas, New York, NY 10010. (212) 989-5800. Irregular. Inquire about price and availability.

Ocean Yearbook. Elisabeth M. Borgese, et al, editors. University of Chicago Press, 5801 Ellis Avenue, Chicago, IL 60637. (312) 702-7700. 1995. $77.00.

ASSOCIATIONS AND PROFESSIONAL SOCIETIES

American Association of State Climatologists. c/o Myron Moinau, University of Idaho, Agriculture Engineering Department, State Climate Service, Moscow, ID 834844-2040.

American Meteorological Society, 45 Beacon Street, Boston, MA 02108-3693. (617) 227-2425. FAX (617) 742-8718. Irregular. Price varies.

Canadian Meteorological and Oceanographic Society, P.O. Box 334, Newmarket, Ontario, L3Y 4X7, Canada. (416) 898-1040. FAX (416) 898-7937.

International Association of Meteorology and Atmospheric Physics. UCAR, P.O. Box 3000, Boulder CO 80307-3000. (303)497-1673.

National Environmental Satellite Data, and Information Service. 2069 Federal Building 4 Room 2060, Washington, DC 20233. (301) 763-71990. FAX (301) 763-4011.

National Weather Association, 6704 Wolke Court, Montgomery, AL 36116-2134. (334) 213-0388. FAX (334) 213-0388.

University Corporation For Atmospheric Research. PO Box 3000, 1850 Table Mesa Drive, Boulder, CO 80307-3000. (303) 497-1673. FAX (303) 497-1654.

Weather Modification Association. PO Box 8116, Fresno CA 93747. (209) 291-8466.

BIBLIOGRAPHIES

Information Sources in the Earth Sciences. David N. Wood, Joan E. Hardy, and Anthony P. Harvey, editors. Bowker-Saur/K.G. Saur. Distributed by R.R. Bowker, 121 Chanlon Road, New Providence, NJ 07974. (800) 521-8110 or (908) 464-6800. Second edition. 1989. $85.00.

ATMOSPHERE

Ency. of Physical Sciences and Engineering Info. Sources

International Bibliography of Meteorology: from the Beginning of Printing to 1889. Oliver L. Fassig. Diane Publishing Co., 600 Upland Avenue, Upland, PA 19015. (610) 499-7415. Reprint edition. 1994. $195.00

Scientific and Technical Books and Serials in Print; An Index To Literature in Science and Technology. R.R. Bowker Co., 205 E. 42nd Street, New York, NY 10017. (800) 521-8110 or (212) 916-1600. 1992. $250.00.

DIRECTORIES AND BIOGRAPHICAL SOURCES

American Men and Women of Science: Physical and Biological Sciences. Eighteenth edition. R.R. Bowker Company, 245 West 17th Street, New York, NY 10011. (800) 521-8810 or (212) 916-1600. $750.00.

Meteorological Services of the World. American Meteorological Society, 45 Beacon Street, Boston, MA 02108- 3693. (617) 227-2425. FAX (617) 742-8718. Annual. $70.00.

National Weather Service offices and Stations. National Oceanic and Atmospheric Administration, Deparatment of Commerce, Silver Spring, MD 20910. (301) 427-7698. Annual. Free

Research Centers Directory. Gale Research, 835 Penobscot Building, Detroit, MI 48226-4094. (800) 877-4253 or (313) 961-2242. 1995. $485.00.

ENCYCLOPEDIAS AND DICTIONARIES

Climates of the States. Bair. Gale Research, 835 Penobscot Building, Detroit, MI 48226-4094. (800) 877-4253 or (313) 961-2242., 5th edition. 1995. $255.00.

Concise Oxford Dictionary of the Earth Sciences. Ailsa Allaby and Michael Allaby, editors. Oxford University Press, Inc., 200 Madison Avenue, New York, NY 10016. (800) 334-4249 or (212) 679-7300. 1990. $42.95.

Encyclopedia of Climatology. Rhodes W.Fairbridge, and John E. Oliver. Chapman & Hall, 1 Penn Plaza, New York, NY 10119. (212) 564-1060. 1986. $115.00. $168.00.

McGraw-Hill Encyclopedia of Science and Technology. McGraw-Hill Publishing Company, 1221 Avenue of the Americas, New York, NY 10020. (800) 262-4729 or (212) 512-3825. Seventh edition. 1992. $1900.00.

Magill's Survey of Science: Earth Science Series. Salem Press, Inc., P.O. Box 1097, Englewood Cliffs, NJ 07632. (800) 221-1592 or (201) 871-3700. Five volumes. 1990. $400.00 for the set.

Smithsonian Meteorological Tables. Books on Demand, 300 North Zeeb Road, Ann Arbor, MI 48106-1346. (313) 761-4700. (800) 521-0600. reprint edition. 199 $153.90.

Water Encyclopedia. Frits van der Leeden, Fred L. Troise, and David K. Todd. Lewis Pubs., CRC Press, Inc., 2000 Corporate Boulevard, NW, Boca Raton, FL 33431. 2nd edition. 1990. $151.95

Weather Satellite Handbook. American Radio Relay League, 225 Main Street, Newington, CT 06111. (203) 666-1541. 1994. $20.00.

GENERAL WORKS

Aerosol-Cloud-Climate Interaction. Peter V. Hobbs. Academic Press, Inc., 6277 Sea Harbor Drive, Orlando, FL. (800) 321-5068. 1993. $65.00.

Atmosphere, Weather and Climate. Roger G. Barry, and Richard J. Chorley. Routledge, Chapman & Hall, Inc., 29 West 35th Street, New York, NY 10001-3291. (212) 244-3336. 1993. $35.00 in paper.

Climate in Human Perspective. A Tribute to Helmut A. Landsberg. F. Baer, et al editors. Kluwer Academic Publishers, 101 Philip Drive, Assinippi Park, Norwell, MA 02061. (617) 871-6600. 1991. $61.50.

Dynamics of Atmospheric Physics. Richard S. Lindzen. Cambridge University Press, 40 West 20th Street, New York, NY 10011-4211. (212) 924-3900. (800) 872-7423. 1990. $47.95. FLUID MECHANICS of the ATMOSPHERE. Robert A. Brown. Academic Press, Inc., 6277 Sea Harbor Drive, Orlando, FL. (800) 321- 5068. 1991. $88.00.

Fundamentals of Atmospheric Dynamics and thermodynamics. C. Riegel. World Scientific Publishing Company, Inc., 1060 Main Street, Unit B, River Edge, NJ 07661. (800) 227-7562 or (201) 487-9655. 1992. $33.00 in paper.

Global and Regional Changes in Atmospheric Composition. Erno Meszaros. Lewis Pubs., CRC Press, Inc., 2000 Corporate Boulevard, NW, Boca Raton, FL 33431. (407) 994-0555. (800) 272-7737. 1993. $59.95.

Greenhouse Effect, Climatic Change & Ecosystems. Bert Bolin and Bo R. Doos, John Wiley and Sons, Inc., 605 Third Avenue, New York, NY 10158. (800) 526-5368 or (212) 850- 6000. 1990. $69.95.

Middle Atmosphere Dynamics. David G. andrews. Academic Press, Inc., 6277 Sea Harbor Drive, Orlando, FL. (800) 321-5068. 1987. $ 44.95 in paper.

Physics of Atmospheres. John T. Houghton. CambridgeUniversity Press, 40 West 20th Street, New York, NY 10011- 4211. (212) 924-3900. (800) 872-7423. 1986. $27.95.

Radiation and Cloud Phenomena in the Atmosphere: Theory, Observation and Modeling. Kuo-Nan Liou. Oxford University Press, 200 Madison Avenue, New York, NY 10016. (212) 725-6000. (800) 334-4249. 1992. $85.00.

Sunsets, Twlights and Evening Skies. Aden B. Meinel and Marjorie P. Mienel. Cambridge University Press, 40 West 20th Street, New York, NY 10011-4211. (212) 924-3900. (800) 872-7423. 1991. $22.95.

Weather Cycles: Real OR Imaginary? William James Burroughs. Cambridge University Press, 40 West 20th Street, New York, NY 10011-4211. (212) 924-3900. 1992. $39.95.

HANDBOOKS AND MANUALS

Author's Guide To the Journals of the American Meteorological Society. American Meteorological Society, 45 Beacon Street,

Boston, MA 02108-3693. (617) 227-2425. FAX (617) 742-8718. 1983. $15.00 in paper.

Handbook of Agricultural Meteorology. John F. Griffin. Oxford University Press, Inc., 200 Madison Avenue, New York, NY 10016. (212) 725-6000. (800) 334-4249. 1994. $85.00.

Handbook in Applied Meteorology. David D. Houghton, editor. John Wiley and Sons, Inc., 605 Third Avenue, New York, NY 10158. (800) 526-5368 or (212) 850-6000. 1985. $168.00.

Weather Almanac. Gale Research, 835 Penobscot Building, Detroit, MI 48226-4094. (313) 961-2242. (800) 877-4253. 7th edition. 1996. $130.00.

Weather Satellite Handbook. American Radio Relay League, 225 Main Street, Newington CT -6111. (203) 666-1541. 1994. $20.00.

ONLINE DATABASES AND CD-ROMS

Climate Assessment Database. National Weather Service, National Meteorological Center, 5200 Auth Road, Suite 101, Camp Springs, MD 20233. (301) 763-8016. Contains daily, weekly and monthly summaries of North American and world climatological data. Also provides five to ten weather forecasts, and 30 to 90 day weather outlook. Subscription required. Inquire as to cost and availability.

CA Search. Chemical Abstracts Service, P.O. Box 3012, Columbus, OH 43210-0012. (614) 447-3600. (800) 848-6533. FAX (614) 447-3709. Very comprehensive guide to worldwide chemical literature and related fields, 1972 to present. Available on BRS,(800) 289-4277, DIALOG, (800) 334-2564, ORBIT (800) 456-7248, and STN International, FIZ Karlsruhe, P.O. Box 2465, W-7500, Karlsruhe 1, Germany, online services. Inquire as to cost and availability.

Compendex Plus. Engineering Information, Inc., 345 East 47th Street, New York, NY 10017. (212) 705-7600 or (800) 221-1044. Contains citations with abstracts to worldwide literature in engineering and technology, from 1970 to present. Available on online BRS,(800) 289-4277, DIALOG, (800) 334-2564, ORBIT (800) 456-7248, and STN International, FIZ Karlsruhe, P.O. Box 2465,W-7500, Karlsruhe 1, Germany, online services. Also available on CD-ROM. Inquire as to cost and availability.

Current Contents Search. Institute for Scientific Information, 3501 Market Street, Philadelphia, PA 19104. (215) 386-0100. FAX (215) 386-6362. Contains citations to articles listed in the table of contents of science and technology journals. Also articles in social sciences and life sciences journals. Available on BRS,(800) 289-4277, DIALOG,(800) 334-2564, online services. Inquire as to cost and availability.

Dissertation Abstracts. University Microfilms International, 300 North Zeeb Road, Ann Arbor, MI 48106. (800) 521-0600 or (313) 761-4700. Scope includes virtually all doctoral dissertations accepted at accredited American institutions from 1861 to present in 252 subject areas. Available on BRS,(800) 289-4277, DIALOG,(800) 334-2564, and OCLC EPIC,(800) 848-5878, online services. Also available on CD-ROM. Inquire as to cost and availability.

INSPEC. Institution of Electrical Engineers, Michael Faraday House, Six Hills Way, Stevenage, Herts. SG1 2AY, England. Telephone: 0438 313311 or FAX 0438 742840. Contains cita-

tions to the worldwide literature of physics, electronics and electrical engineering, computer technology, and related fields. Available on BRS,(800) 289-4277, DIALOG, (800) 334-2564, ORBIT, (800) 456-7248, and STN International, FIZ Karlsruhe, P.O. Box 2465, W-7500, Karlsruhe 1, Germany, online services. Inquire as to cost and availability.

International Aerospace Abstracts. American Institute of Aeronautics and Astronautics, 370 L'Enfant Promenade, S.W., Washington, DC 20024. (202) 646-7400. Contains references and abstracts of journal and monograph literature relating to aerospace science and technology, from 1963 to present. Available through the NASA/RECON system of the National Aeronautics and Space Administration only.

Meteorological and Geoastrophysical Abstracts. American Meteorological Society, 45 Beacon Street, Boston, MA 02108-3693. (617) 227-2425. FAX (617) 742-8718. Contains citations and abstracts to the worldwide literature on significant research in meteorology and geoastrophysics. Related topics include physical oceanography, hydrology, environmental sciences and glaciology. Covers the period 1972 to present. Available on DIALOG,(800) 334-2564, online service. Inquire as to cost and availability.

NTIS Bibliographic Database. National Technical Information Service, 5285 Port Royal Road, Springfield, VA 22161. (703) 487-4929 or FAX (703) 321-8199. Broad coverage of government-sponsored science and technology research reports, 1964 to present. Available on BRS,(800) 289-4277, DIALOG, (800) 334-2564, ORBIT, (800) 456-7248, and STN International, FIZ Karlsruhe, P.O. Box 2465, W-7500, Karlsruhe 1, Germany, online services. Also available on CD-ROM. Inquire as to cost and availability.

Physics Briefs. American Institute of Physics, 335 East 45th Street, New York, NY 10017. (212) 661-9260 or FAX (212) 661-2036. Contains citations with abstracts of the literature of physics and related fields, 1979 to present. Available on the STN International, FIZ Karlsruhe, P.O. Box 2465, W-7500, Karlsruhe 1, Germany, online service. Inquire as to cost and availability.

SCISEARCH. Institute for Scientific Information, 3501 Market Street, Philadelphia, PA 19104. (800) 523-1850 or (215) 386-0100. Broad multidisciplinary title and author index to the international literature of science and technology, 1974 to present. Available on DIALOG and ORBIT online services. Also available on CD-ROM. Inquire as to cost and availability.

WILSONLINE. H.W. Wilson Company, 950 University Avenue, Bronx, NY 10452. (800) 367-6770 or (212) 588-8400. Makes available online versions of the printed H.W. Wilson indexes including Applied Science and Technology Index, Business Periodicals Index, General Science Index, and Readers' Guide to Periodical Literature. Period covered is generally 1983 to present. Available on BRS,(800) 289-4277, DIALOG,(800) 334-2564, and OCLC EPIC,(800) 848-5878, online services. Also available on CD-ROM. Inquire as to cost and availability.

World Climate Disc. Chadwyck-Healey Inc., 1101 King Street, Alexandria, VA 22314. (703) 683-4890. FAX (703) 683-7589. Weather and climate data from approximately 5,000 weather stations worldwide, covering the years 1854 to 1990 on CD-ROM. First edition 1992. Approximately $1200.00 per year with annual updates.

World Weatherdisc. WeatherDisc Associates, Inc., 4584 N.E. 89th Street, Seattle, WA 98115. (206) 524-4314. FAX (206) 543-0308. Meteorological data on CD-ROM which describes

ATMOSPHERE

Ency. of Physical Sciences and Engineering Info. Sources

the climate of the earth today and for the past few hundred years. First edition 1989. Approximately $295.00 per year with biannual updates.

PERIODICALS

Agricultural and Forest Meteorology. Elsevier Science Publishing Company, Inc., 655 Avenue of the Americas, New York, NY 10010. (212) 989-5800. Twenty times per year. $750.00 per year.

American Meteorological Society. BULLETIN. American Meteorological Society, 45 Beacon Street, Boston, MA 02108-3693. (617) 227-2425. FAX (617) 742-8718. Monthly. $60.00 per year.

American Weather Observer. Association of American Weather Observers, 401 Whitney Boulevard, Box 455, Belvedere, IL 61008. (815) 544-5665. FAX (815) 544-6334. Monthly. $21.00 per year.

Atmosphere - Ocean. Canadian Meteorological and Oceanographic Society, P.O. Box 334, Newmarket, Ontario, L3Y 4X7, Canada. (416) 898-1040. FAX (416) 898-7937. Quarterly. $30.00 per year.

Boundary-layer Meteorology: An International Journal of Physical and Biological Processes in the Atmospheric Boundary Layer. Kluwer Academic Publishers, P.O. Box 358, Accord Station, Hingham, MA 02018-0358. (617) 871-6000. Sixteen per year. $785.00 per year.

Climate Change. Kluwer Academic Publishers, P.O. Box 358, Accord Station, Hingham, MA 02018-0358. (617) 871-6000. Six times per year. $327.00 per year.

Dynamics of Atmospheres and Oceans. Elsevier Science Publishing Company, Inc., 655 Avenue of the Americas, New York, NY 10010. (212) 989-5800. Six times per year. $205.00 per year.

Earth. Kalmbach Publishing Company, P.O. Box 1612, Waukesha, WI 53187. (414) 796-0126 or (800) 558-1544. 1991 to present. Bimonthly. $14.95 per year.

International Journal of Climatology. Royal Meteorological Society, distributed by John Wiley and Sons, Inc., 605 Third Avenue, New York, NY 10158. (800) 526-5368 or (212) 850-6000. Eight times per year. $425.00 per year.

JGR: Journal of Geophysical Research: Atmosphere. American Geophysical Union, 2000 Florida Avenue, N.W., Washington, DC 20009. (202) 462-6903. Monthly. $90.00 per year to members.

JGR: Journal of Geophysical Research: Oceans. American Geophysical Union, 2000 Florida Avenue, N.W., Washington, DC 20009. (202) 462-6903. Monthly. $1545.00 per year.

Journal of Applied Meteorology. American Meteorological Society, 45 Beacon Street, Boston, MA 02108-3693. (617) 227-2425. FAX (617) 742-8718. Monthly. $165.00 per year.

Journal of Climate. American Meteorological Society, 45 Beacon Street, Boston, MA 02108-3693. (617) 227-2425. FAX (617) 742-8718. Monthly. $175.00 per year.

Journal of Atmospheric Sciences. American Meteorological Society, 45 Beacon Street, Boston, MA 02108-3693. (617) 227-2425. FAX (617) 742-8718. Semi-monthly. $320.00 per year.

Monthly Weather Review. American Meteorological Society, 45 Beacon Street, Boston, MA 02108-3693. (617) 227-2425. FAX (617) 742-8718. Monthly. $205.00 per year.

National Weather Digest. National Weather Association, 4400 Stamp Road, Room 404, Temple Hills, MD 20748. (301) 899-3784.

Royal Meteorological Society. QUARTERLY JOURNAL. Royal Meteorological Society, 104 Oxford Road, Reading, Berks RG1 7LJ, England. Six times per year. $250.00 per year.

Weather. Royal Meteorological Society, 104 Oxford Road, Reading, Berks RG1 7LJ, England. Monthly. $44.00 per year.

Weatherwise. Heldref Publications, 1319 Eighteenth Street, N.W., Washington, DC 20036-1802. (202) 296-6267. FAX (202) 296-5149. Bi-monthly. $28.00 per year.

RESEARCH CENTERS AND INSTITUTES

Center For Climate Resesarch. Columbia University. Lamont-Doherty Geological Observatory. Palisades, NY 10964. (914) 359-2900.

Center For Climatic Research. University of Wisconsin-Madison, 1225 West Dayton Street, Madison, WI 53706. (608) 262-2839. FAX (608) 262-5964.

Climate Research Institute. Oregon State University. Strand Hall, Room 326, Corvallis, OR 97331-2209.

Cooperative Institute For Mesoscale Meteorological Studies. University of Oklahoma, 401 East Boyd, Norman, OK 73019. (405) 325-3041.

Cooperative Institute For Research in the Atmosphere. Colorado State University. FootHills Campus. Fort Collins, CO 80523. (303) 491-8448.

Florida State Climate Center. Florida State University, 431 Love Building, Tallahassee, FL 32306. (904) 644-3417.

High Plains Climate Center. University of Nebraska-Lincoln, 237 Chase, Lincoln, NE 68583. (402) 472-6706.

Interdisciplinary Center For Aeronomy and Other Atmospheric Sciences. University of Florida, 311 Space Sciences Research Building, Gainsville, FL 32611. (904) 2001.

Joint Institute For Marine and Atmospheric Research. University of Hawaii, 1000 Pope Road, Honolulu, HI 96822. (808) 541-2876. FAX (808) 956-4104

National Center For Atmospheric Research. P.O. Box 3000, Boulder, CO 80307. (303) 497-1000. FAX (303) 497-1654.

ATOM

See: PHYSICS

ATOMIC ENERGY

See: NUCLEAR ENERGY

ATOMIC FISSION

See: FISSION

ATOMIC PHYSICS

See: PHYSICS

ATOMIC POWER

See: NUCLEAR ENERGY

ATOMIC SPECTROSCOPY

See: SPECTROSCOPY

ATOMIC THEORY

See: PHYSICS

AUDIO AMPLIFIER

See: AMPLIFIERS

AUDIO ENGINEERING

See also: ACOUSTICS, AMPLIFIERS, ELECTRICAL ENGINEERING, ELECTRICITY, ELECTRONIC CIRCUITS AND COMPONENTS, ELECTRONIC ENGINEERING

ABSTRACT SERVICES AND INDEXES

Acoustics Abstracts; Parts A & B. Multi-Science Publishing Co. Ltd., Box 176, Avenel, NJ 07001. 1967 to present. Monthly. L265 per year. ISSN: 0001-4974.

Applied Science and Technology Index; A cumulative subject index to English language periodicals in the fields of aeronautics and space science, computer technology, chemistry, construction industry, energy and related areas. H.W. Wilson Co., 950 University Avenue, Bronx, NY 10452. (212) 588-8400.

(800) 367-6770. FAX (718) 590-1617. From 1958 to present. Monthly. Inquire about cost and availability. Also available on CD-ROM andonline. ISSN: 0003-6986.

Current Contents: Engineering, Technology and Applied Sciences. Institute for Scientifi c Information, 3501 Market Street, Philadelphia, PA 19104. (215) 386-0100. FAX (215) 386-6362. 1970 to present. Weekly. $442.00 per year. Also available on CD-ROM and online. Inquire regarding cost and availability. ISSN: 0095-7917.

Electrical and Electronics Abstracts. Institution of Electrical Engineers (IEE), London. Availabl e from INSPEC/IEEE - Institute of Electrical and Electronic Engineers, Box 1331, Hoes Lane, Piscataway, NJ 08855-1331. (908) 562-5549. 1898 to present. Monthly. $2200.00 per year. Also available on CD-ROM and online as INSPEC.

E Ngineering Index Monthly; Indexes and Abstracts the World 's Engineering and Technical Literature. Engineering Information, Inc., Castle Point on the Hudson, Hoboken, NJ 07030. (201) 216-8500. (800) 221-1044. FAX (201) 216-8532. Monthly. $2300.00 per year. Available online as COMPENDEX and also on CD-ROM. ISSN: 0742-1974.

General Science Index. H.W. Wilson Company, 950 University Avenue, Bronx, NY 10452. (212) 588-8400. (800) 367-6770. FAX (718) 590-1617. From 1978 to present. Ten issues per year; quarterly and annual cumulations. Service basis. Available on CD-ROM and online. Inquire regarding cost and availability. ISSN: 0162-1963.

Government Reports Announcements and Index. U. S. National Technical Information Service (NTIS), 5285 Port Royal Road, Springfield, VA 22161. (703) 487-4650. FAX (703) 321-8547. From 1968 to present. Annual. $630.00 per year. Also available online as *NTIS Bibliographic Database* and on CD-ROM.

Index To IEEE Publications. IEEE Service Center, 445 Hoes Lane, Piscataway, NJ 08855-1331. (908) 981-1393. (800) 678-IEEE. FAX (908) 981-9667. 1973 to present. Annual. ISSN: 0099-1368.

Phy sics Abstracts. INSPEC. Section A, Science Abstracts. Institution of Electrical Engineers (IEE), London. Available from: INSPEC/IEEE - Institute of Electrical and Electronic Engineers, Box 1331, Hoes Lane, Piscataway, NJ 08855-1331. (908) 562-5549. 1898 to present. 24 issues per year. $2835.00 per year. Also available online and on CD-ROM. ISSN: 0036-8091.

Physics Briefs (Physikalische Berichte). Information Center for Energy, Physics, Mathematics; German Physical Society. V C H Publishers, Inc., 220 East 23rd Street, New York, NY 10010-4606. (212) 683-8333. 1845 to present. 24 issues per year. $2390.00 per year. Also available online. ISSN: 0179-7434.

Science Citation Index. SCI. Institute for Scientific Information, 3501 Market Street, Philadelphia, PA 19104. (215) 386-0100. (800) 523-1850. FAX (215) 386-2991. 1961 to present. Six issues per year, plus annual cumulation. $11650.00 per year. Also available online and on CD -ROM. Inquire about price and availability. ISSN: 0036-827X.

ANNUAL REVIEWS AND YEARBOOKS

Advances in Electronics and Electron Physics. Academic Press, Inc., 6277 Sea Harbor Drive Orlando FL 32887. (800) 321-5068.

AUDIO ENGINEERING

Ency. of Physical Sciences and Engineering Info. Sources

From 1948 to present. Irregular. Price varies, inquire. ISSN: 0065-2539.

Physical Acoustics. Academic Press, Inc., 6277 Sea Harbor Drive, Orlando, FL 32887. (800) 321-5068. 1964 to present. Irregular. ISSN: 0079-1873.

ASSOCIATIONS AND PROFESSIONAL SOCIETIES

Acoustical Society of America. 500 Sunnyside Boulevard, Woodbury, NY 11797. (516)576-2360.

American Electronics Association. 5201 Great America Way, Suite 520, P.O. 52990, Santa Clara, CA 95056. (408) 987- 4200. FAX (408) 970-8565.

American Institute of Physics. One Physics Ellipse, College Park, MD 20740-3843. (301) 209-3100.

Audio Engineering Society. 60 East 42nd Street, New York, NY 10017. (212) 661-8528.

Creative Audio and Music Electronics Organization. 7753 State Boulevard East, Meridian, MS 39305. (601) 693-5185. FAX (601) 485-7117.

IEEE (institute of Electrical and Electronics Engineers). 345 East 47th Street, New York, NY 10017. (212) 705-7900. FAX (212) 705-4929.

Society of Professional Audio Recording Services. 4300 10th Avenue North, Lake Worth, FL 33461. (407) 641-6648. FAX (407) 642-8263.

BIBLIOGRAPHIES

Science Books & Films. American Association for the Advancement of Science, 1333 H Street NW, Washington, DC 20005. (202) 326-6434. Reviews of print, film and software materials in all sciences. 1965 to present. Nine issues per year. $40.00 per year. ISSN: 0098-342X.

Scientific and Technical Books and Serials in Print; An Index To Literature in Science and Technology. R. R. Bowker. 121 Chanlon Road, New Providence, NJ 07974. (908) 464-6800. FAX (908) 665-3502. (800) 521-8110. 1972 to present. Annual. $299.95 per year. Also available on CD-ROM. ISSN: 0000-054X.

DIRECTORIES AND BIOGRAPHICAL SOURCES

American Men and Women of Science: Physical and Biological Sciences. R. R. Bowker Inc., 121 Chanlon Road, New Providence, NJ 07974. (908) 464-6800. (800) 521-8110. 20th edition. 8 volumes. 1996. $850.00.

E E M - Electronic Engineer's Master. Hearst Business Communications, Inc., 645 Stewart Avenue, Garden City NY 11530. (516) 227-1300. Annual. $90.00 ISSN: 0732-9016.

Engineering Research Centers: Incorporating Electronics Research Centers. Stockton Press, 345 Park Avenue, New York, NY 10010. (212) 689-9200. (800) 221-2123. 4th edition. 1995. $515.00.

IEEE Membership Directory. Institute of Electrical andElectronics Engineers, IEEE Service Center, 445 Hoes Lane, Piscataway, NY 08854. (908) 981-1393. (800) 678-IEEE. FAX (908) 981-9667. 2 volumes. Annual. $190.00. ISSN:

Research Centers Directory. Gale Research Inc., 835 Penobscot Building, Detroit, MI 48226-4094. (313) 961-2242. (800) 877-4253. 20th edition. 1995. $485.00. ISSN: 0080- 1518.

Who's Who in Engineering. Gordon Davis, editor. American Association of Engineering Societies. 1111 19th Street, NY, Suite 608, Washington, DC 20036. (202) 296-2237. (800) 658-8897. 9th edition. 1995. $220.00.

ENCYCLOPEDIAS AND DICTIONARIES

Applied Technical Dictionary: Acoustics. State Mutual Book Service, Ltd. 521 Fifth Avenue, New York, NY 10175. (212) 682-5844. 1986. $75.00.

The Audio Dictionary. Glenn D. White. University of Washington Press, P. O. Box 50096. Seattle, WA 98145-5096. (206) 543-4050. (800) 441-4115. 2nd revised edition. 1991. $19.95 in paper.

Encyclopedia of Applied Physics. George Trigg, editor. VCH Publications, Inc., 220 East 23rd Street, Suite 909, New York, NY 10010-4606. (212) 683-8333. (800) 422-8824. 20 volume set. 1991-. $5990.00.

IEEE Standard Dictionary of Electrical and Electronics Terms. Christopher J. Booth, editor. IEEE Service Center, 445 Hoes Lane, Piscataway, NJ 08855-1331. (908) 981-1393. (800) 678-IEEE. FAX (908) 981-9667. IEEE Standard 100- 1992. 5th edition. 1993. $150.00.

Illustrated Dictionary of Electronics. Stan Gibilsco. TAB Books, P.O. Box 40, Blue Summit, PA 17294-0850. (717) 794- 2191. (800) 233-1128. 7th edition. 1994. $34.95.

McGraw-Hill Encyclopedia of Science and Technology. McGraw- Hill Book Company, Inc., 1221 Avenue of the Americas, New York, NY 10020. (212) 512-2000. (800) 262-4729. 7th edition. 20 volume set. 1992. $1900.00.

GENERAL WORKS

Acoustical Measurements. Leo L. Beranek. Acoustical Society of America, 500 Sunnyside Boulevard, Woodbury, NY 11797. (516)576-2360. 1988. $35.00.

Advanced Audio Production Techniques. Tyree S. Ford. Butterworth-Heinemann, 313 Washington Street, Newton, MA 02158. (618-928-2500. (800) 366-2665. 1993. $24.95 in paper.

Audio Electronics. John L. Hood. Butterworth-Heinemann, 313 Washington Street, Newton, MA 02158. (618-928-2500. (800) 366-2665. 1995. $29.95 in paper.

Basic Acoustics. Donald E. Hall. Krieger Publishing Company, P.O. Box 9542, Melbourne, FL 32902-9542. (407) 724- 9542. 1993. Reprint edition. $61.95.

Introduction to Digital Audio. John Watkinson. Butterworth—Heinemann, 313 Washington Street, Newton, MA 02158. (618-928-2500. (800) 366-2665. 1993. $34.95 in paper.

Mechanics of Sound Recording. Anthony J. Zaza. Prentice- Hall, 113 Sylvan Avenue, Route 9W, Englewood Cliffs, NJ 07632. (201) 592-2000. (800) 922-0579. 1991. $50.00.

Origins in Acoustics: The Science of Sound From Antiquity To the Age of Newton. Books on Demand, 300 North Zeeb Road, Ann Arbor, MI 48106-1346. (313) 761-4700. (800) 521-0600. Reprint edition. $62.50.

The Physics of Sound. Richard E. Berg and David G. Stork. Prentice-Hall, 113 Sylvan Avenue, Route 9W, Englewood Cliffs, NJ 07632. (201) 592-2000. (800) 922-0579. 2nd edition. 1994. $63.00.

Principles of Digital Audio. Ken C. Pohlman. McGraw-Hill Publishing Company, Inc., 1221 Avenue of the Americas, New York, NY 10020. (212) 512-2000. (800) 262-4729. 1995.

Principles of Vibration and Sound. T. D. Rossing and N. H. Fletcher. Springer-Verlag New York, Inc., 175 Fifth Avenue, New York, NY 10010. (212) 460-1500. (800) 777-4643. FAX (212) 473-6272. 1994. $49.95.

Sound Recording and Reproduction. Glyn Alkin. Butterworth-Heinemann, 313 Washington Street, Newton, MA 02158. (618-928-2500. (800) 366-2665. 2nd edition. 1991. $19.95 in paper.

Surface Acoustic Waves and Signal Processing. Michel Feldmann and Jeannine Henaff. Artech House, 685 Canton Street, Norwood, MA 02062. (617) 769-9750. (800) 225-9977. 1989. $89.00.

Vibroacoustical Diagnostics For Machines and Structures. M. F. Dimentberg, et al. John Wiley & Sons, Inc., 605 Third Avenue, New York, NY 10158-0012. (212) 850-6000. (800) 225-5945. 1991. $94.95.

Wonderful Inventions: Motion Pictures, Broadcasting and Recorded Sound. Gordon Press Publications, P.O. Box 459, Bowling Green Station, New York, NY 10004. (718) 624-8419. 1991. $95.00.

HANDBOOKS AND MANUALS

Audio Engineering Handbook. Blair K. Benson, editor. McGraw-Hill Book Company, 1221 Avenue of the Americas, New York, 10020. (212) 512-2000. (800) 628-0004. 1988. $94.50.

Audio Engineer's Reference Book. Michael Talbot-Smith, editor. Butterworth-Heinemann, 313 Washington Street, Newton, MA 02158. (618-928-2500. (800) 366-2665. 1994. $99.95.

CRC Handbook of Chemistry and Physics. David R. Kide, editor. CRC Press, Inc., 2000 Corporate Boulevard, NW, Boca Raton, FL 33431. (407) 994-0555. (800) 272-7737. 77th edition. 1996. $99.95.

Electrical Engineer's Handbook. Donald Christiansen, editor. McGraw Book Company, 1221 Avenue of the Americas, New York, NY 10020. (212) 512-2000. (800) 262-4729. 4th edition. 1996. $110.00.

Electronics Handbook. Jerry C. Whitaker. CRC Press, Inc., 2000 Corporate Boulevard, NW, Boca Raton, FL 33431. (407) 994-0555. (800) 272-7737. 1996. $120.00.

Engineering Tables and Data. A.M. Howatson, P.G. Lund and J.D. Todd. Van Nostrand Reinhold, 115 Fifth Avenue, New York, NY 10003. (212) 254-3232. (800) 926-2665. Second Edition. 1991. $39.95.

Handbook For Sound Engineers: The New Audio Cyclopedia. Glen M. Ballou. Macmillian Publishing Company, Inc,. 200 Old Tappan Road, Old Tappan, NJ 07675. (800) 233-2336. 2nd edition. 1991. $99.95.

Handbook of Acoustical Measurements and Noise Control. Cyril M. Harris. McGraw-Hill Publishing Company, 1221 Avenue of the Americas, New York, NY 10020. (212) 512-2000. (800) 262-4729. 1991. $98.50.

Master Handbook of Acoustics. F. Alton Everest. TAB Books, P.O. Box 40, Blue Summit, PA 17294-0850. (717) 794-2191. (800) 233-1128. 2nd edition. 1988. $19.95.

Sound Recording Handbook. John Woram. Sams, 201 West 103rd Street, Indianapolis, in 46290-1097. (317) 581-3500. (800) 545-5914. 1989. $49.95.

Standard Handbook for Electrical Engineers. Donald Fink. McGraw-Hill Publishing Company, 1221 Avenue of the Americas, New York, NY 10020. (212) 512-2000. (800) 262- 4729. 13th edition. 1996. $110.00.

ONLINE DATABASES AND CD-ROMS

CA Search. Chemical Abstracts Service, P.O. Box 3012, Columbus, OH 43210-0012. (614) 447-3600. (800) 848-6533. FAX (614) 447-3709. Very comprehensive guide to worldwide chemical literature and related fields, 1972 to present. Available on BRS,(800) 289-4277, DIALOG, (800) 334-2564, ORBIT (800) 456-7248, and STN International, FIZ Karlsruhe, P.O. Box 2465, W-7500, Karlsruhe 1, Germany, online services. Inquire as to cost and availability.

Compendex Plus. Engineering Information, Inc., 345 East 47th Street, New York, NY 10017. (212) 705-7600 or (800) 221-1044. Contains citations with abstracts to worldwide literature in engineering and technology, from 1970 to present. Available on online BRS,(800) 289-4277, DIALOG, (800) 334-2564, ORBIT (800) 456-7248, and STN International, FIZ Karlsruhe, P.O. Box 2465, W-7500, Karlsruhe 1, Germany, online services. Also available on CD-ROM. Inquire as to cost and availability.

Current Contents Search. Institute for Scientific Information, 3501 Market Street, Philadelphia, PA 19104. (215) 386-0100. FAX (215) 386-6362. Contains citati ons to articles listed in the table of contents of science and technology journals. Also articles in social sciences and life sciences journals. Available on BRS, (800) 289-4277, DIALOG, (800) 334-2564, online services. Inquire as to cost and availability.

Dissertation Abstracts Online. University Microfilms Internati onal, 300 North Zeeb Road, Ann Arbor, MI 48106. (800) 521-0600 or (313) 761-4700. Scope includes virtual ly all doctoral dissertations accepted at accredited American institutions from 1861 to present in 252 subject areas. Available on BRS, (800) 289-4277, DIALOG, (800) 334-2564, and OCLC EPIC, (800) 848-5878, online services. Also available on CD-ROM. Inquire as to cost andavailability.

INSPEC. Institution of Electrical Engineers, Michael Faraday House, Six Hills WA y, Stevenage, Herts. SG1 2AY , England.

AUDIO ENGINEERING

Ency. of Physical Sciences and Engineering Info. Sources

Telephone: 0438 313311 or FAX 0438 742840. Contains citations to the worldwide literature of physics, electronics and electrical engineering, computer technology, and related fields. Available on BRS, (800) 289-4277, DIALOG, (800) 334-2564, ORBIT, (800) 456-7248, and STN International, FIZ Karlsruhe, P.O. Box 2465, W-7500, Karlsruhe 1, Germany, online services. Inquire as to cost and availability.

NTIS Bibliographic Database. National Technical Information S ervice, 5285 Port Royal Road, Springfield, VA 22161. (703) 487-4929 or FAX (703) 321-8199. Broad coverag e of government-sponsored science and technology research reports, 1964 to present. Available on BRS,(800) 289-4277, DIALOG, (800) 334-2564, ORBIT, (800) 456-7248, and STN International, FIZ Karlsruhe, P.O. Box 2465, W-7500, Karlsruhe 1, Germany, online services. Also available on CD-ROM. Inquire as to cost and availability.

Physics Briefs. American Institute of Physics, 335 East 45th Street , New York, NY 10017. (212) 661-9260 or FAX (212) 661-2036. Contains citations with abstracts of the literature of physics and related fields, 1979 to present. Available on the STN International, FIZ Karlsruhe, P.O. Box 2465, W-7500, Karlsruhe 1, Germany, online service. Inquire as to cost and availability.

Scientific and Technical Books and Serials in Print. R.R. Bowker Inc., 121 Chanlon Road, New Providence, NJ 07974. (800) 521-8110 or (908) 464-6800. List of currently published books and serials in the physical and biological sciences, engineering and technology, with subject, author and titles indexes. Available on ORBIT, (800) 456-7248, online service. Inquire as to cost and availability.

SCISEARCH. Institute for Scientific Information, 3501 Market Street, Philadelphia, PA 19104. (800) 523-1850 or (215) 386-0100. Broad multidisciplinary title and author index to the international literature of science and technology, 1974 to present. Available on DIALOG, (800) 334-2564, and ORBIT, (800) 456-7248, online services. Also available on CD-ROM. Inquire as to cost and availability.

WILSONLINE. H.W. W ilson Company, 950 University Avenue, Bronx, NY 10452. (800) 367-6770 or (212) 588-8400. Makes available onlin e versions of the printed H.W. Wilson indexes includin g Applied Science and Technology Index, Business Periodicals Index, General Science Index, and Readers' Guide to Periodical Literature. Period covered is generally 1983 to present. Available on BRS, (800) 289-4277, DIALOG, (800) 334-2564, and OCLC EPIC, (800) 848-5878, online services. Also available on CD-ROM. Inquire as to cost and availability.

PERIODICALS

Acoustical Society of America Journal. American Instituteof Physics, 500 Sunnyside Boulevard, Woodbury NY 11797-2999. (519) 576-2270. 1929 to present. Monthly. $760.00 per year. ISSN: 0001-4966.

Applied Acoustics. Elsevier Science, 660 White Plains Road, Tarrytown, NY 10591-5153. (914) 524-9200. FAX (914) 333-2444. 1968 to present. Monthly. $770.00 per year. ISSN: 0003-682X.

Audio Engineering Society Journal. Audio Engineering Society, 60 East 42nd Street, New York, NY 10017. (212) 661-8528. 1953 to present. 10 issues per year. $125.00 per year. ISSN: 0004-7554.

DB, The Sound Engineering Magazine. 203 Commack Road, Suite 1010, Commack, NY 1725. (516) 586-6530. 1967 to present. $15.00 per year. ISSN: 0011-7145.

IEEE - Signal Processing Magazine. Institute of Electrical and Electronics Engineers, Inc., Box 1331, 445 Hoes Lane, Piscataway, NJ 08855-1331. (908) 981-0060. 1984 to present. Quarterly. $70.00 per year. ISSN: 1053-5888.

IEEE Transactions On Ultrasonics, Ferroelectrics and Frequency Control. Institute of Electrical and Electronics Engineers, Inc., Box 1331, 445 Hoes Lane, Piscataway, NJ 08855-1331. (908) 981-0060. 1954 to present. Bi-monthly. $230.00 per year. ISSN: 0885-3010.

Journal of Low Frequency Noise and Vibration. Multi-Science Publishing Co. Ltd., Box 176 Avenel, NJ 070001. 1982 to present. Quarterly. L102 per year. ISSN: 0263-0923.

Journal of Sound and Vibration. Academic Press Ltd., 6277 Sea Harbor Drive, Orlando, FL. (800) 321-5068. 1964 to present. 50 issues per year. L1345 per year. ISSN: 0022- 460X.

Ultrasonics. Butterworth-Heinemann, 313 Washington Street, Newton, MA 02158. (618-928-2500. (800) 366-2665. 1963 to present. Bimonthly. L245 per year. ISSN: 0041-624X

RESEARCH CENTERS AND INSTITUTES

Center For Communications and Signal Processing. North Carolina State University. 320 Daniels Hall, Box 7914, Raleigh, NC 27695. (919) 737-3015.

Communications and Signal Processing Laboratory. University of Michigan, 4242 EECS Building, North Campus, Ann Arbor Michigan 48109-2122. (313) 764-5210. FAX (313) 763-1503.

Electronics Research Center. University of Texas at Austin, 132 Engineering Science Building, Austin, TX 78712. (512) 471-3954.

Electrical Engineering Research Laboratories. Purdue University. Electrical Engineering Building, West Lafayette, in 47907. (317) 494-3536. FAX (317) 4944-6440.

Laboratory For Electromagnetic and Electronic Systems. Massachusetts Institute of Technology, 77 Massachusetts Avenue, Cambridge, MA 02139. (617) 253) 4631.

AUTOMATIC CONTROL SYSTEMS

See: AUTOMATION

AUTOMATION

See also: COMPUTER MEMORY, COMPUTER OPERATING SYSTEMS, COMPUTER PROGRAMMING LANGUAGES, COMPUTER VISION, INDUSTRIAL ENGINEERING, PARALLEL COMPUTERS, SOFTWARE, SOFTWARE ENGINEERING

ABSTRACT SERVICES AND INDEXES

ACM Guide To Computing Literature. Association for Computing Machinery, Association for Computing Machinery, 1515 Broadway, 17th Floor, New York, NY 10036-5701. (212) 869-7440. FAX (212) 944-1318. 1964 to present. Monthly. $190.00 per year for non-members.

Applied Science and Technology Index; A cumulative subject index to English language periodicals in the fields of aeronautics and space science, computer technology, chemistry, construction industry, energy and related areas. H.W. Wilson Co., 950 University Avenue, Bronx, NY 10452. (800) 367-6770 or (212) 588-8400. FAX (718) 590-1617. From 1958 to present. Monthly. Inquire about cost and availability. Also available on CD-ROM and online.

Computer Abstracts, MCB University Press Ltd., PO Box 10812, Birmingham, AL 35201-0812. (800) 633-4931. FAX (205) 995-1588. Monthly. Covers computer theory, data, hardware, systems, networks, human-computer interaction, artificial intelligence, as well as applications of computers in aerospace, business, CAD/CAM, cartography, civil engineering, electronics and electrical engineering, industrial engineering, mechanical engineering, medicine, structural engineering, etc. Also available on CD-ROM.

Computer & Control Abstracts (INSPEC). INSPEC/IEEE, Box 1331, 445 Hoes Lane, Piscataway, NJ 08855-1331. (908) 562-5549. Abstracts organized by subjects of international technological information. Monthly. $1455.00 per year. Also available on CD-ROM and online via BRS Online Products, DIALOG Information Services, Orbit Search Service, and STN International.

Computer and Information Systems Abstracts Journal. Cambridge Scientific Abstracts, 7200 Wisconsin Avenue, Bethesda, MD 20814. (301) 961-6750. Fax (301) 961-6720. 1962 to present. Monthly. $1265.00 per year. Also available online via STN International.

Computer Literature Index. Applied Computer Research Inc., Box 82266, Phoenix, AZ 85071-2266. (800) 234-2227. 1971 to present. Quarterly plus annual cumulation. $198.50 per year. Bibliography of books, articles, and reports.

Computing Journal Abstracts. Techgnosis Ltd., Blade House, Battersea Road, Stockport, Cheshire 3AE, England. Telephone 061-442-2639. FAX 061-443-1162. 1969 to present. Monthly. Inquire for price.

Computing Reviews. Association for Computing Machinery, 1515 Broadway, 17th Floor, New York, NY 10036-5701. (212) 869-7440. FAX (212) 944-1318. 1960 to present. Monthly. $130.00 per year for non-members. Also available online via DIALOG Information Services.

Current Contents: Engineering, Technology, and Applied Sciences. Institute for Scientific Information, 3501 Market Street,Philadelphia, PA 19104. (215) 386-0100. FAX (215) 386-6362. 1970 to present. Weekly. $442.00 per year.

Engineering Index Monthly. Engineering Information, Inc., Castle Point on the Hudson, Hoboken, NJ 07030. (800) 221-1044. FAX (201) 216-8532. Monthly. $2300.00 per year. Also available online as COMPENDEX, and also on CD-ROM. Covers chemical engineering, computers, electrical engineering, civil engineering, metals and mining, industrial management, and mechanical engineering.

Science Citation Index. Institute for Scientific Information, 3501 Market Street, Philadelphia, PA 19104. (215) 386-0100. FAX (215) 386-6362. Inquire about availability and cost. Also available on CD-ROM.

ANNUAL REVIEWS AND YEARBOOKS

ACM Monograph Series. Association for Computing Machinery (ACM), 11 West 42nd Street, New York, NY 10036. (212) 869-7440. Fax (212) 869-0481. Irregular. Price varies, inquire.

Advances in Computers. Academic Press, Inc., 1250 Sixth Avenue, San Diego, CA 92101. (619) 231-0926. FAX (619) 699-6715. Irregular. Price varies, inquire.

Advances in Computing Research. JAI Press, Inc., 55 Old Post Road, Number 2, Box 1678, Greenwich, CT 06830. (203) 661-7602. Annual. $68.50 per year.

ASSOCIATIONS AND PROFESSIONAL SOCIETIES

American Federation of Information Processing Societies, 1899 Preston White Drive, Reston, VA 22091-5435. (703) 620-8937.

Association For Computing Machinery. 1515 Broadway, 17th Floor, New York, NY 10036-5701. (212) 869-7440. FAX (212) 944-1318.

Computer and Automation Systems Association of Sme. 1 SME Drive, PO Box 930, Dearborn, MI 48121-0930. (313) 271-1500.

Institute of Electrical and Electronic Engineers. 345 E. 47th Street, New York, NY 10017. (212) 705-7900.

IEEE Computer Society. 1730 Massachusetts Avenue NW, Washington, DC 20036-1992. (202) 371-0101. FAX (202) 728-9614.

IEEE Robotics and Automation Council. 345 E. 47th Street, New York, NY 10017. (212) 705-7900. FAX (212) 705-4929.

Special Interest Group For Design Automation (sigda). c/o James P. Cohoon, PhD., University of Virginia, Thorton Hall, Charlottesville, VA 22903. (804) 982-2210. FAX (804) 982-2214.

BIBLIOGRAPHIES

ACM Guide To Computing Literature. Association for Computing Machinery, 1515 Broadway, 17th Floor, New York, NY 10036-5701. (212) 869-7440. FAX (212) 944-1318. 1964 to present. Annual. $190.00 for non-members. Also available online via DIALOG.

DIRECTORIES AND BIOGRAPHICAL SOURCES

ASIS Handbook and Directory, American Society For Information Science, 8720 Georgia Avenue, Suite 501, Silver Spring, MD 20910-3602. (301) 495-0900. Annual. $100.00.

AUTOMATION

Ency. of Physical Sciences and Engineering Info. Sources

Computing Information Directory. Hildebrandt Inc., PO Box 576, Pullman, WA 99114. 1981 to present. Annual. $199.95 per year.

International Research Centers Directory. Gale Research, 835 Penobscot Building, Detroit, MI 48226-4094. (313) 961-2242. (800) 877-4253. FAX (313) 961-6083. 8th edition. 1995. $410.00.

Research Centers Directory. Gale Research, 835 Penobscot Building, Detroit, MI 48226-4094. (313) 961-2242. (800) 877-4253. FAX (313) 961-6083. $485.00.

Who's Who in Computing. Canadian M I S Database Inc., 268 Lakeshore Road E., Suite 510, Mississauga, ON L5G 1H1, Canada. (905) 271-1601. FAX (905) 271-4522. 1987 to present. Annual. Inquire for price in U.S.

Who's Who in Technology. Gale Research, 835 Penobscot Building, Detroit, MI 48226-4094. (313) 961-2242. (800) 877-4253. FAX (313) 961-6083. 1995. $195.00.

ENCYCLOPEDIAS AND DICTIONARIES

Computer Dictionary. Donald D. Spencer. 4th edition. Camelot Publishing Company, 709 SW 80th Blvd., Gainesville, FL 32607-1537. (904) 331-0952. 1993. $24.95.

Dictionary of Computing. S.M.H. Collin. 2nd edition. Peter Collin Publishing, 8 the Causeway, Teddington, TW11 0HE, England. FAX 0181-943-3386. 1994. Inquire for cost.

Encyclopedia of Computer Science and Engineering. Anthony Ralston & Edwin D. Reilly Jr., editors. 3rd revised edition. Van Nostrand Reinhold, 115 Fifth Avenue, New York, NY 10003. (212) 254-3232 or (800) 926-2665. FAX (212) 254-9499. 1993. $125.00.

Microsoft Press Computer Dictionary: The Comprehensive Standard For Business, School, Library, and Home. 2nd edition. Microsoft Press, One Microsoft Way, Redmond, WA 98052-6399. (206) 936-0055. FAX (206) 823-8101. 1994. $19.95.

GENERAL WORKS

Affordable Automation. Sabrie Soloman. McGraw-Hill Publishing Company, 1221 Avenue of the Americas, New York, NY 10020. (800) 262-4729 or (212) 512-3825. 1996. $59.95.

Computer Automation in Manufacturing: An Introduction. Thomas O. Boucher. Chapman & Hall, 115 Fifth Avenue, New York, NY 10211-0906. 1996. Inquire for price and availability.

Design and Implementation of Intelligent Manufacturing Systems: From Expert Systems, Neural Networks, To Fuzzy Logic. H.R. Parsaei & M. Jamshidi, editors. Prentice Hall, 113 Sylvan Avenue, Route 9W, Englewood Cliffs, NJ 07632. (201) 592-2000 or (800) 922-0579. 1995. $75.00.

Production Planning in Automated Manufacturing. Yves Crama. 2nd revised and enlarged edition. Springer-Verlag, 175 Fifth Avenue, New York, NY 10010. (212) 460-1500 or (800) 777-4643. FAX (212) 473-6272. 1996. $49.95.

HANDBOOKS AND MANUALS

The Automated Factory Handbook: Technology and Management. David Cleland and Bopaya Bidanda. McGraw-Hill Publishing Company, 1221 Avenue of the Americas, New York, NY 10020. (800) 262-4729 or (212) 512-3825. 1990. $69.95.

Handbook of Manufacturing and Automation. Richard C. Dorf and andrew Kusiak, editors. John Wiley and Sons, Inc., 605 ThirdAvenue, New York, NY 10158. (800) 526-5368 or (212) 850-6000. 1994. $135.00.

ONLINE DATABASES AND CD-ROMS

Compendex Plus. Engineering Information, Inc., 345 East 47th Street, New York, NY 10017. (212) 705-7600 or (800) 221-1044. Contains citations with abstracts to worldwide literature in engineering and technology, from 1970 to present. Available on online BRS,(800) 289-4277, DIALOG, (800) 334-2564, ORBIT (800) 456-7248, and STN International, FIZ Karlsruhe, P.O. Box 2465, W-7500, Karlsruhe 1, Germany, online services. Also available on CD-ROM. Inquire as to cost and availability.

Compuscience. FIZ Karlsruhe, D-7514, Eggenstein-Leopoldshafen 2, Germany. Contains citations with abstracts to European and North American literature on computer science, 1972 to present. Available on STN International, FIZ Karlsruhe, P.O. Box 2465, W-7500, Karlsruhe 1, Germany, online service. Inquire as to cost and availability.

Computer and Information Systems Abstracts. Cambridge Scientific Abstracts, 7200 Wisconsin Avenue, Suite 601, Bethesda, MD 20814. (301) 961-6750 or (800) 843-7751. Contains citations to worldwide literature in theoretical and applied computer science and related areas, from 1981 to present.Inquire as to cost and availability.

Computer and Mathematics Search. Institute for Scientific Information, 3501 Market Street, Philadelphia, PA 19104. (215) 386-0100. FAX (215) 386-6362. Covers worldwide literature in computer science and mathematics, from 1980 to present. Available on BRS,(800) 289-4277, online service. Inquire as to cost and availability.

Computer Database. Information Access Company, 362 Lakeside Drive, Foster City, CA 94404. (415) 378-5000 or (800) 227-8431. Contains citations with abstracts to literature from trade journals covering the computer,telecommunications,and electronics industries. Available on the BRS, (800) 289-4277, Compuserve Information Service,(800) 848-8990, and DIALOG,(800) 334-2564, online services. Inquire as to cost and availability.

CSA Engineering. Cambridge Scientific Abstracts, 7200 Wisconsin Avenue, Suite 601, Bethesda, MD 20814. (301) 961-6750 or (800) 843-7751. Contains citations and abstracts of international periodicals and research literature covering all fields of engineering and science and technology,including computer and information science, electronics, mechanical engineering, solid state materials, 1981 to present. Available on BRS,(800) 289-4277, online service.Inquire as to cost and availability.

Current Contents Search. Institute for Scientific Information, 3501 Market Street, Philadelphia, PA 19104. (215) 386-0100. FAX (215) 386-6362. Contains citations to articles listed in the table of contents of science and technology journals. Also articles in social sciences and life sciences journals. Available on

BRS,(800) 289-4277, DIALOG,(800) 334-2564, online services.Inquire as to cost and availability.

Dissertation Abstracts. University Microfilms International, 300 North Zeeb Road, Ann Arbor, MI 48106. (800) 521-0600 or (313) 761-4700. Scope includes virtually all doctoral dissertations accepted at accredited American institutions from 1861 to present in 252 subject areas. Available on BRS,(800) 289-4277, DIALOG,(800) 334-2564, and OCLC EPIC,(800) 848-5878, online services. Also available on CD-ROM. Inquire as to cost and availability.

INSPEC. Institution of Electrical Engineers, Michael Faraday House, Six Hills Way, Stevenage, Herts. SG1 2AY, England. Telephone: 0438 313311 or FAX 0438 742840. Contains citations to the worldwide literature of physics, electronics and electrical engineering, computer technology, and related fields. Available on BRS,(800) 289-4277, DIALOG, (800) 334-2564, ORBIT, (800) 456-7248, and STN International, FIZ Karlsruhe, P.O. Box 2465, W-7500, Karlsruhe 1, Germany, online services. Inquire as to cost and availability.

Mathsci. American Mathematical Society, P.O. Box 6248, Providence, RI 02940. (800) 321-4667 or (401) 455-4000 or FAX (401) 331-3842. Scope includes pure and applied mathematics and related areas of physics, statistics, engineering, computer science, and operations research literature since 1959. Available on DIALOG,(800) 334-2564, online service. Also available on CD-ROM. Inquire as to cost and availability.

NTIS Bibliographic Database. National Technical Information Service, 5285 Port Royal Road, Springfield, VA 22161. (703) 487-4929 or FAX (703) 321-8199. Broad coverage of government-sponsored science and technology research reports, 1964 to present. Available on BRS,(800) 289-4277, DIALOG, (800) 334-2564, ORBIT, (800) 456-7248, and STN International, FIZ Karlsruhe, P.O. Box 2465, W-7500, Karlsruhe 1, Germany, online services. Also available on CD-ROM. Inquire as to cost and availability.

SCISEARCH. Institute for Scientific Information, 3501 Market Street, Philadelphia, PA 19104. (800) 523-1850 or (215) 386-0100. Broad multidisciplinary title and author index to the international literature of science and technology, 1974 to present. Available on DIALOG,(800) 334-2564, and ORBIT,(800) 456-7248, online services. Also available on CD-ROM. Inquire as to cost and availability.

PERIODICALS

ACM Transactions On Computer Systems. Association for Computing Machinery, 1515 Broadway, 17th Floor, New York, NY 10036-5701. (212) 869-7440. FAX (212) 944-1318. 1983 to present. Quarterly. $110.00 for non-members.

Byte. McGraw-Hill Inc., Byte Publications, One Phoenix Mill Lane, Peterborough, NH 03458. (603) 924-9281. FAX (603) 924-2550. 1975 to present. Monthly. $29.95 per year.

Communications of the A C M. Association for ComputingMachinery, 1515 Broadway, 17th Floor, New York, NY 10036-5701. (212) 869-7440. FAX (212) 944-1318. 1958 to present. Monthly. $114.00 for non-members.

Computer Design. PennWell Publishing Company, 1 Technology Park Drive, Westford, MA 01886. (508) 692-0700 or FAX (508) 692-0525. 1962 to present. Semi-monthly. $70.00 per year.

Computer Magazine. IEEE Computer Society, 10662 Vaqueros Circle, Box 3014, Los Alamitos, CA 90720. (714) 821-8380. 1966 to present. Monthly. $49.00 per year.

Computer Technology Review. West World Productions, Inc., 924 Westwood Boulevard, Suite 650, Los Angeles, CA 90024-2927. (213) 208-1335 or FAX (213) 208-1054. 1981 to present. Sixteen times per year. $80.00 per year.

Datamation. Cahners Publishing Company, 275 Washington Street, Newton, MA 02158-1630. (617) 558-4402. 1957 to present. 24 times a year. $75.00 per year.

IEEE Transactions On Computers. IEEE, 345 E. 47th Street, New York, NY 10017. (908) 981-0060. FAX (908) 981-9667. 1952 to present. Monthly. $425.00 for non-members.

Journal of the Association For Computing Machinery. Association for Computing Machinery, 11 West 42nd Street, New York, NY 10036. (212) 869-7440. Fax (212) 869-0481. Quarterly. $85.00 per

RESEARCH CENTERS AND INSTITUTES

Illinois Institute of Technology, Manufacturing Productivity Center. 10 W. 35th Street, Chicago, IL 60616. (312) 567-4800.

Institute For Manufacturing and Automation Research. Hughes Aircraft, 3350 E. Birch Street, Suite 200, Brea, CA 92621. (714) 579-1617.

New York University Robotics Research Laboratory. Courant Institute of Mathematical Sciences, 719 Broadway, 12th floor, New York, NY 10003. (212) 998-3472.

UCLA Center For Manufacturing and Automation Research. School of Engineering and Applied Science, 3531 Boelter Hall, Los Angeles, CA 90024-1597. (213) 825-2212.

AUTOMOBILE TRANSMISSIONS

See: AUTOMOTIVE ENGINEERING

AUTOMOBILES

See: AUTOMOTIVE ENGINEERING

AUTOMOTIVE ENGINEERING

See also: DIESEL ENGINES, ELECTRIC VEHICLES, TRANSPORTATION ENGINEERING

ABSTRACT SERVICES AND INDEXES

Applied Science and Technology Index; A cumulative subject index to English language periodicals in the fields of aeronautics and space science, computer technology, chemistry, construction industry, energy and related areas. H.W. Wilson Co., 950 University Avenue, Bronx, NY 10452. (800) 367-6770 or

(212) 588-8400. FAX (718) 590-1617. From 1958 to present. Monthly. Inquire about cost and availability. Also available on CD-ROM.

Engineering Index Monthly. Engineering Information, Inc., Castle Point on the Hudson, Hoboken, NJ 07030. (800) 221-1044. FAX (212) 832-1857. Monthly. $2200.00 per year. Also available online as COMPENDEX, and also on CD-ROM.

Fuel and Energy Abstracts, Butterworth-heinemann, Linacre House, Jordan Hill, Oxford OX2 8DP, England. Telephone 0865-310366. FAX 0865-310898. 1960 to present. Bimonthly. Inquire for price. A summary of world literature on all technical, scientific, commercial and environmental aspects of fuel and energy.

Physics Abstracts (INSPEC). INSPEC/IEEE, Box 1331, 445 Hoes Lane, Piscataway, NJ 08855-1331. (908) 562-5549. 1898 to present. $2835.00 per year. Also available online. Covers recently published primary research in all areas of physics, including particle, nuclear, fluid, plasma, and solid-state physics, biophysics, geophysics, astrophysics, measurement, and instrumentation.

SAE Global Mobility Database. SAE International, 400 Commonwealth Drive, Warrendale, PA 15096-0001. (412) 776-4970. Bibliographic information and abstracts on mobility engineering technology. Available on CD-ROM and online.

ASSOCIATIONS AND PROFESSIONAL SOCIETIES

Automotive Safety Foundation. 1776 Massachusetts Avenue NW, Suite 500, Washington, DC 20036. (202) 857-1207.

Society of Automotive Engineers (sae International). 400 Commonwealth Drive, Warrendale, PA 15096-0001. (412) 776-4841.

BIBLIOGRAPHIES

Scientific and Technical Books and Serials in Print. R.R. Bowker Inc., 121 Chanlon Road, New Providence, NJ 07974. (800) 521-8110 or (908) 464-6800. List of currently published books and serials in the physical and biological sciences, engineering and technology, with subject, author and titles indexes. Available on ORBIT,(800) 456-7248, online service. Inquire as to cost and availability.

DIRECTORIES AND BIOGRAPHICAL SOURCES

Research Centers Directory. Gale Research, 835 Penobscot Building, Detroit, MI 48226-4094. (313) 961-2242. (800) 877-4253. FAX (313) 961-6083. $485.00.

Scientific and Technical Organizations and Agencies Directory. Gale Research, 835 Penobscot Building, Detroit, MI 48226-4094. (313) 961-2242. (800) 877-4253. FAX (313) 961-6083. 4th edition. 1996. $195.00.

Who's Who in Engineering. American Association of Engineering Societies, 1111 19th Street NW, Suite 608, Washington, DC 20036. (202) 296-2237 or (800) 658-8897. 8th edition. 1991. Inquire for price.

Who's Who in Technology. Gale Research, 835 Penobscot Building, Detroit, MI 48226-4094. (313) 961-2242. (800) 877-4253. FAX (313) 961-6083. 1995. $195.00.

ENCYCLOPEDIAS AND DICTIONARIES

Dictionary For Automotive Engineering. Jean de Coster.K.G. Saur, 121 Chanlon Road, New Providence, NJ 07974. (908) 464-6800 or (800) 521-8110. 1986. Inquire for cost and availability.

Encyclopedia of Physical Science and Technology. Robert A. Meyers, ed. 18 volumes. Academic Press, Inc., 6277 Sea Harbor Drive, Orlando, FL 32887. 1992. $2100.00

McGraw-Hill Dictionary of Scientific and Technical Terms. Sybil P. Parker, ed. 5th edition. McGraw-Hill Publishing Company, 1221 Avenue of the Americas, New York, NY 10020. (800) 262-4729 or (212) 512-3825. 1993. $110.50.

McGraw-Hill Encyclopedia of Engineering. Sybil P. Parker, ed. 2nd edition. McGraw-Hill Publishing Company, 1221 Avenue of the Americas, New York, NY 10020. (800) 262-4729 or (212) 512-3825. 1993. $95.50.

McGraw-Hill Encyclopedia of Science and Technology. Sybil P. Parker, ed. 7th edition. 20 volumes. McGraw-Hill Publishing Company, 1221 Avenue of the Americas, New York, NY 10020. (800) 262-4729 or (212) 512-3825. 1992. $1900.00

SAE Motor Vehicle, Safety and Environmental Terminology. Society of Automotive Engineers, 400 Commonwealth Drive, Warrendale, PA 15096. (412) 776-4841. $19.00.

GENERAL WORKS

Automobile Design: Twelve Great Designers and their Work. Ronald Barker & Anthony Harding, editors. SAE International, 400 Commonwealth Drive, Warrendale, PA 15096-0001.(412) 776-4970. FAX (412) 776-0790. 1992. Inquire for cost.

Automotive Electronics. A. Fletcher. Elsevier Science Inc., 660 White Plains Rd., Tarrytown, NY 10159-5153. (919) 524-9200. FAX (919) 333-2444. 1992. $1070.00.

Automotive Ergonomics. Brian Peacock and Waldemar Karwowski, eds. Taylor and Francis, Inc., 1900 Frost Road, Suite 101, Bristol, PA 19007. (215) 785-5000. FAX (215) 785-5515. 1993. $99.00.

Auto Opium: A Social History of American Automobile Design. David Gartman. Routledge, Chapman and Hall, Inc., 29 West 35th Street, New York, NY 10001-2291. (212) 244-3336 or FAX (212) 563-2269. 1994. Inquire for cost.

HANDBOOKS AND MANUALS

Automotive Handbook. 3rd edition. Robert Bosch GmbH/SAE International, 400 Commonwealth Drive, Warrendale, PA 15096-0001.(412) 776-4970. FAX (412) 776-0790. 1993. $34.00.

SAE Handbook. SAE International, 400 Commonwealth Drive, Warrendale, PA 15096-0001.(412) 776-4970. FAX (412) 776-0790. 1905 to present. Annual. $375.00 for three-volume set.

Standard Handbook of Engineering Calculations. Tyler G. Hicks, ed. 2d ed. McGraw-Hill Publishing Company, 1221 Avenue of the Americas, New York, NY 10020. (800) 262-4729 or (212) 512-3825. 1985. $89.50.

The Wiley Engineer's Desk Reference. Sanford I. Heisler. John Wiley and Sons, Inc., 605 Third Avenue, New York, NY 10158.(800) 526-5368 or (212) 850-6000. 1984. $64.95.

ONLINE DATABASES AND CD-ROMS

Compendex Plus. Engineering Information, Inc., 345 East 47th Street, New York, NY 10017. (212) 705-7600 or (800) 221-1044. Contains citations with abstracts to worldwide literature in engineering and technology, from 1970 to present. Available on online BRS,(800) 289-4277, DIALOG, (800) 334-2564, ORBIT (800) 456-7248, and STN International, FIZ Karlsruhe, P.O. Box 2465, W-7500, Karlsruhe 1, Germany, online services. Also available on CD-ROM. Inquire as to cost and availability.

CSA Engineering. Cambridge Scientific Abstracts, 7200 Wisconsin Avenue, Suite 601, Bethesda, MD 20814. (301) 961-6750 or (800) 843-7751. Contains citations and abstracts of international periodicals and research literature covering all fields of engineering and science and technology,including computer and information science, electronics, mechanical engineering, solid state materials, 1981 to present. Available on BRS,(800) 289-4277, online service.Inquire as to cost and availability.

Current Contents Search. Institute for Scientific Information, 3501 Market Street, Philadelphia, PA 19104. (215) 386-0100. FAX (215) 386-6362. Contains citations to articles listed in the table of contents of science and technology journals. Also articles in social sciences and life sciences journals. Available on BRS,(800) 289-4277, DIALOG,(800) 334-2564, online services. Inquire as to cost and availability.

Dissertation Abstracts. University Microfilms International, 300 North Zeeb Road, Ann Arbor, MI 48106. (800) 521-0600 or (313) 761-4700. Scope includes virtually all doctoral dissertations accepted at accredited American institutions from 1861 to present in 252 subject areas. Available on BRS,(800) 289-4277, DIALOG,(800) 334-2564, and OCLC EPIC,(800) 848-5878, online services. Also available on CD-ROM. Inquire as to cost and availability.

NTIS Bibliographic Database. National Technical Information Service, 5285 Port Royal Road, Springfield, VA 22161. (703) 487-4929 or FAX (703) 321-8199. Broad coverage of government-sponsored science and technology research reports, 1964 to present. Available on BRS,(800) 289-4277, DIALOG, (800) 334-2564, ORBIT, (800) 456-7248, and STN International, FIZ Karlsruhe, P.O. Box 2465, W-7500, Karlsruhe 1, Germany, online services. Also available on CD-ROM. Inquire as to cost and availability.

SAE Global Mobility Database. Society of Automative Engineers (SAE), Electronic Publishing Division, 400 Commonwealth Drive, Warrendale, PA 15098. (412) 776-4841. Contains citations with abstracts to technical papers on automotive and aerospace technology and vehicular-related industries that have been presented at SAE conferences. Covers 1906 to present. Available on ORBIT,(800) 456-7248,online service. Inquire as to cost and availability.

SCISEARCH. Institute for Scientific Information, 3501 Market Street, Philadelphia, PA 19104. (800) 523-1850 or (215) 386-0100. Broad multidisciplinary title and author index to the in-ternational literature of science and technology, 1974 to present. Available on DIALOG,(800) 334-2564, and ORBIT,(800) 456-7248, online services. Also available on CD-ROM. Inquire as to cost and availability.

TRIS (Transportation Research Information). National Academy of Science, 2101 Constitution Avenue, N.W., Washington, DC 20418. (202) 334-3313 or (800) 624-6242. Citations with abstracts of literature on transportation, including air, highway, rail, maritime and other modes from 1968 to present. Available on DIALOG,(800) 334-2564, online service. Inquire as to cost and availability.

WILSONLINE. H.W. Wilson Company, 950 University Avenue, Bronx, NY 10452. (800) 367-6770 or (212) 588-8400. Makes available online versions of the printed H.W. Wilson indexes including Applied Science and Technology Index, Business Periodicals Index, General Science Index, and Readers' Guide to Periodical Literature. Period covered is generally 1983 to present. Available on BRS,(800) 289-4277, DIALOG,(800) 334-2564, and OCLC EPIC,(800) 848-5878, online services. Also available on CD-ROM. Inquire as to cost and availability.

OTHER SOURCES

Federal Motor Vehicle Safety Standards and Regulations. U.S. Department of Transportation, National Highway Traffic Safety Administration. Order from the Superintendent of Documents, U.S. Government Printing office, Washington, DC 20402-9325. (202) 783-3238. 1994. Inquire for cost.

PERIODICALS

Automotive Engineering Magazine. SAE International, 400 Commonwealth Drive, Warrendale, PA 15096-0001. (412) 776-4841. 1905 to present. Monthly. $72.00.

Automotive Engineer. Mechanical Engineering Publications Ltd., Northgate Avenue, Bury St. Edmunds, Suffolk IP32 6BW, England. Telephone 0284-763277. FAX 0284-704006. 1962 to present. 6 times a year. Inquire for cost.

Automotive Industries. Chilton Co., One Chilton Way, Radnor PA 19089. (215) 964-4028. 1895 to present. Monthly. $60.00 per year.

Automotive Week. Automotive Week Publishing Company, Box 3495, Wayne, NJ 07474-3495. (201) 694-7792. 1975 to present. Weekly. $125.00 per year.

Electric Vehicle Progress. Alexander Research & Communications Inc., 215 Park Avenue South, Suite 1301, New York, NY 10003. (212) 228-0246. FAX (212) 228-0376. 1979 to present. Semi-monthly. $387.00 per year.

International Journal of Vehicle Design. Inderscience Enterprises Ltd., World Trade Center Building, 110 Ave. Luis Casai, Case Postale 306, CH-1215 Geneva-Aeroport, Switzerland. 1979 topresent. Bi-monthly. $190.00 per year.

JSAE (Japanese Society of Automotive Engineers) Review. Elsevier Science [journals], 660 White Plains Rd., Tarrytown, NY 10159-5153. (919) 524-9200. FAX (919) 333-2444. 1978 to present. 4 times a year. $168.00 per year.

AUTOMOTIVE ENGINEERING

Ency. of Physical Sciences and Engineering Info. Sources

Proceedings of the Institution of Mechanical Engineers D: Journal of Automobile Engineering. Mechanical Engineering Publications Ltd., Northgate Avenue, Bury St. Edmunds, Suffolk IP32 6BW, England. Telephone 0284-763277. FAX 0284-704006. 1984 to present. Quarterly. $269.00 per year.

SAE Transactions. SAE International, 400 Commonwealth Drive, Warrendale, PA 15096-0001. (412) 776-4841. Annual. $850.00 per year.

RESEARCH CENTERS AND INSTITUTES

Automotive Cooling Systems Institute. PO Box 13966, Research Triangle Park, NC 27709-3966. (919) 549-4800.

Sloan Automotive Laboratory, Massachusetts Institute of Technology. Room 3-340, Cambridge, MA 02139. (617) 253-2243.

Society of Automotive Engineers (sae) International. 400 Commonwealth Drive, Warrendale, PA 15096-0001. (412) 776-4841.

University of Michigan Automotive Engineering Laboratory. Department of Mechanical Engineering & Applied Mechanics, 321 Lay, N. Campus, 1231 Beal, Ann Arbor, MI 48109-2121. (313) 764-4254.

Wayne State University Center For Automotive Research. 5050 Anthony Wayne Drive, Detroit, MI 48202. (313) 577-3887.

AVIONICS

See also: AERONAUTICAL ENGINEERING, AEROSPACE ENGINEERING

ABSTRACT SERVICES AND INDEXES

Aeronautical Engineering. Scientific and Technical Branch, National Aeronautics and Space Administration. National Technical Information Service (NTIS), 5285 Port Royal Road, Springfield, VA 22161. (703) 4878-4650. FAX (703) 321-8547. Monthly. A selection of annotated references to unclassified reports and journal articles that were introduced into the NASA scientific and technical information system and announced in STAR and IAA.

Applied Science and Technology Index; A cumulative subject index to English language periodicals in the fields of aeronautics and space science, computer technology, chemistry, construction industry, energy and related areas. H.W. Wilson Co., 950 University Avenue, Bronx, NY 10452. (800) 367-6770 or (212) 588-8400. FAX (718) 590-1617. From 1958 to present. Monthly. Inquire about cost and availability. Also available on CD-ROM.

Current Contents: Engin, Tech, and Applied Sciences. Institute for Scientific Information, 3501 Market Street, Philadelphia, PA 19104. (215) 386-0100 or (800) 523-1850. FAX (215) 386-6362. Weekly. $360.00 per year.

Engineering Index Monthly. Engineering Information, Inc., Castle Point on the Hudson, Hoboken, NJ 07030. (800) 221-1044. FAX (212) 832-1857. Monthly. $2200.00 per year. Also available onlineas COMPENDEX, and also on CD-ROM.

International Aerospace Abstracts. Technical Information Service, American Institute of Aeronautics and Astronautics, Inc., 555 West 57th St., New York, NY 10019. (212) 247-6500. FAX (212) 582-4861. Semi-monthly. $1295.00 per year.

NASA Patent Abstracts Bibliography. National Aeronautics and Space Administration, Center for Aerospace Information, Box 8757, BWI Airport, Baltimore, MD 21240. (301) 621-0153. Semi-annual. Inquire for price information.

Science Citation Index. Institute for Scientific Information, 3501 Market Street, Philadelphia, PA 19104. (215) 386-0100 or (800) 523-1850. FAX (215) 386-6362. Inquire about availability and cost. Also available on CD-ROM.

Scientific and Technical Aerospace Reports (star). National Aeronautics and Space Administration. NASA Center for Aerospace Information, Box 8757, BWI Airport, Baltimore, MD 21240. (301) 621-0153. Monthly. Inquire about availability and cost. Also available through the NASA online retrieval service (RECON), and through the Aerospace Database through DIALOG.

ANNUAL REVIEWS AND YEARBOOKS

Jane's Airport and ATC Equipment, Jane's Information Group, Dept. DSM, 1340 Braddock Pl., Suite 300, Box 1436, Alexandria, VA 22314-1651. (703) 683-3700. FAX (703) 836-0029. Annual. $225.00.

Jane's Military Communications, Jane's Information Group Dept. DSM, 1340 Braddock Pl., Suite 300, Box 1436, Alexandria, VA 22314-1651. (703) 683-3700. FAX (703) 836-0029. Annual. $245.00. Also available on CD-ROM.

ASSOCIATIONS AND PROFESSIONAL SOCIETIES

Aerospace Industries Association of America. 1250 Eye Street NW, Washington, DC 20005. (202) 371-8400. FAX (202) 371-8573.

Aircraft Electronics Association. PO Box 1963, Independence, MO 64055-0963. (816) 373-6565. FAX (816) 478-3100.

American Institute of Aeronautics and Astronautics, 370 L'enfant Promenade Sw, Washington, DC 20024-2518.

Canadian Aeronautics and Space Institute, 222 Somerset Street West, Suite 601, Ottawa, Ontario, Canada K2p 0j1.

Flight Safety Foundation, Inc., 2200 Wilson Blvd., Suite 500, Arlington, VA 22201-3306. (703) 522-8300. FAX (703) 525-6047.

National Avionics Society. PO Box 5633, Denver, CO 80217-5633.

Society of Automotive Engineers. SAE, Inc., 400 Commonwealth Dr., Warrendale, PA 15096. (412) 776-4841.

BIBLIOGRAPHIES

Aeronautical Engineering: A Continuing Bibliography With. INDEXES. National Aeronautics and Space Administration, Center for Aerospace Information, Box 8757, BWI Airport,

Ency. of Physical Sciences and Engineering Info. Sources

AVIONICS

Baltimore, MD 21240. (301) 621-0153. Inquire for cost and availability.

Scientific and Technical Books and Serials in Print, R.R.Bowker Inc., 121 Chanlon Road, New Providence, NJ 07974. (800) 521-8110 or (908) 464-6800. 3 volumes, annually. $285.00. Also available on CD-ROM.

DIRECTORIES AND BIOGRAPHICAL SOURCES

Research Centers Directory. Gale Research, 835 Penobscot Building, Detroit, MI 48226-4094. (313) 961-2242. (800) 877-4253. FAX (313) 961-6083. $485.00.

Who's Who in Engineering. American Association of Engineering Societies, 1111 19th St. NW, Suite 608, Washington, DC 20036. 7th ed. 1988. $200.00.

World Aviation Directory. McGraw-Hill, Aviation Week Group (NY), 1221 Avenue of the Americas, New York, NY 10020. (215) 237-4112. FAX (215) 586-3232. Semiannual. $250.00.

ENCYCLOPEDIAS AND DICTIONARIES

Dictionary of Aeronautical Terms, Dale Crane, Comp. Aviation Supplies and Academics, Inc., 7005 132nd Place SE, Newcastle, WA 98059. (206) 235-1500 or (800) 426-8338. FAX (206) 235-0128. 1991. $15.95.

Encyclopedia of Physical Science and Technology. Robert A. Meyers, ed. Academic Press, Inc., 6277 Sea Harbor Drive, Orlando, FL 32887. 1992. $2100.00.

Jane's Aerospace Dictionary, B. Guston. 3rd ed. Jane's Information Group, Dept. DSM, 1340 Braddock Pl., Suite 300, Box 1436, Alexandria, VA 22314-1651. (703) 683-3700. FAX (703) 836-0029. 1991. $45.00.

McGraw-Hill Encyclopedia of Science and Technology, Sybil P. Parker, ed. 7th ed. 20 vols. McGraw-Hill Publishing Company, 1221 Avenue of the Americas, New York, NY 10020. (800) 262-4729 or (212) 512-3825. $1900.00

GENERAL WORKS

Aerospace Avionics Systems: A Modern Synthesis. George M. Siouris. Academic Press, Inc., 6277 Sea Harbor Drive, Orlando, FL 32887. 1993. $105.00.

Avionics Systems Design. John R. Newport. CRC Press, Inc., 2000 Corporate Blvd., N.W., Boca Raton, FL 33431. (407) 994-0555 or (800)277-7737. FAX (407) 994-0949. 1994. Inquire for cost and availability.

Manual of Avionics: An Introduction To the Electronics of Civil Aviation. Brian Kendal. 3rd ed. Blackwell Scientific Publishing, Inc., 238 Main Street, Cambridge, MA 02142. (617) 876-7000 or (800) 759-6102. FAX (617) 876-7022. 1993. $49.95.

Trends in Advanced Avionics. Jim Curran. Iowa State University Press, 2121 South State Avenue, Ames, IA 50014-8300. (800) 862-6657. FAX (515) 292-3348. 1992. $35.95.

HANDBOOKS AND MANUALS

Jane's All the World's Aircraft. Jane's Information Group, Dept. DSM, 1340 Braddock Pl., Suite 300, Box 1436, Alexandria, VA 22314-1651. (703) 683-3700. FAX (703) 836-0029. Annual. $245.00. Also available on CD-ROM.

Jane's Avionics. Jane's Information Group, Dept. DSM,1340 Braddock Pl., Suite 300, Box 1436, Alexandria, VA 22314-1651. (703) 683-3700. FAX (703) 836-0029. Annual. $245.00. Also available on CD-ROM.

ONLINE DATABASES AND CD-ROMS

Aerospace Database. American Institute of Aeronautics and Astronautics, 370 L'Enfant Promenade, S.W., Washington, DC 20024. (202) 646-7400. Worldwide published literature on research and development in aerospace and related areas, with abstracts. Covers 1962 to present. Available on DIALOG, (800) 334-2564, online service. Also available on CD-ROM. Inquire as to cost and availability.

Compendex Plus. Engineering Information, Inc., 345 East 47th Street, New York, NY 10017. (212) 705-7600 or (800) 221-1044. Contains citations with abstracts to worldwide literature in engineering and technology, from 1970 to present. Available on online BRS,(800) 289-4277, DIALOG, (800) 334-2564, ORBIT (800) 456-7248, and STN International, FIZ Karlsruhe, P.O. Box 2465, W-7500, Karlsruhe 1, Germany, online services. Also available on CD-ROM. Inquire as to cost and availability.

CSA Engineering. Cambridge Scientific Abstracts, 7200 Wisconsin Avenue, Suite 601, Bethesda, MD 20814. (301) 961-6750 or (800) 843-7751. Contains citations and abstracts of international periodicals and research literature covering all fields of engineering and science and technology,including computer and information science, electronics, mechanical engineering, solid state materials, 1981 to present. Available on BRS,(800) 289-4277, online service.Inquire as to cost and availability.

Current Contents Search. Institute for Scientific Information, 3501 Market Street, Philadelphia, PA 19104. (215) 386-0100. FAX (215) 386-6362. Contains citations to articles listed in the table of contents of science and technology journals. Also articles in social sciences and life sciences journals. Available on BRS,(800) 289-4277, DIALOG,(800) 334-2564, online services. Inquire as to cost and availability.

Dissertation Abstracts. University Microfilms International, 300 North Zeeb Road, Ann Arbor, MI 48106. (800) 521-0600 or (313) 761-4700. Scope includes virtually all doctoral dissertations accepted at accredited American institutions from 1861 to present in 252 subject areas. Available on BRS,(800) 289-4277, DIALOG,(800) 334-2564, and OCLC EPIC,(800) 848-5878, online services. Also available on CD-ROM. Inquire as to cost and availability.

International Aerospace Abstracts. American Institute of Aeronautics and Astronautics, 370 L'Enfant Promenade, S.W., Washington, DC 20024. (202) 646-7400. Contains references and abstracts of journal and monograph literature relating to aerospace science and technology, from 1963 to present. Available through the NASA/RECON system of the National Aeronautics and Space Administration only.

NASA Database. American Institute of Aeronautics andAstronautics, 370 L'Enfant Promenade, S.W., Washington, DC 20024. (202) 646-7400. Citations and abstracts of aeronau-

AVIONICS

Ency. of Physical Sciences and Engineering Info. Sources

tics and astronautics literature, 1962 to present. Also contains citations from STAR, SCIENTIFIC and TECHNICAL AEROSPACE REPORTS, and INTERNATIONAL AEROSPACE ABSTRACTS. Available through NASA/RECON online service. Inquire as to cost and availability.

NTIS Bibliographic Database. National Technical Information Service, 5285 Port Royal Road, Springfield, VA 22161. (703) 487-4929 or FAX (703) 321-8199. Broad coverage of government-sponsored science and technology research reports, 1964 to present. Available on BRS,(800) 289-4277, DIALOG, (800) 334-2564, ORBIT, (800) 456-7248, and STN International, FIZ Karlsruhe, P.O. Box 2465, W-7500, Karlsruhe 1, Germany, online services. Also available on CD-ROM. Inquire as to cost and availability.

SAE Global Mobility Database. Society of Automotive Engineers (SAE), Electronic Publishing Division, 400 Commonwealth Drive, Warrendale, PA 15098. (412) 776-4841. Contains citations with abstracts to technical papers on automotive and aerospace technology and vehicular-related industries that have been presented at SAE conferences. Covers 1906 to present. Available on ORBIT,(800) 456-7248,online service. Inquire as to cost and availability.

Scientific and Technical Books and Serials in Print. R.R. Bowker Inc., 121 Chanlon Road, New Providence, NJ 07974. (800) 521-8110 or (908) 464-6800. List of currently published books and serials in the physical and biological sciences, engineering and technology, with subject, author and titles indexes. Available on ORBIT,(800) 456-7248, online service.Inquire as to cost and availability.

SCISEARCH. Institute for Scientific Information, 3501 Market Street, Philadelphia, PA 19104. (800) 523-1850 or (215) 386-0100. Broad multidisciplinary title and author index to the international literature of science and technology, 1974 to present. Available on DIALOG,(800) 334-2564, and ORBIT,(800) 456-7248, online services. Also available on CD-ROM. Inquire as to cost and availability.

TRIS (Transportation Research Information). National Academy of Science, 2101 Constitution Avenue, N.W., Washington, DC 20418. (202) 334-3313 or (800) 624-6242. Citations with abstracts of literature on transportation, including air, highway, rail, maritime and other modes from 1968 to present. Available on DIALOG,(800) 334-2564, online service. Inquire as to cost and availability.

WILSONLINE. H.W. Wilson Company, 950 University Avenue, Bronx, NY 10452. (800) 367-6770 or (212) 588-8400. Makes available online versions of the printed H.W. Wilson indexes including Applied Science and Technology Index, Business Periodicals Index, General Science Index, and Readers' Guide to Periodical Literature. Period covered is generally 1983 to present. Available on BRS,(800)289-4277, DIALOG,(800) 334-2564, and OCLC EPIC,(800) 848-5878, online services. Also available on CD-ROM. Inquire as to cost and availability.

PERIODICALS

Aeronautical Journal. The Royal Aeronautical Society, 4 Hamilton Place, London W1V 0BQ, England. Monthly except June and August. 1897 to present. Inquire for price information in U.S.

Aerospace America. American Institute of Aeronautics and Astronautics, 370 L'Enfant Promenade SW, Washington, DC 20024-2518. (202) 646-7471. 1932 to present. Monthly. $70 annually (non-members).

Aerospace Daily. McGraw-Hill Inc., Aviation Week Group, 1200 G Street NW, Suite 200, Washington, DC 20005. (202) 383-2350. Five times a week. $1340.00 per year.

Aerospace Engineering. Society of Automotive Engineers, 400 Commonwealth Dr., Warrendale, PA 15096. (412) 776-4841. 1981 to present. Monthly. $48.00 per year.

AIAA Journal. American Institute of Aeronautics and Astronautics, 370 L'Enfant Promenade SW, Washington, DC 20024-2518. (202) 646-7471. 1963 to present. Monthly. $52.00 per year (members) and $435.00 per year (non-members).

Aviation Mechanics Bulletin. Flight Safety Foundation, Inc., 2200 Wilson Blvd., Suite 500, Arlington, VA 22201-3306. (703) 522-8300. FAX (703) 525-6047. 1953 to present. Bimonthly. $15.00 per year (non-members).

Aviation Week and Space Technology. 1221 Avenue of the Americas, New York, NY 10020. (212) 512-2999 or (800) 525-5003. 1916 to present. Weekly. $82.00 per year.

Avionics. PHillips Publishing Inc., 7811 Montrose Rd., Potomac, MD 20854. (301) 340-2100. FAX (301) 309-3847. Monthly. $85.00 per year.

Avionics News Magazine. Aircraft Electronics Association, Box 1981, Independence, MO 64055. (816) 373-6565. FAX (816) 478-3100. 1959 to present. Monthly. Inquire for price.

Avionics Review. Belvoir Publications Inc., 75 Holly Hill Lane, Box 2626, Greenwich, CT 06836-2626. (203) 661-6111. FAX (203) 661-4802. 1990 to present. Monthly. $96.00 per year.

Canadian Aeronautics and Space Journal. Canadian Aeronautics and Space Institute, Suite 818, the National Building, 130 Slater St., Ottawa, Canada K1P 6E2. 1955 to present. Monthly. $75.00 per year (U.S.).

Flying. Hachette Filipacchi Magazines, Inc., 1633 Broadway, New York, NY 10019. (203) 622-2700. FAX (203) 622-2725. 1927 to present. Monthly. $24.00 per year.

IEEE Transactions On Aerospace and Electronic Systems. Institute of Electrical and Electronics Engineers, 445 Hoes Lane, PO Box 1331, Piscataway, NJ 08855-1331. (908) 981-0060. 1965 to present. Quarterly. $10.00 per issue (members), $20.00 per issue (non-members).

Interavia/Aerospace World. Aerospace Media Publishing,Swissair Centre, 31 Route de l'Aeroport, POB Box 437, 1215 Geneva 15, Switzerland. (022) 788 27 88. FAX (022) 788 27 26. 1987 to present. Monthly. $128.00 per year (U.S.).

Journal of Aircraft. American Institute of Aeronautics and Astronautics, 370 L'Enfant Promenade SW, Washington, DC 20024-2518. 1963 to present. Bimonthly. $42.00 (members) and $230.00 (non-members).

Journal of Guidance, Control and Dynamics. American Institute of Aeronautics and Astronautics, 370 L'Enfant Promenade SW, Washington, DC 20024-2518. 1978 to present. Bimonthly. $42.00 (members) and $250.00 (non-members).

B

BALLISTIC MISSILES

See: GUIDED MISSILES

BALLISTICS

ABSTRACT SERVICES AND INDEXES

Applied Science and Technology Index. A cumulative subject index to Englisanguage Periodicals in the Fields of Aeronautics and Space Science, Computer Technology, Chemistry, Construction Industry, Energy and Related Areas. H.W. Wilson Co., 950 University Avenue, Bronx, NY 10452. (800) 367-6770 or (212) 588-8400. FAX (718) 590-1617. From 1958 to present. Monthly. Inquire about cost and availability. Also available on CD-ROM and online.

Computer Abstracts, MCB University Press Ltd., PO Box 10812, Birmingham, AL 35201-0812. (800) 633-4931. FAX (205) 995-1588. Monthly. Covers computer theory, data, hardware, systems, networks, human-computer interaction, artificial intelligence, as well as applications of computers in aerospace, business, CAD/CAM, cartography, civil engineering, electronics and electrical engineering, industrial engineering, mechanical engineering, medicine, structural engineering, etc.

Current Contents: Engineering, Technology, and Applied Sciences. Institute for Scientific Information, 3501 Market Street, Philadelphia, PA 19104. (215) 386-0100 or (800) 523-1850. FAX (215) 386-2991. 1970 to present. Weekly. $442.00 per year.

Engineering Index Monthly. Engineering Information, Inc., Castle Point on the Hudson, Hoboken, NJ 07030. (800) 221-1044. FAX (212) 832-1857. Monthly. $2300.00 per year. Also available online as COMPENDEX, and also on CD-ROM. Covers chemical engineering, computers, electrical engineering, civil engineering, metals and mining, industrial management, and mechanical engineering.

Physical Review Abstracts. American Physical Society, distributed by the American Institute of Physics, 335 East 45th Street, New York, NY 10017 (212) 661-9404 or FAX (516)349-9704. 1970 to present. Semi-monthly. $290.00 per year.

Physics Abstracts. Institute of Electrical Engineers, Michael Faraday House, Six Hill Way, Stevenage, Herts, SG1 2AY, England. Distributed by IEEE, 445 Hoes Lane, Piscataway, NJ 08854. (908) 562-5549. 1898 to present. Monthly. $2700.00 per year. Also available online as INSPEC.

Science Citation Index. Institute for Scientific Information, 3501 Market Street, Philadelphia, PA 19104. (215) 386-0100 or (800) 523-1850. FAX (215) 386-6362. Inquire about availability and cost. Also available on CD-ROM.

ASSOCIATIONS AND PROFESSIONAL SOCIETIES

American Institute of Aeronautics and Astronautics. 370 L'Enfant Promenade, S.W., Washington, DC 20024. (202) 646-7400.

American Society of Crime Laboratory Directors. ExecuSuites, 15200 Shady Grove Road, Rockville, MD 20850. (702) 363-0219. FAX (407) 471-2561.

DIRECTORIES AND BIOGRAPHICAL SOURCES

Directory of Engineering Societies and Related Organizations. Gordon Davis. 13th edition. American Association of Engineering Societies, 1111 19th Street NW, Suite 608, Washington, DC 20036. (202) 296-2237 or (800) 658-8897. 1989. Inquire for price.

International Research Centers Directory. Gale Research, 835 Penobscot Building, Detroit, MI 48226-4094. (313) 961-2242. (800) 347-4253. FAX (313) 961-6083. 8th edition. 1995. $410.00.

Research Centers Directory. Gale Research, 835 Penobscot Building, Detroit, MI 48226-4094. (313) 961-2242. (800) 347-4253. FAX (313) 961-6083. $485.00.

Scientific and Technical Organizations and Agencies Directory. Gale Research, 835 Penobscot Building, Detroit, MI 48226-4094. (313) 961-2242. (800) 347-4253. FAX (313) 961-6083. 4th edition. 1996. $195.00.

Who's Who in Engineering. American Association of Engineering Societies, 1111 19th Street NW, Suite 608, Washington, DC 20036. (202) 296-2237 or (800) 658-8897. 8th edition. 1991. Inquire for price.

BALLISTICS

Ency. of Physical Sciences and Engineering Info. Sources

ENCYCLOPEDIAS AND DICTIONARIES

Encyclopedia of Physical Science and Technology. Robert A. Meyers, ed. 18 volumes. Academic Press, Inc., 6277 Sea Harbor Drive, Orlando, FL 32887. 1992. $2100.00

McGraw-Hill Dictionary of Scientific and Technical Terms. Sybil P. Parker, ed. 5th edition. McGraw-Hill Publishing Company, 1221 Avenue of the Americas, New York, NY 10020. (800) 262-4729 or (212) 512-3825. 1993. $110.50.

McGraw-Hill Encyclopedia of Engineering. Sybil P. Parker, ed. 2nd edition. McGraw-Hill Publishing Company, 1221 Avenue of the Americas, New York, NY 10020. (800) 262-4729 or (212) 512-3825. 1993. $95.50.

McGraw-Hill Encyclopedia of Science and Technology. Sybil P. Parker, ed. 7th edition. 20 volumes. McGraw-Hill Publishing Company, 1221 Avenue of the Americas, New York, NY 10020. (800) 262-4729 or (212) 512-3825. 1992. $1900.00

GENERAL WORKS

ARL Ballistics Research. Steve Taulbee. U.S. Army Research Laboratory, Adelphi, MD. Order from Superintendent of Documents, U.S. Government Printing office, Box 371954, Pittsburgh, PA 15250-7954. (202) 783-3238. FAX (202) 512-2233. 1993. Inquire for price and availability.

Ballistics Simulation. Michael J. Chinni, editor. Society for Computer Simulation International, 4838 Ronson Court Lane, San Diego, CA 92111-1810. (619) 277-3888. 1990. $32.00.

Ballistics Simulation Two. Michael J. Chinni, editor. Society for Computer Simulation International, 4838 Ronson Court Lane, San Diego, CA 92111-1810. (619) 277-3888. 1991. $32.00.

High Velocity Impact Dynamics. Jonas A. Zukas, editor. John Wiley and Sons, Inc., 605 Third Avenue, New York, NY 10158. (800) 526-5368 or (212) 850-6000. 1990. Inquire for cost and availability.

Interior Ballistics of Guns. Herman Krier & Martin Summerfield, editors. American Institute of Aeronautics and Astronautics, 370 L'Enfant Promenade, S.W., Washington, DC 20024. (202) 646-7400. 1979. $86.95.

Modern Practical Ballistics. Arthur Pejsa. 2nd edition. Kenwood Publishing Company, 2120 Kenwood Parkway, Minneapolis, MN 55405-2326. (612) 374-3337. 1991. $19.95.

The Sporting Ballistics Book. Charles W. Matthews. Bill Matthews Inc., PO Box 26727, Lakewood, CO 80226. (303) 922-0055. 1991. Inquire for price and availability.

HANDBOOKS AND MANUALS

Military Ballistics: A Basic Manual. C. L. Farrar & D.W. Leeming. Pergamon Press Inc., Maxwell House, Fairview Park, Elmsford, NY 10523. (914) 592-7700. Fax (914) 592-3625. 1983. $35.00. Elsevier Science Publishing Company, Inc., 655 Avenue of the Americas, New York, NY 10010. (212) 989-5800. 1983. $22.00 (paperback).

Tables of Bullet Performance. Philip Mannes. Wolfe Publishing Company, 6471 Airpark Drive, Prescott, AZ 86301. (602)

445-7810 or (800) 899-7810. FAX (520) 778-5124. 1980. $17.50.

ONLINE DATABASES AND CD-ROM

Compendex Plus. Engineering Information, Inc., 345 East 47th Street, New York, NY 10017. (212) 705-7600 or (800) 221-1044. Contains citations with abstracts to worldwide literature in engineering and technology, from 1970 to present. Available on online BRS,(800) 289-4277, DIALOG, (800) 334-2564, ORBIT (800) 456-7248, and STN International, FIZ Karlsruhe, P.O. Box 2465, W-7500, Karlsruhe 1, Germany, online services. Also available on CD-ROM. Inquire as to cost and availability.

CSA Engineering. Cambridge Scientific Abstracts, 7200 Wisconsin Avenue, Suite 601, Bethesda, MD 20814. (301) 961-6750 or (800) 843-7751. Contains citations and abstracts of international periodicals and research literature covering all fields of engineering and science and technology,including computer and information science, electronics, mechanical engineering, solid state materials, 1981 to present. Available on BRS,(800) 289-4277, online service.Inquire as to cost and availability.

Current Contents Search. Institute for Scientific Information, 3501 Market Street, Philadelphia, PA 19104. (215) 386-0100. FAX (215) 386-6362. Contains citations to articles listed in the table of contents of science and technology journals. Also articles in social sciences and life sciences journals. Available on BRS,(800) 289-4277, DIALOG,(800) 334-2564, online services. Inquire as to cost and availability.

Dissertation Abstracts. University Microfilms International, 300 North Zeeb Road, Ann Arbor, MI 48106. (800)521-0600 or (313) 761-4700. Scope includes virtually all doctoral dissertations accepted at accredited American institutions from 1861 to present in 252 subject areas. Available on BRS,(800) 289-4277, DIALOG,(800) 334-2564, and OCLC EPIC,(800) 848-5878, online services. Also available on CD-ROM. Inquire as to cost and availability.

NTIS Bibliographic Database. National Technical Information Service, 5285 Port Royal Road, Springfield, VA 22161. (703) 487-4929 or FAX (703) 321-8199. Broad coverage of government-sponsored science and technology research reports, 1964 to present. Available on BRS,(800) 289-4277, DIALOG, (800) 334-2564, ORBIT, (800) 456-7248, and STN International, FIZ Karlsruhe, P.O. Box 2465, W-7500, Karlsruhe 1, Germany, online services. Also available on CD-ROM. Inquire as to cost and availability.

Scientific and Technical Books and Serials in Print. R.R. Bowker Inc., 121 Chanlon Road, New Providence, NJ 07974. (800) 521-8110 or (908) 464-6800. List of currently published books and serials in the physical and biological sciences, engineering and technology, with subject, author and titles indexes. Available on ORBIT,(800) 456-7248, online service.Inquire as to cost and availability.

SCISEARCH. Institute for Scientific Information, 3501 Market Street, Philadelphia, PA 19104. (800) 523-1850 or (215) 386-0100. Broad multidisciplinary title and author index to the international literature of science and technology, 1974 to present. Available on DIALOG,(800) 334-2564, and ORBIT,(800) 456-7248, online services. Also available on CD-ROM. Inquire as to cost and availability.

Wilsonline. H.W. Wilson Company, 950 University Avenue, Bronx, NY 10452. (800) 367-6770 or (212) 588-8400. Makes

available online versions of the printed H.W. Wilson indexes including Applied Science and Technology Index, Business Periodicals Index, General Science Index, and Readers' Guide to Periodical Literature. Period covered is generally 1983 to present. Available on BRS,(800) 289-4277, DIALOG,(800) 334-2564, and OCLC EPIC,(800) 848-5878, online services. Also available on CD-ROM. Inquire as to cost and availability.

PERIODICALS

Aerospace America. American Institute of Aeronautics and Astronautics, 370 L'Enfant Promenade SW, Washington, DC 20024-2518. (202) 646-7471. 1932 to present. Monthly. $70 per year (non-members).

American Journal of Physics. American Association of Physics Teachers, 5112 Berwyn Road, College Park, MD 20740. (301) 345-4200. 1933 to present. Monthly. $215.00 per year.

Forensic Science International. Elsevier Science Inc., Box 882, Madison Square Station, New York, NY 10159. (212) 989-5800. 1972 to present. 15 times a year. $672.00 per year.

IBM Journal of Research and Development. IBMCorporation, 500 Columbus Avenue, Thornwood, NY 10394. (914) 742-5850. 1957 to present. Bi-monthly. $45.00 per year.

IEEE Transactions on Magnetics. IEEE, 345 E. 47th Street, New York, NY 10017. (908) 981-0060. FAX (908) 981-9667. 1965 to present. Bi-monthly. $263.00 per year.

Journal of Aircraft. American Institute of Aeronautics and Astronautics, 370 L'Enfant Promenade SW, Washington, DC 20024-2518. Bimonthly. $42.00 (members) and $230.00 (non-members).

Journal of Propulsion and Power. American Institute of Aeronautics and Astronautics, 370 L'Enfant Promenade SW, Washington, DC 20024-2518. Bimonthly. $38.00 (members) and $260.00 (non-members).

RESEARCH CENTERS AND INSTITUTES

U.S. Army Research Laboratory, 2800 Powder Mill Road, Adelphi, MD 20783. (301) 394-1600.

U.S. Army Research Laboratory, Weapons Technology Directorate. Aberdeen Proving Grounds, MD 21005. (410) 678-6244.

BALLOONS AND BALLOONING

See also: AERONAUTICS, AIRSHIPS

ABSTRACT SERVICES AND INDEXES

Applied Science and Technology Index. A cumulative Aeronautics and Space Science, Computer Technology, Chemistry, Construction Industry, Energy and Related Areas. H.W. Wilson Co., 950 University Avenue, Bronx, NY 10452. (800) 367-6770 or (212) 588-8400. FAX (718) 590-1617. From 1958 to present. Monthly. Inquire about cost and availability. Also available on CD-ROM and online.

International Aerospace Abstracts. Technical Information Service, American Institute of Aeronautics and Astronautics, Inc., 555 West 57th St., New York, NY 10019. (212) 247-6500. FAX (212) 582-4861. Semi-monthly. $1295.00 per year.

Scientific and Technical Aerospace Reports (star). National Aeronautics and Space Administration. NASA Center for Aerospace Information, Box 8757, BWI Airport, Baltimore, MD 21240. (301) 621-0153. Monthly. Inquire about availability and cost. ALso available through the NASA online retrieval service (RECON), and through the Aerospace Database through DIALOG.

ASSOCIATIONS AND PROFESSIONAL SOCIETIES

Aeronaut Society. 1241 High Street, Oakland, CA 94601. (510) 261-4222. FAX (510) 261-7908.

Airship Association. 8512 Cedar Street, Silver Spring, MD 20910. (301) 588-7916. FAX (301) 588-2085.

American Institute of Aeronautics and Astronautics, 370 L'enfant Promenade Sw, Washington, DC 20024-2518. (202) 646-7471.

Association of Balloon and Airship Constructors. Box 90864, San Diego, CA 92169. (619) 270-4049.

Balloon Federation of America. PO Box 400, Indianola, IA 50125. (515) 961-8809. FAX (515) 961-3537.

BIBLIOGRAPHIES

Aeronautical Engineering: A Continuing Bibliography With. INDEXES. National Aeronautics and Space Administration, Center for Aerospace Information, Box 8757, BWI Airport, Baltimore, MD 21240. (301) 621-0153. Inquire for cost and availability.

Scientific and Technical Books and Serials in Print. R.R. Bowker Inc., 121 Chanlon Road, New Providence, NJ 07974. (800) 521-8110 or (908) 464-6800. List of currently published books and serials in the physical and biological sciences, engineering and technology, with subject, author and titles indexes. Inquire as to cost and availability.

DIRECTORIES AND BIOGRAPHICAL SOURCES

American Astronautical Society Directory. American Astronautical Society, 6352 Rolling Mill Place, Suite 102, Springfield, VA 22152. Inquire for cost and availability.

Research Centers Directory. Gale Research, 835 Penobscot Building, Detroit, MI 48226-4094. (313) 961-2242. (800) 347-4253. FAX (313) 961-6083. $485.00.

Who's Who in Engineering. American Association of Engineering Societies, 1111 19th Street NW, Suite 608, Washington, DC 20036. (202) 296-2237 or (800) 658-8897. 8th edition. 1991. Inquire for price.

Who's Who of Ballooning: 1783-1983. Rechs Publications, 10298 Skyview Drive, Grass Valley CA, 95949-9596. 1982. $20.00.

BALLOONS AND BALLOONING

Ency. of Physical Sciences and Engineering Info. Sources

World Aviation Directory. McGraw-Hill Aviation Week Group (NY), 1221 Avenue of the Americas, New York, NY 10020. (215) 237-4112. FAX (212) 586-3232. Semiannual. $250.00.

GENERAL WORKS

Aerospace Balloons. E.J. Kirschner. TAB Books, P.O. Box 40, Blue Ridge Summit, PA 17294-0850. (717) 794-2191 or (800) 233-1128. FAX (717) 794-2080. 1985. $12.95.

Ballooning: A Complete Guide To Riding the Winds. Dick Wirth. David McKay Company/Random House, 201 E. 50th Street MD4-6, New York, NY 10022. (212) 751-2600. FAX (212) 872-8026. 1991. $22.50.

Balloon Technology and Observations. W. Riedler & K.M. Torkar. Pergamon Press Inc., Maxwell House, Fairview Park, Elmsford, NY 10523. (914) 592-7700. Fax (914) 592-3625. 1993. $175.00.

The Challengers: A Century of Ballooning. Diane T. Darnall. Hunter Publishing Company, PO Box 9533, Phoenix, AZ 85068. (602) 944-1022. 1990. Write for price and availability information.

The Eagle Aloft: Two Centuries of the Balloon in America. Tom D. Crouch. Smithsonian Institution Press, 4701 L'Enfant Plaza, Suite 7100, Washington, DC 20560. (202) 287-3738. 1983. $62.00.

Hot Air & Gas: The Basics of Balloons. Ed Newman. Greenway Publishing Company, PO Box 148, Roseland, VA 22967. (804) 277-9349. 1992. Inquire for price and availability.

Race To the Stratosphere: Manned Scientific Ballooning in France. David H. DeVorkin. Springer-Verlag, 175 Fifth Avenue, New York, NY 10010. (212) 460-1500 or (800) 777-4643. FAX (212) 473-6272. 1989. $44.00.

HANDBOOKS AND MANUALS

The Ballooning Manual. Bob Howes. Airlife Publishing/Motorbooks International, 729 Prospect Avenue, Osceola, WI 54020. (800) 826-6600. 1991. Inquire for price and availability.

PERIODICALS

Ad Astra. National Space Society, 922 Pennsylvania Avenue SE, Washington, DC 20003-2140. (202) 543-1900. FAX (202) 546-4189. 1989 to present. 6 times a year. $35.00 a year.

Advanced Lighter-Than-Air Review. Airship Association, 8512 Cedar Street, Silver Spring, MD 20910. (301) 588-7916. FAX (301) 588-2085. Quarterly. Inquire for price information.

Aerostation. Association of Balloon and Airship Constructors. Box 90864, San Diego, CA 92169. (619) 270-4049. 1974 to present. Quarterly. Inquire for price and availability.

Ballooning. Balloon Federation of America, PO Box 400, Indianola, IA 50125. (515) 961-8809. FAX (515) 961-3537. 1977 to present. Quarterly. Inquire for price information.

Flight International. Reed Business Publishing, 475 Park Avenue South, 2nd Floor, New York, NY 10016. (212) 679-8888.

FAX (212) 679-9455. 1909 to present. Weekly. $176.00 per year.

RESEARCH CENTERS AND INSTITUTES

National Scientific Ballooning Facility, Box 319, F/m Road, Palestine, TX 75801. (214) 729-0271. FAX (214) 723-8056.

University of Wyoming High Altitude Balloon Research Group, Physics and Astronomy Department, Box 3905, Laramie, WY 82071. (307) 766-4323.

BANACH ALGEBRA

See: ALGEBRA

BAND SPECTRA

See: SPECTROCOPY

BAND THEORY OF SOLIDS

See: SOLID STATE PHYSICS

BAROMETER

See: METEOROLOGY

BARYON

See: PARTICLE PHYSICS

BASES

See: ACIDS AND BASES

BASIC

See:COMPUTER PROGRAMMING LANGUAGES

BATHOLITH

See: VOLCANOLOGY

Ency. of Physical Sciences and Engineering Info. Sources

BATTERIES

BATHYMETRY

See: OCEANOGRAPHY

BATTERIES

See also: ELECTROCHEMISTRY

ABSTRACT SERVICES AND INDEXES

Applied Science and Technology Index. A cumulative subject index to English language periodicals in the fields of aeronautics and space science, computer technology, chemistry, construction industry, energy and related areas. H.W. Wilson Co., 950 University Avenue, Bronx, NY 10452. (212) 588-8400. (800) 367-6770. FAX (718) 590-1617. From 1958 to present. Monthly. Inquire about cost and availability. Also available on CD-ROM and online. ISSN: 0003-6986.

Chemical Abstracts. Chemical Abstracts Service, 2540 Olentangy River Road, Box 3012, Columbus, OH 43210-0012. (614) 447-3600. (800) 848-6538. FAX (614) 447-3713. From 1907 to present. Weekly. $16800.00 per year. Also available on CD-ROM and online. Inquire regarding cost and availability.

Engineering Index Monthly. Engineering Information, Inc., Castle Point on the Hudson, Hoboken, NJ 07030. (201) 216-8500. FAX (201) 216-8532. Monthly. $2200.00 per year. Also available online as COMPENDEX and on CD-ROM. ISSN: 0742-1974.

General Science Index. H.W. Wilson Company, 950 University Avenue, Bronx, NY 10452. (212) 588-8400. (800) 367-6770. FAX (718) 590-1617. From 1978 to present. Ten issues per year; quarterly and annual cumulations. Service basis. Available on CD-ROM and online. Inquire regarding cost and availability. ISSN: 0162-1963.

Physics Abstracts. INSPEC. Section A, Science Abstracts. in stitution of Electrical Engineers (IEE). Available from INSPEC/ IEEE - Institute of Electrical and Electronic Engineers,, Box 1331, 445 Hoes Lane, Piscaway NJ 08855-1331. (908) 562-5549. 1898 to present. 24 issues per year. $2835.00 per year. Also available on CD-ROM and online. ISSN: 0036-8091.

Science Citation Index. SCI. Institute for Scientific Information, 3501 Market Street, Philadelphia, PA 19104. (215) 386-0100. (800) 523-1850. FAX (215) 386-2991. 1961 to present. Six issues per year, plus annual cumulation. $11650.00 per year. Also available online and on CD-ROM. Inquire about price and availability. ISSN: 0036-827X.

ASSOCIATIONS AND PROFESSIONAL SOCIETIES

American Chemical Society. 1155 16th Street, NW, Washington DC 20036. (202) 872-4600.

American Institute of Physics. One Physics Ellipse, College Park, MD 20740-3843. (301) 209-3100.

Association of Consulting Chemists and Chemical Engineers. 40 West 45th Street, New York, NY 10036. (212) 983-3160.

Battery Council International. 401 North Michigan Drive, Chicago, IL 60611. (312) 644-6610.

Electrochemical Society. 10 South Main Street, Pennington, NJ 08534. (609) 737-1902.

Independent Battery Manufacturers Association. 100 Larchwood Drive, Largo FL 34640. (813) 586-1408. FAX (813)586-1400.

DIRECTORIES AND BIOGRAPHICAL SOURCES

American Men and Women of Science: Physical and Biological Sciences. R. R. Bowker Inc., 121 Chanlon Road, New Providence, NJ 07974. (908) 464-6800. (800) 521-8110. 20th edition. 8 volumes. 1996. $850.00.

Electrochemical Society Membership Directory. Electrochemical Society, 10 South Main Street, Pennington, NJ 08534. (609) 737-1902. Annual. Available to members only.

Research Centers Directory. Gale Research, 835 Penobscot Building, Detroit, MI 48226-4094. (313) 961-2242. (800) 347-4253. 20th edition. 1995. $485.00. ISSN: 0080- 1518.

Scientific and Technical Organizations and Agencies Directory. Gale Research, 835 Penobscot Building, Detroit, MI 48226-4094. (313) 961-2242. (800) 347-4253. 4th edition. 1996. $195.00.

Slig Buyers Guide. Independant Battery Manufacturers Association, 100 Larchwood Drive, Largo, FL 34640. (813) 586-1408. FAX (813) 586-1400. Biennial, odd years. $12.00.

Who's Who in Engineering. Gordon Davis, editor. American Society for Engineering Education, 1111 19th Street, NW, Suite 608, Washington, DC 20036. 9th edition. 1995. $220.00.

Who's Who in Technology. Gale Research, 835 Penobscot Building, Detroit, MI 48226-4094. (313) 961-2242. (800) 347-4253. 1995. $195.00.

ENCYCLOPEDIAS AND DICTIONARIES

Encyclopedia of Applied Physics. George Trigg, editor. VCH Publications, Inc., 220 East 23rd Street, Suite 909, New York, NY 10010-4606. (212) 683-8333. (800) 422-8824. 20 volume set. 1991-. $5990.00.

McGraw-Hill Encyclopedia of Science and Technology. McGraw- Hill Book Company, Inc., 1221 Avenue of the Americas, New York, NY 10020. (212) 512-2000. (800) 262-4729. 7th edition. 20 volume set. 1992. $1900.00.

Modern Power Suply and Battery Charger Circuit Encyclopedia. Rudolf F. Graf. McGraw-Hill Book Company, Inc., 1221 Avenue of the Americas, New York, NY 10020. (212) 512- 2000. (800) 262-4729. 1992. $11.95 in paper.

GENERAL WORKS

Battery Hazards and Accident Prevention. S. C. Levy and P. Bro. Plenum Publishing Corp., 233 Spring Street, New York, NY. (212) 620-8000. (800) 2221-9369. FAX (212) 463-0742. 1994. $49.50.

BATTERIES

Ency. of Physical Sciences and Engineering Info. Sources

Fuel Cell Systems. L. J. Blomen and M. N. Bugerwa, editors.

Plenum Publishing Corp., 233 Spring Street, New York, NY. (212) 620-8000. (800) 2221-9369. FAX (212) 463-0742. 1994. $125.00.

Hydrogen Storage Materials, Batteries and Electrochemistry. D. A. Corrigan and S. Srivivasan, editors.Electrochemical Society, 10 South Main Street, Pennington, NJ 08534. (609) 737-1902. 1992. $43.00.

Modern Battery Technology. Clive D. Tuck. Prentice Hall, 113 Sylvan Avenue, Route 9W, Englewood Cliffs, NJ 07632. (201) 592-2000. (800) 922-0579. 1991. $42.00.

Quality and Reliability Methods For Primary Batteries. P. Bro and S. C. Levy. John Wiley & Sons, Inc., 605 Third Avenue, New York, NY 10158-0012. (212) 850-6000. (800) 225-5945. 1990. $122.00.

HANDBOOKS AND MANUALS

Battery Reference Book. T. R. Compton. Butterworth-Heinemann, 313 Washington Street, Newton, MA 02158. (618-928-2500. (800) 366-2665. 1990. $195.00.

Battery Technology Handbook. Kiehne. Marcel Dekker, Inc., 270 Madison Avenue, New York, NY 10016. (212) 696-9000. (800) 228-1160. 1989. $165.00.

CRC Handbook of Chemistry and Physics. David R. Kide, editor. CRC Press, Inc., 2000 Corporate Boulevard, NW, Boca Raton, FL 33431. (407) 994-0555. (800) 272-7737. 77th edition. 1996. $99.95.

Handbook of Batteries. David Linden. McGraw-Hill Publishing Company, Inc., 1221 Avenue of the Americas, New York, NY 10020. (212) 512-2000. (800) 262-4729. 2nd edition. 1995. $125.00.

Handbook of Solid State Batteries and Capacitors. M. Z. Munshi and P. S. Prasad. World Scientific Publishing Company, Inc., 1060 Main Street, River Edge, NJ 07661. (201) 487-9655. (800) 227-7562. 1995. $177.00.

Rechargeable Batteries Applications Handbook. Gates Butterworth-Heinemann, 313 Washington Street, Newton, MA 02158. (618-928-2500. (800) 366-2665. 1992. $49.95.

Standard Handbook For Electrical Engineers. Donald Fink. McGraw-Hill Publishing Company, 1221 Avenue of the Americas, New York, NY 10020. (212) 512-2000. (800) 262- 4729. 13th edition. 1996. $110.00.

ONLINE DATABASES AND CD-ROMS

CA Search. Chemical Abstracts Service, P.O. Box 3012, Columbus, OH 43210-0012. (614) 447-3600. (800) 848-6533. FAX (614) 447-3709. Very comprehensive guide to worldwide chemical literature and related fields, 1972 to present. Available on BRS,(800) 289-4277, DIALOG, (800) 334-2564, ORBIT (800) 456-7248, and STN International, FIZ Karlsruhe, P.O. Box 2465, W-7500, Karlsruhe 1, Germany, online services. Inquire as to cost and availability.

Compendex Plus. Engineering Information, Inc., 345 East 47th Street, New York, NY 10017. (212) 705-7600 or (800) 221-1044. Contains citations with abstracts to worldwide literature in engineering and technology, from 1970 to present. Available on online BRS,(800) 289-4277, DIALOG, (800) 334-2564, ORBIT (800) 456-7248, and STN International, FIZ Karlsruhe, P.O.Box 2465, W-7500, Karlsruhe 1, Germany, online services. Also available on CD-ROM. Inquire as to cost and availability.

INSPEC. Institution of Electrical Engineers, Michael Faraday House, Six Hills Way, Stevenage, Herts. SG1 2AY, England. Telephone: 0438 313311 or FAX 0438 742840. Contains citations to the worldwide literature of physics, electronics and electrical engineering, computer technology, and related fields. Available on BRS, (800) 289-427 7, DIALOG, (800) 334-2564, ORBIT, (800) 456-7248, and STN International, FIZ Karlsruhe, P.O. Box 2465, W-7500, Karlsruhe 1, Germany, online services. Inquire as to cost and availability.

NTIS Bibliographic Database. National Technical Information Service, 5285 Port Royal Road, Springfield, VA 2216 1. (703) 487-4929 or FAX (703) 321-8199. Broad cover age of government-sponsored science and technology research reports, 1964 to present. Available on BRS,(800) 289-4277, DIALOG, (800) 334-2564, ORBIT, (800) 456-7248, and STN International, FIZ Karlsruhe, P.O. Box 2465, W-7500, Karlsruhe 1, Germany, online services. Also available on CD-ROM. Inquire as to cost and availability.

SCISEARCH. Institute for Scientific Information, 3501 MA rket Street, Philadelphia, PA 19104. (800) 523-1850 or (215) 386-0100. Broad multidisciplinary title and author index to the international literature of science and technology, 1974 to present. Available on DIALOG, (800) 334-2564, and ORBIT, (800) 456-7248, online services. Also available on CD-ROM. Inquire as to cost and availability.

Wilsonline. H.W. W ilson Company, 950 University Avenue, Bronx, NY 10452. (800) 367-6770 or (212) 588-8400. Makes available onlin e versions of the printed H.W. Wilson indexes includin g Applied Science and Technology Index, Business Periodicals Index, General Science Index, and Readers' Guide to Periodical Literature. Period covered is generally 1983 to present. Available on BRS, (800) 289-4277, DIALOG, (800) 334-2564, and OCLC EPIC, (800) 848-5878, online services. Also available on CD-ROM. Inquire as to cost and availability.

PERIODICALS

Advanced Battery Technology. Robert Morey Associates, Box 30 Cooperstown, NY 13326-0030. (607) 547-5314. Monthly. $120.00 per year. ISSN: 0001-8627.

Electrochemical Society Journal. Electrochemical Society, 10 South Main Street, Pennington, NJ 08534-2896. (609) 737-1902. 1902 to present. Monthly. $400.00 per year. ISSN: 0013-4651.

Electrochemica Acta. Elsevier Science, 660 White Plains Road, Tarrytown, NY 10591-5151. (914) 524-9200. FAX (914) 333-2444. 18 issues per year. $1650.00 per year. ISSN: 0013- 4686.

Electronic Chemicals News. Chemical Week Associates, 888 Seventh Avenue, New York, NY 10106. (212) 621-4900. 1986 to present. Semi-monthly. $452.00 per year. ISSN: 0886-5671.

Battery Technology Center. Carnegie Mellon University, 4400 Fifth Avenue, Mellon Institute, Pittsburgh, PA 15213-2683. (412) 268-3388. FAX (412) 268-6945.

Electrochemical Analysis Diagnostic Laboratory. Argonne National Laboratory, 9700 South Cass Avenue, Argonne, IL 60439. (708) 972-7764. FAX (708) 972-4176.

Lawrence Berkeley Laboratory. Applied Science Division, 1 Cyclotron Road, Berkeley, CA 94720. (415) 486-5001. FAX (415) 486-5172.

BEACHES

See: COASTS

BEACONS

See: NAVIGATION

BEAM COLUMNS

See: STRUCTURAL ENGINEERING

BEAM FOIL SPECTROSCOPY

See: SPECTROSCOPY

BEAMS

See: STRUCTURAL ENGINEERING

BEARING CAPACITY

See: STRUCTURAL ENGINEERING

BEARINGS AND BALL BEARINGS

ABSTRACTS AND INDEXES

Alloys Index. American Society for Metals, Metals Park, OH 44073. (216) 338-5151. FAX (216) 338-4634. 1974 to present. Monthly. $295.00.

Applied Mechanics Reviews: An Assessment of World Literature in Engineering Sciences. American Society of Mechanical Engineers, 345 East 47th Street, New York, NY 10017. (212) 705-7703. 1948 to present. Monthly. $360.00 per year.

Applied Science and Technology Index. A cumulative subject index to Englisanguage Periodicals in the Fields of Aeronautics and Space Science, Computer Technology, Chemistry, Construction Industry, Energy and Related Areas. H.W. Wilson Co., 950 University Avenue, Bronx, NY 10452. (800) 367-6770 or (212) 588-8400. FAX (718) 590-1617. From 1958 to present. Monthly. Inquire about cost and availability. Also available on CD-ROM.

Current Contents: Enginering, Technology, and Applied Sciences. Institute for Scientific Information, 3501 Market Street, Philadelphia, PA 19104. (215) 386-0100. FAX (215) 386-6362. Weekly. $360.00 per year.

Engineering Index Monthly. Engineering Information, Inc., Castle Point on the Hudson, Hoboken, NJ 07030. (800) 221-1044. FAX (212) 832-1857. Monthly. $2200.00 per year. Also available online as COMPENDEX, and also on CD-ROM.

ISMEC: Mechanical Engineering Abstracts. Cambridge Scientific Abstracts, 7200 Wisconsin Avenue, Suite 601, Bethesda, MD 20814. (301) 961-6750 or (800) 843-7751. Contains citations to the literature in mechanical engineering, industrial andproduction engineering, energy, power, mechanics, devices and related areas, from 1973 to present. Inquire for price and availability. Also available online.

Metals Abstracts and Metals Abstracts Index. American Society for Metals, Metals Park, OH 44073. (216) 338-5151. 1968 to present. Monthly. Abstracts are $1500.00 per year and Index is $460.00 per year.

Physics Abstracts. Institute of Electrical Engineers, Michael Faraday House, Six Hill Way, Stevenage, Herts, SG1 2AY, England. Distributed by IEEE, 445 Hoes Lane, Piscataway, NJ 08854. (908) 562-5549. 1898 to present. Monthly. $2700.00 per year. Also available online as INSPEC.

Tribology and Corrosion Abstracts. Elsevier Science Publishing Company, Inc., 655 Avenue of the Americas, New York, NY 10010. (212) 989-5800. Covers literature and research in friction, wear, lubrication and lubricants, and corrosion problems and protection, 1984 to present. Thirteen times per year. $450.00 per year.

Tribos - Tribology Abstracts. STI Limited, 4 Kings Meadow, Ferry Hinksey Road, Oxford, OX2 ODU, England. Distributed in the United States by Air Science Company, Box 143, Corning NY 14830. (607) 962-5591. Contains citations of literature and research in friction, wear, lubrication and lubricants, and corrosion problems and protection, from 1967 to present. Monthly. $375.00 per year.

ASSOCIATIONS AND PROFESSIONAL SOCIETIES

American Bearing Manufacturing Association. 1101 Connecticut Avenue, Suite 700, Washington, DC 20036. (202) 429-5155. FAX (202) 223-4579.

Ball Manufacturers Engineers Committee. 1101 Connecticut Avenue NW, Suite 700, Washington, DC 20036-4303. (202) 429-5155.

BEARINGS AND BALL BEARINGS

Ency. of Physical Sciences and Engineering Info. Sources

Bearing Specialists Association. Building C, Suite 20, 800 Roosevelt Road, Glen Elyn, IL 60137. (708) 858-3838. FAX (708) 790-3095.

Roller Bearing Engineers Committee. 1101 Connecticut Avenue NW, Suite 700, Washington, DC 20036-4303. (202) 429-1555.

BIBLIOGRAPHIES

Bibliographical Guide To Technology. Two volumes. G.K. Hall & Company, 866 Third Avenue, New York, NY 10022. (212) 702-6789. (800) 257-5755. FAX (212) 605-9350. 1990. $350.00.

DIRECTORIES AND BIBLIOGRAPHIC SOURCES

Research Centers Directory. Gale Research, 835 Penobscot Building, Detroit, MI 48226-4094. (313) 961-2242. (800) 347-4253. FAX (313) 961-6083. $485.00.

Who's Who in Engineering. American Association of Engineering Societies, 1111 19th Street NW, Suite 608, Washington, DC 20036. (202) 296-2237 or (800) 658-8897. 8th edition. 1991. Inquire for price.

ENCYCLOPEDIAS AND DICTIONARIES

Encyclopedia of Physical Science and Technology. Robert A. Meyers, ed. 18 volumes. Academic Press Inc., 6277 Sea Harbor Drive, Orlando, FL 32887. (800) 321-5068. 1992. $2100.00

McGraw-Hill Dictionary of Scientific and Technical Terms. Sybil P. Parker, ed. 5th edition. McGraw-Hill Publishing Company, 1221 Avenue of the Americas, New York, NY 10020. (800) 262-4729 or (212) 512-3825. 1993. $110.50.

McGraw-Hill Encyclopedia of Engineering. Sybil P. Parker, ed. 2nd edition. McGraw-Hill Publishing Company, 1221 Avenue of the Americas, New York, NY 10020. (800) 262-4729 or (212) 512-3825. 1993. $95.50.

McGraw-Hill Encyclopedia of Science and Technology. Sybil P. Parker, ed. 7th edition. 20 volumes. McGraw-Hill Publishing Company, 1221 Avenue of the Americas, New York, NY 10020. (800) 262-4729 or (212) 512-3825. 1992. $1900.00

GENERAL WORKS

Ball and Roller Bearings. P.S. Houghton. Elsevier Applied Science, 660 White Plains Road, Tarrytown, NY 10591-5153. (914) 524-9200. FAX (914) 333-2444. 1976. $147.75.

Ball and Roller Bearings: Theory, Design, and Application. Paul Eschmann, et al. Books on Demand, 300 N. Zeeb Road, Ann Arbor, MI 48106-1346. (313) 761-4700 or (800) 521-0600. 1985. $131.10.

HANDBOOKS AND MANUALS

Bearings: A Tribology Handbook. M.J. Neale, editor. SAE International, 400 Commonwealth Drive, Warrendale, PA 15096-0001. (412) 776-4970. FAX (412) 776-0790. 1993. $39.00.

I.B.I. Guide Bearings. S.H. Friedman. 12th edition. 2 volumes. Interchange Inc., PO Box 16244, St. Louis Park, MN 55416. (612) 929-6669 or (800) 669-6208. $225.00.

ONLINE DATABASES AND CD-ROMS

Compendex Plus. Engineering Information, Inc., 345 East 47th Street, New York, NY 10017. (212) 705-7600 or (800) 221-1044. Contains citations with abstracts to worldwide literature in engineering and technology, from 1970 to present. Available on online BRS,(800) 289-4277, DIALOG, (800) 334-2564, ORBIT (800) 456-7248, and STN International, FIZ Karlsruhe, P.O. Box 2465, W-7500, Karlsruhe 1, Germany, online services. Also available on CD-ROM. Inquire as to cost and availability.

CSA Engineering. Cambridge Scientific Abstracts, 7200 Wisconsin Avenue, Suite 601, Bethesda, MD 20814. (301) 961-6750 or (800) 843-7751. Contains citations and abstracts of international periodicals and research literature covering all fields of engineering and science and technology,including computer and information science, electronics, mechanical engineering, solid state materials, 1981 to present. Available on BRS,(800) 289-4277, online service.Inquire as to cost and availability.

Current Contents Search. Institute for Scientific Information, 3501 Market Street, Philadelphia, PA 19104. (215)386-0100. FAX (215) 386-6362. Contains citations to articles listed in the table of contents of science and technology journals. Also articles in social sciences and life sciences journals. Available on BRS,(800) 289-4277, DIALOG,(800) 334-2564, online services. Inquire as to cost and availability.

ISMEC: Mechanical Engineering Abstracts. Cambridge Scientific Abstracts, 7200 Wisconsin Avenue, Suite 601, Bethesda, MD 20814. (301) 961-6750 or (800) 843-7751. Contains citations to the literature in mechanical engineering, industrial and production engineering, energy, power, mechanics, devices and related areas, from 1973 to present. Available on the DIALOG,(800) 334-2564, online service. Inquire as to cost and availability.

Metadex. Jointly produced by ASM International and the Institute of Materials. Contains more than 925,000 records from the international literature on metals and alloys, concerning properties, processes, materials classes, applications, and metallurgical systems. Updated monthly. Available from ORBIT-QUESTEL (703) 442-0900.

NTIS Bibliographic Database. National Technical Information Service, 5285 Port Royal Road, Springfield, VA 22161. (703) 487-4929 or FAX (703) 321-8199. Broad coverage of government-sponsored science and technology research reports, 1964 to present. Available on BRS,(800) 289-4277, DIALOG, (800) 334-2564, ORBIT, (800) 456-7248, and STN International, FIZ Karlsruhe, P.O. Box 2465, W-7500, Karlsruhe 1, Germany, online services. Also available on CD-ROM. Inquire as to cost and availability.

SCISEARCH. Institute for Scientific Information, 3501 Market Street, Philadelphia, PA 19104. (800) 523-1850 or (215) 386-0100. Broad multidisciplinary title and author index to the international literature of science and technology, 1974 to present. Available on DIALOG,(800) 334-2564, and ORBIT,(800) 456-7248, online services. Also available on CD-ROM. Inquire as to cost and availability.

Wilsonline. H.W. Wilson Company, 950 University Avenue, Bronx, NY 10452. (800) 367-6770 or (212) 588-8400. Makes

available online versions of the printed H.W. Wilson indexes including Applied Science and Technology Index, Business Periodicals Index, General Science Index, and Readers' Guide to Periodical Literature. Period covered is generally 1983 to present. Available on BRS,(800) 289-4277, DIALOG,(800) 334-2564, and OCLC EPIC,(800) 848-5878, online services. Also available on CD-ROM. Inquire as to cost and availability.

PERIODICALS

American Machinist. Penton Publishing, 110 Superior Ave., Cleveland, OH 44114-2543. (216) 696-7000. FAX (216) 696-0177. 1877 to present. Monthly. $65.00 per year.

Journal of Applied Mechanics. American Society of Mechanical Engineers, 345 E. 47th Street, New York, NY 10017.(212) 705-7722. 1935 to present. Quarterly. $120.00.

Journal of Sound & Vibration. Harcourt Brace & Company, Ltd., Foots Cray High Street, Sidcup, Kent DA14 5HP, England. Telephone 44-81-300-3322. FAX 44-81-309-0807. 1964 to present. 50 times a year. Inquire for cost.

Journal of Tribology. American Society of Mechanical Engineers, 345 East 47th Street, New York, NY 10017-2398. (212) 705-7703. 1967 to present. Quarterly. $120.00 per year for non-members.

Lubrication Engineering. Society of Tribologists and Lubrication Engineers, 840 Busse Highway, Park Ridge, IL 60068-2376. (708) 825-5536. FAX (708) 825-1456. 1945 to present. Monthly. $61.00 per year.

Machine Design. Penton Publishing, 110 Superior Ave., Cleveland, OH 44114-2543. 1929 to present. 6 times a year. $180.00 per year.

Shock and Vibration Digest. The Vibration Institute, 6262 S. Kingery Highway, Suite 212, Willowbrook, IL 60514. (708) 654-2254. FAX (708) 654- 2271. 1969 to present. Bi-monthly. $250.00 per year.

Tribology International. Butterworth-Heinemann, Linacre House, Jordan Hill, Oxford OX2 8DP, England. Telephone 0865-310366. FAX 0865-310898. 1968 to present. Bi-monthly. Inquire for cost.

Wear. Elsevier Science Inc., Box 882, Madison Square Station, New York, NY 10159. (212) 989-5800. 1958 to present. 20 times a year. $2585.00.

RESEARCH CENTERS AND INSTITUTES

Gear and Bearing Center. IITRI, 10 W. 35th Street, Chicago, IL 60616-3799. (312) 567-4327. FAX (312) 567-4030.

University of Virginia Center For Magnetic Bearings. MAE Department, Thornton Hall, McCormick Road, Charlottesville, VA 22901. (804) 924-6210. FAX (804) 982-2037.

BEDROCK

See also: CRUST, EARTHQUAKE ENGINEERING, EARTHQUAKES, GEOLOGY, GEOPHYSICS, GEOTECHNICAL

ENGINEERING, PHYSICAL GEOLOGY, PLANETARY SCIENCES, PLATE TECTONICS

ABSTRACT SERVICES AND INDEXES

Bibliography and Index of Geology. American Geological Institute, 4220 King Street, Alexandri a, VA 22302. (703) 379-2480. Fax (703) 379-7563. Monthly. $1295.00 per year. Also available online as GEOREF.

Deep-Sea Research. Part A: Oceanographic Research Papers. Deep-Sea Research. Part B: Oceanographic Literature Review. Pergamon Press, Inc., Maxwell House, Fairview Park, Elmsford, NY 10523. (914) 592-7700. Fax (914) 592-3625. Twelve times per year. $1370 per year for Parts A and B.

General Science Index. H.W. Wilson Co., 950 University Avenue, Bronx, NY 10452. (800) 367-6770 or (212) 588-8400. Inquire about price and availability.

Geological Abstracts. Elsevier - Geo Abstracts, Regency House, 34 Duke Street, Norwich NR3 3AP, England. Monthly. $760.00 per year. Also available online as GEOBASE.

Geophysics Abstracts. Pergamon Press, Inc., Maxwell House, Fairview Park, Elmsford, NY 10523. (914) 592-7700. Fax (914) 592-3625. Twelve times per year. $565.00 per year. Also available in microform.

Mineralogical Abstracts. Mineralogical Society and the Mineralogical Society of America, 41 Queen's Gate, London, SW7 5HR, England. Quarterly. $235.00 per year.

Oceanic Abstracts. Cambridge Scientific Abstracts, 7200 Wisconsin Avenue, Bethesda, MD 20814. (301) 961-6750. Fax (301) 961-6720. Bimonthly. $995.00 per year.

Science Citation Index. Institute for Scientific Information, 3501 Market Street, Philadelphia, PA 19104. (800) 523-1850 or (215) 386-0100. Inquire about price and availability.

ANNUAL REVIEWS AND YEARBOOKS

Advances in Geophysics. Academic Press, Inc., 1250 Sixth Avenue, San Diego, CA 92101. (619) 231-0926. FAX (619) 699-6715. Irregular. Inquire about price and availability.

Annual Review and Earth and Planetary Sciences. Annual Reviews, Inc., 4139 El Camino Way, Palo Alto, CA 94306-0897. (415) 493-4400. Fax (415) 855-9815. Annual. $55.00 per year.

Minerals Yearbook. Bureau of Mines, U.S. Department of the Interior. Available from U.S. Government Printing office, Washington, DC 20402. (202) 783-3238. Annual. Three volumes. $45.00.

ASSOCIATIONS AND PROFESSIONAL SOCIETIES

American Association of Petroleum Geologists. P.O. Box 979, Tulsa, OK 74101. (918) 584-2555.

American Geological Institute. 4220 King Street, Alexandria, VA 22302. (703) 379-2480. Fax (703) 379-7563.

BEDROCK

Ency. of Physical Sciences and Engineering Info. Sources

American Geophysical Union, 2000 Florida Avenue, N.W., Washington, DC 20009. (202) 462-6900.

Association of Engineering Geologists. 323 Boston Post Road, Suite 2D, Sudbury, MA 01776. (508) 443-4639.

Geochemical Society. Department of Terrestrial Magnetism, 5241 Broad Branch Road, N.W., Washington, DC 20015. (202) 686-4370.

Geological Society of America. 3300 Penrose Place, P.O. Box 9140, Boulder, CO 80301-9140. (303) 447-2020.

Society of Exploration Geophysicists. P.O. Box 702740, Tulsa, OK 74170. (918) 493-3516.

Society of Independent Professional Earth ScieNTISts. 4925 Greenville Avenue, Suite 170, Dallas, TX 75206. (214) 363-1780.

BIBLIOGRAPHIES

Geologic Reference Sources: A Subject and Regional Bibliography of Publications and Maps in the Geological Sciences. Dederick C. Ward et al. Scarecrow Press, Inc., 52 Liberty Street, Box 4167, Metuchen, NJ 08840. (800) 537-7107 or (201) 548-8600. 1981. $49.50.

Information Sources in the Earth Sciences. David N. Wood, Joan E. Hardy, and Anthony P. Harvey, editors. Bowker-Saur/K.G. Saur. Distributed by R.R. Bowker, 121 Chanlon Road, New Providence, NJ 07974. (800) 521-8110 or (908) 464-6800. Second edition. 1989. $85.00.

Scientific and Technical Books and Serials in Print; An Index To Literature in Science and Technology. R.R. Bowker Co., 205 E. 42nd Street, New York, NY 10017. (800) 521-8110 or (212) 916-1600. 1992. $250.00.

DIRECTORIES AND BIOGRAPHICAL SOURCES

American Institute of Professional Geologists. Membership Directory. American Institute of Professional Geologists, 7828 Vance Drive, Suite 103, Arvada, CO 80003. (303) 431-0831. Annual. $18.00.

Association of Engineering Geologists. Membership directory. Association of Engineering Geologists, 323 Boston Post Road, Suite 2D, Sudbury, MA 01776. (508) 443-4639. Annual. $18.00.

Directory of Geoscience Departments: United States and Canada. American Geological Institute. 4220 King Street, Alexandria, VA 22302-1507. (703) 379-2480. Fax (703) 379-7563. Twenty-six edition. 1987. $22.00.

Geological Society of America. Membership Directory. Geological Society of America, 3300 Penrose Place, Boulder, CO 80301. (303) 447-2020. Annual. Available to members only.

Research Centers Directory. Gale Research, 835 Penobscot Building, Detroit, MI 48226-4094. (800) 347-4253 or (313) 961-2242. 20th edition, 1992. $485.00.

ENCYCLOPEDIAS AND DICTIONARIES

Dictionary of Geological Terms. American Geological Institute. Doubleday and Company, Inc., 245 Park Avenue, New York, NY 10017. (800) 645-6156 or (212) 953-4561. Third edition. 1984. $12.00 in paper.

Dictionary of Petrology. S.I. Tomkeieff. John Wiley and Sons, Inc., 605 Third Avenue, New York, NY 10158. (800) 526-5368 or (212) 850-6000. 1983. $180.00.

Encyclopedia of Earth System Science. Academic Press, Inc., 1250 Sixth Street, San Diego, CA 92101. (619) 231-0926 or (800) 321-5068. Four volumes, 1992. $950.00.

Encyclopedia of Igneous and Metamorphic Petrology. D.R. Bowes. Van Nostrand Reinhold, 115 Fifth Avenue, New York, NY 10003. (212) 254-3232 or (800) 926-2665. 1989. $125.95.

Encyclopedia of Minerals. Willard L. Roberts, Thomas J. Campbell, and George R. Rapp, editors. Van Nostrand Reinhold, 115 Fifth Avenue, New York, NY 10003. (212) 254-3232 or (800) 926-2665. Second edition. 1990. $100.00.

Facts On File Dictionary of Geology and Geophysics. Dorothy Farris and Donald R. Coates, editors. Facts on File, Inc., 460 Park Avenue South, New York, NY 10016-7382. (800) 322-8755. Fax (800) 683-3633. 1989. $24.95.

Glossary of Geology. Robert L. Bates and Julia A. Jackson. American Geological Institute, 4220 King Street, Alexandria, VA 22032. (703) 379-2480. Third edition. 1987. $75.00.

McGraw-Hill Encyclopedia of Science and Technology. McGraw-Hill Publishing Company, 1221 Avenue of the Americas, New York, NY 10020. (800) 262-4729 or (212) 512-3825. Seventh edition. 1992. $1900.00.

Magill's Survey of Science: Earth Science Series. Salem Press, Inc., P.O. Box 1097, Englewood Cliffs, NJ 07632. (800) 221-1592 or (201) 871-3700. Five volumes. 1990. $400.00 for the set.

GENERAL WORKS

Atlas of Igneous Rocks and their Textures. W.S. Mackenzie, C.H. Donaldson, and C. Guilford. John Wiley and Sons, Incorporated, 605 Third Avenue, New York, NY 10158. (800) 526-5368 or (212) 850-6000. 1982. $29.95.

Bedrock Geology of the World. W.H. Freeman and Company, 41 Madison Avenue, East 26th, 35th Floor, New York, NY 10010. (212) 576-9400. FAX (212) 689-2383. 1995. Inquire for price.

Dynamic Earth: An Introduction To Physical Geology. B.J. Skinner and Stephen C. Porter. John Wiley and Sons, Inc., 605 Third Avenue, New York, NY 10158. (800) 526-5368 or (212) 850-6000. 1990. $51.95.

Plate Tectonics and Crustal Evolution. K.C. Condie. Pergamon Press, Incorporated, Maxwell House, Fairview Park, Elmsford, NY 10523. (914) 592-7700. 1982. $39.95 in paper.

Structural Geology of Rocks and Regions. George H. Davis. John Wiley and Sons, Incorporated, 605 Third Avenue, New

York, NY 10158. (800) 526-5368 or (212) 850-6000. 1984. $39.95.

HANDBOOKS AND MANUALS

Field Mapping For Geology Students. F. Ahmed and D.C. Almond. Unwin Hyman, Inc., 10 East 53rd Street, New York, NY 10022. (212) 207-7626 or (800) 242-7737. 1983. $18.95.

Geology in the Field. R.R. Compton. John Wiley and Sons, Inc., 605 Third Avenue, New York, NY 10158. (800) 526-5368 or (212) 850-6418. 1985. $37.95.

A Manual of Geology For Civil Engineers. John A. Pitts. World Scientific Publishing Company, Inc., 1060 Main Street, Unit B, River Edge, NJ 07661. (800) 227-7562 or (201) 487-9655. 1985. $38.00.

The Mapping of Geological Structures. K.R. McClay. John Wiley and Sons, Inc., 605 Third Avenue, New York, NY 10158. (800) 526-5368 or (212) 850-6000. 1988. $18.95.

ONLINE DATABASES AND CD-ROMS

CA Search. Chemical Abstracts Service, P.O. Box 3012, Columbus, OH 43210-0012. (614) 447-3600. (800) 848-6533. FAX (614) 447-3709. Very comprehensive guide to worldwide chemical literature and related fields, 1972 to present. Available on BRS,(800) 289-4277, DIALOG, (800) 334-2564, ORBIT (800) 456-7248, and STN International, FIZ Karlsruhe, P.O. Box 2465, W-7500, Karlsruhe 1, Germany, online services. Inquire as to cost and availability.

COMPENDEX PLUS. Engineering Information, Inc., 345 East 47th Street, New York, NY 10017. (212) 705-7600 or (800) 221-1044. Contains citations with abstracts to worldwide literature in engineering and technology, from 1970 to present. Available on online BRS,(800) 289-4277, DIALOG, (800) 334-2564, ORBIT (800) 456-7248, and STN International, FIZ Karlsruhe, P.O. Box 2465, W-7500, Karlsruhe 1, Germany, online services. Also available on CD-ROM. Inquire as to cost and availability.

Earth Sciences. U.S. Geological Survey, 12201 Sunrise Valley Drive, Reston, VA 22092-9998. (703) 648-4460. CD-ROM of earth science databases including the U.S. Geological Survey Library, Earth Science Data Directory, and GEOINDEX, citations to published geological maps. $350.00 per year with quarterly updates.

Geobase. Elsevier - Geo Abstracts, Regency House, 34 Duke Street, Norwich NR3 3AP, England. Contains citations to the worldwide earth science literature from 1980 to date. Available on DIALOG,(800) 334-2564, ORBIT (800) 456-7248, online services. Inquire as to cost and availability.

Geomechanics Abstracts. University of London, Imperial College of Science and Technology, Rock Mechanics Information Service, Royal School of Mines, Prince Consort Road, London SW7 2BP England. Telephone: 071-589 5111 or FAX 071-589 6806. Scope includes worldwide literature on engineering geology, hydrogeology, mining, rock mechanics, soil mechanics, and tunnelling, 1977 to present. Available on the ORBIT (800) 456-7248, online service. Inquire as to cost and availability.

GEOREF. American Geological Institute, 4220 King Street, Alexandria, VA 22302. (800) 336-4764 or (703) 379-2480.

Geology and geosciences literature, 1785 to present for North America. Available on DIALOG,(800) 334-2564, ORBIT (800) 456-7248, online services. Also available on CD-ROM. Inquire as to cost and availability.

ICONDA (CIB International Construction Database). Fraunhofer-Gesellschaft, Informationszentrum RAUM und BAU, Nobelstrasse, 12, D-7000, Stuttgart 80, Germany. Contains citations with abstracts to worldwide technical literature on construction and civil engineering, structural engineering, and engineering geology, from 1976 to present. Available on ORBIT, (800) 456-7248, and STN International, FIZ Karlsruhe, P.O. Box 2465, W-7500, Karlsruhe 1, Germany, online services. Inquire as to cost and availability.

Meteorological and Geoastrophysical Abstracts. American Meteorological Society, 45 Beacon Street, Boston, MA 02108-3693. (617) 227-2425. FAX (617) 742-8718. Contains citations and abstracts to the worldwide literature on significant research in meteorology and geoastrophysics. Related topics include physical oceanography, hydrology, environmental sciences and glaciology. Covers the period 1972 to present. Available on DIALOG,(800) 334-2564, online service. Inquire as to cost and availability.

National Information Service For Earthquake Engineering Database. University of California, Berkeley. Contains citations and abstracts to journal articles on earthquakes and earthquake engineering from 1987 to present. Available on BRS,(800) 289-4277, online service. Inquire as to cost and availability.

NTIS Bibliographic Database. National Technical Information Service, 5285 Port Royal Road, Springfield, VA 22161. (703) 487-4929 or FAX (703) 321-8199. Broad coverage of government-sponsored science and technology research reports, 1964 to present. Available on BRS,(800) 289-4277, DIALOG, (800) 334-2564, ORBIT, (800) 456-7248, and STN International, FIZ Karlsruhe, P.O. Box 2465, W-7500, Karlsruhe 1, Germany, online services. Also available on CD-ROM. Inquire as to cost and availability.

Oceanic Abstracts. Cambridge Scientific Abstracts, 7200 Wisconsin Avenue, Suite 601, Bethesda, MD 20814. (301) 961-6750 or (800) 843-7751. Contains citations to worldwide literature on oceanography and marine-related areas, from 1964 to present. Available on DIALOG,(800) 334-2564, online service. Inquire as to cost and availability.

Scientific and Technical Books and Serials in Print. R.R. Bowker Inc., 121 Chanlon Road, New Providence, NJ 07974. (800) 521-8110 or (908) 464-6800. List of currently published books and serials in the physical and biological sciences, engineering and technology, with subject, author and titles indexes. Available on ORBIT,(800) 456-7248, online service. Inquire as to cost and availability.

SCISEARCH. Institute for Scientific Information, 3501 Market Street, Philadelphia, PA 19104. (800) 523-1850 or (215) 386-0100. Broad multidisciplinary title and author index to the international literature of science and technology, 1974 to present. Available on DIALOG,(800) 334-2564, and ORBIT,(800) 456-7248, online services. Also available on CD-ROM. Inquire as to cost and availability.

Tulsa Petroleum Abstracts, 600 South College, Tulsa, OK 74104. (918) 631-2296 or (800) 247-8678. Contains citations with abstracts to the worldwide literature and patents on the exploration, development, and production of petroleum resources, from 1965 to present. Available on DIALOG,(800) 334-2564, and

BEDROCK

Ency. of Physical Sciences and Engineering Info. Sources

ORBIT,(800) 456-7248, online services. Inquire as to cost and availability.

Wilsonline. H.W. Wilson Company, 950 University Avenue, Bronx, NY 10452. (800) 367-6770 or (212) 588-8400. Makes available online versions of the printed H.W. Wilson indexes including Applied Science and Technology Index, BusinessPeriodicals Index, General Science Index, and Readers' Guide to Periodical Literature. Period covered is generally 1983 to present. Available on BRS,(800) 289-4277, DIALOG,(800) 334-2564, and OCLC EPIC,(800) 848-5878, online services. Also available on CD-ROM. Inquire as to cost and availability.

PERIODICALS

Aapg Bulletin. American Association of Petroleum Geologists, P.O. Box 979, Tulsa, OK 7401. (918) 584-2555. Monthly. $120.00 per year.

American Journal of Science. Kline Geology Laboratory, Yale University, Box 6666, Yale Station, New Haven, CT 06511-8130. (203) 432-3131. Ten times per year. $40.00 per year.

Canadian Journal of Earth Sciences. National Research Council of Canada, Research Journals, Ottawa, Ont. K1A OR6, Canada. (613) 993-9084. Monthly. $112.00 per year.

Earth and Planetary Science Letters. Elsevier Science Publishing Company, Inc., 655 Avenue of the Americas, New York, NY 10010. (212) 989-5800. Twenty times per year. $945.00 per year.

Geological Magazine. Cambridge University Press, 40 West 20th Street, New York, NY 10011-4211. (212) 924-3900. Bimonthly. $205.00 per year.

Geological Society. JOURNAL. Geological Society of London. Geological Society Publishing House, Unit 7, Brassmill Enterprise Centre, Brassmill Lane, Bath, Avon BA1 3JN, England. 6 times per year. $432.00.

Geological Society of America Bulletin. P.O. Box 9140, 3300 Penrose Place, Boulder, CO 80301. (303) 447-2020. Fax (303) 447-1133. Monthly. $150.00.

Geology. Geological Society of America. P.O. Box 9140, 3300 Penrose Place, Boulder, CO 80301. (303) 447-2020. Fax (303) 447-1133. Monthly. $120.00 per year.

Geophysics. Society of Exploration Geophysicists, P.O. Box 702740, Tulsa, OK 74170. (918) 493-3516. Monthly. $175.00 per year.

Journal of Geology. University of Chicago Press, 5720 Woodlawn Avenue, Chicago, IL 60637. (312) 753-3347. Bimonthly. $38.00 per year.

Journal of Geophysical Research. American Geophysical Union, 2000 Florida Avenue, N.W., Washington, DC 20009. (202) 462-6903. Monthly. $2170.00 per year.

Journal of Petrology. Oxford University Press, Inc., 200 Madison Avenue, New York, NY 10016. (800) 334-4249 or (212) 679-7300. Bimonthly. $240.00.

Journal of Structural Geology. Pergamon Journals, Inc., Maxwell House, Fairview Park, Elmsford, NY 10523. (914) 592-7700. Ten times per year. $375.00 per year.

Journal of Volcanology and Geothermal Research. Elsevier Science Publishing Company, Inc., 655 Avenue of the Americas, New York, NY 10010. (212) 989-5800. Monthly. $500.00per year.

Lithos; An International Journal of Mineralogy, Petrology, and Geochemistry. Elsevier Science Publishing Company, Inc., 655 Avenue of the Americas, New York, NY 10010. (212) 989-5800. Quarterly. $275.00 per year.

Mountain Geologist. Rocky Mountain Association of Geologists, 4201 West 51st Avenue, Denver, CO 80212-2902. Quarterly. $15.00 per year.

Physics of the Earth and Planetary Interiors. Elsevier Science Publishing Company, Inc., 655 Avenue of the Americas, New York, NY 10010. (212) 989-5800. Twenty-four times per year. $900.00 per year.

Reviews of Geophysics. American Geophysical Union, 2000 Florida Avenue, N.W., Washington, DC 20009. (202) 462-6900. Quarterly. $300.00 per year.

Rock Mechanics and Rock Engineering. Springer-Verlag New York, Inc., 175 Fifth Avenue, New York, NY 10010. (800) 526-7254 or (212) 460-1500. Quarterly. $122.00 per year.

Tectonics. American Geophysical Union, 2000 Florida Avenue, N.W., Washington, DC 20009. (202) 462-6900. Bimonthly. $230.00 per year to individuals.

Tectonophysics; An International Journal of Geotectonics and the Geology and Physics of the Interior of the Earth. Elsevier Science Publishing Company, Inc., 655 Avenue of the Americas, New York, NY 10010. (212) 989-5800. Fifty-six times per year. $1275.00 per year.

RESEARCH CENTERS AND INSTITUTES

California Institute of Technology. Seismological Laboratory, Pasadena, CA 91125. (818) 356-6914.

Carnegie Institution of Washington. Geophysical Laboratory, 5251 Broad Branch Road, N.W., Washington, DC 20015. (202) 686-2410 or FAX (202) 686-2419.

Ocean Drilling Program. Joint Oceanographic Institution Inc., 1755 Massachusetts Avenue, Suite 800, Washington, DC 20036. (202) 232-3900.

U.S. Geological Survey, Geologic Division, National Center, 12201 Sunrise Valley Drive, Reston, VA 22092. (703) 648-6600. The major geological research agency of the federal government conducting research in most areas of pure and applied research in the geosciences.

Ency. of Physical Sciences and Engineering Info. Sources

BENZENE

BENZENE

See also: CARBON, CHEMISTRY, CHEMICAL ENGINEERING, HETEROCYCLIC CHEMISTRY, ORGANIC CHEMISTRY, PLASTICS, PETROLEUM CHEMISTRY

ABSTRACT SERVICES AND INDEXES

Applied Science and Technology Index. A cumulative subject index to Englisanguage Periodicals in the Fields of Aeronautics and Space Science, Computer Technology, Chemistry, Construction Industry, Energy and Related Areas. H.W. Wilson Co., 950 University Avenue, Bronx, NY 10452-9978. (212) 588-8400.(800) 367-6770. FAX (718) 590-1617. From 1958 to present. Monthly. Inquire about cost and availability. Also available online (BRS and Wilsonline) and on CD-ROM. ISSN: 0003-6986.

Chemical Abstracts. Chemical Abstracts Service. 2540 Oleantangy River Road, Box 3012, Columbus, OS 43210-0012. (614) 447-3600. FAX (614) 447-3713. 1907 to present. Weekly. $16,800.00 per year. Available online and on CD- ROM. CA is also available in five section groupings. Inquire regarding cost and availability.

Current Contents: Physical, Chemical and Earth Sciences. Institute For Scientific Information, 3501 Market Street, Philadelphia, PA 19104. (215) 386-0100. (800) 523-1850. Fax (215) 386-2291. 1961 to present. Weekly. $442.00 per year. Also available online (BRS, DIALOG) and on CD-ROM. Inquire regarding cost and availability. ISSN: 0163-2574.

General Science Index. H.W. Wilson Company, 950 University Avenue, Bronx, NY 10452. (212) 588-8400. (800) 367-6770. Fax (718) 590-1617. From 1978 to present. Ten issues per Year; Quarterly and Annual Cumulations. Service basis. Available on Cd-rom and Online. Inquire regarding cost and availability. ISSN: 0162-1963.

Physics Abstracts. INSPEC. Section A, Science Abstracts. Institution of Electrical Engineers (iee), London. available from: INSPEC/IEEE - Institute of Electrical and Electronic engineers, box 1331, hoes lane, Piscataway, NJ 855-1331. (908) 562-5549. 1898 to present. 24 issues per year. $2835.00 per year. Also available online and on CD-ROM. ISSN: 0036-8091.

Science Citation Index. SCI. Institute for Scientific Information, 3501 Market Street, Philadelphia, PA 19104.(215) 386-0100. (800) 523-1850. FAX (215) 386-2991. 1961 to Present. Six issues per year, plus annual cumulation. $11650.00 per year. Also available online and on CD-ROM. Inquire About Price and Availability. ISSN: 0036-827X.

ANNUAL REVIEWS AND YEARBOOKS

Advances in Heterocyclic Chemistry. Alan R. Kitritzky, editor. Academic Press.Inc., 525 B Street, Suite 1900, San Diego, CA 92101-4495. (619) 231-0926. FAX (619) 699-6715. 1963 to Present. Irregular. price varies. ISSN: 0065-2725.

Advances in Physical Organic Chemistry. Academic Press, Inc.,525 B Street, Suite 1900, San Diego, CA 92101-4495. (619) 231-0926. FAX (619) 699-6715. 1963 to present. Irregular. Price Varies. ISSN: 0065-3160.

Studies in Organic Chemistry. Elsevier Science Publishers., Box 882, Madison Square Station, New York, NY 10159. (212) 989-5800. 1973 to present. Irregular. Price varies. ISSN: 0165-3253.

ASSOCIATIONS AND PROFESSIONAL SOCIETIES

American Carbon Society. c/o Carbone of America, 215 Stackpole Street, St. Marys, PS 15857. (814) 781-8410. Fax (814)781-8570.

American Chemical Society. 1155 16th Street, NW, Washington, DC 20036. (202) 872-4600.

American Institute of Chemical Engineers. 345 East 47th Street, New York, NY 10017-2396. (212) 705-7663.

American Oil Chemists' Society. PO Box 3489, Champaign, IL 61826-3489. (217) 359-2344. FAX (217) 351-8091.

Association of official Analytical Chemists. 2200 Wilson Boulevard, Suite 400, Arlington, VA 22001. (703) 522-3032.

Chemical Manufacturers Association. 2501 M Street, N.W., Washington, DC 20037. (202) 887-1182

BIBLIOGRAPHIES

Chemical Information Sources. Gary Wiggins. McGraw-Hill Publishing Company, 1221 Avenue of the Americas, New York, NY 10020. (800) 262-4729 or (212) 512-3825. 1991. $42.50.

Information Sources in Chemistry. R.T. Bottle and J.F.B Rowland, editors. R.R. Bowker Inc., 121 Chanlon Road, New Providence, NJ 07974. (800) 521-8110 or (908) 464-6800. Fourth edition. 1993. $75.00.

Handbooks and Tables in Science and Technology. Russell H. Powell, editor. Oryx Press, 4041 North Central, Suite 700, Phoenix, AZ 85012-3330. (602) 265-2651 or (800) 279-6799. Third edition. 1994. $65.00.

Scientific and Technical Books and Serials in Print; An Index To Literature in Science and Technology. R.R. Bowker Inc., 121 Chanlon Road, New Providence, NJ 07974. (908) 464-6800. (800) 521-8110. FAX (908) 665-3502. 1972 to present. Annual. 4 volumes. 1994. $299.95. Also available on compact disc and online. ISSN: 0000-054X.

DIRECTORIES AND BIOGRAPHICAL SOURCES

American Men and Women of Science. Physical and Biological Sciences. R.R. Bowker Company, 121 Chanlon Road, New Providence, NJ 07974. (908) 464-6800. (800) 521-8110. 20th edition. 8 volumes. 1996. $850.00.

Directory of Chemistry Software 1992. Wendy Warr, Peter Willett, Geoff Downs. American Chemical Society, 1155 16th Street, NW, Washington, DC 20036. (202) 872-4600. 1992. $35.95.

Consulting Services: Chemists and Chemical Engineers. Association of Consulting Chemists and Chemical Engineers, 50 East 41st Street, Suite 92, New York, NY 10017. (212)

BENZENE

Ency. of Physical Sciences and Engineering Info. Sources

684-6255. Biennial. $60.00.

Research Centers Directory. Gale Research Company Inc., 835 Penobscot Building, Detroit, MI 48226-4094. (313) 961-2242. (800) 347-4253. 20th edition. 1995. $485.00. ISSN: 0080-1518.

ENCYCLOPEDIAS AND DICTIONARIES

Academic Press Dictionary of Science and Technology. Christopher Morris, editor. Academic Press, Inc., 1250 Sixth Avenue, San Diego, CA 92101. (619) 231-0926. FAX (619) 699-6715. 1991. $115.00.

Concise Encyclopedia Chemistry. Translated and revised by Mary Eagleson. Walter de Gruyter, Inc., 200 Saw Mill River Road, Hawthorne, New York, 10532. (914) 747-0110 or Fax (914) 747-1326. 1994. $69.95.

Dictionary of Organic Compounds. John Buckingham, et al, editors. Chapman & Hall, 1 Penn Plaza, New York, NY 10119. (212) 564-1060. Fifth edition. 7 volumes. $2550.00 set.

Encyclopedic Dictionary of Science. Candida Hunt and Monica Byles, editors. Facts-on-File, Inc., 460 Park Avenue South, New York, NY 10016-7382. (800) 322-8755. Fax (800) 683-3633. 1988. $32.95.

Encyclopedia of Applied Physics. VCH Publishers, Inc., 303 Northwest 12th Avenue, Deerfield Beach, FL 33442. (800) 367-8249. 1991-. Twenty volumes. $6000.00.

Industrial Chemical thesaurus. Michael Ash and Irene Ash. VCH Publishers, Inc., 220 East 23rd Street, New York, NY 10010-4606. (800) 367-8249. 1992. $295.00.

Kirk-Othmer Encyclopedia of Chemical Technology. John Wiley and Sons, Inc., 605 Third Avenue, New York, NY 10158. (800) 526-5368 or (212) 850-6000. Fourth edition. 1991 - . Twenty-seven volumes. $5400.00.

McGraw-Hill Encyclopedia of Chemistry. Sybil P.Parker, editor. McGraw-Hill Book, Incorporated, 1221 Avenue of the Americas, New York, NY 10020. (212) 997-3675. Second edition. 1993. $95.50.

McGraw-Hill Encyclopedia of Science and Technology. McGraw-Hill Book, Incorporated, 1221 Avenue of the Americas, New York, NY 10020. (212) 997-3675. Seventeenth edition. Twenty volumes. 1992. $1900.00.

GENERAL WORKS

Benzene: Basic and Hazardous Properties. M. Cherimisinoff. Marcel Dekker, Inc., 270 Madison Avenue, New York, NY 10016. (212) 696-9000. (800) 228-1160. 1979. Inquire.

Benzene: Occupational and Environmental Hazards, A Scientific Update. Princeton Science Publishing Co. P.O. Box 2155, Princeton, NJ 08543. (609) 683-4650. 1989. $65.00.

Benzene Toxicity: A Critical Evaluation. Sidney Laskin and Bernard D. Goldstein. Books on Demand, 300 North Zeeb Road, Ann Arbor, MI 48106-1346. (313) 761-4700. (800) 521-0600. Reprint edition. $43.00 in paper.

Benzopyrenes. M. R. Osborne and N. T. Crosby. Cambridge University Press, 40 West 20th Street, New York, NY 10011-4211. (212) 924-3900. (800) 872-7423. 1987.

The Chemical Bond: Structure and Dynamics. Ahmed Zewail. Academic Press, Inc., 1250 Sixth Avenue, San Diego, CA 92101-4311. (619) 231-0926. FAX (619) 699-6715. 1992. $49.95.

Contemporary Heterocyclic Chemistry: Synthesis, Reactions and Applications. G. R. Newkome and W. W. Paudler. John Wiley & Sons, Inc., 605 Third Avenue, New York, NY 10158- 0012. (212) 850-6000. (800) 225-5945. 1982. $99.95.

Halogenated Benzenes: Mutual Solubility of Liquids. Horvath. Franklin Book Co., Inc., Elkins Park, PA 19117. (215) 635-5252. 1985. $120.00.

Organic Chemistry: Its Language and Its State of the Art. Volkan Kisakurek, editor. VCH Publications, Inc., 220 East 23rd Street, Suite 909, New York, NY 10010-4606. (212) 683- 8333. (800) 422-8824. 1993. $117.00.

Organic Reactions: Simplicity and Logic. Pierre Laszlo. *John Wiley & Sons, Inc.*, 605 Third Avenue, New York, NY 10158-0012. (212) 850-6000. (800) 225-5945. 1994. $74.95.

HANDBOOKS AND MANUALS

Chemical Formulary. H. Bennett, editor. Chemical Publishing, Co., Inc. 80 Eighth Avenue, New York, NY 10011. (212 255-1950. Volumes 1 - 30. $60.00 per volume.

Chemometrics: Chemical and Sensory Data. David R. Burgard and James T. Kuznicki. CRC Publishers, Inc., 2000 Corporate Blvd., N.W., Boca Raton, FL 33431. (407) 994-0555 or (800) 333-8300. 1990. $120.00.

Comprehensive Organic Synthesis: Selectivity, Strategy, and Efficiency in Modern Organic Chemical. Barry M. Trost and others, editors. Pergamon Press, Maxwell House, Fairview Park, Elmsford, NY 10523. (914) 592-7700. Fax (914) 592-3625. 1991. Nine volumes. $3900.00.

CRC Handbook of Chemistry and Physics. David R. Kide, editor. CRC Press Inc., 2000 Corporate Blvd., NW, Boca Raton, FL 33431. (407) 994-0555. (800) 333-8300. 77th edition. 1996. $99.95.

CRC Handbook of Data on Organic Compounds. R. C. Weast and M. J. Astle, editors. CRC Press Inc., 2000 Corporate Blvd., NW, Boca Raton, FL 33431. (407) 994-0555. (800) 333-8300. 1985. 2 volumes.

Guide To Basic Chemical Compounds. D.R. Lide, Jr. CRC Publishers, Inc., 2000 Corporate Blvd., N.W., Boca Raton, FL 33431. (407) 994-0555 or (800) 333-8300. 1993. $120.00.

Guide To Iupac Nomenclature of Organic Compounds. J. C. Richer, et al, editors. CRC Press Inc., 2000 Corporate Blvd., NW, Boca Raton, FL 33431. (407) 994-0555. (800) 333-8300. 1993. $39.95.

Handbook of Data and Organic Coumpounds. W Robert C. Weast. CRC Press Inc., 2000 Corporate Blvd., NW, Boca Raton, FL 33431. (407) 994-0555. (800) 333-8300. 2nd edition. 1988. $2655.00.

Handbook of Organic Chemistry. Hans Beyer and Wolfgang Walter. Prentice Hall General Reference & Travel, 15 Columbus Circle, New York, NY 10023. (212) 373-8500. (800) 223-2348. 1995. $96.00.

Handbook of Tables For Organic Compound Identification. Zvi Rappoport. CRC Press Inc., 2000 Corporate Blvd., NW, Boca Raton, FL 33431. (407) 994-0555. (800) 333-8300 1966. $76.95.

Laboratory Handbook of Materials, Equipment, and Techniques. Gary S. Coyne. Prentice Hall (Division of Simon andSchuster), 15 Columbus Circle, New York, NY 10023. (212) 373-8500 or (800) 922-0579. 1992. $45.00.

Lange's Handbook of Chemistry. John A. Dean, editor. McGraw-Hill Publishing Company, 1221 Avenue of the Americas, New York, NY 10020. (800) 262-4729 or (212) 512-3825. Fourteenth edition. 1992. $79.50.

Riegel's Handbook of Industrial Chemistry. James A. Kent, editor. Van Nostrand Reinhold, 115 Fifth Avenue, New York, NY 10003. (212) 254-3232 or (800) 926-2665. Ninth edition. 1992. $114.95.

Tables of Spectral Data For Structure Determination of Organic Compounds. E. Pretsch, et al. Springer-Verlag New York, Inc., 175 Fifth Avenue, New York, NY 10010. (212) 460-1500. (800) 777-4643. FAX (212) 473-6272. 1994. $49.95 in paper.

ONLINE DATABASES AND CD-ROMS

Beilstein Online. Beilstein-Institut fur Literatur der Organischen Chemie, Varrentrapperstrasse, 40-42, D-6000, Frankfurt am Main 90, Germany. Contains data on carbon compounds from the Beilstein Handbook of Organic Chemistry. Structural and factual data for more than 3.4 million heterocyclic, isocyclic, and acyclic compounds included. Covers the period 1830 to present. Available on DIALOG, (800) 334-2564, ORBIT (800) 456-7248, and STN International, FIZ Karlsruhe, P.O. Box 2465, W-7500, Karlsruhe 1, Germany, online services. Inquire as to cost and availability.

CA Search. Chemical Abstracts Service, P.O. Box 3012, Columbus, OH 43210-0012. (614) 447-3600. (800) 848-6533. FAX (614) 447-3709. Very comprehensive guide to worldwide chemical literature and related fields, 1972 to present. Available on BRS,(800) 289-4277, DIALOG, (800) 334-2564, ORBIT (800) 456-7248, and STN International, FIZ Karlsruhe, P.O. Box 2465, W-7500, Karlsruhe 1, Germany, online services. Inquire as to cost and availability.

Chemical Journals of the American Chemical Society. American Chemical Society, 1155 16th Street, N.W., Washington, DC 20036. (202) 872-4381 or (800) 424-6747. Contains complete text of approximately 90,000 articles from 22 primary journals published by the American Chemical Society, from mostly 1982 to present. Available on STN International, FIZ Karlsruhe, P.O. Box 2465, W-7500, Karlsruhe 1, Germany, online service. Inquire as to cost and availability.

Compendex Plus. Engineering Information, Inc., 345 East 47th Street, New York, NY 10017. (212) 705-7600 or (800) 221-1044. Contains citations with abstracts to worldwide literature in engineering and technology, from 1970 to present. Available on online BRS,(800) 289-4277, DIALOG, (800) 334-2564, ORBIT (800) 456-7248, and STN International, FIZ Karlsruhe, P.O. Box 2465, W-7500, Karlsruhe 1, Germany, online services. Also available on CD-ROM. Inquire as to cost and availability.

Current Contents Search. Institute for Scientific Information, 3501 Market Street, Philadelphia, PA 19104. (215) 386-0100. FAX (215) 386-6362. Contains citations to articles listed in the table of contents of science and technology journals. Also articles in social sciences and life sciences journals. Available on BRS,(800) 289-4277, DIALOG,(800) 334-2564, online services. Inquire as to cost and availability.

Dissertation Abstracts. University Microfilms International, 300 North Zeeb Road, Ann Arbor, MI 48106. (800) 521-0600 or (313) 761-4700. Scope includes virtually all doctoral dissertations accepted at accredited American institutions from 1861 to present in 252 subject areas. Available on BRS,(800) 289-4277, DIALOG,(800) 334-2564, and OCLC EPIC,(800) 848-5878, online services. Also available on CD-ROM. Inquire as to cost and availability.

Kirk-Othmer Encyclopedia of Chemical Technology. John Wiley and Sons, Inc., 605 Third Avenue, New York, NY 10158. (800) 526-5368 or (212) 850-6000. Contains the complete text of all chapters in the 27 volume fourth edition of the Kirk-Othmer Encyclopedia of Chemical Technology. 1991. Available on BRS,(800) 289-4277, DIALOG,(800) 334-2564, online services. Inquire as to cost and availability.

NTIS Bibliographic Database. National Technical Information Service, 5285 Port Royal Road, Springfield, VA 22161. (703) 487-4929 or FAX (703) 321-8199. Broad coverage of government-sponsored science and technology research reports, 1964 to present. Available on BRS,(800) 289-4277, DIALOG, (800) 334-2564, ORBIT, (800) 456-7248, and STN International, FIZ Karlsruhe, P.O. Box 2465, W-7500, Karlsruhe 1, Germany, online services. Also available on CD-ROM. Inquire as to cost and availability.

SCISEARCH. Institute for Scientific Information, 3501 Market Street, Philadelphia, PA 19104. (800) 523-1850 or (215) 386-0100. Broad multidisciplinary title and author index to the international literature of science and technology, 1974 to present. Available on DIALOG,(800) 334-2564, and ORBIT,(800) 456-7248, online services. Also available on CD-ROM. Inquire as to cost and availability.

Wilsonline. H.W. Wilson Company, 950 University Avenue, Bronx, NY 10452. (800) 367-6770 or (212) 588-8400. Makes available online versions of the printed H.W. Wilson indexes including Applied Science and Technology Index, Business Periodicals Index, General Science Index, and Readers' Guide to Periodical Literature. Period covered is generally 1983 to present. Available on BRS,(800) 289-4277, DIALOG,(800) 334-2564, and OCLC EPIC,(800) 848-5878, online services. Also available on CD-ROM. Inquire as to cost and availability.

PERIODICALS

American Oil Chemists' Society. JOURNAL. American Oil Chemists' Society, 1608 Broadmoor Drive, Box 3489, Champaign, IL61821-0489. (217) 359-2344. 1917 to present. Monthly. Monthly. $185.00 per year. ISSN: 0003-021X.

Aiche Journal. American Institute of Chemical Engineers, 345 47th Street, New York, NY 10017-2396. (212) 705-7338. Fax (212) 752-3294. 1955 to present. Monthly. $395.00 per year. ISSN: 0001-1541.

Carbon. Elsevier Science Publishing Company, Inc., 660 White Plains Road, Tarrytown, NY 10591-5153. (914-524-9200. FAX

(914) 524-9200. 1963 to present. Bi-monthly. $1045.00 ISSN: 0008-6223.

Chemical Week; Includes Annual Buyers Guide. Chemical *Week Associates,* 888 Seventh Avenue, New York, NY 10106.(212) 621-4900. FAX (212) 621-4949. 1914 to present. Weekly. $99.00 per year. ISSN: 0009-272X.

Chemtech. American Chemical Society, Box 3337, Columbus, OH 43210. (614) 447-3776. (800) 333-9511. FAX (614) 447-3671. 1970 to present. Monthly. $79.00 per year. ISSN: 0009-2703.

Hydrocarbon Processing. Gulf Publishing Company, P.O. Box 2608, Houston, TX 77001. (713) 520-4433. 1922 to present. Monthly. $22.00. ISSN: 0887-0284

Journal of Heterocyclic Chemistry. American Chemical Society, Box 3337, Columbus, OH 43210. (614) 447-3776. (800) 333- 9511. FAX (614) 447-3671. 1964 to present. Bi-monthly. $420.00 per year. ISSN: 0022-152X.

Journal of Organic Chemistry. American Chemical Society, Box 3337, Columbus, OH 43210. (614) 447-3776. (800) 333- 9511. FAX (614) 447-3671. 1936 to present. Bi-weekly. $785.00 per year. ISSN: 0022-3263.

Journal of Polymer Science. Part A: Polymer Chemistry. John Wiley and Sons, Inc., 605 Third Avenue, New York, NY 10158. (800) 526-5368 or (212) 850-6000. 1962 to present. 13 issues per year. $2540.00 per year. ISSN: 9887-624X.

Organometallics. American Chemical Society, Box 3337, Columbus, OH 43210. (614) 447-3776. (800) 333-9511. FAX (614) 447-3671. 1982 to present. Monthly. $850.00 per year. ISSN: 0276-7333.

Tetrahedron. Elsevier Science Publishing Company, Inc., 660 White Plains Road, Tarrytown, NY 10591-5153. (914-524-9200. FAX (914) 524-9200. 1957 to present. Sixty times per year. $5365.00 per year. ISSN: 0040-4020.

Tetrahedron Letters. Elsevier Science Publishing Company, Inc., 660 White Plains Road, Tarrytown, NY 10591-5153. (914-524-9200. FAX (914) 524-9200. 1959 to present. Weekly. $4745.00 per year. ISSN: 0040-4039.

RESEARCH CENTERS AND INSTITUTES

Center For Catalytic Science and Technology. University of Delaware, Departments of Chemical Engineering and Chemistry, Newark, DE 19716. (302) 451-8056. FAX (302) 451-1048.

Chemical Laboratories. Harvard Universiy. OxfordStreet, Cambridge, MA 02138. (617) 495-4283. FAX (617) 496-5618.

Chemistry Laboratories. Rensselaer Polytechnic Institute. Cogswell Laboratory, Troy, NY 12180-3590. (518) 276-8981.

Chemical Research Laboratory. Brown University, Providence, RI 02912. (401) 863-2256.

BENZENE RING

See: ORGANIC CHEMISTRY

BERYLLIUM

See also: ALLOYS, COPPER, METALLURGICAL ENGINEERING, METALLURGY, METALS AND METALWORKING

ABSTRACT SERVICES AND INDEXES

Alloys Index. American Society for Metals, Metals Park, OH 44073. (216) 338-5151. FAX (216) 338-4634. 1974 to present. Monthly. $380.00. Also available on CD-ROM and online via METADEX, STN International, and DIALOG.

Aluminum Industry Abstracts. Aluminum Association, Materials Park, OH 44073. (216) 338-5151. FAX (216) 338-4634. 1968 to present. Monthly. $525.00 per year.

Applied Mechanics Reviews: An Assessment of World Literature in Engineering Sciences. American Society of Mechanical Engineers, 345 East 47th Street, New York, NY 10017. (212) 705-7703. 1948 to present. Monthly. $360.00 per year.

Applied Science and Technology Index. A cumulative subject index to Englisanguage Periodicals in the Fields of Aeronautics and Space Science, Computer Technology, Chemistry, Construction Industry, Energy and Related Areas. H.W. Wilson Co., 950 University Avenue, Bronx, NY 10452. (800) 367-6770 or (212) 588-8400. FAX (718) 590-1617. From 1958 to present. Monthly. Inquire about cost and availability. Also available on CD-ROM.

Corrosion Abstracts; Abstracts of the World's Literature on Corrosion and Corrosion Mitigation. National Association of Corrosion Engineers (NACE), Box 218340, Houston, TX 77218. (713) 492-0535 or FAX (713) 492-8254. 1962 to present. Bi-monthly. $195.00 per year. Also available on CD-ROM.

Current Contents: Engineering, Technology, and Applied Sciences. Institute for Scientific Information, 3501 Market Street, Philadelphia, PA 19104. (215) 386-0100. FAX (215) 386-6362. Weekly. $360.00 per year.

Engineering Index Monthly. Engineering Information, Inc., Castle Point on the Hudson, Hoboken, NJ 07030. (800) 221-1044. FAX (212) 832-1857. Monthly. $2200.00 per year. Also available online as COMPENDEX, and also on CD-ROM.

International Copper Information Bulletin. Copper Development Association, Orchard House, Mutton Lane, Potters Bar, Herts EN6 3AP, England. Telephone 0707-50711. FAX 0707-42769. 1976 to present. 3 times a year. Inquire for cost and availability.

Metals Abstracts and Metals Abstracts Index. American Society for Metals, Metals Park, OH 44073. (216) 338-5151. 1968to present. Monthly. Abstracts are $1925.00 per year and Index is $460.00 per year.

Physics Abstracts. INSPEC/IEEE, Box 1331, 445 Hoes Lane, Piscataway, NJ 08855-1331. (908) 562-5549. 1898 to present. $2835.00 per year. Also available online as INSPEC. Covers recently published primary research in all areas of physics.

ASSOCIATIONS AND PROFESSIONAL SOCIETIES

American Powder Metallurgy Institute. 105 College Road E., Princeton, NJ 08540. (609) 452-7700. (609) 987-8523.

American Society For Metals (asm). Materials Park, OH 44073. (216) 338-5151 or FAX (216) 338-4634.

American Society For Testing and Materials (astm). 1916 Race Street, Philadelphia, PA 19103. (215) 299-5585.

American Society of Mechanical Engineers. 345 East 47th Street, New York, NY 10017-2398. (212) 705-7703.

Metallurgical Society of the Aime (american Institute of Mining, Metallurgical, and Petroleum Engineers). 345 E.47th Street, 14th Floor, New York, NY 10017. (212) 705-7695.

National Association of Corrosion Engineers. P.O. Box 218340, 1440 South Creek, Houston, TX 77218. (713) 492-0535.

DIRECTORIES AND BIOGRAPHICAL SOURCES

Directory of Engineering Societies and Related Organizations. Gordon Davis. 13th edition. American Association of Engineering Societies, 1111 19th Street NW, Suite 608, Washington, DC 20036. (202) 296-2237 or (800) 658-8897. 1989. Inquire for price.

Dun's Industrial Guide: The Metalworking Directory, 1993-94. Dun & Bradstreet Information Services Staff. 3 volumes. Dun & Bradstreet Information Services, 899 Eaton Avenue, Bethlehem, PA 18025. (610) 882-7000 or (800) 526-0651. FAX (610) 882-7269. 1993. $795.00.

International Research Centers Directory. Gale Research, 835 Penobscot Building, Detroit, MI 48226-4094. (313) 961-2242. (800) 347-4253. FAX (313) 961-6083. 8th edition. 1995. $410.00.

Research Centers Directory. Gale Research, 835 Penobscot Building, Detroit, MI 48226-4094. (313) 961-2242. (800) 347-4253. FAX (313) 961-6083. $485.00.

Metallurgical Society of AIME—Membership List. 345 E.47th Street, 14th Floor, New York, NY 10017. (212) 705-7695. Inquire for cost and availability.

Scientific and Technical Organizations and Agencies Directory. Gale Research, 835 Penobscot Building, Detroit, MI 48226-4094. (313) 961-2242. (800) 347-4253. FAX (313) 961-6083. 4th edition. 1996. $195.00.

Who's Who in Engineering. American Association of Engineering Societies, 1111 19th Street NW, Suite 608, Washington, DC 20036. (202) 296-2237 or (800) 658-8897. 8th edition. 1991. Inquire for price.

GENERAL WORKS

Beryllium. Benjamin Petkof. U.S. Bureau of Mines. Inquire for cost and availability from: Superintendent of Documents, U.S. Government Printing office, Box 371954, Pittsburgh, PA 15250-7954. (202) 783-3238. FAX (202) 512-2233. 1979.

Beryllium. Gordon Press Publishers, PO Box 459, Bowling Green Station, New York, NY 10004. (718) 624-8419. 1993. $250.95.

Beryllium: Biomedical and Enviromental Aspects. Milton D. Rossman, et al. Williams & Wilkins, 428 E Preston Street, Baltimore, MD 21202. (800) 527-5597. FAX (410) 528-4422. 1991. Inquire for cost.

Beryllium Science and Technology. Donald Webster, et al. Plenum Publishing Corporation, 233 Spring Street, New York, NY 10013-1578. (212) 620-8000. FAX (212) 463-0742. 1979. Inquire for cost and availability.

HANDBOOKS AND MANUALS

Smithell's Metals Reference Book. E.A. Brandes & G.B. Brook. 7th edition. Butterworth-Heinemann, 313 Washington Street, Newton, MA 02158. (617) 928-2500 or (800) 366-2665. FAX (617) 928-2620. 1992. $250.00.

Woldman's Engineering Alloys. J. Frick, editor. 8th edition. ASM International, Materials Information, Materials Park, OH 44073-0002. (216) 338-5151 or (800) 336-5152. FAX (216) 338-4634. 1994. $175.00.

ONLINE DATABASES AND CD-ROMS

Compendex Plus. Engineering Information, Inc., 345 East 47th Street, New York, NY 10017. (212) 705-7600 or (800) 221-1044. Contains citations with abstracts to worldwide literature in engineering and technology, from 1970 to present. Available on online BRS,(800) 289-4277, DIALOG, (800) 334-2564, ORBIT (800) 456-7248, and STN International, FIZ Karlsruhe, P.O. Box 2465, W-7500, Karlsruhe 1, Germany, online services. Also available on CD-ROM. Inquire as to cost and availability.

Current Contents Search. Institute for Scientific Information, 3501 Market Street, Philadelphia, PA 19104. (215) 386-0100. FAX (215) 386-6362. Contains citations to articles listed in the table of contents of science and technology journals. Also articles in social sciences and life sciences journals. Available on BRS,(800) 289-4277, DIALOG,(800) 334-2564, online services. Inquire as to cost and availability.

NTIS Bibliographic Database. National Technical Information Service, 5285 Port Royal Road, Springfield, VA 22161. (703) 487-4929 or FAX (703) 321-8199. Broad coverage of government-sponsored science and technology research reports, 1964 to present. Available on BRS,(800) 289-4277, DIALOG, (800) 334-2564, ORBIT, (800) 456-7248, and STN International, FIZ Karlsruhe, P.O. Box 2465, W-7500, Karlsruhe 1, Germany, online services. Also available on CD-ROM. Inquire as to cost and availability.

Metadex. Jointly produced by ASM International and theInstitute of Materials. Contains more than 925,000 records from the international literature on metals and alloys, concerning properties, processes, materials classes, applications, and metallurgi-

BERYLLIUM

Ency. of Physical Sciences and Engineering Info. Sources

cal systems. Updated monthly. Available from ORBIT-QUESTEL (703) 442-0900.

NASA Database. American Institute of Aeronautics and Astronautics, 370 L'Enfant Promenade, S.W., Washington, DC 20024. (202) 646-7400. Citations and abstracts of aeronautics and astronautics literature, 1962 to present. Also contains citations from Star, Scientific and Technical Aerospace Reports, and International Aerospace Abstracts. Available through NASA/RECON online service. Inquire as to cost and availability.

Wilsonline. H.W. Wilson Company, 950 University Avenue, Bronx, NY 10452. (800) 367-6770 or (212) 588-8400. Makes available online versions of the printed H.W. Wilson indexes including Applied Science and Technology Index, Business Periodicals Index, General Science Index, and Readers' Guide to Periodical Literature. Period covered is generally 1983 to present. Available on BRS,(800) 289-4277, DIALOG,(800) 334-2564, and OCLC EPIC,(800) 848-5878, online services. Also available on CD-ROM. Inquire as to cost and availability.

PERIODICALS

Alloy Digest. Alloy Digest Inc., 27 Canfield Street, Orange, NJ 07050. (201) 677-9161. 1952 to present. Monthly. $140.00 per year.

J O M (Journal of Metals). Minerals, Metals and Materials Society, 420 Commonwealth Drive, Warrendale, PA 15086. (412) 776-9080. 1949 to present. Monthly. $50.00 per year.

Journal of Applied Mechanics. American Society of Mechanical Engineers, 345 E. 47th Street, New York, NY 10017. (212) 705-7722. 1935 to present. Quarterly. $120.00.

Journal of Engineering For Industry. American Society of Mechanical Engineers, 345 E. 47th Street, New York, NY 10017. (212) 705-7722. 1970 to present. Quarterly. $100.00 per year.

Light Metal Age. Fellom Publishing, 170 S. Spruce Avenue, Suite 120, San Francisco, CA 94080. (415) 588-8832. FAX (415) 588-0901. 1943 to present. Bi-monthly. $35.00 per year.

Metallurgical Transactions A: Physical Metallurgy and Materials Science. ASM International, Materials Park, OH 44073. (216) 338-5151 or FAX (216) 338-4634. 1970 to present. Monthly. $520.00 per year.

Metals Week. McGraw-Hill Publishing Company, 1221 Avenue of the Americas, New York, NY 10020. (800) 262-4729 or (212) 512-3825. 1930 to present. Weekly. $770.00 per year.

Metalworking Digest. Gordon Publications Inc., 301 Gibraltar Drive, Box 350, Morris Plains, NJ 07950-0650. (201) 292-5100. FAX (201) 898-9281. 1968 to present. 12 times a year. $48.00 per year.

RESEARCH CENTERS AND INSTITUTES

Cooperative Program in Metallurgy, Pennsylvania StateUniversity. 208A Steidle Bldg., University Park, PA 16802. (814) 865-5446.

University of Connecticut Institute of Materials Science. U-136, 97 N. Eagleville Road, Storrs, CT 06268. (203) 486-4623. FAX (203) 486-4745.

University of Florida Department of Materials Science and Engineering. Gainesville, FL 32611. (904) 392-1454. FAX (904) 392-6359.

BESSEMER PROCESS

See: STEEL AND STEEL MAKING

BETA RAYS

See: PARTICLE PHYSICS

BETATRON

See: PARTICLE ACCELERATORS

BIDIRECTIONAL TRANSISTOR

See: ELECTRONIC CIRCUITS AND COMPONENTS

BIG BANG THEORY

See also: ASTRONOMY, CELESTIAL MECHANICS, COSMOCHEMISTRY, COSMOLOGY, GALAXIES, INTERSTELLAR MATTER, PLANETS, SOLAR SYSTEM, TELESCOPES

ABSTRACT SERVICES AND INDEXES

Applied Science and Technology Index; A Cumulative Subject Index To English Language Periodicals in the Fields of Aeronautics and Space Science, Computer Technology, Chemistry, Construction Industry, Energy and Related Areas. H.W. Wilson Co., 950 University Avenue, Bronx, NY 10452. (212) 588-8400. (800) 367-6770. FAX (718) 590-1617. From 1958 to present. Monthly. Inquire about cost and availability. Also available on CD-ROM and online. ISSN: 0003-6986.

Astronomy and Astrophysics Abstracts. Springer-Verlag New York, 175 F ifth Avenue, New York, NY 10010. (212) 460-1500. FAX (212) 473-6272. Published for the Astronomisches Rechen-Institut. Comprehensive coverage of all aspects of astronomy, astrophysics and related fields. 1969 to present. Two parts per year. Annual. ISSN: 0067-0022.

General Science Index. H.W. Wilson Company, 950 University Avenue, Bronx, NY 10452. (212) 588-8400. (800) 367-6770. FAX (718) 590-1617. From 1978 to present. Ten issues per year; quarterly and annual cumulations. Service basis. Available on CD-ROM and online. Inquire regarding cost and availability. ISSN: 0162-1963.

Government Reports Announcements and Index. U. S. National Technical Information Service (NTIS), 5285 Port Royal Road, Springfield, VA 22161. (703) 487-4650. FAX (703) 321-8547. From 1968 to present. Annual. $630.00 per year. Also available online as NTIS Bibliographic Database and on CD-ROM.

Meteorological and Geoastrophysical Abstracts. American Meteorological Society, c/o Inforonics, Inc., 550 New town Road, Littleton, MA 01460. (508) 486-8976. FAX (508) 486-0027. Covers literature in environmental sciences, meteorology, astrophysics,hydrology, glaciology, and physical oceanography. From 1950 to present. Monthly. $950.00 per year. Also available on CD-ROM and online. Inquire regarding cost and availability. ISSN: 0026-1130.

NTIS Alerts: Astronomy & Astrophysics. U. S. National Technical Information Service. 5285 Port Royal Road, Springfield, VA 22161. (703) 487-4650. FAX (703) 321-8547. Weekly. $140.00 per year.

Pascal 48: Environnement Cosmique Terrestre, Astronomie Et Geologie Extraterrestre. Centre National de la Recherche Scientifique, Institut de Information Scientifique et Technique, 2 aliee du Parc de Brabois, 54514 Vandoeuvre Les Nancy Cedex, France. TEL 83-50-46-00. FAX: 83-50-46-50. 1985 to present. Ten issues per year. 770 F. Also available on CD-ROM and online.

Physics Abstracts. INSPEC. Section A, Science Abstracts. Institute of Electrical Engineers, London, United Kingdom. Available from: INSPEC/IEEE - Institute of Electrical and Electronic Engineers, Box 1331, Hoes Lane, Piscataway, NJ 08855-1331. (908) 562-5549. 1898 to present. 24 issues per year. $2835.00 per year. Also available online and on CD-ROM. ISSN: 0036-8091.

Science Citation Index. SCI. Institute for Scientific Information, 3501 Market Street, Philadelphia, PA 19104. (215) 386-0100. (800) 523-1850. FAX (215) 386-2991. 1961 to present. Six issues per year, plus annual cumulation. $11650.00 per year. Also available online and on CD-ROM. Inquire about price and availability. ISSN: 0036-827X.

STAR. Scientific and Technical Aerospace Reports. U .S. National Aeronautics and Space Administration. Distributed by U. S. Superintendent of Documents, Washington, DC 20402. 1963 to present. Semi-monthly, with semiannual and annual indexes. $114.00 per year. ISSN: 0036-8741. Also available online and on CD-ROM.

ANNUAL REVIEWS AND YEARBOOKS

Annual Review of Astronomy and Astrophysics; Original Reviews of Critical Literature and Current Developments in Astronomy and Astrophysics. Annual Reviews, Inc., 4139 El Camino Way, Palo Alto, CA 94306-0139. (415) 493-4400. (800) 523-8635. FAX (415) 855-9815. 1963 to date. Annual. $60.00. ISSN: 0066-4146.

Highlights of Astronomy. I. Appenzeller, editor. Kluwer Academic Publishers, Box 358, Accord Station, Hingham, MA 02018-0358. (617) 871-6600. International Astronomical Union Highlights series. ISBN: 07923-3553-8. Triennal. 1995. Price varies.

ASSOCIATIONS AND PROFESSIONAL SOCIETIES

Amateur Astronomers Association. 1010 Park Avenue, New York, NY 10028. (212) 535-2922.

American Association of Variable Star Observers. 23 Birch Street, Cambridge, MA 02138. (617) 354-0484. FAX: (617) 354-0665.

American Astronomical Society. 2000 Florida Avenue NW, Suite 400, Washington, DC 20009. (202) 328-2010. FAX: (202) 234-2560.

American Institute of Physics. One Physics Ellipse, College Park, MD 20740-3843. (301) 209-3100.

Association of University For Research in Astronomy. Suite 701, 1625 Massachusetts Avenue, NW, Washington, DC 20036. (202) 483-2101. FAX (202) 483-2106.

Astronomical League. 2112 Kingfisher Lane East, Rolling Meadows, IL 60008. (708) 398-0562.

Astronomical Society of the Pacific. 390 Ashton Avenue, San Francisco, CA 94112. (415) 337-1100. FAX: (415) 337-5205.

International Astronomical Union. 98bis Boulevard Arago, 75014 Paris, France. TEL 1-43-25-83-58. FAX 1-43-25-26-16.

Royal Astronomical Society of Canada. 136 Dupont Street, Toronto. ON N5R 1V2 Canada. (416) 924-7973. FAX (416) 968-6687.

BIBLIOGRAPHIES

Astronomy and Astrophysics: A Source Guide. Gordon Press, P.O. Box 459, Bowling Green Station, New York, NY 10004. (718) 624-8419. 1991. $250.00.

A Bibliography of Astronomy, 1970-1979. R. A. Sea and S. S. Martin. Libraries Unlimited, Inc., Littleton, CO 80160. 1982.

Science Books and Films. American Association for the Advancement of Science, 1333 H Street NW, Washington, DC 20005. (202) 326-6454. 1965 to present. Nine issues per year. $40.00 per year. ISSN: 0098-342X

Sc Ientific and Technical Books and Serials in Print; An Index To Literature in Science and Technology. R. R. Bowker. 121 Chanlon Road, New Providence, NJ 07974. (908) 464-6800. FAX (908) 665-3502. (800) 521-8110. 1972 to present. Annual. $299.95 per year. Also available on CD-ROM. ISSN: 0000-054X.

A Source Book on Astronomy and Astrophysics, 1900 - 1975. Kenneth Lang and Owen Gingerich, editors. Harvard University Press, 79 Garden Street, Cambridge MA. (617) 495-2600. 1980. $41.95.

DIRECTORIES AND BIOGRAPHICAL SOURCES

American Astronomical Society. Membership Directory. American Astronomical Society. 2000 Florida Avenue NW, Suite 400, Washington, DC 20009. (202) 328-2010. FAX (202) 234-2560 Annual. Included in membership dues. ISSN: 1061- 9038.

American Men and Women of Science: Physical and Biological Sciences. R. R. Bowker Inc., 121 Chanlon Road, New Providence, NJ 07974. (908) 464-6800. (800) 521-8110. 20th edition. 8 volumes. 1996. $850.00.

The Astronomers. Donald Goldsmith. St. Martin's Press, Inc., 175 Fifth Avenue, New York, NY 10010. (212) 674-5151. (800) 221-7945. 1993. $14.95 in paper.

Astronomical Centers of the World. Kevin Krisciunas. Cambridge University Press, 40 West 20th Street, New York, Ny 10011-4211. (212) 924-3900. (800) 872-7423. 1988. $34.95

The Biographical Dictionary of ScieNTISts: Astronomers. D. Abbott, editor. Peter Bedrick Books, Inc., 2112 Broadway, Room 318, New York, NY 10023. (212) 496-0751. 1984.

Directory of Physics and Astronomy Staff Members. American Institute of Physics, One Physics Ellipse, College Park, MD 20740-3843. (301) 209-3100. 1975/76 to present. Annual. $60.00. ISSN: 0361-2228.

Earth and Astronomical Research Centers. Stockton Press, 345 Park Avenue South, New York, NY 10010. 4th edition. 1995. $515.00. ISBN: 1-56169-0967.

Graduate Programs in Physics, Astronomy and Related Fields. 1995-1996. American Institute of Physics, One Physics Ellipse, College Park, MD 20740-3843. (301) 209-3100. 1978 to present. Annual. $45.00. ISSN: 0147-1821.

Research Centers Directory. Gale Research, 835 Penobscot Building, Detroit, MI 48226-4094. (313) 961-2242. (800) 347-4253. 20th edition. 1995. $485.00.

ENCYCLOPEDIAS AND DICTIONARIES

Astronomy From A To Z: A Dictionary of Celestial Objects and Ideas. Charles A. Schweighauser. Illinois Issues, Sangamon State University, K-10, Springfield, IL 62794-9243. (217) 786-9243. 1991. $14.95 in paper.

Concise Dictionary of Astronomy. Jacqueline Mitton. Oxford University Press, Inc., 200 Madison Avenue, New York, NY 10016. (212) 725-6000. (800) 334-4249. 1992. $24.95.

Encyclopedia of Astronomy and Astrophysics. Stephen Maran, editor. Van Nostrand Reinhold, 115 Fifth Avenue, New York, NY 10003. (212) 254-3232. (800) 842-3636. 1992. $129.95.

Facts On File Dictionary of Astronomy. Valerie Illingworth. Facts-on-file, Inc., 460 Park Avenue South, New York, NY 10016-7382. (212) 683-2244. (800) 322-8755. Fax (800) 683-3633. 1994. $27.95.

McGraw-Hill Encyclopedia of Science and Technology. McGraw-Hill Book Company, Inc., 1221 Avenue of the Americas, New York, NY 10020. (212) 512-2000. (800) 262-4729. 7th edition. 20 volume set. 1992. $1900.00.

Oxford Illustrated Encyclopedia. VOLUME 8, the UNIVERSE. Archie Roy, editor. Oxford University Press, Inc., 200 Madison Avenue, New York, NY 10016. (212) 725-6000. (800) 334-4249. 1993. $45.00.

Penguin Dictionary of Astronomy. Jacqueline Mitton. Viking Penguin, 375 Hudson Street, New York, NY 10014-3657. (212) 366-2000. (800) 331-4624. 1993. $12.95 in paper.

Stars, Galaxies, Cosmos. William R. Corliss. Sourcebook Project, P.O. Box 107, Glen Arm, MD 21057. (410) 668-6047. 1987. $17.95.

GENERAL WORKS

Astronomy and Cosmology. John North. W. W. Norton & Company, 500 Fifth Avenue, New York, NY 10110. (212) 354-5500. (800) 223-2584 1990. 1994. $18.95.

The Big Bang: The Creation and Evolution of the Universe. Joseph Silk. W. H. Freeman & Co., 41 Madison Avenue, East 26th Avenue, 35th Floor, New York, NY 10010. (212) 576-0400. Revised edition. 1995.

Cycles of Fire: Stars, Galaxies and the Wonder of Deep Space. William K. Hartmann. Workman Publishing Company, Inc., 708 Broadway, New York, NY 10003. (212) 254-5900. (800) 722-7202. 1988. $15.95 in paper.

Farthest Things in the Universe. Jay M. Pasachoff. Cambridge University Press 40 West 20th Street, New York, NY 10011-4211. (212) 924-3900. (800) 872-7423. 1995. $29.95.

First Three Minutes: A Modern View of the Origin of the Universe. Steven Weinberg. Basic Books, 10 East 53rd Street, New York, NY 10022. (212) 207-7057. (800) 242-7737. 2nd edition; Reprint edition. 1993. $12.00.

Gravitation. Charles W. Misner, et al. W. H. Freeman & Co., 41 Madison Avenue, East 26th Avenue, 35th Floor, New York, NY 10010. (212) 576-0400. 1995.

Hawking on the Big Bang and Black Holes. Stephen W. Hawking. World Scientific Publishing Company, Inc., 1060 Main Street, River Edge, NJ 07661. (201) 487-9655. (800) 227-7562. 1993. $33.00 in paper.

The Left Hand of Creation: The Origin and Evolution of the Expanding Universe. John D. Barrow and Joseph Silk. Oxford University Press, Inc., 200 Madison Avenue, New York, NY 10016. (212) 725-6000. (800) 334-4249. 1994. $23.00.

Masters of Time: How Wormholes, Snakewood and Assults on the Big Bang Have Brought Mystery Back To the Universe. John Boslough. Addison-Wesley Publishing Co., Inc., 1 Jacob Way, Reading, MA 01867. (617) 944-3700. (800) 447-2226. 1992. $wfi

The Moment of Creation: Big Bang Physics From Before the First Millisecond To the Present Universe. James S. Trefil. Simon & Schuster, Inc., 1230 Avenue of the Americas, New York, NY 10020. (212) 698-7000. (800) 223-2348. 1984. $10.95 in paper.

Origin of the Universe. John D. Barrow. Basic Books, 10 East 53rd Street, New York, NY 10022. (212) 207-7057. (800) 242-7737. 1994. $20.00.

The Physics of Stars. A. C. PHillips. John Wiley & Sons, Inc., 605 Third Avenue, New York, NY 10158-0012. (212) 850-6000. (800) 225-5945. 1994. $32.95 in paper.

The Red Limit: The Search For the Edge of the Universe. Timothy Ferris. Morrow, William & Company, 1350 Avenue of the Americas, New York, NY 10019. (212) 261-6500. (800) 843-9389. Revised edition. 1993. $12.95.

Space, Time and Gravity: The theory of the Big Bang and Black Holes. Robert M. Wald. University of Chicago Press, 5801 Ellis

Avenue, Chicago, IL 60637. (312) 702-7700. 2nd edition.1992. $24.95.

Stars and Atoms: From the Big Bang To the Solar System. Stuart Clark. Oxford University Press, Inc., 200 Madison Avenue, New York, NY 10016. (212) 725-6000. (800) 334-4249. 1995. $35.00.

Stephen Hawking's Universe. John Boslough. Morrow, William & Company, 1350 Avenue of the Americas, New York, NY 10019. (212) 261-6500. (800) 843-9389. 1985. $7.95.

The Vindication of the Big Bang: Breakthroughs and Barriers. B. Parker. Plenum Publishing Corp., 233 Spring Street, New York, NY. (212) 620-8000. (800) 2221-9369. 1993. $24.95.

HANDBOOKS AND MANUALS

Astrophysical Data: Planets and Stars. Kenneth R. Lanb. Springer-Verlag New York, Inc., 175 Fifth Avenue, New York, NY 10010. (212) 460-1500. (800) 777-4643. 1993. $59.00.

Burnham's Celestial Handbook: An Observer's Guide To the Universe Beyond the Solar System. Robert Burnham, Jr. Dover Publications, Inc., 180 Varick Street, New York, NY 10014. (212) 255-3755. (800) 223-3130. 3 volumes. 1978. $41.85 for the set.

Cambridge Guide To the Constellations. Michael E. Bakich. Cambridge University Press, 40 West 20th Street, New York, NY 10011-4211. (212) 924-3900. (800) 872-7423. 1995. $69.95.

Encyclopedia of Astronomy and Astrophysics. Stephen Maran, editor. Van Nostrand Reinhold, 115 Fifth Avenue, New York, NY 10003. (212) 254-3232. (800) 842-3636. 1992. $129.95.

Greenwich Guide To Stars, Galaxies and Nebulae. Stuart Malin. Cambridge University Press, 40 West 20th Street, New York, NY 10011-4211. (212) 924-3900. (800) 872-7423. 1990. $10.95.

Handbook of Astronomy, Astrophysics and Geophysics; Volume 2: Galaxies and Cosmology. V. M. Canuto and B. G. Elmegreen. Gordon & Breach Science Publishers, Inc., 50 West 23rd Street, New York, NY 10010. (212) 206-8900. . 1988. $194.00.

Handbook of Space Astronomy and Astrophsyics. Martin V. Zumbeck. Cambridge University Press, 40 West 20th Street, New York, NY 10011-4211. (212) 924-3900. (800) 872-7423. 2nd edition. 1990. $79.95.

Landolt-borenstein Numerical Data and Functional Relationships in Science and Technology: Astronomy, Astrophysics and Space Research. GROUP VI. Springer- Verlag New York, Inc., 175 Fifth Avenue, New York, NY 10010. (212) 460-1500. (800) 777-4643. volumes priced individually.

McGraw-Hill Encyclopedia of Astronomy. Sybil P. Parker and Jay M. Pasachoff, editors. McGraw-Hill Publishing Company, Inc., 1221 Avenue of the Americas, New York, NY 10020. (212) 512-2000. (800) 262-4729. 2nd edition. 1993. $75.00

Observer's Guide To Astronomy. Patrick Martinez, editor. Cambridge University Press, 40 West 20th Street, New York, NY 10011-4211. (212) 924-3900. (800) 872-7423. 2 volumes. 1994. $159.90.

ONLINE DATABASES AND CD-ROM

CA Search. Chemical Abstracts Service, P.O. Box 3012, Columbus, OH 43210-0012. (614) 447-3600. (800) 848-6533. FAX (614) 447-3709. Very comprehensive guide to worldwide chemical literature and related fields, 1972 to present. Available on BRS,(800) 289-4277, DIALOG, (800) 334-2564, ORBIT (800) 456-7248, and STN International, FIZ Karlsruhe, P.O. Box 2465, W-7500, Karlsruhe 1, Germany, online services. Inquire as to cost and availability.

Current Contents Search. Institute for Scientific Information, 3501 Market Street, Philadelphia, PA 19104. (215) 386-0100. FAX (215) 386-6362. Contains citations to articles listed in the table of contents of science and technology journals. Also articles in social scie nces and life sciences journals. Available on BRS, (800) 289-4277, DIALOG, (800) 334-2564, online services. Inquire as to cost and availability.

Dissertation Abstracts Online. University Microfilms Internatio nal, 300 North Zeeb Road, Ann Arbor, MI 48106. (800) 521-0600 or (313) 761-4700. Scope includes virtually all doctoral dissertations accepted at accredited American institutions from 1861 to present in 252 subject areas. Available on BRS, (800) 289-4277, DIALOG, (800) 334-2564, and OCLC EPIC, (800) 848-5878, online services. Also available on CD-ROM. Inquire as to cost and availability.

INSPEC. Institution of Electrical Engineers, Michael Faraday House, Six Hills Way, Stevenage, Herts. SG1 2AY, E ngland. Telephone: 0438 313311 or FAX 0438 7428 40. Contains citations to the worldwide literature of ph ysics, electronics and electrical engineering, computer technology, and related fields. Available on BRS, (800) 289-4277, DIALOG, (800) 334-2564, ORBIT, (800) 456-7248, and STN International, FIZ Karlsruhe, P.O. Box 2465, W-7500, Karlsruhe 1, Germany, online services. Inquire as to cost and availability.

International Aerospace Abstracts. American Institute of Aeronautics and Astronautics, 370 L'Enfant Promenade, S.W., Washington, DC 20024. (202) 646-7400. Contains references and abstracts of journal and monograph literature relating to aerospace science and technology, from 1963 to present. Available through the NASA/RECON system of the National Aeronautics and Space Administration only.

NTIS Bibliographic Database. National Technical Information Service, 5285 Port Royal Road, Springfield, VA 2216 1. (703) 487-4929 or FAX (703) 321-8199. Broad cover age of government-sponsored science and technology research reports, 1964 to present. Available on BRS,(800) 289-4277, DIALOG, (800) 334-2564, ORBIT, (800) 456-7248, and STN International, FIZ Karlsruhe, P.O. Box 2465, W-7500, Karlsruhe 1, Germany, online services. Also available on CD-ROM. Inquire as to cost and availability.

Physics Briefs. American Institute of Physics, 335 East 45th Street , New York, NY 10017. (212) 661-9260 or FAX (212)661-2036. Contains citations with abstracts of the literature of physics and related fields, 1979 to present. Available on the STN International, FIZ Karlsruhe, P.O. Box 2465, W-7500, Karlsruhe 1, Germany, online service. Inquire as to cost and availability.

Scientific and Technical Books and Serials in Print. R.R. B owker Inc., 121 Chanlon Road, New Providence, NJ 07974. (800) 521-8110 or (908) 464-6800. List of currently published books and serials in the physical and biological sciences, engineering and technology, with subject, author and titles indexes. Available

BIG BANG THEORY

Ency. of Physical Sciences and Engineering Info. Sources

on ORBIT, (800) 456-7248, online service. Inquire as to cost and availability.

SCISEARCH. Institute for Scientific Information, 3501 MA rket Street, Philadelphia, PA 19104. (800) 523-1850 or (215) 386-0100. Broad multidisciplinary title and author index to the international literature of science and technology, 1974 to present. Available on DIALOG, (800) 334-2564, and ORBIT, (800) 456-7248, online services. Also available on CD-ROM. Inquire as to cost and availability.

Wilsonline. H.W. W ilson Company, 950 University Avenue, Bronx, NY 10452. (800) 367-6770 or (212) 588-8400. Makes available onlin e versions of the printed H.W. Wilson indexes includin g Applied Science and Technology Index, Business Periodicals Index, General Science Index, and Readers' Guide to Periodical Literature. Period covered is generally 1983 to present. Available on BRS, (800) 289-4277, DIALOG, (800) 334-2564, and OCLC EPIC, (800) 848-5878, online services. Also available on CD-ROM. Inquire as to cost and availability.

OTHER SOURCES

Atlas of Deep-Sky Spendors. Hans Vehrenberg. Sky Publishing Corporation, 49 Bay State Road, Cambridge, MA 02138. (617) 864-7360. (800) 253-0245. 4th edition. 1983. $24.95.

Atlas of the Ultraviolet Sky. Richard C. Henry, Wayne B. Landsman, et al. Johns Hopkins University Press, 2715 North Charles Street, Baltimore, MD 21218-4319. (410) 516-6900. (800) 537-5487. 1988. $70.00.

Atlas of the Universe. Patrick Moore. Rand McNally & Company. 8255 North Central Park, Skokie, IL 60076-2970. (708) 673-9100. 1994. $29.95.

The Bright Star Catalogue. Dorrit Hoffleit and Carlos Jaschek. Yale University Observatory, 260 Whitney Avenue, P.O. Box 6666, New Haven CT 06511. (203) 432-3431. 4th revised edition. 1982. $35.00 in paper.

Cambridge Atlas of Astronomy. Jean Audouze, et al. Cambridge University Press, 40 West 20th Street, New York,

Ny 10011-4211. (212) 924-3900. (800) 872-7423. 3rd edition. 1994. $75.00.

Cambridge Guide To the Constellations. Michael E. Bakich. Cambridge University Press, 40 West 20th Street, New York, NY 10011-4211. (212) 924-3900. (800) 872-7423. 1995. $19.95 in paper.

Cambridge Star Atlas 2000.0. Wil Tirion. Cambridge University Press, 40 West 20th Street, New York, NY 10011- 4211. (212) 924-3900. (800) 872-7423. 2nd edition. 1996. $19.95.

Color Atlas of Galaxies. James Wray. Cambridge University Press, 40 West 20th Street, New York, NY 10011-4211. (212) 924-3900. (800) 872-7423. 1988. $94.95.

Exploring the Southern Sky: A Pictorial Atlas From the European Southern Observatory. S. Lausten, et al. Springer-Verlag, 175 Fifth Avenue, New York, NY 10010. (212) 460-1500. (800) 777-4643. 1987. $59.00.

Observer's Sky Atlas. E. Karkoschka. Springer-Verlag New York, Inc., 175 Fifth Avenue, New York, NY 10010. (212) 460-

1500. (800) 777-4643. 1995. $15.95.

Ngc 2000.0: The Complete New General Catalogue and Index Catalogues of Nebulae and Star Clusters. J. L. E. Dreyer. editor. Cambridge University Press, 40 West 20th Street, New York, NY 10011-4211. (212) 924-3900. (800) 872-7423 1989. $21.95.

Norton's 2000.0 Star Atlas and Reference Handbook. Ian Ridpath, editor. Halsted Press, 605 Third Avenue, New York, 10158-0012. (212) 850-6400. 1989. $49.95.

Observing Handbook and Catalog of Deep-sky Objects. Christian B. Luginbuhl and Brian A. Skiff. Cambridge University Press, 40 West 20th Street, New York, NY 10011- 4211. (212) 924-3900. (800) 872-7423. 1990. $49.95.

Starlist Two Thousand: A Quick Reference Star Catalog For Astronomers. Richard Dibon-Smith. John Wiley & Sons, Inc., 605 Third Avenue, New York, NY 10158-0012. (212) 850-6000. (800) 225-5945. 1992. $29.95.

The Sky: A User's Guide. David H. Levy. Cambridge University Press, 40 West 20th Street, New York, NY 10011- 4211. (212) 924-3900. (800) 872-7423. 1993. $14.95 in paper.

Sky Catalog 2000.0: Volume 1: Stars to Magnitude 8.0. Alan Hirshfeld, et al. Cambridge University Press, 40 West 20th Street, New York, NY 10011-4211. (212) 924-3900. (800) 872-7423. 2nd edition. 1991. $42.95 in paper.

PERIODICALS

Acta Astronomica. Copernicus Foundation for Polish Astronomy, AL Ujazdowskie 4, 00-478 Warsaw, Poland. TEL 48-22-295346. FAX 48-22-294967. 1925 to present. Quarterly. $80.00 per year. ISSN: 0001-5237.

Astronomical Journal. American Institute of Physics, One Physics Ellipse, College Park, MD 20740-3843. (301) 209- 3000. Published for the American Astronomical Society. 1849 to present. Monthly. $280.00 per year. ISSN: 0004- 6256.

Astronomical Society of the Pacific. PUBLICATIONS. Astronomical Society of the Pacific, 390 Ashton Avenue, San Francisco, CA 94112. (415) 337-1100. FAX (415) 337-5205. 1889 to present. Monthly. $175.00 per year. ISSN: 0004- 6280.

Astronomy. Kalmbach Publishing Company, Box 1612, Waukesha, WI 53187-1612. (414) 796-8776. FAX (414) 796-1142. 1973to present. Monthly. $27.00 per year. ISSN: 0091-6358.

Astronomy and Astrophsyics. Springer-Verlag New York, Inc., 44 Hartz Way, Secaucus, NJ 07096-2491. (201) 348-4033. FAX (201) 348-4505. 1969 to date. 36 issues per year in 12 volumes. $1733.00 per year. ISSN: 0004-6361.

Astronomy Reports. American Institute of Physics, One Physics Ellipse, College Park, MD 20740-3843. (301) 209- 3000. English translation of Astronomicheksii Zhurnal. Formerly Soviet Astronomy AJ. 1957 to present. Bimonthly. $1160.00 per year. ISSN: 1063-7729.

Astroparticle Physics. Elsevier Science Publishing Company, Inc., Box 882, Madison Square Station, New York, NY 10159. (212) 989-5800. FAX (212) 633-3990. 1992 to present. Quarterly. $180.00 per year. ISSN: 0927-6505.

Ency. of Physical Sciences and Engineering Info. Sources

BINARY STARS

Astrophysical Journal; An International Review of Astronomy and Astronomical Physics. University of Chicago Press, Journals Division, 5720 South Woodlawn Avenue, Chicago, IL 60637. (312) 753-3347. FAX (312) 753-0811. 1895 to date. three issues per month. $740.00 per year. ISSN: 0004-637X.

Astr Ophysics and Space Science; An International Journal of Cosmic Physics. Kluwer Academic Publishers, Box 358, Accord Station, Hingham, MA 02018-0358. (617) 871-6600. FAX (617) 871-6528. 1968 to present. 24 issues per year. $2544.00 per year. ISSN: 0004-640X.

Celestial Mechanics and Dynamical Astronomy; An International Journal of Space Dynamics. Kluwer Academic Publishers, Box 358, Accord Station, Hingham, MA 02018-0358. (617) 871-6600. FAX (617) 871-6528. 1969 to present. Monthly. $703.50 per year. ISSN: 0923-2958.

Experimental Astronomy; An International Journal on Astronomical Instrumentation and Data Analysis. Kluwer Academic Publishers, Box 358, Accord Station, Hingham, MA 02018-0358. (617) 871-6600. FAX (617) 871-6528. 1989 to present. Quarterly. $161.00 per year. ISSN: 0922-6435.

Mercury. Astronomical Society of the Pacific. 390 Ashton Avenue, San Francisco, CA 94112. (415) 337-1100. FAX: (415) 337-5205. 1972 to present. Bimonthly. $175.00 per year. ISSN: 0047-6773.

Sky and Telescope. Sky Publishing Corporation, Box 9111, Belmont, MA 02178. (617) 864-7360. FAX (617) 864-6117. 1941 to present. Monthly. $27.00 per year. ISSN: 0037-6604.

Space Science Reviews. Kluwer Academic Publishers, Box 358, Accord Station, Hingham, MA 02018-0358. (617) 871-6600. FAX (617) 871-6528. 1962 to present. Sixteen issues per year. $940.00 per year. ISSN: 0038-6308.

Vistas in Astronomy; An International Review Journal. Albert C. Beer, editor. Elsevier Science Publishing Company, Inc., 660 White Plains Road, Terrytown, NY 10591-5153. (914) 524-9200. FAX: (914) 333-2444. 1958 to present. Quarterly. $415.00 per year. ISSN: 0083-6656.

Webb Society Quarterly Journal. Webb Society, 194 Foundry Lane, Freemantle, Southampton, Hampshire SO1 3EQ, England. 1969 to present. Quarterly. $19.50 per year. ISSN: 0043-1680.

RESEARCH CENTERS AND INSTITUTES

Astronomy Department. Yale University. 20 Whitney Avenue. Box 6666, New Haven, CT 06511. (203) 432-3000.

Astronomy Program. University of Maryland, 1207 Computer & Space Sciences Building, College Park, MD 20742. (301) 405-3001. FAX (301) 314-9067.

Center For Astronomy and Space Astrophysics. University of Colorado - Boulder, Campus Box 391, Boulder, CO 80309. (303) 492-4050. (FAX) 492-7178.

Center For Astrophysics and Space Sciences. University of California, San Diego, C-0111, LA Jolla, CA 92093-0111. (619) 534-3460. FAX (619) 534-2294.

Center For Space Science and Astrophysics. Stanford University, 325 Durand Building, Stanford, CA 94305. (415) 723-3592.

Space Sciences Laboratory. University of California, Berkeley, Berkeley, CA 94720. (415) 642-1361. FAX (415) 643-7629.

Space Telescope Science Institute. 3700 San Martin Drive, Baltimore, MD 21218. (301) 338-4700.

Theoretical Astrophysics Program. University of Arizona, Planetary Sciences Department, Tuscon, AZ 85721. (602) 621-6963.

BINARY STARS

See also: ASTRONOMY, BIG BANG, COSMOLOGY, DARK MATTER, GALAXIES, INTERSTELLAR MEDIUM, MILKY WAY, NEUTRON STARS, PULSARS, QUASARS, RADIO ASTRONOMY, STARS, UNIVERSE

ABSTRACT SERVICES AND INDEXES

Applied Science and Technology Index; A Cumulative Subject Index To English Language Periodicals in the Fields of Aeronautics and Space Science, Computer Technology, Chemistry, Construction Industry, Energy and Related Areas. H.W. Wilson Co., 950 University Avenue, Bronx, NY 10452. (212) 588-8400. (800) 367-6770. FAX (718) 590-1617. From 1958 to present. Monthly. Inquire about cost and availability. Also available on CD-ROM and online. ISSN: 0003-6986.

Astronomy and Astrophysics Abstracts. Springer-Verlag New York, 175 Fifth Avenue, New York, NY 10010. (212) 460-1500. FAX (212) 473-6272. Published for the Astronomisches Rechen-Institut. Comprehensive coverage of all aspects of astronomy, astrophysics and related fields. 1969 to present. Two parts per year. Annual. ISSN: 0067-0022.

General Science Index. H.W. Wilson Company, 950 University Avenue, Bronx, NY 10452. (212) 588-8400. (800) 367-6770. FAX (718) 590-1617. From 1978 to present. Ten issues per year; quarterly and annual cumulations. Service basis. Available on CD-ROM and online. Inquire regarding cost and availability. ISSN: 0162-1963.

Government Reports Announcements and Index. U. S. National Technical Information Service (NTIS), 5285 Port Royal Road, Springfield, VA 22161. (703) 487-4650. FAX (703) 321-8547. From 1968 to present. Annual. $630.00 per year. Also available online as NTIS Bibliographic Database and on CD-ROM. ISSN:

Meteorological and Geoastrophysical Abstracts. American Meteorological Society, c/o Inforonics, Inc., 550 New town Road, Littleton, MA 01460. (508) 486-8976. FAX (508) 486-0027. Covers literature in environmental sciences, meteorology, astrophysics, hydrology, glaciology, and physical oceanography. From 1950 to present. Monthly. $950.00 per year. Also available on CD-ROM and online. Inquire regarding cost and availability. ISSN: 0026-1130.

NTIS Alerts: Astronomy & Astrophysics. U. S. National Technical Information Service. 5285 Port Royal Road, Springfield, VA 22161. (703) 487-4650. FAX (703) 321-8547. Weekly. $140.00 per year.

Pascal 48: Environnement Cosmique Terrestre, Astronomie Et Geologie Extraterrestre. Centre National de la Recherche Scientifique, Institut de Information Scientifique et Technique, 2 aliee du Parc de Brabois, 54514 Vandoeuvre Les Nancy Cedex,

BINARY STARS

Ency. of Physical Sciences and Engineering Info. Sources

France. TEL 83-50-46-00. FAX: 83-50-46-50. 1985 to present. Ten issues per year. 770 F. Also available on CD-ROM and online.

Physics Abstracts. INSPEC. Section A, Science Abstracts. Insti tute of Electrical Engineers, London, United Kingdom. Available from: INSPEC/IEEE - Institute of Electrical and Electronic Engineers, Box 1331, Hoes Lane, Piscataway, NJ 08855-1331. (908) 562-5549. 1898 to present. 24 issues per year. $2835.00 per year. ISSN: 0036-8091. Also available online and on CD-ROM.

Physics Briefs (Physikalische Berichte). Information Center for Energy, Physics, Mathematics; German Physical Society.

V C H Publishers, Inc., 220 East 23rd Street, New York, NY 10010-4606. (212) 683-8333. 1845 to present. 24 issues per year. $2390.00 per year. Also available online. ISSN: 0179-7434.

Science Citation Index. SCI. Institute for Scientific Information, 3501 Market Street, Philadelphia, PA 19104. (215) 386-0100. (800) 523-1850. FAX (215) 386-2991. 1961 to present. Six issues per year, plus annual cumulation. $11650.00 per year. Also available online and on CD-ROM. Inquire about price and availability. ISSN: 0036-827X.

STAR. Scientific and Technical Aerospace Reports. U.S. National Aeronautics and Space Administration. Distributed by U. S. Superintendent of Documents, Washington, DC 20402. 1963 to present. Semi-monthly, with semiannual and annual indexes. $114.00 per year. ISSN: 0036-8741. Also available online and on CD-ROM.

ANNUAL REVIEWS AND YEARBOOKS

The Astronomical Almanac. Superintendent of Documents, U. S. Government Printing office, Washington, DC 10402. (202)783-3238. 1981 to present. Supercedes Astronomical Ephemeris and the American Ephermis and Nautical Almanac. Annual. $44.95. ISSN: 0737-6421.

Annual Review of Astronomy and Astrophysics; Original Reviews of Critical Literature and Current Developments in Astronomy and Astrophysics. Annual Reviews, Inc., 4139 El Camino Way, Palo Alto, CA 94306-0139. (415) 493-4400. (800) 523-8635. FAX (415) 855-9815. 1963 to date. Annual. $60.00. ISSN: 0066-4146.

Highlights of Astronomy. I. Appenzeller, editor. Kluwer Academic Publishers, Box 358, Accord Station, Hingham, MA 02018-0358. (617) 871-6600. International Astronomical Union Highlights series. ISBN: 07923-3553-8. Triennal. 1995. Price varies.

ASSOCIATIONS AND PROFESSIONAL SOCIETIES

Amateur Astronomers Association. 1010 Park Avenue, New York, NY 10028. (212) 535-2922.

American Association of Variable Star Observers. 23 Birch Street, Cambridge, MA 02138. (617) 354-0484. FAX: (617) 354-0665.

American Astronomical Society. 2000 Florida Avenue NW, Suite 400, Washington, DC 20009. (202) 328-2010. FAX: (202) 234-2560.

American Institute of Physics. One Physics Ellipse, College Park, MD 20740-3843. (301) 209-3100.

Association of University For Research in Astronomy. Suite 701, 1625 Massachusetts Avenue, NW, Washington, DC 20036. (202) 483-2101. FAX (202) 483-2106.

Astronomical League. 2112 Kingfisher Lane East, Rolling Meadows, IL 60008. (708) 398-0562.

Astronomical Society of the Pacific. 390 Ashton Avenue, San Francisco, CA 94112. (415) 337-1100. FAX: (415) 337-5205.

International Astronomical Union. 98 bis Boulevard Arago, 75014 Paris, France. Tel 1-43-25-83-58. FAX 1-43-25-26-16.

Royal Astronomical Society of Canada. 136 Dupont Street, Toronto. ON N5R 1V2 Canada. (416) 924-7973. FAX (416) 968-6687.

BIBLIOGRAPHIES

Astronomy and Astrophysics: A Source Guide. Gordon Press, P.O. Box 459, Bowling Green Station, New York, NY 10004. (718) 624-8419. 1991. $250.00. .

A Bibliography of Astronomy, 1970-1979. R. A. Sea and S. S. Martin. Libraries Unlimited, Inc., Littleton, CO 80160. 1982.

Catalog of the Naval Observatory Library. Naval Observatory Staff. G. K. Hall & Company, 866 Third Avenue, New York, NY 10022. (212)702-6789. (800) 257-5755. 6 volumes. 1977. $655.00

Science Books and Films. American Association for the Advancement of Science, 1333 H Street NW, Washington, DC 20005.(202) 326-6454. 1965 to present. Nine issues per year. $40.00 per year. ISSN: 0098-342X

Scientific and Technical Books and Serials in Print, 1995. R.R. Bowker Inc., 121 Chanlon Road, New Providence, NJ 07974. (908) 464-6800. (800) 521-8110. 4 volumes. 1994. $299.95. Also available on compact disc and online.

A Source Book on Astronomy and Astrophysics, 1900 - 1975. Kenneth Lang and Owen Gingerich, editors. Harvard University Press, 79 Garden Street, Cambridge MA. (617) 495-2600. 1980. $41.95.

DIRECTORIES AND BIOGRAPHICAL SOURCES

American Men and Women of Science: Physical and Biological Sciences. R. R. Bowker Inc., 121 Chanlon Road, New Providence, NJ 07974. (908) 464-6800. (800) 521-8110. 20th edition. 8 volumes. 1996. $850.00.

The Astronomers. Donald Goldsmith. St. Martin's Press, Inc., 175 Fifth Avenue, New York, NY 10010. (212) 674-5151. (800) 221-7945. 1993. $14.95 in paper.

Astronomical Centers of the World. Kevin Krisciunas. Cambridge University Press, 40 West 20th Street, New York, NY 10011-4211. (212) 924-3900. (800) 872-7423. 1988. $34.95

Ency. of Physical Sciences and Engineering Info. Sources

BINARY STARS

The Biographical Dictionary of Scientists: Astronomers. D. Abbott, editor. Peter Bedrick Books, Inc., 2112 Broadway, Room 318, New York, NY 10023. (212) 496-0751. 1984.

Directory of Physics and Astronomy Staff Members. American Institute of Physics, One Physics Ellipse, College Park, MD 20740-3843. (301) 209-3100. Annual. $45.00.

Earth and Astronomical Research Centers. Stockton Press, 345 Park Avenue South, New York, NY 10010. 4th edition. 1995. $515.00. ISBN: 1-56169-0967.

Graduate Programs in Physics, Astronomy and Related Fields. 1995-1996. American Institute of Physics, One Physics Ellipse, College Park, MD 20740-3843. (301) 209-3100. 1978 to present. Annual. $45.00.

Research Centers Directory. Gale Research, 835 Penobscot Building, Detroit, MI 48226-4094. (313) 961-2242. (800) 347-4253. 20th edition. 1995. $485.00.

ENCYCLOPEDIAS AND DICTIONARIES

Astronomy From A To Z: A Dictionary of Celestial Objects and Ideas. Charles A. Schweighauser. Illinois Issues, Sangamon State University, K-10, Springfield, IL 62794-9243. (217) 786-9243. 1991. $14.95 in paper.

Concise Dictionary of Astronomy. Jacqueline Mitton. Oxford University Press, Inc., 200 Madison Avenue, New York, NY 10016. (212) 725-6000. (800) 334-4249. 1992. $24.95.

Facts On File Dictionary of Astronomy. Valerie Illingworth. Facts-on-file, Inc., 460 Park Avenue South, New York, NY 10016-7382. (212) 683-2244. (800) 322-8755. Fax (800) 683-3633. 1994. $27.95.

Encyclopedia of Astronomy and Astrophysics. StephenMaran, editor. Van Nostrand Reinhold, 115 Fifth Avenue, New York, NY 10003. (212) 254-3232. (800) 842-3636. 1992. $129.95.

L andolt-boernstein Numerical Data and Functional Relationships in Science and Technology: Astronomy, Astrophysics and Space Research; A. O. Madelung and H. H. Boigt, editors. Springer Verlag New York, Inc., 175 Fifth Avenue, New York, NY 10010. (212) 460-1500. (800) 777-4643. 1993. $550.00.

Oxford Illustrated Encyclopedia. Volume 8, the Universe. Archie Roy, editor. Oxford University Press, Inc., 200 Madison Avenue, New York, NY 10016. (212) 725-6000. (800) 334-4249. 1993. $45.00.

McGraw-Hill Encyclopedia of Science and Technology. McGraw-Hill Book Company, Inc., 1221 Avenue of the Americas, New York, NY 10020. (212) 512-2000. (800) 262-4729. 7th edition. 20 volume set. 1992. $1900.00.

Stars, Galaxies, Cosmos. William R. Corliss. Sourcebook Project, P.O. Box 107, Glen Arm, MD 21057. (410) 668-6047. 1987. $17.95.

GENERAL WORKS

Binary and Multiple Systems of Stars. A. H. Batten. Franklin Book Co., Inc., Elkins Park, PA 19117. (215) 635-5252. Reprint edition. 1973. $120.00

Cosmic Questions, Galactic Halos, Cold Dark Matter and the End of Time. Richard Morris. John Wiley & Sons, Inc., 605 Third Avenue, New York, NY 10158-0012. (212) 850-6000. (800) 225-5945. 1993. $24.95.

Cycles of Fire: Stars, Galaxies and the Wonder of Deep Space. William K. Hartmann. Workman Publishing Company, Inc., 708 Broadway, New York, NY 10003. (212) 254-5900. (800) 722-7202. 1988. $15.95 in paper.

Farthest Things in the Universe. Jay M. Pasachoff. Cambridge University Press, 40 West 20th Street, New York, NY 10011-4211. (212) 924-3900. (800) 872-7423. 1995. $29.95.

High-energy Radiation From Magnetized Neutron Stars. Peter Meszaros. University of Chicago Press, 5801 Ellis Avenue, Chicago, IL 60637. (312) 702-7700. 1992. $39.95 in paper.

Interacting Binaries. S. N. Shore, et al. Springer-Verlag New York, Inc., 175 Fifth Avenue, New York, NY 10010. (212) 460-1500. (800) 777-4643. FAX (212) 473-6272. 1994. $59.00.

The Left Hand of Creation: The Origin and Evolution of the Expanding Universe. John D. Barrow and Joseph Silk. Oxford University Press, Inc., 200 Madison Avenue, New York, NY 10016. (212) 725-6000. (800) 334-4249. 1994. $23.00.

Light Curve Modeling of Eclipsing Binary Stars. E. F. Milone. Springer-Verlag New York, Inc., 175 Fifth Avenue, New York, NY 10010. (212) 460-1500. (800) 777-4643. FAX (212) 473-6272. 1992. $65.00.

Interacting Binary Stars. Allen W. Shafter, editor. Astronomical Society of the Pacific, 390 Ashton Avenue, SanFrancisco, CA 94112. (415) 337-1100. 1994. $40.00.

New Frontiers in Binary Research. K. C. Leung and I. S. Nha, editors. Astronomical Society of the Pacific, 390 Ashton Avenue, San Francisco, CA 94112. (415) 337-1100. 1993. $40.00.

New Perspectives on Stellar Pulsation and Pulsating Variable Stars. James M. Nemec and Jaymie M. Matthews. Cambridge U. Press, 40 West 20th Street, New York, NY 10011-4211. (212) 924-3900. 1993. $59.95.

Observational Astrophysics. Robert C. Smith. Cambridge University Press, 40 West 20th Street, New York, NY 10011-4211. (212) 924-3900. (800) 872-7423. 1995. $69.95.

The Physics of Stars. A. C. PHillips. John Wiley & Sons, Inc., 605 Third Avenue, New York, NY 10158-0012. (212) 850-6000. (800) 225-5945. 1994. $32.95 in paper.

Plasma Astrophysics. J. G. Kirk, et al. Springer-Verlag New York, Inc., 175 Fifth Avenue, New York, NY 10010. (212) 460-1500. (800) 777-4643. FAX (212) 473-6272. 1994. $68.00.

Realm of Interacting Binary Stars. J. Schade, editor. Kluwer Academic Publishers, 101 Philip Drive, Assinippi Park, Norwell, MA 02061. (617) 871-6600. 1993. $140.50.

Stars and Atoms: From the Big Bang To the Solar System. Stuart Clark. Oxford University Press, Inc., 200 Madison Avenue, New York, NY 10016. (212) 725-6000. (800) 334-4249. 1995. $35.00.

Stellar and Circumstellar Astrophysics. George Wallerstein and Alberto Noriega-Crespo, editors. Astronomical Society of the Pacific, 390 Ashton Avenue, San Francisco, CA 94112. (415) 337-1100. 1994. $40.00.

Structure and Evolution of Single and Binary Stars. Camiel W. DeLoore and C. Doon. Kluwer Academic Publishers, 101 Philip Drive, Assinippi Park, Norwell, MA 02061. (617) 871-6600. 1992. $75.00 in paper.

X-ray Binaries. Walter H. Lewin, et al, editors. Cambridge University Press, 40 West 20th Street, New York, NY 10011- 4211. (212) 924-3900. (800) 872-7423. 1995. $69.95. 1992.

X-ray Binaries and Recycled Pulsars. E. P. Van den Heuvel and S. A. Rappaport, editors. Kluwer Academic Publishers, 101 Philip Drive, Assinippi Park, Norwell, MA 02061. (617) 871-6600. 1992. $222.00

HANDBOOKS AND MANUALS

Burnham's Celestial Handbook: An Observer's Guide To the Universe Beyond the Solar System. Robert Burnham, Jr. Dover Publications, Inc., 180 Varick Street, New York, NY 10014. (212) 255-3755. (800) 223-3130. 3 volumes. 1978. $41.85 for the set.

Cambridge Guide To the Constellations. Michael E. Bakich. Cambridge University Press, 40 West 20th Street, New York, NY 10011-4211. (212) 924-3900. (800) 872-7423. 1995. $69.95.

Encyclopedia of Astronomy and Astrophysics. Stephen Maran, editor. Van Nostrand Reinhold, 115 Fifth Avenue, New York, NY 10003. (212) 254-3232. (800) 842-3636. 1992. $129.95.

Greenwich Guide To Stars, Galaxies and Nebulae. Stuart Malin. Cambridge University Press, 40 West 20th Street, New York, NY 10011-4211. (212) 924-3900. (800) 872-7423. 1990. $10.95.

Handbook of Astronomy, Astrophysics and Geophysics; Volume 2: Galaxies and Cosmology. V. M. Canuto and B. G. Elmegreen. Gordon & Breach Science Publishers, Inc., P.O. Box 200029, Riverfront Plaza Station, Newark, NJ 07102-0301. (201) 643-7500. 1988. $194.00.

Handbook of Space Astronomy and Astrophsyics. Martin V. Zumbeck. Cambridge University Press, 40 West 20th Street, New York, NY 10011-4211. (212) 924-3900. (800) 872-7423. 2nd edition. 1990. $79.95.

Landolt-borenstein Numerical Data and Functional Relationships in Science and Technology: Astronomy, Astrophysics and Space Research. GROUP VI. Springer- Verlag New York, Inc., 175 Fifth Avenue, New York, NY 10010. (212) 460-1500. (800) 777-4643. volumes priced individually.

McGraw-Hill Encyclopedia of Astronomy. Sybil P. Parker and Jay M. Pasachoff, editors. McGraw-Hill Publishing Company, Inc., 1221 Avenue of the Americas, New York, NY 10020. (212) 512-2000. (800) 262-4729. 2nd edition. 1993. $75.00

National Audobon Society Pocket Guide: Galaxies and Other Deep-sky Objects. Alfred A. Knopf, Inc. 201 E. 50th Street, New York, NY 10022. (800) 733-3000. 1995. $7.95.

Nearby Galaxies Catalog. R. Brent Tully and J. Richard Fisher. Cambridge University Press, 40 West 20th Street, New York,

NY 10011-4211. (212) 924-3900. (800) 872-7423. 1988. $59.95.

Observer's Guide To Astronomy. Patrick Martinez, editor. Cambridge University Press, 40 West 20th Street, New York, NY 10011-4211. (212) 924-3900. (800) 872-7423. 2 volumes. 1994. $159.90.

Properties of Double Stars: A Survey of Parallaxes and Orbits. Leendert Binnenkijk. Books on Demand, 300 North Zeeb Road, Ann Arbor, MI 48106-1346. (313) 761-4700. (800) 521-0600. Reprint edition. $101.50.

ONLINE DATABASES AND CD-ROMS

CA Search. Chemical Abstracts Service, P.O. Box 3012, Colum bus, OH 43210-0012. (614) 447-3600. (800) 848-6533. FAX (614) 447-3709. Very comprehensive guide to worldwide chemical literature and related fields, 1972 to present. Available on BRS,(800) 289-4277, DIALOG, (800) 334-2564, ORBIT (800) 456-7248, and STN International, FIZ Karlsruhe, P.O. Box 2465, W-7500, Karlsruhe 1, Germany, online services. Inquire as to cost and availability.

Current Contents Search. Institute for Scientific Information, 3501 Market Street, Philadelphia, PA 19104. (215) 386-0 100. FAX (215) 386-6362. Contains citations to articles listed in the table of contents of science and technology journals. Also articles in social sciences and life sciencesjournals. Available on BRS, (800) 289-4277, DIALOG, (800) 334-2564, online services. Inquire as to cost and availability.

Dissertation Abstracts Online. University Microfilms Internatio nal, 300 North Zeeb Road, Ann Arbor, MI 48106. (800) 521-0600 or (313) 761-4700. Scope includes virtually all doctoral dissertations accepted at accredited American institutions from 1861 to present in 252 subject areas. Available on BRS, (800) 289-4277, DIALOG, (800) 334-2564, and OCLC EPIC, (800) 848-5878, online services. Also available on CD-ROM. Inquire as to cost and availability.

INSPEC. Institution of Electrical Engineers, Michael Farada y House, Six Hills Way, Stevenage, Herts. SG1 2AY, E ngland. Telephone: 0438 313311 or FAX 0438 742840. Contains citations to the worldwide literature of physics, electronics and electrical engineering, computer technology, and related fields. Available on BRS, (800) 289-4277, DIALOG, (800) 334-2564, ORBIT, (800) 456-7248, and STN International, FIZ Karlsruhe, P.O. Box 2465, W-7500, Karlsruhe 1, Germany, online services. Inquire as to cost and availability.

International Aerospace Abstracts. American Institute of Aeronautics and Astronautics, 370 L'Enfant Promenade, S.W., Washington, DC 20024. (202) 646-7400. Contains references and abstracts of journal and monograph literature relating to aerospace science and technology, from 1963 to present. Available through the NASA/RECON system of the National Aeronautics and Space Administration only.

NTIS Bibliographic Database. National Technical Information Service, 5285 Port Royal Road, Springfield, VA 22161. (703) 487-4929 or FAX (703) 321-8199. Broad coverage of government-sponsored science and technology research reports, 1964 to present. Available on BRS,(800) 289-4277, DIALOG, (800) 334-2564, ORBIT, (800) 456-7248, and STN International, FIZ Karlsruhe, P.O. Box 2465, W-75 00, Karlsruhe 1, Germany, online services. Also available on CD-ROM. Inquire as to cost and availability.

Physics Briefs. American Institute of Physics, 335 East 45th Street , New York, NY 10017. (212) 661-9260 or FAX (212) 661-2036. Contains citations with abstracts of the literature of physics and related fields, 1979 to present. Available on the STN International, FIZ Karlsruhe, P.O. Box 2465, W-7500, Karlsruhe 1, Germany, online service. Inquire as to cost and availability.

Scientific and Technical Books and Serials in Print. R.R. B owker Inc., 121 Chanlon Road, New Providence, NJ 07974. (800) 521-8110 or (908) 464-6800. List of currently published books and serials in the physical and biological sciences, engineering and technology, with subject, author and titles indexes. Available on ORBIT, (800) 456-7248, online service. Inquire as to cost and availability.

SCISEARCH. Institute for Scientific Information, 3501 MA rket Street, Philadelphia, PA 19104. (800) 523-1850 or (215)386-0100. Broad multidisciplinary title and author index to the international literature of science and technology, 1974 to present. Available on DIALOG, (800) 334-2564, and ORBIT, (800) 456-7248, online services. Also available on CD-ROM. Inquire as to cost and availability.

Wilsonline. H.W. W ilson Company, 950 University Avenue, Bronx, NY 10452. (800) 367-6770 or (212) 588-8400. Makes available onlin e versions of the printed H.W. Wilson indexes includin g Applied Science and Technology Index, Business Periodicals Index, General Science Index, and Readers' Guide to Periodical Literature. Period covered is generally 1983 to present. Available on BRS, (800) 289-4277, DIALOG, (800) 334-2564, and OCLC EPIC, (800) 848-5878, online services. Also available on CD-ROM. Inquire as to cost and availability.

OTHER SOURCES

Atlas of Deep-Sky Spendors. Hans Vehrenberg. Sky Publishing Corp., 49 Bay State Road, Cambridge, MA 02138. (617) 864-7360. (800) 253-0245. 4th edition. 1983. $24.95.

Atlas of the Ultraviolet Sky. Richard C. Henry, Wayne B. Landsman, et al. Johns Hopkins University Press, 2715 North Charles Street, Baltimore, MD 21218-4319. (410) 516-6900. (800) 537-5487. 1988. $70.00.

Atlas of the Universe. Patrick Moore. Rand McNally & Company. 8255 North Central Park, Skokie, IL 60076-2970. (708) 673-9100. 1994. $29.95.

Binary Stars: A Pictorial Atlas. Dirk Terrell, et al. Krieger Publishing Company, P.O. Box 9542, Melbourne, FL 32902-9542. (407) 724-9542. . 1992. $42.50.

Cambridge Atlas of Astronomy. Jean Audouze, et al. Cambridge University Press, 40 West 20th Street, New York, NY 10011-4211. (212) 924-3900. (800) 872-7423. 3rd edition. 1994. $75.00.

Carnegie Atlas of Galaxies. Allan Sandage and John Bedke. Carnegie Institution of Washington, 1530 P Street NW, Washington, DC 20005. (202) 387-6411. Two volumes. 1994. $92.00.

Observer's Sky Atlas. E. Karkoschka. Springer-Verlag New York, Inc., 175 Fifth Avenue, New York, NY 10010. (212) 460-1500. (800) 777-4643. 1995. $15.95.

Observing Handbook and Catalog of Deep-sky Objects. Christian B. Luginbuhl and Brian A. Skiff. Cambridge University Press, 40 West 20th Street, New York, NY 10011- 4211. (212) 924-3900. (800) 872-7423. 1990. $49.95.

Photometric Atlas of Northern Bright Galaxies. Kodaira, Keichi, et al editors. Columbia University Press, 562 West 113th Street, New York, NY 10025. (212) 666-1000. (800) 944-8648. 1990. $180.00.

Third Reference Catalogue of Bright Galaxies. Gerard H. DE Vaucouleurs et al, editors. Springer-Verlag New York, Inc., 175 Fifth Avenue, New York, NY 10010. (212) 460-1500. (800) 777-4643. Three volumes. 1991. $198.00 set.

PERIODICALS

Astronomical Journal. American Institute of Physics, One Physics Ellipse, College Park, MD 20740-3843. (301) 209- 3000. Published for the American Astronomical Society. 1849 to present. Monthly. $280.00 per year. ISSN: 0004- 6256.

Astronomical Society of the Pacific. Astronomical Society of the Pacific, 390 Ashton Avenue, San Francisco, CA 94112. (415) 337-1100. FAX (415) 337-5205. 1889 to present. Monthly. $175.00 per year. ISSN: 0004- 6280.

Astronomy. Kalmbach Publishing Company, Box 1612, Waukesha, WI 53187-1612. (414) 796-8776. FAX (414) 796-1142. 1973 to present. Monthly. $27.00 per year. ISSN: 0091-6358.

Astronomy and Astrophsyics. Springer-Verlag New York, Inc., 44 Hartz Way, Secaucus, NJ 07096-2491. (201) 348-4033. FAX (201) 348-4505. 1969 to date. 36 issues per year in 12 volumes. $1733.00 per year. ISSN: 0004-6361.

Astronomy Reports. American Institute of Physics, One Physics Ellipse, College Park, MD 20740-3843. (301) 209- 3000. English translation of Astronomicheksii Zhurnal. Formerly Soviet Astronomy AJ. 1957 to present. Bimonthly. $1160.00 per year. ISSN: 1063-7729.

Astrophysical Journal; An International Review of Astronomy and Astronomical Physics. University of Chicago Press, Journals Division, 5720 South Woodlawn Avenue, Chicago, IL 60637. (312) 753-3347. FAX (312) 753-0811. 1895 to date. three issues per month. $740.00 per year. ISSN: 0004-637X.

Astrophpysics and Space Science; An International Journal of Cosmic Physics. Kluwer Academic Publishers, Box 358, Accord Station, Hingham, MA 02018-0358. (617) 871-6600. FAX (617) 871-6528. 1968 to present. 24 issues per year. $2544.00 per year. ISSN: 0004-640X.

Celestial Mechanics and Dynamical Astronomy; An International Journal of Space Dynamics. Kluwer Academic Publishers, Box 358, Accord Station, Hingham, MA 02018-0358. (617) 871-6600. FAX (617) 871-6528. 1969 to present. Monthly. $703.50 per year. ISSN: 0923-2958.

Experimental Astronomy; An International Journal on Astronomical Instrumentation and Data Analysis. Kluwer Academic Publishers, Box 358, Accord Station, Hingham, MA 02018-0358. (617) 871-6600. FAX (617) 871-6528. 1989 to present. Quarterly. $161.00 per year. ISSN: 0922-6435.

Icarus; International Journal of Solar System Studies. Academic Press, Inc., Journal Division, 525 B Street, Suite 1900, San Diego, CA 92101-4495. (619) 230-1840. FAX (619) 699-6800. 1962 to present. Monthly. $1080.00 per year. ISSN: 0019-1035.

Landolt-Boernstein, Zahlenwerte Und Funktionen Aus Naturwissenschaften Und Technik. Group 6: Astronomy. Springer-Verlag New York, Inc., 175 Fifth Avenue, New York, NY 10010. (212) 460-1500. FAX (212) 473-6272. 1965. Irregular. Price varies. ISSN: 0075-7896.

Mercury. Astronomical Society of the Pacific. 390 Ashton Avenue, San Francisco, CA 94112. (415) 337-1100. FAX: (415) 337-5205. 1972 to present. Bimonthly. $175.00 per year. ISSN: 0047-6773.

Monthly Notices of the Royal Astronomical Society. Blackwell Scientific Publication LT., Osney Mead, Oxford OX2 OEL, England. TEL 0865-240201. FAX 0865-721205. 1827 to present. Fortnightly. $1899.00 per year. ISSN: 0035-8711.

Observatory. Space and Astrophysics Division, Rutherford Appleton Laboratory, Chilton, Didcot, Oxon OX11 OQX England. FAX 0235-445848. 1877 to present. Bimonthly. $42.00 per year.

Sky and Telescope. Sky Publishing Corporation, Box 9111, Belmont, MA 02178. (617) 864-7360. FAX (617) 864-6117. 1941 to present. Monthly. $27.00 per year. ISSN: 0037- 6604.

Space Science Reviews. Kluwer Academic Publishers, Box 358, Accord Station, Hingham, MA 02018- 0358. (617) 871- 6600. FAX (617) 871-6528. 1962 to present. Sixteen issues per year. $940.00 per year. ISSN: 0038-6308.

Vistas in Astronomy; An International Review Journal. Albert C. Beer, editor. Elsevier Science Publishing Company, Inc., 660 White Plains Road, Terrytown, NY 10591- 5153. (914) 524-9200. FAX: (914) 333-2444. 1958 to present. Quarterly. $415.00 per year. ISSN: 0083-6656.

Transactions of the International Astronomical Union. Proceedings of the General Assembly. Kluwer Academic Publishers, Box 358, Accord Station, Hingham, MA 02018- 0358. (617) 871-6600. Issues in two parts: Part A, Reports, and Part B, Proceedings. Triennial. XXI - 1994. Price varies. ISSN: 0080-1372.

Webb Society Quarterly Journal. Webb Society, 194 Foundry Lane, Freemantle, Southampton, Hampshire SO1 3EQ, England. 1969 to present. Quarterly. $19.50 per year. ISSN: 0043- 1680.

RESEARCH CENTERS AND INSTITUTES

Astronomy Program. University of Maryland, 1207 Computer & Space Sciences Building, College Park, MD 20742. (301) 405- 3001. FAX (301) 314-9067.

David Dunlap Observatory. University of Toronto, Box 360, Richmond Hill, ON, Canada L4C 4Y6. (416) 884-9562. FAX (416) 978-3921.

Florida Institute of Technology. OBSERVATORY. Physics and Space Sciences, 150 West University Boulevard, Melbourne, FL 32901. (305) 768-8098.

Lowell Observatory. 1400 West Mars Hill Road. Flagstaff, AZ 86001, (602) 774-3358. FAX (602) 774-6296.

Mount Cuba Astronomical and Observatory, Inc. Hillside Mill Road, P. O. Box 3915, Breenville, DE 19807. (302) 654-6407.

National Radio Astronomy Observatory. Edgemont Road, Charlottesville, VA 22903. (804) 296-0211.

Ritter Astrophysical Research Center. University of Toledo, 2801 West Bancrift Street, Toledo, OH 43606. (419) 537- 2650. FAX (419) 537-2723.

Smithsonian Astrophysical Observatory. 60 Garden Street, Cambridge, MA 02138. (617) 495-7461. FAX (617) 495-7105.

Space Telescope Science Institute. 3700 San Martin Drive, Baltimore, MD 21218. (301) 338-4700.

BIOMEDICAL ENGINEERING

See also: ELECTRICAL ENGINEERING, MECHANICAL ENGINEERING, ROBOTICS

ABSTRACT SERVICES AND INDEXES

Alloys Index. American Society for Metals, Metals Park, OH 44073. (216) 338-5151. FAX (216) 338-4634. 1974 to present. Monthly. $295.00.

Applied Mechanics Reviews. An Assessment of World Literature in Engineering Sciences. American Society of Mechanical Engineers, 345 East 47th Street, New York, NY 10017. (212) 705-7703. 1948 to present. Monthly. $360.00 per year. ISSN: 0003-6900.

Applied Science and Technology Index; a cumulative subject index to English language periodicals in the fields of aeronautics and space science, computer technology, chemistry, construction industry, energy and related areas. H.W. Wilson Co., 950 University Avenue, Bronx, NY 10452. (212) 588-8400. (800) 367-6770. FAX (718) 590-1617. From 1958 to present. Monthly. Inquire about cost and availability. Also available on CD-ROM. ISSN: 0003-6986.

Bioengineering Abstracts. Cambridge Scientific Abstracts, 7200 Wisconsin Avenue, Bethesda, MD 20814. (301) 961-6700. FAX (301) 961-6720. 1974 to present. Monthly. $535.00 per year. ISSN 1068-5693.

Chemical Engineering Abstracts. Royal Society of Chemistry, Information Services, Thomas Graham House, Science Park, Milton Road, Cambridge, CB4 4WF, England. Contains citations, mostly with abstracts, of the worldwide literature on chemical engineering, from 1982 to present. Monthly. $450.00 per year. Also available online.

Current Contents: Engineering, Technology and Applied Sciences. Institute for Scientific Information, 3501 Market Street, Philadelphia, PA 19104. (215) 386-0100. FAX (215) 386-6362. 1970 to present. Weekly. $442.00 per year. Also available on CD-ROM and online. Inquire regarding cost and availability. ISSN 0095-7917.

Current Papers in Electrical and Electronics Engineering. Institution of Electrical Engineers (IEE), London. Distributed by INSPEC/IEEE, Box 1331, 445 Hoes Lane, Piscataway, NJ

08855-1331. (908) 562-5549. 1969 to present. Monthly. $345.00 per year. ISSN: 0011-3778.

Electrical and Electronics Abstracts. Institution of Electrical Engineers (IEE), London. Available from INSPEC/IEEE - Institute of Electrical and Electronic Engineers, Box 1331, 445Hoes Lane, Piscataway, NJ 08854-1331. (908) 562-5549. 1898 to present. Monthly. $2200.00 per year. Also available on CD-ROM and online as INSPEC. ISSN: 0036-8105.

Engineered Materials Abstracts. ASM (American Society of Metals) International, 9639 Kinsman Road, Materials Park, OH 44073. (216) 338-5151 or FAX (216) 338-4634. Covers literature on technical developments in polymer, ceramic, and composite materials and engineering. 1986 to present. Monthly. $1175.00 per year. Also available on CD-ROM. ISSN: 0951-9998.

Engineering Index Monthly; indexes and abstracts the world's engineering and technical literature. Engineering Information, Inc., Castle Point on the Hudson, Hoboken, NJ 07030. (201) 216-8500. (800) 221-1044. FAX (201) 216-8532. Monthly. $2300.00 per year. Available online as COMPENDEX and also on CD-ROM. ISSN: 0742-1974.

General Science Index. H.W. Wilson Company, 950 University Avenue, Bronx, NY 10452. (212) 588-8400. (800) 367-6770. FAX (718) 590-1617. From 1978 to present. Ten issues per year; quarterly and annual cumulations. Service basis. Available on CD-ROM and online. Inquire regarding cost and availability. ISSN: 0162-1963.

Government Reports Announcements and Index. National Technical Information Service (NTIS), 5285 Port Royal Road, Springfield, VA 22161. (703) 487-4650. FAX (703) 321-8547. From 1968 to present. Annual. $630.00 per year.

Science Citation Index. SCI. Institute for Scientific Information, 3501 Market Street, Philadelphia, PA 19104. (215) 386-0100. (800) 523-1850. FAX (215) 386-2991. 1961 to present. Six issues per year, plus annual cumulation. $11650.00 per year. Also available online and on CD-ROM. Inquire about price and availability. ISSN: 0036-827X.

STAR. (Scientific and Technical Aerospace Reports). U.S. National Aeronautics and Space Administration, Scientific and Technical Information Facility, Box 8757, Baltimore-Washington International Airport, MD 21240. (301) 621-0153. Distributed by U. S. Superintendent of Documents, Washington, DC 20402. From 1963 to present. Semimonthly, with semiannual and annual indexes. $114.00 per year. Also available online and on CD-ROM. ISSN: 0036-8741.

World Ceramic Abstracts. Pergamon Press, Maxwell House, Fairview Park, Elmsford, NY 10523. (914) 592-7700. Monthly. 1958 to present. $310.00 per year.

ANNUAL REVIEWS AND YEARBOOKS

Advances in Applied Mechanics. Academic Press, Inc., 1250 Sixth Avenue, San Diego, CA 92101-4311. (619) 231-0926. FAX (619) 699-6715. Irregular. Price varies.

Annual Review of Fluid Mechanics. Annual Reviews, Inc., 4139 Camino Way, P.O. Box 10139, Palo Alto, CA 94303-0897. (800) 523-8635 or (415) 493-4400. Annual. Volume 24, 1992. Price varies.

Annual Review of Materials Science. Annual Reviews, Inc., 4139 Camino Way, P.O. Box 10139, Palo Alto, CA 94303-0897. (800) 523-8635 or (415) 493-4400. Annual. Volume 22, 1992. Price varies.

ASSOCIATIONS AND PROFESSIONAL SOCIETIES

American Institute of Chemical Engineers. 345 East 47th Street, New York, NY 10017-2396. (212) 705-7663.

American Institute of Physics, One Physics Ellipse, College Park, MD 20740-3843. (301) 209-3100.

American Society for Nondestructive Testing, Inc. 1911 Arlington Lane, Columbus, OH 43228-4113. (614) 274-6003.

American Society for Testing and Materials. 1916 Race Street, Philadelphia, PA 19103-1180. (215) 299-5400.

American Society of Mechanical Engineers, 345 East 47th Street, New York, NY 10017-2395. (212) 705-7722.

Association for the Advancement of Medical Instrumentation. 3330 Washington, Boulevard, Suite 400, Arlington, VA 22201. (703) 525-4890. FAX (703) 276-0793.

Biomedical Engineering Society. P.O. Box 2399. Culver City, CA 90231. (310) 618-9322.

IEEE Engineering in Medicine and Biology Society. c/o Institute of Electrical and Electronics Engineers, 345 East 47th Street, New York, NY 10017. (212) 705-7900. FAX (212) 705-4929.

International Conference on Mechanics in Medicine and Biology. University of Michigan, 2150 G.G. Brown Building, Ann Arbor, MI 48109. (313) 764-9910. FAX (313) 747-3170.

Society for Biomaterials. 6518 Walker Street, Suite 215, Minneapolis, MN 55426. (612) 927-8108. FAX (612) 927-8127.

Society of Biomedical Equipment Technicians. 3330 Washington Boulevard, Suite 400. Arlington, VA 22201.(703) 525-4890. FAX (703) 276-0793.

Special Interest Group on Biomedical Computing. (SIGBIO). c/o W. E. Hammond, Duke University Medical Center, Box 2914, Durham, NC 27710. (919) 684-6421. (919) 684-8675.

Special Interest Group for Computers and the Physically Handicapped. (SIGCAPH). Association for Computing Machinery, 1515 Broadway, New York, NY 10036. (212) 869-7440. FAX (212) 302-5826.

BIBLIOGRAPHIES

Biomedical Engineering Index of Modern Information. John C. Bartone. ABBE Publishers Association, 4111 Gallows Road, Annandale, VA 22003. 1990. $39.50.

Scientific and Technical Books and Serials in Print; An index to Literature in Science and Technology. R.R. Bowker Co., 205 E. 42nd Street, New York, NY 10017. (800) 521-8110 or (212) 916-1600. 1992. $250.00.

BIOMEDICAL ENGINEERING

Ency. of Physical Sciences and Engineering Info. Sources

Weather, Health and Biomedicine Subject Analysis Index with Research Bibliography. American Health Research Institue. ABBE Publishers Association, 4111 Gallows Road, Annandale, VA 22003. 1987. $37.50.

DIRECTORIES AND BIOGRAPHICAL SOURCES

American Men and Women of Science: Physical and Biological Sciences. R. R. Bowker Inc., 121 Chanlon Road, New Providence, NJ 07974. (908) 464-6800. (800) 521-8110. 20th edition. 8 volumes. 1996. $850.00.

Biomedical Careers Book. Allan F. Pacela, editor. Quest Publishing Co., 1351 Titan Way, Brea, CA 92621. (714) 738- 6400. FAX (714) 525-6258. 1992. $40.00 in paper.

Biomedical Medical Engineering Society. Membership Directory. Biomedical Medical Engineering Society, PO Box 2399, Culver City, CA 90231. (310) 618-9322. Annual. 1995. $50.00.

Directory of Biomedical & Health Care Grants. Oryx Press, 4041 North Central Avenue, Phoenix, AZ 85012-3397. (602) 265-2651. (800) 279-6799. 9th edition. 1995. $84.50.

Directory of Engineering College Research and Graduate Study, American Society for Engineering Education, Ohio State University, 2070 Neil Avenue, Columbus, OH 43210-1226. Annual. $20.00.

Directory of Engineering Societies and Related Organizations. American Association of Engineering Societies, 1111 19th Street, Suite 608, Washington, DC 20036-3603. (202) 296-2237. Semi-annual. $150.00.

Engineering Research Centers: Incorporating Electronics Research Centers. Stockton Press, 345 Park Avenue, New York, NY 10010. (212) 689-9200. (800) 221-2123. 4th edition. 1995. $515.00.

International Engineering Directory. American Consulting Engineers Council, 1015 15th Street, N.W. Suite 802, Washington, DC 20005-2670. (202) 347-7474. Annual. $10.00.

Libraries, Information Centers, and Databases in Biomedical Sciences: A World Guide. Bettina Bartz, et al, editors. K. G. Saur, 121 Chanlon Road, New Providence, NJ 07974. (908) 464-6800. (800) 521-8110 1991. $170.00.

Research Centers Directory. Gale Research, 835 Penobscot Building, Detroit, MI 48226-4094. (313) 961-2242. (800) 877-4253. 20th edition. 1995. $485.00. ISSN: 0080-1518.

Who's Who in Engineering. Gordon Davis, editor. American Association of Engineering Societies. 1111 19th Street, NY, Suite 608, Washington, DC 20036. (202) 296-2237. (800) 658-8897. 9th edition. 1995. $220.00.

Who's Who in Technology; Gale Research, 835 Penobscot Building, Detroit, MI 48226-4094. (313 961-2242. (800) 521-4253. 7th edition. $195.00. ISN 0-8103-7467-6.

ENCYCLOPEDIAS AND DICTIONARIES

Academic Press Dictionary of Science and Technology. Christopher Morris, editor. Academic Press, Inc., 1250 Sixth Avenue,

San Diego, CA 92101. (619) 231-0926. FAX (619) 699-6715. 1991. $115.00.

Chambers Science and Technology Dictionary. Peter M.B. Walker, editor. Cambridge University Press, 40 West 20th Street, New York, NY 10011-4211. (212) 924-3900. 1988. $39.95.

Encyclopedia of Applied Physics. VCH Publishers, Inc., 303 Northwest 12th Avenue, Deerfield Beach, FL 33442. (800) 367-8249. 1991-. Twenty volumes. $6000.00.

Encyclopedic Dictionary of Science. Candida Hunt and Monica Byles, editors. Facts-on-File, Inc., 460 Park Avenue South, New York, NY 10016-7382. (800) 322-8755. Fax (800) 683-3633. 1988. $32.95.

McGraw-Hill Encyclopedia of Engineering. Sybil P. Parker, editor. McGraw-Hill Book Company, Inc. 1221 Avenue of the Americas, New York, NY 10020. (212) 997-3675. Second edition. 1993. $95.50. ISBN 0-07-051392-9.

McGraw-Hill Encyclopedia of Science and Technology. McGraw-Hill Book, Incorporated, 1221 Avenue of the Americas, New York, NY 10020. (212) 997-3675. Seventh edition. Twenty volumes. 1992. $1900.00.

Medical Engineering Dictionary in Four Languages; English, French, German, and Russian. Roald Albert. IBD LTD, 24 Hudson Street, Kinderhook, NY 12106. (518) 758-1411. (800) 343-3531. 1994. $186.00.

GENERAL WORKS

Biomechanics of Engineering. A. Morecki. Springer-Verlag New York, Inc., 175 Fifth Avenue, New York, NY 10010. (212) 460-1500. (800) 777-4643. FAX (212) 473-6272. 1992. $69.95.

Biomedical Engineering. A. Edward Profio. John Wiley & Sons, Inc., 605 Third Avenue, New York, NY 10158-0012. (212) 850-6000. (800) 225-5945. 1993. $79.95.

Biomedical Engineering Principles: An Introduction to Fluid, Heat and Mass Transport Processes. David Cooney, editor. Marcel Dekker, Inc., 270 Madison Avenue, New York, NY 10016. (212) 696-9000. (800) 228-1160. 12th edition. 1976. $95.00.

Biomedical Instruments: Theory and Design. Academic Press, Inc., 6277 Sea Harbor Drive, Orlando, FL. (800) 321-5068. 1991. $64.95.

Biomedical Signal Processing. Metin Akay. Academic Press, Inc., 6277 Sea Harbor Drive, Orlando, FL. (800) 321-5068. 1994. $49.95.

IEEE Engineering in Medicine and Biology. Annual Conference. Institute of Electrical and Electonics Engineers, 345 East 47th Street, New York, NY 10017. (212) 705-7900. 17th annual conference - 1995.

Mechanical Technology. D. H. Bacon and R. C. Stephens. Industrial Press, Inc., 200 Madison Avenue, New York, NY 10016. (212) 889-6330. 1990. $29.95.

Introduction to Biomedical Equipment Technology. Joseph J. Carr and John M. Brown. 2nd edition. Prentice-Hall, 113 Sylvan Avenue, Route 9W, Englewood Cliffs, NJ 07632. (201) 592-2000. (800) 922-0579. 1993. $79.00.

Understanding the Nervous System: An Engineering Perspective. Sid Deutsch and Alice Deutsch. Institute ofElectrical and Electronics Engineers, 345 East 47th Street, New York, NY 10017. (212) 705-7900. 1993. $34.95 in paper.

HANDBOOKS AND MANUALS

Composite Materials Handbook. Mel M. Schwartz, editor. McGraw-Hill Publishing Company, 1221 Avenue of the Americas, New York, NY 10020. (800) 262-4729 or (212) 512-3825. 1992. $79.50.

CRC Handbook of Chemistry and Physics. David R. Kide, editor. CRC Press Inc., 2000 Corporate Blvd., NW, Boca Raton, FL 33431. (407) 994-0555. (800) 333-8300. 77th edition. 1996. $99.95.

Electrical Engineer's Handbook. Donald Christiansen, editor. McGraw Book Company, 1221 Avenue of the Americas, New York, NY 10020. (212) 512-2000. (800) 262-4729. 4th edition. 1996. $110.00.

Engineering Tables and Data. A.M. Howatson, P.G. Lund and J.D. Todd. Van Nostrand Reinhold, 115 Fifth Avenue, New York, NY 10003. (212) 254-3232 or (800) 926-2665. Second Edition. 1991. $39.95.

Handbook of Biomedical Engineering. Jacob Kline, editor. Academic Press, Inc., 6277 Sea Harbor Drive, Orlando, FL. (800) 321-5068. 1988. $226.00.

Handbook of Clinical Engineering. Barry N. Feinberg, editor. Volume 1. Franklin Book Co., Inc., Elkins Park, PA 19117. (215) 635-5252. 1980. $113.95.

Handbook of Engineering in Medicine and Biology. David G. Fleming and Barry N. Feinberg, editors. Volume 1: General Data; Volume 2: Instruments and Measurements. 1976-78. $150.00 for the two volume set. Volume 1 available from Franklin Note: volume 2 available from CRC Press.

Handbook of Industrial Engineering. Gavriel Salvendy, editor. Institute of Industrial Engineers. Distributed by John Wiley and Sons, Inc., 605 Third Avenue, New York, NY 10158. (800) 526-5368 or (212) 850-6000. Second edition. 1992. $135.00.

Integrated Circuits Handbook. Rector Press, Ltd., 130 Rattlesnake, Leverett, MA 01054-9726. (413) 548-9708. (800) 247-3473. 3 volumes. 1995. $295.00 in paper.

Machinery's Handbook. Robert Green, editor. Industrial Press, Inc. 200 Madison Avenue, New York, NY 10016. Twenty-fourth edition. 1992. $65.00.

Standard Handbook of Engineering Calculations. Tyler G. Hicks. McGraw-Hill Publishing Company, 1221 Avenue of the Americas, New York, NY 10020. (800) 262-4729 or (212) 512-3825. 3rd edition. 1994. $99.50. ISBN 0-07-028812-7.

The Wiley Engineer's Desk Reference: A Guide for the Professional Engineer. Sanford I. Heisler. Wiley-Interscience; distributed by John Wiley & Sons, Inc., 605 Third Avenue, New York, NY 10158. (800) 526-5368 or (212) 850-6000 1984. $79.95. ISBN 0-471-8663206.

ONLINE DATABASES AND CD-ROMS

Aerospace Database. American Institute of Aeronauticsand Astronautics, 370 L'Enfant Promenade, S.W., Washington, DC 20024. (202) 646-7400. Worldwide published literature on research and development in aerospace and related areas, with abstracts. Covers 1962 to present. Available on DIALOG, (800) 334-2564, online service. Also available on CD-ROM. Inquire as to cost and availability.

CA Search. Chemical Abstracts Service, P.O. Box 3012, Columbus, OH 43210-0012. (614) 447-3600. (800) 848-6533. FAX (614) 447-3709. Very comprehensive guide to worldwide chemical literature and related fields, 1972 to present. Available on BRS,(800) 289-4277, DIALOG, (800) 334-2564, ORBIT (800) 456-7248, and STN International, FIZ Karlsruhe, P.O. Box 2465, W-7500, Karlsruhe 1, Germany, online services. Inquire as to cost and availability.

Ceramic Abstracts. American Ceramic Society, Inc., Reference Services, 735 Ceramic Place, Westerville, OH 43081-6136. (614) 890-4700. Citations with abstracts to worldwide literature on ceramic materials and technology, from 1976 to present. Available on DIALOG, (800) 334-2564, ORBIT (800) 456-7248, and STN International, FIZ Karlsruhe, P.O. Box 2465, W-7500, Karlsruhe 1, Germany, online services. Inquire as to cost and availability.

Compendex Plus. Engineering Information, Inc., 345 East 47th Street, New York, NY 10017. (212) 705-7600 or (800) 221-1044. Contains citations with abstracts to worldwide literature in engineering and technology, from 1970 to present. Available on online BRS,(800) 289-4277, DIALOG, (800) 334-2564, ORBIT (800) 456-7248, and STN International, FIZ Karlsruhe, P.O. Box 2465, W-7500, Karlsruhe 1, Germany, online services. Also available on CD-ROM. Inquire as to cost and availability.

Current Contents Search. Institute for Scientific Information, 3501 Market Street, Philadelphia, PA 19104. (215) 386-0100. FAX (215) 386-6362. Contains citations to articles listed in the table of contents of science and technology journals. Also articles in social sciences and life sciences journals. Available on BRS,(800) 289-4277, DIALOG,(800) 334-2564, online services. Inquire as to cost and availability.

Dissertation Abstracts. University Microfilms International, 300 North Zeeb Road, Ann Arbor, MI 48106. (800) 521-0600 or (313) 761-4700. Scope includes virtually all doctoral dissertations accepted at accredited American institutions from 1861 to present in 252 subject areas. Available on BRS,(800) 289-4277, DIALOG,(800) 334-2564, and OCLC EPIC,(800) 848-5878, online services. Also available on CD-ROM. Inquire as to cost and availability.

INSPEC. Institution of Electrical Engineers, Michael Faraday House, Six Hills Way, Stevenage, Herts. SG1 2AY, England. Telephone: 0438 313311 or FAX 0438 742840. Contains citations to the worldwide literature of physics, electronics and electrical engineering, computer technology, and related fields. Availableon BRS,(800) 289-4277, DIALOG, (800) 334-2564, ORBIT, (800) 456-7248, and STN International, FIZ Karlsruhe, P.O. Box 2465, W-7500, Karlsruhe 1, Germany, online services. Inquire as to cost and availability.

International Aerospace Abstracts. American Institute of Aeronautics and Astronautics, 370 L'Enfant Promenade, S.W., Washington, DC 20024. (202) 646-7400. Contains references and abstracts of journal and monograph literature relating to aerospace science and technology, from 1963 to present. Available

BIOMEDICAL ENGINEERING

Ency. of Physical Sciences and Engineering Info. Sources

through the NASA/RECON system of the National Aeronautics and Space Administration only.

ISMEC: Mechanical Engineering Abstracts.Cambridge Scientific Abstracts, 7200 Wisconsin Avenue, Suite 601, Bethesda, MD 20814. (301) 961-6750 or (800) 843-7751. Contains citations to the literature in mechanical engineering, industrial and production engineering, energy, power, mechanics, devices and related areas, from 1973 to present. Available on the DIALOG,(800) 334-2564, online service. Inquire as to cost and availability.

NTIS Bibliographic Database. National Technical Information Service, 5285 Port Royal Road, Springfield, Va 22161. (703) 487-4929 or FAX (703) 321-8199. Broad coverage of government-sponsored science and technology research reports, 1964 to present. Available on BRS,(800) 289-4277, DIALOG, (800) 334-2564, ORBIT, (800) 456-7248, and STN International, FIZ Karlsruhe, P.O. Box 2465, W-7500, Karlsruhe 1, Germany, online services. Also available on CD-ROM. Inquire as to cost and availability.

Physics Briefs. American Institute of Physics, 335 East 45th Street, New York, NY 10017. (212) 661-9260 or FAX (212) 661-2036. Contains citations with abstracts of the literature of physics and related fields, 1979 to present. Available on the STN International, FIZ Karlsruhe, P.O. Box 2465, W-7500, Karlsruhe 1, Germany, online service. Inquire as to cost and availability.

Scientific and Technical Books and Serials in Print. R.R. Bowker Inc., 121 Chanlon Road, New Providence, NJ 07974. (800) 521-8110 or (908) 464-6800. List of currently published books and serials in the physical and biological sciences, engineering and technology, with subject, author and titles indexes. Available on ORBIT,(800) 456-7248, online service.Inquire as to cost and availability.

SCISEARCH. Institute for Scientific Information, 3501 Market Street, Philadelphia, PA 19104. (800) 523-1850 or (215) 386-0100. Broad multidisciplinary title and author index to the international literature of science and technology, 1974 to present. Available on DIALOG,(800) 334-2564, and ORBIT,(800) 456-7248, online services. Also available on CD-ROM. Inquire as to cost and availability.

Wilsonline. H.W. Wilson Company, 950 University Avenue, Bronx, NY 10452. (800) 367-6770 or (212) 588-8400. Makesavailable online versions of the printed H.W. Wilson indexes including *Applied Science and Technology Index, Business Periodicals Index, General Science Index,* and *Readers' Guide to Periodical Literature.* Period covered is generally 1983 to present. Available on BRS,(800) 289-4277, DIALOG,(800) 334-2564, and OCLC EPIC,(800) 848-5878, online services. Also available on CD-ROM. Inquire as to cost and availability.

PERIODICALS

Annals of Biomedical Engineering. Biomedical Engineering Society. PO Box 2399. Culver City, CA 90231. (310) 618-9322. Bimonthly. 1979 to present. $459.00 per year. ISSN: 0090-6964.

Biomedical Instrumentation and Technology. Association for the Advancement of Medical Instrumentation. 3330 Washington Boulevard, Suite 400, Arlington, VA 22201. (703) 525-4890. FAX (703) 276-0793. 1967 to present. Bi-monthly. $86.00 per year. ISSN: 0899-8205.

Journal of Biomedical Materials Research. Society for Biomaterials, 6518 Walker Street, Suite 215, Minneapolis, MN 55426. (612) 927-8108. FAX (612) 927-8127. 1966 to present. Monthly. $880.00. ISSN: 0021-9304.

CRC Critical Reviews in Biomedical Engineering. CRC Publishers, Inc., 2000 Corporate Blvd., N.W., Boca Raton, FL 33431. (407) 994-0555 or (800) 333-8300. 1974 to present. Bi-monthly. $360.00 per year. ISSN: 0278-940X.

IEEE Engineering in Medicine and Biology Magazine. IEEE, 445, Hoes Lane, Piscataway, NJ 08854. (908) 981-0060. 1962 to present. Quarterly. $105.00. ISSN: 0739-5175.

Journal of Biomedical Engineering. Butterworth-Heinemann; subscriptions to: Turpin Transactions, Ltd, Blackhorse Road, Letchworth, Herts SG6 1HN, England. 1979 to present. Bi-monthly. L262 per eyar. ISSN: 0141-5425.

Journal of Micromechanics and Microengineering. IOP Publishing. American Institute of Physics, 335 East 45th Street, New York, NY 10017. (212) 661-9404 or (800) 445-6638. 1991 to present. Quarterly. $3356.00 per year. ISSN: 0960-1317.

RESEARCH CENTERS AND INSTITUTES

Biomedical Engineering Center. Ohio State University, 2300 West Case Road, Columbus, OH 43235. (614) 292-5491.

Center for Biomedical Engineering. New Jersey Institute of Technology, 323 Martin Luther King, Jr., Boulevard, Newark, NJ 07102. (201) 596-3584. FAX (201) 643-3934.

Division of Biomedical Engineering. University of Virginia, Box 337, Medical Center, Charlottesville, VA 22908. (804) 924-5101. FAX (804) 982-3807

Minnesota Rehabilitation Engineering Center. University of Minnesota, 860 Mayo Building, Box 297, Minneapolis, MN 55455. (612) 373-8990.

BLACK HOLES

See also: ASTRONOMY, ASTROPHYSICS, CELESTIAL MECHANICS,COSMOLOGY, GALAXIES, GRAVITATION, INTERSTELLAR MATTER, PLANETS, SOLAR SYSTEM, TELESCOPES

ABSTRACT SERVICES AND INDEXES

Applied Science and Technology Index; a cumulative subject index to English language periodicals in the fields of aeronautics and space science, computer technology, chemistry, construction industry, energy and related areas. H.W. Wilson Co., 950 University Avenue, Bronx, NY 10452. (212) 588-8400. (800) 367-6770. FAX (718) 590-1617. From 1958 to present. Monthly. Inquire about cost and availability. Also available on CD-ROM and online. ISSN: 0003-6986.

Astronomy and Astrophysics Abstracts. Springer-Verlag New York, 175 Fifth Avenue, New York, NY 10010. (212) 460-1500. FAX (212) 473-6272. Published for the Astronomisches Rechen-Institut. Comprehensive coverage of all aspects of astronomy, astrophysics and related fields. 1969 to present. Two parts per year. Annual. ISSN: 0067-0022.

General Science Index. H.W. Wilson Company, 950 University Avenue, Bronx, NY 10452. (212) 588-8400. (800) 367-6770. FAX (718) 590-1617. From 1978 to present. Ten issues per year; quarterly and annual cumulations. Service basis. Available on CD-ROM and online. Inquire regarding cost and availability. ISSN: 0162-1963.

NTIS Alerts: Astronomy & Astrophysics. U. S. National Technical Information Service. 5285 Port Royal Road, Springfield, VA 22161. (703) 487-4650. FAX (703) 321-8547. Weekly. $140.00 per year.

Pascal 48: Environnement Cosmique Terrestre, Astronomie et Geologie Extraterrestre. Centre National de la Recherche Scientifique, Institut de Information Scientifique et Technique, 2 aliee du Parc de Brabois, 54514 Vandoeuvre Les Nancy Cedex, France. TEL 83-50-46-00. FAX: 83-50-46-50. 1985 to present. Ten issues per year. 770 F. Also available on CD-ROM and online.

Physics Abstracts . INSPEC. Section A, Science Abstracts. Institute of Electrical Engineers, London, United Kingdom. Available from: INSPEC/IEEE - Institute of Electrical and Electronic Engineers, Box 1331, Hoes Lane, Piscataway, NJ 08855-1331. (908) 562-5549. 1898 to present. 24 issues per year. $2835.00 per year. ISSN: 0036-8091. Also available online and on CD-ROM.

Science Citation Index (SCI). Institute for Scientific Information, 3501 Market Street, Philadelphia, PA 19104. (215) 386-0100. (800) 523-1850. FAX (215) 386-2991. 1961 to present. Six issues per year, plus annual cumulation. $11650.00 per year. Also available online and on CD-ROM. Inquire about price and availability. ISSN: 0036-827X.

STAR. Scientific and Technical Aerospace Reports. U.S. National Aeronautics and Space Administr ation. Distributed by U. S. Superintendent of Documents, Washington, DC 20402. 1963 to present. Semi-monthly, with semiannual and annual indexes.$114.00 per year. ISSN: 0036-8741. Also available online and on CD-ROM.

ANNUAL REVIEWS AND YEARBOOKS

The Astronomical Almanac. Superintendent of Documents, U. S. Government Printing Office, Washington, DC 10402. (202) 783-3238. 1981 to present. Superceeds Astronomical Ephemeris and the American Ephermis and Nautical Almanac. Annual. $44.95. ISSN: 0737-6421.

Annual Review of Astronomy and Astrophysics; Original reviews of critical literature and current developments in astronomy and astrophysics. Annual Reviews, Inc., 4139 El Camino Way,Palo Alto, CA 94306-0139. (415) 493-4400. (800) 523-8635. FAX (415) 855-9815. 1963 to date. Annual. $60.00. ISSN: 0066-4146.

Highlights of Astronomy. I. Appenzeller, editor. Kluwer Academic Publishers, Box 358, Accord Station, Hingham, MA 02018-0358. (617) 871-6600. International Astronomical Union Highlights series. ISBN 07923-3553-8. Triennal. 1995. Price varies.

ASSOCIATIONS AND PROFESSIONAL SOCIETIES

Amateur Astronomers Association. 1010 Park Avenue, New York, NY 10028. (212) 535-2922.

American Association of Variable *Star* Observers. 23 Birch Street, Cambridge, MA 02138. (617) 354-0484. FAX: (617) 354-0665.

American Astronomical Society. 2000 Florida Avenue NW, Suite 400, Washington, DC 20009. (202) 328-2010. FAX: (202) 659-0134.

American Institute of Physics. One Physics Ellipse, College Park, MD 20740-3843. (301) 209-3100.

Association of University for Research in Astronomy. Suite 701, 1625 Massachusetts Avenue, NW, Washington, DC 20036. (202) 483-2101. FAX (202) 483-2106.

Astronomical League. 2112 Kingfisher Lane East, Rolling Meadows, IL 60008. (708) 398-0562.

Astronomical Society of the Pacific. 390 Ashton Avenue, San Francisco, CA 94112. (415) 337-1100. FAX: (415) 337-5205.

International Astronomical Union. 98bis Boulevard Arago, 75014 Paris, France. TEL 1-43-25-83-58. FAX 1-43-25-26-16.

Royal Astronomical Society of Canada. 136 Dupont Street, Toronto. ON N5R 1V2 Canada. (416) 924-7973. FAX (416) 968-6687.

Universities Research Association, 1111 19th Street NW, Suite 400, Washington, DC 20036. (202) 293-1382.

Webb Society. 1440 South Marmora Avenue, Tuscon, AZ 95713. (602) 628-1077.

BIBLIOGRAPHIES

Astronomy and Astrophysics: A Source Guide. Gordon Press, P.O. Box 459, Bowling Green Station, New York, NY 10004. (718) 624-8419. 1991. $250.00. .

Black Holes: An Annotated Bibliography. Stephen I.Danko. Scarecrow Press, Inc., 52 Liberty Street, Box 4167. Metuchen, NJ 08840. (908) 548-8600. (800) 537-7107. 1985. $27.50.

Catalog of the Naval Observatory Library. Naval Observatory Staff. G. K. Hall & Company, 866 Third Avenue, New York, NY 10022. (212)702-6789. (800) 257-5755. 6 volumes. 1977. $655.00

Science Books and Films. American Association for the Advancement of Science, 1333 H Street NW, Washington, DC 20005. (202) 326-6454. 1965 to present. Nine issues per year. $40.00 per year. ISSN: 0098-342X

Scientific and Technical Books and Serials in Print;

an index to literature in science and technology. R. R. Bowker. 121 Chanlon Road, New Providence, NJ 07974. (908) 464-6800. FAX (908) 665-3502. (800) 521-8110. 1972 to present. Annual. $299.95 per year. Also available on CD-ROM. ISSN: 0000-054X.

Aa Source Book on Astronomy and Astrophysics, 1900 - 1975. Kenneth Lang and Owen Gingerich, editors. Harvard University Press, 79 Garden Street, Cambridge MA. (617) 495-2600. 1980. $41.95.

BLACK HOLES

Ency. of Physical Sciences and Engineering Info. Sources

DIRECTORIES AND BIOGRAPHICAL SOURCES

American Astronomical Society. Membership Directory. American Astronomical Society. 2000 Florida Avenue NW, Suite 400, Washington, DC 20009. (202) 328-2010. FAX (202) 234-2560 Annual. Included in membership dues. ISSN: 1061- 9038.

American Men and Women of Science: Physical and Biological Sciences. R. R. Bowker Inc., 121 Chanlon Road, New Providence, NJ 07974. (908) 464-6800. (800) 521-8110. 20th edition. 8 volumes. 1996. $850.00.

The Astronomers. Donald Goldsmith. St. Martin's Press, Inc., 175 Fifth Avenue, New York, NY 10010. (212) 674-5151. (800) 221-7945. 1993. $14.95 in paper.

Astronomical Centers of the World. Kevin Krisciunas. Cambridge University Press, 40 West 20th Street, New York, NY 10011-4211. (212) 924-3900. (800) 872-7423. 1988. $34.95

The Biographical Dictionary of Scientists: Astronomers. D. Abbott, editor. Peter Bedrick Books, Inc., 2112 Broadway, Room 318, New York, NY 10023. (212) 496-0751. 1984.

Directory of Physics and Astronomy Staff Members. American Institute of Physics, One Physics Ellipse, College Park, MD 20740-3843. (301) 209-3100. 1975/76 to present. Annual. $60.00. ISSN: 0361-2228.

Earth and Astronomical Research Centers. Stockton Press, 345 Park Avenue South, New York, NY 10010. 4th edition. 1995. $515.00. ISBN 1-56169-0967.

Graduate Programs in Physics, Astronomy and Related Fields. 1995-1996. American Institute of Physics, One Physics Ellipse, College Park, MD 20740-3843. (301) 209-3100. 1978 to present. Annual. $45.00. ISSN: 0147-1821.

Research Centers Directory. Gale Research, 835 Penobscot Building, Detroit, MI 48226-4094. (313) 961-2242. (800) 347-4253. 20th edition. 1995. $485.00.

ENCYCLOPEDIAS AND DICTIONARIES

Concise Dictionary of Astronomy. Jacqueline Mitton. Oxford University Press, Inc., 200 Madison Avenue, New York, NY 10016. (212) 725-6000. (800) 334-4249. 1992. $24.95.

Encyclopedia of Astronomy and Astrophysics. Stephen Maran, editor. Van Nostrand Reinhold, 115 Fifth Avenue, New York, NY 10003. (212) 254-3232. (800) 842-3636. 1992. $129.95.

Facts on File Dictionary of Astronomy. Valerie Illingworth. Facts-on-File, Inc., 460 Park Avenue South, New York, NY 10016-7382. (212) 683-2244. (800) 322-8755. Fax (800) 683-3633. 1994. $27.95.

Lan Dolt-boernstein Numerical Data and Functional Relationships in Science and Technology: Astronomy, Astrophysics and Space Research; A. O. Madelung and H. H. Boigt, editors. Springer-Verlag New York, Inc., 175 Fifth Avenue, New York, NY 10010. (212) 460-1500. (800) 777-4643. 1993. $550.00.

Mcgraw-hill Encyclopedia of Science and Technology. McGraw-Hill Book Company, Inc., 1221 Avenue of the Americas, New York, NY 10020. (212) 512-2000. (800) 262-4729. 7th edition. 20 volume set. 1992. $1900.00.

Oxford Illustrated Encyclopedia. Volume 8, The Universe. Archie Roy, editor. Oxford University Press, Inc., 200 Madison Avenue, New York, NY 10016. (212) 725-6000. (800) 334-4249. 1993. $45.00.

Penguin Dictionary of Astronomy. Jacqueline Mitton. Viking Penguin, 375 Hudson Street, New York, NY 10014-3657. (212) 366-2000. (800) 331-4624. 1993. $12.95 in paper.

Stars, Galaxies, Cosmos. William R. Corliss. Sourcebook Project, P.O. Box 107, Glen Arm, MD 21057. (410) 668-6047. 1987. $17.95.

GENERAL WORKS

Astronomer's Universe: Stars, Galaxies and Cosmos. Herbert Friedman. W. W. Norton & Company, 500 Fifth Avenue, New York, NY 10110. (212) 354-5500. (800) 223-2584. 1990. $24.95.

Astronomy and Cosmology. John North. W. W. Norton & Company, 500 Fifth Avenue, New York, NY 10110. (212) 354- 5500. (800) 223-2584 1990. 1994. $18.95.

Black Holes. Jean-Pierre Luminet. Cambridge University Press, 40 West 20th Street, New York, NY 10011-4211. (212) 924-3900. (800) 872-7423. 1992. $19.95 in paper.

Black Holes and the Universe. Igor D. Novikov. Cambridge University Press, 40 West 20th Street, New York, NY 10011-4211. (212) 924-3900. (800) 872-7423. 1990. $19.95 in paper.

Black Holes and Time Warps: Einstein's Outrageous Legacy. Kip S. Thorne. W. W. Norton & Company, 500 Fifth Avenue, New York, NY 10110. (212) 354-5500. (800) 223-2584 1995. $14.95 in paper.

Black Holes, White Dwarfs, and Neutron Stars: The Physics of Compact Objects. Stuart L. Shapiro. John Wiley & Sons, Inc., 605 Third Avenue, New York, NY 10158-0012. (212) 850-6000. (800) 225-5945. 1983. $89.95.

Classical General Relativity. Subrahmanyan Chandrasekhar and Morton D. Hull, editors. Oxford University Press, Inc., 200 Madison Avenue, New York, NY 10016. (212) 725-6000. (800) 334-4249. 1993. $45.00.

Cycles of Fire: Stars, Galaxies and the Wonder of Deep Space. William K. Hartmann. Workman Publishing Company, Inc., 708 Broadway, New York, NY 10003. (212) 254-5900. (800) 722-7202. 1988. $15.95 in paper.

Farthest Things in the Universe. Jay M. Pasachoff. Cambridge University Press 40 West 20th Street, New York, NY 10011-4211. (212) 924-3900. (800) 872-7423. 1995. $29.95.

Gravitation. Charles W. Misner, et al. W. H. Freeman, 41 Madison Avenue, East 26th Avenue, 35th Floor, New York, NY 10010. (212) 576-0400. 1995. $wfi.

Hawking on the Big Bang and Black Holes. Stephen W. Hawking. World Scientific Publishing Company, Inc., 1060 Main

Street, River Edge, NJ 07661. (201) 487-9655. (800) 227-7562. 1993. $33.00 in paper.

Discovering the Cosmos. Robert C. Bless. University Science Books, 55D Gate Five Road, Sausalito, CA 94965. (415) 332-5390. 1995. $58.00.

Universe. William J. Kaufmann. W. H. Freeman & Co., 41 Madison Avenue, East 26th Avenue, 35th Floor, New York, NY 10010. (212) 576-0400. 4th edition. 1995. $54.95.

Introduction to Black Holes. William Kaufmann, et al. Astronomical Society of the Pacific, 390 Ashton Avenue, San Francisco, CA 94112. (415) 337-1100. FAX: (415) 337-5205. 1984. $4.00.

Observational Astrophysics. Robert C. Smith. Cambridge University Press, 40 West 20th Street, New York, NY 10011-4211. (212) 924-3900. (800) 872-7423. 1995. $69.95.

The Physics of Stars. A. C. Phillips. John Wiley & Sons, Inc., 605 Third Avenue, New York, NY 10158-0012. (212) 850-6000. (800) 225-5945. 1994. $32.95 in paper.

Quantum Field Theory in Curved Spacetime and Black Hole Thermodynamics. Robert M. Wald. University of Chicago Press, 5801 Ellis Avenue, Chicago, IL 60637. (312) 702-7700. 1994. $16.95 in paper.

Quasar Astronomy. Daniel W. Weedman. Cambridge University Press, 40 West 20th Street, New York, NY 10011-4211. (212) 924-3900. (800) 872-7423. 1988. $24.95 in paper.

The Red Limit: The Search for the Edge of the Universe. Timothy Ferris. Morrow, William & Company, 1350 Avenue of the Americas, New York, NY 10019. (212) 261-6500. (800) 843-9389. Revised edition. 1993. $12.95.

Space, Time & Gravity: The Theory of the Big Bang and Black Holes. Robert M Wald. University of Chicago Press, 5801 Ellis Avenue, Chicago, IL 60637. (312) 702-7700. 1992. $24.95.

Stars and Atoms: From the Big Bang to the Solar System. Stuart Clark. Oxford University Press, Inc., 200 Madison Avenue, New York, NY 10016. (212) 725-6000. (800) 334-4249. 1995. $35.00.

Stephen Hawking's Universe. John Boslough. Morrow, William & Company, 1350 Avenue of the Americas, New York, NY 10019. (212) 261-6500. (800) 843-9389. 1985. $7.95.

Unveiling the Edge of Time: Black Holes, White Holes, Worm Holes. John Gribbin. Crown Publishing Group, 201 East 50th Street, New York, NY 10022. (212) 751-2600. (800) 726-0600. 1994. $13.00 in paper.

Visual Astronomy of the Deep Sky. Roger N. Clark. Cambridge University Press, 40 West 20th Street, New York, NY 10011-4211. (212) 924-3900. (800) 872-7423. 1991. $39.95.

Workings of the Universe. Time-Life, Inc., 777 Duke Street, Alexandria, VA 22314. (703) 8388-7000. (800) 621-7026. 1991. $25.95.

HANDBOOKS AND MANUALS

Astrophysical Data: Planets and Stars. Kenneth R. Lanb. Springer-Verlag New York, Inc., 175 Fifth Avenue, New York, NY 10010. (212) 460-1500. (800) 777-4643. 1993. $59.00.

Burnham's Celestial Handbook: an Observer's Guide to the Universe Beyond the Solar System. Robert Burnham, Jr. Dover Publications, Inc., 180 Varick Street, New York, NY 10014. (212) 255-3755. (800) 223-3130. 3 volumes. 1978. $41.85 for the set.

Cambridge Guide to the Constellations. Michael E. Bakich. Cambridge University Press, 40 West 20th Street, New York, NY 10011-4211. (212) 924-3900. (800) 872-7423. 1995. $69.95.

Encyclopedia of Astronomy and Astrophysics. Stephen Maran, editor. Van Nostrand Reinhold, 115 Fifth Avenue, New York, NY 10003. (212) 254-3232. (800) 842-3636. 1992. $129.95.

Greenwich Guide to Stars, Galaxies and Nebulae. Stuart Malin. Cambridge University Press, 40 West 20th Street, New York, NY 10011-4211. (212) 924-3900. (800) 872-7423. 1990. $10.95.

Handbook of Astronomy, Astrophysics and Geophysics; Volume 2: Galaxies and Cosmology. V. M. Canuto and B. G. Elmegreen. Gordon & Breach Science Publishers, Inc., 50 West 23rd Street, New York, NY 10010. (212) 206-8900. . 1988. $194.00.

Handbook of Space Astronomy and Astrophsyics. Martin V. Zumbeck. Cambridge University Press, 40 West 20th Street, New York, NY 10011-4211. (212) 924-3900. (800) 872-7423. 2nd edition. 1990. $79.95.

Landolt-Borenstein Numerical Data and Functional Relationships in Science and Technology: Astronomy, Astrophysics and Space Research. GROUP VI. Springer-Verlag New York, Inc., 175 Fifth Avenue, New York, NY 10010. (212) 460-1500. (800) 777-4643. volumes priced individually.

McGraw-hill Encyclopedia of Astronomy. Sybil P. Parker and Jay M. Pasachoff, editors. McGraw-Hill Publishing Company, Inc., 1221 Avenue of the Americas, New York, NY 10020. (212) 512-2000. (800) 262-4729. 2nd edition. 1993. $75.00

Observer's Guide to Astronomy. Patrick Martinez, editor. Cambridge University Press, 40 West 20th Street, New York, NY 10011-4211. (212) 924-3900. (800) 872-7423. 2 volumes. 1994. $159.90.

ONLINE DATABASES AND CD-ROMS

CA Search. Chemical Abstracts Service, P.O. Box 3012, Columbus, OH 43210-0012. (614) 447-3600. (800) 848-6533. FAX (614) 447-3709. Very comprehensive guide to worldwide chemical literature and related fields, 1972 to present. Available on BRS, (800) 289-4277, DIALOG, (800) 334-2564, ORBIT (800) 456-7248, and STN International, FIZ Karlsruhe, P.O. Box 2465, W-7500, Karlsruhe 1, Germany, online services. Inquire as to cost and availability.

Current Contents Search. Institute for Scientific Information, 3501 Market Street, Philadelphia, PA 19104. (215) 386-0100. FAX (215) 386-6362. Contains citations to articles listed in the table of contents of science and technology journals. Also ar-

BLACK HOLES

Ency. of Physical Sciences and Engineering Info. Sources

ticles in social sciences and life sciences journals. Available on BRS, (800) 289-4277, DIALOG, (800) 334-2564, online services. Inquire as to cost and availability.

Dissertation Abstracts Online. University Microfilms International, 300 North Zeeb Road, Ann Arbor, MI 48106. (800) 521-0600 or (313) 761-4700. Scope includes virtually all doctoral dissertations accepted at accredited American institutions from 1861 to present in 252 subject areas. Available on BRS, (800) 289-4277, DIALOG, (800) 334-2564, and OCLC EPIC, (800) 848-5878, online services. Also available on CD-ROM. Inquire as to cost and availability.

INSPEC. Institution of Electrical Engineers, Michael Faraday House, Six Hills Way, Stevenage, Herts. SG1 2AY, England. Telephone: 043 8 313311 or FAX 0438 742840 . Contains citations to the worldwide literature of phys ics, electronics and electrical engineering, computer technology, and related fields. Available on BRS, (800) 289-4277, DIALOG, (800) 334-2564, ORBIT, (800) 456-7248, and STN International, FIZ Karlsruhe, P.O. Box 2465, W-7500, Karlsruhe 1, Germany, online services. Inquire as to cost and availability.

International Aerospace Abstracts. American Institute of Aeronautic s and Astronautics, 370 L'Enfant Promenade, S.W., Washington, DC 20024. (202) 646-7400. Contains references and abstracts of journal and monograph literature relating to aerospace science and technology, from 1963 to present. Available through the NASA/RECON system of the National Aeronautics and Space Administration only.

NTIS Bibliographic Database. National Technical Information Service, 5285 Port Royal Road, Springfield, Va 2216 1. (703) 487-4929 or FAX (703) 321-8199. Broad cover age of government-sponsored science and technology research reports,1964 to present. Available on BRS,(800) 289-4277, DIALOG, (800) 334-2564, ORBIT, (800) 456-7248, and STN International, FIZ Karlsruhe, P.O. Box 2465, W-7500, Karlsruhe 1, Germany, online services. Also available on CD-ROM. Inquire as to cost and availability.

Physics Briefs. American Institute of Physics, 335 East 45th Street , New York, NY 10017. (212) 661-9260 or FAX (212) 661-2036. Contains citations with abstracts of the literature of physics and related fields, 1979 to present. Available on the STN International, FIZ Karlsruhe, P.O. Box 2465, W-7500, Karlsruhe 1, Germany, online service. Inquire as to cost and availability.

Scientific and Technical Books and Serials in Print. R.R. B owker Inc., 121 Chanlon Road, New Providence, NJ 07974. (800) 521-8110 or (908) 464-6800. List of currently published books and serials in the physical and biological sciences, engineering and technology, with subject, author and titles indexes. Available on ORBIT, (800) 456-7248, online service. Inquire as to cost and availability.

SCISearch. Institute for Scientific Information, 3501 M arket Street, Philadelphia, PA 19104. (800) 523-1850 or (215) 386-0100. Broad multidisciplinary title and author index to the international literature of science and technology, 1974 to present. Available on DIALOG, (800) 334-2564, and ORBIT, (800) 456-7248, online services. Also available on CD-ROM. Inquire as to cost and availability.

WilsonLine. H.W. W ilson Company, 950 University Avenue, Bronx, NY 10452. (800) 367-6770 or (212) 588-8400. Makes available onlin e versions of the printed H.W. Wilson indexes includin g Applied Science and Technology Index, Business Periodicals Index, General Science Index, and Readers' Guide

to Periodical Literature. Period covered is generally 1983 to present. Available on BRS, (800) 289-4277, DIALOG, (800) 334-2564, and OCLC EPIC, (800) 848-5878, online services. Also available on CD-ROM. Inquire as to cost and availability.

OTHER SOURCES

Atlas of Deep-sky Spendors. Hans Vehrenberg. Sky Publishing Corporation, 49 Bay State Road, Cambridge, MA 02138. (617) 864-7360. (800) 253-024. 4th edition. 1983. $24.95.

Atlas of the Ultraviolet Sky. Richard C. Henry, Wayne B. Landsman, et al. Johns Hopkins University Press, 2715 North Charles Street, Baltimore, MD 21218-4319. (410) 516-6900. (800) 537-5487. 1988. $70.00.

Atlas of the Universe. Patrick Moore. Rand McNally & Company. 8255 North Central Park, Skokie, IL 60076-2970. (708) 673-9100. 1994. $29.95.

The Bright Star Catalogue. Dorrit Hoffleit and Carlos Jaschek. Yale University Observatory, 260 Whitney Avenue, P.O. Box 6666, New Haven CT 06511. (203) 432-3431. 4th revised edition. 1982. $35.00 in paper.

Cambridge Atlas of Astronomy. Jean Audouze, et al.Cambridge University Press, 40 West 20th Street, New York, NY 10011-4211. (212) 924-3900. (800) 872-7423. 3rd edition. 1994. $75.00.

Cambridge Guide to the Constellations. Michael E. Bakich. Cambridge University Press, 40 West 20th Street, New York, NY 10011-4211. (212) 924-3900. (800) 872-7423. 1995. $19.95 in paper.

Cambridge Star Atlas 2000.0. Wil Tirion. Cambridge University Press, 40 West 20th Street, New York, NY 10011- 4211. (212) 924-3900. (800) 872-7423. 2nd edition. 1996. $19.95.

Color Atlas of Galaxies. James Wray. Cambridge University Press, 40 West 20th Street, New York, NY 10011-4211. (212) 924-3900. (800) 872-7423. 1988. $94.95.

Exploring the Southern Sky: a Pictorial Atlas from the European Southern Observatory. S. Lausten, et al. Springer-Verlag, 175 Fifth Avenue, New York, NY 10010. (212) 460-1500. (800) 777-4643. 1987. $59.00.

Observer's Sky Atlas. E. Karkoschka. Springer-Verlag New York, Inc., 175 Fifth Avenue, New York, NY 10010. (212) 460-1500. (800) 777-4643. 1995. $15.95.

NGC 2000.0: The Complete New General Catalogue and Index Catalogues of Nebulae and Star Clusters. J. L. E. Dreyer. editor. Cambridge University Press, 40 West 20th Street, New York, NY 10011-4211. (212) 924-3900. (800) 872-7423 1989. $21.95.

Norton's 2000.0 Star Atlas and Reference Handbook. Ian Ridpath, editor. Halsted Press, 605 Third Avenue, New York, 10158-0012. (212) 850-6400. 1989. $49.95.

Observing Handbook and Catalog of Deep-sky Objects. Christian B. Luginbuhl and Brian A. Skiff. Cambridge University Press, 40 West 20th Street, New York, NY 10011- 4211. (212) 924-3900. (800) 872-7423. 1990. $49.95.

Starlist Two Thousand: A Quick Reference Star Catalog for Astronomers. Richard Dibon-Smith. John Wiley & Sons, Inc., 605 Third Avenue, New York, NY 10158-0012. (212) 850-6000. (800) 225-5945. 1992. $29.95.

The Sky: A User's Guide. David H. Levy. Cambridge University Press, 40 West 20th Street, New York, NY 10011- 4211. (212) 924-3900. (800) 872-7423. 1993. $14.95 in paper.

Sky Catalog 2000.0: Volume 1: Stars to Magnitude 8.0. Alan Hirshfeld, et al. Cambridge University Press, 40 West 20th Street, New York, NY 10011-4211. (212) 924-3900. (800) 872-7423. 2nd edition. 1991. $42.95 in paper.

PERIODICALS

Astronomical Journal. American Institute of Physics, One Physics Ellipse, College Park, MD 20740-3843. (301) 209- 3000. Published for the American Astronomical Society. 1849 to present. Monthly. $280.00 per year. ISSN: 0004- 6256.

Astronomical Society of the Pacific. Publications. Astronomical Society of the Pacific, 390 Ashton Avenue, San Francisco, CA 94112. (415) 337-1100. FAX (415) 337-5205. 1889 topresent. Monthly. $175.00 per year. ISSN: 0004- 6280.

Astronomy. Kalmbach Publishing Company, Box 1612, Waukesha, WI 53187-1612. (414) 796-8776. FAX (414) 796-1142. 1971 to present. Monthly. $27.00 per year. ISSN: 0091-6358.

Astronomy and Astrophsyics. Springer-Verlag New York, Inc., 44 Hartz Way, Secaucus, NJ 07096-2491. (201) 348-4033. FAX (201) 348-4505. 1969 to date. 36 issues per year in 12 volumes. $1733.00 per year. ISSN: 0004-6361.

Astronomy Reports. American Institute of Physics, One Physics Ellipse, College Park, MD 20740-3843. (301) 209- 3000. English translation of *Astronomicheksii Zhurnal.* Formerly *Soviet Astronomy AJ.* 1957 to present. Bimonthly. $1160.00 per year. ISSN: 1063-7729.

Astroparticle Physics. Elsevier Science Publishing Company, Inc., Box 882, Madison Square Station, New York, NY 10159. (212) 989-5800. FAX (212) 633-3990. 1992 to present. Quarterly. $180.00 per year. ISSN: 0927-6505.

Astrophysical Journal; an international review of astronomy and astronomical physics. University of Chicago Press, Journals Division, 5720 South Woodlawn Avenue, Chicago, IL 60637. (312) 753-3347. FAX (312) 753-0811. 1895 to date. three issues per month. $740.00 per year. ISSN: 0004-637X.

Astrophysics and Space Science; an international journal of cosmic physics. Kluwer Academic Publishers, Box 358, Accord Station, Hingham, MA 02018-0358. (617) 871-6600. FAX (617) 871-6528. 1968 to present. 24 issues per year. $2544.00 per year. ISSN: 0004-640X.

Celestial Mechanics and Dynamical Astronomy; an International Journal of Space Dynamics. Kluwer Academic Publishers, Box 358, Accord Station, Hingham, MA 02018-0358. (617) 871-6600. FAX (617) 871-6528. 1969 to present. Monthly. $703.50 per year. ISSN: 0923-2958.

Experimental Astronomy; an international journal on astronomical instrumentation and data analysis. Kluwer Academic Pub-

lishers, Box 358, Accord Station, Hingham, MA 02018-0358. (617) 871-6600. FAX (617) 871-6528. 1989 to present. Quarterly. $161.00 per year. ISSN: 0922-6435.

ICARUS: International Journal of Solar System Studies. Academic Press, Inc., Journal Division, 525 B Street, Suite 1900, San Diego, CA 92101-4495. (619) 230-1840. FAX (619) 699-6800. 1962 to present. Monthly. $1080.00 per year. ISSN: 0019-1035.

MERCURY. Astronomical Society of the Pacific. 390 Ashton Avenue, San Francisco, CA 94112. (415) 337-1100. FAX: (415) 337-5205. 1972 to present. Bimonthly. $175.00 per year. ISSN: 0047-6773.

Monthly Notices of the Royal Astronomical Society. Blackwell Scientific Publication LT., Osney Mead, Oxford OX2 OEL, England. TEL 0865-240201. FAX 0865-721205. 1827 to present. Fortnightly. $1899.00 per year. ISSN: 0035-8711.

Observatory. c/o Dr. D. J. Stickland, Space andAstrophysics Division, Rutherford Appleton Laboratory, Chilton, Didcot, Oxon OX11 OQX England. FAX 0235-445848. 1877 to present. Bimonthly. $42.00 per year. ISSN: 0029-7704.

Sky and Telescope. Sky Publishing Corporation, Box 9111, Belmont, MA 02178. (617) 864-7360. FAX (617) 864-6117. 1941 to present. Monthly. $27.00 per year. ISSN: 0037- 6604.

Vvistas in Astronomy; an international review journal. Albert C. Beer, editor. Elsevier Science Publishing Company, Inc., 660 White Plains Road, Terrytown, NY 10591- 5153. (914) 524-9200. FAX: (914) 333-2444. 1958 to present. Quarterly. $415.00 per year. ISSN: 0083-6656.

Transactions of the International Astronomical Union. Proceedings of the General Assembly. Kluwer Academic Publishers, Box 358, Accord Station, Hingham, MA 02018- 0358. (617) 871-6600. Issues in two parts: Part A, Reports, and Part B, Proceedings. Triennial. XXI - 1994. Price varies. ISSN: 0080-1372.

Webb Society Quarterly Journal. Webb Society, 194 Foundry Lane, Freemantle, Southampton, Hampshire SO1 3EQ, England. 1969 to present. Quarterly. $19.50 per year. ISSN: 0043- 1680.

RESEARCH CENTERS AND INSTITUTES

Astronomy Program. University of Maryland, 1207 Computer & Space Sciences Building, College Park, Md 20742. (301) 405-3001. FAX (301) 314-9067.

Astrophysical Research Consortium. University of Chicago, Department of Astrophysics, 5640 South Ellis, Chicago, IL 60637. Includes University of Washington, Washington State University, Princeton University and New Mexico State University.

Center for Astronomy and Space Astrophysics. University of Colorado - Boulder, Campus Box 391, Boulder, CO 80309. (303) 492-4050. (FAX) 492-7178.

Center for Astrophysics and Space Sciences. University of California, San Diego, C-0111, La Jolla, CA 92093-0111. (619) 534-3460. FAX (619) 534-2294.

Center for Space Research. Massachusetts Institute of Technology, 77 Massachusetts Avenue, Camridge MA 02139. (617) 253-

BLACK HOLES

Ency. of Physical Sciences and Engineering Info. Sources

7501.

Center for Space Science and Astrophysics. Stanford University, 325 Durand Building, Stanford, CA 94305. (415) 723-3592.

Harvard-Smithsonian Center for Astrophysics. 60 Garden Street, Cambridge, MA 02138. (617) 495-7461. FAX (617) 495-7326.

National Astronomy and Ionosphere Center. Cornell University, Space Sciences Building, Ithaca, NY 14853. (607) 255-3735. (607) 255-8803.

National Radio Astronomy Observatory. Edgemont Road, Charlottesville, VA 22903. (804) 296-0211.

Space Sciences Laboratory. University of California, Berkeley, Berkeley, CA 94720. (415) 642-1361. FAX (415) 643-7629.

Space Telescope Science Institute. 3700 San MartinDrive, Baltimore, MD 21218. (301) 338-4700.

BLAST FURNACES

See: STEEL AND STEEL MAKING

BLASTING

See: EXPLOSIVES

BLIMP

See: AIRSHIPS

BOHR THEORY

See: PHYSICAL CHEMISTRY

BOILERS

See: STEAM ENGINEERING

BOOSTER ROCKETS

See: ROCKETS

BORING

See: PETROLEUM ENGINEERING

BORON

See also: ALLOYS, CHEMISTRY, INORGANIC CHEMISTRY, ORGANIC CHEMISTRY, SILICON, STEEL AND STEEL MAKING

ABSTRACT SERVICES AND INDEXES

Alloys Index. American Society for Metals, Metals Park, OH 44073. (216) 338-5151. FAX (216) 338-4634. 1974 to present. Monthly. $380.00. Also available on CD-ROM and online via METADEX, STN International, and DIALOG.

Applied Mechanics Reviews: An Assessment of World Literature in Engineering Sciences. 1948-present. American Society of Mechanical Engineers, 345 East 47th Street, New York, NY 10017. (212) 705-7703. Monthly. $360.00 per year.

Applied Science and Technology Index; a cumulative subject index to English language periodicals in the fields of aeronautics and space science, computer technology, chemistry, construction industry, energy and related areas. H.W. Wilson Co., 950 University Avenue, Bronx, NY 10452. (800) 367-6770 or (212) 588-8400. FAX (718) 590-1617. From 1958 to present. Monthly. Inquire about cost and availability. Also available on CD-ROM and online.

Chemical Abstracts - Applied Chemistry and Chemical Engineering Sections. Chemical Abstracts Service, 2540 Olentangy River Road, Box 3012, Columbus, OH 43210. (800) 848-6538 or (614) 447-3600, FAX (614) 447-3713. Bi-weekly. $1410.00 per year.

Chemical Abstracts - Organic Chemistry Sections. Chemical Abstracts Service, 2540 Olentangy River Road, Box 3012, Columbus, OH 43210. (800) 848-6538 or (614) 447-3600, FAX (614) 447-3713. Bi-weekly. $1410.00 per year.

Chemical Abstracts - Physical, Inorganic and Analytical Chemistry. Chemical Abstracts Service, 2540 Olentangy River Road, Box 3012, Columbus, OH 43210. (800) 848-6538 or (614) 447-3600, FAX (614) 447-3713. Bi-weekly. $1410.00 per year.

Current Contents: Engineering, Technology, and Applied Sciences. Institute for Scientific Information, 3501 Market Street, Philadelphia, PA 19104. (215) 386-0100. FAX (215)386-6362. 1970 to present. Weekly. $442.00 per year.

Engineering Index Monthly. Engineering Information, Inc., Castle Point on the Hudson, Hoboken, NJ 07030. (800) 221-1044. FAX (212) 832-1857. Monthly. $2300.00 per year. Also available online as COMPENDEX, and also on CD-ROM. Covers chemical engineering, computers, electrical engineering, civil engineering, metals and mining, industrial management, and mechanical engineering.

I.M.M. Abstracts and Index. Institution of Mining and Metallurgy, 44 Portland Pl., London W1N 4BR, England. 071-580-3802. FAX 071-436-5388. Bi-monthly. $364 for non-members.

Manufacturing and Process Engineering Abstracts, Cambridge Scientific Abstracts, 7200 Wisconsin Avenue, Bethesda, MD 20814-4823. (301) 961-6750. FAX (301) 961-6720. 1993 to present. Monthly. $385.00 per year. Covers concurrent engineering, quality control, automated manufacturing, petroleum engineering, oil field operations and equipment, energy management, metallurgy and metallography, foundry practice.

Science Citation Index. Institute for Scientific Information, 3501 Market Street, Philadelphia, PA 19104. (215) 386-0100. FAX (215) 386-6362. Inquire about availability and cost. Also available on CD-ROM.

ASSOCIATIONS AND PROFESSIONAL SOCIETIES

American Chemical Society. 1155 16th Street, N.W., Washington, DC 20036. (202) 872-4381 or (800) 333-9511. FAX (614) 447-3671.

American Iron and Steel Institute. 1101 17th St., NW, Washington, DC 20036-4700. (202) 452-7100. FAX (202) 463-6573.

American Society for Metals (ASM). Materials Park, OH 44073. (216) 338-5151 or FAX (216) 338-4634.

Association of Iron and Steel Engineers. Three Gateway Center, Suite 2350, Pittsburgh, PA 15222. (412) 281-6323.

metallurgical Society of the AIME (American Institute of Mining, Metallurgical, and Petroleum Engineers). 345 E.47th Street, 14th Floor, New York, NY 10017. (212) 705-7695.

DIRECTORIES AND BIOGRAPHICAL SOURCES

American Men & Women of Science, 1995-96. R.R. Bowker Staff, eds. 19th edition. 8 volumes. R.R. Bowker/Reed International Publishing Company, 121 Chanlon Road, New Providence, NJ 07974. (908) 464-6800 or (800) 521-8110. 1995. $850.00

Directory of Engineering Societies and Related Organizations. Gordon Davis. 13th edition. American Association of Engineering Societies, 1111 19th Street NW, Suite 608, Washington, DC 20036. (202) 296-2237 or (800) 658-8897. 1989. Inquire for price.

International Research Centers Directory. Gale Research, 835 Penobscot Building, Detroit, MI 48226-4094. (313) 961-2242. (800) 347-4253. FAX (313) 961-6083. 8th edition. 1995. $410.00.

Research Centers Directory. Annette Piccirelli. 19th edition. 2 volumes. Gale Research, 835 Penobscot Building, Detroit, MI 48226-4094. (313) 961-2242. (800) 347-4253. FAX (313) 961-6083. $470.00.

Scientific and Technical Organizations and Agencies Directory. Gale Research, 835 Penobscot Building, Detroit, MI 48226-4094. (313) 961-2242. (800) 347-4253. FAX (313) 961-6083. 4th edition. 1996. $195.00.

Who's Who in Engineering. American Association of Engineering Societies, 1111 19th Street NW, Suite 608, Washington, DC 20036. (202) 296-2237 or (800) 658-8897. 8th edition. 1991. Inquire for price.

GENERAL WORKS

Boron. Gordon Press Publishers, PO Box 459, Bowling Green Station, New York, NY 10004. (718) 624-8419. 1994. $250.95.

Boron and its Role in Crop Production. Umesh C. Gupta, editor. CRC Press, 2000 Corporate Blvd., N.W., Boca Raton, FL 33431. (407) 994-0555 or (800) 272-7737. FAX (407) 994-0949. 1993. Inquire for cost.

Current Topics in the Chemistry of Boron. George W. Kabalka, editor. CRC Press, 2000 Corporate Blvd., N.W., Boca Raton, FL 33431. (407) 994-0555 or (800) 272-7737. FAX (407) 994-0949. 1994. $110.00.

HANDBOOKS AND MANUALS

CRC Handbook of Chemistry and Physics. David R. Lide, editor. 75th edition. CRC Press, 2000 Corporate Blvd., N.W., Boca Raton, FL 33431. (407) 994-0555 or (800) 272-7737. FAX (407) 994-0949. $99.50.

Metals Handbook. ASM International, Materials Information, Materials Park, OH 44073-0002. (216) 338-5151 or (800) 336-5152. FAX (216) 338-4634. $154.00.

Smithell's Metals Reference Book. E.A. Brandes & G.B. Brook. 7th edition. Butterworth-Heinemann, 313 Washington Street, Newton, MA 02158. (617) 928-2500 or (800) 366-2665. FAX (617) 928-2620. 1992. $250.00.

ONLINE DATABASES AND CD-ROMS

CA Search. Chemical Abstracts Service, P.O. Box 3012, Columbus, OH 43210-0012. (614) 447-3600. (800) 848-6533. FAX (614) 447-3709. Very comprehensive guide to worldwide chemical literature and related fields, 1972 to present. Available on BRS, (800) 289-4277, DIALOG, (800) 334-2564, ORBIT (800) 456-7248, and STN International, FIZ Karlsruhe, P.O. Box 2465, W-7500, Karlsruhe 1, Germany, online services. Inquire as to cost and availability.

COMPENDEX Plus. Engineering Information, Inc., 345 East 47th Street, New York, NY 10017. (212) 705-7600 or (800) 221-1044. Contains citations with abstracts to worldwide literature in engineering and technology, from 1970 to present. Available on online BRS, (800) 289-4277, DIALOG, (800) 334-2564, ORBIT (800) 456-7248, and STN International, FIZ Karlsruhe, P.O.Box 2465, W-7500, Karlsruhe 1, Germany, online services. Also available on CD-ROM. Inquire as to cost and availability.

Dissertation Abstracts. University Microfilms International, 300 North Zeeb Road, Ann Arbor, MI 48106. (800) 521-0600 or (313) 761-4700. Scope includes virtually all doctoral dissertations accepted at accredited American institutions from 1861 to present in 252 subject areas. Available on BRS, (800) 289-4277, DIALOG, (800) 334-2564, and OCLC EPIC, (800) 848-5878, online services. Also available on CD-ROM. Inquire as to cost and availability.

METADEX. Jointly produced by ASM International and the Institute of Materials. Contains more than 925,000 records from the international literature on metals and alloys, concerning properties, processes, materials classes, applications, and metallurgical systems. Updated monthly. Available from ORBIT-QUESTEL (703) 442-0900.

NTIS Bibliographic Database. National Technical Information Service, 5285 Port Royal Road, Springfield, Va 22161. (703) 487-4929 or FAX (703) 321-8199. Broad coverage of government-sponsored science and technology research reports, 1964 to present. Available on BRS, (800) 289-4277, DIALOG, (800) 334-2564, ORBIT, (800) 456-7248, and STN International, FIZ Karlsruhe, P.O. Box 2465, W-7500, Karlsruhe 1, Germany,

BORON

Ency. of Physical Sciences and Engineering Info. Sources

online services. Also available on CD-ROM. Inquire as to cost and availability.

SciSearch. Institute for Scientific Information, 3501 Market Street, Philadelphia, PA 19104. (800) 523-1850 or (215) 386-0100. Broad multidisciplinary title and author index to the international literature of science and technology, 1974 to present. Available on DIALOG,(800) 334-2564, and ORBIT,(800) 456-7248, online services. Also available on CD-ROM. Inquire as to cost and availability.

WILSONLINE. H.W. Wilson Company, 950 University Avenue, Bronx, NY 10452. (800) 367-6770 or (212) 588-8400. Makes available online versions of the printed H.W. Wilson indexes including *Applied Science and Technology Index, Business Periodicals Index, General Science Index,* and *Readers' Guide to Periodical Literature*. Period covered is generally 1983 to present. Available on BRS,(800) 289-4277, DIALOG,(800) 334-2564, and OCLC EPIC,(800) 848-5878, online services. Also available on CD-ROM. Inquire as to cost and availability.

PERIODICALS

American Chemical Society Journal. American Chemical Society, 1155 16th Street, N.W., Washington, DC 20036. (202) 872-4381 or (800) 333-9511. FAX (614) 447-3671. 1879 to present. Bi-weekly. $1055.00 per year for non-members.

Industrial and Engineering Chemistry Research. American Chemical Society, 1155 16th Street, N.W., Washington, DC 20036. (202) 872-4381 or (800) 333-9511. FAX (614) 447-3671. 1962 to present. Monthly. $567.00 per year for non-members.

Inorganic Chemistry. American Chemical Society, 1155 16th Street, N.W., Washington, DC 20036. (202) 872-4381 or (800) 333-9511. FAX (614) 447-3671. 1962 to present. Semi-monthly. $949.00 per year for non-members.

BORON STEEL

See: FERROALLOYS

BOSONS

See: PARTICLE PHYSICS

BOUNDARY LAYER FLOW

See: FLUID MECHANICS

BOUNDARY WAVE

See: FLUID MECHANICS

BRAKES

See: AUTOMOTIVE ENGINEERING

BRASS AND BRONZE

See also: ALLOYS, FERROALLOYS, MATERIALS SCIENCE, METALLURGY

ABSTRACT SERVICES AND INDEXES

Aalloys Index. American Society for Metals, Metals Park, OH 44073. (216) 338-5151. FAX (216) 338-4634. 1974 to present. Monthly. $380.00. Also available on CD-ROM and online via METADEX, STN International, and DIALOG.

Applied Mechanics Reviews: An Assessment of World Literature in Engineering Sciences. 1948-present. American Society of Mechanical Engineers, 345 East 47th Street, New York, NY 10017. (212) 705-7703. Monthly. $360.00 per year.

Applied Science and Technology Index; a cumulative subject index to English language periodicals in the fields of aeronautics and space science, computer technology, chemistry, construction industry, energy and related areas. H.W. Wilson Co., 950 University Avenue, Bronx, NY 10452. (800) 367-6770 or (212) 588-8400. FAX (718) 590-1617. From 1958 to present. Monthly. Inquire about cost and availability. Also available on CD-ROM and online.

Current Contents: Engineering, Technology, and Applied Sciences. Institute for Scientific Information, 3501 Market Street, Philadelphia, PA 19104. (215) 386-0100. FAX (215) 386-6362. 1970 to present. Weekly. $442.00 per year.

Engineering Index Monthly. Engineering Information, Inc., Castle Point on the Hudson, Hoboken, NJ 07030. (800) 221-1044. FAX (212) 832-1857. Monthly. $2300.00 per year. Also available online as COMPENDEX, and also on CD-ROM. Covers chemical engineering, computers, electrical engineering, civil engineering, metals and mining, industrial management, and mechanical engineering.

I.M.M. Abstracts and Index. Institution of Mining and Metallurgy, 44 Portland Pl., London W1N 4BR, England. 071-580-3802. FAX 071-436-5388. Bi-monthly. $364 for non-members.

Manufacturing and Process Engineering Abstracts, Cambridge Scientific Abstracts, 7200 Wisconsin Avenue, Bethesda, MD 20814-4823. (301) 961-6750. FAX (301) 961-6720. 1993 topresent. Monthly. $385.00 per year. Covers concurrent engineering, quality control, automated manufacturing, petroleum engineering, oil field operations and equipment, energy management, metallurgy and metallography, foundry practice.

Metals Abstracts and Metals Abstracts Index. American Society for Metals, Metals Park, OH 44073. (216) 338-5151. 1968 to present. Monthly. Abstracts are $1500.00 per year and Index is $460.00 per year.

ANNUAL REVIEWS AND YEARBOOKS

Annual Review of Materials Science. Annual Reviews, Inc., 4139 El Camino Way, P.O. Box 10139, Palo Alto, CA 94303-0897.

(800) 523-8635 or (415) 493-4400. FAX (415) 855-9815. 1971 to present. Annual. $75.00 per year.

ASSOCIATIONS AND PROFESSIONAL SOCIETIES

American Foundrymen's Society Inc. 505 State Street, Des Plaines, IL 60016. (708) 824-0181.

American Society for Metals. Materials Park, OH 44073. (216) 338-5151 or FAX (216) 338-4634.

Brass and Bronze Ingot Manufacturers. 300 West Washington Street, Suite 1500, Chicago, IL 60606. (312) 236-2715. FAX (312) 236-8772.

Canadian Copper and Brass Development Association. 10 Gateway Blvd., Suite 375, Don Mills, ON M3C 3A1, Canada. (416) 421-0788. FAX (416) 421-8092.

Copper Development Association. 260 Madison Avenue, New York, NY 10016-2401. (212) 251-7200. FAX (203) 251-7234.

Metallurgical Society of the AIME(American Institute of Mining, Metallurgical, and Petroleum Engineers). 345 E.47th Street, 14th Floor, New York, NY 10017. (212) 705-7695.

DIRECTORIES AND BIOGRAPHICAL SOURCES

Dun's Industrial Guide: the Metalworking Directory, 1993-94. Dun & Bradstreet Information Services Staff. 3 volumes. Dun & Bradstreet Information Services, 899 Eaton Avenue, Bethlehem, Pa 18025. (610) 882-7000 or (800) 526-0651. FAX (610) 882-7269. 1993. $795.00.

International Research Centers Directory. Gale Research, 835 Penobscot Building, Detroit, MI 48226-4094. (313) 961-2242. (800) 347-4253. FAX (313) 961-6083. 8th edition. 1995. $410.00.

Research Centers Directory. Annette Piccirelli. 19th edition. 2 volumes. Gale Research, 835 Penobscot Building, Detroit, MI 48226-4094. (313) 961-2242. (800) 347-4253. FAX (313) 961-6083. $470.00.

Scientific and Technical Organizations and Agencies Directory. Gale Research, 835 Penobscot Building, Detroit, MI 48226-4094. (313) 961-2242. (800) 347-4253. FAX (313) 961-6083. 4th edition. 1996. $195.00.

Who's Who in Engineering. American Association of Engineering Societies, 1111 19th Street NW, Suite 608, Washington, DC 20036. (202) 296-2237 or (800) 658-8897. 8thedition. 1991. Inquire for price.

Who's Who in Technology, Amy L. Unterburger, editor. 7th edition. Gale Research, 835 Penobscot Building, Detroit, MI 48226-4094. (313) 961-2242. (800) 347-4253. FAX (313) 961-6083. 1995. $195.00.

ENCYCLOPEDIAS AND DICTIONARIES

Dictionary of Metallurgy and Foundry Technology. Karl Stolzel, editor. Elsevier Science B.V., PO Box 945, Madison Square Station, New York, NY 10159-0945. (212) 633-3650. FAX (212) 633-3680. 1987. Inquire for cost and availability.

Encyclopedia of Physical Science and Technology. Robert A. Meyers, ed. 18 volumes. Academic Press Inc., 6277 Sea Harbor Drive, Orlando, FL 32887. (800) 321-5068. 1992. $2100.00.

McGraw-Hill Dictionary of Scientific and Technical Terms. Sybil P. Parker, ed. 5th edition. McGraw-Hill Publishing Company, 1221 Avenue of the Americas, New York, NY 10020. (800) 262-4729 or (212) 512-3825. 1993. $110.50.

McGraw-Hill Encyclopedia of Engineering. Sybil P. Parker, ed. 2nd edition. McGraw-Hill Publishing Company, 1221 Avenue of the Americas, New York, NY 10020. (800) 262-4729 or (212) 512-3825. 1993. $95.50.

McGraw-Hill Encyclopedia of Science and Technology. Sybil P. Parker, ed. 7th edition. 20 volumes. McGraw-Hill Publishing Company, 1221 Avenue of the Americas, New York, NY 10020. (800) 262-4729 or (212) 512-3825. 1992. $1900.00

GENERAL WORKS

Casting Brass. C.W. Ammen. TAB Books, P.O. Box 40, Blue Ridge Summit, PA 17294-0850. (717) 794-2191 or (800) 233-1128. FAX (717) 794-2080. 1985. $18.95 (hardback) or $14.95 (paperback).

Ferrous and Non-ferrous Foundries. Milos Markovic. Leading Edge Reports, 2171 Jericho Turnpike, No. 200, Commack, NY 11725-2900. (800) 866-4648. FAX (216) 791-0333. 1988. Inquire for cost and availability.

Foundry Processes: Their Chemistry and Physics. S. Katz & C.F. Landefeld, editors. Plenum Publishing Corporation, 233 Spring Street, New York, NY 10013. (212) 620-8000. FAX (212) 463-0742. 1988. $125.00.

Foundry Products and Markets. Leading Edge Reports, 2171 Jericho Turnpike, No. 200, Commack, NY 11725-2900. (800) 866-4648. FAX (216) 791-0333. 1993. $1950.00.

HANDBOOKS AND MANUALS

ASM Handbook. Volume 15: Casting. ASM International Handbook Committee. ASM International, Materials Information, Materials Park, OH 44073-0002. (216) 338-5151 or (800) 336-5152. FAX (216) 338-4634. 1988. $147.00.

The Colouring, Bronzing, and Patination of Metals: A Manual for the Fine Metalworker and Sculptor. Richard Hughes. Watson-Guptill, 1515 Broadway, New York, NY 10036. (212) 536-5121. (800) 451-1741. FAX (212) 536-5359. 1991.

Foundryman's Handbook: Facts, Figures, Formulae. 9th edition. Elsevier Science B.V., PO Box 945, Madison Square Station, New York, NY 10159-0945. (212) 633-3650. FAX (212) 633-3680. 1986. $37.00.

Metals Handbook. ASM International, Materials Information, Materials Park, OH 44073-0002. (216) 338-5151 or (800) 336-5152. FAX (216) 338-4634. $154.00.

Smithell's Metals Reference Book. E.A. Brandes & G.B. Brook. 7th edition. Butterworth-Heinemann, 313 Washington Street, Newton, MA 02158. (617) 928-2500 or (800) 366-2665. FAX (617) 928-2620. 1992. $250.00.

BRASS AND BRONZE

Ency. of Physical Sciences and Engineering Info. Sources

ONLINE DATABASES AND CD-ROMS

COMPENDEX Plus. Engineering Information, Inc., 345 East 47th Street, New York, NY 10017. (212) 705-7600 or (800) 221-1044. Contains citations with abstracts to worldwide literature in engineering and technology, from 1970 to present. Available on online BRS,(800) 289-4277, DIALOG, (800) 334-2564, ORBIT (800) 456-7248, and STN International, FIZ Karlsruhe, P.O. Box 2465, W-7500, Karlsruhe 1, Germany, online services. Also available on CD-ROM. Inquire as to cost and availability.

METADEX. Jointly produced by ASM International and the Institute of Materials. Contains more than 925,000 records from the international literature on metals and alloys, concerning properties, processes, materials classes, applications, and metallurgical systems. Updated monthly. Available from ORBIT-QUESTEL (703) 442-0900.

NASA Database. American Institute of Aeronautics and Astronautics, 370 L'Enfant Promenade, S.W., Washington, DC 20024. (202) 646-7400. Citations and abstracts of aeronautics and astronautics literature, 1962 to present. Also contains citations from *Star, Scientific and Technical Aerospace Reports*, and *International Aerospace Abstracts*. Available through NASA/RECON online service. Inquire as to cost and availability.

NTIS Bibliographic Database. National Technical Information Service, 5285 Port Royal Road, Springfield, Va 22161. (703) 487-4929 or FAX (703) 321-8199. Broad coverage of government-sponsored science and technology research reports, 1964 to present. Available on BRS,(800) 289-4277, DIALOG, (800) 334-2564, ORBIT, (800) 456-7248, and STN International, FIZ Karlsruhe, P.O. Box 2465, W-7500, Karlsruhe 1, Germany, online services. Also available on CD-ROM. Inquire as to cost and availability.

WILSONLINE. H.W. Wilson Company, 950 University Avenue, Bronx, NY 10452. (800) 367-6770 or (212) 588-8400. Makes available online versions of the printed H.W. Wilson indexes including Applied Science and Technology Index, Business Periodicals Index, General Science Index, and Readers' Guide to Periodical Literature. Period covered is generally 1983 to present. Available on BRS,(800) 289-4277, DIALOG,(800) 334-2564, and OCLC EPIC,(800) 848-5878, online services. Also available onCD-ROM. Inquire as to cost and availability.

PERIODICALS

Aalloy Digest. Alloy Digest Inc., 27 Canfield Street, Orange, NJ 07050. (201) 677-9161. 1952 to present. Monthly. $140.00 per year.

Canadian Copper. Canadian Copper & Brass Development Association, 10 Gateway Blvd., Suite 375, Don Mills, ON M3C 3A1, Canada. (416) 421-0788. FAX (416) 421-8092. 1960 to present. Quarterly. Free.

Casting Digest. ASM (American Society of Metals) International, Materials Information, Materials Park, OH 44073. (216) 338-5151 or FAX (216) 338-4634. 1976 to present. Monthly. $1550.00 per year.

Casting World. Continental Communications Inc., Box 1919, Bridgeport, CT 06601-1919. 1969 to present. Quarterly. $20.00 per year.

Foundry Management and Technology. Penton Publishing, 110 Superior Ave., Cleveland, OH 44114-2543. 1892 to present. Monthly. $45.00 per year.

JOM (Journal of Metals). Minerals, Metals and Materials Society, 420 Commonwealth Drive, Warrendale, PA 15086. (412) 776-9080. 1949 to present. Monthly. $50.00 per year.

Metallurgical Transactions A: Physical Metallurgy and Materials Science. ASM International, Materials Park, OH 44073. (216) 338-5151 or FAX (216) 338-4634. 1970 to present. Monthly. $520.00 per year.

Metals Week. McGraw-Hill Publishing Company, 1221 Avenue of the Americas, New York, NY 10020. (800) 262-4729 or (212) 512-3825. 1930 to present. Weekly. $770.00 per year.

Modern Casting. American Foundrymen's Society Inc., 505 State Street, Des Plaines, IL 60016. (708) 824-0181. 1938 to present. Monthly. $45.00 per year.

RESEARCH CENTERS AND INSTITUTES

Texas A&M University Mechanics and Materials Center. Civil Engineering Department, College Station, TX 77843. (409) 845-7512. FAX (409) 845-6156.

University of Connecticut Institute of Materials Science. U-136, 97 N. Eagleville Road, Storrs, CT 06268. (203) 486-4623. FAX (203) 486-4745.

University of Florida Department of Materials Science and Engineering. Gainesville, FL 32611. (904) 392-1454. FAX (904) 392-6359.

University of Wisconsin-Madison, Cast Metals Laboratory. 1509 University Avenue, Madison, WI 53706. (608) 262-2562. FAX (608) 262-6707.

BRAZING

See: WELDING

BREEDER REACTORS

See: NUCLEAR REACTORS

BRICK

See: BUILDING MATERIALS

BRIDGES

See also: CIVIL ENGINEERING, CONCRETE, STRUCTURAL ENGINEERING

ABSTRACT SERVICES AND INDEXES

Abstract Journal in Earthquake Engineering. University of California, Berkeley, Earthquake Engineering Research Center, 1301 South 46th Street, Richmond, CA 94804-4698. (510) 231-9413 or FAX (510) 231-9471. 1972 to present. Annual. $80.00 per year.

Applied Mechanics Reviews: An Assessment of World Literature in Engineering Sciences. American Society of Mechanical Engineers, 345 East 47th Street, New York, NY 10017. (212) 705-7703. 1948 to present. Monthly. $360.00 per year.

Applied Science and Technology Index; a cumulative subject index to English language periodicals in the fields of aeronautics and space science, computer technology, chemistry, construction industry, energy and related areas. H.W. Wilson Co., 950 University Avenue, Bronx, NY 10452. (800) 367-6770 or (212) 588-8400. From 1958 to present. Monthly. Inquire about cost and availability. Also available on CD-ROM.

Concrete Abstracts. American Concrete Institute, 22400 West Seven Mile Road, Detroit, MI 48219-1885. (313) 532-2600 or FAX (313) 538-0655. 1971 to present. Bimonthly. $140.00 per year.

Corrosion Abstracts: Abstracts of the World's Literature on Corrosion and Corrosion Mitigation. National Association of Corrosion Engineers (NACE), Box 218340, Houston, TX 77218. (713) 492-0535 or FAX (713) 492-8254. 1962 to present. Bimonthly. $195.00 per year. Also available on CD-ROM.

Current Contents: Engineering, Technology and Applied Sciences. Institute for Scientific Information, 3501 Market Street, Philadelphia, PA 19104. (215) 386-0100. FAX (215) 386-6362. Weekly. $360.00 per year.

Engineered Materials Abstracts. ASM (American Society of Metals) International, Materials Information, Materials Park, OH 44073. (216) 338-5151 or FAX (216) 338-4634. Covers literature on technical developments in polymer, ceramic, and composite materials and engineering. 1986 to present. Monthly. $1175.00 per year. Also available on CD-ROM.

Engineering Index Monthly. Engineering Information, Inc., Castle Point on the Hudson, Hoboken, NJ 07030. (800) 221-1044. FAX (201) 216-8532. Monthly. $2200.00 per year. Also available online as COMPENDEX, and also on CD-ROM.

General Science Index. H.W. Wilson Company, 950 University Avenue, Bronx, NY 10452. (800) 367-6770 or (212) 588-8400. Inquire as to cost and availability.

Geotechnical Abstracts. Deutsche Gesellschaft fur Erd- und Grundbau e.V., Hohenzollernstrasse 52, 4300 Essen 1, Germany. Also sponsored by International Society for Soil Mechanics and Foundation Engineering. Covers soil mechanics, foundation engineering, and engineering geology. 1970 to present. Monthly.$200.00 per year.

Government Reports Announcements and Index. National Technical Information Service (NTIS), 5285 Port Royal Road, Springfield, VA 22161. (703) 487-4650. 1968 to present. Annual. $525.00 per year. Also available online as NTIS BIBLIOGRAPHIC DATABASE and on CD-ROM.

Metals Abstracts and Metals Abstracts Index. American Society for Metals, Metals Park, OH 44073. (216) 338-5151. 1968 to present. Monthly. Abstracts are $1500.00 per year and Index is $460.00 per year.

Offshore Engineering Abstracts. STI Limited, 4 Kings Meadow, Ferry Hinksey Road, Oxford, OX2 ODU, England. Distributed in the United States by Air Science Company, Box 143, Corning NY 14830. (607) 962-5591. Contains citations on planning, design, and construction of offshore platforms and pipelines, from 1985 to present. Monthly. $400.00 per year.

STAR(Scientific and Technical Aerospace Reports). U.S. National Aeronautics and Space Administration, Scientific and Technical Information Facility, Box 8757, Baltimore-Washington International Airport, MD 21240. (202) 755-2210. Semimonthly, with semiannual and annual indexes. $85.00 per year.

ASSOCIATIONS AND PROFESSIONAL SOCIETIES

American Institute of Steel Construction. 1 East Wacker Drive, Chicago, IL 60601-2001. (312) 670-2400

American Railway Bridge and Building Association. 18154 Harwood Avenue, Homewood, IL 60430. (312) 799-4650.

American Society of Civil Engineers. 345 East 47th Street, New York, NY 10017. (212) 705-7496.

Association for Bridge Construction and Design. Post Office Box 23264, Pittsburgh, PA 15237. (412) 355-5877.

DIRECTORIES AND BIOGRAPHICAL SOURCES

American Society of Civil Engineers. Official Register. American Society of Civil Engineers, 345 East 47th Street, New York, NY 10017. (212) 705-7496. Annual. Free.

Engineers - Civil. American Business Directories, Incorporated, 5707 South 86th Circle, Omaha, NE 68127. (402) 331-7169. Annual. $80.00.

Research Centers Directory. Gale Research, 835 Penobscot Building, Detroit, MI 48226. (800) 347-4253. Twelfth edition. $365.00.

Who's Who in Engineering. Engineers Joint Council, 345 East 47th Street, New York, NY 10017. (212) 705-7010. 1985. $200.00.

Who's Who in Technology Today. Reston Publishing Company, Incorporated, c/o Prentice-Hall, Incorporated, Englewood Cliffs, NJ 07632. (800) 262-6868. Biennial. Five volumes. $425.00. Covers the fields of electronics, computer science, physics, optics, chemistry, biotechnology, mechanics, energy, and earth science.

GENERAL WORKS

Bridge and Highway Maintenance: The Changing Infrastructure. Business Communications Company, Inc., 25 Van Zant St., Norwalk, CT 06855. (203) 853-4266. FAX (203) 853-0348. 1992. $750.00.

Bridge Design: Aesthetics and Developing Technologies. Adele Bacow and Kenneth Kruckemeyer, editors. American Society

BRIDGES

Ency. of Physical Sciences and Engineering Info. Sources

of Civil Engineering, 345 East 47th Street, New York, NY 10017-2398. (212) 705-7520 or (800) 548-2723. 1986. $20.00.

Bridge Engineering: Design, Rehabilitation, and Maintenance of Modern Highway Bridges. Demetrius E. Tonias. McGraw-Hill Publishing Company, 1221 Avenue of the Americas, New York, NY 10020. (800) 262-4729 or (212) 512-3825. 1994. $79.00.

Bridges: Analysis and Design of Reinforced Concrete Bridge Structures. American Concrete Institute, PO Box 19150, Redford Station, Detroit, MI 48219. (313) 532-2600. FAX (213) 538-0655. 1995. $94.50.

Designing Bridges for Improves Appearance: Aesthetic Design Guidelines for Bridge Engineers. Frederick Gottemoeller. Chapman & Hall, 115 Fifth Avenue, New York, NY 10211-0906. 1996. Inquire for price.

Design of Modern Steel Bridges. Sukhen Chatterjee. Sheridan House, Incorporated, 145 Palisade Street, Dobbs Ferry, NY 10522. (914) 693-2410. 1987. $55.00.

HANDBOOKS AND MANUALS

Bridge Analysis Simplified. Baidar Bakht and Leslie G. Jaeger. McGraw-Hill Book Company, 1221 Avenue of the Americas, New York, NY 10020. (800) 628-0004. 1985. $49.95.

CRC Handbook of Tables for Applied Engineering Science. R.E. Bolz and G.L. Tuve, editors. CRC Press, Incorporated, 2000 Corporate Boulevard, NW, Boca Raton, FL 33431. Second edition. 1973. $69.00.

Standard Handbook of Engineering Calculations. Tyler G. Hicks, editor. McGraw-Hill Book Company, 1221 Avenue of the Americas, New York, NY 10020. (212) 512-2000. Second edition. 1984. $59.50.

The Wiley Engineer's Desk Reference. Sanford I. Heisler. John Wiley and Sons, Incorporated, 605 Third Avenue, New York, NY 10158. (800) 526-5368 or (212) 850-6418. 1984. $36.00.

ONLINE DATABASES AND CD-ROMS

Architectural Periodicals Index. Royal Institute of British Architects, British Architectural Library, 66 Portland Place, London, W1N 4AD, England. Citations to worldwide literature on architecture, construction methods and standards, urban planning, and interior design, from 1978 to present. Available on DIALOG, (800) 334-2564, online service. Inquire as to cost and availability.

COMPENDEX Plus. Engineering Information, Inc., 345 East 47th Street, New York, NY 10017. (212) 705-7600 or (800) 221-1044. Contains citations with abstracts to worldwide literature in engineering and technology, from 1970 to present.Available on online BRS,(800) 289-4277, DIALOG, (800) 334-2564, ORBIT (800) 456-7248, and STN International, FIZ Karlsruhe, P.O. Box 2465, W-7500, Karlsruhe 1, Germany, online services. Also available on CD-ROM. Inquire as to cost and availability.

Current Contents Search. Institute for Scientific Information, 3501 Market Street, Philadelphia, PA 19104. (215) 386-0100. FAX (215) 386-6362. Contains citations to articles listed in the table of contents of science and technology journals. Also articles in social sciences and life sciences journals. Available on BRS,(800) 289-4277, DIALOG,(800) 334-2564, online services. Inquire as to cost and availability.

Dissertation Abstracts. University Microfilms International, 300 North Zeeb Road, Ann Arbor, MI 48106. (800) 521-0600 or (313) 761-4700. Scope includes virtually all doctoral dissertations accepted at accredited American institutions from 1861 to present in 252 subject areas. Available on BRS,(800) 289-4277, DIALOG,(800) 334-2564, and OCLC EPIC,(800) 848-5878, online services. Also available on CD-ROM. Inquire as to cost and availability.

Geomechanics Abstracts. University of London, Imperial College of Science and Technology, Rock Mechanics Information Service, Royal School of Mines, Prince Consort Road, London SW7 2BP, England. Telephone: 071-589 5111 or FAX 071-589 6806. Scope includes worldwide literature on engineering geology, hydrogeology, mining, rock mechanics, soil mechanics, and tunnelling, 1977 to present. Available on the ORBIT (800) 456-7248, online service. Inquire as to cost and availability.

ICONDA (CIB International Construction Database). Fraunhofer-Gesellschaft, Informationszentrum RAUM und BAU, Nobelstrasse, 12, D-7000, Stuttgart 80, Germany. Contains citations with abstracts to worldwide technical literature on construction and civil engineering, structural engineering, and engineering geology, from 1976 to present. Available on ORBIT, (800) 456-7248, and STN International, FIZ Karlsruhe, P.O. Box 2465, W-7500, Karlsruhe 1, Germany, online services. Inquire as to cost and availability.

*ISMEC: Mechanical Engineering Abstracts.*Cambridge Scientific Abstracts, 7200 Wisconsin Avenue, Suite 601, Bethesda, MD 20814. (301) 961-6750 or (800) 843-7751. Contains citations to the literature in mechanical engineering, industrial and production engineering, energy, power, mechanics, devices and related areas, from 1973 to present. Available on the DIALOG,(800) 334-2564, online service. Inquire as to cost and availability.

National Information Service for Earthquake Engineering Database. University of California, Berkeley. Contains citations and abstracts to journal articles on earthquakes and earthquake engineering from 1987 to present. Available on BRS,(800) 289-4277, online service. Inquire as to cost and availability.

NTIS Bibliographic Database. National TechnicalInformation Service, 5285 Port Royal Road, Springfield, VA 22161. (703) 487-4929 or FAX (703) 321-8199. Broad coverage of government-sponsored science and technology research reports, 1964 to present. Available on BRS,(800) 289-4277, DIALOG, (800) 334-2564, ORBIT, (800) 456-7248, and STN International, FIZ Karlsruhe, P.O. Box 2465, W-7500, Karlsruhe 1, Germany, online services. Also available on CD-ROM. Inquire as to cost and availability.

Scientific and Technical Books and Serials in Print. R.R. Bowker Inc., 121 Chanlon Road, New Providence, NJ 07974. (800) 521-8110 or (908) 464-6800. List of currently published books and serials in the physical and biological sciences, engineering and technology, with subject, author and titles indexes. Available on ORBIT,(800) 456-7248, online service.Inquire as to cost and availability.

SCISearch. Institute for Scientific Information, 3501 Market Street, Philadelphia, PA 19104. (800) 523-1850 or (215) 386-0100. Broad multidisciplinary title and author index to the in-

ternational literature of science and technology, 1974 to present. Available on DIALOG,(800) 334-2564, and ORBIT,(800) 456-7248, online services. Also available on CD-ROM. Inquire as to cost and availability.

TRIS (Transportation Research Information). National Academy of Science, 2101 Constitution Avenue, N.W., Washington, DC 20418. (202) 334-3313 or (800) 624-6242. Citations with abstracts of literature on transportation, including air, highway, rail, maritime and other modes from 1968 to present. Available on DIALOG,(800) 334-2564, online service. Inquire as to cost and availability.

Who's Who in Technology. Gale Research Inc., 835 Penobscot Building, Detroit, MI 48226-4094. (313) 961-2242. (800) 877-4253. Contains biographical information of contemporary American scientists and engineers. Available on DIALOG,(800) 334-2564, online service. Inquire as to cost and availability.

Wilsonline. H.W. Wilson Company, 950 University Avenue, Bronx, NY 10452. (800) 367-6770 or (212) 588-8400. Makes available online versions of the printed H.W. Wilson indexes including *Applied Science and Technology Index, Business Periodicals Index, General Science Index,* and *Readers' Guide to Periodical Literature.* Period covered is generally 1983 to present. Available on BRS,(800) 289-4277, DIALOG,(800) 334-2564, and OCLC EPIC,(800) 848-5878, online services. Also available on CD-ROM. Inquire as to cost and availability.

PERIODICALS

Journal of Structural Engineering. American Society of Civil Engineers, 345 East 47th Street, New York, NY 10017. (212) 705-7275. Monthly. $140.00 per year.

Journal of Structural Mechanics. Marcel Dekker Journals, 270 Madison Avenue, New York, NY 10016. Quarterly. $75.00 per year.

Structural Engineer. Institution of Structural Engineers, 11 Upper Belgrave Street, London, SW1X 8BH, England. Monthly. $95.00 per year.

RESEARCH CENTERS AND INSTITUTES

New Mexico State University. Center for Transportation Research, College of Engineering, Box 3CE, Las Cruces, NM 88003-0083. (505) 646-3135.

North Carolina State University. Center for Transportation Engineering Studies, Department of Civil Engineering, Box 7908, Raleigh, NC 27695-7908. (919) 737-2331.

University of Maryland. Computer-Aided Design Laboratory, Engineering Research Center, College Park, MD 20742.
<subhead1>BRITISH THERMAL UNIT
See: AIR CONDITIONING

BRITTLENESS

See: MATERIALS SCIENCE

BROADCAST BAND

See: RADIO

BROMINE

See also: CHLORINE, ELEMENTS, FLUORINE, HALIDES, IODINE

ABSTRACT SERVICES AND INDEXES

Applied Science and Technology Index; a cumulative subject index to English language periodicals in the fields of ae ronautics and space science, computer technology, chemis try, construction industry, energy and related areas. H. W. Wilson Co., 950 University Avenue, Bronx, NY 10452-9978. (212) 588-8400. (800) 367-6770. FAX (718) 590-1617. From 1958 to present. Monthly. Inquire about cost and availability. Also available online (BRS and Wilsonline) and on CD-ROM. ISSN: 0003-6986.

Chemical Abstracts. Chemical Abstracts Service. 2540 Olentangy River Road, Box 3012, Columbus, OH 43210-0012. (614) 447-3600. FAX (614) 447-3713. 1907 to present. Weekly. $16,800.00 per year. Available online and on CD- ROM. CA is also available in five section groupings. Inquire regarding cost and availability.

Current Contents: Engineering, Technology and Applied Sciences. Institute for Scientific Information, 3501 Market Street, Philadelphia, PA 19104. (215) 386-0100. FAX (215) 386-2991. From 1970 to present. Weekly. $442.00 per year. Also available online. Inquire regarding cost and availability. ISSN: 0095-7917.

General Science Index. H.W. Wilson Company, 950 University Avenue, Bronx, NY 10452. (212) 588-8400. (800) 367-6770. FAX (718) 590-1617. From 1978 to present. Ten issues per year; quarterly and annual cumulations. Service basis. Available on CD-ROM and online. Inquire regarding cost and availability. ISSN: 0162-1963.

Physics Abstracts. INSPEC. Section A, Science Abstracts. Institution of Electrical Engineers (IEE), London, United Kingdom. Available from: INSPEC/IEEE - Institute of Electrical and Electronic Engineers, Box 1331, 445 Hoes Lane, Piscataway, NJ 08855-1331. (908) 562-5549. 1898 to present. 24 issues per year. $2835.00 per year. ISSN: 0036-8091. Also available online and on CD-ROM. *Science Citation Index.* SCI. Institute for Scientific Information, 3501 Market Street, Philadelphia, PA 19104. (215) 386-0100. (800) 523-1850. FAX (215) 386-2991. 1961 to present. Six issues per year, plus annual cumulation. $11650.00 per year. Also available online and on CD-ROM. Inquire about price and availability. ISSN: 0036-827X.

ASSOCIATIONS AND PROFESSIONAL SOCIETIES

American Chemical Society. 1155 16th Street, NW, Washington, DC 20036. (202) 872-4600.

Association of Consulting Chemists and Chemical Engineers, 50 East 41st Street, Suite 92, New York, NY 10017. (212) 684-6255.

Chemical Manufacturers Association. 2501 M Street, N.W., Washington, DC 20037. (202) 887-1182

Chemical Specialties Manufacturers Association, 1913 Eye Street, NW, Washington, CD 20006, (202) 872-8110. FAX (202) 872-8114.

BIBLIOGRAPHIES

Chemical Information Sources. Gary Wiggins. McGraw-Hill Publishing Company, 1221 Avenue of the Americas, New York, NY 10020. (800) 262-4729 or (212) 512-3825. 1991. $42.50.

Information Sources in Chemistry. R.T. Bottle and J.F.B Rowland, editors. R.R. Bowker Inc., 121 Chanlon Road, New Providence, NJ 07974. (800) 521-8110 or (908) 464-6800. Fourth edition. 1993. $75.00.

Scientific and Technical Books and Serials in Print: An index to Literature in Science and Technology. R.R. Bowker Inc., 121 Chanlon Road, New Providence, NJ 07974. (908) 464-6800. (800) 521-8110. FAX (908) 665-3502. 1972 to present. Annual. 4 volumes. 1994. $299.95. Also available on compact disc and online. ISSN: 0000-054X.

DIRECTORIES AND BIOGRAPHICAL SOURCES

American Men and Women of Science. Physical and Biological Sciences. R.R. Bowker Inc., 121 Chanlon Road, New Providence, NJ 07974. (908) 464-6800. (800) 521-8110. 20th edition. 8 volumes. 1996. $850.00

Chemical Sources International. Mike Desing and Kurt Gandenberger, editors. Chemical Sources International, PO Box 1824, Clemson, SC 29633. (803) 646-7840. 1994. $285.00.

Chemical Week-Buyers Guide Issue. Chemical Week Associates, 888 Seventh Avenue, New York, NY 10106. (212) 621-4900. FAX (212) 621-4949. Annual, October. $50.00.

Directory of Chemistry Software 1992. Wendy Warr, Peter Willett, Geoff Downs. American Chemical Society, 1155 16th Street, NW, Washington, DC 20036. (202) 872-4600. 1992. $35.95.

Research Centers Directory. Gale Research, 835 Penobscot Building, Detroit, MI 48226-4094. (313) 961-2242. (800) 347-4253. 20th edition. 1995. $485.00. ISSN: 0080-1518.

ENCYCLOPEDIAS AND DICTIONARIES

Academic Press Dictionary of Science and Technology. Christopher Morris, editor. Academic Press, Inc., 1250 Sixth Avenue, San Diego, CA 92101. (619) 231-0926. FAX (619) 699-6715. 1991. $115.00.

Concise Encyclopedia Chemistry. Translated and revised by Mary Eagleson. Walter de Gruyter, Inc., 200 Saw Mill River Road, Hawthorne, New York, 10532. (914) 747-0110 or Fax (914) 747-1326. 1994. $69.95.

Encyclopedia of Chemical Processing and Design. McKetta Marcel Dekker, Inc., 270 Madison Avenue, New York, NY 10016. (212) 696-9000. (800) 228-1160. 1976 - . $175.00 per volume. ISSN: .

Grant & Hawks Chemical Dictionary. R. A. Grant. McGraw-Hill Publishing Company, Inc., 1221 Avenue of the Americas,

New York, NY 10020. (212) 512-2000. (800) 262-4729. 5th edition. 1987. $64.50.

Hawley's Condensed Chemical Dictionary. Richard J. Lewis. Van Nostrand Reinhold, 115 Fifth Avenue, New York, NY 10003. (212) 254-3232. (800) 842-3636. 12th edition. 1993. $69.95.

Kirk-Othmer Encyclopedia of Chemical Technology. John Wiley and Sons, Inc., 605 Third Avenue, New York, NY 10158. (800) 526-5368 or (212) 850-6000. Fourth edition. 1991-. Twenty-seven volumes. $5400.00.

McGraw-Hill Encyclopedia of Science and Technology. McGraw-Hill Book, Incorporated, 1221 Avenue of the Americas, New York, NY 10020. (212) 997-3675. Seventeenth editon. 1992. $1900.00.

GENERAL WORKS

Bromine. Gordon Press Publications, P.O. Box 459, Bowling Green Station, New York, NY 10004. (718) 624-8419. Metals and Minerals series. 1993. $250.95.

Bromine Compounds: Chemistry and Applications. D. Price, editor. Elsevier Science Publishing Company, Inc., 655 Avenue of the Americas, New York, NY 10010. (212) 989-5800. 1988. $146.25.

The Chemical Bond: Structure and Dynamics. Ahmed Zewail. Academic Press, Inc., 6277 Sea Harbor Drive, Orlando, FL. (800) 321-5068. 1992. $49.95.

Chemistry of Halides, Pseudohalides and Azides, part 1 and 2. Chemistry of Functional Groups series, supplement no. D. Saul Patai and Zvi Rappaport. Books on Demand, 300 North Zeeb Road, Ann Arbor, MI 48106-1346. (313) 761-4700. (800) 521-0600. 1983. $340.00 set.

The Elements. John Emsley. Oxford University Press, Inc., 200 Madison Avenue, New York, NY 10016. (212) 725-6000. (800) 334-4249. Second edition. 1991. $49,95.

Exploring Chemical Elements and Their Compounds. David L. Heiserman. TAB Books, P.O. Box 40, Blue Summit, PA 17294- 0850. (717 794-2191. (800) 233-1128. 1992. $17.95.

Halogen Chemistry. V. Gutman, editor. Academic Press, Incorporated, 6277 Sea Harbor Drive, Orlando, FL 32821. (800) 321-5068. Three volumes. 1967. Inquire.

HANDBOOKS AND MANUALS

CRC Handbook of Chemistry and Physics. David R. Kide, editor. CRC Press Inc., 2000 Corporate Blvd., NW, Boca Raton, FL 33431. (407) 994-0555. (800) 333-8300. 77th edition. 1996. $99.95.

Guide to Basic Chemical Compounds. D.R. Lide, Jr. CRC Publishers, Inc., 2000 Corporate Blvd., N.W., Boca Raton, FL 33431. (407) 994-0555 or (800) 333-8300. 1993. $120.00.

Handbook of Hazardous Laboratory Chemicals: Information and Disposal. Margaret-Ann Armour. CRC Press, Inc., 2000 Corporate Boulevard, NW, Boca Raton, FL 33431. (407) 994-0555. (800) 272-7737. 1991. $104.95.

Lange's Handbook of Chemistry. John A. Dean, editor. McGraw-Hill Publishing Company, 1221 Avenue of the Americas, New York, NY 10020. (800) 262-4729 or (212) 512-3825. 14th editon. 1992. $79.50.

Riegel's Handbook of Industrial Chemistry. James A. Kent, editor. Van Nostrand Reinhold, 115 Fifth Avenue, New York, NY 10003. (212) 254-3232. (800) 926-2665. 9th edition. 1992. $114.95.

ONLINE DATABASES AND CD-ROMS

CA Search. Chemical Abstracts Service, P.O. Box 3012, Columbus, Oh 43210-0012. (614) 447-3600. (800) 848-6533. FAX (614) 447-3709. Guide to worldwide chemical literature and related fields, 1972 to present. Available on BRS, DIALOG, ORBIT and STN online services. Inquire as to cost and availability.

Chemical Journals of the American Chemical Society. American Chemical Society, 1155 16th Street, N.W., Washington, DC 20036. (202) 872-4381 or (800) 424-6747. Contains complete text of approximately 90,000 articles from 22 primary journals published by the American Chemical Society, from mostly 1982 to present. Available on STN online service. Inquire as to cost and availability.

Dissertation Abstracts. University Microfilms International, 300 North Zeeb Road, Ann Arbor, MI 48106. (800) 521-0600 or (313) 761-4700. Scope includes virtually all doctoral dissertations accepted at accredited American institutions from 1861 to present in 252 subject areas. Available on BRS, DIALOG, and OCLC EPIC online services. Also available on CD-ROM. Inquire as to cost and availability.

Kirk-Othmer Encyclopedia of Chemical Technology. John Wiley and Sons, Inc., 605 Third Avenue, New York, NY 10158. (800) 526-5368 or (212) 850-6000. Contains the complete text of all chapters in the 27 volume fourth edition of the *Kirk-OthmerEncyclopedia of Chemical Technology*. 1991. Available on BRS, DIALOG, online services. Inquire as to cost and availability.

NTIS. National Technical Information Service, 5285 Port Royal Road, Springfield, Va 22161. (703) 487-4929 or FAX (703) 321-8199. Broad coverage of government-sponsored research reports, 1964 to present. Available on BRS, DIALOG, ORBIT and STN online services. Also available on CD-ROM. Inquire as to cost and availability.

SCISearch. Institute for Scientific Information, 3501 Market Street, Philadelphia, PA 19104. (800) 523-1850 or (215) 386-0100. Broad multidisciplinary title and author index to the international literature of science and technology, 1974 to present. Available on DIALOG and ORBIT online services. Also available on CD-ROM. Inquire as to cost and availability.

Wilsonline. H.W. Wilson Company, 950 University Avenue, Bronx, NY 10452. (800) 367-6770 or (212) 588-8400. Makes available online versions of the printed H.W. Wilson indexes including Applied Science and Technology Index, Business Periodicals Index, and Readers' Guide to Periodical Literature. Period covered is generally 1983 to present. Also available on CD-ROM. Inquire as to cost and availability.

Who's Who in Technology. Gale Research, 835 Penobscot Building, Detroit, MI 48226-4094. (313) 961-2242. (800) 347-4253. Contains biographical information of contemporary American scientists and engineers. Available on DIALOG online service. Inquire as to cost and availability.

PERIODICALS

Analytical Chemistry. American Chemical Society, Box 3337, Columbus, OH 43210. (614) 447-3776. 1929 to present. Semimonthly. $415.00 per year. ISSN: 0003-2700.

Angewandte Chemie. VCH Publishers, Inc., 220 East 23rd Street, New York, NY 10010-4606. (212) 683-8333. (800) 367-8249. 1888 to present. 22 issues per year. $840.00 per year. ISSN: 0044-8249.

Chemical Reviews. American Chemical Society, Box 3337, Columbus, OH 43210. (614) 447-3776. 1924 to present. 8 issues per year. $346.00 per year. ISSN: 0009-2665.

Chemical Week; includes annual buyers guide. Chemical Week Associates, 888 Seventh Avenue, New York, NY 10106. (212) 621-4900. FAX (212) 621-4949. 1914 to present. Weekly. $99.00 per year. ISSN: 0009-272X.

CHEMTECH. American Chemical Society. Box 3337, Columbus, OH 43210. (614) 447-3776. 1970 to present. Monthly. $370.00 per year. ISSN: 0009-2703.

Inorganic Chemistry. American Chemical Society, Box 3337, Columbus, OH 43210. (614) 447-3776. 1962 to present. Semimonthly. $500.00 per year. ISSN: 0020-1669.

Journal of the American Chemical Society. American Chemical Society, Box 3337, Columbus, OH 43210. (614) 447-3776. 1879 to present. Biweekly. $1055.00 per year. ISSN: 0002- 7863.

Polyhedron. Elsevier Science Publishing Company, Inc., 660 White Plains Road, Tarrytown, NY 10591-5153. (914-524-9200. FAX (914) 524-9200. 1982 to present. 24 issues per year. $2175.00 per year. ISSN: 0277-5387.

RESEARCH CENTERS AND INSTITUTES

Chemical Laboratories. Harvard University, Oxford Street, Cambridge, MA 02138. (617) 495-4283. FAX (617) 496-5618.

Chemistry Laboratories. Rensselaer Polytechnic Institute, Cogswell Laboratory, Troy, NY 12180-3590. (518) 276-8981.

Research Program in Chemistry and Biochemistry. Southern Illinois University at Carbondale, Carbondale, IL 62901. (618) 453-5721.

Theoretical Chemistry Institute. University of Wisconsin, Madison. 1101 University Avenue, Madison, WI 53706. (608) 262-1511.

University/industry Chemical Research Center. Mississippi State University, Department of Chemistry, P.O. Drawer CH, Mississippi State, MS 39762. (601) 325-3584.

BUILDING CODES

See also: ARCHITECTURAL ENGINEERING, BUILDING MATERIALS, CONSTRUCTION ENGINEERING

ABSTRACT SERVICES AND INDEXES

Abstract Journal in Earthquake Engineering. University of California, Berkeley, Earthquake Engineering Research Center, 1301 South 46th Street, Richmond, CA 94804-4698. (510) 231-9413 or FAX (510) 231-9471. 1972 to present. Annual. $80.00 per year.

Applied Science and Technology Index; a cumulative subject index to English language periodicals in the fields of aeronautics and space science, computer technology, chemistry, construction industry, energy and related areas. H.W. Wilson Co., 950 University Avenue, Bronx, NY 10452. (800) 367-6770 or (212) 588-8400. FAX (718) 590-1617. From 1958 to present. Monthly. Inquire about cost and availability. Also available on CD-ROM and online.

ASCE Combined Annual Combined Index. American Society of Civil Engineers, 345 East 47th Street, New York, NY 10017-2398. (212) 705-7520 or (800) 548-2723. Annual. $48.00.

ASCE Publications Information. 1966-present. American Society of Civil Engineers, 345 East 47th Street, New York, NY 10017-2398. (212) 705-7520 or (800) 548-2723. Bi-monthly. $160.00 per year for non-members.

Civil and Structural Engineering Abstracts, Cambridge Scientific Abstracts, 7200 Wisconsin Avenue, Bethesda, MD 20814-4823. (301) 961-6750. FAX (301) 961-6720. 1993 to present. Monthly. $385.00 per year. Topics covered include structural design, construction equipment and methods, civil defense and military engineering, surveying, highway engineering, maritime and port structures, materials, land reclamation, and soilmechanics.

International Civil Engineering Abstracts. CITIS Ltd., 2 Rosemount Terrace, Blackrock, Dublin, Ireland. Telephone 353-1-2886227. FAX 353-1-885-971. 1974 to present. Monthly. $660.00 per year. Also available on CD-ROM.

NTIS Alerts: Building Industry Technology. U.S. National Technical Information Service, 5825 Port Royal Road, Springfield, VA 22161. (703) 487-4630. FAX (703) 321-8547. Weekly. $135.00 per year.

ASSOCIATIONS AND PROFESSIONAL SOCIETIES

American Association of Building Inspectors. PO Box 273, Occoquan, VA 22125.

American Construction Inspectors Association. PO Box 10579, San Bernardino, CA 92423-0579. (909) 370-3120. FAX (909) 370-2160.

American Institute of Architects. 1735 New York Avenue NW, Washington, DC 20006. (202) 626-7300.

American Institute of Building Design. 16 Wilton Road, Bldg. ED, Westport, CT 06880. (800) 366-2423.

American Society for Testing and Materials (ASTM). 1916 Race Street, Philadelphia, PA 19103. (215) 299-5585.

American Society of Civil Engineers. 345 East 47th Street, New York, NY 10017-2398. (212) 705-7520 or (800) 548-2723.

Building Officials and Code Adminstrators International (BOCAI). 4051 W. Flossmoor Road, Country Club Hills, IL 60478-5795. (708) 799-2300. FAX (708) 799-4981.

International Conference of Building Officials. 5360 Workman Mill Road, Whittier, CA 90601-2298. (310) 699-0541. FAX (310) 692-3853.

National Society of Architectural Engineers. PO Box 395, Lawrence, KS 66044. (913) 864-3434.

DIRECTORIES AND BIOGRAPHICAL SOURCES

Directory of Building Codes and Regulations. National Conference of States on Building Codes and Standards, Inc. NCSBCS, Herndon, VA 22070. 1989. Inquire for availability and price.

ENCYCLOPEDIAS AND DICTIONARIES

Concise Encyclopedia of Building and Construction Materials. Fred Moavenzadeh. Pergamon Press Inc., Maxwell House, Fairview Park, Elmsford, NY 10523. (914) 592-7700. Fax (914) 592-3625. 1990. $210.00.

Construction Glossary: An Encyclopedic Reference and Manual. J. Stewart Stern. 2nd edition. John Wiley and Sons, Inc., 605 Third Avenue, New York, NY 10158. (800) 526-5368 or (212) 850-6000. 1993. $99.95.

Dictionary of Civil Engineering. V.N. Vazirani. South Asia Books, PO Box 502, Columbia, MO 65205. (314) 474-0116. FAX (314) 474-8124. 1992. $20.00.

GENERAL WORKS

Building Code Quick Reference Guide. William J. Brummett and Alec W. Johnson. Professional Publications, 1250 Fifth Ave., Belmont, CA 94002. (415) 593-9119 or (800) 426-1178. 1993. $30.95.

Building Regulations Explained. J. Stephenson. 5th edition. Chapman & Hall, 115 Fifth Avenue, New York, NY 10211-0906. 1995. $85.00.

Code Practice for the Construction of Buildings. Fire Protection Association Staff. State Mutual Book & Periodical Service Ltd., 521 Fifth Ave., 17th floor, New York, NY 10175. (718) 261-1704. FAX (516) 537-0412. 1992. $95.00.

Contractor's Guide to the Building Code. Jack M. Hageman. Revised 2nd edition. Craftsman Book Company, 6058 Corte del Cedro, Carlsbad, CA 92009. (619) 438-7828 or (800) 829-8123. FAX (619) 438-0398. 1990. $28.00.

Uniform Building Code. International Conference of Building Officials, 5360 Workman Mill Road, Whittier, CA 90601-2298. (310) 699-0541. FAX (310) 692-3853. Latest edition 1994. Multi-volume set, inquire for price.

HANDBOOKS AND MANUALS

Building Structural Design Handbook. Richard N. White & Charles G. Salmon, editors. Krieger Publishing Company, P.O. Box 9542, Melbourne, FL 32902-9542. (407) 724-9542. FAX (407) 951-3671. 1987. $95.00.

Civil Engineering Calculations Reference Guide. Tyler G. Hicks. McGraw-Hill Publishing Company, 1221 Avenue of the Americas, New York, NY 10020. (800) 262-4729 or (212) 512-3825. 1987. $46.00.

Civil Engineering Practice. Paul N. Cheremisinoff, et al. 5 volumes. Technomic Publishing Company, Inc., 851 New Holland Avenue, Box 3535, Lancaster, PA 17604. (717) 291-5609 or (800) 233-9936. FAX (717) 295-4538. 1987-88. $199.00.

Construction Inspection Handbook. James J. O'Brien. 3rd edition. Chapman & Hall, 115 Fifth Avenue, New York, NY 10211-0906. 1989. $76.95.

Handbook of Concrete Engineering. Mark Fintel. 2nd edition. Routledge, Chapman and Hall, Inc., 29 West 35th Street, New York, NY 10001-2291. (212) 244-3336 or FAX (212) 563-2269. 1985. $115.00.

Handbook of Mechanics, Materials, and Structures. Alexander Blake, editor. John Wiley and Sons, Inc., 605 Third Avenue, New York, NY 10158. (800) 526-5368 or (212) 850-6000. 1985. $120.00.

Handbook to the Uniform Building Code, 1994. International Conference of Building Officials, 5360 Workman Mill Road, Whittier, CA 90601-2298. (310) 699-0541. FAX (310) 692-3853. 1995. $65.00.

Standard Handbook for Civil Engineers. Frederick S. Merritt, editor. 3rd edition. McGraw-Hill Publishing Company, 1221 Avenue of the Americas, New York, NY 10020. (800) 262-4729 or (212) 512-3825. 1983. $124.50.

Structural Engineering Handbook. Edwin H. Gaylord Jr. and Charles N. Gaylord. 3d ed. McGraw-Hill Publishing Company, 1221 Avenue of the Americas, New York, NY 10020. (800) 262-4729 or (212) 512-3825. 1990. $89.50.

ONLINE DATABASES AND CD-ROMS

Architectural Periodicals Index. Royal Institute of British Architects, British Architectural Library, 66 Portland Place, London, W1N 4AD, England. Citations to worldwide literature on architecture, construction methods and standards, urban planning, and interior design, from 1978 to present. Available on DIALOG, (800) 334-2564, online service. Inquire as to cost and availability.

Compendex Plus. Engineering Information, Inc., 345 East 47th Street, New York, NY 10017. (212) 705-7600 or (800) 221-1044. Contains citations with abstracts to worldwide literature in engineering and technology, from 1970 to present. Available on online BRS, (800) 289-4277, DIALOG, (800) 334-2564, ORBIT (800) 456-7248, and STN International, FIZ Karlsruhe, P.O. Box 2465, W-7500, Karlsruhe 1, Germany, online services. Also available on CD-ROM. Inquire as to cost and availability.

ICONDA (CIB International Construction Database). Fraunhofer-Gesellschaft, Informationszentrum Raum und Bau,

Nobelstrasse, 12, D-7000, Stuttgart 80, Germany. Contains citations with abstracts to worldwide technical literature on construction and civil engineering, structural engineering, and engineering geology, from 1976 to present. Available on ORBIT, (800) 456-7248, and STN International, FIZ Karlsruhe, P.O. Box 2465, W-7500, Karlsruhe 1, Germany, online services. Inquire as to cost and availability.

PERIODICALS

American Concrete Institute Structural Journal. American Concrete Institute, Box 19150, Redford Station, Detroit, MI 48219. (313) 532-2600. FAX (313) 538-0655. 1929 to present. Bi-monthly. $101.00 per year for non-members.

Architecture. BPI Communications Inc., 1515 Broadway, New York, NY 10036. (212) 764-7300. FAX (212) 944-1719. 1913 to present. Monthly. $421.00 per year.

BOCA Bulletin. Building Officials and Code Administrators International, 4051 W. Flossmoor Road, Country Club Hills, IL 60478-5795. (708) 799-2300. FAX (708) 799-4981. Bi-monthly. Inquire for cost.

Building Design and Construction. Cahners Publishing Company (Des Plains), 1350 East Touhy Avenue, Des Plaines, IL 60017-5080. (708) 635-8800. FAX (708) 635-9950. 1950 to present. Monthly. $90.00 per year.

The Building Official. Building Officials and Code Administrators International, 4051 W. Flossmoor Road, Country Club Hills, IL 60478-5795. (708) 799-2300. FAX (708) 799-4981. Monthly. Inquire for cost.

Civil Engineering Asce. American Society of Civil Engineers, 345 East 47th Street, New York, NY 10017-2398. (212) 705-7520 or (800) 548-2723. 1930 to present. Monthly. $85.00 per year.

The Inspector. American Construction Inspectors Association, PO Box 10579, San Bernardino, CA 92423-0579. (909) 370-3120. FAX (909) 370-2160. Quarterly. Inquire for price.

Journal of Structural Engineering. American Society of Civil Engineers, 345 East 47th Street, New York, NY 10017-2398. (212) 705-7520 or (800) 548-2723. 1956 to present. Monthly. $300.00 per year for non-members.

Sstructural Engineer. Structural Engineers Trading Organization Ltd., 11 Upper Belgrave Street, London SW1X 8BH, England. Telephone 071-235-4535. FAX 071-235-4294. 1922 to present. 24 times a year. Inquire for cost.

Structural Engineering International. International Association for Bridge and Structural Engineering, ETH-Hoenggerberg, CH-8093 Zurich, Switzerland. Telephone 01-3772647. FAX 01-3712131. Quarterly. Inquire for cost and availability.

Sstructural Engineering Reviews. Elsevier Science [journals], 660 White Plains Rd., Tarrytown, NY 10159-5153. (919) 524-9200. FAX (919) 333-2444. 1988 to present. 4 times a year. $190.00 per year.

BUILDING CODES

Ency. of Physical Sciences and Engineering Info. Sources

RESEARCH CENTERS AND INSTITUTES

National Institute of Building Sciences. 1201 L Street NW, Suite 400, Washington, DC 20005. (202) 289-7800. FAX (202) 289-1092.

SPECIFICATIONS AND STANDARDS

ASTM Standards in Building Codes, 1992. 28th edition. 4 volumes. American Society for Testing and Materials (ASTM), 1916 Race Street, Philadelphia, PA 19103. (215) 299-5585. $395.00 per set.

Uniform Building Code Standards. International Conference of Building Officials, 5360 Workman Mill Road, Whittier, CA 90601-2298. (310) 699-0541. FAX (310) 692-3853. 1994. $68.75.

BUILDING MATERIALS

See also: ARCHITECTURAL ENGINEERING, BUILDING CODES, CEMENT, CONCRETE, CONSTRUCTION ENGINEERING, STEEL AND STEELMAKING

ABSTRACT SERVICES AND INDEXES

Applied Science and Technology Index; a cumulative subject index to English language periodicals in the fields of aeronautics and space science, computer technology, chemistry, construction industry, energy and related areas. H.W. Wilson Co., 950 University Avenue, Bronx, NY 10452. (800) 367-6770 or (212) 588-8400. FAX (718) 590-1617. From 1958 to present. Monthly. Inquire about cost and availability. Also available on CD-ROM and online.

Civil and Structural Engineering Abstracts, CambridgeScientific Abstracts, 7200 Wisconsin Avenue, Bethesda, MD 20814-4823. (301) 961-6750. FAX (301) 961-6720. 1993 to present. Monthly. $385.00 per year. Topics covered include structural design, construction equipment and methods, civil defense and military engineering, surveying, highway engineering, maritime and port structures, materials, land reclamation, and soil mechanics.

Concrete Abstracts. Covers articles, books, and other publications that report developments in concrete design, construction and technology. American Concrete Institute, PO Box 19150, Detroit, MI 48219. (313) 532-2600. FAX (213) 538-0655. 1972 to present. Bi-monthly. Inquire for price.

Engineering Index Monthly. Engineering Information, Inc., Castle Point on the Hudson, Hoboken, NJ 07030. (800) 221-1044. FAX (212) 832-1857. Monthly. $2300.00 per year. Also available online as COMPENDEX, and also on CD-ROM. Covers chemical engineering, computers, electrical engineering, civil engineering, metals and mining, industrial management, and mechanical engineering.

NTIS Alerts: Building Industry Technology. U.S. National Technical Information Service, 5825 Port Royal Road, Springfield, VA 22161. (703) 487-4630. FAX (703) 321-8547. Weekly. $135.00 per year.

ASSOCIATIONS AND PROFESSIONAL SOCIETIES

American Concrete Institute. PO Box 19150, Redford Station, Detroit, MI 48219. (313) 532-2600. FAX (213) 538-0655.

American Institute of Building Design. 16 Wilton Road, Bldg. ED, Westport, CT 06880. (800) 366-2423.

American Society for Concrete Construction. PO Box 19307, Detroit, MI 48219. (313) 532-2753.

American Institute of Steel Construction, Inc., One East Wacker Drive, Suite 3100, Chicago, IL 60601-2001. (312) 670-2400.

American Society of Civil Engineers. 1015 15th St. NW, Suite 600, Washington, DC 20005. (212) 705-7496.

DIRECTORIES AND BIOGRAPHICAL SOURCES

Aberdeen's Concrete Sourcebook. Aberdeen Group, 426 S. Westgate Street, Addison, IL 60101. (708) 543-0870. FAX (708) 543-3112. Monthly. $30.00 per year.

American Concrete Institute-Directory. American Concrete Institute, PO Box 19150, Detroit, MI 48219. (313) 532-2600. FAX (213) 538-0655. Every 2-3 years. Write or call for price information.

Building Materials Directory 1995. Underwriters Laboratories, 333 Pfingsten Road, Northbook, IL 60062-2096. (708) 272-8800. FAX (312) 272-8129. 1995. $12.50.

Who's Who in Engineering. American Association of Engineering Societies, 1111 19th Street NW, Suite 608, Washington, DC 20036. (202) 296-2237 or (800) 658-8897. 8th edition. 1991. Inquire for price.

ENCYCLOPEDIAS AND DICTIONARIES

Cement and Concrete Terminology. ACI Committee staff. American Concrete Institute, PO Box 19150, Detroit, MI 48219. (313) 532-2600. FAX (213) 538-0655. 1990. $33.95.

*Concise Encyclopedia of Building and Construction Material*S. Fred Moavenzadeh. Pergamon Press Inc., Maxwell House, Fairview Park, Elmsford, NY 10523. (914) 592-7700. Fax (914) 592-3625. 1990. $210.00.

Construction Glossary: an Encyclopedic Reference and Manual. J. Stewart Stern. 2nd edition. John Wiley and Sons, Inc., 605 Third Avenue, New York, NY 10158. (800) 526-5368 or (212) 850-6000. 1993. $99.95.

Construction Materials Reference Book. D.K. Doran. Butterworth-Heinemann, 313 Washington Street, Newton, MA 02158. (617) 928-2500 or (800) 366-2665. FAX (617) 928-2620. 1994. $74.95.

Illustrated Dictionary of Building Materials. Paul Bianchina. John Wiley and Sons, Inc., 605 Third Avenue, New York, NY 10158. (800) 526-5368 or (212) 850-6000. 1993. $42.50 (hardback) and $16.95 (paperback).

GENERAL WORKS

Building Construction Principles, Practices, and Materials. Glenn M. Hardie. Prentice Hall, 113 Sylvan Avenue, Route 9W, Englewood Cliffs, NJ 07632. (201) 592-2000 or (800) 922-0579. 1995. $72.00.

Building Materials Technology: Structural Performance and Environmental Impact. L. Reed Brantley and Ruth T. Brantley. McGraw-Hill Publishing Company, 1221 Avenue of the Americas, New York, NY 10020. (800) 262-4729 or (212) 512-3825. 1995. $56.00.

Building the Future: Innovation in Design, Materials, and Construction. F.K. Garas, editor. E & FN Spon/ Routledge, Chapman and Hall, Inc., 29 West 35th Street, New York, NY 10001-2291. (212) 244-3336 or FAX (212) 563-2269. 1994. Inquire for cost and availability.

Properties of Building Materials. H. J. Eldridge. Reprint edition. Books on Demand, Division of University Microfilms International, 300 North Zeeb Road, Ann Arbor, MI 48106-1346. (313) 761-4700 or (800) 521-0600. Inquire for price and availability.

Twentieth-Century Building Materials: History and Conservation. McGraw-Hill Publishing Company, 1221 Avenue of the Americas, New York, NY 10020. (800) 262-4729 or (212) 512-3825. 1995. Inquire for price.

HANDBOOKS AND MANUALS

ACI Manual of Concrete Practice. 5 volumes. American Concrete Institute, PO Box 19150, Redford Station, Detroit, MI 48219. (313) 532-2600. FAX (213) 538-0655. 1967 to present. Annual. $423.00 per set for non-members.

Concrete Construction Handbook. Joseph J. Waddell & Joseph A. Dobrowolski, editors. 3rd edition. McGraw-HillPublishing Company, 1221 Avenue of the Americas, New York, NY 10020. (800) 262-4729 or (212) 512-3825. 1993. $92.00.

Concrete Manual, 1991 Edition. International Conference of Building Officials, 5360 Workman Mill Road, Whittier, CA 90601-2298. (31) 699-0541 or (800) 423-6587.

Concrete Manual. Materials Engineering Branch, Research and Laboratory Services Division, Denver Office, U.S. Department of the Interior, Bureau of Reclamation. 9th edition. Superintendent of Documents, U.S. Government Printing Office, Box 371954, Pittsburgh, PA 15250-7954. (202) 783-3238. FAX (202) 512-2233. 1992. Inquire for cost and availability.

Handbook of Concrete Engineering. Mark Fintel. 2nd edition. Routledge, Chapman and Hall, Inc., 29 West 35th Street, New York, NY 10001-2291. (212) 244-3336 or FAX (212) 563-2269. 1985. $115.00.

Handbook of Mechanics, Materials, and Structures. Alexander Blake,editor. John Wiley and Sons, Inc., 605 Third Avenue, New York, NY 10158. (800) 526-5368 or (212) 850-6000. 1985. $120.00.

Newnes Construction Materials Pocket Book. D.K. Doran. Butterworth-Heinemann, 313 Washington Street, Newton, MA 02158. (617) 928-2500 or (800) 366-2665. 1995. $28.95.

Structural Design Guide to the AISC Specifications for Buildings. Paul F. Rice, et al. 3rd edition. Books on Demand, Division of University Microfilms International, 300 North Zeeb Road, Ann Arbor, MI 48106-1346. (313) 761-4700 or (800) 521-0600. 1985. $92.00.

ONLINE DATABASES AND CD-ROMS

Architectural Periodicals Index. Royal Institute of British Architects, British Architectural Library, 66 Portland Place, London, W1N 4AD, England. Citations to worldwide literature on architecture, construction methods and standards, urban planning, and interior design, from 1978 to present. Available on DIALOG, (800) 334-2564, online service. Inquire as to cost and availability.

COMPENDEX Plus. Engineering Information, Inc., 345 East 47th Street, New York, NY 10017. (212) 705-7600 or (800) 221-1044. Contains citations with abstracts to worldwide literature in engineering and technology, from 1970 to present. Available on online BRS,(800) 289-4277, DIALOG, (800) 334-2564, ORBIT (800) 456-7248, and STN International, FIZ Karlsruhe, P.O. Box 2465, W-7500, Karlsruhe 1, Germany, online services. Also available on CD-ROM. Inquire as to cost and availability.

ICONDA (Cib International Construction Database). Fraunhofer-Gesellschaft, Informationszentrum RAUM und BAU, Nobelstrasse, 12, D-7000, Stuttgart 80, Germany. Contains citations with abstracts to worldwide technical literature on construction and civil engineering, structural engineering, and engineering geology, from 1976 to present. Available on ORBIT, (800) 456-7248, and STN International, FIZ Karlsruhe, P.O. Box2465, W-7500, Karlsruhe 1, Germany, online services. Inquire as to cost and availability.

NTIS Bibliographic Database. National Technical Information Service, 5285 Port Royal Road, Springfield, Va 22161. (703) 487-4929 or FAX (703) 321-8199. Broad coverage of government-sponsored science and technology research reports, 1964 to present. Available on BRS,(800) 289-4277, DIALOG, (800) 334-2564, ORBIT, (800) 456-7248, and STN International, FIZ Karlsruhe, P.O. Box 2465, W-7500, Karlsruhe 1, Germany, online services. Also available on CD-ROM. Inquire as to cost and availability.

WILSONLINE. H.W. Wilson Company, 950 University Avenue, Bronx, NY 10452. (800) 367-6770 or (212) 588-8400. Makes available online versions of the printed H.W. Wilson indexes including *Applied Science and Technology Index, Business Periodicals Index, General Science Index,* and *Readers' Guide to Periodical Literature.* Period covered is generally 1983 to present. Available on BRS,(800) 289-4277, DIALOG,(800) 334-2564, and OCLC EPIC,(800) 848-5878, online services. Also available on CD-ROM. Inquire as to cost and availability.

PERIODICALS

Aberdeen's Concrete Construction. Aberdeen Group, 426 S. Westgate Street, Addison, IL 60101. (708) 543-0870. FAX (708) 543-3112. 1956 to present. Monthly. $24.00 per year.

ACI Materials Journal. American Concrete Institute, PO Box 19150, Detroit, MI 48219. (313) 532-2600. FAX (213) 538-0655. Monthly. $92.00 per year.

Building Design and Construction. Cahners Publishing Company (Des Plains), 1350 East Touhy Avenue, Des Plaines, IL

BUILDING MATERIALS

Ency. of Physical Sciences and Engineering Info. Sources

60017-5080. (708) 635-8800. FAX (708) 635-9950. 1950 to present. Monthly. $90.00 per year.

Construction Products. Cahners Publishing Company (Des Plaines), 1350 East Touhy Avenue, Des Plaines, IL 60017-5080. (708) 635-8800. 1892 to present. Six times per year. $35.00 per year.

Journal of Materials in Civil Engineering: Properties, Applications, Durability. American Society of Civil Engineers, 345 East 47th Street, New York, NY 10017-2398. (212) 705-7520 or (800) 548-2723. 1989 to present. Quarterly. $120.00 per year.

Journal of Structural Engineering. American Society of Civil Engineers, 345 East 47th Street, New York, NY 10017-2398. (212) 705-7520 or (800) 548-2723. 1956 to present. Monthly. $300.00 per year for non-members.

Journal of the Precast/prestressed Concrete Institute. Precast/ Prestressed Concrete Institute, 175 West Jackson Blvd., Chicago, IL 60604. (312) 786-0300. FAX (312) 786-0353. Bi-monthly. $29.00 per year.

Modern Steel Construction. American Institute of Steel Construction Inc., One East Wacker Drive, Suite 3100, Chicago, IL 60601-2001. (312) 670-2400 or FAX (312) 670-5403. 1944 topresent. Monthly. $30.00 per year.

Steel Construction Today. Kingslea Press Ltd., 137 Newhall Street, Birmingham B3 1SF, England. Telephone 021-236-8112. FAX 021-200-1480. 1987 to present. Bi-monthly. Inquire for cost and availability.

RESEARCH CENTERS AND INSTITUTES

American Concrete Institute. PO Box 19150, Redford Station, Detroit, MI 48219. (313) 532-2600. FAX (213) 538-0655.

American Society for Testing and Materials. 1916 Race Street, Philadelphia, PA 19103. (215) 299-5585.

Construction Industry Institute, University of Texas at Austin, 3208 Red River, Suite 300, Austin, TX 78705-2650. (512) 471-4319.

Council on Tall Buildings and Urban Habitat. Fritz Engineering Laboratory #13, Lehigh University, Bethlehem, PA 18015. (215) 758-3515. FAX (215) 758-4522.

Phil M. Ferguson Structural Engineering Laboratory, University of Texas at Austin. Balcones Research Center, 10100 Burnet Road, Bldg. 24, Austin, TX 78758. (512) 471-3062.

National Association of Home Builders National Research Center, 400 Prince George's Building, Upper Marlboro, MD 20772. (301) 249-4000.

Structural Engineering Laboratory, University of Michigan. 2340 G. G. Brown Bldg., Ann Arbor, MI 48109. (313) 763-3046. FAX (313) 764-4292.

Structural Engineering Materials Laboratory, University of California at BerkeleY. Davis Hall, Berkeley, CA 94720. (415) 642-3464.

Structural Stability Research Council. Fritz Engineering Laboratory, Lehigh University, Bethlehem, PA 18015. (215) 758-3522. FAX (215) 758-4522.

BUOYS

See also: NAVIGATION, OCEAN ENGINEERING, RADAR

ABSTRACT SERVICES AND INDEXES

Applied Science and Technology Index; a cumulative subject index to English language periodicals in the fields of aeronautics and space science, computer technology, chemistry, construction industry, energy and related areas. H.W. Wilson Co., 950 University Avenue, Bronx, NY 10452. (800) 367-6770 or (212) 588-8400. From 1958 to present. Monthly. Inquire about cost and availability. Also available on CD-ROM.

Engineering Index Monthly. Engineering Information, Inc., Castle Point on the Hudson, Hoboken, NJ 07030. (800) 221-1044. FAX (201) 216-8532. Monthly. $2200.00 per year. Also available online as COMPENDEX, and also on CD-ROM.

Government Reports Announcements and Index. National Technical Information Service (NTIS), 5285 Port Royal Road, Springfield, VA 22161. (703) 487-4650. 1968 to present. Annual. $525.00 per year. Also available online as NTIS BIBLIOGRAPHIC DATABASE and on CD-ROM.

Offshore Engineering Abstracts. STI Limited, 4 Kings Meadow, Ferry Hinksey Road, Oxford, OX2 ODU, England. Distributed in the United States by Air Science Company, Box 143, Corning NY 14830. (607) 962-5591. Contains citations on planning, design, and construction of offshore platforms and pipelines, from 1985 to present. Monthly. $400.00 per year.

ASSOCIATIONS AND PROFESSIONAL SOCIETIES

Institute of Navigation. 1025 16th Street, N.W., Suite 104, Washington, DC 20036. (202) 783-4121.

International Omega Association. P.O. Box 2324, 1720 South Eads Street, Arlington, VA 22202-0324. (703) 863-4578.

Permanent International Association of Navigation Congresses, United States Section. 20 Massachusetts Avenue, N.W., Washington, DC 20314-1000. (202) 504-4312.

Society of Naval Architects and Marine Engineers. 601 Pavonia Avenue, Jersey City, NJ 07306. (201) 798-4800.

Wild Goose Association. c/o Nav Com Systems, 7203 Gateway Court, Manassas, VA 22110. (703) 361-0884.

DIRECTORIES AND BIOGRAPHICAL SOURCES

International Research Centers Directory. Gale Research, Detroit, MI 48226. (800) 347-4253. $360.00.

Research Centers Directory. Gale Research, Detroit, MI 48226. (800) 347-4253. $485.00.

Scientific and Technical Organizations and Agencies Directory. Margaret Labash Young, editor. Gale Research, 835 Penobscot Building, Detroit, MI 48226. (800) 347-4253. 2nd edition. 1987. $185.00.

Who's Who in Engineering. Gordon Davis, editor. Hemisphere Publishing Corporation, 79 Madison Avenue, New York, NY 10016-7892. (800) 821-8312. Sixth edition. 1985. $200.00.

GENERAL WORKS

Buoy Engineering. H.O. Berteaux. Books on Demand, 300 North Zeeb Road, Ann Arbor, MI 48106. (313) 761-4700. $98.00 in paper.

Dutton's Navigation and Piloting. Elbert S. Maloney. Naval Institute Press, U.S. Naval Institute, Annapolis, MD 21402. (301) 268-6110. Fourth edition. 1985. $32.95.

Marine Electronic Navigation. S.F. Appleyard. Methuen, Inc., 29 West 35th Street, New York, NY 10001. (212) 244-3336. 1980. $34.95.

Marine Navigation: Piloting and Celestial and Electronic Navigation. Richard R. Hobbs. Naval Institute Press, U.S. Naval Institute, Annapolis, MD 21402. (301) 268-6110. Third edition. 1990. $38.95.

Navigation Control Manual. A.G. Bole and others. Butterworths Publishers, 80 Montvale Avenue, Stoneham, MA 02180. (617) 438-8464 or (800) 366-2665. 1987. $65.00.

Seamanship: Fundamentals for the Deck Officer. David O. Dodge and S.E. Kyriss. Naval Institute Press, U.S. Naval Institute, Annapolis, MD 21402. (301) 268-6110. 1981. $16.95.

Transportation Signal and Control Equipment 1991-1996 Analysis: Traffic Signals, Railroad Grade Crossing Signals, Lighted Buoys and Maritime Beacons, and Airport Lighting Equipment. Dennis Zogbi. World Information Technologies, 24 Woodbine Avenue, Northport, NY 11768-2878. (800) 967-5346. 1991. $1800.00.

ONLINE DATABASES AND CD-ROMS

COMPENDEX Plus. Engineering Information, Inc., 345 East 47th Street, New York, NY 10017. (212) 705-7600 or (800) 221-1044. Contains citations with abstracts to worldwide literature in engineering and technology, from 1970 to present. Available on online BRS, (800) 289-4277, DIALOG, (800) 334-2564, ORBIT (800) 456-7248, and STN International, FIZ Karlsruhe, P.O. Box 2465, W-7500, Karlsruhe 1, Germany, online services. Also available on CD-ROM. Inquire as to cost and availability.

NTIS Bibliographic Database. National Technical Information Service, 5285 Port Royal Road, Springfield, Va 22161. (703) 487-4929 or FAX (703) 321-8199. Broad coverage of government-sponsored science and technology research reports, 1964 to present. Available on BRS, (800) 289-4277, DIALOG, (800) 334-2564, ORBIT, (800) 456-7248, and STN International, FIZ Karlsruhe, P.O. Box 2465, W-7500, Karlsruhe 1, Germany, online services. Also available on CD-ROM. Inquire as to cost and availability.

WilsonLine. H.W. Wilson Company, 950 University Avenue, Bronx, NY 10452. (800) 367-6770 or (212) 588-8400. Makes available online versions of the printed H.W. Wilson indexes including *Applied Science and Technology Index, Business Periodicals Index, General Science Index,* and *Readers' Guide to Periodical Literature.* Period covered is generally 1983 to present. Available on BRS, (800) 289-4277, DIALOG, (800) 334-2564, and OCLC EPIC, (800) 848-5878, online services. Also available on CD-ROM. Inquire as to cost and availability.

PERIODICALS

Journal of Navigation. Royal Institute of Navigation. Cambridge University Press, 32 East 57th Street, New York, NY 10022. (800) 872-7423. 1947 to present. Three per year. $100.00 per year.

Journal of Ship Research. Society of Naval Architects and Marine Engineers, One World Trade Center, Suite 1369, New York, NY 10048. (212) 432-0310. 1957 to present. Quarterly. $40.00 per year.

RESEARCH CENTERS AND INSTITUTES

Charles Stark Draper Laboratory, Inc. 555 Technology Square, Cambridge, MA 02139. (617) 258-1000.

Ocean Engineering Laboratory. University of Washington, Mail Stop FU-10, Mechanical Engineering Building, Seattle, WA 98195. (206) 543-7446.

BUSHINGS

See: BEARINGS AND BALL BEARINGS

C

CABLE

See: WIRE AND CABLE

CABLE TELEVISION

See: TELEVISION

CAD (COMPUTER-AIDED DESIGN)

See also: AUTOMATION, CAM, ROBOTICS

ABSTRACT SERVICES AND INDEXES

ACM Guide To Computing Literature. Association for Computing Machinery, Association for Computing Machinery, 1515 Broadway, 17th Floor, New York, NY 10036-5701. (212) 869-7440. FAX (212) 944-1318. 1964 to present. Monthly. $190.00 per year for non-members.

Applied Science and Technology Index; A Cumulative Subject Index To English Language Periodicals In the Fields of Aeronautics and Space Science, Computer Technology, Chemistry, Construction Industry, Energy and Related Areas. H.W. Wilson Co., 950 University Avenue, Bronx, NY 10452. (800) 367-6770 or (212) 588-8400. FAX (718) 590-1617. From 1958 to present. Monthly. Inquire about cost and availability. Also available on CD-ROM and online.

CAD-CAM Abstracts (formerly *CAD-CAM Profiles*). Techgnosis Ltd., Blade House, Battersea Road, Stockport, Cheshire 3AE, England. Telephone 061-442-2639. FAX 061-443-1162. Six times a year. Inquire for price.

CAD/CAM Abstracts and CAD/CAM Abstracts Annual. R.R. Bowker, Bowker A&I Publishing, 121 Chanlon Road, New Providence, NJ 07974. (908) 771-7714. Fax (908) 771-7725. 1984 to 1992. Ceased publication.

Computer Abstracts, MCB University Press Ltd., PO Box 10812, Birmingham, AL 35201-0812. (800) 633-4931. FAX (205) 995-1588. Monthly. Covers computer theory, data, hardware, systems, networks, human-computer interaction, artificial intelligence, as well as applications of computers in aerospace, business, CAD/CAM, cartography, civil engineering, electronics and electrical engineering, industrial engineering, mechanical engineering, medicine, structural engineering, etc. Also available on CD-ROM.

Computer & Control Abstracts (INSPEC). INSPEC/IEEE, Box 1331, 445 Hoes Lane, Piscataway, NJ 08855-1331. (908) 562-5549. Abstracts organized by subjects of international technological information. Monthly. $1455.00 per year. Also available on CD-ROM and online via BRS Online Products, DIALOG Information Services, Orbit Search Service, and STN International.

Computer and Information Systems Abstracts Journal. Cambridge Scientific Abstracts, 7200 Wisconsin Avenue, Bethesda, MD 20814. (301) 961-6750. Fax (301) 961-6720. 1962 to present. Monthly. $1265.00 per year. Also available online via STN International.

Computer Literature Index. Applied Computer Research Inc., Box 82266, Phoenix, AZ 85071-2266. (800) 234-2227. 1971 to present. Quarterly plus annual cumulation. $198.50 per year.Bibliography of books, articles, and reports.

Computing Journal Abstracts. Techgnosis Ltd., Blade House, Battersea Road, Stockport, Cheshire 3AE, England. Telephone 061-442-2639. FAX 061-443-1162. 1969 to present. Monthly. Inquire for price.

Computing Reviews. Association for Computing Machinery, 1515 Broadway, 17th Floor, New York, NY 10036-5701. (212) 869-7440. FAX (212) 944-1318. 1960 to present. Monthly. $130.00 per year for non-members. Also available online via DIALOG Information Services.

Current Contents: Engineering, Technology, and Applied Sciences. Institute for Scientific Information, 3501 Market Street, Philadelphia, PA 19104. (215) 386-0100. FAX (215) 386-6362. 1970 to present. Weekly. $442.00 per year.

Engineering Index Monthly. Engineering Information, Inc., Castle Point on the Hudson, Hoboken, NJ 07030. (800) 221-1044. FAX (212) 832-1857. Monthly. $2300.00 per year. Also available online as COMPENDEX, and also on CD-ROM. Covers chemical engineering, computers, electrical engineering, civil engineering, metals and mining, industrial management, and mechanical engineering.

Science Citation Index. Institute for Scientific Information, 3501 Market Street, Philadelphia, PA 19104. (215) 386-0100. FAX (215) 386-6362. Inquire about availability and cost. Also available on CD-ROM.

ASSOCIATIONS AND PROFESSIONAL SOCIETIES

American Automatic Control Council. c/o EECS, Northwestern University, 2145 Sheridan Road, Evanston, IL 60208-3118. (708) 491-3641.

Association For Computing Machinery. 1515 Broadway, 17th Floor, New York, NY 10036-5701. (212) 869-7440. FAX (212) 944-1318.

Computer and Automation Systems Association of SME. 1 SME Drive, PO Box 930, Dearborn, MI 48121-0930. (313) 271-1500.

Institute of Electrical and Electronic Engineers. 345 E. 47th Street, New York, NY 10017. (212) 705-7900.

Numerical Control Society/aim Tech. 67 Alexander Drive, PO Box 12277, Research Triangle Park, NC 27709. (815) 399-8700.

BIBLIOGRAPHIES

ACM Guide To Computing Literature. Association for Computing Machinery, 1515 Broadway, 17th Floor, New York, NY 10036-5701. (212) 869-7440. FAX (212) 944-1318. 1964 to present. Annual. $190.00 for non-members. Also available online via DIALOG.

DIRECTORIES AND BIOGRAPHICAL SOURCES

CAD/CAM International Software Directory. TAB Books, P.O. Box 40, Blue Ridge Summit, PA 17294-0850. (717) 794-2191 or (800) 822-8138. FAX (717) 794-2080. 1991. Inquire for cost and availability.

Computer-Aided Design (CAD) Directory. TAB Books, P.O. Box 40, Blue Ridge Summit, PA 17294-0850. (717) 794-2191 or (800) 822-8138. FAX (717) 794-2080. 1991. Inquire for cost and availability.

Computer Graphics World Buyers Guide. PennWell PublishingCompany, Advanced Technology Group, 10 Tara Blvd., 5th floor, Nashua, NH 03062-2801. (603) 891-9111. FAX (603) 891-0539. Annual. $100.00 per year.

Computing Information Directory. Hildebrandt Inc., PO Box 576, Pullman, WA 99114. 1981 to present. Annual. $199.95 per year.

Who's Who In Technology. Gale Research, 835 Penobscot Building, Detroit, MI 48226-4094. (313) 961-2242. (800) 347-4253. FAX (313) 961-6083. 1995. $195.00.

ENCYCLOPEDIAS AND DICTIONARIES

CAD/CAM Dictionary. Preston & Crawford. Marcel Dekker, Inc., 270 Madison Avenue, New York, NY 10016. (212) 696-9000. FAX (212) 685-4540. 1985. $89.75.

CAD/CAM Glossary. CIM Systems, 2425 N. Central, No. 432, Richardson, TX 75080-3549. (214) 437-5171. $19.25.

Computer Dictionary. Donald D. Spencer. 4th edition. Camelot Publishing Company, 709 SW 80th Blvd., Gainesville, FL 32607-1537. (904) 331-0952. 1993. $24.95.

Dictionary of Computing. S.M.H. Collin. 2nd edition. Peter Collin Publishing, 8 the Causeway, Teddington, TW11 0HE, England. FAX 0181-943-3386. 1994. Inquire for cost.

Encyclopedia of Computer Science and Engineering. Anthony Ralston & Edwin D. Reilly Jr., editors. 3rd revised edition. Van Nostrand Reinhold, 115 Fifth Avenue, New York, NY 10003. (212) 254-3232. FAX (212) 254-9499. 1993. $125.00.

Encyclopedic Dictionary of Industrial Automation and Computer Control. David W. South. Prentice Hall, 113 Sylvan Avenue, Route 9W, Englewood Cliffs, NJ 07632. (201) 592-2000 or (800) 922-0579. 1994. Inquire for price and availability.

Microsoft Press Computer Dictionary: the Comprehensive Standard For Business, School, Library, and Home. 2nd edition. Microsoft Press, One Microsoft Way, Redmond, WA 98052-6399. (206) 936-0055. FAX (206) 823-8101. 1994. $19.95.

GENERAL WORKS

Advances In Computer-Aided Design: Computer-Aided Design of VLSI Circuits & Systems, Vol. 1. Alberto Sangiovanni-Vincentelli, editor. Jai Press, 55 Old Post Road No.2, PO Box 1678, Greenwich, CT 06836. (203) 661-7602. FAX (203) 661-0792. 1985. $73.25.

Advances In Computer-Aided Design, Vol. 2. Ibrahim N. Hajj, editor. JAI Press, 55 Old Post Road No.2, PO Box 1678, Greenwich, CT 06836. (203) 661-7602. FAX (203) 661-0792 1988. $73.25.

The Autocad Productivity Book. A.T. Schaefer & J.L. Brittain. Elsevier Science Inc., 660 White Plains Rd., Tarrytown, NY 10159-5153. (919) 524-9200. FAX (919) 333-2444. 1986. Inquire for cost and availability.

CAD/CAM Report. Architecture Technology Corporation. Elsevier Science Inc., 660 White Plains Rd., Tarrytown, NY 10159-5153. (919) 524-9200. FAX (919) 333-2444. 1990. $375.00.

CAD/CAM Theory and Practice. Ibrahim Zeid. McGraw-HillPublishing Company, 1221 Avenue of the Americas, New York, NY 10020. (800) 262-4729 or (212) 512-3825. 1991. Inquire for cost and availability.

Computer-Aided Design and Manufacturing. Faird M.L. Amirouche. Prentice Hall, 113 Sylvan Avenue, Route 9W, Englewood Cliffs, NJ 07632. (201) 592-2000 or (800) 922-0579. 1992. $56.00.

Computer-Assisted Mechanical Design. J. Ed Akin. Prentice Hall, 113 Sylvan Avenue, Route 9W, Englewood Cliffs, NJ 07632. (201) 592-2000 or (800) 922-0579. 1990. $54.00.

Computers In Engineering, 1993. American Society of Mechanical Engineers, 345 East 47th Street, New York, NY 10017-2398. (212) 705-7703. 1993. Inquire for cost and availability.

Making CAD/CAM Work. Richard Halton. Prentice Hall, 113 Sylvan Avenue, Route 9W, Englewood Cliffs, NJ 07632. (201) 592-2000 or (800) 922-0579. 1993. Inquire for price.

HANDBOOKS AND MANUALS

The C4 Handbook: CAD, CAM, CAE, CIM. Carl Machover. TAB Books, P.O. Box 40, Blue Ridge Summit, PA 17294-0850. (717) 794-2191 or (800) 233-1128. FAX (717) 794-2080. 1989. Inquire for price and availability.

McGraw-Hill Computing Essentials, 1993-1994. Timothy J. O'Leary, et al. McGraw-Hill Publishing Company, 1221 Avenue of the Americas, New York, NY 10020. (800) 262-4729 or (212) 512-3825. 1994. Inquire for price and availability.

ONLINE DATABASES AND CD-ROMS

Compendex Plus. Engineering Information, Inc., 345 East 47th Street, New York, NY 10017. (212) 705-7600 or (800) 221-1044. Contains citations with abstracts to worldwide literature in engineering and technology, from 1970 to present. Available on online BRS,(800) 289-4277, DIALOG, (800) 334-2564, ORBIT (800) 456-7248, and STN International, FIZ Karlsruhe, P.O. Box 2465, W-7500, Karlsruhe 1, Germany, online services. Also available on CD-ROM. Inquire as to cost and availability.

CSA Engineering. Cambridge Scientific Abstracts, 7200 Wisconsin Avenue, Suite 601, Bethesda, MD 20814. (301) 961-6750 or (800) 843-7751. Contains citations and abstracts of international periodicals and research literature covering all fields of engineering and science and technology, including computer and information science, electronics, mechanical engineering, solid state materials, 1981 to present. Available on BRS,(800) 289-4277, online service. Inquire as to cost and availability.

Current Contents Search. Institute for Scientific Information, 3501 Market Street, Philadelphia, PA 19104. (215) 386-0100. FAX (215) 386-6362. Contains citations to articles listed in the table of contents of science and technology journals. Also articles in social sciences and life sciences journals. Available on BRS,(800) 289-4277, DIALOG,(800) 334-2564, online services. Inquire as to cost and availability.

Dissertation Abstracts. University MicrofilmsInternational, 300 North Zeeb Road, Ann Arbor, MI 48106. (800) 521-0600 or (313) 761-4700. Scope includes virtually all doctoral dissertations accepted at accredited American institutions from 1861 to present in 252 subject areas. Available on BRS,(800) 289-4277, DIALOG,(800) 334-2564, and OCLC EPIC,(800) 848-5878, online services. Also available on CD-ROM. Inquire as to cost and availability.

NTIS Bibliographic Database. National Technical Information Service, 5285 Port Royal Road, Springfield, Va 22161. (703) 487-4929 or FAX (703) 321-8199. Broad coverage of government-sponsored science and technology research reports, 1964 to present. Available on BRS,(800) 289-4277, DIALOG, (800) 334-2564, ORBIT, (800) 456-7248, and STN International, FIZ Karlsruhe, P.O. Box 2465, W-7500, Karlsruhe 1, Germany, online services. Also available on CD-ROM. Inquire as to cost and availability.

Scisearch. Institute for Scientific Information, 3501 Market Street, Philadelphia, PA 19104. (800) 523-1850 or (215) 386-0100. Broad multi disciplinary title and author index to the international literature of science and technology, 1974 to present.

Available on DIALOG,(800) 334-2564, and ORBIT,(800) 456-7248, online services. Also available on CD-ROM. Inquire as to cost and availability.

WILSONLINE. H.W. Wilson Company, 950 University Avenue, Bronx, NY 10452. (800) 367-6770 or (212) 588-8400. Makes available online versions of the printed H.W. Wilson indexes including Applied Science and Technology Index, Business Periodicals Index, General Science Index, and Readers' Guide to Periodical Literature. Period covered is generally 1983 to present. Available on BRS,(800) 289-4277, DIALOG,(800) 334-2564, and OCLC EPIC,(800) 848-5878, online services. Also available on CD-ROM. Inquire as to cost and availability.

PERIODICALS

ACM Transactions On Graphics. Association for Computing Machinery, 1515 Broadway, 17th Floor, New York, NY 10036-5701. (212) 869-7440. FAX (212) 944-1318. 1982 to present. Quarterly. $110.00.

CADalyst. Advanstar Communications Inc., 7500 Old Oak Road, Cleveland, OH 44130. (216) 826-2839. FAX (216) 891-2726. 1984 to present. 12 times a year. $59.00 per year.

CAD-CAM Update. Worldwide Videotex, Box 3273, Boynton Beach, FL 33424-3273. (407) 738-2276. 1989 to present. Monthly. Inquire for price.

CAD Systems. Kerrwil Publications Ltd., 395 Matheson Blvd. E, Mississauga, Ontario L4Z 2H2, Canada. (905) 890-1846. FAX (905) 890-5769. 1982 to present. Six times a year. $61.00 per year.

CAD User. B.T.C., 24 High Street, Beckenham, Kent BR3 1AY, England. Telephone 081-663-3818. FAX 081-663-3822. 1988 to present. Monthly. Inquire for price.

Computer-aided Design. Butterworth-Heinemann/sub. toTurpin Transactions Ltd., Distribution Centre, Blackhorse Road, Letchworth, Herts SG6 1HN, England. Telephone 0462-672555. 1968 to present. 12 times a year. Inquire for price.

Computer-Aided Design Report. CAD-CAM Publishing Inc., 841 Turquoise Street, Suites D & E, San Diego, CA 92109-1159. (619) 488-0533. FAX (619) 488-6052. 1981 to present. Monthly. $190.00 per year.

Computer-Aided Geometric Design. Elsevier Science, Box 882, Madison Square Station, New York, NY 10159. (212) 989-5800. FAX (212) 633-3990. 1984 to present. Six times a year. $322.00.

Computer Design. PennWell Publishing Company, Advanced Technology Group, 10 Tara Blvd., 5th floor, Nashua, NH 03062-2801. (603) 891-9111. FAX (603) 891-0539. 1962 to present. 12 times a year. $88.00 per year.

Computer Graphics World. PennWell Publishing Company, Advanced Technology Group, 10 Tara Blvd., 5th floor, Nashua, NH 03062-2801. (603) 891-9111. FAX (603) 891-0539. 1978 to present. Monthly. $48.00 per year.

Computers & Graphics. Elsevier Science, 660 White Plains Road, Tarrytown, NY 10159-5153. (919) 524-9200. FAX (919) 333-2444. 1975 to present. Six times a year. $780.00 per year.

CAD (COMPUTER AIDED DESIGN)

Ency. of Physical Sciences and Engineering Info. Sources

IEEE Computer Graphics and Applications. IEEE Computer Society Press, 10662 Los Vaqueros Circle, Box 3014, Los Alamitos, CA 90720-1264. (714) 821-8380. FAX (714) 821-4641. 1981 to present. Bi-monthly. $175.00 per year.

RESEARCH CENTERS AND INSTITUTES

Albert H. Case Center for Computer-Aided Design. Michigan State University, College of Engineering, 112 Engineering Bldg., East Lansing, MI 48824. (517) 355-6453.

Center For Computer-Aided Design and Manufacturing, Utah State University. Mechanical Engineering Department, UMC 4130, Logan, UT 84322-4130. (801) 750-2950.

Computer-Aided Design Laboratory, Massachusetts Institute of Technology. Department of Mechanical Engineering, Bldg. 3, Room 449B, Cambridge, MA 02139. (617) 253-4465.

Computer-Aided Engineering Center, University of Wisconsin—Madison. 1500 Johnson Drive, Madison, WI 53706. (608) 263-3075.

University of Iowa, Center For Computer-Aided Design. 1409 Engineering Bldg., Iowa City, IA 52242. (319) 335-5726. FAX (319) 335-6061.

CADMIUM

See also: ALLOYS, ALUMINUM, BRASS AND BRONZE, COPPER, MATERIALS SCIENCE, METALLURGY

ABSTRACT SERVICES AND INDEXES

Alloys Index. American Society for Metals, Metals Park, OH 44073. (216) 338-5151. FAX (216) 338-4634. 1974 to present. Monthly. $380.00. Also available on CD-ROM and online via METADEX, STN International, and DIALOG.

Aluminum Industry Abstracts. Aluminum Association,Materials Park, OH 44073. (216) 338-5151. FAX (216) 338-4634. 1968 to present. Monthly. $525.00 per year.

Applied Mechanics Reviews: An Assessment of World Literature In Engineering Sciences. 1948-present. American Society of Mechanical Engineers, 345 East 47th Street, New York, NY 10017. (212) 705-7703. Monthly. $360.00 per year.

Applied Science and Technology Index; A Cumulative Subject Index To English Language Periodicals In the Fields of Aeronautics and Space Science, Computer Technology, Chemistry, Construction Industry, Energy and Related Areas. H.W. Wilson Co., 950 University Avenue, Bronx, NY 10452. (800) 367-6770 or (212) 588-8400. FAX (718) 590-1617. From 1958 to present. Monthly. Inquire about cost and availability. Also available on CD-ROM and online.

CADSCAN (formerly *Cadmium Abstracts*). Cadmium Association, 42 Weymouth Street, London W1N 3LQ, England. 1977 to present. Quarterly. Inquire for cost and availability.

Corrosion Abstracts; Abstracts of the World's Literature On Corrosion and Corrosion Mitigation. National Association of Corrosion Engineers (NACE), Box 218340, Houston, TX 77218.

(713) 492-0535 or FAX (713) 492-8254. 1962 to present. Bi-monthly. $195.00 per year. Also available on CD-ROM.

Current Contents: Engineering, Technology, and Applied Sciences. Institute for Scientific Information, 3501 Market Street, Philadelphia, PA 19104. (215) 386-0100. FAX (215) 386-6362. 1970 to present. Weekly. $442.00 per year.

Engineering Index Monthly. Engineering Information, Inc., Castle Point on the Hudson, Hoboken, NJ 07030. (800) 221-1044. FAX (212) 832-1857. Monthly. $2300.00 per year. Also available online as COMPENDEX, and also on CD-ROM. Covers chemical engineering, computers, electrical engineering, civil engineering, metals and mining, industrial management, and mechanical engineering.

International Aerospace Abstracts. Technical Information Service, American Institute of Aeronautics and Astronautics, Inc., 555 West 57th St., New York, NY 10019. (212) 247-6500. FAX (212) 582-4861. Semi-monthly. $1295.00 per year.

Metals Abstracts and Metals Abstracts Index. American Society for Metals, Metals Park, OH 44073. (216) 338-5151. 1968 to present. Monthly. Abstracts are $1500.00 per year and Index is $460.00 per year.

ANNUAL REVIEWS AND YEARBOOKS

Non-Ferrous Metal Data. American Bureau of Metal Statistics, Inc., Box 1405 Plaza Station, 400 Plaza Drive, Secaucus, NJ 07094-0405. (201) 863-6900. FAX (201) 863-6050. 1920 to present. Annual. $350.00 per year.

ASSOCIATIONS AND PROFESSIONAL SOCIETIES

American Society For Metals (ASM). Materials Park, OH 44073. (216) 338-5151 or FAX (216) 338-4634.

Cadmium Association. 42 Weymouth Street, London W1N 3LQ, England.

Cadmium Council, Inc. 12110 Sunset Hills Road, Suite 110, Reston, VA 22090-3223. (703) 709-1400. FAX (703) 709-1402.

Metallurgical Society of the AIME (American Institute of Mining, Metallurgical, and Petroleum Engineers). 345 E.47th Street, 14th Floor, New York, NY 10017. (212) 705-7695.

National Association of Corrosion Engineers. Box 218340, Houston, TX 77218. (713) 492-0535. FAX (713) 492-8254.

DIRECTORIES AND BIOGRAPHICAL SOURCES

Dun's Industrial Guide: the Metalworking Directory, 1993-94. Dun & Bradstreet Information Services Staff. 3 volumes. Dun & Bradstreet Information Services, 899 Eaton Avenue, Bethlehem, Pa 18025. (610) 882-7000 or (800) 526-0651. FAX (610) 882-7269. 1993. $795.00.

International Research Centers Directory. Gale Research, 835 Penobscot Building, Detroit, MI 48226-4094. (313) 961-2242. (800) 347-4253. FAX (313) 961-6083. 8th edition. 1995. $410.00

Materials Performance: Nace Corrosion Engineers' Buyers Guide Issue. National Association of Corrosion Engineers. Box 218340, Houston, TX 77218. (713) 492-0535. FAX (713) 492-8254. Part of subscription to *Materials Performance.*

Research Centers Directory. Gale Research, 835 Penobscot Building, Detroit, MI 48226-4094. (313) 961-2242. (800) 347-4253. FAX (313) 961-6083. $485.00.

Scientific and Technical Organizations and Agencies Directory. Gale Research, 835 Penobscot Building, Detroit, MI 48226-4094. (313) 961-2242. (800) 347-4253. FAX (313) 961-6083. 4th edition. 1996. $195.00.

Who's Who In Engineering. American Association of Engineering Societies, 1111 19th Street NW, Suite 608, Washington, DC 20036. (202) 296-2237 or (800) 658-8897. 8th edition. 1991. Inquire for price.

Who's Who In Technology. Gale Research, 835 Penobscot Building, Detroit, MI 48226-4094. (313) 961-2242. (800) 347-4253. FAX (313) 961-6083. 1995. $195.00.

ENCYCLOPEDIAS AND DICTIONARIES

Encyclopedia of Physical Science and Technology. Robert A. Meyers, ed. 18 volumes. Academic Press Inc., 6277 Sea Harbor Drive, Orlando, FL 32887. (800) 321-5068. 1992. $2100.00.

McGraw-Hill Encyclopedia of Engineering. Sybil P. Parker, ed. 2nd edition. McGraw-Hill Publishing Company, 1221 Avenue of the Americas, New York, NY 10020. (800) 262-4729 or (212) 512-3825. 1993. $95.50.

McGraw-Hill Encyclopedia of Science and Technology. Sybil P. Parker, ed. 7th edition. 20 volumes. McGraw-Hill Publishing Company, 1221 Avenue of the Americas, New York, NY 10020. (800) 262-4729 or (212) 512-3825. 1992. $1900.00

GENERAL WORKS

Cadmium. WHO Task Group on Environmental Health Criteria for Cadmium. World Health Organization, Geneva, Switzerland. 1992. Inquire for price and availability.

Copper, Silver, Gold & Zinc, Cadmium, Mercury Oxides and Hydroxides. T.P. Dirkse. Pergamon Press Inc., Maxwell House, Fairview Park, Elmsford, NY 10523. (914) 592-7700. Fax (914) 592-3625. 1986. $140.00.

HANDBOOKS AND MANUALS

Corrosion Resistant Materials Handbook. D. J. de Renzo, editor. 4th edition. Noyes Data Corporation, 120 Mill Road, Park Ridge, NJ 07656. (201) 391-8484. FAX (201) 391-6833. 1986. $125.00.

Smithell's Metals Reference Book. E.A. Brandes & G.B. Brook. 7th edition. Butterworth-Heinemann, 313 Washington Street, Newton, MA 02158. (617) 928-2500 or (800) 366-2665. FAX (617) 928-2620. 1992. $250.00.

ONLINE DATABASES AND CD-ROMS

Compendex Plus. Engineering Information, Inc., 345 East 47th Street, New York, NY 10017. (212) 705-7600 or (800) 221-1044. Contains citations with abstracts to worldwide literature in engineering and technology, from 1970 to present. Available on online BRS,(800) 289-4277, DIALOG, (800) 334-2564, ORBIT (800) 456-7248, and STN International, FIZ Karlsruhe, P.O. Box 2465, W-7500, Karlsruhe 1, Germany, online services. Also available on CD-ROM. Inquire as to cost and availability.

Current Contents Search. Institute for Scientific Information, 3501 Market Street, Philadelphia, PA 19104. (215) 386-0100. FAX (215) 386-6362. Contains citations to articles listed in the table of contents of science and technology journals. Also articles in social sciences and life sciences journals. Available on BRS,(800) 289-4277, DIALOG,(800) 334-2564, online services. Inquire as to cost and availability.

Metadex. Jointly produced by ASM International and the Institute of Materials. Contains more than 925,000 records from the international literature on metals and alloys, concerning properties, processes, materials classes, applications, and metallurgical systems. Updated monthly. Available from ORBIT-QUESTEL (703) 442-0900.

NASA Database. American Institute of Aeronautics and Astronautics, 370 L'Enfant Promenade, S.W., Washington, DC 20024. (202) 646-7400. Citations and abstracts of aeronautics and astronautics literature, 1962 to present. Inquire as to cost and availability.

NTIS Bibliographic Database. National Technical Information Service, 5285 Port Royal Road, Springfield, Va 22161. (703) 487-4929 or FAX (703) 321-8199. Broad coverage of government-sponsored science and technology research reports, 1964 to present. Available on BRS,(800) 289-4277, DIALOG, (800) 334-2564, ORBIT, (800) 456-7248, and STN International, FIZ Karlsruhe, P.O. Box 2465, W-7500, Karlsruhe 1, Germany, online services. Also available on CD-ROM. Inquire as to cost and availability.

WILSONLINE. H.W. Wilson Company, 950 University Avenue, Bronx, NY 10452. (800) 367-6770 or (212) 588-8400. Makes available online versions of the printed H.W. Wilson indexes including Applied Science and Technology Index, Business Periodicals Index, General Science Index, and Readers' Guide to Periodical Literature. Period covered is generally 1983 to present. Available on BRS,(800) 289-4277, DIALOG,(800) 334-2564, and OCLC EPIC,(800) 848-5878, online services. Also available on CD-ROM. Inquire as to cost and availability.

PERIODICALS

Alloy Digest. Alloy Digest Inc., 27 Canfield Street, Orange, NJ 07050. (201) 677-9161. 1952 to present. Monthly. $140.00 per year.

Corrosion. National Association of Corrosion Engineers, Box 218340, Houston, TX 77218. (713) 492-0535. FAX (713) 492-8254. 1945 to present. $100.00 per year.

Corrosion Prevention and Control. Scientific Surveys Ltd., Box 21, Beaconsfield, Bucks. HP9 1NS, England. Telephone 0494-675139. FAX 0494-670155. 1954 to present. Bi-monthly. $139.00 per year.

Corrosion Science. Elsevier Science [journals], 660 White Plains Rd., Tarrytown, NY 10159-5153. (919) 524-9200. FAX (919) 333-2444. 1961 to present. 12 times a year. $1230.00 per year.

Inorganic Chemistry. American Chemical Society, 1155 16th Street NW Washington, DC 20036. (800) 333-9511. FAX (914) 447-3671. 1962 to present. Semi-monthly. $949.00 per year.

Journal of Alloys and Compounds. Elsevier Science Inc., Box 882, Madison Square Station, New York, NY 10159. (212) 989-5800. 1959 to present. 30 times a year. $3980.00 per year.

J O M (Journal of Metals). Minerals, Metals and Materials Society, 420 Commonwealth Drive, Warrendale, PA 15086. (412) 776-9080. 1949 to present. Monthly. $50.00 per year.

Light Metal Age. Fellom Publishing, 170 S. Spruce Avenue, Suite 120, San Francisco, CA 94080. (415) 588-8832. FAX (415) 588-0901. 1943 to present. Bi-monthly. $35.00 per year.

Materials Performance. National Association of Corrosion Engineers. Box 218340, Houston, TX 77218. (713) 492-0535. FAX (713) 492-8254. 1962 to present. Monthly. $65.00 per year.

Metal Bulletin Monthly. Metal Bulletin Inc., 220 Fifth Avenue, 19th Floor, New York, NY 10001-7781. (800) 638-2525. FAX (212) 213-6273. 1972 to present. Monthly. Inquire for cost.

Metallurgical Transactions A: Physical Metallurgy and Materials Science. ASM International, Materials Park, OH 44073. (216) 338-5151 or FAX (216) 338-4634. 1970 to present. Monthly. $520.00 per year.

Metals Week. McGraw-Hill Publishing Company, Commodity Services Group, 1221 Avenue of the Americas, New York, NY 10020. (800) 262-4729 or (212) 512-3825. 1930 to present. Weekly. $770.00 per year.

Mineral Industry Surveys, Cadmium. U.S. Bureau of Mines,Production and Distribution, Cochrans Mill Rd., Box 18070, Pittsburgh, PA 15236. (412) 892-4411. Quarterly and annually. Inquire for price and availability.

RESEARCH CENTERS AND INSTITUTES

Canada Centre For Mineral and Energy Technology, Materials Technology Laboratories. Energy, Mines & Resources Canada, 568 Booth Street, Ottawa ON, Canada K1A 0G1. (613) 995-8248. FAX (613) 992-8735.

International Lead-Zinc Research Organization. 2525 Meridian Parkway, PO Box 12036, Research Triangle, NC 27709. (919) 361-4647. FAX (919) 361-1957.

Texas A&M University Mechanics and Materials Center. Civil Engineering Department, College Station, TX 77843. (409) 845-7512. FAX (409) 845-6156.

University of Connecticut Institute of Materials Science. U-136, 97 N. Eagleville Road, Storrs, CT 06268. (203) 486-4623. FAX (203) 486-4745.

University of Florida Department of Materials Science and Engineering. Gainesville, FL 32611. (904) 392-1454. FAX (904) 392-6359.

University of Minnesota Corrosion Research Center. 221 Church Street SE, Minneapolis, MN 55455. (612) 625-0717.

SPECIFICATIONS AND STANDARDS

*Annual Book of ASTM Standards, Volume 02.*04. American Society for Testing and Materials (ASTM), 1916 Race Street, Philadelphia, PA 19103. (215) 299-5585. Annual. Inquire for cost and availability.

CAE (COMPUTER-AIDED ENGINEERING)

See also: CAD (COMPUTER-AIDED DESIGN)

ABSTRACT SERVICES AND INDEXES

ACM Guide To Computing Literature. Association for Computing Machinery, Association for Computing Machinery, 1515 Broadway, 17th Floor, New York, NY 10036-5701. (212) 869-7440. FAX (212) 944-1318. 1964 to present. Monthly. $190.00 per year for non-members.

Applied Science and Technology Index; A Cumulative Subject Index To English Language Periodicals In the Fields of Aeronautics and Space Science, Computer Technology, Chemistry, Construction Industry, Energy and Related Areas. H.W. Wilson Co., 950 University Avenue, Bronx, NY 10452. (800) 367-6770 or (212) 588-8400. FAX (718) 590-1617. From 1958 to present. Monthly. Inquire about cost and availability. Also available on CD-ROM and online.

CAD-CAM Abstracts (formerly *CAD-CAM Profiles*). Techgnosis Ltd., Blade House, Battersea Road, Stockport, Cheshire 3AE, England. Telephone 061-442-2639. FAX 061-443-1162. Six times a year. Inquire for price.

CAD/CAM Abstracts and CAD/CAM Abstracts Annual. R.R. Bowker, Bowker A&I Publishing, 121 Chanlon Road, New Providence, NJ 07974. (908) 771-7714. Fax (908) 771-7725. 1984 to 1992. Ceasedpublication.

Computer Abstracts, MCBUniversity Press Ltd., PO Box 10812, Birmingham, AL 35201-0812. (800) 633-4931. FAX (205) 995-1588. Monthly. Covers computer theory, data, hardware, systems, networks, human-computer interaction, artificial intelligence, as well as applications of computers in aerospace, business, CAD/CAM, cartography, civil engineering, electronics and electrical engineering, industrial engineering, mechanical engineering, medicine, structural engineering, etc. Also available on CD-ROM.

Computer & Control Abstracts (inspec). INSPEC/IEEE, Box 1331, 445 Hoes Lane, Piscataway, NJ 08855-1331. (908) 562-5549. Abstracts organized by subjects of international technological information. Monthly. $1455.00 per year. Also available on CD-ROM and online via BRS Online Products, DIALOG Information Services, Orbit Search Service, and STN International.

Computer and Information Systems Abstracts Journal. Cambridge Scientific Abstracts, 7200 Wisconsin Avenue, Bethesda, MD 20814. (301) 961-6750. Fax (301) 961-6720. 1962 to present. $1265.00 per year. Also available online via STN International.

Computer Literature Index. Applied Computer Research Inc., Box 82266, Phoenix, AZ 85071-2266. (800) 234-2227. 1971 to present. Quarterly plus annual cumulation. $198.50 per year. Bibliography of books, articles, and reports.

Computing Journal Abstracts. Techgnosis Ltd., Blade House, Battersea Road, Stockport, Cheshire 3AE, England. Telephone 061-442-2639. FAX 061-443-1162. 1969 to present. Monthly. Inquire for price.

Computing Reviews. Association for Computing Machinery, 1515 Broadway, 17th Floor, New York, NY 10036-5701. (212) 869-7440. FAX (212) 944-1318. 1960 to present. Monthly. $130.00 per year for non-members. Also available online via DIALOG Information Services.

Current Contents: Engineering, Technology, and Applied Sciences. Institute for Scientific Information, 3501 Market Street, Philadelphia, PA 19104. (215) 386-0100. FAX (215) 386-6362. 1970 to present. Weekly. $442.00 per year.

Engineering Index Monthly. Engineering Information, Inc., Castle Point on the Hudson, Hoboken, NJ 07030. (800) 221-1044. FAX (212) 832-1857. Monthly. $2300.00 per year. Also available online as COMPENDEX, and also on CD-ROM. Covers chemical engineering, computers, electrical engineering, civil engineering, metals and mining, industrial management, and mechanical engineering.

Science Citation Index. Institue for Scientific Information, 3501 Market Street, Philadelphia, PA 19104. (215) 386-0100. FAX (215) 386-6362. Inquire about availability and cost. Also available on CD-ROM.

ASSOCIATIONS AND PROFESSIONAL SOCIETIES

American Automatic Control Council. c/o EECS, Northwestern University, 2145 Sheridan Road, Evanston, IL 60208-3118. (708) 491-3641.

Association For Computing Machinery. Association forComputing Machinery, 1515 Broadway, 17th Floor, New York, NY 10036-5701. (212) 869-7440. FAX (212) 944-1318.

Computer and Automation Systems Association of SME. 1 SME Drive, PO Box 930, Dearborn, MI 48121-0930. (313) 271-1500.

Institute of Electrical and Electronic Engineers. 345 E. 47th Street, New York, NY 10017. (212) 705-7900.

Numerical Control Society/aim Tech. 67 Alexander Drive, PO Box 12277, Research Triangle Park, NC 27709. (815) 399-8700.

Society For Computer-aided Engineering. 833 Featherstone Road, Rockford, IL 61107-6301. (815)399-8700. FAX (815) 399-7279.

BIBLIOGRAPHIES

ACM Guide To Computing Literature. Association for Computing Machinery, 1515 Broadway, 17th Floor, New York, NY 10036-5701. (212) 869-7440. FAX (212) 944-1318. 1964 to present. Annual. $190.00 for non-members. Also available online via DIALOG.

DIRECTORIES AND BIOGRAPHICAL SOURCES

CAD/CAM International Software Directory. TAB Books, P.O. Box 40, Blue Ridge Summit, PA 17294-0850. (717) 794-2191 or (800) 233-1128. FAX (717) 794-2080. 1991. Inquire for cost and availability.

Computer-Aided Design (CAD) Directory. TAB Books, P.O. Box 40, Blue Ridge Summit, PA 17294-0850. (717) 794-2191 or (800) 233-1128. FAX (717) 794-2080. 1991. Inquire for cost and availability.

Computer Graphics World Buyers Guide. PennWell Publishing Company, Advanced Technology Group, 10 Tara Blvd., 5th floor, Nashua, NH 03062-2801. (603) 891-9111. FAX (603) 891-0539. Annual. $100.00 per year.

Computing Information Directory. Hildebrandt Inc., PO Box 576, Pullman, WA 99114. 1981 to present. Annual. $199.95 per year.

Who's Who In Technology. Gale Research, 835 Penobscot Building, Detroit, MI 48226-4094. (313) 961-2242. (800) 347-4253. FAX (313) 961-6083. 1995. $195.00.

ENCYCLOPEDIAS AND DICTIONARIES

Computer Dictionary. Donald D. Spencer. 4th edition. Camelot Publishing Company, 709 SW 80th Blvd., Gainesville, FL 32607-1537. (904) 331-0952. (404) 423-9585. 1993. $24.95.

Encyclopedia of Computer Science and Engineering. Anthony Ralston & Edwin D. Reilly Jr., editors. 3rd revised edition. Van Nostrand Reinhold, 115 Fifth Avenue, New York, NY 10003. (212) 254-3232 or FAX (212) 254-9499. 1993. $125.00.

Encyclopedic Dictionary of Industrial Automation and Computer Control. David W. South. Prentice Hall, 113 Sylvan Avenue, Route 9W, Englewood Cliffs, NJ 07632. (201) 592-2000 or (800) 922-0579. 1994. Inquire for price and availability.

Microsoft Press Computer Dictionary: the Comprehensive Standard For Business, School, Library, and Home. 2nd edition. Microsoft Press, One Microsoft Way, Redmond, WA 98052-6399. (206) 936-0055. FAX (206) 823-8101. 1994. $19.95.

GENERAL WORKS

CAE: A Survey of Standards, Trends, and Tools. Stephen A. Ohr. John Wiley and Sons, Inc., 605 Third Avenue, New York, NY 10158. (800) 526-5368 or (212) 850-6000. 1990. Inquire for price.

CAE/Workstations. Architecture Technology Corporation. Elsevier Science Inc., 660 White Plains Rd., Tarrytown, NY 10159-5153. (919) 524-9200. FAX (919) 333-2444. 1991. $410.00.

Computer-Aided Production Engineering. V.C. Venkatesh and J.A. McGeough. Elsevier Science B.V., PO Box 945, Madison Square Station, New York, NY 10159-0945. (212) 633-3650. FAX (212) 633-3680. 1991. $210.50.

Computers In Engineering, 1993. American Society of Mechanical Engineers, 345 East 47th Street, New York, NY 10017-2398. (212) 705-7703. 1993. Inquire for price.

CAM (COMPUTER AIDED MANUFACTURING)

Ency. of Physical Sciences and Engineering Info. Sources

HANDBOOKS AND MANUALS

The C4 Handbook: CAD, CAM, CAE, CIM. Carl Machover. TAB Books, P.O. Box 40, Blue Ridge Summit, PA 17294-0850. (717) 794-2191 or (800) 233-1128. FAX (717) 794-2080.1989. Inquire for cost and availability.

McGraw-Hill Computing Essentials, 1993-1994. Timothy J. O'Leary, et al. McGraw-Hill Publishing Company, 1221 Avenue of the Americas, New York, NY 10020. (800) 262-4729 or (212) 512-3825. 1994. Inquire for price and availability.

ONLINE DATABASES AND CD-ROMS

Compendex Plus. Engineering Information, Inc., 345 East 47th Street, New York, NY 10017. (212) 705-7600 or (800) 221-1044. Contains citations with abstracts to worldwide literature in engineering and technology, from 1970 to present. Available on online BRS,(800) 289-4277, DIALOG, (800) 334-2564, ORBIT (800) 456-7248, and STN International, FIZ Karlsruhe, P.O. Box 2465, W-7500, Karlsruhe 1, Germany, online services. Also available on CD-ROM. Inquire as to cost and availability.

CSA Engineering. Cambridge Scientific Abstracts, 7200 Wisconsin Avenue, Suite 601, Bethesda, MD 20814. (301) 961-6750 or (800) 843-7751. Contains citations and abstracts of international periodicals and research literature covering all fields of engineering and science and technology, including computer and information science, electronics, mechanical engineering, solid state materials, 1981 to present. Available on BRS,(800) 289-4277, online service. Inquire as to cost and availability.

Current Contents Search. Institute for Scientific Information, 3501 Market Street, Philadelphia, PA 19104. (215) 386-0100. FAX (215) 386-6362. Contains citations to articles listed in the table of contents of science and technology journals. Also articles in social sciences and life sciences journals. Available on BRS,(800) 289-4277, DIALOG,(800) 334-2564, online services. Inquire as to cost and availability.

Dissertation Abstracts. University Microfilms International, 300 North Zeeb Road, Ann Arbor, MI 48106. (800) 521-0600 or (313) 761-4700. Scope includes virtually all doctoral dissertations accepted at accredited American institutions from1861 to present in 252 subject areas. Available on BRS,(800) 289-4277, DIALOG,(800) 334-2564, and OCLC EPIC,(800) 848-5878, online services. Also available on CD-ROM. Inquire as to cost and availability.

NTIS Bibliographic Database. National Technical Information Service, 5285 Port Royal Road, Springfield, Va 22161. (703) 487-4929 or FAX (703) 321-8199. Broad coverage of government-sponsored science and technology research reports, 1964 to present. Available on BRS,(800) 289-4277, DIALOG, (800) 334-2564, ORBIT, (800) 456-7248, and STN International, FIZ Karlsruhe, P.O. Box 2465, W-7500, Karlsruhe 1, Germany, online services. Also available on CD-ROM. Inquire as to cost and availability.

Scisearch. Institute for Scientific Information, 3501 Market Street, Philadelphia, PA 19104. (800) 523-1850 or (215) 386-0100. Broad multi disciplinary title and author index to the international literature of science and technology, 1974 to present. Available on DIALOG,(800) 334-2564, and ORBIT,(800) 456-7248, online services. Also available on CD-ROM. Inquire as to cost and availability.

WILSONLINE. H.W. Wilson Company, 950 University Avenue, Bronx, NY 10452. (800) 367-6770 or (212) 588-8400. Makes available online versions of the printed H.W. Wilson indexes including Applied Science and Technology Index, Business Periodicals Index, General Science Index, and Readers' Guide to Periodical Literature. Period covered is generally 1983 to present. Available on BRS,(800) 289-4277, DIALOG,(800) 334-2564, and OCLC EPIC,(800) 848-5878, online services. Also available on CD-ROM. Inquire as to cost and availability.

PERIODICALS

ACM Transactions On Graphics. Association for Computing Machinery, 1515 Broadway, 17th Floor, New York, NY 10036-5701. (212) 869-7440. FAX (212) 944-1318. 1982 to present. Quarterly. $110.00 per year.

CADalyst. Advanstar Communications Inc., 7500 Old Oak Road, Cleveland, OH 44130. (216) 826-2839. FAX (216) 891-2726. 1984 to present. 12 times a year. $59.00 per year.

CAD-CAM Update. Worldwide Videotex, Box 3273, Boynton Beach, FL 33424-3273. (407) 738-2276. 1989 to present. Monthly. Inquire for price.

CAD Systems. Kerrwil Publications Ltd., 395 Matheson Blvd. E, Mississauga, Ontario L4Z 2H2, Canada. (905) 890-1846. FAX (905) 890-5769. 1982 to present. Six times a year. $61.00 per year.

CAD User. B.T.C., 24 High Street, Beckenham, Kent BR3 1AY, England. Telephone 081-663-3818. FAX 081-663-3822. 1988 to present. Monthly. Inquire for price.

Computer-Aided Design. Butterworth-Heinemann/sub. to Turpin Transactions Ltd., Distribution Centre, Blackhorse Road, Letchworth, Herts SG6 1HN, England. Telephone 0462-672555. 1968 to present. 12 times a year. Inquire for price.

Computer-Aided Design Report. CAD-CAM Publishing Inc., 841 Turquoise Street, Suites D & E, San Diego, CA 92109-1159. (619) 488-0533. FAX (619) 488-6052. 1981 to present. Monthly. $190.00 per year.

Computer and Control Engineering Journal. Institute of Electrical Engineers, Michael Faraday House, Six Hill Way, Stevenage, Herts, SG1 2AY, England. Bi-monthly. Inquire for cost.

Computer Design. PennWell Publishing Company, Advanced Technology Group, 10 Tara Blvd., 5th floor, Nashua, NH 03062-2801. (603) 891-9111. FAX (603) 891-0539. 1962 to present. 12 times a year. $88.00 per year.

Computer Graphics World. PennWell Publishing Company, Advanced Technology Group, 10 Tara Blvd., 5th floor, Nashua, NH 03062-2801. (603) 891-9111. FAX (603) 891-0539. 1978 to present. Monthly. $48.00 per year.

Computers & Graphics. Elsevier Science Inc., 660 White Plains Rd., Tarrytown, NY 10159-5153. (919) 524-9200. FAX (919) 333-2444.1975 to present. Six times a year. $780.00 per year.

Computers and Industrial Engineering. Elsevier Science [journals], 660 White Plains Rd., Tarrytown, NY 10159-5153. (919) 524-9200. FAX (919) 333-2444. 1977 to present. 8 times a year. $780.00 per year.

Ency. of Physical Sciences and Engineering Info. Sources

CALCIUM

IEEE Computer Graphics and Applications. IEEE Computer Society Press, 10662 Los Vaqueros Circle, Box 3014, Los Alamitos, CA 90720-1264. (714) 821-8380. FAX (714) 821-4641. 1981 to present. Bi-monthly. $175.00 per year.

Integrated Computer-Aided Engineering. John Wiley & Sons, 605 Third Avenue, New York, NY 10158-0012. (212) 850-6000. FAX (212) 850-6088. 1993 to present. Bi-monthly. $75.00 per year.

RESEARCH CENTERS AND INSTITUTES

Albert H. Case Center for Computer-Aided Design. Michigan State University, College of Engineering, 112 Engineering Bldg., East Lansing, MI 48824. (517) 355-6453.

Center For Computer-Aided Design and Manufacturing, Utah State University. Mechanical Engineering Department, UMC 4130, Logan, UT 84322-4130. (801) 750-2950.

Computer-Aided Design Laboratory, Massachusetts Institute of Technology. Department of Mechanical Engineering, Bldg. 3, Room 449B, Cambridge, MA 02139. (617) 253-4465.

Computer-Aided Engineering Center, University of Wisconsin—Madison. 1500 Johnson Drive, Madison, WI 53706. (608) 263-3075.

University of Iowa, Center For Computer-aided Design. 1409 Engineering Bldg., Iowa City, IA 52242. (319) 335-5726. FAX (319) 335-6061.

University of Kansas Center For Excellence In Computer-aided Systems Engineering. 2219 Irving Hill Road, Lawrence, KS 66045. (913) 864-4896.

CAISSONS

See: CIVIL ENGINEERING

CALCIUM

See also: ANALYTICAL CHEMISTRY, CARBONATES, CHEMISTRY, ELEMENTS, GEOCHEMISTRY, INORGANIC CHEMISTRY, METAL-LURGY

ABSTRACT SERVICES AND INDEXES

Applied Science and Technology Index; A Cumulative Subject Index To English Language Periodicals In the Fields of Aeronautics and Space Science, Computer Technology, Chemistry, Construction Industry, Energy and Related Areas. H.W. Wilson Co., 950 University Avenue, Bronx, NY 10452. (800) 367-6770 or (212) 588-8400. FAX (718) 590-1617. From 1958 to present. Monthly. Inquire about cost and availability. Also available on CD-ROM and online.

Current Contents: Physical, Chemical and Earth Sciences. Institute for Scientific Information, 3501 Market Street, Philadelphia, PA 19104. (215) 386-0100. FAX (215) 386-6362. Weekly. $360.00 per year.

Index To Scientific and Technical Proceedings. Institute for Scientific Information, 3501 Market St., Philadelphia, PA 19104. (215) 386-0100. FAX (215) 386-6362. Monthly. $500.00 per year.

Index To Scientific Reviews. Institute for Scientific Information, 3501 Market St., Philadelphia, PA 19104. (215) 386-0100. FAX (215) 386-6362. Semi-annual.

Science Citation Index. Institue for Scientific Information, 3501 Market Street, Philadelphia, PA 19104. (215) 386-0100. FAX (215) 386-6362. Inquire about availability and cost. Also available on CD-ROM.

ANNUAL REVIEWS AND YEARBOOKS

Calcium and Calcium Compounds: Annual Report. U.S. Department of the Interior, Bureau of Mines. Order from: Superintendent of Documents, U.S. Government Printing office, Box 371954, Pittsburgh, PA 15250-7954. (202) 783-3238. FAX (202) 512-2233. Annual. Inquire for cost and availability.

ASSOCIATIONS AND PROFESSIONAL SOCIETIES

American Chemical Society. 1155 16th Street NW, Washington, DC 20036. (800) 333-9511. FAX (614) 447-3671.

American Institute of Chemical Engineers. 345 East 47th Street, New York, NY 10017-2396. (212) 705-7663.

Metallurgical Society of the AIME (American Institute of Mining, Metallurgical, and Petroleum Engineers). 345 E.47th Street, 14th Floor, New York, NY 10017. (212) 705-7695.

American Society For Metals. ASM International, Materials Park, OH 44073-0002. (216) 338-5151 or (800) 336-5152. FAX (216) 338-4634.

Iron and Steel Society. 410 Commonwealth Drive, Warrendale, PA 15086. (412) 776-9460. FAX (412) 776-0430.

DIRECTORIES AND BIOGRAPHICAL SOURCES

American Men & Women of Science, 1995-96. R.R. Bowker Staff, eds. 19th edition. 8 volumes. R.R. Bowker/Reed International Publishing Company, 121 Chanlon Road, New Providence, NJ 07974.(908) 464-6800 or (800) 521-8110. 1995. $850.00

Directory of Engineering Societies and Related Organizations. Gordon Davis. 13th edition. American Association of Engineering Societies, 1111 19th Street NW, Suite 608, Washington, DC 20036. (202) 296-2237 or (800) 658-8897. 1989. Inquire for price.

International Research Centers Directory. Gale Research, 835 Penobscot Building, Detroit, MI 48226-4094. (313) 961-2242. (800) 347-4253. FAX (313) 961-6083. 20th edition. 1995. $485.00.

Research Centers Directory. Gale Research, 835 Penobscot Building, Detroit, MI 48226-4094. (313) 961-2242. (800) 347-4253. FAX (313) 961-6083. $485.00.

CALCIUM

Ency. of Physical Sciences and Engineering Info. Sources

Scientific and Technical Organizations and Agencies Directory. Gale Research, 835 Penobscot Building, Detroit, MI 48226-4094. (313) 961-2242. (800) 347-4253. FAX (313) 961-6083. 4th edition. 1996. $195.00.

Who's Who In Engineering. American Association of Engineering Societies, 1111 19th Street NW, Suite 608, Washington, DC 20036. (202) 296-2237 or (800) 658-8897. 8th edition. 1991. Inquire for price.

ENCYCLOPEDIAS AND DICTIONARIES

McGraw-hill Encyclopedia of Science and Technology. Sybil P. Parker, ed. 7th edition. 20 volumes. McGraw-Hill Publishing Company, 1221 Avenue of the Americas, New York, NY 10020. (800) 262-4729 or (212) 512-3825. 1992. $1900.00

Thesaurus of Scientific, Technical, and Engineering Terms. Hemisphere Publishing Corporation, 1900 Frost Road, Suite 101, Bristol, PA 19007-1598. (215) 785-5800 or (800) 821-8312. FAX (215) 785-5515. 1987. $173.00.

GENERAL WORKS

Calcium In Human Biology. B.E.C. Nordin, editor. Springer-Verlag, 175 Fifth Avenue, New York, NY 10010. (212) 460-1500 or (800) 777-4643. FAX (212) 473-6272. 1988. $133.00.

Calcium Metallurgy and Technology. C.L. Mantell & Charles Hardy. NY: Reinhold Publishing Corporation, 1945. Out of print.

Metals In Biological Systems. M.J. Kendrick, et al. E. Horwood Publishing, distributed by Prentice Hall, 113 Sylvan Avenue, Route 9W, Englewood Cliffs, NJ 07632. (201) 592-2000 or (800) 922-0579. 1992. Inquire for cost and availability.

Techniques In Calcium Research. M.V. Thomas. Academic Press Inc., 6277 Sea Harbor Drive, Orlando, FL 32887. (800) 321-5068. 1982. $82.00.

HANDBOOKS AND MANUALS

CRC Handbook of Chemistry and Physics. David R. Lide, editor. 75th edition. CRC Press, 2000 Corporate Blvd., N.W., Boca Raton, FL 33431. (407) 994-0555 or (800) 272-7737. FAX (407) 994-0949. $99.50.

Lange's Handbook of Chemistry. John A. Dean. 14th edition. McGraw-Hill Publishing Company, 1221 Avenue of the Americas, New York, NY 10020. (800) 262-4729 or (212) 512-3825. 1992. $79.50.

ONLINE DATABASES AND CD-ROMS

Compendex Plus. Engineering Information, Inc., 345 East 47th Street, New York, NY 10017. (212) 705-7600 or (800) 221-1044. Contains citations with abstracts to worldwide literature in engineering and technology, from 1970 to present. Available on online BRS,(800) 289-4277, DIALOG, (800) 334-2564, ORBIT (800) 456-7248, and STN International, FIZ Karlsruhe, P.O. Box 2465, W-7500, Karlsruhe 1, Germany, online services. Also available on CD-ROM. Inquire as to cost and availability.

Dissertation Abstracts. University Microfilms International, 300 North Zeeb Road, Ann Arbor, MI 48106. (800) 521-0600 or (313) 761-4700. Scope includes virtually all doctoral dissertations accepted at accredited American institutions from 1861 to present in 252 subject areas. Available on BRS,(800) 289-4277, DIALOG,(800) 334-2564, and OCLC EPIC,(800) 848-5878, online services. Also available on CD-ROM. Inquire as to cost and availability.

NTIS Bibliographic Database. National Technical Information Service, 5285 Port Royal Road, Springfield, Va 22161. (703) 487-4929 or FAX (703) 321-8199. Broad coverage of government-sponsored science and technology research reports, 1964 to present. Available on BRS,(800) 289-4277, DIALOG, (800) 334-2564, ORBIT, (800) 456-7248, and STN International, FIZ Karlsruhe, P.O. Box 2465, W-7500, Karlsruhe 1, Germany, online services. Also available on CD-ROM. Inquire as to cost and availability.

Scisearch. Institute for Scientific Information, 3501 Market Street, Philadelphia, PA 19104. (800) 523-1850 or (215) 386-0100. Broad multi disciplinary title and author index to the international literature of science and technology, 1974 to present. Available on DIALOG,(800) 334-2564, and ORBIT,(800) 456-7248, online services. Also available on CD-ROM. Inquire as to cost and availability.

PERIODICALS

American Chemical Society Journal. American Chemical Society, 1155 16th Street NW, Washington, DC 20036. (800) 333-9511. FAX (614) 447-3671. 1879 to present. Bi-weekly. $1055.00 per year for non-members.

Inorganic Chemistry. American Chemical Society, 1155 16th Street NW, Washington, DC 20036. (800) 333-9511. FAX (614) 447-3671. 1962 to present. Semi-monthly. $947.00 per year for non-members.

Inorganica Chimica Acta. Elsevier Science Inc., Box 882, Madison Square Station, New York, NY 10159. (212) 989-5800. 1967 to present. 26 times a year. $3361.00 per year.

Journal of the Chemical Society: Chemical Communications. Royal Society of Chemistry, Information Services, Thomas Graham House, Science Park, Milton Road, Cambridge, CB4 4WF, England. Telephone 0223-420066. FAX 0223-423623. 1965 to present.Semi-monthly. $807.00 per year.

Polyhedron. Elsevier Science [journals], 660 White Plains Rd., Tarrytown, NY 10159-5153. (919) 524-9200. FAX (919) 333-2444. 1982 to present. 24 times a year. $2175.00 per year.

Royal Society of Chemistry Journal: Dalton Transactions. Royal Society of Chemistry, Information Services, Thomas Graham House, Science Park, Milton Road, Cambridge, CB4 4WF, England. 1972 to present. 24 times a year. $1509.00 per year.

RESEARCH CENTERS AND INSTITUTES

Chemical Laboratories, Harvard University. Oxford Street, Cambridge, MA 02138. (617) 495-4283. FAX (617) 495-8824.

University/industry Chemical Research Center, Mississippi State University. Department of Chemistry, PO Drawer CH, Mississippi State, MS 39762. (601) 325-3584. FAX (601) 325-7611.

Ency. of Physical Sciences and Engineering Info. Sources

CALCULUS

CALCULUS

See also: ALGEBRA, DIFFERENTIAL EQUATIONS, GEOMETRY, MATHEMATICS, STATISTICS

ABSTRACT SERVICES AND INDEXES

Applied Mechanics Reviews: An Assessment of World Literature In Engineering Sciences. American Society of Mechanical Engineers, 345 East 47th Street, New York, NY 10017. (212) 705-7703. 1948 to present. Monthly. $360.00 per year. ISSN: 0003-6900.

Applied Science and Technology Index; A Cumulative Subject Index To English Language Periodicals In the Fields of Aeronautics and Space Science, Computer Technology, Chemistry, Construction Industry, Energy and Related Areas. H.W. Wilson Company, 950 University Avenue, Bronx, NY 10452. (212) 588-8400. (800) 367-6770. FAX (718) 590-1617. 1958 to present. Monthly. Available on-line from BRS and WILSONLINE. Also available on CD-ROM. Inquire about cost and availability. ISSN: 0003-6986.

Compactmath - Compact Mathematics Library. Cumulative CD-ROM edition of Zentralblatt fuer Mathematik - Mathematics Abstracts. Springer-Verlag New York, Inc., 44 Hartz Way, Secaucus, NJ 07096-2491. (201) 348-4033. FAX (201) 348-4505. 1993 to present. Annual. Available only on CD-ROM. Inquire about cost and availability. ISSN: 0938-3174.

Compumath Citation Index. Institute for Scientific Information, 3501 Market Street, Philadelphia, PA 19104. (800) 523-1850 or (215) 386-0100. FAX (215) 386-2991. Three times per year. $1955.00. Also available online and on CD-ROM. Inquire regarding cost and availablity. ISSN: 0730-6199.

Current Mathematical Publications. American Mathematical Society. P.O. Box 1571, Annex Station, Providence, RI 02901-9930. (800) 556-7774 or (401) 455-4000. FAX (401) 331-3842. 1969 to present. Seventeen times per year. $377.00 per year. Available online from BRS, DIALOG, European Space Agency. Also available on CD-ROM. Inquire regarding cost and availability. ISSN: 0361-4794.

General Science Index. H.W. Wilson Company, 950 University Avenue, Bronx, NY 10452. (212) 588-8400. (800) 367-6770.FAX (718) 590-1617. From 1978 to present. Ten issues per year; quarterly and annual cumulations. Service basis. Available on CD-ROM and online. Inquire regarding cost and availability. ISSN: 0162-1963.

Index To Mathematical Tables. Alan Fletcher. Addison- Wesley Publishing Co., Inc., 1 Jacob Way, Reading, MA 01867. (617) 944-3700. (800) 447-2226. 2nd edition. 1962. 2 volumes.

Mathematical Reviews: A Review Journal Covering the World Literature of Mathematical Research. American Mathematical Society, P.O. Box 1571, Annex Station, Providence, RI 02901-9930. (800) 556-7774 or (401) 455-4000. 1940 to present. Monthly. $4594.00 per year. Also available via network (MathSciNet), online and on CD-ROM. Inquire regarding cost and availability. ISSN: 0025-5629.

NTIS Alerts: Mathematical Sciences. U.S. National Technical Information Service, 5285 Port Royal Road, Springfield, VA 22161. (703) 487-4650. FAX (703) 321-8547. Weekly. $140.00 per year.

Science Citation Index. SCI. Institute for Scientific Information, 3501 Market Street, Philadelphia, PA 19104. (215) 386-0100. (800) 523-1850. FAX (215) 386-2991. 1961 to present. Six issues per year, plus annual cumulation. $11650.00 per year. Also available online and on CD-ROM. Inquire about price and availability. ISSN: 0036-827X.

Zentralblatt Fuer Mathematik Und Ihre Grenzebiete /mathematics Abstracts. Heidelberger Akademie der Wissenschaften/ Springer-Verlag New York, Inc., 44 Hartz Way, Secaucus, NJ 07096-2491. (201) 348-4033. FAX (201) 348-4505. 1931 to present. 30 issues per year. DM 8340. ISSN: 0044-4235. Also available online and on CD-ROM. Inquire regarding cost and availability

ANNUAL REVIEWS AND YEARBOOKS

Advances In Applied Mathematics. Academic Press, Inc., 6277 Sea Harbor Drive, Orlando, FL 32821-8340. (800) 543-9534. 1980 to present. Irregular. Price varies; inquire. ISSN: 0196-8858.

Applied Mathematical Sciences. Springer-Verlag, 175 Fifth Avenue, New York, NY 10010. (212) 460-2500. FAX (212) 473- 6272. 1972 to present. Irregular. ISSN: 0066-5452.

C B M S -N S F Regional Conference Series In Applied Mathematics. Society for Industrial and Applied Mathematics, 3600 University City Science Center, Philadelphia, PA 19104-2688. (215) 382-9800. FAX (215) 386-7999. 1971 to present. Irregular. ISSN: 0163-9439.

Ergebnisse Der Mathematik Und Ihrer Grenzgebiete. Neue folge. Springer-Verlag New York, Inc. 175 Fifth Avenue, New York, NY 10010. (212) 460-1500. FAX (212) 473-6272. 1953 to present. Irregular. Price varies; inquire. ISSN: 0071-1136.

Lecture Notes In Mathematics. Springer-Verlag New York, Inc. 175 Fifth Avenue, New York, NY 10010. (212) 460-1500. FAX (212) 473-6272. 1964 to present. Irregular. Price varies; inquire. ISSN: 0075-8434.

ASSOCIATIONS AND PROFESSIONAL SOCIETIES

American Mathematical Society. P.O. Box 6248, Providence, RI 02940. (401) 455-4000. FAX (401) 331-3842.

Association For Symbolic Logic. University of Illinois, Department of Mathematics, 1409 West Green Street, Urbana IL 61801. (217) 244-7902. FAX (217) 333-9576.

Association For Women In Mathematics. 4114 Computer and Space Science Building, University of Maryland, College Park, MD 20742-2461. (301) 405-7892.

Mathematical Association of America. 1529 18th Street, NW, Washington, DC 20036. (202) 387-5200. FAX (202) 265-2384.

National Council of Teachers of Mathematics. 1906 Association Drive, Reston, VA 22091. (703) 620-9840. FAX (703) 475-2970.

Society For Industrial and Applied Mathematics. 3600 University City Science Center, Philadelphia, PA 19104-2688. (215) 382-9800. FAX (215) 386-7999.

CALCULUS

Ency. of Physical Sciences and Engineering Info. Sources

Special Interest Group On Numerical Mathematics (signum). c/o ACM. 1515 Broadway, New York, NY 10036-5701. (212) 869-7440. FAX (212) 944-1318.

Special Interest Group For Symbolic and Algebraic Manipulation (SIGSAM). c/o ACM. 1515 Broadway, New York, NY 10036-5701. (212) 869-7440. FAX (212) 944-1318.

BIBLIOGRAPHIES

Bibliography of Early Modern Algebra 1500-1800. Robin E. Rider. University of California. office for History of Science & Technology, 543 Stephens Hall, Berkeley, CA 94720- 2350. (510) 642-4581. 1982. $8.00. ISBN: 0-918102-08-1

Books You Can Count On: Linking Mathematics & Literature. Rachel Griffiths and Margaret Clyne. Heinemann Publ. 361 Hanover Street, Portsmouth, NH 03801-3912. (603) 431-7894. (800) 541-2086. 1991. $15.00. ISBN: 0-435-08322-8.

Mathematical Book Review Index, 1800 1940. Louise S. Grinstein. Garland Publ. Fifth Avenue, Suite 2500, New York, NY 10022-8101. (212) 751-7447. 1992. $72.00. ISBN: 0-8240-4114-3.

Bibliography of Mathematical Works Printed In America Through 1850. Louis C. Karpinski. Supplements 1 & 2, Bernard I. Cohen, editor. Ayer Company Publ. Inc. Lower Mill Road, North Stratford, NH 03590. (603) 922-5105. (800) 282-5413. Reprint edition. 1980. $66.95. ISBN: 0-405-12553-4.

International Mathematical Olympiads 1978-1985. Murray S. Klamkin. National Council of Teachers of Mathematics. 1906 Association Drive, Reston, VA 22091. (703) 620-9840. (800) 235-7566. 1987. $11.75. ISBN: 0-88385-631-X.

Mathematical Journals: An Annotated Guide. Diana F. Liang. Scarecrow Press. Distributed by University Press of America, 4720 Boston Way, Lanham, MD 20706. (800) 462-6420. 1987. $29.50. ISBN: 0-8108-2585-6.

DIRECTORIES AND BIOGRAPHICAL SOURCES

Assistantships and Graduate Fellowships in theMathematical Sciences. American Mathematical Society, P.O. Box 6248, Providence, RI 02940. (401) 455-4000. (800) 321-4267. Annual. $19.00. ISSN: 1040-7650.

Combined Membership List. American Mathematical Society, P.O. Box 6248, Providence, RI 02940. (401) 455-4000. (800) 321-4267. Annual. ISSN: $50.00.

Mathematical Science Professional Directory. American Mathematical Society, P.O. Box 6248, Providence, RI 02940. (401) 455-4000. (800) 321-4267. Annual. $45.00. ISBN: 0- 8218-0173-2.

Research Centers Directory. Gale Research, 835 Penobscot Building, Detroit, MI 48226-4094. (313) 961- 2242. (800) 347-4253. 20th edition. 1995. Annual. $485.00. ISSN: 0080-1518.

World Directory of Mathematicians. G.D. Mostow, editor. International Mathematical Union, Helsinki. 1958 to present. 9th revised edition. 1990. $40.00. Available from: American Mathematical Society, P.O. Box 6248, Providence, RI 02940. (401) 455-4000. ISSN: 0512-2740.

ENCYCLOPEDIAS AND DICTIONARIES

Beyond Numeracy: An Uncommon Dictionary of Mathematics. John A. Paulos. Alfred A. Knopf Inc., 201 East 50th Street, New York, NY 10022. (800) 733-3000. 1991. $21.50. ISBN: 0-394-58640-9.

Concise Oxford Dictionary of Mathematics. Christopher Clapham. Oxford University Press, 200 Madison Avenue, New York, NY 10016. (212) 726-6000. 1990. $10.95. ISBN: 0-19-286103-4.

Encyclopedia of Mathematics. An updated and annotated translation of the Soviet Mathematical Encyclopedia. Michael Hazewinkel, editor Kluwer Academic Publishers, 101 Philip Drive, Assinippi Park, Norwell, MA 02061. (617) 871-6600. 1987 - . 10 volumes. $147.00, volume 1. 1-55608-010-7 set.

Encyclopedia of Mathematics and Its Applications. Addison-Wesley Publishing Co., Inc., 1 Jacob Way, Reading, MA 01867. (617) 944-3700. (800) 447-2226. 1976 to present. irregular. Volumes are individually priced.

Encyclopedic Dictionary of Mathematics. Kiyosi Ito, editor. MIT Press, 55 Hayward Street, Cambridge, MA 02142. (617) 625-8569. 2nd edition, reprinted. 1993. 2 volumes. $70.00. ISBN: 0-262-59020-4.

Harpercollins Dictionary of Mathematics. E. J. Borowski and J. J. Borwein. HarperCollins Publ. 10 East 53rd Street, New York, NY 10022-5299. (800) 331-3761. 1990. $10.95. ISBN: 0-19-286103-4.

Mathematics Dictionary. Robert C. James, editor. Van Nostrand Reinhold, 115 Fifth Avenue, New York, NY 10003. (800)-842-3636. 5th edition. 1992. $42.95. ISBN: 0-442- 00741-8.

Penguin Dictionary of Mathematics. J. Daintith and R. D. Nelson. Viking Penguin, 375 Hudson Street, New York, NY 10014-3657. (800) 331-4624. 1989. $12.00. ISBN: 0-14- 051119-9.

The Words of Mathematics: An Etymological Dictionary of Mathematical Terms Used In English. Steven Schwartzman. Mathematical Association of America, 18th Street NW, Washington, DC20036. (800) 331-1622. 1994. $29.50. ISBN: 0-88385-551.

GENERAL WORKS

Calculus and its Applications. Larry J. Goldstein. Prentice-Hall, 113 Sylvan Avenue, Route 9W, Englewood Cliffs, NJ 07632. (201) 592-2000. (800) 922-0579. 7th edition. 1995. $58.67.

Calculus: One and Several Variables. Robert Ellis and Denny Gulick. SCP: Third World Literature Publishing House, P.O. Box 482, Lithonia, GA 30058-0482. (404) 785-7725. 1990. $40.00.

Calculus Problem Solver. Research and Education Association, 61 Ethel Road, West Piscataway, NJ 08854. (908) 819-8880. Revised edition. 1994. $23.95.

Calculus With Analytical Geometry. Robert Ellis and Dennis Gulick. HB Coll Pubs . 1990. $28.50 in paper. $5.00 for accompanying disk.

A Century of Calculus, 1894-1968. Mathematical Association of America, 1529 18th Street, NW, Washington, DC 20036. (202) 387-5200. FAX (202) 265-2384. Parts I and II. 1992. $39.00 each, in paper.

Excursions In Calculus: An Interplay of the Continuous and the Discrete. Robert M. Young. Mathematical Association of America, 1529 18th Street, NW, Washington, DC 20036. (202) 387-5200. FAX (202) 265-2384. 1992. $42.00.

A First Course In Calculus. Serge A. Lang. Addison-Wesley Publishing Co., Inc., 1 Jacob Way, Reading, MA 01867. (617) 944-3700. (800) 447-2226. 5th edition, reprinted. 1993. $39.95.

Introduction To the Calculus of Variations. Hans Sagan. Revised edition. Dover Publications, Inc., 180 Varick Street, New York, NY 10014. (212) 255-3755. (800) 223- 3130. 1992. $12.95.

Precalculus. Robert Ellis and Dennis Gulick. SCP: Third World Literature Publishing House, P.O. Box 482, Lithonia, GA 30058-0482. (404) 785-7725. 4th edition. 1992. $51.00.

Schaum's Outline of Beginning Calculus. Elliot Mendleson. McGraw-Hill Publishing Company, Inc., 1221 Avenue of the Americas, New York, NY 10020. (212) 512-2000. (800) 262-4729. 1985. $12.95 in paper.

Schaum's Outline of Calculus. Frank Ayres Jr. and Elliot Mendleson. McGraw-Hill Publishing Company, Inc., 1221 Avenue of the Americas, New York, NY 10020. (212) 512-2000. (800) 262-4729. 3rd edition. 1990. $12.95.

Schaum's Three Thousand Solved Problems In Calculus. Elliot Mendleson. McGraw-Hill Publishing Company, Inc., 1221 Avenue of the Americas, New York, NY 10020. (212) 512-2000. (800) 262-4729. Revised edition. 1992. $19.95.

Selected Papers On Precalculus. T. Apostol. Mathematical Association of America, 1529 18th Street, NW, Washington, DC 20036. (202) 387-5200. FAX (202) 265-2384. 199 . $10.00.

Mathematica: A System For Doing Mathematics By Computer. Stephen Wolfram. Addison-Wesley Publishing Company, Inc., 1 Jacob Way, Reading, MA 01867. (617) 944-3700. (800) 447-2226. 2nd edition. 1991. ISBN: 0-201-51502-4.

Writing Mathematics Well. Leonard Gillman. Mathematical Association of America, 18th Street NW, Washington, DC 20036. (800) 331-1622. 1987. $8.00. ISBN: 0-88385-443-0

HANDBOOKS AND MANUALS

Handbook of Boolean Algebras. J. D. Monk editor. Elsevier Science, Inc., 655 Avenue of the Americas, New York, NY 10010. (212) 989-5800. 1989. $72.00. 3 volumes. ISBN: 0- 444-70261-X.

Handbook of Mathematical Sciences. William H. Beyer, editor. CRC Press, 2000 Corporate Boulevard, Boca Raton, FL 33431. (800) 272-7737. 6th edition. 1987. $91.95. ISBN: 0-8493-0656-6.

Handbook of Tables For Probability and Statistics. William H. Beyer, editor. CRC Press, 2000 Corporate Boulevard, Boca Raton, FL 33431. (800) 272-7737. 2nd edition. 1968. $92.95. ISBN: 0-8493-0692-2.

Handbook of Writing For the Mathematical Sciences. Society for Industrial and Applied Mathematics, 1993. $21.50. ISBN: 0-89871-314-5.

Standard Mathematical Tables and Formulas. William H. Beyer, editor. CRC Press, 2000 Corporate Boulevard, Boca Raton, FL 33431. (800) 272-7737. 29th edition. 1991. $41.95. ISBN: 084-930- 6299.

Handbook of Applicable Mathematics. Walter Ledermann, editor. John Wiley & Sons, Inc., 605 Third Avenue, New York, NY 10158-0012. (212) 850-6000. (800) 225-5945. 1985. $1496.00 6 volumes. ISBN: 0-471-90804-5.

Handbook of Applied Mathematics; Selected Results and Methods. Carl E. Pearsen, editor. Chapman & Hall, 1 Penn Plaza, New York NY 10119. (212) 564-1060. 1990. $46.95. ISBN: 0-442-00521-0.

Handbook of Mathematical Functions With Formulas, Graphs, and Mathmetical Tables. Milton Abramowitz and Irene A. Stegun, editors. U.S. National Bureau of Standards. U.S. Government Printing office, Washington, D.C. 10th edition. 1972. $43.00. ISBN: 0-16-000202-8.

Handbook of Mathematics. I. N. Bronshtein and K. A. Semedyayev. Van Nostrand Reinhold, 115 Fifth Avenue, New York, NY 10003. (800)-842-3636. 20th edition. $59.95. ISBN: 0-442-21171-6.

Tables of Integrals, Series and Products: Corrected and Enlarged Edition. I. S. Gradshteyn and I. M. Ryzhik. Academic Press, Inc., 6277 Sea Harbor Drive, Orlando, FL 32821. (800) 321-5068. 5th edition. 1993. $61.00. ISBN: 0-12-294755-X.

ONLINE DATABASES AND CD-ROMS

CA Search. Chemical Abstracts Service, P.O. Box 3012, Columbus, OH 43210 -0012. (614) 447-3600. (800) 848-6533. FAX (614) 447-3709. Very comprehensive guide to worldwide chemical literature and related fields, 1972 to present. Available on BRS,(800) 289-4277, DIALOG, (800) 334-2564, ORBIT (800) 456-7248, and STN International, FIZ Karlsruhe, P.O. Box 2465, W-7500, Karlsruhe 1, Germany, online services. Inquire as to cost and availability.

Compendex Plus. Engineering Information, Inc., 345 East47th Street, New York, NY 1001 7. (212) 705-7 600 or (800) 221-1044. Contains citation s with abstra cts to worldwide literature in engineering and technology, from 1970 to present. Available on online BRS,(800) 289-4277, DIALOG, (800) 334-2564, ORBIT (800) 456-7248, and STN International, FIZ Karlsruhe, P.O. Box 2465, W-7500, Karlsruhe 1, Germany, online services. Also available on CD-ROM. Inquire as to cost and availability.

Dissertation Abstracts. University Microfilms International, 300 North Zeeb Road, Ann Arbor, MI 48106. (800) 521-0600 or (313) 761-4700. Scope includes virtually all doctoral dissertations accepted at accredited American institutions from 1861 to present in 252 subject areas. Available on BRS,(800) 289-4277, DIALOG,(800) 334-2564, and OCLC EPIC,(800) 848-5878, online services. Also available on CD-ROM. Inquire as to cost and availability.

Inspec. Institution of Electrical Engineers, Michael Faraday House, Six Hills Way, Stevenage, Herts. SG1 2AY, England. Telephone: 0438 313311 or FAX 0438 742840. Contains cita-

CALCULUS

Ency. of Physical Sciences and Engineering Info. Sources

tions to the worldwide literature of physics, electronics and electrical engineering, computer technology, and related fields. Available on BRS, (800) 289-4277, DIALOG, (800) 334-2564, ORBIT, (800) 456-7248, and STN International, FIZ Karlsruhe, P.O. Box 2465, W-7500, Karlsruhe 1, Germany, online services. Inquire as to cost and availability.

Mathsci and Mathscinet. American Mathematical Society, P.O. Box 6248, Providence, RI 02940. (800) 321-4667 or (401) 455-4000 or FAX (401) 331-3842. Scope includes pure and applied mathematics and related areas of physics, statistics, engineering, computer science, and operations research literature since 1959. Available on DIALOG, (800) 334-2564, online service. Also available on CD-ROM and MATHSCINET, on the Internet. Inquire as to cost and availability.

NTIS Bibliographic Database. National Technical Information Service, 5285 Port Royal Road, Springfield, Va 22161. (703) 487-4929 or FAX (703) 321-8199. Broad coverage of government-sponsored science and technology research reports, 1964 to present. Available on BRS, (800) 289-4277, DIALOG, (800) 334-2564, ORBIT, (800) 456-7248, and STN International, FIZ Karlsruhe, P.O. Box 2465, W-7500, Karlsruhe 1, Germany, online services. Also available on CD-ROM. Inquire as to cost and availability.

Scientific and Technical Books and Serials In Print. R.R. Bowker Inc., 121 Chanlon Road, New Providence, NJ 07974. (800) 521-8110 or (908) 464-6800. List of currently published books and serials in the physical and biological sciences, engineering and technology, with subject, author and titles indexes. Available on ORBIT, (800) 456-7248, online service. Inquire as to cost and availability.

Scisearch. Institute for Scientific Information, 3501 Market Street, Philadelphia, PA 19104. (800) 523-1850 or (215)386-0100. Broad multidisciplinary title and author index to the international literature of science and technology, 1974 to present. Available on DIALOG, (800) 334-2564, and ORBIT,(800) 456-7248, online services. Also available on CD-ROM. Inquire as to cost and availability.

WILSONLINE. H.W. Wilson Company, 950 University Avenue, Bronx, NY 10452. (800) 367-6770 or (212) 588-8400. Makes available online versions of the printed H.W. Wilson indexes including Applied Science and Technology Index, Business Periodicals Index, General Science Index, and Readers' Guide to Periodical Literature. Period covered is generally 1983 to present. Available on BRS,(800) 289-4277, DIALOG,(800) 334-2564, and OCLC EPIC,(800) 848-5878, online services. Also available on CD-ROM. Inquire as to cost and availability.

PERIODICALS

American Journal of Mathematics. Johns Hopkins University Press, 2715 North Charles Street, Baltimore, Md 21218. (410) 516-6987. FAX (410) 516-6968. 1878 to present. Bimonthly. $162.00 per year. ISSN: 0002-9327.

American Mathematical Monthly. Mathematical Association of America. 1529 Eighteenth Street, NW, Washington, DC 20036. (202) 387-5200. 1894 to present. Ten issues per year. $160.00. ISSN: 0002-9890.

Calculus of Variations and Partial Differential Equations. Springer-Verlag New York, Inc., 44 Hartz Way, Secaucus, NY 07096-2491. (201) 348-4033. FAX (201) 348-4505. 1993 to present. Quarterly. $300.00. ISSN: 0944-2669.

Canadian Applied Mathematical Quarterly. Canadian Mathematical Society. Distributed by Rocky Mountain Mathematics Consortium, Department of Mathematics, Arizona State University, Tempe AZ 85287-1904. 1992 to present. $175.00 per year. ISSN: 1073-1849.

College Mathematics Journal. Mathematical Association of America, 1529 18th Street NW, Washington, DC 20036. (202) 387-5200. 1970 to present. Five issues per year. $98.00. ISSN: 0746-8342.

Communications On Pure and Applied Mathematics. Courant Institute of Mathematical Sciences. Distributed by John Wiley & Sons, Inc. 605 Third Avenue, New York, NY 10158- 0012. (212) 850-6000. FAX (212) 850-6088. 1939 to present. $730.00 per year. ISSN: 0010-3640.

Geometric and Functional Analysis. Birkhauser- Verlag/ Springer-Verlag, Mercedes Distribution Center, 160 Imlay Street, Brooklyn, Ny 11231. 1992 to present. Six issues per year. $278.00. ISSN: 1016-443X.

I M A Journal of Numerical Analysis. Oxford University Press, 2001 Evans Road, Cary, NC 27513. (919) 677-0977. 1981 to present. Quarterly. $270.00 per year. ISSN: 0272- 4979..

Journal of Multivariate Analysis. Academic Press, Inc., 525 B Street, Suite 1900. San Diego, CA 92101-4495. (619) 230- 1840. FAX (619) 699-6800. 1971 to present. Eight issues per year.$654.00 per year. ISSN: 0047-259X.

Mathematical Intelligencer. Springer-Verlag, 175 Fifth Avenue, New York, NY 10010. (212) 460-1500. FAX (212) 473- 6272. (800) 777-4643. 1978 to present. Quarterly. $39.00 per year.

Siam Journal On Applied Mathematics. Society for Industrial and Applied Mathematics, 3600 University City Science Center, Philadelphia, PA 19104-2688. (215) 382- 9800. FAX (215) 386-7999. 1953 to present. Bimonthly. $242.00 per year. ISSN: 0036-1399.

Siam Journal On Mathematical Analysis. Society for Industrial and Applied Mathematics, 3600 University City Science Center, Philadelphia, PA 19104-2688. (215) 382- 9800. FAX (215) 386-7999. 1970 to present. Bimonthly. $352.00 per year. ISSN: 0036-1410.

Siam Journal On Numerical Analysis. Society for Industrial and Applied Mathematics, 3600 University City Science Center, Philadelphia, PA 19104-2688. (215) 382-9800. FAX (215) 386-7999. Bimonthly. $255.00 per year. ISSN: 0036- 1429.

Siam Review. Society for Industrial and Applied Mathematics, 3600 University City Science Center, Philadelphia, PA 19104-2688. (215) 382-9800. 1959 to present. Quarterly. $148.00 per year. ISSN: 0036-1445.

RESEARCH CENTERS AND INSTITUTES

Center For Applied Mathematics. Cornell University. 305 Sage Hall, Ithaca, NY 14853-6201. (607) 255-4335. FAX (607) 255-4336.

Center For Mathematical Science Research. Rutgers University. New Brunswick, NJ 08903. (201) 932-3117. FAX (201) 932-5530.

Center For Mathematical Sciences. University of Wisconsin-Madison. 610 Walnut Street, Madison, WI 53705. (608) 263-2696. FAX (608) 263-2841.

Center For Numerical Analysis. University of Texas at Austin. Austin, TX 78712. (512) 471-1242,

Center For Pure and Applied Mathematics. University of California, Berkeley. 977 Evans Hall, Berkeley, CA 94720. (415) 642-0116. FAX (415) 642-6726.

Institute For Computer Applications In Science and Engineering. Mail Stop 132C, NASA, Langley Research Center, Hampton, VA 23665. (804) 865-2513. FAX (804) 864-6134.

CAM (COMPUTER-AIDED MANUFACTURING)

See also: AUTOMATION, CAD, COMPUTERS, ROBOTICS

ABSTRACT SERVICES AND INDEXES

ACM Guide To Computing Literature. Association for Computing Machinery, Association for Computing Machinery, 1515 Broadway, 17th Floor, New York, NY 10036-5701. (212) 869-7440. FAX (212) 944-1318. 1964 to present. Monthly. $190.00 per year for non-members.

Applied Science and Technology Index; A CumulativeSubject Index To English Language Periodicals In the Fields of Aeronautics and Space Science, Computer Technology, Chemistry, Construction Industry, Energy and Related Areas. H.W. Wilson Co., 950 University Avenue, Bronx, NY 10452. (800) 367-6770 or (212) 588-8400. FAX (718) 590-1617. From 1958 to present. Monthly. Inquire about cost and availability. Also available on CD-ROM and online.

CAD-CAM Abstracts (formerly CAD-CAM Profiles). Techgnosis Ltd., Blade House, Battersea Road, Stockport, Cheshire 3AE, England. Telephone 061-442-2639. FAX 061-443-1162. Six times a year. Inquire for price.

CAD/CAM Abstracts and CAD/CAM Abstracts Annual. R.R. Bowker, Bowker A&I Publishing, 121 Chanlon Road, New Providence, NJ 07974. (908) 771-7714. Fax (908) 771-7725. 1984 to 1992. Ceased publication.

Computer Abstracts, MCB University Press Ltd., PO Box 10812, Birmingham, AL 35201-0812. (800) 633-4931. FAX (205) 995-1588. Monthly. Covers computer theory, data, hardware, systems, networks, human-computer interaction, artificial intelligence, as well as applications of computers in aerospace, business, CAD/CAM, cartography, civil engineering, electronics and electrical engineering, industrial engineering, mechanical engineering, medicine, structural engineering, etc. Also available on CD-ROM.

Computer & Control Abstracts (INSPEC). INSPEC/IEEE, Box 1331, 445 Hoes Lane, Piscataway, NJ 08855-1331. (908) 562-5549. Abstracts organized by subjects of international technological information. Monthly. $1455.00 per year. Also available on CD-ROM and online via BRS Online Products, DIALOG Information Services, Orbit Search Service, and STN International.

Computer and Information Systems Abstracts Journal. Cambridge Scientific Abstracts, 7200 Wisconsin Avenue, Bethesda, MD 20814. (301) 961-6750. Fax (301) 961-6720. 1962 to present. Monthly. $1265.00 per year. Also available online via STN International.

Computer Literature Index. Applied Computer Research Inc., Box 82266, Phoenix, AZ 85071-2266. (800) 234-2227. 1971 to present. Quarterly plus annual cumulation. $198.50 per year. Bibliography of books, articles, and reports.

Computing Journal Abstracts. Techgnosis Ltd., Blade House, Battersea Road, Stockport, Cheshire 3AE, England. Telephone 061-442-2639. FAX 061-443-1162. 1969 to present. Monthly. Inquire for price.

Computing Reviews. Association for Computing Machinery, 1515 Broadway, 17th Floor, New York, NY 10036-5701. (212) 869-7440. FAX (212) 944-1318. 1960 to present. Monthly. $130.00 per year for non-members. Also available online via DIALOG Information Services.

Current Contents: Engineering, Technology, and Applied Sciences. Institute for Scientific Information, 3501 Market Street, Philadelphia, PA 19104. (215) 386-0100. FAX (215) 386-6362. 1970 to present. Weekly. $442.00 per year.

Engineering Index Monthly. Engineering Information, Inc., Castle Point on the Hudson, Hoboken, NJ 07030. (800) 221-1044. FAX (212) 832-1857. Monthly. $2300.00 per year. Also available online as COMPENDEX, and also on CD-ROM. Covers chemical engineering, computers, electrical engineering, civil engineering, metals and mining, industrial management, and mechanical engineering.

Science Citation Index. Institute for Scientific Information, 3501 Market Street, Philadelphia, PA 19104. (215) 386-0100. FAX (215) 386-6362. Inquire about availability and cost. Also available on CD-ROM.

ASSOCIATIONS AND PROFESSIONAL SOCIETIES

American Automatic Control Council. c/o EECS, Northwestern University, 2145 Sheridan Road, Evanston, IL 60208-3118. (708) 491-3641.

Association For Computing Machinery. 1515 Broadway, 17th Floor, New York, NY 10036-5701. (212) 869-7440. FAX (212) 944-1318.

Computer-aided Manufacturing Institute. 1250 East Copeland Road, Suite 500, Arlington, TX 76011. (817) 265-5328. FAX (817) 275-6450.

Computer & Automated Systems Association of Sme. PO Box 930, One SME Drive, Dearborn, MI 48121-0930. (313) 271-1500. FAX (313) 271-2861.

Institute of Electrical and Electronic Engineers. 345 E. 47th Street, New York, NY 10017. (212) 705-7900.

Numerical Control Society/aim Tech. 67 Alexander Drive, PO Box 12277, Research Triangle Park, NC 27709. (815) 399-8700.

Society of Manufacturing Engineers. One SME Drive, Box 930, Dearborn, MI 48121-0930. (313) 271-1500. FAX (313) 271-2861.

CAM (COMPUTER AIDED MANUFACTURING)

Ency. of Physical Sciences and Engineering Info. Sources

BIBLIOGRAPHIES

ACM Guide To Computing Literature. Association for Computing Machinery, 1515 Broadway, 17th Floor, New York, NY 10036-5701. (212) 869-7440. FAX (212) 944-1318. 1964 to present. Annual. $190.00 for non-members. Also available online via DIALOG.

DIRECTORIES AND BIOGRAPHICAL SOURCES

CAD/CAM International Software Directory. TAB Books, P.O. Box 40, Blue Ridge Summit, PA 17294-0850. (717) 794-2191 or (800) 233-1128. FAX (717) 794-2080. 1991. Inquire for cost and availability.

Computer-Aided Design (CAD) Directory. TAB Books, P.O. Box 40, Blue Ridge Summit, PA 17294-0850. (717) 794-2191 or (800) 233-1128. FAX (717) 794-2080. 1991. Inquire for cost and availability.

Computer Graphics World Buyers Guide. PennWell Publishing Company, Advanced Technology Group, 10 Tara Blvd., 5th floor, Nashua, NH 03062-2801. (603) 891-9111. FAX (603) 891-0539. Annual. $100.00 per year.

Computing Information Directory. Hildebrandt Inc., PO Box 576, Pullman, WA 99114. 1981 to present. Annual. $199.95 per year.

Robotics World Directory. Argus Inc., 6151 Powers Ferry Road NW, Atlanta, Georgia 30339-2941. (404) 955-2500. FAX (404)955-0400. Included with subscription to Robotics World. magazine.

Who's Who In Technology. Gale Research, 835 Penobscot Building, Detroit, MI 48226-4094. (313) 961-2242. (800) 347-4253. FAX (313) 961-6083. 1995. $195.00.

ENCYCLOPEDIAS AND DICTIONARIES

CAD/CAM Dictionary. Preston & Crawford. Marcel Dekker, Inc., 270 Madison Avenue, New York, NY 10016. (212) 696-9000. FAX (212) 685-4540. 1985. $89.75.

CAD/CAM Glossary. CIM Systems, 2425 N. Central, No. 432, Richardson, TX 75080-3549. (214) 437-5171. $19.25.

Computer Dictionary. Donald D. Spencer. 4th edition. Camelot Publishing Company, 709 SW 80th Blvd., Gainesville, FL 32607-1537. (904) 331-0952. 1993. $24.95.

Encyclopedia of Computer Science and Engineering. Anthony Ralston & Edwin D. Reilly Jr., editors. 3rd revised edition. Van Nostrand Reinhold, 115 Fifth Avenue, New York, NY 10003. (212) 254-3232 or FAX (212) 254-9499. 1993. $125.00.

Encyclopedic Dictionary of Industrial Automation and Computer Control. David W. South. Prentice Hall, 113 Sylvan Avenue, Route 9W, Englewood Cliffs, NJ 07632. (201) 592-2000 or (800) 922-0579. 1994. Inquire for price and availability.

Microsoft Press Computer Dictionary: the Comprehensive Standard For Business, School, Library, and Home. 2nd edition. Microsoft Press, One Microsoft Way, Redmond, WA 98052-6399. (206) 936-0055. FAX (206) 823-8101. 1994. $19.95.

GENERAL WORKS

Artificial Intelligence In Optimal Design and Manufacturing. Zuomin Dong, editor. Prentice Hall, 113 Sylvan Avenue, Route 9W, Englewood Cliffs, NJ 07632. (201) 592-2000 or (800) 922-0579. 1994. $64.00.

Artificial Intelligence In the Factory. Architecture Technology Corporation. Elsevier Advanced Technology, Elsevier Science Inc., 660 White Plains Rd., Tarrytown, NY 10159-5153. (919) 524-9200. FAX (919) 333-2444. 1990. Inquire for cost and availability.

Better Products Faster: A Practical Guide To Knowledge-based Systems For Manufacturers. William H. VerDuin. Irwin Profession Publishing, 1333 Burr Ridge Parkway, Burr Ridge, IL 60521. (708) 789-4000 or (800) 634-3961. 1995. $40.00.

CAD/CAM: From Principles To Practice. Chris McMahon & Jim Browne. Addison-Wesley Publishing Company, 1 Jacob Way, Reading, MA 01867. (617) 944-3700 or (800) 447-2226. FAX 617-942-1117. 1993. $69.95.

CAD/CAM Report. Architecture Technology Corporation. Elsevier Science Inc., 660 White Plains Rd., Tarrytown, NY 10159-5153. (919) 524-9200. FAX (919) 333-2444. 1990. $375.00.

CAD/CAM theory and Practice. Ibrahim Zeid. McGraw-Hill Publishing Company, 1221 Avenue of the Americas, New York, NY 10020. (800) 262-4729 or (212) 512-3825. 1991. Inquire for cost andavailability.

CIM In Process and Manufacturing Industries. L. Leiviska. Pergamon Press Inc., Maxwell House, Fairview Park, Elmsford, NY 10523. (914) 592-7700. Fax (914) 592-3625. 1993. $83.00.

Computer-Aided Design and Manufacturing. Farid M.L. Amirouche. Prentice Hall, 113 Sylvan Avenue, Route 9W, Englewood Cliffs, NJ 07632. (201) 592-2000 or (800) 922-0579. 1992. $56.00.

Computer-Aided Assembly Planning. A. Delchambre. Chapman and Hall, 29 West 35th Street, New York, NY 10001. (800) 842-3636. FAX (212) 563-2269. 1992. Inquire for cost and availability.

Computer-Aided Design In Manufacturing. David Valliere. Prentice Hall, 113 Sylvan Avenue, Rout 9W, Englewood Cliffs, NJ 07632. (201) 592-2000 or (800) 922-0579. 1990. Inquire for cost and availability.

Computer Integrated Manufacturing and Engineering. W. Rembold, et al. Addison-Wesley Publishing Company, 1 Jacob Way, Reading, MA 01867. (617) 944-3700 or (800) 447-2226. FAX 617-942-1117. 1993. $52.95.

Intelligent Systems In Design and Manufacturing. Cihan H. Dagli & andrew Kusiak, editors. ASME Press, American Society of Mechanical Engineers, 345 East 47th Street, New York, NY 10017-2398. (212) 705-7703. 1994. $77.00.

Making CAD/CAM Work. Richard Halton. Prentice Hall, 113 Sylvan Avenue, Rout 9W, Englewood Cliffs, NJ 07632. (201) 592-2000 or (800) 922-0579. 1993. $32.00.

HANDBOOKS AND MANUALS

The C4 Handbook: CAD, CAM, CAE, CIM. Carl Machover. TAB Books, P.O. Box 40, Blue Ridge Summit, PA 17294-0850. (717) 794-2191 or (800) 233-1128. FAX (717) 794-2080. 1989. Inquire for price and availability.

McGraw-Hill Computing Essentials, 1993-1994. Timothy J. O'Leary, et al. McGraw-Hill Publishing Company, 1221 Avenue of the Americas, New York, NY 10020. (800) 262-4729 or (212) 512-3825. 1994. Inquire for price and availability.

ONLINE DATABASES AND CD-ROMS

Compendex Plus. Engineering Information, Inc., 345 East 47th Street, New York, NY 10017. (212) 705-7600 or (800) 221-1044. Contains citations with abstracts to worldwide literature in engineering and technology, from 1970 to present. Available on online BRS,(800) 289-4277, DIALOG, (800) 334-2564, ORBIT (800) 456-7248, and STN International, FIZ Karlsruhe, P.O. Box 2465, W-7500, Karlsruhe 1, Germany, online services. Also available on CD-ROM. Inquire as to cost and availability.

CSA Engineering. Cambridge Scientific Abstracts, 7200 Wisconsin Avenue, Suite 601, Bethesda, MD 20814. (301) 961-6750 or (800) 843-7751. Contains citations and abstracts of international periodicals and research literature covering all fields of engineering and science and technology,including computer and information science, electronics, mechanical engineering, solid state materials, 1981 to present. Available on BRS,(800) 289-4277, online service.Inquire as to cost and availability.

Current Contents Search. Institute for Scientific Information, 3501 Market Street, Philadelphia, PA 19104. (215) 386-0100. FAX (215) 386-6362. Contains citations to articles listed in the table of contents of science and technology journals. Also articles in social sciences and life sciences journals. Available on BRS,(800) 289-4277, DIALOG,(800) 334-2564, online services. Inquire as to cost and availability.

Dissertation Abstracts. University Microfilms International, 300 North Zeeb Road, Ann Arbor, MI 48106. (800) 521-0600 or (313) 761-4700. Scope includes virtually all doctoral dissertations accepted at accredited American institutions from 1861 to present in 252 subject areas. Available on BRS,(800) 289-4277, DIALOG,(800) 334-2564, and OCLC EPIC,(800) 848-5878, online services. Also available on CD-ROM. Inquire as to cost and availability.

NTIS Bibliographic Database. National Technical Information Service, 5285 Port Royal Road, Springfield, Va 22161. (703) 487-4929 or FAX (703) 321-8199. Broad coverage of government-sponsored science and technology research reports, 1964 to present. Available on BRS,(800) 289-4277, DIALOG, (800) 334-2564, ORBIT, (800) 456-7248, and STN International, FIZ Karlsruhe, P.O. Box 2465, W-7500, Karlsruhe 1, Germany, online services. Also available on CD-ROM. Inquire as to cost and availability.

Scisearch. Institute for Scientific Information, 3501 Market Street, Philadelphia, PA 19104. (800) 523-1850 or (215) 386-0100. Broad multidisciplinary title and author index to the international literature of science and technology, 1974 to present. Available on DIALOG,(800) 334-2564, and ORBIT,(800) 456-7248, online services. Also available on CD-ROM. Inquire as to cost and availability.

WILSONLINE. H.W. Wilson Company, 950 University Avenue, Bronx, NY 10452. (800) 367-6770 or (212) 588-8400. Makes available online versions of the printed H.W. Wilson indexes including Applied Science and Technology Index, Business Periodicals Index, General Science Index, and Readers' Guide to Periodical Literature. Period covered is generally 1983 to present. Available on BRS,(800) 289-4277, DIALOG,(800) 334-2564, and OCLC EPIC,(800) 848-5878, online services. Also available on CD-ROM. Inquire as to cost and availability.

PERIODICALS

Advanced Manufacturing Technology. Technical Insights Inc., 32 N Dean Street, Englewood, NJ 07631. (201) 568-4744. FAX (201) 568-8247. 1980 to present. Monthly. $575.00 per year.

CAD-CAM Update. Worldwide Videotex, Box 3273, Boynton Beach, FL 33424-3273. (407) 738-2276. 1989 to present. Monthly. Inquire for price.

Computer Design. PennWell Publishing Company, Advanced Technology Group, 10 Tara Blvd., 5th floor, Nashua, NH 03062-2801. (603) 891-9111. FAX (603) 891-0539. 1962 to present. 12 times a year. $88.00 per year.

Computer Graphics World. PennWell Publishing Company, Advanced Technology Group, 10 Tara Blvd., 5th floor, Nashua, NH 03062-2801. (603) 891-9111. FAX (603) 891-0539. 1978 to present. Monthly. $48.00 per year.

Computers & Graphics. Elsevier Science, 660 White Plains Rd., Tarrytown, NY 10159-5153. (919) 524-9200. FAX (919) 333-2444. 1975 to present. Six times a year. $780.00 per year.

IEEE Computer Graphics and Applications. IEEE Computer Society Press, 10662 Los Vaqueros Circle, Box 3014, Los Alamitos, CA 90720-1264. (714) 821-8380. FAX (714) 821-4641. 1981 to present. Bi-monthly. $175.00 per year.

Computer-integrated Manufacturing Systems. Butterworth-Heinemann, Linacre House, Jordan Hill, Ocford OX2 8DP, England. Telephone 0865-310366. FAX 0865-310898. 1988 to present. Quarterly. Inquire for cost.

IEEE Robotics and Automation Magazine. IEEE (Institute of Electrical and Electronic Engineers), 445 Hoes Lane, Piscataway, NJ 08854. (908) 562-5545 or (800) 678-IEEE. 1994 to present. Quarterly. $50.00 per year.

Industrial Robot. MCB University Press, 62 Toller Lane, Bradford, W. Yorks. BD8 9BY, England. Telephone 0274-499821. FAX 0274-547143. 1973 to present. 6 times a year. $749.95 per year.

Robotics and Computer-integrated Manufacturing. Elsevier Science [journals], 660 White Plains Rd., Tarrytown, NY 10159-5153. (919) 524-9200. FAX (919) 333-2444. 1984 to present. 4 times a year. $415.00 per year.

Robotics Today. Society of Manufacturing Engineers, One SME Drive, Box 930, Dearborn, MI 48121-0930. (313) 271-1500. FAX (313) 271-2861. 1979 to present. Quarterly. $60.00 per year.

Robotics World. Argus Inc., 6151 Powers Ferry Road NW, Atlanta, Georgia 30339-2941. (404) 955-2500. FAX (404) 955-0400. 1982 to present. 4 times a year. $42.00 per year.

CAM (COMPUTER AIDED MANFACTURING)

Ency. of Physical Sciences and Engineering Info. Sources

RESEARCH CENTERS AND INSTITUTES

Center For Robotics and Advanced Automation. Oakland University, Rochester, MI 48309-4401. (313) 370-2233.

Cornell Manufacturing Engineering and Productivity Program. Cornell University, 100 Engineering & theory Center, Ithaca, NY 14853. (607) 255-8084.

Computer-aided Manufacturing—International, Inc. 1250 E. Copeland Road, Suite 500, Arlington, TX 76011. (817) 860-1654. FAX (817) 275-6450.

Manufacturing Productivity Center, Illinois Institute of Technology. 110 W. 35th Street, Chicago, IL 60616. (312) 567-4800.

University of Wisconsin—madison, Computer-integrated Manufacturing Systems: Design and Simulation Laboratory. Department of Industrial Engineering, 1513 University Avenue, Room 457, Madison, WI 53706. (608) 262-9927.

CAMERAS

See also: AERIAL PHOTOGRAPHY, COLOR, OPTICS, PHOTOGRAMMETRY,PHOTOGRAPHIC FILM, PHOTOGRAPHY, UNDERWATER PHOTOGRAPHY

ABSTRACT SERVICES AND INDEXES

Applied Science and Technology Index; A Cumulative Subject Index To English Language Periodicals In the Fields of Aeronautics and Space Science, Computer Technology, Chemistry, Construction Industry, Energy and Related Areas. H.W. Wilson Co., 950 University Avenue, Bronx, NY 10452. (800) 367-6770 or (212) 588-8400. FAX (718) 590-1617. From 1958 to present. Monthly. Inquire about cost and availability. Also available on CD-ROM and online.

Current Contents: Engineering, Technology, and Applied Sciences. Institute for Scientific Information, 3501 Market Street, Philadelphia, PA 19104. (215) 386-0100. FAX (215) 386-6362. 1970 to present. Weekly. $442.00 per year.

Current Papers In Physics. Institute of Electrical Engineers, Michael Faraday House, Six Hill Way, Stevenage, Herts, SG1 2AY, England. Distributed by IEEE, 445 Hoes Lane, Piscataway, NJ 08854. (908) 562-5549. 1966 to present. Twenty-four times per year. $390.00 per year. Also available online in INSPEC.

Current Physics Index. American Institute of Physics, 335 East 45th Street, New York, NY 10017 (212) 661-9404 or FAX (516) 349-9704. 1975 to present. Quarterly. $625.00 per year.

Engineering Index Monthly. Engineering Information, Inc., Castle Point on the Hudson, Hoboken, NJ 07030. (800) 221-1044. FAX (212) 832-1857. Monthly. $2300.00 per year. Also available online as COMPENDEX, and also on CD-ROM. Covers chemical engineering, computers, electrical engineering, civil engineering, metals and mining, industrial management, and mechanical engineering.

Imaging Abstracts (formerly Photographic Abstracts), Pergamon Press Inc., Maxwell House, Fairview Park, Elmsford, NY 10523. (914) 592-7700. Fax (914) 592-3625. Bimonthly. Inquire for price.

Physics Abstracts. Institute of Electrical Engineers, Michael Faraday House, Six Hill Way, Stevenage, Herts, SG1 2AY, England. Distributed by IEEE, 445 Hoes Lane, Piscataway, NJ 08854. (908) 562-5549. 1898 to present. Monthly. $2700.00 per year. Also available online as INSPEC.

Science Citation Index. Institue for Scientific Information, 3501 Market Street, Philadelphia, PA 19104. (215) 386-0100. FAX (215) 386-6362. Inquire about availability and cost. Also available on CD-ROM.

ASSOCIATIONS AND PROFESSIONAL SOCIETIES

American Society For Photogrammetry and Remote Sensing, 5410 Grosvenor Lane, Suite 210, Bethesda, Md 20814.

American Society of Photographers. Box 3191, Spartanburg, SC 29304-3191. (803) 582-3115. FAX (803) 582-4179.

The International Society For Optical Engineering (spie), Po Box 10, 1000-20th Street, Bellingham, Wa 98227-0010.

Optical Society of America, 2010 Massachusetts Avenue Nw, Washington, Dc 20036. (202) 223-8130.

The Society For Imaging Science and Technology, 7003 Kilworth Lane, Springfield, Va 22151. (703) 642-9090.

Society of Motion Picture and Television Engineers, 595 W. Hartsdale Avenue, White Plains, NY 10607.

Society of Photographic ScieNTISts and Engineers. 7003 Kilworth Lane, Springfield, VA 22151. (703) 642-9090.

Society of Photo-technologists. PO B ox 9634, Denver, CO 80209. (303) 698-1820.

DIRECTORIES AND BIOGRAPHICAL SOURCES

Directory of Engineering Societies and Related Organizations. Gordon Davis. 13th edition. American Association of Engineering Societies, 1111 19th Street NW, Suite 608, Washington, DC 20036. (202) 296-2237 or (800) 658-8897. 1989. Inquire for price.

International Research Centers Directory. Gale Research, 835 Penobscot Building, Detroit, MI 48226-4094. (313) 961-2242. (800) 347-4253. 8th edition. 1995. $410.00

Research Centers Directory. Gale Research, 835 Penobscot Building, Detroit, MI 48226-4094. (313) 961-2242. (800) 347-4253. FAX (313) 961-6083. $485.00.

Scientific and Technical Organizations and Agencies Directory. Gale Research, 835 Penobscot Building, Detroit, MI 48226-4094. (313) 961-2242. (800) 347-4253. FAX (313) 961-6083. 4th edition. 1996. $195.00.

Who's Who In Engineering. American Association of Engineering Societies, 1111 19th Street NW, Suite 608, Washington, DC 20036. (202) 296-2237 or (800) 658-8897. 8th edition. 1991. Inquire for price.

ENCYCLOPEDIAS AND DICTIONARIES

The Focal Encyclopedia of Photography. Leslie Stroebel & Richard D. Zakia, editors. 3rd edition. Butterworth-Heinemann, 313 Washington Street, Newton, MA 02158. (617) 928-2500 or (800) 366-2665. FAX (617) 928-2620. 1993. $125.00.

Thesaurus of Photographic Science and Engineering. Society of Photographic Scientists and Engineers Staff. Books on Demand, 300 N. Zeeb Road, Ann Arbor, MI 48106-1346. (313) 761-4700 or (800) 521-0600. 1968. $34.30.

Thesaurus of Scientific, Technical, and Engineering Terms. Hemisphere Publishing Corporation, 1900 Frost Road, Suite 101, Bristol, PA 19007-1598. (215) 785-5800 or (800) 821-8312. FAX (215) 785-5515. 1987. $173.00.

GENERAL WORKS

Camera Technology. Alastair Jervis. Bookwright/Watts, Franklin Inc., 95 Madison Avenue, New York, NY 10016. (212) 951-2650 or (800) 672-6672. 1991. $12.90.

Camera Technology: the Dark Side of the Lens. Norman Goldberg. Academic Press, Inc., 6277 Sea Harbor Drive, Orlando, FL 32887. (800) 321-5068. 1992. $44.95.

Collecting and Using Classic Cameras. Ivor Matanle. Paperback edition. Thames and Hudson, 500 Fifth Avenue, New York,NY 10110. (212) 354-3763 or FAX (212) 689-0856. 1992. Inquire for cost and availability.

A History of the 35mm Still Camera. Roger Hicks. Focal Press, 80 Montvale Avenue, Stoneham, MA 02180. 1984. Inquire for cost and availability.

Motion Picture Camera and Lighting Equipment: Choice and Technique. David W. Samuelson. 2nd edition. 80 Montvale Avenue, Stoneham, MA 02180. 1987. $19.95.

Science and Technology of Photography. Karlheinz Keller, editor. VCH Publishers, Inc., 303 Northwest 12th Avenue, Deerfield Beach, FL 33442. (800) 422-8824. 1993. Inquire for cost and availability.

The 35mm Film Source Book. Marc Levey. Focal Press, 80 Montvale Avenue, Stoneham, MA 02180. 1992. Inquire for cost and availability.

HANDBOOKS AND MANUALS

Handbook of Photography. Ronald P. Lovell, et al. Delmar Publishers, Division of Thomson Educational Publishing, Inc., 3 Columbia Circle, Box 15015, Albany, NY 12205. (518) 464-3500 or (800) 347-7707. FAX (518) 464-0358. 1993. $27.95.

The New 35mm Handbook: A Complete Course From Basic Techniques To Professional Applications. Michael Freeman. Revised edition. Running Press, 125 S. 22nd Street, Philadelphia, PA 19103-4394. (215) 567-5080 or (800) 345-5359. 1993. $24.95.

The New 35mm Photographer's Handbook. Julian Calder & John Garrett. 2nd revised edition. Crown Publishing Group, 201 E. 50th St., New York, NY 10022. (800) 726-0600 or (800) 733-3000. 1986. Inquire for cost and availability.

ONLINE DATABASES AND CD-ROMS

Compendex Plus. Engineering Information, Inc., 345 East 47th Street, New York, NY 10017. (212) 705-7600 or (800) 221-1044. Contains citations with abstracts to worldwide literature in engineering and technology, from 1970 to present. Available on online BRS,(800) 289-4277, DIALOG, (800) 334-2564, ORBIT (800) 456-7248, and STN International, FIZ Karlsruhe, P.O. Box 2465, W-7500, Karlsruhe 1, Germany, online services. Also available on CD-ROM. Inquire as to cost and availability.

CSA Engineering. Cambridge Scientific Abstracts, 7200 Wisconsin Avenue, Suite 601, Bethesda, MD 20814. (301) 961-6750 or (800) 843-7751. Contains citations and abstracts of international periodicals and research literature covering all fields of engineering and science and technology,including computer and information science, electronics, mechanical engineering, solid state materials, 1981 to present. Available on BRS,(800) 289-4277, online service.Inquire as to cost and availability.

Current Contents Search. Institute for Scientific Information, 3501 Market Street, Philadelphia, PA 19104. (215) 386-0100. FAX (215) 386-6362. Contains citations to articles listed in the table of contents of science and technology journals. Also articles in social sciences and life sciences journals. Availableon BRS,(800) 289-4277, DIALOG,(800) 334-2564, online services. Inquire as to cost and availability.

NTIS Bibliographic Database. National Technical Information Service, 5285 Port Royal Road, Springfield, Va 22161. (703) 487-4929 or FAX (703) 321-8199. Broad coverage of government-sponsored science and technology research reports, 1964 to present. Available on BRS,(800) 289-4277, DIALOG, (800) 334-2564, ORBIT, (800) 456-7248, and STN International, FIZ Karlsruhe, P.O. Box 2465, W-7500, Karlsruhe 1, Germany, online services. Also available on CD-ROM. Inquire as to cost and availability.

Scisearch. Institute for Scientific Information, 3501 Market Street, Philadelphia, PA 19104. (800) 523-1850 or (215) 386-0100. Broad multidisciplinary title and author index to the international literature of science and technology, 1974 to present. Available on DIALOG,(800) 334-2564, and ORBIT,(800) 456-7248, online services. Also available on CD-ROM. Inquire as to cost and availability.

WILSONLINE. H.W. Wilson Company, 950 University Avenue, Bronx, NY 10452. (800) 367-6770 or (212) 588-8400. Makes available online versions of the printed H.W. Wilson indexes including Applied Science and Technology Index, Business Periodicals Index, General Science Index, and Readers' Guide to Periodical Literature. Period covered is generally 1983 to present. Available on BRS,(800) 289-4277, DIALOG,(800) 334-2564, and OCLC EPIC,(800) 848-5878, online services. Also available on CD-ROM. Inquire as to cost and availability.

PERIODICALS

British Journal of Photography. Bouverie Publishing Company Ltd., 147-151 Temple Chambers, Temple Avenue, London EC47 0DT, England. Telephone 071-583-3030. FAX 071-583-4068. 1854 to present. Weekly. Inquire for price and availability.

Industrial Photography. PTN Publishing Corporation, 445 Broad Hollow Road, Melville, NY 11747-4722. (516) 845-2700. FAX (516) 845-7109. 1952 to present. Monthly. $60.00 per year.

Journal of Photographic Science. The Barn, Whitehall, Near Middle Marwood, Barnstaple, N. Devon EX31 4EQ, England. Telephone 0271-72482. FAX 0271-72482. 1952 to present. Bi-monthly. Inquire for price and availability.

Photogrammetric Engineering and Remote Sensing. American Society for Photogrammetry and Remote Sensing, 5410 Grosvenor Lane, Suite 210, Bethesda, MD 20814. (301) 493-0290. FAX (301) 493-0208. 1934-present. Monthly. $160.00 per year.

Psa Journal. Photographic Society of America, 3000 United Founders Blvd., No.103, Oklahoma City, OK 73112-3940. (405) 843-1437. FAX (405) 843-1438. 1934 to present. Monthly. Included with $35.00 annual membership.

Ptn (photographic Trade News). PTN Publishing Corporation, 445 Broad Hollow Road, Suite 21, Melville, NY 11747-4722. (516) 845-2700. FAX (516) 845-7109. 1937 to present. 24times a year. $25.00 per year.

Scientific and Applied Photography and Cinematography. Gordon and Breach Science Publishers, 820 Town Center Drive, Langhorne, PA 19047. (215) 750-2642. FAX (215) 750-6343. 6 times a year. Inquire for price.

RESEARCH CENTERS AND INSTITUTES

George Eastman House, International Museum of Photography and Film. 900 East Avenue, Rochester, NY 14607. (716) 271-3361. FAX (716) 271-3970.

Rochester Institute of Technology, Center For Imaging Science. One Lomb Memorial Drive, Rochester, NY 14623. (716) 475-2774.

University of Rochester, Center For Advanced Optical Technology. Wilmot Building, Rochester, NY 14627. (716) 275-5248.

CANALS

See also: CHANNELS, CIVIL ENGINEERING, DREDGES AND DREDGING, GEOTECHNICAL ENGINEERING, HYDRAULIC ENGINEERING, OCEAN ENGINEERING

ABSTRACT SERVICES AND INDEXES

Applied Science and Technology Index; A Cumulative Subject Index To English Language Periodicals In the Fields of Aeronautics and Space Science, Computer Technology, Chemistry, Construction Industry, Energy and Related Areas. H.W. Wilson Co., 950 University Avenue, Bronx, NY 10452. (800) 367-6770 or (212) 588-8400. FAX (718) 590-1617. From 1958 to present. Monthly. Inquire about cost and availability. Also available on CD-ROM and online.

ASCE Combined Annual Combined Index. American Society of Civil Engineers, 345 East 47th Street, New York, NY 10017-2398. (212) 705-7520 or (800) 548-2723. Annual. $48.00.

ASCE Publications Information. 1966-present. American Society of Civil Engineers, 345 East 47th Street, New York, NY 10017-2398. (212) 705-7520 or (800) 548-2723. Bi-monthly. $160.00 per year for non-members.

Bibliography and Index of Geology. American Geological Institute, 4220 King Street, Alexandria, VA 22302. (703) 379-2480. Fax (703) 379-7563. Monthly. $1295.00 per year. Also available online as GEOREF.

Civil and Structural Engineering Abstracts, Cambridge Scientific Abstracts, 7200 Wisconsin Avenue, Bethesda, Md 20814-4823. (301) 961-6750. FAX (301) 961-6720. 1993 to present. Monthly. $385.00 per year. Topics covered include structural design, construction equipment and methods, civil defense and military engineering, surveying, highway engineering, maritime and port structures, materials, land reclamation, and soil mechanics.

Engineering Index Monthly. Engineering Information, Inc., Castle Point on the Hudson, Hoboken, NJ 07030. (800) 221-1044. FAX (212) 832-1857. Monthly. $2300.00 per year. Also available online as COMPENDEX, and also on CD-ROM. Covers chemical engineering, computers, electrical engineering, civil engineering, metals and mining, industrial management, and mechanical engineering.

Fluid Abstracts: Civil Engineering. Elsevier Science [journals], 660 White Plains Rd., Tarrytown, NY 10159-5153. (919) 524-9200. FAX (919) 333-2444. 1991 to present. Monthly plus annual cumulation. $645.00 per year. Also available online as FLUIDEX. Covers civil engineering applications of fluid mechanics, hydraulics, flow metering and measuring, offshore engineering, environmental hydraulics, and related aspects of wind energy, the atmosphere, and aerodynamics.

Geotechnical Abstracts. Geodex Retrieval Systems, 669 Broadway, Box 279, Sonoma, CA 95476. (707) 939-8476. FAX (707) 996-8734. 1970 to present. Three times a year. $280.00 per year. Covers soil mechanics, foundation engineering, rock mechanics, and engineering geology.

Science Citation Index. Institute for Scientific Information, 3501 Market Street, Philadelphia, PA 19104. (215) 386-0100. FAX (215) 386-6362. Inquire about availability and cost. Also available on CD-ROM.

ASSOCIATIONS AND PROFESSIONAL SOCIETIES

American Canal Society. 809 Rathon Road, York, PA 17403. (717) 843-4035.

American Geological Institute. 4220 King Street, Alexandria, VA (703) 379-2480.

American Society of Civil Engineers. 345 East 47th Street, New York, NY 10017-2398. (212) 705-7520 or (800) 548-2723.

Irrigation Association. 8260 Willow Oak Corp. Drive, Fairfax, VA 22031. (703) 573-3551.

U.S. COMMITTEE ON IRRIGATION and DRAINAGE. 1616 17th Street, No. 483, Denver, CO 80202. (303) 628-5430.

BIBLIOGRAPHIES

Canal Bibliography: With A Primary Emphasis On the United States and Canada. Albright G. Zimmerman. Canal History and

Ency. of Physical Sciences and Engineering Info. Sources

CANALS

Technology Press, 200 S. Delaware Drive, Easton, PA 18044-0877. (610) 250-6700. 1991. $24.95.

DIRECTORIES AND BIOGRAPHICAL SOURCES

Directory of Engineering Societies and Related Organizations. Gordon Davis. 13th edition. American Association of Engineering Societies, 1111 19th Street NW, Suite 608, Washington, DC 20036. (202) 296-2237 or (800) 658-8897. 1989. Inquire for price.

Research Centers Directory. Gale Research, 835 Penobscot Building, Detroit, MI 48226-4094. (313) 961-2242. (800) 347-4253. $470.00

Scientific and Technical Organizations and Agencies Directory. Gale Research, 835 Penobscot Building, Detroit, MI 48226-4094. (313) 961-2242. (800) 347-4253. FAX (313) 961-6083. 4th edition. 1996. $195.00.

Who's Who In Engineering. American Association of Engineering Societies, 1111 19th Street NW, Suite 608, Washington, DC 20036. (202) 296-2237 or (800) 658-8897. 8th edition. 1991. Inquire for price.

ENCYCLOPEDIAS AND DICTIONARIES

Dictionary of Civil Engineering. V.N. Vazirani. South Asia Books, PO Box 502, Columbia, MO 65205. (314) 474-0116. FAX (314) 474-8124. 1992. $20.00.

GENERAL WORKS

Canal and River Levees. P. Peter. Elsevier Science B.V., PO Box 945, Madison Square Station, New York, NY 10159-0945. (212) 633-3650. FAX (212) 633-3680. 1982. $148.75.

Canals For A Nation: the Canal Era In the United States, 1790-1860. University Press of Kentucky, 663 S. Limestone Street, Lexington, KY 40508-4008. (606) 257-8442. 1990. $15.00.

A Hundred Years of the Manchester Ship Canal. Edward Gray. Aurora, Bolton, England. 1994. Inquire for cost and availability.

HANDBOOKS AND MANUALS

Civil Engineering Calculations Reference Guide. Tyler G. Hicks. McGraw-Hill Publishing Company, 1221 Avenue of the Americas, New York, NY 10020. (800) 262-4729 or (212) 512-3825. 1987. $46.00.

Civil Engineering Practice. Paul N. Cheremisinoff, et al. 5 volumes. Technomic Publishing Company, Inc., 851 New Holland Avenue, Box 3535, Lancaster, PA 17604. (717) 291-5609 or (800) 233-9936. FAX (717) 295-4538. 1987-88. $199.00.

Standard Handbook For Civil Engineers. Frederick S. Merritt, editor. 3rd edition. McGraw-Hill Publishing Company, 1221 Avenue of the Americas, New York, NY 10020. (800) 262-4729 or (212) 512-3825. 1983. $15.00.

Surveying Ready-reference Manual. G.O. Stenstrom. McGraw-Hill Publishing Company, 1221 Avenue of the Americas, New York, NY 10020. (800) 262-4729 or (212) 512-3825. 1987. $38.50.

ONLINE DATABASES AND CD-ROMS

Compendex Plus. Engineering Information, Inc., 345 East 47th Street, New York, NY 10017. (212) 705-7600 or (800) 221-1044. Contains citations with abstracts to worldwide literature in engineering and technology, from 1970 to present. Available on online BRS,(800) 289-4277, DIALOG, (800) 334-2564, ORBIT (800) 456-7248, and STN International, FIZ Karlsruhe, P.O. Box 2465, W-7500, Karlsruhe 1, Germany, online services. Also available on CD-ROM. Inquire as to cost and availability.

CSA Engineering. Cambridge Scientific Abstracts, 7200 Wisconsin Avenue, Suite 601, Bethesda, MD 20814. (301) 961-6750 or (800) 843-7751. Contains citations and abstracts of international periodicals and research literature covering all fields of engineering and science and technology,including computer and information science, electronics, mechanical engineering, solid state materials, 1981 to present. Available on BRS,(800) 289-4277, online service.Inquire as to cost and availability.

Current Contents Search. Institute for Scientific Information, 3501 Market Street, Philadelphia, PA 19104. (215) 386-0100. FAX (215) 386-6362. Contains citations to articles listed in the table of contents of science and technology journals. Alsoarticles in social sciences and life sciences journals. Available on BRS,(800) 289-4277, DIALOG,(800) 334-2564, online services. Inquire as to cost and availability.

GEOREF: Online Version of Bibliography and Index of Geology. American Geological Institute, 4220 King Street, Alexandria, VA 22302. (703) 379-2480. Fax (703) 379-7563. Monthly. Inquire for price and availability.

NTIS Bibliographic Database. National Technical Information Service, 5285 Port Royal Road, Springfield, Va 22161. (703) 487-4929 or FAX (703) 321-8199. Broad coverage of government-sponsored science and technology research reports, 1964 to present. Available on BRS,(800) 289-4277, DIALOG, (800) 334-2564, ORBIT, (800) 456-7248, and STN International, FIZ Karlsruhe, P.O. Box 2465, W-7500, Karlsruhe 1, Germany, online services. Also available on CD-ROM. Inquire as to cost and availability.

Scisearch. Institute for Scientific Information, 3501 Market Street, Philadelphia, PA 19104. (800) 523-1850 or (215) 386-0100. Broad multidisciplinary title and author index to the international literature of science and technology, 1974 to present. Available on DIALOG,(800) 334-2564, and ORBIT,(800) 456-7248, online services. Also available on CD-ROM. Inquire as to cost and availability.

WILSONLINE. H.W. Wilson Company, 950 University Avenue, Bronx, NY 10452. (800) 367-6770 or (212) 588-8400. Makes available online versions of the printed H.W. Wilson indexes including Applied Science and Technology Index, Business Periodicals Index, General Science Index, and Readers' Guide to Periodical Literature. Period covered is generally 1983 to present. Available on BRS,(800) 289-4277, DIALOG,(800) 334-2564, and OCLC EPIC,(800) 848-5878, online services. Also available on CD-ROM. Inquire as to cost and availability.

CANALS

Ency. of Physical Sciences and Engineering Info. Sources

PERIODICALS

Civil Engineering. American Society of Civil Engineers, 345 East 47th Street, New York, NY 10017-2398. (212) 705-7520 or (800) 548-2723. 1930 to present. Monthly. $85.00 per year.

Journal of Construction Engineering and Management. American Society of Civil Engineers, Construction Division, 345 East 47th Street, New York, NY 10017-2398. (212) 705-7520 or (800) 548-2723. 1956 to present. Quarterly. $112.00 per year.

Journal of Surveying Engineering. American Society of Civil Engineers, Surveying Engineering Division, 345 East 47th Street, New York, NY 10017-2398. (212) 705-7520 or (800) 548-2723. 1956 to present. Quarterly. $76.00 per year.

Journal of Geotechnical Engineering. American Society of Civil Engineers, Geotechnical Engineering Division, 345 East 47th Street, New York, NY 10017-2398. (212) 705-7520 or (800) 548-2723. 1956 to present. Monthly. $212.00 per year.

Journal of Hydraulic Engineering. American Society of Civil Engineers, Hydraulics Division, 345 East 47th Street, New York, NY 10017-2398. (212) 705-7520 or (800) 548-2723. 1956 to present. Monthly. $200.00 per year.

Journal of Irrigation and Drainage. American Society of Civil Engineers, Irrigation and Drainage Division, 345 East 47th Street, New York, NY 10017-2398. (212) 705-7520 or (800) 548-2723. 1956 to present. Bi-monthly. $136.00 per year.

Journal of Water Resources Planning and Management. American Society of Civil Engineers, Water Resources Planning and Management Division, 345 East 47th Street, New York, NY 10017-2398. (212) 705-7520 or (800) 548-2723. Bi-monthly. $112.00 per year.

Journal of Waterway, Port, Coastal, and Ocean Engineering. American Society of Civil Engineers, Waterway, Port, Coastal, and Ocean Division, 345 East 47th Street, New York, NY 10017-2398. (212) 705-7520 or (800) 548-2723. 1956 to present. Bi-monthly. $100.00 per year.

RESEARCH CENTERS AND INSTITUTES

LSU/GWU National Ports & Waterways Institute. 600 New Hampshire Avenue NW, Suite 475, Washington, DC 20037. (202) 338-9427. FAX (202) 338-9509.

CANTILEVERS

See: STRUCTURAL ENGINEERING

CAPACITORS

See also: BATTERIES, ELECTRICAL ENGINEERING, ELECTRO-CHEMISTRY, ELECTRONIC CIRCUITS AND COMPONENTS, ELECTRONICS, ELECTRONICS ENGINEERING

ABSTRACT SERVICES AND INDEXES

Applied Science and Technology Index; a cumulative subject index to English language periodicals in the fields of aeronautics and space science, computer technology, chemistry, construction industry, energy and related areas. H.W. Wilson Co., 950 University Avenue, Bronx, NY 10452. (212) 588-8400. (800) 367-6770. FAX (718) 590-1617. From 1958 to present. Monthly. Inquire about cost and availability. Also available on CD-ROM and online. ISSN: 0003-6986.

Chemical Abstracts. Chemical Abstracts Service, 2540 Olentangy River Road, Box 3012, Columbus, OH 43210-0012. (614) 447-3600. (800) 848-6538. FAX (614) 447-3713. From 1907 to present. Weekly. $16800.00 per year. Also available on CD-ROM and online. Inquire regarding cost and availability.

Current Papers In Electrical and Electronics Engineering. Institution of Electrical Engineers (IEE), London. Distributed by INSPEC/IEEE, Box 1331, 445 Hoes Lane, Piscataway, NJ 08855-1331. (908) 562-5549. 1969 to present. Monthly. $345.00 per year. ISSN: 0011-3778.

Electrical and Electronics Abstracts. Institute of Electrical Engineers, Michael Faraday House, Six Hill Way, Stevenage, Herts, SG1 2AY, England. Distributed by IEEE, 445 Hoes Lane, Piscataway, NJ 08854. (908) 562-5549. 1898 to present. Monthly. $2200.00 per year. Also available on CD-ROM and online as INSPEC.

Engineering Index Monthly. Engineering Information, Inc., Castle Point on the Hudson, Hoboken, NJ 07030. (201) 216-8500. FAX (201) 216-8532. Monthly. $2200.00 per year. Also available online as COMPENDEX and on CD-ROM. ISSN: 0742-1974.

General Science Index. H.W. Wilson Company, 950 University Avenue, Bronx, NY 10452. (212) 588-8400. (800) 367-6770. FAX (718) 590-1617. From 1978 to present. Ten issues per year; quarterly and annual cumulations. Service basis. Available on CD-ROM and online. Inquire regarding cost and availability. ISSN: 0162-1963.

Index To IEEE Publications. IEEE Service Center, 445 Hoes Lane, Piscataway, NJ 08855-1331. (908) 981-1393. (800) 678-IEEE. FAX (908) 981-9667. 1973 to present. Annual. ISSN: 0099-1368.

Physics Abstracts . INSPEC. Section A, Science Abstracts. Institutio n of Electrical Engineers (IEE). Available from INSPEC/IEEE - Institute of Electrical and Electronic Engineers, Box 1331, 445 Hoes Lane, Piscaway NJ 08855-1331. (908) 562-5549. 1898 to present. 24 issues per year. $2835.00 per year. Also available on CD-ROM and online. ISSN: 0036-8091.

Physics Briefs (Physikalische Berichte). Information Center for Energy, Physics, Mathematics; German Physical Society. V C H Publishers, Inc., 220 East 23rd Street, New York, NY 10010-4606. (212) 683-8333. 1845 to present. 24 issues per year. $2390.00 per year. Also available online. ISSN: 0179-7434.

Science Citation Index. SCI. Institute for Scientific Information, 3501 Market Street, Philadelphia, PA 19104. (215) 386-0100. (800) 523-1850. FAX (215) 386-2991. 1961 to present. Six issues per year, plus annual cumulation. $11650.00 per year. Also available online and on CD-ROM. Inquire about price and availability. ISSN: 0036-827X.

Solid State and Superconductivity Abstracts: Covers theory, Production and Application of Solid State Materials. Cambridge Scientific Abstracts, 7200 Wisconsin Avenue, Bethesda, MD 20824. (301) 961-6750. FAX (301) 961-6720. 1957 to present. Bimonthly. $1320.00 per year. Also available online. ISSN: 0896-5900.

ANNUAL REVIEWS AND YEARBOOKS

Advances in Electrochemistry and Electrochemical Engineering. John Wiley & Sons, Inc., 605 Third Avenue, New York, NY 10158-0012. (212) 850-6000. (800) 225-5945. Irregular. Price varies. inquire.

Advances In Electronics and Electron Physics. Academic Press, Inc., 6277 Sea Harbor Drive, Orlando, FL 32887. (800) 321-5068. From 1948 to present. Irregular. Price varies, inquire. ISSN: 0065-2539.

ASSOCIATIONS AND PROFESSIONAL SOCIETIES

American Chemical Society. 1155 16th Street, NW, Washington DC 20036. (202) 872-4600.

American Electronics Association. 5201 Great America Way, Suite 520, P.O. 52990, Santa Clara, CA 95056. (408) 987- 4200. FAX(408) 970-8565.

American Institute of Physics. One Physics Ellipse, College Park, MD 20740-3843. (301) 209-3100.

Edison Electric Institute. 701 Pennsylvania Avenue NW, Washington, DC 20004-2696. (202) 508-5000, 5454. FAX (202) 508-5360.

Electronics Industries Association. 2500 Wilson Boulevard, Arlington, VA 22201. (703) 907-7500. FAX (202) 457-4985.

IEEE (Institute of Electrical and Electronic Engineers). 345 East 47th Street. New York, NY 10017. (212) 705-7900. FAX (212) 705-4929,

International Society For Hybrid Microelectronics. 1850 Centennial Park Drive, Suite 105, Reston, VA 22091. (703) 758-1060. FAX (703) 758-1066.

DIRECTORIES AND BIOGRAPHICAL SOURCES

American Men and Women of Science: Physical and Biological Sciences. R. R. Bowker Inc., 121 Chanlon Road, New Providence, NJ 07974. (908) 464-6800. (800) 521-8110. 20th edition. 8 volumes. 1996. $850.00.

Directory of Engineering Societies and Related Organizations. American Association of Engineering Societies, 1111 19th Street, Suite 608, Washington, DC 20036-3603. (202) 296-2237. Semi-annual. $150.00.

E E M - Electronic Engineer's Master. Hearst Business Communications, Inc., 645 Stewart Avenue, Garden City NY 11530. (516) 227-1300. Annual. $90.00 ISSN: 0732-9016.

IEEE Membership Directory. Institute of Electrical and Electronics Engineers, IEEE Service Center, 445 Hoes Lane, Picastaway, NY 08854. (908) 981-1393. (800) 678-IEEE. FAX (908) 981-9667. 2 volumes. Annual. $190.00. ISSN:

International Directory of Abbreviations and Acroynms of Electronics, Electrical Engineering, Computer Technology and Information Processing. Peter Wennrich. K. G. Saur, 121 Chanlon Road, New Providence, NJ 07974. (908) 464-6800. (800) 521-8110 2 volumes. 1992. $230.00.

Research Centers Directory. Gale Research Inc., 835 Penobscot Building, Detroit, MI 48226-4094. (313) 961-2242. (800) 347-4253. 20th edition. 1995. $485.00. ISSN:: 0080- 1518.

Scientific and Technical Organizations and Agencies Directory. Gale Research, 835 Penobscot Building, Detroit, MI 48226-4094. (313) 961-2242. (800) 347-4253. 4th edition. 1996. $195.00.

Who's Who In Engineering. Gordon Davis, editor. American Society for Engineering Education, 1111 19th Street, NW, Suite 608, Washington, DC 20036. (202) 296-2237. (800) 658-8897. 9th edition. 1995. $220.00.

Who's Who In Technology. Gale Research, 835 Penobscot Building, Detroit, MI 48226-4094. (313) 961-2242. (800) 347-4253. 7th edition. 1995. $195.00.

ENCYCLOPEDIAS AND DICTIONARIES

Encyclopedia of Applied Physics. George Trigg, editor. VCH Publications, Inc., 220 East 23rd Street, Suite 909, New York, NY 10010-4606. (212) 683-8333. (800) 422-8824. 20 volume set. 1991-. $5990.00.

IEEE Standard Dictionary of Electrical and Electronics Terms. Christopher J. Booth, editor. IEEE Service Center, 445 Hoes Lane, Piscataway, NJ 08855-1331. (908) 981-1393. (800) 678-IEEE. FAX (908) 981-9667. IEEE Standard 100- 1992. 5th edition. 1993. $150.00.

Illustrated Dictionary of Electronics. Stan Gibilsco. TAB Books, P.O. Box 40, Blue Summit, PA 17294-0850. (717) 794- 2191. (800) 233-1128. 7th edition. 1994. $34.95.

McGraw-Hill Encyclopedia of Science and Technology. McGraw- Hill Book Company, Inc., 1221 Avenue of the Americas, New York, NY 10020. (212) 512-2000. (800) 262-4729. 7th edition. 20 volume set. 1992. $1900.00.

Modern Power Supply and Battery Charger Circuit Encyclopedia. Rudolf F. Graf. McGraw-Hill Book Company, Inc., 1221 Avenue of the Americas, New York, NY 10020. (212) 512- 2000. (800) 262-4729. 1992. $11.95 in paper.

GENERAL WORKS

Capacitance, Inductance and Crosstalk Analysis. Charles Walker. Artech House, 685 Canton Street, Norwood, MA 02062. (617) 769-9750. (800) 225-9977. 1990. $69.00.

Dielectric Ceramics: Processing, Properties and Applications. K. M. Nair, et al, editors. American Ceramic Society, 735 Ceramic Place, Westerville, OH 43081. (614) 794- 5824. 1993. $69.00.

Electric Circuits: Principles, Applications and Computer Analysis. David A. Bell. Prentice Hall , 113 Sylvan Avenue, Route 9W, Englewood Cliffs, NJ 07632. (201) 592- 2000. (800) 922-0579. 1993. $72.00.

Electric Fields In Multilayered Structures. Arun Bhattacharyya. Artech House, 685 Canton Street, Norwood, MA 02062. (617) 769-9750. (800) 225-9977. 1993. $69.00.

Exploring Electronic Devices. Mark E. Hazen. SCP: Third World Literature Publishing House, P.O. Box 482, Lithonia, GA 30058-0482. (404) 785-7725. 1991. $50.75.

Switched Capacitor Filters: theory Analysis and Design. Anandmohan, et al. Prentice Hall , 113 Sylvan Avenue, Route 9W, Englewood Cliffs, NJ 07632. (201) 592-2000. (800) 922-0579. 1995. $58.00.

HANDBOOKS AND MANUALS

The Capacitor Handbook. Cletus J. Kaiser. Van Nostrand Reinhold, 115 Fifth Avenue, New York, NY 10003. (212) 254-3232. (800) 842-3636. 1993. $29.95.

CRC Handbook of Chemistry and Physics. David R. Kide, editor. CRC Press, Inc., 2000 Corporate Boulevard, NW, Boca Raton, FL 33431. (407) 994-0555. (800) 272-7737. 77th edition. 1996. $99.95.

Electrical Engineer's Handbook. Donald Christiansen, editor. McGraw Book Company, 1221 Avenue of the Americas, New York, NY 10020. (212) 512-2000. (800) 262-4729. 4th edition. 1996. $110.00.

Handbook of Solid State Batteries and Capacitors. M. Z. Munshi and P. S. Prasad. World Scientific Publishing Company, Inc., 1060 Main Street, River Edge, NJ 07661. (201) 487-9655. (800) 227-7562. 1995. $177.00.

ONLINE DATABASES AND CD-ROMS

Ca Search. Chemical Abstracts Service, P.O. Box 3012, Columbus, OH 43210-0012. (614) 447-3600. (800) 848-6533. FAX (614) 447-3709. Very comprehensive guide to worldwide chemical literature and related fields, 1972 to present. Available on BRS,(800) 289-4277, DIALOG, (800) 334-2564, ORBIT (800) 456-7248, and STN International, FIZ Karlsruhe, P.O. Box 2465, W-7500, Karlsruhe 1, Germany, online services. Inquire as to cost and availability.

Compendex Plus. Engineering Information, Inc., 345 East 47th Street, New York, NY 10017. (212) 705-7600 or (800) 221-1044. Contains citations with abstracts to worldwide literature in engineering and technology, from 1970 to present. Available on online BRS,(800) 289-4277, DIALOG, (800) 334-2564, ORBIT (800) 456-7248, and STN International, FIZ Karlsruhe, P.O. Box 2465, W-7500, Karlsruhe 1, Germany, online services. Also available on CD-ROM. Inquire as to cost and availability.

Dissertation Abstracts. University Microfilms International, 300 North Zeeb Road, Ann Arbor, MI 48106. (800) 521-0600 or (313) 761-4700. Scope includes virtually all doctoral dissertations accepted at accredited American institutions from 1861 to present in 252 subject areas. Available on BRS, (800) 289-4277, DIALOG, (800) 334-2564, and OCLC EPIC, (800) 848-5878,

online services. Also available on CD-ROM. Inquire as to cost and availability.

Inspec. Institution of Electrical Engineers, Michael Faraday House, Six Hills Way, Stevenage, Herts. SG1 2AY, Engl and. Telephone: 0438 313311 or FAX 0438 742840. Contains citations to the worldwide literature of physi cs, electronics and electrical engineering, computer technology, and related fields. Available on BRS, (800) 289-4277, DIALOG, (800) 334-2564, ORBIT, (800) 456-7248, and STN International, FIZ Karlsruhe, P.O. Box 2465, W-7500, Karlsruhe 1, Germany, online services. Inquire as to cost and availability.

NTIS Bibliographic Database. National Technical Information S ervice, 5285 Port Royal Road, Springfield, Va 22161. (703) 487-4929 or FAX (703) 321-8199. Broad coverag e of government-sponsored science and technology research reports, 1964 to present. Available on BRS,(800) 289-4277, DIALOG, (800) 334-2564, ORBIT, (800) 456-7248, and STN International, FIZ Karlsruhe, P.O. Box 2465, W-7500, Karlsruhe 1, Germany, online services. Also available on CD-ROM. Inquire as to cost and availability.

Scisearch. Institute for Scientific Information, 3501 Mar ket Street, Philadelphia, PA 19104. (800) 523-1850 or (215) 386-0100. Broad multidisciplinary title and author index to the international literature of science and technology, 1974 to present. Available on DIALOG, (800) 334-2564, and ORBIT, (800) 456-7248, online services. Also available on CD-ROM. Inquire as to cost and availability.

WILSONLINE. H.W. W ilson Company, 950 University Avenue, Bronx, NY 10452. (800) 367-6770 or (212) 588-8400. Makes available onlin e versions of the printed H.W. Wilson indexes includin g Applied Science and Technology Index, Business Periodicals Index, General Science Index, and Readers' Guide to Periodical Literature. Period covered is generally 1983 to present. Available on BRS, (800) 289-4277, DIALOG, (800) 334-2564, and OCLC EPIC, (800) 848-5878, online services. Also available on CD-ROM. Inquire as to cost and availability.

PERIODICALS

Circuits, Systems, and Signal Processing. Birkhauser; distributed by Springer-Verlag New York, Inc., Box 2485, Secaucus, NH 07096-2491. (201) 348-4033. 1982 to present. Quarterly. $184.00 per year. ISSN: 0278-081X.

Electronic Design. Penton Publishing, San Jose Gateway, Suite 354. 2025 Gateway Place, San Jose, CA 95110. (408) 441-0550. 1952 to present. Biweekly. $95.00 per year. ISSN: 0013-4872.

Electronics. Penton Publishing, San Jose Gateway, Suite 354. 2025 Gateway Place, San Jose, CA 95110. (408) 441- 0550. 1930 to present. Semi-weekly. $98.00. ISSN: 0883- 4989.

IEEE Circuits and Devices Magazine. Institute of Electrical and Electronics Engineers, Inc., Box 1331, 445 Hoes Lane, Piscataway, NJ 08855-1331. (908) 981-0060. 1985 to present. Bi-monthly. $120.00 per year. ISSN: 8755-3996.

IEEE Journal of Solid State Circuits. Institute of Electrical and Electronics Engineers, Inc., Box 1331, 445 Hoes Lane, Piscataway, NJ 08855-1331. (908) 981-0060. 1966 to present. Bi-monthly. $275.00 per year to non-members.

IEEE Transactions On Electron Devices. Institute of

Electrical and Electronics Engineers, Inc., Box 1331, 445 Hoes Lane, Piscataway, NJ 08855-1331. (908) 981-0060. 1952 to present. Monthly. $395.00 per year. ISSN: 0018-9383.

Institute of Electrical and Electronics Engineers. PROCEEDINGS. Institute of Electrical and Electronics Engineers, Inc., Box 1331, 445 Hoes Lane, Piscataway, NJ 08855-1331. (908) 981-0060.

Semiconductor International. Cahners Publishing Company. 1350 East Touhy Avenue, Box 5080, Des Plaines, IL 60017-6080. (708) 635-8800. FAX (708) 390-2770. 1978 to present. 13 issues per year. $84.95 per year.

Solid State Electronics. Elsevier Science, 660 White Plains Road, Tarrytown, NY 10591-5153. (914) 524-9200. FAX (914) 333-2444. 1960 to present. Monthly. $1025.00 per year.

RESEARCH CENTERS AND INSTITUTES

Edison Electric Institute. 701 Pennsylvania Avenue, NW, Washington, DC 20004. (202) 508-5000

Electrical Engineering Research Laboratories. Purdue University, Electrical Engineer4ing Building, West Lafayette, in 47907. (317) 494-3536. FAX (317) 494-6440.

Electronics Research Center. University of Texas at Austin, 132 Engineering Science Building, Austin, TX.

Electronics Research Laboratory. University of California, Berkeley, 243 Cory Hall, Berkeley CA 94720. (415) 642-2301.

Laboratory For Electromagetic and Electronic Systems. Massachusetts Institute of Technology, 77 Massachusetts Avenue, Cambridge, MA 02139.

CARBON

See also: CHEMISTRY, ORGANIC CHEMISTRY, PETROLEUM CHEMISTRY, PETROLEUM ENGINEERING, PLASTICS, POLYMERS

ABSTRACT SERVICES AND INDEXES

A.P.I. Abstracts. American Petroleum Institute, Central Abstracting and Information Services, 275 Seventh Avenue, New York, NY 10001. (212) 366-4040 or FAX (212) 366-4298. Contains citations with abstracts to literature relating to the petroleum refining and the petrochemical industries, from 1964 to present.

Chemical Abstracts - Organic Chemistry Sections. Chemical Abstracts Service, 2540 Olentangy River Road, Box 3012, Columbus, OH 43210. (800) 848-6538 or (614) 447-3600, FAX (614) 447-3713. Bi-weekly. $1410.00 per year.

Current Contents: Physical, Chemical and Earth Sciences. Institute for Scientific Information, 3501 Market Street, Philadelphia, PA 19104. (215) 386-0100. FAX (215) 386-6362. Weekly. $360.00 per year.

General Physics Advance Abstracts. American Institute of Physics, 335 East 45th Street, New York, NY 10017 (212) 661-9404 or FAX (516)349-9704. 1985 to present. Six times per year. $180.00 per year.

Index To Scientific and Technical Proceedings. Institute for Scientific Information, 3501 Market St., Philadelphia, PA 19104. (215) 386-0100. FAX (215) 386-6362. Monthly. $500.00 per year.

Index To Scientific Reviews. Institute for Scientific Information, 3501 Market St., Philadelphia, PA 19104. (215) 386-0100. FAX (215) 386-6362. Semi-annual.

Manufacturing and Process Engineering Abstracts, Cambridge Scientific Abstracts, 7200 Wisconsin Avenue, Bethesda, Md 20814-4823. (301) 961-6750. FAX (301) 961-6720. 1993 to present. Monthly. $385.00 per year. Covers concurrent engineering, quality control, automated manufacturing, petroleum engineering, oil field operations and equipment, energy management, metallurgy and metallography, foundry practice.

Materials Science and Engineering Abstracts, CambridgeScientific Abstracts, 7200 Wisconsin Avenue, Bethesda, Md 20814-4823. (301) 961-6750. FAX (301) 961-6720. 1993 to present. Monthly. $385.00 per year. Focuses on mechanical and physical properties of materials and commercial or industrial applications for materials, methods for strength testing, effects of vibration and other stresses, corrosion and protective coatings, storage and handling, ceramics, composites, metals, wood, plastics, and polymers.

Petroleum Abstracts. 600 South College, Tulsa, OK 74104. (918) 631-2296 or (800) 247-8678. Contains citations with abstracts to the worldwide literature and patents on the exploration, development, and production of petroleum resources, from 1965 to present.

Polymer Blends, Alloys, and Interpenetratin Polymer Network Abstracts. Technomic Publishing Company, Inc., 851 New Holland Avenue, Box 3535, Lancaster, PA 17604. (717) 291-5609. Covers literature in polymer science and product development, from 1987 to present. Monthly. $130.00 per year.

Polymer Contents: International Current Awareness Publication On Polymer Science and Engineering. Elsevier Science Publishing Company, Inc., 655 Avenue of the Americas, New York, NY 10010. (212) 989-5800. 1984 to present. Monthly. $350.00 per year.

Science Citation Index. Institute for Scientific Information, 3501 Market Street, Philadelphia, PA 19104. (215) 386-0100. FAX (215) 386-6362. Inquire about availability and cost. Also available on CD-ROM.

ASSOCIATIONS AND PROFESSIONAL SOCIETIES

American Carbon Society. Carbon of America, 215 Stackpole Street, St. Marys, PA 15857. (814) 781-8410. FAX (814) 781-8570.

American Chemical Society. 1155 16th Street, N.W., Washington, DC 20036. (202) 872-4381 or (800) 333-9511. FAX (614) 447-3671.

American Institute of Chemical Engineers. 345 East 47th Street, New York, NY 10017-2396. (212) 705-7663.

CARBON

Ency. of Physical Sciences and Engineering Info. Sources

American Oil Chemists' Society. PO Box 3489, 1608 Broadmoor Drive, Champaign, IL 61826-3489. (217) 359-2344. FAX (217) 351-8091.

Aoac International (association of official Analytical Chemists). 2200 Wilson Blvd., Suite 400, Arlington, VA 22201-3301. (703) 522-3032. FAX (703) 522-5468.

DIRECTORIES AND BIOGRAPHICAL SOURCES

American Men & Women of Science, 1995-96. R.R. Bowker Staff, eds. 19th edition. 8 volumes. R.R. Bowker/Reed International Publishing Company, 121 Chanlon Road, New Providence, NJ 07974. (908) 464-6800 or (800) 521-8110. 1995. $850.00

Directory of Engineering Societies and Related Organizations. Gordon Davis. 13th edition. American Association of Engineering Societies, 1111 19th Street NW, Suite 608, Washington, DC 20036. (202) 296-2237 or (800) 658-8897. 1989. Inquire for price.

International Research Centers Directory. Gale Research, 835 Penobscot Building, Detroit, MI 48226-4094. (313) 961-2242. (800) 347-4253. FAX (313) 961-6083. 8th edition. 1995. $410.00.

Research Centers Directory. Gale Research, 835 Penobscot Building, Detroit, MI 48226-4094. (313) 961-2242. (800) 347-4253. FAX (313) 961-6083. $485.00.

Scientific and Technical Organizations and Agencies Directory. Gale Research, 835 Penobscot Building, Detroit, MI 48226-4094. (313) 961-2242. (800) 347-4253. FAX (313) 961-6083. 4th edition. 1996. $195.00.

Who's Who In Engineering. American Association of Engineering Societies, 1111 19th Street NW, Suite 608, Washington, DC 20036. (202) 296-2237 or (800) 658-8897. 8th edition. 1991. Inquire for price.

ENCYCLOPEDIAS AND DICTIONARIES

Kirk-Othmer Encyclopedia of Chemical Technology: Bearing Materials To Carbon (volume 4). J.I. Kroschwitz and Mary Howe-Grant, editors. John Wiley and Sons, Inc., 605 Third Avenue, New York, NY 10158. (800) 526-5368 or (212) 850-6000. 1992. $295.00.

MacMillan Encyclopedia of Chemistry. 4 volumes. Macmillan Publishing, 200 Old Tappan Road, Old Tappan, NJ 07675. (800) 223-2336. FAX (800) 445-6991. 1996. $350.00.

Dictionary of Organic Compounds: Tenth Supplement. John B. Buckingham, editor. Chapman and Hall, 29 West 35th Street, New York, NY 10001. (800) 842-3636. FAX (212) 563-2269. 1993. $832.50.

McGraw-Hill Encyclopedia of Chemistry. Sybil P. Parker. Second edition. McGraw-Hill Publishing Company, 1221 Avenue of the Americas, New York, NY 10020. (800) 262-4729 or (212) 512-3825. 1993. $99.50.

GENERAL WORKS

Carbon-carbon Composites. G.M. Savage. Chapman and Hall, 29 West 35th Street, New York, NY 10001. (800) 842-3636. FAX (212) 563-2269. 1993. $77.95.

Carbon: Electrochemical and Physiochemical Properties. K. Kinoshita. John Wiley and Sons, Inc., 605 Third Avenue, New York, NY 10158. (800) 526-5368 or (212) 850-6000. 1988. $150.00.

The Fullerenes: New Horizons For the Chemistry, Physics & Astrophysics of Carbon. H.W. Kroto and D.R. Walton, editors. Cambridge University Press, 40 West 20th Street, New York, NY 10011-4211. (212) 924-3900 or (800) 872-7423. FAX (914) 937-4712. 1993. $24.95.

Introduction To Carbon Science. Harry Marsh, et al. Butterworth-Heinemann, 313 Washington Street, Newton, MA 02158. (617) 928-2500 or (800) 366-2665. FAX (617) 928-2620. 1989. $94.95.

HANDBOOKS AND MANUALS

CRC Handbook of Chemistry and Physics. David R. Lide, editor. 75th edition. CRC Press, 2000 Corporate Blvd., N.W., Boca Raton, FL 33431. (407) 994-0555 or (800) 272-7737. FAX (407) 994-0949. $99.50.

Handbook of Carbon, Graphite, Diamond and Fullerenes: Properties, Processing, and Applications. Hugh O. Pierson. Noyes Publications, 120 Mill Road, Park Ridge, NJ 07656. (201) 391-8484. FAX (201) 391-6833. 1993. $86.00.

Handbook of Organic Chemistry. Hans Beyer and Wolfgang Walter. Prentice Hall, 113 Sylvan Avenue, Route 9W, Englewood Cliffs, NJ 07632. (201) 592-2000 or (800) 922-0579. 1995. $96.00.

ONLINE DATABASES AND CD-ROMS

CA Search. Chemical Abstracts Service, P.O. Box 3012, Columbus, OH 43210-0012. (614) 447-3600. (800) 848-6533. FAX (614) 447-3709. Very comprehensive guide to worldwide chemical literature and related fields, 1972 to present. Available on BRS,(800) 289-4277, DIALOG, (800) 334-2564, ORBIT (800) 456-7248, and STN International, FIZ Karlsruhe, P.O. Box 2465, W-7500, Karlsruhe 1, Germany, online services. Inquire as to cost and availability.

Compendex Plus. Engineering Information, Inc., 345 East 47th Street, New York, NY 10017. (212) 705-7600 or (800) 221-1044. Contains citations with abstracts to worldwide literature in engineering and technology, from 1970 to present. Available on online BRS,(800) 289-4277, DIALOG, (800) 334-2564, ORBIT (800) 456-7248, and STN International, FIZ Karlsruhe, P.O. Box 2465, W-7500, Karlsruhe 1, Germany, online services. Also available on CD-ROM. Inquire as to cost and availability.

Current Contents Search. Institute for Scientific Information, 3501 Market Street, Philadelphia, PA 19104. (215) 386-0100. FAX (215) 386-6362. Contains citations to articles listed in the table of contents of science and technology journals. Also articles in social sciences and life sciences journals. Available on BRS,(800) 289-4277, DIALOG,(800) 334-2564, online services. Inquire as to cost and availability.

Ency. of Physical Sciences and Engineering Info. Sources

CARBONATES

Dissertation Abstracts. University Microfilms International, 300 North Zeeb Road, Ann Arbor, MI 48106. (800) 521-0600 or (313) 761-4700. Scope includes virtually all doctoral dissertations accepted at accredited American institutions from 1861 to present in 252 subject areas. Available on BRS,(800) 289-4277, DIALOG,(800) 334-2564, and OCLC EPIC,(800) 848-5878, online services. Also available on CD-ROM. Inquire as to cost and availability.

NTIS Bibliographic Database. National Technical Information Service, 5285 Port Royal Road, Springfield, Va 22161. (703) 487-4929 or FAX (703) 321-8199. Broad coverage of government-sponsored science and technology research reports, 1964 to present. Available on BRS,(800) 289-4277, DIALOG, (800) 334-2564, ORBIT, (800) 456-7248, and STN International, FIZ Karlsruhe, P.O. Box 2465, W-7500, Karlsruhe 1, Germany, online services. Also available on CD-ROM. Inquire as to cost and availability.

Scisearch. Institute for Scientific Information, 3501 Market Street, Philadelphia, PA 19104. (800) 523-1850 or (215) 386-0100. Broad multidisciplinary title and author index to theinternational literature of science and technology, 1974 to present. Available on DIALOG,(800) 334-2564, and ORBIT,(800) 456-7248, online services. Also available on CD-ROM. Inquire as to cost and availability.

PERIODICALS

American Oil Chemists' Society Journal. American Oil Chemists' Society, PO Box 3489, 1608 Broadmoor Drive, Champaign, IL 61826-3489. (217) 359-2344. FAX (217) 351-8091. 1917 to present. Monthly. $185.00 per year.

Carbon. Elsevier Science [journals], 660 White Plains Rd., Tarrytown, NY 10159-5153. (919) 524-9200. FAX (919) 333-2444. 1963 to present. Eight times a year. $1045.00.

Chemistry and Physics of Carbon: A Series of Advances. Marcel Dekker, Inc., 270 Madison Avenue, New York, NY 10016. (212) 696-9000. FAX (212) 685-4540. 1966 to present. Irregular. Price varies.

Hydrocarbon Processing. Gulf Publishing Company, P.O. Box 2608, Houston, TX. (713) 529-4301 or (800) 231-6275. 1922 to present. Monthly. $22.00 per year.

Journal of Organic Chemistry. American Chemical Society, 1155 16th Street, N.W., Washington, DC 20036. (202) 872-4381 or (800) 333-9511. FAX (614) 447-3671. 1936 to present. Bi-weekly. $785.00 per year for non-members.

Journal of Polymer Science A: Polymer Chemistry. J. Wiley & Sons, Journals, 605 Third Avenue, New York, NY 10158-0012. (212) 692-6000. 1962 to present. 13 times a year. $2540.00 per year (also includes *Journal of Polymer Science B: Polymer Physics*).

Organometallic Chemistry. Royal Society of Chemistry, Thomas Graham House, Science Park, Milton Road, Cambridge CB4 4WF, England. Telephone 0223-420066. FAX 0223-423623. 1971 to present. Annual. price varies.

Studies In Organic Chemistry. Elsevier Science Inc., Box 882, Madison Square Station, New York, NY 10159. (212) 989-5800. FAX (212) 633-3990. 1979 to present. Irregular. Price varies.

Tetrahedron. Elsevier Science [journals], 660 White Plains Rd., Tarrytown, NY 10159-5153. (919) 524-9200. FAX (919) 333-2444. 1957 to present. 60 times a year. $5365.00.

Tetrahedron Letters. Elsevier Science [journals], 660 White Plains Rd., Tarrytown, NY 10159-5153. (919) 524-9200. FAX (919) 333-2444. 1959 to present. 52 times a year. $4745.00.

RESEARCH CENTERS AND INSTITUTES

Mississippi State University, University/industry Chemical Research Center. Department of Chemistry, PO Drawer CH, Mississippi State, MS 39762. (601) 325-3584. FAX (601) 325-7611.

Rensselaer Polytechnic Institute, Chemistry Laboratories. Cogswell Laboratory, Troy, NY 12180-3590. (518) 276-8981.

CARBON-14 DATING

See: RADIOCARBON DATING

CARBONATES

See also: CAVES AND CAVING, GEOLOGY, GEOMORPHOLOGY, KARST,PETROLEUM GEOLOGY, SEDIMENTARY ROCKS

ABSTRACT SERVICES AND INDEXES

Bibliography and Index of Geology. American Geological Institute, 4220 King Street, Alexandria, VA 22302. (703) 379-2480. Fax (703) 379-7563. Monthly. $1295.00 per year. Also available online as GeoRef.

Current Contents: Physical, Chemical and Earth Sciences. Institute for Scientific Information, 3501 Market Street, Philadelphia, PA 19104. (215) 386-0100. FAX (215) 386-6362. Weekly. $360.00 per year.

Geological Abstracts. Elsevier - Geo Abstracts, Regency House, 34 Duke Street, Norwich NR3 3AP, England. Monthly. $760.00 per year. Also available online as GEOBASE.

Geological Society of America. Abstracts with Programs. Geological Society of America. 3300 Penrose Place, P.O. Box 9140, Boulder, CO 80301-9140. (303) 447-2020. Abstracts and programs of the annual conference. Annual. $69.00.

Science Citation Index. Institute for Scientific Information, 3501 Market Street, Philadelphia, PA 19104. (215) 386-0100. FAX (215) 386-6362. Inquire about availability and cost. Also available on CD-ROM.

ASSOCIATIONS AND PROFESSIONAL SOCIETIES

American Geological Institute. 4220 King Street, Alexandria, VA 22302. (703) 379-2480. Fax (703) 379-7563.

Geological Society of America. 3300 Penrose Place, PO Box 9140, Denver, CO 80301-9140. (303) 447-2020.

CARBONATES

Ency. of Physical Sciences and Engineering Info. Sources

National Speleological Society Inc., 2813 Cave Avenue, Huntsville, AL 35810-4431. (205) 852-1300. FAX (205) 851-9241.

Sepm—society For Sedimentary Geology (formerly Society of Economic Paleontologists). 1731 E 71st Street, Tulsa, OK 74136-5108. FAX (918) 493-2093.

DIRECTORIES AND BIOGRAPHICAL SOURCES

American Men & Women of Science, 1995-96. R.R. Bowker Staff, eds. 19th edition. 8 volumes. R.R. Bowker/Reed International Publishing Company, 121 Chanlon Road, New Providence, NJ 07974. (908) 464-6800 or (800) 521-8110. 1995. $850.00

International Research Centers Directory. Gale Research, 835 Penobscot Building, Detroit, MI 48226-4094. (313) 961-2242. (800) 347-4253. FAX (313) 961-6083. 8th edition. 1995. $410.00.

Research Centers Directory. Gale Research, 835 Penobscot Building, Detroit, MI 48226-4094. (313) 961-2242. (800) 347-4253. FAX (313) 961-6083. $485.00.

Scientific and Technical Organizations and Agencies Directory. Gale Research, 835 Penobscot Building, Detroit, MI 48226-4094. (313) 961-2242. (800) 347-4253. FAX (313) 961-6083. 4th edition. 1996. $195.00.

ENCYCLOPEDIAS AND DICTIONARIES

The Encyclopedia of the Solid Earth Sciences. Philip Kearey, editor-in-chief. Blackwell Scientific Publications, 238 Main Street, Cambridge, MA 02142. (617) 876-7000 or (800) 759-6102.FAX (617) 876-7022. 1993. $60.00.

GENERAL WORKS

Cavalcade of Carbonates. John D. Cooper, editor. SEPM—Society for Sedimentary Geology, 1731 E 71st Street, Tulsa, OK 74136-5108. FAX (918) 493-2093. 1989. $11.00.

Geochemical Reference Material Compositions: Rocks, Minerals, Sediments, Soils, Carbonates, Refractories and Ores Used In Research and Industry. P. J. Potts, et al. CRC Press, 2000 Corporate Blvd., N.W., Boca Raton, FL 33431. (407) 994-0555 or (800) 272-7737. FAX (407) 994-0949. 1992. $130.95.

Geochemistry and Sedimentary Carbonates. J.W. Morse, editor. Elsevier Science Inc., Box 882, Madison Square Station, New York, NY 10159. (212) 989-5800. 1990. $113.00 (hardback) or $56.50 (paperback).

HANDBOOKS AND MANUALS

Handbook of Hydrology. David R. Maidment, editor in chief. McGraw-Hill Publishing Company, 1221 Avenue of the Americas, New York, NY 10020. (800) 262-4729 or (212) 512-3825. 1993. $115.00.

ONLINE DATABASES AND CD-ROMS

Current Contents Search. Institute for Scientific Information, 3501 Market Street, Philadelphia, PA 19104. (215) 386-0100. FAX (215) 386-6362. Contains citations to articles listed in the table of contents of science and technology journals. Also articles in social sciences and life sciences journals. Available on BRS,(800) 289-4277, DIALOG,(800) 334-2564, online services. Inquire as to cost and availability.

Dissertation Abstracts. University Microfilms International, 300 North Zeeb Road, Ann Arbor, MI 48106. (800) 521-0600 or (313) 761-4700. Scope includes virtually all doctoral dissertations accepted at accredited American institutions from 1861 to present in 252 subject areas. Available on BRS,(800) 289-4277, DIALOG,(800) 334-2564, and OCLC EPIC,(800) 848-5878, online services. Also available on CD-ROM. Inquire as to cost and availability.

GEOREF: Bibliography and Index of Geology. American Geological Institute, 4220 King Street, Alexandria, VA 22302. (703) 379-2480. Fax (703) 379-7563. Monthly. Inquire for price and availability.

NTIS Bibliographic Database. National Technical Information Service, 5285 Port Royal Road, Springfield, Va 22161. (703) 487-4929 or FAX (703) 321-8199. Broad coverage of government-sponsored science and technology research reports, 1964 to present. Available on BRS,(800) 289-4277, DIALOG, (800) 334-2564, ORBIT, (800) 456-7248, and STN International, FIZ Karlsruhe, P.O. Box 2465, W-7500, Karlsruhe 1, Germany, online services. Also available on CD-ROM. Inquire as to cost and availability.

Scisearch. Institute for Scientific Information, 3501 Market Street, Philadelphia, PA 19104. (800) 523-1850 or (215)386-0100. Broad multidisciplinary title and author index to the international literature of science and technology, 1974 to present. Available on DIALOG,(800) 334-2564, and ORBIT,(800) 456-7248, online services. Also available on CD-ROM. Inquire as to cost and availability.

PERIODICALS

Carbonates and Evaporites. Northeastern Science Foundation Inc., 15 Third Street, Box 746, Troy, NY 12181-0746. (518) 273-3247. 1986 to present. Two times a year. $30 per year (individuals).

Earth Surface Processes and Landforms. J. Wiley & Sons, Journals, 605 Third Avenue, New York, NY 10158-0012. (212) 692-6000. 1976 to present. Nine times a year. $635.00.

Geological Society of America Bulletin. 3300 Penrose Place, PO Box 9140, Denver, CO 80301-9140. (303) 447-2020. 1888 to present. Monthly. $205.00 per year.

Geophysical Research Letters. American Geophysical Union, 2000 Florida Avenue, N.W., Washington, DC 20009. (202) 462-6900. 1974 to present. Semi-monthly. $590.00 per year for nonmembers.

Journal of Geology. University of Chicago Press, Journals Division, 5720 S. Woodlawn Avenue, Chicago, IL 60637. (312) 753-3347. FAX (312) 753-0811. 1893 to present. Bi-monthly. $45.00 per year.

Precambrian Research. Elsevier Science Inc., Box 882, Madison Square Station, New York, NY 10159. (212) 989-5800. FAX (212) 633-3990. 1974 to present. 24 times a year. $1219.00.

Quaternary Research. Academic Press Inc., 525 B Street, Suite 1900, San Diego, CA 92101-4495. (619) 231-0926. FAX (619) 699-6715. 1970 to present. Bi-monthly. $240.00.

Sedimentary Geology. Elsevier Science Inc., Box 882, Madison Square Station, New York, NY 10159. (212) 989-5800. FAX (212) 633-3990. 1967 to present. 24 times a year. $1187.00.

Sedimentology. Blackwell Scientific Publishing, Inc., 3 Cambridge Court, Suite 208, Cambridge, MA 02142. (617) 225-0401. 1952 to present. Six times a year. $434.00.

RESEARCH CENTERS AND INSTITUTES

Bureau of Economic Geology, University of Texas At Austin. Box X, University Station, Austin, TX 78713. (512) 471-7721. FAX (512) 471-0140.

Texas A&M University, Center For Petroleum Reservoir Geology. College of Geosciences, College Station, TX 77843-3115. (409) 845-2460. FAX (409) 845-6162.

New Mexico Bureau of Mines and Mineral Resources. Institute of Mining and Technology, Campus Station, Socorro, NM 87891. (505) 835-5420. FAX (505) 835-5895.

CARBURETORS

See: AUTOMOTIVE ENGINEERING

CARTOGRAPHY

See also: AERIAL PHOTOGRAPHY, GEODESY, PHOTOGRAMMETRY, REMOTE SENSING, SATELLITES, SURVEYING

ABSTRACT SERVICES AND INDEXES

Applied Science and Technology Index; A Cumulative Subject Index To English Language Periodicals In the Fields of Aeronautics and Space Science, Computer Technology, Chemistry, Construction Industry, Energy and Related Areas. H.W. Wilson Co., 950 University Avenue, Bronx, NY 10452. (800) 367-6770 or (212) 588-8400. FAX (718) 590-1617. From 1958 to present. Monthly. Inquire about cost and availability. Also available on CD-ROM and online.

Bibliography and Index of Geology. American Geological Institute, 4220 King Street, Alexandria, VA 22302. (703) 379-2480. Fax (703) 379-7563. Monthly. $1295.00 per year. Also available online as GEOREF.

Current Contents: Engineering, Technology, and Applied Sciences. Institute for Scientific Information, 3501 Market Street, Philadelphia, PA 19104. (215) 386-0100. FAX (215) 386-6362. 1970 to present. Weekly. $442.00 per year.

Current Contents: Physical, Chemical and Earth Sciences. Institute for Scientific Information, 3501 Market Street, Philadelphia, PA 19104. (215) 386-0100. FAX (215) 386-6362. Weekly. $360.00 per year.

Engineering Index Monthly. Engineering Information, Inc., Castle Point on the Hudson, Hoboken, NJ 07030. (800) 221-1044. FAX (212) 832-1857. Monthly. $2300.00 per year. Also available online as COMPENDEX, and also on CD-ROM. Covers chemical engineering, computers, electrical engineering, civil engineering, metals and mining, industrial management, and mechanical engineering.

Geodex Retrieval System For Geotechnical Abstracts. Geodex Retrieval Systems, 669 Broadway, Box 279, Sonoma, CA 95476. (707) 939-8476. FAX (707) 996-8734. 1970 to present. Three times a year. $645.00 per year includes Geotechnical Abstracts, as Geodex Systems-GRS.

Geological Abstracts. Elsevier - Geo Abstracts, Regency House, 34 Duke Street, Norwich NR3 3AP, England. Monthly. $760.00 per year. Also available online as GEOBASE.

Geological Society of America. Abstracts with Programs. Geological Society of America. 3300 Penrose Place, P.O. Box 9140, Boulder, CO 80301-9140. (303) 447-2020. Abstracts and programs of the annual conference. Annual. $69.00.

Geotechnical Abstracts. Geodex Retrieval Systems, 669 Broadway, Box 279, Sonoma, CA 95476. (707) 939-8476. FAX (707) 996-8734. 1970 to present. Three times a year. $280.00 per year. Covers soil mechanics, foundation engineering, rock mechanics, and engineering geology.

Meteorological and Geoastrophysical Abstracts. American Meteorological Society, c/o Inforonics Inc., 550 Newtown Road, Littleton, MA 01460. (508) 486-8976. FAX (508) 486-0027. 1950 to present. Monthly. $950.00.

Remote Sensing of Earth Resources. University of New Mexico Earth Data Analysis Center, 2500 Yale Blvd., Suite 100, Albuquerque, NM 87131-6031. (505) 277-3622. FAX (505) 277-3614. 1974 to present (1991 to present under current title). Quarterly. $175.00 per year.

Science Citation Index. Institute for Scientific Information, 3501 Market Street, Philadelphia, PA 19104. (215) 386-0100. FAX (215) 386-6362. Inquire about availability and cost. Also available on CD-ROM.

ASSOCIATIONS AND PROFESSIONAL SOCIETIES

American Association For Geodetic Surveying. 5410 Grosvenor Lane, Suite 100, Bethesda, MD 20814. (301) 493-0200.

American Cartographic Association. 5410 Grosvenor Lane, Suite 100, Bethesda, MD 20814. (301) 493-0200.

American Congress On Surveying and Mapping. 5410 Grosvenor Lane, Suite 100, Bethesda, MD 20814. (301) 493-0200.

American Society For Photogrammetry and Remote Sensing. 5410 Grosvenor Lane, Suite 210, Bethesda, MD 20814. (301) 493-0290. FAX (301) 493-0208.

Association of American Geographers. 1710 16th Street NW, Washington, DC 20009-3198.

The International Society For Optical Engineering (spie), Po Box 10, 1000-20th Street, Bellingham, Wa 98227-0010.

National Society of Professional Surveyors. 5410 Grosvenor Lane, Suite 100, Bethesda, MD 20814. (301) 493-0200.

North American Cartographic Information Society. PO Box 399, Milwaukee, WI 53201. (414) 229-6282.

Optical Society of America, 2010 Massachusetts Avenue Nw, Washington, Dc 20036. (202) 223-8130.

Society of Photographic ScieNTISts and Engineers. 7003 Kilworth Lane, Springfield, VA 22151. (703) 642-9090.

Society For Imaging Science and Technology. 7003 Kilworth Lane, Springfield, VA 22151. (703) 642-9090. FAX (703) 642-9094.

BIBLIOGRAPHIES

Bibliography of Cartography. Library of Congress, Geography and Map Division. 5 volumes (1973), 2-volume supplement (1980). G.K. Hall, 866 Third Avenue, New York, NY 10022. (212) 702-6789 or (800) 257-5755. Inquire for cost and availability.

Information Sources In Cartography. C.R. Perkins & R.B. Perry, editors. Bowker-Saur, 121 Chanlon Road, New Providence, NJ 07974. (908) 464-6800 or (800) 521-8110. 1990. Inquire for cost and availability.

DIRECTORIES AND BIOGRAPHICAL SOURCES

Directory of Engineering Societies and Related Organizations. Gordon Davis. 13th edition. American Association of Engineering Societies, 1111 19th Street NW, Suite 608, Washington, DC 20036. (202) 296-2237 or (800) 658-8897. 1989. Inquire for price.

International Research Centers Directory. Gale Research, 835 Penobscot Building, Detroit, MI 48226-4094. (313) 961-2242. (800) 347-4253. FAX (313) 961-6083. 8th edition. 1995. $410.00

Research Centers Directory. Gale Research, 835 PenobscotBuilding, Detroit, MI 48226-4094. (313) 961-2242. (800) 347-4253. FAX (313) 961-6083. $485.00.

Scientific and Technical Organizations and Agencies Directory. Gale Research, 835 Penobscot Building, Detroit, MI 48226-4094. (313) 961-2242. (800) 347-4253. FAX (313) 961-6083. 4th edition. 1996. $195.00.

Who's Who In Engineering. American Association of Engineering Societies, 1111 19th Street NW, Suite 608, Washington, DC 20036. (202) 296-2237 or (800) 658-8897. 8th edition. 1991. Inquire for price.

GENERAL WORKS

Advances In Cartography. J.C. Muller. Pergamon Press Inc., Maxwell House, Fairview Park, Elmsford, NY 10523. (914) 592-7700. Fax (914) 592-3625. 1991. $125.00.

Applied Cartography: Introduction To Remote Sensing. Thomas D. Rabenhorst & Paul D. McDermott. Merrill, PO Box 508, Columbus, OH 43216. 1989. Inquire for cost and availability.

Basic Cartography Volume One. R.W. Anson & F.J. Ormeling. 2nd edition. Pergamon Press Inc., Maxwell House, Fairview Park, Elmsford, NY 10523. (914) 592-7700. Fax (914) 592-3625. 1994. $41.50.

Cartographic Design and Production. J.S. Keates. 2nd edition. John Wiley and Sons, Inc., 605 Third Avenue, New York, NY 10158. (800) 526-5368 or (212) 850-6000. 1989. $49.95.

Compendium of Cartographic Techniques. James P. Curran, editor in chief. Elsevier Science Publishing Company, Inc., 655 Avenue of the Americas, New York, NY 10010. (212) 989-5800. 1988. $90.00.

Digital Cartography. Robert G. Cromley. Prentice Hall, 113 Sylvan Avenue, Route 9W, Englewood Cliffs, NJ 07632. (201) 592-2000 or (800) 922-0579. 1992. $50.00.

Geographic Information Systems and Cartographic Modeling. C. Dana Tomlin. Prentice Hall, 113 Sylvan Avenue, Route 9W, Englewood Cliffs, NJ 07632. (201) 592-2000 or (800) 922-0579. 1990. $47.00.

Introduction To thematic Cartography. Judith Tyner. Prentice Hall, 113 Sylvan Avenue, Route 9W, Englewood Cliffs, NJ 07632. (201) 592-2000 or (800) 922-0579. 1992. $44.00.

HANDBOOKS AND MANUALS

Manual of Photogrammetry. Chester C. Slama, ed. 4th edition. American Society for Photogrammetry and Remote Sensing, 5410 Grosvenor Lane, Suite 210, Bethesda, MD 20814. (301) 493-0290. FAX (301) 493-0208. 1980. Inquire for price and availability.

Manual of Remote Sensing. Robert N. Colwell, ed. 2nd edition. American Society for Photogrammetry and Remote Sensing, 5410 Grosvenor Lane, Suite 210, Bethesda, MD 20814. (301) 493-0290. FAX (301) 493-0208. 1983. Inquire for price and availability.

ONLINE DATABASES AND CD-ROMS

Compendex Plus. Engineering Information, Inc., 345 East 47th Street, New York, NY 10017. (212) 705-7600 or (800) 221-1044.Contains citations with abstracts to worldwide literature in engineering and technology, from 1970 to present. Available on online BRS,(800) 289-4277, DIALOG, (800) 334-2564, ORBIT (800) 456-7248, and STN International, FIZ Karlsruhe, P.O. Box 2465, W-7500, Karlsruhe 1, Germany, online services. Also available on CD-ROM. Inquire as to cost and availability.

GEOREF: Bibliography and Index of Geology. American Geological Institute, 4220 King Street, Alexandria, VA 22302. (703) 379-2480. Fax (703) 379-7563. Monthly. Inquire for price and availability.

NTIS Bibliographic Database. National Technical Information Service, 5285 Port Royal Road, Springfield, Va 22161. (703) 487-4929 or FAX (703) 321-8199. Broad coverage of government-sponsored science and technology research reports, 1964 to present. Available on BRS,(800) 289-4277, DIALOG, (800) 334-2564, ORBIT, (800) 456-7248, and STN International, FIZ Karlsruhe, P.O. Box 2465, W-7500, Karlsruhe 1, Germany, online services. Also available on CD-ROM. Inquire as to cost and availability.

Scisearch. Institute for Scientific Information, 3501 Market Street, Philadelphia, PA 19104. (800) 523-1850 or (215) 386-0100. Broad multidisciplinary title and author index to the international literature of science and technology, 1974 to present. Available on DIALOG,(800) 334-2564, and ORBIT,(800) 456-7248, online services. Also available on CD-ROM. Inquire as to cost and availability.

WILSONLINE. H.W. Wilson Company, 950 University Avenue, Bronx, NY 10452. (800) 367-6770 or (212) 588-8400. Makes available online versions of the printed H.W. Wilson indexes including Applied Science and Technology Index, Business Periodicals Index, General Science Index, and Readers' Guide to Periodical Literature. Period covered is generally 1983 to present. Available on BRS,(800) 289-4277, DIALOG,(800) 334-2564, and OCLC EPIC,(800) 848-5878, online services. Also available on CD-ROM. Inquire as to cost and availability.

PERIODICALS

Cartography and Geographic Information Systems. American Congress on Surveying and Mapping, 5410 Grosvenor Lane, Suite 100, Bethesda, MD 20814. (301) 493-0200. 1974 to present. 4 times a year. $85.00 per year.

ACSM Bulletin. American Congress on Surveying and Mapping, 5410 Grosvenor Lane, Suite 100, Bethesda, MD 20814. (301) 493-0200. 1950 to present. Bi-monthly. $75.00 per year.

American Congress On Surveying and Mapping Proceedings. American Congress on Surveying and Mapping, 5410 Grosvenor Lane, Suite 100, Bethesda, MD 20814. (301) 493-0200. 1942 to present. Semi-annual. Price varies.

IEEE Transactions On Geoscience and Remote Sensing. IEEE, 345 E. 47th Street, New York, NY 10017. (908) 981-0060. FAX (908) 981-9667. 1963 to present. Bi-monthly. $188.00 per year.

Journal of Imaging Science and Technology. Society for Imaging Science and Technology, 7003 Kilworth Lane, Springfield, VA 22151. (703) 642-9090. FAX (703) 642-9094. 1950 to present. Bi-monthly. $120.00 per year.

Marine Geodesy. Taylor & Francis, 1900 Frost Road, Suite 101, Bristol, PA 19007. (215) 785-5800. FAX (215) 785-5515. 1977 to present. Quarterly. $141.00 per year.

Photogrammetric Engineering and Remote Sensing. American Society for Photogrammetry and Remote Sensing, 5410 Grosvenor Lane, Suite 210, Bethesda, MD 20814. (301) 493-0290. FAX (301) 493-0208. 1934-present. Monthly. $160.00 per year.

Photogrammetric Record. Photogrammetric Society, Dept. of Photogrammetry & Surveying, University College London, Gower St., London WC1E 6BT England. Telephone 071-387-

7050. FAX 071-380-0453. 1953-present. Semi-annual. $88.00 per year.

Remote Sensing of Environment. Elsevier Science Publishing Company, Inc., 655 Avenue of the Americas, New York, NY 10010. (212) 989-5800. 1969 to present. 12 times a year. $995.00 per year.

Remote Sensing Reviews. Harwood Academic, c/o Gordon & Breach Science Publishers, PO Box 786, Cooper Station, New York, NY 10276. (212) 206-8900. Quarterly. Inquire for cost.

Scientific and Applied Photography and Cinematography. Gordon and Breach Science Publishers, 820 Town Center Drive, Langhorne, PA 19047. (215) 750-2642. FAX (215) 750-6343. 6 times a year. Inquire for price.

RESEARCH CENTERS AND INSTITUTES

Boston University Center For Remote Sensing and Cartography. 725 Commonwealth Avenue, Boston, MA 02215. (617) 353-5081. FAX (617) 353-3200.

Cartographic Center. Ohio University, Porter Hall, Athens, OH 45701. (614) 593-1150.

Michigan State University Center For Remote Sensing. 302 Berkey Hall, East Lansing, MI 48824. (517) 353-7195. FAX (517) 353-1821.

University of Delaware Center For Remote Sensing. College of Marine Studies, Newark, DE 19711. (302) 451-2336.

University of Idaho, Idaho Remote Sensing Research Unit. Moscow, ID 83843. (208) 885-7209.

University of Minnesota Remote Sensing Laboratory. 1530 N. Cleveland Avenue, St. Paul, MN 55108. (612) 624-7764. FAX (612) 625-5212.

University of North Dakota, Institute For Remote Sensing. Geography Department, Grand Forks, ND 58202-8274. (701) 777-4246.

CAST IRON

See also: METALLURGY, STEEL AND STEEL MAKING

ABSTRACT SERVICES AND INDEXES

Alloys Index. American Society for Metals, Metals Park, OH 44073. (216) 338-5151. FAX (216) 338-4634. 1974 to present. Monthly. $380.00. Also available on CD-ROM and online via METADEX,STN International, and DIALOG.

Applied Mechanics Reviews: An Assessment of World Literature In Engineering Sciences. 1948-present. American Society of Mechanical Engineers, 345 East 47th Street, New York, NY 10017. (212) 705-7703. Monthly. $360.00 per year.

Applied Science and Technology Index; A Cumulative Subject Index To English Language Periodicals In the Fields of Aeronautics and Space Science, Computer Technology, Chemistry,

Construction Industry, Energy and Related Areas. H.W. Wilson Co., 950 University Avenue, Bronx, NY 10452. (800) 367-6770 or (212) 588-8400. FAX (718) 590-1617. From 1958 to present. Monthly. Inquire about cost and availability. Also available on CD-ROM and online.

Current Contents: Engineering, Technology, and Applied Sciences. Institute for Scientific Information, 3501 Market Street, Philadelphia, PA 19104. (215) 386-0100. FAX (215) 386-6362. 1970 to present. Weekly. $442.00 per year.

Engineering Index Monthly. Engineering Information, Inc., Castle Point on the Hudson, Hoboken, NJ 07030. (800) 221-1044. FAX (212) 832-1857. Monthly. $2300.00 per year. Also available online as COMPENDEX, and also on CD-ROM. Covers chemical engineering, computers, electrical engineering, civil engineering, metals and mining, industrial management, and mechanical engineering.

I.M.M. Abstracts and Index. Institution of Mining and Metallurgy, 44 Portland Pl., London W1N 4BR, England. 071-580-3802. FAX 071-436-5388. Bi-monthly. $364 for non-members.

Manufacturing and Process Engineering Abstracts, Cambridge Scientific Abstracts, 7200 Wisconsin Avenue, Bethesda, Md 20814-4823. (301) 961-6750. FAX (301) 961-6720. 1993 to present. Monthly. $385.00 per year. Covers concurrent engineering, quality control, automated manufacturing, petroleum engineering, oil field operations and equipment, energy management, metallurgy and metallography, foundry practice.

Metals Abstracts and Metals Abstracts Index. American Society for Metals, Metals Park, OH 44073. (216) 338-5151. 1968 to present. Monthly. Abstracts are $1500.00 per year and Index is $460.00 per year.

ANNUAL REVIEWS AND YEARBOOKS

Annual Review of Materials Science. Annual Reviews Inc., 4139 El Camino Way, PO Box 10139, Palo Alto, CA 94303-0139. (415) 493-4400 or (800) 523-8635. FAX (415) 855-9815. Annual. $75.00 (1994 edition).

ASSOCIATIONS AND PROFESSIONAL SOCIETIES

American Foundrymen's Society Inc. 505 State Street, Des Plaines, IL 60016. (708) 824-0181.

American Iron and Steel Institute. 1101 17th St., NW, Washington, DC 20036-4700. (202) 452-7100. FAX (202) 463-6573.

American Society For Metals. Materials Park, OH 44073. (216) 338-5151 or FAX (216) 338-4634.

Association of Iron and Steel Engineers. Three GatewayCenter, Suite 2350, Pittsburgh, PA 15222. (412) 281-6323.

Casting Industry Suppliers Association. PO Box 280, Greendale, WI 53129. (414) 423-8655.

Iron and Steel Society. 410 Commonwealth Drive, Warrendale, PA 15086. (412) 776-9460. FAX (412) 776-0430.

Metallurgical Society of the Aime (American Institute of Mining, Metallurgical, and Petroleum Engineers). 345 E.47th Street, 14th Floor, New York, NY 10017. (212) 705-7695.

BIBLIOGRAPHIES

Scientific and Technical Books and Serials In Print. R.R. Bowker Inc., 121 Chanlon Road, New Providence, NJ 07974. (800) 521-8110 or (908) 464-6800. List of currently published books and serials in the physical and biological sciences, engineering and technology, with subject, author and titles indexes. Also available on ORBIT,(800) 456-7248, online service. Inquire as to cost and availability.

DIRECTORIES AND BIOGRAPHICAL SOURCES

Dun's Industrial Guide: the Metalworking Directory, 1993-94. Dun & Bradstreet Information Services Staff. 3 volumes. Dun & Bradstreet Information Services, 899 Eaton Avenue, Bethlehem, PA 18025. (610) 882-7000 or (800) 526-0651. FAX (610) 882-7269. 1993. $795.00.

Iron and Steel Works of the World. Metal Bulletin Inc., 220 Fifth Avenue, New York, NY 10001-7781. (212) 213-6202. FAX (212) 213-1870. 1952 to present. Quadrennial. $52.00 for non-members.

Research Centers Directory. Gale Research, 835 Penobscot Building, Detroit, MI 48226-4094. (313) 961-2242. (800) 347-4253. FAX (313) 961-6083. $485.00.

Scientific and Technical Organizations and Agencies Directory. Gale Research, 835 Penobscot Building, Detroit, MI 48226-4094. (313) 961-2242. (800) 347-4253. FAX (313) 961-6083. 4th edition. 1996. $195.00.

Who's Who In Engineering. American Association of Engineering Societies, 1111 19th Street NW, Suite 608, Washington, DC 20036. (202) 296-2237 or (800) 658-8897. 8th edition. 1991. Inquire for price.

Who's Who In Technology. Gale Research, 835 Penobscot Building, Detroit, MI 48226-4094. (313) 961-2242. (800) 347-4253. FAX (313) 961-6083. 1995. $195.00.

ENCYCLOPEDIAS AND DICTIONARIES

Encyclopedia of Physical Science and Technology. Robert A. Meyers, ed. 18 volumes. Academic Press Inc., 6277 Sea Harbor Drive, Orlando, FL 32887. (800) 321-5068. 1992. $2100.00

McGraw-Hill Dictionary of Scientific and Technical Terms. Sybil P. Parker, ed. 5th edition. McGraw-Hill Publishing Company, 1221 Avenue of the Americas, New York, NY 10020. (800) 262-4729 or (212) 512-3825. 1993. $110.50.

McGraw-Hill Encyclopedia of Engineering. Sybil P. Parker, ed. 2nd edition. McGraw-Hill Publishing Company, 1221 Avenue of theAmericas, New York, NY 10020. (800) 262-4729 or (212) 512-3825. 1993. $95.50.

McGraw-Hill Encyclopedia of Science and Technology. Sybil P. Parker, ed. 7th edition. 20 volumes. McGraw-Hill Publishing Company, 1221 Avenue of the Americas, New York, NY 10020. (800) 262-4729 or (212) 512-3825. 1992. $1900.00

GENERAL WORKS

Casting Iron. C.W. Ammen. TAB Books, P.O. Box 40, Blue Ridge Summit, PA 17294-0850. (717) 794-2191 or (800) 233-1128. FAX (717) 794-2080. 1984. Inquire for cost and availability.

The Physical Metallurgy of Cast Iron. I. Minkoff. John Wiley and Sons, Inc., 605 Third Avenue, New York, NY 10158. (800) 526-5368 or (212) 850-6000. 1983. Inquire for cost and availability.

HANDBOOKS AND MANUALS

Asm Handbook Volume 1: Properties and Selection: Irons, Steels, and High-performance Alloys. ASM International Handbook Committee. ASM International, Materials Information, Materials Park, OH 44073-0002. (216) 338-5151 or (800) 336-5152. FAX (216) 338-4634. 1990. $147.00.

Asm Handbook Volume 15: Casting. ASM International Handbook Committee. ASM International, Materials Information, Materials Park, OH 44073-0002. (216) 338-5151 or (800) 336-5152. FAX (216) 338-4634. 1988. $147.00.

Smithell's Metals Reference Book. E.A. Brandes & G.B. Brook. 7th edition. Butterworth-Heinemann, 313 Washington Street, Newton, MA 02158. (617) 928-2500 or (800) 366-2665. FAX (617) 928-2620. 1992. $250.00.

ONLINE DATABASES AND CD-ROMS

Compendex Plus. Engineering Information, Inc., 345 East 47th Street, New York, NY 10017. (212) 705-7600 or (800) 221-1044. Contains citations with abstracts to worldwide literature in engineering and technology, from 1970 to present. Available on online BRS,(800) 289-4277, DIALOG, (800) 334-2564, ORBIT (800) 456-7248, and STN International, FIZ Karlsruhe, P.O. Box 2465, W-7500, Karlsruhe 1, Germany, online services. Also available on CD-ROM. Inquire as to cost and availability.

Metadex. Jointly produced by ASM International and the Institute of Materials. Contains more than 925,000 records from the international literature on metals and alloys, concerning properties, processes, materials classes, applications, and metallurgical systems. Updated monthly. Available from ORBIT-QUESTEL (703) 442-0900.

NTIS Bibliographic Database. National Technical Information Service, 5285 Port Royal Road, Springfield, Va 22161. (703) 487-4929 or FAX (703) 321-8199. Broad coverage of government-sponsored science and technology research reports, 1964 to present. Available on BRS,(800) 289-4277, DIALOG, (800) 334-2564, ORBIT, (800) 456-7248, and STN International, FIZ Karlsruhe, P.O. Box 2465, W-7500, Karlsruhe 1, Germany, online services. Also available on CD-ROM. Inquire as to cost andavailability.

WILSONLINE. H.W. Wilson Company, 950 University Avenue, Bronx, NY 10452. (800) 367-6770 or (212) 588-8400. Makes available online versions of the printed H.W. Wilson indexes including Applied Science and Technology Index, Business Periodicals Index, General Science Index, and Readers' Guide to Periodical Literature. Period covered is generally 1983 to present. Available on BRS,(800) 289-4277, DIALOG,(800) 334-2564, and OCLC EPIC,(800) 848-5878, online services. Also available on CD-ROM. Inquire as to cost and availability.

PERIODICALS

Casting Digest. ASM (American Society of Metals) International, Materials Information, Materials Park, OH 44073. (216) 338-5151 or FAX (216) 338-4634. 1976 to present. Monthly. $1550.00 per year.

Casting World. Continental Communications Inc., Box 1919, Bridgeport, CT 06601-1919. 1969 to present. Quarterly. $20.00 per year.

Foundry Management and Technology. Penton Publishing, 110 Superior Ave., Cleveland, OH 44114-2543. 1892 to present. Monthly. $45.00 per year.

Iron Age. Hitchcock Publishing, 191 S. Gary Avenue, Carol Stream, IL 60188. (708) 462-4641. FAX (708) 462-2205. 1867-1993. Monthly. $55.00 per year.

Iron and Steel Engineer. Association of Iron and Steel Engineers, Three Gateway Center, Suite 2350, Pittsburgh, PA 15222. (412) 281-6323. 1924 to present. Monthly. $50.00 per year.

Iron and Steel International. FMJ International Publications, Ltd., Queensway House, 2 Queensway, Redhill, Surrey RH1 1QS England. Telephone 0737-768611. FAX 0737-761685. Annual. $100.30 per year.

Iron & Steelmaker. Iron and Steel Society, 410 Commonwealth Drive, Warrendale, PA 15086. (412) 776-9460. FAX (412) 776-0430. 1974 to present. Monthly. $50.00 per year.

Ironmaking and Steelmaking. Institute of Materials, 1 Carlton House Terrace, London SW1Y 5DB, England. Telephone 071-839-4071. FAX 071-839-2078. 1974 to present. Bi-monthly. $420.00 per year for non-members.

J O M (journal of Metals). Minerals, Metals and Materials Society, 420 Commonwealth Drive, Warrendale, PA 15086. (412) 776-9080. 1949 to present. Monthly. $50.00 per year.

Metalcaster. American Cast Metals Association, 455 State Street, Des Plaines, IL 60016. (708) 299-9160. FAX (708) 299-3105. 1947 to present. Quarterly. $14.50 per year.

Metallurgical Transactions A: Physical Metallurgy and Materials Science. ASM International, Materials Park, OH 44073. (216) 338-5151 or FAX (216) 338-4634. 1970 to present. Monthly. $520.00 per year.

Metals Week. McGraw-Hill Publishing Company, Commodity Services Group, 1221 Avenue of the Americas, New York, NY 10020.(800) 262-4729 or (212) 512-3825. 1930 to present. Weekly. $770.00 per year.

Modern Casting. American Foundrymen's Society Inc., 505 State Street, Des Plaines, IL 60016. (708) 824-0181. 1938 to present. Monthly. $45.00 per year.

RESEARCH CENTERS AND INSTITUTES

American Iron and Steel Institute. 1101 17th St., NW, Washington, DC 20036-4700. (202) 452-7100. FAX (202) 463-6573.

CAST IRON

Ency. of Physical Sciences and Engineering Info. Sources

Carnegie Mellon University Center For Iron and Steel Making Research. MEMS Department, Pittsburgh, PA 15213. (412) 268-2677.

Colorado School of Mines, Advanced Steel Processing and Products Research Center. Golden, CO 80401. (303) 273-3774.

Iron Casting Research Institute. 2838 Fisher Road, Unit B, Columbus, OH 43204. (614) 275-4201.

University of Connecticut Institute of Materials Science. U-136, 97 N. Eagleville Road, Storrs, CT 06268. (203) 486-4623. FAX (203) 486-4745.

University of Wisconsin—madison, Cast Metals Laboratory. 1509 University Avenue, Madison, WI 53706. (608) 262-2562. FAX (608) 262-6707.

SPECIFICATIONS AND STANDARDS

Annual Book of Astm Standards, Volume 01.02. American Society for Testing and Materials (ASTM), 1916 Race Street, Philadelphia, PA 19103. (215) 299-5585. Annual. Inquire for cost and availability.

CATALYSIS

See also: CHEMICAL ENGINEERING, CHEMISTRY, ORGANIC CHEMISTRY

ABSTRACT SERVICES AND INDEXES

Applied Science and Technology Index; A Cumulative Subject Index To English Language Periodicals In the Fields of Aeronautics and Space Science, Computer Technology, Chemistry, Construction Industry, Energy and Related Areas. H. W. Wilson Co., 950 University Avenue, Bronx, NY 10452-9978. (212) 588-8400. (800) 367-6770. FAX (718) 590-1617. From 1958 to present. Monthly. Inquire about cost and availability. Also available online (BRS and WILSONLINE) and on CD-ROM. ISSN: 0003-6986.

Chemical Abstracts. Chemical Abstracts Service. 2540 Olentangy River Road, Box 3012, Columbus, Oh 43210-0012. (614) 447-3600. FAX (614) 447-3713. 1907 to present. Weekly. $16,800.00 per year. Available online and on CD- Rom. CA is also available in five section groupings. Inquire Regarding Cost and Availability.

Chemical Abstracts - Applied Chemistry and Chemical *Engineering Sections.* Chemical Abstracts Service, 2540 Olentangy River Road, Box 3012, Columbus, Oh 43210. (800) 848-6538 Or (614) 447-3600. FAX (614) 447-3713. 1963 to Present. Biweekly. $2080.00 per year. ISSN: 0090-8363. Also Available Online.

Current Contents: Engineering, Technology and Applied Sciences. Institute for Scientific Information, 3501 Market Street, Philadelphia, PA 19104. (215) 386-0100. FAX (215) 386-2991. From1970 to present. Weekly. $442.00 per year. Also available online. Inquire regarding cost and availability. ISSN: 0095-7917.

Engineering Index Monthly. Engineering Information, Inc., Castle Point on the Hudson, Hoboken, NJ 07030. (201) 216-

8500. (800) 221-1044. FAX (201) 216-8532. Monthly. $2300.00 per year for monthly issues. $1980 per year for annual cumulation. Available on CD-ROM and online as COMPENDEX. ISSN: 0742-1974; 0360-8557.

General Science Index. H.W. Wilson Company, 950 University Avenue, Bronx, NY 10452. (212) 588-8400. (800) 367-6770. FAX (718) 590-1617. From 1978 to present. Ten issues per year; quarterly and annual cumulations. Service basis. Available on CD-ROM and online. Inquire regarding cost and availability. ISSN: 0162-1963.

Process and Chemical Engineering/chemical Engineering Abstracts. Royal Society of Chemistry, Information Services,

thomas graham house, science park, milton road, cambridge,Cb4 4wf, England. Contains citations, mostly with abstracts,of the worldwide literature on chemical engineering; from 1982 To Present. Monthly. $610.00 per year. Also available online. Issn: 0960-5045.

Science Citation Index. SCI. Institute for Scientific Information, 3501 Market Street, Philadelphia, Pa 19104. (215) 386-0100. (800) 523-1850. FAX (215) 386-2991. 1961 To Present. Six issues per year, plus annual Cumulation. $11650.00 per year. Also available online and On Cd-rom. Inquire about price and availability. ISSN: 0036-827x.

Theoretical Chemical Engineering. Royal Society of Chemistry, Information Services, Thomas Graham House, Science Park, Milton Road, Cambridge, Cb4 4wf, England. Covers theoretical chemical engineering, including theory and laboratory Experimentation. 1964 to present. Monthly. $235.00. *Issn: 0960-5053.*

ANNUAL REVIEWS AND YEARBOOKS

Advances in Catalysis. Academic Press, Inc., 6277 Sea Harbor Drive, Orlando, Fl 32887. (800) 321-5068. 1948 to Present. Irregular. ISSN: 0360-0564.

Advances In Chemical Engineering. Academic Press, Inc., 6277 Sea Harbor Drive, Orlando, Fl 32887. (800) 321-5068. 1956 To Present. Irregular. ISSN: 0065-2377.

ASSOCIATIONS AND PROFESSIONAL SOCIETIES

American Chemical Society, 1155 16th Street, Nw, Washington, Dc 20036. (202) 872-4600.

American Institute of Chemical Engineers. 345 East 47th Street, New York, NY 10017. (212) 705-7338.

Association of Consulting Chemists and Chemical Engineers, 40 East 45th Street, New York, Ny 10017. (212) 983-3160. FAX (212) 983-3161.

Catalysis Society of North America. c/o Dr. Michael B. D'Amore. I. I. Dupont Co. Experiment Station, Wilmington DE 19898-9262. (302) 605-2488.

Chemical Manufacturers Association. 2501 M Street, N.W., Washington, DC 20037. (202) 887-1182

BIBLIOGRAPHIES

Chemical Information Sources. Gary Wiggins. McGraw-Hill Publishing Company, 1221 Avenue of the Americas, New York, NY 10020. (800) 262-4729 or (212) 512-3825. 1991. $42.50.

Scientific and Technical Books and Serials In Print; An Index To Literature In Science and Technology. R.R. Bowker

Inc., 121 Chanlon Road, New Providence, NJ 07974. (908) 464-6800. (800) 521-8110. FAX (908) 665-3502. 1972 to present. Annual. 4 volumes. 1994. $299.95. Also available on compact disc and online. ISSN: 0000-054X.

DIRECTORIES AND BIOGRAPHICAL SOURCES

American Institute of Chemical Engineers. DIRECTORY. American Institute of Chemical Engineers. 345 East 47th Street, New York, NY 10017-2396. (212) 705-7663. Annual.

American Men and Women of Science. Physical and Biological Sciences. R.R. Bowker Inc., 121 Chanlon Road, New Providence, NJ 07974. (908) 464-6800. (800) 521-8110. 20th edition. 8 volumes. 1996. $850.00

Consulting Services: Chemists and Chemical Engineers. Association of Consulting Chemists and Chemical Engineers, 40 East 45th Street, New York, NY 10017. (212) 983-3160. FAX (212) 983-3161. Biennial. $60.00.

Directory of Chemistry Software 1992. Wendy Warr, Peter Willett, Geoff Downs. American Chemical Society, 1155 16th Street, NW, Washington, DC 20036. (202) 872-4600. 1992. $35.95.

Research Centers Directory. Gale Research Company, Inc., 835 Penobscot Building, Detroit, MI 48226-4094. (313) 961-2242. (800) 347-4253. 20th edition. 1995. $485.00. ISSN:: 0080-1518.

Who's Who In Engineering. Gordon Davis, editor. American *Association of Engineering Societies. 1111 19th Street,* Suite 608, Washington, DC 20036. (202) 296-2237. (800) 658-8897. 9th edition. 1995. $220.00.

ENCYCLOPEDIAS AND DICTIONARIES

Dictionary of Named Processes in Chemical Technology. Alan E. Comyns. Oxford University Press, Inc., 200 Madison Avenue, New York, NY 10016. (212) 725-6000. (800) 334-4249. 1994. $75.00.

Encyclopedia of Chemical Processing and Design. McKetta Marcel Dekker, Inc., 270 Madison Avenue, New York, NY 10016. (212) 696-9000. (800) 228-1160. 1976 - . $175.00 per

Industrial Chemical thesaurus. Michael Ash and Irene Ash. VCH Publishers, Inc., 220 East 23rd Street, New York, NY 10010-4606. (800) 367-8249. 1992. $295.00.

Kirk-Othmer Encyclopedia of Chemical Technology. John Wiley and Sons, Inc., 605 Third Avenue, New York, NY 10158. (800) 526-5368 or (212) 850-6000. Fourth edition. 1991 - . Twenty-seven volumes. $5400.00.

McGraw-Hill Encyclopedia of Science and Technology. McGraw-Hill Book, Incorporated, 1221 Avenue of the Americas, NewYork, NY 10020. (212) 997-3675. (800) 262-4729. Seventh edition. Twenty volumes. 1992. $1900.00.

Ullman's Encyclopedia of Industrial Chemistry. VCH Publications, Inc., 220 East 23rd Street, Suite 909, New York, NY 10010-4606. (212) 683-8333. (800) 422-8824. 5th edition. 1984 - . Price varies per volume.

GENERAL WORKS

Catalysis: An Integrated Approach To Homogeneous, Heterogeneous and Industrial Catalysis. J. A. Moulijn, editor. Elsevier Science Publishing Company, Inc., 655 Avenue of the Americas, New York, NY 10010. (212) 989-5800. FAX (914) 333-2444. 1993. $184.00

Catalysis At Surfaces. Ian M. Campbell. Chapman & Hall, One Penn Plaza, New York, NY 10119. (212) 564-1060. 1988. $65.00.

Catalysis By Acids and Bases. B. Imelik. Elsevier Science Publishing Company, Inc., 655 Avenue of the Americas, New York, NY 10010. (212) 989-5800. 1985. $141.00.

Catalysis of Organic Reactions. Michael Scaros and Michael L. Prunier, editors. Marcel Dekker, Inc., 270 Madison Avenue, New York, NY 10016. (212) 696-9000. (800) 228- 1160. 1994. $195.00.

Dynamic Processes On Solid Surfaces. Kenzi Tamaru, editor. Plenum Publishing Corp., 233 Spring Street, New York, NY. (212) 620-8000. (800) 2221-9369. FAX (212) 463-0742.. 1993. $95.00.

Industrial Applications of Homgeneous Catalysis. Andre Mortreux and Francis Petit, editors. Kluwer Academic Publishers, 101 Philip Drive, Assinippi Park, Norwell, MA 02061. (617) 871-6600. 1987. $133.00.

Introduction To Surface Chemistry and Catalysis. Gabor A. Jomorjai. John Wiley & Sons, Inc., 605 Third Avenue, New York, NY 10158-0012. (212) 850-6000. (800) 225-5945. 1994. $59.95.

HANDBOOKS AND MANUALS

Catalyst Handbook. Martyn V. Twigg. CRC Press Inc., 2000 Corporate Blvd., NW, Boca Raton, FL 33431. (407) 994-0555. (800) 333-8300. 2nd edition. 1989. $104.00.

Comprehensive Organic Synthesis: Selectivity, Strategy, and Efficiency In Modern Organic Chemical. Barry M. Trost and others, editors. Pergamon Press, Maxwell House, Fairview Park, Elmsford, NY 10523. (914) 592-7700. Fax (914) 592- 3625. 1991. Nine volumes. $3900.00.

CRC Handbook of Chemistry and Physics. David R. Kide, editor. CRC Press Inc., 2000 Corporate Blvd., NW, Boca Raton, FL 33431. (407) 994-0555. (800) 333-8300. 77th edition. 1996. $99.95.

Guide To Basic Chemical Compounds. D.R. Lide, Jr. CRC Publishers, Inc., 2000 Corporate Blvd., N.W., Boca Raton, FL 33431. (407) 994-0555 or (800) 333-8300. 1993. $120.00.

CATALYSIS

Ency. of Physical Sciences and Engineering Info. Sources

Handbook of Homogeneous Catalysis. B. Cornils, editor. VCH Publications, Inc., 220 East 23rd Street, Suite 909, New York, NY 10010-4606. (212) 683-8333. (800) 422-8824. 1996. $345.00.

Lange's Handbook of Chemistry. John A. Dean, editor.McGraw-Hill Publishing Company, 1221 Avenue of the Americas, New York, NY 10020. (212) 512-2000. (800) 262-4729. 14th edition. 1996. $99.95.

Riegel's Handbook of Industrial Chemistry. James A. Kent, editor. Van Nostrand Reinhold, 115 Fifth Avenue, New York, NY 10003. (212) 254-3232 or (800) 926-2665. Ninth edition. 1992. $114.95.

ONLINE DATABASES AND CD-ROMS

CA Search. Chemical Abstracts Service, P.O. Box 3012, Columbus, Oh 43210-0012. (614) 447-3600. (800) 848-6533. FAX (614) 447-3709. Very comprehensive guide to worldwide chemical literature and related fields. 1972 to present. Available on BRS,(800) 289-4277, DIALOG, (800) 334-2564, ORBIT (800) 456-7248, and STN International, FIZ Karlsruhe, P.O. Box 2465, W-7500, Karlsruhe 1, Germany, online services. Inquire as to cost and availability.

Chemical Journals of the American Chemical Society. American Chemical Society, 1155 16th Street, N.W., Washington, DC 20036. (202) 872-4381 or (800) 424-6747. Contains complete text of approximately 90,000 articles from 22 primary journals published by the American Chemical Society, from mostly 1982 to present. Available on STN International, FIZ Karlsruhe, P.O. Box 2465, W-7500, Karlsruhe 1, Germany, online service. Inquire as to cost and availability.

Compendex Plus. Engineering Information, Inc., 345 East 47th Street, New York, NY 10017. (212) 705-7600 or (800) 221-1044. Contains citations with abstracts to worldwide literature in engineering and technology, from 1970 to present. Available on online BRS,(800) 289-4277, DIALOG, (800) 334-2564, ORBIT (800) 456-7248, and STN International, FIZ Karlsruhe, P.O. Box 2465, W-7500, Karlsruhe 1, Germany, online services. Also available on CD-ROM. Inquire as to cost and availability.

Current Contents Search. Institute for Scientific Information, 3501 Market Street, Philadelphia, PA 19104. (215) 386-0100. FAX (215) 386-6362. Contains citations to articles listed in the table of contents of science and technology journals. Also articles in social sciences and life sciences journals. Available on BRS,(800) 289-4277, DIALOG,(800) 334-2564, online services. Inquire as to cost and availability.

Dissertation Abstracts. University Microfilms International, 300 North Zeeb Road, Ann Arbor, MI 48106. (800) 521-0600 or (313) 761-4700. Scope includes virtually all doctoral dissertations accepted at accredited American institutions from 1861 to present in 252 subject areas. Available on BRS, (800) 289-4277, DIALOG, (800) 334-2564, and OCLC EPIC, (800) 848-5878, online services. Also available on CD-ROM. Inquire as to cost and availability.

Kirk-Othmer Encyclopedia of Chemical Technology. John Wiley and Sons, Inc., 605 Third Avenue, New York, NY 10158. (800) 526-5368 or (212) 850-6000. Contains the complete text of all chapters in the 27 volume fourth edition of the *Kirk-OthmerEncyclopedia of Chemical Technology.* 1991. Available on BRS,(800) 289-4277, DIALOG,(800) 334-2564, online services. Inquire as to cost and availability.

NTIS Bibliographic Database. National Technical Information Service, 5285 Port Royal Road, Springfield, Va 22161. (703) 487-4929 or FAX (703) 321-8199. Broad coverage of government-sponsored science and technology research reports, 1964 to present. Available on BRS,(800) 289-4277, DIALOG, (800) 334-2564, ORBIT, (800) 456-7248, and STN International, FIZ Karlsruhe, P.O. Box 2465, W-7500, Karlsruhe 1, Germany, online services. Also available on CD-ROM. Inquire as to cost and availability.

Scisearch. Institute for Scientific Information, 3501 Market Street, Philadelphia, PA 19104. (800) 523-1850 or (215) 386-0100. Broad multidisciplinary title and author index to the international literature of science and technology, 1974 to present. Available on DIALOG, (800) 334-2564, and ORBIT, (800) 456-7248, online services. Also available on CD-ROM. Inquire as to cost and availability.

WILSONLINE. H.W. Wilson Company, 950 University Avenue, Bronx, NY 10452. (800) 367-6770 or (212) 588-8400. Makes available online versions of the printed H.W. Wilson indexes including Applied Science and Technology Index, Business Periodicals Index, General Science Index, and Readers' Guide to Periodical Literature. Period covered is generally 1983 to present. Available on BRS, (800) 289-4277, DIALOG, (800) 334-2564, and OCLC EPIC, (800) 848-5878, online services. Also available on CD-ROM. Inquire as to cost and availability.

PERIODICALS

AICHE Journal. American Institute of Chemical Engineers. 345 East 47th Street, New York, NY 10017. (212) 705-7663. FAX (212) 752-3294. 1955 to present. Monthly. $395.00 per year. ISSN: 0001-1541.

American Laboratory. American Laboratory, 30 Controls Drive, Box 870, Shelton, CT 06484-6111. (203) 926-9300. FAX (203) 926-9310. 1968 to present. Monthly. $150.00 per year. ISSN: 0044-7749.

Analytical Chemistry. American Chemical Society, Box 3337, Columbus, OH 43210. (614) 447-3776. (800) 333-9511. FAX (614) 447-3671. 1929 to present. Monthly. $415.00 per year. ISSN: 0003-2700.

Analytical Methods and Instrumentation. John Wiley and Sons, Inc., 605 Third Avenue, New York, NY 10158. (800) 526-5368 or (212) 850-6000. 1992 to present. Bimonthly. $250.00 per year. ISSN: 1063-5246.

Applied Catalysis A: General. Elsevier Science Inc., Box 882, Madison Square Station, New York, NY 10159-0882. (212) 989-5800. FAX (212) 633-3990. 1981 to present. 28 issues per year. $2921 per year. $3295 in combination with section B Environmental. ISSN: 0926-860X.

Catalysis Reviews: Science and Engineering. Marcel Dekker, Inc., 270 Madison Avenue, New York, NY 10016. (212) 696-9000. FAX (212) 685-4540. 1967 to present. Quarterly. $435.00 per year. ISSN: 0161-4940.

Chemical Engineering Progress. American Institute of Chemical Engineers, 345 East 47th Street, New York, NY 10017. (212) 705-7338. FAX (212) 752-3294. 1947 to present. Monthly. $75.00 per year. ISSN: 0360-7275.

Chemical Week; Includes Annual Buyers Guide. Chemical Week Associates, 888 Seventh Avenue, New York, NY 10106. (212) 621-4900. FAX (212) 621-4949. 1914 to present. Weekly. $99.00 per year. ISSN: 0009-272X.

Chemtech. American Chemical Society, Box 3337, Columbus, OH 43210. (614) 447-3776. (800) 227-5558. 1970 to present. Monthly. $370.00 per year. ISSN: 0009-2703.

Journal of Organic Chemistry. American Chemical Society, Box 3337, Columbus, OH 43210. (614) 447-3776. (800) 227-5558. 1936 to present. Bi-Weekly. $785.00 per year. ISSN: 0022-3263.

RESEARCH CENTERS AND INSTITUTES

Center for Catalytic Science and Technology. University of Delaware. Departments of Chemical Engineering and Chemistry, Newark, DE 19716. (302) 451-8056. (302) 451- 1048.

Center For Catalysis. University of Florida, Gainsville, FL 32601. (904) 392-6043.

Hoke S. Greene Laboratory of Catalysis. University of Cincinnati, Department of Chemistry, Cincinnati, OH 45221. (513) 556-9253.

Institute For Pure and Applied Physical Sciences. University of California, San Diego, 9500 Gillman Drive, La Jolla, CA 92093-0075. (619) 534-4748. FAX (619) 534-0173.

CATALYTIC CONVERTER

See: AIR POLLUTION

CATHODE RAY TUBE

See: COMPUTER COMMUNICATIONS

CAVES AND CAVING

See also: GEOLOGY, GEOMORPHOLOGY, KARST, SEDIMENTOLOGY

ABSTRACT SERVICES AND INDEXES

Bibliography and Index of Geology. American Geological Institute, 4220 King Street, Alexandria, VA 22302. (703) 379-2480. Fax (703) 379-7563. Monthly. $1295.00 per year. Also available online as GeoRef.

Current Contents: Physical, Chemical and Earth Sciences. Institute for Scientific Information, 3501 Market Street, Philadelphia, PA 19104. (215) 386-0100. FAX (215) 386-6362. Weekly. $360.00 per year.

Geological Abstracts. Elsevier - Geo Abstracts, Regency House, 34 Duke Street, Norwich NR3 3AP, England. Monthly. $760.00 per year. Also available online as GEOBASE.

Geological Society of America. Abstracts with Programs. Geological Society of America. 3300 Penrose Place, P.O. Box 9140,Boulder, CO 80301-9140. (303) 447-2020. Abstracts and programs of the annual conference. Annual. $69.00.

Science Citation Index. Institute for Scientific Information, 3501 Market Street, Philadelphia, PA 19104. (215) 386-0100. FAX (215) 386-6362. Inquire about availability and cost. Also available on CD-ROM.

ASSOCIATIONS AND PROFESSIONAL SOCIETIES

British Cave Research Association. c/0 Bryan Ellis, 20 Woodland Avenue, Westonzoyland, Bridgwater, Somerset TA7 0LQ, England.

Cave Research Foundation. 4074 W Redwing Street, Tucson, AZ 85741. (602) 744-2243.

Geological Society of America. 3300 Penrose Place, PO Box 9140, Denver, CO 80301-9140. (303) 447-2020.

National Caves Association. Rt.9, Box 106, McMinnville, TV 37110. (615) 668-3925. FAX (615) 668-3988.

National Speleological Society Inc., 2813 Cave Avenue, Huntsville, AL 35810-4431. (205) 852-1300. FAX (205) 851-9241.

DIRECTORIES AND BIOGRAPHICAL SOURCES

American Men & Women of Science, 1995-96. R.R. Bowker Staff, eds. 19th edition. 8 volumes. R.R. Bowker/Reed International Publishing Company, 121 Chanlon Road, New Providence, NJ 07974. (908) 464-6800 or (800) 521-8110. 1995. $850.00

Caves and Caverns: National Caves Association Directory. National Caves Association, Rt.9, Box 106, McMinnville, TV 37110. (615) 668-3925. FAX (615) 668-3988. Annual. 1973 to date. Free.

Directory of Engineering Societies and Related Organizations. Gordon Davis. 13th edition. American Association of Engineering Societies, 1111 19th Street NW, Suite 608, Washington, DC 20036. (202) 296-2237 or (800) 658-8897. 1989. Inquire for price.

International Research Centers Directory. Gale Research, 835 Penobscot Building, Detroit, MI 48226-4094. (313) 961-2242. (800) 347-4253. FAX (313) 961-6083. 8th edition. 1995. $410.00

National Speleological Society Membership List. 2813 Cave Avenue, Huntsville, AL 35810-4431. (205) 852-1300. FAX (205) 851-9241. Inquire for cost and availability.

Research Centers Directory. Gale Research, 835 Penobscot Building, Detroit, MI 48226-4094. (313) 961-2242. (800) 347-4253. FAX (313) 961-6083. $485.00.

Scientific and Technical Organizations and Agencies Directory. Gale Research, 835 Penobscot Building, Detroit, MI 48226-4094. (313) 961-2242. (800) 347-4253. FAX (313) 961-6083. 4th edition. 1996. $195.00.

CAVES AND CAVING

Ency. of Physical Sciences and Engineering Info. Sources

ENCYCLOPEDIAS AND DICTIONARIES

The Encyclopedia of the Solid Earth Sciences. Philip Kearey, editor-in-chief. Blackwell Scientific Publications, 238 Main Street, Cambridge, MA 02142. (617) 876-7000 or (800) 759-6102. FAX (617) 876-7022. 1993. $60.00.

GENERAL WORKS

Atlas of the Great Caves of the World. Paul Courbon, et al. Cave Books, 756 Harvard Avenue, St. Louis, MO 63130. (314) 862-7646. FAX (314) 935-7349. 1989. $30.00 (hardback) or $20.00 (paperback).

Caving Expeditions. Dick Willis, editor. State Mutual Book & Periodical Service Ltd., 521 Fifth Ave., 17th floor, New York, NY 10175. (718) 261-1704. FAX (516) 537-0412. 1993. $30.00.

Caving In America: the Story of the National Speleological Society. Paul H. Damon, editor. National Speleological Society Inc., 2813 Cave Avenue, Huntsville, AL 35810-4431. (205) 852-1300. FAX (205) 851-9241.

Caving Practice and Equipment. David Judson. Menasha Ridge Press Inc., 3169 Cahaba Heights Road, Birmingham, AL 35243. (800) 247-9437. FAX (205) 967-0580. 1991. $29.95.

Introduction To Speleology. Arthur N. Palmer and Kathleen H. Lavoie. Cave Books, 756 Harvard Avenue, St. Louis, MO 63130. (314) 862-7646. FAX (314) 935-7349. 1996. $35.00 (hardback) or $12.95 (paperback).

HANDBOOKS AND MANUALS

Caves and Caving: A Handbook and Guide To American Caves. Don Jacobson and Lee Stral. Harbor House Publishers Inc., 221 Water Street, Boyne City, MI 49712. (616) 582-2814. 1987. $7.95.

ONLINE DATABASES AND CD-ROMS

Earth Sciences. U.S. Geological Survey, 12201 Sunrise Valley Drive, Reston, VA 22092-9998. (703) 648-4460. CD-ROM of earth science databases including the U.S. Geological Survey Library, Earth Science Data Directory, and GEOINDEX, citations to published geological maps. $350.00 per year with quarterly updates.

Geoarchive. Geosystems, P.O. Box 1024, Westminster, London, England, SW1 P 2JL. Citations to literature on geoscience, 1969 to present. Inquire as to online cost and availability.

Geobase. Elsevier - Geo Abstracts, Regency House, 34 Duke Street, Norwich NR3 3AP, England. Contains citations to the worldwide earth science literature from 1980 to date. Available on DIALOG, ORBIT online services. Inquire as to cost and availability.

GEOREF: Bibliography and Index of Geology. American Geological Institute, 4220 King Street, Alexandria, VA 22302. (703) 379-2480. Fax (703) 379-7563. Monthly. Inquire for price and availability.

NTIS Bibliographic Database. National Technical Information Service, 5285 Port Royal Road, Springfield, Va 22161. (703)

487-4929 or FAX (703) 321-8199. Broad coverage of government-sponsored science and technology research reports, 1964 to present. Available on BRS,(800) 289-4277, DIALOG, (800) 334-2564, ORBIT, (800) 456-7248, and STN International, FIZ Karlsruhe, P.O. Box 2465, W-7500, Karlsruhe 1, Germany, online services. Also available on CD-ROM. Inquire as to cost and availability.

Scisearch. Institute for Scientific Information, 3501 Market Street, Philadelphia, PA 19104. (800) 523-1850 or (215)386-0100. Broad multidisciplinary title and author index to the international literature of science and technology, 1974 to present. Available on DIALOG,(800) 334-2564, and ORBIT,(800) 456-7248, online services. Also available on CD-ROM. Inquire as to cost and availability.

WILSONLINE. H.W. Wilson Company, 950 University Avenue, Bronx, NY 10452. (800) 367-6770 or (212) 588-8400. Makes available online versions of the printed H.W. Wilson indexes including Applied Science and Technology Index, Business Periodicals Index, General Science Index, and Readers' Guide to Periodical Literature. Period covered is generally 1983 to present. Available on BRS,(800) 289-4277, DIALOG,(800) 334-2564, and OCLC EPIC,(800) 848-5878, online services. Also available on CD-ROM. Inquire as to cost and availability.

PERIODICALS

Cave Geology. National Speleological Society, Section of Cave Geology and Geography, 542 Glenn Roard, State College, PA 16803. (814) 237-3187. Irregular. Price varies.

Cave Science: Transactions of the British Cave Research Association. British Cave Research Association, c/0 Bryan Ellis, 20 Woodland Avenue, Westonzoyland, Bridgwater, Somerset TA7 0LQ, England. 1974 to date. Irregular. Inquire for price in U.S.

Caves and Caving. British Cave Research Association, c/0 Bryan Ellis, 20 Woodland Avenue, Westonzoyland, Bridgwater, Somerset TA7 0LQ, England. 1973 to date. Quarterly. Inqire for price in U.S.

Earth Surface Processes and Landforms. J. Wiley & Sons, Journals, 605 Third Avenue, New York, NY 10158-0012. (212) 692-6000. 1976 to present. Nine times a year. $635.00.

Geological Society of America Bulletin. 3300 Penrose Place, PO Box 9140, Denver, CO 80301-9140. (303) 447-2020. 1888 to present. Monthly. $205.00 per year.

Journal of Geology. University of Chicago Press, Journals Division, 5720 S. Woodlawn Avenue, Chicago, IL 60637. (312) 753-3347. FAX (312) 753-0811. 1893 to present. Bi-monthly. $45.00 per year.

N S S Bulletin. National Speleological Society, 2813 Cave Avenue, Huntsville, AL 35810-4431. (205) 852-1300. FAX (205) 851-9241. 1940 to present. Semi-annual. $18.00 for non-members.

N S S News. National Speleological Society, 2813 Cave Avenue, Huntsville, AL 35810-4431. (205) 852-1300. FAX (205) 851-9241. 1943 to present. Monthly. $18.00 for non-members.

Sedimentary Geology. Elsevier Science Inc., Box 882, Madison Square Station, New York, NY 10159. (212) 989-5800. FAX (212) 633-3990. 1967 to present. 24 times a year. $1187.00.

Sedimentology. Blackwell Scientific Publishing, Inc., 3 Cambridge Court, Suite 208, Cambridge, MA 02142. (617) 225-0401. 1952 to present. Six times a year. $434.00.

Studies In Speleology. William Pengelly Cave Studies Trust Ltd., 107 andover Road, Newbury, Berks. RG14 6JH, England.1964 to present. Annual. Inquire for price in U.S.

RESEARCH CENTERS AND INSTITUTES

Cave Research Foundation. 4074 W Redwing Street, Tucson, AZ 85741. (602) 744-2243.

Western Kentucky University Center For Cave and Karst Studies. Department of Geography and Geology, Bowling Green, KY 42101. (502) 745-5989.

CELESTIAL MECHANICS

See also: ASTRONOMY, ASTROPHYSICS, COSMOLOGY, GALAXIES, INTERSTELLAR MATTER, PLANETS, SOLAR SYSTEM, TELESCOPES

ABSTRACT SERVICES AND INDEXES

Applied Science and Technology Index; A Cumulative Subject Index To English Language Periodicals In the Fields of Aeronautics and Space Science, Computer Technology, Ch Emistry, Construction Industry, Energy and Related Areas. H.W. Wilson Co., 950 University Avenue, Bronx, NY 10452. (212) 588-8400. (800) 367-6770. FAX (718) 590-1617. From 1958 to present. Monthly. Inquire about cost and availability. Also available on CD-ROM and online. ISSN: 0003-6986.

Astronomy and Astrophysics Abstracts. Springer-Verlag New York, 175 Fifth Avenue, New York, NY 10010. (212) 460-1500. FAX (212) 473-6272. Published for the Astronomisches Rechen-Institut. Comprehensive coverage of all aspects of astronomy, astrophysics and related fields. 1969 to present. Two parts per year. Annual. ISSN: 0067-0022.

General Science Index. H.W. Wilson Company, 950 University Avenue, Bronx, NY 10452. (212) 588-8400. (800) 367-6770. FAX (718) 590-1617. From 1978 to present. Ten issues per year; quarterly and annual cumulations. Service basis. Available on CD-ROM and online. Inquire regarding cost and availability. ISSN: 0162-1963.

Government Reports Announcements and Index. U. S. National Technical Information Service (NTIS), 5285 Port Royal Road, Springfield, VA 22161. (703) 487-4650. FAX (703) 321-8547. From 1968 to present. Annual. $630.00 per year. Also available online as *NTIS Bibliographic Database* and on CD-ROM. ISSN:

NTIS Alerts: Astronomy & Astrophysics. U. S. National Technical Information Service. 5285 Port Royal Road, Springfield, VA 22161. (703) 487-4650. FAX (703) 321-8547. Weekly. $140.00 per year.

Pascal 48: Environnement Cosmique Terrestre, Astronomie Et Geologie Extraterrestre. Centre National de la Recherche Scientifique, Institut de Information Scientifique et Technique, 2 aliee du Parc de Brabois, 54514 Vandoeuvre Les Nancy Cedex, France. TEL 83-50-46-00. FAX: 83-50-46-50. 1985 to present.

Ten issues per year. 770 F. Also available on CD-ROM and online.

Physics Abstracts . INSPEC. Section A, Science Abstracts. Institute of Electrical Engineers, London, United Kingdom. Available from: INSPEC/IEEE - Institute of Electrical and Electronic Engineers, Box 1331, Hoes Lane, Piscataway, NJ08855-1331. (908) 562-5549. 1898 to present. 24 issues per year. $2835.00 per year. ISSN: 0036-8091. Also available online and on CD-ROM.

Science Citation Index. SCI. Institute for Scientific Information, 3501 Market Street, Philadelphia, PA 19104. (215) 386-0100. (800) 523-1850. FAX (215) 386-2991. 1961 to present. Six issues per year, plus annual cumulation. $11650.00 per year. Also available online and on CD-ROM. Inquire about price and availability. ISSN: 0036-827X.

S T A R. Scientific and Technical Aerospace Reports. U.S. National Aeronautics and Space Administration. Distributed by U. S. Superintendent of Documents, Washington, DC 20402. 1963 to present. Semi-monthly, with semiannual and annual indexes. $114.00 per year. ISSN: 0036-8741. Also available online and on CD-ROM.

ANNUAL REVIEWS AND YEARBOOKS

The Astronomical Almanac. Superintendent of Documents, U. S. Government Printing office, Washington, DC 10402. (202) 783-3238. 1981 to present. Supercedes Astronomical Ephemeris and the American Ephermis and Nautical Almanac. Annual. $44.95. ISSN: 0737-6421.

Annual Review of Astronomy and Astrophysics; Original Reviews of Crit Ical Literature and Current Developments In Astronomy and Astrophysics. Annual Reviews, Inc., 4139 El Camino Way, Palo Alto, CA 94306-0139. (415) 493-4400. (800) 523-8635. FAX (415) 855-9815. 1963 to date. Annual. $60.00. ISSN:: 0066-4146.

Annual Review of Earth and Planetary Sciences. Annual Reviews, Inc., 4139 El Camino Way, Palo Alto CA 94306-0139. (415) 403-4400. (800) 523-8635. Fax (415) 855-9815. 1973 to date. Annual. $62.00. ISSN:: 0084-6597.

Highlights of Astronomy. I. Appenzeller, editor. Kluwer Academic Publishers, Box 358, Accord Station, Hingham, MA 02018-0358. (617) 871-6600. International Astronomical Union Highlights series. ISBN: 07923-3553-8. Triennal. 1995. Price varies.

ASSOCIATIONS AND PROFESSIONAL SOCIETIES

Amateur Astronomers Association. 1010 Park Avenue, New York, NY 10028. (212) 535-2922.

American Association of Variable Star Observers. 23 Birch Street, Cambridge, MA 02138. (617) 354-0484. FAX: (617) 354-0665.

American Astronomical Society. 2000 Florida Avenue NW, Suite 400, Washington, DC 20009. (202) 328-2010. FAX: (202) 234-2560.

CELESTIAL MECHANICS

Ency. of Physical Sciences and Engineering Info. Sources

American Geophysical Union. 2000 Florida Avenue, NW, Washington, DC 20009. (202) 462-6900. (800) 966-AGU1. FAX (202) 328-0566.

American Institute of Physics. One Physics Ellipse, College Park, MD 20740-3843. (301) 209-3100.

Astronomical League. 2112 Kingfisher Lane East, Rolling Meadows, IL 60008. (708) 398-0562.

Astronomical Society of the Pacific. 390 Ashton Avenue,San Francisco, CA 94112. (415) 337-1100. FAX: (415) 337-5205.

Royal Astronomical Society of Canada. 136 Dupont Street, Toronto. ON N5R 1V2 Canada. (416) 924-7973. FAX (416) 968-6687.

Webb Society. 1440 South Marmora Avenue, Tuscon, AZ 95713. (602) 628-1077.

BIBLIOGRAPHIES

Astronomy and Astrophysics: A Source Guide. Gordon Press, P.O. Box 459, Bowling Green Station, New York, NY 10004. (718) 624-8419. 1991. $250.00. .

A Bibliography of Astronomy, 1970-1979. R. A. Sea and S. S. Martin. Libraries Unlimited, Inc., Littleton, CO 80160. 1982.

Catalog of the Naval Observatory Library. Naval Observatory Staff. G. K. Hall & Company, 866 Third Avenue, New York, NY 10022. (212)702-6789. (800) 257-5755. 6 volumes. 1977. $655.00

Science Books and Films. American Association for the Advancement of Science, 1333 H Street NW, Washington, DC 20005. (202) 326-6454. 1965 to present. Nine issues per year. $40.00 per year. ISSN: 0098-342X

Scientific and Technical Books and Serials In Print; An Index To Literature In Science and Technology. R. R. Bowker. 121 Chanlon Road, New Providence, NJ 07974. (908) 464-6800. FAX (908) 665-3502. (800) 521-8110. 1972 to present. Annual. $299.95 per year. Also available on CD-ROM. ISSN: 0000-054X.

A Source Book On Astronomy and Astrophysics, 1900 - 1975. Kenneth Lang and Owen Gingerich, editors. Harvard University Press, 79 Garden Street, Cambridge MA. (617) 495-2600. 1980. $41.95.

DIRECTORIES AND BIOGRAPHICAL SOURCES

American Astronomical Society. MEMBERSHIP DIRECTORY. American Astronomical Society. 2000 Florida Avenue NW, Suite 400, Washington, DC 20009. (202) 328-2010. FAX (202) 234-2560 Annual. Included in membership dues. ISSN:: 1061-9038.

American Men and Women of Science: Physical and Biological Sciences. R. R. Bowker Inc., 121 Chanlon Road, New Providence, NJ 07974. (908) 464-6800. (800) 521-8110. 20th edition. 8 volumes. 1996. $850.00.

The Astronomers. Donald Goldsmith. St. Martin's Press, Inc., 175 Fifth Avenue, New York, NY 10010. (212) 674-5151. (800) 221-7945. 1993. $14.95 in paper.

Astronomical Centers of the World. Kevin Krisciunas. Cambridge University Press, 40 West 20th Street, New York,

Ny 10011-4211. (212) 924-3900. (800) 872-7423. 1988. $34.95

The Biographical Dictionary of Scientists: Astronomers. D. Abbott, editor. Peter Bedrick Books, Inc., 2112 Broadway, Room 318, New York, NY 10023. (212) 496-0751. 1984.

Directory of Physics and Astronomy Staff Members. American Institute of Physics, One Physics Ellipse, College Park, MD 20740-3843. (301) 209-3100. 1975/76 to present. Annual. $60.00. ISSN:: 0361-2228.

Earth and Astronomical Research Centers. Stockton Press, 345 Park Avenue South, New York, NY 10010. 4th edition. 1995. $515.00. ISBN: 1-56169-0967.

Graduate Programs In Physics, Astronomy and Related Fields. 1995-1996. American Institute of Physics, One Physics Ellipse, College Park, MD 20740-3843. (301) 209-3100. 1978 to present. Annual. $45.00. ISSN:: 0147-1821.

Research Centers Directory. Gale Research Inc., 835 Penobscot Building, Detroit, MI 48226-4094. (313) 961-2242. (800) 347-4253. 20th edition. 1995. $485.00.

ENCYCLOPEDIAS AND DICTIONARIES

Astronomy From A To Z: A Dictionary of Celestial Objects and Ideas. Charles A. Schweighauser. Illinois Issues, Sangamon State University, K-10, Springfield, IL 62794-9243. (217) 786-9243. 1991. $14.95 in paper.

Concise Dictionary of Astronomy. Jacqueline Mitton. Oxford University Press, Inc., 200 Madison Avenue, New York, NY 10016. (212) 725-6000. (800) 334-4249. 1992. $24.95.

Dictionary of Astronomical Names. Adrian Room. Routledge, Chapman and Hall, Inc., 29 West 35th Street, New York, NY 10001-3291. (212) 244-3336. 1988. $27.50.

Encyclopedia of Astronomy and Astrophysics. Stephen Maran, editor. Van Nostrand Reinhold, 115 Fifth Avenue, New York, NY 10003. (212) 254-3232. (800) 842-3636. 1992. $129.95.

Facts On File Dictionary of Astronomy. Valerie Illingworth. *Facts-on-file, Inc.*, 460 Park Avenue South, New York, NY 10016-7382. (212) 683-2244. (800) 322-8755. Fax (800) 683-3633. 1994. $27.95.

McGraw-Hill Encyclopedia of Science and Technology. McGraw- Hill Book Company, Inc., 1221 Avenue of the Americas, New York, NY 10020. (212) 512-2000. (800) 262-4729. 7th edition. 20 volume set. 1992. $1900.00.

Oxford Illustrated Encyclopedia. Volume 8, the Universe. Archie Roy, editor. Oxford University Press, Inc., 200 Madison Avenue, New York, NY 10016. (212) 725-6000. (800) 334-4249. 1993. $45.00.

Penguin Dictionary of Astronomy. Jacqueline Mitton. Viking Penguin, 375 Hudson Street, New York, NY 10014-3657.

(212) 366-2000. (800) 331-4624. 1993. $12.95 in paper.

Stars, Galaxies, Cosmos. William R. Corliss. Sourcebook Project, P.O. Box 107, Glen Arm, MD 21057. (410) 668-6047. 1987. $17.95.

GENERAL WORKS

Adventures In Celestial Mechanics: A First Course In the theory of Orbits. Victor G. Szebehely. University of Texas Press, P. O. Box 7819, Austin, TX 78713-7819. (512) 471- 7233. (800) 252-3206. . 1989. $30.00.

Analytical Techniques of Celestial Mechanics. Victor A. Brunberg. Springer-Verlag New York, Inc., 175 Fifth Avenue, New York, NY 10010. (212) 460-1500. (800) 777-4643. 1995. $79.00.

Astronomer's Universe: Stars, Galaxies and Cosmos. Herbert Friedman. W. W. Norton & Company, 500 Fifth Avenue, New York, NY 10110. (212) 354-5500. (800) 223-2584. 1990. $24.95.

Astrophysics Today. A. G. Cameron, editor. American

Institute of Physics, One Physics Ellipse, College Park, Md 20740-3843. (301) 209-3100., 1984. $39.00 in paper.

Celestial Navigation. Arthur Davies. Trafalgar, P.O. Box 257, North Pomfret, VT 05053. (802) 457-1911. (800) 423- 4525. 1993. $29.95.

Compendium of Practical Astronomy. Gunter D. Roth, editor. Springer-Verlag New York, Inc., 175 Fifth Avenue, New York, NY 10010. (212) 460-1500. (800) 777-4643. 3 volumes. Revised edition. 1994. $125.00.

Cycles of Fire: Stars, Galaxies and the Wonder of Deep Space. William K. Hartmann. Workman Publishing Company, Inc., 708 Broadway, New York, NY 10003. (212) 254-5900. (800) 722-7202. 1988. $15.95 in paper.

Essential Relativistics Celestial Mechanics. V. A. Brumberg. IOP Publishing, Public Ledger Building, Suite 1035, Independence Square, Philadelphia, PA 19106. (215) 627-0880. (800) 358-4677. 1991. $118.00.

Foundations of Celestial Mechanics. George W. Collins. Pachart Publishing House, 1130- San Lucas Circle, Tucson, AZ 85704. (520) 297-6760. 1989. $27.00 in paper.

Fundamentals of Celestial Mechanics. Danby . Willmann- Bell, Inc., P.O. Box 35025, Richmond, VA 23235. (804)-*320-7016.* 2nd edition. 1988. $23.95.

Gravitation. Charles W. Misner, et al. W. H. Freeman & Co., 41 Madison Avenue, East 26th Avenue, 35th Floor, New York, NY 10010. (212) 576-0400. 1995. .

Instability, Chaos and Predictability In Celestial Mechanics and Stellar Dynamics. K. B. Bhatnagar, Nova Science Pubs. Inc., 6080 Jericho Turnpike, Suite 207, Commack, NY 11725- 2808. (516) 499-3103. 1986. $145.00.

An Introduction To Astronomy: Stars, Galaxies and the Universe. Gerald D. Waxman. Kendall-Hunt Publishing Co., 4050 Westmark Drive, P.O. Box 1840 Dubuque, IA 52004-1840. (319) 589-1000. (800) 228-0810. 1992. $28.95.

An Introduction To Celestial Mechanics. Forrest R. Moulton, Dover Publications, Inc., 180 Varick Street, New York, NY 10014. (212) 255-3755. (800) 223-3130. 1984. $10.95 in paper.

Newton's Clock: Chaos In the Solar System. Ivars Peterson. W. H. Freeman, 41 Madison Avenue, East 26th Avenue, 35th Floor, New York, NY 10010. (212) 576-0400. 1995. .

Observational Astrophysics. Robert C. Smith. Cambridge University Press, 40 West 20th Street, New York, NY 10011- 4211. (212) 924-3900. (800) 872-7423. 1995. $69.95.

Origin of the Universe. John D. Barrow. Basic Books, 10 East 53rd Street, New York, NY 10022. (212) 207-7057. (800) 242-7737. 1994. $20.00.

The Physics of Stars. A. C. Phillips. John Wiley & Sons, Inc., 605 Third Avenue, New York, NY 10158-0012. (212) 850- 6000. (800) 225-5945. 1994. $32.95 in paper.

Relativity In Astronomy, Celestial Mecanics and Geodesy. M. H. Soffel. Springer-Verlag New York, Inc., 175 Fifth Avenue, New York, NY 10010. (212) 460-1500. (800) 777-4643. 1989. $69.50.

Stars and Atoms: From the Big Bang To the Solar System. Stuart Clark. Oxford University Press, Inc., 200 Madison Avenue, New York, NY 10016. (212) 725-6000. (800) 334-4249. 1995. $35.00.

Workings of the Universe. Time-Life, Inc., 777 Duke Street, Alexandria, VA 22314. (703) 8388-7000. (800) 621-7026. . 1991. $25.95.

Universe. William J. Kaufmann. W. H. Freeman & Co., 41 Madison Avenue, East 26th Avenue, 35th Floor, New York, NY 10010. (212) 576-0400. 4th edition. 1995. $54.95.

HANDBOOKS AND MANUALS

Astrophysical Data: Planets and Stars. Kenneth R. Lanb. Springer-Verlag New York, Inc., 175 Fifth Avenue, New York, NY 10010. (212) 460-1500. (800) 777-4643. 1993. $59.00.

Burnham's Celestial Handbook: An Observer's Guide To the Universe Beyond the Solar System. Robert Burnham, Jr. Dover Publications, Inc., 180 Varick Street, New York, NY 10014. (212) 255-3755. (800) 223-3130. 3 volumes. 1978. $41.85 for the set.

Cambridge Guide To the Constellations. Michael E. Bakich. Cambridge University Press, 40 West 20th Street, New York, NY 10011-4211. (212) 924-3900. (800) 872-7423. 1995. $69.95.

Encyclopedia of Astronomy and Astrophysics. Stephen Maran, editor. Van Nostrand Reinhold, 115 Fifth Avenue, New York, NY 10003. (212) 254-3232. (800) 842-3636. 1992. $129.95.

CELESTIAL MECHANICS

Ency. of Physical Sciences and Engineering Info. Sources

Greenwich Guide To Stars, Galaxies and Nebulae. Stuart Malin. Cambridge University Press, 40 West 20th Street, New York, NY 10011-4211. (212) 924-3900. (800) 872-7423. 1990. $10.95.

Handbook of Astronomy, Astrophysics and Geophysics; Volume 2: Galaxies and Cosmology. V. M. Canuto and B. G. Elmegreen. Gordon & Breach Science Publishers, Inc., P.O.Box 200029, Riverfront Plaza Station, Newark, NJ 07102-0301. (201) 643-7500. 1988. $194.00.

Handbook of Space Astronomy and Astrophsyics. Martin V. Zumbeck. Cambridge University Press, 40 West 20th Street, New York, NY 10011-4211. (212) 924-3900. (800) 872-7423. 2nd edition. 1990. $79.95.

Landolt-borenstein Numerical Data and Functional Relationships In Science and Technology: Astronomy, Astrophysics and Space Research. GROUP VI. Springer- Verlag New York, Inc., 175 Fifth Avenue, New York, NY 10010. (212) 460-1500. (800) 777-4643. volumes priced individually.

McGraw-Hill Encyclopedia of Astronomy. Sybil P. Parker and Jay M. Pasachoff, editors. McGraw-Hill Publishing Company, Inc., 1221 Avenue of the Americas, New York, NY 10020. (212) 512-2000. (800) 262-4729. 2nd edition. 1993. $75.00

Observer's Guide To Astronomy. Patrick Martinez, editor. Cambridge University Press, 40 West 20th Street, New York, NY 10011-4211. (212) 924-3900. (800) 872-7423. 2 volumes. 1994. $159.90.

ONLINE DATABASES AND CD-ROMS

CA Search. Chemical Abstracts Service, P.O. Box 3012, Columbu s, OH 43210-0012. (614) 447-3600. (800) 848-6533. FAX (614) 447-3709. Very comprehensive guide to worldwide chemical literature and related fields, 1972 to present. Available on BRS,(800) 289-4277, DIALOG, (800) 334-2564, OR-BIT (800) 456-7248, and STN International, FIZ Karlsruhe, P.O. Box 2465, W-7500, Karlsruhe 1, Germany, online services. Inquire as to cost and availability.

Current Contents Search. Institute for Scientific Information, 3501 Market Street, Philadelphia, PA 19104. (215) 386-0100 . FAX (215) 386-6362. Contains citations to articles listed in the table of contents of science and technology journals. Also articles in social sciences and life sciences journals. Available on BRS, (800) 289-4277, DIALOG, (800) 334-2564, online services. Inquire as to cost and availability.

Dissertation Abstracts Online. University Microfilms Internatio nal, 300 North Zeeb Road, Ann Arbor, MI 48106. (800) 521-0600 or (313) 761-4700. Scope includes virtually all doctoral dissertations accepted at accredited American institutions from 1861 to present in 252 subject areas. Available on BRS, (800) 289-4277, DIALOG, (800) 334-2564, and OCLC EPIC, (800) 848-5878, online services. Also available on CD-ROM. Inquire as to cost and availability.

Inspec. In stitution of Electrical Engineers, Michael Faraday H ouse, Six Hills Way, Stevenage, Herts. SG1 2AY, Englan d. Telephone: 0438 313311 or FAX 0438 742840. Contains citations to the worldwide literature of physics, electronics and electrical engineering, computer technology, and related fields. Available on BRS, (800) 289-4277, DIALOG, (800) 334-2564, ORBIT, (800) 456-7248, and STN International, FIZ Karlsruhe, P.O. Box 2465, W-7500, Karlsruhe 1, Germany, online services. Inquire as to cost and availability.

International Aerospace Abstracts. American Institute of Aeronautic s and Astronautics, 370 L'Enfant Promenade, S.W., Washington, DC 20024. (202) 646-7400. Contains references and abstracts of journal and monograph literature relating to aerospace science and technology, from 1963 to present. Available through the NASA/RECON system of the National Aeronautics and Space Administration only.

NTIS Bibliographic Database. National Technical Information Service, 5285 Port Royal Road, Springfield, Va 22161 . (703) 487-4929 or FAX (703) 321-8199. Broad covera ge of government-sponsored science and technology research reports, 1964 to present. Available on BRS,(800) 289-4277, DIALOG, (800) 334-2564, ORBIT (800) 456-7248, and STN International, FIZ Karlsruhe, P.O. Box 2465, W-7500, Karlsruhe 1, Germany, onlineservices. Also available on CD-ROM. Inquire as to cost and availability.

Physics Briefs. American Institute of Physics, 335 East 45th Street , New York, NY 10017. (212) 661-9260 or FAX (212) 661-2036. Contains citations with abstracts of the literature of physics and related fields, 1979 to present. Available on the STN International, FIZ Karlsruhe, P.O. Box 2465, W-7500, Karlsruhe 1, Germany, online service. Inquire as to cost and availability.

Scientific and Technical Books and Serials In Print. R.R. Bowker Inc., 121 Chanlon Road, New Providence, NJ 07974. (800) 521-8110 or (908) 464-6800. List of currently published books and serials in the physical and biological sciences, engineering and technology, with subject, author and titles indexes. Available on ORBIT, (800) 456-7248, online service. Inquire as to cost and availability.

Scisear Ch. Institute for Scientific Information, 3501 Market Street, Philadelphia, PA 19104. (800) 523-1850 or (215) 386-0100. Broad multidisciplinary title and author index to the international literature of science and technology, 1974 to present. Available on DIALOG, (800) 334-2564, and ORBIT, (800) 456-7248, online services. Also available on CD-ROM. Inquire as to cost and availability.

WILSONLINE. H.W. W ilson Company, 950 University Avenue, Bronx, NY 10452. (800) 367-6770 or (212) 588-8400. Makes available onlin e versions of the printed H.W. Wilson indexes includin g Applied Science and Technology Index, Business Periodicals Index, General Science Index, and Readers' Guide to Periodical Literature. Period covered is generally 1983 to present. Available on BRS, (800) 289-4277, DIALOG, (800) 334-2564, and OCLC EPIC, (800) 848-5878, online services. Also available on CD-ROM. Inquire as to cost and availability.

OTHER SOURCES

Atlas of the Universe. Patrick Moore. Rand McNally & Company. 8255 North Central Park, Skokie, IL 60076-2970. (708) 673-9100. 1994. $29.95.

Cambridge Atlas of Astronomy. Jean Audouze, et al. Cambridge University Press, 40 West 20th Street, New York,

Ny 10011-4211. (212) 924-3900. (800) 872-7423. 3rd edition. 1994. $75.00.

Cambridge Guide To the Constellations. Michael E. Bakich. Cambridge University Press, 40 West 20th Street, New York, NY 10011-4211. (212) 924-3900. (800) 872-7423. 1995. $19.95 in paper.

Observer's Sky Atlas. E. Karkoschka. Springer-Verlag New York, Inc., 175 Fifth Avenue, New York, NY 10010. (212) 460-1500. (800) 777-4643. 1995. $15.95.

The Sky: A User's Guide. David H. Levy. Cambridge University Press, 40 West 20th Street, New York, NY 10011- 4211. (212) 924-3900. (800) 872-7423. 1993. $14.95 in paper.

PERIODICALS

ACTA Astronomica. Copernicus Foundation for PolishAstronomy, AL Ujazdowskie 4, 00-478 Warsaw, Poland. TEL 48- 22-295346. FAX 48-22-294967. 1925 to present. Quarterly. $80.00 per year. ISSN: 0001-5237.

Astronomical Journal. American Institute of Physics, One Physics Ellipse, College Park, MD 20740-3843. (301) 209- 3000. Published for the American Astronomical Society. 1849 to present. Monthly. $280.00 per year. ISSN: 0004- 6256.

Astronomical Society of the Pacific Publications. Astronomical Society of the Pacific, 390 Ashton Avenue, San Francisco, CA 94112. (415) 337-1100. FAX (415) 337-5205. 1889 to present. Monthly. $175.00 per year. ISSN: 0004- 6280.

Astronomy. Kalmbach Publishing Company, Box 1612, Waukesha, WI 53187-1612. (414) 796-8776. FAX (414) 796-1142. 1973 to present. Monthly. $27.00 per year. ISSN: 0091-6358.

Astronomy and Astrophsyics. Springer-Verlag New York, Inc., 44 Hartz Way, Secaucus, NJ 07096-2491. (201) 348-4033. FAX (201) 348-4505. 1969 to date. 36 issues per year in 12 volumes. $1733.00 per year. ISSN: 0004-6361.

Astronomy Reports. American Institute of Physics, One Physics Ellipse, College Park, MD 20740-3843. (301) 209- 3000. English translation of Astronomicheksii Zhurnal. Formerly Soviet Astronomy AJ. 1957 to present. Bimonthly. $1160.00 per year. ISSN: 1063-7729.

Astrophysical Journal; An International Review of Astronomy and Astronomical Physics. University of Chicago Press, Journals Division, 5720 South Woodlawn Avenue, Chicago, IL 60637. (312) 753-3347. FAX (312) 753-0811. 1895 to date. three issues per month. $740.00 per year. ISSN: 0004-637X.

Astrophys ICS and Space Science; An International Journal of Cosmic Physics. Kluwer Academic Publishers, Box 358, Accord Station, Hingham, MA 02018-0358. (617) 871-6600. FAX (617) 871-6528. 1968 to present. 24 issues per year. $2544.00 per year. ISSN: 0004-640X.

Celestial Mechanics and Dynamical Astronomy; An International Journal of Space Dynamics. Kluwer Academic Publishers, Box 358, Accord Station, Hingham, MA 02018-0358. (617) 871-6600. FAX (617) 871-6528. 1969 to present. Monthly. $703.50 per year. ISSN: 0923-2958.

Ciel Et Terre. Societe Royale Belge d'Astronomie, de Meteorologie et de Physique du Globe, 3 avenue Circulaire, Uccle, Brusels, Belgium. FAX 32-2-374-9822. 1880 to present. Bimonthly. 1600 BEF per year. ISSN: 0009-6709.

Icarus; International Journal of Solar System Studies. Academic Press, Inc., Journal Division, 525 B Street, Suite 1900, San Di-

ego, CA 92101-4495. (619) 230-1840. FAX (619) 699-6800. 1962 to present. Monthly. $1080.00 per year. ISSN: 0019-1035.

Mercury. Astronomical Society of the Pacific. 390 Ashton Avenue, San Francisco, CA 94112. (415) 337-1100. FAX: (415) 337-5205. 1972 to present. Bimonthly. $175.00 per year. ISSN: 0047-6773.

Monthly Notices of the Royal Astronomical Society. Blackwell Scientific Publication LT., Osney Mead, Oxford OX2 OEL, England. TEL 0865-240201. FAX 0865-721205. 1827 to present. Fortnightly. $1899.00 per year. ISSN: 0035-8711.

Sky and Telescope. Sky Publishing Corporation, Box 9111, Belmont, MA 02178. (617) 864-7360. FAX (617) 864-6117. 1941 to present. Monthly. $27.00 per year. ISSN: 0037- 6604.

Space Science Reviews. Kluwer Academic Publishers, Box 358, Accord Station, Hingham, MA 02018- 0358. (617) 871- 6600. FAX (617) 871-6528. 1962 to present. Sixteen issues per year. $940.00 per year. ISSN: 0038-6308.

Vistas In Astronomy; An International Review Journal. Albert C. Beer, editor. Elsevier Science Publishing Company, Inc., 660 White Plains Road, Terrytown, NY 10591- 5153. (914) 524-9200. FAX: (914) 333-2444. 1958 to present. Quarterly. $415.00 per year. ISSN:: 0083-6656.

Transactions of the International Astronomical Union. Proceedings of the General Assembly. Kluwer Academic Publishers, Box 358, Accord Station, Hingham, MA 02018- 0358. (617) 871-6600. Issues in two parts: Part A, Reports, and Part B, Proceedings. Triennial. XXI - 1994. Price varies. ISSN: 0080-1372.

Webb Society Quarterly Journal. Webb Society, 194 Foundry Lane, Freemantle, Southampton, Hampshire SO1 3EQ, England. 1969 to present. Quarterly. $19.50 per year. ISSN: 0043- 1680.

Astronomy Program. University of Maryland, 1207 Computer & Space Sciences Building, College Park, Md 20742. (301) 405-3001. FAX (301) 314-9067.

Institute For Astronomy. University of Hawaii at Manoa, 2680 Woodlawn Drive, Honolulu, HI 96822. (808) 956-8312. FAX (808) 988-2790.

Joint Institute For Laboratory Astrophysics. University of Colorado - Boulder, Boulder CO 80309-0440. (303) 492-7789. FAX (303) 492-5235.

Lowell Observatory. 1400 West Mars Hill Road, Flagstaff, AZ 86001. (602) 774-3358. FAX (602) 774-6296.

Lunar and Planetary Institute. 3303 NASA Road One, Houston, TX 77058-4399. (713) 486-2139. FAX (713) 4 86-2162.

National Astronomy and Ionosphere Center. Cornell University, Space Sciences Building, Ithaca, NY 14853. (607) 255-3735. (607) 255-8803.

Smithsonian Center For Astrophysics. Harvard University, 60 Garden Street, Cambridge, MA 02138. (617) 495-7461. FAX (617) 495-7326.

Space Sciences Laboratory. University of California, Berkeley, Berkeley, CA 94720. (415) 642-1361. FAX (415) 643-7629.

CELESTIAL NAVIGATION

See: NAVIGATION

CEMENT

See also: BUILDING MATERIALS, CONCRETE

ABSTRACT SERVICES AND INDEXES

Applied Science and Technology Index; A Cumulative Subject Index To English Language Periodicals In the Fields of Aeronautics and Space Science, Computer Technology, Chemistry, Construction Industry, Energy and Related Areas. H.W. Wilson Co., 950 University Avenue, Bronx, NY 10452. (800) 367-6770 or (212) 588-8400. FAX (718) 590-1617. From 1958 to present. Monthly. Inquire about cost and availability. Also available on CD-ROM and online.

Concrete Abstracts. Covers articles, books, and other publications that report developments in concrete design, construction and technology. American Concrete Institute, PO Box 19150, Detroit, MI 48219. (313) 532-2600. FAX (213) 538-0655. 1972 to present. Bi-monthly. Inquire for price.

Engineering Index Monthly. Engineering Information, Inc., Castle Point on the Hudson, Hoboken, NJ 07030. (800) 221-1044. FAX (212) 832-1857. Monthly. $2300.00 per year. Also available online as COMPENDEX, and also on CD-ROM. Covers chemical engineering, computers, electrical engineering, civil engineering, metals and mining, industrial management, and mechanical engineering.

ANNUAL REVIEWS AND YEARBOOKS

U.S.Department of the Interior, Bureau of Mines, Minerals Yearbook. Superintendent of Documents, U.S. Government Printing office, Box 371954, Pittsburgh, PA 15250-7954. (202) 783-3238. FAX (202) 512-2233. 1932 to present. Annual. Price varies.

ASSOCIATIONS AND PROFESSIONAL SOCIETIES

American Concrete Institute. PO Box 19150, Redford Station, Detroit, MI 48219. (313) 532-2600. FAX (213) 538-0655.

American Concrete Pavement Association. 3800 N. Wilke Road, Suite 490, Arlington Heights, IL 60004. (708) 394-5577. FAX (708) 394-5610.

American Portland Cement Alliance. 1212 New York Avenue NW, Suite 500, Washington, DC 20005. (202) 408-9494. FAX (202) 408-9392.

American Society For Concrete Construction. 1902 Techny Court, Northbrook, IL 60062. (312) 291-0270.

American Society of Civil Engineers. 345 East 47th Street, New York, NY 10017-2398. (212) 705-7520 or (800) 548-2723.

National Concrete Masonry Association. 2302 Horse Pen Road, Herndon, VA 22071-3406. (703) 713-1900. FAX (703) 713-1910.

National Precast Concrete Association. 10333 North Meridian Street, Suite 272, Indianapolis, in 46290-9500. (800) 366-7731. FAX (317) 571-0041.

National Ready Mixed Concrete Association. 900 Spring Street, Silver Spring, MD 20910. (301) 587-1400. FAX (301) 585-4219.

Portland Cement Association. 5420 Old Orchard Road, Skokie, IL 60077. (708) 966-6200. FAX (708) 966-8389.

Precast/prestressed Concrete Institute. 175 West Jackson Blvd., Chicago, IL 60604. (312) 786-0300. FAX (312) 786-0353.

Reinforced Concrete Research Council. 205 N. Mathews Avenue, Urbana, IL 61801. (217) 333-7384.

DIRECTORIES AND BIOGRAPHICAL SOURCES

American Concrete Institute—directory. American Concrete Institute, PO Box 19150, Detroit, MI 48219. (313) 532-2600. FAX (213) 538-0655. Every 2-3 years. Write or call for price information.

National Precast Concrete Association Directory of Members and Precast Products. National Precast Concrete Association, 10333 North Meridian Street, Suite 272, Indianapolis, in 46290-9500. (800) 366-7731. FAX (317) 571-0041. Annual. $295.00.

Who's Who In Engineering. American Association of Engineering Societies, 1111 19th Street NW, Suite 608, Washington, DC 20036. (202) 296-2237 or (800) 658-8897. 8th edition. 1991. Inquire for price.

World Cement Directory. European Cement Association/Cimeurope, Rue d'Arlon 55, B-1040 Brussels, Belgium. Telephone 32-2-2341011. FAX 32-2-2304720. 1958 to present. Quinquennial. $600.00.

ENCYCLOPEDIAS AND DICTIONARIES

Cement and Concrete Terminology. American Concrete Institute Staff. American Concrete Institute, Box 19150, Redford Station, Detroit, MI 48219. (313) 532-2600. FAX (313) 538-0655. 1990. $33.95.

Concise Encyclopedia of Building and Construction Materials. Fred Moavenzadeh. Pergamon Press Inc., Maxwell House, Fairview Park, Elmsford, NY 10523. (914) 592-7700. Fax (914) 592-3625. 1990. $210.00.

Encyclopedia of Physical Science and Technology. Robert A. Meyers, ed. 18 volumes. Academic Press Inc., 6277 Sea Harbor Drive, Orlando, FL 32887. (800) 321-5068. 1992. $2100.00

McGraw-Hill Dictionary of Scientific and Technical Terms. Sybil P. Parker, ed. 5th edition. McGraw-Hill Publishing Company, 1221 Avenue of the Americas, New York, NY 10020. (800) 262-4729 or (212) 512-3825. 1993. $110.50.

McGraw-Hill Encyclopedia of Engineering. Sybil P. Parker, ed. 2nd edition. McGraw-Hill Publishing Company, 1221 Avenue of the Americas, New York, NY 10020. (800) 262-4729 or (212) 512-3825. 1993. $95.50.

McGraw-Hill Encyclopedia of Science and Technology. Sybil P. Parker, ed. 7th edition. 20 volumes. McGraw-Hill Publishing Company, 1221 Avenue of the Americas, New York, NY 10020. (800) 262-4729 or (212) 512-3825. 1992. $1900.00

GENERAL WORKS

Cement Chemistry. H.F.W. Taylor. Academic Press, Inc., 6277 Sea Harbor Drive, Orlando, FL 32887. (800) 321-5068. 1990. $120.00.

Cement Manufacture and Use. P.W. Brown, editor. ASCE, 345 East 47th Street, New York, NY 10017-2398. (212) 705-7520 or (800) 548-2723. 1994. Inquire for cost and availability.

Cement Technology. E.M. Gartner, H. Uchikawa. American Ceramic Society, 735 Ceramic Place, Westerville, OH 43081. (614) 890-4700. FAX (614) 899-6109. 1994. Inquire for cost and availability.

Concrete Technology. A.M. Neville, J.J. Brooks. John Wiley and Sons, Inc., 605 Third Avenue, New York, NY 10158. (800) 526-5368 or (212) 850-6000. 1987. $59.95.

Concrete Technology. G.R. White. 3rd edition. Delmar Publishers, Division of Thomson Educational Publishing, Inc., 3 Columbia Circle, Box 15015, Albany, NY 12205. (518) 464-3500 or (800) 347-7707. FAX (518) 464-0358. 1991. $21.00.

Concrete Technology: Past, Present, & Future. American Concrete Institute, PO Box 19150, Redford Station, Detroit, MI 48219. (313) 532-2600. FAX (213) 538-0655. 1994. $50.00.

HANDBOOKS AND MANUALS

ACI Manual of Concrete Practice. 5 volumes. American Concrete Institute, PO Box 19150, Redford Station, Detroit, MI 48219. (313) 532-2600. FAX (213) 538-0655. 1967 to present. Annual. $423.00 per set for non-members.

Concrete Construction Handbook. Joseph J. Waddell & Jpseh A. Dobrowolski, editors. 3rd edition. McGraw-Hill Publishing Company, 1221 Avenue of the Americas, New York, NY 10020. (800) 262-4729 or (212) 512-3825. 1993. $92.00.

Concrete Manual. Materials Engineering Branch, Research and Laboratory Services Division, Denver office, U.S. Department of the Interior, Bureau of Reclamation. 9th edition. Superintendent of Documents, U.S. Government Printing office, Box 371954, Pittsburgh, PA 15250-7954. (202) 783-3238. FAX (202) 512-2233. 1992. Inquire for cost and availability.

Handbook of Concrete Engineering. Mark Fintel, editor. 2nd edition. Van Nostrand Reinhold, 115 Fifth Avenue, New York, NY 10003. (212) 254-3232 or FAX (212) 254-9499. 1985. $115.00.

Structural Design Guide To the Asci Specifications For Buildings. Paul F. Rice, et al. 3rd edition. Books on Demand, 300 N. Zeeb Road, Ann Arbor, MI 48106-1346. (313) 761-4700 or (800) 521-0600. 1985. $92.00.

ONLINE DATABASES AND CD-ROMS

Compendex Plus. Engineering Information, Inc., 345 East 47th Street, New York, NY 10017. (212) 705-7600 or (800) 221-1044. Contains citations with abstracts to worldwide literature in engineering and technology, from 1970 to present. Available on online BRS,(800) 289-4277, DIALOG, (800) 334-2564, ORBIT (800) 456-7248, and STN International, FIZ Karlsruhe, P.O. Box 2465, W-7500, Karlsruhe 1, Germany, online services. Also available on CD-ROM. Inquire as to cost and availability.

NTIS Bibliographic Database. National Technical Information Service, 5285 Port Royal Road, Springfield, Va 22161. (703) 487-4929 or FAX (703) 321-8199. Broad coverage of government-sponsored science and technology research reports, 1964 to present. Available on BRS,(800) 289-4277, DIALOG, (800)334-2564, ORBIT, (800) 456-7248, and STN International, FIZ Karlsruhe, P.O. Box 2465, W-7500, Karlsruhe 1, Germany, online services. Also available on CD-ROM. Inquire as to cost and availability.

WILSONLINE. H.W. Wilson Company, 950 University Avenue, Bronx, NY 10452. (800) 367-6770 or (212) 588-8400. Makes available online versions of the printed H.W. Wilson indexes including Applied Science and Technology Index, Business Periodicals Index, General Science Index, and Readers' Guide to Periodical Literature. Period covered is generally 1983 to present. Available on BRS,(800) 289-4277, DIALOG,(800) 334-2564, and OCLC EPIC,(800) 848-5878, online services. Also available on CD-ROM. Inquire as to cost and availability.

PERIODICALS

ACI Materials Journal. American Concrete Institute, PO Box 19150, Detroit, MI 48219. (313) 532-2600. FAX (213) 538-0655. Monthly. $101.00 per year.

ACI Structural Journal. American Concrete Institute, PO Box 19150, Detroit, MI 48219. (313) 532-2600. FAX (213) 538-0655. Bi-monthly. $101.00 per year.

Cement and Concrete Research. Elsevier Science [journals], 660 White Plains Rd., Tarrytown, NY 10159-5153. (919) 524-9200. FAX (919) 333-2444. 1971 to present. 8 times a year. $590.00 per year.

Cement and Concrete Composites. Elsevier Science [journals], 660 White Plains Rd., Tarrytown, NY 10159-5153. (919) 524-9200. FAX (919) 333-2444. 1979 to present. 4 times a year. $280.00 per year.

Cement, Concrete and Aggregates. American Society for Testing and Materials, 1916 Race Street, Philadelphia, PA 19103. (215) 299-5585. 1979-present. Semi-annual. $48.00 per year for non-members.

Cement, Quarry & Mineral Aggregate Newsletter. National Safety Council, Periodicals Department, 1121 Spring Lake Drive, Itasca, IL 60143. (708) 285-1121. Bi-monthly. $19.00 per year.

Concrete. The Concrete Society, Framewood Road, Wexham, Slough, Berks. SL3 6PJ England. Telephone 0753-662226. FAX 0753-662126. 1966 to present. Bi-monthly. $75.00 per year.

Concrete International: Design and Construction. American Concrete Institute, Box 19150, Redford Station, Detroit, MI

CEMENT

Ency. of Physical Sciences and Engineering Info. Sources

48219. (313) 532-2600. FAX (313) 538-0655. 1979 to present. Monthly. $101.00 per year.

Concrete Masonry News. National Concrete Masonry Association, 2302 Horse Pen Road, Herndon, VA 22071-3406. (703) 713-1900. FAX (703) 713-1910. 1962 to present. Monthly. Inquire for cost and availability.

Journal of Ferrocement. International Ferrocement Information Center, Asian Institute of Technology, PO Box 2754, Bangkok 10501, Thailand. Telephone 5160110. FAX 66-2-5162126. 1976 to present. Quarterly. $96.00 per year.

Journal of the Precast/prestressed Concrete Institute. Precast/ Prestressed Concrete Institute, 175 West Jackson Blvd., Chicago, IL 60604. (312) 786-0300. FAX (312) 786-0353. Bi-monthly. $29.00 per year.

Magazine of Concrete Research. Thomas Telford Services Ltd., Thomas Telford House, 1 Heron Quay, London E14 4JD, England. Telephone 071-987-6999. FAX 071-538-4101. 1949 to present. Quarterly. Inquire for cost.

Rock Products. McLean Hunter Publishing Company, 29 N. Wacker Drive, Chicago, IL 60606. (312) 726-2802. FAX (312) 726-2574. 1897 to present. Monthly. $31.25 per year. RESEARCH CENTERS

American Concrete Institute. PO Box 19150, Redford Station, Detroit, MI 48219. (313) 532-2600. FAX (213) 538-0655.

American Society For Testing and Materials. 1916 Race Street, Philadelphia, PA 19103. (215) 299-5585.

Center For Advanced Cement-based Materials. Northwestern University, 1800 Ridge Avenue, Evanston, IL 60208-4400. (708) 491-8569.

Center For Cement Composite Materials. University of Illinois, 202 Ceramics Bldg., 105 S. Goodwin Avenue, Urbana, IL 61801. (217) 333-6960. FAX (217) 244-7750.

Center For Concrete and Geomaterials. Northwestern University, Technological Institute, Room 2474, 2145 Sheridan Road, Evanston, IL 60208. (708) 491-3858.

Concrete Materials Research Council. PO Box 19150, Detroit, MI 48219. (313) 532-2600.

Concrete Research Laboratory. University of Michigan, G.G. Brown Laboratory, Ann Arbor, MI 48109. (313) 764-9660.

National Concrete Masonry Association Research and Development Laboratory. National Concrete Masonry Association, 2302 Horse Pen Road, Herndon, VA 22071-3406. (703) 713-1900. FAX (703) 713-1910.

Portland Cement Association. 5420 Old Orchard Road, Skokie, IL 60077. (708) 966-6200. FAX (708) 966-8389.

SPECIFICATIONS AND STANDARDS

*Annual Book of ASTM Standards, Volume 04.*01. American Society for Testing and Materials (ASTM), 1916 Race Street, Philadelphia, PA 19103. (215) 299-5585. Annual. Inquire for cost and availability.

CERAMIC ENGINEERING

See also: CERAMICS, CLAY, GLASS, MATERIALS SCIENCE

ABSTRACT SERVICES AND INDEXES

Applied Science and Technology Index; A Cumulative Subject Index To English Language Periodicals In the Fields of Aeronautics and Space Science, Computer Technology, Chemistry, Construction Industry, Energy and Related Areas. H.W. Wilson Co., 950 University Avenue, Bronx, NY 10452. (800) 367-6770 or (212) 588-8400. FAX (718) 590-1617. From 1958 to present. Monthly. Inquire about cost and availability. Also available on CD-ROM andonline.

C2C Abstracts: Japan—Materials Science. Scan C2C, 1001 Pennsylvania Ave., NW, No. 1300, Washington, DC 20024-2505. (800) 525-3865. FAX (202) 863-3855. 1990 to present. Monthly. $200.00 per year. Contains abstracts of articles from Japanese scientific, business and technical journals on aggregates, coatings, composites, fibers and whiskers, powder technology, processing, refractories, and wood products. Also available on CD-ROM and online.

CA Selects: Ceramic Materials (Journals). Chemical Abstracts Service, 2450 Olentagy River Road, Box 3012, Columbus, OH 43210-0012. (614) 447-3600. FAX (614) 447-3713. 1988 to present. Semi-weekly. $210.00 per year.

Ceramic Abstracts. American Ceramic Society, Inc., 735 Ceramic Place, Westerville, OH 43081. (614) 890-4700. FAX (614) 899-6109. Bi-monthly. 1922 to present. $138.00 per year for non-members. Also available on CD-ROM, and online through DIALOG, ORBIT, and STN.

Engineered Materials Abstracts. ASM (American Society of Metals) International, Materials Information, Materials Park, OH 44073. (216) 338-5151 or FAX (216) 338-4634. Covers literature on technical developments in polymer, ceramic, and composite materials and engineering. 1986 to present. Monthly. $1200.00 per year. Also available on CD-ROM and online. Covers published materials on technical developments in polymer, ceramic, and composite materials in an engineering environment.

Engineering Index Monthly. Engineering Information, Inc., Castle Point on the Hudson, Hoboken, NJ 07030. (800) 221-1044. FAX (212) 832-1857. Monthly. $2300.00 per year. Also available online as COMPENDEX, and also on CD-ROM. Covers chemical engineering, computers, electrical engineering, civil engineering, metals and mining, industrial management, and mechanical engineering.

Materials Science and Engineering Abstracts, Cambridge Scientific Abstracts, 7200 Wisconsin Avenue, Bethesda, Md 20814-4823. (301) 961-6750. FAX (301) 961-6720. 1993 to present. Monthly. $385.00 per year. Focuses on mechanical and physical properties of materials and commercial or industrial applications for materials, methods for strength testing, effects of vibration and other stresses, corrosion and protective coatings, storage and handling, ceramics, composites, metals, wood, plastics, and polymers.

Materials Science Citation Index. Institute for Scientific Information, 3501 Market Street, Philadelphia, PA 19104. (215) 386-0100. FAX (215) 386-6362. Bi-monthly. $975.00 per year. Also available on CD-ROM.

Physics Abstracts. Institute of Electrical Engineers, Michael Faraday House, Six Hill Way, Stevenage, Herts, SG1 2AY, England. Distributed by IEEE, 445 Hoes Lane, Piscataway, NJ 08854. (908) 562-5549. 1898 to present. Monthly. $2700.00 per year. Also available online as INSPEC.

World Ceramics Abstracts. Ceram Research Ltd., Queens Road, Penkhull, Stoke-on-Trent ST4 7LQ, England. 1958 to present. Monthly. Inquire for price and availability. Also available online through ORBIT.

ASSOCIATIONS AND PROFESSIONAL SOCIETIES

American Ceramic Society. 735 Ceramic Place, Westerville, OH 43081. (614) 890-4700. FAX (614) 899-6109.

Ceramic Educational Council. 735 Ceramic Place, Westerville, OH 43081. (614) 890-4700. FAX (614) 899-6109.

Ceramic Manufacturers Association. 1100-H Brandywine Blvd., PO Box 2188, Zanesville, OH 43702-2188. (614) 452-4541.

International Ceramic Association. PO Box 39, Glen Burnie, MD 21061. (410) 923-3425.

Keramos. Route 4, Box 4400, Kennewick, WA 99337. (509) 627-3907.

Materials Research Society. 9800 McKnight Road, Pittsburgh, PA 15237. (412) 367-3003. FAX (412) 367-4373.

National Institute of Ceramic Engineers. 735 Ceramic Place, Westerville, OH 43081. (614) 890-4700.

Technical Ceramics Manufacturers Association. 25 N. Broadway, Tarrytown, NY 10591. (914) 332-0040.

BIBLIOGRAPHIES

A Bibliography of Ceramics and Glass. Larry L. Hench & B.A. McEldowney. American Ceramic Society, 735 Ceramic Place, Westerville, OH 43081. (614) 890-4700. FAX (614) 899-6109. 1976. $15.00.

DIRECTORIES AND BIOGRAPHICAL SOURCES

American Men & Women of Science, 1995-96. R.R. Bowker Staff, eds. 19th edition. 8 volumes. R.R. Bowker/Reed International Publishing Company, 121 Chanlon Road, New Providence, NJ 07974. (908) 464-6800 or (800) 521-8110. 1995. $850.00.

Ceramic Industry Data Book Buyers Guide. Business News Publishing Company, 755 W. Big Beaver Road, Suite 1000, Troy, MI 48084. (800) 837-1037. FAX (313) 362-0317. 1922 to present. Annual. $25.00 per year. Included as part of subscription to Ceramic Industry periodical (see below).

Guide To Engineering Materials Producers. J.C. Bittence. ASM International, Materials Information, Materials Park, OH 44073-0002. (216) 338-5151 or (800) 336-5152. FAX (216) 338-4634. 1993. $141.00

International Ceramic Directory. London & Sheffield Publishing Company Ltd., 5 Pond Street, Hampstead, London NW3 2PN, England. Telephone 01-794-0800. FAX 01-794-0411. 1984 to present. Every 5 years. Inquire for price and availability.

International Research Centers Directory. Gale Research, 835 Penobscot Building, Detroit, MI 48226-4094. (313) 961-2242. (800) 347-4253. 8th edition. 1995. $410.00

Materials Research Centers: A World Directory of Organizations and Programs In Material Science. 5th edition. Gale Research, 835 Penobscot Building, Detroit, MI 48226-4094. (313)961-2242. (800) 347-4253. FAX (313) 961-6083. 1993. $495.00.

Research Centers Directory. Gale Research, 835 Penobscot Building, Detroit, MI 48226-4094. (313) 961-2242. (800) 347-4253. FAX (313) 961-6083. $485.00.

Scientific and Technical Organizations and Agencies Directory. Gale Research, 835 Penobscot Building, Detroit, MI 48226-4094. (313) 961-2242. (800) 347-4253. FAX (313) 961-6083. 4th edition. 1996. $195.00.

Who's Who In Engineering. American Association of Engineering Societies, 1111 19th Street NW, Suite 608, Washington, DC 20036. (202) 296-2237 or (800) 658-8897. 8th edition. 1991. Inquire for price.

Who's Who In Technology. Gale Research, 835 Penobscot Building, Detroit, MI 48226-4094. (313) 961-2242. (800) 347-4253. FAX (313) 961-6083. 1995. $195.00.

ENCYCLOPEDIAS AND DICTIONARIES

Ceramic Glossary. Walter W. Perkins, editor. American Ceramic Society, 735 Ceramic Place, Westerville, OH 43081. (614) 890-4700. FAX (614) 899-6109. 1984. $12.00.

Concise Encyclopedia of Advanced Ceramic Materials. R.J. Brook, ed. MIT Press, 55 Hayward Street, Cambridge, MA 02142. (617) 253-2884 or (800) 356-0343. FAX (617) 253-1709. 1990. $150.00.

Dictionary of Ceramic Science and Engineering. Ian J. McColm, editor. 2nd edition. Plenum Publishing Corporation, 233 Spring Street, New York, NY 10013. (212) 620-8000. FAX (212) 463-0742. 1994. $75.00.

Encyclopedia of Glass, Ceramics, and Cement. Martin Grayson, editor. John Wiley and Sons, Inc., 605 Third Avenue, New York, NY 10158. (800) 526-5368 or (212) 850-6000. 1985. Inquire for cost and availability.

Encyclopedia of Physical Science and Technology. Robert A. Meyers, ed. 18 volumes. Academic Press Inc., 6277 Sea Harbor Drive, Orlando, FL 32887. (800) 321-5068. 1992. $2100.00.

McGraw-Hill Encyclopedia of Engineering. Sybil P. Parker, ed. 2nd edition. McGraw-Hill Publishing Company, 1221 Avenue of the Americas, New York, NY 10020. (800) 262-4729 or (212) 512-3825. 1993. $95.50.

McGraw-Hill Encyclopedia of Science and Technology. Sybil P. Parker, ed. 7th edition. 20 volumes. McGraw-Hill Publishing

CERAMIC ENGINEERING

Ency. of Physical Sciences and Engineering Info. Sources

Company, 1221 Avenue of the Americas, New York, NY 10020. (800) 262-4729 or (212) 512-3825. 1992. $1900.00

GENERAL WORKS

Ceramics In Energy Applications. Institute of Energy. Pergamon Press Inc., Maxwell House, Fairview Park, Elmsford, NY 10523. (914) 592-7700. Fax (914) 592-3625. 1994. $130.00.

Ceramics Today—Tomorrow's Ceramics. P. Vincenzini. 4 volumes. Elsevier Science B.V., PO Box 945, Madison Square Station, New York, NY 10159-0945. (212) 633-3650. FAX (212) 633-3680. 1991. $1037.50.

Engineering Ceramics: Fabrication Science & Technology. D.P. Thompson, editor. Institute of Materials/Ashgate Publishing Co., Old Post Rd., Brookfield, VT 05036. (802) 276-3162. 1993. $115.00.

Fundamentals of Ceramics Engineering. P. Vincenzini, ed. Elsevier Science Publishing Company, Inc., 655 Avenue of the Americas, New York, NY 10010. (212) 989-5800. 1991. $110.00.

Modern Ceramic Engineering. David W. Richerson. 2d ed. Marcel Dekker, Inc., 270 Madison Avenue, New York, NY 10016. (212) 696-9000. FAX (212) 685-4540. 1992. $150.00.

HANDBOOKS AND MANUALS

ASM Engineered Materials Handbook Volume 4: Ceramics and Glasses. ASM International, Materials Information, Materials Park, OH 44073-0002. (216) 338-5151 or (800) 336-5152. FAX (216) 338-4634. 1995. $147.00.

ASM Engineered Materials Reference Book. Michael L. Bauccio, editor. 2nd edition. ASM International, Materials Information, Materials Park, OH 44073-0002. (216) 338-5151 or (800) 336-5152. FAX (216) 338-4634. 1993. $111.00.

Materials Handbook: An Encyclopedia For Managers, Technical Professionals, Purchasing and Production Managers, Technicians, Supervisors, and Foremen. George S. Brady & Henry R. Clauser. 13th revised edition. McGraw Hill Publishing Company, 1221 Avenue of the Americas, New York, NY 10020. (800) 262-4729 or (212) 512-3825. 1991. $79.50.

ONLINE DATABASES AND CD-ROMS

Compendex Plus. Engineering Information, Inc., 345 East 47th Street, New York, NY 10017. (212) 705-7600 or (800) 221-1044. Contains citations with abstracts to worldwide literature in engineering and technology, from 1970 to present. Available on online BRS,(800) 289-4277, DIALOG, (800) 334-2564, ORBIT (800) 456-7248, and STN International, FIZ Karlsruhe, P.O. Box 2465, W-7500, Karlsruhe 1, Germany, online services. Also available on CD-ROM. Inquire as to cost and availability.

CSA Engineering. Cambridge Scientific Abstracts, 7200 Wisconsin Avenue, Suite 601, Bethesda, MD 20814. (301) 961-6750 or (800) 843-7751. Contains citations and abstracts of international periodicals and research literature covering all fields of engineering and science and technology,including computer and information science, electronics, mechanical engineering,

solid state materials, 1981 to present. Available on BRS,(800) 289-4277, online service.Inquire as to cost and availability.

Current Contents Search. Institute for Scientific Information, 3501 Market Street, Philadelphia, PA 19104. (215) 386-0100. FAX (215) 386-6362. Contains citations to articles listed in the table of contents of science and technology journals. Also articles in social sciences and life sciences journals. Available on BRS,(800) 289-4277, DIALOG,(800) 334-2564, online services. Inquire as to cost and availability.

Dissertation Abstracts. University Microfilms International, 300 North Zeeb Road, Ann Arbor, MI 48106. (800)521-0600 or (313) 761-4700. Scope includes virtually all doctoral dissertations accepted at accredited American institutions from 1861 to present in 252 subject areas. Available on BRS,(800) 289-4277, DIALOG,(800) 334-2564, and OCLC EPIC,(800) 848-5878, online services. Also available on CD-ROM. Inquire as to cost and availability.

NTIS Bibliographic Database. National Technical Information Service, 5285 Port Royal Road, Springfield, Va 22161. (703) 487-4929 or FAX (703) 321-8199. Broad coverage of government-sponsored science and technology research reports, 1964 to present. Available on BRS,(800) 289-4277, DIALOG, (800) 334-2564, ORBIT, (800) 456-7248, and STN International, FIZ Karlsruhe, P.O. Box 2465, W-7500, Karlsruhe 1, Germany, online services. Also available on CD-ROM. Inquire as to cost and availability.

Scisearch. Institute for Scientific Information, 3501 Market Street, Philadelphia, PA 19104. (800) 523-1850 or (215) 386-0100. Broad multidisciplinary title and author index to the international literature of science and technology, 1974 to present. Available on DIALOG,(800) 334-2564, and ORBIT,(800) 456-7248, online services. Also available on CD-ROM. Inquire as to cost and availability.

WILSONLINE. H.W. Wilson Company, 950 University Avenue, Bronx, NY 10452. (800) 367-6770 or (212) 588-8400. Makes available online versions of the printed H.W. Wilson indexes including Applied Science and Technology Index, Business Periodicals Index, General Science Index, and Readers' Guide to Periodical Literature. Period covered is generally 1983 to present. Available on BRS,(800) 289-4277, DIALOG,(800) 334-2564, and OCLC EPIC,(800) 848-5878, online services. Also available on CD-ROM. Inquire as to cost and availability.

PERIODICALS

American Ceramic Society Bulletin. American Ceramic Society, 735 Ceramic Place, Westerville, OH 43081. (614) 890-4700. FAX (614) 899-6109. 1922 to present. Monthly. $28.00 per year for non-members.

American Ceramic Society Journal. American Ceramic Society, 735 Ceramic Place, Westerville, OH 43081. (614) 890-4700. FAX (614) 899-6109. 1899 to present. Monthly. $175.00 per year for non-members.

Canadian Ceramic Society Journal. Canadian Ceramic Society, 2175 Sheppard Avenue E., Suite 100, Willowdale, Ontario M2J 1W8, Canada. (416) 491-2886. FAX (416) 491-1670. 1928 to present. Annual. Inquire for cost.

Ceramic Engineering and Science Proceedings. American Ceramic Society, Inc., 735 Ceramic Place, Westerville, OH 43081.

(614) 890-4700. FAX (614) 899-6109. 1980 to present. Bi-monthly. $80.00 per year.

Ceramic Industry. Business News Publishing Company, 755 W. Big Beaver Road, Suite 1000, Troy, MI 48084. (313) 362-3700. FAX(313) 362-0317. 1923 to present. 13 times a year. $53.00 per year.

Ceramics International. Elsevier Science [journals], 660 White Plains Rd., Tarrytown, NY 10159-5153. (919) 524-9200. FAX (919) 333-2444. 1974 to present. 6 times a year. $530.00 per year.

Glass and Ceramics. Plenum Publishing Corporation, 233 Spring Street, New York, NY 10013. (212) 620-8000. FAX (212) 463-0742. 1956 to present. Monthly. $1215.00 per year.

High Tech Ceramics News. Business Communications Company Inc., 25 Van Zant Street, Suite 13, Norwalk, CT 06855. (203) 853-4266. FAX (203) 853-0348. 1989 to present. Monthly. $325.00 per year.

Journal of Materials Research. Materials Research Society, 9800 McKnight Road, Pittsburgh, PA 15237. (412) 367-3003. FAX (412) 367-4373. 1986 to present. Monthly. $440.00 per year.

Materials Engineering. Penton Publishing, 110 Superior Ave., Cleveland, OH 44114-2543. 1929 to present. Monthly. $50.00 per year.

Materials Evaluation. American Society for Nondestructive Testing, 1711 Arlington Lane, Box 28518, Columbus, OH 43228-0158. (614) 274-6003. FAX (614) 274-6899. 1942 to present. Monthly. $85.00 per year.

World Ceramics and Refractories. London & Sheffield Publishing Company Ltd., 5 Pond Street, Hampstead, London NW3 2PN, England. Telephone 01-794-0800. FAX 01-794-0411. 1990 to present. Bi-monthly. Inquire for price.

RESEARCH CENTERS AND INSTITUTES

Alfred University Center For Advanced Ceramic Technology. New York State College of Ceramics, Alfred, NY 14802. (607) 871-2486. FAX (607) 871-3469.

Center For Ceramics Research. Rutgers University, College of Engineering, Brett & Bowser, PO Box 909, Piscataway, NJ 08855-0909. (201) 932-4817.

Center For Materials Research. Ohio State University, 4108 Smith Laboratory, 174 W. 18th Avenue, Columbus, OH 43210. (614) 292-1133. FAX (614) 292-7557.

Center For Materials Science and Engineering. Massachusetts Institute of Technology, Building 13, Room 2090, 77 Massachusetts Avenue, Cambridge, MA 02139. (617) 253-6701.

Edward Orton Jr. CERAMIC FOUNDATION. 6991 Old 3C Highway, PO Box 460, Westerville, OH 43081. (614) 895-2663. FAX (614) 895-5610.

Materials Research Laboratory. University of Illinois, Urbana, IL 61801. (217) 333-1370.

Materials Science Center. Cornell University, Clark Hall of Science, Ithaca, NY 14853-2501. (607) 255-4272. FAX (607) 255-6428.

SPECIFICATIONS AND STANDARDS

Annual Book of ASTM Standards, Volume 15.02. American Society for Testing and Materials (ASTM), 1916 Race Street, Philadelphia, PA 19103. (215) 299-5585. Annual. Inquire for costand availability.

CERAMICS

See also: CERAMIC ENGINEERING, CLAY, GLASS, MATERIALS SCIENCE

ABSTRACT SERVICES AND INDEXES

Applied Science and Technology Index; A Cumulative Subject Index To English Language Periodicals In the Fields of Aeronautics and Space Science, Computer Technology, Chemistry, Construction Industry, Energy and Related Areas. H.W. Wilson Co., 950 University Avenue, Bronx, NY 10452. (800) 367-6770 or (212) 588-8400. FAX (718) 590-1617. From 1958 to present. Monthly. Inquire about cost and availability. Also available on CD-ROM and online.

C2C Abstracts: Japan—Materials Science. Scan C2C, 1001 Pennsylvania Ave., NW, No. 1300, Washington, DC 20024-2505. (800) 525-3865. FAX (202) 863-3855. 1990 to present. Monthly. $200.00 per year. Contains abstracts of articles from Japanese scientific, business and technical journals on aggregates, coatings, composites, fibers and whiskers, powder technology, processing, refractories, and wood products. Also available on CD-ROM and online.

CA Selects: Ceramic Materials (Journals). Chemical Abstracts Service, 2450 Olentagy River Road, Box 3012, Columbus, OH 43210-0012. (614) 447-3600. FAX (614) 447-3713. 1988 to present. Semi-weekly. $210.00 per year.

Ceramic Abstracts. American Ceramic Society, Inc., 735 Ceramic Place, Westerville, OH 43081. (614) 890-4700. FAX (614) 899-6109. Bi-monthly. 1922 to present. $138.00 per year for non-members. Also available on CD-ROM, and online through DIALOG, ORBIT, and STN.

Engineered Materials Abstracts. ASM (American Society of Metals) International, Materials Information, Materials Park, OH 44073. (216) 338-5151 or FAX (216) 338-4634. Covers literature on technical developments in polymer, ceramic, and composite materials and engineering. 1986 to present. Monthly. $1200.00 per year. Also available on CD-ROM and online. Covers published materials on technical developments in polymer, ceramic, and composite materials in an engineering environment.

Materials Science and Engineering Abstracts, Cambridge Scientific Abstracts, 7200 Wisconsin Avenue, Bethesda, Md 20814-4823. (301) 961-6750. FAX (301) 961-6720. 1993 to present. Monthly. $385.00 per year. Focuses on mechanical and physical properties of materials and commercial or industrial applications for materials, methods for strength testing, effects of vibration and other stresses, corrosion and protective coatings,

CERAMICS

Ency. of Physical Sciences and Engineering Info. Sources

storage and handling, ceramics, composites, metals, wood, plastics, and polymers.

Materials Science Citation Index. Institute for Scientific Information, 3501 Market Street, Philadelphia, PA 19104. (215) 386-0100. FAX (215) 386-6362. Bi-monthly. $975.00 per year. Also available on CD-ROM.

Physics Abstracts. Institute of Electrical Engineers, Michael Faraday House, Six Hill Way, Stevenage, Herts, SG1 2AY, England. Distributed by IEEE, 445 Hoes Lane, Piscataway, NJ 08854. (908) 562-5549. 1898 to present. Monthly. $2700.00 per year. Also available online as INSPEC.

World Ceramics Abstracts. Ceram Research Ltd., Queens Road, Penkhull, Stoke-on-Trent ST4 7LQ, England. 1958 to present. Monthly. Inquire for price and availability. Also available online through ORBIT.

ASSOCIATIONS AND PROFESSIONAL SOCIETIES

American Ceramic Society. 735 Ceramic Place, Westerville, OH 43081. (614) 890-4700. FAX (614) 899-6109.

Ceramic Educational Council. 735 Ceramic Place, Westerville, OH 43081. (614) 890-4700. FAX (614) 899-6109.

Ceramic Manufacturers Association. 1100-H Brandywine Blvd., PO Box 2188, Zanesville, OH 43702-2188. (614) 452-4541.

International Ceramic Association. PO Box 39, Glen Burnie, MD 21061. (410) 923-3425.

Keramos. Route 4, Box 4400, Kennewick, WA 99337. (509) 627-3907.

Materials Research Society. 9800 McKnight Road, Pittsburgh, PA 15237. (412) 367-3003. FAX (412) 367-4373.

National Institute of Ceramic Engineers. 735 Ceramic Place, Westerville, OH 43081. (614) 890-4700.

Technical Ceramics Manufacturers Association. 25 N. Broadway, Tarrytown, NY 10591. (914) 332-0040.

BIBLIOGRAPHIES

A Bibliography of Ceramics and Glass. Larry L. Hench & B.A. McEldowney. American Ceramic Society, 735 Ceramic Place, Westerville, OH 43081. (614) 890-4700. FAX (614) 899-6109. 1976. $15.00.

DIRECTORIES AND BIOGRAPHICAL SOURCES

American Men & Women of Science, 1995-96. R.R. Bowker Staff, eds. 19th edition. 8 volumes. R.R. Bowker/Reed International Publishing Company, 121 Chanlon Road, New Providence, NJ 07974. (908) 464-6800 or (800) 521-8110. 1995. $850.00.

Ceramic Industry Data Book Buyers Guide. Business News Publishing Company, 755 W. Big Beaver Road, Suite 1000, Troy, MI 48084. (313) 362-3700. FAX (313) 362-0317. 1922

to present. Annual. $25.00 per year. Included as part of subscription to CERAMIC INDUSTRY periodical (see below).

Guide To Engineering Materials Producers. J.C. Bittence. ASM International, Materials Information, Materials Park, OH 44073-0002. (216) 338-5151 or (800) 336-5152. FAX (216) 338-4634. 1993. $141.00

International Ceramic Directory. London & Sheffield Publishing Company Ltd., 5 Pond Street, Hampstead, London NW3 2PN, England. Telephone 01-794-0800. FAX 01-794-0411. 1984 to present. Every 5 years. Inquire for price and availability.

International Research Centers Directory. Gale Research, 835 Penobscot Building, Detroit, MI 48226-4094. (313) 961-2242. (800) 347-4253. FAX (313) 961-6083. 8th edition. 1995. $410.00

Materials Research Centers: A World Directory of Organizations and Programs In Material Science. 5th edition. Gale Research, 835 Penobscot Building, Detroit, MI 48226-4094. (313) 961-2242. (800) 347-4253. FAX (313) 961-6083. 1993. $495.00.

Research Centers Directory. Gale Research, 835 Penobscot Building, Detroit, MI 48226-4094. (313) 961-2242. (800) 347-4253. FAX (313) 961-6083. $485.00.

Scientific and Technical Organizations and Agencies Directory. Gale Research, 835 Penobscot Building, Detroit, MI 48226-4094. (313) 961-2242. (800) 347-4253. FAX (313) 961-6083. 4th edition. 1996. $195.00.

Who's Who In Engineering. American Association of Engineering Societies, 1111 19th Street NW, Suite 608, Washington, DC 20036. (202) 296-2237 or (800) 658-8897. 8th edition. 1991. Inquire for price.

Who's Who In Technology. Gale Research, 835 Penobscot Building, Detroit, MI 48226-4094. (313) 961-2242. (800) 347-4253. FAX (313) 961-6083. 1995. $195.00.

ENCYCLOPEDIAS AND DICTIONARIES

Ceramic Glossary. Walter W. Perkins, editor. American Ceramic Society, 735 Ceramic Place, Westerville, OH 43081. (614) 890-4700. FAX (614) 899-6109. 1984. $12.00.

Concise Encyclopedia of Advanced Ceramic Materials. R.J. Brook, ed. MIT Press, 55 Hayward Street, Cambridge, MA 02142. (617) 253-2884 or (800) 356-0343. FAX (617) 253-1709. 1990. $150.00.

Dictionary of Ceramic Science & Engineering. Ian J. McColm. 2nd edition. Plenum Publishing Corporation, 233 Spring Street, New York, NY 10013. (212) 620-8000. FAX (212) 463-0742. 1994. $75.00.

Encyclopedia of Glass, Ceramics, and Cement. Martin Grayson, editor. John Wiley and Sons, Inc., 605 Third Avenue, New York, NY 10158. (800) 526-5368 or (212) 850-6000. 1985. Inquire for cost and availability.

Encyclopedia of Physical Science and Technology. Robert A. Meyers, ed. 18 volumes. Academic Press Inc., 6277 Sea Harbor Drive, Orlando, FL 32887. (800) 321-5068. 1992. $2100.00

McGraw-Hill Encyclopedia of Engineering. Sybil P. Parker, ed. 2nd edition. McGraw-Hill Publishing Company, 1221 Avenue of the Americas, New York, NY 10020. (800) 262-4729 or (212) 512-3825. 1993. $95.50.

McGraw-Hill Encyclopedia of Science and Technology. Sybil P. Parker, ed. 7th edition. 20 volumes. McGraw-Hill Publishing Company, 1221 Avenue of the Americas, New York, NY 10020. (800) 262-4729 or (212) 512-3825. 1992. $1900.00

GENERAL WORKS

Ceramics In Energy Applications. Institute of Energy. Pergamon Press Inc., Maxwell House, Fairview Park, Elmsford, NY 10523. (914) 592-7700. Fax (914) 592-3625. 1994. $130.00.

Ceramics Today—Tomorrow's Ceramics. P. Vincenzini. 4 volumes. Elsevier Science B.V., PO Box 945, Madison Square Station, New York, NY 10159-0945. (212) 633-3650. FAX (212) 633-3680. 1991. $1037.50.

A Concise Introduction To Ceramics. George C. Phillips. Van Nostrand Reinhold, 115 Fifth Avenue, New York, NY 10003. (212) 254-3232 or FAX (212) 254-9499. 1991. $49.95.

Fundamentals of Ceramics Engineering. P. Vincenzini, ed. Elsevier Science Publishing Company, Inc., 655 Avenue of the Americas, New York, NY 10010. (212) 989-5800. 1991. $110.00.

Introduction To the Principles of Ceramic Processing. James S. Reed. John Wiley and Sons, Inc., 605 Third Avenue, New York, NY 10158. (800) 526-5368 or (212) 850-6000. 1988. Inquire for cost.

Modern Ceramic Engineering. David W. Richerson. 2d ed. Marcel Dekker, Inc., 270 Madison Avenue, New York, NY 10016. (212) 696-9000. FAX (212) 685-4540. 1992. $150.00.

HANDBOOKS AND MANUALS

ASM Engineered Materials Handbook Volume 4: Ceramics and Glasses. ASM International, Materials Information, Materials Park, OH 44073-0002. (216) 338-5151 or (800) 336-5152. FAX (216) 338-4634. 1995. $147.00.

ASM Engineered Materials Reference Book. Michael L. Bauccio, editor. 2nd edition. ASM International, Materials Information, Materials Park, OH 44073-0002. (216) 338-5151 or (800) 336-5152. FAX (216) 338-4634. 1993. $111.00.

Materials Handbook: An Encyclopedia For Managers, Technical Professionals, Purchasing and Production Managers, Technicians, Supervisors, and Foremen. George S. Brady & Henry R. Clauser. 13th revised edition. McGraw Hill Publishing Company, 1221 Avenue of the Americas, New York, NY 10020. (800) 262-4729 or (212) 512-3825. 1991. $79.50.

ONLINE DATABASES AND CD-ROMS

Compendex Plus. Engineering Information, Inc., 345 East 47th Street, New York, NY 10017. (212) 705-7600 or (800) 221-1044. Contains citations with abstracts to worldwide literature in engineering and technology, from 1970 to present. Available on online BRS,(800) 289-4277, DIALOG, (800) 334-2564,

ORBIT (800) 456-7248, and STN International, FIZ Karlsruhe, P.O. Box 2465, W-7500, Karlsruhe 1, Germany, online services. Also available on CD-ROM. Inquire as to cost and availability.

CSA Engineering. Cambridge Scientific Abstracts, 7200 Wisconsin Avenue, Suite 601, Bethesda, MD 20814. (301) 961-6750 or (800) 843-7751. Contains citations and abstracts of international periodicals and research literature covering all fields of engineering and science and technology,including computer and information science, electronics, mechanical engineering, solid state materials, 1981 to present. Available on BRS,(800) 289-4277, online service.Inquire as to cost and availability.

Current Contents Search. Institute for ScientificInformation, 3501 Market Street, Philadelphia, PA 19104. (215) 386-0100. FAX (215) 386-6362. Contains citations to articles listed in the table of contents of science and technology journals. Also articles in social sciences and life sciences journals. Available on BRS,(800) 289-4277, DIALOG,(800) 334-2564, online services. Inquire as to cost and availability.

Dissertation Abstracts. University Microfilms International, 300 North Zeeb Road, Ann Arbor, MI 48106. (800) 521-0600 or (313) 761-4700. Scope includes virtually all doctoral dissertations accepted at accredited American institutions from 1861 to present in 252 subject areas. Available on BRS,(800) 289-4277, DIALOG,(800) 334-2564, and OCLC EPIC,(800) 848-5878, online services. Also available on CD-ROM. Inquire as to cost and availability.

NTIS Bibliographic Database. National Technical Information Service, 5285 Port Royal Road, Springfield, Va 22161. (703) 487-4929 or FAX (703) 321-8199. Broad coverage of government-sponsored science and technology research reports, 1964 to present. Available on BRS,(800) 289-4277, DIALOG, (800) 334-2564, ORBIT, (800) 456-7248, and STN International, FIZ Karlsruhe, P.O. Box 2465, W-7500, Karlsruhe 1, Germany, online services. Also available on CD-ROM. Inquire as to cost and availability.

Scisearch. Institute for Scientific Information, 3501 Market Street, Philadelphia, PA 19104. (800) 523-1850 or (215) 386-0100. Broad multidisciplinary title and author index to the international literature of science and technology, 1974 to present. Available on DIALOG,(800) 334-2564, and ORBIT,(800) 456-7248, online services. Also available on CD-ROM. Inquire as to cost and availability.

WILSONLINE. H.W. Wilson Company, 950 University Avenue, Bronx, NY 10452. (800) 367-6770 or (212) 588-8400. Makes available online versions of the printed H.W. Wilson indexes including Applied Science and Technology Index, Business Periodicals Index, General Science Index, and Readers' Guide to Periodical Literature. Period covered is generally 1983 to present. Available on BRS,(800) 289-4277, DIALOG,(800) 334-2564, and OCLC EPIC,(800) 848-5878, online services. Also available on CD-ROM. Inquire as to cost and availability.

PERIODICALS

Advanced Ceramics Report. Elsevier Science [journals], 660 White Plains Rd., Tarrytown, NY 10159-5153. (919) 524-9200. FAX (919) 333-2444. 1986 to present. Monthly. $425.00 per year.

American Ceramic Society Bulletin. American Ceramic Society, 735 Ceramic Place, Westerville, OH 43081. (614) 890-4700.

FAX (614) 899-6109. 1922 to present. Monthly. $28.00 per year for non-members.

American Ceramic Society Journal. American Ceramic Society, 735 Ceramic Place, Westerville, OH 43081. (614) 890-4700. FAX (614) 899-6109. 1899 to present. Monthly. $175.00 per year fornon-members.

Canadian Ceramic Society Journal. Canadian Ceramic Society, 2175 Sheppard Avenue E., Suite 100, Willowdale, Ontario M2J 1W8, Canada. (416) 491-2886. FAX (416) 491-1670. 1928 to present. Annual. Inquire for cost.

Ceramic Engineering and Science Proceedings. American Ceramic Society, Inc., 735 Ceramic Place, Westerville, OH 43081. (614) 890-4700. FAX (614) 899-6109. 1980 to present. Bimonthly. $80.00 per year.

Ceramic Industry. Business News Publishing Company, 755 W. Big Beaver Road, Suite 1000, Troy, MI 48084. (313) 362-3700. FAX (313) 362-0317. 1923 to present. 13 times a year. $53.00 per year.

Ceramics International. Elsevier Science [journals], 660 White Plains Rd., Tarrytown, NY 10159-5153. (919) 524-9200. FAX (919) 333-2444. 1974 to present. 6 times a year. $530.00 per year.

Glass and Ceramics. Plenum Publishing Corporation, 233 Spring Street, New York, NY 10013. (212) 620-8000. FAX (212) 463-0742. 1956 to present. Monthly. $1215.00 per year.

High Tech Ceramics News. Business Communications Company Inc., 25 Van Zant Street, Suite 13, Norwalk, CT 06855. (203) 853-4266. FAX (203) 853-0348. 1989 to present. Monthly. $325.00 per year.

Journal of Materials Research. Materials Research Society, 9800 McKnight Road, Pittsburgh, PA 15237. (412) 367-3003. FAX (412) 367-4373. 1986 to present. Monthly. $440.00 per year.

Materials Engineering. Penton Publishing, 110 Superior Ave., Cleveland, OH 44114-2543. 1929 to present. Monthly. $50.00 per year.

Materials Evaluation. American Society for Nondestructive Testing, 1711 Arlington Lane, Box 28518, Columbus, OH 43228-0158. (614) 274-6003. FAX (614) 274-6899. 1942 to present. Monthly. $85.00 per year.

World Ceramics and Refractories. London & Sheffield Publishing Company Ltd., 5 Pond Street, Hampstead, London NW3 2PN, England. Telephone 01-794-0800. FAX 01-794-0411. 1990 to present. Bi-monthly. Inquire for price.

RESEARCH CENTERS AND INSTITUTES

Alfred University Center For Advanced Ceramic Technology. New York State College of Ceramics, Alfred, NY 14802. (607) 871-2486. FAX (607) 871-3469.

Center For Ceramics Research. Rutgers University, College of Engineering, Brett & Bowser, PO Box 909, Piscataway, NJ 08855-0909. (201) 932-4817.

Center For Materials Research. Ohio State University, 4108 Smith Laboratory, 174 W. 18th Avenue, Columbus, OH 43210. (614) 292-1133. FAX (614) 292-7557.

Center For Materials Science and Engineering. Massachusetts Institute of Technology, Building 13, Room 2090, 77 Massachusetts Avenue, Cambridge, MA 02139. (617) 253-6701.

Edward Orton Jr. CERAMIC FOUNDATION. 6991 Old 3C Highway,PO Box 460, Westerville, OH 43081. (614) 895-2663. FAX (614) 895-5610.

Materials Research Laboratory. University of Illinois, Urbana, IL 61801. (217) 333-1370.

Materials Science Center. Cornell University, Clark Hall of Science, Ithaca, NY 14853-2501. (607) 255-4272. FAX (607) 255-6428.

SPECIFICATIONS AND STANDARDS

*Annual Book of ASTM Standards, Volume 15.*02. American Society for Testing and Materials (ASTM), 1916 Race Street, Philadelphia, PA 19103. (215) 299-5585. Annual. Inquire for cost and availability.

CHAIN REACTION

See: NUCLEAR ENERGY

CHANNELS

See also: CANALS, CIVIL ENGINEERING, DREDGES AND DREDGING, GEOTECHNICAL ENGINEERING, HYDRAULIC ENGINEERING, OCEAN ENGINEERING

ABSTRACT SERVICES AND INDEXES

Applied Science and Technology Index; A Cumulative Subject Index To English Language Periodicals In the Fields of Aeronautics and Space Science, Computer Technology, Chemistry, Construction Industry, Energy and Related Areas. H.W. Wilson Co., 950 University Avenue, Bronx, NY 10452. (800) 367-6770 or (212) 588-8400. FAX (718) 590-1617. From 1958 to present. Monthly. Inquire about cost and availability. Also available on CD-ROM and online.

ASCE Combined Annual Combined Index. American Society of Civil Engineers, 345 East 47th Street, New York, NY 10017-2398. (212) 705-7520 or (800) 548-2723. Annual. $48.00.

ASCE Publications Information. 1966-present. American Society of Civil Engineers, 345 East 47th Street, New York, NY 10017-2398. (212) 705-7520 or (800) 548-2723. Bi-monthly. $160.00 per year for non-members.

Bibliography and Index of Geology. American Geological Institute, 4220 King Street, Alexandria, VA 22302. (703) 379-2480. Fax (703) 379-7563. Monthly. $1295.00 per year. Also available online as GeoRef.

Civil and Structural Engineering Abstracts, Cambridge Scientific Abstracts, 7200 Wisconsin Avenue, Bethesda, Md 20814-4823. (301) 961-6750. FAX (301) 961-6720. 1993 to present. Monthly. $385.00 per year. Topics covered include structural design, construction equipment and methods, civil defense and military engineering, surveying, highway engineering, maritime and port structures, materials, land reclamation, and soil mechanics.

Engineering Index Monthly. Engineering Information, Inc., Castle Point on the Hudson, Hoboken, NJ 07030. (800) 221-1044. FAX (212) 832-1857. Monthly. $2300.00 per year. Also available online as COMPENDEX, and also on CD-ROM. Covers chemical engineering, computers, electrical engineering, civil engineering, metals and mining, industrial management, and mechanical engineering.

Environmental Engineering Abstracts, Cambridge Scientific Abstracts, 7200 Wisconsin Avenue, Bethesda, Md 20814-4823. (301) 961-6750. FAX (301) 961-6720. Monthly. Covers hazardous materials, environmental impact and protection, treatment of sewage and industrial wastes, hydroelectric power, tidal and wind power, artic and tropical engineering.

Fluid Abstracts: Civil Engineering. Elsevier Science [journals], 660 White Plains Rd., Tarrytown, NY 10159-5153. (919) 524-9200. FAX (919) 333-2444. 1991 to present. Monthly plus annual cumulation. $645.00 per year. Also available online as FLUIDEX. Covers civil engineering applications of fluid mechanics, hydraulics, flow metering and measuring, offshore engineering, environmental hydraulics, and related aspects of wind energy, the atmosphere, and aerodynamics.

Science Citation Index. Institute for Scientific Information, 3501 Market Street, Philadelphia, PA 19104. (215) 386-0100. FAX (215) 386-6362. Inquire about availability and cost. Also available on CD-ROM.

ASSOCIATIONS AND PROFESSIONAL SOCIETIES

American Geological Institute. 4220 King Street, Alexandria, VA 22302. (703) 379-2480. Fax (703) 379-7563.

American Society of Civil Engineers. 345 East 47th Street, New York, NY 10017-2398. (212) 705-7520 or (800) 548-2723.

DIRECTORIES AND BIOGRAPHICAL RESOURCES

Directory of Engineering Societies and Related Organizations. Gordon Davis. 13th edition. American Association of Engineering Societies, 1111 19th Street NW, Suite 608, Washington, DC 20036. (202) 296-2237 or (800) 658-8897. 1989. Inquire for price.

International Research Centers Directory. Gale Research, 835 Penobscot Building, Detroit, MI 48226-4094. (313) 961-2242. (800) 347-4253. FAX (313) 961-6083. 8th edition. 1995. $410.00

Research Centers Directory. Gale Research, 835 Penobscot Building, Detroit, MI 48226-4094. (313) 961-2242. (800) 347-4253. FAX (313) 961-6083. $485.00.

Scientific and Technical Organizations and Agencies Directory. Gale Research, 835 Penobscot Building, Detroit, MI 48226-

4094. (313) 961-2242. (800) 347-4253. FAX (313) 961-6083. 4th edition. 1996. $195.00.

Who's Who In Engineering. American Association of Engineering Societies, 1111 19th Street NW, Suite 608, Washington, DC 20036. (202) 296-2237 or (800) 658-8897. 8th edition. 1991. Inquire for price.

ENCYCLOPEDIAS AND DICTIONARIES

Dictionary of Civil Engineering. V.N. Vazirani. South Asia Books, PO Box 502, Columbia, MO 65205. (314) 474-0116. FAX (314) 474-8124. 1992. $20.00.

McGraw-Hill Encyclopedia of Engineering. Sybil P. Parker, ed. 2nd edition. McGraw-Hill Publishing Company, 1221 Avenue of the Americas, New York, NY 10020. (800) 262-4729 or (212) 512-3825.1993. $95.50.

GENERAL WORKS

Open-Channel Hydraulics. Richard H. French. McGraw-Hill Publishing Company, 1221 Avenue of the Americas, New York, NY 10020. (800) 262-4729 or (212) 512-3825. 1985. Inquire for cost and availability.

Report On Ship Channel Design. Task Committee on Ship Channel Design of the Committee on Waterways, Waterway, Port, Coastal, and Ocean Division, ASCE. American Society of Civil Engineers, 345 East 47th Street, New York, NY 10017-2398. (212) 705-7520 or (800) 548-2723. 1993. Inquire for cost and availability.

Transient Flow In Pipes, Open Channels, and Sewers. J.A. Fox. Halsted Press (Division of John Wiley and Sons, Inc.), 605 Third Avenue, New York, NY 10158. (800) 526-5368 or (212) 850-6418. 1989. Inquire for cost and availability.

HANDBOOKS AND MANUALS

Civil Engineering Calculations Reference Guide. Tyler G. Hicks. McGraw-Hill Publishing Company, 1221 Avenue of the Americas, New York, NY 10020. (800) 262-4729 or (212) 512-3825. 1987. $46.00.

Civil Engineering Practice. Paul N. Cheremisinoff, et al. 5 volumes. Technomic Publishing Company, Inc., 851 New Holland Avenue, Box 3535, Lancaster, PA 17604. (717) 291-5609 or (800) 233-9936. FAX (717) 295-4538. 1987-88. $199.00.

Handbook of Coastal and Ocean Engineering Volume 3: Harbors, Navigational Channels, Estuaries and Environmental Effects. John B. Herbich, editor. Gulf Publishing Company, P.O. Box 2608, Houston, TX. (713) 529-4301 or (800) 231-6275. FAX (713) 525-9647. 1992. $195.00.

Standard Handbook For Civil Engineers. Frederick S. Merritt, editor. 3rd edition. McGraw-Hill Publishing Company, 1221 Avenue of the Americas, New York, NY 10020. (800) 262-4729 or (212) 512-3825. 1983. $124.50.

Surveying Ready-reference Manual. G.O. Stenstrom. McGraw-Hill Publishing Company, 1221 Avenue of the Americas, New York, NY 10020. (800) 262-4729 or (212) 512-3825. 1987. $38.50.

Tables For the Hydraulic Design of Pipes, Sewers, and Channels. H.R. Wallingford & D.I.H. Barr. 6th edition. Thomas Telford Services Ltd., Thomas Telford House, 1 Heron Quay, London E14 4JD, England. Telephone 071-987-6999. FAX 071-538-4101. 1994. Inquire for cost and availability.

ONLINE DATABASES AND CD-ROMS

Compendex Plus. Engineering Information, Inc., 345 East 47th Street, New York, NY 10017. (212) 705-7600 or (800) 221-1044. Contains citations with abstracts to worldwide literature in engineering and technology, from 1970 to present. Available on online BRS,(800) 289-4277, DIALOG, (800) 334-2564, ORBIT (800) 456-7248, and STN International, FIZ Karlsruhe, P.O. Box 2465, W-7500, Karlsruhe 1, Germany, online services. Also available on CD-ROM. Inquire as to cost and availability.

CSA Engineering. Cambridge Scientific Abstracts, 7200 Wisconsin Avenue, Suite 601, Bethesda, MD 20814. (301) 961-6750 or (800) 843-7751. Contains citations and abstracts of international periodicals and research literature covering all fields of engineering and science and technology,including computer and information science, electronics, mechanical engineering, solid state materials, 1981 to present. Available on BRS,(800) 289-4277, online service.Inquire as to cost and availability.

Current Contents Search. Institute for Scientific Information, 3501 Market Street, Philadelphia, PA 19104. (215) 386-0100. FAX (215) 386-6362. Contains citations to articles listed in the table of contents of science and technology journals. Also articles in social sciences and life sciences journals. Available on BRS,(800) 289-4277, DIALOG,(800) 334-2564, online services. Inquire as to cost and availability.

GEOREF: Bibliography and Index of Geology. American Geological Institute, 4220 King Street, Alexandria, VA 22302. (703) 379-2480. Fax (703) 379-7563. Monthly. Inquire for price and availability.

NTIS Bibliographic Database. National Technical Information Service, 5285 Port Royal Road, Springfield, Va 22161. (703) 487-4929 or FAX (703) 321-8199. Broad coverage of government-sponsored science and technology research reports, 1964 to present. Available on BRS,(800) 289-4277, DIALOG, (800) 334-2564, ORBIT, (800) 456-7248, and STN International, FIZ Karlsruhe, P.O. Box 2465, W-7500, Karlsruhe 1, Germany, online services. Also available on CD-ROM. Inquire as to cost and availability.

Scisearch. Institute for Scientific Information, 3501 Market Street, Philadelphia, PA 19104. (800) 523-1850 or (215) 386-0100. Broad multidisciplinary title and author index to the international literature of science and technology, 1974 to present. Available on DIALOG,(800) 334-2564, and ORBIT,(800) 456-7248, online services. Also available on CD-ROM. Inquire as to cost and availability.

WILSONLINE. H.W. Wilson Company, 950 University Avenue, Bronx, NY 10452. (800) 367-6770 or (212) 588-8400. Makes available online versions of the printed H.W. Wilson indexes including Applied Science and Technology Index, Business Periodicals Index, General Science Index, and Readers' Guide to Periodical Literature. Period covered is generally 1983 to present. Available on BRS,(800) 289-4277, DIALOG,(800) 334-2564, and OCLC EPIC,(800) 848-5878, online services. Also available on CD-ROM. Inquire as to cost and availability.

PERIODICALS

Civil Engineering. American Society of Civil Engineers, 345 East 47th Street, New York, NY 10017-2398. (212) 705-7520 or (800) 548-2723. 1930 to present. Monthly. $85.00 per year.

Journal of Construction Engineering and Management. American Society of Civil Engineers, Construction Division, 345East 47th Street, New York, NY 10017-2398. (212) 705-7520 or (800) 548-2723. 1956 to present. Quarterly. $112.00 per year.

Journal of Geotechnical Engineering. American Society of Civil Engineers, Geotechnical Engineering Division, 345 East 47th Street, New York, NY 10017-2398. (212) 705-7520 or (800) 548-2723. 1956 to present. Monthly. $212.00 per year.

Journal of Hydraulic Engineering. American Society of Civil Engineers, Hydraulics Division, 345 East 47th Street, New York, NY 10017-2398. (212) 705-7520 or (800) 548-2723. 1956 to present. Monthly. $200.00 per year.

Journal of Irrigation and Drainage. American Society of Civil Engineers, Irrigation and Drainage Division, 345 East 47th Street, New York, NY 10017-2398. (212) 705-7520 or (800) 548-2723. 1956 to present. Bi-monthly. $136.00 per year.

Journal of Water Resources Planning and Management. American Society of Civil Engineers, Water Resources Planning and Management Division, 345 East 47th Street, New York, NY 10017-2398. (212) 705-7520 or (800) 548-2723. Bi-monthly. $112.00 per year.

Journal of Waterway, Port, Coastal, and Ocean Engineering. American Society of Civil Engineers, Waterway, Port, Coastal, and Ocean Division, 345 East 47th Street, New York, NY 10017-2398. (212) 705-7520 or (800) 548-2723. 1956 to present. Bi-monthly. $100.00 per year.

RESEARCH CENTERS AND INSTITUTES

LSU/GWU National Ports & Waterways Institute. 600 New Hampshire Avenue NW, Suite 475, Washington, DC 20037. (202) 338-9427. FAX (202) 338-9509.

CHAOS THEORY

See also: ASTRONOMY, COSMOLOGY, MATHEMATICS, MATTER, MECHANICS, TIME, UNIVERSE

ABSTRACT SERVICES AND INDEXES

Applied Science and Technology Index; a cumulative subject index to English language periodicals in the fields of aeronautics and space science, computer technology, ch emistry, construction industry, energy and related areas. H.W. Wilson Co., 950 University Avenue, Bronx, NY 10452. (212) 588-8400. (800) 367-6770. FAX (718) 590-1617. From 1958 to present. Monthly. Inquire about cost and availability. Also available on CD-ROM and online. ISSN: 0003-6986.

Chemical Abstracts. Chemical Abstracts Service, 2540 Olentangy River Road, Box 3012, Columbus, OH 43210-0012. (614) 447-3600. (800) 848-6538. FAX (614) 447-3713. From

Ency. of Physical Sciences and Engineering Info. Sources

CHAOS THEORY

1907 to present. Weekly. $16800.00 per year. Also available on CD-ROM and online. Inquire regarding cost and availability.

Current Contents: Physical, Chemical and Earth Sciences. Institute for Scientific Information, 3501 Market Street, Philadelphia, PA 19104. (215) 386-0100. FAX (215) 386-2291. From 1961 to present. Weekly. $442.00 per year. Also available on CD-ROM and online. Inquire regarding cost and availability. ISSN: 0163-2574.

Current Papers In Physics. Institute of Electrical Engineers, Michael Faraday House, Six Hill Way, Stevenage, Herts, SG1 2AY, England. Distributed by INSPEC/IEEE, Box 1331, 445 Hoes Lane, Piscataway, NJ 08855-1331. (908) 562- 5549. 1966 to present. Fortnightly. $410.00 per year. ISSN: 0011-3786.

General Science Index. H.W. Wilson Company, 950 University Avenue, Bronx, NY 10452. (212) 588-8400. (800) 367-6770. FAX (718) 590-1617. From 1978 to present. Ten issues per year; quarterly and annual cumulations. Service basis. Available on CD-ROM and online. Inquire regarding cost and availability. ISSN: 0162-1963.

Government Reports Announcements and Index. U. S. National Technical Information Service (NTIS), 5285 Port Royal Road, Springfield, VA 22161. (703) 487-4650. FAX (703) 321-8547. From 1968 to present. Annual. $630.00 per year. Also available online as *NTIS Bibliographic Database* and on CD-ROM. ISSN:

NTIS Alerts: Physics. U. S. National Technical Information Service. 5285 Port Royal Road, Springfield, VA 22161. (703) 487-4650. FAX (703) 321-8547. Weekly. $140.00 per year.

Pascal 10: Mechanique, Acoustique Et Transpert De Chaleur. Centre National De La Recherche Scientifique, Institue De L'information Scientifique Et Technique, 2 Aliee Du Parc De Brabois, 54514 Vandoeuvre-les Nancy Cedex, France. TEC 83-50-46-00. FAX 83-50-46-50. 1984 to present. Ten issues per year. 1530 F per year. Also available on CD-ROM and online. ISSN: 1136-5107.

Physics Abstracts. INSPEC. Section A, Science Abstracts. In stitution of Electrical Engineers (IEE). Available from INSPEC/ IEEE - Institute of Electrical and Electronic Engineers,, Box 1331, 445 Hoes Lane, Piscaway NJ 08855-1331. (908) 562-5549. 1898 to present. 24 issues per year. $2835.00 per year. Also available on CD-ROM and online. ISSN: 0036-8091.

Physics Briefs (Physikalische Berichte). Information Center for Energy, Physics, Mathematics; German Physical Society. V C H Publishers, Inc., 220 East 23rd Street, New York, NY 10010-4606. (212) 683-8333. 1845 to present. 24 issues per year. $2390.00 per year. Also available online. ISSN: 0179-7434.

Science Citation Index. SCI. Institute for Scientific Information, 3501 Market Street, Philadelphia, PA 19104. (215) 386-0100. (800) 523-1850. FAX (215) 386-2991. 1961 to present. Six issues per year, plus annual cumulation. $11650.00 per year. Also available online and on CD-ROM. Inquire about price and availability. ISSN: 0036-827X.

Solid State and Superconductivity Abstracts: Covers theory, Production and Application of Solid State Materials. Cambridge Scientific Abstracts, 7200 Wisconsin Avenue, Bethesda, MD 20824. (301) 961-6750. FAX (301) 961-6720. 1957 to present. Bimonthly. $1320.00 per year. Also available online. ISSN: 0896-5900.

ANNUAL REVIEWS AND YEARBOOKS

Advances in Atomic, Molecular and Optical Physics. Academic Press, Inc., 6277 Sea Harbor Drive, Orlando, FL. (800) 321-5068. 1965 to present. Irregular. ISSN: 1049-250X

Advances In Physics. Taylor & Francis, Ltd., Rankine Road, Basingstoke, Hants RG24 8PR, England. 0256-840366. FAX 0256-47938. 1952 to present. Bimonthly. $511.00 per year. ISSN: 0001-8732.

ASSOCIATIONS AND PROFESSIONAL SOCIETIES

American Institute of Physics. One Physics Ellipse, College Park, MD 20740-3843. (301) 209-3100.

American Mathematical Society. P.O. Box 6248, Providence, RI 02940. (401) 455-4000. FAX (401) 331-3842.

American Physical Society. One Physics Ellipse, College Park, MD 20740-3843. (301) 209-3200. FAX (301) 209-0865.

Universities Research Association, 1111 19th Street Nw, Suite 400, Washington, Dc 20036. (202) 293-1382.

BIBLIOGRAPHIES

Bibliography On Chaos. S. Y Zhang, editor. World Scientific Publishing Company, Inc., 1060 Main Street, River Edge, NJ 07661. (201) 487-9655. (800) 227-7562. 1991. $40.00 in paper.

Science Books and Films. American Association for the Advancement of Science, 1333 H Street NW, Washington, DC 20005. (202) 326-6454. 1965 to present. Nine issues per year. $40.00 per year. ISSN: 0098-342X

Scientific and Technical Books and Serials In Print, 1995. R.R. Bowker Inc., 121 Chanlon Road, New Providence, NJ 07974. (908) 464-6800. (800) 521-8110. 4 volumes. 1994. $299.95. Also available on compact disc and online.

DIRECTORIES AND BIOGRAPHICAL SOURCES

American Men and Women of Science: Physical and Biological Sciences. R. R. Bowker Inc., 121 Chanlon Road, New Providence, NJ 07974. (908) 464-6800. (800) 521-8110. 20th edition. 8 volumes. 1996. $850.00.

American Physical Society. Membership Directory Bulletin ISSUE. American Physical Society, One Physics Ellipse, College Park, MD 20740-3843. (301) 209-3200. FAX (301) 209- 0865. Biennial. $50.00.

The *Biographical Dictionary of Scientists: Physicists*. D. Abbott, editor. Peter Bedrick Books, Inc., 2112 Broadway, Room 318, New York, NY 10023. (212) 496-0751. 1984.

Directory of Physics and Astronomy Staff Members. American Institute of Physics, One Physics Ellipse, College Park, MD 20740-3843. (301) 209-3100. Annual. $45.00.

CHAOS THEORY

Ency. of Physical Sciences and Engineering Info. Sources

Graduate Programs In Physics, Astronomy and Related Fields. 1995-1996. American Institute of Physics, One Physics Ellipse, College Park, MD 20740-3843. (301) 209-3100. Annual. $45.00.

Legends In their Own Time: A Century of American Physical ScieNTISts. Anthony Serafini. Plenum Publishing Corp., 233 Spring Street, New York, NY. (212) 620-8000. (800) 2221-9369. FAX (212)463-0742.. 1993. $27.50.

The Nobel Prize Winners. Frank N. Magill, editor. Salem Press, Inc., P.O. Box 1097, Englewood Cliffs, NJ 07632. (201) 871-3700. (800) 221-1592. 3 volumes. 1989. $210.00.

World Guide To Scientific Associations and Learned Societies. K. G. Saur Inc., 121 Chanlon Road, New Providence, NJ 07974. (908) 464-6800. (800) 521-8110. 5th edition. 1990. $225.00.

ENCYCLOPEDIAS AND DICTIONARIES

A Concise Dictionary of Physics. Oxford University Press, Inc., 200 Madison Avenue, New York, NY 10016. (212) 725- 6000. (800) 334-4249. 1990. $10.95.

Dictionary of Effects and Phenomena In Physics. Joachim Schubert. VCH Publications, Inc., 220 East 23rd Street, Suite 909, New York, NY 10010-4606. (212) 683-8333. (800) 422-8824. 1987. $35.00.

Dictionary of the Physical Sciences: Terms, Formulas, Data. Cesare Emiliani. Oxford University Press, Inc., 200 Madison Avenue, New York, NY 10016. (212) 725-6000. (800) 334-4249. 1989. $19.95.

Encyclopedia of Applied Physics. George Trigg, editor. VCH Publications, Inc., 220 East 23rd Street, Suite 909, New York, NY 10010-4606. (212) 683-8333. (800) 422-8824. 20 volume set. 1991-. $5990.00.

Encyclopedia of Physics. Robert M. Besancon. Chapman & Hall, 1 Penn Plaza, New York, NY 10119. (212) 564-1060. 3rd edition. 1990. $54.95 in paper.

Facts On File Dictionary of Physics. John Daintith, editor. Facts-on-File, Inc., 460 Park Avenue South, New York, NY 10016-7382. (212) 683-2244. (800) 322-8755. Fax (800) 683- 3633. Revised edition. 1990. $12.95.

Encyclopedia of Physical Science and Technology. Academic Press, Inc., 6277 Sea Harbor Drive, Orlando, FL. (800) 321-5068. 2nd edition. 18 volume set. 1992. $2625.00

McGraw-Hill Encyclopedia of Science and Technology. McGraw- Hill Book Company, Inc., 1221 Avenue of the Americas, New York, NY 10020. (212) 512-2000. (800) 262-4729. 7th edition. 20 volume set. 1992. $1900.00.

Penguin Dictionary of Physics. Valerie Illingworth, editor. Viking Penguin, 375 Hudson Street, New York, NY 10014-3657. (212) 366-2000. (800) 331-4624 2nd edition. 1991.

GENERAL WORKS

Applied Chaos. Hong H. Kim and John Stringer. John Wiley & Sons, Inc., 605 Third Avenue, New York, NY 10158-0012. (212) 850-6000. (800) 225-5945.. 1992. $84.95.

Bifurcation and Chaos: Theory and Applications. J. Awrejcewicz, editor. Springer-Verlag New York, Inc., 175 Fifth Avenue, New York, NY 10010. (212) 460-1500. (800) 777-4643. FAX (212) 473-6272. 1994. $84.00.

Chaos and Fractals: New Frontiers of Science. Heinz-Otto Peitgen, et al. Springer-Verlag New York, Inc., 175 Fifth Avenue,New York, NY 10010. (212) 460-1500. (800) 777- 4643. FAX (212) 473-6272. 1993. $49.00.

Chaos Under Control: the Art and Science of Complexity. David P and Michael Frame. W. H. Freeman & Co., 41

Madison Avenue, East 26th Avenue, 35th Floor, New York, Ny 10010. (212) 576-0400 1995. $24.95 in paper.

Cosmic Catastrophies. C. P. Chapman and D. Morrison, editors. Plenum Publishing Corp., 233 Spring Street, New York, NY. (212) 620-8000. (800) 2221-9369. 1989. $22.95.

Exploring Chaos: A Guide To the New Science of Disorder. Nina Hall, editor. W. W. Norton & Company, 500 Fifth

Avenue, New York, Ny 10110. (212) 354-5500. (800) 223-2584 1994. $13.00 in paper.

Fractals In Physics: Essays In Honor of Benoit B. MandELBROT. A. Aharony and Jens Feder, editors. Elsevier Science Publishing Company, Inc., 655 Avenue of the Americas, New York, NY 10010. (212) 989-5800. FAX (914) 333-2444. 1990. $45.25 in paper.

Fundamental Particles and Forces. Richard A. Carrigan, Jr., and W. Peter Trower. W. H. Freeman & Co., 41 Madison Avenue, East 26th Avenue, 35th Floor, New York, NY 10010. (212) 576-0400. 1995. $13.95.

Second Law: Energy, Chaos and Form. Peter Atkins. W. H. Freeman & Co., 41 Madison Avenue, East 26th Avenue, 35th Floor, New York, NY 10010. (212) 576-0400. 1995. $19.95

In Paper.

Selected Papers On Optical Chaos. F. T. Arecchi and R. G. Harison, editors. SPIE - International Society for Optical Engineering, 1000 20th Street, Bellingham, WA 98225. (360) 676-3290. 1993. .

Solutions and Chaos. I. Antoniou and F. J. Lambert, editors. Springer-Verlag New York, Inc., 175 Fifth Avenue, New York, NY 10010. (212) 460-1500. (800) 777-4643. FAX (212) 473-6272. 1991. $69.00 in paper.

HANDBOOKS AND MANUALS

Chemical and Physical Data. Arthur M. James and M. P. Lord, Van Nostrand Reinhold, 115 Fifth Avenue, New York, NY 10003. (212) 254-3232. (800) 842-3636. 1992. $59.95.

CRC Handbook of Chemistry and Physics. David R. Kide, editor. CRC Press, Inc., 2000 Corporate Boulevard, NW, Boca Raton, FL 33431. (407) 994-0555. (800) 272-7737. 77th edition. 1996. $99.95.

Handbook of Physical Quantities. Igor S. Grigoriev and Evgenil Z. Melikhov, editors. CRC Press, Inc., 2000 Corporate Boulevard, NW, Boca Raton, FL 33431. (407) 994-0555. (800) 272-7737. 1995. $99.00.

Landolt-Borenstein Numerical Data and Functional Relationships In Science and Technology: Nuclear Particle and Physics. GROUP I. Springer-Verlag New York, Inc., 175 Fifth Avenue, New York, NY 10010. (212) 460-1500. (800) 777-4643. volumes priced individually.

McGraw-Hill Encyclopedia of Physics. Sybil P. Parker, editor. McGraw-Hill Publishing Company, Inc., 1221 Avenue of the Americas, New York, NY 10020. (212) 512-2000. (800) 262-4729. 2nd edition. 1993. $95.50

A Physicist's Desk Reference. Herbert L. anderson, editor. American Institute of Physics, One Physics Ellipse, College Park, MD 20740-3843. (301) 209-3100. 2nd edition. 1989. $70.00.

Physics Problem Solver. Research and Education Association Staff. Research & Education Association, 61 Ethel Road, West Piscataway, NJ 08854. (908) 819-8880. Revised edition. 1994. $23.95 in paper.

ONLINE DATABASES AND CD-ROMS

CA Search. Chemical Abstracts Service, P.O. Box 3012, Columbus, OH 43210-0012. (614) 447-3600. (800) 848-6533. FAX (614) 447-3709. Very comprehensive guide to worldwide chemical literature and related fields, 1972 to present. Available on BRS, (800) 289-4277, DIALOG, (800) 334-2564, ORBIT (800) 456-7248, and STN International, FIZ Karlsruhe, P.O. Box 2465, W-7500, Karlsruhe 1, Germany, online services. Inquire as to cost and availability.

Current Contents Search. Institute for Scientific Information, 3501 Market Street, Philadelphia, PA 19104. (215) 386-0100. FAX (215) 386-6362. Contains citations to articles listed in the table of contents of science and technology journals. Also articles in social sciences and life sciences journals. Available on BRS, (800) 289-4277, DIALOG, (800) 334-2564, online services. Inquire as to cost and availability.

Dissertation Abstracts Online. University Microfilms International, 300 North Zeeb Road, Ann Arbor, MI 48106. (800) 521-0600 or (313) 761-4700. Scope includes virtually all doctoral dissertations accepted at accredited American institutions from 1861 to present in 252 subject areas. Available on BRS, (800) 289-4277, DIALOG, (800) 334-2564, and OCLC EPIC, (800) 848-5878, online services. Also available on CD-ROM. Inquire as to cost and availability.

Inspec. Institution of Electrical Engineers, Michael

Faraday House, S Ix Hills Way, Stevenage, Herts. SG1 2AY, Englan d. Telephone: 0438 313311 or FAX 0438 742840. Contains citations to the worldwide literature of physics, electronics and electrical engineering, computer technology, and related fields. Available on BRS, (800) 289-4277, DIALOG, (800) 334-2564, ORBIT, (800) 456-7248, and STN International, FIZ Karlsruhe, P.O. Box 2465, W-7500, Karlsruhe 1, Germany, online services. Inquire as to cost and availability.

Mathsci. American Mathematical Society, P.O. Box 6248, Providence, RI 02940. (800) 321-4667 or (401) 455-4000 or FAX (401) 331-3842. Scope includes pure and applied mathematics and related areas of physics, statistics, engineering, computer science, and operations research literature since 1959. Avail-

able on DIALOG, (800) 334-2564, online service. Also available on CD-ROM. Inquire as to cost and availability.

NTIS Bibliographic Database. National Technical Information Service, 5285 Port Royal Road, Springfield, Va 22161. (703) 487-4929 or FAX (703) 321-8199. Broad covera ge of government-sponsored science and technology research reports, 1964 to present. Available on BRS, (800) 289-4277, DIALOG, (800) 334-2564, ORBIT, (800) 456-7248, and STN International, FIZ Karlsruhe, P.O. Box 2465, W-7500, Karlsruhe 1, Germany, online services. Also available on CD-ROM. Inquire as to cost and availability.

Physics Briefs. American Institute of Physics, 335 East 45th Street, New York, NY 10017. (212) 661-9260 or FAX (212) 661-2036. Contains citations with abstracts of the literature of physics and related fields, 1979 to present. Available on the STN International, FIZ Karlsruhe, P.O. Box 2465, W-7500, Karlsruhe 1, Germany, online service. Inquire as to cost and availability.

Scientific and Technical Books and Serials In Print. R.R. Bowker Inc., 121 Chanlon Road, New Providence, NJ 07974. (800) 521-8110 or (908) 464-6800. List of currently published books and serials in the physical and biological sciences, engineering and technology, with subject, author and titles indexes. Available on ORBIT, (800) 456-7248, online service. Inquire as to cost and availability.

Scisearch. Institute for Scientific Information, 3501 Market Street, Philadelphia, PA 19104. (800) 523-1850 or (215) 386-0100. Broad multidisciplinary title and author index to the international literature of science and technology, 1974 to present. Available on DIALOG, (800) 334-2564, and ORBIT, (800) 456-7248, online services. Also available on CD-ROM. Inquire as to cost and availability.

WILSONLINE. H.W. W ilson Company, 950 University Avenue, Bronx, NY 10452. (800) 367-6770 or (212) 588-8400. Makes available onlin e versions of the printed H.W. Wilson indexes includin g Applied Science and Technology Index, Business Periodicals Index, General Science Index, and Readers' Guide to Periodical Literature. Period covered is generally 1983 to present. Available on BRS, (800) 289-4277, DIALOG, (800) 334-2564, and OCLC EPIC, (800) 848-5878, online services. Also available on CD-ROM. Inquire as to cost and availability.

PERIODICALS

American Journal of Physics. American Association of Physics Teachers, 5112 Berywn Road, College Park, MD 20740. (301) 345-4200. 1933 to present. Monthly. $215.00 per year. ISSN: 0002-9505.

Chaos; An Interdisciplinary Journal of Nonlinear Science. American Institute of Physics, One Physics Ellipse, College Park, Md 20740-3843. (301) 209-3000. 1991 to present. Quarterly. $225.00 per year. ISSN: 1054-1500.

Chaos, Solitons and Fractals: Applications In Science and Engineering. Elsevier Science Publishing Company, Inc., 655 Avenue of the Americas, New York, NY 10010. (212) 989-5800. FAX (212)633-3990. 1991 to present. Monthly. $830.00 per year. ISSN: 0960-0779.

Contemporary Physics. Taylor & Francis, Ltd, Rankine Road, Basingstoke, Hants RG24 8PR, England. TEL 0256-840366.

CHAOS THEORY

Ency. of Physical Sciences and Engineering Info. Sources

FAX 0256-479438. 1959 to present. Bimonthly. $318.00 per year. ISSN: 0010-7514.

International Journal of Bifurcations and Chaos In Applied Sciences and Engineering. World Scientific Publishing Co., 1060 Main Street, Suite 18, River Edge, NJ 07661. (800) 227-7562. 1991 to present. Bi-monthly. $560.00 per year. ISSN: 0218-1274.

International Journal of Modern Physics A. World Scientific Publishing Co., 1060 Main Street, Suite 18, River Edge, NJ 07661. (800) 227-7562. 1986 to present. 32 issues per year. $1710.00 per year. ISSN: 0217-751X.

Journal of Physics. A: Mathematical and General. I O P Publishing. U. S. orders to: American Institute of Physics, 500 Sunnyside Boulevard, Woodbury, NY 11797-2999. (516) 576-2200. 1968 to present. Twenty-four issues per year. $2487.00 per year. ISSN: 0305-4470.

Physica D (Nonlinear Phenomena). Elsevier Science Publishing Company, Inc., 655 Avenue of the Americas, New York, NY 10010. (212) 989-5800. FAX (212) 633-3990. 1980

To Present. 36 issues per year. $2048.00 per year. ISSN: 0167-2789.

Physical Review A (General Physics). American Institute of Physics, One Physics Ellipse, College Park, MD 20740-3843. (301) 209-3000. 1970 to present. Monthly. $990.00 per year. ISSN: 1050-2947.

Physics Today. American Institute of Physics, One Physics Ellipse, College Park, MD 20740-3843. (301) 209-3000. 1948 to present. Monthly. $140.00 per year. ISSN: 0031- 9228.

Reports On Progress In Physics. I O P Publishing. U. S. orders to: American Institute of Physics, 500 Sunnyside Boulevard, Woodbury, NY 11797-2999. (516) 576-2200. 1934

To Present. Monthly. $1214.00 per year. ISSN: 0034-4885.

Reviews of Modern Physics. American Institute of Physics, One Physics Ellipse, College Park, MD 20740-3843. (301) 209-3000. 1929 to present. Quarterly. $300.00 per year. ISSN: 0034-6861.

RESEARCH CENTERS AND INSTITUTES

Laboratory For Plasma Research. University of Maryland, 1201W Energy Research Building, College Park, Md 20742. (301) 405-3516. FAX (301) 314-9437.

CHEMICAL BONDS

See also: CHEMISTRY, CRYSTALLOGRAPHY, IONIZATION, PHYSICAL CHEMISTRY

ABSTRACT SERVICES AND INDEXES

Applied Science and Technology Index; a cumulative subject ind ex to English language periodicals in the fields of aeronautics and space science, computer technology, chemi stry, construction industry, energy and related areas. H. W. Wilson

Co.,950 University Avenue, Bronx, NY 10452-9978. (212) 588-8400. (800) 367-6770. FAX (718) 590-1617. From 1958 to present. Monthly. Inquire about cost and availability. Also available online (BRS and WILSONLINE) and on CD-ROM. ISSN: 0003-6986.

Chemical Abstracts. Chemical Abstracts Service. 2540 Olentangy River Road, Box 3012, Columbus, Oh 43210-0012. (614) 447-3600. FAX (614) 447-3713. 1907 to present. Weekly. $16,800.00 per year. Available online and on CD- Rom. CA is also available in five section groupings. Inquire Regarding Cost and Availability.

Current Contents: Physical, Chemical and Earth Sciences. Institute For Scientific Information, 3501 Market Street, Philadelphia, Pa 19104. (215) 386-0100. FAX (215) 386-2291. From 1961 To Present. Weekly. $442.00 per year. Also Available On Cd-rom and Online. Inquire regarding cost and availability. ISSN: 0163-2574.

General Science Index. H.W. Wilson Company, 950 University Avenue, Bronx, NY 10452. (212) 588-8400. (800) 367-6770. Fax (718) 590-1617. From 1978 to present. Ten issues per Year; Quarterly and Annual Cumulations. Service basis. Available On Cd-rom and Online. Inquire regarding cost and availability. ISSN: 0162-1963.

Physics Abstracts. INSPEC. Section A, Science Abstracts. institution of electrical engineers (iee), London, United Kingdom. Available from: INSPEC/IEEE - Institute of Electrical and Electronic Engineers, Box 1331, 445 Hoes Lane, Piscataway, Nj 08855-1331. (908) 562-5549. 1898 to present. 24 Issues Per Year. $2835.00 per year. ISSN: 0036-8091. Also Available Online and On Cd-rom.

Science Citation Index. SCI. Institute for Scientific Information, 3501 Market Street, Philadelphia, Pa 19104. (215) 386-0100. (800) 523-1850. FAX (215) 386-2991. 1961 To Present. Six issues per year, plus annual Cumulation. $11650.00 per year. Also available online and On Cd-rom. Inquire about price and availability. 0036-827x.

ANNUAL REVIEWS AND YEARBOOKS

Annual Review of Physical Chemistry. Annual Reviews, Inc., 4139 El Camino Way, Palo Alto, CA 94306-0897. (415) 493-4400. Fax (415) 855-9815. 1951 to present. Annual. $55.00 per year. ISSN: 0066-426X.

ASSOCIATIONS AND PROFESSIONAL SOCIETIES

American Chemical Society, 1155 16th Street, Nw, Washington, Dc 20036. (202) 872-4600.

American Institute of Chemists. 501 Wythe Street, Alexandria, Va 22314-1917. (703) 836-2090. FAX (703) 836- 2091.

Association of Consulting Chemists and Chemical Engineers, 50 East 41st Street, Suite 92, New York, Ny 10017. (212) 684-6255.

Electrochemical Society. 10 South Main Street, Pennington, NJ 08534. (609) 737-1902.

BIBLIOGRAPHIES

Chemical Information. H. R. Collier, editor. Springer- Verlag New York, Inc., 175 Fifth Avenue, New York, NY 10010. (212) 460-1500. (800) 777-4643. 1990. $91.00.

Chemical Information Sources. Gary Wiggins. McGraw-Hill Publishing Company, 1221 Avenue of the Americas, New York, NY 10020. (800) 262-4729 or (212) 512-3825. 1991. $42.50.

Information Sources In Chemistry. R.T. Bottle and J.F.B Rowland, editors. R.R. Bowker Inc., 121 Chanlon Road, New Providence, NJ 07974. (800) 521-8110 or (908) 464-6800. Fourth edition. 1993. $75.00.

Scientific and Technical Books and Serials In Print; An Index To Literature In Science and Technology. R.R. Bowker Inc., 121 Chanlon Road, New Providence, NJ 07974. (908) 464-6800. (800) 521-8110. FAX (908) 665-3502. 1972 to present. Annual. 4 volumes. 1994. $299.95. Also available on compact disc and online. ISSN: 0000-054X.

DIRECTORIES AND BIOGRAPHICAL SOURCES

American Institute of Chemists - Professional Directory. American Institute of Chemists, 501 Wythe Street, Alexandria, VA 22314-1917. (703) 836-2090. FAX (703) 836- 2091. Annual. $65.00.

American Men and Women of Science. Physical and Biological Sciences. R.R. Bowker Inc., 121 Chanlon Road, New Providence, NJ 07974. (908) 464-6800. (800) 521-8110. 20th edition. 8 volumes. 1996. $850.00

Chemical Sources International. Mike Desing and Kurt Gandenberger, editors. Chemical Sources International, PO Box 1824, Clemson, SC 29633. (803) 646-7840. 1994. $285.00.

Chemical Week — Buyers Guide Issue. Chemical Week Associates, 888 Seventh Avenue, New York, NY 10106. (212) 621-4900. FAX (212) 621-4949. Annual, October. $50.00.

Directory of Chemistry Software 1992. Wendy Warr, Peter Willett, Geoff Downs. American Chemical Society, 1155 16th Street, NW, Washington, DC 20036. (202) 872-4600. 1992. $35.95.

Research Centers Directory. Gale Research., 835 Penobscot Building, Detroit, Mi 48226-4094. (313) 961-2242. (800) 347-4253. 20th edition. 1995. $485.00. ISSN: 0080-1518.

ENCYCLOPEDIAS AND DICTIONARIES

Academic Press Dictionary of Science and Technology. Christopher Morris, editor. Academic Press, Inc., 1250 Sixth Avenue, San Diego, CA 92101. (619) 231-0926. FAX (619) 699-6715. 1991. $115.00.

Concise Encyclopedia Chemistry. Translated and revised by Mary Eagleson. Walter de Gruyter, Inc., 200 Saw Mill River Road, Hawthorne, New York, 10532. (914) 747-0110 or Fax (914) 747-1326. 1994. $69.95.

Dictionary of Chemical Names and Synomyms. Philip A. Howard and Michael W. Neil. Lewis Pub.,/ CRC Press, Inc., 2000 Cor-

porate Boulevard, NW, Boca Raton, FL 33431. (407) 994-0555.(800) 272-7737. 1991. $149.95.

Dictionary of Physical Chemistry. Satish Anand and Raj Kumar. South Asia Publications, PO Box 502 Collumbus, MO 65205. (314) 474-0116. 1990. $23.50.

Encyclopedic Dictionary of Science. Candida Hunt and Monica Byles, editors. Facts-on-File, Inc., 460 Park Avenue South, New York, NY 10016-7382. (800) 322-8755. Fax (800) 683-3633. 1988. $32.95.

Encyclopedia of Applied Physics. VCH Publishers, Inc., 303 Northwest 12th Avenue, Deerfield Beach, FL 33442. (800) 367-8249. 1991-. Twenty volumes. $6000.00

Encyclopedia of Chemical Processing and Design. McKetta. Dekker, Inc., 270 Madison Avenue, New York, NY 10016. (212) 696-9000. (800) 228-1160. 1976 - . $175.00 per volume. ISSN: .

Grant & Hawks Chemical Dictionary. R. A. Grant. McGraw-Hill Publishing Company, Inc., 1221 Avenue of the Americas, New York, NY 10020. (212) 512-2000. (800) 262-4729. 5th edition. 1987. $64.50.

Hawley's Condensed Chemical Dictionary. Richard J. Lewis. Van Nostrand Reinhold, 115 Fifth Avenue, New York, NY 10003. (212) 254-3232. (800) 842-3636. 12th edition. 1993. $69.95.

Industrial Chemical thesaurus. Michael Ash and Irene Ash. VCH Publishers, Inc., 220 East 23rd Street, New York, NY 10010-4606. (800) 367-8249. 1992. $295.00.

Kirk-othmer Encyclopedia of Chemical Technology. John Wiley and Sons, Inc., 605 Third Avenue, New York, NY 10158. (800) 526-5368 or (212) 850-6000. Fourth edition. 1991-. Twenty-seven volumes. $5400.00.

McGraw-Hill Encyclopedia of Chemistry. Sybil P.Parker, editor. McGraw-Hill Book, Incorporated, 1221 Avenue of the Americas, New York, NY 10020. (212) 997-3675. Second edition. 1993. $95.50.

McGraw-Hill Encyclopedia of Science and Technology. McGraw-Hill Book, Incorporated, 1221 Avenue of the Americas, New York, NY 10020. (212) 997-3675. Seventeenth edition. Twenty volumes. 1992. $1900.00.

GENERAL WORKS

Bonding and Strucure: Structural Principles in Inorganic and Organic Chemistry. N. W. Alcock. Prentice-Hall, 113 Sylvan Avenue, Route 9W, Englewood Cliffs, NJ 07632. (201) 592-2000. (800) 922-0579. 1991. $37.50.

The Chemical Bond. John N. Murrell, et al. John Wiley & Sons, Inc., 605 Third Avenue, New York, NY 10158-0012. (212) 850-6000. (800) 225-5945. 1985. $34.95 in paper.

The Chemical Bond: Structure and Dynamics. Ahmed Zewail. Academic Press, Inc., 1250 Sixth Avenue, San Diego, CA 92101-4311. (619) 231-0926. FAX (619) 699-6715. 1992. $54.95

CHEMICAL BONDS

Ency. of Physical Sciences and Engineering Info. Sources

Chemical Bonding In Solids and Fluids. Mark Ladd. Prentice-Hall, 113 Sylvan Avenue, Route 9W, Englewood Cliffs, NJ 07632. (201) 592-2000. (800) 922-0579. 1995. $55.00.

Chemical Structure and Bonding. Roger L. Dekock and Harry B. Gray. 2nd edition. 1989. University Science Books, 55D Gate Five Road, Sausalito, CA 94965. (415) 332-5390. 2nd edition. 1989. $36.50.

Clusters of Atoms and Molecules: theory, Experiment and Clusters of Atoms. Hellmut Haberland, editor. Springer- Verlag, New York, Inc., 175 Fifth Avenue, New York, NY 10010. (212) 460-1500. (800) 777-4643. FAX (212) 473- 6272. 1995. $89.00.

Dynamics of the Liquid State. Umberto Balucani and Marco Zoppi. Oxford University Press, Inc., 200 Madison Avenue, New York, NY 10016. (212) 725-6000. (800) 334-4249. 1995. $95.00.

Electronic Structure of Materials. Adrian Sutton. Oxford University Press, Inc., 200 Madison Avenue, New York, NY 10016. (212) 725-6000. (800) 334-4249. 1993. $26.50 in paper.

The Elements. John Emsley. Oxford University Press, Inc., 200 Madison Avenue, New York, NY 10016. (212) 725-6000. (800) 334-4249. Second edition. 1991. $49,95.

The Elements of Physical Chemistry. P. W. Atkins. W. H. Freeman, 41 Madison Avenue, East 26th Avenue, 35th Floor, New York, NY 10010. (212) 576-0400. 1995. $42.95.

Exploring Chemical Elements and their Compounds. David L.Heiserman. TAB Books, P.O. Box 40, Blue Summit, PA 17294- 0850. (717) 794-2191. (800) 822-8138. 1992. $17.95.

Fundamentals of Structural Chemistry. G. D. Zhou. World Scientific Publishing Company, Inc., 1060 Main Street, River Edge, NJ 07661. (201) 487-9655. (800) 227-7562. 1993. $86.00.

Gases, Liquids and Solids: and Other States of Matter. D. Tabor. Cambridge University Press, 40 West 20th Street, New York, NY 10011-4211. (212) 924-3900. 1992. $80.00.

Multiple Bonds Between Metal Atoms. F. Albert Cotton and Richard A. Walton. Oxford University Press, Inc., 200 Madison Avenue, New York, NY 10016. (212) 725-6000. (800) 334-4249. 2nd edition. 1993. $95.00.

HANDBOOKS AND MANUALS

Chemical Formulary. H. Bennett, editor. Chemical Publishing Company, Inc., 80 Eighth Avenue, New York, NY 10011. (212) 255-1950. Volumes 1-30. $70.00 per volume.

Chemometrics: Chemical and Sensory Data. David R. Burgard and James T. Kuznicki. CRC Publishers, Inc., 2000 Corporate Blvd., N.W., Boca Raton, FL 33431. (407) 994-0555 or (800) 333-8300. 1990. $120.00.

Chemical Reference Handbook. Gordon Press, PO Box 459, Bowling Green Station, New York, NY 10004. (718) 624-8419. 1995. $260.00.

CRC Handbook of Chemistry and Physics. David R. Kide, editor. CRC Press Inc., 2000 Corporate Blvd., NW, Boca Raton,

FL 33431. (407) 994-0555. (800) 333-8300. 77th edition. 1996. $99.95.

Guide To Basic Chemical Compounds. D.R. Lide, Jr. CRC Publishers, Inc., 2000 Corporate Blvd., N.W., Boca Raton, FL 33431. (407) 994-0555. (800) 333-8300. 1993. $120.00.

Improving Safety In the Chemical Laboratory: A Practical Guide. Jay A. Young, editor. John Wiley and Sons, Inc., 605 Third Avenue, New York, NY 10158. (800) 526-5368 or (212) 850-6000. Second edition. 1991. $75.00.

Laboratory Handbook of Materials, Equipment, and Techniques. Gary S. Coyne. Prentice Hall (Division of Simon and Schuster), 15 Columbus Circle, New York, NY 10023. (212) 373-8500 or (800) 922-0579. 1992. $45.00.

Lange's Handbook of Chemistry. John A. Dean, editor. McGraw-Hill Publishing Company, 1221 Avenue of the Americas, New York, NY 10020. (800) 262-4729 or (212) 512-3825. Fourteenth edition. 1992. $79.50.

Riegel's Handbook of Industrial Chemistry. James A. Kent, editor. Van Nostrand Reinhold, 115 Fifth Avenue, New York, NY 10003. (212) 254-3232 or (800) 926-2665. Ninth edition. 1992. $114.95.

ONLINE DATABASES AND CD-ROMS

Analytical Abstracts Online. Royal Society of Chemistry, Information Services, Thomas Graham House, Science Park, Milton Road, Cambridge, CB4 4WF, England. Contains citations, mostly with abstracts, of the worldwide literature on analytical chemistry, from 1980 to present. Available on DIALOG, (800) 334-2564, and STN International, FIZ Karlsruhe, P.O. Box 2465, W-7500, Karlsruhe 1, Germany, online services. Inquire as to cost and availability.

CA Search. Chemical Abstracts Service, P.O. Box 3012, Columbus, OH 43210-0012. (614) 447-3600. (800) 848-6533. FAX (614) 447-3709. Very comprehensive guide to worldwide chemical literature and related fields, 1972 to present. Available on BRS,(800) 289-4277, DIALOG, (800) 334-2564, ORBIT (800) 456-7248, and STN International, FIZ Karlsruhe, P.O. Box 2465, W-7500, Karlsruhe 1, Germany, online services. Inquire as to cost and availability.

Chemical Journals of the American Chemical Society. American Chemical Society, 1155 16th Street, N.W., Washington, DC 20036. (202) 872-4381 or (800) 424-6747. Contains complete text of approximately 90,000 articles from 22 primary journals published by the American Chemical Society, from mostly 1982 to present. Available on STN International, FIZ Karlsruhe, P.O. Box 2465, W-7500, Karlsruhe 1, Germany, online service. Inquire as to cost and availability.

Compendex Plus. Engineering Information, Inc., 345 East 47th Street, New York, NY 10017. (212) 705-7600 or (800) 221-1044. Contains citations with abstracts to worldwide literature in engineering and technology, from 1970 to present. Available on online BRS,(800) 289-4277, DIALOG, (800) 334-2564, ORBIT (800) 456-7248, and STN International, FIZ Karlsruhe, P.O. Box 2465, W-7500, Karlsruhe 1, Germany, online services. Also available on CD-ROM. Inquire as to cost and availability.

Current Contents Search. Institute for Scientific Information, 3501 Market Street, Philadelphia, PA 19104. (215) 386-0100.

FAX (215) 386-6362. Contains citations to articles listedin the table of contents of science and technology journals. Also articles in social sciences and life sciences journals. Available on BRS,(800) 289-4277, DIALOG,(800) 334-2564, online services. Inquire as to cost and availability.

Dissertation Abstracts. University Microfilms International, 300 North Zeeb Road, Ann Arbor, MI 48106. (800) 521-0600 or (313) 761-4700. Scope includes virtually all doctoral dissertations accepted at accredited American institutions from 1861 to present in 252 subject areas. Available on BRS,(800) 289-4277, DIALOG,(800) 334-2564, and OCLC EPIC,(800) 848-5878, online services. Also available on CD-ROM. Inquire as to cost and availability.

Gmelin Database. Gmelin-Institut fur Anorganische Chemie und Grenzgebiete, Varrentrapperstrasse, 40-42, Carl-Bosch-Haus, D-6000, Frankfurt am Main 90, Germany. Contains structural and factual data relating to inorganic and organometallic chemistry. Provides data from the Gmelin Handbook of Inorganic and Organometallic Chemistry. Covers the period 1817 to 1975; 1988-89. Available on STN International, FIZ Karlsruhe, P.O. Box 2465, W-7500, Karlsruhe 1, Germany, online service. Inquire as to cost and availability.

Kirk-othmer Encyclopedia of Chemical Technology. John Wiley and Sons, Inc., 605 Third Avenue, New York, NY 10158. (800) 526-5368 or (212) 850-6000. Contains the complete text of all chapters in the 27 volume fourth edition of the KIRK-OTHMER ENCYCLOPEDIA of CHEMICAL TECHNOLOGY. 1991. Available on BRS,(800) 289-4277, DIALOG,(800) 334-2564, online services. Inquire as to cost and availability.

NTIS Bibliographic Database. National Technical Information Service, 5285 Port Royal Road, Springfield, Va 22161. (703) 487-4929 or FAX (703) 321-8199. Broad coverage of government-sponsored science and technology research reports, 1964 to present. Available on BRS,(800) 289-4277, DIALOG, (800) 334-2564, ORBIT, (800) 456-7248, and STN International, FIZ Karlsruhe, P.O. Box 2465, W-7500, Karlsruhe 1, Germany, online services. Also available on CD-ROM. Inquire as to cost and availability.

Scisearch. Institute for Scientific Information, 3501 Market Street, Philadelphia, PA 19104. (800) 523-1850 or (215) 386-0100. Broad multidisciplinary title and author index to the international literature of science and technology, 1974 to present. Available on DIALOG,(800) 334-2564, and ORBIT,(800) 456-7248, online services. Also available on CD-ROM. Inquire as to cost and availability.

WILSONLINE. H.W. Wilson Company, 950 University Avenue, Bronx, NY 10452. (800) 367-6770 or (212) 588-8400. Makes available online versions of the printed H.W. Wilson indexes including Applied Science and Technology Index, Business Periodicals Index, General Science Index, and Readers' Guide to Periodical Literature. Period covered is generally 1983 to present. Available on BRS,(800)289-4277, DIALOG,(800) 334-2564, and OCLC EPIC,(800) 848-5878, online services. Also available on CD-ROM. Inquire as to cost and availability.

PERIODICALS

American Laboratory. American Laboratory, 30 Controls Drive, Box 870, Shelton, CT 06484-6111. (203) 926-9300. FAX (203) 926-9310. 1968 to present. Monthly. $210.00 per year. ISSN: 0044-7749.

Analytical Chemistry. American Chemical Society, Box *3337, Columbus, Oh 43210. (614) 447-3776. 1929 to Present.* Semimonthly. $415.00 per year. ISSN: 0003-2700.

Analytical Methods and Instrumentation. John Wiley and Sons, Inc., 605 Third Avenue, New York, NY 10158. (800) 526-5368 or (212) 850-6000. Bimonthly. $250.00 per year. ISSN: 1063-5246.

Angewandte Chemie. VCH Publishers, Inc., 220 East 23rd Street, New York, NY 10010-4606. (212) 683-8333. (800) 367-8249. 1888 to present. 22 issues per year. $840.00 per Year. ISSN: 0044-8249.

Chemical Physics. Elsevier Science Publishing Co., Inc., Box 882, Madison Square Station, New York, NY 10159. (212) 989-5800. FAX (212) 633-3990. 1973 to present. 33 issues per year in 11 volumes. $2759.00 per year. ISSN: 0301-0104.

Chemical Physics Letters. Elsevier Science Inc., Publishing Co., Inc., Box 882, Madison Square Station, New York, NY 10159. (212) 989-5800. (212) 633-3990. 1967 to present. 102 issues per year in 17 volumes. $5063.00 per year. ISSN: 0009-2614.

Chemical Reviews. American Chemical Society, Box 3337, Columbus, Oh 43210. (614) 447-3776. 1924 to present. 8 Issues Per Year. $346.00 per year. ISSN: 0009-2665.

Chemical Week; Includes Annual Buyers Guide. Chemical Week Associates, 888 Seventh Avenue, New York, Ny 10106. (212) 621-4900. FAX (212) 621-4949. 1914 to present. Weekly. $99.00 per year. ISSN: 0009-272X.

Chemtech. American Chemical Society. Box 3337, Columbus, OH 43210. (614) 447-3776. 1970 to present. Monthly. $370.00 per year. ISSN: 0009-2703.

Inorganic Chemistry. American Chemical Society, Box 3337, Columbus, Oh 43210. (614) 447-3776. 1962 to Present. Semimonthly. $500.00 per year. ISSN: 0020-1669.

International Journal of Chemical Kinetics. John Wiley and Sons, Inc., 605 Third Avenue, New York, NY 10158. (800) 526-5368 or (212) 850-6000. 1968 to present. Monthly. $730.00 per year. ISSN: 0538- 8066.

Journal of the American Chemical Society. American Chemical Society, Box 3337, Columbus, OH 43210. (614) 447-3776. 1879 to present. Biweekly. $1055.00 per year. ISSN: 0002-7863.

Journal of Organic Chemistry. American Chemical Society, Box 3337, Columbus, Oh 43210. (614) 447-3776. (800) 333- 9511. FAX (614) 447-3671. 1936 to present. Bi-weekly. $785.00 per year. ISSN: 0022-3263.

Journal of Physical Chemistry. American Chemical Society, Box 3337, Columbus, Oh 43210. (614) 447-3776. (800) 333-9511. FAX (614) 447-3671. 1896 to present. Weekly. $1140.00 per year to non-members. ISSN: 0894-3230.

RESEARCH CENTERS AND INSTITUTES

Barnett Institute of Chemical Analysis and Materials Science. Northeastern University, 360 Huntington Avenue, Boston, MA 02115. (617) 437-2864.

Chemical Laboratories. Harvard University, Oxford Street, Cambridge, MA 02138. (617) 495-4283. FAX (617) 496-5618.

Chemistry Laboratories. Rensselaer Polytechnic Institute, Cogswell Laboratory, Troy, NY 12180-3590. (518) 276-8981.

Laboratory For Research On the Structure of Matter.

University of Pennsylvania, 3231 Walnut Street, Philadelphia, Pa 19104. (215) 898-8571.

Research Program In Chemistry and Biochemistry. Southern Illinois University At Carbondale, Carbondale, Il 62901. (618) 453-5721.

Theoretical Chemistry Institute. University of Wisconsin, Madison. 1101 University Avenue, Madison, WI 53706. (608) 262-1511.

CHEMICAL ENGINEERING

See also: AEROSOLS, CATALYSIS, CHEMISTRY, METALLURGY, ORGANIC CHEMISTRY

ABSTRACT SERVICES AND INDEXES

Applied Science and Technology Index; A Cumulative Subject Index To English Language Periodicals In the Fields of Aeron Autics and Space Science, Computer Technology, Chemistry , Construction Industry, Energy and Related Areas. H. W. Wilson Co., 950 University Avenue, Bronx, NY 10452-9978. (212) 588-8400. (800) 367-6770. FAX (718) 590-1617. From 1958 to present. Monthly. Inquire about cost and availability. Also available online (BRS and WILSONLINE) and on CD-ROM. ISSN: 0003-6986.

Chemical Abstracts. Chemical Abstracts Service. 2540 Olentangy River Road, Box 3012, Columbus, Oh 43210-0012. (614) 447-3600. FAX (614) 447-3713. 1907 to present. Weekly. $16,800.00 per year. Available online and on CD-ROM. CA is also available in five section groupings. Inquire Regarding Cost and Availability.

Current Contents : Physical, Chemical and Earth Sciences. Institute for Scientific Information, 3501 Market Street, Philadelphia, PA 19104. (215) 386-0100. (800) 523-1850. FAX (215) 386-2291. 1961 to present. Weekly. $442.00 per year. Also available online (BRS, DIALOG) and on CD-ROM. Inquire regarding cost and availability. ISSN: 0163-2574.

Engineered Materials Abstracts. ASM (American Society of Metals) International, Materials Information, Materials Park, OH 44073. (216) 338-5151 or FAX (216) 338-4634. Covers literature on technical developments in polymer, ceramic, and composite materials and engineering. 1986 to present. Monthly. $1175.00 per year. Also available on CD-ROM.

Engineering Index Monthly. Engineering Information, Inc.,Castle Point on the Hudson, Hoboken, NJ 07030. (201) 216-8500. (800) 221-1044. FAX (201) 216-8532. Monthly. $2300.00 per year for monthly issues. $1980 per year for annual cumulation. Available on CD-ROM and online as COMPENDEX. ISSN: 0742-1974; 0360-8557.

General Science Index. H.W. Wilson Company, 950 University Avenue, Bronx, NY 10452. (212) 588-8400. (800) 367-6770. FAX (718) 590-1617. From 1978 to present. Ten issues per year; quarterly and annual cumulations. Service basis. Available on CD-ROM and online. Inquire regarding cost and availability. ISSN: 0162-1963.

Physics Abstracts. INSPEC. Section A, Science Abstracts. Institution of Electrical Engineers (IEE), London, United Kingdom. Available from: INSPEC/IEEE - Institute of Electrical and Electronic Engineers, Box 1331, 445 Hoes Lane, Piscataway, NJ 08855-1331. (908) 562-5549. 1898 to present. 24 issues per year. $2835.00 per year. ISSN: 0036-8091. Also available online and on CD-ROM.

Polymer Contents: International Current Awareness Publication On Polymer Science and Engineering. Elsevier Science Publishing Company, Inc., 655 Avenue of the Americas, New York, NY 10010. (212) 989-5800. 1984 to present. Monthly. $350.00 per year.

Science Citation Index. SCI. Institute for Scientific Information, 3501 Market Street, Philadelphia, Pa 19104. (215) 386-0100. (800) 523-1850. FAX (215) 386-2991. 1961 To Present. Six issues per year, plus annual cumulation. $11650.00 per year. Also available online and on CD-ROM. Inquire about price and availability. ISSN: 0036-827X.

ANNUAL REVIEWS AND YEARBOOKS

Advances In Catalysis. Academic Press, Inc., 6277 Sea Harbor Drive, Orlando, FL 32887. (800) 321-5068. 1948 to present. Irregular. ISSN: 0360-0564.

Advances In Chemical Engineering. Academic Press, Inc., 6277 Sea Harbor Drive, Orlando, FL 32887. (800) 321-5068. 1956 to present. Irregular. ISSN: 0065-2377.

ASSOCIATIONS AND PROFESSIONAL SOCIETIES

American Chemical Society, 1155 16th Street, Nw, Washington, Dc 20036. (202) 872-4600.

American Institute of Chemical Engineers. 345 East 47th Street, New York, NY 10017. (212) 705-7338.

Association of Consulting Chemists and Chemical Engineers, 50 East 41st Street, Suite 92, New York, Ny 10017. (212) 684-6255.

Association of official Analytical Chemists. 2200 Wilson Blvd, Suite 400, Arlington, VA 22001. (703) 522-3032.

Catalysis Society of North America. c/o Dr. Michael B. D'Amore. I. I. Dupont Co. Experiment Station, Wilmington DE 19898-9262. (302) 605-2488.

Chemical Manufacturers Association. 2501 M Street, N.W., Washington, DC 20037. (202) 887-1182

BIBLIOGRAPHIES

Chemical Information Sources. Gary Wiggins. McGraw-Hill Publishing Company, 1221 Avenue of the Americas, New York, NY 10020. (212) 512-2000. (800) 262-4729. 1991. $42.50.

Handbooks and Tables In Science and Technology. Russell H. Powell, editor. Oryx Press, 4041 North Central, Suite 700, Phoenix, AZ 85012-3330. (602) 265-2651 or (800) 279-6799. Third edition. 1994. $65.00.

Information Sources In Chemistry. R.T. Bottle and J.F.B Rowland, editors. R.R. Bowker Inc., 121 Chanlon Road, New Providence, NJ 07974. (800) 521-8110 or (908) 464-6800. Fourth edition. 1993. $75.00.

Scientific and Technical Books and Serials In Print; An Index To Literature In Science and Technology. R.R. Bowker Inc., 121 Chanlon Road, New Providence, NJ 07974. (908) 464-6800. (800) 521-8110. FAX (908) 665-3502. 1972 to present. Annual. 4 volumes. 1994. $299.95. Also available on compact disc and online. ISSN: 0000-054X.

DIRECTORIES AND BIOGRAPHICAL SOURCES

American Institute of Chemical Engineers Directory. American Institute of Chemical Engineers. 345 East 47th Street, New York, NY 10017-2396. (212) 705-7663. Annual.

American Men and Women of Science. Physical and Biological Sciences. Fifteenth edition. R.R. Bowker Inc., 121 Chanlon Road, New Providence, NJ 07974. (908) 464-6800. (800) 521-8110. 20th edition. 8 volumes. 1996. $850.00

Chemical Engineering Faculties. American Institute of Chemical Engineers. 345 East 47th Street, New York, NY 10017-2396. (212) 705-7663. Annual. $75.00.

Directory of Chemistry Software 1992. Wendy Warr, Peter Willett, Geoff Downs. American Chemical Society, 1155 16th Street, NW, Washington, DC 20036. (202) 872-4600. 1992. $35.95.

Engineering Research Centers: Incorporating Electronics Research Centers. Stockton Press, 345 Park Avenue, New York, NY 10010. (212) 689-9200. (800) 221-2123. 4th edition. 1995. $515.00.

Research Centers Directory. Gale Research Company, Inc., 835 Penobscot Building, Detroit, MI 48226-4094. (313) 961-2242. (800) 347-4253. 20th edition. 1995. $485.00. ISSN:: 0080-1518.

Who's Who In Engineering. Gordon Davis, editor. American Association of Engineering Societies. 1111 19th Street, NY, Suite 608, Washington, Dc 20036. (202) 296-2237. (800) 658-8897. 9th edition. 1995. $220.00.

ENCYCLOPEDIAS AND DICTIONARIES

Dictionary of Named Processes In Chemical Technology. Alan E. Comyns. Oxford University Press, Inc., 200 Madison Avenue, New York, NY 10016. (212) 725-6000. (800) 334-4249. 1994. $75.00.

Encyclopedia of Applied Physics. VCH Publishers, Inc., 303 Northwest 12th Avenue, Deerfield Beach, FL 33442. (800) 367-8249. 1991-. Twenty volumes. $6000.00.

Encyclopedia of Chemical Processing and Design. McKetta Marcel Dekker, Inc., 270 Madison Avenue, New York, NY 10016. (212)696-9000. (800) 228-1160. 1976 - . $175.00 per volume. ISSN:

Illustrated Chemistry Laboratory Terminology. Gerbert W. Ockerman. CRC Press, Inc., 2000 Corporate Boulevard, NW, Boca Raton, FL 33431. (407) 994-0555. (800) 272-7737. 1991. $32.95.

Industrial Chemical thesaurus. Michael Ash and Irene Ash. VCH Publishers, Inc., 220 East 23rd Street, New York, NY 10010-4606. (800) 367-8249. 1992. $295.00.

Kirk-othmer Encyclopedia of Chemical Technology. John Wiley and Sons, Inc., 605 Third Avenue, New York, NY 10158. (800) 526-5368 or (212) 850-6000. Fourth edition. 1991 - . Twenty-seven volumes. $5400.00.

McGraw-Hill Encyclopedia of Science and Technology. McGraw-Hill Book, Incorporated, 1221 Avenue of the Americas, New York, NY 10020. (212) 997-3675. (800) 262-4729. Seventh edition. Twenty volumes. 1992. $1900.00.

Ullman's Encyclopedia of Industrial Chemistry. VCH Publications, Inc., 220 East 23rd Street, Suite 909, New York, NY 10010-4606. (212) 683-8333. (800) 422-8824. 5th edition. 1984 - . Price varies per volume.

GENERAL WORKS

Applied Mathematics and Modeling For Chemical Engineers. Richard G. Rice. John Wiley & Sons, Inc., 605 Third Avenue, New York, NY 10158-0012. (212) 850-6000. (800) 225-5945.

Catalysis: An Integrated Approach To Homogeneous, Heterogeneous and Industrial Catalysis. J. A. Moulijn, editor. Elsevier Science Publishing Company, Inc., 655 Avenue of the Americas, New York, NY 10010. (212) 989-5800. FAX (914) 333-2444. 1993. $184.00

Catalysis of Organic Reactions. Michael Scaros and Michael L. Prunier, editors. Marcel Dekker, Inc., 270 Madison Avenue, New York, NY 10016. (212) 696-9000. (800) 228- 1160. 1994. $195.00.

Chemical Engineering. J. M. Coulson and J. F. Richardson, editors. Elsevier Science Publishing Company, Inc., 655 Avenue of the Americas, New York, NY 10010. (212) 989-5800. FAX (914) 333-2444. volumes 1-5. 1990. price varies.

Chemical Process Design. Robin Smith. McGraw-Hill Publishing Company, Inc., 1221 Avenue of the Americas, New York, NY 10020. (212) 512-2000. (800) 262-4729. 1995. $65.00.

Chemical Reactor Analysis and Design. G. F. Froment and Kenneth B. Bischoff. John Wiley & Sons, Inc., 605 Third Avenue, New York, NY 10158-0012. (212) 850-6000. (800) 225-5945. 2nd edition. 1990.

Complete Chemical Engineer. Robert B, Barat and Worbent Elliot. Kendell-Hunt. Publishing Co., 4050 Westmark Drive, P.O. Box 1840 Dubuque, IA 52004-1840. (319) 589-1000. (800) 228-0810. 1993. $37.44 in paper.

Dynamic Processes On Solid Surfaces. Kenzi Tamaru, editor. Plenum Publishing Corp., 233 Spring Street, New York, NY.(212) 620-8000. (800) 2221-9369. FAX (212) 463-0742.. 1993. $95.00.

Industrial Applications of Homgeneous Catalysis. andre Mortreux and Francis Petit, editors. Kluwer Academic Publish-

CHEMICAL ENGINEERING

Ency. of Physical Sciences and Engineering Info. Sources

ers, 101 Philip Drive, Assinippi Park, Norwell, MA 02061. (617) 871-6600. 1987. $133.00.

Introduction To Surface Chemistry and Catalysis. Gabor A. Jomorjai. John Wiley & Sons, Inc., 605 Third Avenue, New York, NY 10158-0012. (212) 850-6000. (800) 225-5945. 1994. $59.95.

Unit Operations In Chemical Engineering. Warren L. McCabe. McGraw-Hill Publishing Company, Inc., 1221 Avenue of the Americas, New York, NY 10020. (212) 512-2000. (800) 262-4729. 5th edition. 1993.

HANDBOOKS AND MANUALS

Catalyst Handbook. Martyn V. Twigg. CRC Press Inc., 2000 Corporate Blvd., NW, Boca Raton, FL 33431. (407) 994-0555. (800) 333-8300. 2nd edition. 1989. $104.00.

Chemical Engineering Reference Manual. Randall N. Robinson. Professional Publications, Inc., 1250 Fifth Avenue, Belmont, CA 94002. (415) 593-0110. (800) 426-1178. 1988. $49.95.

Chemical Formulary. H. Bennett, editor. Chemical Publishing, Co., Inc. 80 Eighth Avenue, New York, NY 10011. (212 255-1950. Volumes 1 - 30. $60.00 per volume.

Comprehensive Organic Synthesis: Selectivity, Strategy, and Efficiency In Modern Organic Chemical. Barry M. Trost and others, editors. Pergamon Press, Maxwell House, Fairview Park, Elmsford, NY 10523. (914) 592-7700. Fax (914) 592- 3625. 1991. Nine volumes. $3900.00.

CRC Handbook of Chemistry and Physics. David R. Kide, editor. CRC Press Inc., 2000 Corporate Blvd., NW, Boca Raton, FL 33431. (407) 994-0555. (800) 333-8300. 77th edition. 1996. $99.95.

Guide To Basic Chemical Compounds. D.R. Lide, Jr. CRC Publishers, Inc., 2000 Corporate Blvd., N.W., Boca Raton, FL 33431. (407) 994-0555 or (800) 333-8300. 1993. $120.00.

Handbook of Chemical Engineering Calculations. Nicholas P. Chopey and Tyler G. Hicks, editors. McGraw-Hill Publishing Company, Inc., 1221 Avenue of the Americas, New York, NY 10020. (212) 512-2000. (800) 262-4729. 2nd edition. 1993. $69.50.

Handbook of Homogeneous Catalysis. B. Cornils, editor. VCH Publications, Inc., 220 East 23rd Street, Suite 909, New York, NY 10010-4606. (212) 683-8333. (800) 422-8824. 1996. $345.00.

Handbook of Laboratory Safety. A. Keith Furr. CRC Press, Inc., 2000 Corporate Boulevard, NW, Boca Raton, FL 33431. (407) 994-0555. (800) 272-7737. 1995. 125.00.

Handbook of Industrial Chemical Additives. Michael Ash and Irene Ash. VCH Publications, Inc., 220 East 23rd Street, Suite 909, New York, NY 10010-4606. (212) 683-8333. (800) 422-8824. 1991. $195.00.

Laboratory Handbook of Materials, Equipment, and Techniques. Gary S. Coyne. Prentice Hall (Division of Simon and Schuster), 15 Columbus Circle, New York, NY 10023. (212) 373-8500or (800) 922-0579. 1992. $45.00.

Lange's Handbook of Chemistry. John A. Dean, editor. McGraw-Hill Publishing Company, 1221 Avenue of the Americas, New York, NY 10020. (800) 262-4729 or (212) 512-3825. Fourteenth editon. 1992. $79.50.

Perry's Chemical Engineers' Handbook. Robert H. Perry and Donald W. Green, editors. McGraw-Hill Publishing Company, Inc., 1221 Avenue of the Americas, New York, NY 10020. (212) 512-2000. (800) 262-4729. 6th edition. 1996. $129.50.

Riegel's Handbook of Industrial Chemistry. James A. Kent, editor. Van Nostrand Reinhold, 115 Fifth Avenue, New York, NY 10003. (212) 254-3232 or (800) 926-2665. Ninth edition.

1992. $114.95.

ONLINE DATABASES AND CD-ROMS

Beilstein Online. Beilstein-Institut fur Literatur der Organischen Chemie, Varrentrapperstrasse, 40-42, D-6000, Frankfurt am Main 90, Germany. Contains data on carbon compounds from the Beilstein Handbook of Organic Chemistry. Structural and factual data for more than 3.4 million heterocyclic, isocyclic, and acyclic compounds included. Covers the period 1830 to present. Available on DIALOG, (800) 334-2564, ORBIT (800) 456-7248, and STN International, FIZ Karlsruhe, P.O. Box 2465, W-7500, Karlsruhe 1, Germany, online services. Inquire as to cost and availability.

CA Search. Chemical Abstracts Service, P.O. Box 3012, Columbus, Oh 43210-0012. (614) 447-3600. (800) 848-6533. FAX (614) 447-3709. Very comprehensive guide to worldwide chemical literature and related fields. 1972 to present. Available on BRS, (800) 289-4277, DIALOG, (800) 334-2564, ORBIT (800) 456-7248, and STN International, FIZ Karlsruhe, P.O. Box 2465, W-7500, Karlsruhe 1, Germany, online services. Inquire as to cost and availability.

Chemical Journals of the American Chemical Society. American Chemical Society, 1155 16th Street, N.W., Washington, DC 20036. (202) 872-4381 or (800) 424-6747. Contains complete text of approximately 90,000 articles from 22 primary journals published by the American Chemical Society, from mostly 1982 to present. Available on STN International, FIZ Karlsruhe, P.O. Box 2465, W-7500, Karlsruhe 1, Germany, online service. Inquire as to cost and availability.

Compendex Plus. Engineering Information, Inc., 345 East 47th Street, New York, NY 10017. (212) 705-7600 or (800) 221-1044. Contains citations with abstracts to worldwide literature in engineering and technology, from 1970 to present. Available on online BRS, (800) 289-4277, DIALOG, (800) 334-2564, ORBIT (800) 456-7248, and STN International, FIZ Karlsruhe, P.O. Box 2465, W-7500, Karlsruhe 1, Germany, online services. Also available on CD-ROM. Inquire as to cost and availability.

Current Contents Search. Institute for Scientific Information, 3501 Market Street, Philadelphia, PA 19104. (215) 386-0100. FAX (215) 386-6362. Contains citations to articles listed in the table of contents of science and technology journals. Alsoarticles in social sciences and life sciences journals. Available on BRS, (800) 289-4277, DIALOG, (800) 334-2564, online services. Inquire as to cost and availability.

Dissertation Abstracts. University Microfilms International, 300 North Zeeb Road, Ann Arbor, MI 48106. (800) 521-0600 or (313) 761-4700. Scope includes virtually all doctoral dissertations accepted at accredited American institutions from 1861 to

present in 252 subject areas. Available on BRS, (800) 289-4277, DIALOG, (800) 334-2564, and OCLC EPIC, (800) 848-5878, online services. Also available on CD-ROM. Inquire as to cost and availability.

Engineered Materials Abstracts. Materials Information, ASM International, Materials Park, OH 44073. (216) 338-5151. Contains citations with abstracts of worldwide literatur e in the development, processing, and production of cera mic, composite, and polymeric materials for engineering uses.Available on DIALOG, (800) 334-2564, ORBIT (800) 456-7248, and STN International, FIZ Karlsruhe, P.O. Box 2465, W-7500, Karlsruhe 1, Germany, online services. Inquire as to cost and availability.

NTIS Bibliographic Database. National Technical Information Service, 5285 Port Royal Road, Springfield, Va 22161. (703) 487-4929 or FAX (703) 321-8199. Broad coverage of government-sponsored science and technology research reports, 1964 to present. Available on BRS,(800) 289-4277, DIALOG, (800) 334-2564, ORBIT, (800) 456-7248, and STN International, FIZ Karlsruhe, P.O. Box 2465, W-7500, Karlsruhe 1, Germany, online services. Also available on CD-ROM. Inquire as to cost and availability.

Gmelin Database. Gmelin-Institut fur Anorganische Chemie und Grenzgebiete, Varrentrapperstrasse, 40-42, Carl-Bosch-Haus, D-6000, Frankfurt am Main 90, Germany. Contains structural and factual data relating to inorganic and organometallic chemistry. Provides data from the Gmelin Handbook of Inorganic and Organometallic Chemistry. Covers the period 1817 to 1975; 1988-89. Available on STN International, FIZ Karlsruhe, P.O. Box 2465, W-7500, Karlsruhe 1, Germany, online service. Inquire as to cost and availability.

Kirk-Othmer Encyclopedia of Chemical Technology. John Wiley and Sons, Inc., 605 Third Avenue, New York, NY 10158. (800) 526-5368 or (212) 850-6000. Contains the complete text of all chapters in the 27 volume fourth edition of the KIRK-OTHMER ENCYCLOPEDIA of CHEMICAL TECHNOLOGY. 1991. Available on BRS,(800) 289-4277, DIALOG,(800) 334-2564, online services. Inquire as to cost and availability.

NTIS Bibliographic Database. National Technical Information Service, 5285 Port Royal Road, Springfield, Va 22161. (703) 487-4929 or FAX (703) 321-8199. Broad coverage of government-sponsored science and technology research reports, 1964 to present. Available on BRS,(800) 289-4277, DIALOG, (800) 334-2564, ORBIT, (800) 456-7248, and STN International, FIZKarlsruhe, P.O. Box 2465, W-7500, Karlsruhe 1, Germany, online services. Also available on CD-ROM. Inquire as to cost and availability.

Scisearch. Institute for Scientific Information, 3501 Market Street, Philadelphia, PA 19104. (800) 523-1850 or (215) 386-0100. Broad multidisciplinary title and author index to the international literature of science and technology, 1974 to present. Available on DIALOG, (800) 334-2564, and ORBIT, (800) 456-7248, online services. Also available on CD-ROM. Inquire as to cost and availability.

WILSONLINE. H.W. Wilson Company, 950 University Avenue, Bronx, NY 10452. (800) 367-6770 or (212) 588-8400. Makes available online versions of the printed H.W. Wilson indexes including Applied Science and Technology Index, Business Periodicals Index, General Science Index, and Readers' Guide to Periodical Literature. Period covered is generally 1983 to present. Available on BRS, (800) 289-4277, DIALOG, (800) 334-2564, and OCLC EPIC, (800) 848-5878, online services. Also available on CD-ROM. Inquire as to cost and availability.

PERIODICALS

AIChe Journal. American Institute of Chemical Engineers. 345 East 47th Street, New York, NY 10017. (212) 705-7338.

1955 To Present. Monthly. $295.00 per year.

Catalysis Reviews: Science and Engineering. Marcel Dekker, Inc., 270 Madison Avenue, New York, NY 10016. (212) 696-9000. FAX (212) 685-4540. 1967 to present. Quarterly. $400.00 per year.

Chemical Week. Chemical Week Associates, 810 7th Avenue, New York, NY 10019. Weekly. $30.00 per year.

Chemtech. American Chemical Society, Drawer 1734, Atlanta, GA 30301-1734. (800) 227-5558. Monthly. $300.00 per year.

RESEARCH CENTERS AND INSTITUTES

Institute For Systems Design and Optimization. Kansas State University. Department of Chemical Engineering, Manhattan, KS 66506. (913) 532-5584.

Engineering Research Program. Pennsylvania State University, 101 Hammond Building, University Park, PA 16802.

Chemical Laboratories. Harvard Universiy. Oxford Street, Cambridge, MA 02138. (617) 495-4283. FAX (617) 496-5618.

Chemistry Laboratories. Rensselaer Polytechnic Institute. Cogswell Laboratory, Troy, NY 12180-3590. (518) 276-8981.

University/industry Chemical Research Center. Mississippi State University, Department of Chemistry, P.O. Drawer CH, Mississippi State, MS 39762. (601) 325-3584.

CHEMICAL EQUILIBRIUM

see: CHEMICAL ENGINEERING

CHEMICAL FORMULAS

See:CHEMISTRY

CHEMICAL MICROSCOPY

See:MICROSCOPY

CHEMICAL SEPARATION TECHNIQUES

See: ANALYTICAL CHEMISTRY

CHEMISTRY

Ency. of Physical Sciences and Engineering Info. Sources

CHEMISTRY

See also: INORGANIC CHEMISTRY, ORGANIC CHEMISTRY, PHYSICAL CHEMISTRY

ABSTRACT SERVICES AND INDEXES

Analytical A Bstracts. Royal Society of Chemistry, Information Services, Thomas Graham House, Science Park, Milton Road, Cambridge, CB4 4WF, England. Contains citations, mostly with abstracts, of the worldwide literature on analytical chemistry, from 1954 to present. Monthly. $636.00 per year. Also available online.

Applied Science and Technology Index; A Cumulative Subject Index To Engli Sh Language Periodicals In the Fields of Aeronautics and Space Science, Computer Technology, Chemistry, Construction Industry, Energy and Related Areas. H. W. Wilson Co., 950 University Avenue, Bronx, NY 10452-9978. (212) 588-8400. (800) 367-6770. FAX (718) 590-1617. From 1958 to present. Monthly. Inquire about cost and availability. Also available online (BRS and WILSONLINE) and on CD-ROM. ISSN: 0003-6986.

Chemical Abstracts. Chemical Abstracts Service. 2540 Olentangy River Road, Box 3012, Columbus, Oh 43210-0012. (614) 447-3600. FAX (614) 447-3713. 1907 to present. Weekly. $16,800.00 per year. Available online and on CD- Rom. CA is also available in five section groupings. Inquire Regarding Cost and Availability.

Current Contents: Physical, Chemical and Earth Sciences. Institute For Scientific Information, 3501 Market Street, Philadelphia, Pa 19104. (215) 386-0100. FAX (215) 386-2291. From 1961 To Present. Weekly. $442.00 per year. Also Available On Cd-rom and Online. Inquire regarding cost and availability. ISSN: 0163-2574.

Engineering Index Monthly. Engineering Information, Inc., Castle Point on the Hudson, Hoboken, NJ 07030. (201) 216-8500. (800) 221-1044. FAX (201) 216-8532. Monthly. $2300.00 per year for monthly issues. $1980 per year for annual cumulation. Available on CD-ROM and online as COMPENDEX. ISSN: 0742-1974; 0360-8557.

General Science Index. H.W. Wilson Company, 950 University Avenue, Bronx, NY 10452. (212) 588-8400. (800) 367-6770. Fax (718) 590-1617. From 1978 to present. Ten issues per Year; Quarterly and Annual Cumulations. Service basis. Available On Cd-rom and Online. Inquire regarding cost and availability. ISSN: 0162-1963.

Physics Abstracts. INSPEC. Section A, Science Abstracts. Institution of Electrical Engineers (IEE), london, united Kingdom. Available from: INSPEC/IEEE - Institute of Electrical and Electronic Engineers, Box 1331, 445 Hoes Lane, Piscataway, NJ 08855-1331. (908) 562-5549. 1898 to present. 24 Issues Per Year. $2835.00 per year. ISSN: 0036-8091. Also Available Online and On CD-ROM.

Process and Chemical Engineering/Chemical EngineeringAbstracts. Royal Society of Chemistry, Information services, thomas graham house, science park, milton road, Cambridge, Cb4 4wf, England. Contains citations, mostly with abstracts, of the worldwide literature on chemical Engineering; From 1982 To Present. Monthly. $610.00 per Year. Also available online. ISSN: 0960-5045.

Science Citation Index. SCI. Institute for Scientific Information, 3501 Market Street, Philadelphia, Pa 19104. (215) 386-0100. (800) 523-1850. FAX (215) 386-2991. 1961 To Present. Six issues per year, plus annual Cumulation. $11650.00 per year. Also available online and On Cd-rom. Inquire about price and availability. 0036-827x.

ANNUAL REVIEWS AND YEARBOOKS

Advances in Catalysis. Academic Press, Inc., 6277 Sea Harbor Drive, Orlando, Fl 32887. (800) 321-5068. 1948 to Present. Irregular. ISSN: 0360-0564.

Advances In Heterocyclic Chemistry. Alan R. Kitritzky, editor. Academic Press.Inc., 525 B Street, Suite 1900, San Diego, CA 92101-4495. (619) 231-0926. FAX (619) 699-6715. 1963 to present. Irregular. price varies. ISSN: 0065-2725.

Annual Review of Physical Chemistry. Annual Reviews, Inc., 4139 El Camino Way, Palo Alto, CA 94306-0897. (415) 493-4400. Fax (415) 855-9815. 1951 to present. Annual. $55.00 per year. ISSN: 0066-426X

ASSOCIATIONS AND PROFESSIONAL SOCIETIES

American Chemical Society, 1155 16th Street, Nw, Washington, Dc 20036. (202) 872-4600.

American Institute of Chemists. 501 Wythe Street, Alexandria, Va 22314-1917. (703) 836-2090. FAX (703) 836- 2091.

Association of Consulting Chemists and Chemical Engineers, 50 East 41st Street, Suite 92, New York, Ny 10017. (212) 684-6255.

Chemical Manufacturers Association. 2501 M Street, N.W., Washington, DC 20037. (202) 887-1182

Electrochemical Society. 10 South Main Street, Pennington, Nj 08534. (609) 737-1902.

BIBLIOGRAPHIES

Chemical Information. H. R. Collier, editor. Springer- Verlag New York, Inc., 175 Fifth Avenue, New York, NY 10010. (212) 460-1500. (800) 777-4643. 1990. $91.00.

Chemical Information Sources. Gary Wiggins. McGraw-Hill Publishing Company, 1221 Avenue of the Americas, New York, NY 10020. (800) 262-4729 or (212) 512-3825. 1991. $42.50.

Information Sources In Chemistry. R.T. Bottle and J.F.B Rowland, editors. R.R. Bowker Inc., 121 Chanlon Road, New Providence, NJ 07974. (800) 521-8110 or (908) 464-6800. Fourth edition. 1993. $75.00.

Scientific and Technical Books and Serials In Print; An Index To Literature In Science and Technology. R.R. Bowker Inc., 121 Chanlon Road, New Providence, NJ 07974. (908) 464-6800. (800) 521-8110. FAX (908) 665-3502. 1972 to present. Annual. 4 volumes.1994. $299.95. Also available on compact disc and online. ISSN: 0000-054X.

DIRECTORIES AND BIOGRAPHICAL SOURCES

American Institute of Chemists - Professional Directory. American Institute of Chemists, 501 Wythe Street, Alexandria, VA 22314-1917. (703) 836-2090. FAX (703) 836- 2091. Annual. $65.00.

American Men and Women of Science. Physical and Biological Sciences. R.R. Bowker Inc., 121 Chanlon Road, New Providence, NJ 07974. (908) 464-6800. (800) 521-8110. 20th edition. 8 volumes. 1996. $850.00

Chemical Sources International. Mike Desing and Kurt Gandenberger, editors. Chemical Sources International, PO Box 1824, Clemson, SC 29633. (803) 646-7840. 1994. $285.00.

Chemical Week — Buyers Guide Issue. Chemical Week Associates, 888 Seventh Avenue, New York, NY 10106. (212) 621-4900. FAX (212) 621-4949. Annual, October. $50.00.

Directory of Chemistry Software 1992. Wendy Warr, Peter Willett, Geoff Downs. American Chemical Society, 1155 16th Street, NW, Washington, DC 20036. (202) 872-4600. 1992. $35.95.

Research Centers Directory. Gale Research, 835 Penobscot Building, Detroit, MI 48226-4094. (313) 961- 2242. (800) 347-4253. 20th edition. 1995. $485.00. Issn:: 0080-1518.

ENCYCLOPEDIAS AND DICTIONARIES

Academic Press Dictionary of Science and Technology. Christopher Morris, editor. Academic Press, Inc., 1250 Sixth Avenue, San Diego, CA 92101. (619) 231-0926. FAX (619) 699-6715. 1991. $115.00.

Concise Encyclopedia Chemistry. Translated and revised by Mary Eagleson. Walter de Gruyter, Inc., 200 Saw Mill River Road, Hawthorne, New York, 10532. (914) 747-0110 or Fax (914) 747-1326. 1994. $69.95.

Dictionary of Chemical Names and Synomyms. Philip A. Howard and Michael W. Neil. Lewis Pub.,/ CRC Press, Inc., 2000 Corporate Boulevard, NW, Boca Raton, FL 33431. (407) 994- 0555. (800) 272-7737. 1991. $149.95.

Dictionary of Physical Chemistry. Satish Anand and Raj Kumar. South Asia Publications, PO Box 502 Collumbus, MO 65205. (314) 474-0116. 1990. $23.50.

Encyclopedic Dictionary of Science. Candida Hunt and Monica Byles, editors. Facts-on-File, Inc., 460 Park Avenue South, New York, NY 10016-7382. (800) 322-8755. Fax (800) 683-3633. 1988. $32.95.

Encyclopedia of Applied Physics. VCH Publishers, Inc., 303 Northwest 12th Avenue, Deerfield Beach, FL 33442. (800) 367-8249. 1991-. Twenty volumes. $6000.00.

Encyclopedia of Chemical Processing and Design. McKetta. Dekker, Inc., 270 Madison Avenue, New York, NY 10016. (212) 696-9000. (800) 228-1160. 1976 - . $175.00 per volume. ISSN:

Grant & Hawks Chemical Dictionary. R. A. Grant. McGraw-Hill Publishing Company, Inc., 1221 Avenue of the Americas,

NewYork, NY 10020. (212) 512-2000. (800) 262-4729. 5th edition. 1987. $64.50.

Hawley's Condensed Chemical Dictionary. Richard J. Lewis. Van Nostrand Reinhold, 115 Fifth Avenue, New York, NY 10003. (212) 254-3232. (800) 842-3636. 12th edition. 1993. $69.95.

Industrial Chemical thesaurus. Michael Ash and Irene Ash. VCH Publishers, Inc., 220 East 23rd Street, New York, NY 10010-4606. (800) 367-8249. 1992. $295.00.

Kirk-othmer Encyclopedia of Chemical Technology. John Wiley and Sons, Inc., 605 Third Avenue, New York, NY 10158. (800) 526-5368 or (212) 850-6000. Fourth edition. 1991-. Twenty-seven volumes. $5400.00.

McGraw-Hill Encyclopedia of Chemistry. Sybil P.Parker, editor. McGraw-Hill Book, Incorporated, 1221 Avenue of the Americas, New York, NY 10020. (212) 997-3675. Second edition. 1993. $95.50.

McGraw-Hill Encyclopedia of Science and Technology. McGraw-Hill Book, Incorporated, 1221 Avenue of the Americas, New York, NY 10020. (212) 997-3675. Seventeenth edition. Twenty volumes. 1992. $1900.00.

GENERAL WORKS

Basic Physical Chemistry for the Physical Sciences. Peter V. Hobbs. Cambridge University Press, 40 West 20th Street, New York, NY 10011-4211. (212) 924-3900. (800) 872-7423. 1995. $49.95.

The Chemical Bond: Structure and Dynamics. Ahmed Zewail. Academic Press, Inc., 1250 Sixth Avenue, San Diego, CA 92101-4311. (619) 231-0926. FAX (619) 699-6715. 1992. $49.95

Chemistry: the Study of Matter and Its Changes. James E. Brady and John R. Holum. John Wiley & Sons, Inc., 605 Third Avenue, New York, NY 10158-0012. (212) 850-6000. (800) 225-5945. 1993. $ wfi

Concepts and Models of Inorganic Chemistry. B. E. Douglas, et al. John Wiley & Sons, Inc., 605 Third Avenue, New York, NY 10158-0012. (212) 850-6000. (800) 225-5945. 3rd edition. 1994. $59.95.

The Development of Chemical Principles. Cooper H. Langford and Ralph A. Beebe. Dover Publications, Inc., 180 Varick Street, New York, NY 10014. (212) 255-3755. (800) 223- 3130. 1995. $9.95.

The Elements. John Emsley. Oxford University Press, Inc., 200 Madison Avenue, New York, NY 10016. (800) 334-4249 or (212) 679-7300. Second edition. 1991. $49,95.

The Elements of Physical Chemistry. P. W. Atkins. W. H. Freeman & Co., 41 Madison Avenue, East 26th Avenue, 35th Floor, New York, NY 10010. (212) 576-0400. 1995. $42.95.

Exploring Chemical Elements and their Compounds. David L.Heiserman. TAB Books, P.O. Box 40, Blue Summit, PA 17294-0850. (717)794-2191. (800) 822-8138. 1992. $17.95.

CHEMISTRY

Ency. of Physical Sciences and Engineering Info. Sources

From Caveman To Chemist. Hugh W. Salzberg. American Chemical Society, 1155 16th Street, N.W., Washington, DC 20036.(202) 872-4381. (800) 424-6747. 1991. $24.95.

The History of Chemistry. John Hudson. Routledge, Chapman and Hall, Inc., 29 West 35th Street, New York, NY 10001-2291. (212) 244-3336. 1992. $59.95.

Household Chemicals and Emergency First Aid. Charles R. Foden and Jack L. Weddell. Lewis Pub., CRC Press, Inc., 2000 Corporate Boulevard, NW, Boca Raton, FL 33431. (407) 994-0555. (800) 272-7737. 1992. $84.95.

Organic Reactions: Simplicity and Logic. Pierre Laszlo. John Wiley & Sons, Inc., 605 Third Avenue, New York, NY 10158-0012. (212) 850-6000. (800) 225-5945. 1994. $74.95.

Periodic Table of the Elements. D.E.D. Electronic Publishing, 10306 East Live Oak Avenue, Arcadia, CA 91007. Computer software. 1992. $19.95.

Physical Methods For Chemists. Russell S. Drago. SCP: Third World Literature Publishing House, P.O. Box 482, Lithonia, GA 30058-0482. (404) 785-7725. 1992. $63.00.

HANDBOOKS AND MANUALS

Chemical Formulary. H. Bennett, editor. Chemical Publishing Company, Inc., 80 Eighth Avenue, New York, NY 10011. (212) 255-1950. Volumes 1-30. $70.00 per volume.

Chemometrics: Chemical and Sensory Data. David R. Burgard and James T. Kuznicki. CRC Publishers, Inc., 2000 Corporate Blvd., N.W., Boca Raton, FL 33431. (407) 994-0555 or (800) 333-8300. 1990. $120.00.

Chemical Reference Handbook. Gordon Press, PO Box 459, Bowling Green Station, New York, NY 10004. (718) 624-8419. 1995. $260.00.

CRC Handbook of Chemistry and Physics. David R. Kide, editor. CRC Press Inc., 2000 Corporate Blvd., NW, Boca Raton, FL 33431. (407) 994-0555. (800) 333-8300. 77th edition. 1996. $99.95.

Guide To Basic Chemical Compounds. D.R. Lide, Jr. CRC Publishers, Inc., 2000 Corporate Blvd., N.W., Boca Raton, FL 33431. (407) 994-0555. (800) 333-8300. 1993. $120.00.

Improving Safety In the Chemical Laboratory: A Practical Guide. Jay A. Young, editor. John Wiley and Sons, Inc., 605 Third Avenue, New York, NY 10158. (800) 526-5368 or (212) 850-6000. Second edition. 1991. $75.00.

Laboratory Handbook of Materials, Equipment, and Techniques. Gary S. Coyne. Prentice Hall (Division of Simon and Schuster), 15 Columbus Circle, New York, NY 10023. (212) 373-8500 or (800) 922-0579. 1992. $45.00.

Lange's Handbook of Chemistry. John A. Dean, editor. McGraw-Hill Publishing Company, 1221 Avenue of the Americas, New York, NY 10020. (800) 262-4729 or (212) 512-3825. Fourteenth edition. 1992. $79.50.

Riegel's Handbook of Industrial Chemistry. James A. Kent, editor. Van Nostrand Reinhold, 115 Fifth Avenue, New York, NY

10003. (212) 254-3232 or (800) 926-2665. Ninth edition. 1992. $114.95.

ONLINE DATABASES AND CD-ROMS

Analytical Abstracts Online. Royal Society of Chemistry, Information Services, Thomas Graham House, Science Park, Milton Road, Cambridge, CB4 4WF, England. Contains citations, mostly with abstracts, of the worldwide literature on analytical chemistry, from 1980 to present. Available on DIALOG, (800) 334-2564, and STN International, FIZ Karlsruhe, P.O. Box 2465, W-7500, Karlsruhe 1, Germany, online services. Inquire as to cost and availability.

CA Search. Chemical Abstracts Service, P.O. Box 3012, Columbus, OH 43210-0012. (614) 447-3600. (800) 848-6533. FAX (614) 447-3709. Very comprehensive guide to worldwide chemical literature and related fields, 1972 to present. Available on BRS,(800) 289-4277, DIALOG, (800) 334-2564, ORBIT (800) 456-7248, and STN International, FIZ Karlsruhe, P.O. Box 2465, W-7500, Karlsruhe 1, Germany, online services. Inquire as to cost and availability.

Chemical Journals of the American Chemical Society. American Chemical Society, 1155 16th Street, N.W., Washington, DC 20036. (202) 872-4381 or (800) 424-6747. Contains complete text of approximately 90,000 articles from 22 primary journals published by the American Chemical Society, from mostly 1982 to present. Available on STN International, FIZ Karlsruhe, P.O. Box 2465, W-7500, Karlsruhe 1, Germany, online service. Inquire as to cost and availability.

Compendex Plus. Engineering Information, Inc., 345 East 47th Street, New York, NY 10017. (212) 705-7600 or (800) 221-1044. Contains citations with abstracts to worldwide literature in engineering and technology, from 1970 to present. Available on online BRS,(800) 289-4277, DIALOG, (800) 334-2564, ORBIT (800) 456-7248, and STN International, FIZ Karlsruhe, P.O. Box 2465, W-7500, Karlsruhe 1, Germany, online services. Also available on CD-ROM. Inquire as to cost and availability.

Current Contents Search. Institute for Scientific Information, 3501 Market Street, Philadelphia, PA 19104. (215) 386-0100. FAX (215) 386-6362. Contains citations to articles listed in the table of contents of science and technology journals. Also articles in social sciences and life sciences journals. Available on BRS,(800) 289-4277, DIALOG,(800) 334-2564, online services. Inquire as to cost and availability.

Dissertation Abstracts. University Microfilms International, 300 North Zeeb Road, Ann Arbor, MI 48106. (800) 521-0600 or (313) 761-4700. Scope includes virtually all doctoral dissertations accepted at accredited American institutions from 1861 to present in 252 subject areas. Available on BRS,(800) 289-4277, DIALOG,(800) 334-2564, and OCLC EPIC,(800) 848-5878, online services. Also available on CD-ROM. Inquire as to cost and availability.

Gmelin Database. Gmelin-Institut fur Anorganische Chemie und Grenzgebiete, Varrentrapperstrasse, 40-42, Carl-Bosch-Haus, D-6000, Frankfurt am Main 90, Germany. Contains structural and factual data relating to inorganic and organometallic chemistry.Provides data from the Gmelin Handbook of Inorganic and Organometallic Chemistry. Covers the period 1817 to 1975; 1988-89. Available on STN International, FIZ Karlsruhe, P.O. Box 2465, W-7500, Karlsruhe 1, Germany, online service. Inquire as to cost and availability.

Kirk-Othmer Encyclopedia of Chemical Technology. John Wiley and Sons, Inc., 605 Third Avenue, New York, NY 10158. (800) 526-5368 or (212) 850-6000. Contains the complete text of all chapters in the 27 volume fourth edition of the *Kirk-Othmer Encyclopedia of Chemical Technology.* 1991. Available on BRS,(800) 289-4277, DIALOG,(800) 334-2564, online services. Inquire as to cost and availability.

NTIS Bibliographic Database. National Technical Information Service, 5285 Port Royal Road, Springfield, Va 22161. (703) 487-4929 or FAX (703) 321-8199. Broad coverage of government-sponsored science and technology research reports, 1964 to present. Available on BRS,(800) 289-4277, DIALOG, (800) 334-2564, ORBIT, (800) 456-7248, and STN International, FIZ Karlsruhe, P.O. Box 2465, W-7500, Karlsruhe 1, Germany, online services. Also available on CD-ROM. Inquire as to cost and availability.

Scisearch. Institute for Scientific Information, 3501 Market Street, Philadelphia, PA 19104. (800) 523-1850 or (215) 386-0100. Broad multidisciplinary title and author index to the international literature of science and technology, 1974 to present. Available on DIALOG,(800) 334-2564, and ORBIT,(800) 456-7248, online services. Also available on CD-ROM. Inquire as to cost and availability.

WILSONLINE. H.W. Wilson Company, 950 University Avenue, Bronx, NY 10452. (800) 367-6770 or (212) 588-8400. Makes available online versions of the printed H.W. Wilson indexes including Applied Science and Technology Index, Business Periodicals Index, General Science Index, and Readers' Guide to Periodical Literature. Period covered is generally 1983 to present. Available on BRS,(800) 289-4277, DIALOG,(800) 334-2564, and OCLC EPIC,(800) 848-5878, online services. Also available on CD-ROM. Inquire as to cost and availability.

PERIODICALS

American Laboratory. American Laboratory, 30 Controls Drive, Box 870, Shelton, CT 06484-6111. (203) 926-9300. FAX (203) 926-9310. 1968 to present. Monthly. $210.00 per year. ISSN: 0044-7749.

Analytical Chemistry. American Chemical Society, Box 3337, Columbus, Oh 43210. (614) 447-3776. 1929 to Present. Semi-monthly. $415.00 per year. ISSN: 0003-2700.

Analytical Methods and Instrumentation. John Wiley and Sons, Inc., 605 Third Avenue, New York, NY 10158. (800) 526-5368 or (212) 850-6000. Bimonthly. $250.00 per year. *Issn: 1063-5246.*

Angewandte Chemie. VCH Publishers, Inc., 220 East 23rd Street, New York, Ny 10010-4606. (212) 683-8333. (800) 367-8249.1888 to present. 22 issues per year. $840.00 Per Year. ISSN: 0044-8249.

Biochemistry. American Chemical Society, Box 3337, Columbus, OH 43210. (614) 447-3776. Bi-weekly. $

Chemical Physics. Elsevier Science Publishing Co., Inc., Box 882, Madison Square Station, New York, NY 10159. (212) 989-5800. FAX (212) 633-3990. 1973 to present. 33 issues per year in 11 volumes. $2759.00 per year. ISSN: 0301-0104.

Chemical Physics Letters. Elsevier Science Inc., Publishing Co., Inc., Box 882, Madison Square Station, New York, NY 10159.

(212) 989-5800. (212) 633-3990. 1967 to present. 102 issues per year in 17 volumes. $5063.00 per year. ISSN: 0009-2614.

Chemical Reviews. American Chemical Society, Box 3337, Columbus, Oh 43210. (614) 447-3776. 1924 to present. 8 Issues Per Year. $346.00 per year. ISSN: 0009-2665.

Chemical Week; Includes Annual Buyers Guide. Chemical *(212) 621-4900.* FAX (212) 621 Week Associates, 888 Seventh Avenue, New York, Ny 10106. -4949. 1914 to present. *Weekly.* $99.00 per year. ISSN: 0009-272X.

Chemtech. American Chemical Society. Box 3337, Columbus, *Oh 43210.* (614) 447-3776. 1970 to present. Monthly. *$370.00* per year. ISSN: 0009-2703.

Inorganic Chemistry. American Chemical Society, Box 3337, Columbus, Oh 43210. (614) 447-3776. 1962 to Present. Semi-monthly. $500.00 per year. ISSN: 0020-1669. *International Journal of Chemical Kinetics.* John Wiley and Sons, Inc., 605 Third Avenue, New York, NY 10158. (800) 526-5368 or (212) 850-6000. 1968 to present. Monthly. $730.00 per year. ISSN: 0538- 8066.

Journal of the American Chemical Society. American Chemical Society, Box 3337, Columbus, OH 43210. (614) 447-3776. 1879 to present. Biweekly. $1055.00 per year. ISSN: 0002-7863.

Journal of Organic Chemistry. American Chemical Society, *Box 3337, Columbus, Oh 43210.* (614) 447-3776. (800) 333- *9511.* FAX (614) 447-3671. 1936 to present. Bi-weekly. *$785.00* per year. ISSN: 0022-3263.

Journal of Physical Chemistry. American Chemical Society, *Box 3337, Columbus, Oh 43210.* (614) 447-3776. (800) 333- *9511.* FAX (614) 447-3671. 1896 to present. Weekly. *$1140.00* per year to non-members. ISSN: 0894-3230.

RESEARCH CENTERS AND INSTITUTES

Barnett Institute of Chemical Analysis and Materials Science. Northeastern University, 360 Huntington Avenue, Boston, MA 02115. (617) 437-2864.

Chemical Laboratories. Harvard University, Oxford Street, Cambridge, Ma 02138. (617) 495-4283. FAX (617) 496-5618.

Chemistry Laboratories. Rensselaer Polytechnic Institute, Cogswell Laboratory, Troy, NY 12180-3590. (518) 276-8981.

Laboratory For Research On the Structure of Matter. University of Pennsylvania, 3231 Walnut Street, Philadelphia, Pa 19104. (215) 898-8571.

Research Program In Chemistry and Biochemistry. Southern Illinois University At Carbondale, Carbondale, Il 62901. (618) 453-5721.

Theoretical Chemistry Institute. University of Wisconsin, Madison. 1101 University Avenue, Madison, WI 53706. (608) 262-1511.

CHLORINE

Ency. of Physical Sciences and Engineering Info. Sources

CHIPS

See: INTEGRATED CIRCUITS

CHLORINE

See also: BROMINE, ELEMENTS, FLUORINE, HALIDES, IODINE

ABSTRACT SERVICES AND INDEXES

Applied Science and Technology Index; a cumulative subject index to English language periodicals in the fields of aeronautics and space science, computer tec hnology, chemistry, construction industry, energy and related areas. H.W. Wilson Co., 950 University Avenue, Bronx, NY 10452. (212) 588-8400. (800) 367-6770. FAX (718) 590-1617. From 1958 to present. Monthly. Inquire about cost and availability. Also available on CD-ROM and online. ISSN: 0003-6986.

Chemical Abstracts. Chemical Abstracts Service. 2540 Olentangy River Road, Box 3012, Columbus, Oh 43210-0012. (614) 447-3600. FAX (614) 447-3713. 1907 to present. Weekly. $16,800.00 per year. Available online and on CD- Rom. CA is also available in five section groupings. Inquire Regarding Cost and Availability.

Current Contents: Physical, Chemical and Earth Sciences. Institute for Scientific Information, 3501 Market Street, Philadelphia, PA 19104. (215) 386-0100. FAX (215) 386-6362. Weekly. $360.00 per year.

General Science Index. H.W. Wilson Company, 950 University Avenue, Bronx, NY 10452. (212) 588-8400. (800) 367-6770. FAX (718) 590-1617. From 1978 to present. Ten issues per year; quarterly and annual cumulations. Service basis. Available on CD-ROM and online. Inquire regarding cost and availability. ISSN: 0162-1963.

Physics Abstracts. INSPEC. Section A, Science Abstracts. Institution of Electrical Engineers (IEE), London. Available from: INSPEC/IEEE - Institute of Electrical and Electronic Engineers, Box 1331, Hoes Lane, Piscataway, NJ 08855-1331. (908) 562-5549. 1898 to present. 24 issues per year. $2835.00 per year. Also available online and on CD-ROM. ISSN: 0036-8091.

Science Citation Index. SCI. Institute for Scientific Information, 3501 Market Street, Philadelphia, PA 19104. (215) 386-0100. (800) 523-1850. FAX (215) 386-2991. 1961 to present. Six issues per year, plus annual cumulation. $11650.00 per year. Also available online and on CD-ROM. Inquire about price and availability. ISSN: 0036-827X.

ASSOCIATIONS AND PROFESSIONAL SOCIETIES

American Chemical Society. 1155 16th Street, NW, Washington, DC 20036. (202) 872-4600.

Association of Consulting Chemists and Chemical Engineers, 295 Madison Avenue, 27th Floor, New York, Ny 10017.(212) 983-3160. FAX (212) 983-3161.

Chemical Manufacturers Association. 2501 M Street, N.W., Washington, DC 20037. (202) 887-1100. FAX (202) 887-1237.

Chemical Specialties Manufacturers Association, 1913 Eye Street, NW, Washington, DC 20006, (202) 872-8110. FAX (202) 872-8114.

Chlorine Institute, 2001 L Street, Nw, No. 506, Washington, DC 20036. (202) 775-2790. FAX (202) 223-7725.

BIBLIOGRAPHIES

Chemical Information Sources. Gary Wiggins. McGraw-Hill Publishing Company, 1221 Avenue of the Americas, New York, NY 10020. (800) 262-4729 or (212) 512-3825. 1991. $42.50.

Chlorine: An Annotated Bibliography. Ralph G. Smith. Chlorine Institute, 2001 L. Street, No. 506, Washington, DC 20036. (202) 775-2790. FAX (202) 223-7725. 1971. $12.00.

Information Sources In Chemistry. R.T. Bottle and J.F.B Rowland, editors. R.R. Bowker Inc., 121 Chanlon Road, New Providence, NJ 07974. (800) 521-8110 or (908) 464-6800. Fourth edition. 1993. $75.00.

Scientific A Nd Technical Books and Serials In Print; An Index To Literature In Science and Technology. R.R. Bowker Inc., 121 Chanlon Road, New Providence, NJ 07974. (908) 464-6800. (800) 521-8110. FAX (908) 665-3502. 1972 to present. Annual. 4 volumes. 1994. $299.95. Also available on compact disc and online. ISSN: 0000-054X.

DIRECTORIES AND BIOGRAPHICAL SOURCES

American Men and Women of Science. Physical and Biological Sciences. R.R. Bowker Company, 121 Chanlon Road, New Providence, NJ 07974. (908) 464-6800. (800) 521-8110. 20th edition. 8 volumes. 1996. $850.00.

Chemical Week — Buyers Guide Issue. Chemical Week Associates, 888 Seventh Avenue, New York, NY 10106. (212) 621-4900. FAX (212) 621-4949. Annual, October. $50.00.

Consulting Services: Chemists and Chemical Engineers. Association of Consulting Chemists and Chemical Engineers, 50 East 41st Street, Suite 92, New York, NY 10017. (212)

684-6255. Biennial. $60.00.

Directory of Chemistry Software 1992. Wendy Warr, Peter Willett, Geoff Downs. American Chemical Society, 1155 16th Street, NW, Washington, DC 20036. (202) 872-4600. 1992. $35.95.

Research Centers Directory. Gale Research Company Inc., 835 Penobscot Building, Detroit, MI 48226-4094. (313) 961-2242. (800) 347-4253. 20th edition. 1995. $485.00. ISSN:: 0080-1518.

ENCYCLOPEDIAS AND DICTIONARIES

Academic Press Dictionary of Science and Technology. Christopher Morris, editor. Academic Press, Inc., 1250 Sixth Avenue, San Diego, CA 92101. (619) 231-0926. FAX (619) 699-6715. 1991. $115.00.

Dictionary of Named Processes In Chemical Technology. Alan E. Comyns. Oxford University Press, Inc., 200 Madison

Avenue,New York, NY 10016. (212) 725-6000. (800) 334-4249. 1994. $75.00.

Consulting Services: Chemists and Chemical Engineers. Association of Consulting Chemists and Chemical Engineers, 50 East 41st Street, Suite 92, New York, NY 10017. (212) 684-6255. Biennial. $60.00.

Encyclopedia of Chemical Processing and Design. McKetta Marcel Dekker, Inc., 270 Madison Avenue, New York, NY 10016. (212) 696-9000. (800) 228-1160. 1976 - . $175.00 per volume. ISSN: .

Kirk-Othmer Encyclopedia of Chemical Technology. John Wiley and Sons, Inc., 605 Third Avenue, New York, NY 10158. (800) 526-5368 or (212) 850-6000. Fourth edition. 1991 - . Twenty-seven volumes. $5400.00.

McGraw-Hill Encyclopedia of Science and Technology. McGraw-Hill Book, Incorporated, 1221 Avenue of the Americas, New York, NY 10020. (212) 997-3675. (800) 262-4729. Seventh edition. Twenty volumes. 1992. $1900.00.

Ullman's Encyclopedia of Industrial Chemistry. VCH Publications, Inc., 220 East 23rd Street, Suite 909, New York, NY 10010-4606. (212) 683-8333. (800) 422-8824. 5th edition. 1984 - . Price varies per volume.

GENERAL WORKS

Bioreclamation of Chlorinated Compounds. Lewis Pubs., CRC Press, Inc., 2000 Corporate Boulevard, NW, Boca Raton, FL 33431. (407) 994-0555. (800) 272-7737. 1994. $79.95.

The Chemical Bond: Structure and Dynamics. Ahmed Zewail. Academic Press, Inc., 6277 Sea Harbor Drive, Orlando, FL. (800) 321-5068. 1992. $49.95.

Chemistry of Halides, Pseudohalides and Azides, Part 1 and 2. Chemistry of Functional Groups series, supplement no. D. Saul Patai and Zvi Rappaport. Books on Demand, 300 North Zeeb Road, Ann Arbor, MI 48106-1346. (313) 761-4700. (800) 521-0600. 1983. $340.00 set.

Chemistry of the Elements. N. N. Greenwood and A. Earnshaw. Pergamon Press, Inc., Maxwell House, Fairview Park, Elmsford, NY 10523. (914) 592-7700. Fax (914) 592-3625. 1984. $143.00.

Chlorinated Dioxins and Furans: Analytical Methods. : Exposed Populations. 1995. Lewis., CRC Press, Inc., 2000 Corporate Boulevard, NW, Boca Raton, FL 33431. (407) 994- 0555. (800) 272-7737. wfi.

Chlorine: Its Manufacture, Properties and Uses. J. S. Sconse, editor. Krieger Publishing Company, P.O. Box 9542, Melbourne, FL 32902-9542. (407) 724-9542. 1972. $90.50.

The Elements. John Emsley. Oxford University Press, Inc., 200 Madison Avenue, New York, NY 10016. (212) 725-6000. (800) 334-4249. Second edition. 1991. $49,95.

Exploring Chemical Elements and their Compounds. David L.Heiserman. TAB Books, P.O. Box 40, Blue Summit, PA 17294- 0850. (717 794-2191. (800) 233-1128. 1992. $17.95.

Halogen Chemistry. V. Gutman, editor. Academic Press,Incorporated, 6277 Sea Harbor Drive, Orlando, FL 32821. (800) 321-5068. Three volumes. 1967. Inquire

Internatonal thermodynamic Tables of the Fluid State and Chlorine. S. Angus, et al, editors. Franklin Book Co., Inc., Elkins Park, PA 19117. (215) 635-5252. 1987. $77.00.

Properties of Chlorine In Si Units. Chlorine Institute, 2001 L. Street, No. 506, Washington, DC 20036. (202) 775- 2790. FAX (202) 223-7725. 1986. $27.00 in paper.

Sulfur Dioxide, Chlorine, Flourine and Chlorine Oxides. A. S. Young. Franklin Book Co., Inc., Elkins Park, PA 19117. (215) 635-5252. 1983. $204.00

Tocicology of Halogenated Hydrocarbons; Health and Ecological Effects. M. A. Khan and R. H. Stanton, editors. Pergamon Press, Inc., Maxwell House, Fairview Park, Elmsford, NY 10523. (914) 592-7700. Fax (914) 592-3625. 1981. $72.50.

HANDBOOKS AND MANUALS

Chemical Formulary. H. Bennett, editor. Chemical Publishing, Co., Inc. 80 Eighth Avenue, New York, NY 10011. (212 255-1950. Volumes 1 - 30. $60.00 per volume.

CRC Handbook of Chemistry and Physics. David R. Kide, editor. CRC Press Inc., 2000 Corporate Blvd., NW, Boca Raton, FL 33431. (407) 994-0555. (800) 333-8300. 77th edition. 1996. $99.95.

Guide To Basic Chemical Compounds. D.R. Lide, Jr. CRC Publishers, Inc., 2000 Corporate Blvd., N.W., Boca Raton, FL 33431. (407) 994-0555 or (800) 333-8300. 1993. $120.00.

Lange's Handbook of Chemistry. John A. Dean, editor. McGraw-Hill Publishing Company, 1221 Avenue of the Americas, New York, NY 10020. (800) 262-4729 or (212) 512-3825. 14th edition. 1996. $99.95.

Riegel's Handbook of Industrial Chemistry. James A. Kent, editor. Van Nostrand Reinhold, 115 Fifth Avenue, New York, NY 10003. (212) 254-3232. (800) 926-2665. Ninth edition. 1992. $114.95.

ONLINE DATABASES AND CD-ROMS

CA Search. Chemical Abstracts Service, P.O. Box 3012, Columbus, OH 43210-0012. (614) 447-3600. (800) 848-6533. FAX (614) 447-3709. Very comprehensive guide to worldwide chemical literature and related fields, 1972 to present. Available on BRS,(800) 289-4277, DIALOG, (800) 334-2564, ORBIT (800) 456-7248, and STN International, FIZ Karlsruhe, P.O. Box 2465, W-7500, Karlsruhe 1, Germany, online services. Inquire as to cost and availability.

Chemical Journals of the American Chemical Society. American Chemical Society, 1155 16th Street, N.W., Washington, DC 20036. (202) 872-4381 or (800) 424-6747. Contains complete text of approximately 90,000 articles from 22 primary journals published by the American Chemical Society, from mostly 1982 to present. Available on STN International, FIZ Karlsruhe, P.O. Box 2465, W-7500, Karlsruhe 1, Germany, online service. Inquire as to cost and availability.

CHLORINE

Ency. of Physical Sciences and Engineering Info. Sources

Current Contents Search. Institute for ScientificInformation, 3501 Market Street, Philadelphia, PA 19104. (215) 386-0100. FAX (215) 386-6362. Contains citations to articles listed in the table of contents of science and technology journals. Also articles in social sciences and life sciences journals. Available on BRS,(800) 289-4277, DIALOG,(800) 334-2564, online services. Inquire as to cost and availability.

Dissertation Abstracts. University Microfilms International, 300 North Zeeb Road, Ann Arbor, MI 48106. (800) 521-0600 or (313) 761-4700. Scope includes virtually all doctoral dissertations accepted at accredited American institutions from 1861 to present in 252 subject areas. Available on BRS,(800) 289-4277, DIALOG,(800) 334-2564, and OCLC EPIC,(800) 848-5878, online services. Also available on CD-ROM. Inquire as to cost and availability.

Gmelin Datab Ase. Gmelin-Institut fur Anorganische Chemie und Grenzgebiete, Varrentrapperstrasse, 40-42, Carl-Bosch-Haus, D-6000, Frankfurt am Main 90, Germany. Contains structural and factual data relating to inorganic and organometallic chemistry. Provides data from the Gmelin Handbook of Inorganic and Organometal lic Chemistry. Covers the period 1817 to 1975; 1988-89. Available on STN International, FIZ Karlsruhe, P.O. Box 2465, W-7500, Karlsruhe 1, Germany, online service. Inquire as to cost and availability.

NTIS. Na tional Technical Information Service, 5285 Port Royal Road, Springfield, Va 22161. (703) 487-4929 or FAX (703) 321-8199. Broad coverage of government-sponsored research reports, 1964 to present. Available on BRS, DIALOG, ORBIT and STN online services. Also available on CD-ROM. Inquire as to cost and availability.

Scisearch. Institute for Scientific Information, 3501 Market Street, Philadelphia, PA 19104. (800) 523-1850 or (215) 386-0100. Broad multidisciplinary title and author index to the international literature of science and technology, 1974 to present. Available on DIALOG and ORBIT online services. Also available on CD-ROM. Inquire as to cost and availability.

WILSONLINE . H.W. Wilson Company, 950 University Avenue, Bronx, NY 10452. (800) 367-6770 or (212) 588-8400. Make s available online versions of the printed H.W. Wilson indexes including Applied Science and Technology Index, Business Periodicals Index, and Readers' Guide to Periodical Literature. Period covered is generally 1983 to present. Also available on CD-ROM. Inquire as to cost and availability.

Who's Who In Technology. Gale Research Inc., 835 Penobscot Building, Detroit, MI 48226-4094. (313) 961-2242. (800) 347-4253. Contains biographical information of contemporary American scieNTISts and engineers. Available on DIALOG online service. Inquire as to cost and availability.

PERIODICALS

Analytical Chemistry. American Chemical Society, Box *3337, Columbus, Oh 43210.* (614) 447-3776. 1929 to *Present.* Semimonthly. $415.00 per year. ISSN: 0003-2700.

Angewandte Chemie. VCH Publishers, Inc., 220 East 23rd Street, New York, NY 10010-4606. (212) 683-8333. (800) 367-8249. 1888 to present. 22 issues per year. $840.00 per year. ISSN: 0044-8249.

Chemical Reviews. American Chemical Society, Box 3337, Columbus, Oh 43210. (614) 447-3776. 1924 to present. 8 issues per year. $346.00 per year. ISSN: 0009-2665.

Chemical Week; Includes Annual Buyers Guide. Chemical Week Associates, 888 Seventh Avenue, New York, Ny 10106. (212) 621-4900. FAX (212) 621-4949. 1914 to present. Weekly. $99.00 per year. ISSN: 0009-272X.

Chemtech. American Chemical Society. Box 3337, Columbus, OH 43210. (614) 447-3776. 1970 to present. Monthly. *$370.00* per year. ISSN: 0009-2703.

Inorganic Chemistry. American Chemical Society, Box 3337, Columbus, Oh 43210. (614) 447-3776. 1962 to Present. Semimonthly. $500.00 per year. ISSN: 0020-1669.

Journal of the American Chemical Society. American Chemical Society, Box 3337, Columbus, OH 43210. (614) 447-3776. 1879 to present. Biweekly. $1055.00 per year. ISSN: 0002- 7863. *Polyhedron.* Elsevier Science Publishing Company, Inc., 660 White Plains Road, Tarrytown, NY 10591-5153. (914-524-9200. FAX (914) 524-9200. 1982 to present. 24 issues per year. $2175.00 per year. ISSN: 0277-5387.

RESEARCH CENTERS AND INSTITUTES

Chemical Laboratories. Harvard Universiy. Oxford Street, Cambridge, Ma 02138. (617) 495-4283. FAX (617) 496-5618.

Chemistry Laboratories. Rensselaer Polytechnic Institute.Cogswell Laboratory, Troy, Ny 12180-3590. (518) 276-8981.

Research Program In Chemistry and Biochemistry. Southern Illinois University At Carbondale, Carbondale, Il 62901. (618) 453-5721.

Theoretical Chemistry Institute. University of Wisconsin, Madison, 1101 University Avenue, Madison, Wi 53706. (608)262-1511.

University/Industry Chemical Research Center. Mississippi State University, Department of Chemistry, P.O. Drawer CH, Mississippi State, MS 39762. (601) 325-3584.

University/Industry Chemical Research Center. Mississippi State University, Department of Chemistry, P.O. Drawer CH, Mississippi State, MS 39762. (601) 325-3584.

CHROMATOGRAPHY

See also: ANALYTICAL CHEMISTRY, CHEMICAL ENGINEERING, CHEMISTRY, ORGANIC CHEMISTRY, SPECTROSCOPY

ABSTRACT SERVICES AND INDEXES

Applied Science and Technology Index; a cumulative subject index to English language periodicals in the fields of ae ronautics and space science, computer technology, chemis try, construction industry, energy and related areas. H. W. Wilson Co., 950 University Avenue, Bronx, NY 10452-9978. (212) 588-8400. (800)367-6770. FAX (718) 590-1617. From 1958 to present. Monthly. Inquire about cost and availability. Also available online (BRS and WILSONLINE) and on CD-ROM. ISSN: 0003-6986.

Chemical Abstracts. Chemical Abstracts Service. 2540 Olentangy River Road, Box 3012, Columbus, Oh 43210-0012. (614) 447-3600. FAX (614) 447-3713. 1907 to present. Weekly. $16,800.00 per year. Available online and on CD- Rom. CA is also available in five section groupings. Inquire Regarding Cost and Availability.

Chemical Abstracts - Applied Chemistry and Chemical Engineering Sections. Chemical Abstracts Service, 2540 Olentangy River Road, Box 3012, Columbus, OH 43210. (800) 848-6538 or (614) 447-3600, FAX (614) 447-3713. Bi-weekly. $1410.00 per year.

Chemical Abstracts - Organic Chemistry Sections. Chemical Abstracts Service, 2540 Olentangy River Road, Box 3012, Columbus, OH 43210. (800) 848-6538 or (614) 447-3600, FAX (614) 447-3713. Bi-weekly. $1410.00 per year.

Chemical Abstracts - Physical, Inorganic and Analytical Chemistry. Chemical Abstracts Service, 2540 Olentangy River Road, Box 3012, Columbus, OH 43210. (800) 848-6538 or (614) 447-3600, FAX (614) 447-3713. Bi-weekly. $1410.00 per year.

Chromataography Abstracts. Elsevier Science, 660 White Plains Road, Tarrytown, NY 10591-5153. (914) 524-9200. FAX (914) 333-2444. 1958 to present. 10 issues per year. $1040.00 per year. ISSN: 0268-6287.

Current Contents : Physical, Chemical and Earth Sciences. Institute for Scientific Information, 3501 Market Street, Philadelphia, PA 19104. (215) 386-0100. (800) 523-1850. FAX (215) 386-2291. 1961 to present. Weekly. $442.00 per year. Also available online (BRS, DIALOG) and on CD-ROM. Inquire regarding cost and availability. ISSN: 0163-2574.

General Science In Dex. H.W. Wilson Company, 950 University Avenue, Bronx, NY 10452. (212) 588-8400. (800) 367-6770. FAX (718) 590-1617. From 1978 to present. Ten issues per year; quarterly and annual cumulations. Service basis. Available on CD-ROM and online. Inquire regarding cost and availability. ISSN: 0162-1963.

Process and Chemical Engineering/chemical Engineering Abstracts. Royal Society of Chemistry, Information Services, Thomas Graham house, science park, milton road, cambridge, CB4 4WF, England. Contains citations, mostly with abstracts, of the worldwide literature on chemical engineering; from 1982 To Present. Monthly. $610.00 per year. Also available online. ISSN: 0960-5045.

Science Citation Index. SCI. Institute for Scientific Information, 3501 Market Street, Philadelphia, Pa 19104. (215) 386-0100. (800) 523-1850. FAX (215) 386-2991. 1961 To Present. Six issues per year, plus annual Cumulation. $11650.00 per year. Also available online and On Cd-rom. Inquire about price and availability. ISSN: 0036-827x.

Theoretical Chemical Engineering. Royal Society of Chemistry, Information Services, Thomas Graham House, Science Park,Milton Road, Cambridge, CB4 4WF, England. Covers theoretical chemical engineering, including theory and laboratory

experimentation. 1964 to present. Monthly. $235.00. ISSN: 0960-5053.

ANNUAL REVIEWS AND YEARBOOKS

Advances in Chromatography. Marcel Dekker, Inc., 270 Madison Avenue, New York, NY 10016. (212) 696-9000. FAX (212) 685-4540. Irregular. Price varies.

Chromatographic Science Series. Marcel Dekker Inc., 270 Madison Avenue, New York, NY 10016. (212) 696-9000. FAX (212) 685-4540. Irregular. Price varies.

ASSOCIATIONS AND PROFESSIONAL SOCIETIES

American Chemical Society, 1155 16th Street, Nw, Washington, Dc 20036. (202) 872-4600.

Association of Consulting Chemists and Chemical Engineers, 40 East 45th Street, New York, Ny 10017. (212) 983-3160. FAX (212) 983-3161.

Association of official Analytical Chemists. 2200 Wilson Blvd, Suite 400, Arlington, VA 22001. (703) 522-3032.

BIBLIOGRAPHIES

Chemical Information Sources. Gary Wiggins. McGraw-Hill Publishing Company, 1221 Avenue of the Americas, New York, NY 10020. (800) 262-4729 or (212) 512-3825. 1991. $42.50.

Information Sources In Chemistry. R.T. Bottle and J.F.B Rowland, editors. R.R. Bowker Inc., 121 Chanlon Road, New Providence, NJ 07974. (800) 521-8110 or (908) 464-6800. Fourth edition. 1993. $75.00.

DIRECTORIES AND BIOGRAPHICAL SOURCES

American Men and Women of Science. Physical and Biological Sciences. R.R. Bowker Inc., 121 Chanlon Road, New Providence, NJ 07974. (908) 464-6800. (800) 521-8110. 20th edition. 8 volumes. 1996. $850.00

Consulting Services: Chemists and Chemical Engineers. Association of Consulting Chemists and Chemical Engineers, 40 East 45th Street, New York, NY 10017. (212) 983-3160. FAX (212) 983-3161. Annual.

Directory of Chemistry Software 1992. Wendy Warr, Peter Willett, Geoff Downs. American Chemical Society, 1155 16th Street, NW, Washington, DC 20036. (202) 872-4600. 1992. $35.95.

Research Centers Directory. Gale Research Company, Inc., 835 Penobscot Building, Detroit, MI 48226-4094. (313) 961-2242. (800) 347-4253. 20th edition. 1995. $485.00. ISSN:: 0080-1518.

ENCYCLOPEDIAS AND DICTIONARIES

Concise Chemical and Technical Dictionary. H. Bennett. *Chemical Publishing, Co.,* Inc. 80 Eighth Avenue, New York, NY 10011. (212 255-1950. 4th edition. 1986. $160.00.

CHROMATOGRAPHY

Ency. of Physical Sciences and Engineering Info. Sources

Concise Encyclopedia of Chemical Technology. Kirk-Othmer. *John Wiley & Sons, Inc.*, 605 Third Avenue, New York, NY 10158-0012. (212) 850-6000. (800) 225-5945. 3rd edition. 1985. $215.00.

Dictionary of Analytical Reagents. R. Lobinski, editor. Routledge, Chapman and Hall, Inc., 29 West 35th Street, New York, NY 10001-2291. (212) 244-3336. 1993. $1000.00.

Dictionary of Chemical Names and Synomyms. Philip H. Howard and Michael W. Neal. Lewis Publishers, 2000 Corporate Boulevard, NW, Boca Raton, FL 33431. (407) 994-0555. (800) 272-7737. 1992. $151.95.

Dictionary of Chromatography. Ronald C. Denney. John Wiley & Sons, Inc., 605 Third Avenue, New York, NY 10158-0012. (212) 850-6000. (800) 225-5945. 2nd edition. 1982.

Encyclopedic Dictionary of Chemical Technology. D. Noether and Herman Noether. VCH Publishers, Inc., 220 East 23rd Street, Suite 909, New York, NY 10010-4606. (212) 683-8333. (800) 422-8824. 1993. $59.50.

Hawley's Condensed Chemical Dictionary. Van Nostrand Reinhold, 115 Fifth Avenue, New York, NY 10003. (212) 254-3232. (800) 842-3636. 13th edition. 1996. $99.95.

Industrial Chemical thesaurus. Michael Ash and Irene Ash. VCH Publishers, Inc., 220 East 23rd Street, New York, NY 10010-4606. (800) 367-8249. 1992. $295.00.

Kirk-OthmerEncyclopedia of Chemical Technology. John Wiley and Sons, Inc., 605 Third Avenue, New York, NY 10158. (800) 526-5368 or (212) 850-6000. Fourth edition. 1991-. Twenty-seven volumes. $5400.00.

McGraw-hill Encyclopedia of Science and Technology. McGraw-Hill Book, Incorporated, 1221 Avenue of the Americas, New York, NY 10020. (212) 997-3675. Seventeenth editon. 1992. $1900.00.

GENERAL WORKS

Chromatography: Concepts and Contrasts. James M. MIller. John Wiley & Sons, Inc., 605 Third Avenue, New York, NY 10158-0012. (212) 850-6000. (800) 225-5945. 1988. $69.95.

Chromatography For Organic Chemistry. Michael Lederer. John Wiley & Sons, Inc., 605 Third Avenue, New York, NY 10158-0012. (212) 850-6000. (800) 225-5945. 1994. $49.95.

Fundamentals of Preparative and Nonlinear Chromatography. Georges, Guiochon, et al, editors. Academic Press, Inc., 6277 Sea Harbor Drive, Orlando, FL. (800) 321-5068. 1994. $95.00.

The Periodic Table For Chromatographers. Michael Lederer. John Wiley & Sons, Inc., 605 Third Avenue, New York, NY 10158-0012. (212) 850-6000. (800) 225-5945. 1992. $185.00

Principles and Practice of Modern Chromatography. K. Robards, et al. Academic Press, Inc., 6277 Sea Harbor Drive, Orlando, FL. (800) 321-5068. 1994. $55.00.

Process Chromatography: A Practical Guide. Gail K. Sofer and L. E. Nystrom, editors. Academic Press, Inc., 6277 Sea Harbor Drive, Orlando, FL. (800) 321-5068. 1989. $54.00. *Unified*

Separation Science. J. Giddings. John Wiley & Sons, Inc., 605 Third Avenue, New York, NY 10158-0012. (212) 850-6000. (800) 225-5945. 1991. $69.95.

HANDBOOKS AND MANUALS

Annual Book of Astm Standards. Analytical Methods -Spectroscopy, Chromatography. volume 14.01. ASTM-American Society For Testing and Materials, 1916 Race Street, Philadelphia, Pa 19103. (215) 299-5419. 1996.

CRC Handbook of Chromatography: Inorganics. M. Qureshi, editor. CRC Press, Inc., 2000 Corporate Boulevard, NW, Boca Raton, FL 33431. (407) 994-0555. (800) 272-7737. 1986.

Guide To Basic Chemical Compounds. D.R. Lide, Jr. CRC Press, Inc., 2000 Corporate Blvd., N.W., Boca Raton, FL 33431. (407) 994-0555 or (800) 333-8300. 1993. $120.00.

Handbook of Chromatography: General Data and Principles. Gunter Zweig and Joseph Sherma, editors. CRC Press, Inc., 2000 Corporate Boulevard, NW, Boca Raton, FL 33431. (407) 994-0555. (800) 272-7737. 1973. $151.95.

Handbook of Derivatives For Chromatography. Karl and John Halket, editors. John Wiley & Sons, Inc., 605 Third Avenue, New York, NY 10158-0012. (212) 850-6000. (800) 225-5945. 2nd edition. 1993. $99.95.

Handbook of Thin-layer Chromatography. Joseph Sherma and Fried, editors. Marcel Dekker, Inc., 270 Madison Avenue, New York, NY 10016. (212) 696-9000. (800) 228-1160. 1996. $225.00

Laboratory Handbook of Materials, Equipment, and Techniques. Gary S. Coyne. Prentice Hall (Division of Simon and Schuster), 15 Columbus Circle, New York, NY 10023. (212) 373-8500 or (800) 922-0579. 1992. $45.00.

Lange's Handbook of Chemistry. John A. Dean, editor. McGraw-Hill Publishing Company, 1221 Avenue of the Americas, New York, NY 10020. (800) 262-4729 or (212) 512-3825. Fourteenth editon. 1992. $79.50.

ONLINE DATABASES AND CD-ROMS

Analytical Abstracts Online. Royal Society of Chemistry, Information Services, Thomas Graham House, Science Park, Milton Road, Cambridge, CB4 4WF, England. Contains citations, mostly with abstracts, of the worldwide literature on analytical chemistry, from 1980 to present. Available on DIALOG and STN online services. Inquire as to cost and availability.

Ca Search. Chemical Abstracts Service, P.O. Box 3012, Columbus, Oh 43210-0012. (614) 447-3600. (800) 848-6533. FAX (614) 447-3709. Guide to worldwide chemical literature and related fields, 1972 to present. Available on BRS, DIALOG, ORBIT and STN online services. Inquire as to cost and availability.

Chemical Journals of the American Chemical Society. American Chemical Society, 1155 16th Street, N.W., Washington, DC 20036. (202) 872-4381 or (800) 424-6747. Contains complete text of approximately 90,000 articles from 22 pr imary journals published by the American Chemical Society, from mostly 1982 to present. Available on STN International, FIZ Karlsruhe, P.O.

Box 2465, W-7500, Karlsruhe 1, Germany, online service. Inquire as to cost and availability.

Current Contents Search. Institute for Scientific Information, 3501 Market Street, Philadelphia, PA 19104. (215)386-0100. FAX (215) 386-6362. Contains citations to articles listed in the table of contents of science and technology journals. Also articles in social sciences and life sciences journals. Available on BRS,(800) 289-4277, DIALOG,(800) 334-2564, online services. Inquire as to cost and availability.

Dissertation Abstracts. University Microfilms International, 300 North Zeeb Road, Ann Arbor, MI 48106. (800) 521-0600 or (313) 761-4700. Scope includes virtually all doctoral dissertations accepted at accredited American institutions from 1861 to present in 252 subject areas. Available on BRS, DIALOG, and OCLC EPIC online services. Also available on CD-ROM. Inquire as to cost and availability.

NTIS. National Technical Information Service, 5285 Port Royal Road, Springfield, Va 22161. (703) 487-4929 or FAX (703) 321-8199. Broad coverage of government-sponsored research reports, 1964 to present. Available on BRS, DIALOG, ORBIT and STN online services. Also available on CD-ROM. Inquire as to cost and availability.

Scisearch. Institute for Scientific Information, 3501 Market Street, Philadelphia, PA 19104. (800) 523-1850 or (215) 386-0100. Broad multidisciplinary title and author index to the international literature of science and technology, 1974 to present. Available on DIALOG and ORBIT online services. Also available on CD-ROM. Inquire as to cost and availability.

WILSONLINE. H.W. Wilson Company, 950 University Avenue, Bronx, NY 10452. (800) 367-6770 or (212) 588-8400. Makes available online versions of the printed H.W. Wilson indexes including Applied Science and Technology Index, Business Periodicals Index, and Readers' Guide to Periodical Literature. Period covered is generally 1983 to present. Also available on CD-ROM. Inquire as to cost and availability.

PERIODICALS

American Laboratory. American Laboratory, 30 Controls Drive, Box 870, Shelton, CT 06484-6111. (203) 926-9300. FAX (203) 926-9310. Monthly. $150.00 per year. ISSN: 0044-7749.

Analytical Chemistry. American Chemical Society, Box 3337, Columbus, OH 43210. (614) 447-3776. (800) 333-9511. FAX (614) 447-3671. 1929 to present. Monthly. $415.00 per year. ISSN: 0003-2700.

Analytical Methods and Instrumentation. John Wiley and Sons, Inc., 605 Third Avenue, New York, NY 10158. (800) 526-5368 or (212) 850-6000. 1992 to present. Bimonthly. $250.00 per year. ISSN: 1063-5246.

A O A C International Journal. Association of offical Analytical Chemists. 2200 Wilson Blvd, Suite 400, Arlington, VA 22001. (703) 522-3032. 1917 to present. Bimonthly. $160.00 per year. ISSN: 0066-961X.

Journal of Chromatographic Science. Preston Publications, Inc., 7800 Merrimac Avenue, Box 48312, Niles, IL 60714. (708) 965-0566. FAX (708) 965-7639. 1964 to present. Monthly. $190.00 peryear. ISSN: 0021-9665.

Journal of Chromatography. Elsevier Science Inc., Box 882, Madison Square Station, NY 10159-0882. (212) 989-5899. FAX (212) 633-3990. 1958 to present. 62 issues per year in 31 volumes. $4994.00 per year. ISSN: 0021-9673.

Separation and Purification Methods. Marcel Dekker, Inc., 270 Madison Avenue, New York, NY 10016. (212) 696-9000. FAX (212) 685-4540. Subscriptions to: Box 5017 Monticello, NY 12701. 1972 to present. 2 issues per year. $275.00 per year. ISSN: 0369-2540.

Separation Science and Technology. Marcel Dekker, Inc., 270 Madison Avenue, New York, NY 10016. (212) 696-9000. FAX (212) 685-4540. Subscriptions to: Box 5017 Monticello, NY 12701. 1966 to present. 18 issues per year. $1450.00 per year. ISSN: 0149-6395.

RESEARCH CENTERS AND INSTITUTES

Analytical Chemistry Laboratory. U. S. Department of agriculture. Beltsville Agricultural Research Center - East, Building 306, Beltsville, MD 20705. (301) 344-2495.

Barnett Institute of Chemical Analysis and Materials Sci-ence. Northeastern Universiy, 360 Huntington Avenue, Boston, MA 02115. (617) 437-2864.

Center For Chemical Characterization and Analysis. Texas A & M University, College Station, TX 77843. (409) 845- 2341.

Center For Process Analytical Chemistry. University of Washington, BG-10, Seattle, WA 98195. (206) 545-2326.

Chromatography-mass Spectrometry Facility. University of Missouri- Columbia, Room 4, Agriculture Building, Columbia, MO 65211. (314) 882-2608.

Major Analytical Instruments Facility. Case Western Reserve University Millis Science Center, Cleveland, OH 44106. (216) 368-2589.

CHROMIUM

See also: ALLOYS, FERROALLOYS, MATERIALS SCIENCE, METALLURGY, STEEL AND STEEL MAKING

ABSTRACT SERVICES AND INDEXES

Alloys Index. American Society for Metals, Metals Park, OH 44073. (216) 338-5151. FAX (216) 338-4634. 1974 to present. Monthly. $380.00. Also available on CD-ROM and online via METADEX, STN International, and DIALOG.

Applied Mechanics Reviews: An Assessment of World Literature In Engineering Sciences. 1948-present. American Society of Mechanical Engineers, 345 East 47th Street, New York, NY 10017. (212) 705-7703. Monthly. $360.00 per year.

Applied Science and Technology Index; A Cumulative Subject Index To English Language Periodicals In the Fields of Aeronautics and Space Science, Computer Technology, Chemistry, Construction Industry, Energy and Related Areas. H.W. Wilson Co., 950 University Avenue, Bronx, NY 10452. (800) 367-6770 or (212) 588-8400. FAX (718) 590-1617. From 1958 to present.

CHROMIUM

Ency. of Physical Sciences and Engineering Info. Sources

Monthly. Inquire about cost and availability. Also available on CD-ROM and online.

Chemical Abstracts - Applied Chemistry and Chemical Engineering Sections. Chemical Abstracts Service, 2540 Olentangy River Road, Box 3012, Columbus, OH 43210. (800) 848-6538 or (614) 447-3600, FAX (614) 447-3713. Bi-weekly. $1410.00 per year.

Current Contents: Engineering, Technology, and Applied Sciences. Institute for Scientific Information, 3501 Market Street, Philadelphia, PA 19104. (215) 386-0100. FAX (215) 386-6362. 1970 to present. Weekly. $442.00 per year.

Engineering Index Monthly. Engineering Information, Inc., Castle Point on the Hudson, Hoboken, NJ 07030. (800) 221-1044. FAX (212) 832-1857. Monthly. $2300.00 per year. Also available online as COMPENDEX, and also on CD-ROM. Covers chemical engineering, computers, electrical engineering, civil engineering, metals and mining, industrial management, and mechanical engineering.

Metals Abstracts and Metals Abstracts Index. American Society for Metals, Metals Park, OH 44073. (216) 338-5151. 1968 to present. Monthly. Abstracts are $1500.00 per year and Index is $460.00 per year.

ANNUAL REVIEWS AND YEARBOOKS

Annual Review of Materials Science. Annual Reviews Inc., 4139 El Camino Way, PO Box 10139, Palo Alto, CA 94303-0139. (415) 493-4400 or (800) 523-8635. FAX (415) 855-9815. Annual. $75.00 (1994 edition).

ASSOCIATIONS AND PROFESSIONAL SOCIETIES

American Foundrymen's Society Inc. 505 State Street, Des Plaines, IL 60016. (708) 824-0181.

American Iron and Steel Institute. 1101 17th St., NW, Washington, DC 20036-4700. (202) 452-7100. FAX (202) 463-6573.

American Society For Metals (asm). Materials Park, OH 44073. (216) 338-5151 or FAX (216) 338-4634.

Metallurgical Society of the Aime (american Institute of Mining, Metallurgical, and Petroleum Engineers). 345 E.47th Street, 14th Floor, New York, NY 10017. (212) 705-7695.

BIBLIOGRAPHIES

Scientific and Technical Books and Serials In Print. R.R. Bowker Inc., 121 Chanlon Road, New Providence, NJ 07974. (800) 521-8110 or (908) 464-6800. List of currently published books and serials in the physical and biological sciences, engineering and technology, with subject, author and titles indexes. Inquire for price.

DIRECTORIES AND BIOGRAPHICAL SOURCES

Dun's Industrial Guide: the Metalworking Directory, 1993-94. Dun & Bradstreet Information Services Staff. 3 volumes. Dun & Bradstreet Information Services, 899 Eaton Avenue,

Bethlehem, Pa 18025. (610) 882-7000 or (800) 526-0651. FAX (610) 882-7269. 1993. $795.00

International Research Centers Directory. Gale Research, 835 Penobscot Building, Detroit, MI 48226-4094. (313) 961-2242. (800) 347-4253. FAX (313) 961-6083. 8th edition. 1995. $410.00

Metal Finishing Guidebook and Directory. Elsevier SciencePublishing Company, Inc., 655 Avenue of the Americas, New York, NY 10010. (212) 989-5800. Annual. Included with subscription to *Metal Finishing* periodical (see below).

Research Centers Directory. Gale Research, 835 Penobscot Building, Detroit, MI 48226-4094. (313) 961-2242. (800) 347-4253. FAX (313) 961-6083. $485.00.

Scientific and Technical Organizations and Agencies Directory. Gale Research, 835 Penobscot Building, Detroit, MI 48226-4094. (313) 961-2242. (800) 347-4253. FAX (313) 961-6083. 4th edition. 1996. $195.00.

Who's Who In Engineering. American Association of Engineering Societies, 1111 19th Street NW, Suite 608, Washington, DC 20036. (202) 296-2237 or (800) 658-8897. 8th edition. 1991. Inquire for price.

ENCYCLOPEDIAS AND DICTIONARIES

Encyclopedia of Materials Science and Engineering. M.B. Bever and Robert W. Cahn, editors. 8 volumes (1986) and 3 supplementary volumes (1988-1993). Pergamon/Elsevier Science Inc., 660 White Plains Road, Tarrytown, NY 10591-5153. (914) 524-9200. FAX (914) 333-2444. $3900.00 for 11-volume set.

Encyclopedia of Physical Science and Technology. Robert A. Meyers, ed. 18 volumes. Academic Press Inc., 6277 Sea Harbor Drive, Orlando, FL 32887. (800) 321-5068. 1992. $2100.00.

McGraw-hill Encyclopedia of Science and Technology. Sybil P. Parker, ed. 7th edition. 20 volumes. McGraw-Hill Publishing Company, 1221 Avenue of the Americas, New York, NY 10020. (800) 262-4729 or (212) 512-3825. 1992. $1900.00

GENERAL WORKS

Nickel and Chromium Plating. J.K. Dennis and T.E. Such. 3rd edition. Woodhead Publishing Company and ASM International/ ASM International, Materials Information, Materials Park, OH 44073-0002. (216) 338-5151 or (800) 336-5152. FAX (216) 338-4634. 1993. $128.00.

HANDBOOKS AND MANUALS

ASM Handbook Volume 1: Properties and Selection: Irons, Steels, and High-performance Alloys. ASM International Handbook Committee. ASM International, Materials Information, Materials Park, OH 44073-0002. (216) 338-5151 or (800) 336-5152. FAX (216) 338-4634. 1990. $147.00.

Asm Handbook Volume 3: Alloy Phase Diagrams. ASM International Handbook Committee. ASM International, Materials Information, Materials Park, OH 44073-0002. (216) 338-5151 or (800) 336-5152. FAX (216) 338-4634. 1992. $147.00.

Ency. of Physical Sciences and Engineering Info. Sources

CHROMIUM

Smithell's Metals Reference Book. E.A. Brandes & G.B. Brook. 7th edition. Butterworth-Heinemann, 313 Washington Street, Newton, MA 02158. (617) 928-2500 or (800) 366-2665. FAX (617) 928-2620. 1992. $250.00.

ONLINE DATABASES AND CD-ROMS

Compendex Plus. Engineering Information, Inc., 345 East 47th Street, New York, NY 10017. (212) 705-7600 or (800) 221-1044. Contains citations with abstracts to worldwide literature inengineering and technology, from 1970 to present. Available on online BRS,(800) 289-4277, DIALOG, (800) 334-2564, ORBIT (800) 456-7248, and STN International, FIZ Karlsruhe, P.O. Box 2465, W-7500, Karlsruhe 1, Germany, online services. Also available on CD-ROM. Inquire as to cost and availability.

CSA Engineering. Cambridge Scientific Abstracts, 7200 Wisconsin Avenue, Suite 601, Bethesda, MD 20814. (301) 961-6750 or (800) 843-7751. Contains citations and abstracts of international periodicals and research literature covering all fields of engineering and science and technology,including computer and information science, electronics, mechanical engineering, solid state materials, 1981 to present. Available on BRS,(800) 289-4277, online service.Inquire as to cost and availability.

Current Contents Search. Institute for Scientific Information, 3501 Market Street, Philadelphia, PA 19104. (215) 386-0100. FAX (215) 386-6362. Contains citations to articles listed in the table of contents of science and technology journals. Also articles in social sciences and life sciences journals. Available on BRS,(800) 289-4277, DIALOG,(800) 334-2564, online services. Inquire as to cost and availability.

Metadex. Jointly produced by ASM International and the Institute of Materials. Contains more than 925,000 records from the international literature on metals and alloys, concerning properties, processes, materials classes, applications, and metallurgical systems. Updated monthly. Available from ORBIT-QUESTEL (703) 442-0900.

NASA Database. American Institute of Aeronautics and Astronautics, 370 L'Enfant Promenade, S.W., Washington, DC 20024. (202) 646-7400. Citations and abstracts of aeronautics and astronautics literature, 1962 to present. Inquire as to cost and availability.

NTIS Bibliographic Database. National Technical Information Service, 5285 Port Royal Road, Springfield, Va 22161. (703) 487-4929 or FAX (703) 321-8199. Broad coverage of government-sponsored science and technology research reports, 1964 to present. Available on BRS,(800) 289-4277, DIALOG, (800) 334-2564, ORBIT, (800) 456-7248, and STN International, FIZ Karlsruhe, P.O. Box 2465, W-7500, Karlsruhe 1, Germany, online services. Also available on CD-ROM. Inquire as to cost and availability.

WILSONLINE. H.W. Wilson Company, 950 University Avenue, Bronx, NY 10452. (800) 367-6770 or (212) 588-8400. Makes available online versions of the printed H.W. Wilson indexes including Applied Science and Technology Index, Business Periodicals Index, General Science Index, and Readers' Guide to Periodical Literature. Period covered is generally 1983 to present. Available on BRS,(800) 289-4277, DIALOG,(800) 334-2564, and OCLC EPIC,(800) 848-5878, online services. Also available on CD-ROM. Inquire as to cost andavailability.

PERIODICALS

Alloy Digest. Alloy Digest Inc., 27 Canfield Street, Orange, NJ 07050. (201) 677-9161. 1952 to present. Monthly. $140.00 per year.

Inorganic Chemistry. American Chemical Society, 1155 16th Street NW< Washington, DC 20036. (800) 333-9511. FAX (914) 447-3671. 1962 to present. Semi-monthly. $949.00 per year.

J O M (journal of Metals). Minerals, Metals and Materials Society, 420 Commonwealth Drive, Warrendale, PA 15086. (412) 776-9080. 1949 to present. Monthly. $50.00 per year.

Journal of Phase Equilibria (formerly Bulletin of Alloy Phase Diagrams). ASM International, Materials Park, OH 44073-0002. (216) 338-5151. FAX (216) 338-4634. 6 times a year. $658.00 per year for non-members.

Metal Bulletin Monthly. Metal Bulletin Inc., 220 Fifth Avenue, 19th Floor, New York, NY 10001-7781. (800) 638-2525. FAX (212) 213-6273. 1972 to present. Monthly. Inquire for cost.

Metal Finishing. Elsevier Science Publishing Company, Inc., 655 Avenue of the Americas, New York, NY 10010. (212) 989-5800. 1903 to present. 12 times a year. $40.00 per year.

Metallurgical Transactions A: Physical Metallurgy and Materials Science. ASM International, Materials Park, OH 44073. (216) 338-5151 or FAX (216) 338-4634. 1970 to present. Monthly. $520.00 per year.

Metal Science and Heat Treatment. Plenum Publishing Corporation, 233 Spring Street, New York, NY 10013. (212) 620-8000. FAX (212) 463-0742. Monthly. $1215.00 per year.

Modern Metals. Delta Communications, Inc./ Reed Elsevier Inc., 455 N Cityfront Plaza Drive, 24th Floor, Chicago, IL 60611. (312) 222-2000. FAX (312) 222-2026. 1945 to present. Monthly. $50.00 per year.

Mineral Industry Surveys, Chromium. U.S. Bureau of Mines, Production and Distribution, Cochrans Mill Rd., Box 18070, Pittsburgh, PA 15236. (412) 892-4411. Monthly. Inquire for price and availability.

*U.*S. Department of the Interior, Bureau of Mine Annual Reports, Commodities. Superintendent of Documents, U.S. Government Printing office, Box 371954, Pittsburgh, PA 15250-7954. (202) 783-3238. FAX (202) 512-2233. Annual. Price varies.

RESEARCH CENTERS AND INSTITUTES

American Iron and Steel Institute. 1101 17th St., NW, Washington, DC 20036-4700. (202) 452-7100. FAX (202) 463-6573.

Carnegie Mellon University Center For Iron and Steel Making Research. MEMS Department, Pittsburgh, PA 15213. (412)268-2677.

Colorado School of Mines, Advanced Steel Processing and Products Research Center. Golden, CO 80401. (303) 273-3774.

Canada Centre For Mineral and Energy Technology, Materials Technology Laboratories. Energy, Mines & Resources Canada,

CHROMIUM

Ency. of Physical Sciences and Engineering Info. Sources

568 Booth Street, Ottawa ON, Canada K1A 0G1. (613) 995-8248. FAX (613) 992-8735.

Texas A&m University Mechanics and Materials Center. Civil Engineering Department, College Station, TX 77843. (409) 845-7512. FAX (409) 845-6156.

University of Connecticut Institute of Materials Science. U-136, 97 N. Eagleville Road, Storrs, CT 06268. (203) 486-4623. FAX (203) 486-4745.

University of Florida Department of Materials Science and Engineering. Gainesville, FL 32611. (904) 392-1454. FAX (904) 392-6359.

University of Minnesota Corrosion Research Center. 221 Church Street SE, Minneapolis, MN 55455. (612) 625-0717.

University of Wisconsin—madison, Cast Metals Laboratory. 1509 University Avenue, Madison, WI 53706. (608) 262-2562. FAX (608) 262-6707.

CHROMIUM STEEL

See: STEEL and STEEL MAKING

CINDERBLOCKS

See: BUILDING MATERIALS

CINEMATOGRAPHY

See: PHOTOGRAPHY

CIPHERS

See: COMPUTER SECURITY

CIRCUIT BREAKERS

See also: ELECTRICAL ENGINEERING, ELECTRICIY, ELECTRO-MAGNETISM, ELECTRONIC CIRCUITS AND COMPONENTS, ELECTRONICS, ELECTRONICS ENGINEERING, HIGH VOLTAGE, RELAYS

ABSTRACT SERVICES AND INDEXES

Applied Science and Technology Index; A Cumulative Subject Index To English Language Periodicals In the Fields of Aeronautics and Space Science, Computer Technology, Chemistry, Construction Industry, Energy and Related Areas. H. W. Wilson Co., 950 University Avenue, Bronx, NY 10452. (212) 588-8400. (800) 367-6770. FAX (718) 590-1617. From 1958 to present. Monthly. Inquire about cost and availability. Also available on CD-ROM and online. ISSN: 0003-6986.

Current Contents: Engineering, Technology and Applied Sciences. Institute for Scientific Information, 3501 Market Street, Philadelphia, PA 19104. (215) 386-0100. FAX (215) 386-6362. 1970 to present. Weekly. $442.00 per year. Also available on CD-ROM and online. Inquire regarding cost and availability. ISSN: 0095-7917.

Current Papers In Electrical and Electronics Engineering. Institution of Electrical Engineers (IEE), London. Distributed by INSPEC/IEEE, Box 1331, 445 Hoes Lane, Piscataway, NJ 08855-1331.(908) 562-5549. 1969 to present. Monthly. $345.00 per year. ISSN: 0011-3778.

Electric Power Industry Abstracts. Edison Electric Institute. c/o Utility Data Institute, 2011 I Atreet, NW, Suite 700, Washington, DC 20006. 1975 to present. Bimonthly. Inquire as to cost and availability.

Electrical and Electronics Abstracts. I nstitution of Electrical Engineers (IEE), London. Available from INSPEC/IEEE - Institute of Electrical and Electronic Engineers, Box 1331, Hoes Lane, Piscataway, NJ 08855-1331. (908) 562-5549. 1898 to present. Monthly. $2200.00 per year. Also available on CD-ROM and online as INSPEC.

Engineering Index Monthly; Indexes and Abstracts the World's Engineering and Technical Literature. Engineering Information, Inc., Castle Point on the Hudson, Hoboken, NJ 07030. (201) 216-8500. (800) 221-1044. FAX (201) 216-8532. Monthly. $2300.00 per year. Available online as COMPENDEX and also on CD-ROM. ISSN: 0742-1974.

General Science Index. H.W. Wilson Company, 950 University Avenue, Bronx, NY 10452. (212) 588-8400. (800) 367-6770. FAX (718) 590-1617. From 1978 to present. Ten issues per year; quarterly and annual cumulations. Service basis. Available on CD-ROM and online. Inquire regarding cost and availability. ISSN: 0162-1963.

Index To IEEE Publications. IEEE Service Center, 445 Hoes Lane, Piscataway, NJ 08855-1331. (908) 981-1393. (800) 678-IEEE. FAX (908) 981-9667. 1973 to present. Annual. ISSN: 0099-1368.

Physics Abstracts . INSPEC. Section A, Science Abstracts. Institution of Electrical Engineers (IEE), London. Available from: INSPEC/IEEE - Institute of Electrical and Electronic Engineers, Box 1331, Hoes Lane, Piscataway, NJ 08855-1331. (908) 562-5549. 1898 to present. 24 issues per year. $2835.00 per year. Also available online and on CD-ROM. ISSN: 0036-8091.

ASSOCIATIONS AND PROFESSIONAL SOCIETIES

American Electronics Association. 5201 Great America Way, Suite 520, P.O. 52990, Santa Clara, CA 95056. (408) 987- 4200. FAX (408) 970-8565.

American Institute of Physics. One Physics Ellipse, College Park, MD 20740-3843. (301) 209-3100.

Edison Electric Institute. 701 Pennsylvania Avenue NW, Washington, DC 20004-2696. (202) 508-5000, 5454. FAX (202) 508-5360.

Electric Power Research Institute. 3412 Hillview Avenue, Palo Alto, CA 94304. (415) 855-2000. FAX (415) 855-2954.

IEEE (institute of Electrical and Electronic Engineers). 345 East 47th Street. New York, NY 10017. (212) 705-7900. FAX (212) 705-4929.

National Electrical Manufacturers Association. 1300 North 17th Street, Suite 1847, Rosslyn VA 22209. (703) 841-3200. FAX (703) 841-3300.

Utility Data Institute, 1200 G Street Ny, Suite 250, Washington, Dc 20005-3802. (202) 942-8788. (800) 486-3660.

DIRECTORIES AND BIOGRAPHICAL SOURCES

American Men and Women of Science: Physical and Biological Sciences. R. R. Bowker Inc., 121 Chanlon Road, New Providence, NJ 07974. (908) 464-6800. (800) 521-8110. 20th edition. 8 volumes. 1996. $850.00.

Directory of Engineering Societies and Related Organizations. American Association of Engineering Societies, 1111 19th Street, Suite 608, Washington, DC 20036-3603. (202) 296-2237. Semi-annual. $150.00.

E E M - Electronic Engineer's Master. Hearst Business Communications, Inc., 645 Stewart Avenue, Garden City NY 11530. (516) 227-1300. ISSN: 0732-9016.

Engineering Research Centers: Incorporating Electronics Research Centers. Stockton Press, 345 Park Avenue, New York, NY 10010. (212) 689-9200. (800) 221-2123. 4th edition. 1995. $515.00.

IEEE Membership Directory. Institute of Electrical and Electronics Engineers, IEEE Service Center, 445 Hoes Lane, Piscataway, NY 08854. (908) 981-1393. (800) 678-IEEE. FAX (908) 981-9667. 2 volumes. Annual. $190.00. ISSN:

International Directory of Abbreviations and Acroynms of Electronics, Electrical Engineering, Computer Technology and Information Processing. Peter Wennrich. K. G. Saur, 121 Chanlon Road, New Providence, NJ 07974. (908) 464-6800. (800) 521-8110 2 volumes. 1992. $230.00.

Research Centers Directory. Gale Research Company Inc., 835 Penobscot Building, Detroit, MI 48226-4094. (313) 961-2242. (800) 347-4253. 20th edition. 1995. $485.00. ISSN:: 0080-1518.

Who's Who In Engineering. Gordon Davis, editor. American Association of Engineering Societies. 1111 19th Street, NY, Suite 608, Washington, DC 20036. (202) 296-2237. (800) 658-8897. 9th edition. 1995. $220.00.

ENCYCLOPEDIAS AND DICTIONARIES

Chambers Science and Technology Dictionary. Peter M.B. Walker, editor. Cambridge University Press, 40 West 20th Street, New York, NY 10011-4211. (212) 924-3900. 1988. $39.95.

IEEE Standard Dictionary of Electrical and Electronics Terms. Christopher J. Booth, editor. IEEE Service Center, 445 Hoes Lane, Piscataway, NJ 08855-1331. (908) 981-1393. (800) 678-IEEE. FAX (908) 981-9667. IEEE Standard 100- 1992. 5th edition. 1993. $150.00.

Illustrated Dictionary of Electronics. Stan Gibilsco. TAB Books, P.O. Box 40, Blue Summit, PA 17294-0850. (717) 794- 2191. (800) 233-1128. 7th edition. 1994. $34.95.

McGraw-hill Electronics Dictionary. J. Markus. McGraw-Hill Book Company, Inc., 1221 Avenue of the Americas, New York, NY 10020. (212) 997-3675. 5th edition. 1994. $49.95.

McGraw-hill Encyclopedia of Engineering. Sybil P. Parker, editor. McGraw-Hill Book Company, Inc. 1221 Avenue of the Americas,New York, NY 10020. (212) 997-3675. Second edition. 1993. $95.50. ISBN: 0-07-051392-9.

GENERAL WORKS

Electric and Electronic Engineering for ScieNTISts and Engineers. K. A. Krishnamurthy. Halsted Press, 605 Third Avenue, New York, 10158-0012. (212) 850-6400. 1994. $48.95.

Electric Circuits: Principles, Applications and Ocmputer Analysis. David A. Bell. Prentice Hall , 113 Sylvan Avenue, Route 9W, Englewood Cliffs, NJ 07632. (201) 592- 2000. (800) 922-0579. 1993. $72.00.

Electrical and Mathematical Analysis: Encyclopedic Textbook and Reference Handbook. Ernest Joerg. Kendall-Hunt Publishing Co., 4050 Westmark Drive, P.O. Box 1840 Dubuque, IA 52004-1840. (319) 589-1000. (800) 228-0810. 1993. $34.95 in paper.

High-frequency Switching Power Supplies: theory and Design. George C. Chryssis. McGraw-Hill Publishing Company, Inc., 1221 Avenue of the Americas, New York, NY 10020. (212) 512- 2000. (800) 262-4729. 2nd edition. 1989. $48.00.

High-voltage Engineering and Testing. H. M. Ryan, editor. Institution of Electrical Engineers; IEE/INSPEC, Box 1331, 445 Hoes Lane, Piscataway, NJ 08855-1331. (908) 981-0060. 1994. $95.00.

An Introduction To Applied Electromagetism. Christos Christopoulos. John Wiley & Sons, Inc., 605 Third Avenue, New York, NY 10158-0012. (212) 850-6000. (800) 225-5945. 1990. $59.95.

Protective Relaying theory and Applications. Walter Elmore, editor. Marcel Dekker, Inc., 270 Madison Avenue, New York, NY 10016. (212) 696-9000. (800) 228-1160. 1994. $99.75.

Schaum's Outline of Electrical Technology. Milton Kaufman and Peter Brooks. McGraw-Hill Publishing Company, Inc., 1221 Avenue of the Americas, New York, NY 10020. (212) 512- 2000. (800) 262-4729. 2nd edition. 1996. .

Switching Phenomena In High-voltage Circuit Breakers. Kunio Nakanishi, editor. Marcel Dekker, Inc., 270 Madison Avenue, New York, NY 10016. (212) 696-9000. (800) 228-1160. 1991. $125.00.

HANDBOOKS AND MANUALS

Active Electronic Component Handbook. Charles A. Harper and Harold C. Jones. McGraw-Hill Book Company, 1221 Avenue of the Americas, New York, NY 10020. (212) 512-2000. (800) 262-4729. 2nd edition. 1996. $79.50.

CIRCUIT BREAKERS

Ency. of Physical Sciences and Engineering Info. Sources

Electrical Engineer's Handbook. Donald Christiansen, editor. McGraw Book Company, 1221 Avenue of the Americas, New York, NY 10020. (212) 512-2000. (800) 262-4729. 4th edition. 1996. $110.00.

Handbook of Electrical and Electronic Technology. Curtis Johnson. Prentice Hall , 113 Sylvan Avenue, Route 9W, Englewood Cliffs, NJ 07632. (201) 592-2000. (800) 922- 0579. 1996. $88.00.

Industrial Electronics and Systems Handbook. Irwin, CRC Press, Inc., 2000 Corporate Boulevard, NW, Boca Raton, FL

33431. (407) 994-0555. (800) 272-7737. 1996. $129.95.

McGraw-hill's National Electrical Code Handbook. J. F.McPartland. McGraw-Hill Publishing Company, 1221 Avenue of the Americas, New York, NY 10020. (212) 512-2000. (800) 262-4729. 22nd edition. 1996. $65.00.

Standard Handbook For Electrical Engineers. Donald Fink. McGraw-Hill Publishing Company, 1221 Avenue of the Americas, New York, NY 10020. (212) 512-2000. (800) 262- 4729. 13th edition. 1996. $110.00. ONLINE DATA BASES and CD-ROM

Compendex plus. Engineering Information, Inc., 345 East 47th Street, New York, NY 10017. (212) 705-7600 or (800) 221-1044. Contains citations with abstracts to worldwide literature in engineering and technology, from 1970 to present. Available on online BRS,(800) 289-4277, DIALOG, (800) 334-2564, ORBIT (800) 456-7248, and STN International, FIZ Karlsruhe, P.O. Box 2465, W-7500, Karlsruhe 1, Germany, online services. Also available on CD-ROM. Inquire as to cost and availability.

Current Contents Search. Institute for Scientific Information, 3501 Market Street, Philadelphia, PA 19104. (215) 386-0100. FAX (215) 386-6362. Contains citations to articles listed in the table of contents of science and technology journals. Also articles in social sciences and life sciences journals. Available on BRS,(800) 289-4277, DIALOG,(800) 334-2564, online services. Inquire as to cost and availability.

Dissertation Abstracts Online. University Microfilms International, 300 North Zeeb Road, Ann Arbor, MI 48106. (800) 521-0600 or (313) 761-4700. Scope includes virtually all doctoral dissertations accepted at accredited American institutions from 1861 to present in 252 subject areas. Available on BRS, (800) 289-4277, DIALOG, (800) 334-2564, and OCLC EPIC,(800) 848-5878, online services. Also available on CD-ROM. Inquire as to cost and availability.

Inspec. Institution of Electrical Engineers, Michael Faraday House, Six Hills Way, Stevenage, Herts. SG1 2AY, England. Telephone: 0438 313311 or FAX 0438 742840. Contains citations to the worldwide literature of physics, electronics and electrical engineering, computer technology, and related fields. Available on BRS, (800) 289-4277, DIALOG, (800) 334-2564, ORBIT, (800) 456-7248, and STN International, FIZ Karlsruhe, P.O. Box 2465, W-7500, Karlsruhe 1, Germany, online services. Inquire as to cost and availability.

NTIS Bibliographic Database. National Technical Information Service, 5285 Port Royal Road, Springfield, Va 22161. (703) 487-4929 or FAX (703) 321-8199. Broad coverage of government-sponsored science and technology research reports, 1964 to present. Available on BRS,(800) 289-4277, DIALOG, (800) 334-2564, ORBIT, (800) 456-7248, and STN International, FIZ

Karlsruhe, P.O. Box 2465, W-7500, Karlsruhe 1, Germany, online services. Also available on CD-ROM. Inquire as to cost and availability.

WILSONLINE. H.W. Wilson Company, 950 University Avenue, Bronx, NY 10452. (800) 367-6770 or (212) 588-8400. Makes availableonline versions of the printed H.W. Wilson indexes including Applied Science and Technology Index, Business Periodicals Index, General Science Index, and Readers' Guide to Periodical Literature. Period covered is generally 1983 to present. Available on BRS,(800) 289-4277, DIALOG,(800) 334-2564, and OCLC EPIC,(800) 848-5878, online services. Also available on CD-ROM. Inquire as to cost and availability.

PERIODICALS

Circuits, Systems, and Signal Processing. Birkhauser, 675 Massachusetts Avenue, Cambridge, MA 02139-3309. FAX (201) 348-4505. 1982 to present. Quarterly. $184.00 per year. ISSN: 0278-081X.

Electric Light and Power. PennWell Publishing Co. 1421 South Sheridan Road, Box 1260, Tulsa OK 74101. (835-3161. FAX (918) 831-9497. 1922 to present. Monthly. $42.00 per year. ISSN:

Electric Machines and Power Systems. Taylor & Francis, 1900 Frost Road, Suite 101, Bristol, PA 19007-1598. (800) 821- 8312, FAX (215) 785-5515. 1976 to present. Monthly. $355.00 per year. ISSN: 0731-356X.

Electrical World. McGraw-Hill, Inc., Box 513 Hightstown, NJ 08520. (212) 512-3288. 1874 to present. Monthly. $55.00 per year. ISSN: 0013-4457.

Electronic Design. Penton Publishing, San Jose Gateway, Suite 354. 2025 Gateway Place, San Jose, CA 95110. (408) 441-0550. 1952 to present. Fortnightly. $95.00. ISSN: 0013-4872.

Electronics. Penton Publishing, San Jose Gateway, Suite 354. 2025 Gateway Place, San Jose, CA 95110. (408) 441-0550. 1930 to present. Semi-weekly. $98.00 per year. ISSN: 0883-4989.

IEEE Circuits and Devices Magazine. Institute of Electrical and Electronics Engineers, Inc., Box 1331, 445 Hoes Lane, Piscataway, NJ 08855-1331. (908) 981-0060. 1985 to present. Bi-monthly. $120.00 per year. ISSN: 8755-3996.

IEEE Power Engineering Review. Institute of Electrical and Electronics Engineers, Inc., Box 1331, 445 Hoes Lane, Piscataway, NJ 08855-1331. (908) 981-0060. 1981 to present. Monthly. $115.00 per year. ISSN: 0272-1724.

IEEE Spectrum. Institute of Electrical and Electronics Engineers, Inc., Box 1331, 445 Hoes Lane, Piscataway, NJ 08855-1331. (908) 981-0060. 1964 to present. Monthly. $157.00. ISSN: 0018-9235.

IEEE Transactions On Power Delivery. Institute of Electrical and Electronics Engineers, Inc., Box 1331, 445 Hoes Lane, Piscataway, NJ 08855-1331. (908) 981-0060. 1986 to present. Quarterly. $170.00 per year. ISSN: 0885-8977.

IEEE Transactions On Signal Processing. Institute of Electrical and Electronics Engineers, Inc., Box 1331, 445 Hoes Lane, Piscataway, NJ 08855-1331. (908) 981-0060. 1951 To Present. Monthly. $400.00. ISSN: 1053-587X.

Institute of Electrical and Electronics Engineers. Proceedings. Institute of Electrical and Electronics Engineers, Inc., Box 1331, 445 Hoes Lane, Piscataway, NJ 08855-1331. (908)981-0060. 1913 to present. Monthly. $275.00. ISSN: 0018-9219.

RESEARCH CENTERS AND INSTITUTES

Electric Power Institute. Texas A & M University, Department of Electrical Engineering, College Station, TX 77843. (409)845-1423.

Electroscience Laboratory. Ohio State University, 1320 Kinnear Road, Columbus, Ohio 43212. (614) 292-7981, FAX (614) 292-7297.

Electrical Engineering Research Laboratories. Purdue University. Electrical Engineering Building, West Lafayette, in 47907. (317) 494-3536. FAX (317) 494-6440.

Electronics Design Center. Case Western Reserve University, Bingham Building, Cleveland, OH 44106. (216) 368-2934. (216) 368-8738.

Laboratory For Electromagentic and Electronic Systems. Massachusetts Institute of Technology. 77 Massachussetts Avenue, Cambridge, MA 02139. (617) 253-4631.

Systems and Techniques Laboratory. Georgia Institute of Technology, Georgia Tech Research Institute, Atlanta, GA 30332. (404) 528-7010.

Weber Research Institute. Polytechnic University. Route 110, Farmingdale, NY 11735. (516) 755-4250, FAX (516) 755-4404.

CIVIL ENGINEERING

See also: CONSTRUCTION ENGINEERING, ENVIRONMENTAL ENGINEERING, GEOTECHNICAL ENGINEERING, HIGHWAY ENGINEERING, OCEAN ENGINEERING, STRUCTURAL ENGINEERING, SURVEYING, TRANSPORTATION

ABSTRACT SERVICES AND INDEXES

Abstract Journal In Earthquake Engineering. University of California, Berkeley, Earthquake Engineering Research Center, 1301 South 46th Street, Richmond, CA 94804-4698. (510) 231-9413 or FAX (510) 231-9471. 1972 to present. Annual. $80.00 per year.

Applied Mechanics Reviews: An Assessment of World Literature In Engineering Sciences. American Society of Mechanical Engineers, 345 East 47th Street, New York, NY 10017. (212) 705-7703. 1948 to present. Monthly. $360.00 per year.

Applied Science and Technology Index; A Cumulative Subject Index To English Language Periodicals In the Fields of Aeronautics and Space Science, Computer Technology, Chemistry, Construction Industry, Energy and Related Areas. H.W. Wilson Co., 950 University Avenue, Bronx, NY 10452. (800) 367-6770 or (212) 588-8400. From 1958 to present. Monthly. Inquire about cost and availability. Also available on CD-ROM.

Bibliography and Index of Geology. American Geological Institute, 4220 King Street, Alexandria, VA 22302. (703) 379-

2480. Fax (703) 379-7563. Monthly. $1295.00 per year. Also available online as GeoRef.

Bibliography of Economic Geology. Geosystems, Box 40, Didcot Oxon Ox11 9BX, England. Bimonthly. $150.00 per year. Also available online as GeoArchive.

Concrete Abstracts. American Concrete Institute, 22400 West Seven Mile Road, Detroit, MI 48219-1885. (313) 532-2600 or FAX (313) 538-0655. 1971 to present. Bimonthly. $140.00 per year.

Current Contents: Engineering, Technology and Applied Sciences. Institute for Scientific Information, 3501 Market Street, Philadelphia, PA 19104. (215) 386-0100. FAX (215) 386-6362. Weekly. $360.00 per year.

Engineered Materials Abstracts. ASM (American Society of Metals) International, Materials Information, Materials Park, OH 44073. (216) 338-5151 or FAX (216) 338-4634. Covers literature on technical developments in polymer, ceramic, and composite materials and engineering. 1986 to present. Monthly. $1175.00 per year. Also available on CD-ROM.

Engineering Index Monthly. Engineering Information, Inc., Castle Point on the Hudson, Hoboken, NJ 07030. (800) 221-1044. FAX (201) 216-8532. Monthly. $2200.00 per year. Also available online as COMPENDEX, and also on CD-ROM.

Environmental Abstracts. R.R. Bowker Inc., 121 Chanlon Road, New Providence, NJ 07974. (800) 521-8110 or (908) 464-6800. Contains citations, most with abstracts, to literature relating to environmental issues and problems, from 1971 to present. Inquire as to cost and availability.

Fluid Power Abstracts. STI Limited, 4 Kings Meadow, Ferry Hinksey Road, Oxford OX2 ODU, England. Distributed by Air Science Company, Box 143, Corning, NY 14830. (607) 962-5591. Covers oil hydraulic and pneumatic power transmission and control, engineering and design. 1964 to present. Bi-monthly. $260.00 per year.

General Science Index. H.W. Wilson Company, 950 University Avenue, Bronx, NY 10452. (800) 367-6770 or (212) 588-8400. Inquire as to cost and availability.

Geological Abstracts. Elsevier - Geo Abstracts, Regency House, 34 Duke Street, Norwich NR3 3AP, England. Monthly. $760.00 per year. Also available online as GEOBASE.

Geotechnical Abstracts. Deutsche Gesellschaft fur Erd- und Grundbau e.V., Hohenzollernstrasse 52, 4300 Essen 1, Germany. Also sponsored by International Society for Soil Mechanics and Foundation Engineering. Covers soil mechanics, foundation engineering, and engineering geology. 1970 to present. Monthly. $200.00 per year.

Government Reports Announcements and Index. National Technical Information Service (NTIS), 5285 Port Royal Road, Springfield, VA 22161. (703) 487-4650. 1968 to present. Annual. $525.00 per year.

Hydro-Abstracts (formerly *Water Resources Abstracts*). HydroScience Press, 2527 Jackson Street, N.E., Minneapolis, MN 55418. (612) 781-9081. Monthly. $150.00 per year.

CIVIL ENGINEERING

Ency. of Physical Sciences and Engineering Info. Sources

ISMEC Bulletin (Information Service In Mechanical Engineering). Cambridge Scientific Abstracts, 5161 River Road, Bethesda, MD 20816. (301) 951-1400. 1973 to present. Monthly. $750.00 per year.

offshore Engineering Abstracts. STI Limited, 4 Kings Meadow, Ferry Hinksey Road, Oxford, OX2 ODU, England. Distributed in the United States by Air Science Company, Box 143, Corning NY 14830. (607) 962-5591. Contains citations on planning, design, and construction of offshore platforms and pipelines, from 1985 to present. Monthly. $400.00 per year.

ANNUAL REVIEWS AND YEARBOOKS

Advances In Applied Mechanics. Academic Press, Inc., 1250 Sixth Avenue, San Diego, CA 92101-4311. (619) 231-0926. FAX (619) 699-6715. Irregular. Price varies, inquire.

Developments In Civil Engineering. Elsevier Science Publishing Company, Inc., 655 Avenue of the Americas, New York, NY 10010. (212) 989-5800. 1979 to present. Irregular. Inquire about price.

Developments In Geotechnical Engineering. Elsevier Science Publishing Company, Inc., 655 Avenue of the Americas, New York, NY 10010. (212) 989-5800. 1972 to present. Irregular. Inquire about price.

ASSOCIATIONS AND PROFESSIONAL SOCIETIES

American Association of Engineering Societies, 1111 19th Street, Suite 608, Washington, Dc 20036-3603. (202) 296-2237.

American Concrete Institute, 22400 West Seven Mile Road, Detroit, Mi 48219-1885. (313) 532-2600 or FAX (313) 538-0655.

American Congress On Surveying and Mapping, 5410 Grosvenor Lane, Bethesda, Md 20814-2199. (301) 493-0200 or FAX (301) 493-8245.

American Consulting Engineers Council, 1015 15th Street, N.W. Suite 802, Washington, DC 20005-2670. (202) 347-7474.

American Institute of Steel Construction, Inc., One East Wacker Drive, Suite 3100, Chicago, IL 60601-2001. (312) 670-2400.

American Iron and Steel Institute, 1101 17th Street, Washington, Dc 20036-4700. (202) 452-7265 or FAX (202) 463-6573.

American National Standards Institute. 1430 Broadway, New York, NY 10018-3363. (212) 354-4916.

American Public Works Association, 1313 East 60th Street, Chicago, Il 60637-2881. (312) 667-2200.

American Society For Engineering Education, 2070 Neil Avenue, Columbus, Oh 43210-1226. (614) 292-2893.

American Society For Nondestructive Testing, Inc. 1911 Arlington Lane, Columbus, OH 43228-4113. (614) 274-6003.

American Society For Testing and Materials. 1916 Race Street, Philadelphia, PA 19103-1180. (215) 299-5400.

American Society of Civil Engineers. 345 East 47th Street, New York, NY 10017-2398. (212) 705-7388 or FAX (212) 980-4681.

American Society of Mechanical Engineers, 345 East 47th Street, New York, Ny 10017-2395. (212) 705-7722.

American Society of Naval Engineers, 1452 Duke Street, Alexandria, Va 22314-3458. (703) 836-6727.

American Society of Test Engineers, 1050 CommonwealthAvenue, Suite 200, Boston, Ma 02215-1109. (617) 734-4473.

Association of Engineering Geologists. 323 Boston Post Road, Suite 2D, Sudbury, MA 01776. (508) 443-4639.

National Academy of Engineering, 2101 Constitution Avenue, N.W., Washington, DC 20418-0001. (202) 334-1657.

National Society of Black Engineers. 344 Commerce Street, Alexandria, VA 22314. (703) 549-2207.

National Society of Professional Engineers, 1420 King Street, Alexandria, Va 22314-2750. (703) 684-2835.

Society of American Military Engineers. 607 Prince Street, Alexandria, VA 22314-3117. (703) 549-3800.

BIBLIOGRAPHIES

Information Sources In Energy Technology. L.J. Anthony, editor. Butterworths Publishers, 80 Montvale Avenue, Stoneham, MA 02180. (617) 438-8464 or (800) 366-2665. 1988. $105.00.

Scientific and Technical Books and Serials In Print; An Index To Literature In Science and Technology. R.R. Bowker Co., 205 E. 42nd Street, New York, NY 10017. (800) 521-8110 or (212) 916-1600. 1992. $250.00.

DIRECTORIES AND BIOGRAPHICAL SOURCES

Association of Engineering Geologists. Membership directory. Association of Engineering Geologists, 323 Boston Post Road, Suite 2D, Sudbury, MA 01776. (508) 443-4639. Annual. $18.00.

Directory of Engineering College Research and Graduate Study, American Society For Engineering Education, Ohio State University, 2070 Neil Avenue, Columbus, Oh 43210-1226. Annual. $20.00.

Directory of Engineering Societies and Related Organizations. American Association of Engineering Societies, 1111 19th Street, Suite 608, Washington, DC 20036-3603. (202) 296-2237. Semi-annual. $150.00.

Directory of Engineers In Private Practice. National Society of Professional Engineers, 1420 King Street, Alexandria, VA 22314-2750. (703) 684-2835. Biennial. $30.00.

Engineering Research Centers. Gale Research Company, Detroit, MI 48226. (800) 521-0707. 1984. $400.00.

International Engineering Directory. American Consulting Engineers Council, 1015 15th Street, N.W. Suite 802, Washington, DC 20005-2670. (202) 347-7474. Annual. $10.00.

Research Centers Directory. Gale Research Company, Detroit, MI 48226. Seventeenth edition, 1992. (800) 521-0707. $400.00.

ENCYCLOPEDIAS AND DICTIONARIES

Academic Press Dictionary of Science and Technology. Christopher Morris, editor. Academic Press, Inc., 1250 Sixth Avenue, San Diego, CA 92101. (619) 231-0926. FAX (619) 699-6715. 1991. $115.00.

Chambers Science and Technology Dictionary. Peter M.B. Walker, editor. Cambridge University Press, 40 West 20th Street, New York, NY 10011-4211. (212) 924-3900. 1988. $39.95.

Construction Glossary: An Encyclopedic Reference and Manual. J.Stewart Stein. John Wiley and Sons, Inc., 605 Third Avenue, New York, NY 10158. (800) 526-5368 or (212) 850-6000. Second edition. 1993. $95.00.

El Thesaurus. Jessica L. Milstead, editor. Engineering Information, Inc., Castle Point on the Hudson, Hoboken, NJ 07030. (800) 221-1044. FAX (201) 216-8532. 1992. $130.00

Magill's Survey of Science. Applied Science. Frank N. Magill, editor. Salem Press, Inc., P.O. Box 1097, Englewood Cliffs, NJ 07632. (800) 221-1592 or (201) 871-3700. Six volumes. 1993. $475.00.

McGraw-hill Encyclopedia of Science and Technology. McGraw-Hill Book, Incorporated, 1221 Avenue of the Americas, New York, NY 10020. (212) 997-3675. Seventh edition. Twenty volumes. 1992. $1900.00.

GENERAL WORKS

The American Civil Engineer: 19852-1974: the History, Traditions, and Development of the Asce. William H. Wisely. American Society of Civil Engineers, 345 East 47th Street, New York, NY 10017-2398. (212) 705-7520 or (800) 548-2723. FAX (212) 705-7712. 1974. $20.00.

Civil Engineering For the Community. Dennis A. Randolph. American Society of Civil Engineers, 345 East 47th Street, New York, NY 10017-2398. (212) 705-7520 or (800) 548-2723. FAX (212) 705-7712. 1993. $20.00

Civil Engineering Materials. Shan Somayji. Prentice Hall, 113 Sylvan Avenue, Route 9W, Englewood Cliffs, NJ 07632. (201) 592-2000 or (800) 922-0579. 1994. $67.00.

Civil Engineering Procedure. Institution of Civil Engineers. 5th edition. Thomas Telford Services Ltd., Thomas Telford House, 1 Heron Quay, London E14 4JD, England. Telephone 071-987-6999. FAX 071-538-4101. 1996. Inquire for cost.

Design and Construct of Complex Civil Engineering Systems. A. J. DeRidder. Coronet Books, 311 Bainbridge Street, Philadelphia, PA 19147. (215) 925-5083. FAX (215) 925-1912. 1994. $97.50.

HANDBOOKS AND MANUALS

Civil Engineering Handbook. W.F. Chen, editor. CRC Press, 2000 Corporate Blvd., N.W., Boca Raton, FL 33431. (407) 994-0555 or (800) 272-7737. FAX (407) 994-0949. 1995. $99.95.

Civil Engineering Reference Manual. Michael R. Lindeburg. Professional Publications, 1250 Fifth Ave., Belmont, CA 94002. (415) 593-9119 or (800) 426-1178. 1992. $51.95.

Engineering Tables and Data. A.M. Howatson, P.G. Lund and J.D. Todd. Van Nostrand Reinhold, 115 Fifth Avenue, New York, NY 10003. (212) 254-3232 or (800) 926-2665. Second Edition. 1991. $39.95.

A Manual of Geology For Civil Engineers. John A. Pitts. World Scientific Publishing Company, Inc., 1060 Main Street, Unit B, River Edge, NJ 07661. (800) 227-7562 or (201) 487-9655. 1985.$38.00.

ONLINE DATABASES AND CD-ROMS

Architectural Periodicals Index. Royal Institute of British Architects, British Architectural Library, 66 Portland Place, London, W1N 4AD, England. Citations to worldwide literature on architecture, construction methods and standards, urban planning, and interior design, from 1978 to present. Available on DIALOG, (800) 334-2564, online service. Inquire as to cost and availability.

CA Search. Chemical Abstracts Service, P.O. Box 3012, Columbus, OH 43210-0012. (614) 447-3600. (800) 848-6533. FAX (614) 447-3709. Very comprehensive guide to worldwide chemical literature and related fields, 1972 to present. Available on BRS,(800) 289-4277, DIALOG, (800) 334-2564, ORBIT (800) 456-7248, and STN International, FIZ Karlsruhe, P.O. Box 2465, W-7500, Karlsruhe 1, Germany, online services. Inquire as to cost and availability.

Compendex Plus. Engineering Information, Inc., 345 East 47th Street, New York, NY 10017. (212) 705-7600 or (800) 221-1044. Contains citations with abstracts to worldwide literature in engineering and technology, from 1970 to present. Available on online BRS,(800) 289-4277, DIALOG, (800) 334-2564, ORBIT (800) 456-7248, and STN International, FIZ Karlsruhe, P.O. Box 2465, W-7500, Karlsruhe 1, Germany, online services. Also available on CD-ROM. Inquire as to cost and availability.

Current Contents Search. Institute for Scientific Information, 3501 Market Street, Philadelphia, PA 19104. (215) 386-0100. FAX (215) 386-6362. Contains citations to articles listed in the table of contents of science and technology journals. Also articles in social sciences and life sciences journals. Available on BRS,(800) 289-4277, DIALOG,(800) 334-2564, online services. Inquire as to cost and availability.

Dissertation Abstracts. University Microfilms International, 300 North Zeeb Road, Ann Arbor, MI 48106. (800) 521-0600 or (313) 761-4700. Scope includes virtually all doctoral dissertations accepted at accredited American institutions from 1861 to present in 252 subject areas. Available on BRS,(800) 289-4277, DIALOG,(800) 334-2564, and OCLC EPIC,(800) 848-5878, online services. Also available on CD-ROM. Inquire as to cost and availability.

Geomechanics Abstracts. University of London, Imperial College of Science and Technology, Rock Mechanics Information Service, Royal School of Mines, Prince Consort Road, London SW7 2BP, England. Telephone: 071-589 5111 or FAX 071-589 6806. Scope includes worldwide literature on engineering geology, hydrogeology, mining, rock mechanics, soil mechanics, and tunnelling, 1977 to present. Available on the ORBIT (800) 456-7248, online service. Inquire as to cost and availability.

GEOREF. American Geological Institute, 4220 King Street, Alexandria, VA 22302. (800) 336-4764 or (703) 379-2480. Geology and geosciences literature, 1785 to present for North America.Available on DIALOG,(800) 334-2564, ORBIT (800) 456-7248, online services. Also available on CD-ROM. Inquire as to cost and availability.

ICONDA (cib International Construction Database). Fraunhofer-Gesellschaft, Informationszentrum RAUM und BAU, Nobelstrasse, 12, D-7000, Stuttgart 80, Germany. Contains citations with abstracts to worldwide technical literature on construction and civil engineering, structural engineering, and engineering geology, from 1976 to present. Available on ORBIT, (800) 456-7248, and STN International, FIZ Karlsruhe, P.O. Box 2465, W-7500, Karlsruhe 1, Germany, online services. Inquire as to cost and availability.

National Information Service For Earthquake Engineering Database. University of California, Berkeley. Contains citations and abstracts to journal articles on earthquakes and earthquake engineering from 1987 to present. Available on BRS,(800) 289-4277, online service. Inquire as to cost and availability.

NTIS Bibliographic Database. National Technical Information Service, 5285 Port Royal Road, Springfield, Va 22161. (703) 487-4929 or FAX (703) 321-8199. Broad coverage of government-sponsored science and technology research reports, 1964 to present. Available on BRS,(800) 289-4277, DIALOG, (800) 334-2564, ORBIT, (800) 456-7248, and STN International, FIZ Karlsruhe, P.O. Box 2465, W-7500, Karlsruhe 1, Germany, online services. Also available on CD-ROM. Inquire as to cost and availability.

Nuclear Science Abstracts. U.S. Department of Energy, office of Scientific and Technical Energy, P.O. Box 62, Oak Ridge, TN 37831. (615) 576-1189. Contains citations with abstracts to literature in all fields of nuclear science and energy. 1948 to 1976. Available on DIALOG,(800) 334-2564, online service. Inquire as to cost and availability.

Scisearch. Institute for Scientific Information, 3501 Market Street, Philadelphia, PA 19104. (800) 523-1850 or (215) 386-0100. Broad multidisciplinary title and author index to the international literature of science and technology, 1974 to present. Available on DIALOG,(800) 334-2564, and ORBIT,(800) 456-7248, online services. Also available on CD-ROM. Inquire as to cost and availability.

Tris (transportation Research Information). National Academy of Science, 2101 Constitution Avenue, N.W., Washington, DC 20418. (202) 334-3313 or (800) 624-6242. Citations with abstracts of literature on transportation, including air, highway, rail, maritime and other modes from 1968 to present. Available on DIALOG,(800) 334-2564, online service. Inquire as to cost and availability.

Wasteinfo. United Kingdom Atomic Energy Authority, Building 7.12, Harwell Laboratory, Didcot, Oxon, OX11 ORA, England. Contains citations to worldwide literature on non-nuclear waste management, from 1973 to present. Available on ORBIT (800)456-7248, online service. Also available on CD-ROM. Inquire as to cost and availability.

Water Resources Abstracts (wra). U.S. Geological Survey, Water Resources Scientific Information Center, 12201 Sunrise Valley Drive, Reston, VA 22092-9998. (703) 648-4460. Contains citations with abstracts to scientific and technical literature on the water-sources-related aspects of the physical, social and life sciences, from 1968 to present. Available on DIALOG,(800) 334-

2564, and ORBIT,(800) 456-7248, online services. Inquire as to cost and availability.

Waternet. American Water Works Association, Technical Library. Available on DIALOG online services. Citations to literature on water quality, water utility management, water analysis, water pollution, and related areas, 1971 to present. Available on DIALOG,(800) 334-2564, online service. Inquire as to cost and availability.

Who's Who In Technology. Gale Research Inc., 835 Penobscot Building, Detroit, MI 48226-4094. (313) 961-2242. (800) 347-4253. Contains biographical information of contemporary American scieNTISts and engineers. Available on DIALOG,(800) 334-2564, online service. Inquire as to cost and availability.

WILSONLINE. H.W. Wilson Company, 950 University Avenue, Bronx, NY 10452. (800) 367-6770 or (212) 588-8400. Makes available online versions of the printed H.W. Wilson indexes including Applied Science and Technology Index, Business Periodicals Index, General Science Index, and Readers' Guide to Periodical Literature. Period covered is generally 1983 to present. Available on BRS,(800) 289-4277, DIALOG,(800) 334-2564, and OCLC EPIC,(800) 848-5878, online services. Also available on CD-ROM. Inquire as to cost and availability.

PERIODICALS

A.P.W.A. Reporter. American Public Works Association, 1313 East 60th Street, Chicago, IL 60637-2881. (312) 667-2200. 1934 to present. Monthly.

Cement, Concrete, and Aggregates. American Society for Testing and Materials (ASTM), 1916 Race Street, Philadelphia, PA 19103. (215) 299-5585. 1979 to present. Semi-annual. $48.00 per year.

*Civil Engineering A.*S.C.E. American Society of Civil Engineers, 345 East 47th Street, New York, NY 10017-2398. (212) 705-7520 or (800) 548-2723. 1930 to present. Monthly. $85.00 per year.

Civil Engineering Systems. Gordon and Breach Science Publishers, 270 Eighth Avenue, New York, NY 10011. (212) 206-8900 or FAX (212) 645-2459. 1983 to present. Quarterly. $850.00 per year.

Coastal Engineering: An International Journal For Coastal, Harbour, and offshore Engineers. Elsevier Science Publishing Company, Inc., 655 Avenue of the Americas, New York, NY 10010. (212) 989-5800. 1977 to present. Monthly. Inquire aboutprice.

Coastal Engineering Research Council. PROCEEDINGS. American Society of Civil Engineers, 345 East 47th Street, New York, NY 10017-2398. (212) 705-7520 or (800) 548-2723. 1954 to present. Biennial. Price varies.

Computer Applications In Engineering Education. Wiley-Interscience, division of John Wiley and Sons, Inc., 605 Third Avenue, New York, NY 10158. (800) 526-5368 or (212) 850-6000. 1992 to present. Bimonthly. $195.00 per year.

Construction Products. Cahners Publishing Company (Des Plaines), 1350 East Touhy Avenue, Des Plaines, IL 60017-5080.

(708) 635-8800. 1892 to present. Six times per year. $35.00 per year.

E.N.R. McGraw-Hill Publishing Company, 1221 Avenue of the Americas, New York, NY 10020. (800) 262-4729 or (212) 512-3825. 1874 to present. Weekly. $65.00 per year.

Earthquake Engineering and Structural Dynamics. John Wiley and Sons, Inc., 605 Third Avenue, New York, NY 10158. (800) 526-5368 or (212) 850-6000. 1972 to present. Monthly. $800.00 per year.

Engineering and Mining Journal. Maclean Hunter Publishing Company, 29 North Wacker Drive, Chicago, IL 60606-3298. (312) 726-2802. 1866 to present. Monthly. $35.00 per year.

Engineering Geology. Elsevier Science Publishing Company, Inc., 655 Avenue of the Americas, New York, NY 10010. (212) 989-5800. 1965 to present. Monthly. $800.00 per year.

Engineering Journal. American Institute of Steel Construction, Inc., One East Wacker Drive, Suite 3100, Chicago, IL 60601-2001. (312) 670-2400 or FAX (312) 670-5403. 1964 to present. Quarterly. $15.00 per year.

Engineering Structures: the Journal of Earthquake, Wind and Ocean Engineering. Butterworths Publishers, 80 Montvale Avenue, Stoneham, MA 02180. (617) 438-8464 or (800) 366-2665. 1979 to present. Bimonthly. $470.00 per year.

Engineer's Digest. Intertec Publishing Corporation, 9221 Quivira Street, Overland Park, KS 66215. (913) 888-4664. 1973 to present. Monthly. $20.00 per year.

Geotechnical Testing Journal. American Society for Testing and Materials (ASTM), 1916 Race Street, Philadelphia, PA 19103. (215) 299-5585. 1978 to present. Quarterly. $89.00 per year.

International Journal For Numerical and Analytical Methods In Geomechanics. John Wiley and Sons, Inc., 605 Third Avenue, New York, NY 10158. (800) 526-5368 or (212) 850-6000. 1977 to present. Monthly. $825.00 per year.

Journal of Aerospace Engineering. American Society of Civil Engineers, 345 East 47th Street, New York, NY 10017-2398. (212) 705-7520 or (800) 548-2723. 1988 to present. Quarterly. $108.00 per year.

Journal of Construction Engineering and Management. American Society of Civil Engineers, 345 East 47th Street, New York, NY 10017-2398. (212) 705-7520 or (800) 548-2723. 1956 to present. Quarterly. $104.00 per year.

Journal of Energy Engineering. American Society of Civil Engineers, 345 East 47th Street, New York, NY 10017-2398. (212) 705-7520 or (800) 548-2723. 1956 to present. Irregular. $84.00 per year.

Journal of Engineering Mechanics. American Society of Civil Engineers, 345 East 47th Street, New York, NY 10017-2398. (212) 705-7520 or (800) 548-2723. 1956 to present. Monthly. $284.00 per year.

Journal of Geotechnical Engineering. American Society of Civil Engineers, 345 East 47th Street, New York, NY 10017-2398. (212) 705-7520 or (800) 548-2723. 1956 to present. Monthly. $198.00 per year.

Journal of Hydraulic Engineering. American Society of Civil Engineers, 345 East 47th Street, New York, NY 10017-2398. (212) 705-7520 or (800) 548-2723. 1956 to present. Monthly. $198.00 per year.

Journal of Irrigation and Drainage. American Society of Civil Engineers, 345 East 47th Street, New York, NY 10017-2398. (212) 705-7520 or (800) 548-2723. 1956 to present. Bi-monthly. $124.00 per year.

Journal of Materials In Civil Engineering: Properties, Applications, Durability. American Society of Civil Engineers, 345 East 47th Street, New York, NY 10017-2398. (212) 705-7520 or (800) 548-2723. 1989 to present. Quarterly. $120.00 per year.

Journal of Performance of Constructed Facilities. American Society of Civil Engineers, 345 East 47th Street, New York, NY 10017-2398. (212) 705-7520 or (800) 548-2723. 1987 to present. Quarterly. $80.00 per year.

Journal of Structural Engineering. American Society of Civil Engineers, 345 East 47th Street, New York, NY 10017-2398. (212) 705-7520 or (800) 548-2723. 1956 to present. Monthly. $292.00 per year.

Journal of Surveying Engineering. American Society of Civil Engineers, 345 East 47th Street, New York, NY 10017-2398. (212) 705-7520 or (800) 548-2723. 1956 to present. Quarterly. $92.00 per year.

Journal of Water Resources Planning and Management. American Society of Civil Engineers, 345 East 47th Street, New York, NY 10017-2398. (212) 705-7520 or (800) 548-2723. 1956 to present. Bi-monthly. $104.00 per year.

Journal of Waterway, Port, Coastal, and Ocean Engineering. American Society of Civil Engineers, 345 East 47th Street, New York, NY 10017-2398. (212) 705-7520 or (800) 548-2723. 1956 to present. Bi-monthly. $98.00 per year.

Military Engineer. Society of American Military Engineers. 607 Prince Street, Alexandria, VA 22314-3117. (703) 549-3800. 1920 to present. Seven times per year. $24.00 per year.

Public Roads. U.S. Federal Highway Administration, 400 Seventh Street, N.W., Washington, DC 20590. (703) 285-2104. 1918 to present. Quarterly. $12.00 per year.

Reliability Engineering and System Safety. Elsevier Science Publishing Company, Inc., 655 Avenue of the Americas, New York, NY 10010. (212) 989-5800. 1980 to present. Monthly. $775.00 per year.

Roads and Bridges. Scranton Gillette Communications, Inc., 380 East Northwest Highway, Des Plaines, IL 60016. (708) 298-6622. 1892 to present. Monthly. $20.00 per year.

Rock Mechanics and Rock Engineering. Springer-Verlag New York, Inc., 175 Fifth Avenue, New York, NY 10010. (800) 526-7254 or (212) 460-1500. Quarterly. $122.00 per year.

Structural Engineering Review. Pergamon Press Inc., Maxwell House, Fairview Park, Elmsford, NY 10523. (914) 592-7700. Fax (914) 592-3625. 1988 to present. Quarterly. $185.00 per year.

CIVIL ENGINEERING

Ency. of Physical Sciences and Engineering Info. Sources

World Tunnelling. Mining Journal Limited, 60 Worship Street, London EC2A 2HD, England. 1988 to present. Ten times per year. $80.00 per year.

RESEARCH CENTERS AND INSTITUTES

American Concrete Institute, 22400 West Seven Mile Road, Detroit, Mi 48219-1885. (313) 532-2600 or FAX (313) 538-0655.

American Iron and Steel Institute, 1101 17th Street, Washington, Dc 20036-4700. (202) 452-7265 or FAX (202) 463-6573.

Construction Industry Institute, University of Texas At Austin, 3208 Red River, Suite 300, Austin, Tx 78705-2650. (512) 471-4319.

Engineering Societies Library, United Engineering Center, 345 East 47th Street, New York, Ny 10017. (212) 705-7611.

National Association of Home Builders National Research Center, 400 Prince George's Building, Upper Marlboro, Md 20772. (301) 249-4000.

National Research Council Library, 2101 Constitution Avenue, N.W., Washington, DC 20418. (202) 334-2125.

National Research Council of Canada. Institute for Mechanical Engineering, Montreal Road, Ottawa, ON, Canada K1A OR6. (613) 993-2436.

New Mexico State University. Engineering Research Center Box 3449, Las Cruces, NM 88003. (505) 646-3422 or FAX (505) 646-3549.

University of California, Berkeley. office of Research Services. 232 Hearst Memorial Mining Building, Berkeley, CA 94720. (415) 643-6291.

University of Texas At Austin. Bureau of Engineering Research. Cockrell Hall 10.340, Austin, TX 78712-1080. (512) 471-4325.

CLAY

See also: CERAMIC ENGINEERING, CERAMICS, GLASS

ABSTRACT SERVICES AND INDEXES

Applied Science and Technology Index; A Cumulative Subject Index To English Language Periodicals In the Fields of Aeronautics and Space Science, Computer Technology, Chemistry,Construction Industry, Energy and Related Areas. H.W. Wilson Co., 950 University Avenue, Bronx, NY 10452. (800) 367-6770 or (212) 588-8400. FAX (718) 590-1617. From 1958 to present. Monthly. Inquire about cost and availability. Also available on CD-ROM and online.

Bibliography and Index of Geology. American Geological Institute, 4220 King Street, Alexandria, VA 22302. (703) 379-2480. Fax (703) 379-7563. Monthly. $1295.00 per year. Also available online as GeoRef.

CA Selects: Ceramic Materials (Journals). Chemical Abstracts Service, 2450 Olentagy River Road, Box 3012, Columbus, OH 43210-0012. (614) 447-3600. FAX (614) 447-3713. 1988 to present. Semi-weekly. $210.00 per year.

Ceramic Abstracts. American Ceramic Society, Inc., 757 Brooksedge Plaza Drive, Westerville, OH 43081-2821. (614) 890-6136. Bi-monthly. 1922 to present. $120.00 per year.

Engineered Materials Abstracts. ASM (American Society of Metals) International, Materials Information, Materials Park, OH 44073. (216) 338-5151 or FAX (216) 338-4634. Covers literature on technical developments in polymer, ceramic, and composite materials and engineering. 1986 to present. Monthly. $1200.00 per year. Also available on CD-ROM and online. Covers published materials on technical developments in polymer, ceramic, and composite materials in an engineering environment.

Geotechnical Abstracts. Geodex Retrieval Systems, 669 Broadway, Box 279, Sonoma, CA 95476. (707) 939-8476. FAX (707) 996-8734. 1970 to present. Three times a year. $280.00 per year. Covers soil mechanics, foundation engineering, rock mechanics, and engineering geology.

Materials Science and Engineering Abstracts, Cambridge Scientific Abstracts, 7200 Wisconsin Avenue, Bethesda, Md 20814-4823. (301) 961-6750. FAX (301) 961-6720. 1993 to present. Monthly. $385.00 per year. Focuses on mechanical and physical properties of materials and commercial or industrial applications for materials, methods for strength testing, effects of vibration and other stresses, corrosion and protective coatings, storage and handling, ceramics, composites, metals, wood, plastics, and polymers.

World Ceramic Abstracts. Pergamon Press, Maxwell House, Fairview Park, Elmsford, NY 10523. (914) 592-7700. Monthly. 1958 to present. $310.00 per year.

ANNUAL REVIEWS AND YEARBOOKS

Mineral Industry Surveys. CLAYS. U.S. Bureau of Mines, Production and Distribution, Cochrans Mill Road, Box 18070, Pittsburgh, PA 15236. (412) 892-4411. Annual. Inquire for cost and availability.

ASSOCIATIONS AND PROFESSIONAL SOCIETIES

Ceramic Manufacturers Association. 1100-H Brandywine Blvd., PO Box 2188, Zanesville, OH 43702-2188. (614) 452-4541.

Clay Minerals Society. 540 22nd Street, Boulder, CO80302-7911. (303) 444-6405.

Materials Research Society. 9800 McKnight Road, Pittsburgh, PA 15237. (412) 367-3003. FAX (412) 367-4373.

National Institute of Ceramic Engineers. 735 Ceramic Place, Westerville, OH 43081. (614) 890-4700.

Society of Economic Geologists. 5808 S. Rapp Street, Suite 209, Littleton, CO 80120-1942. (303) 797-0332. FAX (303) 797-0417.

DIRECTORIES AND BIOGRAPHICAL SOURCES

American Men & Women of Science, 1995-96. R.R. Bowker Staff, eds. 19th edition. 8 volumes. R.R. Bowker/Reed International Publishing Company, 121 Chanlon Road, New Providence, NJ 07974. (908) 464-6800 or (800) 521-8110. 1995. $850.00

Ceramic Industry Data Book Buyers Guide. Business News Publishing Company, 755 W. Big Beaver Road, Suite 1000, Troy, MI 48084. (313) 362-3700. FAX (313) 362-0317. 1922 to present. Annual. $25.00 per year. Included as part of subscription to *Ceramic Industry* periodical.

Guide To Engineering Materials Producers. J.C. Bittence. ASM International, Materials Information, Materials Park, OH 44073-0002. (216) 338-5151 or (800) 336-5152. FAX (216) 338-4634. 1993. $141.00

International Ceramic Directory. London & Sheffield Publishing Company Ltd., 5 Pond Street, Hampstead, London NW3 2PN, England. Telephone 01-794-0800. FAX 01-794-0411. 1984 to present. Every 5 years. Inquire for price and availability.

International Research Centers Directory. Gale Research, 835 Penobscot Building, Detroit, MI 48226-4094. (313) 961-2242. (800) 347-4253. FAX (313) 961-6083. 8th edition. 1995. $410.00.

Research Centers Directory. Gale Research, 835 Penobscot Building, Detroit, MI 48226-4094. (313) 961-2242. (800) 347-4253. FAX (313) 961-6083. $485.00.

Scientific and Technical Organizations and Agencies Directory. Gale Research, 835 Penobscot Building, Detroit, MI 48226-4094. (313) 961-2242. (800) 347-4253. FAX (313) 961-6083. 4th edition. 1996. $195.00.

Who's Who In Technology. Gale Research, 835 Penobscot Building, Detroit, MI 48226-4094. (313) 961-2242. (800) 347-4253. FAX (313) 961-6083. 1995. $195.00.

ENCYCLOPEDIAS AND DICTIONARIES

Dictionary of Ceramic Science and Engineering. Ian J. McColm, editor. 2nd edition. Plenum Publishing Corporation, 233 Spring Street, New York, NY 10013. (212) 620-8000. FAX (212) 463-0742. 1994. $75.00.

Encyclopedia of Glass, Ceramics, and Cement. Martin Grayson, editor. John Wiley and Sons, Inc., 605 Third Avenue, New York, NY 10158. (800) 526-5368 or (212) 850-6000. 1985. Inquire for cost and availability.

Encyclopedia of Physical Science and Technology. Robert A. Meyers, ed. 18 volumes. Academic Press Inc., 6277 Sea HarborDrive, Orlando, FL 32887. (800) 321-5068. 1992. $2100.00.

GENERAL WORKS

Chemistry and Physics of Clays. Rex W. Grimshaw. Reprint edition. TechBooks, 4012 Williamsburg Court, Fairfax, VA 22032-1139. (703) 352-0001. 1996. $81.00.

Chemistry of Clays and Clay Minerals. A.C. Newman, editor. John Wiley and Sons, Inc., 605 Third Avenue, New York, NY 10158. (800) 526-5368 or (212) 850-6000. 1987. $156.95.

Clay In Engineering Geology. J.E. Gillott. Second revised edition. Elsevier Science Inc., Box 882, Madison Square Station, New York, NY 10159. (212) 989-5800. FAX (212) 633-3990. 1987. $77.00.

Clay Mineralogy. Wilson. Van Nostrand Reinhold, 115 Fifth Avenue, New York, NY 10003. (212) 254-3232 or FAX (212) 254-9499. 1991. $105.00.

Clays. Gordon Press Publishers, PO Box 459, Bowling Green Station, New York, NY 10004. (718) 624-8419. 1993. $250.95.

Clays, Muds and Shales. C.E. Weaver. Elsevier Science Inc., Box 882, Madison Square Station, New York, NY 10159. (212) 989-5800. FAX (212) 633-3990. 1990. $143.75.

Introduction To Clay Minerals: Chemistry, Origins, Uses and Environmental Significance. Bruce Velde. Chapman and Hall, 29 West 35th Street, New York, NY 10001. (800) 842-3636. FAX (212) 563-2269. 1992. $39.95.

HANDBOOKS AND MANUALS

Data Handbook For Clay Minerals and Other Non-metallic Minerals. H. Van Olphen and J.J. Fripiat, editors. Reprint edition. Franklin Book Company, Inc., 7804 Montgomery Ave., Elkins, PA 19117. (215) 635-5252. FAX (215) 635-6155. 1979. $144.00.

ONLINE DATABASES AND CD-ROMS

CA Search. Chemical Abstracts Service, P.O. Box 3012, Columbus, OH 43210-0012. (614) 447-3600. (800) 848-6533. FAX (614) 447-3709. Very comprehensive guide to worldwide chemical literature and related fields, 1972 to present. Available on BRS,(800) 289-4277, DIALOG, (800) 334-2564, ORBIT (800) 456-7248, and STN International, FIZ Karlsruhe, P.O. Box 2465, W-7500, Karlsruhe 1, Germany, online services. Inquire as to cost and availability.

Geoarchive. Geosystems, P.O. Box 1024, Westminster, London, England, SW1 P 2JL. Citations to literature on geoscience, 1969 to present. Inquire as to online cost and availability.

Geobase. Elsevier - Geo Abstracts, Regency House, 34 Duke Street, Norwich NR3 3AP, England. Contains citations to the worldwide earth science literature from 1980 to date. Available on DIALOG, ORBIT online services. Inquire as to cost and availability.

GEOREF: Bibliography and Index of Geology. American Geological Institute, 4220 King Street, Alexandria, VA 22302. (703) 379-2480. Fax (703) 379-7563. Monthly. Inquire for price and availability.

Scisearch. Institute for Scientific Information, 3501 Market Street, Philadelphia, PA 19104. (800) 523-1850 or (215)386-0100. Broad multidisciplinary title and author index to the international literature of science and technology, 1974 to present. Available on DIALOG,(800) 334-2564, and ORBIT,(800) 456-7248, online services. Also available on CD-ROM. Inquire as to cost and availability.

CLAY

Ency. of Physical Sciences and Engineering Info. Sources

PERIODICALS

Clay Minerals. Mineralogical Society, 41 Queens Gate, London SW7 5HR, England. Telephone 071-584-7516. FAX 071-823-8021. 1947 to present. $150.00 per year.

Clay Minerals Society News. Clay Minerals Society, 540 22nd Street, Boulder, CO 80302-7911. (303) 444-6405. Quarterly. $15.00 per year.

Clays and Clay Minerals. Clay Minerals Society, 540 22nd Street, Boulder, CO 80302-7911. (303) 444-6405. 1968 to present. Bi-monthly. $35.00 per year.

Clay Technology. Sherwood House, Holt Lane, Matlock, Derbyshite, DE4 3LY, England. Telephone 0629-583941. FAX 0629-580479. Six times year. Inquire for price.

Economic Geology and the Bulletin of Economic Geologists. Economic Geology Publishing Company, 101 Vowell Hall, University of Texas, El Paso, TX 79968. FAX (915) 544-7416. 1906 to present. Eight times a year. $65.00.

RESEARCH CENTERS AND INSTITUTES

Alfred University Center For Advanced Ceramic Technology. New York State College of Ceramics, Alfred, NY 14802. (607) 871-2486. FAX (607) 871-3469.

Center For Ceramics Research. Rutgers University, College of Engineering, Brett & Bowser, PO Box 909, Piscataway, NJ 08855-0909. (201) 932-4817.

Geotechnical/civil Engineering Materials Research Laboratories. Materials Laboratory/Sprangler Geotechnical Laboratory, Iowa State University, Ames, IA 50011. (515) 294-9470.

CLIMATOLOGY

See also: DESERTIFICATION, DROUGHT, METEOROLOGY, WEATHER

ABSTRACT SERVICES AND INDEXES

Applied Science and Technology Index; A Cumulative Subject Index To English Language Periodicals In the Fields of Aeronautics and Space Science, Computer Technology, Chemistry, Construction Industry, Energy and Related Areas. H.W. Wilson Co., 950 U niversity Avenue, Bronx, NY 10452-9978. (212) 588-8400. (800) 367-6770. FAX (718) 590-1617. From 1958 to present. Monthly. Inquire about cost and availability. Also available online (BRS and WILSONLINE) and on CD-ROM. ISSN: 0003-6986.

Bibliography and Index of Geology. American Geological Institute, 4220 King Street, Alexandria, VA 22302-1507. (703) 379-2480. FAX (703) 379-7563. From 1969 to present. Monthly. $1295.00 Per Year. ISSN: 0098-2784. Also available as georef online (CISTI, DIALOG, ORBIT, STN) and on Cd-rom. Inquire about price and availability.

Chemical Abstracts. Chemical Abstracts Service. 2540 Olentangy river road, box 3012, Columbus, OH 43210-0012. (614) 447-3600. FAX (614) 447-3713. 1907 To Present. Weekly.

$16,800.00 per year. Available Online and On Cd-rom. CA is also Available In Five Section Groupings. Inquire regarding Cost and Availability.

>*Current Contents: Physical and Chemical Science.* Institute for Scientific Information, 3501 Marketstreet, Philadelphia, PA 19104. (215) 386-0100. (800) 523-1850. Fax (215) 386-2291. 1961 to Present. Weekly. $442.00 per Year. Alsoavailable Online (Brs, Dialog) and on Cd-rom. Inquire Regarding Cost and Availability. ISSN: 0163-2574.

Deep-sea Research. Part A: Oceanographic Research Papers. Deep-sea Research. Part B: Oceanographic Literature Review. Pergamon Press, Inc., Maxwell House, Fairview Park, Elmsford, NY 10523. (914) 592-7700. Fax (914) 592-3625. Twelve times per year. $2000.00 per year for Parts A and B. Oceanographic Literature Review also available on CD-ROM. Inquire about price and availability

General Science Index. H.W. Wilson Co., 950 University Avenue, Bronx, NY 10452. (212) 588-8400. (800) 367-6770. FAX (718) 590-1617. From 1978 to present. Ten issues per year; quarterly and annual cumulations. Service basis. Available on CD-ROM and online. Inquire regarding cost and availability. ISSN: 0162-1963.

Government Reports Announcements and Index. National Technical Information Service (NTIS), 5285 Port Royal Road, Springfield, VA 22161. (703) 487-4650. 1968 to present. Annual. $525.00 per year. Also available online as NTIS BIBLIOGRAPHIC DATABASE and on CD-ROM.

Meteorological and Geoastrophysical Abstracts. American Meteorological Society, c/o Inforonics, Inc., 550 Newtown Road, Box 458, Littleton, MA 01460. (508) 486-8976. FAX (508) 486-0027. Covers literature in environmental sciences, meteorology, astrophysics, hydrology, glaciology, and physical oceanography. 1950 to present. Monthly. $950.00 per year. ISSN: 0026-1130. Also available online (DIALOG) and on CD-ROM.

Oceanic Abstracts. Cambridge Scientific Abstracts, 7200 Wisconsin Avenue, Bethesda, MD 20814. (301) 961-6750. Fax (301) 961-6720. Bimonthly. $995.00 per year.

Physics Abstracts. Institute of Electrical Engineers, London, United Kingdom. Available from: Institute of Electrical and Electronic Engineers (IEEE), Box 1331, 445 Hoes Lane, Piscataway, NJ 08855-1331. (908) 562-5549. 1898 to present. 24 issues per year. $2835.00 per year. ISSN: 0036-8091. Also available online and on CD-ROM.

Science Citation Index. Institute for Scientific Information, 3501 Market Street, Philadelphia, PA 19104. (215) 386-0100. (800) 523-1850. FAX (215) 386-2991. 1961 to present. Six issues per year, plus annual cumulation. $11650.00 per year. Also available online and on CD-ROM. Inquire about price and availability. ISSN: 0036-827X.

Selected Water Resources Abstracts. U.S. GeologicalSurvey. Distributed by National Technical Information Service (NTIS), 5285 Port Royal Road, Springfield, VA 22161. (703) 487-4650. Monthly. $115.00 per year.

ANNUAL REVIEWS AND YEARBOOKS

American Meteorological Society. Meteorological Monographs. American Meteorological Society, 45 Beacon Street, Boston,

MA 02108-3693. (617) 227-2425. FAX (617) 742-8718. Irregular. Price varies.

Annual Review and Earth and Planetary Sciences. Annual Reviews, Inc., 4139 El Camino Way, Palo Alto, CA 94306-0897. (415) 493-4400. Fax (415) 855-9815. Annual. $55.00 per year.

Developments In Atmospheric Science Series. Elsevier Science Publishing Company, Inc., 655 Avenue of the Americas, New York, NY 10010. (212) 989-5800. Irregular. Inquire about price and availability.

Ocean Yearbook. Elisabeth M. Borgese, et al, editors. University of Chicago Press, 5801 Ellis Avenue, Chicago, IL 60637. (312) 702-7700. 1995. $77.00.

ASSOCIATIONS AND PROFESSIONAL SOCIETIES

American Association of State Climatologists. c/o Myron Moinau, University of Idaho, Agriculture Engineering Department, State Climate Service, Moscow, ID 834844-2040.

American Meteorological Society, 45 Beacon Street, Boston, Ma 02108-3693. (617) 227-2425. FAX (617) 742-8718. Irregular. Price varies.

Canadian Meteorological and Oceanographic Society, P.O. Box 334, Newmarket, Ontario, L3Y 4X7, Canada. (416) 898-1040. FAX (416) 898-7937.

International Association of Meteorology and Atmospheric Physics. UCAR, P.O. Box 3000, Boulder CO 80307-3000. (303) 497-1673.

National Environmental Satellite Data, and Information Service. 2069 Federal Building 4 Room 2060, Washington, DC 20233. (301) 763-71990. FAX (301) 763-4011.

National Weather Association, 6704 Wolke Court, Montgomery, AL 36116-2134. (334) 213-0388. FAX (334) 213-0388.

University Corporation For Atmospheric Research. PO Box 3000, 1850 Table Mesa Drive, Boulder, CO 80307-3000. (303) 497-1673. FAX (303) 497-1654.

Weather Modification Association. PO Box 8116, Fresno CA 93747. (209) 291-8466.

BIBLIOGRAPHIES

Information Sources in the Earth Sciences. David N. Wood, Joan E. Hardy, and Anthony P. Harvey, editors. Bowker-Saur/K.G. Saur. Distributed by R.R. Bowker, 121 Chanlon Road, New Providence, NJ 07974. (800) 521-8110 or (908) 464-6800. Second edition. 1989. $85.00.

International Bibliography of Meteorology: From the Beginning of Printing To 1889. Oliver L. Fassig. Diane Publishing Co., 600 Upland Avenue, Upland, PA 19015. (610) 499-7415. Reprintedition. 1994. $195.00

Scientific and Technical Books and Serials In Print; An Index To Literature In Science and Technology. R.R. Bowker Co., 205 E. 42nd Street, New York, NY 10017. (800) 521-8110 or (212) 916-1600. 1992. $250.00.

DIRECTORIES AND BIOGRAPHICAL SOURCES

American Men and Women of Science: Physical and Biological Sciences. Eighteenth edition. R.R. Bowker Company, 245 West 17th Street, New York, NY 10011. (800) 521-8810 or (212) 916-1600. $750.00.

Meteorological Services of the World. American Meteorological Society, 45 Beacon Street, Boston, MA 02108- 3693. (617) 227-2425. FAX (617) 742-8718. Annual. $70.00.

National Weather Service offices and Stations. National Oceanic and Atmospheric Administration, Deparatment of Commerce, Silver Spring, MD 20910. (301) 427-7698. Annual. Free

Research Centers Directory. Gale Research, 835 Penobscot Building, Detroit, MI 48226-4094. (800) 347-4253 or (313) 961-2242. 17th edition, 1992. $400.00.

ENCYCLOPEDIAS AND DICTIONARIES

Climates of the States. Gale Research, 835 Penobscot Building, Detroit, MI 48226-4094. (800) 347-4253 OR (313) 961-2242., 5th edition. 1995. $255.00.

Concise Oxford Dictionary of the Earth Sciences. Ailsa Allaby and Michael Allaby, editors. Oxford University Press, Inc., 200 Madison Avenue, New York, NY 10016. (800) 334-4249 or (212) 679-7300. 1990. $42.95.

Encyclopedia of Climatology. Rhodes W.Fairbridge, and John E. Oliver. Chapman & Hall, 1 Penn Plaza, New York, NY 10119. (212) 564-1060. 1986. $115.00.

*$168.*00.

McGraw-Hill Encyclopedia of Science and Technology. McGraw-Hill Publishing Company, 1221 Avenue of the Americas, New York, NY 10020. (800) 262-4729 or (212) 512-3825. Seventh edition. 1992. $1900.00.

Magill's Survey of Science: Earth Science Series. Salem Press, Inc., P.O. Box 1097, Englewood Cliffs, NJ 07632. (800) 221-1592 or (201) 871-3700. Five volumes. 1990. $400.00 for the set.

Smithsonian Meteorological Tables. Books on Demand, 300 North Zeeb Road, Ann Arbor, MI 48106-1346. (313) 761-4700. (800) 521-0600. reprint edition. 199 $153.90.

Water Encyclopedia. Frits van der Leeden, Fred L. Troise, and David K. Todd. Lewis Pubs., CRC Press, Inc., 2000 Corporate Boulevard, NW, Boca Raton, FL 33431. 2nd edition. 1990. $ 151.95

Weather Satellite Handbook. American Radio Relay League, 225 Main Street, Newington, CT 06111. (203) 666-1541. 1994. $20.00.

GENERAL WORKS

Aerosol-cloud-climate Interaction. Peter V. Hobbs. Academic Press, Inc., 6277 Sea Harbor Drive, Orlando, FL. (800)321-5068. 1993. $65.00.

CLIMATOLOGY

Ency. of Physical Sciences and Engineering Info. Sources

Atmosphere, Weather and Climate. Roger G. Barry, and

Richard J. Chorley. Routledge, Chapman & Hall, Inc., 29 West 35th Street, New York, NY 10001-3291. (212) 244-3336. 1993, $35.00 in paper.

Climate Change. John T. Houghton, editor. Cambridge University Press, 40 West 20th Street, New York, NY 10011- 4211. (212) 924-3900. (800) 872-7423. 1992. $19.95.

Climate In Human Perspective. A Tribute to Helmut A. Landsberg. F. Baer, et al editors. Kluwer Academic Publishers, 101 Philip Drive, Assinippi Park, Norwell, MA 02061. (617) 871-6600. 1991. $61.50.

General Climatalogy. H. Flohn, and H. E. Landsberg, editors. World Survey of Climatalogy, volume 2. Elsevier Science Publishing Company, Inc., 655 Avenue of the Americas, New York, NY 10010. (212) 989-5800. FAX (914) 333-2444. 1970, $133.00.

Greenhouse Effect, Climatic Change & Ecosystems. Bert Bolin and Bo R. Doos, John Wiley and Sons, Inc., 605 Third Avenue, New York, Ny 10158. (800) 526-5368 or (212) 850- 6000. 1990. $69.95.

Ocean-atmosphere Interaction & Climate Modeling. Boris A. Kagan. Cambridge University Press, 40 West 20th Street, New York, NY 10011-4211. (212) 924-3900. (800) 872-7423. 1995, $79.95.

Physics of Climate. J. P. Peixoto. American Institute of Physics, One Physics Ellipse, College Park, MD 20740-3843. (301) 209-3100. 1992. $45.00.

The Urban Climate. Helmut Landsberg. Academic Press, Inc., 6277 Sea Harbor Drive, Orlando, FL. (800) 321-5068. 1981. $65.00

Fundamentals of Atmospheric Dynamics and thermodynamics. C. Riegel. World Scientific Publishing Company, Inc., 1060 Main Street, Unit B, River Edge, NJ 07661. (800) 227-7562 or (201) 487-9655. 1992. $33.00 in paper.

Paleoclimatology. Thomas J. Crowley and Gerard R. North. Oxford University Press, Inc., 200 Madison Avenue, New York, NY 10016. (800) 334-4249 or (212) 679-7300. 1991. $59.95.

Weather Companion: An Album of Meteorological History, Science, Legend, and Folklore. Gary Lockhart. John Wiley and Sons, Inc., 605 Third Avenue, New York, NY 10158. (800) 526-5368 or (212) 850-6000. 1988. $12.95.

Weather Cycles: Real Or Imaginary? William James Burroughs. Cambridge University Press, 40 West 20th Street, New York, NY 10011-4211. (212) 924-3900. 1992. $39.95.

HANDBOOKS AND MANUALS

Author's Guide To the Journals of the American Meteorological Society. American Meteorological Society, 45 Beacon Street, Boston, MA 02108-3693. (617) 227-2425. FAX (617) 742-8718. 1983. $15.00 in paper.

Basic Meteorology Lab Manual. Thomas A. Leavy. Allegheny Press, P.O. Box 220, Elgin, PA 19413. (814) 664-8504. Secondedition. 1969. $7.95.

Handbook of Agricultural Meteorology. John F. Griffin. Oxford University Press, Inc., 200 Madison Avenue, New York, NY 10016. (212) 725-6000. (800) 334-4249. 1994. $85.00.

Handbook In Applied Meteorology. David D. Houghton, editor. John Wiley and Sons, Inc., 605 Third Avenue, New York, NY 10158. (800) 526-5368 or (212) 850-6000. 1985. $168.00.

Weather Almanac. Gale Research, 835 Penobscot Building, Detroit, MI 48226-4094. (313) 961-2242. (800) 347-4253. 7th edition. 1996. $130.00.

ONLINE DATABASES AND CD-ROMS

Climate Assessment Database. National Weather Service, National Meteorological Center, 5200 Auth Road, Suite 101, Camp Springs, MD 20233. (301) 763-8016. Contains daily, weekly and monthly summaries of North American and world climatological data. Also provides five to ten weather forecasts, and 30 to 90 day weather outlook. Subscription required. Inquire as to cost and availability.

CA Search. Chemical Abstracts Service, P.O. Box 3012, Columbus, OH 43210-0012. (614) 447-3600. (800) 848-6533. FAX (614) 447-3709. Very comprehensive guide to worldwide chemical literature and related fields, 1972 to present. Available on BRS,(800) 289-4277, DIALOG, (800) 334-2564, ORBIT (800) 456-7248, and STN International, FIZ Karlsruhe, P.O. Box 2465, W-7500, Karlsruhe 1, Germany, online services. Inquire as to cost and availability.

Compendex Plus. Engineering Information, Inc., 345 East 47th Street, New York, NY 10017. (212) 705-7600 or (800) 221-1044. Contains citations with abstracts to worldwide literature in engineering and technology, from 1970 to present. Available on online BRS,(800) 289-4277, DIALOG, (800) 334-2564, ORBIT (800) 456-7248, and STN International, FIZ Karlsruhe, P.O. Box 2465, W-7500, Karlsruhe 1, Germany, online services. Also available on CD-ROM. Inquire as to cost and availability.

Current Contents Search. Institute for Scientific Information, 3501 Market Street, Philadelphia, PA 19104. (215) 386-0100. FAX (215) 386-6362. Contains citations to articles listed in the table of contents of science and technology journals. Also articles in social sciences and life sciences journals. Available on BRS,(800) 289-4277, DIALOG,(800) 334-2564, online services. Inquire as to cost and availability.

Dissertation Abstracts. University Microfilms International, 300 North Zeeb Road, Ann Arbor, MI 48106. (800) 521-0600 or (313) 761-4700. Scope includes virtually all doctoral dissertations accepted at accredited American institutions from 1861 to present in 252 subject areas. Available on BRS,(800) 289-4277, DIALOG,(800) 334-2564, and OCLC EPIC,(800) 848-5878, online services. Also available on CD-ROM. Inquire as to cost and availability.

Inspec. Institution of Electrical Engineers, Michael Faraday House, Six Hills Way, Stevenage, Herts. SG1 2AY, England.Telephone: 0438 313311 or FAX 0438 742840. Contains citations to the worldwide literature of physics, electronics and electrical engineering, computer technology, and related fields. Available on BRS,(800) 289-4277, DIALOG, (800) 334-2564, ORBIT, (800) 456-7248, and STN International, FIZ

Karlsruhe, P.O. Box 2465, W-7500, Karlsruhe 1, Germany, online services. Inquire as to cost and availability.

International Aerospace Abstracts. American Institute of Aeronautics and Astronautics, 370 L'Enfant Promenade, S.W., Washington, DC 20024. (202) 646-7400. Contains references and abstracts of journal and monograph literature relating to aerospace science and technology, from 1963 to present. Available through the NASA/RECON system of the National Aeronautics and Space Administration only.

Meteorological and Geoastrophysical Abstracts. American Meteorological Society, 45 Beacon Street, Boston, MA 02108-3693. (617) 227-2425. FAX (617) 742-8718. Contains citations and abstracts to the worldwide literature on significant research in meteorology and geoastrophysics. Related topics include physical oceanography, hydrology, environmental sciences and glaciology. Covers the period 1972 to present. Available on DIALOG, (800) 334-2564, online service. Inquire as to cost and availability.

NTIS Bibliographic Database. National Technical Information Service, 5285 Port Royal Road, Springfield, Va 22161. (703) 487-4929 or FAX (703) 321-8199. Broad coverage of government-sponsored science and technology research reports, 1964 to present. Available on BRS, (800) 289-4277, DIALOG, (800) 334-2564, ORBIT, (800) 456-7248, and STN International, FIZ Karlsruhe, P.O. Box 2465, W-7500, Karlsruhe 1, Germany, online services. Also available on CD-ROM. Inquire as to cost and availability.

Physics Briefs. American Institute of Physics, 335 East 45th Street, New York, NY 10017. (212) 661-9260 or FAX (212) 661-2036. Contains citations with abstracts of the literature of physics and related fields, 1979 to present. Available on the STN International, FIZ Karlsruhe, P.O. Box 2465, W-7500, Karlsruhe 1, Germany, online service. Inquire as to cost and availability.

Scisearch. Institute for Scientific Information, 3501 Market Street, Philadelphia, PA 19104. (800) 523-1850 or (215) 386-0100. Broad multidisciplinary title and author index to the international literature of science and technology, 1974 to present. Available on DIALOG and ORBIT online services. Also available on CD-ROM. Inquire as to cost and availability.

WILSONLINE. H.W. Wilson Company, 950 University Avenue, Bronx, NY 10452. (800) 367-6770 or (212) 588-8400. Makes available online versions of the printed H.W. Wilson indexes including Applied Science and Technology Index, Business Periodicals Index, General Science Index, and Readers' Guide to Periodical Literature. Period covered is generally 1983 to present. Available on BRS, (800) 289-4277, DIALOG, (800) 334-2564, and OCLC EPIC, (800) 848-5878, online services. Also available on CD-ROM. Inquire as to cost and availability.

World Climate Disc. Chadwyck-Healey Inc., 1101 King Street, Alexandria, VA 22314. (703) 683-4890. FAX (703) 683-7589. Weather and climate data from approximately 5,000 weather stations worldwide, covering the years 1854 to 1990 on CD-ROM. First edition 1992. Approximately $1200.00 per year with annual updates.

World Weatherdisc. WeatherDisc Associates, Inc., 4584 N.E. 89th Street, Seattle, WA 98115. (206) 524-4314. FAX (206) 543-0308. Meteorological data on CD-ROM which describes the climate of the earth today and for the past few hundred years. First edition 1989. Approximately $295.00 per year with biannual updates.

PERIODICALS

Agricultural and Forest Meteorology. Elsevier Science Publishing Company, Inc., 655 Avenue of the Americas, New York, NY 10010. (212) 989-5800. Twenty times per year. $750.00 per year.

American Meteorological Society. BULLETIN. American Meteorological Society, 45 Beacon Street, Boston, MA 02108-3693. (617) 227-2425. FAX (617) 742-8718. Monthly. $60.00 per year.

American Weather Observer. Association of American Weather Observers, 401 Whitney Boulevard, Box 455, Belvedere, IL 61008. (815) 544-5665. FAX (815) 544-6334. Monthly. $21.00 per year.

Atmosphere - Ocean. Canadian Meteorological and Oceanographic Society, P.O. Box 334, Newmarket, Ontario, L3Y 4X7, Canada. (416) 898-1040. FAX (416) 898-7937. Quarterly. $30.00 per year.

Boundary-layer Meteorology: An International Journal of Physical and Biological Processes In the Atmospheric Boundary Layer. Kluwer Academic Publishers, P.O. Box 358, Accord Station, Hingham, MA 02018-0358. (617) 871-6000. Sixteen per year. $785.00 per year.

Climate Change. Kluwer Academic Publishers, P.O. Box 358, Accord Station, Hingham, MA 02018-0358. (617) 871-6000. Six times per year. $327.00 per year.

Dynamics of Atmospheres and Oceans. Elsevier Science Publishing Company, Inc., 655 Avenue of the Americas, New York, NY 10010. (212) 989-5800. Six times per year. $205.00 per year.

Earth. Kalmbach Publishing Company, P.O. Box 1612, Waukesha, WI 53187. (414) 796-0126 or (800) 558-1544. 1991 to present. Bimonthly. $14.95 per year.

International Journal of Climatology. Royal Meteorological Society, distributed by John Wiley and Sons, Inc., 605 Third Avenue, New York, NY 10158. (800) 526-5368 or (212) 850-6000. Eight times per year. $425.00 per year.

Jgr: Journal of Geophysical Research: Atmosphere. American Geophysical Union, 2000 Florida Avenue, N.W., Washington, DC 20009. (202) 462-6903. Monthly. $90.00 per year to members.

Jgr: Journal of Geophysical Research: Oceans. American Geophysical Union, 2000 Florida Avenue, N.W., Washington, DC 20009. (202) 462-6903. Monthly. $1545.00 per year.

Journal of Applied Meteorology. American Meteorological Society, 45 Beacon Street, Boston, MA 02108-3693. (617) 227-2425. FAX (617) 742-8718. Monthly. $165.00 per year.

Journal of Climate. American Meteorological Society, 45 Beacon Street, Boston, MA 02108-3693. (617) 227-2425. FAX (617) 742-8718. Monthly. $175.00 per year.

Journal of Atmospheric Sciences. American Meteorological Society, 45 Beacon Street, Boston, MA 02108-3693. (617) 227-2425. FAX (617) 742-8718. Semi-monthly. $320.00 per year.

Monthly Weather Review. American Meteorological Society, 45 Beacon Street, Boston, MA 02108-3693. (617) 227-2425. FAX (617) 742-8718. Monthly. $205.00 per year.

National Weather Digest. National Weather Association, 4400 Stamp Road, Room 404, Temple Hills, MD 20748. (301) 899-3784.

Royal Meteorological Society. Quarterly Journal. Royal Meteorological Society, 104 Oxford Road, Reading, Berks RG1 7LJ, England. Six times per year. $250.00 per year.

Weather. Royal Meteorological Society, 104 Oxford Road, Reading, Berks RG1 7LJ, England. Monthly. $44.00 per year.

Weatherwise. Heldref Publications, 1319 Eighteenth Street, N.W., Washington, DC 20036-1802. (202) 296-6267. FAX (202) 296-5149. Bi-monthly. $28.00 per year.

RESEARCH CENTERS AND INSTITUTES

Center For Climate Resesarch. Columbia University. Lamont-Doherty Geological Observatory. Palisades, NY 10964. (914) 359-2900.

Center For Climatic Research. University of Wisconsin-Madison, 1225 West Dayton Street, Madison, WI 53706. (608) 262-2839. FAX (608) 262-5964.

Climate Research Institute. Oregon State University. Strand Hall, Room 326, Corvallis, Or 97331-2209.

Cooperative Institute For Research In the Atmosphere. Colorado State University. Foothills Campus. Fort Collins, CO 80523. (303) 491-8448.

Florida State Climate Center. Florida State University, 431 Love Building, Tallahassee, FL 32306. (904) 644-3417.

High Plains Climate Center. University of Nebraska-Lincoln, 237 Chase, Lincoln, NE 68583. (402) 472-6706.

National Center For Atmospheric Research. P.O. Box 3000, Boulder, CO 80307. (303) 497-1000. FAX (303) 497-1654.

CLOUDS

See also: AEROSOLS, FOG, METEOROLOGY, WEATHER MODIFICATION

ABSTRACT SERVICES AND INDEXES

Applied Science and Technology Index; A Cumulative Subject Index To English Language Periodicals In the Fields of Aeronautics and Space Science, Computer Technology , Chemistry, Construction Industry, Energy and Related Areas. H.W. Wilson Co., 950 University Avenue, Bronx, NY 10452. (212) 588-8400. (800) 367-6770. FAX (718) 590-1617. From 1958 to present. Monthly.Inquire about cost and availability. Also available online (BRS and WILSONLINE) and on CD-ROM. ISSN: 0003-6986.

Deep-sea Research. Part A: Oceanographic Research Papers.

Deep-sea Research. Part B: Oceanographic Literature Review. Pergamon Press, Inc., Maxwell House, Fairview Park, Elmsford, NY 10523. (914) 592-7700. Fax (914) 592-3625. 1953 To Present. Twelve times per year. $2000.00 per year for Parts A and B. Oceanographic Literature Review also available on CD-ROM. Inquire about price and availability.

General Science Index. H.W. Wilson Co., 950 University Avenue, Bronx, NY 10452. (212) 588-8400. (800) 367-6770. FAX (718) 590-1617. From 1978 to present. Ten issues per year; quarterly and annual cumulations. Service basis. Available on CD-ROM and online. Inquire regarding cost and availability. ISSN: 0162-1963.

Government Reports Announcements and Index. National Technical Information Service (NTIS), 5285 Port Royal Road, Springfield, VA 22161. (703) 487-4650. FAX (703) 321-8547. From 1968 to present. Annual. $630.00 per year. Also available online as NTIS BIBLIOGRAPHIC DATABASE and on CD-ROM. ISSN:

Meteorological and Geoastrophysical Abstracts. American Meteorological Society, c/o Inforonics, Inc., 550 Newtown Road, Box 458, Littleton, MA 01460. (508) 486-8976. FAX (508) 486-0027. Covers literature in environmental sciences, meteorology, astrophysics, hydrology, glaciology, and physical oceanography. 1950 to present. Monthly. $950.00 per year. ISSN: 0026-1130. Also available online (DIALOG) and on CD-ROM.

Oceanic Abstracts. Cambridge Scientific Abstracts, 7200 Wisconsin Avenue, Bethesda, MD 20814. (301) 961-6750. Fax (301) 961-6720. Bimonthly. $995.00 per year.

Science Citation Index. SCI. Institute for Scientific Information, 3501 Market Street, Philadelphia, PA 19104. (215) 386-0100. (800) 523-1850. FAX (215) 386-2991. 1961 to present. Six issues per year, plus annual cumulation. $11650.00 per year. Also available online and on CD-ROM. Inquire about price and availability. ISSN: 0036-827X.

Selected Water Resources Abstracts. U.S. Geological Survey. Distributed by National Technical Information Service (NTIS), 5285 Port Royal Road, Springfield, VA 22161. (703) 487-4650. Monthly. $115.00 per year.

ANNUAL REVIEWS AND YEARBOOKS

Developments In Atmospheric Science Series. Elsevier Science Publishing Company, Inc., 655 Avenue of the Americas, New York, NY 10010. (212) 989-5800. Irregular. Inquire about price and availability.

ASSOCIATIONS AND PROFESSIONAL SOCIETIES

American Meteorological Society, 45 Beacon Street, Boston, Ma 02108-3693. (617) 227-2425. FAX (617) 742-8718.

National Environmental Satellite Data, and Information Service. 2069 Federal Building 4 Room 2060, Washington, DC 20233. (301) 763-71990. FAX (301) 763-4011.

National Weather Association, 6704 Wolke Court, Montgomery, AL 36116-2134. (334) 213-0388. FAX (334) 213-0388.

University Corporation For Atmospheric Research. PO Box 3000, 1850 Table Mesa Drive, Boulder, CO 80307-3000. (303) 497-1673. FAX (303) 497-1654.

Weather Modification Association. PO Box 8116, Fresno CA 93747. (209) 291-8466.

BIBLIOGRAPHIES

Information Sources In the Earth Sciences. David N. Wood, Joan E. Hardy, and Anthony P. Harvey, editors. Bowker-Saur/ K.G. Saur. Distributed by R.R. Bowker, 121 Chanlon Road, New Providence, NJ 07974. (800) 521-8110 or (908) 464-6800. Second edition. 1989. $85.00.

Scientific and Technical Books and Serials In Print; An Index To Literature In Science and Technology. R.R. Bowker Co., 205 E. 42nd Street, New York, NY 10017. (800) 521-8110 or (212) 916-1600. 1992. $250.00.

DIRECTORIES AND BIOGRAPHICAL SOURCES

American Meteorological Society. Professional Directory. American Meteorological Society. 45 Beacon Street, Boston,

Ma 02108. (617) 227-2425. Included in each issue of the Bulletin of the Society.

Meteorological Services of the World. American Meteorological Society, 45 Beacon Street, Boston, MA 02108- 3693. (617) 227-2425. FAX (617) 742-8718. Annual. $70.00.

National Weather Service offices and Stations. National Oceanic and Atmospheric Administration, Deparatment of Commerce, Silver Spring, MD 20910. (301) 427-7698. Annual. Free

Research Centers Directory. Gale Research, 835 Penobscot Building, Detroit, MI 48226-4094. (800) 347-4253 or (313) 961-2242. 17th edition, 1992. $400.00.

ENCYCLOPEDIAS AND DICTIONARIES

Concise Oxford Dictionary of the Earth Sciences. Ailsa Allaby and Michael Allaby, editors. Oxford University Press, Inc., 200 Madison Avenue, New York, NY 10016. (800) 334-4249 or (212) 679-7300. 1990. $42.95.

McGraw-hill Encyclopedia of Ocean and Atmospheric Sciences. Sybil P. Parker, editor. McGraw-Hill Publishing Company, 1221 Avenue of the Americas, New York, NY 10020. (800) 262-4729 or (212) 512-3825. 1979. $79.95.

McGraw-hill Encyclopedia of Science and Technology. McGraw-Hill Publishing Company, 1221 Avenue of the Americas, New York, NY 10020. (800) 262-4729 or (212) 512-3825. Seventh edition. 1992. $1900.00.

GENERAL WORKS

The Atmosphere: An Introduction To Meteorology. Frederick K. Lutgens and Edward J. Tarbuck. Prentice Hall (Division of Simon and Schuster), 15 Columbus Circle, New York, NY 10023. (212) 373-8500 or (800) 922-0579. Fifth edition. 1991. $49.00.

Cloud Dynamics. Robert A. Houze. Academic Press, Inc.,1250 Sixth Avenue, San Diego, CA 92101-4311. (619) 231-0926. FAX (619) 699-6715. 1993. $149.00.

Clouds and Storms. Audubon Society Staff. Alfred A. Knopf, Inc. 201 E. 50th Street, New York, NY 10022. (800) 733-3000. 1995. $7.99.

Clouds, Rain, and Rainmaking. Basil J. Mason. Books on Demand, Division of University Microfilms International, 300 North Zeeb Road, Ann Arbor, MI 48106-1346. (313) 761-4700 or (800) 521-0600. 1974. $49.50.

Clouds: their Formation, Optical Properties and Effects. Peter V. Hobbs and Adarch Deepak, editors. Academic Press, Inc., 1250 Sixth Avenue, San Diego, CA 92101-4311. (619) 231-0926. FAX (619) 699-6715. 1981. $90.00.

Effective Cloud Cover Variations. Olivi Karner and Sirje Keevallik. A. Deepak Publishing, 101 Research Drive, Hampton, VA 23666. (703) 356-5249. 1993. $60.00.

Introduction To Boundary Layer Meteorology. Roland B. Stull. Kluwer Academic Publishers, P.O. Box 358, Accord Station, Hingham, MA 02018-0358. (617) 871-6000. 1988. $145.00.

Mesoscale Meteorology and Forecasting. Peter S. Ray, editor. American Meteorological Society, 45 Beacon Street, Boston, MA 02108-3693. (617) 227-2425. FAX (617) 742-8718. 1986. $50.00.

Meteorology and Environmental Sciences. R. Guzzi and others. World Scientific Publishing Company, Inc., 1060 Main Street, Unit B, River Edge, NJ 07661. (800) 227-7562 or (201) 487-9655. 1990. $120.00.

Meteorology: the Atmosphere and the Science of Weather. Joseph M. Moran. Macmillan Publishing, 866 3rd Avenue, New York, NY 10024. (212) 689-9140. Third edition. 1991. $49.95.

Physics of Rainclouds. Neville H. Fletcher. Books on Demand, Division of University Microfilms International, 300 North Zeeb Road, Ann Arbor, MI 48106-1346. (313) 761-4700 or (800) 521-0600. $101.00 in paper.

Severe and Unusual Weather. Joe R. Eagleman. Trimedia Publishing Company, 12008 West 87th Street, Suite 117, Lenexa, KS 66215. (913) 599-0505. Second edition. 1990. $41.95.

Storm and Cloud Dynamics. William R. Cotton and Richard A. Anthes. Academic Press, Inc., 1250 Sixth Avenue, San Diego, CA 92101-4311. (619) 231-0926. FAX (619) 699-6715. 1992. $69.95.

HANDBOOKS AND MANUALS

Author's Guide To the Journals of the American Meteorological Society. American Meteorological Society, 45 Beacon Street, Boston, MA 02108-3693. (617) 227-2425. FAX (617) 742-8718. 1983. $15.00 in paper.

Handbook In Applied Meteorology. David D. Houghton, editor. John Wiley and Sons, Inc., 605 Third Avenue, New York, NY 10158. (800) 526-5368 or (212) 850-6000. 1985. $114.00.

CLOUDS

Ency. of Physical Sciences and Engineering Info. Sources

International Cloud Atlas. World Meteorological Organization; American Meteorological Society. 45 Beacon Street, Boston, MA 02108-3693. (617) 227-2425. FAX (617) 742-8718. Volume2-photographs. 1987. $83.00.

Weather Companion: An Album of Meteorological History, Science, Legend, and Folklore. Gary Lockhart. John Wiley and Sons, Inc., 605 Third Avenue, New York, NY 10158. (800) 526-5368 or (212) 850-6000. 1988. $12.95.

ONLINE DATABASES AND CD-ROMS

Climate Assessment Database. National Weather Service, National Meteorological Center, 5200 Auth Road, Suite 101, Camp Springs, MD 20233. (301) 763-8016. Contains daily, weekly and monthly summaries of North American and world climatological data. Also provides five to ten weather forecasts, and 30 to 90 day weather outlook. Subscription required. Inquire as to cost and availability.

Ca Search. Chemical Abstracts Service, P.O. Box 3012, Columbus, OH 43210-0012. (614) 447-3600. (800) 848-6533. FAX (614) 447-3709. Very comprehensive guide to worldwide chemical literature and related fields, 1972 to present. Available on BRS,(800) 289-4277, DIALOG, (800) 334-2564, ORBIT (800) 456-7248, and STN International, FIZ Karlsruhe, P.O. Box 2465, W-7500, Karlsruhe 1, Germany, online services. Inquire as to cost and availability.

Compendex Plus. Engineering Information, Inc., 345 East 47th Street, New York, NY 10017. (212) 705-7600 or (800) 221-1044. Contains citations with abstracts to worldwide literature in engineering and technology, from 1970 to present. Available on online BRS,(800) 289-4277, DIALOG, (800) 334-2564, ORBIT (800) 456-7248, and STN International, FIZ Karlsruhe, P.O. Box 2465, W-7500, Karlsruhe 1, Germany, online services. Also available on CD-ROM. Inquire as to cost and availability.

Current Contents Search. Institute for Scientific Information, 3501 Market Street, Philadelphia, PA 19104. (215) 386-0100. FAX (215) 386-6362. Contains citations to articles listed in the table of contents of science and technology journals. Also articles in social sciences and life sciences journals. Available on BRS,(800) 289-4277, DIALOG,(800) 334-2564, online services. Inquire as to cost and availability.

Dissertation Abstracts. University Microfilms International, 300 North Zeeb Road, Ann Arbor, MI 48106. (800) 521-0600 or (313) 761-4700. Scope includes virtually all doctoral dissertations accepted at accredited American institutions from 1861 to present in 252 subject areas. Available on BRS,(800) 289-4277, DIALOG,(800) 334-2564, and OCLC EPIC,(800) 848-5878, online services. Also available on CD-ROM. Inquire as to cost and availability.

Inspec. Institution of Electrical Engineers, Michael Faraday House, Six Hills Way, Stevenage, Herts. SG1 2AY, England. Telephone: 0438 313311 or FAX 0438 742840. Contains citations to the worldwide literature of physics, electronics and electrical engineering, computer technology, and related fields. Available on BRS,(800) 289-4277, DIALOG, (800) 334-2564, ORBIT, (800) 456-7248, and STN International, FIZ Karlsruhe, P.O. Box 2465, W-7500,Karlsruhe 1, Germany, online services. Inquire as to cost and availability.

International Aerospace Abstracts. American Institute of Aeronautics and Astronautics, 370 L'Enfant Promenade, S.W., Washington, DC 20024. (202) 646-7400. Contains references and abstracts of journal and monograph literature relating to aerospace science and technology, from 1963 to present. Available through the NASA/RECON system of the National Aeronautics and Space Administration only.

Meteorological and Geoastrophysical Abstracts. American Meteorological Society, 45 Beacon Street, Boston, MA 02108-3693. (617) 227-2425. FAX (617) 742-8718. Contains citations and abstracts to the worldwide literature on significant research in meteorology and geoastrophysics. Related topics include physical oceanography, hydrology, environmental sciences and glaciology. Covers the period 1972 to present. Available on DIALOG,(800) 334-2564, online service. Inquire as to cost and availability.

NTIS Bibliographic Database. National Technical Information Service, 5285 Port Royal Road, Springfield, Va 22161. (703) 487-4929 or FAX (703) 321-8199. Broad coverage of government-sponsored science and technology research reports, 1964 to present. Available on BRS,(800) 289-4277, DIALOG, (800) 334-2564, ORBIT, (800) 456-7248, and STN International, FIZ Karlsruhe, P.O. Box 2465, W-7500, Karlsruhe 1, Germany, online services. Also available on CD-ROM. Inquire as to cost and availability.

Scisearch. Institute for Scientific Information, 3501 Market Street, Philadelphia, PA 19104. (800) 523-1850 or (215) 386-0100. Broad multidisciplinary title and author index to the international literature of science and technology, 1974 to present. Available on DIALOG and ORBIT online services. Also available on CD-ROM. Inquire as to cost and availability.

WILSONLINE. H.W. Wilson Company, 950 University Avenue, Bronx, NY 10452. (800) 367-6770 or (212) 588-8400. Makes available online versions of the printed H.W. Wilson indexes including Applied Science and Technology Index, Business Periodicals Index, General Science Index, and Readers' Guide to Periodical Literature. Period covered is generally 1983 to present. Available on BRS,(800) 289-4277, DIALOG,(800) 334-2564, and OCLC EPIC,(800) 848-5878, online services. Also available on CD-ROM. Inquire as to cost and availability.

World Climate Disc. Chadwyck-Healey Inc., 1101 King Street, Alexandria, VA 22314. (703) 683-4890. FAX (703) 683-7589. Weather and climate data from approximately 5,000 weather stations worldwide, covering the years 1854 to 1990 on CD-ROM. First edition 1992. Approximately $1200.00 per year with annual updates.

World Weatherdisc. WeatherDisc Associates, Inc., 4584 N.E. 89th Street, Seattle, WA 98115. (206) 524-4314. FAX (206) 543-0308. Meteorological data on CD-ROM which describes the climate of the earth today and for the past few hundred years. Firstedition 1989. Approximately $295.00 per year with biannual updates.

PERIODICALS

Agricultural and Forest Meteorology. Elsevier Science Publishing Company, Inc., 655 Avenue of the Americas, New York, NY 10010. (212) 989-5800. Twenty times per year. $750.00 per year.

American Meteorological Society. BULLETIN. American Meteorological Society, 45 Beacon Street, Boston, MA 02108-3693. (617) 227-2425. FAX (617) 742-8718. Monthly. $60.00 per year.

American Meteorological Society. METEOROLOGICAL MONOGRAPHS. American Meteorological Society, 45 Beacon Street, Boston, MA 02108-3693. (617) 227-2425. FAX (617) 742-8718. Irregular. Price varies.

American Weather Observer. Association of American Weather Observers, 401 Whitney Boulevard, Box 455, Belvedere, IL 61008. (815) 544-5665. FAX (815) 544-6334. Monthly. $21.00 per year.

Atmosphere - Ocean. Canadian Meteorological and Oceanographic Society, P.O. Box 334, Newmarket, Ontario, L3Y 4X7, Canada. (416) 898-1040. FAX (416) 898-7937. Quarterly. $30.00 per year.

Boundary-layer Meteorology: An International Journal of Physical and Biological Processes In the Atmospheric Boundary Layer. Kluwer Academic Publishers, P.O. Box 358, Accord Station, Hingham, MA 02018-0358. (617) 871-6000. Sixteen per year. $785.00 per year.

Climate Change. Kluwer Academic Publishers, P.O. Box 358, Accord Station, Hingham, MA 02018-0358. (617) 871-6000. Six times per year. $327.00 per year.

Dynamics of Atmospheres and Oceans. Elsevier Science Publishing Company, Inc., 655 Avenue of the Americas, New York, NY 10010. (212) 989-5800. Six times per year. $205.00 per year.

International Journal of Climatology. Royal Meteorological Society, distributed by John Wiley and Sons, Inc., 605 Third Avenue, New York, NY 10158. (800) 526-5368 or (212) 850-6000. Eight times per year. $425.00 per year.

JGR: Journal of Geophysical Research: Atmosphere. American Geophysical Union, 2000 Florida Avenue, N.W., Washington, DC 20009. (202) 462-6903. Monthly. $90.00 per year to members.

JGR: Journal of Geophysical Research: Oceans. American Geophysical Union, 2000 Florida Avenue, N.W., Washington, DC 20009. (202) 462-6903. Monthly. $1545.00 per year.

Journal of Applied Meteorology. American Meteorological Society, 45 Beacon Street, Boston, MA 02108-3693. (617) 227-2425. FAX (617) 742-8718. Monthly. $165.00 per year.

Journal of Climate. American Meteorological Society, 45 Beacon Street, Boston, MA 02108-3693. (617) 227-2425. FAX (617) 742-8718. Monthly. $175.00 per year.

Journal of Atmospheric Sciences. American Meteorological Society, 45 Beacon Street, Boston, MA 02108-3693. (617) 227-2425. FAX (617) 742-8718. Semi-monthly. $320.00 per year.

Monthly Weather Review. American Meteorological Society, 45 Beacon Street, Boston, MA 02108-3693. (617) 227-2425. FAX (617) 742-8718. Monthly. $205.00 per year.

National Weather Digest. National Weather Association, 4400 Stamp Road, Room 404, Temple Hills, MD 20748. (301) 899-3784.

Royal Meteorological Society. QUARTERLY JOURNAL. Royal Meteorological Society, 104 Oxford Road, Reading, Berks RG1 7LJ, England. Six times per year. $250.00 per year.

Weather. Royal Meteorological Society, 104 Oxford Road, Reading, Berks RG1 7LJ, England. Monthly. $44.00 per year.

Weatherwise. Heldref Publications, 1319 Eighteenth Street, N.W., Washington, DC 20036-1802. (202) 296-6267. FAX (202) 296-5149. Bi-monthly. $28.00 per year.

RESEARCH CENTERS AND INSTITUTES

Center for Cloud Physics Research. University of Missouri - Rolla, Rolla, MO 65401. (314) 341-4332.

Cloud Simulation and Aerosol Laboratory. Colorado State University, Fort Collins, CO 80523. (303) 491-8667. FAX (303) 491-8449.

Cooperative Institute For Mesoscale Meteorological Studies. University of Oklahoma, 401 East Boyd, Norman, OK 73019. (405) 325-3041.

Institute of Atmospheric Physics. University of Arizona, Tucson, AZ 85721. (602) 626-6831.

National Center For Atmospheric Research. P.O. Box 3000, Boulder, CO 80307. (303) 496-1000. FAX (303) 497-1654.

CLOTH

See: TEXTILES

COAL

See also: COAL GASIFICATION AND LIQUEFACTION, COAL MINING, GEOLOGY

ABSTRACT SERVICES AND INDEXES

Applied Science and Technology Index; A Cumulative Subject Index To English Language Periodicals In the Fields of Aeronautics and Space Science, Computer Technology, Chemistry, Construction Industry, Energy and Related Areas. H.W. Wilson Co., 950 University Avenue, Bronx, NY 10452. (800) 367-6770 or (212) 588-8400. FAX (718) 590-1617. From 1958 to present. Monthly. Inquire about cost and availability. Also available on CD-ROM.

Bibliography and Index of Geology. American Geological Institute, 4220 King Street, Alexandria, VA 22302. (703) 379-2480. Fax (703) 379-7563. Monthly. $1295.00 per year. Also available online as GeoRef.

Civil and Structural Engineering Abstracts, Cambridge Scientific Abstracts, 7200 Wisconsin Avenue, Bethesda, MD 20814-4823. (301) 961-6750. FAX (301) 961-6720. Monthly. Topics covered include structural design, construction equipment and methods, civil defense and military engineering, surveying, highway engineering, maritime and port structures, materials, land reclamation, and soil mechanics.

Energy Research Abstracts. 1976-present. U.S. Department of Energy, office of Scientific and Technical Information, Box 62,

COAL

Ency. of Physical Sciences and Engineering Info. Sources

Oak Ridge, TN 37831. (615) 574-0733. Subscriptions: Superintendent of Documents, U.S. Government Printing office, Box 371954, Pittsburgh, PA 15250-7954. (202) 783-3238. FAX (202) 512-2233. Monthly. $164.00 per year. Abstracts all the scientific and technical reports, journal articles, conference proceedings, patents, theses, and monographs sponsored by the U.S. Energy Research and Development Administration.

Engineering Index Monthly. Engineering Information, Inc., Castle Point on the Hudson, Hoboken, NJ 07030. (800) 221-1044. FAX (212) 832-1857. Monthly. $2200.00 per year. Also available online as COMPENDEX, and also on CD-ROM.

Fuel and Energy Abstracts. Butterworth-Heinemann, Linacre House, Jordan Hill, Ocford OX2 8DP, England. Telephone 0865-310366. FAX 0865-310898. Summary of world literature on all scientific, technical, commercial, and environmental aspects of fuel and energy. 1960 to present. Six times a year. Inquire for price in U.S.

Geological Society of America. ABSTRACTS WITH PROGRAMS. Geological Society of America. 3300 Penrose Place, P.O. Box 9140, Boulder, CO 80301-9140. (303) 447-2020. Abstracts and programs of the annual conference. Annual. $69.00.

Geotechnical Abstracts. Deutsche Gesellschaft fure Erd- und Grundbau e.V., Hohenzollernstrasse 52, 4300 Essen 1, Germany. Also sponsored by International Society for Soil Mechanics and Foundation Engineering. Covers soil mechanics, foundation engineering, and engineering geology. 1970 to present. Monthly. $200.00 per year.

International Joural of Rock Mechanics and Mining Sciences and Geomechanics Abstracts. Elsevier Science, 660 White Plains Rd., Tarrytown, NY 10591-5153. (914) 524-9200. FAX (914) 333-2444. Six times a year. $845 per year.

I.M.M. Abstracts and Index. Institution of Mining and Metallurgy, 44 Portland Pl., London W1N 4BR, England. 071-580-3802. FAX 071-436-5388. Bi-monthly. $364 for non-members.

ANNUAL REVIEWS AND YEARBOOKS

Coal Data. National Coal Association, 1130 17th Street NW, Washington, DC 20036. (202) 463-2640. Annual. $75.00.

Coal Facts. National Coal Association, 1130 17th Street NW, Washington, DC 20036. (202) 463-2640. 1948 to present. Annual. $15.00.

Coal Production. U.S. Department of Energy, Energy Information Administration, National Energy Information Center, El-231, Room 1F-048, Forrestal Bldg., 1000 Independence Avenue SW, Washington, DC 20585. (202) 586-8800. 1976 to present. Annual. $7.50.

ASSOCIATIONS AND PROFESSIONAL SOCIETIES

American Institute of Mining, Metallurgical & Petroleum Engineers, 345 East 47th Street, New York, Ny 10017.

American Mining Congress, 1920 N St., NW, Suite 300, Washington, DC 20036-1662. (202) 861-2800. FAX (202) 861-7535.

Mineralogical Society of America, 1130 17th St. NW, Suite 330, Washington, DC 20036. (202) 775-4344. FAX (202) 775-0018.

The Minerals, Metals & Materials Society, 410 Commonwealth Drive, Warrendale, Pa 15086.

Mining and Metallurgical Society of America, 275 Madison Ave., Room 2301, New York, NY 10016. (212) 684-4150.

National Coal Association. 1130 17th Street NW, Washington, DC 20036. (202) 463-2625. FAX (202) 463-6152.

Open Pit Mining Association. c/o Consolidated Coal Company, Consol Plaza, Pittsburgh, PA 15241. (412) 831-4440.

Society For Mining, Metallurgy and Exploration, Inc., 8307 Shaffer Parkway, PO Box 625002, Littleton, CO 80162-5002. (303) 973-9550. FAX (303) 973-3845.

Society of Mining Engineers of Aime, Caller Number D, Littleton, Co 80127. (303) 973-9550.

DIRECTORIES AND BIOGRAPHICAL SOURCES

American Mines Handbook. Southam Magazine Group, 1450 Don Mills Rd., Don Mills, ON M3B 2X7, Canada. (416) 442-2004. FAX (416) 442-2261. Annual. Inquire for price information.

Coal Mine Directory 1995. Maclean Hunter Publishing Company, 29 North Wacker Drive, Chicago, IL 60606-3298. (312) 726-2802. 1995. $140.00.

Directory of Engineering Societies and Related Organizations. Gordon Davis. 13th edition. American Association of Engineering Societies, 1111 19th Street NW, Suite 608, Washington, DC 20036. (202) 296-2237 or (800) 658-8897. 1989. Inquire for price.

E & MJ International Directory of Mining. Maclean Hunter Publishing Company, Mining Information Services, 29 N. Wacker Dr., Chicago, IL 60606. (312) 726-2802. FAX (312) 726-2574. Annual. $120.00.

Research Centers Directory. Gale Research, 835 Penobscot Building, Detroit, MI 48226-4094. (313) 961-2242. (800) 347-4253. FAX (313) 961-6083. $485.00.

Who's Who In Technology. Gale Research, 835 Penobscot Building, Detroit, MI 48226-4094. (313) 961-2242. (800) 347-4253. FAX (313) 961-6083. 1995. $195.00.

ENCYCLOPEDIAS AND DICTIONARIES

The Encyclopedia of the Solid Earth Sciences. Philip Kearey, editor-in-chief. Blackwell Scientific Publications, 238 Main Street, Cambridge, MA 02142. (617) 876-7000 or (800) 759-6102. FAX (617) 876-7022. 1993. $60.00.

Encyclopedia of Physical Science and Technology. Robert A. Meyers, ed. 18 volumes. Academic Press Inc., 6277 Sea Harbor Drive, Orlando, FL 32887. (800) 321-5068. 1992. $2100.00

McGraw-hill Encyclopedia of Science and Technology. Sybil P. Parker, ed. 7th edition. 20 volumes. McGraw-Hill Publishing

Company, 1221 Avenue of the Americas, New York, NY 10020. (800)262-4729 or (212) 512-3825. 1992. $1900.00

GENERAL WORKS

Changing Structure of the U.S. Coal Industry: An Update. Diane Publishing Company, 600 Upland Avenue, Upland, PA 19015. (610) 499-7415. FAX (610) 499-7429. 1994. $45.00.

The Chemistry and Technology of Coal. Speight. Second expanded and revised edition. Marcel Dekker, Inc., 270 Madison Avenue, New York, NY 10016. (212) 696-9000. FAX (212) 685-4540. 1994. $195.00.

Clean and Efficient Use of Coal: the New Era For Low-rank Coal. Organization for Economic Cooperation and Development, 2001 L St. NW, Washington, DC 20036. (202) 785-6323 or (800) 456-6323. FAX (202) 785-0350. 1994. $97.00.

Coal and Methane Industry International. Rector Press Ltd., 130 Rattlesnake, Leverett, MA 01054-9726. (800) 247-3473. FAX (413) 367-2853. 1994. $125.00.

Coal: Classification, Coalification, Mineralogy, Trace-element Chemistry, and Oil & Gas Potential. B. Alpern and P.C. Lyon, editors. Elsevier Science Inc., Box 882, Madison Square Station, New York, NY 10159. (212) 989-5800. 1990. $172.00.

Coal Production. Gordon Press Publishers, PO Box 459, Bowling Green Station, New York, NY 10004. (718) 624-8419. 1995. $250.00.

Coal Science. Leo J. Lynch, et al., editors. American Chemical Society, 1155 16th Street, N.W., Washington, DC 20036. (202) 872-4381 or (800) 333-9511. FAX (614) 447-3671. 1991. $77.95.

An Introduction To Coal Technology. Herbert Berkowitz. 2nd edition. Academic Press Inc., 6277 Sea Harbor Drive, Orlando, FL 32887. (800) 321-5068. 1994. $69.00.

Holmes' Principles of Physical Geology. Donald Duff. 4th edition. Chapman and Hall, 29 West 35th Street, New York, NY 10001. (800) 842-3636. FAX (212) 563-2269. 1992. $39.95.

United States Coal Industry 1970-1990: Two Decades of Change. Gordon Press Publishers, PO Box 459, Bowling Green Station, New York, NY 10004. (718) 624-8419. 1995. $250.00.

HANDBOOKS AND MANUALS

Handbook of Practical Coal Geology. Larry Thomas. John Wiley and Sons, Inc., 605 Third Avenue, New York, NY 10158. (800) 526-5368 or (212) 850-6000. 1992. $69.95.

1994 Keystone Coal Industry Manual. Maclean Hunter Publishing Company, 29 North Wacker Drive, Chicago, IL 60606-3298. (312) 726-2802. FAX (312) 726-2574. 1993. $190.00.

ONLINE DATABASES AND CD-ROMS

Ca Search. Chemical Abstracts Service, P.O. Box 3012, Columbus, OH 43210-0012. (614) 447-3600. (800) 848-6533. FAX (614) 447-3709. Very comprehensive guide to worldwide chemical literature and related fields, 1972 to present. Available on

BRS,(800) 289-4277, DIALOG, (800) 334-2564, ORBIT (800) 456-7248, and STN International, FIZ Karlsruhe, P.O. Box 2465, W-7500, Karlsruhe 1, Germany, online services. Inquire as to cost and availability.

Compendex Plus. Engineering Information, Inc., 345 East 47th Street, New York, NY 10017. (212) 705-7600 or (800) 221-1044. Contains citations with abstracts to worldwide literature in engineering and technology, from 1970 to present. Available on online BRS,(800) 289-4277, DIALOG, (800) 334-2564, ORBIT (800) 456-7248, and STN International, FIZ Karlsruhe, P.O. Box 2465, W-7500, Karlsruhe 1, Germany, online services. Also available on CD-ROM. Inquire as to cost and availability.

CSA Engineering. Cambridge Scientific Abstracts, 7200 Wisconsin Avenue, Suite 601, Bethesda, MD 20814. (301) 961-6750 or (800) 843-7751. Contains citations and abstracts of international periodicals and research literature covering all fields of engineering and science and technology, including computer and information science, electronics, mechanical engineering, solid state materials, 1981 to present. Available on BRS,(800) 289-4277, online service.Inquire as to cost and availability.

Current Contents Search. Institute for Scientific Information, 3501 Market Street, Philadelphia, PA 19104. (215) 386-0100. FAX (215) 386-6362. Contains citations to articles listed in the table of contents of science and technology journals. Also articles in social sciences and life sciences journals. Available on BRS,(800) 289-4277, DIALOG,(800) 334-2564, online services. Inquire as to cost and availability.

Dissertation Abstracts. University Microfilms International, 300 North Zeeb Road, Ann Arbor, MI 48106. (800) 521-0600 or (313) 761-4700. Scope includes virtually all doctoral dissertations accepted at accredited American institutions from 1861 to present in 252 subject areas. Available on BRS,(800) 289-4277, DIALOG,(800) 334-2564, and OCLC EPIC,(800) 848-5878, online services. Also available on CD-ROM. Inquire as to cost and availability.

Geoarchive. Geosystems, P.O. Box 1024, Westminster, London, England, SW1 P 2JL. Citations to literature on geoscience, 1969 to present. Inquire as to online cost and availability.

Geobase. Elsevier - Geo Abstracts, Regency House, 34 Duke Street, Norwich NR3 3AP, England. Contains citations to the worldwide earth science literature from 1980 to date. Available on DIALOG, ORBIT online services. Inquire as to cost and availability.

GEOREF: Bibliography and Index of Geology. American Geological Institute, 4220 King Street, Alexandria, VA 22302. (703) 379-2480. Fax (703) 379-7563. Monthly. Inquire for price and availability.

NTIS Bibliographic Database. National Technical Information Service, 5285 Port Royal Road, Springfield, Va 22161. (703) 487-4929 or FAX (703) 321-8199. Broad coverage of government-sponsored science and technology research reports, 1964 to present. Available on BRS,(800) 289-4277, DIALOG, (800) 334-2564, ORBIT, (800) 456-7248, and STN International, FIZ Karlsruhe, P.O. Box 2465, W-7500, Karlsruhe 1, Germany, online services. Also available on CD-ROM. Inquire as to cost andavailability.

COAL

Ency. of Physical Sciences and Engineering Info. Sources

PERIODICALS

Coal. Maclean Hunter Publishing Company, 29 North Wacker Drive, Chicago, IL 60606-3298. (312) 726-2802. FAX (312) 726-2574. 1964 to present. Monthly. $31.25 per year.

Coal Outlook. Pasha Publications Inc., 1616 N. Ft. Myer Drive, Suite 1000, Arlington, VA 22209-3107. (703) 528-1244. FAX (703) 528-1253. 1975 to present. Weekly. $745.00 per year.

Coal Science and Technology. Elsevier Science Inc., Box 882, Madison Square Station, New York, NY 10159. (212) 989-5800. FAX (212) 633-3990. 1981 to present. Irregular. Price varies.

Coal Voice. National Coal Association, 1130 17th Street NW, Washington, DC 20036. (202) 463-2640. 1978 to present. Bi-monthly. $25.00.

Coal Week. McGraw-Hill Inc., Energy & Business Newsletters, 1221 Avenue of the Americas, New York, NY 10020. (800) 262-4729 or (212) 512-3825. 1975 to present. Weekly. $912.00 per year.

Fuel Processing Technology. Elsevier Science Inc., Box 882 Madison Square Station, New York, NY 10159. (212) 989-5800. FAX (212) 633-3990. 1978 to present. 12 times a year. $781.00 per year.

RESEARCH CENTERS AND INSTITUTES

Center For Coal Science. 313 TCNW, Western Kentucky University, Bowling Green, KY 42101. (502) 745-6244. FAX (502) 745-6293.

Center For Research On Sulfur In Coal (crsc). Coal Development Park, Suite 200, PO Box 8, Carterville, IL 62918-0008. (618) 985-3500.

Coal and Lignite Research Laboratory. Texas A&M University, Civil Engineering Department, College Station, TX 77843-3136. (409) 845-5229. FAX (409) 845-6156.

Iowa State University, Iowa State Mining and Mineral Resources Research Institute. 111 Metals Development, Ames, IA 50011. (515) 294-7936. FAX (515) 294-3226.

Pennsylvania State University Cooperative Program In Coal Research. 517A Deike Bldg., University Park, PA 16802. (814) 865-6544.

Pennsylvania State University Energy and Fuels Research Center. 513 Deike Bldg., University Park, PA 16802. (814) 865-6544.

Southern Illinois University At Carbondale, Coal Extraction and Utilization Research Center. Carbondale, IL 62901. (618) 536-5521. FAX (618) 453-7346.

University of Kentucky Center For Applied Energy Research. 3572 Iron Works Pike, Lexington, KY 40511. (606) 257-0305. FAX (606) 0220.

University of North Dakota Fuels & Process Chemistry Research Institute. Box 8213, University Station, Grand Forks, ND 58202. (701) 777-5000. FAX (701) 777-5181.

University of Utah Center For Advanced Coal Technology. Department of Fuels Engineering, 306 William Browning Bldg., SaltLake City, UT 84112-1183. (801) 581-7187.

SPECIFICATIONS AND STANDARDS

Annual Book of ASTM Standards Vol. 05.05: Gaseous Fuels; Coal & Coke. American Society for Testing and Materials (ASTM), 1916 Race Street, Philadelphia, PA 19103. (215) 299-5585. See latest edition.

COAL GASIFICATION AND LIQUEFACTION

See also: COAL, COAL MINING

ABSTRACT SERVICES AND INDEXES

Applied Science and Technology Index; A Cumulative Subject Index To English Language Periodicals In the Fields of Aeronautics and Space Science, Computer Technology, Chemistry, Construction Industry, Energy and Related Areas. H.W. Wilson Co., 950 University Avenue, Bronx, NY 10452. (800) 367-6770 or (212) 588-8400. FAX (718) 590-1617. From 1958 to present. Monthly. Inquire about cost and availability. Also available on CD-ROM.

Energy Research Abstracts. 1976-present. U.S. Department of Energy, office of Scientific and Technical Information, Box 62, Oak Ridge, TN 37831. (615) 574-0733. Subscriptions: Superintendent of Documents, U.S. Government Printing office, Box 371954, Pittsburgh, PA 15250-7954. (202) 783-3238. FAX (202) 512-2233. Monthly. $164.00 per year. Abstracts all the scientific and technical reports, journal articles, conference proceedings, patents, theses, and monographs sponsored by the U.S. Energy Research and Development Administration.

Engineering Index Monthly. Engineering Information, Inc., Castle Point on the Hudson, Hoboken, NJ 07030. (800) 221-1044. FAX (212) 832-1857. Monthly. $2200.00 per year. Also available online as COMPENDEX, and also on CD-ROM.

Fuel and Energy Abstracts. Butterworth-Heinemann, Linacre House, Jordan Hill, Ocford OX2 8DP, England. Telephone 0865-310366. FAX 0865-310898. Summary of world literature on all scientific, technical, commercial, and environmental aspects of fuel and energy. 1960 to present. Six times a year. Inquire for price in U.S.

ASSOCIATIONS AND PROFESSIONAL SOCIETIES

Coal and Slurry Technology Association. 1156 15th Street NW, Suite 525, Washington, DC 20005. (202) 796-1133. FAX (202) 223-3504.

National Coal Association. 1130 17th Street NW, Washington, DC 20036. (202) 463-2640.

DIRECTORIES AND BIOGRAPHICAL SOURCES

American Mines Handbook. Southam Magazine Group, 1450 Don Mills Rd., Don Mills, ON M3B 2X7, Canada. (416) 442-

2004. FAX (416) 442-2261. Annual. Inquire for price information.

Directory of Engineering Societies and Related Organizations. Gordon Davis. 13th edition. American Association of Engineering Societies, 1111 19th Street NW, Suite 608, Washington, DC 20036. (202) 296-2237 or (800) 658-8897. 1989. Inquire for price.

1994 Coal Mine Directory. Maclean Hunter Publishing Company, 29 North Wacker Drive, Chicago, IL 60606-3298. (312) 726-2802. FAX (312) 726-2574. 1993. $120.00.

Research Centers Directory. Gale Research, 835 Penobscot Building, Detroit, MI 48226-4094. (313) 961-2242. (800) 347-4253. FAX (313) 961-6083. $485.00.

Who's Who In Technology. Gale Research, 835 Penobscot Building, Detroit, MI 48226-4094. (313) 961-2242. (800) 347-4253. FAX (313) 961-6083. 1995. $195.00.

ENCYCLOPEDIAS AND DICTIONARIES

Encyclopedia of Physical Science and Technology. Robert A. Meyers, ed. 18 volumes. Academic Press Inc., 6277 Sea Harbor Drive, Orlando, FL 32887. (800) 321-5068. 1992. $2100.00.

McGraw-hill Encyclopedia of Science and Technology. Sybil P. Parker, ed. 7th edition. 20 volumes. McGraw-Hill Publishing Company, 1221 Avenue of the Americas, New York, NY 10020. (800) 262-4729 or (212) 512-3825. 1992. $1900.00

GENERAL WORKS

Coal Gasification: Direct Applications and Synthesis of Chemicals and Fuels. S.S. Penner, editor. Elsevier Science Inc., Box 882, Madison Square Station, New York, NY 10159. (212) 989-5800. 1987. $75.00.

Coal Gasification: Existing Processes and New Developments. H.D. Schilling and B. Bonn. Graham & Trotman Inc., 101 Philip Drive, Norwell, MA 02061-1677. (617) 871-6600. FAX (617) 871-6528. 1990. $56.00.

Coal Liquefaction Fundamentals. D. D. Whitehurst, editor. American Chemical Society, 1155 16th Street, N.W., Washington, DC 20036. (202) 872-4381 or (800) 333-9511. FAX (614) 447-3671. 1980. $49.95.

Fundamentals of Coal Combustion and Gasification. L. Douglas Smoot, editor. Elsevier Science Inc., Box 882, Madison Square Station, New York, NY 10159. (212) 989-5800. 1992. Inquire for cost and availability.

How To Produce Methanol From Coal. Emil Supp. Springer-Verlag, 175 Fifth Avenue, New York, NY 10010. (212) 460-1500. FAX (212) 473-6272. 1990. Inquire for cost and availability.

HANDBOOKS AND MANUALS

1994 Keystone Coal Industry Manual. Maclean Hunter Publishing Company, 29 North Wacker Drive, Chicago, IL 60606-3298. (312) 726-2802. FAX (312) 726-2574. 1993. $190.00.

ONLINE DATABASES AND CD-ROMS

Ca Search. Chemical Abstracts Service, P.O. Box 3012, Columbus, OH 43210-0012. (614) 447-3600. (800) 848-6533. FAX (614) 447-3709. Very comprehensive guide to worldwide chemical literature and related fields, 1972 to present. Available on BRS,(800) 289-4277, DIALOG, (800) 334-2564, ORBIT (800) 456-7248, and STN International, FIZ Karlsruhe, P.O. Box 2465, W-7500, Karlsruhe 1, Germany, online services. Inquire as to cost and availability.

Compendex Plus. Engineering Information, Inc., 345 East 47th Street, New York, NY 10017. (212) 705-7600 or (800) 221-1044. Contains citations with abstracts to worldwide literature in engineering and technology, from 1970 to present. Available on online BRS,(800) 289-4277, DIALOG, (800) 334-2564, ORBIT (800) 456-7248, and STN International, FIZ Karlsruhe, P.O. Box 2465, W-7500, Karlsruhe 1, Germany, online services. Also available on CD-ROM. Inquire as to cost and availability.

Csa Engineering. Cambridge Scientific Abstracts, 7200 Wisconsin Avenue, Suite 601, Bethesda, MD 20814. (301) 961-6750 or (800) 843-7751. Contains citations and abstracts of international periodicals and research literature covering all fields of engineering and science and technology, including computer and information science, electronics, mechanical engineering, solid state materials, 1981 to present. Available on BRS,(800) 289-4277, online service.Inquire as to cost and availability.

Current Contents Search. Institute for Scientific Information, 3501 Market Street, Philadelphia, PA 19104. (215) 386-0100. FAX (215) 386-6362. Contains citations to articles listed in the table of contents of science and technology journals. Also articles in social sciences and life sciences journals. Available on BRS,(800) 289-4277, DIALOG,(800) 334-2564, online services. Inquire as to cost and availability.

Dissertation Abstracts. University Microfilms International, 300 North Zeeb Road, Ann Arbor, MI 48106. (800) 521-0600 or (313) 761-4700. Scope includes virtually all doctoral dissertations accepted at accredited American institutions from 1861 to present in 252 subject areas. Available on BRS,(800) 289-4277, DIALOG,(800) 334-2564, and OCLC EPIC,(800) 848-5878, online services. Also available on CD-ROM. Inquire as to cost and availability.

NTIS Bibliographic Database. National Technical Information Service, 5285 Port Royal Road, Springfield, Va 22161. (703) 487-4929 or FAX (703) 321-8199. Broad coverage of government-sponsored science and technology research reports, 1964 to present. Available on BRS,(800) 289-4277, DIALOG, (800) 334-2564, ORBIT, (800) 456-7248, and STN International, FIZ Karlsruhe, P.O. Box 2465, W-7500, Karlsruhe 1, Germany, online services. Also available on CD-ROM. Inquire as to cost and availability.

PERIODICALS

Coal. Maclean Hunter Publishing Company, 29 North Wacker Drive, Chicago, IL 60606-3298. (312) 726-2802. FAX (312) 726-2574. 1964 to present. Monthly. $31.25 per year.

Coal Outloook. Pasha Publications Inc., 1616 N. Ft. Myer Drive, Suite 1000, Arlington, VA 22209-3107. (703) 528-1244. FAX (703) 528-1253. 1975 to present. Weekly. $745.00 per year.

Coal Science and Technology. Elsevier Science Inc., Box 882, Madison Square Station, New York, NY 10159. (212) 989-5800.

FAX (212) 633-3990. 1981 to present. Irregular. Price varies.

Coal Week. McGraw-Hill Inc., Energy & Business Newsletters, 1221 Avenue of the Americas, New York, NY 10020. (800) 262-4729 or (212) 512-3825. 1975 to present. Weekly. $912.00 per year.

Fuel Processing Technology. Elsevier Science Inc., Box 882 Madison Square Station, New York, NY 10159. (212) 989-5800. FAX (212) 633-3990. 1978 to present. 12 times a year. $781.00 per year.

RESEARCH CENTERS AND INSTITUTES

Center For Coal Science. 313 TCNW, Western Kentucky University, Bowling Green, KY 42101. (502) 745-6244. FAX (502) 745-6293.

City College of City University of New York, Clean Fuels Institute. 104th Street and Convent Avenue, New York, NY 10031. (212) 690-8136.

Empire State Electric Energy Research Corporation. 1155 Avenue of the Americas, New York, NY 10036. (212) 302-1212.

Montana State University Engineering Experiment Station. Bozeman, MT 59717. (406) 994-2272. FAX (406) 994-6292.

Oklahoma State University, University Center For Energy Research. Room 005 Sciences East, Life Sciences E., Stillwater, OK 74078. (405) 744-5700.

Pennsylvania State University Combustion Laboratory. 404 Academic Activities Bldg., University Park, PA 16802. (814) 865-5752.

Pennsylvania State University Energy and Fuels Research Center. 513 Deike Bldg., University Park, PA 16802. (814) 865-6544.

Purdue University Coal Research Center. Potter Engineering Center, West Lafayette, in 47906. (317) 494-7037. FAX (317) 494-2351.

Syracuse University Institute For Energy Research. 103 College Place, Syracuse, NY 13244-4010. (315) 423-3353.

University of Colorado—boulder, Engineering Research Center. Campus Box 423, Boulder, CO 80309. (303) 492-7427.

University of North Dakota Energy & Environmental Research Center. Box 8213, University Station, Grand Forks, ND 58202. (701) 777-5000. FAX (701) 777-5181.

University of Utah Center For Advanced Coal Technology. Department of Fuels Engineering, 306 William Browning Bldg., Salt Lake City, UT 84112-1183. (801) 581-7187.

COAL MINING

See also: COAL, COAL GASIFICATION AND LIQUIFICATION, MINING ENGINEERING

ABSTRACT SERVICES AND INDEXES

Applied Mechanics Reviews: An Assessment of World Literature In Engineering Sciences. American Society of Mechanical Engineers, 345 East 47th Street, New York, NY 10017. (212) 705-7703. 1948 to present. Monthly. $360.00 per year.

Applied Science and Technology Index; A Cumulative Subject Index To English Language Periodicals In the Fields of Aeronautics and Space Science, Computer Technology, Chemistry,Construction Industry, Energy and Related Areas. H.W. Wilson Co., 950 University Avenue, Bronx, NY 10452. (800) 367-6770 or (212) 588-8400. FAX (718) 590-1617.From 1958 to present. Monthly. Inquire about cost and availability. Also available on CD-ROM.

Bibliography and Index of Geology. American Geological Institute, 4220 King Street, Alexandria, VA 22302. (703) 379-2480. Fax (703) 379-7563. Monthly. $1295.00 per year. Also available online as GeoRef.

Civil and Structural Engineering Abstracts, Cambridge Scientific Abstracts, 7200 Wisconsin Avenue, Bethesda, Md 20814-4823. (301) 961-6750. FAX (301) 961-6720. Monthly. Topics covered include structural design, construction equipment and methods, civil defense and military engineering, surveying, highway engineering, maritime and port structures, materials, land reclamation, and soil mechanics.

Engineering Index Monthly. Engineering Information, Inc., Castle Point on the Hudson, Hoboken, NJ 07030. (800) 221-1044. FAX (212) 832-1857. Monthly. $2200.00 per year. Also available online as COMPENDEX, and also on CD-ROM.

Geological Society of America. ABSTRACTS WITH PROGRAMS. Geological Society of America. 3300 Penrose Place, P.O. Box 9140, Boulder, CO 80301-9140. (303) 447-2020. Abstracts and programs of the annual conference. Annual. $69.00.

Geotechnical Abstracts. Deutsche Gesellschaft fure Erd- und Grundbau e.V., Hohenzollernstrasse 52, 4300 Essen 1, Germany. Also sponsored by International Society for Soil Mechanics and Foundation Engineering. Covers soil mechanics, foundation engineering, and engineering geology. 1970 to present. Monthly. $200.00 per year.

International Joural of Rock Mechanics and Mining Sciences and Geomechanics Abstracts. Elsevier Science, 660 White Plains Rd., Tarrytown, NY 10591-5153. (914) 524-9200. FAX (914) 333-2444. Six times a year. $845 per year.

I.M.M. Abstracts and Index. Institution of Mining and Metallurgy, 44 Portland Pl., London W1N 4BR, England. 071-580-3802. FAX 071-436-5388. Bi-monthly. $364 for non-members.

ASSOCIATIONS AND PROFESSIONAL SOCIETIES

American Institute of Mining, Metallurgical & Petroleum Engineers, 345 East 47th Street, New York, Ny 10017.

American Mining Congress, 1920 N St., NW, Suite 300, Washington, DC 20036-1662. (202) 861-2800. FAX (202) 861-7535.

American Society of Civil Engineers, 345 East 47th Street, New York, Ny 10017.

Mineralogical Society of America, 1130 17th St. NW, Suite 330, Washington, DC 20036. (202) 775-4344. FAX (202) 775-0018.

The Minerals, Metals & Materials Society, 410 Commonwealth Drive, Warrendale, Pa 15086.

Mining and Metallurgical Society of America, 275 Madison Ave., Room 2301, New York, NY 10016. (212) 684-4150.

National Coal Association. 1130 17th Street NW, Washington, DC 20036. (202) 463-2625. FAX (202) 463-6152.

Open Pit Mining Association. c/o Consolidated Coal Company, Consol Plaza, Pittsburgh, PA 15241. (412) 831-4440.

Society For Mining, Metallurgy and Exploration, Inc., 8307 Shaffer Parkway, PO Box 625002, Littleton, CO 80162-5002. (303) 973-9550. FAX (303) 973-3845.

Society of Mining Engineers of Aime, Caller Number D, Littleton, Co 80127. (303) 973-9550.

DIRECTORIES AND BIOGRAPHICAL SOURCES

American Mines Handbook. Southam Magazine Group, 1450 Don Mills Rd., Don Mills, ON M3B 2X7, Canada. (416) 442-2004. FAX (416) 442-2261. Annual. Inquire for price information.

Coal Mine Directory 1995. Maclean Hunter Publishing Company, 29 North Wacker Drive, Chicago, IL 60606-3298. (312) 726-2802. FAX (312) 726-2574. 1995. $140.00.

Directory of Engineering Societies and Related Organizations. Gordon Davis. 13th edition. American Association of Engineering Societies, 1111 19th Street NW, Suite 608, Washington, DC 20036. (202) 296-2237 or (800) 658-8897. 1989. Inquire for price.

E & Mj International Directory of Mining. Maclean Hunter Publishing Company, Mining Information Services, 29 N. Wacker Dr., Chicago, IL 60606. (312) 726-2802. FAX (312) 726-2574. Annual. $120.00.

Financial Times International Year Books: Mining. St. James Press, 425 N. Michigan Ave., Chicago, IL 60611. Annual. Inquire for price.

Mining Directory: Mines & Mining Equipment and Service Companies Worldwide (year). Don Nelson Publications Ltd., PPO Box 193, Barnet, Herts EN4 8LP, England. Telephone 081-368-5534. FAX 081-368-7010. Annual. Inquire for price.

Research Centers Directory. Gale Research, 835 Penobscot Building, Detroit, MI 48226-4094. (313) 961-2242. (800) 347-4253. FAX (313) 961-6083. $485.00.

Who's Who In Engineering. American Association of Engineering Societies, 1111 19th Street NW, Suite 608, Washington, DC 20036. (202) 296-2237 or (800) 658-8897. 8th edition. 1991. Inquire for price.

ENCYCLOPEDIAS AND DICTIONARIES

The Contractors' Dictionary of Equipment, Tools, and Techniques: For Civil Engineering, Construction, Forestry, Open-pit Mining, and Public Works. L.F. Webster. John Wiley and Sons, Inc., 605 Third Avenue, New York, NY 10158. (800) 526-5368 or (212) 850-6000. 1995. Inquire for price.

Dictionary of Mining & Processing. Helmut Schmidt. Collets/State Mutual Book & Periodical Service Ltd., 521 Fifth Ave., 17th floor, New York, NY 10175. (718) 261-1704. FAX (516) 537-0412. 1980. $99.00.

The Encyclopedia of the Solid Earth Sciences. Philip Kearey, editor-in-chief. Blackwell Scientific Publications, 238 Main Street, Cambridge, MA 02142. (617) 876-7000 or (800) 759-6102. FAX (617) 876-7022. 1993. $60.00.

GENERAL WORKS

Coal and Methane Industry International. Rector Press Ltd., 130 Rattlesnake, Leverett, MA 01054-9726. (800) 247-3473. FAX (413) 367-2853. 1994. $125.00.

Coal Production. Gordon Press Publishers, PO Box 459, Bowling Green Station, New York, NY 10004. (718) 624-8419. 1995. $250.00.

Elements of Practical Coal Mining. Douglas F. Crickmer and David A. Zegeer, editors. 2nd reprint edition. Books on Demand, Division of University Microfilms International, 300 North Zeeb Road, Ann Arbor, MI 48106-1346. (313) 761-4700 or (800) 521-0600. 1996. $180.00.

Logistics of Underground Coal Mining. James J. Hanslovan & Richard G. Visovsky. Noyes Publications, 120 Mill Road, Park Ridge, NJ 07656. (201) 391-8484. FAX (201) 391-6833. 1984. $32.00.

United States Coal Industry 1970-1990: Two Decades of Change. Gordon Press Publishers, PO Box 459, Bowling Green Station, New York, NY 10004. (718) 624-8419. 1995. $250.00.

HANDBOOKS AND MANUALS

Handbook of Practical Coal Geology. Larry Thomas. John Wiley and Sons, Inc., 605 Third Avenue, New York, NY 10158. (800) 526-5368 or (212) 850-6000. 1992. $69.95.

1994 Keystone Coal Industry Manual. Maclean Hunter Publishing Company, 29 North Wacker Drive, Chicago, IL 60606-3298. (312) 726-2802. FAX (312) 726-2574. 1993. $190.00.

Underground Mining Methods Handbook. William J. Hustrulid. Society for Mining, Metallurgy and Exploration, Inc., 8307 Shaffer Parkway, PO Box 625002, Littleton, CO 80162-5002. (303) 973-9550. FAX (303) 973-3845. 1982. $120.00.

ONLINE DATABASES AND CD-ROMS

Compendex Plus. Engineering Information, Inc., 345 East 47th Street, New York, NY 10017. (212) 705-7600 or (800) 221-1044. Contains citations with abstracts to worldwide literature in engineering and technology, from 1970 to present. Available on online BRS, (800) 289-4277, DIALOG, (800) 334-2564,

COAL MINING

Ency. of Physical Sciences and Engineering Info. Sources

ORBIT (800) 456-7248, and STN International, FIZ Karlsruhe, P.O. Box 2465, W-7500, Karlsruhe 1, Germany, online services. Also available on CD-ROM. Inquire as to cost and availability.

CSA Engineering. Cambridge Scientific Abstracts, 7200 Wisconsin Avenue, Suite 601, Bethesda, MD 20814. (301) 961-6750 or (800) 843-7751. Contains citations and abstracts of international periodicals and research literature covering all fields of engineering and science and technology, including computer and information science, electronics, mechanical engineering, solid state materials, 1981 to present. Available on BRS,(800) 289-4277, online service. Inquire as to cost and availability.

Current Contents Search. Institute for Scientific Information, 3501 Market Street, Philadelphia, PA 19104. (215)386-0100. FAX (215) 386-6362. Contains citations to articles listed in the table of contents of science and technology journals. Also articles in social sciences and life sciences journals. Available on BRS,(800) 289-4277, DIALOG,(800) 334-2564, online services. Inquire as to cost and availability.

Dissertation Abstracts. University Microfilms International, 300 North Zeeb Road, Ann Arbor, MI 48106. (800) 521-0600 or (313) 761-4700. Scope includes virtually all doctoral dissertations accepted at accredited American institutions from 1861 to present in 252 subject areas. Available on BRS,(800) 289-4277, DIALOG,(800) 334-2564, and OCLC EPIC,(800) 848-5878, online services. Also available on CD-ROM. Inquire as to cost and availability.

GEOREF: Bibliography and Index of Geology. American Geological Institute, 4220 King Street, Alexandria, VA 22302. (703) 379-2480. Fax (703) 379-7563. Monthly. Inquire for price and availability.

NTIS Bibliographic Database. National Technical Information Service, 5285 Port Royal Road, Springfield, Va 22161. (703) 487-4929 or FAX (703) 321-8199. Broad coverage of government-sponsored science and technology research reports, 1964 to present. Available on BRS,(800) 289-4277, DIALOG, (800) 334-2564, ORBIT, (800) 456-7248, and STN International, FIZ Karlsruhe, P.O. Box 2465, W-7500, Karlsruhe 1, Germany, online services. Also available on CD-ROM. Inquire as to cost and availability.

Scisearch. Institute for Scientific Information, 3501 Market Street, Philadelphia, PA 19104. (800) 523-1850 or (215) 386-0100. Broad multidisciplinary title and author index to the international literature of science and technology, 1974 to present. Available on DIALOG,(800) 334-2564, and ORBIT,(800) 456-7248, online services. Also available on CD-ROM. Inquire as to cost and availability.

WILSONLINE. H.W. Wilson Company, 950 University Avenue, Bronx, NY 10452. (800) 367-6770 or (212) 588-8400. Makes available online versions of the printed H.W. Wilson indexes including Applied Science and Technology Index, Business Periodicals Index, General Science Index, and Readers' Guide to Periodical Literature. Period covered is generally 1983 to present. Available on BRS,(800) 289-4277, DIALOG,(800) 334-2564, and OCLC EPIC,(800) 848-5878, online services. Also available on CD-ROM. Inquire as to cost and availability.

PERIODICALS

Amc Journal. American Mining Congress, 1920 N St., NW, Suite 300, Washington, DC 20036-1662. (202) 861-2800. FAX (202) 861-7535. Monthly. $36 per year.

Advances In Mining Science and Technology. Elsevier Science Inc., Box 882, Madison Square Station, New York, NY 10159. (212) 989-5800. Irregular. Inquire for price.

Canadian Mining Journal. Southam Magazine Group, 1450 DonMills Rd., Don Mills, ON M3B 2X7, Canada. (416) 442-2004. FAX (416) 442-2261. Bi-monthly. $46.00 per year.

CIM Bulletin. Canadian Institute of Mining, Metallurgy, and Petroleum, Xerox Tower, 3400 de Maisonneuve Blvd. W., Suite 1210, Montreal PQH3Z 3B8, Canada. (514) 939-2710. FAX (514) 939-2714. 1898 to present. Monthly. $105.00 per year.

Coal. Maclean Hunter Publishing Company, 29 North Wacker Drive, Chicago, IL 60606-3298. (312) 726-2802. FAX (312) 726-2574. 1964 to present. Monthly. $31.25 per year.

Coal Outlook. Pasha Publications Inc., 1616 N. Ft. Myer Drive, Suite 1000, Arlington, VA 22209-3107. (703) 528-1244. FAX (703) 528-1253. 1975 to present. Weekly. $745.00 per year.

Coal Science and Technology. Elsevier Science Inc., Box 882, Madison Square Station, New York, NY 10159. (212) 989-5800. FAX (212) 633-3990. 1981 to present. Irregular. Price varies.

Coal Voice. National Coal Association, 1130 17th Street NW, Washington, DC 20036. (202) 463-2640. 1978 to present. Bimonthly. $25.00.

Coal Week. McGraw-Hill Inc., Energy & Business Newsletters, 1221 Avenue of the Americas, New York, NY 10020. (800) 262-4729 or (212) 512-3825. 1975 to present. Weekly. $912.00 per year.

Colorado School of Mines Quarterly. Colorado School of Mines Press, 1500 Illinois, Golden, CO 80401-1887. (303) 273-3000. FAX (303) 273-3310. Quarterly. $50.00 per year.

Engineering & Mining Journal. Maclean Hunter Publishing Company, 29 N. Wacker Dr., Chicago, IL 60606. (312) 726-2802. FAX (312) 726-2574. Monthly. $60.00 per year.

Institution of Mining and Metallurgy. TRANSACTIONS. SECTION A: MINING INDUSTRY. Institution of Mining and Metallurgy, 44 Portland Pl., London W1N 4BR, England. 071-580-3802. FAX 071-436-5388. Three times a year. $126.00 a year.

Mine and Quarry. IML Group, Blair House, High St., Tonbridge, Kent TN(1BQ, England. Telephone 0732-359990. FAX 0732-770049. Monthly. Inquire for price.

Mine & Quarry Trader. Allied Publications, 7355 N. Woodland, Box 603, Indianapolis, in 46206-0603. (317) 297-5500. FAX (317) 299-1356. Montly. $18.00 per year.

Mining Engineer. Institution of Mining Engineers, Danum House, South Parade, Doncaster DN1 2DY, England. Phone 0302-320486. FAX 0302-340554. Monthly. Inquire for price.

Mining Engineering. Society for Mining, Metallurgy, and Exploration, Inc., 8307 Shaffer Parkway, PO Box 625002, Littleton, CO 80162-5002. (303) 973-9550. FAX (303) 973-3845. Monthly. $100.00 per year.

Pit and Quarry. Advanstar Communications Inc., 7500 Old Oak Blvd., Cleveland, OH 44130. (216) 826-2839. FAX (216) 891-2726. 1916 to present. Monthly. $35.00 per year.

World Mining Equipment. Metal Bulletin, Inc., 220 Fifth Ave., New York, NY 10001-7781. (212) 213-6202. FAX (212) 213-1870.11 times a year. Inquire for price.

RESEARCH CENTERS AND INSTITUTES

Coal/mining Productivity Research Center. Ohio University, Department of Industrial Systems Engineering, 280 Stocker Court, Athens, OH 45701-2979. (614) 593-1548.

Mining & Excavation Research Institute. 1825 K St NW, Suite 218, Washington, DC 20006-1202. (202) 785-3756. FAX (202) 429-9417.

Southern Illinois University At Carbondale, Illinois Mining and Mineral Resources Research Institute. Coal Extraction and Utilization Research Center, Carbondale, IL 62901. (618) 536-6637. FAX (618) 453-7455.

University of Kentucky Institute For Mining and Minerals Research. 233 Mining and Mineral Resources Bldg., Lexington, KY 40506. (606) 257-8636. FAX (606) 258-1049.

University of Missouri—rolla, Missouri Mining and Mineral Resources Research Institute. 305 McNutt Hall, Rolla, MO 65401. (314) 341-4153. FAX (314) 341-4192.

COASTAL ENGINEERING

See also: CIVIL ENGINEERING, COASTS, CONTINENTAL MARGINS, OCEAN ENGINEERING

ABSTRACT SERVICES AND INDEXES

Applied Science and Technology Index; A Cumulative Subject Index To English Language Periodicals In the Fields of Aeronautics and Space Science, Computer Technology, Chemistry, Construction Industry, Energy and Related Areas. H.W. Wilson Co., 950 University Avenue, Bronx, NY 10452. (800) 367-6770 or (212) 588-8400. FAX (718) 590-1617. From 1958 to present. Monthly. Inquire about cost and availability. Also available on CD-ROM and online.

ASCE Combined Annual Combined Index. American Society of Civil Engineers, 345 East 47th Street, New York, NY 10017-2398. (212) 705-7520 or (800) 548-2723. Annual. $48.00.

ASCE Publications Information. 1966-present. American Society of Civil Engineers, 345 East 47th Street, New York, NY 10017-2398. (212) 705-7520 or (800) 548-2723. Bi-monthly. $160.00 per year for non-members.

Bibliography and Index of Geology. American Geological Institute, 4220 King Street, Alexandria, VA 22302. (703) 379-2480. Fax (703) 379-7563. Monthly. $1295.00 per year. Also available online as GEOREF.

Civil and Structural Engineering Abstracts, Cambridge Scientific Abstracts, 7200 Wisconsin Avenue, Bethesda, Md 20814-4823. (301) 961-6750. FAX (301) 961-6720. 1993 to present. Monthly. $385.00 per year. Topics covered include structural design, construction equipment and methods, civil defense and military engineering, surveying, highway engineering, maritime

and port structures, materials, land reclamation, and soil mechanics.

Engineering Index Monthly. Engineering Information, Inc., Castle Point on the Hudson, Hoboken, NJ 07030. (800) 221-1044. FAX(212) 832-1857. Monthly. $2300.00 per year. Also available online as COMPENDEX, and also on CD-ROM. Covers chemical engineering, computers, electrical engineering, civil engineering, metals and mining, industrial management, and mechanical engineering.

Environmental Engineering Abstracts, Cambridge Scientific Abstracts, 7200 Wisconsin Avenue, Bethesda, Md 20814-4823. (301) 961-6750. FAX (301) 961-6720. Monthly. Covers hazardous materials, environmental impact and protection, treatment of sewage and industrial wastes, hydroelectric power, tidal and wind power, artic and tropical engineering.

Fluid Abstracts: Civil Engineering. Elsevier Science [journals], 660 White Plains Rd., Tarrytown, NY 10159-5153. (919) 524-9200. FAX (919) 333-2444. 1991 to present. Monthly plus annual cumulation. $645.00 per year. Also available online as FLUIDEX. Covers civil engineering applications of fluid mechanics, hydraulics, flow metering and measuring, offshore engineering, environmental hydraulics, and related aspects of wind energy, the atmosphere, and aerodynamics.

Oceanic Abstracts. Cambridge Scientific Abstracts, 7200 Wisconsin Avenue, Bethesda, MD 20814. (301) 961-6750. Fax (301) 961-6720. Bimonthly. $995.00 per year.

Science Citation Index. Institute for Scientific Information, 3501 Market Street, Philadelphia, PA 19104. (215) 386-0100. FAX (215) 386-6362. Inquire about availability and cost. Also available on CD-ROM.

ASSOCIATIONS AND PROFESSIONAL SOCIETIES

American Society of Civil Engineers. 345 East 47th Street, New York, NY 10017-2398. (212) 705-7520 or (800) 548-2723.

Coastal Engineering Research Council. 4911 Bay Oaks Court, College Station, TX 77845. (409) 690-0306.

The Coastal Society. PO Box 2081, Gloucester, MA 01930-2081. (508) 281-9209.

Estuarine Research Federation. 490 Chippingwood Drive, Suite 2, Port Republic, MD 20676. (301) 855-1876. FAX (301) 586-9226.

Geological Society of America. 3300 Penrose Place, PO Box 9140, Denver, CO 80301-9140. (303) 447-2020.

DIRECTORIES AND BIOGRAPHICAL SOURCES

Directory of Engineering Societies and Related Organizations. Gordon Davis. 13th edition. American Association of Engineering Societies, 1111 19th Street NW, Suite 608, Washington, DC 20036. (202) 296-2237 or (800) 658-8897. 1989. Inquire for price.

International Research Centers Directory. Gale Research, 835 Penobscot Building, Detroit, MI 48226-4094. (313) 961-2242. (800) 347-4253. FAX (313) 961-6083. 8th edition. 1995. $410.00

COASTAL ENGINEERING

Ency. of Physical Sciences and Engineering Info. Sources

Research Centers Directory. Gale Research, 835 Penobscot Building, Detroit, MI 48226-4094. (313) 961-2242. (800) 347-4253. FAX (313) 961-6083. $485.00.

Scientific and Technical Organizations and AgenciesDirectory. Gale Research, 835 Penobscot Building, Detroit, MI 48226-4094. (313) 961-2242. (800) 347-4253. FAX (313) 961-6083. 4th edition. 1996. $195.00.

Who's Who In Engineering. American Association of Engineering Societies, 1111 19th Street NW, Suite 608, Washington, DC 20036. (202) 296-2237 or (800) 658-8897. 8th edition. 1991. Inquire for price.

ENCYCLOPEDIAS AND DICTIONARIES

Dictionary of Civil Engineering. V.N. Vazirani. South Asia Books, Box 502, Columbia, MO 65205. (314) 474-0116. FAX (314) 474-8124. 1992. $20.00.

Encyclopedia of Physical Science and Technology. Robert A. Meyers, ed. 18 volumes. Academic Press Inc., 6277 Sea Harbor Drive, Orlando, FL 32887. (800) 321-5068. 1992. $2100.00

GENERAL WORKS

Advances In Coastal and Ocean Engineering. Philip L-F Liu. World Scientific Publishing Company, Inc., 1060 Main Street, Unit B, River Edge, NJ 07661. (800) 227-7562. FAX (201) 487-9656. 1995. $86.00.

Coastal Engineering: Waves, Beaches, Wave Structure Interactions. T. Sawaragi. Elsevier Science Inc., Box 882, Madison Square Station, New York, NY 10159. (212) 989-5800. 1995. Inquire for cost and availability.

Coastal Management. Institute of Civil Engineers Staff. American Society of Civil Engineers, 345 East 47th Street, New York, NY 10017-2398. (212) 705-7520 or (800) 548-2723. 1989. $95.00.

Coastal Stabilization: Innovative Concepts. Richard Silvester and John R.C. Hsu. Prentice Hall, 113 Sylvan Avenue, Route 9W, Englewood Cliffs, NJ 07632. (201) 592-2000 or (800) 922-0579. 1993. $98.00.

Physical Models and Laboratory Techniques In Coastal Engineering. Steven A. Hughes. World Scientific Publishing Company, Inc., 1060 Main Street, Unit B, River Edge, NJ 07661. (800) 227-7562. FAX (201) 487-9656. 1993. $97.00 (hardback) or $48.00 (paperback).

HANDBOOKS AND MANUALS

Coastal, Estuarial, and Harbour Engineer's Reference Book. M.B. Abbott and W.A. Price. E & FN Spon/ Chapman and Hall, 115 Fifth Avenue, New York, NY 10211-0906. 1994. $115.95.

Coastal Zone Management Handbook. John R. Clark. Lewis Publishers, 121 S. Main St., PO Box 519, Chelsea, MI 48118. (313) 475-8619 or (800) 272-7737. 1995. $89.95.

Handbook of Coastal and Ocean Engineering Volume 1: Wave Phenomena and Coastal Structures. John B. Herbich, editor.

Gulf Publishing Company, P.O. Box 2608, Houston, TX. (713) 529-4301 or (800) 231-6275. 1990. $195.00.

Handbook of Coastal and Ocean Engineering Volume 2: offshore Structures, Marine Foundations, and Sediment Processes. John B. Herbich, editor. Gulf Publishing Company, P.O. Box 2608,Houston, TX. (713) 529-4301 or (800) 231-6275. 1990. $195.00.

Handbook of Coastal and Ocean Engineering Volume 3: Harbors, Navigational Channels, Estuaries and Environmental Effects. John B. Herbich, editor. Gulf Publishing Company, P.O. Box 2608, Houston, TX. (713) 529-4301 or (800) 231-6275. 1992. $195.00.

ONLINE DATABASES AND CD-ROMS

Compendex Plus. Engineering Information, Inc., 345 East 47th Street, New York, NY 10017. (212) 705-7600 or (800) 221-1044. Contains citations with abstracts to worldwide literature in engineering and technology, from 1970 to present. Available on online BRS,(800) 289-4277, DIALOG, (800) 334-2564, ORBIT (800) 456-7248, and STN International, FIZ Karlsruhe, P.O. Box 2465, W-7500, Karlsruhe 1, Germany, online services. Also available on CD-ROM. Inquire as to cost and availability.

Current Contents Search. Institute for Scientific Information, 3501 Market Street, Philadelphia, PA 19104. (215) 386-0100. FAX (215) 386-6362. Contains citations to articles listed in the table of contents of science and technology journals. Also articles in social sciences and life sciences journals. Available on BRS,(800) 289-4277, DIALOG,(800) 334-2564, online services. Inquire as to cost and availability.

Dissertation Abstracts. University Microfilms International, 300 North Zeeb Road, Ann Arbor, MI 48106. (800) 521-0600 or (313) 761-4700. Scope includes virtually all doctoral dissertations accepted at accredited American institutions from 1861 to present in 252 subject areas. Available on BRS,(800) 289-4277, DIALOG,(800) 334-2564, and OCLC EPIC,(800) 848-5878, online services. Also available on CD-ROM. Inquire as to cost and availability.

Earth Sciences. U.S. Geological Survey, 12201 Sunrise Valley Drive, Reston, VA 22092-9998. (703) 648-4460. CD-ROM of earth science databases including the U.S. Geological Survey Library, Earth Science Data Directory, and GEOINDEX, citations to published geological maps. $350.00 per year with quarterly updates.

Geoarchive. Geosystems, P.O. Box 1024, Westminster, London, England, SW1 P 2JL. Citations to literature on geoscience, 1969 to present. Inquire as to online cost and availability.

Geobase. Elsevier - Geo Abstracts, Regency House, 34 Duke Street, Norwich NR3 3AP, England. Contains citations to the worldwide earth science literature from 1980 to date. Available on DIALOG, ORBIT online services. Inquire as to cost and availability.

GEOREF: Bibliography and Index of Geology. American Geological Institute, 4220 King Street, Alexandria, VA 22302. (703) 379-2480. Fax (703) 379-7563. Monthly. Inquire for price and availability.

NTIS Bibliographic Database. National Technical Information Service, 5285 Port Royal Road, Springfield, Va 22161. (703)

487-4929 or FAX (703) 321-8199. Broad coverage of government-sponsored science and technology research reports, 1964 to present. Available on BRS,(800) 289-4277, DIALOG, (800)334-2564, ORBIT, (800) 456-7248, and STN International, FIZ Karlsruhe, P.O. Box 2465, W-7500, Karlsruhe 1, Germany, online services. Also available on CD-ROM. Inquire as to cost and availability.

Scisearch. Institute for Scientific Information, 3501 Market Street, Philadelphia, PA 19104. (800) 523-1850 or (215) 386-0100. Broad multidisciplinary title and author index to the international literature of science and technology, 1974 to present. Available on DIALOG,(800) 334-2564, and ORBIT,(800) 456-7248, online services. Also available on CD-ROM. Inquire as to cost and availability.

WILSONLINE. H.W. Wilson Company, 950 University Avenue, Bronx, NY 10452. (800) 367-6770 or (212) 588-8400. Makes available online versions of the printed H.W. Wilson indexes including Applied Science and Technology Index, Business Periodicals Index, General Science Index, and Readers' Guide to Periodical Literature. Period covered is generally 1983 to present. Available on BRS,(800) 289-4277, DIALOG,(800) 334-2564, and OCLC EPIC,(800) 848-5878, online services. Also available on CD-ROM. Inquire as to cost and availability.

PERIODICALS

Coastal Engineering. Elsevier Science Inc., Box 882, Madison Square Station, New York, NY 10159. (212) 989-5800. 1977 to present. 12 times a year. $642.00 per year.

Coastal Research. Florida State University Geology Department, Tallahassee, FL 32306. (904) 644-5860. FAX (904) 644-4214. 1962 to present. 3 times a year. $6.00 per year.

Coastal Management. Taylor & Francis, 1900 Frost Road, Suite 101, Bristol, PA 19007. (215) 785-5800. FAX (215) 785-5515. 1973 to present. Quarterly. $140.00 per year.

Coastwatch. North Carolina Sea Grant College Program, Box 8605, North Carolina State University, Raleigh, NC 27695-8605. (919) 515-2454. FAX (919) 515-7095. 1970 to present. Bi-monthly. $12.00 per year.

Estuarine Coastal and Shelf Science. Academic Press Ltd./ Harcourt Brace & Company, Ltd., Foots Cray High Street, Sidcup, Kent DA14 5HP, England. Telephone 44-81-300-3322. FAX 44-81-309-0807. 1973 to present. Monthly. Inquire for cost.

Journal of Coastal Research. Coastal Education & Research Foundation, 4310 N.E. 25th Avenue, Ft. Lauderdale, FL 33308. (305) 565-1051. FAX (305) 565-1051. 1985 to present. Quarterly. $58.00.

Journal of Waterway, Port, Coastal, and Ocean Engineering. American Society of Civil Engineers, Waterway, Port, Coastal, and Ocean Division, 345 East 47th Street, New York, NY 10017-2398. (212) 705-7520 or (800) 548-2723. 1956 to present. Bi-monthly. $100.00 per year.

Ocean and Coastal Management. Elsevier Science [journals], 660 White Plains Rd., Tarrytown, NY 10159-5153. (919) 524-9200. FAX (919) 333-2444. 1973 to present. Twelve times a year. $565.00.

Ocean Engineering. Elsevier Science [journals], 660 White Plains Rd., Tarrytown, NY 10159-5153. (919) 524-9200. FAX (919) 333-2444. 1968 to present. 8 times a year. $595.00 per year.

RESEARCH CENTERS AND INSTITUTES

California Institute of Technology, W.m. Keck Laboratory of Hydraulics and Water Resources. 138-78, Pasadena, CA 91125. (818) 356-4404. FAX (818) 356-2940.

Coastal Education & Research Foundation. 4310 N.E. 25th Avenue, Ft. Lauderdale, FL 33308. (305) 565-1051. FAX (305) 565-1051.

Massachusetts Institute of Technology, Ralph M. PARSONS LABORATORY. Rm 48-311, 15 Vassar Street, Cambridge, MA 02139. (617) 253-2117. FAX (617) 258-8850.

Massachusetts Institute of Technology, Sea Grant College Program. 29 Main Street, Cambridge, MA 02139. (617) 253-7041. FAX (617) 253-8000.

Purdue University, Great Lakes Coastal Research Laboratory. c/o School of Civil Engineering, West Lafayette, in 47907. (317) 494-3713.

Stevens Institute of Technology, Davidson Laboratory. Castle Point Station, Hoboken, NJ 07030. (201) 420-5300. FAX (201) 420-5714.

University of Florida Coastal and Oceanographic Engineering Laboratory. 336 Weil Hall, Gainesville, FL 32611. (904) 392-1436. FAX (904) 392-3466.

COASTS

See also: CIVIL ENGINEERING, COASTAL ENGINEERING, CONTI-NENTAL MARGINS, OCEAN ENGINEERING, OCEANOGRAPHY

ABSTRACT SERVICES AND INDEXES

Applied Science and Technology Index; A Cumulative Subject Index To English Language Periodicals In the Fields of Aeronautics and Space Science, Computer Technology, Chemistry, Construction Industry, Energy and Related Areas. H.W. Wilson Co., 950 University Avenue, Bronx, NY 10452. (800) 367-6770 or (212) 588-8400. FAX (718) 590-1617. From 1958 to present. Monthly. Inquire about cost and availability. Also available on CD-ROM and online.

ASCE Combined Annual Combined Index. American Society of Civil Engineers, 345 East 47th Street, New York, NY 10017-2398. (212) 705-7520 or (800) 548-2723. Annual. $48.00.

ASCE Publications Information. 1966-present. American Society of Civil Engineers, 345 East 47th Street, New York, NY 10017-2398. (212) 705-7520 or (800) 548-2723. Bi-monthly. $160.00 per year for non-members.

Bibliography and Index of Geology. American Geological Institute, 4220 King Street, Alexandria, VA 22302. (703) 379-

COASTS

Ency. of Physical Sciences and Engineering Info. Sources

2480. Fax (703) 379-7563. Monthly. $1295.00 per year. Also available online as GEOREF.

Civil and Structural Engineering Abstracts, Cambridge Scientific Abstracts, 7200 Wisconsin Avenue, Bethesda, Md20814-4823. (301) 961-6750. FAX (301) 961-6720. 1993 to present. Monthly. $385.00 per year. Topics covered include structural design, construction equipment and methods, civil defense and military engineering, surveying, highway engineering, maritime and port structures, materials, land reclamation, and soil mechanics.

Engineering Index Monthly. Engineering Information, Inc., Castle Point on the Hudson, Hoboken, NJ 07030. (800) 221-1044. FAX (212) 832-1857. Monthly. $2300.00 per year. Also available online as COMPENDEX, and also on CD-ROM. Covers chemical engineering, computers, electrical engineering, civil engineering, metals and mining, industrial management, and mechanical engineering.

Environmental Engineering Abstracts, Cambridge Scientific Abstracts, 7200 Wisconsin Avenue, Bethesda, Md 20814-4823. (301) 961-6750. FAX (301) 961-6720. Monthly. Covers hazardous materials, environmental impact and protection, treatment of sewage and industrial wastes, hydroelectric power, tidal and wind power, artic and tropical engineering.

Fluid Abstracts: Civil Engineering. Elsevier Science [journals], 660 White Plains Rd., Tarrytown, NY 10159-5153. (919) 524-9200. FAX (919) 333-2444. 1991 to present. Monthly plus annual cumulation. $645.00 per year. Also available online as FLUIDEX. Covers civil engineering applications of fluid mechanics, hydraulics, flow metering and measuring, offshore engineering, environmental hydraulics, and related aspects of wind energy, the atmosphere, and aerodynamics.

Meteorological and Geoastrophysical Abstracts. American Meteorological Society, c/o Inforonics Inc., 550 Newtown Road, Littleton, MA 01460. (508) 486-8976. FAX (508) 486-0027. 1950 to present. Monthly. $950.00 per year. Current abstracts of books, reports, research papers, and miscellaneous literature on environmental sciences, meteorology, astrophysics, hydrology, glaciology, and physical oceanography.

Oceanic Abstracts. Cambridge Scientific Abstracts, 7200 Wisconsin Avenue, Bethesda, MD 20814. (301) 961-6750. Fax (301) 961-6720. Bimonthly. $995.00 per year.

Science Citation Index. Institute for Scientific Information, 3501 Market Street, Philadelphia, PA 19104. (215) 386-0100. FAX (215) 386-6362. Inquire about availability and cost. Also available on CD-ROM.

ASSOCIATIONS AND PROFESSIONAL SOCIETIES

American Society of Civil Engineers. 345 East 47th Street, New York, NY 10017-2398. (212) 705-7520 or (800) 548-2723.

Coastal Engineering Research Council. 4911 Bay Oaks Court, College Station, TX 77845. (409) 690-0306.

The Coastal Society. PO Box 2081, Gloucester, MA 01930-2081. (508) 281-9209.

Estuarine Research Federation. 490 Chippingwood Drive, Suite 2, Port Republic, MD 20676. (301) 855-1876. FAX (301) 586-9226.

Geological Society of America. 3300 Penrose Place, PO Box9140, Denver, CO 80301-9140. (303) 447-2020.

DIRECTORIES AND BIOGRAPHICAL SOURCES

Directory of Engineering Societies and Related Organizations. Gordon Davis. 13th edition. American Association of Engineering Societies, 1111 19th Street NW, Suite 608, Washington, DC 20036. (202) 296-2237 or (800) 658-8897. 1989. Inquire for price.

International Research Centers Directory. Gale Research, 835 Penobscot Building, Detroit, MI 48226-4094. (313) 961-2242. (800) 347-4253. FAX (313) 961-6083. 8th edition. 1995. $410.00

Research Centers Directory. Gale Research, 835 Penobscot Building, Detroit, MI 48226-4094. (313) 961-2242. (800) 347-4253. FAX (313) 961-6083. $485.00.

Scientific and Technical Organizations and Agencies Directory. Gale Research, 835 Penobscot Building, Detroit, MI 48226-4094. (313) 961-2242. (800) 347-4253. FAX (313) 961-6083. 4th edition. 1996. $195.00.

Who's Who In Engineering. American Association of Engineering Societies, 1111 19th Street NW, Suite 608, Washington, DC 20036. (202) 296-2237 or (800) 658-8897. 8th edition. 1991. Inquire for price.

ENCYCLOPEDIAS AND DICTIONARIES

Dictionary of Civil Engineering. V.N. Vazirani. South Asia Books, Box 502, Columbia, MO 65205. (314) 474-0116. FAX (314) 474-8124. 1992. $20.00.

Encyclopedia of Physical Science and Technology. Robert A. Meyers, ed. 18 volumes. Academic Press Inc., 6277 Sea Harbor Drive, Orlando, FL 32887. (800) 321-5068. 1992. $2100.00.

GENERAL WORKS

Coastal Management. Institute of Civil Engineers Staff. American Society of Civil Engineers, 345 East 47th Street, New York, NY 10017-2398. (212) 705-7520 or (800) 548-2723. 1989. $95.00.

Coastal Stabilization: Innovative Concepts. Richard Silvester and John R.C. Hsu. Prentice Hall, 113 Sylvan Avenue, Route 9W, Englewood Cliffs, NJ 07632. (201) 592-2000 or (800) 922-0579. 1993. $98.00.

Coasts. Richard A. Davis. Prentice Hall, 113 Sylvan Avenue, Route 9W, Englewood Cliffs, NJ 07632. (201) 592-2000 or (800) 922-0579. 1995. $25.33.

Coasts: Erosion and Sedimentation. H. Bremer and K.M. Clayton, editors. Lubrecht & Cramer Ltd., 38 Country Route 48, Forestburgh, NY 12777-6400. (914) 794-8539. FAX (914) 791-7575. 1989. $88.00.

The Evolving Coasts. Richard A. Davis. W.H. Freeman and Company, 41 Madison Avenue, East 26th, 35th Floor, New York, NY 10010. (212) 576-9400. FAX (212) 689-2383. 1995. $32.95.

Ency. of Physical Sciences and Engineering Info. Sources

COASTS

HANDBOOKS AND MANUALS

Coastal, Estuarial, and Harbour Engineer's Reference Book. M.B. Abbott & W.A. Price, editors. E & FN Spon/ Routledge,Chapman and Hall, Inc., 29 West 35th Street, New York, NY 10001-2291. (212) 244-3336 or FAX (212) 563-2269. 1994. Inquire for cost and availability.

Handbook of Coastal and Ocean Engineering Volume 1: Wave Phenomena and Coastal Structures. John B. Herbich, editor. Gulf Publishing Company, P.O. Box 2608, Houston, TX. (713) 529-4301 or (800) 231-6275. 1990. $195.00.

Handbook of Coastal and Ocean Engineering Volume 2: offshore Structures, Marine Foundations, and Sediment Processes. John B. Herbich, editor. Gulf Publishing Company, P.O. Box 2608, Houston, TX. (713) 529-4301 or (800) 231-6275. 1990. $195.00.

Handbook of Coastal and Ocean Engineering Volume 3: Harbors, Navigational Channels, Estuaries and Environmental Effects. John B. Herbich, editor. Gulf Publishing Company, P.O. Box 2608, Houston, TX. (713) 529-4301 or (800) 231-6275. 1992. $195.00.

ONLINE DATABASES AND CD-ROMS

Compendex Plus. Engineering Information, Inc., 345 East 47th Street, New York, NY 10017. (212) 705-7600 or (800) 221-1044. Contains citations with abstracts to worldwide literature in engineering and technology, from 1970 to present. Available on online BRS,(800) 289-4277, DIALOG, (800) 334-2564, ORBIT (800) 456-7248, and STN International, FIZ Karlsruhe, P.O. Box 2465, W-7500, Karlsruhe 1, Germany, online services. Also available on CD-ROM. Inquire as to cost and availability.

Current Contents Search. Institute for Scientific Information, 3501 Market Street, Philadelphia, PA 19104. (215) 386-0100. FAX (215) 386-6362. Contains citations to articles listed in the table of contents of science and technology journals. Also articles in social sciences and life sciences journals. Available on BRS,(800) 289-4277, DIALOG,(800) 334-2564, online services. Inquire as to cost and availability.

Dissertation Abstracts. University Microfilms International, 300 North Zeeb Road, Ann Arbor, MI 48106. (800) 521-0600 or (313) 761-4700. Scope includes virtually all doctoral dissertations accepted at accredited American institutions from 1861 to present in 252 subject areas. Available on BRS,(800) 289-4277, DIALOG,(800) 334-2564, and OCLC EPIC,(800) 848-5878, online services. Also available on CD-ROM. Inquire as to cost and availability.

Earth Sciences. U.S. Geological Survey, 12201 Sunrise Valley Drive, Reston, VA 22092-9998. (703) 648-4460. CD-ROM of earth science databases including the U.S. Geological Survey Library, Earth Science Data Directory, and GEOINDEX, citations to published geological maps. $350.00 per year with quarterly updates.

Geoarchive. Geosystems, P.O. Box 1024, Westminster, London, England, SW1 P 2JL. Citations to literature on geoscience, 1969 to present. Inquire as to online cost and availability.

Geobase. Elsevier - Geo Abstracts, Regency House, 34 Duke Street, Norwich NR3 3AP, England. Contains citations to the worldwide earth science literature from 1980 to date. Available

onDIALOG, ORBIT online services. Inquire as to cost and availability.

GEOREF: Bibliography and Index of Geology. American Geological Institute, 4220 King Street, Alexandria, VA 22302. (703) 379-2480. Fax (703) 379-7563. Monthly. Inquire for price and availability.

NTIS Bibliographic Database. National Technical Information Service, 5285 Port Royal Road, Springfield, Va 22161. (703) 487-4929 or FAX (703) 321-8199. Broad coverage of government-sponsored science and technology research reports, 1964 to present. Available on BRS,(800) 289-4277, DIALOG, (800) 334-2564, ORBIT, (800) 456-7248, and STN International, FIZ Karlsruhe, P.O. Box 2465, W-7500, Karlsruhe 1, Germany, online services. Also available on CD-ROM. Inquire as to cost and availability.

Scisearch. Institute for Scientific Information, 3501 Market Street, Philadelphia, PA 19104. (800) 523-1850 or (215) 386-0100. Broad multidisciplinary title and author index to the international literature of science and technology, 1974 to present. Available on DIALOG,(800) 334-2564, and ORBIT,(800) 456-7248, online services. Also available on CD-ROM. Inquire as to cost and availability.

WILSONLINE. H.W. Wilson Company, 950 University Avenue, Bronx, NY 10452. (800) 367-6770 or (212) 588-8400. Makes available online versions of the printed H.W. Wilson indexes including Applied Science and Technology Index, Business Periodicals Index, General Science Index, and Readers' Guide to Periodical Literature. Period covered is generally 1983 to present. Available on BRS,(800) 289-4277, DIALOG,(800) 334-2564, and OCLC EPIC,(800) 848-5878, online services. Also available on CD-ROM. Inquire as to cost and availability.

PERIODICALS

Coastal Engineering. Elsevier Science Inc., Box 882, Madison Square Station, New York, NY 10159. (212) 989-5800. 1977 to present. 12 times a year. $642.00 per year.

Coastal Research. Florida State University Geology Department, Tallahassee, FL 32306. (904) 644-5860. FAX (904) 644-4214. 1962 to present. 3 times a year. $6.00 per year.

Coastal Management. Taylor & Francis, 1900 Frost Road, Suite 101, Bristol, PA 19007. (215) 785-5800. FAX (215) 785-5515. 1973 to present. Quarterly. $140.00 per year.

Coastwatch. North Carolina Sea Grant College Program, Box 8605, North Carolina State University, Raleigh, NC 27695-8605. (919) 515-2454. FAX (919) 515-7095. 1970 to present. Bimonthly. $12.00 per year.

Estuarine Coastal and Shelf Science. Academic Press Ltd./ Harcourt Brace & Company, Ltd., Foots Cray High Street, Sidcup, Kent DA14 5HP, England. Telephone 44-81-300-3322. FAX 44-81-309-0807. 1973 to present. Monthly. Inquire for cost.

Journal of Coastal Research. Coastal Education & Research Foundation, 4310 N.E. 25th Avenue, Ft. Lauderdale, FL 33308. (305)565-1051. FAX (305) 565-1051. 1985 to present. Quarterly. $58.00.

COASTS

Ency. of Physical Sciences and Engineering Info. Sources

Journal of Waterway, Port, Coastal, and Ocean Engineering. American Society of Civil Engineers, Waterway, Port, Coastal, and Ocean Division, 345 East 47th Street, New York, NY 10017-2398. (212) 705-7520 or (800) 548-2723. 1956 to present. Bimonthly. $100.00 per year.

Ocean and Coastal Management. Elsevier Science [journals], 660 White Plains Rd., Tarrytown, NY 10159-5153. (919) 524-9200. FAX (919) 333-2444. 1973 to present. Twelve times a year. $565.00.

Ocean Engineering. Elsevier Science [journals], 660 White Plains Rd., Tarrytown, NY 10159-5153. (919) 524-9200. FAX (919) 333-2444. 1968 to present. 8 times a year. $595.00 per year.

RESEARCH CENTERS AND INSTITUTES

Coastal and Oceanographic Engineering Laboratory. University of Florida, 336 Weil Hall, Gainesville, FL 32611. (904) 392-1436. FAX 9904) 392-3466.

Coastal Education & Research Foundation. 4310 N.E. 25th Avenue, Ft. Lauderdale, FL 33308. (305) 565-1051. FAX (305) 565-1051.

Coastal Research Center. Woods Hole Oceanographic Institution, Woods Hole, MA 02543. (508) 548-1400. FAX (508) 457-2172.

Louisiana State University, Coastal Studies Institute. Coastal Studies Bldg., Baton Rouge, LA 70803. (504) 388-5301. FAX (504) 388-2520.

Purdue University, Great Lakes Coastal Research Laboratory. c/o School of Civil Engineering, West Lafayette, in 47907. (317) 494-3713.

University of California—san Diego, Scripps Institute of Oceanography, Center For Coastal Studies. 9500 Gilman Drive, La Jolla, CA 92093-0209. (619) 534-4333. FAX (619) 534-0300.

University of California, California Sea Grant College Program. 9500 Gilman Drive, La Jolla, CA 92093-0209. (619) 534-4440. FAX (619) 534-2231.

University of Delaware Sea Grant College Program. 196 S. College Avenue, Newark, DE 19716. (302) 451-8182. FAX (302) 451-6838.

COATINGS

See also: CEMENT, CHEMICAL ENGINEERING, CHEMISTRY, ORGANIC CHEMISTRY, PLASTICS, TEXTILES

ABSTRACT SERVICES AND INDEXES

Adhesives Abstracts. R A P R A Technology Ltd., Shawbury, Shrewsbury, Shropshire Sy4 4nr, England. TEL 0939-250383. Fax 0939-251118. 1988 to present. Monthly. L195 per year. *Issn: 0891-7760.*

Applied Science and Technology Index; A Cumulative Subject Index To English Language Periodicals In the Fields of Aeron Autics and Space Science, Computer Technology, Chemistry , Construction Industry, Energy and Related Areas. H. W. Wilson Co.,950 University Avenue, Bronx, NY 10452-9978. (212) 588-8400. (800) 367-6770. FAX (718) 590-1617. From 1958 to present. Monthly. Inquire about cost and availability. Also available online (BRS and WILSONLINE) and on CD-ROM. ISSN: 0003-6986.

Chemical Abstracts. Chemical Abstracts Service. 2540 Olentangy River Road, Box 3012, Columbus, OH 43210-0012. (614) 447-3600. FAX (614) 447-3713. 1907 to present. Weekly. $16,800.00 per year. Available online and on CD- Rom. CA is also available in five section groupings. Inquire Regarding Cost and Availability.

Current Contents: Engineering, Technology and Applied Sciences. Institute for Scientific Information, 3501 Market Street, Philadelphia, PA 19104. (215) 386-0100. FAX (215) 386-2991. From 1970 to present. Weekly. $442.00 per year. Also available online. Inquire regarding cost and availability. ISSN: 0095-7917.

Engineering Index Monthly. Engineering Information, Inc., Castle Point on the Hudson, Hoboken, NJ 07030. (201) 216-8500. (800) 221-1044. FAX (201) 216-8532. Monthly. $2300.00 per year for monthly issues. $1980 per year for annual cumulation. Available on CD-ROM and online as COMPENDEX. ISSN: 0742-1974; 0360-8557.

General Science Index. H.W. Wilson Company, 950 University Avenue, Bronx, NY 10452. (212) 588-8400. (800) 367-6770. FAX (718) 590-1617. From 1978 to present. Ten issues per year; quarterly and annual cumulations. Service basis. Available on CD-ROM and online. Inquire regarding cost and availability. ISSN: 0162-1963.

Institute of Paper Science and Technology Abstract Bulletin. Institute of Paper Science and Technology, 500 10th Street, NW, Atlanta, GA 30318. (404) 853-9500. 1930 to present. Monthly. Price varies. ISSN: 1047-2088.

Process and Chemical Engineering/chemical Engineering Abstracts. Royal Society of Chemistry, Information Services, thomas graham house, science park, milton road, cambridge, Cb4 4wf, England. Contains citations, mostly with abstracts, of the worldwide literature on chemical engineering; from 1982 *To Present.* Monthly. $610.00 per year. Also available online. *Issn: 0960-5045.*

Science Citation Index. SCI. Institute for Scientific *Information, 3501 Market Street, Philadelphia, Pa 19104. (215) 386-0100.* (800) 523-1850. FAX (215) 386-2991. *1961 To Present.* Six issues per year, plus annual *Cumulation.* $11650.00 per year. Also available online and *On Cd-rom.* Inquire about price and availability. *0036-827x.*

Theoretical Chemical Engineering. Royal Society of Chemistry, Information Services, Thomas Graham House, Science Park, Milton Road, Cambridge, Cb4 4wf, England. Covers theoretical chemical engineering, including theory and laboratory Experimentation. 1964 to present. Monthly. $235.00. Issn: 0960-5053.

ASSOCIATIONS AND PROFESSIONAL SOCIETIES

Adhesive and Sealant Council. 1627 K Street NW, Suite 1000, Washington, CD 20006-1707. (202) 452-1500. FAX (202)452-1501.

Adhesives Manufacturers Association. 401 North Michigan Avenue, Chicago, IL 60611. (312) 644-6610. FAX (312) 321-6869.

Adhesion Society. c/o F. James Boerio, University of Cincinnati, Department of Materials Science, Cincinnati, OH 45221-0012. (203) 486-4629.

American Chemical Society, 1155 16th Street, Nw, Washington, Dc 20036. (202) 872-4600.

American Electroplaters and Surface Finishers Society. 12644 Research Parkway, Orlando, FL 32826. (407) 281-6441. FAX (407)281-6446.

American Institute of Chemical Engineers. 345 East 47th Street, New York, NY 10017. (212) 705-7338.

Association For Finishing Processes of the Society of Manufacturing Engineers. 1 SME Drive, PO Box 930, Dearborn, MI 48121-0930. (313) 271-1500. FAX (313) 271-2861.

Association of Consulting Chemists and Chemical Engineers, 50 East 41st Street, Suite 92, New York, Ny 10017. (212) 684-6255.

Chemical Manufacturers Association. 2501 M Street, N.W., Washington, DC 20037. (202) 887-1182.

Federation of Societies For Coating Technology. 429 Norristown Road, Blue Bell, PA 19422. (215) 940-0777. FAX (215) 940-0292.

National Paint and Coating Association. 1500 Rhode Island Avenue NW, Washington, DC 20005. (202) 462-6272.

Pressure Senstitve Tape Council. 401 North Michigan Avenue, Chicago IL 60611-4267. (312) 644-6610. FAX (312) 527-6640.

Steel Structures Painting Council. 40 24th Street, Pittsburgh, PA 15222-4643. (412) 291-2331. FAX (412) 281-9992.

BIBLIOGRAPHIES

Chemical Information Sources. Gary Wiggins. McGraw-Hill Publishing Company, 1221 Avenue of the Americas, New York, NY 10020. (800) 262-4729 or (212) 512-3825. 1991. $42.50.

Scientific and Technical Books and Serials In Print; An Index To Literature In Science and Technology. R.R. Bowker Inc., 121 Chanlon Road, New Providence, NJ 07974. (908) 464-6800. (800) 521-8110. FAX (908) 665-3502. 1972 to present. Annual. 4 volumes. 1994. $299.95. Also available on compact disc and online. ISSN: 0000-054X.

DIRECTORIES AND BIOGRAPHICAL SOURCES

Adhesives. International Plastics Selector, Inc., 9889 Willow Creek Road, San Diego, CA 92126. (619) 578-3910. Annual. $100.00

Adhesives Age Directory. Argus Inc., 6151 Powers Perry Road, NW, Atlanta, GA 30339-2941. (404) 955-2500. FAX (404) 955-0400, 1957 to present. Annual. $51.95. ISSN: 0001-821X.

American Men and Women of Science. Physical and Biological Sciences. R.R. Bowker Inc., 121 Chanlon Road, New Providence, NJ 07974. (908) 464-6800. (800) 521-8110. 20th edition.8 volumes. 1996. $850.00

Directory of Chemistry Software 1992. Wendy Warr, Peter Willett, Geoff Downs. American Chemical Society, 1155 16th Street, NW, Washington, DC 20036. (202) 872-4600. 1992. $35.95.

Research Centers Directory. Gale Research, 835 Penobscot Building, Detroit, MI 48226-4094. (313) 961-2242. (800) 347-4253. 20th edition. 1995. $485.00. ISSN:: 0080-1518.

Who's Who In Engineering. Gordon Davis, editor. American Association of Engineering Societies. 1111 19th Street, NY, Suite 608, Washington, DC 20036. (202) 296-2237. (800) 658-8897. 9th edition. 1995. $220.00.

ENCYCLOPEDIAS AND DICTIONARIES

Encyclopedia of Chemical Processing and Design. McKetta Marcel Dekker, Inc., 270 Madison Avenue, New York, NY 10016. (212) 696-9000. (800) 228-1160. 1976 - . $175.00 per volume. ISSN:

Industrial Chemical thesaurus. Michael Ash and Irene Ash. VCH Publishers, Inc., 220 East 23rd Street, New York, NY 10010-4606. (800) 367-8249. 1992. $295.00.

Illustrated Chemistry Laboratory Terminology. Gerbert W. Ockerman. CRC Press, Inc., 2000 Corporate Boulevard, NW, Boca Raton, FL 33431. (407) 994-0555. (800) 272-7737. 1991. $32.95.

Kirk-OthmerEncyclopedia of Chemical Technology. John Wiley and Sons, Inc., 605 Third Avenue, New York, NY 10158. (800) 526-5368 or (212) 850-6000. Fourth edition. 1991 - . Twenty-seven volumes. $5400.00.

McGraw-hill Encyclopedia of Science and Technology. McGraw-Hill Book, Incorporated, 1221 Avenue of the Americas, New York, NY 10020. (212) 997-3675. (800) 262-4729. Seventh edition. Twenty volumes. 1992. $1900.00.

Ullman's Encyclopedia of Industrial Chemistry. VCH Publications, Inc., 220 East 23rd Street, Suite 909, New York, NY 10010-4606. (212) 683-8333. (800) 422-8824. 5th edition. 1984 - . Price varies per volume. ISSN:

GENERAL WORKS

Adhesive Bonding. L. H. Lee, editor. Plenum Publishing Corp., 233 Spring Street, New York, NY. (212) 620-8000. (800) 2221-9369. FAX (212) 463-0742. . 1991. $115.00.

COATINGS

Ency. of Physical Sciences and Engineering Info. Sources

Adhesives and Adhesive Bonding: theoretical and Practical Aspects. George Epstein, editor. T-C Pr CA 1990. $59.00.

Adhesives and Sealants. American Society for Metals, Volume 3 of Engineered Materials Handbook. ASM International 9639 Kinsman Road, Materials Park, OH 44073. (216) 238-5151. 1990. $153.00.

Adhesives From Renewable Resources. Richard W. Hemmingway, et al, editors. American Chemical Society, 1155 16th Street, NW, Washington, DC 20036. (202) 872-4600. (800) 333-9511. FAX (614) 447-3671. . 1989. $99.95.

Construction and Structural Adhesives and Sealants. Ernest W. Flick. Noyes Data Corporation, 120 Mill Road, Park Ridge, NJ 07656. (201) 391-8484. 1988. $78.00.

Industrial Applications of Adhesive Bonding. M. M. Sadek, editor. Elsevier Science Publishing Company, Inc., 655 Avenue of the Americas, New York, NY 10010. (212) 989-5800. FAX (914) 333-2444. 1987. $72.00.

Science and Technology of Building Seals, Sealants, Glazing and Waterproofing. Charles J. Parise, editor. American Society for Testing and Materials, 1916 Race Street, Philadelphia, PA 19103. (215) 299-5419. STP 1168. 1992. $61.00.

Structural Adhesives: Chemistry and Technology. S. R. Hartshorn, editor. Plenum Publishing Corp., 233 Spring Street, New York, NY. (212) 620-8000. (800) 2221-9369. FAX (212) 463-0742. 1986. $115.00.

Technology of Wood Bonding: Principles In Practice. Alan A. Marra. Chapman & Hall, 1 Penn Plaza, New York, NY 10119. (212) 564-1060. 1992. $59.95.

HANDBOOKS AND MANUALS

Adhesives Technology Handbook. Arthur H. Landrock, Noyes Data Corporation, 120 Mill Road, Park Ridge, NJ 07656. (201) 391-8484. 1986. $64.00.

Chemical Formulary. H. Bennett, editor. Chem Publishing, Co., Inc. 80 Eighth Avenue, New York, NY 10011. (212 255- 1950. volumes 1-30. $70.00 per volume.

Comprehensive Organic Synthesis: Selectivity, Strategy, and Efficiency In Modern Organic Chemical. Barry M. Trost and others, editors. Pergamon Press, Maxwell House, Fairview Park, Elmsford, NY 10523. (914) 592-7700. Fax (914) 592-3625. 1991. Nine volumes. $3900.00.

CRC Handbook of Chemistry and Physics. David R. Kide, editor. CRC Press Inc., 2000 Corporate Blvd., NW, Boca Raton, FL 33431. (407) 994-0555. (800) 333-8300. 77th edition. 1996. $99.95.

Guide To Basic Chemical Compounds. D.R. Lide, Jr. CRC Publishers, Inc., 2000 Corporate Blvd., N.W., Boca Raton, FL 33431. (407) 994-0555 or (800) 333-8300. 1993. $120.00.

Handbook of Adhesion. D. E. Packham, editor. Halsted Press, 605 Third Avenue, New York, 10158-0012. (212) 850- 6400. 1993. $250.00.

Handbook of Adhesives. Irving Skeist. Chapman & Hall, 1 Penn Plaza, New York, NY 10119. (212) 564-1060. 1989. $105.00

Handbook of Adhesive Bonded Structural Repair. Raymond F. Wegman and Thomas R. Tullos. Noyes Data Corporation, 120 Mill Road, Park Ridge, NJ 07656. (201) 391-8484. 1992. $48.00.

Handbook of Adhesive Raw Materials. Ernest W. Flick. Noyes Data Corporation, 120 Mill Road, Park Ridge, NJ 07656. (201) 391-8484. 2nd edition. 1989. $69.00.

Handbook of Adhesive Technology. A. Pizzi, editor. Marcel Dekker, Inc., 270 Madison Avenue, New York, NY 10016. (212) 696-9000. (800) 228-1160. 1994. $195.00.

Handbook of Pressure-sensitive Adhesive Technology. Satas. Chapman & Hall, 1 Penn Plaza, New York, NY 10119. (212) 564-1060. 2nd edition. 1989. $105.00.

Lange's Handbook of Chemistry. John A. Dean, editor. McGraw-Hill Publishing Company, 1221 Avenue of the Americas, New York, NY 10020. (800) 262-4729 or (212) 512-3825. Fourteenth editon. 1992. $79.50.

Perry's Chemical Engineers' Handbook. Robert H. Perry and Donald W. Green, editors. McGraw-Hill Publishing Company, Inc., 1221 Avenue of the Americas, New York, NY 10020. (212) 512-2000. (800) 262-4729. 6th edition. 1996. $129.50.

ONLINE DATABASES AND CD-ROMS

Ca Search. Chemical Abstracts Service, P.O. Box 3012, Columbus, OH 43210-0012. (614) 447-3600. (800) 848-6533. FAX (614) 447-3709. Very comprehensive guide to worldwide chemical literature and related fields, 1972 to present. Available on BRS,(800) 289-4277, DIALOG, (800) 334-2564, ORBIT (800) 456-7248, and STN International, FIZ Karlsruhe, P.O. Box 2465, W-7500, Karlsruhe 1, Germany, online services. Inquire as to cost and availability.

Chemical Journals of the American Chemical Society. American Chemical Society, 1155 16th Street, N.W., Washington, DC 20036. (202) 872-4381 or (800) 424-6747. Contains complete text of approximately 90,000 articles from 22 primary journals published by the American Chemical Society, from mostly 1982 to present. Available on STN International, FIZ Karlsruhe, P.O. Box 2465, W-7500, Karlsruhe 1, Germany, online service. Inquire as to cost and availability.

Dissertation Abstracts. University Microfilms International, 300 North Zeeb Road, Ann Arbor, MI 48106. (800) 521-0600 or (313) 761-4700. Scope includes virtually all doctoral dissertations accepted at accredited American institutions from 1861 to present in 252 subject areas. Available on BRS, (800) 289-4277, DIALOG, (800) 334-2564, and OCLC EPIC, (800) 848-5878, online services. Also available on CD-ROM. Inquire as to cost and availability.

NTIS Bibliographic Database. National Technical Information Service, 5285 Port Royal Road, Springfield, Va 22161. (703) 487-4929 or FAX (703) 321-8199. Broad coverage of government-sponsored science and technology research reports, 1964 to present. Available on BRS,(800) 289-4277, DIALOG, (800) 334-2564, ORBIT, (800) 456-7248, and STN International, FIZ Karlsruhe, P.O. Box 2465, W-7500, Karlsruhe 1, Germany, online services. Also available on CD-ROM. Inquire as to cost and availability.

Scisearch. Institute for Scientific Information, 3501 Market Street, Philadelphia, PA 19104. (800) 523-1850 or (215) 386-0100. Broad multidisciplinary title and author index to the international literature of science and technology, 1974 to present. Available on DIALOG, (800) 334-2564, and ORBIT, (800) 456-7248, online services. Also available on CD-ROM. Inquire as to cost and availability.

WILSONLINE. H.W. Wilson Company, 950 University Avenue, Bronx, NY 10452. (800) 367-6770 or (212) 588-8400. Makes availableonline versions of the printed H.W. Wilson indexes including Applied Science and Technology Index, Business Periodicals Index, General Science Index, and Readers' Guide to Periodical Literature. Period covered is generally 1983 to present. Available on BRS, (800) 289-4277, DIALOG, (800) 334-2564, and OCLC EPIC, (800) 848-5878, online services. Also available on CD-ROM. Inquire as to cost and availability.

PERIODICALS

Adhesives Age. Argus Inc., 6151 Powers Perry Road, NW, Atlanta, GA 30339-2941. (404) 955-2500. FAX (404) 955- 0400, 1957 to present. Monthly, plus annual directory. $52.00 per year. ISSN: 0001-821X.

Chemical Week; Includes Annual Buyers Guide. Chemical Week Associates, 888 Seventh Avenue, New York, NY 10106. (212) 621-4900. FAX (212) 621-4949. 1914 to present. Weekly. $99.00 per year. ISSN: 0009-272X.

Chemtech. American Chemical Society, Box 3337, Columbus, OH 43210. (614) 447-3776. 1970 to present. Monthly. $370.00 per year. ISSN: 0009-2703.

International Journal of Adhesion and Adhesives. Butterworth-Heinemann. Subscriptions to: Turpin Transactions Ltd., Distributin Centre, Blackhorse Road, Letchworth, Herts SG6 1HN, England. TEL 0462-672555. 1980 to present. Quarterly. L180 per year. ISSN: 0143-7496.

Journal of Adhesion Science and Technology. V S P, P.O. Box 346, 3700 AH Zeist, Netherlands. TEL 31-3404-25790. FAX 31-3404-32081. 1987 to present. Monthly. DM 1280 per year. ISSN: 0169-4243.

Journal of Coatings Technology. Federation of Societies for Coating Technology, 429 Norristown Road, Blue Bell, PA 19422. (215) 940-0777. FAX (215) 940-0292. Monthly. Included in membership dues. ISSN: 0361-3473.

Metal Finishing. Elsevier Science Inc., 655 Avenue of the Americas, New York, NY 10010. (212) 989-5800. FAX (212) 633-3990. 1903 to present. Monthly, plus 2 supplements. $40.00 per year. ISSN: 0026-0576.

Modern Paint and Coatings. Argus Inc., 6151 Powers Ferry Road, NW, Atlanta, GA 30339-2941. (404) 955-2500. FAX (404) 955-0400. 1910 to present. Monthly. $49.00. ISSN: 0098-7786.

Plating and Surface Finishing. American Electroplaters and Surface Finishers Society, Inc., 12644 Research Parkway, Orlando, FL 32826-3298. (407) 281-6441. FAX (407) 281- 6446. 1909 to present. Monthly. $35.00. ISSN: 0369-3164.

RESEARCH CENTERS AND INSTITUTES

Center for Adhesion and Sealant Sciences. Virginia Institute for Material Systems, Virginia Polytechnic Institute and State University, 121 Patton Hall, Blacksburg, VA 24060. (703) 961-4565.

Laboratory for Surface Science and Technology. University of Maine, Sawyer Research Center, Orono, ME 044469. (207) 581-2254.FAX (207) 581-2255.

Paint Research Associates. 430 West Forest, Ypsilanti, MI 48197. (313) 483-3401. FAX (313) 487-8755.

Science and Technology Laboratory Group. Gerogia Institute of Technology, Georgia Tech Research Institute, Atlanta, GA 30332. (404) 894-3530. (404) 894-6285.

Steel Structures Painting Council. 4400 Fifth Avenue, Pittsburgh, PA 15213-2683. (412) 268-3327. FAX (412) 268

COBALT

See also: ALLOYS, ELEMENTS, FERROALLOYS, INORGANIC CHEMISTRY, METALLURGICAL ENGINEERING, METALS AND METALWORKING

ABSTRACT SERVICES AND INDEXES

Alloys Index. American Society for Metals, Metals Park, OH 44073. (216) 338-5151. FAX (216) 338-4634. 1974 to present. Monthly. $380.00. Also available on CD-ROM and online via METADEX, STN International, and DIALOG.

Aluminum Industry Abstracts. Aluminum Association, Materials Park, OH 44073. (216) 338-5151. FAX (216) 338-4634. 1968 to present. Monthly. $525.00 per year.

Applied Science and Technology Index; A Cumulative Subject Index To English Language Periodicals In the Fields of Aeronautics and Space Science, Computer Technology, Chemistry, Construction Industry, Energy and Related Areas. H.W. Wilson Co., 950 University Avenue, Bronx, NY 10452. (800) 367-6770 or (212) 588-8400. FAX (718) 590-1617. From 1958 to present. Monthly. Inquire about cost and availability. Also available on CD-ROM and online.

Chemical Abstracts - Applied Chemistry and Chemical Engineering Sections. Chemical Abstracts Service, 2540 Olentangy River Road, Box 3012, Columbus, OH 43210. (800) 848-6538 or (614) 447-3600, FAX (614) 447-3713. Bi-weekly. $1410.00 per year.

Current Contents: Engineering, Technology, and Applied Sciences. Institute for Scientific Information, 3501 Market Street, Philadelphia, PA 19104. (215) 386-0100. FAX (215) 386-6263. 1970 to present. Weekly. $442.00 per year.

Engineering Index Monthly. Engineering Information, Inc., Castle Point on the Hudson, Hoboken, NJ 07030. (800) 221-1044. FAX (212) 832-1857. Monthly. $2300.00 per year. Also available online as COMPENDEX, and also on CD-ROM. Covers chemical engineering, computers, electrical engineering, civil engineering, metals and mining, industrial management, and mechanical engineering.

COBALT

Ency. of Physical Sciences and Engineering Info. Sources

Leadscan (formerly *Lead Abstracts*), Lead Development Association, 42 Weymouth Street, London W1x 3lq, England. 1958 to present. Quarterly. Inquire for cost.

Mechanical Engineering Abstracts (formerly *Ismec),* Cambridge Scientific Abstracts, 7200 Wisconsin Avenue, Bethesda, Md 20814-4823. (301) 961-6750. FAX (301) 961-6720. 1967 to present.Monthly. $895.00 per year. Summarizes world literature in mechanical engineering, production engineering, and engineering management. Also available online.

Metals Abstracts and Metals Abstracts Index. American Society for Metals, Metals Park, OH 44073. (216) 338-5151. 1968 to present. Monthly. Abstracts are $1500.00 per year and Index is $460.00 per year.

Physics Abstracts. Institute of Electrical Engineers, Michael Faraday House, Six Hill Way, Stevenage, Herts, SG1 2AY, England. Distributed by IEEE, 445 Hoes Lane, Piscataway, NJ 08854. (908) 562-5549. 1898 to present. Monthly. $2700.00 per year. Also available online as INSPEC.

Science Citation Index. Institute for Scientific Information, 3501 Market Street, Philadelphia, PA 19104. (215) 386-0100. FAX (215) 386-6362. Inquire about availability and cost. Also available on CD-ROM.

ANNUAL REVIEWS AND YEARBOOKS

U.S. *Department of the Interior, Bureau of Mines Annual Reports, Commodities.* Superintendent of Documents, U.S. Government Printing office, Box 371954, Pittsburgh, PA 15250-7954. (202) 783-3238. FAX (202) 512-2233. Annual. Price varies.

U.S. *Department of the Interior, Bureau of Mines, Minerals Yearbook.* Superintendent of Documents, U.S. Government Printing office, Box 371954, Pittsburgh, PA 15250-7954. (202) 783-3238. FAX (202) 512-2233. 1932 to present. Annual. Price varies.

ASSOCIATIONS AND PROFESSIONAL SOCIETIES

American Powder Metallurgy Institute. 105 College Road E., Princeton, NJ 08540. (609) 452-7700.

American Society For Metals (asm). Materials Park, OH 44073. (216) 338-5151 or FAX (216) 338-4634.

American Society For Testing and Materials (astm). 1916 Race Street, Philadelphia, PA 19103. (215) 299-5585.

American Society of Mechanical Engineers. 345 East 47th Street, New York, NY 10017-2398. (212) 705-7703.

Association of Iron and Steel Engineers. 3 Gateway Center, Suite 2350, Pittsburgh, PA 15222. (412) 281-6323.

Metallurgical Society of the Aime (american Institute of Mining, Metallurgical, and Petroleum Engineers). 345 E.47th Street, 14th Floor, New York, NY 10017. (212) 705-7695.

National Association of Corrosion Engineers. Box 218340, Houston, TX 77218. (713) 492-0535. FAX (713) 492-8254.

DIRECTORIES AND BIOGRAPHICAL SOURCES

Dun's Industrial Guide: the Metalworking Directory, 1993-94. Dun & Bradstreet Information Services Staff. 3 volumes. Dun & Bradstreet Information Services, 899 Eaton Avenue, Bethlehem, Pa 18025. (610) 882-7000 or (800) 526-0651. FAX (610) 882-7269. 1993. $795.00

Metallurgical Society of Aime Membership List. Metallurgical Society of AIME, 345 E.47th Street, 14th Floor, NewYork, NY 10017. (212) 705-7695. Inquire for cost and availability.

Directory of Engineering Societies and Related Organizations. Gordon Davis. 13th edition. American Association of Engineering Societies, 1111 19th Street NW, Suite 608, Washington, DC 20036. (202) 296-2237 or (800) 658-8897. 1989. Inquire for price.

International Research Centers Directory. Gale Research, 835 Penobscot Building, Detroit, MI 48226-4094. (313) 961-2242. (800) 347-4253. 8th edition. 1995. $410.00

Research Centers Directory. Gale Research, 835 Penobscot Building, Detroit, MI 48226-4094. (313) 961-2242. (800) 347-4253. FAX (313) 961-6083. $485.00.

Scientific and Technical Organizations and Agencies Directory. Gale Research, 835 Penobscot Building, Detroit, MI 48226-4094. (313) 961-2242. (800) 347-4253. FAX (313) 961-6083. 4th edition. 1996. $195.00.

Who's Who In Engineering. American Association of Engineering Societies, 1111 19th Street NW, Suite 608, Washington, DC 20036. (202) 296-2237 or (800) 658-8897. 8th edition. 1991. Inquire for price.

ENCYCLOPEDIAS AND DICTIONARIES

McGraw-hill Encyclopedia of Science and Technology. Sybil P. Parker, ed. 7th edition. 20 volumes. McGraw-Hill Publishing Company, 1221 Avenue of the Americas, New York, NY 10020. (800) 262-4729 or (212) 512-3825. 1992. $1900.00

Thesaurus of Scientific, Technical, and Engineering Terms. Hemisphere Publishing Corporation, 1900 Frost Road, Suite 101, Bristol, PA 19007-1598. (215) 785-5800 or (800) 821-8312. FAX (215) 785-5515. 1987. $173.00.

GENERAL WORKS

Cobalt and Its Alloys. W. Betteridge. Halsted Press (Division of John Wiley and Sons, Inc.), 605 Third Avenue, New York, NY 10158. (800) 526-5368 or (212) 850-6418. 1982. Inquire for cost and availability.

Cobalt Availability: A Minerals Availability Appraisal. David G. Willard & Donald I. Bleiwas. U.S. Department of the Interior, Bureau of Mines. Order from Superintendent of Documents, U.S. Government Printing office, Box 371954, Pittsburgh, PA 15250-7954. (202) 783-3238. FAX (202) 512-2233. 1991. Inquire for cost and availability.

Cobalt Facts. Rector Press Ltd., 130 Rattlesnake, Leverett, MA 01054-9726. (413) 548-9708 or (800) 247-3473. FAX (413) 367-2853. 1994. $75.00.

Cobalt In Electronic Technology. Rector Press Ltd., 130 Rattlesnake, Leverett, MA 01054-9726. (413) 548-9708 or (800) 247-3473. FAX (413) 367-2853. 1994. $75.00.

Cobalt In Medicine, Agriculture, & the Environment. Rector Press Ltd., 130 Rattlesnake, Leverett, MA 01054-9726. (413) 548-9708 or (800) 247-3473. FAX (413) 367-2853. 1994. $75.00.

Cobalt In Superalloys. Rector Press Ltd., 130Rattlesnake, Leverett, MA 01054-9726. (413) 548-9708 or (800) 247-3473. FAX (413) 367-2853. 1994. $75.00.

International Strategic Minerals Summary Report, Cobalt. Richard N. Crockett, et al. U.S. Geological Survey Circular 930-F. Order from Superintendent of Documents, U.S. Government Printing office, Box 371954, Pittsburgh, PA 15250-7954. (202) 783-3238. FAX (202) 512-2233. 1987. Inquire for cost and availability.

HANDBOOKS AND MANUALS

Metals Handbook. ASM International, Materials Information, Materials Park, OH 44073-0002. (216) 338-5151 or (800) 336-5152. FAX (216) 338-4634. $154.00.

Smithell's Metals Reference Book. E.A. Brandes & G.B. Brook. 7th edition. Butterworth-Heinemann, 313 Washington Street, Newton, MA 02158. (617) 928-2500 or (800) 366-2665. FAX (617) 928-2620. 1992. $250.00.

Woldman's Engineering Alloys. J. Frick, editor. 8th edition. ASM International, Materials Information, Materials Park, OH 44073-0002. (216) 338-5151 or (800) 336-5152. FAX (216) 338-4634. 1994. $175.00.

ONLINE DATABASES AND CD-ROMS

Compendex Plus. Engineering Information, Inc., 345 East 47th Street, New York, NY 10017. (212) 705-7600 or (800) 221-1044. Contains citations with abstracts to worldwide literature in engineering and technology, from 1970 to present. Available on online BRS,(800) 289-4277, DIALOG, (800) 334-2564, ORBIT (800) 456-7248, and STN International, FIZ Karlsruhe, P.O. Box 2465, W-7500, Karlsruhe 1, Germany, online services. Also available on CD-ROM. Inquire as to cost and availability.

CSA Engineering. Cambridge Scientific Abstracts, 7200 Wisconsin Avenue, Suite 601, Bethesda, MD 20814. (301) 961-6750 or (800) 843-7751. Contains citations and abstracts of international periodicals and research literature covering all fields of engineering and science and technology,including computer and information science, electronics, mechanical engineering, solid state materials, 1981 to present. Available on BRS,(800) 289-4277, online service.Inquire as to cost and availability.

Current Contents Search. Institute for Scientific Information, 3501 Market Street, Philadelphia, PA 19104. (215) 386-0100. FAX (215) 386-6362. Contains citations to articles listed in the table of contents of science and technology journals. Also articles in social sciences and life sciences journals. Available on BRS,(800) 289-4277, DIALOG,(800) 334-2564, online services. Inquire as to cost and availability.

Ismec: Mechanical Engineering Abstracts. Cambridge Scientific Abstracts, 7200 Wisconsin Avenue, Suite 601, Bethesda,

MD 20814. (301) 961-6750 or (800) 843-7751. Contains citations to the literature in mechanical engineering, industrial and production engineering, energy, power, mechanics, devices and related areas, from 1973 to present. Available on the DIALOG,(800) 334-2564, online service. Inquire as to cost and availability.

Metadex. Jointly produced by ASM International and the Institute of Materials. Contains more than 925,000 records from the international literature on metals and alloys, concerning properties, processes, materials classes, applications, and metallurgical systems. Updated monthly. Available from ORBIT-QUESTEL (703) 442-0900.

NTIS Bibliographic Database. National Technical Information Service, 5285 Port Royal Road, Springfield, Va 22161. (703) 487-4929 or FAX (703) 321-8199. Broad coverage of government-sponsored science and technology research reports, 1964 to present. Available on BRS,(800) 289-4277, DIALOG, (800) 334-2564, ORBIT, (800) 456-7248, and STN International, FIZ Karlsruhe, P.O. Box 2465, W-7500, Karlsruhe 1, Germany, online services. Also available on CD-ROM. Inquire as to cost and availability.

Scisearch. Institute for Scientific Information, 3501 Market Street, Philadelphia, PA 19104. (800) 523-1850 or (215) 386-0100. Broad multidisciplinary title and author index to the international literature of science and technology, 1974 to present. Available on DIALOG,(800) 334-2564, and ORBIT,(800) 456-7248, online services. Also available on CD-ROM. Inquire as to cost and availability.

WILSONLINE. H.W. Wilson Company, 950 University Avenue, Bronx, NY 10452. (800) 367-6770 or (212) 588-8400. Makes available online versions of the printed H.W. Wilson indexes including Applied Science and Technology Index, Business Periodicals Index, General Science Index, and Readers' Guide to Periodical Literature. Period covered is generally 1983 to present. Available on BRS,(800) 289-4277, DIALOG,(800) 334-2564, and OCLC EPIC,(800) 848-5878, online services. Also available on CD-ROM. Inquire as to cost and availability.

PERIODICALS

Alloy Digest. Alloy Digest Inc., 27 Canfield Street, Orange, NJ 07050. (201) 677-9161. 1952 to present. Monthly. $140.00 per year.

Inorganic Chemistry. American Chemical Society, 1155 16th Street NW< Washington, DC 20036. (800) 333-9511. FAX (914) 447-3671. 1962 to present. Semi-monthly. $949.00 per year.

Iron and Steel Engineer. Association of Iron and Steel Engineers, Three Gateway Center, Suite 2350, Pittsburgh, PA 15222. (412) 281-6323. 1924 to present. Monthly. $50.00 per year.

Journal of Alloys and Compounds. Elsevier (Mad Sq). 1959 to present. 30 times a year. $3980.00 per year.

J O M (Journal of Metals). Minerals, Metals and Materials Society, 420 Commonwealth Drive, Warrendale, PA 15086. (412) 776-9080. 1949 to present. Monthly. $50.00 per year.

Metal Bulletin Monthly. Metal Bulletin Inc., 220 Fifth Avenue, 19th Floor, New York, NY. 10001-7781. (800) 638-2525. FAX (212) 213-6273. 1972 to present. Monthly. Inquire for cost.

Metal Finishing. Elsevier Science Publishing Company,Inc., 655 Avenue of the Americas, New York, NY 10010. (212) 989-5800. 1903 to present. 12 times a year. $40.00 per year.

Metallurgical Transactions A: Physical Metallurgy and Materials Science. ASM International, Materials Park, OH 44073. (216) 338-5151 or FAX (216) 338-4634. 1970 to present. Monthly. $520.00 per year.

Metals Week. McGraw-Hill Publishing Company, Commodity Services Group, 1221 Avenue of the Americas, New York, NY 10020. (800) 262-4729 or (212) 512-3825. 1930 to present. Weekly. $770.00 per year.

Metalworking Digest. Gordon Publishing, Division of Elsevier Communications, 301 Gibraltar Drive, Box 650, Morris Plains, NJ 07950-3406. (201) 292-5100. FAX (201) 898-9281. 1968 to present. 12 times a year. $48.00 per year.

Mineral Industry Surveys, Cobalt. U.S. Bureau of Mines, Production and Distribution, Cochrans Mill Rd., Box 18070, Pittsburgh, PA 15236. (412) 892-4411. Monthly. Inquire for price and availability.

RESEARCH CENTERS AND INSTITUTES

Cooperative Program In Metallurgy, Pennsylvania State University. 208A Steidle Bldg., University Park, PA 16802. (814) 865-5446.

University of Connecticut Institute of Materials Science. U-136, 97 N. Eagleville Road, Storrs, CT 06268. (203) 486-4623. FAX (203) 486-4745.

University of Florida Department of Materials Science and Engineering. Gainesville, FL 32611. (904) 392-1454. FAX (904) 392-6359.

SPECIFICATIONS AND STANDARDS

*Annual Book of Astm Standards, Volume 02.*04. American Society for Testing and Materials (ASTM), 1916 Race Street, Philadelphia, PA 19103. (215) 299-5585. Annual. Inquire for cost and availability.

COBOL

See: COMPUTER PROGRAMMING LANGUAGES

CODES AND CIPHERS

See: COMPUTER SECURITY

COFFERDAM

See: DAMS

COKE AND COKING

See: STEEL AND STEEL MAKING

COLLOIDS

See Also: ANALYTICAL CHEMISTRY, CATALYSIS, CHEMICAL BONDING, CHEMICAL ENGINEERING, CHEMISTRY, ELECTRO-CHEMISTRY, PHOTOCHEMISTRY, PHYSICAL CHEMISTRY, POLYMERS, SURFACE CHEMISTRY

ABSTRACT SERVICES AND INDEXES

Applied Science and Technology Index; A Cumulative Subject Index To English Language Periodicals In the Fields of Aeron Autics and Space Science, Computer Technology, Chemistry , Construction Industry, Energy and Related Areas. H. W. Wilson Co.,950 University Avenue, Bronx, NY 10452-9978. (212) 588-8400. (800) 367-6770. FAX (718) 590-1617. From 1958 to present. Monthly. Inquire about cost and availability. Also available online (BRS and WILSONLINE) and on CD-ROM. ISSN: 0003-6986.

Chemical Abstracts. Chemical Abstracts Service. 2540 Oleantangy River Road, Box 3012, Columbus, OS 43210-0012. (614) 447-3600. FAX (614) 447-3713. 1907 to present. Weekly. $16,800.00 per year. Available online and on CD- ROM. CA is also available in five section groupings. Inquire regarding cost and availability.

Current Contents : Physical, Chemical and Earth Sciences. Institute for Scientific Information, 3501 Market Street, Philadelphia, PA 19104. (215) 386-0100. (800) 523-1850. FAX (215) 386-2291. 1961 to present. Weekly. $442.00 per year. Also available online (BRS, DIALOG) and on CD-ROM. Inquire regarding cost and availability. ISSN: 0163-2574.

General Science in Dex. H.W. Wilson Company, 950 University Avenue, Bronx, NY 10452. (212) 588-8400. (800) 367-6770. FAX (718) 590-1617. From 1978 to present. Ten issues per year; quarterly and annual cumulations. Service basis. Available on CD-ROM and online. Inquire regarding cost and availability. ISSN: 0162-1963.

Physics Abstracts. INSPEC. Section A, Science Abstracts. Institution of Electrical Engineers (IEE), London, United Kingdom. Available from: INSPEC/IEEE - Institute of Electrical and Electronic Engineers, Box 1331, 445 Hoes Lane, Piscataway, NJ 08855-1331. (908) 562-5549. 1898 to present. 24 issues per year. $2835.00 per year. ISSN: 0036-8091. Also available online and on CD-ROM.

Process and Chemical Engineering/chemical Engineering Abstracts. Royal Society of Chemistry, Information Services, Thomas Graham House, Science Park, Milton Road, Cambridge, CB4 4WF, England. Contains citations, mostly with abstracts, of the worldwide literature on chemical engineering; from 1982 to present. Monthly. $610.00 per year. Also available online. ISSN: 0960-5045.

Science Citation Index. SCI. Institute for Scientific Information, 3501 Market Street, Philadelphia, PA 19104. (215) 386-0100. (800) 523-1850. FAX (215) 386-2991. 1961 to present. Six issues per year, plus annual cumulation. $11650.00 per year. Also available online and on CD-ROM. Inquire about price and availability. ISSN: 0036-827X.

Theoretical Chemical Engineering. Royal Society of Chemistry, Information Services, Thomas Graham House, Science Park, Milton Road, Cambridge, CB4 4WF, England. Covers theoretical chemical engineering, including theory and laboratory experimentation. 1964 to present. Monthly. $235.00. ISSN: 0960-5053.

ANNUAL REVIEWS AND YEARBOOKS

Advances In Photochemistry. John Wiley and Sons, Inc., 605 Third Avenue, New York, NY 10158. (800) 526-5368 or (212) 850-6000.1963 to present. Irregular, price varies.

Advances In Physical Organic Chemistry. Academic Press, Inc., 1250 Sixth Avenue, San Diego, CA 92101-4311. (619) 231-0926. FAX (619) 699-6715. 1963 to present. Irregular, price varies. *Annual Review of Physical Chemistry.* Annual Reviews, Inc., 4139 El Camino Way, Palo Alto, CA 94306-0897. (415) 493-4400. Fax (415) 855-9815. 1951 to present. Annual. $55.00 per year.

Progress In Colloid and Polymer Science Series. Springer-Verlag, 175 Fifth Avenue, New York, NY 10010. (212) 460-1500. Irregular; price varies.

Progress In Surface and Membrane Science. Academic Press, Inc., 1250 Sixth Avenue, San Diego, CA 92101-4311. (619) 231-0926. FAX (619) 699-6715. 1971 to present. Irregular, price varies.

Recent Progress In Surface Membrane Science. Academic Press, Inc., 1250 Sixth Avenue, San Diego, CA 92101-4311. (619) 231-0926. FAX (619) 699-6715. 1964 to present. Irregular, price varies.

ASSOCIATIONS AND PROFESSIONAL SOCIETIES

American Chemical Society, 1155 16th Street, Nw, Washington, Dc 20036. (202) 872-4600.

American Institute of Physics, 335 East 45th Street, New York, Ny 10017 (212) 661-9404 Or Fax (516)349-9704.

Association of Consulting Chemists and Chemical Engineers, 50 East 41st Street, Suite 92, New York, Ny 10017. (212) 684-6255.

Chemical Manufacturers Association. 2501 M Street, N.W., Washington, DC 20037. (202) 887-1182

BIBLIOGRAPHIES

Chemical Information Sources. Gary Wiggins. McGraw-Hill Publishing Company, 1221 Avenue of the Americas, New York, NY 10020. (800) 262-4729 or (212) 512-3825. 1991. $42.50.

Scientific and Technical Books and Serials In Print; An Index To Literature In Science and Technology. R.R. Bowker Inc., 121 Chanlon Road, New Providence, NJ 07974. (908) 464-6800. (800) 521-8110. FAX (908) 665-3502. 1972 to present. Annual. 4 volumes. 1994. $299.95. Also available on compact disc and online. ISSN: 0000-054X.

DIRECTORIES AND BIOGRAPHICAL SOURCES

American Men and Women of Science. Physical and Biological Sciences. R.R. Bowker Inc., 121 Chanlon Road, New Providence, NJ 07974. (908) 464-6800. (800) 521-8110. 20th edition. 8 volumes. 1996. $850.00

Chemical Week - Buyers Guide. Chemical Week Associates, 888 Seventh Avenue, New York, NY 10106. (212) 621-4900. FAX (212) 621-4949. Annual. $50.00. ISSN: 0009-272X.

Consulting Services: Chemists and Chemical Engineers. Association of Consulting Chemists and Chemical Engineers, 40 East 45th Street, New York, NY 10017. (212) 983-3160. FAX (212) 983-3161. Biennial. $60.00.

Research Centers Directory. Gale Research Company,Inc., 835 Penobscot Building, Detroit, MI 48226-4094. (313) 961-2242.(800) 347-4253. 20th edition. 1995. $485.00. ISSN:: 0080-1518.

Who's Who In Engineering. Gordon Davis, editor. American Association of Engineering Societies. 1111 19th Street, NY, Suite 608, Washington, DC 20036. (202) 296-2237. (800) 658-8897. 9th edition. 1995. $220.00.

ENCYCLOPEDIAS AND DICTIONARIES

Dictionary of Organic Compounds. John B. Buckingham, edior. Chapman and Hall, 1 Penn Plaza, New York, NY 10119. (212) 564-1060. 1993. 10th supplement. $832.50.

Encyclopedia of Chemical Processing and Design. McKetta Marcel Dekker, Inc., 270 Madison Avenue, New York, NY 10016. (212) 696-9000. (800) 228-1160. 1976 - . $175.00 per volume. ISSN:

Industrial Chemical thesaurus. Michael Ash and Irene Ash. VCH Publishers, Inc., 220 East 23rd Street, New York, NY 10010-4606. (800) 367-8249. 1992. $295.00.

Kirk-Othmer Encyclopedia of Chemical Technology. John Wiley and Sons, Inc., 605 Third Avenue, New York, NY 10158. (800) 526-5368 or (212) 850 -6000. Fourth edition. 1991 - . Twenty-seven volumes. $5400.00.

Language of Colloid and Interface Science: A Dictionary of Terms. ZLaurier L. Schramm. American Chemical Society, 1155 16th Street, NW, Washington, DC 20036. (202) 872-4600. 1993. $39.95.

McGraw-hill Encyclopedia of Chemistry. Sybil P.Parker, editor. McGraw-Hill Book, Incorporated, 1221 Avenue of the Americas, New York, NY 10020. (212) 997-3675. Second edition. 1993. $95.50.

GENERAL WORKS

Colloidal Systems and Interfaces. Sydney Ross and Ian Morris. John Wiley and Sons, Inc., 605 Third Avenue, New York, NY 10158. (800) 526-5368 or (212) 850-6000. 1988. $110.00.

Colloidal Dispersions. W. B. Russel, et al. Cambridge University Press, 40 West 20th Street, New York, NY 10011-4211. (212) 924-3900. (800) 872-7423. 1992. $42.95.

COLLOIDS

Ency. of Physical Sciences and Engineering Info. Sources

The Colloidal Domain: Where Physics, Chemistry, Biology and Technology Meet. D. Fennell Evans, and Hakan Wennerstrom. VCH Publishers, Inc., 220 East 23rd Street, Suite 909, New York, NY 10010-4606. (212) 683-8333. (800) 422-8824.1994. wfi.

Concepts In Manipulation of Groundwater Colloids For Environmental Restoration. John F. McCarthy, et al,editors. Lewis Pubs/CRC Press, Inc., 2000 Corporate Boulevard, NW, Boca Raton, FL 33431. (407) 994-0555. (800)272-7737. 1993, $79.95.

Disperse Systems, Interfaces and Membranes. K. Hummel, and J. Schurz, editors. Springer-Verlag New York, Inc., 175 Fifth Avenue, New York, NY 10010. (212) 460-1500. (800)777-4643. FAX (212) 473-6272. 1989. $119.00.

Dispersants, Solvents, and Solubilizers. Michael Ash and Irene Ash, editors. Chem Publishing, Co., Inc. 80 Eighth Avenue, New York, NY 10011. (212 255-1950. 1988. $60.00.

Electrochemistry of Colloids and Dispersions. Raymond Mackay and John Texter, editors. VCH Publishers, Inc., 303Northwest 12th Avenue, Deerfield Beach, FL 33442. (800)367-8249. 1992. $115.00.

Foundations of Colloid Science. Robert J. Hunter. Oxford University Press, Inc., 200 Madison Avenue, New York, NY 10016. (212) 725-6000. (800) 334-4249. 2 volumes. 1992.$75.00 in paper.

Introduction To Colloid and Surface Chemistry. Duncan J. Shaw. Butterworth-Heinemann, 313 Washington Street, Newton,MA 02158. (618-928-2500. (800) 366-2665. 1992. $31.95.

*Percolation and Fractals In Colloid and Interface Science.*A. V. Neimark. World Scientific Publishing Company, Inc., 1060 Main Street, River Edge, NJ 07661. (201) 487-9655.(800) 227-7562. 1995. $48.00.

Polymeric Layers. P. Wunsche, editor. Springer-Verlag, New York, Inc., 175 Fifth Avenue, New York, NY 10010. (212)460-1500. (800) 777-4643. FAX (212) 473-6272. 1992.$87.00.

Surface and Colloid Chemistry In Natural Waters and Water Treatment. R. Beckett, editor. Plenum Publishing Corporation, 233 Spring Street, New York, NY 10013. (212) 620-8000. FAX (212) 463-0742. 1990. $65.00.

Surfaces, Interfaces and Colloids: Principles and Applications. Drew Myers. VCH Publishers, Inc., 303 Northwest 12th Avenue, Deerfield Beach, FL 33442. (800) 367-8249. 1991. $49.50.

Theory of Stability of Colloids and Thin Films. V. B. Derjaguin. Plenum Publishing Corp., 233 Spring Street, New York, NY. (212) 620-8000. (800) 2221-9369. FAX (212)463-0742. 1989. $89.50.

HANDBOOKS AND MANUALS

Chemometrics: Chemical and Sensory Data. David R. Burgard and James T. Kuznicki. CRC Publishers, Inc., 2000 Corporate Blvd., N.W., Boca Raton, FL 33431. (407) 994-0555 or (800) 333-8300. 1990. $120.00.

CRC Handbook of Chemistry and Physics. David R. Kide, editor. CRC Press Inc., 2000 Corporate Blvd., NW, Boca Raton,

FL 33431. (407) 994-0555. (800) 333-8300. 77th edition. 1996. $99.95.

Lange's Handbook of Chemistry. John A. Dean, editor. McGraw-Hill Publishing Company, 1221 Avenue of the Americas, New York, NY 10020. (800) 262-4729 or (212) 512-3825. Fourteenth edition. 1992. $79.50.

Perry's Chemical Engineers' Handbook. Robert H. Perry and Donald W. Green, editors. McGraw-Hill Publishing Company, Inc., 1221 Avenue of the Americas, New York, NY 10020. (212) 512-2000. (800) 262-4729. 6th edition. 1996. $129.50.

Riegel's Handbook of Industrial Chemistry. James A. Kent, editor. Van Nostrand Reinhold, 115 Fifth Avenue, New York, NY 10003. (212) 254-3232 or (800) 926-2665. Ninth edition. 1992. $114.95.

ONLINE DATABASES AND CD-ROMS

Analytical Abstracts Online. Royal Society of Chemistry, Information Services, Thomas Graham House, Science Park, Milton Road, Cambridge, CB4 4WF, England. Contains citations, mostly withabstracts, of the worldwide literature on analytical chemistry, from 1980 to present. Available on DIALOG and STN online services. Inquire as to cost and availability.

Ca Search. Chemical Abstracts Service, P.O. Box 3012, Columbus, Oh 43210-0012. (614) 447-3600. (800) 848-6533. FAX (614) 447-3709. Guide to worldwide chemical literature and related fields, 1972 to present. Available on BRS, DIALOG, ORBIT and STN online services. Inquire as to cost and availability.

Chemical Journals of the American Chemical Society. American Chemical Society, 1155 16th Street, N.W., Washington, DC 20036. (202) 872-4381 or (800) 424-6747. Contains complete text of approximately 90,000 articles from 22 primary journals published by the American Chemical Society, from mostly 1982 to present. Available on STN online service. Inquire as to cost and availability.

Dissertation Abstracts. University Microfilms International, 300 North Zeeb Road, Ann Arbor, MI 48106. (800) 521-0600 or (313) 761-4700. Scope includes virtually all doctoral dissertations accepted at accredited American institutions from 1861 to present in 252 subject areas. Available on BRS, DIALOG, and OCLC EPIC online services. Also available on CD-ROM. Inquire as to cost and availability.

Inspec. Institution of Electrical Engineers, Michael Faraday House, Six Hills Way, Stevenage, Herts. SG1 2AY, England. Telephone: 0438 313311 or FAX 0438 742840. Contains citations to the worldwide literature of physics, electronics and electrical engineering, computer technology, and related fields. Available on BRS, DIALOG, ORBIT and STN online services. Inquire as to cost and availability.

Kirk-othmer Encyclopedia of Chemical Technology. John Wiley and Sons, Inc., 605 Third Avenue, New York, NY 10158. (800) 526-5368 or (212) 850-6000. Contains the complete text of all chapters in the 27 volume fourth edition of the KIRK-OTHMER ENCYCLOPEDIA of CHEMICAL TECHNOLOGY. 1991. Available on BRS, DIALOG, online services. Inquire as to cost and availability.

NTIS. National Technical Information Service, 5285 Port Royal Road, Springfield, Va 22161. (703) 487-4929 or FAX (703) 321-8199. Broad coverage of government-sponsored research reports, 1964 to present. Available on BRS, DIALOG, ORBIT and STN online services. Also available on CD-ROM. Inquire as to cost and availability.

SCISEARCH. Institute for Scientific Information, 3501 Market Street, Philadelphia, PA 19104. (800) 523-1850 or (215) 386-0100. Broad multidisciplinary title and author index to the international literature of science and technology, 1974 to present. Available on DIALOG and ORBIT online services. Also available on CD-ROM. Inquire as to cost and availability.

WILSONLINE. H.W. Wilson Company, 950 University Avenue, Bronx, NY 10452. (800) 367-6770 or (212) 588-8400. Makes available online versions of the printed H.W. Wilson indexes includingApplied Science and Technology Index, Business Periodicals Index, and Readers' Guide to Periodical Literature. Period covered is generally 1983 to present. Also available on CD-ROM. Inquire as to cost and availability.

Who's Who In Technology. Gale Research Inc., 835 Penobscot Building, Detroit, MI 48226-4094. (313) 961-2242. (800) 347-4253. Contains biographical information of contemporary American scieNTISts and engineers. Available on DIALOG online service. Inquire as to cost and availability.

PERIODICALS

Advances In Colloid and Interface Science. Elsevier Science Publishing Company, Inc., 655 Avenue of the Americas, New York, NY 10010. (212) 989-5800. 1967 to present. Twenty-four times per year. $1900.00 per year. ISSN: 0001-8686.

Analytical Chemistry. American Chemical Society, Box 3337, Columbus, OH 43210. (614) 447-3776. (800) 333-9511. FAX (614) 447-3671. 1929 to present. Monthly. $415.00 per year. ISSN: 0003-2700.

Analytical Methods and Instrumentation. John Wiley and Sons, Inc., 605 Third Avenue, New York, NY 10158. (800) 526-5368 or (212) 850-6000. Bimonthly.1 992 to present. Bimonthly. $250.00 per year. ISSN: 1063-5246.

Chemical Physics. Elsevier Science Publishing Company, Inc., 655 Avenue of the Americas, New York, NY 10010. (212) 989-5800. 1973 to present. 33 issues per year. $2759.00 per year. ISSN: 0301-0104.

Chemical Week; Includes Annual Buyers Guide. Chemical Week Associates, 888 Seventh Avenue, New York, NY 10106. (212) 621-4900. FAX (212) 621-4949. 1914 to present. Weekly. $99.00 per year. ISSN: 0009-272X

Chemtech. American Chemical Society, Box 3337, Columbus, OH 43210. (614) 447-3776. 1970 to present. Monthly. $370.00 per year. ISSN: 0009-2703.

Colloids and Surfaces A: Physicochemical and Engineering Aspects. Elsevier Science Publishing Company, Inc., 655 Avenue of the Americas, New York, NY 10010. (212) 989-5800. 1980 to present. Thirty-six times per year. $2364.00 per year. ISSN: 0927-7757.

Colloids and Surfaces B: Biointerfaces. Elsevier Science Publishing Company, Inc., 655 Avenue of the Americas, New York,

NY 10010. (212) 989-5800. 1980 to present. Six times per year. $300.00 per year. ISSN: 0927-7765.

Inorganic Chemistry. American Chemical Society, 1155 16th Street, N.W., Washington, DC 20036. (800) 424-6747. FAX (614) 447-3671. 1962 to present. Biweekly. $949.00 per year. ISSN: 0020-1669.

Journal of Colloid and Interface Science. Academic Press, Inc., 1250 Sixth Avenue, San Diego, CA 92101-4311. (619) 231-0926. FAX (619) 699-6715. Subscriptions to: Box 620000, Orlando, FL 32891. (800) 543-9534. 1946 to present. Fourteen times per year. $1302.00 per year. ISSN: 0021-9797

Journal of Dispersion Science and Technology. Marcel Dekker, Inc., 270 Madison Avenue, New York, NY 10016. (212) 696-9000. FAX (212) 685-4540. Subscriptions to: Box 5017 Monticello, NY 12701. 1980 to present. Six times per year. $445.00 per year. ISSN: 0193-2691.

Journal of Membrane Science. Elsevier Science Publishing Company, Inc., 655 Avenue of the Americas, New York, NY 10010. (212) 989-5800. Subscriptions to: Box 882, Madison Square Station, NY 10159-0882. 1977 to present. 36 times per year. $2731.00 per year. ISSN: 0376-7388.

Journal of Organic Chemistry. American Chemical Society, Box 3337, Columbus, OH 43210. (614) 447-3776. (800) 227-5558. 1936 to present. Bi-Weekly. $785.00 per year. ISSN: 0022-3263.

Journal of Photochemistry and Photobiology, A: Chemistry. Elsevier Science Publishing Company, Inc., 655 Avenue of the Americas, New York, NY 10010. (212) 989-5800. 1972 to present. Twenty-four times per year. $1200.00 per year. ISSN: 1010-6030.

Journal of Physical Chemistry. American Chemical Society, Box 3337, Columbus, OH 43210. (614) 447-3776. (800) 227-5558. 1896 to present. Weekly. $1140.00 per year. ISSN: 0022-3654.

Journal of Solution Chemistry. Plenum Publishing Corporation, 233 Spring Street, New York, NY 10013. (212) 620-8000. FAX (212) 463-0742. 1972 to present. Monthly. $560.00 per year. ISSN: 0095-9782.

RESEARCH CENTERS AND INSTITUTES

Chemistry Laboratories. Rensselaer Polytechnic Institute, Cogswell Laboratory, Troy, NY 12180-3590. (518) 276-8981.

Center For Colloidal and Interfacial Dynamics. University of Texas at Arlington, Box 19065, Chemistry Department, Arlington, TX 76091. (817) 273-3803.

Institute of Colloid and Surface Science. Clarkson University. Potsdam, NY 13699-5660. (315) 268-3820. FAX (315) 268-3841.

Zettlemoyer Center For Surface Studies. Lehigh University, Sinclair Laboratory, Building 7, Bethleham, PA 18015. (215) 758-3571.

COLOR

Ency. of Physical Sciences and Engineering Info. Sources

COLOR

See also: CAMERA, DYES AND DYEING, HOLOGRAPHY, OPTICS, PAINTS AND PIGMENTS, PHOTOCHEMISTRY, PHOTOGRAPHIC FILM, PHOTOGRAPHY, SPECTROSCOPY

ABSTRACT SERVICES AND INDEXES

Applied Science and Technology Index; a cumulative subject index to English language periodicals in the fields of aeronautics and space science, computer technology, ch emistry, construction industry, energy and related areas. H.W. Wilson Co., 950 University Avenue, Bronx, NY 10452. (212) 588-8400. (800) 367-6770. FAX (718) 590-1617. From 1958 to present. Monthly. Inquire about cost and availability. Also available on CD-ROM and online. ISSN: 0003-6986.

Chemical Abstracts. Chemical Abstracts Service, 2540Olentangy River Road, Box 3012, Columbus, OH 43210-0012. (614) 447-3600. (800) 848-6538. FAX (614) 447-3713. From 1907 to present. Weekly. $16800.00 per year. Also available on CD-ROM and online. Inquire regarding cost and availability.

Engineering Index Monthly. Engineering Information, Inc., Castle Point on the Hudson, Hoboken, NJ 07030. (201) 216-8500. FAX (201) 216-8532. Monthly. $2200.00 per year. Also available online as COMPENDEX and on CD-ROM. ISSN: 0742-1974.

General Science Index. H.W. Wilson Company, 950 University Avenue, Bronx, NY 10452. (212) 588-8400. (800) 367-6770. FAX (718) 590-1617. From 1978 to present. Ten issues per year; quarterly and annual cumulations. Service basis. Available on CD-ROM and online. Inquire regarding cost and availability. ISSN: 0162-1963.

Imagining Abstracts. Royal Photographic Society of Great Britain. Order from: Elsevier Science Publishers, 660 White Plains Road, Tarrytown, NY 10591-5153. (914) 524-9200. 1921 to present. Bi-monthly. $460.00 per year. ISSN: 0896-100X.

Physics Abstracts. INSPEC. Section A, Science Abstracts. In stitution of Electrical Engineers (IEE). Available from INSPEC/ IEEE - Institute of Electrical and Electronic Engineers,, Box 1331, 445 Hoes Lane, Piscaway NJ 08855-1331. (908) 562-5549. 1898 to present. 24 issues per year. $2835.00 per year. Also available on CD-ROM and online. ISSN: 0036-8091.

Physics Briefs (Physikalische Berichte). Information Center for Energy, Physics, Mathematics; German Physical Society. V C H Publishers, Inc., 220 East 23rd Street, New York, NY 10010-4606. (212) 683-8333. 1845 to present. 24 issues per year. $2390.00 per year. Also available online. ISSN: 0179-7434.

Science Citati On Index. SCI. Institute for Scientific Information, 3501 Market Street, Philadelphia, PA 19104. (215) 386-0100. (800) 523-1850. FAX (215) 386-2991. 1961 to present. Six issues per year, plus annual cumulation. $11650.00 per year. Also available online and on CD-ROM. Inquire about price and availability. ISSN: 0036-827X.

ANNUAL REVIEWS AND YEARBOOKS

Progress In Optics. Elsevier Science Inc., 660 White Plains Road, Tarrytown, NY 10591-5153. (914) 524-9200. FAX (914) 333-2444. 1961 to present. Irregular. Price varies. ISSN:: 0079-6638.

ASSOCIATIONS AND PROFESSIONAL SOCIETIES

American Institute of Physics. One Physics Ellipse, College Park, MD 20740-3843. (301) 209-3100.

Optical Society of America. 2010 Massachusetts Avenue NW, Washington, DC 20036. (202) 223-8130. FAX (202) 223-1096.

Society For Imaging Science and Technology. 7003 Kilworth Lane, Springfield, VA 33252. (703) 64209090. FAX (703) 642-9094.

Spie - the International Society For Optical Engineering. PO Box 10, Bellingham, WA 98227. (206) 676-3290. FAX (206) 647- 1445.

BIBLIOGRAPHIES

Science Books and Films. American Association for the Advancement of Science, 1333 H Street NW, Washington, DC 20005. (202) 326-6454. Reviews of print, film and software materials in all sciences. 1965 to present. Nine issues per year. $40.00 per year. ISSN: 0098-342X

Scientific and Technical Books and Serials in Print, 1995; An Index To Literature In Science and Technology. R.R. Bowker Inc., 121 Chanlon Road, New Providence, NJ 07974. (908) 464-6800. (800) 521-8110. 4 volumes. 1972 to present. Annual. $299.95. Also available on compact disc and online. ISSN:: 0000-054X.

DIRECTORIES AND BIOGRAPHICAL SOURCES

American Men and Women of Science: Physical and Biological Sciences. R. R. Bowker Inc., 121 Chanlon Road, New Providence, NJ 07974. (908) 464-6800. (800) 521-8110. 20th edition. 8 volumes. 1996. $850.00.

International Directory of Engineering Societies and Related Organizations. American Society for Engineering Education, 1818 N Street NW, Suite 600, Washington, DC 20036. (202) 331-3526. 15th edition. 1996. $185.00.

Research Centers Directory. Gale Research Inc., 835 Penobscot Building, Detroit, MI 48226-4094. (313) 961-2242. (800) 347-4253. 20th edition. 1995. $485.00. ISSN:: 0080- 1518.

Scientific and Technical Organizations and Agencies Directory. Gale Research, 835 Penobscot Building, Detroit, MI 48226-4094. (313) 961-2242. (800) 347-4253. 4th edition. 1996. $195.00.

Who's Who In Engineering. Gordon Davis, editor. American Society for Engineering Education, 1111 19th Street, NW, Suite 608, Washington, DC 20036. 9th edition. 1995. $220.00.

Who's Who In Technology. Gale Research, 835 Penobscot Building, Detroit, MI 48226-4094. (313) 961-2242. (800) 347-4253. 7th edition. 1995. $195.00.

ENCYCLOPEDIAS AND DICTIONARIES

Dictionary of the Physical Sciences: Terms, Formulas, Data. Cesare Emiliani. Oxford University Press, Inc., 200 Madison

Ency. of Physical Sciences and Engineering Info. Sources

COLOR

Avenue, New York, NY 10016. (212) 725-6000. (800) 334-4249. 1989. $19.95.

Encyclopedia of Applied Physics. George Trigg, editor. VCH Publications, Inc., 220 East 23rd Street, Suite 909, New York, NY 10010-4606. (212) 683-8333. (800) 422-8824. 20 volume set. 1991-. $5990.00.

Encyclopedia of Physics. Robert M. Besancon. Chapman & Hall, 1 Penn Plaza, New York, NY 10119. (212) 564-1060. 3rd edition. 1990. $54.95 in paper.

Encyclopedia of Lasers and Optical Technology. Robert A. Meyers, editor. Academic Press, Inc., 6277 Sea Harbor Drive, Orlando, FL. (800) 321-5068. 1991. $92.00.

Encyclopedia of Physical Science and Technology. Academic Press, Inc., 6277 Sea Harbor Drive, Orlando, FL. (800) 321-5068.2nd edition. 18 volume set. 1992. $2625.00.

Glossary of Optical Terminology. Thomas K. Farrell. Butterworth-Heinemann, 313 Washington Avenue, Newton, MA 02158. (617) 928-2500. (800) 366-2665. 1985. $24.95.

McGraw-hill Encyclopedia of Science and Technology. McGraw- Hill Book Company, Inc., 1221 Avenue of the Americas, New York, NY 10020. (212) 512-2000. (800) 262-4729. 7th edition. 20 volume set. 1992. $1900.00.

Thesaurus of Photographic Science and Engineering. Society of Photographic ScieNTISts and Engineers. Books on Demand, 300 North Zeeb Road, Ann Arbor, MI 48106-1346. (313) 761-4700. (800) 521-0600.

GENERAL WORKS

Applied Optics and Optical Design. A. E. Conrady. Dover Publications, Inc., 180 Varick Street, New York, NY 10014. (212) 255-3755. (800) 223-3130. 2 volumes. Reprint edition. 1992. $10.95.

Applied Optics and Optical Engineering; A Comprehensive Treatise. Rudolf Kingslake, et al, editors. Academic Press 9 volumes. 1965- . Volumes individually priced.

Color and Color Vision. Paul L. Pease, editor. American Association of Physics Teachers, One Physics Ellipse, College Park Md 20740-3842. (301) 209-3300. 1982. $18.00.

Color and Light In Nature. David K. Lynch and William Livingston. Cambridge University Press, 40 West 20th Street, New York, NY 10011-4211. (212) 924-3900. (800) 872-7423. 1995. $69.95.

Color Science: Concepts and Methods, Quantitative Data and Formulae. Gunter Wyszecki and W. S. Stiles. John Wiley & Sons, Inc., 605 Third Avenue, New York, NY 10158-0012. (212) 850-6000. (800) 225-5945. 1982. $210.00.

Introduction To Classical and Modern Optics. Jurgen R. Meyer-Arendt. Prentice-Hall, 113 Sylvan Avenue, Route 9W, Englewood Cliffs, NJ 07632. (201) 592-2000. (800) 922- 0579. 4th editon. 1994. $70.00.

Principles of Color Technology. Fred W. Billmeyer Jr., and Max Saltzman. John Wiley & Sons, Inc., 605 Third Avenue, New York, NY 10158-0012. (212) 850-6000. (800) 225-5945. 2nd edition. 1981. $69.95.

Theory of Colours. Johann Wolfgang Von Goethe. MIT Press, 55 Hayward Street, Cambridge, MA 02142. (617) 253-8569. 1970. $18.50.

HANDBOOKS AND MANUALS

Handbook of Optics. Optical Society of America. McGraw- Hill Publishing Company, Inc., 1221 Avenue of the Americas, New York, NY 10020. (212) 512-2000. (800) 262- 4729. 2 volumes. 2nd edition. 1993. $99.50 each.

Handbook of Photographic Science and Engineering. Society for Imaging Science and technology, 7003 Kilworth Lane, Springfield, VA 33252. (703) 64209090. FAX (703) 642- 9094. Inquire.

Handbook of Tables of Functions For Applied Optics. LeoLevi, editor. Franklin Book Co., Inc., Elkins Park, PA 19117. (215) 635-5252. CRC Press Reprints. 1974. $152.95.

Kodak Professional Photoguide. Carolyn Grimes, editor. Eastman Kodak Company, 343 State Street, Rochester, NY 14640. (716) 724-4000. 1986.

McGraw-hill Encyclopedia of Physics. Sybil P. Parker, editor. McGraw-Hill Publishing Company, Inc., 1221 Avenue of the Americas, New York, NY 10020. (212) 512-2000. (800) 262-4729. 2nd edition. 1993. $95.50

ONLINE DATABASES AND CD-ROMS

CA Search. Chemical Abstracts Service, P.O. Box 3012, Columbu s, OH 43210-0012. (614) 447-3600. (800) 848-6533. FAX (614) 447-3709. Very comprehensive guide to worldwide chemical literature and related fields, 1972 to present. Available on BRS,(800) 289-4277, DIALOG, (800) 334-2564, ORBIT (800) 456-7248, and STN International, FIZ Karlsruhe, P.O. Box 2465, W-7500, Karlsruhe 1, Germany, online services. Inquire as to cost and availability.

Compendex Plus. Engineering Information, Inc., 345 East 47th Street, New York, NY 10017. (212) 705-7600 or (800) 221-1044. Contains citations with abstracts to worldwide literature in engineering and technology, from 1970 to present. Available on online BRS,(800) 289-4277, DIALOG, (800) 334-2564, ORBIT (800) 456-7248, and STN International, FIZ Karlsruhe, P.O. Box 2465, W-7500, Karlsruhe 1, Germany, online services. Also available on CD-ROM. Inquire as to cost and availability.

Dissert Ation Abstracts Online. University Microfilms Internatio nal, 300 North Zeeb Road, Ann Arbor, MI 48106. (800) 521-0600 or (313) 761-4700. Scope includes virtually all doctoral dissertations accepted at accredited American institutions from 1861 to present in 252 subject areas. Available on BRS, (800) 289-4277, DIALOG, (800) 334-2564, and OCLC EPIC, (800) 848-5878, online services. Also available on CD-ROM. Inquire as to cost and availability.

Inspec. Institution of Electrical Engineers, Michael Faraday House, S ix Hills Way, Stevenage, Herts. SG1 2AY, Englan d. Telephone: 0438 313311 or FAX 0438 742840. Contains citations to the worldwide literature of physics, electronics and electrical engineering, computer technology, and related fields.

Available on BRS, (800) 289-4277, DIALOG, (800) 334-2564, ORBIT, (800) 456-7248, and STN International, FIZ Karlsruhe, P.O. Box 2465, W-7500, Karlsruhe 1, Germany, online services. Inquire as to cost and availability.

NTIS Bibli Ographic Database. National Technical Information Service, 5285 Port Royal Road, Springfield, Va 22161 . (703) 487-4929 or FAX (703) 321-8199. Broad covera ge of government-sponsored science and technology research reports, 1964 to present. Available on BRS,(800) 289-4277, DIALOG, (800) 334-2564, ORBIT, (800) 456-7248, and STN International, FIZ Karlsruhe, P.O. Box 2465, W-7500, Karlsruhe 1, Germany, online services. Also available on CD-ROM. Inquire as to cost andavailability.

Physics Briefs. American Institute of Physics, 335 East 45th Street , New York, NY 10017. (212) 661-9260 or FAX (212) 661-2036. Contains citations with abstracts of the literature of physics and related fields, 1979 to present. Available on the STN International, FIZ Karlsruhe, P.O. Box 2465, W-7500, Karlsruhe 1, Germany, online service. Inquire as to cost and availability.

Scisearch. Institute for Scientific Information, 3501 Market Street, Philadelphia, PA 19104. (800) 523-1850 or (215) 386-0100. Broad multidisciplinary title and author index to the international literature of science and technology, 1974 to present. Available on DIALOG, (800) 334-2564, and ORBIT, (800) 456-7248, online services. Also available on CD-ROM. Inquire as to cost and availability.

WILSONLINE. H.W. W ilson Company, 950 University Avenue, Bronx, NY 10452. (800) 367-6770 or (212) 588-8400. Makes available online versions of the printed H.W. Wilson indexes including Applied Science and Technology Index, Business Periodicals Index, General Science Index, and Readers' Guide to Periodical Literature. Period covered is generally 1983 to present. Available on BRS, (800) 289-4277, DIALOG, (800) 334-2564, and OCLC EPIC, (800) 848-5878, online services. Also available on CD-ROM. Inquire as to cost and availability.

PERIODICALS

Applied Optics. Optical Society of America. 2010 Massachusetts Avenue NW, Washington, DC 20036-1023. (202) 223-8130. 1962 to present. 36 issues per year. $1090.00 per year. ISSN: 0003-6935.

Applied Spectroscopy. Society for Applied Spectroscopy, 198 Thomas Johnson Drive, Suite 2, Frederick, MD 21702. (301) 694-8122. FAX (301) 694-6860. 1946 to present. Monthly. $180.00 per year. ISSN: 0003-7028.

British Journal of Photography. Bouverie Publishing Company, Ltd., 147-151 Temple Chambers, Temple Avenue, London, EC47 ODT, England 1854 to present. Weekly. L45.50 per year. ISSN: 0007-1196.

Fiber and Integrated Optics. Taylor and Francis, 1900 Frost Road, Suite 101. Bristol, PA 19007. (215) 785-5800. FAX (215) 785-5515. 1977 to present. Quarterly. $180.00 per year. ISSN: 0146-8030.

Industrial Photography. PTN Publishing Corporation, 445 Broad Hollow Road, Suite 21, Melville, NY 11747-4722. (516) 845-2700. 1952 to present. Monthly. $60.00 per year. ISSN: 0019-85954.

International Journal of Nonlinear Optical Physics. World Scientific Publishing Co., 1060 Main Street, Suite 18, River Edge, NJ 07661. (800) 227-7562. 1992 to present. Quarterly. $348.00 per year. ISSN: 0218-1991.

Journal of Imaging Science and Technology. Society for Imaging Science and Technology. 7003 Kilworth Lane, Springfield, VA 33252. (703) 642-9090. FAX (703) 642- 9094. 1950 to present.Bi-monthly. $120.00 per year. ISSN: 1062-3701.

Journal of Lightwave Technology. Institute of Electrical and Electronics Engineers, 445 Hoes Lane, PO Box 1331, Piscataway, NJ 08855. (908) 562-3892. FAX (908) 562-8434. 1983 to present. Monthly. $425.00 per year. ISSN: 0733- 8724.

Journal of the Optical Society of America. Part A: Optics and Image Science. Optical Society of America, Inc., 2010 Massachusetts Avenue, NW, Washington, DC 20036-1023. (202) 223-8130. 1917 to present. Monthly. $610.00 per year. ISSN: 0740-3232.

Journal of the Optical Society of America. Part B: Optical Physics. Optical Society of America, Inc., 2010 Massachusetts Avenue, NW, Washington, DC 20036-1023. (202) 223-8130. 1917 to present. Monthly. $610.00 per year. ISSN: 0740-3224.

Laser Focus World. PennWell Publishing Co., 10 Tara Boulevard, Nashua, NH 03062-2891. (603) 891-0123. FAX (603) 891-0574. 1964 to present. Monthly. $104.00 per year. ISSN: 1043-8092.

Optical Engineering. SPIE - International Society for Optical Engineering, Box 10, 1000 20th Street, Bellingham, WA 98227-0010. (206) 676-3290. FAX (206) 647-1445. 1962 to present. Monthly. $170.00 per year. ISSN: 0091-3286.

Optics and Photonics News. Optical Society of America, Inc., 2010 Massachusetts Avenue, NW, Washington, DC 20036-1023. (202) 223-8130. 1975 to present. Monthly. $99.00 per year. ISSN: 1047-6938.

Optics Communications. Elsevier Science Inc., Box 882, Madison Square Station, New York, NY 10159. (212) 989-5800. FAX (212) 633-4990. 1969 to present. 54 issues per year. $2121.00. per year. ISSN: 0030-4018.

Optics Letters. Optical Society of America, Inc., 2010 Massachusetts Avenue, NW, Washington, DC 20036-1023. (202) 223-8130. 1977 to present. Semi-Monthly. $625.00 per year. ISSN: 0146-9592.

Spie - International Society For Optical Engineering. Proceedings. SPIE - International Society for Optical Engineering, Box 10, 1000 20th Street, Bellingham, WA 98227- 0010. (206) 676-3290. FAX (206) 647-1445. 1963 to present. Approximately 50 numbers per year. Indiv idually priced.

RESEARCH CENTERS AND INSTITUTES

Institute of Optics. University of Rochester, Wilmot Building, Rochester, NY 14627. (716) 275-5248.

Center For Applied Optics. University of Alabama in Huntsville, Research Institute Building, Huntsville, AL 25899. (205) 895-6102.

Center For Applied Optics. University of Texas at Dallas, P.O. Box 830688, Richardson, TX 75803-0688. (214) 690-2868. FAX (214) 690-2848.

Center For Imaging Science. Rochester Institute of Technology, One Lomb Memorial Drive, Rochester, NY 14623. (716) 475-2774.

Center For Research In Electro-optics and Lasers. University of Central Florida, 12424 Research Parkway, Orlando, FL 32826. (407) 658-6800. FAX (407) 658-6880.

Institute For Physical Science and Technology. University of Maryland, 4203 Computer and Space Sciences Building, College Park, MD 20742. (301) 454-4874. FAX (301) 314- 9363.

Optical Sciences Center. University of Arizona, Tucson, AZ 85721. (602) 621-6997.

COLOR FILM

See: PHOTOGRAPHIC FILM

COLOR TELEVISION

See: TELEVISION

COLORIMETRY

See: COLOR

COLUMNS

See: STRUCTURAL ENGINEERING

COMBINATORIAL THEORY

See: MATHEMATICS

COMBUSTION

See also: CARBON, CHEMISTRY, CHEMICAL ENGINEERING, THERMOCHEMISTRY

ABSTRACT SERVICES AND INDEXES

Analytical Abstracts. Royal Society of Chemistry, Information Services, Thomas Graham House, Science Park, Milton Road, Cambridge, CB4 4WF, England. Contains citations, mostly with abstracts, of the worldwide literature on analytical chemistry, from 1954 to present. Monthly. $636.00 per year. Also available online.

Applied Science and Technology Index; A Cumulative Subject Index To English Language Periodicals In the Fields of Aeronautics and Space Science, Computer Technology, Chemistry, Construction Industry, Energy and Related Areas. H.W. Wilson Co., 950 University Avenue, Bronx, NY 10452. (212) 588-8400. (800) 367-6770. FAX (718) 590-1617. From 1958 to present. Monthly. Inquire about cost and availability. Also available on CD-ROM and online. ISSN: 0003-6986.

Chemical Abstracts. Chemical Abstracts Service. 2540 Oleantangy River Road, Box 3012, Columbus, OS 43210-0012. (614) 447-3600. FAX (614) 447-3713. 1907 to present. Weekly. $16,800.00 per year. Available online and on CD- ROM. CA is also available in five section groupings. Inquire regarding cost and availability.

Current Contents : Physical, Chemical and Earth Sciences. Institute for Scientific Information, 3501 Market Street, Philadelphia, PA 19104. (215) 386-0100. (800) 523-1850. FAX (215) 386-2291. 1961 to present. Weekly. $442.00 per year. Also available online (BRS, DIALOG) and on CD-ROM. Inquire regarding cost and availability. ISSN: 0163-2574.

Engineering Index Monthly; Indexes and Abstracts the World's Engineering and Technical Literature. Engineering Information, Inc., Castle Point on the Hudson, Hoboken, NJ 07030. (201) 216-8500. (800) 221-1044. FAX (201) 216-8532. Monthly. $2300.00 per year. Available online as COMPENDEX and also on CD-ROM. ISSN: 0742-1974.

General Science Index. H.W. Wilson Company, 950 University Avenue, Bronx, NY 10452. (212) 588-8400. (800) 367-6770. FAX (718) 590-1617. From 1978 to present. Ten issues per year; quarterly and annual cumulations. Service basis. Available on CD-ROM and online. Inquire regarding cost and availability. ISSN: 0162-1963.

Government Reports Announcements and Index. National Technical Information Service (NTIS), 5285 Port Royal Road, Springfield, VA 22161. (703) 487-4650. FAX (703) 321-8547. From 1968 to present. Annual. $630.00 per year. Also available online as *NTIS Bibliographic Database* and on CD-ROM.

Physics Abstracts. INSPEC. Section A, Science Abstracts . Institution of Electrical Engineers (IEE), London. Available from: INSPEC/IEEE - Institute of Electrical and Electronic Engineers, Box 1331, Hoes Lane, Piscataway, NJ 08855-1331. (908) 562-5549. 1898 to present. 24 issues per year. $2835.00 per year. Also available online and on CD-ROM. ISSN: 0036-8091.

Science Citation Index. SCI. Institute for Scientific Information, 3501 Market Street, Philadelphia, PA 19104. (215) 386-0100. (800) 523-1850. FAX (215) 386-2991. 1961 to present. Six issues per year, plus annual cumulation. $11650.00 per year. Also available online and on CD-ROM. Inquire about price and availability. ISSN: 0036-827X.

ANNUAL REVIEWS AND YEARBOOKS

Advances In Catalysis. Academic Press, Inc., 1250 Sixth Avenue, San Diego, CA 92101-4311. (619) 231-0926. FAX (619) 699-6715. 1948 to present. Irregular. Price varies.

Annual Review of Physical Chemistry. Annual Reviews, Inc., 4139 El Camino Way, Palo Alto, CA 94306-0897. (415) 493-4400. Fax (415) 855-9815. 1951 to present. Annual. $55.00 per year.

COMBUSTION

Ency. of Physical Sciences and Engineering Info. Sources

ASSOCIATIONS AND PROFESSIONAL SOCIETIES

American Chemical Society, 1155 16th Street, NW, Washington, DC 20036. (202) 872-4600.

American Institute of Chemical Engineers. 345 East 47th Street, New York, NY 10017-2396. (212) 705-7663.

American Institute of Physics, 335 East 45th Street, New York, Ny 10017 (212) 661-9404 Or Fax (516)349-9704.

Association of Consulting Chemists and Chemical Engineers, 50 East 41st Street, Suite 92, New York, Ny 10017. (212) 684-6255.

Association of official Analytical Chemists, 2200 Wilson Boulevard, Suite 400, Arlington, Va 22001. (703) 522-3032.

Chemical Manufacturers Association. 2501 M Street, N.W., Washington, DC 20037. (202) 887-1182

Combustion Institute. 5001 Baum Boulevard, Pittsburgh,PA 15213-1851. (412) 687-1366. (412) 687-0340.

BIBLIOGRAPHIES

Chemical Information Sources. Gary Wiggins. McGraw-Hill Publishing Company, 1221 Avenue of the Americas, New York, NY 10020. (800) 262-4729 or (212) 512-3825. 1991. $42.50.

Handbooks and Tables In Science and Technology. Russell H. Powell, editor. Oryx Press, 4041 North Central, Suite 700, Phoenix, AZ 85012-3330. (602) 265-2651 or (800) 279-6799. Third edition. 1994. $65.00.

Information Sources In Chemistry. R.T. Bottle and J.F.B Rowland, editors. R.R. Bowker Inc., 121 Chanlon Road, New Providence, NJ 07974. (800) 521-8110 or (908) 464-6800. Fourth edition. 1993. $75.00.

Scientific and Technical Books and Serials In Print; An Index To Literature In Science and Technology. R.R. Bowker Inc., 121 Chanlon Road, New Providence, NJ 07974. (908) 464-6800. (800) 521-8110. FAX (908) 665-3502. 1972 to present. Annual. 4 volumes. 1994. $299.95. Also available on compact disc and online. ISSN: 0000-054X.

DIRECTORIES AND BIOGRAPHICAL SOURCES

American Institute of Chemists - Professional Directory. American Institute of Chemists, 501 Wythe Street, Alexandria, VA 22314-1917. (703) 836-2090. FAX (703) 836- 2091. Annual. $65.00.

American Men and Women of Science. Physical and Biological Sciences. R.R. Bowker Company, 121 Chanlon Road, New Providence, NJ 07974. (908) 464-6800. (800) 521-8110. 20th edition. 8 volumes. 1996. $850.00.

Chemical Week — Buyers Guide Issue. Chemical Week Associates, 888 Seventh Avenue, New York, NY 10106. (212) 621-4900. FAX (212) 621-4949. Annual, October. $50.00.

Directory of Chemistry Software 1992. Wendy Warr, Peter Willett, Geoff Downs. American Chemical Society, 1155 16th

Street, NW, Washington, DC 20036. (202) 872-4600. 1992. $35.95.

Research Centers Directory. Gale Research, 835 Penobscot Building, Detroit, MI 48226-4094. (313) 961-2242. (800) 347-4253. 20th edition. 1995. $485.00. ISSN:: 0080-1518.

ENCYCLOPEDIAS AND DICTIONARIES

Academic Press Dictionary of Science and Technology. Christopher Morris, editor. Academic Press, Inc., 1250 Sixth Avenue, San Diego, CA 92101. (619) 231-0926. FAX (619) 699-6715. 1991. $115.00.

Concise Encyclopedia Chemistry. Translated and revised by Mary Eagleson. Walter de Gruyter, Inc., 200 Saw Mill River Road, Hawthorne, New York, 10532. (914) 747-0110 or Fax (914) 747-1326. 1994. $69.95.

Dictionary of Analytical Reagents. R. Lobinski, editor. Routledge, Chapman and Hall, Inc., 29 West 35th Street, New York, NY 10001-2291. (212) 244-3336. 1993. $1000.00.

Encyclopedic Dictionary of Science. Candida Hunt and Monica Byles, editors. Facts-on-File, Inc., 460 Park Avenue South, New York, NY 10016-7382. (800) 322-8755. Fax (800) 683-3633. 1988. $32.95.

Encyclopedia of Applied Physics. VCH Publishers, Inc., 303 Northwest 12th Avenue, Deerfield Beach, FL 33442. (800) 367-8249. 1991 - . Twenty volumes. $6000.00.

Encyclopedia of Chemical Processing and Design. McKetta Marcel Dekker, Inc., 270 Madison Avenue, New York, NY 10016. (212) 696-9000. (800) 228-1160. 1976 - . $175.00 per volume. ISSN: .

Grant & Hawks Chemical Dictionary. R. A. Grant. McGraw-Hill Publishing Company, Inc., 1221 Avenue of the Americas, New York, NY 10020. (212) 512-2000. (800) 262-4729. 5th edition. 1987. $64.50.

Hawley's Condensed Chemical Dictionary. Richard J. Lewis. Van Nostrand Reinhold, 115 Fifth Avenue, New York, NY 10003. (212) 254-3232. (800) 842-3636. 12th edition. 1993. $69.95.

Kirk-Othmer Encyclopedia of Chemical Technology. John Wiley and Sons, Inc., 605 Third Avenue, New York, NY 10158. (800) 526-5368 or (212) 850-6000. Fourth edition. 1991 - . Twenty-seven volumes. $5400.00.

McGraw-Hill Encyclopedia of Chemistry. Sybil P.Parker, editor. McGraw-Hill Book, Incorporated, 1221 Avenue of the Americas, New York, NY 10020. (212) 997-3675. Second edition. 1993. $95.50.

McGraw-Hill Encyclopedia of Science and Technology. McGraw-Hill Book, Incorporated, 1221 Avenue of the Americas, New York, NY 10020. (212) 997-3675. Seventeenth edition. Twenty volumes. 1992. $1900.00.

GENERAL WORKS

Applied Combustion. Eugene L. Keating. Marcel Dekker, Inc., 270 Madison Avenue, New York, NY 10016. (212) 696- 9000. (800) 228-1160. 1993. $165.00.

Chemistry of Hydrocarbon Combustion. David J. Hucknall. Chapman & Hall, Inc., 29 West 35th Street, New York, NY 10001- 2291. (212) 244-3336. 1985. $85.00.

Combustion Engineering and Gas Utilization. British Gas Staff. 3rd edition. Chapman & Hall, Inc., 29 West 35th Street, New York, NY 10001-2291. (212) 244-3336 . 1992. $145.95.

Combustion Toxicology. Shayne C. Gad and Rosland C. anderson. Telford Press, P.O. Box 287, West Caldwell, NJ 07006. (201) 228-7744. 1990. $47.50.

Dynamics of Gaseous Combustion. Kuhl et al. American Institute of Aeronautics and Astronautics, 370 L'Enfant Promenade, S.W., Washington, DC 20024. (202) 646-7400. (800) 682-2422. 1993. $99.95.

Environmental Sensing and Combustion Diagnostics. L. W. Burgess. SPIE - International Society for Optical Engineering, 1000 20th Street, Bellingham, WA 98227. (206) 676-3290. FAX (206) 647-1445. 1991. $53.00.

Flame and Combustion. J. N. Bradley. Chapman & Hall, Inc., 29 West 35th Street, New York, NY 10001-2291. (212) 244-3336 2nd edition. 1985. $35.00 in paper.

Fuel Combustion: A Source Book. M. Bartok. John Wiley & Sons, Inc., 605 Third Avenue, New York, NY 10158-0012. (212) 850-6000. (800) 225-5945. 1991. $145.00.

Gases, Liquids and Solids: and Other States of Matter. D. Tabor. Cambridge University Press, 40 West 20th Street, New York, NY 10011-4211. (212) 924-3900. 1992. $80.00.

Heat Transfer In Fire and Combustion Systems. B. Farouk, et al. American Society of Mechanical Engineers, 345 East 47th Street, New York, NY 10017. (212) 705-7722. (800) 843- 2763. 1993. $50.00.

Introduction To Combustion Concepts and Applications. Stephen R. Turns. McGraw-Hill Publishing Company, Inc., 1221 Avenue of the Americas, New York, NY 10020. (212) 512- 2000. 1996. wfi.

Organic Reactions: Simplicity and Logic. Pierre Laszlo, John Wiley & Sons, Inc., 605 Third Avenue, New York, NY 10158-0012. (212) 850-6000. (800) 225-5945. 1994. $74.95.

HANDBOOKS AND MANUALS

Chemical Formulary. H. Bennett, editor. Chemical Publishing Company, Inc., 80 Eighth Avenue, New York, NY 10011. (212) 255-1950. Volumes 1-30. $70.00 per volume.

Chemometrics: Chemical and Sensory Data. David R. Burgard and James T. Kuznicki. CRC Publishers, Inc., 2000 Corporate Blvd., N.W., Boca Raton, FL 33431. (407) 994-0555 or (800) 333-8300. 1990. $120.00.

Chemical Reference Handbook. Gordon Press Publications, PO Box 459, Bowling Green Station, New York, NY 10004. (718) 624-8419. 1995. $260.00.

Chemical Safety International Reference Manual. Mervyn Richardson, editor. VCH Publications, Inc., 220 East 23rd Street, Suite 909, New York, NY 10010-4606. (212) 683-8333. (800) 422-8824. 1994. $145.50.

Comprehensive Guide To To the Hazardous Properties of Chemical Substances. Pradyot Patnaik. Van Nostrand Reinhold, 115 Fifth Avenue, New York, NY 10003. (212) 254- 3232. (800) 842-3636. 1992. $99.95.

CRC Handbook of Chemistry and Physics. David R. Kide, editor. CRC Press Inc., 2000 Corporate Blvd., NW, Boca Raton, FL 33431. (407) 994-0555. (800) 333-8300. 77th edition. 1996. $99.95.

Desk Reference For Organic Chemists. Michael B. East and David J. Ager. Krieger Publishing Company, P.O. Box 9542, Melbourne, FL 32902-9542. (407) 724-9542. 1995. $64.50.

Guide To Basic Chemical Compounds. D.R. Lide, Jr. CRC Publishers, Inc., 2000 Corporate Blvd., N.W., Boca Raton, FL 33431. (407) 994-0555 or (800) 333-8300. 1993. $120.00.

Gas Turbine Handbook of Combustion. Rector Press, Ltd., 130 Rattlesnake, Leverett, MA 01054-9726. (413) 548-9708. (800)247-3473. 1994. $295.00.

Handbook of Organic Chemistry. Hans Beyer and Wolfgang Walter. Prentice-Hall General Reference & Travel, 15 Columbus Circle, New York, NY 10023. (212) 373-8500. (800) 223-2348. 1995. $96.00.

Improving Safety In the Chemical Laboratory: A Practical Guide. Jay A. Young, editor. John Wiley and Sons, Inc., 605 Third Avenue, New York, NY 10158. (800) 526-5368 or (212) 850-6000. Second edition. 1991. $75.00.

Laboratory Handbook of Materials, Equipment, and Techniques. Gary S. Coyne. Prentice Hall (Division of Simon and Schuster), 15 Columbus Circle, New York, NY 10023. (212) 373-8500 or (800) 922-0579. 1992. $45.00.

Lange's Handbook of Chemistry. John A. Dean, editor. McGraw-Hill Publishing Company, 1221 Avenue of the Americas, New York, NY 10020. (800) 262-4729 or (212) 512-3825. Fourteenth edition. 1992. $79.50.

North American Combustion Handbook. Combustion fuels, stoichiometry, heat transfer, fluid flow. North American Manufacturing Co. 4455 East 71st Street, Cleveland, OH 44105. (216) 271-6000. 3rd edition. vol.I, 1986. $30.00. vol. II. 1995. $19.88.

Riegel's Handbook of Industrial Chemistry. James A. Kent, editor. Van Nostrand Reinhold, 115 Fifth Avenue, New York, NY 10003. (212) 254-3232 or (800) 926-2665. Ninth edition. 1992. $114.95.

ONLINE DATABASES AND CD-ROMS

Analytical Abstracts Online. Royal Society of Chemistry, Information Services, Thomas Graham House, Science Park, Milton Road, Cambridge, CB4 4WF, England. Contains citations, mostly with abstracts, of the worldwide literature on ana-

COMBUSTION

Ency. of Physical Sciences and Engineering Info. Sources

lytical chemistry, from 1980 to present. Available on DIALOG, (800) 334-2564, and STN International, FIZ Karlsruhe, P.O. Box 2465, W-7500, Karlsruhe 1, Germany, online services. Inquire as to cost and availability.

Beilstein Online. Beilstein-Institut fur Literatur der Organischen Chemie, Varrentrapperstrasse, 40-42, D-6000, Frankfurt am Main 90, Germany. Contains data on carbon compounds from the Beilstein Handbook of Organic Chemistry. Structural and factual data for more than 3.4 million heterocyclic, isocyclic, and acyclic compounds included. Covers the period 1830 to present. Available on DIALOG, (800) 334-2564, ORBIT (800) 456-7248, and STN International, FIZ Karlsruhe, P.O. Box 2465, W-7500, Karlsruhe 1, Germany, online services. Inquire as to cost and availability.

Ca Search. Chemical Abstracts Service, P.O. Box 3012, Columbus, OH 43210-0012. (614) 447-3600. (800) 848-6533. FAX (614) 447-3709. Very comprehensive guide to worldwide chemical literature and related fields, 1972 to present. Available on BRS,(800) 289-4277, DIALOG, (800) 334-2564, ORBIT (800) 456-7248, and STN International, FIZ Karlsruhe, P.O. Box 2465, W-7500, Karlsruhe 1, Germany, online services. Inquire as to cost and availability.

Chemical Journals of the American Chemical Society. American Chemical Society, 1155 16th Street, N.W., Washington, DC 20036. (202) 872-4381 or (800) 424-6747. Contains complete text of approximately 90,000 articles from 22 primary journals published by the American Chemical Society, from mostly 1982 to present. Available on STN International, FIZ Karlsruhe, P.O. Box 2465, W-7500, Karlsruhe 1, Germany, online service. Inquire as to cost and availability.

Compendex Plus. Engineering Information, Inc., 345 East 47th Street, New York, NY 10017. (212) 705-7600 or (800) 221-1044. Contains citations with abstracts to worldwide literature in engineering and technology, from 1970 to present. Available on online BRS,(800) 289-4277, DIALOG, (800) 334-2564, ORBIT (800) 456-7248, and STN International, FIZ Karlsruhe, P.O. Box 2465, W-7500, Karlsruhe 1, Germany, online services. Also available on CD-ROM. Inquire as to cost and availability.

Current Contents Search. Institute for Scientific Information, 3501 Market Street, Philadelphia, PA 19104. (215) 386-0100. FAX (215) 386-6362. Contains citations to articles listed in the table of contents of science and technology journals. Also articles in social sciences and life sciences journals. Available on BRS,(800) 289-4277, DIALOG,(800) 334-2564, online services. Inquire as to cost and availability.

Dissertation Abstracts. University Microfilms International, 300 North Zeeb Road, Ann Arbor, MI 48106. (800) 521-0600 or (313) 761-4700. Scope includes virtually all doctoral dissertations accepted at accredited American institutions from 1861 to present in 252 subject areas. Available on BRS,(800) 289-4277, DIALOG,(800) 334-2564, and OCLC EPIC,(800) 848-5878, online services. Also available on CD-ROM. Inquire as to cost and availability.

Engineered Materials Abstracts. Materials Information, ASM International, Materials Park, OH 44073. (216) 338-5151. Contains citations with abstracts of worldwide literature in the development, processing, and production of ceramic, composite, and polymeric materials for engineering uses. Available on DIALOG, (800) 334-2564, ORBIT (800) 456-7248, and STN International, FIZ Karlsruhe, P.O. Box 2465, W-7500, Karlsruhe 1, Germany, online services. Inquire as to cost and availability.

Gmelin Database. Gmelin-Institut fur Anorganische Chemie und Grenzgebiete, Varrentrapperstrasse, 40-42, Carl-Bosch-Haus, D-6000, Frankfurt am Main 90, Germany. Contains structural and factual data relating to inorganic and organometallic chemistry. Provides data from the Gmelin Handbook of Inorganic and Organometallic Chemistry. Covers the period 1817 to 1975; 1988-89. Available on STN International, FIZ Karlsruhe, P.O. Box 2465, W-7500, Karlsruhe 1, Germany, online service. Inquire as to cost and availability.

Kirk-othmer Encyclopedia of Chemical Technology. John Wiley and Sons, Inc., 605 Third Avenue, New York, NY 10158. (800) 526-5368 or (212) 850-6000. Contains the complete text of allchapters in the 27 volume fourth edition of the KIRK-OTHMER ENCYCLOPEDIA of CHEMICAL TECHNOLOGY. 1991. Available on BRS,(800) 289-4277, DIALOG,(800) 334-2564, online services. Inquire as to cost and availability.

NTIS Bibliographic Database. National Technical Information Service, 5285 Port Royal Road, Springfield, Va 22161. (703) 487-4929 or FAX (703) 321-8199. Broad coverage of government-sponsored science and technology research reports, 1964 to present. Available on BRS,(800) 289-4277, DIALOG, (800) 334-2564, ORBIT, (800) 456-7248, and STN International, FIZ Karlsruhe, P.O. Box 2465, W-7500, Karlsruhe 1, Germany, online services. Also available on CD-ROM. Inquire as to cost and availability.

Paperchem. Institute of Paper Science and Technology, 575 14th Street, N.W., Atlanta, GA 30318. (404) 853-9500. A database of citations with abstracts to the worldwide scientific and technical literature on pulp, paper, and board industries and processes, from 1967 to present. Available on DIALOG,(800) 334-2564, online service. Inquire as to cost and availability.

Physics Briefs. American Institute of Physics, 335 East 45th Street, New York, NY 10017. (212) 661-9260 or FAX (212) 661-2036. Contains citations with abstracts of the literature of physics and related fields, 1979 to present. Available on the STN International, FIZ Karlsruhe, P.O. Box 2465, W-7500, Karlsruhe 1, Germany, online service. Inquire as to cost and availability.

SCISEARCH. Institute for Scientific Information, 3501 Market Street, Philadelphia, PA 19104. (800) 523-1850 or (215) 386-0100. Broad multidisciplinary title and author index to the international literature of science and technology, 1974 to present. Available on DIALOG,(800) 334-2564, and ORBIT,(800) 456-7248, online services. Also available on CD-ROM. Inquire as to cost and availability.

WILSONLINE. H.W. Wilson Company, 950 University Avenue, Bronx, NY 10452. (800) 367-6770 or (212) 588-8400. Makes available online versions of the printed H.W. Wilson indexes including Applied Science and Technology Index, Business Periodicals Index, General Science Index, and Readers' Guide to Periodical Literature. Period covered is generally 1983 to present. Available on BRS,(800) 289-4277, DIALOG,(800) 334-2564, and OCLC EPIC,(800) 848-5878, online services. Also available on CD-ROM. Inquire as to cost and availability.

PERIODICALS

AIChe Journal. American Institute of Chemical Engineers, 345 47th Street, New York, NY 10017-2396. (212) 705-7338. FAX (212) 752-3294. 1955 to present. Monthly. $395.00 per year. ISSN: 0001-1541.

Analytical Chemistry. American Chemical Society, Box 3337, Columbus, OH 43210. (614) 447-3776. 1929 to present. Monthly. $415.00 per year. ISSN: 0003-2700.

Analytical Methods and Instrumentation. John Wiley andSons, Inc., 605 Third Avenue, New York, NY 10158. (800) 526-5368 or (212) 850-6000. Bimonthly. $250.00 per year. ISSN: 1063-5246.

Chemical Week; Includes Annual Buyers Guide. Chemical Week Associates, 888 Seventh Avenue, New York, NY 10106. (212) 621-4900. FAX (212) 621-4949. 1914 to present. Weekly. $99.00 per year. ISSN: 0009-272X.

Chemtech. American Chemical Society, Box 3337, Columbus, OH 43210. (614) 447-3776. (800) 333-9511. FAX (614) 447-3671. 1970 to present. Monthly. $79.00 per year. ISSN: 0009-2703.

Combustion and Flame. Elsevier Science Publishing Company, Inc., 655 Avenue of the Americas, New York, NY 10010. (212) 989-5800. 1963 to present. 16 issues per year. $725.00 per year. ISSN: 0012-2180.

F A M - Fire and Materials. John Wiley & Sons, Ltd., Baffins Lane, Chichester, Sussex, PO19 1UP, England. 1976 to present. Bi-monthly. $495.00 per year. ISSN: 0308-0501.

Journal of Nondestructive Evaluation. Plenum Publishing Corporation, 233 Spring Street, New York, NY 10013. (212) 620-8000. FAX (212) 463-0742. 1980 to present. Quarterly. $230.00 per year. ISSN: 0195-9298.

Journal of Physical Chemistry. American Chemical Society, Box 3337, Columbus, OH 43210. (614) 447-3776. 1896 to present. Weekly. $1140.00 per year to non-members. ISSN: 0022-3654.

RESEARCH CENTERS AND INSTITUTES

Center for Energy and Combustion Research. University of California, San Diego, PO Box 109, JaJolla, CA 92037. (619) 534-4285.

Combustion and Fluid Dynamics Laboratory. University of Wisconsin - Madison, Department of Mechanical /engineering, 1513 University Avenue, Madison, WI 53706. (608) 263-5963. FAX (608) 262-6707.

Combustion Laboratory. Ohio State University, 206 West 18th Avenue, Columbus, OH 43210. (614) 292-0403.

Combustion Laboratory. Pennsylvania State University, 404 Academic Activities Building, University Park, PA 16802. (814) 863-3264. (814) 863-7432.

Lawrence Berkeley Laboratory, Chemical Sciences Division. One Cyclotron Road, Building 66, Berkeley, CA 94720. (510) 486-6062. FAX (510) 486-4995.

COMETS

See also: ASTROGEOLOGY, ASTEROIDS, ASTRONOMY, METEORITES, NAMES OF INDIVIDUAL PLANETS, PLANETARY SCIENCES, SATELLITES (NATURAL), SOLAR SYSTEM

ABSTRACT SERVICES AND INDEXES

Applied Science and Technology Index; A Cumulative Subject Index To English Language Periodicals In the Fields of Aeronautics and Space Science, Computer Technology, Chemistry, Construction Industry, Energy and Related Areas. H.W. Wilson Co., 950 University Avenue, Bronx, NY 10452. (212) 588-8400. (800) 367-6770. FAX (718) 590-1617. From 1958 to present. Monthly.Inquire about cost and availability. Also available on CD-ROM and online. ISSN: 0003-6986.

Astronomy and Astrophysics Abstracts. Springer-Verlag New York, 175 F ifth Avenue, New York, NY 10010. (212) 460-1500. FAX (212) 473-6272. Published for the Astronomisches Rechen-Institut. Comprehensive coverage of all aspects of astronomy, astrophysics and related fields. 1969 to present. Two parts per year. Annual. ISSN: 0067-0022.

General Science In Dex. H.W. Wilson Company, 950 University Avenue, Bronx, NY 10452. (212) 588-8400. (800) 367-6770. FAX (718) 590-1617. From 1978 to present. Ten issues per year; quarterly and annual cumulations. Service basis. Available on CD-ROM and online. Inquire regarding cost and availability. ISSN: 0162-1963.

Govern Ment Reports Announcements and Index. U. S. National Technical Information Service (NTIS), 5285 Port Royal Road, Springfield, VA 22161. (703) 487-4650. FAX (703) 321-8547. From 1968 to present. Annual. $630.00 per year. Also available online as *NTIS Bibliographic Database* and on CD-ROM. ISSN:

Meteorological and Geoastrophysical Abstracts. American M eteorological Society, c/o Inforonics, Inc., 550 Newtown Road, Littleton, MA 01460. (508) 486-8976. FAX (508) 486- 0027. Covers literature in environmental sciences, meteorology, astrophysics, hydrology, glaciology, and physical oceanography. From 1950 to present. Monthly. $950.00 per year. Also available on CD-ROM and online. Inquire regarding cost and availability. ISSN: 0026-1130.

NTIS Alerts: Astronomy & Astrophysics. U. S. National Technical Information Service. 5285 Port Royal Road, Springfield, VA 22161. (703) 487-4650. FAX (703) 321-8547. Weekly. $140.00 per year.

Pascal 48: Environnement Cosmique Terrestre, Astronomie Et Geologie Extraterrestre. Centre National de la Recherche Scientifique, Institut de Information Scientifique et Technique, 2 aliee du Parc de Brabois, 54514 Vandoeuvre Les Nancy Cedex, France. TEL 83-50-46-00. FAX: 83-50-46-50. 1985 to present. Ten issues per year. 770 F. Also available on CD-ROM and online.

Physics Abstracts . INSPEC. Section A, Science Abstracts. Institute of Electrical Engineers, London, United Kingdom. Available from: INSPEC/IEEE - Institute of Electrical and Electronic Engineers, Box 1331, Hoes Lane, Piscataway, NJ 08855-1331. (908) 562-5549. 1898 to present. 24 issues per year. $2835.00 per year. ISSN: 0036-8091. Also available online and on CD-ROM.

Physics Briefs (Physikalische Berichte). Information Cen ter for Energy, Physics, Mathematics; German Physical Society. V C H Publishers, Inc., 220 East 23rd Street, New York, NY 10010-4606. (212) 683-8333. 1845 to present. 24 issues per year. $2390.00 per year. Also available online. ISSN: 0179-7434.

COMETS

Ency. of Physical Sciences and Engineering Info. Sources

Science Citation Index. SCI. Institute for Scientific Information, 3501 Market Street, Philadelphia, PA 19104. (215)386-0100. (800) 523-1850. FAX (215) 386-2991. 1961 to present. Six issues per year, plus annual cumulation. $11650.00 per year. Also available online and on CD-ROM. Inquire about price and availability. ISSN: 0036-827X.

S T A R. Scientific and Technical Aerospace Reports. U.S. National Aeronautics and Space Administration. Distributed by U. S. Superintendent of Documents, Washington, DC 20402. 1963 to present. Semi-monthly, with semiannual and annual indexes. $114.00 per year. ISSN: 0036-8741. Also available online and on CD-ROM.

ANNUAL REVIEWS AND YEARBOOKS

The Astronomical Almanac. Superintendent of Documents, U. S. Government Printing office, Washington, DC 10402. (202) 783-3238. 1981 to present. Supercedes Astronomical Ephemeris. Annual. $44.95. ISSN::

Annual Review of Astronomy and Astrophysics. Original reviews of critical literature and current developments in astronomy and astrophysics. Annual Reviews, Inc., 4139 El Camino Way, Palo Alto, CA 94303-0139. (415) 493-4400. (800) 523-8635. Fax (415) 855-9815. 1963 to date. Annual. $60.00. ISSN:: 0066-4146.

Annual Review of Earth and Planetary Sciences. Annual Reviews, Inc., 4139 El Camino Way, Palo Alto, CA 94303-0139. (415) 493-4400. (800) 523-8635. Fax (415) 855-9815. 1973 to date. Annual. $62.00. ISSN:: 0084-6597.

Proceedings of the Lunar and Planetary Science Conference. National Technical Information Service (NTIS), 5285 Port Royal Road, Springfield, VA 22161. (703) 487-4650. 1970 to date. Annual.

ASSOCIATIONS AND PROFESSIONAL SOCIETIES

American Astronomical Society. 2000 Florida Avenue NW, Suite 400, Washington, DC 20009. (202) 328-2010. FAX\; (202) 234-2560.

American Geophysical Union. 2000 Florida Avenue NW, Washington, DC 20009. (202) 462-6900. (800) 966-AGU1. FAX (202) 328-0566.

American Institute of Physics. 1 Physics Ellipse, College Park, MD 20740-3843. (301) 209-3100.

Association of Universities For Research In Astronomy, Inc. (AURA). Suite 701, 1625 Massachusetts Avenue, NW, Washington, DC 20036. (202) 483-2101. FAX (202) 483-2106.

Astronomical Society of the Pacific. 390 Ashton Avenue, San Francisco, CA 94112. (415) 337-1100. FAX: (415) 337-5205.

Geological Association of America. 3300 Penrose Place, PO Box 9140. Boulder, CO 80301-9140. (303) 447-2020. FAX (303) 447-1133.

IEEE Geoscience and Remote Sensing Society. c/o Institute of Electrical and Electronics Engineers. 345 East 47th Street, New York, NY 10017. (212) 705-7900. FAX (212) 705- 4929.

Meteoritical Society. University of Massachusetts, 125 Marston Hall, Amherst, MA 01003. (413) 545-0300. (413) 545-0724.

Planetary Society. 65 North Catalina Avenue, Pasadena, CA 91106. (818) 793-5100. (800) WOW-MARS. FAX (818) 793-5528.

BIBLIOGRAPHIES

A Bibliography of Astronomy, 1970 - 1979. R. A. Sea and S. S. Martin. Libraries Unlimited, Inc., Littleton, CO 80160. 1982. $37.50.

Halley's Comet: A Bibliography. Ruth Freitag. Library of Congress, U. S. Government Printing office, Washington, D.C. 20402. (202) 783-3238. References to Comet Halley from 1495 to 1983, with annotations. 1984. $26.00.

Halley's Comet, Seventeen Fifty-five To Nineteen Eighty- Four: A Bibliography. Bruce Morton, compiler. Greenwood Press, 1600 Larkin Street, No. 104, San Francisco, CA 94109. (415) 928-4142. 1985. $36.95.

DIRECTORIES AND BIOGRAPHICAL SOURCES

American Astronomical Society. Membership Directory. American Astronomical Society. 2000 Florida Avenue NW, Suite 400, Washington, DC 20009. (202) 328-2010. FAX (202) 234-2560 Annual. Included in membership dues. ISSN:: 1061- 9038.

American Men and Women of Science: Physical and Biological Sciences. R. R. Bowker Inc., 121 Chanlon Road, New Providence, NJ 07974. (908) 464-6800. (800) 521-8110. 20th edition. 8 volumes. 1996. $850.00.

The Astronomers. Donald Goldsmith. St. Martin's Press, Inc., 175 Fifth Avenue, New York, NY 10010. (212) 674-5151. (800) 221-7945. 1993. $14.95 in paper.

Astronomical Centers of the World. Kevin Krisciunas. Cambridge University Press, 40 West 20th Street, New York, NY 10011-4211. (212) 924-3900. (800) 872-7423. 1988. $34.95

Directory of Physics and Astronomy Staff Members. American Institute of Physics, One Physics Ellipse, College Park, MD 20740-3843. (301) 209-3100. 1975/76 to present. Annual. $60.00. ISSN:: 0361-2228.

Earth and Astronomical Research Centers. Stockton Press, 345 Park Avenue South, New York, NY 10010. 4th edition. 1995. $515.00. ISBN: 1-56169-0967.

Graduate Programs In Physics, Astronomy and Related Fields. 1995-1996. American Institute of Physics, One Physics Ellipse, College Park, MD 20740-3843. (301) 209-3100. 1978 to present. Annual. $45.00. ISSN:: 0147-1821.

ENCYCLOPEDIAS AND DICTIONARIES

Concise Dictionary of Astronomy. Jacqueline Mitton. Oxford University Press, Inc., 200 Madison Avenue, New York, NY 10016. (212) 725-6000. (800) 334-4249. 1992. $24.95.

Encyclopedia of Astronomy and Astrophysics. Stephen Maran, editor. Van Nostrand Reinhold, 115 Fifth Avenue, New York, NY 10003. (212) 254-3232. (800) 842-3636. 1992. $129.95.

McGraw-Hill Encyclopedia of Science and Technology. McGraw- Hill Book Company, Inc., 1221 Avenue of the Americas, New York, NY 10020. (212) 512-2000. (800) 262-4729. 7th edition. 20 volume set. 1992. $1900.00.

Moons of the Solar System: An Illustrated Encyclopedia. John Stewart. McFarland & Company, Inc., Box 611, Jefferson, MC 28640. (910) 246-4460. (800) 253-2187. 1991. $49.95.

New Guide To the Planets. Patrick Moore. Trans-Atlantic Publications, Inc., 311 Bainbridge Street, Philadelphia, PA 19147. (215) 925-5083. 1993. $37.50.

Stars and Planets: The Sierra Club Guide To Sky Watching and Direction Finding. W. S. Kals. The Sierra Press, 4988 Gold Leaf Drive, Mariposa, CA 95338. (209) 966-5071. (800) 745-2631. $15.00.

GENERAL WORKS

Comet Halley: Investigations, Results, Interpretations. Mason. Prentice-Hall, 113 Sylvan Avenue, Route 9W, Englewood Cliffs, NJ 07632. (201) 592-2000. (800) 922- 0579. 1990. $120.00.

Comet Halley: Once In A Lifetime. Mark Littmann and Donald Yeomans. American Chemical Society, 1155 16th Street, NW, Washington, DC 20036. (202) 872-4600. (800) 333-9511. FAX (614) 447-3671. 1985. $21.95.

Cometary Plasma Processes. A. D. Johnstone, editor. American Geophysical Society, 2000 Florida Avenue, N.W., Washington, DC 20009. (202) 462-6903. (800) 966-2481. . Geophysical Monograph Series, volume 61. 1991. $56.00.

Cometary theory In Fifteenth-century Europe. Jane L. Jervis. Kluwer Academic Publishers, 101 Philip Drive, Assinippi Park, Norwell, MA 02061. (617) 871-6600. 1985. $39.99.

Comets: A Chronological History of Observation, Science, Myth and Folklore. Donald K. Yeomans. John Wiley & Sons, Inc., 605 Third Avenue, New York, NY 10158-0012. (212) 850- 6000. (800) 225-5945. 1991. $35.00.

Comets: Speculation and Discovery. Nigel Calder. Dover Publications, Inc., 180 Varick Street, New York, NY 10014. (212) 255-3755. (800) 223-3130. Reprint edition. 1994. $8.95.

Comets, Asteroids and Meteorites. Time-Life Inc., 777 Duke Street, Alexandria, VA 22314. (703) 8388-7000. (800) 621-7026. Revised edition. 1992. $18.95.

Comets In the Post-halley Era. R. L. Newburn, Jr., et al, editors. Kluwer Academic Publishers, 101 Philip Drive, Assinippi Park, Norwell, MA 02061. (617) 871-6600. Two volumes. 1991. $61.50 in paper.

Fire and Ice: A History of Comets In Art. Roberta J. Olsen. Walker & Company, 435 Hudson Street, New York, NY 10014. (212) 737-8300. (800) 289-2553. 1985. $24.95.

Hazards Due To Comets and Asteroids. Tom Gehrels, editor. University of Arizona Press, 1230 North Park Avenue, Number 102, Tucson, AZ 85719. (520) 621-1441. (800) 426-3797. 1994. $75.00.

Impact Jupiter: the Crash of Comet Shoemaker-levy 9. David H. Levy. Plenum Publishing Corp., 233 Spring Street, New York, NY. (212) 620-8000. (800) 2221-9369. 1995. $25.95.

Introduction To Comets. John C. Brandt and Robert D. Chapman. Cambridge University Press, 40 West 20th Street, New York, NY 10011-4211. (212) 924-3900. (800) 872-7423.. 1982. $27.95.

Physics and Chemistry of Comets. W. F. Huebner, et al,editors. Springer-Verlag New York, Inc., 175 Fifth Avenue, New York, NY 10010. (212) 460-1500. (800) 777-4643. 1990. $59.00.

Observing Comets, Asteroids, Meteors and the Zodiacal Light. Stephen J. Edberg and David H. Levy. Cambridge University Press, 40 West 20th Street, New York, NY 10011-4211. (212) 924-3900. (800) 872-7423. 1994. $29.95.

The Origin of Comets. M. E. Bailey, et al. Pergamon Press, Inc., Maxwell House, Fairview Park, Elmsford, NY 10523. (914) 592-7700. Fax (914) 592-3625. 1990. $42.00 in paper.

Physical Processes In Comets, Stars and Active Galaxies. W. Hillebrandt, et al, editors. Springer-Verlag New York, Inc., 175 Fifth Avenue, New York, NY 10010. (212) 460- 1500. (800) 777-4643. 1987. $47.00.

Quest For Comets: and Explsive Trail of Beauty and Danger. D. H. Levy. Plenum Publishing Corp., 233 Spring Street, New York, NY. (212) 620-8000. (800) 2221-9369. 1994. $23.95.

Rendezvous in Space: the Science of Comets. John C. Brandt. W. H. Freeman, 41 Madison Avenue, East 26th Avenue, 35th Floor, New York, NY 10010. (212) 576-0400. 1995. $23.95.

Rogue Astroids and Doomsday Comets. Duncan Stasiuk. John Wiley & Sons, Inc., 605 Third Avenue, New York, NY 10158-0012. (212) 850-6000. (800) 225-5945. 1995. $24.95.

The Three Big Bangs: Comet Crashes, Exploding Stars and the Creation of the Universe. Philip M. Dauber and Richard A. Muller. Addison-Wesley Publishing Co., Inc., 1 Jacob Way, Reading, MA 01867. (617) 944-3700. (800) 447-2226. 1996. $25.00.

HANDBOOKS AND MANUALS

Astrophysical Data: Planets and Stars. Kenneth R. Lanb. Springer-Verlag New York, Inc., 175 Fifth Avenue, New York, NY 10010. (212) 460-1500. (800) 777-4643. 1993. $59.00.

New Guide To the Planets. Patrick Moore. Trans-Atlantic Publications, Inc., 311 Bainbridge Street, Philadelphia, PA 19147. (215) 925-5083. 1993. $37.50.

Planet Observer's Handbook. Fred W. Price. Cambridge University Press, 40 West 20th Street, New York, NY 10011- 4211. (212) 924-3900. (800) 872-7423. 1994. $34.95.

Planets and their Moons. Aububon Society Staff. Alfred A. Knopf, Inc. 201 E. 50th Street, New York, NY 10022. (800) 733-3000. 1995. $7.99.

COMETS

Ency. of Physical Sciences and Engineering Info. Sources

Tables of Planetary Phenomena. Neil F. Michelsen. ACS Publications, PO Box 34487, San Diego, CA 92163-4487. (619) 492-9919. (800) 888-9983 2nd revised edition. 1993. $24.95.

ONLINE DATABASES AND CD-ROMS

CA Search. Chemical Abstracts Service, P.O. Box 3012, Columbu s, OH 43210-0012. (614) 447-3600. (800) 848-6533. FAX (614) 447-3709. Very comprehensive guide to worldwide chemical literatureand related fields, 1972 to present. Available on BRS,(800) 289-4277, DIALOG, (800) 334-2564, ORBIT (800) 456-7248, and STN International, FIZ Karlsruhe, P.O. Box 2465, W-7500, Karlsruhe 1, Germany, online services. Inquire as to cost and availability.

Current Contents Search. Institute for Scientific Information, 3501 Market Street, Philadelphia, PA 19104. (215) 386-0100 . FAX (215) 386-6362. Contains citations to articles listed in the table of contents of science and technology journals. Also articles in social sciences and life sciences journals. Available on BRS, (800) 289-4277, DIALOG, (800) 334-2564, online services. Inquire as to cost and availability.

Dissertation Abstracts Online. University Microfilms Internatio nal, 300 North Zeeb Road, Ann Arbor, MI 48106. (800) 521-0600 or (313) 761-4700. Scope includes virtually all doctoral dissertations accepted at accredited American institutions from 1861 to present in 252 subject areas. Available on BRS, (800) 289-4277, DIALOG, (800) 334-2564, and OCLC EPIC, (800) 848-5878, online services. Also available on CD-ROM. Inquire as to cost and availability.

Inspec. In stitution of Electrical Engineers, Michael Faraday Hou se, Six Hills Way, Stevenage, Herts. SG1 2AY, Englan d. Telephone: 0438 313311 or FAX 0438 742840. Cont ains citations to the worldwide literature of physics, electronics and electrical engineering, computer technology, and related fields. Available on BRS, (800) 289-4277, DIALOG, (800) 334-2564, ORBIT, (800) 456-7248, and STN International, FIZ Karlsruhe, P.O. Box 2465, W-7500, Karlsruhe 1, Germany, online services. Inquire as to cost and availability.

International Aerospace Abstracts. American Institute of Aeronautics and Astronautics, 370 L'Enfant Promenade, S.W., Washington, DC 20024. (202) 646-7400. Contains references a nd abstracts of journal and monograph literature relating to aerospace science and technology, from 1963 to present. Available through the NASA/RECON system of the National Aeronautics and Space Administration only.

NTIS Bibliographic Database. National Technical Information Service, 5285 Port Royal Road, Springfield, Va 22161 . (703) 487-4929 or FAX (703) 321-8199. Broad covera ge of government-sponsored science and technology research reports, 1964 to present. Available on BRS,(800) 289-4277, DIALOG, (800) 334-2564, ORBIT, (800) 456-7248, and STN International, FIZ Karlsruhe, P.O. Box 2465, W-7500, Karlsruhe 1, Germany, online services. Also available on CD-ROM. Inquire as to cost and availability.

Physics Briefs. American Institute of Physics, 335 East 45th Street , New York, NY 10017. (212) 661-9260 or FAX (212) 661-2036. Contains citations with abstracts of the literature of physics and related fields, 1979 to present. Available on the STN International, FIZ Karlsruhe, P.O. Box 2465, W-7500, Karlsruhe 1, Germany, online service. Inquire as to cost and availability.

Scientific and Technical Books and Serials In Print. R.R. Bowker Inc., 121 Chanlon Road, New Providence, NJ 07974. (800) 521-8110 or (908) 464-6800. List of currently published books and serials in the physical and biological sciences, engineering and technology, with subject, author and titles indexes. Available on ORBIT, (800) 456-7248, online service. Inquire as to cost and availability.

Scisearch. Institute for Scientific Information, 3501 Market Street, Philadelphia, PA 19104. (800) 523-1850 or (215) 386-0100. Broad multidisciplinary title and author index to the international literature of science and technology, 1974 to present. Available on DIALOG, (800) 334-2564, and ORBIT, (800) 456-7248, online services. Also available on CD-ROM. Inquire as to cost and availability.

WILSONLINE. H.W. W ilson Company, 950 University Avenue, Bronx, NY 10452. (800) 367-6770 or (212) 588-8400. Makes available onlin e versions of the printed H.W. Wilson indexes includin g Applied Science and Technology Index, Business Periodicals Index, General Science Index, and Readers' Guide to Periodical Literature. Period covered is generally 1983 to present. Available on BRS, (800) 289-4277, DIALOG, (800) 334-2564, and OCLC EPIC, (800) 848-5878, online services. Also available on CD-ROM. Inquire as to cost and availability.

OTHER SOURCES

Atlas of the Planets. Vincent DeCallatay and Audouin Dollfus. Books on Demand, 300 North Zeeb Road, Ann Arbor, MI 48106-1346. (313) 761-4700. (800) 521-0600. $45.60.

Atlas of the Solar System. B. Yenne. Simon & Schuster, Inc. 1230 Avenue of the Americas, New York, NY 10020. (212) 698-7000. (800) 223-2348. 1987. $12.98.

Chronology of Eclipses and Comets. AD1 - 1OOO. D. J. Schove. Boydell and Brewer, Inc., P.O. Box 41026, Rochester, NY 14604-4126. (716-275-0419. 1985. $63.00.

Isophotometric Atlas of Comets. W. Hoegner and N. Richter. Springer-Verlag, Inc., 175 Fifth Avenue, New York, NY 10010. (212) 460-1500. (800) 777-4643. 2 parts. 1980. $71.00; $59.00.

Planetary, Lunar and Solar Positions Six Hundred-one B.C. to A.D. One at Five-day and Ten-day Intervals. Bryant Tuckerman. American Philosophical Society, 104 South 5th Street, Philadelphia, PA 19106-3387. (215) 440-3400. FAX (215) 440-3436. Memoirs Series, volume 56. 1979. $20.00.

Planetary, Lunar and Solar Positions, A.D.2 to A.D. 1649 at Five Day and Ten Day Intervals. Bryant Tuckerman. American Philosophical Society, 104 South 5th Street, Philadelphia, PA 19106-3387. (215) 440-3400. FAX (215)440-3436. Memoirs Series, volume 59. 1964. $30.00.

Planetary, Lunar and Solar Positions, 1650 - 1805. Owen Gingerich and Barbara L. Welther. American Philosophical Society, 104 South 5th Street, Philadelphia, PA 19106-3387. (215) 440-3400.FAX (215)440-3436. Memoirs Series, volume 59S. 1983. $20.00.

The Sky: A User's Guide. David H. Levy. Cambridge University Press, 40 West 20th Street, New York, NY 10011- 4211. (212) 924-3900. (800) 872-7423. 1993. $14.95.

Ency. of Physical Sciences and Engineering Info. Sources

COMETS

The View from Space: Photographic Exploration of the Planets. Merton Davies and Bruce C. Murray. Columbia University Press, 562 West 113th Street, New York, NY 10025. (212) 666-1000. (800) 944-8648. 1973. $17.50.

PERIODICALS

Astronomical Journal. American Institute of Physics, One Physics Ellipse, College Park, MD 20740-3843. (301) 209- 3000. Published for the American Astronomical Society. 1849 to present. Monthly. $280.00 per year. ISSN: 0004- 6256.

Astronomical Society of the Pacific. PUBLICATIONS. Astronomical Society of the Pacific, 390 Ashton Avenue, San Francisco, CA 94112. (415) 337-1100. FAX (415) 337-5205. 1889 to present. Monthly. $175.00 per year. ISSN: 0004- 6280.

Astronomy. Kalmbach Publishing Company, Box 1612, Waukesha, WI 53187-1612. (414) 796-8776. FAX (414) 796-1142. 1973 to present. Monthly. $27.00 per year. ISSN: 0091-6358.

Astronomy and Astrophysics. Springer-Verlag New York, Inc., 44 Hartz Way, Secaucus, NJ 07096-2491. (201) 348-4033. FAX (201) 348-4505. 1969 to date. 36 issues per year in 12 volumes. $1733.00 per year. ISSN: 0004-6361.

Celestial Mechanics and Dynamical Astronomy; An International Journal of Space Dynamics. Kluwer Academic Publishers, Box 358, Accord Station, Hingham, MA 02018-0358. (617) 871-6600. FAX (617) 871-6528. 1969 to present. Monthly. $703.50 per year. ISSN: 0923-2958.

Earth, Moon and Planets; An International Journal of Comparative Planetology. Kluwer Academic Publishers, Box 358, Accord Station, Hingham, MA 02018-0358. (617) 871- 6600. FAX (617) 871-6528. 1969 to present. Monthly. $840.00 per year. ISSN: 0167-9295.

Geochemica Et Cosmochimica Acta. Elsevier Science. 660 White Plains Road, Tarrytown, NY 10591-5153. (914) 524- 9200. FAX (914) 333-2444. 1950 to the present. Biweekly. $895.00 per year. ISSN: 0016-7037.

Icarus; International Journal of Solar System Studies. Academic Press, Inc., Journal Division, 525 B Street, Suite 1900, San Diego, CA 92101-4495. (619) 230-1840. FAX (619) 699-6800. 1962 to present. Monthly. $1080.00 per year. ISSN: 0019-1035.

J G R: Journal of Geophysical Research: Planets. American Geophysical Union, 2000 Florida Avenue, NW, Washington, CD 20009. (202) 462-6900. FAX (202) 328-0566. 1991 to present. Monthly. $597.00 per year. ISSN: 0148-0227.

Lunar and Planetary Information Bulletin. Lunar and Planetary Institute. 3600 Bay Area Boulevard, Houston, TX 77058-1113. (713) 486-2175. FAX (713) 486-2125. 1970 to present. Quarterly. Free. Also available online.

Mercury. Astronomical Society of the Pacific. 390 Ashton Avenue, San Francisco, CA 94112. (415) 337-1100. FAX: (415) 337-5205. 1972 to present. Bimonthly. $175.00 per year. ISSN: 0047-6773.

Meteoritics. Meteoritical Society, Department of Chemistry, University of Arkansas, Fayetteville, AR 72701. (501) 575-

7625. FAX (501) 575-7778. 1953 to present. Bimonthly. $210.00 per year. ISSN: 0026-1114.

Monthly Notices of the Royal Astronomical Society. Blackwell Scientific Publications, LT., Osney Mead, Oxford OX2 OEL, England. TEL 0865-240201. FAX 0865-721205. 1827 to present. Fortnightly. $1899.00 per year. ISSN: 0035- 8711.

Planetary and Space Science. Elsevier Science Publishing Company, Inc., 660 White Plains Road, Terrytown, NY 10591- 5153. (914) 524-9200. FAX: (914) 333-2444. 1959 to present. Monthly. $1355.00 per year. ISSN: 0032-0633.

Planetary Report. Planetary Society. 65 North Catalina, Pasadena, CA 91106-2301. (818) 793-5100. FAX (818) 793- 5528. 1980 to present. $25.00 per year. ISSN: 0736-3680.

Sky and Telescope. Sky Publishing Corporation, Box 9111, Belmont, MA 02178. (617) 864-7360. FAX (617) 864-6117. 1941 to present. Monthly. $27.00 per year. ISSN: 0037- 6604.

Vistas In Astronomy; An International Review Journal. Albert C. Beer, editor. Elsevier Science Publishing Company, Inc., 660 White Plains Road, Terrytown, NY 10591- 5153. (914) 524-9200. FAX: (914) 333-2444. 1958 to present. Quarterly. $415.00 per year. ISSN:: 0083-6656.

RESEARCH CENTERS AND INSTITUTES.

Center For Radiophyscs and Space Research. Cornell University, Space Sciences Building, Ithaca NY 14853. (607) 255-4341. FAX (607) 255-9888.

Center For Space Science and Astrophysics. Stanford University. 325 Durand Building, Stanford, CA 04305. (415) 723-3582.

Center For Space Sciences. University of Texas at Dallas, P. O. Box 830688, MS_F022, Richardson, TX 75083-0688. (214) 690-2851. FAX (214) 690-2848.

Earth and Planetary Remote Sensing Laboratory. Washington University, Department of Earth and Planetary Sciences. Campus Box 1169, 1 Brookings Drive, St. Louis, MO 63130. (314) 889-5679. FAX (314) 889-5799.

Institute For Astronomy. University of Hawaii at Manoa, 2680 Woodlawn Drive, Honolulu, HI 96822. (808) 956-8312. FAX (808) 988-2790.

Institute For Terrestrial and Planetary Atmospheres. State University of New York at Stony Brook, Stony Brook, NY 11794-2300. (516) 632-6170. FAX (516) 632-6251.

Institute of Geophysics and Planetary Physics. University of California, Riverside, CA 92521. (714) 787-4503. FAX (714) 787-4529.

Laboratory For Atmospheric and Space Physics. University of Colorado - Boulder, Boulder, CO 80309-0392. (303) 492-7677. (303) 492-6946.

Laboratory For Planetary Geology. Arizona State University, Department of Geology, Tempe, AZ 85281. (601) 965-7029.

Laboratory For Planetary Studies. Cornell University, Space Sciences Building, Ithaca, NY 14853. (607) 255-4971.

COMETS

Ency. of Physical Sciences and Engineering Info. Sources

Lunar and Planetary Institute. 3303 NASA Road One, Houston, TX 77058-4399. (713) 486-2139. FAX 713-496-2162.

Lunar and Planetary Laboratory. University of Arizona, Tucson, AZ 85721. (602) 621-6963. FAX (602) 621-4933.

Mcdonnell Center for the Space Sciences. Washington University, Box 1105, One Brookings Drive, St. Louis MO 63130-4899. (314) 889-6255. FAX (314) 889-6219.

Smithsonian Astrophysical Observatory. 60 Garden Street, Cambridge, MA 02138. (617) 495-7461. FAX (617) 495-7105.

COMMUNICATIONS SATELLITES

See: SATELLITES (ARTIFICIAL)

COMPASSES

See also: AVIONICS, BUOYS, GUIDANCE SYSTEMS, GYROSCOPES, NAVIGATION, RADAR

ABSTRACT SERVICES AND INDEXES

Applied Science and Technology Index; A Cumulative Subject Index To English Language Periodicals In the Fields of Aeronautics and Space Science, Computer Technology, Chemistry, Construction Industry, Energy and Related Areas. H.W. Wilson Co., 950 University Avenue, Bronx, NY 10452. (800) 367-6770 or (212) 588-8400. FAX (718) 590-1617. From 1958 to present. Monthly. Inquire about cost and availability. Also available on CD-ROM and online.

Engineering Index Monthly. Engineering Information, Inc., Castle Point on the Hudson, Hoboken, NJ 07030. (800) 221-1044. FAX (212) 832-1857. Monthly. $2300.00 per year. Also available online as COMPENDEX, and also on CD-ROM. Covers chemical engineering, computers, electrical engineering, civil engineering, metals and mining, industrial management, and mechanical engineering.

International Aerospace Abstracts. Technical Information Service, American Institute of Aeronautics and Astronautics, Inc., 555 West 57th St., New York, NY 10019. (212) 247-6500. FAX (212) 582-4861. Semi-monthly. $1295.00 per year.

ASSOCIATIONS AND PROFESSIONAL SOCIETIES

Institute of Navigation. 1800 Diagonal Road, Suite 480, Alexandria, VA 22314-2840.

International Navigation Association, Po Box 2324, Arlington, Va 22202-0324. (604) 276-4626.

Permanent International Association of Navigation Congresses, United States Section. CECW-PK, 20 Massachusetss Avenue NW, Washington, DC 20314-1000. (202) 504-4312.

Wild Goose Association. Operations, 150 S. Plains Road, the Plains, OH 45780. (614) 797-2081.

DIRECTORIES AND BIOGRAPHICAL SOURCES

Directory of Engineering Societies and Related Organizations. Gordon Davis. 13th edition. American Association of Engineering Societies, 1111 19th Street NW, Suite 608, Washington, DC 20036. (202) 296-2237 or (800) 658-8897. 1989. Inquire for price.

Research Centers Directory. Gale Research, 835 Penobscot Building, Detroit, MI 48226-4094. (313) 961-2242. (800) 347-4253. FAX (313) 961-6083. $485.00.

Who's Who In Engineering. American Association of Engineering Societies, 1111 19th Street NW, Suite 608, Washington, DC 20036. (202) 296-2237 or (800) 658-8897. 8th edition. 1991. Inquire for price.

ENCYCLOPEDIAS AND DICTIONARIES

The Norton Encyclopedic Dictionary of Navigation. David F. Tver. W.W. Norton & Co., Inc., 500 Fifth Ave, New York, NY 11010. (800) 223-2584. FAX (800) 233-2588. 1987. Inquire for cost and availability.

GENERAL WORKS

Maps & Compasses. Percy W. Blandford. 2nd edition. TAB Books, P.O. Box 40, Blue Ridge Summit, PA 17294-0850. (717)794-2191 or (800) 822-8138. 1991. $22.95 (hardback), $14.95 (paperback).

New Explorer's Guide To Maps & Compasses. Percy W. Blandford. TAB Books, P.O. Box 40, Blue Ridge Summit, PA 17294-0850. (717) 794-2191 or (800) 233-1128. FAX (717) 794-2080. 1992. $15.95 (hardback), $7.95 (paperback).

Thataway: the Story of the Magnetic Compass. Bill Eyster. New York: A.S. Barnes, 1970. Out of print.

HANDBOOKS AND MANUALS

Handbook of Magnetic Compass Adjustment. U.S. Defense Mapping Agency, Hydrographic/Topographic Center. 4th edition. Oorder from: Superintendent of Documents, U.S. Government Printing office, Box 371954, Pittsburgh, PA 15250-7954. (202) 783-3238. FAX (202) 512-2233. 1980. Inquire for cost and availability.

Navigational & Surveying Instruments: Industry and Trade Summary. Sundar A. Shetty. Diane Publishing Company, 600 Upland Avenue, Upland, PA 19015. (215) 499-7415. FAX (610) 499-7429. 1994. $40.00.

ONLINE DATABASES AND CD-ROMS

Aerospace Database. American Institute of Aeronautics and Astronautics, 370 L'Enfant Promenade, S.W., Washington, DC 20024. (202) 646-7400. Worldwide published literature on research and development in aerospace and related areas, with abstracts. Covers 1962 to present. Available on DIALOG, (800) 334-2564, online service. Also available on CD-ROM. Inquire as to cost and availability.

Compendex Plus. Engineering Information, Inc., 345 East 47th Street, New York, NY 10017. (212) 705-7600 or (800) 221-1044. Contains citations with abstracts to worldwide literature in engineering and technology, from 1970 to present. Available ononline BRS,(800) 289-4277, DIALOG, (800) 334-2564, ORBIT (800) 456-7248, and STN International, FIZ Karlsruhe, P.O. Box 2465, W-7500, Karlsruhe 1, Germany, online services. Also available on CD-ROM. Inquire as to cost and availability.

International Aerospace Abstracts. American Institute of Aeronautics and Astronautics, 370 L'Enfant Promenade, S.W., Washington, DC 20024. (202) 646-7400. Contains references and abstracts of journal and monograph literature relating to aerospace science and technology, from 1963 to present. Available through the NASA/RECON system of the National Aeronautics and Space Administration only.

NASA Database. American Institute of Aeronautics and Astronautics, 370 L'Enfant Promenade, S.W., Washington, DC 20024. (202) 646-7400. Citations and abstracts of aeronautics and astronautics literature, 1962 to present. Also contains citations from *STAR, Scientific and Technical Aerospace Reports,* and *International Aerospace Abstracts.* Available through NASA/RECON online service. Inquire as to cost and availability.

NTIS Bibliographic Database. National Technical Information Service, 5285 Port Royal Road, Springfield, Va 22161. (703) 487-4929 or FAX (703) 321-8199. Broad coverage of government-sponsored science and technology research reports, 1964 to present. Available on BRS,(800) 289-4277, DIALOG, (800) 334-2564, ORBIT, (800) 456-7248, and STN International, FIZ Karlsruhe, P.O. Box 2465, W-7500, Karlsruhe 1, Germany, online services. Also available on CD-ROM. Inquire as to cost and availability.

SAE Global Mobility Database. Society of Automative Engineers (SAE), Electronic Publishing Division, 400 Commonwealth Drive, Warrendale, PA 15098. (412) 776-4841. Contains citations with abstracts to technical papers on automotive and aerospace technology and vehicular-related industries that have been presented at SAE conferences. Covers 1906 to present. Available on ORBIT,(800) 456-7248,online service. Inquire as to cost and availability.

Scisearch. Institute for Scientific Information, 3501 Market Street, Philadelphia, PA 19104. (800) 523-1850 or (215) 386-0100. Broad multidisciplinary title and author index to the international literature of science and technology, 1974 to present. Available on DIALOG,(800) 334-2564, and ORBIT,(800) 456-7248, online services. Also available on CD-ROM. Inquire as to cost and availability.

TRIS (Transportation Research Information). National Academy of Science, 2101 Constitution Avenue, N.W., Washington, DC 20418. (202) 334-3313 or (800) 624-6242. Citations with abstracts of literature on transportation, including air, highway, rail, maritime and other modes from 1968 to present. Available on DIALOG,(800) 334-2564, online service. Inquire as to cost and availability.

WILSONLINE. H.W. Wilson Company, 950 University Avenue,Bronx, NY 10452. (800) 367-6770 or (212) 588-8400. Makes available online versions of the printed H.W. Wilson indexes including Applied Science and Technology Index, Business Periodicals Index, General Science Index, and Readers' Guide to Periodical Literature. Period covered is generally 1983 to present. Available on BRS,(800) 289-4277, DIALOG,(800) 334-2564, and OCLC EPIC,(800) 848-5878, online services. Also available on CD-ROM. Inquire as to cost and availability.

PERIODICALS

Institute of Navigation, Proceedings of Annual Meeting. Institute of Navigation, 1800 Diagonal Road, Suite 480, Alexandria, VA 22314-2840. Annual. $80.00 per year for non-members.

Journal of Navigation. Cambridge University Press, Journals Department, 40 West 20th Street, New York, NY 10011-4211. (212) 924-3900. 1947 to present. Three times a year. $165.00 per year.

Journal of Ship Research. Society of Naval Architects and Marine Engineers, 601 Pavonia Avenue, Suite 400, Jersey City, NJ 07306. (201) 798-4800. FAX (201) 798-4975. 1957 to present. Quarterly. Inquire for price.

COMPOSITES

See also: MATERIALS SCIENCE

ABSTRACT SERVICES AND INDEXES

Applied Mechanics Reviews: An Assessment of World Literature In Engineering Sciences. 1948-present. American Society of Mechanical Engineers, 345 East 47th Street, New York, NY 10017. (212) 705-7703. Monthly. $360.00 per year.

Applied Science and Technology Index; A Cumulative Subject Index To English Language Periodicals In the Fields of Aeronautics and Space Science, Computer Technology, Chemistry, Construction Industry, Energy and Related Areas. H.W. Wilson Co., 950 University Avenue, Bronx, NY 10452. (800) 367-6770 or (212) 588-8400. FAX (718) 590-1617. From 1958 to present. Monthly. Inquire about cost and availability. Also available on CD-ROM and online.

C2C Abstracts: Japan—Materials Science. Scan C2C, 1001 Pennsylvania Ave., NW, No. 1300, Washington, DC 20024-2505. (800) 525-3865. FAX (202) 863-3855. 1990 to present. Monthly. $200.00 per year. Contains abstracts of articles from Japanese scientific, business and technical journals on aggregates, coatings, composites, fibers and whiskers, powder technology, processing, refractories, and wood products. Also available on CD-ROM and online.

Chemical Abstracts - Applied Chemistry and Chemical Engineering Sections. Chemical Abstracts Service, 2540 Olentangy River Road, Box 3012, Columbus, OH 43210. (800) 848-6538 or (614) 447-3600, FAX (614) 447-3713. Bi-weekly. $1410.00 per year.

Current Contents: Engineering, Technology, and Applied Sciences. Institute for Scientific Information, 3501 Market Street, Philadelphia, PA 19104. (215) 386-0100. FAX (215) 386-6362. 1970 topresent. Weekly. $442.00 per year.

Engineered Materials Abstracts. ASM (American Society for Metals) International, Materials Information, Materials Park, OH 44073. (216) 338-5151 or FAX (216) 338-4634. Covers literature on technical developments in polymer, ceramic, and composite materials and engineering. 1986 to present. Monthly. $1200.00 per year. Also available on CD-ROM and online. Covers published materials on technical developments in polymer, ceramic, and composite materials in an engineering environment.

COMPOSITES

Ency. of Physical Sciences and Engineering Info. Sources

Engineering Index Monthly. Engineering Information, Inc., Castle Point on the Hudson, Hoboken, NJ 07030. (800) 221-1044. FAX (212) 832-1857. Monthly. $2300.00 per year. Also available online as COMPENDEX, and also on CD-ROM. Covers chemical engineering, computers, electrical engineering, civil engineering, metals and mining, industrial management, and mechanical engineering.

International Aerospace Abstracts. Technical Information Service, American Institute of Aeronautics and Astronautics, Inc., 555 West 57th St., New York, NY 10019. (212) 247-6500. FAX (212) 582-4861. Semi-monthly. $1295.00 per year.

Materials Science and Engineering Abstracts, Cambridge Scientific Abstracts, 7200 Wisconsin Avenue, Bethesda, Md 20814-4823. (301) 961-6750. FAX (301) 961-6720. 1993 to present. Monthly. $385.00 per year. Focuses on mechanical and physical properties of materials and commercial or industrial applications for materials, methods for strength testing, effects of vibration and other stresses, corrosion and protective coatings, storage and handling, ceramics, composites, metals, wood, plastics, and polymers.

Materials Science Citation Index. Institute for Scientific Information, 3501 Market Street, Philadelphia, PA 19104. (215) 386-0100. FAX (215) 386-6262 Bi-monthly. $975.00 per year. Also available on CD-ROM.

Metals Abstracts and Metals Abstracts Index. American Society for Metals, Metals Park, OH 44073. (216) 338-5151. 1968 to present. Monthly. Abstracts are $1500.00 per year and Index is $460.00 per year.

Scientific and Technical Aerospace Reports (star). National Aeronautics and Space Administration. NASA Center for Aerospace Information, Box 8757, BWI Airport, Baltimore, MD 21240. (301) 621-0153. Monthly. Inquire about availability and cost. ALso available through the NASA online retrieval service (RECON), and through the Aerospace Database through DIALOG.

ASSOCIATIONS AND PROFESSIONAL SOCIETIES

American Society For Testing and Materials (ASTM). 1916 Race Street, Philadelphia, PA 19103. (215) 299-5585.

American Society For Metals International. Materials Park, OH 44073. (216) 338-5151 or FAX (216) 338-4634.

Composites Fabricators Association. 1735 N. Lynn Street, Suite 950, Arlington, VA 22209. (703) 524-3332.

Composites Manufacturing Association of the Society ofManufacturing Engineers. One SME Drive, PO Box 930, Dearborn, MI 48121-0930. (313) 271-1500. FAX (313) 271-2861.

Society For the Advancement of Material and Process Engineering. PO Box 2459, Covina, CA 91722.

SPI Composites Institute. 355 Lexington Avenue, New York, NY 10017. (212) 351-5410. FAX (212) 370-1731.

Suppliers of Advanced Composite Materials Association. 1600 Wilson Blvd., Suite 1008, Arlington, VA 22209. (703) 841-1556.

BIBLIOGRAPHIES

Scientific and Technical Books and Serials In Print. R.R. Bowker Inc., 121 Chanlon Road, New Providence, NJ 07974. (800) 521-8110 or (908) 464-6800. List of currently published books and serials in the physical and biological sciences, engineering and technology, with subject, author and titles indexes. Inquire for price of latest edition. Also available online on ORBIT service, (800) 456-7248.

DIRECTORIES AND BIOGRAPHICAL SOURCES

Directory of Engineering Societies and Related Organizations. Gordon Davis. 13th edition. American Association of Engineering Societies, 1111 19th Street NW, Suite 608, Washington, DC 20036. (202) 296-2237 or (800) 658-8897. 1989. Inquire for price.

International Research Centers Directory. Gale Research, 835 Penobscot Building, Detroit, MI 48226-4094. (313) 961-2242. (800) 347-4253. FAX (313) 961-6083. 8th edition. 1995. $410.00

Materials Research Centers: A World Directory of Organizations and Programmes In Materials Science. 5th edition. Gale Research, 835 Penobscot Building, Detroit, MI 48226-4094. (313) 961-2242. (800) 347-4253. FAX (313) 961-6083. 1993. $495.00.

Research Centers Directory. Gale Research, 835 Penobscot Building, Detroit, MI 48226-4094. (313) 961-2242. (800) 347-4253. FAX (313) 961-6083. $485.00.

Scientific and Technical Organizations and Agencies Directory. Gale Research, 835 Penobscot Building, Detroit, MI 48226-4094. (313) 961-2242. (800) 347-4253. FAX (313) 961-6083. 4th edition. 1996. $195.00.

Who's Who In Engineering. American Association of Engineering Societies, 1111 19th Street NW, Suite 608, Washington, DC 20036. (202) 296-2237 or (800) 658-8897. 8th edition. 1991. Inquire for price.

Who's Who In Technology. Gale Research, 835 Penobscot Building, Detroit, MI 48226-4094. (313) 961-2242. (800) 347-4253. FAX (313) 961-6083. 1995. $195.00.

ENCYCLOPEDIAS AND DICTIONARIES

Concise Encyclopedia of Composite Materials. Anthony Kelly, editor. Revised edition. Pergamon Press Inc., Maxwell House, Fairview Park, Elmsford, NY 10523. (914) 592-7700. Fax (914) 592-3625. 1994. $57.50.

Dictionary of Composite Materials Technology. Stuart M. Lee. Technomic Publishing Company, Inc., 851 New Holland Avenue,Box 3535, Lancaster, PA 17604. (717) 291-5609 or (800) 233-9936. FAX (717) 295-4538. 1989. Inquire for cost.

Encyclopedia of Composite Materials and Components. Martin Grayson, editor. John Wiley & Sons, 605 Third Avenue, New York, NY 10158-0012. (212) 692-6000. 1983. $225.00.

Encyclopedia of Physical Science and Technology. Robert A. Meyers, ed. 18 volumes. Academic Press Inc., 6277 Sea Harbor Drive, Orlando, FL 32887. 1992. $2100.00

International Encyclopedia of Composites. Stuart M. Lee, editor. 6 volumes. VCH Publishers, Inc., 303 Northwest 12th Avenue, Deerfield Beach, FL 33442. (800) 422-8824. 1990-1992. Subscription, $235.00.

McGraw-Hill Dictionary of Scientific and Technical Terms. Sybil P. Parker, ed. 5th edition. McGraw-Hill Publishing Company, 1221 Avenue of the Americas, New York, NY 10020. (800) 262-4729 or (212) 512-3825. 1993. $110.50.

McGraw-Hill Encyclopedia of Engineering. Sybil P. Parker, ed. 2nd edition. McGraw-Hill Publishing Company, 1221 Avenue of the Americas, New York, NY 10020. (800) 262-4729 or (212) 512-3825. 1993. $95.50.

McGraw-Hill Encyclopedia of Science and Technology. Sybil P. Parker, ed. 7th edition. 20 volumes. McGraw-Hill Publishing Company, 1221 Avenue of the Americas, New York, NY 10020. (800) 262-4729 or (212) 512-3825. 1992. $1900.00

GENERAL WORKS

Advanced Polymer Composites: Principles and Applications. Bor Z. Jang. ASM International, Materials Park, OH 44073-0002. (216) 338-5151 or (800) 336-5152. FAX (216) 338-4634. 1993. $98.00.

Characterization of Composite Materials. Hatsuo Ishida, editor. Butterworth-Heinemann, 313 Washington Street, Newton, MA 02158. (617) 928-2500 or (800) 366-2665. FAX (617) 928-2620. 1994. $59.95.

Composite Materials. A.T. Di Benedetto, L. Nicolais, and R. Watanabe. North-Holland/Elsevier Science B.V., PO Box 945, Madison Square Station, New York, NY 10159-0945. (212) 633-3650. FAX (212) 633-3680. 1992. $148.25.

Composite Materials: Engineering and Science. F.L. Matthews and R.D. Rawlings. Routledge, Chapman and Hall, Inc., 29 West 35th Street, New York, NY 10001-2291. (212) 244-3336 or FAX (212) 563-2269.1994. Inquire for cost and availability.

Composite Materials Technology: Processes and Properties. P.K. Mallick, S. Newman, editors. Hanser-Gardner Publications, 6600 Clough Pike, Cincinnati, OH 45244-4090. (513) 527-8977 or (800) 950-8977. 1990. $89.95.

Engineering Plastics and Composites. William A. Woishnis, editor. ASM International, Materials Park, OH 44073-0002. (216) 338-5151 or (800) 336-5152. FAX (216) 338-4634. 1993. Inquire for cost.

Fundamentals of Metal Matrix Composites. S. Suresh, et al., editors. Butterworth-Heinemann, 313 Washington Street, Newton,MA 02158. (617) 928-2500 or (800) 366-2665. 1993. $115.00.

Principles of Composite Material Mechanics. Ronald F. Gibson. McGraw-Hill Publishing Company, 1221 Avenue of the Americas, New York, NY 10020. (800) 262-4729 or (212) 512-3825. 1993. Inquire for cost and availability.

HANDBOOKS AND MANUALS

Asm Engineered Materials Handbook Volume 1: Composites. ASM International, Materials Information, Materials Park, OH 44073-0002. (216) 338-5151 or (800) 336-5152. FAX (216) 338-4634. 1995. $147.00.

Composite Materials Handbook. Melvin M. Schwartz. 2nd edition. McGraw-Hill Publishing Company, 1221 Avenue of the Americas, New York, NY 10020. (800) 262-4729 or (212) 512-3825. 1992. $79.50.

Handbook of Ceramics and Composites. Nicholas P. Cheremisinoff, editor. Marcel Dekker, Inc., 270 Madison Avenue, New York, NY 10016. (212) 696-9000. FAX (212) 685-4540. Volume 1 (1990), Volume 2 (1991). $210.00 each.

Reference Book For Composites Technology. Stuart M. Lee, editor. 2 volumes. Technomic Publishing Company, Inc., 851 New Holland Avenue, Box 3535, Lancaster, PA 17604. (717) 291-5609 or (800) 233-9936. FAX (717) 295-4538. 1989. $39.00 (volume 1), $29.00 (volume 2).

ONLINE DATABASES AND CD-ROMS

Compendex Plus. Engineering Information, Inc., 345 East 47th Street, New York, NY 10017. (212) 705-7600 or (800) 221-1044. Contains citations with abstracts to worldwide literature in engineering and technology, from 1970 to present. Available on online BRS,(800) 289-4277, DIALOG, (800) 334-2564, ORBIT (800) 456-7248, and STN International, FIZ Karlsruhe, P.O. Box 2465, W-7500, Karlsruhe 1, Germany, online services. Also available on CD-ROM. Inquire as to cost and availability.

CSA Engineering. Cambridge Scientific Abstracts, 7200 Wisconsin Avenue, Suite 601, Bethesda, MD 20814. (301) 961-6750 or (800) 843-7751. Contains citations and abstracts of international periodicals and research literature covering all fields of engineering and science and technology,including computer and information science, electronics, mechanical engineering, solid state materials, 1981 to present. Available on BRS,(800) 289-4277, online service.Inquire as to cost and availability.

Current Contents Search. Institute for Scientific Information, 3501 Market Street, Philadelphia, PA 19104. (215) 386-0100. FAX (215) 386-6362. Contains citations to articles listed in the table of contents of science and technology journals. Also articles in social sciences and life sciences journals. Available on BRS,(800) 289-4277, DIALOG,(800) 334-2564, online services. Inquire as to cost and availability.

International Aerospace Abstracts. American Institute of Aeronautics and Astronautics, 370 L'Enfant Promenade, S.W., Washington, DC 20024. (202) 646-7400. Contains references andabstracts of journal and monograph literature relating to aerospace science and technology, from 1963 to present. Available through the NASA/RECON system of the National Aeronautics and Space Administration only.

NASA Database. American Institute of Aeronautics and Astronautics, 370 L'Enfant Promenade, S.W., Washington, DC 20024. (202) 646-7400. Citations and abstracts of aeronautics and astronautics literature, 1962 to present. Also contains citations from *STAR, Scientific and Technical Aerospace Reports,* and *International Aerospace Abstracts.* Available through NASA/RECON online service. Inquire as to cost and availability.

NTIS Bibliographic Database. National Technical Information Service, 5285 Port Royal Road, Springfield, Va 22161. (703) 487-4929 or FAX (703) 321-8199. Broad coverage of government-sponsored science and technology research reports, 1964 to present. Available on BRS,(800) 289-4277, DIALOG, (800)

COMPOSITES

Ency. of Physical Sciences and Engineering Info. Sources

334-2564, ORBIT, (800) 456-7248, and STN International, FIZ Karlsruhe, P.O. Box 2465, W-7500, Karlsruhe 1, Germany, online services. Also available on CD-ROM. Inquire as to cost and availability.

SCISEARCH. Institute for Scientific Information, 3501 Market Street, Philadelphia, PA 19104. (800) 523-1850 or (215) 386-0100. Broad multidisciplinary title and author index to the international literature of science and technology, 1974 to present. Available on DIALOG,(800) 334-2564, and ORBIT,(800) 456-7248, online services. Also available on CD-ROM. Inquire as to cost and availability.

WILSONLINE. H.W. Wilson Company, 950 University Avenue, Bronx, NY 10452. (800) 367-6770 or (212) 588-8400. Makes available online versions of the printed H.W. Wilson indexes including Applied Science and Technology Index, Business Periodicals Index, General Science Index, and Readers' Guide to Periodical Literature. Period covered is generally 1983 to present. Available on BRS,(800) 289-4277, DIALOG,(800) 334-2564, and OCLC EPIC,(800) 848-5878, online services. Also available on CD-ROM. Inquire as to cost and availability.

PERIODICALS

Aviation Week and Space Technology. 1221 Avenue of the Americas, New York, NY 10020. (212) 512-2999 or (800) 525-5003. 1916 to present. Weekly. $82.00 per year.

Composites. Butterworth-Heinemann, Linacre House, Jordan Hill, Ocford OX2 8DP, England. Telephone 0865-310366. FAX 0865-310898. 1969 to present. 8 times a year. Inquire for cost.

Composites Science and Technology. Elsevier Science [journals], 660 White Plains Rd., Tarrytown, NY 10159-5153. (919) 524-9200. FAX (919) 333-2444. 1968 to present. 12 times a year. $1285.00 per year.

Journal of Applied Polymer Science. J. Wiley & Sons, Journals, 605 Third Avenue, New York, NY 10158-0012. (212) 692-6000. 1956 to present. 48 times a year. $3530.00 per year.

Journal of Composite Materials. Technomic Publishing Company, Inc., 851 New Holland Avenue, Box 3535, Lancaster, PA 17604. (717) 291-5609. 1967 to present. 18 times a year. $795.00 per year.

Journal of Composites Technology and Research. American Society for Testing and Materials (ASTM), 1916 Race Street, Philadelphia, PA 19103. (215) 299-5585. 1979 to present. Quarterly. $95.00 per year for non-members.

Journal of Elastomers and Plastics. Technomic Publishing Company, Inc., 851 New Holland Avenue, Box 3535, Lancaster, PA 17604. (717) 291-5609. 1969 to present. Quarterly. $205.00 per year.

Materials Engineering. Penton Publishing, 110 Superior Ave., Cleveland, OH 44114-2543. 1929 to present. Monthly. $50.00 per year.

RESEARCH CENTERS AND INSTITUTES

Michigan State University Composite Materials and Structures Center. College of Engineering, East Lansing, MI 48824. (517) 353-5466.

University of Florida Center For Studies of Advanced Structural Composites. Engineering Sciences Department, 231 Aerospace Bldg., Gainesville, FL 32611. (904) 392-0961.

COMPUTER COMMUNICATIONS

See also: AUTOMATION, COMPUTER MEMORY, COMPUTER OPERATING SYSTEMS, COMPUTER PROGRAMMING, COMPUTER PROGRAMMING LANGUAGES, COMPUTER VISION, PARALLEL COMPUTING

ABSTRACT SERVICES AND INDEXES

ACM Guide To Computing Literature. Association for Computing Machinery, 11 West 42nd Street, New York, NY 10036. (212) 869-7440. Fax (212) 869-0481. 1964 to present. Annual. $175.00 per year.

Applied Science and Technology Index; A Cumulative Subject Index To English Language Periodicals In the Fields of Aeronautics and Space Science, Computer Technology, Chemistry, Construction Industry, Energy and Related Areas. H.W. Wilson Co., 950 University Avenue, Bronx, NY 10452. (800) 367-6770 or (212) 588-8400. FAX (718) 590-1617. From 1958 to present. Monthly. Inquire about cost and availability. Also available on CD-ROM and online.

Computer Abstracts, MCBUniversity Press Ltd., PO Box 10812, Birmingham, AL 35201-0812. (800) 633-4931. FAX (205) 995-1588. Monthly. Covers computer theory, data, hardware, systems, networks, human-computer interaction, artificial intelligence, as well as applications of computers in aerospace, business, CAD/CAM, cartography, civil engineering, electronics and electrical engineering, industrial engineering, mechanical engineering, medicine, structural engineering, etc.

Computer and Information Systems Abstracts Journal. Cambridge Scientific Abstracts, 7200 Wisconsin Avenue, Bethesda, MD 20814. (301) 961-6750. Fax (301) 961-6720. 1962 to present.Monthly. $1035 per year.

Computing Reviews, Association For Computing Machinery, 11 West 42nd Street, New York, Ny 10036. (212) 869-7440. Fax (212) 869-0481. Monthly.

Engineering Index Monthly. Engineering Information, Inc., Castle Point on the Hudson, Hoboken, NJ 07030. (800) 221-1044. FAX (212) 832-1857. Monthly. $2300.00 per year. Also available online as COMPENDEX, and also on CD-ROM. Covers chemical engineering, computers, electrical engineering, civil engineering, metals and mining, industrial management, and mechanical engineering.

Key Abstracts—Computer Communications and Storage. INSPEC/IEEE, Box 1331, 445 Hoes Lane, Piscataway, NJ 08855-1331. (908) 562-5549. 1987 to present. Monthly. $160.00 per year. Covers multiprocessor systems, storage media, optical disc, interfaces, networks, and network equipment.

Science Citation Index. Institute for Scientific Information, 3501 Market Street, Philadelphia, PA 19104. (215) 386-0100. FAX (215) 386-6362. Inquire about availability and cost. Also available on CD-ROM.

ANNUAL REVIEWS AND YEARBOOKS

ACM Monograph Series. Association for Computing Machinery (ACM), 11 West 42nd Street, New York, NY 10036. (212) 869-7440. Fax (212) 869-0481. Irregular. Price varies, inquire.

Advances In Computers. Academic Press Inc., 6277 Sea Harbor Drive, Orlando, FL 32887. (800) 321-5068. Irregular. Price varies, inquire.

Advances In Computing Research. JAI Press, Inc., 55 Old Post Road, Number 2, Box 1678, Greenwich, CT 06830. (203) 661-7602. FAX (203) 661-0792. Annual. $68.50 per year.

ASSOCIATIONS AND PROFESSIONAL SOCIETIES

American Federation of Information Processing Societies, 1899 Preston White Drive, Reston, Va 22091-5435. (703) 620-8937.

Association For Computing Machinery. 1515 Broadway, 17th Floor, New York, NY 10036-5701. (212) 869-7440. FAX (212) 944-1318.

Computer and Automation Systems Association of Sme. 1 SME Drive, PO Box 930, Dearborn, MI 48121-0930. (313) 271-1500.

Computer & Communications Industry Association. 666 11th Street NW, Suite 600, Washington, DC 20001-4542. (202) 783-0070. FAX (202) 783-0534.

Institute of Electrical and Electronic Engineers. 345 E. 47th Street, New York, NY 10017. (212) 705-7900.

Society For Computer Simulation. Box 17900, San Diego, CA 92177-7900. (619) 277-3888. FAX (619) 277-3930.

Special Interest Group For Data Communications. c/o ACM, 1515 Broadway, New York, NY 10036. (212) 626-0611. FAX (212) 302-5826.

BIBLIOGRAPHIES

ACM Guide To Computing Literature. Association for Computing Machinery, 1515 Broadway, 17th Floor, New York, NY 10036-5701. (212) 869-7440. FAX (212) 944-1318. 1964 to present. Annual. $190.00 for non-members. Also available online via DIALOG.

DIRECTORIES AND BIOGRAPHICAL SOURCES

ACM Adminstrative Directory of College and University Computer Science-data Processing Programs and Computer Facilities. Association for Computing Machinery, 1515 Broadway, 17th Floor, New York, NY 10036-5701. (212) 869-7440. FAX (212) 944-1318. Biennial. $17.00 for non-members.

American Men & Women of Science, 1995-96. R.R. Bowker Staff, eds. 19th edition. 8 volumes. R.R. Bowker/Reed International Publishing Company, 121 Chanlon Road, New Providence, NJ 07974. (908) 464-6800 or (800) 521-8110. 1995. $850.00

ASIS Handbook and Directory, American Society For Information Science, 8720 Georgia Avenue, Suite 501, Silver Spring, Md 20910-3602. (301) 495-0900. Annual. $100.00.

Computing Information Directory. Hildebrandt Inc., PO Box 576, Pullman, WA 99114. 1981 to present. Annual. $199.95 per year.

Data Communications Buyers Guide. McGraw-Hill Publishing Company, 1221 Avenue of the Americas, New York, NY 10020. (800) 262-4729 or (212) 512-3825. 1978 to present. Twice a year. $75.00.

International Research Centers Directory. Gale Research, 835 Penobscot Building, Detroit, MI 48226-4094. (313) 961-2242. (800) 347-4253. FAX (313) 961-6083. 8th edition. 1995. $410.00

Research Centers Directory. Gale Research, 835 Penobscot Building, Detroit, MI 48226-4094. (313) 961-2242. (800) 347-4253. FAX (313) 961-6083. $485.00.

Who's Who In Computing. Canadian M I S Database Inc., 268 Lakeshore Road E., Suite 510, Mississauga, ON L5G 1H1, Canada. (905) 271-1601. FAX (905) 271-4522. 1987 to present. Annual. Inquire for price in U.S.

Who's Who In Technology. Gale Research, 835 Penobscot Building, Detroit, MI 48226-4094. (313) 961-2242. (800) 347-4253. FAX (313) 961-6083. 1995. $195.00.

ENCYCLOPEDIAS AND DICTIONARIES

Computer Dictionary. Donald D. Spencer. 4th edition. Camelot Publishing Company, 709 SW 80th Blvd., Gainesville, FL 32607-1537. (904) 331-0952. 1993. $24.95.

Data Communications and Networking Dictionary. T.D. Pardoe & R. P. Wenig. CBM Books, 1300 Virginia Drive, Suite 400, Ft. Washington, PA 19034. (800) 285-1755. FAX (215) 643-8099. 1992. $12.00.

Dictionary of Computing. S.M.H. Collin. 2nd edition. Peter Collin Publishing, 8 the Causeway, Teddington, TW11 0HE, England. FAX 0181-943-3386. 1994. Inquire for cost.

Encyclopedia of Computer Science and Engineering. Anthony Ralston & Edwin D. Reilly Jr., editors. 3rd revised edition. Van Nostrand Reinhold, 115 Fifth Avenue, New York, NY 10003. (212) 254-3232 or FAX (212) 254-9499. 1993. $125.00.

Microsoft Press Computer Dictionary: the Comprehensive Standard For Business, School, Library, and Home. 2nd edition. Microsoft Press, One Microsoft Way, Redmond, WA 98052-6399. (206) 936-0055. FAX (206) 823-8101. 1994. $19.95.

Networking and Communications Desk Reference. Ken Sochats. Sams/ Prentice-Hall Computer Publishing, 201 W. 103rd Street, Indianapolis, in 46290-1097. (800) 545-5914. FAX (317) 876-9371. 1992. $19.95

GENERAL WORKS

Contemporary Data Communications: A Practical Approach. AL Schroeder and Emilio Ramos. Macmillan Publishing, 200

Old Tappan Road, Old Tappan, NJ 07675. (800) 223-2336. FAX (800) 445-6991. 1994. Inquire for price.

Data and Computer Communications. William Stallings. Fourth edition. Macmillan Publishing, 200 Old Tappan Road, Old Tappan, NJ 07675. (800) 223-2336. FAX (800) 445-6991. 1994. Inquire for price.

Data Communications and Computer Networks: An OSI Approach. Curt M. White. Boyd & Fraser Publishing Co., One Corporate Pl, 55 Ferncroft Road, Danvers, MA 01923. (800) 225-3782. FAX (508) 777-9068. 1994. Inquire for cost and availability.

Data Communications, Computer Networks, and Open Systems. Fred Halsall. Fourth edition. Addison-Wesley Publishing Company, 1 Jacob Way, Reading, MA 01867. (617) 944-3700 or (800) 447-2226. FAX 617-942-1117. 1996. $55.95.

Understanding Data Communications and Networks. William A. Shay. PWS Publishers, 20 Park Plaza, Boston, MA 02116. (617) 542-3377. FAX (617) 338-6134. 1995. $54.95.

HANDBOOKS AND MANUALS

Computer Engineering Handbook. C.H. Chen, editor. McGraw-Hill Publishing Company, 1221 Avenue of the Americas, New York, NY 10020. (800) 262-4729 or (212) 512-3825. 1992. $79.50.

Handbook of Data Communications. National Computing Center staff. Blackwell Publications, 238 Main Street, Cambridge, MA 02142. (617) 547-7110 or (800) 216-2522. FAX (802) 878-1102. 1995. Inquire for cost and availability.

McGraw-Hill Computing Essentials, 1993-1994. Timothy J. O'Leary, et al. McGraw-Hill Publishing Company, 1221 Avenue of the Americas, New York, NY 10020. (800) 262-4729 or (212) 512-3825. 1994. Inquire for price and availability.

McGraw-Hill LAN Communications Handbook. Fred Simonds. McGraw-Hill Publishing Company, 1221 Avenue of the Americas, New York, NY 10020. (800) 262-4729 or (212) 512-3825. 1994. $54.50.

ONLINE DATABASES AND CD-ROMS

Compendex Plus. Engineering Information, Inc., 345 East 47th Street, New York, NY 10017. (212) 705-7600 or (800) 221-1044. Contains citations with abstracts to worldwide literature in engineering and technology, from 1970 to present. Available on online BRS,(800) 289-4277, DIALOG, (800) 334-2564, ORBIT (800) 456-7248, and STN International, FIZ Karlsruhe, P.O. Box 2465, W-7500, Karlsruhe 1, Germany, online services. Also available on CD-ROM. Inquire as to cost and availability.

Compuscience. FIZ Karlsruhe, D-7514,Eggenstein-Leopoldshafen 2, Germany. Contains citations with abstracts to European and North American literature on computer science, 1972 to present. Available on STN International, FIZ Karlsruhe, P.O. Box 2465, W-7500, Karlsruhe 1, Germany, online service. Inquire as to cost and availability.

Computer and Information Systems Abstracts. Cambridge Scientific Abstracts, 7200 Wisconsin Avenue, Suite 601, Bethesda, MD 20814. (301) 961-6750 or (800) 843-7751. Contains cita-

tions to worldwide literature in theoretical and applied computer science and related areas, from 1981 to present.Inquire as to cost and availability.

Computer and Mathematics Search. Institute for Scientific Information, 3501 Market Street, Philadelphia, PA 19104. (215) 386-0100. FAX (215) 386-6362. Covers worldwide literature in computer science and mathematics, from 1980 to present. Available on BRS,(800) 289-4277, online service. Inquire as to cost and availability.

Computer Database. Information Access Company, 362 Lakeside Drive, Foster City, CA 94404. (415) 378-5000 or (800) 227-8431. Contains citations with abstracts to literature from trade journals covering the computer,telecommunications,and electronics industries. Available on the BRS, (800) 289-4277, Compuserve Information Service,(800) 848-8990, and DIALOG,(800) 334-2564, online services. Inquire as to cost and availability.

CSA Engineering. Cambridge Scientific Abstracts, 7200 Wisconsin Avenue, Suite 601, Bethesda, MD 20814. (301) 961-6750 or (800) 843-7751. Contains citations and abstracts of international periodicals and research literature covering all fields of engineering and science and technology,including computer and information science, electronics, mechanical engineering, solid state materials, 1981 to present. Available on BRS,(800) 289-4277, online service.Inquire as to cost and availability.

Current Contents Search. Institute for Scientific Information, 3501 Market Street, Philadelphia, PA 19104. (215) 386-0100. FAX (215) 386-6362. Contains citations to articles listed in the table of contents of science and technology journals. Also articles in social sciences and life sciences journals. Available on BRS,(800) 289-4277, DIALOG,(800) 334-2564, online services. Inquire as to cost and availability.

Dissertation Abstracts. University Microfilms International, 300 North Zeeb Road, Ann Arbor, MI 48106. (800) 521-0600 or (313) 761-4700. Scope includes virtually all doctoral dissertations accepted at accredited American institutions from 1861 to present in 252 subject areas. Available on BRS,(800) 289-4277, DIALOG,(800) 334-2564, and OCLC EPIC,(800) 848-5878, online services. Also available on CD-ROM. Inquire as to cost and availability.

Inspec. Institution of Electrical Engineers, Michael Faraday House, Six Hills Way, Stevenage, Herts. SG1 2AY, England. Telephone: 0438 313311 or FAX 0438 742840. Contains citations tothe worldwide literature of physics, electronics and electrical engineering, computer technology, and related fields. Available on BRS,(800) 289-4277, DIALOG, (800) 334-2564, ORBIT, (800) 456-7248, and STN International, FIZ Karlsruhe, P.O. Box 2465, W-7500, Karlsruhe 1, Germany, online services. Inquire as to cost and availability.

Mathsci. American Mathematical Society, P.O. Box 6248, Providence, RI 02940. (800) 321-4667 or (401) 455-4000 or FAX (401) 331-3842. Scope includes pure and applied mathematics and related areas of physics, statistics, engineering, computer science, and operations research literature since 1959. Available on DIALOG,(800) 334-2564, online service. Also available on CD-ROM. Inquire as to cost and availability.

Microcomputer Index. Learned Information, Inc., 143 Old Marlton Pike, Medford, NJ 08055-8707. (609) 654-6266. Contains citations with abstracts to articles on the use and application of microcomputers and software, from 1981 to date. Available on DIALOG,(800) 334-2564, online service. Inquire as to cost and availability.

NTIS Bibliographic Database. National Technical Information Service, 5285 Port Royal Road, Springfield, Va 22161. (703) 487-4929 or FAX (703) 321-8199. Broad coverage of government-sponsored science and technology research reports, 1964 to present. Available on BRS,(800) 289-4277, DIALOG, (800) 334-2564, ORBIT, (800) 456-7248, and STN International, FIZ Karlsruhe, P.O. Box 2465, W-7500, Karlsruhe 1, Germany, online services. Also available on CD-ROM. Inquire as to cost and availability.

SCISEARCH. Institute for Scientific Information, 3501 Market Street, Philadelphia, PA 19104. (800) 523-1850 or (215) 386-0100. Broad multidisciplinary title and author index to the international literature of science and technology, 1974 to present. Available on DIALOG,(800) 334-2564, and ORBIT,(800) 456-7248, online services. Also available on CD-ROM. Inquire as to cost and availability.

PERIODICALS

ACM Transactions On Computer Systems. Association for Computing Machinery, 11 West 42nd Street, New York, NY 10036. (212) 869-7440. Fax (212) 869-0481. 1964 to present. Annual. $180.00 per year.

Communications of ACM (Association For Computing Machinery) Association For Computing Machinery, 1515 Broadway, 17th Floor, New York, Ny 10036-5701. (212) 869-7440. FAX (212) 944-1318. 1958 to present. Monthly. $114.00 for non-members.

Computer Communications. Butterworth-Heinemann, Linacre House, Jordan Hill, Ocford OX2 8DP, England. Telephone 0865-310366. FAX 0865-310898. 1978 to present. Twelve times a year. Inquire for price in U.S.

Computer Communications Review. Association for Computing Machinery, Special Interest Group on Data Communications, 1515 Broadway, 17th Floor, New York, NY 10036-5701. (212) 869-7440. FAX (212) 944-1318. 1978 to present. Quarterly. $37.00.

Computer Networks and ISDN Systems. Elsevier Science Inc., Box 882, Madison Square Station, New York, NY 10159. (212) 989-5800. FAX (212) 633-3990. 1976 to present. 12 times a year. $579.00.

Data Communications. McGraw-Hill Publishing Company, 1221 Avenue of the Americas, New York, NY 10020. (800) 262-4729 or (212) 512-3825. 1972 to present. Monthly. $95.00 per year.

Datamation. Cahners Publishing Company, 275 Washington Street, Newton, MA 02158-1630. (617) 558-4402. 1957 to present. 24 times a year. $75.00 per year.

IEEE—ACM Transactions On Networking. IEEE, 345 E. 47th Street, New York, NY 10017. (908) 981-0060. FAX (908) 981-9667. 1993 to present. Six times a year. $235.00.

IEEE Network. IEEE, 345 E. 47th Street, New York, NY 10017. (908) 981-0060. FAX (908) 981-9667. 1987 to present. Bi-monthly. $115.00 per year for non-members.

International Journal of Network Management. John Wiley and Sons, Inc., 605 Third Avenue, New York, NY 10158. (800) 526-5368 or (212) 850-6000. 1991 to present. Quarterly. $150.00 per year.

Journal of High-speed Networks. I O S Press, Box 10558, Burke, VA 22009-0559. (703) 323-5554. FAX (703) 250-7405. 1992 to present. Four times a year. $199.00 per year.

Journal of Network and Systems Management. Plenum Publishing Corporation, 233 Spring Street, New York, NY 10013-1578. (212) 620-8000. FAX (212) 463-0742. 1993 to present. Quarterly. $145.00 per year.

L A N. Miller Freeman Inc., 600 Harrison Street, San Francisco, CA 94107. (415) 905-2200. FAX (415) 905-2232. 1986 to present. Monthly. $19.97 per year.

Solid State Communications. Pergamon Press Inc., Maxwell House, Fairview Park, Elmsford, NY 10523. (914) 592-7700. Fax (914) 592-3625. 1963 to present. Weekly. $165.00 per year.

RESEARCH CENTERS AND INSTITUTES

Harvard University Center for Research in Computing Technology. Aiken Computation Laboratory, 33 Oxford Street, Cambridge, MA 02138. (617) 495-4117. FAX (617) 495-9837.

Ohio State University Computer and Information Science Research Center. 2036 Neil Avenue Mall, Columbus, OH 43210. (614) 292-6374. FAX (614) 292-9021.

SPECIFICATIONS AND STANDARDS

ADP and Telecommunications Standards Index. U.S. General Services Administration. Order from: Superintendent of Documents, U.S. Government Printing office, Box 371954, Pittsburgh, PA 15250-7954. (202) 783-3238. FAX (202) 512-2233. Semi-annual. $8.00.

COMPUTER GRAPHICS

See also: AUTOMATION, COMPUTER MEMORY, COMPUTER OPERATING SYSTEMS, COMPUTER PROGRAMMING, COMPUTER PROGRAMMING LANGUAGES, COMPUTER VISION, PARALLEL COMPUTERS

ABSTRACT SERVICES AND INDEXES

ACM Guide To Computing Literature. Association for Computing Machinery, Association for Computing Machinery, 1515 Broadway, 17th Floor, New York, NY 10036-5701. (212) 869-7440. FAX (212) 944-1318. 1964 to present. Monthly. $190.00 per year for non-members.

Applied Science and Technology Index; A Cumulative Subject Index To English Language Periodicals In the Fields of Aeronautics and Space Science, Computer Technology, Chemistry, Construction Industry, Energy and Related Areas. H.W. Wilson Co., 950 University Avenue, Bronx, NY 10452. (800) 367-6770 or (212) 588-8400. FAX (718) 590-1617. From 1958 to present. Monthly. Inquire about cost and availability. Also available on CD-ROM and online.

COMPUTER GRAPHICS

Ency. of Physical Sciences and Engineering Info. Sources

Computer Abstracts, MCBUniversity Press Ltd., PO Box 10812, Birmingham, AL 35201-0812. (800) 633-4931. FAX (205) 995-1588. Monthly. Covers computer theory, data, hardware, systems, networks, human-computer interaction, artificial intelligence, as well as applications of computers in aerospace, business, CAD/CAM, cartography, civil engineering, electronics and electrical engineering, industrial engineering, mechanical engineering, medicine, structural engineering, etc. Also available on CD-ROM. 1957 to present. Monthly. $1399.95 per year.

Computer & Control Abstracts (inspec). INSPEC/IEEE, Box 1331, 445 Hoes Lane, Piscataway, NJ 08855-1331. (908) 562-5549. Abstracts organized by subjects of international technological information. 1966 to present. Monthly. $1455.00 per year. Also available on CD-ROM and online via BRS Online Products, DIALOG Information Services, Orbit Search Service, and STN International.

Computer and Information Systems Abstracts Journal. Cambridge Scientific Abstracts, 7200 Wisconsin Avenue, Bethesda, MD 20814. (301) 961-6750. Fax (301) 961-6720. 1962 to present. 1962 to present. Monthly. $1265.00 per year. Also available online via STN International.

Computer Literature Index. Applied Computer Research Inc., Box 82266, Phoenix, AZ 85071-2266. (800) 234-2227. 1971 to present. Quarterly plus annual cumulation. $198.50 per year. Bibliography of books, articles, and reports.

Computing Journal Abstracts. Techgnosis Ltd., Blade House, Battersea Road, Stockport, Cheshire 3AE, England. Telephone 061-442-2639. FAX 061-443-1162. 1969 to present. Monthly. Inquire for price.

Computing Reviews. Association for Computing Machinery, 1515 Broadway, 17th Floor, New York, NY 10036-5701. (212) 869-7440. FAX (212) 944-1318. 1960 to present. Monthly. $130.00 per year for non-members. Also available online via DIALOG Information Services.

Current Contents: Engineering, Technology, and Applied Sciences. Institute for Scientific Information, 3501 Market Street, Philadelphia, PA 19104. (215) 386-0100. FAX (215) 386-2991. 1970 topresent. Weekly. $442.00 per year.

Engineering Index Monthly. Engineering Information, Inc., Castle Point on the Hudson, Hoboken, NJ 07030. (800) 221-1044. FAX (212) 832-1857. Monthly. $2300.00 per year. Also available online as COMPENDEX, and also on CD-ROM. Covers chemical engineering, computers, electrical engineering, civil engineering, metals and mining, industrial management, and mechanical engineering.

Science Citation Index. Institute for Scientific Information, 3501 Market Street, Philadelphia, PA 19104. (215) 386-0100. FAX (215) 386-6362. Inquire about availability and cost. Also available on CD-ROM.

ANNUAL REVIEWS AND YEARBOOKS

ACM Monograph Series. Association for Computing Machinery (ACM), 11 West 42nd Street, New York, NY 10036. (212) 869-7440. Fax (212) 869-0481. Irregular. Price varies, inquire.

Advances In Computers. Academic Press Inc., 6277 Sea Harbor Drive, Orlando, FL 32887. (800) 321-5068. Irregular. Price varies, inquire.

Advances In Computing Research. JAI Press, Inc., 55 Old Post Road, Number 2, Box 1678, Greenwich, CT 06830. (203) 661-7602. FAX (203) 661-0792. Annual. $68.50 per year.

ASSOCIATIONS AND PROFESSIONAL SOCIETIES

American Federation of Information Processing Societies, 1899 Preston White Drive, Reston, Va 22091-5435. (703) 620-8937.

Association For Computing Machinery. 1515 Broadway, 17th Floor, New York, NY 10036-5701. (212) 869-7440. FAX (212) 944-1318.

Computer and Automation Systems Association of Sme. 1 SME Drive, PO Box 930, Dearborn, MI 48121-0930. (313) 271-1500.

Institute of Electrical and Electronic Engineers. 345 E. 47th Street, New York, NY 10017. (212) 705-7900.

National Computer Graphics Association. 2722 Merrilee Drive, Suite 200, Fairfax, VA 22031-4499. (703) 698-9600. FAX (703) 560-2752.

Society For Computer Simulation International, 4838 Ronson Court Lane, San Diego, Ca 92111-1810. (619) 277-3888.

Special Interest Group On Computer Graphics. c/o ACM, 1515 Broadway, New York, NY 10036. (212) 869-7440. FAX (212) 869-0481.

World Computer Graphics Association. 5201 Leesburg Pike, Suite 201, Falls Church, VA 22041-3203. (703) 578-0301. FAX (703) 578-3386.

BIBLIOGRAPHIES

ACM Guide To Computing Literature. Association for Computing Machinery, 1515 Broadway, 17th Floor, New York, NY 10036-5701. (212) 869-7440. FAX (212) 944-1318. 1964 to present. Annual. $190.00 for non-members. Also available online via DIALOG.

DIRECTORIES AND BIOGRAPHICAL SOURCES

ACM Adminstrative Directory of College and University Computer Science-Data Processing Programs and Computer Facilities. Association for Computing Machinery, 1515 Broadway, 17th Floor, NewYork, NY 10036-5701. (212) 869-7440. FAX (212) 944-1318. Biennial. $17.00 for non-members.

American Men & Women of Science, 1995-96. R.R. Bowker Staff, eds. 19th edition. 8 volumes. R.R. Bowker/Reed International Publishing Company, 121 Chanlon Road, New Providence, NJ 07974. (908) 464-6800 or (800) 521-8110. 1995. $850.00

ASIS Handbook and Directory, American Society For Information Science, 8720 Georgia Avenue, Suite 501, Silver Spring, Md 20910-3602. (301) 495-0900. Annual. $100.00.

Computer Graphics World Buyers Guide. PennWell Publishing Company, Advanced Technology Group, 10 Tara Blvd., 5th floor, Nashua, NH 03062-2801. (603) 891-9111. FAX (603) 891-0539. Annual. $100.00 per year.

Computing Information Directory. Hildebrandt Inc., PO Box 576, Pullman, WA 99114. 1981 to present. Annual. $199.95 per year.

International Research Centers Directory. Gale Research, 835 Penobscot Building, Detroit, MI 48226-4094. (313) 961-2242. (800) 347-4253. FAX (313) 961-6083. 8th edition. 1995. $410.00

Research Centers Directory. Gale Research, 835 Penobscot Building, Detroit, MI 48226-4094. (313) 961-2242. (800) 347-4253. FAX (313) 961-6083. $485.00.

Who's Who In Computing. Canadian M I S Database Inc., 268 Lakeshore Road E., Suite 510, Mississauga, ON L5G 1H1, Canada. (905) 271-1601. FAX (905) 271-4522. 1987 to present. Annual. Inquire for price in U.S.

Who's Who In Technology. Gale Research, 835 Penobscot Building, Detroit, MI 48226-4094. (313) 961-2242. (800) 347-4253. FAX (313) 961-6083. 1995. $195.00.

ENCYCLOPEDIAS AND DICTIONARIES

Computer Dictionary. Donald D. Spencer. 4th edition. Camelot Publishing Company, 709 SW 80th Blvd., Gainesville, FL 32607-1537. (904) 331-0952. 1993. $24.95.

The Dictionary of Computer Graphics Technology. Roy Latham. Second edition. Springer-Verlag, 44 Hartz Way, Secaucus, NJ 07096-2491. (201) 348-4033. FAX (201) 348-4505. 1995. $19.00.

Dictionary of Computing. S.M.H. Collin. 2nd edition. Peter Collin Publishing, 8 The Causeway, Teddington, TW11 0HE, England. FAX 0181-943-3386. 1994. Inquire for cost.

Encyclopedia of Computer Science and Engineering. Anthony Ralston & Edwin D. Reilly Jr., editors. 3rd revised edition. Van Nostrand Reinhold, 115 Fifth Avenue, New York, NY 10003. (212) 254-3232. 1993. $125.00.

Microsoft Press Computer Dictionary: the Comprehensive Standard For Business, School, Library, and Home. 2nd edition. Microsoft Press, One Microsoft Way, Redmond, WA 98052-6399. (206) 936-0055. FAX (206) 823-8101. 1994. $19.95.

GENERAL WORKS

Computer-Aided Graphics and Design. Daniel L. Ryan. Third revised and expanded edition. Marcel Dekker, Inc., 270 Madison Avenue, New York, NY 10016. (212) 696-9000. FAX (212) 685-4540. 1994. $59.75.

Computer Graphics For Industrial Applications. Richard S. Shirley. Prentice Hall, 113 Sylvan Avenue, Route 9W, Englewood Cliffs, NJ 07632. (201) 592-2000 or (800) 922-0579. 1992. $65.00

Computer Graphics Fundamentals. James L. Pomeroy. Palms & Rhodes Publishing Company, 4274 S. Hazelcrest Drive, Las Vegas, NV 89121. (702) 456-4274. 1993. $48.00.

Computer Graphics: Principles and Practice. James D. Foley, et al. Second edition. Addison-Wesley Publishing Company, 1 Jacob Way, Reading, MA 01867. (617) 944-3700 or (800) 447-2226. FAX 617-942-1117. 1996. Inquire for price.

From Pixels To Animation: An Introduction To Graphics Programming. James A. Farrell. Academic Press Inc., 6277 Sea Harbor Drive, Orlando, FL 32887. (800) 321-5068. 1994. $39.95.

Introduction To Computer Graphics. James D. Foley. Addison-Wesley Publishing Company, 1 Jacob Way, Reading, MA 01867. (617) 944-3700 or (800) 447-2226. FAX 617-942-1117. 1994. $45.95.

HANDBOOKS AND MANUALS

McGraw-Hill Computing Essentials, 1993-1994. Timothy J. O'Leary, et al. McGraw-Hill Publishing Company, 1221 Avenue of the Americas, New York, NY 10020. (800) 262-4729 or (212) 512-3825. 1994. Inquire for price and availability.

ONLINE DATABASES AND CD-ROMS

Compendex Plus. Engineering Information, Inc., 345 East 47th Street, New York, NY 10017. (212) 705-7600 or (800) 221-1044. Contains citations with abstracts to worldwide literature in engineering and technology, from 1970 to present. Available on online BRS, (800) 289-4277, DIALOG, (800) 334-2564, ORBIT (800) 456-7248, and STN International, FIZ Karlsruhe, P.O. Box 2465, W-7500, Karlsruhe 1, Germany, online services. Also available on CD-ROM. Inquire as to cost and availability.

Compuscience. FIZ Karlsruhe, D-7514, Eggenstein-Leopoldshafen 2, Germany. Contains citations with abstracts to European and North American literature on computer science, 1972 to present. Available on STN International, FIZ Karlsruhe, P.O. Box 2465, W-7500, Karlsruhe 1, Germany, online service. Inquire as to cost and availability.

Computer and Information Systems Abstracts. Cambridge Scientific Abstracts, 7200 Wisconsin Avenue, Suite 601, Bethesda, MD 20814. (301) 961-6750 or (800) 843-7751. Contains citations to worldwide literature in theoretical and applied computer science and related areas, from 1981 to present. Inquire as to cost and availability.

Computer and Mathematics Search. Institute for Scientific Information, 3501 Market Street, Philadelphia, PA 19104. (215) 386-0100. FAX (215) 386-6362. Covers worldwide literature in computer science and mathematics, from 1980 to present. Available on BRS, (800) 289-4277, online service. Inquire as to cost and availability.

Computer Database. Information Access Company, 362 Lakeside Drive, Foster City, CA 94404. (415) 378-5000 or (800) 227-8431. Contains citations with abstracts to literature from trade journals covering the computer, telecommunications, and electronics industries. Available on the BRS, (800) 289-4277, Compuserve Information Service, (800) 848-8990, and DIALOG, (800) 334-2564, online services. Inquire as to cost and availability.

CSA Engineering. Cambridge Scientific Abstracts, 7200 Wisconsin Avenue, Suite 601, Bethesda, MD 20814. (301) 961-6750 or (800) 843-7751. Contains citations and abstracts of international periodicals and research literature covering all fields of engineering and science and technology, including computer and information science, electronics, mechanical engineering,

solid state materials, 1981 to present. Available on BRS,(800) 289-4277, online service.Inquire as to cost and availability.

Dissertation Abstracts. University Microfilms International, 300 North Zeeb Road, Ann Arbor, MI 48106. (800) 521-0600 or (313) 761-4700. Scope includes virtually all doctoral dissertations accepted at accredited American institutions from 1861 to present in 252 subject areas. Available on BRS,(800) 289-4277, DIALOG,(800) 334-2564, and OCLC EPIC,(800) 848-5878, online services. Also available on CD-ROM. Inquire as to cost and availability.

Mathsci. American Mathematical Society, P.O. Box 6248, Providence, RI 02940. (800) 321-4667 or (401) 455-4000 or FAX (401) 331-3842. Scope includes pure and applied mathematics and related areas of physics, statistics, engineering, computer science, and operations research literature since 1959. Available on DIALOG,(800) 334-2564, online service. Also available on CD-ROM. Inquire as to cost and availability.

NTIS Bibliographic Database. National Technical Information Service, 5285 Port Royal Road, Springfield, Va 22161. (703) 487-4929 or FAX (703) 321-8199. Broad coverage of government-sponsored science and technology research reports, 1964 to present. Available on BRS,(800) 289-4277, DIALOG, (800) 334-2564, ORBIT, (800) 456-7248, and STN International, FIZ Karlsruhe, P.O. Box 2465, W-7500, Karlsruhe 1, Germany, online services. Also available on CD-ROM. Inquire as to cost and availability.

SCISEARCH. Institute for Scientific Information, 3501 Market Street, Philadelphia, PA 19104. (800) 523-1850 or (215) 386-0100. Broad multidisciplinary title and author index to the international literature of science and technology, 1974 to present. Available on DIALOG,(800) 334-2564, and ORBIT,(800) 456-7248, online services. Also available on CD-ROM. Inquire as to cost and availability.

PERIODICALS

ACM Transactions On Computer Systems. Association for Computing Machinery, 11 West 42nd Street, New York, NY 10036. (212) 869-7440. Fax (212) 869-0481. 1964 to present. Annual. $180.00 per year.

ACM Transactions On Graphics. Association for Computing Machinery, 1515 Broadway, 17th Floor, New York, NY 10036-5701. (212) 869-7440. FAX (212) 944-1318. 1982 to present. Quarterly. $110.00.

CAD-CAM Update. Worldwide Videotex, Box 3273, Boynton Beach, FL 33424-3273. (407) 738-2276. 1989 to present. Monthly. Inquire for price.

CAD Systems. Kerrwil Publications Ltd., 395 Matheson Blvd. E, Mississauga, Ontario L4Z 2H2, Canada. (905) 890-1846. FAX (905) 890-5769. 1982 to present. Six times a year. $61.00 per year.

CAD User. B.T.C., 24 High Street, Beckenham, Kent BR3 1AY, England. Telephone 081-663-3818. FAX 081-663-3822. 1988 to present. Monthly. Inquire for price.

CG Professional. Graphic Channels Inc., 1714 Stockton Street, San Francisco, CA 94133. (415) 956-5350. 1989 to present. $89.00 per year.

Computer-Aided Design. Butterworth-Heinemann/sub. to Turpin Transactions Ltd., Distribution Centre, Blackhorse Road, Letchworth, Herts SG6 1HN, England. Telephone 0462-672555. 1968 to present. 12 times a year. Inquire for price.

Computer-Aided Design Report. CAD-CAM Publishing Inc., 841 Turquoise Street, Suites D & E, San Diego, CA 92109-1159. (619) 488-0533. FAX (619) 488-6052. 1981 to present. Monthly. $190.00 per year.

Computer-Aided Geometric Design. Elsevier Science, Box 882, Madison Square Station, New York, NY 10159. (212) 989-5800. FAX (212) 633-3990. 1984 to present. Six times a year. $322.00.

Computer Design. PennWell Publishing Company, Advanced Technology Group, 10 Tara Blvd., 5th floor, Nashua, NH 03062-2801. (603) 891-9111. FAX (603) 891-0539. 1962 to present. 12 times a year. $88.00 per year.

Computer Graphics Forum. Basil Blackwell Ltd., 108 Cowley Road, Oxford OX4 1JF, England. Telephone 0865-791100. FAX 0865-791347. 1982 to present. Five times a year. $305.00 per year.

Computer Graphics World. PennWell Publishing Company, Advanced Technology Group, 10 Tara Blvd., 5th floor, Nashua, NH 03062-2801. (603) 891-9111. FAX (603) 891-0539. 1978 to present. Monthly. $48.00 per year.

Computers & Graphics. Elsevier Science, 660 White Plains Road, Tarrytown, NY 10159-5153. (919) 524-9200. FAX (919) 333-2444. 1975 to present. Six times a year. $780.00 per year.

IEEE Computer Graphics and Applications. IEEE Computer Society Press, 10662 Los Vaqueros Circle, Box 3014, Los Alamitos, CA 90720-1264. (714) 821-8380. FAX (714) 821-4641. 1981 to present. Bi-monthly. $175.00 per year.

International Journal of Computer Vision. Kluwer Academic Publishers, P.O. Box 358, Accord Station, Hingham, MA 02018-0358. (617) 871-6000. 1987 to present. Six times a year. $314.00 per year.

Journal of Visual Communication and Image Representation. Academic Press Inc., Journal Division, 525 B Street, Suite 1900, San Diego, CA 92101-4495. (619) 230-1940. FAX (619) 699-6800. 1990 to present. Quarterly. $158.00 per year.

Multimedia Week. Phillips Business Information Inc., 1201 Seven Locks Road, Potomac, MD 20854. (301) 424-3338. FAX (301) 309-3847. 1992 to present. 50 times a year. $597.00 per year.

Pattern Recognition Letters. Elsevier Science Inc., Box 882, Madison Square Station, New York, NY 10159. (212) 989-5800. FAX (212) 633-3990. 1983 to present. 12 times a year. $798.00 per year.

Visual Computer. Springer-Verlag, 44 Hartz Way, Secaucus, NJ 07096-2491. (201) 348-4033. FAX (201) 348-4505. 1985 to present. Eight times a year. $484.00 per year.

RESEARCH CENTERS AND INSTITUTES

Cornell University Program of Computer Graphics. 120 Rand Hall, Ithaca, NY 14853-5501. (607) 255-4880.

Harvard University Center For Research In Computing Technology. Aiken Computation Laboratory, 33 Oxford Street, Cambridge, MA 02138. (617) 495-4117. FAX (617) 495-9837.

New York Institute of Technology, Computer Graphics Laboratory. Old Westbury, NY 11568. (516) 686-7644.

Ohio State University Computer Graphics Research Group. Advanced Computing Center for Arts and Design, 1224 Kinnear Road, Columbus, OH 43212. (614) 292-3416.

Syracuse University Advanced Graphics Research Laboratory. 215 Machinery Hall, Syracuse, NY 13244-1260. (315) 443-3525.

COMPUTER HACKERS

See: COMPUTER SECURITY

COMPUTER MEMORY AND STORAGE

See also: AUTOMATION, COMPUTER OPERATING SYSTEMS, COMPUTER PROGRAMMING, COMPUTER PROGRAMMING LANGUAGES, PARALLEL COMPUTERS

ABSTRACT SERVICES AND INDEXES

ACM Guide To Computing Literature. Association for Computing Machinery, Association for Computing Machinery, 1515 Broadway, 17th Floor, New York, NY 10036-5701. (212) 869-7440. FAX (212) 944-1318. 1964 to present. Monthly. $190.00 per year for non-members.

Applied Science and Technology Index; A Cumulative Subject Index To English Language Periodicals In the Fields of Aeronautics and Space Science, Computer Technology, Chemistry, Construction Industry, Energy and Related Areas. H.W. Wilson Co., 950 University Avenue, Bronx, NY 10452. (800) 367-6770 or (212) 588-8400. FAX (718) 590-1617. From 1958 to present. Monthly. Inquire about cost and availability. Also available on CD-ROM and online.

Computer Abstracts, MCBUniversity Press Ltd., PO Box 10812, Birmingham, AL 35201-0812. (800) 633-4931. FAX (205) 995-1588. Monthly. Covers computer theory, data, hardware, systems, networks, human-computer interaction, artificial intelligence, aswell as applications of computers in aerospace, business, CAD/CAM, cartography, civil engineering, electronics and electrical engineering, industrial engineering, mechanical engineering, medicine, structural engineering, etc. Also available on CD-ROM.

Computer & Control Abstracts (inspec). INSPEC/IEEE, Box 1331, 445 Hoes Lane, Piscataway, NJ 08855-1331. (908) 562-5549. Abstracts organized by subjects of international technological information. Monthly. $1455.00 per year. Also available on CD-ROM and online via BRS Online Products, DIALOG Information Services, Orbit Search Service, and STN International.

Computer and Information Systems Abstracts Journal. Cambridge Scientific Abstracts, 7200 Wisconsin Avenue, Bethesda, MD 20814. (301) 961-6750. Fax (301) 961-6720. 1962 to present. Monthly. $1265.00 per year. Also available online via STN International.

Computer Literature Index. Applied Computer Research Inc., Box 82266, Phoenix, AZ 85071-2266. (800) 234-2227. 1971 to present. Quarterly plus annual cumulation. $198.50 per year. Bibliography of books, articles, and reports.

Computing Journal Abstracts. Techgnosis Ltd., Blade House, Battersea Road, Stockport, Cheshire 3AE, England. Telephone 061-442-2639. FAX 061-443-1162. 1969 to present. Monthly. Inquire for price.

Computing Reviews. Association for Computing Machinery, 1515 Broadway, 17th Floor, New York, NY 10036-5701. (212) 869-7440. FAX (212) 944-1318. 1960 to present. Monthly. $130.00 per year for non-members. Also available online via DIALOG Information Services.

Current Contents: Engineering, Technology, and Applied Sciences. Institute for Scientific Information, 3501 Market Street, Philadelphia, PA 19104. (215) 386-0100. FAX (215) 386-6362. 1970 to present. Weekly. $442.00 per year.

Engineering Index Monthly. Engineering Information, Inc., Castle Point on the Hudson, Hoboken, NJ 07030. (800) 221-1044. FAX (201) 216-8532. Monthly. $2300.00 per year. Also available online as COMPENDEX, and also on CD-ROM. Covers chemical engineering, computers, electrical engineering, civil engineering, metals and mining, industrial management, and mechanical engineering.

Key Abstracts—Computer Communications and Storage. INSPEC/IEEE, Box 1331, 445 Hoes Lane, Piscataway, NJ 08855-1331. (908) 562-5549. 1987 to present. Monthly. $160.00 per year. Covers multiprocessor systems, storage media, optical disc, interfaces, networks, and network equipment.

Science Citation Index. Institute for Scientific Information, 3501 Market Street, Philadelphia, PA 19104. (215) 386-0100. FAX (215) 386-6362. Inquire about availability and cost. Also available on CD-ROM.

ANNUAL REVIEWS AND YEARBOOKS

ACM Monograph Series. Association for Computing Machinery (ACM), 11 West 42nd Street, New York, NY 10036. (212) 869-7440. Fax (212) 869-0481. Irregular. Price varies, inquire.

Advances In Computers. Academic Press, Inc., 1250 Sixth Avenue, San Diego, CA 92101. (619) 231-0926. FAX (619) 699-6715. Irregular. Price varies, inquire.

Advances In Computing Research. JAI Press, Inc., 55 Old Post Road, Number 2, Box 1678, Greenwich, CT 06830. (203) 661-7602. Annual. $68.50 per year.

ASSOCIATIONS AND PROFESSIONAL SOCIETIES

American Federation of Information Processing Societies, 1899 Preston White Drive, Reston, Va 22091-5435. (703) 620-8937.

Association For Computing Machinery. 1515 Broadway, 17th Floor, New York, NY 10036-5701. (212) 869-7440. FAX (212) 944-1318.

COMPUTER MEMORY AND STORAGE

Ency. of Physical Sciences and Engineering Info. Sources

Computer and Automation Systems Association of Sme. 1 SME Drive, PO Box 930, Dearborn, MI 48121-0930. (313) 271-1500.

Institute of Electrical and Electronic Engineers. 345 E. 47th Street, New York, NY 10017. (212) 705-7900.

IEEE Computer Society. 1730 Massachusetts Avenue NW, Washington, DC 20036-1992. (202) 371-0101. FAX (202) 728-9614.

Special Interest Group For Information Retrieval. c/o ACM, 1515 Broadway, New York, NY 10036. (212) 869-7440. FAX (212) 869-0481.

BIBLIOGRAPHIES

ACM Guide To Computing Literature. Association for Computing Machinery, 1515 Broadway, 17th Floor, New York, NY 10036-5701. (212) 869-7440. FAX (212) 944-1318. 1964 to present. Annual. $190.00 for non-members. Also available online via DIALOG.

DIRECTORIES AND BIOGRAPHICAL SOURCES

ACM Adminstrative Directory of College and University Computer Science-data Processing Programs and Computer Facilities. Association for Computing Machinery, 1515 Broadway, 17th Floor, New York, NY 10036-5701. (212) 869-7440. FAX (212) 944-1318. Biennial. $17.00 for non-members.

ASIS Handbook and Directory, American Society For Information Science, 8720 Georgia Avenue, Suite 501, Silver Spring, Md 20910-3602. (301) 495-0900. Annual. $100.00.

Computing Information Directory. Hildebrandt Inc., PO Box 576, Pullman, WA 99114. 1981 to present. Annual. $199.95 per year.

International Research Centers Directory. Gale Research, 835 Penobscot Building, Detroit, MI 48226-4094. (313) 961-2242. (800) 347-4253. FAX (313) 961-6083. 8th edition. 1995. $410.00.

Research Centers Directory. Gale Research, 835 Penobscot Building, Detroit, MI 48226-4094. (313) 961-2242. (800) 347-4253. FAX (313) 961-6083. $485.00.

Who's Who In Computing. Canadian M I S Database Inc., 268 Lakeshore Road E., Suite 510, Mississauga, ON L5G 1H1, Canada. (905) 271-1601. FAX (905) 271-4522. 1987 to present. Annual. Inquire for price in U.S.

Who's Who In Technology. Gale Research, 835 Penobscot Building, Detroit, MI 48226-4094. (313) 961-2242. (800) 347-4253. FAX (313) 961-6083. 1995. $195.00.

ENCYCLOPEDIAS AND DICTIONARIES

Computer Dictionary. Donald D. Spencer. 4th edition. Camelot Publishing Company, 709 SW 80th Blvd., Gainesville, FL 32607-1537. (904) 331-0952. 1993. $24.95.

Dictionary of Computing. S.M.H. Collin. 2nd edition. Peter Collin Publishing, 8 the Causeway, Teddington, TW11 0HE, England. FAX 0181-943-3386. 1994. Inquire for cost.

Encyclopedia of Computer Science and Engineering. Anthony Ralston & Edwin D. Reilly Jr., editors. 3rd revised edition. Van Nostrand Reinhold, 115 Fifth Avenue, New York, NY 10003. (212) 254-3232 or (800) 926-2665. FAX (212) 254-9499. 1993. $125.00.

Microsoft Press Computer Dictionary: the Comprehensive Standard For Business, School, Library, and Home. 2nd edition. Microsoft Press, One Microsoft Way, Redmond, WA 98052-6399. (206) 936-0055. FAX (206) 823-8101. 1994. $19.95.

GENERAL WORKS

Art of Data Recording. John Watkinson. Butterworth-Heinemann, 313 Washington Street, Newton, MA 02158. (617) 928-2500 or (800) 366-2665. FAX (617) 928-2620. 1994. $69.95.

The Cache Memory Book. Jim Handy. Academic Press, Inc., 6277 Sea Harbor Drive, Orlando, FL 32887. 1993. $44.95.

Mass Storage Guide. Rector Press Ltd., 130 Rattlesnake, Leverett, MA 01054-9726. (800) 247-3473. FAX (413) 367-2853. 1995. $125.00.

New Optical Storage Technology: Including Multimedia, Cd-rom, and Optical Drives. John A. McCormick. Second edition. Irwin Professional Publishing, 1333 Burr Ridge Parkway, Burr Ridge, IL 60521. (800) 634-3961. FAX (800) 926-9495. 1993. $30.00.

Welcome To Memory Management. Phillip Robinson. Henry Holt and Company, Inc., 115 West 18th Street, New York, NY 10011. (212) 886-9200 or (800) 488-5233. FAX (212) 633-0748. 1994. $19.95

HANDBOOKS AND MANUALS

Memory Data Handbook. Rector Press Ltd., 130 Rattlesnake, Leverett, MA 01054-9726. (800) 247-3473. FAX (413) 367-2853. 1995. $125.00.

ONLINE DATABASES AND CD-ROMS

Compendex Plus. Engineering Information, Inc., 345 East 47th Street, New York, NY 10017. (212) 705-7600 or (800) 221-1044. Contains citations with abstracts to worldwide literature in engineering and technology, from 1970 to present. Available on online BRS,(800) 289-4277, DIALOG, (800) 334-2564, ORBIT (800) 456-7248, and STN International, FIZ Karlsruhe, P.O. Box 2465, W-7500, Karlsruhe 1, Germany, online services. Also available on CD-ROM. Inquire as to cost and availability.

Compuscience. FIZ Karlsruhe, D-7514, Eggenstein-Leopoldshafen 2, Germany. Contains citations with abstracts to European and North American literature on computer science, 1972 to present. Available on STN International, FIZ Karlsruhe, P.O. Box 2465, W-7500, Karlsruhe 1, Germany, online service. Inquire as to cost and availability.

Computer and Information Systems Abstracts. CambridgeScientific Abstracts, 7200 Wisconsin Avenue, Suite 601, Bethesda, MD 20814. (301) 961-6750 or (800) 843-7751. Contains citations to worldwide literature in theoretical and applied computer science and related areas, from 1981 to present.Inquire as to cost and availability.

Computer and Mathematics Search. Institute for Scientific Information, 3501 Market Street, Philadelphia, PA 19104. (215) 386-0100. FAX (215) 386-6362. Covers worldwide literature in computer science and mathematics, from 1980 to present. Available on BRS,(800) 289-4277, online service. Inquire as to cost and availability.

Computer Database. Information Access Company, 362 Lakeside Drive, Foster City, CA 94404. (415) 378-5000 or (800) 227-8431. Contains citations with abstracts to literature from trade journals covering the computer,telecommunications,and electronics industries. Available on the BRS, (800) 289-4277, Compuserve Information Service,(800) 848-8990, and DIALOG,(800) 334-2564, online services. Inquire as to cost and availability.

CSA Engineering. Cambridge Scientific Abstracts, 7200 Wisconsin Avenue, Suite 601, Bethesda, MD 20814. (301) 961-6750 or (800) 843-7751. Contains citations and abstracts of international periodicals and research literature covering all fields of engineering and science and technology,including computer and information science, electronics, mechanical engineering, solid state materials, 1981 to present. Available on BRS,(800) 289-4277, online service.Inquire as to cost and availability.

Current Contents Search. Institute for Scientific Information, 3501 Market Street, Philadelphia, PA 19104. (215) 386-0100. FAX (215) 386-6362. Contains citations to articles listed in the table of contents of science and technology journals. Also articles in social sciences and life sciences journals. Available on BRS,(800) 289-4277, DIALOG,(800) 334-2564, online services. Inquire as to cost and availability.

Dissertation Abstracts. University Microfilms International, 300 North Zeeb Road, Ann Arbor, MI 48106. (800) 521-0600 or (313) 761-4700. Scope includes virtually all doctoral dissertations accepted at accredited American institutions from 1861 to present in 252 subject areas. Available on BRS,(800) 289-4277, DIALOG,(800) 334-2564, and OCLC EPIC,(800) 848-5878, online services. Also available on CD-ROM. Inquire as to cost and availability.

Information Science Abstracts. IFI/Plenum Data Company, 302 Swann Avenue, Alexandria, VA 22301. (703) 683-1085 or (800) 368-3093. Contains citations with abstracts to worldwide literature on information science, storage, retrieval and utilization, from 1966 to present.Available on DIALOG,(800) 334-2564, online service. Inquire as to cost and availability.

Inspec. Institution of Electrical Engineers, Michael Faraday House, Six Hills Way, Stevenage, Herts. SG1 2AY, England. Telephone: 0438 313311 or FAX 0438 742840. Contains citations tothe worldwide literature of physics, electronics and electrical engineering, computer technology, and related fields. Available on BRS,(800) 289-4277, DIALOG, (800) 334-2564, ORBIT, (800) 456-7248, and STN International, FIZ Karlsruhe, P.O. Box 2465, W-7500, Karlsruhe 1, Germany, online services. Inquire as to cost and availability.

Mathsci. American Mathematical Society, P.O. Box 6248, Providence, RI 02940. (800) 321-4667 or (401) 455-4000 or FAX (401) 331-3842. Scope includes pure and applied mathematics and related areas of physics, statistics, engineering, computer science, and operations research literature since 1959. Available on DIALOG,(800) 334-2564, online service. Also available on CD-ROM. Inquire as to cost and availability.

NTIS Bibliographic Database. National Technical Information Service, 5285 Port Royal Road, Springfield, Va 22161. (703) 487-4929 or FAX (703) 321-8199. Broad coverage of government-sponsored science and technology research reports, 1964 to present. Available on BRS,(800) 289-4277, DIALOG, (800) 334-2564, ORBIT, (800) 456-7248, and STN International, FIZ Karlsruhe, P.O. Box 2465, W-7500, Karlsruhe 1, Germany, online services. Also available on CD-ROM. Inquire as to cost and availability.

SCISEARCH. Institute for Scientific Information, 3501 Market Street, Philadelphia, PA 19104. (800) 523-1850 or (215) 386-0100. Broad multidisciplinary title and author index to the international literature of science and technology, 1974 to present. Available on DIALOG,(800) 334-2564, and ORBIT,(800) 456-7248, online services. Also available on CD-ROM. Inquire as to cost and availability.

WILSONLINE. H.W. Wilson Company, 950 University Avenue, Bronx, NY 10452. (800) 367-6770 or (212) 588-8400. Makes available online versions of the printed H.W. Wilson indexes including Applied Science and Technology Index, Business Periodicals Index, General Science Index, and Readers' Guide to Periodical Literature. Period covered is generally 1983 to present. Available on BRS,(800) 289-4277, DIALOG,(800) 334-2564, and OCLC EPIC,(800) 848-5878, online services. Also available on CD-ROM. Inquire as to cost and availability.

PERIODICALS

ACM Transactions On Computer Systems. Association for Computing Machinery, 1515 Broadway, 17th Floor, New York, NY 10036-5701. (212) 869-7440. FAX (212) 944-1318. 1983 to present. Quarterly. $110.00 for non-members.

Byte. McGraw-Hill Inc., Byte Publications, One Phoenix Mill Lane, Peterborough, NH 03458. (603) 924-9281. FAX (603) 924-2550. 1975 to present. Monthly. $29.95 per year.

Communications of the A C M. Association for Computing Machinery, 1515 Broadway, 17th Floor, New York, NY 10036-5701. (212) 869-7440. FAX (212) 944-1318. 1958 to present. Monthly. $114.00 for non-members.

Computer Design. PennWell Publishing Company, 1 Technology Park Drive, Westford, MA 01886. (508) 692-0700 or FAX (508) 692-0525. 1962 to present. Semi-monthly. $70.00 per year.

Computer Journal. Cambridge University Press, 40 West 20th Street, New York, NY 10011-4211. (212) 924-3900. Bimonthly. $315.00 per year.

Computer Magazine. IEEE Computer Society, 10662 Vaqueros Circle, Box 3014, Los Alamitos, CA 90720. (714) 821-8380. 1966 to present. Monthly. $49.00 per year.

Computer Technology Review. West World Productions, Inc., 924 Westwood Boulevard, Suite 650, Los Angeles, CA 90024-2927. (213) 208-1335 or FAX (213) 208-1054. 1981 to present. Sixteen times per year. $80.00 per year.

Datamation. Cahners Publishing Company, 275 Washington Street, Newton, MA 02158-1630. (617) 558-4402. 1957 to present. 24 times a year. $75.00 per year.

COMPUTER MEMORY AND STORAGE

Ency. of Physical Sciences and Engineering Info. Sources

IEEE Transactions On Computers. IEEE, 345 E. 47th Street, New York, NY 10017. (908) 981-0060. FAX (908) 981-9667. 1952 to present. Monthly. $425.00 for non-members.

Journal of the Association For Computing Machinery. Association for Computing Machinery, 11 West 42nd Street, New York, NY 10036. (212) 869-7440. Fax (212) 869-0481. Quarterly. $85.00 per

RESEARCH CENTERS AND INSTITUTES

Carnegie Mellon University, Data Storage Systems Center. Schenley Park, Pittsburgh, PA 15213. (412) 268-6600. FAX (412) 268-3497.

University of California—San Diego, Center For Magnetic Recording Research. Mail Code R-001, La Jolla, CA 92093. (619) 534-6210.

University of California—Santa Barbara, Center For Information Processing Research. Department of Electrical and Computer Engineering, Santa Barbara, CA 93106. (805) 893-2037.

COMPUTER OPERATING SYSTEMS

See also: ARTIFICIAL INTELLIGENCE, COMPUTER PROGRAMMING, COMPUTER PROGRAMMING LANGUAGES, PARALLEL COMPUTERS

ABSTRACT SERVICES AND INDEXES

ACM Guide To Computing Literature. Association for Computing Machinery, Association for Computing Machinery, 1515 Broadway, 17th Floor, New York, NY 10036-5701. (212) 869-7440. FAX (212) 944-1318. 1964 to present. Monthly. $190.00 per year for non-members.

Applied Science and Technology Index; A Cumulative Subject Index To English Language Periodicals In the Fields of Aeronautics and Space Science, Computer Technology, Chemistry, Construction Industry, Energy and Related Areas. H.W. Wilson Co., 950 University Avenue, Bronx, NY 10452. (800) 367-6770 or (212) 588-8400. FAX (718) 590-1617. From 1958 to present. Monthly. Inquire about cost and availability. Also available on CD-ROM and online.

Artificial Intelligence Abstracts. R.R. Bowker, BowkerA&I Publishing, 121 Chanlon Road, New Providence, NJ 07974. (908) 771-7714. Fax (908) 771-7725. 1985 to present. Monthly. $495.00 per year.

Computer Abstracts, MCBUniversity Press Ltd., PO Box 10812, Birmingham, AL 35201-0812. (800) 633-4931. FAX (205) 995-1588. Monthly. Covers computer theory, data, hardware, systems, networks, human-computer interaction, artificial intelligence, as well as applications of computers in aerospace, business, CAD/CAM, cartography, civil engineering, electronics and electrical engineering, industrial engineering, mechanical engineering, medicine, structural engineering, etc. Also available on CD-ROM.

Computer & Control Abstracts (inspec). INSPEC/IEEE, Box 1331, 445 Hoes Lane, Piscataway, NJ 08855-1331. (908) 562-5549. Abstracts organized by subjects of international techno-logical information. Monthly. $1455.00 per year. Also available on CD-ROM and online via BRS Online Products, DIALOG Information Services, Orbit Search Service, and STN International.

Computer and Information Systems Abstracts Journal. Cambridge Scientific Abstracts, 7200 Wisconsin Avenue, Bethesda, MD 20814. (301) 961-6750. Fax (301) 961-6720. 1962 to present. Monthly. $1265.00 per year. Also available online via STN International.

Computer Literature Index. Applied Computer Research Inc., Box 82266, Phoenix, AZ 85071-2266. (800) 234-2227. 1971 to present. Quarterly plus annual cumulation. $198.50 per year. Bibliography of books, articles, and reports.

Computing Journal Abstracts. Techgnosis Ltd., Blade House, Battersea Road, Stockport, Cheshire 3AE, England. Telephone 061-442-2639. FAX 061-443-1162. 1969 to present. Monthly. Inquire for price.

Computing Reviews. Association for Computing Machinery, 1515 Broadway, 17th Floor, New York, NY 10036-5701. (212) 869-7440. FAX (212) 944-1318. 1960 to present. Monthly. $130.00 per year for non-members. Also available online via DIALOG Information Services.

Current Contents: Engineering, Technology, and Applied Sciences. Institute for Scientific Information, 3501 Market Street, Philadelphia, PA 19104. (215) 386-0100. FAX (215) 386-6362. 1970 to present. Weekly. $442.00 per year.

Engineering Index Monthly. Engineering Information, Inc., Castle Point on the Hudson, Hoboken, NJ 07030. (800) 221-1044. FAX (212) 832-1857. Monthly. $2300.00 per year. Also available online as COMPENDEX, and also on CD-ROM. Covers chemical engineering, computers, electrical engineering, civil engineering, metals and mining, industrial management, and mechanical engineering.

Index To Scientific and Technical Proceedings. Institute for Scientific Information, 3501 Market St., Philadelphia, PA 19104. (215) 386-0100. FAX (215) 386-6362. Monthly. $500.00 per year.

Index To Scientific Reviews. Institute for Scientific Information, 3501 Market St., Philadelphia, PA 19104. (215)386-0100. FAX (215) 386-6362. Semi-annual.

Microcomputer Index. Learned Information, Inc., 143 Old Marlton Pike, Medford, NJ 08055. (609) 654-6266 or FAX (609) 654-4309. 1980 to present. Bimonthly. $160.00 per year.

Physics Abstracts. Institute of Electrical Engineers, Michael Faraday House, Six Hill Way, Stevenage, Herts, SG1 2AY, England. Distributed by IEEE, 445 Hoes Lane, Piscataway, NJ 08854. (908) 562-5549. 1898 to present. Monthly. $2700.00 per year. Also available online as INSPEC.

Science Citation Index. Institute for Scientific Information, 3501 Market Street, Philadelphia, PA 19104. (215) 386-0100. FAX (215) 386-6362. Inquire about availability and cost. Also available on CD-ROM.

ANNUAL REVIEWS AND YEARBOOKS

ACM Monograph Series. Association for Computing Machinery (ACM), 11 West 42nd Street, New York, NY 10036. (212) 869-7440. Fax (212) 869-0481. Irregular. Price varies, inquire.

Advances In Computers. Academic Press Inc., 6277 Sea Harbor Drive, Orlando, FL 32887. Irregular. Price varies, inquire.

Advances In Computing Research. JAI Press, Inc., 55 Old Post Road, Number 2, Box 1678, Greenwich, CT 06830. (203) 661-7602. FAX (203) 661-0792. Annual. $68.50 per year.

Computer Science and Scientific Computing. Academic Press Inc., 6277 Sea Harbor Drive, Orlando, FL 32887. Irregular. Price varies, inquire.

ASSOCIATIONS AND PROFESSIONAL SOCIETIES

American Association For Artificial Intelligence, 445 Burgess Drive, Menlo Park, Ca 94025-3442. (415) 328-3123.

American Federation of Information Processing Societies, 1899 Preston White Drive, Reston, Va 22091-5435. (703) 620-8937.

Association For Computing Machinery. 1515 Broadway, 17th Floor, New York, NY 10036-5701. (212) 869-7440. FAX (212) 944-1318.

Computer and Automation Systems Association of Sme. 1 SME Drive, PO Box 930, Dearborn, MI 48121-0930. (313) 271-1500.

Institute of Electrical and Electronic Engineers. 345 E. 47th Street, New York, NY 10017. (212) 705-7900.

IEEE Computer Society. 1730 Massachusetts Avenue NW, Washington, DC 20036-1992. (202) 371-0101. FAX (202) 728-9614.

Society For Computer Simulation International, 4838 Ronson Court Lane, San Diego, Ca 92111-1810. (619) 277-3888.

Special Interest Group On Operating Systems. c/o ACM, 1515 Broadway, New York, NY 10036. (212) 869-7440. FAX (212) 869-0481.

BIBLIOGRAPHIES

ACM Guide To Computing Literature. Association for Computing Machinery, 1515 Broadway, 17th Floor, New York, NY 10036-5701. (212) 869-7440. FAX (212) 944-1318. 1964 to present. Annual. $190.00 for non-members. Also available online via DIALOG.

DIRECTORIES AND BIOGRAPHICAL SOURCES

ACM Adminstrative Directory of College and UniversityComputer Science-data Processing Programs and Computer Facilities. Association for Computing Machinery, 1515 Broadway, 17th Floor, New York, NY 10036-5701. (212) 869-7440. FAX (212) 944-1318. Biennial. $17.00 for non-members.

American Men & Women of Science, 1995-96. R.R. Bowker Staff, eds. 19th edition. 8 volumes. R.R. Bowker/Reed International Publishing Company, 121 Chanlon Road, New Providence, NJ 07974. (908) 464-6800 or (800) 521-8110. 1995. $850.00

ASIS Handbook and Directory, American Society For Information Science, 8720 Georgia Avenue, Suite 501, Silver Spring, Md 20910-3602. (301) 495-0900. Annual. $100.00.

Computing Information Directory. Hildebrandt Inc., PO Box 576, Pullman, WA 99114. 1981 to present. Annual. $199.95 per year.

Directory of Engineering Societies and Related Organizations. Gordon Davis. 13th edition. American Association of Engineering Societies, 1111 19th Street NW, Suite 608, Washington, DC 20036. (202) 296-2237 or (800) 658-8897. 1989. Inquire for price.

International Research Centers Directory. Gale Research, 835 Penobscot Building, Detroit, MI 48226-4094. (313) 961-2242. (800) 347-4253. 8th edition. 1995. $410.00

Research Centers Directory. Gale Research, 835 Penobscot Building, Detroit, MI 48226-4094. (313) 961-2242. (800) 347-4253. FAX (313) 961-6083. $485.00.

Scientific and Technical Organizations and Agencies Directory. Gale Research, 835 Penobscot Building, Detroit, MI 48226-4094. (313) 961-2242. (800) 347-4253. FAX (313) 961-6083. 4th edition. 1996. $195.00.

Who's Who In Computing. Canadian M I S Database Inc., 268 Lakeshore Road E., Suite 510, Mississauga, ON L5G 1H1, Canada. (905) 271-1601. FAX (905) 271-4522. 1987 to present. Annual. Inquire for price in U.S.

Who's Who In Technology. Gale Research, 835 Penobscot Building, Detroit, MI 48226-4094. (313) 961-2242. (800) 347-4253. FAX (313) 961-6083. 1995. $195.00.

ENCYCLOPEDIAS AND DICTIONARIES

Computer Dictionary. Donald D. Spencer. 4th edition. Camelot Publishing Company, 709 SW 80th Blvd., Gainesville, FL 32607-1537. (904) 331-0952. 1993. $24.95.

Dictionary of Computing. S.M.H. Collin. 2nd edition. Peter Collin Publishing, 8 the Causeway, Teddington, TW11 0HE, England. FAX 0181-943-3386. 1994. Inquire for cost.

Encyclopedia of Computer Science and Engineering. Anthony Ralston & Edwin D. Reilly Jr., editors. 3rd revised edition. Van Nostrand Reinhold, 115 Fifth Avenue, New York, NY 10003. (212) 254-3232 or FAX (212) 254-9499. 1993. $125.00.

Microsoft Press Computer Dictionary: the Comprehensive Standard For Business, School, Library, and Home. 2nd edition. Microsoft Press, One Microsoft Way, Redmond, WA 98052-6399. (206) 936-0055. FAX (206) 823-8101. 1994. $19.95.

COMPUTER OPERATING SYSTEMS

Ency. of Physical Sciences and Engineering Info. Sources

GENERAL WORKS

Distributed Operating Systems. Andrew S. Tanenbaum. Prentice-Hall, 113 Sylvan Avenue, Route 9W, Englewood Cliffs, NJ 07632. (201) 592-2000 or (800) 922-0579. 1994. $57.00.

Fundamentals of Operating Systems. A.M. Lister & R.D. Eager. Fifth edition. Springer-Verlag, 175 Fifth Avenue, New York, NY 10010. (212) 460-1500. FAX (212) 473-6272. 1993. $29.00.

Modern Operating Systems. Andrew S. Tanenbaum. Prentice-Hall, 113 Sylvan Avenue, Route 9W, Englewood Cliffs, NJ 07632. (201) 592-2000 or (800) 922-0579. 1992. $63.00.

Network Training Guide: Networking Technologies. Debra Niedermiller-Chaffins. Third edition. New Riders Publishing, 201 W. 103rd Street, Indianapolis, in 46290. (317) 581-3500 or (800) 428-5331. FAX (317) 581-3550. 1994. $70.00.

One Thousand Questions and Answers About Unix Systems. Amis Majidimehr. Prentice-Hall, 113 Sylvan Avenue, Route 9W, Englewood Cliffs, NJ 07632. (201) 592-2000 or (800) 922-0579. 1994. $22.50.

Operating Systems. William Stallings. Macmillan Publishing, 200 Old Tappan Road, Old Tappan, NJ 07675. (800) 223-2336. FAX (800) 445-6991. 1994. $65.00.

Operating Systems: A Practical Approach. Robert Switzer. Prentice Hall, 113 Sylvan Avenue, Route 9W, Englewood Cliffs, NJ 07632. (201) 592-2000 or (800) 922-0579. 1993. $50.00.

HANDBOOKS AND MANUALS

The Operating Systems Handbook: Unix, Open Vms, Os-400, Vm & Mvs. Bob DuCharme. McGraw-Hill Publishing Company, 1221 Avenue of the Americas, New York, NY 10020. (800) 262-4729 or (212) 512-3825. 1994. $49.50.

ONLINE DATA BASES AND CD-ROMS

Compendex Plus. Engineering Information, Inc., 345 East 47th Street, New York, NY 10017. (212) 705-7600 or (800) 221-1044. Contains citations with abstracts to worldwide literature in engineering and technology, from 1970 to present. Available on online BRS,(800) 289-4277, DIALOG, (800) 334-2564, ORBIT (800) 456-7248, and STN International, FIZ Karlsruhe, P.O. Box 2465, W-7500, Karlsruhe 1, Germany, online services. Also available on CD-ROM. Inquire as to cost and availability.

Compuscience. FIZ Karlsruhe, D-7514, Eggenstein-Leopoldshafen 2, Germany. Contains citations with abstracts to European and North American literature on computer science, 1972 to present. Available on STN International, FIZ Karlsruhe, P.O. Box 2465, W-7500, Karlsruhe 1, Germany, online service. Inquire as to cost and availability.

Computer and Information Systems Abstracts. Cambridge Scientific Abstracts, 7200 Wisconsin Avenue, Suite 601, Bethesda, MD 20814. (301) 961-6750 or (800) 843-7751. Contains citations to worldwide literature in theoretical and applied computer science and related areas, from 1981 to present.Inquire as to cost and availability.

Computer and Mathematics Search. Institute for Scientific Information, 3501 Market Street, Philadelphia, PA 19104. (215) 386-0100. FAX (215) 386-6362. Covers worldwide literature in computer science and mathematics, from 1980 to present. Available on BRS,(800) 289-4277, online service. Inquire as to cost and availability.

Computer Database. Information Access Company, 362 Lakeside Drive, Foster City, CA 94404. (415) 378-5000 or (800) 227-8431. Contains citations with abstracts to literature from trade journals covering the computer,telecommunications,and electronics industries. Available on the BRS, (800) 289-4277, Compuserve Information Service,(800) 848-8990, and DIALOG,(800) 334-2564, online services. Inquire as to cost and availability.

CSA Engineering. Cambridge Scientific Abstracts, 7200 Wisconsin Avenue, Suite 601, Bethesda, MD 20814. (301) 961-6750 or (800) 843-7751. Contains citations and abstracts of international periodicals and research literature covering all fields of engineering and science and technology,including computer and information science, electronics, mechanical engineering, solid state materials, 1981 to present. Available on BRS,(800) 289-4277, online service.Inquire as to cost and availability.

Current Contents Search. Institute for Scientific Information, 3501 Market Street, Philadelphia, PA 19104. (215) 386-0100. FAX (215) 386-6362. Contains citations to articles listed in the table of contents of science and technology journals. Also articles in social sciences and life sciences journals. Available on BRS,(800) 289-4277, DIALOG,(800) 334-2564, online services. Inquire as to cost and availability.

Dissertation Abstracts. University Microfilms International, 300 North Zeeb Road, Ann Arbor, MI 48106. (800) 521-0600 or (313) 761-4700. Scope includes virtually all doctoral dissertations accepted at accredited American institutions from 1861 to present in 252 subject areas. Available on BRS,(800) 289-4277, DIALOG,(800) 334-2564, and OCLC EPIC,(800) 848-5878, online services. Also available on CD-ROM. Inquire as to cost and availability.

Inspec. Institution of Electrical Engineers, Michael Faraday House, Six Hills Way, Stevenage, Herts. SG1 2AY, England. Telephone: 0438 313311 or FAX 0438 742840. Contains citations to the worldwide literature of physics, electronics and electrical engineering, computer technology, and related fields. Available on BRS,(800) 289-4277, DIALOG, (800) 334-2564, ORBIT, (800) 456-7248, and STN International, FIZ Karlsruhe, P.O. Box 2465, W-7500, Karlsruhe 1, Germany, online services. Inquire as to cost and availability.

NTIS Bibliographic Database. National Technical Information Service, 5285 Port Royal Road, Springfield, Va 22161. (703) 487-4929 or FAX (703) 321-8199. Broad coverage of government-sponsored science and technology research reports, 1964 to present. Available on BRS,(800) 289-4277, DIALOG, (800)334-2564, ORBIT, (800) 456-7248, and STN International, FIZ Karlsruhe, P.O. Box 2465, W-7500, Karlsruhe 1, Germany, online services. Also available on CD-ROM. Inquire as to cost and availability.

SCISEARCH. Institute for Scientific Information, 3501 Market Street, Philadelphia, PA 19104. (800) 523-1850 or (215) 386-0100. Broad multidisciplinary title and author index to the international literature of science and technology, 1974 to present. Available on DIALOG,(800) 334-2564, and ORBIT,(800) 456-7248, online services. Also available on CD-ROM. Inquire as to cost and availability.

Supertech. Bowker A & I Publishing, 121 Chanlon Road, New Providence, NJ 07974. (800) 521-8110 or (908) 464-6800. Contains citations to the world's published and unpublished literature in the fields of artificial intelligence, biotechnology, computer-aided design and manufacturing, robotics, and telecommunications. Covers the period 1973 to present. Available on DIALOG,(800) 334-2564, and ORBIT,(800) 456-7248, online services. Inquire as to cost and availability.

WILSONLINE. H.W. Wilson Company, 950 University Avenue, Bronx, NY 10452. (800) 367-6770 or (212) 588-8400. Makes available online versions of the printed H.W. Wilson indexes including Applied Science and Technology Index, Business Periodicals Index, General Science Index, and Readers' Guide to Periodical Literature. Period covered is generally 1983 to present. Available on BRS,(800) 289-4277, DIALOG,(800) 334-2564, and OCLC EPIC,(800) 848-5878, online services. Also available on CD-ROM. Inquire as to cost and availability.

PERIODICALS

ACM Transactions On Computer Systems. Association for Computing Machinery, 1515 Broadway, 17th Floor, New York, NY 10036-5701. (212) 869-7440. FAX (212) 944-1318. 1983 to present. Quarterly. $110.00 for non-members.

ACM Transactions On Programming Languages and Systems. Association for Computing Machinery, 1515 Broadway, 17th Floor, New York, NY 10036-5701. (212) 869-7440. FAX (212) 944-1318. 1979 to present. Quarterly. $105.00 for non-members.

Byte. McGraw-Hill Inc., Byte Publications, One Phoenix Mill Lane, Peterborough, NH 03458. (603) 924-9281. FAX (603) 924-2550. 1975 to present. Monthly. $29.95 per year.

Communications of the A C M. Association for Computing Machinery, 1515 Broadway, 17th Floor, New York, NY 10036-5701. (212) 869-7440. FAX (212) 944-1318. 1958 to present. Monthly. $114.00 for non-members.

Computer Technology Review. West World Productions, Inc., 924 Westwood Boulevard, Suite 650, Los Angeles, CA 90024-2927. (213) 208-1335 or FAX (213) 208-1054. 1981 to present. Sixteen times per year. $80.00 per year.

Computing Systems. MIT Press, 55 Hayward Street, Cambridge, MA 02142. (617) 253-2884 or (800) 356-0343. 1988 topresent. Quarterly. $60.00 for non-members of Usenix Association.

IEEE Transactions On Computers. IEEE, 345 E. 47th Street, New York, NY 10017. (908) 981-0060. FAX (908) 981-9667. 1952 to present. Monthly. $425.00 for non-members.

IEEE Transactions On Software Engineering. IEEE, 345 E. 47th Street, New York, NY 10017. (908) 981-0060. FAX (908) 981-9667. 1975 to present. Monthly. $400.00 per year for non-members.

International Journal of Systems Automation: Research and Applications. Ablex Publishing Corporation, 355 Chestnut Street, Norwood, NJ 07648. (201) 767-8450 or FAX (201) 767-6717. 1991 to present. Quarterly. $100.00 per year.

Journal of Computer and Software Engineering. Ablex Publishing Corporation, 355 Chestnut Street, Norwood, NJ 07648.

(201) 767-8450 or FAX (201) 767-6717. 1993 to present. Quarterly. $110.00 per year.

Journal of Computer and System Sciences. Academic Press Inc., 525 B Street, Suite 1900, San Diego, CA 92101-4495. (619) 231-0926. FAX (619) 699-6715. 1967 to present. Bi-monthly. $520.00 per year.

Journal of Systems Science and Systems Engineering. International Academic Publishers, Xiwai Dajie, Beijing Exhibition Centre, Beijing 100044, People's Republic of China. 1992 to present. Quarterly. $140.00 per year.

Letters On Programming Languages and Systems. Association for Computing Machinery, 1515 Broadway, 17th Floor, New York, NY 10036-5701. (212) 869-7440. FAX (212) 944-1318. Irregular. $105.00 for non-members.

Unix Review. Miller Freeman Inc., 600 Harrison Street, San Francisco, CA 94107. (415) 905-2200. FAX (415) 905-2232. 1983 to present. Monthly. $55.00 per year.

Unixworld. McGraw-Hill Publishing Company, 1221 Avenue of the Americas, New York, NY 10020. (800) 262-4729 or (212) 512-3825. 1984 to present. Monthly. $18.00 per year.

RESEARCH CENTERS AND INSTITUTES

Harvard University Center For Research In Computing Technology. Aiken Computation Laboratory, 33 Oxford Street, Cambridge, MA 02138. (617) 495-4117. FAX (617) 495-9837.

COMPUTER PROGRAMMING

See also: ARTIFICIAL INTELLIGENCE, COMPUTER OPERATING SYSTEMS, COMPUTER PROGRAMMING LANGUAGES, PARALLEL COMPUTERS

ABSTRACT SERVICES AND INDEXES

ACM Guide To Computing Literature. Association for Computing Machinery, Association for Computing Machinery, 1515 Broadway, 17th Floor, New York, NY 10036-5701. (212) 869-7440. FAX (212) 944-1318. 1964 to present. Monthly. $190.00 per year for non-members.

Applied Science and Technology Index; A Cumulative Subject Index To English Language Periodicals In the Fields of Aeronautics and Space Science, Computer Technology, Chemistry,Construction Industry, Energy and Related Areas. H.W. Wilson Co., 950 University Avenue, Bronx, NY 10452. (800) 367-6770 or (212) 588-8400. FAX (718) 590-1617. From 1958 to present. Monthly. Inquire about cost and availability. Also available on CD-ROM and online.

Computer Abstracts, MCBUniversity Press Ltd., PO Box 10812, Birmingham, AL 35201-0812. (800) 633-4931. FAX (205) 995-1588. Monthly. Covers computer theory, data, hardware, systems, networks, human-computer interaction, artificial intelligence, as well as applications of computers in aerospace, business, CAD/CAM, cartography, civil engineering, electronics and electrical engineering, industrial engineering, mechanical engineering, medicine, structural engineering, etc. Also available on CD-ROM.

COMPUTER PROGRAMMING

Ency. of Physical Sciences and Engineering Info. Sources

Computer & Control Abstracts (inspec). INSPEC/IEEE, Box 1331, 445 Hoes Lane, Piscataway, NJ 08855-1331. (908) 562-5549. Abstracts organized by subjects of international technological information. Monthly. $1455.00 per year. Also available on CD-ROM and online via BRS Online Products, DIALOG Information Services, Orbit Search Service, and STN International.

Computer and Information Systems Abstracts Journal. Cambridge Scientific Abstracts, 7200 Wisconsin Avenue, Bethesda, MD 20814. (301) 961-6750. Fax (301) 961-6720. 1962 to present. Monthly. $1265.00 per year. Also available online via STN International.

Computer Literature Index. Applied Computer Research Inc., Box 82266, Phoenix, AZ 85071-2266. (800) 234-2227. 1971 to present. Quarterly plus annual cumulation. $198.50 per year. Bibliography of books, articles, and reports.

Computing Journal Abstracts. Techgnosis Ltd., Blade House, Battersea Road, Stockport, Cheshire 3AE, England. Telephone 061-442-2639. FAX 061-443-1162. 1969 to present. Monthly. Inquire for price.

Computing Reviews. Association for Computing Machinery, 1515 Broadway, 17th Floor, New York, NY 10036-5701. (212) 869-7440. FAX (212) 944-1318. 1960 to present. Monthly. $130.00 per year for non-members. Also available online via DIALOG Information Services.

Current Contents: Engineering, Technology, and Applied Sciences. Institute for Scientific Information, 3501 Market Street, Philadelphia, PA 19104. (215) 386-0100. FAX (215) 386-6362. 1970 to present. Weekly. $442.00 per year.

Engineering Index Monthly. Engineering Information, Inc., Castle Point on the Hudson, Hoboken, NJ 07030. (800) 221-1044. FAX (212) 832-1857. Monthly. $2300.00 per year. Also available online as COMPENDEX, and also on CD-ROM. Covers chemical engineering, computers, electrical engineering, civil engineering, metals and mining, industrial management, and mechanical engineering.

Index To Scientific and Technical Proceedings. Institute for Scientific Information, 3501 Market St., Philadelphia, PA 19104. (215) 386-0100. FAX (215) 386-6362. Monthly. $500.00 per year.

Index To Scientific Reviews. Institute for Scientific Information, 3501 Market St., Philadelphia, PA 19104. (215) 386-0100. FAX (215) 386-6362. Semi-annual.

Key Abstracts—software Engineering. INSPEC/IEEE, IEEE (Institute of Electrical and Electronic Engineers), 445 Hoes Lane, Piscataway, NJ 08854. (908) 562-5545 or (800) 678-IEEE. 1987 to present. Monthly. $160.00 per year.

Physics Abstracts. Institute of Electrical Engineers, Michael Faraday House, Six Hill Way, Stevenage, Herts, SG1 2AY, England. Distributed by IEEE, 445 Hoes Lane, Piscataway, NJ 08854. (908) 562-5549. 1898 to present. Monthly. $2700.00 per year. Also available online as INSPEC.

Science Citation Index. Institute for Scientific Information, 3501 Market Street, Philadelphia, PA 19104. (215) 386-0100. FAX (215) 386-6362. Inquire about availability and cost. Also available on CD-ROM.

ANNUAL REVIEWS AND YEARBOOKS

ACM Monograph Series. Association for Computing Machinery (ACM), 11 West 42nd Street, New York, NY 10036. (212) 869-7440. Fax (212) 869-0481. Irregular. Price varies, inquire.

Advances In Computers. Academic Press Inc., 6277 Sea Harbor Drive, Orlando, FL 32887. Irregular. Price varies, inquire.

Advances In Computing Research. JAI Press, Inc., 55 Old Post Road, Number 2, Box 1678, Greenwich, CT 06830. (203) 661-7602. FAX (203) 661-0792. Annual. $68.50 per year.

Annual Review In Automatic Programming. Elsevier Science [journals], 660 White Plains Rd., Tarrytown, NY 10159-5153. (919) 524-9200. FAX (919) 333-2444. 1960 to present. Twice a year. $185.00.

ASSOCIATIONS AND PROFESSIONAL SOCIETIES

American Federation of Information Processing Societies, 1899 Preston White Drive, Reston, Va 22091-5435. (703) 620-8937.

Association for Computing Machinery. 1515 Broadway, 17th Floor, New York, NY 10036-5701. (212) 869-7440. FAX (212) 944-1318.

Computer and Automation Systems Association of Sme. 1 SME Drive, PO Box 930, Dearborn, MI 48121-0930. (313) 271-1500.

Computing Research Association. 1875 Connecticut Avenue NW, Suite 718, Washington, DC 20009-5728. (202) 234-2111. FAX (202) 667-1066.

IEEE Computer Society. 1730 Massachusetts Avenue NW, Washington, DC 20036-1992. (202) 371-0101. FAX (202) 728-9614.

Society For Computer Simulation International, 4838 Ronson Court Lane, San Diego, Ca 92111-1810. (619) 277-3888.

Special Interest Group On Programming Languages. c/o ACM, 1515 Broadway, New York, NY 10036. (212) 869-7440. FAX (212) 869-0481.

Special Interest Group On Software Engineering. c/o ACM, 1515 Broadway, New York, NY 10036. (212) 626-0611. FAX (212) 302-5826.

BIBLIOGRAPHIES

ACM Guide To Computing Literature. Association for Computing Machinery, 1515 Broadway, 17th Floor, New York, NY 10036-5701. (212) 869-7440. FAX (212) 944-1318. 1964 to present. Annual. $190.00 for non-members. Also available online via DIALOG.

DIRECTORIES AND BIOGRAPHICAL SOURCES

ACM Adminstrative Directory of College and University Computer Science-Data Processing Programs and Computer Facilities. Association for Computing Machinery, 1515 Broadway, 17th Floor, New York, NY 10036-5701. (212) 869-7440. FAX (212) 944-1318. Biennial. $17.00 for non-members.

American Men & Women of Science, 1995-96. R.R. Bowker Staff, eds. 19th edition. 8 volumes. R.R. Bowker/Reed International Publishing Company, 121 Chanlon Road, New Providence, NJ 07974. (908) 464-6800 or (800) 521-8110. 1995. $850.00

ASIS Handbook and Directory, American Society For Information Science, 8720 Georgia Avenue, Suite 501, Silver Spring, Md 20910-3602. (301) 495-0900. Annual. $100.00.

Computing Information Directory. Hildebrandt Inc., PO Box 576, Pullman, WA 99114. 1981 to present. Annual. $199.95 per year.

Directory of Engineering Societies and Related Organizations. Gordon Davis. 13th edition. American Association of Engineering Societies, 1111 19th Street NW, Suite 608, Washington, DC 20036. (202) 296-2237 or (800) 658-8897. 1989. Inquire for price.

International Research Centers Directory. Gale Research, 835 Penobscot Building, Detroit, MI 48226-4094. (313) 961-2242. (800) 347-4253. FAX (313) 961-6083. 8th edition. 1995. $410.00

Research Centers Directory. Gale Research, 835 Penobscot Building, Detroit, MI 48226-4094. (313) 961-2242. (800) 347-4253. FAX (313) 961-6083. $485.00.

Scientific and Technical Organizations and Agencies Directory. Gale Research, 835 Penobscot Building, Detroit, MI 48226-4094. (313) 961-2242. (800) 347-4253. FAX (313) 961-6083. 4th edition. 1996. $195.00.

Who's Who In Computing. Canadian M I S Database Inc., 268 Lakeshore Road E., Suite 510, Mississauga, ON L5G 1H1, Canada. (905) 271-1601. FAX (905) 271-4522. 1987 to present. Annual. Inquire for price in U.S.

Who's Who In Technology. Gale Research, 835 Penobscot Building, Detroit, MI 48226-4094. (313) 961-2242. (800) 347-4253. FAX (313) 961-6083. 1995. $195.00.

ENCYCLOPEDIAS AND DICTIONARIES

Computer Dictionary. Donald D. Spencer. 4th edition. Camelot Publishing Company, 709 SW 80th Blvd., Gainesville, FL 32607-1537. (904) 331-0952. 1993. $24.95.

Dictionary of Computing. S.M.H. Collin. 2nd edition. Peter Collin Publishing, 8 the Causeway, Teddington, TW11 0HE, England. FAX 0181-943-3386. 1994. Inquire for cost.

Encyclopedia of Computer Science and Engineering. Anthony Ralston & Edwin D. Reilly Jr., editors. 3rd revised edition. VanNostrand Reinhold, 115 Fifth Avenue, New York, NY 10003. (212) 254-3232 or FAX (212) 254-9499. 1993. $125.00.

Microsoft Press Computer Dictionary: the Comprehensive Standard For Business, School, Library, and Home. 2nd edition. Microsoft Press, One Microsoft Way, Redmond, WA 98052-6399. (206) 936-0055. FAX (206) 823-8101. 1994. $19.95.

GENERAL WORKS

Automata & Formal Languages: An Introduction. Dean Kelley. Prentice Hall, 113 Sylvan Avenue, Route 9W, Englewood Cliffs, NJ 07632. (201) 592-2000 or (800) 922-0579. 1995. $52.00.

Introduction To Computer Science: Programming, Problem Solving & Data Structures. Douglas W. Nance & Thomas L. Naps. 3rd edition. West Publishing Company, 610 Opperman Drive, PO Box 64526, St. Paul, MN 55164-0526. (800) 328-9424. 1995. $65.75.

Introduction To Computer Science and Programming. Liia Vilms. Harpercollins Publishers, Inc., 10 E. 53rd Street, New York, NY 10022-5299. (800) 331-3761. FAX (212) 207-7145. 1994. $13.00.

Parallel Programming: An Introduction. Thomas B. Unl. Prentice Hall, 113 Sylvan Avenue, Route 9W, Englewood Cliffs, NJ 07632. (201) 592-2000 or (800) 922-0579. 1993. $40.00.

Programming Language Essentials. Henri Bal. Addison-Wesley Publishing Company, 1 Jacob Way, Reading, MA 01867. (617) 944-3700 or (800) 447-2226. FAX 617-942-1117. 1994. $26.95.

Programming Languages: Principles and Practice. Louden. PWS Publishers, 20 Park Plaza, Boston, MA 02116. (617) 542-3377. FAX (617) 338-6134. 1993. $55.95.

ONLINE DATABASES AND CD-ROMS

Compendex Plus. Engineering Information, Inc., 345 East 47th Street, New York, NY 10017. (212) 705-7600 or (800) 221-1044. Contains citations with abstracts to worldwide literature in engineering and technology, from 1970 to present. Available on online BRS,(800) 289-4277, DIALOG, (800) 334-2564, ORBIT (800) 456-7248, and STN International, FIZ Karlsruhe, P.O. Box 2465, W-7500, Karlsruhe 1, Germany, online services. Also available on CD-ROM. Inquire as to cost and availability.

Compuscience. FIZ Karlsruhe, D-7514, Eggenstein-Leopoldshafen 2, Germany. Contains citations with abstracts to European and North American literature on computer science, 1972 to present. Available on STN International, FIZ Karlsruhe, P.O. Box 2465, W-7500, Karlsruhe 1, Germany, online service. Inquire as to cost and availability.

Computer and Information Systems Abstracts. Cambridge Scientific Abstracts, 7200 Wisconsin Avenue, Suite 601, Bethesda, MD 20814. (301) 961-6750 or (800) 843-7751. Contains citations to worldwide literature in theoretical and applied computer science and related areas, from 1981 to present.Inquire as to cost and availability.

Computer and Mathematics Search. Institute for Scientific Information, 3501 Market Street, Philadelphia, PA 19104. (215) 386-0100. FAX (215) 386-6362. Covers worldwide literature incomputer science and mathematics, from 1980 to present. Available on BRS,(800) 289-4277, online service. Inquire as to cost and availability.

Computer Database. Information Access Company, 362 Lakeside Drive, Foster City, CA 94404. (415) 378-5000 or (800) 227-8431. Contains citations with abstracts to literature from trade journals covering the computer,telecommunications,and electronics industries. Available on the BRS, (800) 289-4277, Compuserve Information Service,(800) 848-8990, and DIA-

COMPUTER PROGRAMMING

Ency. of Physical Sciences and Engineering Info. Sources

LOG,(800) 334-2564, online services. Inquire as to cost and availability.

Dissertation Abstracts. University Microfilms International, 300 North Zeeb Road, Ann Arbor, MI 48106. (800) 521-0600 or (313) 761-4700. Scope includes virtually all doctoral dissertations accepted at accredited American institutions from 1861 to present in 252 subject areas. Available on BRS,(800) 289-4277, DIALOG,(800) 334-2564, and OCLC EPIC,(800) 848-5878, online services. Also available on CD-ROM. Inquire as to cost and availability.

Inspec. Institution of Electrical Engineers, Michael Faraday House, Six Hills Way, Stevenage, Herts. SG1 2AY, England. Telephone: 0438 313311 or FAX 0438 742840. Contains citations to the worldwide literature of physics, electronics and electrical engineering, computer technology, and related fields. Available on BRS,(800) 289-4277, DIALOG, (800) 334-2564, ORBIT, (800) 456-7248, and STN International, FIZ Karlsruhe, P.O. Box 2465, W-7500, Karlsruhe 1, Germany, online services. Inquire as to cost and availability.

Mathsci. American Mathematical Society, P.O. Box 6248, Providence, RI 02940. (800) 321-4667 or (401) 455-4000 or FAX (401) 331-3842. Scope includes pure and applied mathematics and related areas of physics, statistics, engineering, computer science, and operations research literature since 1959. Available on DIALOG,(800) 334-2564, online service. Also available on CD-ROM. Inquire as to cost and availability.

NTIS Bibliographic Database. National Technical Information Service, 5285 Port Royal Road, Springfield, Va 22161. (703) 487-4929 or FAX (703) 321-8199. Broad coverage of government-sponsored science and technology research reports, 1964 to present. Available on BRS,(800) 289-4277, DIALOG, (800) 334-2564, ORBIT, (800) 456-7248, and STN International, FIZ Karlsruhe, P.O. Box 2465, W-7500, Karlsruhe 1, Germany, online services. Also available on CD-ROM. Inquire as to cost and availability.

SCISEARCH. Institute for Scientific Information, 3501 Market Street, Philadelphia, PA 19104. (800) 523-1850 or (215) 386-0100. Broad multidisciplinary title and author index to the international literature of science and technology, 1974 to present. Available on DIALOG,(800) 334-2564, and ORBIT,(800) 456-7248, online services. Also available on CD-ROM. Inquire as to cost and availability.

WILSONLINE. H.W. Wilson Company, 950 University Avenue, Bronx, NY 10452. (800) 367-6770 or (212) 588-8400. Makes available online versions of the printed H.W. Wilson indexes including Applied Science and Technology Index, Business Periodicals Index, General Science Index, and Readers' Guide to Periodical Literature. Period covered is generally 1983 to present. Available on BRS,(800) 289-4277, DIALOG,(800) 334-2564, and OCLC EPIC,(800) 848-5878, online services. Also available on CD-ROM. Inquire as to cost and availability.

PERIODICALS

ACM Transactions On Programming Languages and Systems. Association for Computing Machinery, 1515 Broadway, 17th Floor, New York, NY 10036-5701. (212) 869-7440. FAX (212) 944-1318. 1979 to present. Quarterly. $105.00 for non-members.

ACM Transactions On Software Engineering and Methodology. Association for Computing Machinery, 11 West 42nd Street,

New York, NY 10036. (212) 869-7440. Fax (212) 869-0481. 1991 to present. Quarterly. $85.00 per year.

Byte. McGraw-Hill Inc., Byte Publications, One Phoenix Mill Lane, Peterborough, NH 03458. (603) 924-9281. FAX (603) 924-2550. 1975 to present. Monthly. $29.95 per year.

Communications of the A C M. Association for Computing Machinery, 1515 Broadway, 17th Floor, New York, NY 10036-5701. (212) 869-7440. FAX (212) 944-1318. 1958 to present. Monthly. $114.00 for non-members.

Computer Languages. Elsevier Science [journals], 660 White Plains Rd., Tarrytown, NY 10159-5153. (919) 524-9200. FAX (919) 333-2444. 1976 to present. Four times a year. $445.00 per year.

Dr. Dobb's Journal of Software Tools. M and T Publishing, Inc. 501 Galveston Drive, Redwood City, CA 94063. (415) 366-3600. 1976 to present. Monthly. $30.00 per year.

IEEE Software. IEEE Computer Society Press, 10662 Vaqueros Circle, Box 3014, Los Alamitos, CA 90720. (714) 821-8380. 1984 to present. Bi-monthly. $175.00 per year.

IEEE Transactions On Software Engineering. IEEE, 345 E. 47th Street, New York, NY 10017. (908) 981-0060. FAX (908) 981-9667. 1975 to present. Monthly. $400.00 per year for non-members.

Journal of Computer and Software Engineering. Ablex Publishing Corporation, 355 Chestnut Street, Norwood, NJ 07648. (201) 767-8450 or FAX (201) 767-6717. 1993 to present. Quarterly. $110.00 per year.

Journal of Functional Programming. Cambridge University Press, Journals Department, 40 West 20th Street, New York, NY 10011-4211. (212) 924-3900. 1991 to present. Quarterly. Inquire for price.

Journal of Logic Programming. Elsevier Science Publishing Company, Inc., 655 Avenue of the Americas, New York, NY 10010. (212) 989-5800. 1984 to present. Twelve times a year. $545.00 peryear (institutions).

Journal of Object-Oriented Programming. Sigs Publications Inc., 588 Broadway, Suite 604, New York, NY 10012-5408. (212) 274-0640. FAX (212) 274-0646. 1988 to present. Nine times a year. $59.00 per year.

Journal of Programming Languages. Chapman & Hall, Journals Promotion Department, One Penn Plaza, 41st Floor, New York, NY 10019. (212) 564-1060. FAX (212) 564-1505. 1992 to present. Four times a year. $210.00 per year.

Letters On Programming Languages and Systems. Association for Computing Machinery, 1515 Broadway, 17th Floor, New York, NY 10036-5701. (212) 869-7440. FAX (212) 944-1318. Irregular. $105.00 for non-members.

Programmer's Journal. Oakley Publishing Company, 150 North 4th Street, Springfield, OR 97477. (503) 747-0800. 1983 to present. Bimonthly. $20.00 per year.

Science of Computer Programming. Elsevier Science Inc., Box 882, Madison Square Station, New York, NY 10159. (212) 989-

5800. FAX (212) 633-3990. 1981 to present. Six times a year. $411.00 per year.

Software Engineering. Auerbach Publishers, 1 Penn Plaza, New York, NY 10119-0118. 1989 to present. Bimonthly. $145.00 per year.

Software: Practice and Experience. John Wiley and Sons, Inc., 605 Third Avenue, New York, NY 10158. (800) 526-5368 or (212) 850-6000. 1971 to present. Monthly. $440.00 per year.

Unix Review. Miller Freeman Inc., 600 Harrison Street, San Francisco, CA 94107. (415) 905-2200. FAX (415) 905-2232. 1983 to present. Monthly. $55.00 per year.

Unixworld. McGraw-Hill Publishing Company, 1221 Avenue of the Americas, New York, NY 10020. (800) 262-4729 or (212) 512-3825. 1984 to present. Monthly. $18.00 per year.

RESEARCH CENTERS AND INSTITUTES

Harvard University Center For Research In Computing Technology. Aiken Computational Laboratory, 33 Oxford Street, Cambridge, MA 022138. (617) 495-4117. FAX (617) 495-9837.

Ohio State University Computer and Information Science Research Center. 2036 Neil Avenue Mall, Columbus, OH 43210. (614) 292-6374. FAX (614) 292-9021.

Rice University, Computer and Information Technology Institute. PO Box 1892, Houston, TX 77251. (713) 527-6077. FAX (713) 285-5136.

University of Texas At Austin, Computation Center. Austin, TX 78712. (512) 471-3241.

COMPUTER PROGRAMMING LANGUAGES

See also: ARTIFICIAL INTELLIGENCE, COMPUTER OPERATING SYSTEMS, COMPUTER PROGRAMMING, PARALLEL COMPUTERS

ABSTRACT SERVICES AND INDEXES

ACM Guide To Computing Literature. Association for Computing Machinery, Association for Computing Machinery, 1515Broadway, 17th Floor, New York, NY 10036-5701. (212) 869-7440. FAX (212) 944-1318. 1964 to present. Monthly. $190.00 per year for non-members.

Applied Science and Technology Index; A Cumulative Subject Index To English Language Periodicals In the Fields of Aeronautics and Space Science, Computer Technology, Chemistry, Construction Industry, Energy and Related Areas. H.W. Wilson Co., 950 University Avenue, Bronx, NY 10452. (800) 367-6770 or (212) 588-8400. FAX (718) 590-1617. From 1958 to present. Monthly. Inquire about cost and availability. Also available on CD-ROM and online.

Computer Abstracts, MCBUniversity Press Ltd., PO Box 10812, Birmingham, AL 35201-0812. (800) 633-4931. FAX (205) 995-1588. Monthly. Covers computer theory, data, hardware, systems, networks, human-computer interaction, artificial intelligence, as well as applications of computers in aerospace, business, CAD/CAM, cartography, civil engineering, electronics and electrical engineering, industrial engineering, mechanical engineering, medicine, structural engineering, etc. Also available on CD-ROM.

Computer & Control Abstracts (inspec). INSPEC/IEEE, Box 1331, 445 Hoes Lane, Piscataway, NJ 08855-1331. (908) 562-5549. Abstracts organized by subjects of international technological information. Monthly. $1455.00 per year. Also available on CD-ROM and online via BRS Online Products, DIALOG Information Services, Orbit Search Service, and STN International.

Computer and Information Systems Abstracts Journal. Cambridge Scientific Abstracts, 7200 Wisconsin Avenue, Bethesda, MD 20814. (301) 961-6750. Fax (301) 961-6720. 1962 to present. Monthly. $1265.00 per year. Also available online via STN International.

Computer Literature Index. Applied Computer Research Inc., Box 82266, Phoenix, AZ 85071-2266. (800) 234-2227. 1971 to present. Quarterly plus annual cumulation. $198.50 per year. Bibliography of books, articles, and reports.

Computing Journal Abstracts. Techgnosis Ltd., Blade House, Battersea Road, Stockport, Cheshire 3AE, England. Telephone 061-442-2639. FAX 061-443-1162. 1969 to present. Monthly. Inquire for price.

Computing Reviews. Association for Computing Machinery, 1515 Broadway, 17th Floor, New York, NY 10036-5701. (212) 869-7440. FAX (212) 944-1318. 1960 to present. Monthly. $130.00 per year for non-members. Also available online via DIALOG Information Services.

Current Contents: Engineering, Technology, and Applied Sciences. Institute for Scientific Information, 3501 Market Street, Philadelphia, PA 19104. (215) 386-0100. FAX (215) 386-6362. 1970 to present. Weekly. $442.00 per year.

Engineering Index Monthly. Engineering Information, Inc., Castle Point on the Hudson, Hoboken, NJ 07030. (800) 221-1044. FAX (212) 832-1857. Monthly. $2300.00 per year. Also available online as COMPENDEX, and also on CD-ROM. Covers chemical engineering, computers, electrical engineering, civil engineering, metals and mining, industrial management, and mechanical engineering.

Index To Scientific and Technical Proceedings. Institute for Scientific Information, 3501 Market St., Philadelphia, PA 19104. (215) 386-0100. FAX (215) 386-6362. Monthly. $500.00 per year.

Index To Scientific Reviews. Institute for Scientific Information, 3501 Market St., Philadelphia, PA 19104. (215) 386-0100. FAX (215) 386-6362. Semi-annual.

Key Abstracts—Software Engineering. INSPEC/IEEE, IEEE (Institute of Electrical and Electronic Engineers), 445 Hoes Lane, Piscataway, NJ 08854. (908) 562-5545 or (800) 678-IEEE. 1987 to present. Monthly. $160.00 per year.

Physics Abstracts. Institute of Electrical Engineers, Michael Faraday House, Six Hill Way, Stevenage, Herts, SG1 2AY, England. Distributed by IEEE, 445 Hoes Lane, Piscataway, NJ 08854. (908) 562-5549. 1898 to present. Monthly. $2700.00 per year. Also available online as INSPEC.

COMPUTER PROGRAMMING LANGUAGES

Ency. of Physical Sciences and Engineering Info. Sources

ANNUAL REVIEWS AND YEARBOOKS

Advances In Computers. Academic Press Inc., 6277 Sea Harbor Drive, Orlando, FL 32887. 1960 to present. Irregular. Inquire for price.

Annual Review In Automatic Programming. Elsevier Science [journals], 660 White Plains Rd., Tarrytown, NY 10159-5153. (919) 524-9200. FAX (919) 333-2444. 1960 to present. Twice a year. $185.00.

ASSOCIATIONS AND PROFESSIONAL SOCIETIES

American Federation of Information Processing Societies, 1899 Preston White Drive, Reston, Va 22091-5435. (703) 620-8937.

Association For Computational Linguistics. Bellcore, MRE2A379, 445 South Street, Box 1910, Morristown, NJ 07960-1910. (201) 829-4312. FAX (201) 829-5981.

Association For Computing Machinery. 1515 Broadway, 17th Floor, New York, NY 10036-5701. (212) 869-7440. FAX (212) 944-1318.

Computer and Automation Systems Association of Sme. 1 SME Drive, PO Box 930, Dearborn, MI 48121-0930. (313) 271-1500.

Computing Research Association. 1875 Connecticut Avenue NW, Suite 718, Washington, DC 20009-5728. (202) 234-2111. FAX (202) 667-1066.

IEEE Computer Society. 1730 Massachusetts Avenue NW, Washington, DC 20036-1992. (202) 371-0101. FAX (202) 728-9614.

Special Interest Group On Programming Languages. c/o ACM, 1515 Broadway, New York, NY 10036. (212) 869-7440. FAX (212) 869-0481.

BIBLIOGRAPHIES

ACM Guide To Computing Literature. Association forComputing Machinery, 1515 Broadway, 17th Floor, New York, NY 10036-5701. (212) 869-7440. FAX (212) 944-1318. 1964 to present. Annual. $190.00 for non-members. Also available online via DIALOG.

DIRECTORIES AND BIOGRAPHICAL SOURCES

ACM Adminstrative Directory of College and University Computer Science-data Processing Programs and Computer Facilities. Association for Computing Machinery, 1515 Broadway, 17th Floor, New York, NY 10036-5701. (212) 869-7440. FAX (212) 944-1318. Biennial. $17.00 for non-members.

American Men & Women of Science, 1995-96. R.R. Bowker Staff, eds. 19th edition. 8 volumes. R.R. Bowker/Reed International Publishing Company, 121 Chanlon Road, New Providence, NJ 07974. (908) 464-6800 or (800) 521-8110. 1995. $850.00

Asis Handbook and Directory, American Society For Information Science, 8720 Georgia Avenue, Suite 501, Silver Spring, Md 20910-3602. (301) 495-0900. Annual. $100.00.

Computing Information Directory. Hildebrandt Inc., PO Box 576, Pullman, WA 99114. 1981 to present. Annual. $199.95 per year.

Directory of Engineering Societies and Related Organizations. Gordon Davis. 13th edition. American Association of Engineering Societies, 1111 19th Street NW, Suite 608, Washington, DC 20036. (202) 296-2237 or (800) 658-8897. 1989. Inquire for price.

International Research Centers Directory. Gale Research, 835 Penobscot Building, Detroit, MI 48226-4094. (313) 961-2242. (800) 347-4253. FAX (313) 961-6083. 8th edition. 1995. $410.00

Research Centers Directory. Gale Research, 835 Penobscot Building, Detroit, MI 48226-4094. (313) 961-2242. (800) 347-4253. FAX (313) 961-6083. $485.00.

Scientific and Technical Organizations and Agencies Directory. Gale Research, 835 Penobscot Building, Detroit, MI 48226-4094. (313) 961-2242. (800) 347-4253. FAX (313) 961-6083. 4th edition. 1996. $195.00.

Who's Who In Computing. Canadian M I S Database Inc., 268 Lakeshore Road E., Suite 510, Mississauga, ON L5G 1H1, Canada. (905) 271-1601. FAX (905) 271-4522. 1987 to present. Annual. Inquire for price in U.S.

Who's Who In Technology. Gale Research, 835 Penobscot Building, Detroit, MI 48226-4094. (313) 961-2242. (800) 347-4253. FAX (313) 961-6083. 1995. $195.00.

ENCYCLOPEDIAS AND DICTIONARIES

Computer Dictionary. Donald D. Spencer. 4th edition. Camelot Publishing Company, 709 SW 80th Blvd., Gainesville, FL 32607-1537. (904) 331-0952. 1993. $24.95.

Dictionary of Computing. S.M.H. Collin. 2nd edition. Peter Collin Publishing, 8 the Causeway, Teddington, TW11 0HE, England. FAX 0181-943-3386. 1994. Inquire for cost.

Encyclopedia of Computer Science and Engineering. Anthony Ralston & Edwin D. Reilly Jr., editors. 3rd revised edition. Van Nostrand Reinhold, 115 Fifth Avenue, New York, NY 10003. (212)254-3232 or FAX (212) 254-9499. 1993. $125.00.

Microsoft Press Computer Dictionary: the Comprehensive Standard For Business, School, Library, and Home. 2nd edition. Microsoft Press, One Microsoft Way, Redmond, WA 98052-6399. (206) 936-0055. FAX (206) 823-8101. 1994. $19.95.

GENERAL WORKS

Automata & Formal Languages: An Introduction. Dean Kelley. Prentice Hall, 113 Sylvan Avenue, Route 9W, Englewood Cliffs, NJ 07632. (201) 592-2000 or (800) 922-0579. 1995. $52.00.

Fundamentals of Programming Languages. Ellis Horowitz. Second edition. W.H. Freeman and Company, 41 Madison Avenue, East 26th, 35th Floor, New York, NY 10010. (212) 576-9400. FAX (212) 689-2383. 1995. $46.95.

Programming Language Essentials. Henri Bal. Addison-Wesley Publishing Company, 1 Jacob Way, Reading, MA 01867. (617) 944-3700 or (800) 447-2226. FAX 617-942-1117. 1994. $26.95.

Programming Languages: Principles and Practice. Louden. PWS Publishers, 20 Park Plaza, Boston, MA 02116. (617) 542-3377. FAX (617) 338-6134. 1993. $55.95.

Programming Language theory and Its Implementation. M. Gordon. Prentice Hall, 113 Sylvan Avenue, Route 9W, Englewood Cliffs, NJ 07632. (201) 592-2000 or (800) 922-0579. 1992. $30.00.

ONLINE DATABASES AND CD-ROMS

Compendex Plus. Engineering Information, Inc., 345 East 47th Street, New York, NY 10017. (212) 705-7600 or (800) 221-1044. Contains citations with abstracts to worldwide literature in engineering and technology, from 1970 to present. Available on online BRS,(800) 289-4277, DIALOG, (800) 334-2564, ORBIT (800) 456-7248, and STN International, FIZ Karlsruhe, P.O. Box 2465, W-7500, Karlsruhe 1, Germany, online services. Also available on CD-ROM. Inquire as to cost and availability.

Compuscience. FIZ Karlsruhe, D-7514, Eggenstein-Leopoldshafen 2, Germany. Contains citations with abstracts to European and North American literature on computer science, 1972 to present. Available on STN International, FIZ Karlsruhe, P.O. Box 2465, W-7500, Karlsruhe 1, Germany, online service. Inquire as to cost and availability.

Computer and Information Systems Abstracts. Cambridge Scientific Abstracts, 7200 Wisconsin Avenue, Suite 601, Bethesda, MD 20814. (301) 961-6750 or (800) 843-7751. Contains citations to worldwide literature in theoretical and applied computer science and related areas, from 1981 to present.Inquire as to cost and availability.

Computer and Mathematics Search. Institute for Scientific Information, 3501 Market Street, Philadelphia, PA 19104. (215) 386-0100. FAX (215) 386-6362. Covers worldwide literature in computer science and mathematics, from 1980 to present. Available on BRS,(800) 289-4277, online service. Inquire as to cost and availability.

Computer Database. Information Access Company, 362Lakeside Drive, Foster City, CA 94404. (415) 378-5000 or (800) 227-8431. Contains citations with abstracts to literature from trade journals covering the computer,telecommunications,and electronics industries. Available on the BRS, (800) 289-4277, Compuserve Information Service,(800) 848-8990, and DIALOG,(800) 334-2564, online services. Inquire as to cost and availability.

CSA Engineering. Cambridge Scientific Abstracts, 7200 Wisconsin Avenue, Suite 601, Bethesda, MD 20814. (301) 961-6750 or (800) 843-7751. Contains citations and abstracts of international periodicals and research literature covering all fields of engineering and science and technology,including computer and information science, electronics, mechanical engineering, solid state materials, 1981 to present. Available on BRS,(800) 289-4277, online service.Inquire as to cost and availability.

Current Contents Search. Institute for Scientific Information, 3501 Market Street, Philadelphia, PA 19104. (215) 386-0100. FAX (215) 386-6362. Contains citations to articles listed in the table of contents of science and technology journals. Also articles in social sciences and life sciences journals. Available on

BRS,(800) 289-4277, DIALOG,(800) 334-2564, online services. Inquire as to cost and availability.

Dissertation Abstracts. University Microfilms International, 300 North Zeeb Road, Ann Arbor, MI 48106. (800) 521-0600 or (313) 761-4700. Scope includes virtually all doctoral dissertations accepted at accredited American institutions from 1861 to present in 252 subject areas. Available on BRS,(800) 289-4277, DIALOG,(800) 334-2564, and OCLC EPIC,(800) 848-5878, online services. Also available on CD-ROM. Inquire as to cost and availability.

NTIS Bibliographic Database. National Technical Information Service, 5285 Port Royal Road, Springfield, Va 22161. (703) 487-4929 or FAX (703) 321-8199. Broad coverage of government-sponsored science and technology research reports, 1964 to present. Available on BRS,(800) 289-4277, DIALOG, (800) 334-2564, ORBIT, (800) 456-7248, and STN International, FIZ Karlsruhe, P.O. Box 2465, W-7500, Karlsruhe 1, Germany, online services. Also available on CD-ROM. Inquire as to cost and availability.

SCISEARCH. Institute for Scientific Information, 3501 Market Street, Philadelphia, PA 19104. (800) 523-1850 or (215) 386-0100. Broad multidisciplinary title and author index to the international literature of science and technology, 1974 to present. Available on DIALOG,(800) 334-2564, and ORBIT,(800) 456-7248, online services. Also available on CD-ROM. Inquire as to cost and availability.

WILSONLINE. H.W. Wilson Company, 950 University Avenue, Bronx, NY 10452. (800) 367-6770 or (212) 588-8400. Makes available online versions of the printed H.W. Wilson indexes including Applied Science and Technology Index, Business Periodicals Index, General Science Index, and Readers' Guide to Periodical Literature.Period covered is generally 1983 to present. Available on BRS,(800) 289-4277, DIALOG,(800) 334-2564, and OCLC EPIC,(800) 848-5878, online services. Also available on CD-ROM. Inquire as to cost and availability.

PERIODICALS

ACM Transactions On Programming Languages and Systems. Association for Computing Machinery, 1515 Broadway, 17th Floor, New York, NY 10036-5701. (212) 869-7440. FAX (212) 944-1318. 1979 to present. Quarterly. $105.00 for non-members.

ACM Transactions On Software Engineering and Methodology. Association for Computing Machinery, 11 West 42nd Street, New York, NY 10036. (212) 869-7440. Fax (212) 869-0481. 1991 to present. Quarterly. $85.00 per year.

Byte. McGraw-Hill Inc., Byte Publications, One Phoenix Mill Lane, Peterborough, NH 03458. (603) 924-9281. FAX (603) 924-2550. 1975 to present. Monthly. $29.95 per year.

Communications of the ACM. Association for Computing Machinery, 1515 Broadway, 17th Floor, New York, NY 10036-5701. (212) 869-7440. FAX (212) 944-1318. 1958 to present. Monthly. $114.00 for non-members.

Computer Languages. Elsevier Science [journals], 660 White Plains Rd., Tarrytown, NY 10159-5153. (919) 524-9200. FAX (919) 333-2444. 1976 to present. Four times a year. $445.00 per year.

Dr. DOBB'S JOURNAL of SofTWARE TOOLS. M and T Publishing, Inc. 501 Galveston Drive, Redwood City, CA 94063. (415) 366-3600. 1976 to present. Monthly. $30.00 per year.

IEEE Software. IEEE Computer Society Press, 10662 Vaqueros Circle, Box 3014, Los Alamitos, CA 90720. (714) 821-8380. 1984 to present. Bi-monthly. $175.00 per year.

IEEE Transactions On Software Engineering. IEEE, 345 E. 47th Street, New York, NY 10017. (908) 981-0060. FAX (908) 981-9667. 1975 to present. Monthly. $400.00 per year for non-members.

Journal of Computer and Software Engineering. Ablex Publishing Corporation, 355 Chestnut Street, Norwood, NJ 07648. (201) 767-8450 or FAX (201) 767-6717. 1993 to present. Quarterly. $110.00 per year.

Journal of Functional Programming. Cambridge University Press, Journals Department, 40 West 20th Street, New York, NY 10011-4211. (212) 924-3900. 1991 to present. Quarterly. Inquire for price.

Journal of Logic Programming. Elsevier Science Publishing Company, Inc., 655 Avenue of the Americas, New York, NY 10010. (212) 989-5800. 1984 to present. Twelve times a year. $545.00 per year (institutions).

Journal of Object-oriented Programming. Sigs Publications Inc., 588 Broadway, Suite 604, New York, NY 10012-5408. (212) 274-0640. FAX (212) 274-0646. 1988 to present. Nine times a year. $59.00 per year.

Journal of Programming Languages. Chapman & Hall, Journals Promotion Department, One Penn Plaza, 41st Floor, New York, NY 10019. (212) 564-1060. FAX (212) 564-1505. 1992 to present. Four times a year. $210.00 per year.

Letters On Programming Languages and Systems. Association for Computing Machinery, 1515 Broadway, 17th Floor, New York, NY 10036-5701. (212) 869-7440. FAX (212) 944-1318. Irregular. $105.00 for non-members.

Programmer's Journal. Oakley Publishing Company, 150 North 4th Street, Springfield, OR 97477. (503) 747-0800. 1983 to present. Bimonthly. $20.00 per year.

Science of Computer Programming. Elsevier Science Inc., Box 882, Madison Square Station, New York, NY 10159. (212) 989-5800. FAX (212) 633-3990. 1981 to present. Six times a year. $411.00 per year.

Software Engineering. Auerbach Publishers, 1 Penn Plaza, New York, NY 10119-0118. 1989 to present. Bimonthly. $145.00 per year.

Software: Practice and Experience. John Wiley and Sons, Inc., 605 Third Avenue, New York, NY 10158. (800) 526-5368 or (212) 850-6000. 1971 to present. Monthly. $440.00 per year.

Unix Review. Miller Freeman Inc., 600 Harrison Street, San Francisco, CA 94107. (415) 905-2200. FAX (415) 905-2232. 1983 to present. Monthly. $55.00 per year.

Unixworld. McGraw-Hill Publishing Company, 1221 Avenue of the Americas, New York, NY 10020. (800) 262-4729 or (212) 512-3825. 1984 to present. Monthly. $18.00 per year.

RESEARCH CENTERS AND INSTITUTES

Harvard University Center For Research In Computing Technology. Aiken Computational Laboratory, 33 Oxford Street, Cambridge, MA 02138. (617) 495-4117. FAX (617) 495-9837.

Ohio State University Computer and Information Science Research Center. 2036 Neil Avenue Mall, Columbus, OH 43210. (614) 292-6374. FAX (614) 292-9021.

University of Texas At Austin, Computation Center. Austin, TX 78712. (512) 471-3241.

COMPUTER SECURITY

See also: AUTOMATION, COMPUTER MEMORY, COMPUTER OPERATING SYSTEMS, COMPUTER PROGRAMMING

ABSTRACT SERVICES AND INDEXES

ACM Guide To Computing Literature. Association for Computing Machinery, Association for Computing Machinery, 1515 Broadway, 17th Floor, New York, NY 10036-5701. (212) 869-7440. FAX (212) 944-1318. 1964 to present. Monthly. $190.00 per year for non-members.

Applied Science and Technology Index; A Cumulative Subject Index To English Language Periodicals In the Fields of Aeronautics and Space Science, Computer Technology, Chemistry, Construction Industry, Energy and Related Areas. H.W. Wilson Co., 950 University Avenue, Bronx, NY 10452. (800) 367-6770 or (212)588-8400. FAX (718) 590-1617. From 1958 to present. Monthly. Inquire about cost and availability. Also available on CD-ROM and online.

Computer Abstracts, MCBUniversity Press Ltd., PO Box 10812, Birmingham, AL 35201-0812. (800) 633-4931. FAX (205) 995-1588. Monthly. Covers computer theory, data, hardware, systems, networks, human-computer interaction, artificial intelligence, as well as applications of computers in aerospace, business, CAD/CAM, cartography, civil engineering, electronics and electrical engineering, industrial engineering, mechanical engineering, medicine, structural engineering, etc. Also available on CD-ROM.

Computer & Control Abstracts (inspec). INSPEC/IEEE, Box 1331, 445 Hoes Lane, Piscataway, NJ 08855-1331. (908) 562-5549. Abstracts organized by subjects of international technological information. Monthly. $1455.00 per year. Also available on CD-ROM and online via BRS Online Products, DIALOG Information Services, Orbit Search Service, and STN International.

Computer and Information Systems Abstracts Journal. Cambridge Scientific Abstracts, 7200 Wisconsin Avenue, Bethesda, MD 20814. (301) 961-6750. Fax (301) 961-6720. 1962 to present. Monthly. $1265.00 per year. Also available online via STN International.

Computer Literature Index. Applied Computer Research Inc., Box 82266, Phoenix, AZ 85071-2266. (800) 234-2227. 1971 to present. Quarterly plus annual cumulation. $198.50 per year. Bibliography of books, articles, and reports.

Computer Security and Privacy Abstracts. Techgnosis Ltd., Blade House, Battersea Road, Stockport, Cheshire 3AE, England. Telephone 061-442-2639. FAX 061-443-1162. Six times a year. Inquire for price.

Computing Journal Abstracts. Techgnosis Ltd., Blade House, Battersea Road, Stockport, Cheshire 3AE, England. Telephone 061-442-2639. FAX 061-443-1162. 1969 to present. Monthly. Inquire for price.

Computing Reviews. Association for Computing Machinery, 1515 Broadway, 17th Floor, New York, NY 10036-5701. (212) 869-7440. FAX (212) 944-1318. 1960 to present. Monthly. $130.00 per year for non-members. Also available online via DIALOG Information Services.

Current Contents: Engineering, Technology, and Applied Sciences. Institute for Scientific Information, 3501 Market Street, Philadelphia, PA 19104. (215) 386-0100. FAX (215) 386-6362. 1970 to present. Weekly. $442.00 per year.

Engineering Index Monthly. Engineering Information, Inc., Castle Point on the Hudson, Hoboken, NJ 07030. (800) 221-1044. FAX (212) 832-1857. Monthly. $2300.00 per year. Also available online as COMPENDEX, and also on CD-ROM. Covers chemical engineering, computers, electrical engineering, civil engineering, metals and mining, industrial management, and mechanical engineering.

Index To Scientific and Technical Proceedings. Institute for Scientific Information, 3501 Market St., Philadelphia, PA19104. (215) 386-0100. FAX (215) 386-6362. Monthly. $500.00 per year.

Index To Scientific Reviews. Institute for Scientific Information, 3501 Market St., Philadelphia, PA 19104. (215) 386-0100. FAX (215) 386-6362. Semi-annual.

Physics Abstracts. Institute of Electrical Engineers, Michael Faraday House, Six Hill Way, Stevenage, Herts, SG1 2AY, England. Distributed by IEEE, 445 Hoes Lane, Piscataway, NJ 08854. (908) 562-5549. 1898 to present. Monthly. $2700.00 per year. Also available online as INSPEC.

Science Citation Index. Institute for Scientific Information, 3501 Market Street, Philadelphia, PA 19104. (215) 386-0100. FAX (215) 386-6362. Inquire about availability and cost. Also available on CD-ROM.

ANNUAL REVIEWS AND YEARBOOKS

ACM Monograph Series. Association for Computing Machinery (ACM), 11 West 42nd Street, New York, NY 10036. (212) 869-7440. Fax (212) 869-0481. Irregular. Price varies, inquire.

Advances In Computers. Academic Press Inc., 6277 Sea Harbor Drive, Orlando, FL 32887. (800) 321-5068. Irregular. Price varies, inquire.

Advances In Computing Research. JAI Press, Inc., 55 Old Post Road, Number 2, Box 1678, Greenwich, CT 06830. (203) 661-7602. FAX (203) 661-0792. Annual. $68.50 per year.

ASSOCIATIONS AND PROFESSIONAL SOCIETIES

American Federation of Information Processing Societies, 1899 Preston White Drive, Reston, Va 22091-5435. (703) 620-8937.

Association For Computing Machinery. 1515 Broadway, 17th Floor, New York, NY 10036-5701. (212) 869-7440. FAX (212) 944-1318.

Computer and Automation Systems Association of Sme. 1 SME Drive, PO Box 930, Dearborn, MI 48121-0930. (313) 271-1500.

Computer Security Institute. 600 Harrison Street, San Francisco, CA 94107. (415) 905-2026. FAX (415) 905-2218.

IEEE Computer Society. 1730 Massachusetts Avenue NW, Washington, DC 20036-1992. (202) 371-0101. FAX (202) 728-9614.

Information Systems Security Association. 401 N. Michigan Avenue, Chicago, IL 60611. (312) 644-6610. FAX (312) 321-6869.

National Computer Security Association. 10 S. Courthouse Avenue, Carlisle, PA 17013. (717) 258-1816. FAX (717) 243-8642.

Special Group On Security, Audit and Control. c/o ACM, 1515 Broadway, New York, NY 10036. (212) 869-7440. FAX (212) 869-0481.

BIBLIOGRAPHIES

ACM Guide To Computing Literature. Association for Computing Machinery, 1515 Broadway, 17th Floor, New York, NY 10036-5701. (212) 869-7440. FAX (212) 944-1318. 1964 to present. Annual. $190.00 for non-members. Also available online via DIALOG.

DIRECTORIES AND BIOGRAPHICAL SOURCES

American Men & Women of Science, 1995-96. R.R. Bowker Staff, eds. 19th edition. 8 volumes. R.R. Bowker/Reed InternationalPublishing Company, 121 Chanlon Road, New Providence, NJ 07974. (908) 464-6800 or (800) 521-8110. 1995. $850.00

Computer Security Buyers Guide. Computer Security Institute, 600 Harrison Street, San Francisco, CA 94107. (415) 905-2370. Annual. Inquire for price.

Computing Information Directory. Hildebrandt Inc., PO Box 576, Pullman, WA 99114. 1981 to present. Annual. $199.95 per year.

Directory of Engineering Societies and Related Organizations. Gordon Davis. 13th edition. American Association of Engineering Societies, 1111 19th Street NW, Suite 608, Washington, DC 20036. (202) 296-2237 or (800) 658-8897. 1989. Inquire for price.

International Research Centers Directory. Gale Research, 835 Penobscot Building, Detroit, MI 48226-4094. (313) 961-2242. (800) 347-4253. FAX (313) 961-6083. 8th edition. 1995. $410.00

COMPUTER SECURITY

Ency. of Physical Sciences and Engineering Info. Sources

Research Centers Directory. Gale Research, 835 Penobscot Building, Detroit, MI 48226-4094. (313) 961-2242. (800) 347-4253. FAX (313) 961-6083. $485.00.

Scientific and Technical Organizations and Agencies Directory. Gale Research, 835 Penobscot Building, Detroit, MI 48226-4094. (313) 961-2242. (800) 347-4253. FAX (313) 961-6083. 4th edition. 1996. $195.00.

Who's Who In Technology. Gale Research, 835 Penobscot Building, Detroit, MI 48226-4094. (313) 961-2242. (800) 347-4253. FAX (313) 961-6083. 1995. $195.00.

ENCYCLOPEDIAS AND DICTIONARIES

Computer Dictionary. Donald D. Spencer. 4th edition. Camelot Publishing Company, 709 SW 80th Blvd., Gainesville, FL 32607-1537. (904) 331-0952. 1993. $24.95.

Dictionary of Computing. S.M.H. Collin. 2nd edition. Peter Collin Publishing, 8 the Causeway, Teddington, TW11 0HE, England. FAX 0181-943-3386. 1994. Inquire for cost.

Encyclopedia of Computer Science and Engineering. Anthony Ralston & Edwin D. Reilly Jr., editors. 3rd revised edition. Van Nostrand Reinhold, 115 Fifth Avenue, New York, NY 10003. (212) 254-3232 or FAX (212) 254-9499. 1993. $125.00.

Microsoft Press Computer Dictionary: the Comprehensive Standard For Business, School, Library, and Home. 2nd edition. Microsoft Press, One Microsoft Way, Redmond, WA 98052-6399. (206) 936-0055. FAX (206) 823-8101. 1994. $19.95.

GENERAL WORKS

Computer Security. John M. Carroll. 3rd edition. Butterworth-Heinemann, 313 Washington Street, Newton, MA 02158. (617) 928-2500 or (800) 366-2665. FAX (617) 928-2620. 1995. $49.95.

Computer System & Network Security. Gregory B. White, et al. CRC Publishers, Inc., 2000 Corporate Blvd., N.W., Boca Raton, FL 33431. (407) 994-0555 or (800) 272-7737. FAX (407) 994-0949. 1995. $59.95.

Fundamentals of Computer Security Technology. Edward Amoroso. Prentice Hall, 113 Sylvan Avenue, Route 9W, EnglewoodCliffs, NJ 07632. (201) 592-2000 or (800) 922-0579. 1994. $52.00.

Network and Internetwork Security: Principles and Practice. William Stallings. Macmillan Publishing, 200 Old Tappan Road, Old Tappan, NJ 07675. (800) 223-2336. FAX (800) 445-6991. 1995. $55.00.

Network Security: Data and Voice Communications. Fred Simonds. McGraw-Hill Publishing Company, 1221 Avenue of the Americas, New York, NY 10020. (800) 262-4729 or (212) 512-3825. 1995. $60.00 (hardbound) and $39.95 (paperbound).

Novell's Guide To Network Security. David J. Clarke & Rene Mendoza. Sybex, Inc. 2021 Challenger Drive, Alameda, CA 94501. (800) 227-2346. FAX (510) 523-2373. 1995. $44.99.

The Ultimate Computer Security Survey. James Schwab & Ken D. Biery. Butterworth-Heinemann, 313 Washington Street, New-

ton, MA 02158. (617) 928-2500 or (800) 366-2665. FAX (617) 928-2620. 1995. $59.95.

HANDBOOKS AND MANUALS

Computer Security Handbook. Arthur E. Hutt, et al., editors. 3rd edition. John Wiley and Sons, Inc., 605 Third Avenue, New York, NY 10158. (800) 526-5368 or (212) 850-6000. Inquire for cost and availability.

Internet World's Internet Security Handbook. Mecklermedia Corporation, 20 Ketchum Street, Westport, CT 06880. (203) 226-6967. FAX (203) 454-5840. 1995. $29.95.

ONLINE DATABASES AND CD-ROMS

Compendex Plus. Engineering Information, Inc., 345 East 47th Street, New York, NY 10017. (212) 705-7600 or (800) 221-1044. Contains citations with abstracts to worldwide literature in engineering and technology, from 1970 to present. Available on online BRS,(800) 289-4277, DIALOG, (800) 334-2564, ORBIT (800) 456-7248, and STN International, FIZ Karlsruhe, P.O. Box 2465, W-7500, Karlsruhe 1, Germany, online services. Also available on CD-ROM. Inquire as to cost and availability.

Compuscience. FIZ Karlsruhe, D-7514, Eggenstein-Leopoldshafen 2, Germany. Contains citations with abstracts to European and North American literature on computer science, 1972 to present. Available on STN International, FIZ Karlsruhe, P.O. Box 2465, W-7500, Karlsruhe 1, Germany, online service. Inquire as to cost and availability.

Computer and Information Systems Abstracts. Cambridge Scientific Abstracts, 7200 Wisconsin Avenue, Suite 601, Bethesda, MD 20814. (301) 961-6750 or (800) 843-7751. Contains citations to worldwide literature in theoretical and applied computer science and related areas, from 1981 to present.Inquire as to cost and availability.

Computer and Mathematics Search. Institute for Scientific Information, 3501 Market Street, Philadelphia, PA 19104. (215) 386-0100. FAX (215) 386-6362. Covers worldwide literature in computer science and mathematics, from 1980 to present. Available on BRS,(800) 289-4277, online service. Inquire as to cost andavailability.

Computer Database. Information Access Company, 362 Lakeside Drive, Foster City, CA 94404. (415) 378-5000 or (800) 227-8431. Contains citations with abstracts to literature from trade journals covering the computer,telecommunications,and electronics industries. Available on the BRS, (800) 289-4277, Compuserve Information Service,(800) 848-8990, and DIALOG,(800) 334-2564, online services. Inquire as to cost and availability.

CSA Engineering. Cambridge Scientific Abstracts, 7200 Wisconsin Avenue, Suite 601, Bethesda, MD 20814. (301) 961-6750 or (800) 843-7751. Contains citations and abstracts of international periodicals and research literature covering all fields of engineering and science and technology,including computer and information science, electronics, mechanical engineering, solid state materials, 1981 to present. Available on BRS,(800) 289-4277, online service.Inquire as to cost and availability.

Current Contents Search. Institute for Scientific Information, 3501 Market Street, Philadelphia, PA 19104. (215) 386-0100. FAX (215) 386-6362. Contains citations to articles listed in the table of contents of science and technology journals. Also articles in social sciences and life sciences journals. Available on BRS,(800) 289-4277, DIALOG,(800) 334-2564, online services. Inquire as to cost and availability.

Dissertation Abstracts. University Microfilms International, 300 North Zeeb Road, Ann Arbor, MI 48106. (800) 521-0600 or (313) 761-4700. Scope includes virtually all doctoral dissertations accepted at accredited American institutions from 1861 to present in 252 subject areas. Available on BRS,(800) 289-4277, DIALOG,(800) 334-2564, and OCLC EPIC,(800) 848-5878, online services. Also available on CD-ROM. Inquire as to cost and availability.

Inspec. Institution of Electrical Engineers, Michael Faraday House, Six Hills Way, Stevenage, Herts. SG1 2AY, England. Telephone: 0438 313311 or FAX 0438 742840. Contains citations to the worldwide literature of physics, electronics and electrical engineering, computer technology, and related fields. Available on BRS,(800) 289-4277, DIALOG, (800) 334-2564, ORBIT, (800) 456-7248, and STN International, FIZ Karlsruhe, P.O. Box 2465, W-7500, Karlsruhe 1, Germany, online services. Inquire as to cost and availability.

NTIS Bibliographic Database. National Technical Information Service, 5285 Port Royal Road, Springfield, Va 22161. (703) 487-4929 or FAX (703) 321-8199. Broad coverage of government-sponsored science and technology research reports, 1964 to present. Available on BRS,(800) 289-4277, DIALOG, (800) 334-2564, ORBIT, (800) 456-7248, and STN International, FIZ Karlsruhe, P.O. Box 2465, W-7500, Karlsruhe 1, Germany, online services. Also available on CD-ROM. Inquire as to cost and availability.

SCISEARCH. Institute for Scientific Information, 3501Market Street, Philadelphia, PA 19104. (800) 523-1850 or (215) 386-0100. Broad multidisciplinary title and author index to the international literature of science and technology, 1974 to present. Available on DIALOG,(800) 334-2564, and ORBIT,(800) 456-7248, online services. Also available on CD-ROM. Inquire as to cost and availability.

PERIODICALS

ACM Transactions On Computer Systems. Association for Computing Machinery, 11 West 42nd Street, New York, NY 10036. (212) 869-7440. Fax (212) 869-0481. 1964 to present. Annual. $180.00 per year.

Annual Conference On Computer Security Applications. IEEE Computer Society Press, 10662 Vaqueros Circle, Box 3014, Los Alamitos, CA 90720. (714) 821-8380. 1989 to present. Annual. Price varies.

Byte. Byte Publications, 1 Phoenix Mill Lane, Peterborough, NH 03458-1465. (603) 924-9281. 1975 to present. Monthly. $21.00 per year.

Communications of the Acm. Association for Computing Machinery, 11 West 42nd Street, New York, NY 10036. (212) 869-7440. Fax (212) 869-0481. 1958 to present. Monthly. $105.00 per year.

Computer Design. PennWell Publishing Company, 1 Technology Park Drive, Westford, MA 01886. (508) 692-0700 or FAX (508) 692-0525. 1962 to present. Semi-monthly. $70.00 per year.

Computer Magazine. IEEE Computer Society, 10662 Vaqueros Circle, Box 3014, Los Alamitos, CA 90720. (714) 821-8380. 1966 to present. Monthly. $49.00 per year.

Computer Fraud and Security Bulletin. Elsevier Science [journals], 660 White Plains Rd., Tarrytown, NY 10159-5153. (919) 524-9200. FAX (919) 333-2444. 1978 to present. 12 times a year. $355.00 per year.

Computer Security Digest. Computer Protection Systems Inc., 150 N. Main Street, Plymouth, MI 48170-1236. (313) 459-8787. FAX (313) 459-2720. 1983 to present. Monthly. $125.00 per year.

Computer Security Journal. Computer Security Institute, 600 Harrison Street, San Francisco, CA 94107. (415) 905-2370. fax (415) 905-2234. 1981 to present. Semi-annual. $85.00 per year.

Computer Technology Review. West World Productions, Inc., 924 Westwood Boulevard, Suite 650, Los Angeles, CA 90024-2927. (213) 208-1335 or FAX (213) 208-1054. 1981 to present. Sixteen times per year. $80.00 per year.

Computers and Security. Elsevier Science [journals], 660 White Plains Rd., Tarrytown, NY 10159-5153. (919) 524-9200. FAX (919) 333-2444. 1982 to present. Eight times a year. $325.00 per year.

Datamation. R.R. Bowker Inc., 121 Chanlon Road, New Providence, NJ 07974. (800) 521-8110 or (908) 464-6800. 1957 to present. Biweekly. $40.00 per year.

IEEE Symposium On Research In Security and Privacy: Proceedings. IEEE Computer Society Press, 10662 Vaqueros Circle,Box 3014, Los Alamitos, CA 90720. (714) 821-8380. FAX (714) 821-4641. 1980 to present. Annual. Inquire for price.

Journal of Computer Security. IOS Press, Box 10558, Burke, VA 22009-0558. (703) 323-5554. FAX (703) 250-4705. 1992 to present. Quarterly. $161.00 per year.

National Computer Security Association Newsletter. National Computer Security Association, 10 S. Courthouse Avenue, Carlisle, PA 17013. (717) 258-1816. FAX (717) 243-8642. Six times a year. $25.00 per year for non-members.

RESEARCH CENTERS AND INSTITUTES

Florida Atlantic University, Center For Computer Security and Fault Tolerance. Department of Computer Engineering, Boca Raton, FL 33431. (407) 367-3466. FAX (407) 367-2659.

COMPUTER SCIENCE

ABSTRACT SERVICES AND INDEXES

ACM Guide To Computing Literature. Association for Computing Machinery, 11 West 42nd Street, New York, NY 10036. (212) 869-7440. Fax (212) 869-0481. 1964 to present. Annual. $190.00 per year.

COMPUTER SCIENCE

Ency. of Physical Sciences and Engineering Info. Sources

Applied Mechanics Reviews: An Assessment of World Literature In Engineering Sciences. American Society of Mechanical Engineers, 345 East 47th Street, New York, NY 10017. (212) 705-7703. 1948 to present. Monthly. $360.00 per year.

Applied Science and Technology Index; A Cumulative Subject Index To English Language Periodicals In the Fields of Aeronautics and Space Science, Computer Technology, Chemistry, Construction Industry, Energy and Related Areas. H.W. Wilson Co., 950 University Avenue, Bronx, NY 10452. (800) 367-6770 or (212) 588-8400. FAX (718) 590-1617. From 1958 to present. Monthly. Inquire about cost and availability. Also available on CD-ROM.

Artificial Intelligence Abstracts. R.R. Bowker, Bowker A&I Publishing, 121 Chanlon Road, New Providence, NJ 07974. (908) 771-7714. Fax (908) 771-7725. 1985 to present. Monthly. $495.00 per year.

CAD/CAM Abstracts. R.R. Bowker, Bowker A&I Publishing, 121 Chanlon Road, New Providence, NJ 07974. (908) 771-7714. Fax (908) 771-7725. 1984 to present. Monthly. $495.00 per year.

Computer and Control Abstracts. IEEE (Institute of Electrical and Electronic Engineers), 445 Hoes Lane, Piscataway, NJ 08854. (908) 562-5545 or (800) 678-IEEE. 1969 to present. Monthly. $1385.00 per year.

Computer and Information Systems Abstracts Journal. Cambridge Scientific Abstracts, 7200 Wisconsin Avenue, Bethesda, MD 20814. (301) 961-6750. Fax (301) 961-6720. 1962 to present. Monthly. $1155.00 per year.

Computer Literature Index. Applied Computer Research, Inc., Box 82266, Phoenix, AZ 85071-2266. (800) 234-2227. 1971 to present. Monthly. $195.00.

Computing Reviews. Association for Computing Machinery, 11 West 42nd Street, New York, NY 10036. (212) 869-7440. Fax (212)869-0481. 1960 to present. Monthly. $130.00 per year.

Electrical and Electronics Abstracts. Institute of Electrical Engineers, Michael Faraday House, Six Hill Way, Stevenage, Herts, SG1 2AY, England. Distributed by IEEE, 445 Hoes Lane, Piscataway, NJ 08854. (908) 562-5549. 1898 to present. Monthly. $2200.00 per year. Also available online as INSPEC.

Engineering Index Monthly. Engineering Information, Inc., Castle Point on the Hudson, Hoboken, NJ 07030. (800) 221-1044. FAX (212) 832-1857. Monthly. $2200.00 per year. Also available online as COMPENDEX, and also on CD-ROM.

Government Reports Announcements and Index. National Technical Information Service (NTIS), 5285 Port Royal Road, Springfield, VA 22161. (703) 487-4650. 1968 to present. Annual. $525.00 per year. Also available online as NTIS BIBLIOGRAPHIC DATABASE and on CD-ROM.

Microcomputer Index. Learned Information, Inc., 143 Old Marlton Pike, Medford, NJ 08055. (609) 654-6266 or FAX (609) 654-4309. 1980 to present. Bimonthly. $160.00 per year.

Physics Abstracts. Institute of Electrical Engineers, London, United Kingdom. Available from: Institute of Electrical and Electronic Engineers (IEEE), 345 East 47th Street, New York, NY 10017. (212) 705-7900.

Physics Briefs. American Institute of Physics, 335 East 45th Street, New York, NY 10017 (212) 661-9404 or FAX (516)349-9704. 1980 to present. Six times per year. $2600.00 per year.

Science Citation Index. Institute for Scientific Information, 3501 Market Street, Philadelphia, PA 19104. (215) 386-0100. FAX (215) 386-6362. Inquire about availability and cost. Also available on CD-ROM.

Star. (SCIENTIFIC and TECHNICAL AEROSPACE REPORTS). U.S. National Aeronautics and Space Administration, Scientific and Technical Information Facility, Box 8757, Baltimore-Washington International Airport, Baltimore, MD 21240. (202) 755-2210. Semimonthly, with semiannual and annual indexes. $85.00 per year.

ANNUAL REVIEWS AND YEARBOOKS

ACM Monograph Series. Association for Computing Machinery (ACM), 11 West 42nd Street, New York, NY 10036. (212) 869-7440. Fax (212) 869-0481. Irregular. Price varies, inquire.

Advances In Computers. Academic Press Inc., 6277 Sea Harbor Drive, Orlando, FL 32887. Irregular. Price varies, inquire.

Advances In Computing Research. JAI Press, Inc., 55 Old Post Road, Number 2, Box 1678, Greenwich, CT 06830. (203) 661-7602. FAX (203) 661-0792. Annual. $68.50 per year.

Advances In Electronics and Electron Physics. Academic Press Inc., 6277 Sea Harbor Drive, Orlando, FL 32887. Irregular. Price varies, inquire.

Computer Science and Scientific Computing. Academic Press Inc., 6277 Sea Harbor Drive, Orlando, FL 32887. Irregular. Price varies, inquire.

Proceedings, Association For Computational Linguistics, 445 South Street, Morristown, Nj 07960. (201) 829-4312. 1962 to present. Annual. $25.00 per year.

Solid State Physics - Advances In Research and Applications. Academic Press Inc., 6277 Sea Harbor Drive, Orlando, FL 32887. Irregular. Price varies, inquire

ASSOCIATIONS AND PROFESSIONAL SOCIETIES

American Association For Artificial Intelligence, 445 Burgess Drive, Menlo Park, Ca 94025-3442. (415) 328-3123.

American Federation of Information Processing Societies, 1899 Preston White Drive, Reston, Va 22091-5435. (703) 620-8937.

American Institute of Physics, 335 East 45th Street, New York, Ny 10017 (212) 661-9404 Or Fax (516)349-9704.

American Physical Society, 1 Research Road, Box 1000, Ridge, Ny 10961-2701. (516) 924-5533 or FAX (516) 924-5294.

American Society For Information Science, 8720 Georgia Avenue, Suite 501, Silver Spring, Md 20910-3602. (301) 495-0900.

Association For Computational Linguistics, 445 South Street, Morristown, Nj 07960. (201) 829-4312.

Association For Information Management, 2026-c Opitz Boulevard, Woodbridge, Va 22191. (703) 490-4246.

Boston Computer Society (bcs), Inc., One Center Plaza, Boston, MA 02108-1883.

Society For Computer Simulation International, 4838 Ronson Court Lane, San Diego, Ca 92111-1810. (619) 277-3888.

Society For Industrial and Applied Mathematics (siam), 3600 University City Science Center, Philadelphia, Pa 19104. (215) 382-2688.

BIBLIOGRAPHIES

Scientific and Technical Books In Print; An Index To Literature In Science and Technology. R.R. Bowker Company, 205 East 42nd Street, New York, NY 10017. (800) 521-8110 or (212) 916-1600. 1992. $250.00.

DIRECTORIES AND BIOGRAPHICAL SOURCES

American Men and Women of Science: Physical and Biological Sciences. Eighteenth edition. R.R. Bowker Company, 245 West 17th Street, New York, NY 10011. (800) 521-8810 or (212) 916-1600. $750.00.

ASIS Handbook and Directory, American Society For Information Science, 8720 Georgia Avenue, Suite 501, Silver Spring, Md 20910-3602. (301) 495-0900. Annual. $100.00.

Research Centers Directory. Gale Research, 835 Penobscot Building, Detroit, MI 48226-4094. (313) 961-2242. (800) 347-4253. FAX (313) 961-6083. 1992. (800) 521-0707. $400.00.

Who's Who In Information Management. Association for Information Management, 2026-C Opitz Boulevard, Woodbridge, VA 22191. (703) 490-4246. Annual. Inquire about price and availability.

ENCYCLOPEDIAS AND DICTIONARIES

Academic Press Dictionary of Science and Technology. Christopher Morris, editor. Academic Press Inc., 6277 Sea Harbor Drive, Orlando, FL 32887. 1991. $115.00.

Dictionary of Scientific Literacy. Richard Brennan. John Wiley and Sons, Inc., 605 Third Avenue, New York, NY 10158. (800) 526-5368 or (212) 850-6000. 1991. $22.95.

Encyclopedia of Computer Science. Anthony Ralston and Edwin D. Reilly, editors. Van Nostrand Reinhold, 115 Fifth Avenue, New York, NY 10003. (212) 254-3232 or FAX (212) 254-9499. Third edition. 1993. $99.95.

Macmillan Encyclopedia of Computers. Gary G. Bitter, editor. Macmillan Publishing, 200 Old Tappan Road, Old Tappan, NJ 07675. (800) 223-2336. FAX (800) 445-6991. 1992. $150.00.

McGraw-Hill Encyclopedia of Science and Technology. McGraw-Hill Book, Incorporated, 1221 Avenue of the Americas, New York, NY 10020. (212) 997-3675. Seventh edition, 20 volumes. 1992. $1900.00.

Prentice Hall's Illustrated Dictionary of Computing. Jonar C. Nader. Prentice Hall, 113 Sylvan Avenue, Route 9W, Englewood Cliffs, NJ 07632. (201) 592-2000 or (800) 922-0579. 1992. $24.95.

Software Encyclopedia. R.R. Bowker Inc., 121 Chanlon Road, New Providence, NJ 07974. (800) 521-8110 or (908) 464-6800. Annual. $195.00.

GENERAL WORKS

The Age of Intelligent Machines. Raymond Kurzweil. MIT Press, 55 Hayward Street, Cambridge, MA 02142. (617) 253-2884 or (800) 356-0343. FAX (617) 253-1709. 1990. $29.95.

Applied Artificial Intelligence. K.Warwick and P.Peregrinus, editors. IEEE (Institute of Electrical and Electronic Engineers), 445 Hoes Lane, Piscataway, NJ 08854. (908) 562-5545 or (800) 678-IEEE. 1991. $59.00.

Computers and the Imagination: Visual Adventures Beyond the Edge. Clifford A. Pickover. St. Martin's Press, Inc., 175 Fifth Avenue, New York, NY 10010. (212) 674-5151 or (800) 221-7945. FAX (212) 420-9314. 1991. $29.95.

Computing For Scientists and Engineers: A Workbook of Analysis, Numerics, and Applications. William J. Thompson. Wiley-Interscience, John Wiley and Sons, Inc., 605 Third Avenue, New York, NY 10158. (800) 526-5368 or (212) 850-6000. 1992. $54.95.

Computing the Future: A Broader Agenda For Computer Science and Engineering. Juris Hartmanis and Herbert Lin, editors. National Academy Press, Division of the National Academy of Science, 2101 Constitution Avenue, N.W., Lockbox 285, Washington, DC 20055. (800) 624-6242. FAX (202) 334-2793. 1992. $24.95.

Creating Artificial Life: Computer Modeling Experiments. E. Rietman. TAB Books, P.O. Box 40, Blue Ridge Summit, PA 17294-0850. (717) 794-2191 or (800) 233-1128. FAX (717) 794-2080. 1993. $29.95.

Desktop Computers In Perspective. Richard A. Henle. Oxford University Press, Inc., 200 Madison Avenue, New York, NY 10016. (800) 334-4249 or (212) 679-7300. FAX (212) 726-6446. 1992. $24.95.

Designing Intelligence: A Framework For Smart Systems. Steven H. Kim. Oxford University Press, Inc., 200 Madison Avenue, New York, NY 10016. (800) 334-4249 or (212) 679-7300. FAX (212) 726-6446. 1991. $40.00.

Dynamic Programming: A Practical Introduction. E. Horwood, distributed by Prentice Hall, 113 Sylvan Avenue, Route 9W, Englewood Cliffs, NJ 07632. (201) 592-2000 or (800) 922-0579. 1991. $45.00.

Expert Systems: A Practical Introduction. Peter S. Sell. John Wiley and Sons, Inc., 605 Third Avenue, New York, NY 10158. (800) 526-5368 or (212) 850-6000. 1985. $54.95.

Fuzzy Logics: Exploring Expert Systems. S.H. Hashim. TAB Books, P.O. Box 40, Blue Ridge Summit, PA 17294-0850. (717) 794-2191 or (800) 233-1128. FAX (717) 794-2080. 1993. $32.95.

COMPUTER SCIENCE

Ency. of Physical Sciences and Engineering Info. Sources

In Our Own Image: Building An Artificial Person. Maureen Caudill. Oxford University Press, Inc., 200 Madison Avenue, New York, NY 10016. (800) 334-4249 or (212) 679-7300. 1992. $23.00.

Introduction To Artificial Intelligence and Expert Systems. Presented by Mark S. Fox. IEEE (Institute of Electrical and Electronic Engineers), 445 Hoes Lane, Piscataway, NJ 08854. (908) 562-5545 or (800) 678-IEEE. Videotape. 1991. $109.00.

Introduction To Parallel Computing. Ted G. Lewis. Prentice Hall, 113 Sylvan Avenue, Route 9W, Englewood Cliffs, NJ 07632. (201) 592-2000 or (800) 922-0579. 1992. $48.00.

Introduction To Parallel Programming. K. Mani Chandy. Jones and Bartlett Publishers, Inc., 1 Exeter Plaza, Boston, MA 02116. (800) 832-0034 or (617) 482-3900. FAX (617) 859-7675. 1992. $25.00.

Made-up Minds: A Constructivist Approach To Artificial Intelligence. Gary L. Drescher. MIT Press, 55 Hayward Street, Cambridge, MA 02142. (617) 253-2884 or (800) 356-0343. FAX (617) 253-1709. 1991. $27.50.

Object-oriented Databases. J.G. Hughes. Prentice Hall, 113 Sylvan Avenue, Route 9W, Englewood Cliffs, NJ 07632. (201) 592-2000 or (800) 922-0579. 1991. $40.00.

Pattern Recognition and Machine Learning. Yuichiro Anzai. Academic Press Inc., 6277 Sea Harbor Drive, Orlando, FL 32887. 1992. $74.95.

Past, Present, Parallel: A Survey of Available Parallel Computer Systems. Arthur Trew and Greg Wilson. Springer-Verlag, 175 Fifth Avenue, New York, NY 10010. (212) 460-1500. 1991. $69.00.

Signal Processing, Image Processing, and Pattern Recognition. Stephen P. Banks. Prentice Hall, 113 Sylvan Avenue, Route 9W, Englewood Cliffs, NJ 07632. (201) 592-2000 or (800) 922-0579. 1991. $53.00.

The Social Impact of Computers. Richard S. Rosenberg. Academic Press, Inc., Academic Press Inc., 6277 Sea Harbor Drive, Orlando, FL 32887. (800) 321-5068. 1992. $39.95.

Software Engineering. Robert S. Arnold, editor. IEEE (Institute of Electrical and Electronic Engineers), 445 Hoes Lane, Piscataway, NJ 08854. (908) 562-5549. 1993. $75.00.

Software Engineering: A Holistic View. Bruce I. Blum. Oxford University Press, Inc., 200 Madison Avenue, New York, NY 10016. (800) 334-4249 or (212) 679-7300. 1992. $49.95.

Virtual Reality: Through the New Looking Glass. Ken Pimentel. McGraw-Hill Publishing Company, 1221 Avenue of the Americas, New York, NY 10020. (800) 262-4729 or (212) 512-3825. 1993. $22.95.

Vision, Instruction, and Action. David Chapman. MIT Press, 55 Hayward Street, Cambridge, MA 02142. (617) 253-2884 or (800) 356-0343. FAX (617) 253-1709. 1991. $35.00.

What Computers Still Can't Do: A Critique of Artificial Reason. Hubert L. Dreyfus. MIT Press, 55 Hayward Street, Cambridge, MA 02142. (617) 253-2884 or (800) 356-0343. FAX (617) 253-1709. 1992. $13.95.

HANDBOOKS AND MANUALS

Computer Engineering Handbook. C.H. Chen, editor. McGraw-Hill Publishing Company, 1221 Avenue of the Americas, New York, NY 10020. (800) 262-4729 or (212) 512-3825. 1992. $79.50. <subpara2>Online Databases and CD-ROMs

Compendex Plus. Engineering Information, Inc., 345 East 47th Street, New York, NY 10017. (212) 705-7600 or (800) 221-1044. Contains citations with abstracts to worldwide literature in engineering and technology, from 1970 to present. Available on online BRS,(800) 289-4277, DIALOG, (800) 334-2564, ORBIT (800) 456-7248, and STN International, FIZ Karlsruhe, P.O. Box 2465, W-7500, Karlsruhe 1, Germany, online services. Also available on CD-ROM. Inquire as to cost and availability.

Compuscience. FIZ Karlsruhe, D-7514, Eggenstein-Leopoldshafen 2, Germany. Contains citations with abstracts to European and North American literature on computer science, 1972 to present. Available on STN International, FIZ Karlsruhe, P.O. Box 2465, W-7500, Karlsruhe 1, Germany, online service. Inquire as to cost and availability.

Computer and Information Systems Abstracts. Cambridge Scientific Abstracts, 7200 Wisconsin Avenue, Suite 601, Bethesda, MD 20814. (301) 961-6750 or (800) 843-7751. Contains citations to worldwide literature in theoretical and applied computer science and related areas, from 1981 to present.Inquire as to cost and availability.

Computer and Mathematics Search. Institute for Scientific Information, 3501 Market Street, Philadelphia, PA 19104. (215) 386-0100. FAX (215) 386-6362. Covers worldwide literature in computer science and mathematics, from 1980 to present. Available on BRS,(800) 289-4277, online service. Inquire as to cost and availability.

Computer Database. Information Access Company, 362 Lakeside Drive, Foster City, CA 94404. (415) 378-5000 or (800)227-8431. Contains citations with abstracts to literature from trade journals covering the computer,telecommunications,and electronics industries. Available on the BRS, (800) 289-4277, Compuserve Information Service,(800) 848-8990, and DIALOG,(800) 334-2564, online services. Inquire as to cost and availability.

Current Contents Search. Institute for Scientific Information, 3501 Market Street, Philadelphia, PA 19104. (215) 386-0100. FAX (215) 386-6362. Contains citations to articles listed in the table of contents of science and technology journals. Also articles in social sciences and life sciences journals. Available on BRS,(800) 289-4277, DIALOG,(800) 334-2564, online services. Inquire as to cost and availability.

Dissertation Abstracts. University Microfilms International, 300 North Zeeb Road, Ann Arbor, MI 48106. (800) 521-0600 or (313) 761-4700. Scope includes virtually all doctoral dissertations accepted at accredited American institutions from 1861 to present in 252 subject areas. Available on BRS,(800) 289-4277, DIALOG,(800) 334-2564, and OCLC EPIC,(800) 848-5878, online services. Also available on CD-ROM. Inquire as to cost and availability.

Information Science Abstracts. IFI/Plenum Data Company, 302 Swann Avenue, Alexandria, VA 22301. (703) 683-1085 or (800) 368-3093. Contains citations with abstracts to worldwide literature on information science, storage, retrieval and utilization, from 1966 to present.Available on DIALOG,(800) 334-2564, online service. Inquire as to cost and availability.

INSPEC. Institution of Electrical Engineers, Michael Faraday House, Six Hills Way, Stevenage, Herts. SG1 2AY, England. Telephone: 0438 313311 or FAX 0438 742840. Contains citations to the worldwide literature of physics, electronics and electrical engineering, computer technology, and related fields. Available on BRS,(800) 289-4277, DIALOG, (800) 334-2564, ORBIT, (800) 456-7248, and STN International, FIZ Karlsruhe, P.O. Box 2465, W-7500, Karlsruhe 1, Germany, online services. Inquire as to cost and availability.

Mathsci. American Mathematical Society, P.O. Box 6248, Providence, RI 02940. (800) 321-4667 or (401) 455-4000 or FAX (401) 331-3842. Scope includes pure and applied mathematics and related areas of physics, statistics, engineering, computer science, and operations research literature since 1959. Available on DIALOG,(800) 334-2564, online service. Also available on CD-ROM. Inquire as to cost and availability.

Microcomputer Index. Learned Information, Inc., 143 Old Marlton Pike, Medford, NJ 08055-8707. (609) 654-6266. Contains citations with abstracts to articles on the use and application of microcomputers and software, from 1981 to date. Available on DIALOG,(800) 334-2564, online service. Inquire as to cost and availability.

NTIS Bibliographic Database. National Technical Information Service, 5285 Port Royal Road, Springfield, Va 22161.(703) 487-4929 or FAX (703) 321-8199. Broad coverage of government-sponsored science and technology research reports, 1964 to present. Available on BRS,(800) 289-4277, DIALOG, (800) 334-2564, ORBIT, (800) 456-7248, and STN International, FIZ Karlsruhe, P.O. Box 2465, W-7500, Karlsruhe 1, Germany, online services. Also available on CD-ROM. Inquire as to cost and availability.

Physics Briefs. American Institute of Physics, 335 East 45th Street, New York, NY 10017. (212) 661-9260 or FAX (212) 661-2036. Contains citations with abstracts of the literature of physics and related fields, 1979 to present. Available on the STN International, FIZ Karlsruhe, P.O. Box 2465, W-7500, Karlsruhe 1, Germany, online service. Inquire as to cost and availability.

Scientific and Technical Books and Serials In Print. R.R. Bowker Inc., 121 Chanlon Road, New Providence, NJ 07974. (800) 521-8110 or (908) 464-6800. List of currently published books and serials in the physical and biological sciences, engineering and technology, with subject, author and titles indexes. Available on ORBIT,(800) 456-7248, online service.Inquire as to cost and availability.

SCISEARCH. Institute for Scientific Information, 3501 Market Street, Philadelphia, PA 19104. (800) 523-1850 or (215) 386-0100. Broad multidisciplinary title and author index to the international literature of science and technology, 1974 to present. Available on DIALOG,(800) 334-2564, and ORBIT,(800) 456-7248, online services. Also available on CD-ROM. Inquire as to cost and availability.

Spin (searchable Physics Information Notices). American Institute of Physics, 335 East 45th Street, New York, NY 10017. (212) 661-9260 or FAX (212) 661-2036. Citations with abstracts to United States journal articles, Russian and Chinese journals in translation, and conference proceedings in physics and astronomy that have been published by the American Institute of Physics, 1975 to present. Available on DIALOG,(800) 334-2564, online service. Inquire as to cost and availability.

Supertech. Bowker A & I Publishing, 121 Chanlon Road, New Providence, NJ 07974. (800) 521-8110 or (908) 464-6800. Contains citations to the world's published and unpublished literature in the fields of artificial intelligence, biotechnology, computer-aided design and manufacturing, robotics, and telecommunications. Covers the period 1973 to present. Available on DIALOG,(800) 334-2564, and ORBIT,(800) 456-7248, online services. Inquire as to cost and availability.

Who's Who in Technology. Gale Research Inc., 835 Penobscot Building, Detroit, MI 48226-4094. (313) 961-2242. (800) 347-4253. Contains biographical information of contemporary American scientists and engineers. Available on DIALOG,(800) 334-2564, online service. Inquire as to cost and availability.

WILSONLINE. H.W. Wilson Company, 950 University Avenue, Bronx, NY 10452. (800) 367-6770 or (212) 588-8400. Makes availableonline versions of the printed H.W. Wilson indexes including Applied Science and Technology Index, Business Periodicals Index, General Science Index, and Readers' Guide to Periodical Literature. Period covered is generally 1983 to present. Available on BRS,(800) 289-4277, DIALOG,(800) 334-2564, and OCLC EPIC,(800) 848-5878, online services. Also available on CD-ROM. Inquire as to cost and availability.

PERIODICALS

ACM Transactions On Computer Systems. Association for Computing Machinery, 11 West 42nd Street, New York, NY 10036. (212) 869-7440. Fax (212) 869-0481. 1964 to present. Annual. $180.00 per year.

ACM Transactions On Database Systems. Association for Computing Machinery, 11 West 42nd Street, New York, NY 10036. (212) 869-7440. Fax (212) 869-0481. 1976 to present. Quarterly. $100.00 per year.

ACM Transactions On Graphics. Association for Computing Machinery, 11 West 42nd Street, New York, NY 10036. (212) 869-7440. Fax (212) 869-0481. 1980 to present. Quarterly. $100.00 per year.

ACM Transactions On Information Systems. Association for Computing Machinery, 11 West 42nd Street, New York, NY 10036. (212) 869-7440. Fax (212) 869-0481. 1979 to present. Quarterly. $95.00 per year.

ACM Transactions On Programming Languages and Systems. Association for Computing Machinery, 11 West 42nd Street, New York, NY 10036. (212) 869-7440. Fax (212) 869-0481. 1979 to present. Quarterly. $95.00 per year.

ACM Transactions On Software Engineering and Methodology. Association for Computing Machinery, 11 West 42nd Street, New York, NY 10036. (212) 869-7440. Fax (212) 869-0481. 1991 to present. Quarterly. $85.00 per year.

AI Magazine. American Association for Artificial Intelligence, 445 Burgess Drive, Menlo Park, CA 94025-3442. (415) 328-3123. 1980 to present. Quarterly. $40.00 per year.

Analog Integrated Circuits and Signal Processing. Kluwer Academic Publishers, P.O. Box 358, Accord Station, Hingham, MA 02018-0358. (617) 871-6000. 1991 to present. Quarterly. $175.00 per year.

COMPUTER SCIENCE

Ency. of Physical Sciences and Engineering Info. Sources

Annals of the History of Computing. American Federation of Information Processing Societies, 1899 Preston White Drive, Reston, VA 22091-5435. (703) 620-8937. 1979 to present. Quarterly. $25.00 per year.

Applied Intelligence: The International Journal of Artificial Intelligence, Neural Networks and Complex Problem Solving Technologies. Kluwer Academic Publishers, P.O. Box 358, Accord Station, Hingham, MA 02018-0358. (617) 871-6000. 1991 to present. Quarterly. $150.00 per year.

Automatica. Pergamon Press Inc., Maxwell House, Fairview Park, Elmsford, NY 10523. (914) 592-7700. Fax (914) 592-3625. 1963 to present. Bimonthly. $50.00 per year.

Byte. Byte Publications, 1 Phoenix Mill Lane, Peterborough, NH 03458-1465. (603) 924-9281. 1975 to present. Monthly. $21.00 per year.

Communications of the ACM. Association for Computing Machinery, 11 West 42nd Street, New York, NY 10036. (212) 869-7440. Fax (212) 869-0481. 1958 to present. Monthly. $105.00 per year.

Computer Book Review. Computer Book Review, 735 Ekekela Place, Honolulu, HI 96817. 1983 to present. Bimonthly. $30.00 per year.

Computer Design. PennWell Publishing Company, 1 Technology Park Drive, Westford, MA 01886. (508) 692-0700 or FAX (508) 692-0525. 1962 to present. Semi-monthly. $70.00 per year.

Computer Journal. Cambridge University Press, 40 West 20th Street, New York, NY 10011-4211. (212) 924-3900. Bimonthly. $315.00 per year.

Computer Magazine. IEEE Computer Society, 10662 Vaqueros Circle, Box 3014, Los Alamitos, CA 90720. (714) 821-8380. 1966 to present. Monthly. $49.00 per year.

Computer Technology Review. West World Productions, Inc., 924 Westwood Boulevard, Suite 650, Los Angeles, CA 90024-2927. (213) 208-1335 or FAX (213) 208-1054. 1981 to present. Sixteen times per year. $80.00 per year.

Computers In Physics. American Institute of Physics, 335 East 45th Street, New York, NY 10017 (212) 661-9404 or FAX (516)349-9704. Bimonthly. $135.00 per year.

Computing Surveys. Association for Computing Machinery, 11 West 42nd Street, New York, NY 10036. (212) 869-7440. Fax (212) 869-0481. 1969 to present. Quarterly. $85.00 per year.

Computing Systems. University of California at Berkeley Press, 2120 Berkeley Way, Berkeley, CA 94720. (415) 642-4191 or FAX (415) 643-7127. 1988 to present. Quarterly. $40.00 per year.

Datamation. R.R. Bowker Inc., 121 Chanlon Road, New Providence, NJ 07974. (800) 521-8110 or (908) 464-6800. 1957 to present. Biweekly. $40.00 per year.

Dr. Dobb's Journal of Software Tools. M and T Publishing, Inc. 501 Galveston Drive, Redwood City, CA 94063. (415) 366-3600. 1976 to present. Monthly. $30.00 per year.

Expert Systems With Applications: An International Journal. Pergamon Press Inc., Maxwell House, Fairview Park, Elmsford, NY 10523. (914) 592-7700. Fax (914) 592-3625. 1990 to present. Quarterly. $125.00 per year.

Fractals. World Scientific Publishing Company, Inc., 1060 Main Street, Unit B, River Edge, NJ 07661. (800) 227-7562 or (201) 487-9655. 1993 to present. Quarterly. $176.00 per year.

IEEE Computer Graphics and Applications. IEEE Computer Society, 10662 Vaqueros Circle, Box 3014, Los Alamitos, CA 90720. (714) 821-8380. Bimonthly. $120.00 per year.

IEEE Expert. IEEE Computer Society, 10662 Vaqueros Circle, Box 3014, Los Alamitos, CA 90720. (714) 821-8380. 1986 to present. Bimonthly. $120.00 per year.

IEEE Software. IEEE Computer Society, 10662 Vaqueros Circle, Box 3014, Los Alamitos, CA 90720. (714) 821-8380. 1984 to present. Bimonthly. $95.00 per year.

IEEE Transactions On Computers. IEEE Computer Society, 10662 Vaqueros Circle, Box 3014, Los Alamitos, CA 90720. (714) 821-8380. 1952 to present. Monthly. $335.00 per year.

International Journal of Computer Simulation. Ablex Publishing Corporation, 355 Chestnut Street, Norwood, NJ 07648. (201) 767-8450 or FAX (201) 767-6717. 1991 to present. Quarterly. $100.00 per year.

International Journal of Network Management. John Wiley and Sons, Inc., 605 Third Avenue, New York, NY 10158. (800) 526-5368 or (212) 850-6000. 1991 to present. Quarterly. $150.00 per year.

International Journal of Systems Automation: Research and Applications. Ablex Publishing Corporation, 355 Chestnut Street, Norwood, NJ 07648. (201) 767-8450 or FAX (201) 767-6717. 1991 to present. Quarterly. $100.00 per year.

Journal of Computer and Software Engineering. Ablex Publishing Corporation, 355 Chestnut Street, Norwood, NJ 07648. (201) 767-8450 or FAX (201) 767-6717. 1993 to present. Quarterly. $110.00 per year.

Journal of Functional Programming. Cambridge University Press, 40 West 20th Street, New York, NY 10011-4211. (212) 924-3900. 1991 to present. Semi-annual. $110.00 per year.

Journal of Mathematical Imaging and Vision. Kluwer Academic Publishers, P.O. Box 358, Accord Station, Hingham, MA 02018-0358. (617) 871-6000. 1991 to present. Quarterly. $150.00 per year.

Journal of Parallel and Distributed Computing. Academic Press, Inc., 1250 Sixth Avenue, San Diego, CA 92101. (619) 231-0926. FAX (619) 699-6715. 1984 to present. Monthly. $325.00 per year.

Journal of the Association For Computing Machinery. Association for Computing Machinery, 11 West 42nd Street, New York, NY 10036. (212) 869-7440. Fax (212) 869-0481. Quarterly. $85.00 per

Parallel Processing Letters. World Scientific Publishing Company, Inc., 1060 Main Street, Unit B, River Edge, NJ 07661.

(800) 227-7562 or (201) 487-9655. 1991 to present. Quarterly. $70.00 per year.

Presence: Teleoperators and Virtual Environments. MIT Press, 55 Hayward Street, Cambridge, MA 02142. (617) 253-2884 or (800) 356-0343. 1992 to present. Quarterly. $120.00 per year.

Programmer's Journal. Oakley Publishing Company, 150 North 4th Street, Springfield, OR 97477. (503) 747-0800. 1983 to present. Bimonthly. $20.00 per year.

Robotics Today. Society Manufacturing Engineers, One SME Drive, Box 930, Dearborn, MI 48128-1490. (313) 271-1500 or FAX (313) 271-2861. 1979 to present. Quarterly. $60.00 per year.

Robotics World. Communications Channels, Inc., 6255 Barfield Road, Atlanta, GA 30328-4369. (404) 256-9800 or FAX (404) 256-3116. 1982 to present. Bimonthly. $36.00 per year.

Scientific Computing and Automation. Gordon Publishing, Division of Elsevier Communications, 301 Gibraltar Drive, Morris Plains, NJ 07950-3406. (201) 292-5100. 1984 to present. Monthly. $60.00 per year.

Siam Journal On Computing. Society for Industrial and Applied Mathematics, 3600 University City Science Center, Philadelphia, PA 19104. (215) 382-2688. 1972 to present. Bimonthly. $50.00 per year.

Simulation. Society for Computer Simulation International, 4838 Ronson Court Lane, San Diego, CA 92111-1810. (619) 277-3888. 1963 to present. Monthly. $50.00 per year.

Software Engineering. Auerbach Publishers, 1 Penn Plaza, New York, NY 10119-0118. 1989 to present. Bimonthly. $145.00 per year.

Software: Practice and Experience. John Wiley and Sons, Inc., 605 Third Avenue, New York, NY 10158. (800) 526-5368 or (212) 850-6000. 1971 to present. Monthly. $440.00 per year.

Solid State Communications. Pergamon Press Inc., Maxwell House, Fairview Park, Elmsford, NY 10523. (914) 592-7700. Fax (914) 592-3625. 1963 to present. Weekly. $165.00 per year.

Solid State Electronics. Pergamon Press Inc., Maxwell House, Fairview Park, Elmsford, NY 10523. (914) 592-7700. Fax (914) 592-3625. 1960 to present. Monthly. $125.00 per year.

Supercomputing Review. Supercomputing Review, 8445 Camino Santa Fe, San Diego, CA 92121. (619) 452-4242. Monthly. $55.00 per year.

Systems Integration (Mini-micro Systems). Cahners Publishing Company, 275 Washington Street, Newton, MA 02158-1630. (617) 558-4402. 1968 to present. Sixteen times per year. $65.00 per year.

Unix Review. Miller Freeman Publications, 600 Harrison Street, San Francisco, CA 94107-1370. (415) 397-1881. 1983 to present. Monthly. $35.00 per year.

Unixworld. McGraw-Hill Publishing Company, 1221 Avenue of the Americas, New York, NY 10020. (800) 262-4729 or (212) 512-3825. 1984 to present. Monthly. $18.00 per year.

Visual Computer. Springer-Verlag, 175 Fifth Avenue, New York, NY 10010. (212) 460-1500. Quarterly. $75.00 per year.

RESEARCH CENTERS AND INSTITUTES

At & T Bell Laboratories, Libraries and Information Systems Center, 600 Mountain Avenue, Murray Hill, Nj 07974. (908) 582-4840.

Computer and Information Technology Institute, Rice University, P.O. Box 1892, Houston, TX 77251-1892. (713) 527-6077.

Information Technology Research Centre. University of Toronto, D.L. Pratt Building, Room 286, 6 King's College Road, Toronto, ON, Canada M5S 1A1. (416) 978-7203.

Institute For Computer Applications In Science and Engineering. Mail Stop 132C, NASA Langley Research Center, Hampton, VA 23665. (804) 864-2174.

New York University, Courant Institute of Mathematical Sciences, 251 Mercer Street, New York, Ny 10012. (212) 998-3000.

Systems Research Center. University of Maryland, A.V. Williams Building, Room 115, College Park, MD 20742. (301) 405-6632.

COMPUTER VISION

See also: AUTOMATION, COMPUTER MEMORY, COMPUTER OPERATING SYSTEMS, COMPUTER PROGRAMMING, COMPUTER PROGRAMMING LANGUAGES, PARALLEL COMPUTING, ROBOTICS

ABSTRACT SERVICES AND INDEXES

ACM Guide To Computing Literature. Association for Computing Machinery, Association for Computing Machinery, 1515 Broadway, 17th Floor, New York, NY 10036-5701. (212) 869-7440. FAX (212) 944-1318. 1964 to present. Monthly. $190.00 per year for non-members.

Applied Science and Technology Index; A Cumulative Subject Index To English Language Periodicals In the Fields of Aeronautics and Space Science, Computer Technology, Chemistry, Construction Industry, Energy and Related Areas. H.W. Wilson Co., 950 University Avenue, Bronx, NY 10452. (800) 367-6770 or (212) 588-8400. FAX (718) 590-1617. From 1958 to present. Monthly. Inquire about cost and availability. Also available on CD-ROM and online.

Computer Abstracts, MCBUniversity Press Ltd., PO Box 10812, Birmingham, AL 35201-0812. (800) 633-4931. FAX (205) 995-1588. Monthly. Covers computer theory, data, hardware, systems, networks, human-computer interaction, artificial intelligence, as well as applications of computers in aerospace, business, CAD/CAM, cartography, civil engineering, electronics and electrical engineering, industrial engineering, mechanical engineering, medicine, structural engineering, etc. Also available on CD-ROM.

Computer & Control Abstracts (inspec). INSPEC/IEEE, Box 1331, 445 Hoes Lane, Piscataway, NJ 08855-1331. (908) 562-5549. Abstracts organized by subjects of international technological information. Monthly. $1455.00 per year. Also available on CD-ROM and online via BRS Online Products, DIA-LOG Information Services, Orbit Search Service, and STN International.

Computer and Information Systems Abstracts Journal. Cambridge Scientific Abstracts, 7200 Wisconsin Avenue, Bethesda, MD 20814. (301) 961-6750. Fax (301) 961-6720. 1962 to present. Monthly. $1265.00 per year. Also available online via STN International.

Computer Literature Index. Applied Computer Research Inc., Box 82266, Phoenix, AZ 85071-2266. (800) 234-2227. 1971 to present. Quarterly plus annual cumulation. $198.50 per year. Bibliography of books, articles, and reports.

Computing Journal Abstracts. Techgnosis Ltd., Blade House, Battersea Road, Stockport, Cheshire 3AE, England. Telephone061-442-2639. FAX 061-443-1162. 1969 to present. Monthly. Inquire for price.

Computing Reviews. Association for Computing Machinery, 1515 Broadway, 17th Floor, New York, NY 10036-5701. (212) 869-7440. FAX (212) 944-1318. 1960 to present. Monthly. $130.00 per year for non-members. Also available online via DIALOG Information Services.

Current Contents: Engineering, Technology, and Applied Sciences. Institute for Scientific Information, 3501 Market Street, Philadelphia, PA 19104. (215) 386-0100. FAX (215) 386-6362. 1970 to present. Weekly. $442.00 per year.

Engineering Index Monthly. Engineering Information, Inc., Castle Point on the Hudson, Hoboken, NJ 07030. (800) 221-1044. FAX (212) 832-1857. Monthly. $2300.00 per year. Also available online as COMPENDEX, and also on CD-ROM. Covers chemical engineering, computers, electrical engineering, civil engineering, metals and mining, industrial management, and mechanical engineering.

Index To Scientific and Technical Proceedings. Institute for Scientific Information, 3501 Market St., Philadelphia, PA 19104. (215) 386-0100. FAX (215) 386-6362. Monthly. $500.00 per year.

Index To Scientific Reviews. Institute for Scientific Information, 3501 Market St., Philadelphia, PA 19104. (215) 386-0100. FAX (215) 386-6362. Semi-annual.

Physics Abstracts. Institute of Electrical Engineers, Michael Faraday House, Six Hill Way, Stevenage, Herts, SG1 2AY, England. Distributed by IEEE, 445 Hoes Lane, Piscataway, NJ 08854. (908) 562-5549. 1898 to present. Monthly. $2700.00 per year. Also available online as INSPEC.

Science Citation Index. Institute for Scientific Information, 3501 Market Street, Philadelphia, PA 19104. (215) 386-0100. FAX (215) 386-6362. Inquire about availability and cost. Also available on CD-ROM.

ANNUAL REVIEWS AND YEARBOOKS

ACM Monograph Series. Association for Computing Machinery (ACM), 11 West 42nd Street, New York, NY 10036. (212) 869-7440. Fax (212) 869-0481. Irregular. Price varies, inquire.

Advances In Computers. Academic Press Inc., 6277 Sea Harbor Drive, Orlando, FL 32887. (800) 321-5068. Irregular. Price varies, inquire.

Advances In Computing Research. JAI Press, Inc., 55 Old Post Road, Number 2, Box 1678, Greenwich, CT 06830. (203) 661-7602. FAX (203) 661-0792. Annual. $68.50 per year.

ASSOCIATIONS AND PROFESSIONAL SOCIETIES

American Federation of Information Processing Societies, 1899 Preston White Drive, Reston, Va 22091-5435. (703) 620-8937.

Association For Computing Machinery. 1515 Broadway, 17th Floor, New York, NY 10036-5701. (212) 869-7440. FAX (212) 944-1318.

Computer and Automation Systems Association of Sme. 1 SME Drive, PO Box 930, Dearborn, MI 48121-0930. (313) 271-1500.

IEEE Computer Society. 1730 Massachusetts Avenue NW, Washington, DC 20036-1992. (202) 371-0101. FAX (202) 728-9614.

Society For Computer Simulation International, 4838 Ronson Court Lane, San Diego, Ca 92111-1810. (619) 277-3888.

BIBLIOGRAPHIES

ACM Guide To Computing Literature. Association for Computing Machinery, 1515 Broadway, 17th Floor, New York, NY 10036-5701. (212) 869-7440. FAX (212) 944-1318. 1964 to present. Annual. $190.00 for non-members. Also available online via DIALOG.

DIRECTORIES AND BIOGRAPHICAL SOURCES

American Men & Women of Science, 1995-96. R.R. Bowker Staff, eds. 19th edition. 8 volumes. R.R. Bowker/Reed International Publishing Company, 121 Chanlon Road, New Providence, NJ 07974. (908) 464-6800 or (800) 521-8110. 1995. $850.00

Computing Information Directory. Hildebrandt Inc., PO Box 576, Pullman, WA 99114. 1981 to present. Annual. $199.95 per year.

Directory of Engineering Societies and Related Organizations. Gordon Davis. 13th edition. American Association of Engineering Societies, 1111 19th Street NW, Suite 608, Washington, DC 20036. (202) 296-2237 or (800) 658-8897. 1989. Inquire for price.

International Research Centers Directory. Gale Research, 835 Penobscot Building, Detroit, MI 48226-4094. (313) 961-2242. (800) 347-4253. FAX (313) 961-6083. 8th edition. 1995. $410.00

Research Centers Directory. Gale Research, 835 Penobscot Building, Detroit, MI 48226-4094. (313) 961-2242. (800) 347-4253. FAX (313) 961-6083. $485.00.

Scientific and Technical Organizations and Agencies Directory. Gale Research, 835 Penobscot Building, Detroit, MI 48226-

Ency. of Physical Sciences and Engineering Info. Sources

COMPUTER VISION

4094. (313) 961-2242. (800) 347-4253. FAX (313) 961-6083. 4th edition. 1996. $195.00.

Who's Who In Technology. Gale Research, 835 Penobscot Building, Detroit, MI 48226-4094. (313) 961-2242. (800) 347-4253. FAX (313) 961-6083. 1995. $195.00.

ENCYCLOPEDIAS AND DICTIONARIES

Computer Dictionary. Donald D. Spencer. 4th edition. Camelot Publishing Company, 709 SW 80th Blvd., Gainesville, FL 32607-1537. (904) 331-0952. 1993. $24.95.

Dictionary of Computing. S.M.H. Collin. 2nd edition. Peter Collin Publishing, 8 the Causeway, Teddington, TW11 0HE, England. FAX 0181-943-3386. 1994. Inquire for cost.

Encyclopedia of Computer Science and Engineering. Anthony Ralston & Edwin D. Reilly Jr., editors. 3rd revised edition. Van Nostrand Reinhold, 115 Fifth Avenue, New York, NY 10003. (212) 254-3232 or FAX (212) 254-9499. 1993. $125.00.

Microsoft Press Computer Dictionary: the Comprehensive Standard For Business, School, Library, and Home. 2nd edition. Microsoft Press, One Microsoft Way, Redmond, WA 98052-6399. (206) 936-0055. FAX (206) 823-8101. 1994. $19.95.

GENERAL WORKS

Computer and Robot Vision. Robert M. Haralick & Linda G.Shapiro. Addison-Wesley Publishing Company, 1 Jacob Way, Reading, MA 01867. (617) 944-3700 or (800) 447-2226. FAX 617-942-1117. 1993. $62.95.

Computer Vision: Systems, theory & Applications. A. Basu & X.B. Li. World Scientific Publishing Company, Inc., 1060 Main Street, Unit B, River Edge, NJ 07661. (800) 227-7562. FAX (201) 487-9656. 1993. $109.00.

A Guided Tour of Computer Vision. Vishjit S. Nalwa. Addison-Wesley Publishing Company, 1 Jacob Way, Reading, MA 01867. (617) 944-3700 or (800) 447-2226. FAX 617-942-1117. 1993. $34.95.

Research In Computer and Robot Vision. Colin Archibald & Paul Kwok, editors. World Scientific Publishing Company, Inc., 1060 Main Street, Unit B, River Edge, NJ 07661. (800) 227-7562. FAX (201) 487-9656. 1995. $109.00.

Vision As Process: Basic Research On Computer Vision Systems. J.L. Crowley, editor. Springer-Verlag, 175 Fifth Avenue, New York, NY 10010. (212) 460-1500. FAX (212) 473-6272. 1994. $89.00.

HANDBOOKS AND MANUALS

Handbook On Pattern Recognition and Computer Vision. C.H. Chen, et al. World Scientific Publishing Company, Inc., 1060 Main Street, Unit B, River Edge, NJ 07661. (800) 227-7562. FAX (201) 487-9656. 1993. $84.00.

ONLINE DATABASES AND CD-ROMS

Compendex Plus. Engineering Information, Inc., 345 East 47th Street, New York, NY 10017. (212) 705-7600 or (800) 221-1044. Contains citations with abstracts to worldwide literature in engineering and technology, from 1970 to present. Available on online BRS,(800) 289-4277, DIALOG, (800) 334-2564, ORBIT (800) 456-7248, and STN International, FIZ Karlsruhe, P.O. Box 2465, W-7500, Karlsruhe 1, Germany, online services. Also available on CD-ROM. Inquire as to cost and availability.

Compuscience. FIZ Karlsruhe, D-7514, Eggenstein-Leopoldshafen 2, Germany. Contains citations with abstracts to European and North American literature on computer science, 1972 to present. Available on STN International, FIZ Karlsruhe, P.O. Box 2465, W-7500, Karlsruhe 1, Germany, online service. Inquire as to cost and availability.

Computer and Information Systems Abstracts. Cambridge Scientific Abstracts, 7200 Wisconsin Avenue, Suite 601, Bethesda, MD 20814. (301) 961-6750 or (800) 843-7751. Contains citations to worldwide literature in theoretical and applied computer science and related areas, from 1981 to present.Inquire as to cost and availability.

Computer and Mathematics Search. Institute for Scientific Information, 3501 Market Street, Philadelphia, PA 19104. (215) 386-0100. FAX (215) 386-6362. Covers worldwide literature in computer science and mathematics, from 1980 to present. Available on BRS,(800) 289-4277, online service. Inquire as to cost and availability.

Computer Database. Information Access Company, 362 Lakeside Drive, Foster City, CA 94404. (415) 378-5000 or (800) 227-8431. Contains citations with abstracts to literature from trade journals covering the computer,telecommunications,and electronics industries. Available on the BRS, (800) 289-4277, Compuserve Information Service,(800) 848-8990, and DIALOG,(800) 334-2564, online services. Inquire as to cost and availability.

CSA Engineering. Cambridge Scientific Abstracts, 7200 Wisconsin Avenue, Suite 601, Bethesda, MD 20814. (301) 961-6750 or (800) 843-7751. Contains citations and abstracts of international periodicals and research literature covering all fields of engineering and science and technology,including computer and information science, electronics, mechanical engineering, solid state materials, 1981 to present. Available on BRS,(800) 289-4277, online service.Inquire as to cost and availability.

Current Contents Search. Institute for Scientific Information, 3501 Market Street, Philadelphia, PA 19104. (215) 386-0100. FAX (215) 386-6362. Contains citations to articles listed in the table of contents of science and technology journals. Also articles in social sciences and life sciences journals. Available on BRS,(800) 289-4277, DIALOG,(800) 334-2564, online services. Inquire as to cost and availability.

Dissertation Abstracts. University Microfilms International, 300 North Zeeb Road, Ann Arbor, MI 48106. (800) 521-0600 or (313) 761-4700. Scope includes virtually all doctoral dissertations accepted at accredited American institutions from 1861 to present in 252 subject areas. Available on BRS,(800) 289-4277, DIALOG,(800) 334-2564, and OCLC EPIC,(800) 848-5878, online services. Also available on CD-ROM. Inquire as to cost and availability.

COMPUTER VISION

Ency. of Physical Sciences and Engineering Info. Sources

Inspec. Institution of Electrical Engineers, Michael Faraday House, Six Hills Way, Stevenage, Herts. SG1 2AY, England. Telephone: 0438 313311 or FAX 0438 742840. Contains citations to the worldwide literature of physics, electronics and electrical engineering, computer technology, and related fields. Available on BRS,(800) 289-4277, DIALOG, (800) 334-2564, ORBIT, (800) 456-7248, and STN International, FIZ Karlsruhe, P.O. Box 2465, W-7500, Karlsruhe 1, Germany, online services. Inquire as to cost and availability.

Mathsci. American Mathematical Society, P.O. Box 6248, Providence, RI 02940. (800) 321-4667 or (401) 455-4000 or FAX (401) 331-3842. Scope includes pure and applied mathematics and related areas of physics, statistics, engineering, computer science, and operations research literature since 1959. Available on DIALOG,(800) 334-2564, online service. Also available on CD-ROM. Inquire as to cost and availability.

NTIS Bibliographic Database. National Technical Information Service, 5285 Port Royal Road, Springfield, Va 22161. (703) 487-4929 or FAX (703) 321-8199. Broad coverage of government-sponsored science and technology research reports, 1964to present. Available on BRS,(800) 289-4277, DIALOG, (800) 334-2564, ORBIT, (800) 456-7248, and STN International, FIZ Karlsruhe, P.O. Box 2465, W-7500, Karlsruhe 1, Germany, online services. Also available on CD-ROM. Inquire as to cost and availability.

SCISEARCH. Institute for Scientific Information, 3501 Market Street, Philadelphia, PA 19104. (800) 523-1850 or (215) 386-0100. Broad multidisciplinary title and author index to the international literature of science and technology, 1974 to present. Available on DIALOG,(800) 334-2564, and ORBIT,(800) 456-7248, online services. Also available on CD-ROM. Inquire as to cost and availability.

WILSONLINE. H.W. Wilson Company, 950 University Avenue, Bronx, NY 10452. (800) 367-6770 or (212) 588-8400. Makes available online versions of the printed H.W. Wilson indexes including Applied Science and Technology Index, Business Periodicals Index, General Science Index, and Readers' Guide to Periodical Literature. Period covered is generally 1983 to present. Available on BRS,(800) 289-4277, DIALOG,(800) 334-2564, and OCLC EPIC,(800) 848-5878, online services. Also available on CD-ROM. Inquire as to cost and availability.

PERIODICALS

ACM Transactions On Computer Systems. Association for Computing Machinery, 11 West 42nd Street, New York, NY 10036. (212) 869-7440. Fax (212) 869-0481. 1964 to present. Annual. $180.00 per year.

Computer Vision and Image Understanding: Cviu (formerly Cvgip: Image Understanding). Academic Press Inc., 525 B Street, Suite 1900, San Diego, CA 92101-4495. (619) 231-0926. FAX (619) 699-6715. 1969 to present (1995 to present under current title). Inquire for price.

Graphical Models and Image Processing: Gmip (formerly Cvgip: Graphical Models and Image Processing). Academic Press Inc., 525 B Street, Suite 1900, San Diego, CA 92101-4495. (619) 231-0926. FAX (619) 699-6715. 1969 to present (1995 to present under current title). Inquire for price.

IEEE Computer Graphics and Applications. IEEE Computer Society, 10662 Vaqueros Circle, Box 3014, Los Alamitos, CA 90720. (714) 821-8380. Bimonthly. $120.00 per year.

Image and Vision Computing. Butterworth-Heinemann, Linacre House, Jordan Hill, Ocford OX2 8DP, England. Telephone 0865-310366. FAX 0865-310898. 1982 to present. Ten times a year. Inquire for price.

International Journal of Computer Vision. Kluwer Academic Publishers, P.O. Box 358, Accord Station, Hingham, MA 02018-0358. (617) 871-6000. 1987 to present. Six times a year. $314.00.

Journal of Mathematical Imaging and Vision. Kluwer Academic Publishers, P.O. Box 358, Accord Station, Hingham, MA 02018-0358. (617) 871-6000. 1991 to present. Quarterly. $150.00 per year.

Machine Vision and Applications. Springer-Verlag, 44 Hartz Way, Secaucus, NJ 07096-2491. (201) 348-4033. FAX (201) 348-4505. 1988 to present. Four times a year. $174.00.

Visual Computer. Springer-Verlag, 175 Fifth Avenue, New York, NY 10010. (212) 460-1500. Quarterly. $75.00 per year.

RESEARCH CENTERS AND INSTITUTES

Harvard University Center For Research In Computing Technology. Aiken Computation Laboratory, 33 Oxford Street, Cambridge, MA 02138. (617) 495-4117. FAX (617) 495-9837.

Rice University Computer and Information Technology Institute. PO Box 1892, Houston, TX 77251. (713) 527-6077. FAX (713) 285-5136.

University of Maryland Computer Vision Laboratory. Center for Automation Research, College Park, MD 20742. (301) 405-4526.

University of Texas At Austin, Computer and Vision Research Center. College of Engineering, Austin, TX 78712. (512) 471-3259. FAX (512) 471-5532.

CONCRETE

See also: BUILDING MATERIALS, CEMENT, PORTLAND CEMENT

ABSTRACT SERVICES AND INDEXES

Applied Science and Technology Index; A Cumulative Subject Index To English Language Periodicals In the Fields of Aeronautics and Space Science, Computer Technology, Chemistry, Construction Industry, Energy and Related Areas. H.W. Wilson Co., 950 University Avenue, Bronx, NY 10452. (800) 367-6770 or (212) 588-8400. FAX (718) 590-1617. From 1958 to present. Monthly. Inquire about cost and availability. Also available on CD-ROM and online.

Concrete Abstracts. Covers articles, books, and other publications that report developments in concrete design, construction and technology. American Concrete Institute, PO Box 19150, Detroit, MI 48219. (313) 532-2600. FAX (213) 538-0655. 1972 to present. Bi-monthly. Inquire for price.

Engineering Index Monthly. Engineering Information, Inc., Castle Point on the Hudson, Hoboken, NJ 07030. (800) 221-1044. FAX (212) 832-1857. Monthly. $2300.00 per year. Also available online as COMPENDEX, and also on CD-ROM. Covers chemical engineering, computers, electrical engineering, civil engineering, metals and mining, industrial management, and mechanical engineering.

ANNUAL REVIEWS AND YEARBOOKS

U.S. Department of the Interior, Bureau of Mines, Minerals Yearbook. Superintendent of Documents, U.S. Government Printing office, Box 371954, Pittsburgh, PA 15250-7954. (202) 783-3238. FAX (202) 512-2233. 1932 to present. Annual. Price varies.

ASSOCIATIONS AND PROFESSIONAL SOCIETIES

American Concrete Institute. PO Box 19150, Redford Station, Detroit, MI 48219. (313) 532-2600. FAX (213) 538-0655.

American Concrete Pavement Association. 3800 N. Wilke Road, Suite 490, Arlington Heights, IL 60004. (708) 394-5577. FAX(708) 394-5610.

American Society For Concrete Construction. PO Box 19307, Detroit, MI 48219. (313) 532-2753.

American Society of Civil Engineers. 1015 15th St. NW, Suite 600, Washington, DC 20005. (212) 705-7496.

Concrete Industry Board Inc. 2622 Sunset Hills, Escondido, CA 92025.

National Concrete Masonry Association. 2302 Horse Pen Road, Herndon, VA 22071-3406. (703) 713-1900. FAX (703) 713-1910.

National Precast Concrete Association. 10333 North Meridian Street, Suite 272, Indianapolis, in 46290-9500. (800) 366-7731. FAX (317) 571-0041.

National Ready Mixed Concrete Association. 900 Spring Street, Silver Spring, MD 20910. (301) 587-1400. FAX (301) 585-4219.

Precast/prestressed Concrete Institute. 175 West Jackson Blvd., Chicago, IL 60604. (312) 786-0300. FAX (312) 786-0353.

Reinforced Concrete Research Council. 205 N. Mathews Avenue, Urbana, IL 61801. (217) 333-7384.

DIRECTORIES AND BIOGRAPHICAL SOURCES

Aberdeen's Concrete Sourcebook. Aberdeen Group, 426 S. Westgate Street, Addison, IL 60101. (708) 543-0870. FAX (708) 543-3112. Monthly. $30.00 per year.

American Concrete Institute—Directory. American Concrete Institute, PO Box 19150, Detroit, MI 48219. (313) 532-2600. FAX (213) 538-0655. Every 2-3 years. Write or call for price information.

National Precast Concrete Association Directory of Members and Precast Products. National Precast Concrete Association, 10333 North Meridian Street, Suite 272, Indianapolis, in 46290-9500. (800) 366-7731. FAX (317) 571-0041. Annual. $295.00.

Who's Who In Engineering. American Association of Engineering Societies, 1111 19th Street NW, Suite 608, Washington, DC 20036. (202) 296-2237 or (800) 658-8897. 8th edition. 1991. Inquire for price.

World Cement Directory. European Cement Association/Cimeurope, Rue d'Arlon 55, B-1040 Brussels, Belgium. Telephone 32-2-2341011. FAX 32-2-2304720. 1958 to present. Quinquennial. $600.00.

ENCYCLOPEDIAS AND DICTIONARIES

Cement and Concrete Terminology. ACI Committee staff. American Concrete Institute, PO Box 19150, Detroit, MI 48219. (313) 532-2600. FAX (213) 538-0655. 1990. $33.95.

Encyclopedia of Physical Science and Technology. Robert A. Meyers, ed. 18 volumes. Academic Press Inc., 6277 Sea Harbor Drive, Orlando, FL 32887. (800) 321-5068. 1992. $2100.00

McGraw-Hill Dictionary of Scientific and Technical Terms. Sybil P. Parker, ed. 5th edition. McGraw-Hill Publishing Company, 1221 Avenue of the Americas, New York, NY 10020. (800) 262-4729 or (212) 512-3825. 1993. $110.50.

McGraw-Hill Encyclopedia of Engineering. Sybil P. Parker, ed. 2nd edition. McGraw-Hill Publishing Company, 1221 Avenue of the Americas, New York, NY 10020. (800) 262-4729 or (212) 512-3825. 1993. $95.50.

McGraw-Hill Encyclopedia of Science and Technology. Sybil P. Parker, ed. 7th edition. 20 volumes. McGraw-Hill Publishing Company, 1221 Avenue of the Americas, New York, NY 10020. (800) 262-4729 or (212) 512-3825. 1992. $1900.00

GENERAL WORKS

Concrete Technology. A.M. Neville, J.J. Brooks. John Wiley and Sons, Inc., 605 Third Avenue, New York, NY 10158. (800) 526-5368 or (212) 850-6000. 1987. $59.95.

Concrete Technology. G.R. White. 3rd edition. Delmar Publishers, Division of Thomson Educational Publishing, Inc., 3 Columbia Circle, Box 15015, Albany, NY 12205. (518) 464-3500 or (800) 347-7707. FAX (518) 464-0358. 1991. $21.00.

Concrete Technology: Past, Present, & Future. American Concrete Institute, PO Box 19150, Redford Station, Detroit, MI 48219. (313) 532-2600. FAX (213) 538-0655. 1994. $50.00.

Design and Control of Concrete Mixtures. Steven H. Kosmatka and William C. Panarese. 13th edition. Portland Cement Association, 5420 Old Orchard Road, Skokie, IL 60077-1083. (708) 966-6200. $33.00.

HANDBOOKS AND MANUALS

ACI Manual of Concrete Practice. 5 volumes. American Concrete Institute, PO Box 19150, Redford Station, Detroit, MI 48219. (313) 532-2600. FAX (213) 538-0655. 1967 to present. Annual. $423.00 per set for non-members.

Concrete Construction Handbook. Joseph J. Waddell & Joseph A. Dobrowolski, editors. 3rd edition. McGraw-Hill Publishing

CONCRETE

Ency. of Physical Sciences and Engineering Info. Sources

Company, 1221 Avenue of the Americas, New York, NY 10020. (800) 262-4729 or (212) 512-3825. 1993. $92.00.

Concrete Manual, 1991 Edition. International Conference of Building officials, 5360 Workman Mill Road, Whittier, CA 90601-2298. (31) 699-0541 or (800) 423-6587.

Concrete Manual. Materials Engineering Branch, Research and Laboratory Services Division, Denver office, U.S. Department of the Interior, Bureau of Reclamation. 9th edition. Superintendent of Documents, U.S. Government Printing office, Box 371954, Pittsburgh, PA 15250-7954. (202) 783-3238. FAX (202) 512-2233. 1992. Inquire for cost and availability.

Handbook of Concrete Engineering. Mark Fintel. 2nd edition. Routledge, Chapman and Hall, Inc., 29 West 35th Street, New York, NY 10001-2291. (212) 244-3336 or FAX (212) 563-2269. 1985. $115.00.

Means Concrete Cost Data. R.S. Means Company Inc., 100 Construction Plaza, Box 800, Kingston, MA 02364-0800. (800) 334-3509. FAX (617) 585-7466. 1982 to present. Annual. $74.95.

Structural Design Guide To the AISC Specifications For Buildings. Paul F. Rice, et al. 3rd edition. Books on Demand, Division of University Microfilms International, 300 North Zeeb Road, Ann Arbor, MI 48106-1346. (313) 761-4700 or (800) 521-0600. 1985. $92.00.

ONLINE DATABASES AND CD-ROMS

Compendex Plus. Engineering Information, Inc., 345 East 47th Street, New York, NY 10017. (212) 705-7600 or (800) 221-1044. Contains citations with abstracts to worldwide literature in engineering and technology, from 1970 to present. Available on online BRS,(800) 289-4277, DIALOG, (800) 334-2564, ORBIT (800) 456-7248, and STN International, FIZ Karlsruhe, P.O. Box 2465, W-7500, Karlsruhe 1, Germany, online services. Also available on CD-ROM. Inquire as to cost and availability.

NTIS Bibliographic Database. National Technical Information Service, 5285 Port Royal Road, Springfield, Va 22161. (703) 487-4929 or FAX (703) 321-8199. Broad coverage of government-sponsored science and technology research reports, 1964 to present. Available on BRS,(800) 289-4277, DIALOG, (800) 334-2564, ORBIT, (800) 456-7248, and STN International, FIZ Karlsruhe, P.O. Box 2465, W-7500, Karlsruhe 1, Germany, online services. Also available on CD-ROM. Inquire as to cost and availability.

WILSONLINE. H.W. Wilson Company, 950 University Avenue, Bronx, NY 10452. (800) 367-6770 or (212) 588-8400. Makes available online versions of the printed H.W. Wilson indexes including Applied Science and Technology Index, Business Periodicals Index, General Science Index, and Readers' Guide to Periodical Literature. Period covered is generally 1983 to present. Available on BRS,(800) 289-4277, DIALOG,(800) 334-2564, and OCLC EPIC,(800) 848-5878, online services. Also available on CD-ROM. Inquire as to cost and availability.

PERIODICALS

Aberdeen's Concrete Construction. Aberdeen Group, 426 S. Westgate Street, Addison, IL 60101. (708) 543-0870. FAX (708) 543-3112. 1956 to present. Monthly. $24.00 per year.

Aberdeen's Concrete Journal. Aberdeen Group, 426 S. Westgate Street, Addison, IL 60101. (708) 543-0870. FAX (708) 543-3112. 1983 to present. Monthly. Inquire for cost.

ACI Materials Journal. American Concrete Institute, PO Box 19150, Detroit, MI 48219. (313) 532-2600. FAX (213) 538-0655. Monthly. $92.00 per year.

ACI Structural Journal. American Concrete Institute, PO Box 19150, Detroit, MI 48219. (313) 532-2600. FAX (213) 538-0655. Bi-monthly. $92.00 per year.

Cement and Concrete Composites. Elsevier Science [journals], 660 White Plains Rd., Tarrytown, NY 10159-5153. (919) 524-9200. FAX (919) 333-2444. 1979 to present. 4 times a year. $280.00 per year.

Cement and Concrete Research. Elsevier Science [journals], 660 White Plains Rd., Tarrytown, NY 10159-5153. (919) 524-9200. FAX (919) 333-2444. 1971 to present. 8 times a year.$590.00 per year.

Cement, Concrete and Aggregates. American Society for Testing and Materials, 1916 Race Street, Philadelphia, PA 19103. (215) 299-5585. 1979-present. Semi-annual. $48.00 per year for non-members.

Concrete. The Concrete Society, Framewood Road, Wexham, Slough, Berks. SL3 6PJ England. Telephone 0753-662226. FAX 0753-662126. 1966 to present. Bi-monthly. $75.00 per year.

Concrete International: Design and Construction. American Concrete Institute, Box 19150, Redford Station, Detroit, MI 48219. (313) 532-2600. FAX (313) 538-0655. 1979 to present. Monthly. $101.00 per year.

Concrete Masonry News. National Concrete Masonry Association, 2302 Horse Pen Road, Herndon, VA 22071-3406. (703) 713-1900. FAX (703) 713-1910. 1962 to present. Monthly. Inquire for cost and availability.

Concrete Products. Maclean Hunter Publishing Company, 29 North Wacker Drive, Chicago, IL 60606-3298. (312) 726-2802. 1903 to present. Monthly. $31.25 per year.

Journal of Ferrocement. International Ferrocement Information Center, Asian Institute of Technology, PO Box 2754, Bangkok 10501, Thailand. Telephone 5160110. FAX 66-2-5162126. 1976 to present. Quarterly. $96.00 per year.

Journal of Structural Engineering. American Society of Civil Engineers, 345 East 47th Street, New York, NY 10017-2398. (212) 705-7520 or (800) 548-2723. 1956 to present. Monthly. $300.00 per year for non-members.

Journal of the Precast/Prestressed Concrete Institute. Precast/ Prestressed Concrete Institute, 175 West Jackson Blvd., Chicago, IL 60604. (312) 786-0300. FAX (312) 786-0353. Bi-monthly. $29.00 per year.

Magazine of Concrete Research. Thomas Telford Services Ltd., Thomas Telford House, 1 Heron Quay, London E14 4JD, England. Telephone 071-987-6999. FAX 071-538-4101. 1949 to present. Quarterly. Inquire for cost.

RESEARCH CENTERS AND INSTITUTES

American Concrete Institute. PO Box 19150, Redford Station, Detroit, MI 48219. (313) 532-2600. FAX (213) 538-0655.

American Society For Testing and Materials. 1916 Race Street, Philadelphia, PA 19103. (215) 299-5585.

Center For Advanced Cement-based Materials. Northwestern University, 1800 Ridge Avenue, Evanston, IL 60208-4400. (708) 491-8569.

Center For Cement Composite Materials. University of Illinois, 202 Ceramics Bldg., 105 S. Goodwin Avenue, Urbana, IL 61801. (217) 333-6960. FAX (217) 244-7750.

Center For Concrete and Geomaterials. Northwestern University, Technological Institute, Room 2474, 2145 Sheridan Road, Evanston, IL 60208. (708) 491-3858.

Concrete Materials Research Council. PO Box 19150, Detroit, MI 48219. (313) 532-2600.

Concrete Research Laboratory. University of Michigan, G.G. Brown Laboratory, Ann Arbor, MI 48109. (313) 764-9660.

National Concrete Masonry Association Research and Development Laboratory. National Concrete Masonry Association, 2302 Horse Pen Road, Herndon, VA 22071-3406. (703) 713-1900. FAX (703) 713-1910.

Portland Cement Association. 5420 Old Orchard Road, Skokie, IL 60077. (708) 966-6200. FAX (708) 966-8389.

SPECIFICATIONS AND STANDARDS

*Annual Book of Astm Standards, Volume 04.*02. American Society for Testing and Materials (ASTM), 1916 Race Street, Philadelphia, PA 19103. (215) 299-5585. Annual. Inquire for cost and availability.

CONDUCTION

See: HEAT TRANSFER

CONSTELLATIONS

See: STARS

CONSTRUCTION ENGINEERING

See also: BRIDGES, BUILDING MATERIALS, CIVIL ENGINEERING, CONCRETE, EARTHQUAKES, EARTHQUAKE ENGINEERING, GEOTECHNICAL ENGINEERING, HIGHWAY ENGINEERING, OCEAN ENGINEERING, ROCK MECHANICS, SOIL MECHANICS, STRUCTURAL ENGINEERING

ABSTRACT SERVICES AND INDEXES

Abstract Journal In Earthquake Engineering. University of California, Berkeley, Earthquake Engineering Research Center, 1301 South 46th Street, Richmond, CA 94804-4698. (510) 231-9413 or FAX (510) 231-9471. 1972 to present. Annual. $80.00 per year.

Applied Science and Technology Index; A Cumulative Subject Index To English Language Periodicals In the Fields of Aeronautics and Space Science, Computer Technology, Chemistry, Construction Industry, Energy and Related Areas. H.W. Wilson Co., 950 University Avenue, Bronx, NY 10452. (800) 367-6770 or (212) 588-8400. From 1958 to present. Monthly. Inquire about cost and availability. Also available on CD-ROM.

Bibliography and Index of Geology. American Geological Institute, 4220 King Street, Alexandria, VA 22302. (703) 379-2480. Fax (703) 379-7563. Monthly. $1295.00 per year. Also available online as GeoRef.

Concrete Abstracts. American Concrete Institute, 22400 West Seven Mile Road, Detroit, MI 48219-1885. (313) 532-2600 or FAX (313) 538-0655. 1971 to present. Bimonthly. $140.00 per year.

Current Contents: Engineering, Technology and Applied Sciences. Institute for Scientific Information, 3501 Market Street, Philadelphia, PA 19104. (215) 386-0100. FAX (215) 386-6362. Weekly. $360.00 per year.

Engineering Index Monthly. Engineering Information, Inc., Castle Point on the Hudson, Hoboken, NJ 07030. (800) 221-1044. FAX (201) 216-8532. Monthly. $2200.00 per year. Also available online as COMPENDEX, and also on CD-ROM.

General Science Index. H.W. Wilson Company, 950 University Avenue, Bronx, NY 10452. (800) 367-6770 or (212) 588-8400. Inquire as to cost and availability.

Geotechnical Abstracts. Deutsche Gesellschaft fur Erd- und Grundbau e.V., Hohenzollernstrasse 52, 4300 Essen 1, Germany. Also sponsored by International Society for Soil Mechanics and Foundation Engineering. Covers soil mechanics, foundation engineering, and engineering geology. 1970 to present. Monthly. $200.00 per year.

Government Reports Announcements and Index. National Technical Information Service (NTIS), 5285 Port Royal Road, Springfield, VA 22161. (703) 487-4650. 1968 to present. Annual. $525.00 per year. Also available online as NTIS BIBLIOGRAPHIC DATABASE and on CD-ROM.

Hydro-Abstracts (formerly *Water Resources Abstracts*). HydroScience Press, 2527 Jackson Street, N.E., Minneapolis, MN 55418. (612) 781-9081. Monthly. $150.00 per year.

Offshore Engineering Abstracts. STI Limited, 4 Kings Meadow, Ferry Hinksey Road, Oxford, OX2 ODU, England. Distributed in the United States by Air Science Company, Box 143, Corning NY 14830. (607) 962-5591. Contains citations on planning, design, and construction of offshore platforms and pipelines, from 1985 to present. Monthly. $400.00 per year.

CONSTRUCTION ENGINEERING

Ency. of Physical Sciences and Engineering Info. Sources

ASSOCIATIONS AND PROFESSIONAL SOCIETIES

American Association of Engineering Societies, 1111 19th Street, Suite 608, Washington, Dc 20036-3603. (202) 296-2237.

American Concrete Institute, 22400 West Seven Mile Road, Detroit, Mi 48219-1885. (313) 532-2600 or FAX (313) 538-0655.

American Congress On Surveying and Mapping, 5410 Grosvenor Lane, Bethesda, Md 20814-2199. (301) 493-0200 or FAX (301) 493-8245.

American Consulting Engineers Council, 1015 15th Street, N.W. Suite 802, Washington, DC 20005-2670. (202) 347-7474.

American Institute of Steel Construction, Inc., One East Wacker Drive, Suite 3100, Chicago, IL 60601-2001. (312) 670-2400.

American Iron and Steel Institute, 1101 17th Street, Washington, Dc 20036-4700. (202) 452-7265 or FAX (202) 463-6573.

American National Standards Institute. 1430 Broadway, New York, NY 10018-3363. (212) 354-4916.

American Society For Testing and Materials. 1916 Race Street, Philadelphia, PA 19103-1180. (215) 299-5400.

American Society of Civil Engineers. 345 East 47th Street, New York, NY 10017-2398. (212) 705-7388 or FAX (212) 980-4681.

American Society of Mechanical Engineers, 345 East 47th Street, New York, Ny 10017-2395. (212) 705-7722.

American Society of Test Engineers, 1050 Commonwealth Avenue, Suite 200, Boston, Ma 02215-1109. (617) 734-4473.

Association of Engineering Geologists. 323 Boston Post Road, Suite 2D, Sudbury, MA 01776. (508) 443-4639.

National Academy of Engineering, 2101 Constitution Avenue, N.W., Washington, DC 20418-0001. (202) 334-1657.

National Society of Black Engineers. 344 Commerce Street, Alexandria, VA 22314. (703) 549-2207.

National Society of Professional Engineers, 1420 King Street, Alexandria, Va.22314-2750. (703) 684-2835.

Society of American Military Engineers. 607 Prince Street, Alexandria, VA 22314-3117. (703) 549-3800.

BIBLIOGRAPHIES

Information Sources In Energy Technology. L.J. Anthony, editor. Butterworths Publishers, 80 Montvale Avenue, Stoneham, MA 02180. (617) 438-8464 or (800) 366-2665. 1988. $105.00.

Scientific and Technical Books and Serials In Print; An Index To Literature In Science and Technology. R.R. Bowker Co., 205 E. 42nd Street, New York, NY 10017. (800) 521-8110 or (212) 916-1600. 1992. $250.00.

DIRECTORIES AND BIOGRAPHICAL SOURCES

Association of Engineering Geologists. Membership directory. Association of Engineering Geologists, 323 Boston Post Road, Suite 2D, Sudbury, MA 01776. (508) 443-4639. Annual. $18.00.

Directory of Engineering College Research and Graduate Study, American Society For Engineering Education, Ohio State University, 2070 Neil Avenue, Columbus, Oh 43210-1226. Annual. $20.00.

Directory of Engineering Societies and Related Organizations. American Association of Engineering Societies, 1111 19th Street, Suite 608, Washington, DC 20036-3603. (202) 296-2237. Semi-annual. $150.00.

Directory of Engineers In Private Practice. National Society of Professional Engineers, 1420 King Street, Alexandria, VA 22314-2750. (703) 684-2835. Biennial. $30.00.

Research Centers Directory. Gale Research, Detroit, MI 48226. 20th edition, 1995. (800) 347-4253. $485.00.

ENCYCLOPEDIAS AND DICTIONARIES

Construction Glossary: An Encyclopedic Reference and Manual. J.Stewart Stein. John Wiley and Sons, Inc., 605 Third Avenue, New York, NY 10158. (800) 526-5368 or (212) 850-6000. Second edition. 1993. $95.00.

The Contractor's Dictionary of Equipment: Tools and Construction Techniques. John Wiley and Sons, Inc., 605 Third Avenue, New York, NY 10158. (800) 526-5368 or (212) 850-6000. 1995. $69.95.

EI thesaurus. Jessica L. Milstead, editor. Engineering Information, Inc., Castle Point on the Hudson, Hoboken, NJ 07030. (800) 221-1044. FAX (201) 216-8532. 1992. $130.00

GENERAL WORKS

Applied Structural Steel Design. L. Spiegel and G.F. Limbrunner. Prentice-Hall Publishing, Inc., Englewood Cliffs, NJ 07632. (800) 562-0245. 1986. $34.95.

Building Construction, Drafting and Design. John Molnar.Van Nostrand Reinhold Company, Inc., 135 West 50th Street, New York, NY 10020. (800) 543-2681. 1986. $32.95.

Design of Reinforced Concrete. Samuel E. French. Prentice-Hall Publishing, Inc., Englewood Cliffs, NJ 07632. (800) 562-0245. 1987. $25.00.

Elementary Theory of Structures. Yu H. Yuan. Prentice-Hall Publishing, Inc., Englewood Cliffs, NJ 07632. (800) 562-0245. Third edition. 1988. $42.95.

Foundation Design and Construction. M.J. Tomlinson. John Wiley and Sons, Inc., 605 Third Avenue, New York, NY 10158. (800) 526-5368. Fifth edition. 1986. $45.95.

Highway Engineering. Paul H. Paquette and Randor J. Paquette. John Wiley and Sons, Inc., 605 Third Avenue, New York, NY 10158. (800) 526-5368. Fifth edition. 1986. $45.00.

Highways: Highway Engineering. C.A. O'Flaherty. Edward Arnold Publishers, Limited, 300 North Charles Street, Baltimore, MD 21201. (301) 539-1529. Volume 2. Third edition. 1987. $29.95 in paper.

Materials For Civil and Highway Engineers. K.N. Derucher and G.P. Korgiatis. Prentice-Hall Publishing, Inc., Englewood Cliffs, NJ 07632. (800) 562-0245. Second edition. 1988. $42.50.

Microcomputer-Aided Engineering: Structural Dynamics. Mario Paz. Van Nostrand Reinhold Company, Inc., 135 West 50th Street, New York, NY 10020. (800) 543-2681. 1986. $49.95.

Structural Stability. Wai-Fah Chen and E.M. Lui. Elsevier Science Publishing Company, Inc., 52 Vanderbilt Avenue, New York, NY 10017. (212) 370-5520. 1987. $49.50.

HANDBOOKS AND MANUALS

Engineering Tables and Data. A.M. Howatson, P.G. Lund and J.D. Todd. Van Nostrand Reinhold, 115 Fifth Avenue, New York, NY 10003. (212) 254-3232 or (800) 926-2665. Second Edition. 1991. $39.95.

A Manual of Geology For Civil Engineers. John A. Pitts. World Scientific Publishing Company, Inc., 1060 Main Street, Unit B, River Edge, NJ 07661. (800) 227-7562 or (201) 487-9655. 1985. $38.00.

ONLINE DATABASES AND CD-ROMS

Architectural Periodicals Index. Royal Institute of British Architects, British Architectural Library, 66 Portland Place, London, W1N 4AD, England. Citations to worldwide literature on architecture, construction methods and standards, urban planning, and interior design, from 1978 to present. Available on DIALOG, (800) 334-2564, online service. Inquire as to cost and availability.

CA Search. Chemical Abstracts Service, P.O. Box 3012, Columbus, OH 43210-0012. (614) 447-3600. (800) 848-6533. FAX (614) 447-3709. Very comprehensive guide to worldwide chemical literature and related fields, 1972 to present. Available on BRS,(800) 289-4277, DIALOG, (800) 334-2564, ORBIT (800) 456-7248, and STN International, FIZ Karlsruhe, P.O. Box 2465, W-7500, Karlsruhe 1,Germany, online services. Inquire as to cost and availability.

Compendex Plus. Engineering Information, Inc., 345 East 47th Street, New York, NY 10017. (212) 705-7600 or (800) 221-1044. Contains citations with abstracts to worldwide literature in engineering and technology, from 1970 to present. Available on online BRS,(800) 289-4277, DIALOG, (800) 334-2564, ORBIT (800) 456-7248, and STN International, FIZ Karlsruhe, P.O. Box 2465, W-7500, Karlsruhe 1, Germany, online services. Also available on CD-ROM. Inquire as to cost and availability.

Current Contents Search. Institute for Scientific Information, 3501 Market Street, Philadelphia, PA 19104. (215) 386-0100. FAX (215) 386-6362. Contains citations to articles listed in the table of contents of science and technology journals. Also articles in social sciences and life sciences journals. Available on BRS,(800) 289-4277, DIALOG,(800) 334-2564, online services. Inquire as to cost and availability.

Dissertation Abstracts. University Microfilms International, 300 North Zeeb Road, Ann Arbor, MI 48106. (800) 521-0600 or (313) 761-4700. Scope includes virtually all doctoral dissertations accepted at accredited American institutions from 1861 to present in 252 subject areas. Available on BRS,(800) 289-4277, DIALOG,(800) 334-2564, and OCLC EPIC,(800) 848-5878, online services. Also available on CD-ROM. Inquire as to cost and availability.

Geomechanics Abstracts. University of London, Imperial College of Science and Technology, Rock Mechanics Information Service, Royal School of Mines, Prince Consort Road, London SW7 2BP, England. Telephone: 071-589 5111 or FAX 071-589 6806. Scope includes worldwide literature on engineering geology, hydrogeology, mining, rock mechanics, soil mechanics, and tunnelling, 1977 to present. Available on the ORBIT (800) 456-7248, online service. Inquire as to cost and availability.

GEOREF. American Geological Institute, 4220 King Street, Alexandria, VA 22302. (800) 336-4764 or (703) 379-2480. Geology and geosciences literature, 1785 to present for North America. Available on DIALOG,(800) 334-2564, ORBIT (800) 456-7248, online services. Also available on CD-ROM. Inquire as to cost and availability.

ICONDA (CIB International Construction Database). Fraunhofer-Gesellschaft, Informationszentrum RAUM und BAU, Nobelstrasse, 12, D-7000, Stuttgart 80, Germany. Contains citations with abstracts to worldwide technical literature on construction and civil engineering, structural engineering, and engineering geology, from 1976 to present. Available on ORBIT, (800) 456-7248, and STN International, FIZ Karlsruhe, P.O. Box 2465, W-7500, Karlsruhe 1, Germany, online services. Inquire as to cost and availability.

National Information Service For Earthquake Engineering Database. University of California, Berkeley. Contains citations and abstracts to journal articles on earthquakes and earthquakeengineering from 1987 to present. Available on BRS,(800) 289-4277, online service. Inquire as to cost and availability.

NTIS Bibliographic Database. National Technical Information Service, 5285 Port Royal Road, Springfield, Va 22161. (703) 487-4929 or FAX (703) 321-8199. Broad coverage of government-sponsored science and technology research reports, 1964 to present. Available on BRS,(800) 289-4277, DIALOG, (800) 334-2564, ORBIT, (800) 456-7248, and STN International, FIZ Karlsruhe, P.O. Box 2465, W-7500, Karlsruhe 1, Germany, online services. Also available on CD-ROM. Inquire as to cost and availability.

Nuclear Science Abstracts. U.S. Department of Energy, office of Scientific and Technical Energy, P.O. Box 62, Oak Ridge, TN 37831. (615) 576-1189. Contains citations with abstracts to literature in all fields of nuclear science and energy. 1948 to 1976. Available on DIALOG,(800) 334-2564, online service. Inquire as to cost and availability.

SCISEARCH. Institute for Scientific Information, 3501 Market Street, Philadelphia, PA 19104. (800) 523-1850 or (215) 386-0100. Broad multidisciplinary title and author index to the international literature of science and technology, 1974 to present. Available on DIALOG,(800) 334-2564, and ORBIT,(800) 456-7248, online services. Also available on CD-ROM. Inquire as to cost and availability.

TRIS (Transportation Research Information). National Academy of Science, 2101 Constitution Avenue, N.W., Washington,

DC 20418. (202) 334-3313 or (800) 624-6242. Citations with abstracts of literature on transportation, including air, highway, rail, maritime and other modes from 1968 to present. Available on DIALOG,(800) 334-2564, online service. Inquire as to cost and availability.

Wasteinfo. United Kingdom Atomic Energy Authority, Building 7.12, Harwell Laboratory, Didcot, Oxon, OX11 ORA, England. Contains citations to worldwide literature on non-nuclear waste management, from 1973 to present. Available on ORBIT (800) 456-7248, online service. Also available on CD-ROM. Inquire as to cost and availability.

Water Resources Abstracts (wra). U.S. Geological Survey, Water Resources Scientific Information Center, 12201 Sunrise Valley Drive, Reston, VA 22092-9998. (703) 648-4460. Contains citations with abstracts to scientific and technical literature on the water-sources-related aspects of the physical, social and life sciences, from 1968 to present. Available on DIALOG,(800) 334-2564, and ORBIT,(800) 456-7248, online services. Inquire as to cost and availability.

Waternet. American Water Works Association, Technical Library. Available on DIALOG online services. Citations to literature on water quality, water utility management, water analysis, water pollution, and related areas, 1971 to present. Available on DIALOG,(800) 334-2564, online service. Inquire as tocost and availability.

WILSONLINE. H.W. Wilson Company, 950 University Avenue, Bronx, NY 10452. (800) 367-6770 or (212) 588-8400. Makes available online versions of the printed H.W. Wilson indexes including Applied Science and Technology Index, Business Periodicals Index, General Science Index, and Readers' Guide to Periodical Literature. Period covered is generally 1983 to present. Available on BRS,(800) 289-4277, DIALOG,(800) 334-2564, and OCLC EPIC,(800) 848-5878, online services. Also available on CD-ROM. Inquire as to cost and availability.

PERIODICALS

A.P.W.A. Reporter. American Public Works Association, 1313 East 60th Street, Chicago, IL 60637-2881. (312) 667-2200. 1934 to present. Monthly.

Cement, Concrete, and Aggregates. American Society for Testing and Materials (ASTM), 1916 Race Street, Philadelphia, PA 19103. (215) 299-5585. 1979 to present. Semi-annual. $48.00 per year.

*Civil Engineering A.*S.C.E. American Society of Civil Engineers, 345 East 47th Street, New York, NY 10017-2398. (212) 705-7520 or (800) 548-2723. 1930 to present. Monthly. $85.00 per year.

Civil Engineering Systems. Gordon and Breach Science Publishers, 270 Eighth Avenue, New York, NY 10011. (212) 206-8900 or FAX (212) 645-2459. 1983 to present. Quarterly. $850.00 per year.

Coastal Engineering: An International Journal For Coastal, Harbour, and offshore Engineers. Elsevier Science Publishing Company, Inc., 655 Avenue of the Americas, New York, NY 10010. (212) 989-5800. 1977 to present. Monthly. Inquire about price.

Coastal Engineering Research Council. PROCEEDINGS. American Society of Civil Engineers, 345 East 47th Street, New York, NY 10017-2398. (212) 705-7520 or (800) 548-2723. 1954 to present. Biennial. Price varies.

Construction Products. Cahners Publishing Company (Des Plaines), 1350 East Touhy Avenue, Des Plaines, IL 60017-5080. (708) 635-8800. 1892 to present. Six times per year. $35.00 per year.

E.N.R. McGraw-Hill Publishing Company, 1221 Avenue of the Americas, New York, NY 10020. (800) 262-4729 or (212) 512-3825. 1874 to present. Weekly. $65.00 per year.

Earthquake Engineering and Structural Dynamics. John Wiley and Sons, Inc., 605 Third Avenue, New York, NY 10158. (800) 526-5368 or (212) 850-6000. 1972 to present. Monthly. $800.00 per year.

Engineering and Mining Journal. Maclean Hunter Publishing Company, 29 North Wacker Drive, Chicago, IL 60606-3298. (312) 726-2802. 1866 to present. Monthly. $35.00 per year.

Engineering Geology. Elsevier Science Publishing Company, Inc., 655 Avenue of the Americas, New York, NY 10010. (212)989-5800. 1965 to present. Monthly. $800.00 per year.

Engineering Journal. American Institute of Steel Construction, Inc., One East Wacker Drive, Suite 3100, Chicago, IL 60601-2001. (312) 670-2400 or FAX (312) 670-5403. 1964 to present. Quarterly. $15.00 per year.

Engineering Structures: the Journal of Earthquake, Wind and Ocean Engineering. Butterworths Publishers, 80 Montvale Avenue, Stoneham, MA 02180. (617) 438-8464 or (800) 366-2665. 1979 to present. Bimonthly. $470.00 per year.

Geotechnical Testing Journal. American Society for Testing and Materials (ASTM), 1916 Race Street, Philadelphia, PA 19103. (215) 299-5585. 1978 to present. Quarterly. $89.00 per year.

International Journal For Numerical and Analytical Methods In Geomechanics. John Wiley and Sons, Inc., 605 Third Avenue, New York, NY 10158. (800) 526-5368 or (212) 850-6000. 1977 to present. Monthly. $825.00 per year.

Journal of Construction Engineering and Management. American Society of Civil Engineers, 345 East 47th Street, New York, NY 10017-2398. (212) 705-7520 or (800) 548-2723. 1956 to present. Quarterly. $104.00 per year.

Journal of Geotechnical Engineering. American Society of Civil Engineers, 345 East 47th Street, New York, NY 10017-2398. (212) 705-7520 or (800) 548-2723. 1956 to present. Monthly. $198.00 per year.

Journal of Hydraulic Engineering. American Society of Civil Engineers, 345 East 47th Street, New York, NY 10017-2398. (212) 705-7520 or (800) 548-2723. 1956 to present. Monthly. $198.00 per year.

Journal of Irrigation and Drainage. American Society of Civil Engineers, 345 East 47th Street, New York, NY 10017-2398. (212) 705-7520 or (800) 548-2723. 1956 to present. Bi-monthly. $124.00 per year.

Journal of Materials In Civil Engineering: Properties, Applications, Durability. American Society of Civil Engineers, 345 East 47th Street, New York, NY 10017-2398. (212) 705-7520 or (800) 548-2723. 1989 to present. Quarterly. $120.00 per year.

Journal of Performance of Constructed Facilities. American Society of Civil Engineers, 345 East 47th Street, New York, NY 10017-2398. (212) 705-7520 or (800) 548-2723. 1987 to present. Quarterly. $80.00 per year.

Journal of Structural Engineering. American Society of Civil Engineers, 345 East 47th Street, New York, NY 10017-2398. (212) 705-7520 or (800) 548-2723. 1956 to present. Monthly. $292.00 per year.

Journal of Surveying Engineering. American Society of Civil Engineers, 345 East 47th Street, New York, NY 10017-2398. (212) 705-7520 or (800) 548-2723. 1956 to present. Quarterly. $92.00 per year.

Journal of Waterway, Port, Coastal, and Ocean Engineering. American Society of Civil Engineers, 345 East 47thStreet, New York, NY 10017-2398. (212) 705-7520 or (800) 548-2723. 1956 to present. Bi-monthly. $98.00 per year.

MILITARY ENGINEER. Society of American Military Engineers. 607 Prince Street, Alexandria, VA 22314-3117. (703) 549-3800. 1920 to present. Seven times per year. $24.00 per year.

Public Roads. U.S. Federal Highway Administration, 400 Seventh Street, N.W., Washington, DC 20590. (703) 285-2104. 1918 to present. Quarterly. $12.00 per year.

Roads and Bridges. Scranton Gillette Communications, Inc., 380 East Northwest Highway, Des Plaines, IL 60016. (708) 298-6622. 1892 to present. Monthly. $20.00 per year.

Rock Mechanics and Rock Engineering. Springer-Verlag New York, Inc., 175 Fifth Avenue, New York, NY 10010. (800) 526-7254 or (212) 460-1500. Quarterly. $122.00 per year.

Structural Engineering Review. Pergamon Press Inc., Maxwell House, Fairview Park, Elmsford, NY 10523. (914) 592-7700. Fax (914) 592-3625. 1988 to present. Quarterly. $185.00 per year.

World Tunnelling. Mining Journal Limited, 60 Worship Street, London EC2A 2HD, England. 1988 to present. Ten times per year. $80.00 per year.

RESEARCH CENTERS AND INSTITUTES

American Concrete Institute, 22400 West Seven Mile Road, Detroit, Mi 48219-1885. (313) 532-2600 or FAX (313) 538-0655.

American Iron and Steel Institute, 1101 17th Street, Washington, Dc 20036-4700. (202) 452-7265 or FAX (202) 463-6573.

Construction Industry Institute, University of Texas At Austin, 3208 Red River, Suite 300, Austin, Tx 78705-2650. (512) 471-4319.

Engineering Societies Library, United Engineering Center, 345 East 47th Street, New York, Ny 10017. (212) 705-7611.

National Association of Home Builders National Research Center, 400 Prince George's Building, Upper Marlboro, Md 20772. (301) 249-4000.

University of Texas At Austin. Bureau of Engineering Research. Cockrell Hall 10.340, Austin, TX 78712-1080. (512) 471-4325.

CONTINENTAL DRIFT

See also: CRUST, GEOPHYSICS, PHYSICAL GEOLOGY, PLATE TECTONICS, SEISMOLOGY

ABSTRACT SERVICES AND INDEXES

Bibliography and Index of Geology. American Geological Institute, 4220 King Street, Alexandria, VA 22302. (703) 379-2480. Fax (703) 379-7563. Monthly. $1295.00 per year. Also available online as GeoRef.

Deep-Sea Research. Part A: Oceanographic Research Papers. Deep-Sea Research. Part B: Oceanographic Literature Review. Pergamon Press, Inc., Maxwell House, Fairview Park, Elmsford, NY 10523. (914) 592-7700. Fax (914) 592-3625. Twelve times per year. $1370 per year for Parts A and B.

Geological Abstracts. Elsevier - Geo Abstracts, RegencyHouse, 34 Duke Street, Norwich NR3 3AP, England. Monthly. $760.00 per year. Also available online as GEOBASE.

Geological Society of America. Abstracts with Programs. Geological Society of America. 3300 Penrose Place, P.O. Box 9140, Boulder, CO 80301-9140. (303) 447-2020. Abstracts and programs of the annual conference. Annual. $69.00.

Geophysics Abstracts. Pergamon Press, Inc., Maxwell House, Fairview Park, Elmsford, NY 10523. (914) 592-7700. Fax (914) 592-3625. Twelve times per year. $565.00 per year. Also available in microform.

Oceanic Abstracts. Cambridge Scientific Abstracts, 7200 Wisconsin Avenue, Bethesda, MD 20814. (301) 961-6750. Fax (301) 961-6720. Bimonthly. $995.00 per year.

ANNUAL REVIEWS AND YEARBOOKS

Advances In Geophysics. Academic Press, Inc., 1250 Sixth Avenue, San Diego, CA 92101. (619) 231-0926. FAX (619) 699-6715. Irregular. Inquire about price and availability.

Annual Review and Earth and Planetary Sciences. Annual Reviews, Inc., 4139 El Camino Way, Palo Alto, CA 94306-0897. (415) 493-4400. Fax (415) 855-9815. Annual. $55.00 per year.

ASSOCIATIONS AND PROFESSIONAL SOCIETIES

American Geophysical Union, 2000 Florida Avenue, N.W., Washington, DC 20009. (202) 462-6900.

American Institute of Professional Geologists. 7828 Vance Drive, Suite 103, Arvada, CO 80003. (303) 431-0831.

CONTINENTAL DRIFT

Ency. of Physical Sciences and Engineering Info. Sources

Association of Engineering Geologists. 323 Boston Post Road, Suite 2D, Sudbury, MA 01776. (508) 443-4639.

Geochemical Society. Department of Terrestrial Magnetism, 5241 Broad Branch Road, N.W., Washington, DC 20015. (202) 686-4370.

Geological Society of America. 3300 Penrose Place, P.O. Box 9140, Boulder, CO 80301-9140. (303) 447-2020.

Society of Economic Geologists. P.O. Box 571, Golden, CO 80402. (303) 277-1716.

Society of Exploration Geophysicists. P.O. Box 702740, Tulsa, OK 74170. (918) 493-3516.

BIBLIOGRAPHIES

Geologic Reference Sources: A Subject and Regional Bibliography of Publications and Maps In the Geological Sciences. Dederick C. Ward et al. Scarecrow Press, Inc., 52 Liberty Street, Box 4167, Metuchen, NJ 08840. (800) 537-7107 or (201) 548-8600. 1981. $49.50.

Information Sources In the Earth Sciences. David N. Wood, Joan E. Hardy, and Anthony P. Harvey, editors. Bowker-Saur/K.G. Saur. Distributed by R.R. Bowker, 121 Chanlon Road, New Providence, NJ 07974. (800) 521-8110 or (908) 464-6800. Second edition. 1989. $85.00.

Remote Sensing of Earth Resources: A Quarterly Bibliography. University of New Mexico, Technology Application Center, 2808 Central Avenue, S.E., Albuquerque, NM 87131-6031. (505) 277-3622. Fax (505) 277-3614. $150.00 per year.

DIRECTORIES AND BIOGRAPHICAL SOURCES

Research Centers Directory. Gale Research, 835 Penobscot Building, Detroit, MI 48226-4094. (800) 347-4253 or (313) 961-2242. 20th edition, 1995. $485.00.

ENCYCLOPEDIAS AND DICTIONARIES

Dictionary of Geological Terms. American Geological Institute. Doubleday and Company, Inc., 245 Park Avenue, New York, NY 10017. (800) 645-6156 or (212) 953-4561. Third edition. 1984. $12.00 in paper.

Glossary of Geology. Robert L. Bates and Julia A. Jackson. American Geological Institute, 4220 King Street, Alexandria, VA 22032. (703) 379-2480. Third edition. 1987. $75.00.

GENERAL WORKS

Behavior of the Earth: Continental and Seafloor Mobility. Claude Allegre. Harvard University Press, 79 Garden Street, Cambridge, MA 02138. (617) 495-2600. 1988. $39.95

Continental Drify. Russell Banks. Harpercollins Publishers, Inc., 10 E. 53rd Street, New York, NY 10022-5299. (800) 331-3761. FAX (212) 207-7145. 1994. $12.00.

Continental Drift. AL Young. University of California Press, 2120 Berkeley Way, Berkeley, CA 94720. (800) 777-4726. FAX (510) 642-7127. 1996. $12.95.

Continents In Motion. Walter Sullivan. 3rd edition. American Institute of Physics, AIP Press, 500 Sunnyside Blvd., Woodbury, NY 11797-2999. (800) 809-2247. FAX (516) 576-2223. 1990. $50.00 (hardbound), $25.00 (paperback).

Dynamic Earth: An Introduction To Physical Geology. B.J. Skinner and Stephen C. Porter. John Wiley and Sons, Inc., 605 Third Avenue, New York, NY 10158. (800) 526-5368 or (212) 850-6000. 1990. $51.95.

The Evolving Continents. B.F. Windley. 3rd edition. John Wiley and Sons, Inc., 605 Third Avenue, New York, NY 10158. (800) 526-5368 or (212) 850-6000. 1995. $54.95.

New Views On An Old Planet: A History of Global Change. Tjeerd H. Van andel. Cambridge University Press, 40 West 20th Street, New York, NY 10011-4211. (212) 924-3900 or (800) 872-7423. FAX (914) 937-4712. 1994. $59.95 (hardbound), $24.95 (paperbound).

HANDBOOKS AND MANUALS

Field Mapping For Geology Students. F. Ahmed and D.C. Almond. Unwin Hyman, Inc., 10 East 53rd Street, New York, NY 10022. (212) 207-7626 or (800) 242-7737. 1983. $18.95.

Geology In the Field. R.R. Compton. John Wiley and Sons, Inc., 605 Third Avenue, New York, NY 10158. (800) 526-5368 or (212) 850-6418. 1985. $37.95.

The Mapping of Geological Structures. K.R. McClay. John Wiley and Sons, Inc., 605 Third Avenue, New York, NY 10158. (800) 526-5368 or (212) 850-6000. 1988. $18.95.

ONLINE DATABASES AND CD-ROMS

Earth Sciences. U.S. Geological Survey, 12201 Sunrise Valley Drive, Reston, VA 22092-9998. (703) 648-4460. CD-ROM ofearth science databases including the U.S. Geological Survey Library, Earth Science Data Directory, and GEOINDEX, citations to published geological maps. $350.00 per year with quarterly updates.

Geoarchive. Geosystems, P.O. Box 1024, Westminster, London, England, SW1 P 2JL. Citations to literature on geoscience, 1969 to present. Inquire as to online cost and availability.

Geobase. Elsevier - Geo Abstracts, Regency House, 34 Duke Street, Norwich NR3 3AP, England. Contains citations to the worldwide earth science literature from 1980 to date. Available on DIALOG, ORBIT online services. Inquire as to cost and availability.

Geomechanics Abstracts. University of London, Imperial College of Science and Technology, Rock Mechanics Information Service, Royal School of Mines, Prince Consort Road, London SW7 2BP, England. Telephone: 071-589 5111 or FAX 071-589 6806. Scope includes worldwide literature on engineering geology, hydrogeology, mining, rock mechanics, soil mechanics, and tunnelling, 1977 to present. Available on the ORBIT online service. Inquire as to cost and availability.

GEOREF. American Geological Institute, 4220 King Street, Alexandria, VA 22302. (800) 336-4764 or (703) 379-2480. Geology and geosciences literature, 1785 to present for North America. Available on DIALOG, ORBIT online services. Also available on CD-ROM. Inquire as to cost and availability.

PERIODICALS

American Journal of Science. Kline Geology Laboratory, Yale University, Box 6666, Yale Station, New Haven, CT 06511-8130. (203) 432-3131. Ten times per year. $40.00 per year.

Continental Shelf Research. Pergamon Press, Maxwell House, Fairview Park, Elmsford, NY 10523. (914) 592-7700. Fax (914) 592-3625. Monthly. $420.00.

Earth and Planetary Science Letters. Elsevier Science Publishing Company, Inc., 655 Avenue of the Americas, New York, NY 10010. (212) 989-5800. Twenty times per year. $945.00 per year.

Geological Society Journal. Geological Society of London. Geological Society Publishing House, Unit 7, Brassmill Enterprise Centre, Brassmill Lane, Bath, Avon BA1 3JN, England. 6 times per year. $432.00.

Geological Society of America Bulletin. P.O. Box 9140, 3300 Penrose Place, Boulder, CO 80301. (303) 447-2020. Fax (303) 447-1133. Monthly. $150.00.

Geology. Geological Society of America. P.O. Box 9140, 3300 Penrose Place, Boulder, CO 80301. (303) 447-2020. Fax (303) 447-1133. Monthly. $120.00 per year.

Geophysics. Society of Exploration Geophysicists, P.O. Box 702740, Tulsa, OK 74170. (918) 493-3516. Monthly. $175.00 per year.

Journal of Geology. University of Chicago Press, 5720 Woodlawn Avenue, Chicago, IL 60637. (312) 753-3347. Bimonthly. $38.00 per year.

Journal of Geophysical Research. American GeophysicalUnion, 2000 Florida Avenue, N.W., Washington, DC 20009. (202) 462-6903. Monthly. $2170.00 per year.

Journal of Structural Geology. Pergamon Journals, Inc., Maxwell House, Fairview Park, Elmsford, NY 10523. (914) 592-7700. Ten times per year. $375.00 per year.

Marine Geology. Elsevier Science Publishing Company, Inc., 655 Avenue of the Americas, New York, NY 10010. (212) 989-5800. Twenty-fours times per year. $995.00 per year.

Marine Geophysical Researches. Kluwer Academic Publishers, P.O. Box 358, Hingham, MA 02018-0358. (617) 871-6000. Quarterly. $237.50 per year.

Reviews of Geophysics. American Geophysical Union, 2000 Florida Avenue, N.W., Washington, DC 20009. (202) 462-6900. Quarterly. $300.00 per year.

Tectonics. American Geophysical Union, 2000 Florida Avenue, N.W., Washington, DC 20009. (202) 462-6900. Bimonthly. $230.00 per year to individuals.

Tectonophysics; An International Journal of Geotectonics and the Geology and Physics of the Interior of the Earth. Elsevier Science Publishing Company, Inc., 655 Avenue of the Americas, New York, NY 10010. (212) 989-5800. Fifty-six times per year. $1275.00 per year.

RESEARCH CENTERS AND INSTITUTES

Cornell University. Institute for the Study of Continents. 3122 Snee Hall, Ithaca, NY 14853-1504. (607) 255-3474.

Goddard Space Flight Center. Crustal Dynamics Project. National Aeronautics and Space Administration, Greenbelt, MD 20771. (301) 286-8809. Fax (301) 286-4943.

Smithsonian Institution. Center for Earth and Planetary Studies. National Air and Space Museum, Sixth Street and Independence Avenue, S.W., Washington, DC 20560. (202) 356-1424. Fax (202) 786-2262.

CONTINENTAL MARGINS

See also: CIVIL ENGINEERING, COASTAL ENGINEERING, COASTS, GEOLOGY, OCEAN ENGINEERING, PHYSICAL GEOLOGY

ABSTRACT SERVICES AND INDEXES

ASCE Combined Annual Combined Index. American Society of Civil Engineers, 345 East 47th Street, New York, NY 10017-2398. (212) 705-7520 or (800) 548-2723. Annual. $48.00.

ASCE Publications Information. 1966-present. American Society of Civil Engineers, 345 East 47th Street, New York, NY 10017-2398. (212) 705-7520 or (800) 548-2723. Bi-monthly. $160.00 per year for non-members.

Bibliography and Index of Geology. American Geological Institute, 4220 King Street, Alexandria, VA 22302. (703) 379-2480. Fax (703) 379-7563. Monthly. $1295.00 per year. Also available online as GEOREF.

Civil and Structural Engineering Abstracts, Cambridge Scientific Abstracts, 7200 Wisconsin Avenue, Bethesda, Md 20814-4823. (301) 961-6750. FAX (301) 961-6720. 1993 to present. Monthly. $385.00 per year. Topics covered include structuraldesign, construction equipment and methods, civil defense and military engineering, surveying, highway engineering, maritime and port structures, materials, land reclamation, and soil mechanics.

Meteorological and Geoastrophysical Abstracts. American Meteorological Society, c/o Inforonics Inc., 550 Newtown Road, Littleton, MA 01460. (508) 486-8976. FAX (508) 486-0027. 1950 to present. Monthly. $950.00 per year. Current abstracts of books, reports, research papers, and miscellaneous literature on environmental sciences, meteorology, astrophysics, hydrology, glaciology, and physical oceanography.

Oceanic Abstracts. Cambridge Scientific Abstracts, 7200 Wisconsin Avenue, Bethesda, MD 20814. (301) 961-6750. Fax (301) 961-6720. Bimonthly. $995.00 per year.

Science Citation Index. Institute for Scientific Information, 3501 Market Street, Philadelphia, PA 19104. (215) 386-0100. FAX

CONTINENTAL MARGINS

Ency. of Physical Sciences and Engineering Info. Sources

(215) 386-6362. Inquire about availability and cost. Also available on CD-ROM.

ASSOCIATIONS AND PROFESSIONAL SOCIETIES

American Society of Civil Engineers. 345 East 47th Street, New York, NY 10017. (212) 705-7420.

Coastal Engineering Research Council. 207 East Bay Street, Suite 311, Charleston, SC 29401. (803) 723-4864.

The Coastal Society. 5410 Grosvenor Lane, Suite 110, Bethesda, MD 20814. (301) 897-8616.

Geological Society of America. 3300 Penrose Place, Boulder, CO 80301. (303) 447-2020.

DIRECTORIES AND BIOGRAPHICAL SOURCES

Directory of Engineering Societies and Related Organizations. Gordon Davis. 13th edition. American Association of Engineering Societies, 1111 19th Street NW, Suite 608, Washington, DC 20036. (202) 296-2237 or (800) 658-8897. 1989. Inquire for price.

International Research Centers Directory. Gale Research, 835 Penobscot Building, Detroit, MI 48226-4094. (313) 961-2242. (800) 347-4253. FAX (313) 961-6083. 8th edition. 1995. $410.00

Research Centers Directory. Gale Research, 835 Penobscot Building, Detroit, MI 48226-4094. (313) 961-2242. (800) 347-4253. FAX (313) 961-6083. $485.00.

Scientific and Technical Organizations and Agencies Directory. Gale Research, 835 Penobscot Building, Detroit, MI 48226-4094. (313) 961-2242. (800) 347-4253. FAX (313) 961-6083. 4th edition. 1996. $195.00.

GENERAL WORKS

*The Atlantic Continental Margin: U.*S. R.E. Sheridan and J.A. Grow, editors. Geological Society of America, 3300 Penrose Place, PO Box 9140, Denver, CO 80301-9140. (303) 447-2020. 1988. $49.50.

Continental Drift. James H. Shea, editor. Van Nostrand Reinhold Company, Inc., 135 West 50th Street, New York, NY 10020. (800) 543-2681. 1985. $49.50.

Geology and Geophysics of Continental Margins. Joel S. Watkins, et al., editors. American Association of Petroleum Geologists, PO Box 979, Tulsa, OK 74101-0979. (918) 584-2555 or (800) 364-2274. FAX (918) 584-0469. 1992. $29.00.

The Ocean Basins and Margins. A.E.M. Nairn, editor. Plenum Publishing Corporation, 233 Spring Street, New York, NY 10013. (800) 221-9369. 1973-1985. Eight volumes. Price varies, inquire.

Physical Oceanography of Coastal Waters. K.F. Bowden. John Wiley and Sons, Inc., 605 Third Avenue, New York, NY 10158. (800) 526-5368. 1984. $74.95.

The World's Coastline. E.C. Bird and M.L. Schwartz, editors. Van Nostrand Reinhold Company, Inc., 135 West 50th Street, New York, NY 10020. (800) 543-2681. 1985. $99.95.

HANDBOOKS AND MANUALS

Coastal, Estuarial, and Harbour Engineer's Reference Book. M.B. Abbott & W.A. Price, editors. E & FN Spon/ Routledge, Chapman and Hall, Inc., 29 West 35th Street, New York, NY 10001-2291. (212) 244-3336 or FAX (212) 563-2269. 1994. Inquire for cost and availability.

Handbook of Coastal and Ocean Engineering Volume 1: Wave Phenomena and Coastal Structures. John B. Herbich, editor. Gulf Publishing Company, P.O. Box 2608, Houston, TX. (713) 529-4301 or (800) 231-6275. 1990. $195.00.

Handbook of Coastal and Ocean Engineering Volume 2: Offshore Structures, Marine Foundations, and Sediment Processes. John B. Herbich, editor. Gulf Publishing Company, P.O. Box 2608, Houston, TX. (713) 529-4301 or (800) 231-6275. 1990. $195.00.

Handbook of Coastal and Ocean Engineering Volume 3: Harbors, Navigational Channels, Estuaries and Environmental Effects. John B. Herbich, editor. Gulf Publishing Company, P.O. Box 2608, Houston, TX. (713) 529-4301 or (800) 231-6275. 1992. $195.00.

ONLINE DATABASES AND CD-ROMS

Earth Sciences. U.S. Geological Survey, 12201 Sunrise Valley Drive, Reston, VA 22092-9998. (703) 648-4460. CD-ROM of earth science databases including the U.S. Geological Survey Library, Earth Science Data Directory, and GEOINDEX, citations to published geological maps. $350.00 per year with quarterly updates.

Geoarchive. Geosystems, P.O. Box 1024, Westminster, London, England, SW1 P 2JL. Citations to literature on geoscience, 1969 to present. Inquire as to online cost and availability.

Geobase. Elsevier - Geo Abstracts, Regency House, 34 Duke Street, Norwich NR3 3AP, England. Contains citations to the worldwide earth science literature from 1980 to date. Available on DIALOG, ORBIT online services. Inquire as to cost and availability.

Geomechanics Abstracts. University of London, Imperial College of Science and Technology, Rock Mechanics Information Service, Royal School of Mines, Prince Consort Road, London SW7 2BP, England. Telephone: 071-589 5111 or FAX 071-589 6806. Scope includes worldwide literature on engineering geology, hydrogeology, mining, rock mechanics, soil mechanics, and tunnelling, 1977 to present. Available on the ORBIT (800) 456-7248, online service. Inquire as to cost and availability.

GEOREF: Bibliography and Index of Geology. American Geological Institute, 4220 King Street, Alexandria, VA 22302. (703) 379-2480. Fax (703) 379-7563. Monthly. Inquire for price and availability.

PERIODICALS

Coastal Engineering: An International Journal For Coastal, Harbour, and offshore Engineers. Elsevier Science Publishing Company, Inc., 52 Vanderbilt Avenue, New York, NY 10017. (212) 370-5520. 1977 to present. Quarterly. $95.00 per year.

Coastal Research. Florida State University, Geology Department, Tallahassee, FL 32306. (904) 644-5860. 1962 to present. Three times per year. $5.00 per year.

Coastwatch. University of North Carolina, Sea Grant College Program, Box 8605, North Carolina State University, Raleigh, NC 27695-8605. (919) 737-2454. 1970 to present. Monthly. Free.

Continental Shelf Research. Pergamon Press, Inc., Maxwell House, Fairview Park, Elmsford, NY 10523. (914) 592-7700. 1982 to present. Bimonthly. $1200.00 per year. Includes Oceanographic Literature Review.

Estuarine, Coastal and Shelf Science. Academic Press, Inc., 6277 Sea Harbor Drive, Orlando, FL 32821. (800) 321-5068. 1973 to present. Monthly. $315.00 per year.

Geological Magazine. Cambridge University Press, Journals Department, 40 West 20th Street, New York, NY 10011-4211. (212) 924-3900. 1864 to present. Bi-monthly. $263.00 per year.

Geological Society Journal. Geological Society Publishing House, Unit 7, Brassmill Enterprise Centre, Brassmill Lane, Bath BA1 3JN, England. Telephone 0225-445046. FAX 0225-442836. 1845 to present. Six times a year. $524.00.

Geology. Geological Society of America, 3300 Penrose Place, PO Box 9140, Denver, CO 80301-9140. (303) 447-2020. 1973 to present. Monthly. $170.00 per year for non-members.

Journal of Coastal Research. Coastal Education & Research Foundation, 4310 N.E. 25th Avenue, Ft. Lauderdale, FL 33308. (305) 565-1051. FAX (305) 565-1051. 1985 to present. Quarterly. $58.00.

Journal of Geology. University of Chicago Press, Journals Division, 5720 S. Woodlawn Avenue, Chicago, IL 60637. (312) 753-3347. FAX (312) 753-0811. 1893 to present. Bi-monthly. $45.00 per year.

Journal of Waterway, Port, Coastal, and Ocean Engineering. American Society of Civil Engineers, Waterway, Port, Coastal, and Ocean Division, 345 East 47th Street, New York, NY 10017-2398. (212) 705-7520 or (800) 548-2723. 1956 to present. Bimonthly. $100.00 per year.

Ocean Engineering. Pergamon Press, Inc., Maxwell House, Fairview Park, Elmsford, NY 10523. (914) 592-7700. 1968 to present.Bimonthly. $200.00 per year.

Ocean Science and Engineering. Marcel Dekker Inc., 270 Madison Avenue, New York, NY 10016. (800) 228-1160. 1974 to present. Quarterly. $165.00 per year.

RESEARCH CENTERS AND INSTITUTES

Coastal Research Center. Woods Hole Oceanographic Institution, Woods Hole, MA 02543. (508) 548-1400. FAX (508) 457-2172.

Earth Resources Institute. Texas A&M University, College Station, TX 77843. (409) 845-3651.

Geological Research Division. University of California, San Diego, A-020, Scripps Institution of Oceanography, La Jolla, CA 92093. (619) 534-1830.

Institute For the Study of the Continents. Cornell University, 3120 Snee Hall, Ithaca, NY 14853-1504. (607) 255-3474.

Louisiana State University, Coastal Studies Institute. Coastal Studies Bldg., Baton Rouge, LA 70803. (504) 388-5301. FAX (504) 388-2520.

CONTROL SYSTEMS

See: AUTOMATION

CONTROL THEORY

See also: AUTOMATION

ABSTRACT SERVICES AND INDEXES

ACM Guide To Computing Literature. Association for Computing Machinery, Association for Computing Machinery, 1515 Broadway, 17th Floor, New York, NY 10036-5701. (212) 869-7440. FAX (212) 944-1318. 1964 to present. Monthly. $190.00 per year for non-members.

Applied Science and Technology Index; A Cumulative Subject Index To English Language Periodicals In the Fields of Aeronautics and Space Science, Computer Technology, Chemistry, Construction Industry, Energy and Related Areas. H.W. Wilson Co., 950 University Avenue, Bronx, NY 10452. (800) 367-6770 or (212) 588-8400. FAX (718) 590-1617. From 1958 to present. Monthly. Inquire about cost and availability. Also available on CD-ROM and online.

Computer Abstracts, MCB University Press Ltd., PO Box 10812, Birmingham, AL 35201-0812. (800) 633-4931. FAX (205) 995-1588. Monthly. Covers computer theory, data, hardware, systems, networks, human-computer interaction, artificial intelligence, as well as applications of computers in aerospace, business, CAD/CAM, cartography, civil engineering, electronics and electrical engineering, industrial engineering, mechanical engineering, medicine, structural engineering, etc. Also available on CD-ROM.

Computer & Control Abstracts (inspec). INSPEC/IEEE, Box 1331, 445 Hoes Lane, Piscataway, NJ 08855-1331. (908) 562-5549. Abstracts organized by subjects of international technological information. Monthly. $1455.00 per year. Also available on CD-ROM and online via BRS Online Products, DIALOG Information Services, Orbit Search Service, and STN International.

CONTROL THEORY

Ency. of Physical Sciences and Engineering Info. Sources

Computer and Information Systems Abstracts Journal. Cambridge Scientific Abstracts, 7200 Wisconsin Avenue, Bethesda, MD 20814. (301) 961-6750. Fax (301) 961-6720. 1962 to present. Monthly. $1265.00 per year. Also available online via STN International.

Computer Literature Index. Applied Computer Research Inc., Box 82266, Phoenix, AZ 85071-2266. (800) 234-2227. 1971 to present. Quarterly plus annual cumulation. $198.50 per year. Bibliography of books, articles, and reports.

Computing Journal Abstracts. Techgnosis Ltd., Blade House, Battersea Road, Stockport, Cheshire 3AE, England. Telephone 061-442-2639. FAX 061-443-1162. 1969 to present. Monthly. Inquire for price.

Computing Reviews. Association for Computing Machinery, 1515 Broadway, 17th Floor, New York, NY 10036-5701. (212) 869-7440. FAX (212) 944-1318. 1960 to present. Monthly. $130.00 per year for non-members. Also available online via DIALOG Information Services.

Current Contents: Engineering, Technology, and Applied Sciences. Institute for Scientific Information, 3501 Market Street, Philadelphia, PA 19104. (215) 386-0100. FAX (215) 386-6362. 1970 to present. Weekly. $442.00 per year.

Engineering Index Monthly. Engineering Information, Inc., Castle Point on the Hudson, Hoboken, NJ 07030. (800) 221-1044. FAX (201) 216-8532. Monthly. $2300.00 per year. Also available online as COMPENDEX, and also on CD-ROM. Covers chemical engineering, computers, electrical engineering, civil engineering, metals and mining, industrial management, and mechanical engineering.

Science Citation Index. Institute for Scientific Information, 3501 Market Street, Philadelphia, PA 19104. (215) 386-0100. FAX (215) 386-6362. Inquire about availability and cost. Also available on CD-ROM.

ASSOCIATIONS AND PROFESSIONAL SOCIETIES

Association For Computing Machinery. 1515 Broadway, 17th Floor, New York, NY 10036-5701. (212) 869-7440. FAX (212) 944-1318.

Computer and Automation Systems Association of Sme. 1 SME Drive, PO Box 930, Dearborn, MI 48121-0930. (313) 271-1500.

Institute of Electrical and Electronic Engineers. 345 E. 47th Street, New York, NY 10017. (212) 705-7900.

IEEE Computer Society. 1730 Massachusetts Avenue NW, Washington, DC 20036-1992. (202) 371-0101. FAX (202) 728-9614.

IEEE Robotics and Automation Council. 345 E. 47th Street, New York, NY 10017. (212) 705-7900. FAX (212) 705-4929.

BIBLIOGRAPHIES

ACM Guide To Computing Literature. Association for Computing Machinery, 1515 Broadway, 17th Floor, New York, NY 10036-5701. (212) 869-7440. FAX (212) 944-1318. 1964 to present. Annual. $190.00 for non-members. Also available online via DIALOG.

DIRECTORIES AND BIOGRAPHICAL SOURCES

American Men and Women of Science: Physical and Biological Sciences. Eighteenth edition. R.R. Bowker Company, 245 West 17th Street, New York, NY 10011. (800) 521-8810 or (212) 916-1600. $750.00.

Asis Handbook and Directory, American Society For Information Science, 8720 Georgia Avenue, Suite 501, Silver Spring, Md 20910-3602. (301) 495-0900. Annual. $100.00.

Computing Information Directory. Hildebrandt Inc., PO Box 576, Pullman, WA 99114. 1981 to present. Annual. $199.95 per year.

International Research Centers Directory. Gale Research, 835 Penobscot Building, Detroit, MI 48226-4094. (313) 961-2242. (800) 347-4253. FAX (313) 961-6083. 8th edition. 1995. $410.00.

Research Centers Directory. Gale Research, 835 Penobscot Building, Detroit, MI 48226-4094. (313) 961-2242. (800) 347-4253. FAX (313) 961-6083. $485.00.

Who's Who In Computing. Canadian M I S Database Inc., 268 Lakeshore Road E., Suite 510, Mississauga, ON L5G 1H1, Canada. (905) 271-1601. FAX (905) 271-4522. 1987 to present. Annual. Inquire for price in U.S.

Who's Who In Technology. Gale Research, 835 Penobscot Building, Detroit, MI 48226-4094. (313) 961-2242. (800) 347-4253. FAX (313) 961-6083. 1995. $195.00.

ENCYCLOPEDIAS AND DICTIONARIES

Encyclopedia of Computer Science and Engineering. Anthony Ralston & Edwin D. Reilly Jr., editors. 3rd revised edition. Van Nostrand Reinhold, 115 Fifth Avenue, New York, NY 10003. (212) 254-3232 or (800) 926-2665. FAX (212) 254-9499. 1993. $125.00.

Macmillan Encyclopedia of Computers. Gary G. Bitter, editor. Macmillan Publishing, 200 Old Tappan Road, Old Tappan, NJ 07675. (800) 223-2336. FAX (800) 445-6991. 1992. $150.00.

McGraw-Hill Encyclopedia of Science and Technology. McGraw-Hill Book, Incorporated, 1221 Avenue of the Americas, New York, NY 10020. (212) 997-3675. Seventh edition, 20 volumes. 1992. $1900.00.

Prentice Hall's Illustrated Dictionary of Computing. Jonar C. Nader. Prentice Hall, 113 Sylvan Avenue, Route 9W, Englewood Cliffs, NJ 07632. (201) 592-2000 or (800) 922-0579. 1992. $24.95.

GENERAL WORKS

Automatic Control Systems. Benjamin C. Kuo. Sixth edition. Prentice Hall, 113 Sylvan Avenue, Route 9W, Englewood Cliffs, NJ 07632. (201) 592-2000 or (800) 922-0579. 1990. $76.00.

Introduction To Control theory. O.L. Jacobs. 2nd edition. Oxford University Press, Inc., 198 Madison Avenue, New York, NY 10016-4314. (212) 726-6000. FAX (212) 726-6446. 1994. $75.00 (hardbound), $33.95 (paperback).

Modern Control System theory. M. Gopal. 2nd edition. Halsted Press (Division of John Wiley and Sons, Inc.), 605 Third Avenue, New York, NY 10158-0012. (212) 850-6400. FAX (212) 850-6816. 1993. $56.95.

Modern Control System theory and Design. Stanley M. Shinners. John Wiley and Sons, Inc., 605 Third Avenue, New York, NY10158. (800) 526-5368 or (212) 850-6000. 1992. $69.95.

Schaum's Outline of Feedback and Control Systems. J.J. DiStefano III, et al. 2nd edition. McGraw-Hill Publishing Company, 1221 Avenue of the Americas, New York, NY 10020. (800) 262-4729 or (212) 512-3825. 1990. $13.95.

ONLINE DATABASES AND CD-ROMS

Compendex Plus. Engineering Information, Inc., 345 East 47th Street, New York, NY 10017. (212) 705-7600 or (800) 221-1044. Contains citations with abstracts to worldwide literature in engineering and technology, from 1970 to present. Available on online BRS,(800) 289-4277, DIALOG, (800) 334-2564, ORBIT (800) 456-7248, and STN International, FIZ Karlsruhe, P.O. Box 2465, W-7500, Karlsruhe 1, Germany, online services. Also available on CD-ROM. Inquire as to cost and availability.

Compuscience. FIZ Karlsruhe, D-7514, Eggenstein-Leopoldshafen 2, Germany. Contains citations with abstracts to European and North American literature on computer science, 1972 to present. Available on STN International, FIZ Karlsruhe, P.O. Box 2465, W-7500, Karlsruhe 1, Germany, online service. Inquire as to cost and availability.

Computer and Information Systems Abstracts. Cambridge Scientific Abstracts, 7200 Wisconsin Avenue, Suite 601, Bethesda, MD 20814. (301) 961-6750 or (800) 843-7751. Contains citations to worldwide literature in theoretical and applied computer science and related areas, from 1981 to present.Inquire as to cost and availability.

Computer and Mathematics Search. Institute for Scientific Information, 3501 Market Street, Philadelphia, PA 19104. (215) 386-0100. FAX (215) 386-6362. Covers worldwide literature in computer science and mathematics, from 1980 to present. Available on BRS,(800) 289-4277, online service. Inquire as to cost and availability.

Computer Database. Information Access Company, 362 Lakeside Drive, Foster City, CA 94404. (415) 378-5000 or (800) 227-8431. Contains citations with abstracts to literature from trade journals covering the computer,telecommunications,and electronics industries. Available on the BRS, (800) 289-4277, Compuserve Information Service,(800) 848-8990, and DIA-LOG,(800) 334-2564, online services. Inquire as to cost and availability.

CSA Engineering. Cambridge Scientific Abstracts, 7200 Wisconsin Avenue, Suite 601, Bethesda, MD 20814. (301) 961-6750 or (800) 843-7751. Contains citations and abstracts of international periodicals and research literature covering all fields of engineering and science and technology,including computer and information science, electronics, mechanical engineering, solid state materials, 1981 to present. Available on BRS,(800) 289-4277, online service.Inquire as to cost and availability.

Current Contents Search. Institute for Scientific Information, 3501 Market Street, Philadelphia, PA 19104. (215) 386-0100. FAX (215) 386-6362. Contains citations to articles listedin the table of contents of science and technology journals. Also ar-

ticles in social sciences and life sciences journals. Available on BRS,(800) 289-4277, DIALOG,(800) 334-2564, online services. Inquire as to cost and availability.

Dissertation Abstracts. University Microfilms International, 300 North Zeeb Road, Ann Arbor, MI 48106. (800) 521-0600 or (313) 761-4700. Scope includes virtually all doctoral dissertations accepted at accredited American institutions from 1861 to present in 252 subject areas. Available on BRS,(800) 289-4277, DIALOG,(800) 334-2564, and OCLC EPIC,(800) 848-5878, online services. Also available on CD-ROM. Inquire as to cost and availability.

Inspec. Institution of Electrical Engineers, Michael Faraday House, Six Hills Way, Stevenage, Herts. SG1 2AY, England. Telephone: 0438 313311 or FAX 0438 742840. Contains citations to the worldwide literature of physics, electronics and electrical engineering, computer technology, and related fields. Available on BRS,(800) 289-4277, DIALOG, (800) 334-2564, ORBIT, (800) 456-7248, and STN International, FIZ Karlsruhe, P.O. Box 2465, W-7500, Karlsruhe 1, Germany, online services. Inquire as to cost and availability.

Mathsci. American Mathematical Society, P.O. Box 6248, Providence, RI 02940. (800) 321-4667 or (401) 455-4000 or FAX (401) 331-3842. Scope includes pure and applied mathematics and related areas of physics, statistics, engineering, computer science, and operations research literature since 1959. Available on DIALOG,(800) 334-2564, online service. Also available on CD-ROM. Inquire as to cost and availability.

NTIS Bibliographic Database. National Technical Information Service, 5285 Port Royal Road, Springfield, Va 22161. (703) 487-4929 or FAX (703) 321-8199. Broad coverage of government-sponsored science and technology research reports, 1964 to present. Available on BRS,(800) 289-4277, DIALOG, (800) 334-2564, ORBIT, (800) 456-7248, and STN International, FIZ Karlsruhe, P.O. Box 2465, W-7500, Karlsruhe 1, Germany, online services. Also available on CD-ROM. Inquire as to cost and availability.

SCISEARCH. Institute for Scientific Information, 3501 Market Street, Philadelphia, PA 19104. (800) 523-1850 or (215) 386-0100. Broad multidisciplinary title and author index to the international literature of science and technology, 1974 to present. Available on DIALOG,(800) 334-2564, and ORBIT,(800) 456-7248, online services. Also available on CD-ROM. Inquire as to cost and availability.

PERIODICALS

Automatica. Elsevier Science [journals], 660 White Plains Rd., Tarrytown, NY 10159-5153. (919) 524-9200. FAX (919) 333-2444. 1963 to present. 12 times a year. $985.00.

Control and Computers. International Assocation of Science and Technology for Development, 4500 16th Avenue NW, Suite80, Calgary, AB T3B OM6, Canada. (403) 288-1195. FAX (403) 247-6851. 1972 to present. Three times a year. $172.50.

Control and Dynamic Systems. Academic Press Inc., 6277 Sea Harbor Drive, Orlando, FL 32887. (800) 321-5068. Annual. Inquire for price.

IEE Proceedings D: Control Theory and Applications. Order from: IEEE (Institute of Electrical and Electronic Engineers), 445 Hoes Lane, Piscataway, NJ 08854. (908) 562-5545 or (800) 678-IEEE. 1980 to present. Bi-monthly. Inquire for price.

CONTROL THEORY

Ency. of Physical Sciences and Engineering Info. Sources

Information and Computation. Academic Press Inc., 6277 Sea Harbor Drive, Orlando, FL 32887. (800) 321-5068. 1957 to present. 16 times a year. $895.00.

RESEARCH CENTERS AND INSTITUTES

University of Colorado—boulder, Engineering Research Center. Campus Box 423, Boulder, CO 80309. (303) 492-7427.

University of Illinois, Decision and Control Laboratory. Coordinated Sciences Laboratory, 1101 W. Springfield Avenue, Urbana, IL 61801. (217) 333-0282. FAX (217) 244-1764.

Yale University, Center For Systems Science. Department of Electrical Engineering, PO Box 2157, New Haven, CT 06520. (203) 432-4296.

CONVECTION

See: HEAT TRANSFER

COOLANTS

See also: AIR CONDITIONING, AUTOMOTIVE ENGINEERING, COOLING TOWERS, REFRIGERATION

ABSTRACT SERVICES AND INDEXES

Applied Science and Technology Index; A Cumulative Subject Index To English Language Periodicals In the Fields of Aeronautics and Space Science, Computer Technology, Chemistry, Construction Industry, Energy and Related Areas. H.W. Wilson Co., 950 University Avenue, Bronx, NY 10452. (800) 367-6770 or (212) 588-8400. FAX (718) 590-1617. From 1958 to present. Monthly. Inquire about cost and availability. Also available on CD-ROM.

Engineering Index Monthly. Engineering Information, Inc., Castle Point on the Hudson, Hoboken, NJ 07030. (800) 221-1044. FAX (212) 832-1857. Monthly. $2200.00 per year. Also available online as COMPENDEX, and also on CD-ROM.

Physics Abstracts (Inspec). INSPEC/IEEE, Box 1331, 445 Hoes Lane, Piscataway, NJ 08855-1331. (908) 562-5549. 1898 to present. $2835.00 per year. Also available online. Covers recently published primary research in all areas of physics, including particle, nuclear, fluid, plasma, and solid-state physics, biophysics, geophysics, astrophysics, measurement, and instrumentation.

Sae Global Mobility Database. SAE International, 400 Commonwealth Drive, Warrendale, PA 15096-0001. (412) 776-4970. Bibliographic information and abstracts on mobility engineering technology. Available on CD-ROM and online.

ASSOCIATIONS AND PROFESSIONAL SOCIETIES

Automotive Cooling Systems Institute. PO Box 13966, Research Triangle Park, NC 27709-3966. (919) 549-4800.

National Automotive Radiator Service Association. PO Box 97, East Greenville, PA 18041. (215) 541-4500. FAX (215) 679-4977.

Society of Automotive Engineers (sae International). 400 Commonwealth Drive, Warrendale, PA 15096-0001. (412) 776-4841.

DIRECTORIES AND BIOGRAPHICAL SOURCES

Research Centers Directory. Gale Research, 835 Penobscot Building, Detroit, MI 48226-4094. (313) 961-2242. (800) 347-4253. FAX (313) 961-6083. $485.00.

Scientific and Technical Organizations and Agencies Directory. Gale Research, 835 Penobscot Building, Detroit, MI 48226-4094. (313) 961-2242. (800) 347-4253. FAX (313) 961-6083. 4th edition. 1996. $195.00.

Who's Who In Engineering. American Association of Engineering Societies, 1111 19th Street NW, Suite 608, Washington, DC 20036. (202) 296-2237 or (800) 658-8897. 8th edition. 1991. Inquire for price.

Who's Who In Technology. Gale Research, 835 Penobscot Building, Detroit, MI 48226-4094. (313) 961-2242. (800) 347-4253. FAX (313) 961-6083. 1995. $195.00.

ENCYCLOPEDIAS AND DICTIONARIES

Dictionary For Automotive Engineering. Jean de Coster. K.G. Saur, 121 Chanlon Road, New Providence, NJ 07974. (908) 464-6800 or (800) 521-8110. 1986. Inquire for cost and availability.

Encyclopedia of Physical Science and Technology. Robert A. Meyers, ed. 18 volumes. Academic Press Inc., 6277 Sea Harbor Drive, Orlando, FL 32887. (800) 321-5068. 1992. $2100.00

McGraw-Hill Dictionary of Scientific and Technical Terms. Sybil P. Parker, ed. 5th edition. McGraw-Hill Publishing Company, 1221 Avenue of the Americas, New York, NY 10020. (800) 262-4729 or (212) 512-3825. 1993. $110.50.

McGraw-Hill Encyclopedia of Engineering. Sybil P. Parker, ed. 2nd edition. McGraw-Hill Publishing Company, 1221 Avenue of the Americas, New York, NY 10020. (800) 262-4729 or (212) 512-3825. 1993. $95.50.

McGraw-Hill Encyclopedia of Science and Technology. Sybil P. Parker, ed. 7th edition. 20 volumes. McGraw-Hill Publishing Company, 1221 Avenue of the Americas, New York, NY 10020. (800) 262-4729 or (212) 512-3825. 1992. $1900.00

SAE Motor Vehicle, Safety and Environmental Terminology. Society of Automotive Engineers, 400 Commonwealth Drive, Warrendale, PA 15096. (412) 776-4841. $19.00.

GENERAL WORKS

Automotive and Heavy-Duty Engine Coolant Recycling By Distillation: Technology Evaluation Report. Arun R. Gavasker. U.S. Environmental Protection Agency, office of Research and Development, Risk Reduction Engineering Laboratory, Cincinnati, OH 45202. 1992. Inquire for cost and availability.

Automotive and Heavy-Duty Engine Coolant Recycling By Distillation: Project Summary. Arun R. Gavasker. U.S. Environmental Protection Agency, office of Research and Development, Risk Reduction Engineering Laboratory, Cincinnati, OH 45202. 1992. Inquire for cost and availability.

Engine Coolants and Cooling System Components. SAE International, 400 Commonwealth Dr., Warrendale, PA 15096. (412) 776-4841. 1996. $55.00 for non-members.

Engine Coolants, Cooling System Materials, and Components. Society of Automotive Engineers, 400 Commonwealth Dr., Warrendale, PA 15096. (412) 776-4841. 1993. $56.00.

Principles of Engine Cooling Systems, Components, and Maintenance. 4th edition. Society of Automotive Engineers, 400 Commonwealth Dr., Warrendale, PA 15096. (412) 776-4841. 1991. $45.00.

HANDBOOKS AND MANUALS

Automotive Handbook. 3rd edition. Robert Bosch GmbH/SAE International, 400 Commonwealth Drive, Warrendale, PA 15096-0001.(412) 776-4970. FAX (412) 776-0790. 1993. $34.00.

Manual On Selection and Use of Engine Coolants and Cooling System Chemicals. Joseph A. Lima and George R. Otterman, editors. 3rd edition. American Society for Testing and Materials (ASTM), 1916 Race Street, Philadelphia, PA 19103. (215) 299-5585. 1989. $12.00.

SAE Handbook. SAE International, 400 Commonwealth Drive, Warrendale, PA 15096-0001.(412) 776-4970. FAX (412) 776-0790. 1905 to present. Annual. $375.00 for three-volume set.

ONLINE DATABASES AND CD-ROMS

Compendex Plus. Engineering Information, Inc., 345 East 47th Street, New York, NY 10017. (212) 705-7600 or (800) 221-1044. Contains citations with abstracts to worldwide literature in engineering and technology, from 1970 to present. Available on online BRS,(800) 289-4277, DIALOG, (800) 334-2564, ORBIT (800) 456-7248, and STN International, FIZ Karlsruhe, P.O. Box 2465, W-7500, Karlsruhe 1, Germany, online services. Also available on CD-ROM. Inquire as to cost and availability.

CSA Engineering. Cambridge Scientific Abstracts, 7200 Wisconsin Avenue, Suite 601, Bethesda, MD 20814. (301) 961-6750 or (800) 843-7751. Contains citations and abstracts of international periodicals and research literature covering all fields of engineering and science and technology,including computer and information science, electronics, mechanical engineering, solid state materials, 1981 to present. Available on BRS,(800) 289-4277, online service.Inquire as to cost and availability.

Current Contents Search. Institute for Scientific Information, 3501 Market Street, Philadelphia, PA 19104. (215) 386-0100. FAX (215) 386-6362. Contains citations to articles listed in the table of contents of science and technology journals. Also articles in social sciences and life sciences journals. Available on BRS,(800) 289-4277, DIALOG,(800) 334-2564, online services.Inquire as to cost and availability.

Dissertation Abstracts. University Microfilms International, 300 North Zeeb Road, Ann Arbor, MI 48106. (800) 521-0600 or (313) 761-4700. Scope includes virtually all doctoral dissertations accepted at accredited American institutions from 1861 to present in 252 subject areas. Available on BRS,(800) 289-4277, DIALOG,(800) 334-2564, and OCLC EPIC,(800) 848-5878, online services. Also available on CD-ROM. Inquire as to cost and availability.

NTIS Bibliographic Database. National Technical Information Service, 5285 Port Royal Road, Springfield, Va 22161. (703) 487-4929 or FAX (703) 321-8199. Broad coverage of government-sponsored science and technology research reports, 1964 to present. Available on BRS,(800) 289-4277, DIALOG, (800) 334-2564, ORBIT, (800) 456-7248, and STN International, FIZ Karlsruhe, P.O. Box 2465, W-7500, Karlsruhe 1, Germany, online services. Also available on CD-ROM. Inquire as to cost and availability.

SAE Global Mobility Database. Society of Automative Engineers (SAE), Electronic Publishing Division, 400 Commonwealth Drive, Warrendale, PA 15098. (412) 776-4841. Contains citations with abstracts to technical papers on automotive and aerospace technology and vehicular-related industries that have been presented at SAE conferences. Covers 1906 to present. Available on ORBIT,(800) 456-7248,online service. Inquire as to cost and availability.

Scientific and Technical Books and Serials In Print. R.R. Bowker Inc., 121 Chanlon Road, New Providence, NJ 07974. (800) 521-8110 or (908) 464-6800. List of currently published books and serials in the physical and biological sciences, engineering and technology, with subject, author and titles indexes. Available on ORBIT,(800) 456-7248, online service.Inquire as to cost and availability.

SCISEARCH. Institute for Scientific Information, 3501 Market Street, Philadelphia, PA 19104. (800) 523-1850 or (215) 386-0100. Broad multidisciplinary title and author index to the international literature of science and technology, 1974 to present. Available on DIALOG,(800) 334-2564, and ORBIT,(800) 456-7248, online services. Also available on CD-ROM. Inquire as to cost and availability.

TRIS (Transportation Research Information). National Academy of Science, 2101 Constitution Avenue, N.W., Washington, DC 20418. (202) 334-3313 or (800) 624-6242. Citations with abstracts of literature on transportation, including air, highway, rail, maritime and other modes from 1968 to present. Available on DIALOG,(800) 334-2564, online service. Inquire as to cost and availability.

WILSONLINE. H.W. Wilson Company, 950 University Avenue, Bronx, NY 10452. (800) 367-6770 or (212) 588-8400. Makes available online versions of the printed H.W. Wilson indexes includingApplied Science and Technology Index, Business Periodicals Index, General Science Index, and Readers' Guide to Periodical Literature. Period covered is generally 1983 to present. Available on BRS,(800) 289-4277, DIALOG,(800) 334-2564, and OCLC EPIC,(800) 848-5878, online services. Also available on CD-ROM. Inquire as to cost and availability.

PERIODICALS

ASHRAE Journal: Heating, Refrigerating, Air Conditioning, and Ventilating. American Society of Heating, Refrigeration and Air Conditioning Engineers, 1791 Tullie Circle NE, Atlanta, GA 30329. (404) 636-8400. FAX (404) 321-5478. 1914 to present. Monthly. $49.00 per year for non-members.

COOLANTS

Ency. of Physical Sciences and Engineering Info. Sources

Automotive Engineering Magazine. SAE International, 400 Commonwealth Drive, Warrendale, PA 15096-0001. (412) 776-4841. 1905 to present. Monthly. $72.00.

Automotive Cooling Journal. National Automotive Radiator Service Association, Box 97, E. Greenville, PA 18041. (215) 541-4500. FAX (215) 679-4977. 1956 to present. Monthly. $25.00 per year.

Lubrication Engineering. Society of Tribologists and Lubrication Engineers, 840 Busse Highway, Park Ridge, IL 60068-2376. (708) 825-5536. FAX (708) 825-1456. 1945 to present. Monthly. $61.00 per year.

Manufacturing Engineering. Society of Manufacturing Engineers, One SME Drive, Box 930, Dearborn, MI 48121-0930. (313) 271-1500. FAX (313) 271-2861. 1932 to present. Monthly. $60.00 per year.

SAE Transactions. SAE International, 400 Commonwealth Drive, Warrendale, PA 15096-0001. (412) 776-4841. Annual. $850.00 per year.

Thermal Engineering. Interperiodica, Box 1831, Birmingham, AL 35201. (800) 633-4931. FAX (205) 995-1588. 1964 to present. Monthly. $875.00 per year.

RESEARCH CENTERS AND INSTITUTES

Automotive Cooling Systems Institute. PO Box 13966, Research Triangle Park, NC 27709-3966. (919) 549-4800.

Sloan Automotive Laboratory, Massachusetts Institute of Technology. Room 3-340, Cambridge, MA 02139. (617) 253-2243.

Society of Automotive Engineers (SAE) International. 400 Commonwealth Drive, Warrendale, PA 15096-0001. (412) 776-4841.

University of Michigan Automotive Engineering Laboratory. Department of Mechanical Engineering & Applied Mechanics, 321 Lay, N. Campus, 1231 Beal, Ann Arbor, MI 48109-2121. (313) 764-4254.

Wayne State University Center For Automotive Research. 5050 Anthony Wayne Drive, Detroit, MI 48202. (313) 577-3887.

SPECIFICATIONS AND STANDARDS

*Annual Book of ASTM Standards, Volume 15.*05. American Society for Testing and Materials (ASTM), 1916 Race Street, Philadelphia, PA 19103. (215) 299-5585. Annual. Inquire for cost and availability.

COOLING TOWERS

See also: AIR CONDITIONING, HEAT TRANSFER, MECHANICAL ENGINEERING, NUCLEAR ENERGY, PRESSURE VESSELS, theRMODYNAMICS

ABSTRACT SERVICES AND INDEXES

Applied Mechanics Reviews: An Assessment of World Literature In Engineering Sciences. 1948-present. American Society

of Mechanical Engineers, 345 East 47th Street, New York, NY 10017. (212) 705-7703. Monthly. $360.00 per year.

Applied Science and Technology Index; A Cumulative Subject Index To English Language Periodicals In the Fields of Aeronautics and Space Science, Computer Technology, Chemistry, Construction Industry, Energy and Related Areas. H.W. Wilson Co., 950 University Avenue, Bronx, NY 10452. (800) 367-6770 or (212) 588-8400. From 1958 to present. Monthly. Inquire about cost and availability. Also available on CD-ROM and online.

Asce Combined Annual Combined Index. American Society of Civil Engineers, 345 East 47th Street, New York, NY 10017-2398. (212) 705-7520 or (800) 548-2723. Annual. $48.00.

Asce Publications Information. 1966-present. American Society of Civil Engineers, 345 East 47th Street, New York, NY 10017-2398. (212) 705-7520 or (800) 548-2723. Bi-monthly. $160.00 per year for non-members.

Chemical Abstracts - Applied Chemistry and Chemical Engineering Sections. Chemical Abstracts Service, 2540 Olentangy River Road, Box 3012, Columbus, OH 43210. (800) 848-6538 or (614) 447-3600, FAX (614) 447-3713. Bi-weekly. $1410.00 per year.

Current Contents: Engineering, Technology, and Applied Sciences. Institute for Scientific Information, 3501 Market Street, Philadelphia, PA 19104. (215) 386-0100. FAX (215) 386-2991. 1970 to present. Weekly. $442.00 per year.

Engineering Index Monthly. Engineering Information, Inc., Castle Point on the Hudson, Hoboken, NJ 07030. (800) 221-1044. FAX (201) 216-8532. Monthly. $2300.00 per year. Also available online as COMPENDEX, and also on CD-ROM. Covers chemical engineering, computers, electrical engineering, civil engineering, metals and mining, industrial management, and mechanical engineering.

Mechanical Engineering Abstracts (formerly *Ismec*), Cambridge Scientific Abstracts, 7200 Wisconsin Avenue, Bethesda, Md 20814-4823. (301) 961-6750. FAX (301) 961-6720. 1967 to present. Monthly. $895.00 per year. Summarizes world literature in mechanical engineering, production engineering, and engineering management. Also available online.

ASSOCIATIONS AND PROFESSIONAL SOCIETIES

American Institute of Chemical Engineers. 345 East 47th Street, New York, NY 10017-2396. (212) 705-7663.

American Society of Civil Engineers. 345 East 47th Street, New York, NY 10017-2398. (212) 705-7520 or (800) 548-2723.

American Society of Mechanical Engineers. 345 East 47th Street, New York, NY 10017-2398. (212) 705-7703.

Cooling Tower Institute. PO Box 73383, Houston, TX 77273. (713) 583-4087.

DIRECTORIES AND BIOGRAPHICAL SOURCES

Directory of Engineering Societies and Related Organizations. Gordon Davis. 13th edition. American Association of Engineering Societies, 1111 19th Street NW, Suite 608, Washington,

DC 20036. (202) 296-2237 or (800) 658-8897. 1989. Inquire for price.

International Research Centers Directory. Gale Research, 835 Penobscot Building, Detroit, MI 48226-4094. (313) 961-2242. (800) 347-4253. 8th edition. 1995. $410.00

Research Centers Directory. Gale Research, 835 Penobscot Building, Detroit, MI 48226-4094. (313) 961-2242. (800) 347-4253. $470.00

Scientific and Technical Organizations and Agencies Directory. Gale Research, 835 Penobscot Building, Detroit, MI 48226-4094. (313) 961-2242. (800) 347-4253. 4th edition. 1996. $195.00.

Who's Who In Engineering. American Association of Engineering Societies, 1111 19th Street NW, Suite 608, Washington, DC 20036. (202) 296-2237 or (800) 658-8897. 8th edition. 1991. Inquire for price.

GENERAL WORKS

Cooling Tower Technology: Maintenance, Upgrading, and Rebuilding. Robert Burger. 3rd edition. Fairmont Press, 700 Indian Trail, Lilburn, GA 30247. (404) 925-9388. 1994. Inquire for cost and availability.

Cooling Tower Technology. Robert Burger. 2d ed. Fairmont Press, 700 Indian Trail, Lilburn, GA 30247. (404) 925-9388. 1990. $89.00.

Cooling Towers. Richard K. Miller and Marcia E. Rupnow. Future Technology Surveys, 700 Indian Trial, Lilburn, GA 30247. (404) 717-0779. 1991. $200.00.

Cooling Towers: Principles and Practice. G. B. Hill, E.J. Pring, and Peter D. Osborn. Butterworths Publishers, 80 Montvale Avenue, Stoneham, MA 02180. (617) 438-8464 or (800) 366-2665. 1990. $105.00.

Cooling Towers: Selection, Design & Practice. Nicholas P. Cheremisinoff & Paul N. Cheremisinoff. SciTech Publishers Inc., PO Box 987, Matawan, NJ 07747. (201) 583-6132. 1989. $39.95.

HANDBOOKS AND MANUALS

Handbook of Heat Transfer Applications. CRC Publishers, Inc., 2000 Corporate Blvd., N.W., Boca Raton, FL 33431. (407) 994-0555 or (800) 333-8300. 1993. Inquire for cost and availability.

Heat Exchanger Sourcebook. J.W. Palen. Hemisphere Publishing Corporation, 1900 Frost Road, Suite 101, Bristol, PA 19007-1598. (215) 785-5800 or FAX (215) 785-5515. 1986. $81.00.

ONLINE DATABASES AND CD-ROMS

Compendex Plus. Engineering Information, Inc., 345 East 47th Street, New York, NY 10017. (212) 705-7600 or (800) 221-1044.Contains citations with abstracts to worldwide literature in engineering and technology, from 1970 to present. Available on online BRS,(800) 289-4277, DIALOG, (800) 334-2564, ORBIT (800) 456-7248, and STN International, FIZ Karlsruhe, P.O. Box 2465, W-7500, Karlsruhe 1, Germany, online services. Also available on CD-ROM. Inquire as to cost and availability.

CSA Engineering. Cambridge Scientific Abstracts, 7200 Wisconsin Avenue, Suite 601, Bethesda, MD 20814. (301) 961-6750 or (800) 843-7751. Contains citations and abstracts of international periodicals and research literature covering all fields of engineering and science and technology,including computer and information science, electronics, mechanical engineering, solid state materials, 1981 to present. Available on BRS,(800) 289-4277, online service.Inquire as to cost and availability.

Current Contents Search. Institute for Scientific Information, 3501 Market Street, Philadelphia, PA 19104. (215) 386-0100. FAX (215) 386-6362. Contains citations to articles listed in the table of contents of science and technology journals. Also articles in social sciences and life sciences journals. Available on BRS,(800) 289-4277, DIALOG,(800) 334-2564, online services. Inquire as to cost and availability.

Ismec: Mechanical Engineering Abstracts. Cambridge Scientific Abstracts, 7200 Wisconsin Avenue, Suite 601, Bethesda, MD 20814. (301) 961-6750 or (800) 843-7751. Contains citations to the literature in mechanical engineering, industrial and production engineering, energy, power, mechanics, devices and related areas, from 1973 to present. Available on the DIALOG,(800) 334-2564, online service. Inquire as to cost and availability.

NTIS Bibliographic Database. National Technical Information Service, 5285 Port Royal Road, Springfield, Va 22161. (703) 487-4929 or FAX (703) 321-8199. Broad coverage of government-sponsored science and technology research reports, 1964 to present. Available on BRS,(800) 289-4277, DIALOG, (800) 334-2564, ORBIT, (800) 456-7248, and STN International, FIZ Karlsruhe, P.O. Box 2465, W-7500, Karlsruhe 1, Germany, online services. Also available on CD-ROM. Inquire as to cost and availability.

SCISEARCH. Institute for Scientific Information, 3501 Market Street, Philadelphia, PA 19104. (800) 523-1850 or (215) 386-0100. Broad multidisciplinary title and author index to the international literature of science and technology, 1974 to present. Available on DIALOG,(800) 334-2564, and ORBIT,(800) 456-7248, online services. Also available on CD-ROM. Inquire as to cost and availability.

WILSONLINE. H.W. Wilson Company, 950 University Avenue, Bronx, NY 10452. (800) 367-6770 or (212) 588-8400. Makes available online versions of the printed H.W. Wilson indexes including Applied Science and Technology Index, Business Periodicals Index, General Science Index, and Readers' Guide to Periodical Literature. Period covered is generally 1983 to present. Available on BRS,(800)289-4277, DIALOG,(800) 334-2564, and OCLC EPIC,(800) 848-5878, online services. Also available on CD-ROM. Inquire as to cost and availability.

PERIODICALS

Ashrae Journal: Heating, Refrigerating, Air Conditioning, and Ventilating. American Society of Heating, Refrigeration and Air Conditioning Engineers, 1791 Tullie Circle NE, Atlanta, GA 30329. (404) 636-8400. FAX (404) 321-5478. 1914 to present. Monthly. $49.00 per year for non-members.

CTI Journal. Cooling Tower Institute, PO Box 73383, Houston, TX 77273. (713) 583-4087. FAX (713) 537-1721. Semi-annual. Inquire for cost.

COOLING TOWERS

Ency. of Physical Sciences and Engineering Info. Sources

CTI News. Cooling Tower Institute, PO Box 73383, Houston, TX 77273. (713) 583-4087. FAX (713) 537-1721. Quarterly. Inquire for cost.

Engineering Structures. Butterworth-Heinemann, Linacre House, Jordan Hill, Oxford OX2 8DP, England. Telephone 0865-310366. FAX 0865-310898. 1979 to present. Bi-monthly. Inquire for cost.

Heat Transfer Engineering. Taylor and Francis, Inc., 79 Madison Avenue, Suite 1106, New York, NY 10016. (212) 725-1999 or (800) 821-8312. 1979 to present. Quarterly. $176.00 per year.

Journal of Heat Transfer. American Society of Mechanical Engineers, 345 E. 47th Street, New York, NY 10017. (212) 705-7722. 1970 to present. Quarterly. $155.00 per year.

Journal of Structural Engineering. American Society of Civil Engineers, 345 East 47th Street, New York, NY 10017-2398. (212) 705-7520 or (800) 548-2723. 1956 to present. Monthly. $300.00 per year for non-members.

Power. McGraw-Hill Publishing Company, 1221 Avenue of the Americas, New York, NY 10020. (800) 262-4729 or (212) 512-3825. 1882 to present. Monthly. Inquire for price.

Power Engineering. PennWell Publishing Company, Tulsa 1896 to present. Monthly. $38.00 per year.

RESEARCH CENTERS AND INSTITUTES

Cooling Tower Institute. PO Box 73383, Houston, TX 77273. (713) 583-4087.

Heat Transfer Laboratory. Massachusetts Institute of Technology, 77 Massachusetts Avenue, Cambridge, MA 02139. (716) 253-2248.

Heat Transfer Research. University of Wisconsin—Milwaukee, PO Box 784, Milwaukee, WI 53201. (414) 963-5001. FAX (414) 229-6958.

Heat Transfer Research Facility. Columbia University, 632 W. 125th Street, New York, NY 10027. (212) 280-4163. FAX (212) 678-5279.

COPPER

See also: BRASS AND BRONZE, MATERIALS SCIENCE, METALLURGY

ABSTRACT SERVICES AND INDEXES

Alloys Index. American Society for Metals, Metals Park, OH 44073. (216) 338-5151. FAX (216) 338-4634. 1974 to present.Monthly. $380.00. Also available on CD-ROM and online via METADEX, STN International, and DIALOG.

Applied Science and Technology Index; A Cumulative Subject Index To English Language Periodicals In the Fields of Aeronautics and Space Science, Computer Technology, Chemistry, Construction Industry, Energy and Related Areas. H.W. Wilson Co., 950 University Avenue, Bronx, NY 10452. (800) 367-6770

or (212) 588-8400. FAX (718) 590-1617. From 1958 to present. Monthly. Inquire about cost and availability. Also available on CD-ROM.

Current Contents: Engineering, Technology, and Applied Sciences. Institute for Scientific Information, 3501 Market Street, Philadelphia, PA 19104. (215) 386-0100. FAX (215) 386-6362. Weekly. $360.00 per year.

Engineering Index Monthly. Engineering Information, Inc., Castle Point on the Hudson, Hoboken, NJ 07030. (800) 221-1044. FAX (212) 832-1857. Monthly. $2200.00 per year. Also available online as COMPENDEX, and also on CD-ROM.

International Copper Information Bulletin. Copper Development Association, Orchard House, Mutton Lane, Potters Bar, Herts EN6 3AP, England. Telephone 0707-50711. FAX 0707-42769. 1976 to present. 3 times a year. Inquire for cost and availability.

Metals Abstracts and Metals Abstracts Index. American Society for Metals, Metals Park, OH 44073. (216) 338-5151. 1968 to present. Monthly. Abstracts are $1500.00 per year and Index is $460.00 per year.

ANNUAL REVIEWS AND YEARBOOKS

Annual Review of Materials Science. Annual Reviews Inc., 4139 El Camino Way, PO Box 10139, Palo Alto, CA 94303-0139. (415) 493-4400 or (800) 523-8635. FAX (415) 855-9815. Annual. $75.00 (1994 edition).

Non-ferrous Metal Data. American Bureau of Metal Statistics, Inc., Box 1405 Plaza Station, 400 Plaza Drive, Secaucus, NJ 07094-0405. (201) 863-6900. FAX (201) 863-6050. 1920 to present. Annual. $350.00 per year.

ASSOCIATIONS AND PROFESSIONAL SOCIETIES

American Society for Metals (ASM). Materials Park, OH 44073. (216) 338-5151 or FAX (216) 338-4634.

Canadian Copper and Brass Development Association. 10 Gateway Blvd., Suite 375, Don Mills, ON M3C 3A1, Canada. (416) 421-0788. FAX (416) 421-8092.

Copper Development Association. 260 Madison Avenue, New York, NY 10016-2401. (212) 251-7200. FAX (203) 251-7234.

Metallurgical Society of the Aime (american Institute of Mining, Metallurgical, and Petroleum Engineers). 345 E.47th Street, 14th Floor, New York, NY 10017. (212) 705-7695.

BIBLIOGRAPHIES

Scientific and Technical Books In Print. R.R. Bowker Inc., 121 Chanlon Road, New Providence, NJ 07974. (800) 521-8110 or (908) 464-6800. List of currently published books and serials in the physical and biological sciences, engineering and technology,with subject, author and titles indexes. Also available on ORBIT, (800) 456-7248, online service. Inquire as to cost and availability.

Ency. of Physical Sciences and Engineering Info. Sources

COPPER

DIRECTORIES AND BIOGRAPHICAL SOURCES

Dun's Industrial Guide: the Metalworking Directory, 1993-94. Dun & Bradstreet Information Services Staff. 3 volumes. Dun & Bradstreet Information Services, 899 Eaton Avenue, Bethlehem, Pa 18025. (610) 882-7000 or (800) 526-0651. FAX (610) 882-7269. 1993. $795.00.

International Research Centers Directory. Gale Research, 835 Penobscot Building, Detroit, MI 48226-4094. (313) 961-2242. (800) 347-4253. FAX (313) 961-6083. 8th edition. 1995. $410.00

Research Centers Directory. Gale Research, 835 Penobscot Building, Detroit, MI 48226-4094. (313) 961-2242. (800) 347-4253. FAX (313) 961-6083. $485.00.

Scientific and Technical Organizations and Agencies Directory. Gale Research, 835 Penobscot Building, Detroit, MI 48226-4094. (313) 961-2242. (800) 347-4253. FAX (313) 961-6083. 4th edition. 1996. $195.00.

Who's Who In Engineering. American Association of Engineering Societies, 1111 19th Street NW, Suite 608, Washington, DC 20036. (202) 296-2237 or (800) 658-8897. 8th edition. 1991. Inquire for price.

Who's Who In Technology. Gale Research, 835 Penobscot Building, Detroit, MI 48226-4094. (313) 961-2242. (800) 347-4253. FAX (313) 961-6083. 1995. $195.00.

World Copper Databook. Metal Bulletin Inc., 220 Fifth Ave, New York, NY 10001-7781. (212) 213-6202. FAX (212) 213-1870. 1974 to present. Irregular. $191.00 per issue.

ENCYLOPEDIAS AND DICTIONARIES

Encyclopedia of Physical Science and Technology. Robert A. Meyers, ed. 18 volumes. Academic Press Inc., 6277 Sea Harbor Drive, Orlando, FL 32887. (800) 321-5068. 1992. $2100.00

McGraw-Hill Dictionary of Scientific and Technical Terms. Sybil P. Parker, ed. 5th edition. McGraw-Hill Publishing Company, 1221 Avenue of the Americas, New York, NY 10020. (800) 262-4729 or (212) 512-3825. 1993. $110.50.

McGraw-Hill Encyclopedia of Engineering. Sybil P. Parker, ed. 2nd edition. McGraw-Hill Publishing Company, 1221 Avenue of the Americas, New York, NY 10020. (800) 262-4729 or (212) 512-3825. 1993. $95.50.

McGraw-Hill Encyclopedia of Science and Technology. Sybil P. Parker, ed. 7th edition. 20 volumes. McGraw-Hill Publishing Company, 1221 Avenue of the Americas, New York, NY 10020. (800) 262-4729 or (212) 512-3825. 1992. $1900.00

Thesaurus of Terms On Copper Technology. 7th edition. Copper Development Association, 260 Madison Avenue, New York, NY 10016-2401. (212) 251-7200. FAX (203) 251-7234. $40.00.

GENERAL WORKS

Copper '91. P.J. Mackey, et al. Pergamon Press Inc.,Maxwell House, Fairview Park, Elmsford, NY 10523. (914) 592-7700. Fax (914) 592-3625. 1992. $955.00.

Copper, Silver, Gold & Zinc, Cadmium, Mercury Oxides and Hydroxides. T.P. Dirkse. Pergamon Press Inc., Maxwell House, Fairview Park, Elmsford, NY 10523. (914) 592-7700. Fax (914) 592-3625. 1986. $140.00.

HANDBOOKS AND MANUALS

CRC Handbook of Chemistry and Physics. David R. Lide, editor. 75th edition. CRC Press, 2000 Corporate Blvd., N.W., Boca Raton, FL 33431. (407) 994-0555 or (800) 272-7737. FAX (407) 994-0949. $99.50.

Smithell's Metals Reference Book. E.A. Brandes & G.B. Brook. 7th edition. Butterworth-Heinemann, 313 Washington Street, Newton, MA 02158. (617) 928-2500 or (800) 366-2665. FAX (617) 928-2620. 1992. $250.00.

ONLINE DATABASES AND CD-ROMS

Compendex Plus. Engineering Information, Inc., 345 East 47th Street, New York, NY 10017. (212) 705-7600 or (800) 221-1044. Contains citations with abstracts to worldwide literature in engineering and technology, from 1970 to present. Available on online BRS,(800) 289-4277, DIALOG, (800) 334-2564, ORBIT (800) 456-7248, and STN International, FIZ Karlsruhe, P.O. Box 2465, W-7500, Karlsruhe 1, Germany, online services. Also available on CD-ROM. Inquire as to cost and availability.

Current Contents Search. Institute for Scientific Information, 3501 Market Street, Philadelphia, PA 19104. (215) 386-0100. FAX (215) 386-6362. Contains citations to articles listed in the table of contents of science and technology journals. Also articles in social sciences and life sciences journals. Available on BRS,(800) 289-4277, DIALOG,(800) 334-2564, online services. Inquire as to cost and availability.

Metadex. Jointly produced by ASM International and the Institute of Materials. Contains more than 925,000 records from the international literature on metals and alloys, concerning properties, processes, materials classes, applications, and metallurgical systems. Updated monthly. Available from ORBIT-QUESTEL (703) 442-0900.

NASA Database. American Institute of Aeronautics and Astronautics, 370 L'Enfant Promenade, S.W., Washington, DC 20024. (202) 646-7400. Citations and abstracts of aeronautics and astronautics literature, 1962 to present. Also contains citations from *STAR, Scientific and Technical Aerospace Reports,* and *International Aerospace Abstracts.* Available through NASA/RECON online service. Inquire as to cost and availability.

NTIS Bibliographic Database. National Technical Information Service, 5285 Port Royal Road, Springfield, Va 22161. (703) 487-4929 or FAX (703) 321-8199. Broad coverage of government-sponsored science and technology research reports, 1964 to present. Available on BRS,(800) 289-4277, DIALOG, (800) 334-2564, ORBIT, (800) 456-7248, and STN International, FIZKarlsruhe, P.O. Box 2465, W-7500, Karlsruhe 1, Germany, online services. Also available on CD-ROM. Inquire as to cost and availability.

WILSONLINE. H.W. Wilson Company, 950 University Avenue, Bronx, NY 10452. (800) 367-6770 or (212) 588-8400. Makes available online versions of the printed H.W. Wilson indexes including Applied Science and Technology Index, Business Periodicals Index, General Science Index, and Readers' Guide to Periodical Literature. Period covered is generally 1983 to

COPPER

Ency. of Physical Sciences and Engineering Info. Sources

present. Available on BRS,(800) 289-4277, DIALOG,(800) 334-2564, and OCLC EPIC,(800) 848-5878, online services. Also available on CD-ROM. Inquire as to cost and availability.

PERIODICALS

Alloy Digest. Alloy Digest Inc., 27 Canfield Street, Orange, NJ 07050. (201) 677-9161. 1952 to present. Monthly. $140.00 per year.

Canadian Copper. Canadian Copper & Brass Development Association, 10 Gateway Blvd., Suite 375, Don Mills, ON M3C 3A1, Canada. (416) 421-0788. FAX (416) 421-8092. 1960 to present. Quarterly. Free.

Copper Topics. Copper Development Association, 260 Madison Avenue, New York, NY 10016-2401. (212) 251-7200. FAX (203) 251-7234. 1968 to present. Quarterly. Free.

J O M (Journal of Metals). Minerals, Metals and Materials Society, 420 Commonwealth Drive, Warrendale, PA 15086. (412) 776-9080. 1949 to present. Monthly. $50.00 per year.

Light Metal Age. Fellom Publishing, 170 S. Spruce Avenue, Suite 120, San Francisco, CA 94080. (415) 588-8832. FAX (415) 588-0901. 1943 to present. Bi-monthly. $35.00 per year.

Metal Bulletin Monthly. Metal Bulletin Inc., 220 Fifth Avenue, 19th Floor, New York, NY 10001-7781. (800) 638-2525. FAX (212) 213-6273. 1972 to present. Monthly. Inquire for cost.

Metallurgical Transactions A: Physical Metallurgy and Materials Science. ASM International, Materials Park, OH 44073. (216) 338-5151 or FAX (216) 338-4634. 1970 to present. Monthly. $520.00 per year.

Metals Week. McGraw-Hill Publishing Company, Commodity Services Group, 1221 Avenue of the Americas, New York, NY 10020. (800) 262-4729 or (212) 512-3825. 1930 to present. Weekly. $770.00 per year.

Mineral Industry Surveys, Copper In the United States. U.S. Bureau of Mines, Production and Distribution, Cochrans Mill Rd., Box 18070, Pittsburgh, PA 15236. (412) 892-4411. Monthly. Inquire for price and availability.

Wire Industry. Publex International Ltd., 110 Station Road E., Oxted, Surrey RH8 0QA, England. Telephone 0883-717755. FAX 0883-714554. 1934 to present. Monthly. Inquire for cost.

Wire Industry News. Business Information Services, 7 Hampden Road, Stafford Springs, CT 06076-31404. (203) 684-5347. (203) 684-9158. 1973 to present. Bi-weekly. $250.00 per year.

Wire Journal International. Wire Journal Inc., 1570 Boston Post Road, Box H, Guilford, CT 06437. (203) 453-2777. FAX (203) 453-8384. 1968 to present. Monthly. $60.00 per year.

RESEARCH CENTERS AND INSTITUTES

Canada Centre For Mineral and Energy Technology, Materials Technology Laboratories. Energy, Mines & Resources Canada, 568 Booth Street, Ottawa ON, Canada K1A 0G1. (613) 995-8248. FAX (613) 992-8735.

International Copper Research Association Ltd. 708 Third Avenue, New York, NY 10017. (212) 697-9355. FAX (212) 697-5417.

New Mexico Institute of Mining and Technology, Mining & Mineral Resources Research Institute. Campus Station, Socorro, NM 87801. (505) 835-5210.

University of Connecticut Institute of Materials Science. U-136, 97 N. Eagleville Road, Storrs, CT 06268. (203) 486-4623. FAX (203) 486-4745.

University of Florida Center For Applied thermodynamics and Corrosion. Department of Materials Science and Engineering, Gainesville, FL 32611. (904) 392-8163.

SPECIFICATIONS AND STANDARDS

*Annual Book of ASTM Standards, Volume O2.*01. American Society for Testing and Materials (ASTM), 1916 Race Street, Philadelphia, PA 19103. (215) 299-5585. Annual. Inquire for cost and availability.

CORONA

See: SUN

CORROSION

See also: ALLOYS, ALUMINUM, CADMIUM, ELECTROCHEMISTRY, FERROALLOYS, MATERIALS SCIENCE

ABSTRACT SERVICES AND INDEXES

Alloys Index. American Society for Metals, Metals Park, OH 44073. (216) 338-5151. FAX (216) 338-4634. 1974 to present. Monthly. $380.00. Also available on CD-ROM and online via METADEX, STN International, and DIALOG.

Aluminum Industry Abstracts. Aluminum Association, Materials Park, OH 44073. (216) 338-5151. FAX (216) 338-4634. 1968 to present. Monthly. $525.00 per year.

A.P.I. Abstracts. American Petroleum Institute, Central Abstracting and Information Services, 275 Seventh Avenue, New York, NY 10001. (212) 366-4040 or FAX (212) 366-4298. Contains citations with abstracts to literature relating to the petroleum refining and the petrochemical industries, from 1964 to present.

Applied Mechanics Reviews: An Assessment of World Literature In Engineering Sciences. 1948-present. American Society of Mechanical Engineers, 345 East 47th Street, New York, NY 10017. (212) 705-7703. Monthly. $360.00 per year.

Applied Science and Technology Index; A Cumulative Subject Index To English Language Periodicals In the Fields of Aeronautics and Space Science, Computer Technology, Chemistry, Construction Industry, Energy and Related Areas. H.W. Wilson Co.,950 University Avenue, Bronx, NY 10452. (800) 367-6770 or (212) 588-8400. FAX (718) 590-1617. From 1958 to present.

Ency. of Physical Sciences and Engineering Info. Sources

CORROSION

Monthly. Inquire about cost and availability. Also available on CD-ROM and online.

Chemical Abstracts - Applied Chemistry and Chemical Engineering Sections. Chemical Abstracts Service, 2540 Olentangy River Road, Box 3012, Columbus, OH 43210. (800) 848-6538 or (614) 447-3600, FAX (614) 447-3713. Bi-weekly. $1410.00 per year.

Corrosion Abstracts; Abstracts of the World's Literature On Corrosion and Corrosion Mitigation. National Association of Corrosion Engineers (NACE), Box 218340, Houston, TX 77218. (713) 492-0535 or FAX (713) 492-8254. 1962 to present. Bi-monthly. $195.00 per year. Also available on CD-ROM.

Current Contents: Engineering, Technology, and Applied Sciences. Institute for Scientific Information, 3501 Market Street, Philadelphia, PA 19104. (215) 386-0100. FAX (215) 386-6362. 1970 to present. Weekly. $442.00 per year.

Engineering Index Monthly. Engineering Information, Inc., Castle Point on the Hudson, Hoboken, NJ 07030. (800) 221-1044. FAX (212) 832-1857. Monthly. $2300.00 per year. Also available online as COMPENDEX, and also on CD-ROM. Covers chemical engineering, computers, electrical engineering, civil engineering, metals and mining, industrial management, and mechanical engineering.

International Aerospace Abstracts. Technical Information Service, American Institute of Aeronautics and Astronautics, Inc., 555 West 57th St., New York, NY 10019. (212) 247-6500. FAX (212) 582-4861. Semi-monthly. $1295.00 per year.

Materials Science and Engineering Abstracts, Cambridge Scientific Abstracts, 7200 Wisconsin Avenue, Bethesda, Md 20814-4823. (301) 961-6750. FAX (301) 961-6720. 1993 to present. Monthly. $385.00 per year. Focuses on mechanical and physical properties of materials and commercial or industrial applications for materials, methods for strength testing, effects of vibration and other stresses, corrosion and protective coatings, storage and handling, ceramics, composites, metals, wood, plastics, and polymers.

Materials Science Citation Index. Institute for Scientific Information, 3501 Market Street, Philadelphia, PA 19104. (215) 386-0100. FAX (215) 386-6362. Bi-monthly. $975.00 per year. Also available on CD-ROM.

Metals Abstracts and Metals Abstracts Index. American Society for Metals, Metals Park, OH 44073. (216) 338-5151. 1968 to present. Monthly. Abstracts are $1500.00 per year and Index is $460.00 per year.

Petroleum Abstracts. 600 South College, Tulsa, OK 74104. (918) 631-2296 or (800) 247-8678. Contains citations with abstracts to the worldwide literature and patents on the exploration, development, and production of petroleum resources, from 1965 to present.

Tribology and Corrosion Abstracts. Elsevier SciencePublishing Company, Inc., 655 Avenue of the Americas, New York, NY 10010. (212) 989-5800. Covers literature and research in friction, wear, lubrication and lubricants, and corrosion problems and protection, 1984 to present. Thirteen times per year. $450.00 per year.

Tribos - Tribology Abstracts. STI Limited, 4 Kings Meadow, Ferry Hinksey Road, Oxford, OX2 ODU, England. Distributed in the United States by Air Science Company, Box 143, Corning NY 14830. (607) 962-5591. Contains citations of literature and research in friction, wear, lubrication and lubricants, and corrosion problems and protection, from 1967 to present. Monthly. $375.00 per year.

ASSOCIATIONS AND PROFESSIONAL SOCIETIES

Aluminum Association. 900 19th Street NW, Suite 300, Washington, DC 20006. (202) 862-5100. FAX (202) 862-5164.

American Society For Metals. Materials Park, OH 44073. (216) 338-5151 or FAX (216) 338-4634.

Metallurgical Society of the Aime (american Institute of Mining, Metallurgical, and Petroleum Engineers). 345 E.47th Street, 14th Floor, New York, NY 10017. (212) 705-7695.

National Association of Corrosion Engineers. Box 218340, Houston, TX 77218. (713) 492-0535. FAX (713) 492-8254.

DIRECTORIES AND BIOGRAPHICAL SOURCES

Corrosion and Coatings Buyer's Guide. George Warman Publications Ltd., PO Box 3847, Cape Town 8000, South Africa. Telephone 27-21-24-5320. FAX 27-21-26-1332. Annual. Included with subscription to periodical *Corrosion and Coatings.*

Dun's Industrial Guide: the Metalworking Directory, 1993-94. Dun & Bradstreet Information Services Staff. 3 volumes. Dun & Bradstreet Information Services, 899 Eaton Avenue, Bethlehem, Pa 18025. (610) 882-7000 or (800) 526-0651. FAX (610) 882-7269. 1993. $795.00.

International Research Centers Directory. Gale Research, 835 Penobscot Building, Detroit, MI 48226-4094. (313) 961-2242. (800) 347-4253. FAX (313) 961-6083. 8th edition. 1995. $410.00

Research Centers Directory. Gale Research, 835 Penobscot Building, Detroit, MI 48226-4094. (313) 961-2242. (800) 347-4253. FAX (313) 961-6083. $485.00.

Scientific and Technical Organizations and Agencies Directory. Gale Research, 835 Penobscot Building, Detroit, MI 48226-4094. (313) 961-2242. (800) 347-4253. FAX (313) 961-6083. 4th edition. 1996. $195.00.

Who's Who In Engineering. American Association of Engineering Societies, 1111 19th Street NW, Suite 608, Washington, DC 20036. (202) 296-2237 or (800) 658-8897. 8th edition. 1991. Inquire for price.

Who's Who In Technology. Gale Research, 835 Penobscot Building, Detroit, MI 48226-4094. (313) 961-2242. (800) 347-4253. FAX (313) 961-6083. 1995. $195.00.

ENCYCLOPEDIAS AND DICTIONARIES

Dictionary of Corrosion and Corrosion Control. Helmut Gross, editor in chief. Elsevier Science Publishing Company, Inc., 655 Avenue of the Americas, New York, NY 10010. (212) 989-5800. 1986. $133.50.

CORROSION

Ency. of Physical Sciences and Engineering Info. Sources

Encyclopedia of Physical Science and Technology. Robert A. Meyers, ed. 18 volumes. Academic Press Inc., 6277 Sea Harbor Drive, Orlando, FL 32887. (800) 321-5068. 1992. $2100.00.

A Glossary of Corrosion-Related Terms Used In Science and Industry. Mark S. Vukasovich. SAE International, 400 Commonwealth Drive, Warrendale, PA 15096-0001. (412) 776-4970. FAX (412) 776-0790. 1995. $55.00.

McGraw-Hill Encyclopedia of Engineering. Sybil P. Parker, ed. 2nd edition. McGraw-Hill Publishing Company, 1221 Avenue of the Americas, New York, NY 10020. (800) 262-4729 or (212) 512-3825. 1993. $95.50.

McGraw-Hill Encyclopedia of Science and Technology. Sybil P. Parker, ed. 7th edition. 20 volumes. McGraw-Hill Publishing Company, 1221 Avenue of the Americas, New York, NY 10020. (800) 262-4729 or (212) 512-3825. 1992. $1900.00

GENERAL WORKS

Basic Corrosion Technology For Scientists and Engineers. Einar Mattsson. Halsted Press (Division of John Wiley and Sons, Inc.), 605 Third Avenue, New York, NY 10158. (800) 526-5368 or (212) 850-6418. 1989. Inquire for cost and availability.

The Chemical Engineering Guide To Corrosion. Richard W. Greene, editor. McGraw-Hill Publishing Company, 1221 Avenue of the Americas, New York, NY 10020. (800) 262-4729 or (212) 512-3825. 1986. Inquire for cost and availability.

Corrosion Atlas. E.D.D. During. Elsevier Science B.V., PO Box 945, Madison Square Station, New York, NY 10159-0945. (212) 633-3650. FAX (212) 633-3680. 1991. $620.50.

Corrosion Control. Samuel A. Bradford. Chapman and Hall, 29 West 35th Street, New York, NY 10001. (800) 842-3636. FAX (212) 563-2269. 1992. $59.95.

Corrosion Engineering. Mars G. Fontana. 3rd edition. McGraw-Hill Publishing Company, 1221 Avenue of the Americas, New York, NY 10020. (800) 262-4729 or (212) 512-3825. 1986. Inquire for cost and availability.

Corrosion For Students of Science and Engineering. Kenneth R. Trethewey & John Chamberlain. Halsted Press/John Wiley and Sons, Inc., 605 Third Avenue, New York, NY 10158. (800) 526-5368 or (212) 850-6000. 1988. $53.95.

Corrosion: Industrial Problems, Treatment & Control Techniques. V. Ashworth. Pergamon Press Inc., Maxwell House, Fairview Park, Elmsford, NY 10523. (914) 592-7700. Fax (914) 592-3625. 1987. $305.00.

Corrosion Inhibitors: An Industrial Guide. Ernest W. Flick. 2nd edition. Noyes Publications, 120 Mill Road, Park Ridge, NJ 07656. (201) 391-8484. FAX (201) 391-6833. 1993. $54.00.

Materials Selection For Corrosion Control. Sohan L.Chawla & Rasjeshwar K. Gupta. ASM International, Materials Park, OH 44073. (216) 338-5151 or FAX (216) 338-4634. 1993. $122.00.

Principles and Prevention of Corrosion. Denny A. Jones. Macmillan Publishing, 200 Old Tappan Road, Old Tappan, NJ 07675. (800) 223-2336. FAX (800) 445-6991. 1992. Inquire for cost and availability.

HANDBOOKS AND MANUALS

The Anti-Corrosion Methods and Materials Handbook and Directory. Sawell Publications Ltd., 127 Stanstead Road, London SE223 1JE, England. Annual. Included with subscription to periodical ANTI-CORROSION METHODS and MATERIALS (see below).

ASM Handbook Volume 13: Corrosion. ASM International Handbook Committee. ASM International, Materials Information, Materials Park, OH 44073-0002. (216) 338-5151 or (800) 336-5152. FAX (216) 338-4634. 1987. $147.00.

Corrosion Resistance Tables: Metals, Nonmetals, Coatings, Mortars, Plastics, Elastomers and Linings, and Fabrics. Philip A. Schweitzer. 3rd revised and expanded edition. Marcel Dekker, Inc., 270 Madison Avenue, New York, NY 10016. (212) 696-9000. FAX (212) 685-4540. 1991. In two parts, $145.00 each.

Handbook of Corrosion Data. Bruce D. Craig, D. anderson, editors. 2nd edition. ASM International, Materials Information, Materials Park, OH 44073. (216) 338-5151 or FAX (216) 338-4634. 1994. $188.00.

ONLINE DATABASES AND CD-ROMS

Compendex Plus. Engineering Information, Inc., 345 East 47th Street, New York, NY 10017. (212) 705-7600 or (800) 221-1044. Contains citations with abstracts to worldwide literature in engineering and technology, from 1970 to present. Available on online BRS,(800) 289-4277, DIALOG, (800) 334-2564, ORBIT (800) 456-7248, and STN International, FIZ Karlsruhe, P.O. Box 2465, W-7500, Karlsruhe 1, Germany, online services. Also available on CD-ROM. Inquire as to cost and availability.

Current Contents Search. Institute for Scientific Information, 3501 Market Street, Philadelphia, PA 19104. (215) 386-0100. FAX (215) 386-6362. Contains citations to articles listed in the table of contents of science and technology journals. Also articles in social sciences and life sciences journals. Available on BRS,(800) 289-4277, DIALOG,(800) 334-2564, online services. Inquire as to cost and availability.

Metadex. Jointly produced by ASM International and the Institute of Materials. Contains more than 925,000 records from the international literature on metals and alloys, concerning properties, processes, materials classes, applications, and metallurgical systems. Updated monthly. Available from ORBIT-QUESTEL (703) 442-0900.

NASA Database. American Institute of Aeronautics and Astronautics, 370 L'Enfant Promenade, S.W., Washington, DC 20024. (202) 646-7400. Citations and abstracts of aeronautics and astronautics literature, 1962 to present. Also contains citationsfrom *STAR, Scientific and Technical Aerospace Reports,* and *International Aerospace Abstracts.* Available through NASA/RECON online service. Inquire as to cost and availability.

NTIS Bibliographic Database. National Technical Information Service, 5285 Port Royal Road, Springfield, Va 22161. (703) 487-4929 or FAX (703) 321-8199. Broad coverage of government-sponsored science and technology research reports, 1964 to present. Available on BRS,(800) 289-4277, DIALOG, (800) 334-2564, ORBIT, (800) 456-7248, and STN International, FIZ Karlsruhe, P.O. Box 2465, W-7500, Karlsruhe 1, Germany, online services. Also available on CD-ROM. Inquire as to cost and availability.

WILSONLINE. H.W. Wilson Company, 950 University Avenue, Bronx, NY 10452. (800) 367-6770 or (212) 588-8400. Makes available online versions of the printed H.W. Wilson indexes including Applied Science and Technology Index, Business Periodicals Index, General Science Index, and Readers' Guide to Periodical Literature. Period covered is generally 1983 to present. Available on BRS,(800) 289-4277, DIALOG,(800) 334-2564, and OCLC EPIC,(800) 848-5878, online services. Also available on CD-ROM. Inquire as to cost and availability.

PERIODICALS

Alloy Digest. Alloy Digest Inc., 27 Canfield Street, Orange, NJ 07050. (201) 677-9161. 1952 to present. Monthly. $140.00 per year.

Anti-corrosion Methods and Materials. Sawell Publications Ltd., 127 Stanstead Road, London SE223 1JE, England. 1954 to present. Monthly. $94.00 per year.

British Corrosion Journal. Institute of Materials, 1 Carlton House Terrace, London SW1Y 5DB, England. Telephone 071-839-4071. FAX 071-839-2078. 1965 to present. Quarterly. $296.00 per year.

Corrosion. National Association of Corrosion Engineers, Box 218340, Houston, TX 77218. (713) 492-0535. FAX (713) 492-8254. 1945 to present. $100.00 per year.

Corrosion and Coatings. George Warman Publications Ltd., PO Box 3847, Cape Town 8000, South Africa. Telephone 27-21-24-5320. FAX 27-21-26-1332. 1973 to present. Monthly. Inquire for cost.

Corrosion Engineering (japan). Allerton Press Incorporated, 150 Fifth Avenue, New York, NY 10011. (212) 924-3950. FAX (212) 463-9684. 1987 to present. Monthly. $625.00 per year.

Corrosion Prevention and Control. Scientific Surveys Ltd., Box 21, Beaconsfield, Bucks. HP9 1NS, England. Telephone 0494-675139. FAX 0494-670155. 1954 to present. Bi-monthly. $139.00 per year.

Corrosion Science. Elsevier Science [journals], 660 White Plains Rd., Tarrytown, NY 10159-5153. (919) 524-9200. FAX (919) 333-2444. 1961 to present. 12 times a year. $1230.00 per year.

Industrial Corrosion. Impact Company Publications, Meden House, 55 Old Road, Leighton Buzzard, Beds. LU7 7RB, England.Telephone 44-535-370013. FAX 44-535-382487. 1982 to present. Bi-monthly. Inquire for cost.

J O M (Journal of Metals). Minerals, Metals and Materials Society, 420 Commonwealth Drive, Warrendale, PA 15086. (412) 776-9080. 1949 to present. Monthly. $50.00 per year.

Light Metal Age. Fellom Publishing, 170 S. Spruce Avenue, Suite 120, San Francisco, CA 94080. (415) 588-8832. FAX (415) 588-0901. 1943 to present. Bi-monthly. $35.00 per year.

Materials Performance. National Association of Corrosion Engineers, Box 218340, Houston, TX 77218. (713) 492-0535. FAX (713) 492-8254. 1962 to present. Monthly. $65.00 per year.

Metallurgical Transactions A: Physical Metallurgy and Materials Science. ASM International, Materials Park, OH 44073.

(216) 338-5151 or FAX (216) 338-4634. 1970 to present. Monthly. $520.00 per year.

Metals Week. McGraw-Hill Publishing Company, 1221 Avenue of the Americas, New York, NY 10020. (800) 262-4729 or (212) 512-3825. 1930 to present. Weekly. $770.00 per year.

Oxidation of Metals. Plenum Publishing Corporation, 233 Spring Street, New York, NY 10013. (212) 620-8000. FAX (212) 463-0742. 1969 to present. Monthly. $675.00 per year.

RESEARCH CENTERS AND INSTITUTES

Ohio State University Fontana Corrosion Center. 116 W. 19th Avenue, Columbus, OH 43210. (614) 292-0290.

Texas A&M University Corrosion Engineering Laboratory. Materials Engineering Department, College Station, TX 77843. (409) 845-7512. FAX (409) 845-2944.

University of Florida Center For Applied Thermodynamics and Corrosion. Department of Materials Science and Engineering, Gainesville, FL 32611. (904) 392-8163.

University of Minnesota Corrosion Research Center. 221 Church Street SE, Minneapolis, MN 55455. (612) 625-0717.

SPECIFICATIONS AND STANDARDS

Annual Book of ASTM Standards, Volume 03.02. American Society for Testing and Materials (ASTM), 1916 Race Street, Philadelphia, PA 19103. (215) 299-5585. Annual. Inquire for cost and availability.

COSMIC RAYS

See also: ASTRONOMY, ASTROPHYSICS, CELESTIAL MECHANICS, COSMOLOGY, GALAXIES, INTERSTELLAR MATTER, PARTICLE PHYSICS, PHYSICS

ABSTRACT SERVICES AND INDEXES

Applied Science and Technology Index; A Cumulative Subject Index To English Language Periodicals In the Fields of Aeronautics and Space Science, Computer Technology, Chemistry, Construction Industry, Energy and Related Areas. H.W. Wilson Co., 950 University Avenue, Bronx, NY 10452. (212) 588-8400. (800) 367-6770. FAX (718) 590-1617. From 1958 to present. Monthly. Inquire about cost and availability. Also available on CD-ROM and online. ISSN: 0003-6986.

Astronomy and Astrophysics Abstracts. Springer-Verlag NewYork, 175 F ifth Avenue, New York, NY 10010. (212) 460-1500. FAX (212) 473-6272. Published for the Astronomisches Rechen-Institut. Comprehensive coverage of all aspects of astronomy, astrophysics and related fields. 1969 to present. Two parts per year. Annual. ISSN: 0067-0022.

General Science Index. H.W. Wilson Company, 950 University Avenue, Bronx, NY 10452. (212) 588-8400. (800) 367-6770. FAX (718) 590-1617. From 1978 to present. Ten issues per year; quarterly and annual cumulations. Service basis. Available on

COSMIC RAYS

Ency. of Physical Sciences and Engineering Info. Sources

CD-ROM and online. Inquire regarding cost and availability. ISSN: 0162-1963.

Government Reports Announcements and Index. U. S. National Technical Information Service (NTIS), 5285 Port Royal Road, Sp ringfield, VA 22161. (703) 487-4650. FAX (703) 321-8547. From 1968 to present. Annual. $630.00 per year. Also available online as *NTIS Bibliographic Database* and on CD-ROM. ISSN:

Meteorological and Geoastrophysical Abstracts. American M eteorological Society, c/o Inforonics, Inc., 550 Newtown Road, Littleton, MA 01460. (508) 486-8976. FAX (508) 486- 0027. Covers literature in environmental sciences, meteorology, astrophysics, hydrology, glaciology, and physical oceanography. From 1950 to present. Monthly. $950.00 per year. Also available on CD-ROM and online. Inquire regarding cost and availability. ISSN: 0026-1130.

NTIS Alerts: Astronomy & Astrophysics. U. S. National Technical Information Service. 5285 Port Royal Road, Springfield, VA 22161. (703) 487-4650. FAX (703) 321-8547. Weekly. $140.00 per year.

Pascal 48: Environnement Cosmique Terrestre, Astronomie Et Geologie Extraterrestre. Centre National de la Recherche Scientifique, Institut de Information Scientifique et Technique, 2 aliee du Parc de Brabois, 54514 Vandoeuvre Les Nancy Cedex, France. TEL 83-50-46-00. FAX: 83-50-46-50. 1985 to present. Ten issues per year. 770 F. Also available on CD-ROM and online.

Physics Abstracts . INSPEC. Section A, Science Abstracts. Insti tute of Electrical Engineers, London, United Kingdom. Available from: INSPEC/IEEE - Institute of Electrical and Electronic Engineers, Box 1331, Hoes Lane, Piscataway, NJ 08855-1331. (908) 562-5549. 1898 to present. 24 issues per year. $2835.00 per year. ISSN: 0036-8091. Also available online and on CD-ROM.

Science Citation Index. SCI. Institute for Scientific Information, 3501 Market Street, Philadelphia, PA 19104. (215) 386-0100. (800) 523-1850. FAX (215) 386-2991. 1961 to present. Six issues per year, plus annual cumulation. $11650.00 per year. Also available online and on CD-ROM. Inquire about price and availability. ISSN: 0036-827X.

S T A R. SCIENTIFIC and TECHNICAL AEROSPACE REPORTS. U.S. National Aeronautics and Space Administration. Distributed by U. S. Superintendent of Documents, Washington, DC 20402. 1963 to present. Semi-monthly, with semiannual and annual indexes. $114.00 per year.ISSN: 0036-8741. Also available online and on CD-ROM.

ANNUAL REVIEWS AND YEARBOOKS

Annual Review of Astronomy and Astrophysics; Original reviews of critical literature and current developments in astronomy and astrophysics. Annual Reviews, Inc., 4139 El Camino Way, Palo Alto, CA 94306-0139. (415) 493-4400. (800) 523-8635. FAX (415) 855-9815. 1963 to date. Annual. $60.00. ISSN:: 0066-4146.

Highlights of Astronomy. I. Appenzeller, editor. Kluwer Academic Publishers, Box 358, Accord Station, Hingham, MA 02018-0358. (617) 871-6600. International Astronomical Union Highlights series. ISBN: 07923-3553-8. Triennal. 1995. Price varies.

ASSOCIATIONS AND PROFESSIONAL SOCIETIES

American Association of Variable Star Observers. 23 Birch Street, Cambridge, MA 02138. (617) 354-0484. FAX: (617) 354-0665.

American Astronomical Society. 2000 Florida Avenue NW, Suite 400, Washington, DC 20009. (202) 328-2010. FAX: (202) 234-2560.

American Geophysical Union. 2000 Florida Avenue, NW, Washington, DC 20009. (202) 462-6900. (800) 966-AGU1. FAX (202) 328-0566.

American Institute of Physics. One Physics Ellipse, College Park, MD 20740-3843. (301) 209-3100.

Association of University For Research In Astronomy. Suite 701, 1625 Massachusetts Avenue, NW, Washington, DC 20036. (202) 483-2101. FAX (202) 483-2106.

Astronomical League. 2112 Kingfisher Lane East, Rolling Meadows, IL 60008. (708) 398-0562.

Astronomical Society of the Pacific. 390 Ashton Avenue, San Francisco, CA 94112. (415) 337-1100. FAX: (415) 337-5205.

Joint Institute for Laboratory Astrophysics. University of Colorado, Box 440, Boulder, CO 80309-0440. (303) 492-7780. FAX (303) 492-5235.

Royal Astronomical Society of Canada. 136 Dupont Street, Toronto. ON N5R 1V2 Canada. (416) 924-7973. FAX (416) 968-6687.

Universities Research Association, 1111 19th Street Nw, Suite 400, Washington, Dc 20036. (202) 293-1382.

BIBLIOGRAPHIES

Astronomy and Astrophysics: A Source Guide. Gordon Press, P.O. Box 459, Bowling Green Station, New York, NY 10004. (718) 624-8419. 1991. $250.00. .

A Bibliography of Astronomy, 1970-1979. R. A. Sea and S. S. Martin. Libraries Unlimited, Inc., Littleton, CO 80160. 1982.

Catalog of the Naval Observatory Library. Naval Observatory Staff. G. K. Hall & Company, 866 Third Avenue, New York, NY 10022. (212)702-6789. (800) 257-5755. 6 volumes. 1977. $655.00

Science Books and Films. American Association for the Advancement of Science, 1333 H Street NW, Washington, DC 20005. (202) 326-6454. 1965 to present. Nine issues per year. $40.00 peryear. ISSN: 0098-342X

Scientific and Technical Books and Serials In Print; An Index To Literature In Science and Technology. R. R. Bowker. 121 Chanlon Road, New Providence, NJ 07974. (908) 464-6800. FAX (908) 665-3502. (800) 521-8110. 1972 to present. Annual. $299.95 per year. Also available on CD-ROM. ISSN: 0000-054X.

A Source Book On Astronomy and Astrophysics, 1900 - 1975. Kenneth Lang and Owen Gingerich, editors. Harvard University Press, 79 Garden Street, Cambridge MA. (617) 495-2600. 1980. $41.95.

DIRECTORIES AND BIOGRAPHICAL SOURCES

American Astronomical Society. MEMBERSHIP DIRECTORY. American Astronomical Society. 2000 Florida Avenue NW, Suite 400, Washington, DC 20009. (202) 328-2010. FAX (202) 234-2560 Annual. Included in membership dues. ISSN:: 1061-9038.

American Men and Women of Science: Physical and Biological Sciences. R. R. Bowker Inc., 121 Chanlon Road, New Providence, NJ 07974. (908) 464-6800. (800) 521-8110. 20th edition. 8 volumes. 1996. $850.00.

The Astronomers. Donald Goldsmith. St. Martin's Press, Inc., 175 Fifth Avenue, New York, NY 10010. (212) 674-5151. (800) 221-7945. 1993. $14.95 in paper.

Astronomical Centers of the World. Kevin Krisciunas. Cambridge University Press, 40 West 20th Street, New York, NY 10011-4211. (212) 924-3900. (800) 872-7423. 1988. $34.95

The Biographical Dictionary of Scientists: Astronomers. D. Abbott, editor. Peter Bedrick Books, Inc., 2112 Broadway, Room 318, New York, NY 10023. (212) 496-0751. 1984.

Directory of Physics and Astronomy Staff Members. American Institute of Physics, One Physics Ellipse, College Park, MD 20740-3843. (301) 209-3100. 1975/76 to present. Annual. $60.00. ISSN:: 0361-2228.

Earth and Astronomical Research Centers. Stockton Press, 345 Park Avenue South, New York, NY 10010. 4th edition. 1995. $515.00. ISBN: 1-56169-0967.

Graduate Programs In Physics, Astronomy and Related Fields. 1995-1996. American Institute of Physics, One Physics Ellipse, College Park, MD 20740-3843. (301) 209-3100. 1978 to present. Annual. $45.00. ISSN:: 0147-1821.

Research Centers Directory. Gale Research, 835 Penobscot Building, Detroit, MI 48226-4094. (313) 961-2242. (800) 347-4253. 20th edition. 1995. $485.00.

ENCYCLOPEDIAS AND DICTIONARIES

Astronomy From A To Z: A Dictionary of Celestial Objects and Ideas. Charles A. Schweighauser. Illinois Issues, Sangamon State University, K-10, Springfield, IL 62794-9243. (217) 786-9243. 1991. $14.95 in paper.

Concise Dictionary of Astronomy. Jacqueline Mitton. Oxford University Press, Inc., 200 Madison Avenue, New York, NY 10016. (212) 725-6000. (800) 334-4249. 1992. $24.95.

Encyclopedia of Astronomy and Astrophysics. Stephen Maran, editor. Van Nostrand Reinhold, 115 Fifth Avenue, New York, NY 10003. (212) 254-3232. (800) 842-3636. 1992. $129.95.

Facts On File Dictionary of Astronomy. Valerie Illingworth. Facts On File, Inc., 460 Park Avenue South, New York, NY 10016-7382. (212) 683-2244. (800) 322-8755. Fax (800) 683-3633. 1994. $27.95.

mcgraw-Hill Encyclopedia of Science and Technology. McGraw-Hill Book Company, Inc., 1221 Avenue of the Americas, New York, NY 10020. (212) 512-2000. (800) 262-4729. 7th edition. 20 volume set. 1992. $1900.00.

Oxford Illustrated Encyclopedia. Volume 8, the Universe. Archie Roy, editor. Oxford University Press, Inc., 200 Madison Avenue, New York, NY 10016. (212) 725-6000. (800) 334-4249. 1993. $45.00.

Penguin Dictionary of Astronomy. Jacqueline Mitton. Viking Penguin, 375 Hudson Street, New York, NY 10014-3657.

(212) 366-2000. (800) 331-4624. 1993. $12.95 in paper.

Stars, Galaxies, Cosmos. William R. Corliss. Sourcebook Project, P.O. Box 107, Glen Arm, MD 21057. (410) 668-6047. 1987. $17.95.

GENERAL WORKS

Astronomer's Universe: Stars, Galaxies and COSMOS. Herbert Friedman. W. W. Norton & Company, 500 Fifth Avenue, New York, NY 10110. (212) 354-5500. (800) 223-2584. 1990. $24.95.

Astrophysics of Cosmic Rays. V. L. Ginsburg. Coronet Books, 311 Bainbridge Street, Philadelphia, PA 19147. (210) 925-5083. 1970. $28.50.

Cosmic Radiation In Contemporary Astrophysics. Maurice M. Shapiro, editor. Kluwer Academic Publishers, 101 Philip Drive, Assinippi Park, Norwell, MA 02061. (617) 871-6600. 1985. $101.50.

Cosmic Rays. Michael W. Friedlander. Harvard University Press, 79 Garden Street, Cambridge MA. (617) 495-2600. 1990. $14.95 in paper.

Cosmic Rays, Supernovae and the Interstellar Medium.

Maurice M. Shapiro, et al, editors. Kluwer Academic Publishers, 101 Philip Drive, Assinippi Park, Norwell, MA 02061. (617) 871-6600. 1991. $112.50.

Cosmic Rays, the Sun and Geomagnetism: the Works of Scott E. FORBUSH. James A. Van Allen, editor. American Geophysical Union, 2000 Florida Avenue, N.W., Washington, DC 20009. (202) 462-6903. (800) 966-2481. 1993. $48.50.

Early History of Cosmic Ray Studies: Personal Reminiscemces With Old Photographs. Yataro Sekido and Harry Elliot. Kluwer Academic Publishers, 101 Philip Drive, Assinippi Park, Norwell, MA 02061. (617) 871-6600. 1985. $202.50.

Frontiers of Fundamental Physics. M. Barone and F. Selleri, editors. Plenum Publishing Corp., 233 Spring Street, New York, NY. (212) 620-8000. (800) 2221-9369. 1994. $145.00.

Gravitation. Charles W. Misner, et al. W. H. Freeman, 41 Madison Avenue, East 26th Avenue, 35th Floor, New York, NY 10010. (212) 576-0400. 1995.

Gravitation, Cosmology and Cosmic-ray Physics. National Research Council. National Academy Press, 2101 Constitution

COSMIC RAYS

Ency. of Physical Sciences and Engineering Info. Sources

Avenue NW, Washington, DC 20055. (202) 334-3313. (800) 624-6242. 1986. $24.50.

An Introduction To Astronomy: Stars, Galaxies and the Universe. Gerald D. Waxman. Kendall-Hunt Publishing Co., 4050 Westmark Drive, P.O. Box 1840 Dubuque, IA 52004-1840. (319) 589-1000. (800) 228-0810. 1992. $28.95.

Origin of Cosmic Rays. Giancarlo Setti, et al, editors. Kluwer Academic Publishers, 101 Philip Drive, Assinippi Park, Norwell, MA 02061. (617) 871-6600. 1981. $42.50 in paper.

The Secret of Life: Cosmic Rays and Radiations and Radiations of Living Beings and Electromagnetic Waves. G. Lakhovsky. Gordon Press Publications, P.O. Box 459, Bowling Green Station, - New York, NY 10004. (718) 624-8419. 1991. $79.95.

Universe. William J. Kaufmann. W. H. Freeman & Co., 41 Madison Avenue, East 26th Avenue, 35th Floor, New York, NY 10010. (212) 576-0400. 4th edition. 1995. $54.95.

HANDBOOKS AND MANUALS

Astrophysical Data: Planets and Stars. Kenneth R. Lanb. Springer-Verlag New York, Inc., 175 Fifth Avenue, New York, NY 10010. (212) 460-1500. (800) 777-4643. 1993. $59.00.

Encyclopedia of Astronomy and Astrophysics. Stephen Maran, editor. Van Nostrand Reinhold, 115 Fifth Avenue, New York, NY 10003. (212) 254-3232. (800) 842-3636. 1992. $129.95.

Handbook of Space Astronomy and Astrophsyics. Martin V. Zumbeck. Cambridge University Press, 40 West 20th Street, New York, NY 10011-4211. (212) 924-3900. (800) 872-7423. 2nd edition. 1990. $79.95.

Landolt-Borenstein Numerical Data and Functional Relationships In Science and Technology: Astronomy, Astrophysics and Space Research. Group VI. Springer-Verlag New York, Inc., 175 Fifth Avenue, New York, NY 10010. (212) 460-1500. (800) 777-4643. FAX (212) 473-6272. 1965-. Irregular. Price varies. ISSN: 0075-7896.

McGraw-Hill Encyclopedia of Astronomy. Sybil P. Parker and Jay M. Pasachoff, editors. McGraw-Hill Publishing Company, Inc., 1221 Avenue of the Americas, New York, NY 10020. (212) 512-2000. (800) 262-4729. 2nd edition. 1993. $75.00

Observer's Guide To Astronomy. Patrick Martinez, editor. Cambridge University Press, 40 West 20th Street, New York, NY 10011-4211. (212) 924-3900. (800) 872-7423. 2 volumes. 1994. $159.90.

ONLINE DATABASES AND CD-ROMS

CA Search. Chemical Abstracts Service, P.O. Box 3012, Columbu s, OH 43210-0012. (614) 447-3600. (800) 848-6533. FAX (614) 447-3709. Very comprehensive guide to worldwide chemical literature and related fields, 1972 to present. Available on BRS,(800)289-4277, DIALOG, (800) 334-2564, ORBIT (800) 456-7248, and STN International, FIZ Karlsruhe, P.O. Box 2465, W-7500, Karlsruhe 1, Germany, online services. Inquire as to cost and availability.

Current Contents Search. Institute for Scientific Information, 3501 Market Street, Philadelphia, PA 19104. (215) 386-0100 .

FAX (215) 386-6362. Contains citations to articles listed in the table of contents of science and technology journals. Also articles in social sciences and life sciences journals. Available on BRS, (800) 289-4277, DIALOG, (800) 334-2564, online services. Inquire as to cost and availability.

Dissertation Abstracts Online. University Microfilms International, 300 North Zeeb Road, Ann Arbor, MI 48106. (800) 521-0600 or (313) 761-4700. Scope includes virtually all doctoral dissertations accepted at accredited American institutions from 1861 to present in 252 subject areas. Available on BRS, (800) 289-4277, DIALOG, (800) 33 4-2564, and OCLC EPIC, (800) 848-5878, online services. Also available on CD-ROM. Inquire as to cost and availability.

Inspec. In stitution of Electrical Engineers, Michael Faraday H ouse, Six Hills Way, Stevenage, Herts. SG1 2AY, Englan d. Telephone: 0438 313311 or FAX 0438 742840. Contains citations to the worldwide literature of physics , electronics and electrical engineering, computer technology, and related fields. Available on BRS, (800) 289-4277, DIALOG, (800) 334-2564, ORBIT, (800) 456-7248, and STN International, FIZ Karlsruhe, P.O. Box 2465, W-7500, Karlsruhe 1, Germany, online services. Inquire as to cost and availability.

International Aerospace Abstracts. American Institute of Aeronautic s and Astronautics, 370 L'Enfant Promenade, S.W., Washington, DC 20024. (202) 646-7400. Contains references and abstracts of journal and monograph literature relating to aerospace science and technology, from 1963 to present. Available through the NASA/RECON system of the National Aeronautics and Space Administration only.

NTIS Bibliographic Database. National Technical Information Service, 5285 Port Royal Road, Springfield, Va 22161 . (703) 487-4929 or FAX (703) 321-8199. Broad covera ge of government-sponsored science and technology research reports, 1964 to present. Available on BRS,(800) 289-4277, DIALOG, (800) 334-2564, ORBIT, (800) 456-7248, and STN International, FIZ Karlsruhe, P.O. Box 2465, W-7500, Karlsruhe 1, Germany, online services. Also available on CD-ROM. Inquire as to cost and availability.

Physics Briefs. American Institute of Physics, 335 East 45th Street , New York, NY 10017. (212) 661-9260 or FAX (212) 661-2036. Contains citations with abstracts of the literature of physics and related fields, 1979 to present. Available on the STN International, FIZ Karlsruhe, P.O. Box 2465, W-7500, Karlsruhe 1, Germany, online service. Inquire as to cost and availability.

Scientific and Technical Books and Serials In Print. R.R.Bowker Inc., 121 Chanlon Road, New Providence, NJ 07974. (800) 521-8110 or (908) 464-6800. List of currently published books and serials in the physical and biological sciences, engineering and technology, with subject, author and titles indexes. Available on ORBIT, (800) 456-7248, online service. Inquire as to cost and availability.

Scisearch. Institute for Scientific Information, 3501 Market Street, Philadelphia, PA 19104. (800) 523-1850 or (215) 386-0100. Broad multidisciplinary title and author index to the international literature of science and technology, 1974 to present. Available on DIALOG, (800) 334-2564, and ORBIT, (800) 456-7248, online services. Also available on CD-ROM. Inquire as to cost and availability.

WILSONLINE. H.W. W ilson Company, 950 University Avenue, Bronx, NY 10452. (800) 367-6770 or (212) 588-8400. Makes available onlin e versions of the printed H.W. Wilson indexes includin g Applied Science and Technology Index, Business Periodicals Index, General Science Index, and Readers' Guide to Periodical Literature. Period covered is generally 1983 to present. Available on BRS, (800) 289-4277, DIALOG, (800) 334-2564, and OCLC EPIC, (800) 848-5878, online services. Also available on CD-ROM. Inquire as to cost and availability.

OTHER SOURCES

Atlas of Deep-Sky Spendors. Hans Vehrenberg. Sky Publishing Corp., 49 Bay State Road, Cambridge, MA 02138. (617) 864-7360. (800) 253-0245. 4th edition. 1983. $24.95.

Atlas of the Ultraviolet Sky. Richard C. Henry, Wayne B. Landsman, et al. Johns Hopkins University Press, 2715 North Charles Street, Baltimore, MD 21218-4319. (410) 516-6900. (800) 537-5487. 1988. $70.00.

Atlas of the Universe. Patrick Moore. Rand McNally & Company. 8255 North Central Park, Skokie, IL 60076-2970. (708) 673-9100. 1994. $29.95.

Cambridge Atlas of Astronomy. Jean Audouze, et al. Cambridge University Press, 40 West 20th Street, New York,NY 10011-4211. (212) 924-3900. (800) 872-7423. 3rd edition. 1994. $75.00.

PERIODICALS

Acta Astronomica. Copernicus Foundation for Polish Astronomy, AL Ujazdowskie 4, 00-478 Warsaw, Poland. TEL 48-22-295346. FAX 48-22-294967. 1925 to present. Quarterly. $80.00 per year. ISSN: 0001-5237.

Astronomical Journal. American Institute of Physics, One Physics Ellipse, College Park, MD 20740-3843. (301) 209- 3000. Published for the American Astronomical Society. 1849 to present. Monthly. $280.00 per year. ISSN: 0004- 6256.

Astronomical Society of the Pacific. PUBLICATIONS. Astronomical Society of the Pacific, 390 Ashton Avenue, San Francisco, CA 94112. (415) 337-1100. FAX (415) 337-5205. 1889 to present. Monthly. $175.00 per year. ISSN: 0004- 6280.

Astronomy. Kalmbach Publishing Company, Box 1612, Waukesha, WI 53187-1612. (414) 796-8776. FAX (414) 796-1142. 1973 to present. Monthly. $27.00 per year. ISSN: 0091-6358.

Astronomy and Astrophsyics. Springer-Verlag New York, Inc., 44 Hartz Way, Secaucus, NJ 07096-2491. (201) 348-4033. FAX (201) 348-4505. 1969 to date. 36 issues per year in 12 volumes. $1733.00 per year. ISSN: 0004-6361.

Astronomy Reports. American Institute of Physics, One Physics Ellipse, College Park, MD 20740-3843. (301) 209- 3000. English translation of Astronomicheksii Zhurnal. Formerly Soviet Astronomy AJ. 1957 to present. Bimonthly. $1160.00 per year. ISSN: 1063-7729.

Astroparticle Physics. Elsevier Science Publishing Company, Inc., Box 882, Madison Square Station, New York, NY 10159. (212) 989-5800. FAX (212) 633-3990. 1992 to present. Quarterly. $180.00 per year. ISSN: 0927-6505.

Astrophysical Journal; An International Review of Astronomy and Astronomical Physics. University of Chicago Press, Journals Division, 5720 South Woodlawn Avenue, Chicago, IL 60637. (312) 753-3347. FAX (312) 753-0811. 1895 to date. three issues per month. $740.00 per year. ISSN: 0004-637X.

Astrophysics and Space Science; An International Journal of Cosmic Physics. Kluwer Academic Publishers, Box 358, Accord Station, Hingham, MA 02018-0358. (617) 871-6600. FAX (617) 871-6528. 1968 to present. 24 issues per year. $2544.00 per year. ISSN: 0004-640X.

Experimental Astronomy; An International Journal On Astronomical Instrumentation and Data Analysis. Kluwer Academic Publishers, Box 358, Accord Station, Hingham, MA 02018-0358. (617) 871-6600. FAX (617) 871-6528. 1989 to present. Quarterly. $161.00 per year. ISSN: 0922-6435.

Icarus; International Journal of Solar System Studies. Academic Press, Inc., Journal Division, 525 B Street, Suite 1900, San Diego, CA 92101-4495. (619) 230-1840. FAX (619) 699-6800. 1962 to present. Monthly. $1080.00 per year. ISSN: 0019-1035.

Mercury. Astronomical Society of the Pacific. 390 Ashton Avenue, San Francisco, CA 94112. (415) 337-1100. FAX: (415) 337-5205. 1972 to present. Bimonthly. $175.00 per year. ISSN: 0047-6773.

Monthly Notices of the Royal Astronomical Society. Blackwell Scientific Publication LT., Osney Mead, Oxford OX2 OEL, England. TEL 0865-240201. FAX 0865-721205. 1827 to present. Fortnightly. $1899.00 per year. ISSN: 0035-8711.

Sky and Telescope. Sky Publishing Corporation, Box 9111, Belmont, MA 02178. (617) 864-7360. FAX (617) 864-6117. 1941 to present. Monthly. $27.00 per year. ISSN: 0037- 6604.

Space Science Reviews. Kluwer Academic Publishers, Box 358, Accord Station, Hingham, MA 02018- 0358. (617) 871- 6600. FAX (617) 871-6528. 1962 to present. Sixteen issues per year. $940.00 per year. ISSN: 0038-6308.

Transactions of the International Astronomical Union. Proceedings of the General Assembly. Kluwer Academic Publishers, Box 358, Accord Station, Hingham, MA 02018- 0358. (617) 871-6600. Issues in two parts: Part A, Reports, and Part B, Proceedings. Triennial. XXI - 1994. Price varies. ISSN: 0080-1372.

Vistas In Astronomy; An International Review Journal. Albert C. Beer, editor. Elsevier Science Publishing Company, Inc., 660 White Plains Road, Terrytown, NY 10591- 5153. (914) 524-9200. FAX: (914) 333-2444. 1958 to present. Quarterly. $415.00 per year. ISSN:: 0083-6656.

RESEARCH CENTERS AND INSTITUTES

Astronomy Program. University of Maryland, 1207 Computer & Space Sciences Building, College Park, Md 20742. (301) 405-3001. FAX (301) 314-9067.

Center For Astrophysics and Space Sciences. University of California, San Diego, C-0111, La Jolla, CA 92093-0111. (619) 534-3460. FAX (619) 534-2294.

Cosmic Ray Research Project. University of Utah. 201 James Fletcher Building, Salt Lake City, UT 84112. (801) 581- 6930. FAX (801) 581-4801.

COSMIC RAYS

Ency. of Physical Sciences and Engineering Info. Sources

Joint Institute For Laboratory Astrophysics. University of Colorado - Boulder, Boulder CO 80309-0440. (303) 492-7789. FAX (303) 492-5235.

Lunar and Planetary Laboratory. University of Arizona, Tucson, AZ 85721. (602) 621-6963. FAX (602) 621-4933.

Smithsonian Center For Astrophysics. Harvard University, 60 Garden Street, Cambridge, MA 02138. (617) 495-7461. FAX (617) 495-7326.

Space Sciences Laboratory. University of California, Berkeley, Berkeley, CA 94720. (415) 642-1361. FAX (415) 643-7629.

Theoretical Astrophysics Program. University of Arizona, Planetary Sciences Department, Tuscon, AZ 85721. (602) 621- 6963.

COSMOCHEMISTRY

See also: ASTRONOMY, BIG BANG, COSMOLOGY, DARK MATTER, GALAXIES, PLANETS, SOLAR SYSTEM, UNIVERSE

ABSTRACT SERVICES AND INDEXES

Applied Science and Technology Index; A Cumulative Subject Index To English Language Periodicals In the Fields of Aeronautics and Space Science, Computer Technology, Ch Emistry, Construction Industry, Energy and Related Areas. H.W. Wilson Co., 950 University Avenue, Bronx, NY 10452. (212) 588-8400. (800) 367-6770. FAX (718) 590-1617. From 1958 to present. Monthly. Inquire about cost and availability. Also available on CD-ROM and online. ISSN: 0003-6986.

Astronomy and Astrophysics Abstracts. Springer-Verlag New York, 175 Fifth Avenue, New York, NY 10010. (212) 460-1500. FAX (212) 473-6272. Published for the Astronomisches Rechen-Institut. Comprehensive coverage of all aspects of astronomy, astrophysics and related fields. 1969 to present. Two parts per year. Annual. ISSN: 0067-0022.

Chemical Abstracts. Chemical Abstracts Service, 2540 Olentangy River Road, Box 3012, Columbus, OH 43210-0012. (614) 447-3600. (800) 848-6538. FAX (614) 447-3713. From 1907 to present. Weekly. $16800.00. Available online and on CD-ROM. CA is also available in five section groupings. Inquire regarding cost and availability.

General Science Index. H.W. Wilson Company, 950 University Avenue, Bronx, NY 10452. (212) 588-8400. (800) 367-6770. FAX (718) 590-1617. From 1978 to present. Ten issues per year; quarterly and annual cumulations. Service basis. Available on CD-ROM and online. Inquire regarding cost and availability. ISSN: 0162-1963.

Government Reports Announcements and Index. U. S. National Technical Information Service (NTIS), 5285 Port Royal Road, Springfield, VA 22161. (703) 487-4650. FAX (703) 321-8547. From 1968 to present. Annual. $630.00 per year. Also available online as *NTIS Bibliographic Database* and on CD-ROM.

NTIS Alerts: Astronomy & Astrophysics. U. S. National Technical Information Service. 5285 Port Royal Road, Springfield, VA 22161. (703) 487-4650. FAX (703) 321-8547. Weekly. $140.00 per year.

Pascal 48: Environnement Cosmique Terrestre, Astronomie Et Geologie Extraterrestre. Centre National de la Recherche Scientifique, Institut de Information Scientifique et Technique, 2 aliee du Parc de Brabois, 54514 Vandoeuvre Les Nancy Cedex, France. TEL 83-50-46-00. FAX: 83-50-46-50. 1985 to present. Ten issues per year. 770 F. Also available on CD-ROM and online.

Physics Abstracts. INSPEC. Section A, Science Abstracts. Institute of Electrical Engineers, London, United Kingdom. Available from: INSPEC/IEEE - Institute of Electrical and Electronic Engineers, Box 1331, Hoes Lane, Piscataway, NJ 08855-1331. (908) 562-5549. 1898 to present. 24 issues per year. $2835.00 per year. ISSN: 0036-8091. Also available online and on CD-ROM.

Physics Briefs (physikalische Berichte). Information Center for Energy, Physics, Mathematics; German Physical Society. V C H Publishers, Inc., 220 East 23rd Street, New York, NY 10010-4606. (212) 683-8333. 1845 to present. 24 issues per year. $2390.00 per year. Also available online. ISSN: 0179-7434.

Science Citati On Index. SCI. Institute for Scientific Information, 3501 Market Street, Philadelphia, PA 19104. (215) 386-0100. (800) 523-1850. FAX (215) 386-2991. 1961 to present. Six issues per year, plus annual cumulation. $11650.00 per year. Also available online and on CD-ROM. Inquire about price and availability. ISSN: 0036-827X.

S T A R . Scientific and Technical Aerospace Reports. U.S. National Aeronautics and Space Administration. Distributed by U. S. Superintendent of Documents, Washington, DC 20402. 1963 to present. Semi-monthly, with semiannual and annual indexes. $114.00 per year. ISSN: 0036-8741. Also available online and on CD-ROM.

ANNUAL REVIEWS AND YEARBOOKS

Annual Review of Astronomy and Astrophysics; Original Reviews of Critical Literature and Current Developments In Astronomy and Astrophysics. Annual Reviews, Inc., 4139 El Camino Way, Palo Alto, CA 94306-0139. (415) 493-4400. (800) 523-8635. FAX (415) 855-9815. 1963 to date. Annual. $60.00. ISSN:: 0066-4146.

Highlights of Astronomy. I. Appenzeller, editor. Kluwer Academic Publishers, Box 358, Accord Station, Hingham, MA 02018-0358. (617) 871-6600. International Astronomical Union Highlights series. ISBN: 07923-3553-8. Triennal. 1995. Price varies.

ASSOCIATIONS AND PROFESSIONAL SOCIETIES

American Association of Variable Star Observers. 23 Birch Street, Cambridge, MA 02138. (617) 354-0484. FAX: (617) 354-0665.

American Astronomical Society. 2000 Florida Avenue NW, Suite 400, Washington, DC 20009. (202) 328-2010. FAX: (202) 234-2560.

American Institute of Physics. One Physics Ellipse, College Park, MD 20740-3843. (301) 209-3100.

Association of University For Research In Astronomy. Suite 701, 1625 Massachusetts Avenue, NW, Washington, DC 20036. (202) 483-2101. FAX (202) 483-2106.

Astronomical League. 2112 Kingfisher Lane East, Rolling Meadows, IL 60008. (708) 398-0562.

Astronomical Society of the Pacific. 390 Ashton Avenue, San Francisco, CA 94112. (415) 337-1100. FAX: (415) 337-5205.

International Astronomical Union. 98 bis Boulevard Arago, F-75014 Paris, France. Tel 1-43-25-83-58. FAX 1-43-25-26-16.

National Council For Geocosmic Research. PO Box 1220, Dunkirk, MD 20754. (301) 855-2747. (800) PTOLEMY. FAX (410) 257-2824.

Royal Astronomical Society of Canada. 136 Dupont Street, Toronto. ON N5R 1V2 Canada. (416) 924-7973. FAX (416) 968-6687.

BIBLIOGRAPHIES

Astronomy and Astrophysics: A Source Guide. Gordon Press, P.O. Box 459, Bowling Green Station, New York, NY 10004. (718) 624-8419. 1991. $250.00. .

A Bibliography of Astronomy, 1970-1979. R. A. Sea and S. S. Martin. Libraries Unlimited, Inc., Littleton, CO 80160. 1982.

Catalog of the Naval Observatory Library. Naval Observatory Staff. G. K. Hall & Company, 866 Third Avenue, New York, NY 10022. (212)702-6789. (800) 257-5755. 6 volumes. 1977. $655.00

Science Books and Films. American Association for the Advancement of Science, 1333 H Street NW, Washington, DC 20005. (202) 326-6454. 1965 to present. Nine issues per year. $40.00 per year. ISSN: 0098-342X

Scientific and Technical Books and Serials In Print, 1995. R.R. Bowker Inc., 121 Chanlon Road, New Providence, NJ 07974. (908) 464-6800. (800) 521-8110. 4 volumes. 1994. $299.95. Alsoavailable on compact disc and online.

A Source Book On Astronomy and Astrophysics, 1900 - 1975. Kenneth Lang and Owen Gingerich, editors. Harvard University Press, 79 Garden Street, Cambridge MA. (617) 495-2600. 1980. $41.95.

DIRECTORIES AND BIOGRAPHICAL SOURCES

American Men and Women of Science: Physical and Biological Sciences. R. R. Bowker Inc., 121 Chanlon Road, New Providence, NJ 07974. (908) 464-6800. (800) 521-8110. 20th edition. 8 volumes. 1996. $850.00.

The Astronomers. Donald Goldsmith. St. Martin's Press, Inc., 175 Fifth Avenue, New York, NY 10010. (212) 674-5151. (800) 221-7945. 1993. $14.95 in paper.

Astronomical Centers of the World. Kevin Krisciunas. Cambridge University Press, 40 West 20th Street, New York,

NY 10011-4211. (212) 924-3900. (800) 872-7423. 1988. $34.95

The Biographical Dictionary of ScieNTISts: Astronomers. D. Abbott, editor. Peter Bedrick Books, Inc., 2112 Broadway, Room 318, New York, NY 10023. (212) 496-0751. 1984.

Directory of Physics and Astronomy Staff Members. American Institute of Physics, One Physics Ellipse, College Park, MD 20740-3843. (301) 209-3100. Annual. $45.00.

Earth and Astronomical Research Centers. Stockton Press, 345 Park Avenue South, New York, NY 10010. 4th edition. 1995. $515.00. ISBN: 1-56169-0967.

Graduate Programs In Physics, Astronomy and Related Fields. 1995-1996. American Institute of Physics, One Physics Ellipse, College Park, MD 20740-3843. (301) 209-3100. 1978 to present. Annual. $45.00.

Research Centers Directory. Gale Research, 835 Penobscot Building, Detroit, MI 48226-4094. (313) 961-2242. (800) 347-4253. 20th edition. 1995. $485.00.

ENCYCLOPEDIAS AND DICTIONARIES

Concise Dictionary of Astronomy. Jacqueline Mitton. Oxford University Press, Inc., 200 Madison Avenue, New York, NY 10016. (212) 725-6000. (800) 334-4249. 1992. $24.95.

Encyclopedia of Astronomy and Astrophysics. Stephen Maran, editor. Van Nostrand Reinhold, 115 Fifth Avenue, New York, NY 10003. (212) 254-3232. (800) 842-3636. 1992. $129.95.

Encyclopedia of Cosmology: Historical, Philosophical and Scientific Foundations of Modern Cosmology. Norris S. Hetherington, editor. Garland Publishing, Inc., 717 Fifth Avenue, Suite 2500, New York, NY 10022-8102. (212) 751- 7447. (800) 627-6273. 1993. $125.00.

Oxford Illustrated Encyclopedia. Volume 8, the Universe. Archie Roy, editor. Oxford University Press, Inc., 200 Madison Avenue, New York, NY 10016. (212) 725-6000. (800) 334-4249. 1993. $45.00.

McGraw-Hill Encyclopedia of Science and Technology. McGraw- Hill Book Company, Inc., 1221 Avenue of the Americas, NewYork, NY 10020. (212) 512-2000. (800) 262-4729. 7th edition. 20 volume set. 1992. $1900.00.

Stars, Galaxies, Cosmos. William R. Corliss. Sourcebook Project, P.O. Box 107, Glen Arm, MD 21057. (410) 668-6047. 1987. $17.95.

GENERAL WORKS

Atoms, Stars, and Nebulae. Lawrence H. Aller. Cambridge University Press, 40 West 20th Street, New York, NY 10011- 4211. (212) 924-3900. (800) 872-7423. 1991. $34.95 in paper.

Cosmochemistry and the Origin of Life. Cyril Ponnamperuma.. Kluwer Academic Publishers, 101 Philip Drive, Assinippi Park, Norwell, MA 02061. (617) 871-6600. 1983. $136.50.

The Cosmos. Time-Life, Inc., 777 Duke Street, Alexandria,

VA 22314. (703) 8388-7000. (800) 621-7026. Voyage Through the Universe series. 1989. $24.50.

COSMOCHEMISTRY

Ency. of Physical Sciences and Engineering Info. Sources

Cycles of Fire: Stars, Galaxies and the Wonder of Deep Space. William K. Hartmann. Workman Publishing Company, Inc., 708 Broadway, New York, NY 10003. (212) 254-5900. (800) 722-7202. 1988. $15.95 in paper.

Dust and Chemistry In Astronomy. D. A. Williams and T. J. Miller. IOP Publishing, Public Ledger Building, Suite 1035, Independence Square, Philadelphia, PA 19106. (215) 627-0880. (800) 358-4677. 1993. $149.50.

The First Three Minutes: A Modern View of the Origin of the Universe. Steven Weinberg. Basic Books, 10 East 53rd Street, New York, NY 10022. (212) 207-7057. (800) 242-7737. 2nd edition; Reprint edition. 1993. $12.00.

An Introduction To Cosmochemistry. Charles R. Crowley. Cambridge University Press, 40 West 20th Street, New York, NY 10011-4211. (212) 924-3900. (800) 872-7423. 1995. $29.95 in paper.

The Left Hand of Creation: the Origin and Evolution of the Expanding Universe. John D. Barrow and Joseph Silk. Oxford University Press, Inc., 200 Madison Avenue, New York, NY 10016. (212) 725-6000. (800) 334-4249. 1994. $23.00.

Noble Gas and High Temperature Chemistry. Springer-Verlag, New York, Inc., 175 Fifth Avenue, New York, NY 10010. (212) 460-1500. (800) 777-4643. 1990. $96.00.

Origin of the Universe. John D. Barrow. Basic Basic Books, 10 East 53rd Street, New York, NY 10022. (212) 207-7057. (800) 242-7737. 1994. $20.00.

The Physics of Stars. A. C. Phillips. John Wiley & Sons, Inc., 605 Third Avenue, New York, NY 10158-0012. (212) 850-6000. (800) 225-5945. 1994. $32.95 in paper.

Resources of Near-earth Space. John S. Lewis, et al. University of Arizona Press, 1230 North Park Avenue, Number 102, Tucson, AZ 85719. (520) 621-1441. (800) 426-3797. 1993. $75.00.

Stars and Atoms: From the Big Bang To the Solar System. Stuart Clark. Oxford University Press, Inc., 200 Madison Avenue, New York, NY 10016. (212) 725-6000. (800) 334-4249. 1995. $35.00.

The Theory of Cosmic Grains. Fred Hoyle and Nalin C. Wickramasinghe. Kluwer Academic Publishers, 101 Philip Drive, Assinippi Park, Norwell, MA 02061. (617) 871-6600. 1991. $87.00.

HANDBOOKS AND MANUALS

Encyclopedia of Astronomy and Astrophysics. Stephen Maran, editor. Van Nostrand Reinhold, 115 Fifth Avenue, New York, NY 10003. (212) 254-3232. (800) 842-3636. 1992. $129.95.

Handbook of Astronomy, Astrophysics and Geophysics; Volume 2: Galaxies and Cosmology. V. M. Canuto and B. G. Elmegreen. Gordon & Breach Science Publishers, Inc., 50 West 23rd Street, New York, NY 10010. (212) 206-8900 . 1988. $194.00.

Handbook of Space Astronomy and Astrophsyics. Martin V. Zumbeck. Cambridge University Press, 40 West 20th Street, New York, NY 10011-4211. (212) 924-3900. (800) 872-7423. 2nd edition. 1990. $79.95.

Landolt-Borenstein Numerical Data and Functional Relationships In Science and Technology: Astronomy, Astrophysics and Space Research. Group VI. Springer-Verlag New York, Inc., 175 Fifth Avenue, New York, NY 10010. (212) 460-1500. (800) 777-4643. FAX (212) 473-6272. 1965. Irregular. Price varies. ISSN: 0075-7896.

McGraw-Hill Encyclopedia of Astronomy. Sybil P. Parker and Jay M. Pasachoff, editors. McGraw-Hill Publishing Company, Inc., 1221 Avenue of the Americas, New York, NY 10020. (212) 512-2000. (800) 262-4729. 2nd edition. 1993. $75.00

Observer's Guide To Astronomy. Patrick Martinez, editor. Cambridge University Press, 40 West 20th Street, New York, NY 10011-4211. (212) 924-3900. (800) 872-7423. 2 volumes. 1994. $159.90.

ONLINE DATABASES AND CD-ROMS

CA Search. Chemical Abstracts Service, P.O. Box 3012, Columbu s, OH 43210-0012. (614) 447-3600. (800) 848-6533. FAX (614) 447-3709. Very comprehensive guide to worldwide chemical literature and related fields, 1972 to present. Available on BRS,(800) 289-4277, DIALOG, (800) 334-2564, ORBIT (800) 456-7248, and STN International, FIZ Karlsruhe, P.O. Box 2465, W-7500, Karlsruhe 1, Germany, online services. Inquire as to cost and availability.

Current Contents Search. Institute for Scientific Information, 3501 Market Street, Philadelphia, PA 19104. (215) 386-0100 . FAX (215) 386-6362. Contains citations to articles listed in the table of contents of science and technology journals. Also articles in social sciences and life sciences journals. Available on BRS, (800) 289-4277, DIALOG, (800) 334-2564, online services. Inquire as to cost and availability.

Dissertation Abstracts Online. University Microfilms Internatio nal, 300 North Zeeb Road, Ann Arbor, MI 48106. (800) 521-0600 or (313) 761-4700. Scope includes virtually all doctoral dissertations accepted at accredited American institutions from 1861 to present in 252 subject areas. Available on BRS, (800) 289-4277, DIALOG, (800) 334-2564, and OCLC EPIC, (800) 848-5878, online services. Also available on CD-ROM. Inquire as to cost andavailability.

Inspec. In stitution of Electrical Engineers, Michael Faraday H ouse, Six Hills Way, Stevenage, Herts. SG1 2AY, Englan d. Telephone: 0438 313311 or FAX 0438 742840. Contains citations to the worldwide literature of physics , electronics and electrical engineering, computer technology, and related fields. Available on BRS, (800) 289-4277, DIALOG, (800) 334-2564, ORBIT, (800) 456-7248, and STN International, FIZ Karlsruhe, P.O. Box 2465, W-7500, Karlsruhe 1, Germany, online services. Inquire as to cost and availability.

NTIS Bibliographic Database. National Technical Information Service, 5285 Port Royal Road, Springfield, Va 22161 . (703) 487-4929 or FAX (703) 321-8199. Broad covera ge of government-sponsored science and technology research reports, 1964 to present. Available on BRS,(800) 289-4277, DIALOG, (800) 334-2564, ORBIT, (800) 456-7248, and STN International, FIZ Karlsruhe, P.O. Box 2465, W-7500, Karlsruhe 1, Germany, online services. Also available on CD-ROM. Inquire as to cost and availability.

Physics Briefs. American Institute of Physics, 335 East 45th Street , New York, NY 10017. (212) 661-9260 or FAX (212) 661-2036. Contains citations with abstracts of the literature of physics and related fields, 1979 to present. Available on the STN International, FIZ Karlsruhe, P.O. Box 2465, W-7500, Karlsruhe 1, Germany, online service. Inquire as to cost and availability.

Scientifi C and Technical Books and Serials In Print. R.R. Bowker Inc., 121 Chanlon Road, New Providence, NJ 07974. (800) 521-8110 or (908) 464-6800. List of currently published books and serials in the physical and biological sciences, engineering and technology, with subject, author and titles indexes. Available on ORBIT, (800) 456-7248, online service. Inquire as to cost and availability.

SCISEARCH. Institute for Scientific Information, 3501 Market Street, Philadelphia, PA 19104. (800) 523-1850 or (215) 3 86-0100. Broad multidisciplinary title and author index to the international literature of science and technology, 1974 to present. Available on DIALOG, (800) 334-2564, and ORBIT, (800) 456-7248, online services. Also available on CD-ROM. Inquire as to cost and availability.

WILSONLINE. H.W. W ilson Company, 950 University Avenue, Bronx, NY 10452. (800) 367-6770 or (212) 588-8400. Makes available onlin e versions of the printed H.W. Wilson indexes includin g Applied Science and Technology Index, Business Periodicals Index, General Science Index, and Readers' Guide to Periodical Literature. Period covered is generally 1983 to present. Available on BRS, (800) 289-4277, DIALOG, (800) 334-2564, and OCLC EPIC, (800) 848-5878, online services. Also available on CD-ROM. Inquire as to cost and availability.

OTHER SOURCES

Atlas of the Universe. Patrick Moore. Rand McNally &Company. 8255 North Central Park, Skokie, IL 60076-2970. (708) 673-9100. 1994. $29.95.

Cambridge Atlas of Astronomy. Jean Audouze, et al. Cambridge University Press, 40 West 20th Street, New York, NY 10011-4211. (212) 924-3900. (800) 872-7423. 3rd edition. 1994. $75.00.

Color Atlas of Galaxies. James Wray. Cambridge University Press, 40 West 20th Street, New York, NY 10011-4211. (212) 924-3900. (800) 872-7423. 1988. $94.95.

Observing Handbook and Catalog of Deep-sky Objects. Christian B. Luginbuhl and Brian A. Skiff. Cambridge University Press, 40 West 20th Street, New York, NY 10011- 4211. (212) 924-3900. (800) 872-7423. 1990. $49.95.

PERIODICALS

Astronomical Journal. American Institute of Physics, One Physics Ellipse, College Park, MD 20740-3843. (301) 209- 3000. Published for the American Astronomical Society. 1849 to present. Monthly. $280.00 per year. ISSN: 0004- 6256.

Astronomical Society of the Pacific. Publications. Astronomical Society of the Pacific, 390 Ashton Avenue, San Francisco, CA 94112. (415) 337-1100. FAX (415) 337-5205. 1889 to present. Monthly. $175.00 per year. ISSN: 0004- 6280.

Astronomy. Kalmbach Publishing Company, Box 1612, Waukesha, WI 53187-1612. (414) 796-8776. FAX (414) 796-

1142. 1973 to present. Monthly. $27.00 per year. ISSN: 0091-6358.

Astronomy and Astrophsyics. Springer-Verlag New York, Inc., 44 Hartz Way, Secaucus, NJ 07096-2491. (201) 348-4033. FAX (201) 348-4505. 1969 to date. 36 issues per year in 12 volumes. $1733.00 per year. ISSN: 0004-6361.

Astronomy Reports. American Institute of Physics, One Physics Ellipse, College Park, MD 20740-3843. (301) 209- 3000. English translation of Astronomicheksii Zhurnal. Formerly Soviet Astronomy AJ. 1957 to present. Bimonthly. $1160.00 per year. ISSN: 1063-7729.

Astrophysical Journal; An International Review of Astronomy and Astronomical Physics. University of Chicago Press, Journals Division, 5720 South Woodlawn Avenue, Chicago, IL 60637. (312) 753-3347. FAX (312) 753-0811. 1895 to date. three issues per month. $740.00 per year. ISSN: 0004-637X.

Astrophys Ics and Space Science; An International Journal of Cosmic Physics. Kluwer Academic Publishers, Box 358, Accord Station, Hingham, MA 02018-0358. (617) 871-6600. FAX (617) 871-6528. 1968 to present. 24 issues per year. $2544.00 per year. ISSN: 0004-640X.

Ciel Et Terre. Societe Royale Belge d'Astronomie, de Meteorologie et de Physique du Globe, 3 avenue Circulaire, Uccle, Brusels, Belgium. FAX 32-2-374-9822. 1880 to present. Bimonthly. 1600 BEF per year. ISSN: 0009-6709.

Geochemica Et Cosmochimica Acta. Elsevier Science. 660 White Plains Road, Tarrytown, NY 10591-5153. (914) 524- 9200. FAX(914) 333-2444. 1950 to the present. Biweekly. $895.00 per year. ISSN: 0016-7037.

Icarus; International Journal of Solar System Studies. Academic Press, Inc., Journal Division, 525 B Street, Suite 1900, San Diego, CA 92101-4495. (619) 230-1840. FAX (619) 699-6800. 1962 to present. Monthly. $1080.00 per year. ISSN: 0019-1035.

Mercury. Astronomical Society of the Pacific. 390 Ashton Avenue, San Francisco, CA 94112. (415) 337-1100. FAX: (415) 337-5205. 1972 to present. Bimonthly. $175.00 per year. ISSN: 0047-6773.

Monthly Notices of the Royal Astronomical Society. Blackwell Scientific Publication LT., Osney Mead, Oxford OX2 OEL, England. TEL 0865-240201. FAX 0865-721205. 1827 to present. Fortnightly. $1899.00 per year. ISSN: 0035-8711.

Sky and Telescope. Sky Publishing Corporation, Box 9111, Belmont, MA 02178. (617) 864-7360. FAX (617) 864-6117. 1941 to present. Monthly. $27.00 per year. ISSN: 0037- 6604.

Vistas In Astronomy; An International Review Journal. Albert C. Beer, editor. Elsevier Science Publishing Company, Inc., 660 White Plains Road, Terrytown, NY 10591- 5153. (914) 524-9200. FAX: (914) 333-2444. 1958 to present. Quarterly. $415.00 per year. ISSN:: 0083-6656.

Transactio Ns of the International Astronomical Union. Proceedings of the General Assembly. Kluwer Academic Publishers, Box 358, Accord Station, Hingham, MA 02018- 0358. (617) 871-6600. Issues in two parts: Part A, Reports, and Part B, Proceedings. Triennial. XXI - 1994. Price varies. ISSN: 0080-1372.

COSMOCHEMISTRY

Ency. of Physical Sciences and Engineering Info. Sources

Webb Society Quarterly Journal. Webb Society, 194 Foundry Lane, Freemantle, Southampton, Hampshire SO1 3EQ, England. 1969 to present. Quarterly. $19.50 per year. ISSN: 0043-1680.

RESEARCH CENTERS AND INSTITUTES

Astronomy Program. University of Maryland, 1207 Computer & Space Sciences Building, College Park, Md 20742. (301) 405-3001. FAX (301) 314-9067.

Carnegie Institution of Washington Observatories, 813 Santa Barbara Street, Pasadena, Ca 91101. (818) 577-1122. FAX (818) 759-8136.

Center For Radiophysics and Space Research. Cornell University. Space Sciences Building, Ithaca, NY 14853. (607) 255-4341. FAX (607) 255-9888.

Center For Space Science and Astrophysics. Stanford University, 325 Durand Building, Stanford, CA 94305. (415) 723-3592.

Joint Institute For Laboratory Astrophysics. University of Colorado - Boulder, Boulder CO 80309-0440. (303) 492-7789. FAX (303) 492-5235.

Space Telescope Science Institute. 3700 San Martin Drive, Baltimore, MD 21218. (301) 338-4700.

Theoretical Astrophysics Program. University of Arizona, Planetary Sciences Department, Tuscon, AZ 85721. (602) 621-6963.

COSMOGENY

See: COSMOLOGY

COSMOLOGY

See also: ASTRONOMY, ASTROPHYSICS, BIG BANG, CELESTIAL MECHANICS, COSMOCHEMISTRY, DARK MATTER, GALAXIES, PLANETS, SOLAR SYSTEM

ABSTRACT SERVICES AND INDEXES

Applied Science and Technology Index; A Cumulative Subject Index To English Language Periodicals In the Fields of Aeronautics and Space Science, Computer Technology, Ch Emistry, Construction Industry, Energy and Related Areas. H.W. Wilson Co., 950 University Avenue, Bronx, NY 10452. (212) 588-8400. (800) 367-6770. FAX (718) 590-1617. From 1958 to present. Monthly. Inquire about cost and availability. Also available on CD-ROM and online. ISSN: 0003-6986.

Astronomy and Astrophysics Abstracts. Springer-Verlag, 175 Fifth Avenue, New York, NY 10010. (212) 460-1500. FAX (212) 473-6272. From 1969 to present. Irregular. Price varies; inquire. Also available on CD-ROM. ISSN:: 0067-0022. ISSN: 0067-0022.

General Science Index. H.W. Wilson Company, 950 University Avenue, Bronx, NY 10452. (212) 588-8400. (800) 367-6770. FAX (718) 590-1617. From 1978 to present. Ten issues per year;

quarterly and annual cumulations. Service basis. Available on CD-ROM and online. Inquire regarding cost and availability. ISSN: 0162-196X

Government Reports Announcements and Index. U. S. National Technical Information Service (NTIS), 5285 Port Royal Road, Springfield, VA 22161. (703) 487-4650. FAX (703) 321-8547. From 1968 to present. Annual. $630.00 per year. Also available online as *NTIS Bibliographic Database* and on CD-ROM. ISSN:

International Aerospace Abstracts. American Institute of Aeronautics and Astronautics. Technical Information Service, 555 West 57th Street, New York 10019. (212) 247-6500. FAX (212) 582-496 1. 1961 to present. Semi-monthly with annual index. $1295.00 per year. Also available online and on CD-ROM. ISSN: 0020-5842.

Meteorological and Geoastrophysical Abstracts. American M eteorological Society, c/o Inforonics, Inc., 550 Newtown Road, Littleton, MA 01460. (508) 486-8976. FAX (508) 486-0027. Covers literature in environmental sciences, meteorology, astrophysics, hydrology, glaciology, and physical oceanography. From 1950 to present. Monthly. $950.00 per year. Also available on CD-ROM and online. Inquire regarding cost and availability. ISSN: 0026-1130.

NTIS Alerts: Astronomy & Astrophysics. U. S. National Technical Information Service. 5285 Port Royal Road, Springfield, VA 22161. (703) 487-4650. FAX (703) 321-8547. Weekly. $140.00 per year.

Physics Abstracts . INSPEC. Section A, Science Abstracts. Institution of Electrical Engineers, London, United Kingdom. Available from: INSPEC/IEEE - Institute of Electrical andElectronic Engineers, Box 1331, Hoes Lane, Piscataway, NJ 08855-1331. (908) 562-5549. 1898 to present. 24 issues per year. $2835.00 per year. Also available on CD-ROM and online. ISSN: 0036-8091.

Physics Briefs (Physikalische Berichte). Information Cen ter for Energy, Physics, Mathematics; German Physical Society. V C H Publishers, Inc., 220 East 23rd Street, New York, NY 10010-4606. (212) 683-8333. 1845 to present. 24 issues per year. $2390.00 per year. Also available online. ISSN: 0179-7434.

Science Citation Index. SCI. Institute for Scientific Information, 3501 Market Street, Philadelphia, PA 19104. (215) 386-0100. (800) 523-1850. FAX (215) 386-2991. 1961 to present. Six issues per year, plus annual cumulation. $11650.00 per year. Also available online and on CD-ROM. Inquire about price and availability. ISSN: 0036-827X.

S T A R. Scientific and Technical Aerospace Reports. U.S. National Aeronautics and Space Administration. Distributed by U. S. Superintendent of Documents, Washington, DC 20402. 1963 to present. Semi-monthly, with semiannual and annual indexes. $114.00 per year. ISSN: 0036-8741. Also available online and on CD-ROM.

ANNUAL REVIEWS AND YEARBOOKS

The *Astronomical Almanac.* Superintendent of Documents, U. S. Government Printing office, Washington, DC 10402. (202) 783-3238. 1981 to present. Supercedes Astronomical Ephemeris and the American Ephermis and Nautical Almanac. Annual. $44.95. ISSN: 0737-6421.

Annual Review of Astronomy and Astrophysics; Original Reviews of Critical Literature and Current Developments In Astronomy and Astrophysics. Annual Reviews, Inc., 4139 El Camino Way, Palo Alto, CA 94306-0139. (415) 493-4400. (800) 523-8635. FAX (415) 855-9815. 1963 to date. Annual. $60.00. ISSN:: 0066-4146.

Annual Review of Earth and Planetary Sciences. Annual Reviews, Inc., 4139 El Camino Way, Palo Alto CA 94306-0139. (415) 403-4400. (800) 523-8635. Fax (415) 855-9815. 1973 to date. Annual. $62.00. ISSN:: 0084-6597.

Highlights of Astronomy. I. Appenzeller, editor. Kluwer Academic Publishers, Box 358, Accord Station, Hingham, MA 02018-0358. (617) 871-6600. International Astronomical Union Highlights series. ISBN: 07923-3553-8. Triennal. 1995. Price varies.

ASSOCIATIONS AND PROFESSIONAL SOCIETIES

Amateur Astronomers Association. 1010 Park Avenue, New York, NY 10028. (212) 535-2922.

American Association of Variable Star Observers. 23 Birch Street, Cambridge, MA 02138. (617) 354-0484. FAX: (617) 354-0665.

American Astronomical Society. 2000 Florida Avenue NW, Suite 400, Washington, DC 20009. (202) 328-2010. FAX: (202) 234-2560.

American Institute of Physics. One Physics Ellipse, College Park, MD 20740-3843. (301) 209-3100.

Association of University For Research In Astronomy. Suite 701, 1625 Massachusetts Avenue, NW, Washington, DC 20036. (202) 483-2101. FAX (202) 483-2106.

Astronomical League. 2112 Kingfisher Lane East, Rolling Meadows, IL 60008. (708) 398-0562.

Astronomical Society of the Pacific. 390 Ashton Avenue, San Francisco, CA 94112. (415) 337-1100. FAX: (415) 337-5205.

International Astronomical Union. 98 bis Boulevard Arago, F-75014 Paris, France. Tel 1-43-25-83-58. FAX 1-43-25-26-16. Joint Institute for Laboratory Astrophyics. University of Colorado, Box 440, Boulder, CO 80309-0440. (303) 492-7780. FAX (303) 492-5235.

National Council For Geocosmic Research. PO Box 1220, Dunkirk, MD 20754. (301) 855-2747. (800) PTOLEMY. FAX (410) 257-2824.

Royal Astronomical Society of Canada. 136 Dupont Street, Toronto. ON N5R 1V2 Canada. (416) 924-7973. FAX (416) 968-6687.

BIBLIOGRAPHIES

Astronomy and Astrophysics: A Source Guide. Gordon Press, P.O. Box 459, Bowling Green Station, New York, NY 10004. (718) 624-8419. 1991. $250.00. .

A Bibliography of Astronomy, 1970-1979. R. A. Sea and S. S. Martin. Libraries Unlimited, Inc., Littleton, CO 80160. 1982.

Catalog of the Naval Observatory Library. Naval Observatory Staff. G. K. Hall & Company, 866 Third Avenue, New York, NY 10022. (212)702-6789. (800) 257-5755. 6 volumes. 1977. $655.00

Science Books and Films. American Association for the Advancement of Science, 1333 H Street NW, Washington, DC 20005. (202) 326-6454. 1965 to present. Nine issues per year. $40.00 per year. ISSN: 0098-342X

Scientific and Technical Books and Serials In Print, 1995. R.R. Bowker Inc., 121 Chanlon Road, New Providence, NJ 07974. (908) 464-6800. (800) 521-8110. 4 volumes. 1994. $299.95. Also available on compact disc and online.

A Source Book On Astronomy and Astrophysics, 1900 - 1975. Kenneth Lang and Owen Gingerich, editors. Harvard University Press, 79 Garden Street, Cambridge MA. (617) 495-2600. 1980. $41.95.

DIRECTORIES AND BIOGRAPHICAL SOURCES

American Astronomical Society. Membership Directory. American Astronomical Society. 2000 Florida Avenue NW, Suite 400, Washington, DC 20009. (202) 328-2010. FAX (202) 234-2560 Annual. Included in membership dues. ISSN:: 1061- 9038.

American Men and Women of Science: Physical and Biological Sciences. R. R. Bowker Inc., 121 Chanlon Road, New Providence, NJ 07974. (908) 464-6800. (800) 521-8110. 20th edition. 8 volumes. 1996. $850.00.

The Astronomers. Donald Goldsmith. St. Martin's Press, Inc., 175 Fifth Avenue, New York, NY 10010. (212) 674-5151. (800) 221-7945. 1993. $14.95 in paper.

Astronomical Centers of the World. Kevin Krisciunas. Cambridge University Press, 40 West 20th Street, New York, NY 10011-4211. (212) 924-3900. (800) 872-7423. 1988. $34.95

The Biographical Dictionary of Scientists: Astronomers. D. Abbott, editor. Peter Bedrick Books, Inc., 2112 Broadway, Room 318, New York, NY 10023. (212) 496-0751. 1984.

Directory of Physics and Astronomy Staff Members. American Institute of Physics, One Physics Ellipse, College Park, MD 20740-3843. (301) 209-3100. Annual. $45.00.

Earth and Astronomical Research Centers. Stockton Press, 345 Park Avenue South, New York, NY 10010. 4th edition. 1995. $515.00. ISBN: 1-56169-0967.

Graduate Programs In Physics, Astronomy and Related Fields. 1995-1996. American Institute of Physics, One Physics Ellipse, College Park, MD 20740-3843. (301) 209-3100. Annual. $45.00.

Research Centers Directory. Gale Research, 835 Penobscot Building, Detroit, MI 48226-4094. (313) 961-2242. (800) 347-4253. 20th edition. 1995. $485.00.

COSMOLOGY

Ency. of Physical Sciences and Engineering Info. Sources

ENCYCLOPEDIAS AND DICTIONARIES

Astronomy from A to Z: A Dictionary of Celestial Objects and Ideas. Charles A. Schweighauser. Illinois Issues, Sangamon State University, K-10, Springfield, IL 62794-9243. (217) 786-9243. 1991. $14.95 in paper.

Concise Dictionary of Astronomy. Jacqueline Mitton. Oxford University Press, Inc., 200 Madison Avenue, New York, NY 10016. (212) 725-6000. (800) 334-4249. 1992. $24.95.

Dictionary of Astronomical Names. Adrian Room. Routledge, Chapman and Hall, Inc., 29 West 35th Street, New York, NY 10001-3291. (212) 244-3336. 1988. $27.50.

Facts On File Dictionary of Astronomy. Valerie Illingworth. Facts On File, Inc., 460 Park Avenue South, New York, NY 10016-7382. (212) 683-2244. (800) 322-8755. Fax (800) 683-3633. 1994. $27.95.

Encyclopedia of Astronomy and Astrophysics. Stephen Maran, editor. Van Nostrand Reinhold . 1992. $129.95.

Encyclopedia of Cosmology: Historical, Philosophical and Scientific Foundations of Modern Cosmology. Norris S. Hetherington, editor. Garland. 1993. $125.00.

Oxford Illustrated Encyclopedia. Volume 8, the Universe. Archie Roy, editor. Oxford University Press, Inc., 200 Madison Avenue, New York, NY 10016. (212) 725-6000. (800) 334-4249. 1993. $45.00.

McGraw-Hill Encyclopedia of Science and Technology. McGraw-Hill Book Company, Inc., 1221 Avenue of the Americas, New York, NY 10020. (212) 512-2000. (800) 262-4729. 7th edition. 20 volume set. 1992. $1900.00.

Penguin Dictionary of Astronomy. Jacqueline Mitton. Viking Penguin, 375 Hudson Street, New York, NY 10014-3657.

(212) 366-2000. (800) 331-4624. 1993. $12.95 in paper.

Stars, Galaxies, Cosmos. William R. Corliss. SourcebookProject, P.O. Box 107, Glen Arm, MD 21057. (410) 668-6047. 1987. $17.95.

GENERAL WORKS

Brief History of Time: From the Big Bang To Black Holes. Stephen W. Hawking. Bantam Books, Inc., 1540 Broadway, New York, NY 10036-4094. (212) 354-6500. (800) 223-6834. 1990. $13.95 in paper.

Black Holes and Baby Universes and Other Essays. Stephen W. Hawking. Bantam Books, Inc., 1540 Broadway, New York, NY 10036-4094. (212) 354-6500. (800) 223-6834. 1993. $21.95.

Cosmic Questions, Galactic Halos, Cold Dark Matter and the End of Time. Richard Morris. John Wiley & Sons, Inc., 605 Third Avenue, New York, NY 10158-0012. (212) 850-6000. (800) 225- 5945. 1993. $24.95.

Cosmology and Particle Physics. David Lindley, et al, editors. American Association of Physics Teachers, One Physics Ellipse, College Park Md 20740-3842. (301) 209- 3300. 1991. $18.00.

Cycles of Fire: Stars, Galaxies and the Wonder of Deep Space. William K. Hartmann. Workman Publishing Company, Inc., 708 Broadway, New York, NY 10003. (212) 254-5900. (800) 722-7202. 1988. $15.95 in paper.

Farthest Things In the Universe. Jay M. Pasachoff. Cambridge University Press 40 West 20th Street, New York, NY 10011-4211. (212) 924-3900. (800) 872-7423. 1995. $29.95.

The Left Hand of Creation: the Origin and Evolution of the Expanding Universe. John D. Barrow and Joseph Silk. Oxford University Press, Inc., 200 Madison Avenue, New York, NY 10016. (212) 725-6000. (800) 334-4249. 1994. $23.00.

The New Cosmos. A. Unsold and B. Baschek. Springer-Verlag York, Inc., 175 Fifth Avenue, New York, NY 10010. (212) 460-1500. (800) 777-4643 . 4th revised edition. 1991. $59.00.

Observational Astrophysics. Robert C. Smith. Cambridge University Press, 40 West 20th Street, New York, NY 10011- 4211. (212) 924-3900. (800) 872-7423. 1995. $69.95.

Origin of the Universe. John D. Barrow. Basic Books, 10 East 53rd Street, New York, NY 10022. (212) 207-7057. (800) 242-7737. 1994. $20.00.

The Red Limit: the Search For the Edge of the Universe. Timothy Ferris. Morrow, William & Company, 1350 Avenue of the Americas, New York, NY 10019. (212) 261-6500. (800) 843-9389. Revised edition. 1993. $12.95.

Stars and Atoms: From the Big Bang To the Solar System. Stuart Clark. Oxford University Press, Inc., 200 Madison Avenue, New York, NY 10016. (212) 725-6000. (800) 334-4249. 1995. $35.00.

Stephen Hawking's Universe. John Boslough. Morrow, William & Company, 1350 Avenue of the Americas, New York, NY 10019. (212) 261-6500. (800) 843-9389. 1985. $7.95.

Universe. William J. Kaufmann. W. H. Freeman & Co., 41 Madison Avenue, East 26th Avenue, 35th Floor, New York, NY 10010. (212) 576-0400. 4th edition. 1995. $54.95.

Workings of the Universe. Time-Life, Inc., 777 Duke Street, Alexandria, VA 22314. (703) 8388-7000. (800) 621-7026. . 1991. $25.95.

HANDBOOKS AND MANUALS

Burnham's Celestial Handbook: An Observer's Guide To the Universe Beyond the Solar System. Robert Burnham, Jr. Dover Publications, Inc., 180 Varick Street, New York, NY 10014. (212) 255-3755. (800) 223-3130. 3 volumes. 1978. $41.85 for the set.

Cambridge Guide To the Constellations. Michael E. Bakich. Cambridge University Press, 40 West 20th Street, New York, NY 10011-4211. (212) 924-3900. (800) 872-7423. 1995. $69.95.

Encyclopedia of Astronomy and Astrophysics. Stephen Maran, editor. Van Nostrand Reinhold, 115 Fifth Avenue, New York, NY 10003. (212) 254-3232. (800) 842-3636. 1992. $129.95.

Greenwich Guide To Stars, Galaxies and Nebulae. Stuart Malin. Cambridge University Press, 40 West 20th Street, New York, NY 10011-4211. (212) 924-3900. (800) 872-7423. 1990. $10.95.

Handbook of Space Astronomy and Astrophsyics. Martin V. Zumbeck. Cambridge University Press, 40 West 20th Street, New York, NY 10011-4211. (212) 924-3900. (800) 872-7423. 2nd edition. 1990. $79.95.

Landolt-Borenstein Numerical Data and Functional Relationships In Science and Technology: Astronomy, Astrophysics and Space Research. Group VI. Springer-Verlag New York, Inc., 175 Fifth Avenue, New York, NY 10010. (212) 460-1500. (800) 777-4643. volumes priced individually.

McGraw-Hill Encyclopedia of Astronomy. Sybil P. Parker and Jay M. Pasachoff, editors. McGraw-Hill Publishing Company, Inc., 1221 Avenue of the Americas, New York, NY 10020. (212) 512-2000. (800) 262-4729. 2nd edition. 1993. $75.00

Observer's Guide To Astronomy. Patrick Martinez, editor. Cambridge University Press, 40 West 20th Street, New York, NY 10011-4211. (212) 924-3900. (800) 872-7423. 2 volumes. 1994. $159.90.

ONLINE DATABASES AND CD-ROMS

CA Search. Chemical Abstracts Service, P.O. Box 3012, Columbu s, OH 43210-0012. (614) 447-3600. (800) 848-6533. FAX (614) 447-3709. Very comprehensive guide to worldwide chemical literature and related fields, 1972 to present. Available on BRS,(800) 289-4277, DIALOG, (800) 334-2564, OR-BIT (800) 456-7248, and STN International, FIZ Karlsruhe, P.O. Box 2465, W-7500, Karlsruhe 1, Germany, online services. Inquire as to cost and availability.

Current Contents Search. Institute for Scientific Information, 3501 Market Street, Philadelphia, PA 19104. (215) 386-0100 . FAX (215) 386-6362. Contains citations to articles listed in the table of contents of science and technology journals. Also articles in social sciences and life sciences journals.Available on BRS, (800) 289-4277, DIALOG, (800) 334-2564, online services. Inquire as to cost and availability.

Dissertation Abstracts Online. University Microfilms Internatio nal, 300 North Zeeb Road, Ann Arbor, MI 48106. (800) 521-0600 or (313) 761-4700. Scope includes virtually all doctoral dissertations accepted at accredited American institutions from 1861 to present in 252 subject areas. Available on BRS, (800) 289-4277, DIALOG, (800) 334-2564, and OCLC EPIC, (800) 848-5878, online services. Also available on CD-ROM. Inquire as to cost and availability.

Inspec. In stitution of Electrical Engineers, Michael Faraday H ouse, Six Hills Way, Stevenage, Herts. SG1 2AY, Englan d. Telephone: 0438 313311 or FAX 0438 742840. Contains citations to the worldwide literature of physics , electronics and electrical engineering, computer technology, and related fields. Available on BRS, (800) 289-4277, DIALOG, (800) 334-2564, ORBIT, (800) 456-7248, and STN International, FIZ Karlsruhe, P.O. Box 2465, W-7500, Karlsruhe 1, Germany, online services. Inquire as to cost and availability.

International Aerospace Abstracts. American Institute of Aeronautic s and Astronautics, 370 L'Enfant Promenade, S.W., Washington, DC 20024. (202) 646-7400. Contains references and abstracts of journal and monograph literature relating to aerospace science and technology, from 1963 to present. Available

through the NASA/RECON system of the National Aeronautics and Space Administration only.

NTIS Bibliographic Database. National Technical Information Service, 5285 Port Royal Roa d, Springfield, Va 22161 . (703) 487-4929 or FAX (703) 321-8199. Broad covera ge of government-sponsored science and technology research reports, 1964 to present. Available on BRS,(800) 289-4277, DIALOG, (800) 334-2564, ORBIT, (800) 456-7248, and STN International, FIZ Karlsruhe, P.O. Box 2465, W-7500, Karlsruhe 1, Germany, online services. Also available on CD-ROM. Inquire as to cost and availability.

Physics Briefs. American Institute of Physics, 335 East 45th Street , New York, NY 10017. (212) 661-9260 or FAX (212) 661-2036. Contains citations with abstracts of the literature of physics and related fields, 1979 to present. Available on the STN International, FIZ Karlsruhe, P.O. Box 2465, W-7500, Karlsruhe 1, Germany, online service. Inquire as to cost and availability.

Scientifi C and Technical Books and Serials In Print. R.R. Bowker Inc., 121 Chanlon Road, New Providence, NJ 07974. (800) 521-8110 or (908) 464-6800. List of currently published books and serials in the physical and biological sciences, engineering and technology, with subject, author and titles indexes. Available on ORBIT, (800) 456-7248, online service. Inquire as to cost and availability.

Scisearch. Institute for Scientific Information, 3501 Market Street, Philadelphia, PA 19104. (800) 523-1850 or (215)386-0100. Broad multidisciplinary title and author index to the international literature of science and technology, 1974 to present. Available on DIALOG, (800) 334-2564, and ORBIT, (800) 456-7248, online services. Also available on CD-ROM. Inquire as to cost and availability.

WILSONLINE. H.W. W ilson Company, 950 University Avenue, Bronx, NY 10452. (800) 367-6770 or (212) 588-8400. Makes available onlin e versions of the printed H.W. Wilson indexes includin g Applied Science and Technology Index, Business Periodicals Index, General Science Index, and Readers' Guide to Periodical Literature. Period covered is generally 1983 to present. Available on BRS, (800) 289-4277, DIALOG, (800) 334-2564, and OCLC EPIC, (800) 848-5878, online services. Also available on CD-ROM. Inquire as to cost and availability.

OTHER SOURCES

Atlas of Deep-sky Spendors. Hans Vehrenberg. Sky Publishing Corp., 49 Bay State Road, Cambridge, MA 02138. (617) 864-7360. (800) 253-0245. 4th edition. 1983. $24.95.

Atlas of the Ultraviolet Sky. Richard C. Henry, Wayne B. Landsman, et al. Johns Hopkins University Press, 2715 North Charles Street, Baltimore, MD 21218-4319. (410) 516-6900. (800) 537-5487. 1988. $70.00.

Atlas of the Universe. Patrick Moore. Rand McNally & Company. 8255 North Central Park, Skokie, IL 60076-2970. (708) 673-9100. 1994. $29.95.

Cambridge Atlas of Astronomy. Jean Audouze, et al. Cambridge University Press, 40 West 20th Street, New York,NY 10011-4211. (212) 924-3900. (800) 872-7423. 3rd edition. 1994. $75.00.

COSMOLOGY

Ency. of Physical Sciences and Engineering Info. Sources

Cambridge Guide To the Constellations. Michael E. Bakich. Cambridge University Press, 40 West 20th Street, New York, NY 10011-4211. (212) 924-3900. (800) 872-7423. 1995. $19.95 in paper.

Cambridge Star Atlas 2000.0. Wil Tirion. Cambridge University Press, 40 West 20th Street, New York, NY 10011- 4211. (212) 924-3900. (800) 872-7423. 2nd edition. 1996. $19.95.

Color Atlas of Galaxies. James Wray. Cambridge University Press, 40 West 20th Street, New York, NY 10011-4211. (212) 924-3900. (800) 872-7423. 1988. $94.95.

Observer's Sky Atlas. E. Karkoschka. Springer-Verlag New York, Inc., 175 Fifth Avenue, New York, NY 10010. (212) 460-1500. (800) 777-4643. 1995. $15.95.

Observing Handbook and Catalog of Deep-sky Objects. Christian B. Luginbuhl and Brian A. Skiff. Cambridge University Press, 40 West 20th Street, New York, NY 10011- 4211. (212) 924-3900. (800) 872-7423. 1990. $49.95.

The Sky: A User's Guide. David H. Levy. Cambridge University Press, 40 West 20th Street, New York, NY 10011- 4211. (212) 924-3900. (800) 872-7423. 1993. $14.95 in paper.

PERIODICALS

Acta Astronomica. Copernicus Foundation for Polish Astronomy, AL Ujazdowskie 4, 00-478 Warsaw, Poland. TEL 48-22-295346. FAX 48-22-294967. 1925 to present. Quarterly. $80.00 per year. ISSN: 0001-5237.

Astronomical Journal. American Institute of Physics, One Physics Ellipse, College Park, MD 20740-3843. (301) 209- 3000. Published for the American Astronomical Society. 1849 to present. Monthly. $280.00 per year. ISSN: 0004- 6256.

Astronomical Society of the Pacific. PUBLICATIONS. Astronomical Society of the Pacific, 390 Ashton Avenue, San Francisco, CA 94112. (415) 337-1100. FAX (415) 337-5205. 1889 to present. Monthly. $175.00 per year. ISSN: 0004- 6280.

Astronomy. Kalmbach Publishing Company, Box 1612, Waukesha, WI 53187-1612. (414) 796-8776. FAX (414) 796-1142. 1973 to present. Monthly. $27.00 per year. ISSN: 0091-6358.

Astronomy and Astrophsyics. Springer-Verlag New York, Inc., 44 Hartz Way, Secaucus, NJ 07096-2491. (201) 348-4033. FAX (201) 348-4505. 1969 to date. 36 issues per year in 12 volumes. $1733.00 per year. ISSN: 0004-6361.

Astronomy Reports. American Institute of Physics, One Physics Ellipse, College Park, MD 20740-3843. (301) 209- 3000. English translation of Astronomicheksii Zhurnal. Formerly Soviet Astronomy AJ. 1957 to present. Bimonthly. $1160.00 per year. ISSN: 1063-7729.

Astroparticle Physics. Elsevier Science Publishing Company, Inc., Box 882, Madison Square Station, New York, NY 10159. (212) 989-5800. FAX (212) 633-3990. 1992 to present. Quarterly. $180.00 per year. ISSN: 0927-6505.

Astrophysical Journal; An International Review of Astronomy and Astronomical Physics. University of Chicago Pre ss, Journals Division, 5720 South Woodlawn Avenue, Chicago, IL 60637. (312) 753-3347. FAX (312) 753-0811. 1895 to date. three issues per month. $740.00 per year. ISSN: 0004-637X.

Astrophpy Sics and Space Science; An International Journal of Cosmic Physics. Kluwer Academic Publishers, Box 358, Accord Station, Hingham, MA 02018-0358. (617) 871-6600. FAX (617) 871-6528. 1968 to present. 24 issues per year. $2544.00 per year. ISSN: 0004-640X.

Celestial Mechanics and Dynamical Astronomy; An International Journal of Space Dynamics. Kluwer Academic Publishers, Box 358, Accord Station, Hingham, MA 02018-0358. (617) 871-6600. FAX (617) 871-6528. 1969 to present. Monthly. $703.50 per year. ISSN: 0923-2958.

ICARUS; international journal of solar system studies. Academic Press, Inc., Journal Division, 525 B Street, Suite 1900, San Diego, CA 92101-4495. (619) 230-1840. FAX (619) 699-6800. 1962 to present. Monthly. $1080.00 per year. ISSN: 0019-1035.

Mercury. Astronomical Society of the Pacific. 390 Ashton Avenue, San Francisco, CA 94112. (415) 337-1100. FAX: (415) 337-5205. 1972 to present. Bimonthly. $175.00 per year. ISSN: 0047-6773.

Monthly Notices of the Royal Astronomical Society. Blackwell Scientific Publication LT., Osney Mead, Oxford OX2 OEL, England. TEL 0865-240201. FAX 0865-721205. 1827 to present. Fortnightly. $1899.00 per year. ISSN: 0035-8711.

Observatory. c/o Dr. D. J. Stickland, Space and Astrophysics Division, Rutherford Appleton Laboratory, Chilton, Didcot, Oxon OX11 OQX England. FAX 0235-445848. 1877 to present. Bimonthly. $42.00 per year. ISSN: 0029-7704.

Sky and Telescope. Sky Publishing Corporation, Box 9111, Belmont, MA 02178. (617) 864-7360. FAX (617) 864-6117. 1941 to present. Monthly. $27.00 per year. ISSN: 0037- 6604.

Space Science Reviews. Kluwer Academic Publishers, Box 358, Accord Station, Hingham, MA 02018- 0358. (617) 871- 6600. FAX (617) 871-6528. 1962 to present. Sixteen issues per year. $940.00 per year. ISSN: 0038-6308.

Vistas In Astronomy; An International Review Journal. Albert C. Beer, editor. Elsevier Science Publishing Company, Inc., 660 White Plains Road, Terrytown, NY 10591- 5153. (914) 524-9200. FAX: (914) 333-2444. 1958 to present. Quarterly. $415.00 per year. ISSN:: 0083-6656.

Transactio Ns of the International Astronomical Union. Proceedings of the General Assembly. Kluwer Academic Publishers, Box 358, Accord Station, Hingham, MA 02018- 0358. (617) 871-6600. Issues in two parts: Part A, Reports, and Part B, Proceedings. Triennial. XXI - 1994. Price varies. ISSN: 0080-1372.

Webb Society Quarterly Journal. Webb Society, 194 Foundry Lane, Freemantle, Southampton, Hampshire SO1 3EQ, England. 1969 to present. Quarterly. $19.50 per year. ISSN: 0043- 1680.

RESEARCH CENTERS AND INSTITUTES

Astronomy Program. University of Maryland, 1207 Computer & Space Sciences Building, College Park, Md 20742. (301) 405-3001. FAX (301) 314-9067.

Center For Astronomy and Space Astrophysics. University of Colorado - Boulder, Campus Box 391, Boulder, CO 80309. (303) 492-4050. (FAX) 492-7178.

Center For Space Science and Astrophysics. Stanford University, 325 Durand Building, Stanford, CA 94305. (415) 723-3592.

Joint Institute For Laboratory Astrophysics. University of Colorado - Boulder, Boulder CO 80309-0440. (303) 492-7789. FAX (303) 492-5235.

National Astronomy and Ionosphere Center. Cornell University, Space Sciences Building, Ithaca, NY 14853. (607) 255-3735. (607) 255-8803.

Smithsonian Center For Astrophysics. Harvard University, 60 Garden Street, Cambridge, MA 02138. (617) 495-7461. FAX (617)495-7326.

Space Sciences Laboratory. University of California, Berkeley, Berkeley, CA 94720. (415) 642-1361. FAX (415) 643-7629.

Theoretical Astrophysics Program. University of Arizona, Planetary Sciences Department, Tuscon, AZ 85721. (602) 621- 6963.

COULOMETRIC ANALYSIS

See: ANALYTICAL CHEMISTRY

COVALENT BOND

See: CHEMICAL BONDING

CRACKING

See: CHEMICAL ENGINEERING

CRATERS

See: MOON

CRITICAL MASS

See: NUCLEAR ENERGY

CRUISE MISSILE

See: GUIDED MISSILES

CRUST

See also: GEOPHYSICS, IGNEOUS ROCKS, METAMORPHIC ROCKS, MINERALOGY, PETROLOGY, PHYSICAL GEOLOGY, PLATE TECTONICS, SEDIMENTARY ROCKS, SEISMOLOGY, VOLCANOLOGY

ABSTRACT SERVICES AND INDEXES

Bibliography and Index of Geology. American Geological Institute, 4220 King Street, Alexandria, VA 22302. (703) 379-2480. Fax (703) 379-7563. Monthly. $1295.00 per year. Also available online as GeoRef.

Bibliography of Economic Geology. Geosystems, Box 40, Didcot Oxon Ox11 9BX, England. Bimonthly. $150.00 per year. Also available online as GeoArchive.

Chemical Abstracts - Applied Chemistry and Chemical Engineering Sections. Chemical Abstracts Service, 2540 Olentangy River Road, Box 3012, Columbus, OH 43210. (800) 848-6538 or (614) 447-3600, FAX (614) 447-3713. Bi-weekly. $1410.00 per year. Also available online as CA (Chemical Abstracts).

Geological Abstracts. Elsevier - Geo Abstracts, Regency House, 34 Duke Street, Norwich NR3 3AP, England. Monthly. $760.00 per year. Also available online as GEOBASE.

Geological Society of America. Abstracts with programs. Geological Society of America. 3300 Penrose Place, P.O. Box 9140, Boulder, CO 80301-9140. (303) 447-2020. Abstracts and programs of the annual conference. Annual. $69.00.

Geophysics Abstracts. Pergamon Press, Inc., Maxwell House, Fairview Park, Elmsford, NY 10523. (914) 592-7700. Fax (914) 592-3625. Twelve times per year. $565.00 per year. Also available in microform.

Mineralogical Abstracts. Mineralogical Society and the Mineralogical Society of America, 41 Queen's Gate, London, SW7 5HR, England. Quarterly. $235.00 per year.

Oceanic Abstracts. Cambridge Scientific Abstracts, 7200 Wisconsin Avenue, Bethesda, MD 20814. (301) 961-6750. Fax (301) 961-6720. Bimonthly. $995.00 per year.

ANNUAL REVIEWS AND YEARBOOKS

Advances In Geophysics. Academic Press, Inc., 1250 Sixth Avenue, San Diego, CA 92101. (619) 231-0926. FAX (619) 699-6715. Irregular. Inquire about price and availability.

Annual Review and Earth and Planetary Sciences. Annual Reviews, Inc., 4139 El Camino Way, Palo Alto, CA 94306-0897. (415) 493-4400. Fax (415) 855-9815. Annual. $55.00 per year.

ASSOCIATIONS AND PROFESSIONAL SOCIETIES

American Geophysical Union, 2000 Florida Avenue, N.W., Washington, DC 20009. (202) 462-6900.

American Institute of Professional Geologists. 7828 Vance Drive, Suite 103, Arvada, CO 80003. (303) 431-0831.

CRUST

Ency. of Physical Sciences and Engineering Info. Sources

Association of Engineering Geologists. 323 Boston Post Road, Suite 2D, Sudbury, MA 01776. (508) 443-4639.

Geological Society of America. 3300 Penrose Place, P.O. Box 9140, Boulder, CO 80301-9140. (303) 447-2020.

Society of Economic Geologists. P.O. Box 571, Golden, CO 80402. (303) 277-1716.

Society of Exploration Geophysicists. P.O. Box 702740, Tulsa, OK 74170. (918) 493-3516.

BIBLIOGRAPHIES

Geologic Reference Sources: A Subject and Regional Bibliography of Publications and Maps In the Geological Sciences. Dederick C. Ward et al. Scarecrow Press, Inc., 52 Liberty Street, Box 4167, Metuchen, NJ 08840. (800) 537-7107 or (201) 548-8600. 1981. $49.50.

Information Sources In the Earth Sciences. David N. Wood, Joan E. Hardy, and Anthony P. Harvey, editors. Bowker-Saur/K.G. Saur. Distributed by R.R. Bowker, 121 Chanlon Road, New Providence, NJ 07974. (800) 521-8110 or (908) 464-6800. Second edition. 1989. $85.00.

Remote Sensing of Earth Resources: A Quarterly Bibliography. University of New Mexico, Technology Application Center, 2808 Central Avenue, S.E., Albuquerque, NM 87131-6031. (505) 277-3622. Fax (505) 277-3614. $150.00 per year.

DIRECTORIES AND BIOGRAPHICAL SOURCES

American Institute of Professional Geologists. Membership Directory. American Institute of Professional Geologists, 7828 Vance Drive, Suite 103, Arvada, CO 80003. (303) 431-0831. Annual. $18.00.

Association of Engineering Geologists. Membership directory. Association of Engineering Geologists, 323 Boston PostRoad, Suite 2D, Sudbury, MA 01776. (508) 443-4639. Annual. $18.00.

Geological Society of America. Membership Directory. Geological Society of America, 3300 Penrose Place, Boulder, CO 80301. (303) 447-2020. Annual. Available to members only.

Research Centers Directory. Gale Research, 835 Penobscot Building, Detroit, MI 48226-4094. (800) 347-4253 or (313) 961-2242. 17th edition, 1992. $400.00.

ENCYCLOPEDIAS AND DICTIONARIES

Dictionary of Geological Terms. American Geological Institute. Doubleday and Company, Inc., 245 Park Avenue, New York, NY 10017. (800) 645-6156 or (212) 953-4561. Third edition. 1984. $12.00 in paper.

Glossary of Geology. Robert L. Bates and Julia A. Jackson. American Geological Institute, 4220 King Street, Alexandria, VA 22032. (703) 379-2480. Third edition. 1987. $75.00.

GENERAL WORKS

The Deep Interior of the Earth. Jack A. Jacobs. Chapman & Hall, 115 Fifth Avenue, New York, NY 10211-0906. 1992. $27.50.

Dynamic Earth: An Introduction To Physical Geology. B.J. Skinner and Stephen C. Porter. John Wiley and Sons, Inc., 605 Third Avenue, New York, NY 10158. (800) 526-5368 or (212) 850-6000. 1990. $51.95.

Earth Materials and Earth Processes: An Introduction. Lynn Fichter, et al. Macmillan Publishing, 866 3rd Avenue, New York, NY 10024. (212) 689-9140. Third edition. 1991. $19.50.

Evolution of the Earth. Robert H. Dott, Jr. and Roger L. Batten. McGraw-Hill Publishing Company, 1221 Avenue of the Americas, New York, NY 10020. (800) 262-4729 or (212) 512-3825. Fourth edition. 1988. $41.00.

Exposed Cross-sections of the Continental Crust. Matthew H. Salisbury and David M. Fountain, editors. Kluwer Academic Publishers, P.O. Box 358, Hingham, MA 02018-0358. (617) 871-6000. 1990. $185.00.

Properties & Processes of Earth's Lower Crust. R.F. Mereu, et al., eds. American Geophysical Union, 2000 Florida Avenue NW, Washington, DC 20009. (800) 966-2481. FAX (202) 328-0566. 1989 $32.00.

Rock Cycle: Understanding the Earth's Crust. J.R. Blueford. Math/Science Nucleus, 4009 Pestana Place, Fremont, CA 94538-6301. (510) 490-6284. FAX (510) 490-0256. 1992. $22.95.

HANDBOOKS AND MANUALS

Field Mapping For Geology Students. F. Ahmed and D.C. Almond. Unwin Hyman, Inc., 10 East 53rd Street, New York, NY 10022. (212) 207-7626 or (800) 242-7737. 1983. $18.95.

Geology In the Field. R.R. Compton. John Wiley and Sons, Inc., 605 Third Avenue, New York, NY 10158. (800) 526-5368 or (212) 850-6418. 1985. $37.95.

The Mapping of Geological Structures. K.R. McClay. John Wiley and Sons, Inc., 605 Third Avenue, New York, NY 10158. (800) 526-5368 or (212) 850-6000. 1988. $18.95.

ONLINE DATABASES AND CD-ROM

Earth Sciences. U.S. Geological Survey, 12201 Sunrise Valley Drive, Reston, VA 22092-9998. (703) 648-4460. CD-ROM of earth science databases including the U.S. Geological Survey Library, Earth Science Data Directory, and GEOINDEX, citations to published geological maps. $350.00 per year with quarterly updates.

Geoarchive. Geosystems, P.O. Box 1024, Westminster, London, England, SW1 P 2JL. Citations to literature on geoscience, 1969 to present. Inquire as to online cost and availability.

Geobase. Elsevier - Geo Abstracts, Regency House, 34 Duke Street, Norwich NR3 3AP, England. Contains citations to the worldwide earth science literature from 1980 to date. Available on DIALOG, ORBIT online services. Inquire as to cost and availability.

Ency. of Physical Sciences and Engineering Info. Sources

CRYOGENICS

Geomechanics Abstracts. University of London, Imperial College of Science and Technology, Rock Mechanics Information Service, Royal School of Mines, Prince Consort Road, London SW7 2BP, England. Telephone: 071-589 5111 or FAX 071-589 6806. Scope includes worldwide literature on engineering geology, hydrogeology, mining, rock mechanics, soil mechanics, and tunnelling, 1977 to present. Available on the ORBIT online service. Inquire as to cost and availability.

GEOREF. American Geological Institute, 4220 King Street, Alexandria, VA 22302. (800) 336-4764 or (703) 379-2480. Geology and geosciences literature, 1785 to present for North America. Available on DIALOG, ORBIT online services. Also available on CD-ROM. Inquire as to cost and availability.

PERIODICALS

American Journal of Science. Kline Geology Laboratory, Yale University, Box 6666, Yale Station, New Haven, CT 06511-8130. (203) 432-3131. Ten times per year. $40.00 per year.

Canadian Journal of Earth Sciences. National Research Council of Canada, Research Journals, Ottawa, Ont. K1A OR6, Canada. (613) 993-9084. Monthly. $112.00 per year.

Earth and Planetary Science Letters. Elsevier Science Publishing Company, Inc., 655 Avenue of the Americas, New York, NY 10010. (212) 989-5800. Twenty times per year. $945.00 per year.

Economic Geology and the Bulletin of the Society of Economic Geologists. Society of Economic Geologists, Economic Geology Publishing Company, 101 Vowell Hall, University of Texas at El Paso, El Paso, TX 79968. 8 times per year. $51.00.

Geochimica Et Cosmochimica Acta. Pergamon Journals, Inc., Maxwell House, Fairview Park, Elmsford, NY 10523. (914) 592-7700. Fax (914) 592-3625. Monthly. $580.00 per year.

Geological Society. JOURNAL. Geological Society of London. Geological Society Publishing House, Unit 7, Brassmill Enterprise Centre, Brassmill Lane, Bath, Avon BA1 3JN, England. 6 times per year. $432.00.

Geological Society of America Bulletin. P.O. Box 9140, 3300 Penrose Place, Boulder, CO 80301. (303) 447-2020. Fax (303) 447-1133. Monthly. $150.00.

Geology. Geological Society of America. P.O. Box 9140, 3300 Penrose Place, Boulder, CO 80301. (303) 447-2020. Fax (303) 447-1133. Monthly. $120.00 per year.

Geophysics. Society of Exploration Geophysicists, P.O. Box 702740, Tulsa, OK 74170. (918) 493-3516. Monthly. $175.00 per year.

Journal of Geology. University of Chicago Press, 5720 Woodlawn Avenue, Chicago, IL 60637. (312) 753-3347.

Bimonthly. $38.00 per year.

Journal of Geophysical Research. American Geophysical Union, 2000 Florida Avenue, N.W., Washington, DC 20009. (202) 462-6903. Monthly. $2170.00 per year.

Journal of Structural Geology. Pergamon Journals, Inc., Maxwell House, Fairview Park, Elmsford, NY 10523. (914) 592-7700. Ten times per year. $375.00 per year.

Reviews of Geophysics. American Geophysical Union, 2000 Florida Avenue, N.W., Washington, DC 20009. (202) 462-6900. Quarterly. $300.00 per year.

Tectonics. American Geophysical Union, 2000 Florida Avenue, N.W., Washington, DC 20009. (202) 462-6900. Bimonthly. $230.00 per year to individuals.

Tectonophysics; An International Journal of Geotectonics and the Geology and Physics of the Interior of the Earth. Elsevier Science Publishing Company, Inc., 655 Avenue of the Americas, New York, NY 10010. (212) 989-5800. Fifty-six times per year. $1275.00 per year.

RESEARCH CENTERS AND INSTITUTES

Carnegie Institution of Washington. Geophysical Laboratory. 5251 Broad Branch Road, NW, Washington, DC 20015. (202) 686-2410 or FAX (202) 686-2419.

Cornell University. Institute for the Study of Continents. 3122 Snee Hall, Ithaca, NY 14853-1504. (607) 255-3474.

Goddard Space Flight Center. Crustal Dynamics Project. National Aeronautics and Space Administration, Greenbelt, MD 20771. (301) 286-8809. Fax (301) 286-4943.

Smithsonian Institution. Center for Earth and Planetary Studies. National Air and Space Museum, Sixth Street and Independence Avenue, S.W., Washington, DC 20560. (202) 356-1424. Fax (202) 786-2262.

CRYOGENIC ENGINEERING

See: CRYOGENICS

CRYOGENICS

See also: CHEMICAL ENGINEERING, FLUID MECHNICS, HEAT TRANSFER, PHYSICAL CHEMISTRY, PHYSICS, THERMODYNAMICS

ABSTRACT SERVICES AND INDEXES

Applied Mechanics Reviews: An Assessment of World Literature In Engineering Sciences. American Society of Mechanical Engineers, 345 East 47th Street, New York, NY 10017. (212) 705-7703. 1948 to present. Monthly. $360.00 ISSN: 0003-6900.

Applied Sci Ence and Technology Index; A CumulativeSubject Ind Ex To English Language Periodicals In the Fields of Aeronautics and Space Science, Computer Technology, Chemi Stry, Construction Industry, Energy and Related Areas. H. W. Wil-

CRYOGENICS

Ency. of Physical Sciences and Engineering Info. Sources

son Co., 950 University Avenue, Bronx, NY 10452-9978. (212) 588-8400. (800) 367-6770. FAX (718) 590-1617. From 1958 to present. Monthly. Inquire about cost and availability. Also available online (BRS and WILSONLINE) and on CD-ROM. ISSN: 0003-6986.

Chem Ical Abs Tracts. Chemical Abstracts Service. 2540 Oleantangy River Road, Box 3012, Columbus, OS 43210-0012. (614) 447-3600. FAX (614) 447-3713. 1907 to present. Weekly. $16,800.00 per year. Available online and on CD- ROM. CA is also available in five section groupings. Inquire regarding cost and availability.

Current Contents: Physical, Chemical and Earth Sciences. Institute for Scientific Information, 3501 Market Street, Philadelphia, PA 19104. (215) 386-0100. (800) 523-1850. FAX (215) 386-2291. 1961 to present. Weekly. $442.00 per year. Also available online (BRS, DIALOG) and on CD-ROM. Inquire regarding cost and availability. ISSN: 0163-2574.

Engineering Index Monthly. Engineering Information, Inc., Castle Point on the Hudson, Hoboken, NJ 07030. (201) 216-8500. (800) 221-1044. FAX (201) 216-8532. Monthly. $2300.00 per year for monthly issues. $1980 per year for annual cumulation. Available on CD-ROM and online as COMPENDEX. ISSN: 0742-1974; 0360-8557.

General Science Index. H.W. Wilson and Company, 950 University Av enue, Bronx, NY 10452. (212) 588-8400. (800) 367-6770. FAX (718) 590-1617. From 1978 to present. Ten issues per year; quarterly and annual cumulations. Service basis. Available on CD-ROM and online. Inquire regarding cost and availability. ISSN: 0162-1963.

Ismec: Mechanical E Ngineering Abstracts. Cambridge Scientific Abstracts, 72 00 Wisconsin Avenue, Suite 601, Bethesda, MD 20814. (301) 961-6750 or (800) 843-7751. Contains citations to the literature in mechanical engineering, industrial and production engineering, energy, power, mechanics, devices and related areas, from 1973 to present. Available on the DIALOG,(800) 334-2564, online servic e. Inquire as to cost and availability.

Physics Abstracts. INSPEC. Section A, Science Abstracts. I nstitution of Electrical Engineers (IEE), London. Available from: INSPEC/IEEE - Institute of Electrical and Electronic Engineers, Box 1331, Hoes Lane, Piscataway, NJ 08855-1331. (908) 562-5549. 1898 to present. 24 issues per year. $2835.00 per year. Also available online and on CD-ROM. ISSN: 0036-8091.

Process and Chemical Engineering/chemical Engineerin G Abstracts. Royal Society of Chemistry, Informati on Services, Thomas Graham House, Science Park, Milton Road, Cambridge, CB4 4WF, England. Contains citations, mostly with abstracts, of the worldwide literature on chemical engineering; from 1982 to present. Monthly. $610.00 per year. Also available online. ISSN: 0960-5045.

Science Citation Index. SCI. Institute for Scientific Information, 3501 Market Street, Philadelphia, PA 19104. (215) 386-0100. (800) 523-1850. FAX (215) 386-2991. 1961 to present. Six issues per year, plus annual cumulation. $11650.00 per year. Also available online and on CD-ROM. Inquire about price and availability. ISSN: 0036-827X.

Theoretical Chemical Engineering. Royal Society of Chemistry, Information Services, Thomas Graham House, Science Park, Milton Road, Cambridge, CB4 4WF, England. Covers theoretical chemical engineering, including theory and laboratory experimentation. 1964 to present. Monthly. $235.00. ISSN: 0960-5053.

ANNUAL REVIEWS AND YEARBOOKS

Progress In Low Temperature Physics. Elsevier Science Inc., 655 Avenue of the Americas, New York, NY 10010. (212) 989-5800. Subscriptions to: Box 882, Madison Square Station, NY 10159-0882. 1955 to present. Irregular; price varies. ISSN: 0079-6417.

ASSOCIATIONS AND PROFESSIONAL SOCIETIES

American Chemical Society. 1155 16th Street, N.W., Washington, DC 20036. (202) 872-4600.

American Institute of Chemical Engineers. 345 East 47th Street, New York, NY 10017. (212) 705-7657. FAX (212) 752-3294.

Association of Consulting Chemists and Chemical Engineers, 50 East 41st Street, Suite 92, New York, Ny 10017. (212) 684-6255.

American Society of Mechanical Engineers. 345 East 47th Street, New York, NY 10017. (212) 705-7722. FAX (212) 705- 7739.

Cryogenic Engineering Conference. c/o Dr. P. Kittel, MS 244-10, NASA/AMES, Moffett Field, CA 94035-1000. (415) 604-4297. FAX (415) 604-0673.

Cryogenic Society of America. c/o Huget Advertising, 1033 South Boulevard, Oak Park, IL 60302. (708) 383-6220. FAX (708) 383-9337.

International Institute of Ammonia Refrigeration. 111 East Wacker Drive, Chicago, IL 60601. (312) 644-6610.

DIRECTORIES AND BIOGRAPHICAL SOURCES

American Institute of Chemical Engineers. Directory. American Institute of Chemical Engineers. 345 East 47th Street, New York, NY 10017-2396. (212) 705-7663. Annual.

American Men and Women of Science. Physical and Biological Sciences. R.R. Bowker Inc., 121 Chanlon Road, New Providence, NJ 07974. (908) 464-6800. (800) 521-8110. 20th edition. 8 volumes. 1996. $850.00

Directory of Low Temperature Research and Development In Europe. P. C. McDonald, editor. IOP Publishing, Public Ledger Building, Suite 1035, Independence Square, Philadelphia, PA 19106. (215) 627-0880. (800) 358-4677. 7th edition. 1992. $102.00.

Engineering Research Centers: Incorporating Electronics Research Centers. Stockton Press, 345 Park Avenue, New York, NY 10010. (212) 689-9200. (800) 221-2123. 4th edition. 1995. $515.00.

International Directory of Engineering Societies and Related Organizations. American Society for Engineering Education, 1818 N Street NW, Suite 600, Washington, DC 20036. (202) 331-3526. 15th edition. 1996. $185.00.

International Engineering Directory. American Consulting Engineers Council, 1015 15th Street, N.W. Suite 802, Washington, DC 20005-2670. (202) 347-7474. Annual. $10.00

Research Centers Directory. Gale Research Company, Inc., 835 Penobscot Building, Detroit, MI 48226-4094. (313) 961-2242. (800) 347-4253. 20th edition. 1995. $485.00. ISSN:: 0080-1518.

Scientific and Technical Organizations and Agencies Directory. Gale Research, 835 Penobscot Building, Detroit, MI 48226-4094. (313) 961-2242. (800) 347-4253. 4th edition. 1996. $195.00.

Who's Who In Engineering. Gordon Davis, editor. American Association of Engineering Societies. 1111 19th Street, NY, Suite 608, Washington, DC 20036. (202) 296-2237. (800) 658-8897. 9th edition. 1995. $220.00.

Who's Who In Technology. Gale Research, 835 Penobscot Building, Detroit, MI 48226-4094. (313 961-2242. (800) 521-4253. 7th edition. 1995. $195.00. ISN 0-8103-7467-6.

ENCYCLOPEDIAS AND DICTIONARIES

Chambers Science and Technology Dictionary. Peter M.B. Walker, editor. Cambridge University Press, 40 West 20th Street, New York, NY 10011-4211. (212) 924-3900. 1988. $39.95.

A Concise Dictionary of Physics. Oxford University Press, Inc., 200 Madison Avenue, New York, NY 10016. (212) 725- 6000. (800) 334-4249. 1990. $10.95.

Concise Encyclopedia of Magentic and Superconducting Materials. J. E. Evetts, editor. Elsevier Science Publishing Company, Inc., 655 Avenue of the Americas, New York, NY 10010. (212) 989-5800. FAX (914) 333-2444. 1992. $290.00.

Dictionary of Named Processes In Chemical Technology. Alan E. Comyns. Oxford University Press, Inc., 200 Madison Avenue, New York, NY 10016. (212) 725-6000. (800) 334-4249. 1994. $75.00.

Encyclopedia of Applied Physics. VCH Publishers, Inc., 303 Northwest 12th Avenue, Deerfield Beach, FL 33442. (800) 367-8249. 1991 - . Twenty volumes. $6000.00.

Encyclopedia of Physical Science and Technology. Academic Press, Inc., 6277 Sea Harbor Drive, Orlando, FL. (800) 321-5068. 2nd edition. 18 volume set. 1992. $2625.00.

Kirk-othmer Encyclopedia of Chemical Technology. John Wiley and Sons, Inc., 605 Third Avenue, New York, NY 10158. (800) 526-5368 or (212) 850-6000. Fourth edition. 1991 - . Twenty-seven volumes. $5400.00.

Ullman's Encyclopedia of Industrial Chemistry. VCHPublications, Inc., 220 East 23rd Street, Suite 909, New York, NY 10010-4606. (212) 683-8333. (800) 422-8824. 5thedition. 1984 - . Price varies per volume.

GENERAL WORKS

Applications of Cryogenic Technology. J. P. Kelley, editor. Plenum Publishing Corp., 233 Spring Street, New York, NY. (212) 620-8000. (800) 2221-9369. FAX (212) 463-0742. 1991. $89.50.

Applied thermodyanmics For Engineering Technologies. Thomas D. Eastop and A. McConkey. Halsted Press, 605 Third Avenue, New York, 10158-0012. (212) 850-6400. 5th edition. 1993. $57.95.

Cryogenic Process Engineering. Klaus D. Timmerhaus. Plenum Publishing Corp., 233 Spring Street, New York, NY. (212) 620-8000. (800) 2221-9369. FAX (212) 463-0742. 1989.

Cryogenic Processes and Machinery. Leonard A. Wenzel, editor. American Institute of Chemical Engineers, 345 East 47th Street, New York, NY 10017. (212) 705-7657. FAX (212) 752-3294. 1993.

Experimental Low Temperature Physics. Tony Kent . American Institute of PHysics, One Physics Ellipse, College Park, MD 20740-3843. (301) 209-3100. 1991. $55.00.

Matter and Methods At Low Temperatures. F. Pobll. Springer-Verlag New York, Inc., 175 Fifth Avenue, New York, NY 10010. (212) 460-1500. (800) 777-4643. FAX (212) 473- 6272. 1992. $79.00.

Principles of Modern thermodynamics. B. M. Roy. IOP Publishing, Public Ledger Building, Suite 1035, Independence Square, Philadelphia, PA 19106. (215) 627-0880. (800) 358-4677. 1995. $100.00.

Superconducting Devices. Steven T. Ruggiero and David A. Rudman. Academic Press, Inc., 6277 Sea Harbor Drive, Orlando, FL. (800) 321-5068. 1990. $73.00.

Superconductors: Conquering Technology's New Frontier. R. Simon and andrew Smith. Plenum Publishing Corp., 233 Spring Street, New York, NY. (212) 620- 8000. (800) 2221-9369. FAX (212) 463-0742. 1988. $23.95.

Superconductivity and Cryoelectronics. W. Krech, et al, editors. World Scientific Publishing Company, Inc., 1060 Main Street, River Edge, NJ 07661. (201) 487-9655. (800) 227-7562. 1991. $61.00.

Superconductivity: the New Revolution? Gianfranco Vidali. Cambridge University Press, 40 West 20th Street, New York, NY 10011-4211. (212) 924-3900. (800) 872-7423. 1993. $17.95 in paper.

Temperatures Very Low and Very High. Mark W. Zemansky. Dover Publications, Inc., 180 Varick Street, New York, NY 10014. (212) 255-3755. (800) 223-3130. Reprint edition. 1981. $4.95.

Thermodynamics and the Design, Analysis and Improvement of Energy Systems. H. J. Richter, editor. ASME - American Society of Mechanical Engineers, 345 East 47th Street, New York, NY 10017. (212) 705-7722. (800) 843-2763. 1993. $80.00.

HANDBOOKS AND MANUALS

Chemical Engineering Reference Manual. Randall N. Robinson. Professional Publications, Inc., 1250 Fifth Avenue, Belmont, CA 94002. (415) 593-0110. (800) 426-1178. 1988. $49.95.

Gas Tables: thermodynamics Properties of Air, Products of Combustion, and Component Gases. Joseph H. Keenan, et al.

CRYOGENICS

Ency. of Physical Sciences and Engineering Info. Sources

Krieger Publishing Company, P.O. Box 9542, Melbourne, FL 32902-9542. (407) 724-9542. 2nd edition; reprinted. 1992. 64.50.

Handbook of Applied thermodynamics. David A. Palmer, editor. CRC Press, Inc., 2000 Corporate Boulevard, NW, Boca Raton, FL 33431. (407) 994-0555. (800) 272-7737. 1987. $216.95.

Perry's Chemical Engineers' Handbook. Robert H. Perry and Donald W. Green, editors. McGraw-Hill Publishing Company, Inc., 1221 Avenue of the Americas, New York, NY 10020. (212) 512-2000. (800) 262-4729. 6th edition. 1996. $129.50.

ONLINE DATA BASES AND CD-ROMS

CA SEARCH. Chemical Abstracts Service, P.O. Box 3012, Columbus, OH 43210-0012. (614) 447-3600. (800) 848-6533. FAX (614) 447-3709. Very comprehensive guide to worldwide chemical literature and related fields, 1972 to present. Available on BRS,(800) 289-4277, DIALOG, (800) 334-2564, ORBIT (800) 456-7248, and STN International, FIZ Karlsruhe, P.O. Box 2465, W-7500, Karlsruhe 1, Germany, online services. Inquire as to cost and availability.

Compendex Plus. Engineering Information, Inc., 345 East 47th Street, New York, NY 10017. (212) 705-7600 or (800) 221-1044. Contains citations with abstracts to worldwide literature in engineering and technology, from 1970 to present. Available on online BRS,(800) 289-4277, DIALOG, (800) 334-2564, ORBIT (800) 456-7248, and STN International, FIZ Karlsruhe, P.O. Box 2465, W-7500, Karlsruhe 1, Germany, online services. Also available on CD-ROM. Inquire as to cost and availability.

Dissertation Abstracts. University Microfilms International, 300 North Zeeb Road, Ann Arbor, MI 48106. (800) 521-0600 or (313) 761-4700. Scope includes virtually all doctoral dissertations accepted at accredited American institutions from 1861 to present in 252 subject areas. Available on BRS, (800) 289-4277, DIALOG, (800) 334-2564, and OCLC EPIC, (800) 848-5878, online services. Also available on CD-ROM. Inquire as to cost and availability.

Inspec. Instit ution of Electrical Engineers, Michael Faraday House , Six Hills Way, Stevenage, Herts. SG1 2AY, England. Telephone: 0438 313311 or FAX 0438 742840. Con tains citations to the worldwide literature of physics, electronics and electrical engineering, computer technology, and related fields. Available on BRS, (800) 289-4277, DIALOG, (800) 334-2564, ORBIT, (800) 456-7248, and STN International, FIZ Karlsruhe, P.O. Box 2465, W-7500, Karlsruhe 1, Germany, online services. Inquire as to cost and availability.

NTIS. Natio nal Technical Information Service, 5285 Port Royal Road, Sp ringfield, Va 22161. (703) 487-4929 or FAX (703) 321-8199. Broad coverage of government-sponsored research reports, 1964 to present. Available on BRS, DIALOG, ORBIT and STN online services. Also available on CD-ROM. Inquire as to cost and availability.

Scisearch. Institute for Scientific Information, 3501 Market Street, Philadelphia, PA 19104. (800) 523-1850 or (215) 386-0100. Broad multidisciplinary title and author index to the international literature of science and technology, 1974 to present. Available on BRS, (800) 289-4277, DIALOG, (800) 334-2564, ORBIT, (800) 456-7248, and STN International, FIZ Karlsruhe, P.O. Box 2465, W-7500, Karlsruhe 1, Germany online services. Also available on CD-ROM. Inquire as to cost and availability.

WILSONLINE. H.W. Wilson Company, 950 University Avenue, Bronx, NY 10452. (800) 367-6770 or (212) 588-8400. Makes available online versions of the printed H.W. Wilson indexes including Applied Science and Technology Index, Business Periodicals Index, and Readers' Guide to Periodical Literature. Period covered is generally 1983 to present. Also available on CD-ROM. Inquire as to cost and availability.

PERIODICALS

Cryogas International. J. R. Campbell & Associates, Inc., 5 Militia Drive, Lexington, MA 02173. (617) 863-9411. FAX (617)863-9411. 1963 to present. Monthly. $150.00. ISSN: 1052-0139.

Cryogenics: the International Journal of Low Temperature Engineering and Resarch. Butterworth-Heinemann; Turpin Transactions Ltd., Distribution Center, Blackhorse, Road, Letchworth, Herts SG6 1HN, England. Tel 0462-672555.

Experimental thermal and Fluid Science. Elsevier Science Inc., 655 Avenue of the Americas, New York, NY 10010. (212) 989-5800. Subscriptions to: Box 882, Madison Square Station, NY 10159-0882. 1988 to present. 8 issues per year. $484.00 per year. ISSN: 0894-1777.

Heat Transfer Engineering. Taylor and Francis, 1900 Frost Road, Suite 101, Bristol PA 190077-1598. (800) 821-8312. 1979 to present. Quarterly. $176 per year. ISSN: 0145-7632.

International Communications in Heat and Mass Transfer. Elsevier Science, 660 White Plains Road, Tarrytown NY 10591-5151. (914) 524-9200. FAX (914) 333-2444. 1974 to present. Bi-monthly. $430.00 per year. Combined subscription with International Journal of Heat and Mass Transfer; $2070.00. ISSN: 0735-1933.

International Journal of Heat and Fluid Flow. Butterworth-Heinemann, 313 Washington Street, Newton MA 02158. (617) 928- 2500. (800) 366-2665. 1979 to present. Bi-monthly. $450.00 per year. ISSN: 0142-727X.

Journal of Heat Transfer. American Society of Mechanical Engineers, 345 East 47th Street, New York, NY 10017. (212) 705-7722. volume 92, 1970 to present. Quarterly. $155.00 per year. ISSN: 0022-1481.

Journal of Low Temperature Physics. Plenum Publishing Corp. 233 Spring Street, New York, NY 10013-1578. (212) 620-8000. FAX (212) 463-0742. 1969 to present. 24 issues per year in 3 volumes. $955.00 per year. ISSN: 0022-2291.

Journal of Superconductivity. Plenum Publishing Corp. 233 Spring Street, New York, NY 10013-1578. (212) 620-8000. FAX (212)463-0742. 1988 to present. Bi-monthly. $275.00 per year. ISSN: 0896-1107.

Journal of Thermophysics and Heat Transfer. American Institute of Aeronautics and Astronautics, 370 L'Enfant Prominade SW, Washington, DC 20024. (202) 646-7400. 1987 to present. Quarterly. $165.00 per year. ISSN: 0887-8722.

Numerical Heat Transfer: An International Journal of Computation and Methodology. Taylor and Francis, 1900 Frost Road, Suite 101, Bristol, PA 19007-1598. (800) 821-8312. FAX (215) 785-5515. 1978 to present. 8 issues per year. Part A: Applications, $530.00 per year; ISSN: 1040-7782. Part B: Fundamentals, $330.00 per year; ISSN: 1040-7790.

RESEARCH CENTERS AND INSTITUTES

Applied Superconductivity Research Center. University of Wisconsin - Madison, Engineering Research Building, 1500 Johnson Drive, Madison, WI 53706. (608) 263-5026. (608) 263- 1087.

Center For Thermodynamics. Brigham Young University, 226 ESC, Provo, UT 84602. (801) 378-3668.

Heat Transfer Laboratory. Massachusetts Institute of Technology, 77 Massachusetts Avenue, Cambridge, MA 02139. (716) 253-2248.

Thermodynamics Research Center. Texas A&M University, Texas Engineering Experiment Station, College Station, TX 77843- 3111. (409) 845-4940.

CRYPTOGRAPHY AND CRYPTOLOGY

See: COMPUTER SECURITY

CRYSTAL OPTICS

See: CRYSTALLOGRAPHY

CRYSTALLOGRAPHY

See also: CHEMISTRY, GEOCHEMISTRY, GEOLOGY, LIQUID CRYSTALS, MINERALOGY, SOLID STATE CHEMISTRY, SOLID STATE PHYSICS

ABSTRACT SERVICES AND INDEXES

Bibliography and Index of Geology. American Geological Institute, 4220 King Street, Alexandria, VA 22302. (703) 379-2480. Fax (703) 379-7563. Monthly. $1295.00 per year. Also available online as GeoRef.

Chemical Abstracts - Applied Chemistry and Chemical Engineering Sections. Chemical Abstracts Service, 2540 Olentangy River Road, Box 3012, Columbus, OH 43210. (800) 848-6538 or (614) 447-3600, FAX (614) 447-3713. Bi-weekly. $1410.00 per year.

Current Contents: Physical, Chemical and Earth Sciences. Institute for Scientific Information, 3501 Market Street, Philadelphia, PA 19104. (215) 386-0100. FAX (215) 386-6362. Weekly. $360.00 per year.

Current Papers In Physics. Institute of Electrical Engineers, Michael Faraday House, Six Hill Way, Stevenage, Herts, SG1 2AY, England. Distributed by IEEE, 445 Hoes Lane, Piscataway, NJ 08854. (908) 562-5549. 1966 to present. Twenty-four times per year. $390.00 per year. Also available online in INSPEC.

Current Physics Index. American Institute of Physics, 335East 45th Street, New York, NY 10017 (212) 661-9404 or FAX (516)349-9704. 1975 to present. Quarterly. $625.00 per year.

Engineering Index Monthly. Engineering Information, Inc., Castle Point on the Hudson, Hoboken, NJ 07030. (800) 221-1044. FAX (212) 832-1857. Monthly. $2300.00 per year. Also available online as COMPENDEX, and also on CD-ROM. Covers chemical engineering, computers, electrical engineering, civil engineering, metals and mining, industrial management, and mechanical engineering.

Geological Abstracts. Elsevier - Geo Abstracts, Regency House, 34 Duke Street, Norwich NR3 3AP, England. Monthly. $760.00 per year. Also available online as GEOBASE.

Geological Society of America. Abstracts with programs. Geological Society of America. 3300 Penrose Place, P.O. Box 9140, Boulder, CO 80301-9140. (303) 447-2020. Abstracts and programs of the annual conference. Annual. $69.00.

Index To Scientific Reviews. Institute for Scientific Information, 3501 Market St., Philadelphia, PA 19104. (215) 386-0100. FAX (215) 386-6362. Semi-annual. $960.00 per year.

Mineralogical Abstracts. Mineralogical Society and the Mineralogical Society of America, 41 Queen's Gate, London, SW7 5HR, England. Quarterly. $235.00 per year.

Physical Review Abstracts. American Physical Society, distributed by the American Institute of Physics, 335 East 45th Street, New York, NY 10017 (212) 661-9404 or FAX (516) 349-9704. 1970 to present. Semi-monthly. $290.00 per year.

Physics Abstracts. Institute of Electrical Engineers, Michael Faraday House, Six Hill Way, Stevenage, Herts, SG1 2AY, England. Distributed by IEEE, 445 Hoes Lane, Piscataway, NJ 08854. (908) 562-5549. 1898 to present. Monthly. $2700.00 per year. Also available online as INSPEC.

Physics Briefs. American Institute of Physics, 335 East 45th Street, New York, NY 10017 (212) 661-9404 or FAX (516)349-9704. 1980 to present. Six times per year. $2600.00 per year.

Science Citation Index. Institute for Scientific Information, 3501 Market Street, Philadelphia, PA 19104. (215) 386-0100. FAX (215) 386-6362. Inquire about availability and cost. Also available on CD-ROM.

Solid State and Superconductivity Abstracts. Cambridge Scientific Abstracts, 5161 River Road, Bethesda, MD 20816. (301) 951-1400. 1957 to present. Bi-monthly. $995.00 per year.

ASSOCIATIONS AND PROFESSIONAL SOCIETIES

American Association For Crystal Growth. PO Box 3233, Thousand Oaks, CA 91359-0233. (805) 492-7047.

American Crystallographic Association. PO Box 96, Ellicott Station, Buffalo, NY 14205-0096. (716) 856-9600.

American Federation of Mineralogical Societies. PO Box 26523, Oklahoma City, OK 73126-0523. (405) 324-7870.

American Geological Institute. 4220 King Street, Alexandria, VA 22302. (703) 379-2480. Fax (703) 379-7563.

CRYSTALLOGRAPHY

Ency. of Physical Sciences and Engineering Info. Sources

Geological Society of America. 3300 Penrose Place, P.O. Box 9140, Boulder, CO 80301-9140. (303) 447-2020.

Mineralogical Society of America. 1130 17th Street NW, Washington, DC 20036. (202) 775-4344. FAX (202) 775-0018.

DIRECTORIES AND BIOGRAPHICAL SOURCES

American Men & Women of Science, 1995-96. R.R. Bowker Staff, eds. 19th edition. 8 volumes. R.R. Bowker/Reed International Publishing Company, 121 Chanlon Road, New Providence, NJ 07974. (908) 464-6800 or (800) 521-8110. 1995. $850.00

Geological Society of America Membership Directory. 3300 Penrose Place, P.O. Box 9140, Boulder, CO 80301-9140. (303) 447-2020. Inquire for availability.

International Research Centers Directory. Gale Research, 835 Penobscot Building, Detroit, MI 48226-4094. (313) 961-2242. (800) 347-4253. FAX (313) 961-6083. 8th edition. 1995. $410.00

Research Centers Directory. Gale Research, 835 Penobscot Building, Detroit, MI 48226-4094. (313) 961-2242. (800) 347-4253. FAX (313) 961-6083. $485.00.

Scientific and Technical Organizations and Agencies Directory. Gale Research, 835 Penobscot Building, Detroit, MI 48226-4094. (313) 961-2242. (800) 347-4253. FAX (313) 961-6083. 4th edition. 1996. $195.00.

ENCYCLOPEDIAS AND DICTIONARIES

Dictionary of Geological Terms. American Geological Institute staff. Doubleday & Co. Inc., 1540 Broadway, New York, NY 10036-4094. (212) 354-6500 or (800) 223-6834. 1976. $6.95.

Encyclopedia of Physical Science and Technology. Robert A. Meyers, ed. 18 volumes. Academic Press Inc., 6277 Sea Harbor Drive, Orlando, FL 32887. (800) 321-5068. 1992. $2100.00

The Encyclopedia of the Solid Earth Sciences. Philip Kearey, editor-in-chief. Blackwell Scientific Publications, 238 Main Street, Cambridge, MA 02142. (617) 876-7000 or (800) 759-6102. FAX (617) 876-7022. 1993. $60.00.

McGraw-hill Dictionary of Scientific and Technical Terms. Sybil P. Parker, ed. 5th edition. McGraw-Hill Publishing Company, 1221 Avenue of the Americas, New York, NY 10020. (800) 262-4729 or (212) 512-3825. 1993. $110.50.

McGraw-hill Encyclopedia of Science and Technology. Sybil P. Parker, ed. 7th edition. 20 volumes. McGraw-Hill Publishing Company, 1221 Avenue of the Americas, New York, NY 10020. (800) 262-4729 or (212) 512-3825. 1992. $1900.00

GENERAL WORKS

Introduction To Crystallography. C. Hammond. Rev. ed. Oxford University Press, Inc., 198 Madison Avenue, New York, NY 10016-4314. (212) 726-6000. FAX (212) 726-6446. 1992. $19.95.

Introduction To Crystallography. Donald Sands. Dover Publications, Inc., 180 Varick Street, New York, NY 10014. (212) 255-3755 or (800) 223-3130. FAX (212) 626-9670. 1993. $6.95.

Introduction To Mineral Sciences. A. Putnis. Cambridge University Press, 40 West 20th Street, New York, NY 10011-4211.(212) 924-3900 or (800) 872-7423. FAX (914) 937-4712. 1992. $100.00 (hardback), $39.95 (paperback).

Modern Crystallography. B.K. Vainshtein, editor. Springer-Verlag, 175 Fifth Avenue, New York, NY 10010. (212) 460-1500. FAX (212) 473-6272. 1994. $79.00.

Structure of Crystals. B.K. Vainshtein. 2nd edition. Springer-Verlag, 175 Fifth Avenue, New York, NY 10010. (212) 460-1500. FAX (212) 473-6272. 1995. Inquire for cost and availability.

HANDBOOKS AND MANUALS

CRC Handbook of Chemistry and Physics. David R. Lide, editor. 75th edition. CRC Press, 2000 Corporate Blvd., N.W., Boca Raton, FL 33431. (407) 994-0555 or (800) 272-7737. FAX (407) 994-0949. $99.50.

Handbook of Crystallography. A.G. Jackson. Springer-Verlag, 175 Fifth Avenue, New York, NY 10010. (212) 460-1500. 1991. $49.50.

Manual of Mineralogy. Cornelius Klein. 21st edition. John Wiley and Sons, Inc., 605 Third Avenue, New York, NY 10158. (800) 526-5368 or (212) 850-6000. 1993. Inquire for cost and availability.

Mineral Physics and Crystallography: A Handbook of Physical Constants. American Geophysical Union, 2000 Florida Avenue, N.W., Washington, DC 20009. (202) 462-6900. 1995. Inquire for cost and availability. (Subhead2>Online Databases and CD-ROMS

Compendex Plus. Engineering Information, Inc., 345 East 47th Street, New York, NY 10017. (212) 705-7600 or (800) 221-1044. Contains citations with abstracts to worldwide literature in engineering and technology, from 1970 to present. Available on online BRS,(800) 289-4277, DIALOG, (800) 334-2564, ORBIT (800) 456-7248, and STN International, FIZ Karlsruhe, P.O. Box 2465, W-7500, Karlsruhe 1, Germany, online services. Also available on CD-ROM. Inquire as to cost and availability.

CSA Engineering. Cambridge Scientific Abstracts, 7200 Wisconsin Avenue, Suite 601, Bethesda, MD 20814. (301) 961-6750 or (800) 843-7751. Contains citations and abstracts of international periodicals and research literature covering all fields of engineering and science and technology,including computer and information science, electronics, mechanical engineering, solid state materials, 1981 to present. Available on BRS,(800) 289-4277, online service.Inquire as to cost and availability.

Current Contents Search. Institute for Scientific Information, 3501 Market Street, Philadelphia, PA 19104. (215) 386-0100. FAX (215) 386-6362. Contains citations to articles listed in the table of contents of science and technology journals. Also articles in social sciences and life sciences journals. Available on BRS,(800) 289-4277, DIALOG,(800) 334-2564, online services. Inquire as to cost and availability.

Dissertation Abstracts. University Microfilms International, 300 North Zeeb Road, Ann Arbor, MI 48106. (800)521-0600 or (313) 761-4700. Scope includes virtually all doctoral dissertations accepted at accredited American institutions from 1861 to present in 252 subject areas. Available on BRS,(800) 289-4277, DIA-LOG,(800) 334-2564, and OCLC EPIC,(800) 848-5878, online services. Also available on CD-ROM. Inquire as to cost and availability.

GEOREF: Bibliography and Index of Geology. American Geological Institute, 4220 King Street, Alexandria, VA 22302. (703) 379-2480. Fax (703) 379-7563. Monthly. Inquire for price and availability.

NTIS Bibliographic Database. National Technical Information Service, 5285 Port Royal Road, Springfield, Va 22161. (703) 487-4929 or FAX (703) 321-8199. Broad coverage of government-sponsored science and technology research reports, 1964 to present. Available on BRS,(800) 289-4277, DIALOG, (800) 334-2564, ORBIT, (800) 456-7248, and STN International, FIZ Karlsruhe, P.O. Box 2465, W-7500, Karlsruhe 1, Germany, online services. Also available on CD-ROM. Inquire as to cost and availability.

Scisearch. Institute for Scientific Information, 3501 Market Street, Philadelphia, PA 19104. (800) 523-1850 or (215) 386-0100. Broad multidisciplinary title and author index to the international literature of science and technology, 1974 to present. Available on DIALOG,(800) 334-2564, and ORBIT,(800) 456-7248, online services. Also available on CD-ROM. Inquire as to cost and availability.

WILSONLINE. H.W. Wilson Company, 950 University Avenue, Bronx, NY 10452. (800) 367-6770 or (212) 588-8400. Makes available online versions of the printed H.W. Wilson indexes including Applied Science and Technology Index, Business Periodicals Index, General Science Index, and Readers' Guide to Periodical Literature. Period covered is generally 1983 to present. Available on BRS,(800) 289-4277, DIALOG,(800) 334-2564, and OCLC EPIC,(800) 848-5878, online services. Also available on CD-ROM. Inquire as to cost and availability.

PERIODICALS

Journal of Applied Crystallography. Munksgaard International Publications Ltd., 35 Noerre Soegade, PO Box 2148 DK-1016, Copenhagen K, Denmark. Telephone 33-127030. FAX 33-129387. 1968 to present. Bi-monthly. Inquire for cost.

American Mineralogist. Mineralogical Society of America, 1130 17th Street NW, Washington, DC 20036. (202) 775-4344. 1916 to present. Bi-monthly. $225.00 per year.

Geological Society of America Bulletin. Geological Society of America, 3300 Penrose Place, P.O. Box 9140, Boulder, CO 80301-9140. (303) 447-2020. FAX (303) 447-1133. 1888 to present. Monthly. $25.00 per year.

Journal of Geology. University of Chicago Press, Journals Division, 5720 S. Woodlawn Avenue, Chicago, IL 60637. (312) 753-3347. FAX (312) 753-0811. 1893 to present. Bi-monthly. $45.00per year.

Mineralium Deposita. Springer-Verlag, 44 Hartz Way, Secaucus, NJ 07096-2491. 1966 to present. 6 times a year. $405.00 per year.

Rocks and Minerals. Heldref Publications, 1919 18th Street NW, Washington, DC 20036-1802. (202) 296-6267. FAX (202) 296-5149. 1926 to present. Bi-monthly. $37.00 per year.

RESEARCH CENTERS AND INSTITUTES

Carnegie Institution of Washington Geophysical Laboratory.

(202) 686-2410.

Liquid Crystal Institute. Kent State University, Kent, OH 44242. (216) 672-2654. FAX (216) 672-2796.

University of Connecticut Institute of Materials Science. U-136, 97 N. Eagleville Road, Storrs, CT 06268. (203) 486-4623. FAX (203) 486-4745.

CUMULUS CLOUDS

See: CLOUDS

CURRENTS

See: METEOROLOGY

CYBERNETICS

See also: AUTOMATION, COMPUTER MEMORY, COMPUTER OPERATING SYSTEMS, COMPUTER PROGRAMMING, COMPUTER VISION, PARALLEL COMPUTERS, SOFTWARE

ABSTRACT SERVICES AND INDEXES

ACM Guide To Computing Literature. Association for Computing Machinery, Association for Computing Machinery, 1515 Broadway, 17th Floor, New York, NY 10036-5701. (212) 869-7440. FAX (212) 944-1318. 1964 to present. Monthly. $190.00 per year for non-members.

Applied Science and Technology Index; A Cumulative Subject Index To English Language Periodicals In the Fields of Aeronautics and Space Science, Computer Technology, Chemistry, Construction Industry, Energy and Related Areas. H.W. Wilson Co., 950 University Avenue, Bronx, NY 10452. (800) 367-6770 or (212) 588-8400. FAX (718) 590-1617. From 1958 to present. Monthly. Inquire about cost and availability. Also available on CD-ROM and online.

Computer Abstracts, MCBUniversity Press Ltd., PO Box 10812, Birmingham, AL 35201-0812. (800) 633-4931. FAX (205) 995-1588. Monthly. Covers computer theory, data, hardware, systems, networks, human-computer interaction, artificial intelligence, as well as applications of computers in aerospace, business, CAD/CAM, cartography, civil engineering, electronics and electrical engineering, industrial engineering, mechanical engineering, medicine, structural engineering, etc. Also available on CD-ROM.

CYBERNETICS

Ency. of Physical Sciences and Engineering Info. Sources

Computer & Control Abstracts (inspec). INSPEC/IEEE, Box 1331, 445 Hoes Lane, Piscataway, NJ 08855-1331. (908) 562-5549. Abstracts organized by subjects of international technological information. Monthly. $1455.00 per year. Also available on CD-ROMand online via BRS Online Products, DIALOG Information Services, Orbit Search Service, and STN International.

Computer and Information Systems Abstracts Journal. Cambridge Scientific Abstracts, 7200 Wisconsin Avenue, Bethesda, MD 20814. (301) 961-6750. Fax (301) 961-6720. 1962 to present. Monthly. $1265.00 per year. Also available online via STN International.

Computer Literature Index. Applied Computer Research Inc., Box 82266, Phoenix, AZ 85071-2266. (800) 234-2227. 1971 to present. Quarterly plus annual cumulation. $198.50 per year. Bibliography of books, articles, and reports.

Computing Journal Abstracts. Techgnosis Ltd., Blade House, Battersea Road, Stockport, Cheshire 3AE, England. Telephone 061-442-2639. FAX 061-443-1162. 1969 to present. Monthly. Inquire for price.

Computing Reviews. Association for Computing Machinery, 1515 Broadway, 17th Floor, New York, NY 10036-5701. (212) 869-7440. FAX (212) 944-1318. 1960 to present. Monthly. $130.00 per year for non-members. Also available online via DIALOG Information Services.

Current Contents: Engineering, Technology, and Applied Sciences. Institute for Scientific Information, 3501 Market Street, Philadelphia, PA 19104. (215) 386-0100. FAX (215) 386-6362. 1970 to present. Weekly. $442.00 per year.

Engineering Index Monthly. Engineering Information, Inc., Castle Point on the Hudson, Hoboken, NJ 07030. (800) 221-1044. FAX (201) 216-8532. Monthly. $2300.00 per year. Also available online as COMPENDEX, and also on CD-ROM. Covers chemical engineering, computers, electrical engineering, civil engineering, metals and mining, industrial management, and mechanical engineering.

Science Citation Index. Institute for Scientific Information, 3501 Market Street, Philadelphia, PA 19104. (215) 386-0100. FAX (215) 386-6362. Inquire about availability and cost. Also available on CD-ROM.

ANNUAL REVIEWS AND YEARBOOKS

ACM Monograph Series. Association for Computing Machinery (ACM), 11 West 42nd Street, New York, NY 10036. (212) 869-7440. Fax (212) 869-0481. Irregular. Price varies, inquire.

Advances In Computers. Academic Press, Inc., 1250 Sixth Avenue, San Diego, CA 92101. (619) 231-0926. FAX (619) 699-6715. Irregular. Price varies, inquire.

Advances In Computing Research. JAI Press, Inc., 55 Old Post Road, Number 2, Box 1678, Greenwich, CT 06830. (203) 661-7602. Annual. $68.50 per year.

ASSOCIATIONS AND PROFESSIONAL SOCIETIES

American Federation of Information Processing Societies, 1899 Preston White Drive, Reston, Va 22091-5435. (703) 620-8937.

American Society For Cybernetics. Old Dominion University, Department of Engineering Management, Norfolk, VA 23529-0248.

Association For Computing Machinery. 1515 Broadway, 17thFloor, New York, NY 10036-5701. (212) 869-7440. FAX (212) 944-1318.

Computer and Automation Systems Association of Sme. 1 SME Drive, PO Box 930, Dearborn, MI 48121-0930. (313) 271-1500.

IEEE Computer Society. 1730 Massachusetts Avenue NW, Washington, DC 20036-1992. (202) 371-0101. FAX (202) 728-9614.

BIBLIOGRAPHIES

ACM Guide To Computing Literature. Association for Computing Machinery, 1515 Broadway, 17th Floor, New York, NY 10036-5701. (212) 869-7440. FAX (212) 944-1318. 1964 to present. Annual. $190.00 for non-members. Also available online via DIALOG.

DIRECTORIES AND BIOGRAPHICAL SOURCES

Asis Handbook and Directory, American Society For Information Science, 8720 Georgia Avenue, Suite 501, Silver Spring, Md 20910-3602. (301) 495-0900. Annual. $100.00.

Computing Information Directory. Hildebrandt Inc., PO Box 576, Pullman, WA 99114. 1981 to present. Annual. $199.95 per year.

International Research Centers Directory. Gale Research, 835 Penobscot Building, Detroit, MI 48226-4094. (313) 961-2242. (800) 347-4253. FAX (313) 961-6083. 8th edition. 1995. $410.00.

Research Centers Directory. Gale Research, 835 Penobscot Building, Detroit, MI 48226-4094. (313) 961-2242. (800) 347-4253. FAX (313) 961-6083. $485.00.

Who's Who In Computing. Canadian M I S Database Inc., 268 Lakeshore Road E., Suite 510, Mississauga, ON L5G 1H1, Canada. (905) 271-1601. FAX (905) 271-4522. 1987 to present. Annual. Inquire for price in U.S.

Who's Who In Technology. Gale Research, 835 Penobscot Building, Detroit, MI 48226-4094. (313) 961-2242. (800) 347-4253. FAX (313) 961-6083. 1995. $195.00.

ENCYCLOPEDIAS AND DICTIONARIES

Computer Dictionary. Donald D. Spencer. 4th edition. Camelot Publishing Company, 709 SW 80th Blvd., Gainesville, FL 32607-1537. (904) 331-0952. 1993. $24.95.

The Cyberspace Lexicon: An Illustrated Dictionary of Terms From Multimedia To Virtual Reality. Bob Cotton & Richard Oliver. Chronicle Books, 275 Fifth Street, San Francisco, CA 94103. (800) 722-6657. 1994. $29.95.

Dictionary of Computing. S.M.H. Collin. 2nd edition. Peter Collin Publishing, 8 the Causeway, Teddington, TW11 0HE, England. FAX 0181-943-3386. 1994. Inquire for cost.

Encyclopedia of Computer Science and Engineering. Anthony Ralston & Edwin D. Reilly Jr., editors. 3rd revised edition. Van Nostrand Reinhold, 115 Fifth Avenue, New York, NY 10003. (212) 254-3232 or (800) 926-2665. FAX (212) 254-9499. 1993. $125.00.

Microsoft Press Computer Dictionary: the Comprehensive Standard For Business, School, Library, and Home. 2nd edition. Microsoft Press, One Microsoft Way, Redmond, WA 98052-6399. (206) 936-0055. FAX (206) 823-8101. 1994. $19.95.

GENERAL WORKS

Analytical Robotics & Mechatronics. Wolfram Stader. McGraw-Hill Publishing Company, 1221 Avenue of the Americas, New York, NY 10020. (800) 262-4729 or (212) 512-3825. 1995. $52.95.

Cybernetics and Applied Systems. Marcel Dekker, Inc., 270 Madison Avenue, New York, NY 10016. (212) 696-9000. FAX (212) 685-4540. 1992. $150.00.

Introducing Cyberspace. Joanna Buick. Totem Graphics Inc., 6200F Capitol Blvd., Tumwater, WA 98501. (360) 352-1851. FAX (360) 352-2554. 1995. Inquire for price.

Neural Networks As Cybernetic Systems. Holk Cruse. Thieme Medical Publishers, 381 Park Avenue S., Suite 1501, New York, NY 10016. (800) 782-3488. FAX (212) 683-5088.

Virtual Realities and their Discontents. Robert Markley, editor. Johns Hopkins University Press, 2715 North Charles St., Baltimore, MD 21218. (410) 516-6900 or (800) 537-5487. FAX (410) 516-6998. 1996. $38.50.

ONLINE DATA BASES AND CD-ROMS

Compendex Plus. Engineering Information, Inc., 345 East 47th Street, New York, NY 10017. (212) 705-7600 or (800) 221-1044. Contains citations with abstracts to worldwide literature in engineering and technology, from 1970 to present. Available on online BRS, (800) 289-4277, DIALOG, (800) 334-2564, ORBIT (800) 456-7248, and STN International, FIZ Karlsruhe, P.O. Box 2465, W-7500, Karlsruhe 1, Germany, online services. Also available on CD-ROM. Inquire as to cost and availability.

Compuscience. FIZ Karlsruhe, D-7514, Eggenstein-Leopoldshafen 2, Germany. Contains citations with abstracts to European and North American literature on computer science, 1972 to present. Available on STN International, FIZ Karlsruhe, P.O. Box 2465, W-7500, Karlsruhe 1, Germany, online service. Inquire as to cost and availability.

Computer and Information Systems Abstracts. Cambridge Scientific Abstracts, 7200 Wisconsin Avenue, Suite 601, Bethesda, MD 20814. (301) 961-6750 or (800) 843-7751. Contains citations to worldwide literature in theoretical and applied computer science and related areas, from 1981 to present. Inquire as to cost and availability.

Computer and Mathematics Search. Institute for Scientific Information, 3501 Market Street, Philadelphia, PA 19104. (215) 386-0100. FAX (215) 386-6362. Covers worldwide literature in computer science and mathematics, from 1980 to present. Available on BRS, (800) 289-4277, online service. Inquire as to cost and availability.

Computer Database. Information Access Company, 362 Lakeside Drive, Foster City, CA 94404. (415) 378-5000 or (800) 227-8431. Contains citations with abstracts to literature from trade journals covering the computer, telecommunications, and electronics industries. Available on the BRS, (800) 289-4277, Compuserve Information Service, (800) 848-8990, and DIA-LOG, (800) 334-2564, online services. Inquire as to cost and availability.

CSA Engineering. Cambridge Scientific Abstracts, 7200 Wisconsin Avenue, Suite 601, Bethesda, MD 20814. (301) 961-6750 or (800) 843-7751. Contains citations and abstracts of international periodicals and research literature covering all fields of engineering and science and technology, including computer and information science, electronics, mechanical engineering, solid state materials, 1981 to present. Available on BRS, (800) 289-4277, online service. Inquire as to cost and availability.

Current Contents Search. Institute for Scientific Information, 3501 Market Street, Philadelphia, PA 19104. (215) 386-0100. FAX (215) 386-6362. Contains citations to articles listed in the table of contents of science and technology journals. Also articles in social sciences and life sciences journals. Available on BRS, (800) 289-4277, DIALOG, (800) 334-2564, online services. Inquire as to cost and availability.

Dissertation Abstracts. University Microfilms International, 300 North Zeeb Road, Ann Arbor, MI 48106. (800) 521-0600 or (313) 761-4700. Scope includes virtually all doctoral dissertations accepted at accredited American institutions from 1861 to present in 252 subject areas. Available on BRS, (800) 289-4277, DIALOG, (800) 334-2564, and OCLC EPIC, (800) 848-5878, online services. Also available on CD-ROM. Inquire as to cost and availability.

Inspec. Institution of Electrical Engineers, Michael Faraday House, Six Hills Way, Stevenage, Herts. SG1 2AY, England. Telephone: 0438 313311 or FAX 0438 742840. Contains citations to the worldwide literature of physics, electronics and electrical engineering, computer technology, and related fields. Available on BRS, (800) 289-4277, DIALOG, (800) 334-2564, ORBIT, (800) 456-7248, and STN International, FIZ Karlsruhe, P.O. Box 2465, W-7500, Karlsruhe 1, Germany, online services. Inquire as to cost and availability.

Mathsci. American Mathematical Society, P.O. Box 6248, Providence, RI 02940. (800) 321-4667 or (401) 455-4000 or FAX (401) 331-3842. Scope includes pure and applied mathematics and related areas of physics, statistics, engineering, computer science, and operations research literature since 1959. Available on DIALOG, (800) 334-2564, online service. Also available on CD-ROM. Inquire as to cost and availability.

NTIS Bibliographic Database. National Technical Information Service, 5285 Port Royal Road, Springfield, Va 22161. (703) 487-4929 or FAX (703) 321-8199. Broad coverage of government-sponsored science and technology research reports, 1964 to present. Available on BRS, (800) 289-4277, DIALOG, (800) 334-2564, ORBIT, (800) 456-7248, and STN International, FIZ Karlsruhe, P.O. Box 2465, W-7500, Karlsruhe 1, Germany, online services. Also available on CD-ROM. Inquire as to cost and availability.

Scisearch. Institute for Scientific Information, 3501 Market Street, Philadelphia, PA 19104. (800) 523-1850 or (215) 386-0100. Broad multidisciplinary title and author index to the international literature of science and technology, 1974 to present. Available on DIALOG, (800) 334-2564, and OR-

CYBERNETICS

Ency. of Physical Sciences and Engineering Info. Sources

BIT,(800) 456-7248, online services. Also available on CD-ROM. Inquire as to cost and availability.

PERIODICALS

ACM Transactions On Computer Systems. Association for Computing Machinery, 1515 Broadway, 17th Floor, New York, NY 10036-5701. (212) 869-7440. FAX (212) 944-1318. 1983 to present. Quarterly. $110.00 for non-members.

Communications of the A C M. Association for Computing Machinery, 1515 Broadway, 17th Floor, New York, NY 10036-5701. (212) 869-7440. FAX (212) 944-1318. 1958 to present. Monthly. $114.00 for non-members.

Computer Magazine. IEEE Computer Society, 10662 Vaqueros Circle, Box 3014, Los Alamitos, CA 90720. (714) 821-8380. 1966 to present. Monthly. $49.00 per year.

Cybernetic. American Society for Cybernetics, Old Dominion University, Department of Engineering Management, Norfolk, VA 23529-0248. 1967 to present. Annual. $25.00 per year.

Cybernetics and Systems. Taylor & Francis, 1900 Frost Road, Suite 101, Bristol, PA 19007. (215) 785-5800. FAX (215) 785-5515. 1971 to present. Bi-monthly. $462.00 per year.

IEEE Transactions On Computers. IEEE, 345 E. 47th Street, New York, NY 10017. (908) 981-0060. FAX (908) 981-9667. 1952 to present. Monthly. $425.00 for non-members.

IEEE Transactions On Systems, Man and Cybernetics. IEEE, 345 E. 47th Street, New York, NY 10017. (908) 981-0060. FAX (908) 981-9667. 1945 to present. Bi-monthly. $210.00 per year for non-members.

International Journal of Human-computer Studies. Harcourt Brace & Company, Ltd., Foots Cray High Street, Sidcup, Kent DA14 5HP, England. Telephone 44-81-300-3322. FAX 44-81-309-0807. 1969 to present. Monthly. Inquire for price.

Journal of the Association For Computing Machinery. Association for Computing Machinery, 11 West 42nd Street, New York, NY 10036. (212) 869-7440. Fax (212) 869-0481. Quarterly. $85.00 per year.

RESEARCH CENTERS AND INSTITUTES

Colorado State University, Center For Cybernetic Communications Research. Room B305, Engineering Research Center, Fort Collins, CO 80523. (303) 491-8535.

Purdue University, Purdue Laboratory For Applied Industrial Control. 334 A.A. Potter Engineering Center, West Lafayette, in 47907. (317) 494-7434.

University of Texas At Austin, Center For Cybernetic Studies. CBA 5.202, Austin, TX 78712-1177. (512) 471-1821. FAX (512) 471-7847.

CYCLONES

See also: HURRICANES, METEOROLOGY, TORNADOS

ABSTRACT SERVICES AND INDEXES

Applied Science and Technology Index; A Cumulative Subject Index To English Language Periodicals In the Fields of Aeronautics and Space Science, Computer Technology, Che Mistry, Construction Industry, Energy and Related Areas. H.W. Wilson Co., 950 University Avenue, Bronx, NY 10452-9978. (212) 588-8400. (800) 367-6770. FAX (718) 590-1617. From 1958 to present. Monthly. Inquire about cost and availability. Also available online (BRS and WILSONLINE) and on CD-ROM. ISSN: 0003-6986.

Deep-Sea Research. Part A: Oceanographic Research Papers. Deep-Sea Research. Part B: Oceanographic Literature Review. Pergamon Press, Inc., Maxwell House, Fairview Park, Elmsford, NY 10523. (914) 592-7700. Fax (914) 592-3625.1953 to present. Twelve times per year. $2000.00 per year

for Parts A and B. Oceanographic Literature Review also

available on CD-ROM. Inquire about price and availability.

General Science Index. H.W. Wilson Co., 950 University Avenue, Bronx, NY 10452. (212) 588-8400. (800) 367-6770. FAX (718) 590-1617. From 1978 to present. Ten issues per year; quarterly and annual cumulations. Service basis. Available on CD-ROM and online. Inquire regarding cost and availability. ISSN: 0162-1963.

Government Reports Announcements and Index. National Technical Information Service (NTIS), 5285 Port Royal Road, Springfield, VA 22161. (703) 487-4650. FAX (703) 321-8547. From 1968 to present. Annual. $630.00 per year. Also available online as NTIS BIBLIOGRAPHIC DATABASE and on CD-ROM. ISSN:

Meteorological and Geoastrophysical Abstracts. American Meteorological Society, c/o Inforonics, Inc., 550 Newtown Road, Box 458, Littleton, MA 01460. (508) 486-8976. FAX (508) 486-0027. Covers literature in environmental sciences, meteorology, astrophysics, hydrology, glaciology, and physical oceanography. 1950 to present. Monthly. $950.00 per year. ISSN: 0026-1130. Also available online (DIALOG) and on CD-ROM.

Science Citation Index. SCI. Institute for Scientific Information, 3501 Market Street, Philadelphia, PA 19104. (215) 386-0100. (800) 523-1850. FAX (215) 386-2991. 1961 to present. Six issues per year, plus annual cumulation. $11650.00 per year. Also available online and on CD-ROM. Inquire about price and availability. ISSN: 0036-827X.

Selected Water Resources Abstracts. U.S. Geological Survey. Distributed by National Technical Information Service (NTIS), 5285 Port Royal Road, Springfield, VA 22161. (703) 487-4650. Monthly. $115.00 per year.

ANNUAL REVIEWS AND YEARBOOKS

American Meteorological Society. METEOROLOGICAL MONOGRAPHS. American Meteorological Society, 45 Beacon Street, Boston, MA 02108-3693. (617) 227-2425. FAX (617) 742-8718. Irregular. Price varies.

Annual Review of Earth and Planetary Sciences. Annual Reviews, Inc., 4139 El Camino Way, Palo Alto, CA 94306-0897. (415)493-4400. Fax (415) 855-9815. Annual. $55.00 per year.

Developments In Atmospheric Science Series. Elsevier Science Publishing Company, Inc., 655 Avenue of the Americas, New York, NY 10010. (212) 989-5800. Irregular. Inquire about price and availability.

Ocean Yearbook. Elisabeth M. Borgese, et al, editors. University of Chicago Press, 5801 Ellis Avenue, Chicago, IL 60637. (312) 702-7700. 1995. $77.00.

ASSOCIATIONS AND PROFESSIONAL SOCIETIES

American Association of State Climatologists. c/o Myron Moinau, University of Idaho, Agriculture Engineering Department, State Climate Service, Moscow, ID 834844-2040.

American Meteorological Society, 45 Beacon Street, Boston, Ma 02108-3693. (617) 227-2425. FAX (617) 742-8718. Irregular. Price varies.

Association of American Weather Observers, 401 Whitney Boulevard, Box 455, Belvedere, Il 61008. (815) 544-5665. FAX (815) 544-6334.

International Association of Meteorology and Atmospheric Physics. UCAR, P.O. Box 3000, Boulder CO 80307-3000. (303) 497-1673.

International Association of Severe Weather Specialists. c/o Warren Faidley, PO Box 31808, Tuscon, AZ 85751. (520) 751-9964. FAX (520) 751-1185.

National Environmental Satellite Data, and Information Service. 2069 Federal Building 4 Room 2060, Washington, DC 20233. (301) 763-71990. FAX (301) 763-4011.

National Weather Association, 6704 Wolke Court, Montgomery, AL 36116-2134. (334) 213-0388. FAX (334) 213-0388.

University Corporation For Atmospheric Research. PO Box 3000, 1850 Table Mesa Drive, Boulder, CO 80307-3000. (303) 497-1673. FAX (303) 497-1654.

Weather Modification Association. PO Box 8116, Fresno CA 93747. (209) 291-8466.

BIBLIOGRAPHIES

Information Sources in the Earth Sciences. David N. Wood, Joan E. Hardy, and Anthony P. Harvey, editors. Bowker-Saur/K.G. Saur. Distributed by R.R. Bowker, 121 Chanlon Road, New Providence, NJ 07974. (800) 521-8110 or (908) 464-6800. Second edition. 1989. $85.00.

International Bibliography of Meteorology: from the Beginning of Printing to 1889. Oliver L. Fassig. Diane Publishing Co., 600 Upland Avenue, Upland, PA 19015. (610) 499-7415. Reprint edition. 1994. $195.00

Scientific and Technical Books and Serials In Print; An Index To Literature In Science and Technology. R.R. Bowker Co.,

205 E. 42nd Street, New York, NY 10017. (800) 521-8110 or (212) 916-1600. 1992. $250.00.

DIRECTORIES AND BIOGRAPHICAL SOURCES

American Men and Women of Science: Physical andBiological Sciences. Eighteenth edition. R.R. Bowker Company, 245 West 17th Street, New York, NY 10011. (800) 521-8810 or (212) 916-1600. $750.00.

American Meteorological Society Professional Directory. American Meteorological Society. 45 Beacon Street, Boston, MA 02108. (617) 227-2425. Included in each issue of the Bulletin of the Society.

Meteorological Services of the World. American Meteorological Society, 45 Beacon Street, Boston, MA 02108- 3693. (617) 227-2425. FAX (617) 742-8718. Annual. $70.00.

National Weather Service Offices and Stations. National Oceanic and Atmospheric Administration, Deparatment of Commerce, Silver Spring, MD 20910. (301) 427-7698. Annual. Free

Research Centers Directory. Gale Research, 835 Penobscot Building, Detroit, MI 48226-4094. (313) 961-2242. (800) 347-4253. 20th edition. 1995. $485.00. ISSN:: 0080-1518.

ENCYCLOPEDIAS AND DICTIONARIES

Climates of the States. Bair. Gale Research, 835 Penobscot Building, Detroit, MI 48226-4094. (313) 961-2242. (800) 347-4253. 5th edition. 1995. $255.00.

Concise Oxford Dictionary of the Earth Sciences. Ailsa Allaby and Michael Allaby, editors. Oxford University Press, Inc., 200 Madison Avenue, New York, NY 10016. (800) 334-4249 or (212) 679-7300. 1990. $42.95.

Encyclopedia of Climatology. Rhodes W. Fairbridge, and John E. Oliver. Chapman & Hall, 1 Penn Plaza, New York, NY 10119. (212) 564-1060. 1986. $115.00.

McGraw-Hill Encyclopedia of Science and Technology. McGraw-Hill Publishing Company, 1221 Avenue of the Americas, New York, NY 10020. (800) 262-4729 or (212) 512-3825. Seventh edition. 1992. $1900.00.

Smithsonian Meteorological Tables. Books on Demand, 300 North Zeeb Road, Ann Arbor, MI 48106-1346. (313) 761-4700. (800) 521-0600. reprint edition. $153.90.

Water Encyclopedia. Frits van der Leeden, Fred L. Troise, and David K. Todd. Chelsea Press, 15 East 26th Street, New York, NY 10010. (212) 889-8095. 1990.

Weather Satellite Handbook. American Radio Relay League, 225 Main Street, Newington, CT 06111. (203) 666-1541 1994. $20.00.

GENERAL WORKS

Experimental Investigation and Numerical Modeling of Cyclones for Application at High Temperatures. J. G. Bernard. Coronet Books, 311 Bainbridge Street, Philadelphia PA 19147. (215) 925-5083. 1992. $57.50.

CYCLONES

Ency. of Physical Sciences and Engineering Info. Sources

Extratropical Cyclones. Chester Newton and E. O. Holopainen, editors. American Meteorological Society, 45 Beacon Street, Boston, MA 02108-3693. (617) 227-2425. FAX (617) 742-8718. 1990. $65.00.

The Thermal theory of Cyclones: A History of Meteorological Thought In the Nineteenth Century. Gisela Kutzbach. American Meteorological Society, 45 Beacon Street, Boston,

MA 02108-3693. (617) 227-2425. FAX (617) 742-8718. 1979. $65.00.

Tropical Cyclones of the Pacific. S. S. Visher. Periodicals Service, Co. 11 Main Street, Germantown, NY 12526. (518) 537-4700. (800) 223- 8323. Reprint edition. 1974. $25.00

Tropical Cyclones: their Evolution, Structure and Effects. Richard A. Anthen. American Meteorological Society, 45 Beacon Street, Boston, MA 02108-3693. (617) 227-2425. FAX (617) 742-8718. 1982. $40.00

Ocean-Atmosphere Interaction & Climate Modeling. Boris A. Kagan. Cambridge University Press, 40 West 20th Street, New York, NY 10011-4211. (212) 924-3900. (800) 872-7423. 1995. $79.95.

Physics of Climate. J. P. Peixoto. American Institute of Physics, One Physics Ellipse, College Park, MD 20740-3843. (301) 209-3100. 1992. $45.00.

World Weather Guide. E. A. Pearce and Gordon Smith. Times Books/Random House, Inc., 201 East 50th Street, New York, NY 10022. (212) 751-2600. (800) 733-3000. 1991. $19.50.

Severe and Unusual Weather. Joe R. Eagleman. Trimedia Publishing Company, 12008 West 87th Street, Suite 117, Lenexa, KS 66215. (913) 599-0505. Second edition. 1990. $41.95.

HANDBOOKS AND MANUALS

Author's Guide To the Journals of the American Meteorological Society. American Meteorological Society, 45 Beacon Street, Boston, MA 02108-3693. (617) 227-2425. FAX (617) 742-8718. 1983. $15.00 in paper.

Handbook In Applied Meteorology. David D. Houghton, editor. John Wiley and Sons, Inc., 605 Third Avenue, New York, NY 10158. (800) 526-5368 or (212) 850-6000. 1985. $168.00.

ONLINE DATABASES AND CD-ROMS

Climate Assessment Database. National Weather Service, National Meteorological Center, 5200 Auth Road, Suite 101, Camp Springs, MD 20233. (301) 763-8016. Contains daily, weekly and monthly summaries of North American and world climatological data. Also provides five to ten weather forecasts, and 30 to 90 day weather outlook. Subscription required. Inquire as to cost and availability.

Current Contents Search. Institute for Scientific Information, 3501 Market Street, Philadelphia, PA 19104. (215) 386-0100. FAX (215) 386-6362. Contains citations to articles listed in the table of contents of science and technology journals. Also articles in social sciences and life sciences journals. Available on BRS,(800) 289-4277, DIALOG,(800) 334-2564, online services. Inquire as to cost and availability.

Dissertation Abstracts. University Microfilms International, 300 North Zeeb Road, Ann Arbor, MI 48106. (800) 521-0600 or (313) 761-4700. Scope includes virtually all doctoral dissertations accepted at accredited American institutions from 1861 to present in 252 subject areas. Available on BRS,(800)289-4277, DIALOG,(800) 334-2564, and OCLC EPIC,(800) 848-5878, online services. Also available on CD-ROM. Inquire as to cost and availability.

Meteorological and Geoastrophysical Abstracts. American Meteorological Society, 45 Beacon Street, Boston, MA 02108-3693. (617) 227-2425. FAX (617) 742-8718. Contains citations and abstracts to the worldwide literature on significant research in meteorology and geoastrophysics. Related topics include physical oceanography, hydrology, environmental sciences and glaciology. Covers the period 1972 to present. Available on DIALOG,(800) 334-2564, online service. Inquire as to cost and availability.

NTIS Bibliographic Database. National Technical Information Service, 5285 Port Royal Road, Springfield, Va 22161. (703) 487-4929 or FAX (703) 321-8199. Broad coverage of government-sponsored science and technology research reports, 1964 to present. Available on BRS,(800) 289-4277, DIALOG, (800) 334-2564, ORBIT, (800) 456-7248, and STN International, FIZ Karlsruhe, P.O. Box 2465, W-7500, Karlsruhe 1, Germany, online services. Also available on CD-ROM. Inquire as to cost and availability.

Scisearch. Institute for Scientific Information, 3501 Market Street, Philadelphia, PA 19104. (800) 523-1850 or (215) 386-0100. Broad multidisciplinary title and author index to the international literature of science and technology, 1974 to present. Available on DIALOG and ORBIT online services. Also available on CD-ROM. Inquire as to cost and availability.

WILSONLINE. H.W. Wilson Company, 950 University Avenue, Bronx, NY 10452. (800) 367-6770 or (212) 588-8400. Makes available online versions of the printed H.W. Wilson indexes including Applied Science and Technology Index, Business Periodicals Index, General Science Index, and Readers' Guide to Periodical Literature. Period covered is generally 1983 to present. Available on BRS,(800) 289-4277, DIALOG,(800) 334-2564, and OCLC EPIC,(800) 848-5878, online services. Also available on CD-ROM. Inquire as to cost and availability.

World Climate Disc. Chadwyck-Healey Inc., 1101 King Street, Alexandria, VA 22314. (703) 683-4890. FAX (703) 683-7589. Weather and climate data from approximately 5,000 weather stations worldwide, covering the years 1854 to 1990 on CD-ROM. First edition 1992. Approximately $1200.00 per year with annual updates.

World Weatherdisc. WeatherDisc Associates, Inc., 4584 N.E. 89th Street, Seattle, WA 98115. (206) 524-4314. FAX (206) 543-0308. Meteorological data on CD-ROM which describes the climate of the earth today and for the past few hundred years. First edition 1989. Approximately $295.00 per year with biannual updates.

PERIODICALS

Agricultural and Forest Meteorology. Elsevier Science Publishing Company, Inc., 655 Avenue of the Americas, New York, NY 10010. (212) 989-5800. Twenty times per year. $750.00 per year.

Ency. of Physical Sciences and Engineering Info. Sources

CYCLONES

American Meteorological Society. BULLETIN. AmericanMeteorological Society, 45 Beacon Street, Boston, MA 02108-3693. (617) 227-2425. FAX (617) 742-8718. Monthly. $60.00 per year.

American Weather Observer. Association of American Weather Observers, 401 Whitney Boulevard, Box 455, Belvedere, IL 61008. (815) 544-5665. FAX (815) 544-6334. Monthly. $21.00 per year.

Atmosphere - Ocean. Canadian Meteorological and Oceanographic Society, P.O. Box 334, Newmarket, Ontario, L3Y 4X7, Canada. (416) 898-1040. FAX (416) 898-7937. Quarterly. $30.00 per year.

Dynamics of Atmospheres and Oceans. Elsevier Science Publishing Company, Inc., 655 Avenue of the Americas, New York, NY 10010. (212) 989-5800. Six times per year. $205.00 per year.

JGR: Journal of Geophysical Research: Atmosphere. American Geophysical Union, 2000 Florida Avenue, N.W., Washington, DC 20009. (202) 462-6903. Monthly. $90.00 per year to members.

JGR: Journal of Geophysical Research: Oceans. American Geophysical Union, 2000 Florida Avenue, N.W., Washington, DC 20009. (202) 462-6903. Monthly. $1545.00 per year.

Journal of the Atmospheric Sciences. American Meteorological Society, 45 Beacon Street, Boston, MA 02108-3693. (617) 227-2425. FAX (617) 742-8718. Semi-monthly. $320.00 per year.

Monthly Weather Review. American Meteorological Society, 45 Beacon Street, Boston, MA 02108-3693. (617) 227-2425. FAX (617) 742-8718. Monthly. $205.00 per year.

National Weather Digest. National Weather Association, 4400 Stamp Road, Room 404, Temple Hills, MD 20748. (301) 899-3784.

Weather. Royal Meteorological Society, 104 Oxford Road, Reading, Berks RG1 7LJ, England. Monthly. $44.00 per year.

Weatherwise. Heldref Publications, 1319 Eighteenth Street, N.W., Washington, DC 20036-1802. (202) 296-6267. FAX (202) 296-5149. Bi-monthly. $28.00 per year.

RESEARCH CENTERS AND INSTITUTES

National Center for Atmospheric Research. Box 3000, Boulder, CO 80307. (303) 497-1000. FAX (303) 497-1654. *National Severe Storms Laboratory.* 1313 Halley Circle, Norman, OK 73069. (406) 360-3620. FAX (405) 360-0472.

Remote Sensing Laboratory. University of Miami, P.O. Box 248003, Coral Gables, FL 33124. (305) 284-3881. FAX (305) 284-4792.

Atlantic Oceanographic and Meteorological Laboratory. 4301 Rickenbacker Causeway, Miami, FL 33149. (305) 361-4300.

Climate Analysis Center. National Weather Service, 5200 Auth Road, Camp Springs, MD 20746. (301) 763-8167.

Cooperative Institute for Mesoscale Meteorological Studies. University of Oklahoma, 401 East Boyd, Norman, OK 73019. (405) 325-3041.

University Corporation For Atmospheric Research. P.O. Box 3000, Boulder, CO 80307-3000. Consortium of 58 universities with doctoral programs in atmospheric and/or oceanic sciences. (303) 497-1650. FAX (303) 497-1654.

CYCLOTRON

See: PARTICLE ACCELERATORS

D

DAMS

See also: CHANNELS, CIVIL ENGINEERING, FLOOD CONTROL, GEOTECHNICAL ENGINEERING, HYDRAULIC ENGINEERING, HYDROELECTRIC POWER

ABSTRACT SERVICES AND INDEXES

Applied Science and Technology Index. H.W. Wilson Co., 950 University Avenue, Bronx, NY 10452. (800) 367-6770 or (212) 588-8400. Fax: (718) 590-1617. From 1958 to present. Monthly. Inquire about cost and availability. Also available on CD-ROM and online. A cumulative subject index to English language periodicals in the fields of aeronautics and space science, computer technology, chemistry, construction industry, energy and related areas.

ASCE Combined Annual Combined Index. American Society of Civil Engineers, 345 East 47th Street, New York, NY 10017-2398. (212) 705-7520 or (800) 548-2723. Annual. $48.00.

ASCE Publications Information. American Society of Civil Engineers, 345 East 47th Street, New York, NY 10017-2398. (212) 705-7520 or (800) 548-2723. 1966-present. Bi-monthly. $160.00 per year for non-members.

Civil and Structural Engineering Abstracts. Cambridge Scientific Abstracts, 7200 Wisconsin Avenue, Bethesda, MD 20814-4823. (301) 961-6750. Fax: (301) 961-6720. 1993 to present. Monthly. $385.00 per year. Topics covered include structural design, construction equipment and methods, civil defense and military engineering, surveying, highway engineering, maritime and port structures, materials, land reclamation, and soil mechanics.

Current Contents: Engineering, Technology, and Applied Sciences. Institute for Scientific Information, 3501 Market Street, Philadelphia, PA 19104. (215) 386-0100. Fax: (215) 386-6362. 1970 to present. Weekly. $442.00 per year.

Engineering Index Monthly. Engineering Information, Inc., Castle Point on the Hudson, Hoboken, NJ 07030. (800) 221-1044. Fax: (212) 832-1857. Monthly. $2300.00 per year. Also available online as COMPENDEX, and also on CD-ROM. Covers chemical engineering, computers, electrical engineering, civil engineering, metals and mining, industrial management, and mechanical engineering.

Environmental Engineering Abstracts. Cambridge Scientific Abstracts, 7200 Wisconsin Avenue, Bethesda, MD 20814-4823. (301) 961-6750. Fax: (301) 961-6720. Monthly. Covers hazardous materials, environmental impact and protection, treatment of sewage and industrial wastes, hydroelectric power, tidal and wind power, arctic and tropical engineering.

Fluid Abstracts: Civil Engineering. Elsevier Science [journals], 660 White Plains Rd., Tarrytown, NY 10159-5153. (919) 524-9200. Fax: (919) 333-2444. 1991 to present. Monthly plus annual cumulation. $645.00 per year. Also available online as FLUIDEX. Covers civil engineering applications of fluid mechanics, hydraulics, flow metering and measuring, offshore engineering, environmental hydraulics, and related aspects of wind energy, the atmosphere, and aerodynamics.

Geotechnical Abstracts. Geodex Retrieval Systems, 669 Broadway, P.O. Box 279, Sonoma, CA 95476. (707) 939-8476. Fax: (707) 996-8734. 1970 to present. Three times a year. $280.00 per year. Covers soil mechanics, foundation engineering, rock mechanics, and engineering geology.

Index to Scientific and Technical Proceedings. Institute for Scientific Information, 3501 Market St., Philadelphia, PA 19104. (215) 386-0100. Fax: (215) 386-6362. Monthly. $500.00 per year.

ASSOCIATION AND PROFESSIONAL SOCIETIES

American Society of Civil Engineers. 345 East 47th Street, New York, NY 10017-2398. (212) 705-7520 or (800) 548-2723.

Association of State Dam Safety Officials. 450 Old East Vine, Lexington, KY 40507. (606) 257-5140.

United States Committee on Large Dams. 1616 17th Street, No. 483, Denver, CO 80202. (303) 628-5430.

DIRECTORIES AND BIOGRAPHICAL SOURCES

Directory of Engineering Societies and Related Organizations. Gordon Davis. 13th edition. American Association of Engineering Societies, 1111 19th Street NW, Suite 608, Washington, DC 20036. (202) 296-2237 or (800) 658-8897. 1989. Inquire for price.

DAMS

Ency. of Physical Sciences and Engineering Info. Sources

Research Centers Directory. Gale Research, 835 Penobscot Building, Detroit, MI 48226-4094. (313) 961-2242. (800) 877-4253. Fax: (313) 961-6083. 20th edition. 2 volumes. $485.00.

Scientific and Technical Organizations and Agencies Directory. Gale Research, 835 Penobscot Building, Detroit, MI 48226-4094. (313) 961-2242. (800) 877-4253. Fax: (313) 961-6083. 4th edition. 1996. $195.00.

Who's Who in Engineering. American Association of Engineering Societies, 1111 19th Street NW, Suite 608, Washington, DC 20036. (202) 296-2237 or (800) 658-8897. 8th edition. 1991. Inquire as to price.

ENCYCLOPEDIAS AND DICTIONARIES

Dictionary of Civil Engineering. V.N. Vazirani. South Asia Books, P.O. Box 502, Columbia, MO 65205. (314) 474-0116. Fax: (314) 474-8124. 1992. $20.00.

McGraw-Hill Encyclopedia of Engineering. Sybil P. Parker, ed. 2nd edition. McGraw-Hill Publishing Company, 1221 Avenue of the Americas, New York, NY 10020. (800) 262-4729 or (212) 512-3825. 1993. $95.50.

GENERAL WORKS

Civil Engineering Guidelines for Planning and Designing Hydroelectric Developments. Volume One: Planning, Design of Dams and Related Features. American Society of Civil Engineers, 345 East 47th Street, New York, NY 10017-2398. (212) 705-7520 or (800) 548-2723. 1989. $75.00.

Design of Small Dams. Gordon Press Publishers, P.O. Box 459, Bowling Green Station, New York, NY 10004. (718) 624-8419. 1992. $489.95.

Geotechnical Engineering of Embankment Dams. Robin Fell, et al. A.A. Balkema, Ashgate Publishing Co., Old Post Rd., Brookfield, VT 05036. (802) 276-3162. Fax: (802) 276-3837. 1992. $90.00.

Retaining and Flood Walls. American Society of Civil Engineers, 345 East 47th Street, New York, NY 10017-2398. (212) 705-7520 or (800) 548-2723. 1994. Inquire as to cost and availability.

HANDBOOKS AND MANUALS

Civil Engineering Calculations Reference Guide. Tyler G. Hicks. McGraw-Hill Publishing Company, 1221 Avenue of the Americas, New York, NY 10020. (800) 262-4729 or (212) 512-3825. 1987. $46.00.

Civil Engineering Practice. Paul N. Cheremisinoff, et al. 5 volumes. Technomic Publishing Company, Inc., 851 New Holland Avenue, P.O. Box 3535, Lancaster, PA 17604. (717) 291-5609 or (800) 233-9936. Fax: (717) 295-4538. 1987-88. $199.00.

Concrete Construction Handbook. Joseph J. Waddell & Joseph A. Dobrowolski, editors. 3rd edition. McGraw-Hill Publishing Company, 1221 Avenue of the Americas, New York, NY 10020. (800) 262-4729 or (212) 512-3825. 1993. $92.00.

Handbook of Concrete Engineering. Mark Fintel. 2nd edition. Routledge, Chapman and Hall, Inc., 29 West 35th Street, New York, NY 10001-2291. (212) 244-3336 or Fax: (212) 563-2269. 1985. $115.00.

Handbook of Mechanics, Materials, and Structures. Alexander Blake, editor. John Wiley and Sons, Inc., 605 Third Avenue, New York, NY 10158. (800) 526-5368 or (212) 850-6000. 1985. $120.00.

ONLINE DATABASES AND CD-ROMS

Compendex plus. Engineering Information, Inc., 345 East 47th Street, New York, NY 10017. (212) 705-7600 or (800) 221-1044. Contains citations with abstracts to worldwide literature in engineering and technology, from 1970 to present. Available on online BRS,(800) 289-4277, DIALOG, (800) 334-2564, ORBIT (800) 456-7248, and STN International, FIZ Karlsruhe, P.O. Box 2465, W-7500, Karlsruhe 1, Germany, online services. Also available on CD-ROM. Inquire as to cost and availability.

CSA Engineering. Cambridge Scientific Abstracts, 7200 Wisconsin Avenue, Suite 601, Bethesda, MD 20814. (301) 961-6750 or (800) 843-7751. Contains citations and abstracts of international periodicals and research literature covering all fields of engineering and science and technology,including computer and information science, electronics, mechanical engineering, solid state materials, 1981 to present. Available on BRS,(800) 289-4277, online service.Inquire as to cost and availability.

GEOREF: Bibliography and Index of Geology. American Geological Institute, 4220 King Street, Alexandria, VA 22302. (703) 379-2480. Fax (703) 379-7563. Monthly. Inquire about price and availability.

NTIS Bibliographic Database. National Technical Information Service, 5285 Port Royal Road, Springfield, VA 22161. (703) 487-4929 or Fax: (703) 321-8199. Broad coverage of government-sponsored science and technology research reports, 1964 to present. Available on BRS,(800) 289-4277, DIALOG, (800) 334-2564, ORBIT, (800) 456-7248, and STN International, FIZ Karlsruhe, P.O. Box 2465, W-7500, Karlsruhe 1, Germany, online services. Also available on CD-ROM. Inquire as to cost and availability.

Wilsonline. H.W. Wilson Company, 950 University Avenue, Bronx, NY 10452. (800) 367-6770 or (212) 588-8400. Makes available online versions of the printed H.W. Wilson indexes including *Applied Science and Technology Index, Business Periodicals Index, General Science Index,* and *Readers' Guide to Periodical Literature.* Period covered is generally 1983 to present. Available on BRS,(800) 289-4277, DIALOG,(800) 334-2564, and OCLC EPIC,(800) 848-5878, online services. Also available on CD-ROM. Inquire as to cost and availability.

PERIODICALS

ACI Structural Journal. American Concrete Institute, P.O. Box 19150, Detroit, MI 48219. (313) 532-2600. Fax: (213) 538-0655. Bi-monthly. $101.00 per year.

Civil Engineering. American Society of Civil Engineers, 345 East 47th Street, New York, NY 10017-2398. (212) 705-7520 or (800) 548-2723. 1930 to present. Monthly. $85.00 per year.

Concrete. The Concrete Society, Framewood Road, Wexham, Slough, Berks. SL3 6PJ England. Telephone 0753-662226. Fax: 0753-662126. 1966 to present. Bi-monthly. $75.00 per year.

Construction Products. Gordon Publications/ Cahners-Reed Elsevier, 301 Gilbraltar Drive, P.O. Box 650, Morris Plains, NJ 07950. (201) 252-5100. Fax: (201) 539-3476. 1892 to present. 6 times a year. $40.00 per year.

Journal of Construction Engineering and Management. American Society of Civil Engineers, Construction Division, 345 East 47th Street, New York, NY 10017-2398. (212) 705-7520 or (800) 548-2723. 1956 to present. Quarterly. $112.00 per year.

Journal of Engineering Mechanics. American Society of Civil Engineers, Engineering Mechanics Division, 345 East 47th Street, New York, NY 10017-2398. (212) 705-7520 or (800) 548-2723. 1956 to present. Monthly. $284.00 per year.

Journal of Geotechnical Engineering. American Society of Civil Engineers, Geotechnical Engineering Division, 345 East 47th Street, New York, NY 10017-2398. (212) 705-7520 or (800) 548-2723. 1956 to present. Monthly. $212.00 per year.

Journal of Hydraulic Engineering. American Society of Civil Engineers, Hydraulics Division, 345 East 47th Street, New York, NY 10017-2398. (212) 705-7520 or (800) 548-2723. 1956 to present. Monthly. $200.00 per year.

Journal of Irrigation and Drainage. American Society of Civil Engineers, Irrigation and Drainage Division, 345 East 47th Street, New York, NY 10017-2398. (212) 705-7520 or (800) 548-2723. 1956 to present. Bi-monthly. $136.00 per year.

Journal of Structural Engineering. American Society of Civil Engineers, 345 East 47th Street, New York, NY 10017-2398. (212) 705-7520 or (800) 548-2723. 1956 to present. Monthly. $300.00 per year for non-members.

Journal of Surveying Engineering. American Society of Civil Engineers, 345 East 47th Street, New York, NY 10017-2398. (212) 705-7520 or (800) 548-2723. 1956 to present. Quarterly. $76.00 per year.

Journal of Water Resources Planning and Management. American Society of Civil Engineers, Water Resources Planning and Management Division, 345 East 47th Street, New York, NY 10017-2398. (212) 705-7520 or (800) 548-2723. Bi-monthly. $112.00 per year.

Structural Engineer. Structural Engineers Trading Organization Ltd., 11 Upper Belgrave Street, London SW1X 8BH, England. Telephone 071-235-4535. Fax: 071-235-4294. 1922 to present. 24 times a year. Inquire for cost.

Structural Engineering International. International Association for Bridge and Structural Engineering, ETH-Hoenggerberg, CH-8093 Zurich, Switzerland. Telephone 01-3772647. Fax: 01-3712131. Quarterly. Inquire as to cost and availability.

Structural Engineering Reviews. Elsevier Science [journals], 660 White Plains Rd., Tarrytown, NY 10159-5153. (919) 524-9200. Fax: (919) 333-2444. 1988 to present. 4 times a year. $190.00 per year.

DARK MATTER

See also: ASTRONOMY, ASTROPHYSICS, BIG BANG, COSMO-CHEMISTRY, GALAXIES, INTERSTELLAR MATTER, PLANETS, SOLAR SYSTEM, UNIVERSE

ABSTRACT SERVICES AND INDEXES

Applied Science and Technology Index. H.W. Wilson Co., 950 University Avenue, Bronx, NY 10452. (212) 588-8400. (800) 367-6770. Fax: (718) 590-1617. From 1958 to present. Monthly. Inquire about cost and availability. Also available on CD-ROM and online. A cumulative subject ind ex to English language periodicals in the fields of aeronautics and space science, computer technology,chemistry, construction industry, energy and related areas. ISSN: 0003-6986.

Astronomy and Astrophysics Abstracts. Springer-Verlag New York, 175 Fifth Avenue, New York, NY 10010. (212) 460-1500.Fax: (212) 473-6272. Published for the Astronomisches Rechen-Institut. Comprehensive coverage of all aspects of astronomy, astrophysics and related fields. 1969 to present. Two parts per year. Annual. ISSN: 0067-0022.

Chemical Abstracts. Chemical Abstracts Service, 2540 Olentangy River Road, P.O. Box 3012, Columbus, OH 43210-0012. (614) 447-3600. (800) 848-6538. Fax: (614) 447-3713. From 1907 to present. Weekly. $16800.00. Available online and on CD-ROM. CA is also available in five section groupings. Inquire regarding cost and availability.

General Science Index. H.W. Wilson Company, 950 University Avenue, Bronx, NY 10452. (212) 588-8400. (800) 367-6770. Fax: (718) 590-1617. From 1978 to present. Ten issues per year. quarterly and annual cumulations. Service basis. Available on CD-ROM and online. Inquire regarding cost and availability. ISSN: 0162-1963.

Government Reports Announcements and Index. U.S. National Technical Information Service (NTIS), 5285 Port Royal Road, Springfield, VA 22161. (703) 487-4650. Fax: (703) 321-8547.From 1968 to present. Annual. $630.00 per year. Also available online as NTIS Bibliographic Database and on CD-ROM.

NTIS ALERTS: ASTRONOMY & ASTROPHYSICS. U.S. National Technical Information Service. 5285 Port Royal Road, Springfield, VA 22161. (703) 487-4650. Fax: (703) 321-8547.Weekly. $140.00 per year.

Pascal 48: Environnement Cosmique Terrestre, Astronomie et Geologie Extraterrestre. Centre National de la Recherche Scientifique, Institut de Information Scientifique et Technique, 2 aliee du Parc de Brabois, 54514 Vandoeuvre Les Nancy Cedex, France. TEL 83-50-46-00. Fax: 83-50-46-50. 1985 to present. Ten issues per year. F770. Also available on CD-ROM and online.

Physics Abstracts. Inspec. Section A, Science Abstracts. Institute of Electrical Engineers, London, United Kingdom. Available from: INSPEC/IEEE-Institute of Electrical and Electronic Engineers, P.O. Box 1331, Hoes Lane, Piscataway, NJ 08855-1331. (908) 562-5549. 1898 to present. 24 issues per year. $2835.00 per year. ISSN: 0036-8091. Also available online and on CD-ROM.

SCIENCE CITATION INDEX (SCI). Institute for Scientific Information, 3501 Market Street, Philadelphia, PA 19104. (215)

DARK MATTER

Ency. of Physical Sciences and Engineering Info. Sources

386-0100. (800) 523-1850. Fax: (215) 386-2991. 1961 to present. Six issues per year, plus annual cumulation. $11650.00 per year. Also available online and on CD-ROM. Inquire about price and availability. ISSN: 0036-827X.

STAR. Scientific and Technical Aerospace Reports. U.S. National Aeronautics and Space Administration. Distributed by U.S. Superintendent of Documents, Washington, DC 20402. 1963 to present. Semi-monthly, with semiannual and annual indexes. $114.00 per year. ISSN: 0036-8741. Also available online and on CD-ROM.

ANNUAL REVIEWS AND YEARBOOKS

Annual Review of Astronomy and Astrophysics. Annual Reviews, Inc., 4139 El Camino Way, Palo Alto, CA 94306-0139. (415) 493-4400. (800) 523-8635. Fax: (415) 855-9815. Original reviews of critical literature and current developments in astronomy and astrophysics.1963 to date. Annual. $60.00. ISSN: 0066-4146.

Highlights of Astronomy. I. Appenzeller, editor. Kluwer Academic Publishers, P.O. Box 358, Accord Station, Hingham, MA 02018-0358. (617) 871-6600. International Astronomical Union Highlights series. ISBN: 07923-3553-8. Triennal. 1995. Price varies.

ASSOCIATION AND PROFESSIONAL SOCIETIES

American Association of Variable Star Observers. 23 Birch Street, Cambridge, MA 02138. (617) 354-0484. Fax: (617) 354-0665.

American Astronomical Society. 2000 Florida Avenue NW, Suite 400, Washington, DC 20009. (202) 328-2010. Fax: (202) 234-2560.

American Institute of Physics. One Physics Ellipse, College Park, MD 20740-3843. (301) 209-3100.

Association of University for Research in Astronomy. 1625 Massachusetts Avenue, NW, Suite 701, Washington, DC 20036. (202) 483-2101. Fax: (202) 483-2106.

Astronomical League. 2112 Kingfisher Lane East, Rolling Meadows, IL 60008. (708) 398-0562.

Astronomical Society of the Pacific. 390 Ashton Avenue, San Francisco, CA 94112. (415) 337-1100. Fax: (415) 337-5205.

International Astronomical Union. 98 bis Boulevard Arago, F-75014 Paris, France. Tel 1-43-25-83-58. Fax: 1-43-25-26-16.

Joint Institute for Laboratory Astrophysics. University of Colorado, P.O. Box 440, Boulder, CO 80309-0440. (303) 492-7780. Fax: (303) 492-5235.

National Council for Geocosmic Research. P.O. Box 1220, Dunkirk, MD 20754. (301) 855-2747. (800) PTOLEMY. Fax: (410) 257-2824.

Royal Astronomical Society of Canada. 136 Dupont Street, Toronto, ON N5R 1V2 Canada. (416) 924-7973. Fax: (416) 968-6687.

BIBLIOGRAPHIES

Astronomy and Astrophysics: a Source Guide. Gordon Press, P.O. Box 459, Bowling Green Station, New York, NY 10004. (718) 624-8419. 1991. $250.00.

A Bibliography of Astronomy, 1970-1979. R. A. Sea and S. S. Martin. Libraries Unlimited, Inc., Littleton, CO 80160. 1982.

Catalog of the Naval Observatory Library. Naval Observatory Staff. G.K. Hall & Company, 866 Third Avenue, New York, NY 10022. (212)702-6789. (800) 257-5755. 6 volumes. 1977. $655.00

Science Books and Films. American Association for the Advancement of Science, 1333 H Street NW, Washington, DC 20005. (202) 326-6454. 1965 to present. Nine issues per year. $40.00 per year. ISSN: 0098-342X

Scientific and Technical Books and Serials in Print, 1995. R.R. Bowker Inc., 121 Chanlon Road, New Providence, NJ 07974. (908) 464-6800. (800) 521-8110. 4 volumes. 1994. $299.95. Also available on compact disc and online.

A Source Book on Astronomy and Astrophysics, 1900-1975. Kenneth Lang and Owen Gingerich, editors. Harvard University Press, 79 Garden Street, Cambridge MA. (617) 495-2600. 1980. $41.95.

DIRECTORIES AND BIOGRAPHICAL SOURCES

Aamerican Men and Women of Science: Physical and Biological Sciences. R. R. Bowker Inc., 121 Chanlon Road, New Providence, NJ 07974. (908) 464-6800. (800) 521-8110. 20th edition. 8 volumes. 1996. $850.00.

The Astronomers. Donald Goldsmith. St. Martin's Press, Inc., 175 Fifth Avenue, New York, NY 10010. (212) 674-5151.(800) 221-7945. 1993. $14.95 in paper.

Astronomical Centers of the World. Kevin Krisciunas. Cambridge University Press, 40 West 20th Street, New York, NY 10011-4211. (212) 924-3900. (800) 872-7423. 1988.$34.95

The Biographical Dictionary of Scientists: Astronomers. D.Abbott, editor. Peter Bedrick Books, Inc., 2112 Broadway, Room 318, New York, NY 10023. (212) 496-0751. 1984.

Directory of Physics and Astronomy Staff Members. American Institute of Physics, One Physics Ellipse, College Park, MD20740-3843. (301) 209-3100. Annual. $45.00.

Earth and Astronomical Research Centers. Stockton Press, 345 Park Avenue South, New York, NY 10010. 4th edition. 1995. $515.00. ISBN: 1-56169-0967.

Graduate Programs in Physics, Astronomy and Related Fields, 1995-1996. American Institute of Physics, One Physics Ellipse, College Park, MD 20740-3843. (301) 209-3100. 1978 to present. Annual. $45.00.

Research Centers Directory. Gale Research, 835 Penobscot Building, Detroit, MI 48226-4094. (313) 961-2242.(800) 877-4253. 20th edition. 1995. $485.00.

ENCYCLOPEDIAS AND DICTIONARIES

Concise Dictionary of Astronomy. Jacqueline Mitton. Oxford University Press, Inc., 200 Madison Avenue, New York, NY 10016. (212) 725-6000. (800) 334-4249. 1992. $24.95.

Facts on File Dictionary of Astronomy. Valerie Illingworth. Facts On File, Inc., 460 Park Avenue South, New York, NY 10016-7382. (212) 683-2244. (800) 322-8755. Fax: (800) 683-3633. 1994. $27.95.

Encyclopedia of Astronomy and Astrophysics. Stephen Maran, editor. Van Nostrand Reinhold, 115 Fifth Avenue, New York, NY 10003. (212) 254-3232. (800) 842-3636. 1992. $129.95.

Encyclopedia of Cosmology: Historical, Philosophical and Scientific Foundations of Modern Cosmology. Norris S. Hetherington, editor. Garland Publishing, Inc., 717 Fifth Avenue, Suite 2500, New York, NY 10022-8102. (212) 751-7447. (800) 627-6273. 1993. $125.00.

Oxford Illustrated Encyclopedia. Volume 8, The Universe. Archie Roy, editor. Oxford University Press, Inc., 200 Madison Avenue, New York, NY 10016. (212) 725-6000. (800) 334-4249. 1993. $45.00.

McGraw-Hill Encyclopedia of Science and Technology. McGraw-Hill Book Company, Inc., 1221 Avenue of the Americas, New York, NY 10020. (212) 512-2000. (800) 262-4729. 7th edition. 20 volume set. 1992. $1900.00.

Penguin Dictionary of Astronomy. Jacqueline Mitton. Viking Penguin, 375 Hudson Street, New York, NY 10014-3657. (212) 366-2000. (800) 331-4624. 1993. $12.95 in paper.

Stars, Galaxies, Cosmos. William R. Corliss. Sourcebook Project, P.O. Box 107, Glen Arm, MD 21057. (410) 668-6047. 1987. $17.95.

GENERAL WORKS

Atoms, Stars, and Nebulae. Lawrence H. Aller. Cambridge University Press, 40 West 20th Street, New York, NY 10011-4211. (212) 924-3900. (800) 872-7423. 1991. $34.95 in paper.

Bblinded by the Light: New Theories about the Sun and the Search for Dark Matter. John Gribbin. Crown Publishing Group, East 50th Street, New York, NY 10022. (212) 751-2600. (800) 726-0600. 1991. $20.00

Cosmic Questions, Galactic Halos, Cold Dark Matter and the End of Time. Richard Morris. John Wiley & Sons, Inc., 605 Third Avenue, New York, NY 10158-0012. (212) 850-6000. (800) 225-5945. 1993. $24.95.

The Cosmos. Time-Life, Inc., 777 Duke Street, Alexandria, VA 22314. (703) 8388-7000. (800) 621-7026. Voyage Through the Universe series. 1989. $24.50.

Cycles of Fire: Stars, Galaxies and the Wonder of Deepspace. William K. Hartmann. Workman Publishing Company, Inc., 708 Broadway, New York, NY 10003. (212) 254-5900. (800) 722-7202. 1988. $15.95 in paper.

Farthest Things in the Universe. Jay M. Pasachoff. Cambridge University Press 40 West 20th Street, New York, NY10011-4211. (212) 924-3900. (800) 872-7423. 1995. $29.95.

The First Three Minutes: A Modern View of the Origin of the Universe. Steven Weinberg. Basic Books, 10 East 53rd Street, New York, NY 10022. (212) 207-7057. (800) 242-7737. 2nd edition; Reprint edition. 1993. $12.00.

The Left Hand of Creation: The Origin and Evolution of the Expanding Universe. John D. Barrow and Joseph Silk. Oxford University Press, Inc., 200 Madison Avenue, New York, NY 10016. (212) 725-6000. (800) 334-4249. 1994. $23.00.

Modern Cosmology and the Dark Matter Problem. Dennis W. Sciama. Cambridge University Press 40 West 20th Street, NewYork, NY 10011-4211. (212) 924-3900. (800) 872-7423.1994. $29.95 in paper.

Origin of the Universe. John D. Barrow. Basic Books, 10 East 53rd Street, New York, NY 10022. (212) 207-7057. (800) 242-7737. 1994. $20.00.

Particle Physics and Cosmology: Dark Matter. M. Srednicki, editor. Elsevier Science Publishing Company, Inc., 655 Avenue of the Americas, New York, NY 10010. (212) 989-5800. 1990. $92.50.

The Physics of Stars. A.C. Phillips. John Wiley & Sons, Inc., 605 Third Avenue, New York, NY 10158-0012. (212) 850-6000. (800) 225-5945. 1994. $32.95 in paper.

The Shadows of Creation: Dark Matter and the Structure of the Universe. Michael Riordan and David Schramm. W.H. Freeman & Co., 41 Madison Avenue, East 26th Avenue, 35th Floor, New York, NY 10010. (212) 576-0400. 1995. $19.95.

Space, Time and Gravity: the Theory of the Big Bang and Black Holes. Robert M. Wald. University of Chicago Press, 5801 Ellis Avenue, Chicago, IL 60637. (312) 702-7700. 1992. $24.95.

Stars and Atoms: from the Big Bang to the Solar System. Stuart Clark. Oxford University Press, Inc., 200 Madison Avenue, New York, NY 10016. (212) 725-6000. (800) 334-4249.1995. $35.00.

HANDBOOKS AND MANUALS

Burnham's Celestial Handbook: an Observer's Guide to the Universe Beyond the Solar System. Robert Burnham, Jr. Dover Publications, Inc., 180 Varick Street, New York, NY 10014. (212) 255-3755. (800) 223-3130. 3 volumes. 1978. $41.85 for the set.

Cambridge Guide to the Constellations. Michael E. Bakich. Cambridge University Press, 40 West 20th Street, New York, NY 10011-4211. (212) 924-3900. (800) 872-7423. 1995.$69.95.

Encyclopedia of Astronomy and Astrophysics. Stephen Maran, editor. Van Nostrand Reinhold, 115 Fifth Avenue, New York,NY 10003. (212) 254-3232. (800) 842-3636. 1992. $129.95.

Greenwich Guide to Stars, Galaxies and Nebulae. Stuart Malin. Cambridge University Press, 40 West 20th Street, New York, NY 10011-4211. (212) 924-3900. (800) 872-7423. 1990. $10.95.

DARK MATTER

Ency. of Physical Sciences and Engineering Info. Sources

Handbook of Astronomy, Astrophysics and Geophysics. Volume 2: Galaxies and Cosmology. V. M. Canuto and B. G. Elmegreen. Gordon & Breach Science Publishers, Inc., 50 West 23rd Street, New York, NY 10010. (212) 206-8900 . 1988. $194.00.

Handbook of Space Astronomy and Astrophysics. Martin V. Zumbeck. Cambridge University Press, 40 West 20th Street, New York, NY 10011-4211. (212) 924-3900. (800) 872-7423. 2nd edition. 1990. $79.95.

Landolt-borenstein Numerical Data and Functional Relationships in Science and Technology: Astronomy, Astrophysics and Space Research. Group VI. Springer-Verlag New York, Inc., 175 Fifth Avenue, New York, NY 10010.(212) 460-1500. (800) 777-4643. Fax: (212) 473-6272. 1965. Irregular. Volumes priced individually. ISSN: 0075-7896.

Observer's Guide to Astronomy. Patrick Martinez, editor. Cambridge University Press, 40 West 20th Street, New York,NY 10011-4211. (212) 924-3900. (800) 872-7423. 2 volumes.1994. $159.90.

Observing Handbook and Catalog of Deep-sky Objects. Christian B. Luginbuhl and Brian A. Skiff. Cambridge University Press, 40 West 20th Street, New York, NY 10011- 4211. (212) 924-3900. (800) 872-7423. 1990. $49.95.

ONLINE DATABASES AND CD-ROMS

CA Search. Chemical Abstracts Service, P.O. Box 3012, Columbus, OH 43210-0012. (614) 447-3600. (800) 848-6533. Fax: (614) 447-3709. Very comprehensive guide to worldwide chemical literature and related fields, 1972 to present. Available on BRS,(800) 289-4277, DIALOG, (800) 334-2564, ORBIT (800) 456-7248, and STN International, FIZ Karlsruhe, P.O. Box 2465, W-7500, Karlsruhe 1, Germany, online services. Inquire as to cost and availability.

Current Contents Search. Institute for Scientific Information, 3501 Market Street, Philadelphia, PA 19104. (215) 386-0100. Fax: (215) 386-6362. Contains citations to articles listed in the table of contents of science and technology journals. Also articles in social sciences and life sciences journals. Available on BRS, (800) 289-4277, DIALOG, (800) 334-2564, online services. Inquire as to cost and availability.

Dissertation Abstracts Online. University Microfilms International, 300 North Zeeb Road, Ann Arbor, MI 48106. (800) 521-0600 or (313) 761-4700. Scope includes virtually all doctoral dissertations accepted at accredited American institutions from 1861 to present in 252 subject areas. Available on BRS, (800) 289-4277, DIALOG, (800) 334-2564, and OCLC EPIC, (800) 848-5878, online services. Also available on CD-ROM. Inquire as to cost and availability.

INSPEC. Institution of Electrical Engineers, Michael Faraday House, Six Hills Way, Stevenage, Herts. SG1 2AY, England. Telephone: 0438 313311 or Fax: 0438 742840. Contains citations to the worldwide literature of physics, electronics and electrical engineering, computer technology, and related fields. Available on BRS, (800) 289-4277, DIALOG, (800) 334-2564, ORBIT, (800) 456-7248, and STN International, FIZ Karlsruhe, P.O. Box 2465, W-7500, Karlsruhe 1, Germany, online services. Inquire as to cost and availability.

International Aerospace Abstracts. American Institute of Aeronautics and Astronautics, 370 L'Enfant Promenade, SW, Washington, DC 20024. (202) 646-7400. Contains references and abstracts of journal and monograph literature relating to aerospace science and technology, from 1963 to present. Available through the NASA/RECON system of the National Aeronautics and Space Administration only.

NTIS Bibliographic Database. National Technical Information Service, 5285 Port Royal Road, Springfield, VA 22161. (703) 487-4929 or Fax: (703) 321-8199. Broad coverage of government-sponsored science and technology research reports, 1964 to present. Available on BRS,(800) 289-4277, DIALOG, (800) 334-2564, ORBIT, (800) 456-7248, and STN International, FIZ Karlsruhe, P.O. Box 2465, W-7500, Karlsruhe 1, Germany, online services. Also available on CD-ROM. Inquire as to cost and availability.

Physics Briefs. American Institute of Physics, 335 East 45th Street , New York, NY 10017. (212) 661-9260 or Fax:(212) 661-2036. Contains citations with abstracts of the literature of physics and related fields, 1979 to present. Available on the STN International, FIZ Karlsruhe, P.O. Box 2465, W-7500, Karlsruhe 1, Germany, online service. Inquire as to cost and availability.

Scientific and Technical Books and Serials in Print. R.R. Bowker Inc., 121 Chanlon Road, New Providence, NJ 07974. (800) 521-8110 or (908) 464-6800. List of currently published books and serials in the physical and biological sciences, engineering and technology, with subject, author and titles indexes. Available on ORBIT, (800) 456-7248, online service. Inquire as to cost and availability.

SciSearch. Institute for Scientific Information, 3501 Market Street, Philadelphia, PA 19104. (800) 523-1850 or (215) 386-0100. Broad multidisciplinary title and author index to the international literature of science and technology, 1974 to present. Available on DIALOG, (800) 334-2564, and ORBIT, (800) 456-7248, online services. Also available on CD-ROM. Inquire as to cost and availability.

Wilsonline. H.W. Wilson Company, 950 University Avenue, Bronx, NY 10452. (800) 367-6770 or (212) 588-8400. Makes available online versions of the printed H.W. Wilson indexes. 1983 to present. Available on BRS, (800) 289-4277, DIALOG, (800) 334-2564, and OCLC EPIC, (800) 848-5878, online services. Also available on CD-ROM. Inquire as to cost and availability.

OTHER SOURCES

Atlas of Deep-Sky Splendors. Hans Vehrenberg. Sky Publishing Corporation, 49 Bay State Road, Cambridge, MA 02138. (617) 864-7360. (800) 253-0245. 4th edition. 1983. $24.95.

Atlas of the Ultraviolet Sky. Richard C. Henry, Wayne B. Landsman, et al. Johns Hopkins University Press, 2715 North Charles Street, Baltimore, MD 21218-4319. (410) 516-6900. (800) 537-5487. 1988. $70.00.

Atlas of the Universe. Patrick Moore. Rand McNally & Company. 8255 North Central Park, Skokie, IL 60076-2970. (708) 673-9100. 1994. $29.95.

Cambridge Atlas of Astronomy. Jean Audouze, et al. Cambridge University Press, 40 West 20th Street, New York, NY 10011-4211. (212) 924-3900. (800) 872-7423. 3rd edition. 1994. $75.00.

Color Atlas of Galaxies. James Wray. Cambridge University Press, 40 West 20th Street, New York, NY 10011-4211. (212) 924-3900. (800) 872-7423. 1988. $94.95.

Observer's Sky Atlas. E. Karkoschka. Springer-Verlag New York, Inc., 175 Fifth Avenue, New York, NY 10010. (212) 460-1500. (800) 777-4643. 1995. $15.95.

The Sky: A User's Guide. David H. Levy. Cambridge University Press, 40 West 20th Street, New York, NY 10011- 4211. (212) 924-3900. (800) 872-7423. 1993. $14.95 in paper.

PERIODICALS

ACTA Astronomica. Copernicus Foundation for Polish Astronomy, AL Ujazdowskie 4, 00-478 Warsaw, Poland. TEL 48-22-295346. Fax: 48-22-294967. 1925 to present. Quarterly. $80.00 per year. ISSN: 0001-5237.

Astronomical Journal. American Institute of Physics, One Physics Ellipse, College Park, MD 20740-3843. (301) 209-3000. Published for the American Astronomical Society.1849 to present. Monthly. $280.00 per year. ISSN: 0004-6256.

Astronomical Society of the Pacific. Publications. Astronomical Society of the Pacific, 390 Ashton Avenue, San Francisco, CA 94112. (415) 337-1100. Fax: (415) 337-5205. 1889 to present. Monthly. $175.00 per year. ISSN: 0004-6280.

Astronomy. Kalmbach Publishing Company, P.O. Box 1612, Waukesha,WI 53187-1612. (414) 796-8776. Fax: (414) 796-1142. 1973 to present. Monthly. $27.00 per year. ISSN: 0091-6358.

Astronomy and Astrophysics. Springer-Verlag New York, Inc.,44 Hartz Way, Secaucus, NJ 07096-2491. (201) 348-4033. Fax: (201) 348-4505. 1969 to date. 36 issues per year in 12 volumes. $1733.00 per year. ISSN: 0004-6361.

Astronomy Reports. American Institute of Physics, One Physics Ellipse, College Park, MD 20740-3843. (301) 209-3000. English translation of *Astronomicheksii Zhurnal.* Formerly *Soviet Astronomy AJ.* 1957 to present. Bimonthly. $1160.00 per year. ISSN: 1063-7729.

Astroparticle Physics. Elsevier Science Publishing Company, Inc., P.O. Box 882, Madison Square Station, New York, NY10159. (212) 989-5800. Fax: (212) 633-3990. 1992 to present. Quarterly. $180.00 per year. ISSN: 0927-6505.

Astrophysical Journal. University of Chicago Press,Journals Division, 5720 South Woodlawn Avenue, Chicago, IL 60637. (312) 753-3347. Fax: (312) 753-0811. 1895 to date. An international review of astronomy and astronomical physics. Three issues per month. $740.00 per year. ISSN: 0004-637X.

Astrophysics and Space Science. Kluwer ACADemic Publishers, P.O. Box 358, Accord Station, Hingham, MA 02018-0358. (617) 871-6600. Fax: (617)871-6528. An international journal of cosmic physics. 1968 to present. 24 issues per year. $2544.00 per year. ISSN: 0004-640X.

Celestial Mechanics and Dynamical Astronomy. Kluwer Academic Publishers, P.O. Box 358, Accord Station, Hingham, MA 02018-0358.(617) 871-6600. Fax: (617) 871-6528. An international journal of space dynamics. 1969 to present. Monthly. $703.50 per year. ISSN: 0923-2958.

Experimental Astronomy. Kluwer ACADemic Publishers, P.O. Box 358, Accord Station, Hingham, MA 02018-0358. (617) 871-6600. Fax: (617) 871-6528. An international journal on astronomical instrumentation and data analysis. 1989 to present. Quarterly. $161.00 per year. ISSN: 0922-6435.

Mercury. Astronomical Society of the Pacific. 390 Ashton Avenue, San Francisco, CA 94112. (415) 337-1100. Fax: (415) 337-5205. 1972 to present. Bimonthly. $175.00 per year. ISSN: 0047-6773.

Monthly Notices of the Royal Astronomical Society. Blackwell Scientific Publication LT., Osney Mead, Oxford OX2OEL, England. TEL 0865-240201. Fax: 0865-721205. 1827 to present. Fortnightly. $1899.00 per year. ISSN: 0035-8711.

Planetary and Space Science. Elsevier Science Publishing Company, Inc., 660 White Plains Road, Terrytown, NY 10591- 5153. (914) 524-9200. Fax: (914) 333-2444. 1959 to present. Monthly. $1355.00 per year. ISSN: 0032-0633.

Sky and Telescope. Sky Publishing Corporation, P.O. Box 9111, Belmont, MA 02178. (617) 864-7360. Fax: (617) 864-6117. 1941 to present. Monthly. $27.00 per year. ISSN: 0037-6604.

Vistas in Astronomy. Albert C. Beer, editor. Elsevier Science Publishing Company, Inc., 660 White Plains Road, Terrytown, NY 10591- 5153. (914) 524-9200. Fax: (914) 333-2444. An international review journal. 1958 to present. Quarterly. $415.00 per year. ISSN: 0083-6656.

Transactions of the International Astronomical Union. Proceedings of the General Assembly. Kluwer ACADemic Publishers, P.O. Box 358, Accord Station, Hingham, MA 02018-0358. (617) 871-6600. Issues in two parts: Part A, Reports, and Part B, Proceedings. Triennial. XXI-1994. Price varies. ISSN: 0080-1372.

Webb Society Quarterly Journal. Webb Society, 194 Foundry Lane, Freemantle, Southampton, Hampshire SO1 3EQ, England. 1969 to present. Quarterly. $19.50 per year. ISSN: 0043-1680.

RESEARCH CENTERS AND INSTITUTES

Astronomy Program. University of Maryland, 1207 Computer &Space Sciences Building, College Park, MD 20742. (301) 405-3001. Fax: (301) 314-9067.

Carnegie Institution of Washington Observatories, 813 Santa Barbara Street, Pasadena, CA 91101. (818) 577-1122. Fax: (818) 759-8136.

Center for Particle Astrophysics. University of California, Berkeley, 301 LaConte Hall, Berleley, CA 94720. (415) 642- 4705. Fax: (415) 643-8497.

Center for Space Research. Massachusetts Institute of Technology, 77 Massachusetts Avenue, Camridge MA 02139.(617) 253-7501.

Institute for Astronomy. University of Hawaii at Manoa. 2680 Woodlawn Drive, Honolulu, HI 96822. (808) 956-8312. Fax: (808) 988-2790.

Joint Institute for Laboratory Astrophysics. University of Colorado-Boulder, Boulder CO 80309-0440. (303) 492-7789. Fax: (303) 492-5235.

DARK MATTER

Ency. of Physical Sciences and Engineering Info. Sources

National Radio Astronomy Observatory. Edgemont Road, Charlottesville, VA 22903. (804) 296-0211.

Smithsonian Center for Astrophysics. Harvard University, 60 Garden Street, Cambridge, MA 02138. (617) 495-7461. Fax: (617) 495-7326.

Space Telescope Science Institute. 3700 San Martin Drive, Baltimore, MD 21218. (301) 338-4700.

Theoretical Astrophysics Program. University of Arizona, Planetary Sciences Department, Tuscon, AZ 85721. (602) 621-6963.

Yerkes Observatory. University of Chicago, Williams Bay, WI53191. (414) 245-5555.

DECIMAL SYSTEM

See:MATHEMATICS

DECISION THEORY

See:PROBABILITY

DECODING

See:COMPUTER SECURITY

DEEP SEA PLATFORM

See:MARINE ENGINEERING

DEEP SEA TRENCHES

See:OCEANOGRAPHY

DEPRESSIONS

See:METEOROLOGY

DESERTIFICATION

See also: CLIMATOLOGY, DROUGHT, METEOROLOGY

ABSTRACT SERVICES AND INDEXES

Applied Science and Technology Index. H.W. Wilson Co., 950 University Avenue, Bronx, NY 10452. (800) 367-6770 or (212) 588-8400. Fax: (718) 590-1617. A cumulative subject index to English language periodicals in the fields of aeronautics and space science, computer technology, chemistry, construction

industry, energy and related areas. From 1958 to present. Monthly. Inquire about cost and availability. Also available on CD-ROM and online.

Current Contents: Physical, Chemical and Earth Sciences. Institute for Scientific Information, 3501 Market Street, Philadelphia, PA 19104. (215) 386-0100. Fax: (215) 386-6362. Weekly. $360.00 per year.

Environmental Abstracts. R.R. Bowker Inc., 121 Chanlon Road, New Providence, NJ 07974. (800) 521-8110 or (908) 464-6800. Contains citations to literature relating to environmental issues and problems, from 1971 to present. Inquire as to cost and availability.

Environmental Engineering Abstracts. Cambridge Scientific Abstracts, 7200 Wisconsin Avenue, Bethesda, MD 20814-4823. (301) 961-6750. Fax: (301) 961-6720. Monthly. Covers hazardous materials, environmental impact and protection, treatment of sewage and industrial wastes, hydroelectric power, tidal and wind power, arctic and tropical engineering.

Meteorological and Geoastrophysical Abstracts. American Meteorological Society, c/o Inforonics Inc., 550 Newtown Road, Littleton, MA 01460. (508) 486-8976. Fax: (508) 486-0027. 1950 to present. Monthly. $950.00 per year. Current abstracts of books, reports, research papers, and miscellaneous literature on environmental sciences, meteorology, astrophysics, hydrology, glaciology, and physical oceanography.

Science Citation Index. Institute for Scientific Information, 3501 Market Street, Philadelphia, PA 19104. (215) 386-0100. Fax: (215) 386-6362. Inquire about availability and cost. Also available on CD-ROM.

ASSOCIATION AND PROFESSIONAL SOCIETIES

American Association of State Climatologists. State Climatology Office, 5325 Borlaugh Hall, University of Minnesota, St. Paul, MN 55108. (612) 296-4214. Fax: (612) 625-2208.

American Meteorological Society. 45 Beacon Street, Boston, MA 02108-3693. (617) 227-2425. Fax: (617) 742-8718.

Convention to Combat Desertification. Interim Secretariat: Geneva Executive Centre, C.P. 76, 1219 Chatelaine, Switzerland. Fax: (41-22) 979-9030/1.

IUCN-The World Conservation Union. Rue Mauverney 28, CH-1196 Gland, Switzerland. Telephone 41-22-999-0001. Fax: 41-22-999-0002.

National Weather Service. National Meteorological Center, 5200 Auth Road, Suite 101, Camp Springs, MD 20233. (301) 763-8016.

United Nations Environment Programme. P.O. Box 30552, Nairobi, Kenya. Telephone 254-2-621234. Fax: 254-2-623926.

Weather Modification Association. P.O. Box 8116, Fresno, CA 93747. (209) 434-3486.

World Conservation Monitoring Centre. 219 Huntingdon Road, Cambridge CB3 0DL, United Kingdom. Telephone 44 (0) 1223-277314. Fax: 44 (0) 1223-277136.

Worldwide Fund for Nature International. World Conservation Centre, Avenue du Mont Blanc, CH-1196 Gland, Switzerland.

DIRECTORIES AND BIOGRAPHICAL SOURCES

American Men & Women of Science. R.R. Bowker Staff, ed. 19th edition. 8 volumes. R.R. Bowker/Reed International Publishing Company, 121 Chanlon Road, New Providence, NJ 07974. (908) 464-6800 or (800) 521-8110. 1995. $850.00

Directory of Engineering Societies and Related Organizations. Gordon Davis. 13th edition. American Association of Engineering Societies, 1111 19th Street NW, Suite 608, Washington, DC 20036. (202) 296-2237 or (800) 658-8897. 1989. Inquire as to price.

International Research Centers Directory. Gale Research, 835 Penobscot Building, Detroit, MI 48226-4094. (313) 961-2242. (800) 347-4253. Fax: (313) 961-6083. 8th edition. 1995. $410.00

Research Centers Directory. 20th edition. 2 volumes. Gale Research, 835 Penobscot Building, Detroit, MI 48226-4094. (313) 961-2242. (800) 347-4253. Fax: (313) 961-6083. $485.00.

Scientific and Technical Organizations and Agencies Directory. Gale Research, 835 Penobscot Building, Detroit, MI 48226-4094. (313) 961-2242. (800) 347-4253. Fax: (313) 961-6083. 4th edition. 1996. $195.00.

Who's Who in Technology. Gale Research, 835 Penobscot Building, Detroit, MI 48226-4094. (313) 961-2242. (800) 347-4253. Fax: (313) 961-6083. 7th edition. 1995. $195.00.

ENCYCLOPEDIAS AND DICTIONARIES

The Encyclopedia of Climatology. Rhodes W. Fairbridge and John E. Oliver. Chapman and Hall, 29 West 35th Street, New York, NY 10001. (800) 842-3636. Fax: (212) 563-2269. 1986. $115.00.

Encyclopedia of Physical Science and Technology. Robert A. Meyers, ed. Academic Press, Inc., 6277 Sea Harbor Drive, Orlando, FL 32887. (800) 321-5068. 18 volumes. 1992. $2100.00

McGraw-Hill Dictionary of Scientific and Technical Terms. Sybil P. Parker, ed. 5th edition. McGraw-Hill Publishing Company, 1221 Avenue of the Americas, New York, NY 10020. (800) 262-4729 or (212) 512-3825. 1993. $110.50.

GENERAL WORKS

Desertification: Exploding the Myth. David S.G. Thomas, Nicholas J. Middleton. John Wiley and Sons, Inc., 605 Third Avenue, New York, NY 10158. (800) 526-5368 or (212) 850-6000. 1994. $54.95.

Desertification: Monitoring and Control. A.K. Tewari. State Mutual Book & Periodical Service Ltd., 521 Fifth Ave., 17th floor, New York, NY 10175. (718) 261-1704. Fax: (516) 537-0412. 1988. $135.00.

Desertification: Natural Background and Human Mismanagement. M. Mainguet. Springer-Verlag, 175 Fifth Avenue, New York, NY 10010. (212) 460-1500 or (800) 777-4643. Fax: (212) 473-6272. 1994. $69.00.

Desertification of the World and Their Control. T.S. Chouhan. State Mutual Book & Periodical Service Ltd., 521 Fifth Ave., 17th floor, New York, NY 10175. (718) 261-1704. Fax: (516) 537-0412. 1993. $225.00.

Interactions of Desertification and Climate. Martin A. Williams, editor. Routledge, Chapman and Hall, Inc., 29 West 35th Street, New York, NY 10001-2291. (212) 244-3336 or Fax: (212) 563-2269. 1995. Inquire for price.

World Atlas of Desertification. United Nations Environment Program staff. Halsted Press (Division of John Wiley and Sons, Inc.), 605 Third Avenue, New York, NY 10158. (800) 526-5368 or (212) 850-6418. 1995. $149.00.

ONLINE DATABASES AND CD-ROMS

Current Contents Search. Institute for Scientific Information, 3501 Market Street, Philadelphia, PA 19104. (215) 386-0100. Fax: (215) 386-6362. Contains citations to articles listed in the table of contents of science and technology journals. Also articles in social sciences and life sciences journals. Available on BRS, (800) 289-4277, DIALOG, (800) 334-2564, online services. Inquire as to cost and availability.

Dissertation Abstracts. University Microfilms International, 300 North Zeeb Road, Ann Arbor, MI 48106. (800) 521-0600 or (313) 761-4700. Scope includes virtually all doctoral dissertations accepted at accredited American institutions from 1861 to present in 252 subject areas. Available on BRS, (800) 289-4277, DIALOG, (800) 334-2564, and OCLC EPIC, (800) 848-5878, online services. Also available on CD-ROM. Inquire as to cost and availability.

NTIS Bibliographic Database. National Technical Information Service, 5285 Port Royal Road, Springfield, VA 22161. (703) 487-4929 or Fax: (703) 321-8199. Broad coverage of government-sponsored science and technology research reports, 1964 to present. Available on BRS, (800) 289-4277, DIALOG, (800) 334-2564, ORBIT, (800) 456-7248, and STN International, FIZ Karlsruhe, P.O. Box 2465, W-7500, Karlsruhe 1, Germany, online services. Also available on CD-ROM. Inquire as to cost and availability.

SCISEARCH. Institute for Scientific Information, 3501 Market Street, Philadelphia, PA 19104. (800) 523-1850 or (215) 386-0100. Broad multidisciplinary title and author index to the international literature of science and technology, 1974 to present. Available on DIALOG, (800) 334-2564, and ORBIT, (800) 456-7248, online services. Also available on CD-ROM. Inquire as to cost and availability.

WILSONLINE. H.W. Wilson Company, 950 University Avenue, Bronx, NY 10452. (800) 367-6770 or (212) 588-8400. Makes available online versions of the printed H.W. Wilson indexes including Applied Science and Technology Index, Business Periodicals Index, General Science Index, and Readers' Guide to Periodical Literature. Period covered is generally 1983 to present. Available on BRS, (800) 289-4277, DIALOG, (800) 334-2564, and OCLC EPIC, (800) 848-5878, online services. Also available on CD-ROM. Inquire as to cost and availability.

PERIODICALS

Agricultural and Forest Meteorology. Elsevier Science Inc., P.O. Box 882, Madison Square Station, New York, NY 10159. (212)

DESERTIFICATION

Ency. of Physical Sciences and Engineering Info. Sources

989-5800. Fax: (212) 633-3990. 1964 to present. 20 times a year. $1070.00.

American Meteorological Society Bulletin. 45 Beacon Street, Boston, MA 02108-3693. (617) 227-2425. Fax: (617) 742-8718. 1920 to present. Monthly. $60.00.

Climatic Change. Kluwer Academic Publishers, P.O. Box 358, Accord Station, Hingham, MA 02018-0358. (617) 871-6000. 1977 to present. 12 times a year. $589.50.

Journal of Applied Meteorology. American Meteorological Society, 45 Beacon Street, Boston, MA 02108-3693. (617) 227-2425. Fax: (617) 742-8718. 1962 to present. Monthly. $200.00.

Journal of Climate. American Meteorological Society, 45 Beacon Street, Boston, MA 02108-3693. (617) 227-2425. Fax: (617) 742-8718. 1986 to present. Monthly. $190.00.

Journal of the Atmospheric Sciences. American Meteorological Society, 45 Beacon Street, Boston, MA 02108-3693. (617) 227-2425. Fax: (617) 742-8718. 1944 to present. Semi-monthly. $350.00.

Monthly Weather Review. American Meteorological Society, 45 Beacon Street, Boston, MA 02108-3693. (617) 227-2425. Fax: (617) 742-8718. 1872 to present. Monthly. $295.00.

RESEARCH CENTERS AND INSTITUTES

World Resources Institute. 1709 New York Avenue NW, Suite, 700, Washington, DC 20006. (202) 638-6300. Fax: (202) 638-0036.

DESSICATION

See:INDUSTRIAL ENGINEERING

DESIGN ENGINEERING

See:CAD

DETERGENTS

See also: CHEMICAL ENGINEERING, CHEMISTRY

ABSTRACT SERVICES AND INDEXES

Applied Science and Technology Index. H.W. Wilson Co., 950 University Avenue, Bronx, NY 10452-9978. (212) 588-8400. (800) 367-6770. Fax: (718) 590-1617. A cumulative subject index to English language periodicals in the fields of aeronautics and space science, computer technology, chemistry, construction industry, energy and related areas. From 1958 to present. Monthly. Inquire about cost and availability. Also available online (BRS and Wilsonline) and on CD-ROM. ISSN: 0003-6986.

Chemical Abstracts. Chemical Abstracts Service. 2540 Olentangy River Road, P.O. Box 3012, Columbus, OH 43210-0012. (614) 447-3600. Fax: (614) 447-3713. 1907 to present. Weekly. $16,800.00 per year. Available online and on CD-ROM. CA is also available in five section groupings. Inquire regarding cost and availability.

Current Contents: Physical, Chemical and Earth Sciences. Institute for Scientific Information, 3501 Market Street, Philadelphia, PA 19104. (215) 386-0100. (800) 523-1850. Fax: (215) 386-2291. 1961 to present. Weekly. $442.00 per year. Also available online (BRS, DIALOG) and on CD-ROM. Inquire regarding cost and availability. ISSN: 0163-2574.

Engineering Index Monthly. Engineering Information, Inc., Castle Point on the Hudson, Hoboken, NJ 07030. (201) 216-8500. (800) 221-1044. Fax: (201) 216-8532. Monthly. $2300.00 per year for monthly issues. $1980 per year for annual cumulation. Available on CD-ROM and online as COMPENDEX. ISSN: 0742-1974; 0360-8557.

General Science Index. H.W. Wilson Company, 950 University Avenue, Bronx, NY 10452. (212) 588-8400. (800) 367-6770. Fax: (718) 590-1617. From 1978 to present. Ten issues per year. Quarterly and annual cumulations. Service basis. Available on CD-ROM and online. Inquire regarding cost and availability. ISSN: 0162-1963.

Pphysics Abstracts. Inspec. Section A, Science Abstracts. Institution of Electrical Engineers (IEE), London, United Kingdom. Available from: INSPEC/IEEE - Institute of Electrical and Electronic Engineers, P.O. Box 1331, 445 Hoes Lane, Piscataway, NJ 08855-1331. (908) 562-5549. 1898 to present. 24 issues per year. $2835.00 per year. ISSN: 0036-8091. Also available online and on CD-ROM.

Process and Chemical Engineering/Chemical Engineering Abstracts. Royal Society of Chemistry, Information Services, Thomas Graham House, Science Park, Milton Road, Cambridge, CB4 4WF, England. Contains citations, mostly with abstracts, of the worldwide literature on chemical engineering. from 1982 to present. Monthly. $610.00 per year. Also available online. ISSN: 0960-5045.

Science Citation Index (SCI). Institute for Scientific Information, 3501 Market Street, Philadelphia, PA 19104. (215) 386-0100. (800) 523-1850. Fax: (215) 386-2991. 1961 to present. Six issues per year, plus annual cumulation. $11650.00 per year. Also available online and on CD-ROM. Inquire about price and availability. ISSN:0036-827X.

Theoretical Chemical Engineering. Royal Society of Chemistry, Information Services, Thomas Graham House, Science Park, Milton Road, Cambridge, CB4 4WF, England. Covers theoretical chemical engineering, including theory and laboratory experimentation. 1964 to present. Monthly. $235.00. ISSN: 0960-5053.

ASSOCIATIONS AND PROFESSIONAL SOCIETIES

American Chemical Society, 1155 16th Street, NW, Washington, DC 20036. (202) 872-4600.

American Institute of Chemical Engineers. 345 East 47th Street, New York, NY 10017. (212) 705-7338.

Association of Consulting Chemists and Chemical Engineers, 50 East 41st Street, Suite 92, New York, NY 10017. (212) 684-6255.

Association of Official Analytical Chemists. 2200 Wilson Blvd, Suite 400, Arlington, VA 22001. (703) 522-3032.

Institute for Polyacrylate Absorbants. 1100 New York Avenue, no. 1090, Washington, DC 20005. (202) 414-4190. Fax: (202) 289-8584.

Soap and Detergent Association. 475 Park Avenue South, New York, NY 10016. (212) 725-1262. Fax: (212) 213-0685.

BIBLIOGRAPHIES

Chemical Information Sources. Gary Wiggins. McGraw-Hill Publishing Company, 1221 Avenue of the Americas, New York, NY 10020. (800) 262-4729 or (212) 512-3825. 1991. $42.50.

Scientific and Technical Books and Serials in Print. An index to Literature in Science and Technology. R.R. Bowker, Inc., 121 Chanlon Road, New Providence, NJ 07974. (908)464-6800. (800) 521-8110. Fax: (908) 665-3502. 1972 to present. Annual. 4 volumes. 1994. $299.95. Also available on compact disc and online. ISSN: 0000-054X.

DIRECTORIES AND BIOGRAPHICAL SOURCES

American Institute of Chemical Engineers. Directory. American Institute of Chemical Engineers. 345 East 47th Street, New York, NY 10017-2396. (212) 705-7663. Annual.

American Men and Women of Science. Physical and Biological Sciences. R.R. Bowker, Inc., 121 Chanlon Road, New Providence, NJ 07974. (908) 464-6800. (800) 521-8110. 20th edition. 8 volumes. 1996. $850.00

Consulting Services: Chemists and Chemical Engineers. Association of Consulting Chemists and Chemical Engineers, 40 East 45th Street, New York, NY 10017. (212) 983-3160. Fax: (212) 983-3161. Biennial. $60.00.

Directory of Chemistry Software 1992. Wendy Warr, Peter Willett, Geoff Downs. American Chemical Society, 1155 16th Street, NW, Washington, DC 20036. (202) 872-4600. 1992. $35.95.

Research Centers Directory. Gale Research, 835 Penobscot Building, Detroit, MI 48226-4094. (313) 961-2242. (800) 347-4253. 20th edition. 1995. $485.00. ISSN: 0080-1518.

Who's Who in Engineering. Gordon Davis, editor. American Association of Engineering Societies. 1111 19th Street, NY, Suite 608, Washington, DC 20036. (202) 296-2237. (800) 658-8897. 9th edition. 1995. $220.00.

ENCYCLOPEDIAS AND DICTIONARIES

Condensed Encyclopedia of Surfactants. Michael Ash and Irene Ash, ed. Chemical Publishing Company, Inc., 80 Eighth Avenue, New York, NY 10011. (212) 255-1950. 1989. $125.00.

Dictionary of Named Processes in Chemical Technology. Alan E. Comyns. Oxford University Press, Inc., 200 Madison Av-

enue, New York, NY 10016. (212) 725-6000. (800) 334-4249.1994. $75.00.

Dictionary of Surfactants. K. Siekamnn. Springer-Verlag New York, Inc., 175 Fifth Avenue, New York, NY 10010. (212) 460-1500. (800) 777-4643. 1987. $95.00 in paper.

Encyclopedia of Chemical Processing and Design. Marcel Dekker, Inc., 270 Madison Avenue, New York, NY 10016.(212) 696-9000. (800) 228-1160. 1976 - . $175.00 per volume.

Industrial Chemical Thesaurus. Michael Ash and Irene Ash. VCH Publishers, Inc., 220 East 23rd Street, New York, NY 10010-4606. (800) 367-8249. 1992. $295.00.

Kirk-Othmer Encyclopedia of Chemical Technology. John Wiley and Sons, Inc., 605 Third Avenue, New York, NY 10158. (800) 526-5368 or (212) 850-6000. Fourth edition. 1991-. 27 volumes. $5400.00.

McGraw-Hill Encyclopedia of Science and Technology. McGraw-Hill Book, Inc., 1221 Avenue of the Americas, New York, NY 10020. (212) 997-3675. (800) 262-4729. Seventh edition. 20 volumes. 1992. $1900.00.

Ullman's Encyclopedia of Industrial Chemistry. VCH Publications, Inc., 220 East 23rd Street, Suite 909, New York, NY 10010-4606. (212) 683-8333. (800) 422-8824. 5th edition. 1984 - . Price varies per volume.

GENERAL WORKS

Advanced Cleaning Product Formulations: Household, Industrial, Automotive. Ernest W. Flick. Noyes Data Corporation, 120 Mill Road, Park Ridge, NJ 07656. (201) 391-8484. 1989. $56.00.

Biodegradable Detergents and the Environment. William McGucken. 1991. $38.50. Texas A & M University Press, Lewis Street, John Lindsey Building, College Station, TX 77843-4354. (409) 845-1436. (800) 826-8911. 1991. $38.50.

Tthe Consumer's Guide to Household Compounds. Robert J.Palma, Sr. and Mark Espenscheid. Prometheus Books, 59 John Glenn Drive, Amherst, NY 14228-2197. (716) 691-0133. (800)421-0351. 1992. $22.95.

Detergents and Textile Washing: Principles and Practice. Gunter Jakobi and Albrecht Lohr. VCH Publications, Inc., 220 East 23rd Street, Suite 909, New York, NY 10010-4606. (212) 683-8333. (800) 422-8824. 1988. $90.00.

Formulary of Detergents and Other Cleaning Agents. M. Ash and I. Ash. Chemical Publishing Company, Inc., 80 Eighth Avenue, New York, NY 10011. (212) 255-1950. 1980. $60.00.

Industrial Surfactants: An Industrial Guide. Ernest W.Flick. Noyes Data Corporation, 120 Mill Road, Park Ridge, NJ 07656. (201) 391-8484. 1993. $86.00.

Introduction to Surfactant Analysis. D.C. Cullum, ed. Routledge Chapman & Hall, Inc., 29 West 35th Street, New York, NY 10001-3291. (212) 244-3336. 1994. $199.95.

Polymer Additives for High Performing Detergents. Paolo Zini. Technomic Publishing Co., 852 New Holland Avenue, Lancaster,

DETERGENTS

Ency. of Physical Sciences and Engineering Info. Sources

PA 17604 (717-291-5609. (800) 233-9936. 1994. $65.00 in paper.

Surfactant Aggregation. J. H. Clint. Chapman & Hall, 1 Penn Plaza, New York, NY 10119. (212) 564-1060. 1991.$117.50.

Surfactant-Enhanced Subsurface Remediation: Emerging Technologies. David A. Sabatini, et al. American Chemical Society, 1155 16th Street, NW, Washington, DC 20036. (202) 872-4600. (800) 333-9511. Fax: (614) 447-3671. 1995. $84.95.

Surfactants and Interfacial Phenomena. Milton Rosen. John Wiley & Sons, Inc., 605 Third Avenue, New York, NY 10158-0012. (212) 850-6000. (800) 225-5945. 2nd edition. 1989. $79.95.

Synthetic Detergents. Alfred S. Davidsohn. Halsted Press, 605 Third Avenue, New York, 10158-0012. (212) 850-6400. 1987. $115.00.

HANDBOOKS AND MANUALS

Chemical Formulary. H. Bennett, editor. Chemical Publishing, Co., Inc. 80 Eighth Avenue, New York, NY 10011.(212 255-1950. 30 volumes. $60.00 per volume.

CRC Handbook of Chemistry and Physics. David R. Kide, editor. CRC Press Inc., 2000 Corporate Blvd., NW, Boca Raton, FL 33431. (407) 994-0555. (800) 333-8300. 77th edition. 1996. $99.95.

Guide to Basic Chemical Compounds. D.R. Lide, Jr. CRC Publishers, Inc., 2000 Corporate Blvd., NW, Boca Raton, FL 33431. (407) 994-0555 or (800) 333-8300. 1993. $120.00.

Handbook for Detergent Formulators. K. Robert Lange, editor. Hanser-Gardner Publications, 6600 Clough Pike, Cincinnati, OH 45244-4090. (513) 527-8977. (800) 950-8977. 1994.

Hhandbook of Industrial Surfactants. Irene Ash and Michael Ash. Ashgate Publishing Co., Old Post Road, Brookfield, VT 05036. (802) 276-3162. 1993. $250.00.

Handbook of Surfactants. M.R. Porter. Routledge Chapman & Hall, Inc., 29 West 35th Street, New York, NY 10001-3291. (212) 244-3336. 1991. $99.50.

ONLINE DATABASES AND CD-ROMS

Analytical Abstracts Online. Royal Society of Chemistry, Information Services, Thomas Graham House, Science Park, Milton Road, Cambridge, CB4 4WF, England. Contains citations, mostly with abstracts, of the worldwide literature on analytical chemistry, from 1980 to present. Available on DIALOG and STN online services. Inquire as to cost and availability.

CA Search. Chemical Abstracts Service, P.O. Box 3012, Columbus, OH 43210-0012. (614) 447-3600. (800) 848-6533. Fax: (614) 447-3709. Guide to worldwide chemical literature and related fields, 1972 to present. Available on BRS, DIALOG, ORBIT and STN online services. Inquire as to cost and availability.

Dissertation Abstracts. University Microfilms International, 300 North Zeeb Road, Ann Arbor, MI 48106. (800) 521-0600 or (313) 761-4700. Scope includes virtually all doctoral dissertations accepted at accredited American institutions from 1861 to present in 252 subject areas. Available on BRS, DIALOG, and OCLC EPIC online services. Also available on CD-ROM. Inquire as to cost and availability.

INSPEC. Institution of Electrical Engineers, Michael Faraday House, Six Hills Way, Stevenage, Herts. SG1 2AY, England. Telephone: 0438 313311. Fax: 0438 742840. Contains citations to the worldwide literature of physics, electronics and electrical engineering, computer technology, and related fields. Available on BRS, DIALOG, ORBIT and STN online services. Inquire as to cost and availability.

Kirk-Othmer Encyclopedia of Chemical Technology. John Wiley and Sons, Inc., 605 Third Avenue, New York, NY 10158. (800) 526-5368 or (212) 850-6000. Contains the complete text of all chapters in the 27 volume fourth edition of the *Kirk-Othmer Encyclopedia of Chemical Technology.* 1991. Available on BRS,(800) 289-4277, DIALOG,(800) 334-2564, online services. Inquire as to cost and availability.

NTIS. National Technical Information Service, 5285 Port Royal Road, Springfield, VA 22161. (703) 487-4929 or Fax: (703) 321-8199. Broad coverage of government-sponsored research reports, 1964 to present. Available on BRS, DIALOG, ORBIT and STN online services. Also available on CD-ROM. Inquire as to cost and availability.

SCISEARCH. Institute for Scientific Information, 3501 Market Street, Philadelphia, PA 19104. (800) 523-1850 or (215) 386-0100. Broad multidisciplinary title and author index to the international literature of science and technology, 1974 to present. Available on DIALOG and ORBIT online services. Also available on CD-ROM. Inquire as to cost and availability.

WILSONLINE. H.W. Wilson Company, 950 University Avenue, Bronx, NY 10452. (800) 367-6770 or (212) 588-8400. Makes available online versions of the printed H.W. Wilson indexes including Applied Science and Technology Index, Business Periodicals Index, and Readers' Guide to Periodical Literature. Period covered is generally 1983 to present. Also available on CD-ROM. Inquire as to cost and availability.

PERIODICALS

American Laboratory. American Laboratory, 30 Controls Drive, P.O. Box 870, Shelton, CT 06484-6111. (203) 926-9300. Fax: (203) 926-9310. 1968 to present. Monthly. $150.00 per year. ISSN: 0044-7749.

Analytical Chemistry. American Chemical Society, P.O. Box 3337, Columbus, OH 43210. (614) 447-3776. (800) 333-9511. Fax: (614) 447-3671. 1929 to present. Monthly. $415.00 per year. ISSN: 0003-2700.

Analytical Methods and Instrumentation. John Wiley and Sons, Inc., 605 Third Avenue, New York, NY 10158. (800) 526-5368 or (212) 850-6000. 1992 to present. Bimonthly. $250.00 per year. ISSN: 1063-5246.

Association of Official Analytical Chemists Journal. Association of Offical Analytical Chemists. 2200 Wilson Blvd, Suite 400, Arlington, VA 22001. (703) 522-3032. Bimonthly. $180.00 per year.

Chemical Week. Includes annual buyers guide. Chemical Week Associates, 888 Seventh Avenue, New York, NY 10106. (212)

621-4900. Fax: (212) 621-4949. 1914 to present. Weekly. $99.00 per year. ISSN: 0009-272X.

CHEMTECH. American Chemical Society. P.O. Box 3337, Columbus, OH 43210. (614) 447-3776. 1970 to present. Monthly. $370.00 per year. ISSN: 0009-2703.

International Laboratory. International Scientific Communications, Inc., 30 Controls Drive, Shelton, CT 06484-0870. (203) 926-9300. 1971 to present. Nine times per year. $1200 per year. ISSN: 0010-2164.

Journal of Dispersion Science and Technology. Marcel Dekker, Inc., 270 Madison Avenue, New York, NY 10016. (212) 696-9000. Fax: (212) 685-4540. Subscriptions to: P.O. Box 5017 Monticello, NY 12701. 1980 to present. Six times per year. $445.00 per year. ISSN: 0193-2691.

RESEARCH CENTERS AND INSTITUTES

International Fabricare Institute. 12251 Tech Road, Montgomery Industrial Park, Silver Spring, MD 20904. (301) 622-1900. Fax: (301) 236-9320.

Texas Research Center for Laundry and Drycleaning. Texas Woman's University, P.O. Box 22509, TWU Station, Denton, TX 76204. (817) 898-2670. Fax: (817) 898-3198.

DETONATIONS

See:EXPLOSIVES

DEUTERIUM

See:HYDROGEN

DIAMONDS, INDUSTRIAL

See also: MACHINING

ABSTRACT SERVICES AND INDEXES

Applied Science and Technology Index. H.W. Wilson Co., 950 University Avenue, Bronx, NY 10452. (800) 367-6770 or (212) 588-8400. Fax: (718) 590-1617. A cumulative subject index to English language periodicals in the fields of aeronautics and space science, computer technology, chemistry, construction industry, energy and related areas.From 1958 to present. Monthly. Inquire about cost and availability. Also available on CD-ROM and online.

Current Contents: Engineering, Technology, and Applied Sciences. Institute for Scientific Information, 3501 Market Street, Philadelphia, PA 19104. (215) 386-0100. Fax: (215) 386-6362. 1970 to present. Weekly. $442.00 per year.

Engineering Index Monthly. Engineering Information, Inc., Castle Point on the Hudson, Hoboken, NJ 07030. (800) 221-1044. Fax: (212) 832-1857. Monthly. $2300.00 per year. Also

available online as COMPENDEX, and also on CD-ROM. Covers chemical engineering, computers, electrical engineering, civil engineering, metals and mining, industrial management, and mechanical engineering.

Mineralogical Abstracts. Mineralogical Society and the Mineralogical Society of America, 41 Queen's Gate, London, SW7 5HR, England. Quarterly. $235.00 per year.

Science Citation Index. Institute for Scientific Information, 3501 Market Street, Philadelphia, PA 19104. (215) 386-0100. Fax: (215) 386-6362. Inquire about availability and cost. Also available on CD-ROM.

ASSOCIATIONS AND PROFESSIONAL SOCIETIES

American Society of Mechanical Engineers. 345 East 47th Street, New York, NY 10017-2398. (212) 705-7703.

Diamond Wheel Manufacturers Institute. 30200 Detroit Road, Cleveland, OH 44145-1967. (216) 899-0010.

Drilling Equipment Manufacturers Association. 3008 Millwood Avenue, Columbia, SC 29205. (803) 252-5646. Fax: (803) 765-0860.

Industrial Diamond Association of America. 355 Lexington Avenue, New York, NY 10017. (212) 661-4261. Fax: (212) 370-9047.

BIBLIOGRAPHIES

Scientific and Technical Books and Serials in Print. R.R. Bowker Inc., 121 Chanlon Road, New Providence, NJ 07974. (800) 521-8110 or (908) 464-6800. List of currently published books and serials in the physical and biological sciences, engineering and technology, with subject, author and titles indexes. Inquire for cost of latest edition. Also available on ORBIT online service, (800) 456-7248.

DIRECTORIES AND BIOGRAPHICAL SOURCES

BCC Diamond Directory 1992: A Comprehensive Directory of the Companies Involved in Industrial Diamond & CBN Technology. Business Communications Company, Inc., 25 Van Zant St., Norwalk, CT 06855. (203) 853-4266. Fax: (203) 853-0348. 1992. $575.00

Industrial Diamonds & CBN: Materials, Films & Products: U.S. Directory and Market Guide. Business Communications Company, Inc., 25 Van Zant St., Norwalk, CT 06855. (203) 853-4266. Fax: (203) 853-0348. 1992. $575.00

Who's Who in Engineering. American Association of Engineering Societies, 1111 19th Street NW, Suite 608, Washington, DC 20036. (202) 296-2237 or (800) 658-8897. 8th edition. 1991. Inquire for price.

ENCYCLOPEDIAS AND DICTIONARIES

Encyclopedia of Physical Science and Technology. Robert A. Meyers, ed. Academic Press Inc., 6277 Sea Harbor Drive, Orlando, FL 32887. (800) 321-5068. 18 volumes. 1992. $2100.00

McGraw-Hill Dictionary of Scientific and Technical Terms. Sybil P. Parker, ed. 5th edition. McGraw-Hill Publishing Company, 1221 Avenue of the Americas, New York, NY 10020. (800) 262-4729 or (212) 512-3825. 1993. $110.50.

McGraw-Hill Encyclopedia of Engineering. Sybil P. Parker, ed. McGraw-Hill Publishing Company, 1221 Avenue of the Americas, New York, NY 10020. (800) 262-4729 or (212) 512-3825. 2nd edition. 1993. $95.50.

McGraw-Hill Encyclopedia of Science and Technology. Sybil P. Parker, ed. McGraw-Hill Publishing Company, 1221 Avenue of the Americas, New York, NY 10020. (800) 262-4729 or (212) 512-3825. 7th edition. 20 volumes. 1992. $1900.00

Thesaurus of Scientific, Technical, and Engineering Terms. Hemisphere Publishing Corporation, 1900 Frost Road, Suite 101, Bristol, PA 19007-1598. (215) 785-5800 or (800) 821-8312. Fax: (215) 785-5515. 1987. $173.00.

GENERAL WORKS

Diamond: Electronic Properties and Applications. Lawrence S. Pan. Kluwer Academic Publishers, P.O. Box 358, Accord Station, Hingham, MA 02018-0358. (617) 871-6000. Fax: (617) 871-6528. 1994. Inquire about cost and availability.

Diamonds: Exploration, Sampling and Evaluation. Prospectors and Developers Association of Canada, Toronto, Ontario, Canada. 1993. Write for price and availability.

Practical Uses of Diamonds. A. Bakon and A. Szymanski. E. Horwood Publishing, distributed by Prentice Hall, 113 Sylvan Avenue, Route 9W, Englewood Cliffs, NJ 07632. (201) 592-2000 or (800) 922-0579. 1993. Inquire about price.

Properties & Applications of Diamonds. J. Wilks & E.M. Wilks. 3rd ed. Butterworth-Heinemann, 313 Washington Street, Newton, MA 02158. (617) 928-2500 or (800) 366-2665. Fax: (617) 928-2620. 1991. $179.00

ONLINE DATABASES AND CD-ROMS

COMPENDEX PLUS. Engineering Information, Inc., 345 East 47th Street, New York, NY 10017. (212) 705-7600 or (800) 221-1044. Contains citations with abstracts to worldwide literature in engineering and technology, from 1970 to present. Available on online BRS,(800) 289-4277, DIALOG, (800) 334-2564, ORBIT (800) 456-7248, and STN International, FIZ Karlsruhe, P.O. Box 2465, W-7500, Karlsruhe 1, Germany, online services. Also available on CD-ROM. Inquire as to cost and availability.

CSA Engineering. Cambridge Scientific Abstracts, 7200 Wisconsin Avenue, Suite 601, Bethesda, MD 20814. (301) 961-6750 or (800) 843-7751. Contains citations and abstracts of international periodicals and research literature covering all fields of engineering and science and technology,including computer and information science, electronics, mechanical engineering, solid state materials, 1981 to present. Available on BRS,(800) 289-4277, online service. Inquire as to cost and availability.

CURRENT CONTENTS SEARCH. Institute for Scientific Information, 3501 Market Street, Philadelphia, PA 19104. (215) 386-0100. Fax: (215) 386-6362. Contains citations to articles listed in the table of contents of science and technology journals. Also articles in social sciences and life sciences journals. Available

on BRS,(800) 289-4277, DIALOG,(800) 334-2564, online services. Inquire as to cost and availability.

ISMEC: Mechanical Engineering Abstracts. Cambridge Scientific Abstracts, 7200 Wisconsin Avenue, Suite 601, Bethesda, MD 20814. (301) 961-6750 or (800) 843-7751. Contains citations to the literature in mechanical engineering, industrial and production engineering, energy, power, mechanics, devices and related areas, from 1973 to present. Available on the DIALOG,(800) 334-2564, online service. Inquire as to cost and availability.

NTIS Bibliographic Database. National Technical Information Service, 5285 Port Royal Road, Springfield, VA 22161. (703) 487-4929 or Fax: (703) 321-8199. Broad coverage of government-sponsored science and technology research reports, 1964 to present. Available on BRS,(800) 289-4277, DIALOG, (800) 334-2564, ORBIT, (800) 456-7248, and STN International, FIZ Karlsruhe, P.O. Box 2465, W-7500, Karlsruhe 1, Germany, online services. Also available on CD-ROM. Inquire as to cost and availability.

PERIODICALS

Diamond and Related Materials. Elsevier Science Inc., P.O. Box 882, Madison Square Station, New York, NY 10159. (212) 989-5800. 1991 to present. 12 times a year. $799.00 per year.

Diamond Depositions. Superconductivity Publications, Sunset Plaza, 710 Easton Avenue, Suite C, Somerset, NJ 08873-1855. (908) 846-2002. Fax: (908) 846-2050. 1990 to present. 10 times a year. $132.00 per year.

Finer Points. Industrial Diamond Association of America, 355 Lexington Avenue, New York, NY 10017. (212) 661-4261. Fax: (212) 370-9047. 4 times a year. Inquire for cost.

Industrial Diamond Review. De Beers Industrial Diamond Division Pty. Ltd., Charters, SunningHill, Ascot, Berks. SL5 9PX, England. Fax: 0344-28188. 1940 to present. Bi-monthly. Inquire for cost and availability.

Industrial Minerals. Metal Bulletin Inc., 220 Fifth Avenue, 19th Floor, New York, NY 10001-7781. (800) 638-2525. Fax: (212) 213-6273. 1967 to present. Monthly. Inquire for cost.

Journal of Vacuum Science and Technology A: Vacuum, Surfaces & Films. American Institute of Physics, One Physics Ellipse, College Park, MD 20740. (301) 209-3000. 1964 to present. 6 times a year. $530.00 per year.

Journal of Vacuum Science and Technology B: Microelectronics & Nanometer Structures. American Institute of Physics, One Physics Ellipse, College Park, MD 20740. (301) 209-3000. 1964 to present. 6 times a year. $430.00 per year.

Mineral Industry Surveys, Gem Stones. U.S. Bureau of Mines, Production and Distribution, Cochrans Mill Rd., P.O. Box 18070, Pittsburgh, PA 15236. (412) 892-4411. Annually. Inquire for price and availability.

Materials Science and Technology. Institute of Materials, 1 Carlton House Terrace, London SW1Y 5DB, England. Telephone 071-839-4071. 1985 to present. Monthly. $37.00 per year.

Mining Engineering. Society for Mining, Metallurgy, and Exploration, Inc., 8307 Shaffer Parkway, P.O. Box 625002,

Littleton, CO 80162-5002. (303) 973-9550. Fax: (303) 973-3845. Monthly. $100.00 per year.

Precision Engineering. Butterworths-Heinemann/Reed Elsevier, 313 Washington Street, Newton, MA 02158. 1979 to present. Quarterly. $340.00.

RESEARCH CENTERS AND INSTITUTES

Wayne State University Diamond Laboratory. Institute for Manufacturing Research, 277 Physics Bldg., Detroit, MI 48202. (313) 577-0846. Fax: (313) 577-7743.

DIELECTRICS

See also: ELECTRICAL ENGINEERING, ELECTROMAGNETISM, ELECTRONICS, ELECTRONICS ENGINEERING, MATERIALS SCIENCE, PIEZOELECTRICS, SOLID STATE PHYSICS

ABSTRACT SERVICES AND INDEXES

Applied Mechanics Reviews. An assessment of world literature in engineering sciences. American Society of Mechanical Engineers, 345 East 47th Street, New York, NY 10017. (212)705-7703. 1948 to present. Monthly. $360.00 ISSN: 0003-6900.

Applied Science and Technology Index. H.W. Wilson Co., 950 University Avenue, Bronx, NY 10452. (212) 588-8400. (800) 367-6770. Fax: (718) 590-1617. A cumulative subject index to English language periodicals in the fields of aeronautics and space science, computer technology,chemistry, construction industry, energy and related areas. From 1958 to present. Monthly. Inquire about cost and availability. Also available on CD-ROM and online. ISSN: 0003-6986.

Current Contents: Engineering, Technology and Applied Sciences. Institute for Scientific Information, 3501 Market Street, Philadelphia, PA 19104. (215) 386-0100. Fax: (215) 386-6362. 1970 to present. Weekly. $442.00 per year. Also available on CD-ROM and online. Inquire regarding cost and availability. ISSN: 0095-7917.

Current Papers in Electrical and Electronics Engineering. Institution of Electrical Engineers (IEE), London. Distributed by INSPEC/IEEE, P.O. Box 1331, 445 Hoes Lane, Piscataway, NJ 08855-1331. (908) 562-5549. 1969 to present. Monthly. $345.00 per year. ISSN: 0011-3778.

Electric Power Industry Abstracts. Edison Electric Institute, c/o Utility Data Institute, 2011 I Street, NW, Suite 700, Washington, DC 20006. 1975 to present.Bimonthly. Inquire as to cost and availability.

Electrical and Electronics Abstracts. Institution of Electrical Engineers (IEE), London. Available from INSPEC/IEEE - Institute of Electrical and Electronic Engineers, P.O. Box 1331, Hoes Lane, Piscataway, NJ 08855-1331. (908) 562-5549. 1898 to present. Monthly. $2200.00 per year. Also available on CD-ROM and online as INSPEC.

Engineering Index Monthly. Indexes and abstracts the world's engineering and technical literature. Engineering Information, Inc., Castle Point on the Hudson, Hoboken, NJ 07030. (201) 216-8500. (800) 221-1044. Fax: (201) 216-8532. Monthly.

$2300.00 per year. Available online as COMPENDEX and also on CD-ROM. ISSN: 0742-1974.

General Science Index. H.W. Wilson Company, 950 University Avenue, Bronx, NY 10452. (212) 588-8400. (800) 367-6770. Fax: (718) 590-1617. From 1978 to present. Ten issues per year. quarterly and annual cumulations. Service basis. Available on CD-ROM and online. Inquire regarding cost and availability. ISSN: 0162-1963.

Government Reports Announcements and Index. U. S. National Technical Information Service (NTIS), 5285 Port Royal Road, Springfield, VA 22161. (703) 487-4650. Fax: (703) 321-8547. From 1968 to present. Annual. $630.00 per year. Also available online as NTIS Bibliographic Database and on CD-ROM.

Index to Ieee Publications. IEEE Service Center, 445 Hoes Lane, Piscataway, NJ 08855-1331. (908) 981-1393. (800) 678-IEEE. Fax: (908) 981-9667. 1973 to present. Annual. ISSN: 0099-1368.

Physics Abstracts. INSPEC. Section A, Science Abstracts. Institution of Electrical Engineers (IEE), London. Available from: INSPEC/IEEE - Institute of Electrical and Electronic Engineers, P.O. Box 1331, Hoes Lane, Piscataway, NJ 08855-1331. (908) 562-5549. 1898 to present. 24 issues per year. $2835.00 per year. Also available online and on CD-ROM. ISSN: 0036-8091.

PHYSICS BRIEFS (Physikalische Berichte). Information Center for Energy, Physics, Mathematics. German Physical Society. V C H Publishers, Inc., 220 East 23rd Street, New York, NY 10010-4606. (212) 683-8333. 1845 to present. 24 issues per year. $2390.00 per year. Also available online. ISSN: 0179-7434.

Science Citation Index (SCI). Institute for Scientific Information, 3501 Market Street, Philadelphia, PA 19104. (215) 386-0100. (800) 523-1850. Fax: (215) 386-2991. 1961 to present. Six issues per year, plus annual cumulation. $11650.00 per year. Also available online and on CD-ROM. Inquire about price and availability. ISSN: 0036-827X.

Solid State and Superconductivity Abstracts. Cambridge Scientific Abstracts, 7200 Wisconsin Avenue, Bethesda, MD 20824. (301) 961-6750. Fax: (301) 961-6720. Covers theory, production and application of solid state materials. 1957 to present. Bimonthly. $1320.00 per year. Also available online. ISSN: 0896-5900.

ANNUAL REVIEWS AND YEARBOOKS

Advances in Electronics and Electron Physics. Academic Press, Inc., 6277 Sea Harbor Drive, Orlando, FL 32887. (800) 321-5068. From 1948 to present. Irregular. Price varies. ISSN: 0065-2539.

Critical Reviews in Solid State and Materials Sciences. CRC Press, Inc., 2000 Corporate Boulevard, NW, Boca Raton, FL 33431. (407) 994-0555. (800) 272-7737. Fax: (407) 998- 9784. 1970 to present. Bimonthly. $265.00 per year. ISSN: 1040-8436.

Engineering Dielectrics Series. American Society for Testing and Materials, 1916 Race Street, Philadelphia, PA 19103. (215) 299-5419. Volume 3, 1993. Irregular. Inquire.

DIELECTRICS

Ency. of Physical Sciences and Engineering Info. Sources

ASSOCIATIONS AND PROFESSIONAL SOCIETIES

American Association of Engineering Societies, 1111 19th Street, Suite 608, Washington, DC 20036-3603. (202) 296-2237. Fax: (202) 296-1151.

American Electronics Association. 5201 Great America Way, Suite 520, P.O. 52990, Santa Clara, CA 95056. (408) 987-4200. Fax: (408) 970-8565.

American Institute of Physics. One Physics Ellipse, College Park, MD 20740-3843. (301) 209-3100.

American Society for Testing and Materials. 1916 Race Street, Philadelphia, PA 19103-1180. (215) 299-5400.

Edison Electric Institute. 701 Pennsylvania Avenue NW, Washington, DC 20004-2696. (202) 508-5000, 5454. Fax: (202)508-5360.

Electric Power Research Institute. 3412 Hillview Avenue, Palo Alto, CA 94304. (415) 855-2000. Fax: (415) 855-2954.

Electronics Industries Association. 2500 Wilson Boulevard, Arlington, VA 22201. (703) 907-7500. Fax: (202) 457-4985.

IEEE Dielectrics and Electrical Insulation Society. c/o Institute of Electrical and Electronic Engineers. 345 East 47th Street. New York, NY 10017. (212) 705-7900. Fax: (212)705-4929.

Institute of Electrical and Electronics Engineers. 345 East 47th Street. New York, NY 10017. (212) 705-7900. Fax: (212) 705-4929.

National Electrical Manufacturers Association. 1300 North 17th Street, Suite 1847, Rosslyn, VA 22209. (703) 841-3200. Fax: (703) 841-3300.

DIRECTORIES AND BIOGRAPHICAL SOURCES

American Electronics Association. Directory. 5201 Great America Way, Suite 520, P.O. 52990, Santa Clara, CA 95056. (408) 987-4200. Fax: (408) 970-8565. Annual. $175.00.

American Men and Women of Science. Physical and Biological Sciences. R. R. Bowker Inc., 121 Chanlon Road, New Providence, NJ 07974. (908) 464-6800. (800) 521-8110. 20th edition. 8 volumes. 1996. $850.00.

Directory of Engineering Societies and Related Organizations. American Association of Engineering Societies, 1111 19th Street, Suite 608, Washington, DC 20036-3603. (202) 296-2237. Semi-annual. $150.00.

EEM-Electronic Engineer's Master. Hearst Business Communications, Inc., 645 Stewart Avenue, Garden City NY 11530. (516) 227-1300. ISSN: 0732-9016.

Electrical and Electronics Trades Directory: The Blue Book. Institution of Electrical Engineers. c/o IEEE Service Center, 445 Hoes Lane, Piscataway, NY 08854. (908) 981-1393. (800) 678-IEEE. Fax: (908) 981-9667. Annual. 1995. $140.00

Engineering Research Centers: Incorporating Electronics Research Centers. Stockton Press, 345 Park Avenue, New York,

NY 10010. (212) 689-9200. (800) 221-2123. 4th edition. 1995. $515.00.

IEEE Membership Directory. Institute of Electrical and Electronics Engineers, IEEE Service Center, 445 Hoes Lane, Piscataway, NY 08854. (908) 981-1393. (800) 678-IEEE. Fax:(908) 981-9667. 2 volumes. Annual. $190.00. ISSN:

International Directory of Abbreviations and Acronyms of Electronics, Electrical Engineering, Computer Technology and Information Processing. Peter Wennrich. K. G. Saur, 121 Chanlon Road, New Providence, NJ 07974. (908) 464-6800. (800) 521-8110 2 volumes. 1992. $230.00.

International Engineering Directory. American Consulting Engineers Council, 1015 15th Street, N.W. Suite 802, Washington, DC 20005-2670. (202) 347-7474. Annual. $10.00.

Research Centers Directory. Gale Research, 835 Penobscot Building, Detroit, MI 48226-4094. (313) 961-2242.(800) 347-4253. 20th edition. 1995. $485.00. ISSN: 0080-1518.

Who's Who in Engineering. Gordon Davis, editor. American Association of Engineering Societies. 1111 19th Street, NY, Suite 608, Washington, DC 20036. (202) 296-2237. (800) 658-8897. 9th edition. 1995. $220.00.

Who's Who in Technology. Biographies and Index. Amy L.Unterburger, editor. Gale Research, 835 Penobscot Building, Detroit, MI 48226-4094. (313 961-2242. (800) 347-4253. 7th edition. 1995. $195.00. ISBN 0-8103-7467-6.

ENCYCLOPEDIAS AND DICTIONARIES

Chambers Science and Technology Dictionary. Peter M.B. Walker, editor. Cambridge University Press, 40 West 20th Street, New York, NY 10011-4211. (212) 924-3900. 1988. $39.95.

Encyclopedia of Applied Physics. George Trigg, editor. VCH Publications, Inc., 220 East 23rd Street, Suite 909, New York, NY 10010-4606. (212) 683-8333. (800) 422-8824. 20 volume set. 1991- . $5990.00.

IEEE Standard Dictionary of Electrical and Electronics Terms. Christopher J. Booth, editor. IEEE Service Center,445 Hoes Lane, Piscataway, NJ 08855-1331. (908) 981-1393. (800) 678-IEEE. Fax: (908) 981-9667. IEEE Standard 100- 1992. 5th edition. 1993. $150.00.

Illustrated Dictionary of Electronics. Stan Gibilsco. TAB Books, P.O. Box 40, Blue Summit, PA 17294-0850. (717) 794-2191. (800) 233-1128. 7th edition. 1994. $34.95.

McGraw-Hill Electronics Dictionary. J. Markus. McGraw-Hill Book Company, Inc., 1221 Avenue of the Americas, New York,NY 10020. (212) 997-3675. 5th edition. 1994. $49.95.

McGraw-Hill Encyclopedia of Engineering. Sybil P. Parker, editor. McGraw-Hill Book Company, Inc. 1221 Avenue of the Americas, New York, NY 10020. (212) 997-3675. Second edition. 1993. $95.50. ISBN: 0-07-051392-9.

McGraw-Hill Encyclopedia of Science and Technology. McGraw-Hill Book, Incorporated, 1221 Avenue of the Americas, New York, NY 10020. (212) 997-3675. (800) 262-4729. Seventh edition. Twenty volumes. 1992. $1900.00.

Ency. of Physical Sciences and Engineering Info. Sources

DIELECTRICS

GENERAL WORKS

Dielectric Ceramics: Processing, Properties and Applications. K. M. Nair, et al, editors. American Ceramic Society, 735 Ceramic Place, Westerville, OH 43081. (614) 794-5824. 1993. $69.00.

Dielectrics and Waves. Arthur von Hippel. Artech House, 685 Canton Street, Norwood, MA 02062. (617) 769-9750. (800) 225-9977. 2 volumes. 1994. $110.00.

Electrets. G. M. Sessler. Springer-Verlag New York, Inc., 175 Fifth Avenue, New York, NY 10010. (212) 460-1500. (800) 777-4643. Fax: (212) 473-6272. 2nd enlarged edition. 1987. $69.00 in paper.

Electrical Insulating Liquids. R. Bartnikas, editor. American Society for Testing and Materials, 1916 Race Street, Philadelphia, PA 19103. (215) 299-5419. 1993. $83.00.

Electromagnetic Fields in Multilayered Structures. Arun Bhattacharyya. Artech House, 685 Canton Street, Norwood, MA 02062. (617) 769-9750. (800) 225-9977.

Engineering Dielectrics. Volume 1: Corona Measurement and Interpretation. ASTM STP 669. R. Bartnikas and E. J. McMahon, editors. American Society for Testing and Materials, 1916 Race Street, Philadelphia, PA 19103. (215) 299-5419. 1979. $49.00.

Exploring Electronic Devices. Mark E. Hazen. SCP: Third World Literature Publishing House, P.O. Box 482, Lithonia, GA 30058-0482. (404) 785-7725. 1991. $50.75.

Ferroelectric Materials and Their Applications. Xu Yuhuan. Elsevier Science Publishing Company, Inc., 655 Avenue of the Americas, New York, NY 10010. (212) 989-5800. Fax: (914) 333-2444. 1991. $171.50.

IEEE International Conference on Properties and Applications of Dielectric Materials. Institute of Electrical and Electronic Engineers (IEEE), 345 East 47th Street, New York,NY 10017. (212) 705-7900. 5th conference held 1995. Irregular. Inquire.

Theory of Dielectric Optical Waveguides. Dietrich Marcuse. Academic Press, Inc., 6277 Sea Harbor Drive, Orlando, FL. (800) 321-5068. 2nd edition. 1991. $69.96.

HANDBOOKS AND MANUALS

Active Electronic Component Handbook. Charles A. Harper and Harold C. Jones. McGraw-Hill Book Company, 1221 Avenue of the Americas, New York, NY 10020. (212) 512-2000. (800) 262-4729. 2nd edition. 1996. $79.50.

The Capacitor Handbook. Cletus J. Kaiser. Van Nostrand Reinhold, 115 Fifth Avenue, New York, NY 10003. (212) 254-3232. (800) 842-3636. 1993. $29.95.

Dielectric Properties of Binary Solutions: a Data Handbook. Ya Y. Akhadov. Franklin Book Co., Inc., Elkins Park, PA 19117. (215) 635-5252. 1980. $200.00.

EDN Designer's Companion. Ian Hickman and Bill Travis, editors. Butterworth-Heinemann, 313 Washington Street,Newton, MA 02158. (618-928-2500. (800) 366-2665. 1994. $29.95.

Electrical Engineering Handbook. Richard C. Dorf, editor. CRC Press, Inc., 2000 Corporate Boulevard, NW, Boca Raton, FL 33431. (407) 994-0555. (800) 272-7737. 1993. $99.95.

Handbook of Electrical and Electronic Technology. Curtis Johnson. Prentice Hall , 113 Sylvan Avenue, Route 9W, Englewood Cliffs, NJ 07632. (201) 592-2000. (800) 922-0579. 1996. $88.00.

Handbook of Solid State Batteries and Capacitors. M. Z. Munshi and P. S. Prasad. World Scientific Publishing Company, Inc., 1060 Main Street, River Edge, NJ 07661. (201)487-9655. (800) 227-7562. 1995. $177.00.

Standard Handbook for Electrical Engineers. Donald Fink. McGraw-Hill Publishing Company, 1221 Avenue of the Americas, New York, NY 10020. (212) 512-2000. (800) 262-4729. 13th edition. 1996. $110.00.

ONLINE DATABASES AND CD-ROMS

CA Search. Chemical Abstracts Service, P.O. Box 3012, Columbus, OH 43210-0012. (614) 447-3600. (800) 848-6533. Fax: (614) 447-3709. Very comprehensive guide to worldwide chemical literature and related fields, 1972 to present. Available on BRS,(800) 289-4277, DIALOG, (800) 334-2564, ORBIT (800) 456-7248, and STN International, FIZ Karlsruhe, P.O. Box 2465, W-7500, Karlsruhe 1, Germany, online services. Inquire as to cost and availability.

COMPENDEX PLUS. Engineering Information, Inc., 345 East 47th Street, New York, NY 10017. (212) 705-7600 or (800) 221-1044. Contains citations with abstracts to worldwide literature in engineering and technology, from 1970 to present. Available on online BRS,(800) 289-4277, DIALOG, (800) 334-2564, ORBIT (800) 456-7248, and STN International, FIZ Karlsruhe, P.O. Box 2465, W-7500, Karlsruhe 1, Germany, online services. Also available on CD-ROM. Inquire as to cost and availability.

Current Contents Search. Institute for Scientific Information, 3501 Market Street, Philadelphia, PA 19104. (215) 386-0100. Fax: (215) 386-6362. Contains citations to articles listed in the table of contents of science and technology journals. Also articles in social sciences and life sciences journals. Available on BRS,(800) 289-4277, DIALOG,(800) 334-2564, online services. Inquire as to cost and availability.

Dissertation Abstracts Online. University Microfilms International, 300 North Zeeb Road, Ann Arbor, MI 48106. (800) 521-0600 or (313) 761-4700. Scope includes virtually all doctoral dissertations accepted at accredited American institutions from 1861 to present in 252 subject areas. Available on BRS, (800) 289-4277, DIALOG, (800) 334-2564, and OCLC EPIC,(800) 848-5878, online services. Also available on CD-ROM. Inquire as to cost and availability.

INSPEC. Institution of Electrical Engineers, Michael Faraday House, Six Hills Way, Stevenage, Herts. SG1 2AY, England. Telephone: 0438 313311 or Fax: 0438 742840. Contains citations to the worldwide literature of physics, electronics and electrical engineering, computer technology, and related fields. Available on BRS, (800) 289-4277, DIALOG, (800) 334-2564, ORBIT, (800) 456-7248, and STN International, FIZ Karlsruhe, P.O. Box 2465, W-7500, Karlsruhe 1, Germany, online services. Inquire as to cost and availability.

NTIS Bibliographic Database. National Technical Information Service, 5285 Port Royal Road, Springfield, VA 22161. (703)

487-4929 or Fax: (703) 321-8199. Broad coverage of government-sponsored science and technology research reports, 1964 to present. Available on BRS,(800) 289-4277, DIALOG, (800) 334-2564, ORBIT, (800) 456-7248, and STN International, FIZ Karlsruhe, P.O. Box 2465, W-7500, Karlsruhe 1, Germany, online services. Also available on CD-ROM. Inquire as to cost and availability.

Physics Briefs. American Institute of Physics, 335 East 45th Street , New York, NY 10017. (212) 661-9260 or Fax:(212) 661-2036. Contains citations with abstracts of the literature of physics and related fields, 1979 to present. Available on the STN International, FIZ Karlsruhe, P.O. Box 2465, W-7500, Karlsruhe 1, Germany, online service. Inquire as to cost and availability.

SCIsearch. Institute for Scientific Information, 3501 Market Street, Philadelphia, PA 19104. (800) 523-1850 or (215) 386-0100. Broad multidisciplinary title and author index to the international literature of science and technology, 1974 to present. Available on DIALOG,(800) 334-2564, and ORBIT,(800) 456-7248, online services. Also available on CD-ROM. Inquire as to cost and availability.

WILSONLINE. H.W. Wilson Company, 950 University Avenue, Bronx, NY 10452. (800) 367-6770 or (212) 588-8400. Makes available online versions of the printed H.W. Wilson indexes including Applied Science and Technology Index, Business Periodicals Index, General Science Index, and Readers' Guide to Periodical Literature. Period covered is generally 1983 to present. Available on BRS,(800) 289-4277, DIALOG,(800) 334-2564, and OCLC EPIC,(800) 848-5878, online services. Also available on CD-ROM. Inquire as to cost and availability.

PERIODICALS

Electrical World. McGraw-Hill, Inc., P.O. Box 513 Hightstown, NJ 08520. (212) 512-3288. 1874 to present. Monthly. $55.00 per year. ISSN: 0013-4457.

Electronic Design. Penton Publishing, San Jose Gateway, Suite 354. 2025 Gateway Place, San Jose, CA 95110. (408) 441-0550. 1952 to present. Fortnightly. $95.00. ISSN: 0013-4872.

Electronic Engineering Times. CMP Publications, Inc., 600 Community Drice, Manhasset, NY 11030. (516) 562-5000. Fax: (516) 562-5325. 1972 to present. Weekly. $159.00 per year. ISSN: 0192-1541.

Electronics. Penton Publishing, San Jose Gateway,Suite 354. 2025 Gateway Place, San Jose, CA 95110. (408) 441-0550. 1930 to present. Semi-weekly. $98.00 per year. ISSN: 0883-4989.

Electronics World and Wireless World. Reed Business Publishing, Ltd., Quadrant House, The Quadrant, Sutton, Surrey SM2 5AS, England. 1911 to present. Monthly. L35. ISSN: 0266-3244.

IEEE Circuits and Devices Magazine. Institute of Electrical and Electronics Engineers, Inc., P.O. Box 1331, 445 Hoes Lane, Piscataway, NJ 08855-1331. (908) 981-0060. 1985 to present. Bi-monthly. $120.00 per year. ISSN: 8755-3996.

IEEE Journal of Solid State Circuits. Institute of Electrical and Electronics Engineers, Inc., P.O. Box 1331, 445 Hoes Lane,

Piscataway, NJ 08855-1331. (908) 981-0060. 1966 to present. Bi-monthly. $275.00 per year. ISSN: 0018-9200.

IEEE Spectrum. Institute of Electrical and Electronics Engineers, Inc., P.O. Box 1331, 445 Hoes Lane, Piscataway, NJ 08855-1331. (908) 981-0060. 1964 to present. Monthly. $157.00 per year. ISSN: 0018-9235.

IEEE Transactions on Electron Devices. Institute of Electrical and Electronics Engineers, Inc., P.O. Box 1331, 445 Hoes Lane, Piscataway, NJ 08855-1331. (908) 981-0060. 1952 to present. Monthly. $395.00 per year. ISSN: 0018-9393.

Institute of Electrical and Electronics Engineers. Proceedings. Institute of Electrical and Electronics Engineers, Inc., P.O. Box 1331, 445 Hoes Lane, Piscataway, NJ 08855-1331. (908) 981-0060. 1913 to present. Monthly. $275.00. ISSN: 0018-9219.

IEEE Transactions on Ultrasonics, Ferroelectrics and Frequency Control. Institute of Electrical and Electronics Engineers, Inc., P.O. Box 1331, 445 Hoes Lane, Piscataway, NJ 08855-1331. (908) 981-0060. 1954 to present. Bi-monthly. $230.00 per year. ISSN: 0885-3010.

Solid State Electronics. Elsevier Science, 660 White Plains Road, Tarrytown, NY 10591-5153. (914) 524-9200. Fax: (914)333-2444. 1960 to present. Monthly. $1025.00 per year. ISSN: 0038-1101.

RESEARCH CENTERS AND INSTITUTES

Alabama Microelectronics Science and Technology Center.Auburn University, 200 Broun Hall, Auburn University, AL 36849. (205) 844-1871. (205) 844-1809.

Center for Integrated Systems. Stanford University, School of Engineering, Stanford, CA 94305-4028. (415) 723-9038.

Edison Electric Institute. 701 Pennsylvania Avenue, NW, Washington, DC 20004-2696. (202) 508-5000, 5454. Fax: (202)508-5360. capacitors

Electrical Engineering Research Laboratories. Purdue University. Electrical Engineering Building, West Lafayette, IN 47907. (317) 494-3536. Fax: (317) 494-6440.

Electrical Power Research Institute. 3412 Hillview Avenue, P.O. Box 10412, Palo Alto, CA 94303. (415) 855-2000. Fax: (415) 855-2954.

Electronics Research Laboratory. University of California, Berkeley, 253 Cory Hall, Berkeley, CA 94720. (415) 642- 2301.

Engineering Research Center. University of Maryland, Clark School of Engineering, Potomac Building, Room 2104. (301) 405-3906. Fax: (301-403-4105.

Solid State Electronics Laboratory. North Carolina State University, 432 Daniels Hall, Raleigh, NC 27695. (919) 737-2336.

DIESEL ENGINES

See also: AUTOMOTIVE ENGINEERING, ENGINES

ABSTRACT SERVICES AND INDEXES

Applied Science and Technology Index. H.W. Wilson Co., 950 University Avenue, Bronx, NY 10452. (800) 367-6770 or (212) 588-8400. Fax: (718) 590-1617. A cumulative subject index to English language periodicals in the fields of aeronautics and space science, computer technology, chemistry, construction industry, energy and related areas. From 1958 to present. Monthly. Inquire about cost and availability. Also available on CD-ROM and online.

Engineering Index Monthly. Engineering Information, Inc., Castle Point on the Hudson, Hoboken, NJ 07030. (800) 221-1044. Fax: (212) 832-1857. Monthly. $2300.00 per year. Also available online as COMPENDEX, and also on CD-ROM. Covers chemical engineering, computers, electrical engineering, civil engineering, metals and mining, industrial management, and mechanical engineering.

Fuel and Energy Abstracts. Butterworth-Heinemann, Linacre House, Jordan Hill, Oxford OX2 8DP, England. Telephone 0865-310366. Fax: 0865-310898. 1960 to present. Bimonthly. Inquire for price. A summary of world literature on all technical, scientific, commercial and environmental aspects of fuel and energy.

ASSOCIATIONS AND PROFESSIONAL SOCIETIES

Association of Diesel Specialists. 9140 Ward Parkway, Kansas City, MO 64114. (816) 444-3500. Fax: (816) 444-0330.

Engine Manufacturers Association. 401 N. Michigan Avenue, Chicago, IL 60611. (312) 644-6610.

Society of Automotive Engineers. 400 Commonwealth Drive, Warrendale, PA 15096. (412) 776-4841.

DIRECTORIES AND BIOGRAPHICAL SOURCES

Diesel and Gas Turbine Worldwide Catalog. Diesel and Gas Turbine Publications, 13555 Bishop's Court, Brookfield, WI 53005-6286. (914) 784-9177. Fax: (914) 784-9177. 1935 to present. Annual. $75.00 per year.

Marine Engineers Review Directory of Marine Diesel Engines. Learned Information Inc., 143 Old Marlton Pike, Medford, NJ 08055-8750. (607) 654-6266. Fax: (607) 654-4309. 1988 to present. Annual. Inquire for price.

Research Centers Directory. Gale Research, 835 Penobscot Building, Detroit, MI 48226-4094. (313) 961-2242. (800) 347-4253. Fax: (313) 961-6083. 20th edition. 2 volumes. $485.00.

Scientific and Technical Organizations and Agencies Directory. Gale Research, 835 Penobscot Building, Detroit, MI 48226-4094. (313) 961-2242. (800) 347-4253. Fax: (313) 961-6083. 4th edition. 1996. $195.00.

Who's Who in Engineering. American Association of Engineering Societies, 1111 19th Street NW, Suite 608, Washington, DC 20036. (202) 296-2237 or (800) 658-8897. 8th edition. 1991. Inquire for price.

Who's Who in Technology. Gale Research, 835 Penobscot Building, Detroit, MI 48226-4094. (313) 961-2242. (800) 347-4253. Fax: (313) 961-6083. 1995. $195.00.

ENCYCLOPEDIAS AND DICTIONARIES

Encyclopedia of Physical Science and Technology. Robert A. Meyers, ed. 18 volumes. ACADemic Press Inc., 6277 Sea Harbor Drive, Orlando, FL 32887. (800) 321-5068. 1992. $2100.00.

McGraw-Hill Dictionary of Scientific and Technical Terms. Sybil P. Parker, ed. 5th edition. McGraw-Hill Publishing Company, 1221 Avenue of the Americas, New York, NY 10020. (800) 262-4729 or (212) 512-3825. 1993. $110.50.

McGraw-Hill Encyclopedia of Engineering. Sybil P. Parker, ed. 2nd edition. McGraw-Hill Publishing Company, 1221 Avenue of the Americas, New York, NY 10020. (800) 262-4729 or (212) 512-3825. 1993. $95.50.

McGraw-Hill Encyclopedia of Science and Technology. Sybil P. Parker, ed. 7th edition. 20 volumes. McGraw-Hill Publishing Company, 1221 Avenue of the Americas, New York, NY 10020. (800) 262-4729 or (212) 512-3825. 1992. $1900.00

Thesaurus of Scientific, Technical, and Engineering Terms. Hemisphere Publishing Corporation, 1900 Frost Road, Suite 101, Bristol, PA 19007-1598. (215) 785-5800 or (800) 821-8312. Fax: (215) 785-5515. 1987. $173.00.

GENERAL WORKS

Analysis of New Diesel Engine and Component Design. SAE International, 400 Commonwealth Drive, Warrendale, PA 15096. (412) 776-4970. Fax: (412) 776-0790. 1995. $36.00.

Diesel Engine Design for the 1990s. David R. Merrion. SAE International, 400 Commonwealth Drive, Warrendale, PA 15096. (412) 776-4970. Fax: (412) 776-0790. 1994. $15.00.

Diesel Engine Maintenance. Peter F. Caplen. Trafalgar Square, P.O. Box 257, North Pomfret, VT 05053. (800) 423-4525. Fax: (802) 457-1913. 1993. $19.95.

Diesel Engines. A. J. Wharton. 3nd ed. Butterworth-Heinemann, 313 Washington Street, Newton, MA 02158. (617) 928-2500 or (800) 366-2665. Fax: (617) 928-2620. 1991. $24.95.

Diesel's Engine Volume One: From Conception to 1918. Lyle Cummins. Carnot Press/ SAE International, 400 Commonwealth Drive, Warrendale, PA 15096. (412) 776-4970. Fax: (412) 776-0790. 1993. $55.00.

New Developments in Diesel Engines and Components. SAE International, 400 Commonwealth Drive, Warrendale, PA 15096. (412) 776-4970. Fax: (412) 776-0790. 1994. $29.00.

New Diesel Engines. SAE International, 400 Commonwealth Drive, Warrendale, PA 15096. (412) 776-4970. Fax: (412) 776-0790. 1985. $32.00.

Nnew Diesel Engines and Components. SAE International, 400 Commonwealth Drive, Warrendale, PA 15096. (412) 776-4970. Fax: (412) 776-0790. 1993. $39.00.

DIESEL ENGINES

Ency. of Physical Sciences and Engineering Info. Sources

Wind-Diesel Systems: A Guide to the Technology and Its Implementation. Ray Hunter & George Elliot, editors. Cambridge University Press, 40 West 20th Street, New York, NY 10011-4211. (212) 924-3900 or (800) 872-7423. Fax: (914) 937-4712. 1994. $59.95.

HANDBOOKS AND MANUALS

Automotive Handbook. 3rd edition. SAE International, 400 Commonwealth Dr, Warrendale, PA 15096-0001. (412) 776-4970. Fax: (412) 776-0790. 1993. $34.00.

Diesel Engineering Handbook. Business Journals, 50 Day Street, P.O. Box 5550, South Norwalk, CT 06856. (203) 853-6015. $32.00.

Diesel Engine Manual. P.O. Black, et al. 4th edition. Macmillan Publishing, 200 Old Tappan Road, Old Tappan, NJ 07675. (800) 223-2336. Fax: (800) 445-6991. 1983. $22.50.

Diesel Generator Handbook. L.L.J. Mahon. Butterworth-Heinemann, 313 Washington Street, Newton MA 02158. (617) 928-2500 or (800) 366-2665. Fax: (617) 928-2620. 1992. $195.00.

Diesel Plant Operations Handbook. Clive T. James. McGraw-Hill Publishing Company, 1221 Avenue of the Americas, New York, NY 10020. (800) 262-4729 or (212) 512-3825. 1990. $69.50.

SAE Handbook. SAE International, 400 Commonwealth Dr., Warrendale, PA 15096. (412) 776-4841. 1905 to present. Annual $375.00 for three-volume set.

Standard Handbook of Engineering Calculations. Tyler G. Hicks, ed. 2nd ed. McGraw-Hill Publishing Company, 1221 Avenue of the Americas, New York, NY 10020. (800) 262-4729 or (212) 512-3825. 1985. $89.50.

The Wiley Engineer's Desk Reference. Sanford I. Heisler. John Wiley and Sons, Inc., 605 Third Avenue, New York, NY 10158. (800) 526-5368 or (212) 850-6000. 1984. $64.95.

ONLINE DATABASES AND CD-ROMS

COMPENDEX PLUS. Engineering Information, Inc., 345 East 47th Street, New York, NY 10017. (212) 705-7600 or (800) 221-1044. Contains citations with abstracts to worldwide literature in engineering and technology, from 1970 to present. Available on online BRS,(800) 289-4277, DIALOG, (800) 334-2564, ORBIT (800) 456-7248, and STN International, FIZ Karlsruhe, P.O. Box 2465, W-7500, Karlsruhe 1, Germany, online services. Also available on CD-ROM. Inquire as to cost and availability.

CSA Engineering. Cambridge Scientific Abstracts, 7200 Wisconsin Avenue, Suite 601, Bethesda, MD 20814. (301) 961-6750 or (800) 843-7751. Contains citations and abstracts of international periodicals and research literature covering all fields of engineering and science and technology,including computer and information science, electronics, mechanical engineering, solid state materials, 1981 to present. Available on BRS,(800) 289-4277, online service.Inquire as to cost and availability.

Current Contents Search. Institute for Scientific Information, 3501 Market Street, Philadelphia, PA 19104. (215) 386-0100. Fax: (215) 386-6362. Contains citations to articles listed in the table of contents of science and technology journals. Also articles in social sciences and life sciences journals. Available on BRS,(800) 289-4277, DIALOG,(800) 334-2564, online services. Inquire as to cost and availability.

NTIS Bibliographic Database. National Technical Information Service, 5285 Port Royal Road, Springfield, VA 22161. (703) 487-4929 or Fax: (703) 321-8199. Broad coverage of government-sponsored science and technology research reports, 1964 to present. Available on BRS,(800) 289-4277, DIALOG, (800) 334-2564, ORBIT, (800) 456-7248, and STN International, FIZ Karlsruhe, P.O. Box 2465, W-7500, Karlsruhe 1, Germany, online services. Also available on CD-ROM. Inquire as to cost and availability.

SAE Global Mobility Database. Society of Automotive Engineers (SAE), Electronic Publishing Division, 400 Commonwealth Drive, Warrendale, PA 15098. (412) 776-4841. Contains citations with abstracts to technical papers on automotive and aerospace technology and vehicular-related industries that have been presented at SAE conferences. Covers 1906 to present. Available on ORBIT,(800) 456-7248,online service. Inquire as to cost and availability.

TRIS (Transportation Research Information). National Academy of Science, 2101 Constitution Avenue, N.W., Washington, DC 20418. (202) 334-3313 or (800) 624-6242. Citations with abstracts of literature on transportation, including air, highway, rail, maritime and other modes from 1968 to present. Available on DIALOG,(800) 334-2564, online service. Inquire as to cost and availability.

WILSONLINE. H.W. Wilson Company, 950 University Avenue, Bronx, NY 10452. (800) 367-6770 or (212) 588-8400. Makes available online versions of the printed H.W. Wilson indexes including Applied Science and Technology Index, Business Periodicals Index, General Science Index, and Readers' Guide to Periodical Literature. Period covered is generally 1983 to present. Available on BRS,(800) 289-4277, DIALOG,(800) 334-2564, and OCLC EPIC,(800) 848-5878, online services. Also available on CD-ROM. Inquire as to cost and availability.

PERIODICALS

Automotive Engineering Magazine. SAE International, 400 Commonwealth Dr., Warrendale, PA 15096. (412) 776-4841. 1905 to present. Monthly. $72.00 per year.

Diesel and Gas Turbine Worldwide. Diesel and Gas Turbine Publications, 13555 Bishop's Court, Brookfield, WI 53005-6286. (914) 784-9177. Fax: (914) 784-9177. 1969 to present. 10 times a year. $55.00 per year.

Diesel Progress Engines and Drives. Diesel and Gas Turbine Publications, 13555 Bishop's Court, Brookfield, WI 53005-6286. (914) 784-9177. Fax: (914) 784-9177. 1935 to present. Monthly. $60.00 per year.

High-Speed Diesels and Drives. Diesel and Gas Turbine Publications, 13555 Bishop's Court, Brookfield, WI 53005-6286. (914) 784-9177. Fax: (914) 784-9177. 1982 to present. 9 times a year. $50.00 per year.

Transactions of the Institution of Diesel and Gas Turbine Engineers. Mechanical Engineering Publications Ltd., Northgate Avenue, Bury St. Edmunds, Suffolk IP32 6BW, England. Telephone 0284-763277. Fax: 0284-704006. 6 times a year. Inquire for cost.

Ency. of Physical Sciences and Engineering Info. Sources

DIFFERENTIAL EQUATIONS

RESEARCH CENTERS AND INSTITUTES

Coordinating Research Council Inc. 219 Perimeter Center Parkway, Atlanta, GA 30346. (404) 396-3400.

Sloan Automotive Laboratory, Massachusetts Institute of Technology. Room 3-340, Cambridge, MA 02139. (617) 253-2243.

University of Michigan Automotive Engineering Laboratory. Department of Mechanical Engineering & Applied Mechanics, 321 Lay, N. Campus, 1231 Beal, Ann Arbor, MI 48109-2121. (313) 764-4254.

Wayne State University Center for Automotive Research. 5050 Anthony Wayne Drive, Detroit, MI 48202. (313) 577-3887.

DIFFERENTIAL CALCULUS

See:CALCULUS

DIFFERENTIAL EQUATIONS

See also: ALGEBRA, CALCULUS, ALGEBRAIC GEOMETRY, GEOMETRY, MATHEMATICS, STATISTICS

ABSTRACT SERVICES AND INDEXES

Applied Mechanics Reviews. An assessment of world literature in engineering sciences. American Society of Mechanical Engineers, 345 East 47th Street, New York, NY 10017. (212)705-7703. 1948 to present. Monthly. $360.00 per year. ISSN: 0003-6900.

Applied Science and Technology Index. A cumulative subject index to English language periodicals in the fields of aeronautics and space science, computer technology, chemistry, construction industry, energy and related areas. H.W. Wilson Company, 950 University Avenue, Bronx, NY 10452. (800) 367-6770. Fax: (718) 590-1617. 1958 to present. Monthly. Available on-line from BRS and Wilsonline. Also available on CD-ROM. Inquire about cost and availability. ISSN: 0003-6986.

COMPACTMATH-Compact Mathematics Library. Cumulative CD-ROM edition of *Zentralblatt fuer Mathematik-Mathematics Abstracts*. Springer-Verlag New York, Inc., 44 Hartz Way, Secaucus, NJ 07096-2491. (201) 348-4033. Fax: (201) 348-4505. 1993 to present. Annual. Available only on CD-ROM. Inquire about cost and availability. ISSN: 0938-3174.

COMPUMATH Citation Index. Institute for Scientific Information, 3501 Market Street, Philadelphia, PA 19104. (800) 523-1850 or (215) 386-0100. Fax: (215) 386-2991. Three times per year. $1955.00. Also available online and on CD-ROM. Inquire regarding cost and availability. ISSN: 0730-6199.

Current Mathematical Publications. American Mathematical Society. P.O. Box 1571, Annex Station, Providence, RI 02901-9930. (800) 556-7774 or (401) 455-4000. Fax: (401) 331-3842. 1969 to present. Seventeen times per year. $377.00 per year. Available online from BRS, DIALOG, European Space Agency. Also available on CD-ROM. Inquire regarding cost and availability. ISSN: 0361-4794.

General Science Index. H.W. Wilson Company, 950 University Avenue, Bronx, NY 10452. (212) 588-8400. (800) 367-6770. Fax: (718) 590-1617. From 1978 to present. Ten issues per year. quarterly and annual cumulations. Service basis. Available on CD-ROM and online. Inquire regarding cost and availability. ISSN: 0162-1963.

Index to Mathematical Tables. Alan Fletcher. Addison-Wesley Publishing Co., Inc., 1 Jacob Way, Reading, MA 01867.(617) 944-3700. (800) 447-2226. 2nd edition. 1962. 2 volumes.

Mathematical Reviews. A Review journal covering the world literature of mathematical research. American Mathematical Society, P.O. Box 1571, Annex Station, Providence, RI 02901-9930. (800) 556-7774 or (401) 455-4000. 1940 to present. Monthly. $4594.00 per year. Also available online (BRS, DIALOG, ESA), and on CD-ROM. Inquire regarding cost and availability. ISSN: 0025-5629.

NTIS Alerts: Mathematical Sciences. U.S. National Technical Information Service, 5285 Port Royal Road, Springfield, VA 22161. (703) 487-4650. Fax: (703) 321-8547.Weekly. $140.00 per year.

science Citation Index. Institute for Scientific Information. 3501 Market Street, Philadelphia, PA 19104. (215) 386-0100. (800) 523-1850. Fax: (215) 386-2991. 1961 to present. Six issues per year, plus annual cumulation. $11650.00 per year. Also available online and on CD-ROM. Inquire about price and availability. ISSN: 0036-827X.

Zentr Alblatt Fuer Mathematik und Ihre Grenzebiete/Mathematics Abstracts. Heidelberger Akademie der Wissenschaften/ Springer-Verlag New York, Inc., 44 Hartz Way, Seacacus, NJ 07096-2491. (201) 348-4033. Fax: (201) 348-4505. 1931 to present. 30 issues per year. DM 8340. ISSN: 0044-4235. Also available online and on CD-ROM. Inquire regarding cost and availability.

ANNUAL REVIEWS AND YEARBOOKS

Advances in Applied Mathematics. ACADemic Press, Inc., 6277 Sea Harbor Drive, Orlando, FL 32821-8340. (800) 543-9534. 1980 to present. Irregular. Price varies. ISSN: 0196-8858.

Applied Mathematical Sciences. Springer-Verlag, 175 Fifth Avenue, New York, NY 10010. (212) 460-2500. Fax: (212) 473-6272. 1972 to present. Irregular. ISSN: 0066-5452.

CBMS-NSF Regional Conference Series in Applied Mathematics. Society for Industrial and Applied Mathematics, 3600 University City Science Center, Philadelphia, PA 19104-2688. (215) 382-9800. Fax: (215) 386-7999. 1971 to present. Irregular. ISSN: 0163-9439.

Ergebnisse Der Mathematik Und Ihrer Grenzgebiete. Neuefolge. Springer-Verlag New York, Inc. 175 Fifth Avenue, New York, NY 10010. (212) 460-1500. Fax: (212) 473-6272. 1953 to present. Irregular. Price varies. ISSN: 0071-1136.

Lecture Notes in Mathematics. Springer-Verlag New York, Inc. 175 Fifth Avenue, New York, NY 10010. (212) 460-1500.Fax: (212) 473-6272. 1964 to present. Irregular. Price varies. inquire. ISSN: 0075-8434.

DIFFERENTIAL EQUATIONS

Ency. of Physical Sciences and Engineering Info. Sources

ASSOCIATIONS AND PROFESSIONAL SOCIETIES

American Mathematical Society. P.O. Box 6248, Providence, RI 02940. (401) 455-4000. Fax: (401) 331-3842.

Association for Symbolic Logic. University of Illinois, Department of Mathematics, 1409 West Green Street, Urbana IL 61801. (217) 244-7902. Fax: (217) 333-9576.

Association for Women in Mathematics. 4114 Computer and Space Science Building, University of Maryland, College Park, MD 20742-2461. (301) 405-7892.

Mathematical Association of America. 1529 18th Street, NW, Washington, DC 20036. (202) 387-5200. Fax: (202) 265-2384.

National Council of Teachers of Mathematics. 1906 Association Drive, Reston, VA 22091. (703) 620-9840. Fax: (703) 475-2970.

Society for Industrial and Applied Mathematics. 3600 University City Science Center, Philadelphia, PA 19104-2688.(215) 382-9800. Fax: (215) 386-7999.

Special Interest Group on Numerical Mathematics (SIGNUM). c/o ACM. 1515 Broadway, New York, NY 10036-5701. (212) 869-7440. Fax: (212) 944-1318.

Special Interest Group for Symbolic and Algebraic Manipulation (SIGSAM). c/o ACM. 1515 Broadway, New York,NY 10036-5701. (212) 869-7440. Fax: (212) 944-1318.

BIBLIOGRAPHIES

Books You Can Count On: Linking Mathematics & Literature. Rachel Griffiths and Margaret Clyne. Heinemann Publishing, 361 Hanover Street, Portsmouth, NH 03801-3912. (603) 431-7894. (800) 541-2086. 1991. $15.00. ISBN: 0-435-08322-8.

Bibliography of Mathematical Works Printed in America Through 1850. Louis C. Karpinski. Supplements 1 & 2, Bernard I. Cohen, editor. Ayer Company Publishing, Inc. Lower Mill Road, North Stratford, NH 03590. (603) 922-5105. (800) 282-5413. Reprint edition. 1980. $66.95. ISBN: 0-405-12553-4.

Mathematical Book Review Index, 1800-1940. Louise S. Grinstein. Garland Publishing, Fifth Avenue, Suite 2500, New York, NY 10022-8101. (212) 751-7447. 1992. $72.00. ISBN: 0-8240-4114-3.

International Mathematical Olympiads 1978-1985. Murray S.Klamkin. National Council of Teachers of Mathematics. 1906 Association Drive, Reston, VA 22091. (703) 620-9840. (800) 235-7566. 1987. $11.75. ISBN: 0-88385-631-X.

Mathematical Journals: An Annotated Guide. Diana F. Liang. Scarecrow Press. Distributed by University Press of America, 4720 Boston Way, Lanham, MD 20706. (800) 462-6420.1987. $29.50. ISBN: 0-8108-2585-6.

DIRECTORIES AND BIOGRAPHICAL SOURCES

Assistantships and Graduate Fellowships in the Mathematical Sciences. American Mathematical Society, P.O. Box 6248, Provi-

dence, RI 02940. (401) 455-4000. (800) 321-4267.Annual. $19.00. ISSN: 1040-7650.

Combined Membership List. American Mathematical Society, P.O. Box 6248, Providence, RI 02940. (401) 455-4000. (800)321-4267. Annual. ISSN: 0569-6461.

Mathematical Science Professional Directory. American Mathematical Society, P.O. Box 6248, Providence, RI 02940. Annual. $45.00. ISBN: 0-8218-0173-2.

Research Centers Directory. Gale Research, 835 Penobscot Building, Detroit, MI 48226. (313) 961-2242. (800) 347-4253. 20th edition. 1995. Annual. $485.00. ISSN: 0080-1518.

Women of Mathematics. A biobibliographic sourcebook. Louise S. Grinstein and Paul J. Campbell, editors. Greenwood Press. 1987.

World Directory of Mathematicians. G.D. Mostow, editor. International Mathematical Union, Helsinki. 1958 to present. 9th revised edition. 1990. $40.00. Available from: American Mathematical Society, P.O. Box 6248, Providence, RI 02940. (401) 455-4000. ISSN: 0512-2740.

ENCYCLOPEDIAS AND DICTIONARIES

Beyond Numeracy: An Uncommon Dictionary of Mathematics. John A. Paulos. Alfred A. Knopf Inc., 201 East 50th Street,New York, NY 10022. (800) 733-3000. 1991. $21.50. ISBN: 0-394-58640-9.

Concise Oxford Dictionary of Mathematics. Christopher Clapham. Oxford University Press, 200 Madison Avenue, New York, NY 10016. (212) 726-6000. 1990. $10.95. ISBN: 0-19-286103-4.

Encyclopedia of Mathematics. An updated and annotated translation of the Soviet Mathematical Encyclopedia. Michael Hazewinkel, editor. Kluwer ACADemic Publishers. 101 Philip Drive, Assinippi Park, Norwell, MA 02061. (617) 871-6600. 1987 - . 10 volumes. $147.00, volume 1.

Encyclopedia of Mathematics and its Applications. Addison-Wesley Publishing Co., Inc., 1 Jacob Way, Reading, MA 01867.(617) 944-3700. (800) 447-2226. 1976 to present.irregular. Volumes are individually priced.

Encyclopedic Dictionary of Mathematics. Kiyosi Ito, editor. MIT Press, 55 Hayward Street, Cambridge, MA 02142. (617) 625-8569. 2nd edition, reprinted. 1993. 2 volumes. $70.00. ISBN: 0-262-59020-4.

HarperCollins Dictionary of Mathematics. E. J. Borowski and J. J. Borwein. HarperCollins, 10 East 53rd Street, New York, NY 10022-5299. (800) 331-3761. 1990. $10.95. ISBN: 0-19-286103-4.

Mathematics Dictionary. Robert C. James, editor. Van Nostrand Reinhold, 115 Fifth Avenue, New York, NY 10003. (800)-842-3636. 5th edition. 1992. $42.95. ISBN: 0-442-00741-8.

Penguin Dictionary of Mathematics. J. Daintith and R. D. Nelson. Viking Penguin, 375 Hudson Street, New York, NY 10014-3657. (800) 331-4624. 1989. $12.00. ISBN: 0-14-051119-9.

The Words of Mathematics: An Etymological Dictionary of Mathematical Terms Used in English. Steven Schwartzman. Mathematical Association of America, 18th Street NW, Washington, DC 20036. (800) 331-1622. 1994. $29.50. ISBN: 0-88385-551.

GENERAL WORKS

Differential Equations with Applications to Mathematical Physics. William F. Ames, et al, editors. ACADemic Press, Inc., 6277 Sea Harbor Drive, Orlando, FL. (800) 321-5068. 1993. $69.95.

Differential Equations and Dynamical Systems. L. M. Perko. Springer-Verlag New York, Inc., 175 Fifth Avenue, New York, NY 10010. (212) 460-1500. (800) 777-4643. Fax: (212) 473-6272. 1993. $39.95.

Differential Equations Problem Solver. Research and Education Association, 61 Ethel Road, West Piscataway, NJ 08854. (908) 819-8880. Revised edition. 1994. $29.95 in paper.

First Course in Differential Equations with Applications. Dennis G. Zill. PWS Publications, 20 Park Plaza, Boston, MA 02116. (617) 542-3377. (800) 354-9706. 5th edition. 1993. $64.95.

Fundamentals of Mathematics. H. Behnke, et al. MIT Press, 55 Howard Street, Cambridge, MA 02142. (617) 625-8569. Volume 1, Foundations of Mathematics: Real Number Systems and Algebra. Volume 2, Geometry. Volume 3, Analysis. 1984. reprint edition. $75.00. ISBN: 0-685-07931-7.

Linear Algebra and Differential Equations. Charles G. Cullen. PWS Publications, 20 Park Plaza, Boston, MA 02116. (617) 542-3377. (800) 354-9706. 1991. $65.95.

Modeling with Ordinary Differential Equations. T. P. Dreyer. CRC Press, Inc., 2000 Corporate Boulevard, NW, Boca Raton, FL 33431. (407) 994-0555. (800) 272-7737. 1993. $35.95.

Numerical Methods for Physics. Alejandro Garcia. Prentice Hall, 113 Sylvan Avenue, Route 9W, Englewood Cliffs, NJ 07632. (201) 592-2000. (800) 922-0579. 1994. $67.00.

Numerical Solution of Partial Differential Equations. K. W.Morton and D. F. Mayers. Cambridge University Press, 40 West 20th Street, New York, NY 10011-4211. (212) 924-3900. (800) 872-7423. 1995. $22.95.

Schaum's Outline of Differential Equations. Richard Bronson. McGraw-Hill Publishing Company, Inc., 1221 Avenue of the Americas, New York, NY 10020. (212) 512-2000. (800) 262-4729. 1993. $12.95 in paper.

Writing Mathematics Well. Leonard Gillman. Mathematical Association of America, 18th Street NW, Washington, DC 20036. (800) 331-1622. 1987. $8.00. ISBN: 0-88385-443-0.

HANDBOOKS AND MANUALS

Handbook of Applied Mathematics. Selected Results and Methods. Carl E. Pearsen, editor. Chapman & Hall, 1 Penn Plaza, New York NY 10119. (212) 564-1060. 1990. $46.95. ISBN: 0-442-00521-0.

Handbook of Boolean Algebras. J. D. Monk, editor. Elsevier Science, Inc., 655 Avenue of the Americas, New York, NY 10010. (212) 989-5800. 1989. $72.00. 3 volumes.

Handbook of Lie Group Analysis of Differential Equations. Nail H. Ibragimov. CRC Press, 2000 Corporate Boulevard, Boca Raton, FL 33431. (800) 272-7737. 1993. $85.00.

Hhandbook of Mathematical Functions with Formulas, Graphs, and Mathematical Tables. Milton Abramowitz and Irene A.Stegun, editors. U.S. National Bureau of Standards. U.S. Government Printing Office, Washington, DC. 10th edition. 1972. $43.00. ISBN: 0-16-000202-8.

Handbook of Mathematical Sciences. William H. Beyer, editor. CRC Press, 2000 Corporate Boulevard, Boca Raton, FL 33431. (800) 272-7737. 6th edition. 1987. $91.95. ISBN: 0-8493-0656-6.

Handbook of Mathematics. I. N. Bronshtein and K. A. Semedyayev. Van Nostrand Reinhold, 115 Fifth Avenue, New York, NY 10003. (800)-842-3636. 20th edition. $59.95.ISBN: 0-442-21171-6.

Handbook of Writing for the Mathematical Sciences. Society for Industrial and Applied Mathematics, 3600 University City Science Center, Philadelphia, PA 19104-2688. (215) 382-9800. Fax: (215) 386-7999. 1993. $21.50.

Standard Mathematical Tables and Formulas. William H. Beyer, editor. CRC Press, 2000 Corporate Boulevard, Boca Raton, FL 33431. (800) 272-7737. 29th edition. 1991.$41.95. ISBN: 084-930-6299.

Tables of Integrals, Series and Products. Corrected and enlarged edition. I. S. Gradshteyn and I. M. Ryzhik. Academic Press, Inc., 6277 Sea Harbor Drive, Orlando, FL 32821. (800) 321-5068. 5th edition. 1993. $61.00. ISBN: 0-12-294755-X.

ONLINE DATABASES AND CD-ROMS

CA Search. Chemical Abstracts Service, P.O. Box 3012, Columbus, OH 43210-0012. (614) 447-3600. (800) 848-6533. FA X (614) 447-3709. Very comprehensive guide to worldwide chemical literature and related fields, 1972 to present. Available on BRS,(800) 289-4277, DIALOG, (800) 334-2564, OR-BIT (800) 456-7248, and STN International, FIZ Karlsruhe, P.O. Box 2465, W-7500, Karlsruhe 1, Germany, online services. Inquire as to cost and availability.

COMPENDEX Plus. Engineering Information, Inc., 345 East 47th Street, New York, NY 10017. (212) 705-7600 or (800) 221-1044. Contains citations with abstracts to worldwide literature in engineering and technology, from 1970 to present. Available on online BRS, (800) 289-4277, DIALOG, (800) 334-2564, ORBIT (800) 456-7248, and STN International, FIZ Karlsruhe, P.O. Box 2465, W-7500, Karlsruhe 1, Germany, online services. Also available on CD-ROM. Inquire as to cost and availability.

Dissertation Abstracts. University Microfilms International, 300 North Zeeb Road, Ann Arbor, MI 48106. (800) 521-0600 or (313) 761-4700. Scope includes virtually all doctoral dissertations accepted at accredited American institutions from 1861 to present in 252 subject areas. Available on BRS,(800) 289-4277, DIALOG,(800) 334-2564, and OCLC EPIC,(800) 848-5878,

DIFFERENTIAL EQUATIONS

Ency. of Physical Sciences and Engineering Info. Sources

online services. Also available on CD-ROM. Inquire as to cost and availability.

INSPEC. Institution of Electrical Engineers, Michael Faraday House, Six Hills Way, Stevenage, Herts. SG1 2AY, England. Telephone: 0438 313311 or Fax: 0438 742840. Contains citations to the worldwide literature of physics, electronics and electrical engineering, computer technology, and related fields. Available on BRS,(800) 289-4277, DIALOG, (800) 334-2564, ORBIT, (800) 456-7248, and STN International, FIZ Karlsruhe, P.O. Box 2465, W-7500, Karlsruhe 1, Germany, online services. Inquire as to cost and availability.

MATHSCI. American Mathematical Society, P.O. Box 6248, Providence, RI 02940. (800) 321-4667 or (401) 455-4000 or Fax: (401) 331-3842. Scope includes pure and applied mathematics and related areas of physics, statistics, engineering, computer science, and operations research literature since 1959. Available on DIALOG,(800) 334-2564, online service. Also available on CD-ROM and MATHSCINET, on the Internet. Inquire as to cost and availability.

NTIS Bibliographic Database. National Technical Information Service, 5285 Port Royal Road, Springfield, VA 22161. (703) 487-4929 or Fax: (703) 321-8199. Broad coverage of government-sponsored science and technology research reports, 1964 to present. Available on BRS,(800) 289-4277, DIALOG, (800) 334-2564, ORBIT, (800) 456-7248, and STN International, FIZ Karlsruhe, P.O. Box 2465, W-7500, Karlsruhe 1, Germany, online services. Also available on CD-ROM. Inquire as to cost and availability.

Scientific and Technical Books and Serials in Print. R.R. Bowker Inc., 121 Chanlon Road, New Providence, NJ 07974. (800) 521-8110 or (908) 464-6800. List of currently published books and serials in the physical and biological sciences, engineering and technology, with subject, author and titles indexes. Available on ORBIT,(800) 456-7248, online service.Inquire as to cost and availability.

SCISearch. Institute for Scientific Information, 3501 Market Street, Philadelphia, PA 19104. (800) 523-1850 or (215) 386-0100. Broad multidisciplinary title and author index to the international literature of science and technology, 1974 to present. Available on DIALOG,(800) 334-2564, and ORBIT,(800) 456-7248, online services. Also available on CD-ROM. Inquire as to cost and availability.

WILSONLINE. H.W. Wilson Company, 950 University Avenue, Bronx, NY 10452. (800) 367-6770 or (212) 588-8400. Makes available online versions of the printed H.W. Wilson indexes including Applied Science and Technology Index, Business Periodicals Index, General Science Index, and Readers' Guide to Periodical Literature. Period covered is generally 1983 to present. Available on BRS,(800) 289-4277, DIALOG,(800) 334-2564, and OCLC EPIC,(800) 848-5878, online services. Also available on CD-ROM. Inquire as to cost and availability.

PERIODICALS

American Journal of Mathematics. Johns Hopkins University Press, 2715 North Charles Street, Baltimore, MD 21218. (410) 516-6987. Fax: (410) 516-6968. 1878 to present. Bimonthly. $162.00 per year. ISSN: 0002-9327.

American Mathematical Monthly. Mathematical Association of America. 1529 Eighteenth Street, NW, Washington, DC 20036.

(202) 387-5200. 1894 to present. Ten issues per year.$160.00. ISSN: 0002-9890.

Calculus of Variations and Partial Differential Equations. Springer-Verlag New York, Inc., 44 Hartz Way, Secaucus, NY 07096-2491. (201) 348-4033. Fax: (201) 348-4505. 1993 to present. Quarterly. $300.00. ISSN: 0944-2669.

Canadian Applied Mathematical Quarterly. Canadian Mathematical Society. Distributed by Rocky Mountain Mathematics Consortium, Department of Mathematics, Arizona State University, Tempe, AZ 85287-1904. 1992 to present. $175.00 per year. ISSN: 1073-1849.

College Mathematics Journal. Mathematical Association of America, 1529 18th Street NW, Washington, DC 20036. (202) 387-5200. 1970 to present. Five issues per year. $98.00. ISSN: 0746-8342.

Differential and Integral Equations. Ohio University, Department of Mathematics, Athens, OH 45701. (614) 593-1268. 1988 to date. Quarterly. $156.00 per year. ISSN: 0893-4983.

Differential Equations. English translation of *Differentsial'nye Uravneniya.* Plenum Publishing Corp.,Consultants Bureau, 233 Spring Street, New York, NY 10013- 1578. (212) 620-8468. Fax: (212) 463-0742. 1965 to present. Monthly. $1250.00 per year. ISSN: 0012-2661.

Ergodic Theory and Dynamical Systems. London Mathematical Society/Cambridge University Press, 40 West 20th Street, New York, NY 10011. (212) 924-3900. Fax: (212) 691-3239. 1981 to present. Quarterly. $339.00 per year. ISSN: 0143-3857.

Journal of Dynamics and Differential Equations. Plenum Publishing Corp., 233 Spring Street, New York, NY 10013-1578. (212) 620-8000. Fax: (212) 463-0742. 1989 to present. Quarterly. $180.00 per year. ISSN: 1040-7294.

Journal of Geometric Analysis. CRC Press, Inc., 2000 Corporate Boulevard, NW, Boca Raton, FL 33431. (407) 994- 0555. Fax: (407) 997-0949. 1991 to present. Bimonthly. $224.00 per year. ISSN: 1050-6926.

Mathematical Intelligencer. Springer-Verlag Journals, 175 Fifth Avenue, New York, NY 10010. (212) 460-1500. Fax:(212) 473-6272. 1978 to present. Quarterly. $39.00 per year. ISSN: 0343-6993.

Mathematics Magazine. Mathematical Association of America, 1529 18th Street, NW, Washington, DC 20036. (202) 387-5200.1926 to present. Bimonthly. $80.00 per year. ISSN: 0025-570X.

Numerical Methods for Partial Differential Equations. John Wiley & Sons, Inc., 605 Third Avenue, New York, NY 10158-0012. (212) 850-6000. Fax: (212) 850-6088. 1985 to present. Bimonthly. $430.00 per year. ISSN: 0749-159X.

SIAM Journal on Applied Mathematics. Society for Industrial and Applied Mathematics, 3600 University City Science Center, Philadelphia, PA 19104-2688. (215) 382-9800. Fax: (215) 386-7999. 1953 to present. Bimonthly. $242.00 per year. ISSN: 0036-1399.

SIAM Journal on Mathematical Analysis. Society for Industrial and Applied Mathematics, 3600 University City Science Cen-

Ency. of Physical Sciences and Engineering Info. Sources

DIODE

ter, Philadelphia, PA 19104-2688. (215) 382- 9800. Fax: (215) 386-7999. 1970 to present. Bimonthly. $352.00 per year. ISSN: 0036-1410.

SIAM Review. Society for Industrial and Applied Mathematics, 3600 University City Science Center, Philadelphia, PA 19104-2688. (215) 382-9800. 1959 to present. Quarterly. $148.00 per year. ISSN: 0036-1445.

RESEARCH CENTERS AND INSTITUTES

Center for Applied Mathematics. University of Florida, 201 Walker Hall, Gainsville, FL 32611. (904) 392-0281.

Center for Mathematical Sciences. University of Wisconsin-Madison. 610 Walnut Street, Madison, WI 53705. (608) 263-2696. Fax: (608) 263-2841.

Center for Numerical Analysis. University of Texas at Austin, Austin, TX 78712. (512) 471-1242.

Center for Pure and Applied Mathematics. University of California, Berkeley. 977 Evans Hall, Berkeley, CA 94720. (415) 642-0116. Fax: (415) 642-6726.

Courant Institute of Mathematical Sciences. New York University, 251 Mercer Street, New York, NY 10012. (212) 998-3000.

Institute for Mathematics and Its Applications. University of Minnesota, 514 Vincent Hall, 206 Church Street, SE, Minneapolis, MN 55455. (612) 624-6066. Fax: (612) 626-7370.

Institute for Physical Science and Technology. University of Maryland. College Park, MD 20742. (301) 405-4874. Fax: (301) 314-9363.

Institute of Applied Mathematics. University of Missouri- Rolla, Rolla, MO 65401. (314) 341-4641.

Lefschetz Center for Dynamical Systems. Brown University, 182 George Street, Providence, RI 02912. (401) 863-2358.

DIFFRACTION GRATING

See:SPECTROSCOPY

DIGITAL CIRCUITS

See:ELECTRONIC CIRCUITS AND COMPONENTS

DIKES

See:DAMS

DIODE

See also: ELECTRICAL ENGINEERING, ELECTRICITY, ELECTRONIC CIRCUITS AND COMPONENTS, ELECTRONICS, ELECTRONICS ENGINEERING, MICROELECTRONICS

ABSTRACT SERVICES AND INDEXES

Applied Science and Technology Index. A cumulative subject index to English language periodicals in the fields of aeronautics and space science, computer technology, chemistry, construction industry, energy and related areas. H.W. Wilson Co., 950 University Avenue, Bronx, NY 10452. (212) 588-8400. (800) 367-6770. Fax: (718) 590-1617. From 1958 to present. Monthly. Inquire about cost and availability. Also available on CD-ROM and online. ISSN: 0003-6986.

Chemical Abstracts. Chemical Abstracts Service, 2540 Olentangy River Road, P.O. Box 3012, Columbus, OH 43210-0012. (614) 447-3600. (800) 848-6538. Fax: (614) 447-3713. From 1907 to present. Weekly. $16800.00 per year. Also available on CD-ROM and online. Inquire regarding cost and availability.

Current Papers in Electrical and Electronics Engineering. Institution of Electrical Engineers (IEE), London.Distributed by INSPEC/IEEE, P.O. Box 1331, 445 Hoes Lane, Piscataway, NJ 08855-1331. (908) 562-5549. 1969 to present. Monthly. $345.00 per year. ISSN: 0011-3778.

Electrical and Electronics Abstracts. Science Abstracts Section B. Institute of Electrical Engineers, Michael Faraday House, Six Hill Way, Stevenage, Herts, SG1 2AY,England. Distributed by IEEE, 445 Hoes Lane, Piscataway, NJ 08854. (908) 562-5549. 1898 to present. Monthly. $2200.00 per year. Also available on CD-ROM and online as INSPEC. ISSN: 0036-8105.

Engineering Index Monthly. Engineering Information, Inc., Castle Point on the Hudson, Hoboken, NJ 07030. (201) 216-8500. Fax: (201) 216-8532. Monthly. $2200.00 per year. Also available online as COMPENDEX and on CD-ROM. ISSN: 0742-1974.

General Science Index. H.W. Wilson Company, 950 University Avenue, Bronx, NY 10452. (212) 588-8400. (800) 367-6770. Fax: (718) 590-1617. From 1978 to present. Ten issues per year. quarterly and annual cumulations. Service basis. Available on CD-ROM and online. Inquire regarding cost and availability. ISSN: 0162-1963.

Index to IEEE Publications. IEEE Service Center, 445 Hoes Lane, Piscataway, NJ 08855-1331. (908) 981-1393. (800) 678-IEEE. Fax: (908) 981-9667. 1973 to present. Annual. ISSN: 0099-1368.

Physics Abstracts. INSPEC. Section A, Science Abstracts. Institution of Electrical Engineers (IEE). Available from INSPEC/IEEE - Institute of Electrical and Electronic Engineers, P.O. Box 1331, 445 Hoes Lane, Piscaway NJ 08855-1331. (908) 562-5549. 1898 to present. 24 issues per year. $2835.00 per year. Also available on CD-ROM and online. ISSN: 0036-8091.

Physics Briefs (Physikalische Berichte). Information Center for Energy, Physics, Mathematics. German Physical Society. VCH Publishers, Inc., 220 East 23rd Street, New York, NY 10010-4606. (212) 683-8333. 1845 to present. 24 issues per year. $2390.00 per year. Also available online. ISSN: 0179-7434.

DIODE

Ency. of Physical Sciences and Engineering Info. Sources

Science Citation Index (SCI). Institute for Scientific Information, 3501 Market Street, Philadelphia, PA 19104. (215) 386-0100. (800) 523-1850. Fax: (215) 386-2991. 1961 to present. Six issues per year, plus annual cumulation. $11650.00 per year. Also available online and on CD-ROM. Inquire about price and availability. ISSN: 0036-827X.

Solid State and Superconductivity Abstracts. Covers theory, production and application of solid state materials. Cambridge Scientific Abstracts, 7200 Wisconsin Avenue, Bethesda, MD 20824. (301) 961-6750. Fax: (301) 961-6720. 1957 to present. Bimonthly. $1320.00 per year. Also available online. ISSN: 0896-5900.

ANNUAL REVIEWS AND YEARBOOKS

Advances in Electronics and Electron Physics. Academic Press, Inc., 6277 Sea Harbor Drive, Orlando, FL 32887. (800) 321-5068. From 1948 to present. Irregular. Price varies, inquire. ISSN: 0065-2539.

ASSOCIATIONS AND PROFESSIONAL SOCIETIES

American Electronics Association. 5201 Great America Way, Suite 520, P.O. 52990, Santa Clara, CA 95056. (408) 987-4200. Fax: (408) 970-8565.

American Institute of Physics. One Physics Ellipse, College Park, MD 20740-3843. (301) 209-3100.

Edison Electric Institute. 701 Pennsylvania Avenue NW, Washington, DC 20004-2696. (202) 508-5000, 5454. Fax: (202) 508-5360.

Electronics Industries Association. 2500 Wilson Boulevard, Arlington, VA 22201. (703) 907-7500. Fax: (202) 457-4985.

IEEE (Institute of Electrical and Electronic Engineers). 345 East 47th Street. New York, NY 10017. (212) 705-7900. Fax: (212) 705-4929.

International Society for Hybrid Microelectronics. 1850 Centennial Park Drive, Suite 105, Reston, VA 22091. (703) 758-1060. Fax: (703) 758-1066.

DIRECTORIES AND BIOGRAPHICAL SOURCES

American Men and Women of Science. Physical and Biological Sciences. R. R. Bowker Inc., 121 Chanlon Road, New Providence, NJ 07974. (908) 464-6800. (800) 521-8110. 20th edition. 8 volumes. 1996. $850.00.

Directory of Engineering Societies and Related Organizations. American Association of Engineering Societies, 1111 19th Street, Suite 608, Washington, DC 20036-3603. (202) 296-2237. Semi-annual. $150.00.

EEM-Electronic Engineer's Master. Hearst Business Communications, Inc., 645 Stewart Avenue, Garden City NY 11530. (516) 227-1300. Annual. $90.00 ISSN: 0732-9016.

Engineering Research Centers: Incorporating Electronics Research Centers. Stockton Press, 345 Park Avenue, New York, NY 10010. (212) 689-9200. (800) 221-2123. 4th edition. 1995. $515.00.

IC Master. Hearst Business Communications, Inc., 645 Stewart Avenue, Garden City, NY 11530. (516) 227-1300.1975 to present. Annual. $170.00 per year. ISSN: 0894-6809.

IEEE Membership Directory. Institute of Electrical and Electronics Engineers, IEEE Service Center, 445 Hoes Lane, Picastaway, NJ 08854. (908) 981-1393. (800) 678-IEEE. Fax:(908) 981-9667. 2 volumes. Annual. $190.00.

International Directory of Abbreviations and Acronyms of Electronics, Electrical Engineering, Computer Technology and Information Processing. Peter Wennrich. K. G. Saur, 121 Chanlon Road, New Providence, NJ 07974. (908) 464-6800. (800) 521-8110 2 volumes. 1992. $230.00.

International Engineering Directory. American Consulting Engineers Council, 1015 15th Street, N.W. Suite 802, Washington, DC 20005-2670. (202) 347-7474. Annual. $10.00.

Research Centers Directory. Gale Research, 835 Penobscot Building, Detroit, MI 48226-4094. (313) 961-2242.(800) 347-4253. 20th edition. 1995. $485.00. ISSN: 0080-1518.

Scientific and Technical Organizations and Agencies Directory. Gale Research, 835 Penobscot Building, Detroit, MI 48226-4094. (313) 961-2242. (800) 347-4253. 4th edition. 1996. $195.00.

Who's Who in Engineering. Gordon Davis, editor. American Society for Engineering Education, 1111 19th Street, NW, Suite 608, Washington, DC 20036. (202) 296-2237. (800) 658-8897. 9th edition. 1995. $220.00.

Who's Who in Technology. Gale Research, 835 Penobscot Building, Detroit, MI 48226-4094. (313) 961-2242. (800) 347-4253. 1995. $195.00.

ENCYCLOPEDIAS AND DICTIONARIES

Encyclopedia of Applied Physics. George Trigg, editor. VCH Publications, Inc., 220 East 23rd Street, Suite 909, New York, NY 10010-4606. (212) 683-8333. (800) 422-8824. 20 volume set. 1991-. $5990.00.

Encyclopedia of Integrated Circuits. Walter A. Buchsbaum. Prentice Hall, 113 Sylvan Avenue, Route 9W, Englewood Cliffs, NJ 07632. (201) 592-2000. (800) 922-0579. 1987. $39.95.

IEEE Standard Dictionary of Electrical and Electronics Terms. Christopher J. Booth, editor. IEEE Service Center, 445 Hoes Lane, Piscataway, NJ 08855-1331. (908) 981-1393. (800) 678-IEEE. Fax: (908) 981-9667. IEEE Standard 100- 1992. 5th edition. 1993. $150.00.

Illustrated Dictionary of Electronics. Stan Gibilsco. TAB Books, P.O. Box 40, Blue Summit, PA 17294-0850. (717) 794- 2191. (800) 233-1128. 7th edition. 1994. $34.95.

International Encyclopedia of Integrated Circuits. Arthur A. Seidman. TAB Books, P.O. Box 40, Blue Summit, PA 17294-0850. (717) 794-2191. (800) 233-1128. 1991. $75.00.

McGraw-Hill Encyclopedia of Science and Technology. McGraw-Hill Book Company, Inc., 1221 Avenue of the Americas, New York, NY 10020. (212) 512-2000. (800) 262-4729. 7th edition. 20 volume set. 1992. $1900.00.

GENERAL WORKS

Diode Laser Arrays. Dan Botez and Don R. Scifres, editors. Cambridge University Press, 40 West 20th Street, New York, NY 10011-4211. (212) 924-3900. (800) 872-7423. 1994. $100.00.

Electric Circuits: Principles, Applications and Computer Analysis. David A. Bell. Prentice Hall , 113 Sylvan Avenue, Route 9W, Englewood Cliffs, NJ 07632. (201) 592- 2000. (800) 922-0579. 1993. $72.00.

The Experimenter's Guide to Solid State Diodes. Robert J. Traister. Prentice-Hall, 113 Sylvan Avenue, Route 9W, Englewood Cliffs, NJ 07632. (201) 592-2000. (800) 922- 0579. 1986. $16.00.

Exploring Electronic Devices. Mark E. Hazen. SCP: Third World Literature Publishing House, P.O. Box 482, Lithonia, GA 30058-0482. (404) 785-7725. 1991. $50.75.

Luminescence and the Light Emitting Diode. E. W. Williams, and R. Hall. Franklin Book Co., Inc., Elkins Park, PA 19117. (215) 635-5252. 1978. $110.00.

The PN Junction Diode. Gerald W. Neudeck. Addison-Wesley Publishing Co., Inc., 1 Jacob Way, Reading, MA 01867. (617)944-3700. (800) 447-2226. . 2nd edition. 1989. $20.50.

Variable Capacitance Diodes. Kenneth E. Mortenson. Books on Demand, 300 North Zeeb Road, Ann Arbor, MI 48106-1346. (313) 761-4700. (800) 521-0600. Reprint edition. $40.50.

HANDBOOKS AND MANUALS

Electrical Engineer's Handbook. Donald Christiansen, editor. McGraw-Hill Book Company, 1221 Avenue of the Americas, New York, NY 10020. (212) 512-2000. (800) 262-4729. 4th edition. 1996. $110.00.

Handbook of Electrical and Electronic Technology. Curtis Johnson. Prentice Hall , 113 Sylvan Avenue, Route 9W, Englewood Cliffs, NJ 07632. (201) 592-2000. (800) 922- 0579. 1996. $88.00.

Handbook of Quality Integrated Circuit Manufacturing. Robert Zurich. ACADemic Press, Inc., 6277 Sea Harbor Drive,Orlando, FL. (800) 321-5068. 1991. $59.95.

Integrated Circuits Handbook. Rector Press, Ltd., 130 Rattlesnake, Leverett, MA 01054-9726. (413) 548-9708. (800) 247-3473. 3 volumes. 1995. $295.00 in paper.

Standard Handbook for Electrical Engineers. Donald Fink. McGraw-Hill Publishing Company, 1221 Avenue of the Americas, New York, NY 10020. (212) 512-2000. (800) 262- 4729. 13th edition. 1996. $110.00.

ONLINE DATABASES AND CD-ROMS

CA Search. Chemical Abstracts Service, P.O. Box 3012, Columbus, OH 43210-0012. (614) 447-3600. (800) 848-6533. Fax: (614) 447-3709. Very comprehensive guide to worldwide chemical literature and related fields, 1972 to present. Available on BRS,(800) 289-4277, DIALOG, (800) 334-2564, ORBIT (800) 456-7248, and STN International, FIZ Karlsruhe, P.O. Box 2465,

W-7500, Karlsruhe 1, Germany, online services. Inquire as to cost and availability.

COMPENDEX Plus. Engineering Information, Inc., 345 East 47th Street, New York, NY 10017. (212) 705-7600 or (800) 221-1044. Contains citations with abstracts to worldwide literature in engineering and technology, from 1970 to present. Available on online BRS,(800) 289-4277, DIALOG, (800) 334-2564, ORBIT (800) 456-7248, and STN International, FIZ Karlsruhe, P.O. Box 2465, W-7500, Karlsruhe 1, Germany, online services. Also available on CD-ROM. Inquire as to cost and availability.

Dissertation Abstracts. University Microfilms International, 300 North Zeeb Road, Ann Arbor, MI 48106. (800) 521-0600 or (313) 761-4700. Scope includes virtually all doctoral dissertations accepted at accredited American institutions from 1861 to present in 252 subject areas. Available on BRS, (800) 289-4277, DIALOG, (800) 334-2564, and OCLC EPIC, (800) 848-5878, online services. Also available on CD-ROM. Inquire as to cost and availability.

INSPEC. Institution of Electrical Engineers, Michael Faraday House, Six Hills Way, Stevenage, Herts. SG1 2AY, England. Telephone: 0438 313311 or Fax: 0438 742840 . Contains citations to the worldwide literature of physics, electronics and electrical engineering, computer technology, and related fields. Available on BRS, (800) 289-4277, DIALOG, (800) 334-2564, ORBIT, (800) 456-7248, and STN International, FIZ Karlsruhe, P.O. Box 2465, W-7500, Karlsruhe 1, Germany, online services. Inquire as to cost and availability.

NTIS Bibliographic Database. National Technical Information Service, 5285 Port Royal Road, Springfield, VA 22161. (703) 487-4929 or Fax: (703) 321-8199. Broad coverage of government-sponsored science and technology research reports, 1964 to present. Available on BRS,(800) 289-4277, DIALOG, (800) 334-2564, ORBIT, (800) 456-7248, and STN International, FIZ Karlsruhe, P.O. Box 2465, W-7500, Karlsruhe 1, Germany, online services. Also available on CD-ROM. Inquire as to cost and availability.

SCISearch. Institute for Scientific Information, 3501 Market Street, Philadelphia, PA 19104. (800) 523-1850 or (215) 386-0100. Broad multidisciplinary title and author index to the international literature of science and technology, 1974 to present. Available on DIALOG, (800) 334-2564, and ORBIT, (800) 456-7248, online services. Also available on CD-ROM. Inquire as to cost and availability.

WILSONLINE. H.W. Wilson Company, 950 University Avenue, Bronx, NY 10452. (800) 367-6770 or (212) 588-8400. Makes available online versions of the printed H.W. Wilson indexes. 1983 to present. Available on BRS, (800) 289-4277, DIALOG, (800) 334-2564, and OCLC EPIC, (800) 848-5878, online services. Also available on CD-ROM. Inquire as to cost and availability.

PERIODICALS

Electronic Design. Penton Publishing, San Jose Gateway, Suite 354. 2025 Gateway Place, San Jose, CA 95110. (408) 441-0550. 1952 to present. Biweekly. $95.00 per year. ISSN: 0013-4872.

ELECTRONIC ENGINEERING TIMES. CMP Publications, Inc., 600 Community Drice, Manhasset, NY 11030. (516) 562-5000. Fax: (516) 562-5325. 1972 to present. Weekly. $159.00 per year. ISSN: 0192-1541.

DIODE

Ency. of Physical Sciences and Engineering Info. Sources

Electronics. Penton Publishing, San Jose Gateway, Suite 354. 2025 Gateway Place, San Jose, CA 95110. (408) 441- 0550. 1930 to present. Semi-weekly. $98.00. ISSN: 0883- 4989.

IEEE Circuits and Devices Magazine. Institute of Electrical and Electronics Engineers, Inc., P.O. Box 1331, 445 Hoes Lane, Piscataway, NJ 08855-1331. (908) 981-0060. 1985 to present. Bi-monthly. $120.00 per year. ISSN: 8755-3996.

IEEE Journal of Solid State Circuits. Institute of Electrical and Electronics Engineers, Inc., P.O. Box 1331, 445 Hoes Lane, Piscataway, NJ 08855-1331. (908) 981-0060. 1966 to present. Bi-monthly. $275.00 per year to non-members.

IEEE Transactions on Electron Devices. Institute of Electrical and Electronics Engineers, Inc., P.O. Box 1331, 445 Hoes Lane, Piscataway, NJ 08855-1331. (908) 981-0060. 1952 to present. Monthly. $395.00 per year. ISSN: 0018-9383.

Institute of Electrical and Electronics Engineers. Proceedings. Institute of Electrical and Electronics Engineers, Inc., P.O. Box 1331, 445 Hoes Lane, Piscataway, NJ 08855-1331. (908) 981-0060.

Microelectronic Engineering. An interdisciplinary journal of semiconductor manufacturing technology. Elsevier Science, 660 White Plains Road, Tarrytown, NY 10591-5153. (914) 524-9200. Fax: (914) 333-2444. 1983 to present. 16 issues per year. $845.00 per year. ISSN: 0167-9317.

Microelectronics and Reliability. Elsevier Science, 660 White Plains Road, Tarrytown, NY 10591-5153. (914) 524-9200. Fax: (914) 333-2444. 1962 to present. 15 issues per year. $1145.00 per year. ISSN: 0026-2714.

Microelectronics Journal. Elsevier Science, 660 White Plains Road, Tarrytown, NY 10591-5153. (914) 524-9200. Fax:(914) 333-2444. 1967 to present. 8 issues per year. $455.00 per year. ISSN: 0026-2692.

Semiconductor International. Cahners Publishing Company. 1350 East Touhy Avenue, P.O. Box 5080, Des Plaines, IL 60017-6080. (708) 635-8800. Fax: (708) 390-2770. 1978 to present. 13 issues per year. $84.95 per year.

Solid State Electronics. Elsevier Science, 660 White Plains Road, Tarrytown, NY 10591-5153. (914) 524-9200. Fax:(914) 333-2444. 1960 to present. Monthly. $1025.00 per year.

RESEARCH CENTERS AND INSTITUTES

Alabama Microelectronics Science and Technology Center. Auburn University, 200 Broun Hall, Auburn University, AL 36849. (205) 844-1871. (205) 844-1809.

Edison Electric Institute. 701 Pennsylvania Avenue, NW, Washington, DC 20004. (202) 508-5000

Electrical Engineering Research Laboratories. Purdue University, Electrical Engineering Building, West Lafayette, IN 47907. (317) 494-3536. Fax: (317) 494-6440.

Electronics Research Center. University of Texas at Austin,132 Engineering Science Building, Austin, TX.

Electronics Research Laboratory. University of California, Berkeley, 243 Cory Hall, Berkeley CA 94720. (415) 642-2301.

Engineering Research Center. University of Maryland, Clark School of Engineering, Potomac Building, Room 2104. (301) 405-3906. Fax: (301-403-4105.

Laboratory for Electromagnetic and Electronic Systems. Massachusetts Institute of Technology, 77 Massachusetts Avenue, Cambridge, MA 02139.

DIRECT CURRENT

See:ELECTRICITY

DIRIGIBLE

See:AIRSHIPS

DISPLAY TERMINALS

See:COMPUTER COMMUNICATIONS

DISTILLATION

See also: ALCOHOL, ANALYTICAL CHEMISTRY, CHEMICAL ENGINEERING, CHEMISTRY, COAL GASIFICATION, ORGANIC CHEMISTRY

ABSTRACT SERVICES AND INDEXES

Applied Science and Technology Index. A cumulative subject index to English language periodicals in the fields of aeronautics and space science, computer technology, chemistry, construction industry, energy and related areas. H. W. Wilson Co., 950 University Avenue, Bronx, NY 10452-9978. (212) 588-8400. (800) 367-6770. Fax: (718) 590-1617. From 1958 to present. Monthly. Inquire about cost and availability. Also available online (BRS and Wilsonline) and on CD-ROM. ISSN: 0003-6986.

Chemical Abstracts. Chemical Abstracts Service. 2540 Olentangy River Road, P.O. Box 3012, Columbus, OH 43210-0012. (614) 447-3600. Fax: (614) 447-3713. 1907 to present. Weekly. $16,800.00 per year. Available online and on CD-ROM. CA is also available in five section groupings. Inquire regarding cost and availability.

Current Contents: Physical, Chemical and Earth Sciences. Institute for Scientific Information, 3501 Market Street, Philadelphia, PA 19104. (215) 386-0100. (800) 523-1850.Fax: (215) 386-2291. 1961 to present. Weekly. $442.00 per year. Also available online (BRS, DIALOG) and on CD-ROM. Inquire regarding cost and availability. ISSN: 0163-2574.

General Science Index. H.W. Wilson Company, 950 University Avenue, Bronx, NY 10452. (212) 588-8400. (800) 367-6770. Fax: (718) 590-1617. From 1978 to present. Ten issues per year.

quarterly and annual cumulations. Service basis. Available on CD-ROM and online. Inquire regarding cost and availability. ISSN: 0162-1963.

Process and Chemical Engineering/Chemical Engineering Abstracts. Royal Society of Chemistry, Information Services, Thomas Graham House, Science Park, Milton Road, Cambridge, CB4 4WF, England. Contains citations, mostly with abstracts, of the worldwide literature on chemical engineering. from 1982 to present. Monthly. $610.00 per year. Also available online. ISSN: 0960-5045.

Science Citation Index. Institute for Scientific Information, 3501 Market Street, Philadelphia, PA 19104. (215) 386-0100. (800) 523-1850. Fax: (215) 386-2991. 1961 to present. Six issues per year, plus annual cumulation. $11650.00 per year. Also available online and on CD-ROM. Inquire about price and availability. ISSN:0036-827X.

Theoretical Chemical Engineering. Royal Society of Chemistry, Information Services, Thomas Graham House, Science Park, Milton Road, Cambridge, CB4 4WF, England. Covers theoretical chemical engineering, including theory and laboratory experimentation. 1964 to present. Monthly. $235.00. ISSN: 0960-5053.

ANNUAL REVIEWS AND YEARBOOKS

Advances in Physical Organic Chemistry. Academic Press, Inc., 1250 Sixth Avenue, San Diego, CA 92101. (619) 231-0926. Fax: (619) 699-6715.1963 to present. Irregular. Price varies, inquire.

Annual Review of Physical Chemistry. Annual Reviews, Inc., 4139 El Camino Way, Palo Alto, CA 94306-0897. (415) 493-4400. Fax (415) 855-9815. Annual. $55.00 per year.

The Essential Oils. Individual Essential Oils of the Plant Families. Ernest Guenther, et al. editors. Krieger Publishing Company, P.O. Box 9542, Melbourne, FL 32902-9542. (407) 724-9542. 6 volumes. 1992. $495.00 set.

Studies in Organic Chemistry. Marcel Dekker, Inc., 270 Madison Avenue, New York, NY 10016. (212) 696-9000. Fax: (212) 685-4540. 1973 to present. Irregular. Price varies.

ASSOCIATIONS AND PROFESSIONAL SOCIETIES

American Chemical Society, 1155 16th Street, NW, Washington, DC 20036. (202) 872-4600.

American Institute of Chemical Engineers. 345 East 47th Street, New York, NY 10017. (212) 705-7338.

Association of Consulting Chemists and Chemical Engineers, 50 East 41st Street, Suite 92, New York, NY 10017. (212) 684-6255.

Chemical Manufacturers Association. 2501 M Street, N.W., Washington, DC 20037. (202) 887-1182

BIBLIOGRAPHIES

Chemical Information Sources. Gary Wiggins. McGraw-Hill Publishing Company, 1221 Avenue of the Americas, New York, NY 10020. (800) 262-4729 or (212) 512-3825. 1991. $42.50.

International Bibliography of Studies on Alcohol. Sarah S. Jordy, compiler. Rutgers Center of Alcohol Studies Publications, P.O. Box 969, Piscataway, NJ 08855. 3 volumes. 1901-1960. 1981. $150.00 set.

Scientific and Technical Books and Serials in Print. An index to Literature in Science and Technology. R.R. Bowker Co., 121 Chanlon Road, New Providence, NJ 07974. (908) 464-6800. (800) 521-8110. Fax: (908) 665-3502. 1972 to present. Annual. 4 volumes. 1994. $299.95. Also available on compact disc and online. ISSN: 0000-054X.

DIRECTORIES AND BIOGRAPHICAL SOURCES

American Institute of Chemical Engineers. Directory. American Institute of Chemical Engineers. 345 East 47th Street, New York, NY 10017-2396. (212) 705-7663. Annual.

American Men and Women of Science. Physical and Biological Sciences. R. R. Bowker, 121 Chanlon Road, New Providence, NJ 07974. (908) 464-6800. (800) 521-8110. 20th edition. 8 volumes. 1996. $850.00

Consulting Services: Chemists and Chemical Engineers. Association of Consulting Chemists and Chemical Engineers, 40 East 45th Street, New York, NY 10017. (212) 983-3160. Fax: (212) 983-3161. Biennial. $60.00.

Directory of Chemistry Software 1992. Wendy Warr, Peter Willett, Geoff Downs. American Chemical Society, 1155 16th Street, NW, Washington, DC 20036. (202) 872-4600. 1992. $35.95.

Directory of Essential Oils. Wanda Sellar. Beekman Publishers, Inc., P.O. Box 888, Woodstock, NY 12498. (914) 679-2300. Fax: (914) 679-2301. 2nd edition. 19 . $26.95 in paper.

Engineering Research Centers: Incorporating Electronics Research Centers. Stockton Press, 345 Park Avenue, NewYork, NY 10010. (212) 689-9200. (800) 221-2123. 4th edition. 1995. $515.00.

Research Centers Directory. Gale Research, 835 Penobscot Building, Detroit, MI 48226-4094. (313) 961-2242. (800) 347-4253. 20th edition. 1995. $485.00. ISSN: 0080-1518.

Who's Who in Engineering. Gordon Davis, editor. American Association of Engineering Societies. 1111 19th Street, NY, Suite 608, Washington, DC 20036. (202) 296-2237. (800) 658-8897. 9th edition. 1995. $220.00.

ENCYCLOPEDIAS AND DICTIONARIES

Dictionary of Organic Compounds. John B. Buckingham, editor. Chapman and Hall, 1 Penn Plaza, New York, NY 10119. (212) 564-1060. 1993. 10th supplement. $832.50.

Encyclopedia of Chemical Processing and Design. McKetta Marcel Dekker, Inc., 270 Madison Avenue, New York, NY

DISTILLATION

Ency. of Physical Sciences and Engineering Info. Sources

10016.(212) 696-9000. (800) 228-1160. 1976- . $175.00 per volume. ISSN:

Industrial Chemical Thesaurus. Michael Ash and Irene Ash. VCH Publishers, Inc., 220 East 23rd Street, New York, NY 10010-4606. (800) 367-8249. 1992. $295.00.

Kirk-Othmer Encyclopedia of Chemical Technology. John Wiley and Sons, Inc., 605 Third Avenue, New York, NY 10158. (800) 526-5368 or (212) 850-6000. Fourth edition. 1991 - . 27 volumes. $5400.00.

McGraw-Hill Encyclopedia of Science and Technology. McGraw-Hill Book, Inc., 1221 Avenue of the Americas, New York, NY 10020. (212) 997-3675. 17th edition. 1992. $1900.00.

Ullman's Encyclopedia of Industrial Chemistry. VCH Publications, Inc., 220 East 23rd Street, Suite 909, New York, NY 10010-4606. (212) 683-8333. (800) 422-8824. 5th edition. 1984 - . Price varies per volume.

GENERAL WORKS

Alcohol Fuels: Impacts from Increased Use of Ethanol Blended Fuels. Diane Publishing Co., 600 Upland Avenue, Upland, PA 19015. (610) 499-7415. 1993. $45.00.

Alcohols and Derivatives. J. Tremolieres, editor. Franklin Book Co., Inc., Elkins Park, PA 19117. (215) 635-5252. 2 volumes. 1970. $128.00 set.

Chemical Process Design. Robin Smith. McGraw-Hill Publishing Company, Inc., 1221 Avenue of the Americas,New York, NY 10020. (212) 512-2000. (800) 262-4729. 1995. $65.00.

Corn into Alcohol: Its Chemistry. Gordon Press Publications, P.O. Box 459, Bowling Green Station, New York, NY 10004. (718) 624-8419. 1991. $69.00.

Distillation and Absorption. G. G. Haselden, et al, editors. Hemisphere Publishing Corp., 1900 Frost Road,Suite 101, Bristol PA 19007. (215) 785-5800. (800) 821- 8312. 2 volumes. 1988. $275.00 set.

Distillation Design. Henry Z. Kister. McGraw-Hill Publishing Company, Inc., 1221 Avenue of the Americas, New York, NY 10020. (212) 512-2000. (800) 262-4729. 1992.$75.00.

Distillation Design in Practice. L. M. Rose. Elsevier Science Publishing Company, Inc., 655 Avenue of the Americas, New York, NY 10010. (212) 989-5800. Fax: (914) 333-2444. 1985. $37.50.

Increasing the Efficiency of Ethanol Production Through the Use of a Membrane Technology. Diane Publishing Co., 600 Upland Avenue, Upland, PA 19015. (610) 499-7415. 1994. $40.00 in paper.

Multicomponent Distillation. J. M. Calo and E. J. Henley, editors. American Institute of Chemical Engineers, 345 East 47th Street, New York, NY 10017. (212) 705-7663. Fax: (212) 752-3294. 1981. $44.00.

Solar Distillation. M. A. Malik, et al. Franklin Book Co.,Inc., Elkins Park, PA 19117. (215) 635-5252. 1982. $82.00.

HANDBOOKS AND MANUALS

Comprehensive Organic Synthesis: Selectivity, Strategy, and Efficiency in Modern Organic Chemical. Barry M. Trost and others, editors. Pergamon Press, Maxwell House, Fairview Park, Elmsford, NY 10523. (914) 592-7700. Fax (914) 592-3625. 1991. Nine volumes. $3900.00.

CRC Handbook of Chemistry and Physics. Robert C. Weast, editor. CRC Publishers, Inc., 2000 Corporate Blvd., N.W., Boca Raton, FL 33431. (407) 994-0555 or (800) 333-8300. Seventy-third edition. 1992. $100.00.

Guide to Basic Chemical Compounds. D.R. Lide, Jr. CRC Publishers, Inc., 2000 Corporate Blvd., N.W., Boca Raton, FL 33431. (407) 994-0555 or (800) 333-8300. 1993. $120.00.

Handbook for Alcohol Distillers. Gordon Press Publications,P.O. Box 459, Bowling Green Station, New York, NY 10004. (718) 624-8419. 1991. $250.00.

Handbook of Laboratory Distillation. E. Krell, editor. Elsevier Science Publishing Company, Inc., 655 Avenue of the Americas, New York, NY 10010. (212) 989-5800. Fax: (914) 333-2444. 2nd edition. 1982. $186.50.

Lange's Handbook of Chemistry. John A. Dean, editor. McGraw-Hill Publishing Company, 1221 Avenue of the Americas, New York, NY 10020. (800) 262-4729 or (212) 512-3825. Fourteenth edition. 1992. $79.50.

Perry's Chemical Engineers' Handbook. Robert H. Perry and Donald W. Green, editors. McGraw-Hill Publishing Company, Inc., 1221 Avenue of the Americas, New York, NY 10020.(212) 512-2000. (800) 262-4729. 6th edition. 1996. $129.50.

ONLINE DATABASES AND CD-ROMS

CA SEARCH. Chemical Abstracts Service, P.O. Box 3012, Columbus, OH 43210-0012. (614) 447-3600. (800) 848-6533. Fax: (614) 447-3709. Very comprehensive guide to worldwide chemical literature and related fields, 1972 to present. Available on BRS,(800) 289-4277, DIALOG, (800) 334-2564, ORBIT (800) 456-7248, and STN International, FIZ Karlsruhe, P.O. Box 2465, W-7500, Karlsruhe 1, Germany, online services. Inquire as to cost and availability.

Chemical Journals of the American Chemical Society. American Chemical Society, 1155 16th Street, N.W., Washington, DC 20036. (202) 872-4381 or (800) 424-6747. Contains complete text of approximately 90,000 articles from 22 primary journals published by the American Chemical Society, from mostly 1982 to present. Available on STN International, FIZ Karlsruhe, P.O. Box 2465, W-7500, Karlsruhe 1, Germany, online service. Inquire as to cost and availability.

Kirk-Othmer Encyclopedia of Chemical Technology. John Wiley and Sons, Inc., 605 Third Avenue, New York, NY 10158. (800)526-5368 or (212) 850-6000. Contains the complete text of all chapters in the 27 volume fourth edition of the *Kirk-Othmer Encyclopedia of Chemical Technology.* 1991. Available on BRS,(800) 289-4277, DIALOG,(800) 334-2564, online services. Inquire as to cost and availability.

NTIS Bibliographic Database. National Technical Information Service, 5285 Port Royal Road, Springfield, VA 22161. (703) 487-4929 or Fax: (703) 321-8199. Broad coverage of govern-

Ency. of Physical Sciences and Engineering Info. Sources

DISTILLATION

ment-sponsored science and technology research reports, 1964 to present. Available on BRS,(800) 289-4277, DIALOG, (800) 334-2564, ORBIT, (800) 456-7248, and STN International, FIZ Karlsruhe, P.O. Box 2465, W-7500, Karlsruhe 1, Germany, online services. Also available on CD-ROM. Inquire as to cost and availability.

SCISearch. Institute for Scientific Information, 3501 Market Street, Philadelphia, PA 19104. (800) 523-1850 or (215) 386-0100. Broad multidisciplinary title and author index to the international literature of science and technology, 1974 to present. Available on DIALOG,(800) 334-2564, and ORBIT,(800) 456-7248, online services. Also available on CD-ROM. Inquire as to cost and availability.

WILSONLINE. H.W. Wilson Company, 950 University Avenue, Bronx, NY 10452. (800) 367-6770 or (212) 588-8400. Makes available online versions of the printed H.W. Wilson indexes including *Applied Science and Technology Index, Business Periodicals Index,* and *Readers' Guide to Periodical Literature.* Period covered is generally 1983 to present. Also available on CD-ROM. Inquire as to cost and availability.

PERIODICALS

AICHE Journal. American Institute of Chemical Engineers.345 East 47th Street, New York, NY 10017. (212) 705-7663. Fax: (212) 752-3294. 1955 to present. Monthly. $395.00 per year. ISSN: 0001-1541.

American Laboratory. American Laboratory, 30 Controls Drive, P.O. Box 870, Shelton, CT 06484-6111. (203) 926-9300. Fax: (203) 926-9310. 1968 to present. Monthly. $150.00 per year. ISSN: 0044-7749.

Analytical Chemistry. American Chemical Society, P.O. Box 3337, Columbus, OH 43210. (614) 447-3776. (800) 333-9511. Fax: (614) 447-3671. 1929 to present. Monthly. $415.00 per year. ISSN: 0003-2700.

Analytical Methods and Instrumentation. John Wiley and Sons, Inc., 605 Third Avenue, New York, NY 10158. (800) 526-5368 or (212) 850-6000. 1992 to present. Bimonthly. $250.00 per year. ISSN: 1063-5246.

Chemical and Engineering News. American Chemical Society, 1155 16th Street, NW, Washington, DC 20036. (202) 872-4600. 1923 to present. Weekly. $115.00 per year. ISSN: 0009-2347.

Chemical Engineering. The chemical and process industries journal. Institution of Chemical Engineers/Taylor and Francis, Ltd., 1900 Frost Road, Suite 101, Bristol, PA 19007-1598. (215) 785-5800. 1923 to present. Fortnightly. $176.00 per year. ISSN: 0302-0797.

Chemical Engineering Progress. American Institute of Chemical Engineers, 345 East 47th Street, New York, NY 10017. (212) 705-7338. Fax: (212) 752-3294. 1947 to present. Monthly. $75.00 per year. ISSN: 0360-7275.

Chemical Week. Chemical Week Associates, 888 Seventh Avenue, New York, NY 10106. (212) 621-4900. Fax: (212) 621-4949. 1914 to present. Weekly. $99.00 per year. ISSN: 0009-272X.

Chemtech. American Chemical Society. P.O. Box 3337, Columbus, OH 43210. (614) 447-3776. 1970 to present. Monthly. $370.00 per year. ISSN: 0009-2703.

Gas Separation and Purification. Butterworth-Heinemann, 313 Washington Street, Newton, MA 02158. (618-928-2500. (800) 366-2665. 1987 to present. Quarterly. L185. ISSN: 0950- 4214.

Industrial and Engineering Chemistry Research. American Chemical Society, P.O. Box 3337, Columbus, OH 43210. (614) 447-3776. (800) 333-9511. Fax: (614) 447-3671. 1962 to present. Monthly. $567.00 per year. ISSN: 0888-5885.

Journal of Membrane Science. Elsevier Science Publishing Company, Inc., 655 Avenue of the Americas, New York, NY 10010. (212) 989-5800. Subscriptions to: P.O. Box 882, Madison Square Station, NY 10159-0882. 1977 to present. 36 times per year. $2731.00 per year. ISSN: 0376-7388.

Journal of Organic Chemistry. American Chemical Society, P.O. Box 3337, Columbus, OH 43210. (614) 447-3776. (800) 227-5558. 1936 to present. Bi-Weekly. $785.00 per year. ISSN: 0022-3263.

Separation Science and Technology. Marcel Dekker, Inc., 270 Madison Avenue, New York, NY 10016. (212) 696-9000. (800) 228-1160. 1966 to present. 18 issues per year.$1450.00 per year. ISSN: 0149-6395.

RESEARCH CENTERS AND INSTITUTES

Center for Catalytic Science and Technology. University of Delaware. Departments of Chemical Engineering and Chemistry, Newark, DE 19716. (302) 451-8056. (302) 451- 1048.

Center for Separation Science. University of Arizona, Building 20, Tucson, AZ 85721. (602) 621-4952. Fax: (602) 621-4189.

Separations Research Program. University of Texas at Austin, CPE Building 4, Austin, TX 78712-01062. (512) 471-7792. Fax: (512) 471-1720.

University/industry Chemical Research Center. Mississippi State University, Department of Chemistry, P.O. Drawer CH, Mississippi State, MS 39762. (601) 325-3584.

DOLBY SYSTEM

See: AUDIO ENGINEERING

DOMES

See: STRUCTURAL ENGINEERING

DOPING

See: SEMICONDUCTORS

DREDGES AND DREDGING

Ency. of Physical Sciences and Engineering Info. Sources

DOPPLER RADAR

See:RADAR

DOSIMETRY

See:RADIATION

DRAFTING

See:CAD

DRAG

See:AERODYNAMICS

DRAWBRIDGES

See:BRIDGES

DREDGES AND DREDGING

See also: CHANNELS, CIVIL ENGINEERING, COASTAL ENGINEER-
ING, COASTS, GEOTECHNICAL ENGINEERING, NAVIGATION,
OCEAN ENGINEERING

ABSTRACT SERVICES AND INDEXES

Applied Science and Technology Index. A cumulative subject
index to English language periodicals in the fields of aeronautics and space science, computer technology, chemistry, construction industry, energy and related areas. H.W. Wilson Co.,
950 University Avenue, Bronx, NY 10452. (800) 367-6770 or
(212) 588-8400. Fax: (718) 590-1617. From 1958 to present.
Monthly. Inquire about cost and availability. Also available on
CD-ROM and online.

ASCE Combined Annual Combined Index. American Society of
Civil Engineers, 345 East 47th Street, New York, NY 10017-
2398. (212) 705-7520 or (800) 548-2723. Annual. $48.00.

ASCE Publications Information. 1966-present. American Society of Civil Engineers, 345 East 47th Street, New York, NY
10017-2398. (212) 705-7520 or (800) 548-2723. Bi-monthly.
$160.00 per year for non-members.

Bibliography and Index of Geology. American Geological Institute, 4220 King Street, Alexandria, VA 22302. (703) 379-
2480. Fax (703) 379-7563. Monthly. $1295.00 per year. Also
available online as GeoRef.

Civil and Structural Engineering Abstracts. Cambridge Scientific Abstracts, 7200 Wisconsin Avenue, Bethesda, MD 20814-
4823. (301) 961-6750. Fax: (301) 961-6720. 1993 to present.
Monthly. $385.00 per year. Topics covered include structural
design, construction equipment and methods, civil defense and

military engineering, surveying, highway engineering, maritime
and port structures, materials, land reclamation, and soil mechanics.

*Current Contents: Engineering, Technology, and Applied Sci-
ences.* Institute for Scientific Information, 3501 Market Street,
Philadelphia, PA 19104. (215) 386-0100. Fax: (215) 386-6362.
1970 to present. Weekly. $442.00 per year.

Engineering Index Monthly. Engineering Information, Inc.,
Castle Point on the Hudson, Hoboken, NJ 07030. (800) 221-
1044. Fax: (212) 832-1857. Monthly. $2300.00 per year. Also
available online as COMPENDEX, and also on CD-ROM. Covers chemical engineering, computers, electrical engineering, civil
engineering, metals and mining, industrial management, and
mechanical engineering.

Fluid Abstracts: Civil Engineering. Elsevier Science [journals],
660 White Plains Rd., Tarrytown, NY 10159-5153. (919) 524-
9200. Fax: (919) 333-2444. 1991 to present. Monthly plus annual cumulation. $645.00 per year. Also available online as
FLUIDEX. Covers civil engineering applications of fluid mechanics, hydraulics, flow metering and measuring, offshore engineering, environmental hydraulics, and related aspects of wind
energy, the atmosphere, and aerodynamics.

Geodex Retrieval System for Geotechnical Abstracts. Geodex
Retrieval Systems, 669 Broadway, P.O. Box 279, Sonoma, CA
95476. (707) 939-8476. Fax: (707) 996-8734. 1970 to present.
Three times a year. $645.00 per year includes Geotechnical
Abstracts, as Geodex Systems-GRS.

Geotechnical Abstracts. Geodex Retrieval Systems, 669 Broadway, P.O. Box 279, Sonoma, CA 95476. (707) 939-8476. Fax:
(707) 996-8734. 1970 to present. Three times a year. $280.00
per year. Covers soil mechanics, foundation engineering, rock
mechanics, and engineering geology.

Oceanic Abstracts. Cambridge Scientific Abstracts, 7200 Wisconsin Avenue, Bethesda, MD 20814. (301) 961-6750. Fax (301)
961-6720. Bimonthly. $995.00 per year.

Science Citation Index. Institute for Scientific Information, 3501
Market Street, Philadelphia, PA 19104. (215) 386-0100. Fax:
(215) 386-6362. Inquire about availability and cost. Also available on CD-ROM.

ASSOCIATIONS AND PROFESSIONAL SOCIETIES

American Society of Civil Engineers. 345 East 47th Street, New
York, NY 10017-2398. (212) 705-7520 or (800) 548-2723.

Coastal Engineering Research Council. 4911 Bay Oaks Court,
College Station, TX 77845. (409) 690-0306.

The Coastal Society. P.O. Box 2081, Gloucester, MA 01930-
2081. (508) 281-9209.

Dredging Contractors of America. 1733 King Street, Suite 300,
Alexandria, VA 22314-2720. (703) 548-8300. Fax: (703) 548-
0421.

Dredging Industry Size Standard Committee. c/o Joe R. Reeder,
Patton, Boggs & Blow, 2550 M Street NW, Suite 900, Washington, DC 20037. (202) 457-6436.

Geological Society of America. 3300 Penrose Place, P.O. Box 9140, Denver, CO 80301-9140. (303) 447-2020.

World Organization of Dredging Associations. P.O. Box 4562, Vancouver, WA 98662. (503) 224-9087.

DIRECTORIES AND BIOGRAPHICAL SOURCES

Directory of Engineering Societies and Related Organizations. Gordon Davis. 13th edition. American Association of Engineering Societies, 1111 19th Street NW, Suite 608, Washington, DC 20036. (202) 296-2237 or (800) 658-8897. 1989. Inquire for price.

Research Centers Directory. Gale Research, 835 Penobscot Building, Detroit, MI 48226-4094. (313) 961-2242. (800) 347-4253. Fax: (313) 961-6083. 1995. $485.00.

Scientific and Technical Organizations and Agencies Directory. Gale Research, 835 Penobscot Building, Detroit, MI 48226-4094. (313) 961-2242. (800) 347-4253. Fax: (313) 961-6083. 4th edition. 1996. $195.00.

Who's Who in Engineering. American Association of Engineering Societies, 1111 19th Street NW, Suite 608, Washington, DC 20036. (202) 296-2237 or (800) 658-8897. 8th edition. 1991. Inquire for price.

ENCYCLOPEDIAS AND DICTIONARIES

McGraw-Hill Encyclopedia of Engineering. Sybil P. Parker, ed. 2nd edition. McGraw-Hill Publishing Company, 1221 Avenue of the Americas, New York, NY 10020. (800) 262-4729 or (212) 512-3825. 1993. $95.50.

GENERAL WORKS

Dredging Coastal Ports: An Assessment of the Issues. National Research Council. National Academy Press, Division of the National Academy of Science, 2101 Constitution Avenue, N.W., Lockbox 285, Washington, DC 20055. (800) 624-6242. Fax: (202) 334-2793. 1985. $21.50.

Fundamentals of Hydraulic Dredging. Thomas M. Turner. Cornell Maritime Press Inc., P.O. Box 456, Centreville, MD 21617. (410) 758-1075 or (800) 638-7641. Fax: (410) 758-6849. 1984. $20.00.

Hydraulic Dredging: Principles, Equipment, Processes, Methods. John Huston. John Huston Inc., 514 Santa Monica, Corpus Christi, TX 78411. (512) 853-6512. 1986. $89.00.

HANDBOOKS AND MANUALS

Complete Manual of Dredging. Frederick Schmidt. McGraw-Hill Publishing Company, 1221 Avenue of the Americas, New York, NY 10020. (800) 262-4729 or (212) 512-3825. 1988. $49.95.

Handbook of Dredging Engineering. John B. Herbich, ed. McGraw-Hill Publishing Company, 1221 Avenue of the Americas, New York, NY 10020. (800) 262-4729 or (212) 512-3825. 1992. $59.50.

ONLINE DATABASES AND CD-ROMS

Compendex Plus. Engineering Information, Inc., 345 East 47th Street, New York, NY 10017. (212) 705-7600 or (800) 221-1044. Contains citations with abstracts to worldwide literature in engineering and technology, from 1970 to present. Available on online BRS,(800) 289-4277, DIALOG, (800) 334-2564, ORBIT (800) 456-7248, and STN International, FIZ Karlsruhe, P.O. Box 2465, W-7500, Karlsruhe 1, Germany, online services. Also available on CD-ROM. Inquire as to cost and availability.

CSA Engineering. Cambridge Scientific Abstracts, 7200 Wisconsin Avenue, Suite 601, Bethesda, MD 20814. (301) 961-6750 or (800) 843-7751. Contains citations and abstracts of international periodicals and research literature covering all fields of engineering and science and technology, including computer and information science, electronics, mechanical engineering, solid state materials, 1981 to present. Available on BRS,(800) 289-4277, online service. Inquire as to cost and availability.

Current Contents Search. Institute for Scientific Information, 3501 Market Street, Philadelphia, PA 19104. (215) 386-0100. Fax: (215) 386-6362. Contains citations to articles listed in the table of contents of science and technology journals. Also articles in social sciences and life sciences journals. Available on BRS,(800) 289-4277, DIALOG,(800) 334-2564, online services. Inquire as to cost and availability.

GEOREF: Bibliography and Index of Geology. American Geological Institute, 4220 King Street, Alexandria, VA 22302. (703) 379-2480. Fax (703) 379-7563. Monthly. Inquire as to price and availability.

NTIS Bibliographic Database. National Technical Information Service, 5285 Port Royal Road, Springfield, VA 22161. (703) 487-4929 or Fax: (703) 321-8199. Broad coverage of government-sponsored science and technology research reports, 1964 to present. Available on BRS,(800) 289-4277, DIALOG, (800) 334-2564, ORBIT, (800) 456-7248, and STN International, FIZ Karlsruhe, P.O. Box 2465, W-7500, Karlsruhe 1, Germany, online services. Also available on CD-ROM. Inquire as to cost and availability.

SciSearch. Institute for Scientific Information, 3501 Market Street, Philadelphia, PA 19104. (800) 523-1850 or (215) 386-0100. Broad multidisciplinary title and author index to the international literature of science and technology, 1974 to present. Available on DIALOG,(800) 334-2564, and ORBIT,(800) 456-7248, online services. Also available on CD-ROM. Inquire as to cost and availability.

WILSONLINE. H.W. Wilson Company, 950 University Avenue, Bronx, NY 10452. (800) 367-6770 or (212) 588-8400. Makes available online versions of the printed H.W. Wilson indexes including Applied Science and Technology Index, Business Periodicals Index, General Science Index, and Readers' Guide to Periodical Literature. Period covered is generally 1983 to present. Available on BRS,(800) 289-4277, DIALOG,(800) 334-2564, and OCLC EPIC,(800) 848-5878, online services. Also available on CD-ROM. Inquire as to cost and availability.

PERIODICALS

Coastal Engineering. Elsevier Science Inc., P.O. Box 882, Madison Square Station, New York, NY 10159. (212) 989-5800. 1977 to present. 12 times a year. $642.00 per year.

DREDGES AND DREDGING

Ency. of Physical Sciences and Engineering Info. Sources

Coastal Research. Florida State University Geology Department, Tallahassee, FL 32306. (904) 644-5860. Fax: (904) 644-4214. 1962 to present. 3 times a year. $6.00 per year.

Coastal Management. Taylor & Francis, 1900 Frost Road, Suite 101, Bristol, PA 19007. (215) 785-5800. Fax: (215) 785-5515. 1973 to present. Quarterly. $140.00 per year.

Coastwatch. North Carolina Sea Grant College Program, P.O. Box 8605, North Carolina State University, Raleigh, NC 27695-8605. (919) 515-2454. Fax: (919) 515-7095. 1970 to present. Bi-monthly. $12.00 per year.

Estuarine Coastal and Shelf Science. ACADemic Press Ltd./ Harcourt Brace & Company, Ltd., Foots Cray High Street, Sidcup, Kent DA14 5HP, England. Telephone 44-81-300-3322. Fax: 44-81-309-0807. 1973 to present. Monthly. Inquire for cost.

Journal of Waterway, Port, Coastal, and Ocean Engineering. American Society of Civil Engineers, Waterway, Port, Coastal, and Ocean Division, 345 East 47th Street, New York, NY 10017-2398. (212) 705-7520 or (800) 548-2723. 1956 to present. Bi-monthly. $100.00 per year.

Ocean Engineering. Elsevier Science [journals], 660 White Plains Rd., Tarrytown, NY 10159-5153. (919) 524-9200. Fax: (919) 333-2444. 1968 to present. 8 times a year. $595.00 per year.

RESEARCH CENTERS AND INSTITUTES

Center for Dredging Studies, Texas A&M University. Texas Engineering Experiment Station, College Station, TX 77843-3136. (409) 845-4517.

Coastal and Oceanographic Engineering Laboratory, University of Florida. 336 Weil Hall, Gainesville, FL 32611. (904) 392-1436. Fax: (904) 392-3466.

J.K.K. Look Laboratory of Oceanographic Engineering, University of Hawaii. 811 Olomehani Street, Honolulu, HI 96813. (808) 533-6412. Fax: (818) 537-5607.

DRILLING

See:PETROLEUM ENGINEERING

DROUGHT

See also: CLIMATE, DESERTIFICATION, METEOROLOGY, WATER RESOURCES

ABSTRACT SERVICES AND INDEXES

Applied Science and Technology Index. H.W. Wilson Co., 950 University Avenue, Bronx, NY 10452-9978. (212) 588-8400. (800) 367-6770. Fax: (718) 590-1617. From 1958 to present. Monthly. Inquire about cost and availability. Also available online (BRS and Wilsonline) and on CD-ROM. A cumulative subject index to English language periodicals in the fields of aeronautics and space science, computer technology, chemis-

try, construction industry, energy and related areas. ISSN: 0003-6986.

General Science Index. H.W. Wilson Co., 950 University Avenue, Bronx, NY 10452. (212) 588-8400. (800) 367-6770. Fax: (718) 590-1617. From 1978 to present. Ten issues per year. Quarterly and annual cumulations. Service basis. Available on CD-ROM and online. Inquire regarding cost and availability. ISSN: 0162-1963.

Meteorological and Geoastrophysical Abstracts. American Meteorological Society, c/o Inforonics, Inc., 550 Newtown Road, P.O. Box 458, Littleton, MA 01460. (508) 486-8976. Fax: (508) 486-0027. Covers literature in environmental sciences, meteorology, astrophysics, hydrology, glaciology, and physical oceanography. 1950 to present. Monthly. $950.00 per year. ISSN: 0026-1130. Also available online (DIALOG) and on CD-ROM.

Science Citation Index (SCI). Institute for Scientific Information, 3501 Market Street, Philadelphia, PA 19104. (215) 386-0100. (800) 523-1850. Fax: (215) 386-2991. 1961 to present. Six issues per year, plus annual cumulation. $11650.00 per year. Also available online and on CD-ROM. Inquire about price and availability. ISSN: 0036-827X.

Selected Water Resources Abstracts. U.S. Geological Survey. Distributed by National Technical Information Service (NTIS), 5285 Port Royal Road, Springfield, VA 22161. (703) 487-4650. Monthly. $115.00 per year.

ANNUAL REVIEWS AND YEARBOOKS

American Meteorological Society. Meteorological Monographs. American Meteorological Society, 45 Beacon Street, Boston, MA 02108-3693. (617) 227-2425. Fax: (617) 742-8718. Irregular. Price varies.

Annual Review of Earth and Planetary Sciences. Annual Reviews, Inc., 4139 El Camino Way, Palo Alto, CA 94306-0897. (415) 493-4400. Fax (415) 855-9815. Annual. $55.00 per year.

Developments in Atmospheric Science Series. Elsevier Science Publishing Company, Inc., 655 Avenue of the Americas, New York, NY 10010. (212) 989-5800. Irregular. Inquire about price and availability.

ASSOCIATIONS AND PROFESSIONAL SOCIETIES

American Association of State Climatologists. c/o Myron Moinau, University of Idaho, Agriculture Engineering Department, State Climate Service, Moscow, ID 834844-2040

American Meteorological Society, 45 Beacon Street, Boston, MA 02108-3693. (617) 227-2425. Fax: (617) 742-8718.

National Environmental Satellite Data, and Information Service. 2069 Federal Building 4, Room 2060, Washington, DC 20233. (301) 763-71990. Fax: (301) 763-4011.

National Weather Association, 6704 Wolke Court, Montgomery, AL 36116-2134. (334) 213-0388. Fax: (334) 213-0388.

University Corporation for Atmospheric Research. P.O. Box 3000 1850 Table Mesa Drive, Boulder, CO 80307. (303) 497-1000. Fax: (303) 497-1654

Weather Modification Association. P.O. Box 8116, Fresno CA 93747. (209) 291-8466.

BIBLIOGRAPHIES

Information Sources in the Earth Sciences. David N. Wood, Joan E. Hardy, and Anthony P. Harvey, editors. Bowker-Saur/K.G. Saur. Distributed by R.R. Bowker, 121 Chanlon Road, New Providence, NJ 07974. (800) 521-8110 or (908) 464-6800. Second edition. 1989. $85.00.

Scientific and Technical Books and Serials in Print. An index to Literature in Science and Technology. R.R. Bowker Co., 205 E. 42nd Street, New York, NY 10017. (800) 521-8110 or (212) 916-1600. 1992. $250.00.

DIRECTORIES AND BIOGRAPHICAL SOURCES

American Men and Women of Science: *Physical and Biological Sciences*. Eighteenth edition. R.R. Bowker Company, 245 West 17th Street, New York, NY 10011. (800) 521-8810 or (212) 916-1600. $750.00.

American Meteorological Society. Professional Directory. American Meteorological Society. 45 Beacon Street, Boston, MA 02108. (617) 227-2425. Included in each issue of the Bulletin of the Society.

Meteorological Services of the World. American Meteorological Society, 45 Beacon Street, Boston, MA 02108- 3693. (617) 227-2425. Fax: (617) 742-8718. Annual. $70.00.

National Weather Service Offices and Stations. National Oceanic and Atmospheric Administration, Department of Commerce, Silver Spring, MD 20910. (301) 427-7698. Annual. Free.

Research Centers Directory. Gale Research, 835 Penobscot Building, Detroit, MI 48226-4094. (313) 961-2242. (800) 347-4253. 20th edition. 1995. $485.00. ISSN: 0080-1518.

ENCYCLOPEDIAS AND DICTIONARIES

Climates of the States. Gale Research, 835 Penobscot Building, Detroit, MI 48226-4094. (313) 961-2242.(800) 347-4253. 5th edition. 1995. $255.00.

Concise Oxford Dictionary of the Earth Sciences. Ailsa Allaby and Michael Allaby, editors. Oxford University Press, Inc., 200 Madison Avenue, New York, NY 10016. (800) 334-4249 or (212) 679-7300. 1990. $42.95.

Encyclopedia of Climatology. Rhodes W. Fairbridge, and John E. Oliver. Chapman & Hall, 1 Penn Plaza, New York, NY 10119. (212) 564-1060. 1986. $115.00.

McGraw-Hill Encyclopedia of Science and Technology. McGraw-Hill Publishing Company, 1221 Avenue of the Americas, New York, NY 10020. (800) 262-4729 or (212) 512-3825. Seventh edition. 1992. $1900.00.

Smithsonian Meteorological Tables. Books on Demand, 300 North Zeeb Road, Ann Arbor, MI 48106-1346. (313) 761-4700. (800) 521-0600. reprint edition. $153.90.

Water Encyclopedia. Frits van der Leeden, Fred L. Troise, and David K. Todd. Lewis Pubs., CRC Press, Inc., 2000 Corporate Boulevard, NW, Boca Raton, FL 33431. 2nd edition.1990. $ 151.95

GENERAL WORKS

Coping with Seasonality and Drought. Martha A. Chen. Sage. 1991. $29.95.

Drought Assessment, Management and Planning: Theory and Case Studies. Donald A Wilhite. Kluwer ACADemic Publishers, 101 Philip Drive, Assinippi Park, Norwell, MA 02061. (617) 871-6600. 1993. $99.95.

Drought Follows the Plow: Cultivating Marginal Areas. Michael H. Glantz. Cambridge University Press, 40 West 20th Street, New York, NY 10011-4211. (212) 924-3900. (800) 872-7423. 1994. $19.95 in paper.

Drought in the Great Plains: Research on Impacts and Strategies. Norman J. Rosenberg. US Committee on Irrigation, 1616 17th Street, Suite 483, Denver, CO 80202. (303) 628-5430. 1990. $40.00.

Drought, Its Causes and Effects. Ivan R. Tannehill. Books on Demand, 300 North Zeeb Road, Ann Arbor, MI 48106-1346. (313) 761-4700. (800) 521-0600. reprint edition. $79.00.

The Dust Bowl: An Agricultural and Social History. R. Douglas Hurt. Nelson-Hall, Inc., 111 North Canal Street, Chicago, IL 60606. (312) 930-9446. 1981. $30.95.

Human Encounter with Drought. P. K. Nayak and Anil Mahajan. Reliance Publishing Co., 380 Steinwehr Avenue, Gettysburg, PA 17325. (717) 334-1103. 1991. $30.00.

Rooted in Dust: Surviving Drought and Depression in Southwestern Kansas. Pamela Riney-Kehrberg. University Press of Kansas, 2501 West 15th Street, Lawrence KS 66049. (913) 864-4154. 1994. $25.00.

Social Aspects of Sustainable Dryland Managment. Daniel Stiles. John Wiley & Sons, Inc., 605 Third Avenue, NewYork, NY 10158-0012. (212) 850-6000. (800) 225-5945. 1995. $74.95.

Unstable Agriculture and Droughts. H. Rao. Asia Press, P.O. Box 5446, Berkeley, CA 94705-0446. 9510) 849-2042. 1988. $25.00

HANDBOOKS AND MANUALS

Handbook of Agricultural Meteorology. John F. Griffin. Oxford University Press,Inc., 200 Madison Avenue, New York, NY 10016. (212) 725-6000. (800) 334-4249. 1994. $85.00.

Handbook of Applied Meteorology. David D. Houghton, editor. John Wiley & Sons, Inc., 605 Third Avenue, New York, NY10158-0012. (212) 850-6000. (800) 225-5945. 1985.

Weather Satellite Handbook. American Radio Relay League, 225 Main Street, Newington, CT 06111. (203) 666-1541 1994. $20.00. $168.00.

DROUGHT

Ency. of Physical Sciences and Engineering Info. Sources

ONLINE DATABASES AND CD-ROMS

Climate Assessment Database. National Weather Service, National Meteorological Center, 5200 Auth Road, Suite 101, Camp Springs, MD 20233. (301) 763-8016. Contains daily, weekly and monthly summaries of North American and world climatological data. Also provides five to ten weather forecasts, and 30 to 90 day weather outlook. Subscription required. Inquire as to cost and availability.

Current Contents Search. Institute for Scientific Information, 3501 Market Street, Philadelphia, PA 19104. (215) 386-0100. Fax: (215) 386-6362. Contains citations to articles listed in the table of contents of science and technology journals. Also articles in social sciences and life sciences journals. Available on BRS,(800) 289-4277, DIALOG,(800) 334-2564, online services. Inquire as to cost and availability.

Dissertation Abstracts. University Microfilms International, 300 North Zeeb Road, Ann Arbor, MI 48106. (800) 521-0600 or (313) 761-4700. Scope includes virtually all doctoral dissertations accepted at accredited American institutions from 1861 to present in 252 subject areas. Available on BRS,(800) 289-4277, DIALOG,(800) 334-2564, and OCLC EPIC,(800) 848-5878, online services. Also available on CD-ROM. Inquire as to cost and availability.

Meteorological and Geoastrophysical Abstracts. American Meteorological Society, 45 Beacon Street, Boston, MA 02108-3693. (617) 227-2425. Fax: (617) 742-8718. Contains citations and abstracts to the worldwide literature on significant research in meteorology and geoastrophysics. Related topics include physical oceanography, hydrology, environmental sciences and glaciology. Covers the period 1972 to present. Available on DIALOG,(800) 334-2564, online service. Inquire as to cost and availability.

NTIS BIBLIOGRAPHIC DATABASE. National Technical Information Service, 5285 Port Royal Road, Springfield, VA 22161. (703) 487-4929 or Fax: (703) 321-8199. Broad coverage of government-sponsored science and technology research reports, 1964 to present. Available on BRS,(800) 289-4277, DIALOG, (800) 334-2564, ORBIT, (800) 456-7248, and STN International, FIZ Karlsruhe, P.O. Box 2465, W-7500, Karlsruhe 1, Germany, online services. Also available on CD-ROM. Inquire as to cost and availability.

SciSearch. Institute for Scientific Information, 3501 Market Street, Philadelphia, PA 19104. (800) 523-1850 or (215) 386-0100. Broad multidisciplinary title and author index to the international literature of science and technology, 1974 to present. Available on DIALOG and ORBIT online services. Also available on CD-ROM. Inquire as to cost and availability.

WILSONLINE. H.W. Wilson Company, 950 University Avenue, Bronx, NY 10452. (800) 367-6770 or (212) 588-8400. Makes available online versions of the printed H.W. Wilson indexes including Applied Science and Technology Index, Business Periodicals Index, General Science Index, and Readers' Guide to Periodical Literature. Period covered is generally 1983 to present. Available on BRS,(800) 289-4277, DIALOG,(800) 334-2564, and OCLC EPIC,(800) 848-5878, online services. Also available on CD-ROM. Inquire as to cost and availability.

World Climate Disc. Chadwyck-Healey Inc., 1101 King Street, Alexandria, VA 22314. (703) 683-4890. Fax: (703) 683-7589. Weather and climate data from approximately 5,000 weather stations worldwide, covering the years 1854 to 1990 on CD-ROM. First edition 1992. Approximately $1200.00 per year with annual updates.

World Weatherdisc. WeatherDisc Associates, Inc., 4584 N.E. 89th Street, Seattle, WA 98115. (206) 524-4314. Fax: (206) 543-0308. Meteorological data on CD-ROM which describes the climate of the earth today and for the past few hundred years. First edition, 1989. Approximately $295.00 per year with biannual updates.

PERIODICALS

Agricultural and Forest Meteorology. Elsevier Science Publishing Company, Inc., 655 Avenue of the Americas, New York, NY 10010. (212) 989-5800. Twenty times per year. $750.00 per year.

American Meteorological Society. Bulletin. American Meteorological Society, 45 Beacon Street, Boston, MA 02108-3693. (617) 227-2425. Fax: (617) 742-8718. Monthly. $60.00 per year.

Climate Change. Kluwer Academic Publishers, P.O. Box 358, Accord Station, Hingham, MA 02018-0358. (617) 871-6000. Six times per year. $327.00 per year.

Earth. Kalmbach Publishing Company, P.O. Box 1612, Waukesha, WI 53187. (414) 796-0126 or (800) 558-1544. 1991 to present. Bimonthly. $14.95 per year.

International Journal of Climatology. Royal Meteorological Society, distributed by John Wiley and Sons, Inc., 605 Third Avenue, New York, NY 10158. (800) 526-5368 or (212) 850-6000. Eight times per year. $425.00 per year.

JGR: Journal of Geophysical Research: Atmosphere. American Geophysical Union, 2000 Florida Avenue, NW, Washington, DC 20009. (202) 462-6903. Monthly. $90.00 per year to members.

Journal of Applied Meteorology. American Meteorological Society, 45 Beacon Street, Boston, MA 02108-3693. (617) 227-2425. Fax: (617) 742-8718. Monthly. $165.00 per year.

Journal of Climate. American Meteorological Society, 45 Beacon Street, Boston, MA 02108-3693. (617) 227-2425. Fax: (617) 742-8718. Monthly. $175.00 per year.

Journal of the Atmospheric Sciences. American Meteorological Society, 45 Beacon Street, Boston, MA 02108-3693. (617) 227-2425. Fax: (617) 742-8718. Semi-monthly. $320.00 per year.

Monthly Weather Review. American Meteorological Society, 45 Beacon Street, Boston, MA 02108-3693. (617) 227-2425. Fax: (617) 742-8718. Monthly. $205.00 per year.

National Weather Digest. National Weather Association, 4400 Stamp Road, Room 404, Temple Hills, MD 20748. (301) 899-3784.

Weather. Royal Meteorological Society, 104 Oxford Road, Reading, Berks RG1 7LJ, England. Monthly. $44.00 per year.

Weatherwise. Heldref Publications, 1319 Eighteenth Street, N.W., Washington, DC 20036-1802. (202) 296-6267. Fax: (202) 296-5149. Bi-monthly. $28.00 per year.

RESEARCH CENTERS AND INSTITUTES

Center for Climate Research. Columbia University. Lamont-Doherty Geological Observatory. Palisades, NY 10964. (914)359-2900.

Center for Climatic Research. University of Delaware, Department of Geography, Newark, DE 19716. (302) 451-8998.

Center for Climatic Research. University of Wisconsin-Madison, 1225 West Dayton Street, Madison, WI 53706. (608) 262-2839. Fax: (608) 262-5964.

High Plains Climate Center. University of Nebraska-Lincoln, 237 Chase, Lincoln, NE 68583. (402) 472-6706.

National Center for Atmospheric Research. P.O. Box 3000, Boulder, CO 80307. (303) 497-1000. Fax: (303) 497-1654.

DRYING

See:INDUSTRIAL ENGINEERING

DUST STORMS

See also: AIR POLLUTION, CLIMATOLOGY, DESERTIFICATION, DROUGHT, METEOROLOGY, VOLCANOS, WATER RESOURCES,

ABSTRACT SERVICES AND INDEXES

Applied Science and Technology Index. H.W. Wilson Co., 950 University Avenue, Bronx, NY 10452-9978. (212) 588-8400. (800) 367-6770. Fax: (718) 590-1617. From 1958 to present. Monthly. Inquire about cost and availability. Also available online (BRS and Wilsonline) and on CD-ROM. A cumulative subject index to English language periodicals in the fields of aeronautics and space science, computer technology, chemistry, construction industry, energy and related areas. ISSN: 0003-6986.

Bibliography and Index of Geology. American Geological Institute, 4220 King Street, Alexandria, VA 22302-1507. (703) 379-2480. Fax: (703) 379-7563. From 1969 to present. Monthly. $1295.00 per year. ISSN: 0098-2784. Also available as GEOREF online (CISTI, DIALOG, Orbit, STN) and on CD-ROM. Inquire about price and availability.

General Science Index. H.W. Wilson Co., 950 University Avenue, Bronx, NY 10452. (212) 588-8400. (800) 367-6770. Fax: (718) 590-1617. From 1978 to present. Ten issues per year. Quarterly and annual cumulations. Service basis. Available on CD-ROM and online. Inquire regarding cost and availability. ISSN: 0162-1963.

Meteorological and Geoastrophysical Abstracts. American Meteorological Society, c/o Inforonics, Inc., 550 Newtown Road, P.O. Box 458, Littleton, MA 01460. (508) 486-8976. Fax: (508) 486-0027. Covers literature in environmental sciences, meteorology, astrophysics, hydrology, glaciology, and physical oceanography. 1950 to present. Monthly. $950.00 per year. ISSN: 0026-1130. Also available online (DIALOG) and on CD-ROM.

Science Citation Index (SCI*).* Institute for Scientific Information, 3501 Market Street, Philadelphia, PA 19104. (215) 386-0100. (800) 523-1850. Fax: (215) 386-2991. 1961 to present. Six issues per year, plus annual cumulation. $11650.00 per year. Also available online and on CD-ROM. Inquire about price and availability. ISSN: 0036-827X.

Selected Water Resources Abstracts. U.S. Geological Survey. Distributed by National Technical Information Service (NTIS), 5285 Port Royal Road, Springfield, VA 22161. (703) 487-4650. Monthly. $115.00 per year.

ANNUAL REVIEWS AND YEARBOOKS

American Meteorological Society. Meteorological Monographs. American Meteorological Society, 45 Beacon Street, Boston, MA 02108-3693. (617) 227-2425. Fax: (617) 742-8718. Irregular. Price varies.

Annual Review of Earth and Planetary Sciences. Annual Reviews, Inc., 4139 El Camino Way, Palo Alto, CA 94306-0897. (415) 493-4400. Fax (415) 855-9815. Annual. $55.00 per year.

Developments in Atmospheric Science Series. Elsevier Science Publishing Company, Inc., 655 Avenue of the Americas, New York, NY 10010. (212) 989-5800. Irregular. Inquire about price and availability.

ASSOCIATIONS AND PROFESSIONAL SOCIETIES

American Association of State Climatologists. c/o Myron Moinau, University of Idaho, Agriculture Engineering Department, State Climate Service, Moscow, ID 834844-2040.

American Meteorological Society, 45 Beacon Street, Boston, MA 02108-3693. (617) 227-2425. Fax: (617) 742-8718. Irregular. Price varies.

International Association of Meteorology and Atmospheric Physics. UCAR, P.O. Box 3000, Boulder CO 80307-3000. (303) 497-1673.

National Environmental Satellite Data, and Information Service. 2069 Federal Building 4, Room 2060, Washington, DC 20233. (301) 763-71990. Fax: (301) 763-4011.

National Weather Association, 6704 Wolke Court, Montgomery, AL 36116-2134. (334) 213-0388. Fax: (334) 213-0388.

University Corporation for Atmospheric Research. P.O. Box 3000, 1850 Table Mesa Drive, Boulder, CO 80307-3000. (303) 497-1673. Fax: (303) 497-1654.

Weather Modification Association. P.O. Box 8116, Fresno CA 93747. (209) 291-8466.

BIBLIOGRAPHIES

Information Sources in the Earth Sciences. David N. Wood, Joan E. Hardy, and Anthony P. Harvey, editors. Bowker-Saur/K.G. Saur. Distributed by R.R. Bowker, 121 Chanlon Road, New Providence, NJ 07974. (800) 521-8110 or (908) 464-6800. Second edition. 1989. $85.00.

DUST STORMS

Ency. of Physical Sciences and Engineering Info. Sources

Scientific and Technical Books and Serials in Print. An index to Literature in Science and Technology. R.R. Bowker Co., 205 E. 42nd Street, New York, NY 10017. (800) 521-8110 or (212) 916-1600. 1992. $250.00.

DIRECTORIES AND BIOGRAPHICAL SOURCES

American Men and Women of Science: Physical and Biological Sciences. 18th edition. R.R. Bowker Company, 245 West 17th Street, New York, NY 10011. (800) 521-8810 or (212) 916-1600. $750.00.

American Meteorological Society. Professional Directory. American Meteorological Society. 45 Beacon Street, Boston, MA 02108. (617) 227-2425. Included in each issue of the Bulletin of the Society.

Meteorological Services of the World. American Meteorological Society, 45 Beacon Street, Boston, MA 02108- 3693. (617) 227-2425. Fax: (617) 742-8718. Annual. $70.00.

National Weather Service Offices and Stations. National Oceanic and Atmospheric Administration, Department of Commerce, Silver Spring, MD 20910. (301) 427-7698. Annual. Free

Research Centers Directory. Gale Research, 835 Penobscot Building, Detroit, MI 48226-4094. (313) 961-2242. (800) 347-4253. 20th edition. 1995. $485.00. ISSN: 0080-1518.

ENCYCLOPEDIAS AND DICTIONARIES

Climates of the States. Bair. Gale Research, 835 Penobscot Building, Detroit, MI 48226-4094. (313) 961-2242.(800) 347-4253. 5th edition. 1995. $255.00.

Concise Oxford Dictionary of the Earth Sciences. Ailsa Allaby and Michael Allaby, editors. Oxford University Press, Inc., 200 Madison Avenue, New York, NY 10016. (800) 334-4249 or (212) 679-7300. 1990. $42.95.

Encyclopedia of Climatology. Rhodes W. Fairbridge, and John E. Oliver. Chapman & Hall, 1 Penn Plaza, New York, NY 10119. (212) 564-1060. 1986. $115.00.

McGraw-Hill Encyclopedia of Science and Technology. McGraw-Hill Publishing Company, 1221 Avenue of the Americas, New York, NY 10020. (800) 262-4729 or (212) 512-3825. Seventh edition. 1992. $1900.00.

Smithsonian Meteorological Tables. Books on Demand, 300 North Zeeb Road, Ann Arbor, MI 48106-1346. (313) 761-4700. (800) 521-0600. reprint edition. $153.90.

Water Encyclopedia. Frits van der Leeden, Fred L. Troise, and David K. Todd. Chelsea Press, 15 East 26th Street, New York, NY 10010. (212) 889-8095.1990.

Weather Satellite Handbook. American Radio Relay League, 225 Main Street, Newington, CT 06111. (203) 666-1541. 1994. $20.00.

GENERAL WORKS

The Dust Bowl: An Agricultural and Social History. Douglas R. Hurt. Nelson-Hall, Inc., 111 North Canal Street,Chicago,IL 60606. (312) 930-9446. 1981. $22.95.

Aerosol-Cloud-Climate Interaction. Peter V. Hobbs. Academic Press, Inc., 6277 Sea Harbor Drive, Orlando, FL. (800) 321-5068. 1993. $65.00.

Drought Follows the Plow: Cultivating Marginal Areas. Michael H. Glantz. Cambridge University Press, 40 West 20th Street, New York, NY 10011-4211. (212) 924-3900. (800) 872-7423. 1994. $39.95.

Weather Cycles: Real or Imaginary? William James Burroughs. Cambridge University Press, 40 West 20th Street, New York, NY 10011-4211. (212) 924-3900. 1992. $39.95.

HANDBOOKS AND MANUALS

Author's Guide to the Journals of the American Meteorological Society. American Meteorological Society, 45 Beacon Street, Boston, MA 02108-3693. (617) 227-2425. Fax: (617) 742-8718. 1983. $15.00 in paper.

Handbook of Agricultural Meteorology. John F. Griffin.Oxford University Press, Inc., 200 Madison Avenue, New York, NY 10016. (212) 725-6000. (800) 334-4249. 1994. $85.00.

Handbook of Applied Meteorology. David D. Houghton, editor. John Wiley & Sons, Inc., 605 Third Avenue, New York, NY 10158-0012. (212) 850-6000. (800) 225-5945. 1985. $168.00.

ONLINE DATABASES AND CD-ROMS

Climate Assessment Database. National Weather Service, National Meteorological Center, 5200 Auth Road, Suite 101, Camp Springs, MD 20233. (301) 763-8016. Contains daily, weekly and monthly summaries of North American and world climatological data. Also provides five to ten weather forecasts, and 30 to 90 day weather outlook. Subscription required. Inquire as to cost and availability.

Current Contents Search. Institute for Scientific Information, 3501 Market Street, Philadelphia, PA 19104. (215) 386-0100. Fax: (215) 386-6362. Contains citations to articles listed in the table of contents of science and technology journals. Also articles in social sciences and life sciences journals. Available on BRS,(800) 289-4277, DIALOG,(800) 334-2564, online services. Inquire as to cost and availability.

Dissertation Abstracts. University Microfilms International, 300 North Zeeb Road, Ann Arbor, MI 48106. (800) 521-0600 or (313) 761-4700. Scope includes virtually all doctoral dissertations accepted at accredited American institutions from 1861 to present in 252 subject areas. Available on BRS,(800) 289-4277, DIALOG,(800) 334-2564, and OCLC EPIC,(800) 848-5878, online services. Also available on CD-ROM. Inquire as to cost and availability.

Inspec. Institution of Electrical Engineers, Michael Faraday House, Six Hills Way, Stevenage, Herts. SG1 2AY, England. Telephone: 0438 313311 or Fax: 0438 742840. Contains citations to the worldwide literature of physics, electronics and electrical engineering, computer technology, and related fields. Available on BRS,(800) 289-4277, DIALOG, (800) 334-2564, ORBIT, (800) 456-7248, and STN International, FIZ Karlsruhe, P.O. Box 2465, W-7500, Karlsruhe 1, Germany, online services. Inquire as to cost and availability.

NTIS Bibliographic Database. National Technical Information Service, 5285 Port Royal Road, Springfield, VA 22161. (703) 487-4929 or Fax: (703) 321-8199. Broad coverage of government-sponsored science and technology research reports. 1964 to present. Available on BRS, (800) 289-4277, DIALOG, (800) 334-2564, ORBIT, (800) 456-7248, and STN International, FIZ Karlsruhe, P.O. Box 2465, W-7500, Karlsruhe 1, Germany, online services. Also available on CD-ROM. Inquire as to cost and availability.

SciSearch. Institute for Scientific Information, 3501 Market Street, Philadelphia, PA 19104. (800) 523-1850 or (215) 386-0100. Broad multidisciplinary title and author index to the international literature of science and technology, 1974 to present. Available on DIALOG and ORBIT online services. Also available on CD-ROM. Inquire as to cost and availability.

WILSONLINE. H.W. Wilson Company, 950 University Avenue, Bronx, NY 10452. (800) 367-6770 or (212) 588-8400. Makes available online versions of the printed H.W. Wilson indexes including Applied Science and Technology Index, Business Periodicals Index, General Science Index, and Readers' Guide to Periodical Literature. Period covered is generally 1983 to present. Available on BRS,(800) 289-4277, DIALOG,(800) 334-2564, and OCLC EPIC,(800) 848-5878, online services. Also available on CD-ROM. Inquire as to cost and availability.

World Climate Disc. Chadwyck-Healey Inc., 1101 King Street, Alexandria, VA 22314. (703) 683-4890. Fax: (703) 683-7589. Weather and climate data from approximately 5,000 weather stations worldwide, covering the years 1854 to 1990 on CD-ROM. First edition 1992. Approximately $1200.00 per year with annual updates.

World WeatherDisc. WeatherDisc Associates, Inc., 4584 N.E. 89th Street, Seattle, WA 98115. (206) 524-4314. Fax: (206) 543-0308. Meteorological data on CD-ROM which describes the climate of the earth today and for the past few hundred years. First edition, 1989. Approximately $295.00 per year with biannual updates.

PERIODICALS

Agricultural and Forest Meteorology. Elsevier Science Publishing Company, Inc., 655 Avenue of the Americas, New York, NY 10010. (212) 989-5800. Twenty times per year. $750.00 per year.

American Meteorological Society. Bulletin. American Meteorological Society, 45 Beacon Street, Boston, MA 02108-3693. (617) 227-2425. Fax: (617) 742-8718. Monthly. $60.00 per year.

Climate Change. Kluwer Academic Publishers, P.O. Box 358, Accord Station, Hingham, MA 02018-0358. (617) 871-6000. Six times per year. $327.00 per year.

Earth. Kalmbach Publishing Company, P.O. Box 1612, Waukesha, WI 53187. (414) 796-0126 or (800) 558-1544. 1991 to present. Bimonthly. $14.95 per year.

International Journal of Climatology. Royal Meteorological Society, distributed by John Wiley and Sons, Inc., 605 Third Avenue, New York, NY 10158. (800) 526-5368 or (212) 850-6000. Eight times per year. $425.00 per year.

JGR: Journal of Geophysical Research: Atmosphere. American Geophysical Union, 2000 Florida Avenue, NW, Washington, DC 20009. (202) 462-6903. Monthly. $90.00 per year to members.

Journal of Applied Meteorology. American Meteorological Society, 45 Beacon Street, Boston, MA 02108-3693. (617) 227-2425. Fax: (617) 742-8718. Monthly. $165.00 per year.

Journal of Climate. American Meteorological Society, 45 Beacon Street, Boston, MA 02108-3693. (617) 227-2425. Fax: (617) 742-8718. Monthly. $175.00 per year.

Journal of the Atmospheric Sciences. American Meteorological Society, 45 Beacon Street, Boston, MA 02108-3693. (617) 227-2425. Fax: (617) 742-8718. Semi-monthly. $320.00 per year.

Monthly Weather Review. American Meteorological Society, 45 Beacon Street, Boston, MA 02108-3693. (617) 227-2425. Fax: (617) 742-8718. Monthly. $205.00 per year.

National Weather Digest. National Weather Association, 4400 Stamp Road, Room 404, Temple Hills, MD 20748. (301) 899-3784.

Weather. Royal Meteorological Society, 104 Oxford Road, Reading, Berks RG1 7LJ, England. Monthly. $44.00 per year.

Weatherwise. Heldref Publications, 1319 Eighteenth Street, N.W., Washington, DC 20036-1802. (202) 296-6267. Fax: (202) 296-5149. Bi-monthly. $28.00 per year.

RESEARCH CENTERS AND INSTITUTES

Center for Climate Research. Columbia University. Lamont-Doherty Geological Observatory. Palisades, NY 10964. (914) 359-2900.

Center for Climatic Research. University of Delaware, Department of Geography, Newark, DE 19716. (302) 451-8998.

Center for Climatic Research. University of Wisconsin-Madison, 1225 West Dayton Street, Madison, WI 53706. (608) 262-2839. Fax: (608) 262-5964.

Climate Research Institute. Oregon State University. Strand Hall, Room 326, Corvallis, OR 97331-2209.

High Plains Climate Center. University of Nebraska-Lincoln, 237 Chase, Lincoln, NE 68583. (402) 472-6706.

National Center for Atmospheric Research. P.O. Box 3000, Boulder, CO 80307. (303) 497-1000. Fax: (303) 497-1654.

National Severe Storms Forecast Center. 601 East 12th Street, Kansas City, MO 64106. (816) 374-5922.

DYES AND DYEING

Ency. of Physical Sciences and Engineering Info. Sources

DWARF GALAXY

See:GALAXIES

DYES AND DYEING

See also: LEAD, ORGANIC CHEMISTRY, PAINTS AND PIGMENTS, SURFACE CHEMISTRY, TEXTILES

ABSTRACT SERVICES AND INDEXES

Applied Science and Technology Index. H. W. Wilson Co., 950 University Avenue, Bronx, NY 10452-9978. (212) 588-8400. (800) 367-6770. Fax: (718) 590-1617. From 1958 to present. Monthly. Inquire about cost and availability. Also available online (BRS and Wilsonline) and on CD-ROM. A cumulative subject index to English language periodicals in the fields of aeronautics and space science, computer technology, chemistry, construction industry, energy and related areas. ISSN: 0003-6986.

Chemical Abstracts. Chemical Abstracts Service, 2540 Olentangy River Road, P.O. Box 3012, Columbus, OH 43210-0012. (614) 447-3600. Fax: (614) 447-3713. 1907 to present. Weekly. $16,800.00 per year. Available online and on CD-ROM. CA is also available in five section groupings. Inquire regarding cost and availability.

Current Contents: Physical, Chemical and Earth Sciences. Institute for Scientific Information, 3501 Market Street, Philadelphia, PA 19104. (215) 386-0100. (800) 523-1850. Fax: (215) 386-2291. 1961 to present. Weekly. $442.00 per year. Also available online (BRS, DIALOG) and on CD-ROM. Inquire regarding cost and availability. ISSN: 0163-2574.

Engineering Index Monthly. Engineering Information, Inc., Castle Point on the Hudson, Hoboken, NJ 07030. (201) 216-8500. (800) 221-1044. Fax: (201) 216-8532. Monthly. $2300.00 per year for monthly issues. $1980 per year for annual cumulation. Available on CD-ROM and online as COMPENDEX. ISSN: 0742-1974. 0360-8557.

General Science Index. H.W. Wilson Company, 950 University Avenue, Bronx, NY 10452. (212) 588-8400. (800) 367-6770. Fax: (718) 590-1617. From 1978 to present. Ten issues per year. quarterly and annual cumulations. Service basis. Available on CD-ROM and online. Inquire regarding cost and availability. ISSN: 0162-1963.

Physics Abstracts. Inspec. Section A, Science Abstracts. Institution of Electrical Engineers (IEE), London, United Kingdom. Available from: INSPEC/IEEE-Institute of Electrical and Electronic Engineers, P.O. Box 1331, 445 Hoes Lane, Piscataway, NJ 08855-1331. (908) 562-5549. 1898 to present. 24 issues per year. $2835.00 per year. ISSN: 0036-8091. Also available online and on CD-ROM.

Process and Chemical Engineering/chemical Engineering Abstracts. Royal Society of Chemistry, Information Services, Thomas Graham House, Science Park, Milton Road, Cambridge, CB4 4WF, England. Contains citations, mostly with abstracts, of the worldwide literature on chemical engineering. from 1982 to present. Monthly. $610.00 per year. Also available online. ISSN: 0960-5045.

Science Citation Index (SCI). Institute for Scientific Information, 3501 Market Street, Philadelphia, PA 19104. (215) 386-0100. (800) 523-1850. Fax: (215) 386-2991. 1961 to present. Six issues per year, plus annual cumulation. $11650.00 per year. Also available online and on CD-ROM. Inquire about price and availability. ISSN: 0036-827X.

Theoretical Chemical Engineering. Royal Society of Chemistry, Information Services, Thomas Graham House, Science Park, Milton Road, Cambridge, CB4 4WF, England. Covers theoretical chemical engineering, including theory and laboratory experimentation. 1964 to present. Monthly. $235.00. ISSN: 0960-5053.

ASSOCIATIONS AND PROFESSIONAL SOCIETIES

American Association of Textile Chemists and Colorists. P.O. Box 12215, Research Triangle Park, NC 27709-2215. (919) 549-8141. Fax: (919) 549-8933.

American Chemical Society, 1155 16th Street, NW, Washington, DC 20036. (202) 872-4600.

American Institute of Chemical Engineers. 345 East 47th Street, New York, NY 10017. (212) 705-7338.

American Society for Testing and Materials. 1916 Race Street, Philadelphia, PA 19103-1187. (215) 29905400. Fax: (215) 977-9679.

Association of Consulting Chemists and Chemical Engineers, 50 East 41st Street, Suite 92, New York, NY 10017. (212) 684-6255.

Chemical Manufacturers Association. 2501 M Street, NW, Washington, DC 20037. (202) 887-1182

Color Pigments Manufacturers Association. P.O. Box 20839, Alexandria, VA 22320. (703) 684-4044. Fax: (703) 684-1795.

Federation of Societies for Coating Technology. 429 Norristown Road, Blue Bell, PA 19422. (215) 940-0777. Fax: (215) 940-0292.

National Paint and Coating Association. 1500 Rhode Island Avenue NW, Washington, DC 20005. (202) 462-6272.

Steel Structures Painting Council. 40 24th Street,Pittsburgh, PA 15222-4643. (412) 291-2331. Fax: (412) 281-9992.

TAPPI. Technical Association of the Pulp and Paper Industry. Technology Park, P.O. Box 105113, Atlanta, GA 30348-5113.(770) 446-1400. Fax: (770) 446-6947.

BIBLIOGRAPHIES

Chemical Information Sources. Gary Wiggins. McGraw-Hill Publishing Company, 1221 Avenue of the Americas, New York, NY 10020. (212) 512-2000. (800) 262-4729. 1991. $42.50.

Information Sources in Chemistry. R.T. Bottle and J.F.B Rowland, editors. R.R. Bowker Inc., 121 Chanlon Road, New Providence, NJ 07974. (800) 521-8110 or (908) 464-6800. Fourth edition. 1993. $75.00.

DIRECTORIES AND BIOGRAPHICAL SOURCES

American Institute of Chemical Engineers Directory. American Institute of Chemical Engineers. 345 East 47th Street, New York, NY 10017-2396. (212) 705-7663. Annual.

American Men and Women of Science. Physical and Biological Sciences. R.R. Bowker Inc., 121 Chanlon Road, New Providence, NJ 07974. (908) 464-6800. (800) 521-8110. 20th edition. 8 volumes. 1996. $850.00

Consulting Services: Chemists and Chemical Engineers. Association of Consulting Chemists and Chemical Engineers, 40 East 45th Street, New York, NY 10017. (212) 983-3160. Fax: (212) 983-3161. Biennial. $60.00.

Research Centers Directory. Gale Research, 835 Penobscot Building, Detroit, MI 48226-4094. (313) 961- 2242. (800) 347-4253. 20th edition. 1995. $485.00.

ENCYCLOPEDIAS AND DICTIONARIES

Dictionary of Named Processes in Chemical Technology. Alan E. Comyns. Oxford University Press, Inc., 200 Madison Avenue, New York, NY 10016. (212) 725-6000. (800) 334-4249.

Encyclopedia of Chemical Processing and Design. McKetta Marcel Dekker, Inc., 270 Madison Avenue, New York, NY 10016.(212) 696-9000. (800) 228-1160. 1976 - . $175.00 per volume.

Illustrated Chemistry Laboratory Terminology. Gerbert W. Ockerman. CRC Press, Inc., 2000 Corporate Boulevard, NW, Boca Raton, FL 33431. (407) 994-0555. (800) 272-7737.1991. $32.95.

Industrial Chemical Thesaurus. Michael Ash and Irene Ash. VCH Publishers, Inc., 220 East 23rd Street, New York, NY 1994. $75.00. 10010-4606. (800) 367-8249. 1992. $295.00.

Kirk-Othmer Encyclopedia of Chemical Technology. John Wiley and Sons, Inc., 605 Third Avenue, New York, NY 10158. (800) 526-5368 or (212) 850-6000. Fourth edition. 1991- . 27 volumes. $5400.00.

McGraw-Hill Encyclopedia of Chemistry. Sybil P.Parker, editor. McGraw-Hill Book, Incorporated, 1221 Avenue of the Americas, New York, NY 10020. (212) 997-3675. Second edition. 1993. $95.50.

Paint Manufacturing Industry: Guides to Pollution Prevention. Diane Publishing Co., 600 Upland Avenue, Upland, PA 19015. (610) 499-7415. 1993. $45.00.

Painting Materials: a Short Encyclopedia. Rutherford J. Gettens and George L. Stoug. Dover Publications, Inc., 180 Varick Street, New York, NY 10014. (212) 255-3755. (800) 223-3130. 1965. $6.95.

Ullman's Encyclopedia of Industrial Chemistry. VCH Publications, Inc., 220 East 23rd Street, Suite 909, New York, NY 10010-4606. (212) 683-8333. (800) 422-8824. 5th edition. 1984 - . Price varies per volume.

GENERAL WORKS

Basics of Dyeing and Finishing. American Association of Textile Chemists and Colorists (AATCC). P.O. Box 12215, Research Triangle Park, NC 27709-2215. (919) 549-8141. Fax: (919) 549-8933. 1993. $65.00.

Chemistry and Application of Dyes. D. R. Waring and G. Hallas, editors. Plenum Publishing Corp., 233 Spring Street, New York, NY. (212) 620-8000. (800) 2221-9369. Fax: (212) 463-0742. 1990. $105.00.

Chemistry of Synthetic Dyes. K. Venkataraman. Academic Press Inc., 6277 Sea Harbor Drive, Orlando, FL. (800) 321-5068. 8 volumes. 1971.

Color Chemistry: Synthesis, Properties and Applications of Organic Dyes and Pigments. Heinrich Zollinger. VCH Publications, Inc., 220 East 23rd Street, Suite 909, New York, NY 10010-4606. (212) 683-8333. (800) 422-8824. 2nd revised edition. 1991. $150.00.

The Dyer's Companion. Elijah Bemiss. Dover Publications, Inc., 180 Varick Street, New York, NY 10014. (212) 255-3755. (800) 223-3130. 1973. $7.95.

Fundamentals of the Chemistry and Application of Dyes. Raul Rys and H. Zollinger. Books on Demand, 300 North Zeeb Road, Ann Arbor, MI 48106-1346. (313) 761-4700. (800) 521-0600. $58.00.

Industrial Color Technology. Ruth M. Johnston and Max Saltzman, editors. American Chemical Society, 1155 16th Street, NW, Washington, DC 20036. (202) 872-4600. (800) 333-9511. Fax: (614) 447-3671. 1972. $21.95.

Industrial Inorganic Pigments. Gunter Buxbaum. editor. VCH Publications, Inc., 220 East 23rd Street, Suite 909, NewYork, NY 10010-4606. (212) 683-8333. (800) 422-8824. 1993. $92.00.

Industrial Organic Pigments: Production, Properties, Applications. Willy Herbst and Kaus Hunger. VCH Publications, Inc., 220 East 23rd Street, Suite 909, New York, NY 10010-4606. (212) 683-8333. (800) 422-8824. 1993. $195.00.

Infrared Absorbing Dyes. M. Matsuoka, editor. Plenum Publishing Corp., 233 Spring Street, New York, NY. (212) 620-8000. (800) 2221-9369. Fax: (212) 463-0742. 1990. $69.50.

Master Dyers to the World: Technique and Trade in Early Indian Dyed Cotton Textiles. Mattiebelle Gittinger. Textile Museum, 2320 S Street, NW, Washington, DC 20008. (202) 667-0441. 1982. $20.00 in paper.

The Rainbow Makers: The Origins of the Synthetic Dyestuffs Industry in Western Europe. Anthony S. Travis. Lehigh University Press. c/o Associated University Presses, 440 Forsgate Drive, Cranbury, NJ 08512. (609) 655-4770.1993. $49.50.

Textile Processing and Properties: Preparation, Dyeing, Finishing and Performance. Tyrone L. Vigo. Elsevier Science Publishing Company, Inc., 655 Avenue of the Americas, New York, NY 10010. (212) 989-5800. Fax: (914) 333-2444. 1994. $265.75.

DYES AND DYEING

Ency. of Physical Sciences and Engineering Info. Sources

HANDBOOKS AND MANUALS

Chemical Engineering Reference Manual. Randall N. Robinson.Professional Publications, Inc., 1250 Fifth Avenue, Belmont,CA 94002. (415) 593-0110. (800) 426-1178. 1988. $49.95.

Chemical Formulary. H. Bennett, editor. Chemical Publishing, Co., Inc. 80 Eighth Avenue, New York, NY 10011.(212 255-1950. Volumes 1-30. $60.00 per volume.

CRC Handbook of Chemistry and Physics. David R. Kide, editor. CRC Press Inc., 2000 Corporate Blvd., NW, BocaRaton, FL 33431. (407) 994-0555. (800) 333-8300. 77th edition. 1996. $99.95.

Guide to Basic Chemical Compounds. D.R. Lide, Jr. CRC Publishers, Inc., 2000 Corporate Blvd., N.W., Boca Raton, FL 33431. (407) 994-0555 or (800) 333-8300. 1993. $120.00.

Handbook of Chemical Engineering Calculations. Nicholas P. Chopey and Tyler G. Hicks, editors. McGraw-Hill Publishing Company, Inc., 1221 Avenue of the Americas, New York, NY 10020. (212) 512-2000. (800) 262-4729. 2nd edition. 1993. $69.50.

Handbook of Paint Raw Materials. Ernest w. Flich. Noyes Data Corporation, 120 Mill Road, Park Ridge, NJ 07656. (201) 391-8484. 1989. $98.00.

Lange's Handbook of Chemistry. John A. Dean, editor. McGraw-Hill Publishing Company, 1221 Avenue of the Americas, New York, NY 10020. (800) 262-4729 or (212) 512-3825. Fourteenth edition. 1992. $79.50.

Pigment Handbook. Peter A. Lewis. John Wiley & Sons, Inc.,605 Third Avenue, New York, NY 10158-0012. (212) 850-6000. (800) 225-5945. 3 volumes. 1988. $608.00 set.

Riegel's Handbook of Industrial Chemistry. James A. Kent, editor. Van Nostrand Reinhold, 115 Fifth Avenue, New York, NY 10003. (212) 254-3232 or (800) 926-2665. Ninth edition. 1992. $114.95.

ONLINE DATABASES AND CD-ROMS

Analytical Abstracts Online. Royal Society of Chemistry, Information Services, Thomas Graham House, Science Park, Milton Road, Cambridge, CB4 4WF, England. Contains citations, mostly with abstracts, of the worldwide literature on analytical chemistry, from 1980 to present. Available on DIALOG and STN online services. Inquire as to cost and availability.

CA Search. Chemical Abstracts Service, P.O. Box 3012, Columbus, OH 43210-0012. (614) 447-3600. (800) 848-6533. Fax: (614) 447-3709. Guide to worldwide chemical literature and related fields, 1972 to present. Available on BRS, DIALOG, ORBIT and STN online services. Inquire as to cost and availability.

Chemical Journals of the American Chemical Society. American Chemical Society, 1155 16th Street, N.W., Washington, DC 20036. (202) 872-4381 or (800) 424-6747. Contains complete text of approximately 90,000 articles from 22 primary journals published by the American Chemical Society, from mostly 1982 to present. Available on STN online service. Inquire as to cost and availability.

Dissertation Abstracts. University Microfilms International, 300 North Zeeb Road, Ann Arbor, MI 48106. (800) 521-0600 or (313) 761-4700. Scope includes virtually all doctoral dissertations accepted at accredited American institutions from 1861 to present in 252 subject areas. Available on BRS, DIALOG, and OCLC EPIC online services. Also available on CD-ROM. Inquire as to cost and availability.

INSPEC. Institution of Electrical Engineers, Michael Faraday House, Six Hills Way, Stevenage, Herts. SG1 2AY, England. Telephone: 0438 313311 or Fax: 0438 742840. Contains citations to the worldwide literature of physics, electronics and electrical engineering, computer technology, and related fields. Available on BRS, DIALOG, ORBIT and STN online services. Inquire as to cost and availability.

Kirk-Othmer Encyclopedia of Chemical Technology. John Wiley and Sons, Inc., 605 Third Avenue, New York, NY 10158. (800) 526-5368 or (212) 850-6000. Contains the complete text of all chapters in the 27 volume fourth edition of the *Kirk-Othmer Encyclopedia of Chemical Technology.* 1991. Available on BRS, DIALOG, online services. Inquire as to cost and availability.

NTIS. National Technical Information Service, 5285 Port Royal Road, Springfield, VA 22161. (703) 487-4929 or Fax: (703) 321-8199. Broad coverage of government-sponsored research reports, 1964 to present. Available on BRS, DIALOG, ORBIT and STN online services. Also available on CD-ROM. Inquire as to cost and availability.

SciSearch. Institute for Scientific Information, 3501 Market Street, Philadelphia, PA 19104. (800) 523-1850 or (215) 386-0100. Broad multidisciplinary title and author index to the international literature of science and technology, 1974 to present. Available on DIALOG and ORBIT online services. Also available on CD-ROM. Inquire as to cost and availability.

WILSONLINE. H.W. Wilson Company, 950 University Avenue, Bronx, NY 10452. (800) 367-6770 or (212) 588-8400. Makes available online versions of the printed H.W. Wilson indexes including *Applied Science and Technology Index, Business Periodicals Index,* and *Readers' Guide to Periodical Literature.* Period covered is generally 1983 to present. Also available on CD-ROM. Inquire as to cost and availability.

Who's Who in Technology. Gale Research, 835 Penobscot Building, Detroit, MI 48226-4094. (313) 961-2242. (800) 347-4253. Contains biographical information of contemporary American scieNTISts and engineers. Available on DIALOG online service. Inquire as to cost and availability.

PERIODICALS

AICHE Journal. American Institute of Chemical Engineers.345 East 47th Street, New York, NY 10017. (212) 705-7663. Fax: (212) 752-3294. 1955 to present. Monthly. $395.00 per year. ISSN: 0001-1541.

American Dyestuff Reporter. SAF International Publications,Promenade A, Suite 2, Harmon Cove Towers, Secucus, NJ 70904.(201) 867-9230. Fax: (201) 867-9230. 1917 to present. Monthly. $17.00 per year. ISSN: 0002-8266.

Analytical Chemistry. American Chemical Society, P.O. Box 3337, Columbus, OH 43210. (614) 447-3776. (800) 333-9511. Fax: (614) 447-3671. 1929 to present. Monthly. $415.00 per year. ISSN: 0003-2700.

Chemical Engineering Progress. American Institute of Chemical Engineers, 345 East 47th Street, New York, NY 10017. (212) 705-7338. Fax: (212) 752-3294. 1947 to present. Monthly. $75.00 per year. ISSN: 0360-7275.

Chemical Week. includes annual buyers guide. Chemical Week Associates, 888 Seventh Avenue, New York, NY 10106. (212) 621-4900. Fax: (212) 621-4949. 1914 to present. Weekly. $99.00 per year. ISSN: 0009-272X.

CHEMTECH. American Chemical Society. P.O. Box 3337, Columbus, OH 43210. (614) 447-3776. 1970 to present. Monthly. $370.00 per year. ISSN: 0009-2703.

Industrial and Engineering Chemistry Research. American Chemical Society, P.O. Box 3337, Columbus, OH 43210. (614) 447-3776. (800) 333-9511. Fax: (614) 447-3671. 1962 to present. Monthly. $567.00 per year. ISSN: 0888-5885.

Journal of Dispersion Science and Technology. Marcel Dekker, Inc., 270 Madison Avenue, New York, NY 10016. (212) 696-9000. Fax: (212) 685-4540. Subscriptions to: P.O. Box 5017 Monticello, NY 12701. 1980 to present. Six times per year. $445.00 per year. ISSN: 0193-2691.

Journal of Physical Chemistry. American Chemical Society, P.O. Box 3337, Columbus, OH 43210. (614) 447-3776. (800) 227-5558. 1896 to present. Weekly. $1140.00 per year. ISSN: 0022-3654.

Textile Research Journal. Textile Research Institute, 601 Prospect Avenue, P.O. Box 625, Princeton, NJ 08542. (609) 924-3150. Fax: (609) 683-7836. 1930 to present. Monthly. $170.00 per year. ISSN: 0040-5175.

RESEARCH CENTERS AND INSTITUTES

American Association of Textile Chemists and Colorists. P.O. Box 12215, Research Triangle Park, MC 27709-2215. (919) 549-8141. Fax: (919) 549-8933.

Dye Chemistry Laboratory. North Carolina State University, Department of Textile Engineering, Chemistry and Science, School of Textiles, P.O. Box 8302, Raleigh, NC 27695. (919) 737-2551.

International Fabricare Institute. 12251 Tech Road, Montgomery Industrial Park, Silver Spring, MD 20904. (301) 622-1900. Fax: (301) 236-9320.

Texas Research Center for Laundry and Drycleaning. Texas Woman's University, P.O. Box 22509, TWU Station, Denton, TX 76204. (817) 898-2670. Fax: (817) 898-3198.

DYNAMICS

See:MECHANICS

DYNAMOS

See:GENERATORS

E

EARTH

See also: ASTROGEOLOGY, ASTEROIDS, ASTRONOMY, COMETS, METEORITES, NAMES OF INDIVIDUAL PLANETS, PLANETARY SCIENCES,SATELLITES (NATURAL), SEISMOLOGY, SOLAR SYSTEM, TEKTITES, VOLCANOLOGY

ABSTRACT SERVICES AND INDEXES

Applied Science and Technology Index; A Cumulative Subject Index To English Language Periodicals in the Fields of Aeronautics and Space Science, Computer Technology, Chemistry, Construction Industry, Energy and Related Areas. H.W. Wilson Co., 950 University Avenue, Bronx, NY 10452. (212) 588-8400. (800) 367-6770. FAX (718) 590-1617. From 1958 to present. Monthly. Inquire about cost and availability. Also available on CD-ROM and online. ISSN: 0003-6986.

Astronomy and Astrophysics Abstracts. Springer-Verlag New York, 175 Fifth Avenue, New York, NY 10010. (212) 460-1500. FAX (212) 473-6272. Published for the Astronomisches Rechen-Institut. Comprehensive coverage of all aspects of astronomy, astrophysics and related fields. 1969 to present. Two parts per year. Annual. ISSN: 0067-0022.

Bibliography and Index of Geology. American Geological Institute, 4220 King Street, Alexandria, VA 22302-1507. (703) 379-2480. FAX (703) 379-7563. From 1969 to present. Monthly. $1295.00 per year. ISSN: 0098-2784. Also available as GEOREF online (CISTI, DIALOG, Orbit, STN) and on CD-ROM. Inquire about price and availability.

Chemical Abstracts. Chemical abstracts Service. 2540 Oleantangy River Road, Box 3012, Columbus, OS 43210-0012. (614) 447-3600. (800) 848-6538. FAX (614) 447-3713. 1907 to present. Weekly. $16,800.00 per year. Available online and on CD-ROM. CA is also available in five section groupings. Inquire regarding cost and availability. ISSN:

Current Contents: Physical, Chemical and Earth Sciences. Institute for Scientific Information, 3501 Market Street, Philadelphia, PA 19104. (215) 386-0100. (800) 523-1850. FAX (215) 386-2291. 1961 to present. Weekly. $442.00 per year. Also available online (BRS, DIALOG) and on CD-ROM. Inquire regarding cost and availability. ISSN: 0163-2574.

General Science Index. H.W. Wilson Company, 950 University Avenue, Bronx, NY 10452. (212) 588-8400. (800) 367-6770. FAX (718) 590-1617. From 1978 to present. Ten issues per year;

quarterly and annual cumulations. Service basis. Available on CD-ROM and online. Inquire regarding cost and availability. ISSN: 0162-1963.

Government Reports Announcements and Index. U. S. National Technical Information Service (NTIS), 5285 Port Royal Road, Springfield, VA 22161. (703) 487-4650. FAX (703) 321-8547. From 1968 to present. Annual. $630.00 per year. Also available online as *NTIS Bibliographic Database* and on CD-ROM.

Meteorological and Geoastrophysical Abstracts. American Meteorological Society, c/o Inforonics, Inc., 550 Newtown Road, Littleton, MA 01460. (508) 486-8976. FAX (508) 486-0027. Covers literature in environmental sciences, meteorology, astrophysics, hydrology, glaciology, and physical oceanography. From 1950 to present. Monthly. $950.00 per year. Also available on CD-ROM and online. Inquire regarding cost and availability. ISSN: 0026-1130.

NTIS Alerts: Astronomy & Astrophysics. U. S. National Technical Information Service. 5285 Port Royal Road, Springfield, VA 22161. (703) 487-4650. FAX (703) 321-8547. Weekly. $140.00 per year.

Pascal 48: Environnement Cosmique Terrestre, Astronomie Et Geologie Extraterrestre. Centre National de la Recherche Scientifique, Institut de Information Scientifique et Technique, 2 aliee du Parc de Brabois, 54514 Vandoeuvre Les Nancy Cedex, France. TEL 83-50-46-00. FAX: 83-50-46-50. 1985 to present. Ten issues per year. 770 F. Also available on CD-ROM and online.

Physics Abstracts. INSPEC. Section A, Science Abstracts. Institute of Electrical Engineers, London, United Kingdom. Available from: INSPEC/IEEE - Institute of Electrical and Electronic Engineers, Box 1331, Hoes Lane, Piscataway, NJ 08855-1331. (908) 562-5549. 1898 to present. 24 issues per year. $2835.00 per year. ISSN: 0036-8091. Also available online and on CD-ROM.

Physics Briefs (Physikalische Berichte). Information Center for Energy, Physics, Mathematics; German Physical Society. V C H Publishers, Inc., 220 East 23rd Street, New York, NY 10010-4606. (212) 683-8333. 1845 to present. 24 issues per year. $2390.00 per year. Also available online. ISSN: 0179-7434.

Science Citation Index. SCI. Institute for Scientific Information, 3501 Market Street, Philadelphia, PA 19104. (215) 386-0100. (800) 523-1850. FAX (215) 386-2991. 1961 to present.

Six issues per year, plus annual cumulation. $11650.00 per year. Also available online and on CD-ROM. Inquire about price and availability. ISSN: 0036-827X.

STAR. Scientific and Technical Aerospace Reports. U.S. National Aeronautics and Space Administration. Distributed by U. S. Superintendent of Documents, Washington, DC 20402. 1963 to present. Semi-monthly, with semiannual and annual indexes. $114.00 per year. ISSN: 0036-8741. Also available online and on CD-ROM.

ANNUAL REVIEWS AND YEARBOOKS

The Astronomical Almanac. Superintendent of Documents, U. S. Government Printing office, Washington, DC 10402. (202) 783-3238. 1981 to present. Supercedes AstronomicalEphemeris. Annual. $44.95. ISSN: 0737-6421.

Annual Review of Astronomy and Astrophysics. Original reviews of critical literature and current developments in astronomy and astrophysics. Annual Reviews, Inc., 4139 El Camino Way, Palo Alto, CA 94303-0139. (415) 493-4400. (800) 523-8635. Fax (415) 855-9815. 1963 to date. Annual. $60.00. ISSN: 0066-4146.

Annual Review of Earth and Planetary Sciences. Annual Reviews, Inc., 4139 El Camino Way, Palo Alto, CA 94303-0139. (415) 493-4400. (800) 523-8635. Fax (415) 855-9815. 1973 to date. Annual. $62.00. ISSN: 0084-6597.

Proceedings of the Lunar and Planetary Science Conference. National Technical Information Service (NTIS), 5285 Port Royal Road, Springfield, VA 22161. (703) 487-4650. 1970 to date. Annual.

ASSOCIATIONS AND PROFESSIONAL SOCIETIES

Amateur Astronomers Association. 1010 Park Avenue, New York, NY 10028. (212) 535-2922.

American Astronomical Society. 2000 Florida Avenue NW, Suite 400, Washington, DC 20009. (202) 328-2010. FAX (202) 234-2560.

American Geophysical Union. 2000 Florida Avenue NW, Washington, DC 20009. (202) 462-6900. (800) 966-AGU1. FAX (202) 328-0566.

American Institute of Physics. 1 Physics Ellipse, College Park, MD 20740-3843. (301) 209-3100.

Association of Universities For Research in Astronomy, Inc. (AURA). Suite 701, 1625 Massachusetts Avenue, NW, Washington, DC 20036. (202) 483-2101. FAX (202) 483-2106.

Astronomical Society of the Pacific. 390 Ashton Avenue, San Francisco, CA 94112. (415) 337-1100. FAX: (415) 337-5205.

Geological Society of America. 3300 Penrose Place, PO Box 9140. Boulder, CO 80301-9140. (303) 447-2020. FAX (303) 447-1133.

IEEE Geoscience and Remote Sensing Society. c/o Instituteof Electrical and Electronics Engineers. 345 East 47th Street, New York, NY 10017. (212) 705-7900. FAX (212) 705-4929.

Meteoritical Society. University of Massachusetts, 125 Marston Hall, Amherst, MA 01003. (413) 545-0300. (413) 545-0724.

Planetary Society. 65 North Catalina Avenue, Pasadena, CA 91106. (818) 793-5100. (800) WOW-MARS. FAX (818) 793-5528.

BIBLIOGRAPHIES

Astronomy and Astrophysics: A Source Guide. Gordon Press, P.O. Box 459, Bowling Green Station, New York, NY 10004. (718) 624-8419. 1991. $250.00.

A Bibliography of Astronomy, 1970 - 1979. R. A. Sea and S. S. Martin. Libraries Unlimited, Inc., Littleton, CO 80160. 1982. $37.50.

Scientific and Technical Books and Serials in Print, 1995. R.R. Bowker Inc., 121 Chanlon Road, New Providence, NJ 07974. (908) 464-6800. (800) 521-8110. 4 volumes. 1994. $299.95. Also available on compact disc and online.

A Source Book On Astronomy and Astrophysics, 1900 - 1975. Kenneth Lang and Owen Gingerich, editors. Harvard University Press, 79 Garden Street, Cambridge MA. (617) 495-2600. 1980. $41.95.

DIRECTORIES AND BIOGRAPHICAL SOURCES

American Astronomical Society. Membership Directory.

American Men and Women of Science: Physical and Biologicalsciences. R. R. Bowker Inc., 121 Chanlon Road, New Providence, NJ 07974. (908) 464-6800. (800) 521-8110. 20th edition. 8 volumes. 1996. $850.00.

The Astronomers. Donald Goldsmith. St. Martin's Press, Inc., 175 Fifth Avenue, New York, NY 10010. (212) 674-5151. (800) 221-7945. 1993. $14.95 in paper.

Astronomical Centers of the World. Kevin Krisciunas. Cambridge University Press, 40 West 20th Street, New York, NY 10011-4211. (212) 924-3900. (800) 872-7423. 1988. $34.95

Directory of Physics and Astronomy Staff Members. American Institute of Physics, One Physics Ellipse, College Park, MD 20740-3843. (301) 209-3100. Annual. $45.00.

Earth and Astronomical Research Centers. Stockton Press, 345 Park Avenue South, New York, NY 10010. 4th edition. 1995. $515.00. ISBN: 1-56169-0967.

Graduate Programs in Physics, Astronomy and Related Fields. 1995-1996. American Institute of Physics, One Physics Ellipse, College Park, MD 20740-3843. (301) 209-3100. 1978 to present. Annual. $45.00.

Research Centers Directory. Gale Research Inc., 835 Penobscot Building, Detroit, MI 48226-4094. (313) 961-2242. (800) 347-4253. 20th edition. 1995. $485.00.

ENCYCLOPEDIAS AND DICTIONARIES

Concise Dictionary of Astronomy. Jacqueline Mitton. Oxford University Press, Inc., 200 Madison Avenue, New York, NY 10016. (212) 725-6000. (800) 334-4249. 1992. $24.95.

Encyclopedia of Astronomy and Astrophysics. Stephen Maran, editor. Van Nostrand Reinhold, 115 Fifth Avenue, New York, NY 10003. (212) 254-3232. (800) 842-3636. 1992. $129.95. lead>*McGraw-Hill Encyclopedia of Science and Technology.* McGraw-Hill Book Company, Inc., 1221 Avenue of the Americas, New York, NY 10020. (212) 512-2000. (800) 262-4729. 7th edition. 20 volume set. 1992. $1900.00.

Moons of the Solar System: An Illustrated Encyclopedia. John Stewart. McFarland & Company, Inc., Box 611, Jefferson, MC 28640. (910) 246-4460. (800) 253-2187. 1991. $49.95.

New Guide To the Planets. Patirck Moore. Trans-Atlantic Publications, Inc., 311 Bainbridge Street, Philadelphia, PA 19147. (215) 925-5083. 1993. $37.50.

Stars and Planets: The Sierra Club Guide To Sky Watching and Direction Finding. W. S. Kals. The Sierra Press, 4988 Gold Leaf Drive, Mariposa, CA 95338. (209) 966-5071. (800) 745-2631. $15.00.

GENERAL WORKS

Asimov's Guide To Earth and Space. Isaac Asimov. Random House, Inc., 201 East 50th Street, New York, NY 10022. (212) 751-2600. (800) 733-3000. 1991. $19.50.

Birth of the Earth. David E. Fisher. Columbia University Press, 562 West 113th Street, New York, NY 10025. (212) 666-1000. (800) 944-8648. 1988. $16.50 in paper.

The Blue Planet: An Introduction To Earth System Science. Brian J. Skinner and Stephen C. Porter. John Wiley & Sons, Inc., 605 Third Avenue, New York, NY 10158-0012. (212) 850-6000. (800) 225-5945. 1994. $wfi

Chemistry and Physics of the Terrestrial Planets. Surendra K. Saxwna. Springer-Verlag New York, Inc., 175 Fifth Avenue, New York, NY 10010. (212) 460-1500. (800) 777-4643. 1986. $109.00.

Earth, Moon and Planets. Fred L. Whipple. Harvard University Press, 79 Garden Street, Cambridge MA. (617) 495-2600. 3rd edition. 1968. $29.00.

Earth: The Water Planet. Jack Gartrell, et al. National Science Teachers Association, 1840 Wilson Boulevard, Arlington, VA 22201. (703) 243-71000. (800) 722-6782. 1992. $18.50.

Evolution of the Earth and Planets. E. Takahashi, et al, editors. American Geophysical Union, 2000 Florida Avenue, N.W., Washington, DC 20009. (202) 462-6903. (800) 966-2481. 1993. $28.00.

Exploring the Planets. W. Kenneth Hamblin and Eric H. Christiansen. Macmillan Publishing Company, Inc,. 200 Old Tappan Road, Old Tappan, NJ 07675. (800) 233-2336. 2nd edition. 1995. $62.00.

The Geology of Multi-Ring Impact Basins: The Moon and Other Planets. P. D. Spudis. Cambridge University Press, 40 West 20th Street, New York, NY 10011-4211. (212) 924-3900. (800) 872-7423. Planetary Sciences Series. 1993. $59.95.

Geology of the Terrestrial Planets. Michael H. Carr. U. S. Government Printing office, Superintendent of Documents, Washington, DC 20402-9325. (202) 783-3238. 1985. $16.00

A History of Arabic Astronomy: Planetary theories During the Golden Age of Islam. George Saliba. New York University Press, 70 Washington Square South, New York, NY 10012. (212) 998-2575. (800) 996-6987. 1994. $45.00.

An Introduction To Cosmochemistry. Charles R. Cowley. Cambridge University Press, 40 West 20th Street, New York, NY 10011-4211. (212) 924-3900. (800) 872-7423. 1995. $29.95 in paper.

Meteorites: An Introduction. Fritz Heide. Springer-Verlag New York, Inc., 175 Fifth Avenue, New York, NY 10010. (212) 460-1500. (800) 777-4643. 1995. $24.00 in paper.

Meteorites and the Origin of Planets. John A. Wood. Books on Demand, 300 North Zeeb Road, Ann Arbor, MI 48106-1346. (313) 761-4700. (800) 521-0600. Reprint edition. $35.00 in paper.

Moons and Planets. William K. Hartmann. Wadsworth Publishing Co., 10 Davis Drive, Belmont, CA 94002. (415) 595-2350. (800) 354-9706. 3rd edition. 1993. $49.95.

Music of the Heavens: Kepler's Harmonic Astronomy. Bruce Stephenson. Princeton University Press. 41 William Street, Princeton, NJ 08540. (609) 258-4900. (800) 777-4726. 1994. $39.50.

The Near Planets. Time-Life Books Editors. Time-Life, Inc., 777 Duke Street, Alexandria, VA 22314. (703) 8388-7000. (800) 621-7026. Voyage through the Universe Series. 1992. $ wfi.

Planet Earth. Jill Bailey and Catherine Thomas, editors. Oxford University Press Inc., 200 Madison Avenue, New York, NY 10016. (212) 725-6000. (800) 334-4249. 1993. $30.00.

Planetary Nebulae. Hynes . Willmann-Bell, Inc., P.O. Box 35025, Richmond, VA 23235. (804)-320-7016. 1991. $24.95.

Planetary Landscapes. Ronald Greeley. Chapman & Hall, 1 Penn Plaza, New York, NY 10119. (212) 564-1060. Div. of Routledge. 2nd edition. 1994. $49.95.

The Planetary System. Tobias Owen and David Morrison. Addison-Wesley Publishing Co., Inc., 1 Jacob Way, Reading, MA 01867. (617) 944-3700. (800) 447-2226. 1988. $49.50.

Planets and Their Atmospheres. John S. Lewis and Ronald G. Primm. Academic Press Inc., 6277 Sea Harbor Drive, Orlando, FL. (800) 321-5068. 1983. $65.00 in paper.

*Proceedings of Lunar and Planetary Science, Vol.*22. Graham Ryder and Virgil L. Sharpton, editors. Lunar and Planetary Institute, 3600 Bay Area Boulevard, Houston, TX 77058. (713) 486-2143. 1992. $50.00.

EARTH

Ency. of Physical Sciences and Engineering Info. Sources

Resources of Near-earth Space. John S. Lewis, et al. University of Arizona Press, 1230 North Park Avenue, Number 102, Tucson, AZ 85719. (520) 621-1441. (800) 426-3797. 1993. $75.00.

Solar and Planetary Dynamos. M. R. Proctor, et al, editors. Cambridge University Press, 40 West 20th Street, New York, NY 10011-4211. (212) 924-3900. (800) 872-7423. Publications of the Newton Institue. 1994. $49.95.

Space Technology and Planetary Astronomy. Joseph N. Tatarewicz. Indiana University Pres, 601 North Morton Street, Bloomington, in 47404-3797. (812) 855-6804. (800) 842-6796. 1990. $29.95.

Story of the Earth. Peter Cattermole and Patrick Moore. Cambridge University Press, 40 West 20th Street, New York, NY 10011-4211. (212) 924-3900. (800) 872-7423. 1986. $5.95 in paper.

Sun and Earth. Herbert Friedman. W. H. Freeman & Co., 41 Madison Avenue, East 26th Avenue, 35th Floor, New York, NY 10010. (212) 576-0400. 1995. Revised edition. $32.95.

Terraforming: Engineering Planetary Environments. Martyn J. Fogg. Society of Automotive Engineers (SAE), 400 Commonwealth Drive, Warrendale, PA 15096. (412) 776-4841. 1995. $49.00.

The Third Planet. Time-Life Inc., 777 Duke Street, Alexandria, VA 22314. (703) 8388-7000. (800) 621-7026.. Voyage through the Universe series. 1989. $24.60.

This Island Earth. Oran W. Nicks, editor. U. S. Government Printing office, Superintendent of Documents, Washington, DC 20402-9325. (202) 783-3238. NASA-SP 250. 1987. $12.00.

Worlds Apart: A Textbook in Planetary Sciences. Guy Consolmagno. Prentice-Hall, 113 Sylvan Avenue, Route 9W, Englewood Cliffs, NJ 07632. (201) 592-2000. (800) 922-0579. 1994. $62.00.

HANDBOOKS AND MANUALS

Astrophysical Data: Planets and Stars. Kenneth R. Lanb. Springer-Verlag New York, Inc., 175 Fifth Avenue, New York, NY 10010. (212) 460-1500. (800) 777-4643. 1993. $59.00.

Field Guide to the Stars and Planets. Jay M. Pasachoff and Donald H. Menzel. Houghton Mifflin Co., 222 Berkeley Street, Boston, MA 02116. (617) 351-5000. (800) 225-3362. Revised edition. 1992. $24.95.

New Guide To the Planets. Patrick Moore. Trans-Atlantic Publications, Inc., 311 Bainbridge Street, Philadelphia, PA 19147. (215) 925-5083. 1993. $37.50.

Planet Observer's Handbook. Fred W. Price. Cambridge University Press, 40 West 20th Street, New York, NY 10011-4211. (212) 924-3900. (800) 872-7423. 1994. $34.95.

Planets and their Moons. Aububon Society Staff. Alfred A. Knopf, Inc. 201 E. 50th Street, New York, NY 10022. (800) 733-3000. 1995. $7.99.

Tables of Planetary Phenomena. Neil F. Michelsen. ACS Publications, PO Box 34487, San Diego, CA 92163-4487. (619) 492-9919. (800) 888-9983. 2nd revised edition. 1993. $24.95.

ONLINE DATABASES AND CD-ROMS

CA Search. Chemical Abstracts Service, P.O. Box 3012, Columbus, OH 43210-0012. (614) 447-3600. (800) 848-6533. FAX (614) 447-3709. Very comprehensive guide to worldwide chemical literature and related fields, 1972 to present. Available on BRS,(800) 289-4277, DIALOG, (800) 334-2564, ORBIT (800) 456-7248, and STN International, FIZ Karlsruhe, P.O. Box 2465, W-7500, Karlsruhe 1, Germany, online services. Inquire as to cost and availability.

Current Contents Search. Institute for Scientific Information, 3501 Market Street, Philadelphia, PA 19104. (215) 386-0100. FAX (215) 386-6362. Contains citations to articles listed in the table of contents of science and technology journals. Also articles in social sciences and life sciences journals. Available on BRS, (800) 289-4277, DIALOG, (800) 334-2564, online services. Inquire as to cost and availability.

Dissertation Abstracts Online. University Microfilms International, 300 North Zeeb Road, Ann Arbor, MI 48106. (800) 521-0600 or (313) 761-4700. Scope includes virtually all doctoral dissertations accepted at accredited American institutions from 1861 to present in 252 subject areas. Available on BRS, (800) 289-4277, DIALOG, (800) 334-2564, and OCLC EPIC, (800) 848-5878, online services. Also available on CD-ROM. Inquire as to cost and availability.

INSPEC. Institution of Electrical Engineers, Michael Faraday House, Six Hills Way, Stevenage, Herts. SG1 2AY, England. Telephone: 0438 313311 or FAX 0438 742840. Contains citations to the worldwide literature of physics, electronics and electrical engineering, computer technology, and related fields. Available on BRS, (800) 289-4277, DIALOG, (800) 334-2564, ORBIT, (800) 456-7248, and STN International, FIZ Karlsruhe, P.O. Box 2465, W-7500, Karlsruhe 1, Germany, online services. Inquire as to cost and availability.

International Aerospace Abstracts. American Institute ofAeronautics and Astronautics, 370 L'Enfant Promenade, S.W., Washington, DC 20024. (202) 646-7400. Contains references and abstracts of journal and monograph literature relating to aerospace science and technology, from 1963 to present. Available through the NASA/RECON system of the National Aeronautics and Space Administration only.

NTIS Bibliographic Database. National Technical Information Service, 5285 Port Royal Road, Springfield, VA 22161. (703) 487-4929 or FAX (703) 321-8199. Broad coverage of government-sponsored science and technology research reports, 1964 to present. Available on BRS,(800) 289-4277, DIALOG, (800) 334-2564, ORBIT, (800) 456-7248, and STN International, FIZ Karlsruhe, P.O. Box 2465, W-7500, Karlsruhe 1, Germany, online services. Also available on CD-ROM. Inquire as to cost and availability.

Physics Briefs. American Institute of Physics, 335 East 45th Street, New York, NY 10017. (212) 661-9260 or FAX (212) 661-2036. Contains citations with abstracts of the literature of physics and related fields, 1979 to present. Available on the STN International, FIZ Karlsruhe, P.O. Box 2465, W-7500, Karlsruhe 1, Germany, online service. Inquire as to cost and availability.

Scientific and Technical Books and Serials in Print. R.R. Bowker Inc., 121 Chanlon Road, New Providence, NJ 07974. (800) 521-8110 or (908) 464-6800. List of currently published books and serials in the physical and biological sciences, engineering and technology, with subject, author and titles indexes. Available on ORBIT, (800) 456-7248, online service. Inquire as to cost and availability.

SCISEARCH. Institute for Scientific Information, 3501 Market Street, Philadelphia, PA 19104. (800) 523-1850 or (215) 386-0100. Broad multidisciplinary title and author index to the international literature of science and technology, 1974 to present. Available on DIALOG, (800) 334-2564, and ORBIT, (800) 456-7248, online services. Also available on CD-ROM. Inquire as to cost and availability.

WILSONLINE. H.W. Wilson Company, 950 University Avenue, Bronx, NY 10452. (800) 367-6770 or (212) 588-8400. Makes available online versions of the printed H.W. Wilson indexes including Applied Science and Technology Index, Business Periodicals Index, General Science Index, and Readers' Guide to Periodical Literature. Period covered is generally 1983 to present. Available on BRS, (800) 289-4277, DIALOG, (800) 334-2564, and OCLC EPIC, (800) 848-5878, online services. Also available on CD-ROM. Inquire as to cost and availability.

PERIODICALS

Astronomical Journal. American Institute of Physics, One Physics Ellipse, College Park, MD 20740-3843. (301) 209-3000. Published for the American Astronomical Society. 1849 to present. Monthly. $280.00 per year. ISSN: 0004-6256.

Astronomical Society of the Pacific. Publications. Astronomical Society of the Pacific, 390 Ashton Avenue, San Francisco, CA 94112. (415) 337-1100. FAX (415) 337-5205. 1889 to present. Monthly. $175.00 per year. ISSN: 0004-6280.

Astronomy. Kalmbach Publishing Company, Box 1612, Waukesha, WI 53187-1612. (414) 796-8776. FAX (414) 796-1142. 1973 to present. Monthly. $27.00 per year. ISSN: 0091-6358.

Astronomy and Astrophysics. Springer-Verlag New York, Inc., 44 Hartz Way, Secaucus, NJ 07096-2491. (201) 348-4033. FAX (201) 348-4505. 1969 to date. 36 issues per year in 12 volumes. $1733.00 per year. ISSN: 0004-6361.

Earth, Moon and Planets: An International Journal of Comparative Planetology. Kluwer Academic Publishers, Box 358, Accord Station, Hingham, MA 02018-0358. (617) 871-6600. FAX (617) 871-6528. 1969 to present. Monthly. $840.00 per year. ISSN: 0167-9295.

Geochemica Et Cosmochimica Acta. Elsevier Science. 660 White Plains Road, Tarrytown, NY 10591-5153. (914) 524-9200. FAX (914) 333-2444. 1950 to the present. Biweekly. $895.00 per year. ISSN: 0016-7037.

Icarus; International Journal of Solar System Studies. Academic Press, Inc., Journal Division, 525 B Street, Suite 1900, San Diego, CA 92101-4495. (619) 230-1840. FAX (619) 699-6800. 1962 to present. Monthly. $1080.00 per year. ISSN: 0019-1035.

J G R: Journal of Geophysical Research: Planets. American Geophysical Union, 2000 Florida Avenue, NW, Washington,

CD 20009. (202) 462-6900. FAX (202) 328-0566. 1991 to present. Monthly. $597.00 per year. ISSN: 0148-0227.

Lunar and Planetary Information Bulletin. Lunar and Planetary Institute. 3600 Bay Area Boulevard, Houston, TX 77058-1113. (713) 486-2175. FAX (713) 486-2125. 1970 to present. Quarterly. Free. Also available online.

Mercury. Astronomical Society of the Pacific. 390 Ashton Avenue, San Francisco, CA 94112. (415) 337-1100. FAX: (415) 337-5205. 1972 to present. Bimonthly. $175.00 per year. ISSN: 0047-6773.

Meteoritics. Meteoritical Society, Department of Chemistry, University of Arkansas, Fayetteville, AR 72701. (501) 575-7625. FAX (501) 575-7778. 1953 to present. Bimonthly. $210.00 per year. ISSN: 0026-1114.

Planetary and Space Science. Elsevier Science Publishing Company, Inc., 660 White Plains Road, Terrytown, NY 10591-5153. (914) 524-9200. FAX: (914) 333-2444. 1959 to present. Monthly. $1355.00 per year. ISSN: 0032-0633.

Planetary Report. Planetary Society. 65 North Catalina, Pasadena, CA 91106-2301. (818) 793-5100. FAX (818) 793-5528. 1980 to present. $25.00 per year. ISSN: 0736-3680.

Sky and Telescope. Sky Publishing Corporation, Box 9111, Belmont, MA 02178. (617) 864-7360. FAX (617) 864-6117. 1941 to present. Monthly. $27.00 per year. ISSN: 0037-6604.

RESEARCH CENTERS AND INSTITUTES

Center For Radiophyscs and Space Research. Cornell University, Space Sciences Building, Ithaca NY 14853. (607) 255-4341. FAX (607) 255-9888.

Center For Space Science and Astrophysics. Stanford University. 325 Durand Building, Stanford, CA 04305. (415) 723-3582.

Center For Space Sciences. University of Texas at Dallas, P. O. Box 830688, MS_F022, Richardson, TX 75083-0688. (214) 690-2851. FAX (214) 690-2848.

Earth and Planetary Remote Sensing Laboratory. Washington University, Department of Earth and Planetary Sciences. Campus Box 1169, 1 Brookings Drive, St. Louis, MO 63130. (314) 889-5679. FAX (314) 889-5799.

Institute For Astronomy. University of Hawaii at Manoa, 2680 Woodlawn Drive, Honolulu, HI 96822. (808) 956-8312. FAX (808) 988-2790.

Institute For Terrestrial and Planetary Atmospheres. State University of New York at Stony Brook, Stony Brook, NY 11794-2300. (516) 632-6170. FAX (516) 632-6251.

Institute of Geophysics and Planetary Physics. University of California, Riverside, CA 92521. (714) 787-4503. FAX (714) 787-4529.

Laboratory For Atmospheric and Space Physics. University of Colorado - Boulder, Boulder, CO 80309-0392. (303) 492-7677. (303) 492-6946.

EARTH

Ency. of Physical Sciences and Engineering Info. Sources

Laboratory For Planetary Geology. Arizona State University, Department of Geology, Tempe, AZ 85281. (601) 965-7029.

Laboratory For Planetary Studies. Cornell University, Space Sciences Building, Ithaca, NY 14853. (607) 255-4971.

Lunar and Planetary Institute. 3303 NASA Road One, Houston, TX 77058-4399. (713) 486-2139. FAX 713-496-2162.

Lunar and Planetary Laboratory. University of Arizona, Tucson, AZ 85721. (602) 621-6963. FAX (602) 621-4933.

McDonnell Center for the Space Sciences. Washington University, Box 1105, One Brookings Drive, St. Louis MO 63130-4899. (314) 889-6255. FAX (314) 889-6219.

Smithsonian Astrophysical Observatory. 60 Garden Street, Cambridge, MA 02138. (617) 495-7461. FAX (617) 495-7105.

EARTH SCIENCES

See: GEOLOGY, METEOROLOGY, OCEANOGRAPHY

EARTHQUAKE ENGINEERING

See also: CIVIL ENGINEERING, CONSTRUCTION ENGINEERING, EARTHQUAKES, GEOPHYSICS, GEOTECHNICAL ENGINEERING, ROCK MECHANICS, STRUCTURAL ENGINEERING

ABSTRACT SERVICES AND INDEXES

Abstract Journal in Earthquake Engineering. University of California, Berkeley, Earthquake Engineering Research Center, 1301 South 46th Street, Richmond, CA 94804-4698. (510) 231-9413 or FAX (510) 231-9471. 1972 to present. Annual. $80.00 per year.

Applied Science and Technology Index; A Cumulative Subject Index To English Language Periodicals in the Fields of Aeronautics and Space Science, Computer Technology, Chemistry, Construction Industry, Energy and Related Areas. H.W. Wilson Co., 950 University Avenue, Bronx, NY 10452. (800) 367-6770 or (212) 588-8400. FAX (718) 590-1617. From 1958 to present. Monthly. Inquire about cost and availability. Also available on CD-ROM and online.

ASCE Combined Annual Combined Index. American Society of Civil Engineers, 345 East 47th Street, New York, NY 10017-2398. (212) 705-7520 or (800) 548-2723. Annual. $48.00.

ASCE Publications Information. 1966-present. American Society of Civil Engineers, 345 East 47th Street, New York, NY 10017-2398. (212) 705-7520 or (800) 548-2723. Bi-monthly. $160.00 per year for non-members.

Bibliography and Index of Geology. American Geological Institute, 4220 King Street, Alexandria, VA 22302. (703) 379-2480. Fax (703) 379-7563. Monthly. $1295.00 per year. Also available online as GeoRef.

Civil and Structural Engineering Abstracts, Cambridge Scientific Abstracts, 7200 Wisconsin Avenue, Bethesda, MD 20814-4823. (301) 961-6750. FAX (301) 961-6720. 1993 to present. Monthly. $385.00 per year. Topics covered include structural design, construction equipment and methods, civil defense and military engineering, surveying, highway engineering, maritime and port structures, materials, land reclamation, and soil mechanics.

Computer Abstracts, MCBUniversity Press Ltd., PO Box 10812, Birmingham, AL 35201-0812. (800) 633-4931. FAX (205) 995-1588. Monthly. Covers computer theory, data, hardware, systems, networks, human-computer interaction, artificial intelligence, as well as applications of computers in aerospace, business, CAD/CAM, cartography, civil engineering, electronics and electrical engineering, industrial engineering, mechanical engineering, medicine, structural engineering, etc. 1957 to present. Monthly. $1399.95 per year.

Current Contents: Engineering, Technology, and Applied Sciences. Institute for Scientific Information, 3501 Market Street, Philadelphia, PA 19104. (215) 386-0100. FAX (215) 386-6362. 1970 to present. Weekly. $442.00 per year.

Engineering Index Monthly. Engineering Information, Inc., Castle Point on the Hudson, Hoboken, NJ 07030. (800) 221-1044. FAX (212) 832-1857. Monthly. $2300.00 per year. Also available online as COMPENDEX, and also on CD-ROM. Covers chemical engineering, computers, electrical engineering, civil engineering, metals and mining, industrial management, and mechanical engineering.

Geodex Retrieval System For Geotechnical Abstracts. Geodex Retrieval Systems, 669 Broadway, Box 279, Sonoma, CA 95476. (707) 939-8476. FAX (707) 996-8734. 1970 to present. Three times a year. $645.00 per year includes Geotechnical Abstracts, as Geodex Systems-GRS.

Geological Abstracts. Elsevier - Geo Abstracts, Regency House, 34 Duke Street, Norwich NR3 3AP, England. Monthly. $760.00 per year. Also available online as GEOBASE.

Geological Society of America. Abstracts with Programs. Geological Society of America. 3300 Penrose Place, P.O. Box 9140, Boulder, CO 80301-9140. (303) 447-2020. Abstracts and programs of the annual conference. Annual. $69.00.

Geotechnical Abstracts. Geodex Retrieval Systems, 669 Broadway, Box 279, Sonoma, CA 95476. (707) 939-8476. FAX (707) 996-8734. 1970 to present. Three times a year. $280.00 per year. Covers soil mechanics, foundation engineering, rock mechanics, and engineering geology.

Index To Scientific and Technical Proceedings. Institute for Scientific Information, 3501 Market St., Philadelphia, PA 19104. (215) 386-0100. FAX (215) 386-6362. Monthly. $500.00 per year.

Index To Scientific Reviews. Institute for Scientific Information, 3501 Market St., Philadelphia, PA 19104. (215) 386-0100. FAX (215) 386-6362. Semi-annual.

International Civil Engineering Abstracts. CITIS Ltd., 2 Rosemount Terrace, Blackrock, Dublin, Ireland. Telephone 353-1-2886227. FAX 353-1-885-971. 1974 to present. Monthly. $660.00 per year. Also available on CD-ROM.

Meteorological and Geoastrophysical Abstracts. American Meteorological Society, c/o Inforonics Inc., 550 Newtown Road, Littleton, MA 01460. (508) 486-8976. FAX (508) 486-0027. 1950 to present. Monthly. $950.00 per year. Current abstracts of books, reports, research papers, and miscellaneous literature on environmental sciences, meteorology, astrophysics, hydrology, glaciology, and physical oceanography.

NTIS Alerts: Building Industry Technology. U.S. National Technical Information Service, 5825 Port Royal Road, Springfield, VA 22161. (703) 487-4630. FAX (703) 321-8547. Weekly. $135.00 per year.

Oceanic Abstracts. Cambridge Scientific Abstracts, 7200 Wisconsin Avenue, Bethesda, MD 20814. (301) 961-6750. Fax (301) 961-6720. Bimonthly. $995.00 per year.

Physics Abstracts. Institute of Electrical Engineers, Michael Faraday House, Six Hill Way, Stevenage, Herts, SG1 2AY, England. Distributed by IEEE, 445 Hoes Lane, Piscataway, NJ 08854. (908) 562-5549. 1898 to present. Monthly. $2700.00 per year. Also available online as INSPEC.

Physics Briefs. American Institute of Physics, 335 East 45th Street, New York, NY 10017 (212) 661-9404 or FAX (516)349-9704. 1980 to present. Six times per year. $2600.00 per year.

Science Citation Index. Institute for Scientific Information, 3501 Market Street, Philadelphia, PA 19104. (215) 386-0100. FAX (215) 386-6362. Inquire about availability and cost. Also available on CD-ROM.

ANNUAL REVIEWS AND YEARBOOKS

International Institute of Seismology and Earthquake Engineering Bulletin. Building Research Institute—Ministry of Construction, 1 Tatehara, Tsukuba-City, Ibaraki Prefecture 305, Japan. 1964 to present. Annual. Inquire for cost.

International Institute of Seismology and Earthquake Engineering Yearbook. Building Research Institute—Ministry of Construction, 1 Tatehara, Tsukuba-city, Ibaraki Prefecture 305, Japan. 1964 to present. Biennial. Inquire for cost.

ASSOCIATIONS AND PROFESSIONAL SOCIETIES

American Geological Institute. 4220 King Street, Alexandria, VA 22302. (703) 379-2480. Fax (703) 379-7563.

Earthquake Engineering Research Institute. 499 14th Street, Suite 320, Oakland, CA 94612-1902. (510) 451-0905.

Geological Society of America. 3300 Penrose Place, PO Box 9140, Denver, CO 80301-9140. (303) 447-2020.

International Institute of Seismology and Earthquake Engineering. Building Research Institute—Ministry of Construction, 1 Tatehara, Tsukuba-city, Ibaraki Prefecture 305, Japan.

International Seismological Centre. Pipers Lane, Thatcham, Newbury, Berks. RG13 4NS, England. Telephone 0635-61022. FAX 0635-72351.

Seismological Society of America. 201 Plaza Professional Bldg., El Cerrito, CA 94530. (510) 525-5474. FAX (510) 525-7204.

DIRECTORIES AND BIOGRAPHICAL SOURCES

Directory of Engineering Societies and Related Organizations. Gordon Davis. 13th edition. American Association of Engineering Societies, 1111 19th Street NW, Suite 608, Washington, DC 20036. (202) 296-2237 or (800) 658-8897. 1989. Inquire for price.

International Research Centers Directory. Gale Research, 835 Penobscot Building, Detroit, MI 48226-4094. (313) 961-2242. (800) 347-4253. FAX (313) 961-6083. 8th edition. 1995. $410.00

Research Centers Directory. Gale Research, 835 Penobscot Building, Detroit, MI 48226-4094. (313) 961-2242. (800) 347-4253. FAX (313) 961-6083. $485.00.

Scientific and Technical Organizations and Agencies Directory. Gale Research, 835 Penobscot Building, Detroit, MI 48226-4094. (313) 961-2242. (800) 347-4253. FAX (313) 961-6083. 4th edition. 1996. $195.00.

Who's Who in Engineering. American Association of Engineering Societies, 1111 19th Street NW, Suite 608, Washington, DC 20036. (202) 296-2237 or (800) 658-8897. 8th edition. 1991. Inquire for price.

ENCYCLOPEDIAS AND DICTIONARIES

Dictionary of Civil Engineering. V.N. Vazirani. South Asia Books, Box 502, Columbia, MO 65205. (314) 474-0116. FAX (314) 474-8124. 1992. $20.00.

Dictionary of Structural Engineering. Cyril M. Harris. McGraw-Hill Publishing Company, 1221 Avenue of the Americas, New York, NY 10020. (800) 262-4729 or (212) 512-3825. 1992. $14.95.

Encyclopedia of Physical Science and Technology. Robert A. Meyers, ed. 18 volumes. Academic Press Inc., 6277 Sea Harbor Drive, Orlando, FL 32887. (800) 321-5068. 1992. $2100.00.

McGraw-Hill Dictionary of Scientific and Technical Terms. Sybil P. Parker, ed. 5th edition. McGraw-Hill Publishing Company, 1221 Avenue of the Americas, New York, NY 10020. (800) 262-4729 or (212) 512-3825. 1993. $110.50.

McGraw-Hill Encyclopedia of Engineering. Sybil P. Parker, ed. 2nd edition. McGraw-Hill Publishing Company, 1221 Avenue of the Americas, New York, NY 10020. (800) 262-4729 or (212) 512-3825. 1993. $95.50.

GENERAL WORKS

Dynamics of Structures: Theory and Applications To Earthquake Engineering. Anil K. Chopra. Prentice Hall, 113 Sylvan Avenue, Route 9W, Englewood Cliffs, NJ 07632. (201) 592-2000 or (800) 922-0579. 1995. Inquire for cost.

Earthquake Engineering. Y-X. Hu, S-C. Liu, W. Dong. Chapman & Hall, 115 Fifth Avenue, New York, NY 10211-0906. 1996. $139.00.

Earthquakes and Earthquake Engineering: A Source Guide. Gordon Press Publishers, PO Box 459, Bowling Green Station, New York, NY 10004. (718) 624-8419. 1991. $250.00.

EARTHQUAKE ENGINEERING

Ency. of Physical Sciences and Engineering Info. Sources

Fundamentals of Earthquake-resistant Construction. Ellis L. Krinitzsky, et al. John Wiley and Sons, Inc., 605 Third Avenue, New York, NY 10158. (800) 526-5368 or (212) 850-6000. 1992. $64.95.

Geotechnical Earthquake Engineering. Steven L. Kramer. Prentice Hall, 113 Sylvan Avenue, Route 9W, Englewood Cliffs, NJ 07632. (201) 592-2000 or (800) 922-0579. 1994. $74.00.

Seismic Design of Building Structures: A Professional's Introduction To Earthquake Forces and Design Details. Michael R. Lindeburg. Sixth edition. Professional Publications, 1250 Fifth Ave., Belmont, CA 94002. (415) 593-9119 or (800) 426-1178. 1994. $21.95.

HANDBOOKS AND MANUALS

Building Structural Design Handbook. Richard N. White & Charles G. Salmon, editors. Krieger Publishing Company, P.O. Box 9542, Melbourne, FL 32902-9542. (407) 724-9542. FAX (407) 951-3671. 1987. $95.00.

Civil Engineering Calculations Reference Guide. Tyler G. Hicks. McGraw-Hill Publishing Company, 1221 Avenue of the Americas, New York, NY 10020. (800) 262-4729 or (212) 512-3825. 1987. $46.00.

Civil Engineering Practice. Paul N. Cheremisinoff, et al. 5 volumes. Technomic Publishing Company, Inc., 851 New Holland Avenue, Box 3535, Lancaster, PA 17604. (717) 291-5609 or (800) 233-9936. FAX (717) 295-4538. 1987-88. $199.00.

Handbook of Mechanics, Materials, and Structures. Alexander Blake, editor. John Wiley and Sons, Inc., 605 Third Avenue, New York, NY 10158. (800) 526-5368 or (212) 850-6000. 1985. $120.00.

International Handbook of Earthquake Engineering: Codes, Programs & Examples. Mario Paz, editor. Chapman and Hall, 115 Fifth Avenue, New York, NY 10211-0906. 1994. $84.95.

Seismic Design Handbook. Farzad Naeim. Chapman and Hall, 115 Fifth Avenue, New York, NY 10211-0906. 1989. $102.95.

Standard Handbook For Civil Engineers. Frederick S. Merritt, editor. 3rd edition. McGraw-Hill Publishing Company, 1221 Avenue of the Americas, New York, NY 10020. (800) 262-4729 or (212) 512-3825. 1983. $124.50.

Structural Engineering Handbook. Edwin H. Gaylord Jr. and Charles N. Gaylord. 3d ed. McGraw-Hill Publishing Company, 1221 Avenue of the Americas, New York, NY 10020. (800) 262-4729 or (212) 512-3825. 1990. $89.50.

ONLINE DATABASES AND CD-ROMS

Architectural Periodicals Index. Royal Institute of British Architects, British Architectural Library, 66 Portland Place, London, W1N 4AD, England. Citations to worldwide literature on architecture, construction methods and standards, urban planning, and interior design, from 1978 to present. Available on DIALOG, (800) 334-2564, online service. Inquire as to cost and availability.

Compendex Plus. Engineering Information, Inc., 345 East 47th Street, New York, NY 10017. (212) 705-7600 or (800) 221-

1044. Contains citations with abstracts to worldwide literature in engineering and technology, from 1970 to present. Available on online BRS,(800) 289-4277, DIALOG, (800) 334-2564, ORBIT (800) 456-7248, and STN International, FIZ Karlsruhe, P.O. Box 2465, W-7500, Karlsruhe 1, Germany, online services. Also available on CD-ROM. Inquire as to cost and availability.

CSA Engineering. Cambridge Scientific Abstracts, 7200 Wisconsin Avenue, Suite 601, Bethesda, MD 20814. (301) 961-6750 or (800) 843-7751. Contains citations and abstracts of international periodicals and research literature covering all fields of engineering and science and technology,including computer and information science, electronics, mechanical engineering, solid state materials, 1981 to present. Available on BRS,(800) 289-4277, online service.Inquire as to cost and availability.

Earth Sciences. U.S. Geological Survey, 12201 Sunrise Valley Drive, Reston, VA 22092-9998. (703) 648-4460. CD-ROM of earth science databases including the U.S. Geological Survey Library, Earth Science Data Directory, and GEOINDEX, citations to published geological maps. $350.00 per year with quarterly updates.

Geoarchive. Geosystems, P.O. Box 1024, Westminster, London, England, SW1 P 2JL. Citations to literature on geoscience, 1969 to present. Inquire as to online cost and availability.

Geobase. Elsevier - Geo Abstracts, Regency House, 34 Duke Street, Norwich NR3 3AP, England. Contains citations to the worldwide earth science literature from 1980 to date. Available on DIALOG, ORBIT online services. Inquire as to cost and availability.

Georef: Bibliography and Index of Geology. American Geological Institute, 4220 King Street, Alexandria, VA 22302. (703) 379-2480. Fax (703) 379-7563. Monthly. Inquire for price and availability.

ICONDA (CIB International Construction Database). Fraunhofer-Gesellschaft, Informationszentrum RAUM und BAU, Nobelstrasse, 12, D-7000, Stuttgart 80, Germany. Contains citations with abstracts to worldwide technical literature on construction and civil engineering, structural engineering, and engineering geology, from 1976 to present. Available on ORBIT, (800) 456-7248, and STN International, FIZ Karlsruhe, P.O. Box 2465, W-7500, Karlsruhe 1, Germany, online services. Inquire as to cost and availability.

National Information Service for Earthquake Engineering Database. University of California, Berkeley. Contains citations and abstracts to journal articles on earthquakes and earthquake engineering from 1987 to present. Available on BRS,(800) 289-4277, online service. Inquire as to cost and availability.

NTIS Bibliographic Database. National Technical Information Service, 5285 Port Royal Road, Springfield, VA 22161. (703) 487-4929 or FAX (703) 321-8199. Broad coverage of government-sponsored science and technology research reports, 1964 to present. Available on BRS,(800) 289-4277, DIALOG, (800) 334-2564, ORBIT, (800) 456-7248, and STN International, FIZ Karlsruhe, P.O. Box 2465, W-7500, Karlsruhe 1, Germany, online services. Also available on CD-ROM. Inquire as to cost and availability.

SCISEARCH. Institute for Scientific Information, 3501 Market Street, Philadelphia, PA 19104. (800) 523-1850 or (215) 386-0100. Broad multidisciplinary title and author index to the international literature of science and technology, 1974 to present. Available on DIALOG,(800) 334-2564, and ORBIT,(800) 456-

7248, online services. Also available on CD-ROM. Inquire as to cost and availability.

PERIODICALS

Disasters. Basil Blackwell Inc., 238 Main Street, Cambridge, MA 02142. (617) 547-7110. FAX (617) 547-0789. 1977 to present. $89.50.

Earthquake Engineering and Structural Dynamics. J. Wiley & Sons Journals, Baffins Lane, Chichester, Sussex PO19 1UD, England. Telephone 0243-779773. FAX 0243-775878. 1972 to present. 12 times a year. $995.00 per year.

Earthquake Spectra. Earthquake Engineering Research Institute, 499 14th Street, Suite 320, Oakland, CA 94612-1902. (510) 451-0905. 1984 to present. Quarterly. $120.00.

Geophysical and Astrophysical Fluid Dynamics. Gordon and Breach Science Publishers, 820 Town Center Drive, Langhorne, PA 19047. (215) 750-2642. FAX (215) 750-6343. 1970 to present. 28 times a year. Inquire for price.

Geophysical Research Letters. American Geophysical Union, 2000 Florida Avenue, N.W., Washington, DC 20009. (202) 462-6900. 1974 to present. Semi-monthly. $590.00 per year for non-members.

Journal of Volcanology and Geothermal Research. Elsevier Science Inc., Box 882, Madison Square Station, New York, NY 10159. (212) 989-5800. FAX (212) 633-3990. 1976 to present. 24 times a year. $1155.00.

Seismological Society of America Bulletin. 201 Plaza Professional Bldg., El Cerrito, CA 94530. (510) 525-5474. FAX (510) 525-7204. 1911 to present. Bi-monthly. $135.00.

Soil Dynamics and Earthquake Engineering. Elsevier Science [journals], 660 White Plains Rd., Tarrytown, NY 10159-5153. (919) 524-9200. FAX (919) 333-2444. 1981 to present. Six times a year. $515.00.

RESEARCH CENTERS AND INSTITUTES

Earthquake Engineering Research Institute. 499 14th Street, Suite 320, Oakland, CA 94612-1902. (510) 451-0905.

International Institute of Seismology and Earthquake Engineering. Building Research Institute—Ministry of Construction, 1 Tatehara, Tsukuba-city, Ibaraki Prefecture 305, Japan.

State University of New York—Buffalo, National Center for Earthquake Engineering Research. Red Jacket Quadrangle, Buffalo, NY 14261. (716) 636-3391. FAX (716) 636-3399.

University of California—Berkeley, Earthquake Engineering Research Center. 1301 S. 46th Street, Richmond, CA 94804-4698. (415) 231-9554. FAX (415) 231-9471.

University of Michigan, Structural Engineering Laboratory. 2340 G.G. Brown Bldg., Ann Arbor, MI 48109. (313) 763-3046. FAX (313) 764-4292.

EARTHQUAKE PREDICTION

See: PREDICTION

EARTHQUAKES

See also: EARTHQUAKE ENGINEERING, GEOLOGY, GEOPHYSICS, GEOTECHNICAL ENGINEERING, OCEANOGRAPHY, SEISMOLOGY, VOLCANOLOGY

ABSTRACT SERVICES AND INDEXES

Abstract Journal in Earthquake Engineering. University of California, Berkeley, Earthquake Engineering Research Center, 1301 South 46th Street, Richmond, CA 94804-4698. (510) 231-9413 or FAX (510) 231-9471. 1972 to present. Annual. $80.00 per year.

Bibliography and Index of Geology. American Geological Institute, 4220 King Street, Alexandria, VA 22302. (703) 379-2480. Fax (703) 379-7563. Monthly. $1295.00 per year. Also available online as GeoRef.

Current Contents: Physical, Chemical and Earth Sciences. Institute for Scientific Information, 3501 Market Street, Philadelphia, PA 19104. (215) 386-0100. FAX (215) 386-6362. Weekly. $360.00 per year.

Geological Abstracts. Elsevier - Geo Abstracts, Regency House, 34 Duke Street, Norwich NR3 3AP, England. Monthly. $760.00 per year. Also available online as GEOBASE.

Geological Society of America. Abstracts with Programs. Geological Society of America. 3300 Penrose Place, P.O. Box 9140, Boulder, CO 80301-9140. (303) 447-2020. Abstracts and programs of the annual conference. Annual. $69.00.

Index To Scientific and Technical Proceedings. Institute for Scientific Information, 3501 Market St., Philadelphia, PA 19104. (215) 386-0100. FAX (215) 386-6362. Monthly. $500.00 per year.

Index To Scientific Reviews. Institute for Scientific Information, 3501 Market St., Philadelphia, PA 19104. (215) 386-0100. FAX (215) 386-6362. Semi-annual.

Meteorological and Geoastrophysical Abstracts. American Meteorological Society, c/o Inforonics Inc., 550 Newtown Road, Littleton, MA 01460. (508) 486-8976. FAX (508) 486-0027. 1950 to present. Monthly. $950.00 per year. Current abstracts of books, reports, research papers, and miscellaneous literature on environmental sciences, meteorology, astrophysics, hydrology, glaciology, and physical oceanography.

Oceanic Abstracts. Cambridge Scientific Abstracts, 7200 Wisconsin Avenue, Bethesda, MD 20814. (301) 961-6750. Fax (301) 961-6720. Bimonthly. $995.00 per year.

Physics Abstracts. Institute of Electrical Engineers, Michael Faraday House, Six Hill Way, Stevenage, Herts, SG1 2AY, England. Distributed by IEEE, 445 Hoes Lane, Piscataway, NJ 08854. (908) 562-5549. 1898 to present. Monthly. $2700.00 per year. Also available online as INSPEC.

EARTHQUAKES

Ency. of Physical Sciences and Engineering Info. Sources

Physics Briefs. American Institute of Physics, 335 East 45th Street, New York, NY 10017 (212) 661-9404 or FAX (516)349-9704. 1980 to present. Six times per year. $2600.00 per year.

Science Citation Index. Institute for Scientific Information, 3501 Market Street, Philadelphia, PA 19104. (215) 386-0100. FAX (215) 386-6362. Inquire about availability and cost. Also available on CD-ROM.

ANNUAL REVIEWS AND YEARBOOKS

International Institute of Seismology and Earthquake Engineering Bulletin. Building Research Institute—Ministry of Construction, 1 Tatehara, Tsukuba-city, Ibaraki Prefecture 305, Japan. 1964 to present. Annual. Inquire for cost.

International Institute of Seismology and Earthquake Engineering Yearbook. Building Research Institute—Ministry of Construction, 1 Tatehara, Tsukuba-city, Ibaraki Prefecture 305, Japan. 1964 to present. Biennial. Inquire for cost.

ASSOCIATIONS AND PROFESSIONAL SOCIETIES

American Geological Institute. 4220 King Street, Alexandria, VA 22302. (703) 379-2480. Fax (703) 379-7563.

American Geophysical Union. 2000 Florida Avenue, N.W., Washington, DC 20009. (202) 462-6900. FAX (202) 328-0566.

American Institute of Physics. One Physics Ellipse, College Park, MD 20740. (301) 209-3000.

Earthquake Engineering Research Institute. 499 14th Street, Suite 320, Oakland, CA 94612-1902. (510) 451-0905.

Geological Society of America. 3300 Penrose Place, PO Box 9140, Denver, CO 80301-9140. (303) 447-2020.

International Institute of Seismology and Earthquake Engineering. Building Research Institute—Ministry of Construction, 1 Tatehara, Tsukuba-city, Ibaraki Prefecture 305, Japan.

International Seismological Centre. Pipers Lane, Thatcham, Newbury, Berks. RG13 4NS, England. Telephone 0635-61022. FAX 0635-72351.

Seismological Society of America. 201 Plaza Professional Bldg., El Cerrito, CA 94530. (510) 525-5474. FAX (510) 525-7204.

DIRECTORIES AND BIOGRAPHICAL SOURCES

American Men & Women of Science, 1995-96. R.R. Bowker Staff, eds. 19th edition. 8 volumes. R.R. Bowker/Reed International Publishing Company, 121 Chanlon Road, New Providence, NJ 07974. (908) 464-6800 or (800) 521-8110. 1995. $850.00

Directory of Engineering Societies and Related Organizations. Gordon Davis. 13th edition. American Association of Engineering Societies, 1111 19th Street NW, Suite 608, Washington, DC 20036. (202) 296-2237 or (800) 658-8897. 1989. Inquire for price.

International Research Centers Directory. Gale Research, 835 Penobscot Building, Detroit, MI 48226-4094. (313) 961-2242. (800) 347-4253. FAX (313) 961-6083. 8th edition. 1995. $410.00

Research Centers Directory. Gale Research, 835 Penobscot Building, Detroit, MI 48226-4094. (313) 961-2242. (800) 347-4253. FAX (313) 961-6083. $485.00.

Scientific and Technical Organizations and Agencies Directory. Gale Research, 835 Penobscot Building, Detroit, MI 48226-4094. (313) 961-2242. (800) 347-4253. FAX (313) 961-6083. 4th edition. 1996. $195.00.

Who's Who in Engineering. American Association of Engineering Societies, 1111 19th Street NW, Suite 608, Washington, DC 20036. (202) 296-2237 or (800) 658-8897. 8th edition. 1991. Inquire for price.

ENCYCLOPEDIAS AND DICTIONARIES

Encyclopedia of Earthquakes and Volcanoes. David Ritchie. Facts-on-File, Inc., 460 Park Avenue South, New York, NY 10016-7382. (800) 322-8755. Fax (212) 213-4578. 1994. $40.00.

The Encyclopedia of the Solid Earth Sciences. Philip Kearey, editor-in-chief. Blackwell Scientific Publications, 238 Main Street, Cambridge, MA 02142. (617) 876-7000 or (800) 759-6102. FAX (617) 876-7022. 1993. $60.00.

GENERAL WORKS

Earthquakes. Bruce A. Bolt. 3rd edition. W.H. Freeman and Company, 41 Madison Avenue, East 26th, 35th Floor, New York, NY 10010. (212) 576-9400. FAX (212) 689-2383. 1995. $16.95.

Earthquakes and Geological Discovery. Bruce A. Bolt. W.H. Freeman and Company, 41 Madison Avenue, East 26th, 35th Floor, New York, NY 10010. (212) 576-9400. FAX (212) 689-2383. 1995. $32.95.

Earthquake Prediction. Haroun Tazieff. McGraw-Hill Publishing Company, 1221 Avenue of the Americas, New York, NY 10020. (800) 262-4729 or (212) 512-3825. 1992. $11.95.

Earthquakes and Earthquake Engineering: A Source Guide. Gordon Press Publishers, PO Box 459, Bowling Green Station, New York, NY 10004. (718) 624-8419. 1991. $250.00.

Fundamentals of Earthquake Prediction. Cinna Lomnitz. John Wiley and Sons, Inc., 605 Third Avenue, New York, NY 10158. (800) 526-5368 or (212) 850-6000. 1994. $79.95.

Quakes, Eruptions and Other Geologic Cataclysms. Jon Erickson. Facts-on-File, Inc., 460 Park Avenue South, New York, NY 10016-7382. (800) 322-8755. Fax (212) 213-4578. 1994. $24.95.

ONLINE DATABASES AND CD-ROMS

Compendex Plus. Engineering Information, Inc., 345 East 47th Street, New York, NY 10017. (212) 705-7600 or (800) 221-1044. Contains citations with abstracts to worldwide literature in engineering and technology, from 1970 to present. Available on online BRS,(800) 289-4277, DIALOG, (800) 334-2564, ORBIT (800) 456-7248, and STN International, FIZ Karlsruhe, P.O. Box 2465, W-7500, Karlsruhe 1, Germany, online services. Also available on CD-ROM. Inquire as to cost and availability.

Earth Sciences. U.S. Geological Survey, 12201 Sunrise Valley Drive, Reston, VA 22092-9998. (703) 648-4460. CD-ROM of earth science databases including the U.S. Geological Survey Library, Earth Science Data Directory, and GEOINDEX, citations to published geological maps. $350.00 per year with quarterly updates.

Geoarchive. Geosystems, P.O. Box 1024, Westminster, London, England, SW1 P 2JL. Citations to literature on geoscience, 1969 to present. Inquire as to online cost and availability.

Geobase. Elsevier - Geo Abstracts, Regency House, 34 Duke Street, Norwich NR3 3AP, England. Contains citations to the worldwide earth science literature from 1980 to date. Available on DIALOG, ORBIT online services. Inquire as to cost and availability.

Georef: Bibliography and Index of Geology. American Geological Institute, 4220 King Street, Alexandria, VA 22302. (703) 379-2480. Fax (703) 379-7563. Monthly. Inquire for price and availability.

NTIS Bibliographic Database. National Technical Information Service, 5285 Port Royal Road, Springfield, VA 22161. (703) 487-4929 or FAX (703) 321-8199. Broad coverage of government-sponsored science and technology research reports, 1964 to present. Available on BRS,(800) 289-4277, DIALOG, (800) 334-2564, ORBIT, (800) 456-7248, and STN International, FIZ Karlsruhe, P.O. Box 2465, W-7500, Karlsruhe 1, Germany, online services. Also available on CD-ROM. Inquire as to cost and availability.

SCISEARCH. Institute for Scientific Information, 3501 Market Street, Philadelphia, PA 19104. (800) 523-1850 or (215) 386-0100. Broad multidisciplinary title and author index to the international literature of science and technology, 1974 to present. Available on DIALOG,(800) 334-2564, and ORBIT,(800) 456-7248, online services. Also available on CD-ROM. Inquire as to cost and availability.

PERIODICALS

Disasters. Basil Blackwell Inc., 238 Main Street, Cambridge, MA 02142. (617) 547-7110. FAX (617) 547-0789. 1977 to present. $89.50.

Earthquake History of the United States. National Oceanic and Atmospheric Administration, National Geophysical Data Center, 325 Broadway, Boulder, CO 80303-3328. (303) 497-6419. 1928 to present. Irregular (approximately every five years). Price varies. Also available on CD-ROM.

Earthquake Spectra. Earthquake Engineering Research Institute, 499 14th Street, Suite 320, Oakland, CA 94612-1902. (510) 451-0905. 1984 to present. Quarterly. $120.00.

Geophysical and Astrophysical Fluid Dynamics. Gordon and Breach Science Publishers, 820 Town Center Drive, Langhorne, PA 19047. (215) 750-2642. FAX (215) 750-6343. 1970 to present. 28 times a year. Inquire for price.

Geophysical Research Letters. American Geophysical Union, 2000 Florida Avenue, N.W., Washington, DC 20009. (202) 462-6900. 1974 to present. Semi-monthly. $590.00 per year for non-members.

International Seismological Centre Bulletin. Pipers Lane, Thatcham, Newbury, Berks. RG13 4NS, England. Telephone 0635-61022. FAX 0635-72351. 1964 to present. Monthly. Inquire for price in U.S.

International Seismological Centre Regional Catalogue of Earthquakes. Pipers Lane, Thatcham, Newbury, Berks. RG13 4NS, England. Telephone 0635-61022. FAX 0635-72351. 1964 to present. Semi-annual. Inquire for price in U.S.

Journal of Volcanology and Geothermal Research. Elsevier Science Inc., Box 882, Madison Square Station, New York, NY 10159. (212) 989-5800. FAX (212) 633-3990. 1976 to present. 24 times a year. $1155.00.

Seismological Research Letters. Seismological Society of America, 201 Plaza Professional Bldg., El Cerrito, CA 94530. (510) 525-5474. FAX (510) 525-7204. 1929 to present. Quarterly. $15.00.

Seismological Society of America Bulletin. 201 Plaza Professional Bldg., El Cerrito, CA 94530. (510) 525-5474. FAX (510) 525-7204. 1911 to present. Bi-monthly. $135.00.

Tectonophysics. Elsevier Science Inc., Box 882, Madison Square Station, New York, NY 10159. (212) 989-5800. FAX (212) 633-3990. 1964 to present. 56 times a year. $2429.00.

Volcanology and Seismology. Gordon and Breach Science Publishers, 820 Town Center Drive, Langhorne, PA 19047. (215) 750-2642. FAX (215) 750-6343. Twelve times a year. Inquire for cost.

RESEARCH CENTERS AND INSTITUTES

California Institute of Technology, Seismological Laboratory. Pasadena, CA 91125. (818() 356-6914.

Center For Earthquake Research and Information (ceri). Memphis State University, Memphis, TN 38152. (901) 678-2007. FAX (901) 323-2857.

Earthquake Engineering Research Institute. 499 14th Street, Suite 320, Oakland, CA 94612-1902. (510) 451-0905.

Incorporated Research Institutions For Seismology. Suite 1440, 1616 N. Fort Myer Drive, Arlington, VA 22209. (702) 524-6222. (703) 527-7256.

International Institute of Seismology and Earthquake Engineering. Building Research Institute—Ministry of Construction, 1 Tatehara, Tsukuba-city, Ibaraki Prefecture 305, Japan.

International Seismological Centre. Pipers Lane, Thatcham, Newbury, Berks. RG13 4NS, England. Telephone 0635-61022. FAX 0635-72351.

University of California—Santa Cruz, Charles F. Richter Seismological Laboratory. Institute of Tectonics, Santa Cruz, CA 95064. (408) 459-4137.

ECLIPSES

Ency. of Physical Sciences and Engineering Info. Sources

ECHO

See: RADAR, SONAR

ECLIPSES

See also: ASTRONOMY, COSMOLOGY, STARS

ABSTRACT SERVICES AND INDEXES

Applied Science and Technology Index; A Cumulative Subject Index To English Language Periodicals in the Fields of Aeronautics and Space Science, Computer Technology, Chemistry, Construction Industry, Energy and Related Areas. H.W. Wilson Co., 950 University Avenue, Bronx, NY 10452. (212) 588-8400. (800) 367-6770. FAX (718) 590-1617. From 1958 to present. Monthly. Inquire about cost and availability. Also available on CD-ROM and online. ISSN: 0003-6986.

Astronomy and Astrophysics Abstracts. Springer-Verlag New York, 175 Fifth Avenue, New York, NY 10010. (212) 460-1500. FAX (212) 473-6272. Published for the Astronomisches Rechen-Institut. Comprehensive coverage of all aspects of astronomy, astrophysics and related fields. 1969 to present. Two parts per year. Annual. ISSN: 0067-0022.

General Science Index. H.W. Wilson Company, 950 University Avenue, Bronx, NY 10452. (212) 588-8400. (800) 367-6770. FAX (718) 590-1617. From 1978 to present. Ten issues per year; quarterly and annual cumulations. Service basis. Available on CD-ROM and online. Inquire regarding cost and availability. ISSN: 0162-1963.

Government Reports Announcements and Index. U. S. National Technical Information Service (NTIS), 5285 Port Royal Road, Springfield, VA 22161. (703) 487-4650. FAX (703) 321-8547. From 1968 to present. Annual. $630.00 per year. Also available online as *NTIS Bibliographic Database* and on CD-ROM. ISSN:

NTIS Alerts: Astronomy & Astrophysics. U. S. National Technical Information Service. 5285 Port Royal Road, Springfield, VA 22161. (703) 487-4650. FAX (703) 321-8547. Weekly. $140.00 per year.

Pascal 48: Environnement Cosmique Terrestre, Astronomie Et Geologie Extraterrestre. Centre National de la Recherche Scientifique, Institut de Information Scientifique et Technique, 2 aliee du Parc de Brabois, 54514 Vandoeuvre Les Nancy Cedex, France. TEL 83-50-46-00. FAX: 83-50-46-50. 1985 to present. Ten issues per year. 770 F. Also available on CD-ROM and online.

Physics Abstracts. INSPEC. Section A, Science Abstracts. Institute of Electrical Engineers, London, United Kingdom. Available from: INSPEC/IEEE - Institute of Electrical and Electronic Engineers, Box 1331, Hoes Lane, Piscataway, NJ 08855-1331. (908) 562-5549. 1898 to present. 24 issues per year. $2835.00 per year. ISSN: 0036-8091. Also available online and on CD-ROM.

Science Citation Index. SCI. Institute for Scientific Information, 3501 Market Street, Philadelphia, PA 19104. (215) 386-0100. (800) 523-1850. FAX (215) 386-2991. 1961 to present. Six issues per year, plus annual cumulation. $11650.00 per year. Also available online and on CD-ROM. Inquire about price and availability. ISSN: 0036-827X.

STAR. Scientific and Technical Aerospace Reports. U.S. National Aeronautics and Space Administration. Distributed by U. S. Superintendent of Documents, Washington, DC 20402. 1963 to present. Semi-monthly, with semiannual and annual indexes. $114.00 per year. ISSN: 0036-8741. Also available online and on CD-ROM.

ANNUAL REVIEWS AND YEARBOOKS

The Astronomical Almanac. Superintendent of Documents, U. S. Government Printing office, Washington, DC 10402. (202) 783-3238. 1981 to present. Supercedes AstronomicalEphemeris and the American Ephermis and Nautical Almanac. Annual. $44.95. ISSN: 0737-6421.

Annual Review of Astronomy and Astrophysics; Original Reviews of Critical Literature and Current Developments in Astronomy and Astrophysics. Annual Reviews, Inc., 4139 El Camino Way, Palo Alto, CA 94306-0139. (415) 493-4400. (800) 523-8635. FAX (415) 855-9815. 1963 to date. Annual. $60.00. ISSN: 0066-4146.

Annual Review of Earth and Planetary Sciences. Annual Reviews, Inc., 4139 El Camino Way, Palo Alto CA 94306-0139. (415) 403-4400. (800) 523-8635. Fax (415) 855-9815. 1973 to date. Annual. $62.00. ISSN: 0084-6597.

Highlights of Astronomy. I. Appenzeller, editor. Kluwer Academic Publishers, Box 358, Accord Station, Hingham, MA 02018-0358. (617) 871-6600. International Astronomical Union Highlights series. ISBN: 07923-3553-8. Triennal. 1995. Price varies.

ASSOCIATIONS AND PROFESSIONAL SOCIETIES

Amateur Astronomers Association. 1010 Park Avenue, New York, NY 10028. (212) 535-2922.

American Astronomical Society. 2000 Florida Avenue NW, Suite 400, Washington, DC 20009. (202) 328-2010. FAX: (202) 234-2560.

American Institute of Physics. One Physics Ellipse, College Park, MD 20740-3843. (301) 209-3100.

American Lunar Society. P. O. Box 209, East Pittsburgh, PA 15112. (608) 837-6054.

Astronomical League. 2112 Kingfisher Lane East, Rolling Meadows, IL 60008. (708) 398-0562.

Astronomical Society of the Pacific. 390 Ashton Avenue, San Francisco, CA 94112. (415) 337-1100. FAX: (415) 337-5205.

Royal Astronomical Society of Canada. 136 Dupont Street, Toronto. ON N5R 1V2 Canada. (416) 924-7973. FAX (416) 968-6687.

BIBLIOGRAPHIES

Astronomy and Astrophysics: A Source Guide. Gordon Press, P.O. Box 459, Bowling Green Station, New York, NY 10004. (718) 624-8419. 1991. $250.00. .

A Bibliography of Astronomy, 1970-1979. R. A. Sea and S. S. Martin. Libraries Unlimited, Inc., Littleton, CO 80160. 1982.

Catalog of the Naval Observatory Library. Naval Observatory Staff. G. K. Hall & Company, 866 Third Avenue, New York, NY 10022. (212)702-6789. (800) 257-5755. 6 volumes. 1977. $655.00

Science Books and Films. American Association for the Advancement of Science, 1333 H Street NW, Washington, DC 20005. (202) 326-6454. 1965 to present. Nine issues per year. $40.00 per year. ISSN: 0098-342X

Scientific and Technical Books and Serials in Print, 1995. R.R. Bowker Inc., 121 Chanlon Road, New Providence, NJ 07974. (908) 464-6800. (800) 521-8110. 4 volumes. 1994. $299.95. Also available on compact disc and online.

A Source Book On Astronomy and Astrophysics, 1900 - 1975. Kenneth Lang and Owen Gingerich, editors. Harvard University Press, 79 Garden Street, Cambridge MA. (617) 495-2600. 1980. $41.95.

DIRECTORIES AND BIOGRAPHICAL SOURCES

American Men and Women of Science: Physical and Biologicalsciences. R. R. Bowker Inc., 121 Chanlon Road, New Providence, NJ 07974. (908) 464-6800. (800) 521-8110. 20th edition. 8 volumes. 1996. $850.00.

The Astronomers. Donald Goldsmith. St. Martin's Press, Inc., 175 Fifth Avenue, New York, NY 10010. (212) 674-5151. (800) 221-7945. 1993. $14.95 in paper.

Astronomical Centers of the World. Kevin Krisciunas. Cambridge University Press, 40 West 20th Street, New York, NY 10011-4211. (212) 924-3900. (800) 872-7423. 1988. $34.95

The Biographical Dictionary of Scientists: Astronomers. D. Abbott, editor. Peter Bedrick Books, Inc., 2112 Broadway, Room 318, New York, NY 10023. (212) 496-0751. 1984.

Directory of Physics and Astronomy Staff Members. American Institute of Physics, One Physics Ellipse, College Park, MD 20740-3843. (301) 209-3100. Annual. $45.00.

Earth and Astronomical Research Centers. Stockton Press, 345 Park Avenue South, New York, NY 10010. 4th edition. 1995. $515.00. ISBN: 1-56169-0967.

Graduate Programs in Physics, Astronomy and Related Fields. 1995-1996. American Institute of Physics, One Physics Ellipse, College Park, MD 20740-3843. (301) 209-3100. 1978 to present. Annual. $45.00.

Research Centers Directory. Gale Research Inc., 835 Penobscot Building, Detroit, MI 48226-4094. (313) 961-2242. (800) 347-4253. 20th edition. 1995. $485.00.

ENCYCLOPEDIAS AND DICTIONARIES

Astronomy From A To Z: A Dictionary of Celestial Objects and Ideas. Charles A. Schweighauser. Illinois Issues, Sangamon State University, K-10, Springfield, IL 62794-9243. (217) 786-9243. 1991. $14.95 in paper.

Concise Dictionary of Astronomy. Jacqueline Mitton. Oxford University Press, Inc., 200 Madison Avenue, New York, NY 10016. (212) 725-6000. (800) 334-4249. 1992. $24.95.

Facts On File Dictionary of Astronomy. Valerie Illingworth. Facts-on-File, Inc., 460 Park Avenue South, New York, NY 10016-7382. (212) 683-2244. (800) 322-8755. Fax (800) 683-3633. 1994. $27.95.

Encyclopedia of Astronomy and Astrophysics. Stephen Maran, editor. Van Nostrand Reinhold, 115 Fifth Avenue, New York, NY 10003. (212) 254-3232. (800) 842-3636. 1992. $129.95.

Oxford Illustrated Encyclopedia. Volume 8, *The Universe.* Archie Roy, editor. Oxford University Press, Inc., 200 Madison Avenue, New York, NY 10016. (212) 725-6000. (800) 334-4249. 1993. $45.00.

McGraw-Hill Encyclopedia of Science and Technology. McGraw-Hill Book Company, Inc., 1221 Avenue of the Americas, New York, NY 10020. (212) 512-2000. (800) 262-4729. 7th edition. 20 volume set. 1992. $1900.00.

Stars, Galaxies, Cosmos. William R. Corliss. Sourcebook Project, P.O. Box 107, Glen Arm, MD 21057. (410) 668-6047. 1987. $17.95.

GENERAL WORKS

The Cambridge Eclipse Photography Guide: How and Where To Observe and Photograph Solar and Lunar Eclipses. Jay M. Pasachoff and Michael A. Covington. Cambridge University Press, 40 West 20th Street, New York, NY 10011-4211. (212) 924-3900. (800) 872-7423. 1993. $16.95.

Canon of Lunar Eclipses. Bao-Lin Liu and Alan D. Fiala. Willmann-Bell, Inc., P.O. Box 35025, Richmond, VA 23235. (804) 320-7016. 1989. $19.95. 1992. $24.95.

Chasing the Shadow: An Observer's Guide To Eclipses. Joel Harris and Richard Talcott. Kalmbach Publishing Company, P.O. Box 1612, Waukesha, WI 53187. (414) 796-0126. (800) 446-5489. 1994. $18.95.

Chronology of Eclipses and Comets A.D. 1-1000. D. J. Schove. Boydell and Brewer, Inc., P.O.Box 41026, Rochester, NY 14604-4126. (716) 275-0419. 1985. $63.00.

Elements of Solar Eclipses, 1951-2200. Meeus. Willmann-Bell, Inc., P.O. Box 35025, Richmond, VA 23235. (804) 320-7016. 1989. $19.95.

Fifty-year Canon of Solar Eclipses, 1986-2035. Fred Espenak. Sky Publishing Corp., 49 Bay State Road, Cambridge, MA 02138. (617) 864-7360. (800) 253-0245. Reprint edition. 1987. $24.95.

HANDBOOKS AND MANUALS

Cambridge Guide to the Constellations. Michael E. Bakich. Cambridge University Press, 40 West 20th Street, New York, NY 10011-4211. (212) 924-3900. (800) 872-7423. 1995. $69.95.

Encyclopedia of Astronomy and Astrophysics. Stephen Maran, editor. Van Nostrand Reinhold, 115 Fifth Avenue, New York, NY 10003. (212) 254-3232. (800) 842-3636. 1992. $129.95.

ECLIPSES

Ency. of Physical Sciences and Engineering Info. Sources

Handbook of Astronomy, Astrophysics and Geophysics; Volume 2: Galaxies and Cosmology. V. M. Canuto and B. G. Elmegreen. Gordon & Breach Science Publishers, Inc., P.O. Box 200029, Riverfront Plaza Station, Newark, NJ 07102-0301. (201) 643-7500. 1988. $194.00.

Handbook of Space Astronomy and Astrophysics. Martin V. Zumbeck. Cambridge University Press, 40 West 20th Street, New York, NY 10011-4211. (212) 924-3900. (800) 872-7423. 2nd edition. 1990. $79.95.

Mcgraw-Hill Encyclopedia of Astronomy. Sybil P. Parker and Jay M. Pasachoff, editors. McGraw-Hill Publishing Company, Inc., 1221 Avenue of the Americas, New York, NY 10020. (212) 512-2000. (800) 262-4729. 2nd edition. 1993. $75.00

Observer's Guide To Astronomy. Patrick Martinez, editor. Cambridge University Press, 40 West 20th Street, New York, NY 10011-4211. (212) 924-3900. (800) 872-7423. 2 volumes. 1994. $159.90.

ONLINE DATABASES AND CD-ROMS

CA Search. Chemical Abstracts Service, P.O. Box 3012, Columbus, OH 43210-0012. (614) 447-3600. (800) 848-6533. FAX (614) 447-3709. Very comprehensive guide to worldwide chemical literature and related fields, 1972 to present. Available on BRS,(800) 289-4277, DIALOG, (800) 334-2564, ORBIT (800) 456-7248, and STN International, FIZ Karlsruhe, P.O. Box 2465, W-7500, Karlsruhe 1, Germany, online services. Inquire as to cost and availability.

Current Contents Search. Institute for Scientific Information, 3501 Market Street, Philadelphia, PA 19104. (215) 386-0100. FAX (215) 386-6362. Contains citations to articles listed in the table of contents of science and technology journals. Also articles in social sciences and life sciences journals. Available on BRS, (800) 289-4277, DIALOG, (800) 334-2564, online services. Inquire as to cost and availability.

Dissertation Abstracts Online. University Microfilms International, 300 North Zeeb Road, Ann Arbor, MI 48106. (800) 521-0600 or (313) 761-4700. Scope includes virtually all doctoral dissertations accepted at accredited American institutions from 1861 to present in 252 subject areas. Available on BRS, (800) 289-4277, DIALOG, (800) 334-2564, and OCLC EPIC, (800) 848-5878, online services. Also available on CD-ROM. Inquire as to cost and availability.

INSPEC. Institution of Electrical Engineers, Michael Faraday House, Six Hills Way, Stevenage, Herts. SG1 2AY, England. Telephone: 0438 313311 or FAX 0438 742840. Contains citations to the worldwide literature of physics, electronics and electrical engineering, computer technology, and related fields. Available on BRS, (800) 289-4277, DIALOG, (800) 334-2564, ORBIT, (800) 456-7248, and STN International, FIZ Karlsruhe, P.O. Box 2465, W-7500, Karlsruhe 1, Germany, online services. Inquire as to cost and availability.

International Aerospace Abstracts. American Institute ofAeronautics and Astronautics, 370 L'Enfant Promenade, S.W., Washington, DC 20024. (202) 646-7400. Contains references and abstracts of journal and monograph literature relating to aerospace science and technology, from 1963 to present. Available through the NASA/RECON system of the National Aeronautics and Space Administration only.

NTIS Bibliographic Database. National Technical Information Service, 5285 Port Royal Road, Springfield, VA 22161. (703) 487-4929 or FAX (703) 321-8199. Broad coverage of government-sponsored science and technology research reports, 1964 to present. Available on BRS,(800) 289-4277, DIALOG, (800) 334-2564, ORBIT, (800) 456-7248, and STN International, FIZ Karlsruhe, P.O. Box 2465, W-7500, Karlsruhe 1, Germany, online services. Also available on CD-ROM. Inquire as to cost and availability.

Physics Briefs. American Institute of Physics, 335 East 45th Street, New York, NY 10017. (212) 661-9260 or FAX (212) 661-2036. Contains citations with abstracts of the literature of physics and related fields, 1979 to present. Available on the STN International, FIZ Karlsruhe, P.O. Box 2465, W-7500, Karlsruhe 1, Germany, online service. Inquire as to cost and availability.

Scientific and Technical Books and Serials in Print. R.R. Bowker Inc., 121 Chanlon Road, New Providence, NJ 07974. (800) 521-8110 or (908) 464-6800. List of currently published books and serials in the physical and biological sciences, engineering and technology, with subject, author and titles indexes. Available on ORBIT, (800) 456-7248, online service. Inquire as to cost and availability.

SCISEARCH. Institute for Scientific Information, 3501 Market Street, Philadelphia, PA 19104. (800) 523-1850 or (215) 386-0100. Broad multidisciplinary title and author index to the international literature of science and technology, 1974 to present. Available on DIALOG, (800) 334-2564, and ORBIT, (800) 456-7248, online services. Also available on CD-ROM. Inquire as to cost and availability.

WILSONLINE. H.W. Wilson Company, 950 University Avenue, Bronx, NY 10452. (800) 367-6770 or (212) 588-8400. Makes available online versions of the printed H.W. Wilson indexes including Applied Science and Technology Index, Business Periodicals Index, General Science Index, and Readers' Guide to Periodical Literature. Period covered is generally 1983 to present. Available on BRS, (800) 289-4277, DIALOG, (800) 334-2564, and OCLC EPIC, (800) 848-5878, online services. Also available on CD-ROM. Inquire as to cost and availability.

PERIODICALS

Astronomical Journal. American Institute of Physics, One Physics Ellipse, College Park, MD 20740-3843. (301) 209-3000. Published for the American Astronomical Society. 1849 to present. Monthly. $280.00 per year. ISSN: 0004-6256.

Astronomical Society of the Pacific. Publications. Astronomical Society of the Pacific, 390 Ashton Avenue, San Francisco, CA 94112. (415) 337-1100. FAX (415) 337-5205. 1889 to present. Monthly. $175.00 per year. ISSN: 0004-6280.

Astronomy. Kalmbach Publishing Company, Box 1612, Waukesha, WI 53187-1612. (414) 796-8776. FAX (414) 796-1142. 1973 to present. Monthly. $27.00 per year. ISSN: 0091-6358.

Astronomy and Astrophysics. Springer-Verlag New York, Inc., 44 Hartz Way, Secaucus, NJ 07096-2491. (201) 348-4033. FAX (201) 348-4505. 1969 to date. 36 issues per year in 12 volumes. $1733.00 per year. ISSN: 0004-6361.

Celestial Mechanics and Dynamical Astronomy; An International Journal of Space Dynamics. Kluwer Academic Publishers, Box 358, Accord Station, Hingham, MA 02018-0358. (617)

871-6600. FAX (617) 871-6528. 1969 to present. Monthly. $703.50 per year. ISSN: 0923-2958.

Ciel Et Terre. Societe Royale Belge d'Astronomie, de Meteorologie et de Physique du Globe, 3 avenue Circulaire, Uccle, Brusels, Belgium. FAX 32-2-374-9822. 1880 to present. Bimonthly. 1600 BEF per year. ISSN: 0009-6709.

Earth, Moon and Planets: An International Journal of Comparative Planetology. Kluwer Academic Publishers, Box 358, Accord Station, Hingham, MA 02018-0358. (617) 871-6600. FAX (617) 871-6528. 1969 to present. Monthly. $840.00 per year. ISSN: 0167-9295.

Experimental Astronomy; An International Journal On Astronomical Instrumentation and Data Analysis. Kluwer Academic Publishers, Box 358, Accord Station, Hingham, MA 02018-0358. (617) 871-6600. FAX (617) 871-6528. 1989 to present. Quarterly. $161.00 per year. ISSN: 0922-6435.

Icarus: International Journal of Solar System Studies. Academic Press, Inc., Journal Division, 525 B Street, Suite 1900, San Diego, CA 92101-4495. (619) 230-1840. FAX (619) 699-6800. 1962 to present. Monthly. $1080.00 per year. ISSN: 0019-1035.

Lunar and Planetary Information Bulletin. Lunar and Planetary Institute. 3600 Bay Area Boulevard, Houston, TX 77058-1113. (713) 486-2175. FAX (713) 486-2125. 1970 to present. Quarterly. Free. Also available online.

Mercury. Astronomical Society of the Pacific. 390 Ashton Avenue, San Francisco, CA 94112. (415) 337-1100. FAX: (415) 337-5205. 1972 to present. Bimonthly. $175.00 per year. ISSN: 0047-6773.

Monthly Notices of the Royal Astronomical Society. Blackwell Scientific Publication LT., Osney Mead, Oxford OX2 OEL, England. TEL 0865-240201. FAX 0865-721205. 1827 to present. Fortnightly. $1899.00 per year. ISSN: 0035-8711.

Observatory. c/o Dr. D. J. Stickland Space and Astrophysics Division, Rutherford Appleton Laboratory, Chilton, Didcot, Oxon OX11 OQX England. FAX 0235-445848. 1877 to present. Bimonthly. $42.00 per year. ISSN: 0029-7704.

Planetary and Space Science. Elsevier Science Publishing Company, Inc., 660 White Plains Road, Terrytown, NY 10591-5153. (914) 524-9200. FAX: (914) 333-2444. 1959 to present. Monthly. $1355.00 per year. ISSN: 0032-0633.

Sky and Telescope. Sky Publishing Corporation, Box 9111,Belmont, MA 02178. (617) 864-7360. FAX (617) 864-6117. 1941 to present. Monthly. $27.00 per year. ISSN: 0037-6604.

Space Science Reviews. Kluwer Academic Publishers, Box 358, Accord Station, Hingham, MA 02018-0358. (617) 871-6600. FAX (617) 871-6528. 1962 to present. Sixteen issues per year. $940.00 per year. ISSN: 0038-6308.

Vistas in Astronomy; An International Review Journal. Albert C. Beer, editor. Elsevier Science Publishing Company, Inc., 660 White Plains Road, Terrytown, NY 10591-5153. (914) 524-9200. FAX: (914) 333-2444. 1958 to present. Quarterly. $415.00 per year. ISSN: 0083-6656.

Webb Society Quarterly Journal. Webb Society, 194 Foundry Lane, Freemantle, Southampton, Hampshire SO1 3EQ, England. 1969 to present. Quarterly. $19.50 per year. ISSN: 0043-1680.

Webb Society Quarterly Journal. Webb Society, 194 Foundry Lane, Freemantle, Southampton, Hampshire SO1 3EQ, England. 1969 to present. Quarterly. $19.50 per year. ISSN: 0043-1680.

RESEARCH CENTERS AND INSTITUTES

Carnegie Institution of Washington Observatories, 813 Santa Barbara Street, Pasadena, CA 91101. (818) 577-1122. FAX (818) 759-8136.

Institute For Astronomy. University of Hawaii at Manoa. 2680 Woodlawn Drive, Honolulu, HI 96822. (808) 956-8312. FAX (808) 988-2790.

Lowell Observatory. 1400 West Mars Hill Road. Flagstaff, AZ 86001, (602) 774-3358. FAX (602) 774-6296.

Lunar and Planetary Institute. 3303 NASA Road One, Houston, TX 77058-4399. (713) 486-2139. FAX (713) 486-2162.

Lunar and Planetary Laboratory. University of Arizona, Tucson, AZ 85721. (602) 621-6963.

McDonald Observatory. University of Texas at Austin. P.O. Box 1337, Fort Davis TX 79734. (915) 426-3263.

National Astronomy and Ionosphere Center. Cornell University, Space Sciences Building, Ithaca, NY 14853. (607) 255-3735. (607) 255-8803.

National Optical Astronomy Observatories. 950 North Cherry Avenue, Tucson, AZ 85719. (602) 325-9282. FAX (602) 325-9360.

EFFLUENT

See: ENVIRONMENTAL ENGINEERING

EL NINO

See: OCEANOGRAPHY

ELASTICITY

See: MECHANICS

ELECTRIC CURRENT

See: ELECTRICITY

ELECTRIC MOTORS

Ency. of Physical Sciences and Engineering Info. Sources

ELECTRIC MOTORS

See also: ELECTRIC POWER ENGINEERING, ELECTRICAL CODES, ELECTRICAL ENGINEERING, ELECTRICITY, ELECTROMAGENTISM, GENERATORS, HYDROELECTRIC POWER

ABSTRACT SERVICES AND INDEXES

Applied Mechanics Reviews: An Assessment of World Literature in Engineering Sciences. American Society of Mechanical Engineers, 345 East 47th Street, New York, NY 10017. (212) 705-7703. 1948 to present. Monthly. $360.00 ISSN: 0003-6900.

Applied Science and Technology Index; A Cumulative Subject Index To English Language Periodicals in the Fields of Aeronautics and Space Science, Computer Technology, Chemistry, Construction Industry, Energy and Related Areas. H.W. Wilson Co., 950 University Avenue, Bronx, NY 10452. (212) 588-8400. (800) 367-6770. FAX (718) 590-1617. From 1958 to present. Monthly. Inquire about cost and availability. Also available on CD-ROM and online. ISSN: 0003-6986.

Current Contents: Engineering, Technology and Applied Sciences. Institute for Scientific Information, 3501 Market Street, Philadelphia, PA 19104. (215) 386-0100. FAX (215) 386-6362. 1970 to present. Weekly. $442.00 per year. Also available on CD-ROM and online. Inquire regarding cost and availability. ISSN: 0095-7917.

Current Papers in Electrical and Electronics Engineering. Institution of Electrical Engineers (IEE), London. Distributed by INSPEC/IEEE, Box 1331, 445 Hoes Lane, Piscataway, NJ 08855-1331. (908) 562-5549. 1969 to present. Monthly. $345.00 per year. ISSN: 0011-3778.

Electric Power Industry Abstracts. Edison Electric Institute, c/o Utility Data Institute, 2011 I Atreet, NW, Suite 700, Washington, DC 20006. 1975 to present. Bimonthly. Inquire as to cost and availability.

Electrical and Electronics Abstracts. Institution of Electrical Engineers (IEE), London. Available from INSPEC/IEEE - Institute of Electrical and Electronic Engineers, Box 1331, Hoes Lane, Piscataway, NJ 08855-1331. (908) 562-5549. 1898 to present. Monthly. $2200.00 per year. Also available on CD-ROM and online as INSPEC. ISSN: 0036-8105.

Engineering Index Monthly; Indexes and Abstracts the World's Engineering and Technical Literature. Engineering Information, Inc., Castle Point on the Hudson, Hoboken, NJ 07030. (201) 216-8500. (800) 221-1044. FAX (201) 216-8532. Monthly. $2300.00 per year. Available online as COMPENDEX and also on CD-ROM. ISSN: 0742-1974.

General Science Index. H.W. Wilson Company, 950 University Avenue, Bronx, NY 10452. (212) 588-8400. (800) 367-6770. FAX (718) 590-1617. From 1978 to present. Ten issues per year; quarterly and annual cumulations. Service basis. Available on CD-ROM and online. Inquire regarding cost and availability. ISSN: 0162-1963.

Government Reports Announcements and Index. U. S. National Technical Information Service (NTIS), 5285 Port Royal Road, Springfield, VA 22161. (703) 487-4650. FAX (703) 321-8547. From 1968 to present. Annual. $630.00 per year. Also available online as *NTIS Bibliographic Database* and on CD-ROM. ISSN:

Index To IEEE Publications. IEEE Service Center, 445 Hoes Lane, Piscataway, NJ 08855-1331. (908) 981-1393. (800) 678-IEEE. FAX (908) 981-9667. 1973 to present. Annual. ISSN: 0099-1368.

Physics Abstracts. INSPEC. Section A, Science Abstracts. Institution of Electrical Engineers (IEE), London. Available from: INSPEC/IEEE - Institute of Electrical and Electronic Engineers, Box 1331, Hoes Lane, Piscataway, NJ 08855-1331. (908) 562-5549. 1898 to present. 24 issues per year. $2835.00 per year. Also available online and on CD-ROM. ISSN: 0036-8091.

Physics Briefs (Physikalische Berichte). Information Center for Energy, Physics, Mathematics; German Physical Society. V C H Publishers, Inc., 220 East 23rd Street, New York, NY 10010-4606. (212) 683-8333. 1845 to present. 24 issues per year. $2390.00 per year. Also available online. ISSN: 0179-7434.

Science Citation Index. SCI. Institute for Scientific Information, 3501 Market Street, Philadelphia, PA 19104. (215) 386-0100. (800) 523-1850. FAX (215) 386-2991. 1961 to present. Six issues per year, plus annual cumulation. $11650.00 per year. Also available online and on CD-ROM. Inquire about price and availability. ISSN: 0036-827X.

ANNUAL REVIEWS AND YEARBOOKS

Advances in Applied Mechanics. Academic Press, Inc., 6277 Sea Harbor Drive, Orlando, FL 32887. (619) 231-0926. (800) 321-5068. FAX (619) 699-6715. 1948 to present. Irregular. Price varies, inquire. ISSN: 0065-2156.

ASSOCIATIONS AND PROFESSIONAL SOCIETIES

American Electronics Association. 5201 Great America Way, Suite 520, P.O. 52990, Santa Clara, CA 95056. (408) 987-4200. FAX (408) 970-8565.

Edison Electric Institute. 701 Pennsylvania Avenue NW, Washington, DC 20004-2696. (202) 508-5000, 5454. FAX (202) 508-5360.

Electric Power Research Institute. 3412 Hillview Avenue, Palo Alto, CA 94304. (415) 855-2000. FAX (415) 855-2954.

Electronics Industries Association. 2500 Wilson Boulevard, Arlington, VA 22201. (703) 907-7500. FAX (202) 457-4985.

IEEE (institute of Electrical and Electronic Engineers). 345 East 47th Street. New York, NY 10017. (212) 705-7900. FAX (212) 705-4929.

IEEE Power Engineering Society. c/o Institute of Electrical and Electronic Engineers. 345 East 47th Street, New York, NY 10017. (212) 705-7900. FAX (212) 705-4929.

National Association of Power Engineers. 1 Springfield Street, Chicopee, MA 01013. (413) 592-6273. FAX (413) 592-1998.

National Electrical Manufacturers Association. 1300 North 17th Street, Suite 1847, Rosslyn VA 22209. (703) 841-3200. FAX (703) 841-3300.

SMMA - The Association For Electric Motors, their Control and Application. 4 Hollis Street, PO Box 378, Sherborn, MA 01770. (508) 655-4409. FAX (508) 651-3920.

Ency. of Physical Sciences and Engineering Info. Sources

ELECTRIC MOTORS

DIRECTORIES AND BIOGRAPHICAL SOURCES

American Electronics Association. Directory. American Electronics Association, 5201 Great America Way, Suite 520, P.O. 52990, Santa Clara, CA 95056. (408) 987-4200. FAX (408) 970-8565. Annual. $175.00.

American Men and Women of Science: Physical and Biological Sciences. R. R. Bowker Inc., 121 Chanlon Road, New Providence, NJ 07974. (908) 464-6800. (800) 521-8110. 20th edition. 8 volumes. 1996. $850.00.

Directory of Engineering Societies and Related Organizations. American Association of Engineering Societies, 1111 19th Street, Suite 608, Washington, DC 20036-3603. (202) 296-2237. Semi-annual. $150.00.

E E M - Electronic Engineer's Master. Hearst Business Communications, Inc., 645 Stewart Avenue, Garden City NY 11530. (516) 227-1300. ISSN: 0732-9016.

Electrical and Electronic Trades Directory: The Blue Book. Institution of Electrical Engineers, Michael Faraday House, Six Hills Way, Stevenage, Herts. SG1 2AY, England. Telephone: 0438 313311 or FAX 0438 742840. 1995. $140.00

Electrical World Directory of Electrical Utilities. McGraw-Hill Book Company, Inc., 1221 Avenue of the Americas, New York, NY 10020. (212) 997-3675. Annual.

Engineering Research Centers: Incorporating Electronics Research Centers. Stockton Press, 345 Park Avenue, New York, NY 10010. (212) 689-9200. (800) 221-2123. 4th edition. 1995. $515.00.

IEEE Membership Directory. Institute of Electrical and Electronics Engineers, IEEE Service Center, 445 Hoes Lane, Piscataway, NY 08854. (908) 981-1393. (800) 678-IEEE. FAX (908) 981-9667. 2 volumes. Annual. $190.00.

International Engineering Directory. American Consulting Engineers Council, 1015 15th Street, N.W. Suite 802, Washington, DC 20005-2670. (202) 347-7474. Annual. $10.00.

Research Centers Directory. Gale Research Company Inc., 835 Penobscot Building, Detroit, MI 48226-4094. (313) 961-2242. (800) 347-4253. 20th edition. 1995. $485.00. ISSN: 0080-1518.

Scientific and Technical Organizations and Agencies Directory. Gale Research, 835 Penobscot Building, Detroit, MI 48226-4094. (313) 961-2242. (800) 347-4253. 4th edition. 1996. $195.00.

Who's Who in Electric Transmission and Distribution. Utility Data Institute, 1200 G Street NY, Suite 250, Washington, DC 20005-3802. (202) 942-8788. (800) 486-3660. 1995. $150.00.

Who's Who in Engineering. Gordon Davis, editor. American Association of Engineering Societies. 1111 19th Street, NY, Suite 608, Washington, DC 20036. (202) 296-2237. (800) 658-8897. 9th edition. 1995. $220.00.

Who's Who in Technology. Gale Research, 835 Penobscot Building, Detroit, MI 48226-4094. (313 961-2242. (800) 521-4253. 7th edition. 1995. $195.00. ISN 0-8103-7467-6.

ENCYCLOPEDIAS AND DICTIONARIES

Chambers Science and Technology Dictionary. Peter M.B. Walker, editor. Cambridge University Press, 40 West 20th Street, New York, NY 10011-4211. (212) 924-3900. 1988. $39.95.

IEEE Standard Dictionary of Electrical and Electronics Terms. Christopher J. Booth, editor. IEEE Service Center, 445 Hoes Lane, Piscataway, NJ 08855-1331. (908) 981-1393. (800) 678-IEEE. FAX (908) 981-9667. IEEE Standard 100-1992. 5th edition. 1993. $150.00.

Illustrated Dictionary of Electronics. Stan Gibilsco. TAB Books, P.O. Box 40, Blue Summit, PA 17294-0850. (717) 794-2191. (800) 233-1128. 7th edition. 1994. $34.95.

McGraw-Hill Electronics Dictionary. J. Markus. McGraw-Hill Book Company, Inc., 1221 Avenue of the Americas, New York, NY 10020. (212) 997-3675. 5th edition. 1994. $49.95.

McGraw-Hill Encyclopedia of Science and Technology. McGraw-Hill Book, Incorporated, 1221 Avenue of the Americas, New York, NY 10020. (212) 997-3675. (800) 262-4729. Seventh edition. Twenty volumes. 1992. $1900.00.

GENERAL WORKS

Basic Electrical and Electronic Engineering. E. C. Bell and R. W. Whitehead. Blackwell Scientific Publications, Inc., 238 Main Street, Cambridge MA 02142. (617) 876-7000. (800) 759-6102. 4th revised edition. 1993.

Basic Electrical Power and Machines. David Bradley. Chapman & Hall, 1 Penn Plaza, New York, NY 10119. (212) 564-1060. 1994. $30.95.

Design of Small Electrical Machines. H. S. Hamdi. John Wiley & Sons, Inc., 605 Third Avenue, New York, NY 10158-0012. (212) 850-6000. (800) 225-5945. 1994. $67.95

Electric Motors and Motor Controls. Jeffrey J. Keljik. Delmar Publishing, 3 Columbia Circle, Box 15015, Albany, NY 12212. (518) 464-3500. (800) 347-7707. 1995. $35.95.

Electrical Machines, Drives and Power Systems. Theodore Wildi. Prentice Hall, 113 Sylvan Avenue, Route 9W, Englewood Cliffs, NJ 07632. (201) 592-2000. (800) 922-0579. 2nd edition. 1990. $79.75.

Electrical Power Engineering. Henslay W. Kabisama. McGraw-Hill Publishing Company, Inc., 1221 Avenue of the Americas, New York, NY 10020. (212) 512-2000. (800) 262-4729. 1993. $55.00.

An Introduction To Applied Electromagetism. Christos Christopoulos. John Wiley & Sons, Inc., 605 Third Avenue, New York, NY 10158-0012. (212) 850-6000. (800) 225-5945. 1990. $59.95.

Linear Induction Drives. Jacek F. Gieras. Oxford University Press, Inc., 200 Madison Avenue, New York, NY 10016. (212) 725-6000. (800) 334-4249. 1994. $ 75.00

Magneto-hydro-dynamic Electrical Power Generation. Hugo K. Messerie. John Wiley & Sons, Inc., 605 Third Avenue, New

ELECTRIC MOTORS

Ency. of Physical Sciences and Engineering Info. Sources

York, NY 10158-0012. (212) 850-6000. (800) 225-5945. 1994. $34.95 in paper.

Schaum's Electric Power Systems. Syed A. Masar. McGraw-Hill Publishing Company, Inc., 1221 Avenue of the Americas, New York, NY 10020. (212) 512-2000. (800) 262-4729. 1990. $10.95 in paper.

Troubleshooting Electric Motors. Thomas E. Proctor and G. A. Mazur. American Technical Pubs, Inc., 1155 West 175th Street, Homewood, IL 60430. (708) 957-1100. (800) 323-3471. 1993. $26.00.

HANDBOOKS AND MANUALS

American Electricians Handbook. Terrell Croft and Wilford I. Summers. McGraw-Hill Book Company, 1221 Avenue of the Americas, New York, NY 10020. (212) 512-2000. (800) 262-4729. 12th edition. 1992. $77.50.

Complete Handbook of Electric Motor Controls. John Traister. Prentice Hall, 113 Sylvan Avenue, Route 9W, Englewood Cliffs, NJ 07632. (201) 592-2000. (800) 922-0579. 2nd edition. 1994. $69.00.

Electrical Engineer's Handbook. Donald Christiansen, editor. McGraw Book Company, 1221 Avenue of the Americas, New York, NY 10020. (212) 512-2000. (800) 262-4729. 4th edition. 1996. $110.00.

Electronics Handbook. Jerry C. Whitaker. CRC Press, Inc., 2000 Corporate Boulevard, NW, Boca Raton, FL 33431. (407) 994-0555. (800) 272-7737. 1996. $120.00.

Handbook of Electric Motors. John E. Traister. Prentice-Hall, 113 Sylvan Avenue, Route 9W, Englewood Cliffs, NJ 07632. (201) 592-2000. (800) 922-0579. 1992. $64.00.

Handbook of Electrical and Electronic Technology. Curtis Johnson. Prentice Hall , 113 Sylvan Avenue, Route 9W, Englewood Cliffs, NJ 07632. (201) 592-2000. (800) 922-0579. 1996. $88.00.

Handbook of Electronics Manufacturing Engineering. Bernard S. Matisoff. Chapman & Hall, 1 Penn Plaza, New York, NY 10119. (212) 564-1060. 1996.

Standard Handbook For Electrical Engineers. Donald Fink. McGraw-Hill Publishing Company, 1221 Avenue of the Americas, New York, NY 10020. (212) 512-2000. (800) 262-4729. 13th edition. 1996. $110.00.

ONLINE DATABASES AND CD-ROMS

CA Search. Chemical Abstracts Service, P.O. Box 3012, Columbus, OH 43210-0012. (614) 447-3600. (800) 848-6533. FAX (614) 447-3709. Very comprehensive guide to worldwide chemical literature and related fields, 1972 to present. Available on BRS,(800) 289-4277, DIALOG, (800) 334-2564, ORBIT (800) 456-7248, and STN International, FIZ Karlsruhe, P.O. Box 2465, W-7500, Karlsruhe 1, Germany, online services. Inquire as to cost and availability.

COMPENDEX PLUS. Engineering Information, Inc., 345 East 47th Street, New York, NY 10017. (212) 705-7600 or (800) 221-1044. Contains citations with abstracts to worldwide literature in engineering and technology, from 1970 to present. Available on online BRS,(800) 289-4277, DIALOG, (800) 334-2564, ORBIT (800) 456-7248, and STN International, FIZ Karlsruhe, P.O. Box 2465, W-7500, Karlsruhe 1, Germany, online services. Also available on CD-ROM. Inquire as to cost and availability.

Current Contents Search. Institute for Scientific Information, 3501 Market Street, Philadelphia, PA 19104. (215) 386-0100. FAX (215) 386-6362. Contains citations to articles listed in the table of contents of science and technology journals. Also articles in social sciences and life sciences journals. Available on BRS,(800) 289-4277, DIALOG,(800) 334-2564, online services. Inquire as to cost and availability.

Dissertation Abstracts Online. University Microfilms International, 300 North Zeeb Road, Ann Arbor, MI 48106. (800) 521-0600 or (313) 761-4700. Scope includes virtually all doctoral dissertations accepted at accredited American institutions from 1861 to present in 252 subject areas. Available on BRS, (800) 289-4277, DIALOG, (800) 334-2564, and OCLC EPIC,(800) 848-5878, online services. Also available on CD-ROM. Inquire as to cost and availability.

INSPEC. Institution of Electrical Engineers, Michael Faraday House, Six Hills Way, Stevenage, Herts. SG1 2AY, England. Telephone: 0438 313311 or FAX 0438 742840. Contains citations to the worldwide literature of physics, electronics and electrical engineering, computer technology, and related fields. Available on BRS, (800) 289-4277, DIALOG, (800) 334-2564, ORBIT, (800) 456-7248, and STN International, FIZ Karlsruhe, P.O. Box 2465, W-7500, Karlsruhe 1, Germany, online services. Inquire as to cost and availability.

NTIS Bibliographic Database. National Technical Information Service, 5285 Port Royal Road, Springfield, VA 22161. (703) 487-4929 or FAX (703) 321-8199. Broad coverage of government-sponsored science and technology research reports, 1964 to present. Available on BRS,(800) 289-4277, DIALOG, (800) 334-2564, ORBIT, (800) 456-7248, and STN International, FIZ Karlsruhe, P.O. Box 2465, W-7500, Karlsruhe 1, Germany, online services. Also available on CD-ROM. Inquire as to cost and availability.

Physics Briefs. American Institute of Physics, 335 East 45th Street, New York, NY 10017. (212) 661-9260 or FAX (212) 661-2036. Contains citations with abstracts of the literature of physics and related fields, 1979 to present. Available on the STN International, FIZ Karlsruhe, P.O. Box 2465, W-7500, Karlsruhe 1, Germany, online service. Inquire as to cost and availability.

Scientific and Technical Books and Serials in Print. R.R. Bowker Inc., 121 Chanlon Road, New Providence, NJ 07974. (800) 521-8110 or (908) 464-6800. List of currently published books and serials in the physical and biological sciences, engineering and technology, with subject, author and titles indexes. Available on ORBIT,(800) 456-7248, online service.Inquire as to cost and availability.

SCISEARCH. Institute for Scientific Information, 3501 Market Street, Philadelphia, PA 19104. (800) 523-1850 or (215) 386-0100. Broad multidisciplinary title and author index to the international literature of science and technology, 1974 to present. Available on DIALOG,(800) 334-2564, and ORBIT,(800) 456-

7248, online services. Also available on CD-ROM. Inquire as to cost and availability.

WILSONLINE. H.W. Wilson Company, 950 University Avenue, Bronx, NY 10452. (800) 367-6770 or (212) 588-8400. Makes available online versions of the printed H.W. Wilson indexes including Applied Science and Technology Index, Business Periodicals Index, General Science Index, and Readers' Guide to Periodical Literature. Period covered is generally 1983 to present. Available on BRS,(800) 289-4277, DIALOG,(800) 334-2564, and OCLC EPIC,(800) 848-5878, online services. Also available on CD-ROM. Inquire as to cost and availability.

PERIODICALS

Electric Light and Power. PennWell Publishing Co. 1421 South Sheridan Road, Box 1260, Tulsa OK 74101. (835-3161. FAX (918) 831-9497. 1922 to present. Monthly. $42.00 per year.

Electric Machines and Power Systems. Taylor & Francis, 1900 Frost Road, Suite 101, Bristol, PA 19007-1598. (800) 821-8312, FAX (215) 785-5515. 1976 to present. Monthly. $355.00 per year. ISSN: 0731-356X.

Electrical World. McGraw-Hill, Inc., Box 513 Hightstown, NJ 08520. (212) 512-3288. 1874 to present. Monthly. $55.00 per year. ISSN: 0013-4457.

Electronic Design. Penton Publishing, San Jose Gateway, Suite 354. 2025 Gateway Place, San Jose, CA 95110. (408) 441-0550. 1952 to present. Fortnightly. $95.00. ISSN: 0013-4872.

Electronics. Penton Publishing, San Jose Gateway, Suite 354. 2025 Gateway Place, San Jose, CA 95110. (408) 441-0550. 1930 to present. Semi-weekly. $98.00 per year. ISSN: 0883-4989.

IEEE Power Engineering Review. Institute of Electrical and Electronics Engineers, Inc., Box 1331, 445 Hoes Lane, Piscataway, NJ 08855-1331. (908) 981-0060. 1981 to present. Monthly. $115.00 per year. ISSN: 0272-1724.

IEEE Spectrum. Institute of Electrical and Electronics Engineers, Inc., Box 1331, 445 Hoes Lane, Piscataway, NJ 08855-1331. (908) 981-0060. 1964 to present. Monthly. $157.00. ISSN: 0018-9235.

IEEE Transactions On Power Delivery. Institute of Electrical and Electronics Engineers, Inc., Box 1331, 445 Hoes Lane, Piscataway, NJ 08855-1331. (908) 981-0060. 1986 to present. Quarterly. $170.00 per year. ISSN: 0885-8977.

IEE Proceedings: Electric Power Applications. Institution of Electrical Engineers (IEE), London. Available from INSPEC/IEEE - Institute of Electrical and Electronic Engineers, Box 1331, Hoes Lane, Piscataway, NJ 08855-1331. (908) 562-5549. 1980 to present. Bi-monthly. ISSN: 0143-7038.

Institute of Electrical and Electronics Engineers. Proceedings. Institute of Electrical and Electronics Engineers, Inc., Box 1331, 445 Hoes Lane, Piscataway, NJ 08855-1331. (908) 981-0060. 1913 to present. Monthly. $275.00. ISSN: 0018-9219.

Power Engineering (Tulsa). PennWell Publishing Co. 1421 South Sheridan Road, Box 1260, Tulsa OK 74101. (835-3161. FAX (918) 831-9497. 1896 to present. Monthly. $28.00 per year. ISSN: 0032-5961

Power Engineering Journal. Institution of Electrical Engineers (IEE), London. Available from INSPEC/IEEE- Institute of Electrical and Electronic Engineers, Box 1331, 445 Hoes Lane, Piscataway, NJ 08855-1331. (908) 562-5549. 1987 to present. Bi-monthly. L85 per year. ISSN: 0950-3366.

RESEARCH CENTERS AND INSTITUTES

Edison Electric Institute. 701 Pennsylvania Avenue, NW, Washington, DC 20004-2696. (202) 508-5000, 5454. FAX (202) 508-5360.

Electroscience Laboratory. Ohio State University, 1320 Kinnear Road, Columbus, Ohio 43212. (614) 292-7981, FAX (614) 292-7297.

Electrical Engineering Research Laboratories. Purdue University. Electrical Engineering Building, West Lafayette, in 47907. (317) 494-3536. FAX (317) 494-6440.

Electrical Power Research Institute. 3412 Hillview Avenue, PO Box 10412, Palo Alto, CA 94303. (415) 855-2000. FAX (415) 855-2954.

Laboratory For Electromagnetic and Electronic Systems. Massachusetts Institute of Technology, 77 Massachusetts Avenue, Cambridge, MA 02139. (617) 253-4631. FAX (617) 258-6774.

Weber Research Institute. Polytechnic University. Route 110, Farmingdale, NY 11735. (516) 755-4250, FAX (516) 755-4404.

ELECTRIC POWER ENGINEERING

See also: ELECTRIC MOTORS, ELECTRICAL ENGINEERING, ELECTRICITY, ELECTROMAGENTISM, GENERATORS, HYDRO-ELECTRIC POWER

ABSTRACT SERVICES AND INDEXES

Applied Science and Technology Index; A Cumulative Subject Index To English Language Periodicals in the Fields of Aeronautics and Space Science, Computer Technology, Chemistry, Construction Industry, Energy and Related Areas. H.W. Wilson Co., 950 University Avenue, Bronx, NY 10452. (212) 588-8400. (800) 367-6770. FAX (718) 590-1617. From 1958 to present. Monthly. Inquire about cost and availability. Also available on CD-ROM and online. ISSN: 0003-6986.

Current Contents: Engineering, Technology and Applied Sciences. Institute for Scientific Information, 3501 Market Street, Philadelphia, PA 19104. (215) 386-0100. FAX (215) 386-6362. 1970 to present. Weekly. $442.00 per year. Also available on CD-ROM and online. Inquire regarding cost and availability. ISSN: 0095-7917.

Current Papers in Electrical and Electronics Engineering. Institution of Electrical Engineers (IEE), London. Distributed by INSPEC/IEEE, Box 1331, 445 Hoes Lane, Piscataway, NJ 08855-1331. (908) 562-5549. 1969 to present. Monthly. $345.00 per year. ISSN: 0011-3778.

Electric Power Industry Abstracts. Edison Electric Institute, c/o Utility Data Institute, 2011 I Atreet, NW, Suite 700, Washing-

ELECTRIC POWER ENGINEERING

Ency. of Physical Sciences and Engineering Info. Sources

ton, DC 20006. 1975 to present. Bimonthly. Inquire as to cost and availability.

Electrical and Electronics Abstracts. Institution of Electrical Engineers (IEE), London. Available from INSPEC/IEEE - Institute of Electrical and Electronic Engineers, Box 1331, Hoes Lane, Piscataway, NJ 08855-1331. (908) 562-5549. 1898 to present. Monthly. $2200.00 per year. Also available on CD-ROM and online as INSPEC.

Engineering Index Monthly; Indexes and Abstracts the World's Engineering and Technical Literature. Engineering Information, Inc., Castle Point on the Hudson, Hoboken, NJ 07030. (201) 216-8500. (800) 221-1044. FAX (201) 216-8532. Monthly. $2300.00 per year. Available online as COMPENDEX and also on CD-ROM. ISSN: 0742-1974.

General Science Index. H.W. Wilson Company, 950 University Avenue, Bronx, NY 10452. (212) 588-8400. (800) 367-6770. FAX (718) 590-1617. From 1978 to present. Ten issues per year; quarterly and annual cumulations. Service basis. Available on CD-ROM and online. Inquire regarding cost and availability. ISSN: 0162-1963.

Government Reports Announcements and Index. U. S. National Technical Information Service (NTIS), 5285 Port Royal Road, Springfield, VA 22161. (703) 487-4650. FAX (703) 321-8547. From 1968 to present. Annual. $630.00 per year. Also available online as NTIS BIBLIOGRAPHIC DATABASE and on CD-ROM.

Index To IEEE Publications. IEEE Service Center, 445 Hoes Lane, Piscataway, NJ 08855-1331. (908) 981-1393. (800) 678-IEEE. FAX (908) 981-9667. 1973 to present. Annual. ISSN: 0099-1368.

Physics Abstracts. INSPEC. Section A, Science Abstracts. Institution of Electrical Engineers (IEE), London. Available from: INSPEC/IEEE - Institute of Electrical and Electronic Engineers, Box 1331, Hoes Lane, Piscataway, NJ 08855-1331. (908) 562-5549. 1898 to present. 24 issues per year. $2835.00 per year. Also available online and on CD-ROM. ISSN: 0036-8091.

Physics Briefs (Physikalische Berichte). Information Center for Energy, Physics, Mathematics; German Physical Society. V C H Publishers, Inc., 220 East 23rd Street, New York, NY 10010-4606. (212) 683-8333. 1845 to present. 24 issues per year. $2390.00 per year. Also available online. ISSN: 0179-7434.

Science Citation Index. SCI. Institute for Scientific Information, 3501 Market Street, Philadelphia, PA 19104. (215) 386-0100. (800) 523-1850. FAX (215) 386-2991. 1961 to present. Six issues per year, plus annual cumulation. $11650.00 per year. Also available online and on CD-ROM. Inquire about price and availability. ISSN: 0036-827X.

ASSOCIATIONS AND PROFESSIONAL SOCIETIES

American Electronics Association. 5201 Great America Way, Suite 520, P.O. 52990, Santa Clara, CA 95056. (408) 987-4200. FAX (408) 970-8565.

Edison Electric Institute. 701 Pennsylvania Avenue NW, Washington, DC 20004-2696. (202) 508-5000, 5454. FAX (202) 508-5360.

Electric Power Research Institute. 3412 Hillview Avenue, Palo Alto, CA 94304. (415) 855-2000. FAX (415) 855-2954.

Electronics Industries Association. 2500 Wilson Boulevard, Arlington, VA 22201. (703) 907-7500. FAX (202) 457-4985.

IEEE (Institute of Electrical and Electronic Engineers). 345 East 47th Street. New York, NY 10017. (212) 705-7900. FAX (212) 705-4929.

IEEE Power Engineering Society. c/o Institute of Electrical and Electronic Engineers. 345 East 47th Street, New York, NY 10017. (212) 705-7900. FAX (212) 705-4929.

National Association of Power Engineers. 1 Springfield Street, Chicopee, MA 01013. (413) 592-6273. FAX (413) 592-1998.

DIRECTORIES AND BIOGRAPHICAL SOURCES

American Men and Women of Science: Physical and Biological Sciences. R. R. Bowker Inc., 121 Chanlon Road, New Providence, NJ 07974. (908) 464-6800. (800) 521-8110. 20th edition. 8 volumes. 1996. $850.00.

Directory of Engineering Societies and Related Organizations. American Association of Engineering Societies, 1111 19th Street, Suite 608, Washington, DC 20036-3603. (202) 296-2237. Semi-annual. $150.00.

E E M - Electronic Engineer's Master. Hearst Business Communications, Inc., 645 Stewart Avenue, Garden City NY 11530. (516) 227-1300. ISSN: 0732-9016.

Electrical and Electronic Trades Directory: The Blue Book. Institution of Electrical Engineers, Michael Faraday House, Six Hills Way, Stevenage, Herts. SG1 2AY, England. Telephone: 0438 313311 or FAX 0438 742840. 1995. $140.00

Electrical World Directory of Electrical Utilities. McGraw-Hill Book Company, Inc., 1221 Avenue of the Americas, New York, NY 10020. (212) 997-3675. Annual.

Engineering Research Centers: Incorporating Electronics Research Centers. Stockton Press, 345 Park Avenue, New York, NY 10010. (212) 689-9200. (800) 221-2123. 4th edition. 1995. $515.00.

IEEE Membership Directory. Institute of Electrical and Electronics Engineers, IEEE Service Center, 445 Hoes Lane, Piscataway, NY 08854. (908) 981-1393. (800) 678-IEEE. FAX (908) 981-9667. 2 volumes. Annual. $190.00.

Research Centers Directory. Gale Research Company Inc., 835 Penobscot Building, Detroit, MI 48226-4094. (313) 961-2242. (800) 347-4253. 20th edition. 1995. $485.00. ISSN: 0080-1518.

Who's Who in Electric Transmission and Distribution. Utility Data Institute, 1200 G Street NY, Suite 250, Washington, DC 20005-3802. (202) 942-8788. (800) 486-3660. 1995. $150.00.

Who's Who in Engineering. Gordon Davis, editor. American Association of Engineering Societies. 1111 19th Street, NY, Suite 608, Washington, DC 20036. (202) 296-2237. (800) 658-8897. 9th edition. 1995. $220.00.

Who's Who in Technology. Gale Research, 835 Penobscot Building, Detroit, MI 48226-4094. (313 961-2242. (800) 521-4253. 7th edition. 1995. $195.00. ISBN 0-8103-7467-6.

ENCYCLOPEDIAS AND DICTIONARIES

Chambers Science and Technology Dictionary. Peter M.B. Walker, editor. Cambridge University Press, 40 West 20th Street, New York, NY 10011-4211. (212) 924-3900. 1988. $39.95.

IEEE Standard Dictionary of Electrical and Electronics Terms. Christopher J. Booth, editor. IEEE Service Center, 445 Hoes Lane, Piscataway, NJ 08855-1331. (908) 981-1393. (800) 678-IEEE. FAX (908) 981-9667. IEEE Standard 100-1992. 5th edition. 1993. $150.00.

Illustrated Dictionary of Electronics. Stan Gibilsco. TAB Books, P.O. Box 40, Blue Summit, PA 17294-0850. (717) 794-2191. (800) 233-1128. 7th edition. 1994. $34.95.

McGraw-Hill Electronics Dictionary. J. Markus. McGraw-Hill Book Company, Inc., 1221 Avenue of the Americas, New York, NY 10020. (212) 997-3675. 5th edition. 1994. $49.95.

McGraw-Hill Encyclopedia of Science and Technology. McGraw-Hill Book, Incorporated, 1221 Avenue of the Americas, New York, NY 10020. (212) 997-3675. (800) 262-4729. Seventh edition. Twenty volumes. 1992. $1900.00.

GENERAL WORKS

Basic Electrical and Electronic Engineering. E. C. Bell and R. W. Whitehead. Blackwell Scientific Publications, Inc., 238 Main Street, Cambridge MA 02142. (617) 876-7000. (800) 759-6102. 4th revised edition. 1993.

Electrical and Mathematical Analysis: Encyclopedic Textbook and Reference Handbook. Ernest Joerg. Kendall-Hunt Publishing Co., 4050 Westmark Drive, P.O. Box 1840 Dubuque, IA 52004-1840. (319) 589-1000. (800) 228-0810. 1993. $34.95 in paper.

Electrical Machines, Drives and Power Systems. Theodore Wildi. Prentice Hall, 113 Sylvan Avenue, Route 9W, Englewood Cliffs, NJ 07632. (201) 592-2000. (800) 922-0579. 2nd edition. 1990. $79.75.

Electrical Power Engineering. Henslay W. Kabisama. McGraw-Hill Publishing Company, Inc., 1221 Avenue of the Americas, New York, NY 10020. (212) 512-2000. (800) 262-4729. 1993. $55.00.

Electricity, Economics and Planning. T. W. Berrie. Institution of Electrical Engineers; IEE/INSPEC, Box 1331, 445 Hoes Lane, Piscataway, NJ 08855-1331. (908) 981-0060. 1992. $82.00.

High-voltage Engineering and Testing. H. M. Ryan, editor. Institution of Electrical Engineers; IEE/INSPEC, Box 1331, 445 Hoes Lane, Piscataway, NJ 08855-1331. (908) 981-0060. 1994. $95.00.

An Introduction To Applied Electromagetism. Christos Christopoulos. John Wiley & Sons, Inc., 605 Third Avenue, New York, NY 10158-0012. (212) 850-6000. (800) 225-5945. 1990. $59.95.

Magneto-Hydro-Dynamic Electrical Power Generation. Hugo K. Messerie. John Wiley & Sons, Inc., 605 Third Avenue, New York, NY 10158-0012. (212) 850-6000. (800) 225-5945. 1994. $34.95 in paper.

Power Supplies, Switching Regulators, Inverters and Converters. Irving M. Gottlieb. TAB Books, P.O. Box 40, Blue Summit, PA 17294-0850. (717) 794-2191. (800) 233-1128. 1993. Reprint edition. $32.95.

Power System Control and Stability. Paul M. anderson and Azia A. Fouad. IEEE - Institute of Electrical and Electronic Engineers, Box 1331, Hoes Lane, Piscataway, NJ 08855-1331. (908) 562-5549. 1993. $49.95.

Schaum's Electric Power Systems. Syed A. Masar. McGraw-Hill Publishing Company, Inc., 1221 Avenue of the Americas, New York, NY 10020. (212) 512-2000. (800) 262-4729. 1990. $10.95 in paper.

Simulation and Control of Electrical Power Stations. J. B. Knowles. John Wiley & Sons, Inc., 605 Third Avenue, New York, NY 10158-0012. (212) 850-6000. (800) 225-5945. 1991. $167.00.

Uninterruptible Power Supplies. J. Platts and J. D. St. Aubyn. Institution of Electrical Engineers; IEE/INSPEC, Box 1331, 445 Hoes Lane, Piscataway, NJ 08855-1331. (908) 981-0060. 1992. $70.00.

HANDBOOKS AND MANUALS

AC Power Systems Handbook. Jerry Whitaker.CRC Press, Inc., 2000 Corporate Boulevard, NW, Boca Raton, FL 33431. (407) 994-0555. (800) 272-7737. 1991. $55.00.

Active Electronic Component Handbook. Charles A. Harper and Harold C. Jones. McGraw-Hill Book Company, 1221 Avenue of the Americas, New York, NY 10020. (212) 512-2000. (800) 262-4729. 2nd edition. 1996. $79.50.

Electrical Engineering Handbook. Richard C. Dorf, editor. CRC Press, Inc., 2000 Corporate Boulevard, NW, Boca Raton, FL 33431. (407) 994-0555. (800) 272-7737. 1993. $99.95.

Electrical Engineer's Handbook. Donald Christiansen, editor. McGraw Book Company, 1221 Avenue of the Americas, New York, NY 10020. (212) 512-2000. (800) 262-4729. 4th edition. 1996. $110.00.

Electrical Power Systems Safety Handbook. John CADick. McGraw-Hill Book Company, 1221 Avenue of the Americas, New York, NY 10020. (212) 512-2000. 1994. $54.50.

Electronics Handbook. Jerry C. Whitaker. CRC Press, Inc., 2000 Corporate Boulevard, NW, Boca Raton, FL 33431. (407) 994-0555. (800) 272-7737. 1996. $120.00.

Handbook of Electrical and Electronic Technology. Curtis Johnson. Prentice Hall , 113 Sylvan Avenue, Route 9W, Englewood Cliffs, NJ 07632. (201) 592-2000. (800) 922-0579. 1996. $88.00.

Handbook of Solid State Batteries and Capacitors. M. Z. Munshi and P. S. Prasad. World Scientific Publishing Company, Inc., 1060 Main Street, River Edge, NJ 07661. (201) 487-9655. (800) 227-7562. 1995. $177.00.

McGraw-Hill's Handbook of Electrical Construction Calculations. J. F. McPartland. McGraw-Hill Publishing Company, 1221 Avenue of the Americas, New York, NY 10020. (212) 512-2000. (800) 262-4729. 1996. $54.50.

McGraw-Hill's National Electrical Code Handbook. J. F. McPartland. McGraw-Hill Publishing Company, 1221 Avenue of the Americas, New York, NY 10020. (212) 512-2000. (800) 262-4729. 22nd edition. 1996. $65.00.

Power Systems Handbook: Design, Operation and Maintenance. O. C. Seevers. Prentice Hall , 113 Sylvan Avenue, Route 9W, Englewood Cliffs, NJ 07632. (201) 592-2000. (800) 922-0579. 1991. $58.00.

Standard Handbook For Electrical Engineers. Donald Fink. McGraw-Hill Publishing Company, 1221 Avenue of the Americas, New York, NY 10020. (212) 512-2000. (800) 262-4729. 13th edition. 1996. $110.00.

Standard Handbook of Engineering Calculations. Tyler G. Hicks. McGraw-Hill Publishing Company, 1221 Avenue of the Americas, New York, NY 10020. (212) 512-2000. (800) 262-4729. 3rd edition. 1994. $99.50. ISBN: 0-07-028812-7.

Underground Systems Reference Book. Edison Electric Institute, 701 Pennsylvania Avenue NW, Washington, DC 20004-2696. (202) 508-5000, 5454. FAX (202) 508-5360. 1955. $14.50.

ONLINE DATABASES AND CD-ROMS

CA Search. Chemical Abstracts Service, P.O. Box 3012, Columbus, OH 43210-0012. (614) 447-3600. (800) 848-6533. FAX (614) 447-3709. Very comprehensive guide to worldwide chemical literature and related fields, 1972 to present. Available on BRS,(800) 289-4277, DIALOG, (800) 334-2564, ORBIT (800) 456-7248, and STN International, FIZ Karlsruhe, P.O. Box 2465, W-7500, Karlsruhe 1, Germany, online services. Inquire as to cost and availability.

Compendex Plus. Engineering Information, Inc., 345 East 47th Street, New York, NY 10017. (212) 705-7600 or (800) 221-1044. Contains citations with abstracts to worldwide literature in engineering and technology, from 1970 to present. Available on online BRS,(800) 289-4277, DIALOG, (800) 334-2564, ORBIT (800) 456-7248, and STN International, FIZ Karlsruhe, P.O. Box 2465, W-7500, Karlsruhe 1, Germany, online services. Also available on CD-ROM. Inquire as to cost and availability.

Current Contents Search. Institute for Scientific Information, 3501 Market Street, Philadelphia, PA 19104. (215) 386-0100. FAX (215) 386-6362. Contains citations to articles listed in the table of contents of science and technology journals. Also articles in social sciences and life sciences journals. Available on BRS,(800) 289-4277, DIALOG,(800) 334-2564, online services. Inquire as to cost and availability.

Dissertation Abstracts Online. University Microfilms International, 300 North Zeeb Road, Ann Arbor, MI 48106. (800) 521-0600 or (313) 761-4700. Scope includes virtually all doctoral dissertations accepted at accredited American institutions from 1861 to present in 252 subject areas. Available on BRS, (800) 289-4277, DIALOG, (800) 334-2564, and OCLC EPIC,(800) 848-5878, online services. Also available on CD-ROM. Inquire as to cost and availability.

INSPEC. Institution of Electrical Engineers, Michael Faraday House, Six Hills Way, Stevenage, Herts. SG1 2AY, England. Telephone: 0438 313311 or FAX 0438 742840. Contains citations to the worldwide literature of physics, electronics and electrical engineering, computer technology, and related fields. Available on BRS, (800) 289-4277, DIALOG, (800) 334-2564, ORBIT, (800) 456-7248, and STN International, FIZ Karlsruhe, P.O. Box 2465, W-7500, Karlsruhe 1, Germany, online services. Inquire as to cost and availability.

NTIS Bibliographic Database. National Technical Information Service, 5285 Port Royal Road, Springfield, VA 22161. (703) 487-4929 or FAX (703) 321-8199. Broad coverage of government-sponsored science and technology research reports, 1964 to present. Available on BRS,(800) 289-4277, DIALOG, (800) 334-2564, ORBIT, (800) 456-7248, and STN International, FIZ Karlsruhe, P.O. Box 2465, W-7500, Karlsruhe 1, Germany, online services. Also available on CD-ROM. Inquire as to cost and availability.

Physics Briefs. American Institute of Physics, 335 East 45th Street, New York, NY 10017. (212) 661-9260 or FAX (212) 661-2036. Contains citations with abstracts of the literature of physics and related fields, 1979 to present. Available on the STN International, FIZ Karlsruhe, P.O. Box 2465, W-7500, Karlsruhe 1, Germany, online service. Inquire as to cost and availability.

Scientific and Technical Books and Serials in Print. R.R. Bowker Inc., 121 Chanlon Road, New Providence, NJ 07974. (800) 521-8110 or (908) 464-6800. List of currently published books and serials in the physical and biological sciences, engineering and technology, with subject, author and titles indexes. Available on ORBIT,(800) 456-7248, online service.Inquire as to cost and availability.

SCISEARCH. Institute for Scientific Information, 3501 Market Street, Philadelphia, PA 19104. (800) 523-1850 or (215) 386-0100. Broad multidisciplinary title and author index to the international literature of science and technology, 1974 to present. Available on DIALOG,(800) 334-2564, and ORBIT,(800) 456-7248, online services. Also available on CD-ROM. Inquire as to cost and availability.

WILSONLINE. H.W. Wilson Company, 950 University Avenue, Bronx, NY 10452. (800) 367-6770 or (212) 588-8400. Makes available online versions of the printed H.W. Wilson indexes including Applied Science and Technology Index, Business Periodicals Index, General Science Index, and Readers' Guide to Periodical Literature. Period covered is generally 1983 to present. Available on BRS,(800) 289-4277, DIALOG,(800) 334-2564, and OCLC EPIC,(800) 848-5878, online services. Also available on CD-ROM. Inquire as to cost and availability.

PERIODICALS

Electric Light and Power. PennWell Publishing Co. 1421 South Sheridan Road, Box 1260, Tulsa OK 74101. (835-3161. FAX (918) 831-9497. 1922 to present. Monthly. $42.00 per year.

Electric Machines and Power Systems. Taylor & Francis, 1900 Frost Road, Suite 101, Bristol, PA 19007-1598. (800) 821-8312, FAX (215) 785-5515. 1976 to present. Monthly. $355.00 per year. ISSN: 0731-356X.

Electrical World. McGraw-Hill, Inc., Box 513 Hightstown, NJ 08520. (212) 512-3288. 1874 to present. Monthly. $55.00 per year. ISSN: 0013-4457.

Electronic Design. Penton Publishing, San Jose Gateway, Suite 354. 2025 Gateway Place, San Jose, CA 95110. (408) 441-0550. 1952 to present. Fortnightly. $95.00. ISSN: 0013-4872.

Electronics. Penton Publishing, San Jose Gateway, Suite 354. 2025 Gateway Place, San Jose, CA 95110. (408) 441-0550. 1930 to present. Semi-weekly. $98.00 per year. ISSN: 0883-4989.

IEEE Power Engineering Review. Institute of Electrical and Electronics Engineers, Inc., Box 1331, 445 Hoes Lane, Piscataway, NJ 08855-1331. (908) 981-0060. 1981 to present. Monthly. $115.00 per year. ISSN: 0272-1724.

IEEE Spectrum. Institute of Electrical and Electronics Engineers, Inc., Box 1331, 445 Hoes Lane, Piscataway, NJ 08855-1331. (908) 981-0060. 1964 to present. Monthly. $157.00. ISSN: 0018-9235.

IEEE Transactions On Power Delivery. Institute of Electrical and Electronics Engineers, Inc., Box 1331, 445 Hoes Lane, Piscataway, NJ 08855-1331. (908) 981-0060. 1986 to present. Quarterly. $170.00 per year. ISSN: 0885-8977.

IEE Proceedings: Electric Power Applications. Institution of Electrical Engineers (IEE), London. Available from INSPEC/IEEE - Institute of Electrical and Electronic Engineers, Box 1331, Hoes Lane, Piscataway, NJ 08855-1331. (908) 562-5549. 1980 to present. Bi-monthly. $ ISSN:

Institute of Electrical and Electronics Engineers. Proceedings. Institute of Electrical and Electronics Engineers, Inc., Box 1331, 445 Hoes Lane, Piscataway, NJ 08855-1331. (908) 981-0060. 1913 to present. Monthly. $275.00. ISSN: 0018-9219.

Power Engineering (Tulsa). PennWell Publishing Co. 1421 South Sheridan Road, Box 1260, Tulsa OK 74101. (835-3161. FAX (918) 831-9497. 1896 to present. Monthly. $28.00 per year. ISSN: 0032-5961

Power Engineering Journal. Institution of Electrical Engineers (IEE), London. Available from INSPEC/IEEE- Institute of Electrical and Electronic Engineers, Box 1331, 445 Hoes Lane, Piscataway, NJ 08855-1331. (908) 562-5549. 1987 to present. Bi-monthly. L85 per year. ISSN: 0950-3366.

RESEARCH CENTERS AND INSTITUTES

Edison Electric Institute. 701 Pennsylvania Avenue, NW, Washington, DC 20004-2696. (202) 508-5000, 5454. FAX (202) 508-5360.

Electroscience Laboratory. Ohio State University, 1320 Kinnear Road, Columbus, Ohio 43212. (614) 292-7981, FAX (614) 292-7297.

Electrical Engineering Research Laboratories. Purdue University. Electrical Engineering Building, West Lafayette, in 47907. (317) 494-3536. FAX (317) 494-6440.

Electrical Power Research Institute. 3412 Hillview Avenue, PO Box 10412, Palo Alto, CA 94303. (415) 855-2000. FAX (415) 855-2954.

Laboratory For Electromagnetic and Electronic Systems. Massachusetts Institute of Technology, 77 Massachusetts Avenue, Cambridge, MA 02139. (617) 253-4631. FAX (617) 258-6774.

Weber Research Institute. Polytechnic University. Route 110, Farmingdale, NY 11735. (516) 755-4250, FAX (516) 755-4404.

ELECTRIC VEHICLES

See also: AUTOMOTIVE ENGINEERING

ABSTRACT SERVICES AND INDEXES

Applied Science and Technology Index; A Cumulative Subject Index To English Language Periodicals in the Fields of Aeronautics and Space Science, Computer Technology, Chemistry, Construction Industry, Energy and Related Areas. H.W. Wilson Co., 950 University Avenue, Bronx, NY 10452. (800) 367-6770 or (212) 588-8400. FAX (718) 590-1617. From 1958 to present. Monthly. Inquire about cost and availability. Also available on CD-ROM.

Engineering Index Monthly. Engineering Information, Inc., Castle Point on the Hudson, Hoboken, NJ 07030. (800) 221-1044. FAX (212) 832-1857. Monthly. $2200.00 per year. Also available online as COMPENDEX, and also on CD-ROM.

ASSOCIATIONS AND PROFESSIONAL SOCIETIES

Electric Car Association. 1249 Lane Street, Belmont, CA 94002. (415) 591-6698.

Electric Vehicle Association of the Americas. 20823 Stevens Creek Blvd., Suite 440, Cupertino, CA 95014. (408) 253-5262.

Society of Automotive Engineers (sae). 400 Commonwealth Drive, Warrendale, PA 15096. (412) 776-4841.

BIBLIOGRAPHIES

Scientific and Technical Books and Serials in Print. R.R. Bowker Inc., 121 Chanlon Road, New Providence, NJ 07974. (800) 521-8110 or (908) 464-6800. List of currently published books and serials in the physical and biological sciences, engineering and technology, with subject, author and titles indexes. Inquire as to cost and availability. Also available on ORBIT online service, (800) 456-7248.

DIRECTORIES AND BIOGRAPHICAL SOURCES

Electric Vehicle Directory, 1992. Philip Terpstra. Spirit Publishing, PO Box 23417, Tucscon, AZ 85734. (602) 822-2030. 1991. $11.00.

Research Centers Directory. Gale Research, 835 Penobscot Building, Detroit, MI 48226-4094. (313) 961-2242. (800) 347-4253. FAX (313) 961-6083. $485.00.

Scientific and Technical Organizations and Agencies Directory. Gale Research, 835 Penobscot Building, Detroit, MI 48226-

ELECTRIC VEHICLES

Ency. of Physical Sciences and Engineering Info. Sources

4094. (313) 961-2242. (800) 347-4253. FAX (313) 961-6083. 4th edition. 1996. $195.00.

Who's Who in Engineering. American Association of Engineering Societies, 1111 19th Street NW, Suite 608, Washington, DC 20036. (202) 296-2237 or (800) 658-8897. 8th edition. 1991. Inquire for price.

Who's Who in Technology. Gale Research, 835 Penobscot Building, Detroit, MI 48226-4094. (313) 961-2242. (800) 347-4253. FAX (313) 961-6083. 1995. $195.00.

ENCYCLOPEDIAS AND DICTIONARIES

Encyclopedia of Physical Science and Technology. Robert A. Meyers, ed. 18 volumes. Academic Press Inc., 6277 Sea Harbor Drive, Orlando, FL 32887. (800) 321-5068. 1992. $2100.00

McGraw-Hill Dictionary of Scientific and Technical Terms. Sybil P. Parker, ed. 5th edition. McGraw-Hill Publishing Company, 1221 Avenue of the Americas, New York, NY 10020. (800) 262-4729 or (212) 512-3825. 1993. $110.50.

McGraw-Hill Encyclopedia of Engineering. Sybil P. Parker, ed. 2nd edition. McGraw-Hill Publishing Company, 1221 Avenue of the Americas, New York, NY 10020. (800) 262-4729 or (212) 512-3825. 1993. $95.50.

McGraw-Hill Encyclopedia of Science and Technology. Sybil P. Parker, ed. 7th edition. 20 volumes. McGraw-Hill Publishing Company, 1221 Avenue of the Americas, New York, NY 10020. (800) 262-4729 or (212) 512-3825. 1992. $1900.00

SAE Motor Vehicle, Safety and Environmental Terminology. Society of Automotive Engineers, 400 Commonwealth Drive, Warrendale, PA 15096. (412) 776-4841. $19.00.

GENERAL WORKS

Alternative Cars in the 21st Century. Robert Q. Riley. Society of Automotive Engineers, 400 Commonwealth Dr., Warrendale, PA 15096. (412) 776-4841. 1995. $39.00.

Building the E-Motive Industry. Scott A. Cronk. Society of Automotive Engineers, 400 Commonwealth Dr., Warrendale, PA 15096. (412) 776-4841. 1995. $35.00.

Design Innovations in Electric and Hybrid Electric Vehicles. Society of Automotive Engineers, 400 Commonwealth Dr., Warrendale, PA 15096. (412) 776-4841. 1995. $49.00.

Electric Vehicle Design and Development. Society of Automotive Engineers, 400 Commonwealth Dr., Warrendale, PA 15096. (412) 776-4841. 1991. $34.00

Electric Vehicle Structures and Components. Philip Terpstra. Spirit Publishing, PO Box 23417, Tucscon, AZ 85734. (602) 822-2030. 1993. $10.00.

Electric Vehicles: A Decade of Transition. Society of Automotive Engineers, 400 Commonwealth Dr., Warrendale, PA 15096. (412) 776-4841. 1992. $74.00.

Electric Vehicles: Technology, Performance and Potential. Organization for Economic Cooperation and Development, 2001

L St. NW, Washington, DC 20036. (202) 785-6323 or (800) 456-6323. FAX (202) 785-0350. 1993. Inquire for cost and availability.

Solo: Life With An Electric Car. Noel Perrin. W.W. Norton & Co., Inc., 500 Fifth Ave, New York, NY 11010. (800) 223-2584. FAX (800) 233-2588. 1992. $18.95.

The Urban Electric Vehicle. Organization for Economic Cooperation and Development, 2001 L St. NW, Washington, DC 20036. (202) 785-6323 or (800) 456-6323. FAX (202) 785-0350. 1992. $72.00

HANDBOOKS AND MANUALS

Automotive Handbook. 3rd edition. Robert Bosch GmbH/SAE International, 400 Commonwealth Drive, Warrendale, PA 15096-0001.(412) 776-4970. FAX (412) 776-0790. 1993. $34.00.

SAE Handbook. SAE International, 400 Commonwealth Drive, Warrendale, PA 15096-0001.(412) 776-4970. FAX (412) 776-0790. 1905 to present. Annual. $375.00 for three-volume set.

Standard Handbook of Engineering Calculations. Tyler G. Hicks, ed. 2d ed. McGraw-Hill Publishing Company, 1221 Avenue of the Americas, New York, NY 10020. (800) 262-4729 or (212) 512-3825. 1985. $89.50.

The Wiley Engineer's Desk Reference. Sanford I. Heisler. John Wiley and Sons, Inc., 605 Third Avenue, New York, NY 10158. (800) 526-5368 or (212) 850-6000. 1984. $64.95.

ONLINE DATABASES AND CD-ROMS

Compendex Plus. Engineering Information, Inc., 345 East 47th Street, New York, NY 10017. (212) 705-7600 or (800) 221-1044. Contains citations with abstracts to worldwide literature in engineering and technology, from 1970 to present. Available on online BRS,(800) 289-4277, DIALOG, (800) 334-2564, ORBIT (800) 456-7248, and STN International, FIZ Karlsruhe, P.O. Box 2465, W-7500, Karlsruhe 1, Germany, online services. Also available on CD-ROM. Inquire as to cost and availability.

CSA Engineering. Cambridge Scientific Abstracts, 7200 Wisconsin Avenue, Suite 601, Bethesda, MD 20814. (301) 961-6750 or (800) 843-7751. Contains citations and abstracts of international periodicals and research literature covering all fields of engineering and science and technology,including computer and information science, electronics, mechanical engineering, solid state materials, 1981 to present. Available on BRS,(800) 289-4277, online service.Inquire as to cost and availability.

Current Contents Search. Institute for Scientific Information, 3501 Market Street, Philadelphia, PA 19104. (215) 386-0100. FAX (215) 386-6362. Contains citations to articles listed in the table of contents of science and technology journals. Also articles in social sciences and life sciences journals. Available on BRS,(800) 289-4277, DIALOG,(800) 334-2564, online services. Inquire as to cost and availability.

Dissertation Abstracts. University Microfilms International, 300 North Zeeb Road, Ann Arbor, MI 48106. (800) 521-0600 or (313) 761-4700. Scope includes virtually all doctoral dissertations accepted at accredited American institutions from 1861 to present in 252 subject areas. Available on BRS,(800) 289-4277,

DIALOG,(800) 334-2564, and OCLC EPIC,(800) 848-5878, online services. Also available on CD-ROM. Inquire as to cost and availability.

NTIS Bibliographic Database. National Technical Information Service, 5285 Port Royal Road, Springfield, VA 22161. (703) 487-4929 or FAX (703) 321-8199. Broad coverage of government-sponsored science and technology research reports, 1964 to present. Available on BRS,(800) 289-4277, DIALOG, (800) 334-2564, ORBIT, (800) 456-7248, and STN International, FIZ Karlsruhe, P.O. Box 2465, W-7500, Karlsruhe 1, Germany, online services. Also available on CD-ROM. Inquire as to cost and availability.

SAE Global Mobility Database. Society of Automative Engineers (SAE), Electronic Publishing Division, 400 Commonwealth Drive, Warrendale, PA 15098. (412) 776-4841. Contains citations with abstracts to technical papers on automotive and aerospace technology and vehicular-related industries that have been presented at SAE conferences. Covers 1906 to present. Available on ORBIT,(800) 456-7248,online service. Inquire as to cost and availability.

Scientific and Technical Books and Serials in Print. R.R. Bowker Inc., 121 Chanlon Road, New Providence, NJ 07974. (800) 521-8110 or (908) 464-6800. List of currently published books and serials in the physical and biological sciences, engineering and technology, with subject, author and titles indexes. Available on ORBIT,(800) 456-7248, online service.Inquire as to cost and availability.

SCISEARCH. Institute for Scientific Information, 3501 Market Street, Philadelphia, PA 19104. (800) 523-1850 or (215) 386-0100. Broad multidisciplinary title and author index to the international literature of science and technology, 1974 to present. Available on DIALOG,(800) 334-2564, and ORBIT,(800) 456-7248, online services. Also available on CD-ROM. Inquire as to cost and availability.

TRIS (Transportation Research Information). National Academy of Science, 2101 Constitution Avenue, N.W., Washington, DC 20418. (202) 334-3313 or (800) 624-6242. Citations with abstracts of literature on transportation, including air, highway, rail, maritime and other modes from 1968 to present. Available on DIALOG,(800) 334-2564, online service. Inquire as to cost and availability.

WILSONLINE. H.W. Wilson Company, 950 University Avenue, Bronx, NY 10452. (800) 367-6770 or (212) 588-8400. Makes available online versions of the printed H.W. Wilson indexes including Applied Science and Technology Index, Business Periodicals Index, General Science Index, and Readers' Guide to Periodical Literature. Period covered is generally 1983 to present. Available on BRS,(800) 289-4277, DIALOG,(800) 334-2564, and OCLC EPIC,(800) 848-5878, online services. Also available on CD-ROM. Inquire as to cost and availability.

PERIODICALS

Automotive Engineering Magazine. SAE International, 400 Commonwealth Drive, Warrendale, PA 15096-0001. (412) 776-4841. 1905 to present. Monthly. $72.00.

Automotive Engineer. Mechanical Engineering Publications Ltd., Northgate Avenue, Bury St. Edmunds, Suffolk IP32 6BW, England. Telephone 0284-763277. FAX 0284-704006. 1962 to present. 6 times a year. Inquire for cost.

Automotive Industries. Chilton Co., One Chilton Way, Radnor PA 19089. (215) 964-4028. 1895 to present. Monthly. $60.00 per year.

Automotive Week. Automotive Week Publishing Company, Box 3495, Wayne, NJ 07474-3495. (201) 694-7792. 1975 to present. Weekly. $125.00 per year.

Electric Vehicle Progress. Alexander Research & Communications Inc., 215 Park Avenue South, Suite 1301, New York, NY 10003. (212) 228-0246. FAX (212) 228-0376. 1979 to present. Semi-monthly. $387.00 per year.

International Journal of Vehicle Design. Inderscience Enterprises Ltd., World Trade Center Building, 110 Ave. Luis Casai, Case Postale 306, CH-1215 Geneva-Aeroport, Switzerland. 1979 to present. Bi-monthly. $190.00 per year.

JSAE (Japanese Society of Automotive Engineers) Review. Elsevier Science [journals], 660 White Plains Rd., Tarrytown, NY 10159-5153. (919) 524-9200. FAX (919) 333-2444. 1978 to present. 4 times a year. $168.00 per year.

Proceedings of the Institution of Mechanical Engineers D: Journal of Automobile Engineering. Mechanical Engineering Publications Ltd., Northgate Avenue, Bury St. Edmunds, Suffolk IP32 6BW, England. Telephone 0284-763277. FAX 0284-704006. 1984 to present. Quarterly. $269.00 per year.

SAE Transactions. SAE International, 400 Commonwealth Drive, Warrendale, PA 15096-0001. (412) 776-4841. Annual. $850.00 per year.

RESEARCH CENTERS AND INSTITUTES

Society of Automotive Engineers (sae) International. 400 Commonwealth Drive, Warrendale, PA 15096-0001. (412) 776-4841.

U.S. Department of Energy, Office of Transporation Systems, Electric and Hybrid Propulsion Division. Mail Stop 5G-030, 1000 Independence Avenue SW, Washington, DC 20585. (202) 586-2198. FAX (202) 586-1600.

University of Michigan Automotive Engineering Laboratory. Department of Mechanical Engineering & Applied Mechanics, 321 Lay, N. Campus, 1231 Beal, Ann Arbor, MI 48109-2121. (313) 764-4254.

Wayne State University Center For Automotive Research. 5050 Anthony Wayne Drive, Detroit, MI 48202. (313) 577-3887.

ELECTRICAL CODES

See: BUILDING CODES, WIRING ELECTRICAL

ELECTRICAL ENGINEERING

Ency. of Physical Sciences and Engineering Info. Sources

ELECTRICAL ENGINEERING

See also: ELECTROMAGNETISM, ELECTRONIC CIRCUITS and COMPONENTS, ELECTRICITY, ELECTRONICS, ELECTRONICS ENGINEERING, MICROWAVES, RADIO SIGNAL PROCESSING

ABSTRACT SERVICES AND INDEXES

Applied Science and Technology Index; A Cumulative Subject Index To English Language Periodicals in the Fields of Aeronautics and Space Science, Computer Technology, Chemistry, Construction Industry, Energy and Related Areas. H.W. Wilson Co., 950 University Avenue, Bronx, NY 10452. (212) 588-8400. (800) 367-6770. FAX (718) 590-1617. From 1958 to present. Monthly. Inquire about cost and availability. Also available on CD-ROM and online. ISSN: 0003-6986.

Current Contents: Engineering, Technology and Applied Sciences. Institute for Scientific Information, 3501 Market Street, Philadelphia, PA 19104. (215) 386-0100. FAX (215) 386-6362. Weekly. $360.00 per year.

Current Papers in Electrical and Electronics Engineering. Institution of Electrical Engineers (IEE), London. Distributed by INSPEC/IEEE, Box 1331, 445 Hoes Lane, Piscataway, NJ 08855-1331. (908) 562-5549. 1969 to present. Monthly. $345.00 per year. ISSN: 0011-3778.

Electrical and Electronics Abstracts. Institute of Electrical Engineers, Michael Faraday House, Six Hill Way, Stevenage, Herts, SG1 2AY, England. Distributed by INSPEC/IEEE, 445 Hoes Lane, Piscataway, NJ 08854. (908) 562-5549. 1898 to present. Monthly. $2200.00 per year. Also available on CD-ROM and online as INSPEC. ISSN: 0036-8105.

Engineered Materials Abstracts. ASM (American Society of Metals) International, Materials Information, Materials Park, OH 44073. (216) 338-5151 or FAX (216) 338-4634. Covers literature on technical developments in polymer, ceramic, and composite materials and engineering. 1986 to present. Monthly. $1175.00 per year. Also available on CD-ROM.

Engineering Index Monthly; Indexes and Abstracts the World's Engineering and Technical Literature. Engineering Information, Inc., Castle Point on the Hudson, Hoboken, NJ 07030. (201) 216-8500. (800) 221-1044. FAX (201) 216-8532. Monthly. $2300.00 per year. Available online as COMPENDEX and also on CD-ROM. ISSN: 0742-1974.

General Science Index. H.W. Wilson Company, 950 University Avenue, Bronx, NY 10452. (212) 588-8400. (800) 367-6770. FAX (718) 590-1617. From 1978 to present. Ten issues per year; quarterly and annual cumulations. Service basis. Available on CD-ROM and online. Inquire regarding cost and availability. ISSN: 0162-1963.

Government Reports Announcements and Index. U. S. National Technical Information Service (NTIS), 5285 Port Royal Road, Springfield, VA 22161. (703) 487-4650. FAX (703) 321-8547. From 1968 to present. Annual. $630.00 per year. Also available online as NTIS BIBLIOGRAPHIC DATABASE and on CD-ROM. ISSN:

Index To IEEE Publications. IEEE Service Center, 445 Hoes Lane, Piscataway, NJ 08855-1331. (908) 981-1393. (800) 678-IEEE. FAX (908) 981-9667. 1973 to present. Annual. ISSN: 0099-1368.

Physics Abstracts. INSPEC. Section A, Science Abstracts. Institution of Electrical Engineers (IEE), London. Available from: INSPEC/IEEE - Institute of Electrical and Electronic Engineers, Box 1331, Hoes Lane, Piscataway, NJ 08855-1331. (908) 562-5549. 1898 to present. 24 issues per year. $2835.00 per year. Also available online and on CD-ROM. ISSN: 0036-8091.

Physics Briefs (Physikalische Berichte). Information Center for Energy, Physics, Mathematics; German Physical Society. V C H Publishers, Inc., 220 East 23rd Street, New York, NY 10010-4606. (212) 683-8333. 1845 to present. 24 issues per year. $2390.00 per year. Also available online. ISSN: 0179-7434.

Science Citation Index. SCI. Institute for Scientific Information, 3501 Market Street, Philadelphia, PA 19104. (215) 386-0100. (800) 523-1850. FAX (215) 386-2991. 1961 to present. Six issues per year, plus annual cumulation. $11650.00 per year. Also available online and on CD-ROM. Inquire about price and availability. ISSN: 0036-827X.

Solid State and Superconductivity Abstracts: Covers theory, Production and Application of Solid State Materials. Cambridge Scientific Abstracts, 7200 Wisconsin Avenue, Bethesda, MD 20824. (301) 961-6750. FAX (301) 961-6720. 1957 to present. Bimonthly. $1320.00 per year. Also available online. ISSN: 0896-5900.

STAR. Scientific and Technical Aerospace Reports. U.S. National Aeronautics and Space Administration, Scientific and Technical Information Facility, Box 8757, Baltimore-Washington International Airport, MD 21240. (301) 621-0153. Distributed by U. S. Superintendent of Documents, Washington, DC 20402. From 1963 to present. Semimonthly, with semiannual and annual indexes. $114.00 per year. Also available online and on CD-ROM. ISSN: 0036-8741.

ANNUAL REVIEWS AND YEARBOOKS

Advances in Electronics and Electron Physics. Academic Press, Inc., 1250 Sixth Avenue, San Diego, CA 92101-4311. (619) 231-0926. FAX (619) 699-6715. Irregular. Price varies, inquire.

ASSOCIATIONS AND PROFESSIONAL SOCIETIES

American Association of Engineering Societies, 1111 19th Street, Suite 608, Washington, DC 20036-3603. (202) 296-2237. FAX (202) 296-1151.

American Institute of Physics. 1 Physics Ellipse, College Park, MD 20740-3843. (301) 209-3100.

Association of Old Crows (Electronic Warfare). 1000 North Payne Street, Alexandria, VA 22314. (703) 549-1600.

Edison Electric Institute. 701 Pennsylvania Avenue NW, Washington, DC 20004-2696. (202) 508-5000, 5454. FAX (202) 508-5360.

Electronics Industries Association. 2500 Wilson Boulevard, Arlington, VA 22201. (703) 907-7500. FAX (202) 457-4985.

Equipment Manufacturers Institute. 10 South Riverside Plaza, Suite 1220, Chicago, IL 60606. (312) 321-1470. FAX (312) 321-1480.

IEEE (institute of Electrical and Electronic Engineers). 345 East 47th Street. New York, NY 10017. (212) 705-7900. FAX (212) 705-4929,

International Society For Hybrid Microelectronics. 1850 Centennial Park Drive, Suite 105, Reston, VA 22091. (703) 758-1060. FAX (703) 758-1066.

BIBLIOGRAPHIES

Bibliographical History of Electricity and Magnetism. Paul F. Mottelay, editor. Ayer Company Pubs., Inc., Lower Mill Road, North Stratford, NH 03590. (603) 922-5105. (800) 282-5413. Reprint edition. 1975. $52.95.

Science Books and Films. American Association for the Advancement of Science, 1333 H Street NW, Washington, DC 20005. (202) 326-6454. 1965 to present. Nine issues per year. $40.00 per year. ISSN: 0098-342X.

Scientific and Technical Books and Serials in Print; An Index To Literature in Science and Technology. R.R. Bowker Inc., 121 Chanlon Road, New Providence, NJ 07974. (908) 464-6800. (800) 521-8110. FAX (908) 665-3502. 1972 to present. Annual. 4 volumes. 1994. $299.95. Also available on compact disc and online. ISSN: 0000-054X.

DIRECTORIES AND BIOGRAPHICAL SOURCES

American Electronics Association. Directory. American Electronics Association, 5201 Great America Way, Suite 520, P.O. 52990, Santa Clara, CA 95056. (408) 987-4200. FAX (408) 970-8565. Annual. $175.00.

American Men and Women of Science: Physical and Biological Sciences. R. R. Bowker Inc., 121 Chanlon Road, New Providence, NJ 07974. (908) 464-6800. (800) 521-8110. 20th edition. 8 volumes. 1996. $850.00.

Directory of Engineering College Research and Graduate Study. American Society for Engineering Education, Ohio State University, 2070 Neil Avenue, Columbus, OH 43210-1226. Annual. $20.00.

Directory of Engineering Societies and Related Organizations. American Association of Engineering Societies, 1111 19th Street, Suite 608, Washington, DC 20036-3603. (202) 296-2237. Semi-annual. $150.00.

Directory of Engineers in Private Practice. National Society of Professional Engineers, 1420 King Street, Alexandria, VA 22314-2750. (703) 684-2835. Biennial. $30.00.

EEM: Electronic Engieers Master Catalog. Herst Business Communications, Inc., 645 Stewart Avenue, Garden City, NY 11530. (516) 227-1300. FAX (516) 227-1901. 38th edition. 1996. Annual. $90.00. ISSN: 0732-9016.

Electrical and Electronic Trades Directory: The Blue Book. Institution of Electrical Engineers, Michael Faraday House, Six Hills Way, Stevenage, Herts. SG1 2AY, England. Telephone: 0438 313311 or FAX 0438 742840. 1995. $140.00

Electrical World Directory of Electrical Utilities. McGraw-Hill Book Company, Inc., 1221 Avenue of the Americas, New York, NY 10020. (212) 997-3675. Annual.

Engineering Research Centers: Incorporating Electronics Research Centers. Stockton Press, 345 Park Avenue, New York, NY 10010. (212) 689-9200. (800) 221-2123. 4th edition. 1995. $515.00.

IEEE Membership Directory. Institute of Electrical and Electronics Engineers, IEEE Service Center, 445 Hoes Lane, Piscataway, NY 08854. (908) 981-1393. (800) 678-IEEE. FAX (908) 981-9667. 2 volumes. Annual. $190.00. ISSN:

International Directory of Abbreviations and Acroynms of Electronics, Electrical Engineering, Computer Technology and Information Processing. Peter Wennrich. K. G. Saur, 121 Chanlon Road, New Providence, NJ 07974. (908) 464-6800. (800) 521-8110 2 volumes. 1992. $230.00.

International Engineering Directory. American Consulting Engineers Council, 1015 15th Street, N.W. Suite 802, Washington, DC 20005-2670. (202) 347-7474. Annual. $10.00.

Research Centers Directory. Gale Research Company Inc., 835 Penobscot Building, Detroit, MI 48226-4094. (313) 961-2242. (800) 347-4253. 20th edition. 1995. $485.00.

Who's Who in Engineering. Gordon Davis, editor. American Association of Engineering Societies. 1111 19th Street, NY, Suite 608, Washington, DC 20036. (202) 296-2237. (800) 658-8897. 9th edition. 1995. $220.00.

Who's Who in Technology. Gale Research, 835 Penobscot Building, Detroit, MI 48226-4094. (313 961-2242. (800) 521-4253. 7th edition. 1995. $195.00. ISN 0-8103-7467-6.

ENCYCLOPEDIAS AND DICTIONARIES

Communications Standards Dictionary. Martin H. Weik. Chapman & Hall, 1 Penn Plaza, New York, NY 10119. (212) 564-1060 3rd edition. 1996. $69.95.

Dictionary of Communications Technology: Terms, Definitions, and Appreviations. Gilbert Held. John Wiley & Sons, Inc., 605 Third Avenue, New York, NY 10158-0012. (212) 850-6000. (800) 225-5945. 1995. $49.95.

Electrical and Electronic Technologies: A Chronology of Events and Inventors From 1940 To 1980. Henry B. O. Davis. Scarecrow Press 1985. $35.00.

Encyclopedia of Applied Physics. George Trigg, editor. VCH Publications, Inc., 220 East 23rd Street, Suite 909, New York, NY 10010-4606. (212) 683-8333. (800) 422-8824. 20 volume set. 1991 - . $5990.00.

IEEE Standard Dictionary of Electrical and Electronics Terms. Christopher J. Booth, editor. IEEE Service Center, 445 Hoes Lane, Piscataway, NJ 08855-1331. (908) 981-1393. (800) 678-IEEE. FAX (908) 981-9667. IEEE Standard 100-1992. 5th edition. 1993. $150.00.

Illustrated Dictionary of Electronics. Stan Gibilsco. TAB Books, P.O. Box 40, Blue Summit, PA 17294-0850. (717) 794-2191. (800) 233-1128. 7th edition. 1994. $34.95.

McGraw-Hill Circuit Encyclopedia and Troubleshooting Guide. John Lenk, editor. McGraw-Hill Book Company, Inc. 1221 Avenue of the Americas, New York, NY 10020. (212) 997-3675 volume 1-3. 1993 - . $36.95 each, in paper.

ELECTRICAL ENGINEERING

Ency. of Physical Sciences and Engineering Info. Sources

McGraw-Hill Electronics Dictionary. J. Markus. McGraw-Hill Book Company, Inc., 1221 Avenue of the Americas, New York, NY 10020. (212) 997-3675. 5th edition. 1994. $49.95.

McGraw-Hill Encyclopedia of Science and Technology. McGraw-Hill Book, Incorporated, 1221 Avenue of the Americas, New York, NY 10020. (212) 997-3675. (800) 262-4729. Seventh edition. Twenty volumes. 1992. $1900.00.

Telecommunications Engineer's Reference Book. Fraidoon Mazda, editor. Butterworth Heinmann, 313 Washington Street, Newton, MA 02158. (618-928-2500. (800) 366-2665. 1993. $130.00.

GENERAL WORKS

Alternating Current Fundamentals. John R. Duff and Stephen L. Herman. Delmar Publications, 3 Columbia Circle, Box 15015, Albany, NY 12212. (518) 464-3500. (800) 347-7707. 1995.

Electric and Electronic Engineering For Scientists and Engineers. K. A. Krishnamurthy. Halsted Press, 605 Third Avenue, New York, 10158-0012. (212) 850-6400. 1994. $48.95.

Electricity and Electronics: A Survey. Dale R. Patrick and Stephen W. Fardo. Prentice Hall , 113 Sylvan Avenue, Route 9W, Englewood Cliffs, NJ 07632. (201) 592-2000. (800) 922-0579. 3rd edition. 1995. $68.00.

High-Voltage Engineering: Theory and Practice. Khalifa.

Marcel Dekker, Inc., 270 Madison Avenue, New York, NY 10016. (212) 696-9000. (800) 228-1160. 1989. $125.00.

Industrial Electricity. John M. Nadon, et al. Delmar Publications, 3 Columbia Circle, Box 15015, Albany, NY 12212. (518) 464-3500. (800) 347-7707. 1994. $37.95.

VHF and UHF Antennas. R. A. Burberry. Institution of Electrical Engineers, Michael Faraday House, Six Hills Way, Stevenage, Herts. SG1 2AY, England. Telephone: 0438 313311 or FAX 0438 742840. 1992. $95.00

HANDBOOKS AND MANUALS

American Electricians' Handbook. Terrell Croft and Wilford I. Summers, editors. McGraw-Hill Publishing Company, Inc., 1221 Avenue of the Americas, New York, NY 10020. (212) 512-2000. (800) 262-4729. 12th editon. 1992. $74.50.

Antenna Handbook. Y. T. Lo and S. W. Lee, editors. Van Nostrand Reinhold, 115 Fifth Avenue, New York, NY 10003. (212) 254-3232. (800) 842-3636. Four volumes: Fundamentals and mathematical techniques, Antenna theory, Applications, Related topics. 1993. $150.95.

Circuits and Filters Handbook. Wai-Kai Chen, editor-in-chief. CRC Press Inc., 2000 Corporate Blvd., N.W., Boca Raton, FL 33431. (407) 994-0555. (800) 333-8300. Published in cooperation with IEEE Press. 1995. $129.95.

Electrical and Electronics Graphic and Letter Symbols and Reference Designations Standards Collection. Institute of Electrical and Electronics Engineers, Inc., C-2, 1993. $75.00.

Electrical Engineering Handbook. Richard Dorf, editor. CRC Press, Inc., 2000 Corporate Boulevard, NW, Boca Raton, FL 33431. (407) 994-0555. (800) 272-7737. 1993. $99.95.

Electrical Engineer's Handbook. Donald Christiansen, editor. McGraw Publishing Company Inc., 1221 Avenue of the Americas, New York, NY 10020. (212) 512-2000. (800) 262-4729 4th edition. 1996. $110.00.

Handbook of Practical Electrical Design. Joseph F. McPartland and Brian J. McPartland. McGraw-Hill Publishing Company Inc., 1221 Avenue of the Americas, New York, NY 10020. (212) 512-2000. (800) 262-4729 2nd edition. 1995. $62.50.

National Electrical Safety Code. Institute of Electrical and Electronic Engineers (IEEE), 345 East 47th Street, New York, NY 10017. (212) 705-7900. C2-1993. 1992. $86.00.

Rapid Electrical Estimating and Pricing. C. Kenneth Kolstad, and Gerald V. Kohnert. McGraw-Hill Publishing Company Inc., 1221 Avenue of the Americas, New York, NY 10020. (212) 512-2000. (800) 262-4729 5th editon. 1993. $65.00

Standard Handbook For Electrical Engineers. Donald G. Fink and H. Wayne Beaty, editors. McGraw-Hill Publishing Company, 1221 Avenue of the Americas, New York, NY 10020. (212) 512-2000. (800) 262-4729. 13th edition. 1996. $110.00.

Standard Handbook of Engineering Calculations. Tyler G. Hicks. McGraw-Hill Publishing Company, 1221 Avenue of the Americas, New York, NY 10020. (800) 262-4729 or (212) 512-3825. 3rd edition. 1994. $99.50. ISBN: 0-07-028812-7.

ONLINE DATABASES AND CD-ROMS

CA Search. Chemical Abstracts Service, P.O. Box 3012, Columbus, OH 43210-0012. (614) 447-3600. (800) 848-6533. FAX (614) 447-3709. Very comprehensive guide to worldwide chemical literature and related fields, 1972 to present. Available on BRS,(800) 289-4277, DIALOG, (800) 334-2564, ORBIT (800) 456-7248, and STN International, FIZ Karlsruhe, P.O. Box 2465, W-7500, Karlsruhe 1, Germany, online services. Inquire as to cost and availability.

Compendex Plus. Engineering Information, Inc., 345 East 47th Street, New York, NY 10017. (212) 705-7600 or (800) 221-1044. Contains citations with abstracts to worldwide literature in engineering and technology, from 1970 to present. Available on online BRS,(800) 289-4277, DIALOG, (800) 334-2564, ORBIT (800) 456-7248, and STN International, FIZ Karlsruhe, P.O. Box 2465, W-7500, Karlsruhe 1, Germany, online services. Also available on CD-ROM. Inquire as to cost and availability.

Current Contents Search. Institute for Scientific Information, 3501 Market Street, Philadelphia, PA 19104. (215) 386-0100. FAX (215) 386-6362. Contains citations to articles listed in the table of contents of science and technology journals. Also articles in social sciences and life sciences journals. Available on BRS,(800) 289-4277, DIALOG,(800) 334-2564, online services. Inquire as to cost and availability.

Dissertation Abstracts. University Microfilms International, 300 North Zeeb Road, Ann Arbor, MI 48106. (800) 521-0600 or (313) 761-4700. Scope includes virtually all doctoral dissertations accepted at accredited American institutions from 1861 to present in 252 subject areas. Available on BRS,(800) 289-4277, DIALOG,(800) 334-2564, and OCLC EPIC,(800) 848-5878,

online services. Also available on CD-ROM. Inquire as to cost and availability.

INSPEC. Institution of Electrical Engineers, Michael Faraday House, Six Hills Way, Stevenage, Herts. SG1 2AY, England. Telephone: 0438 313311 or FAX 0438 742840. Contains citations to the worldwide literature of physics, electronics and electrical engineering, computer technology, and related fields. Available on BRS,(800) 289-4277, DIALOG, (800) 334-2564, ORBIT, (800) 456-7248, and STN International, FIZ Karlsruhe, P.O. Box 2465, W-7500, Karlsruhe 1, Germany, online services. Inquire as to cost and availability.

NTIS Bibliographic Database. National Technical Information Service, 5285 Port Royal Road, Springfield, VA 22161. (703) 487-4929 or FAX (703) 321-8199. Broad coverage of government-sponsored science and technology research reports, 1964 to present. Available on BRS,(800) 289-4277, DIALOG, (800) 334-2564, ORBIT, (800) 456-7248, and STN International, FIZ Karlsruhe, P.O. Box 2465, W-7500, Karlsruhe 1, Germany, online services. Also available on CD-ROM. Inquire as to cost and availability.

Scientific and Technical Books and Serials in Print. R.R. Bowker Inc., 121 Chanlon Road, New Providence, NJ 07974. (800) 521-8110 or (908) 464-6800. List of currently published books and serials in the physical and biological sciences, engineering and technology, with subject, author and titles indexes. Available on ORBIT,(800) 456-7248, online service.Inquire as to cost and availability.

SCISEARCH. Institute for Scientific Information, 3501 Market Street, Philadelphia, PA 19104. (800) 523-1850 or (215) 386-0100. Broad multidisciplinary title and author index to the international literature of science and technology, 1974 to present. Available on DIALOG,(800) 334-2564, and ORBIT,(800) 456-7248, online services. Also available on CD-ROM. Inquire as to cost and availability.

WILSONLINE. H.W. Wilson Company, 950 University Avenue, Bronx, NY 10452. (800) 367-6770 or (212) 588-8400. Makes available online versions of the printed H.W. Wilson indexes including Applied Science and Technology Index, Business Periodicals Index, General Science Index, and Readers' Guide to Periodical Literature. Period covered is generally 1983 to present. Available on BRS,(800) 289-4277, DIALOG,(800) 334-2564, and OCLC EPIC,(800) 848-5878, online services. Also available on CD-ROM. Inquire as to cost and availability.

PERIODICALS

Chaos: An Interdisciplinary Journal. American Institute of Physics, 335 East 45th Street, New York, NY 10017 (212) 661-9404 or FAX (516)349-9704. 1991 to present. Quarterly. $160.00 per year.

Circuits, Systems, and Signal Processing. Birkhauser, 675 Massachusetts Avenue, Cambridge, MA 02139-3309. FAX (201) 348-4505. 1982 to present. Quarterly. $184.00 per year. ISSN: 0278-081X.

CRC Critical Reviews in Solid State and Materials Science. CRC Publishers, Inc., 2000 Corporate Blvd., N.W., Boca Raton, FL 33431. (407) 994-0555 or (800) 333-8300. Quarterly. $295.00 per year.

Electrical World. McGraw-Hill, Inc., Box 513 Hightstown, NJ 08520. (212) 512-3288. 1874 to present. Monthly. $55.00 per year. ISSN: 0013-4457.

Electronic Design. Penton Publishing, San Jose Gateway, Suite 354. 2025 Gateway Place, San Jose, CA 95110. (408) 441-0550. 1952 to present. Fortnightly. $95.00. ISSN: 0013-4872.

Electronics. Penton Publishing, San Jose Gateway, Suite 354. 2025 Gateway Place, San Jose, CA 95110. (408) 441-0550. 1930 to present. Semi-weekly. $98.00 per year. ISSN: 0883-4989.

IEEE Circuits and Devices Magazine. Institute of Electrical and Electronics Engineers, Inc., Box 1331, 445 Hoes Lane, Piscataway, NJ 08855-1331. (908) 981-0060. 1985 to present. Bi-monthly. $120.00 per year. ISSN: 8755-3996.

IEEE Spectrum. Institute of Electrical and Electronics Engineers, Inc., Box 1331, 445 Hoes Lane, Piscataway, NJ 08855-1331. (908) 981-0060. 1964 to present. Monthly. $157.00. ISSN: 0018-9235.

IEEE Transactions On Circuits and Systems. Part 1: Fundamental theory and Applications. Institute of Electrical and Electronics Engineers, Inc., Box 1331, 445 Hoes Lane, Piscataway, NJ 08855-1331. (908) 981-0060. 1952 to present. Monthly. $241.00 per year. ISSN: 1057-7122.

IEEE Transactions On Electron Devices. Institute of Electrical and Electronics Engineers, Inc., Box 1331, 445 Hoes Lane, Piscataway, NJ 08855-1331. (908) 981-0060. 1952 to present. Monthly. $395.00 per year. ISSN: 0018-9393.

IEEE Transactions On Power Delivery. Institute of Electrical and Electronics Engineers, Inc., Box 1331, 445 Hoes Lane, Piscataway, NJ 08855-1331. (908) 981-0060. 1986 to present. Quarterly. $170.00 per year. ISSN: 0885-8977.

Institute of Electrical and Electronics Engineers. Proceedings. Institute of Electrical and Electronics Engineers, Inc., Box 1331, 445 Hoes Lane, Piscataway, NJ 08855-1331. (908) 981-0060. 1913 to present. Monthly. $275.00. ISSN: 0018-9219.

International Journal of Microwave and Millimeter-wave Computer Aided Engineering. John Wiley and Sons, Inc., 605 Third Avenue, New York, NY 10158. (800) 526-5368 or (212) 850-6000. 1991 to present. Quarterly. $175.00 per year.

RESEARCH CENTERS AND INSTITUTES

Edison Electric Institute. 701 Pennsylvania Avenue, NW, Washington, DC 20004. (202) 508-5000

Electrical Engineering Research Laboratories. Purdue University. Electrical Engineering Building, West Lafayette, in 47907. (317) 494-3536. FAX (317) 494-6440.

Electrical Power Institute. Texas A and M University, Department of Electrical Engineering, College Station, TX 77843.

Electrical Power Research Institute. 3412 Hillview Avenue, PO Box 10412, Palo Alto, CA 94303. (415) 855-2000. FAX (415) 855-2954.

ELECTRICAL ENGINEERING

Ency. of Physical Sciences and Engineering Info. Sources

Laboratory For Electromagnetic and Electronic Systems. Massachusetts Institute of Technology, 77 Massachusetts Avenue, Cambridge, MA 02139. (617) 253) 4631.

Electroscience Laboratory. Ohio State University, 1320 Kinnear Road, Columbus, Ohio 43212. (614) 292-7981, FAX (614) 292-7297.

Weber Research Institute. Polytechnic University. Route 110, Farmingdale, NY 11735. (516) 755-4250, FAX (516) 755-4404.

ELECTRICAL IMPEDANCE

See: ELECTRICITY

ELECTRICAL INSULATION

See: ELECTRICITY

ELECTRICAL RESISTANCE

See: ELECTRICITY

ELECTRICITY

See also: ELECTRIC POWER ENGINEERING, ELECTRICAL ENGINEERING, ELECTROCHEMISTRY, ELECTROMAGENTISM, ELECTRONICS, ELECTRONICS ENGINEERING

ABSTRACT SERVICES AND INDEXES

Applied Science and Technology Index; A Cumulative Subject Index To English Language Periodicals in the Fields of Aeronautics and Space Science, Computer Technology, Chemistry, Construction Industry, Energy and Related Areas. H.W. Wilson Co., 950 University Avenue, Bronx, NY 10452. (212) 588-8400. (800) 367-6770. FAX (718) 590-1617. From 1958 to present. Monthly. Inquire about cost and availability. Also available on CD-ROM and online. ISSN: 0003-6986.

Current Contents: Engineering, Technology and Applied Sciences. Institute for Scientific Information, 3501 Market Street, Philadelphia, PA 19104. (215) 386-0100. FAX (215) 386-6362. 1970 to present. Weekly. $442.00 per year. Also available on CD-ROM and online. Inquire regarding cost and availability. ISSN: 0095-7917.

Current Papers in Electrical and Electronics Engineering. Institution of Electrical Engineers (IEE), London. Distributed by INSPEC/IEEE, Box 1331, 445 Hoes Lane, Piscataway, NJ 08855-1331. (908) 562-5549. 1969 to present. Monthly. $345.00 per year. ISSN: 0011-3778.

Electric Power Industry Abstracts. Edison Electic Institute, c/o Utlity Data Institute, 2011 I Street, NW Suite 700, Washington, DC 20006. 1975 to present. Bimonthly. Inquire as to cost and availability.

Electrical and Electronics Abstracts. Institution of Electrical Engineers (IEE), London. Available from INSPEC/IEEE - Institute of Electrical and Electronic Engineers, Box 1331, Hoes Lane, Piscataway, NJ 08855-1331. (908) 562-5549. 1898 to present. Monthly. $2200.00 per year. Also available on CD-ROM and online as INSPEC.

Engineering Index Monthly; Indexes and Abstracts the World's Engineering and Technical Literature. Engineering Information, Inc., Castle Point on the Hudson, Hoboken, NJ 07030. (201) 216-8500. (800) 221-1044. FAX (201) 216-8532. Monthly. $2300.00 per year. Available online as COMPENDEX and also on CD-ROM. ISSN: 0742-1974.

General Science Index. H.W. Wilson Company, 950 University Avenue, Bronx, NY 10452. (212) 588-8400. (800) 367-6770. FAX (718) 590-1617. From 1978 to present. Ten issues per year; quarterly and annual cumulations. Service basis. Available on CD-ROM and online. Inquire regarding cost and availability. ISSN: 0162-1963.

Government Reports Announcements and Index. U. S. National Technical Information Service (NTIS), 5285 Port Royal Road, Springfield, VA 22161. (703) 487-4650. FAX (703) 321-8547. From 1968 to present. Annual. $630.00 per year. Also available online as *NTIS Bibliographic Database* and on CD-ROM.

Index To IEEE Publications. IEEE Service Center, 445 Hoes Lane, Piscataway, NJ 08855-1331. (908) 981-1393. (800) 678-IEEE. FAX (908) 981-9667. 1973 to present. Annual. ISSN: 0099-1368.

Physics Abstracts. INSPEC. Section A, Science Abstracts. Institution of Electrical Engineers (IEE), London. Available from: INSPEC/IEEE - Institute of Electrical and Electronic Engineers, Box 1331, Hoes Lane, Piscataway, NJ 08855-1331. (908) 562-5549. 1898 to present. 24 issues per year. $2835.00 per year. Also available online and on CD-ROM. ISSN: 0036-8091.

Physics Briefs (Physikalische Berichte). Information Center for Energy, Physics, Mathematics; German Physical Society. V C H Publishers, Inc., 220 East 23rd Street, New York, NY 10010-4606. (212) 683-8333. 1845 to present. 24 issues per year. $2390.00 per year. Also available online. ISSN: 0179-7434.

Science Citation Index. SCI. Institute for Scientific Information, 3501 Market Street, Philadelphia, PA 19104. (215) 386-0100. (800) 523-1850. FAX (215) 386-2991. 1961 to present. Six issues per year, plus annual cumulation. $11650.00 per year. Also available online and on CD-ROM. Inquire about price and availability. ISSN: 0036-827X.

ANNUAL REVIEWS AND YEARBOOKS

Advances in Electronics and Electron Physics. Academic Press, Inc., 6277 Sea Harbor Drive, Orlando, FL 32887. (800) 321-5068. From 1948 to present. Irregular. Price varies, inquire. ISSN: 0065-2539.

Critical Reviews in Solid State and Materials Sciences. CRC Press, Inc., 2000 Corporate Boulevard, NW, Boca Raton, FL 33431. (407) 994-0555. (800) 272-7737. FAX (407) 998-9784. 1970 to present. Bimonthly. $265.00 per year. ISSN: 1040-8436.

ASSOCIATIONS AND PROFESSIONAL SOCIETIES

American Electronics Association. 5201 Great America Way, Suite 520, P.O. 52990, Santa Clara, CA 95056. (408) 987-4200. FAX (408) 970-8565.

American Institute of Physics. One Physics Ellipse, College Park, MD 20740-3843. (301) 209-3100.

Edison Electric Institute. 701 Pennsylvania Avenue NW, Washington, DC 20004-2696. (202) 508-5000, 5454. FAX (202) 508-5360.

Electric Power Research Institute. 3412 Hillview Avenue, Palo Alto, CA 94304. (415) 855-2000. FAX (415) 855-2954.

Electronics Industries Association. 2500 Wilson Boulevard, Arlington, VA 22201. (703) 907-7500. FAX (202) 457-4985.

IEEE (institute of Electrical and Electronic Engineers). 345 East 47th Street. New York, NY 10017. (212) 705-7900. FAX (212) 705-4929.

IEEE Power Engineering Society. c/o Institute of Electrical and Electronic Engineers. 345 East 47th Street, New York, NY 10017. (212) 705-7900. FAX (212) 705-4929.

National Electrical Manufacturers Association. 1300 North 17th Street, Suite 1847, Rosslyn VA 22209. (703) 841-3200. FAX (703) 841-3300.

BIBLIOGRAPHIES

Bibliographical History of Electricity and Magnetism. Paul F. Mottelay, editor. Ayer Company Pubs., Inc., Lower Mill Road, North Stratford, NH 03590. (603) 922-5105. (800) 282-5413. Reprint edition. 1975. $52.95.

Science Books and Films. American Association for the Advancement of Science, 1333 H Street NW, Washington, DC 20005. (202) 326-6454. 1965 to present. Nine issues per year. $40.00 per year. ISSN: 0098-342X.

Scientific and Technical Books and Serials in Print; An Index To Literature in Science and Technology. R.R. Bowker Inc., 121 Chanlon Road, New Providence, NJ 07974. (908) 464-6800. (800) 521-8110. FAX (908) 665-3502. 1972 to present. Annual. 4 volumes. 1994. $299.95. Also available on compact disc and online. ISSN: 0000-054X.

DIRECTORIES AND BIOGRAPHICAL SOURCES

American Men and Women of Science: Physical and Biologicalsciences. R. R. Bowker Inc., 121 Chanlon Road, New Providence, NJ 07974. (908) 464-6800. (800) 521-8110. 20th edition. 8 volumes. 1996. $850.00.

Directory of Engineering College Research and Graduate Study. American Society for Engineering Education, Ohio State University, 2070 Neil Avenue, Columbus, OH 43210-1226. Annual. $20.00.

Directory of Engineering Societies and Related Organizations. American Association of Engineering Societies, 1111 19th Street, Suite 608, Washington, DC 20036-3603. (202) 296-2237. Semi-annual. $150.00.

Directory of Engineers in Private Practice. National Society of Professional Engineers, 1420 King Street, Alexandria, VA 22314-2750. (703) 684-2835. Biennial. $30.00.

E E M - Electronic Engineer's Master. Hearst Business Communications, Inc., 645 Stewart Avenue, Garden City NY 11530. (516) 227-1300. ISSN: 0732-9016.

Electrical and Electronic Trades Directory: The Blue Book. Institution of Electrical Engineers, Michael Faraday House, Six Hills Way, Stevenage, Herts. SG1 2AY, England. Telephone: 0438 313311 or FAX 0438 742840. 1995. $140.00

Electrical World Directory of Electrical Utilities. McGraw-Hill Book Company, Inc., 1221 Avenue of the Americas, New York, NY 10020. (212) 997-3675. Annual.

Engineering Research Centers: Incorporating Electronics Research Centers. Stockton Press, 345 Park Avenue, New York, NY 10010. (212) 689-9200. (800) 221-2123. 4th edition. 1995. $515.00.

IEEE Membership Directory. Institute of Electrical and Electronics Engineers, IEEE Service Center, 445 Hoes Lane, Piscataway, NY 08854. (908) 981-1393. (800) 678-IEEE. FAX (908) 981-9667. 2 volumes. Annual. $190.00. ISSN:

International Directory of Abbreviations and Acroynms of Electronics, Electrical Engineering, Computer Technology and Information Processing. Peter Wennrich. K. G. Saur, 121 Chanlon Road, New Providence, NJ 07974. (908) 464-6800. (800) 521-8110 2 volumes. 1992. $230.00.

International Engineering Directory. American Consulting Engineers Council, 1015 15th Street, N.W. Suite 802, Washington, DC 20005-2670. (202) 347-7474. Annual. $10.00.

Research Centers Directory. Gale Research Company Inc., 835 Penobscot Building, Detroit, MI 48226-4094. (313) 961-2242. (800) 347-4253. 20th edition. 1995. $485.00. ISSN: 0080-1518.

Who's Who in Electric Transmission and Distribution. Utility Data Institute, 1200 G Street NY, Suite 250, Washington, DC 20005-3802. (202) 942-8788. (800) 486-3660. 1995. $150.00.

Who's Who in Engineering. Gordon Davis, editor. American Association of Engineering Societies. 1111 19th Street, NY, Suite 608, Washington, DC 20036. (202) 296-2237. (800) 658-8897. 9th edition. 1995. $220.00.

Who's Who in Technology. Gale Research, 835 Penobscot Building, Detroit, MI 48226-4094. (313 961-2242. (800) 521-4253. 7th edition. 1995. $195.00. ISN 0-8103-7467-6.

World Guide to Scientific Associations and Learned Societies. K. G. Saur, 121 Chanlon Road, New Providence, NJ 07974. (908) 464-6800. (800) 521-8110. 5th edition. 1990. $225.00.

ENCYCLOPEDIAS AND DICTIONARIES

Chambers Science and Technology Dictionary. Peter M.B. Walker, editor. Cambridge University Press, 40 West 20th Street, New York, NY 10011-4211. (212) 924-3900. 1988. $39.95.

ELECTRICITY

Ency. of Physical Sciences and Engineering Info. Sources

Encyclopedia of Applied Physics. George Trigg, editor. VCH Publications, Inc., 220 East 23rd Street, Suite 909, New York, NY 10010-4606. (212) 683-8333. (800) 422-8824. 20 volume set. 1991 - . $5990.00.

IEEE Standard Dictionary of Electrical and Electronics Terms. Christopher J. Booth, editor. IEEE Service Center, 445 Hoes Lane, Piscataway, NJ 08855-1331. (908) 981-1393. (800) 678-IEEE. FAX (908) 981-9667. IEEE Standard 100-1992. 5th edition. 1993. $150.00.

Illustrated Dictionary of Electronics. Stan Gibilsco. TAB Books, P.O. Box 40, Blue Summit, PA 17294-0850. (717) 794-2191. (800) 233-1128. 7th edition. 1994. $34.95.

International Encyclopedia of Integrated Circuits. Arthur A. Seidman. TAB Books, P.O. Box 40, Blue Summit, PA 17294-0850. (717) 794-2191. (800) 233-1128. 1991. $75.00.

McGraw-Hill Electronics Dictionary. J. Markus. McGraw-Hill Book Company, Inc., 1221 Avenue of the Americas, New York, NY 10020. (212) 997-3675. 5th edition. 1994. $49.95.

McGraw-Hill Encyclopedia of Science and Technology. McGraw-Hill Book, Incorporated, 1221 Avenue of the Americas, New York, NY 10020. (212) 997-3675. (800) 262-4729. Seventh edition. Twenty volumes. 1992. $1900.00.

GENERAL WORKS

Basic Electrical and Electronic Engineering. E. C. Bell and R. W. Whitehead. Blackwell Scientific Publications, Inc., 238 Main Street, Cambridge MA 02142. (617) 876-7000. (800) 759-6102. 4th revised edition. 1993.

Berkeley Physics Course: Electricity and Magnetism. E. M. Purcell. McGraw-Hill Publishing Company, Inc., 1221 Avenue of the Americas, New York, NY 10020. (212) 512-2000. (800) 262-4729. 2nd edition. volume 2. 1985.

Classical theory of Electricity and Magnetism. A. K. Roychawdjuri. Oxford University Press, Inc., 200 Madison Avenue, New York, NY 10016. (212) 725-6000. (800) 334-4249. 1991. $35.00.

Design and Technology of Integrated Circuits. Donald DeCogan. John Wiley & Sons, Inc., 605 Third Avenue, New York, NY 10158-0012. (212) 850-6000. (800) 225-5945. 1990. $75.00.

Electric and Electronic Engineering For Scientists and Engineers. K. A. Krishnamurthy. Halsted Press, 605 Third Avenue, New York, 10158-0012. (212) 850-6400. 1994. $48.95.

Electricity and Electronics For the Microcomputer Age. Stanley L. Rosen. Prentice Hall , 113 Sylvan Avenue, Route 9W, Englewood Cliffs, NJ 07632. (201) 592-2000. (800) 922-0579. 1987. $43.95.

Electricity: Principles and Applications. Richard J. Fowler. McGraw-Hill Publishing Company, Inc., 1221 Avenue of the Americas, New York, NY 10020. (212) 512-2000. (800) 262-4729. 3rd edition. 1989. $35.00.

Exploring Electronic Devices. Mark E. Hazen. SCP: Third World Literature Publishing House, P.O. Box 482, Lithonia, GA 30058-0482. (404) 785-7725. 1991. $50.75.

High-voltage Engineering and Testing. H. M. Ryan, editor. Institution of Electrical Engineers; IEE/INSPEC, Box 1331, 445 Hoes Lane, Piscataway, NJ 08855-1331. (908) 981-0060. 1994. $95.00.

An Introduction To Applied Electromagetism. Christos Christopoulos. John Wiley & Sons, Inc., 605 Third Avenue, New York, NY 10158-0012. (212) 850-6000. (800) 225-5945. 1990. $59.95.

Introduction To Vlsi Technology. T. E. Price. Prentice Hall, 113 Sylvan Avenue, Route 9W, Englewood Cliffs, NJ 07632. (201) 592-2000. (800) 922-0579. 1994. $65.00.

PRACTICAL ELECTRICITY. Robert G. Middleton. Macmillan Publishing Company, Inc,. 200 Old Tappan Road, Old Tappan, NJ 07675. (800) 233-2336. 4th edition. 1988. $22.50.

HANDBOOKS AND MANUALS

Active Electronic Component Handbook. Charles A. Harper and Harold C. Jones. McGraw-Hill Book Company, 1221 Avenue of the Americas, New York, NY 10020. (212) 512-2000. (800) 262-4729. 2nd edition. 1996. $79.50.

Electrical Engineering Handbook. Richard C. Dorf, editor. CRC Press, Inc., 2000 Corporate Boulevard, NW, Boca Raton, FL 33431. (407) 994-0555. (800) 272-7737. 1993. $99.95.

Electrical Engineer's Handbook. Donald Christiansen, editor. McGraw Book Company Inc., 1221 Avenue of the Americas, New York, NY 10020. (212) 512-2000. (800) 262-4729. 4th edition. 1996. $110.00.

Electronics Handbook. Jerry C. Whitaker. CRC Press, Inc., 2000 Corporate Boulevard, NW, Boca Raton, FL 33431. (407) 994-0555. (800) 272-7737. 1996. $120.00.

Engineering Tables and Data. A.M. Howatson, P.G. Lund and J.D. Todd. Van Nostrand Reinhold, 115 Fifth Avenue, New York, NY 10003. (212) 254-3232. (800) 926-2665. Second Edition. 1991. $39.95.

Handbook of Electrical and Electronic Technology. Curtis Johnson. Prentice Hall , 113 Sylvan Avenue, Route 9W, Englewood Cliffs, NJ 07632. (201) 592-2000. (800) 922-0579. 1996. $88.00.

Industrial Electronics and Systems Handbook. Irwin, CRC Press, Inc., 2000 Corporate Boulevard, NW, Boca Raton, FL

33431. (407) 994-0555. (800) 272-7737. 1996. $129.95.

Newnes Electrical Pocket Book. E. A. Reeves, editor. Butterworth-Heinemann, 313 Washington Street, Newton, MA 02158. (618-928-2500. (800) 366-2665. 21th edition. 1992. $32.95 in paper.

Standard Handbook For Electrical Engineers. Donald Fink. McGraw-Hill Publishing Company, 1221 Avenue of the Americas, New York, NY 10020. (212) 512-2000. (800) 262-4729. 13th edition. 1996. $110.00.

ONLINE DATABASES AND CD-ROMS

CA Search. Chemical Abstracts Service, P.O. Box 3012, Columbus, OH 43210-0012. (614) 447-3600. (800) 848-6533. FAX (614) 447-3709. Very comprehensive guide to worldwide chemical literature and related fields, 1972 to present. Available on BRS,(800) 289-4277, DIALOG, (800) 334-2564, ORBIT (800) 456-7248, and STN International, FIZ Karlsruhe, P.O. Box 2465, W-7500, Karlsruhe 1, Germany, online services. Inquire as to cost and availability.

Compendex Plus. Engineering Information, Inc., 345 East 47th Street, New York, NY 10017. (212) 705-7600 or (800) 221-1044. Contains citations with abstracts to worldwide literature in engineering and technology, from 1970 to present. Available on online BRS,(800) 289-4277, DIALOG, (800) 334-2564, ORBIT (800) 456-7248, and STN International, FIZ Karlsruhe, P.O. Box 2465, W-7500, Karlsruhe 1, Germany, online services. Also available on CD-ROM. Inquire as to cost and availability.

Current Contents Search. Institute for Scientific Information, 3501 Market Street, Philadelphia, PA 19104. (215) 386-0100. FAX (215) 386-6362. Contains citations to articles listed in the table of contents of science and technology journals. Also articles in social sciences and life sciences journals. Available on BRS,(800) 289-4277, DIALOG,(800) 334-2564, online services. Inquire as to cost and availability.

Dissertation Abstracts Online. University Microfilms International, 300 North Zeeb Road, Ann Arbor, MI 48106. (800) 521-0600 or (313) 761-4700. Scope includes virtually all doctoral dissertations accepted at accredited American institutions from 1861 to present in 252 subject areas. Available on BRS, (800) 289-4277, DIALOG, (800) 334-2564, and OCLC EPIC,(800) 848-5878, online services. Also available on CD-ROM. Inquire as to cost and availability.

INSPEC. Institution of Electrical Engineers, Michael Faraday House, Six Hills Way, Stevenage, Herts. SG1 2AY, England. Telephone: 0438 313311 or FAX 0438 742840. Contains citations to the worldwide literature of physics, electronics and electrical engineering, computer technology, and related fields. Available on BRS, (800) 289-4277, DIALOG, (800) 334-2564, ORBIT, (800) 456-7248, and STN International, FIZ Karlsruhe, P.O. Box 2465, W-7500, Karlsruhe 1, Germany, online services. Inquire as to cost and availability.

NTIS Bibliographic Database. National Technical Information Service, 5285 Port Royal Road, Springfield, VA 22161. (703) 487-4929 or FAX (703) 321-8199. Broad coverage of government-sponsored science and technology research reports, 1964 to present. Available on BRS,(800) 289-4277, DIALOG, (800) 334-2564, ORBIT, (800) 456-7248, and STN International, FIZ Karlsruhe, P.O. Box 2465, W-7500, Karlsruhe 1, Germany, online services. Also available on CD-ROM. Inquire as to cost and availability.

Physics Briefs. American Institute of Physics, 335 East 45th Street, New York, NY 10017. (212) 661-9260 or FAX (212) 661-2036. Contains citations with abstracts of the literature of physics and related fields, 1979 to present. Available on the STN International, FIZ Karlsruhe, P.O. Box 2465, W-7500, Karlsruhe 1, Germany, online service. Inquire as to cost and availability.

Scientific and Technical Books and Serials in Print. R.R. Bowker Inc., 121 Chanlon Road, New Providence, NJ 07974. (800) 521-8110 or (908) 464-6800. List of currently published books and serials in the physical and biological sciences, engineering and technology, with subject, author and titles indexes. Available

on ORBIT,(800) 456-7248, online service.Inquire as to cost and availability.

SCISEARCH. Institute for Scientific Information, 3501 Market Street, Philadelphia, PA 19104. (800) 523-1850 or (215) 386-0100. Broad multidisciplinary title and author index to the international literature of science and technology, 1974 to present. Available on DIALOG,(800) 334-2564, and ORBIT,(800) 456-7248, online services. Also available on CD-ROM. Inquire as to cost and availability.

WILSONLINE. H.W. Wilson Company, 950 University Avenue, Bronx, NY 10452. (800) 367-6770 or (212) 588-8400. Makes available online versions of the printed H.W. Wilson indexes including Applied Science and Technology Index, Business Periodicals Index, General Science Index, and Readers' Guide to Periodical Literature. Period covered is generally 1983 to present. Available on BRS,(800) 289-4277, DIALOG,(800) 334-2564, and OCLC EPIC,(800) 848-5878, online services. Also available on CD-ROM. Inquire as to cost and availability.

PERIODICALS

Electric Light and Power. PennWell Publishing Co. 1421 South Sheridan Road, Box 1260, Tulsa OK 74101. (835-3161. FAX (918) 831-9497. 1922 to present. Monthly. $42.00 per year.

Electric Machines and Power Systems. Taylor & Francis, 1900 Frost Road, Suite 101, Bristol, PA 19007-1598. (800) 821-8312, FAX (215) 785-5515. 1976 to present. Monthly. $355.00 per year. ISSN: 0731-356X.

Electrical World. McGraw-Hill, Inc., Box 513 Hightstown, NJ 08520. (212) 512-3288. 1874 to present. Monthly. $55.00 per year. ISSN: 0013-4457.

Electronic Design. Penton Publishing, San Jose Gateway, Suite 354. 2025 Gateway Place, San Jose, CA 95110. (408) 441-0550. 1952 to present. Fortnightly. $95.00. ISSN: 0013-4872.

Electronics. Penton Publishing, San Jose Gateway, Suite 354. 2025 Gateway Place, San Jose, CA 95110. (408) 441-0550. 1930 to present. Semi-weekly. $98.00 per year. ISSN: 0883-4989.

IEEE Power Engineering Review. Institute of Electrical and Electronics Engineers, Inc., Box 1331, 445 Hoes Lane, Piscataway, NJ 08855-1331. (908) 981-0060. 1981 to present. Monthly. $115.00 per year. ISSN: 0272-1724.

IEEE Spectrum. Institute of Electrical and Electronics Engineers, Inc., Box 1331, 445 Hoes Lane, Piscataway, NJ 08855-1331. (908) 981-0060. 1964 to present. Monthly. $157.00. ISSN: 0018-9235.

IEEE Transactions On Power Delivery. Institute of Electrical and Electronics Engineers, Inc., Box 1331, 445 Hoes Lane, Piscataway, NJ 08855-1331. (908) 981-0060. 1986 to present. Quarterly. $170.00 per year. ISSN: 0885-8977.

Institute of Electrical and Electronics Engineers. Proceedings. Institute of Electrical and Electronics Engineers, Inc., Box 1331, 445 Hoes Lane, Piscataway, NJ 08855-1331. (908) 981-0060. 1913 to present. Monthly. $275.00. ISSN: 0018-9219.

Power Engineering (Tulsa). PennWell Publishing Co. 1421 South Sheridan Road, Box 1260, Tulsa OK 74101. (835-3161.

ELECTRICITY

Ency. of Physical Sciences and Engineering Info. Sources

FAX (918) 831-9497. 1896 to present. Monthly. $28.00 per year. ISSN: 0032-5961

RESEARCH CENTERS AND INSTITUTES

Edison Electric Institute. 701 Pennsylvania Avenue, NW, Washington, DC 20004-2696. (202) 508-5000, 5454. FAX (202) 508-5360.

Electroscience Laboratory. Ohio State University, 1320 Kinnear Road, Columbus, Ohio 43212. (614) 292-7981, FAX (614) 292-7297.

Electrical Engineering Research Laboratories. Purdue University. Electrical Engineering Building, West Lafayette, in 47907. (317) 494-3536. FAX (317) 494-6440.

Electrical Power Research Institute. 3412 Hillview Avenue, PO Box 10412, Palo Alto, CA 94303. (415) 855-2000. FAX (415) 855-2954.

Laboratory For Electromagnetic and Electronic Systems. Massachusetts Institute of Technology, 77 Massachusetts Avenue, Cambridge, MA 02139. (617) 253-4631. FAX (617) 258-6774.

ELECTROACOUSTICS

See: ACOUSTICS

ELECTROCHEMISTRY

See also: BATTERIES

ABSTRACT SERVICES AND INDEXES

Applied Science and Technology Index; A Cumulative Subject Index To English Language Periodicals in the Fields of Aeronautics and Space Science, Computer Technology, Chemistry, Construction Industry, Energy and Related Areas. H.W. Wilson Co., 950 University Avenue, Bronx, NY 10452. (212) 588-8400. (800) 367-6770. FAX (718) 590-1617. From 1958 to present. Monthly. Inquire about cost and availability. Also available on CD-ROM and online. ISSN: 0003-6986.

Chemical Abstracts. Chemical Abstracts Service, 2540 Olentangy River Road, Box 3012, Columbus, OH 43210-0012. (614) 447-3600. (800) 848-6538. FAX (614) 447-3713. From 1907 to present. Weekly. $16800.00 per year. Also available on CD-ROM and online. Inquire regarding cost and availability.

Engineering Index Monthly. Engineering Information, Inc., Castle Point on the Hudson, Hoboken, NJ 07030. (201) 216-8500. FAX (201) 216-8532. Monthly. $2200.00 per year. Also available online as COMPENDEX and on CD-ROM. ISSN: 0742-1974.

General Science Index. H.W. Wilson Company, 950 University Avenue, Bronx, NY 10452. (212) 588-8400. (800) 367-6770. FAX (718) 590-1617. From 1978 to present. Ten issues per year; quarterly and annual cumulations. Service basis. Available on

CD-ROM and online. Inquire regarding cost and availability. ISSN: 0162-1963.

Physics Abstracts. INSPEC. Section A, Science Abstracts. Institution of Electrical Engineers (IEE). Available from INSPEC/IEEE - Institute of Electrical and Electronic Engineers,, Box 1331, 445 Hoes Lane, Piscaway NJ 08855-1331. (908) 562-5549. 1898 to present. 24 issues per year. $2835.00 per year. Also available on CD-ROM and online. ISSN: 0036-8091.

Science Citation Index. SCI. Institute for Scientific Information, 3501 Market Street, Philadelphia, PA 19104. (215) 386-0100. (800) 523-1850. FAX (215) 386-2991. 1961 to present. Six issues per year, plus annual cumulation. $11650.00 per year. Also available online and on CD-ROM. Inquire about price and availability. ISSN: 0036-827X.

ASSOCIATIONS AND PROFESSIONAL SOCIETIES

American Chemical Society. 1155 16th Street, NW, Washington DC 20036. (202) 872-4600.

American Institute of Physics. One Physics Ellipse, College Park, MD 20740-3843. (301) 209-3100.

Association of Consulting Chemists and Chemical Engineers. 40 West 45th Street, New York, NY 10036. (212) 983-3160.

Battery Council International. 401 North Michigan Drive, Chicago, IL 60611. (312) 644-6610.

Electrochemical Society. 10 South Main Street, Pennington, NJ 08534. (609) 737-1902.

BIBLIOGRAPHY

Electrochemical Synthesis of Inorganic Compounds: A Bibliography. Zolan Nagy, editor. Plenum Publishing Corp., 233 Spring Street, New York, NY. (212) 620-8000. (800) 2221-9369. FAX (212) 463-0742. 1985. $125.00.

DIRECTORIES AND BIOGRAPHICAL SOURCES

American Men and Women of Science: Physical and Biological Sciences. R. R. Bowker Inc., 121 Chanlon Road, New Providence, NJ 07974. (908) 464-6800. (800) 521-8110. 20th edition. 8 volumes. 1996. $850.00.

Electrochemical Society Membership Directory. Electrochemical Society, 10 South Main Street, Pennington, NJ 08534. (609) 737-1902. Annual. Available to members only.

Research Centers Directory. Gale Research Inc., 835 Penobscot Building, Detroit, MI 48226-4094. (313) 961-2242. (800) 347-4253. 20th edition. 1995. $485.00. ISSN: 0080-1518.

Scientific and Technical Organizations and Agencies Directory. Gale Research, 835 Penobscot Building, Detroit, MI 48226-4094. (313) 961-2242. (800) 347-4253. 4th edition. 1996. $195.00.

Who's Who in Engineering. Gordon Davis, editor. American Society for Engineering Education, 1111 19th Street, NW, Suite 608, Washington, DC 20036. 9th edition. 1995. $220.00.

Who's Who in Technology. Gale Research, 835 Penobscot Building, Detroit, MI 48226-4094. (313) 961-2242. (800) 347-4253. 7th edition. 1995. $195.00.

ENCYCLOPEDIAS AND DICTIONARIES

Dictionary of Electrochemistry. D. B. Hibbert and A. M. James. John Wiley & Sons, Inc., 605 Third Avenue, New York, NY 10158-0012. (212) 850-6000. (800) 225-5945. 2nd edition. 1985.

Encyclopedia of Applied Physics. George Trigg, editor. VCH Publications, Inc., 220 East 23rd Street, Suite 909, New York, NY 10010-4606. (212) 683-8333. (800) 422-8824. 20 volume set. 1991-. $5990.00.

McGraw-Hill Encyclopedia of Science and Technology. McGraw-Hill Book Company, Inc., 1221 Avenue of the Americas, New York, NY 10020. (212) 512-2000. (800) 262-4729. 7th edition. 20 volume set. 1992. $1900.00.

Modern Power Suply and Battery Charger Circuit Encyclopedia. Rudolf F. Graf. McGraw-Hill Book Company, Inc., 1221 Avenue of the Americas, New York, NY 10020. (212) 512-2000. (800) 262-4729. 1992. $11.95 in paper.

GENERAL WORKS

Battery Hazards and Accident Prevention. S. C. Levy and P. Bro. Plenum Publishing Corp., 233 Spring Street, New York, NY. (212) 620-8000. (800) 2221-9369. FAX (212) 463-0742. 1994. $49.50.

Electrochemical Cell Design. R. White, editor. Plenum Publishing Corp., 233 Spring Street, New York, NY. (212) 620-8000. (800) 2221-9369. FAX (212) 463-0742. 1984. $95.00.

Electrochemical Engineering Principles. Geoffrey A. Prentice. Prentice Hall, 113 Sylvan Avenue, Route 9W, Englewood Cliffs, NJ 07632. (201) 592-2000. (800) 922-0579. 1990. $78.00.

Electrochemical Reaction Engineering. K. Scott. Academic Press, Inc., 6277 Sea Harbor Drive, Orlando, FL. (800) 321-5068. 1991. $143.00.

Fundamentals of Electrochemical Science. Keith Oldham and Jan C. Myland. Academic Press, Inc., 6277 Sea Harbor Drive, Orlando, FL. (800) 321-5068. 1993. $49.95.

Fuel Cell Systems. L. J. Blomen and M. N. Bugerwa, editors. Plenum Publishing Corp., 233 Spring Street, New York, NY. (212) 620-8000. (800) 2221-9369. FAX (212) 463-0742. 1994. $125.00.

Hydrogen Storage Materials, Batteries and Electrochemistry. D. A. Corrigan and S. Srivivasan, editors. Electrochemical Society, 10 South Main Street, Pennington, NJ 08534. (609) 737-1902. 1992. $43.00.

Modern Battery Technology. Clive D. Tuck. Prentice Hall, 113 Sylvan Avenue, Route 9W, Englewood Cliffs, NJ 07632. (201) 592-2000. (800) 922-0579. 1991. $42.00.

HANDBOOKS AND MANUALS

Battery Reference Book. T. R. Compton. Butterworth-Heinemann, 313 Washington Street, Newton, MA 02158. (618-928-2500. (800) 366-2665. 1990. $195.00.

Battery Technology Handbook. Kiehne. Marcel Dekker, Inc., 270 Madison Avenue, New York, NY 10016. (212) 696-9000. (800) 228-1160. 1989. $165.00.

CRC Handbook of Chemistry and Physics. David R. Kide, editor. CRC Press, Inc., 2000 Corporate Boulevard, NW, Boca Raton, FL 33431. (407) 994-0555. (800) 272-7737. 77th edition. 1996. $99.95.

Handbook of Batteries. David Linden. McGraw-Hill Publishing Company, Inc., 1221 Avenue of the Americas, New York, NY 10020. (212) 512-2000. (800) 262-4729. 2nd edition. 1995. $125.00.

Handbook of Solid State Batteries and Capacitors. M. Z. Munshi and P. S. Prasad. World Scientific Publishing Company, Inc., 1060 Main Street, River Edge, NJ 07661. (201) 487-9655. (800) 227-7562. 1995. $177.00.

Photovoltaic Engineering Handbook. F. Lasnier and T. Ang. IOP Publishing, Public Ledger Building, Suite 1035, Independence Square, Philadelphia, PA 19106. (215) 627-0880. (800) 358-4677. 1990. $176.00.

Rechargeable Batteries Applications Handbook. Gates Butterworth-Heinemann, 313 Washington Street, Newton, MA 02158. (618-928-2500. (800) 366-2665. 1992. $49.95.

ONLINE DATABASES

CA Search. Chemical Abstracts Service, P.O. Box 3012, Columbus, OH 43210-0012. (614) 447-3600. (800) 848-6533. FAX (614) 447-3709. Very comprehensive guide to worldwide chemical literature and related fields, 1972 to present. Available on BRS,(800) 289-4277, DIALOG, (800) 334-2564, ORBIT (800) 456-7248, and STN International, FIZ Karlsruhe, P.O. Box 2465, W-7500, Karlsruhe 1, Germany, online services. Inquire as to cost and availability.

Compendex Plus. Engineering Information, Inc., 345 East 47th Street, New York, NY 10017. (212) 705-7600 or (800) 221-1044. Contains citations with abstracts to worldwide literature in engineering and technology, from 1970 to present. Available on online BRS,(800) 289-4277, DIALOG, (800) 334-2564, ORBIT (800) 456-7248, and STN International, FIZ Karlsruhe, P.O. Box 2465, W-7500, Karlsruhe 1, Germany, online services. Also available on CD-ROM. Inquire as to cost and availability.

INSPEC. Institution of Electrical Engineers, Michael

Faraday House, Six Hills Way, Stevenage, Herts. SG1 2AY, England. Telephone: 0438 313311 or FAX 0438 742840. Contains citations to the worldwide literature of physics, electronics and electrical engineering, computer technology, and related fields. Available on BRS, (800) 289-4277, DIALOG, (800) 334-2564, ORBIT, (800) 456-7248, and STN International, FIZ Karlsruhe, P.O. Box 2465, W-7500, Karlsruhe 1, Germany, online services. Inquire as to cost and availability.

NTIS Bibliographic Database. National Technical Information Service, 5285 Port Royal Road, Springfield, VA 22161. (703)

ELECTROCHEMISTRY

Ency. of Physical Sciences and Engineering Info. Sources

487-4929 or FAX (703) 321-8199. Broad coverage of government-sponsored science and technology research reports, 1964 to present. Available on BRS,(800) 289-4277, DIALOG, (800) 334-2564, ORBIT, (800) 456-7248, and STN International, FIZ Karlsruhe, P.O. Box 2465, W-7500, Karlsruhe 1, Germany, online services. Also available on CD-ROM. Inquire as to cost and availability.

SCISEARCH. Institute for Scientific Information, 3501 Market Street, Philadelphia, PA 19104. (800) 523-1850 or (215) 386-0100. Broad multidisciplinary title and author index to the international literature of science and technology, 1974 to present. Available on DIALOG, (800) 334-2564, and ORBIT, (800) 456-7248, online services. Also available on CD-ROM. Inquire as to cost and availability.

WILSONLINE. H.W. Wilson Company, 950 University Avenue, Bronx, NY 10452. (800) 367-6770 or (212) 588-8400. Makes available online versions of the printed H.W. Wilson indexes including Applied Science and Technology Index, Business Periodicals Index, General Science Index, and Readers' Guide to Periodical Literature. Period covered is generally 1983 to present. Available on BRS, (800) 289-4277, DIALOG, (800) 334-2564, and OCLC EPIC, (800) 848-5878, online services. Also available on CD-ROM. Inquire as to cost and availability.

PERIODICALS

Electrochemical Society Journal. Electrochemical Society, 10 South Main Street, Pennington, NJ 08534-2896. (609) 737-1902. 1902 to present. Monthly. $400.00 per year. ISSN: 0013-4651.

Electrochemica Acta. Elsevier Science, 660 White Plains Road, Tarrytown, NY 10591-5151. (914) 524-9200. FAX (914) 333-2444. 18 issues per year. $1650.00 per year. ISSN: 0013-4686.

Electronic Chemicals News. Chemical Week Associates, 888 Seventh Avenue, New York, NY 10106. (212) 621-4900. 1986 to present. Semi-monthly. $452.00 per year. ISSN: 0886-5671.

RESEARCH CENTERS AND INSTITUTES

Battery Technology Center. Carnegie Mellon University, 4400 Fifth Avenue, Mellon Institute, Pittsburgh, PA 15213-2683. (412) 268-3388. FAX (412) 268-6945.

Case Center For Electrochemical Science. Case Western Reserve University. Cleveland, OH 44106. (216) 368-3626. FAX (216) 368-4874.

Electrochemical Analysis Diagnostic Laboratory. Argonne National Laboratory, 9700 South Cass Avenue, Argonne, IL 60439. (708) 972-7764. FAX (708) 972-4176.

Electrochemistry and Enthalpimetric Analysis Laboratory. Pennsylvania State University, 428B Davey Laboratory Building, University Park, PA 16802. (814) 865-2022. FAX (814) 865-3314.

Laboratory For Electrochemical Research. University of Texas at Austin, Austin TX 78712. (512) 471-3761.

International Lead-zinc Research Organization. 2525 Meridian Parkway, PO Box 12036, Research Triangle, NC 27709. (919) 361-4647. (919) 361-1957.

Lawrence Berkeley Laboratory. Applied Science Division, 1 Cyclotron Road, Berkeley, CA 94720. (415) 486-5001. FAX (415) 486-5172.

ELECTRODES

See: ELECTROCHEMISTRY

ELECTRODYNAMICS

See: ELECTRICITY

ELECTROLYSIS

See: ELECTROCHEMISTRY

ELECTROLYTES

See: ELECTROCHEMISTRY

ELECTROMAGNETISM

See also: ELECTRIC MOTORS, ELECTRIC POWER ENGINEERING, ELECTRICAL ENGINEERING, ELECTRICITY, GENERATORS, PHYSICS

ABSTRACT SERVICES AND INDEXES

Applied Mechanics Reviews: an Assessment of World Literature in Engineering Sciences. American Society of Mechanical Engineers, 345 East 47th Street, New York, NY 10017. (212) 705-7703. 1948 to present. Monthly. $360.00 ISSN: 0003-6900.

Applied Science and Technology Index: A Cumulative Subject Index To English Language Periodicals in the Fields of Aeronautics and Space Science, Computer Technology, Chemistry, Construction Industry, Energy and Related Areas. H.W. Wilson Co., 950 University Avenue, Bronx, NY 10452. (212) 588-8400. (800) 367-6770. FAX (718) 590-1617. From 1958 to present. Monthly. Inquire about cost and availability. Also available on CD-ROM and online. ISSN: 0003-6986.

Current Contents: Engineering, Technology and Applied Sciences. Institute for Scientific Information, 3501 Market Street, Philadelphia, PA 19104. (215) 386-0100. FAX (215) 386-6362. 1970 to present. Weekly. $442.00 per year. Also available on CD-ROM and online. Inquire regarding cost and availability. ISSN: 0095-7917.

Current Papers in Electrical and Electronics Engineering. Institution of Electrical Engineers (IEE), London. Distributed by INSPEC/IEEE, Box 1331, 445 Hoes Lane, Piscataway, NJ 08855-1331. (908) 562-5549. 1969 to present. Monthly. $345.00 per year. ISSN: 0011-3778.

Electric Power Industry Abstracts. Edison Electric Institute, c/o Utility Data Institute, 2011 I Atreet, NW, Suite 700, Washington, DC 20006. 1975 to present. Bimonthly. Inquire as to cost and availability.

Electrical and Electronics Abstracts. Institution of Electrical Engineers (IEE), London. Available from INSPEC/IEEE - Institute of Electrical and Electronic Engineers, Box 1331, Hoes Lane, Piscataway, NJ 08855-1331. (908) 562-5549. 1898 to present. Monthly. $2200.00 per year. Also available on CD-ROM and online as INSPEC. ISSN: 0036-8105.

Engineering Index Monthly; Indexes and Abstracts the World's Engineering and Technical Literature. Engineering Information, Inc., Castle Point on the Hudson, Hoboken, NJ 07030. (201) 216-8500. (800) 221-1044. FAX (201) 216-8532. Monthly. $2300.00 per year. Available online as COMPENDEX and also on CD-ROM. ISSN: 0742-1974.

General Science Index. H.W. Wilson Company, 950 University Avenue, Bronx, NY 10452. (212) 588-8400. (800) 367-6770. FAX (718) 590-1617. From 1978 to present. Ten issues per year; quarterly and annual cumulations. Service basis. Available on CD-ROM and online. Inquire regarding cost and availability. ISSN: 0162-1963.

Government Reports Announcements and Index. U. S. National Technical Information Service (NTIS), 5285 Port Royal Road, Springfield, VA 22161. (703) 487-4650. FAX (703) 321-8547. From 1968 to present. Annual. $630.00 per year. Also available online as *NTIS Bibliographic Database* and on CD-ROM.

Index To IEEE Publications. IEEE Service Center, 445 Hoes Lane, Piscataway, NJ 08855-1331. (908) 981-1393. (800) 678-IEEE. FAX (908) 981-9667. 1973 to present. Annual. ISSN: 0099-1368.

Physics Abstracts. INSPEC. Section A, Science Abstracts. Institution of Electrical Engineers (IEE), London. Available from: INSPEC/IEEE - Institute of Electrical and Electronic Engineers, Box 1331, Hoes Lane, Piscataway, NJ 08855-1331. (908) 562-5549. 1898 to present. 24 issues per year. $2835.00 per year. Also available online and on CD-ROM. ISSN: 0036-8091.

Physics Briefs (Physikalische Berichte). Information Center for Energy, Physics, Mathematics; German Physical Society. V C H Publishers, Inc., 220 East 23rd Street, New York, NY 10010-4606. (212) 683-8333. 1845 to present. 24 issues per year. $2390.00 per year. Also available online. ISSN: 0179-7434.

Science Citation Index. SCI. Institute for Scientific Information, 3501 Market Street, Philadelphia, PA 19104. (215) 386-0100. (800) 523-1850. FAX (215) 386-2991. 1961 to present. Six issues per year, plus annual cumulation. $11650.00 per year. Also available online and on CD-ROM. Inquire about price and availability. ISSN: 0036-827X.

ANNUAL REVIEWS AND YEARBOOKS

Advances in Applied Mechanics. Academic Press, Inc., 6277 Sea Harbor Drive, Orlando, FL 32887. (619) 231-0926. (800) 321-5068. FAX (619) 699-6715. 1948 to present. Irregular. Price varies, inquire. ISSN: 0065-2156.

Advances in Electronics and Electron Physics. Academic Press, Inc., 6277 Sea Harbor Drive, Orlando, FL 32887. (800) 321-

5068. From 1948 to present. Irregular. Price varies, inquire. ISSN: 0065-2539.

ASSOCIATIONS AND PROFESSIONAL SOCIETIES

American Electronics Association. 5201 Great America Way, Suite 520, P.O. 52990, Santa Clara, CA 95056. (408) 987-4200. FAX (408) 970-8565.

American Institute of Physics. One Physics Ellipse, College Park, MD 20740-3843. (301) 209-3100.

Edison Electric Institute. 701 Pennsylvania Avenue NW, Washington, DC 20004-2696. (202) 508-5000, 5454. FAX (202) 508-5360.

Electric Power Research Institute. 3412 Hillview Avenue, Palo Alto, CA 94304. (415) 855-2000. FAX (415) 855-2954.

Electromagnetic Energy Association. 1255 23rd Street NW Suite 850, Washington, DC 20017, (202) 452-1070.

Electronics Industries Association. 2500 Wilson Boulevard, Arlington, VA 22201. (703) 907-7500. FAX (202) 457-4985.

IEEE (institute of Electrical and Electronic Engineers). 345 East 47th Street. New York, NY 10017. (212) 705-7900. FAX (212) 705-4929.

IEEE Power Engineering Society. c/o Institute of Electrical and Electronic Engineers. 345 East 47th Street, New York, NY 10017. (212) 705-7900. FAX (212) 705-4929.

National Association of Power Engineers. 1 Springfield Street, Chicopee, MA 01013. (413) 592-6273. FAX (413) 592-1998.

National Electrical Manufacturers Association. 1300 North 17th Street, Suite 1847, Rosslyn VA 22209. (703) 841-3200. FAX (703) 841-3300.

SMMA--The Association For Electric Motors. 4 Hollis Street, PO Box 378, Sherborn, MA 01770. (508) 655-4409.

DIRECTORIES AND BIOGRAPHICAL SOURCES

American Electronics Association. Directory. American Electronics Association, 5201 Great America Way, Suite 520, P.O. 52990, Santa Clara, CA 95056. (408) 987-4200. FAX (408) 970-8565. Annual. $175.00.

American Men and Women of Science: Physical and Biological Sciences. R. R. Bowker Inc., 121 Chanlon Road, New Providence, NJ 07974. (908) 464-6800. (800) 521-8110. 20th edition. 8 volumes. 1996. $850.00.

Directory of Engineering Societies and Related Organizations. American Association of Engineering Societies, 1111 19th Street, Suite 608, Washington, DC 20036-3603. (202) 296-2237. Semi-annual. $150.00.

E E M - Electronic Engineer's Master. Hearst Business Communications, Inc., 645 Stewart Avenue, Garden City NY 11530. (516) 227-1300. ISSN: 0732-9016.

ELECTROMAGNETISM

Ency. of Physical Sciences and Engineering Info. Sources

Electrical and Electronic Trades Directory: The Blue Book. Institution of Electrical Engineers, Michael Faraday House, Six Hills Way, Stevenage, Herts. SG1 2AY, England. Telephone: 0438 313311 or FAX 0438 742840. 1995. $140.00

Engineering Research Centers: Incorporating Electronics Research Centers. Stockton Press, 345 Park Avenue, New York, NY 10010. (212) 689-9200. (800) 221-2123. 4th edition. 1995. $515.00.

IEEE Membership Directory. Institute of Electrical and Electronics Engineers, IEEE Service Center, 445 Hoes Lane, Piscataway, NY 08854. (908) 981-1393. (800) 678-IEEE. FAX (908) 981-9667. 2 volumes. Annual. $190.00.

International Engineering Directory. American Consulting Engineers Council, 1015 15th Street, N.W. Suite 802, Washington, DC 20005-2670. (202) 347-7474. Annual. $10.00.

Research Centers Directory. Gale Research Company Inc., 835 Penobscot Building, Detroit, MI 48226-4094. (313) 961-2242. (800) 347-4253. 20th edition. 1995. $485.00. ISSN: 0080-1518.

Scientific and Technical Organizations and Agencies Directory. Gale Research, 835 Penobscot Building, Detroit, MI 48226-4094. (313) 961-2242. (800) 347-4253. 4th edition. 1996. $195.00.

Who's Who in Electric Transmission and Distribution. Utility Data Institute, 1200 G Street NY, Suite 250, Washington, DC 20005-3802. (202) 942-8788. (800) 486-3660. 1995. $150.00.

Who's Who in Engineering. Gordon Davis, editor. American Association of Engineering Societies. 1111 19th Street, NY, Suite 608, Washington, DC 20036. (202) 296-2237. (800) 658-8897. 9th edition. 1995. $220.00.

Who's Who in Technology. Gale Research, 835 Penobscot Building, Detroit, MI 48226-4094. (313 961-2242. (800) 521-4253. 7th edition. 1995. $195.00. ISN 0-8103-7467-6.

ENCYCLOPEDIAS AND DICTIONARIES

Chambers Science and Technology Dictionary. Peter M.B. Walker, editor. Cambridge University Press, 40 West 20th Street, New York, NY 10011-4211. (212) 924-3900. 1988. $39.95.

Encyclopedia of Applied Physics. George Trigg, editor. VCH Publications, Inc., 220 East 23rd Street, Suite 909, New York, NY 10010-4606. (212) 683-8333. (800) 422-8824. 20 volume set. 1991 - . $5990.00.

IEEE Standard Dictionary of Electrical and Electronics Terms. Christopher J. Booth, editor. IEEE Service Center, 445 Hoes Lane, Piscataway, NJ 08855-1331. (908) 981-1393. (800) 678-IEEE. FAX (908) 981-9667. IEEE Standard 100-1992. 5th edition. 1993. $150.00.

Illustrated Dictionary of Electronics. Stan Gibilsco. TAB Books, P.O. Box 40, Blue Summit, PA 17294-0850. (717) 794-2191. (800) 233-1128. 7th edition. 1994. $34.95.

McGraw-Hill Electronics Dictionary. J. Markus. McGraw-Hill Book Company, Inc., 1221 Avenue of the Americas, New York, NY 10020. (212) 997-3675. 5th edition. 1994. $49.95.

McGraw-Hill Encyclopedia of Science and Technology. McGraw-Hill Book, Incorporated, 1221 Avenue of the Americas, New York, NY 10020. (212) 997-3675. (800) 262-4729. Seventh edition. Twenty volumes. 1992. $1900.00.

GENERAL WORKS

Basic Electrical and Electronic Engineering. E. C. Bell and R. W. Whitehead. Blackwell Scientific Publications, Inc., 238 Main Street, Cambridge MA 02142. (617) 876-7000. (800) 759-6102. 4th revised edition. 1993.

Basic Electrical Power and Machines. David Bradley. Chapman & Hall, 1 Penn Plaza, New York, NY 10119. (212) 564-1060. 1994. $30.95.

Classical Electromagnetic Radiation. James B. Marion and Mark A. Heald. SCP: Third World Literature Publishing House, P.O. Box 482, Lithonia, GA 30058-0482. (404) 785-7725. 2nd edition. 1980. $28.50 in paper.

Design of Small Electrical Machines. H. S. Hamdi. John Wiley & Sons, Inc., 605 Third Avenue, New York, NY 10158-0012. (212) 850-6000. (800) 225-5945. 1994. $67.95

Electrical Machines, Drives and Power Systems. Theodore Wildi. Prentice Hall, 113 Sylvan Avenue, Route 9W, Englewood Cliffs, NJ 07632. (201) 592-2000. (800) 922-0579. 2nd edition. 1990. $79.75.

Electrical Power Engineering. Henslay W. Kabisama. McGraw-Hill Publishing Company, Inc., 1221 Avenue of the Americas, New York, NY 10020. (212) 512-2000. (800) 262-4729. 1993. $55.00.

Electromagnetics and Computer Techniques. Rector Press, Ltd., 130 Rattlesnake, Leverett, MA 01054-9726. (413) 548-9708. (800) 247-3473. 1994, $195.00 in paper.

Electromagnetism For Electronic Engineers. R. G. Carter. Chapman & Hall, 1 Penn Plaza, New York, NY 10119. (212) 564-1060. 2nd edition. 1992. $24.95 in paper.

Elements of Engineering Electromagnetism. N. N. Rao. Prentice Hall , 113 Sylvan Avenue, Route 9W, Englewood Cliffs, NJ 07632. (201) 592-2000. (800) 922-0579. 4th edition. 1993. $74.00.

Essentials of Electromagnetism. David E. Dugdale. American Institute of Physics. One Physics Ellipse, College Park, MD 20740-3843. (301) 209-3100. 1993. $55.00.

An *Introduction to Applied Electromagetism.* Christos Christopoulos. John Wiley & Sons, Inc., 605 Third Avenue, New York, NY 10158-0012. (212) 850-6000. (800) 225-5945. 1990. $59.95.

Magneto-hydro-dynamic Electrical Power Generation. Hugo K. Messerie. John Wiley & Sons, Inc., 605 Third Avenue, New York, NY 10158-0012. (212) 850-6000. (800) 225-5945. 1994. $34.95 in paper.

Schaum's Outline of Electromagnetism. Joseph Edminster. McGraw-Hill Publishing Company, Inc., 1221 Avenue of the Americas, New York, NY 10020. (212) 512-2000. (800) 262-4729. 1993. $12.95 in paper.

Ency. of Physical Sciences and Engineering Info. Sources

ELECTROMAGNETISM

HANDBOOKS AND MANUALS

American Electricians Handbook. Terrell Croft and Wilford I. Summers. McGraw-Hill Book Company, 1221 Avenue of the Americas, New York, NY 10020. (212) 512-2000. (800) 262-4729. 12th edition. 1992. $77.50.

Complete Handbook of Electric Motor Controls. John Traister. Prentice Hall, 113 Sylvan Avenue, Route 9W, Englewood Cliffs, NJ 07632. (201) 592-2000. (800) 922-0579. 2nd edition. 1994. $69.00.

Electrical Engineer's Handbook. Donald Christiansen, editor. McGraw Book Company, 1221 Avenue of the Americas, New York, NY 10020. (212) 512-2000. (800) 262-4729. 4th edition. 1996. $110.00.

Electronics Handbook. Jerry C. Whitaker. CRC Press, Inc., 2000 Corporate Boulevard, NW, Boca Raton, FL 33431. (407) 994-0555. (800) 272-7737. 1996. $120.00.

Handbook of Antenna Design. A. W. Rudge, et al, editors.

Institution of Electrical Engineers, Michael Faraday House, Six Hills Way, Stevenage, Herts. SG1 2AY, England. Telephone: 0438 313311 or FAX 0438 742840. Electromagnetic waves series, 15 and 16. 2 volumes in 1. 1986. $165.00.

Handbook of Electric Motors. John E. Traister. Prentice-Hall, 113 Sylvan Avenue, Route 9W, Englewood Cliffs, NJ 07632. (201) 592-2000. (800) 922-0579. 1992. $64.00.

Handbook of Electrical and Electronic Technology. Curtis Johnson. Prentice Hall , 113 Sylvan Avenue, Route 9W, Englewood Cliffs, NJ 07632. (201) 592-2000. (800) 922-0579. 1996. $88.00.

Handbook of Electronics Manufacturing Engineering. Bernard S. Matisoff. Chapman & Hall, 1 Penn Plaza, New York, NY 10119. (212) 564-1060. 1996.

Standard Handbook For Electrical Engineers. Donald Fink. McGraw-Hill Publishing Company, 1221 Avenue of the Americas, New York, NY 10020. (212) 512-2000. (800) 262-4729. 13th edition. 1996. $110.00.

ONLINE DATABASES AND CD-ROMS

CA Search. Chemical Abstracts Service, P.O. Box 3012, Columbus, OH 43210-0012. (614) 447-3600. (800) 848-6533. FAX (614) 447-3709. Very comprehensive guide to worldwide chemical literature and related fields, 1972 to present. Available on BRS,(800) 289-4277, DIALOG, (800) 334-2564, ORBIT (800) 456-7248, and STN International, FIZ Karlsruhe, P.O. Box 2465, W-7500, Karlsruhe 1, Germany, online services. Inquire as to cost and availability.

Compendex Plus. Engineering Information, Inc., 345 East 47th Street, New York, NY 10017. (212) 705-7600 or (800) 221-1044. Contains citations with abstracts to worldwide literature in engineering and technology, from 1970 to present. Available on online BRS,(800) 289-4277, DIALOG, (800) 334-2564, ORBIT (800) 456-7248, and STN International, FIZ Karlsruhe, P.O. Box 2465, W-7500, Karlsruhe 1, Germany, online services. Also available on CD-ROM. Inquire as to cost and availability.

Current Contents Search. Institute for Scientific Information, 3501 Market Street, Philadelphia, PA 19104. (215) 386-0100. FAX (215) 386-6362. Contains citations to articles listed in the table of contents of science and technology journals. Also articles in social sciences and life sciences journals. Available on BRS,(800) 289-4277, DIALOG,(800) 334-2564, online services. Inquire as to cost and availability.

Dissertation Abstracts Online. University Microfilms International, 300 North Zeeb Road, Ann Arbor, MI 48106. (800) 521-0600 or (313) 761-4700. Scope includes virtually all doctoral dissertations accepted at accredited American institutions from 1861 to present in 252 subject areas. Available on BRS, (800) 289-4277, DIALOG, (800) 334-2564, and OCLC EPIC,(800) 848-5878, online services. Also available on CD-ROM. Inquire as to cost and availability.

INSPEC. Institution of Electrical Engineers, Michael Faraday House, Six Hills Way, Stevenage, Herts. SG1 2AY, England. Telephone: 0438 313311 or FAX 0438 742840. Contains citations to the worldwide literature of physics, electronics and electrical engineering, computer technology, and related fields. Available on BRS, (800) 289-4277, DIALOG, (800) 334-2564, ORBIT, (800) 456-7248, and STN International, FIZ Karlsruhe, P.O. Box 2465, W-7500, Karlsruhe 1, Germany, online services. Inquire as to cost and availability.

NTIS Bibliographic Database. National Technical Information Service, 5285 Port Royal Road, Springfield, VA 22161. (703) 487-4929 or FAX (703) 321-8199. Broad coverage of government-sponsored science and technology research reports, 1964 to present. Available on BRS,(800) 289-4277, DIALOG, (800) 334-2564, ORBIT, (800) 456-7248, and STN International, FIZ Karlsruhe, P.O. Box 2465, W-7500, Karlsruhe 1, Germany, online services. Also available on CD-ROM. Inquire as to cost and availability.

Physics Briefs. American Institute of Physics, 335 East 45th Street, New York, NY 10017. (212) 661-9260 or FAX (212) 661-2036. Contains citations with abstracts of the literature of physics and related fields, 1979 to present. Available on the STN International, FIZ Karlsruhe, P.O. Box 2465, W-7500, Karlsruhe 1, Germany, online service. Inquire as to cost and availability.

Scientific and Technical Books and Serials in Print. R.R. Bowker Inc., 121 Chanlon Road, New Providence, NJ 07974. (800) 521-8110 or (908) 464-6800. List of currently published books and serials in the physical and biological sciences, engineering and technology, with subject, author and titles indexes. Available on ORBIT,(800) 456-7248, online service.Inquire as to cost and availability.

SCISEARCH. Institute for Scientific Information, 3501 Market Street, Philadelphia, PA 19104. (800) 523-1850 or (215) 386-0100. Broad multidisciplinary title and author index to the international literature of science and technology, 1974 to present. Available on DIALOG,(800) 334-2564, and ORBIT,(800) 456-7248, online services. Also available on CD-ROM. Inquire as to cost and availability.

WILSONLINE. H.W. Wilson Company, 950 University Avenue, Bronx, NY 10452. (800) 367-6770 or (212) 588-8400. Makes available online versions of the printed H.W. Wilson indexes including Applied Science and Technology Index, Business Periodicals Index, General Science Index, and Readers' Guide to Periodical Literature. Period covered is generally 1983 to present. Available on BRS,(800) 289-4277, DIALOG,(800) 334-2564, and OCLC EPIC,(800) 848-5878, online services. Also available on CD-ROM. Inquire as to cost and availability.

ELECTROMAGNETISM

Ency. of Physical Sciences and Engineering Info. Sources

PERIODICALS

Electric Light and Power. PennWell Publishing Co. 1421 South Sheridan Road, Box 1260, Tulsa OK 74101. (835-3161. FAX (918) 831-9497. 1922 to present. Monthly. $42.00 per year.

Electric Machines and Power Systems. Taylor & Francis, 1900 Frost Road, Suite 101, Bristol, PA 19007-1598. (800) 821-8312, FAX (215) 785-5515. 1976 to present. Monthly. $355.00 per year. ISSN: 0731-356X.

Electrical World. McGraw-Hill, Inc., Box 513 Hightstown, NJ 08520. (212) 512-3288. 1874 to present. Monthly. $55.00 per year. ISSN: 0013-4457.

Electronic Design. Penton Publishing, San Jose Gateway, Suite 354. 2025 Gateway Place, San Jose, CA 95110. (408) 441-0550. 1952 to present. Fortnightly. $95.00. ISSN: 0013-4872.

Electronics. Penton Publishing, San Jose Gateway, Suite 354. 2025 Gateway Place, San Jose, CA 95110. (408) 441-0550. 1930 to present. Semi-weekly. $98.00 per year. ISSN: 0883-4989.

IEEE Power Engineering Review. Institute of Electrical and Electronics Engineers, Inc., Box 1331, 445 Hoes Lane, Piscataway, NJ 08855-1331. (908) 981-0060. 1981 to present. Monthly. $115.00 per year. ISSN: 0272-1724.

IEEE Transactions On Electromagnetic Compatability. Institute of Electrical and Electronics Engineers, Inc., Box 1331, 445 Hoes Lane, Piscataway, NJ 08855-1331. (908) 981-0060. 1959 to present. Quarterly. $93.00 per year. ISSN: 0018-9375.

IEEE Transactions On Magentics. Institute of Electrical and Electronics Engineers, Inc., Box 1331, 445 Hoes Lane, Piscataway, NJ 08855-1331. (908) 981-0060. 1965 to present. Bi-monthly. $263.00 per year. ISSN: 0018-9464.

IEE Proceedings: Electric Power Applications. Institution of Electrical Engineers (IEE), London. Available from INSPEC/ IEEE - Institute of Electrical and Electronic Engineers, Box 1331, Hoes Lane, Piscataway, NJ 08855-1331. (908) 562-5549. 1980 to present. Bi-monthly. ISSN: 0143-7038.

Institute of Electrical and Electronics Engineers. PROCEEDINGS. Institute of Electrical and Electronics Engineers, Inc., Box 1331, 445 Hoes Lane, Piscataway, NJ 08855-1331. (908) 981-0060. 1913 to present. Monthly. $275.00. ISSN: 0018-9219.

RESEARCH CENTERS AND INSTITUTES

Edison Electric Institute. 701 Pennsylvania Avenue, NW, Washington, DC 20004-2696. (202) 508-5000, 5454. FAX (202) 508-5360.

Electroscience Laboratory. Ohio State University, 1320 Kinnear Road, Columbus, Ohio 43212. (614) 292-7981, FAX (614) 292-7297.

Electrical Engineering Research Laboratories. Purdue University. Electrical Engineering Building, West Lafayette, in 47907. (317) 494-3536. FAX (317) 494-6440.

Electrical Power Research Institute. 3412 Hillview Avenue, PO Box 10412, Palo Alto, CA 94303. (415) 855-2000. FAX (415) 855-2954.

Laboratory For Electromagnetic and Electronic Systems. Massachusetts Institute of Technology, 77 Massachusetts Avenue, Cambridge, MA 02139. (617) 253-4631. FAX (617) 258-6774.

Weber Research Institute. Polytechnic University. Route 110, Farmingdale, NY 11735. (516) 755-4250, FAX (516) 755-4404.

ELECTRON LENS

See: ELECTRON OPTICS

ELECTRON MICROSCOPY

See also: MICROSCOPY, OPTICS

ABSTRACT SERVICES AND INDEXES

Applied Science and Technology Index: a cumulative subject index to English language periodicals in the fields of aeronautics and space science, computer technology, chemistry, construction industry, energy and related areas. H.W. Wilson Co., 950 University Avenue, Bronx, NY 10452. (212) 588-8400. (800) 367-6770. FAX (718) 590-1617. From 1958 to present. Monthly. Inquire about cost and availability. Also available on CD-ROM and online. ISSN: 0003-6986.

Bibliography and Index of Geology. American Geological Institute, 4220 King Street, Alexandria, VA 22302-1507.(703) 379-2480. FAX (703) 379-7563. From 1969 to present. Monthly. $1295.00 per year. Also available as GEOREF online (CISTI, DIALOG, Orbit, STN) and on CD-ROM. Inquire about price and availability. ISSN: 0098-2784.

Chemical Abstracts. Chemical Abstracts Service, 2540 Olentangy River Road, Box 3012, Columbus, OH 43210-0012. (614) 447-3600. (800) 848-6538. FAX (614) 447-3713. From 1907 to present. Weekly. $16800.00 per year. Also available on CD-ROM and online. Inquire regarding cost and availability.

Current Contents: Engineering, Technology and Applied Sciences. Institute for Scientific Information, 3501 Market Street, Philadelphia, PA 19104. (215) 386-0100. FAX (215) 386-2291. From 1961 to present. Weekly. $442.00 per year. Also available on CD-ROM and online. Inquire regarding cost and availability. ISSN: 0095-7917.

Engineering Index Monthly. Engineering Information, Inc., Castle Point on the Hudson, Hoboken, NJ 07030. (201) 216-8500. FAX (201) 216-8532. Monthly. $2200.00 per year. Also available online as COMPENDEX and on CD-ROM. ISSN: 0742-1974.

General Science Index. H.W. Wilson Company, 950 University Avenue, Bronx, NY 10452. (212) 588-8400. (800) 367-6770. FAX (718) 590-1617. From 1978 to present. Ten issues per year; quarterly and annual cumulations. Service basis. Available on CD-ROM and online. Inquire regarding cost and availability. ISSN: 0162-1963.

NTIS Alerts: Physics. U. S. National Technical Information Service. 5285 Port Royal Road, Springfield, VA 22161. (703) 487-4650. FAX (703) 321-8547. Weekly. $140.00 per year.

Physics Abstracts. INSPEC. Section A, Science Abstracts. Institution of Electrical Engineers (IEE). Available from INSPEC/ IEEE - Institute of Electrical and Electronic Engineers,, Box 1331, 445 Hoes Lane, Piscaway NJ 08855-1331. (908) 562-5549. 1898 to present. 24 issues per year. $2835.00 per year. Also available on CD-ROM and online. ISSN: 0036-8091.

Physics Briefs (Physikalische Berichte). Information Center for Energy, Physics, Mathematics; German Physical Society. V C H Publishers, Inc., 220 East 23rd Street, New York, NY 10010-4606. (212) 683-8333. 1845 to present. 24 issues per year. $2390.00 per year. Also available online. ISSN: 0179-7434.

Science Citation Index. SCI. Institute for Scientific Information, 3501 Market Street, Philadelphia, PA 19104. (215) 386-0100. (800) 523-1850. FAX (215) 386-2991. 1961 to present. Six issues per year, plus annual cumulation. $11650.00 per year. Also available online and on CD-ROM. Inquire about price and availability. ISSN: 0036-827X.

ANNUAL REVIEWS AND YEARBOOKS

Advances in Optical and Electron Microscopy. Academic Press, Inc., 6277 Sea Harbor Drive, Orlando, FL. (800) 321-5068. 1966 to present. Irregular. ISSN: 0065-3012.

ASSOCIATIONS AND PROFESSIONAL SOCIETIES

American Institute of Physics. One Physics Ellipse, College Park, MD 20740-3843. (301) 209-3100.

IEEE Lasers and Electro-optics Society. c/o IEEE, 345 East 47th Street, New York, NY 10017. 445 Hoes Lane, PO Box 1331, Piscataway, NJ 08855. (908) 562-3892. FAX (908) 562-8434.

Microbeam Analysis Society. c/o VCH Publishers, Inc., 303 NW 12th Avenue, Deerfield Beach, FL 33442-1788. (800) 367-8249. FAX (305) 428-8201.

Microscopy Society of America. PO Box MSA, Woods Hole, MA 02543. (508) 540-7630. FAX (508) 548-9053.

Optical Society of America. 2010 Massachusetts Avenue NW, Washington, DC 20036. (202) 223-8130. FAX (202) 223-1096.

Spie - the International For Optical Engineering. PO Box 10, Bellingham, WA 98227. (206) 676-3290. FAX (206) 647-1445.

DIRECTORIES AND BIOGRAPHICAL SOURCES

American Men and Women of Science: *Physical and Biologicalsciences.* R. R. Bowker Inc., 121 Chanlon Road, New Providence, NJ 07974. (908) 464-6800. (800) 521-8110. 20th edition. 8 volumes. 1996. $850.00.

International Directory of Engineering Societies and Related Organizations. American Society for Engineering Education, 1818 N Street NW, Suite 600, Washington, DC 20036. (202) 331-3526. 15th edition. 1996. $185.00.

Research Centers Directory. Gale Research Inc., 835 Penobscot Building, Detroit, MI 48226-4094. (313) 961-2242. (800) 347-4253. 20th edition. 1995. $485.00. ISSN: 0080-1518.

Scientific and Technical Organizations and Agencies Directory. Gale Research, 835 Penobscot Building, Detroit, MI 48226-4094. (313) 961-2242. (800) 347-4253. 4th edition. 1996. $195.00.

Who's Who in Engineering. Gordon Davis, editor. American Society for Engineering Education, 1111 19th Street, NW, Suite 608, Washington, DC 20036. 9th edition. 1995. $220.00.

Who's Who in Technology. Gale Research, 835 Penobscot Building, Detroit, MI 48226-4094. (313) 961-2242. (800) 347-4253. 7th edition. 1995. $195.00.

ENCYCLOPEDIAS AND DICTIONARIES

Encyclopedia of Applied Physics. George Trigg, editor. VCH Publications, Inc., 220 East 23rd Street, Suite 909, New York, NY 10010-4606. (212) 683-8333. (800) 422-8824. 20 volume set. 1991-. $5990.00.

Encyclopedia of Lasers and Optical Technology. Robert A. Meyers, editor. Academic Press, Inc., 6277 Sea Harbor Drive, Orlando, FL. (800) 321-5068. 1991. $92.00.

Glossary of Optical Terminology. Thomas K. Farrell. Butterworth-Heinemann, 313 Washington Avenue, Newton, MA 02158. (617) 928-2500. (800) 366-2665. 1985. $24.95.

Mcgraw-Hill Encyclopedia of Science and Technology. McGraw-Hill Book Company, Inc., 1221 Avenue of the Americas, New York, NY 10020. (212) 512-2000. (800) 262-4729. 7th edition. 20 volume set. 1992. $1900.00.

GENERAL WORKS

Applied Optics and Optical Design. A. E. Conrady. Dover Publications, Inc., 180 Varick Street, New York, NY 10014. (212) 255-3755. (800) 223-3130. 2 volumes. Reprint edition. 1992. $10.95.

Applied Optics and Optical Engineering: A Comprehensive Treatise. Rudolf Kingslake, et al, editors. Academic Press 9 volumes. 1965- . Volumes individually priced.

Electron Beam Analysis of Materials. M. H. Loretto. Chapman & Hall, 1 Penn Plaza, New York, NY 10119. (212) 564-1060. 2nd edition. 1993.

Electron Microscopy and Analysis. P. Goodhew and F. Humphreys, editors. Taylor and Francis Inc., 1900 Frost Road, Suite 101, Bristol PA 19007-1598. (215) 785-5800. 2nd edition. 1988. $39.50 in paper.

High-resolution Transmission Electron Microscopy and Associated Technologies. Peter Buseck, et al. Oxford University Press, Inc., 200 Madison Avenue, New York, NY 10016. (212) 725-6000. (800) 334-4249. 1989. $75.00.

Introduction To Classical and Modern Optics. Jurgen R. Meyer-Arendt. Prentice-Hall, 113 Sylvan Avenue, Route 9W, Englewood Cliffs, NJ 07632. (201) 592-2000. (800) 922-0579. 4th editon. 1994. $70.00.

Practical Electron Microscopy: A Beginner's Illustrated Guide. Elaine Hunter. Cambridge University Press, 40 West 20th Street,

ELECTRON MICROSCOPY

Ency. of Physical Sciences and Engineering Info. Sources

New York, NY 10011-4211. (212) 924-3900. (800) 872-7423. 1993. 34.95 in paper.

HANDBOOKS AND MANUALS

CRC Handbook of Chemistry and Physics. David R. Kide, editor. CRC Press, Inc., 2000 Corporate Boulevard, NW, Boca Raton, FL 33431. (407) 994-0555. (800) 272-7737. 77th edition. 1996. $99.95.

Handbook of Optics. Optical Society of America. McGraw-Hill Publishing Company, Inc., 1221 Avenue of the Americas, New York, NY 10020. (212) 512-2000. (800) 262-4729. 2 volumes. 2nd edition. 1993. $99.50 each.

ONLINE DATABASES AND CD-ROMS

CA Search. Chemical Abstracts Service, P.O. Box 3012, Columbus, OH 43210-0012. (614) 447-3600. (800) 848-6533. FAX (614) 447-3709. Very comprehensive guide to worldwide chemical literature and related fields, 1972 to present. Available on BRS,(800) 289-4277, DIALOG, (800) 334-2564, ORBIT (800) 456-7248, and STN International, FIZ Karlsruhe, P.O. Box 2465, W-7500, Karlsruhe 1, Germany, online services. Inquire as to cost and availability.

Compendex Plus. Engineering Information, Inc., 345 East 47th Street, New York, NY 10017. (212) 705-7600 or (800) 221-1044. Contains citations with abstracts to worldwide literature in engineering and technology, from 1970 to present. Available on online BRS,(800) 289-4277, DIALOG, (800) 334-2564, ORBIT (800) 456-7248, and STN International, FIZ Karlsruhe, P.O. Box 2465, W-7500, Karlsruhe 1, Germany, online services. Also available on CD-ROM. Inquire as to cost and availability.

Current Contents Search. Institute for Scientific Information, 3501 Market Street, Philadelphia, PA 19104. (215) 386-0100. FAX (215) 386-6362. Contains citations to articles listed in the table of contents of science and technology journals. Also articles in social sciences and life sciences journals. Available on BRS, (800) 289-4277, DIALOG, (800) 334-2564, online services. Inquire as to cost and availability.

Dissertation Abstracts Online. University Microfilms International, 300 North Zeeb Road, Ann Arbor, MI 48106. (800) 521-0600 or (313) 761-4700. Scope includes virtually all doctoral dissertations accepted at accredited American institutions from 1861 to present in 252 subject areas. Available on BRS, (800) 289-4277, DIALOG, (800) 334-2564, and OCLC EPIC, (800) 848-5878, online services. Also available on CD-ROM. Inquire as to cost and availability.

Georef. American Geological Institute, 4220 King Street, Alexandria, VA 22302. (800) 336-4764 or (703) 379-2480. Geology and geosciences literature, 1785 to present for North America. Available on DIALOG,(800) 334-2564, ORBIT (800) 456-7248, online services. Also available on CD-ROM. Inquire as to cost and availability.

INSPEC. Institution of Electrical Engineers, Michael

Faraday House, Six Hills Way, Stevenage, Herts. SG1 2AY, England. Telephone: 0438 313311 or FAX 0438 742840. Contains citations to the worldwide literature of physics, electronics and electrical engineering, computer technology, and related fields. Available on BRS, (800) 289-4277, DIALOG, (800) 334-2564, ORBIT, (800) 456-7248, and STN International, FIZ

Karlsruhe, P.O. Box 2465, W-7500, Karlsruhe 1, Germany, online services. Inquire as to cost and availability.

NTIS Bibliographic Database. National Technical Information Service, 5285 Port Royal Road, Springfield, VA 22161. (703) 487-4929 or FAX (703) 321-8199. Broad coverage of government-sponsored science and technology research reports, 1964 to present. Available on BRS,(800) 289-4277, DIALOG, (800) 334-2564, ORBIT, (800) 456-7248, and STN International, FIZ Karlsruhe, P.O. Box 2465, W-7500, Karlsruhe 1, Germany, online services. Also available on CD-ROM. Inquire as to cost and availability.

Physics Briefs. American Institute of Physics, 335 East 45th Street, New York, NY 10017. (212) 661-9260 or FAX (212) 661-2036. Contains citations with abstracts of the literature of physics and related fields, 1979 to present. Available on the STN International, FIZ Karlsruhe, P.O. Box 2465, W-7500, Karlsruhe 1, Germany, online service. Inquire as to cost and availability.

SCISEARCH. Institute for Scientific Information, 3501 Market Street, Philadelphia, PA 19104. (800) 523-1850 or (215) 386-0100. Broad multidisciplinary title and author index to the international literature of science and technology, 1974 to present. Available on DIALOG, (800) 334-2564, and ORBIT, (800) 456-7248, online services. Also available on CD-ROM. Inquire as to cost and availability.

WILSONLINE. H.W. Wilson Company, 950 University Avenue, Bronx, NY 10452. (800) 367-6770 or (212) 588-8400. Makes available online versions of the printed H.W. Wilson indexes including Applied Science and Technology Index, Business Periodicals Index, General Science Index, and Readers' Guide to Periodical Literature. Period covered is generally 1983 to present. Available on BRS, (800) 289-4277, DIALOG, (800) 334-2564, and OCLC EPIC, (800) 848-5878, online services. Also available on CD-ROM. Inquire as to cost and availability.

PERIODICALS

Journal of Electron Microscopy. Journal of Electron Microscopy, c/o Business Center for Academic Societies Japan, 5-16-9 Honkomagome, Bunkyo-ku, Tokyo 113, Japan. TEL 03-5814-5811. FAX 03-5814-5823. 1953 to present. Bimonthly. $200.00 per year. ISSN: 0022-0744.

Journal of Microscopy. Blackwell Scientific Publications, Inc., 52 Beacon Street, Boston, MA 02108. (800) 325-4177. 1878 to present. Monthly. $552.00 per year. ISSN: 0022-2720.

Journal of Microscopy Research and Technique. John Wiley & Sons, Inc., Journal, 605 Third Avenue, New York, NY 10158. (212) 850-6000. FAX (212) 850-6088. 1984 to present. Sixteen issues per year. $898.00 per year. ISSN: 1059-910X.

Optical Engineering. SPIE - International Society for Optical Engineering, Box 10, 1000 20th Street, Bellingham, WA 98227-0010. (206) 676-3290. FAX (206) 647-1445. 1962

to present. Monthly. $170.00 per year. ISSN: 0091-3286.

Scanning Electron Microscopy. Scanning Microscopy International, Inc., Box 66507, AMF O'Hare, Chicago, IL 60666. (708) 529-6677. FAX (708) 980-6698. 1968 to present. Quarterly. $164.00 per year. ISSN: 0891-7035.

Ency. of Physical Sciences and Engineering Info. Sources

ELECTRON OPTICS

SPIE - International Society For Optical Engineering. PROCEEDINGS. SPIE - International Society for Optical Engineering, Box 10, 1000 20th Street, Bellingham, WA 98227-0010. (206) 676-3290. FAX (206) 647-1445. 1963 to present. Approximately 50 numbers per year. Individually priced.

RESEARCH CENTERS AND INSTITUTES

Center For Electron Microscopy. University of Illinois, 74 Bevier Hall, 905 S. Goodwin Avenue, Urbana, IL 61801. (217) 333-2108. FAX (217) 333-1656.

Center For High Resolution Electron Microscopy. Arizona State University, Center for Solid State Science, Tempe, AZ 85287-1704. (602) 965-6459. FAX (602) 965-7954.

Electron Microbeam Analysis Laboratory. University of Michigan, Main Campus, 2501 C.C. Little Building, Ann Arbor MI 48109-1063. (313) 936-1555.

National Center For Electron Microscopy. Lawrence Berkeley Laboratory, 1 Cyclotron Road, Berkeley, CA 94720. (415) 486-5006. FAX (415) 486-5888.

ELECTRON OPTICS

See also: ELECTRON MICROSCOPY, ELECTRON SPECTROSCOPY, OPTICS, OPTOELECTRONICS

ABSTRACT SERVICES AND INDEXES

Applied Science and Technology Index; a cumulative subject index to English language periodicals in the fields of aeronautics and space science, computer technology, chemistry, construction industry, energy and related areas. H.W. Wilson Co., 950 University Avenue, Bronx, NY 10452. (212) 588-8400. (800) 367-6770. FAX (718) 590-1617. From 1958 to present. Monthly. Inquire about cost and availability. Also available on CD-ROM and online. ISSN: 0003-6986.

Chemical Abstracts. Chemical Abstracts Service, 2540 Olentangy River Road, Box 3012, Columbus, OH 43210-0012. (614) 447-3600. (800) 848-6538. FAX (614) 447-3713. From 1907 to present. Weekly. $16800.00 per year. Also available on CD-ROM and online. Inquire regarding cost and availability. ISSN:

Current Contents: Engineering, Technology and Applied Sciences. Institute for Scientific Information, 3501 Market Street, Philadelphia, PA 19104. (215) 386-0100. FAX (215) 386-2291. From 1961 to present. Weekly. $442.00 per year. Also available on CD-ROM and online. Inquire regarding cost and availability. ISSN: 0095-7917.

Engineering Index Monthly. Engineering Information, Inc., Castle Point on the Hudson, Hoboken, NJ 07030. (201) 216-8500. FAX (201) 216-8532. Monthly. $2200.00 per year. Also available online as COMPENDEX and on CD-ROM. ISSN: 0742-1974.

General Science Index. H.W. Wilson Company, 950 University Avenue, Bronx, NY 10452. (212) 588-8400. (800) 367-6770. FAX (718) 590-1617. From 1978 to present. Ten issues per year; quarterly and annual cumulations. Service basis. Available on CD-ROM and online. Inquire regarding cost and availability. ISSN: 0162-1963.

NTIS Alerts: Physics. U. S. National Technical Information Service. 5285 Port Royal Road, Springfield, VA 22161. (703) 487-4650. FAX (703) 321-8547. Weekly. $140.00 per year.

Physics Abstracts. INSPEC. Section A, Science Abstracts. Institution of Electrical Engineers (IEE). Available from INSPEC/ IEEE - Institute of Electrical and Electronic Engineers,, Box 1331, 445 Hoes Lane, Piscaway NJ 08855-1331. (908) 562-5549. 1898 to present. 24 issues per year. $2835.00 per year. Also available on CD-ROM and online. ISSN: 0036-8091.

Physics Briefs (Physikalische Berichte). Information Center for Energy, Physics, Mathematics; German Physical Society. V C H Publishers, Inc., 220 East 23rd Street, New York, NY 10010-4606. (212) 683-8333. 1845 to present. 24 issues per year. $2390.00 per year. Also available online. ISSN: 0179-7434.

Science Citation Index. SCI. Institute for Scientific Information, 3501 Market Street, Philadelphia, PA 19104. (215) 386-0100. (800) 523-1850. FAX (215) 386-2991. 1961 to present. Six issues per year, plus annual cumulation. $11650.00 per year. Also available online and on CD-ROM. Inquire about price and availability. ISSN: 0036-827X.

ANNUAL REVIEWS AND YEARBOOKS

Advances in Optical and Electron Microscopy. Academic Press, Inc., 6277 Sea Harbor Drive, Orlando, FL. (800) 321-5068. 1966 to present. Irregular. ISSN: 0065-3012.

ASSOCIATIONS AND PROFESSIONAL SOCIETIES

American Institute of Physics. One Physics Ellipse, College Park, MD 20740-3843. (301) 209-3100.

IEEE Lasers and Electro-optics Society. c/o IEEE, 345 East 47th Street, New York, NY 10017. 445 Hoes Lane, PO Box 1331, Piscataway, NJ 08855. (908) 562-3892. FAX (908) 562-8434.

Microbeam Analysis Society. c/o VCH Publishers, Inc., 303 NW 12th Avenue, Deerfield Beach, FL 33442-1788. (800) 367-8249. FAX (305) 428-8201.

Microscopy Society of America. PO Box MSA, Woods Hole, MA 02543. (508) 540-7630. FAX (508) 548-9053.

Optical Society of America. 2010 Massachusetts Avenue NW, Washington, DC 20036. (202) 223-8130. FAX (202) 223-1096.

Spie - the International Society For Optical Engineering. PO Box 10, Bellingham, WA 98227. (206) 676-3290. FAX (206) 647-1445.

DIRECTORIES AND BIOGRAPHICAL SOURCES

American Men and Women of Science: Physical and Biological Sciences. R. R. Bowker Inc., 121 Chanlon Road, New Providence, NJ 07974. (908) 464-6800. (800) 521-8110. 20th edition. 8 volumes. 1996. $850.00.

International Directory of Engineering Societies and Related Organizations. American Society for Engineering Education,

1818 N Street NW, Suite 600, Washington, DC 20036. (202) 331-3526. 15th edition. 1996. $185.00.

Research Centers Directory. Gale Research Inc., 835 Penobscot Building, Detroit, MI 48226-4094. (313) 961-2242. (800) 347-4253. 20th edition. 1995. $485.00. ISSN: 0080-1518.

Scientific and Technical Organizations and Agencies Directory. Gale Research, 835 Penobscot Building, Detroit, MI 48226-4094. (313) 961-2242. (800) 347-4253. 4th edition. 1996. $195.00.

SPIE - the International Society For Optical Engineering. Annual Membership Directory. PO Box 10, Bellingham, WA 98227. (206) 676-3290. FAX (206) 647-1445. Annual. Inquire.

Who's Who in Engineering. Gordon Davis, editor. American Society for Engineering Education, 1111 19th Street, NW, Suite 608, Washington, DC 20036. 9th edition. 1995. $220.00.

Who's Who in Technology. Gale Research, 835 Penobscot Building, Detroit, MI 48226-4094. (313) 961-2242. (800) 347-4253. 7th edition. 1995. $195.00.

ENCYCLOPEDIAS AND DICTIONARIES

Encyclopedia of Applied Physics. George Trigg, editor. VCH Publications, Inc., 220 East 23rd Street, Suite 909, New York, NY 10010-4606. (212) 683-8333. (800) 422-8824. 20 volume set. 1991-. $5990.00.

Encyclopedial of Lasers and Optical Technology. Robert A. Meyers, editor. Academic Press, Inc., 6277 Sea Harbor Drive, Orlando, FL. (800) 321-5068. 1991. $92.00.

Glossary of Optical Terminology. Thomas K. Farrell. Butterworth-Heinemann, 313 Washington Avenue, Newton, MA 02158. (617) 928-2500. (800) 366-2665. 1985. $24.95.

McGraw-Hill Encyclopedia of Science and Technology. McGraw-Hill Book Company, Inc., 1221 Avenue of the Americas, New

York, NY 10020. (212) 512-2000. (800) 262-4729. 7th edition. 20 volume set. 1992. $1900.00.

GENERAL WORKS

Applied Optics and Optical Design. A. E. Conrady. Dover Publications, Inc., 180 Varick Street, New York, NY 10014. (212) 255-3755. (800) 223-3130. 2 volumes. Reprint edition. 1992. $10.95.

Applied Optics and Optical Engineering; A Comprehensive Treatise. Rudolf Kingslake, et al, editors. Academic Press 9 volumes. 1965- . Volumes individually priced.

Basis of Electron Optics. David A. DeWolf. John Wiley &Sons, Inc., 605 Third Avenue, New York, NY 10158-0012. (212) 850-6000. (800) 225-5945. 1990. $99.95.

Crystals and Light: an Introduction to Optical Crystallography. Elizabeth A. Wood. Dover Publications, Inc., 180 Varick Street, New York, NY 10014. (212) 255-3755. (800) 223-3130. 1977. $5.95 in paper.

Electron Beam Analysis of Materials. M. H. Loretto. Chapman & Hall, 1 Penn Plaza, New York, NY 10119. (212) 564-1060. 2nd edition. 1993.

Electron Microscopy and Analysis. P. Goodhew and F. Humphreys, editors. Taylor and Francis Inc., 1900 Frost Road, Suite 101, Bristol PA 19007-1598. (215) 785-5800. 2nd edition. 1988. $39.50 in paper.

High-resolution Transmission Electron Microscopy and Associated Technologies. Peter Buseck, et al. Oxford University Press, Inc., 200 Madison Avenue, New York, NY 10016. (212) 725-6000. (800) 334-4249. 1989. $75.00.

Introduction To Classical and Modern Optics. Jurgen R. Meyer-Arendt. Prentice-Hall, 113 Sylvan Avenue, Route 9W, Englewood Cliffs, NJ 07632. (201) 592-2000. (800) 922-0579. 4th editon. 1994. $70.00.

Magnetism and Optics of Molecular Crystals J. W. Rohleder and R. W. Munn. John Wiley & Sons, Inc., 605 Third Avenue, New York, NY 10158-0012. (212) 850-6000. (800) 225-5945. 1992. $79.95.

Optical Electronics. Anmon Yariv. SCP: Third World Literature Publishing House, P.O. Box 482, Lithonia, GA30058-0482. (404) 785-7725. 4th edition. 1991. $67.00.

Practical Electron Microscopy: A Beginner's Illustrated Guide. Elaine Hunter. Cambridge University Press, 40 West 20th Street, New York, NY 10011-4211. (212) 924-3900. (800) 872-7423. 1993. 34.95 in paper.

Principles of Electron Optics. Peter W. Hawkes. ACADemicPress, Inc., 6277 Sea Harbor Drive, Orlando, FL. (800) 321-5068. 3 volumes. 1989. $108.00 ea.

HANDBOOKS AND MANUALS

CRC Handbook of Chemistry and Physics. David R. Kide, editor. CRC Press, Inc., 2000 Corporate Boulevard, NW, Boca Raton, FL 33431. (407) 994-0555. (800) 272-7737. 77th edition. 1996. $99.95.

Handbook of Microwave and Optical Components. Kai Chang, John Wiley & Sons, Inc., 605 Third Avenue, New York, NY 10158-0012. (212) 850-6000. (800) 225-5945. 4 volumes. 1991. $324.00 set.

Handbook of Optics. Optical Society of America. McGraw-Hill Publishing Company, Inc., 1221 Avenue of the Americas, New York, NY 10020. (212) 512-2000. (800) 262-4729. 2 volumes. 2nd edition. 1993. $99.50 each.

ONLINE DATABASES AND CD-ROMS

CA Search. Chemical Abstracts Service, P.O. Box 3012, Columbus, OH 43210-0012. (614) 447-3600. (800) 848-6533. FAX (614) 447-3709. Very comprehensive guide to worldwide chemical literature and related fields, 1972 to present. Available on BRS,(800) 289-4277, DIALOG, (800) 334-2564, ORBIT (800) 456-7248, and STN International, FIZ Karlsruhe, P.O. Box 2465, W-7500, Karlsruhe 1, Germany, online services. Inquire as to cost and availability.

Ency. of Physical Sciences and Engineering Info. Sources

ELECTRON OPTICS

Compendex Plus. Engineering Information, Inc., 345 East 47th Street, New York, NY 10017. (212) 705-7600 or (800) 221-1044. Contains citations with abstracts to worldwide literature in engineering and technology, from 1970 to present. Available on online BRS,(800) 289-4277, DIALOG, (800) 334-2564, ORBIT (800) 456-7248, and STN International, FIZ Karlsruhe, P.O. Box 2465, W-7500, Karlsruhe 1, Germany, online services. Also available on CD-ROM. Inquire as to cost and availability.

Current Contents Search. Institute for Scientific Information, 3501 Market Street, Philadelphia, PA 19104. (215) 386-0100. FAX (215) 386-6362. Contains citations to articles listed in the table of contents of science and technology journals. Also articles in social sciences and life sciences journals. Available on BRS, (800) 289-4277, DIALOG, (800) 334-2564, online services. Inquire as to cost and availability.

Dissertation Abstracts Online. University Microfilms International, 300 North Zeeb Road, Ann Arbor, MI 48106. (800) 521-0600 or (313) 761-4700. Scope includes virtually all doctoral dissertations accepted at accredited American institutions from 1861 to present in 252 subject areas. Available on BRS, (800) 289-4277, DIALOG, (800) 334-2564, and OCLC EPIC, (800) 848-5878, online services. Also available on CD-ROM. Inquire as to cost and availability.

Georef. American Geological Institute, 4220 King Street, Alexandria, VA 22302. (800) 336-4764 or (703) 379-2480. Geology and geosciences literature, 1785 to present for North America. Available on DIALOG,(800) 334-2564, ORBIT (800) 456-7248, online services. Also available on CD-ROM. Inquire as to cost and availability.

INSPEC. Institution of Electrical Engineers, Michael Faraday House, Six Hills Way, Stevenage, Herts. SG1 2AY, England. Telephone: 0438 313311 or FAX 0438 742840. Contains citations to the worldwide literature of physics, electronics and electrical engineering, computer technology, and related fields. Available on BRS, (800) 289-4277, DIALOG, (800) 334-2564, ORBIT, (800) 456-7248, and STN International, FIZ Karlsruhe, P.O. Box 2465, W-7500, Karlsruhe 1, Germany, online services. Inquire as to cost and availability.

NTIS Bibliographic Database. National Technical Information Service, 5285 Port Royal Road, Springfield, VA 22161. (703) 487-4929 or FAX (703) 321-8199. Broad coverage of government-sponsored science and technology research reports, 1964 to present. Available on BRS,(800) 289-4277, DIALOG, (800) 334-2564, ORBIT, (800) 456-7248, and STN International, FIZ Karlsruhe, P.O. Box 2465, W-7500, Karlsruhe 1, Germany, online services. Also available on CD-ROM. Inquire as to cost and availability.

Physics Briefs. American Institute of Physics, 335 East 45th Street, New York, NY 10017. (212) 661-9260 or FAX (212) 661-2036. Contains citations with abstracts of the literature of physics and related fields, 1979 to present. Available on the STN International, FIZ Karlsruhe, P.O. Box 2465, W-7500, Karlsruhe 1, Germany, online service. Inquire as to cost and availability.

SCISEARCH. Institute for Scientific Information, 3501 Market Street, Philadelphia, PA 19104. (800) 523-1850 or (215) 386-0100. Broad multidisciplinary title and author index to the international literature of science and technology, 1974 to present. Available on DIALOG, (800) 334-2564, and ORBIT, (800) 456-7248, online services. Also available on CD-ROM. Inquire as to cost and availability.

WILSONLINE. H.W. Wilson Company, 950 University Avenue, Bronx, NY 10452. (800) 367-6770 or (212) 588-8400. Makes available online versions of the printed H.W. Wilson indexes including Applied Science and Technology Index, Business Periodicals Index, General Science Index, and Readers' Guide to Periodical Literature. Period covered is generally 1983 to present. Available on BRS, (800) 289-4277, DIALOG, (800) 334-2564, and OCLC EPIC, (800) 848-5878, online services. Also available on CD-ROM. Inquire as to cost and availability.

PERIODICALS

Applied Optics. Optical Society of America. 2010 Massachusetts Avenue NW, Washington, DC 20036-1023. (202)223-8130. 1962 to present. 36 issues per year. $1090.00 per year. ISSN: 0003-6935.

Electro-optics. PennWell Publishing Co., 119 Russell Street, Littleton, MA 01460. (617) 486-9501. 1969 to present. Twenty issues per year. $ ISSN: 0745-5003.

Journal of Lightwave Technology. Institute of Electrical and Electronics Engineers, 445 Hoes Lane, PO Box 1331, Piscataway, NJ 08855. (908) 562-3892. FAX (908) 562-8434. 1983 to present. Monthly. $425.00 per year. ISSN: 0733-8724.

Optical Engineering. SPIE - International Society for Optical Engineering, Box 10, 1000 20th Street, Bellingham,

WA 98227-0010. (206) 676-3290. FAX (206) 647-1445. 1962 to present. Monthly. $170.00 per year. ISSN: 0091-3286.

Scanning Electron Microscopy. Scanning Microscopy International, Inc., Box 66507, AMF O'Hare, Chicago, IL 60666. (708) 529-6677. FAX (708) 980-6698. 1968 to present. Quarterly. $164.00 per year. ISSN: 0891-7035.

SPIE - International Society For Optical Engineering. PROCEEDINGS. SPIE - International Society for Optical Engineering, Box 10, 1000 20th Street, Bellingham, WA 98227-0010. (206) 676-3290. FAX (206) 647-1445. 1963 to present. Approximately 50 numbers per year. Individually priced.

RESEARCH CENTERS AND INSTITUTES

Institute of Optics. University of Rochester, Wilmot Building, Rochester, NY 14627. (716) 275-5248.

Center For Applied Optics. University of Alabama in Huntsville, Research Institute Building, Huntsville, AL 25899. (205) 895-6102.

Center For Applied Optics. University of Texas at Dallas, P.O. Box 830688, Richardson, TX 75803-0688. (214) 690-2868. FAX (214) 690-2848.

Center For Imaging Science. Rochester Institute of Technology, One Lomb Memorial Drive, Rochester, NY 14623. (716) 475-2774.

Center For Research in Electro-optics and Lasers. University of Central Florida, 12424 Research Parkway, Orlando, FL 32826. (407) 658-6800. FAX (407) 658-6880.

ELECTRON OPTICS

Ency. of Physical Sciences and Engineering Info. Sources

Institute For Physical Science and Technology. University of Maryland, 4203 Computer and Space Sciences Building, College Park, MD 20742. (301) 454-4874. FAX (301) 314-9363.

Optical Sciences Center. University of Arizona, Tucson, AZ 85721. (602) 621-6997.

ELECTRON SPECTROSCOPY

See also: ANALYTICAL CHEMISTRY, MASS SPECTROMETRY, NUCLEAR MAGNETIC RESONANCE, SPECTROSCOPY

ABSTRACT SERVICES AND INDEXES

Applied Science and Technology Index: A Cumulative Subject Index to English Language Periodicals in the Fields of Aeronautics and Space Science, Computer Technology, Chemistry, Construction Industry, Energy and Related Areas. H.W. Wilson Co., 950 University Avenue, Bronx, NY 10452. (212) 588-8400. (800) 367-6770. FAX (718) 590-1617. From 1958 to present. Monthly. Inquire about cost and availability. Also available on CD-ROM and online. ISSN: 0003-6986.

Chemical Abstracts. Chemical Abstracts Service. 2540 Olentangy River Road, Box 3012, Columbus, OH 43210-0012.(614) 447-3600. FAX (614) 447-3713. 1907 to present. Weekly. $16,800.00 per year. Available online and on CD-ROM. CA is also available in five section groupings. Inquire regarding cost and availability.

Current Contents: Physical, Chemical and Earth Sciences. Institute for Scientific Information, 3501 Market Street, Philadelphia, PA 19104. (215) 386-0100. FAX (215) 386-2291. From 1961 to present. Weekly. $442.00 per year. Also available on CD-ROM and online. Inquire regarding cost andavailability. ISSN: 0163-2574.

Engineering Index Monthly; Indexes and Abstracts the World's Engineering and Technical Literature. Engineering Information, Inc., Castle Point on the Hudson, Hoboken, NJ 07030. (201) 216-8500. (800) 221-1044. FAX (201) 216-8532. Monthly. $2300.00 per year. Available online as COMPENDEX and also on CD-ROM. ISSN: 0742-1974.

General Science Index. H.W. Wilson Company, 950 University Avenue, Bronx, NY 10452. (212) 588-8400. (800) 367-6770. FAX (718) 590-1617. From 1978 to present. Ten issues per year; quarterly and annual cumulations. Service basis. Available on CD-ROM and online. Inquire regarding cost and availability. ISSN: 0162-1963.

Physics Abstracts. INSPEC. Section A, Science Abstracts. Institution of Electrical Engineers (IEE), London. Available from: INSPEC/IEEE - Institute of Electrical and Electronic Engineers, Box 1331, Hoes Lane, Piscataway, NJ 08855-1331. (908) 562-5549. 1898 to present. 24 issues per year. $2835.00 per year. Also available online and on CD-ROM. ISSN: 0036-8091.

Physics Briefs (Physikalische Berichte). Information Centerfor Energy, Physics, Mathematics; German Physical Society. V C H Publishers, Inc., 220 East 23rd Street, New York, NY 10010-4606. (212) 683-8333. 1845 to present. 24 issues per year. $2390.00 per year. Also available online. ISSN: 0179-7434.

Science Citation Index. SCI. Institute for Scientific Information, 3501 Market Street, Philadelphia, PA 19104. (215) 386-0100. (800) 523-1850. FAX (215) 386-2991. 1961 to present. Six issues per year, plus annual cumulation. $11650.00 per year. Also available online and on CD-ROM. Inquire about price and availability. ISSN: 0036-827X.

ASSOCIATIONS AND PROFESSIONAL SOCIETIES

American Chemical Society. 1155 16th Street NW, Washington, DC 20036. (202) 872-4600.

American Institute of Physics. One Physics Ellipse, College Park, MD 20740-3843. (301) 209-3100. FAX (415) 855-2954.

American Society For Mass Spectroscopy. 1201 Don Diego, Santa Fe, NM 87505. (505) 989-4517.

Coblentz Society. c/o David W. Schiering, Perkin-Elmer Corporation, 761 Main Avenue, Norwalk, CT 06859-0240. (203) 761-2915. FAX (203) 761-2842.

Federation of Analytical Chemistry and Spectroscopoy Societies. c/o Jo Ann Brown, 201-B Broadway Street, Frederick, MD 21701. (301) 846-4797. FAX (301) 694-6860.

Society For Applied Spectroscopy. 201B Broadway Street, Frederick, MD 21701. (301) 694-8122. FAX (301) 694-6860.

DIRECTORIES AND BIOGRAPHICAL SOURCES

American Men and Women of Science: Physical and Biological Sciences. R. R. Bowker Inc., 121 Chanlon Road, New Providence, NJ 07974. (908) 464-6800. (800) 521-8110. 20th edition. 8 volumes. 1996. $850.00.

Engineering Research Centers: Incorporating Electronics Research Centers. Stockton Press, 345 Park Avenue, New York, NY 10010. (212) 689-9200. (800) 221-2123. 4th edition. 1995. $515.00.

International Engineering Directory. American Consulting Engineers Council, 1015 15th Street, N.W. Suite 802, Washington, DC 20005-2670. (202) 347-7474. Annual. $10.00.

Research Centers Directory. Gale Research Company Inc., 835Penobscot Building, Detroit, MI 48226-4094. (313) 961-2242. (800) 347-4253. 20th edition. 1995. $485.00. ISSN: 0080 1518.

Scientific and Technical Organizations and Agencies Directory. Gale Research, 835 Penobscot Building, Detroit, MI 48226-4094.(313) 961-2242. (800) 347-4253. 4th edition. 1996. $195.00.

Society For Applied Spectroscopy Membership Directory. Society for Applied Spectroscopy, 201B Broadway Street, Frederick, MD 21701. (301) 694-8122. FAX (301) 694-6860. Irregular.

ENCYCLOPEDIAS AND DICTIONARIES

Chambers Nuclear Energy and Radiation Dictionary. Peter M.B. Walker, editor. Cambridge University Press, 40 West 20th Street, New York, NY 10011-4211. (212) 924-3900. 1992. $40.00.

Encyclopedia of Applied Physics. George Trigg, editor. VCH Publications, Inc., 220 East 23rd Street, Suite 909, New

York, NY 10010-4606. (212) 683-8333. (800) 422-8824. 20 volume set. 1991 - . $5990.00.

McGraw-Hill Encyclopedia of Science and Technology. McGraw-Hill Book, Incorporated, 1221 Avenue of the Americas, New York, NY 10020. (212) 997-3675. (800) 262-4729. Seventh edition. Twenty volumes. 1992. $1900.00.

GENERAL WORKS

Electron Energy Loss Spectrometers. Harald Ibach. Springer-Verlag New York, Inc., 175 Fifth Avenue, New York, NY 10010. (212) 460-1500. (800) 777-4643. FAX (212) 473-6272. 1991. $64.00.

Energy Levels in Atoms and Molecules. W. Graham Richards and Peter Scott. Oxford University Press, Inc., 200 Madison Avenue, New York, NY 10016. (212) 725-6000. (800) 334-4249. 1995. $9.95.

Engineering Applications of Correlation and Spectral Analysis. Julius Bendat and Allan G. Piersol. John Wiley & Sons, Inc., 605 Third Avenue, New York, NY 10158-0012. (212) 850-6000. (800) 225-5945.

Introduction to Surface Analysis by Electron Spectroscopy. John F. Watts. Oxford University Press, Inc., 200 Madison Avenue, New York, NY 10016. (212) 725-6000. (800) 334-4249. 1990. $23.95 in paper.

Photoelectron Spectroscopy: Principe and Application. Stefan Hufner. Springer-Verlag New York, Inc., 175 Fifth Avenue, New York, NY 10010. (212) 460-1500. (800) 777-4643. FAX (212) 473-6272. 1994. $99.00.

Principles of Electron Tunneling Spectroscopy. E. L. Wolf. Oxford University Press, Inc., 200 Madison Avenue, New York, NY 10016. (212) 725-6000. (800) 334-4249. Reprint editon. 1989. $59.95 in paper.

Quantum Optics and Spectroscopy. J. Fiutak, et al, editors. Nova Science Publishers, Inc., 6080 Jericho Turnpike, Suite 207, Commack, NY 11725-2808. (516) 499-3103. 1993. $83.00.

Solid State Phenomena As Seen By Muons, Protons and Excited Nuclei. Erik B. Karlsson. Oxford University Press, Inc., 200 Madison Avenue, New York, NY 10016. (212) 725-6000. (800) 334-4249.

Symmetry and Spectroscopy. Daniel C. Harris. Dover Publications, Inc., 180 Varick Street, New York, NY 10014. (212) 255-3755. (800) 223-3130. 1989. $14.95.

HANDBOOKS AND MANUALS

CRC HandBOOK of CHEMISTRY and PHYSICS. David R. Kide, editor. CRC Press Inc., 2000 Corporate Blvd., NW, Boca Raton, FL 33431. (407) 994-0555. (800) 333-8300. 77th edition. 1996. $99.95.

Handbook of Nuclear Spectrometry. Juhnai Kantele. Academic Press, Inc., 6277 Sea Harbor Drive, Orlando, FL. (800) 321-5068. 1995. $79.95.

Practical Handbook of Spectroscopy. James W. Robinson. CRC Press, Inc., 2000 Corporate Boulevard, NW, Boca Raton, FL 33431. (407) 994-0555. (800) 272-7737. 1991. $96.50.

ONLINE DATABASES AND CD-ROMS

CA SEARCH. Chemical Abstracts Service, P.O. Box 3012, Columbus, OH 43210-0012. (614) 447-3600. (800) 848-6533. FAX (614) 447-3709. Very comprehensive guide to worldwide chemical literature and related fields, 1972 to present. Available on BRS,(800) 289-4277, DIALOG, (800) 334-2564, ORBIT (800) 456-7248, and STN International, FIZ Karlsruhe, P.O. Box 2465, W-7500, Karlsruhe 1, Germany, online services. Inquire as to cost and availability.

Current Contents Search. Institute for Scientific Information, 3501 Market Street, Philadelphia, PA 19104. (215) 386-0100. FAX (215) 386-6362. Contains citations to articles listed in the table of contents of science and technology journals. Also articles in social sciences and life sciences journals. Available on BRS,(800) 289-4277, DIALOG,(800) 334-2564, online services. Inquire as to cost and availability.

Dissertation Abstracts Online. University Microfilms International, 300 North Zeeb Road, Ann Arbor, MI 48106. (800) 521-0600 or (313) 761-4700. Scope includes virtually all doctoral dissertations accepted at accredited American institutions from 1861 to present in 252 subject areas. Available on BRS, (800) 289-4277, DIALOG, (800) 334-2564, and OCLC EPIC,(800) 848-5878, online services. Also available on CD-ROM. Inquire as to cost and availability.

INSPEC. Institution of Electrical Engineers, Michael Faraday House, Six Hills Way, Stevenage, Herts. SG1 2AY, England. Telephone: 0438 313311 or FAX 0438 742840. Contains citations to the worldwide literature of physics, electronics and electrical engineering, computer technology, and related fields. Available on BRS, (800) 289-4277, DIALOG, (800) 334-2564, ORBIT, (800) 456-7248, and STN International, FIZ Karlsruhe, P.O. Box 2465, W-7500, Karlsruhe 1, Germany, online services. Inquire as to cost and availability.

NTIS Bibliographic Database. National Technical Information Service, 5285 Port Royal Road, Springfield, VA 22161. (703) 487-4929 or FAX (703) 321-8199. Broad coverage of government-sponsored science and technology research reports, 1964 to present. Available on BRS,(800) 289-4277, DIALOG, (800) 334-2564, ORBIT, (800) 456-7248, and STN International, FIZ Karlsruhe, P.O. Box 2465, W-7500, Karlsruhe 1, Germany, online services. Also available on CD-ROM. Inquire as to cost and availability.

Physics Briefs. American Institute of Physics, 335 East 45th Street, New York, NY 10017. (212) 661-9260 or FAX (212) 661-2036. Contains citations with abstracts of the literature of physics and related fields, 1979 to present. Available on the STN International, FIZ Karlsruhe, P.O. Box 2465, W-7500, Karlsruhe 1, Germany, online service. Inquire as to cost and availability.

SCISEARCH. Institute for Scientific Information, 3501 Market Street, Philadelphia, PA 19104. (800) 523-1850 or (215) 386-0100. Broad multidisciplinary title and author index to the international literature of science and technology, 1974 to present. Available on DIALOG,(800) 334-2564, and ORBIT,(800) 456-7248, online services. Also available on CD-ROM. Inquire as to cost and availability.

ELECTRON SPECTROSCOPY

Ency. of Physical Sciences and Engineering Info. Sources

WILSONLINE. H.W. Wilson Company, 950 University Avenue, Bronx, NY 10452. (800) 367-6770 or (212) 588-8400. Makes available online versions of the printed H.W. Wilson indexes including Applied Science and Technology Index, Business Periodicals Index, General Science Index, and Readers' Guide to Periodical Literature. Period covered is generally 1983 to present. Available on BRS,(800) 289-4277, DIALOG,(800) 334-2564, and OCLC EPIC,(800) 848-5878, online services. Also available on CD-ROM. Inquire as to cost and availability.

PERIODICALS

Applied Spectroscopy. Society for Applied Spectroscopy, 198 Thomas Johnson Drive, Suite 2, Frederick, MD 21702. (301) 694-8122. 1946 to present. Monthly. $180.00 per year. ISSN: 0003-7028.

Applied Spectroscopy Reviews. Marcel Dekker Journals, 270 Madison Avenue, New York, NY 10016. (212) 696-9000. FAX (212) 685-4540. 1964 to present. Quarterly. $395.00 per year. ISSN: 0570-4928.

International Journal of Mass Spectrometry and Ion Processes. Elsevier Science Inc., Publishing Co., Inc., Box 882, Madison Square Station, New York, NY 10159. (212) 989-5800. (212) 633-3990. 1968 to present. 36 issues per year in 12 volumes. $2504.00 per year. ISSN: 0168-1176.

Journal of Electron Spectroscopy and Related Phenomena. Elsevier Science Inc., Publishing Co., Inc., Box 882, Madison Square Station, New York, NY 10159. (212) 989-5800. (212) 633-3990. 1972 to present. 20 issues per year. $1030.00 per year. ISSN: 0368-2048.

Journal of Molecular Spectroscopy. Academic Press, Inc., 525 B Street, Suite 1900, San Diego, CA 92101-4495. (619) 230-1840. 1957 to present. Monthly. $1216.00 per year. ISSN: 0022-2852.

Journal of Quantiative Spectroscopy and Radiative Transfer. Elsevier Science Inc., Publishing Co., Inc., Box 882, Madison Square Station, New York, NY 10159. (212) 989-5800. (212) 633-3990. 1961 to present. Monthly. $1815.00 per year. ISSN: 0022-4073.

Mass Spectrometry Reviews. John Wiley and Sons, Inc., 605 Third Avenue, New York, NY 10158. (800) 526-5368 or (212) 850-6000. 1982 to present. Bi-monthly. $330.00 per year. ISSN: 0277-7037.

Spectrochemica Acta A: Molecular Spectroscopy. Elsevier Science Inc., Publishing Co., Inc., Box 882, Madison Square Station, New York, NY 10159. (212) 989-5800. (212) 633-3990. 1939 to present. 14 issues per year. $1270.00 per year. ISSN: 0584-8539.

Spectrochemica Acta B: Atomic Spectroscopy. Elsevier Science Inc., Publishing Co., Inc., Box 882, Madison Square Station, New York, NY 10159. (212) 989-5800. (212) 633-3990. 1939 to present. 14 issues per year. $1445.00 per year. ISSN: 0584-8547.

Spectroscopy Letters. Marcel Dekker Journals, 270 Madison Avenue, New York, NY 10016. (212) 696-9000. FAX (212) 685-4540. 1985 to present. 9 issues per year. $59.00 per year. ISSN: 0887-6703.

RESEARCH CENTERS AND INSTITUTES

Atomic Physics and Chemical Physics Laboratory. University of Nevada - Reno, Department of Physics, Reno, NV 89507. (702) 784-4920.

George R. Harrison Spectroscopy Laboratory. Massachusetts Institute of Technology, 77 Massachusetts Avenue, Cambridge, MA 02139. (617) 253-7700.

Institute For Physical Science and Technology. University of Maryland, College Park, MD 20742. (301) 454-2636.

Institute of Optics. University of Rochester, Rochester, NY 14627. (716) 275-5428.

Nuclear Magnetic Resonance Facility. University of Missouri-Columbia, Chemical Building, Columbia, MO 65211. (314) 882-7725. FAX (314) 882-2754.

South Carolina Nuclear Magnetic Resonance Laboratory. University of South Carolina, Department of Chemistry, Columbia, SC 29208. (803) 777-7341.

ELECTRONIC CIRCUITS AND COMPONENTS

See also: ELECTRICAL ENGINEERING, ELECTRICITY, ELECTRON-ICS ENGINEERING, MICROELECTRICS

ABSTRACT SERVICES AND INDEXES

Applied Science and Technology Index; A Cumulative Subject Index To English Language Periodicals in the Fields of Aeronautics and Space Science, Computer Technology, Chemistry, Construction Industry, Energy and Related Areas. H.W. Wilson Co., 950 University Avenue, Bronx, NY 10452. (212) 588-8400. (800) 367-6770. FAX (718) 590-1617. From 1958 to present. Monthly. Inquire about cost and availability. Also available on CD-ROM and online. ISSN: 0003-6986.

Current Contents: Engineering, Technology and Applied Sciences. Institute for Scientific Information, 3501 Market Street, Philadelphia, PA 19104. (215) 386-0100. FAX (215) 386-6362. 1970 to present. Weekly. $442.00 per year. Also available on CD-ROM and online. Inquire regarding cost and availability. ISSN: 0095-7917.

Current Papers in Electrical and Electronics Engineering. Institution of Electrical Engineers (IEE), London. Distributed by INSPEC/IEEE, Box 1331, 445 Hoes Lane, Piscataway, NJ 08855-1331. (908) 562-5549. 1969 to present. Monthly. $345.00 per year. ISSN: 0011-3778.

Electrical and Electronics Abstracts. Institution of Electrical Engineers (IEE), London. Available from INSPEC/IEEE - Institute of Electrical and Electronic Engineers, Box 1331, Hoes Lane, Piscataway, NJ 08855-1331. (908) 562-5549. 1898 to present. Monthly. $2200.00 per year. Also available on CD-ROM and online as INSPEC.

Engineering Index Monthly: Indexes and Abstracts the World's Engineering and Technical Literature. Engineering Information, Inc., Castle Point on the Hudson, Hoboken, NJ 07030. (201) 216-8500. (800) 221-1044. FAX (201) 216-8532. Monthly.

$2300.00 per year. Available online as COMPENDEX and also on CD-ROM. ISSN: 0742-1974.

General Science Index. H.W. Wilson Company, 950 University Avenue, Bronx, NY 10452. (212) 588-8400. (800) 367-6770. FAX (718) 590-1617. From 1978 to present. Ten issues per year; quarterly and annual cumulations. Service basis. Available on CD-ROM and online. Inquire regarding cost and availability. ISSN: 0162-1963.

Government Reports Announcements and Index. U. S. National Technical Information Service (NTIS), 5285 Port Royal Road, Springfield, VA 22161. (703) 487-4650. FAX (703) 321-8547. From 1968 to present. Annual. $630.00 per year. Also available online as NTIS BIBLIOGRAPHIC DATABASE and on CD-ROM. ISSN:

Index To IEEE Publications. IEEE Service Center, 445 Hoes Lane, Piscataway, NJ 08855-1331. (908) 981-1393. (800) 678-IEEE. FAX (908) 981-9667. 1973 to present. Annual. ISSN: 0099-1368.

Physics Abstracts. INSPEC. Section A, Science Abstracts. Institution of Electrical Engineers (IEE), London. Available from: INSPEC/IEEE - Institute of Electrical and Electronic Engineers, Box 1331, Hoes Lane, Piscataway, NJ 08855-1331. (908) 562-5549. 1898 to present. 24 issues per year. $2835.00 per year. Also available online and on CD-ROM. ISSN: 0036-8091.

Physics Briefs (Physikalische Berichte). Information Center for Energy, Physics, Mathematics; German Physical Society. V C H Publishers, Inc., 220 East 23rd Street, New York, NY 10010-4606. (212) 683-8333. 1845 to present. 24 issues per year. $2390.00 per year. Also available online. ISSN: 0179-7434.

Science Citation Index. SCI. Institute for Scientific Information, 3501 Market Street, Philadelphia, PA 19104. (215) 386-0100. (800) 523-1850. FAX (215) 386-2991. 1961 to present. Six issues per year, plus annual cumulation. $11650.00 per year. Also available online and on CD-ROM. Inquire about price and availability. ISSN: 0036-827X.

Solid State and Superconductivity Abstracts: Covers theory, Production and Application of Solid State Materials. Cambridge Scientific Abstracts, 7200 Wisconsin Avenue, Bethesda, MD 20824. (301) 961-6750. FAX (301) 961-6720. 1957 to present. Bimonthly. $1320.00 per year. Also available online. ISSN: 0896-5900.

ANNUAL REVIEWS AND YEARBOOKS

Advances in Electronics and Electron Physics. Academic Press, Inc., 6277 Sea Harbor Drive, Orlando, FL 32887. (800) 321-5068. From 1948 to present. Irregular. Price varies, inquire. ISSN: 0065-2539.

Critical Reviews in Solid State and Materials Sciences. CRC Press, Inc., 2000 Corporate Boulevard, NW, Boca Raton, FL 33431. (407) 994-0555. (800) 272-7737. FAX (407) 998-9784. 1970 to present. Bimonthly. $265.00 per year. ISSN: 1040-8436.

ASSOCIATIONS AND PROFESSIONAL SOCIETIES

American Association of Engineering Societies, 1111 19th Street, Suite 608, Washington, DC 20036-3603. (202) 296-2237. FAX (202) 296-1151.

American Electronics Association. 5201 Great America Way, Suite 520, P.O. 52990, Santa Clara, CA 95056. (408) 987-4200. FAX (408) 970-8565.

American Institute of Physics. One Physics Ellipse, College Park, MD 20740-3843. (301) 209-3100.

Edison Electric Institute. 701 Pennsylvania Avenue NW, Washington, DC 20004-2696. (202) 508-5000, 5454. FAX (202) 508-5360.

Electric Power Research Institute. 3412 Hillview Avenue, Palo Alto, CA 94304. (415) 855-2000. FAX (415) 855-2954.

Electronics Industries Association. 2500 Wilson Boulevard, Arlington, VA 22201. (703) 907-7500. FAX (202) 457-4985.

IEEE Circuits and Systems Society. c/o Institute of Electrical and Electronic Engineers. 345 East 47th Street. New York, NY 10017. (212) 705-7900. FAX (212) 705-4929.

IEEE Solid-state Circuits Council. c/o Institute of Electrical and Electronic Engineers. 345 East 47th Street. New York, NY 10017. (212) 705-7900. FAX (212) 705-4929.

Institute For Interconnecting and Packaging Electronic Electronic Circuits. 2215 Sanders Road, Northbrook, IL 60062-6135. (708) 509-9700. (708) 509-9798.

Institute of Electrical and Electronics Engineers. 345 East 47th Street. New York, NY 10017. (212) 705-7900. FAX (212) 705-4929.

International Society for Hydrid Microelectronics. 1850 Centennial Park Drive, Suite 105, Reston, VA 22091. (703) 758-1060. FAX (703) 758-1066.

National Electrical Manufacturers Association. 1300 North 17th Street, Suite 1847, Rosslyn VA 22209. (703) 841-3200. FAX (703) 841-3300.

DIRECTORIES AND BIOGRAPHICAL SOURCES

American Electronics Association. Directory. 5201 Great America Way, Suite 520, P.O. 52990, Santa Clara, CA 95056. (408) 987-4200. FAX (408) 970-8565. Annual. $175.00.

American Men and Women of Science: Physical and Biological Sciences. R. R. Bowker Inc., 121 Chanlon Road, New Providence, NJ 07974. (908) 464-6800. (800) 521-8110. 20th edition. 8 volumes. 1996. $850.00.

Directory of Engineering Societies and Related Organizations. American Association of Engineering Societies, 1111 19th Street, Suite 608, Washington, DC 20036-3603. (202) 296-2237. Semi-annual. $150.00.

E E M - Electronic Engineer's Master. Hearst Business Communications, Inc., 645 Stewart Avenue, Garden City NY 11530. (516) 227-1300. ISSN: 0732-9016.

Electrical and Electronics Trades Directory: The Blue Book. Institution of Electrical Engineers. c.o IEEE Service Center, 445 Hoes Lane, Piscataway, NY 08854. (908) 981-1393. (800) 678-IEEE. FAX (908) 981-9667. Annual. 1995. $140.00

ELECTRONIC CIRCUITS AND COMPONENTS

Ency. of Physical Sciences and Engineering Info. Sources

Engineering Research Centers: Incorporating Electronics Research Centers. Stockton Press, 345 Park Avenue, New York, NY 10010. (212) 689-9200. (800) 221-2123. 4th edition. 1995. $515.00.

Ic Master. Hearst Business Communications, Inc., 645 Stewart Avenue, Garden City, NY 11530. (516) 227-1300. 1975 to present. Annual. $170.00 per year. ISSN: 0894-6809.

IEEE Membership Directory. Institute of Electrical and Electronics Engineers, IEEE Service Center, 445 Hoes Lane, Piscataway, NY 08854. (908) 981-1393. (800) 678-IEEE. FAX (908) 981-9667. 2 volumes. Annual. $190.00. ISSN:

International Directory of Abbreviations and Acroynms of Electronics, Electrical Engineering, Computer Technology and Information Processing. Peter Wennrich. K. G. Saur, 121 Chanlon Road, New Providence, NJ 07974. (908) 464-6800. (800) 521-8110 2 volumes. 1992. $230.00.

International Engineering Directory. American Consulting Engineers Council, 1015 15th Street, N.W. Suite 802, Washington, DC 20005-2670. (202) 347-7474. Annual. $10.00.

Research Centers Directory. Gale Research Company Inc., 835 Penobscot Building, Detroit, MI 48226-4094. (313) 961-2242. (800) 347-4253. 20th edition. 1995. $485.00. ISSN: 0080-1518.

Who's Who in Engineering. Gordon Davis, editor. American Association of Engineering Societies. 1111 19th Street, NY, Suite 608, Washington, DC 20036. (202) 296-2237. (800) 658-8897. 9th edition. 1995. $220.00.

Who's Who in Technology. Gale Research, 835 Penobscot Building, Detroit, MI 48226-4094. (313 961-2242. (800) 521-4253. 7th edition. 1995. $195.00. ISN 0-8103-7467-6.

ENCYCLOPEDIAS AND DICTIONARIES

Chambers Science and Technology Dictionary. Peter M.B. Walker, editor. Cambridge University Press, 40 West 20th Street, New York, NY 10011-4211. (212) 924-3900. 1988. $39.95.

Encyclopedia of Applied Physics. George Trigg, editor. VCH Publications, Inc., 220 East 23rd Street, Suite 909, New York, NY 10010-4606. (212) 683-8333. (800) 422-8824. 20 volume set. 1991 - . $5990.00.

Encyclopedia of Integrated Circuits. Walter A. Buchsbaum. Prentice Hall, 113 Sylvan Avenue, Route 9W, Englewood Cliffs, NJ 07632. (201) 592-2000. (800) 922-0579. 1987. $39.95.

IEEE Standard Dictionary of Electrical and Electronics Terms. Christopher J. Booth, editor. IEEE Service Center, 445 Hoes Lane, Piscataway, NJ 08855-1331. (908) 981-1393. (800) 678-IEEE. FAX (908) 981-9667. IEEE Standard 100-1992. 5th edition. 1993. $150.00.

Illustrated Dictionary of Electronics. Stan Gibilsco. TAB Books, P.O. Box 40, Blue Summit, PA 17294-0850. (717) 794-2191. (800) 233-1128. 7th edition. 1994. $34.95.

Illustrated Encyclopedic Dictionary of Electronic Circuits. John Douglas-Young. Prentice Hall, 113 Sylvan Avenue, Route 9W, Englewood Cliffs, NJ 07632. (201) 592-2000. (800) 922-0579. 1983. $32.95.

International Encyclopedia of Integrated Circuits. Arthur

A. Seidman. TAB Books, P.O. Box 40, Blue Summit, PA 17294-0850. (717) 794-2191. (800) 233-1128. 1991. $75.00.

McGraw-Hill Electronics Dictionary. J. Markus. McGraw-Hill Book Company, Inc., 1221 Avenue of the Americas, New York, NY 10020. (212) 997-3675. 5th edition. 1994. $49.95.

McGraw-Hill Encyclopedia of Engineering. Sybil P. Parker, editor. McGraw-Hill Book Company, Inc. 1221 Avenue of the Americas, New York, NY 10020. (212) 997-3675. Second edition. 1993. $95.50. ISBN: 0-07-051392-9.

McGraw-Hill Encyclopedia of Science and Technology. McGraw-Hill Book, Incorporated, 1221 Avenue of the Americas, New York, NY 10020. (212) 997-3675. (800) 262-4729. Seventh edition. Twenty volumes. 1992. $1900.00.

GENERAL WORKS

Design and Technology of Integrated Circuits. Donald DeCogan. John Wiley & Sons, Inc., 605 Third Avenue, New York, NY 10158-0012. (212) 850-6000. (800) 225-5945. 1990. $75.00.

Diode Lasers and Photonic Integrated Circuits. David J. Comer. John Wiley & Sons, Inc., 605 Third Avenue, New York, NY 10158-0012. (212) 850-6000. (800) 225-5945. 1995. $64.95.

Computer Hardware Diagnostics For Engineers. Ronald E. Howland. McGraw-Hill Publishing Company, Inc., 1221 Avenue of the Americas, New York, NY 10020. (212) 512-2000. (800) 262-4729. 1995. $55.00.

Electric Circuits: Principles, Applications and Computer Analysis. David A. Bell. Prentice Hall, 113 Sylvan Avenue, Route 9W, Englewood Cliffs, NJ 07632. (201) 592-2000. (800) 922-0579. 5th edition. 1995. $72.00.

Exploring Electronic Devices. Mark E. Hazen. SCP: Third World Literature Publishing House, P.O. Box 482, Lithonia, GA 30058-0482. (404) 785-7725. 1991. $50.75.

Integrated Circuits and Microprocessors. Robin C. Holland. Elsevier Science Publishing Company, Inc., 655 Avenue of the Americas, New York, NY 10010. (212) 989-5800. FAX (914) 333-2444. 1986. $30.00 in paper.

Introduction To Electronics Design. Fred H. Mitchell and P. H. Mitchell, Sr. Prentice Hall, 113 Sylvan Avenue, Route 9W, Englewood Cliffs, NJ 07632. (201) 592-2000. (800) 922-0579. 2nd edition. 1991. $81.00

Introduction To Fields and Circuits. Gordon Lancaster. Oxford University Press, Inc., 200 Madison Avenue, New York, NY 10016. (212) 725-6000. (800) 334-4249. 1992 $39.95.

Introduction To Vlsi Techology. T. E. Price. Prentice Hall, 113 Sylvan Avenue, Route 9W, Englewood Cliffs, NJ 07632. (201) 592-2000. (800) 922-0579. 1994. $60.00.

Mastering IC Circuits. Joseph J. Carr. TAB Books, P.O. Box 40, Blue Summit, PA 17294-0850. (717) 794-2191. (800) 233-1128. 1991. $32.95.

Operational Amplifiers, Integrated and Hybrid Circuits. George B. Rutkowski. Prentice Hall, 113 Sylvan Avenue, Route 9W, Englewood Cliffs, NJ 07632. (201) 592-2000. (800) 922-0579. 1993. $87.95.

Simplified Design of Linear Power Supplies. John D. Lenk. Butterworth-Heinemann, 313 Washington Street, Newton, MA 02158. (618-928-2500. (800) 366-2665. 1994. $29.95.

Transistors: Fundamentals For the Integrated-circuit Engineer. R. M. Warner Jr., and B. L. Grung. Krieger Publishing Company, P.O. Box 9542, Melbourne, FL 32902-9542. (407) 724-9542. 1990. $72.50.

HANDBOOKS AND MANUALS

Active Electronic Component Handbook. Charles A. Harper and Harold C. Jones. McGraw-Hill Book Company, 1221 Avenue of the Americas, New York, NY 10020. (212) 512-2000. (800) 262-4729. 2nd edition. 1996. $79.50.

Edn Designer's Companion. Ian Hickman and Bill Travis, editors. Butterworth-Heinemann, 313 Washington Street, Newton, MA 02158. (618-928-2500. (800) 366-2665. 1994. $29.95.

Electronics Circuits Handbook. M. Tooley. Butterworth-Heinemann, 313 Washington Street, Newton, MA 02158. (618-928-2500. (800) 366-2665. 1993. $49.95.

Electronics Handbook. Jerry C. Whitaker. CRC Press, Inc., 2000 Corporate Boulevard, NW, Boca Raton, FL 33431. (407) 994-0555. (800) 272-7737. 1996. $120.00.

Handbook of Electrical and Electronic Technology. Curtis Johnson. Prentice Hall , 113 Sylvan Avenue, Route 9W, Englewood Cliffs, NJ 07632. (201) 592-2000. (800) 922-0579. 1996. $88.00.

Handbook of Quality Integrated Circuit Manufacturing. Robert Zurich. Academic Press, Inc., 6277 Sea Harbor Drive, Orlando, FL. (800) 321-5068. 1991. $59.95.

Handbook of Vlsi Chip Design and Expert Systems. A. F. Schwartz. Academic Press, Inc., 6277 Sea Harbor Drive, Orlando, FL. (800) 321-5068.

Integrated Circuits Handbook. Rector Press, Ltd., 130 Rattlesnake, Leverett, MA 01054-9726. (413) 548-9708. (800) 247-3473. 3 volumes. 1995. $295.00 in paper.

Standard Handbook For Electrical Engineers. Donald Fink. McGraw-Hill Publishing Company, 1221 Avenue of the Americas, New York, NY 10020. (212) 512-2000. (800) 262-4729. 13th edition. 1996. $110.00.

ONLINE DATABASES AND CD-ROMS

Compendex Plus. Engineering Information, Inc., 345 East 47th Street, New York, NY 10017. (212) 705-7600 or (800) 221-1044. Contains citations with abstracts to worldwide literature in engineering and technology, from 1970 to present. Available on online BRS,(800) 289-4277, DIALOG, (800) 334-2564, ORBIT (800) 456-7248, and STN International, FIZ Karlsruhe, P.O. Box 2465, W-7500, Karlsruhe 1, Germany, online services. Also available on CD-ROM. Inquire as to cost and availability.

Current Contents Search. Institute for Scientific Information, 3501 Market Street, Philadelphia, PA 19104. (215) 386-0100. FAX (215) 386-6362. Contains citations to articles listed in the table of contents of science and technology journals. Also articles in social sciences and life sciences journals. Available on BRS,(800) 289-4277, DIALOG,(800) 334-2564, online services. Inquire as to cost and availability.

Dissertation Abstracts Online. University Microfilms International, 300 North Zeeb Road, Ann Arbor, MI 48106. (800) 521-0600 or (313) 761-4700. Scope includes virtually all doctoral dissertations accepted at accredited American institutions from 1861 to present in 252 subject areas. Available on BRS, (800) 289-4277, DIALOG, (800) 334-2564, and OCLC EPIC,(800) 848-5878, online services. Also available on CD-ROM. Inquire as to cost and availability.

INSPEC. Institution of Electrical Engineers, Michael Faraday House, Six Hills Way, Stevenage, Herts. SG1 2AY, England. Telephone: 0438 313311 or FAX 0438 742840. Contains citations to the worldwide literature of physics, electronics and electrical engineering, computer technology, and related fields. Available on BRS, (800) 289-4277, DIALOG, (800) 334-2564, ORBIT, (800) 456-7248, and STN International, FIZ Karlsruhe, P.O. Box 2465, W-7500, Karlsruhe 1, Germany, online services. Inquire as to cost and availability.

NTIS Bibliographic Database. National Technical Information Service, 5285 Port Royal Road, Springfield, VA 22161. (703) 487-4929 or FAX (703) 321-8199. Broad coverage of government-sponsored science and technology research reports, 1964 to present. Available on BRS,(800) 289-4277, DIALOG, (800) 334-2564, ORBIT, (800) 456-7248, and STN International, FIZ Karlsruhe, P.O. Box 2465, W-7500, Karlsruhe 1, Germany, online services. Also available on CD-ROM. Inquire as to cost and availability.

Physics Briefs. American Institute of Physics, 335 East 45th Street, New York, NY 10017. (212) 661-9260 or FAX (212) 661-2036. Contains citations with abstracts of the literature of physics and related fields, 1979 to present. Available on the STN International, FIZ Karlsruhe, P.O. Box 2465, W-7500, Karlsruhe 1, Germany, online service. Inquire as to cost and availability.

SCISEARCH. Institute for Scientific Information, 3501 Market Street, Philadelphia, PA 19104. (800) 523-1850 or (215) 386-0100. Broad multidisciplinary title and author index to the international literature of science and technology, 1974 to present. Available on DIALOG,(800) 334-2564, and ORBIT,(800) 456-7248, online services. Also available on CD-ROM. Inquire as to cost and availability.

WILSONLINE. H.W. Wilson Company, 950 University Avenue, Bronx, NY 10452. (800) 367-6770 or (212) 588-8400. Makes available online versions of the printed H.W. Wilson indexes including Applied Science and Technology Index, Business Periodicals Index, General Science Index, and Readers' Guide to Periodical Literature. Period covered is generally 1983 to present. Available on BRS,(800) 289-4277, DIALOG,(800) 334-2564, and OCLC EPIC,(800) 848-5878, online services. Also available on CD-ROM. Inquire as to cost and availability.

PERIODICALS

Circuits, Systems, and Signal Processing. Birkhauser, 675 Massachusetts Avenue, Cambridge, MA 02139-3309. FAX (201) 348-4505. 1982 to present. Quarterly. $184.00 per year. ISSN: 0278-081X.

ELECTRONIC CIRCUITS AND COMPONENTS

Ency. of Physical Sciences and Engineering Info. Sources

Electrical World. McGraw-Hill, Inc., Box 513 Hightstown, NJ 08520. (212) 512-3288. 1874 to present. Monthly. $55.00 per year. ISSN: 0013-4457.

Electronic Design. Penton Publishing, San Jose Gateway, Suite 354. 2025 Gateway Place, San Jose, CA 95110. (408) 441-0550. 1952 to present. Fortnightly. $95.00. ISSN: 0013-4872.

Electronic Engineering Times. CMP Publications, Inc., 600 Community Drice, Manhasset, NY 11030. (516) 562-5000. FAX (516) 562-5325. 1972 to present. Weekly. $159.00 per year. ISSN: 0192-1541.

Electronics. Penton Publishing, San Jose Gateway, Suite 354. 2025 Gateway Place, San Jose, CA 95110. (408) 441-0550. 1930 to present. Semi-weekly. $98.00 per year. ISSN: 0883-4989.

Electronics World and Wireless World. Reed Business Publishing, Ltd., Quadrant House, the Quadrant, Sutton, Surrey SM2 5AS, England. 1911 to present. Monthly. L35. ISSN: 0266-3244.

IEEE Circuits and Devices Magazine. Institute of Electrical and Electronics Engineers, Inc., Box 1331, 445 Hoes Lane, Piscataway, NJ 08855-1331. (908) 981-0060. 1985 to present. Bi-monthly. $120.00 per year. ISSN: 8755-3996.

IEEE Journal of Solid State Circuits. Institute of Electrical and Electronics Engineers, Inc., Box 1331, 445 Hoes Lane, Piscataway, NJ 08855-1331. (908) 981-0060. 1966 to present. Bi-monthly. $275.00 per year. ISSN: 0018-9200.

IEEE Spectrum. Institute of Electrical and Electronics Engineers, Inc., Box 1331, 445 Hoes Lane, Piscataway, NJ 08855-1331. (908) 981-0060. 1964 to present. Monthly. $157.00 per year. ISSN: 0018-9235.

IEEE Transactions On Circuits and Systems. Part 1:Fundamental theory and Applications. Institute of Electrical and Electronics Engineers, Inc., Box 1331, 445 Hoes Lane, Piscataway, NJ 08855-1331. (908) 981-0060. 1952 to present. Monthly. $241.00 per year. ISSN: 1057-7122.

IEEE Transactions On Electron Devices. Institute of Electrical and Electronics Engineers, Inc., Box 1331, 445 Hoes Lane, Piscataway, NJ 08855-1331. (908) 981-0060. 1952 to present. Monthly. $395.00 per year. ISSN: 0018-9393.

Institute of Electrical and Electronics Engineers. Proceedings. Institute of Electrical and Electronics Engineers, Inc., Box 1331, 445 Hoes Lane, Piscataway, NJ 08855-1331. (908) 981-0060. 1913 to present. Monthly. $275.00. ISSN: 0018-9219.

Semiconductor International. Cahners Publishing Co., 44 Cook Street, Denver, CO 80206. (708) 635-8000. FAX (708) 390-2770. (800) 662-7776. 1978 to present. 13 issues per year. $84.95 per year. ISSN: 0163-3767.

Solid State Electronics. Elsevier Science, 660 White Plains Road, Tarrytown, NY 10591-5153. (914) 524-9200. FAX (914) 333-2444. 1960 to present. Monthly. $1025.00 per year. ISSN: 0038-1101.

RESEARCH CENTERS AND INSTITUTES

Alabama Microelectronics Science and Technology Center. Auburn University, 200 Broun Hall, Auburn University, AL 36849. (205) 844-1871. (205) 844-1809.

Center For Integrated Sensors and Circuits. University of Michigan, 1246 EECS Building, Ann Arbor, MI 48109-2122. (313) 764-3346. FAX (313) 747-1781.

Center For Integrated Circuit Packaging. Lehigh University, Sherman Fairchild Laboratory, Building 161. Bethlehem, PA 18015. (215) 758-4409. FAX (215) 758-4561.

Center For Integrated Systems. Stanford University, School of Engineering, Stanford, CA 94305-4028. (415) 723-9038.

Electrical Engineering Research Laboratories. Purdue University. Electrical Engineering Building, West Lafayette, in 47907. (317) 494-3536. FAX (317) 494-6440.

Electronics Research Laboratory. University of California, Berkeley, 253 Cory Hall, Berkeley, CA 94720. (415) 642-2301.

Engineering Research Center. University of Maryland, Clark School of Engineering, Potomac Building, Room 2104. (301) 405-3906. FAX (301) 403-4105.

Solid State Electronics Laboratory. North Carolina State University, 432 Daniels Hall, Raleigh, NC 27695. (919) 737-2336.

ELECTRONIC MAIL

See: COMPUTER COMMUNICATIONS

ELECTRONIC SECURITY SYSTEMS

See: ALARM SYSTEMS

ELECTRONICS

See also: ELECTRICAL ENGINEERING, ELECTROMAGNETISM, ELECTRONIC CIRCUITS and COMPONENTS, ELECTRONICS ENGINEERING, MICROWAVES, RADIO, SIGNAL PROCESSING, SOLID STATE PHYSICS

ABSTRACT SERVICES AND INDEXES

Applied Science and Technology Index: A Cumulative Subject Index to English Language Periodicals in the Fields of Aeronautics and Space Science, Computer Technology,chemistry, Construction Industry, Energy and Related Areas. H.W. Wilson Co., 950 University Avenue, Bronx, NY 10452. (212) 588-8400. (800) 367-6770. FAX (718) 590-1617. From 1958 to present. Monthly. Inquire about cost and availability. Also available on CD-ROM and online. ISSN: 0003-6986.

Current Contents: Engineering, Technology and Applied Sciences. Institute for Scientific Information, 3501 Market Street, Philadelphia, PA 19104. (215) 386-0100. FAX (215) 386-6362. 1970 to present. Weekly. $442.00 per year. Also available on CD-ROM and online. Inquire regarding cost and availability. ISSN: 0095-7917.

Current Papers in Electrical and Electronics Engineering. Institution of Electrical Engineers (IEE), London. Distributed by

INSPEC/IEEE, Box 1331, 445 Hoes Lane, Piscataway, NJ 08855-1331. (908) 562-5549. 1969 to present. Monthly. $345.00 per year. ISSN: 0011-3778.

Electrical and Electronics Abstracts. Institute of Electrical Engineers, (IEE), London. Available from INSPEC/IEEE - Institute of Electrical and Electronic Engineers, Box 1331, Hoes Lane, Piscataway, NJ 08855-1331. (908) 562-5549. 1898 to present. Monthly. $2200.00 per year. Also available on CD-ROM and online as INSPEC. ISSN: 0036-8105.

Engineering Index Monthly: Indexes and Abstracts the World's Engineering and Technical Literature. Engineering Information, Inc., Castle Point on the Hudson, Hoboken, NJ 07030. (201) 216-8500. (800) 221-1044. FAX (201) 216-8532. Monthly. $2300.00 per year. Available online as COMPENDEX and also on CD-ROM. ISSN: 0742-1974.

General Science Index. H.W. Wilson Company, 950 University Avenue, Bronx, NY 10452. (212) 588-8400. (800) 367-6770. FAX (718) 590-1617. From 1978 to present. Ten issues per year; quarterly and annual cumulations. Service basis. Available on CD-ROM and online. Inquire regarding cost and availability. ISSN: 0162-1963.

Government Reports Announcements and Index. U. S. National Technical Information Service (NTIS), 5285 Port Royal Road, Springfield, VA 22161. (703) 487-4650. FAX (703) 321-8547. From 1968 to present. Annual. $630.00 per year. Also available online as *NTIS Bibliographic Database* and on CD-ROM.

Index To IEEE Publications. IEEE Service Center, 445 Hoes Lane, Piscataway, NJ 08855-1331. (908) 981-1393. (800) 678-IEEE. FAX (908) 981-9667. 1973 to present. Annual. $ ISSN: 0099-1368.

NTIS Alerts: Electrotechnology. U. S. National Technical Information Service, 5285 Port Royal Road, Springfield, VA 22161. (703) 487-4630. FAX (703) 321-8547. Weekly. $135.00 per year.

Pascal 20: Electronique Et Telecommunications. Centre National de la Recherche Scientifique, Institut de I'information Scientific et Technique, 2 allee de Parc de Brabois, 54514 Vandoeuvre-Les Nancy Cedex, France. TEL 83-50-46-00. FAX 83-50-46-50. 1984 to present. 10 issues per year. 1295 F per year. ISSN: 1146-5352.

Physics Abstracts. INSPEC. Section A, Science Abstracts. Institute of Electrical Engineers, London, United Kingdom. Available from: INSPEC/IEEE - Institute of Electrical and Electronic Engineers, Box 1331, Hoes Lane, Piscataway, NJ 08855-1331. (908) 562-5549. 1898 to present. 24 issues per year. $2835.00 per year. ISSN: 0036-8091. Also available online and on CD-ROM.

Physics Briefs (Physikalische Berichte). Information Center for Energy, Physics, Mathematics; German Physical Society. V C H Publishers, Inc., 220 East 23rd Street, New York, NY 10010-4606. (212) 683-8333. 1845 to present. 24 issues per year. $2390.00 per year. Also available online. ISSN: 0179-7434.

Science Citation Index. SCI. Institute for Scientific Information, 3501 Market Street, Philadelphia, PA 19104. (215) 386-0100. (800) 523-1850. FAX (215) 386-2991. 1961 to present. Six issues per year, plus annual cumulation. $11650.00 per year. Also available online and on CD-ROM. Inquire about price and availability. ISSN: 0036-827X.

Solid State and Superconductivity Abstracts: Covers theory, Production and Application of Solid State Materials. Cambridge Scientific Abstracts, 7200 Wisconsin Avenue, Bethesda, MD 20824. (301) 961-6750. FAX (301) 961-6720. 1957 to present. Bimonthly. $1320.00 per year. Also available online. ISSN: 0896-5900.

STAR. (Scientific and Technical Aerospace Reports). U.S. National Aeronautics and Space Administration, Scientific and Technical Information Facility, Box 8757, Baltimore-Washington International Airport, MD 21240. (301) 621-0153. From 1963 to present. Semimonthly, with semiannual and annual indexes. $114.00 per year. Also available online. ISSN: 0036-8741.

ANNUAL REVIEWS AND YEARBOOKS

Advances in Electronics and Electron Physics. Academic Press, Inc., 1250 Sixth Avenue, San Diego, CA 92101-4311. (619) 231-0926. FAX (619) 699-6715. Irregular. Price varies, inquire.

ASSOCIATIONS AND PROFESSIONAL SOCIETIES

American Electronics Association. 5201 Great America Way, Suite 520, P.O. 52990, Santa Clara, CA 95056. (408) 987-4200. FAX (408) 970-8565.

American Institute of Physics. One Physics Ellipse, College Park, MD 20740-3843. (301) 209-3100.

Edison Electric Institute. 701 Pennsylvania Avenue NW, Washington, DC 20004-2696. (202) 508-5000, 5454. FAX (202) 508-5360.

Electric Power Research Institute. 3412 Hillview Avenue, Palo Alto, CA 94304. (415) 855-2000. FAX (415) 855-2954.

Electronics Industries Association. 2500 Wilson Boulevard, Arlington, VA 22201. (703) 907-7500. FAX (202) 457-4985.

IEEE (institute of Electrical and Electronic Engineers). 345 East 47th Street. New York, NY 10017. (212) 705-7900. FAX (212) 705-4929.

International Society For Hybrid Microelectronics. 1850 Centennial Park Drive, Suite 105, Reston, VA 22091. (703) 758-1060. FAX (703) 758-1066.

National Electrical Manufacturers Association. 1300 North 17th Street, Suite 1847, Rosslyn VA 22209. (703) 841-3200. FAX (703) 841-3300.

DIRECTORIES AND BIOGRAPHICAL SOURCES

American Men and Women of Science: Physical and Biological Sciences. R. R. Bowker Inc., 121 Chanlon Road, New Providence, NJ 07974. (908) 464-6800. (800) 521-8110. 20th edition. 8 volumes. 1996. $850.00.

Directory of Engineering Societies and Related Organizations. American Association of Engineering Societies, 1111 19th Street, Suite 608, Washington, DC 20036-3603. (202) 296-2237. Semi-annual. $150.00.

ELECTRONICS

Ency. of Physical Sciences and Engineering Info. Sources

Directory of Engineers in Private Practice. National Society of Professional Engineers, 1420 King Street, Alexandria, VA 22314-2750. (703) 684-2835. Biennial. $30.00.

E E M - Electronic Engineer's Master. Hearst Business Communications, Inc., 645 Stewart Avenue, Garden City NY 11530. (516) 227-1300. Annual. $90.00 ISSN: 0732-9016.

Electrical and Electronic Trades Directory: The Blue Book. Institution of Electrical Engineers, Michael Faraday House, Six Hills Way, Stevenage, Herts. SG1 2AY, England. Telephone: 0438 313311 or FAX 0438 742840. 1995. $140.00

Electronics Manufacturers Directory. Harris InfoSource, 2057 Aurora Road, Twinsburg, OH 44087-1999. (216) 425-9000. FAX (216) 425-7150. 1995. $323.50

Engineering Research Centers: Incorporating Electronics Research Centers. Stockton Press, 345 Park Avenue, New York, NY 10010. (212) 689-9200. (800) 221-2123. 4th edition. 1995. $515.00.

IEEE Membership Directory. Institute of Electrical and Electronics Engineers, IEEE Service Center, 445 Hoes Lane, Piscataway, NY 08854. (908) 981-1393. (800) 678-IEEE. FAX (908) 981-9667. 2 volumes. Annual. $190.00.

International Directory of Abbreviations and Acroynms of Electronics, Electrical Engineering, Computer Technology and Information Processing. Peter Wennrich. K. G. Saur, 121 Chanlon Road, New Providence, NJ 07974. (908) 464-6800. (800) 521-8110 2 volumes. 1992. $230.00.

International Engineering Directory. American Consulting Engineers Council, 1015 15th Street, N.W. Suite 802, Washington, DC 20005-2670. (202) 347-7474. Annual. $10.00.

Research Centers Directory. Gale Research Inc., 835 Penobscot Building, Detroit, MI 48226-4094. (313) 961-2242. (800) 347-4253. 20th edition. 1995. $485.00. ISSN: 0080-1518.

Who's Who in Engineering. Gordon Davis, editor. American Association of Engineering Societies. 1111 19th Street, NY, Suite 608, Washington, DC 20036. (202) 296-2237. (800) 658-8897. 9th edition. 1995. $220.00.

Who's Who in Technology. Gale Research, 835 Penobscot Building, Detroit, MI 48226-4094. (313 961-2242. (800) 521-4253. 7th edition. 1995. $195.00. ISN 0-8103-7467-6.

ENCYCLOPEDIAS AND DICTIONARIES

IEEE Standard Dictionary of Electrical and Electronics Terms. Christopher J. Booth, editor. IEEE Service Center, 445 Hoes Lane, Piscataway, NJ 08855-1331. (908) 981-1393. (800) 678-IEEE. FAX (908) 981-9667. IEEE Standard 100-1992. 5th edition. 1993. $150.00.

Illustrated Dictionary of Electronics. Stan Gibilsco. TAB Books, P.O. Box 40, Blue Summit, PA 17294-0850. (717) 794-2191. (800) 233-1128. 7th edition. 1994. $34.95.

McGraw-Hill Electronics Dictionary. J. Markus. McGraw-Hill Book Company, Inc., 1221 Avenue of the Americas, New York, NY 10020. (212) 997-3675. 5th edition. 1994. $49.95.

McGraw-Hill Encyclopedia of Engineering. Sybil P. Parker, editor. McGraw-Hill Book Company, Inc. 1221 Avenue of the Americas, New York, NY 10020. (212) 997-3675. Second edition. 1993. $95.50. ISBN: 0-07-051392-9.

McGraw-Hill Encyclopedia of Science and Technology. McGraw-Hill Book, Incorporated, 1221 Avenue of the Americas, New York, NY 10020. (212) 997-3675. (800) 262-4729. Seventh edition. Twenty volumes. 1992. $1900.00.

GENERAL WORKS

The Art of Electronics. Paul Horowitz and Winfield Hill. Cambridge University Press, 40 West 20th Street, New York, NY 10011-4211. (212) 924-3900. (800) 872-7423. 1989. $59.95.

Basic Electronics For Scientists. James J. Brophy. McGraw-Hill Book, Incorporated, 1221 Avenue of the Americas, New York, NY 10020. (212) 997-3675. (800) 262-4729. 1990.

Electric and Electronic Engineering For Scientists and Engineers. K. A. Krishnamurthy. Halsted Press, 605 Third Avenue, New York, 10158-0012. (212) 850-6400. 1944. $48.95.

Electrical and Electronic Technologies: A Chronology of Events and Inventors From 1940 To 1980. Henry B. Davis. Scarecrow Press, Inc., 52 Liberty Street, Box 4167, Metuchen, NJ 08840. (908) 548-8600. 1985. $35.00.

Electromagnetism For Electronic Engineers. R. G. Carter. Chapman & Hall, 1 Penn Plaza, New York, NY 10119. (212) 564-1060. 1992. $24.95.

Electronics Problem Solver. Research and Education Association, 61 Ethel Road, West Piscataway, NJ 08854. (908) 819-8880. Revised edition. 1994. $ 29.95.

Exploring Electronic Devices. Mark E. Hazen. SCP: Third World Literature Publishing House, P.O. Box 482, Lithonia, GA 30058-0482. (404) 785-7725. 1991. $50.75.

Fundamentals of Electronic Communication Systems. Wayne Tomasi. Prentice-Hall, 113 Sylvan Avenue, Route 9W, Englewood Cliffs, NJ 07632. (201) 592-2000. (800) 922-0579. 1993. $72.00.

High-voltage Engineering and Testing. H. M. Ryan, editor. Institution of Electrical Engineers; IEE/INSPEC, Box 1331, 445 Hoes Lane, Piscataway, NJ 08855-1331. (908) 981-0060. 1994. $95.00.

An Introduction To Applied Electromagetism. Christos Christopoulos. John Wiley & Sons, Inc., 605 Third Avenue, New York, NY 10158-0012. (212) 850-6000. (800) 225-5945. 1990. $59.95.

Introductory AC-DC Circuits. Nigel P. Cook. Prentice-Hall, 113 Sylvan Avenue, Route 9W, Englewood Cliffs, NJ 07632. (201) 592-2000. (800) 922-0579. 1995. $66.00.

Schaum's Outline of Electrical Technology. Milton Kaufman and Peter Brooks. McGraw-Hill Publishing Company, Inc., 1221 Avenue of the Americas, New York, NY 10020. (212) 512-2000. (800) 262-4729. 2nd edition. 1996.

HANDBOOKS AND MANUALS

Active Electronic Component Handbook. Charles A. Harper and Harold C. Jones. McGraw-Hill Book Company, 1221 Avenue of the Americas, New York, NY 10020. (212) 512-2000. (800) 262-4729. 2nd edition. 1996. $79.50.

Electrical Engineering Handbook. Richard C. Dorf, editor. CRC Press, Inc., 2000 Corporate Boulevard, NW, Boca Raton, FL 33431. (407) 994-0555. (800) 272-7737. 1993. $99.95.

Electronic Engineer's Handbook. Donald G. Fink and Donald Christiansen, editors. McGraw-Hill Book Company, 1221 Avenue of the Americas, New York, NY 10020. (212) 512-2000. (800) 262-4729. 3rd edition. 1989. $105.00.

Electronics Handbook. Jerry C. Whitaker. CRC Press, Inc., 2000 Corporate Boulevard, NW, Boca Raton, FL 33431. (407) 994-0555. (800) 272-7737. 1996. $120.00.

Engineering Tables and Data. A.M. Howatson, P.G. Lund and J.D. Todd. Van Nostrand Reinhold, 115 Fifth Avenue, New York, NY 10003. (212) 254-3232. (800) 926-2665. Second Edition. 1991. $39.95.

Handbook of Electrical and Electronic Technology. Curtis Johnson. Prentice Hall, 113 Sylvan Avenue, Route 9W, Englewood Cliffs, NJ 07632. (201) 592-2000. (800) 922-0579. 1996. $88.00.

Handbook of Solid State Batteries and Capacitors. M. Z. Munshi and P. S. Prasad. World Scientific Publishing Company, Inc., 1060 Main Street, River Edge, NJ 07661. (201) 487-9655. (800) 227-7562. 1995. $177.00.

Industrial Electronics and Systems Handbook. Irwin, CRC Press, Inc., 2000 Corporate Boulevard, NW, Boca Raton, FL 33431. (407) 994-0555. (800) 272-7737. 1996. $129.95.

Standard Handbook For Electrical Engineers. Donald Fink. McGraw-Hill Publishing Company, 1221 Avenue of the Americas, New York, NY 10020. (212) 512-2000. (800) 262-4729. 13th edition. 1996. $110.00.

Standard Handbook of Engineering Calculations. Tyler G. Hicks. McGraw-Hill Publishing Company, 1221 Avenue of the Americas, New York, NY 10020. (212) 512-2000. (800) 262-4729. 3rd edition. 1994. $99.50. ISBN: 0-07-028812-7.

ONLINE DATABASES AND CD-ROMS

CA SEARCH. Chemical Abstracts Service, P.O. Box 3012, Columbus, OH 43210-0012. (614) 447-3600. (800) 848-6533. FAX (614) 447-3709. Very comprehensive guide to worldwide chemical literature and related fields, 1972 to present. Available on BRS,(800) 289-4277, DIALOG, (800) 334-2564, ORBIT (800) 456-7248, and STN International, FIZ Karlsruhe, P.O. Box 2465, W-7500, Karlsruhe 1, Germany, online services. Inquire as to cost and availability.

Compendex Plus. Engineering Information, Inc., 345 East 47th Street, New York, NY 10017. (212) 705-7600 or (800) 221-1044. Contains citations with abstracts to worldwide literature in engineering and technology, from 1970 to present. Available on online BRS,(800) 289-4277, DIALOG, (800) 334-2564, ORBIT (800) 456-7248, and STN International, FIZ Karlsruhe,

P.O. Box 2465, W-7500, Karlsruhe 1, Germany, online services. Also available on CD-ROM. Inquire as to cost and availability.

Current Contents Search. Institute for Scientific Information, 3501 Market Street, Philadelphia, PA 19104. (215) 386-0100. FAX (215) 386-6362. Contains citations to articles listed in the table of contents of science and technology journals. Also articles in social sciences and life sciences journals. Available on BRS,(800) 289-4277, DIALOG,(800) 334-2564, online services. Inquire as to cost and availability.

Dissertation Abstracts. University Microfilms International, 300 North Zeeb Road, Ann Arbor, MI 48106. (800) 521-0600 or (313) 761-4700. Scope includes virtually all doctoral dissertations accepted at accredited American institutions from 1861 to present in 252 subject areas. Available on BRS,(800) 289-4277, DIALOG,(800) 334-2564, and OCLC EPIC,(800) 848-5878, online services. Also available on CD-ROM. Inquire as to cost and availability.

INSPEC. Institution of Electrical Engineers, Michael Faraday House, Six Hills Way, Stevenage, Herts. SG1 2AY, England. Telephone: 0438 313311 or FAX 0438 742840. Contains citations to the worldwide literature of physics, electronics and electrical engineering, computer technology, and related fields. Available on BRS,(800) 289-4277, DIALOG, (800) 334-2564, ORBIT, (800) 456-7248, and STN International, FIZ Karlsruhe, P.O. Box 2465, W-7500, Karlsruhe 1, Germany, online services. Inquire as to cost and availability.

NTIS Bibliographic Database. National Technical Information Service, 5285 Port Royal Road, Springfield, VA 22161. (703) 487-4929 or FAX (703) 321-8199. Broad coverage of government-sponsored science and technology research reports, 1964 to present. Available on BRS,(800) 289-4277, DIALOG, (800) 334-2564, ORBIT, (800) 456-7248, and STN International, FIZ Karlsruhe, P.O. Box 2465, W-7500, Karlsruhe 1, Germany, online services. Also available on CD-ROM. Inquire as to cost and availability.

Scientific and Technical Books and Serials in Print. R.R. Bowker Inc., 121 Chanlon Road, New Providence, NJ 07974. (800) 521-8110 or (908) 464-6800. List of currently published books and serials in the physical and biological sciences, engineering and technology, with subject, author and titles indexes. Available on ORBIT,(800) 456-7248, online service.Inquire as to cost and availability.

SCISEARCH. Institute for Scientific Information, 3501 Market Street, Philadelphia, PA 19104. (800) 523-1850 or (215) 386-0100. Broad multidisciplinary title and author index to the international literature of science and technology, 1974 to present. Available on DIALOG,(800) 334-2564, and ORBIT,(800) 456-7248, online services. Also available on CD-ROM. Inquire as to cost and availability.

WILSONLINE. H.W. Wilson Company, 950 University Avenue, Bronx, NY 10452. (800) 367-6770 or (212) 588-8400. Makes available online versions of the printed H.W. Wilson indexes including Applied Science and Technology Index, Business Periodicals Index, General Science Index, and Readers' Guide to Periodical Literature. Period covered is generally 1983 to present. Available on BRS,(800) 289-4277, DIALOG,(800) 334-2564, and OCLC EPIC,(800) 848-5878, online services. Also available on CD-ROM. Inquire as to cost and availability.

PERIODICALS

Circuits, Systems, and Signal Processing. Birkhauser, 675 Massachusetts Avenue, Cambridge, MA 02139-3309. FAX (201) 348-4505. 1982 to present. Quarterly. $184.00 per year. ISSN: 0278-081X.

E D N Magazine. Cahners Publishing Company, 44 Cook Street, Denver CO 80206. (800) 662-7776. 1956 to present. 26 issues per year. $120.00 per year.

Electrical World. McGraw-Hill, Inc., Box 513 Hightstown, NJ 08520. (212) 512-3288. 1874 to present. Monthly. $55.00 per year. ISSN: 0013-4457.

Electronic Design. Penton Publishing, San Jose Gateway, Suite 354. 2025 Gateway Place, San Jose, CA 95110. (408) 441-0550. 1952 to present. Fortnightly. $95.00. ISSN: 0013-4872.

Electronic Engineering Times. CMP Publications, Inc., 600 Community Drice, Manhasset, NY 11030. (516) 562-5000. FAX (516) 562-5325. 1972 to present. Weekly. $159.00 per year. ISSN: 0192-1541.

Electronics. Penton Publishing, San Jose Gateway, Suite 354. 2025 Gateway Place, San Jose, CA 95110. (408) 441-0550. 1930 to present. Semi-weekly. $98.00 per year. ISSN: 0883-4989.

Electronics Now. Gernsback Publications, Inc., 500-B Bi-County Boulevard, Farmingdale, NY 11735. (516) 293-3000. FAX (516) 293-3115. 1929 to present. Monthly. $19.95. ISSN: 0033-7862.

Electronics World and Wireless World. Reed Business Publishing, Ltd., Quadrant House, the Quadrant, Sutton, Surrey SM2 5AS, England. 1911 to present. Monthly. L35. ISSN: 0266-3244.

IEEE Circuits and Devices Magazine. Institute of Electrical and Electronics Engineers, Inc., Box 1331, 445 Hoes Lane, Piscataway, NJ 08855-1331. (908) 981-0060. 1985 to present. Bi-monthly. $120.00 per year. ISSN: 8755-3996.

IEEE Journal of Quantum Electronics. Institute of Electrical and Electronics Engineers, Inc., Box 1331, 445 Hoes Lane, Piscataway, NJ 08855-1331. (908) 981-0060. 1965 to present. Monthly. $535.00 per year. ISSN: 0018-9197.

IEEE Spectrum. Institute of Electrical and Electronics Engineers, Inc., Box 1331, 445 Hoes Lane, Piscataway, NJ 08855-1331. (908) 981-0060. 1964 to present. Monthly. $157.00. ISSN: 0018-9235.

IEEE Transactions On Electron Devices. Institute of Electrical and Electronics Engineers, Inc., Box 1331, 445 Hoes Lane, Piscataway, NJ 08855-1331. (908) 981-0060. 1952 to present. Monthly. $395.00 per year. ISSN: 0018-938

International Journal of High Speed Electronics. World Scientific Publishing Co., 1060 Main Street, Suite 18, River Edge, NJ 07661. (800) 227-7562. 1990 to present. Quarterly. $270.00 per year. ISSN: 0129-1564.

Institute of Electrical and Electronics Engineers. PROCEEDINGS. Institute of Electrical and Electronics Engineers, Inc., Box 1331, 445 Hoes Lane, Piscataway, NJ 08855-1331. (908) 981-0060. 1913 to present. Monthly. $275.00. ISSN: 0018-9219.

Optical and Quantum Electronics. Chapman & Hall, One Penn Plaza, New York, NY 10019. (212) 564-1060. FAX (212) 564-1505. 1969 to present. Monthly. $657.00 per year. ISSN: 0306-8919.

Semiconductor International. Cahners Publishing Co., 44 Cook Street, Denver, CO 80206. (708) 635-8000. FAX (708) 390-2770. (800) 662-7776. 1978 to present. 13 issues per year. $84.95. ISSN: 0163-3767.

Semiconductor Science and Technology. IOP, 500 Sunnyside Boulevard, Woodbury, NY 11808-2999. (516) 576-2200. 1986 to present. Monthly. $1238.00 per year. ISSN: 0268-1242.

Solid State Technology. PennWell Publishing Co., 10 Tara Boulevard, Nausha, NH 03062-2801. (603) 891-0123. FAX (609) 891-0597. 1958 to present. Monthly. $110.00 per year. ISSN: 0038-111X.

RESEARCH CENTERS AND INSTITUTES

Edison Electric Institute. 701 Pennsylvania Avenue, NW, Washington, DC 20004-2696. (202) 508-5000, 5454. FAX (202) 508-5360.

Electroscience Laboratory. Ohio State University, 1320 Kinnear Road, Columbus, Ohio 43212. (614) 292-7981, FAX (614) 292-7297.

Electrical Engineering Research Laboratories. Purdue University. Electrical Engineering Building, West Lafayette, in 47907. (317) 494-3536. FAX (317) 494-6440.

Electronics Research Center. University of Texas at Austin, 132 Engineering Science Building, Austin, TX 78712. (512) 471-3954.

Electronics Research Laboratory. University of California, Berkeley, 253 Cory Hall, Berkeley, CA 94720. (415) 642-2301.

Laboratory For Electromagnetic and Electronic Systems. Massachusetts Institute of Technology, 77 Massachusetts Avenue, Cambridge, MA 02139. (617) 253) 4631.

ELECTRONICS ENGINEERING

See also: ELECTROMAGNETISM, ELECTRONIC CIRCUITS AND COMPONENTS, ELECTRICITY, ELECTRONICS, MICROWAVES, RADIO SIGNAL PROCESSING

ABSTRACT SERVICES AND INDEXES

Applied Science and Technology Index; A Cumulative Subject Index To English Language Periodicals in the Fields of Aeronautics and Space Science, Computer Technology, Chemistry, Construction Industry, Energy and Related Areas. H.W. Wilson Co., 950 University Avenue, Bronx, NY 10452. (212) 588-8400. (800) 367-6770. FAX (718) 590-1617. From 1958 to present. Monthly. Inquire about cost and availability. Also available on CD-ROM and online. ISSN: 0003-6986.

Current Contents: Engineering, Technology and Applied Sciences. Institute for Scientific Information, 3501 Market Street, Philadelphia, PA 19104. (215) 386-0100. FAX (215) 386-6362.

1970 to present. Weekly. $442.00 per year. Also available on CD-ROM and online. Inquireregarding cost and availability. ISSN: 0095-7917.

Current Papers in Electrical and Electronics Engineering. Institution of Electrical Engineers (IEE), London. Distributed by INSPEC/IEEE, Box 1331, 445 Hoes Lane, Piscataway, NJ 08855-1331. (908) 562-5549. 1969 to present. Monthly. $345.00 per year. ISSN: 0011-3778.

Electrical and Electronics Abstracts. Institute of Electrical Engineers, (IEE), London. Available from INSPEC/IEEE - Institute of Electrical and Electronic Engineers, Box 1331, Hoes Lane, Piscataway, NJ 08855-1331. (908) 562-5549. 1898 to present. Monthly. $2200.00 per year. Also available on CD-ROM and online as INSPEC. ISSN: 0036-8105.

Engineered Materials Abstracts. ASM (American Society of Metals) International, 9639 Kinsman Road, Materials Park, OH 44073. (216) 338-5151 or FAX (216) 338-4634. Covers literature on technical developments in polymer, ceramic, and composite materials and engineering. 1986 to present. Monthly. $1175.00 per year. Also available on CD-ROM. ISSN: 0951-9998.

Engineering Index Monthly: Indexes and Abstracts the World's Engineering and Technical Literature. Engineering Information, Inc., Castle Point on the Hudson, Hoboken, NJ 07030. (201) 216-8500. (800) 221-1044. FAX (201) 216-8532. Monthly. $2300.00 per year. Available online as COMPENDEX and also on CD-ROM. ISSN: 0742-1974.

General Science Index. H.W. Wilson Company, 950 University Avenue, Bronx, NY 10452. (212) 588-8400. (800) 367-6770. FAX (718) 590-1617. From 1978 to present. Ten issues per year; quarterly and annual cumulations. Service basis. Available on CD-ROM and online. Inquire regarding cost and availability. ISSN: 0162-1963.

Government Reports Announcements and Index. U. S. National Technical Information Service (NTIS), 5285 Port Royal Road, Springfield, VA 22161. (703) 487-4650. FAX (703) 321-8547. From 1968 to present. Annual. $630.00 per year. Also available online as *NTIS Bibliographic Database* and on CD-ROM. ISSN:

Index To IEEE Publications. IEEE Service Center, 445 Hoes Lane, Piscataway, NJ 08855-1331. (908) 981-1393. (800) 678-IEEE. FAX (908) 981-9667. 1973 to present. Annual. ISSN: 0099-1368.

Physics Abstracts. INSPEC. Section A, Science Abstracts. Institution of Electrical Engineers (IEE), London. Available from: INSPEC/IEEE - Institute of Electrical and Electronic Engineers, Box 1331, Hoes Lane, Piscataway, NJ 08855-1331. (908) 562-5549. 1898 to present. 24 issues per year. $2835.00 per year. Also available online and on CD-ROM. ISSN: 0036-8091.

Physics Briefs (Physikalische Berichte). Information Center for Energy, Physics, Mathematics; German Physical Society. V C H Publishers, Inc., 220 East 23rd Street, New York, NY 10010-4606. (212) 683-8333. 1845 to present. 24 issues per year. $2390.00 per year. Also available online. ISSN: 0179-7434.

Science Citation Index. SCI. Institute for Scientific Information, 3501 Market Street, Philadelphia, PA 19104. (215) 386-0100. (800) 523-1850. FAX (215) 386-2991. 1961 to present. Six issues per year, plus annual cumulation. $11650.00 per year. Also

available online and on CD-ROM. Inquire about price and availability. ISSN: 0036-827X.

Solid State and Superconductivity Abstracts: Covers theory, Production and Application of Solid State Materials. Cambridge Scientific Abstracts, 7200 Wisconsin Avenue, Bethesda, MD 20824. (301) 961-6750. FAX (301) 961-6720. 1957 to present. Bimonthly. $1320.00 per year. Also available online. ISSN: 0896-5900.

STAR. Scientific and Technical Aerospace Reports. U.S. National Aeronautics and Space Administration, Scientific and Technical Information Facility, Box 8757, Baltimore-Washington International Airport, MD 21240. (301) 621-0153. Distributed by U. S. Superintendent of Documents, Washington, DC 20402. From 1963 to present. Semimonthly, with semiannual and annual indexes. $114.00 per year. Also available online and on CD-ROM. ISSN: 0036-8741.

ANNUAL REVIEWS AND YEARBOOKS

Advances in Electronics and Electron Physics. Academic Press, Inc., 1250 Sixth Avenue, San Diego, CA 92101-4311. (619) 231-0926. FAX (619) 699-6715. Irregular. Price varies, inquire.

ASSOCIATIONS AND PROFESSIONAL SOCIETIES

American Association of Engineering Societies, 1111 19th Street, Suite 608, Washington, DC 20036-3603. (202) 296-2237. FAX (202) 296-1151.

American Institute of Physics. 1 Physics Ellipse, College Park, MD 20740-3843. (301) 209-3100.

Association of Old Crows (electronic Warfare). 1000 North Payne Street, Alexandria, VA 22314. (703) 549-1600.

Edison Electric Institute. 701 Pennsylvania Avenue NW, Washington, DC 20004-2696. (202) 508-5000, 5454. FAX (202) 508-5360.

Electronics Industries Association. 2500 Wilson Boulevard, Arlington, VA 22201. (703) 907-7500. FAX (202) 457-4985.

Equipment Manufacturers Institute. 10 South Riverside Plaza, Suite 1220, Chicago, IL 60606. (312) 321-1470. FAX (312) 321-1480.

IEEE (institute of Electrical and Electronic Engineers). 345 East 47th Street. New York, NY 10017. (212) 705-7900. FAX (212) 705-4929,

International Society For Hybrid Microelectronics. 1850 Centennial Park Drive, Suite 105, Reston, VA 22091. (703) 758-1060. FAX (703) 758-1066.

BIBLIOGRAPHIES

Bibliographical History of Electricity and Magnetism. Paul F. Mottelay, editor. Ayer Company Pubs., Inc., Lower Mill Road, North Stratford, NH 03590. (603) 922-5105. (800) 282-5413. Reprint edition. 1975. $52.95.

Science Books and Films. American Association for the Advancement of Science, 1333 H Street NW, Washington, DC

ELECTRONICS ENGINEERING

Ency. of Physical Sciences and Engineering Info. Sources

20005. (202) 326-6454. 1965 to present. Nine issues per year. $40.00 per year. ISSN: 0098-342X.

Scientific and Technical Books and Serials in Print; An Index To Literature in Science and Technology. R.R. Bowker Inc., 121 Chanlon Road, New Providence, NJ 07974. (908) 464-6800. (800) 521-8110. FAX (908) 665-3502. 1972 to present. Annual. 4 volumes. 1994. $299.95. Also available on compact disc and online. ISSN: 0000-054X.

DIRECTORIES AND BIOGRAPHICAL SOURCES

American Association of Engineering Societies, 1111 19th Street, Suite 608, Washington, DC 20036-3603. (202) 296-2237. FAX (202) 296-1151.

American Electronics Association. 5201 Great America Way, Suite 520, P.O. 52990, Santa Clara, CA 95056. (408) 987-4200. FAX (408) 970-8565.

American Men and Women of Science: Physical and Biological Sciences. R. R. Bowker Inc., 121 Chanlon Road, New Providence, NJ 07974. (908) 464-6800. (800) 521-8110. 20th edition. 8 volumes. 1996. $850.00.

Directory of Engineering College Research and Graduate Study. American Society for Engineering Education, Ohio State University, 2070 Neil Avenue, Columbus, OH 43210-1226. Annual. $20.00.

Directory of Engineering Societies and Related Organizations. American Association of Engineering Societies, 1111 19th Street, Suite 608, Washington, DC 20036-3603. (202) 296-2237. Semi-annual. $150.00.

Directory of Engineers in Private Practice. National Society of Professional Engineers, 1420 King Street, Alexandria, VA 22314-2750. (703) 684-2835. Biennial. $30.00.

EEM: Electronic Engieers Master Catalog. Herst Business Communications, Inc., 645 Stewart Avenue, Garden City, NY 11530. (516) 227-1300. FAX (516) 227-1901. 38th edition. 1996. Annual. $90.00. ISSN: 0732-9016.

Electrical and Electronic Trades Directory: The Blue Book. Institution of Electrical Engineers, Michael Faraday House, Six Hills Way, Stevenage, Herts. SG1 2AY, England. Telephone: 0438 313311 or FAX 0438 742840. 1995. $140.00

Engineering Research Centers: Incorporating Electronics Research Centers. Stockton Press, 345 Park Avenue, New York, NY 10010. (212) 689-9200. (800) 221-2123. 4th edition. 1995. $515.00.

IEEE Membership Directory. Institute of Electrical and Electronics Engineers, IEEE Service Center, 445 Hoes Lane, Piscataway, NY 08854. (908) 981-1393. (800) 678-IEEE. FAX (908) 981-9667. 2 volumes. Annual. $190.00. ISSN:

International Directory of Abbreviations and Acroynms of Electronics, Electrical Engineering, Computer Technology and Information Processing. Peter Wennrich. K. G. Saur, 121 Chanlon Road, New Providence, NJ 07974. (908) 464-6800. (800) 521-8110 2 volumes. 1992. $230.00.

International Engineering Directory. American Consulting Engineers Council, 1015 15th Street, N.W. Suite 802, Washing-ton, DC 20005-2670. (202) 347-7474. Annual. $10.00.

Research Centers Directory. Gale Research Inc., 835 Penobscot Building, Detroit, MI 48226-4094. (313) 961-2242. (800) 347-4253. 20th edition. 1995. $485.00.

Scientific and Technical Organizations and Agencies Directory. Gale Research, 835 Penobscot Building, Detroit, MI 48226-4094. (313) 961-2242. (800) 347-4253. 4th edition. 1996. $195.00.

Who's Who in Engineering. Gordon Davis, editor. American Association of Engineering Societies. 1111 19th Street, NY, Suite 608, Washington, DC 20036. (202) 296-2237. (800) 658-8897. 9th edition. 1995. $220.00.

Who's Who in Technology. Gale Research, 835 Penobscot Building, Detroit, MI 48226-4094. (313 961-2242. (800) 521-4253. 7th edition. 1995. $195.00. ISN 0-8103-7467-6.

ENCYCLOPEDIAS AND DICTIONARIES

Academic Press Dictionary of Science and Technology. Christopher Morris, editor. Academic Press, Inc., 1250 Sixth Avenue, San Diego, CA 92101. (619) 231-0926. FAX (619) 699-6715. 1991. $115.00.

Communications Standards Dictinary. Martin H. Weik. Chapman & Hall, 1 Penn Plaza, New York, NY 10119. (212) 564-1060 3rd edition. 1996. $69.95.

Dictionary of Communications Technology; Terms, Definitions, and Appreviations. Gilbert Held. John Wiley & Sons, Inc., 605 Third Avenue, New York, NY 10158-0012. (212) 850-6000. (800) 225-5945. 2nd edition. 1995. $49.95.

Electrical and Electronic Technologies: A Chronology of Events and Inventors From 1940 To 1980. Henry B. O. Davis. Scarecrow Press 1985. $35.00.

Encyclopedia of Applied Physics. George Trigg, editor. VCH Publications, Inc., 220 East 23rd Street, Suite 909, New York, NY 10010-4606. (212) 683-8333. (800) 422-8824. 20 volume set. 1991 - . $5990.00.

IEEE Standard Dictionary of Electrical and Electronics Terms. Christopher J. Booth, editor. IEEE Service Center, 445 Hoes Lane, Piscataway, NJ 08855-1331. (908) 981-1393. (800) 678-IEEE. FAX (908) 981-9667. IEEE Standard 100-1992. 5th edition. 1993. $150.00.

Illustrated Dictionary of Electronics. Stan Gibilsco. TAB Books, P.O. Box 40, Blue Summit, PA 17294-0850. (717) 794-2191. (800) 233-1128. 7th edition. 1994. $34.95.

Illustrated Encyclopedic Dictionary of Electronic Circuits. John Douglas-Young. Prentice Hall , 113 Sylvan Avenue, Route 9W, Englewood Cliffs, NJ 07632. (201) 592-2000. (800) 922-0579. 1983. $32.95.

Parat Index of Acronyms and Abbreviations in Electrical and Electronic Enginering. H. D. Bunge, editor. VCH Publications, Inc., 220 East 23rd Street, Suite 909, New York, NY 10010-4606. (212) 683-8333. (800) 422-8824. 1989. $195.00

McGraw-Hill Circuit Encyclopedia and Troubleshooting Guide. John Lenk, editor. McGraw-Hill Book Company, Inc. 1221 Avenue of the Americas, New York, NY 10020. (212) 997-3675 Volume 1-3. 1993. $36.95 each.

McGraw-Hill Electronics Dictionary. J. Markus. McGraw-Hill Book Company, Inc., 1221 Avenue of the Americas, New York, NY 10020. (212) 997-3675. 5th edition. 1994. $49.95.

McGraw-Hill Encyclopedia of Science and Technology. McGraw-Hill Book, Incorporated, 1221 Avenue of the Americas, New York, NY 10020. (212) 997-3675. (800) 262-4729. Seventh edition. Twenty volumes. 1992. $1900.00.

Telecommunications Engineer's Reference Book. Fraidoon Mazda, editor. Butterworth Heinmann, 313 Washington Street, Newton, MA 02158. (618-928-2500. (800) 366-2665. 1993. $130.00.

GENERAL WORKS

Alternating Current Fundamentals. John R. Duff and Stephen L. Herman. Delmar Publications, 3 Columbia Circle, Box 15015, Albany, NY 12212. (518) 464-3500. (800) 347-7707. 1995.

Electric and Electronic Engineering For Scientists and Engineers. K. A. Krishnamurthy. Halsted Press, 605 Third Avenue, New York, 10158-0012. (212) 850-6400. 1994. $48.95.

Electricity and Electronics: A Survey. Dale R. Patrick and Stephen W. Fardo. Prentice Hall , 113 Sylvan Avenue, Route 9W, Englewood Cliffs, NJ 07632. (201) 592-2000. (800) 922-0579. 3rd edition. 1995. $68.00.

Electronics Problem Solver. Research and Education Association, 61 Ethel Road, West Piscataway, NJ 08854. (908) 819-8880. Revised edition. 1994. $29.95 in paper.

High-voltage Engineering: Theory and Practice. Khalifa. Marcel Dekker, Inc., 270 Madison Avenue, New York, NY 10016. (212) 696-9000. (800) 228-1160. 1989. $125.00.

Industrial Electricity. John M. Nadon, et al. Delmar Publications, 3 Columbia Circle, Box 15015, Albany, NY 12212. (518) 464-3500. (800) 347-7707. 1994. $37.95.

Vhf and Uhf Antennas. R. A. Burberry. Institution of Electrical Engineers, Michael Faraday House, Six Hills Way, Stevenage, Herts. SG1 2AY, England. Telephone: 0438 313311

or FAX 0438 742840. 1992. $95.00

HANDBOOKS AND MANUALS

Active Electronic Component Handbook. Charles A. Harper and Harold C. Jones. McGraw-Hill Book Company, 1221 Avenue of the Americas, New York, NY 10020. (212) 512-2000. (800) 262-4729. 2nd edition. 1996. $79.50.

American Electricians' Handbook. Terrell Croft and Wilford I. Summers, editors. McGraw-Hill Publishing Company, Inc., 1221 Avenue of the Americas, New York, NY 10020. (212) 512-2000. (800) 262-4729. 12th editon. 1992. $74.50.

Antenna Handbook. Y. T. Lo and S. W. Lee, editors. Van Nostrand Reinhold, 115 Fifth Avenue, New York, NY 10003. (212) 254-3232. (800) 842-3636. Four volumes: Fundamentals and mathematical techniques, Antenna theory, Applications, Related topics. 1993. $150.95.

Audio Engineer's Reference Book. Michael Talbot-Smith, editor. Butterworth-Heinemann, 313 Washington Street, Newton, MA 02158. (618-928-2500. (800) 366-2665. 1994. $99.95.

Circuits and Filters Handbook. Wai-Kai Chen, editor-in-chief. CRC Press Inc., 2000 Corporate Blvd., N.W., Boca Raton, FL 33431. (407) 994-0555. (800) 333-8300. Published in cooperation with IEEE Press. 1995. $129.95.

Electrical and Electronics Graphic and Letter Symbols and Reference Designations Standards Collection. Institute of Electrical and Electronics Engineers, Inc., 1992.

Electrical Engineering Handbook. Richard Dorf, editor. CRC Press, Inc., 2000 Corporate Boulevard, NW, Boca Raton, FL 33431. (407) 994-0555. (800) 272-7737. 1993. $99.95.

Electronic Engineers Handbook. Donald G. Fink, and Donald Christiansen, editors. McGraw-Hill Book Company, 1221 Avenue of the Americas, New York, NY 10020. (212) 512-2000. (800) 262-4729. 3rd edition. 1989. $105.00.

Electrical Engineer's Handbook. Donald Christiansen, editor. McGraw Publishing Company Inc., 1221 Avenue of the Americas, New York, NY 10020. (212) 512-2000. (800) 262-4729. 4th edition. 1996. $110.00.

Electronics Handbook. Jerry C. Whitaker. CRC Press, Inc., 2000 Corporate Boulevard, NW, Boca Raton, FL 33431. (407) 994-0555. (800) 272-7737. 1996. $120.00.

Engineering Tables and Data. A.M. Howatson, P.G. Lund and J.D. Todd. Van Nostrand Reinhold, 115 Fifth Avenue, New York, NY 10003. (212) 254-3232. (800) 926-2665. Second Edition. 1991. $39.95.

Handbook of Electrical and Electronic Technology. Curtis Johnson. Prentice Hall , 113 Sylvan Avenue, Route 9W, Englewood Cliffs, NJ 07632. (201) 592-2000. (800) 922-0579. 1996. $88.00.

Handbook of Electronics Manufacturing Engineering. Bernard S. Matisoff. Chapman & Hall, 1 Penn Plaza, New York, NY 10119. (212) 564-1060. 1996.

Handbook of Practical Electrical Design. Joseph F. McPartland and Brian J. McPartland. McGraw-Hill Publishing Company Inc., 1221 Avenue of the Americas, New York, NY 10020. (212) 512-2000. (800) 262-4729 2nd edition. 1995.

Industrial Electronics and Systems Handbook. Irwin, CRC Press, Inc., 2000 Corporate Boulevard, NW, Boca Raton, FL

33431. (407) 994-0555. (800) 272-7737. 1996. $129.95.

National Electrical Safety Code. Institute of Electrical and Electronic Engineers (IEEE), 345 East 47th Street, New York, NY 10017. (212) 705-7900. C2-1993. 1992.

Rapid Electrical Estimating and Pricing. C. Kenneth Kolstad, and Gerald V. Kohnert. McGraw-Hill Publishing Company Inc.,

ELECTRONICS ENGINEERING

Ency. of Physical Sciences and Engineering Info. Sources

1221 Avenue of the Americas, New York, NY 10020. (212) 512-2000. (800) 262-4729 5th editon. 1993.

Standard Handbook For Electrical Engineers. Donald G. Fink and H. Wayne Beaty, editors. McGraw-Hill Publishing Company, 1221 Avenue of the Americas, New York, NY 10020. (212) 512-2000. (800) 262-4729. 13th edition. 1996. $110.00.

Standard Handbook of Engineering Calculations. Tyler G. Hicks. McGraw-Hill Publishing Company, 1221 Avenue of the Americas, New York, NY 10020. (800) 262-4729 or (212) 512-3825. 3rd edition. 1994. $99.50. ISBN: 0-07-028812-7.

ONLINE DATABASES AND CD-ROMS

CA Search. Chemical Abstracts Service, P.O. Box 3012, Columbus, OH 43210-0012. (614) 447-3600. (800) 848-6533. FAX (614) 447-3709. Very comprehensive guide to worldwide chemical literature and related fields, 1972 to present. Available on BRS,(800) 289-4277, DIALOG, (800) 334-2564, ORBIT (800) 456-7248, and STN International, FIZ Karlsruhe, P.O. Box 2465, W-7500, Karlsruhe 1, Germany, online services. Inquire as to cost and availability.

Compendex Plus. Engineering Information, Inc., 345 East 47th Street, New York, NY 10017. (212) 705-7600 or (800) 221-1044. Contains citations with abstracts to worldwide literature in engineering and technology, from 1970 to present. Available on online BRS,(800) 289-4277, DIALOG, (800) 334-2564, ORBIT (800) 456-7248, and STN International, FIZ Karlsruhe, P.O. Box 2465, W-7500, Karlsruhe 1, Germany, online services. Also available on CD-ROM. Inquire as to cost and availability.

Current Contents Search. Institute for Scientific Information, 3501 Market Street, Philadelphia, PA 19104. (215) 386-0100. FAX (215) 386-6362. Contains citations to articles listed in the table of contents of science and technology journals. Also articles in social sciences and life sciences journals. Available on BRS,(800) 289-4277, DIALOG,(800) 334-2564, online services. Inquire as to cost and availability.

Dissertation Abstracts. University Microfilms International, 300 North Zeeb Road, Ann Arbor, MI 48106. (800) 521-0600 or (313) 761-4700. Scope includes virtually all doctoral dissertations accepted at accredited American institutions from 1861 to present in 252 subject areas. Available on BRS,(800) 289-4277, DIALOG,(800) 334-2564, and OCLC EPIC,(800) 848-5878, online services. Also available on CD-ROM. Inquire as to cost and availability.

INSPEC. Institution of Electrical Engineers, Michael Faraday House, Six Hills Way, Stevenage, Herts. SG1 2AY, England. Telephone: 0438 313311 or FAX 0438 742840. Contains citations to the worldwide literature of physics, electronics and electrical engineering, computer technology, and related fields. Available on BRS,(800) 289-4277, DIALOG, (800) 334-2564, ORBIT, (800) 456-7248, and STN International, FIZ Karlsruhe, P.O. Box 2465, W-7500, Karlsruhe 1, Germany, online services. Inquire as to cost and availability.

NTIS Bibliographic Database. National Technical Information Service, 5285 Port Royal Road, Springfield, VA 22161. (703) 487-4929 or FAX (703) 321-8199. Broad coverage of government-sponsored science and technology research reports, 1964 to present. Available on BRS,(800) 289-4277, DIALOG, (800) 334-2564, ORBIT, (800) 456-7248, and STN International, FIZ Karlsruhe, P.O. Box 2465, W-7500, Karlsruhe 1, Germany,

online services. Also available on CD-ROM. Inquire as to cost and availability.

Scientific and Technical Books and Serials in Print. R.R. Bowker Inc., 121 Chanlon Road, New Providence, NJ 07974. (800) 521-8110 or (908) 464-6800. List of currently published books and serials in the physical and biological sciences, engineering and technology, with subject, author and titles indexes. Available on ORBIT,(800) 456-7248, online service.Inquire as to cost and availability.

SCISEARCH. Institute for Scientific Information, 3501 Market Street, Philadelphia, PA 19104. (800) 523-1850 or (215) 386-0100. Broad multidisciplinary title and author index to the international literature of science and technology, 1974 to present. Available on DIALOG,(800) 334-2564, and ORBIT,(800) 456-7248, online services. Also available on CD-ROM. Inquire as to cost and availability.

WILSONLINE. H.W. Wilson Company, 950 University Avenue, Bronx, NY 10452. (800) 367-6770 or (212) 588-8400. Makes available online versions of the printed H.W. Wilson indexes including Applied Science and Technology Index, Business Periodicals Index, General Science Index, and Readers' Guide to Periodical Literature. Period covered is generally 1983 to present. Available on BRS,(800) 289-4277, DIALOG,(800) 334-2564, and OCLC EPIC,(800) 848-5878, online services. Also available on CD-ROM. Inquire as to cost and availability.

PERIODICALS

Audio Engineering Society. Journal. Audio Engineering Society, 60 East 42nd Street, New York, NY 10017. (212) 661-8528. 1953 to present. 10 issues per year. $125.00 per year. ISSN: 0004-7554.

Chaos: An Interdisciplinary Journal. American Institute of Physics, 335 East 45th Street, New York, NY 10017 (212) 661-9404 or FAX (516)349-9704. 1991 to present. Quarterly. $160.00 per year.

Circuits, Systems, and Signal Processing. Birkhauser, 675 Massachusetts Avenue, Cambridge, MA 02139-3309. FAX (201) 348-4505. 1982 to present. Quarterly. $184.00 per year. ISSN: 0278-081X.

CRC Critical Reviews in Solid State and Materials Science. CRC Publishers, Inc., 2000 Corporate Blvd., N.W., Boca Raton, FL 33431. (407) 994-0555 or (800) 333-8300. Quarterly. $295.00 per year.

E D N Magazine. Cahners Publishing Company, 44 Cook Street, Denver CO 80206. (800) 662-7776. 1956 to present. 26 issues per year. $120.00 per year.

Electrical World. McGraw-Hill, Inc., Box 513 Hightstown, NJ 08520. (212) 512-3288. 1874 to present. Monthly. $55.00 per year. ISSN: 0013-4457.

Electronic Design. Penton Publishing, San Jose Gateway, Suite 354. 2025 Gateway Place, San Jose, CA 95110. (408) 441-0550. 1952 to present. Fortnightly. $95.00. ISSN: 0013-4872.

Electronic Engineering Times. CMP Publications, Inc., 600 Community Drice, Manhasset, NY 11030. (516) 562-5000. FAX (516) 562-5325. 1972 to present. Weekly. $159.00 per year. ISSN: 0192-1541.

Electronics. Penton Publishing, San Jose Gateway, Suite 354. 2025 Gateway Place, San Jose, CA 95110. (408) 441-0550. 1930 to present. Semi-weekly. $98.00 per year. ISSN: 0883-4989.

Electronics Now. Gernsback Publications, Inc., 500-B Bi-County Boulevard, Farmingdale, NY 11735. (516) 293-3000. FAX (516) 293-3115. 1929 to present. Monthly. $19.95. ISSN: 0033-7862.883-4989.

IEEE Circuits and Devices Magazine. Institute of Electrical and Electronics Engineers, Inc., Box 1331, 445 Hoes Lane, Piscataway, NJ 08855-1331. (908) 981-0060. 1985 to present. Bi-monthly. $120.00 per year. ISSN: 8755-3996.

IEEE Journal of Quantum Electronics. Institute of Electrical and Electronics Engineers, Inc., Box 1331, 445 Hoes Lane, Piscataway, NJ 08855-1331. (908) 981-0060. 1965 to present. Monthly. $535.00 per year. ISSN: 0018-9197.

IEEE Spectrum. Institute of Electrical and Electronics Engineers, Inc., Box 1331, 445 Hoes Lane, Piscataway, NJ 08855-1331. (908) 981-0060. 1964 to present. Monthly. $157.00. ISSN: 0018-9235.

IEEE Transactions On Circuits and Systems. Part 1: Fundamental theory and Applications. Institute of Electrical and Electronics Engineers, Inc., Box 1331, 445 Hoes Lane, Piscataway, NJ 08855-1331. (908) 981-0060. 1952 to present. Monthly. $241.00 per year. ISSN: 1057-7122.

IEEE Transactions On Electron Devices. Institute of Electrical and Electronics Engineers, Inc., Box 1331, 445 Hoes Lane, Piscataway, NJ 08855-1331. (908) 981-0060. 1952 to present. Monthly. $395.00 per year. ISSN: 0018-9393.

IEEE Transactions On Power Delivery. Institute of Electrical and Electronics Engineers, Inc., Box 1331, 445 Hoes Lane, Piscataway, NJ 08855-1331. (908) 981-0060. 1986 to present. Quarterly. $170.00 per year. ISSN: 0885-8977.

Institute of Electrical and Electronics Engineers. Proceedings. Institute of Electrical and Electronics Engineers, Inc., Box 1331, 445 Hoes Lane, Piscataway, NJ 08855-1331. (908) 981-0060. 1913 to present. Monthly. $275.00. ISSN: 0018-9219.

International Journal of High Speed Electronics. World Scientific Publishing Co., 1060 Main Street, Suite 18, River Edge, NJ 07661. (800) 227-7562. 1990 to present. Quarterly. $270.00 per year. ISSN: 0129-1564.

International Journal of Microwave and Millimeter-wave Computer Aided Engineering. John Wiley and Sons, Inc., 605 Third Avenue, New York, NY 10158. (800) 526-5368 or (212) 850-6000. 1991 to present. Quarterly. $175.00 per year.

Microwaves and Rf. Penton Publishing Co., Circulation Department, 1100 Superior Avenue, Cleveland, OH 44114. (201) 393-6060. FAX (201) 393-6297. From 1962 to present. Monthly. $60.00. ISSN: 0745-2992.

Optical and Quantum Electronics. Chapman & Hall, One Penn Plaza, New York, NY 10019. (212) 564-1060. FAX (212) 564-1505. 1969 to present. Monthly. $657.00 per year. ISSN: 0306-8919.

Solid State Electronics. Elsevier Science, 660 White Plains Road, Tarrytown, NY 10591-5153. (914) 524-9200. FAX (914) 333-2444. 1960 to present. Monthly. $1025.00 per year. ISSN: 0038-1101.

RESEARCH CENTERS AND INSTITUTES

Electrical Engineering Research Laboratories. Purdue University. Electrical Engineering Building, West Lafayette, in 47907. (317) 494-3536. FAX (317) 494-6440.

Electrical Power Research Institute. 3412 Hillview Avenue, PO Box 10412, Palo Alto, CA 94303. (415) 855-2000. FAX (415) 855-2954.

Electroscience Laboratory. Ohio State University, 1320 Kinnear Road, Columbus, Ohio 43212. (614) 292-7981, FAX (614) 292-7297.

Electronics Design Center. Case Western Reserve University, Bingham Building, Cleveland, OH 44106. (216) 368-2934. (216) 368-8738.

Electronics Research Laboratory. University of California, Berkeley, 253 Cory Hall, Berkeley, CA 94720. (415) 642-2301.

Electronics Research Center. University of Texas at Austin, 132 Engineering Science Building, Austin, TX 78712. (512) 471-3954.

Engineering Research Center. University of Maryland, Clark School of Engineering, Potomac Building, Room 2104. (301) 405-3906. FAX (301) 403-4105.

Laboratory For Electromagnetic and Electronic Systems. Massachusetts Institute of Technology, 77 Massachusetts Avenue, Cambridge, MA 02139. (617) 253) 4631.

Solid State Electronics Laboratory. North Carolina State University, 432 Daniels Hall, Raleigh, NC 27695. (919) 737-2336.

Weber Research Institute. Polytechnic University. Route 110, Farmingdale, NY 11735. (516) 755-4250, FAX (516) 755-4404.

ELECTROPHOTOGRAPHY

ABSTRACT SERVICES AND INDEXES

Applied Science and Technology Index; A Cumulative Subject Index To English Language Periodicals in the Fields of Aeronautics and Space Science, Computer Technology, Chemistry, Construction Industry, Energy and Related Areas. H.W. Wilson Co., 950 University Avenue, Bronx, NY 10452. (800) 367-6770 or (212) 588-8400. FAX (718) 590-1617. From 1958 to present. Monthly. Inquire about cost and availability. Also available on CD-ROM and online.

Current Contents: Engineering, Technology, and Applied Sciences. Institute for Scientific Information, 3501 Market Street, Philadelphia, PA 19104. (215) 386-0100. FAX (215) 386-6362. 1970 to present. Weekly. $442.00 per year.

Current Papers in Physics. Institute of Electrical Engineers, Michael Faraday House, Six Hill Way, Stevenage, Herts, SG1

2AY, England. Distributed by IEEE, 445 Hoes Lane, Piscataway, NJ 08854. (908) 562-5549. 1966 to present. Twenty-four times per year. $390.00 per year. Also available online in INSPEC.

Current Physics Index. American Institute of Physics, 335 East 45th Street, New York, NY 10017 (212) 661-9404 or FAX (516)349-9704. 1975 to present. Quarterly. $625.00 per year.

Engineering Index Monthly. Engineering Information, Inc., Castle Point on the Hudson, Hoboken, NJ 07030. (800) 221-1044. FAX (212) 832-1857. Monthly. $2300.00 per year. Also available online as COMPENDEX, and also on CD-ROM. Covers chemical engineering, computers, electrical engineering, civil engineering, metals and mining, industrial management, and mechanical engineering.

*Imaging Abstracts (*formerly *Photographic Abstracts),* Pergamon Press Inc., Maxwell House, Fairview Park, Elmsford, NY 10523. (914) 592-7700. Fax (914) 592-3625. Bimonthly. Inquire for price.

Physics Abstracts. Institute of Electrical Engineers, Michael Faraday House, Six Hill Way, Stevenage, Herts, SG1 2AY, England. Distributed by IEEE, 445 Hoes Lane, Piscataway, NJ 08854. (908) 562-5549. 1898 to present. Monthly. $2700.00 per year. Also available online as INSPEC.

Science Citation Index. Institue for Scientific Information, 3501 Market Street, Philadelphia, PA 19104. (215) 386-0100. FAX (215) 386-6362. Inquire about availability and cost. Also available on CD-ROM.

ASSOCIATIONS AND PROFESSIONAL SOCIETIES

The International Society For Optical Engineering (spie), Po Box 10, 1000-20th Street, Bellingham, WA 98227-0010.

Optical Society of America, 2010 Massachusetts Avenue Nw, Washington, DC 20036. (202) 223-8130.

Society of Motion Picture and Television Engineers, 595 W. Hartsdale Avenue, White Plains, NY 10607.

The Society For Imaging Science and Technology, 7003 Kilworth Lane, Springfield, VA 22151. (703) 642-9090.

DIRECTORIES AND BIOGRAPHICAL SOURCES

Directory of Engineering Societies and Related Organizations. Gordon Davis. 13th edition. American Association of Engineering Societies, 1111 19th Street NW, Suite 608, Washington, DC 20036. (202) 296-2237 or (800) 658-8897. 1989. Inquire for price.

International Research Centers Directory. Gale Research, 835 Penobscot Building, Detroit, MI 48226-4094. (313) 961-2242. (800) 347-4253. FAX (313) 961-6083. 8th edition. 1995. $410.00

Research Centers Directory. Gale Research, 835 Penobscot Building, Detroit, MI 48226-4094. (313) 961-2242. (800) 347-4253. FAX (313) 961-6083. $485.00.

Scientific and Technical Organizations and Agencies Directory. Gale Research, 835 Penobscot Building, Detroit, MI 48226-

4094. (313) 961-2242. (800) 347-4253. FAX (313) 961-6083. 4th edition. 1996. $195.00.

Who's Who in Engineering. American Association of Engineering Societies, 1111 19th Street NW, Suite 608, Washington, DC 20036. (202) 296-2237 or (800) 658-8897. 8th edition. 1991. Inquire for price.

ENCYCLOPEDIAS AND DICTIONARIES

Thesaurus of Scientific, Technical, and Engineering Terms. Hemisphere Publishing Corporation, 1900 Frost Road, Suite 101, Bristol, PA 19007-1598. (215) 785-5800 or (800) 821-8312. FAX (215) 785-5515. 1987. $173.00.

GENERAL WORKS

Camera-Sensitive Electrophotography. Helmut Keiss. Focal Press, 80 Montvale Avenue, Stoneham, MA 02180. 1980. Inquire for cost and availability.

Electronic Photography. John J. Larish. TAB Books, P.O. Box 40, Blue Ridge Summit, PA 17294-0850. (717) 794-2191 or (800) 233-1128. FAX (717) 794-2080. 1990. Inquire for cost and availability.

Electrophotography and Development Physics. L.B. Schein. 2d ed. Springer-Verlag, 175 Fifth Avenue, New York, NY 10010. (212) 460-1500 or (800) 777-4643. FAX (212) 473-6272. 1992. $52.00.

Science and Technology of Photography. Karlheinz Keller, editor. VCH Publishers, Inc., 303 Northwest 12th Avenue, Deerfield Beach, FL 33442. (800) 422-8824. 1993. Inquire for cost and availability.

HANDBOOKS AND MANUALS

Handbook of Imaging Materials. Arthur S. Diamond, editor. Marcel Dekker, Inc., 270 Madison Avenue, New York, NY 10016. (212) 696-9000. FAX (212) 685-4540. 1991. $190.00.

ONLINE DATABASES AND CD-ROMS

Compendex Plus. Engineering Information, Inc., 345 East 47th Street, New York, NY 10017. (212) 705-7600 or (800) 221-1044. Contains citations with abstracts to worldwide literature in engineering and technology, from 1970 to present. Available on online BRS,(800) 289-4277, DIALOG, (800) 334-2564, ORBIT (800) 456-7248, and STN International, FIZ Karlsruhe, P.O. Box 2465, W-7500, Karlsruhe 1, Germany, online services. Also available on CD-ROM. Inquire as to cost and availability.

CSA Engineering. Cambridge Scientific Abstracts, 7200 Wisconsin Avenue, Suite 601, Bethesda, MD 20814. (301) 961-6750 or (800) 843-7751. Contains citations and abstracts of international periodicals and research literature covering all fields of engineering and science and technology,including computer and information science, electronics, mechanical engineering, solid state materials, 1981 to present. Available on BRS,(800) 289-4277, online service.Inquire as to cost and availability.

Current Contents Search. Institute for Scientific Information, 3501 Market Street, Philadelphia, PA 19104. (215) 386-0100.

FAX (215) 386-6362. Contains citations to articles listed in the table of contents of science and technology journals. Also articles in social sciences and life sciences journals. Available on BRS,(800) 289-4277, DIALOG,(800) 334-2564, online services. Inquire as to cost and availability.

NTIS Bibliographic Database. National Technical Information Service, 5285 Port Royal Road, Springfield, VA 22161. (703) 487-4929 or FAX (703) 321-8199. Broad coverage of government-sponsored science and technology research reports, 1964 to present. Available on BRS,(800) 289-4277, DIALOG, (800) 334-2564, ORBIT, (800) 456-7248, and STN International, FIZ Karlsruhe, P.O. Box 2465, W-7500, Karlsruhe 1, Germany, online services. Also available on CD-ROM. Inquire as to cost and availability.

SCISEARCH. Institute for Scientific Information, 3501 Market Street, Philadelphia, PA 19104. (800) 523-1850 or (215) 386-0100. Broad multidisciplinary title and author index to the international literature of science and technology, 1974 to present. Available on DIALOG,(800) 334-2564, and ORBIT,(800) 456-7248, online services. Also available on CD-ROM. Inquire as to cost and availability.

WILSONLINE. H.W. Wilson Company, 950 University Avenue, Bronx, NY 10452. (800) 367-6770 or (212) 588-8400. Makes available online versions of the printed H.W. Wilson indexes including Applied Science and Technology Index, Business Periodicals Index, General Science Index, and Readers' Guide to Periodical Literature. Period covered is generally 1983 to present. Available on BRS,(800) 289-4277, DIALOG,(800) 334-2564, and OCLC EPIC,(800) 848-5878, online services. Also available on CD-ROM. Inquire as to cost and availability.

PERIODICALS

British Journal of Photography. Bouverie Publishing Company Ltd., 147-151 Temple Chambers, Temple Avenue, London EC47 0DT, England. Telephone 071-583-3030. FAX 071-583-4068. 1854 to present. Weekly. Inquire for price and availability.

Electronic Photography News. Photofinishing News Inc., 10915 Bonita Beach Road, Bonita Springs, FL 33923. (813) 992-4421. FAX (813) 992-6328. 1986 to present. Monthly. $90.00 per year.

Electrophotography. Society of Electrophotography of Japan, c/o Tokyo Institute of Polytechnics, 2-9-5 Honcho, Nakano-Ku, Tokyo 164, Japan. Telephone 03-3373-9576. FAX 03-3372-4414. 1959 to present. Quarterly. Inquire for cost and availability.

Journal of Photographic Science. The Barn, Whitehall, Near Middle Marwood, Barnstaple, N. Devon EX31 4EQ, England. Telephone 0271-72482. FAX 0271-72482. 1952 to present. Bi-monthly. Inquire for price and availability.

Scientific and Applied Photography and Cinematography. Gordon and Breach Science Publishers, 820 Town Center Drive, Langhorne, PA 19047. (215) 750-2642. FAX (215) 750-6343. 6 times a year. Inquire for price.

RESEARCH CENTERS AND INSTITUTES

Image Permanence Institute. Rochester Institute of Technology, RIT City Center, 50 W. Main Street, Rochester, NY 14614. (716) 475-5199. FAX (716) 475-5097.

Rochester Institute of Technology, Center For Imaging Science. One Lomb Memorial Drive, Rochester, NY 14623. (716) 475-2774.

University of Rochester, Center For Advanced Optical Technology. Wilmot Building, Rochester, NY 14627. (716) 275-5248.

ELECTROPLATING

See: ELECTROCHEMISTRY

ELECTROSTATICS

See: ELECTRICITY

ELEMENTARY PARTICLES

See: PARTICLE PHYSCIS

ELEMENTS (CHEMICAL)

See also: ANALYTICAL CHEMISTRY, CHEMISTRY. COSMOCHEMISTRY, ELECTROCHEMISTRY, GEOCHEMISTRY, INORGANIC CHEMISTRY, ORGANIC CHEMISTRY, PHYSICAL CHEMISTRY, AND NAMES OF INDIVIDUAL ELEMENTS.

ABSTRACT SERVICES AND INDEXES

Analytical Abstracts. Royal Society of Chemistry, Information Services, Thomas Graham House, Science Park, Milton Road, Cambridge, CB4 4WF, England. Contains citations, mostly with abstracts, of the worldwide literature on analytical chemistry, from 1954 to present. Monthly. $636.00 per year. Also available online.

Applied Science and Technology Index; A Cumulative Subject Index To English Language Periodicals in the Fields of Aeronautics and Space Science, Computer Technology, Chemistry, Construction Industry, Energy and Related Areas. H. W. Wilson Co., 950 University Avenue, Bronx, NY 10452-9978. (212) 588-8400. (800) 367-6770. FAX (718) 590-1617. From 1958 to present. Monthly. Inquire about cost and availability. Also available online (BRS and WILSONLINE) and on CD-ROM. ISSN: 0003-6986.

Chemical Abstracts. Chemical Abstracts Service. Olentangy River Road, Box 3012, Columbus, OH 43210-0012. (614) 447-3600. FAX (614) 447-3713. 1907 to present. Weekly. $16,800.00 per year. Available online and on CD-ROM. CA is also available in five section groupings.Inquire regarding cost and availability.

Current Contents: Physical, Chemical and Earth Sciences. Institute for Scientific Information, 3501 Market Street, Philadelphia, PA 19104. (215) 386-0100. FAX (215) 386-2291. From 1961 to present. Weekly. $442.00 per year. Also available on CD-ROM and online. Inquire regarding cost and availability. ISSN: 0163-2574.

ELEMENTS (CHEMICAL)

Ency. of Physical Sciences and Engineering Info. Sources

Engineering Index Monthly. Engineering Information, Inc., Castle Point on the Hudson, Hoboken, NJ 07030. (800) 221-1044. FAX (201) 216-8532. Monthly. $2200.00 per year. Also available online as COMPENDEX, and also on CD-ROM.

General Science Index. H.W. Wilson Company, 950 University Avenue, Bronx, NY 10452. (212) 588-8400. (800) 367-6770. FAX (718) 590-1617. From 1978 to present. Ten issues per year; quarterly and annual cumulations. Service basis. Available on CD-ROM and online. Inquire regarding cost and availability. ISSN: 0162-1963.

Physics Abstracts. INSPEC. Section A, Science Abstracts. Institution of Electrical Engineers (IEE), London, UnitedKingdom. Available from: INSPEC/IEEE - Institute of Electrical and Electronic Engineers, Box 1331, 445 Hoes Lane, Piscataway, NJ 08855-1331. (908) 562-5549. 1898 to present 24 issues per year. $2835.00 per year. ISSN: 0036-8091. Also available online and on CD-ROM.

Science Citation Index. SCI. Institute for Scientific Information, 3501 Market Street, Philadelphia, PA 19104. (215) 386-0100. (800) 523-1850. FAX (215) 386-2991. 1961 to present. Six issues per year, plus annual cumulation. $11650.00 per year. Also available online and on CD-ROM. Inquire about price and availability. ISSN: 0036-827X.

ANNUAL REVIEWS AND YEARBOOKS

Advances in Inorganic and Radiochemistry. Academic Press, Inc., 1250 Sixth Avenue, San Diego, CA 92101. (619) 231-0926. FAX (619) 699-6715. 1959 to present. Price varies. Inquire.

ASSOCIATIONS AND PROFESSIONAL SOCIETIES

American Chemical Society, 1155 16th Street, Nw, Washington, DC 20036. (202) 872-4600.

American Institute of Chemical Engineers. 345 East 47th Street, New York, NY 10017-2396. (212) 705-7663. (202) 752-3294.

American Institute of Chemists. 7315 Wisconson Avenue, Bethesda, MD 20814. (301) 652-2447. FAX (301) 657-3549.

American Institute of Physics, 335 East 45th Street, New York, NY 10017 (212) 661-9404 or Fax (516)349-9704.

Association of Consulting Chemists and Chemical Engineers, 50 East 41st Street, Suite 92, New York, NY 10017. (212) 684-6255.

Chemical Manufacturers Association. 2501 M Street, N.W., Washington, DC 20037. (202) 887-1100. FAX (202) 202-1237.

Electrochemical Society. 10 South Main Street, Pennington, NJ 08534. (609) 737-1902.

BIBLIOGRAPHIES

Chemical Information. H. R. Collier, editor. Springer-Verlag New York, Inc., 175 Fifth Avenue, New York, NY 10010. (212) 460-1500. (800) 777-4643. 1990. $91.00.

Chemical Information Sources. Gary Wiggins. McGraw-Hill Publishing Company, 1221 Avenue of the Americas, New York, NY 10020. (212) 512-2000. (800) 262-4729. 1991. $42.50.

Handbooks and Tables in Science and Technology. Russell H. Powell, editor. Oryx Press, 4041 North Central, Suite 700, Phoenix, AZ 85012-3330. (602) 265-2651 or (800) 279-6799. Third edition. 1994. $65.00.

Information Sources in Chemistry. R.T. Bottle and J.F.B Rowland, editors. R.R. Bowker Inc., 121 Chanlon Road, New Providence, NJ 07974. (908) 464-6800. (800) 521-8110. Fourth edition. 1993. $75.00.

Scientific and Technical Books and Serials in Print; An Index To Literature in Science and Technology. R.R. Bowker Inc., 121 Chanlon Road, New Providence, NJ 07974. (908) 464-6800. (800) 521-8110. FAX (908) 665-3502. 1972 to present. Annual. 4 volumes. 1994. $299.95. Also available on compact disc and online. ISSN: 0000-054X.

DIRECTORIES AND BIOGRAPHICAL SOURCES

American Institute of Chemists - Professional Directory. American Institute of Chemists, 501 Wythe Street, Alexandria, VA 22314-1917. (703) 836-2090. FAX (703) 836- 2091. Annual. $65.00.

American Men and Women of Science. Physical and Biological Sciences. R.R. Bowker Inc., 121 Chanlon Road, New Providence, NJ 07974. (908) 464-6800. (800) 521-8110. 20th edition. 8 volumes. 1996. $850.00

Chemical Week — Buyers Guide Issue. Chemical Week Associates, 888 Seventh Avenue, New York, NY 10106. (212) 621-4900. FAX (212) 621-4949. Annual, October. $50.00.

Chemical Sources International. Mike Desing and Kurt Gandenberger, editors. Chemical Sources International, Inc., P.O. Box 1824, Clemson, SC 29633. (803) 646-7840. 1994. $285.00.

Directory of Chemistry Software 1992. Wendy Warr, Peter Willett, Geoff Downs. American Chemical Society, 1155 16th Street, NW, Washington, DC 20036. (202) 872-4600. 1992. $35.95.

Research Centers Directory. Gale Research Company,Inc., 835 Penobscot Building, Detroit, MI 48226-4094. (313) 961-2242. (800) 347-4253. 20th edition. 1995. $485.00. ISSN: 0080-1518.

ENCYCLOPEDIAS AND DICTIONARIES

Academic Press Dictionary of Science and Technology. Christopher Morris, editor. Academic Press, Inc., 6277 Sea Harbor Drive, Orlando, FL. (800) 321-5068. 1991. $115.00.

Concise Encyclopedia Chemistry. Translated and revised by Mary Eagleson. Walter de Gruyter, Inc., 200 Saw Mill River Road, Hawthorne, New York, 10532. (914) 747-0110. Fax (914) 747-1326. 1994. $69.95.

Concise Encyclopedia of Chemical Technology. John Wiley & Sons, Inc., 605 Third Avenue, New York, NY 10158-0012. (212) 850-6000. (800) 225-5945. 2nd edition. 1989. $99.95.

Ency. of Physical Sciences and Engineering Info. Sources

ELEMENTS (CHEMICAL)

Dictionary of Chemical Names and Synomyms. Philip A. Howard and Michael W. Neil. Lewis Pub.,/ CRC Press, Inc., 2000 Corporate Boulevard, NW, Boca Raton, FL 33431. (407) 994- 0555. (800) 272-7737. 1991. $149.95.

Encyclopedia of Applied Physics. VCH Publications, Inc., 220 East 23rd Street, Suite 909, New York, NY 10010-4606. (212) 683-8333. (800) 422-8824. 1991-. Twenty volumes. $6000.00.

Gardner's Chemical Synomyns and Trade Names. Michael Ash and Irene Ash. Ashgate Publishing Co., Old Post Road, Brookfield, VT 05036. (802) 276-3162 10th edition. 1994. $195.00.

Grant & Hawks Chemical Dictionary. R. A. Grant. McGraw-Hill Publishing Company, 1221 Avenue of the Americas, New York, NY 10020. (212) 512-2000. (800) 262-4729. 5th edition. 1987. $64.50.

Hawley's Condensed Chemical Dictionary. Richard J. Lewis. Van Nostrand Reinhold, 115 Fifth Avenue, New York, NY 10003. (212) 254-3232. (800) 842-3636. 12th edition. 1993. $69.95.

Kirk-Othmer Encyclopedia of Chemical Technology. John Wiley and Sons, Inc., 605 Third Avenue, New York, NY 10158-0012. (212) 850-6000. (800) 225-5945. Fourth edition. 1991- . Twenty-seven volumes. $5400.00.

McGraw-Hill Encyclopedia of Chemistry. Sybil P.Parker, editor. McGraw-Hill Book, Incorporated, 1221 Avenue of the Americas, New York, NY 10020. (212) 512-2000. (800) 262-4729. Second edition. 1993. $95.50.

McGraw-Hill Encyclopedia of Science and Technology. McGraw- Hill Book, Incorporated, 1221 Avenue of the Americas, New York, NY 10020. (212) 512-2000. (800) 262-4729. Seventh edition. Twenty volumes. 1992. $1900.00.

Mcketta's Encyclopedia of Chemical Processes and Design. Marcel Dekker, Inc., 270 Madison Avenue, New York, NY 10016. (212) 696-9000. (800) 228-1160., 1979-

GENERAL WORKS

The Chemical Elements: Chemistry, Physical Properties and Uses in Science and Industry. Liam P. Roche. Prentice Hall, 113 Sylvan Avenue, Route 9W, Englewood Cliffs, NJ 07632. (201) 592-2000. (800) 922-0579. 1992.

Discovery of the Elements. Mary E. Weeks. Books on Demand, 300 North Zeeb Road, Ann Arbor, MI 48106-1346. (313) 761-4700. (800) 521-0600. 7th revised edition. Reprint. $180.00 in paper.

Elements: A Computer File. P. Thomas Vernier. CRC Publishers, Inc., 2000 Corporate Blvd., N.W., Boca Raton, FL 33431. (407) 994-0555. (800) 272-7737. IBM-PC compatible. 1992. $39.95.

The Elements. John Emsley. Oxford University Press, Inc., 200 Madison Avenue, New York, NY 10016. (212) 725-6000. (800) 334-4249. Second edition. 1991. $24.95 in paper.

The Elements: Their Origin, Abundance, and Distribution. P. A. Cox. Oxford University Press, Inc., 200 Madison Avenue,

New York, NY 10016. (212) 725-6000. (800) 334-4249. 1989. $19.95 in paper.

Exploring Chemical Elements and their Compounds. David L.Heiserman. TAB Books, P.O. Box 40, Blue Summit, PA 17294-0850. (717) 794-2191. (800) 233-1128. 1992. $17.95 in paper.

Origin and Evolution of the Elements. S. Kubono and T. Kajino. World Scientific Publishing Company, Inc., 1060 Main Street, River Edge, NJ 07661. (201) 487-9655. (800) 227-7562. 1993. $95.95.

The Origin of the Chemical Elements. Roger J. Tayler. Taylor & Francis, Inc., 1900 Frost Road, Suite 101, Bristol PA 19007-1598. (215) 785-5800. Revised edition. 1977. $26.00 in paper.

The Periodic Kingdom: A Journey Into the Land of the Chemical Elements. P. W. Arkins. Basic Books, 10 East 53rd Street, New York, NY 10022. (212) 207-7057. (800) 242- 7737. 1995. $20.00.

Periodic Table of the Elements. D.E.D. Electronic Publishing, 10306 East Live Oak Avenue, Arcadia, CA 91007. Computer software. 1992. $19.95.

Standard Methods of Chemical Analysis: The Elements. N. Howell Furman, editor. Krieger Publishing Company, P.O. Box 9542, Melbourne, FL 32902-9542. (407) 724-9542. Volume I. 6th edition. 1975. Reprint edition. $142.25.

HANDBOOKS AND MANUALS

Chemical Formulary. H. Bennett, editor. Chem Publishing Company, Inc., 80 Eighth Avenue, New York, NY 10011. (212) 255-1950. Volumes 1-30. $70.00 per volume.

Chemical Reference Handbook. Gordon Press Publications, P.O. Box 459, Bowling Green Station, New York, NY 10004. (718) 624-8419. 1995. $260.00.

CRC Handbook of Chemistry and Physics. David R. Kide, editor. CRC Press Inc., 2000 Corporate Blvd., NW, Boca Raton, FL 33431. (407) 994-0555. (800) 333-8300. 77th edition. 1996. $99.95.

Guide To Basic Chemical Compounds. D.R. Lide, Jr. CRC Publishers, Inc., 2000 Corporate Blvd., N.W., Boca Raton, FL 33431. (407) 994-0555 or (800) 333-8300. 1993. $120.00.

Lange's Handbook of Chemistry. John A. Dean, editor. McGraw-Hill Publishing Company, 1221 Avenue of the Americas, New York, NY 10020. (800) 262-4729 or (212) 512-3825. Fourteenth edition. 1992. $79.50.

Riegel's Handbook of Industrial Chemistry. James A. Kent, editor. Van Nostrand Reinhold, 115 Fifth Avenue, New York, NY 10003. (212) 254-3232 or (800) 926-2665. Ninth edition. 1992. $114.95.

ONLINE DATABASES AND CD-ROMS

Analytical Abstracts Online. Royal Society of Chemistry, Information Services, Thomas Graham House, Science Park, Milton Road, Cambridge, CB4 4WF, England. Contains citations, mostly with abstracts, of the worldwide literature on ana-

ELEMENTS (CHEMICAL)

Ency. of Physical Sciences and Engineering Info. Sources

lytical chemistry, from 1980 to present. Available on DIALOG, (800) 334-2564, and STN International, FIZ Karlsruhe, P.O. Box 2465, W-7500, Karlsruhe 1, Germany, online services. Inquire as to cost and availability.

CA Search. Chemical Abstracts Service, P.O. Box 3012, Columbus, OH 43210-0012. (614) 447-3600. (800) 848-6533. FAX (614) 447-3709. Very comprehensive guide to worldwide chemical literature and related fields, 1972 to present. Available on BRS,(800) 289-4277, DIALOG, (800) 334-2564, ORBIT (800) 456-7248, and STN International, FIZ Karlsruhe, P.O. Box 2465, W-7500, Karlsruhe 1, Germany, online services. Inquire as to cost and availability.

Chemical Journals of the American Chemical Society. American Chemical Society, 1155 16th Street, N.W., Washington, DC 20036. (202) 872-4381 or (800) 424-6747. Contains complete text of approximately 90,000 articles from 22 primary journals published by the American Chemical Society, from mostly 1982 to present. Available on STN International, FIZ Karlsruhe, P.O. Box 2465, W-7500, Karlsruhe 1, Germany, online service. Inquire as to cost and availability.

Compendex Plus. Engineering Information, Inc., 345 East 47th Street, New York, NY 10017. (212) 705-7600 or (800) 221-1044. Contains citations with abstracts to worldwide literature in engineering and technology, from 1970 to present. Available on online BRS,(800) 289-4277, DIALOG, (800) 334-2564, ORBIT (800) 456-7248, and STN International, FIZ Karlsruhe, P.O. Box 2465, W-7500, Karlsruhe 1, Germany, online services. Also available on CD-ROM. Inquire as to cost and availability.

Current Contents Search. Institute for Scientific Information, 3501 Market Street, Philadelphia, PA 19104. (215) 386-0100. FAX (215) 386-6362. Contains citations to articles listed in the table of contents of science and technology journals. Also articles in social sciences and life sciences journals. Available on BRS,(800) 289-4277, DIALOG,(800) 334-2564, online services. Inquire as to cost and availability.

Dissertation Abstracts. University Microfilms International, 300 North Zeeb Road, Ann Arbor, MI 48106. (800) 521-0600 or (313) 761-4700. Scope includes virtually all doctoral dissertations accepted at accredited American institutions from 1861 to present in 252 subject areas. Available on BRS,(800) 289-4277, DIALOG,(800) 334-2564, and OCLC EPIC,(800) 848-5878, online services. Also available on CD-ROM. Inquire as to cost and availability.

Gmelin Database. Gmelin-Institut fur Anorganische Chemie und Grenzgebiete, Varrentrapperstrasse, 40-42, Carl-Bosch-Haus, D-6000, Frankfurt am Main 90, Germany. Contains structural and factual data relating to inorganic and organometallic chemistry. Provides data from the Gmelin Handbook of Inorganic and Organometallic Chemistry. Covers the period 1817 to 1975; 1988-89. Available on STN International, FIZ Karlsruhe, P.O. Box 2465, W-7500, Karlsruhe 1, Germany, online service. Inquire as to cost and availability.

Kirk-Othmer Encyclopedia of Chemical Technology. John Wiley and Sons, Inc., 605 Third Avenue, New York, NY 10158. (800) 526-5368 or (212) 850-6000. Contains the complete text of all chapters in the 27 volume fourth edition of the KIRK-Othmer ENCYCLOPEDIA of CHEMICAL TECHNOLOGY. 1991. Available on BRS,(800) 289-4277, DIALOG,(800) 334-2564, online services. Inquire as to cost and availability.

Metals Datafile. Materials Information, ASM International, Materials Park, OH 44073. (216) 338-5151.Contains designa-

tion and specification numbers for ferrous and nonferrous metals and alloys. Covers the period 1982 to present. Available on ORBIT,(800) 456-7248, and STN International, FIZ Karlsruhe, P.O. Box 2465, W-7500, Karlsruhe 1, Germany, online services. Inquire as to cost and availability.

NTIS Bibliographic Database. National Technical Information Service, 5285 Port Royal Road, Springfield, VA 22161. (703) 487-4929 or FAX (703) 321-8199. Broad coverage of government-sponsored science and technology research reports, 1964 to present. Available on BRS,(800) 289-4277, DIALOG, (800) 334-2564, ORBIT, (800) 456-7248, and STN International, FIZ Karlsruhe, P.O. Box 2465, W-7500, Karlsruhe 1, Germany, online services. Also available on CD-ROM. Inquire as to cost and availability.

Physics Briefs. American Institute of Physics, 335 East 45th Street, New York, NY 10017. (212) 661-9260 or FAX (212) 661-2036. Contains citations with abstracts of the literature of physics and related fields, 1979 to present. Available on the STN International, FIZ Karlsruhe, P.O. Box 2465, W-7500, Karlsruhe 1, Germany, online service. Inquire as to cost and availability.

Scientific and Technical Books and Serials in Print. R.R. Bowker Inc., 121 Chanlon Road, New Providence, NJ 07974. (800) 521-8110 or (908) 464-6800. List of currently published books and serials in the physical and biological sciences, engineering and technology, with subject, author and titles indexes. Available on ORBIT,(800) 456-7248, online service. Inquire as to cost and availability.

SCISEARCH. Institute for Scientific Information, 3501 Market Street, Philadelphia, PA 19104. (800) 523-1850 or (215) 386-0100. Broad multidisciplinary title and author index to the international literature of science and technology, 1974 to present. Available on DIALOG,(800) 334-2564, and ORBIT,(800) 456-7248, online services. Also available on CD-ROM. Inquire as to cost and availability.

WILSONLINE. H.W. Wilson Company, 950 University Avenue, Bronx, NY 10452. (800) 367-6770 or (212) 588-8400. Makes available online versions of the printed H.W. Wilson indexes including Applied Science and Technology Index, Business Periodicals Index, General Science Index, and Readers' Guide to Periodical Literature. Period covered is generally 1983 to present. Available on BRS,(800) 289-4277, DIALOG,(800) 334-2564, and OCLC EPIC,(800) 848-5878, online services. Also available on CD-ROM. Inquire as to cost and availability.

PERIODICALS

Analytical Chemistry. American Chemical Society, Box 3337, Columbus, OH 43210. (614) 447-3776. 1929 to present. Semimonthly. $415.00 per year. ISSN: 0003-2700.

Angewandte Chemie. VCH Publishers, Inc., 220 East 23rd Street, New York, NY 10010-4606. (212) 683-8333. (800) 367-8249. 1888 to present. 22 issues per year. $840.00 per year. ISSN: 0044-8249.

Chemical Reviews. American Chemical Society, Box 3337, Columbus, OH 43210. (614) 447-3776. 1924 to present. 8 issues per year. $346.00 per year. ISSN: 0009-2665.

Chemical Week; Includes Annual Buyers Guide. Chemical Week Associates, 888 Seventh Avenue, New York, NY 10106. (212) 621-4900. FAX (212) 621-4949. 1914 to present. Weekly. $99.00 per year. ISSN: 0009-272X.

Chemtech. American Chemical Society. Box 3337, Columbus, OH 43210. (614) 447-3776. 1970 to present. Monthly. $370.00 per year. ISSN: 0009-2703.

Inorganic Chemistry. American Chemical Society, Box 3337, Columbus, OH 43210. (614) 447-3776. 1962 to present. Semimonthly. $500.00 per year. ISSN: 0020-1669.

Journal of the American Chemical Society. American Chemical Society, Box 3337, Columbus, OH 43210. (614) 447-3776. 1879 to present. Biweekly. $1055.00 per year. ISSN: 0002-7863.

Polyhedron. Elsevier Science Publishing Company, Inc., 660 White Plains Road, Tarrytown, NY 10591-5153. (914-524-9200. FAX (914) 524-9200. 1982 to present. 24 issues per year. $2175.00 per year. ISSN: 0277-5387.

RESEARCH CENTERS AND INSTITUTES

Chemical Laboratories. Harvard University, Oxford Street, Cambridge, MA 02138. (617) 495-4283. FAX (617) 496-5618.

Chemistry Laboratories. Rensselaer Polytechnic Institute, Cogswell Laboratory, Troy, NY 12180-3590. (518) 276-8981.

Lawrence Berkeley Laboratory, Chemical Sciences Division. One Cyclotron Road, Building 66, Berkeley, CA 94720. (510) 486-6062. FAX (510) 486-4995.

Theoretical Chemistry Institute. University of Wisconsin, Madison. 1101 University Avenue, Madison, WI 53706. (608) 262-1511.

University/industry Chemical Research Center. Mississippi State University, Department of Chemistry, P.O. Drawer CH, Mississippi State, MS 39762. (601) 325-3584.

ELLIPTICAL GALAXIES

See: GALAXIES

EMISSION ELECTRON MICROSCOPES

See: ELECTRON MICROSCOPY

EMISSION SPECTROSCOPY

See: SPECTROSCOPY

EMULSIONS

See: COLLOIDS

ENERGY

See also: COAL, ENERGY POWER ENGINEERING, FUSION, GEOTHERMAL POWER, HYDROELECTRIC POWER, HYDROGEN, NATURAL GAS, NUCLEAR ENERGY, PETROLEUM, SOLAR ENERGY, SYNTHETIC FUELS

ABSTRACT SERVICES AND INDEXES

Applied Science and Technology Index; A Cumulative Subject Index To English Language Periodicals in the Fields of Aeronautics and Space Science, Computer Technology, Chemistry, Construction Industry, Energy and Related Areas. H. W. Wilson Co., 950 University Avenue, Bronx, NY 10452-9978. (212) 588-8400. (800) 367-6770. FAX (718) 590-1617. From 1958 to present. Monthly. Inquire about cost and availability. Also available online (BRS and WILSONLINE) and on CD-ROM. ISSN: 0003-6986.

Chemical Abstracts. Chemical Abstracts Service. 2540 Olentangy River Road, Box 3012, Columbus, OH 43210-0012. (614) 447-3600. FAX (614) 447-3713. 1907 to present. Weekly. $16,800.00 per year. Available online and on CD-ROM. CA is also available in five section groupings. Inquire regarding cost and availability.

Current C Ontents: Engineering, Technology and Applied Sciences. Institute for Scientific Information, 3501 Market Street, Philadelphia, PA 19104. (215) 386-0100. FAX (215) 386-2991. From 1970 to present. Weekly. $442.00 per year. Also available online. Inquire regarding cost and availability. ISSN: 0095-7917.

Energy Research Abstracts. U. S. Department of Energy, office of Scientific and Technical Information, Box 62, Oak Ridge, TN 37831. (615) 674-0733. Subscriptions to: U. S. Government Printing office, Box 371954, Pittsburgh, PA 15250-7954. (202) 783-3238. 1976 to present. Monthly. $164.00 per year. ISSN: 0160-3604.

Engineering Index Monthly. Engineering Information, Inc., Castle Point on the Hudson, Hoboken, NJ 07030. (201) 216-8500. (800) 221-1044. FAX (201) 216-8532. Monthly. $2300.00 per year for monthly issues. $1980 per year for annual cumulation. Available on CD-ROM and online as COMPENDEX. ISSN: 0742-1974; 0360-8557.

General Science in Dex. H.W. Wilson Company, 950 University Avenue, Bronx, NY 10452. (212) 588-8400. (800) 367-6770. FAX (718) 590-1617. From 1978 to present. Ten issues per year; quarterly and annual cumulations. Service basis. Available on CD-ROM and online. Inquire regarding cost and availability. ISSN: 0162-1963.

Index To IEEE Publications. IEEE Service Center, 445 Hoes Lane, Piscataway, NJ 08855-1331. (908) 981-1393. (800) 678-IEEE. FAX (908) 981-9667. 1973 to present. Annual. ISSN: 0099-1368.

ISMEC: Mechanical Engineering Abstracts. Cambridge Scientific Abstracts, 7200 Wisconsin Avenue, Suite 601, Bethesda, MD 20814. (301) 961-6750 or (800) 843-7751. Contains citations to the literature in mechanical engineering, industrial and production engineering, energy, power, mechanics, devices and related areas, from 1973 to present. Available on the DIALOG,(800) 334-2564, online service. Inquire as to cost and availability.

ENERGY

Ency. of Physical Sciences and Engineering Info. Sources

N T I S Alerts: Energy. U.S. National Technical Information Service, 5285 Port Royal Road, Springfield, VA 22161. (703) 487-4929. FAX (703) 321-8547. Weekly. $160.00 per year.

Physics Abstracts. INSPEC. Section A, Science Abstracts. Institution of Electrical Engineers (IEE), London, United Kingdom. Available from: INSPEC/IEEE - Institute of Electrical and Electronic Engineers, Box 1331, 445 Hoes Lane, Piscataway, NJ 08855-1331. (908) 562-5549. 1898 to present. 24 issues per year. $2835.00 per year. ISSN: 0036-8091. Also available online and on CD-ROM.

Physics Briefs (Physikalische Berichte). Information Center for Energy, Physics, Mathematics; German Physical Society. V C H Publishers, Inc., 220 East 23rd Street, New York, NY 10010-4606. (212) 683-8333. 1845 to present. 24 issues per year. $2390.00 per year. Also available online. ISSN: 0179-7434.

Process and Chemical Engineering/chemical Engineering Abstracts. Royal Society of Chemistry, Information Services,Thomas Graham House, Science Park, Milton Road, Cambridge, CB4 4WF, England. Contains citations, mostly with abstracts,of the worldwide literature on chemical engineering; from 1982 to present. Monthly. $610.00 per year. Also available online. ISSN: 0960-5045.

Science Citation Index. SCI. Institute for Scientific Information, 3501 Market Street, Philadelphia, PA 19104. (215) 386-0100. (800) 523-1850. FAX (215) 386-2991. 1961 to present. Six issues per year, plus annual cumulation. $11650.00 per year. Also available online and on CD-ROM. Inquire about price and availability. ISSN: 0036-827X.

ANNUAL REVIEWS AND YEARBOOKS

Advances in Chemical Engineering. Academic Press, Inc., 6277 Sea Harbor Drive, Orlando, FL 32887. (800) 321-5068. 1956 to present. Irregular. ISSN: 0065-2377.

Advances in Energy Systems and Technology. Academic Press, Inc., 6277 Sea Harbor Drive, Orlando, FL 32887. (800) 321-5068. 1979 to present. Irregular. Inquire.

Advances in Nuclear Science and Technology. Plenum Publishing Corp., 233 Spring Street, New York, NY. (212) 620-8000. (800) 2221-9369. FAX (212) 463-0742. 1977 to present. Irregular.

Annual Review of Energy and the Environment. Annual Reviews, Inc., 4139 El Camino Way, Palo Alto, CA 94303-0139. (415) 943-4400. (800) 523-8635. 1976 to present. Annual. $71.00. ISSN: 1056-3466.

ASSOCIATIONS AND PROFESSIONAL SOCIETIES

American Institute of Chemical Engineers. 345 East 47th Street, New York, NY 10017. (212) 705-7338.

American Institute of Physics. One Physics Ellipse, College Park, MD 20740-3843. (301) 209-3100.

American Wind Energy Associaton. 122 C Street NW, Washington, DC 20001. (202) 383-2500.

Association of Energy Engineers, 4025 Pleasantdale Road, Suite 340, Atlanta, GA 30340. (770) 447-5083.

Edison Electric Institute, 701 Pennsylvania Avenue Nw, Washington, DC 20004-2696. (202) 508-5000.

IEEE (institute of Electrical and Electronics Engineers) 345 East 47th Street, New York, NY 10017. (212) 705-7900. FAX

212) 705-4929.

BIBLIOGRAPHIES

Information Sources in Chemistry. R.T. Bottle and J.F.B Rowland, editors. R.R. Bowker Inc., 121 Chanlon Road, New Providence, NJ 07974. (800) 521-8110 or (908) 464-6800. Fourth edition. 1993. $75.00.

Handbooks and Tables in Science and Technology. Russell H. Powell, editor. Oryx Press, 4041 North Central, Suite 700, Phoenix, AZ 85012-3330. (602) 265-2651 or (800) 279-6799. Third edition. 1994. $65.00.

Information Sources in Engineering. Peter Hicks, editor. Bowker-Saur, 121 Chanlon Road, New Providence, NJ 07974. (800) 521-8110 or (908) 464-6800. 3rd edition. 1996. $125.00.

Science Books and Films. American Association for the Advancement of Science, 1333 H Street NW, Washington, DC 20005. (202) 326-6454. 1965 to present. Nine issues per year. $40.00 per year. ISSN: 0098-342X.

Scientific and Technical Books and Serials in Print; An Index To Literature in Science and Technology. R.R. Bowker Inc., 121 Chanlon Road, New Providence, NJ 07974. (908) 464-6800. (800) 521-8110. FAX (908) 665-3502. 1972 to present. Annual. 4 volumes. 1994. $299.95. Also available on compact disc and online. ISSN: 0000-054X.

DIRECTORIES AND BIOGRAPHICAL SOURCES

American Institute of Chemical Engineers. Directory. American Institute of Chemical Engineers. 345 East 47th Street, New York, NY 10017-2396. (212) 705-7663. Annual.

American Men and Women of Science. Physical and Biological Sciences. Fifteenth edition. R.R. Bowker Inc., 121 Chanlon Road, New Providence, NJ 07974. (908) 464-6800. (800) 521-8110. 20th edition. 8 volumes. 1996. $850.00

Energy and Nuclear Science International Who's Who. Gale Research, 835 Penobscot Building, Detroit, MI 48226- 4094. (313) 961-2242. (800) 347-4253. Triennial. 1993. $325.00.

Energy Information Centers Directory. U. S. Council for Energy Awareness, 1776 I Street, NW, Suite 400, Washington, DC 20006. (202) 293-0770. Triennial. No charge.

Engineering Research Centers: Incorporating Electronics Research Centers. Stockton Press, 345 Park Avenue, New York, NY 10010. (212) 689-9200. (800) 221-2123. 4th edition. 1995. $515.00.

Research Centers Directory. Gale Research Company, Inc., 835 Penobscot Building, Detroit, MI 48226-4094. (313) 961-2242. (800) 347-4253. 20th edition. 1995. $485.00. ISSN: 0080-1518.

Who's Who in Engineering. Gordon Davis, editor. American Association of Engineering Societies. 1111 19th Street, NY, Suite 608, Washington, DC 20036. (202) 296-2237. (800) 658-8897. 9th edition. 1995. $220.00.

Who's Who in Technology. Gale Research, 835 Penobscot Building, Detroit, MI 48226-4094. (313 961-2242. (800) 521-4253. 7th edition. 1995. $195.00. ISBN 0-8103-7467-6.

World Energy and Nuclear Directory. Gale Research Company,Inc., 835 Penobscot Building, Detroit, MI 48226-4094. (313) 961-2242. (800) 347-4253. Biennial, even years. $450.00.

ENCYCLOPEDIAS AND DICTIONARIES

Encyclopedia of Applied Physics. VCH Publishers, Inc., 303 Northwest 12th Avenue, Deerfield Beach, FL 33442. (800) 367-8249. 1991-. Twenty volumes. $6000.00.

Encyclopedia of Chemical Processing and Design. McKetta Marcel Dekker, Inc., 270 Madison Avenue, New York, NY 10016. (212) 696-9000. (800) 228-1160. 1976 - . $175.00 per volume. ISSN:

Kirk-Othmer Encyclopedia of Chemical Technology. John Wiley and Sons, Inc., 605 Third Avenue, New York, NY 10158. (800) 526-5368 or (212) 850-6000. Fourth edition. 1991 - . Twenty-seven volumes. $5400.00.

McGraw-Hill Encyclopedia of Science and Technology. McGraw-Hill Book, Incorporated, 1221 Avenue of the Americas, New York, NY 10020. (212) 997-3675. (800) 262-4729. Seventh edition. Twenty volumes. 1992. $1900.00.

Ullman's Encyclopedia of Industrial Chemistry. VCH Publications, Inc., 220 East 23rd Street, Suite 909, New York, NY 10010-4606. (212) 683-8333. (800) 422-8824. 5th edition. 1984 - . Price varies per volume. ISSN:

GENERAL WORKS

Alternative Fuels: Alcohols, Hydrogen, Natural Gas and Propane. Society of Automotive Engineers (SAE), 400 Commonwealth Drive, Warrendale, PA 15096. (412) 776-4841. 1993. $57.00

Clean Fuels: Progress and Experiences of Demonstration Programs. Soc. Auto Engineers. Society of Automotive Engineers (SAE), 400 Commonwealth Drive, Warrendale, PA 15096. (412) 776-4841. SP 985. 1992. $35.00.

Combustion of Solid Fuels and Wastes. David A. Tillman. Academic Press, Inc., 6277 Sea Harbor Drive, Orlando, FL. (800) 321-5068. 1991. $60.00

Combustion of Synthetic Fuels. William Bartok, editor. American Chemical Society, 1155 16th Street, NW, Washington, DC 20036. (202) 872-4600. (800) 333-9511. FAX (614) 447- 3671. 1983. $38.95.

Electricity Supply: Efforts To Develop Solar and Wind Energy. Diane Publishing Co., 600 Upland Avenue, Upland, PA 19015. (610) 499-7415. $60.00 in paper.

Energy Conversion: Systems, Flow Physics and Engineering. Reiner Decher. Oxford University Press, Inc., 200 Madison Avenue, New York, NY 10016. (212) 725-6000. (800) 334-4249. 1994. $69.95.

Energy Efficiency: Challenges and Trends For Electric Utilities. Diane Publishing Co., 600 Upland Avenue, Upland, PA 19015. (610) 499-74151994. $60.00.

Fuels and Combustion. Samir Sarkar. Apt Books, 25 Brixham Road, Eliot, ME 03903-1209. 2nd edition 1990. $20.00 in paper

Nuclear Engineering: Theory and Technology of Commercial Nuclear Power. Ronald A. Knief. Hemisphere Publishing Corp., 1900 Frost Road, Suite 101, Bristol PA 19007. (215) 785-5800. (800) 821-8312. 2nd edition. 1992. $156.00.

Nuclear Power, Nuclear Fuel Cycle and Waste Management: Status and Trends 1992. UNIPUB, 4511-F Assembly Drive, Lanham, MD 20706-4391. (301) 459-7666. (800) 274-4888. 1992. $19.00.

Synthetic Fuel Technology Development in the United States: A Retrospective Assessment. Michael Crow, et al. Greenwood Press, 88 Post Road West, Box 5507, Westport, CT 06881. (203) 226-3571. (800) 225-5800. 1988. $55.00.

HANDBOOKS AND MANUALS

Automotive Fuels Handbook. Keith Owen and Trevor Coley. Society of Automotive Engineers (SAE), 400 Commonwealth Drive, Warrendale, PA 15096. (412) 776-4841. 2nd edition. 1995. $129.00.

Biomass Handbook. Osamu Katini and C. W. Hall, editors. Gordon and Breach Science Publishers, Inc., P.O.Box 200029, Riverfront Plaza Station, Newark, NJ 07102-0301. (201) 643-7500 1989. $419.00

Energy Engineering Handbook. Fairmont Press Staff. Prentice-Hall, 113 Sylvan Avenue, Route 9W, Englewood Cliffs, NJ 07632. (201) 592-2000. (800) 922-0579. 3rd edition. 1995. $75.00.

Fuel Cells: A Handbook. J. M. Hirschenfoffer Business/Technology Information Services, P. O. Box 574, Orinda, CA 94563. (510) 299-1829. (800) 808-8811. 1996. $115.00.

Fuel Science and Technology Handbook. James G. Speight. Marcel Dekker, Inc., 270 Madison Avenue, New York, NY 10016. (212) 696-9000. (800) 228-1160. 1990. $250.00.

Perry's Chemical Engineers' Handbook. Robert H. Perry and Donald W. Green, editors. McGraw-Hill Publishing Company, Inc., 1221 Avenue of the Americas, New York, NY 10020. (212) 512-2000. (800) 262-4729. 6th edition. 1996. $129.50.

Practical Handbook of Processing and Recycling of Municipal Waste. A. G. Manser and Alan A. Keeling. Lewis Pubs., CRC Press, Inc., 2000 Corporate Boulevard, NW, Boca Raton, FL 33431. (407) 994-0555. (800) 272-7737. 1996. $69.95.

Riegel's Handbook of Industrial Chemistry. James A. Kent, editor. Van Nostrand Reinhold, 115 Fifth Avenue, New York, NY 10003. (212) 254-3232 or (800) 926-2665. Ninth edition. 1992. $114.95.

ENERGY

Ency. of Physical Sciences and Engineering Info. Sources

Sourcebook of Methods of Analysis For Biomass and Biomass Conversion Processes. T. A. Milne, et al. Elsevier Science Publishing Company, Inc., 655 Avenue of the Americas, New York, NY 10010. (212) 989-5800. FAX (914) 333-2444. 1990. 93.75.

ONLINE DATABASES AND CD-ROMS

CA Search. Chemical Abstracts Service, P.O. Box 3012, Columbus, OH 43210-0012. (614) 447-3600. (800) 848-6533. FAX (614) 447-3709. Very comprehensive guide to worldwide chemical literature and related fields. 1972 to present. Available on BRS,(800) 289-4277, DIALOG, (800) 334-2564, ORBIT (800) 456-7248, and STN International, FIZ Karlsruhe, P.O. Box 2465, W-7500, Karlsruhe 1, Germany, online services. Inquire as to cost and availability.

Compendex Plus. Engineering Information, Inc., 345 East 47th Street, New York, NY 10017. (212) 705-7600 or (800) 221-1044. Contains citations with abstracts to worldwide literature in engineering and technology, from 1970 to present. Available on online BRS,(800) 289-4277, DIALOG, (800) 334-2564, ORBIT (800) 456-7248, and STN International, FIZ Karlsruhe, P.O. Box 2465, W-7500, Karlsruhe 1, Germany, online services. Also available on CD-ROM. Inquire as to cost and availability.

Current Contents Search. Institute for Scientific Information, 3501 Market Street, Philadelphia, PA 19104. (215) 386-0100. FAX (215) 386-6362. Contains citations to articles listed in the table of contents of science and technology journals. Also articles in social sciences and life sciences journals. Available on BRS,(800) 289-4277, DIALOG,(800) 334-2564, online services. Inquire as to cost and availability.

Dissertation Abstracts. University Microfilms International, 300 North Zeeb Road, Ann Arbor, MI 48106. (800) 521-0600 or (313) 761-4700. Scope includes virtually all doctoral dissertations accepted at accredited American institutions from 1861 to present in 252 subject areas. Available on BRS, (800) 289-4277, DIALOG, (800) 334-2564, and OCLC EPIC, (800) 848-5878, online services. Also available on CD-ROM. Inquire as to cost and availability.

INSPEC. Institution of Electrical Engineers, Michael Faraday House, Six Hills Way, Stevenage, Herts. SG1 2AY, England. Telephone: 0438 313311 or FAX 0438 742840. Contains citations to the worldwide literature of physics, electronics and electrical engineering, computer technology, and related fields. Available on BRS, (800) 289-4277, DIALOG, (800) 334-2564, ORBIT, (800) 456-7248, and STN International, FIZ Karlsruhe, P.O. Box 2465, W-7500, Karlsruhe 1, Germany, online services. Inquire as to cost and availability.

Kirk-Othmer Encyclopedia of Chemical Technology. John Wiley and Sons, Inc., 605 Third Avenue, New York, NY 10158. (800) 526-5368 or (212) 850-6000. Contains the complete text of all chapters in the 27 volume fourth edition of the KIRK-Othmer *Encyclopedia of Chemical Technology.* 1991. Available on BRS,(800) 289-4277, DIALOG,(800) 334-2564, online services. Inquire as to cost and availability.

NTIS Bibliographic Database. National Technical Information Service, 5285 Port Royal Road, Springfield, VA 22161. (703) 487-4929 or FAX (703) 321-8199. Broad coverage of government-sponsored science and technology research reports, 1964 to present. Available on BRS,(800) 289-4277, DIALOG, (800) 334-2564, ORBIT, (800) 456-7248, and STN International, FIZ

Karlsruhe, P.O. Box 2465, W-7500, Karlsruhe 1, Germany, online services. Also available on CD-ROM. Inquire as to cost and availability.

Physics Briefs. American Institute of Physics, 335 East 45th Street, New York, NY 10017. (212) 661-9260 or FAX (212) 661-2036. Contains citations with abstracts of the literature of physics and related fields, 1979 to present. Available on the STN International, FIZ Karlsruhe, P.O. Box 2465, W-7500, Karlsruhe 1, Germany, online service. Inquire as to cost and availability.

SCISEARCH. Institute for Scientific Information, 3501 Market Street, Philadelphia, PA 19104. (800) 523-1850 or (215) 386-0100. Broad multidisciplinary title and author index to the international literature of science and technology, 1974 to present. Available on DIALOG, (800) 334-2564, and ORBIT, (800) 456-7248, online services. Also available on CD-ROM. Inquire as to cost and availability.

WILSONLINE. H.W. Wilson Company, 950 University Avenue, Bronx, NY 10452. (800) 367-6770 or (212) 588-8400. Makes available online versions of the printed H.W. Wilson indexes including Applied Science and Technology Index, Business Periodicals Index, General Science Index, and Readers' Guide to Periodical Literature. Period covered is generally 1983 to present. Available on BRS, (800) 289-4277, DIALOG, (800) 334-2564, and OCLC EPIC, (800) 848-5878, online services. Also available on CD-ROM. Inquire as to cost and availability.

Who's Who in Technology. Gale Research Inc., 835 Penobscot Building, Detroit, MI 48226-4094. (313) 961-2242. (800) 347-4253. Contains biographical information of contemporary American scieNTISts and engineers. Available on DIALOG,(800) 334-2564, online service. Inquire as to cost and availability.

PERIODICALS

Aiche Journal. American Institute of Chemical Engineers. 345 East 47th Street, New York, NY 10017. (212) 705-7338. 1955 to present. Monthly. $295.00 per year.

Coal and Synfuels Technology. Pasha Publicaitons, Inc., 1616 North Fort Myer Drive, Arlington, VA 22209-3107. (703) 528-1244. 1979 to present. Weekly. $790.00 per year. ISSN: 0883-9735.

Energy and Fuels. American Chemical Society, Box 3337, Columbus, OH 43210. (614) 447-3774. 1987 to present. Bimonthly. $345.00 per year. ISSN: 0887-0624.

Energy Engineering. Association of Energy Engineers. Available from Fairmont Press, 700 Indian Trail, Lilburn, GA 30324. (404) 925-9388. 1904 to present. Bimonthly. $99.00 per year. ISSN: 0199-8595.

Energy Sources. Taylor and Francis, 1900 Forst Road, Stuie 101, Bristol, PA 19007-1598. (215) 785-5800. 1973 to present. Quarterly. $225.00. ISSN: 0090-8312.

International Journal of Solar Energy. Harwood Academic Publishers, 820 Town Center Drive, Langhorne, PA 19047. (215) 750-2642. 1982 to present. 8 issues per year. 254 ECU per year. ISSN: 0142-5919.

Journal of Energy Resources Technology. American Society of Mechanical Engineers, 345 East 47th Street, New York, NY

10017. (212) 705-7703. 1979 to present. Quarterly. $100.00 per year. ISSN: 0195-0738.

Power Engineering (Tulsa). PenWell Publishing Co., 1421 Sheridan Road, Tulsa, OK 74101. (918) 8335-3161. 1896 to present. Monthly. $38.00 per year. ISSN: 0032-5961.

Solar Energy. Elsevier Science Publishing Company, Inc., 660 White Plains Road, Tarrytown, NY 10591-5133. (914) 524-9200. 1957 to present. Monthly. $600.00 per year. ISSN: 0038-092X.

Wind Engineering. Multi-Science Publishing Company, Ltd., 107 High Street, Brentwood, Essex, CM14 4RX, England. Subscriptions to: Box 176 Avenel, NJ 07001. 1977 to present. Bimonthly. L135 per year. ISSN: 0309-524X.

RESEARCH CENTERS AND INSTITUTES

Center For Applied Energy Research. University of Kentucky, 3572 Iron Works Pike, Lexington, KY (606) 257-0305. FAX (606) 257-0220.

Center For Energy Studies. University of Texas at Austin, 10100 Burnet Road, Building 133, Austin, TX 78758. (512) 471-4496. FAX (512) 471-0781.

Energy Research Laboratories. Canada Centre for Mineral and Energy Technology, 555 Booth Street, Ottawa, ON Canada KIA 0G1. (613) 996-8201. FAX (613) 995-9584.

Institute of Gas Technology. 3424 South State Street, Chicago, IL 60616. (312) 567-3650. FAX (312) 567-5209.

Oak Ridge National Laboratory. U. S. Department of Energy, P. O. Box 2008, Oak Ridge, TN 37831-6255. (615) 576-2900. FAX (615) 576-2912.

ENGINEERING DESIGN

See: CAD

ENGINEERING GEOLOGY

See: GEOTECHNICAL ENGINEERING

ENGINES

See also: AUTOMOTIVE ENGINEERING, DIESEL ENGINES

ABSTRACT SERVICES AND INDEXES

Applied Science and Technology Index; A Cumulative Subject Index To English Language Periodicals in the Fields of Aeronautics and Space Science, Computer Technology, Chemistry, Construction Industry, Energy and Related Areas. H.W. Wilson Co., 950 University Avenue, Bronx, NY 10452. (800) 367-6770 or (212) 588-8400. FAX (718) 590-1617. From 1958 to

present. Monthly. Inquire about cost and availability. Also available on CD-ROM and online.

Engineering Index Monthly. Engineering Information, Inc., Castle Point on the Hudson, Hoboken, NJ 07030. (800) 221-1044. FAX (212) 832-1857. Monthly. $2300.00 per year. Also available online as COMPENDEX, and also on CD-ROM. Covers chemical engineering, computers, electrical engineering, civil engineering, metals and mining, industrial management, and mechanical engineering.

ASSOCIATIONS AND PROFESSIONAL SOCIETIES

Association of Diesel Specialists. 9140 Ward Parkway, Kansas City, MO 64114. (816) 444-3500.

Association of Marine Engine Manufacturers. 3050 K Street NW, Suite 145, Washington, DC 20007. (202) 944-4980.

Engine Manufacturers Association. 401 N. Michigan Avenue, Chicago, IL 60611. (312) 644-6610.

Engine Manufacturers Committee. c/o M. Molish II, Speed Code S22C, Allison Gas Turbine Division, PO Box 420, Indianapolis, in 46206-0420. (317) 230-2639.

Society of Automotive Engineers (sae International). 400 Commonwealth Drive, Warrendale, PA 15096. (412) 776-4841.

DIRECTORIES AND BIOGRAPHICAL SOURCES

Marine Engineers Review Directory of Marine Diesel Engines. Learned Information Inc., 143 Old Marlton Pike, Medford, NJ 08055-8750. (607) 654-6266. FAX (607) 654-4309. 1988 to present. Annual. Inquire for price.

Research Centers Directory. Gale Research, 835 Penobscot Building, Detroit, MI 48226-4094. (313) 961-2242. (800) 347-4253. FAX (313) 961-6083. $485.00.

Scientific and Technical Organizations and Agencies Directory. Gale Research, 835 Penobscot Building, Detroit, MI 48226-4094. (313) 961-2242. (800) 347-4253. FAX (313) 961-6083. 4th edition. 1996. $195.00.

Who's Who in Engineering. American Association of Engineering Societies, 1111 19th Street NW, Suite 608, Washington, DC 20036. (202) 296-2237 or (800) 658-8897. 8th edition. 1991. Inquire for price.

Who's Who in Technology. Gale Research, 835 Penobscot Building, Detroit, MI 48226-4094. (313) 961-2242. (800) 347-4253. FAX (313) 961-6083. 1995. $195.00.

ENCYCLOPEDIAS AND DICTIONARIES

Encyclopedia of Physical Science and Technology. Robert A. Meyers, ed. 18 volumes. Academic Press Inc., 6277 Sea Harbor Drive, Orlando, FL 32887. (800) 321-5068. 1992. $2100.00.

McGraw-Hill Dictionary of Scientific and Technical Terms. Sybil P. Parker, ed. 5th edition. McGraw-Hill Publishing Company, 1221 Avenue of the Americas, New York, NY 10020. (800) 262-4729 or (212) 512-3825. 1993. $110.50.

ENGINES

Ency. of Physical Sciences and Engineering Info. Sources

McGraw-Hill Encyclopedia of Engineering. Sybil P. Parker, ed. 2nd edition. McGraw-Hill Publishing Company, 1221 Avenue of the Americas, New York, NY 10020. (800) 262-4729 or (212) 512-3825. 1993. $95.50.

McGraw-Hill Encyclopedia of Science and Technology. Sybil P. Parker, ed. 7th edition. 20 volumes. McGraw-Hill Publishing Company, 1221 Avenue of the Americas, New York, NY 10020. (800) 262-4729 or (212) 512-3825. 1992. $1900.00

Thesaurus of Scientific, Technical, and Engineering Terms. Hemisphere Publishing Corporation, 1900 Frost Road, Suite 101, Bristol, PA 19007-1598. (215) 785-5800 or (800) 821-8312. FAX (215) 785-5515. 1987. $173.00.

GENERAL WORKS

Advanced Engine Technology. Heinz Heisler. SAE International, 400 Commonwealth Drive, Warrendale, PA 15096-0001. (412) 776-4970. FAX (412) 776-0790. 1995. $49.00 for non-members.

Diesel Engines. A. J. Wharton. 3d ed. Butterworths Publishers, 80 Montvale Avenue, Stoneham, MA 02180. (617) 438-8464 or (800) 366-2665. 1991. $24.95.

Introduction To Internal Combustion Engines. Richard Stone. 2nd edition. SAE International, 400 Commonwealth Drive, Warrendale, PA 15096-0001. (412) 776-4970. FAX (412) 776-0790. 1993. $39.00.

Internal Combustion Engine Fundamentals. John B. Heywood. McGraw-Hill Publishing Company, 1221 Avenue of the Americas, New York, NY 10020. (800) 262-4729 or (212) 512-3825. 1988. $69.00.

The Internal Combustion Engine in theory and Practice Volume One: Thermodynamics, Fluid Flow, Performance. Charles Fayette Taylor. 2nd edition. MIT Press, 55 Hayward Street, Cambridge, MA 02142. (617) 253-2884 or (800) 356-0343. FAX (617) 253-1709. 1985. $39.00.

The Internal Combustion Engine in theory and Practice Volume Two: Combustion, Fuels, Materials, Design. Charles Fayette Taylor. 2nd edition. MIT Press, 55 Hayward Street, Cambridge, MA 02142. (617) 253-2884 or (800) 356-0343. FAX (617) 253-1709. 1985. $39.00.

Internal Fire: The Internal Combustion Engine 1673-1900. Lyle Cummins. 2nd edition. SAE International, 400 Commonwealth Drive, Warrendale, PA 15096-0001. (412) 776-4970. FAX (412) 776-0790. 1989. $35.00.

Rotary Engine Design: Analysis and Developments. SAE International, 400 Commonwealth Drive, Warrendale, PA 15096-0001. (412) 776-4970. FAX (412) 776-0790. 1989. $42.00.

HANDBOOKS AND MANUALS

Automotive Handbook. 3rd edition. Robert Bosch GmbH/SAE International, 400 Commonwealth Drive, Warrendale, PA 15096-0001.(412) 776-4970. FAX (412) 776-0790. 1993. $29.00.

Diesel Engineering Handbook. Business Journals, 50 Day Street, PO Box 5550, South Norwalk, CT 06856. (203) 853-6015. $32.00.

Diesel Engine Manual. P.O. Black, et al. 4th edition. Macmillan Publishing, 200 Old Tappan Road, Old Tappan, NJ 07675. (800) 223-2336. FAX (800) 445-6991. 1983. $22.50.

Standard Handbook of Engineering Calculations. Tyler G. Hicks, ed. 2d ed. McGraw-Hill Publishing Company, 1221 Avenue of the Americas, New York, NY 10020. (800) 262-4729 or (212) 512-3825. 1985. $89.50.

The Wiley Engineer's Desk Reference. Sanford I. Heisler. John Wiley and Sons, Inc., 605 Third Avenue, New York, NY 10158. (800) 526-5368 or (212) 850-6000. 1984. $64.95.

ONLINE DATABASES AND CD-ROMS

Compendex Plus. Engineering Information, Inc., 345 East 47th Street, New York, NY 10017. (212) 705-7600 or (800) 221-1044. Contains citations with abstracts to worldwide literature in engineering and technology, from 1970 to present. Available on online BRS,(800) 289-4277, DIALOG, (800) 334-2564, ORBIT (800) 456-7248, and STN International, FIZ Karlsruhe, P.O. Box 2465, W-7500, Karlsruhe 1, Germany, online services. Also available on CD-ROM. Inquire as to cost and availability.

CSA Engineering. Cambridge Scientific Abstracts, 7200 Wisconsin Avenue, Suite 601, Bethesda, MD 20814. (301) 961-6750 or (800) 843-7751. Contains citations and abstracts of international periodicals and research literature covering all fields of engineering and science and technology,including computer and information science, electronics, mechanical engineering, solid state materials, 1981 to present. Available on BRS,(800) 289-4277, online service.Inquire as to cost and availability.

Current Contents Search. Institute for Scientific Information, 3501 Market Street, Philadelphia, PA 19104. (215) 386-0100. FAX (215) 386-6362. Contains citations to articles listed in the table of contents of science and technology journals. Also articles in social sciences and life sciences journals. Available on BRS,(800) 289-4277, DIALOG,(800) 334-2564, online services. Inquire as to cost and availability.

NTIS Bibliographic Database. National Technical Information Service, 5285 Port Royal Road, Springfield, VA 22161. (703) 487-4929 or FAX (703) 321-8199. Broad coverage of government-sponsored science and technology research reports, 1964 to present. Available on BRS,(800) 289-4277, DIALOG, (800) 334-2564, ORBIT, (800) 456-7248, and STN International, FIZ Karlsruhe, P.O. Box 2465, W-7500, Karlsruhe 1, Germany, online services. Also available on CD-ROM. Inquire as to cost and availability.

SAE Global Mobility Database. Society of Automative Engineers (SAE), Electronic Publishing Division, 400 Commonwealth Drive, Warrendale, PA 15098. (412) 776-4841. Contains citations with abstracts to technical papers on automotive and aerospace technology and vehicular-related industries that have been presented at SAE conferences. Covers 1906 to present. Available on ORBIT,(800) 456-7248,online service. Inquire as to cost and availability.

WILSONLINE. H.W. Wilson Company, 950 University Avenue, Bronx, NY 10452. (800) 367-6770 or (212) 588-8400. Makes available online versions of the printed H.W. Wilson indexes including Applied Science and Technology Index, Business

Periodicals Index, General Science Index, and Readers' Guide to Periodical Literature. Period covered is generally 1983 to present. Available on BRS,(800) 289-4277, DIALOG,(800) 334-2564, and OCLC EPIC,(800) 848-5878, online services. Also available on CD-ROM. Inquire as to cost and availability.

PERIODICALS

Aerospace Engineering. Society of Automotive Engineers, 400 Commonwealth Dr., Warrendale, PA 15096. (412) 776-4841. 1981 to present. Monthly. $48.00 per year.

Automotive Engineering Magazine. SAE International, 400 Commonwealth Drive, Warrendale, PA 15096-0001. (412) 776-4841. 1905 to present. Monthly. $72.00.

Aviation Week and Space Technology. 1221 Avenue of the Americas, New York, NY 10020. (212) 512-2999 or (800) 525-5003. 1916 to present. Weekly. $82.00 per year.

Design News. Cahners Publishing Company, 275 Washington Street, Newton, MA 02158-1630. (617) 558-4402. 1946 to present. 24 times a year. $95.00 per year.

Diesel and Gas Turbine Worldwide. Diesel and Gas Turbine Publications, 13555 Bishop's Court, Brookfield, WI 53005-6286. (914) 784-9177. FAX (914) 784-9177. 1969 to present. 10 times a year. $55.00 per year.

Diesel Progress Engines and Drives. Diesel and Gas Turbine Publications, 13555 Bishop's Court, Brookfield, WI 53005-6286. (914) 784-9177. FAX (914) 784-9177. 1935 to present. Monthly. $60.00 per year.

International Journal of Vehicle Design. Inderscience Enterprises Ltd., World Trade Center Building, 110 Ave. Luis Casai, Case Postale 306, CH-1215 Geneva-Aeroport, Switzerland. 1979 to present. Bi-monthly. $190.00 per year.

Journal of Applied Mechanics. American Society of Mechanical Engineers, 345 East 47th Street, New York, NY 10017-2398. (212) 705-7703. 1935 to present. Quarterly. $120.00 per year.

Journal of Engineering For Gas Turbines and Power. American Society of Mechanical Engineers, 345 East 47th Street, New York, NY 10017-2398. (212) 705-7703. Quarterly. Inquire for price.

Journal of Mechanical Design. American Society of Mechanical Engineers, 22 Law Drive, Box 2300, Fairfield, NJ 07007-2300. (800) 321-2633. 1978 to present. Quarterly. $100.00 per year.

Machine Design. Penton Publishing, 110 Superior Ave., Cleveland, OH 44114-2543. 1929 to present. 28 times a year. $180.00 per year.

SAE Transactions. SAE International, 400 Commonwealth Drive, Warrendale, PA 15096-0001. (412) 776-4841. Annual. $850.00 per year.

RESEARCH CENTERS AND INSTITUTES

Machinery & Engine Technology Laboratory. Institute for Mechanical Engineering, Bldg. M-7, Montreal Road, Ottawa, ON, Canada K1A 0R6. (613) 993-2425. FAX (613) 957-3281.

Sloan Automotive Laboratory, Massachusetts Institute of Technology. Room 3-340, Cambridge, MA 02139. (617) 253-2243.

University of Michigan Automotive Engineering Laboratory. Department of Mechanical Engineering & Applied Mechanics, 321 Lay, N. Campus, 1231 Beal, Ann Arbor, MI 48109-2121. (313) 764-4254.

University of Nebraska—lincoln, Center For Engine Technology. 255 WSEC, Lincoln, NE 68588-0525. (402) 472-2375. FAX (402) 472-1465.

Wayne State University Center For Automotive Research. 5050 Anthony Wayne Drive, Detroit, MI 48202. (313) 577-3887.

ENTROPY

See: TheRMODYNAMICS

ENVIRONMENTAL ENGINEERING

See also: AIR POLLUTION, GROUND WATER POLLUTION,POLLUTION, SOLID WASTE DISPOSAL, SLUDGE, SEWAGE TREATMENT, WATER POLLUTION, WATER TREATMENT

ABSTRACT SERVICES AND INDEXES

Applied Science and Technology Index; A Cumulative Subject Index To English Language Periodicals in the Fields of Aeronautics and Space Science, Computer Technology, Chemistry, Construction Industry, Energy and Related Areas. H.W. Wilson Co., 950 University Avenue, Bronx, NY 10452. (800) 367-6770 or (212) 588-8400. From 1958 to present. Monthly. Inquire about cost and availability. Also available on CD-ROM and online.

Civil and Structural Engineering Abstracts, Cambridge Scientific Abstracts, 7200 Wisconsin Avenue, Bethesda, MD 20814-4823. (301) 961-6750. FAX (301) 961-6720. 1993 to present. Monthly. $385.00 per year. Topics covered include structural design, construction equipment and methods, civil defense and military engineering, surveying, highway engineering, maritime and port structures, materials, land reclamation, and soil mechanics.

Current Contents: Engineering, Technology, and Applied Sciences. Institute for Scientific Information, 3501 Market Street, Philadelphia, PA 19104. (215) 386-0100. FAX (215) 386-2991. 1970 to present. Weekly. $442.00 per year.

Engineering Index Monthly. Engineering Information, Inc., Castle Point on the Hudson, Hoboken, NJ 07030. (800) 221-1044. FAX (201) 216-8532. Monthly. $2300.00 per year. Also available online as COMPENDEX, and also on CD-ROM. Covers chemical engineering, computers, electrical engineering, civil engineering, metals and mining, industrial management, and mechanical engineering.

Environmental Abstracts. R.R. Bowker Inc., 121 Chanlon Road, New Providence, NJ 07974. (800) 521-8110 or (908) 464-6800. Contains citations, most with abstracts, to literature relating to environmental issues and problems, from 1971 to present. Inquire as to cost and availability.

Environmental Engineering Abstracts, Cambridge Scientific Abstracts, 7200 Wisconsin Avenue, Bethesda, MD 20814-4823. (301) 961-6750. FAX (301) 961-6720. Monthly. Covers hazardous materials, environmental impact and protection, treatment of sewage and industrial wastes, hydroelectric power, tidal and wind power, arctic and tropical engineering.

Meteorological and Geoastrophysical Abstracts. American Meteorological Society, c/o Inforonics Inc., 550 Newtown Road, Littleton, MA 01460. (508) 486-8976. FAX (508) 486-0027. 1950 to present. Monthly. $950.00 per year. Current abstracts of books, reports, research papers, and miscellaneous literature on environmental sciences, meteorology, astrophysics, hydrology, glaciology, and physical oceanography.

Pollution Abstracts, Cambridge Scientific Abstracts, 7200 Wisconsin Avenue, Bethesda, MD 20814-4823. (301) 961-6750. FAX (301) 961-6720. Inquire for cost and availability.

Science Citation Index. Institute for Scientific Information, 3501 Market Street, Philadelphia, PA 19104. (215) 386-0100. FAX (215) 386-6362. Inquire about availability and cost. Also available on CD-ROM.

ASSOCIATIONS AND PROFESSIONAL SOCIETIES

American Academy of Environmental Engineers. 130 Holiday Ct., Suite 100, Annapolis, MD 21401. (410) 266-3311. FAX (410) 266-7653.

American Society of Civil Engineers. 345 East 47th Street, New York, NY 10017-2398. (212) 705-7520 or (800) 548-2723.

American Water Resources Association. 5410 Grosvenor Lane, Suite 220, Bethesda, MD 20814-2192. (301) 493-8600. FAX (301) 493-5844.

American Water Works Association. 6666 W. Quincy Avenue, Denver, CO 80235. (303) 794-7711.

Association of Ground Water Scientists and Engineers. A division of the National Water Well Association, 6375 Riverside Drive, Dublin, OH 43017. (614) 761-1711. FAX (614) 761-3446.

National Association of Environmental Professionals. 5165 MacArthur Blvd. NW, PO Box 9400, Washington, DC 20016-3315. (202) 966-1500. FAX (202) 966-1977.

National Ground Water Association, 6375 Riverside Drive, Dublin, OH 43017. (614) 761-1711. FAX (614) 761-3446.

National Water Resources Association. 3800 N. Fairfax Drive, Suite 4, Arlington, VA 22203-1703. (703) 524-1544. FAX (703) 524-1548.

National Water Well Association. 6375 Riverside Drive, Dublin, OH 43017. (614) 761-1711. FAX (614) 761-3446.

Water Quality Association. 4151 Naperville Road, Lisle, IL 60532. (708) 505-0160. FAX (708) 505-9637.

DIRECTORIES AND BIOGRAPHICAL SOURCES

Citation Who's Who Environmental Registry, 1992. Citation Directories Ltd., Inc. staff. 1992. Citation Directories Ltd., 1003 Central Avenue, PO Box 1036, Fort Dodge, IA 50501. (800) 848-9059. FAX (515) 386-2092. 1992. $239.95.

Directory of Engineering Societies and Related Organizations. Gordon Davis. 13th edition. American Association of Engineering Societies, 1111 19th Street NW, Suite 608, Washington, DC 20036. (202) 296-2237 or (800) 658-8897. 1989. Inquire for price.

Directory of Engineering and Technology Certification Programs 1995-1996. American Academy of Environmental Engineers, 130 Holiday Ct., Suite 100, Annapolis, MD 21401. (410) 266-3311. FAX (410) 266-7653. 1996. Inquire for price and availability.

Environmental Engineering Selection Guide 1995: A Directory To Engineering Firms and Educational Institutions Employing Board Certified Specialists. American Academy of Environmental Engineers, 130 Holiday Ct., Suite 100, Annapolis, MD 21401. (410) 266-3311. FAX (410) 266-7653. 1995. Inquire for price and availability.

International Research Centers Directory. Gale Research, 835 Penobscot Building, Detroit, MI 48226-4094. (313) 961-2242. (800) 347-4253. 8th edition. 1995. $410.00

Research Centers Directory. Gale Research, 835 Penobscot Building, Detroit, MI 48226-4094. (313) 961-2242. (800) 347-4253. $470.00

Scientific and Technical Organizations and Agencies Directory. Gale Research, 835 Penobscot Building, Detroit, MI 48226-4094. (313) 961-2242. (800) 347-4253. 4th edition. 1996. $195.00.

The Stevens Environmental Sourcebook. Stevens Publishing, 225 N. New Road, Waco TX 76710. (800) 727-7573. FAX (817) 751-5190. 1995. Inquire for price and availability.

Who's Who in Engineering. American Association of Engineering Societies, 1111 19th Street NW, Suite 608, Washington, DC 20036. (202) 296-2237 or (800) 658-8897. 8th edition. 1991. Inquire for price.

ENCYCLOPEDIAS AND DICTIONARIES

Concise Dictionary of Environmental Engineering. Tom M. Pankratz. CRC Press, 2000 Corporate Blvd., N.W., Boca Raton, FL 33431. (407) 994-0555 or (800) 272-7737. FAX (407) 994-0949. 1996. Inquire for price and availability.

Dictionary of Environmental Science and Technology. andrew Porteous. 2nd edition. John Wiley and Sons, Inc., 605 Third Avenue, New York, NY 10158. (800) 526-5368 or (212) 850-6000. 1996. Inquire for price and availability.

McGraw-Hill Dictionary of Scientific and Technical Terms. Sybil P. Parker, ed. 5th edition. McGraw-Hill Publishing Company, 1221 Avenue of the Americas, New York, NY 10020. (800) 262-4729 or (212) 512-3825. 1993. $110.50.

McGraw-Hill Encyclopedia of Engineering. Sybil P. Parker, ed. 2nd edition. McGraw-Hill Publishing Company, 1221 Avenue of the Americas, New York, NY 10020. (800) 262-4729 or (212) 512-3825. 1993. $95.50.

Thesaurus of Scientific, Technical, and Engineering Terms. Hemisphere Publishing Corporation, 1900 Frost Road, Suite 101, Bristol, PA 19007-1598. (215) 785-5800 or FAX (215) 785-5515. 1987. $173.00.

GENERAL WORKS

Basic Environmental Engineering. Henry R. Bungay. 2d edition. BiLine Associates, 2 Fairlawn Lane, Troy, NY 12180. (518) 274-2945. 1992. $38.00. Includes a diskette with computer programs.

Civil and Environmental Engineering Systems. Charles ReVelle. Prentice Hall, 113 Sylvan Avenue, Route 9W, Englewood Cliffs, NJ 07632. (201) 592-2000 or (800) 922-0579. 1996. $81.00.

Environmental Technologies and Trends. R.K. Jain. Springer-Verlag, 175 Fifth Avenue, New York, NY 10010. (212) 460-1500 or (800) 777-4643. FAX (212) 473-6272. 1996. Inquire for price and availability.

EPA Environmental Engineering Sourcebook. J.R. Boulding. Ann Arbor Press Inc., PO Box 310, Chelsea, MI 48118. (800) 858-5299. FAX (313) 475-8852. 1996. $59.95.

Recycling and Resource Recovery Engineering: Principles of Waste Processing. Richard I. Stessel. Springer-Verlag, 175 Fifth Avenue, New York, NY 10010. (212) 460-1500 or (800) 777-4643. FAX (212) 473-6272. 1996. Inquire for cost and availability.

HANDBOOKS AND MANUALS

Environmental Science and Technology Handbook. Porter C. Knowles, et al. Government Institutes Inc., 4 Research Place, Suite 200, Rockville, MD 20850. (301) 921-2355. 1994. $75.00.

Handbook of Environmental Health and Safety: Principles and Practices. Herman Koren and Michael S. Bisesi. 3d edition. Lewis Publishers, 2000 Corporate Blvd. NW, Boca Raton, FL 33431. (407) 994-0555 or (800) 272-7737. 1995. $85.00.

ONLINE DATABASES AND CD-ROMS

Compendex Plus. Engineering Information, Inc., 345 East 47th Street, New York, NY 10017. (212) 705-7600 or (800) 221-1044. Contains citations with abstracts to worldwide literature in engineering and technology, from 1970 to present. Available on online BRS,(800) 289-4277, DIALOG, (800) 334-2564, ORBIT (800) 456-7248, and STN International, FIZ Karlsruhe, P.O. Box 2465, W-7500, Karlsruhe 1, Germany, online services. Also available on CD-ROM. Inquire as to cost and availability.

CSA Engineering. Cambridge Scientific Abstracts, 7200 Wisconsin Avenue, Suite 601, Bethesda, MD 20814. (301) 961-6750 or (800) 843-7751. Contains citations and abstracts of international periodicals and research literature covering all fields of engineering and science and technology,including computer and information science, electronics, mechanical engineering,

solid state materials, 1981 to present. Available on BRS,(800) 289-4277, online service.Inquire as to cost and availability.

Current Contents Search. Institute for Scientific Information, 3501 Market Street, Philadelphia, PA 19104. (215) 386-0100. FAX (215) 386-6362. Contains citations to articles listed in the table of contents of science and technology journals. Also articles in social sciences and life sciences journals. Available on BRS,(800) 289-4277, DIALOG,(800) 334-2564, online services. Inquire as to cost and availability.

Dissertation Abstracts. University Microfilms International, 300 North Zeeb Road, Ann Arbor, MI 48106. (800) 521-0600 or (313) 761-4700. Scope includes virtually all doctoral dissertations accepted at accredited American institutions from 1861 to present in 252 subject areas. Available on BRS,(800) 289-4277, DIALOG,(800) 334-2564, and OCLC EPIC,(800) 848-5878, online services. Also available on CD-ROM. Inquire as to cost and availability.

NTIS Bibliographic Database. National Technical Information Service, 5285 Port Royal Road, Springfield, VA 22161. (703) 487-4929 or FAX (703) 321-8199. Broad coverage of government-sponsored science and technology research reports, 1964 to present. Available on BRS,(800) 289-4277, DIALOG, (800) 334-2564, ORBIT, (800) 456-7248, and STN International, FIZ Karlsruhe, P.O. Box 2465, W-7500, Karlsruhe 1, Germany, online services. Also available on CD-ROM. Inquire as to cost and availability.

SCISEARCH. Institute for Scientific Information, 3501 Market Street, Philadelphia, PA 19104. (800) 523-1850 or (215) 386-0100. Broad multidisciplinary title and author index to the international literature of science and technology, 1974 to present. Available on DIALOG,(800) 334-2564, and ORBIT,(800) 456-7248, online services. Also available on CD-ROM. Inquire as to cost and availability.

PERIODICALS

American Water Works Association Journal. American Water Works Association, 6666 W. Quincy Avenue, Denver, CO 80235. (303) 794-7711. 1914 to present. Monthly. $85.00 to libraries and governmental agencies only.

Environmental Engineering. Mechanical Engineering Publications Ltd., Northgate Avenue, Bury St. Edmunds, Suffolk IP32 6BW, England. Telephone 0284-763277. FAX 0284-704006. 1972 to present. Quarterly. Inquire for price.

Environmental Pollution. Elsevier Science [journals], 660 White Plains Rd., Tarrytown, NY 10159-5153. (919) 524-9200. FAX (919) 333-2444. 1967 to present. 1970 to present. 12 times a year. $1250.00 per year.

Environmental Science and Technology. American Chemical Society, 1155 16th Street, N.W., Washington, DC 20036. (202) 872-4381 or (800) 333-9511. FAX (614) 447-3671. 1967 to present. Monthly. $89.00 per year for non-members.

Ground Water. Ground Water Publishing Company, National Well Water Association, 6375 Riverside Drive, Dublin, OH 43017. (614) 761-1711. FAX (614) 761-3446. 1963 to present. Bi-monthly. $90.00 per year.

Journal of Environmental Engineering. American Society of Civil Engineers, Environmental Engineering Division, 345 East

ENVIRONMENTAL ENGINEERING

Ency. of Physical Sciences and Engineering Info. Sources

47th Street, New York, NY 10017-2398. (212) 705-7520 or (800) 548-2723. FAX (212) 705-7712. 1956 to present. Bi-monthly. $136.00 per year for non-members.

Journal of Environmental Science and Health, Part A: Environmental Science and Engineering. Marcel Dekker, Inc., 270 Madison Avenue, New York, NY 10016. (212) 696-9000. FAX (212) 685-4540. 1968 to present. 10 times a year. $725.00 per year.

Journal of Water Resources Planning and Management. American Society of Civil Engineers, Water Resources Planning and Management Division, 345 East 47th Street, New York, NY 10017-2398. (212) 705-7520 or (800) 548-2723. Bi-monthly. $112.00 per year.

Pollution Engineering. Cahners Publishing Company (Des Plains), 1350 East Touhy Avenue, Des Plaines, IL 60017-5080. (800) 323-4958. FAX (708) 390-2779. 1969 to prsent. 13 times a year. $70.00 per year.

Water, Air and Soil Pollution. Kluwer Academic Publishers, P.O. Box 358, Accord Station, Hingham, MA 02018-0358. (617) 871-6000. FAX (617) 871-6528. 1971 to present. 28 times a year. $1519.00.

Water Resources Bulletin. American Water Resources Association, 5410 Grosvenor Lane, Suite 220, Bethesda, MD 20814-2192. (301) 493-8600. FAX (301) 493-5844. 1965 to present. Bi-monthly. $115.00.

Water Resources Research. American Geophysical Union, 2000 Florida Avenue, N.W., Washington, DC 20009. (202) 462-6900. 1965 to present. Monthly. $660.00 for non-members.

World Water and Environmental Engineer. Thomas Telford Services Ltd., Thomas Telford House, 1 Heron Quay, London E14 4JD, England. Telephone 071-987-6999. FAX 071-538-4101. 1978 to present. Monthly. Inquire for price in U.S.

RESEARCH CENTERS AND INSTITUTES

Center For Energy and Environmental Studies. Carnegie Mellon University, Schenley Park, Pittsburgh, PA 15213. (412) 268-5897.

Pennsylvania State University Environmental Engineering Laboratory. 212 Sackett Bldg., University Park, PA 16802. (814) 863-4385. FAX (814) 863-7304.

University of Arizona Environmental Engineering Laboratory. Civil Engineering Department, Room 206, Tucson, AZ 85721. (602) 621-6586.

University of Central Florida, Environmental Systems Engineering Institute. Department of Civil Engineering and Environmental Science, PO Box 25000, Orlando, FL 32816. (305) 275-2785.

University of Florida Environmental Engineering Research Center. College of Engineering, 217 Black Hall, Gainesville, FL 32611. (904) 392-0841. FAX (904) 392-3076.

University of Illinois, Advanced Environmental Control Technology Research Center. 3230 Newmark C.E. Laboratory, 205 N. Mathews Avenue, Urbana, IL 61801. (217) 333-3822. FAX (217) 333-9464.

EPOXY RESINS

See: PLASTICS

EPSILON MESON

See: PARTICLE PHYSICS

EQUATIONS

See: ALGEBRA, MAtheMATICS

ERGONOMICS

See also: CAD (COMPUTER-AIDED DESIGN), INDUSTRIAL ENGINEERING, MECHANICAL ENGINEERING

ABSTRACT SERVICES AND INDEXES

Applied Mechanics Reviews: An Assessment of World Literature in Engineering Sciences. 1948-present. American Society of Mechanical Engineers, 345 East 47th Street, New York, NY 10017. (212) 705-7703. Monthly. $360.00 per year.

Applied Science and Technology Index; A Cumulative Subject Index To English Language Periodicals in the Fields of Aeronautics and Space Science, Computer Technology, Chemistry, Construction Industry, Energy and Related Areas. H.W. Wilson Co., 950 University Avenue, Bronx, NY 10452. (800) 367-6770 or (212) 588-8400. FAX (718) 590-1617. From 1958 to present. Monthly. Inquire about cost and availability. Also available on CD-ROM and online.

Current Contents: Engineering, Technology, and Applied Sciences. Institute for Scientific Information, 3501 Market Street, Philadelphia, PA 19104. (215) 386-0100. FAX (215) 386-6362. 1970 to present. Weekly. $442.00 per year.

Engineering Index Monthly. Engineering Information, Inc., Castle Point on the Hudson, Hoboken, NJ 07030. (800) 221-1044. FAX (212) 832-1857. Monthly. $2300.00 per year. Also available online as COMPENDEX, and also on CD-ROM. Covers chemical engineering, computers, electrical engineering, civil engineering, metals and mining, industrial management, and mechanical engineering.

Ergonomics Abstracts. Taylor & Francis Ltd., 242 Cherry Street, Philadephia, PA 19106. 1968 to present. Bi-monthly. $718.00 per year.

Index To Scientific and Technical Proceedings. Institute for Scientific Information, 3501 Market St., Philadelphia, PA 19104. (215) 386-0100. FAX (215) 386-6362. Monthly. $500.00 per year.

Index To Scientific Reviews. Institute for Scientific Information, 3501 Market St., Philadelphia, PA 19104. (215) 386-0100. FAX (215) 386-6362. Semi-annual.

Mechanical Engineering Abstracts (formerly *ISMEC*), Cambridge Scientific Abstracts, 7200 Wisconsin Avenue, Bethesda,

Ency. of Physical Sciences and Engineering Info. Sources

ERGONOMICS

MD 20814-4823. (301) 961-6750. FAX (301) 961-6720. 1967 to present. Monthly. $895.00 per year. Summarizes world literature in mechanical engineering, production engineering, and engineering management. Also available online.

Science Citation Index. Institute for Scientific Information, 3501 Market Street, Philadelphia, PA 19104. (215) 386-0100. FAX (215) 386-6362. Inquire about availability and cost. Also available on CD-ROM.

ASSOCIATIONS AND PROFESSIONAL SOCIETIES

American Society of Mechanical Engineers. 345 East 47th Street, New York, NY 10017-2398. (212) 705-7703.

The Ergonomics Society. Devonshire House, Devonshire Square, Loughborough LE11 3DW, United Kingdom. Telephone 44-1509-234904. FAX 44-1509-234904.

Human Factors and Ergonomics Society. Box 1369, Santa Monica, CA 90406-1369. (310) 394-1811. FAX (310) 394-2410.

The International Ergonomics Association. c/o Prof. ir. Pieter Rookmaaker, Netherlands Railways, SE ARBO/Ergonomics, PO Box 2025, 3500 HA Utrecht, the Netherlands. Telephone 31-30-354455. FAX 31-30-357639.

DIRECTORIES AND BIOGRAPHICAL SOURCES

American Men & Women of Science, 1995-96. R.R. Bowker Staff, eds. 19th edition. 8 volumes. R.R. Bowker/Reed International Publishing Company, 121 Chanlon Road, New Providence, NJ 07974. (908) 464-6800 or (800) 521-8110. 1995. $850.00

Directory of Engineering Societies and Related Organizations. Gordon Davis. 13th edition. American Association of Engineering Societies, 1111 19th Street NW, Suite 608, Washington, DC 20036. (202) 296-2237 or (800) 658-8897. 1989. Inquire for price.

International Research Centers Directory. Gale Research, 835 Penobscot Building, Detroit, MI 48226-4094. (313) 961-2242. (800) 347-4253. FAX (313) 961-6083. 8th edition. 1995. $410.00.

Research Centers Directory. Gale Research, 835 Penobscot Building, Detroit, MI 48226-4094. (313) 961-2242. (800) 347-4253. FAX (313) 961-6083. $485.00.

Scientific and Technical Organizations and Agencies Directory. Gale Research, 835 Penobscot Building, Detroit, MI 48226-4094. (313) 961-2242. (800) 347-4253. FAX (313) 961-6083. 4th edition. 1996. $195.00.

Who's Who in Engineering. American Association of Engineering Societies, 1111 19th Street NW, Suite 608, Washington, DC 20036. (202) 296-2237 or (800) 658-8897. 8th edition. 1991. Inquire for price.

ENCYCLOPEDIAS AND DICTIONARIES

Information Sources in Engineering. Ken Mildren & Peter Hicks, editors. 3rd edition. Bowker-Saur, 121 Chanlon Road, New

Providence, NJ 07974. (800) 521-8110. FAX (908) 665-6707. 1995. $100.00.

McGraw-Hill Dictionary of Scientific and Technical Terms. Sybil P. Parker, ed. 5th edition. McGraw-Hill Publishing Company, 1221 Avenue of the Americas, New York, NY 10020. (800) 262-4729 or (212) 512-3825. 1993. $110.50.

McGraw-Hill Encyclopedia of Engineering. Sybil P. Parker, ed. 2nd edition. McGraw-Hill Publishing Company, 1221 Avenue of the Americas, New York, NY 10020. (800) 262-4729 or (212) 512-3825. 1993. $95.50.

GENERAL WORKS

Engineering Ergonomics. Aaron J. Saber. Prentice Hall, 113 Sylvan Avenue, Route 9W, Englewood Cliffs, NJ 07632. (201) 592-2000 or (800) 922-0579. 1993. $50.00

Ergonomics At Work. David A. Osborne. 3rd edition. John Wiley and Sons, Inc., 605 Third Avenue, New York, NY 10158. (800) 526-5368 or (212) 850-6000. 1995. $39.95.

Ergonomics: Practical Guide. H. Panjwani. Van Nostrand Reinhold, 115 Fifth Avenue, New York, NY 10003. (212) 254-3232 or FAX (212) 254-9499. 1993. Inquire for cost and availability.

A Guide To the Ergonomics of Manufacturing. Martin Helander. Taylor & Francis, 1900 Frost Road, Suite 101, Bristol, PA 19007. (215) 785-5800. FAX (215) 785-5515. 1995. Inquire for cost and availability.

Introduction To Ergonomics. R.S. Bridger. McGraw-Hill Publishing Company, 1221 Avenue of the Americas, New York, NY 10020. (800) 262-4729 or (212) 512-3825. 1995. Inquire for cost and availability.

Step By Step Guide To Industrial Ergonomics. R. Webb. Van Nostrand Reinhold, 115 Fifth Avenue, New York, NY 10003. (212) 254-3232 or FAX (212) 254-9499. 1995. Inquire for cost and availability.

Work Design: Industrial Ergonomics. Stephen Konz. 4th edition. Publishing Horizons Inc., 8233 Via Paseo del Norte, Suite F400, Scottsdale, AZ 85258. (602) 991-7881. FAX (602) 991-4770. 1995. $65.00

HANDBOOKS AND MANUALS

Applied Ergonomics Handbook. Mike Burke. Lewis Publishers, 121 S. Main St., PO Box 519, Chelsea, MI 48118. (313) 475-8619 or (800) 272-7737. 1991. $59.95.

Artificial Intelligence, Knowledge Engineering, Expert Systems, Natural Language, Human Factors Ergonomics, Man-Machine Interface Design: Handbook of Human-Computer Interaction. M. Helander, editor. Elsevier Science Inc., Box 882, Madison Square Station, New York, NY 10159. (212) 989-5800. FAX (212) 633-3990. 1993. $251.00 (hardbound), $83.00 (paperbound).

Ergonomics Handbook. Rector Press Ltd., 130 Rattlesnake, Leverett, MA 01054-9726. (800) 247-3473. FAX (413) 367-2853. 1994. $75.00.

ERGONOMICS

Ency. of Physical Sciences and Engineering Info. Sources

Ergonomics Process Manual. Erik Roy, editor. Genium Publishing Corporation, 1 Genium Plaza, Schenectady, NY 12304. (800) 243-6486. FAX (518) 377-1891. 1993. $50.00.

Handbook of Ergonomic and Human Factors Tables. Jon Weimer. Prentice Hall, 113 Sylvan Avenue, Route 9W, Englewood Cliffs, NJ 07632. (201) 592-2000 or (800) 922-0579. 1993. $76.00.

ONLINE DATABASES AND CD-ROMS

Compendex Plus. Engineering Information, Inc., 345 East 47th Street, New York, NY 10017. (212) 705-7600 or (800) 221-1044. Contains citations with abstracts to worldwide literature in engineering and technology, from 1970 to present. Available on online BRS,(800) 289-4277, DIALOG, (800) 334-2564, ORBIT (800) 456-7248, and STN International, FIZ Karlsruhe, P.O. Box 2465, W-7500, Karlsruhe 1, Germany, online services. Also available on CD-ROM. Inquire as to cost and availability.

CSA Engineering. Cambridge Scientific Abstracts, 7200 Wisconsin Avenue, Suite 601, Bethesda, MD 20814. (301) 961-6750 or (800) 843-7751. Contains citations and abstracts of international periodicals and research literature covering all fields of engineering and science and technology,including computer and information science, electronics, mechanical engineering, solid state materials, 1981 to present. Available on BRS,(800) 289-4277, online service.Inquire as to cost and availability.

Current Contents Search. Institute for Scientific Information, 3501 Market Street, Philadelphia, PA 19104. (215) 386-0100. FAX (215) 386-6362. Contains citations to articles listed in the table of contents of science and technology journals. Also articles in social sciences and life sciences journals. Available on BRS,(800) 289-4277, DIALOG,(800) 334-2564, online services. Inquire as to cost and availability.

Dissertation Abstracts. University Microfilms International, 300 North Zeeb Road, Ann Arbor, MI 48106. (800) 521-0600 or (313) 761-4700. Scope includes virtually all doctoral dissertations accepted at accredited American institutions from 1861 to present in 252 subject areas. Available on BRS,(800) 289-4277, DIALOG,(800) 334-2564, and OCLC EPIC,(800) 848-5878, online services. Also available on CD-ROM. Inquire as to cost and availability.

ISMEC: Mechanical Engineering Abstracts. Cambridge Scientific Abstracts, 7200 Wisconsin Avenue, Suite 601, Bethesda, MD 20814. (301) 961-6750 or (800) 843-7751. Contains citations to the literature in mechanical engineering, industrial and production engineering, energy, power, mechanics, devices and related areas, from 1973 to present. Available on the DIALOG,(800) 334-2564, online service. Inquire as to cost and availability.

NTIS Bibliographic Database. National Technical Information Service, 5285 Port Royal Road, Springfield, VA 22161. (703) 487-4929 or FAX (703) 321-8199. Broad coverage of government-sponsored science and technology research reports, 1964 to present. Available on BRS,(800) 289-4277, DIALOG, (800) 334-2564, ORBIT, (800) 456-7248, and STN International, FIZ Karlsruhe, P.O. Box 2465, W-7500, Karlsruhe 1, Germany, online services. Also available on CD-ROM. Inquire as to cost and availability.

SCISEARCH. Institute for Scientific Information, 3501 Market Street, Philadelphia, PA 19104. (800) 523-1850 or (215) 386-0100. Broad multidisciplinary title and author index to the international literature of science and technology, 1974 to present. Available on DIALOG,(800) 334-2564, and ORBIT,(800) 456-7248, online services. Also available on CD-ROM. Inquire as to cost and availability.

WILSONLINE. H.W. Wilson Company, 950 University Avenue, Bronx, NY 10452. (800) 367-6770 or (212) 588-8400. Makes available online versions of the printed H.W. Wilson indexes including Applied Science and Technology Index, Business Periodicals Index, General Science Index, and Readers' Guide to Periodical Literature. Period covered is generally 1983 to present. Available on BRS,(800) 289-4277, DIALOG,(800) 334-2564, and OCLC EPIC,(800) 848-5878, online services. Also available on CD-ROM. Inquire as to cost and availability.

PERIODICALS

Advances in Human Factors—ergonomics. Elsevier Science Inc., Box 882, Madison Square Station, New York, NY 10159. (212) 989-5800. FAX (212) 633-3990. 1984 to present. Irregular. Price varies.

Advances in Human-computer Interaction. Ablex Publishing Corporation, 355 Chestnut Street, Norwood, NJ 07648. (201) 767-8450 or FAX (201) 767-6717. 1985 to present. Irregular. Price varies.

Applied Ergonomics. Butterworth-Heinemann, Linacre House, Jordan Hill, Ocford OX2 8DP, England. Telephone 0865-310366. FAX 0865-310898. 1969 to present. Bi-monthly. Inquire for cost.

Ergonomics: The official Publication of the Ergonomics Research Society. Taylor & Francis Ltd., 4 John Street, London WC1N 2ET, England. Telephone 0256-840366. FAX 0256-479438. 1957 to present. Monthly. $635.00.

Human Factors. Human Factors and Ergonomics Society, Box 1369, Santa Monica, CA 90406-1369. (310) 394-1811. FAX (310) 394-2410. Quarterly. $115.00 for non-members.

Human Factors and Ergonomics Society Bulletin. Human Factors and Ergonomics Society, Box 1369, Santa Monica, CA 90406-1369. (310) 394-1811. FAX (310) 394-2410. 1958 to present. Monthly. $36.00 for non-members.

International Journal of Industrial Ergonomics. Elsevier Science Inc., Box 882, Madison Square Station, New York, NY 10159. (212) 989-5800. FAX (212) 633-3990. 1986 to present. Eight times a year. $388.00.

International Journal of Human Computer Studies. Harcourt Brace & Company, Ltd., Foots Cray High Street, Sidcup, Kent DA14 5HP, England. Telephone 44-81-300-3322. FAX 44-81-309-0807. 1969 to present. Monthly. Inquire for cost.

RESEARCH CENTERS AND INSTITUTES

Kansas State University Human Factors/ergonomics Laboratory. Department of Industrial Engineering, Manhattan, KS 66506. (913) 532-5606. FAX (913) 532-7810.

University of Cincinnati Ergonomics and Engineering Controls Research Laboratory. Cincinnati, OH 45221. (513) 556-2652.

Ency. of Physical Sciences and Engineering Info. Sources

EROSION

University of Louisville Center For Industrial Ergonomics. James B. Speed Scientific School, Louisville, KY 40292. (502) 588-7173. FAX (502) 588-7397.

University of Michigan Center For Ergonomics. 1205 Beal Street, I.O.E. Bldg., Ann Arbor, MI 48109-2117. (313) 763-2243. FAX (313) 764-3451.

EROSION

See also: GEOLOGY, SOIL SCIENCE

ABSTRACT SERVICES AND INDEXES

Biological and Agricultural Index. H.W. Wilson Company, 950 University Avenue, Bronx, NY 10452. (800) 367-6770. FAX (718) 590-1617. 1964 to present. Monthly. Inquire for price. Also available online and on CD-ROM.

Civil and Structural Engineering Abstracts, Cambridge Scientific Abstracts, 7200 Wisconsin Avenue, Bethesda, MD 20814-4823. (301) 961-6750. FAX (301) 961-6720. 1993 to present. Monthly. $385.00 per year. Topics covered include structural design, construction equipment and methods, civil defense and military engineering, surveying, highway engineering, maritime and port structures, materials, land reclamation, and soil mechanics.

Current Contents: Physical, Chemical and Earth Sciences. Institute for Scientific Information, 3501 Market Street, Philadelphia, PA 19104. (215) 386-0100. FAX (215) 386-6362. Weekly. $360.00 per year.

Environmental Abstracts. R.R. Bowker Inc., 121 Chanlon Road, New Providence, NJ 07974. (800) 521-8110 or (908) 464-6800. Contains citations, most with abstracts, to literature relating to environmental issues and problems, from 1971 to present. Inquire as to cost and availability.

Environmental Engineering Abstracts, Cambridge Scientific Abstracts, 7200 Wisconsin Avenue, Bethesda, MD 20814-4823. (301) 961-6750. FAX (301) 961-6720. Monthly. Covers hazardous materials, environmental impact and protection, treatment of sewage and industrial wastes, hydroelectric power, tidal and wind power, artic and tropical engineering.

Selected Water Resources Abstracts. U.S. Geological Survey, Water Resources Scientific Information Center, 425 National Center, Reston, VA 22092. (703) 648-6820. Monthly. $115.00 per year. Also available on CD-ROM.

ASSOCIATIONS AND PROFESSIONAL SOCIETIES

American Geological Institute. 4220 King Street, Alexandria, VA 22302. (703) 379-2480. Fax (703) 379-7563.

American Geophysical Union. 2000 Florida Avenue, N.W., Washington, DC 20009. (202) 462-6900. FAX (202) 328-0566.

Geological Society of America. 3300 Penrose Place, PO Box 9140, Denver, CO 80301-9140. (303) 447-2020. FAX (303) 447-1133.

International Erosion Control Association. PO Box 4904, Steamboat Springs, CO 80477-4904. (303) 879-3010. FAX (303) 879-8563.

International Research and Training Center On Erosion and Sedimentation (irtces). 20 Chegongzhuang Xilu, PO Box 366, Beijing, People's Republic of China. Telephone 1-8413372. FAX 1-8411174.

Soil and Water Conservation Society of America. 7515 Northeast Ankeny Road, Ankeny, IA 50021. (515) 289-2331. FAX (515) 289-1227.

DIRECTORIES AND BIOGRAPHICAL SOURCES

American Men & Women of Science, 1995-96. R.R. Bowker Staff, eds. 19th edition. 8 volumes. R.R. Bowker/Reed International Publishing Company, 121 Chanlon Road, New Providence, NJ 07974. (908) 464-6800 or (800) 521-8110. 1995. $850.00

Conservation Directory. National Wildlife Federation, 1400 16th Street NW, Washington, DC 20036-2266. (202) 797-6800. FAX (703) 442-7332. 1955 to present. Annual. $20.00.

Directory of Engineering Societies and Related Organizations. Gordon Davis. 13th edition. American Association of Engineering Societies, 1111 19th Street NW, Suite 608, Washington, DC 20036. (202) 296-2237 or (800) 658-8897. 1989. Inquire for price.

International Research Centers Directory. Gale Research, 835 Penobscot Building, Detroit, MI 48226-4094. (313) 961-2242. (800) 347-4253. FAX (313) 961-6083. 8th edition. 1995. $410.00

Research Centers Directory. Gale Research, 835 Penobscot Building, Detroit, MI 48226-4094. (313) 961-2242. (800) 347-4253. FAX (313) 961-6083. $485.00.

Scientific and Technical Organizations and Agencies Directory. Gale Research, 835 Penobscot Building, Detroit, MI 48226-4094. (313) 961-2242. (800) 347-4253. FAX (313) 961-6083. 4th edition. 1996. $195.00.

ENCYCLOPEDIAS AND DICTIONARIES

Dictionary of Civil Engineering. V.N. Vazirani. South Asia Books, Box 502, Columbia, MO 65205. (314) 474-0116. FAX (314) 474-8124. 1992. $20.00.

The Encyclopedia of the Solid Earth Sciences. Philip Kearey, editor-in-chief. Blackwell Scientific Publications, 238 Main Street, Cambridge, MA 02142. (617) 876-7000 or (800) 759-6102. FAX (617) 876-7022. 1993. $60.00.

Encyclopedia of Physical Science and Technology. Robert A. Meyers, ed. 18 volumes. Academic Press Inc., 6277 Sea Harbor Drive, Orlando, FL 32887. (800) 321-5068. 1992. $2100.00.

GENERAL WORKS

Coasts: Erosion and Sedimentation. H. Bremer and K.M. Clayton, editors. Lubrecht & Cramer Ltd., 38 Country Route

EROSION

Ency. of Physical Sciences and Engineering Info. Sources

48, Forestburgh, NY 12777-6400. (914) 794-8539. FAX (914) 791-7575. 1989. $88.00.

Erosion and Sedimentation. Pierre Y. Julien. Cambridge University Press, 40 West 20th Street, New York, NY 10011-4211. (212) 924-3900 or (800) 872-7423. FAX (914) 937-4712. 1995. $54.95.

Field Measurement of Soil Erosion and Runoff. N.W. Hudson. UNIPUB, 4611-F Assembly Drive, Lanham, MD 20706-4391. (800) 274-4888. FAX (301) 459-0056. 1993. $17.00

Soil Erosion: A Research Guide. Gordon Press Publishers, PO Box 459, Bowling Green Station, New York, NY 10004. (718) 624-8419. 1991. $75.00.

Soil Erosion, Conservation, and Rehabilitation. Menachem Agassi, editor. Marcel Dekker, Inc., 270 Madison Avenue, New York, NY 10016. (212) 696-9000. FAX (212) 685-4540. 1995. Inquire for price.

Soil Erosion Research Methods. Rattan Lal, editor. St. Lucie Press, 100 E. Linton Boulevard, Suite 403B, Delray Beach, FL 33483. (407) 274-9906. FAX (407) 274-9927. 1994. $42.50.

World Soil Erosion and Conservation. David Pimental, editor. Cambridge University Press, 40 West 20th Street, New York, NY 10011-4211. (212) 924-3900 or (800) 872-7423. FAX (914) 937-4712. 1993. $99.95.

HANDBOOKS AND MANUALS

Erosion and Sediment Control Handbook. Association of Bay Area Governments, PO Box 2050, Oakland, CA 94604-2050. (510) 464-7900. FAX (510) 464-7970. 1986. $65.00.

Handbook of Hydrology. David R. Maidment, editor in chief. McGraw-Hill Publishing Company, 1221 Avenue of the Americas, New York, NY 10020. (800) 262-4729 or (212) 512-3825. 1993. $115.00.

ONLINE DATABASES AND CD-ROMS

Current Contents Search. Institute for Scientific Information, 3501 Market Street, Philadelphia, PA 19104. (215) 386-0100. FAX (215) 386-6362. Contains citations to articles listed in the table of contents of science and technology journals. Also articles in social sciences and life sciences journals. Available on BRS, (800) 289-4277, DIALOG, (800) 334-2564, online services. Inquire as to cost and availability.

Dissertation Abstracts. University Microfilms International, 300 North Zeeb Road, Ann Arbor, MI 48106. (800) 521-0600 or (313) 761-4700. Scope includes virtually all doctoral dissertations accepted at accredited American institutions from 1861 to present in 252 subject areas. Available on BRS, (800) 289-4277, DIALOG, (800) 334-2564, and OCLC EPIC, (800) 848-5878, online services. Also available on CD-ROM. Inquire as to cost and availability.

Earth Sciences. U.S. Geological Survey, 12201 Sunrise Valley Drive, Reston, VA 22092-9998. (703) 648-4460. CD-ROM of earth science databases including the U.S. Geological Survey Library, Earth Science Data Directory, and GEOINDEX, citations to published geological maps. $350.00 per year with quarterly updates.

Geoarchive. Geosystems, P.O. Box 1024, Westminster, London, England, SW1 P 2JL. Citations to literature on geoscience, 1969 to present. Inquire as to online cost and availability.

Geobase. Elsevier - Geo Abstracts, Regency House, 34 Duke Street, Norwich NR3 3AP, England. Contains citations to the worldwide earth science literature from 1980 to date. Available on DIALOG, ORBIT online services. Inquire as to cost and availability.

Georef: Bibliography and Index of Geology. American Geological Institute, 4220 King Street, Alexandria, VA 22302. (703) 379-2480. Fax (703) 379-7563. Monthly. Inquire for price and availability.

NTIS Bibliographic Database. National Technical Information Service, 5285 Port Royal Road, Springfield, VA 22161. (703) 487-4929 or FAX (703) 321-8199. Broad coverage of government-sponsored science and technology research reports, 1964 to present. Available on BRS, (800) 289-4277, DIALOG, (800) 334-2564, ORBIT, (800) 456-7248, and STN International, FIZ Karlsruhe, P.O. Box 2465, W-7500, Karlsruhe 1, Germany, online services. Also available on CD-ROM. Inquire as to cost and availability.

SCISEARCH. Institute for Scientific Information, 3501 Market Street, Philadelphia, PA 19104. (800) 523-1850 or (215) 386-0100. Broad multidisciplinary title and author index to the international literature of science and technology, 1974 to present. Available on DIALOG, (800) 334-2564, and ORBIT, (800) 456-7248, online services. Also available on CD-ROM. Inquire as to cost and availability.

PERIODICALS

American Journal of Science. Box 6666, Yale Station, New Haven, CT 06511-8130. (203) 432-3131. FAX (203) 432-5668. 1818 to present. Monthly except July and August. $40.00 per year.

Geochimica Et Cosmochimica Acta. Elsevier Science [journals], 660 White Plains Rd., Tarrytown, NY 10159-5153. (919) 524-9200. FAX (919) 333-2444. 1950 to present. Twenty-four times a year. $895.00.

Geological Magazine. Cambridge University Press, Journals Department, 40 West 20th Street, New York, NY 10011-4211. (212) 924-3900. 1864 to present. Bi-monthly. $263.00 per year.

Geological Society Journal. Geological Society Publishing House, Unit 7, Brassmill Enterprise Centre, Brassmill Lane, Bath BA1 3JN, England. Telephone 0225-445046. FAX 0225-442836. 1845 to present. Six times a year. $524.00.

Geological Society of America Bulletin. 3300 Penrose Place, PO Box 9140, Denver, CO 80301-9140. (303) 447-2020. 1888 to present. Monthly. $205.00 per year.

Geology. Geological Society of America, 3300 Penrose Place, PO Box 9140, Denver, CO 80301-9140. (303) 447-2020. 1973 to present. Monthly. $170.00 per year for non-members.

Journal of Geology. University of Chicago Press, Journals Division, 5720 S. Woodlawn Avenue, Chicago, IL 60637. (312) 753-3347. FAX (312) 753-0811. 1893 to present. Bi-monthly. $45.00 per year.

Journal of Soil and Water Conservation. Soil Conservation Society of America, 7515 Northeast Ankeny Road, Ankeny, IA 50021. (515) 289-2331. FAX (515) 289-1227. 1946 to present. Bi-monthly. $39.00.

Land Degradation and Rehabilitation. J. Wiley & Sons, Journals, 605 Third Avenue, New York, NY 10158-0012. (212) 692-6000. 1989 to present. Quarterly. $165.00.

NASLR Newsletter. National Association of State Land Reclamationists, 459B Carlisle Drive, Herndon, VA 22070. (703) 709-8654. FAX (703) 709-8655. Quarterly. Inquire for price and availability.

Professional Geologist. American Institute of Professional Geologists, 7828 Vance Drive, Suite 103, Arvada, CO 80003. (303) 431-0831. FAX (303) 431-1332. 1964 to present. Monthly. $25.00.

Restoration and Management Notes. University of Wisconsin Press, Journals Division, 114 N. Murray Street, Madison, WI 53715. (608) 262-4952. FAX (608) 262-7560. 1981 to present. Semi-annual. $18.00.

Soil Technology. Elsevier Science Inc., Box 882, Madison Square Station, New York, NY 10159. (212) 989-5800. FAX (212) 633-3990. 1988 to present. Quarterly. $176.00.

RESEARCH CENTERS AND INSTITUTES

North Central Soil Conservation Laboratory. N. Iowa Avenue, Morris, MN 56267. (612) 589-3411. FAX (612) 589-3787.

USDA National Soil Erosion Research Laboratory. Purdue University, SOIL Bldg., West Lafayette, in 47907. (317) 494-8673. FAX (317) 494-5948.

USDA Northern Plains Soil & Water Research Center. Montana State University, PO Box 1109, Sidney, MT 59270. (406) 482-2020.

USDA Soil & Water Management Research Unit. Route 1, 3793 N., 3600 E., Kimberly, ID 83341. (208) 423-5582.

USDA Wind Erosion Research Unit. Room 105-B, E. Waters Hall, Kansas State University, Manhattan, KS 66506. (913) 532-6807. FAX (913) 776-0962.

ETHYL ALCOHOL

See: ALCOHOL

EUCLIDEAN GEOMETRY

See: GEOMETRY

EXPANSION JOINTS

See: JOINTS

EXPERIMENTAL DESIGN

See: CAD

EXPERT SYSTEMS

See also: ARTIFICIAL INTELLIGENCE

ABSTRACT SERVICES AND INDEXES

ACM Guide To Computing Literature. Association for Computing Machinery, 11 West 42nd Street, New York, NY 10036. (212) 869-7440. Fax (212) 869-0481. 1964 to present. Annual. $175.00 per year.

Applied Science and Technology Index; A Cumulative Subject Index To English Language Periodicals in the Fields of Aeronautics and Space Science, Computer Technology, Chemistry, Construction Industry, Energy and Related Areas. H.W. Wilson Co., 950 University Avenue, Bronx, NY 10452. (800) 367-6770 or (212) 588-8400. FAX (718) 590-1617. From 1958 to present. Monthly. Inquire about cost and availability. Also available on CD-ROM and online.

Artificial Intelligence Abstracts. R.R. Bowker, Bowker A&I Publishing, 121 Chanlon Road, New Providence, NJ 07974. (908) 771-7714. Fax (908) 771-7725. 1985 to present. Monthly. $495.00 per year.

CAD/CAM Abstracts. R.R. Bowker, Bowker A&I Publishing, 121 Chanlon Road, New Providence, NJ 07974. (908) 771-7714. Fax (908) 771-7725. 1984 to present. Monthly. $495.00 per year.

Computer Abstracts, MCB University Press Ltd., PO Box 10812, Birmingham, AL 35201-0812. (800) 633-4931. FAX (205) 995-1588. Monthly. Covers computer theory, data, hardware, systems, networks, human-computer interaction, artificial intelligence, as well as applications of computers in aerospace, business, CAD/CAM, cartography, civil engineering, electronics and electrical engineering, industrial engineering, mechanical engineering, medicine, structural engineering, etc. 1957 to present. Monthly. $1399.95 per year.

Computer and Control Abstracts (aka INSPEC). Order from: INSPEC/IEEE, Box 1331, 445 Hoes Lane, Piscataway, NJ 08855-1331. (908) 562-5549. Section C: 1966 to present. Monthly. $1455.00 per year. Abstracts of international technical information.

Computer and Information Systems Abstracts Journal. Cambridge Scientific Abstracts, 7200 Wisconsin Avenue, Bethesda, MD 20814. (301) 961-6750. Fax (301) 961-6720. 1962 to present. Monthly. $1035 per year.

Computer Literature Index. Applied Computer Research Inc., Box 82266, Phoenix, AZ 85071-2266. (800) 234-2227. 1971 to present. Quarterly plus annual cumulation. $198.50 per year. Bibliography of books, articles, and reports.

Computing Journal Abstracts. Techgnosis Ltd., Blade House, Battersea Road, Stockport, Cheshire 3AE, England. Telephone 061-442-2639. FAX 061-443-1162. 1969 to present. Monthly. Inquire for price.

EXPERT SYSTEMS

Ency. of Physical Sciences and Engineering Info. Sources

Computing Reviews, Association For Computing Machinery, 11 West 42nd Street, New York, NY 10036. (212) 869-7440. Fax (212) 869-0481. Monthly.

Science Citation Index. Institute for Scientific Information, 3501 Market Street, Philadelphia, PA 19104. (215) 386-0100. FAX (215) 386-6362. Inquire about availability and cost. Also available on CD-ROM.

ASSOCIATIONS AND PROFESSIONAL SOCIETIES

American Association For Artificial Intelligence. 445 Burgess Drive, Menlo Park, CA 94025-3496. (415) 328-3123. FAX (415) 321-4457.

Association For Computing Machinery. 1515 Broadway, New York, NY 10036. (212) 869-7440. FAX (212) 869-0481.

Association For Intelligent Systems Technology. 2-212 Center for Science and Technology, Syracuse, NY 13244. (315) 443-2143. FAX (315) 443-1865.

Society For Machine Intelligence. 100 Farnsworth, Detroit, MI 48202. (313) 832-5400. FAX (313) 832-5920.

Special Interest Group On Artificial Intelligence. c/o ACM, 1515 Broadway, New York, NY 10036. (212) 869-7440. FAX (212) 869-0481.

BIBLIOGRAPHIES

ACM Guide To Computing Literature. Association for Computing Machinery, 1515 Broadway, 17th Floor, New York, NY 10036-5701. (212) 869-7440. FAX (212) 944-1318. 1964 to present. Annual. $190.00 for non-members. Also available online via DIALOG.

DIRECTORIES AND BIOGRAPHICAL SOURCES

The AI Directory. American Association for Artificial Intelligence, 445 Burgess Drive, Menlo Park, CA 94025. (415) 328-3123. FAX (415) 321-4457. 1985 to present (1990 to present under current title). Annual. Inquire for cost.

American Men & Women of Science, 1995-96. R.R. Bowker Staff, eds. 19th edition. 8 volumes. R.R. Bowker/Reed International Publishing Company, 121 Chanlon Road, New Providence, NJ 07974. (908) 464-6800 or (800) 521-8110. 1995. $850.00

ASIS Handbook and Directory, American Society For Information Science, 8720 Georgia Avenue, Suite 501, Silver Spring, MD 20910-3602. (301) 495-0900. Annual. $100.00.

Computing Information Directory. Hildebrandt Inc., Box 285, Colville, WA 99114. (509) 684-2324. FAX (509) 684-2324. 1981 to present. Annual. $199.95 per year.

Directory of Engineering Societies and Related Organizations. Gordon Davis. 13th edition. American Association of Engineering Societies, 1111 19th Street NW, Suite 608, Washington, DC 20036. (202) 296-2237 or (800) 658-8897. 1989. Inquire for price.

International Research Centers Directory. Gale Research, 835 Penobscot Building, Detroit, MI 48226-4094. (313) 961-2242 or (800) 347-4253. FAX (313) 961-6083. 8th edition. 1995. $410.00.

Research Centers Directory. Gale Research, 835 Penobscot Building, Detroit, MI 48226-4094. (313) 961-2242. (800) 347-4253. FAX (313) 961-6083. $485.00.

Scientific and Technical Organizations and Agencies Directory. Gale Research, 835 Penobscot Building, Detroit, MI 48226-4094. (313) 961-2242. (800) 347-4253. FAX (313) 961-6083. 4th edition. 1996. $195.00.

Who's Who in Computing. Canadian M I S Database Inc., 268 Lakeshore Road E., Suite 510, Mississauga, ON L5G 1H1, Canada. (905) 271-1601. FAX (905) 271-4522. 1987 to present. Annual. Inquire for price in U.S.

Who's Who in Technology. Gale Research, 835 Penobscot Building, Detroit, MI 48226-4094. (313) 961-2242. (800) 347-4253. FAX (313) 961-6083. 1995. $195.00.

ENCYCLOPEDIAS AND DICTIONARIES

Computer Dictionary. Donald D. Spencer. 4th edition. Camelot Publishing Company, 709 SW 80th Blvd., Gainesville, FL 32607-1537. (904) 331-0952. 1993. $24.95.

Dictionary of Computing. S.M.H. Collin. 2nd edition. Peter Collin Publishing, 8 the Causeway, Teddington, TW11 0HE, England. FAX 0181-943-3386. 1994. Inquire for cost.

Encyclopedia of Artificial Intelligence. Stuart C. Shapiro, editor. Second edition. John Wiley and Sons, Inc., 605 Third Avenue, New York, NY 10158. (800) 526-5368 or (212) 850-6000. 1992. $297.50.

Encyclopedia of Computer Science and Engineering. Anthony Ralston & Edwin D. Reilly Jr., editors. 3rd revised edition. Van Nostrand Reinhold, 115 Fifth Avenue, New York, NY 10003. (212) 254-3232. FAX (212) 254-9499. 1993. $125.00.

The McGraw-Hill Illustrated Encyclopedia of Robotics and Artificial Intelligence. Stan Gonilisco. McGraw-Hill Publishing Company, 1221 Avenue of the Americas, New York, NY 10020. (800) 262-4729 or (212) 512-3825. 1994. $34.95 (hardbound), $24.95 (paperbound).

GENERAL WORKS

Artificial Intelligence and Expert Systems For Engineers. C.S. Krishnamoorthy. CRC Press, 2000 Corporate Blvd., N.W., Boca Raton, FL 33431. (407) 994-0555 or (800) 272-7737. FAX (407) 994-0949. 1996. $96.00.

Computers As Assistants: A New Generation of Support Systems. Peter Hoschka, editor. L. Erlbaum Associates, 10 Industrial Avenue, Mahwah, NJ 07430-2262. (800) 926-6579. 1996. $25.00.

Design and Implementation of Intelligent Manufacturing Systems: From Expert Systems, Neural Networks, To Fuzzy Logic. H.R. Parsaei & M. Jamshidi, editors. Prentice Hall, 113 Sylvan Avenue, Route 9W, Englewood Cliffs, NJ 07632. (201) 592-2000 or (800) 922-0579. 1995. $75.00.

Expert Systems in Engineering Applications. Spryos G. Tzafestas. Springer-Verlag, 175 Fifth Avenue, New York, NY 10010. (212) 460-1500 or (800) 777-4643. FAX (212) 473-6272. 1993. $199.95.

Industrial and Engineering Applications of Expert Systems. Frank D. Anger, editor. Gordon and Breach Science Publishers, 820 Town Center Drive, Langhorne, PA 19047. (215) 750-2642. FAX (215) 750-6343. 1994. $105.00.

Introduction To Knowledge Systems. Mark Stefik. Morgan Kaufmann Inc., 340 Pine Street, 6th Floor, San Francisco, CA 94104. (800) 745-7323. FAX (415) 982-2665. 1995. $69.95.

Knowledge Based Systems Usage. Lise Land, editor. McGraw-Hill Publishing Company, 1221 Avenue of the Americas, New York, NY 10020. (800) 262-4729 or (212) 512-3825. 1995. Inquire for price.

HANDBOOKS AND MANUALS

Artificial Intelligence, Knowledge Engineering, Expert Systems, Natural Language, Human Factors Ergonomics, Man-Machine Interface Design: Handbook of Human-Computer Interaction. M. Helander, editor. Elsevier Science Publishing Company, Inc., 655 Avenue of the Americas, New York, NY 10010. (212) 989-5800. 1993. $251.00 (hardbound), $83.00 (paperbound).

ONLINE DATABASES AND CD-ROMS

Compuscience. FIZ Karlsruhe, D-7514, Eggenstein-Leopoldshafen 2, Germany. Contains citations with abstracts to European and North American literature on computer science, 1972 to present. Available on STN International, FIZ Karlsruhe, P.O. Box 2465, W-7500, Karlsruhe 1, Germany, online service. Inquire as to cost and availability.

Computer and Information Systems Abstracts. Cambridge Scientific Abstracts, 7200 Wisconsin Avenue, Suite 601, Bethesda, MD 20814. (301) 961-6750 or (800) 843-7751. Contains citations to worldwide literature in theoretical and applied computer science and related areas, from 1981 to present.Inquire as to cost and availability.

Computer and Mathematics Search. Institute for Scientific Information, 3501 Market Street, Philadelphia, PA 19104. (215) 386-0100. FAX (215) 386-6362. Covers worldwide literature in computer science and mathematics, from 1980 to present. Available on BRS,(800) 289-4277, online service. Inquire as to cost and availability.

Computer Database. Information Access Company, 362 Lakeside Drive, Foster City, CA 94404. (415) 378-5000 or (800) 227-8431. Contains citations with abstracts to literature from trade journals covering the computer,telecommunications,and electronics industries. Available on the BRS, (800) 289-4277, Compuserve Information Service,(800) 848-8990, and DIALOG,(800) 334-2564, online services. Inquire as to cost and availability.

Current Contents Search. Institute for Scientific Information, 3501 Market Street, Philadelphia, PA 19104. (215) 386-0100. FAX (215) 386-6362. Contains citations to articles listed in the table of contents of science and technology journals. Also articles in social sciences and life sciences journals. Available on BRS,(800) 289-4277, DIALOG,(800) 334-2564, online services. Inquire as to cost and availability.

Dissertation Abstracts. University Microfilms International, 300 North Zeeb Road, Ann Arbor, MI 48106. (800) 521-0600 or (313) 761-4700. Scope includes virtually all doctoral dissertations accepted at accredited American institutions from 1861 to present in 252 subject areas. Available on BRS,(800) 289-4277, DIALOG,(800) 334-2564, and OCLC EPIC,(800) 848-5878, online services. Also available on CD-ROM. Inquire as to cost and availability.

INSPEC. Institution of Electrical Engineers, Michael Faraday House, Six Hills Way, Stevenage, Herts. SG1 2AY, England. Telephone: 0438 313311 or FAX 0438 742840. Contains citations to the worldwide literature of physics, electronics and electrical engineering, computer technology, and related fields. Available on BRS,(800) 289-4277, DIALOG, (800) 334-2564, ORBIT, (800) 456-7248, and STN International, FIZ Karlsruhe, P.O. Box 2465, W-7500, Karlsruhe 1, Germany, online services. Inquire as to cost and availability.

NTIS Bibliographic Database. National Technical Information Service, 5285 Port Royal Road, Springfield, VA 22161. (703) 487-4929 or FAX (703) 321-8199. Broad coverage of government-sponsored science and technology research reports, 1964 to present. Available on BRS,(800) 289-4277, DIALOG, (800) 334-2564, ORBIT, (800) 456-7248, and STN International, FIZ Karlsruhe, P.O. Box 2465, W-7500, Karlsruhe 1, Germany, online services. Also available on CD-ROM. Inquire as to cost and availability.

SCISEARCH. Institute for Scientific Information, 3501 Market Street, Philadelphia, PA 19104. (800) 523-1850 or (215) 386-0100. Broad multidisciplinary title and author index to the international literature of science and technology, 1974 to present. Available on DIALOG,(800) 334-2564, and ORBIT,(800) 456-7248, online services. Also available on CD-ROM. Inquire as to cost and availability.

Supertech. Bowker A & I Publishing, 121 Chanlon Road, New Providence, NJ 07974. (800) 521-8110 or (908) 464-6800. Contains citations to the world's published and unpublished literature in the fields of artificial intelligence, biotechnology, computer-aided design and manufacturing, robotics, and telecommunications. Covers the period 1973 to present. Available on DIALOG,(800) 334-2564, and ORBIT,(800) 456-7248, online services. Inquire as to cost and availability.

WILSONLINE. H.W. Wilson Company, 950 University Avenue, Bronx, NY 10452. (800) 367-6770 or (212) 588-8400. Makes available online versions of the printed H.W. Wilson indexes including Applied Science and Technology Index, Business Periodicals Index, General Science Index, and Readers' Guide to Periodical Literature. Period covered is generally 1983 to present. Available on BRS,(800) 289-4277, DIALOG,(800) 334-2564, and OCLC EPIC,(800) 848-5878, online services. Also available on CD-ROM. Inquire as to cost and availability.

PERIODICALS

AI Applications. University of Idaho, Box 3066, Moscow, ID 93943. (208) 885-7033. FAX (208) 885-6226. 1987 to present. Three times a year. $37.00.

AI Expert. Miller Freeman Inc., 600 Harrison Street, San Francisco, CA 94107. (415) 905-2200. FAX (415) 905-2232. 1986 to present. Monthly. $34.00.

EXPERT SYSTEMS

Ency. of Physical Sciences and Engineering Info. Sources

AI Magazine. American Association for Artificial Intelligence, 445 Burgess Drive, Menlo Park, CA 94025. (415) 328-3123. FAX (415) 321-4457. 1979 to present. Quarterly. $40.00.

Applied Artificial Intelligence: AAI. Taylor & Francis, 1900 Frost Road, Suite 101, Bristol, PA 19007. (215) 785-5800. FAX (215) 785-5515. 1987 to present. Quarterly. $190.00.

Artificial Intelligence. Elsevier Science Inc., Box 882, Madison Square Station, New York, NY 10159. (212) 989-5800. FAX (212) 633-3990. 1970 to present. 14 times a year. $1279.00.

The Artificial Intelligence Review. Kluwer Academic Publishers, P.O. Box 358, Accord Station, Hingham, MA 02018-0358. (617) 871-6000. 1986 to present. Six times a year. $233.50.

International Journal of Pattern Recognition and Artificial Intelligence. World Scientific Publishing Company, Inc., 1060 Main Street, Unit B, River Edge, NJ 07661. (800) 227-7562. FAX (201) 487-9656. 1987 to present. Six times a year. $350.00.

Law, Computers and Artificial Intelligence. Triangle Journals Ltd., PO Box 65, Wallingford, Oxfordshire, OX10 0YG, England. Telephone 0491-838013. FAX 0491-834968. 1992 to present. Three times a year. Inquire for cost.

Robotica. Cambridge University Press, Journals Department, 40 W 20th Street, New York, NY 10011-4211. (212) 924-3900. 1983 to present. Bi-monthly. $119.00.

Sigart Bulletin. Special Interest Group on Artificial Intelliegence, c/o ACM, 1515 Broadway, New York, NY 10036. (212) 869-7440. FAX (212) 869-0481. 1990 to present. Quarterly. $40.00 for non-members.

RESEARCH CENTERS AND INSTITUTES

Center For Machine Intelligence. 2001 Commowealth Blvd., Ann Arbor, MI 48105. (313) 995-0900.

Stanford University, Center For Integrated Facility Engineering. Terman Engineering Center, Stanford, CA 94305-4020. (415) 725-6486. FAX (415) 723-4806.

University of Michigan Artificial Intelligence Laboratory. 1101 Beal Avenue, Ann Arbor, MI 48109-2110. (313) 764-8505. Usc Expert Systems Laboratory. PO Box 30041, Long Beach, CA 90853-0041. (213) 439-7021.

EXPLOSIVES

See also: CHEMICAL ENGINEERING, FIRE PROTECTION, FUELS, PROPULSION SYSTEMS, PROPELLANTS, ROCKETS, SAFETY ENGINEERING

ABSTRACT SERVICES AND INDEXES

Applied Science and Technology Index; A Cumulative Subject Index To English Language Periodicals in the Fields of Aeronautics and Space Science, Computer Technology, Chemistry, Construction Industry, Energy and Related Areas. H. W. Wilson Co., 950 University Avenue, Bronx, NY 10452-9978. (212) 588-8400. (800) 367-6770. FAX (718) 590-1617. From 1958 to present. Monthly. Inquire about cost and availability. Also

available online (BRS and WILSONLINE) and on CD-ROM. ISSN: 0003-6986.

Chemical Abstracts. Chemical Abstracts Service. 2540 Olentangy River Road, Box 3012, Columbus, OH 43210-0012. (614) 447-3600. FAX (614) 447-3713. 1907 to present. Weekly. $16,800.00 per year. Available online and on CD-ROM. CA is also available in five section groupings. Inquire regarding cost and availability.

Chemical Abstracts - Applied Chemistry and Chemical Engineering Sections. Chemical Abstracts Service, 2540 Olentangy River Road, Box 3012, Columbus, OH 43210. (800) 848-6538 or (614) 447-3600, FAX (614) 447-3713. Bi-weekly. $1410.00 per year. Monthly. $450.00 per year. Also available online.

Current Contents: Engineering, Technology and Applied Sciences. Institute for Scientific Information, 3501 Market Street, Philadelphia, PA 19104. (215) 386-0100. FAX (215) 386-2991. From 1970 to present. Weekly. $442.00 per year. Also available online. Inquire regarding cost and availability. ISSN: 0095-7917.

Engineering Index Monthly. Engineering Information, Inc., Castle Point on the Hudson, Hoboken, NJ 07030. (201) 216-8500. (800) 221-1044. FAX (201) 216-8532. Monthly. $2300.00 per year for monthly issues. $1980 per year for annual cumulation. Available on CD-ROM and online as COMPENDEX. ISSN: 0742-1974; 0360-8557.

General Science Index. H.W. Wilson and Company, 950 University Avenue, Bronx, NY 10452. (212) 588-8400. (800) 367-6770. FAX (718) 590-1617. From 1978 to present. Ten issues per year; quarterly and annual cumulations. Service basis. Available on CD-ROM and online. Inquire regarding cost and availability. ISSN: 0162-1963.

Process and Chemical Engineering/Chemical Engineering Abstracts. Royal Society of Chemistry, Information Services, Thomas Graham House, Science Park, Milton Road, Cambridge, CB4 4WF, England. Contains citations, mostly with abstracts, of the worldwide literature on chemical engineering; from 1982 to present. Monthly. $610.00 per year. Also available online. ISSN: 0960-5045.

Science Citation Index. SCI. Institute for Scientific Information, 3501 Market Street, Philadelphia, PA 19104. (215) 386-0100. (800) 523-1850. FAX (215) 386-2991. 1961 to present. Six issues per year, plus annual cumulation. $11650.00 per year. Also available online and on CD-ROM. Inquire about price and availability. ISSN: 0036-827X.

Theoretical Chemical Engineering. Royal Society of Chemistry, Information Services, Thomas Graham House, Science Park, Milton Road, Cambridge, CB4 4WF, England. Covers theoreticalchemical engineering, including theory and laboratory experimentation. 1964 to present. Monthly. $235.00. ISSN: 0960-5053.

ANNUAL REVIEWS AND YEARBOOKS

Advances in Chemical Engineering. Academic Press, Inc., 6277 Sea Harbor Drive, Orlando, FL 32887. (800) 321-5068. 1956 to present. Irregular. ISSN: 0065-2377.

Explosives and Blasting Technicals. International Society of Explosives Engineers, 29100 Aurora Road, Cleveland, OH 44139- 1800. (216) 349-4004. FAX (216) 349-3788.

ASSOCIATIONS AND PROFESSIONAL SOCIETIES

American Astronautical Society. 6352 Rolling Mill Place, Suite 102, Springfield, VA 22152. (703) 866-0020. (703) 866- 3526. American Chemical Society. 1155 16th Street, N.W., Washington, DC 20036. (202) 872-4600.

American Institute of Aeronautics and Astronautics. 370 L'Enfant Promenade SW, Washington, DC 20024. (202) 646-7400. FAX (202) 646-7508.

American Institute of Chemical Engineers. 345 East 47th Street, New York, NY 10017. (212) 705-7657. FAX (212) 752-3294.

American Pyrotechnics Association. PO Box 213, Chestertown, MD 21620. (410) 778-6825.

Association of Consulting Chemists and Chemical Engineers, 50 East 41st Street, Suite 92, New York, NY 10017. (212) 684-6255.

Association of official Analytical Chemists. 1111 North 19th Street, Suite 210, Arlington, VA 22209. (703) 522-3032.

Institute of Makers of Explosives. 1120 19th Street NW, Suite 310, Washington, CD 20036. (202) 429-9280. (202) 293-2420.

International Allociation of Bomb Technicians and Investigators. PO Box 8629, Naples, FL 33941. (941) 353- 6843. FAX (941) 353-6841.

International Society of Explosives Engineers. 29100 Aurora Road, Cleveland, OH 44139-1800. (216) 349-4004. FAX (216) 349-3788.

Pyrotechnics Guild International. 221 Spring Avenue, Lutherville, MD 21093. (410) 560-0513.

DIRECTORIES AND BIOGRAPHICAL SOURCES

American Men and Women of Science. Physical and Biological Sciences. Fifteenth edition. R.R. Bowker Company, 205 East 42nd Street, New York, NY 10017. (800) 521-8110 or (212) 916-1600.

ARI Directory. Agricultural Research Institute, 9650 Rockville Pike, Bethesda, MD 20841-3998. (301) 530-7122. $5.00 per copy.

Research Centers Directory. Gale Research Company, Detroit, MI 48226. Seventeenth edition, 1992. (800) 521-0707. $400.00.

International Society of Explosives Engineers Directory of Member Specialties. 29100 Aurora Road, Cleveland, OH 44139-1800. (216) 349-4004. FAX (216) 349-3788. Annual. Inquire.

Who's Who in Engineering. Gordon Davis, editor. American Association of Engineering Societies. 1111 19th Street, NY, Suite 608, Washington, DC 20036. (202) 296-2237. (800) 658-8897. 9th edition. 1995. $220.00.

ENCYCLOPEDIAS AND DICTIONARIES

Dictionary of Explosives. Arthur Marshall. Gordon Press Publications, P.O. Box 459, Bowling Green Station, New York, NY 10004. (718) 624-8419. 1989. $250.00.

Industrial Chemical Thesaurus. Michael Ash and Irene Ash. VCH Publishers, Inc., 220 East 23rd Street, New York, NY 10010-4606. (800) 367-8249. 1992. $295.00.

Kirk-Othmer Encyclopedia of Chemical Technology. John Wiley and Sons, Inc., 605 Third Avenue, New York, NY 10158. (800) 526-5368 or (212) 850-6000. Fourth edition. 1991 - . Twenty-seven volumes. $5400.00.

GENERAL WORKS

Chemical Rockets, and Flame and Explosives Technology. Richard T. Holzmann. Books on Demand, 300 North Zeeb Road, Ann Arbor, MI 48106-1346. (313) 761-4700. (800) 521-0600. $131.00.

Chemistry of Energetic Materials. George A. Olah and David R. Squire. Academic Press, Inc., 6277 Sea Harbor Drive, Orlando, FL. (800) 321-5068. 1991. $79.00.

Explosives For North American Engineers. C. Gregory. LPS Distribution Center, 52 North LaBombard Road, Lebanon, NH 03766. (603) 448-0037. 3rd edition. 1979. $10.00.

Improvised Batteries and Detonating Devices. Gordon Press Publications, P.O. Box 459, Bowling Green Station, New York, NY 10004. (718) 624-8419. 1991. $75.00.

Modern Methods and Applications in Analysis of Explosives. Jehuda Yinon and Smuel Zitrin. John Wiley & Sons, Inc., 605 Third Avenue, New York, NY 10158-0012. (212) 850-6000. (800) 225-5945. 1993. $125.00.

HANDBOOKS AND MANUALS

CRC Handbook of Chemistry and Physics. Robert C. Weast, editor. CRC Publishers, Inc., 2000 Corporate Blvd., N.W., Boca Raton, FL 33431. (407) 994-0555 or (800) 333-8300. Seventy-third edition. 1992. $100.00.

Lange's Handbook of Chemistry. John A. Dean, editor. McGraw-Hill Publishing Company, 1221 Avenue of the Americas, New York, NY 10020. (800) 262-4729 or (212) 512-3825. Fourteenth editon. 1992. $79.50.

ONLINE DATABASES AND CD-ROMS

CA Search. Chemical Abstracts Service, P.O. Box 3012, Columbus, OH 43210-0012. (614) 447-3600. (800) 848-6533. FAX (614) 447-3709. Guide to worldwide chemical literature and related fields, 1972 to present. Available on BRS, DIALOG, ORBIT and STN online services. Inquire as to cost and availability.

NTIS. National Technical Information Service, 5285 Port Royal Road, Springfield, VA 22161. (703) 487-4929 or FAX (703) 321-8199. Broad coverage of government-sponsored research reports, 1964 to present. Available on BRS, DIALOG, ORBIT and STN online services. Also available on CD-ROM. Inquire as to cost and availability.

SCISEARCH. Institute for Scientific Information, 3501 Market Street, Philadelphia, PA 19104. (800) 523-1850 or (215) 386-0100. Broad multidisciplinary title and author index to the international literature of science and technology, 1974 to present. Available on DIALOG and ORBIT online services. Also available on CD-ROM. Inquire as to cost and availability.

WILSONLINE. H.W. Wilson Company, 950 University Avenue, Bronx, NY 10452. (800) 367-6770 or (212) 588-8400. Makes available online versions of the printed H.W. Wilson indexes including Applied Science and Technology Index, Business Periodicals Index, and Readers' Guide to Periodical Literature. Period covered is generally 1983 to present. Also available on CD-ROM. Inquire as to cost and availability.

PERIODICALS

Inorganic Chemistry. American Chemical Society, 1155 16th Street, N.W., Washington, DC 20036. (800) 424-6747. 1962 to present. Biweekly. $864.00 per year.

Journal of Explosives Engineering. Internatinal Society of Explosives Engineers, 29100 Aurora Road, Cleveland, OH 44139-1800. (216) 349-4004. FAX (216) 349-3788. Bimonthly. $30.00.

Propellants, Explosives, Pyrotechnics. International Pyrotechnics Society. Available from: VCH Publishers, Inc., 220 East 23rd Street, New York, NY 10010-4606. 1976 to present. Bimonthly. $435.00 per year. ISSN: 0721-3115.

Pyrotechnica: Occasional Papers in Pyrotechnics. Pyrotechnica Publications, 2302 Tower Drive, Austin TX 78703. (512) 476-4062. FAX (512) 453-1353. 1977 to present. Annual. $32.00 per year. ISSN: 0272-6521.

RESEARCH CENTERS AND INSTITUTES

Center for Explosives Technology Research. New Mexico Institute of Mining and Technology, Campus Station, Socorro, NM 87801. (505) 835-5733. (505) 835-5680.

Factory Mutural Research Corporation. 1151 Poston-Providence Turnpike, Norwood, MA 02062. (617) 762-4300. (671) 761-9375.

Rock Mechanics and Explosives Research Center. University of Missouri-Rolla, Rolla, MO 654-1. (314) 341-4364. (314) 341-6078.

EXTRACTION PROCESSES

See: METALLURGICAL ENGINEERING

EXTRUSION

See: METALLURGICAL ENGINEERING

F

FABRICS

See: TEXTILES

FACTOR ANALYSIS

See: MATHEMATICS

FAILURE ANALYSIS

See also: AERONAUTICAL ENGINEERING, CIVIL ENGINEERING, FRACTURE MECHANICS, MATERIALS SCIENCE, MECHANICAL ENGINEERING, NONDESTRUCTIVE TESTING, STRUCTURAL ENGINEERING

ABSTRACT SERVICES AND INDEXES

Abstract Journal in Earthquake Engineering. University of California, Berkeley, Earthquake Engineering Research Center, 1301 South 46th Street, Richmond, CA 94804-4698. (510) 231-9413 or FAX (510) 231-9471. 1972 to present. Annual. $80.00 per year.

Applied Science and Technology Index; A Cumulative Subject Index To English Language Periodicals in the Fields of Aeronautics and Space Science, Computer Technology, Chemistry, Construction Industry, Energy and Related Areas. H.W. Wilson Co., 950 University Avenue, Bronx, NY 10452. (800) 367-6770 or (212) 588-8400. FAX (718) 590-1617. From 1958 to present. Monthly. Inquire about cost and availability. Also available on CD-ROM and online.

ASCE Combined Annual Combined Index. American Society of Civil Engineers, 345 East 47th Street, New York, NY 10017-2398. (212) 705-7520 or (800) 548-2723. Annual. $48.00.

ASCE Publications Information. 1966-present. American Society of Civil Engineers, 345 East 47th Street, New York, NY 10017-2398. (212) 705-7520 or (800) 548-2723. Bi-monthly. $160.00 per year for non-members.

Civil and Structural Engineering Abstracts, Cambridge Scientific Abstracts, 7200 Wisconsin Avenue, Bethesda, MD 20814-4823. (301) 961-6750. FAX (301) 961-6720. 1993 to present.

Monthly. $385.00 per year. Topics covered include structural design, construction equipment and methods, civil defense and military engineering, surveying, highway engineering, maritime and port structures, materials, land reclamation, and soil mechanics.

Current Contents: Engineering, Technology, and Applied Sciences. Institute for Scientific Information, 3501 Market Street, Philadelphia, PA 19104. (215) 386-0100. FAX (215) 386-6362. 1970 to present. Weekly. $442.00 per year.

Engineering Index Monthly. Engineering Information, Inc., Castle Point on the Hudson, Hoboken, NJ 07030. (800) 221-1044. FAX (212) 832-1857. Monthly. $2300.00 per year. Also available online as COMPENDEX, and also on CD-ROM. Covers chemical engineering, computers, electrical engineering, civil engineering, metals and mining, industrial management, and mechanical engineering.

Index To Scientific and Technical Proceedings. Institute for Scientific Information, 3501 Market St., Philadelphia, PA 19104. (215) 386-0100. FAX (215) 386-6362. Monthly. $500.00 per year.

Mechanical Engineering Abstracts (formerly ISMEC), Cambridge Scientific Abstracts, 7200 Wisconsin Avenue, Bethesda, MD 20814-4823. (301) 961-6750. FAX (301) 961-6720. 1967 to present. Monthly. $895.00 per year. Summarizes world literature in mechanical engineering, production engineering, and engineering management. Also available online.

Science Citation Index. Institute for Scientific Information, 3501 Market Street, Philadelphia, PA 19104. (215) 386-0100. FAX (215) 386-6362. Inquire about availability and cost. Also available on CD-ROM.

ASSOCIATIONS AND PROFESSIONAL SOCIETIES

American Society For Nondestructive Testing. 1711 Arlingate Lane, Box 28518, Columbus, OH 43228-0158. (614) 274-6003. FAX (614) 274-6899.

American Society For Testing and Materials. 1916 Race Street, Philadelphia, PA 19103. (215) 299-5585.

American Society of Civil Engineers. 345 East 47th Street, New York, NY 10017-2398. (212) 705-7520 or (800) 548-2723.

American Society of Mechanical Engineers. 345 East 47th Street, New York, NY 10017-2398. (212) 705-7703.

DIRECTORIES AND BIOGRAPHICAL SOURCES

Directory of Engineering Societies and Related Organizations. Gordon Davis. 13th edition. American Association of Engineering Societies, 1111 19th Street NW, Suite 608, Washington, DC 20036. (202) 296-2237 or (800) 658-8897. 1989. Inquire for price.

International Research Centers Directory. Gale Research, , 835 Penobscot Building, Detroit, MI 48226-4094. (313) 961-2242. (800) 347-4253. FAX (313) 961-6083. 8th edition. 1995. $410.00

Research Centers Directory. Gale Research, , 835 Penobscot Building, Detroit, MI 48226-4094. (313) 961-2242. (800) 347-4253. FAX (313) 961-6083. $485.00.

Scientific and Technical Organizations and Agencies Directory. Gale Research, , 835 Penobscot Building, Detroit, MI 48226-4094. (313) 961-2242. (800) 347-4253. FAX (313) 961-6083. 4th edition. 1996. $195.00.

Who's Who in Engineering. American Association of Engineering Societies, 1111 19th Street NW, Suite 608, Washington, DC 20036. (202) 296-2237 or (800) 658-8897. 8th edition. 1991. Inquire for price.

ENCYCLOPEDIAS AND DICTIONARIES

Encyclopedia of Physical Science and Technology. Robert A. Meyers, ed. 18 volumes. Academic Press Inc., 6277 Sea Harbor Drive, Orlando, FL 32887. (800) 321-5068. 1992. $2100.00

McGraw-Hill Encyclopedia of Engineering. Sybil P. Parker, ed. 2nd edition. McGraw-Hill Publishing Company, 1221 Avenue of the Americas, New York, NY 10020. (800) 262-4729 or (212) 512-3825. 1993. $95.50.

McGraw-Hill Encyclopedia of Science and Technology. Sybil P. Parker, ed. 7th edition. 20 volumes. McGraw-Hill Publishing Company, 1221 Avenue of the Americas, New York, NY 10020. (800) 262-4729 or (212) 512-3825. 1992. $1900.00

Thesaurus of Scientific, Technical, and Engineering Terms. Hemisphere Publishing Corporation, 1900 Frost Road, Suite 101, Bristol, PA 19007-1598. (215) 785-5800 or (800) 821-8312. FAX (215) 785-5515. 1987. $173.00.

GENERAL WORKS

Component Failures: Maintenance and Repair. M.J. Neale, editor. SAE International, 400 Commonwealth Drive, Warrendale, PA 15096-0001. (412) 776-4970. FAX (412) 776-0790. 1995. $39.00.

Engineering Materials 3: Materials Failure Analysis. David Rayner Hunkin Jones. Pergamon Press Inc., Maxwell House, Fairview Park, Elmsford, NY 10523. (914) 592-7700. Fax (914) 592-3625. 1993. $96.01 (hardbound) and $32.01 (paperback).

Failure Analysis: Techniques and Applications. J. Ivan Dixon, editor. ASM International, Materials Information, Materials Park,

OH 44073. (216) 338-5151 or FAX (216) 338-4634. 1992. $103.00.

Handbook of Case Histories in Failure Analysis. Khelfa Esakul, ed. ASM International, Materials Information, Materials Park, OH 44073. (216) 338-5151 or FAX (216) 338-4634. 1992. Inquire for cost and availability.

Machinery Failure Analysis and Troubleshooting. Heniz P. Bloch and Fred K. Gertner. 2d ed. Gulf Publishing Company, P.O. Box 2608, Houston, TX. (713) 529-4301 or (800) 231-6275. 1994. Write for information.

Metallurgical Failure Analysis. Charlie R. Brooks. McGraw-Hill Publishing Company, 1221 Avenue of the Americas, New York, NY 10020. (800) 262-4729 or (212) 512-3825. 1993. $49.95.

Microelectronic Failure Analysis. Thomas W. Lee and Seshu V. Pabbisetty, eds. ASM (American Society of Metals) International, Materials Information, Materials Park, OH 44073. (216) 338-5151 or FAX (216) 338-4634. 1993. Write for information.

HANDBOOKS AND MANUALS

Asm Handbook Volume 11: Failure Analysis and Prevention. ASM International Handbook Committee. ASM International, Materials Information, Materials Park, OH 44073-0002. (216) 338-5151 or (800) 336-5152. FAX (216) 338-4634. 1986. $147.00.

Handbook of Mechanics, Materials, and Structures. Alexander Blake, editor. John Wiley and Sons, Inc., 605 Third Avenue, New York, NY 10158. (800) 526-5368 or (212) 850-6000. 1985. $120.00.

Marks' Standard Handbook For Mechanical Engineers. E.A. Avallone and T. Baumeister III. 9th edition. McGraw-Hill Publishing Company, 1221 Avenue of the Americas, New York, NY 10020. (800) 262-4729 or (212) 512-3825. 1987. $133.00.

ONLINE DATABASES AND CD-ROMS

Compendex Plus. Engineering Information, Inc., 345 East 47th Street, New York, NY 10017. (212) 705-7600 or (800) 221-1044. Contains citations with abstracts to worldwide literature in engineering and technology, from 1970 to present. Available on online BRS,(800) 289-4277, DIALOG, (800) 334-2564, ORBIT (800) 456-7248, and STN International, FIZ Karlsruhe, P.O. Box 2465, W-7500, Karlsruhe 1, Germany, online services. Also available on CD-ROM. Inquire as to cost and availability.

CSA Engineering. Cambridge Scientific Abstracts, 7200 Wisconsin Avenue, Suite 601, Bethesda, MD 20814. (301) 961-6750 or (800) 843-7751. Contains citations and abstracts of international periodicals and research literature covering all fields of engineering and science and technology,including computer and information science, electronics, mechanical engineering, solid state materials, 1981 to present. Available on BRS,(800) 289-4277, online service.Inquire as to cost and availability.

Current Contents Search. Institute for Scientific Information, 3501 Market Street, Philadelphia, PA 19104. (215) 386-0100. FAX (215) 386-6362. Contains citations to articles listed in the table of contents of science and technology journals. Also ar-

ticles in social sciences and life sciences journals. Available on BRS,(800) 289-4277, DIALOG,(800) 334-2564, online services. Inquire as to cost and availability.

Dissertation Abstracts. University Microfilms International, 300 North Zeeb Road, Ann Arbor, MI 48106. (800) 521-0600 or (313) 761-4700. Scope includes virtually all doctoral dissertations accepted at accredited American institutions from 1861 to present in 252 subject areas. Available on BRS,(800) 289-4277, DIALOG,(800) 334-2564, and OCLC EPIC,(800) 848-5878, online services. Also available on CD-ROM. Inquire as to cost and availability.

ISMEC: Mechanical Engineering Abstracts. Cambridge Scientific Abstracts, 7200 Wisconsin Avenue, Suite 601, Bethesda, MD 20814. (301) 961-6750 or (800) 843-7751. Contains citations to the literature in mechanical engineering, industrial and production engineering, energy, power, mechanics, devices and related areas, from 1973 to present. Available on the DIALOG,(800) 334-2564, online service. Inquire as to cost and availability.

NTIS Bibliographic Database. National Technical Information Service, 5285 Port Royal Road, Springfield, VA 22161. (703) 487-4929 or FAX (703) 321-8199. Broad coverage of government-sponsored science and technology research reports, 1964 to present. Available on BRS,(800) 289-4277, DIALOG, (800) 334-2564, ORBIT, (800) 456-7248, and STN International, FIZ Karlsruhe, P.O. Box 2465, W-7500, Karlsruhe 1, Germany, online services. Also available on CD-ROM. Inquire as to cost and availability.

SCISEARCH. Institute for Scientific Information, 3501 Market Street, Philadelphia, PA 19104. (800) 523-1850 or (215) 386-0100. Broad multidisciplinary title and author index to the international literature of science and technology, 1974 to present. Available on DIALOG,(800) 334-2564, and ORBIT,(800) 456-7248, online services. Also available on CD-ROM. Inquire as to cost and availability.

WILSONLINE. H.W. Wilson Company, 950 University Avenue, Bronx, NY 10452. (800) 367-6770 or (212) 588-8400. Makes available online versions of the printed H.W. Wilson indexes including Applied Science and Technology Index, Business Periodicals Index, General Science Index, and Readers' Guide to Periodical Literature. Period covered is generally 1983 to present. Available on BRS,(800) 289-4277, DIALOG,(800) 334-2564, and OCLC EPIC,(800) 848-5878, online services. Also available on CD-ROM. Inquire as to cost and availability.

PERIODICALS

Journal of Engineering Mechanics. American Society of Civil Engineers, Engineering Mechanics Division, 345 East 47th Street, New York, NY 10017-2398. (212) 705-7520 or (800) 548-2723. 1956 to present. Monthly. $284.00 per year.

Journal of Structural Engineering. American Society of Civil Engineers, 345 East 47th Street, New York, NY 10017-2398. (212) 705-7520 or (800) 548-2723. 1956 to present. Monthly. $300.00 per year for non-members.

Journal of Testing and Evaluation. American Society for Testing and Materials (ASTM), 1916 Race Street, Philadelphia, PA 19103. (215) 299-5585. FAX (215) 977-9679. 1966 to present. Bi-monthly. $103.00 per year for non-members.

Materials Engineering. Penton Publishing, 110 Superior Ave., Cleveland, OH 44114-2543. 1929 to present. Monthly. $50.00 per year.

Materials Evaluation. American Society for Nondestructive Testing, 1711 Arlingate Lane, Box 28518, Columbus, OH 43228-0158. (614) 274-6003. FAX (614) 274-6899. 1942 to present. Monthly. $85.00 per year.

SAMPE Journal. Society for the Advancement of Material and Process Engineering, Covina, CA 91724. (818) 331-0610. FAX (818) 332-8929. 1965 to present. Bi-monthly. $65.00 per year.

RESEARCH CENTERS AND INSTITUTES

Center For Quality Engineering and Failure Prevention. Northwestern University, Technological Institute, Evanston, IL 60208. (708) 491-5527. FAX (708) 491-5227.

Structural Engineering Materials Laboratory, University of California At Berkeley. Davis Hall, Berkeley, CA 94720. (415) 642-3464.

Structural Stability Research Council. Fritz Engineering Laboratory, Lehigh University, Bethlehem, PA 18015. (215) 758-3522. FAX (215) 758-4522.

University of Wisconsin—madison, Center For Structural and Materials Testing. 1415 Johnson Street, Madison, WI 53706. (608) 262-3205.

FALLOUT

See: RADIATION

FANJET

See: JET PROPULSION

FATIGUE

See: FAILURE ANALYSIS

FAULTS

See: PHYSICAL GEOLOGY

FEEDBACK CONTROL SYSTEMS

See: AUTOMATION

FEEDFORWARD CONTROL SYSTEMS

See: AUTOMATION

FERROALLOYS

See also: ALLOYS, CAST IRON, MATERIALS SCIENCE, STEEL and STEELMAKING

ABSTRACT SERVICES AND INDEXES

Alloys Index. American Society for Metals, Metals Park, OH 44073. (216) 338-5151. FAX (216) 338-4634. 1974 to present. Monthly. $380.00. Also available on CD-ROM and online via METADEX, STN International, and DIALOG.

Applied Mechanics Reviews: An Assessment of World Literature in Engineering Sciences. 1948-present. American Society of Mechanical Engineers, 345 East 47th Street, New York, NY 10017. (212) 705-7703. Monthly. $360.00 per year.

Applied Science and Technology Index; A Cumulative Subject Index To English Language Periodicals in the Fields of Aeronautics and Space Science, Computer Technology, Chemistry, Construction Industry, Energy and Related Areas. H.W. Wilson Co., 950 University Avenue, Bronx, NY 10452. (800) 367-6770 or (212) 588-8400. FAX (718) 590-1617. From 1958 to present. Monthly. Inquire about cost and availability. Also available on CD-ROM and online.

Chemical Abstracts - Applied Chemistry and Chemical Engineering Sections. Chemical Abstracts Service, 2540 Olentangy River Road, Box 3012, Columbus, OH 43210. (800) 848-6538 or (614) 447-3600, FAX (614) 447-3713. Bi-weekly. $1410.00 per year.

Current Contents: Engineering, Technology, and Applied Sciences. Institute for Scientific Information, 3501 Market Street, Philadelphia, PA 19104. (215) 386-0100. FAX (215) 386-6362. 1970 to present. Weekly. $442.00 per year.

Engineering Index Monthly. Engineering Information, Inc., Castle Point on the Hudson, Hoboken, NJ 07030. (800) 221-1044. FAX (212) 832-1857. Monthly. $2300.00 per year. Also available online as COMPENDEX, and also on CD-ROM. Covers chemical engineering, computers, electrical engineering, civil engineering, metals and mining, industrial management, and mechanical engineering.

*I.*M.M. ABSTRACTS and INDEX. Institution of Mining and Metallurgy, 44 Portland Pl., London W1N 4BR, England. 071-580-3802. FAX 071-436-5388. Bi-monthly. $364 for non-members.

Manufacturing and Process Engineering Abstracts, Cambridge Scientific Abstracts, 7200 Wisconsin Avenue, Bethesda, MD 20814-4823. (301) 961-6750. FAX (301) 961-6720. 1993 to present. Monthly. $385.00 per year. Covers concurrent engineering, quality control, automated manufacturing, petroleum engineering, oil field operations and equipment, energy management, metallurgy and metallography, foundry practice.

Metals Abstracts and Metals Abstracts Index. American Society for Metals, Metals Park, OH 44073. (216) 338-5151. 1968 to present. Monthly. Abstracts are $1500.00 per year and Index is $460.00 per year.

ANNUAL REVIEWS AND YEARBOOKS

Annual Review of Materials Science. Annual Reviews Inc., 4139 El Camino Way, PO Box 10139, Palo Alto, CA 94303-0139. (415) 493-4400 or (800) 523-8635. FAX (415) 855-9815. Annual. $75.00 (1994 edition).

Minerals Yearbook: Ferroalloys. U.S. Bureau of Mines. Order from: Superintendent of Documents, U.S. Government Printing office, Box 371954, Pittsburgh, PA 15250-7954. (202) 783-3238. FAX (202) 512-2233. Annual. Inquire for cost and availability.

ASSOCIATIONS AND PROFESSIONAL SOCIETIES

American Foundrymen's Society Inc. 505 State Street, Des Plaines, IL 60016. (708) 824-0181.

American Iron and Steel Institute. 1101 17th St., NW, Washington, DC 20036-4700. (202) 452-7100. FAX (202) 463-6573.

American Society For Metals (asm). Materials Park, OH 44073. (216) 338-5151 or FAX (216) 338-4634.

Ferroalloys Association. 900 2nd Street NE, Suite 306, Washington, DC 20002-3557. (202) 842-0292. FAX (202) 842-4840.

Iron and Steel Society. 410 Commonwealth Drive, Warrendale, PA 15086. (412) 776-9460. FAX (412) 776-0430.

Metallurgical Society of the Aime (american Institute of Mining, Metallurgical, and Petroleum Engineers). 345 E.47th Street, 14th Floor, New York, NY 10017. (212) 705-7695.

BIBLIOGRAPHIES

Scientific and Technical Books in Print. R.R. Bowker Inc., 121 Chanlon Road, New Providence, NJ 07974. (800) 521-8110 or (908) 464-6800. List of currently published books and serials in the physical and biological sciences, engineering and technology, with subject, author and titles indexes. Also available on ORBIT, (800) 456-7248, online service. Inquire as to cost and availability.

DIRECTORIES AND BIOGRAPHICAL SOURCES

Dun's Industrial Guide: The Metalworking Directory, 1993-94. Dun & Bradstreet Information Services Staff. 3 volumes. Dun & Bradstreet Information Services, 899 Eaton Avenue, Bethlehem, PA 18025. (610) 882-7000 or (800) 526-0651. FAX (610) 882-7269. 1993. $795.00.

Ferro-alloy Directory & Databook. Metal Bulletin Inc., 220 Fifth Avenue, New York, NY 10001-7781. (212) 213-6202. FAX (212) 213-1870. 1992. Inquire for cost and availability.

International Research Centers Directory. Gale Research, , 835 Penobscot Building, Detroit, MI 48226-4094. (313) 961-2242. (800) 347-4253. FAX (313) 961-6083. 8th edition. 1995. $410.00

Iron and Steel Works of the World. Metal Bulletin Inc., 220 Fifth Avenue, New York, NY 10001-7781. (212) 213-6202. FAX (212) 213-1870. 1952 to present. Quadrennial. $52.00 for non-members.

Research Centers Directory. Gale Research, , 835 Penobscot Building, Detroit, MI 48226-4094. (313) 961-2242. (800) 347-4253. FAX (313) 961-6083. $485.00.

Scientific and Technical Organizations and Agencies Directory. Gale Research, , 835 Penobscot Building, Detroit, MI 48226-4094. (313) 961-2242. (800) 347-4253. FAX (313) 961-6083. 4th edition. 1996. $195.00.

Who's Who in Engineering. American Association of Engineering Societies, 1111 19th Street NW, Suite 608, Washington, DC 20036. (202) 296-2237 or (800) 658-8897. 8th edition. 1991. Inquire for price.

Who's Who in Technology. Gale Research, , 835 Penobscot Building, Detroit, MI 48226-4094. (313) 961-2242. (800) 347-4253. FAX (313) 961-6083. 1995. $195.00.

ENCYCLOPEDIAS AND DICTIONARIES

Encyclopedia of Physical Science and Technology. Robert A. Meyers, ed. 18 volumes. Academic Press Inc., 6277 Sea Harbor Drive, Orlando, FL 32887. (800) 321-5068. 1992. $2100.00.

McGraw-Hill Dictionary of Scientific and Technical Terms. Sybil P. Parker, ed. 5th edition. McGraw-Hill Publishing Company, 1221 Avenue of the Americas, New York, NY 10020. (800) 262-4729 or (212) 512-3825. 1993. $110.50.

McGraw-Hill Encyclopedia of Engineering. Sybil P. Parker, ed. 2nd edition. McGraw-Hill Publishing Company, 1221 Avenue of the Americas, New York, NY 10020. (800) 262-4729 or (212) 512-3825. 1993. $95.50.

McGraw-Hill Encyclopedia of Science and Technology. Sybil P. Parker, ed. 7th edition. 20 volumes. McGraw-Hill Publishing Company, 1221 Avenue of the Americas, New York, NY 10020. (800) 262-4729 or (212) 512-3825. 1992. $1900.00

GENERAL WORKS

Ferroalloys and Other Additives To Liquid Iron and Steel—stp 739. Lampman & Peters, editors. American Society for Testing and Materials (ASTM), 1916 Race Street, Philadelphia, PA 19103. (215) 299-5585. 1981. $24.75.

Phase Diagrams of Binary Iron Alloys. H. Okamoto. ASM International, Materials Information, Materials Park, OH 44073-0002. (216) 338-5151 or (800) 336-5152. FAX (216) 338-4634. 1993. $301.00.

HANDBOOKS AND MANUALS

Asm Handbook Volume 1: Properties and Selection: Irons, Steels, and High-performance Alloys. ASM International Handbook Committee. ASM International, Materials Information, Materials Park, OH 44073-0002. (216) 338-5151 or (800) 336-5152. FAX (216) 338-4634. 1990. $147.00.

Asm Handbook Volume 3: Alloy Phase Diagrams. ASM International Handbook Committee. ASM International, Materials Information, Materials Park, OH 44073-0002. (216) 338-5151 or (800) 336-5152. FAX (216) 338-4634. 1992. $147.00.

CRC Handbook of Chemistry and Physics. David R. Lide, editor. 75th edition. CRC Press, 2000 Corporate Blvd., N.W., Boca Raton, FL 33431. (407) 994-0555 or (800) 272-7737. FAX (407) 994-0949. $99.50.

Properties of Selected Ferrous Alloying Elements. Y.S. Touloukian & C.Y. Ho. Hemisphere Publishing Corporation, 1900 Frost Road, Suite 101, Bristol, PA 19007-1598. (215) 785-5800 or (800) 821-8312. FAX (215) 785-5515. 1989. $121.00.

Smithell's Metals Reference Book. E.A. Brandes & G.B. Brook. 7th edition. Butterworth-Heinemann, 313 Washington Street, Newton, MA 02158. (617) 928-2500 or (800) 366-2665. FAX (617) 928-2620. 1992. $250.00.

Woldman's Engineering Alloys. J. Frick, editor. 8th edition. ASM International, Materials Information, Materials Park, OH 44073-0002. (216) 338-5151 or (800) 336-5152. FAX (216) 338-4634. 1994. $175.00.

ONLINE DATABASES AND CD-ROMS

Compendex Plus. Engineering Information, Inc., 345 East 47th Street, New York, NY 10017. (212) 705-7600 or (800) 221-1044. Contains citations with abstracts to worldwide literature in engineering and technology, from 1970 to present. Available on online BRS,(800) 289-4277, DIALOG, (800) 334-2564, ORBIT (800) 456-7248, and STN International, FIZ Karlsruhe, P.O. Box 2465, W-7500, Karlsruhe 1, Germany, online services. Also available on CD-ROM. Inquire as to cost and availability.

Current Contents Search. Institute for Scientific Information, 3501 Market Street, Philadelphia, PA 19104. (215) 386-0100. FAX (215) 386-6362. Contains citations to articles listed in the table of contents of science and technology journals. Also articles in social sciences and life sciences journals. Available on BRS,(800) 289-4277, DIALOG,(800) 334-2564, online services. Inquire as to cost and availability.

Metadex. Jointly produced by ASM International and the Institute of Materials. Contains more than 925,000 records from the international literature on metals and alloys, concerning properties, processes, materials classes, applications, and metallurgical systems. Updated monthly. Available from ORBIT-QUESTEL (703) 442-0900.

NASA Database. American Institute of Aeronautics and Astronautics, 370 L'Enfant Promenade, S.W., Washington, DC 20024. (202) 646-7400. Citations and abstracts of aeronautics and astronautics literature, 1962 to present. Also contains citations from STAR, SCIENTIFIC and TECHNICAL AEROSPACE REPORTS, and INTERNATIONAL AEROSPACE ABSTRACTS. Available through NASA/RECON online service. Inquire as to cost and availability.

NTIS Bibliographic Database. National Technical Information Service, 5285 Port Royal Road, Springfield, VA 22161. (703) 487-4929 or FAX (703) 321-8199. Broad coverage of government-sponsored science and technology research reports, 1964 to present. Available on BRS,(800) 289-4277, DIALOG, (800) 334-2564, ORBIT, (800) 456-7248, and STN International, FIZ Karlsruhe, P.O. Box 2465, W-7500, Karlsruhe 1, Germany, online services. Also available on CD-ROM. Inquire as to cost and availability.

WILSONLINE. H.W. Wilson Company, 950 University Avenue, Bronx, NY 10452. (800) 367-6770 or (212) 588-8400. Makes available online versions of the printed H.W. Wilson indexes including Applied Science and Technology Index, Business Periodicals Index, General Science Index, and Readers' Guide to Periodical Literature. Period covered is generally 1983 to present. Available on BRS,(800) 289-4277, DIALOG,(800) 334-2564, and OCLC EPIC,(800) 848-5878, online services. Also available on CD-ROM. Inquire as to cost and availability.

PERIODICALS

Alloy Digest. Alloy Digest Inc., 27 Canfield Street, Orange, NJ 07050. (201) 677-9161. 1952 to present. Monthly. $140.00 per year.

International Journal of Powder Metallurgy. American Powder Metallurgy Institute, 105 College Road E., Princeton, NJ 08540. (609) 452-7700. (609) 987-8523. 1965 to present. Quarterly. $70.00 per year.

Iron Age. Hitchcock Publishing, 191 S. Gary Avenue, Carol Stream, IL 60188. (708) 462-4641. FAX (708) 462-2205. 1867-1993. Monthly. $55.00 per year.

Iron and Steel Engineer. Association of Iron and Steel Engineers, Three Gateway Center, Suite 2350, Pittsburgh, PA 15222. (412) 281-6323. 1924 to present. Monthly. $50.00 per year.

Iron and Steel International. FMJ International Publications, Ltd., Queensway House, 2 Queensway, RedHill, Surrey RH1 1QS England. Telephone 0737-768611. FAX 0737-761685. Annual. $100.30 per year.

Iron & Steelmaker. Iron and Steel Society, 410 Commonwealth Drive, Warrendale, PA 15086. (412) 776-9460. FAX (412) 776-0430. 1974 to present. Monthly. $50.00 per year.

Ironmaking and Steelmaking. Institute of Materials, 1 Carlton House Terrace, London SW1Y 5DB, England. Telephone 071-839-4071. FAX 071-839-2078. 1974 to present. Bi-monthly. $420.00 per year for non-members.

Journal of Phase Equilibria (formerly Bulletin of Alloy Phase Diagrams). ASM International, Materials Park, OH 44073-0002. (216) 338-5151. FAX (216) 338-4634. 6 times a year. $658.00 per year for non-members.

J O M (journal of Metals). Minerals, Metals and Materials Society, 420 Commonwealth Drive, Warrendale, PA 15086. (412) 776-9080. 1949 to present. Monthly. $50.00 per year.

Metallurgical Transactions A: Physical Metallurgy and Materials Science. ASM International, Materials Park, OH 44073. (216) 338-5151 or FAX (216) 338-4634. 1970 to present. Monthly. $520.00 per year.

Metals Week. McGraw-Hill Publishing Company, Commodity Services Group, 1221 Avenue of the Americas, New York, NY 10020. (800) 262-4729 or (212) 512-3825. 1930 to present. Weekly. $770.00 per year.

RESEARCH CENTERS AND INSTITUTES

American Iron and Steel Institute. 1101 17th St., NW, Washington, DC 20036-4700. (202) 452-7100. FAX (202) 463-6573.

Carnegie Mellon University Center For Iron and Steel Making Research. MEMS Department, Pittsburgh, PA 15213. (412) 268-2677.

Colorado School of Mines, Advanced Steel Processing and Products Research Center. Golden, CO 80401. (303) 273-3774.

Canada Centre For Mineral and Energy Technology, Materials Technology Laboratories. Energy, Mines & Resources Canada, 568 Booth Street, Ottawa ON, Canada K1A 0G1. (613) 995-8248. FAX (613) 992-8735.

Texas A&m University Mechanics and Materials Center. Civil Engineering Department, College Station, TX 77843. (409) 845-7512. FAX (409) 845-6156.

University of Connecticut Institute of Materials Science. U-136, 97 N. Eagleville Road, Storrs, CT 06268. (203) 486-4623. FAX (203) 486-4745.

University of Florida Department of Materials Science and Engineering. Gainesville, FL 32611. (904) 392-1454. FAX (904) 392-6359.

University of Minnesota Corrosion Research Center. 221 Church Street SE, Minneapolis, MN 55455. (612) 625-0717.

SPECIFICATIONS AND STANDARDS

*Annual Book of Astm Standards, Volume 01.*02. American Society for Testing and Materials (ASTM), 1916 Race Street, Philadelphia, PA 19103. (215) 299-5585. Annual. Inquire for cost and availability.

FERROELECTRICS

See: ELECTRICITY, ELECTROMAGNETISM

FIBER OPTICS

See also: HOLOGRAPHY, OPTICAL COMMUNICATIONS, OPTICAL DISKS, OPTOELECTRONICS, PHYSICS,

ABSTRACT SERVICES AND INDEXES

Applied Science and Technology Index; a cumulative subject index to English language periodicals in the fields of aeronautics and space science, computer technology, chemistry, construction industry, energy and related areas. H.W. Wilson Co., 950 University Avenue, Bronx, NY 10452. (212) 588-8400. (800) 367-6770. FAX (718) 590-1617. From 1958 to present. Monthly. Inquire about cost and availability. Also available on CD-ROM and online. ISSN: 0003-6986.

Chemical Abstracts. Chemical Abstracts Service, 2540 Olentangy River Road, Box 3012, Columbus, OH 43210-0012. (614) 447-3600. (800) 848-6538. FAX (614) 447-3713. From 1907 to present. Weekly. $16800.00 per year. Also available on CD-ROM and online. Inquire regarding cost and availability. ISSN:

Current Contents: Engineering, Technology and Applied Sciences. Institute for Scientific Information, 3501 Market Street, Philadelphia, PA 19104. (215) 386-0100. FAX (215) 386-2291. From 1961 to present. Weekly. $442.00 per year. Also available on CD-ROM and online. Inquire regarding cost and availability. ISSN: 0095-7917.

Current Papers in Physics. Institute of Electrical Engineers, Michael Faraday House, Six Hill Way, Stevenage, Herts, SG1 2AY, England. Distributed by INSPEC/IEEE, Box 1331, 445 Hoes Lane, Piscataway, NJ 08855-1331. (908) 562-5549. 1966 to present. Fortnightly. $410.00 per year. ISSN: 0011-3786.

Engineering Index Monthly. Engineering Information, Inc., Castle Point on the Hudson, Hoboken, NJ 07030. (201) 216-8500. FAX (201) 216-8532. Monthly. $2200.00 per year. Also available online as COMPENDEX and on CD-ROM. ISSN: 0742-1974.

General Science Index. H.W. Wilson Company, 950 University Avenue, Bronx, NY 10452. (212) 588-8400. (800) 367-6770. FAX (718) 590-1617. From 1978 to present. Ten issues per year; quarterly and annual cumulations. Service basis. Available on CD-ROM and online. Inquire regarding cost and availability. ISSN: 0162-1963.

Government Reports Announcements and Index. U. S. National Technical Information Service (NTIS), 5285 Port Royal Road, Springfield, VA 22161. (703) 487-4650. FAX (703) 321-8547. From 1968 to present. Annual. $630.00 per year. Also available online as NTIS BIBLIOGRAPHIC DATABASE and on CD-ROM.

NTIS Alerts: Physics. U. S. National Technical Information Service. 5285 Port Royal Road, Springfield, VA 22161. (703) 487-4650. FAX (703) 321-8547. Weekly. $140.00 per year.

Pascal 10: Mechanique, Acoustique Et Transport DE Chaleur. Centre National de la Recherche Scientifique, Institue de l'Information Scientifique et Technique, 2 aliee du Parc de Brabois, 54514 Vandoeuvre-Les Nancy Cedex, France. TEC 83-50-46-00. FAX 83-50-46-50. 1984 to present. Ten issues per year. 1530 F per year. Also available on CD-ROM and online. ISSN: 1136-5107.

Physics Abstracts. INSPEC. Section A, Science Abstracts. Institution of Electrical Engineers (IEE). Available from INSPEC/IEEE - Institute of Electrical and Electronic Engineers,, Box 1331, 445 Hoes Lane, Piscaway NJ 08855-1331.(908) 562-5549. 1898 to present. 24 issues per year. $2835.00 per year. Also available on CD-ROM and online. ISSN: 0036-8091.

Physics Briefs (Physikalische Berichte). Information Center for Energy, Physics, Mathematics; German Physical Society. V C H Publishers, Inc., 220 East 23rd Street, New York, NY 10010-4606. (212) 683-8333. 1845 to present. 24 issues per year. $2390.00 per year. Also available online. ISSN: 0179-7434.

Science Citation Index. SCI. Institute for Scientific Information, 3501 Market Street, Philadelphia, PA 19104. (215) 386-0100. (800) 523-1850. FAX (215) 386-2991. 1961 to present. Six issues per year, plus annual cumulation. $11650.00 per year. Also available online and on CD-ROM. Inquire about price and availability. ISSN: 0036-827X.

ANNUAL REVIEWS AND YEARBOOKS

Advances in Atomic, Molecular and Optical Physics. Academic Press, Inc., 6277 Sea Harbor Drive, Orlando, FL. (800) 321-5068. 1965 to present. Irregular. ISSN: 1049-250X.

Advances in Electronics and Electron Physics. Academic Press, Inc., 6277 Sea Harbor Drive, Orlando, FL. (800) 321- 5068. From 1948 to present. Irregular. ISSN: 0065-2539.

Advances in Physics. Taylor & Francis, Ltd., Rankine Road, Basingstoke, Hants RG24 8PR, England. 0256-840366. FAX 0256-47938. 1952 to present. Bimonthly. $511.00 per year. ISSN: 0001-8732.

Progress in Optics. Elsevier Science Inc., 660 White Plains Road, Tarrytown, NY 10591-5153. (914) 524-9200. FAX (914) 333-2444. 1961 to present. Irregular. Price varies. ISSN: 0079-6638.

Progress in Quantum Electronics. Elsevier Science Inc., 660 White Plains Road, Tarrytown, NY 10591-5153. (914) 524-9200. FAX (914) 333-2444. 1969 to present. 5 issues per year. $395.00 per year. ISSN: 0079-6727.

Solid State Physics: Advances in Research and Applications. Academic Press, Inc., 6277 Sea Harbor Drive, Orlando, FL. (800) 321-5068. 1955 to present. Irregular. ISSN: 0081- 1947.

Topics in Applied Physics. Springer-Verlag New York, Inc., 175 Fifth Avenue, New York, NY 10010. (212) 460-1500. 1(800) 777-4643. FAX (212) 473-6272. ISSN: 0303-4216.

ASSOCIATIONS AND PROFESSIONAL SOCIETIES

American Institute of Physics. One Physics Ellipse, College Park, MD 20740-3843. (301) 209-3100.

IEEE Lasers and Electro-optics Society. c/o IEEE, 445 Hoes Lane, PO Box 1331, Piscataway, NJ 08855. (908) 562-3892. FAX (908) 562-8434.

Optical Society of America. 2010 Massachusetts Avenue NW, Washington, DC 20036. (202) 223-8130. FAX (202) 223-1096.

Society For Imaging Science and Technology. 7003 Kilworth Lane, Springfield, VA 33252. (703) 64209090. FAX (703) 642-9094.

Spie - the International For Optical Engineering. PO Box 10, Bellingham, WA 98227. (206) 676-3290. FAX (206) 647- 1445.

BIBLIOGRAPHIES

Science Books and Films. American Association for the Advancement of Science, 1333 H Street NW, Washington, DC 20005. (202) 326-6454. Reviews of print, film and software materials in all sciences. 1965 to present. Nine issues per year. $40.00 per year. ISSN: 0098-342X

Scientific and Technical Books and Serials in Print, 1995; An Index To Literature in Science and Technology. R.R. Bowker Inc., 121 Chanlon Road, New Providence, NJ 07974. (908) 464-6800. (800) 521-8110. 4 volumes. 1972 to present. Annual. $299.95. Also available on compact disc and online. ISSN: 0000-054X.

DIRECTORIES AND BIOGRAPHICAL SOURCES

American Men and Women of Science: Physical and Biological Sciences. R. R. Bowker Inc., 121 Chanlon Road, New Providence, NJ 07974. (908) 464-6800. (800) 521-8110. 20th edition. 8 volumes. 1996. $850.00.

American Physical Society. MEMBERSHIP DIRECTORY BULLETIN ISSUE. American Physical Society, One Physics Ellipse, College Park, MD 20740-3843. (301) 209-3200. FAX (301) 209-0865. Biennial. $50.00

International Directory of Engineering Societies and Related Organizations. American Society for Engineering Education, 1111 19th Street, NW, Suite 608, Washington, DC 20036. 15th edition. 1996. $185.00.

Graduate Programs in Physics, Astronomy and Related Fields. 1995-1996. American Institute of Physics, One Physics Ellipse, College Park, MD 20740-3843. (301) 209-3100. Annual. $45.00. ISSN: 0147-1821.

Optics Education. SPIE, International PO Box 10, Bellingham, WA 98227. (206) 676-3290. FAX (206) 647-1445. Annual. $

Research Centers Directory. Gale Research Inc., 835 Penobscot Building, Detroit, MI 48226-4094. (313) 961-2242. (800) 347-4253. 20th edition. 1995. $485.00. ISSN: 0080- 1518.

Scientific and Technical Organizations and Agencies Directory. Gale Research, , 835 Penobscot Building, Detroit, MI 48226-4094. (313) 961-2242. (800) 347-4253. 4th edition. 1996. $195.00.

Who's Who in Engineering. Gordon Davis, editor. American Society for Engineering Education, 1111 19th Street, NW, Suite 608, Washington, DC 20036. 9th edition. 1995. $220.00.

Who's Who in Technology. Gale Research, 835 Penobscot Building, Detroit, MI 48226-4094. (313) 961-2242. (800) 347-4253. 7th edition. 1995. $195.00.

ENCYCLOPEDIAS AND DICTIONARIES

Dictionary of the Physical Sciences: Terms, Formulas, Data. Cesare Emiliani. Oxford University Press, Inc., 200 Madison Avenue, New York, NY 10016. (212) 725-6000. (800) 334-4249. 1989. $19.95.

Encyclopedia of Applied Physics. George Trigg, editor. VCH Publications, Inc., 220 East 23rd Street, Suite 909, New York, NY 10010-4606. (212) 683-8333. (800) 422-8824. 20 volume set. 1991-. $5990.00.

Encyclopedia of Physics. Robert M. Besancon. Chapman & Hall, 1 Penn Plaza, New York, NY 10119. (212) 564-1060. 3rd edition. 1990. $54.95 in paper.

Encyclopedial of Lasers and Optical Technology. Robert A. Meyers, editor. Academic Press, Inc., 6277 Sea Harbor Drive, Orlando, FL. (800) 321-5068. 1991. $92.00.

Encyclopedia of Physical Science and Technology. Academic Press, Inc., 6277 Sea Harbor Drive, Orlando, FL. (800) 321-5068. 2nd edition. 18 volume set. 1992. $2625.00.

Fiber Optics Standard Dictionary. Martin H. Weik. 2nd edition. Van Nostrand Reinhold, 115 Fifth Avenue, New York, NY 10003. (212) 254-3232. (800) 842-3636.. 2nd edition. 1989. $49.95.

Glossary of Optical Terminology. Thomas K. Farrell. Butterworth-Heinemann, 313 Washington Avenue, Newton, MA 02158. (617) 928-2500. (800) 366-2665. 1985. $24.95.

McGraw-Hill Encyclopedia of Science and Technology. McGraw-Hill Book Company, Inc., 1221 Avenue of the Americas, New York, NY 10020. (212) 512-2000. (800) 262-4729. 7th edition. 20 volume set. 1992. $1900.00.

GENERAL WORKS

Distributed and Multiplexed Fiber Optics Sensors. J. P. Dakin and A. D. Kersey. SPIE - International Society for Optical Engineering, 1000 20th Street, Bellingham, WA 98227. (206) 676-3290. FAX (206) 647-1445. 1992. $53.00.

Fiber-optic Communication Systems. Govind P. Agrawal. John Wiley & Sons, Inc., 605 Third Avenue, New York, NY 10158-0012. (212) 850-6000. (800) 225-5945.

1992. $74.95.

Fiber Optic Sensors: An Introduction For Engineers and Scientists. Iric Udd. John Wiley & Sons, Inc., 605 Third Avenue, New York, NY 10158-0012. (212) 850-6000. (800) 225-5945. 1991. $110.00.

Fiber Optics Communications. Joseph C. Palais. Prentice- Hall, 113 Sylvan Avenue, Route 9W, Englewood Cliffs, NJ 07632. (201) 592-2000. (800) 922-0579. 3rd edition. 1992. $66.00.

Fiber Optics Systems Design. John M Simmons. TAB Books, P.O. Box 40, Blue Summit, PA 17294-0850. (717) 794-2191. (800) 233-1128. 1992. $42.95.

Fundamentals of Optical Fibers. John A. Buck. John Wiley & Sons, Inc., 605 Third Avenue, New York, NY 10158-0012. (212) 850-6000. 1995. $59.95.

Optical Fibre Sensor Technology. Ken Grattan and Beverley Meggitt. Prentice-Hall, 113 Sylvan Avenue, Route 9W, Englewood Cliffs, NJ 07632. (201) 592-2000. (800) 922- 0579. 1994. $33.33.

Optical Guided Waves and Devices. Richard Syms. McGraw-Hill Publishing Company, Inc., 1221 Avenue of the Americas, New York, NY 10020. (212) 512-2000. (800) 262-4729 1992. $55.00.

Optics and Lasers, Including Fibers and Optical Waveguides. M. Young. Springer-Verlag New York, Inc., 175 Fifth Avenue, New York, NY 10010. (212) 460-1500. (800) 777-4643. FAX (212) 473-6272. 4th revised edition. 1993. $34.50.

Principles of Optics, Optoelectronics and Photonics. Alan Billings. Prentice-Hall, 113 Sylvan Avenue, Route 9W, Englewood Cliffs, NJ 07632. (201) 592-2000. (800) 922- 0579. 1994. $40.00 in paper.

The Rewiring of American: The Fiber Optics Revolution. G. David Chaffee. Academic Press, Inc., 6277 Sea Harbor Drive, Orlando, FL. (800) 321-5068. 1987. $54.00.

Technologies of Light. E. U. Kotte, et al, editors. Springer-Verlag New York, Inc., 175 Fifth Avenue, New York, NY 10010. (212) 460-1500. (800) 777-4643. FAX (212) 473- 6272. 1988. $66.00.

Ultrfast Fiber Switching Devices and Systems. Mohammed N. Islam. Cambridge University Press, 40 West 20th Street, New York, NY 10011-4211. (212) 924-3900. (800) 872-7423. 1992. $49.95.

Understanding Fiber Optics. Jeff Hecht Sams, 201 West 103rd Street, Indianapolis, in 46290-1097. (317) 581-3500.(800) 545-5914. 2nd edition. 1993. $26.95.

Understanding Telecommunications and Lightwave Systems. John G. Nellist. Institute of Electrical and Electronic Engineers (IEEE), 345 East 47th Street, New York, NY 10017. (212) 705-7900. 2nd edition. 1995.

HANDBOOKS AND MANUALS

CRC Handbook of Chemistry and Physics. David R. Kide, editor. CRC Press, Inc., 2000 Corporate Boulevard, NW, Boca Raton, FL 33431. (407) 994-0555. (800) 272-7737. 77th edition. 1996. $99.95.

Design Handbook For Optical Fiber Systems. R. P. DePaula and E. Udd. Information Gatekeepers, 214 Harvard Avenue, Boston, MA 02134. (617) 232-3111. (800) 323-1088. $75.00.

FIBER OPTIC COMMUNICATION DESIGN HandBOOK. Robert H. Hoss. Prentice-Hall, 113 Sylvan Avenue, Route 9W, Englewood Cliffs, NJ 07632. (201) 592-2000. (800) 922-0579. 1990. $66.00.

Fiber Optic Communications Handbook. TAB Books, P.O. Box 40, Blue Summit, PA 17294-0850. (717) 794-2191. (800) 233-1128. 2nd edition. 1990. $89.50.

Fiber Optic Metrology and Standards. O. D. Soares. SPIE - International Society for Optical Engineering, 1000 20th Street, Bellingham, WA 98227. (206) 676-3290. FAX (206) 647-1445. 1991. $68.00.

Handbook of Optics. Optical Society of America. McGraw- Hill Publishing Company, Inc., 1221 Avenue of the Americas, New York, NY 10020. (212) 512-2000. (800) 262- 4729. 2 volumes. 2nd edition. 1993. $99.50 each.

Handbook of Physical Quantities. Igor S. Grigoriev and Evgenil Z. Melikhov, editors. CRC Press, Inc., 2000 Corporate Boulevard, NW, Boca Raton, FL 33431. (407) 994- 0555. (800) 272-7737. 1995. $99.00.

Handbook of Tables of Functions For Applied Optics. Leo Levi, editor. Franklin Book Co., Inc., Elkins Park, PA 19117. (215) 635-5252. CRC Press Reprints. 1974. $152.95.

McGraw-Hill Encyclopedia of Physics. Sybil P. Parker, editor. McGraw-Hill Publishing Company, Inc., 1221 Avenue of the Americas, New York, NY 10020. (212) 512-2000. (800) 262-4729. 2nd edition. 1993. $95.50

Newnes Engineering and Physical Science Pocketbook. J. O. Bird and P. J. Chivers. Butterworth-Heinmann, 313 Washington Avenue, Newton, MA 02158. (617) 928-2500. (800) 366-2665. 1993. $49.95.

Optics Problem Solver. Research and Education Association Staff. Research & Education Association, 61 Ethel Road, West Piscataway, NJ 08854. (908) 819-8880. Revised edition. 1994. $29.95 in paper.

ONLINE DATABASES AND CD-ROMS

CA Search. Chemical Abstracts Service, P.O. Box 3012, Columbus, OH 43210-0012. (614) 447-3600. (800) 848-6533. FAX (614) 447-3709. Very comprehensive guide to worldwide chemical literature and related fields, 1972 to present. Available on BRS,(800) 289-4277, DIALOG, (800) 334-2564, ORBIT (800) 456-7248, and STN International, FIZ Karlsruhe, P.O. Box 2465, W-7500, Karlsruhe 1, Germany, online services. Inquire as to cost and availability.

Compendex Plus. Engineering Information, Inc., 345 East 47th Street, New York, NY 10017. (212) 705-7600 or (800) 221-1044. Contains citations with abstracts to worldwide literature in engineering and technology, from 1970 to present. Available on online BRS,(800) 289-4277, DIALOG, (800) 334-2564, ORBIT (800) 456-7248, and STN International, FIZ Karlsruhe, P.O. Box 2465, W-7500, Karlsruhe 1, Germany, online services. Also available on CD-ROM. Inquire as to cost and availability.

Current Contents Search. Institute for Scientific Information, 3501 Market Street, Philadelphia, PA 19104. (215) 386-0100. FAX (215) 386-6362. Contains citations to articles listed in the table of contents of science and technology journals. Also articles in social sciences and life sciences journals. Available on BRS, (800) 289-4277, DIALOG, (800) 334-2564, online services. Inquire as to cost and availability.

Dissertation Abstracts Online. University Microfilms International, 300 North Zeeb Road, Ann Arbor, MI 48106. (800) 521-0600 or (313) 761-4700. Scope includes virtually all doctoral dissertations accepted at accredited American institutions from 1861 to present in 252 subject areas. Available on BRS, (800) 289-4277, DIALOG, (800) 334-2564, and OCLC EPIC, (800) 848-5878, online services. Also available on CD-ROM. Inquire as to cost and availability.

INSPEC. Institution of Electrical Engineers, Michael Faraday House, Six Hills Way, Stevenage, Herts. SG1 2AY, England. Telephone: 0438 313311 or FAX 0438 742840. Contains citations to the worldwide literature of physics, electronics and electrical engineering, computer technology, and related fields. Available on BRS, (800) 289-4277, DIALOG, (800) 334-2564, ORBIT, (800) 456-7248, and STN International, FIZ Karlsruhe, P.O. Box 2465, W-7500, Karlsruhe 1, Germany, online services. Inquire as to cost and availability.

NTIS Bibliographic Database. National Technical Information Service, 5285 Port Royal Road, Springfield, VA 22161. (703) 487-4929 or FAX (703) 321-8199. Broad coverage of government-sponsored science and technology research reports, 1964 to present. Available on BRS,(800) 289-4277, DIALOG, (800) 334-2564, ORBIT, (800) 456-7248, and STN International, FIZ Karlsruhe, P.O. Box 2465, W-7500, Karlsruhe 1, Germany, online services. Also available on CD-ROM. Inquire as to cost and availability.

FEEDFORWARD CONTROL SYSTEMS

Ency. of Physical Sciences and Engineering Info. Sources

Physics Briefs. American Institute of Physics, 335 East 45th Street, New York, NY 10017. (212) 661-9260 or FAX (212) 661-2036. Contains citations with abstracts of the literature of physics and related fields, 1979 to present. Available on the STN International, FIZ Karlsruhe, P.O. Box 2465, W-7500, Karlsruhe 1, Germany, online service. Inquire as to cost and availability.

Scientific and Technical Books and Serials in Print. R.R. Bowker Inc., 121 Chanlon Road, New Providence, NJ 07974. (800) 521-8110 or (908) 464-6800. List of currently published books and serials in the physical and biological sciences, engineering and technology, with subject, author and titles indexes. Available on ORBIT, (800) 456-7248,online service. Inquire as to cost and availability.

SCISEARCH. Institute for Scientific Information, 3501 Market Street, Philadelphia, PA 19104. (800) 523-1850 or (215) 386-0100. Broad multidisciplinary title and author index to the international literature of science and technology, 1974 to present. Available on DIALOG, (800) 334-2564, and ORBIT, (800) 456-7248, online services. Also available on CD-ROM. Inquire as to cost and availability.

WILSONLINE. H.W. Wilson Company, 950 University Avenue, Bronx, NY 10452. (800) 367-6770 or (212) 588-8400. Makes available online versions of the printed H.W. Wilson indexes including Applied Science and Technology Index, Business Periodicals Index, General Science Index, and Readers' Guide to Periodical Literature. Period covered is generally 1983 to present. Available on BRS, (800) 289-4277, DIALOG, (800) 334-2564, and OCLC EPIC, (800) 848-5878, online services.Also available on CD-ROM. Inquire as to cost and availability.

PERIODICALS

Applied Optics. Optical Society of America. 2010 Massachusetts Avenue NW, Washington, DC 20036-1023. (202) 223-8130. 1962 to present. 36 issues per year. $1090.00 per year. ISSN: 0003-6935.

Applied Spectroscopy. Society for Applied Spectroscopy, 198 Thomas Johnson Drive, Suite 2, Frederick, MD 21702. (301) 694-8122. FAX (301) 694-6860. 1946 to present. Monthly. $180.00 per year. ISSN: 0003-7028.

Fiber and Integrated Optics. Taylor and Francis, 1900 Frost Road, Suite 101. Bristol, PA 19007. (215) 785-5800. FAX (215) 785-5515. 1977 to present. Quarterly. $180.00 per year. ISSN: 0146-8030.

International Journal of Nonlinear Optical Physics. World Scientific Publishing Co., 1060 Main Street, Suite 18, River Edge, NJ 07661. (800) 227-7562. 1992 to present. Quarterly. $348.00 per year. ISSN: 0218-1991.

Journal of Lightwave Technology. Institute of Electrical and Electronics Engineers, 445 Hoes Lane, PO Box 1331, Piscataway, NJ 08855. (908) 562-3892. FAX (908) 562-8434. 1983 to present. Monthly. $425.00 per year. ISSN: 0733- 8724.

IEEE Journal of Quantum Electronics. IEEE Lasers and Electro-optics Society/IEEE, 445 Hoes Lane, PO Box 1331, Piscataway, NJ 08855. (908) 981-0060. FAX (908) 981-9667. 1965 to present. Monthly. $535.00 per year. ISSN: 0018- 9197.

Journal of the Optical Society of America. Part A: Optics and Image Science. Optical Society of America, Inc., 2010 Massachusetts Avenue, NW, Washington, DC 20036-1023. (202) 223-8130. 1917 to present. Monthly. $610.00 per year. ISSN: 0740-3232.

Journal of the Optical Society of America. Part B: Optical Physics. Optical Society of America, Inc., 2010 Massachusetts Avenue, NW, Washington, DC 20036-1023. (202)223-8130. 1917 to present. Monthly. $610.00 per year. ISSN: 0740-3224.

Laser Focus World. PennWell Publishing Co., 10 Tara Boulevard, Nashua, NH 03062-2891. (603) 891-0123. FAX(603) 891-0574. 1964 to present. Monthly. $104.00 per year. ISSN: 1043-8092.

Optical Engineering. SPIE - International Society for Optical Engineering, Box 10, 1000 20th Street, Bellingham, WA 98227-0010. (206) 676-3290. FAX (206) 647-1445. 1962 to present. Monthly. $170.00 per year. ISSN: 0091-3286.

Optics and Photonics News. Optical Society of America, Inc., 2010 Massachusetts Avenue, NW, Washington, DC 20036-1023. (202) 223-8130. 1975 to present. Monthly. $99.00 per year. ISSN: 1047-6938.

Optics Communications. Elsevier Science Inc., Box 882, Madison Square Station, New York, NY 10159. (212) 989-5800. FAX (212) 633-4990. 1969 to present. 54 issues per year. $2121.00. per year. ISSN: 0030-4018.

Optics Letters. Optical Society of America, Inc., 2010 Massachusetts Avenue, NW, Washington, DC 20036-1023. (202) 223-8130. 1977 to present. Semi-Monthly. $625.00 per year. ISSN: 0146-9592.

Optics Materials. Elsevier Science Inc., Box 882, Madison Square Station, New York, NY 10159. (212) 989-5800. FAX (212) 633-4990. 1991 to present. Quarterly. $206.00 per year. ISSN: 0925-3467.

Spie - International Society For Optical Engineering. PROCEEDINGS. SPIE - International Society for Optical Engineering, Box 10, 1000 20th Street, Bellingham, WA 98227-0010. (206) 676-3290. FAX (206) 647-1445. 1963 to present. Approximately 50 numbers per year. Individually priced.

RESEARCH CENTERS AND INSTITUTES

Institute of Optics. University of Rochester, Wilmot Building, Rochester, NY 14627. (716) 275-5248.

Center For Applied Optics. University of Alabama in Huntsville, Research Institute Building, Huntsville, AL 25899. (205) 895-6102.

Center For Applied Optics. University of Texas at Dallas, P.O. Box 830688, Richardson, TX 75803-0688. (214) 690-2868. FAX (214) 690-2848.

Center For Imaging Science. Rochester Institute of Technology, One Lomb Memorial Drive, Rochester, NY 14623. (716) 475-2774.

Center For Research in Electro-optics and Lasers.University of Central Florida, 12424 Research Parkway, Orlando, FL 32826. (407) 658-6800. FAX (407) 658-6880.

Institute For Physical Science and Technology. University of Maryland, 4203 Computer and Space Sciences Building, College Park, MD 20742. (301) 454-4874. FAX (301) 314-9363.

Optical Sciences Center. University of Arizona, Tucson, AZ 85721. (602) 621-6997.

FIBRES (TEXTILES)

See: TEXTILES

FIELD-EFFECT TRANSISTORS

See: SEMICONDUCTORS

FIRE PROTECTION ENGINEERING

See also: ALARM SYSTEMS, COMBUSTION, EXPLOSIVES, SAFETY ENGINEERING

ABSTRACT SERVICES AND INDEXES

Applied Science and Technology Index; A Cumulative Subject Index To English Language Periodicals in the Fields of Aeronautics and Space Science, Computer Technology, Chemistry, Construction Industry, Energy and Related Areas. H.W. Wilson Co., 950 University Avenue, Bronx, NY 10452. (800) 367-6770 or (212) 588-8400. FAX (718) 590-1617. From 1958 to present. Monthly. Inquire about cost and availability. Also available on CD-ROM and online.

Chemical Abstracts - Applied Chemistry and Chemical Engineering Sections. Chemical Abstracts Service, 2540 Olentangy River Road, Box 3012, Columbus, OH 43210. (800) 848-6538 or (614) 447-3600, FAX (614) 447-3713. Bi-weekly. $1410.00 per year.

Current Contents: Physical, Chemical and Earth Sciences. Institute for Scientific Information, 3501 Market Street, Philadelphia, PA 19104. (215) 386-0100. FAX (215) 386-6362. Weekly. $360.00 per year.

Engineering Index Monthly. Engineering Information, Inc., Castle Point on the Hudson, Hoboken, NJ 07030. (800) 221-1044. FAX (212) 832-1857. Monthly. $2200.00 per year. Also available online as COMPENDEX, and also on CD-ROM.

Health and Safety Science Abstracts. Cambridge Scientific Abstracts, 7200 Wisconsin Avenue, Bethesda, MD 20814-4823. (301) 961-6750. FAX (301) 961-6720. 1973 to present. 4 times a year. $645.00 per year.

ASSOCIATIONS AND PROFESSIONAL SOCIETIES

American Fire Sprinkler Association. 12959 Jupiter Rd., Suite 142, Dallas, TX 75238. (214) 349-5965. FAX (214) 343-8898.

American Society of Safety Engineers. 1800 E Oakton St., Des Plaines, IL 60018-2187. (708) 692-4121. FAX (708) 296-3769.

Automatic Fire Alarm Association. PO Box 1652, Barrington, IL 60011. (708) 381-6422. FAX (708) 381-0727.

Board of Certified Safety Professionals. 208 Burwash, Savoy, IL 61874. (217) 359-9263.

Fire and Emergency Manufacturers and Services Association. 808 17th Street NW, Suite 200, Washington, DC 20006. (202) 223-9669. FAX (202) 223-9569.

Fire Protection Association. 140 Aldersgate St., London EC1A 4HX, England. Telephone 071-606-3757. FAX 071-600-1487.

Fire Retardant Chemicals Association. 851 New Holland Avenue, Box 3535, Lancaster, PA 17604. (717) 291-5616. FAX (717) 295-4538.

Fire Suppression Systems Association. 5024-R Campbell Blvd., Baltimore, MD 21236-5974. (410) 931-8100. FAX (410) 931-8111.

National Association of Fire Equipment Distributors. 401 N. Michigan Ave., Chicago, IL 60611. (312) 644-6610. FAX (312) 321-6869.

National Burglar & Fire Alarm Association. 7101 Wisconsin Avenue, Bethesda, MD 20814. (301) 907-3202. FAX (301) 907-7897.

National Fire Protection Association. 1 Batterymarch Park, Quincy, MA 02269. (617) 770-3500.

National Fire Sprinkler Association. Robin Hill Corporate Park, Route 22, Box 1000, Patterson, NY 12563. (914) 878-4200. FAX (914) 878-4215.

National Safety Council. 1121 Spring Lake Dr., Itasca, IL 60143. (708) 285-1121.

Society of Fire Protection Engineers. 1 Liberty Square, Boston, MA 02109-4825. (617) 482-0686. FAX (617) 482-8184.

System Safety Society, Inc. Technology Trading Park, 5 Export Dr., Suite A, Sterling, VA 22170-4421. (703) 444-6520.

DIRECTORIES AND BIOGRAPHICAL SOURCES

Consultants' Directory [system Safety Society Directory of Consultants]. System Safety Society, Technology Trading Park, 5 Export Dr., Suite A, Sterling, VA 22170-4421. (703) 444-6520. Periodic. $20.00 per issue.

Directory of Engineering Societies and Related Organizations. Gordon Davis. 13th edition. American Association of Engineering Societies, 1111 19th Street NW, Suite 608, Washington, DC 20036. (202) 296-2237 or (800) 658-8897. 1989. Inquire for price.

Fitech International. FMJ International Publications, Ltd., Queensway House, 2 Queensway, RedHill, Surrey RH1 1QS England. Telephone 0737-768611. FAX 0737-761685. Annual. $82.50.

Research Centers Directory. Gale Research, , 835 Penobscot Building, Detroit, MI 48226-4094. (313) 961-2242. (800) 347-4253. FAX (313) 961-6083. $485.00.

Scientific and Technical Organizations and Agencies Directory. Gale Research, , 835 Penobscot Building, Detroit, MI 48226-4094. (313) 961-2242. (800) 347-4253. FAX (313) 961-6083. Inquire for cost.

Society of Fire Protection Engineers—roster. 1 Liberty Square, Boston, MA 02109-4825. (617) 482-0686. FAX (617) 482-8184. Annual. Write or call for price information.

Who's Who in Engineering. American Association of Engineering Societies, 1111 19th Street NW, Suite 608, Washington, DC 20036. (202) 296-2237 or (800) 658-8897. 8th edition. 1991. Inquire for price.

GENERAL WORKS

Design Against Fire. P. Stollard & L. Johnston, editors. Chapman & Hall, 115 Fifth Avenue, New York, NY 10211-0906. 1994. $44.95.

Fire and Explosion Protection Systems. Michael R. Lindeburg. Professional Publications, 1250 Fifth Ave., Belmont, CA 94002. (415) 593-9119 or (800) 426-1178. 1993. $15.95. 2nd ed. to be published in 1995. Write or call for information.

Fire Alarm Signaling Systems. Richard Bukowski. 2md ed. National Fire Protection Association. 1 Batterymarch Park, Quincy, MA 02269. (617) 770-3500. 1994. Inquire for cost and availability.

Standard For Control Units For Fire-protective Signaling Systems. 7th ed. Underwriters' Laboratories, 333 Pfingsten Rd., Northbrook, IL 60062-2096. (708) 272-8800. FAX (312) 272-8129. 1993. Write or call for price information.

Structural Fire Protection. T.T. Lee, ed. American Society of Civil Engineers, 345 East 47th Street, New York, NY 10017-2398. (212) 705-7520 or (800) 548-2723. 1992. $56.00.

Understanding Necode Calculations. Charles Michael Holt. Delmar Publishers, Division of Thomson Educational Publishing, Inc., 3 Columbia Circle, Box 15015, Albany, NY 12205. (518) 464-3500 or (800) 347-7707. FAX (518) 464-0358. 1994. $26.95.

HANDBOOKS AND MANUALS

Fire and Flammability Handbook. Neil Schultz. Van Nostrand Reinhold, 115 Fifth Avenue, New York, NY 10003. (212) 254-3232 or FAX (212) 254-9499. 1985. Inquire for cost and availability.

Fire Protection Handbook. National Fire Protection Association, 1 Batterymarch Park, Quincy, MA 02269. (617) 770-3500. Quinquennial. $125.00 for non-members.

Boca National Fire Prevention Code. Building officials and Code Administrators International, 4051 W. Flossmoor Rd., Country Club Hills, IL 60477-5795. Triennial. $36.00.

ONLINE DATABASES AND CD-ROMS

Compendex Plus. Engineering Information, Inc., 345 East 47th Street, New York, NY 10017. (212) 705-7600 or (800) 221-1044. Contains citations with abstracts to worldwide literature in engineering and technology, from 1970 to present. Available on online BRS,(800) 289-4277, DIALOG, (800) 334-2564, ORBIT (800) 456-7248, and STN International, FIZ Karlsruhe, P.O. Box 2465, W-7500, Karlsruhe 1, Germany, online services. Also available on CD-ROM. Inquire as to cost and availability.

CSA Engineering. Cambridge Scientific Abstracts, 7200 Wisconsin Avenue, Suite 601, Bethesda, MD 20814. (301) 961-6750 or (800) 843-7751. Contains citations and abstracts of international periodicals and research literature covering all fields of engineering and science and technology,including computer and information science, electronics, mechanical engineering, solid state materials, 1981 to present. Available on BRS,(800) 289-4277, online service.Inquire as to cost and availability.

Current Contents Search. Institute for Scientific Information, 3501 Market Street, Philadelphia, PA 19104. (215) 386-0100. FAX (215) 386-6362. Contains citations to articles listed in the table of contents of science and technology journals. Also articles in social sciences and life sciences journals. Available on BRS,(800) 289-4277, DIALOG,(800) 334-2564, online services. Inquire as to cost and availability.

Dissertation Abstracts. University Microfilms International, 300 North Zeeb Road, Ann Arbor, MI 48106. (800) 521-0600 or (313) 761-4700. Scope includes virtually all doctoral dissertations accepted at accredited American institutions from 1861 to present in 252 subject areas. Available on BRS,(800) 289-4277, DIALOG,(800) 334-2564, and OCLC EPIC,(800) 848-5878, online services. Also available on CD-ROM. Inquire as to cost and availability.

NTIS Bibliographic Database. National Technical Information Service, 5285 Port Royal Road, Springfield, VA 22161. (703) 487-4929 or FAX (703) 321-8199. Broad coverage of government-sponsored science and technology research reports, 1964 to present. Available on BRS,(800) 289-4277, DIALOG, (800) 334-2564, ORBIT, (800) 456-7248, and STN International, FIZ Karlsruhe, P.O. Box 2465, W-7500, Karlsruhe 1, Germany, online services. Also available on CD-ROM. Inquire as to cost and availability.

SCISEARCH. Institute for Scientific Information, 3501 Market Street, Philadelphia, PA 19104. (800) 523-1850 or (215) 386-0100. Broad multidisciplinary title and author index to the international literature of science and technology, 1974 to present. Available on DIALOG,(800) 334-2564, and ORBIT,(800) 456-7248, online services. Also available on CD-ROM. Inquire as to cost and availability.

WILSONLINE. H.W. Wilson Company, 950 University Avenue, Bronx, NY 10452. (800) 367-6770 or (212) 588-8400. Makes available online versions of the printed H.W. Wilson indexes including Applied Science and Technology Index, Business Periodicals Index, General Science Index, and Readers' Guide to Periodical Literature. Period covered is generally 1983 to present. Available on BRS,(800) 289-4277, DIALOG,(800) 334-2564, and OCLC EPIC,(800) 848-5878, online services. Also available on CD-ROM. Inquire as to cost and availability.

PERIODICALS

Accident Analysis and Prevention. Elsevier Science, 660 White Plains Rd., Tarrytown, NY 10159-5153. (919) 524-9200. FAX (919) 333-2444. 6 times a year. $605.00.

Alarm/allarme/alarme. Cerberus AG, CH- 8708 Maennedorf, Switzerland. Telephone 1-922611. FAX 1-922450. 3 times a year. Free.

American Fire Journal. Fire Publications, Inc., c/o J.A. Ackerman Publishing, 9072 E Artesia Blvd., Bellflower, CA 90706. (213) 866-1664. Monthly. $19.95 per year.

Fire. FMJ International Publications, Ltd., Queensway House, 2 Queensway, RedHill, Surrey RH1 1QS England. Telephone 0737-768611. FAX 0737-761685. Monthly. Write for price.

Fire and Flammability Bulletin. Elsevier Science, 660 White Plains Rd., Tarrytown, NY 10159-5153. (919) 524-9200. FAX (919) 333-2444. Monthly. $390.00 per year.

Fire Engineering. Pennwell Publishing Co., Park 80 W. Plaza 2, Saddle Brook, NJ 07662-5812. (201) 845-0800. FAX (201) 845-6275. Monthly. $23.50 per year.

Fire International. FMJ International Publications, Ltd., Queensway House, 2 Queensway, RedHill, Surrey RH1 1QS England. Telephone 0737-768611. FAX 0737-761685. Quarterly. Write for price information.

Fire News. National Fire Protection Association, 1 Batterymarch Park, Quincy, MA 02269. (617) 770-3500. 6 times a year. Free with membership.

Fire Prevention. Fire Protection Association, 140 Aldersgate St., London EC1A 4HX, England. Telephone 071-606-3757. FAX 071-600-1487. 10 times a year. Write for price information.

Fire Safety Engineering. Paramount Publishing Ltd., 17-21 Shenley Rd., Borehamwood, Herts WD6 1RT, England. Telephone 081-207-2598. Bi-monthly. Write for price information.

Fire Safety Journal. Elsevier Science, 660 White Plains Rd., Tarrytown, NY 10159-5153. (919) 524-9200. FAX (919) 333-2444. 8 times a year. $510.00.

Fire Systems. National Association of Fire Equipment Distributors, 401 N. Michigan Ave., Chicago, IL 60611. (312) 644-6610. FAX (312) 321-6869. Twice a year. $10.00.

Fire Technology. National Fire Protection Association, 1 Batterymarch Park, Quincy, MA 02269. (617) 770-3500. Quarterly. $39.50.

Hazard Prevention. System Safety Society, Inc., Technology Trading Park, 5 Export Dr., Suite A, Sterling, VA 22170-4421. (703) 444-6520. Quarterly. $5.00 for non-members.

Industrial Fire Chief. Argus Inc., 6151 Powers Ferry Rd. NW, Atlanta, GA 30339-2941. (404) 955-1500. FAX (404) 955-0400. Bi-monthly. $24.00 a year.

Journal of Fire Protection Engineering. Society of Fire Protection Engineers, 1 Liberty Square, Boston, MA 02109-4825. (617) 482-0686. FAX (617) 482-8184. Quarterly. $170.00 per year.

Journal of Fire Sciences. Technomic Publishing Company, Inc., 851 New Holland Avenue, Box 3535, Lancaster, PA 17604. (717) 291-5609. 1983 to present. Bi-monthly. $295.00 per year.

Journal of Safety Research. Elsevier Science [journals], 660 White Plains Rd., Tarrytown, NY 10159-5153. (919) 524-9200. FAX (919) 333-2444. 1982 to present. 4 times a year. $305.00 per year.

Nfpa Journal. National Fire Protection Association, 1 Batterymarch Park, Quincy, MA 02269. (617) 770-3500. 1991 to present. Bi-monthly. Included with $75.00 annual membership.

Professional Safety. American Society of Safety Engineers, 1800 E Oakton St., Des Plaines, IL 60018-2187. (708) 692-4121. FAX (708) 296-3769. Monthly. $50.00 per year.

Sfpe Bulletin. Society of Fire Protection Engineers, 1 Liberty Square, Boston, MA 02109-4825. (617) 482-0686. FAX (617) 482-8184. Six times a year. Write for price information for non-members (free with membership).

RESEARCH CENTERS AND INSTITUTES

Factory Mutual Research Corporation. 1151 Boston-Providence Turnpike, Norwood, MA 02062. (617) 762-4300. FAX 9617) 762-9375.

Institute For Research in Construction. National Research Council of Canada, Montreal Road, Bldg. M-20, Ottawa, ON Canada K1A0R6. (613) 993-2607. FAX (613) 954-5984.

SPECIFICATIONS AND STANDARDS

*Annual Book of Astm Standards, Volume 04.*07. American Society for Testing and Materials (ASTM), 1916 Race Street, Philadelphia, PA 19103. (215) 299-5585. Annual. Inquire for cost and availability.

FISSION

See also: ENERGY, FUSION, NUCLEAR ENERGY, NUCLEAR ENGINEERING, NUCLEAR PHYSICS, NUCLEAR REACTORS

ABSTRACT SERVICES AND INDEXES

Applied Mechanics Reviews: An assessment of world literature in engineering sciences. American Society of Mechanical Engineers, 345 East 47th Street, New York, NY 10017. (212) 705-7703. 1948 to present. Monthly. $360.00 per year. ISSN: 0003-6900.

Applied Science and Technology Index; A Cumulative Subject Index To English Language Periodicals in the Fields of Aeronautics and Space Science, Computer Technology, Chemistry, Construction Industry, Energy and Related Areas. H. W. Wilson Co., 950 University Avenue, Bronx, NY 10452- 9978. (212) 588-8400. (800) 367-6770. FAX (718) 590-1617. From 1958 to present. Monthly. Inquire about cost and availability. Also available online (BRS and WILSONLINE) and on CD-ROM. ISSN: 0003-6986.

Chemical Abstracts. Chemical Abstracts Service. 2540 Olentangy River Road, Box 3012, Columbus, OH 43210-0012. (614) 447-3600. FAX (614) 447-3713. 1907 to present. Weekly. $16,800.00 per year. Available online and on CD- ROM. CA is

FISSION

Ency. of Physical Sciences and Engineering Info. Sources

also available in five section groupings. Inquire regarding cost and availability.

Current Contents: Engineering, Technology and Applied Sciences. Institute for Scientific Information, 3501 Market Street, Philadelphia, PA 19104. (215) 386-0100. FAX (215) 386-2991. From 1970 to present. Weekly. $442.00 per year. Also available online. Inquire regarding cost and availability. ISSN: 0095-7917.

Engineering Index Monthly. Engineering Information, Inc., Castle Point on the Hudson, Hoboken, NJ 07030. (201) 216-8500. (800) 221-1044. FAX (201) 216-8532. Monthly. $2300.00 per year for monthly issues. $1980 per year for annual cumulation. Available on CD-ROM and online as COMPENDEX. ISSN: 0742-1974; 0360-8557.

General Science Index. H.W. Wilson Company, 950 University Avenue, Bronx, NY 10452. (212) 588-8400. (800) 367-6770. FAX (718) 590-1617. From 1978 to present. Ten issues per year; quarterly and annual cumulations. Service basis. Available on CD-ROM and online. Inquire regarding cost and availability. ISSN: 0162-1963.

*ISMEC: Mechanical Engineering Abstracts.*Cambridge Scientific Abstracts, 7200 Wisconsin Avenue, Suite 601, Bethesda, MD 20814. (301) 961-6750 or (800) 843-7751. Contains citations to the literature in mechanical engineering, industrial and production engineering, energy, power, mechanics, devices and related areas, from 1973 to present. Available on the DIALOG,(800) 334-2564, online service. Inquire as to cost and availability.

NTIS Alerts: Energy. U.S. National Technical Information Service, 5285 Port Royal Road, Springfield, VA 22161. (703) 487-4929. FAX (703) 321-8547. Weekly. $160.00 per year.

Physics Abstracts. INSPEC. Section A, Science Abstracts. Institution of Electrical Engineers (IEE), London, United Kingdom. Available from: INSPEC/IEEE - Institute of Electrical and Electronic Engineers, Box 1331, 445 Hoes Lane, Piscataway, NJ 08855-1331. (908) 562-5549. 1898 to present. 24 issues per year. $2835.00 per year. ISSN: 0036-8091. Also available online and on CD-ROM.

Physics Briefs (Physikalische Berichte). Information Center for Energy, Physics, Mathematics; German Physical Society. V C H Publishers, Inc., 220 East 23rd Street, New York, NY 10010-4606. (212) 683-8333. 1845 to present. 24 issues per year. $2390.00 per year. Also available online. ISSN: 0179-7434.

Science Citation Index. SCI. Institute for Scientific Information, 3501 Market Street, Philadelphia, PA 19104. (215) 386-0100. (800) 523-1850. FAX (215) 386-2991. 1961 to present. Six issues per year, plus annual cumulation. $11650.00 per year. Also available online and on CD-ROM. Inquire about price and availability. ISSN: 0036-827X.

ANNUAL REVIEWS AND YEARBOOKS

Advances in Nuclear Science and Technology. Plenum Publishing Corp., 233 Spring Street, New York, NY. (212) 620-8000. (800) 2221-9369. FAX (212) 463-0742. 1977 to present. Irregular.

ASSOCIATIONS AND PROFESSIONAL SOCIETIES

American Institute of Chemical Engineers. 345 East 47th Street, New York, NY 10017. (212) 705-7338.

American Institute of Physics. One Physics Ellipse, College Park, MD 20740-3843. (301) 209-3100.

American Nuclear Society. 555 North Kensington Avenue, LA Grange Park, IL 60525. (708) 352-6611. (800) NUC-NEWS.

Fusion Power Associates. 2 Professional Drive, Suite 248, Gaithersburg, MD 20879. (301) 258-0545. FAX (301) 975-9869.

Institute of Nuclear Materials Management. 60 Revere Drive, Northbrook, IL 60062. (847) 480-9573. FAX (947) 480-9282.

BIBLIOGRAPHIES

Handbooks and Tables in Science and Technology. Russell H. Powell, editor. Oryx Press, 4041 North Central, Suite 700, Phoenix, AZ 85012-3330. (602) 265-2651 or (800) 279-6799. Third edition. 1994. $65.00.

Information Sources in Engineering. Peter Hicks, editor. Bowker-Saur, 121 Chanlon Road, New Providence, NJ 07974. (800) 521-8110 or (908) 464-6800. 3rd edition. 1996. $125.00.

Scientific and Technical Books and Serials in Print; An Index To Literature in Science and Technology. R.R. Bowker Inc., 121 Chanlon Road, New Providence, NJ 07974. (908) 464-6800. (800) 521-8110. FAX (908) 665-3502. 1972 to present. Annual. 4 volumes. 1994. $299.95. Also available on compact disc and online. ISSN: 0000-054X.

DIRECTORIES AND BIOGRAPHICAL SOURCES

American Institute of Chemical Engineers. DIRECTORY. American Institute of Chemical Engineers. 345 East 47th Street, New York, NY 10017-2396. (212) 705-7663. Annual.

American Men and Women of Science. Physical and Biological Sciences. Fifteenth edition. R.R. Bowker Inc., 121 Chanlon Road, New Providence, NJ 07974. (908) 464-6800. (800) 521-8110. 20th edition. 8 volumes. 1996. $850.00

Energy and Nuclear Science International Who's Who. Gale Research, , 835 Penobscot Building, Detroit, MI 48226- 4094. (313) 961-2242. (800) 347-4253. Triennial. 1993. $325.00.

Energy Information Centers Directory. U. S. Council for Energy Awareness, 1776 I Street, NW, Suite 400, Washington, DC 20006. (202) 293-0770. Triennial. No charge.

Engineering Research Centers: Incorporating Electronics Research Centers. Stockton Press, 345 Park Avenue, New York, NY 10010. (212) 689-9200. (800) 221-2123. 4th edition. 1995. $515.00.

Nuclear News - Directory Issues. American Nuclear Society, 555 North Kensington Avenue, LA Grange Park, IL 60525. (708) 352-6611. (800) NUC-NEWS. March issue contains Buyer's Guide; $65.00. World List of Nuclear Power Plants. Semi-annual in February and August. $15.00 per issue.

Nuclear Reactors Built, Being Built OR Planned in the United States. National Technical Information Service, Springfield, VA 22161. (703) 487-4780. 1989. $17.00.

Research Centers Directory. Gale Research Company, Inc., 835 Penobscot Building, Detroit, MI 48226-4094. (313) 961-2242. (800) 347-4253. 20th edition. 1995. $485.00. ISSN: 0080-1518.

Who's Who in Engineering. Gordon Davis, editor. American Association of Engineering Societies. 1111 19th Street, NY, Suite 608, Washington, DC 20036. (202) 296-2237. (800) 658-8897. 9th edition. 1995. $220.00.

Who's Who in Technology. Gale Research, 835 Penobscot Building, Detroit, MI 48226-4094. (313) 961-2242. (800) 521-4253. 7th edition. 1995. $195.00. ISN 0-8103-7467-6.

World Energy and Nuclear Directory. Gale Research Company, Inc., 835 Penobscot Building, Detroit, MI 48226-4094. (313)

961-2242. (800) 347-4253. Biennial, even years. $450.00.

ENCYCLOPEDIAS AND DICTIONARIES

Dictionary of Named Processes in Chemical Technology. Alan E. Comyns. Oxford University Press, Inc., 200 Madison Avenue, New York, NY 10016. (212) 725-6000. (800) 334-4249. 1994. $75.00.

Dictionary of Nuclear Technology. Pro/Am Music Resources, Inc., 63 Prospect Street, White Plains, New York, 10606. (914) 948-7436. 1986. $245.00.

Encyclopedia of Applied Physics. VCH Publishers, Inc., 303 Northwest 12th Avenue, Deerfield Beach, FL 33442. (800) 367-8249. 1991-. Twenty volumes. $6000.00.

IEEE Nuclear Science Standards Collection. IEEE Standards office, Box 1331, 445 Hoes Lane, Piscataway, NJ 08855-1331. (908) 562-3822. (800) 678-4333. 1991. $90.00 in paper.

Kirk-Othmer Encyclopedia of Chemical Technology. John Wiley and Sons, Inc., 605 Third Avenue, New York, NY 10158. (800) 526-5368 or (212) 850-6000. Fourth edition. 1991 - . Twenty-seven volumes. $5400.00.

McGraw-Hill Encyclopedia of Science and Technology. McGraw-Hill Book, Incorporated, 1221 Avenue of the Americas, New York, NY 10020. (212) 997-3675. (800) 262-4729. Seventh edition. Twenty volumes. 1992. $1900.00.

Ullman's Encyclopedia of Industrial Chemistry. VCH Publications, Inc., 220 East 23rd Street, Suite 909, New York, NY 10010-4606. (212) 683-8333. (800) 422-8824. 5th edition. 1984 - . Price varies per volume. ISSN:

GENERAL WORKS

Atomic and Plasma Material Interaction Processes in Controlled Thermonuclear Fusion. R. K. Janev and H. W. Drawin, editors. Elsevier Science Publishing Company, Inc., 655 Avenue of the Americas, New York, NY 10010. (212) 989- 5800. 1993. $191.50.

Clean Fission. Richard A. Weiss. K & W Publications, P.O. Box 09121, Columbus, OH 43209-0121. (614) 898-2724. 1992. $48.00.

Controlled Nuclear Chain Reactions: The First Fifty Years. American Nuclear Society, 555 North Kensington Avenue, LA Grange Park, IL 60525. (708) 352-6611. (800) NUC-NEWS. 1992. $25.00.

The Discovery of Nuclear Fission. Hans G. Graetzerk and David L. anderson. Ayer Company Pubs., Inc., Lower Mill Road, North Stratford, NH 03590. (603) 922-5105. (800) 282-5413. 1981. $15.95.

Fiftieth Anniversary of Nuclear Fission. L. V. Drapchinsky, editor. Nova Science Publishers, Inc., 6080 Jericho Turnpike, Suite 207, Commack, NY 11725-2808. (516) 499- 3103. 1994. $145.00.

Internal Fission. Richard A. Weiss. K & W Publications, P.O. Box 09121, Columbus, OH 43209-0121. (614) 898-2724.1995. $50.00.

The Nuclear Fission Process. Cryiel Wagemans. CRC Press, Inc., 2000 Corporate Boulevard, NW, Boca Raton, FL

33431. (407) 994-0555. (800) 272-7737. 1991. $245.95.

Physics and Chemistry of Fission. H. Marten and E. Seelinger, editors. Nova Science Publishers, Inc., 6080 Jericho Turnpike, Suite 207, Commack, NY 11725-2808. (516) 499-3103. 1992. $160.00

Physics of Nuclear Fission. N. A. Perfilov and V. p. Eismont. Coronet Books, 311 Bainbridge Street, Philadelphia PA 19147. (215) 925-5083. 1964. $59.50.

The Search For Endless Energy. Robin Herman. Cambridge University Press, 40 West 20th Street, New York, NY 10011-4211. (212) 924-3900. (800) 872-7423. 1990. $24.95.

Legacy of Chernobyl. Zhores A. Medvedev. W. W. Norton & Company, 500 Fifth Avenue, New York, NY 10110. (212) 354-5500. (800) 223-2584. 1992. $10.95 in paper.

Managing Nuclear Accidents: A Model Emergency Response Plan For Power Plants and Communities. Dominic Golding, et al. Westview Press. 5500 Central Avenue, Boulder CO 80301-2847. (303) 444-3541. (800) 456-1995. 1

HANDBOOKS AND MANUALS

CRC Handbook of Chemistry and Physics. David R. Kide, editor. CRC Press Inc., 2000 Corporate Blvd., NW, Boca Raton, FL 33431. (407) 994-0555. (800) 333-8300. 77th edition. 1996. $99.95.

A Guide To Nuclear Power Technology: A Resource For Decision Making. Frank J. Rahn, et al. Krieger Publishing Company, P.O. Box 9542, Melbourne, FL 32902-9542. (407) 724-9542. Reprint editon. 1992. $145.00.

Information Center On Nuclear Standards. American Nuclear Society, 555 North Kensington Avenue, LA Grange Park, IL 60525. (708) 352-6611. Standards for all aspects of the design, construction, operation and maintenance of nuclear power plants.

Nuclear Engineering Databases, Standards and Numerical Analysis. Jack Jedruch. Van Nostrand Reinhold, 115 Fifth Avenue, New York, NY 10003. (212) 254-3232. (800) 842- 3636. 1985.

Nuclear Engineering Sourcebook: A Compilation of Documents For Nuclear Equipment Qualification. IEEE Standards office, Box 1331, 445 Hoes Lane, Piscataway, NJ 08855-1331. (908) 562-3822. (800) 678-4333. 1992.

Summary Description of Design Criteria, Codes, Standards, and Regulatory Provisions Typically Used For the Civil and Structural Design of Nuclear Fuel Cycle Facilities. American Society of Civil Engineers. 345 East 47th Street, New York, NY 10017. (212) 705-7496. (800) 548-2723. 1988. $17.00

ONLINE DATABASES AND CD-ROMS

CA Search. Chemical Abstracts Service, P.O. Box 3012, Columbus, OH 43210-0012. (614) 447-3600. (800) 848-6533. FAX (614) 447-3709. Very comprehensive guide to worldwide chemical literature and related fields. 1972 to present. Available on BRS,(800) 289-4277, DIALOG, (800) 334-2564, ORBIT (800) 456-7248, and STN International, FIZ Karlsruhe, P.O. Box 2465, W-7500, Karlsruhe 1, Germany, online services. Inquire as to cost and availability.

Compendex Plus. Engineering Information, Inc., 345 East 47th Street, New York, NY 10017. (212) 705-7600 or (800) 221-1044. Contains citations with abstracts to worldwide literature in engineering and technology, from 1970 to present. Available on online BRS,(800) 289-4277, DIALOG, (800) 334-2564, ORBIT (800) 456-7248 and STN International, FIZ Karlsruhe, P.O. Box 2465, W-7500, Karlsruhe 1, Germany, online services. Also available on CD-ROM. Inquire as to cost and availability.

Current Contents Search. Institute for Scientific Information, 3501 Market Street, Philadelphia, PA 19104. (215) 386-0100. FAX (215) 386-6362. Contains citations to articles listed in the table of contents of science and technology journals. Also articles in social sciences and life sciences journals. Available on BRS,(800) 289-4277, DIALOG,(800) 334-2564, online services. Inquire as to cost and availability.

Dissertation Abstracts. University Microfilms International, 300 North Zeeb Road, Ann Arbor, MI 48106. (800) 521-0600 or (313) 761-4700. Scope includes virtually all doctoral dissertations accepted at accredited American institutions from 1861 to present in 252 subject areas. Available on BRS, (800) 289-4277, DIALOG, (800) 334-2564, and OCLC EPIC, (800) 848-5878, online services. Also available on CD-ROM. Inquire as to cost and availability.

INSPEC. Institution of Electrical Engineers, Michael Faraday House, Six Hills Way, Stevenage, Herts. SG1 2AY, England. Telephone: 0438 313311 or FAX 0438 742840. Contains citations to the worldwide literature of physics, electronics and electrical engineering, computer technology, and related fields. Available on BRS, (800) 289-4277, DIALOG, (800) 334-2564, ORBIT, (800) 456-7248, and STN International, FIZ Karlsruhe, P.O. Box 2465, W-7500, Karlsruhe 1, Germany, online services. Inquire as to cost and availability.

Kirk-Othmer Encyclopedia of Chemical Technology. John Wiley and Sons, Inc., 605 Third Avenue, New York, NY 10158. (800) 526-5368 or (212) 850-6000. Contains the complete text of all chapters in the 27 volume fourth edition of the KIRK-Othmer ENCYCLOPEDIA of CHEMICAL TECHNOLOGY. 1991.

Available on BRS,(800) 289-4277, DIALOG,(800) 334-2564, online services. Inquire as to cost and availability.

NTIS Bibliographic Database. National Technical Information Service, 5285 Port Royal Road, Springfield, VA 22161. (703) 487-4929 or FAX (703) 321-8199. Broad coverage of government-sponsored science and technology research reports, 1964 to present. Available on BRS,(800) 289-4277, DIALOG, (800) 334-2564, ORBIT, (800) 456-7248, and STN International, FIZ Karlsruhe, P.O. Box 2465, W-7500, Karlsruhe 1, Germany, online services. Also available on CD-ROM. Inquire as to cost and availability.

Nuclear Science Abstracts. U.S. Department of Energy, office of Scientific and Technical Energy, P.O. Box 62, Oak Ridge, TN 37831. (615) 576-1189. Contains citations with abstracts to literature in all fields of nuclear science and energy. 1948 to 1976. Available on DIALOG,(800) 334-2564, online service. Inquire as to cost and availability.

Physics Briefs. American Institute of Physics, 335 East 45th Street, New York, NY 10017. (212) 661-9260 or FAX (212) 661-2036. Contains citations with abstracts of the literature of physics and related fields, 1979 to present. Available on the STN International, FIZ Karlsruhe, P.O. Box 2465, W-7500, Karlsruhe 1, Germany, online service. Inquire as to cost and availability.

SCISEARCH. Institute for Scientific Information, 3501 Market Street, Philadelphia, PA 19104. (800) 523-1850 or (215) 386-0100. Broad multidisciplinary title and author index to the international literature of science and technology, 1974 to present. Available on DIALOG, (800) 334-2564, and ORBIT, (800) 456-7248, online services. Also available on CD-ROM. Inquire as to cost and availability.

Scientific and Technical Books and Serials in Print. R.R. Bowker Inc., 121 Chanlon Road, New Providence, NJ 07974. (800) 521-8110 or (908) 464-6800. List of currently published books and serials in the physical and biological sciences, engineering and technology, with subject, author and titles indexes. Available on ORBIT, (800) 456-7248, online service. Inquire as to cost and availability.

WILSONLINE. H.W. Wilson Company, 950 University Avenue, Bronx, NY 10452. (800) 367-6770 or (212) 588-8400. Makes available online versions of the printed H.W. Wilson indexes including Applied Science and Technology Index, Business Periodicals Index, General Science Index, and Readers' Guide to Periodical Literature. Period covered is generally 1983 to present. Available on BRS, (800) 289-4277, DIALOG, (800) 334-2564, and OCLC EPIC, (800) 848-5878, online services. Also available on CD-ROM. Inquire as to cost and availability.

Who's Who in Technology. Gale Research Inc., 835 Penobscot Building, Detroit, MI 48226-4094. (313) 961-2242. (800) 347-4253. Contains biographical information of contemporary American scieNTISts and engineers. Available on DIALOG,(800) 334-2564, online service. Inquire as to cost and availability.

PERIODICALS

American Nuclear Society Transactions. American Nuclear Society, 555 North Kensington Avenue, LA Grange Park, IL 60525. (708) 352-6611. 1958 to present. 3 issues per year. $340.00 per year. ISSN: 0003-018X

Ency. of Physical Sciences and Engineering Info. Sources

FISSION

Annals of Nuclear Energy. Elsevier Science Publishing Company, Inc., 660 White Plains Road, Tarrytown, NY 10591- 5133. (914) 524-9200. 1974 to present. Monthly. $815.00 per year. ISSN: 0306-4549.

Bulletin of the Atomic Scientists. Educational Foundation for Nuclear Science, 6042 South Kimbark Avenue, Chicago, IL 60637. (312) 702-2555. 1945 to present. 10 issues per year. $30.00 per year. ISSN: 0096-3402.

Fusion Technology. American Nuclear Society, 555 North Kensington Avenue, LA Grange Park, IL 60525. (708) 352-6611. 1981 to present. 8 issues per year. $$485.00 per year. ISSN: 0748-1896.

Journal of Fusion Energy. Plenum Publishing Corporation, 233 Spring Street, New York, NY 10013- 1578. (212) 620- 8000. 1981 to present. Quarterly. $315.00 per year. ISSN: 0164-0313.

Nuclear Engineer. Institution of Nuclear Engineers, 1 Penerley Road, London, SE6 2LQ, England. 1959 to present. Bi-monthly. L77 per year. ISSN: 0262-5091.

Nuclear Engineering and Design. Elsevier Science Publishing Company, Inc., 660 White Plains Road, Tarrytown, NY 10591- 5133. (914) 524-9200. 1965 to present. 24 issues per year. $2748.00 per year.

Nuclear Engineering International. Reed Business Pulbishing Group, Quadrant House, the Quadrant, Sutton, Surrey, SM2 5AS, Englnad. 1956 to present. Monthly. $236.00 per year. ISSN: 0029-5507.

Nuclear Science and Engineering. American Nuclear Society, 555 North Kensington Avenue, LA Grange Park, IL 60525. (708) 352-6611. 1956 to present. Monthly. $405.00 per year. ISSN: 0029-5639.

Nuclear Technology. American Nuclear Society, 555 North Kensington Avenue, LA Grange Park, IL 60525. (708) 352-6611. 1965 to present. Monthly. $440.00 per year. ISSN: 0029-5450.

RESEARCH CENTERS AND INSTITUTES

Assistant Secretary for Nuclear Energy. U. S. Department of Energy, 1000 Independence Avenue, SW, Washington, DC 20585. (202) 252-6450.

Department of Nuclear Engineering and Engineering Physics. University of Wisconsin - Madison, 1500 Johnson Drive, Madison, WI 53706. (608) 263-1648. FAX (608) 262-6707.

Frank H. Neely Nuclear Research Center. Georgia Institute of Technology, 900 Atantic Drive NW, Atlanta, GA 30332. (404) 894-3600.

Institute of Nuclear Science and Engineering. Oregon State University Radiation Center- A100, 35th and Jefferson Streets, Corvallis, OR 97331-5903. (503) 737-2341.

FAX (503) 737-0480.

Radiation Science and Engineering Center. Pennsylvania State University, University park, PA 16802. (814) 865- 6351. FAX (814) 863-4840.

Whiteshell Laboratories. Research Company, Atomic Energy of Canada Limited, Pinawa, MB, Canada ROE ILO. (204) 753-2311.

FAX (204) 753-8404.

FISSION REACTORS

See: NUCLEAR REACTORS

FLAME EMISSION SPECTROSCOPY

See: SPECTROSCOPY

FLAME LASERS

See: LASERS

FLASH WELDING

See: WELDING

FLIGHT

See: AERONAUTICS

FLIGHT DYNAMICS

See: AERODYNAMICS

FLOOD CONTROL

See also: CHANNELS, CIVIL ENGINEERING, DAMS, GEOTECHNICAL ENGINEERING, HYDRAULIC ENGINEERING, HYDRAULICS

ABSTRACT SERVICES AND INDEXES

Applied Science and Technology Index; A Cumulative Subject Index To English Language Periodicals in the Fields of Aeronautics and Space Science, Computer Technology, Chemistry, Construction Industry, Energy and Related Areas. H.W. Wilson Co., 950 University Avenue, Bronx, NY 10452. (800) 367-6770 or (212) 588-8400. FAX (718) 590-1617. From 1958 to present. Monthly. Inquire about cost and availability. Also available on CD-ROM and online.

ASCE Combined Annual Combined Index. American Society of Civil Engineers, 345 East 47th Street, New York, NY 10017-2398. (212) 705-7520 or (800) 548-2723. Annual. $48.00.

FLOOD CONTROL

Ency. of Physical Sciences and Engineering Info. Sources

ASCE Publications Information. 1966-present. American Society of Civil Engineers, 345 East 47th Street, New York, NY 10017-2398. (212) 705-7520 or (800) 548-2723. Bi-monthly. $160.00 per year for non-members.

Civil and Structural Engineering Abstracts, Cambridge Scientific Abstracts, 7200 Wisconsin Avenue, Bethesda, MD 20814-4823. (301) 961-6750. FAX (301) 961-6720. 1993 to present. Monthly. $385.00 per year. Topics covered include structural design, construction equipment and methods, civil defense and military engineering, surveying, highway engineering, maritime and port structures, materials, land reclamation, and soil mechanics.

Engineering Index Monthly. Engineering Information, Inc., Castle Point on the Hudson, Hoboken, NJ 07030. (800) 221-1044. FAX (212) 832-1857. Monthly. $2300.00 per year. Also available online as COMPENDEX, and also on CD-ROM. Covers chemical engineering, computers, electrical engineering, civil engineering, metals and mining, industrial management, and mechanical engineering.

Environmental Abstracts. R.R. Bowker Inc., 121 Chanlon Road, New Providence, NJ 07974. (800) 521-8110 or (908) 464-6800. Contains citations, most with abstracts, to literature relating to environmental issues and problems, from 1971 to present. Inquire as to cost and availability.

Environmental Engineering Abstracts, Cambridge Scientific Abstracts, 7200 Wisconsin Avenue, Bethesda, MD 20814-4823. (301) 961-6750. FAX (301) 961-6720. Monthly. Covers hazardous materials, environmental impact and protection, treatment of sewage and industrial wastes, hydroelectric power, tidal and wind power, artic and tropical engineering.

Fluid Abstracts: Civil Engineering. Elsevier Science [journals], 660 White Plains Rd., Tarrytown, NY 10159-5153. (919) 524-9200. FAX (919) 333-2444. 1991 to present. Monthly plus annual cumulation. $645.00 per year. Also available online as FLUIDEX. Covers civil engineering applications of fluid mechanics, hydraulics, flow metering and measuring, offshore engineering, environmental hydraulics, and related aspects of wind energy, the atmosphere, and aerodynamics.

Hydro-abstracts (formerly Water Resources Abstracts). HydroScience Press, 2527 Jackson Street, N.E., Minneapolis, MN 55418. (612) 781-9081. Monthly. $150.00 per year.

Selected Water Resources Abstracts. U.S. Geological Survey, Water Resources Scientific Information Center, 425 National Center, Reston, VA 22092. (703) 648-6820. Monthly. $115.00 per year. Also available on CD-ROM.

ASSOCIATIONS AND PROFESSIONAL SOCIETIES

American Institute of Hydrology. 3416 University Avenue SE, Minneapolis, MN 55414-3328. (612) 379-1030.

American Society of Civil Engineers. 345 East 47th Street, New York, NY 10017-2398. (212) 705-7520 or (800) 548-2723.

Association of State Floodplain Managers. PO Box 2051, Madison, WI 53701-2051. (608) 266-1926. FAX (608) 264-9200.

National Association of Flood and Storm Water Management Agencies. 1225 Eye Street NW, Suite 300, Washington, DC 20005. (202) 682-3761. FAX (202) 842-0621.

U.S. Commission on Irrigation and Drainage. 1616 17th Street, No. 483, Denver, CO 80202. (303) 628-5430.

BIBLIOGRAPHIES

Flood Damage Prevention: Recent References. Mary Vance. Vance Bibliographies, PO Box 229, Monticello, IL 61856. 1990. Out of print.

DIRECTORIES AND BIOGRAPHICAL SOURCES

Directory of Engineering Societies and Related Organizations. Gordon Davis. 13th edition. American Association of Engineering Societies, 1111 19th Street NW, Suite 608, Washington, DC 20036. (202) 296-2237 or (800) 658-8897. 1989. Inquire for price.

Research Centers Directory. Gale Research, , 835 Penobscot Building, Detroit, MI 48226-4094. (313) 961-2242. (800) 347-4253. FAX (313) 961-6083. $485.00.

Scientific and Technical Organizations and Agencies Directory. Gale Research, , 835 Penobscot Building, Detroit, MI 48226-4094. (313) 961-2242. (800) 347-4253. FAX (313) 961-6083. 4th edition. 1996. $195.00.

Who's Who in Engineering. American Association of Engineering Societies, 1111 19th Street NW, Suite 608, Washington, DC 20036. (202) 296-2237 or (800) 658-8897. 8th edition. 1991. Inquire for price.

ENCYCLOPEDIAS AND DICTIONARIES

Dictionary of Civil Engineering. V.N. Vazirani. South Asia Books, Box 502, Columbia, MO 65205. (314) 474-0116. FAX (314) 474-8124. 1992. $20.00.

McGraw-Hill Encyclopedia of Engineering. Sybil P. Parker, ed. 2nd edition. McGraw-Hill Publishing Company, 1221 Avenue of the Americas, New York, NY 10020. (800) 262-4729 or (212) 512-3825. 1993. $95.50.

GENERAL WORKS

Flood Control & Drainage Engineering. S.N. Ghosh. South Asia Books, Box 502, Columbia, MO 65205. (314) 474-0116. FAX (314) 474-8124. $14.00.

The Flood Control Challenge: Past, Present, and Future. Howard Rosen & Martin Reuss, editors. Public Works Historical Society, Northwestern University Research Park, 1801 Maple Avenue, Chicago, IL 60201-3135. (708) 491-5829. 1988. $20.00.

Floodplain Management in the United States: An Assessment Report. Federal Interagency Floodplain Management Task Force. Federal Emergency Management Agency. Order from: Superintendent of Documents, U.S. Government Printing office, Box 371954, Pittsburgh, PA 15250-7954. (202) 783-3238. FAX (202) 512-2233. 1992. Inquire for cost and availability.

Hydraulic Design of Flood Control Channels: Engineering and Design. Department of the Army, Corps of Engineers, office of the Chief of Engineers. Order from: Superintendent of Docu-

Ency. of Physical Sciences and Engineering Info. Sources

FLOOD CONTROL

ments, U.S. Government Printing office, Box 371954, Pittsburgh, PA 15250-7954. (202) 783-3238. FAX (202) 512-2233. 1994. Inquire for cost and availability.

HANDBOOKS AND MANUALS

Civil Engineering Practice. Paul N. Cheremisinoff, et al. 5 volumes. Technomic Publishing Company, Inc., 851 New Holland Avenue, Box 3535, Lancaster, PA 17604. (717) 291-5609 or (800) 233-9936. FAX (717) 295-4538. 1987-88. $199.00.

Flood Hydrology Manual. Arthur G. Cudworth Jr. U.S. Department of the Interior, Bureau of Reclamation. Order from: Superintendent of Documents, U.S. Government Printing office, Box 371954, Pittsburgh, PA 15250-7954. (202) 783-3238. FAX (202) 512-2233. 1989. Inquire for cost and availability.

Flood Hydrology Manual: Planning, Design, Construction and Operation of Water Control Facilities. Gordon Press Publishers, PO Box 459, Bowling Green Station, New York, NY 10004. (718) 624-8419. 1992. $250.00.

Floodplain Management Handbook. Flood Loss Reduction Associates. U.S. Water Resources Council. Order from: Superintendent of Documents, U.S. Government Printing office, Box 371954, Pittsburgh, PA 15250-7954. (202) 783-3238. FAX (202) 512-2233. 1981. Inquire for cost and availability.

Hydraulic Design of Flood Control Channels: Engineering and Design. Department of the Army, Corps of Engineers, office of the Chief of Engineers. Order from: Superintendent of Documents, U.S. Government Printing office, Box 371954, Pittsburgh, PA 15250-7954. (202) 783-3238. FAX (202) 512-2233. 1994. Inquire for cost and availability.

Standard Handbook For Civil Engineers. Frederick S. Merritt, editor. 3rd edition. McGraw-Hill Publishing Company, 1221 Avenue of the Americas, New York, NY 10020. (800) 262-4729 or (212) 512-3825. 1983. Inquire for cost and availability.

ONLINE DATABASES AND CD-ROMS

Compendex Plus. Engineering Information, Inc., 345 East 47th Street, New York, NY 10017. (212) 705-7600 or (800) 221-1044. Contains citations with abstracts to worldwide literature in engineering and technology, from 1970 to present. Available on online BRS,(800) 289-4277, DIALOG, (800) 334-2564, ORBIT (800) 456-7248, and STN International, FIZ Karlsruhe, P.O. Box 2465, W-7500, Karlsruhe 1, Germany, online services. Also available on CD-ROM. Inquire as to cost and availability.

CSA Engineering. Cambridge Scientific Abstracts, 7200 Wisconsin Avenue, Suite 601, Bethesda, MD 20814. (301) 961-6750 or (800) 843-7751. Contains citations and abstracts of international periodicals and research literature covering all fields of engineering and science and technology,including computer and information science, electronics, mechanical engineering, solid state materials, 1981 to present. Available on BRS,(800) 289-4277, online service.Inquire as to cost and availability.

Current Contents Search. Institute for Scientific Information, 3501 Market Street, Philadelphia, PA 19104. (215) 386-0100. FAX (215) 386-6362. Contains citations to articles listed in the table of contents of science and technology journals. Also articles in social sciences and life sciences journals. Available on BRS,(800) 289-4277, DIALOG,(800) 334-2564, online services. Inquire as to cost and availability.

GEOREF: Bibliography and Index of Geology. American Geological Institute, 4220 King Street, Alexandria, VA 22302. (703) 379-2480. Fax (703) 379-7563. Monthly. Inquire for price and availability.

NTIS Bibliographic Database. National Technical Information Service, 5285 Port Royal Road, Springfield, VA 22161. (703) 487-4929 or FAX (703) 321-8199. Broad coverage of government-sponsored science and technology research reports, 1964 to present. Available on BRS,(800) 289-4277, DIALOG, (800) 334-2564, ORBIT, (800) 456-7248, and STN International, FIZ Karlsruhe, P.O. Box 2465, W-7500, Karlsruhe 1, Germany, online services. Also available on CD-ROM. Inquire as to cost and availability.

SCISEARCH. Institute for Scientific Information, 3501 Market Street, Philadelphia, PA 19104. (800) 523-1850 or (215) 386-0100. Broad multidisciplinary title and author index to the international literature of science and technology, 1974 to present. Available on DIALOG,(800) 334-2564, and ORBIT,(800) 456-7248, online services. Also available on CD-ROM. Inquire as to cost and availability.

WILSONLINE. H.W. Wilson Company, 950 University Avenue, Bronx, NY 10452. (800) 367-6770 or (212) 588-8400. Makes available online versions of the printed H.W. Wilson indexes including Applied Science and Technology Index, Business Periodicals Index, General Science Index, and Readers' Guide to Periodical Literature. Period covered is generally 1983 to present. Available on BRS,(800) 289-4277, DIALOG,(800) 334-2564, and OCLC EPIC,(800) 848-5878, online services. Also available on CD-ROM. Inquire as to cost and availability.

PERIODICALS

Civil Engineering. American Society of Civil Engineers, 345 East 47th Street, New York, NY 10017-2398. (212) 705-7520 or (800) 548-2723. 1930 to present. Monthly. $85.00 per year.

Journal of Construction Engineering and Management. American Society of Civil Engineers, Construction Division, 345 East 47th Street, New York, NY 10017-2398. (212) 705-7520 or (800) 548-2723. 1956 to present. Quarterly. $112.00 per year.

Journal of Geotechnical Engineering. American Society of Civil Engineers, Geotechnical Engineering Division, 345 East 47th Street, New York, NY 10017-2398. (212) 705-7520 or (800) 548-2723. 1956 to present. Monthly. $212.00 per year.

Journal of Hydraulic Engineering. American Society of Civil Engineers, Hydraulics Division, 345 East 47th Street, New York, NY 10017-2398. (212) 705-7520 or (800) 548-2723. 1956 to present. Monthly. $200.00 per year.

Journal of Irrigation and Drainage. American Society of Civil Engineers, Irrigation and Drainage Division, 345 East 47th Street, New York, NY 10017-2398. (212) 705-7520 or (800) 548-2723. 1956 to present. Bi-monthly. $136.00 per year.

Journal of Hydrology. Elsevier Science B.V., PO Box 945, Madison Square Station, New York, NY 10159-0945. (212) 633-3650. FAX (212) 633-3680. 1963 to present. 48 times a year. $2309.00 per year.

Journal of Surveying Engineering. American Society of Civil Engineers, Surveying Engineering Division, 345 East 47th Street,

FLOOD CONTROL

Ency. of Physical Sciences and Engineering Info. Sources

New York, NY 10017-2398. (212) 705-7520 or (800) 548-2723. 1956 to present. Quarterly. $76.00 per year.

Journal of Water Resources Planning and Management. American Society of Civil Engineers, Water Resources Planning and Management Division, 345 East 47th Street, New York, NY 10017-2398. (212) 705-7520 or (800) 548-2723. Bi-monthly. $112.00 per year.

Public Works. Public Works Journal Corporation, 200 S. Broad Street, Ridgewood, NJ 07451. (201) 445-5800. FAX (201) 445-5170. 1896 to present. 13 times a year. $45.00 per year.

RESEARCH CENTERS AND INSTITUTES

W.M. Keck Engineering Laboratory of Hydraulics and Water Resources. California Institute of Technology, 138-78, Pasadena, CA 91125. (818) 356-4404. (818) 356-2940.

FLOPPY DISKS

See: COMPUTER MEMORY and STORAGE

FLOW

See: FLUID MECHANICS

FLOW WELDING

See: WELDING

FLUID DYNAMICS

See also: AERODYNAMICS, FLUID MECHANICS, FLUIDS, HYDRAULIC ENGINEERING, HYDRAULICS, HYDRODYNAMICS

ABSTRACT SERVICES AND INDEXES

Applied Mechanics Reviews: An Assessment of World Literature in Engineering Sciences. American Society of Mechanical Engineers, 345 East 47th Street, New York, NY 10017. (212) 705-7703. 1948 to present. Monthly. $360.00 per year. ISSN: 0003-6900.

Applied Science and Technology Index; A Cumulative Subject Index To English Language Periodicals in the Fields of Aeronautics and Space Science, Computer Technology, Chemistry, Construction Industry, Energy and Related Areas. H.W. Wilson Co., 950 University Avenue, Bronx, NY 10452. (212) 588-8400. (800) 367-6770. FAX (718) 590-1617. From 1958 to present. Monthly. Inquire about cost and availability. Also available on CD-ROM and online. ISSN: 0003-6986.

Chemical Abstracts. Chemical Abstracts Service, 2540 Olentangy River Road, Box 3012, Columbus, OH 43210-0012. (614) 447-3600. (800) 848-6538. FAX (614) 447-3713. From 1907 to present. Weekly. $16800.00 per year. Also available on

CD-ROM and online. Inquire regarding cost and availability. ISSN:

Civil Engineering Hydraulics Abstracts. BHRA Fluid Engineering, Cranfield, Bedford MK43 OAJ, England. Distributed by Learned Information, Incorporated, 143 Old Marlton Pike, Medford NJ 08055. 1968 to present. Monthly.

Current Contents: Engineering, Technology, and Applied Sciences. Institute for Scientific Information, 3501 Market Street, Philadelphia, PA 19104. (215) 386-0100. FAX (215) 386-2291. From 1961 to present. Weekly. $442.00 per year. Also available on CD-ROM and online. Inquire regarding cost and availability. ISSN: 0095-7917.

Current Papers in Physics. Institute of Electrical Engineers, Michael Faraday House, Six Hill Way, Stevenage, Herts, SG1 2AY, England. Distributed by INSPEC/IEEE, Box 1331, 445 Hoes Lane, Piscataway, NJ 08855-1331. (908) 562-5549. 1966 to present. Fortnightly. $410.00 per year. ISSN: 0011-3786.

Engineering Index Monthly. Engineering Information, Inc., Castle Point on the Hudson, Hoboken, NJ 07030. (201) 216-8500. FAX (201) 216-8532. Monthly. $2200.00 per year. Also available online as COMPENDEX and on CD-ROM. ISSN: 0742-1974.

Fluid Power Abstracts. BHRA Fluid Engineering, Cranfield, Bedford MK43 OAJ, England. Distributed by Learned Information, Incorporated, 143 Old Marlton Pike, Medford NJ 08055. 1970 to present. Bimonthly.

General Science Index. H.W. Wilson Company, 950 University Avenue, Bronx, NY 10452. (212) 588-8400. (800) 367-6770. FAX (718) 590-1617. From 1978 to present. Ten issues per year; quarterly and annual cumulations. Service basis. Available on CD-ROM and online. Inquire regarding cost and availability. ISSN: 0162-1963.

Government Reports Announcements and Index. U. S. National Technical Information Service (NTIS), 5285 Port Royal Road, Springfield, VA 22161. (703) 487-4650. FAX (703) 321-8547. From 1968 to present. Annual. $630.00 per year. Also available online as NTIS BIBLIOGRAPHIC DATABASE and on CD-ROM. ISSN:

International Aerospace Abstracts. American Institute of Aeronautics and Astronautics. Technical Information Service, 555 West 57th Street, New York 10019. (212) 247- 6500. FAX (212) 582-4961. 1961 to present. Semi-monthly with annual index. $1295.00 per year. Also available online and on CD-ROM. ISSN:

ISMEC: Mechanical Engineering Abstracts. Cambridge Scientific Abstracts, 7200 Wisconsin Avenue, Suite 601, Bethesda, MD 20814. (301) 961-6750 or (800) 843-7751. Contains citations to the literature in mechanical engineering, industrial and production engineering, energy, power, mechanics, devices and related areas, from 1973 to present. Available on the DIALOG,(800) 334-2564, online service. Inquire as to cost and availability.

NTIS Alerts: Physics. U. S. National Technical Information Service. 5285 Port Royal Road, Springfield, VA 22161. (703) 487-4650. FAX (703) 321-8547. Weekly. $140.00 per year.

Physics Abstracts. INSPEC. Section A, Science Abstracts. Institution of Electrical Engineers (IEE). Available from INSPEC/IEEE - Institute of Electrical and Electronic Engineers,, Box

1331, 445 Hoes Lane, Piscaway NJ 08855-1331. (908) 562-5549. 1898 to present. 24 issues per year. $2835.00 per year. Also available on CD-ROM and online. ISSN: 0036-8091.

Physics Briefs (Physikalische Berichte). Information Center for Energy, Physics, Mathematics; German Physical Society. V C H Publishers, Inc., 220 East 23rd Street, New York, NY 10010-4606. (212) 683-8333. 1845 to present. 24 issues per year. $2390.00 per year. Also available online. ISSN: 0179-7434.

Science Citation Index. SCI. Institute for Scientific Information, 3501 Market Street, Philadelphia, PA 19104. (215) 386-0100. (800) 523-1850. FAX (215) 386-2991. 1961 to present. Six issues per year, plus annual cumulation. $11650.00 per year. Also available online and on CD-ROM. Inquire about price and availability. ISSN: 0036-827X.

ANNUAL REVIEWS AND YEARBOOKS

Advances in Applied Mechanics. Academic Press, Inc., 6277 Sea Harbor Drive, Orlando, FL. (800) 321-5068. 1948 to present. Irregular. ISSN: 0065-2156.

Annual Review of Fluid Mechanics. Annual Reviews, Inc., 4139 Camino Way, P.O. Box 10139, Palo Alto, CA 94303-0897. (800) 523-8635 or (415) 493-4400. 1969 to present. Annual. $47.00 per year. ISSN: 0066-4189.

Topics in Applied Physics. Springer-Verlag New York, Inc., 175 Fifth Avenue, New York, NY 10010. (212) 460-1500. (800) 777-4643. FAX (212) 473-6272. ISSN: 0303-4216.

ASSOCIATIONS AND PROFESSIONAL SOCIETIES

American Institute of Aeronautics and Astronautics. 370 L'Enfant Promenade SW, Washington, DC 20024. (202) 646-7400.FAX (202) 646-7508.

American Institute of Physics. One Physics Ellipse, College Park, MD 20740-3843. (301) 209-3100.

American Society of Civil Engineers. 1015 15th Street NW, Suite 600, Washington, DC 20005. (202) 789-2200.

American Society of Mechanical Engineers, 345 East 47th Street, New York, NY 10017. (212) 705-7722. FAX (212) 705- 7739.

Fluid Controls Institute. P.O. Box 9036, Morristown, NJ 07960. (201) 829-0990.

Fluid Power Society. 2433 North Mayfair Road, Suite 111, Milwaukee WI 5322i6. (414) 257-0910.

Hydraulic Institute. 9 Sylvan Way, Parsippany, NJ 07054-3802. (201) 267-9700. FAX (3\201 267-9055.

International Association For Hydraulics Research. 185 Rotterdamseweg, Box 177, Delft, the Netherlands.

National Fluid Power Association. 3333 North Mayfair Road, Suite 311, Milwaukee WI 53226. (414) 778-3344.

Society of Automotive Engineers, SAE Incorporated, 400 Commonwealth Drive, Warrendale, PA 15096-0001. (412) 776- 4841. FAX (412) 776-5760.

DIRECTORIES AND BIOGRAPHICAL SOURCES

American Institute of Aeronautics and Astronautics. Roster. 370 L'Efant Promenade SW, Washington, DC 20024. (202) 646-7400. FAX (202) 646-7508.

American Men and Women of Science: Physical and Biological Sciences. R. R. Bowker Inc., 121 Chanlon Road, New Providence, NJ 07974. (908) 464-6800. (800) 521-8110. 20th edition. 8 volumes. 1996. $850.00.

American Physical Society. MEMBERSHIP DIRECTORY BULLETIN ISSUE. American Physical Society, One Physics Ellipse, College Park, MD 20740-3843. (301) 209-3200. FAX (301) 209-0865. Biennial. $50.00.

American Society of Civil Engineers - official Register. 1015 15th Street NW, Suite 600, Washington, DC 20005. (202) 789-2200. Annual. No charge.

Automotive Engineering - SAE Membership Directory. SAE Incorporated, 400 Commonwealth Drive, Warrendale, PA 15096. (412) 776-4841. Annual. Included in member dues.

Directory of Physics and Astronomy Staff Members. American Institute of Physics, One Physics Ellipse, College Park, MD 20740-3843. (301) 209-3100. Annual. $45.00.

International Directory of Engineering Societies and Related Organizations. American Society for Engineering Education, 1818 N Street NW, Suite 600, Washington, DC 20036. (202) 331-3526. 15th edition. 1996. $185.00.

International Engineering Directory. American Consulting Engineers Council, 1015 15th Street, N.W. Suite 802, Washington, DC 20005-2670. (202) 347-7474. Annual. $10.00

Research Centers Directory. Gale Research Inc., 835 Penobscot Building, Detroit, MI 48226-4094. (313) 961-2242. (800) 347-4253. 20th edition. 1995. $485.00. ISSN: 0080- 1518.

Scientific and Technical Organizations and Agencies Directory. Gale Research, , 835 Penobscot Building, Detroit, MI 48226-4094. (313) 961-2242. (800) 347-4253. 4th edition. 1996. $195.00.

Who's Who in Engineering. Gordon Davis, editor. American Society for Engineering Education, 1111 19th Street, NW, Suite 608, Washington, DC 20036. 9th edition. 1995. $220.00.

Who's Who in Technology. Gale Research, 835 Penobscot Building, Detroit, MI 48226-4094. (313 961-2242. (800) 521-4253. 7th edition. 1995. $195.00. ISN 0-8103-7467-6.

ENCYCLOPEDIAS AND DICTIONARIES

Chambers Science and Technology Dictionary. Peter M.B. Walker, editor. Cambridge University Press, 40 West 20th Street, New York, NY 10011-4211. (212) 924-3900. 1988. $39.95.

A Concise Dictionary of Physics. Oxford University Press, Inc., 200 Madison Avenue, New York, NY 10016. (212) 725- 6000. (800) 334-4249. 1990. $10.95.

Encyclopedia of Applied Physics. George Trigg, editor. VCH Publications, Inc., 220 East 23rd Street, Suite 909, New York,

NY 10010-4606. (212) 683-8333. (800) 422-8824. 20 volume set. 1991-. $5990.00.

Encyclopedia of Fluid Mechanics. Nicholas P. Cheremisinoff, editor. Gulf Publishing Company, P.O. Box 208, Houston TX 77001. (713) 520-4444. 10 volumes plus supplements. 1986 - $1950.00

Encyclopedia of Physical Science and Technology. Academic Press, Inc., 6277 Sea Harbor Drive, Orlando, FL. (800) 321-5068. 2nd edition. 18 volume set. 1992. $2625.00.

McGraw-Hill Encyclopedia of Science and Technology. McGraw-Hill Book Company, Inc., 1221 Avenue of the Americas, New York, NY 10020. (212) 512-2000. (800) 262-4729. 7th edition. 20 volume set. 1992. $1900.00.

GENERAL WORKS

Astrophysical Fluid Dynamics. Eduardo Battaner. Cambridge University Press, 40 West 20th Street, New York, NY 10011-4211. (212) 924-3900. (800) 872-7423. 1994. $

Computational Fluid Dynamics: An Introduction. J. Von Wendt, editor. Springer-Verlag New York, Inc., 175 Fifth Avenue, New York, NY 10010. (212) 460-1500. (800) 777- 4643. FAX (212) 473-6272. 1992. $89.00.

Elementary Fluid Dynamics. D. J. Acheson. Oxford University Press, Inc., 200 Madison Avenue, New York, NY 10016. (212) 725-6000. (800) 334-4249. 1990. $32.50.

Fundamental Fluid Mechanics. Philip M. Gerhart, et al. Addison-Wesley Publishing Co., Inc., 1 Jacob Way, Reading, MA 01867. (617) 944-3700. (800) 447-2226.. 2nd edition. 1992. $68.95.

Introduction To Geophysical Fluid Dynamics. Benoit Cushman-Roisin. Prentice-Hall , 113 Sylvan Avenue, Route 9W, Englewood Cliffs, NJ 07632. (201) 592-2000. (800) 922-0579 1994. $67.00.

Modern Compressible Flow: With Historical Perspectives. John D. anderson. McGraw-Hill Publishing Company, Inc., 1221 Avenue of the Americas, New York, NY 10020. (212) 512- 2000. (800) 262-4729. 2nd edition. 1990.

Practical Fluid Mechanics For Engineers and Scientists. Nicholas P. Cheremisinoff. Technomic, 851 New Holland Avenue, Box 3535, Lancaster, PA 17604. (717) 291-5609. (800) 233-9936. 1990. $45.00.

Transport Processes in Bubbles, Drops and Particles. D. DeKee and R. P Chabra. Hemisphere Publishing Corp., 1900 Frost Road, Suite 101, Bristol PA 19007. (215) 785-5800. (800) 821-8312. 1991. $73.00

HANDBOOKS AND MANUALS

Applied Fluid Dynamics Handbook. Robert D. Blevins. Krieger Publishing Company, P.O. Box 9542, Melbourne, FL 32902-9542. (407) 724-9542. 1992. $83.50.

Engineers Guide To Fluid Flow. Nicholas P. Cheremisinoff. SciTech Pubs P.O. Box 86. Morganville, NJ 07751-0086. 1989. $39.95.

ONLINE DATABASES AND CD-ROMS

Current Contents Search. Institute for Scientific Information, 3501 Market Street, Philadelphia, PA 19104. (215) 386-0100. FAX (215) 386-6362. Contains citations to articles listed in the table of contents of science and technology journals. Also articles in social sciences and life sciences journals. Available on BRS, (800) 289-4277, DIALOG, (800) 334-2564, online services. Inquire as to cost and availability.

Compendex Plus. Engineering Information, Inc., 345 East 47th Street, New York, NY 10017. (212) 705-7600 or (800) 221-1044. Contains citations with abstracts to worldwide literature in engineering and technology, from 1970 to present. Available on online BRS,(800) 289-4277, DIALOG, (800) 334-2564, ORBIT (800) 456-7248, and STN International, FIZ Karlsruhe, P.O. Box 2465, W-7500, Karlsruhe 1, Germany, online services. Also available on CD-ROM. Inquire as to cost and availability.

Dissertation Abstracts Online. University Microfilms International, 300 North Zeeb Road, Ann Arbor, MI 48106. (800) 521-0600 or (313) 761-4700. Scope includes virtually all doctoral dissertations accepted at accredited American institutions from 1861 to present in 252 subject areas. Available on BRS, (800) 289-4277, DIALOG, (800) 334-2564,and OCLC EPIC, (800) 848-5878, online services. Also available on CD-ROM. Inquire as to cost and availability.

*ISMEC: Mechanical Engineering Abstracts.*Cambridge Scientific Abstracts, 7200 Wisconsin Avenue, Suite 601, Bethesda, MD 20814. (301) 961-6750 or (800) 843-7751. Contains citations to the literature in mechanical engineering, industrial and production engineering, energy, power, mechanics, devices and related areas, from 1973 to present. Available on the DIALOG,(800) 334-2564, online service. Inquire as to cost and availability.

INSPEC. Institution of Electrical Engineers, Michael Faraday House, Six Hills Way, Stevenage, Herts. SG1 2AY, England. Telephone: 0438 313311 or FAX 0438 742840. Contains citations to the worldwide literature of physics, electronics and electrical engineering, computer technology, and related fields. Available on BRS, (800) 289-4277, DIALOG, (800) 334-2564, ORBIT, (800) 456-7248, and STN International, FIZ Karlsruhe, P.O. Box 2465, W-7500, Karlsruhe 1, Germany, online services. Inquire as to cost and availability.

NTIS Bibliographic Database. National Technical Information Service, 5285 Port Royal Road, Springfield, VA 22161. (703) 487-4929 or FAX (703) 321-8199. Broad coverage of government-sponsored science and technology research reports, 1964 to present. Available on BRS,(800) 289-4277, DIALOG, (800) 334-2564, ORBIT, (800) 456-7248, and STN International, FIZ Karlsruhe, P.O. Box 2465, W-7500, Karlsruhe 1, Germany, online services. Also available on CD-ROM. Inquire as to cost and availability.

Physics Briefs. American Institute of Physics, 335 East 45th Street, New York, NY 10017. (212) 661-9260 or FAX (212) 661-2036. Contains citations with abstracts of the literature of physics and related fields, 1979 to present. Available on the STN International, FIZ Karlsruhe, P.O. Box 2465, W-7500, Karlsruhe 1, Germany, online service. Inquire as to cost and availability.

SCISEARCH. Institute for Scientific Information, 3501 Market Street, Philadelphia, PA 19104. (800) 523-1850 or (215) 386-0100. Broad multidisciplinary title and author index to the international literature of science and technology, 1974 to present. Available on DIALOG, (800) 334-2564, and ORBIT, (800) 456-

7248, online services. Also available on CD-ROM. Inquire as to cost and availability.

WILSONLINE. H.W. Wilson Company, 950 University Avenue, Bronx, NY 10452. (800) 367-6770 or (212) 588-8400. Makes available online versions of the printed H.W. Wilson indexes including Applied Science and Technology Index, Business Periodicals Index, General Science Index, and Readers' Guide to Periodical Literature. Period covered is generally 1983 to present. Available on BRS, (800) 289-4277, DIALOG, (800) 334-2564, and OCLC EPIC, (800) 848-5878, online services.Also available on CD-ROM. Inquire as to cost and availability.

PERIODICALS

American Journal of Physics. American Association of Physics Teachers, 5112 Berywn Road, College Park, MD 20740. (301) 345-4200. 1933 to present. Monthly. $215.00 per year. ISSN: 0002-9505.

Archive For Rational Mechanics and Analysis. Springer-Verlag New York, 44 Hartz Way, Secaucus, NJ 07096-2491. (201) 348-4033. 1957 to present. 16 issues per year in 4 volumes. $1238.00. ISSN: 0003-9527.

Experiments in Fluids. Springer-Verlag New York, 44 Hartz Way, Secaucus, NJ 07096-2491. (201) 348-4033. 1983 to present. Monthly. $674.00 per year. ISSN: 0723-4864.

Fluid Mechanics Research. Scripta Publishing Compnay; distributed by John Wiley & Sons, 650 Third Avenue, New York, NY 10158. (212) 852-6000. 1972 to present. Bi-monthly. $956.00 per year. ISSN: 1064-2277.

Hydraulics and Pneumatics. Penton Publishing, 1100 Superior Avenue, Cleveland, OH 44114-2543. (216) 696-7000. 1947 to present. Monthly. $45.00 per year. ISSN: 0018-814X.

Journal of Fluid Control (fluidics Quarterly). Delbridge Publishing Co., Box 160817, Cupertino, CA 95016. (408) 446-3131. 1967 to present. Quarterly. $145.00 per year. ISSN: 8755-8564.

Journal of Fluid Mechanics. Cambridge University Press, 40 West 20th Street, New York, NY 10011. (212) 924-3900. 1956 to present. 24 issues per year. $595.00 per year. ISSN: 0022-1120.

Journal of Fluids Engineering. American Society of Mechanical Engineers, 345 East 47th Street, New York, NY 10017. (212) 705-7722. FAX (212) 705-7739. 1919 to present. Quarterly. $100.00. ISSN: 0098-2202.

Journal of Hydraulic Engineering. American Society of Civil Engineers, 345 East 47th Street, New York, NY 10017. (212) 705-7288. 1956 to present. Monthly. $200.00 per year. ISSN: 0733-9429.

RESEARCH CENTERS AND INSTITUTES

Aeronautical Laboratory. University of Washington, FS-10, Seattle, WA 98105. (206) 543-0439. FAX (206) 543-0217.

Computational Fluid Mechanics Laboratory. University of Arizona, Building #16, Room 312, Tucson, AZ 85721. (602) 621-4423.

Environmental Fluid Mechanics Laboratory. Stanford University. Department of Civil Engineering, Stanford, CA 94305-4020. (415) 723-1825. FAX (415) 725-8662.

Fluid Mechanics Laboratory. Purdue University, School of Mechanical Engineering, West Lafayette, in 47907. (317) 494-5633.

Fluid Properties Research, Inc., School of Chemical Engineering, Georgia Institute of Technology, Atlanta, GA 30332. (405) 894-3098.

Institute of Hydraulic Research. University of Iowa. Iowa City, IA 52242. (319) 353-5236. FAX (319) 335-5238.

Institute of thermo-fluid Engineering and Science. LeHigh University, Building A, Mountain Top Campus, Bethleham, PA 18015. (215) 758-4091. FAX (215) 758-5057.

FLUID MECHANICS

See also: AERODYNAMICS, FLUID DYNAMICS, HYDRAULIC ENGINEERING, HYDRAULICS, HYDRODYNAMICS

ABSTRACT SERVICES AND INDEXES

Applied Mechanics Reviews: An Assessment of World Literature in Engineering Sciences. American Society of Mechanical Engineers, 345 East 47th Street, New York, NY 10017. (212) 705-7703. 1948 to present. Monthly. $360.00 per year.

ISSN: 0003-6900.

Applied Science and Technology Index; A Cumulative Subject Index To English Language Periodicals in the Fields of Aeronautics and Space Science, Computer Technology, Chemistry, Construction Industry, Energy and Related Areas. H.W. Wilson Co., 950 University Avenue, Bronx, NY 10452. (212) 588-8400. (800) 367-6770. FAX (718) 590-1617. From 1958 to present. Monthly. Inquire about cost and availability. Also available on CD-ROM and online. ISSN: 0003-6986.

Chemical Abstracts. Chemical Abstracts Service, 2540 Olentangy River Road, Box 3012, Columbus, OH 43210-0012. (614) 447-3600. (800) 848-6538. FAX (614) 447-3713.From 1907 to present. Weekly. $16800.00 per year. Also available on CD-ROM and online. Inquire regarding cost and availability. ISSN:

Civil Engineering Hydraulics Abstracts. BHRA Fluid Engineering, Cranfield, Bedford MK43 OAJ, England. Distributed by Learned Information, Incorporated, 143 Old Marlton Pike, Medford NJ 08055. 1968 to present. Monthly.

Current Contents: Engineering, Technology, and Applied Sciences. Institute for Scientific Information, 3501 Market Street, Philadelphia, PA 19104. (215) 386-0100. FAX (215) 386-2291. From 1961 to present. Weekly. $442.00 per year. Also available on CD-ROM and online. Inquire regarding cost and availability. ISSN: 0095-7917.

Current Papers in Physics. Institute of Electrical Engineers, Michael Faraday House, Six Hill Way, Stevenage, Herts, SG1 2AY, England. Distributed by INSPEC/IEEE, Box 1331, 445

Hoes Lane, Piscataway, NJ 08855-1331. (908) 562- 5549. 1966 to present. Fortnightly. $410.00 per year. ISSN: 0011-3786.

Engineering Index Monthly. Engineering Information, Inc., Castle Point on the Hudson, Hoboken, NJ 07030. (201) 216-8500. FAX (201) 216-8532. Monthly. $2200.00 per year. Also available online as COMPENDEX and on CD-ROM. ISSN: 0742-1974.

Fluid Power Abstracts. BHRA Fluid Engineering, Cranfield, Bedford MK43 OAJ, England. Distributed by Learned Information, Incorporated, 143 Old Marlton Pike, Medford NJ 08055. 1970 to present. Bimonthly.

General Science Index. H.W. Wilson Company, 950 University Avenue, Bronx, NY 10452. (212) 588-8400. (800) 367-6770. FAX (718) 590-1617. From 1978 to present. Ten issues per year; quarterly and annual cumulations. Service basis. Available on CD-ROM and online. Inquire regarding cost and availability. ISSN: 0162-1963.

Government Reports Announcements and Index. U. S. National Technical Information Service (NTIS), 5285 Port Royal Road, Springfield, VA 22161. (703) 487-4650. FAX (703) 321-8547. From 1968 to present. Annual. $630.00 per year. Also available online as NTIS BIBLIOGRAPHIC DATABASE and on CD-ROM. ISSN:

International Aerospace Abstracts. American Institute of Aeronautics and Astronautics. Technical Information Service, 555 West 57th Street, New York 10019. (212) 247- 6500. FAX (212) 582-4961. 1961 to present. Semi-monthlywith annual index. $1295.00 per year. Also available online and on CD-ROM. ISSN:

*ISMEC: Mechanical Engineering Abstracts.*Cambridge Scientific Abstracts, 7200 Wisconsin Avenue, Suite 601, Bethesda, MD 20814. (301) 961-6750 or (800) 843-7751. Contains citations to the literature in mechanical engineering, industrial and production engineering, energy, power, mechanics, devices and related areas, from 1973 to present. Available on the DIALOG,(800) 334-2564, online service. Inquire as to cost and availability.

Physics Abstracts. INSPEC. Section A, Science Abstracts. Institution of Electrical Engineers (IEE). Available from INSPEC/IEEE - Institute of Electrical and Electronic Engineers,, Box 1331, 445 Hoes Lane, Piscaway NJ 08855-1331. (908) 562-5549. 1898 to present. 24 issues per year. $2835.00 per year. Also available on CD-ROM and online. ISSN: 0036-8091.

Science Citation Index. SCI. Institute for Scientific Information, 3501 Market Street, Philadelphia, PA 19104. (215) 386-0100. (800) 523-1850. FAX (215) 386-2991. 1961 to present. Six issues per year, plus annual cumulation. $11650.00 per year. Also available online and on CD-ROM. Inquire about price and availability. ISSN: 0036-827X.

ANNUAL REVIEWS AND YEARBOOKS

Advances in Applied Mechanics. Academic Press, Inc., 6277 Sea Harbor Drive, Orlando, FL. (800) 321-5068. 1948 to present. Irregular. ISSN: 0065-2156.

Annual Review of Fluid Mechanics. Annual Reviews, Inc., 4139 Camino Way, P.O. Box 10139, Palo Alto, CA 94303-0897. (800) 523-8635 or (415) 493-4400. 1969 to present. Annual. $47.00 per year. ISSN: 0066-4189.

Topics in Applied Physics. Springer-Verlag New York, Inc., 175 Fifth Avenue, New York, NY 10010. (212) 460-1500. (800) 777-4643. FAX (212) 473-6272. ISSN: 0303-4216.

ASSOCIATIONS AND PROFESSIONAL SOCIETIES

American Institute of Aeronautics and Astronautics. 370 L'Enfant Promenade SW, Washington, DC 20024. (202) 646-7400. FAX (202) 646-7508.

American Institute of Physics. One Physics Ellipse, College Park, MD 20740-3843. (301) 209-3100.

American Society of Civil Engineers. 1015 15th Street NW, Suite 600, Washington, DC 20005. (202) 789-2200.

American Society of Mechanical Engineers, 345 East 47th Street, New York, NY 10017. (212) 705-7722. FAX (212) 705-7739.

Fluid Controls Institute. P.O. Box 9036, Morristown, NJ 07960. (201) 829-0990.

Fluid Power Society. 2433 North Mayfair Road, Suite 111, Milwaukee WI 5322i6. (414) 257-0910.

Hydraulic Institute. 9 Sylvan Way, Parsippany, NJ 07054-3802. (201) 267-9700. FAX (201) 267-9055.

International Association For Hydraulics Research. 185 Rotterdamseweg, Box 177, Delft, the Netherlands.

National Fluid Power Association. 3333 North Mayfair Road, Suite 311, Milwaukee WI 53226. (414) 778-3344.

Society of Automtive Engineers, SAE Incorporated, 400 Commonwealth Drive, Warrendale, PA 15096-0001. (412) 776- 4841. FAX (412) 776-5760.

DIRECTORIES AND BIOGRAPHICAL SOURCES

American Men and Women of Science: Physical and Biological Sciences. R. R. Bowker Inc., 121 Chanlon Road, New Providence, NJ 07974. (908) 464-6800. (800) 521-8110. 20th edition. 8 volumes. 1996. $850.00.

American Physical Society. MEMBERSHIP DIRECTORY BULLETIN ISSUE. American Physical Society, One Physics Ellipse, College Park, MD 20740-3843. (301) 209-3200. FAX (301) 209-0865. Biennial. $50.00.

American Society of Civil Engineers - official Register. 1015 15th Street NW, Suite 600, Washington, DC 20005. (202) 789-2200. Annual. No charge.

Automotive Engineering - SAE Membership Directory. SAE Incorporated, 400 Commonwealth Drive, Warrendale, PA 15096. (412) 776-4841. Annual. Included in member dues.

Directory of Physics and Astronomy Staff Members. American Institute of Physics, One Physics Ellipse, College Park, MD 20740-3843. (301) 209-3100. Annual. $45.00.

International Directory of Engineering Societies and Related Organizations. American Society for Engineering Education,

1818 N Street NW, Suite 600, Washington, DC 20036. (202) 331-3526. 15th edition. 1996. $185.00.

International Engineering Directory. American Consulting Engineers Council, 1015 15th Street, N.W. Suite 802, Washington, DC 20005-2670. (202) 347-7474. Annual. $10.00

Research Centers Directory. Gale Research Inc., 835 Penobscot Building, Detroit, MI 48226-4094. (313) 961-2242. (800) 347-4253. 20th edition. 1995. $485.00. ISSN: 0080- 1518.

Scientific and Technical Organizations and Agencies Directory. Gale Research, , 835 Penobscot Building, Detroit, MI 48226-4094. (313) 961-2242. (800) 347-4253. 4th edition. 1996. $195.00.

Who's Who in Engineering. Gordon Davis, editor. American Society for Engineering Education, 1111 19th Street, NW, Suite 608, Washington, DC 20036. 9th edition. 1995. $220.00.

Who's Who in Technology. Gale Research, 835 Penobscot Building, Detroit, MI 48226-4094. (313 961-2242. (800)521-4253. 7th edition. 1995. $195.00. ISN 0-8103-7467-6.

ENCYCLOPEDIAS AND DICTIONARIES

Chambers Science and Technology Dictionary. Peter M.B. Walker, editor. Cambridge University Press, 40 West 20th Street, New York, NY 10011-4211. (212) 924-3900. 1988.$39.95.

A Concise Dictionary of Physics. Oxford University Press, Inc., 200 Madison Avenue, New York, NY 10016. (212) 725- 6000. (800) 334-4249. 1990. $10.95.

Encyclopedia of Applied Physics. George Trigg, editor. VCH Publications, Inc., 220 East 23rd Street, Suite 909, New York, NY 10010-4606. (212) 683-8333. (800) 422-8824. 20 volume set. 1991-. $5990.00.

Encyclopedia of Fluid Mechanics. Nicholas P. Cheremisinoff, editor. Gulf Publishing Company, P.O. Box 208, Houston TX 77001. (713) 520-4444. 10 volumes plus supplements. 1986 - $1950.00.

Encyclopedia of Physical Science and Technology. Academic Press, Inc., 6277 Sea Harbor Drive, Orlando, FL. (800) 321-5068. 2nd edition. 18 volume set. 1992. $2625.00.

McGraw-Hill Encyclopedia of Science and Technology. McGraw-Hill Book Company, Inc., 1221 Avenue of the Americas, New York, NY 10020. (212) 512-2000. (800) 262-4729. 7th edition. 20 volume set. 1992. $1900.00.

GENERAL WORKS

Astrophysical Fluid Dynamics. Eduardo Battaner. Cambridge University Press, 40 West 20th Street, New York, NY 10011-4211. (212) 924-3900. (800) 872-7423. 1994. $

Computational Fluid Dynamics: An Introduction. J. Von Wendt, editor. Springer-Verlag New York, Inc., 175 Fifth Avenue, New York, NY 10010. (212) 460-1500. (800) 777- 4643. FAX (212) 473-6272. 1992. $89.00.

Fluid Mechanics For Engineering Technology. Irving Granet. Prentice-Hall , 113 Sylvan Avenue, Route 9W, Englewood Cliffs, NJ 07632. (201) 592-2000. (800) 922-0579 4th edition. 1995. $67.00.

Fundamental Fluid Mechanics. Philip M. Gerhart, et al. Addison-Wesley Publishing Co., Inc., 1 Jacob Way, Reading, MA 01867. (617) 944-3700. (800) 447-2226.. 2nd edition. 1992. $68.95.

Introduction To Geophysical Fluid Dynamics. Benoit Cushman-Roisin. Prentice-Hall , 113 Sylvan Avenue, Route 9W, Englewood Cliffs, NJ 07632. (201) 592-2000. (800) 922-0579 1994. $67.00.

Introduction To Fluid Mechanics. James A. Fay. MIT Press, 55 Hayward Street, Cambridge, MA 02142. (617) 253-8569. 1994. $49.95.

Modern Compressible Flow: With Historical Perspectives. John D. anderson. McGraw-Hill Publishing Company, Inc., 1221 Avenue of the Americas, New York, NY 10020. (212) 512- 2000. (800) 262-4729. 2nd edition. 1990.

Practical Fluid Mechanics For Engineers and Scientists. Nicholas P. Cheremisinoff. Technomic, 851 New Holland Avenue, Box 3535, Lancaster, PA 17604. (717) 291-5609. (800) 233-9936. 1990. $45.00.

HANDBOOKS AND MANUALS

Applied Fluid Dynamics Handbook. Robert D. Blevins. Krieger Publishing Company, P.O. Box 9542, Melbourne, FL 32902-9542. (407) 724-9542. 1992. $83.50.

Engineers Guide To Fluid Flow. Nicholas P. Cheremisinoff. SciTech Pubs P.O. Box 86. Morganville, NJ 07751-0086. 1989. $39.95.

Handbook of Fluid Dynamics and Fluid Machinery. Allen E. Fuhs and Joseph A. Schetz. John Wiley & Sons, Inc., 605 Third Avenue, New York, NY 10158-0012. (212) 850-6000. (800) 225-5945. 1995.

ONLINE DATABASES AND CD-ROMS

Current Contents Search. Institute for Scientific Information, 3501 Market Street, Philadelphia, PA 19104. (215) 386-0100. FAX (215) 386-6362. Contains citations to articles listed in the table of contents of science and technology journals. Also articles in social sciences and life sciences journals. Available on BRS, (800) 289-4277, DIALOG, (800) 334-2564, online services. Inquire as to cost and availability.

Compendex Plus. Engineering Information, Inc., 345 East 47th Street, New York, NY 10017. (212) 705-7600 or (800) 221-1044. Contains citations with abstracts to worldwide literature in engineering and technology, from 1970 to present. Available on online BRS,(800) 289-4277, DIALOG, (800) 334-2564, ORBIT (800) 456-7248, and STN International, FIZ Karlsruhe, P.O. Box 2465, W-7500, Karlsruhe 1, Germany, online services. Also available on CD-ROM. Inquire as to cost and availability.

Dissertation Abstracts Online. University Microfilms International, 300 North Zeeb Road, Ann Arbor, MI 48106. (800) 521-0600 or (313) 761-4700. Scope includes virtually all doctoral dissertations accepted at accredited American institutions from

1861 to present in 252 subject areas. Available on BRS, (800) 289-4277, DIALOG, (800) 334-2564, and OCLC EPIC, (800) 848-5878, online services. Also available on CD-ROM. Inquire as to cost and availability.

*ISMEC: Mechanical Engineering Abstracts.*Cambridge Scientific Abstracts, 7200 Wisconsin Avenue, Suite 601, Bethesda, MD 20814. (301) 961-6750 or (800) 843-7751. Contains citations to the literature in mechanical engineering, industrial and production engineering, energy, power, mechanics, devices and related areas, from 1973 to present. Available on the DIALOG,(800) 334-2564, online service. Inquire as to cost and availability.

INSPEC. Institution of Electrical Engineers, Michael Faraday House, Six Hills Way, Stevenage, Herts. SG1 2AY, England. Telephone: 0438 313311 or FAX 0438 742840. Contains citations to the worldwide literature of physics, electronics and electrical engineering, computer technology, and related fields. Available on BRS, (800) 289-4277, DIALOG, (800) 334-2564, ORBIT, (800) 456-7248, and STN International, FIZ Karlsruhe, P.O. Box 2465, W-7500, Karlsruhe 1, Germany, online services. Inquire as to cost and availability.

NTIS Bibliographic Database. National Technical Information Service, 5285 Port Royal Road, Springfield, VA 22161. (703) 487-4929 or FAX (703) 321-8199. Broad coverage of government-sponsored science and technology research reports, 1964 to present. Available on BRS,(800) 289-4277, DIALOG, (800) 334-2564, ORBIT, (800) 456-7248, and STN International, FIZ Karlsruhe, P.O. Box 2465, W-7500, Karlsruhe 1, Germany, online services. Also available on CD-ROM. Inquire as to cost and availability.

Physics Briefs. American Institute of Physics, 335 East 45th Street, New York, NY 10017. (212) 661-9260 or FAX (212) 661-2036. Contains citations with abstracts of the literature of physics and related fields, 1979 to present. Available on the STN International, FIZ Karlsruhe, P.O. Box 2465, W-7500, Karlsruhe 1, Germany, online service. Inquire as to cost and availability.

SCISEARCH. Institute for Scientific Information, 3501 Market Street, Philadelphia, PA 19104. (800) 523-1850 or (215) 386-0100. Broad multidisciplinary title and author index to the international literature of science and technology, 1974 to present. Available on DIALOG, (800) 334-2564, and ORBIT, (800) 456-7248, online services. Also available on CD-ROM. Inquire as to cost and availability.

WILSONLINE. H.W. Wilson Company, 950 University Avenue, Bronx, NY 10452. (800) 367-6770 or (212) 588-8400. Makes available online versions of the printed H.W. Wilson indexes including Applied Science and Technology Index, Business Periodicals Index, General Science Index, and Readers' Guide to Periodical Literature. Period covered is generally 1983 to present. Available on BRS, (800) 289-4277, DIALOG, (800) 334-2564, and OCLC EPIC, (800) 848-5878, online services.Also available on CD-ROM. Inquire as to cost and availability.

PERIODICALS

Archive for Rational Mechanics and Analysis. Springer-Verlag New York, 44 Hartz Way, Secaucus, NJ 07096-2491. (201) 348-4033. 1957 to present. 16 issues per year in 4 volumes. $1238.00. ISSN: 0003-9527.

Experiments in Fluids. Springer-Verlag New York, 44 Hartz Way, Secaucus, NJ 07096-2491. (201) 348-4033. 1983 to present. Monthly. $674.00 per year. ISSN: 0723-4864.

Fluid Mechanics Research. Scripta Publishing Compnay; distributed by John Wiley & Sons, 650 Third Avenue, New York, NY 10158. (212) 852-6000. 1972 to present. Bi-monthly. $956.00 per year. ISSN: 1064-2277.

Hydraulics and Pneumatics. Penton Publishing, 1100 Superior Avenue, Cleveland, OH 44114-2543. (216) 696-7000. 1947 to present. Monthly. $45.00 per year. ISSN: 0018-814X.

Journal of Fluid Control (fluidics Quarterly). Delbridge Publishing Co., Box 160817, Cupertino, CA 95016. (408) 446-3131. 1967 to present. Quarterly. $145.00 per year. ISSN: 8755-8564.

Journal of Fluid Mechanics. Cambridge University Press, 40 West 20th Street, New York, NY 10011. (212) 924-3900. 1956 to present. 24 issues per year. $595.00 per year. ISSN: 0022-1120.

Journal of Fluids Engineering. American Society of Mechanical Engineers, 345 East 47th Street, New York, NY 10017. (212) 705-7722. FAX (212) 705-7739. 1919 to present. Quarterly. $100.00. ISSN: 0098-2202.

Journal of Hydraulic Engineering. American Society of Civil Engineers, 345 East 47th Street, New York, NY 10017. (212) 705-7288. 1956 to present. Monthly. $200.00 per year. ISSN: 0733-9429.

RESEARCH CENTERS AND INSTITUTES

Aeronautical Laboratory. University of Washington, FS-10, Seattle, WA 98105. (206) 543-0439. FAX (206) 543-0217.

Computational Fluid Mechanics Laboratory. University of Arizona, Building #16, Room 312, Tucson, AZ 85721. (602) 621-4423.

Environmental Fluid Mechanics Laboratory. Stanford University. Department of Civil Engineering, Stanford, CA 94305-4020. (415) 723-1825. FAX (415) 725-8662.

Fluid Mechanics Laboratory. Purdue University, School of Mechanical Engineering, West Lafayette, in 47907. (317) 494-5633.

Fluid Power Laboratory. Ohio State University, Mechanical Engineering Department, 206 West 18th Avenue, Columbus, OH 43210. (614) 292-9044. FAX (614) 292-3163.

Fluid Power Research Center. Oklahoma State University, 1724 West Tyler, Stillwater, OK 74078. (405) 744-7375.

Institute of Hydraulic Research. University of Iowa. Iowa City, IA 52242. (319) 353-5236. FAX (319) 335-5238.

Institute of thermo-fluid Engineering and Science. LeHigh University, Building A, Mountain Top Campus, Bethleham, PA 18015. (215) 758-4091. FAX (215) 758-5057.

FLUID STATISTICS

See: FLUID MECHANICS

FLUIDS ENGINEERING

See also: FLUID DYNAMICS, FLUID MECHANICS, HYDRAULIC
ENGINEERING, HYDRAULICS, HYDRODYNAMICS

ABSTRACT SERVICES AND INDEXES

Applied Mechanics Reviews: An Assessment of World Literature in Engineering Sciences. 1948-present. American Society of Mechanical Engineers, 345 East 47th Street, New York, NY 10017. (212) 705-7703. Monthly. $360.00 per year.

Applied Science and Technology Index; A Cumulative Subject Index To English Language Periodicals in the Fields of Aeronautics and Space Science, Computer Technology, Chemistry, Construction Industry, Energy and Related Areas. H.W. Wilson Co., 950 University Avenue, Bronx, NY 10452. (800) 367-6770 or (212) 588-8400. FAX (718) 590-1617. From 1958 to present. Monthly. Inquire about cost and availability. Also available on CD-ROM and online.

ASCE Combined Annual Combined Index. American Society of Civil Engineers, 345 East 47th Street, New York, NY 10017-2398. (212) 705-7520 or (800) 548-2723. Annual. $48.00.

ASCE Publications Information. 1966-present. American Society of Civil Engineers, 345 East 47th Street, New York, NY 10017-2398. (212) 705-7520 or (800) 548-2723. Bi-monthly. $160.00 per year for non-members.

Current Contents: Engineering, Technology, and Applied Sciences. Institute for Scientific Information, 3501 Market Street, Philadelphia, PA 19104. (215) 386-0100. FAX (215) 386-6362. 1970 to present. Weekly. $442.00 per year.

Engineering Index Monthly. Engineering Information, Inc., Castle Point on the Hudson, Hoboken, NJ 07030. (800) 221-1044. FAX (212) 832-1857. Monthly. $2300.00 per year. Also available online as COMPENDEX, and also on CD-ROM. Covers chemical engineering, computers, electrical engineering, civil engineering, metals and mining, industrial management, and mechanical engineering.

Environmental Engineering Abstracts, Cambridge Scientific Abstracts, 7200 Wisconsin Avenue, Bethesda, MD 20814-4823. (301) 961-6750. FAX (301) 961-6720. Monthly. Covers hazardous materials, environmental impact and protection, treatment of sewage and industrial wastes, hydroelectric power, tidal and wind power, artic and tropical engineering.

Fluid Power Abstracts. STI Limited, 4 Kings Meadow, Ferry Hinksey Road, Oxford OX2 0DU, England. Distributed by Air Science Company, Box 143, Corning, NY 14830. (607) 962-5591. Covers oil hydraulic and pneumatic power transmission and control, engineering and design. 1964 to present. Bi-monthly. $260.00 per year.

Fluid Abstracts: Civil Engineering. Elsevier Science [journals], 660 White Plains Rd., Tarrytown, NY 10159-5153. (919) 524-9200. FAX (919) 333-2444. 1991 to present. Monthly plus annual cumulation. $645.00 per year. Also available online as FLUIDEX. Covers civil engineering applications of fluid mechanics, hydraulics, flow metering and measuring, offshore engineering, environmental hydraulics, and related aspects of wind energy, the atmosphere, and aerodynamics.

International Aerospace Abstracts. Technical Information Service, American Institute of Aeronautics and Astronautics, Inc., 555 West 57th St., New York, NY 10019. (212) 247-6500. FAX (212) 582-4861. Semi-monthly. $1295.00 per year.

Mechanical Engineering Abstracts (formerly ISMEC), Cambridge Scientific Abstracts, 7200 Wisconsin Avenue, Bethesda, MD 20814-4823. (301) 961-6750. FAX (301) 961-6720. 1967 to present. Monthly. $895.00 per year. Summarizes world literature in mechanical engineering, production engineering, and engineering management. Also available online.

Science Citation Index. Institute for Scientific Information, 3501 Market Street, Philadelphia, PA 19104. (215) 386-0100. FAX (215) 386-6362. Inquire about availability and cost. Also available on CD-ROM.

Scientific and Technical Aerospace Reports (star). National Aeronautics and Space Administration. NASA Center for Aerospace Information, Box 8757, BWI Airport, Baltimore, MD 21240. (301) 621-0153. Monthly. Inquire about availability and cost. ALso available through the NASA online retrieval service (RECON), and through the Aerospace Database through DIALOG.

ASSOCIATIONS AND PROFESSIONAL SOCIETIES

American Institute of Aeronautics and Astronautics. The Aerospace Center, 370 L'Enfant Promenade SW, Washington, DC 20024. (202) 646-7400.

American Society of Civil Engineers. 345 East 47th Street, New York, NY 10017-2398. (212) 705-7520 or (800) 548-2723.

Fluid Controls Institute. PO Box 9036, Morristown, NJ 07960. (201) 829-0990.

Fluid Power Consultants International. PO Box 106, 1000 Grandview Drive, Elm Grove, WI 53122-0106. (414) 782-0410.

Fluid Power Society. 2433 N. Mayfair Road, Suite 111, Milwaukee, WI 53226. (414) 257-0910.

Hydraulic Institute. 9 Sylvan Way, Parsippany, NJ 07054-3802. (201) 267-9700.

International Association For Hydraulic Research. Rotterdamsweg 185, PO Box 177, 2600 MH Delft, Netherlands. Telephone 31-15-569353. FAX 31-51-619674.

National Conference On Fluid Power. 3333 N. Mayfair Road, Suite 111, Milwaukee, WI 53226. (414) 778-3368.

National Fluid Power Association. 3333 N. Mayfair Road, Suite 111, Milwaukee, WI 53226. (414) 778-3344.

Society of Automotive Engineers. 400 Commonwealth Drive, Warrendale, PA 15098. (412) 776-4841.

FLUID STATISTICS

Ency. of Physical Sciences and Engineering Info. Sources

DIRECTORIES AND BIOGRAPHICAL SOURCES

American Astronautical Society Directory. American Astronautical Society, 6352 Rolling Mill Place, Suite 102, Springfield, VA 22152. Inquire for cost and availability.

Fluid Power Society Membership Directory. Fluid Power Society, 2433 N. Mayfair Road, Suite 111, Milwaukee, WI 53226. (414) 257-0910. Inquire for cost and availability.

National Fluid Power Association Membership Directory. National Fluid Power Association, 3333 N. Mayfair Road, Suite 111, Milwaukee, WI 53226. (414) 778-3344. Inquire for cost and availability.

Research Centers Directory. Gale Research, , 835 Penobscot Building, Detroit, MI 48226-4094. (313) 961-2242. (800) 347-4253. FAX (313) 961-6083. $485.00.

Who's Who in Engineering. American Association of Engineering Societies, 1111 19th Street NW, Suite 608, Washington, DC 20036. (202) 296-2237 or (800) 658-8897. 8th edition. 1991. Inquire for price.

ENCYCLOPEDIAS AND DICTIONARIES

Encyclopedia of Fluid Mechanics. 10 volumes plus 3 supplements. Gulf Publishing Company, P.O. Box 2608, Houston, TX. (713) 529-4301 or (800) 231-6275. 1986-90 (vols. 1-10), 1993-94 (supplements). $195.00 each (vols. 1-10), $139.00 each (supplements).

Encyclopedia of Physical Science and Technology. Robert A. Meyers, ed. 18 volumes. Academic Press Inc., 6277 Sea Harbor Drive, Orlando, FL 32887. (800) 321-5068. 1992. $2100.00

McGraw-Hill Dictionary of Scientific and Technical Terms. Sybil P. Parker, ed. 5th edition. McGraw-Hill Publishing Company, 1221 Avenue of the Americas, New York, NY 10020. (800) 262-4729 or (212) 512-3825. 1993. $110.50.

McGraw-Hill Encyclopedia of Engineering. Sybil P. Parker, ed. 2nd edition. McGraw-Hill Publishing Company, 1221 Avenue of the Americas, New York, NY 10020. (800) 262-4729 or (212) 512-3825. 1993. $95.50.

McGraw-Hill Encyclopedia of Science and Technology. Sybil P. Parker, ed. 7th edition. 20 volumes. McGraw-Hill Publishing Company, 1221 Avenue of the Americas, New York, NY 10020. (800) 262-4729 or (212) 512-3825. 1992. $1900.00

Thesaurus of Scientific, Technical, and Engineering Terms. Hemisphere Publishing Corporation, 1900 Frost Road, Suite 101, Bristol, PA 19007-1598. (215) 785-5800 or (800) 821-8312. FAX (215) 785-5515. 1987. $173.00.

GENERAL WORKS

Computational Fluid Dynamics: An Introduction. John F. Wendt, editor. Springer-Verlag, 175 Fifth Avenue, New York, NY 10010. (212) 460-1500 or (800) 777-4643. FAX (212) 473-6272. 1992. $89.00.

Engineering Fluid Mechanics. John A. Roberson & Clayton T. Crowe. 4th edition. Houghton Mifflin, 222 Berkeley Street,

Boston, MA 02116. (617) 351-5000 or (800) 225-3362. FAX (617) 227-5409. 1990. $70.36.

Fluid Power Technology. Robert P. Kokernak. Macmillan Publishing, 200 Old Tappan Road, Old Tappan, NJ 07675. (800) 223-2336. FAX (800) 445-6991. 1994. Inquire for cost and availability.

Introduction To Fluid Mechanics. James A. Fay. MIT Press, 55 Hayward Street, Cambridge, MA 02142. (617) 253-2884 or (800) 356-0343. FAX (617) 253-1709. 1994. $49.95.

Mechanics of Fluids. Irving H. Shames. McGraw-Hill Publishing Company, 1221 Avenue of the Americas, New York, NY 10020. (800) 262-4729 or (212) 512-3825. 1992. Inquire for cost and availability.

HANDBOOKS AND MANUALS

Plant Engineering Magazine's Fluid Power Handbook. Anton H. Hehn. 2 volumes. Gulf Publishing Company, P.O. Box 2608, Houston, TX. (713) 529-4301 or (800) 231-6275. 1993. Inquire for cost and availability.

ONLINE DATABASES AND CD-ROMS

Compendex Plus. Engineering Information, Inc., 345 East 47th Street, New York, NY 10017. (212) 705-7600 or (800) 221-1044. Contains citations with abstracts to worldwide literature in engineering and technology, from 1970 to present. Available on online BRS,(800) 289-4277, DIALOG, (800) 334-2564, ORBIT (800) 456-7248, and STN International, FIZ Karlsruhe, P.O. Box 2465, W-7500, Karlsruhe 1, Germany, online services. Also available on CD-ROM. Inquire as to cost and availability.

CSA Engineering. Cambridge Scientific Abstracts, 7200 Wisconsin Avenue, Suite 601, Bethesda, MD 20814. (301) 961-6750 or (800) 843-7751. Contains citations and abstracts of international periodicals and research literature covering all fields of engineering and science and technology,including computer and information science, electronics, mechanical engineering, solid state materials, 1981 to present. Available on BRS,(800) 289-4277, online service.Inquire as to cost and availability.

Current Contents Search. Institute for Scientific Information, 3501 Market Street, Philadelphia, PA 19104. (215) 386-0100. FAX (215) 386-6362. Contains citations to articles listed in the table of contents of science and technology journals. Also articles in social sciences and life sciences journals. Available on BRS,(800) 289-4277, DIALOG,(800) 334-2564, online services. Inquire as to cost and availability.

International Aerospace Abstracts. American Institute of Aeronautics and Astronautics, 370 L'Enfant Promenade, S.W., Washington, DC 20024. (202) 646-7400. Contains references and abstracts of journal and monograph literature relating to aerospace science and technology, from 1963 to present. Available through the NASA/RECON system of the National Aeronautics and Space Administration only.

ISMEC: Mechanical Engineering Abstracts. Cambridge Scientific Abstracts, 7200 Wisconsin Avenue, Suite 601, Bethesda, MD 20814. (301) 961-6750 or (800) 843-7751. Contains citations to the literature in mechanical engineering, industrial and production engineering, energy, power, mechanics, devices and related areas, from 1973 to present. Available on the DIA-

LOG,(800) 334-2564, online service. Inquire as to cost and availability.

NASA Database. American Institute of Aeronautics and Astronautics, 370 L'Enfant Promenade, S.W., Washington, DC 20024. (202) 646-7400. Citations and abstracts of aeronautics and astronautics literature, 1962 to present. Also contains citations from STAR, Scientific and Technical Aerospace Reports, and *International Aerospace Abstracts.* Available through NASA/RECON online service. Inquire as to cost and availability.

NTIS Bibliographic Database. National Technical Information Service, 5285 Port Royal Road, Springfield, VA 22161. (703) 487-4929 or FAX (703) 321-8199. Broad coverage of government-sponsored science and technology research reports, 1964 to present. Available on BRS,(800) 289-4277, DIALOG, (800) 334-2564, ORBIT, (800) 456-7248, and STN International, FIZ Karlsruhe, P.O. Box 2465, W-7500, Karlsruhe 1, Germany, online services. Also available on CD-ROM. Inquire as to cost and availability.

PERIODICALS

Applied Scientific Research. Kluwer Academic Publishers, P.O. Box 358, Accord Station, Hingham, MA 02018-0358. (617) 871-6000. 1947 to present. 8 times a year. $468.00 per year.

Experiments in Fluids. Springer-Verlag, 44 Hartz Way, Secaucus, NJ 07096-2491. 1983 to present. 12 times a year. $675.00 per year.

Hydraulics and Pneumatics. Penton Publishing, 110 Superior Ave., Cleveland, OH 44114-2543. 1947 to present. Monthly. $45.00 per year.

International Journal of Heat and Fluid Flow. Butterworths-Heinemann/ Reed Elsevier, 313 Washington Street, Newton, MA 02158. (617) 928-2500 or (800) 366-2665. FAX 9617) 928-2610. 1979 to present. 6 times a year. $85.00 per year.

Journal of Fluid Control. Delbridge Publishing Company, Box 160817, Cupertino, CA 95016. (408) 446-3131. FAX (408) 446-3131. 1967 to present. Quarterly. $145.00 per year.

Journal of Fluid Mechanics. Cambridge University Press, 40 West 20th Street, New York, NY 10011-4211. (212) 924-3900. 1956 to present. 24 times a year. $595.00 per year.

Journal of Fluids and Structures. Harcourt Brace & Company, Ltd., Foots Cray High Street, Sidcup, Kent DA14 5HP, England. Telephone 44-81-300-3322. FAX 44-81-309-0807. 1987 to present. 8 times a year. Inquire for cost and availability.

Journal of Fluids Engineering. American Society of Mechanical Engineers, 345 East 47th Street, New York, NY 10017-2398. (212) 705-7703. 1919 to present. Quarterly. $100.00 per year.

Journal of Hydraulic Engineering. American Society of Civil Engineers, Hydraulics Division, 345 East 47th Street, New York, NY 10017-2398. (212) 705-7520 or (800) 548-2723. 1956 to present. Monthly. $200.00 per year.

Journal of Hydraulic Research. International Association for Hydraulic Research, Rotterdamsweg 185, PO Box 177, 2600 MH Delft, Netherlands. Telephone 31-15-569353. FAX 31-51-619674. 1963 to present. 6 times a year. Inquire for cost.

Journal of Non-newtonian Fluid Mechanics. Elsevier Science Inc., Box 882, Madison Square Station, New York, NY 10159. (212) 989-5800. 1976 to present. 15 times a year. $1205.00 per year.

Journal of Supercritical Fluids. Polymer Research Assoc. Inc., 9200 Montgomery Road, Suite 23B, Cincinnati, OH 45242. (513) 891-7030. 1988 to present. Quarterly. $225.00 per year.

RESEARCH CENTERS AND INSTITUTES

Fluid Properties Research, Inc. School of Chemical Engineering, Georgia Institute of Technology, Atlanta, GA 30332. (404) 894-3098. FAX (404) 894-2866.

Lehigh University Institute of thermo-fluid Engineering and Science. Bldg. A, Mountain Top Campus, Bethlehem, PA 18015. (215) 758-4091. FAX (215) 758-5057.

Purdue University Fluid Mechanics Laboratory. School of Mechanical Engineering, West Lafayette, in 47907. (317) 494-5633.

Stanford University Environmental Fluid Mechanics Laboratory. Department of Civil Engineering, Stanford, CA 94305-4020. (415) 723-1925. FAX (415) 725-8662.

University of Arizona Computational Fluid Mechanics Laboratory. Bldg. 16, Room 312, Tucson, AZ 85721. (602) 621-4423.

University of Iowa Institute of Hydraulic Research. Iowa City, IA 52242. (319) 335-5236. FAX (319) 335-5238.

University of Washington Aeronautical Laboratory. FS-10, Seattle, WA 98105. (206) 543-0439. FAX (206) 542-0217.

FLUORESCENT LIGHTING

See: ILLUMINATION

FLUORINE

See also: BROMINE, CHLORINE, ELEMENTS, HALIDES, IODINE

ABSTRACT SERVICES AND INDEXES

Applied Science and Technology Index; a cumulative subject index to English language periodicals in the fields of aeronautics and space science, computer technology, chemistry, construction industry, energy and related areas. H.W. Wilson Co., 950 University Avenue, Bronx, NY 10452. (212) 588-8400. (800) 367-6770. FAX (718) 590-1617. From 1958 to present. Monthly. Inquire about cost and availability. Also available on CD-ROM and online. ISSN: 0003-6986.

Chemical Abstracts. Chemical Abstracts Service. 2540 Olentangy River Road, Box 3012, Columbus, OH 43210-0012. (614) 447-3600. FAX (614) 447-3713. 1907 to present. Weekly. $16,800.00 per year. Available online and on CD- ROM. CA is also available in five section groupings. Inquire regarding cost and availability.

FLUORINE

Ency. of Physical Sciences and Engineering Info. Sources

Current Contents: Physical, Chemical and Earth Sciences. Institute for Scientific Information, 3501 Market Street, Philadelphia, PA 19104. (215) 386-0100. FAX (215) 386-6362. Weekly. $360.00 per year.

General Science Index. H.W. Wilson Company, 950 University Avenue, Bronx, NY 10452. (212) 588-8400. (800) 367-6770. FAX (718) 590-1617. From 1978 to present. Ten issues per year; quarterly and annual cumulations. Service basis. Available on CD-ROM and online. Inquire regarding cost and availability. ISSN: 0162-1963.

Physics Abstracts. INSPEC. Section A, Science Abstracts. Institution of Electrical Engineers (IEE), London. Available from: INSPEC/IEEE - Institute of Electrical and Electronic Engineers, Box 1331, Hoes Lane, Piscataway, NJ 08855-1331. (908) 562-5549. 1898 to present. 24 issues per year. $2835.00 per year. Also available online and on CD-ROM. ISSN: 0036-8091.

Science Citation Index. SCI. Institute for Scientific Information, 3501 Market Street, Philadelphia, PA 19104. (215) 386-0100. (800) 523-1850. FAX (215) 386-2991. 1961 to present. Six issues per year, plus annual cumulation. $11650.00 per year. Also available online and on CD-ROM. Inquire about price and availability. ISSN: 0036-827X.

ASSOCIATIONS AND PROFESSIONAL SOCIETIES

American Chemical Society. 1155 16th Street, NW, Washington, DC 20036. (202) 872-4600.

Association of Consulting Chemists and Chemical Engineers, 295 Madison Avenue, 27th Floor, New York, NY 10017. (212) 983-3160. FAX (212) 983-3161.

Chemical Manufacturers Association. 2501 M Street, N.W., Washington, DC 20037. (202) 887-1100. FAX (202) 887-1237.

Chemical Specialties Manufacturers Association, 1913 Eye Street, Nw, Washington, Cd 20006, (202) 872-8110. FAX (202) 872-8114.

Chlorine Institute, 2001 L Street, Nw, No. 506, Washington, DC 20036. (202) 775-2790. FAX (202) 223-7725.

International Society For Fluoride Research. 8IA Landscape Road, Mount Eden Auckland 4, New Zealand. TEL 96307114. FAX 96307114.

BIBLIOGRAPHIES

Chlorine: an Annotated Bibliography. Ralph G. Smith. Chlorine Institute, 2001 L. Street, No. 506, Washington, DC 20036. (202) 775-2790. FAX (202) 223-7725. 1971. $12.00.

Chemical Information Sources. Gary Wiggins. McGraw-Hill Publishing Company, 1221 Avenue of the Americas, New York, NY 10020. (800) 262-4729 or (212) 512-3825. 1991. $42.50.

Information Sources in Chemistry. R.T. Bottle and J.F.B Rowland, editors. R.R. Bowker Inc., 121 Chanlon Road, New Providence, NJ 07974. (800) 521-8110 or (908) 464-6800. Fourth edition. 1993. $75.00.

Handbooks and Tables in Science and Technology. Russell H. Powell, editor. Oryx Press, 4041 North Central, Suite 700, Phoe-

nix, AZ 85012-3330. (602) 265-2651 or (800) 279-6799. Third edition. 1994. $65.00.

Scientific and Technical Books and Serials in Print; An Index To Literature in Science and Technology. R.R. Bowker Inc., 121 Chanlon Road, New Providence, NJ 07974. (908) 464-6800. (800) 521-8110. FAX (908) 665-3502. 1972 to present. Annual. 4 volumes. 1994. $299.95. Also available on compact disc and online. ISSN: 0000-054X.

DIRECTORIES AND BIOGRAPHICAL SOURCES

American Men and Women of Science. Physical and Biological Sciences. R.R. Bowker Company, 121 Chanlon Road, New Providence, NJ 07974. (908) 464-6800. (800) 521-8110. 20th edition. 8 volumes. 1996. $850.00.

Chemical Week — Buyers Guide Issue. Chemical Week Associates, 888 Seventh Avenue, New York, NY 10106. (212) 621-4900. FAX (212) 621-4949. Annual, October. $50.00.

Consulting Services: Chemists and Chemical Engineers. Association of Consulting Chemists and Chemical Engineers, 50 East 41st Street, Suite 92, New York, NY 10017. (212) 684-6255. Biennial. $60.00.

Directory of Chemistry Software 1992. Wendy Warr, Peter Willett, Geoff Downs. American Chemical Society, 1155 16th Street, NW, Washington, DC 20036. (202) 872-4600. 1992. $35.95.

Research Centers Directory. Gale Research Company Inc., 835 Penobscot Building, Detroit, MI 48226-4094. (313) 961-2242. (800) 347-4253. 20th edition. 1995. $485.00. ISSN: 0080-1518.

ENCYCLOPEDIAS AND DICTIONARIES

Academic Press Dictionary of Science and Technology. Christopher Morris, editor. Academic Press, Inc., 1250 Sixth Avenue, San Diego, CA 92101. (619) 231-0926. FAX (619) 699-6715. 1991. $115.00.

Dictionary of Named Processes in Chemical Technology. Alan E. Comyns. Oxford University Press, Inc., 200 Madison Avenue, New York, NY 10016. (212) 725-6000. (800) 334-4249. 1994. $75.00.

Consulting Services: Chemists and Chemical Engineers. Association of Consulting Chemists and Chemical Engineers, 50 East 41st Street, Suite 92, New York, NY 10017. (212)

684-6255. Biennial. $60.00.

Encyclopedia of Chemical Processing and Design. McKetta Marcel Dekker, Inc., 270 Madison Avenue, New York, NY 10016. (212) 696-9000. (800) 228-1160. 1976 - . $175.00 per volume. ISSN: .

Kirk-Othmer Encyclopedia of Chemical Technology. John Wiley and Sons, Inc., 605 Third Avenue, New York, NY 10158. (800) 526-5368 or (212) 850-6000. Fourth edition. 1991 - . Twenty-seven volumes. $5400.00.

McGraw-Hill Encyclopedia of Science and Technology. McGraw-Hill Book, Incorporated, 1221 Avenue of the Americas, New York, NY 10020. (212) 997-3675. (800) 262-4729. Seventh edition. Twenty volumes. 1992. $1900.00.

Ullman's Encyclopedia of Industrial Chemistry. VCH Publications, Inc., 220 East 23rd Street, Suite 909, New York, NY 10010-4606. (212) 683-8333. (800) 422-8824. 5th edition. 1984 - . Price varies per volume. ISSN:

GENERAL WORKS

Carbon-fluorine Compounds: Chemistry, Biochemistry and Biological Activities. Ciba Foundation. Books on Demand, 300 North Zeeb Road, Ann Arbor, MI 48106-1346. (313) 761-4700. (800) 521-0600. $121.00 in paper.

The Chemical Bond: Structure and Dynamics. Ahmed Zewail. Academic Press, Inc., 6277 Sea Harbor Drive, Orlando, FL. (800) 321-5068. 1992. $49.95.

Chemistry of Halides, Pseudohalides and Azides, Part 1 and 2. Chemistry of Functional Groups series, supplement no. D. Saul Patai and Zvi Rappaport. Books on Demand, 300 North Zeeb Road, Ann Arbor, MI 48106-1346. (313) 761-4700. (800) 521-0600. 1983. $340.00 set.

Chemistry of the Elements. N. N. Greenwood and A. Earnshaw. Pergamon Press, Inc., Maxwell House, Fairview Park, Elmsford, NY 10523. (914) 592-7700. Fax (914) 592-3625. 1984. $143.00.

Chlorinated Dioxins and Furans: Analytical Methods: Exposed Populations. Lewis Pub., CRC Press, Inc., 2000 Corporate Boulevard, NW, Boca Raton, FL 33431. (407) 994-0555. (800) 272-7737. 1995. wfi

The Elements. John Emsley. Oxford University Press, Inc., 200 Madison Avenue, New York, NY 10016. (212) 725-6000. (800) 334-4249. Second edition. 1991. $49,95.

Exploring Chemical Elements and their Compounds. David L.Heiserman. TAB Books, P.O. Box 40, Blue Summit, PA 17294-0850. (717 794-2191. (800) 233-1128. 1992. $17.95.

Fluorine Chemistry: A Comprehensive Treatment. Mary Howe-Grant. John Wiley & Sons, Inc., 605 Third Avenue, New York, NY 10158-0012. (212) 850-6000. (800) 225-5945. 1995. $79.95.

Graphite Fluorides and Carbon-fluorine Compounds. Tsuyoshi Nakajima and Nobuatsu Watanabe. Franklin Book Co., Inc., Elkins Park, PA 19117. (215) 635-5252. 1990. $144.00

Inorganic Fluoride Chemistry: Toward the 21st Century. Joseph S. Thrasher. American Chemical Society, 1155 16th Street, NW, Washington, DC 20036. (202) 872-4600. 1994.$109.95.

Inorganic Solid Fluorides. Paul Hagenmuller, editor. Academic Press, Inc., 6277 Sea Harbor Drive, Orlando, FL. (800) 321-5068. 1985. $190.00

Halogen Chemistry. V. Gutman, editor. Academic Press, Incorporated, 6277 Sea Harbor Drive, Orlando, FL 32821. (800) 321-5068. Three volumes. 1967. Inquire

Sulfur Dioxide, Chlorine, Flourine and Chlorine Oxides. A. S. Young. Franklin Book Co., Inc., Elkins Park, PA 19117. (215) 635-5252. 1983. $204.00

Review of Fluoride Benefits and Risks; Report of the Ad Hoc Subcommittee On Fluoride. Diane Publishing Co., 600 Upland Avenue, Upland, PA 19015. (610) 499-7415. 1994. $65.00 in paper.

Synthetic Fluorine Chemistry. George A. Olah. John Wiley & Sons, Inc., 605 Third Avenue, New York, NY 10158-0012. (212) 850-6000. (800) 225-5945. 1992. $117.00

Toxicology of Halogenated Hydrocarbons; Health and Ecological Effects. M. A. Khan and R. H. Stanton, editors. Pergamon Press, Inc., Maxwell House, Fairview Park, Elmsford, NY 10523. (914) 592-7700. Fax (914) 592-3625. 1981. $72.50.

HANDBOOKS AND MANUALS

Chemical Formulary. H. Bennett, editor. Chemical Publishing, Co., Inc. 80 Eighth Avenue, New York, NY 10011. (212 255-1950. Volumes 1 - 30. $60.00 per volume.

CRC Handbook of Chemistry and Physics. David R. Kide, editor. CRC Press Inc., 2000 Corporate Blvd., NW, Boca Raton, FL 33431. (407) 994-0555. (800) 333-8300. 77th edition. 1996. $99.95.

Guide To Basic Chemical Compounds. D.R. Lide, Jr. CRC Publishers, Inc., 2000 Corporate Blvd., N.W., Boca Raton, FL 33431. (407) 994-0555 or (800) 333-8300. 1993. $120.00.

Lange's Handbook of Chemistry. John A. Dean, editor. McGraw-Hill Publishing Company, 1221 Avenue of the Americas, New York, NY 10020. (212) 512-2000. (800) 262-4729. 14th edition. 1996. $99.95.

Riegel's Handbook of Industrial Chemistry. James A. Kent, editor. Van Nostrand Reinhold, 115 Fifth Avenue, New York, NY 10003. (212) 254-3232. (800) 926-2665. Ninth edition. 1992. $114.95.

ONLINE DATABASES AND CD-ROMS

CA Search. Chemical Abstracts Service, P.O. Box 3012, Columbus, OH 43210-0012. (614) 447-3600. (800) 848-6533. FAX (614) 447-3709. Very comprehensive guide to worldwide chemical literature and related fields, 1972 to present. Available on BRS,(800) 289-4277, DIALOG, (800) 334-2564, ORBIT (800) 456-7248, and STN International, FIZ Karlsruhe, P.O. Box 2465, W-7500, Karlsruhe 1, Germany, online services. Inquire as to cost and availability.

Chemical Journals of the American Chemical Society. American Chemical Society, 1155 16th Street, N.W., Washington, DC 20036. (202) 872-4381 or (800) 424-6747. Contains complete text of approximately 90,000 articles from 22 primary journals published by the American Chemical Society, from mostly 1982 to present. Available on STN International, FIZ Karlsruhe, P.O. Box 2465, W-7500, Karlsruhe 1, Germany, online service. Inquire as to cost and availability.

Current Contents Search. Institute for Scientific Information, 3501 Market Street, Philadelphia, PA 19104. (215) 386-0100. FAX (215) 386-6362. Contains citations to articles listed in the table of contents of science and technology journals. Also articles in social sciences and life sciences journals. Available on BRS,(800) 289-4277, DIALOG,(800) 334-2564, online services. Inquire as to cost and availability.

Dissertation Abstracts. University Microfilms International, 300 North Zeeb Road, Ann Arbor, MI 48106. (800) 521-0600 or (313) 761-4700. Scope includes virtually all doctoral dissertations accepted at accredited American institutions from 1861 to present in 252 subject areas. Available on BRS,(800) 289-4277, DIALOG,(800) 334-2564, and OCLC EPIC,(800) 848-5878, online services. Also available on CD-ROM. Inquire as to cost and availability.

Gmelin Database. Gmelin-Institut fur Anorganische Chemie und Grenzgebiete, Varrentrapperstrasse, 40-42, Carl-Bosch-Haus, D-6000, Frankfurt am Main 90, Germany. Contains structural and factual data relating to inorganic and organometallic chemistry. Provides data from the Gmelin Handbook of Inorganic and Organometallic Chemistry. Covers the period 1817 to 1975; 1988-89. Available on STN International, FIZ Karlsruhe, P.O. Box 2465, W-7500, Karlsruhe 1, Germany, online service. Inquire as to cost and availability.

Kirk-Othmer Encyclopedia of Chemical Technology. John Wiley and Sons, Inc., 605 Third Avenue, New York, NY 10158. (800) 526-5368 or (212) 850-6000. Contains the complete text of all chapters in the 27 volume fourth edition of the *Kirk-othmer Encyclopedia of Chemical Technology.* 1991. Available on BRS,(800) 289-4277, DIALOG,(800) 334-2564, online services. Inquire as to cost and availability.

NTIS Bibliographic Database. National Technical Information Service, 5285 Port Royal Road, Springfield, VA 22161. (703) 487-4929 or FAX (703) 321-8199. Broad coverage of government-sponsored science and technology research reports, 1964 to present. Available on BRS,(800) 289-4277, DIALOG, (800) 334-2564, ORBIT, (800) 456-7248, and STN International, FIZ Karlsruhe, P.O. Box 2465, W-7500, Karlsruhe 1, Germany, online services. Also available on CD-ROM. Inquire as to cost and availability.

SCISEARCH. Institute for Scientific Information, 3501 Market Street, Philadelphia, PA 19104. (800) 523-1850 or (215) 386-0100. Broad multidisciplinary title and author index to the international literature of science and technology, 1974 to present. Available on DIALOG,(800) 334-2564, and ORBIT,(800) 456-7248, online services. Also available on CD-ROM. Inquire as to cost and availability.

WILSONLINE. H.W. Wilson Company, 950 University Avenue, Bronx, NY 10452. (800) 367-6770 or (212) 588-8400. Makes available online versions of the printed H.W. Wilson indexes including Applied Science and Technology Index, Business Periodicals Index, General Science Index, and Readers' Guide to Periodical Literature. Period covered is generally 1983 to present. Available on BRS,(800) 289-4277, DIALOG,(800) 334-2564, and OCLC EPIC,(800) 848-5878, online services. Also available on CD-ROM. Inquire as to cost and availability.

PERIODICALS

Analytical Chemistry. American Chemical Society, Box 3337, Columbus, OH 43210. (614) 447-3776. 1929 to present. Semi-monthly. $415.00 per year. ISSN: 0003-2700.

Angewandte Chemie. VCH Publishers, Inc., 220 East 23rd Street, New York, NY 10010-4606. (212) 683-8333. (800) 367-8249. 1888 to present. 22 issues per year. $840.00 per year. ISSN: 0044-8249.

Chemical Reviews. American Chemical Society, Box 3337, Columbus, OH 43210. (614) 447-3776. 1924 to present. issues per year. $346.00 per year. ISSN: 0009-2665.

Chemical Week; Includes Annual Buyers Guide. Chemical Week Associates, 888 Seventh Avenue, New York, NY 10106. (212) 621-4900. FAX (212) 621-4949. 1914 to present. Weekly. $99.00 per year. ISSN: 0009-272X.

Chemtech. American Chemical Society. Box 3337, Columbus, OH 43210. (614) 447-3776. 1970 to present. Monthly. $370.00 per year. ISSN: 0009-2703.

Fluoride. International Society for Fluoride Research. 8IA Landscape Road, Mount Eden Auckland 4, New Zealand. TEL 96307114. FAX 96307114. Quarterly. $50.00. ISSN: 0015-4725.

Inorganic Chemistry. American Chemical Society, Box 3337, Columbus, OH 43210. (614) 447-3776. 1962 to present. Semi-monthly. $500.00 per year. ISSN: 0020-1669.

Journal of Fluorine Chemistry. Elsevier Science Publisher,Box 882, Madison Square Station, New York, NY 10159. (212) 989-5800. 1971 to present. 15 issues per year. $1327.00 per year. ISSN: 0022- 1139.

Journal of the American Chemical Society. American Chemical Society, Box 3337, Columbus, OH 43210. (614) 447-3776.

1879 to present. Biweekly. $1055.00 per year. ISSN: 0002- 7863.

Polyhedron. Elsevier Science Publishing Company, Inc., 660 White Plains Road, Tarrytown, NY 10591-5153. (914) 524-9200. FAX (914) 524-9200. 1982 to present. 24 issues per year. $2175.00 per year. ISSN: 0277-5387.

RESEARCH CENTERS AND INSTITUTES

Chemical Laboratories. Harvard Universiy. Oxford Street, Cambridge, MA 02138. (617) 495-4283. FAX (617) 496-5618.

Chemistry Laboratories. Rensselaer Polytechnic Institute. Cogswell Laboratory, Troy, NY 12180-3590. (518) 276-8981.

Research Program in Chemistry and Biochemistry. Southern Illinois University at Carbondale, Carbondale, IL 62901. (618) 453-5721.

Theoretical Chemistry Institute. University of Wisconsin, Madison, 1101 University Avenue, Madison, WI 53706. (608) 262-1511.

FLUOROCARBONS
See: HALIDES

FM RADIO
See: RADIO

FOAM

See: PLASTICS

FOCAL LENGTH

See: OPTICS

FOG

See also: AEROSOLS, CLIMATE, CLOUDS, METEOROLOGY, RAIN,
WEATHER MODIFICATION

ABSTRACT SERVICES AND INDEXES

Applied Science and Technology Index; A Cumulative Subject Index To English Language Periodicals in the Fields of Aeronautics and Space Science, Computer Technology, Chemistry, Construction Industry, Energy and Related Areas. H.W. Wilson Co., 950 University Avenue, Bronx, NY 10452 -9978. (212) 588-8400. (800) 367-6770. FAX (718) 590-1617. From 1958 to present. Monthly. Inquire about cost and availability. Also available online (BRS and WILSONLINE) and on CD-ROM. ISSN: 0003-6986.

General Science Index. H.W. Wilson Co., 950 University Avenue, Bronx, NY 10452. (212) 588-8400. (800) 367-6770. FAX (718) 590-1617. From 1978 to present. Ten issues per year; quarterly and annual cumulations. Service basis. Available on CD-ROM and online. Inquire regarding cost and availability. ISSN: 0162-1963.

Government Reports Announcements and Index. National Technical Information Service (NTIS), 5285 Port Royal Road, Springfield, VA 22161. (703) 487-4650. FAX (703) 321-8547. From 1968 to present. Annual. $630.00 per year. Also available online as NTIS BIBLIOGRAPHIC DATABASE and on CD-ROM. ISSN:

Meteorological and Geoastrophysical Abstracts. American Meteorological Society, c/o Inforonics, Inc., 550 Newtown Road, Box 458, Littleton, MA 01460. (508) 486-8976. FAX (508) 486-0027. Covers literature in environmental sciences, meteorology, astrophysics, hydrology, glaciology, and physical oceanography. 1950 to present. Monthly. $950.00 per year. ISSN: 0026-1130. Also available online (DIALOG) and on CD-ROM.

Science Citation Index. SCI. Institute for Scientific Information, 3501 Market Street, Philadelphia, PA 19104. (215) 386-0100. (800) 523-1850. FAX (215) 386-2991. 1961 to present. Six issues per year, plus annual cumulation. $11650.00 per year. Also available online and on CD-ROM. Inquire about price and availability. ISSN: 0036-827X.

Selected Water Resources Abstracts. U.S. Geological Survey. Distributed by National Technical Information Service (NTIS), 5285 Port Royal Road, Springfield, VA 22161. (703) 487-4650. Monthly. $115.00 per year.

ANNUAL REVIEWS AND YEARBOOKS

Developments in Atmospheric Science Series. Elsevier Science Publishing Company, Inc., 655 Avenue of the Americas, New York, NY 10010. (212) 989-5800. Irregular. Inquire about price and availability.

ASSOCIATIONS AND PROFESSIONAL SOCIETIES

American Meteorological Society, 45 Beacon Street, Boston, MA 02108-3693. (617) 227-2425. FAX (617) 742-8718. Irregular. Price varies.

Association of American Weather Observers, 401 Whitney Boulevard, Box 455, Belvedere, IL 61008. (815) 544-5665. FAX (815) 544-6334.

International Association of Meteorology and Atmospheric Physics. c/o UCAR, P.O. Box 3000, Boulder CO 80307-3000. (303) 497-1673.

National Weather Association, 6704 Wolke Court, Montgomery, AL 36116-2134. (334) 213-0388. FAX (334) 213-0388.

University Corporation For Atmospheric Research. PO Box 3000, 1850 Table Mesa Drive, Boulder, CO 80307-3000. (303) 497-1673. FAX (303) 497-1654.

Weather Modification Association. PO Box 26926, Fresno, CA 93729-6926. (209) 434-3486.

BIBLIOGRAPHIES

Information Sources in the Earth Sciences. David N. Wood, Joan E. Hardy, and Anthony P. Harvey, editors. Bowker-Saur/K.G. Saur. Distributed by R.R. Bowker, 121 Chanlon Road, New Providence, NJ 07974. (800) 521-8110 or (908) 464-6800. Second edition. 1989. $85.00.

SCIENTIFIC and TECHNICAL BOOKS and SERIALS in PRINT; An index to Literature in Science and Technology. R.R. Bowker Co., 205 E. 42nd Street, New York, NY 10017. (800) 521-8110 or (212) 916-1600. 1992. $250.00.

DIRECTORIES AND BIOGRAPHICAL SOURCES

American Meteorological Society. PRofESSIONAL DIRECTORY. American Meteorological Society. 45 Beacon Street, Boston, MA 02108. (617) 227-2425. Included in each issue of the Bulletin of the Society.

Meteorological Services of the World. American Meteorological Society, 45 Beacon Street, Boston, MA 02108- 3693. (617) 227-2425. FAX (617) 742-8718. Annual. $70.00.

National Weather Service offices and Stations. National Oceanic and Atmospheric Administration, Deparatment of Commerce, Silver Spring, MD 20910. (301) 427-7698. Annual. Free

Research Centers Directory. Gale Research, , 835 Penobscot Building, Detroit, MI 48226-4094. (313) 961-2242. (800) 347-4253. 20th edition. 1995. $485.00. ISSN: 0080-1518.

ENCYCLOPEDIAS AND DICTIONARIES

Climates of the States. Bair. Gale Research, , 835 Penobscot Building, Detroit, MI 48226-4094. (313) 961-2242. (800) 347-4253. 5th edition. 1995. $255.00.

Concise Oxford Dictionary of the Earth Sciences. Ailsa Allaby and Michael Allaby, editors. Oxford University Press, Inc., 200 Madison Avenue, New York, NY 10016. (800) 334-4249 or (212) 679-7300. 1990. $42.95.

McGraw-Hill Encyclopedia of Science and Technology. McGraw-Hill Publishing Company, 1221 Avenue of the Americas, New York, NY 10020. (800) 262-4729 or (212) 512-3825. Seventh edition. 1992. $1900.00.

GENERAL WORKS

The Atmosphere: an Introduction to Meteorology. Frederick K. Lutgens and Edward J. Tarbuck. Prentice Hall (Division of Simon and Schuster), 15 Columbus Circle, New York, NY 10023. (212) 373-8500 or (800) 922-0579. Fifth edition. 1991. $49.00.

Cloud Dynamics. Robert A. Houze. Academic Press, Inc., 1250 Sixth Avenue, San Diego, CA 92101-4311. (619) 231-0926. FAX (619) 699-6715. 1993. $149.00.

Clouds and Storms. Audubon Society Staff. Alfred A. Knopf, Inc. 201 E. 50th Street, New York, NY 10022. (800) 733-3000. 1995, $7.99.

Highway Fog - Visability Measures and Guidance Systems. Transportation Research Board, National Academy of Sciences, 2101 Constitution Avenue NW, Washington, DC 20055. (202) 334- 3313. (800) 624-6242. 1976. $4.00.

Meteorology: The Atmosphere and the Science of Weather. Joseph M. Moran. Macmillan Publishing, 866 3rd Avenue, New York, NY 10024. (212) 689-9140. Third edition. 1991. $49.95.

Physics of Rainclouds. Neville H. Fletcher. Books on Demand, Division of University Microfilms International, 300 North Zeeb Road, Ann Arbor, MI 48106-1346. (313) 761-4700 or (800) 521-0600. $101.00 in paper.

Severe and Unusual Weather. Joe R. Eagleman. Trimedia Publishing Company, 12008 West 87th Street, Suite 117, Lenexa, KS 66215. (913) 599-0505. Second edition. 1990. $41.95.

Storm and Cloud Dynamics. William R. Cotton and Richard A. Anthes. Academic Press, Inc., 1250 Sixth Avenue, San Diego, CA 92101-4311. (619) 231-0926. FAX (619) 699-6715. 1992. $69.95.

Theory of Fog Condensation. A. G. Amelin. Coronet Books, 311 Bainbridge Street, Philadelphia PA 19147. (215) 925-5083. 1967. $62.50.

HANDBOOKS AND MANUALS

Author's Guide To the Journals of the American Meteorological Society. American Meteorological Society, 45 Beacon Street, Boston, MA 02108-3693. (617) 227-2425. FAX (617) 742-8718. 1983. $15.00 in paper.

Handbook in Applied Meteorology. David D. Houghton, editor. John Wiley and Sons, Inc., 605 Third Avenue, New York, NY 10158. (800) 526-5368 or (212) 850-6000. 1985. $114.00.

ONLINE DATABASES AND CD-ROMS

Climate Assessment Database. National Weather Service, National Meteorological Center, 5200 Auth Road, Suite 101, Camp Springs, MD 20233. (301) 763-8016. Contains daily, weekly and monthly summaries of North American and world climatological data. Also provides five to ten weather forecasts, and 30 to 90 day weather outlook. Subscription required. Inquire as to cost and availability.

Current Contents Search. Institute for Scientific Information, 3501 Market Street, Philadelphia, PA 19104. (215) 386-0100. FAX (215) 386-6362. Contains citations to articles listed in the table of contents of science and technology journals. Also articles in social sciences and life sciences journals. Available on BRS,(800) 289-4277, DIALOG,(800) 334-2564, online services. Inquire as to cost and availability.

Dissertation Abstracts. University Microfilms International, 300 North Zeeb Road, Ann Arbor, MI 48106. (800) 521-0600 or (313) 761-4700. Scope includes virtually all doctoral dissertations accepted at accredited American institutions from 1861 to present in 252 subject areas. Available on BRS,(800) 289-4277, DIALOG,(800) 334-2564, and OCLC EPIC,(800) 848-5878, online services. Also available on CD-ROM. Inquire as to cost and availability.

Meteorological and Geoastrophysical Abstracts. American Meteorological Society, 45 Beacon Street, Boston, MA 02108-3693. (617) 227-2425. FAX (617) 742-8718. Contains citations and abstracts to the worldwide literature on significant research in meteorology and geoastrophysics. Related topics include physical oceanography, hydrology, environmental sciences and glaciology. Covers the period 1972 to present. Available on DIALOG,(800) 334-2564, online service. Inquire as to cost and availability.

NTIS Bibliographic Database. National Technical Information Service, 5285 Port Royal Road, Springfield, VA 22161. (703) 487-4929 or FAX (703) 321-8199. Broad coverage of government-sponsored science and technology research reports, 1964 to present. Available on BRS,(800) 289-4277, DIALOG, (800) 334-2564, ORBIT, (800) 456-7248, and STN International, FIZ Karlsruhe, P.O. Box 2465, W-7500, Karlsruhe 1, Germany, online services. Also available on CD-ROM. Inquire as to cost and availability.

SCISEARCH. Institute for Scientific Information, 3501 Market Street, Philadelphia, PA 19104. (800) 523-1850 or (215) 386-0100. Broad multidisciplinary title and author index to the international literature of science and technology, 1974 to present. Available on DIALOG and ORBIT online services. Also available on CD-ROM. Inquire as to cost and availability.

WILSONLINE. H.W. Wilson Company, 950 University Avenue, Bronx, NY 10452. (800) 367-6770 or (212) 588-8400. Makes available online versions of the printed H.W. Wilson indexes including Applied Science and Technology Index, Business Periodicals Index, General Science Index, and Readers' Guide to Periodical Literature. Period covered is generally 1983 to present. Available on BRS,(800) 289-4277, DIALOG,(800) 334-2564, and OCLC EPIC,(800) 848-5878, online services. Also available on CD-ROM. Inquire as to cost and availability.

World Climate Disc. Chadwyck-Healey Inc., 1101 King Street, Alexandria, VA 22314. (703) 683-4890. FAX (703) 683-7589. Weather and climate data from approximately 5,000 weather stations worldwide, covering the years 1854 to 1990 on CD-

ROM. First edition 1992. Approximately $1200.00 per year with annual updates.

World Weatherdisc. WeatherDisc Associates, Inc., 4584 N.E. 89th Street, Seattle, WA 98115. (206) 524-4314. FAX (206) 543-0308. Meteorological data on CD-ROM which describes the climate of the earth today and for the past few hundred years. First edition 1989. Approximately $295.00 per year with biannual updates.

PERIODICALS

American Meteorological Society. BULLETIN. American Meteorological Society, 45 Beacon Street, Boston, MA 02108-3693. (617) 227-2425. FAX (617) 742-8718. Monthly. $60.00 per year.

American Meteorological Society. METEOROLOGICAL MONOGRAPHS. American Meteorological Society, 45 Beacon Street, Boston, MA 02108-3693. (617) 227-2425. FAX (617) 742-8718. Irregular. Price varies.

Atmosphere - Ocean. Canadian Meteorological and Oceanographic Society, P.O. Box 334, Newmarket, Ontario, L3Y 4X7, Canada. (416) 898-1040. FAX (416) 898-7937. Quarterly. $30.00 per year.

Dynamics of Atmospheres and Oceans. Elsevier Science Publishing Company, Inc., 655 Avenue of the Americas, New York, NY 10010. (212) 989-5800. Six times per year. $205.00 per year.

International Journal of Climatology. Royal Meteorological Society, distributed by John Wiley and Sons, Inc., 605 Third Avenue, New York, NY 10158. (800) 526-5368 or (212) 850-6000. Eight times per year. $425.00 per year.

JGR: Journal of Geophysical Research: Atmosphere. American Geophysical Union, 2000 Florida Avenue, N.W., Washington, DC 20009. (202) 462-6903. Monthly. $90.00 per year to members.

Journal of Applied Meteorology. American Meteorological Society, 45 Beacon Street, Boston, MA 02108-3693. (617) 227-2425. FAX (617) 742-8718. Monthly. $165.00 per year.

Journal of the Atmospheric Sciences. American Meteorological Society, 45 Beacon Street, Boston, MA 02108-3693. (617) 227-2425. FAX (617) 742-8718. Semi-monthly. $320.00 per year.

Monthly Weather Review. American Meteorological Society, 45 Beacon Street, Boston, MA 02108-3693. (617) 227-2425. FAX (617) 742-8718. Monthly. $205.00 per year.

National Weather Digest. National Weather Association, 4400 Stamp Road, Room 404, Temple Hills, MD 20748. (301) 899-3784.

Weather. Royal Meteorological Society, 104 Oxford Road, Reading, Berks RG1 7LJ, England. Monthly. $44.00 per year.

Weatherwise. Heldref Publications, 1319 Eighteenth Street, N.W., Washington, DC 20036-1802. (202) 296-6267. FAX (202) 296-5149. Bi-monthly. $28.00 per year.

RESEARCH CENTERS AND INSTITUTES

Center for Cloud Physics Research. University of Missouri - Rolla, Rolla, MO 65401. (314) 341-4332.

Cloud Simulation and Aerosol Laboratory. Colorado State University, Fort Collins, CO 80523. (303) 491-8667. FAX (303) 491-8449.

Cooperative Institute For Mesoscale Meteorological Studies. University of Oklahoma, 401 East Boyd, Norman, OK 73019. (405) 325-3041.

Institute of Atmospheric Physics. University of Arizona, Tucson, AZ 85721. (602) 626-6831.

National Center For Atmospheric Research. P.O. Box 3000, Boulder, CO 80307. (303) 496-1000.

FOLDS (GEOLOGY)

See: PHYSICAL GEOLOGY

FORTRAN

See: COMPUTER PROGRAMMING LANGUAGES

FOUNDRIES

See also: ALLOYS, CAST IRON, FERROALLOYS, MACHINING, METALLURGICAL ENGINEERING, METALLURGY, METALS and METALWORKING, STEEL and STEEL MAKING

ABSTRACT SERVICES AND INDEXES

Alloys Index. American Society for Metals, Metals Park, OH 44073. (216) 338-5151. FAX (216) 338-4634. 1974 to present. Monthly. $380.00. Also available on CD-ROM and online via METADEX, STN International, and DIALOG.

Applied Science and Technology Index; A Cumulative Subject Index To English Language Periodicals in the Fields of Aeronautics and Space Science, Computer Technology, Chemistry, Construction Industry, Energy and Related Areas. H.W. Wilson Co., 950 University Avenue, Bronx, NY 10452. (800) 367-6770 or (212) 588-8400. FAX (718) 590-1617. From 1958 to present. Monthly. Inquire about cost and availability. Also available on CD-ROM.

Current Contents: Enginering, Technology, and Applied Sciences. Institute for Scientific Information, 3501 Market Street, Philadelphia, PA 19104. (215) 386-0100. FAX (215) 386-6362. Weekly. $360.00 per year.

Engineering Index Monthly. Engineering Information, Inc., Castle Point on the Hudson, Hoboken, NJ 07030. (800) 221-1044. FAX (212) 832-1857. Monthly. $2200.00 per year. Also available online as COMPENDEX, and also on CD-ROM.

FOUNDRIES

Ency. of Physical Sciences and Engineering Info. Sources

Metals Abstracts and Metals Abstracts Index. American Society for Metals, Metals Park, OH 44073. (216) 338-5151. 1968 to present. Monthly. Abstracts are $1500.00 per year and Index is $460.00 per year.

ASSOCIATIONS AND PROFESSIONAL SOCIETIES

American Foundrymen's Society Inc. 505 State Street, Des Plaines, IL 60016. (708) 824-0181.

American Iron and Steel Institute. 1101 17th St., NW, Washington, DC 20036-4700. (202) 452-7100. FAX (202) 463-6573.

American Powder Metallurgy Institute. 105 College Road E., Princeton, NJ 08540. (609) 452-7700.

American Society For Metals (asm). Materials Park, OH 44073. (216) 338-5151 or FAX (216) 338-4634.

Association of Iron and Steel Engineers. Three Gateway Center, Suite 2350, Pittsburgh, PA 15222. (412) 281-6323.

Casting Industry Suppliers Association. PO Box 280, Greendale, WI 53129. (414) 423-8655.

Iron and Steel Society. 410 Commonwealth Drive, Warrendale, PA 15086. (412) 776-9460. FAX (412) 776-0430.

Metallurgical Society of the Aime (american Institute of Mining, Metallurgical, and Petroleum Engineers). 345 E.47th Street, 14th Floor, New York, NY 10017. (212) 705-7695.

National Association of Corrosion Engineers. Box 218340, Houston, TX 77218. (713) 492-0535. FAX (713) 492-8254.

Non-ferrous Founders' Society. 455 State Street, Suite 100, Des Plaines, IL 60016. (708) 299-0950. FAX (708) 299-3598.

Steel Founders Society of America. Cast Metals Federation Bldg., 455 State Street, Des Plaines, IL 60016. (708) 299-9160. FAX (708) 299-3105.

DIRECTORIES AND BIOGRAPHICAL SOURCES

Die Casting Buyer's Guide. Die Casting Industry Publications, 415 Bennett Road, Elk Grove Village, IL 60007. (708) 364-1222. FAX (708) 364-1268. 1990 to present. Annual. $44.95.

Directory of Steel Foundries and Buyer's Guide. Steel Founders Society of America, Cast Metals Federation Bldg., 455 State Street, Des Plaines, IL 60016. (708) 299-9160. FAX (708) 299-3105. Biennial. $70.00 (1995 edition).

Dun's Industrial Guide: The Metalworking Directory, 1993-94. Dun & Bradstreet Information Services Staff. 3 volumes. Dun & Bradstreet Information Services, 899 Eaton Avenue, Bethlehem, PA 18025. (610) 882-7000 or (800) 526-0651. FAX (610) 882-7269. 1993. $795.00.

Foundry Databook & Catalog File. Penton Publishing, 110 Superior Ave., Cleveland, OH 44114-2543. (800) 321-7003. 1970 to present. Annual. $10.00.

Iron and Steel Works of the World. Metal Bulletin Inc., 220 Fifth Avenue, New York, NY 10001-7781. (212) 213-6202.

FAX (212) 213-1870. 1952 to present. Quadrennial. $52.00 for non-members.

Metallurgical Society of Aime Membership List. American Institute of Mining, Metallurgical and Petroleum Engineers, 345 E.47th Street, 14th Floor, New York, NY 10017. (212) 705-7695. Inquire for availability.

1994-95 North American Directory of Non-ferrous Foundries. Non-Ferrous Founders' Society, 455 State Street, Suite 100, Des Plaines, IL 60016. (708) 299-0950. FAX (708) 299-3598. Biennial. Inquire for cost.

Research Centers Directory. Gale Research, , 835 Penobscot Building, Detroit, MI 48226-4094. (313) 961-2242. (800) 347-4253. FAX (313) 961-6083. $485.00.

Scientific and Technical Organizations and Agencies Directory. Gale Research, , 835 Penobscot Building, Detroit, MI 48226-4094. (313) 961-2242. (800) 347-4253. FAX (313) 961-6083. 4th edition. 1996. $195.00.

Who's Who in Engineering. American Association of Engineering Societies, 1111 19th Street NW, Suite 608, Washington, DC 20036. (202) 296-2237 or (800) 658-8897. 8th edition. 1991. Inquire for price.

ENCYCLOPEDIAS AND DICTIONARIES

Dictionary of Metallurgy and Foundry Technology. Karl Stolzel, editor. Elsevier Science B.V., PO Box 945, Madison Square Station, New York, NY 10159-0945. (212) 633-3650. FAX (212) 633-3680. 1987. Inquire for cost and availability.

GENERAL WORKS

Clean Cast Steel Technology. Christopher R. Wanstall. Steel Founders Society of America, Cast Metals Federation Bldg., 455 State Street, Des Plaines, IL 60016. (708) 299-9160. FAX (708) 299-3105. 1994. Inquire for cost and availability.

Ferrous and Non-ferrous Foundries. Milos Markovic. Leading Edge Reports, 2171 Jericho Turnpike, No. 200, Commack, NY 11725-2900. (800) 866-4648. FAX (216) 791-0333. 1988. Inquire for cost and availability.

Foundry Processes: Their Chemistry and Physics. S. Katz & C.F. Landefeld, editors. Plenum Publishing Corporation, 233 Spring Street, New York, NY 10013. (212) 620-8000. FAX (212) 463-0742. 1988. $125.00.

Foundry Products and Markets. Leading Edge Reports, 2171 Jericho Turnpike, No. 200, Commack, NY 11725-2900. (800) 866-4648. FAX (216) 791-0333. 1993. $1950.00.

Troubleshooting the Steel Casting Process. John M. Svoboda & Barbara Linskey, editors. Steel Founders Society of America, Cast Metals Federation Bldg., 455 State Street, Des Plaines, IL 60016. (708) 299-9160. FAX (708) 299-3105. 1987. $45.00.

HANDBOOKS AND MANUALS

Asm Handbook Volume 15: Casting. ASM International Handbook Committee. ASM International, Materials Information,

Materials Park, OH 44073-0002. (216) 338-5151 or (800) 336-5152. FAX (216) 338-4634. 1988. $147.00.

Foundryman's Handbook: Facts, Figures, Formulae. 9th edition. Elsevier Science B.V., PO Box 945, Madison Square Station, New York, NY 10159-0945. (212) 633-3650. FAX (212) 633-3680. 1986. $37.00.

Metals Handbook. ASM International, Materials Information, Materials Park, OH 44073-0002. (216) 338-5151 or (800) 336-5152. FAX (216) 338-4634. $154.00.

Woldman's Engineering Alloys. J. Frick, editor. 8th edition. ASM International, Materials Information, Materials Park, OH 44073-0002. (216) 338-5151 or (800) 336-5152. FAX (216) 338-4634. 1994. $175.00.

ONLINE DATABASES AND CD-ROMS

Compendex Plus. Engineering Information, Inc., 345 East 47th Street, New York, NY 10017. (212) 705-7600 or (800) 221-1044. Contains citations with abstracts to worldwide literature in engineering and technology, from 1970 to present. Available on online BRS,(800) 289-4277, DIALOG, (800) 334-2564, ORBIT (800) 456-7248, and STN International, FIZ Karlsruhe, P.O. Box 2465, W-7500, Karlsruhe 1, Germany, online services. Also available on CD-ROM. Inquire as to cost and availability.

CSA Engineering. Cambridge Scientific Abstracts, 7200 Wisconsin Avenue, Suite 601, Bethesda, MD 20814. (301) 961-6750 or (800) 843-7751. Contains citations and abstracts of international periodicals and research literature covering all fields of engineering and science and technology,including computer and information science, electronics, mechanical engineering, solid state materials, 1981 to present. Available on BRS,(800) 289-4277, online service.Inquire as to cost and availability.

Metadex. Jointly produced by ASM International and the Institute of Materials. Contains more than 925,000 records from the international literature on metals and alloys, concerning properties, processes, materials classes, applications, and metallurgical systems. Updated monthly. Available from ORBIT-QUESTEL (703) 442-0900.

NTIS Bibliographic Database. National Technical Information Service, 5285 Port Royal Road, Springfield, VA 22161. (703) 487-4929 or FAX (703) 321-8199. Broad coverage of government-sponsored science and technology research reports, 1964 to present. Available on BRS,(800) 289-4277, DIALOG, (800) 334-2564, ORBIT, (800) 456-7248, and STN International, FIZ Karlsruhe, P.O. Box 2465, W-7500, Karlsruhe 1, Germany, online services. Also available on CD-ROM. Inquire as to cost and availability.

SCISEARCH. Institute for Scientific Information, 3501 Market Street, Philadelphia, PA 19104. (800) 523-1850 or (215) 386-0100. Broad multidisciplinary title and author index to the international literature of science and technology, 1974 to present. Available on DIALOG,(800) 334-2564, and ORBIT,(800) 456-7248, online services. Also available on CD-ROM. Inquire as to cost and availability.

WILSONLINE. H.W. Wilson Company, 950 University Avenue, Bronx, NY 10452. (800) 367-6770 or (212) 588-8400. Makes available online versions of the printed H.W. Wilson indexes including Applied Science and Technology Index, Business Periodicals Index, General Science Index, and Readers' Guide to Periodical Literature. Period covered is generally 1983 to

present. Available on BRS,(800) 289-4277, DIALOG,(800) 334-2564, and OCLC EPIC,(800) 848-5878, online services. Also available on CD-ROM. Inquire as to cost and availability.

PERIODICALS

Casting Digest. ASM (American Society of Metals) International, Materials Information, Materials Park, OH 44073. (216) 338-5151 or FAX (216) 338-4634. 1976 to present. Monthly. $1550.00 per year.

Casting World. Continental Communications Inc., Box 1919, Bridgeport, CT 06601-1919. 1969 to present. Quarterly. $20.00 per year.

Crucible. American Chemical Society, Pittsburgh Section, c/o H.J. Heinz Company Analytical Laboratory, 1062 Progress Street, Pittsburgh, PA 15212. (412) 261-4300. 1918 to present. Monthly except July-August. $2.00 per year.

Die Casting Engineer. North American Die Casting Association, 9701 W. Higgins Road, Number 880, Rosemont, IL 60018-4721. (708) 292-3600. FAX (708) 292-3620. 1957 to present. Bi-monthly. $48.00 per year.

Die Casting Management. C-K Publishing Inc., Box 247, Wonder Lake, IL 60097-0247. (815) 728-0912. 1983 to present. Bi-monthly. $35.00 per year.

Foundry International. FMJ International Publications, Ltd., Queensway House, 2 Queensway, RedHill, Surrey RH1 1QS England. Telephone 0737-768611. FAX 0737-761685. 1978 to present. Quarterly. $152.00 per year.

Foundry Management and Technology. Penton Publishing, 110 Superior Ave., Cleveland, OH 44114-2543. 1892 to present. Monthly. $45.00 per year.

Foundry Trade Journal. FMJ International Publications, Ltd., Queensway House, 2 Queensway, RedHill, Surrey RH1 1QS England. Telephone 0737-768611. FAX 0737-761685. 1902 to present. Bi-weekly. $232.60 per year.

Foundryman. IBF Publications, Bordsley Hall, the Holloway, Alvechurch, Birmingham B48 7QA, England. Telephone 0527-596101. FAX 0527-596102. 1956 to present. Monthly. Inquire for cost.

International Journal of Rapid Solidification. AB Academic Publishing, PO Box 42, Bicester Oxon., 0X6 7NW, England. Telephone 0869-320949. 1984 to present. 4 times a year. $198.00 per year.

Iron Age. Hitchcock Publishing, 191 S. Gary Avenue, Carol Stream, IL 60188. (708) 462-4641. FAX (708) 462-2205. 1867-1993. Monthly. $55.00 per year.

Iron and Steel Engineer. Association of Iron and Steel Engineers, Three Gateway Center, Suite 2350, Pittsburgh, PA 15222. (412) 281-6323. 1924 to present. Monthly. $50.00 per year.

Iron and Steel International. FMJ International Publications, Ltd., Queensway House, 2 Queensway, RedHill, Surrey RH1 1QS England. Telephone 0737-768611. FAX 0737-761685. Annual. $100.30 per year.

FOUNDRIES

Ency. of Physical Sciences and Engineering Info. Sources

Iron & Steelmaker. Iron and Steel Society, 410 Commonwealth Drive, Warrendale, PA 15086. (412) 776-9460. FAX (412) 776-0430. 1974 to present. Monthly. $50.00 per year.

Ironmaking and Steelmaking. Institute of Materials, 1 Carlton House Terrace, London SW1Y 5DB, England. Telephone 071-839-4071. FAX 071-839-2078. 1974 to present. Bi-monthly. $420.00 per year for non-members.

J O M (journal of Metals). Minerals, Metals and Materials Society, 420 Commonwealth Drive, Warrendale, PA 15086. (412) 776-9080. 1949 to present. Monthly. $50.00 per year.

Metal Casting and Surface Finishing. RALA Information Service Pty. Ltd., 203-205 Darling Street, Balmain N.S.W., Australia. Telephone 02-555-1944. FAX 02-555-1496. 1955 to present. Bi-monthly. Inquire for cost and availability.

Metalcaster. American Cast Metals Association, 455 State Street, Des Plaines, IL 60016. (708) 299-9160. FAX (708) 299-3105. 1947 to present. Quarterly. $14.50 per year.

Metallurgical Transactions A: Physical Metallurgy and Materials Science. ASM International, Materials Park, OH 44073. (216) 338-5151 or FAX (216) 338-4634. 1970 to present. Monthly. $520.00 per year.

Metals Week. McGraw-Hill Publishing Company, Commodity Services Group, 1221 Avenue of the Americas, New York, NY 10020. (800) 262-4729 or (212) 512-3825. 1930 to present. Weekly. $770.00 per year.

Modern Casting. American Foundrymen's Society Inc., 505 State Street, Des Plaines, IL 60016. (708) 824-0181. 1938 to present. Monthly. $45.00 per year.

RESEARCH CENTERS AND INSTITUTES

American Iron and Steel Institute. 1101 17th St., NW, Washington, DC 20036-4700. (202) 452-7100. FAX (202) 463-6573.

Carnegie Mellon University Center For Iron and Steel Making Research. MEMS Department, Pittsburgh, PA 15213. (412) 268-2677.

Colorado School of Mines, Advanced Steel Processing and Products Research Center. Golden, CO 80401. (303) 273-3774.

Cooperative Program in Metallurgy. Pennsylvania State University. 208A Steidle Bldg., University Park, PA 16802. (814) 865-5446.

University of Wisconsin—milwaukee Foundry & Solidification Processing Research Laboratory. PO Box 784, Mechanical Engineering Department, College of Engineering & Applied Sciences, Milwaukee, WI 53201. (414) 229-4987.

FOURIER ANALYSIS

See: MATHEMATICS

FRACTIONATING COLUMNS

See: CHEMICAL ENGINEERING

FRACTALS

See also: ANALYTIC GEOMETRY, CALCULUS, CHAOS, MAtheMATICS, TOPOLOGY

ABSTRACT SERVICES AND INDEXES

Applied Mechanics Reviews: An Assessment of World Literature in Engineering Sciences. American Society of Mechanical Engineers, 345 East 47th Street, New York, NY 10017. (212) 705-7703. 1948 to present. Monthly. $360.00 per year. ISSN: 0003-6900.

Applied Science and Technology Index; A Cumulative Subject Index To English Language Periodicals in the Fields of Aeronautics and Space Science, Computer Technology, Chemistry, Construction Industry, Energy and Related Areas. . H.W. Wilson Company, 950 University Avenue, Bronx, NY 10452. (800) 367-6770. FAX (718) 590-1617. 1958 to present. Monthly. Available on-line from BRS and WILSONLINE. Also available on CD-ROM. Inquire about cost and availability. ISSN: 0003-6986.

Compactmath - Compact Mathematics Library. Cumulative CD-ROM edition of Zentralblatt fuer Mathematik - Mathematics Abstracts. Springer-Verlag New York, Inc. 44 Hartz Way. Secaucus. NJ. 07096-2491. (201) 348-4033. FAX (201) 348-4505. 1993 to present. Annual. Available only on CD- ROM. Inquire regarding cost and availability. ISSN: 0938- 3174.

Compumath Citation Index. Institute for Scientific Information, 3501 Market Street, Philadelphia, PA 19104. (800) 523-1850 or (215)-386-0100. FAX (215) 386-2991. Three times per year. $1955.00. Also available online and on CD-ROM. Inquire regarding cost and availablity. ISSN: 0730-6199.

Current Mathematical Publications. American Mathematical Society. P.O. Box 1571, Annex Station, Providence, RI 02901-9930. (800) 556-7774 or (401) 455-4000. FAX (401) 331-3842. 1969 to present. Seventeen times per year. $377.00 per year. Available online from BRS, DIALOG, European Space Agency. Also available on CD-ROM. Inquire regarding cost and availability. ISSN: 0361-4794.

General Science Index. H.W. Wilson Company, 950 University Avenue, Bronx, NY 10452. (212) 588-8400. (800) 367-6770. FAX (718) 590-1617. From 1978 to present. Ten issues per year; quarterly and annual cumulations. Service basis. Available on CD-ROM and online. Inquire regarding cost and availability. ISSN: 0162-1963.

Mathematical Reviews: A Review Journal Covering the World Literature of Mathematical Research. American Mathematical Society, P.O. Box 1571, Annex Station, Providence, RI 02901-9930. (800) 556-7774 or (401) 455- 4000. 1940 to present. Monthly. $4594.00 per year. Also available via network (MathSciNet), online and on CD-ROM.Inquire regarding cost and availability. ISSN: 0025-5629.

N T I S Alerts: Mathematical Sciences. U.S. National Technical Information Service, 5285 Port Royal Road, Springfield, VA 22161. (703) 487-4650. FAX (703) 321-8547.Weekly. $140.00.

Science Citation Index. Institute for Scientific Information. 3501 Market Street, Philadelphia, PA 19104. (215) 386-0100. (800) 523-1850. FAX (215) 386-2991. 1961 to present. Six issues per year, plus annual cumulation. $11650.00 per year. Also available online and on CD-ROM.Inquire about price and availability. ISSN: 0036-827X.

ANNUAL REVIEWS AND YEARBOOKS

Advances in Applied Mathematics. Academic Press, Inc., 6277 Sea Harbor Drive, Orlando, FL 32821-8340. (800) 5439534.1980 to pesent. Irregular, Price varies; inquire.

Applied Mathematical Sciences. Springer-Verlag, 175 Fifth Avenue, New York, NY 10010. (212) 460-2500. FAX (212) 473-6272. 1972 to present. Irregular. ISSN: 0066-5452.

CBMS-NSF Regional Conference Series in Applied Mathematics. Society for Industrial and Applied Mathematics, 3600 University City Science Center, Philadelphia, PA 19104-2688. (215) 382-9800. FAX (215) 386-7999. 1971 to present. Irregular. ISSN: 0163-9439.

Ergebnisse Der Mathematik Und Ihrer Grenzgebiete. Neue folge. Springer-Verlag New York, Inc. 175 Fifth Avenue, New York, NY 10010. (212) 460-1500. FAX (212) 473-6272. 1953 to present. Irregular. Price varies; inquire. ISSN: 0071-1136.

Lecture Notes in Mathematics. Springer-Verlag New York, Inc. 44 Hartz Way. Secaucus. NJ. 07096-2491. (201) 348-4033. FAX (201) 348-4505. Irregular. Price varies; inquire.

ASSOCIATIONS AND PROFESSIONAL SOCIETIES

American Astronomical Society. 2000 Florida Avenue NW, Suite 400, Washington, DC 20009. (202) 328-2010. FAX: (202) 234-2560.

American Geophysical Union. 2000 Florida Avenue, NW, Washington, DC 20009. (202) 462-6900. (800) 966-AGU1. FAX (202) 328-0566.

American Institute of Physics. One Physics Ellipse, College Park, MD 20740-3843. (301) 209-3100.

American Mathematical Society. P.O. Box 6248, Providence, RI 02940. (401) 455-4000. FAX (401) 331-3842.

American Physical Society. One Physics Ellipse, College Park, MD 20740-3843. (301) 209-3200. FAX (301) 209-0865.

Association For Women in Mathematics. 4114 Computer and Space Science Building, University of Maryland, College Park, MD 20742-2461. (301) 405-7892.

Mathematical Association of America. 1529 18th Street, NW, Washington, DC 20036. (202) 387-5200. FAX (202) 265-2384.

NATIONAL COUNCIL of TEACHERS of MAtheMATICS. 1906 Association Drive, Reston, VA 22091. (703) 620-9840. FAX (703) 475-2970.

Operations Research Society of America. 1314 Guilford Avenue, Baltimore, MD. 21202. (410) 528-4146. FAX (410)528-8556.

Society For Industrial and Applied Mathematics. 3600 University City Science Center, Philadelphia, PA 19104-2688. (215) 382-9800. FAX (215) 386-7999.

BIBLIOGRAPHIES

Bibliography On Chaos. S. Y Zhang, editor. World Scientific Publishing Company, Inc., 1060 Main Street, River Edge, NJ 07661. (201) 487-9655. (800) 227-7562. 1991. $40.00 in paper.

DIRECTORIES AND BIOGRAPHICAL SOURCES

American Men and Women of Science: Physical and Biological Sciences. R. R. Bowker Inc., 121 Chanlon Road, New Providence, NJ 07974. (908) 464-6800. (800) 521-8110. 20th edition. 8 volumes. 1996. $850.00.

American Physical Society. Membership Directory Bulletin Issue. American Physical Society, One Physics Ellipse, College Park, MD 20740-3843. (301) 209-3200. FAX (301) 209- 0865. Biennial. $50.00.

Assistantships and Graduate Fellowships in the Mathematical Sciences. American Mathematical Society, P.O. Box 6248, Providence, RI 02940. (401) 455-4000. (800) 321-4267. Annual. $19.00. ISSN: 1040-7650.

Combined Membership List. American Mathematical Society, P.O. Box 6248, Providence, RI 02940. (401) 455-4000. 800) 321-4267. Annual. $50.00 per year. ISSN: 0560-6461.

Mathematical Science Professional Directory. American Mathematical Society, P.O. Box 6248, Providence, RI 02940. Annual. $45.00. ISBN: 0-8218-0173-2.

Research Centers Directory. Gale Research Company, 835 Penobscot Building, Detroit, MI 48226-4094. (313) 961- 2242. (800) 347-4253. 20th edition. 1995. Annual. $485.00. Issn 0080-1518.

World Directory of Mathematicians. G.D. Mostow, editor. International Mathematical Union, Helsinki. 1958 to present. 9th revised edition. 1990. $40.00. Available from: American Mathematical Society, P.O. Box 6248, Providence, RI 02940. (401) 455-4000. ISSN: 0512-2740.

ENCYCLOPEDIAS AND DICTIONARIES

Concise Oxford Dictionary of Mathematics. Christopher Clapham. Oxford University Press, 200 Madison Avenue, New York, NY 10016. (212) 725-6000. (800) 334-4249. 1990. $10.95. ISBN: 0-19-286103-4.

Encyclopedia of Applied Physics. George Trigg, editor. VCH Publications, Inc., 220 East 23rd Street, Suite 909, New York, NY 10010-4606. (212) 683-8333. (800) 422-8824. 20 volume set. 1991-. $5990.00.

Encclopedia of Mathematics. An updated and annotated translation of the Soviet Mathematical Encyclopedia. Michael Hazewinkel, editor. Kluwer Academic Publishers. 101 Philip Drive, Assinippi Park, Norwell, MA 02061. (617) 871-6600. 1987 - . 10 volumes. $147.00, volume 1. 1-55608-010-7 set.

FRACTALS

Ency. of Physical Sciences and Engineering Info. Sources

Encyclopedia of Mathematics and Its Applications. Addison-Wesley Publishing Co., Inc., 1 Jacob Way, Reading, MA 01867. (617) 944-3700. (800) 447-2226. 1976 to present. irregular. Volumes are individually priced.

Encyclopedic Dictionary of Mathematics. Kiyosi Ito, editor. MIT Press, 55 Hayward Street, Cambridge, MA 02142. (617) 625-8569. 2nd edition, reprinted. 1993. 2 volumes. $70.00. ISBN: 0-262-59020-4.

Mathematics Dictionary. Robert C. James, editor. Van Nostrand Reinhold, 115 Fifth Avenue, New York, NY 10003. (800)-842-3636. 5th edition. 1992. $42.95. ISBN: 0-442- 00741-8.

The Words of Mathematics: An Etymological Dictionary of Mathematical Terms Used in English. Steven Schwartzman. Mathematical Association of America, 18th Street NW, Washington, DC 20036. (800) 331-1622. 1994. $29.50. ISBN: 0-88385-551.

GENERAL WORKS

Applications of Fractals and Chaos. A. J. Crilly. Springer-Verlag New York, Inc., 175 FifthAvenue, New York, NY 10010. (212) 460-1500. (800) 777-4643. 1993. $69.00.

Chaotic and Fractal Dynamics: An Introduction For Applied Scientists and Enginers. Francis C. Moon. John Wiley & Sons, Inc., 605 Third Avenue, New York, NY 10158-0012. (212) 850-6000. (800) 225-5945. 1992. $64.95.

Chaos and Fractals: New Frontiers of Science. Heinz-Otto Peitgen, et al. Springer-Verlag New York, Inc., 175 Fifth Avenue, New York, NY 10010. (212) 460-1500. (800) 777- 4643. 1993. $49.00

Chaos, Dynamics and Fractals: An Algorithmic Approach To Deterministic Chaos. Joseph L. McCauley. Cambridge University Press, 40 West 20th Street, New York, NY 10011- 4211. (212) 924-3900. 1994. $27.95 in paper.

Classics On Fractals. Gerald A. Edger. Addison-Wesley Publishing Co., Inc., 1 Jacob Way, Reading, MA 01867. (617) 944-3700. (800) 447-2226. 1993. $59.95.

Encounters With Chaos. Denny Gulick. McGraw-Hill Publishing Company, 1221 Avenue of the Americas, New York,

NY 10020. (212) 512-2000. (800) 262-4729. 1992. $wfi

Fractal Surfaces. John C. Russ. Plenum Publishing Corp., 233 Spring Street, New York, NY. (212) 620-8000. (800) 2221-9369. 1994. 55.00.

Fractals and Disordered Systems. A. Bunde and S. Havlin, editors. Springer-Verlag New York, Inc., 175 Fifth Avenue, New York, NY 10010. (212) 460-1500. (800) 777- 4643. FAX (212) 473-6272. 1991. $64.50.

Fractals and Dynamic Systems in Geoscience. J. H. Kruhl, editor. Springer-Verlag New York, Inc., 175 Fifth Avenue, New York, NY 10010. (212) 460-1500. (800) 777- 4643. 1994. $99.00.

Fractals Everywhere. Michael F. Barnsley. Academic Press, Inc., 6277 Sea Harbor Drive, Orlando, FL (800) 321-5068. 2nd edition. 1993. $49.95.

Fractals in Geophysics. Benoit Mandelbrot and Christopher H. Scholz. Birkhauser Boston, 675 Massachusetts Avenue, Cambridge, MA 02139. (617) 876-2333. (800) 777-4643. 1990. $34.50.

Fractals in Physics: Essays in Honor of Benoit B. Mandelbrot. A. Aharony and Jens Feder, editors. Elsevier Science Publishing Company, Inc., 655 Avenue of the Americas, New York, NY 10010. (212) 989-5800. FAX (914) 333-2444. 1990. $45.25 in paper.

Fractals, Random Shapes and Point Fields: Methods of Geometrical Statistics. Deitrich Stoyan and Helga Stoyan.

John Wiley & Sons, Inc., 605 Third Avenue, New York, NY 10158-0012. (212) 850-6000. (800) 225-5945. 1994.$59.95.

Fractals: Selected Reprints. Alan J. Hurd, editor. Anerican Association of Physics Teachers, One Physics Ellipse, College Park MD 20740-3842. (301) 209-3300. 1989. $18.00 in paper.

Introduction To Fractals and Chaos. Richard M. Crownover. Jones & Bartlett Pubs., Inc., 1 Exeter Plaza, Boston, MA 02116. (617) 859-3900. (800) 832-0034. 1994. $51.25.

Measure, Topology and Fractal Geometry. Gerald A. Edgar. Addison-Wesley Publishing Co., Inc., 1 Jacob Way, Reading, MA 01867. (617) 944-3700. (800) 447-2226. 1992. $29.95.

Physics and Fractal Structure Structures. Jean-Francois Gouyet. Springer-Verlag New York, Inc., 175 Fifth Avenue, New York, NY 10010. (212) 460-1500. (800) 777- 4643. 1994. $39.00

A Random Walk Through Fractal Dimensions. Brian H. Kaye. VCH Publications, Inc., 220 East 23rd Street, Suite 909, New York, NY 10010-4606. (212) 683-8333. (800) 422-8824.

Statistical Mechanics and Fractals. R. L. Dobrushin and S. Kusukoa. Springer-Verlag New York, Inc., 175 Fifth Avenue, New York, NY 10010. (212) 460-1500. (800) 777- 4643. 1994. $23.00.

Wavelets, Fractals, and Fourier Transforms. M. Farge, et al, editors. Oxford University Press, Inc., 200 Madison Avenue, New York, NY 10016. (212) 725-6000. (800) 334-4249. 1993. $98.00.

HANDBOOKS AND MANUALS

Handbook of Mathematical Sciences. William H. Beyer, editor. CRC Press, 2000 Corporate Boulevard, Boca Raton, FL 33431. (800) 272-7737. 6th edition. 1987. $91.95. ISBN: 0-8493-0656-6.

Handbook of Physical Quantities. Igor S. Grigoriev and Evgenil Z. Melikhov, editors. CRC Press, Inc., 2000 Corporate Boulevard, NW, Boca Raton, FL 33431. (407) 994- 0555. (800) 272-7737. 1995. $99.00.

Handbook of Writing For the Mathematical Sciences. Society for Industrial and Applied Mathematics, 1993. $21.50.

Standard Mathematical Tables and Formulas. William H. Beyer, editor. CRC Press, 2000 Corporate Boulevard, Boca Raton, FL 33431. (800) 272-7737. 29th edition. 1991. $41.95. ISBN: 084-930- 6299.

Handbook of Applicable Mathematics. Walter Ledermann, editor. Wiley 1 Wiley Drive, Somerset, NJ 08875-1272. 1985. $1496.00 6 volumes. ISBN: 0-471-90804-5.

Tables of Integrals, Series and Products: Corrected and Enlarged Edition. I. S. Gradshteyn and I. M. Ryzhik. Academic Press, Inc., 6277 Sea Harbor Drive, Orlando, FL 32821. (800) 321-5068. 5th edition. 1993. $61.00. ISBN: 0-12-294755-X.

ONLINE DATABASES AND CD ROMS

CA Search. Chemical Abstracts Service, P.O. Box 3012, Columbus, OH 43210-0012. (614) 447-3600. (800) 848-6533. FAX (614) 447-3709. Very comprehensive guide to worldwide chemical literature and related fields, 1972 to present. Available on BRS,(800) 289-4277, DIALOG, (800) 334-2564, ORBIT (800) 456-7248, and STN International, FIZ Karlsruhe, P.O. Box 2465, W-7500, Karlsruhe 1, Germany, online services. Inquire as to cost and availability.

Compendex Plus. Engineering Information, Inc., 345 East 47th Street, New York, NY 10017. (212) 705-7600 or (800) 221-1044. Contains citations with abstracts to worldwide literature in engineering and technology, from 1970 to present. Available online BRS,(800) 289-4277, DIALOG, (800) 334-2564, ORBIT (800) 456-7248, and STN International,FIZ Karlsruhe, P.O. Box 2465, W-7500, Karlsruhe 1, Germany, online services. Also available on CD-ROM. Inquire as tocost and availability.

Dissertation Abstracts. University Microfilms International, 300 North Zeeb Road, Ann Arbor, MI 48106. (800) 521-0600 or (313) 761-4700. Scope includes virtually all doctoral dissertations accepted at accredited American institutions from 1861 to present in 252 subject areas. Available on BRS,(800) 289-4277, DIALOG,(800) 334-2564, and OCLC EPIC,(800) 848-5878, online services. Also available on CD-ROM. Inquire as to cost and availability.

INSPEC. Institution of Electrical Engineers, Michael Faraday House, Six Hills Way, Stevenage, Herts. SG1 2AY, England. Telephone: 0438 313311 or FAX 0438 742840. Contains citations to the worldwide literature of physics, electronics and electrical engineering, computer technology, and related fields. Available on BRS,(800) 289-4277, DIALOG, (800) 334-2564, ORBIT, (800) 456-7248, and STN International, FIZ Karlsruhe, P.O. Box 2465, W-7500, Karlsruhe 1, Germany, online services. Inquire as to cost and availability.

Mathsci. American Mathematical Society, P.O. Box 6248, Providence, RI 02940. (800) 321-4667 or (401) 455-4000 or FAX (401) 331-3842. Scope includes pure and applied mathematics and related areas of physics, statistics, engineering, computer science, and operations research literature since 1959. Available on DIALOG,(800) 334-2564, online service. Also available on CD-ROM and MATHSCINET, on the Internet. Inquire as to cost and availability.

NTIS Bibliographic Database. National Technical Information Service, 5285 Port Royal Road, Springfield, VA 22161. (703) 487-4929 or FAX (703) 321-8199. Broad coverage of government-sponsored science and technology research reports, 1964 to present. Available on BRS,(800) 289-4277, DIALOG, (800) 334-2564, ORBIT, (800) 456-7248, and STN International, FIZ

Karlsruhe, P.O. Box 2465, W-7500, Karlsruhe 1, Germany, online services. Also available on CD-ROM. Inquire as to cost and availability.

Scientific and Technical Books and Serials in Print. R.R. Bowker Inc., 121 Chanlon Road, New Providence, NJ 07974. (800) 521-8110 or (908) 464-6800. List of currently published books and serials in the physical and biological sciences, engineering and technology, with subject, author and titles indexes. Available on ORBIT,(800) 456-7248, online service.Inquire as to cost and availability.

SCISEARCH. Institute for Scientific Information, 3501 Market Street, Philadelphia, PA 19104. (800) 523-1850 or (215) 386-0100. Broad multidisciplinary title and author index to the international literature of science and technology, 1974 to present. Available on DIALOG,(800) 334-2564, and ORBIT,(800) 456-7248, online services. Also available on CD-ROM. Inquire as to cost and availability.

WILSONLINE. H.W. Wilson Company, 950 University Avenue, Bronx, NY 10452. (800) 367-6770 or (212) 588-8400. Makes available online versions of the printed H.W. Wilson indexes including Applied Science and Technology Index, Business Periodicals Index, General Science Index, and Readers' Guide to Periodical Literature. Period covered is generally 1983 to present. Available on BRS,(800) 289-4277, DIALOG,(800) 334-2564, and OCLC EPIC,(800) 848-5878, online services. Also available on CD-ROM. Inquire as to cost and availability.

PERIODICALS

American Journal of Mathematics. Johns Hopkins University Press, 2715 North Charles Street, Baltimore, MD 21218. (410) 516-6987. FAX (410) 516-6968. 1878 to present. Bimonthly. $162.00 per year. ISSN: 0002-9327.

American Mathematical Monthly. Mathematical Association of America. 1529 Eighteenth Street, NW, Washington, DC 20036. (202) 387-5200. 1894 to present. Ten issues per year. $160.00. ISSN: 0002-9890.

American Journal of Physics. American Association of Physics Teachers, 5112 Berywn Road, College Park, MD 20740. (301) 345-4200. 1933 to present. Monthly. $215.00 per year. ISSN: 0002-9505.

Chaos; An Interdisciplinary Journal of Nonlinear Science. American Institute of Physics, One Physics Ellipse, College Park, MD 20740-3843. (301) 209-3000. 1991 to present. Quarterly. $225.00 per year. ISSN: 1054-1500.

Chaos, Solitons and Fractals: Applications in Science and Engineering. Elsevier Science Publishing Company, Inc., 655 Avenue of the Americas, New York, NY 10010. (212) 989- 5800. FAX (212) 633-3990. 1991 to present. Monthly. $830.00 per year. ISSN: 0960-0779.

Communications On Pure and Applied Mathematics. Courant Institute of Mathematical Sciences. Distributed by John Wiley & Sons, Inc. 605 Third Avenue, New York, NY 10158- 0012. (212) 850-6000. FAX (212) 850-6088. 1939 to present. $730.00 per year. ISSN: 0010-3640.

Contemporary Physics. Taylor & Francis, Ltd, Rankine Road, Basingstoke, Hants RG24 8PR, England. TEL 0256-840366.

FRACTALS

Ency. of Physical Sciences and Engineering Info. Sources

FAX 0256-479438. 1959 to present. Bimonthly. $318.00 per year. ISSN: 0010-7514.

I M A Journal of Applied Mathematics. Oxford University Press, 2001 Evans Road, Cary, NC 27513. (919) 677-0977. 1981 to present. Bimonthly. $410.00 per year. ISSN: 0272- 4960.

International Journal of Bifurcations and Chaos in Applied Sciences and Engineering. World Scientific Publishing Co., 1060 Main Street, Suite 18, River Edge, NJ 07661. (800) 227-7562. 1991 to present. Bi-monthly. $560.00 per year. ISSN: 0218-1274.

Mathematics Magazine. Mathematical Association of America. 1529 Eighteenth Street, NW, Washington, DC 20036. (202) 387-5200. 1926 to present. Bimonthly. $80.00 per year. ISSN: 0025-570X.

Physics Today. American Institute of Physics, One Physics Ellipse, College Park, MD 20740-3843. (301) 209-3000.1948 to present. Monthly. $140.00 per year. ISSN: 0031- 9228.

Siam Journal On Applied Mathematics. Society for Industrial and Applied Mathematics, 3600 University City Science Center, Philadelphia, PA 19104-2688. (215) 382- 9800. FAX (215) 386-7999. 1953 to present. Bimonthly. $242.00 per year. ISSN: 0036-1399.

Siam Review. Society for Industrial and Applied Mathematics, 3600 University City Science Center, Philadelphia, PA 19104-2688. (215) 382-9800. 1959 to present. Quarterly. $148.00 per year. ISSN: 0036-1445.

RESEARCH CENTERS AND INSTITUTES

Center for Applied Mathematics. Cornell University. 305 Sage Hall, Ithaca, NY 14853-6201. (607) 255-4335. FAX (607) 255-4336.

Center For Mathematical Science Research. Rutgers University. New Brunswick, NJ 08903. (201) 932-3117. FAX (201) 932-5530.

Center For Mathematical Sciences. University of Wisconsin-Madison. 610 Walnut Street, Madison, WI 53705. (608) 263-2696. FAX (608) 263-2841.

Center For Numerical Analysis. University of Texas at Austin. Austin, TX 78712. (512) 471-1242,

Center For Pure and Applied Mathematics. University of California, Berkeley. 977 Evans Hall, Berkeley, CA 94720. (415) 642-0116. FAX (415) 642-6726.

Institute For Computer Applications in Science and Engineering. Mail Stop 132C, NASA, Langley Research Center, Hampton, VA 23665. (804) 865-2513. FAX (804) 864-6134.

FRACTURE MECHANICS

See also: AERONAUTICAL ENGINEERING, CIVIL ENGINEERING, MATERIALS SCIENCE, MECHANICAL ENGINEERING, MECHANICS, NONDESTRUCTIVE TESTING, STRUCTURAL ENGINEERING

ABSTRACT SERVICES AND INDEXES

Abstract Journal in Earthquake Engineering. University of California, Berkeley, Earthquake Engineering Research Center, 1301 South 46th Street, Richmond, CA 94804-4698. (510) 231-9413 or FAX (510) 231-9471. 1972 to present. Annual. $80.00 per year.

Applied Science and Technology Index; A Cumulative Subject Index To English Language Periodicals in the Fields of Aeronautics and Space Science, Computer Technology, Chemistry, Construction Industry, Energy and Related Areas. H.W. Wilson Co., 950 University Avenue, Bronx, NY 10452. (800) 367-6770 or (212) 588-8400. FAX (718) 590-1617. From 1958 to present. Monthly. Inquire about cost and availability. Also available on CD-ROM and online.

ASCE Combined Annual Combined Index. American Society of Civil Engineers, 345 East 47th Street, New York, NY 10017-2398. (212) 705-7520 or (800) 548-2723. Annual. $48.00.

ASCE Publications Information. 1966-present. American Society of Civil Engineers, 345 East 47th Street, New York, NY 10017-2398. (212) 705-7520 or (800) 548-2723. Bi-monthly. $160.00 per year for non-members.

Current Contents: Engineering, Technology, and Applied Sciences. Institute for Scientific Information, 3501 Market Street, Philadelphia, PA 19104. (215) 386-0100. FAX (215) 386-6362. 1970 to present. Weekly. $442.00 per year.

Engineering Index Monthly. Engineering Information, Inc., Castle Point on the Hudson, Hoboken, NJ 07030. (800) 221-1044. FAX (212) 832-1857. Monthly. $2300.00 per year. Also available online as COMPENDEX, and also on CD-ROM. Covers chemical engineering, computers, electrical engineering, civil engineering, metals and mining, industrial management, and mechanical engineering.

Index To Scientific and Technical Proceedings. Institute for Scientific Information, 3501 Market St., Philadelphia, PA 19104. (215) 386-0100. FAX (215) 386-6362. Monthly. $500.00 per year.

International Civil Engineering Abstracts. CITIS Ltd., 2 Rosemount Terrace, Blackrock, Dublin, Ireland. Telephone 353-1-2886227. FAX 353-1-885-971. 1974 to present. Monthly. $660.00 per year. Also available on CD-ROM.

Mechanical Engineering Abstracts (formerly ISMEC), Cambridge Scientific Abstracts, 7200 Wisconsin Avenue, Bethesda, MD 20814-4823. (301) 961-6750. FAX (301) 961-6720. 1967 to present. Monthly. $895.00 per year. Summarizes world literature in mechanical engineering, production engineering, and engineering management. Also available online.

Science Citation Index. Institute for Scientific Information, 3501 Market Street, Philadelphia, PA 19104. (215) 386-0100. FAX (215) 386-6362. Inquire about availability and cost. Also available on CD-ROM.

ASSOCIATIONS AND PROFESSIONAL SOCIETIES

American Society For Nondestructive Testing. 1711 Arlingate Lane, Box 28518, Columbus, OH 43228-0158. (614) 274-6003. FAX (614) 274-6899.

American Society For Testing and Materials. 1916 Race Street, Philadelphia, PA 19103. (215) 299-5585.

American Society of Civil Engineers. 345 East 47th Street, New York, NY 10017-2398. (212) 705-7520 or (800) 548-2723.

American Society of Mechanical Engineers. 345 East 47th Street, New York, NY 10017-2398. (212) 705-7703.

DIRECTORIES AND BIOGRAPHICAL SOURCES

Directory of Engineering Societies and Related Organizations. Gordon Davis. 13th edition. American Association of Engineering Societies, 1111 19th Street NW, Suite 608, Washington, DC 20036. (202) 296-2237 or (800) 658-8897. 1989. Inquire for price.

International Research Centers Directory. Gale Research, , 835 Penobscot Building, Detroit, MI 48226-4094. (313) 961-2242. (800) 347-4253. FAX (313) 961-6083. 8th edition. 1995. $410.00

Research Centers Directory. Gale Research, , 835 Penobscot Building, Detroit, MI 48226-4094. (313) 961-2242. (800) 347-4253. FAX (313) 961-6083. $485.00.

Scientific and Technical Organizations and Agencies Directory. Gale Research, , 835 Penobscot Building, Detroit, MI 48226-4094. (313) 961-2242. (800) 347-4253. FAX (313) 961-6083. 4th edition. 1996. $195.00.

Who's Who in Engineering. American Association of Engineering Societies, 1111 19th Street NW, Suite 608, Washington, DC 20036. (202) 296-2237 or (800) 658-8897. 8th edition. 1991. Inquire for price.

GENERAL WORKS

Advances in Fracture Research. K. Salama, et al. Pergamon Press Inc., Maxwell House, Fairview Park, Elmsford, NY 10523. (914) 592-7700. Fax (914) 592-3625. 1989. $1175.00.

Fracture Mechanics: An Introduction. E.E. Gdoutos. Kluwer Academic Publishers, P.O. Box 358, Accord Station, Hingham, MA 02018-0358. (617) 871-6000. FAX (617) 871-6528. 1993. $140.00.

Fracture Mechanics & Failure Control For Inspectors and Engineers. P.F. Timmins. ASM International, Materials Park, OH 44073-0002. (800) 336-5152. FAX (216) 338-4634. 1994. $40.00.

HANDBOOKS AND MANUALS

Asm Handbook Volume 12: Fractograpy. ASM International Handbook Committee. ASM International, Materials Information, Materials Park, OH 44073-0002. (216) 338-5151 or (800) 336-5152. FAX (216) 338-4634. 1987. $147.00.

Handbook of Mechanics, Materials, and Structures. Alexander Blake, editor. John Wiley and Sons, Inc., 605 Third Avenue, New York, NY 10158. (800) 526-5368 or (212) 850-6000. 1985. $120.00.

Marks' Standard Handbook For Mechanical Engineers. E.A. Avallone and T. Baumeister III. 9th edition. McGraw-Hill Publishing Company, 1221 Avenue of the Americas, New York, NY 10020. (800) 262-4729 or (212) 512-3825. 1987. $133.00.

ONLINE DATABASES AND CD-ROMS

Compendex Plus. Engineering Information, Inc., 345 East 47th Street, New York, NY 10017. (212) 705-7600 or (800) 221-1044. Contains citations with abstracts to worldwide literature in engineering and technology, from 1970 to present. Available on online BRS,(800) 289-4277, DIALOG, (800) 334-2564, ORBIT (800) 456-7248, and STN International, FIZ Karlsruhe, P.O. Box 2465, W-7500, Karlsruhe 1, Germany, online services. Also available on CD-ROM. Inquire as to cost and availability.

CSA Engineering. Cambridge Scientific Abstracts, 7200 Wisconsin Avenue, Suite 601, Bethesda, MD 20814. (301) 961-6750 or (800) 843-7751. Contains citations and abstracts of international periodicals and research literature covering all fields of engineering and science and technology,including computer and information science, electronics, mechanical engineering, solid state materials, 1981 to present. Available on BRS,(800) 289-4277, online service.Inquire as to cost and availability.

Current Contents Search. Institute for Scientific Information, 3501 Market Street, Philadelphia, PA 19104. (215) 386-0100. FAX (215) 386-6362. Contains citations to articles listed in the table of contents of science and technology journals. Also articles in social sciences and life sciences journals. Available on BRS,(800) 289-4277, DIALOG,(800) 334-2564, online services. Inquire as to cost and availability.

Dissertation Abstracts. University Microfilms International, 300 North Zeeb Road, Ann Arbor, MI 48106. (800) 521-0600 or (313) 761-4700. Scope includes virtually all doctoral dissertations accepted at accredited American institutions from 1861 to present in 252 subject areas. Available on BRS,(800) 289-4277, DIALOG,(800) 334-2564, and OCLC EPIC,(800) 848-5878, online services. Also available on CD-ROM. Inquire as to cost and availability.

NTIS Bibliographic Database. National Technical Information Service, 5285 Port Royal Road, Springfield, VA 22161. (703) 487-4929 or FAX (703) 321-8199. Broad coverage of government-sponsored science and technology research reports, 1964 to present. Available on BRS,(800) 289-4277, DIALOG, (800) 334-2564, ORBIT, (800) 456-7248, and STN International, FIZ Karlsruhe, P.O. Box 2465, W-7500, Karlsruhe 1, Germany, online services. Also available on CD-ROM. Inquire as to cost and availability.

SCISEARCH. Institute for Scientific Information, 3501 Market Street, Philadelphia, PA 19104. (800) 523-1850 or (215) 386-0100. Broad multidisciplinary title and author index to the international literature of science and technology, 1974 to present. Available on DIALOG,(800) 334-2564, and ORBIT,(800) 456-7248, online services. Also available on CD-ROM. Inquire as to cost and availability.

WILSONLINE. H.W. Wilson Company, 950 University Avenue, Bronx, NY 10452. (800) 367-6770 or (212) 588-8400. Makes available online versions of the printed H.W. Wilson indexes including Applied Science and Technology Index, Business Periodicals Index, General Science Index, and Readers' Guide to Periodical Literature. Period covered is generally 1983 to present. Available on BRS,(800) 289-4277, DIALOG,(800) 334-

2564, and OCLC EPIC,(800) 848-5878, online services. Also available on CD-ROM. Inquire as to cost and availability.

PERIODICALS

Engineering Fracture Mechanics. Elsevier Science [journals], 660 White Plains Rd., Tarrytown, NY 10159-5153. (919) 524-9200. FAX (919) 333-2444. 1968 to present. 18 times a year. $1945.00 per year.

International Journal of Fracture. Kluwer Academic Publishers, Box 358, Accord Station, Hingham, MA 02018-0358. (617) 871-6600. FAX (617) 871-6528. 1965 to present. 24 times a year. $1572.00 per year.

Journal of Engineering Mechanics. American Society of Civil Engineers, Engineering Mechanics Division, 345 East 47th Street, New York, NY 10017-2398. (212) 705-7520 or (800) 548-2723. 1956 to present. Monthly. $284.00 per year.

Journal of Materials Research. Materials Research Society, 9800 McKnight Road, Pittsburg, PA 15237. (412) 367-3003. FAX (412) 367-4373. 1986 to present. Monthly. $440.00 per year.

Journal of Structural Engineering. American Society of Civil Engineers, 345 East 47th Street, New York, NY 10017-2398. (212) 705-7520 or (800) 548-2723. 1956 to present. Monthly. $300.00 per year for non-members.

Journal of Testing and Evaluation. American Society for Testing and Materials (ASTM), 1916 Race Street, Philadelphia, PA 19103. (215) 299-5585. FAX (215) 977-9679. 1966 to present. Bi-monthly. $103.00 per year for non-members.

Materials Engineering. Penton Publishing, 110 Superior Ave., Cleveland, OH 44114-2543. 1929 to present. Monthly. $50.00 per year.

Materials Evaluation. American Society for Nondestructive Testing, 1711 Arlingate Lane, Box 28518, Columbus, OH 43228-0158. (614) 274-6003. FAX (614) 274-6899. 1942 to present. Monthly. $85.00 per year.

SAMPE Journal. Society for the Advancement of Material and Process Engineering, Covina, CA 91724. (818) 331-0610. FAX (818) 332-8929. 1965 to present. Bi-monthly. $65.00 per year.

RESEARCH CENTERS AND INSTITUTES

Center For Quality Engineering and Failure Prevention. Northwestern University, Technological Institute, Evanston, IL 60208. (708) 491-5527. FAX (708) 491-5227.

Structural Engineering Materials Laboratory, University of California At Berkeley. Davis Hall, Berkeley, CA 94720. (415) 642-3464.

Structural Stability Research Council. Fritz Engineering Laboratory, Lehigh University, Bethlehem, PA 18015. (215) 758-3522. FAX (215) 758-4522.

University of Toronto Structural Integrity, Fatigue, and Fracture Research Laboratory. Department of Mechanical Engineering, 5 King's College Road, Toronto ON Canada M5S 1A4. (416) 978-6721.

University of Wisconsin—madison, Center For Structural and Materials Testing. 1415 Johnson Street, Madison, WI 53706. (608) 262-3205.

SPECIFICATIONS AND STANDARDS

*Annual Book of Astm Standards, Volume 03.*01. American Society for Testing and Materials (ASTM), 1916 Race Street, Philadelphia, PA 19103. (215) 299-5585. Annual. Inquire for cost and availability.

FREQUENCY MODULATION

See: RADIO

FRICTION

See also: LUBRICATION, MACHINERY, MECHANICAL ENGINEERING, MECHANICS,PHYSICS

ABSTRACT SERVICES AND INDEXES

Applied Mechanics Reviews: An Assessment of World Literature in Engineering Sciences. American Society of Mechanical Engineers, 345 East 47th Street, New York, NY 10017. (212) 705-7703. 1948 to present. Monthly. $360.00 per year.

ISSN: 0003-6900.

Applied Science and Technology Index; A Cumulative Subject Index To English Language Periodicals in the Fields of Aeronautics and Space Science, Computer Technology, Chemistry, Construction Industry, Energy and Related Areas. H.W. Wilson Co., 950 University Avenue, Bronx, NY 10452. (212) 588-8400. (800) 367-6770. FAX (718) 590-1617. From 1958 to present. Monthly. Inquire about cost and availability. Also available on CD-ROM and online. ISSN: 0003-6986.

Chemical Abstracts. Chemical Abstracts Service, 2540 Olentangy River Road, Box 3012, Columbus, OH 43210-0012. (614) 447-3600. (800) 848-6538. FAX (614) 447-3713.From 1907 to present. Weekly. $16800.00 per year. Also available on CD-ROM and online. Inquire regarding cost and availability. ISSN:

Current Contents: Engineering, Technology and Applied Sciences. Institute for Scientific Information, 3501 Market Street, Philadelphia, PA 19104. (215) 386-0100. FAX (215) 386-2991. From 1970 to present. Weekly. $442.00 per year. Also available online. Inquire regarding cost and availability. ISSN: 0095-7917.

Current Papers in Physics. Institute of Electrical Engineers, Michael Faraday House, Six Hill Way, Stevenage, Herts, SG1 2AY, England. Distributed by INSPEC/IEEE, Box 1331, 445 Hoes Lane, Piscataway, NJ 08855-1331. (908) 562-5549. 1966 to present. Fortnightly. $410.00 per year. ISSN: 0011-3786.

Engineering Index Monthly. Engineering Information, Inc., Castle Point on the Hudson, Hoboken, NJ 07030. (201) 216-8500. FAX (201) 216-8532. Monthly. $2200.00 per year. Also

Ency. of Physical Sciences and Engineering Info. Sources

FRICTION

available online as COMPENDEX and on CD-ROM. ISSN: 0742-1974.

General Science Index. H.W. Wilson Company, 950 University Avenue, Bronx, NY 10452. (212) 588-8400. (800) 367-6770. FAX (718) 590-1617. From 1978 to present. Ten issues per year; quarterly and annual cumulations. Service basis. Available on CD-ROM and online. Inquire regarding cost and availability. ISSN: 0162-1963.

ISMEC Bulletin (information Service in Mechanical Engineering). Cambridge Scientific Abstracts, 5161 River Road, Bethesda, MD 20816. (301) 951-1400. 1973 to present. Monthly. $750.00 per year.

NTIS Alerts: Physics. U. S. National Technical Information Service. 5285 Port Royal Road, Springfield, VA 22161. (703) 487-4650. FAX (703) 321-8547. Weekly. $140.00 per year.

Physics Abstracts. INSPEC. Section A, Science Abstracts. Institution of Electrical Engineers (IEE). Available from INSPEC/ IEEE - Institute of Electrical and Electronic Engineers,, Box 1331, 445 Hoes Lane, Piscaway NJ 08855-1331. (908) 562-5549. 1898 to present. 24 issues per year. $2835.00 per year. Also available on CD-ROM and online. ISSN: 0036-8091.

Physics Briefs (Physikalische Berichte). Information Center for Energy, Physics, Mathematics; German Physical Society. V C H Publishers, Inc., 220 East 23rd Street, New York, NY 10010-4606. (212) 683-8333. 1845 to present. 24 issues per year. $2390.00 per year. Also available online. ISSN: 0179-7434.

Science Citation Index. SCI. Institute for Scientific Information, 3501 Market Street, Philadelphia, PA 19104. (215) 386-0100. (800) 523-1850. FAX (215) 386-2991. 1961 to present. Six issues per year, plus annual cumulation. $11650.00 per year. Also available online and on CD-ROM.Inquire about price and availability. ISSN: 0036-827X.

ANNUAL REVIEWS AND YEARBOOKS

Advances in Applied Mechanics. Academic Press, Inc., 6277 Sea Harbor Drive, Orlando, FL. (800) 321-5068. 1948 to present. Irregular. ISSN: 0065-2156.

Advances in Atomic, Molecular and Optical Physics. Academic Press, Inc., 6277 Sea Harbor Drive, Orlando, FL. (800) 321-5068. 1965 to present. Irregular. ISSN: 1049-250X.

ASSOCIATIONS AND PROFESSIONAL SOCIETIES

American Institute of Physics. One Physics Ellipse, College Park, MD 20740-3843. (301) 209-3100.

American Physical Society. One Physics Ellipse, College Park, MD 20740-3843. (301) 209-3200. FAX (301) 209-0865.

Asm International (american Society For Metals). 9639 Kinsman, Materials Park, OH 44073-0002. (216) 338-5151. FAX (216) 338-4634.

American Society of Mechanical Engineers, 345 East 47th Street, New York, NY 10017. (212) 705-7722. FAX (212) 705- 7739.

Minerals, Metals, and Materials Society. 420 Commonweatlh Drive, Warrendale, PA 15086. (412) 776-9000. FAX (412) 776-3770.

Nace International. PO Box 218340, Houston, TX 77218. (713) 492-0535. FAX (713) 492-8254.

NATIONAL LUBRICATING GREASE INSTITUTE. 4635 Wyandotte Street, Kansas City, MO 64112. (816) 931-9480. FAX (816) 753-5026.

SAE, International. 400 Commonwealth Drive, Warrendale, PA 15096-0001. (412) 776-4841. FAX (412) 776-5760.

Society of Tribologists and Lubrication Engineers. 840 Busse Highway, Park Ridge, IL 60068-2376. (708) 825-5536. FAX (708) 825-1456.

DIRECTORIES AND BIOGRAPHICAL SOURCES

American Men and Women of Science: Physical and Biological Sciences. R. R. Bowker Inc., 121 Chanlon Road, New Providence, NJ 07974. (908) 464-6800. (800) 521-8110. 20th edition. 8 volumes. 1996. $850.00.

International Directory of Engineering Societies and Related Organizations. American Society for Engineering Education, 1818 N Street NW, Suite 600, Washington, DC 20036. (202) 331-3526. 15th edition. 1996. $185.00.

Membership Roster. Society of Tribologists and Lubrication Engineers. 840 Busse Highway, Park Ridge, IL 60068-2376. (708) 825-5536. FAX (708) 825-1456. Included in membership dues. Annual.

Research Centers Directory. Gale Research Inc., 835 Penobscot Building, Detroit, MI 48226-4094. (313) 961-2242. (800) 347-4253. 20th edition. 1995. $485.00. ISSN: 0080- 1518.

Scientific and Technical Organizations and Agencies Directory. Gale Research, , 835 Penobscot Building, Detroit, MI 48226-4094. (313) 961-2242. (800) 347-4253. 4th edition. 1996. $195.00.

Who's Who in Engineering. Gordon Davis, editor. American Society for Engineering Education, 1111 19th Street, NW, Suite 608, Washington, DC 20036. 9th edition. 1995. $220.00.

ENCYCLOPEDIAS AND DICTIONARIES

Dictionary of the Physical Sciences: Terms, Formulas, Data. Cesare Emiliani. Oxford University Press, Inc., 200 Madison Avenue, New York, NY 10016. (212) 725-6000. (800) 334-4249.1989. $19.95.

Encyclopedia of Applied Physics. George Trigg, editor. VCH Publications, Inc., 220 East 23rd Street, Suite 909, New York, NY 10010-4606. (212) 683-8333. (800) 422-8824. 20 volume set. 1991-. $5990.00.

McGraw-Hill Encyclopedia of Engineering. Sybil P. Parker, editor. McGraw-Hill Book Company, Inc. 1221 Avenue of the Americas, New York, NY 10020. (212) 997-3675. Second edition. 1993. $95.50. ISBN: 0-07-051392-9.

McGraw-Hill Encyclopedia of Science and Technology. McGraw-Hill Book Company, Inc., 1221 Avenue of the Americas, New York, NY 10020. (212) 512-2000. (800) 262-4729. 7th edition. 20 volume set. 1992. $1900.00.

GENERAL WORKS

Adhesion and Friction. M. Grunze, editor. Springer-Verlag New York, Inc., 175 Fifth Avenue, New York, NY 10010. (212) 460-1500. (800) 777-4643. FAX (212) 473-6272. 1989. $57.00.

Engineering Tribology. Williams. Oxford University Press, Inc., 200 Madison Avenue, New York, NY 10016. (212) 725-6000. (800) 334-4249. 1995. $42.50.

Faulting, Friction, and Earthquake Mechanics. Chris J. Marone, editor. Birkhauser Boston, 675 Massachusetts Avenue, Cambridge, MA 02139. (617) 876-2333. (800) 777-4643. 1994. $34.50.

Friction and Wear of Materials. Ernest Rabinowicz. John Wiley & Sons, Inc., 605 Third Avenue, New York, NY 10158-0012. (212) 850-6000. (800) 225-5945. 2nd edition. 1995. $49.95.

Friction Science and Technology. Blau. Marcel Dekker, Inc., 270 Madison Avenue, New York, NY 10016. (212) 696-9000. (800) 228-1160. 1995. $150.00.

Friction, Wear and Lubrication: A Textbook in Trobology. K. C. Ludema. CRC Press, Inc., 2000 Corporate Boulevard, NW, Boca Raton, FL 33431. (407) 994-0555. (800) 272-7737. 1996.

Tribology: Friction and Wear of Engineering Materials. I. M. Hutchings. CRC Press, Inc., 2000 Corporate Boulevard, NW, Boca Raton, FL 33431. (407) 994-0555. (800) 272-7737. 1992. $58.00.

HANDBOOKS AND MANUALS

Compressible Flow Tables For Engineers. James Palmer, Kenneth Ramsden, and Eric Goodger. Scholium International, Inc., 14 Vanderventer Avenue, PO Box 1519, Port Washington, NY 11050-0306. (516) 767-7171. 1987. $21.50.

The CRC - Stle Handbook of Tribology. E. R. Booser. CRC Press, Inc., 2000 Corporate Boulevard, NW, Boca Raton, FL

33431. (407) 994-0555. (800) 272-7737. 1995.

Lubrication: A Tribology Handbook. M. H. Neale. Society of Automotive Engineers (SAE), 400 Commonwealth Drive, Warrendale, PA 15096. (412) 776-4841. 1993. $39.00.

Newnes Engineering and Physical Science Pocketbook. J. O. Bird and P. J. Chivers. Butterworth-Heinmann, 313 Washington Avenue, Newton, MA 02158. (617) 928-2500. (800) 366-2665. 1993. $49.95.

Tables For the Calculation of Friction in Internal Flow. H. R. Wallingford and D. I. Barr. American Society of Civil Engineers, 345 East 47th Street, New York, NY 10017. (212) 705-7496. (800) 548-2723. 1995. $86.40.

The Tribology Handbook. M. H. Neale. Butterworth-Heinemann, 313 Washington Street, Newton, MA 02158. (618-928-2500. (800) 366-2665. 2nd editon. 1996. $145.00.

ONLINE DATABASES AND CD-ROMS

CA Search. Chemical Abstracts Service, P.O. Box 3012, Columbus, OH 43210-0012. (614) 447-3600. (800) 848-6533. FAX (614) 447-3709. Very comprehensive guide to worldwide chemical literature and related fields, 1972 to present. Available on BRS,(800) 289-4277, DIALOG, (800) 334-2564, ORBIT (800) 456-7248, and STN International, FIZ Karlsruhe, P.O. Box 2465, W-7500, Karlsruhe 1, Germany, online services. Inquire as to cost and availability.

Current Contents Search. Institute for Scientific Information, 3501 Market Street, Philadelphia, PA 19104. (215) 386-0100. FAX (215) 386-6362. Contains citations to articles listed in the table of contents of science and technology journals. Also articles in social sciences and life sciences journals. Available on BRS, (800) 289-4277, DIALOG, (800) 334-2564, online services. Inquire as to cost and availability.

Compendex Plus. Engineering Information, Inc., 345 East 47th Street, New York, NY 10017. (212) 705-7600 or (800) 221-1044. Contains citations with abstracts to worldwide literature in engineering and technology, from 1970 to present. Available on online BRS,(800) 289-4277, DIALOG, (800) 334-2564, ORBIT (800) 456-7248, and STN International, FIZ Karlsruhe, P.O. Box 2465, W-7500, Karlsruhe 1, Germany, online services. Also available on CD-ROM. Inquire as to cost and availability.

Dissertation Abstracts Online. University Microfilms International, 300 North Zeeb Road, Ann Arbor, MI 48106. (800) 521-0600 or (313) 761-4700. Scope includes virtually all doctoral dissertations accepted at accredited American institutions from 1861 to present in 252 subject areas. Available on BRS, (800) 289-4277, DIALOG, (800) 334-2564, and OCLC EPIC, (800) 848-5878, online services. Also available on CD-ROM. Inquire as to cost and availability.

INSPEC. Institution of Electrical Engineers, Michael Faraday House, Six Hills Way, Stevenage, Herts. SG1 2AY, England. Telephone: 0438 313311 or FAX 0438 742840. Contains citations to the worldwide literature of physics, electronics and electrical engineering, computer technology, and related fields. Available on BRS, (800) 289-4277, DIALOG, (800) 334-2564, ORBIT, (800) 456-7248, and STN International, FIZ Karlsruhe, P.O. Box 2465, W-7500, Karlsruhe 1, Germany, online services. Inquire as to cost and availability.

NTIS Bibliographic Database. National Technical Information Service, 5285 Port Royal Road, Springfield, VA 22161. (703) 487-4929 or FAX (703) 321-8199. Broad coverage of government-sponsored science and technology research reports, 1964 to present. Available on BRS,(800) 289-4277, DIALOG, (800) 334-2564, ORBIT, (800) 456-7248, and STN International, FIZ Karlsruhe, P.O. Box 2465, W-7500, Karlsruhe 1, Germany, online services. Also available on CD-ROM. Inquire as to cost and availability.

Physics Briefs. American Institute of Physics, 335 East 45th Street, New York, NY 10017. (212) 661-9260 or FAX (212) 661-2036. Contains citations with abstracts of the literature of physics and related fields, 1979 to present. Available on the STN International, FIZ Karlsruhe, P.O. Box 2465, W-7500, Karlsruhe 1, Germany, online service. Inquire as to cost and availability.

SCISEARCH. Institute for Scientific Information, 3501 Market Street, Philadelphia, PA 19104. (800) 523-1850 or (215) 386-0100. Broad multidisciplinary title and author index to the international literature of science and technology, 1974 to present. Available on DIALOG, (800) 334-2564, and ORBIT, (800) 456-7248, online services. Also available on CD-ROM. Inquire as to cost and availability.

WILSONLINE. H.W. Wilson Company, 950 University Avenue, Bronx, NY 10452. (800) 367-6770 or (212) 588-8400. Makes available online versions of the printed H.W. Wilson indexes including Applied Science and Technology Index, Business Periodicals Index, General Science Index, and Readers' Guide to Periodical Literature. Period covered is generally 1983 to present. Available on BRS, (800) 289-4277, DIALOG, (800) 334-2564, and OCLC EPIC, (800) 848-5878, online services. Also available on CD-ROM. Inquire as to cost and availability.

PERIODICALS

Journal of Applied Mechanics. American Society of Mechanical Engineers. 345 East 47th Street, New York, NY 10017. (212) 705-7722. FAX (212) 705-7739. 1935 to present. Quarterly. $120.00 per year. ISSN: 0021-8936.

Journal of Engineering For Industry. American Society of Mechanical Engineers. 345 East 47th Street, New York, NY 10017. (212) 705-7722. FAX (212) 705-7739. 1970 topresent. Quarterly. $100.00 per year. ISSN: 0022-0817.

Journal of Fluid Control. Delbridge Publishing Co., Box 160817, Cupertino, CA 95016. (408) 446-3131. 1967 to present. Quarterly. $145.00 per year. ISSN: 8755-8564.

Journal of Mechanical Design. American Society of Mechanical Engineers. 345 East 47th Street, New York, NY 10017. (212) 705-7722. FAX (212) 705-7739. 1978 to present. Quarterly. $100.00 per year. ISSN: 10500-0472.

Journal of Synthetic Lubrication. Coppin Publishing, Co., PO Box 111 Deal, Dent CT14 6SX, England. 1984 to present. Quarterly. $185.00 per year. ISSN: 0265-6582.

Journal of Tribology. American Society of Mechanical Engineers. 345 East 47th Street, New York, NY 10017. (212) 705-7722. FAX (212) 705-7739. 1967 to present. Quarterly. $120.00 per year. ISSN: 0742-4787.

Lubrication Engineering. Society of Tribologists and Lubrication Engineers, 840 Busse Highway, Park Ridge, IL 60068-2376. (708) 825-5536. FAX(708) 825-1456. 1945 to present. Monthly. $61.00. ISSN: 0024-7154.

Machine Design. Penton Publishing, 1100 Superior Avenue, Cleveland, OH 44114-2543. (216) 696-7000. FAX (216) 696-0177. 1929 to present. 28 issues per year. $180.00 per year. ISSN: 0024-9114.

Mechanical Engineering. American Society of Mechanical Engineers. 345 East 47th Street, New York, NY 10017. (212) 705-7722. FAX (212) 705-7739. 1906 to present. Monthly. $45.00 per year. ISSN: 0025-6501.

N. L. G. I. Spokesman. National Lubrication Grease Institute, 4635 Wyandotte Street, Kansas City, MO 64112. (816) 931-

9480. FAX (816) 753-5026. 1937 to present. Monthly. $20.00 per year. ISSN: 0027-6782.

RESEARCH CENTERS AND INSTITUTES

Center for Engineering Tribology. Northwestern University, Tech Institute, Evanston, IL 60208. (708) 491-3112. FAX (708) 491-4133.

Coordinating Research Council, Inc. 219 Perimeter Center Parkway, Atlanta, GA 30346. (404) 396-3400.

Fluid Power Research Center. Oklahoma State University, 1724 West Tyler, Stillwater, OK 74078. (405) 744-7375.

Institute For Wear Control and Tribology. Rensselaer Polytechnic Institute, Mechanical Engineering Department, Jonsson Engineering Center, Troy, NY 12181. (518) 276-6992. FAX (518) 276-8788.

FRONT (METEOROLOGY)

See: METEOROLOGY

FUELS CELLS

See: ELECTROCHEMISTRY

FUELS

See also: CHEMICAL ENGINEERING, COAL, ENERGY, NATURAL GAS, NUCLEAR ENERGY, PETROLEUM, SYNTHETIC FUELS

ABSTRACT SERVICES AND INDEXES

Applied Science and Technology Index; a cumulative subject index to English language periodicals in the fields of aeronautics and space science, computer technology, chemistry, construction industry, energy and related areas. H. W. Wilson Co., 950 University Avenue, Bronx, NY 10452- 9978. (212) 588-8400. (800) 367-6770. FAX (718) 590-1617. From 1958 to present. Monthly. Inquire about cost and availability. Also available online (BRS and WILSONLINE) and on CD-ROM. ISSN: 0003-6986.

Chemical Abstracts. Chemical Abstracts Service. 2540 Olentangy River Road, Box 3012, Columbus, OH 43210-0012. (614) 447-3600. FAX (614) 447-3713. 1907 to present. Weekly. $16,800.00 per year. Available online and on CD- ROM. CA is also available in five section groupings. Inquire regarding cost and availability.

Current Contents: Engineering, Technology and Applied Sciences. Institute for Scientific Information, 3501 Market Street, Philadelphia, PA 19104. (215) 386-0100. FAX (215) 386-2991. From 1970 to present. Weekly. $442.00 per year. Also available online. Inquire regarding cost and availability. ISSN: 0095-7917.

FUELS

Ency. of Physical Sciences and Engineering Info. Sources

Energy Research Abstracts. U. S. Department of Energy, office of Scientific and Technical Information, Box 62, Oak Ridge, TN 37831. (615) 674-0733. Subscriptions to: U. S. Government Printing office, Box 371954, Pittsburgh, PA 15250-7954. (202) 783-3238. 1976 to present. Monthly. $164.00 per year. ISSN: 0160-3604.

Engineering Index Monthly. Engineering Information, Inc., Castle Point on the Hudson, Hoboken, NJ 07030. (201) 216-8500. (800) 221-1044. FAX (201) 216-8532. Monthly. $2300.00 per year for monthly issues. $1980 per year for annual cumulation. Available on CD-ROM and online as COMPENDEX. ISSN: 0742-1974; 0360-8557.

General Science Index. H.W. Wilson Company, 950 University Avenue, Bronx, NY 10452. (212) 588-8400. (800) 367-6770. FAX (718) 590-1617. From 1978 to present. Ten issues per year; quarterly and annual cumulations. Service basis. Available on CD-ROM and online. Inquire regarding cost and availability. ISSN: 0162-1963.

NTIS Alerts: Energy. U.S. National Technical Information Service, 5285 Port Royal Road, Springfield, VA 22161. (703) 487-4929. FAX (703) 321-8547. Weekly. $160.00 per year.

Physics Abstracts. INSPEC. Section A, Science Abstracts.

Institution of Electrical Engineers (IEE), London, United Kingdom. Available from: INSPEC/IEEE - Institute of Electrical and Electronic Engineers, Box 1331, 445 Hoes Lane, Piscataway, NJ 08855-1331. (908) 562-5549. 1898 to present. 24 issues per year. $2835.00 per year. ISSN: 0036-8091. Also available online and on CD-ROM.

Physics Briefs (Physikalische Berichte). Information Center for Energy, Physics, Mathematics; German Physical Society. V C H Publishers, Inc., 220 East 23rd Street, New York, NY 10010-4606. (212) 683-8333. 1845 to present. 24 issues per year. $2390.00 per year. Also available online. ISSN: 0179-7434.

Process and Chemical Engineering/chemical Engineering Abstracts. Royal Society of Chemistry, Information Services, Thomas Graham House, Science Park, Milton Road, Cambridge, CB4 4WF, England. Contains citations, mostly with abstracts, of the worldwide literature on chemical engineering; from 1982 to present. Monthly. $610.00 per year. Also available online. ISSN: 0960-5045.

Science Citation Index. SCI. Institute for Scientific Information, 3501 Market Street, Philadelphia, PA 19104. (215) 386-0100. (800) 523-1850. FAX (215) 386-2991. 1961 to present. Six issues per year, plus annual cumulation. $11650.00 per year. Also available online and on CD-ROM. Inquire about price and availability. ISSN: 0036-827X.

Theoretical Chemical Engineering. Royal Society of Chemistry, Information Services, Thomas Graham House, Science Park, Milton Road, Cambridge, CB4 4WF, England. Covers theoretical chemical engineering, including theory and laboratory experimentation. 1964 to present. Monthly. $235.00. ISSN: 0960-5053.

ANNUAL REVIEWS AND YEARBOOKS

Advances in Chemical Engineering. Academic Press, Inc., 6277 Sea Harbor Drive, Orlando, FL 32887. (800) 321-5068. 1956 to present. Irregular. ISSN: 0065-2377.

Advances in Energy Systems and Technology. Academic Press, Inc., 6277 Sea Harbor Drive, Orlando, FL 32887. (800) 321-5068. 1979 to present. Irregular. Inquire.

Advances in Nuclear Science and Technology. Plenum Publishing Corp., 233 Spring Street, New York, NY. (212) 620-8000. (800) 2221-9369. FAX (212) 463-0742. 1977 to present. Irregular.

Annual Review of Energy and the Environment. Annual Reviews, Inc., 4139 El Camino Way, Palo Alto, CA 94303-0139. (415) 943-4400.(800) 523-8635. 1976 to present. Annual. $71.00. ISSN: 1056-3466.

ASSOCIATIONS AND PROFESSIONAL SOCIETIES

American Chemical Society, 1155 16th Street, Nw, Washington, DC 20036. (202) 872-4600.

American Institute of Chemical Engineers. 345 East 47th Street, New York, NY 10017. (212) 705-7338.

American Institute of Physics. One Physics Ellipse, College Park, MD 20740-3843. (301) 209-3100.

Association of Energy Engineers, 4025 Pleasantdale Road, Suite 340, Atlanta, GA 30340. (770) 447-5083.

Edison Electric Institute, 701 pennsylvania avenue nw, Washington, DC 20004-2696. (202) 508-5000.

IEEE (institute of Electrical and Electronics Engineers) 345 East 47th Street, New York, NY 10017. (212) 705-7900. FAX 212) 705-4929.

BIBLIOGRAPHIES

Information Sources in Chemistry. R.T. Bottle and J.F.B Rowland, editors. R.R. Bowker Inc., 121 Chanlon Road, New Providence, NJ 07974. (800) 521-8110 or (908) 464-6800. Fourth edition. 1993. $75.00.

Information Sources in Engineering. Peter Hicks, editor. Bowker-Saur, 121 Chanlon Road, New Providence, NJ 07974. (800) 521-8110 or (908) 464-6800. 3rd edition. 1996. $125.00.

Scientific and Technical Books and Serials in Print; An Index To Literature in Science and Technology. R.R. Bowker

Inc., 121 Chanlon Road, New Providence, NJ 07974. (908) 464-6800. (800) 521-8110. FAX (908) 665-3502. 1972 to present. Annual. 4 volumes. 1994. $299.95. Also available on compact disc and online. ISSN: 0000-054X.

DIRECTORIES AND BIOGRAPHICAL SOURCES

American Institute of Chemical Engineers. DIRECTORY. American Institute of Chemical Engineers. 345 East 47th Street, New York, NY 10017-2396. (212) 705-7663. Annual.

American Men and Women of Science. Physical and Biological Sciences. Fifteenth edition. R.R. Bowker Inc., 121 Chanlon Road, New Providence, NJ 07974. (908) 464-6800. (800) 521-8110. 20th edition. 8 volumes. 1996. $850.00

Energy and Nuclear Science International Who's Who. Gale Research, , 835 Penobscot Building, Detroit, MI 48226- 4094. (313) 961-2242. (800) 347-4253. Triennial. 1993. $325.00.

Energy Information Centers Directory. U. S. Council for Energy Awareness, 1776 I Street, NW, Suite 400, Washington, DC 20006. (202) 293-0770. Triennial. No charge.

Engineering Research Centers: Incorporating Electronics Research Centers. Stockton Press, 345 Park Avenue, New York, NY 10010. (212) 689-9200. (800) 221-2123. 4th edition. 1995. $515.00.

Research Centers Directory. Gale Research Company, Inc., 835 Penobscot Building, Detroit, MI 48226-4094. (313) 961-2242. (800) 347-4253. 20th edition. 1995. $485.00. ISSN: 0080-1518.

Who's Who in Engineering. Gordon Davis, editor. American Association of Engineering Societies. 1111 19th Street, NY, Suite 608, Washington, DC 20036. (202) 296-2237. (800) 658-8897. 9th edition. 1995. $220.00.

Who's Who in Technology. Gale Research, 835 Penobscot Building, Detroit, MI 48226-4094. (313 961-2242. (800) 521-4253. 7th edition. 1995. $195.00. ISN 0-8103-7467-6.

World Energy and Nuclear Directory. Gale Research Company,Inc., 835 Penobscot Building, Detroit, MI 48226-4094. (313)

961-2242. (800) 347-4253. Biennial, even years. $450.00.

ENCYCLOPEDIAS AND DICTIONARIES

Encyclopedia of Applied Physics. VCH Publishers, Inc., 303 Northwest 12th Avenue, Deerfield Beach, FL 33442. (800) 367-8249. 1991-. Twenty volumes. $6000.00.

Encyclopedia of Chemical Processing and Design. McKetta Marcel Dekker, Inc., 270 Madison Avenue, New York, NY 10016. (212) 696-9000. (800) 228-1160. 1976 - . $175.00 per volume. ISSN:

Industrial Chemical thesaurus. Michael Ash and Irene Ash. VCH Publishers, Inc., 220 East 23rd Street, New York, NY 10010-4606. (800) 367-8249. 1992. $295.00.

Kirk-Othmer Encyclopedia of Chemical Technology. John Wiley and Sons, Inc., 605 Third Avenue, New York, NY 10158. (800) 526-5368 or (212) 850-6000. Fourth edition. 1991 - . Twenty-seven volumes. $5400.00.

McGraw-Hill Encyclopedia of Science and Technology. McGraw-Hill Book, Incorporated, 1221 Avenue of the Americas, New York, NY 10020. (212) 997-3675. (800) 262-4729. Seventh edition. Twenty volumes. 1992. $1900.00.

Ullman's Encyclopedia of Industrial Chemistry. VCH Publications, Inc., 220 East 23rd Street, Suite 909, New York, NY 10010-4606. (212) 683-8333. (800) 422-8824. 5th edition. 1984 - . Price varies per volume. ISSN:

GENERAL WORKS

Alternative Fuels: Alcohols, Hydrogen, Natural Gas and Propane. Society of Automotive Engineers (SAE), 400 Commonwealth Drive, Warrendale, PA 15096. (412) 776-4841.1993. $57.00

Clean Fuels: Progress and Experiences of Demonstration Programs. Soc. Auto Engineers. Society of Automotive Engineers (SAE), 400 Commonwealth Drive, Warrendale, PA 15096. (412) 776-4841. SP 985. 1992. $35.00.

Combustion of Solid Fuels and Wastes. David A. Tillman. Academic Press, Inc., 6277 Sea Harbor Drive, Orlando, FL. (800) 321-5068. 1991. $60.00

Combustion of Synthetic Fuels. William Bartok, editor. American Chemical Society, 1155 16th Street, NW, Washington, DC 20036. (202) 872-4600. (800) 333-9511. FAX (614) 447- 3671. 1983. $38.95.

Electricity Supply: Efforts To Develop Solar and Wind Energy. Diane Publishing Co., 600 Upland Avenue, Upland, PA 19015. (610) 499-7415. $60.00 in paper.

Energy Conversion: Systems, Flow Physics and Engineering. Reiner Decher. Oxford University Press, Inc., 200 Madison Avenue, New York, NY 10016. (212) 725-6000. (800) 334-4249. 1994. $69.95.

Energy Efficiency: Challenges and Trends For Electric Utilities. Diane Publishing Co., 600 Upland Avenue, Upland, PA 19015. (610) 499-74151994. $60.00.

Fuels and Combustion. Samir Sarkar. Apt Books, 25 Brixham Road, Eliot, ME 03903-1209. 2nd edition 1990. $20.00 in paper

Nuclear Engineering: Theory and Technology of Commercial Nuclear Power. Ronald A. Knief. Hemisphere Publishing Corp., 1900 Frost Road, Suite 101, Bristol PA 19007. (215) 785-5800. (800) 821-8312. 2nd edition. 1992. $156.00.

Nuclear Power, Nuclear Fuel Cycle and Waste Management: Status and Trends 1992. UNIPUB, 4511-F Assembly Drive, Lanham, MD 20706-4391. (301) 459-7666. (800) 274-4888.1992. $19.00.

Synthetic Fuel Technology Development in the United States: A Retrospective Assessment. Michael Crow, et al. Greenwood Press, 88 Post Road West, Box 5507, Westport, CT 06881. (203) 226-3571. (800) 225-5800. 1988. $55.00.

HANDBOOKS AND MANUALS

Automotive Fuels Handbook. Keith Owen and Trevor Coley. Society of Automotive Engineers (SAE), 400 Commonwealth Drive, Warrendale, PA 15096. (412) 776-4841. 2nd edition. 1995. $129.00.

Biomass Handbook. Osamu Katini and C. W. Hall, editors. Gordon and Breach Science Publishers, Inc., P.O.Box 200029, Riverfront Plaza Station, Newark, NJ 07102-0301. (201) 643-7500 1989. $419.00

FUELS

Ency. of Physical Sciences and Engineering Info. Sources

Energy Engineering Handbook. Fairmont Press Staff. Prentice-Hall, 113 Sylvan Avenue, Route 9W, Englewood Cliffs, NJ 07632. (201) 592-2000. (800) 922-0579. 3rd edition. 1995. $75.00.

Fuel Cells: A Handbook. J. M. Hirschenfoffer Business/Technology Information Services, P. O. Box 574, Orinda, CA 94563. (510) 299-1829. (800) 808-8811. 1996. $115.00.

Fuel Science and Technology Handbook. James G. Speight. Marcel Dekker, Inc., 270 Madison Avenue, New York, NY 10016. (212) 696-9000. (800) 228-1160. 1990. $250.00.

Perry's Chemical Engineers' Handbook. Robert H. Perry and Donald W. Green, editors. McGraw-Hill Publishing Company, Inc., 1221 Avenue of the Americas, New York, NY 10020. (212) 512-2000. (800) 262-4729. 6th edition. 1996. $129.50.

Practical Handbook of Processing and Recycling of Municipal Waste. A. G. Manser and Alan A. Keeling. Lewis Pubs., CRC Press, Inc., 2000 Corporate Boulevard, NW, Boca Raton, FL 33431. (407) 994-0555. (800) 272-7737. 1996. $69.95.

RIEGEL'S HandBOOK of INDUSTRIAL CHEMISTRY. James A. Kent, editor. Van Nostrand Reinhold, 115 Fifth Avenue, New York, NY 10003. (212) 254-3232 or (800) 926-2665. Ninth edition.

1992. $114.95.

Sourcebook of Methods of Analysis For Biomass and Biomass Conversion Processes. T. A. Milne, et al. Elsevier Science Publishing Company, Inc., 655 Avenue of the Americas, New York, NY 10010. (212) 989-5800. FAX (914) 333-2444. 1990. 93.75.

ONLINE DATABASES AND CD-ROMS

CA Search. Chemical Abstracts Service, P.O. Box 3012, Columbus, OH 43210-0012. (614) 447-3600. (800) 848-6533. FAX (614) 447-3709. Very comprehensive guide to worldwide chemical literature and related fields. 1972 to present. Available on BRS,(800) 289-4277, DIALOG, (800) 334-2564, ORBIT (800) 456-7248, and STN International, FIZ Karlsruhe, P.O. Box 2465, W-7500, Karlsruhe 1, Germany, online services. Inquire as to cost and availability.

Compendex Plus. Engineering Information, Inc., 345 East 47th Street, New York, NY 10017. (212) 705-7600 or (800) 221-1044. Contains citations with abstracts to worldwide literature in engineering and technology, from 1970 to present. Available on online BRS,(800) 289-4277, DIALOG, (800) 334-2564, ORBIT (800) 456-7248, and STN International, FIZ Karlsruhe, P.O. Box 2465, W-7500, Karlsruhe 1, Germany, online services. Also available on CD-ROM. Inquire as to cost and availability.

Current Contents Search. Institute for Scientific Information, 3501 Market Street, Philadelphia, PA 19104. (215) 386-0100. FAX (215) 386-6362. Contains citations to articles listed in the table of contents of science and technology journals. Also articles in social sciences and life sciences journals. Available on BRS,(800) 289-4277, DIALOG,(800) 334-2564, online services. Inquire as to cost and availability.

Dissertation Abstracts. University Microfilms International, 300 North Zeeb Road, Ann Arbor, MI 48106. (800) 521-0600 or (313) 761-4700. Scope includes virtually all doctoral dissertations accepted at accredited American institutions from 1861 to

present in 252 subject areas. Available on BRS, (800) 289-4277, DIALOG, (800) 334-2564, and OCLC EPIC, (800) 848-5878, online services. Also available on CD-ROM. Inquire as to cost and availability.

INSPEC. Institution of Electrical Engineers, Michael Faraday House, Six Hills Way, Stevenage, Herts. SG1 2AY, England. Telephone: 0438 313311 or FAX 0438 742840. Contains citations to the worldwide literature of physics, electronics and electrical engineering, computer technology, and related fields. Available on BRS, (800) 289-4277, DIALOG, (800) 334-2564, ORBIT, (800) 456-7248, and STN International, FIZ Karlsruhe, P.O. Box 2465, W-7500, Karlsruhe 1, Germany, online services. Inquire as to cost and availability.

Kirk-Othmer Encyclopedia of Chemical Technology. John Wiley and Sons, Inc., 605 Third Avenue, New York, NY 10158. (800) 526-5368 or (212) 850-6000. Contains the complete text of all chapters in the 27 volume fourth edition of the KIRK-Othmer ENCYCLOPEDIA of CHEMICAL TECHNOLOGY. 1991. Available on BRS,(800) 289-4277, DIALOG,(800) 334-2564, online services. Inquire as to cost and availability.

NTIS Bibliographic Database. National Technical Information Service, 5285 Port Royal Road, Springfield, VA 22161. (703) 487-4929 or FAX (703) 321-8199. Broad coverage of government-sponsored science and technology research reports, 1964 to present. Available on BRS,(800) 289-4277, DIALOG, (800) 334-2564, ORBIT, (800) 456-7248, and STN International, FIZ Karlsruhe, P.O. Box 2465, W-7500, Karlsruhe 1, Germany, online services. Also available on CD-ROM. Inquire as to cost and availability.

Physics Briefs. American Institute of Physics, 335 East 45th Street, New York, NY 10017. (212) 661-9260 or FAX (212) 661-2036. Contains citations with abstracts of the literature of physics and related fields, 1979 to present. Available on the STN International, FIZ Karlsruhe, P.O. Box 2465, W-7500, Karlsruhe 1, Germany, online service. Inquire as to cost and availability.

SCISEARCH. Institute for Scientific Information, 3501 Market Street, Philadelphia, PA 19104. (800) 523-1850 or (215) 386-0100. Broad multidisciplinary title and author index to the international literature of science and technology, 1974 to present. Available on DIALOG, (800) 334-2564, and ORBIT, (800) 456-7248, online services. Also available on CD-ROM. Inquire as to cost and availability.

WILSONLINE. H.W. Wilson Company, 950 University Avenue, Bronx, NY 10452. (800) 367-6770 or (212) 588-8400. Makes available online versions of the printed H.W. Wilson indexes including Applied Science and Technology Index, Business Periodicals Index, General Science Index, and Readers' Guide to Periodical Literature. Period covered is generally 1983 to present. Available on BRS, (800) 289-4277, DIALOG, (800) 334-2564, and OCLC EPIC, (800) 848-5878, online services. Also available on CD-ROM. Inquire as to cost and availability.

Who's Who in Technology. Gale Research Inc., 835 Penobscot Building, Detroit, MI 48226-4094. (313) 961-2242. (800) 347-4253. Contains biographical information of contemporary American scieNTISts and engineers. Available on DIALOG,(800) 334-2564, online service. Inquire as to cost and availability.

PERIODICALS

Aiche Journal. American Institute of Chemical Engineers. 345 East 47th Street, New York, NY 10017. (212) 705-7338. 1955 to present. Monthly. $295.00 per year.

Coal and Synfuels Technology. Pasha Publicaitons, Inc., 1616 North Fort Myer Drive, Arlington, VA 22209-3107. (703) 528-1244. 1979 to present. Weekly. $790.00 per year. ISSN: 0883-9735.

Energy and Fuels. American Chemical Society, Box 3337, Columbus, OH 43210. (614) 447-3774. 1987 to present. Bimonthly. $345.00 per year. ISSN: 0887-0624.

Energy Engineering. Association of Energy Engineers. Available from Fairmont Press, 700 Indian Trail, Lilburn, GA 30324. (404) 925-9388. 1904 to present. Bimonthly. $99.00 per year. ISSN: 0199-8595.

Energy Sources. Taylor and Francis, 1900 Forst Road, Stuie 101, Bristol, PA 19007-1598. (215) 785-5800. 1973 to present. Quarterly. $225.00. ISSN: 0090-8312.

International Journal of Solar Energy. Harwood Academic Publishers, 820 Town Center Drive, Langhorne, PA 19047. (215) 750-2642. 1982 to present. 8 issues per year. 254 ECU per year. ISSN: 0142-5919.

Journal of Energy Resources Technology. American Society of Mechanical Engineers, 345 East 47th Street, New York, NY 10017. (212) 705-7703. 1979 to present. Quarterly. $100.00 per year. ISSN: 0195-0738.

Power Engineering (tulsa). PenWell Publishing Co., 1421 Sheridan Road, Tulsa, OK 74101. (918) 8335-3161. 1896 to present. Monthly. $38.00 per year. ISSN: 0032-5961.

RESEARCH CENTERS AND INSTITUTES

Center For Applied Energy Research. University of Kentucky, 3572 Iron Works Pike, Lexington, KY (606) 257-0305. FAX (606) 257-0220.

Center For Energy Studies. University of Texas at Austin, 10100 Burnet Road, Building 133, Austin, TX 78758. (512) 471-4496. FAX (512) 471-0781.

Energy Research Laboratories. Canada Centre for Mineral and Energy Technology, 555 Booth Street, Ottawa, ON Canada KIA 0G1. (613) 996-8201. FAX (613) 995-9584.

Institute of Gas Technology. 3424 South State Street, Chicago, IL 60616. (312) 567-3650. FAX (312) 567-5209.

Oak Ridge National Laboratory. U. S. Department of Energy, P. O. Box 2008, Oak Ridge, TN 37831-6255. (615) 576-2900. FAX (615) 576-2912.

FUNDAMENTAL PARTICLES

See: PARTICLE PHYSICS

FUSES (ELECTRICAL)

See: CIRCUIT BREAKERS

FUSION

See also: ENERGY, FISSION, NUCLEAR ENERGY, NUCLEAR ENGINEERING, NUCLEAR PHYSICS, NUCLEAR REACTORS

ABSTRACT SERVICES AND INDEXES

Applied Mechanics Reviews: An assessment of world literature in engineering sciences. American Society of Mechanical Engineers, 345 East 47th Street, New York, NY 10017. (212) 705-7703. 1948 to present. Monthly. $360.00 per year. ISSN: 0003-6900.

Applied Science and Technology Index; A Cumulative Subject Index To English Language Periodicals in the Fields of Aeronautics and Space Science, Computer Technology, Chemistry, Construction Industry, Energy and Related Areas. H. W. Wilson Co., 950 University Avenue, Bronx, NY 10452- 9978. (212) 588-8400. (800) 367-6770. FAX (718) 590-1617. From 1958 to present. Monthly. Inquire about cost and availability. Also available online (BRS and WILSONLINE) and on CD-ROM. ISSN: 0003-6986.

Chemical Abstracts. Chemical Abstracts Service. 2540 Olentangy River Road, Box 3012, Columbus, OH 43210-0012. (614) 447-3600. FAX (614) 447-3713. 1907 to present. Weekly. $16,800.00 per year. Available online and on CD- ROM. CA is also available in five section groupings. Inquire regarding cost and availability.

Current Contents: Engineering, Technology and Applied Sciences. Institute for Scientific Information, 3501 Market Street, Philadelphia, PA 19104. (215) 386-0100. FAX (215) 386-2991. From 1970 to present. Weekly. $442.00 per year. Also available online. Inquire regarding cost and availability. ISSN: 0095-7917.

Engineering Index Monthly. Engineering Information, Inc., Castle Point on the Hudson, Hoboken, NJ 07030. (201) 216-8500. (800) 221-1044. FAX (201) 216-8532. Monthly. $2300.00 per year for monthly issues. $1980 per year for annual cumulation. Available on CD-ROM and online as COMPENDEX. ISSN: 0742-1974; 0360-8557.

General Science Index. H.W. Wilson Company, 950 University Avenue, Bronx, NY 10452. (212) 588-8400. (800) 367-6770. FAX (718) 590-1617. From 1978 to present. Ten issues per year; quarterly and annual cumulations. Service basis. Available on CD-ROM and online. Inquire regarding cost and

ISMEC: Mechanical Engineering Abstracts. Cambridge Scientific Abstracts, 7200 Wisconsin Avenue, Suite 601, Bethesda, MD 20814. (301) 961-6750 or (800) 843-7751. Contains citations to the literature in mechanical engineering, industrial and production engineering, energy, power, mechanics, devices and related areas, from 1973 to present. Available on the DIALOG,(800) 334-2564, online service. Inquire as to cost and availability.

N T I S Alerts: Energy. U.S. National Technical Information Service, 5285 Port Royal Road, Springfield, VA 22161. (703) 487-4929. FAX (703) 321-8547. Weekly. $160.00 per year.

FUSION

Ency. of Physical Sciences and Engineering Info. Sources

Physics Abstracts. INSPEC. Section A, Science Abstracts. Institution of Electrical Engineers (IEE), London, United Kingdom. Available from: INSPEC/IEEE - Institute of Electrical and Electronic Engineers, Box 1331, 445 Hoes Lane, Piscataway, NJ 08855-1331. (908) 562-5549. 1898 to present. 24 issues per year. $2835.00 per year. ISSN: 0036-8091. Also available online and on CD-ROM.

Physics Briefs (Physikalische Berichte). Information Center for Energy, Physics, Mathematics; German Physical Society. V C H Publishers, Inc., 220 East 23rd Street, New York, NY 10010-4606. (212) 683-8333. 1845 to present. 24 issues per year. $2390.00 per year. Also available online. ISSN: 0179-7434.

Process and Chemical Engineering/chemical Engineering Abstracts. Royal Society of Chemistry, Information Services, Thomas Graham House, Science Park, Milton Road, Cambridge, CB4 4WF, England. Contains citations, mostly with abstracts, of the worldwide literature on chemical engineering; from 1982 to present. Monthly. $610.00 per year. Also available online. ISSN: 0960-5045.

Science Citation Index. SCI. Institute for Scientific Information, 3501 Market Street, Philadelphia, PA 19104. (215) 386-0100. (800) 523-1850. FAX (215) 386-2991. 1961 to present. Six issues per year, plus annual cumulation. $11650.00 per year. Also available online and on CD-ROM. Inquire about price and availability. ISSN: 0036-827X.

ANNUAL REVIEWS AND YEARBOOKS

Advances in Chemical Engineering. Academic Press, Inc., 6277 Sea Harbor Drive, Orlando, FL 32887. (800) 321-5068. 1956 to present. Irregular. ISSN: 0065-2377.

Advances in Nuclear Science and Technology. Plenum Publishing Corp., 233 Spring Street, New York, NY. (212) 620-8000. (800) 2221-9369. FAX (212) 463-0742. 1977 to present. Irregular.

ASSOCIATIONS AND PROFESSIONAL SOCIETIES

American Institute of Chemical Engineers. 345 East 47th Street, New York, NY 10017. (212) 705-7338.

American Institute of Physics. One Physics Ellipse, College Park, MD 20740-3843. (301) 209-3100.

American Nuclear Society. 555 North Kensington Avenue, LA Grange Park, IL 60525. (708) 352-6611. (800) NUC-NEWS.

Fusion Power Associates. 2 Professional Drive, Suite 248, Gaithersburg, MD 20879. (301) 258-0545. FAX (301) 975-9869.

Institute of Nuclear Materials Management. 60 Revere Drive, Northbrook, IL 60062. (847) 480-9573. FAX (947) 480-9282.

BIBLIOGRAPHIES

Handbooks and Tables in Science and Technology. Russell H. Powell, editor. Oryx Press, 4041 North Central, Suite 700, Phoenix, AZ 85012-3330. (602) 265-2651 or (800) 279-6799. Third edition. 1994. $65.00.

Information Sources in Engineering. Peter Hicks, editor. Bowker-Saur, 121 Chanlon Road, New Providence, NJ 07974. (800) 521-8110 or (908) 464-6800. 3rd edition. 1996. $125.00.

Scientific and Technical Books and Serials in Print; An Index To Literature in Science and Technology. R.R. Bowker Inc., 121 Chanlon Road, New Providence, NJ 07974. (908) 464-6800. (800) 521-8110. FAX (908) 665-3502. 1972 to present. Annual. 4 volumes. 1994. $299.95. Also available on compact disc and online. ISSN: 0000-054X.

DIRECTORIES AND BIOGRAPHICAL SOURCES

American Institute of Chemical Engineers. DIRECTORY. American Institute of Chemical Engineers. 345 East 47th Street, New York, NY 10017-2396. (212) 705-7663. Annual.

American Men and Women of Science. Physical and Biological Sciences. Fifteenth edition. R.R. Bowker Inc., 121 Chanlon Road, New Providence, NJ 07974. (908) 464-6800. (800) 521-8110. 20th edition. 8 volumes. 1996. $850.00

Energy and Nuclear Science International Who's Who. Gale Research, 835 Penobscot Building, Detroit, MI 48226-4094. (313) 961-2242. (800) 347-4253. Triennial. 1993. $325.00.

Energy Information Centers Directory. U. S. Council for Energy Awareness, 1776 I Street, NW, Suite 400, Washington, DC 20006. (202) 293-0770. Triennial. No charge.

Engineering Research Centers: Incorporating Electronics Research Centers. Stockton Press, 345 Park Avenue, New York, NY 10010. (212) 689-9200. (800) 221-2123. 4th edition. 1995. $515.00.

Nuclear News - Directory Issues. American Nuclear Society, 555 North Kensington Avenue, LA Grange Park, IL 60525. (708) 352-6611. (800) NUC-NEWS. March issue contains Buyer's Guide; $65.00. World List of Nuclear Power Plants. Semi-annual in February and August. $15.00 per issue.

Nuclear Reactors Built, Being Built OR Planned in the United States. National Technical Information Service, Springfield, VA 22161. (703) 487-4780. 1989. $17.00.

Research Centers Directory. Gale Research Company, Inc., 835 Penobscot Building, Detroit, MI 48226-4094. (313) 961-2242. (800) 347-4253. 20th edition. 1995. $485.00. ISSN: 0080-1518.

Who's Who in Engineering. Gordon Davis, editor. American Association of Engineering Societies. 1111 19th Street, NY, Suite 608, Washington, DC 20036. (202) 296-2237. (800) 658-8897. 9th edition. 1995. $220.00.

Who's Who in Technology. Gale Research, 835 Penobscot Building, Detroit, MI 48226-4094. (313 961-2242. (800) 521-4253. 7th edition. 1995. $195.00. ISN 0-8103-7467-6.

World Energy and Nuclear Directory. Gale Research Company, Inc., 835 Penobscot Building, Detroit, MI 48226-4094. (313)961-2242. (800) 347-4253. Biennial, even years. $450.00.

ENCYCLOPEDIAS AND DICTIONARIES

Dictionary of Named Processes in Chemical Technology. Alan E. Comyns. Oxford University Press, Inc., 200 Madison Avenue, New York, NY 10016. (212) 725-6000. (800) 334-4249. 1994. $75.00.

Dictionary of Nuclear Technology. Pro/Am Music Resources, Inc., 63 Prospect Street, White Plains, New York, 10606. (914) 948-7436. 1986. $245.00.

Encyclopedia of Applied Physics. VCH Publishers, Inc., 303 Northwest 12th Avenue, Deerfield Beach, FL 33442. (800) 367-8249. 1991-. Twenty volumes. $6000.00.

IEEE Nuclear Science Standards Collection. IEEE Standards office, Box 1331, 445 Hoes Lane, Piscataway, NJ 08855-1331. (908) 562-3822. (800) 678-4333. 1991. $90.00 in paper.

Kirk-Othmer Encyclopedia of Chemical Technology. John Wiley and Sons, Inc., 605 Third Avenue, New York, NY 10158. (800) 526-5368 or (212) 850-6000. Fourth edition. 1991 - . Twenty-seven volumes. $5400.00.

McGraw-Hill Encyclopedia of Science and Technology. McGraw-Hill Book, Incorporated, 1221 Avenue of the Americas, New York, NY 10020. (212) 997-3675. (800) 262-4729. Seventh edition. Twenty volumes. 1992. $1900.00.

Ullman's Encyclopedia of Industrial Chemistry. VCH Publications, Inc., 220 East 23rd Street, Suite 909, New York, NY 10010-4606. (212) 683-8333. (800) 422-8824. 5th edition. 1984 - . Price varies per volume. ISSN:

GENERAL WORKS

Atomic and Plasma Material Interaction Processes in Controlled Thermonuclear Fusion. R. K. Janev and H. W. Drawin, editors. Elsevier Science Publishing Company, Inc., 655 Avenue of the Americas, New York, NY 10010. (212) 989- 5800. FAX (914) 333-2444. 1993. $191.50.

Controlled Nuclear Chain Reactions: The First Fifty Years. American Nuclear Society, 555 North Kensington Avenue, LA Grange Park, IL 60525. (708) 352-6611. 1992. $25.00.

Fifieth Anniversary of Nuclear Fission. L. V. Drapchinsky, editor. Nova Science Publishers, Inc., 6080 Jericho Turnpike, Suite 207, Commack, NY 11725-2808. (516) 499- 3103. 1994. $145.00.

Fusion Energy in Space Propulsion. Terry Kammash, editor. American Institute of Aeronautics and Astronautics, 370 L'Enfant Promenade, S.W., Washington, DC 20024. (202) 646-7400. (800) 682-2422. 1995. $84.95.

Fusion Plasma Analysis. Weston M. Stacey. Krieger Publishing Company, P.O. Box 9542, Melbourne, FL 32902-9542. (407) 724-9542. 1992. $54.50.

Legacy of Chernobyl. Zhores A. Medvedev. W. W. Norton & Company, 500 Fifth Avenue, New York, NY 10110. (212) 354-5500. (800) 223-2584. 1992. $10.95 in paper.

Managing Nuclear Accidents: A Model Emergency Response Plan For Power Plants and Communities. Dominic Golding, et al. Westview Press. 5500 Central Avenue, Boulder CO 80301-2847. (303) 444-3541. (800) 456-1995.

The Search For Endless Energy. Robin Herman. Cambridge University Press, 40 West 20th Street, New York, NY 10011-4211. (212) 924-3900. (800) 872-7423. 1990. $24.95.

HANDBOOKS AND MANUALS

CRC Handbook of Chemistry and Physics. David R. Kide, editor. CRC Press Inc., 2000 Corporate Blvd., NW, Boca Raton, FL 33431. (407) 994-0555. (800) 333-8300. 77th edition. 1996. $99.95.

A Guide To Nuclear Power Technology: A Resource For Decision Making. Frank J. Rahn, et al. Krieger Publishing Company, P.O. Box 9542, Melbourne, FL 32902-9542. (407) 724-9542. Reprint editon. 1992. $145.00.

Information Center On Nuclear Standards. American Nuclear Society, 555 North Kensington Avenue, LA Grange Park, IL 60525. (708) 352-6611. Standards for all aspects of the design, construction, operation and maintenance of nuclear power plants.

Nuclear Engineering Databases, Standards and Numerical Analysis. Jack Jedruch. Van Nostrand Reinhold, 115 Fifth Avenue, New York, NY 10003. (212) 254-3232. (800) 842- 3636. 1985.

Nuclear Engineering Sourcebook: A Complilation of Documents For Nuclear Equipment Qualification. IEEE Standards office, Box 1331, 445 Hoes Lane, Piscataway, NJ 08855-1331. (908) 562-3822. (800) 678-4333. 1992.

Summary Description of Design Criteria, Codes, Standards, and Regulatory Provisions Typically Used For the Civil and Structural Design of Nuclear Fuel Cycle Facilities. American Society of Civil Engineers. 345 East 47th Street, New York, NY 10017. (212) 705-7496. (800) 548-2723. 1988. $17.00

ONLINE DATABASES AND CD-ROMS

CA Search. Chemical Abstracts Service, P.O. Box 3012, Columbus, OH 43210-0012. (614) 447-3600. (800) 848-6533. FAX (614) 447-3709. Very comprehensive guide to worldwide chemical literature and related fields. 1972 to present. Available on BRS,(800) 289-4277, DIALOG, (800) 334-2564, ORBIT (800) 456-7248, and STN International, FIZ Karlsruhe, P.O. Box 2465, W-7500, Karlsruhe 1, Germany, online services. Inquire as to cost and availability.

Compendex Plus. Engineering Information, Inc., 345 East 47th Street, New York, NY 10017. (212) 705-7600 or (800) 221-1044. Contains citations with abstracts to worldwide literature in engineering and technology, from 1970 to present. Available on online BRS,(800) 289-4277, DIALOG, (800) 334-2564, ORBIT (800) 456-7248, and STN International, FIZ Karlsruhe, P.O. Box 2465, W-7500, Karlsruhe 1, Germany, online services. Also available on CD-ROM. Inquire as to cost and availability.

Current Contents Search. Institute for Scientific Information, 3501 Market Street, Philadelphia, PA 19104. (215) 386-0100. FAX (215) 386-6362. Contains citations to articles listed in the table of contents of science and technology journals. Also articles in social sciences and life sciences journals. Available on BRS,(800) 289-4277, DIALOG,(800) 334-2564, online services. Inquire as to cost and availability.

FUSION

Ency. of Physical Sciences and Engineering Info. Sources

Dissertation Abstracts. University Microfilms International, 300 North Zeeb Road, Ann Arbor, MI 48106. (800) 521-0600 or (313) 761-4700. Scope includes virtually all doctoral dissertations accepted at accredited American institutions from 1861 to present in 252 subject areas. Available on BRS, (800) 289-4277, DIALOG, (800) 334-2564, and OCLC EPIC, (800) 848-5878, online services. Also available on CD-ROM. Inquire as to cost and availability.

INSPEC. Institution of Electrical Engineers, Michael Faraday House, Six Hills Way, Stevenage, Herts. SG1 2AY, England. Telephone: 0438 313311 or FAX 0438 742840. Contains citations to the worldwide literature of physics, electronics and electrical engineering, computer technology, and related fields. Available on BRS, (800) 289-4277, DIALOG, (800) 334-2564, ORBIT, (800) 456-7248, and STN International, FIZ Karlsruhe, P.O. Box 2465, W-7500, Karlsruhe 1, Germany, online services. Inquire as to cost and availability.

Kirk-Othmer Encyclopedia of Chemical Technology. John Wiley and Sons, Inc., 605 Third Avenue, New York, NY 10158. (800) 526-5368 or (212) 850-6000. Contains the complete text of all chapters in the 27 volume fourth edition of the *Kirk-othmer Encyclopedia of Chemical Technology.* 1991. Available on BRS,(800) 289-4277, DIALOG,(800) 334-2564, online services. Inquire as to cost and availability.

NTIS Bibliographic Database. National Technical Information Service, 5285 Port Royal Road, Springfield, VA 22161. (703) 487-4929 or FAX (703) 321-8199. Broad coverage of government-sponsored science and technology research reports, 1964 to present. Available on BRS,(800) 289-4277, DIALOG, (800) 334-2564, ORBIT, (800) 456-7248, and STN International, FIZ Karlsruhe, P.O. Box 2465, W-7500, Karlsruhe 1, Germany, online services. Also available on CD-ROM. Inquire as to cost and availability.

Nuclear Science Abstracts. U.S. Department of Energy, office of Scientific and Technical Energy, P.O. Box 62, Oak Ridge, TN 37831. (615) 576-1189. Contains citations with abstracts to literature in all fields of nuclear science and energy. 1948 to 1976. Available on DIALOG,(800) 334-2564, online service. Inquire as to cost and availability.

Physics Briefs. American Institute of Physics, 335 East 45th Street, New York, NY 10017. (212) 661-9260 or FAX (212) 661-2036. Contains citations with abstracts of the literature of physics and related fields, 1979 to present. Available on the STN International, FIZ Karlsruhe, P.O. Box 2465, W-7500, Karlsruhe 1, Germany, online service. Inquire as to cost and availability.

SCISEARCH. Institute for Scientific Information, 3501 Market Street, Philadelphia, PA 19104. (800) 523-1850 or (215) 386-0100. Broad multidisciplinary title and author index to the international literature of science and technology, 1974 to present. Available on DIALOG, (800) 334-2564, and ORBIT, (800) 456-7248, online services. Also available on CD-ROM. Inquire as to cost and availability.

Scientific and Technical Books and Serials in Print. R.R. Bowker Inc., 121 Chanlon Road, New Providence, NJ 07974. (800) 521-8110 or (908) 464-6800. List of currently published books and serials in the physical and biological sciences, engineering and technology, with subject, author and titles indexes. Available on ORBIT, (800) 456-7248, online service. Inquire as to cost and availability.

WILSONLINE. H.W. Wilson Company, 950 University Avenue, Bronx, NY 10452. (800) 367-6770 or (212) 588-8400. Makes available online versions of the printed H.W. Wilson indexes including Applied Science and Technology Index, Business Periodicals Index, General Science Index, and Readers' Guide to Periodical Literature. Period covered is generally 1983 to present. Available on BRS, (800) 289-4277, DIALOG, (800) 334-2564, and OCLC EPIC, (800) 848-5878, online services. Also available on CD-ROM. Inquire as to cost and availability.

Who's Who in Technology. Gale Research Inc., 835 Penobscot Building, Detroit, MI 48226-4094. (313) 961-2242. (800) 347-4253. Contains biographical information of contemporary American scieNTISts and engineers. Available on DIALOG,(800) 334-2564, online service. Inquire as to cost and availability.

PERIODICALS

American Nuclear Society Transactions. American Nuclear Society, 555 North Kensington Avenue, LA Grange Park, IL 60525. (708) 352-6611. 1958 to present. 3 issues per year. $340.00 per year. ISSN: 0003-018X

Annals of Nuclear Energy. Elsevier Science Publishing Company, Inc., 660 White Plains Road, Tarrytown, NY 10591- 5133. (914) 524-9200. 1974 to present. Monthly. $815.00 per year. ISSN: 0306-4549.

Bulletin of the Atomic Scientists. Educational Foundation for Nuclear Science, 6042 South Kimbark Avenue, Chicago, IL 60637. (312) 702-2555. 1945 to present. 10 issues per year. $30.00 per year. ISSN: 0096-3402.

Fusion Technology. American Nuclear Society, 555 North Kensington Avenue, LA Grange Park, IL 60525. (708) 352-6611. 1981 to present. 8 issues per year. $$485.00 per year. ISSN: 0748-1896.

Journal of Fusion Energy. Plenum Publishing Corporation, 233 Spring Street, New York, NY 10013- 1578. (212) 620- 8000. 1981 to present. Quarterly. $315.00 per year. ISSN:0164-0313.

Nuclear Engineer. Institution of Nuclear Engineers, 1 Penerley Road, London, SE6 2LQ, England. 1959 to present. Bi-monthly. L77 per year. ISSN: 0262-5091.

Nuclear Engineering and Design. Elsevier Science Publishing Company, Inc., 660 White Plains Road,

Tarrytown, NY 10591-5133. (914) 524-9200. 1965 to present. 24 issues per year. $2748.00 per year.

Nuclear Engineering International. Reed Business Pulbishing Group, Quadrant House, the Quadrant, Sutton, Surrey, SM2 5AS, Englnad. 1956 to present. Monthly. $236.00 per year. ISSN: 0029-5507.

Nuclear Science and Engineering. American Nuclear Society, 555 North Kensington Avenue, LA Grange Park, IL 60525. (708) 352-6611. 1956 to present. Monthly. $405.00 per year. ISSN: 0029-5639.

Nuclear Technology. American Nuclear Society, 555 North Kensington Avenue, LA Grange Park, IL 60525. (708) 352-6611. 1965 to present. Monthly. $440.00 per year. ISSN: 0029-5450.

Ency. of Physical Sciences and Engineering Info. Sources

FUSION

RESEARCH CENTERS AND INSTITUTES

Department of Nuclear Engineering and Engineering Physics. University of Wisconsin - Madison, 1500 Johnson Drive, Madison, WI 53706. (608) 263-1648. FAX (608) 262-6707.

Fusion Research Center. University of Texas at Austin. RLM 11.222, Austin, TX 78712. ((512) 471-3926. FAX (512) 471-8865.

Frank H. Neely Nuclear Research Center. Georgia Institute of Technology, 900 Atantic Drive NW, Atlanta, GA 30332. (404) 894-3600.

Institute of Nuclear Science and Engineering. Oregon State University Radiation Center- A100, 35th and Jefferson Streets, Corvallis, OR 97331-5903. (503) 737-2341. FAX (503) 737-0480.

Institute of Plasma and Fusion Research. University of California, Los Angeles, 44-139 Engineering IV 405 Hilgard Avenue, Los Angeles, CA 90024-1597. (213) 825-1613. FAX (213) 206-4832.

Radiation Science and Engineering Center. Pennsylvania State University, University park, PA 16802. (814) 865- 6351. FAX (814) 863-4840.

Whiteshell Laboratories. Research Company, Atomic Energy of Canada Limited, Pinawa, MB, Canada ROE ILO. (204) 753-2311. FAX (204) 753-8404.

FUSION REACTOR

See: NUCLEAR REACTORS

FUSION WELDING

See: WELDING

G

GALAXIES

See also: ASTRONOMY, BIG BANG, COSMOCHEMISTRY, COSMOLOGY, DARK MATTER, INTERSTELLAR MEDIUM, MILKY WAY, NEBULAE, STARS, UNIVERSE

ABSTRACT SERVICES AND INDEXES

*Applied Science and Technology Index; A Cumulative Subject Index To English Language Periodicals in the Fields ofaeronautics and Space Science, Computer Technology,chemistry, Construction Industry, Energy and Related Areas.*H.W. Wilson Co., 950 University Avenue, Bronx, NY 10452.(212) 588-8400. (800) 367-6770. FAX (718) 590-1617. From1958 to present. Monthly. Inquire about cost andavailability. Also available on CD-ROM and online. ISSN: 0003-6986.

Astronomy and Astrophysics Abstracts. Springer-Verlag New York, 175 Fifth Avenue, New York, NY 10010. (212) 460-1500.FAX (212) 473-6272. Published for the AstronomischesRechen-Institut. Comprehensive coverage of all aspects ofastronomy, astrophysics and related fields. 1969 to present. Two parts per year. Annual. ISSN: 0067-0022.

Chemical Abstracts. Chemical Abstracts Service, 2540Olentangy River Road, Box 3012, Columbus, OH 43210-0012.(614) 447-3600. (800) 848-6538. FAX (614) 447-3713.From 1907 to present. Weekly. $16800.00. Available onlineand on CD-ROM. CA is also available in five section groupings. Inquire regarding cost and availability.

Chemical Abstracts — Physical, Inorganic and Analytical Chemistry. Chemical Abstracts Service, 2540 Olentangy RiverRoad, Box 3012, Columbus, OH 43210-0012. (614) 447-3600.(800) 848-6538. FAX (614) 447-3713. Bi-weekly. $1870.00per year. Also available on CD-ROM and online. Inquireregarding cost and availability.

General Science Index. H.W. Wilson Company, 950 University Avenue, Bronx, NY 10452. (212) 588-8400. (800) 367-6770.FAX (718) 590-1617. From 1978 to present. Ten issues peryear; quarterly and annual cumulations. Service basis.Available on CD-ROM and online. Inquire regarding cost andavailability. ISSN: 0162-1963.

Government Reports Announcements and Index. U. S. National Technical Information Service (NTIS), 5285 Port Royal Road,Springfield, VA 22161. (703) 487-4650. FAX (703) 321-8547.From 1968 to present. Annual. $630.00 per year. Alsoavailable online as *NTIS Bibliographic Database* and on CD- ROM. ISSN:

International Aerospace Abstracts. American Institute of Aeronautics and Astronautics. Technical InformationService, 555 West 57th Street, New York 10019. (212) 247-6500. FAX (212) 582-4961. 1961 to present. Semi-monthlywith annual index. $1295.00 per year. Also available online and on CD-ROM. ISSN: 0020-5842.

Meteorological and Geoastrophysical Abstracts. American Meteorological Society, c/o Inforonics, Inc., 550 NewtownRoad, Littleton, MA 01460. (508) 486-8976. FAX (508) 486-0027. Covers literature in environmental sciences,meteorology, astrophysics, hydrology, glaciology, and physical oceanography. From 1950 to present. Monthly.$950.00 per year. Also available on CD-ROM and online.Inquire regarding cost and availability. ISSN: 0026-1130.

NTIS Alerts: Astronomy & Astrophysics. U. S. National Technical Information Service. 5285 Port Royal Road,Springfield, VA 22161. (703) 487-4650. FAX (703) 321-8547.Weekly. $140.00 per year.

Pascal 48: Environnement Cosmique Terrestre, Astronomie Et Geologie Extraterrestre. Centre National de la RechercheScientifique, Institut de Information Scientifique etTechnique, 2 aliee du Parc de Brabois, 54514 VandoeuvreLesNancy Cedex, France. TEL 83-50-46-00. FAX: 83-50-46-50. 1985 to present. Ten issuesper year. 770 F. Alsoavailable on CD-ROM and online.

Physics Abstracts. INSPEC. Section A, Science Abstracts.Institute of Electrical Engineers, London, United Kingdom.Available from: INSPEC/IEEE - Institute of Electrical andElectronic Engineers, Box 1331, Hoes Lane, Piscataway, NJ08855-1331. (908) 562-5549. 1898 to present. 24 issuesper year. $2835.00 per year. ISSN: 0036-8091. Alsoavailable online and on CD-ROM.

Physics Briefs (Physikalische Berichte). InformationCenter for Energy, Physics, Mathematics; German Physical Society. V C H Publishers, Inc., 220 East 23rd Street, New York, NY 10010-4606. (212) 683-8333. 1845 to present. 24 issues per year. $2390.00 per year. Also available online.ISSN: 0179-7434.

Science Citation Index. SCI. Institute for ScientificInformation, 3501 Market Street, Philadelphia, PA 19104. (215) 386-0100.

GALAXIES

Ency. of Physical Sciences and Engineering Info. Sources

(800) 523-1850. FAX (215) 386-2991. 1961 to present. Six issues per year, plus annual cumulation. $11650.00 per year. Also
available online and on CD-ROM. Inquire about price and availability. ISSN: 0036-827X.

STAR. SCIENTIFIC and TECHNICAL AEROSPACE RE-
PORTS. U.S. National Aeronautics and Space Administration.
Distributed by U. S. Superintendent of Documents, Washing-
ton, DC 20402. 1963 to present. Semi-monthly, with semian-
nual and annual indexes. $114.00 per year. ISSN: 0036-8741.
Also available online and on CD-ROM.

ANNUAL REVIEWS AND YEARBOOKS

The Astronomical Almanac. Superintendent of Documents, U.
S. Government Printing office, Washington, DC 10402. (202)
783-3238. 1981 to present. Supercedes Astronomical Ephemeris and the American Ephermis and Nautical Almanac. Annual.
$44.95. ISSN: 0737-6421.

*Annual Review of Astronomy and Astrophysics; Original Re-
views of Critical Literature and Current Developments in Astronomy and Astrophysics.* Annual Reviews, Inc., 4139 El
Camino Way, Palo Alto, CA 94306-0139. (415) 493-4400. (800)
523-8635. FAX (415) 855-9815. 1963 to date. Annual. $60.00.
ISSN: 0066-4146.

Highlights of Astronomy. I. Appenzeller, editor. Kluwer Academic Publishers, Box 358, Accord Station, Hingham, MA
02018-0358. (617) 871-6600. International Astronomical Union
Highlights series. ISBN: 07923-3553-8. Triennal. 1995. Price
varies.

ASSOCIATIONS AND PROFESSIONAL SOCIETIES

Amateur Astronomers Association. 1010 Park Avenue, New
York, NY 10028. (212) 535-2922.

American Association of Variable Star Observers. 23 Birch
Street, Cambridge, MA 02138. (617) 354-0484. FAX: (617)354-
0665.

American Astronomical Society. 2000 Florida Avenue NW,
Suite 400, Washington, DC 20009. (202) 328-2010. FAX:(202)
234-2560.

American Institute of Physics. One Physics Ellipse, College Park,
MD 20740-3843. (301) 209-3100.

Association of University For Research in Astronomy. Suite 701,
1625 Massachusetts Avenue, NW, Washington, DC 20036. (202)
483-2101. FAX (202) 483-2106.

Astronomical League. 2112 Kingfisher Lane East, Rolling Meadows, IL 60008. (708) 398-0562.

Astronomical Society of the Pacific. 390 Ashton Avenue, San
Francisco, CA 94112. (415) 337-1100. FAX: (415) 337-5205.

International Astronomical Union. 98 bis Boulevard Arago, F-
75014 Paris, France. Tel 1-43-25-83-58. FAX 1-43-25-26-16.

Joint Institute For Laboratory Astrophyics. University of Colorado, Box 440, Boulder, CO 80309-0440. (303) 492-7780. FAX
(303) 492-5235.

National Council For Geocosmic Research. PO Box 1220,
Dunkirk, MD 20754. (301) 855-2747. (800) PTOLEMY. FAX
(410) 257-2824.

Royal Astronomical Society of Canada. 136 Dupont Street,
Toronto. ON N5R 1V2 Canada. (416) 924-7973. FAX (416)
968-6687.

BIBLIOGRAPHIES

Astronomy and Astrophysics: A Source Guide. Gordon Press,
P.O. Box 459, Bowling Green Station, New York, NY 10004.
(718) 624-8419. 1991. $250.00. .

A Bibliography of Astronomy, 1970-1979. R. A. Sea and S. S.
Martin. Libraries Unlimited, Inc., Littleton, CO 80160. 1982.

Catalog of the Naval Observatory Library. Naval Observatory
Staff. G. K. Hall & Company, 866 Third Avenue, New York,
NY 10022. (212)702-6789. (800) 257-5755. 6 volumes. 1977.
$655.00

Science Books and Films. American Association for the Advancement of Science, 1333 H Street NW, Washington, DC
20005. (202) 326-6454. 1965 to present. Nine issues per year.
$40.00 per year. ISSN: 0098-342X

Scientific and Technical Books and Serials in Print, 1995. R.R.
Bowker Inc., 121 Chanlon Road, New Providence, NJ 07974.
(908) 464-6800. (800) 521-8110. 4 volumes. 1994. $299.95.
Also available on compact disc and online.

A Source Book On Astronomy and Astrophysics, 1900 - 1975.
Kenneth Lang and Owen Gingerich, editors. Harvard University Press, 79 Garden Street, Cambridge MA. (617) 495-2600.
1980. $41.95.

DIRECTORIES AND BIOGRAPHICAL SOURCES

*American Men and Women of Science: Physical and Biological
Sciences.* R. R. Bowker Inc., 121 Chanlon Road, New Providence, NJ 07974. (908) 464-6800. (800) 521-8110. 20th edition. 8 volumes. 1996. $850.00.

The Astronomers. Donald Goldsmith. St. Martin's Press, Inc.,
175 Fifth Avenue, New York, NY 10010. (212) 674-5151. (800)
221-7945. 1993. $14.95 in paper.

Astronomical Centers of the World. Kevin Krisciunas. Cambridge University Press, 40 West 20th Street, New York, NY
10011-4211. (212) 924-3900. (800) 872-7423. 1988. $34.95

The Biographical Dictionary of Scientists: Astronomers. D.
Abbott, editor. Peter Bedrick Books, Inc., 2112 Broadway, Room
318, New York, NY 10023. (212) 496-0751. 1984.

Directory of Physics and Astronomy Staff Members. American
Institute of Physics, One Physics Ellipse, College Park, MD
20740-3843. (301) 209-3100. Annual. $45.00.

Earth and Astronomical Research Centers. Stockton Press, 345
Park Avenue South, New York, NY 10010. 4th edition. 1995.
$515.00. ISBN: 1-56169-0967.

Ency. of Physical Sciences and Engineering Info. Sources

GALAXIES

Graduate Programs in Physics, Astronomy and Related Fields. 1995-1996. American Institute of Physics, One Physics Ellipse, College Park, MD 20740-3843. (301) 209-3100. 1978 to present. Annual. $45.00.

Research Centers Directory. Gale Research Inc., 835 Penobscot Building, Detroit, MI 48226-4094. (313) 961-2242. (800) 877-4253. 20th edition. 1995. $485.00.

ENCYCLOPEDIAS AND DICTIONARIES

Astronomy From A To Z: A Dictionary of Celestial Objects and Ideas. Charles A. Schweighauser. Illinois Issues, Sangamon State University, K-10, Springfield, IL 62794-9243. (217) 786-9243. 1991. $14.95 in paper.

Concise Dictionary of Astronomy. Jacqueline Mitton. Oxford University Press, Inc., 200 Madison Avenue, New York, NY 10016. (212) 725-6000. (800) 334-4249. 1992. $24.95.

Facts On File Dictionary of Astronomy. Valerie Illingworth. *Facts-on-file, Inc.,* 460 Park Avenue South, New York, NY 10016-7382. (212) 683-2244. (800) 322-8755. Fax (800) 683-3633. 1994. $27.95.

Encyclopedia of Astronomy and Astrophysics. Stephen Maran, editor. Van Nostrand Reinhold, 115 Fifth Avenue, New York, NY 10003. (212) 254-3232. (800) 842-3636. 1992. $129.95.

Landolt-boernstein Numerical Data and Functional Relationships in Science and Technology: Astronomy, Astrophysics and Space Research; A. O. Madelung and H. H. Boigt, editors. Springer Verlag New York, Inc., 175 Fifth Avenue, New York, NY 10010. (212) 460-1500. (800) 777- 4643. 1993. $550.00.

Oxford Illustrated Encyclopedia. VOLUME 8, the UNIVERSE. Archie Roy, editor. Oxford University Press, Inc., 200 Madison Avenue, New York, NY 10016. (212) 725-6000. (800) 334-4249. 1993. $45.00.

McGraw-Hill Encyclopedia of Science and Technology. McGraw-Hill Book Company, Inc., 1221 Avenue of the Americas, New York, NY 10020. (212) 512-2000. (800) 262-4729. 7th edition. 20 volume set. 1992. $1900.00.

Stars, Galaxies, Cosmos. William R. Corliss. Sourcebook Project, P.O. Box 107, Glen Arm, MD 21057. (410) 668-6047. 1987. $17.95.

GENERAL WORKS

Astronomer's Universe: Stars, Galaxies and Cosmos. Herbert Friedman. W. W. Norton & Company, 500 Fifth Avenue, New York, NY 10110. (212) 354-5500. (800) 223-2584. 1990. $24.95.

Brief History of Time: From the Big Bang To Black Holes. Stephen W. Hawking. Bantam Books, Inc., 1540 Broadway, New York, NY 10036-4094. (212) 354-6500. (800) 223-6934. 1990. $13.95 in paper.

Black Holes and Baby Universes and Other Essays. Stephen W. Hawking. Bantam Books, Inc., 1540 Broadway, New York, NY 10036-4094. (212) 354-6500. (800) 223-6934. 1993. $21.95.

Cosmic Questions, Galactic Halos, Cold Dark Matter and the End of Time. Richard Morris. John Wiley & Sons, Inc., 605 Third Avenue, New York, NY 10158-0012. (212) 850-6000. (800) 225-5945. 1993. $24.95.

Cycles of Fire: Stars, Galaxies and the Wonder of Deep Space. William K. Hartmann. Workman Publishing Company, Inc., 708 Broadway, New York, NY 10003. (212) 254-5900. (800) 722-7202. 1988. $15.95 in paper.

The Discovery of Our Galaxy. Charles A. Whitney. Iowa State University Press, 2121 South State Avenue, Ames, IA 50014-8300. (515) 292-0140. (800) 862-6657. Reprint edition. 1988. $13.95 in paper.

The Early Universe: Facts and Fiction. Gerhard Borner. Springer-Verlag New York, Inc., 175 Fifth Avenue, New York, NY 10010. (212) 460-1500. (800) 777-4643. 1995. $69.00.

The Environment and Evolution of Galaxies. J. Michael Shull, editor. Kluwer Academic Publishers, 101 Philip Drive, Assinippi Park, Norwell, MA 02061. (617) 871-6600. 1993. $41.00.

Evolution of Interstellar Matter and Dynamics of Galaxies. J. Palous et al, editors. Cambridge University Press, 40 West 20th Street, New York, NY 10011-4211. (212) 924-3900. (800) 872-7423. 1992. $69.95.

Farthest Things in the Universe. Jay M. Pasachoff. Cambridge University Press, 40 West 20th Street, New York, NY 10011-4211. (212) 924-3900. (800) 872-7423. 1995. $29.95.

The First Three Minutes: A Modern View of the Origin of the Universe. Steven Weinberg. Basic Books, 10 East 53rd Street, New York, NY 10022. (212) 207-7057. (800) 242-7737. Reprint edition. 1993. $12.00.

Galaxies and the Universe. David J. Eicher, editor. Kalmbach Publishing Company, P.O. Box 1612, Waukesha, WI 53187. (414) 796-0126. (800) 446-5489. Readings from Deep Sky Magazine. 1992. $14.95.

Universe. William J. Kaufmann. W. H. Freeman & Co., 41 Madison Avenue, East 26th Avenue, 35th Floor, New York, NY 10010. (212) 576-0400. 4th edition. 1995. $54.95.

The Left Hand of Creation: The Origin and Evolution of the Expanding Universe. John D. Barrow and Joseph Silk. Oxford University Press, Inc., 200 Madison Avenue, New York, NY 10016. (212) 725-6000. (800) 334-4249. 1994. $23.00.

Man Discovers the Galaxies. Richard Berendzen. Columbia University Press, 562 West 113th Street, New York, NY 10025. (212) 666-1000. (800) 944-8648. 1984. $17.50 in paper.

The Nearest Active Galaxies. J. E. Beckman, editor. Kluwer Academic Publishers, 101 Philip Drive, Assinippi Park, Norwell, MA 02061. (617) 871-6600. 1993. $87.50.

The New Cosmos. A. Unsold and B. Baschek. Springer-Verlag New York, Inc., 175 Fifth Avenue, New York, NY 10010. (212) 460-1500. (800) 777-4643 . 4th revised edition. 1991. $59.00.

Observational Astrophysics. Robert C. Smith. Cambridge University Press, 40 West 20th Street, New York, NY 10011- 4211. (212) 924-3900. (800) 872-7423. 1995. $69.95.

GALAXIES

Ency. of Physical Sciences and Engineering Info. Sources

The Origin and Evolution of Galaxies. B. J. Jones and J. E. Jones. Kluwer Academic Publishers, 101 Philip Drive, Assinippi Park, Norwell, MA 02061. (617) 871-6600. 1982. $52.50 in paper.

The Outer Galaxy. Leo Blitz and F. J. Lockman, editors. Springer-Verlag New York, Inc., 175 Fifth Avenue, New York, NY 10010. (212) 460-1500. (800) 777-4643. 1988. $39.00.

Physical Processes in Comets, Stars and Active Galaxies. W. Hildebrandt, et al editors. Springer-Verlag New York, Inc., 175 Fifth Avenue, New York, NY 10010. (212) 460-1500. (800) 777-4643. 1987. $47.00.

The Physics of Stars. A. C. PHillips. John Wiley & Sons, Inc., 605 Third Avenue, New York, NY 10158-0012. (212) 850- 6000. (800) 225-5945. 1994. $32.95 in paper.

Star Formation, Galaxies and the Interstellar Medium. J. Franco, et al, editors. Cambridge University Press, 40 West 20th Street, New York, NY 10011-4211. (212) 924-3900.(800) 872-7423. 1993. $59.95.

Stars and Atoms: From the Big Bang To the Solar System. Stuart Clark. Oxford University Press, Inc., 200 Madison Avenue, New York, NY 10016. (212) 725-6000. (800) 334-4249. 1995. $35.00.

Stars and Galaxies: Astronomy's Guide To Exploring the Cosmos. Astronomy Magazine Staff. Kalmbach Publishing Company, P.O. Box 1612, Waukesha, WI 53187. (414) 796-0126. (800) 446-5489. 1992. $29.95 in paper.

Structure and Evolution of Galaxies. Roger J. Tayler. Cambridge University Press, 40 West 20th Street, New York, NY 10011-4211. (212) 924-3900. (800) 872-7423. 1993. $49.95. $22.95 in paper.

Universe of Galaxies. Paul W. Hodge, editor. W. H. Freeman & Co., 41 Madison Avenue, East 26th Avenue, 35th Floor, New York, NY 10010. (212) 576-0400. Readings from Scientific American. 1995.

Visual Astronomy of the Deep Sky. Roger N. Clark. Cambridge University Press, 40 West 20th Street, New York, *NY 10011- 4211.* (212) 924-3900. (800) 872-7423. 1991. $39.95.

The World of Galaxies. H. G. Corwin and L. Bettinelli, editors. Springer-Verlag New York, Inc., 175 Fifth Avenue, New York, NY 10010. (212) 460-1500. (800) 777-4643. 1989. $79.00

HANDBOOKS AND MANUALS

Burnham's Celestial Handbook: An Observer's Guide To the Universe Beyond the Solar System. Robert Burnham, Jr. Dover Publications, Inc., 180 Varick Street, New York, NY 10014. (212) 255-3755. (800) 223-3130. 3 volumes. 1978. $41.85 for the set.

Cambridge Guide To the Constellations. Michael E. Bakich. Cambridge University Press, 40 West 20th Street, New York, NY 10011-4211. (212) 924-3900. (800) 872-7423. 1995. $69.95.

Encyclopedia of Astronomy and Astrophysics. Stephen Maran, editor. Van Nostrand Reinhold, 115 Fifth Avenue, New York, NY 10003. (212) 254-3232. (800) 842-3636. 1992. $129.95.

Greenwich Guide To Stars, Galaxies and Nebulae. Stuart Malin. Cambridge University Press, 40 West 20th Street, New York, NY 10011-4211. (212) 924-3900. (800) 872-7423.1990. $10.95.

Handbook of Astronomy, Astrophysics and Geophysics; Volume 2: Galaxies and Cosmology. V. M. Canuto and B. G. Elmegreen. Gordon & Breach Science Publishers, Inc., P.O. Box 200029, Riverfront Plaza Station, Newark, NJ 07102-0301.(201) 643-7500. 1988. $194.00.

Handbook of Space Astronomy and Astrophsyics. Martin V. Zumbeck. Cambridge University Press, 40 West 20th Street, New York, NY 10011-4211. (212) 924-3900. (800) 872-7423.2nd edition. 1990. $79.95.

Landolt-borenstein Numerical Data and Functional Relationships in Science and Technology: Astronomy, Astrophysics and Space Research. GROUP VI. Springer- Verlag New York, Inc., 175 Fifth Avenue, New York, NY 10010. (212) 460-1500. (800) 777-4643. volumes priced individually.

McGraw-Hill Encyclopedia of Astronomy. Sybil P. Parker and Jay M. Pasachoff, editors. McGraw-Hill Publishing Company, Inc., 1221 Avenue of the Americas, New York, NY 10020. (212) 512-2000. (800) 262-4729. 2nd edition. 1993. $75.00

National Audobon Society Pocket Guide: Galaxies and Other Deep-sky Objects. Alfred A. Knopf, Inc. 201 E. 50th Street, New York, NY 10022. (800) 733-3000. 1995. $7.95.

Nearby Galaxies Catalog. R. Brent Tully and J. Richard Fisher. Cambridge University Press, 40 West 20th Street, New York, NY 10011-4211. (212) 924-3900. (800) 872-7423.1988. $59.95.

Observer's Guide To Astronomy. Patrick Martinez, editor. Cambridge University Press, 40 West 20th Street, New York, NY 10011-4211. (212) 924-3900. (800) 872-7423. 2 volumes. 1994. $159.90.

ONLINE DATABASES AND CD-ROMS

CA SEARCH. Chemical Abstracts Service, P.O. Box 3012, Columbus, OH 43210-0012. (614) 447-3600. (800) 848-6533. FAX (614) 447-3709. Very comprehensive guide to worldwide chemical literature and related fields, 1972 to present. Available on BRS,(800) 289-4277, DIALOG, (800) 334-2564, ORBIT (800) 456-7248, and STN International, FIZ Karlsruhe, P.O. Box 2465, W-7500, Karlsruhe 1, Germany, online services. Inquire as to cost and availability.

Current Contents Search. Institute for Scientific Information, 3501 Market Street, Philadelphia, PA 19104. (215) 386-0100. FAX (215) 386-6362. Contains citations to articles listed in the table of contents of science and technology journals. Also articles in social sciences and life sciences journals. Available on BRS, (800) 289-4277, DIALOG, (800) 334-2564, online services. Inquire as to cost and availability.

Dissertation Abstracts Online. University Microfilms International, 300 North Zeeb Road, Ann Arbor, MI 48106. (800) 521-0600 or (313) 761-4700. Scope includes virtually all doctoral dissertations accepted at accredited American institutions from 1861 to present in 252 subject areas. Available on BRS, (800) 289-4277, DIALOG, (800) 334-2564, and OCLC EPIC, (800) 848-5878, online services. Also available on CD-ROM. Inquire as to cost and availability.

INSPEC. Institution of Electrical Engineers, Michael Faraday House, Six Hills Way, Stevenage, Herts. SG1 2AY, England. Telephone: 0438 313311 or FAX 0438 742840. Contains citations to the worldwide literature of physics, electronics and electrical engineering, computer technology, and related fields. Available on BRS, (800) 289-4277, DIALOG, (800) 334-2564, ORBIT, (800) 456-7248, and STN International, FIZ Karlsruhe, P.O. Box 2465, W-7500, Karlsruhe 1, Germany, online services. Inquire as to cost and availability.

International Aerospace Abstracts. American Institute of Aeronautics and Astronautics, 370 L'Enfant Promenade, S.W., Washington, DC 20024. (202) 646-7400. Contains referencesand abstracts of journal and monograph literature relating to aerospace science and technology, from 1963 to present. Available through the NASA/RECON system of the National Aeronautics and Space Administration only.

NTIS Bibliographic Database. National Technical Information Service, 5285 Port Royal Road, Springfield, VA 22161. (703) 487-4929 or FAX (703) 321-8199. Broad coverage of government-sponsored science and technology research reports, 1964 to present. Available on BRS,(800) 289-4277, DIALOG, (800) 334-2564, ORBIT, (800) 456-7248, and STN International, FIZ Karlsruhe, P.O. Box 2465, W-7500, Karlsruhe 1, Germany, online services. Also available on CD-ROM. Inquire as to cost and availability.

Physics Briefs. American Institute of Physics, 335 East 45th Street, New York, NY 10017. (212) 661-9260 or FAX (212) 661-2036. Contains citations with abstracts of the literature of physics and related fields, 1979 to present. Available on the STN International, FIZ Karlsruhe, P.O. Box 2465, W-7500, Karlsruhe 1, Germany, online service. Inquire as to cost and availability.

Scientific and Technical Books and Serials in Print. R.R. Bowker Inc., 121 Chanlon Road, New Providence, NJ 07974. (800) 521-8110 or (908) 464-6800. List of currently published books and serials in the physical and biological sciences, engineering and technology, with subject, author and titles indexes. Available on ORBIT, (800) 456-7248, online service. Inquire as to cost and availability.

SCISEARCH. Institute for Scientific Information, 3501 Market Street, Philadelphia, PA 19104. (800) 523-1850 or (215) 386-0100. Broad multidisciplinary title and author index to the international literature of science and technology, 1974 to present. Available on DIALOG, (800) 334-2564, and ORBIT, (800) 456-7248, online services. Also available on CD-ROM. Inquire as to cost and availability.

WILSONLINE. H.W. Wilson Company, 950 University Avenue, Bronx, NY 10452. (800) 367-6770 or (212) 588-8400. Makes available online versions of the printed H.W. Wilson indexes including Applied Science and Technology Index, Business Periodicals Index, General Science Index, and Readers' Guide to Periodical Literature. Period covered is generally 1983 to present. Available on BRS, (800) 289-4277, DIALOG, (800) 334-2564, and OCLC EPIC, (800) 848-5878, online services. Also available on CD-ROM. Inquire as to cost and availability. subhead2>Other Sources

Atlas of the Andromeda Galaxy. Paul W. Hodge. University of Washington Press, P. O. Box 50096. Seattle, WA 98145- 5096. (206) 543-4050. (800) 441-4115. 1981. $75.00.

Atlas of Deep-sky Spendors. Hans Vehrenberg. Sky Publishing Corp., 49 Bay State Road, Cambridge, MA 02138. (617) 864-7360. (800) 253-0245. 4th edition. 1983.$24.95.

Atlas of the Ultraviolet Sky. Richard C. Henry, Wayne B. Landsman, et al. Johns Hopkins University Press, 2715 North Charles Street, Baltimore, MD 21218-4319. (410) 516-6900. (800) 537-5487. 1988. $70.00.

Atlas of the Universe. Patrick Moore. Rand McNally & Company. 8255 North Central Park, Skokie, IL 60076-2970. (708) 673-9100. 1994. $29.95.

Cambridge Atlas of Astronomy. Jean Audouze, et al. Cambridge University Press, 40 West 20th Street, New York, NY 10011-4211. (212) 924-3900. (800) 872-7423. 3rd edition. 1994. $75.00.

Cambridge Guide To the Constellations. Michael E. Bakich. Cambridge University Press, 40 West 20th Street, New York, NY 10011-4211. (212) 924-3900. (800) 872-7423. 1995. $19.95 in paper.

Carnegie Atlas of Galaxies. Allan Sandage and John Bedke. Carnegie Institution of Washington, 1530 P Street NW, Washington, DC 20005. (202) 387-6411. Two volumes. 1994. $92.00.

Color Atlas of Galaxies. James Wray. Cambridge University Press, 40 West 20th Street, New York, NY 10011-4211. (212) 924-3900. (800) 872-7423. 1988. $94.95.

Nearby Galaxy Atlas. R. Brent Tully and J. Richard Fisher. Cambridge University Press, 40 West 20th Street, New York, NY 10011-4211. (212) 924-3900. (800) 872-7423., 1987. $74.95.

Observer's Sky Atlas. E. Karkoschka. Springer-Verlag New York, Inc., 175 Fifth Avenue, New York, NY 10010. (212) 460-1500. (800) 777-4643. 1995. $15.95.

Observing Handbook and Catalog of Deep-sky Objects. Christian B. Luginbuhl and Brian A. Skiff. Cambridge University Press, 40 West 20th Street, New York, NY 10011-4211. (212) 924-3900. (800) 872-7423. 1990. $49.95.

Photometric Atlas of Northern Bright Galaxies. Kodaira, Keichi, et al editors. Columbia University Press, 562 West 113th Street, New York, NY 10025. (212) 666-1000. (800) 944-8648. 1990. $180.00.

Third Reference Catalogue of Bright Galaxies. Gerard H. DE Vaucouleurs et al, editors. Springer-Verlag New York, Inc., 175 Fifth Avenue, New York, NY 10010. (212) 460-1500. (800) 777-4643. Three volumes. 1991. $198.00 set.

The Sky: A User's Guide. David H. Levy. Cambridge University Press, 40 West 20th Street, New York, NY 10011- 4211. (212) 924-3900. (800) 872-7423. 1993. $14.95 in paper.

PERIODICALS

Astronomical Journal. American Institute of Physics, One Physics Ellipse, College Park, MD 20740-3843. (301) 209- 3000. Published for the American Astronomical Society. 1849 to present. Monthly. $280.00 per year. ISSN: 0004- 6256.

Astronomical Society of the Pacific. PUBLICATIONS. Astronomical Society of the Pacific, 390 Ashton Avenue, San Francisco, CA 94112. (415) 337-1100. FAX (415) 337-5205.1889 to present. Monthly. $175.00 per year. ISSN: 0004- 6280.

GALAXIES

Ency. of Physical Sciences and Engineering Info. Sources

Astronomy. Kalmbach Publishing Company, Box 1612, Waukesha, WI 53187-1612. (414) 796-8776. FAX (414) 796-1142. 1973 to present. Monthly. $27.00 per year. ISSN: 0091-6358.

Astronomy and Astrophsyics. Springer-Verlag New York, Inc., 44 Hartz Way, Secaucus, NJ 07096-2491. (201) 348-4033. FAX (201) 348-4505. 1969 to date. 36 issues per year in 12 volumes. $1733.00 per year. ISSN: 0004-6361.

Astronomy Reports. American Institute of Physics, One Physics Ellipse, College Park, MD 20740-3843. (301) 209-3000. English translation of Astronomicheksii Zhurnal. Formerly Soviet Astronomy AJ. 1957 to present. Bimonthly. $1160.00 per year. ISSN: 1063-7729.

Astrophysical Journal; An International Review of Astronomy and Astronomical Physics. University of Chicago Press, Journals Division, 5720 South Woodlawn Avenue, Chicago, IL 60637. (312) 753-3347. FAX (312) 753-0811. 1895 to date.three issues per month. $740.00 per year. ISSN: 0004-637X.

Astrophpysics and Space Science; An International Journal of Cosmic Physics. Kluwer Academic Publishers, Box 358, Accord Station, Hingham, MA 02018-0358. (617) 871-6600. FAX (617) 871-6528. 1968 to present. 24 issues per year. $2544.00 per year. ISSN: 0004-640X.

Celestial Mechanics and Dynamical Astronomy; An International Journal of Space Dynamics. Kluwer Academic Publishers, Box 358, Accord Station, Hingham, MA 02018-0358. (617) 871-6600. FAX (617) 871-6528. 1969 to present. Monthly. $703.50 per year. ISSN: 0923-2958.

Experimental Astronomy; An International Journal On Astronomical Instrumentation and Data Analysis. Kluwer Academic Publishers, Box 358, Accord Station, Hingham, MA 02018-0358. (617) 871-6600. FAX (617) 871-6528. 1989 to present. Quarterly. $161.00 per year. ISSN: 0922-6435.

Icarus; International Journal of Solar System Studies. Academic Press, Inc., Journal Division, 525 B Street, Suite 1900, San Diego, CA 92101-4495. (619) 230-1840. FAX (619) 699-6800. 1962 to present. Monthly. $1080.00 per year. ISSN: 0019-1035.

Landolt-boernstein, Zahlenwerte Und Funktionen Aus Naturwissenschaften Und Technik. GROUP 6: ASTRONOMY. Springer-Verlag New York, Inc., 175 Fifth Avenue, New York, NY 10010. (212) 460-1500. FAX (212) 473-6272. 1965. Irregular. Price varies. ISSN: 0075-7896.

Mercury. Astronomical Society of the Pacific. 390 Ashton Avenue, San Francisco, CA 94112. (415) 337-1100. FAX: (415) 337-5205. 1972 to present. Bimonthly. $175.00 per year. ISSN: 0047-6773.

Monthly Notices of the Royal Astronomical Society. Blackwell Scientific Publication LT., Osney Mead, Oxford OX2 OEL, England. TEL 0865-240201. FAX 0865-721205. 1827 to present. Fortnightly. $1899.00 per year. ISSN: 0035-8711.

Observatory. Space and Astrophysics Division, Rutherford Appleton Laboratory, Chilton, Didcot, Oxon OX11 OQX England. FAX 0235-445848. 1877 to present. Bimonthly. $42.00 per year.

Sky and Telescope. Sky Publishing Corporation, Box 9111, Belmont, MA 02178. (617) 864-7360. FAX (617) 864-6117.1941 to present. Monthly. $27.00 per year. ISSN: 0037-6604.

Space Science Reviews. Kluwer Academic Publishers, Box 358, Accord Station, Hingham, MA 02018-0358. (617) 871-6600. FAX (617) 871-6528. 1962 to present. Sixteen issues per year. $940.00 per year. ISSN: 0038-6308.

Vistas in Astronomy; An International Review Journal. Albert C. Beer, editor. Elsevier Science Publishing Company, Inc., 660 White Plains Road, Terrytown, NY 10591- 5153. (914) 524-9200. FAX: (914) 333-2444. 1958 to present. Quarterly. $415.00 per year. ISSN: 0083-6656.

Transactions of the International Astronomical Union. Proceedings of the General Assembly. Kluwer Academic Publishers, Box 358, Accord Station, Hingham, MA 02018-0358. (617) 871-6600. Issues in two parts: Part A, Reports, and Part B, Proceedings. Triennial. XXI - 1994. Price varies. ISSN: 0080-1372.

Webb Society Quarterly Journal. Webb Society, 194 Foundry Lane, Freemantle, Southampton, Hampshire SO1 3EQ, England. 1969 to present. Quarterly. $19.50 per year. ISSN: 0043- 1680.

RESEARCH CENTERS AND INSTITUTES

Astronomy Program. University of Maryland, 1207 Computer & Space Sciences Building, College Park, MD 20742. (301) 405- 3001. FAX (301) 314-9067.

Carnegie Institution of Washington Observatories, 813 Santa Barbara Street, Pasadena, CA 91101. (818) 577-1122. FAX (818) 759-8136.

Center For Astronomy and Space Astrophysics. University of Colorado - Boulder, Campus Box 391, Boulder, CO 80309. (303) 492-4050. (FAX) 492-7178.

Center For Space Research. Massachusetts Institute of Technology, 77 Massachusetts Avenue, Camridge MA 02139. (617) 253-7501.

David Dunlap Observatory. University of Toronto. Box 360, Richmond Hill ON, Canada L4C 4Y6. (416) 884-9562. FAX (416) 978-3921.

Lick Observatory. University of California, Santa Cruz, CA 95064. (408) 459-2991. FAX (408) 426-3115.

Lowell Observatory. 1400 West Mars Hill Road. Flagstaff, AZ 86001, (602) 774-3358. FAX (602) 774-6296.

National Astronomy and Ionosphere Center. Cornell University, Space Sciences Building, Ithaca, NY 14853. (607) 255-3735. (607) 255-8803.

National Optical Astronomy Observatories. 950 North Cherry Avenue, Tucson, AZ 85719. (602) 325-9282. FAX (602) 325-9360.

National Radio Astronomy Observatory. Edgemont Road, Charlottesville, VA 22903. (804) 296-0211.

Smithsonian Center For Astrophysics. Harvard University, 60 Garden Street, Cambridge, MA 02138. (617) 495-7461. FAX (617) 495-7326.

Space Telescope Science Institute. 3700 San Martin Drive, Baltimore, MD 21218. (301) 338-4700.

Theoretical Astrophysics Program. University of Arizona, Planetary Sciences Department, Tuscon, AZ 85721. (602) 621- 6963.

GAMMA-RAY ASTRONOMY

See: ASTRONOMY

GAMMA-RAY LASER

See: LASERS

GAMMA-RAY SPECTROSCOPY

See: SPECTROSCOPY

GAMMA RAYS

See also: ASTRONOMY, ASTROPHYSICS, GAMMA RAY ASTRONOMY, PHYSICS

ABSTRACT SERVICES AND INDEXES

Applied Science and Technology Index; A Cumulative Subject Index To English Language Periodicals in the Fields of Aeronautics and Space Science, Computer Technology, Chemistry, Construction Industry, Energy and Related Areas. H.W. Wilson Co., 950 University Avenue, Bronx, NY 10452. (212) 588-8400. (800) 367-6770. FAX (718) 590-1617. From1958 to present. Monthly. Inquire about cost and availability. Also available on CD-ROM and online. ISSN: 0003-6986.

Astronomy and Astrophysics Abstracts. Springer-Verlag New York, 175 Fifth Avenue, New York, NY 10010. (212) 460-1500. FAX (212) 473-6272. Published for the Astronomisches Rechen-Institut. Comprehensive coverage of all aspects of astronomy, astrophysics and related fields. 1969 to present. Two parts per year. Annual. ISSN: 0067-0022.

General Science Index. H.W. Wilson Company, 950 University Avenue, Bronx, NY 10452. (212) 588-8400. (800) 367-6770. FAX (718) 590-1617. From 1978 to present. Ten issues per year; quarterly and annual cumulations. Service basis. Available on CD-ROM and online. Inquire regarding cost and availability. ISSN: 0162-1963.

NTIS Alerts: Astronomy & Astrophysics. U. S. National Technical Information Service. 5285 Port Royal Road, Springfield, VA 22161. (703) 487-4650. FAX (703) 321-8547. Weekly. $140.00 per year.

Physics Abstracts. INSPEC. Section A, Science Abstracts. Institute of Electrical Engineers, London, United Kingdom. Available from: INSPEC/IEEE - Institute of Electrical and Electronic Engineers, Box 1331, Hoes Lane, Piscataway, NJ 08855-1331. (908) 562-5549. 1898 to present. 24 issues per year. $2835.00 per year. ISSN: 0036-8091. Also available online and on CD-ROM.

Physics Briefs (Physikalische Berichte). Information Center for Energy, Physics, Mathematics; German Physical Society. V C H Publishers, Inc., 220 East 23rd Street, New York, NY 10010-4606. (212) 683-8333. 1845 to present. 24 issues per year. $2390.00 per year. Also available online. ISSN: 0179-7434.

Science Citation Index. SCI. Institute for Scientific Information, 3501 Market Street, Philadelphia, PA 19104. (215) 386-0100. (800) 523-1850. FAX (215) 386-2991. 1961 to present. Six issues per year, plus annual cumulation. $11650.00 per year. Also available online and on CD-ROM.Inquire about price and availability. ISSN: 0036-827X.

STAR. Scientific and Technical Aerospace Reports. U.S. National Aeronautics and Space Administration. Distributed by U. S. Superintendent of Documents, Washington, DC 20402. 1963 to present. Semi-monthly, with semiannual and annual indexes. $114.00 per year. ISSN: 0036-8741. Also available online and on CD-ROM.

ANNUAL REVIEWS AND YEARBOOKS

The Astronomical Almanac. Superintendent of Documents, U. S. Government Printing office, Washington, DC 10402. (202) 783-3238. 1981 to present. Supercedes Astronomical Ephemeris and the American Ephermis and Nautical Almanac. Annual. $44.95. ISSN: 0737-6421.

Annual Review of Astronomy and Astrophysics; Original Reviews of Critical Literature and Current Developments in Astronomy and Astrophysics. Annual Reviews, Inc., 4139 El Camino Way, Palo Alto, CA 94306-0139. (415) 493-4400.(800) 523-8635.FAX (415) 855-9815. 1963 to date. Annual. $60.00. ISSN: 0066-4146.

ASSOCIATIONS AND PROFESSIONAL SOCIETIES

American Astronomical Society. 2000 Florida Avenue NW, Suite 400, Washington, DC 20009. (202) 328-2010. FAX:(202) 234-2560.

American Institute of Physics. One Physics Ellipse, College Park, MD 20740-3843. (301) 209-3100.

American Society of Nondestructive Testing. 1711 Arlingate Lane, Columbus, OH 43228-0518. (614) 274-6003. FAX (614)274-6899.

Association of Universities For Research in Astronomy. Suite 701, 1625 Massachusetts Avenue, NW, Washington, DC 20036. (202) 483-2101. FAX (202) 483-2106.

National Council On Radiation Protection and Measurements. 7910 Woodmont Avenue, Suite 1016, Bethesda, MD 10814. (301) 657-2652.

Radiation Research Society. 2021 Spring Road, Suite 600, Oak Brook Road, IL 60521. (708) 571-2881. (708) 571-7837.

BIBLIOGRAPHIES

Astronomy and Astrophysics: A Source Guide. Gordon Press, P.O. Box 459, Bowling Green Station, New York, NY 10004. (718) 624-8419. 1991. $250.00. .

A Bibliography of Astronomy, 1970-1979. R. A. Sea and S. S. Martin. Libraries Unlimited, Inc., Littleton, CO 80160. 1982.

Catalog of the Naval Observatory Library. Naval Observatory Staff. G. K. Hall & Company, 866 Third Avenue, New York, NY 10022. (212)702-6789. (800) 257-5755. 6 volumes. 1977. $655.00

Science Books and Films. American Association for the Advancement of Science, 1333 H Street NW, Washington, DC 20005. (202) 326-6454. 1965 to present. Nine issues per year. $40.00 per year. ISSN: 0098-342X

Scientific and Technical Books and Serials in Print, 1995. R.R. Bowker Inc., 121 Chanlon Road, New Providence, NJ 07974. (908) 464-6800. (800) 521-8110. 4 volumes. 1994. $299.95. Also available on compact disc and online.

A Source Book On Astronomy and Astrophysics, 1900 - 1975. Kenneth Lang and Owen Gingerich, editors. Harvard University Press, 79 Garden Street, Cambridge MA. (617) 495-2600. 1980. $41.95.

DIRECTORIES AND BIOGRAPHICAL SOURCES

American Men and Women of Science: Physical and Biological Sciences. R. R. Bowker Inc., 121 Chanlon Road, New Providence, NJ 07974. (908) 464-6800. (800) 521-8110.20th edition. 8 volumes. 1996. $850.00.

The Astronomers. Donald Goldsmith. St. Martin's Press, Inc., 175 Fifth Avenue, New York, NY 10010. (212) 674-5151. (800) 221-7945. 1993. $14.95 in paper.

Astronomical Centers of the World. Kevin Krisciunas. Cambridge University Press, 40 West 20th Street, New York, NY 10011-4211. (212) 924-3900. (800) 872-7423. 1988. $34.95

The Biographical Dictionary of Scientists: Astronomers. D. Abbott, editor. Peter Bedrick Books, Inc., 2112 Broadway, Room 318, New York, NY 10023. (212) 496-0751. 1984.

Directory of Physics and Astronomy Staff Members. American Institute of Physics, One Physics Ellipse, College Park, MD 20740-3843. (301) 209-3100. Annual. $45.00.

Earth and Astronomical Research Centers. Stockton Press, 345 Park Avenue South, New York, NY 10010. 4th edition. 1995. $515.00. ISBN: 1-56169-0967.

Graduate Programs in Physics, Astronomy and Related Fields. 1995-1996. American Institute of Physics, One Physics Ellipse,

College Park, MD 20740-3843. (301) 209-3100. 1978 to present. Annual. $45.00.

Research Centers Directory. Gale Research Inc., 835 Penobscot Building, Detroit, MI 48226-4094. (313) 961-2242. (800) 877-4253. 20th edition. 1995. $485.00.

ENCYCLOPEDIAS AND DICTIONARIES

Astronomy From A To Z: A Dictionary of Celestial Objects and Ideas. Charles A. Schweighauser. Illinois Issues, Sangamon State University, K-10, Springfield, IL 62794-9243. (217) 786-9243. 1991. $14.95 in paper.

Encyclopedia of Astronomy and Astrophysics. Stephen Maran, editor. Van Nostrand Reinhold, 115 Fifth Avenue, New York, NY 10003. (212) 254-3232. (800) 842-3636. 1992. $129.95.

Landolt-boernstein Numerical Data and Functional Relationships in Science and Technology: Astronomy, Astrophysics and Space Research; A. O. Madelung and H. H. Boigt, editors. Springer Verlag New York, Inc., 175 Fifth Avenue, New York, NY 10010. (212) 460-1500. (800) 777- 4643. 1993. $550.00.

McGraw-Hill Encyclopedia of Science and Technology. McGraw-Hill Book Company, Inc., 1221 Avenue of the Americas, New York, NY 10020. (212) 512-2000. (800) 262-4729. 7th edition. 20 volume set. 1992. $1900.00.

GENERAL WORKS

Cosmic Gamma Rays, Neutrinos, and Related Astrophysics. Maurice M. Shapiro and John Pl Weld, editors. Kluwer Academic Publishers, 101 Philip Drive, Assinippi Park, Norwell, MA 02061. (617) 871-6600. 1989. $207.50.

Gamma and X-ray Spectrometry With Semiconductor Detectors. K. Debertin and R. G. Helmer. Elsevier Science Publishing Company, Inc., 655 Avenue of the Americas, New York, NY 10010. (212) 989-5800. FAX (914) 333-2444. 1988. $92.50.

Gamma Ray Bursts, Observations, Analyses and theories. Cheng Ho, et al, editors. Cambridge University Press, 40 West 20th Street, New York, NY 10011-4211. (212) 924-3900. (800) 872-7423. 1992. $74.95.

High Energy Gamma Ray Astronomy. James Matthews, editor. American Institute of Physics, One Physics Ellipse, College Park, MD 20740-3843. (301) 209-3100. 1991. $85.00.

Observational Astrophysics. Robert C. Smith. Cambridge University Press, 40 West 20th Street, New York, NY 10011- 4211. (212) 924-3900. (800) 872-7423. 1995. $69.95.

The Physics of Stars. A. C. PHillips. John Wiley & Sons, Inc., 605 Third Avenue, New York, NY 10158-0012. (212) 850- 6000. (800) 225-5945. 1994. $32.95 in paper.

Practical Gamma-ray Spectrometry. Gordon Gilmore and John D. Hemingway. John Wiley & Sons, Inc., 605 Third Avenue, New York, NY 10158-0012. (212) 850-6000. (800) 225-5945. 1995. $96.00.

X-ray Binaries and Recycled Pulsars. E. P. Van den Heuvel and S. A. Rappaport, editors. Kluwer Academic Publishers, 101

Ency. of Physical Sciences and Engineering Info. Sources

GAMMA RAYS

Philip Drive, Assinippi Park, Norwell, MA 02061. (617) 871-6600. 1992. $222.00.

HANDBOOKS AND MANUALS

Encyclopedia of Astronomy and Astrophysics. Stephen Maran, editor. Van Nostrand Reinhold, 115 Fifth Avenue, New York, NY 10003. (212) 254-3232. (800) 842-3636. 1992. $129.95.

Handbook of Space Astronomy and Astrophsyics. Martin V. Zumbeck. Cambridge University Press, 40 West 20th Street, New York, NY 10011-4211. (212) 924-3900. (800) 872-7423.2nd edition. 1990. $79.95.

McGraw-Hill Encyclopedia of Astronomy. Sybil P. Parker and Jay M. Pasachoff, editors. McGraw-Hill Publishing Company, Inc., 1221 Avenue of the Americas, New York, NY 10020. (212) 512-2000. (800) 262-4729. 2nd edition. 1993. $75.00

ONLINE DATABASES AND CD-ROMS

CA Search. Chemical Abstracts Service, P.O. Box 3012, Columbus, OH 43210-0012. (614) 447-3600. (800) 848-6533. FAX (614) 447-3709. Very comprehensive guide to worldwide chemical literature and related fields, 1972 to present. Available on BRS,(800) 289-4277, DIALOG, (800) 334-2564, ORBIT (800) 456-7248, and STN International, FIZ Karlsruhe, P.O. Box 2465, W-7500, Karlsruhe 1, Germany, online services. Inquire as to cost and availability.

Dissertation Abstracts Online. University Microfilms International, 300 North Zeeb Road, Ann Arbor, MI 48106. (800) 521-0600 or (313) 761-4700. Scope includes virtually all doctoral dissertations accepted at accredited American institutions from 1861 to present in 252 subject areas. Available on BRS, (800) 289-4277, DIALOG, (800) 334-2564, and OCLC EPIC, (800) 848-5878, online services. Also available on CD-ROM. Inquire as to cost and availability.

INSPEC. Institution of Electrical Engineers, Michael Faraday House, Six Hills Way, Stevenage, Herts. SG1 2AY, England. Telephone: 0438 313311 or FAX 0438 742840. Contains citations to the worldwide literature of physics, electronics and electrical engineering, computer technology, and related fields. Available on BRS, (800) 289-4277, DIALOG, (800) 334-2564, ORBIT, (800) 456-7248, and STN International, FIZ Karlsruhe, P.O. Box 2465, W-7500, Karlsruhe 1, Germany, online services. Inquire as to cost and availability.

NTIS Bibliographic Database. National Technical Information Service, 5285 Port Royal Road, Springfield, VA 22161. (703) 487-4929 or FAX (703) 321-8199. Broad coverage of government-sponsored science and technology research reports, 1964 to present. Available on BRS,(800) 289-4277, DIALOG, (800) 334-2564, ORBIT, (800) 456-7248, and STN International, FIZ Karlsruhe, P.O. Box 2465, W-7500, Karlsruhe 1, Germany, online services. Also available on CD-ROM. Inquire as to cost and availability.

Physics Briefs. American Institute of Physics, 335 East 45th Street, New York, NY 10017. (212) 661-9260 or FAX (212) 661-2036. Contains citations with abstracts of the literature of physics and related fields, 1979 to present. Available on the STN International, FIZ Karlsruhe, P.O. Box 2465, W-7500, Karlsruhe 1, Germany, online service. Inquire as to cost and availability.

SCISEARCH. Institute for Scientific Information, 3501 Market Street, Philadelphia, PA 19104. (800) 523-1850 or (215) 386-0100. Broad multidisciplinary title and author index to the international literature of science and technology, 1974 to present. Available on DIALOG, (800) 334-2564, and ORBIT, (800) 456-7248, online services. Also available on CD-ROM. Inquire as to cost and availability.

WILSONLINE. H.W. Wilson Company, 950 University Avenue, Bronx, NY 10452. (800) 367-6770 or (212) 588-8400. Makes available online versions of the printed H.W. Wilson indexes including Applied Science and Technology Index, Business Periodicals Index, General Science Index, and Readers' Guide to Periodical Literature. Period covered is generally 1983 to present. Available on BRS, (800) 289-4277, DIALOG, (800) 334-2564, and OCLC EPIC, (800) 848-5878, online services. Also available on CD-ROM. Inquire as to cost and availability.

PERIODICALS

Astronomical Journal. American Institute of Physics, One Physics Ellipse, College Park, MD 20740-3843. (301) 209- 3000. Published for the American Astronomical Society. 1849 to present. Monthly. $280.00 per year. ISSN: 0004- 6256.

Astronomical Society of the Pacific. PUBLICATIONS. Astronomical Society of the Pacific, 390 Ashton Avenue, San Francisco, CA 94112. (415) 337-1100. FAX (415) 337-5205. 1889 to present. Monthly. $175.00 per year. ISSN: 0004- 6280.

Astronomy. Kalmbach Publishing Company, Box 1612, Waukesha, WI 53187-1612. (414) 796-8776. FAX (414) 796-1142. 1973

To Present. Monthly. $27.00 per year. ISSN: 0091-6358.

Astrophysical Journal; An International Review of Astronomy and Astronomical Physics. University of Chicago Press, Journals Division, 5720 South Woodlawn Avenue, Chicago, IL 60637. (312) 753-3347. FAX (312) 753-0811. 1895 to date.three issues per month. $740.00 per year. ISSN: 0004-637X.

Mercury. Astronomical Society of the Pacific. 390 Ashton Avenue, San Francisco, CA 94112. (415) 337-1100. FAX: (415) 337-5205. 1972 to present. Bimonthly. $175.00 per year. ISSN: 0047-6773.

Monthly Notices of the Royal Astronomical Society. Blackwell Scientific Publication LT., Osney Mead, Oxford OX2 OEL, England. TEL 0865-240201. FAX 0865-721205. 1827 to present. Fortnightly. $1899.00 per year. ISSN: 0035-8711.

Nuclear Science and Engineering. American Nuclear Society, 555 North Kensington Avenue, LA Grange Park, IL 60525. (708) 352-6611. 1956 to present. Monthly. $405.00. ISSN: 0029- 5639.

Observatory. Space and Astrophysics Division, Rutherford Appleton Laboratory, Chilton, Didcot, Oxon OX11 OQX England. FAX 0235-445848. 1877 to present. Bimonthly. $42.00 per year.

Radiation Research. Kluge Carden Jennings, 853 West Main Street, Charlottesville, VA 22903. 1954 to present. Monthly. $560.00 per year. ISSN: 0033-7587.

Solar Physics. Kluwer Academic Publishers, Box 358, Accord Station, Hingham, MA 02018-0358. (617) 871-6600. FAX (617)

GAMMA RAYS

Ency. of Physical Sciences and Engineering Info. Sources

871-6528. 1967 to present. 14 issues per year. $1595.00 per year. ISSN: 0038-0938.

RESEARCH CENTERS AND INSTITUTES

Center For Astrophysics and Space Physics. University of California, San Diego, C-0111, LA Jolla, CA 92093-0111. (619) 534-3460. FAX (619) 534-2294.

Center For Engineering Applications of Radioisotopes. North Carolina State University, Box 7909, Department of Nuclear Engineering, Raleigh, NC 27695-7909. (919) 737-3378. FAX (919) 737-3928.

Cosmic Ray and Gamma Ray Astronomy Group. University of New Hampshire, Institute for the Study of Earth, Oceans and Space, Science and Engineering Research Building, Durham, NH 03824. (603) 862-3510. FAX (603) 862-4685.

Fred Lawrence Whipple Observatory. P. O. Box 97, Amado, AZ 85645. (602) 670-6741. FAX (602) 670-6779.

National Scientific Balloon Facility. Box 319 F/M Road, 3224, Palestine, TX 75801. (214) 729-0271. FAX (214) 723-8056.

Nuclear Physics Laboratory, University of Washington, Seattle, WA 98195. (206) 543-4080.

Space Astronomy Laboratory. Institute for Space Science and Technology, Inc., 1810 NW 6th Street, Gainesville, FL 32609-3530. (904) 371-4778. FAX (904) 372-5042.

GARBAGE DISPOSAL

See: SOLID WASTE DISPOSAL

GAS CHROMATOGRAPHY

See: CHROMATOGRAPHY

GAS INJECTION WELLS

See: PETROLEUM ENGINEERING

GAS LASER

See: LASERS

GAS MASERS

See: MASERS

GAS PIPELINES

See: NATURAL GAS, PIPELINE TECHNOLOGY

GAS TURBINES

See also: AERONAUTICS, AIRCRAFT, AIRCRAFT ENGINES, AUTOMOTIVE ENGINEERING, ENGINES, JET PROPULSION, TURBINES

ABSTRACT SERVICES AND INDEXES

Aeronautical Engineering. Scientific and Technical Branch, National Aeronautics and Space Administration. National Technical Information Service (NTIS), Springfield, VA 22161. Monthly. A selection of annotated references to unclassified reports and journal articles that were introduced into the NASA scientific and technical information system and announced in STAR and IAA.

Applied Mechanics Reviews: An Assessment of World Literature in Engineering Sciences. 1948-present. American Society of Mechanical Engineers, 345 East 47th Street, New York, NY 10017. (212) 705-7703. Monthly. $360.00 per year.

Applied Science and Technology Index; A Cumulative Subject Index To English Language Periodicals in the Fields of Aeronautics and Space Science, Computer Technology, Chemistry, Construction Industry, Energy and Related Areas. H.W. Wilson Co., 950 University Avenue, Bronx, NY 10452. (800) 367-6770 or (212) 588-8400. FAX (718) 590-1617. From 1958 to present. Monthly. Inquire about cost and availability. Also available on CD-ROM and online.

Current Contents: Engineering, Technology, and Applied Sciences. Institute for Scientific Information, 3501 Market Street, Philadelphia, PA 19104. (215) 386-0100. FAX (215) 386-6362. 1970 to present. Weekly. $442.00 per year.

Engineering Index Monthly. Engineering Information, Inc., Castle Point on the Hudson, Hoboken, NJ 07030. (800) 221-1044. FAX (212) 832-1857. Monthly. $2300.00 per year. Also available online as COMPENDEX, and also on CD-ROM. Covers chemical engineering, computers, electrical engineering, civil engineering, metals and mining, industrial management, and mechanical engineering.

International Aerospace Abstracts. Technical Information Service, American Institute of Aeronautics and Astronautics, Inc., 555 West 57th St., New York, NY 10019. (212) 247-6500. FAX (212) 582-4861. Semi-monthly. $1295.00 per year.

Scientific and Technical Aerospace Reports (star). National Aeronautics and Space Administration. NASA Center for Aerospace Information, Box 8757, BWI Airport, Baltimore, MD 21240. (301) 621-0153. Monthly. Inquire about availability and cost. ALso available through the NASA online retrieval service (RECON), and through the Aerospace Database through DIALOG.

ASSOCIATIONS AND PROFESSIONAL SOCIETIES

American Institute of Aeronautics and Astronautics. 370 L'Enfant Promenade, S.W., Washington, DC 20024. (202) 646-7400.

American Society of Mechanical Engineers. 345 East 47th Street, New York, NY 10017-2398. (212) 705-7703.

International Gas Turbine Institute, Asme. 5801 Peachtree Dunwoody Road, Suite 100, Atlanta, GA 30342-1503. (404) 847-0072.

Society of Automotive Engineers (sae International). 400 Commonwealth Drive, Warrendale, PA 15096. (412) 776-4841.

DIRECTORIES AND BIOGRAPHICAL SOURCES

Diesel and Gas Turbine Worldwide Catalog. Diesel and Gas Turbine Publications, 13555 Bishop's Court, Brookfield, WI 53005-6286. (914) 784-9177. FAX (914) 784-9177. 1935 to present. Annual. $75.00 per year.

Directory of Engineering Societies and Related Organizations. Gordon Davis. 13th edition. American Association of Engineering Societies, 1111 19th Street NW, Suite 608, Washington, DC 20036. (202) 296-2237 or (800) 658-8897. 1989. Inquire for price.

Research Centers Directory. Gale Research, 835 Penobscot Building, Detroit, MI 48226-4094. (313) 961-2242. (800) 877-4253. FAX (313) 961-6083. $485.00.

Scientific and Technical Organizations and Agencies Directory. Gale Research, 835 Penobscot Building, Detroit, MI 48226-4094. (313) 961-2242. (800) 877-4253. FAX (313) 961-6083. 4th edition. 1996. $195.00.

Who's Who in Engineering. American Association of Engineering Societies, 1111 19th Street NW, Suite 608, Washington, DC 20036. (202) 296-2237 or (800) 658-8897. 8th edition. 1991. Inquire for price.

ENCYCLOPEDIAS AND DICTIONARIES

McGraw-Hill Encyclopedia of Science and Technology. Sybil P. Parker, ed. 7th edition. 20 volumes. McGraw-Hill Publishing Company, 1221 Avenue of the Americas, New York, NY 10020. (800) 262-4729 or (212) 512-3825. 1992. $1900.00

Thesaurus of Scientific, Technical, and Engineering Terms. Hemisphere Publishing Corporation, 1900 Frost Road, Suite 101, Bristol, PA 19007-1598. (215) 785-5800 or (800) 821-8312. FAX (215) 785-5515. 1987. $173.00.

GENERAL WORKS

Aircraft Engines and Gas Turbines. Jack L. Kerrebrock. 2d ed. MIT Press, 55 Hayward Street, Cambridge, MA 02142. (617) 253-2884 or (800) 356-0343. FAX (617) 253-1709. 1992. $45.00.

Elements of Gas Turbine Propulsion. Jack D. Mattingly. McGraw-Hill Publishing Company, 1221 Avenue of the Americas, New York, NY 10020. (800) 262-4729 or (212) 512-3825. 1996. Inquire for cost and availability.

Fundamentals of Gas Turbines. William W. Bathie. John Wiley and Sons, Inc., 605 Third Avenue, New York, NY 10158. (800) 526-5368 or (212) 850-6000. 1984. Inquire for cost and availability.

Steam and Gas Turbines. A. Stodola. 2 volumes. 6th edition. Peter Smith Publishing, 5 Lexington Avenue, Magnolia, MA 01930. (508) 525-3562. $48.00.

HANDBOOKS AND MANUALS

Gas Turbine Engineering Handbook. Meherwan P. Boyce. Gulf Publishing Company, P.O. Box 2608, Houston, TX. (713) 529-4301 or (800) 231-6275. 1982. $79.00.

Sawyer's Turbomachinery Maintenance Handbooks. John W. Sawyer & David Japikse, editors. 3rd edition. 3 volumes. Turbomachinery International Publications, Business Journals, 50 Day Street, PO Box 5550, South Norwalk, CT 06856. (203) 853-6015. $248.50.

ONLINE DATABASES AND CD-ROMS

Compendex Plus. Engineering Information, Inc., 345 East 47th Street, New York, NY 10017. (212) 705-7600 or (800) 221-1044. Contains citations with abstracts to worldwide literature in engineering and technology, from 1970 to present. Available on online BRS,(800) 289-4277, DIALOG, (800) 334-2564, ORBIT (800) 456-7248, and STN International, FIZ Karlsruhe, P.O. Box 2465, W-7500, Karlsruhe 1, Germany, online services. Also available on CD-ROM. Inquire as to cost and availability.

NTIS Bibliographic Database. National Technical Information Service, 5285 Port Royal Road, Springfield, VA 22161. (703) 487-4929 or FAX (703) 321-8199. Broad coverage of government-sponsored science and technology research reports, 1964 to present. Available on BRS,(800) 289-4277, DIALOG, (800) 334-2564, ORBIT, (800) 456-7248, and STN International, FIZ Karlsruhe, P.O. Box 2465, W-7500, Karlsruhe 1, Germany, online services. Also available on CD-ROM. Inquire as to cost and availability.

SAE Global Mobility Database. Society of Automotive Engineers (SAE), Electronic Publishing Division, 400 Commonwealth Drive, Warrendale, PA 15098. (412) 776-4841. Contains citations with abstracts to technical papers on automotive and aerospace technology and vehicular-related industries that have been presented at SAE conferences. Covers 1906 to present. Available on ORBIT,(800) 456-7248,online service. Inquire as to cost and availability.

TRIS (Transportation Research Information). National Academy of Science, 2101 Constitution Avenue, N.W., Washington, DC 20418. (202) 334-3313 or (800) 624-6242. Citations with abstracts of literature on transportation, including air, highway, rail, maritime and other modes from 1968 to present. Available on DIALOG,(800) 334-2564, online service. Inquire as to cost and availability.

WILSONLINE. H.W. Wilson Company, 950 University Avenue, Bronx, NY 10452. (800) 367-6770 or (212) 588-8400. Makes available online versions of the printed H.W. Wilson indexes including Applied Science and Technology Index, Business Periodicals Index, General Science Index, and Readers' Guide to Periodical Literature. Period covered is generally 1983 to present. Available on BRS,(800) 289-4277, DIALOG,(800) 334-2564, and OCLC EPIC,(800) 848-5878, online services. Also available on CD-ROM. Inquire as to cost and availability.

GAS TURBINES

Ency. of Physical Sciences and Engineering Info. Sources

PERIODICALS

Diesel and Gas Turbine Worldwide. Diesel and Gas Turbine Publications, 13555 Bishop's Court, Brookfield, WI 53005-6286. (914) 784-9177. FAX (914) 784-9177. 1969 to present. 10 times a year. $55.00 per year.

High-speed Diesels and Drives. Diesel and Gas Turbine Publications, 13555 Bishop's Court, Brookfield, WI 53005-6286. (914) 784-9177. FAX (914) 784-9177. 1982 to present. 9 times a year. $50.00 per year.

Institution of Diesel and Gas Turbine Engineers Transactions. Mechanical Engineering Publications Ltd., Northgate Avenue, Bury St. Edmunds, Suffolk IP32 6BW, England. Telephone 0284-763277. FAX 0284-704006. 6 times a year. Inquire for cost.

International Journal of Turbo and Jet Engines. Freund Publishing House, Ltd., PO Box 35010, Tel Aviv 63150, Israel. 972-2-5662925. FX 972-3-5605335. Quarterly. $190.00 per year.

Journal of Engineering For Gas Turbines and Power. American Society of Mechanical Engineers, 345 E. 47th Street, New York, NY 10017. (212) 705-7722. Quarterly. Inquire for cost.

Journal of Propulsion and Power. American Institute of Aeronautics and Astronautics, 370 L'Enfant Promenade SW, Washington, DC 20024-2518. Bimonthly. $38.00 (members) and $260.00 (non-members).

Journal of Turbomachinery. American Society of Mechanical Engineers, 345 E. 47th Street, New York, NY 10017. (212) 705-7722. Quarterly. $100.00 per year for non-members.

Turbomachinery International. Turbomachinery International Publications, Business Journals Inc., Box 5550, Norwalk, CT 06856. (203) 853-6015. FAX (203) 853-8175. 1959 to present. 7 times a year. $49.00 per year.

RESEARCH CENTERS AND INSTITUTES

International Gas Turbine Institute, Asme. 5801 Peachtree Dunwoody Road, Suite 100, Atlanta, GA 30342-1503. (404) 847-0072.

Machinery and Engine Technology Laboratory. Institute for Mechanical Engineering, Bldg. M-7, Montreal Road, Ottawa, ON, Canada K1A 0R6. (613) 993-2425. FAX (613) 957-3281.

Sloan Automotive Laboratory, Massachusetts Institute of Technology. Room 3-340, Cambridge, MA 02139. (617) 253-2243.

Turbomachinery Laboratory. Texas A&M University, Mechanical Engineering Department, College Station, TX 77843. (409) 845-7417. FAX (409) 845-1835.

Virginia Polytechnic Institute Center For Turbomachinery and Propulsion Research. Department of Mechanical Engineering, Blacksburg, VA 24061-0238. (703) 231-6641. FAX (703) 231-7248.

GASIFICATION
See: COAL GASIFICATION AND LIQUEFACTION

GASOHOL
See:FUELS

GASOLINE
See: FUELS

GEARS AND GEARING
See:MACHINERY

GEMS
See: MINERALOGY

GENERATORS
See Also: ELECTRIC MOTORS, ELECTRIC POWER ENGINEERING, ELECTRICAL ENGINEERING, ELECTRICITY, ELECTROMAGENTISM, GENERATORS, HIGH VOLTAGE, HYDRO-ELECTRIC POWER

ABSTRACT SERVICES AND INDEXES

Applied Mechanics Reviews: An Assessment of World Literature in Engineering Sciences. American Society of Mechanical Engineers, 345 East 47th Street, New York, NY 10017. (212) 705-7703. 1948 to present. Monthly. $360.00 ISSN: 0003-6900.

Applied Science and Technology Index; A Cumulative Subject Index To English Language Periodicals in the Fields of Aeronautics and Space Science, Computer Technology, Chemistry, Construction Industry, Energy and Related Areas.

H.W. Wilson Co., 950 University Avenue, Bronx, NY 10452. (212) 588-8400. (800) 367-6770. FAX (718) 590-1617. From 1958 to present. Monthly. Inquire about cost and availability. Also available on CD-ROM and online. ISSN: 0003-6986.

Current Contents: Engineering, Technology and Applied Sciences. Institute for Scientific Information, 3501 Market Street, Philadelphia, PA 19104. (215) 386-0100. FAX (215)386-6362. 1970 to present. Weekly. $442.00 per year. Also available on CD-ROM and online. Inquire regarding cost and availability. ISSN: 0095-7917.

Current Papers in Electrical and Electronics Engineering. Institution of Electrical Engineers (IEE), London. Distributed by INSPEC/IEEE, Box 1331, 445 Hoes Lane, Piscataway, NJ 08855-1331. (908) 562-5549. 1969 to present. Monthly. $345.00 per year. ISSN: 0011-3778.

Electric Power Industry Abstracts. Edison Electric Institute, c/o Utility Data Institute, 2011 I Atreet, NW, Suite 700, Washington, DC 20006. 1975 to present. Bimonthly. Inquire as to cost and availability.

Electrical and Electronics Abstracts. Institution of Electrical Engineers (IEE), London. Available from INSPEC/IEEE - Institute of Electrical and Electronic Engineers, Box 1331, Hoes Lane, Piscataway, NJ 08855-

1331. (908) 562-5549. 1898 to present. Monthly. $2200.00 per year. Also available on CD-ROM and online as INSPEC.

Issn: 0036-8105.

Engineering Index Monthly; Indexes and Abstracts the World's Engineering and Technical Literature. Engineering Information, Inc., Castle Point on the Hudson, Hoboken, NJ 07030. (201) 216-8500. (800) 221-1044. FAX (201) 216-8532. Monthly. $2300.00 per year. Available online as COMPENDEX and also on CD-ROM. ISSN: 0742-1974.

General Science Index. H.W. Wilson Company, 950 University Avenue, Bronx, NY 10452. (212) 588-8400. (800) 367-6770. FAX (718) 590-1617. From 1978 to present. Ten issues per year; quarterly and annual cumulations. Service basis. Available on CD-ROM and online. Inquire regarding cost and availability. ISSN: 0162-1963.

Government Reports Announcements and Index. U. S. National Technical Information Service (NTIS), 5285 Port Royal Road, Springfield, VA 22161. (703) 487-4650. FAX (703) 321-8547. From 1968 to present. Annual. $630.00 per year. Also available online as *NTIS Bibliographic Database* and on CD-ROM.

Index To IEEE Publications. IEEE Service Center, 445 Hoes Lane, Piscataway, NJ 08855-1331. (908) 981-1393. (800) 678-IEEE. FAX (908) 981-9667. 1973 to present. Annual. ISSN: 0099-1368.

Physics Abstracts. INSPEC. Section A, Science Abstracts. Institution of Electrical Engineers (IEE), London. Available from: INSPEC/IEEE - Institute of Electrical and Electronic Engineers, Box 1331, Hoes Lane, Piscataway, NJ 08855-1331. (908) 562-5549. 1898 to present. 24 issues per year. $2835.00 per year. Also available online and on CD-ROM. ISSN: 0036-8091.

Physics Briefs (Physikalische Berichte). Information Center for Energy, Physics, Mathematics; German Physical Society. V C H Publishers, Inc., 220 East 23rd Street, New York, NY 10010-4606. (212) 683-8333. 1845 to present. 24 issues per year. $2390.00 per year. Also available online. ISSN: 0179-7434.

Science Citation Index. SCI. Institute for Scientific Information, 3501 Market Street, Philadelphia, PA 19104. (215) 386-0100. (800) 523-1850. FAX (215) 386-2991. 1961 to present. Six issues per year, plus annual cumulation. $11650.00 per year. Also available online and on CD-ROM. Inquire about price and availability. ISSN: 0036-827X.

ASSOCIATIONS AND PROFESSIONAL SOCIETIES

American Electronics Association. 5201 Great America Way, Suite 520, P.O. 52990, Santa Clara, CA 95056. (408) 987- 4200. FAX (408) 970-8565.

Edison Electric Institute. 701 Pennsylvania Avenue NW, Washington, DC 20004-2696. (202) 508-5000, 5454. FAX (202) 508-5360.

Electric Power Research Institute. 3412 Hillview Avenue, Palo Alto, CA 94304. (415) 855-2000. FAX (415) 855-2954.

Electronics Industries Association. 2500 Wilson Boulevard, Arlington, VA 22201. (703) 907-7500. FAX (202) 457-4985.

IEEE (institute of Electrical and Electronic Engineers). 345 East 47th Street. New York, NY 10017. (212) 705-7900. FAX (212) 705-4929.

IEEE Power Engineering Society. c/o Institute of Electrical and Electronic Engineers. 345 East 47th Street, New York, NY 10017. (212) 705-7900. FAX (212) 705-4929.

National Association of Power Engineers. 1 Springfield Street, Chicopee, MA 01013. (413) 592-6273. FAX (413) 592- 1998.

National Electrical Manufacturers Association. 1300 North 17th Street, Suite 1847, Rosslyn VA 22209. (703) 841-3200. FAX (703) 841-3300.

SMMA - the Association For Electric Motors, their Control and Application. 4 Hollis Street, PO Box 378, Sherborn, MA 01770. (508) 655-4409. FAX (508) 651-3920.   <subhead2>Directories and Biographical Sources

American Electronics Association. Directory. American Electronics Association, 5201 Great America Way, Suite 520, P.O. 52990, Santa Clara, CA 95056. (408) 987-4200. FAX (408) 970-8565. Annual. $175.00.

American Men and Women of Science: Physical and Biological Sciences. R. R. Bowker Inc., 121 Chanlon Road, New Providence, NJ 07974. (908) 464-6800. (800) 521-8110.20th edition. 8 volumes. 1996. $850.00.

Directory of Engineering Societies and Related Organizations. American Association of Engineering Societies, 1111 19th Street, Suite 608, Washington, DC 20036-3603. (202) 296-2237. Semi-annual. $150.00.

E E M - Electronic Engineer's Master. Hearst Business Communications, Inc., 645 Stewart Avenue, Garden City NY 11530. (516) 227-1300. ISSN: 0732-9016.

Electrical and Electronic Trades Directory: The Blue Book.

Institution of Electrical Engineers, Michael Faraday House, Six Hills Way, Stevenage, Herts. SG1 2AY, England. Telephone: 0438 313311 or FAX 0438 742840. 1995. $140.00

Electrical World Directory of Electrical Utilities. McGraw-Hill Book Company, Inc., 1221 Avenue of the Americas, New York, NY 10020. (212) 997-3675. Annual.

Engineering Research Centers: Incorporating Electronics Research Centers. Stockton Press, 345 Park Avenue, New York, NY 10010. (212) 689-9200. (800) 221-2123. 4th edition. 1995. $515.00.

IEEE Membership Directory. Institute of Electrical and Electronics Engineers, IEEE Service Center, 445 Hoes Lane,

GENERATORS

Ency. of Physical Sciences and Engineering Info. Sources

Piscataway, NY 08854. (908) 981-1393. (800) 678-IEEE. FAX(908) 981-9667. 2 volumes. Annual. $190.00. ISSN:

International Engineering Directory. American Consulting Engineers Council, 1015 15th Street, N.W. Suite 802, Washington, DC 20005-2670. (202) 347-7474. Annual. $10.00.

Research Centers Directory. Gale Research Company Inc., 835 Penobscot Building, Detroit, MI 48226-4094. (313) 961-2242. (800) 877-4253. 20th edition. 1995. $485.00. ISSN: 0080- 1518.

Scientific and Technical Organizations and Agencies Directory. Gale Research, 835 Penobscot Building, Detroit, MI 48226-4094. (313) 961-2242. (800) 877-4253.

4th Edition. 1996. $195.00.

Who's Who in Electric Transmission and Distribution. Utility Data Institute, 1200 G Street NY, Suite 250, Washington, DC 20005-3802. (202) 942-8788. (800) 486-3660. 1995. $150.00.

Who's Who in Engineering. Gordon Davis, editor. American Association of Engineering Societies. 1111 19th Street, NY, Suite 608, Washington, DC 20036. (202) 296-2237. (800) 658-8897. 9th edition. 1995. $220.00.

Who's Who in Technology. Gale Research, 835 Penobscot Building, Detroit, MI 48226-4094. (313 961-2242. (800) 521-4253. 7th edition. 1995. $195.00. ISN 0-8103-7467-6.

ENCYCLOPEDIAS AND DICTIONARIES

Chambers Science and Technology Dictionary. Peter M.B. Walker, editor. Cambridge University Press, 40 West 20th Street, New York, NY 10011-4211. (212) 924-3900. 1988. $39.95.

IEEE Standard Dictionary of Electrical and Electronics Terms. Christopher J. Booth, editor. IEEE Service Center, 445 Hoes Lane, Piscataway, NJ 08855-1331. (908) 981-1393. (800) 678-IEEE. FAX (908) 981-9667. IEEE Standard 100- 1992. 5th edition. 1993. $150.00.

Illustrated Dictionary of Electronics. Stan Gibilsco. TAB Books, P.O. Box 40, Blue Summit, PA 17294-0850. (717) 794- 2191. (800) 233-1128. 7th edition. 1994. $34.95.

McGraw-Hill Electronics Dictionary. J. Markus. McGraw-HillBook Company, Inc., 1221 Avenue of the Americas, New York, NY 10020. (212) 997-3675. 5th edition. 1994. $49.95.

McGraw-Hill Encyclopedia of Science and Technology. McGraw-Hill Book, Incorporated, 1221 Avenue of the Americas, New York, NY 10020. (212) 997-3675. (800) 262-4729. Seventh edition. Twenty volumes. 1992. $1900.00.

GENERAL WORKS

A. C. Generators: Design and Application. Robert L. Ames.*John Wiley & Sons, Inc.*, 605 Third Avenue, New York, NY 10158-0012. (212) 850-6000. (800) 225-5945. 1990.$185.00

Basic Electrical and Electronic Engineering. E. C. Bell and R. W. Whitehead. Blackwell Scientific Publications, Inc., 238 Main Street, Cambridge MA 02142. (617) 876-7000. (800) 759-6102. 4th revised edition. 1993.

Basic Electrical Power and Machines. David Bradley. Chapman & Hall, 1 Penn Plaza, New York, NY 10119. (212)*564-1060.* 1994. $30.95.

Electric Motor and Generator Repair. Gordon Press Publications, P.O. Box 459, Bowling Green Station, New York, NY 10004. (718) 624-8419. 1991. $75.00.

Electric Motors and Motor Controls. Jeffrey J. Keljik. *Delmar Publishing, 3 Columbia Circle, Box 15015, Albany, NY 12212.* (518) 464-3500. (800) 347-7707. 1995. $35.95.

Electrical Machines, Drives and Power Systems. Theodore Wildi. Prentice Hall, 113 Sylvan Avenue, Route 9W, Englewood Cliffs, NJ 07632. (201) 592-2000. (800) 922-0579. 2nd edition. 1990. $79.75. *Electrical Power Engineering.* Henslay W. Kabisama. McGraw-Hill Publishing Company, Inc., 1221 Avenue of the Americas, New York, NY 10020. (212) 512-2000. (800) 262- 4729. 1993. $55.00.

Energy Conversion: Electric Motors and Generators. Raymond S. Ramshaw and R. G. Van Heeswijk. Oxford University Press, Inc., 200 Madison Avenue, New York, NY 10016. (212) 725-6000. (800) 334-4249 1995. $64.50.

High Voltage Engineering. M. S. Naidu and V. Kamaraju. McGraw-Hill Publishing Company, Inc., 1221 Avenue of the Americas, New York, NY 10020. (212) 512-2000. (800) 262-4729. 1995. $55.00.

Magneto-hydro-dynamic Electrical Power Generation. Hugo K. Messerie. John Wiley & Sons, Inc., 605 Third Avenue, New York, NY 10158-0012. (212) 850-6000. (800) 225-5945. 1994. $34.95 in paper.

Monitoring and Diagnosis of Turbine-driven Generators. Aveline J. Gonzalez. Prentice-Hall, 113 Sylvan Avenue, Route 9W, Englewood Cliffs, NJ 07632. (201) 592-2000. (800) 922-0579. 1995. $68.00.

Motors, Generators and Transformers. Multi-Amp Institute Staff. Delmar Publishing, 3 Columbia Circle, Box 15015, Albany, NY 12212. Z (518) 464-3500. (800) 347-7707. 1993. $39.00.

Schaum's Electric Power Systems. Syed A. Masar. McGraw-Hill Publishing Company, Inc., 1221 Avenue of the Americas, New York, NY 10020. (212) 512-2000. (800) 262-4729. 1990. $10.95 in paper.

Transformers and Motors. George Shultz. Sams, 201 West 103rd Street, Indianapolis, in 46290-1097. (317) 581-3500.(800) 545-5914. 1991. $29.95.

HANDBOOKS AND MANUALS

American Electricians Handbook. Terrell Croft and Wilford I. Summers. McGraw-Hill Book Company, 1221 Avenue of the Americas, New York, NY 10020. (212) 512-2000. (800) 262-4729. 12th edition. 1992. $77.50.

Complete Handbook of Electric Motor Controls. John Traister. Prentice Hall, 113 Sylvan Avenue, Route 9W, Englewood Cliffs, NJ 07632. (201) 592-2000. (800) 922-0579. 2nd edition. 1994. $69.00.

Diesel Generator Handbook. L. L. Mahon. Butterworth-Heinemann, 313 Washington Street, Newton, MA 02158. (618-928-2500. (800) 366-2665. 1992. $200.00

Electrical Engineer's Handbook. Donald Christiansen, editor. McGraw Book Company, 1221 Avenue of the Americas, New York, NY 10020. (212) 512-2000. (800) 262-4729. 4th edition. 1996. $110.00.

Handbook of Electric Motors. John E. Traister. Prentice-Hall, 113 Sylvan Avenue, Route 9W, Englewood Cliffs, NJ 07632. (201) 592-2000. (800) 922-0579. 1992. $64.00.

Handbook of Electrical and Electronic Technology. Curtis Johnson. Prentice Hall , 113 Sylvan Avenue, Route 9W, Englewood Cliffs, NJ 07632. (201) 592-2000. (800) 922- 0579. 1996. $88.00.

Handbook of Electronics Manufacturing Engineering. Bernard S. Matisoff. Chapman & Hall, 1 Penn Plaza, New York, NY 10119. (212) 564-1060. 1996.

Handbook of Transformer Design and Applications, William M. Flanagan. McGraw-Hill Publishing Company, Inc., 1221 Avenue of the Americas, New York, NY 10020. (212) 512-2000. (800) 262-4729. 1993. $69.95.

Standard For Safety For Electric Motors and Generators For Use in Hazardous Locations, Ul 674 Underwriters Laboratories, Inc., 333 Pfingsten Road, Northbrook, IL, 60062-2096. (708) 272-8800. 2nd edition. 1989. $75.00.

Standard For Safety For High Voltage Industrial Control Equipment, Ul 347. Underwriters Laboratories, Inc., 333 Pfingsten Road, Northbrook, IL, 60062-2096. (708) 272-8800. 4th edition. 1993. $175.00.

Standard Handbook For Electrical Engineers. Donald Fink. McGraw-Hill Publishing Company, 1221 Avenue of the Americas, New York, NY 10020. (212) 512-2000. (800) 262- 4729. 13th edition. 1996. $110.00.

ONLINE DATABASES AND CD-ROMS

CA Search. Chemical Abstracts Service, P.O. Box 3012, Columbus, OH 43210-0012. (614) 447-3600. (800) 848-6533. FAX (614) 447-3709. Very comprehensive guide to worldwide chemical literature and related fields, 1972 to present. Available on BRS,(800) 289-4277, DIALOG, (800) 334-2564, ORBIT (800) 456-7248, and STN International, FIZ Karlsruhe, P.O. Box 2465, W-7500, Karlsruhe 1, Germany, online services. Inquire as to cost and availability.

Compendex Plus. Engineering Information, Inc., 345 East 47th Street, New York, NY 10017. (212) 705-7600 or (800) 221-1044. Contains citations with abstracts to worldwide literature in engineering and technology, from 1970 to present. Available on online BRS,(800) 289-4277, DIALOG, (800) 334-2564, ORBIT (800) 456-7248, and STN International, FIZ Karlsruhe, P.O. Box 2465, W-7500, Karlsruhe 1, Germany, online services. Also available on CD-ROM. Inquire as to cost and availability.

Current Contents Search. Institute for Scientific Information, 3501 Market Street, Philadelphia, PA 19104. (215) 386-0100. FAX (215) 386-6362. Contains citations to articles listed in the table of contents of science and technology journals. Also articles in social sciences and life sciences journals. Available on

BRS,(800) 289-4277, DIALOG,(800) 334-2564, online services. Inquire as to cost and availability.

Dissertation Abstracts Online. University Microfilms International, 300 North Zeeb Road, Ann Arbor, MI 48106. (800) 521-0600 or (313) 761-4700. Scope includes virtually all doctoral dissertations accepted at accredited American institutions from 1861 to present in 252 subject areas. Available on BRS, (800) 289-4277, DIALOG, (800) 334-2564, and OCLC EPIC,(800) 848-5878, online services. Also available on CD-ROM. Inquire as to cost and availability.

INSPEC. Institution of Electrical Engineers, Michael Faraday House, Six Hills Way, Stevenage, Herts. SG1 2AY, England. Telephone: 0438 313311 or FAX 0438 742840. Contains citations to the worldwide literature of physics, electronics and electrical engineering, computer technology, and related fields. Available on BRS, (800) 289-4277, DIALOG, (800) 334-2564, ORBIT, (800) 456-7248, and STN International, FIZ Karlsruhe, P.O. Box 2465, W-7500, Karlsruhe 1, Germany, online services. Inquire as to cost and availability.

NTIS Bibliographic Database. National Technical Information Service, 5285 Port Royal Road, Springfield, VA 22161. (703) 487-4929 or FAX (703) 321-8199. Broad coverage of government-sponsored science and technology research reports, 1964 to present. Available on BRS,(800) 289-4277, DIALOG, (800) 334-2564, ORBIT, (800) 456-7248, and STN International, FIZ Karlsruhe, P.O. Box 2465, W-7500, Karlsruhe 1, Germany, online services. Also available on CD-ROM. Inquire as to cost and availability.

Physics Briefs. American Institute of Physics, 335 East 45th Street, New York, NY 10017. (212) 661-9260 or FAX (212) 661-2036. Contains citations with abstracts of the literature of physics and related fields, 1979 to present. Available on the STN International, FIZ Karlsruhe, P.O. Box 2465, W-7500, Karlsruhe 1, Germany, online service. Inquire as to cost and availability.

Scientific and Technical Books and Serials in Print. R.R. Bowker Inc., 121 Chanlon Road, New Providence, NJ 07974. (800) 521-8110 or (908) 464-6800. List of currently published books and serials in the physical and biological sciences, engineering and technology, with subject, author and titles indexes. Available on ORBIT,(800) 456-7248, online service.Inquire as to cost and availability.

SCISEARCH. Institute for Scientific Information, 3501 Market Street, Philadelphia, PA 19104. (800) 523-1850 or (215) 386-0100. Broad multidisciplinary title and author index to the international literature of science and technology, 1974 to present. Available on DIALOG,(800) 334-2564, and ORBIT,(800) 456-7248, online services. Also available on CD-ROM. Inquire as to cost and availability.

WILSONLINE. H.W. Wilson Company, 950 University Avenue, Bronx, NY 10452. (800) 367-6770 or (212) 588-8400. Makes available online versions of the printed H.W. Wilson indexes including Applied Science and Technology Index, Business Periodicals Index, General Science Index, and Readers' Guide to Periodical Literature. Period covered is generally 1983 to present. Available on BRS,(800) 289-4277, DIALOG,(800) 334-2564, and OCLC EPIC,(800) 848-5878, online services. Also available on CD-ROM. Inquire as to cost and availability.

GENERATORS

Ency. of Physical Sciences and Engineering Info. Sources

PERIODICALS

Electric Light and Power. Pennwell Publishing Co. 1421 South Sheridan Road, Box 1260, Tulsa OK 74101. (835-3161. FAX (918) 831-9497. 1922 to present. Monthly. $42.00 per year.

Electric Machines and Power Systems. Taylor & Francis, 1900 Frost Road, Suite 101, Bristol, PA 19007-1598. (800) 821-8312, FAX (215) 785-5515. 1976 to present. Monthly. $355.00 per year. ISSN: 0731-356X.

Electrical World. McGraw-Hill, Inc., Box 513 Hightstown, NJ 08520. (212) 512-3288. 1874 to present. Monthly. $55.00 per year. ISSN: 0013-4457.

Electronic Design. Penton Publishing, San Jose Gateway, Suite 354. 2025 Gateway Place, San Jose, CA 95110. (408) 441-0550. 1952 to present. Fortnightly. $95.00. ISSN: 0013-4872.

Electronics. Penton Publishing, San Jose Gateway, Suite 354. 2025 Gateway Place, San Jose, CA 95110. (408) 441-0550. 1930 to present. Semi-weekly. $98.00 per year. ISSN: 0883-4989.

IEEE Power Engineering Review. Institute of Electrical and Electronics Engineers, Inc., Box 1331, 445 Hoes Lane, Piscataway, NJ 08855-1331. (908) 981-0060. 1981 to present. Monthly. $115.00 per year. ISSN: 0272-1724.

IEEE Spectrum. Institute of Electrical and Electronics Engineers, Inc., Box 1331, 445 Hoes Lane, Piscataway, NJ 08855-1331. (908) 981-0060. 1964 to present. Monthly. $157.00. ISSN: 0018-9235.

IEEE Transactions On Power Delivery. Institute of Electrical and Electronics Engineers, Inc., Box 1331, 445 Hoes Lane, Piscataway, NJ 08855-1331. (908) 981-0060. 1986 to present. Quarterly. $170.00 per year. ISSN: 0885-8977.

Iee Proceedings: Electric Power Applications. Institution of Electrical Engineers (IEE), London. Available from INSPEC/ IEEE - Institute of Electrical and Electronic Engineers, Box 1331, Hoes Lane, Piscataway, NJ 08855-1331. (908) 562-5549. 1980 to present. Bi-monthly. $ ISSN:

Institute of Electrical and Electronics Engineers. PROCEEDINGS. Institute of Electrical and Electronics Engineers, Inc., Box 1331, 445 Hoes Lane, Piscataway, NJ 08855-1331. (908) 981-0060. 1913 to present. Monthly. $275.00. ISSN: 0018-9219.

Power Engineering (tulsa). PennWell Publishing Co. 1421 South Sheridan Road, Box 1260, Tulsa OK 74101. (835-3161. FAX (918) 831-9497. 1896 to present. Monthly. $28.00 per year. ISSN: 0032-5961

Power Engineering Journal. Institution of Electrical Engineers (IEE), London. Available from INSPEC/IEEE- Institute of Electrical and Electronic Engineers, Box 1331, 445 Hoes Lane, Piscataway, NJ 08855-1331. (908) 562-5549. *1987* To Present. Bi-monthly. L85 per year. ISSN: 0950-3366.

RESEARCH CENTERS AND INSTITUTES

Edison Electric Institute. 701 Pennsylvania Avenue, NW, Washington, DC 20004-2696. (202) 508-5000, 5454. FAX (202) 508-5360.

Electroscience Laboratory. Ohio State University, 1320 Kinnear Road, Columbus, Ohio 43212. (614) 292-7981, FAX (614) 292-7297.

Electrical Engineering Research Laboratories. Purdue University. Electrical Engineering Building, West Lafayette, in 47907. (317) 494-3536. FAX (317) 494-6440.

Electrical Power Research Institute. 3412 Hillview Avenue, PO Box 10412, Palo Alto, CA 94303. (415) 855-2000. FAX (415) 855-2954.

Laboratory For Electromagnetic and Electronic Systems. Massachusetts Institute of Technology, 77 Massachusetts Avenue, Cambridge, MA 02139. (617) 253-4631. FAX (617) 258-6774.

Weber Research Institute. Polytechnic University. Route 110, Farmingdale, NY 11735. (516) 755-4250, FAX (516) 755-4404.

GEOCHEMISTRY

See also: GEOLOGY, GEOPHYSICS, IGNEOUS ROCKS, METAMORPHIC ROCKS, PETROLEUM GEOLOGY, SEDIMENTARY ROCKS

ABSTRACT SERVICES AND INDEXES

Bibliography and Index of Geology. American Geological Institute, 4220 King Street, Alexandria, VA 22302. (703) 379-2480. Fax (703) 379-7563. Monthly. $1295.00 per year. Also available online as GeoRef.

Chemical Abstracts - Physical, Inorganic and Analytical Chemistry. Chemical Abstracts Service, 2540 Olentangy River Road, Box 3012, Columbus, OH 43210. (800) 848-6538 or (614) 447-3600, FAX (614) 447-3713. Bi-weekly. $1410.00 per year.

Current Contents: Physical, Chemical and Earth Sciences. Institute for Scientific Information, 3501 Market Street, Philadelphia, PA 19104. (215) 386-0100. FAX (215) 386-6362. Weekly. $360.00 per year.

Deep-sea Research. PART A: OCEANOGRAPHIC RESEARCH PAPERS. DEEP-SEA RESEARCH. PART B: OCEANOGRAPHIC LITERATURE REVIEW. Pergamon Press, Inc., Maxwell House, Fairview Park, Elmsford, NY 10523. (914) 592-7700. Fax (914) 592-3625. Twelve times per year. $1370 per year for Parts A and B.

Geological Abstracts. Elsevier - Geo Abstracts, Regency House, 34 Duke Street, Norwich NR3 3AP, England. Monthly. $760.00 per year. Also available online as GEOBASE.

Geological Society of America. Abstracts with Programs. Geological Society of America. 3300 Penrose Place, P.O. Box 9140, Boulder, CO 80301-9140. (303) 447-2020. Abstracts and programs of the annual conference. Annual. $69.00.

Mineralogical Abstracts. Mineralogical Society and the Mineralogical Society of America, 41 Queen's Gate, London, SW7 5HR, England. Quarterly. $235.00 per year.

Oceanic Abstracts. Cambridge Scientific Abstracts, 7200 Wisconsin Avenue, Bethesda, MD 20814. (301) 961-6750. Fax (301) 961-6720. Bimonthly. $995.00 per year.

Science Citation Index. Institute for Scientific Information, 3501 Market Street, Philadelphia, PA 19104. (215) 386-0100. FAX (215) 386-6362. Inquire about availability and cost. Also available on CD-ROM.

ANNUAL REVIEWS AND YEARBOOKS

Annual Review of Earth and Planetary Sciences. Annual Reviews Inc., 4139 El Camino Way, PO Box 10139, Palo Alto, CA 94303-0139. (800) 523-8635 or (415) 493-4400. FAX (415) 855-9815. Annual. $62.00 (1995 edition).

ASSOCIATIONS AND PROFESSIONAL SOCIETIES

American Association of Petroleum Geologists. Box 979, Tulsa, OK 74101. (918) 584-2555. FAX (918) 584-0469.

American Geological Institute. 4220 King Street, Alexandria, VA 22302. (703) 379-2480. Fax (703) 379-7563.

American Geophysical Union. 2000 Florida Avenue, N.W., Washington, DC 20009. (202) 462-6900. FAX (202) 328-0566.

Association of Engineering Geologists. 323 Boston Post Road, Suite 2D, Sudbury, MA 01776. (508) 443-4639.

Geochemical Society. Department of Terrestrial Magnetism, 5241 Broad Branch Road NW, Washington, DC 20015. (202) 686-4387.

Geological Society of America. 3300 Penrose Place, PO Box 9140, Denver, CO 80301-9140. (303) 447-2020. FAX (303) 447-1133.

SEPM (formerly Society For Economic Paleontologists and Mineralogists). 1731 E 71st Street, Tulsa, OK 74136-5108. FAX (918) 493-2093.)

DIRECTORIES AND BIOGRAPHICAL SOURCES

American Men & Women of Science, 1995-96. R.R. Bowker Staff, eds. 19th edition. 8 volumes. R.R. Bowker/Reed International Publishing Company, 121 Chanlon Road, New Providence, NJ 07974. (908) 464-6800 or (800) 521-8110. 1995. $850.00

Directory of Engineering Societies and Related Organizations. Gordon Davis. 13th edition. American Association of Engineering Societies, 1111 19th Street NW, Suite 608, Washington, DC 20036. (202) 296-2237 or (800) 658-8897. 1989. Inquire for price.

International Research Centers Directory. Gale Research, 835 Penobscot Building, Detroit, MI 48226-4094. (313) 961-2242. (800) 877-4253. FAX (313) 961-6083. 8th edition. 1995. $410.00

Research Centers Directory. Gale Research, 835 Penobscot Building, Detroit, MI 48226-4094. (313) 961-2242. (800) 877-4253. FAX (313) 961-6083. $485.00.

Scientific and Technical Organizations and Agencies Directory. Gale Research, 835 Penobscot Building, Detroit, MI 48226-4094. (313) 961-2242. (800) 877-4253. FAX (313) 961-6083. 4th edition. 1996. $195.00.

Who's Who in Technology. Gale Research, 835 Penobscot Building, Detroit, MI 48226-4094. (313) 961-2242. (800) 877-4253. FAX (313) 961-6083. 1995. $195.00.

ENCYCLOPEDIAS AND DICTIONARIES

The Encyclopedia of the Solid Earth Sciences. Philip Kearey, editor-in-chief. Blackwell Scientific Publications, 238 Main Street, Cambridge, MA 02142. (617) 876-7000 or (800) 759-6102. FAX (617) 876-7022. 1993. $60.00.

Encyclopedia of Physical Science and Technology. Robert A. Meyers, ed. 18 volumes. Academic Press Inc., 6277 Sea Harbor Drive, Orlando, FL 32887. (800) 321-5068. 1992. $2100.00.

Illustrated Glossary of Petroleum Geochemistry. Jennifer A. Miles. Reprint edition. Oxford University Press, Inc., 198 Madison Avenue, New York, NY 10016-4314. (212) 726-6000. FAX (212) 726-6446. 1994. $21.95.

GENERAL WORKS

Introduction To Geochemistry. Konrad B. Krauskopf and Dennis K. Bud. 3rd edition. McGraw-Hill Publishing Company, 1221 Avenue of the Americas, New York, NY 10020. (800) 262-4729 or (212) 512-3825. 1994. Inquire for cost and availability.

Introduction To Organic Geochemistry. S.D. Killops and V.J. Killops. Halsted Press (Division of John Wiley and Sons, Inc.), 605 Third Avenue, New York, NY 10158. (800) 526-5368 or (212) 850-6418. 1993. $43.95.

Physics and Chemistry of Earth Materials. Alexandra Navrotsky. Cambridge University Press, 40 West 20th Street, New York, NY 10011-4211. (212) 924-3900 or (800) 872-7423. FAX (914) 937-4712. 1994. $79.95 (hardback) or $34.95 (paperback).

Principles and Applications of Geochemistry. Gunter Faure. The Free Press, division of Simon & Schuster, Inc. 1230 Avenue of the Americas, New York, NY 10020. (212) 698-7000. 1990. $84.00.

Sediments and Environmental Geochemistry. D. Heling, et al., editors. Springer-Verlag, 175 Fifth Avenue, New York, NY 10010. (212) 460-1500 or (800) 777-4643. FAX (212) 473-6272. 1990. $109.00.

Theoretical Geochemistry: Applications of Quantum Mechanics in the Earth and Mineral Sciences. David J. Vaughan and John A. Tossel. Oxford University Press, Inc., 198 Madison Avenue, New York, NY 10016-4314. (212) 726-6000. FAX (212) 726-6446. 1992. $85.00.

ONLINE DATABASES AND CD-ROMS

CA Search. Chemical Abstracts Service, P.O. Box 3012, Columbus, OH 43210-0012. (614) 447-3600. (800) 848-6533. FAX (614) 447-3709. Very comprehensive guide to worldwide chemical literature and related fields, 1972 to present. Available on BRS,(800) 289-4277, DIALOG, (800) 334-2564, ORBIT (800) 456-7248, and STN International, FIZ Karlsruhe, P.O. Box 2465, W-7500, Karlsruhe 1, Germany, online services. Inquire as to cost and availability.

GEOCHEMISTRY

Ency. of Physical Sciences and Engineering Info. Sources

Dissertation Abstracts. University Microfilms International, 300 North Zeeb Road, Ann Arbor, MI 48106. (800) 521-0600 or (313) 761-4700. Scope includes virtually all doctoral dissertations accepted at accredited American institutions from 1861 to present in 252 subject areas. Available on BRS,(800) 289-4277, DIALOG,(800) 334-2564, and OCLC EPIC,(800) 848-5878, online services. Also available on CD-ROM. Inquire as to cost and availability.

Earth Sciences. U.S. Geological Survey, 12201 Sunrise Valley Drive, Reston, VA 22092-9998. (703) 648-4460. CD-ROM of earth science databases including the U.S. Geological Survey Library, Earth Science Data Directory, and GEOINDEX, citations to published geological maps. $350.00 per year with quarterly updates.

Geoarchive. Geosystems, P.O. Box 1024, Westminster, London, England, SW1 P 2JL. Citations to literature on geoscience, 1969 to present. Inquire as to online cost and availability.

Geobase. Elsevier - Geo Abstracts, Regency House, 34 Duke Street, Norwich NR3 3AP, England. Contains citations to the worldwide earth science literature from 1980 to date. Available on DIALOG, ORBIT online services. Inquire as to cost and availability.

GEOREF: Bibliography and Index of Geology. American Geological Institute, 4220 King Street, Alexandria, VA 22302. (703) 379-2480. Fax (703) 379-7563. Monthly. Inquire for price and availability.

NTIS Bibliographic Database. National Technical Information Service, 5285 Port Royal Road, Springfield, VA 22161. (703) 487-4929 or FAX (703) 321-8199. Broad coverage of government-sponsored science and technology research reports, 1964 to present. Available on BRS,(800) 289-4277, DIALOG, (800) 334-2564, ORBIT, (800) 456-7248, and STN International, FIZ Karlsruhe, P.O. Box 2465, W-7500, Karlsruhe 1, Germany, online services. Also available on CD-ROM. Inquire as to cost and availability.

SCISEARCH. Institute for Scientific Information, 3501 Market Street, Philadelphia, PA 19104. (800) 523-1850 or (215) 386-0100. Broad multidisciplinary title and author index to the international literature of science and technology, 1974 to present. Available on DIALOG,(800) 334-2564, and ORBIT,(800) 456-7248, online services. Also available on CD-ROM. Inquire as to cost and availability.

PERIODICALS

A A P G Bulletin. American Association of Petroleum Geologists, Box 979, Tulsa, OK 74101. (918) 584-2555. FAX (918) 584-0469. 1917 to present. Monthly. $135.00.

A A P G Studies in Geology Series. American Association of Petroleum Geologists, Box 979, Tulsa, OK 74101. (918) 584-2555. FAX (918) 584-0469. Irregular. Price varies.

American Journal of Science. Box 6666, Yale Station, New Haven, CT 06511-8130. (203) 432-3131. FAX (203) 432-5668. 1818 to present. Monthly except July and August. $40.00 per year.

Economic Geology and the Bulletin of the Society of Economic Geologists. Economic Geology Publishing Company, 101

Vowell Hall, University of Texas, El Paso, TX 79968. FAX (915) 544-7416. 1906 to present. Eight times a year. $65.00.

Geochemical Journal. Business Center for Academic Societies Japan, 5-16-9 Honkomagome, Bunkyo-ku, Tokyo 113, Japan. FAX 035814-5822. 1966 to present. Bi-monthly. $160.00.

Geochemistry International. J. Wiley & Sons, Journals, 605 Third Avenue, New York, NY 10158-0012. (212) 692-6000. 1964 to present. Monthly. $1150.00.

Geochimica Et Cosmochimica Acta. Elsevier Science [journals], 660 White Plains Rd., Tarrytown, NY 10159-5153. (919) 524-9200. FAX (919) 333-2444. 1950 to present. Twenty-four times a year. $895.00.

Geological Magazine. Cambridge University Press, Journals Department, 40 West 20th Street, New York, NY 10011-4211. (212) 924-3900. 1864 to present. Bi-monthly. $263.00 per year.

Geological Society Journal. Geological Society Publishing House, Unit 7, Brassmill Enterprise Centre, Brassmill Lane, Bath BA1 3JN, England. Telephone 0225-445046. FAX 0225-442836. 1845 to present. Six times a year. $524.00.

Geological Society of America Bulletin. 3300 Penrose Place, PO Box 9140, Denver, CO 80301-9140. (303) 447-2020. 1888 to present. Monthly. $205.00 per year.

Geology. Geological Society of America, 3300 Penrose Place, PO Box 9140, Denver, CO 80301-9140. (303) 447-2020. 1973 to present. Monthly. $170.00 per year for non-members.

Journal of Geology. University of Chicago Press, Journals Division, 5720 S. Woodlawn Avenue, Chicago, IL 60637. (312) 753-3347. FAX (312) 753-0811. 1893 to present. Bi-monthly. $45.00 per year.

Lithos. Elsevier Science Inc., Box 882, Madison Square Station, New York, NY 10159. (212) 989-5800. FAX (212) 633-3990. 1968 to present. Twelve times a year. $529.00.

RESEARCH CENTERS AND INSTITUTES

Carnegie Institution of Washington. Department of Terrestrial Magnetism, 5241 Broad Branch Road NW, Washington, DC 20015. (202) 966-0863. FAX (202) 364-8726.

Harvard University, Hoffman Laboratory of Experimental Geology. Department of Earth & Planetary Sciences, 20 Oxford Street, Cambridge, MA 02138. (617) 495-4215.

University of Arizona Laboratory of Isotope Geochemistry. 136 Gould-Simpson Building, Tucson, AZ 85721. (602) 621-6014. FAX (602) 621-2672.

University of Arizona Laboratory of Organic Geochemistry. Department of Geosciences, Tucson, AZ 85721. (602) 621-6973.

University of California—san Diego, Scripps Institute of Oceanography, Geological Research Division. San Diego, CA 92093-0220. (619) 534-1829. FAX (619) 534-0784.

GEOCHRONOLOGY

See also: GEOLOGY, GEOPHYSICS, PALEONTOLOGY, SEDIMEN-
TARY ROCKS

ABSTRACT SERVICES AND INDEXES

Bibliography and Index of Geology. American Geological Institute, 4220 King Street, Alexandria, VA 22302. (703) 379-2480. Fax (703) 379-7563. Monthly. $1295.00 per year. Also available online as GeoRef.

Current Contents: Physical, Chemical and Earth Sciences. Institute for Scientific Information, 3501 Market Street, Philadelphia, PA 19104. (215) 386-0100. FAX (215) 386-6362. Weekly. $360.00 per year.

Geological Abstracts. Elsevier - Geo Abstracts, Regency House, 34 Duke Street, Norwich NR3 3AP, England. Monthly. $760.00 per year. Also available online as GEOBASE.

Geological Society of America. ABSTRACTS WITH PROGRAMS. Geological Society of America. 3300 Penrose Place, P.O. Box 9140, Boulder, CO 80301-9140. (303) 447-2020. Abstracts and programs of the annual conference. Annual. $69.00.

Mineralogical Abstracts. Mineralogical Society and the Mineralogical Society of America, 41 Queen's Gate, London, SW7 5HR, England. Quarterly. $235.00 per year.

Oceanic Abstracts. Cambridge Scientific Abstracts, 7200 Wisconsin Avenue, Bethesda, MD 20814. (301) 961-6750. Fax (301) 961-6720. Bimonthly. $995.00 per year.

Science Citation Index. Institute for Scientific Information, 3501 Market Street, Philadelphia, PA 19104. (215) 386-0100. FAX (215) 386-6362. Inquire about availability and cost. Also available on CD-ROM.

ANNUAL REVIEWS AND YEARBOOKS

Annual Review of Earth and Planetary Sciences. Annual Reviews Inc., 4139 El Camino Way, PO Box 10139, Palo Alto, CA 94303-0139. (800) 523-8635 or (415) 493-4400. FAX (415) 855-9815. Annual. $62.00 (1995 edition).

ASSOCIATIONS AND PROFESSIONAL SOCIETIES

American Association of Petroleum Geologists. Box 979, Tulsa, OK 74101. (918) 584-2555.

American Geological Institute. 4220 King Street, Alexandria, VA 22302. (703) 379-2480. Fax (703) 379-7563.

American Geophysical Union. 2000 Florida Avenue, N.W., Washington, DC 20009. (202) 462-6900. FAX (202) 328-0566.

Association of Engineering Geologists. 323 Boston Post Road, Suite 2D, Sudbury, MA 01776. (508) 443-4639.

Geological Society of America. 3300 Penrose Place, PO Box 9140, Denver, CO 80301-9140. (303) 447-2020.

SEPM (formerly Society For Economic Paleontologists and Mineralogists). 1731 E 71st Street, Tulsa, OK 74136-5108. FAX (918) 493-2093.)

DIRECTORIES AND BIOGRAPHICAL SOURCES

American Men & Women of Science, 1995-96. R.R. Bowker Staff, eds. 19th edition. 8 volumes. R.R. Bowker/Reed International Publishing Company, 121 Chanlon Road, New Providence, NJ 07974. (908) 464-6800 or (800) 521-8110. 1995. $850.00

Directory of Engineering Societies and Related Organizations. Gordon Davis. 13th edition. American Association of Engineering Societies, 1111 19th Street NW, Suite 608, Washington, DC 20036. (202) 296-2237 or (800) 658-8897. 1989. Inquire for price.

International Research Centers Directory. Gale Research, 835 Penobscot Building, Detroit, MI 48226-4094. (313) 961-2242. (800) 877-4253. FAX (313) 961-6083. 8th edition. 1995. $410.00.

Research Centers Directory. Gale Research, 835 Penobscot Building, Detroit, MI 48226-4094. (313) 961-2242. (800) 877-4253. FAX (313) 961-6083. $485.00.

Scientific and Technical Organizations and Agencies Directory. Gale Research, 835 Penobscot Building, Detroit, MI 48226-4094. (313) 961-2242. (800) 877-4253. FAX (313) 961-6083. 4th edition. 1996. $195.00.

Who's Who in Technology. Gale Research, 835 Penobscot Building, Detroit, MI 48226-4094. (313) 961-2242. (800) 877-4253. FAX (313) 961-6083. 1995. $195.00.

ENCYCLOPEDIAS AND DICTIONARIES

The Encyclopedia of the Solid Earth Sciences. Philip Kearey, editor-in-chief. Blackwell Scientific Publications, 238 Main Street, Cambridge, MA 02142. (617) 876-7000 or (800) 759-6102. FAX (617) 876-7022. 1993. $60.00.

Encyclopedia of Physical Science and Technology. Robert A. Meyers, ed. 18 volumes. Academic Press Inc., 6277 Sea Harbor Drive, Orlando, FL 32887. (800) 321-5068. 1992. $2100.00

Illustrated Glossary of Petroleum Geochemistry. Jennifer A. Miles. Reprint edition. Oxford University Press, Inc., 198 Madison Avenue, New York, NY 10016-4314. (212) 726-6000. FAX (212) 726-6446. 1994. $21.95.

GENERAL WORKS

Absolute Age Determination: Physical and Chemical Dating Methods and their Applications. H. Schleicher and M.A. Geyh. Springer-Verlag, 175 Fifth Avenue, New York, NY 10010. (212) 460-1500 or (800) 777-4643. FAX (212) 473-6272. 1990. $69.00.

The Age of the Earth. G. Brent Dalrymple. Stanford University Press, Stanford, CA 94305-7235. (415) 723-1593. 1991. $55.00.

The Earth's Age and Geochronology. R.M. Farquhar and D. York. Franklin Book Company, 7804 Montgomery Avenue,

GEOCHRONOLOGY

Ency. of Physical Sciences and Engineering Info. Sources

Elkins Park, PA 19117. (215) 635-5252. FAX (215) 635-6155. 1972. $72.00.

Lab Studies in Earth History. James C. Bruce, et al. Fifth edition. William C. Brown Publications, 2460 Kerper Blvd., Dubuque, IA 52001. (800) 338-5578. 1992. Inquire for price.

ONLINE DATABASES AND CD-ROMS

Earth Sciences. U.S. Geological Survey, 12201 Sunrise Valley Drive, Reston, VA 22092-9998. (703) 648-4460. CD-ROM of earth science databases including the U.S. Geological Survey Library, Earth Science Data Directory, and GEOINDEX, citations to published geological maps. $350.00 per year with quarterly updates.

Geoarchive. Geosystems, P.O. Box 1024, Westminster, London, England, SW1 P 2JL. Citations to literature on geoscience, 1969 to present. Inquire as to online cost and availability.

Geobase. Elsevier - Geo Abstracts, Regency House, 34 Duke Street, Norwich NR3 3AP, England. Contains citations to the worldwide earth science literature from 1980 to date. Available on DIALOG, ORBIT online services. Inquire as to cost and availability.

GEOREF: Bibliography and Index of Geology. American Geological Institute, 4220 King Street, Alexandria, VA 22302. (703) 379-2480. Fax (703) 379-7563. Monthly. Inquire for price and availability.

NTIS Bibliographic Database. National Technical Information Service, 5285 Port Royal Road, Springfield, VA 22161. (703) 487-4929 or FAX (703) 321-8199. Broad coverage of government-sponsored science and technology research reports, 1964 to present. Available on BRS,(800) 289-4277, DIALOG, (800) 334-2564, ORBIT, (800) 456-7248, and STN International, FIZ Karlsruhe, P.O. Box 2465, W-7500, Karlsruhe 1, Germany, online services. Also available on CD-ROM. Inquire as to cost and availability.

SCISEARCH. Institute for Scientific Information, 3501 Market Street, Philadelphia, PA 19104. (800) 523-1850 or (215) 386-0100. Broad multidisciplinary title and author index to the international literature of science and technology, 1974 to present. Available on DIALOG,(800) 334-2564, and ORBIT,(800) 456-7248, online services. Also available on CD-ROM. Inquire as to cost and availability.

PERIODICALS

A A P G Bulletin. American Association of Petroleum Geologists, Box 979, Tulsa, OK 74101. (918) 584-2555. FAX (918) 584-0469. 1917 to present. Monthly. $135.00.

A A P G Studies in Geology Series. American Association of Petroleum Geologists, Box 979, Tulsa, OK 74101. (918) 584-2555. FAX (918) 584-0469. Irregular. Price varies.

American Journal of Science. Box 6666, Yale Station, New Haven, CT 06511-8130. (203) 432-3131. FAX (203) 432-5668. 1818 to present. Monthly except July and August. $40.00 per year.

Economic Geology and the Bulletin of the Society of Economic Geologists. Economic Geology Publishing Company, 101

Vowell Hall, University of Texas, El Paso, TX 79968. FAX (915) 544-7416. 1906 to present. $65.00 per year.

Geochimica Et Cosmochimica Acta. Elsevier Science [journals], 660 White Plains Rd., Tarrytown, NY 10159-5153. (919) 524-9200. FAX (919) 333-2444. 1950 to present. 24 times a year. $895.00.

Geological Magazine. Cambridge University Press, Journals Department, 40 West 20th Street, New York, NY 10011-4211. (212) 924-3900. 1864 to present. Bi-monthly. $263.00 per year.

Geological Society Journal. Geological Society Publishing House, Unit 7, Brassmill Enterprise Centre, Brassmill Lane, Bath BA1 3JN, England. Telephone 0225-445046. FAX 0225-442836. 1845 to present. Six times a year. $524.00.

Geological Society of America Bulletin. 3300 Penrose Place, PO Box 9140, Denver, CO 80301-9140. (303) 447-2020. 1888 to present. Monthly. $205.00 per year.

Geology. Geological Society of America, 3300 Penrose Place, PO Box 9140, Denver, CO 80301-9140. (303) 447-2020. 1973 to present. Monthly. $170.00 per year for non-members.

Journal of Geology. University of Chicago Press, Journals Division, 5720 S. Woodlawn Avenue, Chicago, IL 60637. (312) 753-3347. FAX (312) 753-0811. 1893 to present. Bi-monthly. $45.00 per year.

Lithos. Elsevier Science Inc., Box 882, Madison Square Station, New York, NY 10159. (212) 989-5800. FAX (212) 633-3990. 1968 to present. Twelve times a year. $529.00.

RESEARCH CENTERS AND INSTITUTES

Carnegie Institution of Washington. Department of Terrestrial Magnetism, 5241 Broad Branch Road NW, Washington, DC 20015. (202) 966-0863. FAX (202) 364-8726.

Geochronology Center, Institute of Human Origins. 2453 Ridge Road, Berkeley, CA 94709. (415) 644-9200. FAX (415) 845-9453.

University of Alaska—fairbanks, Geochronology Laboratory. Geophysical Institute, C.T. Elvey Building, Fairbanks, AK 99775-0800. (907) 474-5514. FAX (907) 474-7290.

University of Arizona Laboratory of Isotope Geochemistry. 136 Gould-Simpson Building, Tucson, AZ 85721. (602) 621-6014. FAX (602) 621-2672.

University of Colorado—boulder, Center For Geochronological Research. Boulder, CO 80309-0450. (303) 492-6962. (303) 492-6388.

GEODESY

See also: CARTOGRAPHY, REMOTE SENSING, SURVEYING

ABSTRACT SERVICES AND INDEXES

Applied Science and Technology Index; A Cumulative Subject Index To English Language Periodicals in the Fields of Aeronautics and Space Science, Computer Technology, Chemistry, Construction Industry, Energy and Related Areas. H.W. Wilson Co., 950 University Avenue, Bronx, NY 10452. (800) 367-6770 or (212) 588-8400. FAX (718) 590-1617. From 1958 to present. Monthly. Inquire about cost and availability. Also available on CD-ROM and online.

Bibliography and Index of Geology. American Geological Institute, 4220 King Street, Alexandria, VA 22302. (703) 379-2480. Fax (703) 379-7563. Monthly. $1295.00 per year. Also available online as GEOREF.

Current Contents: Engineering, Technology, and Applied Sciences. Institute for Scientific Information, 3501 Market Street, Philadelphia, PA 19104. (215) 386-0100. FAX (215) 386-6362. 1970 to present. Weekly. $442.00 per year.

Current Contents: Physical, Chemical and Earth Sciences. Institute for Scientific Information, 3501 Market Street, Philadelphia, PA 19104. (215) 386-0100. FAX (215) 386-6362. Weekly. $360.00 per year.

Engineering Index Monthly. Engineering Information, Inc., Castle Point on the Hudson, Hoboken, NJ 07030. (800) 221-1044. FAX (212) 832-1857. Monthly. $2300.00 per year. Also available online as COMPENDEX, and also on CD-ROM. Covers chemical engineering, computers, electrical engineering, civil engineering, metals and mining, industrial management, and mechanical engineering.

Geodex Retrieval System For Geotechnical Abstracts. Geodex Retrieval Systems, 669 Broadway, Box 279, Sonoma, CA 95476. (707) 939-8476. FAX (707) 996-8734. 1970 to present. Three times a year. $645.00 per year includes Geotechnical Abstracts, as Geodex Systems-GRS.

Geological Abstracts. Elsevier - Geo Abstracts, Regency House, 34 Duke Street, Norwich NR3 3AP, England. Monthly. $760.00 per year. Also available online as GEOBASE.

Geological Society of America. ABSTRACTS WITH PROGRAMS. Geological Society of America. 3300 Penrose Place, P.O. Box 9140, Boulder, CO 80301-9140. (303) 447-2020. Abstracts and programs of the annual conference. Annual. $69.00.

Geotechnical Abstracts. Geodex Retrieval Systems, 669 Broadway, Box 279, Sonoma, CA 95476. (707) 939-8476. FAX (707) 996-8734. 1970 to present. Three times a year. $280.00 per year. Covers soil mechanics, foundation engineering, rock mechanics, and engineering geology.

Meteorological and Geoastrophysical Abstracts. American Meteorological Society, c/o Inforonics Inc., 550 Newtown Road, Littleton, MA 01460. (508) 486-8976. FAX (508) 486-0027. 1950 to present. Monthly. $950.00.

Remote Sensing of Earth Resources. University of New Mexico Earth Data Analysis Center, 2500 Yale Blvd., Suite 100, Albuquerque, NM 87131-6031. (505) 277-3622. FAX (505) 277-3614. 1974 to present (1991 to present under current title). Quarterly. $175.00 per year.

Science Citation Index. Institute for Scientific Information, 3501 Market Street, Philadelphia, PA 19104. (215) 386-0100. FAX (215) 386-6362. Inquire about availability and cost. Also available on CD-ROM.

ASSOCIATIONS AND PROFESSIONAL SOCIETIES

American Association For Geodetic Surveying. 5410 Grosvenor Lane, Suite 100, Bethesda, MD 20814. (301) 493-0200.

American Cartographic Association. 5410 Grosvenor Lane, Suite 100, Bethesda, MD 20814. (301) 493-0200.

American Congress On Surveying and Mapping. 5410 Grosvenor Lane, Suite 100, Bethesda, MD 20814. (301) 493-0200.

American Society For Photogrammetry and Remote Sensing. 5410 Grosvenor Lane, Suite 210, Bethesda, MD 20814. (301) 493-0290. FAX (301) 493-0208.

Association of American Geographers. 1710 16th Street NW, Washington, DC 20009-3198.

National Society of Professional Surveyors. 5410 Grosvenor Lane, Suite 100, Bethesda, MD 20814. (301) 493-0200.

North American Cartographic Information Society. PO Box 399, Milwaukee, WI 53201. (414) 229-6282.

DIRECTORIES AND BIOGRAPHICAL SOURCES

Directory of Engineering Societies and Related Organizations. Gordon Davis. 13th edition. American Association of Engineering Societies, 1111 19th Street NW, Suite 608, Washington, DC 20036. (202) 296-2237 or (800) 658-8897. 1989. Inquire for price.

International Research Centers Directory. Gale Research, 835 Penobscot Building, Detroit, MI 48226-4094. (313) 961-2242. (800) 877-4253. FAX (313) 961-6083. 8th edition. 1995. $410.00

Research Centers Directory. Gale Research, 835 Penobscot Building, Detroit, MI 48226-4094. (313) 961-2242. (800) 877-4253. FAX (313) 961-6083. $485.00.

Scientific and Technical Organizations and Agencies Directory. Gale Research, 835 Penobscot Building, Detroit, MI 48226-4094. (313) 961-2242. (800) 877-4253. FAX (313) 961-6083. 4th edition. 1996. $195.00.

Who's Who in Engineering. American Association of Engineering Societies, 1111 19th Street NW, Suite 608, Washington, DC 20036. (202) 296-2237 or (800) 658-8897. 8th edition. 1991. Inquire for price.

GENERAL WORKS

Applications of Geodesy To Engineering. K. Linkwitz, et al., eds. Springer-Verlag, 175 Fifth Avenue, New York, NY 10010. (212) 460-1500 or (800) 777-4643. FAX (212) 473-6272. 1993. $158.95.

Geodesy. Wolfgang Torge. Walter DE Gruyter Inc., 200 Saw Mill River Road, Hawthorne, NY 10532. (914) 747-0110. FAX (914) 747-1326. 1991. $44.95.

Geodesy in the Year 2000. National Academy Press, Division of the National Academy of Science, 2101 Constitution Avenue, N.W., Lockbox 285, Washington, DC 20055. (800) 624-6242. FAX (202) 334-2793. 1990. $20.00.

HANDBOOKS AND MANUALS

Manual of Remote Sensing. Robert N. Colwell, ed. 2nd edition. American Society for Photogrammetry and Remote Sensing, 5410 Grosvenor Lane, Suite 210, Bethesda, MD 20814. (301) 493-0290. FAX (301) 493-0208. 1983. Inquire for price and availability.

ONLINE DATABASES AND CD-ROMS

Compendex Plus. Engineering Information, Inc., 345 East 47th Street, New York, NY 10017. (212) 705-7600 or (800) 221-1044. Contains citations with abstracts to worldwide literature in engineering and technology, from 1970 to present. Available on online BRS,(800) 289-4277, DIALOG, (800) 334-2564, ORBIT (800) 456-7248, and STN International, FIZ Karlsruhe, P.O. Box 2465, W-7500, Karlsruhe 1, Germany, online services. Also available on CD-ROM. Inquire as to cost and availability.

GEOREF: Bibliography and Index of Geology. American Geological Institute, 4220 King Street, Alexandria, VA 22302. (703) 379-2480. Fax (703) 379-7563. Monthly. Inquire for price and availability.

NTIS Bibliographic Database. National Technical Information Service, 5285 Port Royal Road, Springfield, VA 22161. (703) 487-4929 or FAX (703) 321-8199. Broad coverage of government-sponsored science and technology research reports, 1964 to present. Available on BRS,(800) 289-4277, DIALOG, (800) 334-2564, ORBIT, (800) 456-7248, and STN International, FIZ Karlsruhe, P.O. Box 2465, W-7500, Karlsruhe 1, Germany, online services. Also available on CD-ROM. Inquire as to cost and availability.

SCISEARCH. Institute for Scientific Information, 3501 Market Street, Philadelphia, PA 19104. (800) 523-1850 or (215) 386-0100. Broad multidisciplinary title and author index to the international literature of science and technology, 1974 to present. Available on DIALOG,(800) 334-2564, and ORBIT,(800) 456-7248, online services. Also available on CD-ROM. Inquire as to cost and availability.

WILSONLINE. H.W. Wilson Company, 950 University Avenue, Bronx, NY 10452. (800) 367-6770 or (212) 588-8400. Makes available online versions of the printed H.W. Wilson indexes including Applied Science and Technology Index, Business Periodicals Index, General Science Index, and Readers' Guide to Periodical Literature. Period covered is generally 1983 to present. Available on BRS,(800) 289-4277, DIALOG,(800) 334-2564, and OCLC EPIC,(800) 848-5878, online services. Also available on CD-ROM. Inquire as to cost and availability.

PERIODICALS

Cartography and Geographic Information Systems. American Congress on Surveying and Mapping, 5410 Grosvenor Lane,

Suite 100, Bethesda, MD 20814. (301) 493-0200. 1974 to present. 4 times a year. $85.00 per year.

Acsm Bulletin. American Congress on Surveying and Mapping, 5410 Grosvenor Lane, Suite 100, Bethesda, MD 20814. (301) 493-0200. 1950 to present. Bi-monthly. $75.00 per year.

American Congress On Surveying and Mapping Proceedings. American Congress on Surveying and Mapping, 5410 Grosvenor Lane, Suite 100, Bethesda, MD 20814. (301) 493-0200. 1942 to present. Semi-annual. Price varies.

IEEE Transactions On Geoscience and Remote Sensing. IEEE, 345 E. 47th Street, New York, NY 10017. (908) 981-0060. FAX (908) 981-9667. 1963 to present. Bi-monthly. $188.00 per year.

Marine Geodesy. Taylor & Francis, 1900 Frost Road, Suite 101, Bristol, PA 19007. (215) 785-5800. FAX (215) 785-5515. 1977 to present. Quarterly. $141.00 per year.

Remote Sensing of Environment. Elsevier Science Publishing Company, Inc., 655 Avenue of the Americas, New York, NY 10010. (212) 989-5800. 1969 to present. 12 times a year. $995.00 per year.

Remote Sensing Reviews. Harwood ACADemic, c/o Gordon & Breach Science Publishers, PO Box 786, Cooper Station, New York, NY 10276. (212) 206-8900. Quarterly. Inquire for cost.

RESEARCH CENTERS AND INSTITUTES

University of Colorado—boulder, Engineering Research Center. Campus Box 423, Boulder, CO 80309. (303) 492-7427.

University of Hawaii At Manoa, Hawaii Institute of Geophysics. 2525 Correa Road, Honolulu, HI 96822. (808) 948-8760.

GEOLOGY

See also: CRUST, GEOCHEMISTRY, GEOCHRONOLOGY, GEOPHYSICS, IGNEOUS ROCKS, METAMORPHIC ROCKS, MINERALOGY, PETROLOGY, PHYSICAL GEOLOGY, SEDIMENTARY ROCKS, SEISMOLOGY, VOLCANOLOGY

ABSTRACT SERVICES AND INDEXES

Bibliography and Index of Geology. American Geological Institute, 4220 King Street, Alexandria, VA 22302. (703) 379-2480. Fax (703) 379-7563. Monthly. $1295.00 per year. Also available online as GeoRef.

Bibliography of Economic Geology. Geosystems, Box 40, Didcot Oxon Ox11 9BX, England. Bimonthly. $150.00 per year. Also available online as GeoArchive.

Chemical Abstracts - Applied Chemistry and Chemical Engineering Sections. Chemical Abstracts Service, 2540 Olentangy River Road, Box 3012, Columbus, OH 43210. (800) 848-6538 or (614) 447-3600, FAX (614) 447-3713. Bi-weekly. $1410.00 per year. Also available online as CA (Chemical Abstracts).

Deep-sea Research. PART A: OCEANOGRAPHIC RESEARCH PAPERS.

General Science Index. H.W. Wilson Co., 950 University Avenue, Bronx, NY 10452. (800) 367-6770 or (212) 588-8400. Inquire about price and availability.

Geological Abstracts. Elsevier - Geo Abstracts, Regency House, 34 Duke Street, Norwich NR3 3AP, England. Monthly. $760.00 per year. Also available online as GEOBASE.

Geological Society of America. ABSTRACTS WITH PROGRAMS. Geological Society of America. 3300 Penrose Place, P.O. Box 9140, Boulder, CO 80301-9140. (303) 447-2020. Abstracts and programs of the annual conference. Annual. $69.00.

Geophysics Abstracts. Pergamon Press, Inc., Maxwell House, Fairview Park, Elmsford, NY 10523. (914) 592-7700. Fax (914) 592-3625. Twelve times per year. $565.00 per year. Also available in microform.

Mineralogical Abstracts. Mineralogical Society and the Mineralogical Society of America, 41 Queen's Gate, London, SW7 5HR, England. Quarterly. $235.00 per year.

Oceanic Abstracts. Cambridge Scientific Abstracts, 7200 Wisconsin Avenue, Bethesda, MD 20814. (301) 961-6750. Fax (301) 961-6720. Bimonthly. $995.00 per year.

Science Citation Index. Institute for Scientific Information, 3501 Market Street, Philadelphia, PA 19104. (800) 523-1850 or (215) 386-0100. Inquire about price and availability.

ANNUAL REVIEWS AND YEARBOOKS

Advances in Geophysics. Academic Press, Inc., 1250 Sixth Avenue, San Diego, CA 92101. (619) 231-0926. FAX (619) 699-6715. Irregular. Inquire about price and availability.

Annual Review and Earth and Planetary Sciences. Annual Reviews, Inc., 4139 El Camino Way, Palo Alto, CA 94306-0897. (415) 493-4400. Fax (415) 855-9815. Annual. $55.00 per year.

Minerals Yearbook. Bureau of Mines, U.S. Department of the Interior. Available from U.S. Government Printing office, Washington, DC 20402. (202) 783-3238. Annual. Three volumes. $45.00.

ASSOCIATIONS AND PROFESSIONAL SOCIETIES

American Association of Petroleum Geologists. P.O. Box 979, Tulsa, OK 74101. (918) 584-2555.

American Geological Institute. 4220 King Street, Alexandria, VA 22302. (703) 379-2480. Fax (703) 379-7563.

American Geophysical Union, 2000 Florida Avenue, N.W., Washington, DC 20009. (202) 462-6900.

American Institute of Professional Geologists. 7828 Vance Drive, Suite 103, Arvada, CO 80003. (303) 431-0831.

Association of Engineering Geologists. 323 Boston Post Road, Suite 2D, Sudbury, MA 01776. (508) 443-4639.

Geochemical Society. Department of Terrestrial Magnetism, 5241 Broad Branch Road, N.W., Washington, DC 20015. (202) 686-4370.

Geological Society of America. 3300 Penrose Place, P.O. Box 9140, Boulder, CO 80301-9140. (303) 447-2020.

Geoscience Information Society. American Geological Institute. 4220 King Street, Alexandria, VA 22302. (703) 379-2480. Fax (703) 379-7563.

Society of Economic Geologists. P.O. Box 571, Golden, CO 80402. (303) 277-1716.

Society of Exploration Geophysicists. P.O. Box 702740, Tulsa, OK 74170. (918) 493-3516.

Society of Independent Professional Earth Scientists. 4925 Greenville Avenue, Suite 170, Dallas, TX 75206. (214) 363-1780.

BIBLIOGRAPHIES

Geologic Reference Sources: A Subject and Regional Bibliography of Publications and Maps in the Geological Sciences. Dederick C. Ward et al. Scarecrow Press, Inc., 52 Liberty Street, Box 4167, Metuchen, NJ 08840. (800) 537-7107 or (201) 548-8600. 1981. $49.50.

Information Sources in the Earth Sciences. David N. Wood, Joan E. Hardy, and Anthony P. Harvey, editors. Bowker-Saur/K.G. Saur. Distributed by R.R. Bowker, 121 Chanlon Road, New Providence, NJ 07974. (800) 521-8110 or (908) 464-6800. Second edition. 1989. $85.00.

Remote Sensing of Earth Resources: A Quarterly Bibliography. University of New Mexico, Technology Application Center, 2808 Central Avenue, S.E., Albuquerque, NM 87131-6031. (505) 277-3622. Fax (505) 277-3614. $150.00 per year.

Scientific and Technical Books and Serials in Print; An Index To Literature in Science and Technology. R.R. Bowker Co., 205 E. 42nd Street, New York, NY 10017. (800) 521-8110 or (212) 916-1600. 1992. $250.00.

DIRECTORIES AND BIOGRAPHICAL SOURCES

American Men and Women of Science: Physical and Biological Sciences. Eighteenth edition. R.R. Bowker Company, 245 West 17th Street, New York, NY 10011. (800) 521-8810 or (212) 916-1600. $750.00.

American Institute of Professional Geologists. Membership Directory. American Institute of Professional Geologists, 7828 Vance Drive, Suite 103, Arvada, CO 80003. (303) 431-0831. Annual. $18.00.

Association of Engineering Geologists. Membership directory. Association of Engineering Geologists, 323 Boston Post Road, Suite 2D, Sudbury, MA 01776. (508) 443-4639. Annual. $18.00.

Directory of Geoscience Departments: United States and Canada. American Geological Institute. 4220 King Street, Alexandria, VA 22302-1507. (703) 379-2480. Fax (703) 379-7563. Twenty-six edition. 1987. $22.00.

GEOLOGY

Ency. of Physical Sciences and Engineering Info. Sources

Geological Society of America. Membership Directory. Geological Society of America, 3300 Penrose Place, Boulder, CO 80301. (303) 447-2020. Annual. Available to members only.

Research Centers Directory. Gale Research, 835 Penobscot Building, Detroit, MI 48226-4094. (800) 877-4253 or (313) 961-2242. 17th edition, 1992. $400.00.

ENCYCLOPEDIAS AND DICTIONARIES

Concise Oxford Dictionary of the Earth Sciences. Ailsa Allaby and Michael Allaby, editors. Oxford University Press, Inc., 200 Madison Avenue, New York, NY 10016. (800) 334-4249 or (212) 679-7300. 1990. $42.95.

Dictionary of Geological Terms. American Geological Institute. Doubleday and Company, Inc., 245 Park Avenue, New York, NY 10017. (800) 645-6156 or (212) 953-4561. Third edition. 1984. $12.00 in paper.

Dictionary of Petrology. S.I. Tomkeieff. John Wiley and Sons, Inc., 605 Third Avenue, New York, NY 10158. (800) 526-5368 or (212) 850-6000. 1983. $180.00.

Encyclopedia of Earth System Science. Academic Press, Inc., 1250 Sixth Street, San Diego, CA 92101. (619) 231-0926 or (800) 321-5068. Four volumes, 1992. $950.00.

Encyclopedia of Igneous and Metamorphic Petrology. D.R. Bowes. Van Nostrand Reinhold, 115 Fifth Avenue, New York, NY 10003. (212) 254-3232 or (800) 926-2665. 1989. $125.95.

Encyclopedia of Minerals. Willard L. Roberts, Thomas J. Campbell, and George R. Rapp, editors. Van Nostrand Reinhold, 115 Fifth Avenue, New York, NY 10003. (212) 254-3232 or (800) 926-2665. Second edition. 1990. $100.00.

Facts On File Dictionary of Geology and Geophysics. Dorothy Farris and Donald R. Coates, editors. Facts on File, Inc., 460 Park Avenue South, New York, NY 10016-7382. (800) 322-8755. Fax (800) 683-3633. 1989. $24.95.

Glossary of Geology. Robert L. Bates and Julia A. Jackson. American Geological Institute, 4220 King Street, Alexandria, VA 22032. (703) 379-2480. Third edition. 1987. $75.00.

McGraw-Hill Encyclopedia of Science and Technology. McGraw-Hill Publishing Company, 1221 Avenue of the Americas, New York, NY 10020. (800) 262-4729 or (212) 512-3825. Seventh edition. 1992. $1900.00.

Magill's Survey of Science: Earth Science Series. Salem Press, Inc., P.O. Box 1097, Englewood Cliffs, NJ 07632. (800) 221-1592 or (201) 871-3700. Five volumes. 1990. $400.00 for the set.

GENERAL WORKS

Dynamic Earth: An Introduction To Physical Geology. B.J. Skinner and Stephen C. Porter. John Wiley and Sons, Inc., 605 Third Avenue, New York, NY 10158. (800) 526-5368 or (212) 850-6000. 1990. $51.95.

Earth and Life Through Time. Steven M. Stanley. W.H. Freeman and Company, 41 Madison Avenue, East 26th, 35th Floor,

New York, NY 10010. (212) 576-9400. Second edition. 1988. $42.95.

Earth Materials and Earth Processes: An Introduction. Lynn Fichter, et al. Macmillan Publishing, 866 3rd Avenue, New York, NY 10024. (212) 689-9140. Third edition. 1991. $19.50.

Earth, Time and Life: An Introduction To Geology. Charles W. Barnes. John Wiley and Sons, Inc., 605 Third Avenue, New York, NY 10158. (800) 526-5368 or (212) 850-6000. Second edition. 1988. $47.95.

Evolution of the Earth. Robert H. Dott, Jr. and Roger L. Batten. McGraw-Hill Publishing Company, 1221 Avenue of the Americas, New York, NY 10020. (800) 262-4729 or (212) 512-3825. Fourth edition. 1988. $41.00.

Gems, Granites, and Gravels: Knowing and Using Rocks and Minerals. R.V. Dietrich and Brian J. Skinner. Cambridge University Press, 40 West 20th Street, New York, NY 10011-4211. (212) 924-3900. 1990. $24.95.

Geology: The Science of A Changing Earth. Ira S. Allison et al. McGraw-Hill Publishing Company, 1221 Avenue of the Americas, New York, NY 10020. (800) 262-4729 or (212) 512-3825. Sixth edition. 1974. $18.95 in paper.

Stratigraphy and Sedimentation. William C. Krumbein and L.L. Sloss. W.H. Freeman and Company, 41 Madison Avenue, East 26th, 35th Floor, New York, NY 10010. (212) 576-9400. Second edition. 1963. $37.95. HandBOOKS and MANUAL

Agi Data Sheets For Geology in the Field, Laboratory, and office. J.T. Dutro, R.V. Dietrich, and R.M. Foose, editors. American Geological Institute, 4220 King Street, Alexandria, VA 223020-1507 (800) 336-4764 or (703) 379-2480. 1989. $24.95.

Field Mapping For Geology Students. F. Ahmed and D.C. Almond. Unwin Hyman, Inc., 10 East 53rd Street, New York, NY 10022. (212) 207-7626 or (800) 242-7737. 1983. $18.95.

Geology in the Field. R.R. Compton. John Wiley and Sons, Inc., 605 Third Avenue, New York, NY 10158. (800) 526-5368 or (212) 850-6418. 1985. $37.95.

A Manual of Geology For Civil Engineers. John A. Pitts. World Scientific Publishing Company, Inc., 1060 Main Street, Unit B, River Edge, NJ 07661. (800) 227-7562 or (201) 487-9655. 1985. $38.00.

The Mapping of Geological Structures. K.R. McClay. John Wiley and Sons, Inc., 605 Third Avenue, New York, NY 10158. (800) 526-5368 or (212) 850-6000. 1988. $18.95.

ONLINE DATABASES AND CD-ROMS

Earth Sciences. U.S. Geological Survey, 12201 Sunrise Valley Drive, Reston, VA 22092-9998. (703) 648-4460. CD-ROM of earth science databases including the U.S. Geological Survey Library, Earth Science Data Directory, and GEOINDEX, citations to published geological maps. $350.00 per year with quarterly updates.

Geoarchive. Geosystems, P.O. Box 1024, Westminster, London, England, SW1 P 2JL. Citations to literature on geoscience, 1969 to present. Inquire as to online cost and availability.

Geobase. Elsevier - Geo Abstracts, Regency House, 34 Duke Street, Norwich NR3 3AP, England. Contains citations to the worldwide earth science literature from 1980 to date. Available on DIALOG, ORBIT online services. Inquire as to cost and availability.

Geomechanics Abstracts. University of London, Imperial College of Science and Technology, Rock Mechanics Information Service, Royal School of Mines, Prince Consort Road, London SW7 2BP, England. Telephone: 071-589 5111 or FAX 071-589 6806. Scope includes worldwide literature on engineering geology, hydrogeology, mining, rock mechanics, soil mechanics, and tunnelling, 1977 to present. Available on the ORBIT online service. Inquire as to cost and availability.

GEOREF. American Geological Institute, 4220 King Street, Alexandria, VA 22302. (800) 336-4764 or (703) 379-2480. Geology and geosciences literature, 1785 to present for North America. Available on DIALOG, ORBIT online services. Also available on CD-ROM. Inquire as to cost and availability.

NTIS Bibliographic Database. National Technical Information Service, 5285 Port Royal Road, Springfield, VA 22161. (703) 487-4630. Broad coverage of government- sponsored research reports, 1964 to present. Available on BRS, DIALOG, ORBIT and STN online services. Also available on CD-ROM. Inquire as to cost and availability.

SCISEARCH. Institute for Scientific Information, 3501 Market Street, Philadelphia, PA 19104. (800) 523-1850 or (215) 386-0100. Broad multidisciplinary title and author index to the international literature of science and technology, 1974 to present. Available on DIALOG and ORBIT online services. Also available on CD-ROM. Inquire as to cost and availability.

PERIODICALS

Aapg Bulletin. American Association of Petroleum Geologists, P.O. Box 979, Tulsa, OK 7401. (918) 584-2555. Monthly. $120.00 per year.

American Journal of Science. Kline Geology Laboratory, Yale University, Box 6666, Yale Station, New Haven, CT 06511-8130. (203) 432-3131. Ten times per year. $40.00 per year.

Canadian Journal of Earth Sciences. National Research Council of Canada, Research Journals, Ottawa, Ont. K1A OR6, Canada. (613) 993-9084. Monthly. $112.00 per year.

Earth and Planetary Science Letters. Elsevier Science Publishing Company, Inc., 655 Avenue of the Americas, New York, NY 10010. (212) 989-5800. Twenty times per year. $945.00 per year.

Economic Geology and the Bulletin of the Society of Economic Geologists. Society of Economic Geologists, Economic Geology Publishing Company, 101 Vowell Hall, University of Texas at El Paso, El Paso, TX 79968. 8 times per year. $51.00.

Geochimica Et Cosmochimica Acta. Pergamon Journals, Inc., Maxwell House, Fairview Park, Elmsford, NY 10523. (914) 592-7700. Fax (914) 592-3625. Monthly. $580.00 per year.

Geoforum. Pergamon Press, Maxwell House, Fairview Park, Elmsford, NY 10523. (914) 592-7700. Fax (914) 592-3625. Quarterly. $305.00 per year.

Geological Magazine. Cambridge University Press, 40 West 20th Street, New York, NY 10011-4211. (212) 924-3900. Bimonthly. $205.00 per year.

Geological Society. JOURNAL. Geological Society of London. Geological Society Publishing House, Unit 7, Brassmill Enterprise Centre, Brassmill Lane, Bath, Avon BA1 3JN, England. 6 times per year. $432.00.

Geological Society of America Bulletin. P.O. Box 9140, 3300 Penrose Place, Boulder, CO 80301. (303) 447-2020. Fax (303) 447-1133. Monthly. $150.00.

Geology. Geological Society of America. P.O. Box 9140, 3300 Penrose Place, Boulder, CO 80301. (303) 447-2020. Fax (303) 447-1133. Monthly. $120.00 per year.

Geophysics. Society of Exploration Geophysicists, P.O. Box 702740, Tulsa, OK 74170. (918) 493-3516. Monthly. $175.00 per year.

Geotimes. American Geological Institute, 4220 King Street, Alexandria, VA 22032. (703) 379-2480. Monthly. $24.95 per year.

Journal of Geology. University of Chicago Press, 5720 Woodlawn Avenue, Chicago, IL 60637. (312) 753-3347. Bimonthly. $38.00 per year.

Journal of Geophysical Research. American Geophysical Union, 2000 Florida Avenue, N.W., Washington, DC 20009. (202) 462-6903. Monthly. $2170.00 per year.

Journal of Petrology. Oxford University Press, Inc., 200 Madison Avenue, New York, NY 10016. (800) 334-4249 or (212) 679-7300. Bimonthly. $240.00.

Journal of Sedimentary Petrology. Society of Economic Paleontologists and Mineralogists, P.O. Box 4756, Tulsa, OK 74159-0756. (918) 743-9765. Bimonthly. $120.00 per year.

Journal of Structural Geology. Pergamon Journals, Inc., Maxwell House, Fairview Park, Elmsford, NY 10523. (914) 592-7700. Ten times per year. $375.00 per year.

Journal of Volcanology and Geothermal Research. Elsevier Science Publishing Company, Inc., 655 Avenue of the Americas, New York, NY 10010. (212) 989-5800. Monthly. $500.00 per year.

Lithos; An International Journal of Mineralogy, Petrology, and Geochemistry. Elsevier Science Publishing Company, Inc., 655 Avenue of the Americas, New York, NY 10010. (212) 989-5800. Quarterly. $275.00 per year.

Mathematical Geology. Plenum Publishing Corporation, 233 Spring Street, New York, NY 10013. (212) 620-8000. Fax (212) 463-0742. Eight times per year. $360.00 per year.

Mountain Geologist. Rocky Mountain Association of Geologists, 4201 West 51st Avenue, Denver, CO 80212-2902. Quarterly. $15.00 per year.

Physics of the Earth and Planetary Interiors. Elsevier Science Publishing Company, Inc., 655 Avenue of the Americas, New York, NY 10010. (212) 989-5800. Twenty-four times per year. $900.00 per year.

Reviews of Geophysics. American Geophysical Union, 2000 Florida Avenue, N.W., Washington, DC 20009. (202) 462-6900. Quarterly. $300.00 per year.

Rock Mechanics and Rock Engineering. Springer-Verlag New York, Inc., 175 Fifth Avenue, New York, NY 10010. (800) 526-7254 or (212) 460-1500. Quarterly. $122.00 per year.

Sedimentary Geology. Elsevier Science Publishing Company, Inc., 655 Avenue of the Americas, New York, NY 10010. (212) 989-5800. Twenty times per year. $901.00 per year.

Sedimentology. Blackwell Scientific Publishing, Inc., 3 Cambridge Court, Suite 208, Cambridge, MA 02142. (617) 225-0401. Six times per year. $370.00 per year.

Tectonics. American Geophysical Union, 2000 Florida Avenue, N.W., Washington, DC 20009. (202) 462-6900. Bimonthly. $230.00 per year to individuals.

Tectonophysics; An International Journal of Geotectonics and the Geology and Physics of the Interior of the Earth. Elsevier Science Publishing Company, Inc., 655 Avenue of the Americas, New York, NY 10010. (212) 989-5800. Fifty-six times per year. $1275.00 per year.

RESEARCH CENTERS AND INSTITUTES

U.S. Geological Survey, Geologic Division, National Center, 12201 Sunrise Valley Drive, Reston, VA 22092. (703) 648-6600. The major geological research agency of the federal government conducting research in most areas of pure and applied research in the geosciences.

Geological Society of America. P.O. Box 9140, 3300 Penrose Place, Boulder, CO 80301. (303) 447-2020.

Society of Economic Geologists. P.O. Box 571, Golden, CO 80402.

Society of Economic Paleontologists and Mineralogists. P.O. Box 4756, Tulsa, OK 74159. (918) 743-9765.

GEOMAGNETISM

See also: GEOLOGY, GEOPHYSICS

ABSTRACT SERVICES AND INDEXES

Bibliography and Index of Geology. American Geological Institute, 4220 King Street, Alexandria, VA 22302. (703) 379-2480. Fax (703) 379-7563. Monthly. $1295.00 per year. Also available online as GeoRef.

Geological Abstracts. Elsevier - Geo Abstracts, Regency House, 34 Duke Street, Norwich NR3 3AP, England. Monthly. $760.00 per year. Also available online as GEOBASE.

Geological Society of America. ABSTRACTS WITH PROGRAMS. Geological Society of America. 3300 Penrose Place, P.O. Box 9140, Boulder, CO 80301-9140. (303) 447-2020. Abstracts and programs of the annual conference. Annual. $69.00.

Geophysics Abstracts. Pergamon Press, Inc., Maxwell House, Fairview Park, Elmsford, NY 10523. (914) 592-7700. Fax (914) 592-3625. Twelve times per year. $565.00 per year. Also available in microform.

Mineralogical Abstracts. Mineralogical Society and the Mineralogical Society of America, 41 Queen's Gate, London, SW7 5HR, England. Quarterly. $235.00 per year.

Oceanic Abstracts. Cambridge Scientific Abstracts, 7200 Wisconsin Avenue, Bethesda, MD 20814. (301) 961-6750. Fax (301) 961-6720. Bimonthly. $995.00 per year.

Science Citation Index. Institute for Scientific Information, 3501 Market Street, Philadelphia, PA 19104. (800) 523-1850 or (215) 386-0100. Inquire about price and availability.

General Science Index. H.W. Wilson Co., 950 University Avenue, Bronx, NY 10452. (800) 367-6770 or (212) 588-8400. Inquire about price and availability.

ANNUAL REVIEWS AND YEARBOOKS

Advances in Geophysics. Academic Press, Inc., 1250 Sixth Avenue, San Diego, CA 92101. (619) 231-0926. FAX (619) 699-6715. Irregular. Inquire about price and availability.

Annual Review and Earth and Planetary Sciences. Annual Reviews, Inc., 4139 El Camino Way, Palo Alto, CA 94306-0897. (415) 493-4400. Fax (415) 855-9815. Annual. $55.00 per year.

ASSOCIATIONS AND PROFESSIONAL SOCIETIES

American Association of Petroleum Geologists. P.O. Box 979, Tulsa, OK 74101. (918) 584-2555.

American Geological Institute. 4220 King Street, Alexandria, VA 22302. (703) 379-2480. Fax (703) 379-7563.

American Geophysical Union, 2000 Florida Avenue, N.W., Washington, DC 20009. (202) 462-6900.

American Institute of Professional Geologists. 7828 Vance Drive, Suite 103, Arvada, CO 80003. (303) 431-0831.

Association of Engineering Geologists. 323 Boston Post Road, Suite 2D, Sudbury, MA 01776. (508) 443-4639.

Geological Society of America. 3300 Penrose Place, P.O. Box 9140, Boulder, CO 80301-9140. (303) 447-2020.

Geoscience Information Society. American Geological Institute. 4220 King Street, Alexandria, VA 22302. (703) 379-2480. Fax (703) 379-7563.

Society of Economic Geologists. P.O. Box 571, Golden, CO 80402. (303) 277-1716.

BIBLIOGRAPHIES

Geologic Reference Sources: A Subject and Regional Bibliography of Publications and Maps in the Geological Sciences. Dederick C. Ward et al. Scarecrow Press, Inc., 52 Liberty Street,

Box 4167, Metuchen, NJ 08840. (800) 537-7107 or (201) 548-8600. 1981. $49.50.

Information Sources in the Earth Sciences. David N. Wood, Joan E. Hardy, and Anthony P. Harvey, editors. Bowker-Saur/K.G. Saur. Distributed by R.R. Bowker, 121 Chanlon Road, New Providence, NJ 07974. (800) 521-8110 or (908) 464-6800. Second edition. 1989. $85.00.

Scientific and Technical Books and Serials in Print; An Index To Literature in Science and Technology. R.R. Bowker Co., 205 E. 42nd Street, New York, NY 10017. (800) 521-8110 or (212) 916-1600. 1992. $250.00. DICTIONARIES and BIOGRAPHICAL SOURCES

American Men and Women of Science: Physical and Biological Sciences. Eighteenth edition. R.R. Bowker Company, 245 West 17th Street, New York, NY 10011. (800) 521-8810 or (212) 916-1600. $750.00.

American Institute of Professional Geologists. Membership Directory. American Institute of Professional Geologists, 7828 Vance Drive, Suite 103, Arvada, CO 80003. (303) 431-0831. Annual. $18.00.

Association of Engineering Geologists. Membership directory. Association of Engineering Geologists, 323 Boston Post Road, Suite 2D, Sudbury, MA 01776. (508) 443-4639. Annual. $18.00.

Directory of Geoscience Departments: United States and Canada. American Geological Institute. 4220 King Street, Alexandria, VA 22302-1507. (703) 379-2480. Fax (703) 379-7563. Twenty-six edition. 1987. $22.00.

Geological Society of America. Membership Directory. Geological Society of America, 3300 Penrose Place, Boulder, CO 80301. (303) 447-2020. Annual. Available to members only.

Research Centers Directory. Gale Research, 835 Penobscot Building, Detroit, MI 48226-4094. (800) 877-4253 or (313) 961-2242. 17th edition, 1992. $400.00.

ENCYCLOPEDIAS AND DICTIONARIES

Concise Oxford Dictionary of the Earth Sciences. Ailsa Allaby and Michael Allaby, editors. Oxford University Press, Inc., 200 Madison Avenue, New York, NY 10016. (800) 334-4249 or (212) 679-7300. 1990. $42.95.

Encyclopedia of Solid Earth Geophysics. David E. James, editor. Van Nostrand Reinhold, 115 Fifth Avenue, New York, NY 10003. (212) 254-3232 or (800) 926-2665. 1989. $124.95.

Glossary of Geology. Robert L. Bates and Julia A. Jackson. American Geological Institute, 4220 King Street, Alexandria, VA 22032. (703) 379-2480. Third edition. 1987. $75.00.

GENERAL WORKS

Electrical Environment: The Earth's Electrical Environment. Rector Press Ltd., 130 Rattlesnake, Leverett, MA 01054-9726. (800) 247-3473. FAX (413) 367-2853. 1995. $145.00.

Foundations of Geomagnetism. George Backus, et al. Cambridge University Press, 40 West 20th Street, New York, NY 10011-

4211. (212) 924-3900 or (800) 872-7423. FAX (914) 937-4712. 1996. $59.95.

Geomagnetism. Jack A. Jacobs, editor. Academic Press Inc., 6277 Sea Harbor Drive, Orlando, FL 32887. (800) 321-5068. 1987-1991. 4 volumes. Price varies, inquire.

Geomagnetism Applications. Wallace H. Campbell. U.S. Geological Survey. Superintendent of Documents, U.S. Government Printing office, Box 371954, Pittsburgh, PA 15250-7954. (202) 783-3238. FAX (202) 512-2233. 1995. Inquire for cost.

ONLINE DATABASES AND CD-ROMS

Earth Sciences. U.S. Geological Survey, 12201 Sunrise Valley Drive, Reston, VA 22092-9998. (703) 648-4460. CD-ROM of earth science databases including the U.S. Geological Survey Library, Earth Science Data Directory, and GEOINDEX, citations to published geological maps. $350.00 per year with quarterly updates.

Geoarchive. Geosystems, P.O. Box 1024, Westminster, London, England, SW1 P 2JL. Citations to literature on geoscience, 1969 to present. Inquire as to online cost and availability.

Geobase. Elsevier - Geo Abstracts, Regency House, 34 Duke Street, Norwich NR3 3AP, England. Contains citations to the worldwide earth science literature from 1980 to date. Available on DIALOG, ORBIT online services. Inquire as to cost and availability.

GEOREF. American Geological Institute, 4220 King Street, Alexandria, VA 22302. (800) 336-4764 or (703) 379-2480. Geology and geosciences literature, 1785 to present for North America. Available on DIALOG, ORBIT online services. Also available on CD-ROM. Inquire as to cost and availability.

NTIS Bibliographic Database. National Technical Information Service, 5285 Port Royal Road, Springfield, VA 22161. (703) 487-4630. Broad coverage of government- sponsored research reports, 1964 to present. Available on BRS, DIALOG, ORBIT and STN online services. Also available on CD-ROM. Inquire as to cost and availability.

SCISEARCH. Institute for Scientific Information, 3501 Market Street, Philadelphia, PA 19104. (800) 523-1850 or (215) 386-0100. Broad multidisciplinary title and author index to the international literature of science and technology, 1974 to present. Available on DIALOG and ORBIT online services. Also available on CD-ROM. Inquire as to cost and availability.

PERIODICALS

Aapg Bulletin. American Association of Petroleum Geologists, P.O. Box 979, Tulsa, OK 7401. (918) 584-2555. Monthly. $120.00 per year.

American Journal of Science. Kline Geology Laboratory, Yale University, Box 6666, Yale Station, New Haven, CT 06511-8130. (203) 432-3131. Ten times per year. $40.00 per year.

Canadian Journal of Earth Sciences. National Research Council of Canada, Research Journals, Ottawa, Ont. K1A OR6, Canada. (613) 993-9084. Monthly. $112.00 per year.

Geological Society. JOURNAL. Geological Society of London. Geological Society Publishing House, Unit 7, Brassmill Enterprise Centre, Brassmill Lane, Bath, Avon BA1 3JN, England. 6 times per year. $432.00.

Geological Society of America Bulletin. P.O. Box 9140, 3300 Penrose Place, Boulder, CO 80301. (303) 447-2020. Fax (303) 447-1133. Monthly. $150.00.

Geology. Geological Society of America. P.O. Box 9140, 3300 Penrose Place, Boulder, CO 80301. (303) 447-2020. Fax (303) 447-1133. Monthly. $120.00 per year.

Geomagnetic Indices Bulletin. NOAA, National Geophysical Data Center, 325 Broadway E-GC4, Boulder, CO 80303-3328. (303) 497-6223.

Geophysical Research Letters. American Geophysical Union, 2000 Florida Avenue, N.W., Washington, DC 20009. (202) 462-6900. Monthly. $480.00 per year.

Geophysics. Society of Exploration Geophysicists, P.O. Box 702740, Tulsa, OK 74170. (918) 493-3516. Monthly. $175.00 per year.

Journal of Geology. University of Chicago Press, 5720 Woodlawn Avenue, Chicago, IL 60637. (312) 753-3347. Bi-monthly. $38.00 per year.

RESEARCH CENTERS AND INSTITUTES

Carnegie Institution of Washington, Department of Terrestrial Magnetism. 5241 Broad Branch Road NW. Washington, DC 20015. (202) 966-0863. FAX (202) 364-8726.

College Magnetic & Seismological Observatory. 800 Yukon Drive, Fairbanks, AK 99775-5160. (907) 479-6146. FAX (907) 479-5663.

University of California, Institute of Geophysics and Planetary Physics. 405 Hilgard Avenue, Los Angeles, CA 90024. (213) 825-1580. FAX (213) 206-3051.

University of Pittsburgh, Earth Magnetism Laboratory. 321 Old Engineering Hall, Pittsburgh, PA 15260. (412) 624-4718.

GEOMETRY

See also: ANALYTIC GEOMETRY, CALCULUS, MATHEMATICS

ABSTRACT SERVICES AND INDEXES

Applied Mechanics Reviews: An Assessment of World Literature in Engineering Sciences. American Society of MechanicalEngineers, 345 East 47th Street, New York, NY 10017. (212) 705-7703. 1948 to present. Monthly. $360.00 per year. ISSN:0003-6900.

Applied Science and Technology Index; A Cumulative Subject Index To English Language Periodicals in the Fields ofaeronautics and Space Science, Computer Technology, Chemistry, Construction Industry, Energy and Related Areas.. H.W. WilsonCompany, 950 University Avenue, Bronx, NY 10452.

(212) 588-8400. (800) 367-6770. FAX (718) 590-1617.1958 topresent. Monthly. Available on-line from BRS and WILSONLINE. Also available on CD-ROM. Inquire about cost and availability. ISSN: 0003-6986.

Compactmath - Compact Mathematics Library. Cumulative CD-ROM edition of Zentralblatt fuer Mathematik - Mathematics Abstracts. Springer-Verlag New York, Inc. 44 Hartz Way. Secaucus. NJ. 07096-2491. (201) 348-4033. FAX (201) 348-4505. 1993 to present. Annual. Available only on CD-ROM. Inquire regarding cost and availability. ISSN: 0938- 3174.

Compumath Citation Index. Institute for Scientific Information, 3501 Market Street, Philadelphia, PA 19104. (800) 523-1850 or (215)-386-0100. FAX (215) 386-2991. Three times per year. $1955.00. Also available online and on CD-ROM. Inquire regarding cost and availablity. ISSN: 0730-6199.

Current Mathematical Publications. American Mathematical Society. P.O. Box 1571, Annex Station, Providence, RI 02901-9930. (800) 556-7774 or (401) 455-4000. FAX (401) 331-3842. 1969 to present. Seventeen times per year.$377.00 per year. Available online from BRS, DIALOG, European Space Agency. Also available on CD-ROM. Inquire regarding cost and availability. ISSN: 0361-4794.

General Science Index. H.W. Wilson Company, 950 University Avenue, Bronx, NY 10452. (212) 588-8400. (800) 367-6770.FAX (718) 590-1617. From 1978 to present. Ten issues peryear; quarterly and annual cumulations. Service basis. Available on CD-ROM and online. Inquire regarding cost and availability. ISSN: 0162-1963.

Index To Mathematical Tables. Alan Fletcher. Addison-Wesley, Publishing Co., Inc., 1 Jacob Way, Reading, MA01867. (617) 944-3700. (800) 447-2226. 2nd edition. 1962. 2volumes.

Mathematical Reviews: A Review Journal Covering the World Literature of Mathematical Research. American Mathematical Society, P.O. Box 1571, Annex Station, Providence, RI 02901-9930. (800) 556-7774 or (401) 455- 4000. 1940 to present. Monthly. $4594.00 per year. Also available via network (MathSciNet), online and on CD-ROM. Inquire regarding cost and availability. ISSN: 0025-5629.

N T I S Alerts: Mathematical Sciences. U.S. National Technical Information Service, 5285 Port Royal Road, Springfield, VA 22161. (703) 487-4650. FAX (703) 321-8547. Weekly. $140.00.

Science Citation Index. Institute for ScientificInformation, 3501 Market Street, Philadelphia, PA 19104. (215) 386-0100. (800) 523-1850. FAX (215) 386-2991. 1961 to present. Six issues per year, plus annual cumulation.$11650.00 per year. Also available online and on CD-ROM. Inquire about price and availability. ISSN: 0036-827X.

Zentralblatt Fuer Mathematik Und Ihre Grenzebiete/mathematics Abstracts. Heidelberger Akademie der Wissenschaften/ Springer-Verlag New York, Inc., 44 Hartz Way, Seacaucus, NJ 07096-2491. (201) 348-4033. FAX (201) 348-4505. 1931 to present. 30 issues per year. DM 8340. ISSN: 0044-4235. Also available online and on CD-ROM. Inquire regarding cost and availability. ANNUAL REVIEWS, YEARBOOKS and MONOGRAPHIC SERIES.

Advances in Applied Mathematics. Academic Press, Inc., 6277 Sea Harbor Drive, Orlando, FL 32821-8340. (800) 5439534.1980 to pesent. Irregular, Price varies; inquire.

Applied Mathematical Sciences. Springer-Verlag, 175 Fifth Avenue, New York, NY 10010. (212) 460-2500. FAX (212) 473-6272. 1972 to present. Irregular. ISSN: 0066-5452.

C B M S -N S F Regional Conference Series in Applied Mathematics. Society for Industrial and Applied Mathematics, 3600 University City Science Center, Philadelphia, PA 19104-2688. (215) 382-9800. FAX (215) 386-7999. 1971 to present. Irregular. ISSN: 0163-9439.

Ergebnisse Der Mathematik Und Ihrer Grenzgebiete. Neue folge. Springer-Verlag New York, Inc. 175 Fifth Avenue, New York, NY 10010. (212) 460-1500. FAX (212) 473-6272. 1953 to present. Irregular. Price varies; inquire. ISSN: 0071-1136.

Lecture Notes in Mathematics. Springer-Verlag New York, Inc. 44 Hartz Way. Secaucus. NJ. 07096-2491. (201) 348-4033. FAX (201) 348-4505. Irregular. Price varies; inquire.

ASSOCIATIONS AND PROFESSIONAL SOCIETIES

American Mathematical Society. P.O. Box 6248, Providence, RI 02940. (401) 455-4000. FAX (401) 331-3842.

Association For Symbolic Logic. University of Illinois, Department of Mathematics, 1409 West Green Street, Urbana IL 61801. (217) 244-7902. FAX (217) 333-9576.

Association For Women in Mathematics. 4114 Computer and Space Science Building, University of Maryland, College Park, MD 20742-2461. (301) 405-7892.

Mathematical Association of America. 1529 18th Street, NW, Washington, DC 20036. (202) 387-5200. FAX (202) 265-2384.

National Council of Teachers of Mathematics. 1906 Association Drive, Reston, VA 22091. (703) 620-9840. FAX (703) 475-2970.

Society For Industrial and Applied Mathematics. 3600 University City Science Center, Philadelphia, PA 19104-2688. (215) 382-9800. FAX (215) 386-7999.

Special Interest Group For Symbolic and Algebraic Manipulation (sigsam). c/o ACM. 1515 Broadway, New York, NY 10036-5701. (212) 869-7440. FAX (212) 944-1318.

BIBLIOGRAPHIES

Books You Can Count On: Linking Mathematics & Literature. Rachel Griffiths and Margaret Clyne. Heinemann Publ. 361 Hanover Street, Portsmouth, NH 03801-3912. (603) 431-7894. (800) 541-2086. 1991. $15.00. ISBN: 0-435-08322-8.

Bibliography of Mathematical Works Printed in America Through 1850. Louis C. Karpinski. Supplements 1 & 2, Bernard I. Cohen, editor. Ayer Company Publ. Inc. Lower Mill Road, North Stratford, NH 03590. (603) 922-5105.(800) 282-5413. Reprint edition. 1980. $66.95. ISBN: 0-405-12553-4.

Bibliograhy of Non-Euclidean Geometry. Duncan Y. Summerville. Chelsea Publishing Inc., 15 East 26th Street, New York, NY 10010. (212) 889-8095. 1960. $29.95.

International Mathematical Olympiads 1978-1985. Murray S. Klamkin. National Council of Teachers of Mathematics. 1906 Association Drive, Reston, VA 22091. (703) 620-9840. (800) 235-7566. 1987. $11.75. ISBN: 0-88385-631-X.

Mathematical Book Review Index, 1800-1940. Louise S. Grinstein. Garland Publ. Fifth Avenue, Suite 2500, New York, NY 10022-8101. (212) 751-7447. 1992. $72.00. ISBN: 0-8240-4114-3.

Mathematical Journals: An Annotated Guide. Diana F. Liang. Scarecrow Press. Distributed by University Press of America, 4720 Boston Way, Lanham, MD 20706. (800) 462-6420. 1987. $29.50. ISBN: 0-8108-2585-6.

DIRECTORIES AND BIOGRAPHICAL SOURCES

Assistantships and Graduate Fellowships in the Mathematical Sciences. American Mathematical Society, P.O. Box 6248, Providence, RI 02940. (401) 455-4000. (800) 321-4267. Annual. $19.00. ISSN: 1040-7650.

Combined Membership List. American Mathematical Society, P.O. Box 6248, Providence, RI 02940. (401) 455-4000. (800) 321-4267. Annual. ISSN: 0569-6461.

Mathematical Science Professional Directory. American Mathematical Society, P.O. Box 6248, Providence, RI 02940. Annual. $45.00. ISBN: 0-8218-0173-2.

Research Centers Directory. Gale Research Company, 835 Penobscot Building, Detroit, MI 48226. (313) 961- 2242. (800) 877-4253. 20th edition. 1995. Annual. $485.00. ISBN: 0-8103-9094-9.

World Directory of Mathematicians. G.D. Mostow, editor. International Mathematical Union, Helsinki. 1958 to present. 9th revised edition. 1990. $40.00. Available from: American Mathematical Society, P.O. Box 6248, Providence, RI 02940. (401) 455-4000. ISSN: 0512-2740.

ENCYCLOPEDIAS AND DICTIONARIES

Beyond Numeracy: An Uncommon Dictionary of Mathematics. John A. Paulos. Alfred A. Knopf Inc., 201 East 50th Street, New York, NY 10022. (800) 733-3000. 1991. $21.50. ISBN: 0-394-58640-9.

Concise Oxford Dictionary of Mathematics. Christopher Clapham. Oxford University Press, 200 Madison Avenue, New York, NY 10016. (212) 726-6000. 1990. $10.95. ISBN: 0-19-286103-4.

Encyclopedia of Mathematics. An updated and annotated translation of the Soviet Mathematical Encyclopedia. Michael Hazewinkel, editor Kluwer Academic Publishers, 101 Philip Drive, Assinippi Park, Norwell, MA 02061. (617) 871-6600. 1987 - . 10 volumes. $147.00, volume 1. 1-55608-010-7 set.

Encyclopedia of Mathematics and Its Applications. Addison-Wesley Publishing Co., Inc., 1 Jacob Way, Reading, MA 01867. (617) 944-3700. (800) 447-2226. 1976 to present. irregular. Volumes are individually priced.

Encyclopedic Dictionary of Mathematics. Kiyosi Ito, editor. MIT Press, 55 Hayward Street, Cambridge, MA 02142. (617) 625-

GEOMETRY

Ency. of Physical Sciences and Engineering Info. Sources

8569. 2nd edition, reprinted. 1993. 2 volumes. $70.00. ISBN: 0-262-59020-4.

Harpercollins Dictionary of Mathematics. E. J. Borowski and J. J. Borwein. HarperCollins Publ. 10 East 53rd Street, New York, NY 10022-5299. (800) 331-3761. 1990. $10.95.ISBN: 0-19-286103-4.

Mathematics Dictionary. Robert C. James, editor. Van Nostrand Reinhold, 115 Fifth Avenue, New York, NY 10003. (800)-842-3636. 5th edition. 1992. $42.95. ISBN: 0-442- 00741-8.

Penguin Dictionary of Curious and Interesting Geometry. David Wells. Viking Penguin, 375 Hudson Street, New York, NY 10014-3657. (800) 331-4624. 1992. $20.00 in paper.

The Words of Mathematics: An Etymological Dictionary of Mathematical Terms Used in English. Steven Schwartzman. Mathematical Association of America, 18th Street NW, Washington, DC 20036. (800) 331-1622. 1994. $29.50.
<subhead2>General Works

Basic Algebraic Geometry. I. R. Shafarevich. Springer-Verlag New York, Inc., 175 Fifth Avenue, New York, NY 10010. (212) 460-1500. (800) 777-4643. FAX (212) 473-6272. 1990. $32.00 in paper.

Computational Algebraic Geometry. F. Eysette and A. Galligo, editors. Birkhauser Boston, 675 Massachusetts Avenue, Cambridge, MA 02139. (617) 876-2333. (800) 777-4643. 1994. $74.50.

Computer-aided Surface Geometry and Design. Adrian Bowyer, editor. Oxford University Press, Inc., 200 Madison Avenue, New York, NY 10016. (212) 725-6000. (800) 334-4249. 1994. $98.00.

Euclidean and Non-euclidean Geometries: Development and History. Marvin J. Greenberg. W. H. Freeman & Co., 41 Madison Avenue, East 26th Avenue, 35th Floor, New York, NY 10010. (212) 576-0400. 1995.

Geometric and Analytic Number theory. E. Hlawka, et al.

Springer-verlag New York, Inc., 175 Fifth Avenue, New York, NY 10010. (212) 460-1500. (800) 777-4643. FAX (212) 473-6272. 1991. $32.00 in paper.

Geometries, Codes and Cryptographpy. G. Longo et al, editors.Springer-verlag New York, Inc., 175 Fifth Avenue, New York, NY 10010. (212) 460-1500. (800) 777-4643. FAX (212) 473-6272. 1990. $47.00 in paper.

Geometry and the Visual Arts. Dan Pedoe. Dover Publications, Inc., 180 Varick Street, New York, NY 10014. (212) 255-3755. (800) 223-3130. Reprint edition. 1983. $7.95.

Geometry Problem Solver. Research & Education Association, 61 Ethel Road, West Piscataway, NJ 08854. (908) 819-8880. 1994; revised edition. $29.95. ISBN: 0-87891-549-4.

Geometry, Topology and Field theory. K. Fukaya, et al. World Scientific Publishing Company, Inc., 1060 Main Street, River Edge, NJ 07661. (201) 487-9655. (800) 227-7562. 1994. $55.00.

Fractals Everywhere. Michael F. Barnsley. Academic Press Inc., 6277 Sea Harbor Drive, Orlando, FL 32887. (800) 321- 5068. 1988. $49.00. ISBN: 0-12-079062-9.

Introduction To Differential Geometry For Engineers. Doolin.

Marcel Dekker, Inc., 270 Madison Avenue, New York, NY 10016. (212) 696-9000. (800) 228-1160. 1990. $55.00.

Kinematic Geometry of Mechanisms. K. H. Hunt. Oxford University Press, Inc., 200 Madison Avenue, New York, NY 10016. (212) 725-6000. (800) 334-4249. 1990. $51.00 in paper.

Mathematica: A System For Doing Mathematics By Computer. Stephen Wolfram. Addison-Wesley Publishing Company, Inc., 1 Jacob Way, Reading, MA 01867. (617) 944-3700. (800) 447-2226. 2nd edition. 1991. ISBN: 0-201-51502-4

Mathematics: Queen and Servant of Science. Eric T. Bell. Mathematical Association of America, 18th Street NW, Washington, DC 20036. (202) 387-5200. (800) 331-1622 1988. $16.50; reprint editon. ISBN: 1-55615-173-X or 0-88385-447-3.

The Visual Mind: Art and Mathematics. Michele Emmer. MIT Press, 55 Hayward Street, Cambridge, MA 02142. (617) 625-8569. 1993. $45.00. ISBN: 0-262-05048-X.

Writing Mathematics Well. Leonard Gillman. Mathematical Association of America, 18th Street NW, Washington, DC 20036. (202) 387-5200. (800) 331-1622. 1987. $8.00. ISBN: 0-88385-443-0.

HANDBOOKS AND MANUALS

Handbook of Mathematical Functions with Formulas, Graphs, and Mathmetical Tables. Milton Abramowitz and Irene A. Stegun, editors. U.S. National Bureau of Standards. U.S. Government Printing office, Washington, D.C. 10th edition. 1972. $43.00. ISBN: 0-16-000202-8.

Handbook of Mathematics. I. N. Bronshtein and K. A. Semedyayev. Van Nostrand Reinhold, 115 Fifth Avenue, New York, NY 10003. (800)-842-3636. 20th edition. $59.95. ISBN: 0-442-21171-6.

Handbook of Writing For the Mathematical Sciences. Society for Industrial and Applied Mathematics, 3600 University City Science Center, Philadelphia, PA 19104-2688. (215) 382-9800. FAX (215) 386-7999. 1993. $21.50. ISBN: 0- 89871-314-5.

Standard Mathematical Tables and Formulas. William H. Beyer, editor. CRC Press, 2000 Corporate Boulevard, Boca Raton, FL 33431. (800) 272-7737. 29th edition. 1991. $41.95. ISBN: 084-930- 6299.

Tables of Integrals, Series and Products: Corrected and Enlarged Edition. I. S. Gradshteyn and I. M. Ryzhik. Academic Press, Inc., 6277 Sea Harbor Drive, Orlando, FL 32821. (800) 321-5068. 5th edition. 1993. $61.00. ISBN: 0-12-294755-X.

ONLINE DATABASES AND CD-ROMS

CA Search. Chemical Abstracts Service, P.O. Box 3012, Columbus, OH 43210-0012. (614) 447-3600. (800) 848-6533. FAX (614) 447-3709. Very comprehensive guide to worldwide chemical literature and related fields, 1972 to present. Available on

BRS,(800) 289-4277, DIALOG, (800) 334-2564, ORBIT (800) 456-7248, and STN International, FIZ Karlsruhe, P.O. Box 2465, W-7500, Karlsruhe 1, Germany, online services. Inquire as to cost and availability.

Compendex Plus. Engineering Information, Inc., 345 East 47th Street, New York, NY 10017. (212) 705-7600 or (800) 221-1044. Contains citations with abstracts to worldwide literature in engineering and technology, from 1970 to present. Available on online BRS,(800) 289-4277, DIALOG, (800) 334-2564, ORBIT (800) 456-7248, and STNInternational, FIZ Karlsruhe, P.O. Box 2465, W-7500, Karlsruhe 1, Germany, online services. Also available on CD-ROM. Inquire as to cost and availability.

Dissertation Abstracts. University Microfilms International, 300 North Zeeb Road, Ann Arbor, MI 48106. (800) 521-0600 or (313) 761-4700. Scope includes virtually all doctoral dissertations accepted at accredited American institutions from 1861 to present in 252 subject areas. Available on BRS,(800) 289-4277, DIALOG,(800) 334-2564, and OCLC EPIC,(800) 848-5878, online services. Also available on CD-ROM. Inquire as to cost and availability.

INSPEC. Institution of Electrical Engineers, Michael Faraday House, Six Hills Way, Stevenage, Herts. SG1 2AY, England. Telephone: 0438 313311 or FAX 0438 742840. Contains citations to the worldwide literature of physics, electronics and electrical engineering, computer technology, and related fields. Available on BRS,(800) 289-4277, DIALOG, (800) 334-2564, ORBIT, (800) 456-7248, and STN International, FIZ Karlsruhe, P.O. Box 2465, W-7500, Karlsruhe 1, Germany, online services. Inquire as to cost and availability.

Mathsci. American Mathematical Society, P.O. Box 6248, Providence, RI 02940. (800) 321-4667 or (401) 455-4000 or FAX (401) 331-3842. Scope includes pure and applied mathematics and related areas of physics, statistics, engineering, computer science, and operations research literature since 1959. Available on DIALOG,(800) 334-2564, online service. Also available on CD-ROM and MATHSCINET, on the Internet. Inquire as to cost and availability.

NTIS Bibliographic Database. National Technical Information Service, 5285 Port Royal Road, Springfield, VA 22161. (703) 487-4929 or FAX (703) 321-8199. Broad coverage of government-sponsored science and technology research reports, 1964 to present. Available on BRS,(800) 289-4277, DIALOG, (800) 334-2564, ORBIT, (800) 456-7248, and STN International, FIZ Karlsruhe, P.O. Box 2465, W-7500, Karlsruhe 1, Germany, online services. Also available on CD-ROM. Inquire as to cost and availability.

Scientific and Technical Books and Serials in Print. R.R. Bowker Inc., 121 Chanlon Road, New Providence, NJ 07974. (800) 521-8110 or (908) 464-6800. List of currently published books and serials in the physical and biological sciences, engineering and technology, with subject, author and titles indexes. Available on ORBIT,(800) 456-7248, online service.Inquire as to cost and availability.

SCISEARCH. Institute for Scientific Information, 3501 Market Street, Philadelphia, PA 19104. (800) 523-1850 or (215) 386-0100. Broad multidisciplinary title and author index to the international literature of science and technology, 1974 to present. Available on DIALOG,(800) 334-2564, and ORBIT,(800) 456-7248, online services. Also available on CD-ROM. Inquire as to cost and availability.

WILSONLINE. H.W. Wilson Company, 950 University Avenue, Bronx, NY 10452. (800) 367-6770 or (212) 588-8400. Makes available online versions of the printed H.W. Wilson indexes including Applied Science and Technology Index, Business Periodicals Index, General Science Index, and Readers' Guide to Periodical Literature. Period covered is generally 1983 to present. Available on BRS,(800) 289-4277, DIALOG,(800) 334-2564, and OCLC EPIC,(800) 848-5878, online services. Also available on CD-ROM. Inquire as to cost and availability. <2>Periodicals

American Journal of Mathematics. Johns Hopkins University Press, 2715 North Charles Street, Baltimore, MD 21218. (410) 516-6987. FAX (410) 516-6968. 1878 to present. Bimonthly. $162.00 per year. ISSN: 0002-9327.

American Mathematical Monthly. Mathematical Association of America. 1529 Eighteenth Street, NW, Washington, DC 20036. (202) 387-5200. 1894 to present. Ten issues per year. $160.00. ISSN: 0002-9890.

Canadian Applied Mathematical Quarterly. Canadian Mathematical Society. Distributed by Rocky Mountain Mathematics Consortium, Department of Mathematics, Arizona State University, Tempe AZ 85287-1904. 1992 to present. Quarterly. $175.00 per year. ISSN: 1073-1849.

Discrete and Computational Geometry. Springer-Verlag New York, Inc., 175 Fifth Avenue, New York, NY 10010. (212) 460-1500. (800) 777-4643. FAX (212) 473-6272. 1986 to present. 8 issues per year. $254.00 per year. ISSN: 0179-5376.

Geometriae Dedicata. Kluwer Academic Publishers, Box 358, Accord Station, Hingham, MA 02018-0358. (617) 871-6600. FAX (617) 871-6528. 1972 to present. 15 issues per year. $872.50 per year. ISSN: 0046-5755.

Geometric and Functional Analysis. Birkhauser-Verlag/ Springer-Verlag New York, Inc., 175 Fifth Avenue, New York, NY 10010. (212) 460-1500. (800) 777-4643. FAX (212) 473-6272. 1992. Bimonthly. $278.00 per year. ISSN: 1016-443X.

I M A Journal of Applied Mathematics. Oxford University Press, 2001 Evans Road, Cary, NC 27513. (919) 677-0977. 1981 to present. Bimonthly. $410.00 per year. ISSN: 0272- 4960.

Journal of Algebraic Geometry. American Mathematical Society, Box 1571, Annex Station, Providence, RI 02901-9930. (401) 455-4000. 1991 to present. Quarterly. $188.00 per year. ISSN: 1056-3911.

Journal of Geometry. Birkhauser Verlag/Springer-Verlag New York, 175 Fifth Avenue, New York, NY 10010. (212) 460-1500. FAX (212) 473-6272. (800) 777-4643. 1971 to present. Bimonthly. $237.00. ISSN: 0047-2468.

Mathematical Intelligencer. Springer-Verlag, 175 Fifth Avenue, New York, NY 10010. (212) 460-1500. FAX (212) 473- 6272. (800) 777-4643. 1978 to present. Quarterly. $39.00 per year. ISSN: -343-6993.

Mathematics Magazine. Mathematical Association of America. 1529 Eighteenth Street, NW, Washington, DC 20036. (202) 387-5200. 1926 to present. Bimonthly. $80.00 per year. ISSN: 0025-570X.

Quarterly Journal of Mechanics and Applied Mathematics. entOxford University Press, 2001 Evans Road, Cary, NC 27513.

GEOMETRY

Ency. of Physical Sciences and Engineering Info. Sources

(919) 677-0977. 1948 to present. Quarterly. $214.00 per year. ISSN: 0033-5614.

Siam Journal On Applied Mathematics. Society for Industrial and Applied Mathematics, 3600 University City Science Center, Philadelphia, PA 19104-2688. (215) 382-9800. FAX (215) 386-7999. 1953 to present. Bimonthly. $242.00 per year. ISSN: 0036-1399.

Siam Journal On Control and Optimization. Society for Industrial and Applied Mathematics, 3600 University City Science Center, Philadelphia, PA 19104-2688. (215) 382-9800. FAX (215) 386-7999. 1963 to present. Bimonthly. $287.00 per year. ISSN: 0036-0129.

Siam Journal On Discrete Mathematics. Society for Industrial and Applied Mathematics, 3600 University City Science Center, Philadelphia, PA 19104-2688. (215) 382-9800. FAX (215) 386-7999. 1980 to present. Quarterly. $220.00 per year. ISSN: 0895-4801.

Siam Journal On Numerical Analysis. Society for Industrial and Applied Mathematics, 3600 University City Science Center, Philadelphia, PA 19104-2688. (215) 382-9800. FAX (215) 386-7999. Bimonthly. $255.00 per year. ISSN: 0036-1429

Siam Review. Society for Industrial and Applied Mathematics, 3600 University City Science Center, Philadelphia, PA 19104-2688. (215) 382-9800. 1959 to present. Quarterly. $148.00 per year. ISSN: 0036-1445.

RESEARCH CENTERS AND INSTITUTES

Center for Applied Mathematics. University of Georgia, Athens, GA 30602. (404) 542-3491.

Center For Mathematical Sciences. University of Wisconsin-Madison. 610 Walnut Street, Madison, WI 53705. (608) 263-2696. FAX (608) 263-2841.

Center For Pure and Applied Mathematics. University of California, Berkeley. 977 Evans Hall, Berkeley, CA 94720. (415) 642-0116. FAX (415) 642-6726.

Design Research Center. Rensselaer Polytechnic Institute, CII 7015, Troy, NY 12180-3590. (518) 266-6751.

Project in Geometry and Physics. University of California, San Diego, Department of Mathematics, C-012, LA Jolla, CA 92093-0112. (619) 534-2218.

GEOMORPHOLOGY

See also: GEOLOGY, IGNEOUS ROCKS, METAMORPHIC ROCKS, PHYSICAL GEOLOGY, SEDIMENTARY ROCKS

ABSTRACT SERVICES AND INDEXES

Bibliography and Index of Geology. American Geological Institute, 4220 King Street, Alexandria, VA 22302. (703) 379-2480. Fax (703) 379-7563. Monthly. $1295.00 per year. Also available online as GeoRef.

Current Contents: Physical, Chemical and Earth Sciences. Institute for Scientific Information, 3501 Market Street, Philadelphia, PA 19104. (215) 386-0100. FAX (215) 386-6362. Weekly. $360.00 per year.

Deep-sea Research. Part A: Oceanographic Research Papers. Deep-sea Research. Part B: Oceanographic Literature Review. Pergamon Press, Inc., Maxwell House, Fairview Park, Elmsford, NY 10523. (914) 592-7700. Fax (914) 592-3625. Twelve times per year. $1370 per year for Parts A and B.

Geological Abstracts. Elsevier - Geo Abstracts, Regency House, 34 Duke Street, Norwich NR3 3AP, England. Monthly. $760.00 per year. Also available online as GEOBASE.

Geological Society of America. Abstracts with Programs. Geological Society of America. 3300 Penrose Place, P.O. Box 9140, Boulder, CO 80301-9140. (303) 447-2020. Abstracts and programs of the annual conference. Annual. $69.00.

Mineralogical Abstracts. Mineralogical Society and the Mineralogical Society of America, 41 Queen's Gate, London, SW7 5HR, England. Quarterly. $235.00 per year.

Oceanic Abstracts. Cambridge Scientific Abstracts, 7200 Wisconsin Avenue, Bethesda, MD 20814. (301) 961-6750. Fax (301) 961-6720. Bimonthly. $995.00 per year.

Science Citation Index. Institute for Scientific Information, 3501 Market Street, Philadelphia, PA 19104. (215) 386-0100. FAX (215) 386-6362. Inquire about availability and cost. Also available on CD-ROM.

ASSOCIATIONS AND PROFESSIONAL SOCIETIES

American Geological Institute. 4220 King Street, Alexandria, VA 22302. (703) 379-2480. Fax (703) 379-7563.

American Geophysical Union, 2000 Florida Avenue, N.W., Washington, DC 20009. (202) 462-6900.

American Institute of Professional Geologists. 7828 Vance Drive, Suite 103, Arvada, CO 80003. (303) 431-0831.

Association of Engineering Geologists. 323 Boston Post Road, Suite 2D, Sudbury, MA 01776. (508) 443-4639.

Geological Society of America. 3300 Penrose Place, P.O. Box 9140, Boulder, CO 80301-9140. (303) 447-2020.

DIRECTORIES AND BIOGRAPHICAL SOURCES

American Men and Women of Science: Physical and Biological Sciences. Eighteenth edition. R.R. Bowker Company, 245 West 17th Street, New York, NY 10011. (800) 521-8810 or (212) 916-1600. $750.00.

American Institute of Professional Geologists. Membership Directory. American Institute of Professional Geologists, 7828 Vance Drive, Suite 103, Arvada, CO 80003. (303) 431-0831. Annual. $18.00.

Directory of Geoscience Departments: United States and Canada. American Geological Institute. 4220 King Street, Alexandria, VA 22302-1507. (703) 379-2480. Fax (703) 379-7563. Twenty-six edition. 1987. $22.00.

Geological Society of America. Membership Directory. Geological Society of America, 3300 Penrose Place, Boulder, CO 80301. (303) 447-2020. Annual. Available to members only.

Research Centers Directory. Gale Research, 835 Penobscot Building, Detroit, MI 48226-4094. (800) 877-4253 or (313) 961-2242. 17th edition, 1992. $400.00.

ENCYCLOPEDIAS AND DICTIONARIES

Concise Oxford Dictionary of the Earth Sciences. Ailsa Allaby and Michael Allaby, editors. Oxford University Press, Inc., 200 Madison Avenue, New York, NY 10016. (800) 334-4249 or (212) 679-7300. 1990. $42.95.

Dictionary of Geological Terms. American Geological Institute. Doubleday and Company, Inc., 245 Park Avenue, New York, NY 10017. (800) 645-6156 or (212) 953-4561. Third edition. 1984. $12.00 in paper.

Encyclopedia of Igneous and Metamorphic Petrology. D.R. Bowes. Van Nostrand Reinhold, 115 Fifth Avenue, New York, NY 10003. (212) 254-3232 or (800) 926-2665. 1989. $125.95.

Glossary of Geology. Robert L. Bates and Julia A. Jackson. American Geological Institute, 4220 King Street, Alexandria, VA 22032. (703) 379-2480. Third edition. 1987. $75.00.

GENERAL WORKS

The Changing Earth: Geomorphological Processes and Time. andre Goudie. Blackwell Scientific Publications, 238 Main Street, Cambridge, MA 02142. (617) 876-7000 or (800) 759-6102. FAX (617) 876-7022. 1995. $78.95.

Coastal Problems: Geomorphology, Ecology & Society At the Coast. Heather Viles & Tom Spencer. Halsted Press (Division of John Wiley and Sons, Inc.), 605 Third Avenue, New York, NY 10158-0012. (212) 850-6400. FAX (212) 850-6816. 1995. $31.95.

Geomorphology & Groundwater. A.G.Brown, editor. John Wiley and Sons, Inc., 605 Third Avenue, New York, NY 10158. (800) 526-5368 or (212) 850-6000. 1995. $125.00.

Soil Geomorphology. Raymond B. Daniels & Richard D. Hammer. John Wiley and Sons, Inc., 605 Third Avenue, New York, NY 10158. (800) 526-5368 or (212) 850-6000. 1992. $79.95.

ONLINE DATABASES AND CD-ROMS

Earth Sciences. U.S. Geological Survey, 12201 Sunrise Valley Drive, Reston, VA 22092-9998. (703) 648-4460. CD-ROM of earth science databases including the U.S. Geological Survey Library, Earth Science Data Directory, and GEOINDEX, citations to published geological maps. $350.00 per year with quarterly updates.

Geoarchive. Geosystems, P.O. Box 1024, Westminster, London, England, SW1 P 2JL. Citations to literature on geoscience, 1969 to present. Inquire as to online cost and availability.

Geobase. Elsevier - Geo Abstracts, Regency House, 34 Duke Street, Norwich NR3 3AP, England. Contains citations to the worldwide earth science literature from 1980 to date. Available on DIALOG, ORBIT online services. Inquire as to cost and availability.

Geomechanics Abstracts. University of London, Imperial College of Science and Technology, Rock Mechanics Information Service, Royal School of Mines, Prince Consort Road, London SW7 2BP, England. Telephone: 071-589 5111 or FAX 071-589 6806. Scope includes worldwide literature on engineering geology, hydrogeology, mining, rock mechanics, soil mechanics, and tunnelling, 1977 to present. Available on the ORBIT online service. Inquire as to cost and availability.

GEOREF. American Geological Institute, 4220 King Street, Alexandria, VA 22302. (800) 336-4764 or (703) 379-2480. Geology and geosciences literature, 1785 to present for North America. Available on DIALOG, ORBIT online services. Also available on CD-ROM. Inquire as to cost and availability.

NTIS Bibliographic Database. National Technical Information Service, 5285 Port Royal Road, Springfield, VA 22161. (703) 487-4630. Broad coverage of government- sponsored research reports, 1964 to present. Available on BRS, DIALOG, ORBIT and STN online services. Also available on CD-ROM. Inquire as to cost and availability.

SCISEARCH. Institute for Scientific Information, 3501 Market Street, Philadelphia, PA 19104. (800) 523-1850 or (215) 386-0100. Broad multidisciplinary title and author index to the international literature of science and technology, 1974 to present. Available on DIALOG and ORBIT online services. Also available on CD-ROM. Inquire as to cost and availability.

PERIODICALS

American Journal of Science. Kline Geology Laboratory, Yale University, Box 6666, Yale Station, New Haven, CT 06511-8130. (203) 432-3131. Ten times per year. $40.00 per year.

Canadian Journal of Earth Sciences. National Research Council of Canada, Research Journals, Ottawa, Ont. K1A OR6, Canada. (613) 993-9084. Monthly. $112.00 per year.

Earth and Planetary Science Letters. Elsevier Science Publishing Company, Inc., 655 Avenue of the Americas, New York, NY 10010. (212) 989-5800. Twenty times per year. $945.00 per year.

Geological Magazine. Cambridge University Press, 40 West 20th Street, New York, NY 10011-4211. (212) 924-3900. Bimonthly. $205.00 per year.

Geological Society. JOURNAL. Geological Society of London. Geological Society Publishing House, Unit 7, Brassmill Enterprise Centre, Brassmill Lane, Bath, Avon BA1 3JN, England. 6 times per year. $432.00.

Geological Society of America Bulletin. P.O. Box 9140, 3300 Penrose Place, Boulder, CO 80301. (303) 447-2020. Fax (303) 447-1133. Monthly. $150.00.

Geology. Geological Society of America. P.O. Box 9140, 3300 Penrose Place, Boulder, CO 80301. (303) 447-2020. Fax (303) 447-1133. Monthly. $120.00 per year.

Geomorphology. Elsevier Science B.V., PO Box 945, Madison Square Station, New York, NY 10160-0757. (212) 633-3650. FAX (212) 633-3680. 1987 to present. 12 times a year. $594.00.

Journal of Geology. University of Chicago Press, 5720 Woodlawn Avenue, Chicago, IL 60637. (312) 753-3347.

Bimonthly. $38.00 per year.

RESEARCH CENTERS AND INSTITUTES

Colorado State University, Hydraulics and Hydromachinery Research Laboratories. Engineering Research Center, Ft. Collins, CO 80523. (303) 491-8404. FAX (303) 491-8671.

Cornell University, Institute For the Study of Continents. 3122 Snee Hall, Ithaca, NY 14853-1504. (607) 255-3474.

Iowa Department of Natural Resources, Geological Survey Bureau. 123 N. Capitol Street, Iowa City, IA 52242. (319) 335-1575. FAX (319) 335-2754.

University of Washington, Quaternary Research Center. AK-60, Seattle, WA 98195. (206) 543-1166. FAX (206) 543-3936.

GEOPHYSICAL ENGINEERING

See: GEOTECHNICAL ENGINEERING

GEOPHYSICS

See also: GEOLOGY, MINERAL EXPLORATION, PHYSICAL GEOLOGY, PLATE TECTONICS, SEISMOLOGY

ABSTRACT SERVICES AND INDEXES

Bibliography and Index of Geology. American Geological Institute, 4220 King Street, Alexandria, VA 22302. (703) 379-2480. Fax (703) 379-7563. Monthly. $1295.00 per year. Also available online as GeoRef.

Bibliography of Economic Geology. Geosystems, Box 40, Didcot Oxon Ox11 9BX, England. Bimonthly. $150.00 per year. Also available online as GeoArchive.

Chemical Abstracts - Applied Chemistry and Chemical Engineering Sections. Chemical Abstracts Service, 2540 Olentangy River Road, Box 3012, Columbus, OH 43210. (800) 848-6538 or (614) 447-3600, FAX (614) 447-3713. Bi-weekly. $1410.00 per year. Also available online as CA (Chemical Abstracts).

Deep-sea Research. PART A: Oceanographic Research Papers. Deep-sea Research. Part B: Oceanographic Literature Review. Pergamon Press, Inc., Maxwell House, Fairview Park, Elmsford, NY 10523. (914) 592-7700. Fax (914) 592-3625. Twelve times per year. $1370 per year for Parts A and B.

Geological Abstracts. Elsevier - Geo Abstracts, Regency House, 34 Duke Street, Norwich NR3 3AP, England. Monthly. $760.00 per year. Also available online as GEOBASE.

Geological Society of America. ABSTRACTS WITH PROGRAMS. Geological Society of America. 3300 Penrose Place, P.O. Box 9140, Boulder, CO 80301-9140. (303) 447-2020. Abstracts and programs of the annual conference. Annual. $69.00.

Geophysics Abstracts. Pergamon Press, Inc., Maxwell House, Fairview Park, Elmsford, NY 10523. (914) 592-7700. Fax (914) 592-3625. Twelve times per year. $565.00 per year. Also available in microform.

Mineralogical Abstracts. Mineralogical Society and the Mineralogical Society of America, 41 Queen's Gate, London, SW7 5HR, England. Quarterly. $235.00 per year.

Oceanic Abstracts. Cambridge Scientific Abstracts, 7200 Wisconsin Avenue, Bethesda, MD 20814. (301) 961-6750. Fax (301) 961-6720. Bimonthly. $995.00 per year.

Science Citation Index. Institute for Scientific Information, 3501 Market Street, Philadelphia, PA 19104. (800) 523-1850 or (215) 386-0100. Inquire about price and availability.

General Science Index. H.W. Wilson Co., 950 University Avenue, Bronx, NY 10452. (800) 367-6770 or (212) 588-8400. Inquire about price and availability.

ANNUAL REVIEWS AND YEARBOOKS

Advances in Geophysics. Academic Press, Inc., 1250 Sixth Avenue, San Diego, CA 92101. (619) 231-0926. FAX (619) 699-6715. Irregular. Inquire about price and availability.

Annual Review and Earth and Planetary Sciences. Annual Reviews, Inc., 4139 El Camino Way, Palo Alto, CA 94306-0897. (415) 493-4400. Fax (415) 855-9815. Annual. $55.00 per year.

ASSOCIATIONS AND PROFESSIONAL SOCIETIES

American Association of Petroleum Geologists. P.O. Box 979, Tulsa, OK 74101. (918) 584-2555.

American Geological Institute. 4220 King Street, Alexandria, VA 22302. (703) 379-2480. Fax (703) 379-7563.

American Geophysical Union, 2000 Florida Avenue, N.W., Washington, DC 20009. (202) 462-6900.

American Institute of Professional Geologists. 7828 Vance Drive, Suite 103, Arvada, CO 80003. (303) 431-0831.

Association of Engineering Geologists. 323 Boston Post Road, Suite 2D, Sudbury, MA 01776. (508) 443-4639.

Geochemical Society. Department of Terrestrial Magnetism, 5241 Broad Branch Road, N.W., Washington, DC 20015. (202) 686-4370.

Geological Society of America. 3300 Penrose Place, P.O. Box 9140, Boulder, CO 80301-9140. (303) 447-2020.

Geoscience Information Society. American Geological Institute. 4220 King Street, Alexandria, VA 22302. (703) 379-2480. Fax (703) 379-7563.

Seismological Society of America. 201 Plaza Professional Building, El Cerrito, CA 94530. (510) 525-5474 or Fax (510) 525-7204

Society of Economic Geologists. P.O. Box 571, Golden, CO 80402. (303) 277-1716.

Society of Exploration Geophysicists. P.O. Box 702740, Tulsa, OK 74170. (918) 493-3516.

Society of Independent Professional Earth Scientists. 4925 Greenville Avenue, Suite 170, Dallas, TX 75206. (214) 363-1780.

BIBLIOGRAPHIES

Geologic Reference Sources: A Subject and Regional Bibliography of Publications and Maps in the Geological Sciences. Dederick C. Ward et al. Scarecrow Press, Inc., 52 Liberty Street, Box 4167, Metuchen, NJ 08840. (800) 537-7107 or (201) 548-8600. 1981. $49.50.

Information Sources in the Earth Sciences. David N. Wood, Joan E. Hardy, and Anthony P. Harvey, editors. Bowker-Saur/K.G. Saur. Distributed by R.R. Bowker, 121 Chanlon Road, New Providence, NJ 07974. (800) 521-8110 or (908) 464-6800. Second edition. 1989. $85.00.

Scientific and Technical Books and Serials in Print; An Index To Literature in Science and Technology. R.R. Bowker Co., 205 E. 42nd Street, New York, NY 10017. (800) 521-8110 or (212) 916-1600. 1992. $250.00. DICTIONARIES and BIOGRAPHICAL SOURCES

American Men and Women of Science: Physical and Biological Sciences. Eighteenth edition. R.R. Bowker Company, 245 West 17th Street, New York, NY 10011. (800) 521-8810 or (212) 916-1600. $750.00.

American Institute of Professional Geologists. Membership Directory. American Institute of Professional Geologists, 7828 Vance Drive, Suite 103, Arvada, CO 80003. (303) 431-0831. Annual. $18.00.

Association of Engineering Geologists. Membership directory. Association of Engineering Geologists, 323 Boston Post Road, Suite 2D, Sudbury, MA 01776. (508) 443-4639. Annual. $18.00.

Directory of Geoscience Departments: United States and Canada. American Geological Institute. 4220 King Street, Alexandria, VA 22302-1507. (703) 379-2480. Fax (703) 379-7563. Twenty-six edition. 1987. $22.00.

Geological Society of America. Membership Directory. Geological Society of America, 3300 Penrose Place, Boulder, CO 80301. (303) 447-2020. Annual. Available to members only.

Research Centers Directory. Gale Research, 835 Penobscot Building, Detroit, MI 48226-4094. (800) 877-4253 or (313) 961-2242. 17th edition, 1992. $400.00.

ENCYCLOPEDIAS AND DICTIONARIES

Concise Oxford Dictionary of the Earth Sciences. Ailsa Allaby and Michael Allaby, editors. Oxford University Press, Inc., 200 Madison Avenue, New York, NY 10016. (800) 334-4249 or (212) 679-7300. 1990. $42.95.

Dictionary of Geological Terms. American Geological Institute. Doubleday and Company, Inc., 245 Park Avenue, New York, NY 10017. (800) 645-6156 or (212) 953-4561. Third edition. 1984. $12.00 in paper.

Encyclopedia of Applied Physics. VCH Publishers, Inc., 220 East 23rd Street, New York, NY 10010. (800) 367-8249. Twenty volumes. 1991 -. $5950.00 for set.

Encyclopedia of Earth System Science. Academic Press, Inc., 1250 Sixth Street, San Diego, CA 92101. (619) 231-0926 or (800) 321-5068. Four volumes, 1992. $950.00.

Encyclopedia of Solid Earth Geophysics. David E. James, editor. Van Nostrand Reinhold, 115 Fifth Avenue, New York, NY 10003. (212) 254-3232 or (800) 926-2665. 1989. $124.95.

Glossary of Geology. Robert L. Bates and Julia A. Jackson. American Geological Institute, 4220 King Street, Alexandria, VA 22032. (703) 379-2480. Third edition. 1987. $75.00.

McGraw-Hill Encyclopedia of Science and Technology. McGraw-Hill Publishing Company, 1221 Avenue of the Americas, New York, NY 10020. (800) 262-4729 or (212) 512-3825. Seventh edition. 1992. $1900.00.

Magill's Survey of Science: Earth Science Series. Salem Press, Inc., P.O. Box 1097, Englewood Cliffs, NJ 07632. (800) 221-1592 or (201) 871-3700. Five volumes. 1990. $400.00 for the set.

GENERAL WORKS

Dynamic Earth: An Introduction To Physical Geology. B.J. Skinner and Stephen C. Porter. John Wiley and Sons, Inc., 605 Third Avenue, New York, NY 10158. (800) 526-5368 or (212) 850-6000. 1990. $51.95.

An Introduction To Applied & Environmental Geophysics. John M. Reynolds. John Wiley and Sons, Inc., 605 Third Avenue, New York, NY 10158. (800) 526-5368 or (212) 850-6000. 1996. Inquire for cost.

Introduction To Geophysical Exploration. P. Keary & M. Brooks. 2nd edition. Blackwell Scientific Publications, 238 Main Street, Cambridge, MA 02142. (617) 876-7000 or (800) 759-6102. FAX (617) 876-7022. 1991. $125.00 (hardbound), $55.00 (paperbound).

Introduction To Geophysical Fluid Dynamics. Benoit Cushman-Roisin. Prentice Hall, 113 Sylvan Avenue, Route 9W, Englewood Cliffs, NJ 07632. (201) 592-2000 or (800) 922-0579. 1994. $72.00.

The Solid Earth: An Introduction To Global Geophysics. C.M. Fowler. Cambridge University Press, 40 West 20th Street, New York, NY 10011-4211. (212) 924-3900 or (800) 872-7423. FAX (914) 937-4712. 1990. $39.95.

Theory of the Earth. Don anderson. Blackwell Scientific Publications, 238 Main Street, Cambridge, MA 02142. (617) 876-7000 or (800) 759-6102. FAX (617) 876-7022. 1990. $39.95.

GEOPHYSICS

Ency. of Physical Sciences and Engineering Info. Sources

HANDBOOKS AND MANUALS

Field Mapping For Geology Students. F. Ahmed and D.C. Almond. Unwin Hyman, Inc., 10 East 53rd Street, New York, NY 10022. (212) 207-7626 or (800) 242-7737. 1983. $18.95.

Geology in the Field. R.R. Compton. John Wiley and Sons, Inc., 605 Third Avenue, New York, NY 10158. (800) 526-5368 or (212) 850-6418. 1985. $37.95.

A Manual of Geology For Civil Engineers. John A. Pitts. World Scientific Publishing Company, Inc., 1060 Main Street, Unit B, River Edge, NJ 07661. (800) 227-7562 or (201) 487-9655. 1985. $38.00.

The Mapping of Geological Structures. K.R. McClay. John Wiley and Sons, Inc., 605 Third Avenue, New York, NY 10158. (800) 526-5368 or (212) 850-6000. 1988. $18.95.

ONLINE DATABASES AND CD-ROMS

Earth Sciences. U.S. Geological Survey, 12201 Sunrise Valley Drive, Reston, VA 22092-9998. (703) 648-4460. CD-ROM of earth science databases including the U.S. Geological Survey Library, Earth Science Data Directory, and GEOINDEX, citations to published geological maps. $350.00 per year with quarterly updates.

Geoarchive. Geosystems, P.O. Box 1024, Westminster, London, England, SW1 P 2JL. Citations to literature on geoscience, 1969 to present. Inquire as to online cost and availability.

Geobase. Elsevier - Geo Abstracts, Regency House, 34 Duke Street, Norwich NR3 3AP, England. Contains citations to the worldwide earth science literature from 1980 to date. Available on DIALOG, ORBIT online services. Inquire as to cost and availability.

GEOREF. American Geological Institute, 4220 King Street, Alexandria, VA 22302. (800) 336-4764 or (703) 379-2480. Geology and geosciences literature, 1785 to present for North America. Available on DIALOG, ORBIT online services. Also available on CD-ROM. Inquire as to cost and availability.

NTIS Bibliographic Database. National Technical Information Service, 5285 Port Royal Road, Springfield, VA 22161. (703) 487-4630. Broad coverage of government- sponsored research reports, 1964 to present. Available on BRS, DIALOG, ORBIT and STN online services. Also available on CD-ROM. Inquire as to cost and availability.

SCISEARCH. Institute for Scientific Information, 3501 Market Street, Philadelphia, PA 19104. (800) 523-1850 or (215) 386-0100. Broad multidisciplinary title and author index to the international literature of science and technology, 1974 to present. Available on DIALOG and ORBIT online services. Also available on CD-ROM. Inquire as to cost and availability.

PERIODICALS

Aapg Bulletin. American Association of Petroleum Geologists, P.O. Box 979, Tulsa, OK 7401. (918) 584-2555. Monthly. $120.00 per year.

American Journal of Science. Kline Geology Laboratory, Yale University, Box 6666, Yale Station, New Haven, CT 06511-8130. (203) 432-3131. Ten times per year. $40.00 per year.

Canadian Journal of Earth Sciences. National Research Council of Canada, Research Journals, Ottawa, Ont. K1A OR6, Canada. (613) 993-9084. Monthly. $112.00 per year.

Earth and Planetary Science Letters. Elsevier Science Publishing Company, Inc., 655 Avenue of the Americas, New York, NY 10010. (212) 989-5800. Twenty times per year. $945.00 per year.

Eos. American Geophysical Union, 2000 Florida Avenue, N.W., Washington, DC 20009. (202) 462-6900. Weekly. $190.00 per year.

Geological Society. JOURNAL. Geological Society of London. Geological Society Publishing House, Unit 7, Brassmill Enterprise Centre, Brassmill Lane, Bath, Avon BA1 3JN, England. 6 times per year. $432.00.

Geological Society of America Bulletin. P.O. Box 9140, 3300 Penrose Place, Boulder, CO 80301. (303) 447-2020. Fax (303) 447-1133. Monthly. $150.00.

Geology. Geological Society of America. P.O. Box 9140, 3300 Penrose Place, Boulder, CO 80301. (303) 447-2020. Fax (303) 447-1133. Monthly. $120.00 per year.

Geophysical Research Letters. American Geophysical Union, 2000 Florida Avenue, N.W., Washington, DC 20009. (202) 462-6900. Monthly. $480.00 per year.

Geophysics. Society of Exploration Geophysicists, P.O. Box 702740, Tulsa, OK 74170. (918) 493-3516. Monthly. $175.00 per year.

Geotimes. American Geological Institute, 4220 King Street, Alexandria, VA 22032. (703) 379-2480. Monthly. $24.95 per year.

JGR: Journal of Geophysical Research: Oceans. American Geophysical Union, 2000 Florida Avenue, N.W., Washington, DC 20009. (202) 462-6903. Monthly. $1545.00 per year.

JGR: Journal of Geophysical Research: Solid Earth. American Geophysical Union, 2000 Florida Avenue, N.W., Washington, DC 20009. (202) 462-6903. Monthly. $1385.00 per year.

JGR: Journal of Geophysical Research: Space Physics. American Geophysical Union, 2000 Florida Avenue, N.W., Washington, DC 20009. (202) 462-6903. Monthly. $1285.00 per year.

Journal of Geology. University of Chicago Press, 5720 Woodlawn Avenue, Chicago, IL 60637. (312) 753-3347. Bimonthly. $38.00 per year.

Journal of Structural Geology. Pergamon Journals, Inc., Maxwell House, Fairview Park, Elmsford, NY 10523. (914) 592-7700. Ten times per year. $375.00 per year.

Journal of Volcanology and Geothermal Research. Elsevier Science Publishing Company, Inc., 655 Avenue of the Americas, New York, NY 10010. (212) 989-5800. Monthly. $500.00 per year.

Reviews of Geophysics. American Geophysical Union, 2000 Florida Avenue, N.W., Washington, DC 20009. (202) 462-6900. Quarterly. $220.00 per year.

Rock Mechanics and Rock Engineering. Springer-Verlag New York, Inc., 175 Fifth Avenue, New York, NY 10010. (800) 526-7254 or (212) 460-1500. Quarterly. $122.00 per year.

Tectonics. American Geophysical Union, 2000 Florida Avenue, N.W., Washington, DC 20009. (202) 462-6900. Bimonthly. $230.00 per year to individuals.

Tectonophysics; An International Journal of Geotectonics and the Geology and Physics of the Interior of the Earth. Elsevier Science Publishing Company, Inc., 655 Avenue of the Americas, New York, NY 10010. (212) 989-5800. Fifty-six times per year. $1275.00 per year.

RESEARCH CENTERS AND INSTITUTES

California Institute of Technology. Seismological Laboratory. Pasadena, CA 91125. (818) 356-6914.

Carnegie Institution of Washington. Geophysical Laboratory. 5251 Broad Branch Road, NW, Washington, DC 20015. (202) 686-2410 or FAX (202) 686-2419.

National Science Foundation. Directorate for Geosciences. Division of Earth Sciences. Experimental and theoretical Geochemistry Program. 1800 G Street, NW, Washington, DC 20550. (202) 357-7498.

University of Alaska Fairbanks. Geophysical Institute. C.T. Elvey Building, Fairbanks, AK 99775-0800. (907) 474-7282. Fax (907) 474-7290.

University of California, Santa Barbara. Institute for Crustal Studies. Santa Barbara, CA 93105. (805) 961-8231.

GEOSTATIONARY SATELLITES

See: SATELLITES (ARTIFICIAL)

GEOTECHNICAL ENGINEERING

See also: CIVIL ENGINEERING, PHYSICAL GEOLOGY, ROCK MECHANICS

ABSTRACT SERVICES AND INDEXES

ASCE Combined Annual Combined Index. American Society of Civil Engineers, 345 East 47th Street, New York, NY 10017-2398. (212) 705-7520 or (800) 548-2723. Annual. $48.00.

ASCE Publications Information. 1966-present. American Society of Civil Engineers, 345 East 47th Street, New York, NY 10017-2398. (212) 705-7520 or (800) 548-2723. Bi-monthly. $160.00 per year for non-members.

Bibliography and Index of Geology. American Geological Institute, 4220 King Street, Alexandria, VA 22302. (703) 379-2480. Fax (703) 379-7563. Monthly. $1295.00 per year. Also available online as GeoRef.

Civil and Structural Engineering Abstracts, Cambridge Scientific Abstracts, 7200 Wisconsin Avenue, Bethesda, MD 20814-4823. (301) 961-6750. FAX (301) 961-6720. 1993 to present. Monthly. $385.00 per year. Topics covered include structural design, construction equipment and methods, civil defense and military engineering, surveying, highway engineering, maritime and port structures, materials, land reclamation, and soil mechanics.

Engineering Index Monthly. Engineering Information, Inc., Castle Point on the Hudson, Hoboken, NJ 07030. (800) 221-1044. FAX (212) 832-1857. Monthly. $2300.00 per year. Also available online as COMPENDEX, and also on CD-ROM. Covers chemical engineering, computers, electrical engineering, civil engineering, metals and mining, industrial management, and mechanical engineering.

Geodex Retrieval System For Geotechnical Abstracts. Geodex Retrieval Systems, 669 Broadway, Box 279, Sonoma, CA 95476. (707) 939-8476. FAX (707) 996-8734. 1970 to present. Three times a year. $645.00 per year includes Geotechnical Abstracts, as Geodex Systems-GRS.

Geological Abstracts. Elsevier - Geo Abstracts, Regency House, 34 Duke Street, Norwich NR3 3AP, England. Monthly. $760.00 per year. Also available online as GEOBASE.

Geological Society of America. ABSTRACTS WITH PROGRAMS. Geological Society of America. 3300 Penrose Place, P.O. Box 9140, Boulder, CO 80301-9140. (303) 447-2020. Abstracts and programs of the annual conference. Annual. $69.00.

Geotechnical Abstracts. Geodex Retrieval Systems, 669 Broadway, Box 279, Sonoma, CA 95476. (707) 939-8476. FAX (707) 996-8734. 1970 to present. Three times a year. $280.00 per year. Covers soil mechanics, foundation engineering, rock mechanics, and engineering geology.

International Journal of Rock Mechanics and Mining Sciences and Geomechanics Abstracts. Elsevier Science, 660 White Plains Rd., Tarrytown, NY 10591-5153. (914) 524-9200. FAX (914) 333-2444. Six times a year. $845 per year.

Mineralogical Abstracts. Mineralogical Society and the Mineralogical Society of America, 41 Queen's Gate, London, SW7 5HR, England. Quarterly. $235.00 per year.

Science Citation Index. Institute for Scientific Information, 3501 Market Street, Philadelphia, PA 19104. (215) 386-0100. FAX (215) 386-6362. Inquire about availability and cost. Also available on CD-ROM.

ANNUAL REVIEWS AND YEARBOOKS

Advances in Geophysics. Academic Press Inc., 6277 Sea Harbor Drive, Orlando, FL 32887. (800) 321-5068. Annual. Inquire for price.

Annual Review of Earth and Planetary Sciences. Annual Reviews Inc., 4139 El Camino Way, PO Box 10139, Palo Alto, CA 94303-0139. (800) 523-8635 or (415) 493-4400. FAX (415) 855-9815. Annual. $62.00 (1995 edition).

ASSOCIATIONS AND PROFESSIONAL SOCIETIES

American Association of Petroleum Geologists. Box 979, Tulsa, OK 74101. (918) 584-2555. FAX (918) 584-0469.

American Geological Institute. 4220 King Street, Alexandria, VA 22302. (703) 379-2480. Fax (703) 379-7563.

American Geophysical Union. 2000 Florida Avenue, N.W., Washington, DC 20009. (202) 462-6900. FAX (202) 328-0566.

American Institute of Professional Geologists. 7828 Vance Drive, Suite 103, Arvada, CO 80003. (303) 431-0831. FAX (303) 431-1332.

American Society of Civil Engineers. 345 East 47th Street, New York, NY 10017-2398. (212) 705-7520 or (800) 548-2723.

Association of Engineering Geologists. 323 Boston Post Road, Suite 2D, Sudbury, MA 01776. (508) 443-4639.

Geological Society of America. 3300 Penrose Place, PO Box 9140, Denver, CO 80301-9140. (303) 447-2020.

International Society for Rock Mechanics. c/o Laboratorio Nacional de Engenharia Civil, Avenida do Brasil 101, Lisbon Codex, Portugal. Telephone 1-8482131. FAX 1-897660.

DIRECTORIES AND BIOGRAPHICAL SOURCES

American Men & Women of Science, 1995-96. R.R. Bowker Staff, eds. 19th edition. 8 volumes. R.R. Bowker/Reed International Publishing Company, 121 Chanlon Road, New Providence, NJ 07974. (908) 464-6800 or (800) 521-8110. 1995. $850.00

Directory of Engineering Societies and Related Organizations. Gordon Davis. 13th edition. American Association of Engineering Societies, 1111 19th Street NW, Suite 608, Washington, DC 20036. (202) 296-2237 or (800) 658-8897. 1989. Inquire for price.

International Research Centers Directory. Gale Research, 835 Penobscot Building, Detroit, MI 48226-4094. (313) 961-2242. (800) 877-4253. FAX (313) 961-6083. 8th edition. 1995. $410.00.

Research Centers Directory. Gale Research, 835 Penobscot Building, Detroit, MI 48226-4094. (313) 961-2242. (800) 877-4253. FAX (313) 961-6083. $485.00.

Scientific and Technical Organizations and Agencies Directory. Gale Research, 835 Penobscot Building, Detroit, MI 48226-4094. (313) 961-2242. (800) 877-4253. FAX (313) 961-6083. 4th edition. 1996. $195.00.

Who's Who in Engineering. American Association of Engineering Societies, 1111 19th Street NW, Suite 608, Washington, DC 20036. (202) 296-2237 or (800) 658-8897. 8th edition. 1991. Inquire for price.

Who's Who in Technology. Gale Research, 835 Penobscot Building, Detroit, MI 48226-4094. (313) 961-2242. (800) 877-4253. FAX (313) 961-6083. 1995. $195.00.

ENCYCLOPEDIAS AND DICTIONARIES

Dictionary of Civil Engineering. V.N. Vazirani. South Asia Books, Box 502, Columbia, MO 65205. (314) 474-0116. FAX (314) 474-8124. 1992. $20.00.

The Encyclopedia of the Solid Earth Sciences. Philip Kearey, editor-in-chief. Blackwell Scientific Publications, 238 Main Street, Cambridge, MA 02142. (617) 876-7000 or (800) 759-6102. FAX (617) 876-7022. 1993. $60.00.

Encyclopedia of Physical Science and Technology. Robert A. Meyers, ed. 18 volumes. Academic Press Inc., 6277 Sea Harbor Drive, Orlando, FL 32887. (800) 321-5068. 1992. $2100.00

Glossary of Rock Engineering. Harith S. Faruqi. French & European Publications, Inc., Rockefeller Center Promenade, 610 Fifth Ave., New York, NY 10020-2479. (212) 581-8810. FAX (212) 265-1094. 1980. $75.00.

Thesaurus of Rock & Soil Mechanics Terms. J.P. Jenkins & A.M. Smith, eds. Franklin Book Company, Inc., 7804 Montgomery Ave., Elkins, PA 19117. (215) 635-5252. FAX (215) 635-6155. 1984. $37.00.

GENERAL WORKS

Geotechnical Engineering. Renato Lancelotta. A.A. Balkema/ Ashgate Publishing Co., Old Post Rd., Brookfield, VT 05036. (802) 276-3162. FAX (802) 276-3837. 1994. $95.00 (hardback) and $60.00 paperback.

Geotechnical Engineering: Emerging Trends in Design and Practice. K.R. Saxena, editor. A.A. Balkema/ Ashgate Publishing Co., Old Post Rd., Brookfield, VT 05036. (802) 276-3162. FAX (802) 276-3837. 1994. $60.00.

Geotechnical Engineering: Foundation Design. John N. Cernica. John Wiley and Sons, Inc., 605 Third Avenue, New York, NY 10158. (800) 526-5368 or (212) 850-6000. 1994. $75.95.

Geotechnical Engineering: Soil Mechanics. John N. Cernica. John Wiley and Sons, Inc., 605 Third Avenue, New York, NY 10158. (800) 526-5368 or (212) 850-6000. 1994. $77.95.

Principles of Geotechnical Engineering. Braja M. Das. Second edition. PWS Publishers, 20 Park Plaza, Boston, MA 02116. (617) 542-3377. FAX (617) 338-6134. 1990. $69.95.

HANDBOOKS AND MANUALS

Geotechnical Engineering Handbook. Harold Carter. Pentech Press, distributed by Chapman and Hall, 29 West 35th Street, New York, NY 10001. (800) 842-3636. FAX (212) 563-2269. 1983. Inquire for price and availability.

Practical Handbook For Underground Rock Mechanics. T.R. Stacey & C.H. Page. LPS Distribution Center, 52 LaBombard Rd., Lebanon, NH 03766. (603) 448-0037. FAX (603) 448-2576. 1986. $38.00.

ONLINE DATABASES AND CD-ROMS

Compendex Plus. Engineering Information, Inc., 345 East 47th Street, New York, NY 10017. (212) 705-7600 or (800) 221-

1044. Contains citations with abstracts to worldwide literature in engineering and technology, from 1970 to present. Available on online BRS,(800) 289-4277, DIALOG, (800) 334-2564, ORBIT (800) 456-7248, and STN International, FIZ Karlsruhe, P.O. Box 2465, W-7500, Karlsruhe 1, Germany, online services. Also available on CD-ROM. Inquire as to cost and availability.

CSA Engineering. Cambridge Scientific Abstracts, 7200 Wisconsin Avenue, Suite 601, Bethesda, MD 20814. (301) 961-6750 or (800) 843-7751. Contains citations and abstracts of international periodicals and research literature covering all fields of engineering and science and technology,including computer and information science, electronics, mechanical engineering, solid state materials, 1981 to present. Available on BRS,(800) 289-4277, online service.Inquire as to cost and availability.

Earth Sciences. U.S. Geological Survey, 12201 Sunrise Valley Drive, Reston, VA 22092-9998. (703) 648-4460. CD-ROM of earth science databases including the U.S. Geological Survey Library, Earth Science Data Directory, and GEOINDEX, citations to published geological maps. $350.00 per year with quarterly updates.

Geoarchive. Geosystems, P.O. Box 1024, Westminster, London, England, SW1 P 2JL. Citations to literature on geoscience, 1969 to present. Inquire as to online cost and availability.

Geobase. Elsevier - Geo Abstracts, Regency House, 34 Duke Street, Norwich NR3 3AP, England. Contains citations to the worldwide earth science literature from 1980 to date. Available on DIALOG, ORBIT online services. Inquire as to cost and availability.

Geomechanics Abstracts. University of London, Imperial College of Science and Technology, Rock Mechanics Information Service, Royal School of Mines, Prince Consort Road, London SW7 2BP, England. Telephone: 071-589 5111 or FAX 071-589 6806. Scope includes worldwide literature on engineering geology, hydrogeology, mining, rock mechanics, soil mechanics, and tunnelling, 1977 to present. Available on the ORBIT (800) 456-7248, online service. Inquire as to cost and availability.

GEOREF: Bibliography and Index of Geology. American Geological Institute, 4220 King Street, Alexandria, VA 22302. (703) 379-2480. Fax (703) 379-7563. Monthly. Inquire for price and availability.

ICONDA (CIB International Construction Database). Fraunhofer-Gesellschaft, Informationszentrum RAUM und BAU, Nobelstrasse, 12, D-7000, Stuttgart 80, Germany. Contains citations with abstracts to worldwide technical literature on construction and civil engineering, structural engineering, and engineering geology, from 1976 to present. Available on ORBIT, (800) 456-7248, and STN International, FIZ Karlsruhe, P.O. Box 2465, W-7500, Karlsruhe 1, Germany, online services. Inquire as to cost and availability.

NTIS Bibliographic Database. National Technical Information Service, 5285 Port Royal Road, Springfield, VA 22161. (703) 487-4929 or FAX (703) 321-8199. Broad coverage of government-sponsored science and technology research reports, 1964 to present. Available on BRS,(800) 289-4277, DIALOG, (800) 334-2564, ORBIT, (800) 456-7248, and STN International, FIZ Karlsruhe, P.O. Box 2465, W-7500, Karlsruhe 1, Germany, online services. Also available on CD-ROM. Inquire as to cost and availability.

SCISEARCH. Institute for Scientific Information, 3501 Market Street, Philadelphia, PA 19104. (800) 523-1850 or (215) 386-

0100. Broad multidisciplinary title and author index to the international literature of science and technology, 1974 to present. Available on DIALOG,(800) 334-2564, and ORBIT,(800) 456-7248, online services. Also available on CD-ROM. Inquire as to cost and availability.

PERIODICALS

American Journal of Science. Box 6666, Yale Station, New Haven, CT 06511-8130. (203) 432-3131. FAX (203) 432-5668. 1818 to present. Monthly except July and August. $40.00 per year.

Geological Magazine. Cambridge University Press, Journals Department, 40 West 20th Street, New York, NY 10011-4211. (212) 924-3900. 1864 to present. Bi-monthly. $263.00 per year.

Geological Society Journal. Geological Society Publishing House, Unit 7, Brassmill Enterprise Centre, Brassmill Lane, Bath BA1 3JN, England. Telephone 0225-445046. FAX 0225-442836. 1845 to present. Six times a year. $524.00.

Geotechnical News. BiTech Publishers Ltd., 173-11860 Hammersmith Way, Richmond BC V7A 5G1, Canada. (604) 277-4250. FAX (604) 277-8125. 1982 to present. Quarterly. $45.00.

Geotechnique International. Thomas Telford Services Ltd., Thomas Telford House, 1 Heron Quay, London E14 4JD, England. Telephone 071-987-6999. FAX 071-538-4101. 1948 to present. Quarterly. Inquire for price in U.S.A.

Geology. Geological Society of America, 3300 Penrose Place, PO Box 9140, Denver, CO 80301-9140. (303) 447-2020. 1973 to present. Monthly. $170.00 per year for non-members.

Journal of Geology. University of Chicago Press, Journals Division, 5720 S. Woodlawn Avenue, Chicago, IL 60637. (312) 753-3347. FAX (312) 753-0811. 1893 to present. Bi-monthly. $45.00 per year.

Journal of Geotechnical Engineering. American Society of Civil Engineers, Geotechnical Engineering Division, 345 East 47th Street, New York, NY 10017-2398. (212) 705-7520 or (800) 548-2723. 1956 to present. Monthly. $212.00 per year.

Journal of Structural Geology. Elsevier Science [journals], 660 White Plains Rd., Tarrytown, NY 10159-5153. (919) 524-9200. FAX (919) 333-2444. 1979 to present. Twelve times a year. $640.00.

Lithos. Elsevier Science Inc., Box 882, Madison Square Station, New York, NY 10159. (212) 989-5800. FAX (212) 633-3990. 1968 to present. Twelve times a year. $529.00.

Mountain Geologist. Rocky Mountain Association of Geologist, 5139 Raleigh Street, Denver, CO 80212-2609. 1964 to present. Quarterly. $15.00.

Professional Geologist. American Institute of Professional Geologists, 7828 Vance Drive, Suite 103, Arvada, CO 80003. (303) 431-0831. FAX (303) 431-1332. 1964 to present. Monthly. $25.00.

Rock Mechanics and Rock Engineering. Springer Verlag/Journal Fulfillment Services, Box 2485, Secaucus, NJ 07096-2491.

GEOSTATIONARY SATELLITES

Ency. of Physical Sciences and Engineering Info. Sources

(800) 777-4643. FAX (201) 348-4505. 1929 to present. Four times a year. $187.00 per year.

Tectonics. American Geophysical Union, 2000 Florida Avenue, N.W., Washington, DC 20009. (202) 462-6900. FAX (202) 328-0566. 1982 to present. Bi-monthly. $330.00 for non-members.

Tectonophysics. Elsevier Science Inc., Box 882, Madison Square Station, New York, NY 10159. (212) 989-5800. FAX (212) 633-3990. 1964 to present. 56 times a year. $2429.00.

RESEARCH CENTERS AND INSTITUTES

Geotechnical Engineering Center. University of Texas at Austin, Department of Civil Engineering, ECJ 9.227, Austin, TX 78712. (512) 471-4929.

Houston Advanced Research Center, Geotechnology Research Institute. 4802 Research Forest Drive, the Woodlands, TX 77381. (713) 363-7915. FAX (713) 363-7914.

Iowa State University, Spangler Geotechnical Laboratory. College of Engineering, Ames, IA 50011. (515) 294-7690. FAX (515) 294-7689.

Massachusetts Institute of Technology—remergence Laboratory. Department of Civil Engineering, Cambridge, MA 02139. (617) 253-3598. FAX (617) 253-6044.

University of Colorado—boulder, Center For Geotechnical Engineering. Campus Box 428, Boulder, CO 80309-0428. (303) 492-7315.

GEOTHERMAL ENGERGY

See also: GEOLOGY, GEOPHYSICS, VOLCANOLOGY

ABSTRACT SERVICES AND INDEXES

Applied Science and Technology Index; A Cumulative Subject Index To English Language Periodicals in the Fields of Aeronautics and Space Science, Computer Technology, Chemistry, Construction Industry, Energy and Related Areas. H.W. Wilson Co., 950 University Avenue, Bronx, NY 10452. (800) 367-6770 or (212) 588-8400. FAX (718) 590-1617. From 1958 to present. Monthly. Inquire about cost and availability. Also available on CD-ROM and online.

Bibliography and Index of Geology. American Geological Institute, 4220 King Street, Alexandria, VA 22302. (703) 379-2480. Fax (703) 379-7563. Monthly. $1295.00 per year. Also available online as GeoRef.

Energy Research Abstracts. 1976-present. U.S. Department of Energy, office of Scientific and Technical Information, Box 62, Oak Ridge, TN 37831. (615) 574-0733. Subscriptions: Superintendent of Documents, U.S. Government Printing office, Box 371954, Pittsburgh, PA 15250-7954. (202) 783-3238. FAX (202) 512-2233. Monthly. $164.00 per year. Abstracts all the scientific and technical reports, journal articles, conference proceedings, patents, theses, and monographs sponsored by the U.S. Energy Research and Development Administration.

Geological Abstracts. Elsevier - Geo Abstracts, Regency House, 34 Duke Street, Norwich NR3 3AP, England. Monthly. $760.00 per year. Also available online as GEOBASE.

Geological Society of America. ABSTRACTS WITH PROGRAMS. Geological Society of America. 3300 Penrose Place, P.O. Box 9140, Boulder, CO 80301-9140. (303) 447-2020. Abstracts and programs of the annual conference. Annual. $69.00.

Mineralogical Abstracts. Mineralogical Society and the Mineralogical Society of America, 41 Queen's Gate, London, SW7 5HR, England. Quarterly. $235.00 per year.

Oceanic Abstracts. Cambridge Scientific Abstracts, 7200 Wisconsin Avenue, Bethesda, MD 20814. (301) 961-6750. Fax (301) 961-6720. Bimonthly. $995.00 per year.

Science Citation Index. Institute for Scientific Information, 3501 Market Street, Philadelphia, PA 19104. (215) 386-0100. FAX (215) 386-6362. Inquire about availability and cost. Also available on CD-ROM.

ANNUAL REVIEWS AND YEARBOOKS

Annual Review of Earth and Planetary Sciences. Annual Reviews Inc., 4139 El Camino Way, PO Box 10139, Palo Alto, CA 94303-0139. (800) 523-8635 or (415) 493-4400. FAX (415) 855-9815. Annual. $62.00 (1995 edition).

ASSOCIATIONS AND PROFESSIONAL SOCIETIES

American Association of Petroleum Geologists. Box 979, Tulsa, OK 74101. (918) 584-2555. FAX (918) 584-0469.

American Geological Institute. 4220 King Street, Alexandria, VA 22302. (703) 379-2480. Fax (703) 379-7563.

American Geophysical Union. 2000 Florida Avenue, N.W., Washington, DC 20009. (202) 462-6900. FAX (202) 328-0566.

American Institute of Professional Geologists. 7828 Vance Drive, Suite 103, Arvada, CO 80003. (303) 431-0831. FAX (303) 431-1332.

Association of Engineering Geologists. 323 Boston Post Road, Suite 2D, Sudbury, MA 01776. (508) 443-4639.

Geological Society of America. 3300 Penrose Place, PO Box 9140, Denver, CO 80301-9140. (303) 447-2020.

Geothermal Resources Council. PO Box 1350, Davis, CA 95617-1350. (916) 758-2360. FAX (916) 758-2839.

SEPM (formerly Society For Economic Paleontologists and Mineralogists). 1731 E 71st Street, Tulsa, OK 74136-5108. FAX (918) 493-2093.)

DIRECTORIES AND BIOGRAPHICAL SOURCES

American Men & Women of Science, 1995-96. R.R. Bowker Staff, eds. 19th edition. 8 volumes. R.R. Bowker/Reed International Publishing Company, 121 Chanlon Road, New Providence, NJ 07974. (908) 464-6800 or (800) 521-8110. 1995. $850.00

American Institute of Professional Geologists Directory. 7828 Vance Drive, Suite 103, Arvada, CO 80003. (303) 431-0831. FAX (303) 431-1332. Annual. Inquire for price and availability.

Directory of Engineering Societies and Related Organizations. Gordon Davis. 13th edition. American Association of Engineering Societies, 1111 19th Street NW, Suite 608, Washington, DC 20036. (202) 296-2237 or (800) 658-8897. 1989. Inquire for price.

Geological Society of America Membership Directory. 3300 Penrose Place, PO Box 9140, Denver, CO 80301-9140. (303) 447-2020. FAX (303) 447-1133. Inquire for price and availability.

International Research Centers Directory. Gale Research, 835 Penobscot Building, Detroit, MI 48226-4094. (313) 961-2242. (800) 877-4253. FAX (313) 961-6083. 8th edition. 1995. $410.00.

Research Centers Directory. Gale Research, 835 Penobscot Building, Detroit, MI 48226-4094. (313) 961-2242. (800) 877-4253. FAX (313) 961-6083. $485.00.

Scientific and Technical Organizations and Agencies Directory. Gale Research, 835 Penobscot Building, Detroit, MI 48226-4094. (313) 961-2242. (800) 877-4253. FAX (313) 961-6083. 4th edition. 1996. $195.00.

Who's Who in Technology. Gale Research, 835 Penobscot Building, Detroit, MI 48226-4094. (313) 961-2242. (800) 877-4253. FAX (313) 961-6083. 1995. $195.00.

ENCYCLOPEDIAS AND DICTIONARIES

The Encyclopedia of the Solid Earth Sciences. Philip Kearey, editor-in-chief. Blackwell Scientific Publications, 238 Main Street, Cambridge, MA 02142. (617) 876-7000 or (800) 759-6102. FAX (617) 876-7022. 1993. $60.00.

GENERAL WORKS

Geothermal Direct Use Engineering and Design Guidebook. Paul J. Lienau and Ben C. Lunis. Oregon Institute of Technology, Geo-Heat Center, 3201 Campus Drive, Klamath Falls, OR 97601. (503) 885-1750. FAX (503) 885-1754. 1991. $20.00.

Geothermal Energy. Mary H. Dickson and Mario Fanelli, editors. John Wiley and Sons, Inc., 605 Third Avenue, New York, NY 10158. (800) 526-5368 or (212) 850-6000. 1995. $31.95.

Geothermal Energy: A Source Guide. Gordon Press Publishers, PO Box 459, Bowling Green Station, New York, NY 10004. (718) 624-8419. 1991. $75.00.

Geothermal Energy: Barriers To the Use of Geothermal Heat Pumps. Victor S. Rezendes. Diane Publishing Company, 600 Upland Avenue, Upland, PA 19015. (610) 499-7415. FAX (610) 499-7429. 1994. $45.00.

Geothermal Reservoir Engineering. Ender Okandan, editor. Kluwer Academic Publishers, P.O. Box 358, Accord Station,

Hingham, MA 02018-0358. (617) 871-6000. FAX (617) 871-6528. 1988. $125.50.

Tapping the Earth's Natural Heat. Wendel A. Duffield, et al. United States Geological Survey/ order from Superintendent of Documents, U.S. Government Printing office, Box 371954, Pittsburgh, PA 15250-7954. (202) 783-3238. FAX (202) 512-2233. 1994. Inquire for price and availability.

ONLINE DATABASES AND CD-ROMS

CA Search. Chemical Abstracts Service, P.O. Box 3012, Columbus, OH 43210-0012. (614) 447-3600. (800) 848-6533. FAX (614) 447-3709. Very comprehensive guide to worldwide chemical literature and related fields, 1972 to present. Available on BRS,(800) 289-4277, DIALOG, (800) 334-2564, ORBIT (800) 456-7248, and STN International, FIZ Karlsruhe, P.O. Box 2465, W-7500, Karlsruhe 1, Germany, online services. Inquire as to cost and availability.

Compendex Plus. Engineering Information, Inc., 345 East 47th Street, New York, NY 10017. (212) 705-7600 or (800) 221-1044. Contains citations with abstracts to worldwide literature in engineering and technology, from 1970 to present. Available on online BRS,(800) 289-4277, DIALOG, (800) 334-2564, ORBIT (800) 456-7248, and STN International, FIZ Karlsruhe, P.O. Box 2465, W-7500, Karlsruhe 1, Germany, online services. Also available on CD-ROM. Inquire as to cost and availability.

Dissertation Abstracts. University Microfilms International, 300 North Zeeb Road, Ann Arbor, MI 48106. (800) 521-0600 or (313) 761-4700. Scope includes virtually all doctoral dissertations accepted at accredited American institutions from 1861 to present in 252 subject areas. Available on BRS,(800) 289-4277, DIALOG,(800) 334-2564, and OCLC EPIC,(800) 848-5878, online services. Also available on CD-ROM. Inquire as to cost and availability.

Earth Sciences. U.S. Geological Survey, 12201 Sunrise Valley Drive, Reston, VA 22092-9998. (703) 648-4460. CD-ROM of earth science databases including the U.S. Geological Survey Library, Earth Science Data Directory, and GEOINDEX, citations to published geological maps. $350.00 per year with quarterly updates.

Geoarchive. Geosystems, P.O. Box 1024, Westminster, London, England, SW1 P 2JL. Citations to literature on geoscience, 1969 to present. Inquire as to online cost and availability.

Geobase. Elsevier - Geo Abstracts, Regency House, 34 Duke Street, Norwich NR3 3AP, England. Contains citations to the worldwide earth science literature from 1980 to date. Available on DIALOG, ORBIT online services. Inquire as to cost and availability.

GEOREF: Bibliography and Index of Geology. American Geological Institute, 4220 King Street, Alexandria, VA 22302. (703) 379-2480. Fax (703) 379-7563. Monthly. Inquire for price and availability.

NTIS Bibliographic Database. National Technical Information Service, 5285 Port Royal Road, Springfield, VA 22161. (703) 487-4929 or FAX (703) 321-8199. Broad coverage of government-sponsored science and technology research reports, 1964 to present. Available on BRS,(800) 289-4277, DIALOG, (800) 334-2564, ORBIT, (800) 456-7248, and STN International, FIZ Karlsruhe, P.O. Box 2465, W-7500, Karlsruhe 1, Germany,

GEOTHERMAL ENERGY

Ency. of Physical Sciences and Engineering Info. Sources

online services. Also available on CD-ROM. Inquire as to cost and availability.

SCISEARCH. Institute for Scientific Information, 3501 Market Street, Philadelphia, PA 19104. (800) 523-1850 or (215) 386-0100. Broad multidisciplinary title and author index to the international literature of science and technology, 1974 to present. Available on DIALOG,(800) 334-2564, and ORBIT,(800) 456-7248, online services. Also available on CD-ROM. Inquire as to cost and availability.

PERIODICALS

American Journal of Science. Box 6666, Yale Station, New Haven, CT 06511-8130. (203) 432-3131. FAX (203) 432-5668. 1818 to present. Monthly except July and August. $40.00 per year.

Economic Geology and the Bulletin of the Society of Economic Geologists. Economic Geology Publishing Company, 101 Vowell Hall, University of Texas, El Paso, TX 79968. FAX (915) 544-7416. 1906 to present. Eight times a year. $65.00.

Geochimica Et Cosmochimica Acta. Elsevier Science [journals], 660 White Plains Rd., Tarrytown, NY 10159-5153. (919) 524-9200. FAX (919) 333-2444. 1950 to present. Twenty-four times a year. $895.00.

Geological Magazine. Cambridge University Press, Journals Department, 40 West 20th Street, New York, NY 10011-4211. (212) 924-3900. 1864 to present. Bi-monthly. $263.00 per year.

Geological Society Journal. Geological Society Publishing House, Unit 7, Brassmill Enterprise Centre, Brassmill Lane, Bath BA1 3JN, England. Telephone 0225-445046. FAX 0225-442836. 1845 to present. Six times a year. $524.00.

Geological Society of America Bulletin. 3300 Penrose Place, PO Box 9140, Denver, CO 80301-9140. (303) 447-2020. 1888 to present. Monthly. $205.00 per year.

Geology. Geological Society of America, 3300 Penrose Place, PO Box 9140, Denver, CO 80301-9140. (303) 447-2020. 1973 to present. Monthly. $170.00 per year for non-members.

Geothermics. Elsevier Science Inc., Box 882, Madison Square Station, New York, NY 10159. (212) 989-5800. FAX (212) 633-3990. 1972 to present. Six times a year. $425.00 per year.

Journal of Geology. University of Chicago Press, Journals Division, 5720 S. Woodlawn Avenue, Chicago, IL 60637. (312) 753-3347. FAX (312) 753-0811. 1893 to present. Bi-monthly. $45.00 per year.

Journal of Volcanology and Geothermal Energy. Elsevier Science Inc., Box 882, Madison Square Station, New York, NY 10159. (212) 989-5800. FAX (212) 633-3990. 1976 to present. 24 times a year. $1155.00.

RESEARCH CENTERS AND INSTITUTES

Geothermal Resources Council. PO Box 1350, Davis, CA 95617-1350. (916) 758-2360. FAX (916) 758-2839.

Oregon Institute of Technology, Geo-heat Center. 3201 Campus Drive, Klamath Falls, OR 97601. (503) 885-1750. FAX (503) 885-1754.

Pacific International Center For High Technology Research. 711 Kapiolani Blvd., Suite 200, Honolulu, HI 96813. (808) 533-7500. FAX (808) 538-0677.

University of California, Institute of Geophysics and Planetary Physics. UC-San Diego, A-025, La Jolla, CA 92093. (619) 534-2870. STandARDS

Annual Book of Astm Standards, Vol. 12.02: NUCLEAR, SOLAR, and GEOtheRMAL ENERGY. American Society for Testing and Materials (ASTM), 1916 Race Street, Philadelphia, PA 19103. (215) 299-5585. Issued annually, but individual standards are not always updated. Check latest year.

GIANT STARS

See: STARS

GLACIOLOGY

See also: GEOLOGY, GEOMORPHOLOGY, IGNEOUS ROCKS, METAMORPHIC ROCKS, PHYSICAL GEOLOGY, SEDIMENTARY ROCKS

ABSTRACT SERVICES AND INDEXES

Bibliography and Index of Geology. American Geological Institute, 4220 King Street, Alexandria, VA 22302. (703) 379-2480. Fax (703) 379-7563. Monthly. $1295.00 per year. Also available online as GeoRef.

Current Contents: Physical, Chemical and Earth Sciences. Institute for Scientific Information, 3501 Market Street, Philadelphia, PA 19104. (215) 386-0100. FAX (215) 386-6362. Weekly. $360.00 per year.

Geological Abstracts. Elsevier - Geo Abstracts, Regency House, 34 Duke Street, Norwich NR3 3AP, England. Monthly. $760.00 per year. Also available online as GEOBASE.

Geological Society of America. ABSTRACTS WITH PROGRAMS. Geological Society of America. 3300 Penrose Place, P.O. Box 9140, Boulder, CO 80301-9140. (303) 447-2020. Abstracts and programs of the annual conference. Annual. $69.00.

Meteorological and Geoastrophysical Abstracts. American Meteorological Society, c/o Inforonics Inc., 550 Newtown Road, Littleton, MA 01460. (508) 486-8976. FAX (508) 486-0027. 1950 to present. Monthly. $950.00 per year. Current abstracts of books, reports, research papers, and miscellaneous literature on environmental sciences, meteorology, astrophysics, hydrology, glaciology, and physical oceanography.

Science Citation Index. Institute for Scientific Information, 3501 Market Street, Philadelphia, PA 19104. (215) 386-0100. FAX (215) 386-6362. Inquire about availability and cost. Also available on CD-ROM.

ASSOCIATIONS AND PROFESSIONAL SOCIETIES

American Geological Institute. 4220 King Street, Alexandria, VA 22302. (703) 379-2480. Fax (703) 379-7563.

American Geophysical Union. 2000 Florida Avenue, N.W., Washington, DC 20009. (202) 462-6900. FAX (202) 328-0566.

American Institute of Professional Geologists. 7828 Vance Drive, Suite 103, Arvada, CO 80003. (303) 431-0831. FAX (303) 431-1332.

Association of Engineering Geologists. 323 Boston Post Road, Suite 2D, Sudbury, MA 01776. (508) 443-4639.

Geological Society of America. 3300 Penrose Place, PO Box 9140, Denver, CO 80301-9140. (303) 447-2020.

DIRECTORIES AND BIOGRAPHICAL SOURCES

American Men & Women of Science, 1995-96. R.R. Bowker Staff, eds. 19th edition. 8 volumes. R.R. Bowker/Reed International Publishing Company, 121 Chanlon Road, New Providence, NJ 07974. (908) 464-6800 or (800) 521-8110. 1995. $850.00

American Institute of Professional Geologists Membership Directory. 7828 Vance Drive, Suite 103, Arvada, CO 80003. (303) 431-0831. FAX (303) 431-1332. Annual. Inquire for price and availability.

Association of Engineering Geologists Directory. 323 Boston Post Road, Suite 2D, Sudbury, MA 01776. (508) 443-4639. Annual. Inquire for price and availability.

Geological Society of America Membership Directory. 3300 Penrose Place, PO Box 9140, Denver, CO 80301-9140. (303) 447-2020. Annual. Inquire for cost and availability.

International Research Centers Directory. Gale Research, 835 Penobscot Building, Detroit, MI 48226-4094. (313) 961-2242. (800) 877-4253. FAX (313) 961-6083. 8th edition. 1995. $410.00

Research Centers Directory. Gale Research, 835 Penobscot Building, Detroit, MI 48226-4094. (313) 961-2242. (800) 877-4253. FAX (313) 961-6083. $485.00.

Scientific and Technical Organizations and Agencies Directory. Gale Research, 835 Penobscot Building, Detroit, MI 48226-4094. (313) 961-2242. (800) 877-4253. FAX (313) 961-6083. 4th edition. 1996. $195.00.

ENCYCLOPEDIAS AND DICTIONARIES

Glossary of Geology. Robert Bates, et al. Reprint of third edition. American Geological Institute, 4220 King Street, Alexandria, VA 22302-1507. (703) 379-2480. FAX (301) 379-7563. 1990. $225.00.

Elsevier's Dictionary of Glaciology: in English, Russian, French & German (with Definitions in English & Russian). V.M. Kotlyakov and N.A. Smolyarova. Elsevier Science Publishing Company, Inc., 655 Avenue of the Americas, New York, NY 10010. (212) 989-5800. 1991. $114.50

The Encyclopedia of the Solid Earth Sciences. Philip Kearey, editor-in-chief. Blackwell Scientific Publications, 238 Main Street, Cambridge, MA 02142. (617) 876-7000 or (800) 759-6102. FAX (617) 876-7022. 1993. $60.00.

Encyclopedia of Physical Science and Technology. Robert A. Meyers, ed. 18 volumes. Academic Press Inc., 6277 Sea Harbor Drive, Orlando, FL 32887. (800) 321-5068. 1992. $2100.00.

McGraw-Hill Dictionary of Scientific and Technical Terms. Sybil P. Parker, ed. 5th edition. McGraw-Hill Publishing Company, 1221 Avenue of the Americas, New York, NY 10020. (800) 262-4729 or (212) 512-3825. 1993. $110.50.

GENERAL WORKS

Glaciers of North America: A Field Guide. Sue A. Ferguson. Fulcrum Publishing, 250 Indiana Street, Suite 350, Golden, CO 80401. (800) 992-2908. FAX (303) 279-7111. 1992. $14.95.

Living Ice: Understanding Glaciers and Glaciation. Robert P. Sharp. Cambridge University Press, 40 West 20th Street, New York, NY 10011-4211. (212) 924-3900 or (800) 872-7423. FAX (914) 937-4712. 1991. $17.95.

Modern Glacial Environments: Processes, Dynamics & Sediments. Menzies. Elsevier Science Publishing Company, Inc., 655 Avenue of the Americas, New York, NY 10010. (212) 989-5800. 1994. $115.01 (hardback) or $42.01 (paperback).

The Physics of Glaciers. W.S.B. Paterson. Third edition. Elsevier Science Publishing Company, Inc., 655 Avenue of the Americas, New York, NY 10010. (212) 989-5800. 1994. $110.00.

ONLINE DATABASES AND CD-ROMS

Earth Sciences. U.S. Geological Survey, 12201 Sunrise Valley Drive, Reston, VA 22092-9998. (703) 648-4460. CD-ROM of earth science databases including the U.S. Geological Survey Library, Earth Science Data Directory, and GEOINDEX, citations to published geological maps. $350.00 per year with quarterly updates.

Geoarchive. Geosystems, P.O. Box 1024, Westminster, London, England, SW1 P 2JL. Citations to literature on geoscience, 1969 to present. Inquire as to online cost and availability.

Geobase. Elsevier - Geo Abstracts, Regency House, 34 Duke Street, Norwich NR3 3AP, England. Contains citations to the worldwide earth science literature from 1980 to date. Available on DIALOG, ORBIT online services. Inquire as to cost and availability.

GEOREF: Bibliography and Index of Geology. American Geological Institute, 4220 King Street, Alexandria, VA 22302. (703) 379-2480. Fax (703) 379-7563. Monthly. Inquire for price and availability.

NTIS Bibliographic Database. National Technical Information Service, 5285 Port Royal Road, Springfield, VA 22161. (703) 487-4929 or FAX (703) 321-8199. Broad coverage of government-sponsored science and technology research reports, 1964 to present. Available on BRS, (800) 289-4277, DIALOG, (800) 334-2564, ORBIT, (800) 456-7248, and STN International, FIZ Karlsruhe, P.O. Box 2465, W-7500, Karlsruhe 1, Germany,

GIANT STARS

Ency. of Physical Sciences and Engineering Info. Sources

online services. Also available on CD-ROM. Inquire as to cost and availability.

SCISEARCH. Institute for Scientific Information, 3501 Market Street, Philadelphia, PA 19104. (800) 523-1850 or (215) 386-0100. Broad multidisciplinary title and author index to the international literature of science and technology, 1974 to present. Available on DIALOG,(800) 334-2564, and ORBIT,(800) 456-7248, online services. Also available on CD-ROM. Inquire as to cost and availability.

PERIODICALS

American Journal of Science. Box 6666, Yale Station, New Haven, CT 06511-8130. (203) 432-3131. FAX (203) 432-5668. 1818 to present. Monthly except July and August. $40.00 per year.

Geological Magazine. Cambridge University Press, Journals Department, 40 West 20th Street, New York, NY 10011-4211. (212) 924-3900. 1864 to present. Bi-monthly. $263.00 per year.

Geological Society Journal. Geological Society Publishing House, Unit 7, Brassmill Enterprise Centre, Brassmill Lane, Bath BA1 3JN, England. Telephone 0225-445046. FAX 0225-442836. 1845 to present. Six times a year. $524.00.

Geological Society of America Bulletin. 3300 Penrose Place, PO Box 9140, Denver, CO 80301-9140. (303) 447-2020. 1888 to present. Monthly. $205.00 per year.

Geology. Geological Society of America, 3300 Penrose Place, PO Box 9140, Denver, CO 80301-9140. (303) 447-2020. 1973 to present. Monthly. $170.00 per year for non-members.

Glaciological Data. World Data Center A for Glaciology (Snow & Ice), National Snow and Ice Data Center, CIRES, Campus Box 449, University of Colorado, Boulder, CO 80309. (303) 492-5171. FAX (303) 492-2468. 1977 to present. Irregular. Inquire for price and availability.

Glaciology and Quaternary Geology. Kluwer Academic Publishers, P.O. Box 358, Accord Station, Hingham, MA 02018-0358. (617) 871-6000. 1985 to present. Irregular. Price varies.

Journal of Geology. University of Chicago Press, Journals Division, 5720 S. Woodlawn Avenue, Chicago, IL 60637. (312) 753-3347. FAX (312) 753-0811. 1893 to present. Bi-monthly. $45.00 per year.

Sedimentary Geology. Elsevier Science Inc., Box 882, Madison Square Station, New York, NY 10159. (212) 989-5800. FAX (212) 633-3990. 1967 to present. 24 times a year. $1187.00.

Sedimentology. Blackwell Scientific Publishing, Inc., 3 Cambridge Court, Suite 208, Cambridge, MA 02142. (617) 225-0401. 1952 to present. Six times a year. $434.00.

RESEARCH CENTERS AND INSTITUTES

Foundation For Glacier and Environmental Research, Inc. 514 E. First Street, Moscow, ID 83843.

University of Alaska Fairbanks, Alaska Quaternary Center (aqc). University of Alaska Museum, 907 Yukon Drive, Fairbanks, AK 99775-1200. (907) 474-7817. FAX (907) 474-5469.

University of Colorado—boulder, Institute of Arctic and Alpine Research (instaar). Campus Box 450, Boulder, CO 80309. (303) 492-6387.

University of Idaho, Juneau Icefield Research Program (jirp). Glaciological and Arctic Sciences Institute, Moscow, ID 83843. (208) 885-6192. FAX (208) 885-5724.

University of New Hampshire, Glacier Research Group. Durham, NH 03824. (603) 862-3146.

University of Washington—madison, Geophysical and Polar Research Center. Weeks Hall, 1215 W. Dayton Street, Madison, WI 53706-1692. (608) 262-1921. FAX (608) 262-0693.

World Data Center A For Glaciology (snow & Ice). National Snow and Ice Data Center, CIRES, Campus Box 449, University of Colorado, Boulder, CO 80309. (303) 492-5171. FAX (303) 492-2468.

GLASS

See also: CERAMIC ENGINEERING, CERAMICS, MATERIALS SCIENCE, OPTICS, SILICON

ABSTRACT SERVICES AND INDEXES

Applied Science and Technology Index; A Cumulative Subject Index To English Language Periodicals in the Fields of Aeronautics and Space Science, Computer Technology, Chemistry, Construction Industry, Energy and Related Areas. H.W. Wilson Co., 950 University Avenue, Bronx, NY 10452. (800) 367-6770 or (212) 588-8400. FAX (718) 590-1617. From 1958 to present. Monthly. Inquire about cost and availability. Also available on CD-ROM and online.

CA Selects: Ceramic Materials (journals). Chemical Abstracts Service, 2450 Olentagy River Road, Box 3012, Columbus, OH 43210-0012. (614) 447-3600. FAX (614) 447-3713. 1988 to present. Semi-weekly. $210.00 per year.

Ceramic Abstracts. American Ceramic Society, Inc., 735 Ceramic Place, Westerville, OH 43081. (614) 890-4700. FAX (614) 899-6109. Bi-monthly. 1922 to present. $138.00 per year for non-members. Also available on CD-ROM, and online through DIALOG, ORBIT, and STN.

Current Contents: Engineering, Technology, and Applied Sciences. Institute for Scientific Information, 3501 Market Street, Philadelphia, PA 19104. (215) 386-0100. FAX (215) 386-6362. 1970 to present. Weekly. $442.00 per year.

Engineered Materials Abstracts. ASM (American Society of Metals) International, Materials Information, Materials Park, OH 44073. (216) 338-5151 or FAX (216) 338-4634. Covers literature on technical developments in polymer, ceramic, and composite materials and engineering. 1986 to present. Monthly. $1200.00 per year. Also available on CD-ROM and online. Covers published materials on technical developments in polymer, ceramic, and composite materials in an engineering environment.

Engineering Index Monthly. Engineering Information, Inc., Castle Point on the Hudson, Hoboken, NJ 07030. (800) 221-1044. FAX (212) 832-1857. Monthly. $2300.00 per year. Also available online as COMPENDEX, and also on CD-ROM. Covers chemical engineering, computers, electrical engineering, civil engineering, metals and mining, industrial management, and mechanical engineering.

Materials Science and Engineering Abstracts, Cambridge Scientific Abstracts, 7200 Wisconsin Avenue, Bethesda, MD 20814-4823. (301) 961-6750. FAX (301) 961-6720. 1993 to present. Monthly. $385.00 per year. Focuses on mechanical and physical properties of materials and commercial or industrial applications for materials, methods for strength testing, effects of vibration and other stresses, corrosion and protective coatings, storage and handling, ceramics, composites, metals, wood, plastics, and polymers.

Materials Science Citation Index. Institute for Scientific Information, 3501 Market Street, Philadelphia, PA 19104. (215) 386-0100. FAX (215) 386-6362. Bi-monthly. $975.00 per year. Also available on CD-ROM.

Physics Abstracts. Institute of Electrical Engineers, Michael Faraday House, Six Hill Way, Stevenage, Herts, SG1 2AY, England. Distributed by IEEE, 445 Hoes Lane, Piscataway, NJ 08854. (908) 562-5549. 1898 to present. Monthly. $2700.00 per year. Also available online as INSPEC.

Science Citation Index. Institute for Scientific Information, 3501 Market Street, Philadelphia, PA 19104. (215) 386-0100. FAX (215) 386-6362. Inquire about availability and cost. Also available on CD-ROM.

World Ceramics Abstracts. Ceram Research Ltd., Queens Road, Penkhull, Stoke-on-Trent ST4 7LQ, England. 1958 to present. Monthly. Inquire for price and availability. Also available online through ORBIT.

ASSOCIATIONS AND PROFESSIONAL SOCIETIES

American Ceramic Society. 735 Ceramic Place, Westerville, OH 43081. (614) 890-4700. FAX (614) 899-6109.

American Society For Nondestructive Testing. 1711 Arlington Lane, Box 28518, Columbus, OH 43228-0158. (614) 274-6003. FAX (614) 274-6899.

American Society For Testing and Materials. 1916 Race Street, Philadelphia, PA 19103. (215) 299-5585.

Ceramic Manufacturers Association. 1100-H Brandywine Blvd., PO Box 2188, Zanesville, OH 43702-2188. (614) 452-4541.

Materials Properties Council. 345 E. 47th Street, New York, NY 10017. (212) 705-7693.

Materials Research Society. 9800 McKnight Road, Pittsburgh, PA 15237. (412) 367-3003. FAX (412) 367-4373.

National Institute of Ceramic Engineers. 735 Ceramic Place, Westerville, OH 43081. (614) 890-4700.

National Materials Advisory Board. 2101 Constitution Avenue NW, Washington, DC 20418. (202) 334-3505.

Society For the Advancement of Material and Process Engineering (sampe). PO Box 889, Covina, CA 91722. (818) 331-0616.

BIBLIOGRAPHIES

A Bibliography of Ceramics and Glass. Larry L. Hench & B.A. McEldowney. American Ceramic Society, 735 Ceramic Place, Westerville, OH 43081. (614) 890-4700. FAX (614) 899-6109. 1976. $15.00.

DIRECTORIES AND BIOGRAPHICAL SOURCES

American Glass Review Glass Factory Directory. Doctorow Communications Inc., 1033 Clifton Avenue, Clifton, NJ 07013. (201) 779-1600. FAX (201) 779-3242. Annual, as part of AMERICAN GLASS REVIEW periodical. $25.00 per year.

Ceramic Industry Data Book Buyers Guide. Business News Publishing Company, 755 W. Big Beaver Road, Suite 1000, Troy, MI 48084. (800) 837-1037. FAX (313) 362-0317. 1922 to present. Annual. $25.00 per year. Included as part of subscription to CERAMIC INDUSTRY periodical.

Glass Industry Annual Directory Issue. Ashlee Publishing Company Inc., 310 Madison Avenue, New York, NY 10017. (212) 682-7681. FAX (212) 697-8331. Annual. Included as part of subscription to *Glass Industry* periodical.

Guide To Engineering Materials Producers. J.C. Bittence. ASM International, Materials Information, Materials Park, OH 44073-0002. (216) 338-5151 or (800) 336-5152. FAX (216) 338-4634. 1993. $141.00

Directory of Engineering Societies and Related Organizations. Gordon Davis. 13th edition. American Association of Engineering Societies, 1111 19th Street NW, Suite 608, Washington, DC 20036. (202) 296-2237 or (800) 658-8897. 1989. Inquire for price.

Research Centers Directory. Gale Research, 835 Penobscot Building, Detroit, MI 48226-4094. (313) 961-2242. (800) 877-4253. FAX (313) 961-6083. $485.00.

Scientific and Technical Organizations and Agencies Directory. Gale Research, 835 Penobscot Building, Detroit, MI 48226-4094. (313) 961-2242. (800) 877-4253. FAX (313) 961-6083. 4th edition. 1996. $195.00.

Who's Who in Engineering. American Association of Engineering Societies, 1111 19th Street NW, Suite 608, Washington, DC 20036. (202) 296-2237 or (800) 658-8897. 8th edition. 1991. Inquire for price.

ENCYCLOPEDIAS AND DICTIONARIES

Ceramic Glossary. Walter W. Perkins, editor. American Ceramic Society, 735 Ceramic Place, Westerville, OH 43081. (614) 890-4700. FAX (614) 899-6109. 1984. $12.00.

Encyclopedia of Glass, Ceramics, and Cement. Martin Grayson, editor. John Wiley and Sons, Inc., 605 Third Avenue, New York, NY 10158. (800) 526-5368 or (212) 850-6000. 1985. Inquire for cost and availability.

Thesaurus of Scientific, Technical, and Engineering Terms. Hemisphere Publishing Corporation, 1900 Frost Road, Suite 101, Bristol, PA 19007-1598. (215) 785-5800 or (800) 821-8312. FAX (215) 785-5515. 1987. $173.00.

GLASS

Ency. of Physical Sciences and Engineering Info. Sources

GENERAL WORKS

Glass: Nature, Structure, and Properties. Horst Scholze. Springer-Verlag, 175 Fifth Avenue, New York, NY 10010. (212) 460-1500 or (800) 777-4643. FAX (212) 473-6272. 1991. $108.00.

Glasses and their Applications. H. Rawson. Ashgate Publishing Company, Old Post Road, Brookfield, VT 05036. (802) 276-3162. FAX (802) 276-3837. 1991. $70.00.

Schott Guide To Glass. H.G. Pfaender. 2nd edition. Chapman & Hall, 115 Fifth Avenue, New York, NY 10211-0906. 1996. $46.95.

Strength and Fracture of Glass and Ceramics. Jaroslav Mencik. Elsevier Science Publishing Company, Inc., 655 Avenue of the Americas, New York, NY 10010. (212) 989-5800. 1992. Inquire for cost and availability.

HANDBOOKS AND MANUALS

Asm Engineered Materials Handbook Volume 4: Ceramics and Glasses. ASM International, Materials Information, Materials Park, OH 44073-0002. (216) 338-5151 or (800) 336-5152. FAX (216) 338-4634. 1995. $147.00.

Asm Engineered Materials Reference Book. Michael L. Bauccio, editor. 2nd edition. ASM International, Materials Information, Materials Park, OH 44073-0002. (216) 338-5151 or (800) 336-5152. FAX (216) 338-4634. 1993. $111.00.

Glass Engineering Handbook. George W. McLellan & Errol B. Shand, technical consultants. McGraw-Hill Publishing Company, 1221 Avenue of the Americas, New York, NY 10020. (800) 262-4729 or (212) 512-3825. 1984. Inquire for cost and availability.

Materials Handbook: An Encyclopedia For Managers, Technical Professionals, Purchasing and Production Managers, Technicians, Supervisors, and Foremen. George S. Brady & Henry R. Clauser. 13th revised edition. McGraw Hill Publishing Company, 1221 Avenue of the Americas, New York, NY 10020. (800) 262-4729 or (212) 512-3825. 1991. $79.50.

ONLINE DATABASES AND CD-ROMS

Compendex Plus. Engineering Information, Inc., 345 East 47th Street, New York, NY 10017. (212) 705-7600 or (800) 221-1044. Contains citations with abstracts to worldwide literature in engineering and technology, from 1970 to present. Available on online BRS,(800) 289-4277, DIALOG, (800) 334-2564, ORBIT (800) 456-7248, and STN International, FIZ Karlsruhe, P.O. Box 2465, W-7500, Karlsruhe 1, Germany, online services. Also available on CD-ROM. Inquire as to cost and availability.

CSA Engineering. Cambridge Scientific Abstracts, 7200 Wisconsin Avenue, Suite 601, Bethesda, MD 20814. (301) 961-6750 or (800) 843-7751. Contains citations and abstracts of international periodicals and research literature covering all fields of engineering and science and technology,including computer and information science, electronics, mechanical engineering, solid state materials, 1981 to present. Available on BRS,(800) 289-4277, online service.Inquire as to cost and availability.

Current Contents Search. Institute for Scientific Information, 3501 Market Street, Philadelphia, PA 19104. (215) 386-0100. FAX (215) 386-6362. Contains citations to articles listed in the table of contents of science and technology journals. Also articles in social sciences and life sciences journals. Available on BRS,(800) 289-4277, DIALOG,(800) 334-2564, online services. Inquire as to cost and availability.

Dissertation Abstracts. University Microfilms International, 300 North Zeeb Road, Ann Arbor, MI 48106. (800) 521-0600 or (313) 761-4700. Scope includes virtually all doctoral dissertations accepted at accredited American institutions from 1861 to present in 252 subject areas. Available on BRS,(800) 289-4277, DIALOG,(800) 334-2564, and OCLC EPIC,(800) 848-5878, online services. Also available on CD-ROM. Inquire as to cost and availability.

NTIS Bibliographic Database. National Technical Information Service, 5285 Port Royal Road, Springfield, VA 22161. (703) 487-4929 or FAX (703) 321-8199. Broad coverage of government-sponsored science and technology research reports, 1964 to present. Available on BRS,(800) 289-4277, DIALOG, (800) 334-2564, ORBIT, (800) 456-7248, and STN International, FIZ Karlsruhe, P.O. Box 2465, W-7500, Karlsruhe 1, Germany, online services. Also available on CD-ROM. Inquire as to cost and availability.

SCISEARCH. Institute for Scientific Information, 3501 Market Street, Philadelphia, PA 19104. (800) 523-1850 or (215) 386-0100. Broad multidisciplinary title and author index to the international literature of science and technology, 1974 to present. Available on DIALOG,(800) 334-2564, and ORBIT,(800) 456-7248, online services. Also available on CD-ROM. Inquire as to cost and availability.

WILSONLINE. H.W. Wilson Company, 950 University Avenue, Bronx, NY 10452. (800) 367-6770 or (212) 588-8400. Makes available online versions of the printed H.W. Wilson indexes including Applied Science and Technology Index, Business Periodicals Index, General Science Index, and Readers' Guide to Periodical Literature. Period covered is generally 1983 to present. Available on BRS,(800) 289-4277, DIALOG,(800) 334-2564, and OCLC EPIC,(800) 848-5878, online services. Also available on CD-ROM. Inquire as to cost and availability.

PERIODICALS

American Ceramic Society Bulletin. American Ceramic Society, 735 Ceramic Place, Westerville, OH 43081. (614) 890-4700. FAX (614) 899-6109. 1922 to present. Monthly. $28.00 per year for non-members.

American Ceramic Society Journal. American Ceramic Society, 735 Ceramic Place, Westerville, OH 43081. (614) 890-4700. FAX (614) 899-6109. 1899 to present. Monthly. $175.00 per year for non-members.

American Glass Review. Doctorow Communications Inc., 1033 Clifton Avenue, Clifton, NJ 07013. (201) 779-1600. FAX (201) 779-3242. 1882 to present. 13 times per year (13th issue is AMERICAN GLASS REVIEW GLASS FACTORY DIRECTORY) $25.00 per year.

Ceramic Engineering and Science Proceedings. American Ceramic Society, Inc., 735 Ceramic Place, Westerville, OH 43081. (614) 890-4700. FAX (614) 899-6109. 1980 to present. Bimonthly. $80.00 per year.

Ceramic Industry. Business News Publishing Company, 755 W. Big Beaver Road, Suite 1000, Troy, MI 48084. (313) 362-3700. FAX (313) 362-0317. 1923 to present. 13 times a year. $53.00 per year.

Ceramics International. Elsevier Science [journals], 660 White Plains Rd., Tarrytown, NY 10159-5153. (919) 524-9200. FAX (919) 333-2444. 1974 to present. 6 times a year. $530.00 per year.

Composites Science and Technology. Elsevier Science [journals], 660 White Plains Rd., Tarrytown, NY 10159-5153. (919) 524-9200. FAX (919) 333-2444. 1968 to present. 12 times a year. $1285.00 per year.

Glass and Ceramics. Plenum Publishing Corporation, 233 Spring Street, New York, NY 10013. (212) 620-8000. FAX (212) 463-0742. 1956 to present. Monthly. $1215.00 per year.

Glass Industry. Ashlee Publishing Company Inc., 310 Madison Avenue, New York, NY 10017. (212) 682-7681. FAX (212) 697-8331. 1920 to present. Monthly. $30.00 per year.

Glass International. FMJ International Publications, Ltd., Queensway House, 2 Queensway, RedHill, Surrey RH1 1QS England. Telephone 0737-768611. FAX 0737-761685. 1978 to present. Quarterly. Inquire for price.

Glass Science and Technology. Elsevier Science Inc., Box 882, Madison Square Station, New York, NY 10159. (212) 989-5800. 1977 to present. Irregular (v.13 is 1993). Price varies.

Glass Technology. Society of Glass Technologists, Thornton, 20 Hallam Gate Road, Sheffield S10 5BT, England. TEL 0742-663168. FAX 0742-665252. 1960 to present. Bi-monthly. Inquire for price.

Journal of Materials Research. Materials Research Society, 9800 McKnight Road, Pittsburgh, PA 15237. (412) 367-3003. FAX (412) 367-4373. 1986 to present. Monthly. $440.00 per year.

Materials Engineering. Penton Publishing, 110 Superior Ave., Cleveland, OH 44114-2543. 1929 to present. Monthly. $50.00 per year.

Materials Evaluation. American Society for Nondestructive Testing, 1711 Arlington Lane, Box 28518, Columbus, OH 43228-0158. (614) 274-6003. FAX (614) 274-6899. 1942 to present. Monthly. $85.00 per year.

Physics and Chemistry of Glasses. Society of Glass Technologists, Thornton, 20 Hallam Gate Road, Sheffield S10 5BT, England. TEL 0742-663168. FAX 0742-665252. 1960 to present. Bi-monthly. Inquire for price.

SAMPE Journal. Society for the Advancement of Material and Process Engineering (SAMPE), SAMPE Journal, Covina, CA 91724. (818) 331-0610. FAX (818) 332-8929. 1965 to present. Bi-monthly. $65.00 per year.

RESEARCH CENTERS AND INSTITUTES

Alfred University Center For Glass Research. New York State College of Ceramics, Alfred, NY 14802. (607) 871-2432. FAX (607) 871-2383.

Center for Materials Research. Ohio State University, 4108 Smith Laboratory, 174 W. 18th Avenue, Columbus, OH 43210. (614) 292-1133. FAX (614) 292-7557.

Center For Materials Science and Engineering. Massachusetts Institute of Technology, Building 13, Room 2090, 77 Massachusetts Avenue, Cambridge, MA 02139. (617) 253-6701.

Glass Technical Institute. 12653 Portada Place, San Diego, CA 92130. (619) 481-1277.

Materials Research Laboratory. University of Illinois, Urbana, IL 61801. (217) 333-1370.

Materials Science Center. Cornell University, Clark Hall of Science, Ithaca, NY 14853-2501. (607) 255-4272. FAX (607) 255-6428.

Texas Tech University Glass Research and Testing Laboratory. Department of Civil Engineering, PO Box 4089, Lubbock, TX 79409-1023. (806) 742-1930. FAX (806) 742-3488.

SPECIFICATIONS AND STANDARDS

*Annual Book of Astm Standards, Volume 15.*02. American Society for Testing and Materials (ASTM), 1916 Race Street, Philadelphia, PA 19103. (215) 299-5585. Annual. Inquire for cost and availability.

GLUES

See: ADHESIVES

GOLD AND GOLD MINING

See also: METALS AND METALWORKING, MINING ENGINEERING

ABSTRACT SERVICES AND INDEXES

Alloys Index. American Society for Metals, Metals Park, OH 44073. (216) 338-5151. FAX (216) 338-4634. 1974 to present. Monthly. $295.00.

Applied Science and Technology Index; A Cumulative Subject Index To English Language Periodicals in the Fields of Aeronautics and Space Science, Computer Technology, Chemistry, Construction Industry, Energy and Related Areas. H.W. Wilson Co., 950 University Avenue, Bronx, NY 10452. (800) 367-6770 or (212) 588-8400. FAX (718) 590-1617. From 1958 to present. Monthly. Inquire about cost and availability. Also available on CD-ROM.

Bibliography and Index of Geology. American Geological Institute, 4220 King Street, Alexandria, VA 22302. (703) 379-2480. Fax (703) 379-7563. Monthly. $1295.00 per year. Also available online as GeoRef.

Current Contents: Engineering, Technology, and Applied Sciences. Institute for Scientific Information, 3501 Market Street, Philadelphia, PA 19104. (215) 386-0100. FAX (215) 386-6362. 1970 to present. Weekly. $442.00 per year.

GLUES

Ency. of Physical Sciences and Engineering Info. Sources

Engineering Index Monthly. Engineering Information, Inc., Castle Point on the Hudson, Hoboken, NJ 07030. (800) 221-1044. FAX (212) 832-1857. Monthly. $2200.00 per year. Also available online as COMPENDEX, and also on CD-ROM.

I.M.M. Abstracts and Index. Institution of Mining and Metallurgy, 44 Portland Pl., London W1N 4BR, England. 071-580-3802. FAX 071-436-5388. Bi-monthly. $364 for non-members.

Metals Abstracts and Metals Abstracts Index. American Society for Metals, Metals Park, OH 44073. (216) 338-5151. 1968 to present. Monthly. Abstracts are $1925.00 per year and Index is $460.00 per year.

Mineralogical Abstracts. Mineralogical Society and the Mineralogical Society of America, 41 Queen's Gate, London, SW7 5HR, England. Quarterly. $235.00 per year.

ANNUAL REVIEWS AND YEARBOOKS

Gold. U.S. Department of the Interior, Bureau of Mines. Annual report. Superintendent of Documents, U.S. Government Printing office, Box 371954, Pittsburgh, PA 15250-7954. (202) 783-3238. FAX (202) 512-2233. Inquire for cost and availability.

World Mine Production of Gold. Gold Institute—Administrative office, 1112 16th Street NE, Suite 240, Washington, DC 20036-4823. (202) 835-0185. FAX (202) 835-0155. 1979 to present. Annual. $50.00 per year.

ASSOCIATIONS AND PROFESSIONAL SOCIETIES

American Institute of Mining, Metallurgical & Petroleum Engineers, 345 East 47th Street, New York, NY 10017.

Asm International, Materials Information, Materials Park, OH 44073-0002. (216) 338-5151 or (800) 336-5152. FAX (216) 338-4634.

Gold Institute. 1026 16th St., NW, Suite 101, Washington, DC 20036. (202) 783-0500.

The Minerals, Metals & Materials Society, 410 Commonwealth Drive, Warrendale, PA 15086.

Society For Mining, Metallurgy and Exploration, Inc., 8307 Shaffer Parkway, PO Box 625002, Littleton, CO 80162-5002.

DIRECTORIES AND BIOGRAPHICAL SOURCES

Dun's Industrial Guide: The Metalworking Directory, 1993-94. Dun & Bradstreet Information Services Staff. 3 volumes. Dun & Bradstreet Information Services, 899 Eaton Avenue, Bethlehem, PA 18025. (610) 882-7000 or (800) 526-0651. FAX (610) 882-7269. 1993. $795.00

International Research Centers Directory. Gale Research, 835 Penobscot Building, Detroit, MI 48226-4094. (313) 961-2242. (800) 877-4253. FAX (313) 961-6083. 8th edition. 1995. $410.00

Research Centers Directory. Gale Research, 835 Penobscot Building, Detroit, MI 48226-4094. (313) 961-2242. (800) 877-4253. FAX (313) 961-6083. $485.00.

Scientific and Technical Organizations and Agencies Directory. Gale Research, 835 Penobscot Building, Detroit, MI 48226-4094. (313) 961-2242. (800) 877-4253. FAX (313) 961-6083. 4th edition. 1996. $195.00.

Who's Who in Engineering. American Association of Engineering Societies, 1111 19th Street NW, Suite 608, Washington, DC 20036. (202) 296-2237 or (800) 658-8897. 8th edition. 1991. Inquire for price.

Who's Who in Technology. Gale Research, 835 Penobscot Building, Detroit, MI 48226-4094. (313) 961-2242. (800) 877-4253. FAX (313) 961-6083. 1995. $195.00.

ENCYCLOPEDIAS AND DICTIONARIES

Encyclopedia of Physical Science and Technology. Robert A. Meyers, ed. 18 volumes. Academic Press Inc., 6277 Sea Harbor Drive, Orlando, FL 32887. (800) 321-5068. 1992. $2100.00

McGraw-Hill Dictionary of Scientific and Technical Terms. Sybil P. Parker, ed. 5th edition. McGraw-Hill Publishing Company, 1221 Avenue of the Americas, New York, NY 10020. (800) 262-4729 or (212) 512-3825. 1993. $110.50.

McGraw-Hill Encyclopedia of Engineering. Sybil P. Parker, ed. 2nd edition. McGraw-Hill Publishing Company, 1221 Avenue of the Americas, New York, NY 10020. (800) 262-4729 or (212) 512-3825. 1993. $95.50.

McGraw-Hill Encyclopedia of Science and Technology. Sybil P. Parker, ed. 7th edition. 20 volumes. McGraw-Hill Publishing Company, 1221 Avenue of the Americas, New York, NY 10020. (800) 262-4729 or (212) 512-3825. 1992. $1900.00

GENERAL WORKS

Gold Miners of the Wild West. Jeff Savage. Enslow Publications, Bloy Street & Ramsey Avenue, Box 777, Hillside, NJ 07205-0777. (908) 964-4116 or (800) 398-2504. 1995. Inquire for cost and availability.

Gold Mining in the 1990s: The Complete Book of Modern Gold Mining Procedure. Dave McCracken. New Era Publications, PO Box 47, Happy Camp, CA 96039. (916) 493-2062. 1993. $19.95.

Precious Dust: The American Gold Rush Era, 1848-1900. Paula Mitchell Marks. William Morrow and Company, Inc., 1350 Avenue of the Americas, New York, NY 10019. (212) 261-6500 or (800) 843-9389. FAX (212) 779-0965. 1994. $25.00.

Solution Mining: Leaching and Fluid Recovery of Materials. Robert W. Bartlett. Gordon and Breach Science Publishers, 820 Town Center Drive, Langhorne, PA 19047. (215) 750-2642. FAX (215) 750-6343. 1992. $65.00.

World Gold: A Minerals Availability Appraisal. United States Dept of the Interior, Bureau of Mines Special Publication SP 24-94. Superintendent of Documents, U.S. Government Printing office, Box 371954, Pittsburgh, PA 15250-7954. (202) 783-3238. FAX (202) 512-2233. 1994. Inquire for cost and availability.

HANDBOOKS AND MANUALS

Metals Handbook. ASM International, Materials Information, Materials Park, OH 44073-0002. (216) 338-5151 or (800) 336-5152. FAX (216) 338-4634. $154.00.

Smithells Metals Reference Book. E.A. Brandes & G.B. Brook. 7th edition. Butterworth-Heinemann, 313 Washington Street, Newton, MA 02158. (617) 928-2500 or (800) 366-2665. FAX (617) 928-2620. 1992. $250.00.

ONLINE DATABASES AND CD-ROMS

Compendex Plus. Engineering Information, Inc., 345 East 47th Street, New York, NY 10017. (212) 705-7600 or (800) 221-1044. Contains citations with abstracts to worldwide literature in engineering and technology, from 1970 to present. Available on online BRS,(800) 289-4277, DIALOG, (800) 334-2564, ORBIT (800) 456-7248, and STN International, FIZ Karlsruhe, P.O. Box 2465, W-7500, Karlsruhe 1, Germany, online services. Also available on CD-ROM. Inquire as to cost and availability.

CSA Engineering. Cambridge Scientific Abstracts, 7200 Wisconsin Avenue, Suite 601, Bethesda, MD 20814. (301) 961-6750 or (800) 843-7751. Contains citations and abstracts of international periodicals and research literature covering all fields of engineering and science and technology,including computer and information science, electronics, mechanical engineering, solid state materials, 1981 to present. Available on BRS,(800) 289-4277, online service.Inquire as to cost and availability.

Current Contents Search. Institute for Scientific Information, 3501 Market Street, Philadelphia, PA 19104. (215) 386-0100. FAX (215) 386-6362. Contains citations to articles listed in the table of contents of science and technology journals. Also articles in social sciences and life sciences journals. Available on BRS,(800) 289-4277, DIALOG,(800) 334-2564, online services. Inquire as to cost and availability.

GEOREF: Bibliography and Index of Geology. American Geological Institute, 4220 King Street, Alexandria, VA 22302. (703) 379-2480. Fax (703) 379-7563. Monthly. Inquire for price and availability.

Metadex. Jointly produced by ASM International and the Institute of Materials. Contains more than 925,000 records from the international literature on metals and alloys, concerning properties, processes, materials classes, applications, and metallurgical systems. Updated monthly. Available from ORBIT-QUESTEL (703) 442-0900.

NASA Database. American Institute of Aeronautics and Astronautics, 370 L'Enfant Promenade, S.W., Washington, DC 20024. (202) 646-7400. Citations and abstracts of aeronautics and astronautics literature, 1962 to present. Also contains citations from STAR, SCIENTIFIC and TECHNICAL AEROSPACE REPORTS, and INTERNATIONAL AEROSPACE ABSTRACTS. Available through NASA/RECON online service. Inquire as to cost and availability.

NTIS Bibliographic Database. National Technical Information Service, 5285 Port Royal Road, Springfield, VA 22161. (703) 487-4929 or FAX (703) 321-8199. Broad coverage of government-sponsored science and technology research reports, 1964 to present. Available on BRS,(800) 289-4277, DIALOG, (800) 334-2564, ORBIT, (800) 456-7248, and STN International, FIZ Karlsruhe, P.O. Box 2465, W-7500, Karlsruhe 1, Germany,

online services. Also available on CD-ROM. Inquire as to cost and availability.

WILSONLINE. H.W. Wilson Company, 950 University Avenue, Bronx, NY 10452. (800) 367-6770 or (212) 588-8400. Makes available online versions of the printed H.W. Wilson indexes including Applied Science and Technology Index, Business Periodicals Index, General Science Index, and Readers' Guide to Periodical Literature. Period covered is generally 1983 to present. Available on BRS,(800) 289-4277, DIALOG,(800) 334-2564, and OCLC EPIC,(800) 848-5878, online services. Also available on CD-ROM. Inquire as to cost and availability.

PERIODICALS

Alloy Digest. Alloy Digest Inc., 27 Canfield Street, Orange, NJ 07050. (201) 677-9161. 1952 to present. Monthly. $140.00 per year.

Engineering and Mining Journal. Maclean Hunter Publishing Company, 29 North Wacker Drive, Chicago, IL 60606-3298. (312) 726-2802. 1866 to present. Monthly. $60.00 per year.

Gold Bulletin. World Gold Council, 1 rue de la Rotisserie, CH-1024, Geneva, Switzerland. Telephone 022-219666. FAX 022-288160. 1968 to present. Quarterly. Free.

The Gold News. Gold Institute, 1026 16th St., NW, Suite 101, Washington, DC 20036. (202) 783-0500. 1976 to present. Bimonthly. $25.00 per year.

J O M (journal of Metals). Minerals, Metals and Materials Society, 420 Commonwealth Drive, Warrendale, PA 15086. (412) 776-9080. 1949 to present. Monthly. $50.00 per year.

Metallurgical Transactions A: Physical Metallurgy and Materials Science. ASM (American Society for Metals) International, Materials Information, Materials Park, OH 44073. (216) 338-5151 or FAX (216) 338-4634. 1970 to present. Monthly. $520.00 per year for non-members.

Metals Week. McGraw-Hill Publishing Company, 1221 Avenue of the Americas, New York, NY 10020. (800) 262-4729 or (212) 512-3825. 1930 to present. Weekly. $770.00 per year.

Mineral Industry Surveys, Gold & Silver. U.S. Bureau of Mines, Production and Distribution, Cochrans Mill Rd., Box 18070, Pittsburgh, PA 15236. (412) 892-4411. Monthly. Inquire for price and availability.

RESEARCH CENTERS AND INSTITUTES

New Mexico Institute of Mining and Technology, Mining & Mineral Resources Research Institute. Campus Station, Socorro, NM 87801. (505) 835-5210.

GRAPH THEORY

See: MATHEMATICS

GRATING

GLUES

Ency. of Physical Sciences and Engineering Info. Sources

See: SPECTROSCOPY

GRAVIMETRIC ANALYSIS

See: ANALYTICAL CHEMISTRY

GRAVITATION

See also: ASTRONOMY, COSMOLOGY, MATTER, MECHANICS, RELATIVITY, TIME, UNIVERSE

ABSTRACT SERVICES AND INDEXES

Applied Science and Technology Index; A Cumulative Subject Index To English Language Periodicals in the Fields of Aeronautics and Space Science, Computer Technology, Chemistry, Construction Industry, Energy and Related Areas. H.W. Wilson Co., 950 University Avenue, Bronx, NY 10452. (212) 588-8400. (800) 367-6770. FAX (718) 590-1617. From 1958 to present. Monthly. Inquire about cost and availability. Also available on CD-ROM and online. ISSN: 0003-6986.

Current Contents: Physical, Chemical and Earth Sciences. Institute for Scientific Information, 3501 Market Street, Philadelphia, PA 19104. (215) 386-0100. FAX (215) 386-2291. From 1961 to present. Weekly. $442.00 per year. Also available on CD-ROM and online. Inquire regarding cost and availability. ISSN: 0163-2574.

Current Papers in Physics. Institute of Electrical Engineers, Michael Faraday House, Six Hill Way, Stevenage, Herts, SG1 2AY, England. Distributed by INSPEC/IEEE, Box 1331, 445 Hoes Lane, Piscataway, NJ 08855-1331. (908) 562-5549. 1966 to present. Fortnightly. $410.00 per year. ISSN: 0011-3786.

General Science Index. H.W. Wilson Company, 950 University Avenue, Bronx, NY 10452. (212) 588-8400. (800) 367-6770. FAX (718) 590-1617. From 1978 to present. Ten issues per year; quarterly and annual cumulations. Service basis. Available on CD-ROM and online. Inquire regarding cost and availability. ISSN: 0162-1963.

NTIS Alerts: Physics. U. S. National Technical Information Service. 5285 Port Royal Road, Springfield, VA 22161. (703) 487-4650. FAX (703) 321-8547. Weekly. $140.00 per year.

Physics Abstracts. INSPEC. Section A, Science Abstracts. Institution of Electrical Engineers (IEE). Available from INSPEC/IEEE - Institute of Electrical and Electronic Engineers,, Box 1331, 445 Hoes Lane, Piscaway NJ 08855-1331. (908) 562-5549. 1898 to present. 24 issues per year. $2835.00 per year. Also available on CD-ROM and online. ISSN: 0036-8091.

Physics Briefs (Physikalische Berichte). Information Center for Energy, Physics, Mathematics; German Physical Society. V C H Publishers, Inc., 220 East 23rd Street, New York, NY 10010-4606. (212) 683-8333. 1845 to present. 24 issues per year. $2390.00 per year. Also available online. ISSN: 0179-7434.

Science Citation Index. SCI. Institute for Scientific Information, 3501 Market Street, Philadelphia, PA 19104. (215) 386-0100. (800) 523-1850. FAX (215) 386-2991. 1961 to present. Six issues per year, plus annual cumulation. $11650.00 per year. Also

available online and on CD-ROM. Inquire about price and availability. ISSN: 0036-827X.

ANNUAL REVIEWS AND YEARBOOKS

Advances in Applied Mechanics. Academic Press, Inc., 6277 Sea Harbor Drive, Orlando, FL. (800) 321-5068. 1948 to present. Irregular. ISSN: 0065-2156.

Advances in Atomic, Molecular and Optical Physics. Academic Press, Inc., 6277 Sea Harbor Drive, Orlando, FL. (800) 321-5068. 1965 to present. Irregular. ISSN: 1049-250X.

Advances in Electronics and Electron Physics. Academic Press, Inc., 6277 Sea Harbor Drive, Orlando, FL. (800) 321-5068. From 1948 to present. Irregular. ISSN: 0065-2539.

Advances in Nuclear Physics. Plenum Publishing Corp., 233 Spring Street, New York, NY. (212) 620-8000. (800) 2221-9369. FAX (212) 463-0742. 1968 to present. Irregular. Price varies. ISSN: 0065-2970.

Advances in Physics. Taylor & Francis, Ltd., Rankine Road, Basingstoke, Hants RG24 8PR, England. 0256-840366. FAX 0256-47938. 1952 to present. Bimonthly. $511.00 per year. ISSN: 0001-8732.

Annual Review of Nuclear and Particle Science. Annual Reviews, Inc., 4139 El Camino Way, Palo Alto, CA 94303-0139. (415) 493-4400. (800) 523-8635. Fax (415) 855-9815. 1952 to present. Annual. $62.00.

Topics in Applied Physics. Springer-Verlag New York, Inc., 175 Fifth Avenue, New York, NY 10010. (212) 460-1500. (800) 777-4643. FAX (212) 473-6272. From 1974 to present. Irregular. ISSN: 0303-4216.

ASSOCIATIONS AND PROFESSIONAL SOCIETIES

American Astronomical Society. 2000 Florida Avenue NW, Suite 400, Washington, DC 20009. (202) 328-2010. FAX: (202) 234-2560.

American Geophysical Union. 2000 Florida Avenue, NW, Washington, DC 20009. (202) 462-6900. (800) 966-AGU1. FAX (202) 328-0566.

American Institute of Physics. One Physics Ellipse, College Park, MD 20740-3843. (301) 209-3100.

American Physical Society. One Physics Ellipse, College Park, MD 20740-3843. (301) 209-3200. FAX (301) 209-0865.

International Society For General Relativity and Gravitation. Universities Research Association, 1111 19th Street NW, Suite 400, Washington, DC 20036. (202) 293-1382.

BIBLIOGRAPHIES

Guide To the Archival Collectins in the Niels Bohr Libary At the American Institute of Physics. Niels Bohr Library Staff. American Institute of Physics, One Physics Ellipse, College Park, MD 20740-3843. (301) 209-3100. 1995. $140.00

Science Books and Films. American Association for the Advancement of Science, 1333 H Street NW, Washington, DC 20005. (202) 326-6454. 1965 to present. Nine issues per year. $40.00 per year. ISSN: 0098-342X

Scientific and Technical Books and Serials in Print, 1995. R.R. Bowker Inc., 121 Chanlon Road, New Providence, NJ 07974. (908) 464-6800. (800) 521-8110. 4 volumes. 1994. $299.95. Also available on compact disc and online.

DIRECTORIES AND BIOGRAPHICAL SOURCES

American Association of Physics Teachers. Directory of members. American Association of Physics Teachers, 5112 Berywn Road, College Park, MD 20740. (301) 345-4200.

American Men and Women of Science: Physical and Biological Sciences. R. R. Bowker Inc., 121 Chanlon Road, New Providence, NJ 07974. (908) 464-6800. (800) 521-8110. 20th edition. 8 volumes. 1996. $850.00.

American Physical Society. MEMBERSHIP DIRECTORY BULLETIN ISSUE. American Physical Society, One Physics Ellipse, College Park, MD 20740-3843. (301) 209-3200. FAX (301) 209-0865. Biennial. $50.00.

The Biographical Dictionary of Scientists: Physicists. D. Abbott, editor. Peter Bedrick Books, Inc., 2112 Broadway, Room 318, New York, NY 10023. (212) 496-0751. 1984.

Directory of Physics and Astronomy Staff Members. American Institute of Physics, One Physics Ellipse, College Park, MD 20740-3843. (301) 209-3100. Annual. $45.00.

Graduate Programs in Physics, Astronomy and Related Fields. 1995-1996. American Institute of Physics, One Physics Ellipse, College Park, MD 20740-3843. (301) 209-3100. Annual. $45.00.

Legends in their Own Time: A Century of American Physical Scientists. Anthony Serafini. Plenum Publishing Corp., 233 Spring Street, New York, NY. (212) 620-8000. (800) 2221-9369. FAX (212) 463-0742.. 1993. $27.50.

The Nobel Prize Winners. Frank N. Magill, editor. Salem Press, Inc., P.O. Box 1097, Englewood Cliffs, NJ 07632. (201) 871-3700. (800) 221-1592. 3 volumes. 1989. $210.00.

Research Centers Directory. Gale Research Inc., 835 Penobscot Building, Detroit, MI 48226-4094. (313) 961-2242. (800) 877-4253. 20th edition. 1995. $485.00.

Scientific and Technical Organizations and Agencies Directory. Gale Research, 835 Penobscot Building, Detroit, MI 48226-4094. (313) 961-2242. (800) 877-4253. 4th edition. 1996. $195.00.

World Guide To Scientific Associations and Learned Societies. K. G. Saur Inc., 121 Chanlon Road, New Providence, NJ 07974. (908) 464-6800. (800) 521-8110. 5th edition.1990. $225.00.

ENCYCLOPEDIAS AND DICTIONARIES

A Concise Dictionary of Physics. Oxford University Press, Inc., 200 Madison Avenue, New York, NY 10016. (212) 725-6000. (800) 334-4249. 1990. $10.95.

Dictionary of Effects and Phenomena in Physics. Joachim Schubert. VCH Publications, Inc., 220 East 23rd Street, Suite 909, New York, NY 10010-4606. (212) 683-8333. (800) 422-8824. 1987. $35.00.

Dictionary of the Physical Sciences: Terms, Formulas, Data. Cesare Emiliani. Oxford University Press, Inc., 200 Madison Avenue, New York, NY 10016. (212) 725-6000. (800) 334-4249. 1989. $19.95.

Encyclopedia of Applied Physics. George Trigg, editor. VCH Publications, Inc., 220 East 23rd Street, Suite 909, New York, NY 10010-4606. (212) 683-8333. (800) 422-8824. 20 volume set. 1991-. $5990.00.

Encyclopedia of Physics. Robert M. Besancon. Chapman & Hall, 1 Penn Plaza, New York, NY 10119. (212) 564-1060. 3rd edition. 1990. $54.95 in paper.

Facts On File Dictionary of Physics. John Daintith, editor. Facts-on-File, Inc., 460 Park Avenue South, New York, NY 10016-7382. (212) 683-2244. (800) 322-8755. Fax (800) 683-3633. Revised edition. 1990. $12.95.

Encyclopedia of Physical Science and Technology. Academic Press, Inc., 6277 Sea Harbor Drive, Orlando, FL. (800) 321-5068. 2nd edition. 18 volume set. 1992. $2625.00.

Landolt-boernstein Numerical Data and Functional Relationships in Science and Technology: Nuclear and Particle Physics. K. H. Hellwege editor. Springer-Verlag New York, Inc., 175 Fifth Avenue, New York, NY 10010. (212) 460-1500. (800) 777-4643. FAX (212) 473-6272. 1993. Volumes priced separately.

McGraw-Hill Encyclopedia of Science and Technology. McGraw-Hill Book Company, Inc., 1221 Avenue of the Americas, New York, NY 10020. (212) 512-2000. (800) 262-4729. 7th edition. 20 volume set. 1992. $1900.00.

Penguin Dictionary of Physics. Valerie Illingworth, editor. Viking Penguin, 375 Hudson Street, New York, NY 10014-3657. (212) 366-2000. (800) 331-4624 2nd edition. 1991.

GENERAL WORKS

Fundamental Particles and Forces. Richard A. Carrigan, Jr., and W. Peter Trower. W. H. Freeman & Co., 41 Madison Avenue, East 26th Avenue, 35th Floor, New York, NY 10010. (212) 576-0400. 1995. $13.95.

Gravitation. Charles W. Misner, et al. W. H. Freeman & Co., 41 Madison Avenue, East 26th Avenue, 35th Floor, New York, NY 10010. (212) 576-0400. 1995.

Gravitation and Inertia. Ignazio Ciufolini and John A. Wheeler. Princeton University Press, 41 William Street, Princeton, NJ 08540. (609) 258-4900. (800) 777-4726. 1998. $49.50.

Lectures On Gravitation. Richard Feyman. Addison-Wesley Publishing Co., Inc., 1 Jacob Way, Reading, MA 01867. (617) 944-3700. (800) 447-2226. 1995. $55.95.

Postprincipia: Gravitation For Physicists and Astronomers. Peter Rastall. World Scientific Publishing Company, Inc., 1060 Main Street, River Edge, NJ 07661. (201) 487-9655. (800) 227-7562. 1991. $48.00.

GRAVITATION

Ency. of Physical Sciences and Engineering Info. Sources

Relativity and Gravitation - Classical and Quantum. J. C. D'Olivo, et al, editors. World Scientific Publishing Company, Inc., 1060 Main Street, River Edge, NJ 07661. (201) 487-9655. (800) 227-7562. 1991. $104.00.

Riddle of Gravitation. Peter G. Bergmann. Dover Publications, Inc., 180 Varick Street, New York, NY 10014. (212) 255-3755. (800) 223-3130. 1995. Reprint edition. 1993. $7.95.

Space, Time and Gravity: The theory of the Big Bang and Black Holes. Robert M. Wald. University of Chicago Press, 5801 Ellis Avenue, Chicago, IL 60637. (312) 702-7700. 2nd edition. 1992. $24.95.

Theory and Experiment in Gravitational Physics. Clifford M. Will. Cambridge University Press, 40 West 20th Street, New York, NY 10011-4211. (212) 924-3900. (800) 872-7423. 1993. Revised edition. $39.95 in paper.

Three Hundred Years of Gravitation. S. W. Hawking and W. Israel, editors. Cambridge University Press, 40 West 20th Street, New York, NY 10011-4211. (212) 924-3900. (800) 872-7423. 1989. $54.95 in paper.

HANDBOOKS AND MANUALS

Chemical and Physical Data. Arthur M. James and M. P. Lord, Van Nostrand Reinhold, 115 Fifth Avenue, New York, NY 10003. (212) 254-3232. (800) 842-3636. 1992. $59.95.

CRC Handbook of Chemistry and Physics. David R. Kide, editor. CRC Press, Inc., 2000 Corporate Boulevard, NW, Boca Raton, FL 33431. (407) 994-0555. (800) 272-7737. 77th edition. 1996. $99.95.

Handbook of Physical Quantities. Igor S. Grigoriev and Evgenil Z. Melikhov, editors. CRC Press, Inc., 2000 Corporate Boulevard, NW, Boca Raton, FL 33431. (407) 994-0555. (800) 272-7737. 1995. $99.00.

Handbook of Space Astronomy and Astrophsyics. Martin V. Zumbeck. Cambridge University Press, 40 West 20th Street, New York, NY 10011-4211. (212) 924-3900. (800) 872-7423. 2nd edition. 1990. $79.95.

Mathematical Methods For Physicists. George B. Arfken. Academic Press, Inc., 6277 Sea Harbor Drive, Orlando, FL. (800) 321-5068. 4th edition. 1995. $40.00.

McGraw-Hill Encyclopedia of Physics. Sybil P. Parker, editor. McGraw-Hill Publishing Company, Inc., 1221 Avenue of the Americas, New York, NY 10020. (212) 512-2000. (800) 262-4729. 2nd edition. 1993. $95.50

A Physicist's Desk Reference. Herbert L. anderson, editor. American Institute of Physics, One Physics Ellipse, College Park, MD 20740-3843. (301) 209-3100. 2nd edition. 1989. $70.00.

Physics Problem Solver. Research and Education Association Staff. Research & Education Association, 61 Ethel Road, West Piscataway, NJ 08854. (908) 819-8880. Revised edition. 1994. $23.95 in paper.

ONLINE DATABASES AND CD-ROMS

Current Contents Search. Institute for Scientific Information, 3501 Market Street, Philadelphia, PA 19104. (215) 386-0100. FAX (215) 386-6362. Contains citations to articles listed in the table of contents of science and technology journals. Also articles in social sciences and life sciences journals. Available on BRS, (800) 289-4277, DIALOG, (800) 334-2564, online services. Inquire as to cost and availability.

Dissertation Abstracts Online. University Microfilms International, 300 North Zeeb Road, Ann Arbor, MI 48106. (800) 521-0600 or (313) 761-4700. Scope includes virtually all doctoral dissertations accepted at accredited American institutions from 1861 to present in 252 subject areas. Available on BRS, (800) 289-4277, DIALOG, (800) 334-2564, and OCLC EPIC, (800) 848-5878, online services. Also available on CD-ROM. Inquire as to cost and availability.

INSPEC. Institution of Electrical Engineers, Michael Faraday House, Six Hills Way, Stevenage, Herts. SG1 2AY, England. Telephone: 0438 313311 or FAX 0438 742840. Contains citations to the worldwide literature of physics, electronics and electrical engineering, computer technology, and related fields. Available on BRS, (800) 289-4277, DIALOG, (800) 334-2564, ORBIT, (800) 456-7248, and STN International, FIZ Karlsruhe, P.O. Box 2465, W-7500, Karlsruhe 1, Germany, online services. Inquire as to cost and availability.

Physics Briefs. American Institute of Physics, 335 East 45th Street, New York, NY 10017. (212) 661-9260 or FAX (212) 661-2036. Contains citations with abstracts of the literature of physics and related fields, 1979 to present. Available on the STN International, FIZ Karlsruhe, P.O. Box 2465, W-7500, Karlsruhe 1, Germany, online service. Inquire as to cost and availability.

Scientific and Technical Books and Serials in Print. R.R. Bowker Inc., 121 Chanlon Road, New Providence, NJ 07974. (800) 521-8110 or (908) 464-6800. List of currently published books and serials in the physical and biological sciences, engineering and technology, with subject, author and titles indexes. Available on ORBIT, (800) 456-7248, online service. Inquire as to cost and availability.

SCISEARCH. Institute for Scientific Information, 3501 Market Street, Philadelphia, PA 19104. (800) 523-1850 or (215) 386-0100. Broad multidisciplinary title and author index to the international literature of science and technology, 1974 to present. Available on DIALOG, (800) 334-2564, and ORBIT, (800) 456-7248, online services. Also available on CD-ROM. Inquire as to cost and availability.

WILSONLINE. H.W. Wilson Company, 950 University Avenue, Bronx, NY 10452. (800) 367-6770 or (212) 588-8400. Makes available online versions of the printed H.W. Wilson indexes including Applied Science and Technology Index, Business Periodicals Index, General Science Index, and Readers' Guide to Periodical Literature. Period covered is generally 1983 to present. Available on BRS, (800) 289-4277, DIALOG, (800) 334-2564, and OCLC EPIC, (800) 848-5878, online services. Also available on CD-ROM. Inquire as to cost and availability.

PERIODICALS

American Journal of Physics. American Association of Physics Teachers, 5112 Berywn Road, College Park, MD 20740. (301) 345-4200. 1933 to present. Monthly. $215.00 per year. ISSN: 0002-9505.

Canadian Journal of Physics. National Research Council of Canada. Research Journals, Ottawa K1A OR6, Canada. (613) 993-9084. FAX (613) 952-7656. 1929 to present. Monthly. $285.00 per year. ISSN: 0008-4204.

Classical and Quantum Gravity. I O P Publishing. U. S. subscriptions to: American Institute of Physics, 500 Sunnyside Boulevard, Woodbury, NY 11797-2999. (516) 575-2270. 1984 to present. Monthly. $1497.00 per year. ISSN: 0264-9381.

Contemporary Physics. Taylor & Francis, Ltd, Rankine Road, Basingstoke, Hants RG24 8PR, England. TEL 0256-840366. FAX 0256-479438. 1959 to present. Bimonthly. $318.00 per year. ISSN: 0010-7514.

European Journal of Physics. I O P Publishing. U. S. subscriptions to: American Institute of Physics, 500 Sunnyside Boulevard, Woodbury, NY 11797-2999. (516) 575-2270. 1980 to present. Bimonthly. $339.00 per year. ISSN: 0143-0807.

General Relativity and Gravitation. Plenum Publishing Corp., 233 Spring Street, New York, NY 10013-1578. (212) 620-8000. FAX (212) 463-0742. 1970 to present. Monthly. $710.00. ISSN: 0001-7701.

International Journal of Modern Physics D: Gravitation, Astrophysics and Cosmology. World Scientific Publishing Co., 1060 Main Street, Suite 18, River Edge, NJ 07661. (800) 227-7562. 1992 to present. Quarterly. $195.00 per year. ISSN: 0218-2718.

Journal of Physics. A: MAtheMATICAL and GENERAL. I O P Publishing. U. S. orders to: American Institute of Physics, 500 Sunnyside Boulevard, Woodbury, NY 11797-2999. (516) 576-2200. 1968 to present. Twenty-four issues per year. $2487.00 per year. ISSN: 0305-4470.

Modern Physics Letters A. World Scientific Publishing Co., 1060 Main Street, Suite 18, River Edge, NJ 07661. (800) 227-7562. 1986 to present. 40 issues per year. $1089.00 per year. ISSN: 0217-7323.

Physical Review D (particles, Fields and Gravitation). American Institute of Physics, One Physics Ellipse, College Park, MD 20740-3843. (301) 209-3000. 1970 to present. Monthly. $1280.00 per year. ISSN: 0556-2821.

Physics and Chemistry of the Earth. Elsevier Science 660 White Plains Road, Tarrytown, NY 10591-5153. (914) 524-9200. FAX (914) 333-2444. 1956 to present. Six issues per year. $550.00. ISSN: 0079-1946.

Physics - Doklady. (translation of Rossiiskaya Adamemiya Nauk. Doklady). American Institute of Physics, One Physics Ellipse, College Park, MD 20740-3843. (301) 209-3000. 1956 to present. Monthly. $1300.00. ISSN: 1063-7753.

Physics Today. American Institute of Physics, One Physics Ellipse, College Park, MD 20740-3843. (301) 209-3000. 1948 to present. Monthly. $140.00 per year. ISSN: 0031-9228.

Reviews of Modern Physics. American Institute of Physics, One Physics Ellipse, College Park, MD 20740-3843. (301) 209-3000. 1929 to present. Quarterly. $300.00 per year. ISSN: 0034-6861.

Royal Society of London, Philosophical Transactions. SERIES A, PHYSICAL SCIENCES and ENGINEERING. Royal Society of London, 6 Carlton House Terrace, London SW1Y 5AG, England. TEL 071-839-5561. FAX 071-976-1837. 1665 to present. Monthly. L485 per year. ISSN: 0962-8428.

RESEARCH CENTERS AND INSTITUTES

Center For Relativity. University of Texas at austin. Department of Physics. RLM 9.224, Austin, TX 78712. (512) 471-1103.

Center For Space Research. Massachusetts Institute of Technology, 77 Massachusetts Avenue, Cambridge MA 02139. (617) 253-7501.

Center For Space Science and Astrophysics. Stanford University, 325 Durand Building, Stanford, CA 94305. (415) 723-3592.

Joint Institute For Laboratory Astrophysics. University of Colorado - Boulder, Boulder CO 80309-0440. (303) 492-7789. FAX (303) 492-5235.

GRAVITATIONAL RED SHIFT

See: GRAVITY

GRAVITY

See: GRAVITATION

GREASE

See: LUBRICATION

GREENHOUSE EFFECT

See also: CLIMATOLOGY, DESERTIFICATION, METEOROLOGY, OZONE

ABSTRACT SERVICES AND INDEXES

Applied Science and Technology Index; A Cumulative Subject Index To English Language Periodicals in the Fields of Aeronautics and Space Science, Computer Technology, Chemistry, Construction Industry, Energy and Related Areas. H.W. Wilson Co., 950 University Avenue, Bronx, NY 10452-9978. (212) 588-8400. (800) 367-6770. FAX (718) 590-1617. From 1958 to present. Monthly. Inquire about cost and availability. Also available online (BRS and WILSONLINE) and on CD-ROM. ISSN: 0003-6986.

Bibliography and Index of Geology. American Geological Institute, 4220 King Street, Alexandria, VA 22302-1507. (703) 379-2480. FAX (703) 379-7563. From 1969 to present. Monthly. $1295.00 per year. ISSN: 0098-2784. Also available as GEOREF online (CISTI, DIALOG, Orbit, STN) and on CD-ROM. Inquire about price and availability.

GREENHOUSE EFFECT

Ency. of Physical Sciences and Engineering Info. Sources

General Science Index. H.W. Wilson Co., 950 University Avenue, Bronx, NY 10452. (212) 588-8400. (800) 367-6770. FAX (718) 590-1617. From 1978 to present. Ten issues per year; quarterly and annual cumulations. Service basis. Available on CD-ROM and online. Inquire regarding cost and availability. ISSN: 0162-1963.

Meteorological and Geoastrophysical Abstracts. American Meteorological Society, c/o Inforonics, Inc., 550 Newtown Road, Box 458, Littleton, MA 01460. (508) 486-8976. FAX (508) 486-0027. Covers literature in environmental sciences, meteorology, astrophysics, hydrology, glaciology, and physical oceanography. 1950 to present. Monthly. $950.00 per year. ISSN: 0026-1130. Also available online (DIALOG) and on CD-ROM.

Oceanic Abstracts. Cambridge Scientific Abstracts, 7200 Wisconsin Avenue, Bethesda, MD 20814. (301) 961-6750. Fax (301) 961-6720. Bimonthly. $995.00 per year.

Science Citation Index. SCI. Institute for Scientific Information, 3501 Market Street, Philadelphia, PA 19104. (215) 386-0100. (800) 523-1850. FAX (215) 386-2991. 1961 to present. Six issues per year, plus annual cumulation. $11650.00 per year. Also available online and on CD-ROM. Inquire about price and availability. ISSN: 0036-827X.

Selected Water Resources Abstracts. U.S. Geological Survey. Distributed by National Technical Information Service (NTIS), 5285 Port Royal Road, Springfield, VA 22161. (703) 487-4650. Monthly. $115.00 per year.

ANNUAL REVIEWS AND YEARBOOKS

American Meteorological Society. METEOROLOGICAL MONOGRAPHS. American Meteorological Society, 45 Beacon Street, Boston, MA 02108-3693. (617) 227-2425. FAX (617) 742-8718. Irregular. Price varies.

Annual Review of Earth and Planetary Sciences. Annual Reviews, Inc., 4139 El Camino Way, Palo Alto, CA 94306-0897. (415) 493-4400. Fax (415) 855-9815. Annual. $55.00 per year.

Developments in Atmospheric Science Series. Elsevier Science Publishing Company, Inc., 655 Avenue of the Americas, New York, NY 10010. (212) 989-5800. Irregular. Inquire about price and availability.

Ocean Yearbook. Elisabeth M. Borgese, et al, editors. University of Chicago Press, 5801 Ellis Avenue, Chicago, IL 60637. (312) 702-7700. 1995. $77.00.

ASSOCIATIONS AND PROFESSIONAL SOCIETIES

American Association of State Climatologists. c/o Myron Moinau, University of Idaho, Agriculture Engineering Department, State Climate Service, Moscow, ID 834844-2040.

American Meteorological Society, 45 Beacon Street, Boston, MA 02108-3693. (617) 227-2425. FAX (617) 742-8718. Irregular. Price varies.

Greenhouse Crisis Foundation, 1130 17th Street Nw, Suite 630, Washington, DC 20036. (202) 460-2823.

International Association of Meteorology and Atmospheric Physics. UCAR, P.O. Box 3000, Boulder CO 80307-3000. (303) 497-1673.

National Weather Association, 6704 Wolke Court, Montgomery, AL 36116-2134. (334) 213-0388. FAX (334) 213-0388.

University Corporation For Atmospheric Research. PO Box 3000, 1850 Table Mesa Drive, Boulder, CO 80307-3000. (303) 497-1673. FAX (303) 497-1654.

Weather Modification Association. PO Box 8116, Fresno CA 93747. (209) 291-8466.

BIBLIOGRAPHIES

The Greenhouse Effect: A Bibliography. Joan Nordquist. Reference & Research Services, 511 Lincols Nstrre, Santa Cruz, CA 95060. (408) 426-4479. 1990. $15.00.

Information Sources in the Earth Sciences. David N. Wood, Joan E. Hardy, and Anthony P. Harvey, editors. Bowker-Saur/K.G. Saur. Distributed by R.R. Bowker, 121 Chanlon Road, New Providence, NJ 07974. (800) 521-8110 or (908) 464-6800. Second edition. 1989. $85.00.

Scientific and Technical Books and Serials in Print; An Index To Literature in Science and Technology. R.R. Bowker Co., 205 E. 42nd Street, New York, NY 10017. (800) 521-8110 or (212) 916-1600. 1992. $250.00.

DIRECTORIES AND BIOGRAPHICAL SOURCES

American Men and Women of Science: Physical and Biological Sciences. Eighteenth edition. R.R. Bowker Company, 245 West 17th Street, New York, NY 10011. (800) 521-8810 or (212) 916-1600. $750.00.

American Meteorological Society. PRofESSIONAL DIRECTORY. American Meteorological Society. 45 Beacon Street, Boston, MA 02108. (617) 227-2425. Included in each issue of the Bulletin of the Society.

Meteorological Services of the World. American Meteorological Society, 45 Beacon Street, Boston, MA 02108-3693. (617) 227-2425. FAX (617) 742-8718. Annual. $70.00.

National Weather Service offices and Stations. National Oceanic and Atmospheric Administration, Deparatment of Commerce, Silver Spring, MD 20910. (301) 427-7698. Annual. Free

Research Centers Directory. Gale Research, 835 Penobscot Building, Detroit, MI 48226-4094. (313) 961-2242. (800) 877-4253. 20th edition. 1995. $485.00. ISSN: 0080-1518.

ENCYCLOPEDIAS AND DICTIONARIES

Climates of the States. Bair. Gale Research, 835 Penobscot Building, Detroit, MI 48226-4094. (313) 961-2242. (800) 877-4253. 5th edition. 1995. $255.00.

Concise Oxford Dictionary of the Earth Sciences. Ailsa Allaby and Michael Allaby, editors. Oxford University Press, Inc., 200 Madison Avenue, New York, NY 10016. (800) 334-4249 or (212) 679-7300. 1990. $42.95.

Encyclopedia of Climatology. Rhodes W. Fairbridge, and John E. Oliver. Chapman & Hall, 1 Penn Plaza, New York, NY 10119. (212) 564-1060. 1986. $115.00.

McGraw-Hill Encyclopedia of Science and Technology. McGraw-Hill Publishing Company, 1221 Avenue of the Americas, New York, NY 10020. (800) 262-4729 or (212) 512-3825. Seventh edition. 1992. $1900.00.

Weather Satellite Handbook. American Radio Relay League, 225 Main Street, Newington, CT 06111. (203) 666-1541 1994. $20.00.

GENERAL WORKS

Climate Change: The Ipcc Scientific Assessment. John T. Houghton, et. al editors. Cambridge University Press, 40 West 20th Street, New York, NY 10011-4211. (212) 924-3900. (800) 872-7423. 1990. $39.95 in paper.

Energy Technologies For Reducing Emissions of Greenhouse Gases. OECD Staff. Organization for Economic Cooperation and Development, 2001 L Street NW, Washington, DC 20036. (202) 785-6323. (800) 456-6323. 1989. $84.00.

Global Warming and Biological Diversity. Robert L. Peters and Thomas E. Lovejoy. Yale University Press, 302 Temple Street, New Haven, CT 06511. (203) 432-0940. 1992. $17.00.

Greenhouse Effect, Sea Level and Drought. Roland Paepe, editor. Kluwer Academic Publishers, 101 Philip Drive, Assinippi Park, Norwell, MA 02061. (617) 871-6600. 1990. $218.50.

Greenhouse Effect: Climatic Change and Ecosystems. Bert Bolin and Bo R. Doos. John Wiley and Sons, Inc., 605 Third Avenue, New York, NY 10158. 1986. $27.50.

Greenhouse Warming: Abatement and Adaptation. Norman J. Rosenberg. Resources for the Future, Inc., 1616 P Street, NW, Washington, DC 20036. (202) 328-5086. 1989. $18.95.

Impact of Carbon Dioxide, Trace Gases and Climate Change On Global Agriculture. B. A. Kimball, editor. American Society of Agronomy, 677 South Segoe Road, Madison, WI 53711-1086. (608) 273-8080. 1990. $15.00.

Land Use and the Causes of Global Warming. W. Neil Adger and Katrina Brown. John Wiley and Sons, Inc., e605 Third Avenue, New York, NY 10158. 1995. $54.95.

Primer of Greenhouse Effect Gases. Donald L. Wuebbles. Lewis Publs., CRC Press, Inc., 2000 Corporate Boulevard, NW, Boca Raton, FL 33431. (407) 994-0555. (800) 272-7737. 1991. $69.95.

Soil Management and the Greenhouse Effect. Ratan Lal, editor. Lewis Publs., CRC Press, Inc., 2000 Corporate Boulevard, NW, Boca Raton, FL 33431. (407) 994-0555. (800) 272-7737. 1995, $79.95.

Soils and Global Change. Rattan Lal, et al editors. Lewis Publs., CRC Press, Inc., 2000 Corporate Boulevard, NW, Boca Raton, FL 33431. (407) 994-0555. (800) 272-7737. 1995, $85.00. HandB00KS and MANUALS

Handbook of Agricultural Meteorology. John F. Griffin. Oxford University Press, 200 Madison Avenue, New York, NY 10016. (212) 725-6000. (800) 334-4249. 1994. $85.00.

Handbook of Applied Meteorology. David D. Houghton, editor. John Wiley & Sons, Inc., 605 Third Avenue, New York, NY 10158-0012. (212) 850-6000. (800) 225-5945. 1985.

ONLINE DATABASES AND CD-ROMS

Climate Assessment Database. National Weather Service, National Meteorological Center, 5200 Auth Road, Suite 101, Camp Springs, MD 20233. (301) 763-8016. Contains daily, weekly and monthly summaries of North American and world climatological data. Also provides five to ten weather forecasts, and 30 to 90 day weather outlook. Subscription required. Inquire as to cost and availability.

Current Contents Search. Institute for Scientific Information, 3501 Market Street, Philadelphia, PA 19104. (215) 386-0100. FAX (215) 386-6362. Contains citations to articles listed in the table of contents of science and technology journals. Also articles in social sciences and life sciences journals. Available on BRS,(800) 289-4277, DIALOG,(800) 334-2564, online services. Inquire as to cost and availability.

Dissertation Abstracts. University Microfilms International, 300 North Zeeb Road, Ann Arbor, MI 48106. (800) 521-0600 or (313) 761-4700. Scope includes virtually all doctoral dissertations accepted at accredited American institutions from 1861 to present in 252 subject areas. Available on BRS,(800) 289-4277, DIALOG,(800) 334-2564, and OCLC EPIC,(800) 848-5878, online services. Also available on CD-ROM. Inquire as to cost and availability.

Meteorological and Geoastrophysical Abstracts. American Meteorological Society, 45 Beacon Street, Boston, MA 02108-3693. (617) 227-2425. FAX (617) 742-8718. Contains citations and abstracts to the worldwide literature on significant research in meteorology and geoastrophysics. Related topics include physical oceanography, hydrology, environmental sciences and glaciology. Covers the period 1972 to present. Available on DIALOG,(800) 334-2564, online service. Inquire as to cost and availability.

NTIS Bibliographic Database. National Technical Information Service, 5285 Port Royal Road, Springfield, VA 22161. (703) 487-4929 or FAX (703) 321-8199. Broad coverage of government-sponsored science and technology research reports, 1964 to present. Available on BRS,(800) 289-4277, DIALOG, (800) 334-2564, ORBIT, (800) 456-7248, and STN International, FIZ Karlsruhe, P.O. Box 2465, W-7500, Karlsruhe 1, Germany, online services. Also available on CD-ROM. Inquire as to cost and availability.

SCISEARCH. Institute for Scientific Information, 3501 Market Street, Philadelphia, PA 19104. (800) 523-1850 or (215) 386-0100. Broad multidisciplinary title and author index to the international literature of science and technology, 1974 to present. Available on DIALOG and ORBIT online services. Also available on CD-ROM. Inquire as to cost and availability.

WILSONLINE. H.W. Wilson Company, 950 University Avenue, Bronx, NY 10452. (800) 367-6770 or (212) 588-8400. Makes available online versions of the printed H.W. Wilson indexes including Applied Science and Technology Index, Business Periodicals Index, General Science Index, and Readers' Guide to Periodical Literature. Period covered is generally 1983 to

GREENHOUSE EFFECT

Ency. of Physical Sciences and Engineering Info. Sources

present. Available on BRS,(800) 289-4277, DIALOG,(800) 334-2564, and OCLC EPIC,(800) 848-5878, online services. Also available on CD-ROM. Inquire as to cost and availability.

World Climate Disc. Chadwyck-Healey Inc., 1101 King Street, Alexandria, VA 22314. (703) 683-4890. FAX (703) 683-7589. Weather and climate data from approximately 5,000 weather stations worldwide, covering the years 1854 to 1990 on CD-ROM. First edition 1992. Approximately $1200.00 per year with annual updates.

World Weatherdisc. WeatherDisc Associates, Inc., 4584 N.E. 89th Street, Seattle, WA 98115. (206) 524-4314. FAX (206) 543-0308. Meteorological data on CD-ROM which describes the climate of the earth today and for the past few hundred years. First edition 1989. Approximately $295.00 per year with biannual updates.

PERIODICALS

Agricultural and Forest Meteorology. Elsevier Science Publishing Company, Inc., 655 Avenue of the Americas, New York, NY 10010. (212) 989-5800. Twenty times per year. $750.00 per year.

American Meteorological Society. BULLETIN. American Meteorological Society, 45 Beacon Street, Boston, MA 02108-3693. (617) 227-2425. FAX (617) 742-8718. Monthly. $60.00 per year.

Atmosphere - Ocean. Canadian Meteorological and Oceanographic Society, P.O. Box 334, Newmarket, Ontario, L3Y 4X7, Canada. (416) 898-1040. FAX (416) 898-7937. Quarterly. $30.00 per year.

Climate Change. Kluwer Academic Publishers, P.O. Box 358, Accord Station, Hingham, MA 02018-0358. (617) 871-6000. Six times per year. $327.00 per year.

International Journal of Climatology. Royal Meteorological Society, distributed by John Wiley and Sons, Inc., 605 Third Avenue, New York, NY 10158. (800) 526-5368 or (212) 850-6000. Eight times per year. $425.00 per year.

JGR: Journal of Geophysical Research: Atmosphere. American Geophysical Union, 2000 Florida Avenue, N.W., Washington, DC 20009. (202) 462-6903. Monthly. $90.00 per year to members.

Journal of Applied Meteorology. American Meteorological Society, 45 Beacon Street, Boston, MA 02108-3693. (617) 227-2425. FAX (617) 742-8718. Monthly. $165.00 per year.

Journal of Climate. American Meteorological Society, 45 Beacon Street, Boston, MA 02108-3693. (617) 227-2425. FAX (617) 742-8718. Monthly. $175.00 per year.

Journal of the Atmospheric Sciences. American Meteorological Society, 45 Beacon Street, Boston, MA 02108-3693. (617) 227-2425. FAX (617) 742-8718. Semi-monthly. $320.00 per year.

Monthly Weather Review. American Meteorological Society, 45 Beacon Street, Boston, MA 02108-3693. (617) 227-2425. FAX (617) 742-8718. Monthly. $205.00 per year.

National Weather Digest. National Weather Association, 4400 Stamp Road, Room 404, Temple Hills, MD 20748. (301) 899-3784.

Weather. Royal Meteorological Society, 104 Oxford Road, Reading, Berks RG1 7LJ, England. Monthly. $44.00 per year.

Weatherwise. Heldref Publications, 1319 Eighteenth Street, N.W., Washington, DC 20036-1802. (202) 296-6267. FAX (202) 296-5149. Bi-monthly. $28.00 per year.

RESEARCH CENTERS AND INSTITUTES

Center For Climate Resesarch. Columbia University. Lamont-Doherty Geological Observatory. Palisades, NY 10964. (914) 359-2900.

Center For Climatic Research. University of Delaware, Department of Geography, Newark, DE 19716. (302) 451-8998.

Center For Climatic Research. University of Wisconsin-Madison, 1225 West Dayton Street, Madison, WI 53706. (608) 262-2839. FAX (608) 262-5964.

Climate Research Institute. Oregon State University. Strand Hall, Room 326, Corvallis, OR 97331-2209.

Florida State Climate Center. Florida State University, 431 Love Building, Tallahassee, FL 32306. (904) 644-3417.

National Center For Atmospheric Research. P.O. Box 3000, Boulder, CO 80307. (303) 497-1000. FAX (303) 497-1654. ++++++

Alaska Sar (synthetic Aperature Radar) Facility. University of Alaska Fairbanks, Geophysical Institute, Fairbanks, AK 99775-0800. (907) 474-7954. = ATLANTIC OCEANOGRAPHIC and METEOROLOGICAL LABORATORY. 4301 Rickenbacker Causeway, Miami, FL 33149. (305) 361-4300.

Atmospheric Sciences Research Center. State University of New York, 100 Fuller Road. Albany NY 12205. (518) 442-3820. FAX (518) 442-3867.

Center For Atmospheric theory and Analysis. University of Colorado-Boulder, Campus Box 391, Boulder CO 80309-0391. (303) 492-6487. FAX (303) 492-0642.

Center For Climate Research. Columbia University, Lamont-Doherty Geological Observatory. Palisades, NY 10964. (914) 359-2900.

Centre For Climate and Global Change Research. McGill University, 805 Sherbrooke Street, West, Montreal, PQ Canada H3A 2K6. (514) 398-3759. (514) 398-6115.

Centre For Climatic Chemistry. Institute of Ocean Sciences. Canadian Department of Fisheries and Oceans, P.O. Box 6000, 9860 West Saanich Road, Sidney, BC, Canada V9L 4B2. (604) 356-6329.

Climatic Research Institute. Oregon State University. Strand Hall, Room 326, Corvallis, OR 97331-2209. (503) 737-5705. = CLIMATE ANALYSIS CENTER. National Weather Service, 5200 Auth Road, Camp Springs, MD 20746. (301) 763-8167. = COOPERATIVE INSTITUTE FOR MESOSCALE METEOROLOGICAL STUDIES. University of Oklahoma, 401 East Boyd, Norman, OK 73019. (405) 325-3041.

Cooperative Institute For Meteorological Satellite Studies. Space Science & Engineering Center, University of Wisconsin-Madison, 1225 West Dayton Street, Madison, WI 53706. (608) 263-7435.

Cooperative Institute For Research in the Atmosphere. Colorado State University, FootHills Campus. Fort Collins, CO 80523. (303) 491-8448.

Florida State Climate Center. Florida State University. 431 Love Building, Tallahassee, FL 32306. (904) 644-3417.

Geophysical Fluid Dynamics Laboratory. Princeton University, P.O. Box 308, Princeton, NJ 08542. (609) 452-6500.

Geophysical Institute. University of Alaska Fairbanks, C.T. Elvey Building, Fairbanks, AK 99770-0800. (907) 474-7282. FAX (907) 474-7290.

Goddard Space Flight Center. Laboratory for Atmospheres, Mail Code 610, Greenbelt, MD 20771. (301) 286-5002. = HURRICANE RESEARCH DIVISION. ATLANTIC OCEANOGRAPHIC and METEOROLOGICAL LABORATORY. 4301 Rickenbacker Causeway, Miami, FL 33149. (305) 361-4400.

Institute of Atmospheric Sciences. South Dakota School of Mines and Technology. 501 East Saint Joseph Street, Rapid City, SD 57701-3995. (605) 394-2291. FAX (605) 394-6131.

Institute of Snow Research. Michigan Technological University. Houghton, MI 49931. (906) 487-2750. FAX (906) 487-2202.

Joint Institute For Marine and Atmospheric Research. University of Hawaii at Manoa, 1000 Pope Road, MSB 312, Honolulu, HI 96822. (808) 956-8083. FAX (808) 956-4104.

National Center For Atmospheric Research. P.O. Box 3000, Boulder, CO 80307. (303) 496-1000. FAX (303) 497-1654.

National Hurricane Center. 1320 South Dixie Highway, Coral Gables, FL 33146. (305) 666-4612.

National Meteorological Center. World Weather Building, Room 101, 5200 Auth Road, Camp Springs, MD 20746. (301) 763-8016.

National Severe Storms Forecast Center. 601 East 12th Street, Kansas City, MO 64106. (816) 374-5922.

National Severe Storms Laboratory. National Oceanic and Atmospheric Administration Environment Research Laboratories, 1313 Halley Circle, Norman, OK 73069. (405) 360-3620. FAX (405) 366-0472.

National Weather Service. 1325 East-West Highway, Silver Spring, MD 20910. (301) 427-7689.

Program in Atmospheric and Ocenic Sciences. Princeton University. P.O. CN 710, Sayre Hall, Forrestal Campus, Princeton, NJ 08544-0710. (609) 258-6571. FAX (609) 258-2850.

Radar Systems and Remote Sensing Laboratory. University of Kansas. 2291 Irving Hill Road, Lawrence, Kansas. 66045-2969. (913) 864-4835. FAX (913) 864-7789.

Radar Weather Observatory. McGill University, Macdonald College P.O. Box 241, Saint Anne de Bellevue, PQ Canada, H9X 1C0. (514) 398-7733. FAX (514) 398-7755.

University Corporation For Atmospheric Research. P.O. Box 3000, Boulder, CO 80307-3000. Consortium of 58 universities with doctoral programs in atmospheric and/or oceanic sciences. (303) 497-1650. FAX (303) 497-1654.

GRINDING AND POLISHING

See: MACHINING

GROUND EFFECT MACHINES

See: AERONAUTICS

GROUNDWATER

See also: HYDROGEOLOGY, HYDROLOGY, HYDRODYNAMICS, WATER RESOURCES

ABSTRACT SERVICES AND INDEXES

Aqualine Abstracts. Water Research Centre, PO Box 85, Frankland Rd., Blagrove, Swindon, Wilts. SN5 8YF, England. 1927 to present. 26 times a year. Inquire for price in U.S. Also available online.

Bibliography and Index of Geology. American Geological Institute, 4220 King Street, Alexandria, VA 22302. (703) 379-2480. Fax (703) 379-7563. Monthly. $1295.00 per year. Also available online as GeoRef.

Current Contents: Physical, Chemical and Earth Sciences. Institute for Scientific Information, 3501 Market Street, Philadelphia, PA 19104. (215) 386-0100. FAX (215) 386-6362. Weekly. $360.00 per year.

Fluid Abstracts: Civil Engineering. Elsevier Science [journals], 660 White Plains Rd., Tarrytown, NY 10159-5153. (919) 524-9200. FAX (919) 333-2444. 1991 to present. Monthly plus annual cumulation. $645.00 per year. Also available online as FLUIDEX. Covers civil engineering applications of fluid mechanics, hydraulics, flow metering and measuring, offshore engineering, environmental hydraulics, and related aspects of wind energy, the atmosphere, and aerodynamics.

Geological Abstracts. Elsevier - Geo Abstracts, Regency House, 34 Duke Street, Norwich NR3 3AP, England. Monthly. $760.00 per year. Also available online as GEOBASE.

Geological Society of America. ABSTRACTS WITH PROGRAMS. Geological Society of America. 3300 Penrose Place, P.O. Box 9140, Boulder, CO 80301-9140. (303) 447-2020. Abstracts and programs of the annual conference. Annual. $69.00.

GROUNDWATER

Ency. of Physical Sciences and Engineering Info. Sources

Hydro-abstracts (formerly Water Resources Abstracts). HydroScience Press, 2527 Jackson Street, N.E., Minneapolis, MN 55418. (612) 781-9081. Monthly. $150.00 per year.

Meteorological and Geoastrophysical Abstracts. American Meteorological Society, c/o Inforonics Inc., 550 Newtown Road, Littleton, MA 01460. (508) 486-8976. FAX (508) 486-0027. 1950 to present. Monthly. $950.00 per year. Current abstracts of books, reports, research papers, and miscellaneous literature on environmental sciences, meteorology, astrophysics, hydrology, glaciology, and physical oceanography.

Science Citation Index. Institute for Scientific Information, 3501 Market Street, Philadelphia, PA 19104. (215) 386-0100. FAX (215) 386-6362. Inquire about availability and cost. Also available on CD-ROM.

Selected Water Resources Abstracts. U.S. Geological Survey, Water Resources Scientific Information Center, 425 National Center, Reston, VA 22092. (703) 648-6820. Monthly. $115.00 per year. Also available on CD-ROM.

ANNUAL REVIEWS AND YEARBOOKS

Advances in Hydrosciences. Academic Press Inc., 6277 Sea Harbor Drive, Orlando, FL 32887. (800) 321-5068. Inquire for cost and availability.

ASSOCIATIONS AND PROFESSIONAL SOCIETIES

American Institute of Hydrology. 3416 University Avenue SE, Minneapolis, MN 55414-3328. (612) 379-1030. FAX (612) 379-0169.

American Public Works Association. 106 West 11th Street, Suite 1800, Kansas City, MO 64015-1806. (816) 472-6100. FAX (816) 472-1610.

American Society of Civil Engineers. 345 East 47th Street, New York, NY 10017-2398. (212) 705-7520 or (800) 548-2723.

American Society of Limnology and Oceanography. School of Oceanography, WB-10, University of Washington, Seattle, WA 98195.

American Water Resources Association. 5410 Grosvenor Lane, Suite 220, Bethesda, MD 20814-2192. (301) 493-8600. FAX (301) 493-5844.

American Water Works Association. 6666 W. Quincy Avenue, Denver, CO 80235. (303) 794-7711. FAX (303) 794-7310.

Association of Ground Water Scientists and Engineers. A division of the National Water Well Association, 6375 Riverside Drive, Dublin, OH 43017. (614) 761-1711. FAX (614) 761-3446.

Ground Water Protection Council. 827 NW 63rd Street, Suite 103, Oklahoma City, OK 73116-7639. (405) 848-0690. FAX (405) 848-0722.

National Ground Water Association, 6375 Riverside Drive, Dublin, OH 43017. (614) 761-1711. FAX (614) 761-3446.

National Water Resources Association. 3800 N. Fairfax Drive, Suite 4, Arlington, VA 22203-1703. (703) 524-1544. FAX (703) 524-1548.

National Water Well Association. 6375 Riverside Drive, Dublin, OH 43017. (614) 761-1711. FAX (614) 761-3446.

Universities Council On Water Resources. 4543 Faner Hall, Department of Geography, Southern Illinois University at Carbondale, Carbondale, IL 62901-4526. (618) 536-7571. FAX (618) 453-2671.

Water Environment Federation. 601 Wythe Street, Alexandria, VA 22314-1994. (703) 684-2400. FAX (703) 684-2492. Formerly the Water Pollution Control Federation.

Water Quality Association. 4151 Naperville Road, Lisle, IL 60532. (708) 505-0160. FAX (708) 505-9637.

BIBLIOGRAPHIES

Geraghty & Miller's Groundwater Bibliography. Fifth edition. Lewis Publishers, 121 S. Main St., PO Box 519, Chelsea, MI 48118. (313) 475-8619 or (800) 272-7737. 1991. $79.95.

DIRECTORIES AND BIOGRAPHICAL SOURCES

Directory of Engineering Societies and Related Organizations. Gordon Davis. 13th edition. American Association of Engineering Societies, 1111 19th Street NW, Suite 608, Washington, DC 20036. (202) 296-2237 or (800) 658-8897. 1989. Inquire for price.

International Research Centers Directory. Gale Research, 835 Penobscot Building, Detroit, MI 48226-4094. (313) 961-2242. (800) 877-4253. FAX (313) 961-6083. 8th edition. 1995. $410.00

Research Centers Directory. Gale Research, 835 Penobscot Building, Detroit, MI 48226-4094. (313) 961-2242. (800) 877-4253. FAX (313) 961-6083. $485.00.

Scientific and Technical Organizations and Agencies Directory. Gale Research, 835 Penobscot Building, Detroit, MI 48226-4094. (313) 961-2242. (800) 877-4253. FAX (313) 961-6083. 4th edition. 1996. $195.00.

ENCYCLOPEDIAS AND DICTIONARIES

The Encyclopedia of the Solid Earth Sciences. Philip Kearey, editor-in-chief. Blackwell Scientific Publications, 238 Main Street, Cambridge, MA 02142. (617) 876-7000 or (800) 759-6102. FAX (617) 876-7022. 1993. $60.00.

Encyclopedia of Physical Science and Technology. Robert A. Meyers, ed. 18 volumes. Academic Press Inc., 6277 Sea Harbor Drive, Orlando, FL 32887. (800) 321-5068. 1992. $2100.00.

McGraw-Hill Dictionary of Scientific and Technical Terms. Sybil P. Parker, ed. 5th edition. McGraw-Hill Publishing Company, 1221 Avenue of the Americas, New York, NY 10020. (800) 262-4729 or (212) 512-3825. 1993. $110.50.

GENERAL WORKS

Geochemistry, Groundwater and Pollution. C.A. Appelo and D. Postma. Ashgate Publishing Co., Old Post Rd., Brookfield, VT 05036. (802) 276-3162. FAX (802) 276-3837. $45.00.

Geomorphology and Groundwater. A.G. Brown, editor. John Wiley and Sons, Inc., 605 Third Avenue, New York, NY 10158. (800) 526-5368 or (212) 850-6000. 1995. $89.95.

Groundwater Contamination and Control. Uri Zoller. Marcel Dekker, Inc., 270 Madison Avenue, New York, NY 10016. (212) 696-9000. FAX (212) 685-4540. 1994. $195.00.

Groundwater Ecology. Janine Gilbert, et al., editors. Academic Press Inc., 6277 Sea Harbor Drive, Orlando, FL 32887. (800) 321-5068. 1994. $74.95.

Hydrology in Practice. E. Shaw. 3rd edition. Chapman & Hall, 115 Fifth Avenue, New York, NY 10211-0906. 1994. $36.95.

Introducing Groundwater. M. Price. 2nd edition. Chapman & Hall, 115 Fifth Avenue, New York, NY 10211-0906. 1996. $23.95.

HANDBOOKS AND MANUALS

Groundwater: Managing the Unseen Resources: A Handbook For States. Edwin H. Clark and Philip J. Cherry. World Wildlife Fund & the Conservation Foundation, 1250 24th Street NW, Washington, DC 20037. (202) 293-4800. FAX (410) 516-6998. 1992. $8.50.

Handbook of Hydrology. David R. Maidment, editor in chief. McGraw-Hill Publishing Company, 1221 Avenue of the Americas, New York, NY 10020. (800) 262-4729 or (212) 512-3825. 1993. $115.00.

Practical Groundwater Hydrology. Michael E. Renz. Lewis Publishers, 121 S. Main St., PO Box 519, Chelsea, MI 48118. (313) 475-8619 or (800) 272-7737. 1995. Inquire for cost and availability.

ONLINE DATABASES AND CD-ROMS

Compendex Plus. Engineering Information, Inc., 345 East 47th Street, New York, NY 10017. (212) 705-7600 or (800) 221-1044. Contains citations with abstracts to worldwide literature in engineering and technology, from 1970 to present. Available on online BRS,(800) 289-4277, DIALOG, (800) 334-2564, ORBIT (800) 456-7248, and STN International, FIZ Karlsruhe, P.O. Box 2465, W-7500, Karlsruhe 1, Germany, online services. Also available on CD-ROM. Inquire as to cost and availability.

Current Contents Search. Institute for Scientific Information, 3501 Market Street, Philadelphia, PA 19104. (215) 386-0100. FAX (215) 386-6362. Contains citations to articles listed in the table of contents of science and technology journals. Also articles in social sciences and life sciences journals. Available on BRS,(800) 289-4277, DIALOG,(800) 334-2564, online services. Inquire as to cost and availability.

Dissertation Abstracts. University Microfilms International, 300 North Zeeb Road, Ann Arbor, MI 48106. (800) 521-0600 or (313) 761-4700. Scope includes virtually all doctoral dissertations accepted at accredited American institutions from 1861 to present in 252 subject areas. Available on BRS,(800) 289-4277, DIALOG,(800) 334-2564, and OCLC EPIC,(800) 848-5878, online services. Also available on CD-ROM. Inquire as to cost and availability.

Earth Sciences. U.S. Geological Survey, 12201 Sunrise Valley Drive, Reston, VA 22092-9998. (703) 648-4460. CD-ROM of earth science databases including the U.S. Geological Survey Library, Earth Science Data Directory, and GEOINDEX, citations to published geological maps. $350.00 per year with quarterly updates.

Geoarchive. Geosystems, P.O. Box 1024, Westminster, London, England, SW1 P 2JL. Citations to literature on geoscience, 1969 to present. Inquire as to online cost and availability.

Geobase. Elsevier - Geo Abstracts, Regency House, 34 Duke Street, Norwich NR3 3AP, England. Contains citations to the worldwide earth science literature from 1980 to date. Available on DIALOG, ORBIT online services. Inquire as to cost and availability.

GEOREF: Bibliography and Index of Geology. American Geological Institute, 4220 King Street, Alexandria, VA 22302. (703) 379-2480. Fax (703) 379-7563. Monthly. Inquire for price and availability.

NTIS Bibliographic Database. National Technical Information Service, 5285 Port Royal Road, Springfield, VA 22161. (703) 487-4929 or FAX (703) 321-8199. Broad coverage of government-sponsored science and technology research reports, 1964 to present. Available on BRS,(800) 289-4277, DIALOG, (800) 334-2564, ORBIT, (800) 456-7248, and STN International, FIZ Karlsruhe, P.O. Box 2465, W-7500, Karlsruhe 1, Germany, online services. Also available on CD-ROM. Inquire as to cost and availability.

Water Resources Abstracts (wra). U.S. Geological Survey, Water Resources Scientific Information Center, 12201 Sunrise Valley Drive, Reston, VA 22092-9998. (703) 648-4460. Contains citations with abstracts to scientific and technical literature on the water-sources-related aspects of the physical, social and life sciences, from 1968 to present. Available on DIALOG,(800) 334-2564, and ORBIT,(800) 456-7248, online services. Inquire as to cost and availability.

Waternet. American Water Works Association, Technical Library. Available on DIALOG online services. Citations to literature on water quality, water utility management, water analysis, water pollution, and related areas, 1971 to present. Available on DIALOG,(800) 334-2564, online service. Inquire as to cost and availability.

PERIODICALS

American Water Works Association Journal. 6666 W. Quincy Avenue, Denver, CO 80235. (303) 794-7711. FAX (303) 794-7310. 1914 to present. Monthly. $85.00 to libraries and government agencies only.

Ground Water. Ground Water Publishing Company, National Well Water Association, 6375 Riverside Drive, Dublin, OH 43017. (614) 761-1711. FAX (614) 761-3446. 1963 to present. Bi-monthly. $90.00 per year.

Ground Water Monitoring and Remediation. Ground Water Publishing Company, National Well Water Association, 6375

GROUNDWATER

Ency. of Physical Sciences and Engineering Info. Sources

Riverside Drive, Dublin, OH 43017. (614) 761-1711. FAX (614) 761-3446. 1981 to present. Quarterly. $30.00 per year.

Journal of Hydrology. Elsevier Science Inc., Box 882, Madison Square Station, New York, NY 10159. (212) 989-5800. FAX (212) 633-3990. 1963 to present. 48 times a year. $2309.00.

Journal of Irrigation and Drainage. American Society of Civil Engineers, Irrigation and Drainage Division, 345 East 47th Street, New York, NY 10017-2398. (212) 705-7520 or (800) 548-2723. 1956 to present. Bi-monthly. $136.00 per year.

Journal of Water Resources Planning and Management. American Society of Civil Engineers, Water Resources Planning and Management Division, 345 East 47th Street, New York, NY 10017-2398. (212) 705-7520 or (800) 548-2723. Bi-monthly. $112.00 per year.

Limnology and Oceanography. American Society of Limnology and Oceanography, School of Oceanography, WB-10, University of Washington, Seattle, WA 98195. 1956 to present. Eight times a year. $160.00 per year.

Water Resources Bulletin. American Water Resources Association, 5410 Grosvenor Lane, Suite 220, Bethesda, MD 20814-2192. (301) 493-8600. FAX (301) 493-5844. 1965 to present. Bi-monthly. $15.00.

Water Resources Research. American Geophysical Union, 2000 Florida Avenue, N.W., Washington, DC 20009. (202) 462-6900. FAX (202) 328-0566. 1965 to present. Monthly. $660.00 for non-members.

RESEARCH CENTERS AND INSTITUTES

National Center For Ground Water Research. University of Oklahoma, 200 Telgar Street, Room 127, Norman, OK 73019-0470. (405) 325-5202.

Texas A&M University, Center For Engineering Geosciences. College Station, TX 77843. (409) 845-3224. FAX (409) 845-6162.

University of Cincinnati, Groundwater Research Center. College of Engineering, Mail Location 18, Cincinnati, OH 45221-0018. (513) 475-2933.

Wright State University, Center For Ground Water Management. 056 Library, Dayton, OH 45435. (513) 873-3648. FAX (513) 873-4106.

GROUNDWATER POLLUTION

See also: GROUNDWATER, HYDROGEOLOGY, HYDROLOGY, HYDRODYNAMICS, WATER POLLUTION, WATER RESOURCES

ABSTRACT SERVICES AND INDEXES

Bibliography and Index of Geology. American Geological Institute, 4220 King Street, Alexandria, VA 22302. (703) 379-2480. Fax (703) 379-7563. Monthly. $1295.00 per year. Also available online as GeoRef.

Current Contents: Physical, Chemical and Earth Sciences. Institute for Scientific Information, 3501 Market Street, Philadelphia, PA 19104. (215) 386-0100. FAX (215) 386-6362. Weekly. $360.00 per year.

Environmental Abstracts. R.R. Bowker Inc., 121 Chanlon Road, New Providence, NJ 07974. (800) 521-8110 or (908) 464-6800. Contains citations, most with abstracts, to literature relating to environmental issues and problems, from 1971 to present. Inquire as to cost and availability.

Fluid Abstracts: Civil Engineering. Elsevier Science [journals], 660 White Plains Rd., Tarrytown, NY 10159-5153. (919) 524-9200. FAX (919) 333-2444. 1991 to present. Monthly plus annual cumulation. $645.00 per year. Also available online as FLUIDEX. Covers civil engineering applications of fluid mechanics, hydraulics, flow metering and measuring, offshore engineering, environmental hydraulics, and related aspects of wind energy, the atmosphere, and aerodynamics.

Geological Abstracts. Elsevier - Geo Abstracts, Regency House, 34 Duke Street, Norwich NR3 3AP, England. Monthly. $760.00 per year. Also available online as GEOBASE.

Geological Society of America. ABSTRACTS WITH PROGRAMS. Geological Society of America. 3300 Penrose Place, P.O. Box 9140, Boulder, CO 80301-9140. (303) 447-2020. Abstracts and programs of the annual conference. Annual. $69.00.

Hydro-abstracts (formerly Water Resources Abstracts). HydroScience Press, 2527 Jackson Street, N.E., Minneapolis, MN 55418. (612) 781-9081. Monthly. $150.00 per year.

Meteorological and Geoastrophysical Abstracts. American Meteorological Society, c/o Inforonics Inc., 550 Newtown Road, Littleton, MA 01460. (508) 486-8976. FAX (508) 486-0027. 1950 to present. Monthly. $950.00 per year. Current abstracts of books, reports, research papers, and miscellaneous literature on environmental sciences, meteorology, astrophysics, hydrology, glaciology, and physical oceanography.

Pollution Abstracts, Cambridge Scientific Abstracts, 7200 Wisconsin Avenue, Bethesda, MD 20814-4823. (301) 961-6750. FAX (301) 961-6720. Inquire for cost and availability.

Science Citation Index. Institute for Scientific Information, 3501 Market Street, Philadelphia, PA 19104. (215) 386-0100. FAX (215) 386-6362. Inquire about availability and cost. Also available on CD-ROM.

Selected Water Resources Abstracts. U.S. Geological Survey, Water Resources Scientific Information Center, 425 National Center, Reston, VA 22092. (703) 648-6820. Monthly. $115.00 per year. Also available on CD-ROM.

ANNUAL REVIEWS AND YEARBOOKS

Advances in Hydrosciences. Academic Press Inc., 6277 Sea Harbor Drive, Orlando, FL 32887. (800) 321-5068. Inquire for cost and availability.

ASSOCIATIONS AND PROFESSIONAL SOCIETIES

American Institute of Hydrology. 3416 University Avenue SE, Minneapolis, MN 55414-3328. (612) 379-1030. FAX (612) 379-0169.

American Public Works Association. 106 West 11th Street, Suite 1800, Kansas City, MO 64015-1806. (816) 472-6100. FAX (816) 472-1610.

American Society of Civil Engineers. 345 East 47th Street, New York, NY 10017-2398. (212) 705-7520 or (800) 548-2723.

American Water Resources Association. 5410 Grosvenor Lane, Suite 220, Bethesda, MD 20814-2192. (301) 493-8600. FAX (301) 493-5844.

American Water Works Association. 6666 W. Quincy Avenue, Denver, CO 80235. (303) 794-7711. FAX (303) 794-7310.

Association of Ground Water Scientists and Engineers. A division of the National Water Well Association, 6375 Riverside Drive, Dublin, OH 43017. (614) 761-1711. FAX (614) 761-3446.

Ground Water Protection Council. 827 NW 63rd Street, Suite 103, Oklahoma City, OK 73116-7639. (405) 848-0690. FAX (405) 848-0722.

National Water Resources Association. 3800 N. Fairfax Drive, Suite 4, Arlington, VA 22203-1703. (703) 524-1544. FAX (703) 524-1548.

National Water Well Association. 6375 Riverside Drive, Dublin, OH 43017. (614) 761-1711. FAX (614) 761-3446.

Water Environment Federation. 601 Wythe Street, Alexandria, VA 22314-1994. (703) 684-2400. FAX (703) 684-2492. Formerly the Water Pollution Control Federation.

Water Quality Association. 4151 Naperville Road, Lisle, IL 60532. (708) 505-0160. FAX (708) 505-9637.

BIBLIOGRAPHIES

Geraghty & Miller's Groundwater Bibliography. Fifth edition. Lewis Publishers, 121 S. Main St., PO Box 519, Chelsea, MI 48118. (313) 475-8619 or (800) 272-7737. 1991. $79.95.

DIRECTORIES AND BIOGRAPHICAL SOURCES

Directory of Engineering Societies and Related Organizations. Gordon Davis. 13th edition. American Association of Engineering Societies, 1111 19th Street NW, Suite 608, Washington, DC 20036. (202) 296-2237 or (800) 658-8897. 1989. Inquire for price.

International Research Centers Directory. Gale Research, 835 Penobscot Building, Detroit, MI 48226-4094. (313) 961-2242. (800) 877-4253. FAX (313) 961-6083. 8th edition. 1995. $410.00

Research Centers Directory. Gale Research, 835 Penobscot Building, Detroit, MI 48226-4094. (313) 961-2242. (800) 877-4253. FAX (313) 961-6083. $485.00.

Scientific and Technical Organizations and Agencies Directory. Gale Research, 835 Penobscot Building, Detroit, MI 48226-4094. (313) 961-2242. (800) 877-4253. FAX (313) 961-6083. 4th edition. 1996. $195.00.

ENCYCLOPEDIAS AND DICTIONARIES

The Encyclopedia of the Solid Earth Sciences. Philip Kearey, editor-in-chief. Blackwell Scientific Publications, 238 Main Street, Cambridge, MA 02142. (617) 876-7000 or (800) 759-6102. FAX (617) 876-7022. 1993. $60.00.

Encyclopedia of Physical Science and Technology. Robert A. Meyers, ed. 18 volumes. Academic Press Inc., 6277 Sea Harbor Drive, Orlando, FL 32887. (800) 321-5068. 1992. $2100.00.

Ground Water Chemicals Desk Reference. Lewis Publishers, 121 S. Main St., PO Box 519, Chelsea, MI 48118. (313) 475-8619 or (800) 272-7737. 1991. $89.95.

McGraw-Hill Dictionary of Scientific and Technical Terms. Sybil P. Parker, ed. 5th edition. McGraw-Hill Publishing Company, 1221 Avenue of the Americas, New York, NY 10020. (800) 262-4729 or (212) 512-3825. 1993. $110.50.

GENERAL WORKS

Geochemistry, Groundwater and Pollution. C.A. Appelo and D. Postma. Ashgate Publishing Co., Old Post Rd., Brookfield, VT 05036. (802) 276-3162. FAX (802) 276-3837. 1993. $45.00.

Groundwater Contamination and Control. Uri Zoller. Marcel Dekker, Inc., 270 Madison Avenue, New York, NY 10016. (212) 696-9000. FAX (212) 685-4540. 1994. $195.00.

Ground Water Contamination, Transport & Remediation. Philip B. Bedient. Prentice Hall, 113 Sylvan Avenue, Route 9W, Englewood Cliffs, NJ 07632. (201) 592-2000 or (800) 922-0579. 1994. $64.00.

Practical Groundwater Geophysics. Hoekstra. Lewis Publishers, 121 S. Main St., PO Box 519, Chelsea, MI 48118. (313) 475-8619 or (800) 272-7737. 1995. Inquire for price.

HANDBOOKS AND MANUALS

Groundwater: Managing the Unseen Resources: A Handbook For States. Edwin H. Clark and Philip J. Cherry. World Wildlife Fund & the Conservation Foundation, 1250 24th Street NW, Washington, DC 20037. (202) 293-4800. FAX (410) 516-6998. 1992. $8.50.

Handbook For the Identification, Location & Investigation of Pollution Sources Affecting Ground Water. G. Oudijk and K. Mujica. National Water Well Association, 6375 Riverside Drive, Dublin, OH 43017. (614) 761-1711. FAX (614) 761-3446. 1989. $22.50.

Handbook of Hydrology. David R. Maidment, editor in chief. McGraw-Hill Publishing Company, 1221 Avenue of the Americas, New York, NY 10020. (800) 262-4729 or (212) 512-3825. 1993. $115.00.

Practical Groundwater Hydrology. Michael E. Renz. Lewis Publishers, 121 S. Main St., PO Box 519, Chelsea, MI 48118.

GROUNDWATER POLLUTION

Ency. of Physical Sciences and Engineering Info. Sources

(313) 475-8619 or (800) 272-7737. 1995. Inquire for cost and availability.

Practical Handbook of Soil, Vadose Zone & Ground Water Contamination: Assessment, Prevention, and Remediation. J. Russell Boulding. Lewis Publishers, 121 S. Main St., PO Box 519, Chelsea, MI 48118. (313) 475-8619 or (800) 272-7737. 1995. $89.95.

ONLINE DATABASES AND CD-ROMS

Compendex Plus. Engineering Information, Inc., 345 East 47th Street, New York, NY 10017. (212) 705-7600 or (800) 221-1044. Contains citations with abstracts to worldwide literature in engineering and technology, from 1970 to present. Available on online BRS,(800) 289-4277, DIALOG, (800) 334-2564, ORBIT (800) 456-7248, and STN International, FIZ Karlsruhe, P.O. Box 2465, W-7500, Karlsruhe 1, Germany, online services. Also available on CD-ROM. Inquire as to cost and availability.

Current Contents Search. Institute for Scientific Information, 3501 Market Street, Philadelphia, PA 19104. (215) 386-0100. FAX (215) 386-6362. Contains citations to articles listed in the table of contents of science and technology journals. Also articles in social sciences and life sciences journals. Available on BRS,(800) 289-4277, DIALOG,(800) 334-2564, online services. Inquire as to cost and availability.

Dissertation Abstracts. University Microfilms International, 300 North Zeeb Road, Ann Arbor, MI 48106. (800) 521-0600 or (313) 761-4700. Scope includes virtually all doctoral dissertations accepted at accredited American institutions from 1861 to present in 252 subject areas. Available on BRS,(800) 289-4277, DIALOG,(800) 334-2564, and OCLC EPIC,(800) 848-5878, online services. Also available on CD-ROM. Inquire as to cost and availability.

Earth Sciences. U.S. Geological Survey, 12201 Sunrise Valley Drive, Reston, VA 22092-9998. (703) 648-4460. CD-ROM of earth science databases including the U.S. Geological Survey Library, Earth Science Data Directory, and GEOINDEX, citations to published geological maps. $350.00 per year with quarterly updates.

Geoarchive. Geosystems, P.O. Box 1024, Westminster, London, England, SW1 P 2JL. Citations to literature on geoscience, 1969 to present. Inquire as to online cost and availability.

Geobase. Elsevier - Geo Abstracts, Regency House, 34 Duke Street, Norwich NR3 3AP, England. Contains citations to the worldwide earth science literature from 1980 to date. Available on DIALOG, ORBIT online services. Inquire as to cost and availability.

GEOREF: Bibliography and Index of Geology. American Geological Institute, 4220 King Street, Alexandria, VA 22302. (703) 379-2480. Fax (703) 379-7563. Monthly. Inquire for price and availability.

NTIS Bibliographic Database. National Technical Information Service, 5285 Port Royal Road, Springfield, VA 22161. (703) 487-4929 or FAX (703) 321-8199. Broad coverage of government-sponsored science and technology research reports, 1964 to present. Available on BRS,(800) 289-4277, DIALOG, (800) 334-2564, ORBIT, (800) 456-7248, and STN International, FIZ Karlsruhe, P.O. Box 2465, W-7500, Karlsruhe 1, Germany, online services. Also available on CD-ROM. Inquire as to cost and availability.

Water Resources Abstracts (wra). U.S. Geological Survey, Water Resources Scientific Information Center, 12201 Sunrise Valley Drive, Reston, VA 22092-9998. (703) 648-4460. Contains citations with abstracts to scientific and technical literature on the water-sources-related aspects of the physical, social and life sciences, from 1968 to present. Available on DIALOG,(800) 334-2564, and ORBIT,(800) 456-7248, online services. Inquire as to cost and availability.

Waternet. American Water Works Association, Technical Library. Available on DIALOG online services. Citations to literature on water quality, water utility management, water analysis, water pollution, and related areas, 1971 to present. Available on DIALOG,(800) 334-2564, online service. Inquire as to cost and availability.

PERIODICALS

American Water Works Association Journal. 6666 W. Quincy Avenue, Denver, CO 80235. (303) 794-7711. FAX (303) 794-7310. 1914 to present. Monthly. $85.00 to libraries and government agencies only.

Ground Water. Ground Water Publishing Company, National Well Water Association, 6375 Riverside Drive, Dublin, OH 43017. (614) 761-1711. FAX (614) 761-3446. 1963 to present. Bi-monthly. $90.00 per year.

Ground Water Monitoring and Remediation. Ground Water Publishing Company, National Well Water Association, 6375 Riverside Drive, Dublin, OH 43017. (614) 761-1711. FAX (614) 761-3446. 1981 to present. Quarterly. $30.00 per year.

Journal of Hydrology. Elsevier Science Inc., Box 882, Madison Square Station, New York, NY 10159. (212) 989-5800. FAX (212) 633-3990. 1963 to present. 48 times a year. $2309.00.

Journal of Irrigation and Drainage. American Society of Civil Engineers, Irrigation and Drainage Division, 345 East 47th Street, New York, NY 10017-2398. (212) 705-7520 or (800) 548-2723. 1956 to present. Bi-monthly. $136.00 per year.

Journal of Water Resources Planning and Management. American Society of Civil Engineers, Water Resources Planning and Management Division, 345 East 47th Street, New York, NY 10017-2398. (212) 705-7520 or (800) 548-2723. Bi-monthly. $112.00 per year.

Water Environment and Technology (formerly Water Pollution Control Federation Journal). Water Environment Federation, 601 Wythe Street, Alexandria, VA 22314-1994. (703) 684-2400. FAX (703) 684-2492. 13 times a year. $144.00.

Water Resources Bulletin. American Water Resources Association, 5410 Grosvenor Lane, Suite 220, Bethesda, MD 20814-2192. (301) 493-8600. FAX (301) 493-5844. 1965 to present. Bi-monthly. $15.00.

Water Resources Research. American Geophysical Union, 2000 Florida Avenue, N.W., Washington, DC 20009. (202) 462-6900. FAX (202) 328-0566. 1965 to present. Monthly. $660.00 for non-members.

RESEARCH CENTERS AND INSTITUTES

Hazardous Substances Research Center For U.S. EPA REGIONS 7 & 8. Department of Chemical Engineering, Durland Hall, Manhattan, KS 66506-5102. (913) 532-5584. FAX (913) 532-7810.

Texas A&m University, Center For Engineering Geosciences. College Station, TX 77843. (409) 845-3224. FAX (409) 845-6162.

University of Oklahoma, Environmental and Ground Water Institute. 200 Felgar Street, Room 127, Norman, OK 73019-0470. (405) 325-5202. FAX (405) 325-7596.

GUIDANCE SYSTEMS

See also: AEROSPACE ENGINEERING, GUIDED MISSILES, ROCKETS

ABSTRACT SERVICES AND INDEXES

Applied Science and Technology Index; A Cumulative Subject Index To English Language Periodicals in the Fields of Aeronautics and Space Science, Computer Technology, Chemistry, Construction Industry, Energy and Related Areas. H.W. Wilson Co., 950 University Avenue, Bronx, NY 10452. (800) 367-6770 or (212) 588-8400. FAX (718) 590-1617. From 1958 to present. Monthly. Inquire about cost and availability. Also available on CD-ROM and online.

Computer Abstracts, MCBUniversity Press Ltd., PO Box 10812, Birmingham, AL 35201-0812. (800) 633-4931. FAX (205) 995-1588. Monthly. Covers computer theory, data, hardware, systems, networks, human-computer interaction, artificial intelligence, as well as applications of computers in aerospace, business, CAD/CAM, cartography, civil engineering, electronics and electrical engineering, industrial engineering, mechanical engineering, medicine, structural engineering, etc.

Current Contents: Engineering, Technology, and Applied Sciences. Institute for Scientific Information, 3501 Market Street, Philadelphia, PA 19104. (215) 386-0100. FAX (215) 386-6362. 1970 to present. Weekly. $442.00 per year.

Engineering Index Monthly. Engineering Information, Inc., Castle Point on the Hudson, Hoboken, NJ 07030. (800) 221-1044. FAX (212) 832-1857. Monthly. $2300.00 per year. Also available online as COMPENDEX, and also on CD-ROM. Covers chemical engineering, computers, electrical engineering, civil engineering, metals and mining, industrial management, and mechanical engineering.

International Aerospace Abstracts. Technical Information Service, American Institute of Aeronautics and Astronautics, Inc., 555 West 57th St., New York, NY 10019. (212) 247-6500. FAX (212) 582-4861. Semi-monthly. $1295.00 per year.

Science Citation Index. Institute for Scientific Information, 3501 Market Street, Philadelphia, PA 19104. (215) 386-0100. FAX (215) 386-6362. Inquire about availability and cost. Also available on CD-ROM.

Scientific and Technical Aerospace Reports (star). National Aeronautics and Space Administration. NASA Center for Aero-

space Information, Box 8757, BWI Airport, Baltimore, MD 21240. (301) 621-0153. Monthly. Inquire about availability and cost. ALso available through the NASA online retrieval service (RECON), and through the Aerospace Database through DIALOG.

ASSOCIATIONS AND PROFESSIONAL SOCIETIES

American Astronautical Society. 6352 Rolling Mill Place, Suite 102, Springfield, VA 22152. (703) 866-0020. FAX (703) 866-3526.

American Institute of Aeronautics and Astronautics. 370 L'Enfant Promenade, S.W., Washington, DC 20024. (202) 646-7400. (202) 646-7400. FAX (202) 646-7508.

National Space Society, 922 Pennsylvania Avenue Se, Washington, DC 20003-2140. (202) 543-1900. FAX (202) 546-4189.

Society of Automotive Engineers. SAE, Inc., 400 Commonwealth Dr., Warrendale, PA 15096. (412) 776-4841.

DIRECTORIES AND BIOGRAPHICAL SOURCES

American Astronautical Society Directory. American Astronautical Society, 6352 Rolling Mill Place, Suite 102, Springfield, VA 22152. Inquire for cost and availability.

American Institute of Aeronautics and Astronautics Roster. American Institute of Aeronautics and Astronautics, 370 L'Enfant Promenade, S.W., Washington, DC 20024. (202) 646-7400. Inquire for availability.

International Research Centers Directory. Gale Research, 835 Penobscot Building, Detroit, MI 48226-4094. (313) 961-2242. (800) 877-4253. FAX (313) 961-6083. 8th edition. 1995. $410.00

Research Centers Directory. Gale Research, 835 Penobscot Building, Detroit, MI 48226-4094. (313) 961-2242. (800) 877-4253. FAX (313) 961-6083. $485.00.

Scientific and Technical Organizations and Agencies Directory. Gale Research, 835 Penobscot Building, Detroit, MI 48226-4094. (313) 961-2242. (800) 877-4253. FAX (313) 961-6083. 4th edition. 1996. $195.00.

Who's Who in Engineering. American Association of Engineering Societies, 1111 19th Street NW, Suite 608, Washington, DC 20036. (202) 296-2237 or (800) 658-8897. 8th edition. 1991. Inquire for price.

GENERAL WORKS

Automatic Control of Aircraft and Missiles. John H. Blakelock. 2d ed. John Wiley and Sons, Inc., 605 Third Avenue, New York, NY 10158. (800) 526-5368 or (212) 850-6000. 1991. $79.95.

Inventing Accuracy: A Historical Sociology of Nuclear Missile Guidance. Donald A. Mackenzie. MIT Press, 55 Hayward Street, Cambridge, MA 02142. (617) 253-2884 or (800) 356-0343. FAX (617) 253-1709. 1990. $35.00.

Modern Navigation Guidance and Control Processing. Ching-Fang Lin. Prentice Hall, 113 Sylvan Avenue, Route 9W,

GUIDANCE SYSTEMS

Ency. of Physical Sciences and Engineering Info. Sources

Englewood Cliffs, NJ 07632. (201) 592-2000 or (800) 922-0579. 1991. $60.00.

Tactical and Strategic Missile Guidance. Paul Zarchan. American Institute of Aeronautics and Astronautics, 370 L'Enfant Promenade, S.W., Washington, DC 20024. (202) 646-7400. 1990. $65.95.

ONLINE DATABASES AND CD-ROMS

Aerospace Database. American Institute of Aeronautics and Astronautics, 370 L'Enfant Promenade, S.W., Washington, DC 20024. (202) 646-7400. Worldwide published literature on research and development in aerospace and related areas, with abstracts. Covers 1962 to present. Available on DIALOG, (800) 334-2564, online service. Also available on CD-ROM. Inquire as to cost and availability.

Compendex Plus. Engineering Information, Inc., 345 East 47th Street, New York, NY 10017. (212) 705-7600 or (800) 221-1044. Contains citations with abstracts to worldwide literature in engineering and technology, from 1970 to present. Available on online BRS,(800) 289-4277, DIALOG, (800) 334-2564, ORBIT (800) 456-7248, and STN International, FIZ Karlsruhe, P.O. Box 2465, W-7500, Karlsruhe 1, Germany, online services. Also available on CD-ROM. Inquire as to cost and availability.

CSA Engineering. Cambridge Scientific Abstracts, 7200 Wisconsin Avenue, Suite 601, Bethesda, MD 20814. (301) 961-6750 or (800) 843-7751. Contains citations and abstracts of international periodicals and research literature covering all fields of engineering and science and technology,including computer and information science, electronics, mechanical engineering, solid state materials, 1981 to present. Available on BRS,(800) 289-4277, online service.Inquire as to cost and availability.

Current Contents Search. Institute for Scientific Information, 3501 Market Street, Philadelphia, PA 19104. (215) 386-0100. FAX (215) 386-6362. Contains citations to articles listed in the table of contents of science and technology journals. Also articles in social sciences and life sciences journals. Available on BRS,(800) 289-4277, DIALOG,(800) 334-2564, online services. Inquire as to cost and availability.

Dissertation Abstracts. University Microfilms International, 300 North Zeeb Road, Ann Arbor, MI 48106. (800) 521-0600 or (313) 761-4700. Scope includes virtually all doctoral dissertations accepted at accredited American institutions from 1861 to present in 252 subject areas. Available on BRS,(800) 289-4277, DIALOG,(800) 334-2564, and OCLC EPIC,(800) 848-5878, online services. Also available on CD-ROM. Inquire as to cost and availability.

International Aerospace Abstracts. American Institute of Aeronautics and Astronautics, 370 L'Enfant Promenade, S.W., Washington, DC 20024. (202) 646-7400. Contains references and abstracts of journal and monograph literature relating to aerospace science and technology, from 1963 to present. Available through the NASA/RECON system of the National Aeronautics and Space Administration only.

NASA Database. American Institute of Aeronautics and Astronautics, 370 L'Enfant Promenade, S.W., Washington, DC 20024. (202) 646-7400. Citations and abstracts of aeronautics and astronautics literature, 1962 to present. Also contains citations from *STAR, Scientific and Technical Aerospace Reports, and International Aerospace Abstracts.* Available through NASA/RECON online service. Inquire as to cost and availability.

NTIS Bibliographic Database. National Technical Information Service, 5285 Port Royal Road, Springfield, VA 22161. (703) 487-4929 or FAX (703) 321-8199. Broad coverage of government-sponsored science and technology research reports, 1964 to present. Available on BRS,(800) 289-4277, DIALOG, (800) 334-2564, ORBIT, (800) 456-7248, and STN International, FIZ Karlsruhe, P.O. Box 2465, W-7500, Karlsruhe 1, Germany, online services. Also available on CD-ROM. Inquire as to cost and availability.

SAE Global Mobility Database. Society of Automative Engineers (SAE), Electronic Publishing Division, 400 Commonwealth Drive, Warrendale, PA 15098. (412) 776-4841. Contains citations with abstracts to technical papers on automotive and aerospace technology and vehicular-related industries that have been presented at SAE conferences. Covers 1906 to present. Available on ORBIT,(800) 456-7248,online service. Inquire as to cost and availability.

SCISEARCH. Institute for Scientific Information, 3501 Market Street, Philadelphia, PA 19104. (800) 523-1850 or (215) 386-0100. Broad multidisciplinary title and author index to the international literature of science and technology, 1974 to present. Available on DIALOG,(800) 334-2564, and ORBIT,(800) 456-7248, online services. Also available on CD-ROM. Inquire as to cost and availability.

TRIS (Transportation Research Information). National Academy of Science, 2101 Constitution Avenue, N.W., Washington, DC 20418. (202) 334-3313 or (800) 624-6242. Citations with abstracts of literature on transportation, including air, highway, rail, maritime and other modes from 1968 to present. Available on DIALOG,(800) 334-2564, online service. Inquire as to cost and availability.

WILSONLINE. H.W. Wilson Company, 950 University Avenue, Bronx, NY 10452. (800) 367-6770 or (212) 588-8400. Makes available online versions of the printed H.W. Wilson indexes including Applied Science and Technology Index, Business Periodicals Index, General Science Index, and Readers' Guide to Periodical Literature. Period covered is generally 1983 to present. Available on BRS,(800) 289-4277, DIALOG,(800) 334-2564, and OCLC EPIC,(800) 848-5878, online services. Also available on CD-ROM. Inquire as to cost and availability.

PERIODICALS

Aerospace America. American Institute of Aeronautics and Astronautics, 370 L'Enfant Promenade SW, Washington, DC 20024-2518. (202) 646-7471. 1932 to present. Monthly. $70 annually (non-members).

Aerospace Engineering. Society of Automotive Engineers, 400 Commonwealth Dr., Warrendale, PA 15096. (412) 776-4841. 1981 to present. Monthly. $48.00 per year.

AIAA Journal. American Institute of Aeronautics and Astronautics, 370 L'Enfant Promenade SW, Washington, DC 20024-2518. (202) 646-7471. 1963 to present. Monthly. $52.00 per year (members) and $435.00 per year (non-members).

Aviation Week and Space Technology. 1221 Avenue of the Americas, New York, NY 10020. (212) 512-2999 or (800) 525-5003. Weekly. $82.00 per year.

IEEE Transactions On Aerospace and Electronic Systems. Institute of Electrical and Electronics Engineers, 445 Hoes Lane, PO Box 1331, Piscataway, NJ 08855-1331. (908) 981-0060.

Quarterly. $10.00 per issue (members), $20.00 per issue (non-members).

Journal of Guidance, Control and Dynamics. American Institute of Aeronautics and Astronautics, 370 L'Enfant Promenade SW, Washington, DC 20024-2518. 1978 to present. Bimonthly. $42.00 (members) and $250.00 (non-members).

Journal of Navigation. 1947 to present. Cambridge University Press, Journals Department, 40 West 20th Street, New York, NY 10011-4211. (212) 924-3900. Three times a year. $165.00 per year.

Journal of Robotic Systems. J. Wiley & Sons, Journals, 605 Third Avenue, New York, NY 10158-0012. (212) 692-6000.1984 to present. 8 times a year. $525.00 per year.

Journal of Spacecraft and Rockets. American Institute of Aeronautics and Astronautics, 370 L'Enfant Promenade SW, Washington, DC 20024-2518. Bimonthly. $33.00 (members) and $215.00 (non-members).

RESEARCH CENTERS AND INSTITUTES

Center For Control and Systems Research, University of Texas At Austin, Aerospace Engineering and Engineering Mechanics, Wrw 408, Austin, TX 78712-1085. (512) 471-4908. FAX (512) 471-3788.

Charles Stark Draper Laboratory Inc., 555 Technology Square, Cambridge, MA 02139. (617) 258-1000.

Goddard Space Flight Center, Greenbelt, MD 20771. (301) 286-7351.

Instrumentation and Control Laboratory, Princeton University, Dept. of MAE, Engineering Quadrangle, Princeton, NJ 08544. (609) 452-5154.

George C. MARSHALL SPACE FLIGHT CENTER, Huntsville, AL 35812. (205) 544-0024.

GUIDED MISSILES

See also: AEROSPACE ENGINEERING, BALLISTICS, ROCKETS

ABSTRACT SERVICES AND INDEXES

Applied Science and Technology Index; A Cumulative Subject Index To English Language Periodicals in the Fields of Aeronautics and Space Science, Computer Technology, Chemistry, Construction Industry, Energy and Related Areas. H.W. Wilson Co., 950 University Avenue, Bronx, NY 10452. (800) 367-6770 or (212) 588-8400. FAX (718) 590-1617. From 1958 to present. Monthly. Inquire about cost and availability. Also available on CD-ROM and online.

Computer Abstracts, MCBUniversity Press Ltd., PO Box 10812, Birmingham, AL 35201-0812. (800) 633-4931. FAX (205) 995-1588. Monthly. Covers computer theory, data, hardware, systems, networks, human-computer interaction, artificial intelligence, as well as applications of computers in aerospace, business, CAD/CAM, cartography, civil engineering, electronics and electrical engineering, industrial engineering, mechanical engineering, medicine, structural engineering, etc.

Engineering Index Monthly. Engineering Information, Inc., Castle Point on the Hudson, Hoboken, NJ 07030. (800) 221-1044. FAX (212) 832-1857. Monthly. $2300.00 per year. Also available online as COMPENDEX, and also on CD-ROM. Covers chemical engineering, computers, electrical engineering, civil engineering, metals and mining, industrial management, and mechanical engineering.

International Aerospace Abstracts. Technical Information Service, American Institute of Aeronautics and Astronautics, Inc., 555 West 57th St., New York, NY 10019. (212) 247-6500. FAX (212) 582-4861. Semi-monthly. $1295.00 per year.

Science Citation Index. Institute for Scientific Information, 3501 Market Street, Philadelphia, PA 19104. (215) 386-0100. FAX (215) 386-6362. Inquire about availability and cost. Also available on CD-ROM.

Scientific and Technical Aerospace Reports (star). National Aeronautics and Space Administration. NASA Center for Aerospace Information, Box 8757, BWI Airport, Baltimore, MD 21240. (301) 621-0153. Monthly. Inquire about availability and cost. ALso available through the NASA online retrieval service (RECON), and through the Aerospace Database through DIALOG.

ASSOCIATIONS AND PROFESSIONAL SOCIETIES

American Astronautical Society. 6352 Rolling Mill Place, Suite 102, Springfield, VA 22152. (703) 866-0020. FAX (703) 866-3526.

American Institute of Aeronautics and Astronautics. The Aerospace Center, 370 L'Enfant Promenade SW, Washington, DC 20024. (202) 646-7400. FAX (202) 646-7508.

National Space Society, 922 Pennsylvania Avenue Se, Washington, DC 20003-2140. (202) 543-1900. FAX (202) 546-4189.

Society of Automotive Engineers. SAE, Inc., 400 Commonwealth Dr., Warrendale, PA 15096. (412) 776-4841.

DIRECTORIES AND BIOGRAPHICAL SOURCES

American Astronautical Society Directory. American Astronautical Society, 6352 Rolling Mill Place, Suite 102, Springfield, VA 22152. Inquire for cost and availability.

American Institute of Aeronautics and Astronautics Roster. American Institute of Aeronautics and Astronautics. 370 L'Enfant Promenade SW, Washington, DC 20024. (202) 646-7400. Inquire for cost and availability.

International Research Centers Directory. Gale Research, 835 Penobscot Building, Detroit, MI 48226-4094. (313) 961-2242. (800) 877-4253. FAX (313) 961-6083. 8th edition. 1995. $410.00

Research Centers Directory. Gale Research, 835 Penobscot Building, Detroit, MI 48226-4094. (313) 961-2242. (800) 877-4253. FAX (313) 961-6083. $485.00.

GUIDED MISSILES

Ency. of Physical Sciences and Engineering Info. Sources

Scientific and Technical Organizations and Agencies Directory.
Gale Research, 835 Penobscot Building, Detroit, MI 48226-
4094. (313) 961-2242. (800) 877-4253. FAX (313) 961-6083.
4th edition. 1996. $195.00.

Who's Who in Engineering. American Association of Engineer-
ing Societies, 1111 19th Street NW, Suite 608, Washington,
DC 20036. (202) 296-2237 or (800) 658-8897. 8th edition. 1991.
Inquire for price.

ENCYCLOPEDIAS AND DICTIONARIES

McGraw-Hill Encyclopedia of Science and Technology. Sybil
P. Parker, ed. 7th edition. 20 volumes. McGraw-Hill Publishing
Company, 1221 Avenue of the Americas, New York, NY 10020.
(800) 262-4729 or (212) 512-3825. 1992. $1900.00

Thesaurus of Scientific, Technical, and Engineering Terms.
Hemisphere Publishing Corporation, 1900 Frost Road, Suite 101,
Bristol, PA 19007-1598. (215) 785-5800 or (800) 821-8312.
FAX (215) 785-5515. 1987. $173.00.

GENERAL WORKS

Doodlebugs and Rockets: The Battle of the Flying Bombs. Bob
Ogley. Froglets Publishing, Kent, England. 1992. Inquire for
price and availability.

*The Early Development of Guided Weapons in the United King-
dom, 1940-1960.* Stephen Robert Twigge. Harwood ACADemic,
c/o Gordon & Breach Science Publishers, 820 Town Center
Drive, Langhorne, PA 19047. (215) 750-2642. FAX (215) 750-
6343. 1993. Inquire for cost and availability.

Icbm: The Making of the Weapon That Changed the World. G.
Harry Stine. Orion Books, Crown Publishing Group, 201 E. 50th
St., New York, NY 10022. (800) 726-0600 or (800) 733-3000.
1991. Inquire for cost and availability.

*U.*S. MISSILE DATA BOOK, 1994. Ted G. Nicholas. Data
Search Associates, Fountain Valley, CA 92708. 1993. Write for
cost and availability.

ONLINE DATABASES AND CD-ROMS

Aerospace Database. American Institute of Aeronautics and
Astronautics, 370 L'Enfant Promenade, S.W., Washington, DC
20024. (202) 646-7400. Worldwide published literature on research and development in aerospace and related areas, with
abstracts. Covers 1962 to present. Available on DIALOG, (800)
334-2564, online service. Also available on CD-ROM. Inquire
as to cost and availability.

Compendex Plus. Engineering Information, Inc., 345 East 47th
Street, New York, NY 10017. (212) 705-7600 or (800) 221-
1044. Contains citations with abstracts to worldwide literature
in engineering and technology, from 1970 to present. Available
on online BRS,(800) 289-4277, DIALOG, (800) 334-2564,
ORBIT (800) 456-7248, and STN International, FIZ Karlsruhe,
P.O. Box 2465, W-7500, Karlsruhe 1, Germany, online services.
Also available on CD-ROM. Inquire as to cost and availability.

CSA Engineering. Cambridge Scientific Abstracts, 7200 Wisconsin Avenue, Suite 601, Bethesda, MD 20814. (301) 961-
6750 or (800) 843-7751. Contains citations and abstracts of in-

ternational periodicals and research literature covering all fields
of engineering and science and technology,including computer
and information science, electronics, mechanical engineering,
solid state materials, 1981 to present. Available on BRS,(800)
289-4277, online service.Inquire as to cost and availability.

Current Contents Search. Institute for Scientific Information,
3501 Market Street, Philadelphia, PA 19104. (215) 386-0100.
FAX (215) 386-6362. Contains citations to articles listed in the
table of contents of science and technology journals. Also articles in social sciences and life sciences journals. Available on
BRS,(800) 289-4277, DIALOG,(800) 334-2564, online services.
Inquire as to cost and availability.

International Aerospace Abstracts. American Institute of Aeronautics and Astronautics, 370 L'Enfant Promenade, S.W., Washington, DC 20024. (202) 646-7400. Contains references and
abstracts of journal and monograph literature relating to aerospace science and technology, from 1963 to present. Available
through the NASA/RECON system of the National Aeronautics
and Space Administration only.

NASA Database. American Institute of Aeronautics and Astronautics, 370 L'Enfant Promenade, S.W., Washington, DC 20024.
(202) 646-7400. Citations and abstracts of aeronautics and astronautics literature, 1962 to present. Also contains citations from
*STAR, Scientific and Technical Aerospace Reports, and Inter-
national Aerospace Abstracts.* Available through NASA/RE-
CON online service. Inquire as to cost and availability.

NTIS Bibliographic Database. National Technical Information
Service, 5285 Port Royal Road, Springfield, VA 22161. (703)
487-4929 or FAX (703) 321-8199. Broad coverage of government-sponsored science and technology research reports, 1964
to present. Available on BRS,(800) 289-4277, DIALOG, (800)
334-2564, ORBIT, (800) 456-7248, and STN International, FIZ
Karlsruhe, P.O. Box 2465, W-7500, Karlsruhe 1, Germany,
online services. Also available on CD-ROM. Inquire as to cost
and availability.

SAE Global Mobility Database. Society of Automative Engineers (SAE), Electronic Publishing Division, 400 Commonwealth Drive, Warrendale, PA 15098. (412) 776-4841. Contains
citations with abstracts to technical papers on automotive and
aerospace technology and vehicular-related industries that have
been presented at SAE conferences. Covers 1906 to present.
Available on ORBIT,(800) 456-7248,online service. Inquire as
to cost and availability.

SCISEARCH. Institute for Scientific Information, 3501 Market
Street, Philadelphia, PA 19104. (800) 523-1850 or (215) 386-
0100. Broad multidisciplinary title and author index to the international literature of science and technology, 1974 to present.
Available on DIALOG,(800) 334-2564, and ORBIT,(800) 456-
7248, online services. Also available on CD-ROM. Inquire as
to cost and availability.

WILSONLINE. H.W. Wilson Company, 950 University Avenue,
Bronx, NY 10452. (800) 367-6770 or (212) 588-8400. Makes
available online versions of the printed H.W. Wilson indexes
including Applied Science and Technology Index, Business
Periodicals Index, General Science Index, and Readers' Guide
to Periodical Literature. Period covered is generally 1983 to
present. Available on BRS,(800) 289-4277, DIALOG,(800) 334-
2564, and OCLC EPIC,(800) 848-5878, online services. Also
available on CD-ROM. Inquire as to cost and availability.

PERIODICALS

Aeronautical Journal. The Royal Aeronautical Society, 4 Hamilton Place, London W1V 0BQ, England. 1897 to present. Monthly except June and August. Inquire for price information in U.S.

Aerospace America. American Institute of Aeronautics and Astronautics, 370 L'Enfant Promenade SW, Washington, DC 20024-2518. (202) 646-7471. 1932 to present. Monthly. $70 annually (non-members).

Aerospace Engineering. Society of Automotive Engineers, 400 Commonwealth Dr., Warrendale, PA 15096. (412) 776-4841. 1981 to present. Monthly. $48.00 per year.

Aerospace Propulsion. McGraw-Hill Inc., Aviation Week Group, 1200 G Street NW, Suite 200, Washington, DC 20005. (202) 383-2350. 1963 to present. 25 times a year. $550.00 a year.

AIAA Journal. American Institute of Aeronautics and Astronautics, 370 L'Enfant Promenade SW, Washington, DC 20024-2518. (202) 646-7471. 1963 to present. Monthly. $52.00 per year (members) and $435.00 per year (non-members).

Aviation Week and Space Technology. 1221 Avenue of the Americas, New York, NY 10020. (212) 512-2999 or (800) 525-5003. Weekly. $82.00 per year.

Journal of Guidance, Control and Dynamics. American Institute of Aeronautics and Astronautics, 370 L'Enfant Promenade SW, Washington, DC 20024-2518. 1978 to present. Bimonthly. $42.00 (members) and $250.00 (non-members).

Journal of Spacecraft and Rockets. American Institute of Aeronautics and Astronautics, 370 L'Enfant Promenade SW, Washington, DC 20024-2518. Bimonthly. $33.00 (members) and $215.00 (non-members).

RESEARCH CENTERS AND INSTITUTES

Center For Control and Systems Research, University of Texas At Austin, Aerospace Engineering and Engineering Mechanics, Wrw 408, Austin, TX 78712-1085. (512) 471-4908. FAX (512) 471-3788.

Goddard Space Flight Center, Greenbelt, MD 20771. (301) 286-7351.

George C. Marshall Space Flight Center, Huntsville, AL 35812. (205) 544-0024.

GUMS AND RESINS

See also: CEMENT, CHEMICAL ENGINEERING, CHEMISTRY, ORGANIC CHEMISTRY, PLASTICS, TEXTILES

ABSTRACT SERVICES AND INDEXES

Adhesives Abstracts. R A P R A Technology Ltd., Shawbury, Shrewsbury, Shropshire SY4 4NR, England. TEL 0939-250383.

FAX 0939-251118. 1988 to present. Monthly. L195 per year. ISSN: 0891-7760.

Applied Science and Technology Index; A Cumulative Subject Index To English Language Periodicals in the Fields of Aeronautics and Space Science, Computer Technology, Chemistry, Construction Industry, Energy and Related Areas. H. W. Wilson Co., 950 University Avenue, Bronx, NY 10452-9978. (212) 588-8400. (800) 367-6770. FAX (718) 590-1617. From 1958 to present. Monthly. Inquire about cost and availability. Also available online (BRS and WILSONLINE) and on CD-ROM. ISSN: 0003-6986.

Chemical Abstracts. Chemical Abstracts Service. 2540 Olentangy River Road, Box 3012, Columbus, OH 43210-0012. (614) 447-3600. FAX (614) 447-3713. 1907 to present. Weekly. $16,800.00 per year. Available online and on CD-ROM. CA is also available in five section groupings.

Current Contents: Engineering, Technology and Applied Sciences. Institute for Scientific Information, 3501 Market Street, Philadelphia, PA 19104. (215) 386-0100. FAX (215) 386-2991. From 1970 to present. Weekly. $442.00 per year. Also available online. Inquire regarding cost and availability. ISSN: 0095-7917.

Engineering Index Monthly. Engineering Information, Inc., Castle Point on the Hudson, Hoboken, NJ 07030. (201) 216-8500. (800) 221-1044. FAX (201) 216-8532. Monthly. $2300.00 per year for monthly issues. $1980 per year for annual cumulation. Available on CD-ROM and online as COMPENDEX. ISSN: 0742-1974; 0360-8557.

General Science Index. H.W. Wilson Company, 950 University Avenue, Bronx, NY 10452. (212) 588-8400. (800) 367-6770. FAX (718) 590-1617. From 1978 to present. Ten issues per year; quarterly and annual cumulations. Service basis. Available on CD-ROM and online. Inquire regarding cost and availability. ISSN: 0162-1963.

Process and Chemical Engineering/chemical Engineering Abstracts. Royal Society of Chemistry, Information Services, Thomas Graham House, Science Park, Milton Road, Cambridge, CB4 4WF, England. Contains citations, mostly with abstracts, of the worldwide literature on chemical engineering; from 1982 to present. Monthly. $610.00 per year. Also available online. ISSN: 0960-5045.

Science Citation Index. SCI. Institute for Scientific Information, 3501 Market Street, Philadelphia, PA 19104. (215) 386-0100. (800) 523-1850. FAX (215) 386-2991. 1961 to present. Six issues per year, plus annual cumulation. $11650.00 per year. Also available online and on CD-ROM. Inquire about price and availability. ISSN: 0036-827X. theORETICAL CHEMICAL ENGINEERING. Royal Society of Chemistry, Information Services, Thomas Graham House, Science Park, Milton Road, Cambridge, CB4 4WF, England. Covers theoretical chemical engineering, including theory and laboratory experimentation. 1964 to present. Monthly. $235.00. ISSN: 0960-5053.

ASSOCIATIONS AND PROFESSIONAL SOCIETIES

American Chemical Society, 1155 16th Street, NW, Washington, DC 20036. (202) 872-4600.

American Institute of Chemical Engineers. 345 East 47th Street, New York, NY 10017. (212) 705-7338.

GUMS AND RESINS

Ency. of Physical Sciences and Engineering Info. Sources

Association of Consulting Chemists and Chemical Engineers, 50 East 41st Street, Suite 92, New York, NY 10017. (212) 684-6255.

Chemical Manufacturers Association. 2501 M Street, N.W., Washington, DC 20037. (202) 887-1182.

Pressure Senstitive Tape Council. 401 North Michigan Avenue, Chicago IL 60611-4267. (312) 644-6610. FAX (312) 527-6640.

Society of Plastics Engineers, Inc., 14 Fairfield Drive, Brookfield, CT 06805-0403. (203) 775-0471. FAX (203) 775-8490.

BIBLIOGRAPHIES

Chemical Information Sources. Gary Wiggins. McGraw-Hill Publishing Company, 1221 Avenue of the Americas, New York, NY 10020. (800) 262-4729 or (212) 512-3825. 1991. $42.50.

Information Sources in Polymers and Plastics. R. T. Adkins. Bowker-Saur, 121 Chanlon Road, New Providence, NJ 07974. (908) 464-6800. (800) 521-8110. 1989. $75.00

Scientific and Technical Books and Serials in Print; An Index To Literature in Science and Technology. R.R. Bowker Inc., 121 Chanlon Road, New Providence, NJ 07974. (908) 464-6800. (800) 521-8110. FAX (908) 665-3502. 1972 to present. Annual. 4 volumes. 1994. $299.95. Also available on compact disc and online. ISSN: 0000-054X.

DIRECTORIES AND BIOGRAPHICAL SOURCES

Adhesives Age Directory. Argus Inc., 6151 Powers Perry Road, NW, Atlanta, GA 30339-2941. (404) 955-2500. FAX (404) 955-0400, 1957 to present. Annual. $51.95. ISSN: 0001-821X.

American Institute of Chemical Engineers. DIRECTORY. American Institute of Chemical Engineers. 345 East 47th Street, New York, NY 10017-2396. (212) 705-7663. Annual.

American Men and Women of Science. Physical and Biological Sciences. R.R. Bowker Inc., 121 Chanlon Road, New Providence, NJ 07974. (908) 464-6800. (800) 521-8110. 20th edition. 8 volumes. 1996. $850.00

Consulting Services: Chemists and Chemical Engineers. Association of Consulting Chemists and Chemical Engineers, 40 East 45th Street, New York, NY 10017. (212) 983-3160. FAX (212) 983-3161. Biennial. $60.00.

Research Centers Directory. Gale Research Company, Inc., 835 Penobscot Building, Detroit, MI 48226-4094. (313) 961-2242. (800) 877-4253. 20th edition. 1995. $485.00. ISSN: 0080-1518.

Who's Who in Engineering. Gordon Davis, editor. American Association of Engineering Societies. 1111 19th Street, NY, Suite 608, Washington, DC 20036. (202) 296-2237. (800) 658-8897. 9th edition. 1995. $220.00.

ENCYCLOPEDIAS AND DICTIONARIES

Concise Encyclopedia of Polymer Science. Herman F. Mark. John Wiley & Sons, Inc., 605 Third Avenue, New York, NY 10158-0012. (212) 850-6000. (800) 225-5945. 1990. $199.00

Encyclopedia of Chemical Processing and Design. McKetta Marcel Dekker, Inc., 270 Madison Avenue, New York, NY 10016. (212) 696-9000. (800) 228-1160. 1976 - . $175.00 per volume. ISSN:

Hawley's Condensed Chemical Dictionary. Van Nostrand Reinhold, 115 Fifth Avenue, New York, NY 10003. (212) 254-3232. (800) 842-3636. 13th edition. 1996. $99.95.

Industrial Chemical Thesaurus. Michael Ash and Irene Ash. VCH Publishers, Inc., 220 East 23rd Street, New York, NY 10010-4606. (800) 367-8249. 1992. $295.00.

Illustrated Chemistry Laboratory Terminology. Gerbert W. Ockerman. CRC Press, Inc., 2000 Corporate Boulevard, NW, Boca Raton, FL 33431. (407) 994-0555. (800) 272-7737. 1991. $32.95.

Kirk-Othmer Encyclopedia of Chemical Technology. John Wiley and Sons, Inc., 605 Third Avenue, New York, NY 10158. (800) 526-5368 or (212) 850-6000. Fourth edition. 1991 - . Twenty-seven volumes. $5400.00.

McGraw-Hill Encyclopedia of Science and Technology. McGraw-Hill Book, Incorporated, 1221 Avenue of the Americas, New York, NY 10020. (212) 997-3675. (800) 262-4729. Seventh edition. Twenty volumes. 1992. $1900.00.

Modern Plastics - Encyclopedia Issue. McGraw-Hill 1221 Avenue of the Americas, New York, NY 10020. (212) 512-6255. Annual; issued in October. $49.95. ISSN: 0085-3518.

Ullman's Encyclopedia of Industrial Chemistry. VCH Publications, Inc., 220 East 23rd Street, Suite 909, New York, NY 10010-4606. (212) 683-8333. (800) 422-8824. 5th edition. 1984 - . Price varies per volume. ISSN:

GENERAL WORKS

Adhesives and Sealants. American Society for Metals, Volume 3 of Engineered Materials Handbook. ASM International 9639 Kinsman Road, Materials Park, OH 44073. (216) 238-5151. 1990. $153.00.

Cyanoacrylate Resins: The Instant Adhesives. Henry Lee. T-C Publications, CA. Reprint edition. 1991. $52.00.

Engineering Resins: An Industrial Guide. Ernest W. Flick. Noyes, 120 Mill Road, Park Ridge, NJ 07656. (201) 391-8484. (212) 354-5500. (800) 223-2584 1989. $94.00.

Industrial Applications of Adhesive Bonding. M. M. Sadek, editor. Elsevier Science Publishing Company, Inc., 655 Avenue of the Americas, New York, NY 10010. (212) 989-5800. FAX (914) 333-2444. 1987. $72.00.

Industrial Gums, Polysaccharides and their Deriavatives. Roy L Whistler and James N. Miller, editors. Academic Press, Inc., 6277 Sea Harbor Drive, Orlando, FL. (800) 321-5068. 3rd edition. 1992. $175.00.

Ion Exhcange Resins. Robert Kunin. Krieger Publishing Company, P.O. Box 9542, Melbourne, FL 32902-9542. (407) 724-9542. 1990. $52.00.

Polymer Chemistry: An Introduction. Raymond B. Seymour and Charles E. Carraher Jr. Marcel Dekker, Inc., 270 Madison Avenue, New York, NY 10016. (212) 696-9000. (800) 228-1160. 3rd edition. 1992. $59.75.

Resins For Aerospace. Clayton A. May. American Chemical Society, 1155 16th Street, NW, Washington, DC 20036. (202) 872-4600. (800) 333-9511. FAX (614) 447-3671. 1980. $54.95.

Resins For Surface Coatings. M. J. Husbands. Scholium Int Volume III, 2nd edition. 1987. $110.00

HANDBOOKS AND MANUALS

Atlas of Polymer and Plastics Analysis. Dieter O. Hummel. VCH Publications, Inc., 220 East 23rd Street, Suite 909, New York, NY 10010-4606. (212) 683-8333. (800) 422-8824. Volume 1, parts A and B. 3rd edition. 1991. $525.00

Catalyst Handbook. Martyn V. Twigg. CRC Press Inc., 2000 Corporate Blvd., NW, Boca Raton, FL 33431. (407) 994-0555. (800) 333-8300. 2nd edition. 1989. $104.00.

Chemical Formulary. H. Bennett, editor. Chem Publishing, Co., Inc. 80 Eighth Avenue, New York, NY 10011. (212 255-1950. volumes 1-30. $70.00 per volume.

CRC Handbook of Chemistry and Physics. David R. Kide, editor. CRC Press Inc., 2000 Corporate Blvd., NW, Boca Raton, FL 33431. (40?-??4-0555. (800) 333-8300. 77th edition. 1996. $99.95.

Data Book of thermoset Resins For Composites. Trevor F. Starr. Elsevier Science Publishing Company, Inc., 655 Avenue of the Americas, New York, NY 10010. (212) 989-5800. FAX (914) 333-2444.

Guide To Basic Chemical Compounds. D.R. Lide, Jr. CRC Publishers, Inc., 2000 Corporate Blvd., N.W., Boca Raton, FL 33431. (407) 994-0555 or (800) 333-8300. 1993. $120.00.

Handbook of Adhesive Technology. A. Pizzi, editor. Marcel Dekker, Inc., 270 Madison Avenue, New York, NY 10016. (212) 696-9000. (800) 228-1160. 1994. $195.00.

Handbook of Plastic Compounds, Elastomers and Resins; An International Guide By Category, Tradename, Composition and Supplier. Michael Ash and Irene Ash, editors. VCH Publishers, Inc., 220 East 23rd Street, Suite 909, New York, NY 10010-4606. (212) 683-8333. (800) 422-8824. 1992. $195.00.

Lange's Handbook of Chemistry. John A. Dean, editor. McGraw-Hill Publishing Company, 1221 Avenue of the Americas, New York, NY 10020. (800) 262-4729 or (212) 512-3825. Fourteenth editon. 1992. $79.50.

Perry's Chemical Engineers' Handbook. Robert H. Perry and Donald W. Green, editors. McGraw-Hill Publishing Company, Inc., 1221 Avenue of the Americas, New York, NY 10020. (212) 512-2000. (800) 262-4729. 6th edition. 1996. $129.50.

ONLINE DATABASES AND CD-ROMS

CA Search. Chemical Abstracts Service, P.O. Box 3012, Columbus, OH 43210-0012. (614) 447-3600. (800) 848-6533. FAX (614) 447-3709. Very comprehensive guide to worldwide chemi-

cal literature and related fields, 1972 to present. Available on BRS,(800) 289-4277, DIALOG, (800) 334-2564, ORBIT (800) 456-7248, and STN International, FIZ Karlsruhe, P.O. Box 2465, W-7500, Karlsruhe 1, Germany, online services. Inquire as to cost and availability.

Chemical Journals of the American Chemical Society. American Chemical Society, 1155 16th Street, N.W., Washington, DC 20036. (202) 872-4381 or (800) 424-6747. Contains complete text of approximately 90,000 articles from 22 primary journals published by the American Chemical Society, from mostly 1982 to present. Available on STN International, FIZ Karlsruhe, P.O. Box 2465, W-7500, Karlsruhe 1, Germany, online service. Inquire as to cost and availability.

Dissertation Abstracts. University Microfilms International, 300 North Zeeb Road, Ann Arbor, MI 48106. (800) 521-0600 or (313) 761-4700. Scope includes virtually all doctoral dissertations accepted at accredited American institutions from 1861 to present in 252 subject areas. Available on BRS, (800) 289-4277, DIALOG, (800) 334-2564, and OCLC EPIC, (800) 848-5878, online services. Also available on CD-ROM. Inquire as to cost and availability.

NTIS Bibliographic Database. National Technical Information Service, 5285 Port Royal Road, Springfield, VA 22161. (703) 487-4929 or FAX (703) 321-8199. Broad coverage of government-sponsored science and technology research reports, 1964 to present. Available on BRS,(800) 289-4277, DIALOG, (800) 334-2564, ORBIT, (800) 456-7248, and STN International, FIZ Karlsruhe, P.O. Box 2465, W-7500, Karlsruhe 1, Germany, online services. Also available on CD-ROM. Inquire as to cost and availability.

SCISEARCH. Institute for Scientific Information, 3501 Market Street, Philadelphia, PA 19104. (800) 523-1850 or (215) 386-0100. Broad multidisciplinary title and author index to the international literature of science and technology, 1974 to present. Available on DIALOG, (800) 334-2564, and ORBIT, (800) 456-7248, online services. Also available on CD-ROM. Inquire as to cost and availability.

WILSONLINE. H.W. Wilson Company, 950 University Avenue, Bronx, NY 10452. (800) 367-6770 or (212) 588-8400. Makes available online versions of the printed H.W. Wilson indexes including Applied Science and Technology Index, Business Periodicals Index, General Science Index, and Readers' Guide to Periodical Literature. Period covered is generally 1983 to present. Available on BRS, (800) 289-4277, DIALOG, (800) 334-2564, and OCLC EPIC, (800) 848-5878, online services. Also available on CD-ROM. Inquire as to cost and availability.

PERIODICALS

Adhesives Age. Argus Inc., 6151 Powers Perry Road, NW, Atlanta, GA 30339-2941. (404) 955-2500. FAX (404) 955-0400, 1957 to present. Monthly, plus annual directory. $52.00 per year. ISSN: 0001-821X.

Chemical Week; Includes Annual Buyers Guide. Chemical Week Associates, 888 Seventh Avenue, New York, NY 10106. (212) 621-4900. FAX (212) 621-4949. 1914 to present. Weekly. $99.00 per year. ISSN: 0009-272X.

Chemtech. American Chemical Society, Box 3337, Columbus, OH 43210. (614) 447-3776. 1970 to present. Monthly. $370.00 per year. ISSN: 0009-2703.

Industrial and Engineering Chemistry Research. American Chemical Society, Box 3337, Columbus, OH 43210. (614) 447-3776. (800) 333-9511. FAX (614) 447-3671. 1962 to present. Monthly. $567.00 per year. ISSN: 0888-5885.

International Journal of Adhesion and Adhesives. Butterworth-Heinemann. Subscriptions to: Turpin Transactions Ltd., Distributin Centre, Blackhorse Road, Letchworth, Herts SG6 1HN, England. TEL 0462-672555. 1980 to present. Quarterly. L180 per year. ISSN: 0143-7496.

Journal of Coatings Technology. Federation of Societies for Coating Technology, 429 Norristown Road, Blue Bell, PA 19422. (215) 940-0777. FAX (215) 940-0292. Monthly. Included in membership dues. ISSN: 0361-8773.

Modern Plastics. McGraw-Hill, Inc., 1221 Avenue of the Americas, New York, NY 10020. (212) 512-6267. FAX (212) 512-6111. 1925 to present. Monthly. $41.75 per year. ISSN: 0026-8275.

Plastics Compounding: For Resin Producers, Formulators and Compounders. Advanstar Communications, Inc., 7500 Old Oak Boulevard, Cleveland, OH 44130. (216) 826-2839. FAX (216) 891-2726. 1977 to present. Bi-monthly. $40.00 per year. ISSN: 0148-9119.

Plastics Engineering. Society of Plastics Engineers, Inc., 14 Fairfield Drive, Brookfield, CT 06805-0403. (203) 775-0471. FAX (203) 775-8490. 1945 to present. Monthly. $50.00 per year. ISSN: 0091-9578.

RESEARCH CENTERS AND INSTITUTES

Institute of Polymer Engineering. University of Akron, Akron, OH 44325-0301. (216) 972-6865. (216) 258-2339.

Plastics Institute of America. 277 Fairfield Road, Suite 100, Fairfield, NJ 07004-1932. (201) 808-5950. FAX (201) 808-5953.

Polymer Processing Institute. Stevens Institute of Technology, Castle Point Station, Hoboken, NJ 07030. (201) 420-5819.

Polymer Research Center. University of Cincinnati, Mail Location 172, Cincinnati, OH 45221. (513) 465-2453.

Polymer Research Laboratory. University of Michigan, 2045 Dow Building, Ann Arbor, MI 48109. (313) 763-2240. FAX (313) 763-4788.

GYROSCOPES

See also: AEROSPACE ENGINEERING, COMPASSES, GUIDANCE SYSTEMS, GUIDED MISSILES, NAVIGATION, ROCKETS

ABSTRACT SERVICES AND INDEXES

Applied Science and Technology Index; A Cumulative Subject Index To English Language Periodicals in the Fields of Aeronautics and Space Science, Computer Technology, Chemistry, Construction Industry, Energy and Related Areas. H.W. Wilson Co., 950 University Avenue, Bronx, NY 10452. (800) 367-6770 or (212) 588-8400. FAX (313) 590-1617. From 1958 to present. Monthly. Inquire about cost and availability. Also available on CD-ROM and online.

Engineering Index Monthly. Engineering Information, Inc., Castle Point on the Hudson, Hoboken, NJ 07030. (800) 221-1044. FAX (212) 832-1857. Monthly. $2300.00 per year. Also available online as COMPENDEX, and also on CD-ROM. Covers chemical engineering, computers, electrical engineering, civil engineering, metals and mining, industrial management, and mechanical engineering.

International Aerospace Abstracts. Technical Information Service, American Institute of Aeronautics and Astronautics, Inc., 555 West 57th St., New York, NY 10019. (212) 247-6500. FAX (212) 582-4861. Semi-monthly. $1295.00 per year.

ASSOCIATIONS AND PROFESSIONAL SOCIETIES

Institute of Navigation. 1800 Diagonal Road, Suite 480, Alexandria, VA 22314-2840.

International Navigation Association, Po Box 2324, Arlington, VA 22202-0324. (604) 276-4626.

Permanent International Association of Navigation Congresses, United States Section. CECW-PK, 20 Massachusetss Avenue NW, Washington, DC 20314-1000. (202) 504-4312.

Wild Goose Association. Operations, 150 S. Plains Road, the Plains, OH 45780. (614) 797-2081.

DIRECTORIES AND BIOGRAPHICAL SOURCES

Directory of Engineering Societies and Related Organizations. Gordon Davis. 13th edition. American Association of Engineering Societies, 1111 19th Street NW, Suite 608, Washington, DC 20036. (202) 296-2237 or (800) 658-8897. 1989. Inquire for price.

Research Centers Directory. Gale Research, 835 Penobscot Building, Detroit, MI 48226-4094. (313) 961-2242. (800) 877-4253. FAX (313) 961-6083. $485.00.

Who's Who in Engineering. American Association of Engineering Societies, 1111 19th Street NW, Suite 608, Washington, DC 20036. (202) 296-2237 or (800) 658-8897. 8th edition. 1991. Inquire for price.

GENERAL WORKS

The Fiber-optic Gyroscope. Herve Lefevre. Artech House, 685 Canton St., Norwood, MA 02062. (617) 769-9750 or (800) 225-9977. FAX (617) 769-6334. 1993. $85.00.

The Gyroscope: Theory and Applications. James B. Scarborough. Interscience, 1958. Out of print.

The Gyroscope Through the Ages. Phoenix, AZ: Sperry, Flight Systems Division, 1971. Out of print.

Gyroscopic theory. G. GreenHill. Chelsea Publishing Company, 15 E. 26th Street, New York, NY 10010. (212) 889-8095. 1966. $32.50.

Gyroscopic theory, Design, and Instrumentation. Walter Wrigley, et al. MIT Press, 55 Hayward Street, Cambridge, MA 02142. (617) 625-8569 or (800) 356-0343. FAX (617) 253-1709. 1969. Inquire for cost and availability.

Mechanics of Gyroscopic Systems. A. Yu Ishlinskii. Coronet Books, 311 Bainbridge Street, Philadelphia, PA 19147. (215) 925-5083. FAX (215) 925-1912. 1965. $79.50.

Navigational & Surveying Instruments: Industry and Trade Summary. Sundar A. Shetty. Diane Publishing Company, 600 Upland Avenue, Upland, PA 19015. (215) 499-7415. FAX (610) 499-7429. 1994. $40.00.

ONLINE DATABASES AND CD-ROMS

Aerospace Database. American Institute of Aeronautics and Astronautics, 370 L'Enfant Promenade, S.W., Washington, DC 20024. (202) 646-7400. Worldwide published literature on research and development in aerospace and related areas, with abstracts. Covers 1962 to present. Available on DIALOG, (800) 334-2564, online service. Also available on CD-ROM. Inquire as to cost and availability.

Compendex Plus. Engineering Information, Inc., 345 East 47th Street, New York, NY 10017. (212) 705-7600 or (800) 221-1044. Contains citations with abstracts to worldwide literature in engineering and technology, from 1970 to present. Available on online BRS,(800) 289-4277, DIALOG, (800) 334-2564, ORBIT (800) 456-7248, and STN International, FIZ Karlsruhe, P.O. Box 2465, W-7500, Karlsruhe 1, Germany, online services. Also available on CD-ROM. Inquire as to cost and availability.

International Aerospace Abstracts. American Institute of Aeronautics and Astronautics, 370 L'Enfant Promenade, S.W., Washington, DC 20024. (202) 646-7400. Contains references and abstracts of journal and monograph literature relating to aerospace science and technology, from 1963 to present. Available through the NASA/RECON system of the National Aeronautics and Space Administration only.

NASA Database. American Institute of Aeronautics and Astronautics, 370 L'Enfant Promenade, S.W., Washington, DC 20024. (202) 646-7400. Citations and abstracts of aeronautics and astronautics literature, 1962 to present. Also contains citations from STAR, SCIENTIFIC and TECHNICAL AEROSPACE REPORTS, and INTERNATIONAL AEROSPACE ABSTRACTS. Available through NASA/RECON online service. Inquire as to cost and availability.

NTIS Bibliographic Database. National Technical Information Service, 5285 Port Royal Road, Springfield, VA 22161. (703) 487-4929 or FAX (703) 321-8199. Broad coverage of government-sponsored science and technology research reports, 1964 to present. Available on BRS,(800) 289-4277, DIALOG, (800) 334-2564, ORBIT, (800) 456-7248, and STN International, FIZ Karlsruhe, P.O. Box 2465, W-7500, Karlsruhe 1, Germany, online services. Also available on CD-ROM. Inquire as to cost and availability.

SAE Global Mobility Database. Society of Automative Engineers (SAE), Electronic Publishing Division, 400 Commonwealth Drive, Warrendale, PA 15098. (412) 776-4841. Contains citations with abstracts to technical papers on automotive and aerospace technology and vehicular-related industries that have been presented at SAE conferences. Covers 1906 to present. Available on ORBIT,(800) 456-7248,online service. Inquire as to cost and availability.

SCISEARCH. Institute for Scientific Information, 3501 Market Street, Philadelphia, PA 19104. (800) 523-1850 or (215) 386-0100. Broad multidisciplinary title and author index to the international literature of science and technology, 1974 to present. Available on DIALOG,(800) 334-2564, and ORBIT,(800) 456-7248, online services. Also available on CD-ROM. Inquire as to cost and availability.

TRIS (Transportation Research Information). National Academy of Science, 2101 Constitution Avenue, N.W., Washington, DC 20418. (202) 334-3313 or (800) 624-6242. Citations with abstracts of literature on transportation, including air, highway, rail, maritime and other modes from 1968 to present. Available on DIALOG,(800) 334-2564, online service. Inquire as to cost and availability.

WILSONLINE. H.W. Wilson Company, 950 University Avenue, Bronx, NY 10452. (800) 367-6770 or (212) 588-8400. Makes available online versions of the printed H.W. Wilson indexes including Applied Science and Technology Index, Business Periodicals Index, General Science Index, and Readers' Guide to Periodical Literature. Period covered is generally 1983 to present. Available on BRS,(800) 289-4277, DIALOG,(800) 334-2564, and OCLC EPIC,(800) 848-5878, online services. Also available on CD-ROM. Inquire as to cost and availability.

PERIODICALS

Aerospace America. American Institute of Aeronautics and Astronautics, 370 L'Enfant Promenade SW, Washington, DC 20024-2518. (202) 646-7471. 1932 to present. Monthly. $70 annually (non-members).

Aerospace Engineering. Society of Automotive Engineers, 400 Commonwealth Dr., Warrendale, PA 15096. (412) 776-4841. 1981 to present. Monthly. $48.00 per year.

AIAA Journal. American Institute of Aeronautics and Astronautics, 370 L'Enfant Promenade SW, Washington, DC 20024-2518. (202) 646-7471. 1963 to present. Monthly. $52.00 per year (members) and $435.00 per year (non-members).

Aviation Week and Space Technology. 1221 Avenue of the Americas, New York, NY 10020. (212) 512-2999 or (800) 525-5003. 1916 to present. Weekly. $82.00 per year.

Journal of Guidance, Control and Dynamics. American Institute of Aeronautics and Astronautics, 370 L'Enfant Promenade SW, Washington, DC 20024-2518. 1978 to present. Bimonthly. $42.00 (members) and $250.00 (non-members).

Journal of Navigation. Cambridge University Press, Journals Department, 40 West 20th Street, New York, NY 10011-4211. (212) 924-3900. 1947 to present. Three times a year. $165.00 per year.

Journal of Ship Research. Society of Naval Architects and Marine Engineers, 601 Pavonia Avenue, Suite 400, Jersey City, NJ 07306. (201) 798-4800. FAX (201) 798-4975. 1957 to present. Quarterly. $80.00 per year.

Mechanical Engineering. American Society of Mechanical Engineers, 345 E. 47th Street, New York, NY 10017. (212) 705-7722. 1906 to present. Monthly. $45.00 for non-members.

GYROSCOPES

Ency. of Physical Sciences and Engineering Info. Sources

Review of Scientific Instruments. American Institute of Physics, One Physics Ellipse, College Park, MD 20740. (301) 209-3000. 1930 to present. Monthly. $800.00 per year.

Sensors: The Journal of Applied Sensing Technology. Helmers Publishing Company Inc., 174 Concord Street, Box 874, Peterborough, NH 03458-0874. (603) 924-9631. FAX (603) 924-2076. 1984 to present. Monthly. $55.00 per year.

RESEARCH CENTERS AND INSTITUTES

Charles Stark Draper Laboratory Inc., 555 Technology Square, Cambridge, MA 02139. (617) 258-1000.

Goddard Space Flight Center, Greenbelt, MD 20771. (301) 286-7351.

Instrumentation and Control Laboratory, Princeton University, Dept. of MAE, Engineering Quadrangle, Princeton, NJ 08544. (609) 452-5154.

George C. Marshall Space Flight Center, Huntsville, AL 35812. (205) 544-0024.

H

HACKERS

See: COMPUTER SECURITY

HADRONS

See: PARTICLE PHYSICS

HAIL

See also: METEOROLOGY, RAIN, THUNDERSTORMS

ABSTRACT SERVICES AND INDEXES

Applied Science and Technology Index; A Cumulative Subject Index To English Language Periodicals in the Fields of Aeronautics and Space Science, Computer Technology, Chemistry, Construction Industry, Energy and Related Areas. H.W. Wilson Co., 950 University Avenue, Bronx, NY 10452-9978. (212) 588-8400. (800) 367-6770. FAX (718) 590-1617. From 1958 to present. Monthly. Inquire about cost and availability. Also available online (BRS and WILSONLINE)and on CD-ROM. ISSN: 0003-6986.

General Science Index. H.W. Wilson Co., 950 University Avenue, Bronx, NY 10452. (212) 588-8400. (800) 367-6770. FAX (718) 590-1617. From 1978 to present. Ten issues per year; quarterly and annual cumulations. Service basis. Available on CD-ROM and online. Inquire regarding cost and availability. ISSN: 0162-1963.

Meteorological and Geoastrophysical Abstracts. American Meteorological Society, c/o Inforonics, Inc., 550 Newtown Road, Box 458, Littleton, MA 01460. (508) 486-8976. FAX (508) 486-0027. Covers literature in environmental sciences, meteorology, astrophysics, hydrology, glaciology, and physical oceanography. 1950 to present. Monthly. $950.00 per year. ISSN: 0026-1130. Also available online (DIALOG) and on CD-ROM.

Science Citation Index. SCI. Institute for Scientific Information, 3501 Market Street, Philadelphia, PA 19104. (215) 386-0100. (800) 523-1850. FAX (215) 386-2991. 1961 to present. Six issues per year, plus annual cumulation. $11650.00 per year. Also

available online and on CD-ROM. Inquire about price and availability. ISSN: 0036-827X.

Selected Water Resources Abstracts. U.S. Geological Survey. Distributed by National Technical Information Service (NTIS), 5285 Port Royal Road, Springfield, VA 22161. (703) 487-4650. Monthly. $115.00 per year.

ANNUAL REVIEWS AND YEARBOOKS

Developments in Atmospheric Science Series. Elsevier Science Publishing Company, Inc., 655 Avenue of the Americas, New York, NY 10010. (212) 989-5800. Irregular. Inquire about price and availability.

ASSOCIATIONS AND PROFESSIONAL SOCIETIES

American Meteorological Society, 45 Beacon Street, Boston, MA 02108-3693. (617) 227-2425. FAX (617) 742-8718.

National Center For Atmospheric Research. P.O. Box 3000, Boulder, CO 80307. (303) 496-1000. FAX (303) 497-1654.

National Severe Storms Forecast Center. 601 East 12th Street, Kansas City, MO 64106. (816) 374-5922.

National Weather Association, 6704 Wolke Court, Montgomery, AL 36116-2134. (334) 213-0388. FAX (334) 213-0388.

University Corporation For Atmospheric Research. PO Box 3000, 1850 Table Mesa Drive, Boulder, CO 80307-3000. (303) 497-1673. FAX (303) 497-1654.

Weather Modification Association. PO Box 8116, Fresno CA 93747. (209) 291-8466.

BIBLIOGRAPHIES

Information Sources in the Earth Sciences. David N. Wood, Joan E. Hardy, and Anthony P. Harvey, editors. Bowker-Saur/ K.G. Saur. Distributed by R.R. Bowker, 121 Chanlon Road, New Providence, NJ 07974. (800) 521-8110 or (908) 464-6800. Second edition. 1989. $85.00.

Scientific and Technical Books and Serials in Print; An Index To Literature in Science and Technology. R.R. Bowker Co., 205 E. 42nd Street, New York, NY 10017. (800) 521-8110 or (212) 916-1600. 1992. $250.00.

DIRECTORIES AND BIOGRAPHICAL SOURCES

American Men and Women of Science: Physical and Biological Sciences. Eighteenth edition. R.R. Bowker Company, 245 West 17th Street, New York, NY 10011. (800) 521-8810 or (212) 916-1600. $750.00.

American Meteorological Society. PRofESSIONAL DIRECTORY. American Meteorological Society. 45 Beacon Street, Boston, MA 02108. (617) 227-2425. Included in each issue of the Bulletin of the Society.

Meteorological Services of the World. American Meteorological Society, 45 Beacon Street, Boston, MA 02108- 3693. (617) 227-2425. FAX (617) 742-8718. Annual. $70.00.

National Weather Service Offices and Stations. National Oceanic and Atmospheric Administration, Deparatment of Commerce, Silver Spring, MD 20910. (301) 427-7698. Annual. Free

Research Centers Directory. Gale Research, 835 Penobscot Building, Detroit, MI 48226-4094. (313) 961-2242. (800) 347-4253. 20th edition. 1995. $485.00. ISSN: 0080-1518.

ENCYCLOPEDIAS AND DICTIONARIES

Climates of the States. Gale Research,835 Penobscot Building, Detroit, MI 48226-4094. (313) 961-2242. (800) 347-4253. 5th edition. 1995. $255.00.

Concise Oxford Dictionary of the Earth Sciences. Ailsa Allaby and Michael Allaby, editors. Oxford University Press, Inc., 200 Madison Avenue, New York, NY 10016. (800) 334-4249 or (212) 679-7300. 1990. $42.95.

Encyclopedia of Climatology. John E. Oliver and Rhodes W. Fairbridge, editors. Van Nostrand Reinhold Inc., 115 Fifth Avenue, New York, NY 10033. (800) 543-2681. 1987. $115.00.

McGraw-Hill Encyclopedia of Ocean and Atmospheric Sciences. Sybil P. Parker, editor. McGraw-Hill Publishing Company, 1221 Avenue of the Americas, New York, NY 10020. (800) 262-4729 or (212) 512-3825. 1979. $79.95.

McGraw-Hill Encyclopedia of Science and Technology. McGraw-Hill Publishing Company, 1221 Avenue of the Americas, New York, NY 10020. (800) 262-4729 or (212) 512-3825. Seventh edition. 1992. $1900.00.

GENERAL WORKS

The Atmosphere: An Introduction To Meteorology. Frederick K. Lutgens and Edward J. Tarbuck. Prentice Hall (Division of Simon and Schuster), 15 Columbus Circle, New York, NY 10023. (212) 373-8500 or (800) 922-0579. Fifth edition. 1991. $49.00.

Cloud Dynamics. Robert A. Houze. Academic Press, Inc., 1250 Sixth Avenue, San Diego, CA 92101-4311. (619) 231-0926. FAX (619) 699-6715. 1993. $149.00.

Formation of Precipitation & Modification of Hail Processes. G. K. Sulakvelidze. Coronet Books, 311 Bainbridge Street, Philadelphia PA 19147. (215) 925-5083. 1967. $58.00.

Hail: A Review of Hail Science and Hail Supression. Brant Foote and Charles A. Knight. American Meteorological Society, 45 Beacon Street, Boston, MA 02108-3693. (617) 227-2425. FAX (617) 742-8718. 1977. $40.00.

Hailstorms & Hailstone Growth. Narayan R. Gokhale. State University of New York Press, State University Plaza, Albany, NY 12246-0001. (518) 472-5000. (800) 666-2211. 1974. $49.50.

Hailstorms of the Central High Plains. Charles Knight and Patrick Squires. Books on Demand 300 North Zeeb Road, Ann Arbor, MI 48106-1346. (313) 761-4700. (800) 521-0600.

$83.50.

Mesoscale Meteorology and Forecasting. Peter S. Ray, editor. American Meteorological Society, 45 Beacon Street, Boston, MA 02108-3693. (617) 227-2425. FAX (617) 742-8718. 1986. $50.00.

Physics of Rainclouds. Neville H. Fletcher. Books on Demand, Division of University Microfilms International, 300 North Zeeb Road, Ann Arbor, MI 48106-1346. (313) 761-4700 or (800) 521-0600. $101.00 in paper.

Severe and Unusual Weather. Joe R. Eagleman. Trimedia Publishing Company, 12008 West 87th Street, Suite 117, Lenexa, KS 66215. (913) 599-0505. Second edition. 1990. $41.95.

Storm and Cloud Dynamics. William R. Cotton and Richard A. Anthes. Academic Press, Inc., 1250 Sixth Avenue, San Diego, CA 92101-4311. (619) 231-0926. FAX (619) 699-6715. 1992. $69.95.

HANDBOOKS AND MANUAL

Author's Guide To the Journals of the American Meteorological Society. American Meteorological Society, 45 Beacon Street, Boston, MA 02108-3693. (617) 227-2425. FAX (617) 742-8718. 1983. $15.00 in paper.

Basic Meteorology Lab Manual. Thomas A. Leavy. Allegheny Press, P.O. Box 220, Elgin, PA 19413. (814) 664-8504. Second edition. 1969. $7.95.

Handbook in Applied Meteorology. David D. Houghton, editor. John Wiley and Sons, Inc., 605 Third Avenue, New York, NY 10158. (800) 526-5368 or (212) 850-6000. 1985. $114.00.

International Cloud Atlas. World Meteorological Organization. American Meteorological Society, 45 Beacon Street, Boston, MA 02108-3693. (617) 227-2425. FAX (617) 742-8718. Volume 2- photographs. 1987. $83.00.

ONLINE DATABASES AND CD-ROMS

Climate Assessment Database. National Weather Service, National Meteorological Center, 5200 Auth Road, Suite 101, Camp Springs, MD 20233. (301) 763-8016. Contains daily, weekly and monthly summaries of North American and world climatological data. Also provides five to ten weather forecasts, and 30 to 90 day weather outlook. Subscription required. Inquire as to cost and availability.

Current Contents Search. Institute for Scientific Information, 3501 Market Street, Philadelphia, PA 19104. (215) 386-0100. FAX (215) 386-6362. Contains citations to articles listed in the table of contents of science and technology journals. Also articles in social sciences and life sciences journals. Available on BRS,(800) 289-4277, DIALOG,(800) 334-2564, online services. Inquire as to cost and availability.

Dissertation Abstracts. University Microfilms International, 300 North Zeeb Road, Ann Arbor, MI 48106. (800) 521-0600 or (313) 761-4700. Scope includes virtually all doctoral dissertations accepted at accredited American institutions from 1861 to present in 252 subject areas. Available on BRS,(800) 289-4277, DIALOG,(800) 334-2564, and OCLC EPIC,(800) 848-5878, online services. Also available on CD-ROM. Inquire as to cost and availability.

Meteorological and Geoastrophysical Abstracts. American Meteorological Society, 45 Beacon Street, Boston, MA 02108-3693. (617) 227-2425. FAX (617) 742-8718. Contains citations and abstracts to the worldwide literature on significant research in meteorology and geoastrophysics. Related topics include physical oceanography, hydrology, environmental sciences and glaciology. Covers the period 1972 to present. Available on DIALOG,(800) 334-2564, online service. Inquire as to cost and availability.

NTIS Bibliographic Database. National Technical Information Service, 5285 Port Royal Road, Springfield, VA 22161. (703) 487-4929 or FAX (703) 321-8199. Broad coverage of government-sponsored science and technology research reports, 1964 to present. Available on BRS,(800) 289-4277, DIALOG, (800) 334-2564, ORBIT, (800) 456-7248, and STN International, FIZ Karlsruhe, P.O. Box 2465, W-7500, Karlsruhe 1, Germany, online services. Also available on CD-ROM. Inquire as to cost and availability.

SCISEARCH. Institute for Scientific Information, 3501 Market Street, Philadelphia, PA 19104. (800) 523-1850 or (215) 386-0100. Broad multidisciplinary title and author index to the international literature of science and technology, 1974 to present. Available on DIALOG and ORBIT online services. Also available on CD-ROM. Inquire as to cost and availability.

WILSONLINE. H.W. Wilson Company, 950 University Avenue, Bronx, NY 10452. (800) 367-6770 or (212) 588-8400. Makes available online versions of the printed H.W. Wilson indexes including Applied Science and Technology Index, Business Periodicals Index, General Science Index, and Readers' Guide to Periodical Literature. Period covered is generally 1983 to present. Available on BRS,(800) 289-4277, DIALOG,(800) 334-2564, and OCLC EPIC,(800) 848-5878, online services. Also available on CD-ROM. Inquire as to cost and availability.

World Climate Disc. Chadwyck-Healey Inc., 1101 King Street, Alexandria, VA 22314. (703) 683-4890. FAX (703) 683-7589. Weather and climate data from approximately 5,000 weather stations worldwide, covering the years 1854 to 1990 on CD-ROM. First edition 1992. Approximately $1200.00 per year with annual updates.

World Weatherdisc. WeatherDisc Associates, Inc., 4584 N.E. 89th Street, Seattle, WA 98115. (206) 524-4314. FAX (206) 543-0308. Meteorological data on CD-ROM which describes the climate of the earth today and for the past few hundred years.

First edition 1989. Approximately $295.00 per year with biannual updates.

PERIODICALS

Agricultural and Forest Meteorology. Elsevier Science Publishing Company, Inc., 655 Avenue of the Americas, New York, NY 10010. (212) 989-5800. Twenty times per year. $750.00 per year.

American Meteorological Society. BULLETIN. American Meteorological Society, 45 Beacon Street, Boston, MA 02108-3693. (617) 227-2425. FAX (617) 742-8718. Monthly. $60.00 per year.

American Meteorological Society. METEOROLOGICAL MONOGRAPHS. American Meteorological Society, 45 Beacon Street, Boston, MA 02108-3693. (617) 227-2425. FAX (617) 742-8718. Irregular. Price varies.

American Weather Observer. Association of American Weather Observers, 401 Whitney Boulevard, Box 455, Belvedere, IL 61008. (815) 544-5665. FAX (815) 544-6334. Monthly. $21.00 per year.

JGR: Journal of Geophysical Research: Atmosphere. American Geophysical Union, 2000 Florida Avenue, N.W., Washington, DC 20009. (202) 462-6903. Monthly. $90.00 per year to members. JOURNAL of APPLIED METEOROLOGY. American Meteorological Society, 45 Beacon Street, Boston, MA 02108-3693. (617) 227-2425. FAX (617) 742-8718. Monthly. $165.00 per year.

Journal of the Atmospheric Sciences. American Meteorological Society, 45 Beacon Street, Boston, MA 02108-3693. (617) 227-2425. FAX (617) 742-8718. Semi-monthly. $320.00 per year.

Monthly Weather Review. American Meteorological Society, 45 Beacon Street, Boston, MA 02108-3693. (617) 227-2425. FAX (617) 742-8718. Monthly. $205.00 per year.

National Weather Digest. National Weather Association, 4400 Stamp Road, Room 404, Temple Hills, MD 20748. (301) 899-3784.

Weather. Royal Meteorological Society, 104 Oxford Road, Reading, Berks RG1 7LJ, England. Monthly. $44.00 per year.

Weatherwise. Heldref Publications, 1319 Eighteenth Street, N.W., Washington, DC 20036-1802. (202) 296-6267. FAX (202) 296-5149. Bi-monthly. $28.00 per year.

RESEARCH CENTERS AND INSTITUTES

Cooperative Institute for Mesoscale Meteorological Studies. University of Oklahoma, 401 East Boyd, Norman, OK 73019. (405) 325-3041.

Goddard Space Flight Center. Laboratory for Atmospheres, Mail Code 610, Greenbelt, MD 20771. (301) 286-5002.

Joint Institute For Marine and Atmospheric Research. University of Hawaii, 1000 Pope Road, Honolulu, HI 96822. (808) 541-2876. FAX (808) 956-4104

HAIL

Ency. of Physical Sciences and Engineering Info. Sources

National Center For Atmospheric Research. P.O. Box 3000, Boulder, CO 80307. (303) 496-1000. FAX (303) 497-1654.

National Meteorological Center. World Weather Building, Room 101, 5200 Auth Road, Camp Springs, MD 20746. (301) 763-8016.

National Severe Storms Forecast Center. 601 East 12th Street, Kansas City, MO 64106. (816) 374-5922.

National Weather Service. 1325 East-West Highway, Silver Spring, MD 20910. (301) 427-7689.

HALF-LIFE

See: RADIATION

HALIDES

See also: BROMINE, CHLORINE, ELEMENTS, FLORINE, IODINE

ABSTRACT SERVICES AND INDEXES

Analytical Abstracts. Royal Society of Chemistry, Information Services, Thomas Graham House, Science Park, Milton Road, Cambridge, CB4 4WF, England. Contains citations, mostly with abstracts, of the worldwide literature on analytical chemistry, from 1954 to present. Monthly. $636.00 per year. Also available online.

Applied Science and Technology Index; A Cumulative Subject Index To English Language Periodicals in the Fields of Aeronautics and Space Science, Computer Technology, Chemistry, Construction Industry, Energy and Related Areas. H.W. Wilson Co., 950 University Avenue, Bronx, NY 10452- 9978. (212) 588-8400. (800) 367-6770. FAX (718) 590-1617.From 1958 to present. Monthly. Inquire about cost and availability. Also available online (BRS and WILSONLINE) and on CD-ROM. ISSN: 0003-6986.

Chemical Abstracts. Chemical Abstracts Service. 2540 Olentangy River Road, Box 3012, Columbus, OH 43210-0012. (614) 447-3600. FAX (614) 447-3713. 1907 to present. Weekly. $16,800.00 per year. Available online and on CD-ROM. CA is also available in five section groupings. Inquire Regarding Cost and Availability.

Current Contents: Physical, Chemical and Earth Sciences. Institute for Scientific Information, 3501 Market Street, Philadelphia, PA 19104. (215) 386-0100. FAX (215) 386-6362. Weekly. $360.00 per year.

Engineering Index Monthly. Engineering Information, Inc., Castle Point on the Hudson, Hoboken, NJ 07030. (800) 221-1044. FAX (201) 216-8532. Monthly. $2200.00 per year. Also available online as COMPENDEX, and also on CD-ROM.

General Science Index. H.W. Wilson Company, 950 University Avenue, Bronx, NY 10452. (212) 588-8400. (800) 367-6770. FAX (718) 590-1617. From 1978 to present. Ten issues per year; quarterly and annual cumulations. Service basis. Available on CD-ROM and online. Inquire regarding cost and availability. ISSN: 0162-1963.

Physics Abstracts. INSPEC. Section A, Science Abstracts. Institution of Electrical Engineers (IEE), London, United Kingdom. Available from: INSPEC/IEEE - Institute of Electrical and Electronic Engineers, Box 1331, 445 Hoes Lane, Piscataway, NJ 08855-1331. (908) 562-5549. 1898 to present. 24 issues per year. $2835.00 per year. ISSN: 0036-8091. Also available online and on CD-ROM.

Science Citation Index. SCI. Institute for Scientific Information, 3501 Market Street, Philadelphia, PA 19104. (215) 386-0100. (800) 523-1850. FAX (215) 386-2991. 1961 to present. Six issues per year, plus annual cumulation. $11650.00 per year. Also available online and on CD-ROM. Inquire about price and availability. ISSN: 0036-827X.

ANNUAL REVIEWS AND YEARBOOKS

Advances in Photochemistry. John Wiley and Sons, Inc., 605 Third Avenue, New York, NY 10158. (800) 526-5368 or (212) 850-6000. 1963 to present. Irregular. Price varies.

Annual Review of Physical Chemistry. Annual Reviews, Inc., 4139 El Camino Way, Palo Alto, CA 94306-0897. (415) 493-4400. Fax (415) 855-9815. 1951 to present. Annual. $55.00 per year.

ASSOCIATIONS AND PROFESSIONAL SOCIETIES

American Chemical Society, 1155 16th Street, Nw, Washington, DC 20036. (202) 872-4600.

American Institute of Chemical Engineers. 345 East 47th Street, New York, NY 10017-2396. (212) 705-7663.

American Institute of Physics, 335 East 45th Street, New York, NY 10017 (212) 661-9404 OR Fax (516)349-9704.

Association of Consulting Chemists and Chemical Engineers, 50 East 41st Street, Suite 92, New York, NY 10017. (212) 684-6255.

Chemical Manufacturers Association. 2501 M Street, N.W., Washington, DC 20037. (202) 887-1182

Chlorine Institute. 2001 L Street NW, Number 506, Washington, DC 20036. (202) 775-2790. FAX (202) 223-7225.

Chlorobenzene Producers Association. 1330 Connecticut Avenue, Nw, Suite 300, Washington, DC 20036. (202) 659-0060. FAX (202) 659-1699.

International Society For Fluoride Research. 8IA Landscape Road, Mount Eden Auckland 4, New Zealand. TEL 96307114. Fax 96307114.

BIBLIOGRAPHIES

Chemical Information Sources. Gary Wiggins. McGraw-Hill Publishing Company, 1221 Avenue of the Americas, New York, NY 10020. (800) 262-4729 or (212) 512-3825. 1991. $42.50.

Handbooks and Tables in Science and Technology. Russell H. Powell, editor. Oryx Press, 4041 North Central, Suite 700, Phoenix, AZ 85012-3330. (602) 265-2651 or (800) 279-6799. Third edition. 1994. $65.00.

Information Sources in Chemistry. R.T. Bottle and J.F.B Rowland, editors. R.R. Bowker Inc., 121 Chanlon Road, New Providence, NJ 07974. (800) 521-8110 or (908) 464-6800. Fourth edition. 1993. $75.00.

Scientific and Technical Books and Serials in Print; An Index To Literature in Science and Technology. R.R. Bowker Inc., 121 Chanlon Road, New Providence, NJ 07974. (908) 464-6800. (800) 521-8110. FAX (908) 665-3502. 1972 to present. Annual. 4 volumes. 1994. $299.95. Also available on compact disc and online. ISSN: 0000-054X.

DIRECTORIES AND BIOGRAPHICAL SOURCES

American Men and Women of Science. Physical and Biological Sciences. R.R. Bowker Company, 121 Chanlon Road, New Providence, NJ 07974. (908) 464-6800. (800) 521-8110. 20th edition. 8 volumes. 1996. $850.00

Chemical Sources International. Mike Desing and Kurt Gandenberger, editors. Chemical Sources International, PO Box 1824, Clemson, SC 29633. (803) 646-7840. 1994. $285.00.

Chemical Week — Buyers Guide Issue. Chemical Week Associates, 888 Seventh Avenue, New York, NY 10106. (212) 621-4900. FAX (212) 621-4949. Annual, October. $50.00.

Consulting Services: Chemists and Chemical Engineers. Association of Consulting Chemists and Chemical Engineers, 50 East 41st Street, Suite 92, New York, NY 10017. (212)

684-6255. Biennial. $60.00.

Directory of Chemistry Software 1992. Wendy Warr, Peter Willett, Geoff Downs. American Chemical Society, 1155 16th Street, NW, Washington, DC 20036. (202) 872-4600. 1992. $35.95.

Research Centers Directory. Gale Research, 835 Penobscot Building, Detroit, MI 48226-4094. (313) 961-2242. (800) 347-4253. 20th edition. 1995. $485.00. ISSN: 0080-1518.

ENCYCLOPEDIAS AND DICTIONARIES

Academic Press Dictionary of Science and Technology. Christopher Morris, editor. Academic Press, Inc., 1250 Sixth Avenue, San Diego, CA 92101. (619) 231-0926. FAX (619) 699-6715. 1991. $115.00.

Concise Encyclopedia Chemistry. Translated and revised by Mary Eagleson. Walter de Gruyter, Inc., 200 Saw Mill River Road, Hawthorne, New York, 10532. (914) 747-0110 or Fax (914) 747-1326. 1994. $69.95.

Dictionary of Named Processes in Chemical Technology. Alan E. Comyns. Oxford University Press, Inc., 200 Madison Avenue, New York, NY 10016. (212) 725-6000. (800) 334-4249. 1994. $75.00.

Encyclopedia of Chemical Processing and Design. McKetta Marcel Dekker, Inc., 270 Madison Avenue, New York, NY

10016. (212) 696-9000. (800) 228-1160. 1976 - . $175.00 per volume. ISSN: .

Gardner's Chemical Synonyns and Trade Names. Michael Ash and Irene Ash. Ashgate Publishing Co., Old Post Road, Brookfield, VT 05036. (802) 276-3162. Co. 10th edition. 1994. $195.00.

Grant & Hawks Chemical Dictionary. R. A. Grant. McGraw-Hill Publishing Company, Inc., 1221 Avenue of the Americas, New York, NY 10020. (212) 512-2000. (800) 262-4729. 5th edition. 1987. $64.50.

Hawley's Condensed Chemical Dictionary. Richard J. Lewis. Van Nostrand Reinhold, 115 Fifth Avenue, New York, NY 10003. (212) 254-3232. (800) 842-3636. 12th edition. 1993. $69.95.

Kirk-Othmer Encyclopedia of Chemical Technology. John Wiley and Sons, Inc., 605 Third Avenue, New York, NY 10158. (800) 526-5368 or (212) 850-6000. Fourth edition. 1991 - . Twenty-seven volumes. $5400.00.

McGraw-Hill Encyclopedia of Chemistry. Sybil P.Parker, editor. McGraw-Hill Book, Incorporated, 1221 Avenue of the Americas, New York, NY 10020. (212) 997-3675. Second edition. 1993. $95.50.

McGraw-Hill Encyclopedia of Science and Technology. McGraw-Hill Book, Incorporated, 1221 Avenue of the Americas, New York, NY 10020. (212) 997-3675. Seventeenth edition. Twenty volumes. 1992. $1900.00.

GENERAL WORKS

Biochemistry of the Elemental Halogens and Inorganic Halides. K. L. Kirk. Plenum Publishing Corp., 233 Spring Street, New York, NY. (212) 620-8000. (800) 2221-9369. 1991. $85.00

The Chemical Bond: Structure and Dynamics. Ahmed Zewail. Academic Press, Inc., 1250 Sixth Avenue, San Diego, CA 92101-4311. (619) 231-0926. FAX (619) 699-6715. 1992. $49.95.

Chemistry of Halides, Pseudohalides, and Azides. Saul Patai and Zvi Rappoport, editors. Books on Demand, 300 North Zeeb Road, Ann Arbor, MI 48106-1346. (313) 761-4700. (800) 521-0600. Parts 1 and 2. $340.00.

Chemistry of the Elements. N. N. Greenwood and A. Earnshaw. Elsevier Science Publishing Company, Inc., 655 Avenue of the Americas, New York, NY 10010. (212) 989-5800. 1984. Inquire.

Elements: A Computer File. P. Thomas Vernier. CRC Publishers, Inc., 2000 Corporate Blvd., N.W., Boca Raton, FL 33431. (407) 994-0555 or (800) 333-8300. IBM-PC compatible. 1992. $39.95.

The Elements. John Emsley. Oxford University Press, Inc., 200 Madison Avenue, New York, NY 10016. (800) 334-4249 or (212) 679-7300. Second edition. 1991. $49,95.

Exploring Chemical Elements and their Compounds. David L.Heiserman. TAB Books, P.O. Box 40, Blue Summit, PA 17294-0850. (717) 794-2191 or (800) 822-8138. 1992. $17.95.

Fluorine Chemistry: A Comprehensive Treatment. Mary Howe-Grant. John Wiley & Sons, Inc., 605 Third Avenue, New York, NY 10158-0012. (212) 850-6000. (800) 225-5945. 1995. $79.95.

The History of Chemistry. John Hudson. Routledge, Chapman and Hall, Inc., 29 West 35th Street, New York, NY 10001-2291. (212) 244-3336. 1992. $59.95.

Periodic Table of the Elements. D.E.D. Electronic Publishing, 10306 East Live Oak Avenue, ArCADia, CA 91007. Computer software. 1992. $19.95.

Photographic Sensitivity: Theory and Mechanisms. Tadaaki Tani. Oxford University Press, Inc., 200 Madison Avenue, New York, NY 10016. (212) 725-6000. (800) 334-4249. . 1995. $69.95.

HANDBOOKS AND MANUALS

Chemical Formulary. H. Bennett, editor. Chemical Publishing, Co., Inc. 80 Eighth Avenue, New York, NY 10011. (212 255-1950. Volumes 1 - 30. $60.00 per volume.

Chemometrics: Chemical and Sensory Data. David R. Burgard and James T. Kuznicki. CRC Publishers, Inc., 2000 Corporate Blvd., N.W., Boca Raton, FL 33431. (407) 994-0555 or (800) 333-8300. 1990. $120.00.

Chemical Reference Handbook. Gordon Press Publications, PO Box 459, Bowling Green Station, New York, NY 10004. (718) 624-8419. 1995. $260.00.

Chemical Safety International Reference Manual. Mervyn Richardson, editor. VCH Publications, Inc., 220 East 23rd Street, Suite 909, New York, NY 10010-4606. (212) 683-8333. (800) 422-8824. 1994. $145.50.

Comprehensive Guide To To the Hazardous Properties of Chemical Substances. Pradyot Patnaik. Van Nostrand Reinhold, 115 Fifth Avenue, New York, NY 10003. (212) 254- 3232. (800) 842-3636. 1992. $99.95.

CRC Handbook of Chemistry and Physics. David R. Kide, editor. CRC Press Inc., 2000 Corporate Blvd., NW, Boca Raton, FL 33431. (407) 994-0555. (800) 333-8300. 77th edition. 1996. $99.95.

Guide To Basic Chemical Compounds. D.R. Lide, Jr. CRC Publishers, Inc., 2000 Corporate Blvd., N.W., Boca Raton, FL 33431. (407) 994-0555 or (800) 333-8300. 1993. $120.00.

Handbook of Hazardous Laboratory Chemicals: Information and Disposal. Margaret-Ann Armour. CRC Press, Inc., 2000 Corporate Boulevard, NW, Boca Raton, FL 33431. (407) 994-0555. (800) 272-7737. 1991. $104.95.

Handbooks and Tables in Science and Technology. Russell H. Powell, editor. Oryx Press, 4041 North Central, Suite 700, Phoenix, AZ 85012-3330. (602) 265-2651 or (800) 279-6799. Third edition. 1994. $65.00.

Improving Safety in the Chemical Laboratory: A Practical Guide. Jay A. Young, editor. John Wiley and Sons, Inc., 605 Third Avenue, New York, NY 10158. (800) 526-5368 or (212) 850-6000. Second edition. 1991. $75.00.

Laboratory Handbook of Materials, Equipment, and Techniques. Gary S. Coyne. Prentice Hall (Division of Simon and Schuster), 15 Columbus Circle, New York, NY 10023. (212) 373-8500 or (800) 922-0579. 1992. $45.00.

Lange's Handbook of Chemistry. John A. Dean, editor. McGraw-Hill Publishing Company, 1221 Avenue of the Americas, New York, NY 10020. (800) 262-4729 or (212) 512-3825. Fourteenth edition. 1992. $79.50.

Riegel's Handbook of Industrial Chemistry. James A. Kent, editor. Van Nostrand Reinhold, 115 Fifth Avenue, New York, NY 10003. (212) 254-3232 or (800) 926-2665. Ninth edition. 1992. $114.95.

ONLINE DATABASES AND CD-ROMS

Analytical Abstracts Online. Royal Society of Chemistry, Information Services, Thomas Graham House, Science Park, Milton Road, Cambridge, CB4 4WF, England. Contains citations, mostly with abstracts, of the worldwide literature on analytical chemistry, from 1980 to present. Available on DIALOG, (800) 334-2564, and STN International, FIZ Karlsruhe, P.O. Box 2465, W-7500, Karlsruhe 1, Germany, online services. Inquire as to cost and availability.

Beilstein Online. Beilstein-Institut fur Literatur der Organischen Chemie, Varrentrapperstrasse, 40-42, D-6000, Frankfurt am Main 90, Germany. Contains data on carbon compounds from the Beilstein Handbook of Organic Chemistry. Structural and factual data for more than 3.4 million heterocyclic, isocyclic, and acyclic compounds included. Covers the period 1830 to present. Available on DIALOG, (800) 334-2564, ORBIT (800) 456-7248, and STN International, FIZ Karlsruhe, P.O. Box 2465, W-7500, Karlsruhe 1, Germany, online services. Inquire as to cost and availability.

CA Search. Chemical Abstracts Service, P.O. Box 3012, Columbus, OH 43210-0012. (614) 447-3600. (800) 848-6533. FAX (614) 447-3709. Very comprehensive guide to worldwide chemical literature and related fields, 1972 to present. Available on BRS,(800) 289-4277, DIALOG, (800) 334-2564, ORBIT (800) 456-7248, and STN International, FIZ Karlsruhe, P.O. Box 2465, W-7500, Karlsruhe 1, Germany, online services. Inquire as to cost and availability.

Chemical Journals of the American Chemical Society. American Chemical Society, 1155 16th Street, N.W., Washington, DC 20036. (202) 872-4381 or (800) 424-6747. Contains complete text of approximately 90,000 articles from 22 primary journals published by the American Chemical Society, from mostly 1982 to present. Available on STN International, FIZ Karlsruhe, P.O. Box 2465, W-7500, Karlsruhe 1, Germany, online service. Inquire as to cost and availability.

Compendex Plus. Engineering Information, Inc., 345 East 47th Street, New York, NY 10017. (212) 705-7600 or (800) 221-1044. Contains citations with abstracts to worldwide literature in engineering and technology, from 1970 to present. Available on online BRS,(800) 289-4277, DIALOG, (800) 334-2564, ORBIT (800) 456-7248, and STN International, FIZ Karlsruhe, P.O. Box 2465, W-7500, Karlsruhe 1, Germany, online services. Also available on CD-ROM. Inquire as to cost and availability.

Current Contents Search. Institute for Scientific Information, 3501 Market Street, Philadelphia, PA 19104. (215) 386-0100. FAX (215) 386-6362. Contains citations to articles listed in the table of contents of science and technology journals. Also ar-

ticles in social sciences and life sciences journals. Available on BRS,(800) 289-4277, DIALOG,(800) 334-2564, online services. Inquire as to cost and availability.

Dissertation Abstracts. University Microfilms International, 300 North Zeeb Road, Ann Arbor, MI 48106. (800) 521-0600 or (313) 761-4700. Scope includes virtually all doctoral dissertations accepted at accredited American institutions from 1861 to present in 252 subject areas. Available on BRS,(800) 289-4277, DIALOG,(800) 334-2564, and OCLC EPIC,(800) 848-5878, online services. Also available on CD-ROM. Inquire as to cost and availability.

Gmelin Database. Gmelin-Institut fur Anorganische Chemie und Grenzgebiete, Varrentrapperstrasse, 40-42, Carl-Bosch-Haus, D-6000, Frankfurt am Main 90, Germany. Contains structural and factual data relating to inorganic and organometallic chemistry. Provides data from the Gmelin Handbook of Inorganic and Organometallic Chemistry. Covers the period 1817 to 1975; 1988-89. Available on STN International, FIZ Karlsruhe, P.O. Box 2465, W-7500, Karlsruhe 1, Germany, online service. Inquire as to cost and availability.

Kirk-Othmer Encyclopedia of Chemical Technology. John Wiley and Sons, Inc., 605 Third Avenue, New York, NY 10158. (800) 526-5368 or (212) 850-6000. Contains the complete text of all chapters in the 27 volume fourth edition of the KIRK-Othmer ENCYCLOPEDIA of CHEMICAL TECHNOLOGY. 1991. Available on BRS,(800) 289-4277, DIALOG,(800) 334-2564, online services. Inquire as to cost and availability.

NASA Database. American Institute of Aeronautics and Astronautics, 370 L'Enfant Promenade, S.W., Washington, DC 20024. (202) 646-7400. Citations and abstracts of aeronautics and astronautics literature, 1962 to present. Also contains citations from STAR, SCIENTIFIC and TECHNICAL AEROSPACE REPORTS, and INTERNATIONAL AEROSPACE ABSTRACTS. Available through NASA/RECON online service. Inquire as to cost and availability.

NTIS Bibliographic Database. National Technical Information Service, 5285 Port Royal Road, Springfield, VA 22161. (703) 487-4929 or FAX (703) 321-8199. Broad coverage of government-sponsored science and technology research reports, 1964 to present. Available on BRS,(800) 289-4277, DIALOG, (800) 334-2564, ORBIT, (800) 456-7248, and STN International, FIZ Karlsruhe, P.O. Box 2465, W-7500, Karlsruhe 1, Germany, online services. Also available on CD-ROM. Inquire as to cost and availability.

Nuclear Science Abstracts. U.S. Department of Energy, office of Scientific and Technical Energy, P.O. Box 62, Oak Ridge, TN 37831. (615) 576-1189. Contains citations with abstracts to literature in all fields of nuclear science and energy. 1948 to 1976. Available on DIALOG,(800) 334-2564, online service. Inquire as to cost and availability.

Physics Briefs. American Institute of Physics, 335 East 45th Street, New York, NY 10017. (212) 661-9260 or FAX (212) 661-2036. Contains citations with abstracts of the literature of physics and related fields, 1979 to present. Available on the STN International, FIZ Karlsruhe, P.O. Box 2465, W-7500, Karlsruhe 1, Germany, online service. Inquire as to cost and availability.

Scientific and Technical Books and Serials in Print. R.R. Bowker Inc., 121 Chanlon Road, New Providence, NJ 07974. (800) 521-8110 or (908) 464-6800. List of currently published books and serials in the physical and biological sciences, engineering and technology, with subject, author and titles indexes. Available

on ORBIT,(800) 456-7248, online service.Inquire as to cost and availability.

SCISEARCH. Institute for Scientific Information, 3501 Market Street, Philadelphia, PA 19104. (800) 523-1850 or (215) 386-0100. Broad multidisciplinary title and author index to the international literature of science and technology, 1974 to present. Available on DIALOG,(800) 334-2564, and ORBIT,(800) 456-7248, online services. Also available on CD-ROM. Inquire as to cost and availability.

WILSONLINE. H.W. Wilson Company, 950 University Avenue, Bronx, NY 10452. (800) 367-6770 or (212) 588-8400. Makes available online versions of the printed H.W. Wilson indexes including Applied Science and Technology Index, Business Periodicals Index, General Science Index, and Readers' Guide to Periodical Literature. Period covered is generally 1983 to present. Available on BRS,(800) 289-4277, DIALOG,(800) 334-2564, and OCLC EPIC,(800) 848-5878, online services. Also available on CD-ROM. Inquire as to cost and availability.

PERIODICALS

Analytical Chemistry. American Chemical Society, Box 3337, Columbus, OH 43210. (614) 447-3776. 1929 to Present. Semimonthly. $415.00 per year. ISSN: 0003-2700

Chemical Week. Includes Annual Buyers Guide. Chemical Week Associates, 888 Seventh Avenue, New York, NY 10106. (212) 621-4900. FAX (212) 621-4949. 1914 to present. Weekly. $99.00 per year. ISSN: 0009-272X.

Chemtech. American Chemical Society. Box 3337, Columbus, OH 43210. (614) 447-3776. 1970 to present. Monthly. $370.00 per year. ISSN: 0009-2703.

Fluoride. International Society for Fluoride Research. 8ia Landscape Road, Mount Eden Auckland 4, New Zealand. TEL 96307114. FAX 96307114. Quarterly. $50.00. ISSN: 0015-4725.

Inorganic Chemistry. American Chemical Society, Box 3337, Columbus, OH 43210. (614) 447-3776. 1962 to present. Semimonthly. $500.00 per year. ISSN: 0020-1669.

Journal of Fluorine Chemistry. Elsevier Science Publisher, Box 882, Madison Square Station, New York, NY 10159. (212) 989-5800. 1971 to present. 15 issues per year. $1327.00 per year. ISSN: 0022-1139.

Journal of the American Chemical Society. American Chemical Society, Box 3337, Columbus, OH 43210. (614) 447-3776. 1879 to present. Biweekly. $1055.00 per year. ISSN: 0002- 7863.

Polyhedron. Elsevier Science Publishing Company, Inc., 660 White Plains Road, Tarrytown, NY 10591-5153. (914-524-9200. FAX (914) 524-9200. 1982 to present. 24 issues per year. $2175.00 per year. ISSN: 0277-5387.

RESEARCH CENTERS AND INSTITUTES

Chemical Laboratories. Harvard University, Oxford Street, Cambridge, MA 02138. (617) 495-4283. FAX (617) 496-5618.

Chemistry Laboratories. Rensselaer Polytechnic Institute. Cogswell Laboratory, Troy, NY 12180-3590. (518) 276-8981.

HALIDES

Ency. of Physical Sciences and Engineering Info. Sources

Lawrence Berkeley Laboratory, Chemical Sciences Division. One Cyclotron Road, Building 66, Berkeley, CA 94720. (510) 486-6062. FAX (510) 486-4995.

Research Program in Chemistry and Biochemistry. Southern Illinois University at Carbondale, Carbondale, IL 62901. (618) 453-5721.

Theoretical Chemistry Institute. University of Wisconsin, Madison. 1101 University Avenue, Madison, WI 53706. (608) 262-1511.

University/industry Chemical Research Center. Mississippi State University, Department of Chemistry, P.O. Drawer CH, Mississippi State, MS 39762. (601) 325-3584.

HAM RADIO

See: AMATEUR RADIO

HARBORS

See also: COASTAL ENGINEERING, DREDGES and DREDGING

ABSTRACT SERVICES AND INDEXES

Applied Science and Technology Index; A Cumulative Subject Index To English Language Periodicals in the Fields of Aeronautics and Space Science, Computer Technology, Chemistry, Construction Industry, Energy and Related Areas. H.W. Wilson Co., 950 University Avenue, Bronx, NY 10452. (800) 367-6770 or (212) 588-8400. FAX (718) 590-1857. From 1958 to present. Monthly. Inquire about cost and availability. Also available on CD-ROM and online.

ASCE Combined Annual Combined Index. American Society of Civil Engineers, 345 East 47th Street, New York, NY 10017-2398. (212) 705-7520 or (800) 548-2723. Annual. $48.00.

ASCE PUBLICATIONS Information. 1966-present. American Society of Civil Engineers, 345 East 47th Street, New York, NY 10017-2398. (212) 705-7520 or (800) 548-2723. Bimonthly. $160.00 per year for non-members.

Bibliography and Index of Geology. American Geological Institute, 4220 King Street, Alexandria, VA 22302. (703) 379-2480. Fax (703) 379-7563. Monthly. $1295.00 per year. Also available online as GEOREF.

Civil and Structural Engineering Abstracts, Cambridge Scientific Abstracts, 7200 Wisconsin Avenue, Bethesda, MD 20814-4823. (301) 961-6750. FAX (301) 961-6720. 1993 to present. Monthly. $385.00 per year. Topics covered include structural design, construction equipment and methods, civil defense and military engineering, surveying, highway engineering, maritime and port structures, materials, land reclamation, and soil mechanics.

Current Contents: Engineering, Technology, and Applied Sciences. Institute for Scientific Information, 3501 Market Street, Philadelphia, PA 19104. (215) 386-0100. FAX (215) 386-6362. 1970 to present. Weekly. $442.00 per year.

Engineering Index Monthly. Engineering Information, Inc., Castle Point on the Hudson, Hoboken, NJ 07030. (800) 221-1044. FAX (212) 832-1857. Monthly. $2300.00 per year. Also available online as COMPENDEX, and also on CD-ROM. Covers chemical engineering, computers, electrical engineering, civil engineering, metals and mining, industrial management, and mechanical engineering.

Environmental Engineering Abstracts, Cambridge Scientific Abstracts, 7200 Wisconsin Avenue, Bethesda, MD 20814-4823. (301) 961-6750. FAX (301) 961-6720. Monthly. Covers hazardous materials, environmental impact and protection, treatment of sewage and industrial wastes, hydroelectric power, tidal and wind power, artic and tropical engineering.

Fluid Abstracts: Civil Engineering. Elsevier Science [journals], 660 White Plains Rd., Tarrytown, NY 10159-5153. (919) 524-9200. FAX (919) 333-2444. 1991 to present. Monthly plus annual cumulation. $645.00 per year. Also available online as FLUIDEX. Covers civil engineering applications of fluid mechanics, hydraulics, flow metering and measuring, offshore engineering, environmental hydraulics, and related aspects of wind energy, the atmosphere, and aerodynamics.

Oceanic Abstracts. Cambridge Scientific Abstracts, 7200 Wisconsin Avenue, Bethesda, MD 20814. (301) 961-6750. Fax (301) 961-6720. Bimonthly. $995.00 per year.

Science Citation Index. Institute for Scientific Information, 3501 Market Street, Philadelphia, PA 19104. (215) 386-0100. FAX (215) 386-6362. Inquire about availability and cost. Also available on CD-ROM.

ASSOCIATIONS AND PROFESSIONAL SOCIETIES

American Society of Civil Engineers. 345 East 47th Street, New York, NY 10017-2398. (212) 705-7520 or (800) 548-2723.

Coastal Engineering Research Council. 4911 Bay Oaks Court, College Station, TX 77845. (409) 690-0306.

The Coastal Society. PO Box 2081, Gloucester, MA 01930-2081. (508) 281-9209.

Geological Society of America. 3300 Penrose Place, PO Box 9140, Denver, CO 80301-9140. (303) 447-2020.

International Association of Ports and Harbors. Kotohira-Kaikan Building, 1-2-8 Toranomon, Minato-Ku, Tokyo 105, Japan. Telephone 3-35914261. FAX 3-35800364.

Society of Marine Port Engineers. PO Box 466, Avenel, NJ 07001-2402. (908) 381-7673. FAX (908) 381-2046.

DIRECTORIES AND BIOGRAPHICAL SOURCES

Directory of Engineering Societies and Related Organizations. Gordon Davis. 13th edition. American Association of Engineering Societies, 1111 19th Street NW, Suite 608, Washington, DC 20036. (202) 296-2237 or (800) 658-8897. 1989. Inquire for price.

Research Centers Directory. Gale Research, 835 Penobscot Building, Detroit, MI 48226-4094. (313) 961-2242. (800) 347-4253. FAX (313) 961-6083. $485.00.

Ency. of Physical Sciences and Engineering Info. Sources

HARBORS

Scientific and Technical Organizations and Agencies Directory. Gale Research, 835 Penobscot Building, Detroit, MI 48226-4094. (313) 961-2242. (800) 347-4253. FAX (313) 961-6083. 4th edition. 1996. $195.00.

Who's Who in Engineering. American Association of Engineering Societies, 1111 19th Street NW, Suite 608, Washington, DC 20036. (202) 296-2237 or (800) 658-8897. 8th edition. 1991. Inquire for price.

GENERAL WORKS

Design of Marine Facilities For the Berthing, Mooring, and Repair of Vessels. John W. Gaythwaite. Chapman and Hall, 29 West 35th Street, New York, NY 10001. (800) 842-3636. FAX (212) 563-2269. 1990. $85.00.

Engineering and Design of Military Ports. Headquarters, Department of the Army/ order from Superintendent of Documents, U.S. Government Printing office, Box 371954, Pittsburgh, PA 15250-7954. (202) 783-3238. FAX (202) 512-2233. 1993. Inquire for cost and availability.

Marinas and Small Craft Harbors. Bruce O. Tobiasson & Ronald C. Kollmeyer. Chapman and Hall, 29 West 35th Street, New York, NY 10001. (800) 842-3636. FAX (212) 563-2269. 1991. $62.95.

Marine Structures Engineering. Gregory Tsinker. Chapman & Hall, 115 Fifth Avenue, New York, NY 10211-0906. 1995. $66.95.

Planning and Design Guidelines For Small Craft Harbors. Task Committee on Marinas 2000. Revised edition. American Society of Civil Engineers, 345 East 47th Street, New York, NY 10017-2398. (212) 705-7520 or (800) 548-2723. 1994. Inquire for price and availability.

HANDBOOKS AND MANUALS

Coastal, Estuarial, and Harbour Engineer's Reference Book. M.B. Abbott & W.A. Price, editors. E & FN Spon/ Routledge, Chapman and Hall, Inc., 115 Fifth Avenue, New York, NY 10211-0906. 1994. $115.95.

Handbook of Coastal and Ocean Engineering Volume 3: Harbors, Navigational Channels, Estuaries and Environmental Effects. John B. Herbich, editor. Gulf Publishing Company, P.O. Box 2608, Houston, TX. (713) 529-4301 or (800) 231-6275. 1992. $195.00.

Handbook of Port and Harbor Engineering. Gregory Tsinker. Chapman and Hall, Inc., 115 Fifth Avenue, New York, NY 10211-0906. 1996. $159.95.

ONLINE DATABASES AND CD-ROMS

Compendex Plus. Engineering Information, Inc., 345 East 47th Street, New York, NY 10017. (212) 705-7600 or (800) 221-1044. Contains citations with abstracts to worldwide literature in engineering and technology, from 1970 to present. Available on online BRS,(800) 289-4277, DIALOG, (800) 334-2564, ORBIT (800) 456-7248, and STN International, FIZ Karlsruhe, P.O. Box 2465, W-7500, Karlsruhe 1, Germany, online services. Also available on CD-ROM. Inquire as to cost and availability.

CSA Engineering. Cambridge Scientific Abstracts, 7200 Wisconsin Avenue, Suite 601, Bethesda, MD 20814. (301) 961-6750 or (800) 843-7751. Contains citations and abstracts of international periodicals and research literature covering all fields of engineering and science and technology,including computer and information science, electronics, mechanical engineering, solid state materials, 1981 to present. Available on BRS,(800) 289-4277, online service.Inquire as to cost and availability.

Current Contents Search. Institute for Scientific Information, 3501 Market Street, Philadelphia, PA 19104. (215) 386-0100. FAX (215) 386-6362. Contains citations to articles listed in the table of contents of science and technology journals. Also articles in social sciences and life sciences journals. Available on BRS,(800) 289-4277, DIALOG,(800) 334-2564, online services. Inquire as to cost and availability.

Dissertation Abstracts. University Microfilms International, 300 North Zeeb Road, Ann Arbor, MI 48106. (800) 521-0600 or (313) 761-4700. Scope includes virtually all doctoral dissertations accepted at accredited American institutions from 1861 to present in 252 subject areas. Available on BRS,(800) 289-4277, DIALOG,(800) 334-2564, and OCLC EPIC,(800) 848-5878, online services. Also available on CD-ROM. Inquire as to cost and availability.

GEOREF: Bibliography and Index of Geology. American Geological Institute, 4220 King Street, Alexandria, VA 22302. (703) 379-2480. Fax (703) 379-7563. Monthly. Inquire for price and availability.

NTIS Bibliographic Database. National Technical Information Service, 5285 Port Royal Road, Springfield, VA 22161. (703) 487-4929 or FAX (703) 321-8199. Broad coverage of government-sponsored science and technology research reports, 1964 to present. Available on BRS,(800) 289-4277, DIALOG, (800) 334-2564, ORBIT, (800) 456-7248, and STN International, FIZ Karlsruhe, P.O. Box 2465, W-7500, Karlsruhe 1, Germany, online services. Also available on CD-ROM. Inquire as to cost and availability.

SCISEARCH. Institute for Scientific Information, 3501 Market Street, Philadelphia, PA 19104. (800) 523-1850 or (215) 386-0100. Broad multidisciplinary title and author index to the international literature of science and technology, 1974 to present. Available on DIALOG,(800) 334-2564, and ORBIT,(800) 456-7248, online services. Also available on CD-ROM. Inquire as to cost and availability.

TRIS (Transportation Research Information). National Academy of Science, 2101 Constitution Avenue, N.W., Washington, DC 20418. (202) 334-3313 or (800) 624-6242. Citations with abstracts of literature on transportation, including air, highway, rail, maritime and other modes from 1968 to present. Available on DIALOG,(800) 334-2564, online service. Inquire as to cost and availability.

WILSONLINE. H.W. Wilson Company, 950 University Avenue, Bronx, NY 10452. (800) 367-6770 or (212) 588-8400. Makes available online versions of the printed H.W. Wilson indexes including Applied Science and Technology Index, Business Periodicals Index, General Science Index, and Readers' Guide to Periodical Literature. Period covered is generally 1983 to present. Available on BRS,(800) 289-4277, DIALOG,(800) 334-2564, and OCLC EPIC,(800) 848-5878, online services. Also available on CD-ROM. Inquire as to cost and availability.

HARBORS

Ency. of Physical Sciences and Engineering Info. Sources

PERIODICALS

Civil Engineering. American Society of Civil Engineers, 345 East 47th Street, New York, NY 10017-2398. (212) 705-7520 or (800) 548-2723. 1930 to present. Monthly. $85.00 per year.

Coastal Engineering. Elsevier Science Inc., Box 882, Madison Square Station, New York, NY 10159. (212) 989-5800. 1977 to present. 12 times a year. $642.00 per year.

Coastal Research. Florida State University Geology Department, Tallahassee, FL 32306. (904) 644-5860. FAX (904) 644-4214. 1962 to present. 3 times a year. $6.00 per year.

Coastal Management. Taylor & Francis, 1900 Frost Road, Suite 101, Bristol, PA 19007. (215) 785-5800. FAX (215) 785-5515. 1973 to present. Quarterly. $140.00 per year.

Coastwatch. North Carolina Sea Grant College Program, Box 8605, North Carolina State University, Raleigh, NC 27695-8605. (919) 515-2454. FAX (919) 515-7095. 1970 to present. Bimonthly. $12.00 per year.

Estuarine Coastal and Shelf Science. Academic Press Ltd./ Harcourt Brace & Company, Ltd., Foots Cray High Street, Sidcup, Kent DA14 5HP, England. Telephone 44-81-300-3322. FAX 44-81-309-0807. 1973 to present. Monthly. Inquire for cost.

Journal of Waterway, Port, Coastal, and Ocean Engineering. American Society of Civil Engineers, Waterway, Port, Coastal, and Ocean Division, 345 East 47th Street, New York, NY 10017-2398. (212) 705-7520 or (800) 548-2723. 1956 to present. Bimonthly. $100.00 per year.

Ocean Engineering. Elsevier Science [journals], 660 White Plains Rd., Tarrytown, NY 10159-5153. (919) 524-9200. FAX (919) 333-2444. 1968 to present. 8 times a year. $595.00 per year.

RESEARCH CENTERS AND INSTITUTES

Coastal and Oceanographic Engineering Laboratory. University of Florida, 336 Weil Hall, Gainesville, FL 32611. (904) 392-1436. FAX 9904) 392-3466.

University of Massachusetts At Boston, Urban Harbors Institute. Harbor Campus, Boston, MA 02125. (617) 287-5570. FAX (617) 265-7173.

HARD DISKS

See: COMPUTER MEMORY AND STORAGE

HAZARDOUS MATERIALS AND WASTES

ABSTRACT SERVICES AND INDEXES

Applied Science and Technology Index; A Cumulative Subject Index To English Language Periodicals in the Fields of Aeronautics and Space Science, Computer Technology, Chemistry, Construction Industry, Energy and Related Areas. H.W. Wilson Co., 950 University Avenue, Bronx, NY 10452. (800) 367-6770 or (212) 588-8400. From 1958 to present. Monthly. Inquire about cost and availability. Also available on CD-ROM and online.

Environmental Abstracts. R.R. Bowker Inc., 121 Chanlon Road, New Providence, NJ 07974. (800) 521-8110 or (908) 464-6800. Contains citations, most with abstracts, to literature relating to environmental issues and problems, from 1971 to present. Inquire as to cost and availability.

Environmental Engineering Abstracts, Cambridge Scientific Abstracts, 7200 Wisconsin Avenue, Bethesda, MD 20814-4823. (301) 961-6750. FAX (301) 961-6720. Monthly. Covers hazardous materials, environmental impact and protection, treatment of sewage and industrial wastes, hydroelectric power, tidal and wind power, arctic and tropical engineering.

Pollution Abstracts, Cambridge Scientific Abstracts, 7200 Wisconsin Avenue, Bethesda, MD 20814-4823. (301) 961-6750. FAX (301) 961-6720. Inquire for cost and availability.

Science Citation Index. Institute for Scientific Information, 3501 Market Street, Philadelphia, PA 19104. (215) 386-0100. FAX (215) 386-6362. Inquire about availability and cost. Also available on CD-ROM.

ASSOCIATIONS AND PROFESSIONAL SOCIETIES

Citizen's Clearing House For Hazardous Wastes. PO Box 6806, Falls Church, VA 22040. (703) 237-2249.

Hazardous Waste Treatment Council. 915 15th Street NW, Fifth floor, Washington, DC 20005. (202) 783-0870. FAX (202) 737-2038.

BIBLIOGRAPHIES

Catalog of Hazardous and Solid Waste Publications. U.S. Environmental Protection Agency, office of Solid Waste and Emergency Response. 8th edition. Superintendent of Documents, U.S. Government Printing office, Box 371954, Pittsburgh, PA 15250-7954. (202) 783-3238. FAX (202) 512-2233. 1995. Inquire for cost and availability.

DIRECTORIES AND BIOGRAPHICAL SOURCES

Directory of Engineering Societies and Related Organizations. Gordon Davis. 13th edition. American Association of Engineering Societies, 1111 19th Street NW, Suite 608, Washington, DC 20036. (202) 296-2237 or (800) 658-8897. 1989. Inquire for price.

Directory of Radioactive Waste officials. Business Publishers Inc., 951 Pershing Drive, Silver Spring, MD 20910-4464. (301) 587-6300. FAX (301) 585-9075. Irregular (second edition, 1992). $89.00.

International Research Centers Directory. Gale Research, 835 Penobscot Building, Detroit, MI 48226-4094. (313) 961-2242. (800) 347-4253. 8th edition. 1995. $410.00

Research Centers Directory. Gale Research,835 Penobscot Building, Detroit, MI 48226-4094. (313) 961-2242. (800) 347-4253. $470.00

Scientific and Technical Organizations and Agencies Directory. Gale Research, 835 Penobscot Building, Detroit, MI 48226-4094. (313) 961-2242. (800) 347-4253. 4th edition. 1996. $195.00.

ENCYCLOPEDIAS AND DICTIONARIES

Encyclopedia of Physical Science and Technology. Robert A. Meyers, ed. 18 volumes. Academic Press, Inc., 1250 Sixth Avenue, San Diego, CA 92101-4311. (619) 231-0926. FAX (619) 699-6715. 1992. $2100.00

McGraw-Hill Dictionary of Scientific and Technical Terms. Sybil P. Parker, ed. 5th edition. McGraw-Hill Publishing Company, 1221 Avenue of the Americas, New York, NY 10020. (800) 262-4729 or (212) 512-3825. 1993. $110.50.

GENERAL WORKS

Basic Hazardous Waste Management. William C. Blackman. 2d edition. Lewis Publishers, 2000 Corporate Blvd. NW, Boca Raton, FL 33431. (407) 994-0555 or (800) 272-7737. 1992. $59.95.

Hazardous Materials and Hazardous Waste Management: A Technical Guide. Gayle Woodside. John Wiley and Sons, Inc., 605 Third Avenue, New York, NY 10158. (800) 526-5368 or (212) 850-6000. 1993. $64.95.

Hazardous Materials and Waste Management. W. Kuhne. Prentice Hall, 113 Sylvan Avenue, Route 9W, Englewood Cliffs, NJ 07632. (201) 592-2000 or (800) 922-0579. 1994. $54.00.

Hazardous Waste Management. Michael LaGrega. McGraw-Hill Publishing Company, 1221 Avenue of the Americas, New York, NY 10020. (800) 262-4729 or (212) 512-3825. 1994. $65.32.

HANDBOOKS AND MANUALS

CRC Handbook of Chemistry and Physics. David R. Lide, editor. 75th edition. CRC Press/Times Mirror Company, 2000 Corporate Blvd. NW, Boca Raton, FL 33431. (407) 994-0555 or (800) 272-7737. $99.50.

Handbook of Pollution and Hazardous Materials Compliance. Nicholas P. Cheremisinoff and Madelyn Graffia. Marcel Dekker, Inc., 270 Madison Avenue, New York, NY 10016. (212) 696-9000. FAX (212) 685-4540. 1996. Inquire for cost.

Hazardous Materials Handbook. Van Nostrand Reinhold, 115 Fifth Avenue, New York, NY 10003. (212) 254-3232. FAX (212) 254-9499. 1996. $300.00.

Hazardous Materials Storage and Handling Handbook. Diane Publishing Company, 600 Upland Avenue, Upland, PA 19015. (610) 499-7415. FAX (610) 499-7429. 1995. $20.00.

Hazardous Waste Handling Pocket Guide. Christine Gorman, editor. Genium Publishing Corporation, 1 Genium Plaza, Schenectady, NY 12304. (800) 243-6486. FAX (518) 377-1891. 1993. $41.80.

Hazardous Waste Management Handbook. Southam Information and Technology Group, 1450 Don Mills Road, Don Mills, ON M3B 2X7, Canada. (416) 445-6641. FAX (416) 442-2200. 1982 to present. Annual. $189.39.

Hazardous Waste Site Operations: A Training Manual For Site Professionals. Michael F. Waxman. John Wiley and Sons, Inc., 605 Third Avenue, New York, NY 10158. (800) 526-5368 or (212) 850-6000. 1996. $59.95.

Radioactive Waste Management Handbook. Harwood ACADemic, c/o Gordon & Breach Science Publishers, 820 Town Center Drive, Langhorne, PA 19047. (215) 750-2642. FAX (215) 750-6343. Inquire for price and availability.

ONLINE DATABASES AND CD-ROMS

CA Search. Chemical Abstracts Service, P.O. Box 3012, Columbus, OH 43210-0012. (614) 447-3600. (800) 848-6533. FAX (614) 447-3709. Very comprehensive guide to worldwide chemical literature and related fields, 1972 to present. Available on BRS,(800) 289-4277, DIALOG, (800) 334-2564, ORBIT (800) 456-7248, and STN International, FIZ Karlsruhe, P.O. Box 2465, W-7500, Karlsruhe 1, Germany, online services. Inquire as to cost and availability.

Compendex Plus. Engineering Information, Inc., 345 East 47th Street, New York, NY 10017. (212) 705-7600 or (800) 221-1044. Contains citations with abstracts to worldwide literature in engineering and technology, from 1970 to present. Available on online BRS,(800) 289-4277, DIALOG, (800) 334-2564, ORBIT (800) 456-7248, and STN International, FIZ Karlsruhe, P.O. Box 2465, W-7500, Karlsruhe 1, Germany, online services. Also available on CD-ROM. Inquire as to cost and availability.

CSA Engineering. Cambridge Scientific Abstracts, 7200 Wisconsin Avenue, Suite 601, Bethesda, MD 20814. (301) 961-6750 or (800) 843-7751. Contains citations and abstracts of international periodicals and research literature covering all fields of engineering and science and technology,including computer and information science, electronics, mechanical engineering, solid state materials, 1981 to present. Available on BRS,(800) 289-4277, online service.Inquire as to cost and availability.

Current Contents Search. Institute for Scientific Information, 3501 Market Street, Philadelphia, PA 19104. (215) 386-0100. FAX (215) 386-6362. Contains citations to articles listed in the table of contents of science and technology journals. Also articles in social sciences and life sciences journals. Available on BRS,(800) 289-4277, DIALOG,(800) 334-2564, online services. Inquire as to cost and availability.

Dissertation Abstracts. University Microfilms International, 300 North Zeeb Road, Ann Arbor, MI 48106. (800) 521-0600 or (313) 761-4700. Scope includes virtually all doctoral dissertations accepted at accredited American institutions from 1861 to present in 252 subject areas. Available on BRS,(800) 289-4277, DIALOG,(800) 334-2564, and OCLC EPIC,(800) 848-5878, online services. Also available on CD-ROM. Inquire as to cost and availability.

NTIS Bibliographic Database. National Technical Information Service, 5285 Port Royal Road, Springfield, VA 22161. (703) 487-4929 or FAX (703) 321-8199. Broad coverage of government-sponsored science and technology research reports, 1964 to present. Available on BRS,(800) 289-4277, DIALOG, (800) 334-2564, ORBIT, (800) 456-7248, and STN International, FIZ Karlsruhe, P.O. Box 2465, W-7500, Karlsruhe 1, Germany,

HAZARDOUS MATERIALS AND WASTES

Ency. of Physical Sciences and Engineering Info. Sources

online services. Also available on CD-ROM. Inquire as to cost and availability.

SCISEARCH. Institute for Scientific Information, 3501 Market Street, Philadelphia, PA 19104. (800) 523-1850 or (215) 386-0100. Broad multidisciplinary title and author index to the international literature of science and technology, 1974 to present. Available on DIALOG,(800) 334-2564, and ORBIT,(800) 456-7248, online services. Also available on CD-ROM. Inquire as to cost and availability.

Wasteinfo. United Kingdom Atomic Energy Authority, Building 7.12, Harwell Laboratory, Didcot, Oxon, OX11 ORA, England. Contains citations to worldwide literature on non-nuclear waste management, from 1973 to present. Available on ORBIT (800) 456-7248, online service. Also available on CD-ROM. Inquire as to cost and availability.

PERIODICALS

EI Digest. Environmental Information Ltd., 4801 W 81st Street, Suite 119, Minneapolis, MN 55437. (612) 831-2473. FAX (612) 831-6550. 1989 to present. Monthly. $1250.00 per year.

Hazardous Materials Control. HMC Resources Institute, 1 Church Street, Suite 120, Rockville, MD 20850-4129. (301) 251-1900. FAX (301) 738-2330. 1988 to present. Bi-monthly. $18.00 per year.

Hazardous Waste and Hazardous Materials. HMC Resources Institute, 1 Church Street, Suite 120, Rockville, MD 20850-4129. (301) 251-1900. FAX (301) 738-2330. 1984 to present. Quarterly. $152.00 per year.

Hazardous Waste Consultant. Elsevier Science [journals], 660 White Plains Rd., Tarrytown, NY 10159-5153. (919) 524-9200. FAX (919) 333-2444. 1983 to present. Bi-monthly. $495.00 per year.

Hazardous Waste News. Business Publishers Inc., 951 Pershing Drive, Silver Spring, MD 20910-4464. (301) 587-6300. FAX (301) 585-9075. 1979 to present. Weekly. $540.00 per year.

Journal of Hazardous Materials. Elsevier Science [journals], 660 White Plains Rd., Tarrytown, NY 10159-5153. (919) 524-9200. FAX (919) 333-2444. 1976 to present. 12 times a year. $899.00 per year.

World Wastes. Argus Inc., 6151 Powers Ferry Road NW, Atlanta, GA 30339-2941. (404) 955-2500. FAX (404) 955-0400. 1957 to present. Monthly. $48.00 per year.

RESEARCH CENTERS AND INSTITUTES

Center For Hazardous Materials Research. University of Pittsburgh Applied Research Center, 320 William Pitt Way, Pittsburgh, PA 15238. (412) 826-5320. FAX (412) 826-5552.

Environmental Defense Fund. 257 Park Avenue S, New York, NY 10010. (212) 505-2100. FAX (212) 505-2375.

Environmental Policy Institute. 218 D Street NE, Washington, DC 20003. (202) 544-2600.

Environmental Research Foundation. PO Box 3541, Princeton, NJ 08543-3541. (609) 683-0707.

Florida State University Center For Biomedical Toxicological Research and Hazardous Waste Management. Bellamy Building, Tallahassee, FL 32306. (904) 644-5524. FAX (904) 574-6704.

Hazardous Materials Control Research Institute. 9300 Columbia Blvd., Silver Spring, MD 20910. (301) 587-9390. FAX (301) 589-0192.

Hazardous Substance Management Research Center. New Jersey Institute of Technology, Newark, NJ 07102. (201) 596-3233.

Hazardous Waste Research and Information Center. One E Hazelwood Drive, Champaign, IL 61820. (217) 333-8940. FAX (217) 333-8944.

Tufts University Center For Environmental Management. Curtis Hall, 474 Boston Avenue, Medford, MA 02155. (617) 381-3486. FAX (617) 381-3084.

University of Alabama, Hazardous Materials Management and Resource Recovery. Department of Chemical Engineering, PO Box 870203, Tuscaloosa, AL 35487-0203. (205) 348-8401.

University of Connecticut Environmental Research Institute For Hazardous Materials and Wastes. 191 Auditorium Road, Box U210, Storrs, CT 06268. (203) 486-4015.

HEAT EXCHANGERS

See also: COOLING TOWERS, FLUID MECHANICS, HEAT TRANSFER, MECHANICAL ENGINEERING, PRESSURE VESSELS, THERMODYNAMICS

ABSTRACT SERVICES AND INDEXES

Applied Mechanics Reviews: An Assessment of World Literature in Engineering Sciences. 1948-present. American Society of Mechanical Engineers, 345 East 47th Street, New York, NY 10017. (212) 705-7703. Monthly. $360.00 per year.

Applied Science and Technology Index; A Cumulative Subject Index To English Language Periodicals in the Fields of Aeronautics and Space Science, Computer Technology, Chemistry, Construction Industry, Energy and Related Areas. H.W. Wilson Co., 950 University Avenue, Bronx, NY 10452. (800) 367-6770 or (212) 588-8400. FAX (718) 590-1617. From 1958 to present. Monthly. Inquire about cost and availability. Also available on CD-ROM and online.

Current Contents: Engineering, Technology, and Applied Sciences. Institute for Scientific Information, 3501 Market Street, Philadelphia, PA 19104. (215) 386-0100. FAX (215) 386-6362. 1970 to present. Weekly. $442.00 per year.

Energy Research Abstracts. 1976-present. U.S. Department of Energy, office of Scientific and Technical Information, Box 62, Oak Ridge, TN 37831. (615) 574-0733. Subscriptions: Superintendent of Documents, U.S. Government Printing office, Box 371954, Pittsburgh, PA 15250-7954. (202) 783-3238. FAX (202) 512-2233.Monthly. $164.00 per year. Abstracts all the scientific and technical reports, journal articles, conference proceedings, patents, theses, and monographs sponsored by the U.S. Energy Research and Development Administration.

Engineering Index Monthly. Engineering Information, Inc., Castle Point on the Hudson, Hoboken, NJ 07030. (800) 221-1044. FAX (212) 832-1857. Monthly. $2300.00 per year. Also available online as COMPENDEX, and also on CD-ROM. Covers chemical engineering, computers, electrical engineering, civil engineering, metals and mining, industrial management, and mechanical engineering.

Physics Abstracts. Institute of Electrical Engineers, Michael Faraday House, Six Hill Way, Stevenage, Herts, SG1 2AY, England. Distributed by IEEE, 445 Hoes Lane, Piscataway, NJ 08854. (908) 562-5549. 1898 to present. Monthly. $2700.00 per year. Also available online as INSPEC.

Physics Briefs. American Institute of Physics, 335 East 45th Street, New York, NY 10017 (212) 661-9404 or FAX (516)349-9704. 1980 to present. Six times per year. $2600.00 per year.

Previews of Heat and Mass Transfer. Elsevier Science [journals], 660 White Plains Rd., Tarrytown, NY 10159-5153. (919) 524-9200. FAX (919) 333-2444. 1974 to present. Six times a year. $265.00 per year. Contains citations with abstracts to the worldwide literature on heat and mass transfer from more than 100 journals.

Science Citation Index. Institute for Scientific Information, 3501 Market Street, Philadelphia, PA 19104. (215) 386-0100. FAX (215) 386-6362. Inquire about availability and cost. Also available on CD-ROM.

ASSOCIATIONS AND PROFESSIONAL SOCIETIES

American Institute of Chemical Engineers. 345 East 47th Street, New York, NY 10017-2396. (212) 705-7663.

American Institute of Physics. One Physics Ellipse, College Park, MD 20740. (301) 209-3000.

American Society of Mechanical Engineers. 345 East 47th Street, New York, NY 10017-2398. (212) 705-7703.

Heat Exchange Institute. 1300 Sumner Avenue, Cleveland, OH 44115-2851. (216) 241-7333. FAX (216) 241-0105.

DIRECTORIES AND BIOGRAPHICAL SOURCES

American Men & Women of Science, 1995-96. R.R. Bowker Staff, eds. 19th edition. 8 volumes. R.R. Bowker/Reed International Publishing Company, 121 Chanlon Road, New Providence, NJ 07974. (908) 464-6800 or (800) 521-8110. 1995. $850.00

Directory of Engineering Societies and Related Organizations. Gordon Davis. 13th edition. American Association of Engineering Societies, 1111 19th Street NW, Suite 608, Washington, DC 20036. (202) 296-2237 or (800) 658-8897. 1989. Inquire for price.

Research Centers Directory. Gale Research,835 Penobscot Building, Detroit, MI 48226-4094. (313) 961-2242. (800) 347-4253. FAX (313) 961-6083. $485.00.

Scientific and Technical Organizations and Agencies Directory. Gale Research, 835 Penobscot Building, Detroit, MI 48226-4094. (313) 961-2242. (800) 347-4253. FAX (313) 961-6083. 4th edition. 1996. $195.00.

Who's Who in Engineering. American Association of Engineering Societies, 1111 19th Street NW, Suite 608, Washington, DC 20036. (202) 296-2237 or (800) 658-8897. 8th edition. 1991. Inquire for price.

ENCYCLOPEDIAS AND DICTIONARIES

Encyclopedia of Physical Science and Technology. Robert A. Meyers, ed. 18 volumes. Academic Press Inc., 6277 Sea Harbor Drive, Orlando, FL 32887. (800) 321-5068. 1992. $2100.00.

McGraw-Hill Encyclopedia of Engineering. Sybil P. Parker, ed. 2nd edition. McGraw-Hill Publishing Company, 1221 Avenue of the Americas, New York, NY 10020. (800) 262-4729 or (212) 512-3825. 1993. $95.50.

McGraw-Hill Encyclopedia of Science and Technology. Sybil P. Parker, ed. 7th edition. 20 volumes. McGraw-Hill Publishing Company, 1221 Avenue of the Americas, New York, NY 10020. (800) 262-4729 or (212) 512-3825. 1992. $1900.00

Thesaurus of Scientific, Technical, and Engineering Terms. Hemisphere Publishing Corporation, 1900 Frost Road, Suite 101, Bristol, PA 19007-1598. (215) 785-5800 or (800) 821-8312. FAX (215) 785-5515. 1987. $173.00.

GENERAL WORKS

Heat Exchangers. Holger Martin. Rev. and updated. Hemisphere Publishing Corporation, 1900 Frost Road, Suite 101, Bristol, PA 19007-1598. (215) 785-5800 or (800) 821-8312. FAX (215) 785-5515. 1992. $49.50.

Heat Transfer Equipment. Nicolas P. Cheremisinoff. Prentice Hall, 113 Sylvan Avenue, Route 9W, Englewood Cliffs, NJ 07632. (201) 592-2000 or (800) 922-0579. 1993. $56.00.

Industrial Heat Exchangers: A Basic Guide. G. Walker. 2nd edition. Hemisphere Publishing Corporation, 1900 Frost Road, Suite 101, Bristol, PA 19007-1598. (215) 785-5800 or (800) 821-8312. FAX (215) 785-5515. 1990. $73.00.

Principles of Enhanced Heat Transfer. Ralph L. Webb. John Wiley and Sons, Inc., 605 Third Avenue, New York, NY 10158. (800) 526-5368 or (212) 850-6000. 1993. Write for information.

HANDBOOKS AND MANUALS

Handbook of Heat Exchanger Calculations. G.F.Hewitt & P.B. Whalley. CRC Press, 2000 Corporate Blvd., N.W., Boca Raton, FL 33431. (407) 994-0555 or (800) 272-7737. FAX (407) 994-0949. 1990. $45.00.

Handbook of Heat Exchanger Design. G.F. Hewitt, editor. Begell House Publishers, 79 Madison Avenue, Suite 1106, New York, NY 10016. (212) 775-1999. FAX (212) 213-8368. 1989. $125.00.

Handbook of Heat Transfer. Kutatkerkadze. CRC Press, 2000 Corporate Blvd., N.W., Boca Raton, FL 33431. (407) 994-0555 or (800) 272-7737. FAX (407) 994-0949. 1993. Inquire for cost and availability.

HEAT EXCHANGERS

Ency. of Physical Sciences and Engineering Info. Sources

Handbook of Heat Transfer Applications. W.M. Rohsenow, et al., editors. McGraw-Hill Publishing Company, 1221 Avenue of the Americas, New York, NY 10020. (800) 262-4729 or (212) 512-3825. 1985. Inquire for cost and availability.

Heat Exchanger Sourcebook. J. W. Palen. Hemisphere Publishing Corporation, 1900 Frost Road, Suite 101, Bristol, PA 19007-1598. (215) 785-5800 or (800) 821-8312. FAX (215) 785-5515. 1986. $81.00.

ONLINE DATABASES AND CD-ROMS

Compendex Plus. Engineering Information, Inc., 345 East 47th Street, New York, NY 10017. (212) 705-7600 or (800) 221-1044. Contains citations with abstracts to worldwide literature in engineering and technology, from 1970 to present. Available on online BRS,(800) 289-4277, DIALOG, (800) 334-2564, ORBIT (800) 456-7248, and STN International, FIZ Karlsruhe, P.O. Box 2465, W-7500, Karlsruhe 1, Germany, online services. Also available on CD-ROM. Inquire as to cost and availability.

CSA Engineering. Cambridge Scientific Abstracts, 7200 Wisconsin Avenue, Suite 601, Bethesda, MD 20814. (301) 961-6750 or (800) 843-7751. Contains citations and abstracts of international periodicals and research literature covering all fields of engineering and science and technology,including computer and information science, electronics, mechanical engineering, solid state materials, 1981 to present. Available on BRS,(800) 289-4277, online service.Inquire as to cost and availability.

Current Contents Search. Institute for Scientific Information, 3501 Market Street, Philadelphia, PA 19104. (215) 386-0100. FAX (215) 386-6362. Contains citations to articles listed in the table of contents of science and technology journals. Also articles in social sciences and life sciences journals. Available on BRS,(800) 289-4277, DIALOG,(800) 334-2564, online services. Inquire as to cost and availability.

Dissertation Abstracts. University Microfilms International, 300 North Zeeb Road, Ann Arbor, MI 48106. (800) 521-0600 or (313) 761-4700. Scope includes virtually all doctoral dissertations accepted at accredited American institutions from 1861 to present in 252 subject areas. Available on BRS,(800) 289-4277, DIALOG,(800) 334-2564, and OCLC EPIC,(800) 848-5878, online services. Also available on CD-ROM. Inquire as to cost and availability.

ISMEC: Mechanical Engineering Abstracts. Cambridge Scientific Abstracts, 7200 Wisconsin Avenue, Suite 601, Bethesda, MD 20814. (301) 961-6750 or (800) 843-7751. Contains citations to the literature in mechanical engineering, industrial and production engineering, energy, power, mechanics, devices and related areas, from 1973 to present. Available on the DIALOG,(800) 334-2564, online service. Inquire as to cost and availability.

NTIS Bibliographic Database. National Technical Information Service, 5285 Port Royal Road, Springfield, VA 22161. (703) 487-4929 or FAX (703) 321-8199. Broad coverage of government-sponsored science and technology research reports, 1964 to present. Available on BRS,(800) 289-4277, DIALOG, (800) 334-2564, ORBIT, (800) 456-7248, and STN International, FIZ Karlsruhe, P.O. Box 2465, W-7500, Karlsruhe 1, Germany, online services. Also available on CD-ROM. Inquire as to cost and availability.

SCISEARCH. Institute for Scientific Information, 3501 Market Street, Philadelphia, PA 19104. (800) 523-1850 or (215) 386-0100. Broad multidisciplinary title and author index to the international literature of science and technology, 1974 to present. Available on DIALOG,(800) 334-2564, and ORBIT,(800) 456-7248, online services. Also available on CD-ROM. Inquire as to cost and availability.

WILSONLINE. H.W. Wilson Company, 950 University Avenue, Bronx, NY 10452. (800) 367-6770 or (212) 588-8400. Makes available online versions of the printed H.W. Wilson indexes including Applied Science and Technology Index, Business Periodicals Index, General Science Index, and Readers' Guide to Periodical Literature. Period covered is generally 1983 to present. Available on BRS,(800) 289-4277, DIALOG,(800) 334-2564, and OCLC EPIC,(800) 848-5878, online services. Also available on CD-ROM. Inquire as to cost and availability.

PERIODICALS

Experimental Thermal and Fluid Science. Elsevier Science Publishing Company, Inc., 655 Avenue of the Americas, New York, NY 10010. (212) 989-5800. 1988 to present. 8 times a year. $484.00 per year.

Heat Transfer Engineering. Taylor and Francis, Inc., 79 Madison Avenue, Suite 1106, New York, NY 10016. (212) 725-1999 or (800) 821-8312. 1979 to present. Quarterly. $176.00 per year.

International Communications in Heat and Mass Transfer. Elsevier Science [journals], 660 White Plains Rd., Tarrytown, NY 10159-5153. (919) 524-9200. FAX (919) 333-2444. 1974 to present. 6 times a year. $430.00 per year.

Journal of Heat Transfer. American Society of Mechanical Engineers, 345 E. 47th Street, New York, NY 10017. (212) 705-7722. 1970 to present. Quarterly. $155.00 per year.

Numerical Heat Transfer A: Applications. Taylor and Francis, Inc., 79 Madison Avenue, Suite 1106, New York, NY 10016. (212) 725-1999 or (800) 821-8312. 1978 to present. 8 times a year. $530.00 per year.

Numerical Heat Transfer B: Fundamentals. Taylor and Francis, Inc., 79 Madison Avenue, Suite 1106, New York, NY 10016. (212) 725-1999 or (800) 821-8312. 1978 to present. 8 times a year. $330.00 per year.

RESEARCH CENTERS AND INSTITUTES

Heat Transfer Laboratory. Massachusetts Institute of Technology, 77 Massachusetts Avenue, Cambridge, MA 02139. (716) 253-2248.

Heat Transfer Research. University of Wisconsin—Milwaukee, PO Box 784, Milwaukee, WI 53201. (414) 963-5001. FAX (414) 229-6958.

Institute of Thermo-fluid Engineering and Science. Lehigh University, Bldg. A, Mountain Top Campus, Bethlehem, PA 18015. (215) 758-4091. FAX (215) 758-5057.

Heat Transfer Research Facility. Columbia University, 632 W. 125th Street, New York, NY 10027. (212) 280-4163. FAX (212) 678-5279.

HEAT PIPES

See also: FLUID MECHANICS, HEAT EXCHANGERS, HEATING and VENTILATING, MECHANICAL ENGINEERING, THERMODYNAM-ICS

ABSTRACT SERVICES AND INDEXES

Applied Science and Technology Index; A Cumulative Subject Index To English Language Periodicals in the Fields of Aeronautics and Space Science, Computer Technology, Chemistry, Construction Industry, Energy and Related Areas. H.W. Wilson Co., 950 University Avenue, Bronx, NY 10452. (800) 367-6770 or (212) 588-8400. FAX (718) 590-1617. From 1958 to present. Monthly. Inquire about cost and availability. Also available on CD-ROM and online.

Computer Abstracts, MCB University Press Ltd., PO Box 10812, Birmingham, AL 35201-0812. (800) 633-4931. FAX (205) 995-1588. Monthly. Covers computer theory, data, hardware, systems, networks, human-computer interaction, artificial intelligence, as well as applications of computers in aerospace, business, CAD/CAM, cartography, civil engineering, electronics and electrical engineering, industrial engineering, mechanical engineering, medicine, structural engineering, etc.

Current Contents: Engineering, Technology, and Applied Sciences. Institute for Scientific Information, 3501 Market Street, Philadelphia, PA 19104. (215) 386-0100. FAX (215) 386-6362. 1970 to present. Weekly. $442.00 per year.

Engineering Index Monthly. Engineering Information, Inc., Castle Point on the Hudson, Hoboken, NJ 07030. (800) 221-1044. FAX (212) 832-1857. Monthly. $2300.00 per year. Also available online as COMPENDEX, and also on CD-ROM. Covers chemical engineering, computers, electrical engineering, civil engineering, metals and mining, industrial management, and mechanical engineering.

International Building Services Abstracts. Building Service Research and Information Association, Old Bracknell Lane W, Bracknell, Berks. RG12 7AH, England. Telephone 0344-426511. FAX 0344-487575. 1966 to present. Bi-monthly. Inquire for price. Also available online. A survey of world literature on mechanical and technical services of buildings, including heating, sanitation, ventilation, air conditioning, lighting, and communications.

ASSOCIATIONS AND PROFESSIONAL SOCIETIES

National Association of Plumbing, Heating, Cooling Contractors. 180 S. Washington Street, Box 6808, Falls Church, VA 22040-1148. (703) 237-8100. FAX (703) 237-7442.

Pipe Fabrication Institute. Box 173, 612 Lenore Avenue, Springdale, PA 15144-1518. (412) 274-4722. FAX (412) 274-4722.

Society of Piping Engineers and Designers. One Main Street, Suite N-723, Houston, TX 77002. (713) 221-8090. FAX (713) 221-8064.

TPF, the Tube & Pipe Fabricators Association International. 833 Featherstone Road, Rockford, IL 61107-6203. (815) 399-8700. FAX (815) 399-7279.

DIRECTORIES AND BIOGRAPHICAL SOURCES

Air Conditioning, Heating and Refrigerating News Directory. Business News Publishing Company, 755 W. Big Beaver Road, Suite 1000, Troy, MI 48084. (313) 362-3700. FAX (313) 362-0317. Included with subscription to AIR CONDITIONING, HEATING and REFRIGERATING NEWS (see below).

Heating, Plumbing and Air Conditioning Buyers' Guide. Cowgate Communications Inc., 1370 Don Mills Road, Suite 300, Don Mills, ON M3B 3N7, Canada. (416) 759-2500. FAX (416) 759-6979. 1923 to present. Annual. $30.00

Research Centers Directory. Gale Research, 835 Penobscot Building, Detroit, MI 48226-4094. (313) 961-2242. (800) 347-4253. FAX (313) 961-6083. $485.00.

Scientific and Technical Organizations and Agencies Directory. Gale Research, 835 Penobscot Building, Detroit, MI 48226-4094. (313) 961-2242. (800) 347-4253. FAX (313) 961-6083. 4th edition. 1996. $195.00.

Who's Who in Engineering. American Association of Engineering Societies, 1111 19th Street NW, Suite 608, Washington, DC 20036. (202) 296-2237 or (800) 658-8897. 8th edition. 1991. Inquire for price.

Who's Who in Technology. Gale Research, 835 Penobscot Building, Detroit, MI 48226-4094. (313) 961-2242. (800) 347-4253. FAX (313) 961-6083. 1995. $195.00.

ENCYCLOPEDIAS AND DICTIONARIES

Encyclopedia of Physical Science and Technology. Robert A. Meyers, ed. 18 volumes. Academic Press Inc., 6277 Sea Harbor Drive, Orlando, FL 32887. (800) 321-5068. 1992. $2100.00

McGraw-Hill Encyclopedia of Engineering. Sybil P. Parker, ed. 2nd edition. McGraw-Hill Publishing Company, 1221 Avenue of the Americas, New York, NY 10020. (800) 262-4729 or (212) 512-3825. 1993. $95.50.

GENERAL WORKS

Design and Technology of Heat Pipes For Cooling and Heat Exchange. Calvin C. Silverstein. Hemisphere Publishing Corporation, 1900 Frost Road, Suite 101, Bristol, PA 19007-1598. (215) 785-5800 or (800) 821-8312. FAX (215) 785-5515. 1992. $99.50.

Heat Pipes. P.D. Dunn and D.A. Reay. 4th edition. Pergamon/Elsevier Science Inc., 660 White Plains Road, Tarrytown, NY 10591-5153. (914) 524-9200. FAX (914) 333-2444. 1994. $88.00.

HANDBOOKS AND MANUALS

Piping Guide: A Reference For the Design & Drafting of Piping Systems. Sherwood. 2nd edition. Syentek Inc., PO Box 26588, San Francisco, CA 94126. (415) 928-0471. 1991. Inquire for cost and availability.

Piping Handbook. Mohinder L. Nayyar, editor in chief. 6th edition. McGraw-Hill Publishing Company, 1221 Avenue of the

HEAT PIPES

Ency. of Physical Sciences and Engineering Info. Sources

Americas, New York, NY 10020. (800) 262-4729 or (212) 512-3825. 1992. $99.50.

ONLINE DATABASES AND CD-ROMS

Compendex Plus. Engineering Information, Inc., 345 East 47th Street, New York, NY 10017. (212) 705-7600 or (800) 221-1044. Contains citations with abstracts to worldwide literature in engineering and technology, from 1970 to present. Available on online BRS,(800) 289-4277, DIALOG, (800) 334-2564, ORBIT (800) 456-7248, and STN International, FIZ Karlsruhe, P.O. Box 2465, W-7500, Karlsruhe 1, Germany, online services. Also available on CD-ROM. Inquire as to cost and availability.

CSA Engineering. Cambridge Scientific Abstracts, 7200 Wisconsin Avenue, Suite 601, Bethesda, MD 20814. (301) 961-6750 or (800) 843-7751. Contains citations and abstracts of international periodicals and research literature covering all fields of engineering and science and technology,including computer and information science, electronics, mechanical engineering, solid state materials, 1981 to present. Available on BRS,(800) 289-4277, online service.Inquire as to cost and availability.

Current Contents Search. Institute for Scientific Information, 3501 Market Street, Philadelphia, PA 19104. (215) 386-0100. FAX (215) 386-6362. Contains citations to articles listed in the table of contents of science and technology journals. Also articles in social sciences and life sciences journals. Available on BRS,(800) 289-4277, DIALOG,(800) 334-2564, online services. Inquire as to cost and availability.

Dissertation Abstracts. University Microfilms International, 300 North Zeeb Road, Ann Arbor, MI 48106. (800) 521-0600 or (313) 761-4700. Scope includes virtually all doctoral dissertations accepted at accredited American institutions from 1861 to present in 252 subject areas. Available on BRS,(800) 289-4277, DIALOG,(800) 334-2564, and OCLC EPIC,(800) 848-5878, online services. Also available on CD-ROM. Inquire as to cost and availability.

ISMEC: Mechanical Engineering Abstracts. Cambridge Scientific Abstracts, 7200 Wisconsin Avenue, Suite 601, Bethesda, MD 20814. (301) 961-6750 or (800) 843-7751. Contains citations to the literature in mechanical engineering, industrial and production engineering, energy, power, mechanics, devices and related areas, from 1973 to present. Available on the DIALOG,(800) 334-2564, online service. Inquire as to cost and availability.

NTIS Bibliographic Database. National Technical Information Service, 5285 Port Royal Road, Springfield, VA 22161. (703) 487-4929 or FAX (703) 321-8199. Broad coverage of government-sponsored science and technology research reports, 1964 to present. Available on BRS,(800) 289-4277, DIALOG, (800) 334-2564, ORBIT, (800) 456-7248, and STN International, FIZ Karlsruhe, P.O. Box 2465, W-7500, Karlsruhe 1, Germany, online services. Also available on CD-ROM. Inquire as to cost and availability.

SCISEARCH. Institute for Scientific Information, 3501 Market Street, Philadelphia, PA 19104. (800) 523-1850 or (215) 386-0100. Broad multidisciplinary title and author index to the international literature of science and technology, 1974 to present. Available on DIALOG,(800) 334-2564, and ORBIT,(800) 456-7248, online services. Also available on CD-ROM. Inquire as to cost and availability.

WILSONLINE. H.W. Wilson Company, 950 University Avenue, Bronx, NY 10452. (800) 367-6770 or (212) 588-8400. Makes available online versions of the printed H.W. Wilson indexes including Applied Science and Technology Index, Business Periodicals Index, General Science Index, and Readers' Guide to Periodical Literature. Period covered is generally 1983 to present. Available on BRS,(800) 289-4277, DIALOG,(800) 334-2564, and OCLC EPIC,(800) 848-5878, online services. Also available on CD-ROM. Inquire as to cost and availability.

PERIODICALS

Air Conditioning, Heating, and Refrigeration News. Business News Publishing Company, 755 W. Big Beaver Road, Suite 1000, Troy, MI 48084. (313) 362-3700. FAX (313) 362-0317. 1926 to present. Weekly. $72.00 per year. Includes the annual directory and statistical summary issues.

ASHRAE Journal: Heating, Refrigerating, Air Conditioning, and Ventilating. American Society of Heating, Refrigeration and Air Conditioning Engineers, 1791 Tullie Circle NE, Atlanta, GA 30329. (404) 636-8400. FAX (404) 321-5478. 1914 to present. Monthly. $49.00 per year for non-members.

ASHRAE Transactions. American Society of Heating, Refrigeration and Air Conditioning Engineers, 1791 Tullie Circle NE, Atlanta, GA 30329. (404) 636-8400. FAX (404) 321-5478. 1895 to present. Semi-annual. $170.00 per volume.

Engineered Systems. Business News Publishing Company, 755 W. Big Beaver Road, Suite 1000, Troy, MI 48084. (313) 362-3700. FAX (313) 362-0317. 1985 to present. 12 times a year. $39.00 per year.

Heating and Air Conditioning Journal. Maclean Hunter Publishing Company, 29 North Wacker Drive, Chicago, IL 60606-3298. (312) 726-2802. 1931 to present. Monthly. Inquire for price and availability.

Heating, Piping, Air Conditioning. Penton Publishing, 110 Superior Ave., Cleveland, OH 44114-2543. 1929 to present. Monthly. $50.00 per year.

Heating, Plumbing, and Air Conditioning. Cowgate Communications Inc., 1370 Don Mills Road, Suite 300, Don Mills, ON M3B 3N7, Canada. (416) 759-2500. FAX (416) 759-6979. 1923 to present. 7 times a year. $43.00 per year.

Indoor Comfort News. Institute of Heating and Air Conditioning Industries, 606 N Larchmont Blvd., Suite 4A, Los Angeles, CA 90004. (213) 467-1158. FAX (213) 461-2588. 1955 to present. Monthly. $12.00 per year.

TPQ, The Tube and Pipe Quarterly. Croydon Group Ltd., 833 Featherstone Road, Rockford, IL 61107-6301. (815) 399-8700. FAX (815) 399-7279. 1990 to present. Quarterly. $15.00 per year.

Tube International. Publex International Ltd., 110 Station Road E., Oxted, Surrey, RH8 0QA, England. Telephone 0883-717755. FAX 0883-714554. 1982 to present. Bi-monthly. Free.

RESEARCH CENTERS AND INSTITUTES

Heat Transfer Laboratory. Massachusetts Institute of Technology, 77 Massachusetts Avenue, Cambridge, MA 02139. (716) 253-2248.

Heat Transfer Research. University of Wisconsin—Milwaukee, PO Box 784, Milwaukee, WI 53201. (414) 963-5001. FAX (414) 229-6958.

Institute of thermo-fluid Engineering and Science. Lehigh University, Bldg. A, Mountain Top Campus, Bethlehem, PA 18015. (215) 758-4091. FAX (215) 758-5057.

Heat Transfer Research Facility. Columbia University, 632 W. 125th Street, New York, NY 10027. (212) 280-4163. FAX (212) 678-5279.

SPECIFICATIONS AND STANDARDS

Annual Book of ASTM Standards, Volume 01.01, Volume 04.06. American Society for Testing and Materials (ASTM), 1916 Race Street, Philadelphia, PA 19103. (215) 299-5585. Annual. Inquire for cost and availability.

HEAT PUMPS

See: HEATING and VENTILATION

HEATR RESISTANT MATERIALS

See: MATERIALS SCIENCE

HEAT TRANSFER

See also: COOLING TOWERS, FLUID MECHANICS, HEAT EXCHANGERS, MECHANICAL ENGINEERING, THERMODYNAMICS

ABSTRACT SERVICES AND INDEXES

Applied Mechanics Reviews: An Assessment of World Literature in Engineering Sciences. 1948-present. American Society of Mechanical Engineers, 345 East 47th Street, New York, NY 10017. (212) 705-7703. Monthly. $360.00 per year.

Applied Science and Technology Index; A Cumulative Subject Index To English Language Periodicals in the Fields of Aeronautics and Space Science, Computer Technology, Chemistry, Construction Industry, Energy and Related Areas. H.W. Wilson Co., 950 University Avenue, Bronx, NY 10452. (800) 367-6770 or (212) 588-8400. FAX (718) 590-1617. From 1958 to present. Monthly. Inquire about cost and availability. Also available on CD-ROM and online.

Current Contents: Engineering, Technology, and Applied Sciences. Institute for Scientific Information, 3501 Market Street, Philadelphia, PA 19104. (215) 386-0100. FAX (215) 386-6362. 1970 to present. Weekly. $442.00 per year.

Energy Research Abstracts. 1976-present. U.S. Department of Energy, office of Scientific and Technical Information, Box 62, Oak Ridge, TN 37831. (615) 574-0733. Subscriptions: Superintendent of Documents, U.S. Government Printing office, Box 371954, Pittsburgh, PA 15250-7954. (202) 783-3238. FAX (202) 512-2233.Monthly. $164.00 per year. Abstracts all the scientific and technical reports, journal articles, conference proceedings, patents, theses, and monographs sponsored by the U.S. Energy Research and Development Administration.

Engineering Index Monthly. Engineering Information, Inc., Castle Point on the Hudson, Hoboken, NJ 07030. (800) 221-1044. FAX (212) 832-1857. Monthly. $2300.00 per year. Also available online as COMPENDEX, and also on CD-ROM. Covers chemical engineering, computers, electrical engineering, civil engineering, metals and mining, industrial management, and mechanical engineering.

Physics Abstracts. Institute of Electrical Engineers, Michael Faraday House, Six Hill Way, Stevenage, Herts, SG1 2AY, England. Distributed by IEEE, 445 Hoes Lane, Piscataway, NJ 08854. (908) 562-5549. 1898 to present. Monthly. $2700.00 per year. Also available online as INSPEC.

Physics Briefs. American Institute of Physics, 335 East 45th Street, New York, NY 10017 (212) 661-9404 or FAX (516)349-9704. 1980 to present. Six times per year. $2600.00 per year.

Previews of Heat and Mass Transfer. Elsevier Science [journals], 660 White Plains Rd., Tarrytown, NY 10159-5153. (919) 524-9200. FAX (919) 333-2444. 1974 to present. Six times a year. $265.00 per year. Contains citations with abstracts to the worldwide literature on heat and mass transfer from more than 100 journals.

Science Citation Index. Institute for Scientific Information, 3501 Market Street, Philadelphia, PA 19104. (215) 386-0100. FAX (215) 386-6362. Inquire about availability and cost. Also available on CD-ROM.

ASSOCIATIONS AND PROFESSIONAL SOCIETIES

American Institute of Physics. One Physics Ellipse, College Park, MD 20740. (301) 209-3000.

American Society of Mechanical Engineers. 345 East 47th Street, New York, NY 10017-2398. (212) 705-7703.

DIRECTORIES AND BIOGRAPHICAL SOURCES

American Men & Women of Science, 1995-96. R.R. Bowker Staff, eds. 19th edition. 8 volumes. R.R. Bowker/Reed International Publishing Company, 121 Chanlon Road, New Providence, NJ 07974. (908) 464-6800 or (800) 521-8110. 1995. $850.00

Directory of Engineering Societies and Related Organizations. Gordon Davis. 13th edition. American Association of Engineering Societies, 1111 19th Street NW, Suite 608, Washington, DC 20036. (202) 296-2237 or (800) 658-8897. 1989. Inquire for price.

Research Centers Directory. Gale Research, 835 Penobscot Building, Detroit, MI 48226-4094. (313) 961-2242. (800) 347-4253. FAX (313) 961-6083. $485.00.

HEAT TRANSFER

Ency. of Physical Sciences and Engineering Info. Sources

Scientific and Technical Organizations and Agencies Directory. Gale Research, 835 Penobscot Building, Detroit, MI 48226-4094. (313) 961-2242. (800) 347-4253. FAX (313) 961-6083. 4th edition. 1996. $195.00.

Who's Who in Engineering. American Association of Engineering Societies, 1111 19th Street NW, Suite 608, Washington, DC 20036. (202) 296-2237 or (800) 658-8897. 8th edition. 1991. Inquire for price.

ENCYCLOPEDIAS AND DICTIONARIES

McGraw-Hill Encyclopedia of Science and Technology. Sybil P. Parker, ed. 7th edition. 20 volumes. McGraw-Hill Publishing Company, 1221 Avenue of the Americas, New York, NY 10020. (800) 262-4729 or (212) 512-3825. 1992. $1900.00

Thesaurus of Scientific, Technical, and Engineering Terms. Hemisphere Publishing Corporation, 1900 Frost Road, Suite 101, Bristol, PA 19007-1598. (215) 785-5800 or (800) 821-8312. FAX (215) 785-5515. 1987. $173.00.

GENERAL WORKS

Introduction To Heat Transfer. Frank P. Incropera. John Wiley and Sons, Inc., 605 Third Avenue, New York, NY 10158. (800) 526-5368 or (212) 850-6000. 1990. $57.95.

Principles of Convective Heat Transfer. M. Kaviany. Springer-Verlag, 175 Fifth Avenue, New York, NY 10010. (212) 460-1500 or (800) 777-4643. FAX (212) 473-6272. 1994. Inquire for cost and availability.

Principles of Enhanced Heat Transfer. Ralph L. Webb. John Wiley and Sons, Inc., 605 Third Avenue, New York, NY 10158. (800) 526-5368 or (212) 850-6000. 1993. Write for information.

HANDBOOKS AND MANUALS

Handbook of Heat Exchanger Calculations. G.F.Hewitt & P.B. Whalley. CRC Press, 2000 Corporate Blvd., N.W., Boca Raton, FL 33431. (407) 994-0555 or (800) 272-7737. FAX (407) 994-0949. 1990. $45.00.

Handbook of Heat Exchanger Design. G.F. Hewitt, editor. Begell House Publishers, 79 Madison Avenue, Suite 1106, New York, NY 10016. (212) 775-1999. 1989. $125.00.

Handbook of Heat Transfer. Kutatkerkadze. CRC Press, 2000 Corporate Blvd., N.W., Boca Raton, FL 33431. (407) 994-0555 or (800) 272-7737. FAX (407) 994-0949. 1993. Inquire for cost and availability.

Handbook of Heat Transfer Applications. W.M. Rohsenow, et al., editors. McGraw-Hill Publishing Company, 1221 Avenue of the Americas, New York, NY 10020. (800) 262-4729 or (212) 512-3825. 1985. Inquire for cost and availability.

Heat Exchanger Sourcebook. J. W. Palen. Hemisphere Publishing Corporation, 1900 Frost Road, Suite 101, Bristol, PA 19007-1598. (215) 785-5800 or (800) 821-8312. FAX (215) 785-5515. 1986. $81.00.

ONLINE DATABASES AND CD-ROMS

Compendex Plus. Engineering Information, Inc., 345 East 47th Street, New York, NY 10017. (212) 705-7600 or (800) 221-1044. Contains citations with abstracts to worldwide literature in engineering and technology, from 1970 to present. Available on online BRS,(800) 289-4277, DIALOG, (800) 334-2564, ORBIT (800) 456-7248, and STN International, FIZ Karlsruhe, P.O. Box 2465, W-7500, Karlsruhe 1, Germany, online services. Also available on CD-ROM. Inquire as to cost and availability.

CSA Engineering. Cambridge Scientific Abstracts, 7200 Wisconsin Avenue, Suite 601, Bethesda, MD 20814. (301) 961-6750 or (800) 843-7751. Contains citations and abstracts of international periodicals and research literature covering all fields of engineering and science and technology,including computer and information science, electronics, mechanical engineering, solid state materials, 1981 to present. Available on BRS,(800) 289-4277, online service.Inquire as to cost and availability.

Current Contents Search. Institute for Scientific Information, 3501 Market Street, Philadelphia, PA 19104. (215) 386-0100. FAX (215) 386-6362. Contains citations to articles listed in the table of contents of science and technology journals. Also articles in social sciences and life sciences journals. Available on BRS,(800) 289-4277, DIALOG,(800) 334-2564, online services. Inquire as to cost and availability.

Dissertation Abstracts. University Microfilms International, 300 North Zeeb Road, Ann Arbor, MI 48106. (800) 521-0600 or (313) 761-4700. Scope includes virtually all doctoral dissertations accepted at accredited American institutions from 1861 to present in 252 subject areas. Available on BRS,(800) 289-4277, DIALOG,(800) 334-2564, and OCLC EPIC,(800) 848-5878, online services. Also available on CD-ROM. Inquire as to cost and availability.

ISMEC: Mechanical Engineering Abstracts. Cambridge Scientific Abstracts, 7200 Wisconsin Avenue, Suite 601, Bethesda, MD 20814. (301) 961-6750 or (800) 843-7751. Contains citations to the literature in mechanical engineering, industrial and production engineering, energy, power, mechanics, devices and related areas, from 1973 to present. Available on the DIALOG,(800) 334-2564, online service. Inquire as to cost and availability.

NTIS Bibliographic Database. National Technical Information Service, 5285 Port Royal Road, Springfield, VA 22161. (703) 487-4929 or FAX (703) 321-8199. Broad coverage of government-sponsored science and technology research reports, 1964 to present. Available on BRS,(800) 289-4277, DIALOG, (800) 334-2564, ORBIT, (800) 456-7248, and STN International, FIZ Karlsruhe, P.O. Box 2465, W-7500, Karlsruhe 1, Germany, online services. Also available on CD-ROM. Inquire as to cost and availability.

SCISEARCH. Institute for Scientific Information, 3501 Market Street, Philadelphia, PA 19104. (800) 523-1850 or (215) 386-0100. Broad multidisciplinary title and author index to the international literature of science and technology, 1974 to present. Available on DIALOG,(800) 334-2564, and ORBIT,(800) 456-7248, online services. Also available on CD-ROM. Inquire as to cost and availability.

WILSONLINE. H.W. Wilson Company, 950 University Avenue, Bronx, NY 10452. (800) 367-6770 or (212) 588-8400. Makes available online versions of the printed H.W. Wilson indexes including Applied Science and Technology Index, Business Periodicals Index, General Science Index, and Readers' Guide

to Periodical Literature. Period covered is generally 1983 to present. Available on BRS,(800) 289-4277, DIALOG,(800) 334-2564, and OCLC EPIC,(800) 848-5878, online services. Also available on CD-ROM. Inquire as to cost and availability.

PERIODICALS

Advances in Heat Transfer. Academic Press, Inc., 1250 Sixth Avenue, San Diego, CA 92101-4311. (619) 231-0926. FAX (619) 699-6715. 1964 to present. Irregular. Inquire for cost and availability.

Experimental Heat Transfer. Taylor & Francis, 1900 Frost Road, Suite 101, Bristol, PA 19007-1598. (800) 821-8312. FAX (215) 785-5515. Quarterly. $182.00 per year.

Heat Transfer Engineering. Taylor and Francis, Inc., 79 Madison Avenue, Suite 1106, New York, NY 10016. (212) 725-1999 or (800) 821-8312. 1979 to present. Quarterly. $176.00 per year.

Heat Transfer Research. John Wiley and Sons, Inc., 605 Third Avenue, New York, NY 10158. (800) 526-5368 or (212) 850-6000. 1969 to present. 8 times a year. $1196.00 per year.

International Journal of Heat and Fluid Flow. Butterworth-Heinemann, 313 Washington Street, Newton, MA 02158. (617) 928-2500 or (800) 366-2665. FAX (617) 928-2610. 1979 to present. 6 times a year. $85.00 per year.

Journal of Heat Transfer. American Society of Mechanical Engineers, 345 E. 47th Street, New York, NY 10017. (212) 705-7722. 1970 to present. Quarterly. $155.00 per year.

Journal of thermophysics and Heat Transfer. American Institute of Aeronautics and Astronautics, 370 L'Enfant Promenade, S.W., Washington, DC 20024. (202) 646-7400. 1987 to present. Quarterly. $165.00 per year.

Numerical Heat Transfer A: Applications. Taylor and Francis, Inc., 79 Madison Avenue, Suite 1106, New York, NY 10016. (212) 725-1999 or (800) 821-8312. 1978 to present. 8 times a year. $530.00 per year.

Numerical Heat Transfer B: Fundamentals. Taylor and Francis, Inc., 79 Madison Avenue, Suite 1106, New York, NY 10016. (212) 725-1999 or (800) 821-8312. 1978 to present. 8 times a year. $330.00 per year.

RESEARCH CENTERS AND INSTITUTES

Heat Transfer Laboratory. Massachusetts Institute of Technology, 77 Massachusetts Avenue, Cambridge, MA 02139. (716) 253-2248.

Heat Transfer Research. University of Wisconsin—Milwaukee, PO Box 784, Milwaukee, WI 53201. (414) 963-5001. FAX (414) 229-6958.

Institute of thermo-fluid Engineering and Science. Lehigh University, Bldg. A, Mountain Top Campus, Bethlehem, PA 18015. (215) 758-4091. FAX (215) 758-5057.

Heat Transfer Research Facility. Columbia University, 632 W. 125th Street, New York, NY 10027. (212) 280-4163. FAX (212) 678-5279.

HEATING AND VENTILATION

See also: AIR CONDITIONING, FLUID MECHANICS, HEAT PIPES, MECHANICAL ENGINEERING

ABSTRACT SERVICES AND INDEXES

Applied Science and Technology Index; A Cumulative Subject Index To English Language Periodicals in the Fields of Aeronautics and Space Science, Computer Technology, Chemistry, Construction Industry, Energy and Related Areas. H.W. Wilson Co., 950 University Avenue, Bronx, NY 10452. (800) 367-6770 or (212) 588-8400. FAX (718) 590-1617. From 1958 to present. Monthly. Inquire about cost and availability. Also available on CD-ROM and online.

Computer Abstracts, MCBUniversity Press Ltd., PO Box 10812, Birmingham, AL 35201-0812. (800) 633-4931. FAX (205) 995-1588. Monthly. Covers computer theory, data, hardware, systems, networks, human-computer interaction, artificial intelligence, as well as applications of computers in aerospace, business, CAD/CAM, cartography, civil engineering, electronics and electrical engineering, industrial engineering, mechanical engineering, medicine, structural engineering, etc.

Current Contents: Engineering, Technology, and Applied Sciences. Institute for Scientific Information, 3501 Market Street, Philadelphia, PA 19104. (215) 386-0100. FAX (215) 386-6362. 1970 to present. Weekly. $442.00 per year.

Engineering Index Monthly. Engineering Information, Inc., Castle Point on the Hudson, Hoboken, NJ 07030. (800) 221-1044. FAX (212) 832-1857. Monthly. $2300.00 per year. Also available online as COMPENDEX, and also on CD-ROM. Covers chemical engineering, computers, electrical engineering, civil engineering, metals and mining, industrial management, and mechanical engineering.

International Building Services Abstracts. Building Service Research and Information Association, Old Bracknell Lane W, Bracknell, Berks. RG12 7AH, England. Telephone 0344-426511. FAX 0344-487575. 1966 to present. Bi-monthly. Inquire for price. Also available online. A survey of world literature on mechanical and technical services of buildings, including heating, sanitation, ventilation, air conditioning, lighting, and communications.

ASSOCIATIONS AND PROFESSIONAL SOCIETIES

American Academy of Environmental Engineers, 132 Holiday Ct., Suite 206, Annapolis, MD 21041.

American Society of Heating, Refrigerating & Air-conditioning Engineers, 1791 Tullie Circle, N.E., Atlanta, GA 30329. (404) 636-8400. FAX (404) 321-5478.

Institute of Heating & Air Conditioning Industries. 606 N Larchmont Blvd., Suite 4A, Los Angeles, CA 90004. (213) 467-1158. FAX (213) 461-2588.

National Association of Plumbing, Heating, Cooling Contractors. 180 S. Washington Street, Box 6808, Falls Church, VA 22040-1148. (703) 237-8100. FAX (703) 237-7442.

HEAT ING AND VENTILATION

Ency. of Physical Sciences and Engineering Info. Sources

DIRECTORIES AND BIOGRAPHICAL SOURCES

Air Conditioning, Heating and Refrigerating News Directory. Business News Publishing Company, 755 W. Big Beaver Road, Suite 1000, Troy, MI 48084. (313) 362-3700. FAX (313) 362-0317. Included with subscription to AIR CONDITIONING, HEATING and REFRIGERATING NEWS (see below).

Heating, Plumbing and Air Conditioning Buyers' Guide. Cowgate Communications Inc., 1370 Don Mills Road, Suite 300, Don Mills, ON M3B 3N7, Canada. (416) 759-2500. FAX (416) 759-6979. 1923 to present. Annual. $30.00

Research Centers Directory. Gale Research, 835 Penobscot Building, Detroit, MI 48226-4094. (313) 961-2242. (800) 347-4253. FAX (313) 961-6083. $485.00.

Scientific and Technical Organizations and Agencies Directory. Gale Research, 835 Penobscot Building, Detroit, MI 48226-4094. (313) 961-2242. (800) 347-4253. FAX (313) 961-6083. 4th edition. 1996. $195.00.

Who's Who in Engineering. American Association of Engineering Societies, 1111 19th Street NW, Suite 608, Washington, DC 20036. (202) 296-2237 or (800) 658-8897. 8th edition. 1991. Inquire for price.

Who's Who in Technology. Gale Research, 835 Penobscot Building, Detroit, MI 48226-4094. (313) 961-2242. (800) 347-4253. FAX (313) 961-6083. 1995. $195.00.

ENCYCLOPEDIAS AND DICTIONARIES

International Dictionary of Heating, Ventilating, and Air Conditioning. REHVA. 2nd edition. Chapman & Hall, 115 Fifth Avenue, New York, NY 10211-0906. 1994. $196.95.

Encyclopedia of Physical Science and Technology. Robert A. Meyers, ed. 18 volumes. Academic Press Inc., 6277 Sea Harbor Drive, Orlando, FL 32887. (800) 321-5068. 1992. $2100.00

McGraw-Hill Encyclopedia of Engineering. Sybil P. Parker, ed. 2nd edition. McGraw-Hill Publishing Company, 1221 Avenue of the Americas, New York, NY 10020. (800) 262-4729 or (212) 512-3825. 1993. $95.50.

GENERAL WORKS

Air Handling System Design. Tseng-Yao Sun. McGraw-Hill Publishing Company, 1221 Avenue of the Americas, New York, NY 10020. (800) 262-4729 or (212) 512-3825. 1994. Inquire for price and availability.

Building Energy Management Systems. G.J. Levermore. Chapman & Hall, 115 Fifth Avenue, New York, NY 10211-0906. 1993. $40.95.

Control Systems For Heating, Ventilating, and Air Conditioning. Roger W. Haines & Douglas C. Hittler. Chapman & Hall, 115 Fifth Avenue, New York, NY 10211-0906. 1993. $57.95.

Indoor Air Quality and Hvac Systems. David W. Bearg. Lewis Publishers, 121 S. Main St., PO Box 519, Chelsea, MI 48118. (313) 475-8619 or (800) 272-7737. 1993. $59.95.

Modern Heating and Ventilating Systems Design. George Clifford. Prentice Hall, 113 Sylvan Avenue, Route 9W, Englewood Cliffs, NJ 07632. (201) 592-2000 or (800) 922-0579. 1993. $58.67.

HANDBOOKS AND MANUALS

Air Conditioning & Heating Manual 1991-93. Chilton Staff, editors. Chilton Co., One Chilton Way, Radnor PA 19089. (215) 964-4028 or (800) 695-1214. FAX (610) 964-4745. 1993. Inquire for price.

Ashrae Handbook: Fundamentals. 4 volumes. American Society of Heating, Refrigeration and Air Conditioning Engineers, 1791 Tullie Circle NE, Atlanta, GA 30329. (404) 636-8400. FAX (404) 321-5478. 1922 to present. Annual. $119.00.

Hvac Engineer's Handbook. Fred Porges. 9th ed. Butterworths Publishers, 80 Montvale Avenue, Stoneham, MA 02180. (617) 438-8464 or (800) 366-2665. 1991. $70.00.

ONLINE DATABASES AND CD-ROMS

Compendex Plus. Engineering Information, Inc., 345 East 47th Street, New York, NY 10017. (212) 705-7600 or (800) 221-1044. Contains citations with abstracts to worldwide literature in engineering and technology, from 1970 to present. Available on online BRS,(800) 289-4277, DIALOG, (800) 334-2564, ORBIT (800) 456-7248, and STN International, FIZ Karlsruhe, P.O. Box 2465, W-7500, Karlsruhe 1, Germany, online services. Also available on CD-ROM. Inquire as to cost and availability.

CSA Engineering. Cambridge Scientific Abstracts, 7200 Wisconsin Avenue, Suite 601, Bethesda, MD 20814. (301) 961-6750 or (800) 843-7751. Contains citations and abstracts of international periodicals and research literature covering all fields of engineering and science and technology, including computer and information science, electronics, mechanical engineering, solid state materials, 1981 to present. Available on BRS,(800) 289-4277, online service. Inquire as to cost and availability.

Current Contents Search. Institute for Scientific Information, 3501 Market Street, Philadelphia, PA 19104. (215) 386-0100. FAX (215) 386-6362. Contains citations to articles listed in the table of contents of science and technology journals. Also articles in social sciences and life sciences journals. Available on BRS,(800) 289-4277, DIALOG,(800) 334-2564, online services. Inquire as to cost and availability.

Dissertation Abstracts. University Microfilms International, 300 North Zeeb Road, Ann Arbor, MI 48106. (800) 521-0600 or (313) 761-4700. Scope includes virtually all doctoral dissertations accepted at accredited American institutions from 1861 to present in 252 subject areas. Available on BRS,(800) 289-4277, DIALOG,(800) 334-2564, and OCLC EPIC,(800) 848-5878, online services. Also available on CD-ROM. Inquire as to cost and availability.

ISMEC: Mechanical Engineering Abstracts. Cambridge Scientific Abstracts, 7200 Wisconsin Avenue, Suite 601, Bethesda, MD 20814. (301) 961-6750 or (800) 843-7751. Contains citations to the literature in mechanical engineering, industrial and production engineering, energy, power, mechanics, devices and related areas, from 1973 to present. Available on the DIALOG,(800) 334-2564, online service. Inquire as to cost and availability.

NTIS Bibliographic Database. National Technical Information Service, 5285 Port Royal Road, Springfield, VA 22161. (703) 487-4929 or FAX (703) 321-8199. Broad coverage of government-sponsored science and technology research reports, 1964 to present. Available on BRS,(800) 289-4277, DIALOG, (800) 334-2564, ORBIT, (800) 456-7248, and STN International, FIZ Karlsruhe, P.O. Box 2465, W-7500, Karlsruhe 1, Germany, online services. Also available on CD-ROM. Inquire as to cost and availability.

SCISEARCH. Institute for Scientific Information, 3501 Market Street, Philadelphia, PA 19104. (800) 523-1850 or (215) 386-0100. Broad multidisciplinary title and author index to the international literature of science and technology, 1974 to present. Available on DIALOG,(800) 334-2564, and ORBIT,(800) 456-7248, online services. Also available on CD-ROM. Inquire as to cost and availability.

WILSONLINE. H.W. Wilson Company, 950 University Avenue, Bronx, NY 10452. (800) 367-6770 or (212) 588-8400. Makes available online versions of the printed H.W. Wilson indexes including Applied Science and Technology Index, Business Periodicals Index, General Science Index, and Readers' Guide to Periodical Literature. Period covered is generally 1983 to present. Available on BRS,(800) 289-4277, DIALOG,(800) 334-2564, and OCLC EPIC,(800) 848-5878, online services. Also available on CD-ROM. Inquire as to cost and availability.

PERIODICALS

Air Conditioning, Heating, and Refrigeration News. Business News Publishing Company, 755 W. Big Beaver Road, Suite 1000, Troy, MI 48084. (313) 362-3700. FAX (313) 362-0317. 1926 to present. Weekly. $72.00 per year. Includes the annual directory and statistical summary issues.

ASHRAE Journal: Heating, Refrigerating, Air Conditioning, and Ventilating. American Society of Heating, Refrigeration and Air Conditioning Engineers, 1791 Tullie Circle NE, Atlanta, GA 30329. (404) 636-8400. FAX (404) 321-5478. 1914 to present. Monthly. $49.00 per year for non-members.

ASHRAE Transactions. American Society of Heating, Refrigeration and Air Conditioning Engineers, 1791 Tullie Circle NE, Atlanta, GA 30329. (404) 636-8400. FAX (404) 321-5478. 1895 to present. Semi-annual. $170.00 per volume.

Engineered Systems. Business News Publishing Company, 755 W. Big Beaver Road, Suite 1000, Troy, MI 48084. (313) 362-3700. FAX (313) 362-0317. 1985 to present. 12 times a year. $39.00 per year.

Heating and Air Conditioning Journal. Maclean Hunter Publishing Company, 29 North Wacker Drive, Chicago, IL 60606-3298. (312) 726-2802. 1931 to present. Monthly. Inquire for price and availability.

Heating, Piping, Air Conditioning. Penton Publishing, 110 Superior Ave., Cleveland, OH 44114-2543. 1929 to present. Monthly. $50.00 per year.

Heating, Plumbing, and Air Conditioning. Cowgate Communications Inc., 1370 Don Mills Road, Suite 300, Don Mills, ON M3B 3N7, Canada. (416) 759-2500. FAX (416) 759-6979. 1923 to present. 7 times a year. $43.00 per year.

Indoor Comfort News. Institute of Heating and Air Conditioning Industries, 606 N Larchmont Blvd., Suite 4A, Los Angeles, CA 90004. (213) 467-1158. FAX (213) 461-2588. 1955 to present. Monthly. $12.00 per year.

Mechanical Engineering. American Society of Mechanical Engineers, 345 E. 47th Street, New York, NY 10017. (212) 705-7722. 1906 to present. Monthly. $45.00 for non-members.

RESEARCH CENTERS AND INSTITUTES

Kansas State University Mechanical Engineering Research Laboratory. Durland Hall, Manhattan, KS 66506. (913) 532-5610.

Lehigh University Institute of thermo-fluid Engineering and Science. Bldg. A, Mountain Top Campus, Bethlehem, PA 18015. (215) 758-4091. FAX (215) 758-5057.

Refrigeration Research Foundation. 7315 Wisconsin Avenue, Bethesda, MD 20814. (301) 652-5674. FAX (301) 652-7269.

Texas A&M University Energy Systems Laboratory. College Station, TX 77843. (409) 845-6402.

SPECIFICATIONS AND STANDARDS

*Annual Book of ASTM Standards, Volume 04.*06. American Society for Testing and Materials (ASTM), 1916 Race Street, Philadelphia, PA 19103. (215) 299-5585. Annual. Inquire for cost and availability.

HEAT METAL ALLOYS

See: ALLOYS

HEAVY WATER

See: HYDROGEN

HELICOPTERS

See also: AERODYNAMICS, AERONAUTICAL ENGINEERING, AERONAUTICS, AEROSPACE ENGINEERING, AIRCRAFT

ABSTRACT SERVICES AND INDEXES

Aeronautical Engineering. Scientific and Technical Branch, National Aeronautics and Space Administration. National Technical Information Service (NTIS), 5285 Port Royal Road, Springfield, VA 22161. (703) 487-4650. FAX (703) 321-8547. Monthly. A selection of annotated references to unclassified reports and journal articles that were introduced into the NASA scientific and technical information system and announced in STAR and IAA.

Alloys Index. American Society for Metals, Metals Park, OH 44073. (216) 338-5151. FAX (216) 338-4634. 1974 to present. Monthly. $295.00.

HELICOPTERS

Ency. of Physical Sciences and Engineering Info. Sources

Applied Mechanics Reviews: An Assessment of World Literature in Engineering Sciences. American Society of Mechanical Engineers, 345 East 47th Street, New York, NY 10017. (212) 705-7703. 1948 to present. Monthly. $360.00 per year.

Applied Science and Technology Index; A Cumulative Subject Index To English Language Periodicals in the Fields of Aeronautics and Space Science, Computer Technology, Chemistry, Construction Industry, Energy and Related Areas. H.W. Wilson Co., 950 University Avenue, Bronx, NY 10452. (800) 367-6770 or (212) 588-8400. FAX (718) 590-1617. From 1958 to present. Monthly. Inquire about cost and availability. Also available on CD-ROM.

Current Contents: Engineering, Technology, and Applied Sciences. Institute for Scientific Information, 3501 Market Street, Philadelphia, PA 19104. (215) 386-0100. FAX (215) 386-6362. Weekly. $360.00 per year.

Engineering Index Monthly. Engineering Information, Inc., Castle Point on the Hudson, Hoboken, NJ 07030. (800) 221-1044. FAX (201) 216-8532. Monthly. $2200.00 per year. Also available online as COMPENDEX, and also on CD-ROM.

International Aerospace Abstracts. Technical Information Service, American Institute of Aeronautics and Astronautics, Inc., 555 West 57th St., New York, NY 10019. (212) 247-6500. FAX (212) 582-4861. Semi-monthly. $1295.00 per year.

Metals Abstracts and Metals Abstracts Index. American Society for Metals, Metals Park, OH 44073. (216) 338-5151. 1968 to present. Monthly. Abstracts are $1500.00 per year and Index is $460.00 per year.

Science Citation Index. Institute for Scientific Information, 3501 Market Street, Philadelphia, PA 19104. (215) 386-0100. FAX (215) 386-6362. Inquire about availability and cost. Also available on CD-ROM.

World Aluminum Abstracts. Aluminum Association, Inc., 818 Connecticut Ave. NW, Washington, DC 20006. (202) 862-5156.

ASSOCIATIONS AND PROFESSIONAL SOCIETIES

American Aviation Historical Society. 2333 Otis St., Santa Ana, CA 92704.

American Helicopter Society, Inc., 217 N. Washington St., Alexandria, VA 22314. (703) 684-6777. FAX (703) 739-9279.

American Institute of Aeronautics and Astronautics, 370 L'enfant Promenade Sw, Washington, DC 20024-2518.

Helicopter Association International. 1619 Duke Street, Alexandria, VA 22314. (703) 683-4646. FAX (703) 683-4745.

Society of Automotive Engineers. SAE, Inc., 400 Commonwealth Dr., Warrendale, PA 15096. (412) 776-4841.

BIBLIOGRAPHIES

Aeronautical Engineering: A Continuing Bibliography With. INDEXES. National Aeronautics and Space Administration, Center for Aerospace Information, Box 8757, BWI Airport, Baltimore, MD 21240. (301) 621-0153. Inquire for cost and availability.

Scientific and Technical Books and Serials in Print, R.R. Bowker Inc., 121 Chanlon Road, New Providence, NJ 07974. (800) 521-8110 or (908) 464-6800. 3 volumes, annually. $285.00. Also available on CD-ROM.

DIRECTORIES AND BIOGRAPHICAL SOURCES

Research Centers Directory. Gale Research, 835 Penobscot Building, Detroit, MI 48226-4094. (313) 961-2242. (800) 347-4253. FAX (313) 961-6083. 20th edition. $485.00.

Who's Who in Engineering. American Association of Engineering Societies, 1111 19th St. NW, Suite 608, Washington, DC 20036. 7th ed. 1988. $200.00.

World Aviation Directory. McGraw-Hill, Aviation Week Group (NY), 1221 Avenue of the Americas, New York, NY 10020. (215) 237-4112. FAX (215) 586-3232. Semiannual. $250.00.

The World Civil Helicopter Market. Gunter Endres. Jane's Information Group, 1340 Braddock Place, Suite 300, Alexandria, VA 22314-1651. (703) 683-3700. FAX (703) 836-0029. 1994. Inquire for cost and availability.

ENCYCLOPEDIAS AND DICTIONARIES

Dictionary of Aeronautical Terms, Dale Crane, Comp. Aviation Supplies and ACADemics, Inc., 7005 132nd Place SE, Newcastle, WA 98059. (206) 235-1500 or (800) 426-8338. FAX (206) 235-0128. 1991. $15.95.

Encyclopedia of Physical Science and Technology. Robert A. Meyers, ed. Academic Press Inc., 6277 Sea Harbor Drive, Orlando, FL 32887. (800) 321-5068. 1992. $2100.00.

Jane's Aerospace Dictionary, B. Guston. 3rd ed. Jane's Information Group, 1340 Braddock Place, Suite 300, Alexandria, VA 22314-1651. (703) 683-3700. FAX (703) 836-0029. 1991. $45.00.

McGraw-Hill Encyclopedia of Science and Technology, Sybil P. Parker, ed. 7th ed. 20 vols. McGraw-Hill Publishing Company, 1221 Avenue of the Americas, New York, NY 10020. (800) 262-4729 or (212) 512-3825. $1900.00

GENERAL WORKS

Basic Helicopter Aerodynamics. J. Seddon. American Institute of Aeronautics and Astronautics, 370 L'Enfant Promenade, S.W., Washington, DC 20024. (202) 646-7400. 1990. $59.95.

The Foundations of Helicopter Flight. Simon Newman. Edward Arnold/Routledge, Chapman and Hall, Inc., 29 West 35th Street, New York, NY 10001-2291. (212) 244-3336 or FAX (212) 563-2269. 1994. Inquire for cost and availability.

The Helicopter: History, Piloting, and How It Flies. John Fay. 4th edition. David & Charles/Sterling Publishing, 387 Park Avenue South, New York, NY 10016-8810. (212) 532-7160 or (800) 367-9692. 1987. Inquire for cost and availability.

Helicopter Performance Stability and Control. Raymond W. Prouty. Krieger Publishing Company, P.O. Box 9542, Melbourne, FL 32902-9542. (407) 724-9542. FAX (407) 951-3671. 1990. $86.50.

Principles of Helicopter Flight. Jean-Pierre Harrison. Pilot Training Publications, PO Box 394, Mountain View, CA 94042-0394. (415) 962-1097. 1993. $19.95.

HANDBOOKS AND MANUALS

Basic Helicopter Handbook. Gordon Press Publishers, PO Box 459, Bowling Green Station, New York, NY 10004. (718) 624-8419. 1991. $79.00.

Jane's All the World's Aircraft. Jane's Information Group, 1340 Braddock Place, Suite 300, Alexandria, VA 22314-1651. (703) 683-3700. FAX (703) 836-0029. Annual. $245.00. Also available on CD-ROM.

ONLINE DATABASES AND CD-ROMS

Aerospace Database. American Institute of Aeronautics and Astronautics, 370 L'Enfant Promenade, S.W., Washington, DC 20024. (202) 646-7400. Worldwide published literature on research and development in aerospace and related areas, with abstracts. Covers 1962 to present. Available on DIALOG, (800) 334-2564, online service. Also available on CD-ROM. Inquire as to cost and availability.

Compendex Plus. Engineering Information, Inc., 345 East 47th Street, New York, NY 10017. (212) 705-7600 or (800) 221-1044. Contains citations with abstracts to worldwide literature in engineering and technology, from 1970 to present. Available on online BRS,(800) 289-4277, DIALOG, (800) 334-2564, ORBIT (800) 456-7248, and STN International, FIZ Karlsruhe, P.O. Box 2465, W-7500, Karlsruhe 1, Germany, online services. Also available on CD-ROM. Inquire as to cost and availability.

CSA Engineering. Cambridge Scientific Abstracts, 7200 Wisconsin Avenue, Suite 601, Bethesda, MD 20814. (301) 961-6750 or (800) 843-7751. Contains citations and abstracts of international periodicals and research literature covering all fields of engineering and science and technology,including computer and information science, electronics, mechanical engineering, solid state materials, 1981 to present. Available on BRS,(800) 289-4277, online service.Inquire as to cost and availability.

Current Contents Search. Institute for Scientific Information, 3501 Market Street, Philadelphia, PA 19104. (215) 386-0100. FAX (215) 386-6362. Contains citations to articles listed in the table of contents of science and technology journals. Also articles in social sciences and life sciences journals. Available on BRS,(800) 289-4277, DIALOG,(800) 334-2564, online services. Inquire as to cost and availability.

Dissertation Abstracts. University Microfilms International, 300 North Zeeb Road, Ann Arbor, MI 48106. (800) 521-0600 or (313) 761-4700. Scope includes virtually all doctoral dissertations accepted at accredited American institutions from 1861 to present in 252 subject areas. Available on BRS,(800) 289-4277, DIALOG,(800) 334-2564, and OCLC EPIC,(800) 848-5878, online services. Also available on CD-ROM. Inquire as to cost and availability.

*ISMEC: Mechanical Engineering Abstracts.*Cambridge Scientific Abstracts, 7200 Wisconsin Avenue, Suite 601, Bethesda, MD 20814. (301) 961-6750 or (800) 843-7751. Contains citations to the literature in mechanical engineering, industrial and production engineering, energy, power, mechanics, devices and related areas, from 1973 to present. Available on the DIA-LOG,(800) 334-2564, online service. Inquire as to cost and availability.

NTIS Bibliographic Database. National Technical Information Service, 5285 Port Royal Road, Springfield, VA 22161. (703) 487-4929 or FAX (703) 321-8199. Broad coverage of government-sponsored science and technology research reports, 1964 to present. Available on BRS,(800) 289-4277, DIALOG, (800) 334-2564, ORBIT, (800) 456-7248, and STN International, FIZ Karlsruhe, P.O. Box 2465, W-7500, Karlsruhe 1, Germany, online services. Also available on CD-ROM. Inquire as to cost and availability.

SAE Global Mobility Database. Society of Automative Engineers (SAE), Electronic Publishing Division, 400 Commonwealth Drive, Warrendale, PA 15098. (412) 776-4841. Contains citations with abstracts to technical papers on automotive and aerospace technology and vehicular-related industries that have been presented at SAE conferences. Covers 1906 to present. Available on ORBIT,(800) 456-7248,online service. Inquire as to cost and availability.

Scientific and Technical Books and Serials in Print. R.R. Bowker Inc., 121 Chanlon Road, New Providence, NJ 07974. (800) 521-8110 or (908) 464-6800. List of currently published books and serials in the physical and biological sciences, engineering and technology, with subject, author and titles indexes. Available on ORBIT,(800) 456-7248, online service.Inquire as to cost and availability.

SCISEARCH. Institute for Scientific Information, 3501 Market Street, Philadelphia, PA 19104. (800) 523-1850 or (215) 386-0100. Broad multidisciplinary title and author index to the international literature of science and technology, 1974 to present. Available on DIALOG,(800) 334-2564, and ORBIT,(800) 456-7248, online services. Also available on CD-ROM. Inquire as to cost and availability.

TRIS (Transportation Research Information). National Academy of Science, 2101 Constitution Avenue, N.W., Washington, DC 20418. (202) 334-3313 or (800) 624-6242. Citations with abstracts of literature on transportation, including air, highway, rail, maritime and other modes from 1968 to present. Available on DIALOG,(800) 334-2564, online service. Inquire as to cost and availability.

WILSONLINE. H.W. Wilson Company, 950 University Avenue, Bronx, NY 10452. (800) 367-6770 or (212) 588-8400. Makes available online versions of the printed H.W. Wilson indexes including Applied Science and Technology Index, Business Periodicals Index, General Science Index, and Readers' Guide to Periodical Literature. Period covered is generally 1983 to present. Available on BRS,(800) 289-4277, DIALOG,(800) 334-2564, and OCLC EPIC,(800) 848-5878, online services. Also available on CD-ROM. Inquire as to cost and availability.

PERIODICALS

Aviation Week and Space Technology. 1221 Avenue of the Americas, New York, NY 10020. (212) 512-2999 or (800) 525-5003. 1916 to present. Weekly. $82.00 per year.

Interavia/Aerospace World. Aerospace Media Publishing, Swissair Centre, 31 Route de l'Aeroport, POB Box 437, 1215 Geneva 15, Switzerland. (022) 788 27 88. FAX (022) 788 27 26. 1987 to present.Monthly. $128.00 per year (U.S.).

International Journal of Turbo and Jet Engines. Freund Publishing House, Ltd., PO Box 35010, Tel Aviv 63150, Israel. 972-2-5662925. FX 972-3-5605335. 1983 to present. Quarterly. $190.00 per year.

Journal of Aircraft. American Institute of Aeronautics and Astronautics, 370 L'Enfant Promenade SW, Washington, DC 20024-2518. 1963 to present. Bimonthly. $42.00 (members) and $230.00 (non-members).

Journal of the American Helicopter Society. American Helicopter Society, Inc., 217 North Washington St., Alexandria, VA 22314. (703) 684-6777. 1956 to present. Quarterly. $35.00 per year.

Rotor and Wing International. PHillips Publishing Inc., 7811 Montrose Rd., Potomac, MD 20854. (301) 340-2100. FAX (301) 309-3847. 1967 to present. Monthly. $49.00 a year.

Vertiflite. American Helicopter Society, 217 N. Washington St., Alexandria, VA 22314. (703) 684-6777. FAX (703) 739-9279. 1955 to present. Bimonthly. $25 per year (members) and $35 per year (non-members).

RESEARCH CENTERS AND INSTITUTES

Aerospace Research Applications Center, Indianapolis Center For Advanced Research, 611 N. Capitol Ave., Indianapolis, in 46204. (317) 262-5003. FAX (317) 262-5044.

Aviation Safety Institute, 6797 N. High St., Suite 316, PO Box 304, Worthington, OH 43085. (614) 885-4242. FAX (614) 885-5891.

Center For Aerospace Technology, Weber State College, Ogden, UT 84408-1805. (801) 626-7272.

Flight Research Laboratory, University of Kansas, Raymond Nichols Hall, Lawrence, KS 66045. (913) 864-3043. FAX (913) 864-7789.

Joint Institute For Advancement of Flight Sciences, George Washington University, Langley Research Center, Mail Stop 269, Hampton, VA 23665. (804) 864-1982. FAX (804) 864-5894.

Machinery and Engine Technology Laboratory, Institute For Mechanical Engineering, Bldg. M-7, Montreal Rd., Ottawa, Ontario, Canada K1A 0R6. (613) 993-2425. (613) 957-3281.

Transportation Research and Education Center, Georgia Institute of Technology, Mason Civil Engineering Bldg., Atlanta, GA 30332. (404) 894-2236. FAX (404) 894-2278.

HELIUM

See also: AIRSHIPS, CRYOGENICS, ELEMENTS, INORGANIC CHEMISTRY,

ABSTRACT SERVICES AND INDEXES

Analytical Abstracts. Royal Society of Chemistry, Information Services, Thomas Graham House, Science Park, Milton Road, Cambridge, CB4 4WF, England. Contains citations, mostly with abstracts, of the worldwide literature on analytical chemistry, from 1954 to present. Monthly. $636.00 per year. Also available online.

Applied Science and Technology Index; A Cumulative Subject Index To English Language Periodicals in the Fields of Aeronautics and Space Science, Computer Technology, Chemistry, Construction Industry, Energy and Related Areas. H. W. Wilson Co., 950 University Avenue, Bronx, NY 10452- 9978. (212) 588-8400. (800) 367-6770. FAX (718) 590-1617. From 1958 to present. Monthly. Inquire about cost and availability. Also available online (BRS and WILSONLINE) and on CD-ROM. ISSN: 0003-6986.

Chemical Abstracts. Chemical Abstracts Service. 2540 Olentangy River Road, Box 3012, Columbus, OH 43210-0012. (614) 447-3600. FAX (614) 447-3713. 1907 to present. Weekly. $16,800.00 per year. Available online and on CD-ROM. CA is also available in five section groupings. Inquire Regarding Cost and Availability.

Current Contents: Physical, Chemical and Earth Sciences. Institute For Scientific Information, 3501 Market Street, Philadelphia, PA 19104. (215) 386-0100. FAX (215) 386-2291. From 1961 To Present. Weekly. $442.00 per year. Also Available On CD-ROM and Online. Inquire regarding cost and availability. ISSN: 0163-2574.

Engineering Index Monthly. Engineering Information, Inc., Castle Point on the Hudson, Hoboken, NJ 07030. (201) 216-8500. (800) 221-1044. FAX (201) 216-8532. Monthly. $2300.00 per year for monthly issues. $1980 per year for annual cumulation. Available on CD-ROM and online as COMPENDEX. ISSN: 0742-1974; 0360-8557.

General Science Index. H.W. Wilson Company, 950 University Avenue, Bronx, NY 10452. (212) 588-8400. (800) 367-6770. Fax (718) 590-1617. From 1978 to present. Ten issues per year; Quarterly and Annual Cumulations. Service basis. Available On CD-ROM and Online. Inquire regarding cost and availability. ISSN: 0162-1963.

Physics Abstracts. INSPEC. Section A, Science Abstracts. Institution of Electrical Engineers (IEE), London, United Kingdom. Available from: INSPEC/IEEE -Institute of Electrical and Electronic Engineers, Box 1331, 445 Hoes Lane, Piscataway, NJ 08855-1331. (908) 562-5549. 1898 to present. 24 Issues Per Year. $2835.00 per year. ISSN: 0036-8091. Also Available Online and on CD-ROM.

Science Citation Index. SCI. Institute for Scientific Information, 3501 Market Street, Philadelphia, PA 19104. (215) 386-0100. (800) 523-1850. FAX (215) 386-2991. 1961 To Present. Six issues per year, plus annual cumulation. $11650.00 per year. Also available online and on CD-ROM. Inquire about price and availability. ISSN: 0036-827x.

ASSOCIATIONS AND PROFESSIONAL SOCIETIES

American Chemical Society, 1155 16th Street, Nw, Washington, DC 20036. (202) 872-4600.

American Institute of Chemists. 7315 Wisconson Avenue,

Bethesda, MD 20814. (301) 652-2447. FAX (301) 657-3549.

Ency. of Physical Sciences and Engineering Info. Sources

HELIUM

American Institute of Physics, 335 East 45th Street, New York, NY 10017 (212) 661-9404 OR Fax (516)349-9704.

American Society For Metals: ASM International, 9639 Kinsman, Materials Park, OH 44073-0002. (216) 338-5151. FAX (216) 338-4634

Cryogenic Sociey of America. c/o Laurie Huget, Huget Advertising, 1033 South Boulevard, Oak Park, IL 60302. (708) 383-6220. FAX (708) 383-9337.

BIBLIOGRAPHIES

Chemical Information. H. R. Collier, editor. Springer- Verlag New York, Inc., 175 Fifth Avenue, New York, NY 10010. (212) 460-1500. (800) 777-4643. 1990. $91.00.

Chemical Information Sources. Gary Wiggins. McGraw-Hill Publishing Company, 1221 Avenue of the Americas, New York, NY 10020. (212) 512-2000. (800) 262-4729. 1991. $42.50.

Information Sources in Chemistry. R.T. Bottle and J.F.B Rowland, editors. R.R. Bowker Inc., 121 Chanlon Road, New Providence, NJ 07974. (908) 464-6800. (800) 521-8110. Fourth edition. 1993. $75.00.

Scientific and Technical Books and Serials in Print; An Index To Literature in Science and Technology. R.R. Bowker Inc., 121 Chanlon Road, New Providence, NJ 07974. (908) 464-6800. (800) 521-8110. FAX (908) 665-3502. 1972 to present. Annual. 4 volumes. 1994. $299.95. Also available on compact disc and online. ISSN: 0000-054X.

DIRECTORIES AND BIOGRAPHICAL SOURCES

American Institute of Chemists - Professional Directory. American Institute of Chemists, 501 Wythe Street, Alexandria, VA 22314-1917. (703) 836-2090. FAX (703) 836-2091. Annual. $65.00.

American Men and Women of Science. Physical and Biological Sciences. R.R. Bowker Inc., 121 Chanlon Road, New Providence, NJ 07974. (908) 464-6800. (800) 521-8110. 20thedition. 8 volumes. 1996. $850.00

Chemical Week — Buyers Guide Issue. Chemical Week Associates, 888 Seventh Avenue, New York, NY 10106. (212) 621-4900. FAX (212) 621-4949. Annual, October. $50.00.

Chemical Sources International. Mike Desing and Kurt Gandenberger, editors. Chemical Sources International, Inc., P.O. Box 1824, Clemson, SC 29633. (803) 646-7840. 1994. $285.00.

Directory of Chemistry Software 1992. Wendy Warr, Peter Willett, Geoff Downs. American Chemical Society, 1155 16th Street, NW, Washington, DC 20036. (202) 872-4600. 1992.$35.95.

Research Centers Directory. Gale Research, 835 Penobscot Building, Detroit, MI 48226-4094. (313) 961- 2242. (800) 347-4253. 20th edition. 1995. $485.00. ISSN: 0080-1518.

ENCYCLOPEDIAS AND DICTIONARIES

Academic Press Dictionary of Science and Technology. Christopher Morris, editor. Academic Press, Inc., 6277 Sea Harbor Drive, Orlando, FL. (800) 321-5068. 1991. $115.00.

Concise Encyclopedia of Chemical Technology. John Wiley & Sons, Inc., 605 Third Avenue, New York, NY 10158-0012. (212) 850-6000. (800) 225-5945. . 2nd edition. 1989. $99.95.

Dictionary of Chemical Names and Synomyms. Philip A. Howard and Michael W. Neil. Lewis Pub.,/ CRC Press, Inc., 2000 Corporate Boulevard, NW, Boca Raton, FL 33431. (407) 994- 0555. (800) 272-7737. 1991. $149.95.

Encyclopedia of Applied Physics. VCH Publications, Inc., 220 East 23rd Street, Suite 909, New York, NY 10010-4606. (212) 683-8333. (800) 422-8824. 1991-. Twenty volumes. $6000.00.

Grant & Hawks Chemical Dictionary. R. A. Grant. McGraw-Hill Publishing Company, 1221 Avenue of the Americas, New York, NY 10020. (212) 512-2000. (800) 262-4729. 5th edition. 1987. $64.50.

Hawley's Condensed Chemical Dictionary. Richard J. Lewis. Van Nostrand Reinhold, 115 Fifth Avenue, New York, NY 10003. (212) 254-3232. (800) 842-3636. 12th edition. 1993. $69.95.

Kirk-Othmer Encyclopedia of Chemical Technology. John Wiley and Sons, Inc., 605 Third Avenue, New York, NY 10158-0012. (212) 850-6000. (800) 225-5945. Fourth edition. 1991- . Twenty-seven volumes. $5400.00.

McGraw-Hill Encyclopedia of Science and Technology. McGraw-Hill Book, Incorporated, 1221 Avenue of the Americas, New York, NY 10020. (212) 512-2000. (800) 262-4729. Seventh edition. Twenty volumes. 1992. $1900.00.

GENERAL WORKS

Excitations in Liquid and Solid Helium. Henry Glyde. Oxford University Press, 200 Madison Avenue, New York, NY 10016. (212) 725-6000. (800) 334-4249. 1995. $98.00.

Helium Cryogenics. Steven W. Van Sciver. Plenum Publishing Corp., 233 Spring Street, New York, NY. (212) 620-8000. (800) 221-9369. FAX (212) 463-0742. 1986. $115.00.

Helium Isotopes in Nature. B. A. Manyrin and I. N. Tolstikhin. Elsevier Science Publishing Company, Inc., 655 Avenue of the Americas, New York, NY 10010. (212) 989-5800. FAX (914) 333-2444. 1984. $79.50.

Liquid and Solid Helium. C. G. Kuper and S. G. Lipson. Coronet Books, 311 Bainbridge Street, Philadelphia PA 19147. (215) 925-5083. 1975. $97.50.

Origin and Evolution of the Elements. S. Kubono and T. Kajino. World Scientific Publishing Company, Inc., 1060 Main Street, River Edge, NJ 07661. (201) 487-9655. (800) 227-7562. 1993. $95.95.

Solid Helium Three. E. R. Dobbs. Oxford University Press, 200 Madison Avenue, New York, NY 10016. (212) 725-6000. (800) 334-4249. 1994. $67.50.

Thermodynamic Properties of Helium. V. V. Sychev. Hemisphere Publishing Corp., 1900 Frost Road, Suite 101, Bristol PA 19007. (215) 785-5800. (800) 821-8312.

1987. $167.00.

Thermodynamic and Thermophysical Properties of Helium. N. V. Tsederberg, et al. Coronet Books, 311 Bainbridge Street, Philadelphia PA 19147. (215) 925-5083. 1971. $69.00.

HANDBOOKS AND MANUALS

Chemical Reference Handbook. Gordon Press Publications, P.O. Box 459, Bowling Green Station, New York, NY 10004. (718) 624-8419. 1995. $260.00.

CRC Handbook of Chemistry and Physics. David R. Kide, editor. CRC Press Inc., 2000 Corporate Blvd., NW, Boca Raton, FL 33431. (407) 994-0555. (800) 333-8300. 77th edition. 1996. $99.95.

Guide To Basic Chemical Compounds. D.R. Lide, Jr. CRC Publishers, Inc., 2000 Corporate Blvd., N.W., Boca Raton, FL 33431. (407) 994-0555 or (800) 333-8300. 1993. $120.00.

Lange's Handbook of Chemistry. John A. Dean, editor. McGraw-Hill Publishing Company, 1221 Avenue of the Americas, New York, NY 10020. (800) 262-4729 or (212) 512-3825. Fourteenth edition. 1992. $79.50.

Riegel's Handbook of Industrial Chemistry. James A. Kent, editor. Van Nostrand Reinhold, 115 Fifth Avenue, New York, NY 10003. (212) 254-3232 or (800) 926-2665. Ninth edition. 1992. $114.95.

ONLINE DATABASES AND CD-ROMS

Analytical Abstracts Online. Royal Society of Chemistry, Information Services, Thomas Graham House, Science Park, Milton Road, Cambridge, CB4 4WF, England. Contains citations, mostly with abstracts, of the worldwide literature on analytical chemistry, from 1980 to present. Available on DIALOG, (800) 334-2564, and STN International, FIZ Karlsruhe, P.O. Box 2465, W-7500, Karlsruhe 1, Germany, online services. Inquire as to cost and availability.

CA Search. Chemical Abstracts Service, P.O. Box 3012, Columbus, OH 43210-0012. (614) 447-3600. (800) 848-6533. FAX (614) 447-3709. Very comprehensive guide to worldwide chemical literature and related fields, 1972 to present. Available on BRS,(800) 289-4277, DIALOG, (800) 334-2564, ORBIT (800) 456-7248, and STN International, FIZ Karlsruhe, P.O. Box 2465, W-7500, Karlsruhe 1, Germany, online services. Inquire as to cost and availability.

Chemical Journals of the American Chemical Society. American Chemical Society, 1155 16th Street, N.W., Washington, DC 20036. (202) 872-4381 or (800) 424-6747. Contains complete text of approximately 90,000 articles from 22 primary journals published by the American Chemical Society, from mostly 1982 to present. Available on STN International, FIZ Karlsruhe, P.O. Box 2465, W-7500, Karlsruhe 1, Germany, online service. Inquire as to cost and availability.

Compendex Plus. Engineering Information, Inc., 345 East 47th Street, New York, NY 10017. (212) 705-7600 or (800) 221-

1044. Contains citations with abstracts to worldwide literature in engineering and technology, from 1970 to present. Available on online BRS,(800) 289-4277, DIALOG, (800) 334-2564, ORBIT (800) 456-7248, and STN International, FIZ Karlsruhe, P.O. Box 2465, W-7500, Karlsruhe 1, Germany, online services. Also available on CD-ROM. Inquire as to cost and availability.

Current Contents Search. Institute for Scientific Information, 3501 Market Street, Philadelphia, PA 19104. (215) 386-0100. FAX (215) 386-6362. Contains citations to articles listed in the table of contents of science and technology journals. Also articles in social sciences and life sciences journals. Available on BRS,(800) 289-4277, DIALOG,(800) 334-2564, online services. Inquire as to cost and availability.

Dissertation Abstracts. University Microfilms International, 300 North Zeeb Road, Ann Arbor, MI 48106. (800) 521-0600 or (313) 761-4700. Scope includes virtually all doctoral dissertations accepted at accredited American institutions from 1861 to present in 252 subject areas. Available on BRS,(800) 289-4277, DIALOG,(800) 334-2564, and OCLC EPIC,(800) 848-5878, online services. Also available on CD-ROM. Inquire as to cost and availability.

Gmelin Database. Gmelin-Institut fur Anorganische Chemie und Grenzgebiete, Varrentrapperstrasse, 40-42, Carl-Bosch-Haus, D-6000, Frankfurt am Main 90, Germany. Contains structural and factual data relating to inorganic and organometallic chemistry. Provides data from the Gmelin Handbook of Inorganic and Organometallic Chemistry. Covers the period 1817 to 1975; 1988-89. Available on STN International, FIZ Karlsruhe, P.O. Box 2465, W-7500, Karlsruhe 1, Germany, online service. Inquire as to cost and availability.

Kirk-Othmer Encyclopedia of Chemical Technology. John Wiley and Sons, Inc., 605 Third Avenue, New York, NY 10158. (800) 526-5368 or (212) 850-6000. Contains the complete text of all chapters in the 27 volume fourth edition of the KIRK-Othmer ENCYCLOPEDIA of CHEMICAL TECHNOLOGY. 1991. Available on BRS,(800) 289-4277, DIALOG,(800) 334-2564, online services. Inquire as to cost and availability.

NTIS Bibliographic Database. National Technical Information Service, 5285 Port Royal Road, Springfield, VA 22161. (703) 487-4929 or FAX (703) 321-8199. Broad coverage of government-sponsored science and technology research reports, 1964 to present. Available on BRS,(800) 289-4277, DIALOG, (800) 334-2564, ORBIT, (800) 456-7248, and STN International, FIZ Karlsruhe, P.O. Box 2465, W-7500, Karlsruhe 1, Germany, online services. Also available on CD-ROM. Inquire as to cost and availability.

Physics Briefs. American Institute of Physics, 335 East 45th Street, New York, NY 10017. (212) 661-9260 or FAX (212) 661-2036. Contains citations with abstracts of the literature of physics and related fields, 1979 to present. Available on the STN International, FIZ Karlsruhe, P.O. Box 2465, W-7500, Karlsruhe 1, Germany, online service. Inquire as to cost and availability.

SCISEARCH. Institute for Scientific Information, 3501 Market Street, Philadelphia, PA 19104. (800) 523-1850 or (215) 386-0100. Broad multidisciplinary title and author index to the international literature of science and technology, 1974 to present. Available on DIALOG,(800) 334-2564, and ORBIT,(800) 456-7248, online services. Also available on CD-ROM. Inquire as to cost and availability.

WILSONLINE. H.W. Wilson Company, 950 University Avenue, Bronx, NY 10452. (800) 367-6770 or (212) 588-8400. Makes

available online versions of the printed H.W. Wilson indexes including Applied Science and Technology Index, Business Periodicals Index, General Science Index, and Readers' Guide to Periodical Literature. Period covered is generally 1983 to present. Available on BRS,(800) 289-4277, DIALOG,(800) 334-2564, and OCLC EPIC,(800) 848-5878, online services. Also available on CD-ROM. Inquire as to cost and availability.

PERIODICALS

Analytical Chemistry. American Chemical Society, Box 3337, Columbus, OH 43210. (614) 447-3776. 1929 to present. Semimonthly. $415.00 per year. ISSN: 0003-2700.

Angewandte Chemie. VCH Publishers, Inc., 220 East 23rd Street, New York, NY 10010-4606. (212) 683-8333. (800) 367-8249. 1888 to present. 22 issues per year. $840.00 per year. ISSN: 0044-8249.

Chemical Reviews. American Chemical Society, Box 3337, Columbus, OH 43210. (614) 447-3776. 1924 to present. 8 issues per year. $346.00 per year. ISSN: 0009-2665.

Chemical Week; Includes Annual Buyers Guide. Chemical Week Associates, 888 Seventh Avenue, New York, NY 10106. (212) 621-4900. FAX (212) 621-4949. 1914 to present. Weekly. $99.00 per year. ISSN: 0009-272X.

Chemtech. American Chemical Society. Box 3337, Columbus, OH 43210. (614) 447-3776. 1970 to present. Monthly. $370.00 per year. ISSN: 0009-2703.

Inorganic Chemistry. American Chemical Society, Box 3337, Columbus, OH 43210. (614) 447-3776. 1962 to present. Semimonthly. $500.00 per year. ISSN: 0020-1669.

Journal of the American Chemical Society. American Chemical Society, Box 3337, Columbus, OH 43210. (614) 447-3776. 1879 to present. Biweekly. $1055.00 per year. ISSN: 0002-7863.

Polyhedron. Elsevier Science Publishing Company, Inc., 660 White Plains Road, Tarrytown, NY 10591-5153. (914-524-9200. FAX (914) 524-9200. 1982 to present. 24 issues per year. $2175.00 per year. ISSN: 0277-5387.

RESEARCH CENTERS AND INSTITUTES

Chemical Laboratories. Harvard University, Oxford Street, Cambridge, MA 02138. (617) 495-4283. FAX (617) 496-5618.

Chemistry Laboratories. Rensselaer Polytechnic Institute, Cogswell Laboratory, Troy, NY 12180-3590. (518) 276-8981.

Lawrence Berkeley Laboratory, Chemical Sciences Division. One Cyclotron Road, Building 66, Berkeley, CA 94720. (510) 486-6062. FAX (510) 486-4995.

Theoretical Chemistry Institute. University of Wisconsin, Madison. 1101 University Avenue, Madison, WI 53706. (608) 262-1511.

University/industry Chemical Research Center. Mississippi State University, Department of Chemistry, P.O. Drawer CH, Mississippi State, MS 39762. (601) 325-3584.

HETEROCYCLIC CHEMISTRY

See also: CARBON, CHEMISTRY, CHEMICAL ENGINEERING, ORGANIC CHEMISTRY, PLASTICS

ABSTRACT SERVICES AND INDEXES

Applied Science and Technology Index; A cumulative subject index to English language periodicals in the fields of aeronautics and space science, computer technology, chemistry, construction industry, energy and related areas. H.W. Wilson Co., 950 University Avenue, Bronx, NY 10452-9978. (212) 588-8400. (800) 367-6770. FAX (718) 590-1617. From 1958 to present. Monthly. Inquire about cost and availability. Also available online (BRS and WILSONLINE) and on CD-ROM. ISSN: 0003-6986.

Chemical Abstracts. Chemical Abstracts Service. 2540 Oleantangy River Road, Box 3012, Columbus, OS 43210-0012. (614) 447-3600. FAX (614) 447-3713. 1907 to present. Weekly. $16,800.00 per year. Available online and on CD-ROM. CA is also available in five section groupings. Inquire regarding cost and availability.

Current Contents: Physical, Chemical and Earth Sciences. Institute For Scientific Information, 3501 Market Street, Philadelphia, PA 19104. (215) 386-0100. (800) 523-1850. Fax (215) 386-2291. 1961 to present. Weekly. $442.00 per year. Also available online (BRS, DIALOG) and on CD-ROM. Inquire regarding cost and availability. ISSN: 0163-2574.

General Science Index. H.W. Wilson Company, 950 University Avenue, Bronx, NY 10452. (212) 588-8400. (800) 367-6770. Fax (718) 590-1617. From 1978 to present. Ten issues per year; Quarterly and Annual Cumulations. Service basis. Available on CD-ROM and Online. Inquire regarding cost and availability. ISSN: 0162-1963.

Physics Abstracts. INSPEC. Section A, Science Abstracts. Institution of Electrical Engineers (IEE), London. available from: inspec/IEEE - institute of electrical and Electronic engineers, box 1331, hoes lane, piscataway, nj 08855-1331. (908) 562-5549. 1898 to present. 24 issues per year. $2835.00 per year. Also available online and on CD-ROM. ISSN: 0036-8091.

Science Citation Index. SCI. Institute for Scientific Information, 3501 Market Street, Philadelphia, PA 19104. (215) 386-0100. (800) 523-1850. FAX (215) 386-2991. 1961 to Present. Six issues per year, plus annual cumulation. $11650.00 per year. Also available online and on CD-ROM. Inquire about price and availability. ISSN: 0036-827X.

ANNUAL REVIEWS AND YEARBOOKS

Advances in Heterocyclic Chemistry. Alan R. Kitritzky, editor. Academic Press.Inc., 525 B Street, Suite 1900, San Diego, CA 92101-4495. (619) 231-0926. FAX (619) 699-6715. 1963 to present. Irregular. price varies. ISSN: 0065-2725.

Advances in Physical Organic Chemistry. Academic Press, Inc., 525 B Street, Suite 1900, San Diego, CA 92101-4495. (619) 231-0926. FAX (619) 699-6715. 1963 to present. Irregular. Price varies. ISSN: 0065-3160.

Studies in Organic Chemistry. Elsevier Science Publishers., Box 882, Madison Square Station, New York, NY 10159. (212) 989-5800. 1973 to present. Irregular. Price varies. ISSN: 0165-3253.

ASSOCIATIONS AND PROFESSIONAL SOCIETIES

American Carbon Society. c/o Carbone of America, 215 Stackpole Street, St. Marys, PS 15857. (814) 781-8410. Fax (814) 781-8570.

American Chemical Society. 1155 16th Street, NW, Washington, DC 20036. (202) 872-4600.

American Institute of Chemical Engineers. 345 East 47th Street, New York, NY 10017-2396. (212) 705-7663.

American Oil Chemists' Society. PO Box 3489, Champaign, IL 61826-3489. (217) 359-2344. FAX (217) 351-8091.

Association of official Analytical Chemists. 2200 Wilson Boulevard, Suite 400, Arlington, VA 22001. (703) 522-3032.

Chemical Manufacturers Association. 2501 M Street, N.W., Washington, DC 20037. (202) 887-1182

BIBLIOGRAPHIES

Chemical Information Sources. Gary Wiggins. McGraw-Hill Publishing Company, 1221 Avenue of the Americas, New York, NY 10020. (800) 262-4729 or (212) 512-3825. 1991. $42.50.

Information Sources in Chemistry. R.T. Bottle and J.F.B Rowland, editors. R.R. Bowker Inc., 121 Chanlon Road, New Providence, NJ 07974. (800) 521-8110 or (908) 464-6800. Fourth edition. 1993. $75.00.

Handbooks and Tables in Science and Technology. Russell H. Powell, editor. Oryx Press, 4041 North Central, Suite 700, Phoenix, AZ 85012-3330. (602) 265-2651 or (800) 279-6799. Third edition. 1994. $65.00.

Organic Reaction Mechanisms: An Annual Survey of Literature. A. C. Knipe and W. E. Watts, editors. John Wiley and Sons, Inc., 605 Third Avenue, New York, NY 10158. (800) 526-5368 or (212) 850-6000. vol. 92, 1992. price varies.

Scientific and Technical Books and Serials in Print. R.R. Bowker Inc., 121 Chanlon Road, New Providence, NJ 07974. (908) 464-6800. (800) 521-8110. FAX (908) 665-3502. 1972 to present. Annual. 4 volumes. 1994. $299.95. Also available on compact disc and online. ISSN: 0000-054X.

DIRECTORIES AND BIOGRAPHICAL SOURCES

American Men and Women of Science. Physical and Biological Sciences. R.R. Bowker Company, 121 Chanlon Road, New Providence, NJ 07974. (908) 464-6800. (800) 521-8110. 20th edition. 8 volumes. 1996. $850.00.

Directory of Chemistry Software 1992. Wendy Warr, Peter Willett, Geoff Downs. American Chemical Society, 1155 16th Street, NW, Washington, DC 20036. (202) 872-4600. 1992. $35.95.

Consulting Services: Chemists and Chemical Engineers. Association of Consulting Chemists and Chemical Engineers, 50 East 41st Street, Suite 92, New York, NY 10017. (212)

684-6255. Biennial. $60.00.

Research Centers Directory. Gale Research Company, 835 Penobscot Building, Detroit, MI 48226-4094. (313) 961-2242. (800) 347-4253. 20th edition. 1995. $485.00. ISSN: 0080-1518.

ENCYCLOPEDIAS AND DICTIONARIES

Academic Press Dictionary of Science and Technology. Christopher Morris, editor. Academic Press, Inc., 1250 Sixth Avenue, San Diego, CA 92101. (619) 231-0926. FAX (619) 699-6715. 1991. $115.00.

Concise Encyclopedia Chemistry. Translated and revised by Mary Eagleson. Walter de Gruyter, Inc., 200 Saw Mill River Road, Hawthorne, New York, 10532. (914) 747-0110 or Fax (914) 747-1326. 1994. $69.95.

Dictionary of Organic Compounds. John Buckingham, et al, editors. Chapman & Hall, 1 Penn Plaza, New York, NY 10119. (212) 564-1060. Fifth edition. 7 volumes. $2550.00 set.

Encyclopedic Dictionary of Science. Candida Hunt and Monica Byles, editors. Facts-on-File, Inc., 460 Park Avenue South, New York, NY 10016-7382. (800) 322-8755. Fax (800) 683-3633. 1988. $32.95.

Encyclopedia of Applied Physics. VCH Publishers, Inc., 303 Northwest 12th Avenue, Deerfield Beach, FL 33442. (800) 367-8249. 1991-. Twenty volumes. $6000.00.

Industrial Chemical thesaurus. Michael Ash and Irene Ash. VCH Publishers, Inc., 220 East 23rd Street, New York, NY 10010-4606. (800) 367-8249. 1992. $295.00.

Kirk-Othmer Encyclopedia of Chemical Technology. John Wiley and Sons, Inc., 605 Third Avenue, New York, NY 10158. (800) 526-5368 or (212) 850-6000. Fourth edition. 1991 - . Twenty-seven volumes. $5400.00.

McGraw-Hill Encyclopedia of Chemistry. Sybil P.Parker, editor. McGraw-Hill Book, Incorporated, 1221 Avenue of the Americas, New York, NY 10020. (212) 997-3675. Second edition. 1993. $95.50.

McGraw-Hill Encyclopedia of Science and Technology. McGraw-Hill Book, Incorporated, 1221 Avenue of the Americas, New York, NY 10020. (212) 997-3675. Seventeenth edition. Twenty volumes. 1992. $1900.00.

GENERAL WORKS

Advanced Organic Chemistry. Frank A. Carey and R. J. Sundberg. Plenum Publishing Corp., 233 Spring Street, New York, NY. (212) 620-8000. (800) 221-9369. FAX (212) 463-0742. 3rd edition. 1990. Part A: Structure and Mechanisms, $32.50; Part B: Reactions and synthesis, $34.50 in paper.

Advanced Organic Chemistry. Jerry March. John Wiley & Sons, Inc., 605 Third Avenue, New York, NY 10158-0012. (212) 850-6000. (800) 225-5945. 4th edition. 1992. $59.95.

The Art of Problem Solving in Organic Chemistry. Miguel E. Alonso. John Wiley & Sons, Inc., 605 Third Avenue, New York, NY 10158-0012. (212) 850-6000. (800) 225-5945. 1987. $64.95.

The Chemical Bond: Structure and Dynamics. Ahmed Zewail. Academic Press, Inc., 1250 Sixth Avenue, San Diego, CA 92101-4311. (619) 231-0926. FAX (619) 699-6715. 1992. $49.95.

Comprehensive Organic Chemistry. Derek H. Barton and W. D. Ollis, editors. Elsevier Science Publishing Company, Inc., 655 Avenue of the Americas, New York, NY 10010. (212) 989-5800. FAX (914) 333-2444. 6 volume set. 1979. $800.00 in paper.

Contemporary Heterocyclic Chemistry: Synthesis, Reactions and Applications. G. R. Newkome and W. W. Paudler. John Wiley & Sons, Inc., 605 Third Avenue, New York, NY 10158-0012. (212) 850-6000. (800) 225-5945. 1982. $99.95.

Desk Reference For Organic Chemists. Michael B. East and David J. Ager. Krieger Publishing Company, P.O. Box 9542, Melbourne, FL 32902-9542. (407) 724-9542. 1995. $64.50.

Organic Acids in Geologic Processes. E. D. Pittman and M. D. Lewan. Springer-Verlag New York, Inc., 175 Fifth Avenue, New York, NY 10010. (212) 460-1500. (800) 777-4643. FAX (212) 473-6272. 1994. $89.00.

Organic Chemistry: Its Language and Its State of the Art. Volkan Kisakurek, editor. VCH Publications, Inc., 220 East 23rd Street, Suite 909, New York, NY 10010-4606. (212) 683-8333. (800) 422-8824. 1993. $117.00.

Organic Reactions: Simplicity and Logic. Pierre Laszlo. John Wiley & Sons, Inc., 605 Third Avenue, New York, NY 10158-0012. (212) 850-6000. (800) 225-5945. 1994.$74.95.

Physical Basis of Organic Chemistry. Howard Maskill. Oxford University Press, Inc., 200 Madison Avenue, New York, NY 10016. (212) 725-6000. (800) 334-4249. 1986. $35.00.

Reaction Mechanisms in Environmental Organic Chemistry. Richard A. Larson and Erick J. Weber. Lewis Pubs., CRC Press, Inc., 2000 Corporate Boulevard, NW, Boca Raton, FL 33431. (407) 994-0555. (800) 272-7737. 1994. $69.95.

HANDBOOKS AND MANUALS

Chemical Formulary. H. Bennett, editor. Chemical Publishing, Co., Inc. 80 Eighth Avenue, New York, NY 10011. (212 255-1950. Volumes 1 - 30. $60.00 per volume.

Chemometrics: Chemical and Sensory Data. David R. Burgard and James T. Kuznicki. CRC Publishers, Inc., 2000 Corporate Blvd., N.W., Boca Raton, FL 33431. (407) 994-0555 or (800) 333-8300. 1990. $120.00.

Comprehensive Organic Synthesis: Selectivity, Strategy, and Efficiency in Modern Organic Chemical. Barry M. Trost and others, editors. Pergamon Press, Maxwell House, Fairview Park, Elmsford, NY 10523. (914) 592-7700. Fax (914) 592-3625. 1991. Nine volumes. $3900.00.

CRC Handbook of Chemistry and Physics. David R. Kide, editor. CRC Press Inc., 2000 Corporate Blvd., NW, Boca Raton, FL 33431. (407) 994-0555. (800) 333-8300. 77th edition. 1996. $99.95.

CRC Handbook of Data On Organic Compounds. R. C. Weast and M. J. Astle, editors. CRC Press Inc., 2000 Corporate Blvd., NW, Boca Raton, FL 33431. (407) 994-0555. (800) 333-8300. 1985. 2 volumes.

Guide To Basic Chemical Compounds. D.R. Lide, Jr. CRC Publishers, Inc., 2000 Corporate Blvd., N.W., Boca Raton, FL 33431. (407) 994-0555 or (800) 333-8300. 1993. $120.00.

Guide to IUPAC Nomenclature of Organic Compounds. J. C. Richer, et al, editors. CRC Press Inc., 2000 Corporate Blvd., NW, Boca Raton, FL 33431. (407) 994-0555. (800) 333-8300. 1993. $39.95.

Handbook of Data and Organic Coumpounds. WRobert C. Weast. CRC Press Inc., 2000 Corporate Blvd., NW, Boca Raton, FL 33431. (407) 994-0555. (800) 333-8300. 2nd edition. 1988. $2655.00.

Handbook of Organic Chemistry. Hans Beyer and Wolfgang Walter. Prentice Hall General Reference & Travel, 15 Columbus Circle, New York, NY 10023. (212) 373-8500. (800) 223-2348. 1995. $96.00.

Handbook of Tables For Organic Compound Identification. Zvi Rappoport. CRC Press Inc., 2000 Corporate Blvd., NW, Boca Raton, FL 33431. (407) 994-0555. (800) 333-8300.. 1966. $76.95.

Laboratory Handbook of Materials, Equipment, and Techniques. Gary S. Coyne. Prentice Hall (Division of Simon and Schuster), 15 Columbus Circle, New York, NY 10023. (212) 373-8500 or (800) 922-0579. 1992. $45.00.

Lange's Handbook of Chemistry. John A. Dean, editor. McGraw-Hill Publishing Company, 1221 Avenue of the Americas, New York, NY 10020. (800) 262-4729 or (212) 512-3825. Fourteenth edition. 1992. $79.50.

Riegel's Handbook of Industrial Chemistry. James A. Kent, editor. Van Nostrand Reinhold, 115 Fifth Avenue, New York, NY 10003. (212) 254-3232 or (800) 926-2665. Ninth edition. 1992. $114.95.

Tables of Spectral Data For Structure Determination of Organic Compounds. E. Pretsch, et al. Springer-Verlag

New York, Inc., 175 Fifth Avenue, New York, NY 10010. (212) 460-1500. (800) 777-4643. FAX (212) 473-6272. 1994. $49.95 in paper.

ONLINE DATABASES AND CD-ROMS

Beilstein Online. Beilstein-Institut fur Literatur der Organischen Chemie, Varrentrapperstrasse, 40-42, D-6000, Frankfurt am Main 90, Germany. Contains data on carbon compounds from the Beilstein Handbook of Organic Chemistry. Structural and factual data for more than 3.4 million heterocyclic, isocyclic, and acyclic compounds included. Covers the period 1830 to present. Available on DIALOG, (800) 334-2564, ORBIT (800) 456-7248, and STN International, FIZ Karlsruhe, P.O. Box 2465, W-7500, Karlsruhe 1, Germany, online services. Inquire as to cost and availability.

CA Search. Chemical Abstracts Service, P.O. Box 3012, Columbus, OH 43210-0012. (614) 447-3600. (800) 848-6533. FAX (614) 447-3709. Very comprehensive guide to worldwide chemical literature and related fields, 1972 to present. Available on BRS,(800) 289-4277, DIALOG, (800) 334-2564, ORBIT (800) 456-7248, and STN International, FIZ Karlsruhe, P.O. Box 2465, W-7500, Karlsruhe 1, Germany, online services. Inquire as to cost and availability.

Chemical Journals of the American Chemical Society. American Chemical Society, 1155 16th Street, N.W., Washington, DC 20036. (202) 872-4381 or (800) 424-6747. Contains complete text of approximately 90,000 articles from 22 primary journals published by the American Chemical Society, from mostly 1982 to present. Available on STN International, FIZ Karlsruhe, P.O. Box 2465, W-7500, Karlsruhe 1, Germany, online service. Inquire as to cost and availability.

Compendex Plus. Engineering Information, Inc., 345 East 47th Street, New York, NY 10017. (212) 705-7600 or (800) 221-1044. Contains citations with abstracts to worldwide literature in engineering and technology, from 1970 to present. Available on online BRS,(800) 289-4277, DIALOG, (800) 334-2564, ORBIT (800) 456-7248, and STN International, FIZ Karlsruhe, P.O. Box 2465, W-7500, Karlsruhe 1, Germany, online services. Also available on CD-ROM. Inquire as to cost and availability.

Current Contents Search. Institute for Scientific Information, 3501 Market Street, Philadelphia, PA 19104. (215) 386-0100. FAX (215) 386-6362. Contains citations to articles listed in the table of contents of science and technology journals. Also articles in social sciences and life sciences journals. Available on BRS,(800) 289-4277, DIALOG,(800) 334-2564, online services. Inquire as to cost and availability.

Dissertation Abstracts. University Microfilms International, 300 North Zeeb Road, Ann Arbor, MI 48106. (800) 521-0600 or (313) 761-4700. Scope includes virtually all doctoral dissertations accepted at accredited American institutions from 1861 to present in 252 subject areas. Available on BRS,(800) 289-4277, DIALOG,(800) 334-2564, and OCLC EPIC,(800) 848-5878, online services. Also available on CD-ROM. Inquire as to cost and availability.

Kirk-Othmer Encyclopedia of Chemical Technology. John Wiley and Sons, Inc., 605 Third Avenue, New York, NY 10158. (800) 526-5368 or (212) 850-6000. Contains the complete text of all chapters in the 27 volume fourth edition of the KIRK-Othmer ENCYCLOPEDIA of CHEMICAL TECHNOLOGY. 1991. Available on BRS,(800) 289-4277, DIALOG,(800) 334-2564, online services. Inquire as to cost and availability.

NTIS Bibliographic Database. National Technical Information Service, 5285 Port Royal Road, Springfield, VA 22161. (703) 487-4929 or FAX (703) 321-8199. Broad coverage of government-sponsored science and technology research reports, 1964 to present. Available on BRS,(800) 289-4277, DIALOG, (800) 334-2564, ORBIT, (800) 456-7248, and STN International, FIZ Karlsruhe, P.O. Box 2465, W-7500, Karlsruhe 1, Germany, online services. Also available on CD-ROM. Inquire as to cost and availability.

SCISEARCH. Institute for Scientific Information, 3501 Market Street, Philadelphia, PA 19104. (800) 523-1850 or (215) 386-0100. Broad multidisciplinary title and author index to the international literature of science and technology, 1974 to present. Available on DIALOG,(800) 334-2564, and ORBIT,(800) 456-7248, online services. Also available on CD-ROM. Inquire as to cost and availability.

WILSONLINE. H.W. Wilson Company, 950 University Avenue, Bronx, NY 10452. (800) 367-6770 or (212) 588-8400. Makes available online versions of the printed H.W. Wilson indexes including Applied Science and Technology Index, Business Periodicals Index, General Science Index, and Readers' Guide to Periodical Literature. Period covered is generally 1983 to present. Available on BRS,(800) 289-4277, DIALOG,(800) 334-

2564, and OCLC EPIC,(800) 848-5878, online services. Also available on CD-ROM. Inquire as to cost and availability.

PERIODICALS

American Oil Chemists' Society. JOURNAL. American Oil Chemists' Society, 1608 Broadmoor Drive, Box 3489, Champaign, IL 61821-0489. (217) 359-2344. 1917 to present. Monthly. Monthly. $185.00 per year. ISSN: 0003-021X.

AICHE Journal. American Institute of Chemical Engineers, 345 47th Street, New York, NY 10017-2396. (212) 705-7338.

FAX (212) 752-3294. 1955 to present. Monthly. $395.00 per year. ISSN: 0001-1541.

Carbon. Elsevier Science Publishing Company, Inc., 660 White Plains Road, Tarrytown, NY 10591-5153. (914-524-9200. FAX (914) 524-9200. 1963 to present. Bi-monthly. $1045.00 ISSN: 0008-6223.

Chemical Week; Includes Annual Buyers Guide. Chemical Week Associates, 888 Seventh Avenue, New York, NY 10106. (212) 621-4900. FAX (212) 621-4949. 1914 to present. Weekly. $99.00 per year. ISSN: 0009-272X.

Chemtech. American Chemical Society, Box 3337, Columbus, OH 43210. (614) 447-3776. (800) 333-9511. FAX (614) 447-3671. 1970 to present. Monthly. $79.00 per year. ISSN: 0009-2703.

Hydrocarbon Processing. Gulf Publishing Company, P.O. Box 2608, Houston, TX 77001. (713) 520-4433. 1922 to present. Monthly. $22.00. ISSN: 0887-0284

Journal of Heterocyclic Chemistry. American Chemical Society, Box 3337, Columbus, OH 43210. (614) 447-3776. (800) 333-9511. FAX (614) 447-3671. 1964 to present. Bi- monthly. $420.00 per year. ISSN: 0022-152X.

Journal of Organic Chemistry. American Chemical Society, Box 3337, Columbus, OH 43210. (614) 447-3776. (800) 333- 9511. FAX (614) 447-3671. 1936 to present. Bi-weekly. $785.00 per year. ISSN: 0022-3263.

Journal of Polymer Science. Part A: Polymer Chemistry. John Wiley and Sons, Inc., 605 Third Avenue, New York, NY 10158. (800) 526-5368 or (212) 850-6000. 1962 to present. 13 issues per year. $2540.00 per year. ISSN: 9887-624X.

Organometallics. American Chemical Society, Box 3337, Columbus, OH 43210. (614) 447-3776. (800) 333-9511. FAX(614) 447-3671. 1982 to present. Monthly. $850.00 per year. ISSN: 0276-7333.

Tetrahedron. Elsevier Science Publishing Company, Inc., 660 White Plains Road, Tarrytown, NY 10591-5153. (914-524-9200.FAX (914) 524-9200. 1957 to present. Sixty times per year. $5365.00 per year. ISSN: 0040-4020.

Tetrahedron Letters. Elsevier Science Publishing Company, Inc., 660 White Plains Road, Tarrytown, NY 10591-5153. (914-524-9200. FAX (914) 524-9200. 1959 to present. Weekly. $4745.00 per year. ISSN: 0040-4039.

RESEARCH CENTERS AND INSTITUTES

Center for Catalytic Science and Technology. University of Delaware, Departments of Chemical Engineering and Chemistry, Newark, DE 19716. (302) 451-8056. FAX (302) 451-1048.

Chemical Laboratories. Harvard Universiy. Oxford Street, Cambridge, MA 02138. (617) 495-4283. FAX (617) 496-5618.

Chemistry Laboratories. Rensselaer Polytechnic Institute. Cogswell Laboratory, Troy, NY 12180-3590. (518) 276-8981.

Chemical Research Laboratory. Brown University, Providence, RI 02912. (401) 863-2256.

University/industry Chemical Research Center. Mississippi State University, Department of Chemistry, P.O. Drawer CH, Mississippi State, MS 39762. (601) 325-3584.

HIGH CARBON STEEL

See: STEEL and STEEL MAKING

HIGH ENERGY PHYSICS

See: PARTICLE PHYSICS

HIGH EXPLOSIVES

See: EXPLOSIVES

HIGH FIDELITY

See: AUDIO ENGINEERING

HIGH TEMPERATURE MATERIALS

See: MATERIALS SCIENCE

HIGH VOLTAGE

See also: ELECTRIC POWER ENGINEERING, ELECTRICAL ENGINEERING, ELECTRICITY, ELECTROMAGENTISM, GENERATORS, HYDROELECTRIC POWER

ABSTRACT SERVICES AND INDEXES

Applied Science and Technology Index; A cumulative subject index to English language periodicals in the fields of aeronautics and space science, computer technology, chemistry, construction industry, energy and related areas.

H.W. Wilson Co., 950 University Avenue, Bronx, NY 10452. (212) 588-8400. (800) 367-6770. FAX (718) 590-1617. From 1958 to present. Monthly. Inquire about cost and availability. Also available on CD-ROM and online. ISSN: 0003-6986.

Current Contents: Engineering, Technology and Applied Sciences. Institute for Scientific Information, 3501 Market Street, Philadelphia, PA 19104. (215) 386-0100. FAX (215)386-6362. 1970 to present. Weekly. $442.00 per year. Also available on CD-ROM and online. Inquire regarding cost and availability. ISSN: 0095-7917.

Current Papers in Electrical and Electronics Engineering. Institution of Electrical Engineers (IEE), London. Distributed by INSPEC/IEEE, Box 1331, 445 Hoes Lane, Piscataway, NJ 08855-1331. (908) 562-5549. 1969 to present. Monthly. $345.00 per year. ISSN: 0011-3778.

Electric Power Industry Abstracts. Edison Electric Institute, c/o Utility Data Institute, 2011 I Atreet, NW, Suite 700, Washington, DC 20006. 1975 to present. Bimonthly. Inquire as to cost and availability.

Electrical and Electronics Abstracts. Institution of Electrical Engineers (IEE), London. Available from INSPEC/IEEE - Institute of Electrical and Electronic Engineers, Box 1331, Hoes Lane, Piscataway, NJ 08855-

1331. (908) 562-5549. 1898 to present. Monthly. $2200.00 per year. Also available on CD-ROM and online as INSPEC.

ISSN: 0036-8105.

Engineering Index Monthly; Indexes and Abstracts the World's Engineering and Technical Literature. Engineering Information, Inc., Castle Point on the Hudson, Hoboken, NJ 07030. (201) 216-8500. (800) 221-1044. FAX (201) 216-8532. Monthly. $2300.00 per year. Available online as COMPENDEX and also on CD-ROM. ISSN: 0742-1974.

General Science Index. H.W. Wilson Company, 950 University Avenue, Bronx, NY 10452. (212) 588-8400. (800) 367-6770. FAX (718) 590-1617. From 1978 to present. Ten issues per year; quarterly and annual cumulations. Service basis. Available on CD-ROM and online. Inquire regarding cost and availability. ISSN: 0162-1963.

Government Reports Announcements and Index. U. S. National Technical Information Service (NTIS), 5285 Port Royal Road, Springfield, VA 22161. (703) 487-4650. FAX (703) 321-8547. From 1968 to present. Annual. $630.00 per year. Also available online as *NTIS Bibliographic Database* and on CD-ROM.

Index To IEEE Publications. IEEE Service Center, 445 Hoes Lane, Piscataway, NJ 08855-1331. (908) 981-1393. (800) 678-IEEE. FAX (908) 981-9667. 1973 to present. Annual. ISSN: 0099-1368.

Physics Abstracts. INSPEC. Section A, Science Abstracts. Institution of Electrical Engineers (IEE), London. Available from: INSPEC/IEEE - Institute of Electrical and Electronic Engineers, Box 1331, Hoes Lane, Piscataway, NJ 08855-1331. (908) 562-5549. 1898 to present. 24 issues per year. $2835.00 per year. Also available online and on CD-ROM. ISSN: 0036-8091.

Physics Briefs (Physikalische Berichte). Information Center for Energy, Physics, Mathematics; German Physical Society. V C H Publishers, Inc., 220 East 23rd Street, New York, NY 10010-

HIGH VOLTAGE

Ency. of Physical Sciences and Engineering Info. Sources

4606. (212) 683-8333. 1845 to present. 24 issues per year. $2390.00 per year. Also available online. ISSN: 0179-7434.

Science Citation Index. SCI. Institute for Scientific Information, 3501 Market Street, Philadelphia, PA 19104. (215) 386-0100. (800) 523-1850. FAX (215) 386-2991. 1961 to present. Six issues per year, plus annual cumulation. $11650.00 per year. Also available online and on CD-ROM. Inquire about price and availability. ISSN: 0036-827X.

ASSOCIATIONS AND PROFESSIONAL SOCIETIES

American Electronics Association. 5201 Great America Way, Suite 520, P.O. 52990, Santa Clara, CA 95056. (408) 987- 4200. FAX (408) 970-8565.

Edison Electric Institute. 701 Pennsylvania Avenue NW, Washington, DC 20004-2696. (202) 508-5000, 5454. FAX (202) 508-5360.

Electric Power Research Institute. 3412 Hillview Avenue, Palo Alto, CA 94304. (415) 855-2000. FAX (415) 855-2954.

Electronics Industries Association. 2500 Wilson Boulevard, Arlington, VA 22201. (703) 907-7500. FAX (202) 457-4985.

IEEE (Institute of Electrical and Electronic Engineers). 345 East 47th Street. New York, NY 10017. (212) 705-7900. FAX (212) 705-4929.

IEEE Power Engineering Society. c/o Institute of Electrical and Electronic Engineers. 345 East 47th Street, New York, NY 10017. (212) 705-7900. FAX (212) 705-4929.

National Association of Power Engineers. 1 Springfield Street, Chicopee, MA 01013. (413) 592-6273. FAX (413) 592-1998.

National Electrical Manufacturers Association. 1300 North 17th Street, Suite 1847, Rosslyn VA 22209. (703) 841-3200. FAX (703) 841-3300.

SMMA - The Association For Electric Motors, their Control and Application. 4 Hollis Street, PO Box 378, Sherborn, MA 01770. (508) 655-4409. FAX (508) 651-3920.

DIRECTORIES AND BIOGRAPHICAL SOURCES

American Electronics Association Directory. American Electronics Association, 5201 Great America Way, Suite 520, P.O. 52990, Santa Clara, CA 95056. (408) 987-4200. FAX (408) 970-8565. Annual. $175.00.

American Men and Women of Science: Physical and Biological Sciences. R. R. Bowker Inc., 121 Chanlon Road, New Providence, NJ 07974. (908) 464-6800. (800) 521-8110. 20th edition. 8 volumes. 1996. $850.00.

Directory of Engineering Societies and Related Organizations. American Association of Engineering Societies, 1111 19th Street, Suite 608, Washington, DC 20036-3603. (202) 296-2237. Semi-annual. $150.00.

E E M - Electronic Engineer's Master. Hearst Business Communications, Inc., 645 Stewart Avenue, Garden City NY 11530. (516) 227-1300. ISSN: 0732-9016.

Electrical and Electronic Trades Directory: The Blue Book. Institution of Electrical Engineers, Michael Faraday House, Six Hills Way, Stevenage, Herts. SG1 2AY, England. Telephone: 0438 313311 or FAX 0438 742840. 1995. $140.00

Electrical World Directory of Electrical Utilities. McGraw-Hill Book Company, Inc., 1221 Avenue of the Americas, New York, NY 10020. (212) 997-3675. Annual.

Engineering Research Centers: Incorporating Electronics Research Centers. Stockton Press, 345 Park Avenue, New York, NY 10010. (212) 689-9200. (800) 221-2123. 4th edition. 1995. $515.00.

IEEE Membership Directory. Institute of Electrical and Electronics Engineers, IEEE Service Center, 445 Hoes Lane, Piscataway, NY 08854. (908) 981-1393. (800) 678-IEEE. FAX(908) 981-9667. 2 volumes. Annual. $190.00. ISSN:

International Engineering Directory. American Consulting Engineers Council, 1015 15th Street, N.W. Suite 802, Washington, DC 20005-2670. (202) 347-7474. Annual. $10.00.

Research Centers Directory. Gale Research, 835 Penobscot Building, Detroit, MI 48226-4094. (313) 961-2242. (800) 347-4253. 20th edition. 1995. $485.00. ISSN: 0080- 1518.

Scientific and Technical Organizations and Agencies Directory. Gale Research, 835 Penobscot Building, Detroit, MI 48226-4094. (313) 961-2242. (800) 347-4253.

4th edition. 1996. $195.00.

Who's Who in Electric Transmission and Distribution. Utility Data Institute, 1200 G Street NY, Suite 250, Washington, DC 20005-3802. (202) 942-8788. (800) 486-3660. 1995. $150.00.

Who's Who in Engineering. Gordon Davis, editor. American Association of Engineering Societies. 1111 19th Street, NY, Suite 608, Washington, DC 20036. (202) 296-2237. (800) 658-8897. 9th edition. 1995. $220.00.

Who's Who in Technology. Gale Research 835 Penobscot Building, Detroit, MI 48226-4094. (313)961-2242. (800) 347-4253. 1995. $195.00. ISN 0-8103-7467-6.

ENCYCLOPEDIAS AND DICTIONARIES

Chambers Science and Technology Dictionary. Peter M.B. Walker, editor. Cambridge University Press, 40 West 20th Street, New York, NY 10011-4211. (212) 924-3900. 1988. $39.95.

IEEE Standard Dictionary of Electrical and Electronics Terms. Christopher J. Booth, editor. IEEE Service Center, 445 Hoes Lane, Piscataway, NJ 08855-1331. (908) 981-1393. (800) 678-IEEE. FAX (908) 981-9667. IEEE Standard 100- 1992. 5th edition. 1993. $150.00.

Illustrated Dictionary of Electronics. Stan Gibilsco. TAB Books, P.O. Box 40, Blue Summit, PA 17294-0850. (717) 794- 2191. (800) 233-1128. 7th edition. 1994. $34.95.

McGraw-Hill Electronics Dictionary. J. Markus. McGraw-Hill Book Company, Inc., 1221 Avenue of the Americas, New York, NY 10020. (212) 997-3675. 5th edition. 1994. $49.95.

Ency. of Physical Sciences and Engineering Info. Sources

HIGH VOLTAGE

McGraw-Hill Encyclopedia of Science and Technology. McGraw-Hill Book, Incorporated, 1221 Avenue of the Americas, New York, NY 10020. (212) 997-3675. (800) 262-4729. Seventh edition. Twenty volumes. 1992. $1900.00.

GENERAL WORKS

Basic Electrical and Electronic Engineering. E. C. Bell and R. W. Whitehead. Blackwell Scientific Publications, Inc., 238 Main Street, Cambridge MA 02142. (617) 876-7000. (800) 759-6102. 4th revised edition. 1993.

Basic Electrical Power and Machines. David Bradley. Chapman & Hall, 1 Penn Plaza, New York, NY 10119. (212) 564-1060. 1994. $30.95.

Electric Motors and Motor Controls. Jeffrey J. Keljik. Delmar Publishing, 3 Columbia Circle, Box 15015, Albany, NY 12212. (518) 464-3500. (800) 347-7707. 1995. $35.95.

Electrical Machines, Drives and Power Systems. Theodore Wildi. Prentice Hall, 113 Sylvan Avenue, Route 9W, Englewood Cliffs, NJ 07632. (201) 592-2000. (800) 922-0579. 2nd edition. 1990. $79.75.

Electrical Power Engineering. Henslay W. Kabisama. McGraw-Hill Publishing Company, Inc., 1221 Avenue of the Americas, New York, NY 10020. (212) 512-2000. (800) 262- 4729. 1993. $55.00.

High Voltage Engineering. M. S. Naidu and V. Kamaraju. McGraw-Hill Publishing Company, Inc., 1221 Avenue of the Americas, New York, NY 10020. (212) 512-2000. (800) 262-4729. 1995. $55.00.

High Voltage Engineering and Testing. H. M. Ryan, editor. Institution of Electrical Engineers, Michael Faraday House, Six Hills Way, Stevenage, Herts. SG1 2AY, England. Telephone: 0438 313311 or FAX 0438 742840; for Peter Peregrinus Press. 1994. $95.00

High-Voltage Engineering: Theory and Practice. Khalifa, editor. Marcel Dekker, Inc., 270 Madison Avenue, New York, NY 10016. (212) 696-9000. (800) 228-1160. 1989. $125.00.

High Voltage Power Equipment Engineering. Pansini and Fairmont Press Staff. Prentice Hall, 113 Sylvan Avenue, Route 9W, Englewood Cliffs, NJ 07632. (201) 592-2000. (800) 922-0579. 1995. $95.00.

Linear Induction Drives. Jacek F. Gieras. Oxford University Press, Inc., 200 Madison Avenue, New York, NY 10016. (212) 725-6000. (800) 334-4249. 1994. $ 75.00

Magneto-Hydro-dynamic Electrical Power Generation. Hugo K. Messerle. John Wiley & Sons, Inc., 605 Third Avenue, New York, NY 10158-0012. (212) 850-6000. (800) 225-5945. 1994. $34.95 in paper.

Schaum's Electric Power Systems. Syed A. Masar. McGraw-Hill Publishing Company, Inc., 1221 Avenue of the Americas, New York, NY 10020. (212) 512-2000. (800) 262-4729. 1990. $10.95 in paper.

HANDBOOKS AND MANUALS

American Electricians Handbook. Terrell Croft and Wilford I. Summers. McGraw-Hill Book Company, 1221 Avenue of the Americas, New York, NY 10020. (212) 512-2000. (800) 262-4729. 12th edition. 1992. $77.50.

Complete Handbook of Electric Motor Controls. John Traister. Prentice Hall, 113 Sylvan Avenue, Route 9W, Englewood Cliffs, NJ 07632. (201) 592-2000. (800) 922-0579. 2nd edition. 1994. $69.00.

Electrical Engineer's Handbook. Donald Christiansen, editor. McGraw Book Company, 1221 Avenue of the Americas, New York, NY 10020. (212) 512-2000. (800) 262-4729. 4th edition. 1996. $110.00.

Electronics Handbook. Jerry C. Whitaker. CRC Press, Inc., 2000 Corporate Boulevard, NW, Boca Raton, FL 33431. (407) 994-0555. (800) 272-7737. 1996. $120.00.

Handbook of Electric Motors. John E. Traister. Prentice- Hall, 113 Sylvan Avenue, Route 9W, Englewood Cliffs, NJ 07632. (201) 592-2000. (800) 922-0579. 1992. $64.00.

Handbook of Electrical and Electronic Technology. Curtis Johnson. Prentice Hall , 113 Sylvan Avenue, Route 9W, Englewood Cliffs, NJ 07632. (201) 592-2000. (800) 922- 0579. 1996. $88.00.

Handbook of Electronics Manufacturing Engineering. Bernard S. Matisoff. Chapman & Hall, 1 Penn Plaza, New York, NY 10119. (212) 564-1060. 1996.

Standard For Safety For High Voltage Industrial Control Equipment, Ul 347. Underwriters Laboratories, Inc., 333 Pfingsten Road, Northbrook, IL, 60062-2096. (708) 272-8800. 4th edition. 1993. $175.00.

Standard Handbook For Electrical Engineers. Donald Fink. McGraw-Hill Publishing Company, 1221 Avenue of the Americas, New York, NY 10020. (212) 512-2000. (800) 262- 4729. 13th edition. 1996. $110.00.

ONLINE DATABASES AND CD-ROMS

CA Search. Chemical Abstracts Service, P.O. Box 3012, Columbus, OH 43210-0012. (614) 447-3600. (800) 848-6533. FAX (614) 447-3709. Very comprehensive guide to worldwide chemical literature and related fields, 1972 to present. Available on BRS,(800) 289-4277, DIALOG, (800) 334-2564, ORBIT (800) 456-7248, and STN International, FIZ Karlsruhe, P.O. Box 2465, W-7500, Karlsruhe 1, Germany, online services. Inquire as to cost and availability.

Compendex Plus. Engineering Information, Inc., 345 East 47th Street, New York, NY 10017. (212) 705-7600 or (800) 221-1044. Contains citations with abstracts to worldwide literature in engineering and technology, from 1970 to present. Available on online BRS,(800) 289-4277, DIALOG, (800) 334-2564, ORBIT (800) 456-7248, and STN International, FIZ Karlsruhe, P.O. Box 2465, W-7500, Karlsruhe 1, Germany, online services. Also available on CD-ROM. Inquire as to cost and availability.

Current Contents Search. Institute for Scientific Information, 3501 Market Street, Philadelphia, PA 19104. (215) 386-0100. FAX (215) 386-6362. Contains citations to articles listed in the

HIGH VOLTAGE

Ency. of Physical Sciences and Engineering Info. Sources

table of contents of science and technology journals. Also articles in social sciences and life sciences journals. Available on BRS,(800) 289-4277, DIALOG,(800) 334-2564, online services. Inquire as to cost and availability.

Dissertation Abstracts Online. University Microfilms International, 300 North Zeeb Road, Ann Arbor, MI 48106. (800) 521-0600 or (313) 761-4700. Scope includes virtually all doctoral dissertations accepted at accredited American institutions from 1861 to present in 252 subject areas. Available on BRS, (800) 289-4277, DIALOG, (800) 334-2564, and OCLC EPIC,(800) 848-5878, online services. Also available on CD-ROM. Inquire as to cost and availability.

INSPEC. Institution of Electrical Engineers, Michael Faraday House, Six Hills Way, Stevenage, Herts. SG1 2AY, England. Telephone: 0438 313311 or FAX 0438 742840. Contains citations to the worldwide literature of physics, electronics and electrical engineering, computer technology, and related fields. Available on BRS, (800) 289-4277, DIALOG, (800) 334-2564, ORBIT, (800) 456-7248, and STN International, FIZ Karlsruhe, P.O. Box 2465, W-7500, Karlsruhe 1, Germany, online services. Inquire as to cost and availability.

NTIS Bibliographic Database. National Technical Information Service, 5285 Port Royal Road, Springfield, VA 22161. (703) 487-4929 or FAX (703) 321-8199. Broad coverage of government-sponsored science and technology research reports, 1964 to present. Available on BRS,(800) 289-4277, DIALOG, (800) 334-2564, ORBIT, (800) 456-7248, and STN International, FIZ Karlsruhe, P.O. Box 2465, W-7500, Karlsruhe 1, Germany, online services. Also available on CD-ROM. Inquire as to cost and availability.

Physics Briefs. American Institute of Physics, 335 East 45th Street, New York, NY 10017. (212) 661-9260 or FAX (212) 661-2036. Contains citations with abstracts of the literature of physics and related fields, 1979 to present. Available on the STN International, FIZ Karlsruhe, P.O. Box 2465, W-7500, Karlsruhe 1, Germany, online service. Inquire as to cost and availability.

Scientific and Technical Books and Serials in Print. R.R. Bowker Inc., 121 Chanlon Road, New Providence, NJ 07974. (800) 521-8110 or (908) 464-6800. List of currently published books and serials in the physical and biological sciences, engineering and technology, with subject, author and titles indexes. Available on ORBIT,(800) 456-7248, online service.Inquire as to cost and availability.

SCISEARCH. Institute for Scientific Information, 3501 Market Street, Philadelphia, PA 19104. (800) 523-1850 or (215) 386-0100. Broad multidisciplinary title and author index to the international literature of science and technology, 1974 to present. Available on DIALOG,(800) 334-2564, and ORBIT,(800) 456-7248, online services. Also available on CD-ROM. Inquire as to cost and availability.

WILSONLINE. H.W. Wilson Company, 950 University Avenue, Bronx, NY 10452. (800) 367-6770 or (212) 588-8400. Makes available online versions of the printed H.W. Wilson indexes including Applied Science and Technology Index, Business Periodicals Index, General Science Index, and Readers' Guide to Periodical Literature. Period covered is generally 1983 to present. Available on BRS,(800) 289-4277, DIALOG,(800) 334-2564, and OCLC EPIC,(800) 848-5878, online services. Also available on CD-ROM. Inquire as to cost and availability.

PERIODICALS

Electric Light and Power. PennWell Publishing Co. 1421 South Sheridan Road, Box 1260, Tulsa OK 74101. (835-3161. FAX (918) 831-9497. 1922 to present. Monthly. $42.00 per year.

Electric Machines and Power Systems. Taylor & Francis, 1900 Frost Road, Suite 101, Bristol, PA 19007-1598. (800) 821-8312, FAX (215) 785-5515. 1976 to present. Monthly. $355.00 per year. ISSN: 0731-356X.

Electrical World. McGraw-Hill, Inc., Box 513 Hightstown, NJ 08520. (212) 512-3288. 1874 to present. Monthly. $55.00 per year. ISSN: 0013-4457.

Electronic Design. Penton Publishing, San Jose Gateway, Suite 354. 2025 Gateway Place, San Jose, CA 95110. (408) 441-0550. 1952 to present. Fortnightly. $95.00. ISSN: 0013-4872.

Electronics. Penton Publishing, San Jose Gateway, Suite 354. 2025 Gateway Place, San Jose, CA 95110. (408) 441-0550. 1930 to present. Semi-weekly. $98.00 per year. ISSN: 0883-4989.

IEEE Power Engineering Review. Institute of Electrical and Electronics Engineers, Inc., Box 1331, 445 Hoes Lane, Piscataway, NJ 08855-1331. (908) 981-0060. 1981 to present. Monthly. $115.00 per year. ISSN: 0272-1724.

IEEE Spectrum. Institute of Electrical and Electronics Engineers, Inc., Box 1331, 445 Hoes Lane, Piscataway, NJ 08855-1331. (908) 981-0060. 1964 to present. Monthly. $157.00. ISSN: 0018-9235.

IEEE Transactions On Power Delivery. Institute of Electrical and Electronics Engineers, Inc., Box 1331, 445 Hoes Lane, Piscataway, NJ 08855-1331. (908) 981-0060. 1986 to present. Quarterly. $170.00 per year. ISSN: 0885-8977.

IEE Proceedings: Electric Power Applications. Institution of Electrical Engineers (IEE), London. Available from INSPEC/IEEE - Institute of Electrical and Electronic Engineers, Box 1331, Hoes Lane, Piscataway, NJ 08855-1331. (908) 562-5549. 1980 to present. Bi-monthly. $ ISSN:

Institute of Electrical and Electronics Engineers Proceedings. Institute of Electrical and Electronics Engineers, Inc., Box 1331, 445 Hoes Lane, Piscataway, NJ 08855-1331. (908) 981-0060. 1913 to present. Monthly. $275.00. ISSN: 0018-9219.

Power Engineering (TULSA). PennWell Publishing Co. 1421 South Sheridan Road, Box 1260, Tulsa OK 74101. (835-3161. FAX (918) 831-9497. 1896 to present. Monthly. $28.00 per year. ISSN: 0032-5961

Power Engineering Journal. Institution of Electrical Engineers (IEE), London. Available from INSPEC/IEEE- Institute of Electrical and Electronic Engineers, Box 1331, 445 Hoes Lane, Piscataway, NJ 08855-1331. (908) 562-5549.

1987 to present. Bi-monthly. L85 per year. ISSN: 0950-3366.

RESEARCH CENTERS AND INSTITUTES

Edison Electric Institute. 701 Pennsylvania Avenue, NW, Washington, DC 20004-2696. (202) 508-5000, 5454. FAX (202) 508-5360.

Electroscience Laboratory. Ohio State University, 1320 Kinnear Road, Columbus, Ohio 43212. (614) 292-7981, FAX (614) 292-7297.

Electrical Engineering Research Laboratories. Purdue University. Electrical Engineering Building, West Lafayette, in 47907. (317) 494-3536. FAX (317) 494-6440.

Electrical Power Research Institute. 3412 Hillview Avenue, PO Box 10412, Palo Alto, CA 94303. (415) 855-2000. FAX (415) 855-2954.

Laboratory For Electromagnetic and Electronic Systems. Massachusetts Institute of Technology, 77 Massachusetts Avenue, Cambridge, MA 02139. (617) 253-4631. FAX (617) 258-6774.

Weber Research Institute. Polytechnic University. Route 110, Farmingdale, NY 11735. (516) 755-4250, FAX (516) 755- 4404.

HIGHWAY ENGINEERING

See also: AGGREGATES, ASPHALT, BRIDGES, CIVIL ENGINEERING, CONCRETE, CONSTRUCTION ENGINEERING, TRANSPORTATION ENGINEERING

ABSTRACT SERVICES AND INDEXES

Applied Science and Technology Index; A Cumulative Subject Index To English Language Periodicals in the Fields of Aeronautics and Space Science, Computer Technology, Chemistry, Construction Industry, Energy and Related Areas. H.W. Wilson Co., 950 University Avenue, Bronx, NY 10452. (800) 367-6770 or (212) 588-8400. FAX (718) 590-1617. From 1958 to present. Monthly. Inquire about cost and availability. Also available on CD-ROM and online.

ASCE Combined Annual Combined Index. American Society of Civil Engineers, 345 East 47th Street, New York, NY 10017-2398. (212) 705-7520 or (800) 548-2723. Annual. $48.00.

ASCE Publications Information. 1966-present. American Society of Civil Engineers, 345 East 47th Street, New York, NY 10017-2398. (212) 705-7520 or (800) 548-2723. Bi-monthly. $160.00 per year for non-members.

Civil and Structural Engineering Abstracts, Cambridge Scientific Abstracts, 7200 Wisconsin Avenue, Bethesda, MD 20814-4823. (301) 961-6750. FAX (301) 961-6720. 1993 to present. Monthly. $385.00 per year. Topics covered include structural design, construction equipment and methods, civil defense and military engineering, surveying, highway engineering, maritime and port structures, materials, land reclamation, and soil mechanics.

Current Contents: Engineering, Technology, and Applied Sciences. Institute for Scientific Information, 3501 Market Street, Philadelphia, PA 19104. (215) 386-0100. FAX (215) 386-6362. 1970 to present. Weekly. $442.00 per year.

Engineering Index Monthly. Engineering Information, Inc., Castle Point on the Hudson, Hoboken, NJ 07030. (800) 221-1044. FAX (212) 832-1857. Monthly. $2300.00 per year. Also available online as COMPENDEX, and also on CD-ROM. Covers chemical engineering, computers, electrical engineering, civil

engineering, metals and mining, industrial management, and mechanical engineering.

Index To Scientific and Technical Proceedings. Institute for Scientific Information, 3501 Market St., Philadelphia, PA 19104. (215) 386-0100. FAX (215) 386-6362. Monthly. $500.00 per year.

Science Citation Index. Institute for Scientific Information, 3501 Market Street, Philadelphia, PA 19104. (215) 386-0100. FAX (215) 386-6362. Inquire about availability and cost. Also available on CD-ROM.

ASSOCIATIONS AND PROFESSIONAL SOCIETIES

American Association of State Highway and Transportation officials. 444 North Capitol Street, Suite 249, Washington, DC 20001. (202) 624-5800. FAX (202) 624-5806.

American Road & Transportation Builders Association. 1010 Massachusetts Avenue NW, Washington, DC 20001. (202) 488-2722. FAX (202) 488-3631.

American Society For Testing and Materials (astm). 1916 Race Street, Philadelphia, PA 19103. (215) 299-5585.

American Society of Civil Engineers. 345 East 47th Street, New York, NY 10017-2398. (212) 705-7520 or (800) 548-2723.

Institute of Transportation Engineers. 525 School Street SW, Suite 410, Washington, DC 20024. (202) 554-8050. FAX (202) 863-5486.

BIBLIOGRAPHIES

Scientific and Technical Books and Serials in Print. R.R. Bowker Inc., 121 Chanlon Road, New Providence, NJ 07974. (800) 521-8110 or (908) 464-6800. List of currently published books and serials in the physical and biological sciences, engineering and technology, with subject, author and titles indexes. Inquire for cost and availability of latest edition. Also available on ORBIT, 800-456-7248, online service.

DIRECTORIES AND BIOGRAPHICAL SOURCES

American Road and Transportation Builders Association—Officials and Engineers Directory. 1010 Massachusetts Avenue NW, Washington, DC 20001. (202) 488-2722. FAX (202) 488-3631. Inquire for cost and availability.

Directory of Engineering Societies and Related Organizations. Gordon Davis. 13th edition. American Association of Engineering Societies, 1111 19th Street NW, Suite 608, Washington, DC 20036. (202) 296-2237 or (800) 658-8897. 1989. Inquire for price.

International Research Centers Directory. Gale Research, 835 Penobscot Building, Detroit, MI 48226-4094. (313) 961-2242. (800) 347-4253. FAX (313) 961-6083. 8th edition. 1995. $410.00.

Public Works Manual. Public Works Journal Corporation, 200 S. Broad Street, Ridgewood, NJ 07451. (201) 445-5800. FAX (201) 445-5170. Annual. Included with subscription to PUBLIC WORKS magazine (see below).

HIGHWAY ENGINEERING

Ency. of Physical Sciences and Engineering Info. Sources

Research Centers Directory. Gale Research, 835 Penobscot Building, Detroit, MI 48226-4094. (313) 961-2242. (800) 347-4253. FAX (313) 961-6083. $485.00.

Scientific and Technical Organizations and Agencies Directory. Gale Research, 835 Penobscot Building, Detroit, MI 48226-4094. (313) 961-2242. (800) 347-4253. FAX (313) 961-6083. 4th edition. 1996. $195.00.

Who's Who in Engineering. American Association of Engineering Societies, 1111 19th Street NW, Suite 608, Washington, DC 20036. (202) 296-2237 or (800) 658-8897. 8th edition. 1991. Inquire for price.

ENCYCLOPEDIAS AND DICTIONARIES

Concise Encyclopedia of Traffic & Transportation Systems. M. Papageorgiou. Pergamon Press Inc., Maxwell House, Fairview Park, Elmsford, NY 10523. (914) 592-7700. Fax (914) 592-3625. 1991. $388.00.

Construction Glossary: An Encyclopedic Reference and Manual. J. Stewart Stern. 2nd edition. John Wiley and Sons, Inc., 605 Third Avenue, New York, NY 10158. (800) 526-5368 or (212) 850-6000. 1993. $99.95.

Dictionary of Civil Engineering. V.N. Vazirani. South Asia Books, PO Box 502, Columbia, MO 65205. (314) 474-0116. FAX (314) 474-8124. 1992. $20.00.

McGraw-Hill Encyclopedia of Engineering. Sybil P. Parker, ed. 2nd edition. McGraw-Hill Publishing Company, 1221 Avenue of the Americas, New York, NY 10020. (800) 262-4729 or (212) 512-3825. 1993. $95.50.

McGraw-Hill Encyclopedia of Science and Technology. Sybil P. Parker, ed. 7th edition. 20 volumes. McGraw-Hill Publishing Company, 1221 Avenue of the Americas, New York, NY 10020. (800) 262-4729 or (212) 512-3825. 1992. $1900.00

Thesaurus of Scientific, Technical, and Engineering Terms. Hemisphere Publishing Corporation, 1900 Frost Road, Suite 101, Bristol, PA 19007-1598. (215) 785-5800 or (800) 821-8312. FAX (215) 785-5515. 1987. $173.00.

GENERAL WORKS

The Geometric Design of Roads. Robin T. Underwood. Macmillan Publishing, 200 Old Tappan Road, Old Tappan, NJ 07675. (800) 223-2336. FAX (800) 445-6991. 1991. Inquire for cost and availability.

Highway and Transportation Engineering and Planning. Gavin Macpherson. John Wiley and Sons, Inc., 605 Third Avenue, New York, NY 10158. (800) 526-5368 or (212) 850-6000. 1993. $49.95.

Highway Construction and Maintenance. John Watson. John Wiley and Sons, Inc., 605 Third Avenue, New York, NY 10158. (800) 526-5368 or (212) 850-6000. 1994. $57.95.

Highway Engineering. P.H. Wright & R.J. Paquette. 5th edition. John Wiley and Sons, Inc., 605 Third Avenue, New York, NY 10158. (800) 526-5368 or (212) 850-6000. 1987. Inquire for cost and availability.

Materials For Civil and Highway Engineers. Kenneth N. Derucher, ed. 2d ed. Prentice Hall, 113 Sylvan Avenue, Route 9W, Englewood Cliffs, NJ 07632. (201) 592-2000 or (800) 922-0579. 1994. $61.00

HANDBOOKS AND MANUALS

Civil Engineering Calculations Reference Guide. Tyler G. Hicks. McGraw-Hill Publishing Company, 1221 Avenue of the Americas, New York, NY 10020. (800) 262-4729 or (212) 512-3825. 1987. $46.00.

Civil Engineering Practice. Paul N. Cheremisinoff, et al. 5 volumes. Technomic Publishing Company, Inc., 851 New Holland Avenue, Box 3535, Lancaster, PA 17604. (717) 291-5609 or (800) 233-9936. FAX (717) 295-4538. 1987-88. $199.00.

Handbook of Concrete Engineering. Mark Fintel. 2nd edition. Routledge, Chapman and Hall, Inc., 29 West 35th Street, New York, NY 10001-2291. (212) 244-3336 or FAX (212) 563-2269. 1985. $115.00.

Handbook of Soil Mechanics Volume 3: Soil Mechanics of Earthworks, Foundations, and Highway Engineering. Elsevier Science Publishing Company, Inc., 655 Avenue of the Americas, New York, NY 10010. (212) 989-5800. $123.00.

Spon's Civil Engineering and Highway Works Price Book 1996. Davis Langdon & Everest. Chapman & Hall, 115 Fifth Avenue, New York, NY 10211-0906. 1996. $156.50.

Standard Handbook For Civil Engineers. Frederick S. Merritt, editor. 3rd edition. McGraw-Hill Publishing Company, 1221 Avenue of the Americas, New York, NY 10020. (800) 262-4729 or (212) 512-3825. 1983. $124.50.

Surveying Ready-Reference Manual. G.O. Stenstrom. McGraw-Hill Publishing Company, 1221 Avenue of the Americas, New York, NY 10020. (800) 262-4729 or (212) 512-3825. 1987. $38.50.

ONLINE DATABASES AND CD-ROMS

Compendex Plus. Engineering Information, Inc., 345 East 47th Street, New York, NY 10017. (212) 705-7600 or (800) 221-1044. Contains citations with abstracts to worldwide literature in engineering and technology, from 1970 to present. Available on online BRS,(800) 289-4277, DIALOG, (800) 334-2564, ORBIT (800) 456-7248, and STN International, FIZ Karlsruhe, P.O. Box 2465, W-7500, Karlsruhe 1, Germany, online services. Also available on CD-ROM. Inquire as to cost and availability.

CSA Engineering. Cambridge Scientific Abstracts, 7200 Wisconsin Avenue, Suite 601, Bethesda, MD 20814. (301) 961-6750 or (800) 843-7751. Contains citations and abstracts of international periodicals and research literature covering all fields of engineering and science and technology, including computer and information science, electronics, mechanical engineering, solid state materials, 1981 to present. Available on BRS,(800) 289-4277, online service. Inquire as to cost and availability.

Current Contents Search. Institute for Scientific Information, 3501 Market Street, Philadelphia, PA 19104. (215) 386-0100. FAX (215) 386-6362. Contains citations to articles listed in the table of contents of science and technology journals. Also articles in social sciences and life sciences journals. Available on

BRS,(800) 289-4277, DIALOG,(800) 334-2564, online services. Inquire as to cost and availability.

Dissertation Abstracts. University Microfilms International, 300 North Zeeb Road, Ann Arbor, MI 48106. (800) 521-0600 or (313) 761-4700. Scope includes virtually all doctoral dissertations accepted at accredited American institutions from 1861 to present in 252 subject areas. Available on BRS,(800) 289-4277, DIALOG,(800) 334-2564, and OCLC EPIC,(800) 848-5878, online services. Also available on CD-ROM. Inquire as to cost and availability.

GEOREF: Bibliography and Index of Geology. American Geological Institute, 4220 King Street, Alexandria, VA 22302. (703) 379-2480. Fax (703) 379-7563. Monthly. Inquire for price and availability.

NTIS Bibliographic Database. National Technical Information Service, 5285 Port Royal Road, Springfield, VA 22161. (703) 487-4929 or FAX (703) 321-8199. Broad coverage of government-sponsored science and technology research reports, 1964 to present. Available on BRS,(800) 289-4277, DIALOG, (800) 334-2564, ORBIT, (800) 456-7248, and STN International, FIZ Karlsruhe, P.O. Box 2465, W-7500, Karlsruhe 1, Germany, online services. Also available on CD-ROM. Inquire as to cost and availability.

SAE Global Mobility Database. Society of Automative Engineers (SAE), Electronic Publishing Division, 400 Commonwealth Drive, Warrendale, PA 15098. (412) 776-4841. Contains citations with abstracts to technical papers on automotive and aerospace technology and vehicular-related industries that have been presented at SAE conferences. Covers 1906 to present. Available on ORBIT,(800) 456-7248,online service. Inquire as to cost and availability.

SCISEARCH. Institute for Scientific Information, 3501 Market Street, Philadelphia, PA 19104. (800) 523-1850 or (215) 386-0100. Broad multidisciplinary title and author index to the international literature of science and technology, 1974 to present. Available on DIALOG,(800) 334-2564, and ORBIT,(800) 456-7248, online services. Also available on CD-ROM. Inquire as to cost and availability.

TRIS (Transportation Research Information). National Academy of Science, 2101 Constitution Avenue, N.W., Washington, DC 20418. (202) 334-3313 or (800) 624-6242. Citations with abstracts of literature on transportation, including air, highway, rail, maritime and other modes from 1968 to present. Available on DIALOG,(800) 334-2564, online service. Inquire as to cost and availability.

WILSONLINE. H.W. Wilson Company, 950 University Avenue, Bronx, NY 10452. (800) 367-6770 or (212) 588-8400. Makes available online versions of the printed H.W. Wilson indexes including Applied Science and Technology Index, Business Periodicals Index, General Science Index, and Readers' Guide to Periodical Literature. Period covered is generally 1983 to present. Available on BRS,(800) 289-4277, DIALOG,(800) 334-2564, and OCLC EPIC,(800) 848-5878, online services. Also available on CD-ROM. Inquire as to cost and availability.

PERIODICALS

Civil Engineering ASCE. American Society of Civil Engineers, 345 East 47th Street, New York, NY 10017-2398. (212) 705-7520 or (800) 548-2723. 1930 to present. Monthly. $85.00 per year.

Concrete. The Concrete Society, Framewood Road, Wexham, Slough, Berks. SL3 6PJ England. Telephone 0753-662226. FAX 0753-662126. 1966 to present. Bi-monthly. $75.00 per year.

Construction Products. Gordon Publications/ Cahners-Reed Elsevier, 301 Gilbraltar Drive, Box 650, Morris Plains, NJ 07950. (201) 252-5100. FAX (201) 539-3476. 1892 to present. 6 times a year. $40.00 per year.

Highway Builder. Naylor Publications Inc., 11350 McCormick Road, Executive Plaza Three, Hunt Valley, MD 21031. (410) 785-2445. 1921 to present. Quarterly. $10.00 per year.

Highways and Transportation. Institution of Highways and Transportation, 3 Lygon Place, Ebury Street, London SW1 0JS, England. Telephone 071-730-5245. FAX 071-730-1628. 1930 to present. Monthly. Inquire for cost.

ITE Journal. Institute of Transportation Engineers, 525 School Street SW, Suite 410, Washington, DC 20024. (202) 554-8050. 1930 to present. Monthly. $50.00 per year.

Journal of Construction Engineering and Management. American Society of Civil Engineers, Construction Division, 345 East 47th Street, New York, NY 10017-2398. (212) 705-7520 or (800) 548-2723. 1956 to present. Quarterly. $112.00 per year.

Journal of Structural Engineering. American Society of Civil Engineers, 345 East 47th Street, New York, NY 10017-2398. (212) 705-7520 or (800) 548-2723. 1956 to present. Monthly. $300.00 per year for non-members.

Journal of Surveying Engineering. 1983-present. American Society of Civil Engineers, 345 East 47th Street, New York, NY 10017-2398. (212) 705-7520 or (800) 548-2723.

Journal of Transportation Engineering. American Society of Civil Engineers, Air Transport, Highway, Pipeline, Urban Transportation Divisions, 345 East 47th Street, New York, NY 10017-2398. (212) 705-7520 or (800) 548-2723. 1969 to present. Bi-monthly. $124.00 per year.

Public Roads. U.S. Federal Highway Administration, 400 7th Street NW, Washington, DC 20590. (703) 285-2104. 1918 to present. Quarterly. $12.00 per year.

Public Works. Public Works Journal Corporation, 200 S. Broad Street, Ridgewood, NJ 07451. (201) 445-5800. FAX (201) 445-5170. 1896 to present. 13 times a year. $45.00 per year.

Roads & Bridges. Scranton Gillette Communications Inc., 380 E. Northwest Highway, Des Plaines, IL 60016. (708) 298-6622. FAX (708) 390-0408. 1892 to present. Monthly. $20.00 per year.

Traffic Engineering & Control. Printerhall Ltd., 29 Newman Street, London W1P 3PE, England. Telephone 071-636-3956. FAX 071-436-7016. 1960 to present. Monthly. $106.00 per year.

Transportation Builder. American Road and Transportation Builders Association, 1010 Massachusetts Avenue NW, Washington, DC 20001. (202) 488-2722. FAX (202) 488-3631. 1923 to present. 6 times a year. $50.00.

Transportation Research A: Policy and Practice. Elsevier Science [journals], 660 White Plains Rd., Tarrytown, NY 10159-5153. (919) 524-9200. FAX (919) 333-2444. 1967 to present. 6 times a year. $405.00 per year ($910.00 for A, B, and C).

HIGHWAY ENGINEERING

Ency. of Physical Sciences and Engineering Info. Sources

Transportation Research B: Methodological. Elsevier Science [journals], 660 White Plains Rd., Tarrytown, NY 10159-5153. (919) 524-9200. FAX (919) 333-2444. 1967 to present. 6 times a year. $405.00 per year ($910.00 for A, B, and C).

Transportation Research C: Emerging Technologies. Elsevier Science [journals], 660 White Plains Rd., Tarrytown, NY 10159-5153. (919) 524-9200. FAX (919) 333-2444. 1993 to present. 4 times a year. $260.00 per year ($910.00 for A, B, and C).

Transportation Research Record. National Research Council, Transportation Research Board, 2101 Constitution Avenue NW, Washington, DC 20418. (202) 334-3218. FAX (202) 334-2519. 1963 to present. Irregular. $865.00 per year.

RESEARCH CENTERS AND INSTITUTES

Highway Research Center. Auburn University, Auburn, AL 36849. (205) 826-5250. FAX (205) 844-2672.

Joint Highway Research Project. Purdue University, Civil Engineering Building, West Lafayette, in 47907. (317) 494-2159. FAX (317) 494-0395.

North Carolina State University, Center For Transportation Engineering Studies. Department of Civil Engineering, Box 7908, Raleigh, NC 27695-7908. (919) 737-2331.

Texas A&m University, Texas Transportation Institute, System Division. College Station, TX 77843-3135. (409) 945-9871.

SPECIFICATIONS AND STANDARDS

Annual Book of ASTM Standards, Volume 04.03. American Society for Testing and Materials (ASTM), 1916 Race Street, Philadelphia, PA 19103. (215) 299-5585. Annual. Inquire for cost and availability.

HOLOGRAPHY

See also: FIBER OPTICS, LASERS, OPTICAL COMMUNICATIONS, OPTICS, OPTOELECTRONICS, PHYSICS,

ABSTRACT SERVICES AND INDEXES

Applied Science and Technology Index; A cumulative subject index to English language periodicals in the fields of aeronautics and space science, computer technology, chemistry, construction industry, energy and related areas. H.W. Wilson Co., 950 University Avenue, Bronx, NY 10452. (212) 588-8400. (800) 367-6770. FAX (718) 590-1617. From 1958 to present. Monthly. Inquire about cost and availability. Also available on CD-ROM and online. ISSN: 0003-6986.

Chemical Abstracts. Chemical Abstracts Service, 2540 Olentangy River Road, Box 3012, Columbus, OH 43210-0012. (614) 447-3600. (800) 848-6538. FAX (614) 447-3713. From 1907 to present. Weekly. $16800.00 per year. Also available on CD-ROM and online. Inquire regarding cost and availability.

Current Contents: Engineering, Technology and Applied Sciences. Institute for Scientific Information, 3501 Market Street,

Philadelphia, PA 19104. (215) 386-0100. FAX (215) 386-2291. From 1961 to present. Weekly. $442.00 per year. Also available on CD-ROM and online. Inquire regarding cost and availability. ISSN: 0095-7917.

Electrical and Electronics Abstracts. Science Abstracts Section B. Institute of Electrical Engineers, Michael Faraday House, Six Hill Way, Stevenage, Herts, SG1 2AY, England. Distributed by IEEE, 445 Hoes Lane, Piscataway, NJ 08854. (908) 562-5549. 1898 to present. Monthly. $2200.00 per year. Also available on CD-ROM and online as INSPEC. ISSN: 0036-8105.

Engineering Index Monthly. Engineering Information, Inc., Castle Point on the Hudson, Hoboken, NJ 07030. (201) 216-8500. FAX (201) 216-8532. Monthly. $2200.00 per year. Also available online as COMPENDEX and on CD-ROM. ISSN: 0742-1974.

General Science Index. H.W. Wilson Company, 950 University Avenue, Bronx, NY 10452. (212) 588-8400. (800) 367-6770. FAX (718) 590-1617. From 1978 to present. Ten issues per year; quarterly and annual cumulations. Service basis. Available on CD-ROM and online. Inquire regarding cost and availability. ISSN: 0162-1963.

Government Reports Announcements and Index. U. S. National Technical Information Service (NTIS), 5285 Port Royal Road, Springfield, VA 22161. (703) 487-4650. FAX (703) 321-8547. From 1968 to present. Annual. $630.00 per year. Also available online as *NTIS Bibliographic Database* and on CD-ROM.

NTIS Alerts: Electrotechnology. U. S. National Technical Information Service. 5285 Port Royal Road, Springfield, VA 22161. (703) 487-4650. FAX (703) 321-8547. Weekly. $140.00 per year.

Physics Abstracts. INSPEC. Section A, Science Abstracts. Institution of Electrical Engineers (IEE). Available from INSPEC/IEEE - Institute of Electrical and Electronic Engineers,, Box 1331, 445 Hoes Lane, Piscaway NJ 08855-1331. (908) 562-5549. 1898 to present. 24 issues per year. $2835.00 per year. Also available on CD-ROM and online. ISSN: 0036-8091.

Physics Briefs (Physikalische Berichte). Information Center for Energy, Physics, Mathematics; German Physical Society. V C H Publishers, Inc., 220 East 23rd Street, New York, NY 10010-4606. (212) 683-8333. 1845 to present. 24 issues per year. $2390.00 per year. Also available online. ISSN: 0179-7434.

Science Citation Index. SCI. Institute for Scientific Information, 3501 Market Street, Philadelphia, PA 19104. (215) 386-0100. (800) 523-1850. FAX (215) 386-2991. 1961 to present. Six issues per year, plus annual cumulation. $11650.00 per year. Also available online and on CD-ROM. Inquire about price and availability. ISSN: 0036-827X.

ANNUAL REVIEWS AND YEARBOOKS

Acoustical Imagining. Plenum Publishing Corp., 233 Spring Street, New York, NY. (212) 620-8000. (800) 2221-9369.

FAX (212) 463-0742. From 1969 to present. Irregular. ISSN: 0270-5117.

Progress in Optics. Elsevier Science Inc., 660 White Plains Road, Tarrytown, NY 10591-5153. (914) 524-9200. FAX (914) 333-2444. 1961 to present. Irregular. Price varies. ISSN: 0079-6638.

Ency. of Physical Sciences and Engineering Info. Sources

HOLOGRAPHY

Springer Series in Optical Sciences. Springer-Verlag New York, Inc., 175 Fifth Avenue, New York, NY 10010. (212) 460-1500. (800) 777-4643. FAX (212) 473-6272. 1976 to present. Irregular. Price varies.

ASSOCIATIONS AND PROFESSIONAL SOCIETIES

American Institute of Physics. One Physics Ellipse, College Park, MD 20740-3843. (301) 209-3100.

IEEE Lasers and Electro-optics Society. c/o IEEE, 445 Hoes Lane, PO Box 1331, Piscataway, NJ 08855. (908) 562-3892. FAX (908) 562-8434.

Institute of Electrical and Electronics Engineers. 345 East 47th Street, New York, NY 10017. (212) 705-7900. FAX (212) 705-4929.

Laser Institute of America. 12424 Research Parkway, Suite 130, Orlando, FL 32926. (407) 380-1553. FAX (407) 380- 5588.

Optical Society of America. 2010 Massachusetts Avenue NW, Washington, DC 20036. (202) 223-8130. FAX (202) 223-1096.

Society For Imaging Science and Technology. 7003 Kilworth Lane, Springfield, VA 22151. (703) 642-9090. FAX (703) 642-9094.

Spie - the International Society For Optical Engineering. PO Box 10, 1000 20th Street, Bellingham, WA 98227. (206) 676-3290. FAX (206) 647-1445.

BIBLIOGRAPHIES

Science Books and Films. American Association for the Advancement of Science, 1333 H Street NW, Washington, DC 20005. (202) 326-6454. Reviews of print, film and software materials in all sciences. 1965 to present. Nine issues per year. $40.00 per year. ISSN: 0098-342X

Scientific and Technical Books and Serials in Print, 1995; An Index To Literature in Science and Technology. R.R. Bowker Inc., 121 Chanlon Road, New Providence, NJ 07974. (908) 464-6800. (800) 521-8110. 4 volumes. 1972 to present. Annual. $299.95. Also available on compact disc and online. ISSN: 0000-054X.

DIRECTORIES AND BIOGRAPHICAL SOURCES

American Men and Women of Science: Physical and Biological Sciences. R. R. Bowker Inc., 121 Chanlon Road, New Providence, NJ 07974. (908) 464-6800. (800) 521-8110.20th edition. 8 volumes. 1996. $850.00.

American Physical Society Membership Directory Bulletin. American Physical Society, One Physics Ellipse, College Park, MD 20740-3843. (301) 209-3200. FAX (301) 209- 0865. Biennial. $50.00.

IEEE Membership Directory. Institute of Electrical and Electronics Engineers. 345 East 47th Street, New York, NY 10017. (212) 705-7900. FAX (212) 705-4929.

International Directory of Engineering Societies and Related Organizations. American Society for Engineering Education,

1818 N Street NW, Suite 600, Washington, DC 20036. (202) 331-3526. 15th edition. 1996. $185.00.

Laser Focus World Buyers Guide. PennWell Publishing Co., 1421 Sheridan Road, Tulsa, OK 74112. (918) 831-9421. Annual. $75.00.

Laser Pioneers. Jeff Hecht. Academic Press, Inc., 6277 Sea Harbor Drive, Orlando, FL. (800) 321-5068. 2nd edition. 1991. $49.95.

Lasers and Optoelectronics Buying Guide. Gordon Publications. P.O. Box 650, Morris Plains, NJ 07950. (908) 292-5100. Annual. $95.00.

Optical Society of America. Member Directory. Optical Society of America. 2010 Massachusetts Avenue NW, Washington, DC 20036-1023. (202) 223-8130. (800) 762-6960. Biennial. $12.00.

Optics Education. SPIE - the International Society for Optical Engineering, PO Box 10, 1000 20th Street, Bellingham, WA 98227. (206) 676-3290. FAX (206) 647-1445. Annual. No charge.

Research Centers Directory. Gale Research Inc., 835 Penobscot Building, Detroit, MI 48226-4094. (313) 961-2242. (800) 347-4253. 20th edition. 1995. $485.00. ISSN: 0080- 1518.

Scientific and Technical Organizations and Agencies Directory. Gale Research, 835 Penobscot Building, Detroit, MI 48226-4094. (313) 961-2242. (800) 347-4253. 4th edition. 1996. $195.00.

Who's Who in Engineering. Gordon Davis, editor. American Society for Engineering Education, 1111 19th Street, NW, Suite 608, Washington, DC 20036. 9th edition. 1995. $220.00.

Who's Who in Technology. Gale Research, 835 Penobscot Building, Detroit, MI 48226-4094. (313) 961-2242. (800) 347-4253. 1995. $195.00.

ENCYCLOPEDIAS AND DICTIONARIES

Encyclopedia of Applied Physics. George Trigg, editor. VCH Publications, Inc., 220 East 23rd Street, Suite 909, New York, NY 10010-4606. (212) 683-8333. (800) 422-8824. 20 volume set. 1991-. $5990.00.

Encyclopedia of Lasers and Optical Technology. Robert A. Meyers, editor. Academic Press, Inc., 6277 Sea Harbor Drive, Orlando, FL. (800) 321-5068. 1991. $92.00.

Encyclopedia of Physical Science and Technology. Academic Press, Inc., 6277 Sea Harbor Drive, Orlando, FL. (800) 321-5068. 2nd edition. 18 volume set. 1992. $2625.00.

Fiber Optics Standard Dictionary. Martin H. Weik. 2nd edition. Van Nostrand Reinhold, 115 Fifth Avenue, New York, NY 10003. (212) 254-3232. (800) 842-3636. 2nd edition. 1989. $49.95.

Glossary of Optical Terminology. Thomas K. Farrell. Butterworth-Heinemann, 313 Washington Avenue, Newton, MA 02158. (617) 928-2500. (800) 366-2665. 1985. $24.95.

McGraw-Hill Encyclopedia of Science and Technology. McGraw-Hill Book Company, Inc., 1221 Avenue of the Americas, New York, NY 10020. (212) 512-2000. (800) 262-4729. 7th edition. 20 volume set. 1992. $1900.00.

GENERAL WORKS

Applied Optics and Optical Engineering; A Comprehensive Treatise. Rudolf Kingslake, et al, editors. Academic Press, Inc., 6277 Sea Harbor Drive, Orlando, FL. (800) 321- 5068. 9 volumes. 1965- . Volumes individually priced.

Coherent Optics. W. Lauterborn, et al. Springer-Verlag New York, Inc., 175 Fifth Avenue, New York, NY 10010. (212) 460-1500. (800) 777-4643. FAX (212) 473-6272. 1995. $49.00.

Electron Holography. A. Tonomura. Springer-Verlag New York, Inc., 175 Fifth Avenue, New York, NY 10010. (212) 460-1500. (800) 777-4643. FAX (212) 473-6272. 1994. $54.00.

Fundamentals of Laser Optics. K. Iga. Plenum Publishing Corp., 233 Spring Street, New York, NY. (212) 620-8000. (800) 2221-9369. FAX (212) 463-0742. 1994. $49.50.

Industrial Applications For Optical Data Processing and Holography. Jean Robbillard and Edgar Conley, editors. CRC Press, Inc., 2000 Corporate Boulevard, NW, Boca Raton, FL

33431. (407) 994-0555. (800) 272-7737. 1992. $85.00.

Lasers and Holography. V. V. Rampal and P. C. Mehta. World Scientific Publishing Company, Inc., 1060 Main Street, River Edge, NJ 07661. (201) 487-9655. (800) 227-7562. 1993. $98.00.

Lasers and Holography: An Introduction To Coherent Optics. Winston E. Kock. Revised edition. Dover Publications, Inc., 180 Varick Street, New York, NY 10014. (212) 255- 3755. (800) 223-3130. 1981. $4.95.

Optics and Lasers: Including Fibers and Optical Waveguides. M. Young. Springer-Verlag New York, Inc., 175 Fifth Avenue, New York, NY 10010. (212) 460-1500. (800) 777-4643. FAX (212) 473-6272. 4th revised edition. 1993. $43.50.

Practical Holography. Graham Saxby. Prentice-Hall, 113 Sylvan Avenue, Route 9W, Englewood Cliffs, NJ 07632. (201) 592-2000. (800) 922-0579 2nd edition. 1994. $58.00 in paper.

Tunable Lasers. L. F. Mollenauer, et al, editors. Springer-Verlag New York, Inc., 175 Fifth Avenue, New York, NY 10010. (212) 460-1500. (800) 777-4643. FAX (212) 473-6272. 2nd edition. 1992. $69.00 in paper.

HANDBOOKS AND MANUALS

Handbook of Optical Holography. H. John Caulfield, editor. Academic Press, Inc., 6277 Sea Harbor Drive, Orlando, FL. (800) 321-5068. 1979. $136.00.

Handbook of Optics. Optical Society of America. McGraw- Hill Publishing Company, Inc., 1221 Avenue of the Americas, New York, NY 10020. (212) 512-2000. (800) 262- 4729. 2 volumes. 2nd edition. 1993. $99.50 each.

Industrial Laser Handbook. M. Levitt and D. Belforte, editors. Springer-Verlag New York, Inc., 175 Fifth Avenue, New York, NY 10010. (212) 460-1500. (800) 777-4643. FAX (212) 473-6272. 1992. $163.00.

Laser Experimenter's Handbook. Delton T. Horn. TAB Books, P.O. Box 40, Blue Summit, PA 17294-0850. (717) 794-2191. (800) 233-1128. 2nd edition. 1988. $21.95.

Laser Handbook. F. T. Areechi, E. D. Schultz-Dubois, et al, editors. Elsevier Science Inc., 660 White Plains Road, Tarrytown, NY 10591-5153. (914) 524-9200. FAX (914) 333- 2444. 6 volumes. 1973 -

Manual of Practical Holography. Graham Saxby. Butterworth-Heinemann, 313 Washington Street, Newton, MA 02158. (618-928- 2500. (800) 366-2665. 1991. $9.95 in paper.

ONLINE DATABASES AND CD-ROMS

CA Search. Chemical Abstracts Service, P.O. Box 3012, Columbus, OH 43210-0012. (614) 447-3600. (800) 848-6533. FAX (614) 447-3709. Very comprehensive guide to worldwide chemical literature and related fields, 1972 to present. Available on BRS,(800) 289-4277, DIALOG, (800) 334-2564, ORBIT (800) 456-7248, and STN International, FIZ Karlsruhe, P.O. Box 2465, W-7500, Karlsruhe 1, Germany, online services. Inquire as to cost and availability.

Compendex Plus. Engineering Information, Inc., 345 East 47th Street, New York, NY 10017. (212) 705-7600 or (800) 221-1044. Contains citations with abstracts to worldwide literature in engineering and technology, from 1970 to present. Available on online BRS,(800) 289-4277, DIALOG, (800) 334-2564, ORBIT (800) 456-7248, and STN International, FIZ Karlsruhe, P.O. Box 2465, W-7500, Karlsruhe 1, Germany, online services. Also available on CD-ROM. Inquire as to cost and availability.

Dissertation Abstracts Online. University Microfilms International, 300 North Zeeb Road, Ann Arbor, MI 48106. (800) 521-0600 or (313) 761-4700. Scope includes virtually all doctoral dissertations accepted at accredited American institutions from 1861 to present in 252 subject areas. Available on BRS, (800) 289-4277, DIALOG, (800) 334-2564, and OCLC EPIC, (800) 848-5878, online services. Also available on CD-ROM. Inquire as to cost and availability.

INSPEC. Institution of Electrical Engineers, Michael

Faraday House, Six Hills Way, Stevenage, Herts. SG1 2AY, England. Telephone: 0438 313311 or FAX 0438 742840. Contains citations to the worldwide literature of physics, electronics and electrical engineering, computer technology, and related fields. Available on BRS, (800) 289-4277, DIALOG, (800) 334-2564, ORBIT, (800) 456-7248, and STN International, FIZ Karlsruhe, P.O. Box 2465, W-7500, Karlsruhe 1, Germany, online services. Inquire as to cost and availability.

NTIS Bibliographic Database. National Technical Information Service, 5285 Port Royal Road, Springfield, VA 22161. (703) 487-4929 or FAX (703) 321-8199. Broad coverage of government-sponsored science and technology research reports, 1964 to present. Available on BRS,(800) 289-4277, DIALOG, (800) 334-2564, ORBIT, (800) 456-7248, and STN International, FIZ Karlsruhe, P.O. Box 2465, W-7500, Karlsruhe 1, Germany, online services. Also available on CD-ROM. Inquire as to cost and availability.

Ency. of Physical Sciences and Engineering Info. Sources

HOLOGRAPHY

Physics Briefs. American Institute of Physics, 335 East 45th Street, New York, NY 10017. (212) 661-9260 or FAX (212) 661-2036. Contains citations with abstracts of the literature of physics and related fields, 1979 to present. Available on the STN International, FIZ Karlsruhe, P.O. Box 2465, W-7500, Karlsruhe 1, Germany, online service. Inquire as to cost and availability.

SCISEARCH. Institute for Scientific Information, 3501 Market Street, Philadelphia, PA 19104. (800) 523-1850 or (215) 386-0100. Broad multidisciplinary title and author index to the international literature of science and technology, 1974 to present. Available on DIALOG, (800) 334-2564, and ORBIT, (800) 456-7248, online services. Also available on CD-ROM. Inquire as to cost and availability.

WILSONLINE. H.W. Wilson Company, 950 University Avenue, Bronx, NY 10452. (800) 367-6770 or (212) 588-8400. Makes available online versions of the printed H.W. Wilson indexes including Applied Science and Technology Index, Business Periodicals Index, General Science Index, and Readers' Guide to Periodical Literature. Period covered is generally 1983 to present. Available on BRS, (800) 289-4277, DIALOG, (800) 334-2564, and OCLC EPIC, (800) 848-5878, online services. Also available on CD-ROM. Inquire as to cost and availability.

PERIODICALS

Applied Optics. Optical Society of America. 2010 Massachusetts Avenue NW, Washington, DC 20036-1023. (202) 223-8130. 1962 to present. 36 issues per year. $1090.00 per year. ISSN: 0003-6935.

International Journal of Nonlinear Optical Physics. World Scientific Publishing Co., 1060 Main Street, Suite 18, River Edge, NJ 07661. (800) 227-7562. 1992 to present. Quarterly. $348.00 per year. ISSN: 0218-1991.

Journal of Applied Physics. American Institute of Physics. One Physics Ellipse, College Park, MD 20740-3843. (301) 209-3100. 1931 to present. Semi-monthly. $1655.00 per year. ISSN: 0021-8979.

Journal of Laser Applications. Laser Institute of America, 12424 Research Parkway, Suite 130, Orlando, FL 32926. (407) 380-1553. FAX (407) 380-5588.

Journal of Lightwave Technology. Institute of Electrical and Electronics Engineers, 445 Hoes Lane, PO Box 1331, Piscataway, NJ 08855. (908) 562-3892. FAX (908) 562-8434. 1983 to present. Monthly. $425.00 per year. ISSN: 0733-8724.

Journal of the Optical Society of America. Part A: Optics and Image Science. Optical Society of America, Inc., 2010 Massachusetts Avenue, NW, Washington, DC 20036-1023. (202)223-8130. 1917 to present. Monthly. $610.00 per year. ISSN: 0740-3232.

Journal of the Optical Society of America. Part B: Optical Physics. Optical Society of America, Inc., 2010 Massachusetts Avenue, NW, Washington, DC 20036-1023. (202) 223-8130. 1917 to present. Monthly. $610.00 per year. ISSN: 0740-3224.

Laser Focus World. PennWell Publishing Co., 10 Tara Boulevard, Nashua, NH 03062-2891. (603) 891-0123. FAX (603) 891-0574. 1964 to present. Monthly. $104.00 per year. ISSN: 1043-8092.

Optical Engineering. SPIE - International Society for Optical Engineering, Box 10, 1000 20th Street, Bellingham, WA 98227-0010. (206) 676-3290. FAX (206) 647-1445. 1962 to present. Monthly. $170.00 per year. ISSN: 0091-3286.

Optics and Lasers in Enginering. Elsevier Science, 660 White Plains Road, Tarrytown, NY 10591-5153. (914) 545-9200. FAX (914) 333-2444. 1980 to present. Ten issues per year. $500.00 per year. ISSN: 0143-8166.

Optics and Photonics News. Optical Society of America, Inc., 2010 Massachusetts Avenue, NW, Washington, DC 20036-1023. (202) 223-8130. 1975 to present. Monthly. $99.00 per year. ISSN: 1047-6938.

Optics Communications. Elsevier Science, 660 White Plains Road, Tarrytown, NY 10591-5153. (914) 545-9200. FAX (914) 333-2444. 1969 to present. 54 issues per year. $2121.00. per year. ISSN: 0030-4018.

Optics Letters. Optical Society of America, Inc., 2010 Massachusetts Avenue, NW, Washington, DC 20036-1023. (202) 223-8130. 1977 to present. Semi-Monthly. $625.00 per year. ISSN: 0146-9592.

SPIE - International Society For Optical Engineering. PROCEEDINGS. SPIE - International Society for Optical Engineering, Box 10, 1000 20th Street, Bellingham, WA 98227-0010. (206) 676-3290. FAX (206) 647-1445. 1963 to present. Approximately 50 numbers per year. Individually priced.

RESEARCH CENTERS AND INSTITUTES

Center for Applied Optics. University of Alabama in Huntsville, Research Institute Building, Huntsville, AL 25899. (205) 895-6102.

Center For Imaging Science. Rochester Institute of Technology, One Lomb Memorial Drive, Rochester, NY 14623. (716) 475-2774.

Center For Laser Science and Engineering. University of Iowa, Iowa City, IA 52242. (319) 335-4580. FAX (319) 335-2951.

Center For Research in Electro-optics and Lasers.University of Central Florida, 12424 Research Parkway, Orlando, FL 32826. (407) 658-6800. FAX (407) 658-6880.

Fiber and Electro-optics Research Center. Virginia Polytechnic Institute and State University, Department of Electrical Engineering, 648 Whittemore, Blacksburg, VA 24061. (703) 231-7203.

HOMING

See: NAVIGATION

HOVERCRAFT

See: AERONAUTICS

HULLS

See: NAVAL ARCHITECTURE

HUMAN ENGINEERING

See: ERGONOMICS

HURRICANES

See also: CYCLONES, JET STREAM, METEOROLOGY, RAIN, TORNADOS, WEATHER FORECASTING

ABSTRACT SERVICES AND INDEXES

Applied Science and Technology Index; A cumulative subject index to English language periodicals in the fields of aeronautics and space science, computer technology, chemistry, construction industry, energy and related areas. H.W. Wilson Co., 950 University Avenue, Bronx, NY 10452. -9978. (212) 588-8400. (800) 367-6770. FAX (718) 590-1617. From 1958 to present. Monthly. Inquire about cost and availability. Also available online (BRS and WILSONLINE) and on CD-ROM. ISSN: 0003-6986.

Deep-Sea Research. Part A: Oceanographic Research Papers. Deep-Sea Research. Part B: Oceanographic Literature Review. Pergamon Press, Inc., Maxwell House, Fairview Park, Elmsford, NY 10523. (914) 592-7700. Fax (914) 592-3625. 1953 to present. Twelve times per year. $2000.00 per year for Parts A and B. Oceanographic Literature Review also available on CD-ROM. Inquire about price and availability.

General Science Index. H.W. Wilson Co., 950 University Avenue, Bronx, NY 10452. (212) 588-8400. (800) 367-6770. FAX (718) 590-1617. From 1978 to present. Ten issues per year; quarterly and annual cumulations. Service basis. Available on CD-ROM and online. Inquire regarding cost and availability. ISSN: 0162-1963.

Government Reports Announcements and Index. National Technical Information Service (NTIS), 5285 Port Royal Road, Springfield, VA 22161. (703) 487-4650. FAX (703) 321-8547. From 1968 to present. Annual. $630.00 per year. Also available online as *NTIS Bibliographic Database* and on CD-ROM. ISSN:

Meteorological and Geoastrophysical Abstracts. American Meteorological Society, c/o Inforonics, Inc., 550 Newtown Road, Box 458, Littleton, MA 01460. (508) 486-8976. FAX (508) 486-0027. Covers literature in environmental sciences, meteorology, astrophysics, hydrology, glaciology, and physical oceanography. 1950 to present. Monthly. $950.00 per year. ISSN: 0026-1130. Also available online (DIALOG) and on CD-ROM.

Oceanic Abstracts. Cambridge Scientific Abstracts, 7200 Wisconsin Avenue, Bethesda, MD 20814. (301) 961-6750. Fax (301) 961-6720. Bimonthly. $995.00 per year.

Science Citation Index. SCI. Institute for Scientific Information, 3501 Market Street, Philadelphia, PA 19104. (215) 386-0100. (800) 523-1850. FAX (215) 386-2991. 1961 to present. Six issues per year, plus annual cumulation. $11650.00 per year. Also

available online and on CD-ROM. Inquire about price and availability. ISSN: 0036-827X.

Selected Water Resources Abstracts. U.S. Geological Survey. Distributed by National Technical Information Service (NTIS), 5285 Port Royal Road, Springfield, VA 22161. (703) 487-4650. Monthly. $115.00 per year.

ANNUAL REVIEWS AND YEARBOOKS

American Meteorological Society. Meteorological Monographs. American Meteorological Society, 45 Beacon Street, Boston, MA 02108-3693. (617) 227-2425. FAX (617) 742-8718. Irregular. Price varies.

Annual Review of Earth and Planetary Sciences. Annual Reviews, Inc., 4139 El Camino Way, Palo Alto, CA 94306-0897. (415) 493-4400. (800) 523-8635. Fax (415) 855-9815. Annual. $55.00 per year.

Developments in Atmospheric Science Series. Elsevier Science Publishing Company, Inc., 655 Avenue of the Americas, New York, NY 10010. (212) 989-5800. Irregular. Inquire about price and availability.

Ocean Yearbook. Elisabeth M. Borgese, et al, editors. University of Chicago Press, 5801 Ellis Avenue, Chicago, IL 60637. (312) 702-7700. 1995. $77.00.

ASSOCIATIONS AND PROFESSIONAL SOCIETIES

American Association of State Climatologists. c/o Myron Moinau, University of Idaho, Agriculture Engineering Department, State Climate Service, Moscow, ID 834844-2040.

American Meteorological Society, 45 Beacon Street, Boston, MA 02108-3693. (617) 227-2425. FAX (617) 742-8718.

Association of American Weather Observers, 401 Whitney Boulevard, Box 455, Belvedere, IL 61008. (815) 544-5665. FAX (815) 544-6334.

International Association of Meteorology and Atmospheric Physics. UCAR, P.O. Box 3000, Boulder CO 80307-3000.

(303) 497-1673.

International Association of Severe Weather Specialists. c/o Warren Faidley, PO Box 31808, Tuscon, AZ 85751. (520) 751-9964. FAX (520) 751-1185.

National Environmental Satellite Data, and Information Service. 2069 Federal Building 4 Room 2060, Washington, DC 20233. (301) 763-71990. FAX (301) 763-4011.

National Weather Association, 6704 Wolke Court, Montgomery, AL 36116-2134. (334) 213-0388. FAX (334) 213-0388.

University Corporation For Atmospheric Research. PO Box 3000, 1850 Table Mesa Drive, Boulder, CO 80307-3000. (303) 497-1673. FAX (303) 497-1654.

Weather Modification Association. PO Box 26926, Fresno, CA 93729-6926. (209) 434-3486.

BIBLIOGRAPHIES

Information Sources in the Earth Sciences. David N. Wood, Joan E. Hardy, and Anthony P. Harvey, editors. Bowker-Saur/K.G. Saur. Distributed by R.R. Bowker, 121 Chanlon Road, New Providence, NJ 07974. (800) 521-8110 or (908) 464-6800. Second edition. 1989. $85.00.

Scientific and Technical Books and Serials in Print; An Index To Literature in Science and Technology. R.R. Bowker Co., 205 E. 42nd Street, New York, NY 10017. (800) 521-8110 or (212) 916-1600. 1992. $250.00.

DIRECTORIES AND BIOGRAPHICAL SOURCES

American Men and Women of Science: Physical and Biological Sciences. Eighteenth edition. R.R. Bowker Company, 245 West 17th Street, New York, NY 10011. (800) 521-8810 or (212) 916-1600. $750.00.

American Meteorological Society. PRofESSIONAL DIRECTORY. American Meteorological Society. 45 Beacon Street, Boston, MA 02108. (617) 227-2425. Included in each issue of the Bulletin of the Society.

Meteorological Services of the World. American Meteorological Society, 45 Beacon Street, Boston, MA 02108- 3693. (617) 227-2425. FAX (617) 742-8718. Annual. $70.00.

National Weather Service offices and Stations. National Oceanic and Atmospheric Administration, Deparatment of Commerce, Silver Spring, MD 20910. (301) 427-7698. Annual. Free

Research Centers Directory. Gale Research, 835 Penobscot Building, Detroit, MI 48226-4094. (313) 961-2242. (800) 347-4253. 20th edition. 1995. $485.00. ISSN: 0080-1518.

ENCYCLOPEDIAS AND DICTIONARIES

Climates of the States. Gale Research, 835 Penobscot Building, Detroit, MI 48226-4094. (313) 961-2242. (800) 347-4253. 5th edition. 1995. $255.00.

Concise Oxford Dictionary of the Earth Sciences. Ailsa Allaby and Michael Allaby, editors. Oxford University Press, Inc., 200 Madison Avenue, New York, NY 10016. (800) 334-4249 or (212) 679-7300. 1990. $42.95.

Encyclopedia of Climatology. Rhodes W. Fairbridge, and John E. Oliver. Chapman & Hall, 1 Penn Plaza, New York, NY 10119. (212) 564-1060. 1986. $115.00.

McGraw-Hill Encyclopedia of Science and Technology. McGraw-Hill Publishing Company, 1221 Avenue of the Americas, New York, NY 10020. (800) 262-4729 or (212) 512-3825. Seventh edition. 1992. $1900.00.

Smithsonian Meteorological Tables. Books on Demand, 300 North Zeeb Road, Ann Arbor, MI 48106-1346. (313) 761-4700. (800) 521-0600. Reprint edition. $153.90.

Weather Satellite Handbook. American Radio Relay League, 225 Main Street, Newington, CT 06111. (203) 666-1541 1994. $20.00.

GENERAL WORKS

Dark Wind: A True Account of Hurricane Gloria's Assault on Fire Island. John Jiler. St. Martin's Press, Inc., 175 Fifth Avenue, New York, NY 10010. (212) 674-5151. (800) 221-7945. 1993. $21.95.

Florida Hurricanes and Tropical Storms: 1871-1993: An Historical Survey. Fred Doehring, et al editors. Sea Grant Publications, Narragansett Bay Campus, Narragansett, RI 02882- 1197. (401) 792-6842. 1994. $5.00.

Early American Hurricanes, 1492-1870. American Meteorological Society, 45 Beacon Street, Boston, MA 02108- 3693. (617) 227-2425. FAX (617) 742-8718. 1990. $20.00.

The Hurricane. Roger A. Pieke. Routledge Chapman & Hall, Inc., 29 West 35th Street, New York, NY 10001-3291. (212) 244-3336. 1990. $59.95

Hurricane andrew, the Public Schools and the Rebuilding of Community. Eugene F. Prevenzo Jr., and Sandra H. Fradd. State University of New York Press, State University Plaza, Albany, NY 12246-0001. (518) 472-5000. (800) 666-2211.1995. $14.95 in paper.

Hurricane Hugo One Year Later. American Society of Civil Engineers. 345 East 47th Street, New York, NY 10017. (212) 705-7496. (800) 548-2723. 1991. $33.00.

Hurricanes. Jerome Gold, et al. Black Heron Press, P. O. Box 95676, Seattle, WA 98145. (206) 363-5210. 1994. $18.50.

Hurricanes, Storms and Tornados: Geographic Characteristics and Geologic Activity. D. Nalivkin. Ashgate Publishing Co., Old Post Road, Brookfield, VT 05036. (802) 276-3162., 1983. $130.00

1001 Questions Answered About Hurricanes, Tornadoes and Other Natural Air Disasters. Barbara Tuffy. Dover Publications, Inc., 180 Varick Street, New York, NY 10014. (212) 255-3755. (800) 223-3130. 1987. $7.95.

HANDBOOKS AND MANUALS

Basic Meteorology Lab Manual. Thomas A. Leavy. Allegheny Press, P.O. Box 220, Elgin, PA 19413. (814) 664-8504. Second edition. 1969. $7.95.

Handbook in Applied Meteorology. David D. Houghton, editor. John Wiley and Sons, Inc., 605 Third Avenue, New York, NY 10158. (800) 526-5368 or (212) 850-6000. 1985. $168.00.

The Hurricane Handbook: A Practical Guide For Residents of the Hurricane Belt. Sharon M. Carpenter and Toni G. Carpenter. Tailored Tours Publications, Inc., Box 22861, Lake Buena Vista, FL 32830. (407) 345-9126. 1993. $9.95.

Hurricane Survival Guide: How To Prepare Your Family and Home For the Next Hurricane. Sun, Sentinal Staff. Contemporary Books, Inc., 2 Prudential Plaza, Suite 1200, Chicago, IL 60601. (312) 540-4500. (800) 621-1918. . 1993. $9.95.

HURRICANES

Ency. of Physical Sciences and Engineering Info. Sources

ONLINE DATABASES AND CD-ROMS

Climate Assessment Database. National Weather Service, National Meteorological Center, 5200 Auth Road, Suite 101, Camp Springs, MD 20233. (301) 763-8016. Contains daily, weekly and monthly summaries of North American and world climatological data. Also provides five to ten weather forecasts, and 30 to 90 day weather outlook. Subscription required. Inquire as to cost and availability.

Current Contents Search. Institute for Scientific Information, 3501 Market Street, Philadelphia, PA 19104. (215) 386-0100. FAX (215) 386-6362. Contains citations to articles listed in the table of contents of science and technology journals. Also articles in social sciences and life sciences journals. Available on BRS,(800) 289-4277, DIALOG,(800) 334-2564, online services. Inquire as to cost and availability.

Dissertation Abstracts. University Microfilms International, 300 North Zeeb Road, Ann Arbor, MI 48106. (800) 521-0600 or (313) 761-4700. Scope includes virtually all doctoral dissertations accepted at accredited American institutions from 1861 to present in 252 subject areas. Available on BRS,(800) 289-4277, DIALOG,(800) 334-2564, and OCLC EPIC,(800) 848-5878, online services. Also available on CD-ROM. Inquire as to cost and availability.

Meteorological and Geoastrophysical Abstracts. American Meteorological Society, 45 Beacon Street, Boston, MA 02108-3693. (617) 227-2425. FAX (617) 742-8718. Contains citations and abstracts to the worldwide literature on significant research in meteorology and geoastrophysics. Related topics include physical oceanography, hydrology, environmental sciences and glaciology. Covers the period 1972 to present. Available on DIALOG,(800) 334-2564, online service. Inquire as to cost and availability.

NTIS Bibliographic Database. National Technical Information Service, 5285 Port Royal Road, Springfield, VA 22161. (703) 487-4929 or FAX (703) 321-8199. Broad coverage of government-sponsored science and technology research reports, 1964 to present. Available on BRS,(800) 289-4277, DIALOG, (800) 334-2564, ORBIT, (800) 456-7248, and STN International, FIZ Karlsruhe, P.O. Box 2465, W-7500, Karlsruhe 1, Germany, online services. Also available on CD-ROM. Inquire as to cost and availability.

SCISEARCH. Institute for Scientific Information, 3501 Market Street, Philadelphia, PA 19104. (800) 523-1850 or (215) 386-0100. Broad multidisciplinary title and author index to the international literature of science and technology, 1974 to present. Available on DIALOG and ORBIT online services. Also available on CD-ROM. Inquire as to cost and availability.

WILSONLINE. H.W. Wilson Company, 950 University Avenue, Bronx, NY 10452. (800) 367-6770 or (212) 588-8400. Makes available online versions of the printed H.W. Wilson indexes including Applied Science and Technology Index, Business Periodicals Index, General Science Index, and Readers' Guide to Periodical Literature. Period covered is generally 1983 to present. Available on BRS,(800) 289-4277, DIALOG,(800) 334-2564, and OCLC EPIC,(800) 848-5878, online services. Also available on CD-ROM. Inquire as to cost and availability.

World Climate Disc. Chadwyck-Healey Inc., 1101 King Street, Alexandria, VA 22314. (703) 683-4890. FAX (703) 683-7589. Weather and climate data from approximately 5,000 weather stations worldwide, covering the years 1854 to 1990 on CD-ROM. First edition 1992. Approximately $1200.00 per year with annual updates.

World Weatherdisc. WeatherDisc Associates, Inc., 4584 N.E. 89th Street, Seattle, WA 98115. (206) 524-4314. FAX (206) 543-0308. Meteorological data on CD-ROM which describes the climate of the earth today and for the past few hundred years. First edition 1989. Approximately $295.00 per year with biannual updates.

PERIODICALS

Agricultural and Forest Meteorology. Elsevier Science Publishing Company, Inc., 655 Avenue of the Americas, New York, NY 10010. (212) 989-5800. Twenty times per year. $750.00 per year.

American Meteorological Society. BULLETIN. American Meteorological Society, 45 Beacon Street, Boston, MA 02108-3693. (617) 227-2425. FAX (617) 742-8718. Monthly. $60.00 per year.

American Weather Observer. Association of American Weather Observers, 401 Whitney Boulevard, Box 455, Belvedere, IL 61008. (815) 544-5665. FAX (815) 544-6334. Monthly. $21.00 per year.

Atmosphere - Ocean. Canadian Meteorological and Oceanographic Society, P.O. Box 334, Newmarket, Ontario, L3Y 4X7, Canada. (416) 898-1040. FAX (416) 898-7937. Quarterly. $30.00 per year.

Boundary-layer Meteorology: An International Journal of Physical and Biological Processes in the Atmospheric Boundary Layer. Kluwer Academic Publishers, P.O. Box 358, Accord Station, Hingham, MA 02018-0358. (617) 871-6000. Sixteen per year. $785.00 per year.

Dynamics of Atmospheres and Oceans. Elsevier Science Publishing Company, Inc., 655 Avenue of the Americas, New York, NY 10010. (212) 989-5800. Six times per year. $205.00 per year.

JGR: Journal of Geophysical Research: Atmosphere. American Geophysical Union, 2000 Florida Avenue, N.W., Washington, DC 20009. (202) 462-6903. Monthly. $90.00 per year to members.

Journal of Applied Meteorology. American Meteorological Society, 45 Beacon Street, Boston, MA 02108-3693. (617) 227-2425. FAX (617) 742-8718. Monthly. $165.00 per year.

Journal of the Atmospheric Sciences. American Meteorological Society, 45 Beacon Street, Boston, MA 02108-3693. (617) 227-2425. FAX (617) 742-8718. Semi-monthly. $320.00 per year.

Monthly Weather Review. American Meteorological Society, 45 Beacon Street, Boston, MA 02108-3693. (617) 227-2425. FAX (617) 742-8718. Monthly. $205.00 per year.

National Weather Digest. National Weather Association, 4400 Stamp Road, Room 404, Temple Hills, MD 20748. (301) 899-3784.

Weather. Royal Meteorological Society, 104 Oxford Road, Reading, Berks RG1 7LJ, England. Monthly. $44.00 per year.

Weatherwise. Heldref Publications, 1319 Eighteenth Street, N.W., Washington, DC 20036-1802. (202) 296-6267. FAX (202) 296-5149. Bi-monthly. $28.00 per year.

RESEARCH CENTERS AND INSTITUTES

Coastal and Oceanographic Engineering Laboratory. University of Florida, 336 Weil Hall, Gainsville, FL 32607. (904) 392-1436.

Cooperative Institute For Mesoscale Meteorological Studies. University of Oklahoma, 401 East Boyd, Norman, OK 73019. (405) 325-3041.

Florida State Climate Center. 431 Love Building, Florida State University, Tallahassee, FL 32306. (904) 644-3417.

Hurricane Research Division. Atlantic Oceanographic and Meteorological Laboratory. 4301 Rickenbacker Causeway, Miami, FL 33149. (305) 361-4400.

National Center For Atmospheric Research. Box 3000, Boulder, CO 80307. (303) 497-1000. FAX (303) 497-1654.

National Hurricane Center. 1320 South Dixie Highway, Coral Gables, FL 33146. (305) 666-4612.

National Severe Storms Laboratory. 1313 Halley Circle, Norman, OK 73069. (406) 360-3620. FAX (405) 360-0472.

Program in Atmospheric and Oceanic Sciences. Princeton University. P.O. CN 710, Sayre Hall, Forrestal Campus, Princeton, NJ 08544-0719. (609) 258-6571. FAX (609) 258- 2850.

Remote Sensing Laboratory. University of Miami, P.O. Box 248003, Coral Gables, FL 33124. (305) 284-3881. FAX (305) 284-4792.

Weather Research Center. 3710 Mt. Vernon, Houston, TX 77006. (713) 529-3076.

HYDRAULIC ENGINEERING

See also: FLUID DYNAMICS, HYDRAULICS, HYDROLOGY, HYDRODYNAMICS

ABSTRACT SERVICES AND INDEXES

Applied Mechanics Reviews: An Assessment of World Literature in Engineering Sciences. 1948-present. American Society of Mechanical Engineers, 345 East 47th Street, New York, NY 10017. (212) 705-7703. Monthly. $360.00 per year.

Applied Science and Technology Index; A Cumulative Subject Index To English Language Periodicals in the Fields of Aeronautics and Space Science, Computer Technology, Chemistry, Construction Industry, Energy and Related Areas. H.W. Wilson Co., 950 University Avenue, Bronx, NY 10452. (800) 367-6770 or (212) 588-8400. FAX (718) 590-1617. From 1958 to present. Monthly. Inquire about cost and availability. Also available on CD-ROM and online.

ASCE Combined Annual Combined Index. American Society of Civil Engineers, 345 East 47th Street, New York, NY 10017-2398. (212) 705-7520 or (800) 548-2723. Annual. $48.00.

ASCE Publications Information. 1966-present. American Society of Civil Engineers, 345 East 47th Street, New York, NY 10017-2398. (212) 705-7520 or (800) 548-2723. Bi-monthly. $160.00 per year for non-members.

Current Contents: Engineering, Technology, and Applied Sciences. Institute for Scientific Information, 3501 Market Street, Philadelphia, PA 19104. (215) 386-0100. FAX (215) 386-6362. 1970 to present. Weekly. $442.00 per year.

Engineering Index Monthly. Engineering Information, Inc., Castle Point on the Hudson, Hoboken, NJ 07030. (800) 221-1044. FAX (212) 832-1857. Monthly. $2300.00 per year. Also available online as COMPENDEX, and also on CD-ROM. Covers chemical engineering, computers, electrical engineering, civil engineering, metals and mining, industrial management, and mechanical engineering.

Fluid Power Abstracts. STI Limited, 4 Kings Meadow, Ferry Hinksey Road, Oxford OX2 ODU, England. Distributed by Air Science Company, Box 143, Corning, NY 14830. (607) 962-5591. Covers oil hydraulic and pneumatic power transmission and control, engineering and design. 1964 to present. Bi-monthly. $260.00 per year.

Fluid Abstracts: Civil Engineering. Elsevier Science [journals], 660 White Plains Rd., Tarrytown, NY 10159-5153. (919) 524-9200. FAX (919) 333-2444. 1991 to present. Monthly plus annual cumulation. $645.00 per year. Also available online as FLUIDEX. Covers civil engineering applications of fluid mechanics, hydraulics, flow metering and measuring, offshore engineering, environmental hydraulics, and related aspects of wind energy, the atmosphere, and aerodynamics.

International Aerospace Abstracts. Technical Information Service, American Institute of Aeronautics and Astronautics, Inc., 555 West 57th St., New York, NY 10019. (212) 247-6500. FAX (212) 582-4861. Semi-monthly. $1295.00 per year.

Mechanical Engineering Abstracts (formerly *ISMEC*), Cambridge Scientific Abstracts, 7200 Wisconsin Avenue, Bethesda, MD 20814-4823. (301) 961-6750. FAX (301) 961-6720. 1967 to present. Monthly. $895.00 per year. Summarizes world literature in mechanical engineering, production engineering, and engineering management. Also available online.

Science Citation Index. Institute for Scientific Information, 3501 Market Street, Philadelphia, PA 19104. (215) 386-0100. FAX (215) 386-6362. Inquire about availability and cost. Also available on CD-ROM.

STAR. Scientific and Technical Aerospace Reports. National Aeronautics and Space Administration. NASA Center for Aerospace Information, Box 8757, BWI Airport, Baltimore, MD 21240. (301) 621-0153. Monthly. Inquire about availability and cost. ALso available through the NASA online retrieval service (RECON), and through the Aerospace Database through DIALOG.

ANNUAL REVIEWS AND YEARBOOKS

Developments in Hydraulic Engineering. Elsevier Science Publishing Company, Inc., 655 Avenue of the Americas, New York,

NY 10010. (212) 989-5800. Inquire for latest volume price and availability.

ASSOCIATIONS AND PROFESSIONAL SOCIETIES

American Institute of Aeronautics and Astronautics. The Aerospace Center, 370 L'Enfant Promenade SW, Washington, DC 20024. (202) 646-7400.

American Society of Civil Engineers. 345 East 47th Street, New York, NY 10017-2398. (212) 705-7520 or (800) 548-2723.

American Society of Mechanical Engineers. 345 East 47th Street, New York, NY 10017-2398. (212) 705-7703.

Fluid Controls Institute. PO Box 9036, Morristown, NJ 07960. (201) 829-0990.

Fluid Power Consultants International. PO Box 106, 1000 Grandview Drive, Elm Grove, WI 53122-0106. (414) 782-0410.

Fluid Power Society. 2433 N. Mayfair Road, Suite 111, Milwaukee, WI 53226. (414) 257-0910.

Hydraulic Institute. 9 Sylvan Way, Parsippany, NJ 07054-3802. (201) 267-9700.

International Association For Hydraulic Research. Rotterdamsweg 185, PO Box 177, 2600 MH Delft, Netherlands. Telephone 31-15-569353. FAX 31-51-619674.

National Conference On Fluid Power. 3333 N. Mayfair Road, Suite 111, Milwaukee, WI 53226. (414) 778-3368.

National Fluid Power Association. 3333 N. Mayfair Road, Suite 111, Milwaukee, WI 53226. (414) 778-3344.

Society of Automotive Engineers. 400 Commonwealth Drive, Warrendale, PA 15096-0001. (412) 776-4841.

DIRECTORIES AND BIOGRAPHICAL SOURCES

American Astronautical Society Directory. American Astronautical Society, 6352 Rolling Mill Place, Suite 102, Springfield, VA 22152. Inquire for cost and availability.

Fluid Power Society Membership Directory. Fluid Power Society, 2433 N. Mayfair Road, Suite 111, Milwaukee, WI 53226. (414) 257-0910. Inquire for cost and availability.

National Fluid Power Association Membership Directory. National Fluid Power Association, 3333 N. Mayfair Road, Suite 111, Milwaukee, WI 53226. (414) 778-3344. Inquire for cost and availability.

Research Centers Directory. Gale Research, 835 Penobscot Building, Detroit, MI 48226-4094. (313) 961-2242. (800) 347-4253. FAX (313) 961-6083. $485.00.

Scientific and Technical Organizations and Agencies Directory. Gale Research, 835 Penobscot Building, Detroit, MI 48226-4094. (313) 961-2242. (800) 347-4253. FAX (313) 961-6083. 4th edition. 1996. $195.00.

Who's Who in Engineering. American Association of Engineering Societies, 1111 19th Street NW, Suite 608, Washington, DC 20036. (202) 296-2237 or (800) 658-8897. 8th edition. 1991. Inquire for price.

ENCYCLOPEDIAS AND DICTIONARIES

Encyclopedia of Fluid Mechanics. 10 volumes plus 3 supplements. Gulf Publishing Company, P.O. Box 2608, Houston, TX. (713) 529-4301 or (800) 231-6275. 1986-90 (vols. 1-10), 1993-94 (supplements). $195.00 each (vols. 1-10), $139.00 each (supplements).

Encyclopedia of Physical Science and Technology. Robert A. Meyers, ed. 18 volumes. Academic Press Inc., 6277 Sea Harbor Drive, Orlando, FL 32887. (800) 321-5068. 1992. $2100.00

McGraw-Hill Dictionary of Scientific and Technical Terms. Sybil P. Parker, ed. 5th edition. McGraw-Hill Publishing Company, 1221 Avenue of the Americas, New York, NY 10020. (800) 262-4729 or (212) 512-3825. 1993. $110.50.

McGraw-Hill Encyclopedia of Engineering. Sybil P. Parker, ed. 2nd edition. McGraw-Hill Publishing Company, 1221 Avenue of the Americas, New York, NY 10020. (800) 262-4729 or (212) 512-3825. 1993. $95.50.

McGraw-Hill Encyclopedia of Science and Technology. Sybil P. Parker, ed. 7th edition. 20 volumes. McGraw-Hill Publishing Company, 1221 Avenue of the Americas, New York, NY 10020. (800) 262-4729 or (212) 512-3825. 1992. $1900.00

Thesaurus of Scientific, Technical, and Engineering Terms. Hemisphere Publishing Corporation, 1900 Frost Road, Suite 101, Bristol, PA 19007-1598. (215) 785-5800 or (800) 821-8312. FAX (215) 785-5515. 1987. $173.00.

GENERAL WORKS

Fundamentals of Hydraulic Engineering. Alan L. Prasuhn. Saunders College Publishing, the Curtis Center, Independence Square W., Philadelphia, PA 19106. (215) 238-7800. 1987. $63.00.

Fundamentals of Hydraulic Engineering Systems. Ned H.C. Hwang. Prentice Hall, 113 Sylvan Avenue, Route 9W, Englewood Cliffs, NJ 07632. (201) 592-2000 or (800) 922-0579. 1987. $63.00.

Hydraulics and Hydraulic Research: A Historical Review. Gunther Garbrecht, editor. A.A. Balkema/ Ashgate Publishing Co., Old Post Rd., Brookfield, VT 05036. (802) 276-3162. FAX (802) 276-3837. 1987. $90.00.

Hydraulic Machines: Turbines and Pumps. Grigor Krivchenko. 2nd edition. Lewis Publishers, 121 S. Main St., PO Box 519, Chelsea, MI 48118. (313) 475-8619 or (800) 272-7737. 1994. $85.00.

Hydraulics in Civil and Environmental Engineering. A.J. Chadwick. 2nd edition. E & FN Spon/ Routledge, Chapman and Hall, Inc., 29 West 35th Street, New York, NY 10001-2291. (212) 244-3336 or FAX (212) 563-2269. 1993. Inquire for cost and availability.

Hydrology and Hydraulic Systems. Ram S. Gupta. Prentice Hall, 113 Sylvan Avenue, Route 9W, Englewood Cliffs, NJ 07632. (201) 592-2000 or (800) 922-0579. 1989. $61.00.

Industrial Hydraulic Systems. D.D. Banks and D.S. Banks. Prentice Hall, 113 Sylvan Avenue, Route 9W, Englewood Cliffs, NJ 07632. (201) 592-2000 or (800) 922-0579. 1988. $46.20.

Water Resources Engineering. Ray K. Linsley, et al. 4th edition. McGraw-Hill Publishing Company, 1221 Avenue of the Americas, New York, NY 10020. (800) 262-4729 or (212) 512-3825. 1992. Inquire for cost and availability.

HANDBOOKS AND MANUALS

Davis's Handbook of Applied Hydraulics. V.J. Zipparro, editor. McGraw-Hill Publishing Company, 1221 Avenue of the Americas, New York, NY 10020. (800) 262-4729 or (212) 512-3825. 1993. $89.50.

Fluid Power Design Handbook. Frank Yeaple. 2nd edition. Marcel Dekker, Inc., 270 Madison Avenue, New York, NY 10016. (212) 696-9000. FAX (212) 685-4540. 1990. Inquire for cost and availability.

Handbook of Hydraulic Engineering. Armando Lencastre. Halsted Press (Division of John Wiley and Sons, Inc.), 605 Third Avenue, New York, NY 10158. (800) 526-5368 or (212) 850-6418. 1987. Inquire for cost and availability.

Handbook of Hydraulics. E.F. Brate. 6th edition. McGraw-Hill Publishing Company, 1221 Avenue of the Americas, New York, NY 10020. (800) 262-4729 or (212) 512-3825. 1976. $95.50.

Handbook of Hydraulic Resistance. 3rd edition. CRC Press, 2000 Corporate Blvd., N.W., Boca Raton, FL 33431. (407) 994-0555 or (800) 272-7737. FAX (407) 994-0949. 1993. $129.95.

Hydraulics Field Manual. Robert O. Parmley. McGraw-Hill Publishing Company, 1221 Avenue of the Americas, New York, NY 10020. (800) 262-4729 or (212) 512-3825. 1992. $42.50.

Plant Engineering Magazine's Fluid Power Handbook. Anton H. Hehn. 2 volumes. Gulf Publishing Company, P.O. Box 2608, Houston, TX. (713) 529-4301 or (800) 231-6275. 1993. Inquire for cost and availability.

The Wiley Engineer's Desk Reference. Sanford I. Heisler. John Wiley and Sons, Inc., 605 Third Avenue, New York, NY 10158. (800) 526-5368 or (212) 850-6000. 1984. $64.95.

ONLINE DATABASES AND CD-ROMS

Aerospace Database. American Institute of Aeronautics and Astronautics, 370 L'Enfant Promenade, S.W., Washington, DC 20024. (202) 646-7400. Worldwide published literature on research and development in aerospace and related areas, with abstracts. Covers 1962 to present. Available on DIALOG, (800) 334-2564, online service. Also available on CD-ROM. Inquire as to cost and availability.

Compendex Plus. Engineering Information, Inc., 345 East 47th Street, New York, NY 10017. (212) 705-7600 or (800) 221-1044. Contains citations with abstracts to worldwide literature in engineering and technology, from 1970 to present. Available on online BRS,(800) 289-4277, DIALOG, (800) 334-2564,

ORBIT (800) 456-7248, and STN International, FIZ Karlsruhe, P.O. Box 2465, W-7500, Karlsruhe 1, Germany, online services. Also available on CD-ROM. Inquire as to cost and availability.

CSA Engineering. Cambridge Scientific Abstracts, 7200 Wisconsin Avenue, Suite 601, Bethesda, MD 20814. (301) 961-6750 or (800) 843-7751. Contains citations and abstracts of international periodicals and research literature covering all fields of engineering and science and technology,including computer and information science, electronics, mechanical engineering, solid state materials, 1981 to present. Available on BRS,(800) 289-4277, online service.Inquire as to cost and availability.

Current Contents Search. Institute for Scientific Information, 3501 Market Street, Philadelphia, PA 19104. (215) 386-0100. FAX (215) 386-6362. Contains citations to articles listed in the table of contents of science and technology journals. Also articles in social sciences and life sciences journals. Available on BRS,(800) 289-4277, DIALOG,(800) 334-2564, online services. Inquire as to cost and availability.

Dissertation Abstracts. University Microfilms International, 300 North Zeeb Road, Ann Arbor, MI 48106. (800) 521-0600 or (313) 761-4700. Scope includes virtually all doctoral dissertations accepted at accredited American institutions from 1861 to present in 252 subject areas. Available on BRS,(800) 289-4277, DIALOG,(800) 334-2564, and OCLC EPIC,(800) 848-5878, online services. Also available on CD-ROM. Inquire as to cost and availability.

GEOREF: Bibliography and Index of Geology. American Geological Institute, 4220 King Street, Alexandria, VA 22302. (703) 379-2480. Fax (703) 379-7563. Monthly. Inquire for price and availability.

International Aerospace Abstracts. American Institute of Aeronautics and Astronautics, 370 L'Enfant Promenade, S.W., Washington, DC 20024. (202) 646-7400. Contains references and abstracts of journal and monograph literature relating to aerospace science and technology, from 1963 to present. Available through the NASA/RECON system of the National Aeronautics and Space Administration only.

ISMEC: Mechanical Engineering Abstracts. Cambridge Scientific Abstracts, 7200 Wisconsin Avenue, Suite 601, Bethesda, MD 20814. (301) 961-6750 or (800) 843-7751. Contains citations to the literature in mechanical engineering, industrial and production engineering, energy, power, mechanics, devices and related areas, from 1973 to present. Available on the DIALOG,(800) 334-2564, online service. Inquire as to cost and availability.

NASA Database. American Institute of Aeronautics and Astronautics, 370 L'Enfant Promenade, S.W., Washington, DC 20024. (202) 646-7400. Citations and abstracts of aeronautics and astronautics literature, 1962 to present. Also contains citations from STAR, SCIENTIFIC and TECHNICAL AEROSPACE REPORTS, and INTERNATIONAL AEROSPACE ABSTRACTS. Available through NASA/RECON online service. Inquire as to cost and availability.

NTIS Bibliographic Database. National Technical Information Service, 5285 Port Royal Road, Springfield, VA 22161. (703) 487-4929 or FAX (703) 321-8199. Broad coverage of government-sponsored science and technology research reports, 1964 to present. Available on BRS,(800) 289-4277, DIALOG, (800) 334-2564, ORBIT, (800) 456-7248, and STN International, FIZ Karlsruhe, P.O. Box 2465, W-7500, Karlsruhe 1, Germany,

online services. Also available on CD-ROM. Inquire as to cost and availability.

SAE Global Mobility Database. Society of Automative Engineers (SAE), Electronic Publishing Division, 400 Commonwealth Drive, Warrendale, PA 15098. (412) 776-4841. Contains citations with abstracts to technical papers on automotive and aerospace technology and vehicular-related industries that have been presented at SAE conferences. Covers 1906 to present. Available on ORBIT,(800) 456-7248,online service. Inquire as to cost and availability.

SCISEARCH. Institute for Scientific Information, 3501 Market Street, Philadelphia, PA 19104. (800) 523-1850 or (215) 386-0100. Broad multidisciplinary title and author index to the international literature of science and technology, 1974 to present. Available on DIALOG,(800) 334-2564, and ORBIT,(800) 456-7248, online services. Also available on CD-ROM. Inquire as to cost and availability.

TRIS (Transportation Research Information). National Academy of Science, 2101 Constitution Avenue, N.W., Washington, DC 20418. (202) 334-3313 or (800) 624-6242. Citations with abstracts of literature on transportation, including air, highway, rail, maritime and other modes from 1968 to present. Available on DIALOG,(800) 334-2564, online service. Inquire as to cost and availability.

WILSONLINE. H.W. Wilson Company, 950 University Avenue, Bronx, NY 10452. (800) 367-6770 or (212) 588-8400. Makes available online versions of the printed H.W. Wilson indexes including Applied Science and Technology Index, Business Periodicals Index, General Science Index, and Readers' Guide to Periodical Literature. Period covered is generally 1983 to present. Available on BRS,(800) 289-4277, DIALOG,(800) 334-2564, and OCLC EPIC,(800) 848-5878, online services. Also available on CD-ROM. Inquire as to cost and availability.

PERIODICALS

Experiments in Fluids. Springer-Verlag, 44 Hartz Way, Secaucus, NJ 07096-2491. 1983 to present. 12 times a year. $675.00 per year.

Fluid Mechanics Research. J. Wiley & Sons, Journals, 605 Third Avenue, New York, NY 10158-0012. (212) 692-6000. 1972 to present. Bi-monthly. $896.00 per year.

Hydraulics and Pneumatics. Penton Publishing, 110 Superior Ave., Cleveland, OH 44114-2543. 1947 to present. Monthly. $45.00 per year.

Journal of Fluid Control. Delbridge Publishing Company, Box 160817, Cupertino, CA 95016. (408) 446-3131. FAX (408) 446-3131. 1967 to present. Quarterly. $145.00 per year.

Journal of Fluid Mechanics. Cambridge University Press, 40 West 20th Street, New York, NY 10011-4211. (212) 924-3900. 1956 to present. 24 times a year. $595.00 per year.

Journal of Fluids Engineering. American Society of Mechanical Engineers, 345 East 47th Street, New York, NY 10017-2398. (212) 705-7703. 1919 to present. Quarterly. $100.00 per year.

Journal of Hydraulic Engineering. American Society of Civil Engineers, Hydraulics Division, 345 East 47th Street, New York,

NY 10017-2398. (212) 705-7520 or (800) 548-2723. 1956 to present. Monthly. $200.00 per year.

Journal of Hydraulic Research. International Association for Hydraulic Research, Rotterdamsweg 185, PO Box 177, 2600 MH Delft, Netherlands. Telephone 31-15-569353. FAX 31-51-619674. 1963 to present. 6 times a year. Inquire for cost.

RESEARCH CENTERS AND INSTITUTES

California Institute of Technology, W.M. Keck Engineering Laboratory of Hydraulics and Water Resources. 138-78, Pasadena, CA 91125. (818) 356-4404. FAX (818) 356-2940.

Colorado State University Hydraulics and Hydromachinery Research Laboratories. Engineering Research Center, Fort Collins, CO 805232. (303) 491-8404. FAX (303) 491-8671.

University of Illinois Hydrosystems Laboratory. Department of Civil Engineering, Urbana, IL 61801. (217) 333-0107.

University of Iowa Institute of Hydraulic Research. Iowa City, IA 52242. (319) 335-5236. (319) 335-5238.

University of Minnesota St. Anthony Falls Hydraulic Laboratory. Mississippi River at 3rd Avenue SE, Minneapolis, MN 55414-2196. (612) 627-4010. FAX (612) 627-4609.

HYDRAULICS

See also: FLUID DYNAMICS, HYDRAULIC ENGINEERING, HYDRO-DYNAMICS

ABSTRACT SERVICES AND INDEXES

Applied Mechanics Reviews: An Assessment of World Literature in Engineering Sciences. 1948-present. American Society of Mechanical Engineers, 345 East 47th Street, New York, NY 10017. (212) 705-7703. Monthly. $360.00 per year.

Applied Science and Technology Index; A Cumulative Subject Index To English Language Periodicals in the Fields of Aeronautics and Space Science, Computer Technology, Chemistry, Construction Industry, Energy and Related Areas. H.W. Wilson Co., 950 University Avenue, Bronx, NY 10452. (800) 367-6770 or (212) 588-8400. FAX (718) 590-1617. From 1958 to present. Monthly. Inquire about cost and availability. Also available on CD-ROM and online.

ASCE Combined Annual Combined Index. American Society of Civil Engineers, 345 East 47th Street, New York, NY 10017-2398. (212) 705-7520 or (800) 548-2723. Annual. $48.00.

ASCE Publications Information. 1966-present. American Society of Civil Engineers, 345 East 47th Street, New York, NY 10017-2398. (212) 705-7520 or (800) 548-2723. Bi-monthly. $160.00 per year for non-members.

Current Contents: Engineering, Technology, and Applied Sciences. Institute for Scientific Information, 3501 Market Street, Philadelphia, PA 19104. (215) 386-0100. FAX (215) 386-6362. 1970 to present. Weekly. $442.00 per year.

Engineering Index Monthly. Engineering Information, Inc., Castle Point on the Hudson, Hoboken, NJ 07030. (800) 221-1044. FAX (212) 832-1857. Monthly. $2300.00 per year. Also available online as COMPENDEX, and also on CD-ROM. Covers chemical engineering, computers, electrical engineering, civil engineering, metals and mining, industrial management, and mechanical engineering.

Fluid Power Abstracts. STI Limited, 4 Kings Meadow, Ferry Hinksey Road, Oxford OX2 ODU, England. Distributed by Air Science Company, Box 143, Corning, NY 14830. (607) 962-5591. Covers oil hydraulic and pneumatic power transmission and control, engineering and design. 1964 to present. Bi-monthly. $260.00 per year.

Fluid Abstracts: Civil Engineering. Elsevier Science [journals], 660 White Plains Rd., Tarrytown, NY 10159-5153. (919) 524-9200. FAX (919) 333-2444. 1991 to present. Monthly plus annual cumulation. $645.00 per year. Also available online as FLUIDEX. Covers civil engineering applications of fluid mechanics, hydraulics, flow metering and measuring, offshore engineering, environmental hydraulics, and related aspects of wind energy, the atmosphere, and aerodynamics.

International Aerospace Abstracts. Technical Information Service, American Institute of Aeronautics and Astronautics, Inc., 555 West 57th St., New York, NY 10019. (212) 247-6500. FAX (212) 582-4861. Semi-monthly. $1295.00 per year.

Mechanical Engineering Abstracts (formerly *ISMEC*), Cambridge Scientific Abstracts, 7200 Wisconsin Avenue, Bethesda, MD 20814-4823. (301) 961-6750. FAX (301) 961-6720. 1967 to present. Monthly. $895.00 per year. Summarizes world literature in mechanical engineering, production engineering, and engineering management. Also available online.

Science Citation Index. Institute for Scientific Information, 3501 Market Street, Philadelphia, PA 19104. (215) 386-0100. FAX (215) 386-6362. Inquire about availability and cost. Also available on CD-ROM.

STAR. Scientific and Technical Aerospace Reports. National Aeronautics and Space Administration. NASA Center for Aerospace Information, Box 8757, BWI Airport, Baltimore, MD 21240. (301) 621-0153. Monthly. Inquire about availability and cost. ALso available through the NASA online retrieval service (RECON), and through the Aerospace Database through DIALOG.

ANNUAL REVIEWS AND YEARBOOKS

Developments in Hydraulic Engineering. Elsevier Science Publishing Company, Inc., 655 Avenue of the Americas, New York, NY 10010. (212) 989-5800. Inquire for latest volume price and availability.

ASSOCIATIONS AND PROFESSIONAL SOCIETIES

American Institute of Aeronautics and Astronautics. The Aerospace Center, 370 L'Enfant Promenade SW, Washington, DC 20024. (202) 646-7400.

American Society of Civil Engineers. 345 East 47th Street, New York, NY 10017-2398. (212) 705-7520 or (800) 548-2723.

American Society of Mechanical Engineers. 345 East 47th Street, New York, NY 10017-2398. (212) 705-7703.

Fluid Controls Institute. PO Box 9036, Morristown, NJ 07960. (201) 829-0990.

Fluid Power Consultants International. PO Box 106, 1000 Grandview Drive, Elm Grove, WI 53122-0106. (414) 782-0410.

Fluid Power Society. 2433 N. Mayfair Road, Suite 111, Milwaukee, WI 53226. (414) 257-0910.

Hydraulic Institute. 9 Sylvan Way, Parsippany, NJ 07054-3802. (201) 267-9700.

International Association For Hydraulic Research. Postbus 177, NL-2600 MH Delft, Netherlands. Telephone 31-15-569353. FAX 31-51-619674.

National Conference On Fluid Power. 3333 N. Mayfair Road, Suite 111, Milwaukee, WI 53226. (414) 778-3368.

National Fluid Power Association. 3333 N. Mayfair Road, Suite 111, Milwaukee, WI 53226. (414) 778-3344.

Society of Automotive Engineers. 400 Commonwealth Drive, Warrendale, PA 15096-0001. (412) 776-4841.

DIRECTORIES AND BIOGRAPHICAL SOURCES

American Astronautical Society Directory. American Astronautical Society, 6352 Rolling Mill Place, Suite 102, Springfield, VA 22152. Inquire for cost and availability.

Fluid Power Society Membership Directory. Fluid Power Society, 2433 N. Mayfair Road, Suite 111, Milwaukee, WI 53226. (414) 257-0910. Inquire for cost and availability.

National Fluid Power Association Membership Directory. National Fluid Power Association, 3333 N. Mayfair Road, Suite 111, Milwaukee, WI 53226. (414) 778-3344. Inquire for cost and availability.

Research Centers Directory. Gale Research, 835 Penobscot Building, Detroit, MI 48226-4094. (313) 961-2242. (800) 347-4253. FAX (313) 961-6083. $485.00.

Scientific and Technical Organizations and Agencies Directory. Gale Research, 835 Penobscot Building, Detroit, MI 48226-4094. (313) 961-2242. (800) 347-4253. FAX (313) 961-6083. 4th edition. 1996. $195.00.

Who's Who in Engineering. American Association of Engineering Societies, 1111 19th Street NW, Suite 608, Washington, DC 20036. (202) 296-2237 or (800) 658-8897. 8th edition. 1991. Inquire for price.

ENCYCLOPEDIAS AND DICTIONARIES

Encyclopedia of Fluid Mechanics. 10 volumes plus 3 supplements. Gulf Publishing Company, P.O. Box 2608, Houston, TX. (713) 529-4301 or (800) 231-6275. 1986-90 (vols. 1-10), 1993-94 (supplements). $195.00 each (vols. 1-10), $139.00 each (supplements).

Encyclopedia of Physical Science and Technology. Robert A. Meyers, ed. 18 volumes. Academic Press Inc., 6277 Sea Harbor Drive, Orlando, FL 32887. (800) 321-5068. 1992. $2100.00

McGraw-Hill Dictionary of Scientific and Technical Terms. Sybil P. Parker, ed. 5th edition. McGraw-Hill Publishing Company, 1221 Avenue of the Americas, New York, NY 10020. (800) 262-4729 or (212) 512-3825. 1993. $110.50.

McGraw-Hill Encyclopedia of Engineering. Sybil P. Parker, ed. 2nd edition. McGraw-Hill Publishing Company, 1221 Avenue of the Americas, New York, NY 10020. (800) 262-4729 or (212) 512-3825. 1993. $95.50.

McGraw-Hill Encyclopedia of Science and Technology. Sybil P. Parker, ed. 7th edition. 20 volumes. McGraw-Hill Publishing Company, 1221 Avenue of the Americas, New York, NY 10020. (800) 262-4729 or (212) 512-3825. 1992. $1900.00

Thesaurus of Scientific, Technical, and Engineering Terms. Hemisphere Publishing Corporation, 1900 Frost Road, Suite 101, Bristol, PA 19007-1598. (215) 785-5800 or (800) 821-8312. FAX (215) 785-5515. 1987. $173.00.

GENERAL WORKS

Fluid Power Technology. F. Don Norvelle. West Publishing Company, 610 Opperman Drive, PO Box 64526, St. Paul, MN 55164-0526. (800) 328-9424. 1995. Inquire for cost and availability.

Hydraulics and Hydraulic Research: A Historical Review. Gunther Garbrecht, editor. A.A. Balkema/ Ashgate Publishing Co., Old Post Rd., Brookfield, VT 05036. (802) 276-3162. FAX (802) 276-3837. 1987. $90.00.

Hydraulic Machines: Turbines and Pumps. Grigor Krivchenko. 2nd edition. Lewis Publishers, 121 S. Main St., PO Box 519, Chelsea, MI 48118. (313) 475-8619 or (800) 272-7737. 1994. $85.00.

Hydraulics in Civil and Environmental Engineering. A.J. Chadwick. 2nd edition. E & FN Spon/ Routledge, Chapman and Hall, Inc., 115 Fifth Avenue, New York, NY 10211-0906. 1994. $38.95.

Industrial Hydraulic Systems. D.D. Banks and D.S. Banks. Prentice Hall, 113 Sylvan Avenue, Route 9W, Englewood Cliffs, NJ 07632. (201) 592-2000 or (800) 922-0579. 1988. $46.20.

HANDBOOKS AND MANUALS

Davis's Handbook of Applied Hydraulics. V.J. Zipparro, editor. McGraw-Hill Publishing Company, 1221 Avenue of the Americas, New York, NY 10020. (800) 262-4729 or (212) 512-3825. 1993. $89.50.

Fluid Power Design Handbook. Frank Yeaple. 2nd edition. Marcel Dekker, Inc., 270 Madison Avenue, New York, NY 10016. (212) 696-9000. FAX (212) 685-4540. 1990. Inquire for cost and availability.

Handbook of Hydraulics. E.F. Brate. 6th edition. McGraw-Hill Publishing Company, 1221 Avenue of the Americas, New York, NY 10020. (800) 262-4729 or (212) 512-3825. 1976. $95.50.

Handbook of Hydraulic Resistance. 3rd edition. CRC Press, 2000 Corporate Blvd., N.W., Boca Raton, FL 33431. (407) 994-0555 or (800) 272-7737. FAX (407) 994-0949. 1993. $129.95.

Handbook of Hydraulic Resistance. I.E. Idelchik, editor. Third edition. Begell House Inc., 79 Madison Avenue, New York, NY 10016. (212) 725-1999. FAX (212) 213-8368. 1994. $138.95.

Hydraulics Field Manual. Robert O. Parmley. McGraw-Hill Publishing Company, 1221 Avenue of the Americas, New York, NY 10020. (800) 262-4729 or (212) 512-3825. 1992. $42.50.

Plant Engineering Magazine's Fluid Power Handbook. Anton H. Hehn. 2 volumes. Gulf Publishing Company, P.O. Box 2608, Houston, TX. (713) 529-4301 or (800) 231-6275. 1993. Inquire for cost and availability.

The Wiley Engineer's Desk Reference. Sanford I. Heisler. John Wiley and Sons, Inc., 605 Third Avenue, New York, NY 10158. (800) 526-5368 or (212) 850-6000. 1984. $64.95.

ONLINE DATABASES AND CD-ROMS

Aerospace Database. American Institute of Aeronautics and Astronautics, 370 L'Enfant Promenade, S.W., Washington, DC 20024. (202) 646-7400. Worldwide published literature on research and development in aerospace and related areas, with abstracts. Covers 1962 to present. Available on DIALOG, (800) 334-2564, online service. Also available on CD-ROM. Inquire as to cost and availability.

Compendex Plus. Engineering Information, Inc., 345 East 47th Street, New York, NY 10017. (212) 705-7600 or (800) 221-1044. Contains citations with abstracts to worldwide literature in engineering and technology, from 1970 to present. Available on online BRS,(800) 289-4277, DIALOG, (800) 334-2564, ORBIT (800) 456-7248, and STN International, FIZ Karlsruhe, P.O. Box 2465, W-7500, Karlsruhe 1, Germany, online services. Also available on CD-ROM. Inquire as to cost and availability.

CSA Engineering. Cambridge Scientific Abstracts, 7200 Wisconsin Avenue, Suite 601, Bethesda, MD 20814. (301) 961-6750 or (800) 843-7751. Contains citations and abstracts of international periodicals and research literature covering all fields of engineering and science and technology,including computer and information science, electronics, mechanical engineering, solid state materials, 1981 to present. Available on BRS,(800) 289-4277, online service.Inquire as to cost and availability.

Current Contents Search. Institute for Scientific Information, 3501 Market Street, Philadelphia, PA 19104. (215) 386-0100. FAX (215) 386-6362. Contains citations to articles listed in the table of contents of science and technology journals. Also articles in social sciences and life sciences journals. Available on BRS,(800) 289-4277, DIALOG,(800) 334-2564, online services. Inquire as to cost and availability.

Dissertation Abstracts. University Microfilms International, 300 North Zeeb Road, Ann Arbor, MI 48106. (800) 521-0600 or (313) 761-4700. Scope includes virtually all doctoral dissertations accepted at accredited American institutions from 1861 to present in 252 subject areas. Available on BRS,(800) 289-4277, DIALOG,(800) 334-2564, and OCLC EPIC,(800) 848-5878, online services. Also available on CD-ROM. Inquire as to cost and availability.

GEOREF: Bibliography and Index of Geology. American Geological Institute, 4220 King Street, Alexandria, VA 22302. (703) 379-2480. Fax (703) 379-7563. Monthly. Inquire for price and availability.

International Aerospace Abstracts. American Institute of Aeronautics and Astronautics, 370 L'Enfant Promenade, S.W., Washington, DC 20024. (202) 646-7400. Contains references and abstracts of journal and monograph literature relating to aerospace science and technology, from 1963 to present. Available through the NASA/RECON system of the National Aeronautics and Space Administration only.

ISMEC: Mechanical Engineering Abstracts. Cambridge Scientific Abstracts, 7200 Wisconsin Avenue, Suite 601, Bethesda, MD 20814. (301) 961-6750 or (800) 843-7751. Contains citations to the literature in mechanical engineering, industrial and production engineering, energy, power, mechanics, devices and related areas, from 1973 to present. Available on the DIALOG,(800) 334-2564, online service. Inquire as to cost and availability.

NASA Database. American Institute of Aeronautics and Astronautics, 370 L'Enfant Promenade, S.W., Washington, DC 20024. (202) 646-7400. Citations and abstracts of aeronautics and astronautics literature, 1962 to present. Also contains citations from STAR, SCIENTIFIC and TECHNICAL AEROSPACE REPORTS, and INTERNATIONAL AEROSPACE ABSTRACTS. Available through NASA/RECON online service. Inquire as to cost and availability.

NTIS Bibliographic Database. National Technical Information Service, 5285 Port Royal Road, Springfield, VA 22161. (703) 487-4929 or FAX (703) 321-8199. Broad coverage of government-sponsored science and technology research reports, 1964 to present. Available on BRS,(800) 289-4277, DIALOG, (800) 334-2564, ORBIT, (800) 456-7248, and STN International, FIZ Karlsruhe, P.O. Box 2465, W-7500, Karlsruhe 1, Germany, online services. Also available on CD-ROM. Inquire as to cost and availability.

SAE Global Mobility Database. Society of Automative Engineers (SAE), Electronic Publishing Division, 400 Commonwealth Drive, Warrendale, PA 15098. (412) 776-4841. Contains citations with abstracts to technical papers on automotive and aerospace technology and vehicular-related industries that have been presented at SAE conferences. Covers 1906 to present. Available on ORBIT,(800) 456-7248,online service. Inquire as to cost and availability.

SCISEARCH. Institute for Scientific Information, 3501 Market Street, Philadelphia, PA 19104. (800) 523-1850 or (215) 386-0100. Broad multidisciplinary title and author index to the international literature of science and technology, 1974 to present. Available on DIALOG,(800) 334-2564, and ORBIT,(800) 456-7248, online services. Also available on CD-ROM. Inquire as to cost and availability.

TRIS (Transportation Research Information). National Academy of Science, 2101 Constitution Avenue, N.W., Washington, DC 20418. (202) 334-3313 or (800) 624-6242. Citations with abstracts of literature on transportation, including air, highway, rail, maritime and other modes from 1968 to present. Available on DIALOG,(800) 334-2564, online service. Inquire as to cost and availability.

WILSONLINE. H.W. Wilson Company, 950 University Avenue, Bronx, NY 10452. (800) 367-6770 or (212) 588-8400. Makes available online versions of the printed H.W. Wilson indexes including Applied Science and Technology Index, Business Periodicals Index, General Science Index, and Readers' Guide to Periodical Literature. Period covered is generally 1983 to present. Available on BRS,(800) 289-4277, DIALOG,(800) 334-2564, and OCLC EPIC,(800) 848-5878, online services. Also available on CD-ROM. Inquire as to cost and availability.

PERIODICALS

Experiments in Fluids. Springer-Verlag, 44 Hartz Way, Secaucus, NJ 07096-2491. 1983 to present. 12 times a year. $675.00 per year.

Fluid Mechanics Research. J. Wiley & Sons, Journals, 605 Third Avenue, New York, NY 10158-0012. (212) 692-6000. 1972 to present. Bi-monthly. $896.00 per year.

Hydraulics and Pneumatics. Penton Publishing, 110 Superior Ave., Cleveland, OH 44114-2543. 1947 to present. Monthly. $45.00 per year.

Journal of Fluid Control. Delbridge Publishing Company, Box 160817, Cupertino, CA 95016. (408) 446-3131. FAX (408) 446-3131. 1967 to present. Quarterly. $145.00 per year.

Journal of Fluid Mechanics. Cambridge University Press, 40 West 20th Street, New York, NY 10011-4211. (212) 924-3900. 1956 to present. 24 times a year. $595.00 per year.

Journal of Fluids Engineering. American Society of Mechanical Engineers, 345 East 47th Street, New York, NY 10017-2398. (212) 705-7703. 1919 to present. Quarterly. $100.00 per year.

Journal of Hydraulic Engineering. American Society of Civil Engineers, Hydraulics Division, 345 East 47th Street, New York, NY 10017-2398. (212) 705-7520 or (800) 548-2723. 1956 to present. Monthly. $200.00 per year.

Journal of Hydraulic Research. International Association for Hydraulic Research, Rotterdamsweg 185, PO Box 177, 2600 MH Delft, Netherlands. Telephone 31-15-569353. FAX 31-51-619674. 1963 to present. 6 times a year. Inquire for cost.

RESEARCH CENTERS AND INSTITUTES

California Institute of Technology, W.M. Keck Engineering Laboratory of Hydraulics and Water Resources. 138-78, Pasadena, CA 91125. (818) 356-4404. FAX (818) 356-2940.

Colorado State University Hydraulics and Hydromachinery Research Laboratories. Engineering Research Center, Fort Collins, CO 805232. (303) 491-8404. FAX (303) 491-8671.

University of Illinois Hydrosystems Laboratory. Department of Civil Engineering, Urbana, IL 61801. (217) 333-0107.

University of Iowa Institute of Hydraulic Research. Iowa City, IA 52242. (319) 335-5236. (319) 335-5238.

University of Minnesota St. Anthony Falls Hydraulic Laboratory. Mississippi River at 3rd Avenue SE, Minneapolis, MN 55414-2196. (612) 627-4010. FAX (612) 627-4609.

HYDRODYNAMICS

Ency. of Physical Sciences and Engineering Info. Sources

HYDROCARBONS

See: PETROLEUM CHEMISTRY

HYDRODYNAMICS

See also: FLUID DYNAMICS, HYDRAULIC ENGINEERING, HYDRAULICS, HYDROLOGY, OCEAN ENGINEERING

ABSTRACT SERVICES AND INDEXES

Applied Mechanics Reviews: An Assessment of World Literature in Engineering Sciences. 1948-present. American Society of Mechanical Engineers, 345 East 47th Street, New York, NY 10017. (212) 705-7703. Monthly. $360.00 per year.

Applied Science and Technology Index; A Cumulative Subject Index To English Language Periodicals in the Fields of Aeronautics and Space Science, Computer Technology, Chemistry, Construction Industry, Energy and Related Areas. H.W. Wilson Co., 950 University Avenue, Bronx, NY 10452. (800) 367-6770 or (212) 588-8400. FAX (718) 590-1617. From 1958 to present. Monthly. Inquire about cost and availability. Also available on CD-ROM and online.

ASCE Combined Annual Combined Index. American Society of Civil Engineers, 345 East 47th Street, New York, NY 10017-2398. (212) 705-7520 or (800) 548-2723. Annual. $48.00.

ASCE Publications Information. 1966-present. American Society of Civil Engineers, 345 East 47th Street, New York, NY 10017-2398. (212) 705-7520 or (800) 548-2723. Bi-monthly. $160.00 per year for non-members.

Current Contents: Engineering, Technology, and Applied Sciences. Institute for Scientific Information, 3501 Market Street, Philadelphia, PA 19104. (215) 386-0100. FAX (215) 386-6362. 1970 to present. Weekly. $442.00 per year.

Engineering Index Monthly. Engineering Information, Inc., Castle Point on the Hudson, Hoboken, NJ 07030. (800) 221-1044. FAX (212) 832-1857. Monthly. $2300.00 per year. Also available online as COMPENDEX, and also on CD-ROM. Covers chemical engineering, computers, electrical engineering, civil engineering, metals and mining, industrial management, and mechanical engineering.

Fluid Power Abstracts. STI Limited, 4 Kings Meadow, Ferry Hinksey Road, Oxford OX2 ODU, England. Distributed by Air Science Company, Box 143, Corning, NY 14830. (607) 962-5591. Covers oil hydraulic and pneumatic power transmission and control, engineering and design. 1964 to present. Bi-monthly. $260.00 per year.

Fluid Abstracts: Civil Engineering. Elsevier Science [journals], 660 White Plains Rd., Tarrytown, NY 10159-5153. (919) 524-9200. FAX (919) 333-2444. 1991 to present. Monthly plus annual cumulation. $645.00 per year. Also available online as FLUIDEX. Covers civil engineering applications of fluid mechanics, hydraulics, flow metering and measuring, offshore engineering, environmental hydraulics, and related aspects of wind energy, the atmosphere, and aerodynamics.

International Aerospace Abstracts. Technical Information Service, American Institute of Aeronautics and Astronautics, Inc., 555 West 57th St., New York, NY 10019. (212) 247-6500. FAX (212) 582-4861. Semi-monthly. $1295.00 per year.

Mechanical Engineering Abstracts (formerly *ISMEC*), Cambridge Scientific Abstracts, 7200 Wisconsin Avenue, Bethesda, MD 20814-4823. (301) 961-6750. FAX (301) 961-6720. 1967 to present. Monthly. $895.00 per year. Summarizes world literature in mechanical engineering, production engineering, and engineering management. Also available online.

Science Citation Index. Institute for Scientific Information, 3501 Market Street, Philadelphia, PA 19104. (215) 386-0100. FAX (215) 386-6362. Inquire about availability and cost. Also available on CD-ROM.

Scientific and Technical Aerospace Reports (star). National Aeronautics and Space Administration. NASA Center for Aerospace Information, Box 8757, BWI Airport, Baltimore, MD 21240. (301) 621-0153. Monthly. Inquire about availability and cost. ALso available through the NASA online retrieval service (RECON), and through the Aerospace Database through DIALOG.

ASSOCIATIONS AND PROFESSIONAL SOCIETIES

American Institute of Aeronautics and Astronautics. The Aerospace Center, 370 L'Enfant Promenade SW, Washington, DC 20024. (202) 646-7400.

American Society of Civil Engineers. 345 East 47th Street, New York, NY 10017-2398. (212) 705-7520 or (800) 548-2723.

American Society of Mechanical Engineers. 345 East 47th Street, New York, NY 10017-2398. (212) 705-7703.

Fluid Controls Institute. PO Box 9036, Morristown, NJ 07960. (201) 829-0990.

Fluid Power Consultants International. PO Box 106, 1000 Grandview Drive, Elm Grove, WI 53122-0106. (414) 782-0410.

Fluid Power Society. 2433 N. Mayfair Road, Suite 111, Milwaukee, WI 53226. (414) 257-0910.

Hydraulic Institute. 9 Sylvan Way, Parsippany, NJ 07054-3802. (201) 267-9700.

International Association For Hydraulic Research. Rotterdamsweg 185, PO Box 177, 2600 MH Delft, Netherlands. Telephone 31-15-569353. FAX 31-51-619674.

National Conference On Fluid Power. 3333 N. Mayfair Road, Suite 111, Milwaukee, WI 53226. (414) 778-3368.

National Fluid Power Association. 3333 N. Mayfair Road, Suite 111, Milwaukee, WI 53226. (414) 778-3344.

DIRECTORIES AND BIOGRAPHICAL SOURCES

American Astronautical Society Directory. American Astronautical Society, 6352 Rolling Mill Place, Suite 102, Springfield, VA 22152. Inquire for cost and availability.

Fluid Power Society Membership Directory. Fluid Power Society, 2433 N. Mayfair Road, Suite 111, Milwaukee, WI 53226. (414) 257-0910. Inquire for cost and availability.

National Fluid Power Association Membership Directory. National Fluid Power Association, 3333 N. Mayfair Road, Suite 111, Milwaukee, WI 53226. (414) 778-3344. Inquire for cost and availability.

Research Centers Directory. Gale Research, 835 Penobscot Building, Detroit, MI 48226-4094. (313) 961-2242. (800) 347-4253. FAX (313) 961-6083. $485.00.

Scientific and Technical Organizations and Agencies Directory. Gale Research, 835 Penobscot Building, Detroit, MI 48226-4094. (313) 961-2242. (800) 347-4253. FAX (313) 961-6083. 4th edition. 1996. $195.00.

Who's Who in Engineering. American Association of Engineering Societies, 1111 19th Street NW, Suite 608, Washington, DC 20036. (202) 296-2237 or (800) 658-8897. 8th edition. 1991. Inquire for price.

ENCYCLOPEDIAS AND DICTIONARIES

Encyclopedia of Fluid Mechanics. 10 volumes plus 3 supplements. Gulf Publishing Company, P.O. Box 2608, Houston, TX. (713) 529-4301 or (800) 231-6275. 1986-90 (vols. 1-10), 1993-94 (supplements). $195.00 each (vols. 1-10), $139.00 each (supplements).

Encyclopedia of Physical Science and Technology. Robert A. Meyers, ed. 18 volumes. Academic Press Inc., 6277 Sea Harbor Drive, Orlando, FL 32887. (800) 321-5068. 1992. $2100.00.

McGraw-Hill Dictionary of Scientific and Technical Terms. Sybil P. Parker, ed. 5th edition. McGraw-Hill Publishing Company, 1221 Avenue of the Americas, New York, NY 10020. (800) 262-4729 or (212) 512-3825. 1993. $110.50.

McGraw-Hill Encyclopedia of Engineering. Sybil P. Parker, ed. 2nd edition. McGraw-Hill Publishing Company, 1221 Avenue of the Americas, New York, NY 10020. (800) 262-4729 or (212) 512-3825. 1993. $95.50.

McGraw-Hill Encyclopedia of Science and Technology. Sybil P. Parker, ed. 7th edition. 20 volumes. McGraw-Hill Publishing Company, 1221 Avenue of the Americas, New York, NY 10020. (800) 262-4729 or (212) 512-3825. 1992. $1900.00

Thesaurus of Scientific, Technical, and Engineering Terms. Hemisphere Publishing Corporation, 1900 Frost Road, Suite 101, Bristol, PA 19007-1598. (215) 785-5800 or (800) 821-8312. FAX (215) 785-5515. 1987. $173.00.

GENERAL WORKS

Advances in Marine Hydrodynamics. M. Ohkusu, editor. Computational Mechanics Publications, 25 Bridge Street, Billerica, MA 01821. (508) 667-5841. FAX (508) 667-7582. 1996. $170.00

Coastal Engineering: Waves, Beaches, Wave Structure Interactions. T. Sawaragi. Elsevier Science Publishing Company, Inc., 655 Avenue of the Americas, New York, NY 10010. (212) 989-5800. 1995. $218.75.

Dynamics of the Liquid State. Umbert Balucani & Marco Zoppi. Oxford University Press, Inc., 198 Madison Avenue, New York, NY 10016-4314. (212) 726-6000. FAX (212) 726-6446. 1995. $95.00.

Hydrodynamics. Lamb. Third edition. Oxford University Press, Inc., 198 Madison Avenue, New York, NY 10016-4314. (212) 726-6000. FAX (212) 726-6446. 1991. $85.00.

HANDBOOKS AND MANUALS

Applied Fluid Dynamics Handbook. Robert D. Blevins, editor. Reprint edition. Krieger Publishing Company, P.O. Box 9542, Melbourne, FL 32902-9542. (407) 724-9542. FAX (407) 951-3671. 1992. $83.50.

Davis's Handbook of Applied Hydraulics. V.J. Zipparro, editor. McGraw-Hill Publishing Company, 1221 Avenue of the Americas, New York, NY 10020. (800) 262-4729 or (212) 512-3825. 1993. $89.50.

Fluid Power Design Handbook. Frank Yeaple. 2nd edition. Marcel Dekker, Inc., 270 Madison Avenue, New York, NY 10016. (212) 696-9000. FAX (212) 685-4540. 1990. Inquire for cost and availability.

Handbook of Fluid Dynamics and Fluid Machinery. Allen E. Fuhs and Joseph A. Schetz, editors. John Wiley and Sons, Inc., 605 Third Avenue, New York, NY 10158. (800) 526-5368 or (212) 850-6000. 1995. Inquire for price

Handbook of Hydraulic Engineering. Armando Lencastre. Halsted Press (Division of John Wiley and Sons, Inc.), 605 Third Avenue, New York, NY 10158. (800) 526-5368 or (212) 850-6418. 1987. Inquire for cost and availability.

Handbook of Hydraulics. E.F. Brate. 6th edition. McGraw-Hill Publishing Company, 1221 Avenue of the Americas, New York, NY 10020. (800) 262-4729 or (212) 512-3825. 1976. $95.50.

The Wiley Engineer's Desk Reference. Sanford I. Heisler. John Wiley and Sons, Inc., 605 Third Avenue, New York, NY 10158. (800) 526-5368 or (212) 850-6000. 1984. $64.95.

ONLINE DATABASES AND CD-ROMS

Compendex Plus. Engineering Information, Inc., 345 East 47th Street, New York, NY 10017. (212) 705-7600 or (800) 221-1044. Contains citations with abstracts to worldwide literature in engineering and technology, from 1970 to present. Available on online BRS,(800) 289-4277, DIALOG, (800) 334-2564, ORBIT (800) 456-7248, and STN International, FIZ Karlsruhe, P.O. Box 2465, W-7500, Karlsruhe 1, Germany, online services. Also available on CD-ROM. Inquire as to cost and availability.

Current Contents Search. Institute for Scientific Information, 3501 Market Street, Philadelphia, PA 19104. (215) 386-0100. FAX (215) 386-6362. Contains citations to articles listed in the table of contents of science and technology journals. Also articles in social sciences and life sciences journals. Available on BRS,(800) 289-4277, DIALOG,(800) 334-2564, online services. Inquire as to cost and availability.

Dissertation Abstracts. University Microfilms International, 300 North Zeeb Road, Ann Arbor, MI 48106. (800) 521-0600 or (313) 761-4700. Scope includes virtually all doctoral disserta-

tions accepted at accredited American institutions from 1861 to present in 252 subject areas. Available on BRS,(800) 289-4277, DIALOG,(800) 334-2564, and OCLC EPIC,(800) 848-5878, online services. Also available on CD-ROM. Inquire as to cost and availability.

International Aerospace Abstracts. American Institute of Aeronautics and Astronautics, 370 L'Enfant Promenade, S.W., Washington, DC 20024. (202) 646-7400. Contains references and abstracts of journal and monograph literature relating to aerospace science and technology, from 1963 to present. Available through the NASA/RECON system of the National Aeronautics and Space Administration only.

ISMEC: Mechanical Engineering Abstracts. Cambridge Scientific Abstracts, 7200 Wisconsin Avenue, Suite 601, Bethesda, MD 20814. (301) 961-6750 or (800) 843-7751. Contains citations to the literature in mechanical engineering, industrial and production engineering, energy, power, mechanics, devices and related areas, from 1973 to present. Available on the DIALOG,(800) 334-2564, online service. Inquire as to cost and availability.

NASA Database. American Institute of Aeronautics and Astronautics, 370 L'Enfant Promenade, S.W., Washington, DC 20024. (202) 646-7400. Citations and abstracts of aeronautics and astronautics literature, 1962 to present. Also contains citations from STAR, SCIENTIFIC and TECHNICAL AEROSPACE REPORTS, and INTERNATIONAL AEROSPACE ABSTRACTS. Available through NASA/RECON online service. Inquire as to cost and availability.

NTIS Bibliographic Database. National Technical Information Service, 5285 Port Royal Road, Springfield, VA 22161. (703) 487-4929 or FAX (703) 321-8199. Broad coverage of government-sponsored science and technology research reports, 1964 to present. Available on BRS,(800) 289-4277, DIALOG, (800) 334-2564, ORBIT, (800) 456-7248, and STN International, FIZ Karlsruhe, P.O. Box 2465, W-7500, Karlsruhe 1, Germany, online services. Also available on CD-ROM. Inquire as to cost and availability.

PERIODICALS

Computers and Fluids. Elsevier Science [journals], 660 White Plains Rd., Tarrytown, NY 10159-5153. (919) 524-9200. FAX (919) 333-2444. 1973 to present. Eight times a year. $985.00 per year.

Experiments in Fluids. Springer-Verlag, 44 Hartz Way, Secaucus, NJ 07096-2491. 1983 to present. 12 times a year. $675.00 per year.

Fluid Dynamics. Plenum Publishing Corporation, Consultants Bureau, 233 Spring Street, New York, NY 10013-1578. (212) 620-8468. FAX (212) 463-0742. 1966 to present. $1195.00.

Fluid Dynamics Research. Elsevier Science Inc., Box 882, Madison Square Station, New York, NY 10159. (212) 989-5800. FAX (212) 633-3990. 1986 to present. Twelve times a year. $458.00.

Hydraulics and Pneumatics. Penton Publishing, 110 Superior Ave., Cleveland, OH 44114-2543. 1947 to present. Monthly. $45.00 per year.

Journal of Fluids Engineering. American Society of Mechanical Engineers, 345 East 47th Street, New York, NY 10017-2398. (212) 705-7703. 1919 to present. Quarterly. $100.00 per year.

Journal of Hydraulic Engineering. American Society of Civil Engineers, Hydraulics Division, 345 East 47th Street, New York, NY 10017-2398. (212) 705-7520 or (800) 548-2723. 1956 to present. Monthly. $200.00 per year.

RESEARCH CENTERS AND INSTITUTES

University of Alaska Fairbanks, Water Research Center. Fairbanks, AK 99775-1760. (907) 474-7350. FAX (907) 474-6087.

University of California—San Diego, Institute For Pure and Applied Physical Sciences. 9500 Gilman Drive, LA Jolla, CA 92093-0075. (619) 534-4748. FAX (619) 534-0173.

University of Iowa, Institute of Hydraulic Research. Iowa City, IA 52242. (319) 335-5236. FAX (319) 335-5238.

University of Minnesota, St. Anthony Falls Hydraulic Laboratory. Mississippi River at 3rd Avenue, Minneapolis, MN 55414-2196. (612) 627-4010. FAX (612) 627-4609.

HYDROELECTRIC POWER

See also: DAMS, ELECTRICAL ENGINEERING

ABSTRACT SERVICES AND INDEXES

Applied Science and Technology Index; A Cumulative Subject Index To English Language Periodicals in the Fields of Aeronautics and Space Science, Computer Technology, Chemistry, Construction Industry, Energy and Related Areas. H.W. Wilson Co., 950 University Avenue, Bronx, NY 10452. (800) 367-6770 or (212) 588-8400. FAX (718) 590-1617. From 1958 to present. Monthly. Inquire about cost and availability. Also available on CD-ROM and online.

ASCE Combined Annual Combined Index. American Society of Civil Engineers, 345 East 47th Street, New York, NY 10017-2398. (212) 705-7520 or (800) 548-2723. Annual. $48.00.

ASCE Publications Information. 1966-present. American Society of Civil Engineers, 345 East 47th Street, New York, NY 10017-2398. (212) 705-7520 or (800) 548-2723. Bi-monthly. $160.00 per year for non-members.

Civil and Structural Engineering Abstracts, Cambridge Scientific Abstracts, 7200 Wisconsin Avenue, Bethesda, MD 20814-4823. (301) 961-6750. FAX (301) 961-6720. 1993 to present. Monthly. $385.00 per year. Topics covered include structural design, construction equipment and methods, civil defense and military engineering, surveying, highway engineering, maritime and port structures, materials, land reclamation, and soil mechanics.

Current Contents: Engineering, Technology, and Applied Sciences. Institute for Scientific Information, 3501 Market Street, Philadelphia, PA 19104. (215) 386-0100. FAX (215) 386-6362. 1970 to present. Weekly. $442.00 per year.

Electrical and Electronics Abstracts. Institute of Electrical Engineers, Michael Faraday House, Six Hill Way, Stevenage, Herts, SG1 2AY, England. Distributed by IEEE, 445 Hoes Lane, Piscataway, NJ 08854. (908) 562-5549. 1898 to present. Monthly. $2200.00 per year. Also available on CD-ROM and online as INSPEC.

Energy Research Abstracts. 1976-present. U.S. Department of Energy, office of Scientific and Technical Information, Box 62, Oak Ridge, TN 37831. (615) 574-0733. Subscriptions: Superintendent of Documents, U.S. Government Printing office, Box 8757, BWI Airport, Baltimore, MD 21240. (301) 621-0153.Monthly. $164.00 per year. Abstracts all the scientific and technical reports, journal articles, conference proceedings, patents, theses, and monographs sponsored by the U.S. Energy Research and Development Administration.

Engineering Index Monthly. Engineering Information, Inc., Castle Point on the Hudson, Hoboken, NJ 07030. (800) 221-1044. FAX (212) 832-1857. Monthly. $2300.00 per year. Also available online as COMPENDEX, and also on CD-ROM. Covers chemical engineering, computers, electrical engineering, civil engineering, metals and mining, industrial management, and mechanical engineering.

Environmental Engineering Abstracts, Cambridge Scientific Abstracts, 7200 Wisconsin Avenue, Bethesda, MD 20814-4823. (301) 961-6750. FAX (301) 961-6720. Monthly. Covers hazardous materials, environmental impact and protection, treatment of sewage and industrial wastes, hydroelectric power, tidal and wind power, artic and tropical engineering.

Science Citation Index. Institute for Scientific Information, 3501 Market Street, Philadelphia, PA 19104. (215) 386-0100. FAX (215) 386-6362. Inquire about availability and cost. Also available on CD-ROM.

ASSOCIATIONS AND PROFESSIONAL SOCIETIES

American Electronics Association. 5201 Great America Parkway, Suite 520, PO Box 54990, Santa Clara, CA 95054-0990. (408) 987-4200.

American Society of Civil Engineers. 345 East 47th Street, New York, NY 10017-2398. (212) 705-7520 or (800) 548-2723.

Edison Electric Institute. 701 Pennsylvania Avenue NW, Washington, DC 20004-2696. (202) 508-5000.

Electric Generation Association. 2715 M Street NW, Suite 150, Washington, DC 20007. (202) 965-1134. FAX (202) 965-1139.

Electric Power Research Institute. Box 10412, Palo Alto, CA 94303-0813. (415) 855-2000. FAX (415) 855-2954.

Institute of Electrical and Electronics Engineers. 345 E. 47th Street, New York, NY 10017. (212) 705-7900.

National Hydropower Association. 555 13th Street NW, Suite 900-E, Washington, DC 20004. (202) 637-9115. FAX (202) 637-8116.

DIRECTORIES AND BIOGRAPHICAL SOURCES

Electric Plant Datapak. Utility Data Institute, 1200 K Street NW, Suite 250, Washington, DC 20005. (202) 942-8788 or (800) 486-3660. FAX (202) 942-8789. 1990 to present. Annual. Issued on diskette. $495.00 per year.

Major Dams, Reservoirs, and Hydroelectric Plants Worldwide. James M. Tilsley. Revised edition. United States Department of the Interior, Bureau of Reclamation. Order from: Superintendent of Documents, U.S. Government Printing office, Box 371954, Pittsburgh, PA 15250-7954. (202) 783-3238. FAX (202) 512-2233. 1983. Inquire for cost and availability.

Directory of Engineering Societies and Related Organizations. Gordon Davis. 13th edition. American Association of Engineering Societies, 1111 19th Street NW, Suite 608, Washington, DC 20036. (202) 296-2237 or (800) 658-8897. 1989. Inquire for price.

International Research Centers Directory. Gale Research, 835 Penobscot Building, Detroit, MI 48226-4094. (313) 961-2242. (800) 347-4253. FAX (313) 961-6083. 8th edition. 1995. $410.00

Research Centers Directory. Gale Research, 835 Penobscot Building, Detroit, MI 48226-4094. (313) 961-2242. (800) 347-4253. FAX (313) 961-6083. $485.00.

Scientific and Technical Organizations and Agencies Directory. Gale Research, 835 Penobscot Building, Detroit, MI 48226-4094. (313) 961-2242. (800) 347-4253. FAX (313) 961-6083. 4th edition. 1996. $195.00.

Water Power and Dam Construction Directory. Reed Business Publishing Group, Quadrant House, the Quadrant, Sutton, Surrey SM2 5AS, England. Telephone 081-652-3369. FAX 081-652-8180. 1989. Inquire for cost and availability.

Who's Who in Engineering. American Association of Engineering Societies, 1111 19th Street NW, Suite 608, Washington, DC 20036. (202) 296-2237 or (800) 658-8897. 8th edition. 1991. Inquire for price.

GENERAL WORKS

Civil Engineering Guidelines For Planning and Designing Hydroelectric Developments. 5 volumes. American Society of Civil Engineers, 345 East 47th Street, New York, NY 10017-2398. (212) 705-7520 or (800) 548-2723. 1989. Prices vary per volume.

Hydropower Engineering. C.C. Warnick, et al. Prentice Hall, 113 Sylvan Avenue, Route 9W, Englewood Cliffs, NJ 07632. (201) 592-2000 or (800) 922-0579. 1984. $65.50.

Lessons Learned From the Design, Construction, and Operation of Hydroelectric Facilities. Task Committee on Lessons Learned from the Design, Construction, and Operation of Hydroelectric Facilities of the Hydropower Committee, Energy Division, ASCE. American Society of Civil Engineers, 345 East 47th Street, New York, NY 10017-2398. (212) 705-7520 or (800) 548-2723. 1994. Inquire for cost and availability.

HANDBOOKS AND MANUALS

The Almanac of Renewable Energy. Richard Golub & Eric Brus. Henry Holt and Company, Inc., 115 West 18th Street, New York, NY 10011. (212) 886-9200 or (800) 488-5233. FAX (212) 633-0748. 1993. $50.00.

HYDROELECTRIC POWER

Ency. of Physical Sciences and Engineering Info. Sources

Handbook of Modern Electronics and Electrical Engineering. Charles Belove, editor in chief. John Wiley and Sons, Inc., 605 Third Avenue, New York, NY 10158. (800) 526-5368 or (212) 850-6000. 1986. Inquire for cost and availability.

Hydropower Engineering Handbook. John S. Gulliver & Roger E.A. Arndt, editors in chief. McGraw-Hill Publishing Company, 1221 Avenue of the Americas, New York, NY 10020. (800) 262-4729 or (212) 512-3825. 1991. $72.50.

Standard Handbook For Electrical Engineers. D.G. Fink & W. Beaty, editors. 14th edition. McGraw-Hill Publishing Company, 1221 Avenue of the Americas, New York, NY 10020. (800) 262-4729 or (212) 512-3825. 1993. $110.50.

ONLINE DATABASES AND CD-ROMS

Compendex Plus. Engineering Information, Inc., 345 East 47th Street, New York, NY 10017. (212) 705-7600 or (800) 221-1044. Contains citations with abstracts to worldwide literature in engineering and technology, from 1970 to present. Available on online BRS,(800) 289-4277, DIALOG, (800) 334-2564, ORBIT (800) 456-7248, and STN International, FIZ Karlsruhe, P.O. Box 2465, W-7500, Karlsruhe 1, Germany, online services. Also available on CD-ROM. Inquire as to cost and availability.

CSA Engineering. Cambridge Scientific Abstracts, 7200 Wisconsin Avenue, Suite 601, Bethesda, MD 20814. (301) 961-6750 or (800) 843-7751. Contains citations and abstracts of international periodicals and research literature covering all fields of engineering and science and technology,including computer and information science, electronics, mechanical engineering, solid state materials, 1981 to present. Available on BRS,(800) 289-4277, online service.Inquire as to cost and availability.

Current Contents Search. Institute for Scientific Information, 3501 Market Street, Philadelphia, PA 19104. (215) 386-0100. FAX (215) 386-6362. Contains citations to articles listed in the table of contents of science and technology journals. Also articles in social sciences and life sciences journals. Available on BRS,(800) 289-4277, DIALOG,(800) 334-2564, online services. Inquire as to cost and availability.

ISMEC: Mechanical Engineering Abstracts. Cambridge Scientific Abstracts, 7200 Wisconsin Avenue, Suite 601, Bethesda, MD 20814. (301) 961-6750 or (800) 843-7751. Contains citations to the literature in mechanical engineering, industrial and production engineering, energy, power, mechanics, devices and related areas, from 1973 to present. Available on the DIALOG,(800) 334-2564, online service. Inquire as to cost and availability.

NTIS Bibliographic Database. National Technical Information Service, 5285 Port Royal Road, Springfield, VA 22161. (703) 487-4929 or FAX (703) 321-8199. Broad coverage of government-sponsored science and technology research reports, 1964 to present. Available on BRS,(800) 289-4277, DIALOG, (800) 334-2564, ORBIT, (800) 456-7248, and STN International, FIZ Karlsruhe, P.O. Box 2465, W-7500, Karlsruhe 1, Germany, online services. Also available on CD-ROM. Inquire as to cost and availability.

SCISEARCH. Institute for Scientific Information, 3501 Market Street, Philadelphia, PA 19104. (800) 523-1850 or (215) 386-0100. Broad multidisciplinary title and author index to the international literature of science and technology, 1974 to present. Available on DIALOG,(800) 334-2564, and ORBIT,(800) 456-

7248, online services. Also available on CD-ROM. Inquire as to cost and availability.

WILSONLINE. H.W. Wilson Company, 950 University Avenue, Bronx, NY 10452. (800) 367-6770 or (212) 588-8400. Makes available online versions of the printed H.W. Wilson indexes including Applied Science and Technology Index, Business Periodicals Index, General Science Index, and Readers' Guide to Periodical Literature. Period covered is generally 1983 to present. Available on BRS,(800) 289-4277, DIALOG,(800) 334-2564, and OCLC EPIC,(800) 848-5878, online services. Also available on CD-ROM. Inquire as to cost and availability.

PERIODICALS

Civil Engineering ASCE. American Society of Civil Engineers, 345 East 47th Street, New York, NY 10017-2398. (212) 705-7520 or (800) 548-2723. 1930 to present. Monthly. $85.00 per year.

Electric Light and Power. PennWell Publishing, 1421 S. Sheridan Road, Box 1260, Tulsa, OK 74101. (918) 835-3161. FAX (918) 835-9497. 1922 to present. Monthly. $42.00 per year.

Electric Machines and Power Systems. Taylor & Francis, 1900 Frost Road, Suite 101, Bristol, PA 19007. (215) 785-5800. FAX (215) 785-5515. 1976 to present. Monthly. $355.00 per year.

Electrical World. McGraw-Hill Publishing Company, 1221 Avenue of the Americas, New York, NY 10020. (800) 262-4729 or (212) 512-3825. 1874 to present. Monthly. $55.00 per year.

IEEE Power Engineering Review. IEEE (Institute of Electrical and Electronic Engineers), 445 Hoes Lane, Piscataway, NJ 08854. (908) 562-5545 or (800) 678-IEEE. 1981 to present. Monthly. $115.00 per year.

IEEE Transactions On Power Delivery. IEEE (Institute of Electrical and Electronic Engineers), 445 Hoes Lane, Piscataway, NJ 08854. (908) 562-5545 or (800) 678-IEEE. 1986 to present. Quarterly. $170.00 per year.

IEEE Transactions On Power Systems. IEEE (Institute of Electrical and Electronic Engineers), 445 Hoes Lane, Piscataway, NJ 08854. (908) 562-5545 or (800) 678-IEEE. 1986 to present. Quarterly. $180.00 per year.

International Water Power and Dam Construction. Reed Business Publishing Group, Quadrant House, the Quadrant, Sutton, Surrey SM2 5AS, England. Telephone 081-652-3111. FAX 081-652-8904. 1949 to present. Monthly. $226.00.

RESEARCH CENTERS AND INSTITUTES

Edison Electric Institute. 701 Pennsylvania Avenue NW, Washington, DC (202) 508-5000.

Electric Power Institute, Texas A&M University. Department of Electricial Engineering, College Station, TX 77843. (409) 845-1423.

Electrical Engineering Research Laboratories, Purdue University. Electrical Engineering Building, West Lafayette, Indiana 47907. (317) 494-3536. FAX (317) 494- 6440.

Ency. of Physical Sciences and Engineering Info. Sources

HYDROGEN

U.S. Department of Energy, Office of Renewable Energy Technologies, Wind/Hydro/Ocean Technologies Division. Mail Stop CE-121, 1000 Independence Avenue SW, Room 511047, Washington, DC 20585. (202) 586-5630. FAX (202) 586-5124.

STATISTICS SOURCES

International Energy Annual. U.S. Department of Energy, Energy Information Administration, 1000 Independence Avenue SW, Washington, DC 20585. (202) 586-8800.

HYDROFOIL CRAFT

See: NAVAL ARCHITECTURE

HYDROGEN

See also: COSMOCHEMISTRY, ELECTROCHEMISTRY, ELEMENTS, GEOHEMISTRY, AND NAMES OF INDIVIDUAL ELEMENTS.

ABSTRACT SERVICES AND INDEXES

Analytical Abstracts. Royal Society of Chemistry, Information Services, Thomas Graham House, Science Park, Milton Road, Cambridge, CB4 4WF, England. Contains citations, mostly with abstracts, of the worldwide literature on analytical chemistry, from 1954 to present. Monthly. $636.00 per year. Also available online.

Applied Science and Technology Index; A Cumulative Subject Index To English Language Periodicals in the Fields of Aeronautics and Space Science, Computer Technology, Chemistry, Construction Industry, Energy and Related Areas. H. W. Wilson Co., 950 University Avenue, Bronx, NY 10452- 9978. (212) 588-8400. (800) 367-6770. FAX (718) 590-1617.From 1958 to present. Monthly. Inquire about cost and availability. Also available online (BRS and WILSONLINE) and on CD-ROM. ISSN: 0003-6986.

Chemical Abstracts. Chemical Abstracts Service. 2540 Olentangy River Road, Box 3012, Columbus, OH 43210-0012. (614) 447-3600. FAX (614) 447-3713. 1907 to present. Weekly. $16,800.00 per year. Available online and on CD-ROM. CA is also available in five section groupings. Inquire regarding cost and availability.

Current Contents: Physical, Chemical and Earth Sciences. Institute for Scientific Information, 3501 Market Street, Philadelphia, PA 19104. (215) 386-0100. FAX (215) 386-2291. From 1961 to present. Weekly. $442.00 per year. Also available on CD-ROM and online. Inquire regarding cost and availability. ISSN: 0163-2574.

Engineering Index Monthly. Engineering Information, Inc., Castle Point on the Hudson, Hoboken, NJ 07030. (201) 216-8500. (800) 221-1044. FAX (201) 216-8532. Monthly. $2300.00 per year for monthly issues. $1980 per year for annual cumulation. Available on CD-ROM and online as COMPENDEX. ISSN: 0742-1974; 0360-8557.

General Science Index. H.W. Wilson Company, 950 University Avenue, Bronx, NY 10452. (212) 588-8400. (800) 367-6770.

FAX (718) 590-1617. From 1978 to present. Ten issues per year; quarterly and annual cumulations. Service basis. Available on CD-ROM and online. Inquire regarding cost and availability. ISSN: 0162-1963.

Physics Abstracts. INSPEC. Section A, Science Abstracts. Institution of Electrical Engineers (IEE), London, United Kingdom. Available from: INSPEC/IEEE -Institute of Electrical and Electronic Engineers, Box 1331, 445 Hoes Lane, Piscataway, NJ 08855-1331. (908) 562-5549. 1898 to present. 24 issues per year. $2835.00 per year. ISSN: 0036-8091. Also available online and on CD-ROM.

Science Citation Index. SCI. Institute for Scientific Information, 3501 Market Street, Philadelphia, PA 19104. (215) 386-0100. (800) 523-1850. FAX (215) 386-2991. 1961 to present. Six issues per year, plus annual cumulation. $11650.00 per year. Also available online and on CD-ROM. Inquire about price and availability. ISSN: 0036-827X.

ASSOCIATIONS AND PROFESSIONAL SOCIETIES

American Chemical Society, 1155 16th Street, Nw, Washington, DC 20036. (202) 872-4600.

American Hydrogen Association. 216 South Clark Drive, Suite 103, Tempe, AZ 852281. (602) 921-0422. FAX (602) 921-6601.

American Institute of Chemists. 7315 Wisconson Avenue, Bethesda, MD 20814. (301) 652-2447. FAX (301) 657-3549.

American Institute of Physics, 335 East 45th Street, New York, NY 10017 (212) 661-9404 OR Fax (516)349-9704.

American Society For Metals: ASM International, 9639 Kinsman, Materials Park, OH 44073-0002. (216) 338-5151. FAX (216) 338-4634

Association of Energy Engineers. 4025 Pleasantdale Road, Suite 420, Atlanta, GA 30340. (770) 447-5083. FAX (770)447-3969.

International Association For Hydrogen Energy. PO Box 248266, Coral Gables, FL 33124. (305) 284-4666. FAX (305) 284-4792.

BIBLIOGRAPHIES

Chemical Information. H. R. Collier, editor. Springer- Verlag New York, Inc., 175 Fifth Avenue, New York, NY 10010. (212) 460-1500. (800) 777-4643. 1990. $91.00.

Chemical Information Sources. Gary Wiggins. McGraw-Hill Publishing Company, 1221 Avenue of the Americas, New York, NY 10020. (212) 512-2000. (800) 262-4729. 1991. $42.50.

Hydrogen Fuels: A Bibliography. ERDA Technical Inforamtion Center Staff. U. S. Department of Energy, National Technical Information Service, 5285 Port Royal Road, Springfield, VA 22161. (703) 487-4650. 1981. $29.50 in paper.

Information Sources in Chemistry. R.T. Bottle and J.F.B Rowland, editors. R.R. Bowker Inc., 121 Chanlon Road, New Providence, NJ 07974. (908) 464-6800. (800) 521-8110. Fourth edition. 1993. $75.00.

Scientific and Technical Books and Serials in Print; An Index To Literature in Science and Technology. R.R. Bowker Inc., 121 Chanlon Road, New Providence, NJ 07974. (908) 464-6800. (800) 521-8110. FAX (908) 665-3502. 1972 to present. Annual. 4 volumes. 1994. $299.95. Also available on compact disc and online. ISSN: 0000-054X.

DIRECTORIES AND BIOGRAPHICAL SOURCES

American Institute of Chemists - Professional Directory. American Institute of Chemists, 501 Wythe Street, Alexandria, VA 22314-1917. (703) 836-2090. FAX (703) 836-2091. Annual. $65.00.

American Men and Women of Science. Physical and Biological Sciences. R.R. Bowker Inc., 121 Chanlon Road, New Providence, NJ 07974. (908) 464-6800. (800) 521-8110. 20th edition. 8 volumes. 1996. $850.00

Chemical Week — Buyers Guide Issue. Chemical Week Associates, 888 Seventh Avenue, New York, NY 10106. (212) 621-4900. FAX (212) 621-4949. Annual, October. $50.00.

Chemical Sources International. Mike Desing and Kurt Gandenberger, editors. Chemical Sources International, Inc., P.O. Box 1824, Clemson, SC 29633. (803) 646-7840. 1994. $285.00.

Directory of Chemistry Software 1992. Wendy Warr, Peter Willett, Geoff Downs. American Chemical Society, 1155 16th Street, NW, Washington, DC 20036. (202) 872-4600. 1992.$35.95.

Hydrogen Energy International Directory. Rector Press, Ltd., 130 Rattlesnake, Leverett, MA 01054-9726. (413) 548-9708. (800) 247-3473. 1994. $395.00.

Research Centers Directory. Gale Research,Inc., 835 Penobscot Building, Detroit, MI 48226-4094. (313) 961-2242. (800) 347-4253. 20th edition. 1995. $485.00. ISSN: 0080-1518.

ENCYCLOPEDIAS AND DICTIONARIES

Academic Press Dictionary of Science and Technology. Christopher Morris, editor. Academic Press, Inc., 6277 Sea Harbor Drive, Orlando, FL. (800) 321-5068. 1991. $115.00.

Encyclopedia of Applied Physics. VCH Publications, Inc., 220 East 23rd Street, Suite 909, New York, NY 10010-4606. (212) 683-8333. (800) 422-8824. 1991-. Twenty volumes. $6000.00.

Grant & Hawks Chemical Dictionary. R. A. Grant. McGraw-Hill Publishing Company, 1221 Avenue of the Americas, New York, NY 10020. (212) 512-2000. (800) 262-4729. 5th edition. 1987. $64.50.

Kirk-Othmer Encyclopedia of Chemical Technology. John Wiley and Sons, Inc., 605 Third Avenue, New York, NY 10158-0012. (212) 850-6000. (800) 225-5945. Fourth edition. 1991-. Twenty-seven volumes. $5400.00.

McGraw-Hill Encyclopedia of Chemistry. Sybil P.Parker, editor. McGraw-Hill Book, Incorporated, 1221 Avenue of the Americas, New York, NY 10020. (212) 512-2000. (800) 262-4729. Second edition. 1993. $95.50.

McGraw-Hill Encyclopedia of Science and Technology. McGraw-Hill Book, Incorporated, 1221 Avenue of the Americas, New York, NY 10020. (212) 512-2000. (800) 262-4729. Seventh edition. Twenty volumes. 1992. $1900.00.

GENERAL WORKS

Exploring Chemical Elements and Their Compounds. David L.Heiserman. TAB Books, P.O. Box 40, Blue Summit, PA 17294-0850. (717) 794-2191. (800) 233-1128. 1992. $17.95 in paper.

Fuel From Water: Energy Independence With Hydrogen.

Michael Peavey. Merit Products Inc., 2307 Tyler Lane, Louisville, KY 40205. (502) 459-2058. Revised edition. 1995. $20.00.

Hydrogen As An Energy Carrier. C. J. Winter and J. Nitsch. Springer-Verlag New York, Inc., 175 Fifth Avenue, New York, NY 10010. (212) 460-1500. (800) 777-4643. 1988. $119.00.

HYDROGEN EMBRITTLEMENT: PREVENTION and CONTROL. Lou Raymond, editor. Special Technical Publication 962. American Society for Testing and Materials, 1916 Race Street, Philadelphia, PA 19103. (215) 299-5419. 1988. $54.00.

Hydrogen in Water-Cooled Nuclear Power Reactors. UNIPUB, 4511-F Assembly Drive, Lanham, MD 20706-4391. (301) 459-7666. (800) 274-4888. 1992. $40.00 in paper.

Hydrogen Power: An Introduction To Hydrogen Energy and Its Applications. L. O. Williams. Franklin Book Co., Inc., Elkins Park, PA 19117. (215) 635-5252. 1980. $76.00.

Hydrogen Production and Marketing. J. G. Santangelo and W. Novis Smith, editors. American Chemical Society, 1155 16th Street, NW, Washington, DC 20036. (202) 872-4600. (800) 333-9511. FAX (614) 447-3671. 1980. $49.95.

Liquid Hydrogen - Fuel of the Future. W. Peschka. Springer-Verlag New York, Inc., 175 Fifth Avenue, New York, NY 10010. (212) 460-1500. (800) 777-4643. 1992. $98.99.

Metal-hydrogen System: Basic Bulk Properties. Yuh Fukai. Springer-Verlag New York, Inc., 175 Fifth Avenue, New York, NY 10010. (212) 460-1500. (800) 777-4643. 1993. $98.00.

MODELING the HYDROGEN BOND. Douglas A. Smith. American Chemical Society, 1155 16th Street, NW, Washington, DC 20036. (202) 872-4600. (800) 333-9511. FAX (614) 447- 3671. 1994. $74.95.

Solar Hydrogen: Moving Beyond Fossil Fuels. Joan M. Ogden and Robert H. Williams. World Resources Institute, 1709 New York Avenue, NW, Washington, DC 20006. (202) 638-6300. 1989. $10.00 in paper.

Spectrum of Atomic Hydrogen: Advances. G. Series. World Scientific Publishing Company, Inc., 1060 Main Street, River Edge, NJ 07661. (201) 487-9655. (800) 227-7562. 1988. $43.00.

Stress Corrosion Cracking and Hydrogen Embrittlement of Iron Base Alloys. NACE International, 1440 South Creek Drive, Houston, TX 77218. (713) 492-0535. NACE Reference Book, No. 5. 1977. $10.00.

HANDBOOKS AND MANUALS

Chemical Reference Handbook. Gordon Press Publications, P.O. Box 459, Bowling Green Station, New York, NY 10004. (718) 624-8419. 1995. $260.00.

CRC Handbook of Chemistry and Physics. David R. Kide, editor. CRC Press Inc., 2000 Corporate Blvd., NW, Boca Raton, FL 33431. (407) 994-0555. (800) 333-8300. 77th edition. 1996. $99.95.

Guide To Basic Chemical Compounds. D.R. Lide, Jr. CRC Publishers, Inc., 2000 Corporate Blvd., N.W., Boca Raton, FL 33431. (407) 994-0555 or (800) 333-8300. 1993. $120.00.

Handbook of Properties of Condensed Phases of Hydrogen and Oxygen. B. I. Verkin. Hemisphere Publishing Corp., 1900 Frost Road, Suite 101, Bristol PA 19007. (215) 785-5800. (800) 821-8312. 1991. $116.00.

Lange's Handbook of Chemistry. John A. Dean, editor. McGraw-Hill Publishing Company, 1221 Avenue of the Americas, New York, NY 10020. (800) 262-4729 or (212) 512-3825. Fourteenth edition. 1992. $79.50.

Riegel's Handbook of Industrial Chemistry. James A. Kent, editor. Van Nostrand Reinhold, 115 Fifth Avenue, New York, NY 10003. (212) 254-3232 or (800) 926-2665. Ninth edition. 1992. $114.95.

ONLINE DATABASES AND CD-ROMS

Analytical Abstracts Online. Royal Society of Chemistry, Information Services, Thomas Graham House, Science Park, Milton Road, Cambridge, CB4 4WF, England. Contains citations, mostly with abstracts, of the worldwide literature on analytical chemistry, from 1980 to present. Available on DIALOG, (800) 334-2564, and STN International, FIZ Karlsruhe, P.O. Box 2465, W-7500, Karlsruhe 1, Germany, online services. Inquire as to cost and availability.

Beilstein Online. Beilstein-Institut fur Literatur der Organischen Chemie, Varrentrapperstrasse, 40-42, D-6000, Frankfurt am Main 90, Germany. Contains data on carbon compounds from the Beilstein Handbook of Organic Chemistry. Structural and factual data for more than 3.4 million heterocyclic, isocyclic, and acyclic compounds included. Covers the period 1830 to present. Available on DIALOG, (800) 334-2564, ORBIT (800) 456-7248, and STN International, FIZ Karlsruhe, P.O. Box 2465, W-7500, Karlsruhe 1, Germany, online services. Inquire as to cost and availability.

CA Search. Chemical Abstracts Service, P.O. Box 3012, Columbus, OH 43210-0012. (614) 447-3600. (800) 848-6533. FAX (614) 447-3709. Very comprehensive guide to worldwide chemical literature and related fields, 1972 to present. Available on BRS,(800) 289-4277, DIALOG, (800) 334-2564, ORBIT (800) 456-7248, and STN International, FIZ Karlsruhe, P.O. Box 2465, W-7500, Karlsruhe 1, Germany, online services. Inquire as to cost and availability.

Chemical Journals of the American Chemical Society. American Chemical Society, 1155 16th Street, N.W., Washington, DC 20036. (202) 872-4381 or (800) 424-6747. Contains complete text of approximately 90,000 articles from 22 primary journals published by the American Chemical Society, from mostly 1982 to present. Available on STN International, FIZ Karlsruhe, P.O.

Box 2465, W-7500, Karlsruhe 1, Germany, online service. Inquire as to cost and availability.

Compendex Plus. Engineering Information, Inc., 345 East 47th Street, New York, NY 10017. (212) 705-7600 or (800) 221-1044. Contains citations with abstracts to worldwide literature in engineering and technology, from 1970 to present. Available on online BRS,(800) 289-4277, DIALOG, (800) 334-2564, ORBIT (800) 456-7248, and STN International, FIZ Karlsruhe, P.O. Box 2465, W-7500, Karlsruhe 1, Germany, online services. Also available on CD-ROM. Inquire as to cost and availability.

Current Contents Search. Institute for Scientific Information, 3501 Market Street, Philadelphia, PA 19104. (215) 386-0100. FAX (215) 386-6362. Contains citations to articles listed in the table of contents of science and technology journals. Also articles in social sciences and life sciences journals. Available on BRS,(800) 289-4277, DIALOG,(800) 334-2564, online services. Inquire as to cost and availability.

Dissertation Abstracts. University Microfilms International, 300 North Zeeb Road, Ann Arbor, MI 48106. (800) 521-0600 or (313) 761-4700. Scope includes virtually all doctoral dissertations accepted at accredited American institutions from 1861 to present in 252 subject areas. Available on BRS,(800) 289-4277, DIALOG,(800) 334-2564, and OCLC EPIC,(800) 848-5878, online services. Also available on CD-ROM. Inquire as to cost and availability.

Gmelin Database. Gmelin-Institut fur Anorganische Chemie und Grenzgebiete, Varrentrapperstrasse, 40-42, Carl-Bosch-Haus, D-6000, Frankfurt am Main 90, Germany. Contains structural and factual data relating to inorganic and organometallic chemistry. Provides data from the Gmelin Handbook of Inorganic and Organometallic Chemistry. Covers the period 1817 to 1975; 1988-89. Available on STN International, FIZ Karlsruhe, P.O. Box 2465, W-7500, Karlsruhe 1, Germany, online service. Inquire as to cost and availability.

Kirk-Othmer Encyclopedia of Chemical Technology. John Wiley and Sons, Inc., 605 Third Avenue, New York, NY 10158. (800) 526-5368 or (212) 850-6000. Contains the complete text of all chapters in the 27 volume fourth edition of the KIRK-Othmer ENCYCLOPEDIA of CHEMICAL TECHNOLOGY. 1991. Available on BRS,(800) 289-4277, DIALOG,(800) 334-2564, online services. Inquire as to cost and availability.

Metals Datafile. Materials Information, ASM International, Materials Park, OH 44073. (216) 338-5151.Contains designation and specification numbers for ferrous and nonferrous metals and alloys. Covers the period 1982 to present. Available on ORBIT,(800) 456-7248, and STN International, FIZ Karlsruhe, P.O. Box 2465, W-7500, Karlsruhe 1, Germany, online services. Inquire as to cost and availability.

NTIS Bibliographic Database. National Technical Information Service, 5285 Port Royal Road, Springfield, VA 22161. (703) 487-4929 or FAX (703) 321-8199. Broad coverage of government-sponsored science and technology research reports, 1964 to present. Available on BRS,(800) 289-4277, DIALOG, (800) 334-2564, ORBIT, (800) 456-7248, and STN International, FIZ Karlsruhe, P.O. Box 2465, W-7500, Karlsruhe 1, Germany, online services. Also available on CD-ROM. Inquire as to cost and availability.

Physics Briefs. American Institute of Physics, 335 East 45th Street, New York, NY 10017. (212) 661-9260 or FAX (212) 661-2036. Contains citations with abstracts of the literature of physics and related fields, 1979 to present. Available on the STN

HYDROGEN

Ency. of Physical Sciences and Engineering Info. Sources

International, FIZ Karlsruhe, P.O. Box 2465, W-7500, Karlsruhe 1, Germany, online service. Inquire as to cost and availability.

SCISEARCH. Institute for Scientific Information, 3501 Market Street, Philadelphia, PA 19104. (800) 523-1850 or (215) 386-0100. Broad multidisciplinary title and author index to the international literature of science and technology, 1974 to present. Available on DIALOG,(800) 334-2564, and ORBIT,(800) 456-7248, online services. Also available on CD-ROM. Inquire as to cost and availability.

WILSONLINE. H.W. Wilson Company, 950 University Avenue, Bronx, NY 10452. (800) 367-6770 or (212) 588-8400. Makes available online versions of the printed H.W. Wilson indexes including Applied Science and Technology Index, Business Periodicals Index, General Science Index, and Readers' Guide to Periodical Literature. Period covered is generally 1983 to present. Available on BRS,(800) 289-4277, DIALOG,(800) 334-2564, and OCLC EPIC,(800) 848-5878, online services. Also available on CD-ROM. Inquire as to cost and availability.

PERIODICALS

Analytical Chemistry. American Chemical Society, Box 3337, Columbus, OH 43210. (614) 447-3776. 1929 to present. Semimonthly. $415.00 per year. ISSN: 0003-2700.

Angewandte Chemie. VCH Publishers, Inc., 220 East 23rd Street, New York, NY 10010-4606. (212) 683-8333. (800) 367-8249. 1888 to present. 22 issues per year. $840.00 per year. ISSN: 0044-8249.

Chemical Reviews. American Chemical Society, Box 3337, Columbus, OH 43210. (614) 447-3776. 1924 to present. 8 issues per year. $346.00 per year. ISSN: 0009-2665.

Chemical Week; Includes Annual Buyers Guide. Chemical Week Associates, 888 Seventh Avenue, New York, NY 10106. (212) 621-4900. FAX (212) 621-4949. 1914 to present. Weekly. $99.00 per year. ISSN: 0009-272X.

Chemtech. American Chemical Society. Box 3337, Columbus, OH 43210. (614) 447-3776. 1970 to present. Monthly.

$370.00 per year. ISSN: 0009-2703.

International Journal of Hydrogen Energy. International Association for Hydrogen Energy. PO Box 248266, Coral Gables, FL 33124. (305) 284-4666. FAX (305) 284-4792. Monthly. $1026.00 per year. ISSN: 0360-3199.

Inorganic Chemistry. American Chemical Society, Box 3337, Columbus, OH 43210. (614) 447-3776. 1962 to present. Semimonthly. $500.00 per year. ISSN: 0020-1669.

Journal of the American Chemical Society. American Chemical Society, Box 3337, Columbus, OH 43210. (614) 447-3776. 1879 to present. Biweekly. $1055.00 per year. ISSN: 0002- 7863.

Polyhedron. Elsevier Science Publishing Company, Inc., 660 White Plains Road, Tarrytown, NY 10591-5153. (914-524-9200. FAX (914) 524-9200. 1982 to present. 24 issues per year. $2175.00 per year. ISSN: 0277-5387.

RESEARCH CENTERS AND INSTITUTES

Chemical Laboratories. Harvard University, Oxford Street, Cambridge, MA 02138. (617) 495-4283. FAX (617) 496-5618.

Chemistry Laboratories. Rensselaer Polytechnic Institute, Cogswell Laboratory, Troy, NY 12180-3590. (518) 276-8981.

Lawrence Berkeley Laboratory, Chemical Sciences Division. One Cyclotron Road, Building 66, Berkeley, CA 94720. (510) 486-6062. FAX (510) 486-4995.

Theoretical Chemistry Institute. University of Wisconsin, Madison. 1101 University Avenue, Madison, WI 53706. (608) 262-1511.

University/industry Chemical Research Center. Mississippi State University, Department of Chemistry, P.O. Drawer CH, Mississippi State, MS 39762. (601) 325-3584.

HYDROGENATION

See also: CATALYSIS, COAL GASIFICATION AND LIQUEFACTION, CHEMISTRY, ELEMENTS, HYDROGEN

ABSTRACT SERVICES AND INDEXES

Applied Science and Technology Index; A cumulative subject index to English language periodicals in the fields of aeronautics and space science, computer technology, chemistry, construction industry, energy and related areas. H. W. Wilson Co., 950 University Avenue, Bronx, NY 10452- 9978. (212) 588-8400. (800) 367-6770. FAX (718) 590-1617.From 1958 to present. Monthly. Inquire about cost and availability. Also available online (BRS and WILSONLINE) and on CD-ROM. ISSN: 0003-6986.

Chemical Abstracts. Chemical Abstracts Service. 2540 Olentangy River Road, Box 3012, Columbus, OH 43210-0012. (614) 447-3600. FAX (614) 447-3713. 1907 to present. Weekly. $16,800.00 per year. Available online and on CD-ROM. CA is also available in five section groupings. Inquire regarding cost and availability.

Current Contents: Physical, Chemical and Earth Sciences. Institute for Scientific Information, 3501 Market Street, Philadelphia, PA 19104. (215) 386-0100. FAX (215) 386-2291. From 1961 to present. Weekly. $442.00 per year. Also available on CD-ROM and online. Inquire regarding cost and availability. ISSN: 0163-2574.

Engineering Index Monthly. Engineering Information, Inc., Castle Point on the Hudson, Hoboken, NJ 07030. (201) 216-8500. (800) 221-1044. FAX (201) 216-8532. Monthly. $2300.00 per year for monthly issues. $1980 per year for annual cumulation. Available on CD-ROM and online as COMPENDEX. ISSN: 0742-1974; 0360-8557.

General Science Index. H.W. Wilson Company, 950 University Avenue, Bronx, NY 10452. (212) 588-8400. (800) 367-6770. FAX (718) 590-1617. From 1978 to present. Ten issues per year; quarterly and annual cumulations. Service basis. Available on CD-ROM and online. Inquire regarding cost and availability. ISSN: 0162-1963.

Physics Abstracts. INSPEC. Section A, Science Abstracts. Institution of Electrical Engineers (IEE), London, United Kingdom. Available from: INSPEC/IEEE -Institute of Electrical and Electronic Engineers, Box 1331, 445 Hoes Lane, Piscataway, NJ 08855-1331. (908) 562-5549. 1898 to present. 24 issues per year. $2835.00 per year. ISSN: 0036-8091. Also available online and on CD-ROM.

Process and Chemical Engineering/chemical Engineering Abstracts. Royal Society of Chemistry, Information Services, Thomas Graham House, Science Park, Milton Road, Cambridge, CB4 4WF, England. Contains citations, mostly with abstracts, of the worldwide literature on chemical engineering; from 1982 to present. Monthly. $610.00 per year. Also available online. ISSN: 0960-5045.

Science Citation Index. SCI. Institute for Scientific Information, 3501 Market Street, Philadelphia, PA 19104. (215) 386-0100. (800) 523-1850. FAX (215) 386-2991. 1961 to present. Six issues per year, plus annual cumulation. $11650.00 per year. Also available online and on CD-ROM. Inquire about price and availability. ISSN: 0036-827X.

ASSOCIATIONS AND PROFESSIONAL SOCIETIES

American Chemical Society, 1155 16th Street, Nw, Washington, DC 20036. (202) 872-4600.

American Hydrogen Association. 216 South Clark Drive, Suite 103, Tempe, AZ 852281. (602) 921-0422. FAX (602) 921-6601.

American Institute of Chemical Engineers. 345 East 47th Street, New York, NY 10017-2396. (212) 705-7663. (202) 752-3294.

American Oil Chemists Society. PO Box 3489, Champaign, IL 61826-3489. (217) 359-2344. FAX (217) 351-8091.

International Association For Hydrogen Energy. PO Box 248266, Coral Gables, FL 33124. (305) 284-4666. FAX (305) 284-4792.

BIBLIOGRAPHIES

Chemical Information. H. R. Collier, editor. Springer- Verlag New York, Inc., 175 Fifth Avenue, New York, NY 10010. (212) 460-1500. (800) 777-4643. 1990. $91.00.

Chemical Information Sources. Gary Wiggins. McGraw-Hill Publishing Company, 1221 Avenue of the Americas, New York, NY 10020. (212) 512-2000. (800) 262-4729. 1991. $42.50.

Information Sources in Chemistry. R.T. Bottle and J.F.B Rowland, editors. R.R. Bowker Inc., 121 Chanlon Road, New Providence, NJ 07974. (908) 464-6800. (800) 521-8110. Fourth edition. 1993. $75.00.

Scientific and Technical Books and Serials in Print; An Index To Literature in Science and Technology. R.R. Bowker Inc., 121 Chanlon Road, New Providence, NJ 07974. (908) 464-6800. (800) 521-8110. FAX (908) 665-3502. 1972 to present. Annual. 4 volumes. 1994. $299.95. Also available on compact disc and online. ISSN: 0000-054X.

DIRECTORIES AND BIOGRAPHICAL SOURCES

American Institute of Chemical Engineers Directory. American Institute of Chemical Engineers. 345 East 47th Street, New York, NY 10017-2396. (212) 705-7663. Annual.

American Institute of Chemists - Professional Directory. American Institute of Chemists, 501 Wythe Street, Alexandria, VA 22314-1917. (703) 836-2090. FAX (703) 836-2091. Annual. $65.00.

American Men and Women of Science. Physical and Biological Sciences. R.R. Bowker Inc., 121 Chanlon Road, New Providence, NJ 07974. (908) 464-6800. (800) 521-8110. 20th edition. 8 volumes. 1996. $850.00

Chemical Week — Buyers Guide Issue. Chemical Week Associates, 888 Seventh Avenue, New York, NY 10106. (212) 621-4900. FAX (212) 621-4949. Annual, October. $50.00.

Chemical Sources International. Mike Desing and Kurt Gandenberger, editors. Chemical Sources International, Inc., P.O. Box 1824, Clemson, SC 29633. (803) 646-7840. 1994. $285.00.

Consulting Services: Chemists and Chemical Engineers. Association of Consulting Chemists and Chemical Engineers, 40 East 45th Street, New York, NY 10017. (212) 983-3160. FAX (212) 983-3161. Biennial. $60.00.

Hydrogen Energy International Directory. Rector Press, Ltd., 130 Rattlesnake, Leverett, MA 01054-9726. (413) 548- 9708. (800) 247-3473. 1994. $395.00.

Research Centers Directory. Gale Research,Inc., 835 Penobscot Building, Detroit, MI 48226-4094. (313) 961-2242. (800) 347-4253. 20th edition. 1995. $485.00. ISSN: 0080-1518.

ENCYCLOPEDIAS AND DICTIONARIES

Academic Press Dictionary of Science and Technology. Christopher Morris, editor. Academic Press, Inc., 6277 Sea Harbor Drive, Orlando, FL. (800) 321-5068. 1991. $115.00.

Dictionary of Named Processes in Chemical Technology. Alan E. Comyns. Oxford University Press, Inc., 200 Madison Avenue, New York, NY 10016. (212) 725-6000. (800) 334-4249. 1994. $75.00.

Encyclopedia of Applied Physics. VCH Publications, Inc., 220 East 23rd Street, Suite 909, New York, NY 10010-4606. (212) 683-8333. (800) 422-8824. 1991-. Twenty volumes. $6000.00.

Gardner's Chemical Synomyns and Trade Names. Michael Ash and Irene Ash. Ashgate Publishing Co., Old Post Road, Brookfield, VT 05036. (802) 276-3162 10th edition. 1994. $195.00.

Grant & Hawks Chemical Dictionary. R. A. Grant. McGraw-Hill Publishing Company, 1221 Avenue of the Americas, New York, NY 10020. (212) 512-2000. (800) 262-4729. 5th edition. 1987. $64.50.

Hawley's Condensed Chemical Dictionary. Richard J. Lewis. Van Nostrand Reinhold, 115 Fifth Avenue, New York, NY 10003. (212) 254-3232. (800) 842-3636. 12th edition. 1993. $69.95.

Industrial Chemical thesaurus. Michael Ash and Irene Ash. VCH Publishers, Inc., 220 East 23rd Street, New York, NY 10010-4606. (212) 683-8333. (800) 422-8824. 1992. $295.00.

Kirk-Othmer Encyclopedia of Chemical Technology. John Wiley and Sons, Inc., 605 Third Avenue, New York, NY 10158-0012. (212) 850-6000. (800) 225-5945. Fourth edition. 1991- . Twenty-seven volumes. $5400.00.

McGraw-Hill Encyclopedia of Science and Technology. McGraw-Hill Book, Incorporated, 1221 Avenue of the Americas, New York, NY 10020. (212) 512-2000. (800) 262-4729. Seventh edition. Twenty volumes. 1992. $1900.00.

Ullman's Encyclopedia of Industrial Chemistry. VCH Publications, Inc., 220 East 23rd Street, Suite 909, New York, NY 10010-4606. (212) 683-8333. (800) 422-8824. 5th edition. 1984 - . Price varies per volume.

GENERAL WORKS

Catalytic Hydrogenation in Organic Synthesis: Procedures and Commentary. Morris Freifelder. Books on Demand, 300 North Zeeb Road, Ann Arbor, MI 48106-1346. (313) 761-4700. (800) 521-0600. $58.50 in paper.

Chemical Process Design. Robin Smith. McGraw-Hill Publishing Company, Inc., 1221 Avenue of the Americas, New York, NY 10020. (212) 512-2000. (800) 262-4729. 1995. $65.00.

Coal Liquefaction Fundamentals. D. D. Whitehurst, editor. American Chemical Society, 1155 16th Street, NW, Washington, DC 20036. (202) 872-4600. (800) 333-9511. FAX (614) 447-3671. 1980. $49.95.

Homogeneous Hydrogenation in Organic Chemistry. F. J. McQullian, editor. Kluwer Academic Publishers, 101 Philip Drive, Assinippi Park, Norwell, MA 02061. (617) 871-6600. 1975. $62.00.

Hydrogen Effects On Materials Behavior. A. W. Thompson, and N. R. Moody, editors. Minerals Metals & Materials Society, 420 Commonwealth Drive, Warrendale, PA 15086. (412) 776-9000. (800) 759-4867. 1989. $20.00.

Hydrogenation Methods. Paul Rylander, editor. Academic Press, Inc., 6277 Sea Harbor Drive, Orlando, FL. (800) 321-5068. 1990. Reprint editon. $55.00.

Hydrogenation of Fats and Oils: Theory and Practice. H. B. Patterson. AOCS Press, PO Box 3489, Champaign, IL 61826-3489. (217) 359-2344. (800) 336-2627. 1994. $75.00.

Hydropressing Catalysts For Heavy Oil and Coal. M. J. Satriana, editor. Noyes, 120 Mill Road, Park Ridge, NJ 07656. (201) 391-8484. 1982. $45.00.

Photogeneration of Hydrogen. A. Harriman and M. A. West, editors. Academic Press, Inc., 6277 Sea Harbor Drive, Orlando, FL. (800) 321-5068. 1983. $84.00.

HANDBOOKS AND MANUALS

Catalyst Handbook. Martyn V. Twigg. CRC Press Inc., 2000 Corporate Blvd., NW, Boca Raton, FL 33431. (407) 994-0555. (800) 333-8300. 2nd edition. 1989. $104.00.

Chemical Reference Handbook. Gordon Press Publications, P.O. Box 459, Bowling Green Station, New York, NY 10004. (718) 624-8419. 1995. $260.00.

CRC Handbook of Chemistry and Physics. David R. Kide, editor. CRC Press Inc., 2000 Corporate Blvd., NW, Boca Raton, FL 33431. (407) 994-0555. (800) 333-8300. 77th edition. 1996. $99.95.

Guide To Basic Chemical Compounds. D.R. Lide, Jr. CRC Publishers, Inc., 2000 Corporate Blvd., N.W., Boca Raton, FL 33431. (407) 994-0555 or (800) 333-8300. 1993. $120.00.

Handbook of Homogeneous Catalysis. B. Cornils, editor. VCH Publications, Inc., 220 East 23rd Street, Suite 909, New York, NY 10010-4606. (212) 683-8333. (800) 422-8824. 1996. $345.00.

Handbook of Properties of Condensed Phases of Hydrogen and Oxygen. B. I. Verkin. Hemisphere Publishing Corp., 1900 Frost Road, Suite 101, Bristol PA 19007. (215) 785-5800. (800) 821-8312. 1991. $116.00.

Lange's Handbook of Chemistry. John A. Dean, editor. McGraw-Hill Publishing Company, 1221 Avenue of the Americas, New York, NY 10020. (800) 262-4729 or (212) 512-3825. Fourteenth edition. 1992. $79.50.

Riegel's Handbook of Industrial Chemistry. James A. Kent, editor. Van Nostrand Reinhold, 115 Fifth Avenue, New York, NY 10003. (212) 254-3232 or (800) 926-2665. Ninth edition. 1992. $114.95.

ONLINE DATABASES AND CD-ROMS

Beilstein Online. Beilstein-Institut fur Literatur der Organischen Chemie, Varrentrappersstrasse, 40-42, D-6000, Frankfurt am Main 90, Germany. Contains data on carbon compounds from the Beilstein Handbook of Organic Chemistry. Structural and factual data for more than 3.4 million heterocyclic, isocyclic, and acyclic compounds included. Covers the period 1830 to present. Available on DIALOG, (800) 334-2564, ORBIT (800) 456-7248, and STN International, FIZ Karlsruhe, P.O. Box 2465, W-7500, Karlsruhe 1, Germany, online services. Inquire as to cost and availability.

CA Search. Chemical Abstracts Service, P.O. Box 3012, Columbus, OH 43210-0012. (614) 447-3600. (800) 848-6533. FAX (614) 447-3709. Very comprehensive guide to worldwide chemical literature and related fields, 1972 to present. Available on BRS,(800) 289-4277, DIALOG, (800) 334-2564, ORBIT (800) 456-7248, and STN International, FIZ Karlsruhe, P.O. Box 2465, W-7500, Karlsruhe 1, Germany, online services. Inquire as to cost and availability.

Chemical Journals of the American Chemical Society. American Chemical Society, 1155 16th Street, N.W., Washington, DC 20036. (202) 872-4381 or (800) 424-6747. Contains complete text of approximately 90,000 articles from 22 primary journals published by the American Chemical Society, from mostly 1982 to present. Available on STN International, FIZ Karlsruhe, P.O. Box 2465, W-7500, Karlsruhe 1, Germany, online service. Inquire as to cost and availability.

Compendex Plus. Engineering Information, Inc., 345 East 47th Street, New York, NY 10017. (212) 705-7600 or (800) 221-1044. Contains citations with abstracts to worldwide literature in engineering and technology, from 1970 to present. Available on online BRS,(800) 289-4277, DIALOG, (800) 334-2564, ORBIT (800) 456-7248, and STN International, FIZ Karlsruhe, P.O. Box 2465, W-7500, Karlsruhe 1, Germany, online services. Also available on CD-ROM. Inquire as to cost and availability.

Current Contents Search. Institute for Scientific Information, 3501 Market Street, Philadelphia, PA 19104. (215) 386-0100. FAX (215) 386-6362. Contains citations to articles listed in the table of contents of science and technology journals. Also articles in social sciences and life sciences journals. Available on BRS,(800) 289-4277, DIALOG,(800) 334-2564, online services. Inquire as to cost and availability.

Dissertation Abstracts. University Microfilms International, 300 North Zeeb Road, Ann Arbor, MI 48106. (800) 521-0600 or (313) 761-4700. Scope includes virtually all doctoral dissertations accepted at accredited American institutions from 1861 to present in 252 subject areas. Available on BRS,(800) 289-4277, DIALOG,(800) 334-2564, and OCLC EPIC,(800) 848-5878, online services. Also available on CD-ROM. Inquire as to cost and availability.

Kirk-Othmer Encyclopedia of Chemical Technology. John Wiley and Sons, Inc., 605 Third Avenue, New York, NY 10158. (800) 526-5368 or (212) 850-6000. Contains the complete text of all chapters in the 27 volume fourth edition of the KIRK-Othmer ENCYCLOPEDIA of CHEMICAL TECHNOLOGY. 1991. Available on BRS,(800) 289-4277, DIALOG,(800) 334-2564, online services. Inquire as to cost and availability.

NTIS Bibliographic Database. National Technical Information Service, 5285 Port Royal Road, Springfield, VA 22161. (703) 487-4929 or FAX (703) 321-8199. Broad coverage of government-sponsored science and technology research reports, 1964 to present. Available on BRS,(800) 289-4277, DIALOG, (800) 334-2564, ORBIT, (800) 456-7248, and STN International, FIZ Karlsruhe, P.O. Box 2465, W-7500, Karlsruhe 1, Germany, online services. Also available on CD-ROM. Inquire as to cost and availability.

SCISEARCH. Institute for Scientific Information, 3501 Market Street, Philadelphia, PA 19104. (800) 523-1850 or (215) 386-0100. Broad multidisciplinary title and author index to the international literature of science and technology, 1974 to present. Available on DIALOG,(800) 334-2564, and ORBIT,(800) 456-7248, online services. Also available on CD-ROM. Inquire as to cost and availability.

WILSONLINE. H.W. Wilson Company, 950 University Avenue, Bronx, NY 10452. (800) 367-6770 or (212) 588-8400. Makes available online versions of the printed H.W. Wilson indexes including Applied Science and Technology Index, Business Periodicals Index, General Science Index, and Readers' Guide to Periodical Literature. Period covered is generally 1983 to present. Available on BRS,(800) 289-4277, DIALOG,(800) 334-2564, and OCLC EPIC,(800) 848-5878, online services. Also available on CD-ROM, Inquire as to cost and availability.

PERIODICALS

AICHE Journal. American Institute of Chemical Engineers. 345 East 47th Street, New York, NY 10017. (212) 705-7338. FAX (212) 752-3294. 1955 to present. Monthly. $395.00 per year. ISSN: 0001-1541.

Applied Catalysis A: General. Elsevier Science Inc., Box 882, Madison Square Station, New York, NY 10159-0882. (212) 989-

5800. FAX (212) 633-3990. 1981 to present. 28 issues per year. $2921 per year. $3295 in combination with section B Environmental. ISSN: 0926-860X.

Catalysis Reviews: Science and Engineering. Marcel Dekker, Inc., 270 Madison Avenue, New York, NY 10016. (212) 696-9000. FAX (212) 685-4540. 1967 to present. Quarterly. $435.00 per year. ISSN: 0161-4940.

Chemical and Engineering News. American Chemical Society, 1155 16th Street, NW, Washington, DC 20036. (202) 872-4600.

1923 to present. Weekly. $115.00 per year. ISSN: 0009- 2347.

Chemical Processing. Putman Publishing Co., 301 East Erie Street, Chicago, IL 60611. (312) 644-2020. 1938 to present. Monthly. $45.00 per year. ISSN: 0009-2630.

Chemical Week; Includes Annual Buyers Guide. Chemical Week Associates, 888 Seventh Avenue, New York, NY 10106. (212) 621-4900. FAX (212) 621-4949. 1914 to present. Weekly. $99.00 per year. ISSN: 0009-272X.

Chemtech. American Chemical Society. Box 3337, Columbus, OH 43210. (614) 447-3776. 1970 to present. Monthly. $370.00 per year. ISSN: 0009-2703.

Industrial and Engineering Chemistry Research. American Chemical Society, Box 3337, Columbus, OH 43210. (614) 447-3776. (800) 333-9511. FAX (614) 447-3671. 1962 to present. Monthly. $567.00 per year. ISSN: 0888-5885.

Hydrocarbon Processing. Gulf Publishing Company, P.O. Box 2608, Houston, TX 77001. (713) 520-4433. 1922 to present. Monthly. $22.00. ISSN: 0887-0284

International Journal of Hydrogen Energy. International Association for Hydrogen Energy. PO Box 248266, Coral Gables, FL 33124. (305) 284-4666. FAX (305) 284-4792. Monthly. $1026.00 per year. ISSN: 0360-3199.

RESEARCH CENTERS AND INSTITUTES

Center For Catalytic Science and Technology. University of Delaware. Departments of Chemical Engineering and Chemistry, Newark, DE 19716. (302) 451-8056. (302) 451- 1048.

Chemical Laboratories. Harvard Universiy. Oxford Street, Cambridge, MA 02138. (617) 495-4283. FAX (617) 496-5618.

Chemistry Laboratories. Rensselaer Polytechnic Institute. Cogswell Laboratory, Troy, NY 12180-3590. (518) 276-8981.

Engineering Research Program. Pennsylvania State University, 101 Hammond Building, University Park, PA 16802.

Institute For Systems Design and Optimization. Kansas State University. Department of Chemical Engineering, Manhattan, KS 66506. (913) 532-5584.

University/industry Chemical Research Center. Mississippi State University, Department of Chemistry, P.O. Drawer CH, Mississippi State, MS 39762. (601) 325-3584.

HYDROGEOLOGY

See also: GROUND WATER, HYDROLOGY, HYDRODYNAMICS, WATER RESOURCES

ABSTRACT SERVICES AND INDEXES

Aqualine Abstracts. Water Research Centre, PO Box 85, Frankland Rd., Blagrove, Swindon, Wilts. SN5 8YF, England. 1927 to present. 26 times a year. Inquire for price in U.S. Also available online.

Bibliography and Index of Geology. American Geological Institute, 4220 King Street, Alexandria, VA 22302. (703) 379-2480. Fax (703) 379-7563. Monthly. $1295.00 per year. Also available online as GEOREF.

Current Contents: Physical, Chemical and Earth Sciences. Institute for Scientific Information, 3501 Market Street, Philadelphia, PA 19104. (215) 386-0100. FAX (215) 386-6362. Weekly. $360.00 per year.

Engineering Index Monthly. Engineering Information, Inc., Castle Point on the Hudson, Hoboken, NJ 07030. (800) 221-1044. FAX (212) 832-1857. Monthly. $2300.00 per year. Also available online as COMPENDEX, and also on CD-ROM. Covers chemical engineering, computers, electrical engineering, civil engineering, metals and mining, industrial management, and mechanical engineering.

Geological Abstracts. Elsevier - Geo Abstracts, Regency House, 34 Duke Street, Norwich NR3 3AP, England. Monthly. $760.00 per year. Also available online as GEOBASE.

Geological Society of America. ABSTRACTS WITH PROGRAMS. Geological Society of America. 3300 Penrose Place, P.O. Box 9140, Boulder, CO 80301-9140. (303) 447-2020. Abstracts and programs of the annual conference. Annual. $69.00.

Hydro-abstracts (formerly Water Resources Abstracts). HydroScience Press, 2527 Jackson Street, N.E., Minneapolis, MN 55418. (612) 781-9081. Monthly. $150.00 per year.

Meteorological and Geoastrophysical Abstracts. American Meteorological Society, c/o Inforonics Inc., 550 Newtown Road, Littleton, MA 01460. (508) 486-8976. FAX (508) 486-0027. 1950 to present. Monthly. $950.00 per year. Current abstracts of books, reports, research papers, and miscellaneous literature on environmental sciences, meteorology, astrophysics, hydrology, glaciology, and physical oceanography.

Science Citation Index. Institute for Scientific Information, 3501 Market Street, Philadelphia, PA 19104. (215) 386-0100. FAX (215) 386-6362. Inquire about availability and cost. Also available on CD-ROM.

Selected Water Resources Abstracts. U.S. Geological Survey, Water Resources Scientific Information Center, 425 National Center, Reston, VA 22092. (703) 648-6820. Monthly. $115.00 per year. Also available on CD-ROM.

ASSOCIATIONS AND PROFESSIONAL SOCIETIES

American Institute of Hydrology. 3416 University Avenue S.E., Minneapolis, MN 55414-3329. (612) 379-1030. FAX (379-0169.

American Society of Civil Engineers. 345 East 47th Street, New York, NY 10017-2398. (212) 705-7520 or (800) 548-2723.

International Association of Hydrogeology. National Rivers Authority, Sapphire E 550, St.sBrook Road, Solihull, W. Midlands B91 1QT, England. Telephone 121-7112324. FAX 121-7112794.

DIRECTORIES AND BIOGRAPHICAL SOURCES

Directory of Engineering Societies and Related Organizations. Gordon Davis. 13th edition. American Association of Engineering Societies, 1111 19th Street NW, Suite 608, Washington, DC 20036. (202) 296-2237 or (800) 658-8897. 1989. Inquire for price.

International Research Centers Directory. Gale Research, 835 Penobscot Building, Detroit, MI 48226-4094. (313) 961-2242. (800) 347-4253. FAX (313) 961-6083. 8th edition. 1995. $410.00

Research Centers Directory. Gale Research, 835 Penobscot Building, Detroit, MI 48226-4094. (313) 961-2242. (800) 347-4253. FAX (313) 961-6083. $485.00.

Scientific and Technical Organizations and Agencies Directory. Gale Research, 835 Penobscot Building, Detroit, MI 48226-4094. (313) 961-2242. (800) 347-4253. FAX (313) 961-6083. 4th edition. 1996. $195.00.

Who's Who in Engineering. American Association of Engineering Societies, 1111 19th Street NW, Suite 608, Washington, DC 20036. (202) 296-2237 or (800) 658-8897. 8th edition. 1991. Inquire for price.

ENCYCLOPEDIAS AND DICTIONARIES

Dictionary of Civil Engineering. V.N. Vazirani. South Asia Books, Box 502, Columbia, MO 65205. (314) 474-0116. FAX (314) 474-8124. 1992. $20.00.

The Encyclopedia of the Solid Earth Sciences. Philip Kearey, editor-in-chief. Blackwell Scientific Publications, 238 Main Street, Cambridge, MA 02142. (617) 876-7000 or (800) 759-6102. FAX (617) 876-7022. 1993. $60.00.

Encyclopedia of Physical Science and Technology. Robert A. Meyers, ed. 18 volumes. Academic Press Inc., 6277 Sea Harbor Drive, Orlando, FL 32887. (800) 321-5068. 1992. $2100.00.

McGraw-Hill Dictionary of Scientific and Technical Terms. Sybil P. Parker, ed. 5th edition. McGraw-Hill Publishing Company, 1221 Avenue of the Americas, New York, NY 10020. (800) 262-4729 or (212) 512-3825. 1993. $110.50.

McGraw-Hill Encyclopedia of Science and Technology. Sybil P. Parker, ed. 7th edition. 20 volumes. McGraw-Hill Publishing Company, 1221 Avenue of the Americas, New York, NY 10020. (800) 262-4729 or (212) 512-3825. 1992. $1900.00

ENCYCLOPEDIAS AND DICTIONARIES

The Encyclopedia of the Solid Earth Sciences. Philip Kearey, editor-in-chief. Blackwell Scientific Publications, 238 Main

Street, Cambridge, MA 02142. (617) 876-7000 or (800) 759-6102. FAX (617) 876-7022. 1993. $60.00.

Encyclopedia of Physical Science and Technology. Robert A. Meyers, ed. 18 volumes. Academic Press Inc., 6277 Sea Harbor Drive, Orlando, FL 32887. (800) 321-5068. 1992. $2100.00.

GENERAL WORKS

Applied Hydrogeology. Charles W. Fetter. 3rd edition. Macmillan Publishing, 200 Old Tappan Road, Old Tappan, NJ 07675. (800) 223-2336. FAX (800) 445-6991. 1994. Inquire for cost and availability.

Applied Hydrogeology For Scientists and Engineers. Sekai Sen. Lewis Publishers, 121 S. Main St., PO Box 519, Chelsea, MI 48118. (313) 475-8619 or (800) 272-7737. 1995. Inquire for cost and availability.

Hydrogeology. Stanley N. Davis and Roger J. DE Wrest. Reprint edition. Krieger Publishing Company, P.O. Box 9542, Melbourne, FL 32902-9542. (407) 724-9542. FAX (407) 951-3671. 1991. $64.95.

Hydrogeology and Engineering Geology. A.M. Galperin, editor. Ashgate Publishing Co., Old Post Rd., Brookfield, VT 05036. (802) 276-3162. FAX (802) 276-3837. 1993. $105.00.

Introduction To Environmental Hydrogeology. Eric Eslinger, et al. SEPM—Society for Sedimentary Geology, 1731 E 71st Street, Tulsa, OK 74136-5108. FAX (918) 493-2093. 1994. $48.00.

HANDBOOKS AND MANUALS

Handbook of Hydrology. David R. Maidment, editor in chief. McGraw-Hill Publishing Company, 1221 Avenue of the Americas, New York, NY 10020. (800) 262-4729 or (212) 512-3825. 1993. $115.00.

Laboratory Manual in Hydrogeology. Keenan Lee. Macmillan Publishing, 200 Old Tappan Road, Old Tappan, NJ 07675. (800) 223-2336. FAX (800) 445-6991. 1994. Inquire for cost and availability.

ONLINE DATABASES AND CD-ROMS

Compendex Plus. Engineering Information, Inc., 345 East 47th Street, New York, NY 10017. (212) 705-7600 or (800) 221-1044. Contains citations with abstracts to worldwide literature in engineering and technology, from 1970 to present. Available on online BRS,(800) 289-4277, DIALOG, (800) 334-2564, ORBIT (800) 456-7248, and STN International, FIZ Karlsruhe, P.O. Box 2465, W-7500, Karlsruhe 1, Germany, online services. Also available on CD-ROM. Inquire as to cost and availability.

Earth Sciences. U.S. Geological Survey, 12201 Sunrise Valley Drive, Reston, VA 22092-9998. (703) 648-4460. CD-ROM of earth science databases including the U.S. Geological Survey Library, Earth Science Data Directory, and GEOINDEX, citations to published geological maps. $350.00 per year with quarterly updates.

Geoarchive. Geosystems, P.O. Box 1024, Westminster, London, England, SW1 P 2JL. Citations to literature on geoscience, 1969 to present. Inquire as to online cost and availability.

Geobase. Elsevier - Geo Abstracts, Regency House, 34 Duke Street, Norwich NR3 3AP, England. Contains citations to the worldwide earth science literature from 1980 to date. Available on DIALOG, ORBIT online services. Inquire as to cost and availability.

Geomechanics Abstracts. University of London, Imperial College of Science and Technology, Rock Mechanics Information Service, Royal School of Mines, Prince Consort Road, London SW7 2BP, England. Telephone: 071-589 5111 or FAX 071-589 6806. Scope includes worldwide literature on engineering geology, hydrogeology, mining, rock mechanics, soil mechanics, and tunnelling, 1977 to present. Available on the ORBIT (800) 456-7248, online service. Inquire as to cost and availability.

GEOREF: Bibliography and Index of Geology. American Geological Institute, 4220 King Street, Alexandria, VA 22302. (703) 379-2480. Fax (703) 379-7563. Monthly. Inquire for price and availability.

NTIS Bibliographic Database. National Technical Information Service, 5285 Port Royal Road, Springfield, VA 22161. (703) 487-4929 or FAX (703) 321-8199. Broad coverage of government-sponsored science and technology research reports, 1964 to present. Available on BRS,(800) 289-4277, DIALOG, (800) 334-2564, ORBIT, (800) 456-7248, and STN International, FIZ Karlsruhe, P.O. Box 2465, W-7500, Karlsruhe 1, Germany, online services. Also available on CD-ROM. Inquire as to cost and availability.

Water Resources Abstracts (WRA). U.S. Geological Survey, Water Resources Scientific Information Center, 12201 Sunrise Valley Drive, Reston, VA 22092-9998. (703) 648-4460. Contains citations with abstracts to scientific and technical literature on the water-sources-related aspects of the physical, social and life sciences, from 1968 to present. Available on DIALOG,(800) 334-2564, and ORBIT,(800) 456-7248, online services. Inquire as to cost and availability.

PERIODICALS

Ground Water. Ground Water Publishing Company, National Well Water Association, 6375 Riverside Drive, Dublin, OH 43017. (614) 761-1711. FAX (614) 761-3446. 1963 to present. Bi-monthly. $90.00 per year.

Hydrological Science and Technology. American Institute of Hydrology, 3416 University Avenue SE, Minneapolis, MN 55414-3328. (612) 379-1030. FAX (612) 379-0169. 1985 to present. Quarterly. $75.00.

Journal of Hydrology. Elsevier Science Inc., Box 882, Madison Square Station, New York, NY 10159. (212) 989-5800. FAX (212) 633-3990. 1963 to present. 48 times a year. $2309.00.

Journal of Irrigation and Drainage. American Society of Civil Engineers, Irrigation and Drainage Division, 345 East 47th Street, New York, NY 10017-2398. (212) 705-7520 or (800) 548-2723. 1956 to present. Bi-monthly. $136.00 per year.

Journal of Water Resources Planning and Management. American Society of Civil Engineers, Water Resources Planning and Management Division, 345 East 47th Street, New York, NY 10017-2398. (212) 705-7520 or (800) 548-2723. Bi-monthly. $112.00 per year.

Limnology and Oceanography. American Society of Limnology and Oceanography, School of Oceanography, WB-10, University of Washington, Seattle, WA 98195. 1956 to present. Eight times a year. $160.00 per year.

Water Resources Bulletin. American Water Resources Association, 5410 Grosvenor Lane, Suite 220, Bethesda, MD 20814-2192. (301) 493-8600. FAX (301) 493-5844. 1965 to present. Bi-monthly. $15.00.

Water Resources Research. American Geophysical Union, 2000 Florida Avenue, N.W., Washington, DC 20009. (202) 462-6900. FAX (202) 328-0566. 1965 to present. Monthly. $660.00 for non-members.

RESEARCH CENTERS AND INSTITUTES

National Center For Ground Water Research. University of Oklahoma, 200 Telgar Street, Room 127, Norman, OK 73019-0470. (405) 325-5202.

University of British Columbia, Interdisciplinary Hydrology Programme. Department of Civil Engineering, Vancouver, BC, Canada V6T 1W3. (604) 228-2826. FAX (694) 228-6901.

University of Cincinnati, Groundwater Research Center. College of Engineering, Mail Location 18, Cincinnati, OH 45221-0018. (513) 475-2933.

HYDROLOGY

See also: WATER, HYDRAULIC ENGINEERING, HYDRAULICS, HYDRODYNAMICS, WATER RESOURCES

ABSTRACT SERVICES AND INDEXES

Aqualine Abstracts. Water Research Centre, PO Box 85, Frankland Rd., Blagrove, Swindon, Wilts. SN5 8YF, England. 1927 to present. 26 times a year. Inquire for price in U.S. Also available online.

Bibliography and Index of Geology. American Geological Institute, 4220 King Street, Alexandria, VA 22302. (703) 379-2480. Fax (703) 379-7563. Monthly. $1295.00 per year. Also available online as GEOREF.

Current Contents: Physical, Chemical and Earth Sciences. Institute for Scientific Information, 3501 Market Street, Philadelphia, PA 19104. (215) 386-0100. FAX (215) 386-6362. Weekly. $360.00 per year.

Fluid Abstracts: Civil Engineering. Elsevier Science [journals], 660 White Plains Rd., Tarrytown, NY 10159-5153. (919) 524-9200. FAX (919) 333-2444. 1991 to present. Monthly plus annual cumulation. $645.00 per year. Also available online as FLUIDEX. Covers civil engineering applications of fluid mechanics, hydraulics, flow metering and measuring, offshore engineering, environmental hydraulics, and related aspects of wind energy, the atmosphere, and aerodynamics.

Geological Abstracts. Elsevier - Geo Abstracts, Regency House, 34 Duke Street, Norwich NR3 3AP, England. Monthly. $760.00 per year. Also available online as GEOBASE.

Geological Society of America. ABSTRACTS WITH PROGRAMS. Geological Society of America. 3300 Penrose Place, P.O. Box 9140, Boulder, CO 80301-9140. (303) 447-2020. Abstracts and programs of the annual conference. Annual. $69.00.

Hydro-Abstracts (formerly Water Resources Abstracts). HydroScience Press, 2527 Jackson Street, N.E., Minneapolis, MN 55418. (612) 781-9081. Monthly. $150.00 per year.

Meteorological and Geoastrophysical Abstracts. American Meteorological Society, c/o Inforonics Inc., 550 Newtown Road, Littleton, MA 01460. (508) 486-8976. FAX (508) 486-0027. 1950 to present. Monthly. $950.00 per year. Current abstracts of books, reports, research papers, and miscellaneous literature on environmental sciences, meteorology, astrophysics, hydrology, glaciology, and physical oceanography.

Science Citation Index. Institute for Scientific Information, 3501 Market Street, Philadelphia, PA 19104. (215) 386-0100. FAX (215) 386-6362. Inquire about availability and cost. Also available on CD-ROM.

Selected Water Resources Abstracts. U.S. Geological Survey, Water Resources Scientific Information Center, 425 National Center, Reston, VA 22092. (703) 648-6820. Monthly. $115.00 per year. Also available on CD-ROM.

ANNUAL REVIEWS AND YEARBOOKS

Advances in Hydrosciences. Academic Press Inc., 6277 Sea Harbor Drive, Orlando, FL 32887. (800) 321-5068. Inquire for cost and availability.

ASSOCIATIONS AND PROFESSIONAL SOCIETIES

American Institute of Hydrology. 3416 University Avenue SE, Minneapolis, MN 55414-3328. (612) 379-1030. FAX (612) 379-0169.

American Water Resources Association. 5410 Grosvenor Lane, Suite 220, Bethesda, MD 20814-2192. (301) 493-8600. FAX (301) 493-5844.

Association of Ground Water Scientists and Engineers. A division of the National Water Well Association, 6375 Riverside Drive, Dublin, OH 43017. (614) 761-1711. FAX (614) 761-3446.

Ground Water Protection Council. 827 NW 63rd Street, Suite 103, Oklahoma City, OK 73116-7639. (405) 848-0690. FAX (405) 848-0722.

National Ground Water Association, 6375 Riverside Drive, Dublin, OH 43017. (614) 761-1711. FAX (614) 761-3446.

National Water Resources Association. 3800 N. Fairfax Drive, Suite 4, Arlington, VA 22203-1703. (703) 524-1544. FAX (703) 524-1548.

Universities Council On Water Resources. 4543 Faner Hall, Department of Geography, Southern Illinois University at Carbondale, Carbondale, IL 62901-4526. (618) 536-7571. FAX (618) 453-2671.

DIRECTORIES AND BIOGRAPHICAL SOURCES

Directory of Engineering Societies and Related Organizations. Gordon Davis. 13th edition. American Association of Engineering Societies, 1111 19th Street NW, Suite 608, Washington, DC 20036. (202) 296-2237 or (800) 658-8897. 1989. Inquire for price.

International Research Centers Directory. Gale Research, 835 Penobscot Building, Detroit, MI 48226-4094. (313) 961-2242. (800) 347-4253. FAX (313) 961-6083. 8th edition. 1995. $410.00

Research Centers Directory. Gale Research, 835 Penobscot Building, Detroit, MI 48226-4094. (313) 961-2242. (800) 347-4253. FAX (313) 961-6083. $485.00.

Scientific and Technical Organizations and Agencies Directory. Gale Research, 835 Penobscot Building, Detroit, MI 48226-4094. (313) 961-2242. (800) 347-4253. FAX (313) 961-6083. 4th edition. 1996. $195.00.

ENCYCLOPEDIAS AND DICTIONARIES

The Encyclopedia of the Solid Earth Sciences. Philip Kearey, editor-in-chief. Blackwell Scientific Publications, 238 Main Street, Cambridge, MA 02142. (617) 876-7000 or (800) 759-6102. FAX (617) 876-7022. 1993. $60.00.

Encyclopedia of Physical Science and Technology. Robert A. Meyers, ed. 18 volumes. Academic Press Inc., 6277 Sea Harbor Drive, Orlando, FL 32887. (800) 321-5068. 1992. $2100.00.

McGraw-Hill Dictionary of Scientific and Technical Terms. Sybil P. Parker, ed. 5th edition. McGraw-Hill Publishing Company, 1221 Avenue of the Americas, New York, NY 10020. (800) 262-4729 or (212) 512-3825. 1993. $110.50.

GENERAL WORKS

Applied Principles of Hydrology. John C. Manning. 2nd edition. Macmillan Publishing, 200 Old Tappan Road, Old Tappan, NJ 07675. (800) 223-2336. FAX (800) 445-6991. 1991. $37.40.

Hydrology: An Introduction To Hydrologic Science. Rafael A. Bras. Addison-Wesley Publishing Company, 1 Jacob Way, Reading, MA 01867. (617) 944-3700 or (800) 447-2226. FAX 617-942-1117. 1990. Inquire for cost and availability.

Hydrology in Practice. Elizabeth M. Shaw. Third edition. Chapman & Hall, 115 Fifth Avenue, New York, NY 10211-0906. 1994. $34.50.

Introduction To Hydrology. Warren Viessman, et al. Fourth edition. Harpercollins Publishers, Inc., 10 E. 53rd Street, New York, NY 10022-5299. (800) 331-3761. FAX (212) 207-7145. 1995. $57.00.

Practical Hydrololology. Jack D. Keen. D'Zign Land Survey and Development, PO Box 1370, Pacifica, CA 94004. 1995. $32.00.

HANDBOOKS AND MANUALS

Groundwater: Managing the Unseen Resources: A Handbook For States. Edwin H. Clark and Philip J. Cherry. World Wildlife Fund & the Conservation Foundation, 1250 24th Street NW, Washington, DC 20037. (202) 293-4800. FAX (410) 516-6998. 1992. $8.50.

Handbook of Hydrology. David R. Maidment, editor in chief. McGraw-Hill Publishing Company, 1221 Avenue of the Americas, New York, NY 10020. (800) 262-4729 or (212) 512-3825. 1993. $115.00.

Practical Groundwater Hydrology. Michael E. Renz. Lewis Publishers, 121 S. Main St., PO Box 519, Chelsea, MI 48118. (313) 475-8619 or (800) 272-7737. 1995. Inquire for cost and availability.

ONLINE DATABASES AND CD-ROMS

Compendex Plus. Engineering Information, Inc., 345 East 47th Street, New York, NY 10017. (212) 705-7600 or (800) 221-1044. Contains citations with abstracts to worldwide literature in engineering and technology, from 1970 to present. Available on online BRS,(800) 289-4277, DIALOG, (800) 334-2564, ORBIT (800) 456-7248, and STN International, FIZ Karlsruhe, P.O. Box 2465, W-7500, Karlsruhe 1, Germany, online services. Also available on CD-ROM. Inquire as to cost and availability.

Current Contents Search. Institute for Scientific Information, 3501 Market Street, Philadelphia, PA 19104. (215) 386-0100. FAX (215) 386-6362. Contains citations to articles listed in the table of contents of science and technology journals. Also articles in social sciences and life sciences journals. Available on BRS,(800) 289-4277, DIALOG,(800) 334-2564, online services. Inquire as to cost and availability.

Dissertation Abstracts. University Microfilms International, 300 North Zeeb Road, Ann Arbor, MI 48106. (800) 521-0600 or (313) 761-4700. Scope includes virtually all doctoral dissertations accepted at accredited American institutions from 1861 to present in 252 subject areas. Available on BRS,(800) 289-4277, DIALOG,(800) 334-2564, and OCLC EPIC,(800) 848-5878, online services. Also available on CD-ROM. Inquire as to cost and availability.

Earth Sciences. U.S. Geological Survey, 12201 Sunrise Valley Drive, Reston, VA 22092-9998. (703) 648-4460. CD-ROM of earth science databases including the U.S. Geological Survey Library, Earth Science Data Directory, and GEOINDEX, citations to published geological maps. $350.00 per year with quarterly updates.

Geoarchive. Geosystems, P.O. Box 1024, Westminster, London, England, SW1 P 2JL. Citations to literature on geoscience, 1969 to present. Inquire as to online cost and availability.

Geobase. Elsevier - Geo Abstracts, Regency House, 34 Duke Street, Norwich NR3 3AP, England. Contains citations to the worldwide earth science literature from 1980 to date. Available on DIALOG, ORBIT online services. Inquire as to cost and availability.

GEOREF: Bibliography and Index of Geology. American Geological Institute, 4220 King Street, Alexandria, VA 22302. (703) 379-2480. Fax (703) 379-7563. Monthly. Inquire for price and availability.

NTIS Bibliographic Database. National Technical Information Service, 5285 Port Royal Road, Springfield, VA 22161. (703) 487-4929 or FAX (703) 321-8199. Broad coverage of government-sponsored science and technology research reports, 1964 to present. Available on BRS,(800) 289-4277, DIALOG, (800) 334-2564, ORBIT, (800) 456-7248, and STN International, FIZ Karlsruhe, P.O. Box 2465, W-7500, Karlsruhe 1, Germany, online services. Also available on CD-ROM. Inquire as to cost and availability.

Water Resources Abstracts (WRA). U.S. Geological Survey, Water Resources Scientific Information Center, 12201 Sunrise Valley Drive, Reston, VA 22092-9998. (703) 648-4460. Contains citations with abstracts to scientific and technical literature on the water-sources-related aspects of the physical, social and life sciences, from 1968 to present. Available on DIALOG,(800) 334-2564, and ORBIT,(800) 456-7248, online services. Inquire as to cost and availability.

Waternet. American Water Works Association, Technical Library. Available on DIALOG online services. Citations to literature on water quality, water utility management, water analysis, water pollution, and related areas, 1971 to present. Available on DIALOG,(800) 334-2564, online service. Inquire as to cost and availability.

PERIODICALS

Ground Water. Ground Water Publishing Company, National Well Water Association, 6375 Riverside Drive, Dublin, OH 43017. (614) 761-1711. FAX (614) 761-3446. 1963 to present. Bi-monthly. $90.00 per year.

Ground Water Monitoring and Remediation. Ground Water Publishing Company, National Well Water Association, 6375 Riverside Drive, Dublin, OH 43017. (614) 761-1711. FAX (614) 761-3446. 1981 to present. Quarterly. $30.00 per year.

Hydrological Science and Technology. American Institute of Hydrology, 3416 University Avenue SE, Minneapolis, MN 55414-3328. (612) 379-1030. FAX (612) 379-0169. 1985 to present. Quarterly. $75.00 per year.

Journal of Hydrology. Elsevier Science Inc., Box 882, Madison Square Station, New York, NY 10159. (212) 989-5800. FAX (212) 633-3990. 1963 to present. 48 times a year. $2309.00.

Journal of Water Resources Planning and Management. American Society of Civil Engineers, Water Resources Planning and Management Division, 345 East 47th Street, New York, NY 10017-2398. (212) 705-7520 or (800) 548-2723. Bi-monthly. $112.00 per year.

Water Resources Research. American Geophysical Union, 2000 Florida Avenue, N.W., Washington, DC 20009. (202) 462-6900. FAX (202) 328-0566. 1965 to present. Monthly. $660.00 for non-members.

RESEARCH CENTERS AND INSTITUTES

Colorado State University, Hydraulics and Hydromachinery Research Laboratories. Engineering Research Center, Fort Collins, CO 80523. (303) 491-8404. FAX (303) 491-8671.

University of Alaska Fairbanks, Water Research Center. Fairbanks, AK 99775-1760. (907) 474-7350. FAX (907) 474-6087.

University of California At Berkeley, Hydraulic Laboratories. 412 O'Brien Hall, Berkeley, CA 94720. (415) 642-6777.

University of Minnesota, St. ANTHONY FALLS HYDRAULIC LABORATORY. Mississippi River at 3rd Avenue, Minneapolis, MN 55414-2196. (612) 627-4010. FAX (612) 627-4609.

HYDROMECHANICS

See: HYDRODYNAMICS

HYDROSTATICS

See: HYDRODYNAMICS

HYDROSTATICS

See: HYDRODYNAMICS

HYPERBOLIC NAVIGATION SYSTEM

See: NAVIGATION

HYPERSONIC FLIGHT

See: AERODYNAMICS

HYPERSONICS

See: ACOUSTICS

I

ICE

See also: CLOUDS, METEOROLOGY, RAIN, WATER RESOURCES, WEATHER FORECASTING

ABSTRACT SERVICES AND INDEXES

*Applied Science and Technology Index; A Cumulative Subject Index To English Language Periodicals in the Fields ofaeronautics and Space Science, Computer Technology,chemistry, Construction Industry, Energy and Related Areas.*H.W. Wilson Co., 950 University Avenue, Bronx, NY 10452. -9978. (212) 588-8400. (800) 367-6770. FAX (718) 590-1617.From 1958 to present. Monthly. Inquire about cost andavailability. Also available online (BRS and WILSONLINE)and on CD-ROM. ISSN: 0003-6986.

General Science Index. H.W. Wilson Co., 950 University Avenue, Bronx, NY 10452. (212) 588-8400. (800) 367-6770. FAX (718) 590-1617. From 1978 to present. Ten issues per year; quarterly and annual cumulations. Service basis. Available on CD-ROM and online. Inquire regarding cost and availability. ISSN: 0162-1963.

Government Reports Announcements and Index. National Technical Information Service (NTIS), 5285 Port Royal Road, Springfield, VA 22161. (703) 487-4650. FAX (703) 321-8547. From 1968 to present. Annual. $630.00 per year. Also available online as *NTIS Bibliographic Database* and on CD-ROM.

Meteorological and Geoastrophysical Abstracts. American Meteorological Society, c/o Inforonics, Inc., 550 Newtown Road, Box 458, Littleton, MA 01460. (508) 486-8976. FAX (508) 486-0027. Covers literature in environmental sciences, meteorology, astrophysics, hydrology, glaciology, and physical oceanography. 1950 to present. Monthly. $950.00 per year. ISSN: 0026-1130. Also available online (DIALOG) and on CD-ROM.

Science Citation Index. SCI. Institute for Scientific Information, 3501 Market Street, Philadelphia, PA 19104. (215) 386-0100. (800) 523-1850. FAX (215) 386-2991. 1961 to present. Six issues per year, plus annual cumulation. $11650.00 per year. Also available online and on CD-ROM. Inquire about price and availability. ISSN: 0036-827X.

Selected Water Resources Abstracts. U.S. Geological Survey. Distributed by National Technical Information Service (NTIS), 5285 Port Royal Road, Springfield, VA 22161. (703) 487-4650. Monthly. $115.00 per year.

ANNUAL REVIEWS AND YEARBOOKS

Annual Review of Earth and Planetary Sciences. Annual Reviews, Inc., 4139 El Camino Way, Palo Alto, CA 94306-0897. (415) 493-4400. (800) 523-8635. FAX (415) 855-9815. Annual. $55.00 per year.

Developments in Atmospheric Science Series. Elsevier Science Publishing Company, Inc., 655 Avenue of the Americas, New York, NY 10010. (212) 989-5800. Irregular. Inquire about price and availability.

ASSOCIATIONS AND PROFESSIONAL SOCIETIES

American Meteorological Society, 45 Beacon Street, Boston, MA 02108-3693. (617) 227-2425. FAX (617) 742-8718.

International Association of Meteorology and Atmospheric Physics. c/o UCAR, P.O. Box 3000, Boulder CO 80307-3000. (303) 497-1673.

International Association of Severe Weather Specialists. c/o Warren Faidley, PO Box 31808, Tuscon, AZ 85751. (520) 751-9964. FAX (520) 751-1185.

National Environmental Satellite Data, and Information Service. 2069 Federal Building 4 Room 2060, Washington, DC 20233. (301) 763-71990. FAX (301) 763-4011.

National Weather Association, 6704 Wolke Court, Montgomery, AL 36116-2134. (334) 213-0388. FAX (334) 213-0388.

University Corporation For Atmospheric Research. PO Box 3000, 1850 Table Mesa Drive, Boulder, CO 80307-3000. (303) 497-1673. FAX (303) 497-1654.

BIBLIOGRAPHIES

Information Sources in the Earth Sciences. David N. Wood, Joan E. Hardy, and Anthony P. Harvey, editors. Bowker-Saur/ K.G. Saur. Distributed by R.R. Bowker, 121 Chanlon Road, New Providence, NJ 07974. (800) 521-8110 or (908) 464-6800. Second edition. 1989. $85.00.

ICE

Ency. of Physical Sciences and Engineering Info. Sources

DIRECTORIES AND BIOGRAPHICAL SOURCES

American Men and Women of Science: Physical and Biological Sciences. Eighteenth edition. R.R. Bowker Company, 245 West 17th Street, New York, NY 10011. (800) 521-8810 or (212) 916-1600. $750.00.

American Meteorological Society. Professional Directory. American Meteorological Society. 45 Beacon Street, Boston, MA 02108. (617) 227-2425. Included in each issue of the Bulletin of the Society.

Meteorological Services of the World. American Meteorological Society, 45 Beacon Street, Boston, MA 02108- 3693. (617) 227-2425. FAX (617) 742-8718. Annual. $70.00.

national Weather Service Offices and Stations. National Oceanic and Atmospheric Administration, Deparatment of Commerce, Silver Spring, MD 20910. (301) 427-7698. Annual. Free

Research Centers Directory. Gale Research, 835 Penobscot Building, Detroit, MI 48226-4094. (313) 961-2242. (800) 347-4253. 20th edition. 1995. $485.00. ISSN: 0080-1518.

ENCYCLOPEDIAS AND DICTIONARIES

Climates of the States. Bair. Gale Research, 835 Penobscot Building, Detroit, MI 48226-4094. (313) 961-2242. (800) 347-4253. 5th edition. 1995. $255.00.

Concise Oxford Dictionary of the Earth Sciences. Ailsa Allaby and Michael Allaby, editors. Oxford University Press, Inc., 200 Madison Avenue, New York, NY 10016. (800) 334-4249 or (212) 679-7300. 1990. $42.95.

Encyclopedia of Climatology. Rhodes W. Fairbridge, and John E. Oliver. Chapman & Hall, 1 Penn Plaza, New York, NY 10119. (212) 564-1060. 1986. $115.00.

McGraw-Hill Encyclopedia of Ocean and Atmospheric Sciences. Sybil P. Parker, editor. McGraw-Hill Publishing Company, 1221 Avenue of the Americas, New York, NY 10020. (800) 262-4729 or (212) 512-3825. 1979. $79.95.

GENERAL WORKS

Cold Climate Hydrometeorology. D.S. Upadhyay. Halsted Press, 605 Third Avenue, New York, NY 10158-0012. (212) 850-6400. (800) 526-5368. 1995. $44.95.

Development of Anti-icing Technology. Robert R. Blackburn. SHRP - Strategic Highway Research Program 2101 Constitution Avenue NW, Washington, DC 20055. (202) 334-3313. (800) 624-6242. 1993. $20.00 in paper.

Dynamics of Snow and Ice Masses. Samuel C. Colbeck. Academic Press, Inc., 6277 Sea Harbor Drive, Orlando FL 32887. (800) 3221-5068. 1980. $118.00.

Ice Destruction Methods and Technology. V. V. Bogorodsky, et al. Kluwer Academic Publishers, 101 Philip Drive, Assinippi Park, Norwell, MA 02061. 1986. $79.00

Ice Technology For Polar Operations. T. K. Murthy, et al, editors. Computational Mech MA 1990. $149.00.

Improved Cutting Edges For Ice Removal. Wilfred A. ___ SHRP-Strategic Highway Research Program, 2101 Constitution Avenue NW, Washington, DC 20055. (202) 334-3313. (800) 624-6242. 1993. $15.00 in paper.

Remote Sensing of Sea Ice and Icebergs. Simon Haykin, et al, editors. John Wiley & Sons, Inc., 605 Third Avenue, New York, NY 10158-0012. (212) 850-6000. (800) 225-5945. 1994. $95.00.

Severe and Unusual Weather. Joe R. Eagleman. Trimedia Publishing Company, 12008 West 87th Street, Suite 117, Lenexa, KS 66215. (913) 599-0505. Second edition. 1990. $41.95.

Snow Crystals. W.A. Bentley and W.J. Humphreys. Dover Publications, Inc., 180 Varick Street, New York, NY 10014. (212) 255-3755 or (800) 223-3130. 1931. $15.95 in paper.

Studies in Ice Physics and Ice Engineering. G. N. Takovlev.

Coronet Books, 311 Bainbridge Street, Philadelphia PA 19147. (215) 925-5083., 1972. $56.00.

HANDBOOKS AND MANUAL

Author's Guide To the Journals of the American Meteorological Society. American Meteorological Society, 45 Beacon Street, Boston, MA 02108-3693. (617) 227-2425. FAX (617) 742-8718. 1983. $15.00 in paper.

Basic Meteorology Lab Manual. Thomas A. Leavy. Allegheny Press, P.O. Box 220, Elgin, PA 19413. (814) 664-8504. Second edition. 1969. $7.95.

Guidelines For the Control of Blowing and Drifting Snow. Ronald D. Tabler. SHRP - Strategic Highway Research Program. 2101 Constitution Avenue NW, Washington, DC 20055. (202) 334-3313. (800) 624-6242. 1994. $20.00.

Handbook of Test Methods For Evaluating Chemical Deicers. Cecil C. Chappelow, et al. SHRP - Strategic Highway Research Program. 2101 Constitution Avenue NW, Washington, DC 20055. (202) 334-3313. (800) 624-6242.1992. $20.00 in paper.

Weather Almanac. Gale Research, 835 Penobscot Building, Detroit, MI 48226-4094. (313) 961-2242. (800) 347- 4253. 7th edition. 1996. $130.00.

Weather Satellite Handbook. American Radio Relay League, 225 Main Street, Newington CT -6111. (203) 666-1541. 1994. $20.00.

ONLINE DATABASES AND CD-ROMS

Climate Assessment Database. National Weather Service, National Meteorological Center, 5200 Auth Road, Suite 101, Camp Springs, MD 20233. (301) 763-8016. Contains daily, weekly and monthly summaries of North American and world climatological data. Also provides five to ten weather forecasts, and 30 to 90 day weather outlook. Subscription required. Inquire as to cost and availability.

Current Contents Search. Institute for Scientific Information, 3501 Market Street, Philadelphia, PA 19104. (215) 386-0100. FAX (215) 386-6362. Contains citations to articles listed in the table of contents of science and technology journals. Also articles in social sciences and life sciences journals. Available on

BRS,(800) 289-4277, DIALOG,(800) 334-2564, online services. Inquire as to cost and availability.

Dissertation Abstracts. University Microfilms International, 300 North Zeeb Road, Ann Arbor, MI 48106. (800) 521-0600 or (313) 761-4700. Scope includes virtually all doctoral dissertations accepted at accredited American institutions from 1861 to present in 252 subject areas. Available on BRS,(800) 289-4277, DIALOG,(800) 334-2564, and OCLC EPIC,(800) 848-5878, online services. Also available on CD-ROM. Inquire as to cost and availability.

Meteorological and Geoastrophysical Abstracts. American Meteorological Society, 45 Beacon Street, Boston, MA 02108-3693. (617) 227-2425. FAX (617) 742-8718. Contains citations and abstracts to the worldwide literature on significant research in meteorology and geoastrophysics. Related topics include physical oceanography, hydrology, environmental sciences and glaciology. Covers the period 1972 to present. Available on DIALOG,(800) 334-2564, online service. Inquire as to cost and availability.

NTIS Bibliographic Database. National Technical Information Service, 5285 Port Royal Road, Springfield, VA 22161. (703) 487-4929 or FAX (703) 321-8199. Broad coverage of government-sponsored science and technology research reports, 1964 to present. Available on BRS,(800) 289-4277, DIALOG, (800) 334-2564, ORBIT, (800) 456-7248, and STN International, FIZ Karlsruhe, P.O. Box 2465, W-7500, Karlsruhe 1, Germany, online services. Also available on CD-ROM. Inquire as to cost and availability.

SCISEARCH. Institute for Scientific Information, 3501 Market Street, Philadelphia, PA 19104. (800) 523-1850 or (215) 386-0100. Broad multidisciplinary title and author index to the international literature of science and technology, 1974 to present. Available on DIALOG and ORBIT online services. Also available on CD-ROM. Inquire as to cost and availability.

WILSONLINE. H.W. Wilson Company, 950 University Avenue, Bronx, NY 10452. (800) 367-6770 or (212) 588-8400. Makes available online versions of the printed H.W. Wilson indexes including Applied Science and Technology Index, Business Periodicals Index, General Science Index, and Readers' Guide to Periodical Literature. Period covered is generally 1983 to present. Available on BRS,(800) 289-4277, DIALOG,(800) 334-2564, and OCLC EPIC,(800) 848-5878, online services. Also available on CD-ROM. Inquire as to cost and availability.

World Climate Disc. Chadwyck-Healey Inc., 1101 King Street, Alexandria, VA 22314. (703) 683-4890. FAX (703) 683-7589. Weather and climate data from approximately 5,000 weather stations worldwide, covering the years 1854 to 1990 on CD-ROM. First edition 1992. Approximately $1200.00 per year with annual updates.

World Weatherdisc. WeatherDisc Associates, Inc., 4584 N.E. 89th Street, Seattle, WA 98115. (206) 524-4314. FAX (206) 543-0308. Meteorological data on CD-ROM which describes the climate of the earth today and for the past few hundred years. First edition 1989. Approximately $295.00 per year with biannual updates.

PERIODICALS

American Meteorological Society. Bulletin. American Meteorological Society, 45 Beacon Street, Boston, MA 02108-3693. (617) 227-2425. FAX (617) 742-8718. Monthly. $60.00 per year.

American Meteorological Society. Meteorological Monographs. American Meteorological Society, 45 Beacon Street, Boston, MA 02108-3693. (617) 227-2425. FAX (617) 742-8718. Irregular. Price varies.

American Weather Observer. Association of American Weather Observers, 401 Whitney Boulevard, Box 455, Belvedere, IL 61008. (815) 544-5665. FAX (815) 544-6334. Monthly. $21.00 per year. *Journal of Applied Meteorology.* American Meteorological Society, 45 Beacon Street, Boston, MA 02108-3693. (617) 227-2425. FAX (617) 742-8718. Monthly. $165.00 per year.

Journal of Climate. American Meteorological Society, 45 Beacon Street, Boston, MA 02108-3693. (617) 227-2425. FAX (617) 742-8718. Monthly. $175.00 per year.

Journal of the Atmospheric Sciences. American Meteorological Society, 45 Beacon Street, Boston, MA 02108-3693. (617) 227-2425. FAX (617) 742-8718. Semi-monthly. $320.00 per year.

Monthly Weather Review. American Meteorological Society, 45 Beacon Street, Boston, MA 02108-3693. (617) 227-2425. FAX (617) 742-8718. Monthly. $205.00 per year.

National Weather Digest. National Weather Association, 4400 Stamp Road, Room 404, Temple Hills, MD 20748. (301) 899-3784.

Weather. Royal Meteorological Society, 104 Oxford Road, Reading, Berks RG1 7LJ, England. Monthly. $44.00 per year.

Weatherwise. Heldref Publications, 1319 Eighteenth Street, N.W., Washington, DC 20036-1802. (202) 296-6267. FAX (202) 296-5149. Bi-monthly. $28.00 per year.

RESEARCH CENTERS AND INSTITUTES

Atmospheric Ice Laboratory. University of Nevada, PO Box 60220, Reno, NV 89506. (702) 972-1676.

Ice Core Laboratory. State University of New York at Buffalo, Department of Geological Sciences, 4240 Ridge Lea Road, Amherst, NY 14226. (716) 831-3054.

Ice Research Laboratory. Dartmouth college, Thayer School of Engineering, Hanover, NH 03755. (603) 646-2888.

Glacier Research Group. University of New Hampshire, Durham, NH 03824. (603) 862-1718.

Climate Analysis Center. National Weather Service, 5200 Auth Road, Camp Springs, MD 20746. (301) 763-8167.

Cooperative Institute For Mesoscale Meteorological Studies. University of Oklahoma, 401 East Boyd, Norman, OK 73019. (405) 325-3041.

Desert Research Institute, Atmospheric Water Resources Laboratory. P.O. Box 60220, Reno NV 89506. (702) 677-3123.

Institute of Snow Research. Michigan Technological University, Houghton, MI 49931. (906) 487-2750. FAX (906) 487-2202.

National Center For Atmospheric Research. P.O. Box 3000, Boulder, CO 80307. (303) 496-1000. FAX (303) 497-1654.

National Snow and Ice Data Center. University of Colorado - Boulder, CB 449, Boulder, CO 80309-0449. (303) 492-5171. FAX (303) 492-2468.

ICE AGES

See also: GEOLOGY, GLACIOLOGY, STRATIGRAPHY

ABSTRACT SERVICES AND INDEXES

Bibliography and Index of Geology. American Geological Institute, 4220 King Street, Alexandria, VA 22302. (703) 379-2480. Fax (703) 379-7563. Monthly. $1295.00 per year. Also available online as GEOREF.

Current Contents: Physical, Chemical and Earth Sciences. Institute for Scientific Information, 3501 Market Street, Philadelphia, PA 19104. (215) 386-0100. FAX (215) 386-6362. Weekly. $360.00 per year.

Geological Abstracts. Elsevier - Geo Abstracts, Regency House, 34 Duke Street, Norwich NR3 3AP, England. Monthly. $760.00 per year. Also available online as GEOBASE.

Geological Society of America. Abstracts with Programs. Geological Society of America. 3300 Penrose Place, P.O. Box 9140, Boulder, CO 80301-9140. (303) 447-2020. Abstracts and programs of the annual conference. Annual. $69.00.

Meteorological and Geoastrophysical Abstracts. American Meteorological Society, c/o Inforonics Inc., 550 Newtown Road, Littleton, MA 01460. (508) 486-8976. FAX (508) 486-0027. 1950 to present. Monthly. $950.00 per year. Current abstracts of books, reports, research papers, and miscellaneous literature on environmental sciences, meteorology, astrophysics, hydrology, glaciology, and physical oceanography.

Science Citation Index. Institute for Scientific Information, 3501 Market Street, Philadelphia, PA 19104. (215) 386-0100. FAX (215) 386-6362. Inquire about availability and cost. Also available on CD-ROM.

ASSOCIATIONS AND PROFESSIONAL SOCIETIES

American Geological Institute. 4220 King Street, Alexandria, VA 22302. (703) 379-2480. Fax (703) 379-7563.

Geological Society of America. 3300 Penrose Place, PO Box 9140, Denver, CO 80301-9140. (303) 447-2020.

DIRECTORIES AND BIOGRAPHICAL SOURCES

American Men & Women of Science, 1995-96. R.R. Bowker Staff, eds. 19th edition. 8 volumes. R.R. Bowker/Reed International Publishing Company, 121 Chanlon Road, New Providence, NJ 07974. (908) 464-6800 or (800) 521-8110. 1995. $850.00

Geological Society of America Membership Directory. 3300 Penrose Place, PO Box 9140, Denver, CO 80301-9140. (303) 447-2020. Annual. Inquire for cost and availability.

International Research Centers Directory. Gale Research, 835 Penobscot Building, Detroit, MI 48226-4094. (313) 961-2242. (800) 347-4253. FAX (313) 961-6083. 8th edition. 1995. $410.00.

Research Centers Directory. Gale Research, 835 Penobscot Building, Detroit, MI 48226-4094. (313) 961-2242. (800) 347-4253. FAX (313) 961-6083. $485.00.

Scientific and Technical Organizations and Agencies Directory. Gale Research, 835 Penobscot Building, Detroit, MI 48226-4094. (313) 961-2242. (800) 347-4253. FAX (313) 961-6083. 4th edition. 1996. $195.00.

ENCYCLOPEDIAS AND DICTIONARIES

Glossary of Geology. Robert Bates, et al. Reprint of third edition. American Geological Institute, 4220 King Street, Alexandria, VA 22302-1507. (703) 379-2480. FAX (301) 379-7563. 1990. $225.00.

The Encyclopedia of the Solid Earth Sciences. Philip Kearey, editor-in-chief. Blackwell Scientific Publications, 238 Main Street, Cambridge, MA 02142. (617) 876-7000 or (800) 759-6102. FAX (617) 876-7022. 1993. $60.00.

GENERAL WORKS

After the Ice Age: The Return of Life To Glaciated North America. E.C. Pielou. University of Chicago Press, 5801 Ellis Avenue, 4th Floor, Chicago, IL 60637. (312) 702-7700. FAX (312) 702-9756. 1991. $24.95 (hardback) and $13.95 (paperback).

Earth's Glacial Record. M. Deynoux, et al., editors. Cambridge University Press, 40 West 20th Street, New York, NY 10011-4211. (212) 924-3900 or (800) 872-7423. FAX (914) 937-4712. 1994. $89.95.

Ice Age Earth: Late Quaternary Geology and Climate. Alastair G. Dawson. Routledge, Chapman and Hall, Inc., 29 West 35th Street, New York, NY 10001-2291. (212) 244-3336 or FAX (212) 563-2269. 1991. $62.50 (hardback) or $25.00 (paperback).

Ice Ages: Past and Future. Jon S. Erickson. TAB Books, P.O. Box 40, Blue Ridge Summit, PA 17294-0850. (717) 794-2191 or (800) 233-1128. FAX (717) 794-2080. 1990. $22.95 (hardback) or $15.95 (paperback).

ONLINE DATABASES AND CD-ROMS

Earth Sciences. U.S. Geological Survey, 12201 Sunrise Valley Drive, Reston, VA 22092-9998. (703) 648-4460. CD-ROM of earth science databases including the U.S. Geological Survey Library, Earth Science Data Directory, and GEOINDEX, citations to published geological maps. $350.00 per year with quarterly updates.

Geoarchive. Geosystems, P.O. Box 1024, Westminster, London, England, SW1 P 2JL. Citations to literature on geoscience, 1969 to present. Inquire as to online cost and availability.

Geobase. Elsevier - Geo Abstracts, Regency House, 34 Duke Street, Norwich NR3 3AP, England. Contains citations to the worldwide earth science literature from 1980 to date. Available on DIALOG, ORBIT online services. Inquire as to cost and availability.

GEOREF: Bibliography and Index of Geology. American Geological Institute, 4220 King Street, Alexandria, VA 22302. (703) 379-2480. Fax (703) 379-7563. Monthly. Inquire for price and availability.

SCISEARCH. Institute for Scientific Information, 3501 Market Street, Philadelphia, PA 19104. (800) 523-1850 or (215) 386-0100. Broad multidisciplinary title and author index to the international literature of science and technology, 1974 to present. Available on DIALOG,(800) 334-2564, and ORBIT,(800) 456-7248, online services. Also available on CD-ROM. Inquire as to cost and availability.

PERIODICALS

American Journal of Science. Box 6666, Yale Station, New Haven, CT 06511-8130. (203) 432-3131. FAX (203) 432-5668. 1818 to present. Monthly except July and August. $40.00 per year.

Geological Magazine. Cambridge University Press, Journals Department, 40 West 20th Street, New York, NY 10011-4211. (212) 924-3900. 1864 to present. Bi-monthly. $263.00 per year.

Geological Society Journal. Geological Society Publishing House, Unit 7, Brassmill Enterprise Centre, Brassmill Lane, Bath BA1 3JN, England. Telephone 0225-445046. FAX 0225-442836. 1845 to present. Six times a year. $524.00.

Geological Society of America Bulletin. 3300 Penrose Place, PO Box 9140, Denver, CO 80301-9140. (303) 447-2020. 1888 to present. Monthly. $205.00 per year.

Geology. Geological Society of America, 3300 Penrose Place, PO Box 9140, Denver, CO 80301-9140. (303) 447-2020. 1973 to present. Monthly. $170.00 per year for non-members.

Glaciological Data. World Data Center A for Glaciology (Snow & Ice), National Snow and Ice Data Center, CIRES, Campus Box 449, University of Colorado, Boulder, CO 80309. (303) 492-5171. FAX (303) 492-2468. 1977 to present. Irregular. Inquire for price and availability.

Glaciology and Quaternary Geology. Kluwer Academic Publishers, P.O. Box 358, Accord Station, Hingham, MA 02018-0358. (617) 871-6000. 1985 to present. Irregular. Price varies.

Journal of Geology. University of Chicago Press, Journals Division, 5720 S. Woodlawn Avenue, Chicago, IL 60637. (312) 753-3347. FAX (312) 753-0811. 1893 to present. Bi-monthly. $45.00 per year.

RESEARCH CENTERS AND INSTITUTES

University of Maine, Institute For Quaternary Studies. Orono, ME 04469. (207) 581-2190.

University of Washington, Quaternary Research Center. AK-60, Seattle, WA 98195. (206) 543-1166. FAX (206) 543-3936.

ICEBERGS

See: ICE

IGNEOUS ROCKS

See also: CRUST, GEOCHEMISTRY, METAMORPHIC ROCKS, MINERALOGY, PETROLOGY, PHYSICAL GEOLOGY, SEISMOLOGY, VOLCANOLOGY

ABSTRACT SERVICES AND INDEXES

Bibliography and Index of Geology. American Geological Institute, 4220 King Street, Alexandria, VA 22302. (703) 379-2480. Fax (703) 379-7563. Monthly. $1295.00 per year. Also available online as GEOREF.

Bibliography of Economic Geology. Geosystems, Box 40, Didcot Oxon Ox11 9BX, England. Bimonthly. $150.00 per year. Also available online as GeoArchive.

Chemical Abstracts - Applied Chemistry and Chemical Engineering Sections. Chemical Abstracts Service, 2540 Olentangy River Road, Box 3012, Columbus, OH 43210. (800) 848-6538 or (614) 447-3600, FAX (614) 447-3713. Bi-weekly. $1410.00 per year. Also available online as CA (Chemical Abstracts).

Deep-sea Research. Part A: Oceanographic Research Papers.

deep-sea Research. Part B: Oceanographic Literature Review. Pergamon Press, Inc., Maxwell House, Fairview Park, Elmsford, NY 10523. (914) 592-7700. Fax (914) 592-3625. Twelve times per year. $1370 per year for Parts A and B.

Geological Abstracts. Elsevier - Geo Abstracts, Regency House, 34 Duke Street, Norwich NR3 3AP, England. Monthly. $760.00 per year. Also available online as GEOBASE.

Geological Society of America. Abstracts with Programs. Geological Society of America. 3300 Penrose Place, P.O. Box 9140, Boulder, CO 80301-9140. (303) 447-2020. Abstracts and programs of the annual conference. Annual. $69.00.

Geophysics Abstracts. Pergamon Press, Inc., Maxwell House, Fairview Park, Elmsford, NY 10523. (914) 592-7700. Fax (914) 592-3625. Twelve times per year. $565.00 per year. Also available in microform.

Mineralogical Abstracts. Mineralogical Society and the Mineralogical Society of America, 41 Queen's Gate, London, SW7 5HR, England. Quarterly. $235.00 per year.

Oceanic Abstracts. Cambridge Scientific Abstracts, 7200 Wisconsin Avenue, Bethesda, MD 20814. (301) 961-6750. Fax (301) 961-6720. Bimonthly. $995.00 per year.

Science Citation Index. Institute for Scientific Information, 3501 Market Street, Philadelphia, PA 19104. (800) 523-1850 or (215) 386-0100. Inquire about price and availability.

General Science Index. H.W. Wilson Co., 950 University Avenue, Bronx, NY 10452. (800) 367-6770 or (212) 588-8400. Inquire about price and availability.

ANNUAL REVIEWS AND YEARBOOKS

Annual Review and Earth and Planetary Sciences. Annual Reviews, Inc., 4139 El Camino Way, Palo Alto, CA 94306-0897. (415) 493-4400. Fax (415) 855-9815. Annual. $55.00 per year.

Minerals Yearbook. Bureau of Mines, U.S. Department of the Interior. Available from U.S. Government Printing office, Washington, DC 20402. (202) 783-3238. Annual. Three volumes. $45.00.

ASSOCIATIONS AND PROFESSIONAL SOCIETIES

American Geological Institute. 4220 King Street, Alexandria, VA 22302. (703) 379-2480. Fax (703) 379-7563.

American Geophysical Union, 2000 Florida Avenue, N.W., Washington, DC 20009. (202) 462-6900.

American Institute of Professional Geologists. 7828 Vance Drive, Suite 103, Arvada, CO 80003. (303) 431-0831.

Association of Engineering Geologists. 323 Boston Post Road, Suite 2D, Sudbury, MA 01776. (508) 443-4639.

Geochemical Society. Department of Terrestrial Magnetism, 5241 Broad Branch Road, N.W., Washington, DC 20015. (202) 686-4370.

Society of Independent Professional Earth Scientists. 4925 Greenville Avenue, Suite 170, Dallas, TX 75206. (214) 363-1780.

BIBLIOGRAPHIES

Geologic Reference Sources: A Subject and Regional Bibliography of Publications and Maps in the Geological Sciences. Dederick C. Ward et al. Scarecrow Press, Inc., 52 Liberty Street, Box 4167, Metuchen, NJ 08840. (800) 537-7107 or (201) 548-8600. 1981. $49.50.

Information Sources in the Earth Sciences. David N. Wood, Joan E. Hardy, and Anthony P. Harvey, editors. Bowker-Saur/K.G. Saur. Distributed by R.R. Bowker, 121 Chanlon Road, New Providence, NJ 07974. (800) 521-8110 or (908) 464-6800. Second edition. 1989. $85.00.

DIRECTORIES AND BIOGRAPHICAL SOURCES

American Men and Women of Science: Physical and Biological Sciences. Eighteenth edition. R.R. Bowker Company, 245 West 17th Street, New York, NY 10011. (800) 521-8810 or (212) 916-1600. $750.00.

American Institute of Professional Geologists. Membership Directory. American Institute of Professional Geologists, 7828 Vance Drive, Suite 103, Arvada, CO 80003. (303) 431-0831. Annual. $18.00.

Directory of Geoscience Departments: United States and Canada. American Geological Institute. 4220 King Street, Alexandria, VA 22302-1507. (703) 379-2480. Fax (703) 379-7563. Twenty-six edition. 1987. $22.00.

Geological Society of America. Membership Directory. Geological Society of America, 3300 Penrose Place, Boulder, CO 80301. (303) 447-2020. Annual. Available to members only.

Research Centers Directory. Gale Research, 835 Penobscot Building, Detroit, MI 48226-4094. (313) 961-2242. (800) 347-4253. $470.00

ENCYCLOPEDIAS AND DICTIONARIES

Concise Oxford Dictionary of the Earth Sciences. Ailsa Allaby and Michael Allaby, editors. Oxford University Press, Inc., 200 Madison Avenue, New York, NY 10016. (800) 334-4249 or (212) 679-7300. 1990. $42.95.

Dictionary of Geological Terms. American Geological Institute. Doubleday and Company, Inc., 245 Park Avenue, New York, NY 10017. (800) 645-6156 or (212) 953-4561. Third edition. 1984. $12.00 in paper.

Dictionary of Petrology. S.I. Tomkeieff. John Wiley and Sons, Inc., 605 Third Avenue, New York, NY 10158. (800) 526-5368 or (212) 850-6000. 1983. $180.00.

Encyclopedia of Igneous and Metamorphic Petrology. D.R. Bowes. Van Nostrand Reinhold, 115 Fifth Avenue, New York, NY 10003. (212) 254-3232 or (800) 926-2665. 1989. $125.95.

Glossary of Geology. Robert L. Bates and Julia A. Jackson. American Geological Institute, 4220 King Street, Alexandria, VA 22032. (703) 379-2480. Third edition. 1987. $75.00.

GENERAL WORKS

Atlas of Igneous Rocks and their Textures. W.S. MacKenzie and others. Halsted Press (Division of John Wiley and Sons, Inc.), 605 Third Avenue, New York, NY 10158. (800) 526-5368 or (212) 850-6418. 1982. $37.95.

The Field Description of Igneous Rocks. R.S. Thorpe & G.C. Brown. Reprint edition. John Wiley and Sons, Inc., 605 Third Avenue, New York, NY 10158. (800) 526-5368 or (212) 850-6000. 1993. Inquire for price.

Holmes' Principles of Physical Geology. P. McL. D. Duff. 4th edition. Chapman & Hall, 115 Fifth Avenue, New York, NY 10211-0906. 1992. $39.95.

Igneous and Metamorphic Rocks Under the Microscope. David Shelley. Chapman & Hall, 115 Fifth Avenue, New York, NY 10211-0906. 1992. $45.00.

Igneous Petrology. Anthony Hall. Halsted Press (Division of John Wiley and Sons, Inc.), 605 Third Avenue, New York, NY 10158. (800) 526-5368 or (212) 850-6418. 1987. $51.95.

Origins of Igneous Rocks. Paul C. Hess. Harvard University Press, 79 Garden Street, Cambridge, MA 02138. (617) 495-2600. 1989. $77.50.

Petrology: The Study of Igneous, Sedimentary & Metamorphic Rocks. Loren A. Raymond. William C. Brown Publications, 2460 Kerper Blvd., Dubuque, IA 52001. (800) 338-5578. 1994. Inquire for cost and availability. HandBOOKS and MANUAL

Agi Data Sheets For Geology in the Field, Laboratory, and office. J.T. Dutro, R.V. Dietrich, and R.M. Foose, editors. American Geological Institute, 4220 King Street, Alexandria, VA 223020-1507 (800) 336-4764 or (703) 379-2480. 1989. $24.95.

Field Mapping For Geology Students. F. Ahmed and D.C. Almond. Unwin Hyman, Inc., 10 East 53rd Street, New York, NY 10022. (212) 207-7626 or (800) 242-7737. 1983. $18.95.

Geology in the Field. R.R. Compton. John Wiley and Sons, Inc., 605 Third Avenue, New York, NY 10158. (800) 526-5368 or (212) 850-6418. 1985. $37.95.

A Manual of Geology For Civil Engineers. John A. Pitts. World Scientific Publishing Company, Inc., 1060 Main Street, Unit B, River Edge, NJ 07661. (800) 227-7562 or (201) 487-9655. 1985. $38.00.

The Mapping of Geological Structures. K.R. McClay. John Wiley and Sons, Inc., 605 Third Avenue, New York, NY 10158. (800) 526-5368 or (212) 850-6000. 1988. $18.95.

ONLINE DATABASES AND CD-ROMS

Earth Sciences. U.S. Geological Survey, 12201 Sunrise Valley Drive, Reston, VA 22092-9998. (703) 648-4460. CD-ROM of earth science databases including the U.S. Geological Survey Library, Earth Science Data Directory, and GEOINDEX, citations to published geological maps. $350.00 per year with quarterly updates.

Geoarchive. Geosystems, P.O. Box 1024, Westminster, London, England, SW1 P 2JL. Citations to literature on geoscience, 1969 to present. Inquire as to online cost and availability.

Geobase. Elsevier - Geo Abstracts, Regency House, 34 Duke Street, Norwich NR3 3AP, England. Contains citations to the worldwide earth science literature from 1980 to date. Available on DIALOG, ORBIT online services. Inquire as to cost and availability.

Geomechanics Abstracts. University of London, Imperial College of Science and Technology, Rock Mechanics Information Service, Royal School of Mines, Prince Consort Road, London SW7 2BP, England. Telephone: 071-589 5111 or FAX 071-589 6806. Scope includes worldwide literature on engineering geology, hydrogeology, mining, rock mechanics, soil mechanics, and tunnelling, 1977 to present. Available on the ORBIT online service. Inquire as to cost and availability.

GEOREF. American Geological Institute, 4220 King Street, Alexandria, VA 22302. (800) 336-4764 or (703) 379-2480. Geology and geosciences literature, 1785 to present for North America. Available on DIALOG, ORBIT online services. Also available on CD-ROM. Inquire as to cost and availability.

NTIS Bibliographic Database. National Technical Information Service, 5285 Port Royal Road, Springfield, VA 22161. (703) 487-4630. Broad coverage of government-sponsored research reports, 1964 to present. Available on BRS, DIALOG, ORBIT and STN online services. Also available on CD-ROM. Inquire as to cost and availability.

SCISEARCH. Institute for Scientific Information, 3501 Market Street, Philadelphia, PA 19104. (800) 523-1850 or (215) 386-0100. Broad multidisciplinary title and author index to the international literature of science and technology, 1974 to present.

Available on DIALOG and ORBIT online services. Also available on CD-ROM. Inquire as to cost and availability.

PERIODICALS

American Journal of Science. Kline Geology Laboratory, Yale University, Box 6666, Yale Station, New Haven, CT 06511-8130. (203) 432-3131. Ten times per year. $40.00 per year.

Canadian Journal of Earth Sciences. National Research Council of Canada, Research Journals, Ottawa, Ont. K1A OR6, Canada. (613) 993-9084. Monthly. $112.00 per year.

Earth and Planetary Science Letters. Elsevier Science Publishing Company, Inc., 655 Avenue of the Americas, New York, NY 10010. (212) 989-5800. Twenty times per year. $945.00 per year.

Geochimica Et Cosmochimica Acta. Pergamon Journals, Inc., Maxwell House, Fairview Park, Elmsford, NY 10523. (914) 592-7700. Fax (914) 592-3625. Monthly. $580.00 per year.

Geoforum. Pergamon Press, Maxwell House, Fairview Park, Elmsford, NY 10523. (914) 592-7700. Fax (914) 592-3625. Quarterly. $305.00 per year.

Geological Magazine. Cambridge University Press, 40 West 20th Street, New York, NY 10011-4211. (212) 924-3900. Bimonthly. $205.00 per year.

Geological Society. Journal. Geological Society of London. Geological Society Publishing House, Unit 7, Brassmill Enterprise Centre, Brassmill Lane, Bath, Avon BA1 3JN, England. 6 times per year. $432.00.

Geological Society of America Bulletin. P.O. Box 9140, 3300 Penrose Place, Boulder, CO 80301. (303) 447-2020. Fax (303) 447-1133. Monthly. $150.00.

Geology. Geological Society of America. P.O. Box 9140, 3300 Penrose Place, Boulder, CO 80301. (303) 447-2020. Fax (303) 447-1133. Monthly. $120.00 per year.

Geophysics. Society of Exploration Geophysicists, P.O. Box 702740, Tulsa, OK 74170. (918) 493-3516. Monthly. $175.00 per year.

Journal of Geology. University of Chicago Press, 5720 Woodlawn Avenue, Chicago, IL 60637. (312) 753-3347. Bimonthly. $38.00 per year.

Journal of Geophysical Research. American Geophysical Union, 2000 Florida Avenue, N.W., Washington, DC 20009. (202) 462-6903. Monthly. $2170.00 per year.

Journal of Petrology. Oxford University Press, Inc., 200 Madison Avenue, New York, NY 10016. (800) 334-4249 or (212) 679-7300. Bimonthly. $240.00.

Journal of Structural Geology. Pergamon Journals, Inc., Maxwell House, Fairview Park, Elmsford, NY 10523. (914) 592-7700. Ten times per year. $375.00 per year.

Journal of Volcanology and Geothermal Research. Elsevier Science Publishing Company, Inc., 655 Avenue of the Ameri-

IGNEOUS ROCKS

Ency. of Physical Sciences and Engineering Info. Sources

cas, New York, NY 10010. (212) 989-5800. Monthly. $500.00 per year.

LITHOS; an international journal of mineralogy, petrology, and geochemistry. Elsevier Science Publishing Company, Inc., 655 Avenue of the Americas, New York, NY 10010. (212) 989-5800. Quarterly. $275.00 per year.

Physics of the Earth and Planetary Interiors. Elsevier Science Publishing Company, Inc., 655 Avenue of the Americas, New York, NY 10010. (212) 989-5800. Twenty-four times per year. $900.00 per year.

Reviews of Geophysics. American Geophysical Union, 2000 Florida Avenue, N.W., Washington, DC 20009. (202) 462-6900. Quarterly. $300.00 per year.

Rock Mechanics and Rock Engineering. Springer-Verlag New York, Inc., 175 Fifth Avenue, New York, NY 10010. (800) 526-7254 or (212) 460-1500. Quarterly. $122.00 per year.

Tectonics. American Geophysical Union, 2000 Florida Avenue, N.W., Washington, DC 20009. (202) 462-6900. Bimonthly. $230.00 per year to individuals.

Tectonophysics; An International Journal of Geotectonics and the Geology and Physics of the Interior of the Earth. Elsevier Science Publishing Company, Inc., 655 Avenue of the Americas, New York, NY 10010. (212) 989-5800. Fifty-six times per year. $1275.00 per year.

RESEARCH CENTERS AND INSTITUTES

Geological Society of America. P.O. Box 9140, 3300 Penrose Place, Boulder, CO 80301. (303) 447-2020.

U.S. Geological Survey, Geologic Division, National Center, 12201 Sunrise Valley Drive, Reston, VA 22092. (703) 648-6600. The major geological research agency of the federal government conducting research in most areas of pure and applied research in the geosciences.

University of California, San Diego. Scripps Institution of Oceanography, Geological Research Division. San Diego, CA 92093-0220. (619) 534-1829. FAX (619) 534-0784.

University of Hawaii At Manoa. Hawaii Institute of Geophysics. 2525 Correa Road, Honolulu, HI 96822. (808) 948-8760.

ILLUMINATION

See also: ELECTRICAL ENGINEERING, POWER ENGINEERING

ABSTRACT SERVICES AND INDEXES

Applied Science and Technology Index; A Cumulative Subject Index To English Language Periodicals in the Fields of Aeronautics and Space Science, Computer Technology, Chemistry, Construction Industry, Energy and Related Areas.

H.W. Wilson Co., 950 University Avenue, Bronx, NY 10452. (212) 588-8400. (800) 367-6770. FAX (718) 590-1617. From

1958 to present. Monthly. Inquire about cost and availability. Also available on CD-ROM and online. ISSN: 0003-6986.

Current Contents: Engineering, Technology and Applied Sciences. Institute for Scientific Information, 3501 Market Street, Philadelphia, PA 19104. (215) 386-0100. FAX (215) 386-6362. 1970 to present. Weekly. $442.00 per year. Also available on CD-ROM and online. Inquire regarding cost and availability. ISSN: 0095-7917.

Current Papers in Electrical and Electronics Engineering. Institution of Electrical Engineers (IEE), London. Distributed by INSPEC/IEEE, Box 1331, 445 Hoes Lane, Piscataway, NJ 08855-1331. (908) 562-5549. 1969 to present. Monthly. $345.00 per year. ISSN: 0011-3778.

Electric Power Industry Abstracts. Edison Electric Institute, c/o Utility Data Institute, 2011 I Street, NW, Suite 700, Washington, DC 20006. 1975 to present. Bimonthly. Inquire as to cost and availability.

Electrical and Electronics Abstracts. Institution of Electrical Engineers (IEE), London. Available from INSPEC/IEEE - Institute of Electrical and Electronic Engineers, Box 1331, Hoes Lane, Piscataway, NJ 08855-1331. (908) 562-5549. 1898 to present. Monthly. $2200.00 per year. Also available on CD-ROM and online as INSPEC.

Engineering Index Monthly; Indexes and Abstracts the World's Engineering and Technical Literature. Engineering Information, Inc., Castle Point on the Hudson, Hoboken, NJ 07030. (201) 216-8500. (800) 221-1044. FAX (201) 216-8532. Monthly. $2300.00 per year. Available online as COMPENDEX and also on CD-ROM. ISSN: 0742-1974.

General Science Index. H.W. Wilson Company, 950 University Avenue, Bronx, NY 10452. (212) 588-8400. (800) 367-6770. FAX (718) 590-1617. From 1978 to present. Ten issues per year; quarterly and annual cumulations. Service basis. Available on CD-ROM and online. Inquire regarding cost and availability. ISSN: 0162-1963.

Government Reports Announcements and Index. U. S. National Technical Information Service (NTIS), 5285 Port Royal Road, Springfield, VA 22161. (703) 487-4650. FAX (703) 321-8547. From 1968 to present. Annual. $630.00 per year. Also available online as *NTIS Bibliographic Database* and on CD-ROM.

Index To IEEE Publications. IEEE Service Center, 445 Hoes Lane, Piscataway, NJ 08855-1331. (908) 981-1393. (800) 678-IEEE. FAX (908) 981-9667. 1973 to present. Annual. ISSN: 0099-1368.

Physics Abstracts. INSPEC. Section A, Science Abstracts. Institution of Electrical Engineers (IEE), London. Available from: INSPEC/IEEE - Institute of Electrical and Electronic Engineers, Box 1331, Hoes Lane, Piscataway, NJ 08855-1331. (908) 562-5549. 1898 to present. 24 issues per year. $2835.00 per year. Also available online and on CD-ROM. ISSN: 0036-8091.

Physics Briefs (Physikalische Berichte). Information Center for Energy, Physics, Mathematics; German Physical Society. V C H Publishers, Inc., 220 East 23rd Street, New York, NY 10010-4606. (212) 683-8333. 1845 to present. 24 issues per year. $2390.00 per year. Also available online. ISSN: 0179-7434.

Science Citation Index. SCI. Institute for Scientific Information, 3501 Market Street, Philadelphia, PA 19104. (215) 386-0100.

(800) 523-1850. FAX (215) 386-2991. 1961 to present. Six issues per year, plus annual cumulation. $11650.00 per year. Also
available online and on CD-ROM.Inquire about price and availability. ISSN: 0036-827X.

ASSOCIATIONS AND PROFESSIONAL SOCIETIES

American Electronics Association. 5201 Great America Way,
Suite 520, P.O. 52990, Santa Clara, CA 95056. (408) 987-
4200. FAX (408) 970-8565.

American Institute of Physics. One Physics Ellipse, College Park,
MD 20740-3843. (301) 209-3100.

American Lighting Association. PO Box 580168, 2050
Stemmons Freeway, Dallas, TX 75258, Dallas, TX 75258. (214)
698- 9898. FAX (214) 698-9899.

Edison Electric Institute. 701 Pennsylvania Avenue NW, Washington, DC 20004-2696. (202) 508-5000, 5454. FAX (202) 508-
5360.

Electronics Industries Association. 2500 Wilson Boulevard,
Arlington, VA 22201. (703) 907-7500. FAX (202) 457-4985.

IEEE (Institute of Electrical and Electronic Engineers). 345 East
47th Street. New York, NY 10017. (212) 705-7900. FAX (212)
705-4929.

Illuminating Engineering Society of North America. 120 Wall
Street, 17th Floor, New York, NY 10005-4001. (212) 248- 5000.
FAX (212) 248-5017.

DIRECTORIES AND BIOGRAPHICAL SOURCES

*American Men and Women of Science: Physical and Biological
Sciences.* R. R. Bowker Inc., 121 Chanlon Road, New Providence, NJ 07974. (908) 464-6800. (800) 521-8110. 20th edition. 8 volumes. 1996. $850.00.

Directory of Engineering Societies and Related Organizations.
American Association of Engineering Societies, 1111 19th
Street, Suite 608, Washington, DC 20036-3603. (202) 296-2237.
Semi-annual. $150.00.

E E M - Electronic Engineer's Master. Hearst Business Communications, Inc., 645 Stewart Avenue, Garden City NY 11530.
(516) 227-1300. ISSN: 0732-9016.

Electrical and Electronic Trades Directory: The Blue Book. Institution of Electrical Engineers, Michael Faraday House, Six
Hills Way, Stevenage, Herts. SG1 2AY, England. Telephone:
0438 313311 or FAX 0438 742840. 1995. $140.00

*Engineering Research Centers: Incorporating Electronics Re-
search Centers.* Stockton Press, 345 Park Avenue, New York,
NY 10010. (212) 689-9200. (800) 221-2123. 4th edition. 1995.
$515.00.

IEEE Membership Directory. Institute of Electrical and Electronics Engineers, IEEE Service Center, 445 Hoes Lane,

Piscataway, NY 08854. (908) 981-1393. (800) 678-IEEE. FAX
(908) 981-9667. 2 volumes. Annual. $190.00.

Illuminating Engineering Society of North America - Membership Guide and Directory. Harris Publishing, 3 Barker Avenue,
White Plains, NY 10601. (914) 946-7500. Quadrennial. Members only. $30.00.

Lighting Dimensions - Directory Issue. Lighting Dimensions
Associates, 135 Fifth Aveue, New York, NY 10010. (212) 677-
5997. Annual, December. $10.00.

Research Centers Directory. Gale Research Company Inc., 835
Penobscot Building, Detroit, MI 48226-4094. (313) 961-2242.
(800) 347-4253. 20th edition. 1995. $485.00. ISSN: 0080- 1518.

Who's Who in Engineering. Gordon Davis, editor. American
Association of Engineering Societies. 1111 19th Street, NY,
Suite 608, Washington, DC 20036. (202) 296-2237. (800) 658-
8897. 9th edition. 1995. $220.00.

Who's Who in Technology. Gale Research, 835 Penobscot Building, Detroit, MI 48226-4094. (313 961-2242. (800) 521-4253.
7th edition. 1995. $195.00. ISN 0-8103-7467-6.

ENCYCLOPEDIAS AND DICTIONARIES

Chambers Science and Technology Dictionary. Peter M.B.
Walker, editor. Cambridge University Press, 40 West 20th Street,
New York, NY 10011-4211. (212) 924-3900. 1988. $39.95.

IEEE Standard Dictionary of Electrical and Electronics Terms.
Christopher J. Booth, editor. IEEE Service Center, 445 Hoes
Lane, Piscataway, NJ 08855-1331. (908) 981-1393. (800) 678-
IEEE. FAX (908) 981-9667. IEEE Standard 100- 1992. 5th edi-
tion. 1993. $150.00.

Illustrated Dictionary of Electronics. Stan Gibilsco. TAB Books,
P.O. Box 40, Blue Summit, PA 17294-0850. (717) 794- 2191.
(800) 233-1128. 7th edition. 1994. $34.95.

McGraw-Hill Electronics Dictionary. J. Markus. McGraw-Hill
Book Company, Inc., 1221 Avenue of the Americas, New York,
NY 10020. (212) 997-3675. 5th edition. 1994. $49.95.

McGraw-Hill Encyclopedia of Science and Technology.
McGraw-Hill Book, Incorporated, 1221 Avenue of the Ameri-
cas, New York, NY 10020. (212) 997-3675. (800) 262-4729.
Seventh edition. Twenty volumes. 1992. $1900.00.

Nomenclature and Definitions For Illuminating Engineering.
Illuminating Engineering Society of North America, 120 Wall
Street, 17th Floor, New York, NY 10005-4001. (212)248-5000.
FAX (212) 248-5017. Revised edition, ANSI standard. 1986.
$30.00.

GENERAL WORKS

Applied Illumination Engineering. Jack L. Lindsey, Prentice-
Hall, 113 Sylvan Avenue, Route 9W, Englewood Cliffs, NJ
07632. (201) 592-2000. (800) 922-0579. 1991. $69.00.

*Chemistry of Artificial Lighting Devices, Lamps, Phosphors, and
Chatode Ray Tubes.* R. C. Ropp. Elsevier Science Publishing
Company, Inc., 655 Avenue of the Americas, New York, NY
10010. (212) 989-5800. FAX (914) 333-2444. 1993.

ILLUMINATION

Ency. of Physical Sciences and Engineering Info. Sources

Color and Light in Man-made Environments. R. Mahnke. Van Nostrand Reinhold, 115 Fifth Avenue, New York, NY 10003. (212) 254-3232. (800) 842-3636. 1993. $39.95 in paper.

Edison's Electric Light: Biography of An Invention. Robert Friedel, et al. Rutgers University Press, 109 Church Street, New Brunswick, NY 08901. (908) 932-7764. (800) 446-9323. 1987. $16.95 in paper.

Effective Industrial Illuminating Systems. Kao Chen, Fairmont Press, Inc., 700 Indian Trail, Liburn, GA 30247. (404) 925-9388. 1994.

Lighting By Design: A Technical Guide. Brian Fitt and Joe Thornley. Butterworth-Heinemann, 313 Washington Street, Newton, MA 02158. (618-928-2500. (800) 366-2665. 1993. $44.95

Lighting Efficiency Applications. Albert Thumann. Fairmont Press, Inc., 700 Indian Trail, Liburn, GA 30247. (404) 925-9388. 2nd edition. 1991. $67.00.

Lighting For Energy Efficient Luminous Environment. Ronald N. Helms and M. Clay Belcher. Prentice-Hall, 113 Sylvan Avenue, Route 9W, Englewood Cliffs, NJ 07632. (201) 592-2000. (800) 922-0579. 1991. $84.00.

HANDBOOKS AND MANUALS

Active Electronic Component Handbook. Charles A. Harper and Harold C. Jones. McGraw-Hill Book Company, 1221 Avenue of the Americas, New York, NY 10020. (212) 512-2000. (800) 262-4729. 2nd edition. 1996. $79.50.

Electrical Engineer's Handbook. Donald Christiansen, editor. McGraw Book Company, 1221 Avenue of the Americas, New York, NY 10020. (212) 512-2000. (800) 262-4729. 4th edition. 1996. $110.00.

Handbook of Electrical and Electronic Technology. Curtis Johnson. Prentice Hall , 113 Sylvan Avenue, Route 9W, Englewood Cliffs, NJ 07632. (201) 592-2000. (800) 922-0579. 1996. $88.00.

Iesna Lighting Handbook: Reference and Applications. Mark Rea, editor. Illuminating Engineering Society of North America, 120 Wall Street, 17th Floor, New York, NY 10005- 4001. (212) 248-5000. FAX (212) 248-5017. Revised edition, ANSI standard. 1986. $30.00. 1993. $398.00.

Lighting Design Handbook. Lee H. Watson. McGraw-Hill Publishing Company, Inc., 1221 Avenue of the Americas, New York, NY 10020. (212) 512-2000. (800) 262-4729. 1990. $72.50.

Lighting Management Handbook. Craig Dilouie. Fairmont Press, Inc., 700 Indian Trail, Liburn, GA 30247. (404) 925- 9388. 1994. $69.00.

McGraw-Hill's Handbook of Electrical Construction Calculations. J. F. McPartland. McGraw-Hill Publishing Company, 1221 Avenue of the Americas, New York, NY 10020. (212) 512-2000. (800) 262-4729. 1996. $54.50.

McGraw-Hill's National Electrical Code Handbook. J. F. McPartland. McGraw-Hill Publishing Company, 1221 Avenue of the Americas, New York, NY 10020. (212) 512-2000. (800) 262-4729. 22nd edition. 1996. $65.00.

Power Systems Handbook: Design, Operation and Maintenance. O. C. Seevers. Prentice Hall , 113 Sylvan Avenue, Route 9W, Englewood Cliffs, NJ 07632. (201) 592-2000. (800) 922- 0579. 1991. $58.00.

Standard Handbook For Electrical Engineers. Donald Fink. McGraw-Hill Publishing Company, 1221 Avenue of the Americas, New York, NY 10020. (212) 512-2000. (800) 262- 4729. 13th edition. 1996. $110.00.

Standard Handbook of Engineering Calculations. Tyler G.

Hicks. McGraw-Hill Publishing Company, 1221 Avenue of the Americas, New York, NY 10020. (212) 512-2000. (800) 262-4729. 3rd edition. 1994. $99.50. ISBN: 0-07-028812-7.

Underground Systems Reference Book. Edison Electric Institute, 701 Pennsylvania Avenue NW, Washington, DC 20004-2696. (202) 508-5000, 5454. FAX (202) 508-5360.

1955. $14.50.

ONLINE DATABASES AND CD-ROMS

CA Search. Chemical Abstracts Service, P.O. Box 3012, Columbus, OH 43210-0012. (614) 447-3600. (800) 848-6533. FAX (614) 447-3709. Very comprehensive guide to worldwide chemical literature and related fields, 1972 to present. Available on BRS,(800) 289-4277, DIALOG, (800) 334-2564, ORBIT (800) 456-7248, and STN International, FIZ Karlsruhe, P.O. Box 2465, W-7500, Karlsruhe 1, Germany, online services. Inquire as to cost and availability.

Compendex Plus. Engineering Information, Inc., 345 East 47th Street, New York, NY 10017. (212) 705-7600 or (800) 221-1044. Contains citations with abstracts to worldwide literature in engineering and technology, from 1970 to present. Available on online BRS,(800) 289-4277, DIALOG, (800) 334-2564, ORBIT (800) 456-7248, and STN International, FIZ Karlsruhe, P.O. Box 2465, W-7500, Karlsruhe 1, Germany, online services. Also available on CD-ROM. Inquire as to cost and availability.

Current Contents Search. Institute for Scientific Information, 3501 Market Street, Philadelphia, PA 19104. (215) 386-0100. FAX (215) 386-6362. Contains citations to articles listed in the table of contents of science and technology journals. Also articles in social sciences and life sciences journals. Available on BRS,(800) 289-4277, DIALOG,(800) 334-2564, online services. Inquire as to cost and availability.

Dissertation Abstracts Online. University Microfilms International, 300 North Zeeb Road, Ann Arbor, MI 48106. (800) 521-0600 or (313) 761-4700. Scope includes virtually all doctoral dissertations accepted at accredited American institutions from 1861 to present in 252 subject areas. Available on BRS, (800) 289-4277, DIALOG, (800) 334-2564, and OCLC EPIC,(800) 848-5878, online services. Also available on CD-ROM. Inquire as to cost and availability.

INSPEC. Institution of Electrical Engineers, Michael Faraday House, Six Hills Way, Stevenage, Herts. SG1 2AY, England. Telephone: 0438 313311 or FAX 0438 742840. Contains citations to the worldwide literature of physics, electronics and electrical engineering, computer technology, and related fields. Available on BRS, (800) 289-4277, DIALOG, (800) 334-2564, ORBIT, (800) 456-7248, and STN International, FIZ Karlsruhe,

P.O. Box 2465, W-7500, Karlsruhe 1, Germany, online services. Inquire as to cost and availability.

NTIS Bibliographic Database. National Technical Information Service, 5285 Port Royal Road, Springfield, VA 22161. (703) 487-4929 or FAX (703) 321-8199. Broad coverage of government-sponsored science and technology research reports, 1964 to present. Available on BRS,(800) 289-4277, DIALOG, (800) 334-2564, ORBIT, (800) 456-7248, and STN International, FIZ Karlsruhe, P.O. Box 2465, W-7500, Karlsruhe 1, Germany, online services. Also available on CD-ROM. Inquire as to cost and availability.

Physics Briefs. American Institute of Physics, 335 East 45th Street, New York, NY 10017. (212) 661-9260 or FAX (212) 661-2036. Contains citations with abstracts of the literature of physics and related fields, 1979 to present. Available on the STN International, FIZ Karlsruhe, P.O. Box 2465, W-7500, Karlsruhe 1, Germany, online service. Inquire as to cost and availability.

Scientific and Technical Books and Serials in Print. R.R. Bowker Inc., 121 Chanlon Road, New Providence, NJ 07974. (800) 521-8110 or (908) 464-6800. List of currently published books and serials in the physical and biological sciences, engineering and technology, with subject, author and titles indexes. Available on ORBIT,(800) 456-7248, online service.Inquire as to cost and availability.

SCISEARCH. Institute for Scientific Information, 3501 Market Street, Philadelphia, PA 19104. (800) 523-1850 or (215) 386-0100. Broad multidisciplinary title and author index to the international literature of science and technology, 1974 to present. Available on DIALOG,(800) 334-2564, and ORBIT,(800) 456-7248, online services. Also available on CD-ROM. Inquire as to cost and availability.

WILSONLINE. H.W. Wilson Company, 950 University Avenue, Bronx, NY 10452. (800) 367-6770 or (212) 588-8400. Makes available online versions of the printed H.W. Wilson indexes including Applied Science and Technology Index, Business Periodicals Index, General Science Index, and Readers' Guide to Periodical Literature. Period covered is generally 1983 to present. Available on BRS,(800) 289-4277, DIALOG,(800) 334-2564, and OCLC EPIC,(800) 848-5878, online services. Also available on CD-ROM. Inquire as to cost and availability.

PERIODICALS

Electric Light and Power. PennWell Publishing Co. 1421 South Sheridan Road, Box 1260, Tulsa OK 74101. (835-3161. FAX (918) 831-9497. 1922 to present. Monthly. $42.00 per year.

Electric Machines and Power Systems. Taylor & Francis, 1900 Frost Road, Suite 101, Bristol, PA 19007-1598. (800) 821-8312, FAX (215) 785-5515. 1976 to present. Monthly. $355.00 per year. ISSN: 0731-356X.

Electrical World. McGraw-Hill, Inc., Box 513 Hightstown, NJ 08520. (212) 512-3288. 1874 to present. Monthly. $55.00 per year. ISSN: 0013-4457.

Electronic Design. Penton Publishing, San Jose Gateway, Suite 354. 2025 Gateway Place, San Jose, CA 95110. (408) 441-0550. 1952 to present. Fortnightly. $95.00. ISSN: 0013-4872.

Electronics. Penton Publishing, San Jose Gateway, Suite 354. 2025 Gateway Place, San Jose, CA 95110. (408) 441-0550. 1930 to present. Semi-weekly. $98.00 per year. ISSN: 0883-4989.

IEEE SPECTRUM. Institute of Electrical and Electronics Engineers, Inc., Box 1331, 445 Hoes Lane, Piscataway, NJ 08855-1331. (908) 981-0060. 1964 to present. Monthly. $157.00. ISSN: 0018-9235.

IEEE Transactions On Power Delivery. Institute of Electrical and Electronics Engineers, Inc., Box 1331, 445 Hoes Lane, Piscataway, NJ 08855-1331. (908) 981-0060. 1986to present. Quarterly. $170.00 per year. ISSN: 0885-8977.

Institute of Electrical and Electronics Engineers. Proceedings. Institute of Electrical and Electronics Engineers, Inc., Box 1331, 445 Hoes Lane, Piscataway, NJ 08855-1331. (908) 981-0060. 1913 to present. Monthly. $275.00. ISSN: 0018-9219.

Journal of the Illuminating Engineering Society. Illuminating Engineering Society of North America, 120 Wall Street, 17th Floor, New York, NY 10005-4001. (212) 248-5000. FAX (212) 248-5017. 1971 to present. Semiannual. $195.00 per year. ISSN: 0099-4480.

L D and A: Lighting Design and Applicaton. Illuminating Engineering Society of North America, 120 Wall Street, 17th Floor, New York, NY 10005-4001. (212) 248-5000. FAX (212) 248-5017. 1906 to present. MOnthly. $39.00 per year. ISSN: 0360-6325.

RESEARCH CENTERS AND INSTITUTES

Edison Electric Institute. 701 Pennsylvania Avenue, NW, Washington, DC 20004-2696. (202) 508-5000, 5454. FAX (202) 508-5360.

Electrical Engineering Research Laboratories. Purdue University. Electrical Engineering Building, West Lafayette, in 47907. (317) 494-3536. FAX (317) 494-6440.

Lighting Research Institute. 120 Wall Street, 17th Floor, New York, NY 10005-4001. (212) 248-5014. FAX (212) 248-5019.

office of Building Research. Lousiana State University, School of Architecture, Baton Rouge, LA 70803. (504) 388-6885.

IMAGE ANALYSIS

See: CAD

IMAGE PROCESSING

See: CAD

INDUSTRIAL ENGINEERING

See also: CHEMICAL ENGINEERING, CONTROL SYSTEMS, DESIGN ENGINEERING, MATERIALS HANDLING, MECHANICAL ENGINEERING, QUALITY CONTROL ENGINEERING, ROBOTICS

ABSTRACT SERVICES AND INDEXES

Alloys Index. American Society for Metals, Metals Park, OH 44073. (216) 338-5151. FAX (216) 338-4634. 1974 to present. Monthly. $295.00.

Analytical Abstracts. Royal Society of Chemistry, Information Services, Thomas Graham House, Science Park, Milton Road, Cambridge, CB4 4WF, England. Contains citations, mostly with abstracts, of the worldwide literature on analytical chemistry, from 1954 to present. Monthly. $636.00 per year. Also available online.

A.P.I. Abstracts. American Petroleum Institute, Central Abstracting and Information Services, 275 Seventh Avenue, New York, NY 10001. (212) 366-4040 or FAX (212) 366-4298. Contains citations with abstracts to literature relating to the petroleum refining and the petrochemical industries, from 1964 to present.

Applied Mechanics Reviews: An Assessment of World Literature in Engineering Sciences. American Society of Mechanical Engineers, 345 East 47th Street, New York, NY 10017. (212) 705-7703. 1948 to present. Monthly. $360.00 per year.

Applied Science and Technology Index; A Cumulative Subject Index To English Language Periodicals in the Fields of Aeronautics and Space Science, Computer Technology, Chemistry, Construction Industry, Energy and Related Areas. H.W. Wilson Co., 950 University Avenue, Bronx, NY 10452. (800) 367-6770 or (212) 588-8400. From 1958 to present. Monthly. Inquire about cost and availability. Also available on CD-ROM.

CAD/cam Abstracts. R.R. Bowker, Bowker A&I Publishing, 121 Chanlon Road, New Providence, NJ 07974. (908) 771-7714. Fax (908) 771-7725. 1984 to present. Monthly. $495.00 per year.

Ceramic Abstracts. American Ceramic Society, Inc., 757 Brooksedge Plaza Drive, Westerville, OH 43081-2821. (614) 890-6136. Bi-monthly. 1922 to present. $120.00 per year.

Chemical Abstracts - Physical, Inorganic and Analytical Chemistry. Chemical Abstracts Service, 2540 Olentangy River Road, Box 3012, Columbus, OH 43210. (800) 848-6538 or (614) 447-3600, FAX (614) 447-3713. Bi-weekly. $1410.00 per year.

Chemical Engineering Abstracts. Royal Society of Chemistry, Information Services, Thomas Graham House, Science Park, Milton Road, Cambridge, CB4 4WF, England. Contains citations, mostly with abstracts, of the worldwide literature on chemical engineering, from 1982 to present. Monthly. $450.00 per year. Also available online.

Corrosion Abstracts; Abstracts of the World's Literature On Corrosion and Corrosion Mitigation. National Association of Corrosion Engineers (NACE), Box 218340, Houston, TX 77218. (713) 492-0535 or FAX (713) 492-8254. 1962 to present. Bi-monthly. $195.00 per year. Also available on CD-ROM.

Current Contents: Engineering, Technology and Applied Sciences. Institute for Scientific Information, 3501 Market Street, Philadelphia, PA 19104. (215) 386-0100. FAX (215) 386-6362. Weekly. $360.00 per year.

Current Contents: Physical, Chemical and Earth Sciences. Institute for Scientific Information, 3501 Market Street, Philadelphia, PA 19104. (215) 386-0100. FAX (215) 386-6362. Weekly. $360.00 per year.

Electrical and Electronics Abstracts. Institute of Electrical Engineers, Michael Faraday House, Six Hill Way, Stevenage, Herts, SG1 2AY, England. Distributed by IEEE, 445 Hoes Lane, Piscataway, NJ 08854. (908) 562-5549. 1898 to present. Monthly. $2200.00 per year. Also available online as INSPEC.

Engineered Materials Abstracts. ASM (American Society of Metals) International, Materials Information, Materials Park, OH 44073. (216) 338-5151 or FAX (216) 338-4634. Covers literature on technical developments in polymer, ceramic, and composite materials and engineering. 1986 to present. Monthly. $1175.00 per year. Also available on CD-ROM.

Engineering Index Monthly. Engineering Information, Inc., Castle Point on the Hudson, Hoboken, NJ 07030. (800) 221-1044. FAX (201) 216-8532. Monthly. $2200.00 per year. Also available online as COMPENDEX, and also on CD-ROM.

Fluid Power Abstracts. STI Limited, 4 Kings Meadow, Ferry Hinksey Road, Oxford OX2 ODU, England. Distributed by Air Science Company, Box 143, Corning, NY 14830. (607) 962-5591. Covers oil hydraulic and pneumatic power transmission and control, engineering and design. 1964 to present. Bi-monthly. $260.00 per year.

ISMEC Bulletin (Information Service in Mechanical Engineering). Cambridge Scientific Abstracts, 5161 River Road, Bethesda, MD 20816. (301) 951-1400. 1973 to present. Monthly. $750.00 per year.

Literature Abstracts: Catalysts and Catalysis. American Petroleum Institute, Central Abstracting and Information Services, 275 Seventh Avenue, New York, NY 10001. (212) 366-4040 or FAX (212) 366-4298. Contains citations with abstracts to literature relating to catalysts used in petroleum refining and the petrochemical industries, from 1985 to present. Weekly. $160.00 per year. Also available online.

Metals Abstracts and Metals Abstracts Index. American Society for Metals, Metals Park, OH 44073. (216) 338-5151. 1968 to present. Monthly. Abstracts are $1500.00 per year and Index is $460.00 per year.

Petroleum Abstracts. 600 South College, Tulsa, OK 74104. (918) 631-2296 or (800) 247-8678. Contains citations with abstracts to the worldwide literature and patents on the exploration, development, and production of petroleum resources, from 1965 to present.

Polymer Blends, Alloys, and Interpenetration Polymer Network Abstracts. Technomic Publishing Company, Inc., 851 New Holland Avenue, Box 3535, Lancaster, PA 17604. (717) 291-5609. Covers literature in polymer science and product development, from 1987 to present. Monthly. $130.00 per year.

Polymer Contents: International Current Awareness Publication on Polymer Science and Engineering. Elsevier Science Publishing Company, Inc., 655 Avenue of the Americas, New York, NY 10010. (212) 989-5800. 1984 to present. Monthly. $350.00 per year.

Previews of Heat and Mass Transfer. Pergamon Press, Maxwell House, Fairview Park, Elmsford, NY 10523. (914) 592-7700. Fax (914) 592-3625. Contains citations with abstracts to the worldwide literature on heat and mass transfer from more than 100 journals, 1974 to present. Six times per year. $230.00 per year.

Science Citation Index. Institute for Scientific Information, 3501 Market Street, Philadelphia, PA 19104. (215) 386-0100. FAX (215) 386-6362. Inquire about availability and cost. Also available on CD-ROM.

Tribology and Corrosion Abstracts. Elsevier Science Publishing Company, Inc., 655 Avenue of the Americas, New York, NY 10010. (212) 989-5800. Covers literature and research in friction, wear, lubrication and lubricants, and corrosion problems and protection, 1984 to present. Thirteen times per year. $450.00 per year.

Tribos - Tribology Abstracts. STI Limited, 4 Kings Meadow, Ferry Hinksey Road, Oxford, OX2 ODU, England. Distributed in the United States by Air Science Company, Box 143, Corning NY 14830. (607) 962-5591. Contains citations of literature and research in friction, wear, lubrication and lubricants, and corrosion problems and protection, from 1967 to present. Monthly. $375.00 per year.

World Ceramic Abstracts. Pergamon Press, Maxwell House, Fairview Park, Elmsford, NY 10523. (914) 592-7700. Monthly. 1958 to present. $310.00 per year.

ANNUAL REVIEWS AND YEARBOOKS

Materials Handling Engineering Handbook. Penton-IPC, 1100 Superior Avenue, Cleveland, OH 44114. (216) 696-7000. Annual. Inquire.

ASSOCIATIONS AND PROFESSIONAL SOCIETIES

American Institute of Plant Engineers. 3975 Erie Avenue, Cincinnati, OH 45208. (513) 561-6000.

American Production and Inventory Control Society. 500 West Annandale Road, Falls Church, VA 22046. (703) 237-8344.

American Society For Quality Control. 230 West Wells Street, Suite 700, Milwaukee, WI 53203. (414) 272-8575.

American Society of Mechanical Engineers. 345 East 47th Street, New York, NY 10017. (212) 705-7703.

Industrial Designers Society of America. 1360 Beverly Road, McLean, VA 22101. (703) 556-0919.

Institute of Industrial Engineers. 25 Technology Park/Atlanta, Norcross, GA 30092. (404) 449-0460.

International Association of Quality Circles. 801-B West Eighth Street, Suite 301, Cincinnati, OH 45203. (513) 381-1959.

Society of American Value Engineers. 221 North LaSalle Street, Suite 2026, Chicago, IL 60601. (312) 346-3265.

Society of Logistics Engineers. 303 Williams Avenue, Suite 922, Huntsville, AL 35801. (205) 539-3800.

Society of Manufacturing Engineers. One SME Drive, Box 930, Dearborn, MI 48121. (313) 271-1500.

Society of Packing and Handling Engineers. Reston International Center, Reston, VA 22091. (703) 620-9380.

DIRECTORIES AND BIOGRAPHICAL SOURCES

Directory of Engineering Societies and Related Organizations. American Association of Engineering Societies, 1111 19th Street, Suite 608, Washington, DC 20036-3603. (202) 296-2237. Semi-annual. $150.00.

Directory of Engineers in Private Practice. National Society of Professional Engineers, 1420 King Street, Alexandria, VA 22314-2750. (703) 684-2835. Biennial. $30.00.

Engineering Research Centers. Gale Research Company, Detroit, MI 48226. (800) 521-0707. 1984. $400.00.

International Engineering Directory. American Consulting Engineers Council, 1015 15th Street, N.W. Suite 802, Washington, DC 20005-2670. (202) 347-7474. Annual. $10.00.

Plant Engineering Directory. Technical Publishing Company, 1301 South Grove Avenue, Chicago, IL 60010. (312) 381-1840. Annual. $35.00.

Society of Logistics Engineers Membership Directory. 303 Williams Avenue, Suite 922, Huntsville, AL 35801. (205) 539-3800. Annual. Inquire.

Society of Packing and Handling Engineers Directory. Reston International Center, Reston, VA 22091. (703) 620-9380. Annual. Inquire.

ENCYCLOPEDIAS AND DICTIONARIES

Ei thesaurus. Jessica L. Milstead, editor. Engineering Information, Inc., Castle Point on the Hudson, Hoboken, NJ 07030. (800) 221-1044. FAX (201) 216-8532. 1992. $130.00

McGraw-Hill Encyclopedia of Science and Technology. McGraw-Hill Book, Incorporated, 1221 Avenue of the Americas, New York, NY 10020. (212) 997-3675. Seventh edition. Twenty volumes. 1992. $1900.00.

GENERAL WORKS

Industrial Energy Management: Principles and Applications. Giovanni Petrecca. Kluwer Academic Publishers, P.O. Box 358, Accord Station, Hingham, MA 02018-0358. (617) 871-6000. FAX (617) 871-6528. 1993. $143.00.

Industrial Engineering and Management: A New Perspective. Philip E. Hicks. 2nd edition. McGraw-Hill Publishing Company, 1221 Avenue of the Americas, New York, NY 10020. (800) 262-4729 or (212) 512-3825. 1994. Inquire for price.

Integrated Manufacturing: Strategy, Planning, and Implementation. E.R.G. Gerelle and J. Stark. McGraw-Hill Book Company, 1221 Avenue of the Americas, New York, NY 10020. (212) 512-2000. 1988. $34.50.

Introduction To Industrial and Systems Engineering. Wayne C. Turner, et al. 3rd edition. Prentice Hall, 113 Sylvan Avenue, Route 9W, Englewood Cliffs, NJ 07632. (201) 592-2000 or (800) 922-0579. 1992. $78.00.

Manufacturing Engineering: Principles of Optimization. Daniel T. Koenig. Hemisphere Publishing Corporation, 79 Madison

Avenue, New York, NY 10016-7892. (800) 821-8312. 1987. $45.75.

Principles of Engineering Organization. Stephen Wearne. 2nd edition. Thomas Telford Services Ltd., Thomas Telford House, 1 Heron Quay, London E14 4JD, England. Telephone 071-987-6999. FAX 071-538-4101. 1993. Inquire for cost and availability.

HANDBOOKS AND MANUALS

Computer-integrated Manufacturing Handbook. E. Telcholz and J.N. Orr. McGraw-Hill Book Company, 1221 Avenue of the Americas, New York, NY 10020. (212) 512-2000. 1987. $59.95.

Engineering Tables and Data. A.M. Howatson, P.G. Lund and J.D. Todd. Van Nostrand Reinhold, 115 Fifth Avenue, New York, NY 10003. (212) 254-3232 or (800) 926-2665. Second Edition. 1991. $39.95.

Handbook of Industrial Engineering. Gavriel Salvendy, editor. Institute of Industrial Engineers. Distributed by John Wiley and Sons, Inc., 605 Third Avenue, New York, NY 10158. (800) 526-5368 or (212) 850-6000. Second edition. 1992. $135.00.

Maintenance Engineering Handbook. Lindley R. Higgins. McGraw-Hill Book Company, 1221 Avenue of the Americas, New York, NY 10020. (212) 512-2000. 1987. $79.50.

Production Handbook. J.A. White. John Wiley and Sons, Inc., 605 Third Avenue, New York, NY 10158. (800) 526-5368. 4th edition. 1986. $78.50.

ONLINE DATABASES AND CD-ROMS

Aerospace Database. American Institute of Aeronautics and Astronautics, 370 L'Enfant Promenade, S.W., Washington, DC 20024. (202) 646-7400. Worldwide published literature on research and development in aerospace and related areas, with abstracts. Covers 1962 to present. Available on DIALOG, (800) 334-2564, online service. Also available on CD-ROM. Inquire as to cost and availability.

Analytical Abstracts Online. Royal Society of Chemistry, Information Services, Thomas Graham House, Science Park, Milton Road, Cambridge, CB4 4WF, England. Contains citations, mostly with abstracts, of the worldwide literature on analytical chemistry, from 1980 to present. Available on DIALOG, (800) 334-2564, and STN International, FIZ Karlsruhe, P.O. Box 2465, W-7500, Karlsruhe 1, Germany, online services. Inquire as to cost and availability.

Apilit. American Petroleum Institute, Central Abstracting and Information Services, 275 Seventh Avenue, New York, NY 10001. (212) 366-4040. Contains citations with abstracts to literature relating to the petroleum refining and the petrochemical industries, from 1964 to present. Available on BRS,(800) 289-4277, DIALOG, (800) 334-2564, ORBIT (800) 456-7248, and STN International, FIZ Karlsruhe, P.O. Box 2465, W-7500, Karlsruhe 1, Germany, online services. Inquire as to cost and availability.

CA Search. Chemical Abstracts Service, P.O. Box 3012, Columbus, OH 43210-0012. (614) 447-3600. (800) 848-6533. FAX (614) 447-3709. Very comprehensive guide to worldwide chemical literature and related fields, 1972 to present. Available on

BRS,(800) 289-4277, DIALOG, (800) 334-2564, ORBIT (800) 456-7248, and STN International, FIZ Karlsruhe, P.O. Box 2465, W-7500, Karlsruhe 1, Germany, online services. Inquire as to cost and availability.

Ceramic Abstracts. American Ceramic Society, Inc., Reference Services, 735 Ceramic Place, Westerville, OH 43081-6136. (614) 890-4700. Citations with abstracts to worldwide literature on ceramic materials and technology, from 1976 to present. Available on DIALOG, (800) 334-2564, ORBIT (800) 456-7248, and STN International, FIZ Karlsruhe, P.O. Box 2465, W-7500, Karlsruhe 1, Germany, online services. Inquire as to cost and availability.

Compendex Plus. Engineering Information, Inc., 345 East 47th Street, New York, NY 10017. (212) 705-7600 or (800) 221-1044. Contains citations with abstracts to worldwide literature in engineering and technology, from 1970 to present. Available on online BRS,(800) 289-4277, DIALOG, (800) 334-2564, ORBIT (800) 456-7248, and STN International, FIZ Karlsruhe, P.O. Box 2465, W-7500, Karlsruhe 1, Germany, online services. Also available on CD-ROM. Inquire as to cost and availability.

Current Contents Search. Institute for Scientific Information, 3501 Market Street, Philadelphia, PA 19104. (215) 386-0100. FAX (215) 386-6362. Contains citations to articles listed in the table of contents of science and technology journals. Also articles in social sciences and life sciences journals. Available on BRS,(800) 289-4277, DIALOG,(800) 334-2564, online services. Inquire as to cost and availability.

International Aerospace Abstracts. American Institute of Aeronautics and Astronautics, 370 L'Enfant Promenade, S.W., Washington, DC 20024. (202) 646-7400. Contains references and abstracts of journal and monograph literature relating to aerospace science and technology, from 1963 to present. Available through the NASA/RECON system of the National Aeronautics and Space Administration only.

*ISMEC: Mechanical Engineering Abstracts.*Cambridge Scientific Abstracts, 7200 Wisconsin Avenue, Suite 601, Bethesda, MD 20814. (301) 961-6750 or (800) 843-7751. Contains citations to the literature in mechanical engineering, industrial and production engineering, energy, power, mechanics, devices and related areas, from 1973 to present. Available on the DIALOG,(800) 334-2564, online service. Inquire as to cost and availability.

NTIS Bibliographic Database. National Technical Information Service, 5285 Port Royal Road, Springfield, VA 22161. (703) 487-4929 or FAX (703) 321-8199. Broad coverage of government-sponsored science and technology research reports, 1964 to present. Available on BRS,(800) 289-4277, DIALOG, (800) 334-2564, ORBIT, (800) 456-7248, and STN International, FIZ Karlsruhe, P.O. Box 2465, W-7500, Karlsruhe 1, Germany, online services. Also available on CD-ROM. Inquire as to cost and availability.

Supertech. Bowker A & I Publishing, 121 Chanlon Road, New Providence, NJ 07974. (800) 521-8110 or (908) 464-6800. Contains citations to the world's published and unpublished literature in the fields of artificial intelligence, biotechnology, computer-aided design and manufacturing, robotics, and telecommunications. Covers the period 1973 to present. Available on DIALOG,(800) 334-2564, and ORBIT,(800) 456-7248, online services. Inquire as to cost and availability.

Tulsa. Petroleum Abstracts, 600 South College, Tulsa, OK 74104. (918) 631-2296 or (800) 247-8678. Contains citations with abstracts to the worldwide literature and patents on the exploration, development, and production of petroleum resources, from 1965 to present. Available on DIALOG,(800) 334-2564, and ORBIT,(800) 456-7248, online services. Inquire as to cost and availability.

OTHER SOURCES

What Every Engineer Should Know About Engineering Sources. Marcel Dekker Inc., 270 Madison Avenue, New York, NY 10016. (800) 228-1160. 1984. $24.95.

PERIODICALS

Advanced Manufacturing Processes. Marcel Dekker Inc., 270 Madison Avenue, New York, NY 10016. (800) 228-1160. 1986 to present. Quarterly. $185.00 per year.

Advanced Manufacturing Technology. Technical Insights, Inc., Box 1304, Fort Lee, NJ 07024. (201) 568-4744. 1977 to present. Biweekly. $520.00 per year.

American Industry. Publications for Industry, 21 Russell Woods Road, Great Neck, NY 11021. (516) 487-0990. 1946 to present. Monthly. $40.00 per year.

Assembly Engineering. Hitchcock Publishing Company, 25 West 550 Geneva Road, Wheaton, IL 60188. (312) 665-1000. 1958 to present. Monthly. $95.00 per year.

IEEE Transactions On Industry Applications. Institute of Electrical and Electronics Engineers. IEEE Service Center, 445 Hoes Lane, Piscataway, NJ 08854. 1965 to present. Bimonthly. $185.00 per year.

IIE Transactions: Industrial Engineering Research and Development. Insturture of Industrial Engineers, 25 Technology Park/ Atlanta, Norcross, GA 30092. (404) 449-0460. 1969 to present. Quarterly. $85.00 per year.

Journal of Engineering For Industry. American Society of Mechanical Engineers, 345 East 47th Street, New York, NY 10017. (212) 705-7703. 1928 to present. Quarterly. $180.00 per year.

Journal of Quality Technology. American Society for Quality Control, 230 West Wells Street, Suite 700, Milwaukee, WI 53203. (414) 272-8575. 1969 to present. Quarterly. $20.00 per year.

Journal of Operations Management. American Production and Inventory Control Society, 500 West Annandale Road, Falls Church, VA 22046. (703) 237-8344. 1980 to present. Quarterly. $35.00 per year.

Manufacturing Engineering. Society of Manufactruing Engineers, One SME Drive, Box 930, Dearborn, MI 48121. (313) 271-1500. 1932 to present. Monthly. $90.00 per year.

Material Handling Engineering. Penton-IPC, 1100 Superior Avenue, Cleveland, OH 44114. (216) 696-7000. 1945 to present. Monthly. $95.00 per year.

Materials Engineering. Penton-IPC, 1100 Superior Avenue, Cleveland, OH 44114. (216)696-7000. 1929 to present. Monthly. $80.00 per year.

Plant Engineering Magazine. Technical Publishing Company, 1301 South Grove Avenue, Chicago, IL 60010. (312) 381-1840. 1947 to present. Semimonthly. $100.00 per year.

Production. Production Publishing, 44 East Long Lake Road, Bloomfield Hills, MI 48303. (313) 647-8400. 1934 to present. Monthly. $75.00 per year.

Production and Inventory Management Journal. American Production and Inventory Control Society, 500 West Annandale Road, Falls Church, VA 22046. (703) 237-8344. 1960 to present. Quarterly. $60.00 per year.

Quality Assurance. Hitchcock Publishing Company, 25 West 550 Geneva Road, Wheaton, IL 60188. (312) 665-1000. 1962 to present. Monthly. $85.00 per year.

RESEARCH CENTERS AND INSTITUTES

Advanced Manufacturing Center. Cleveland State University, Euclid Avenue at East 24th Street, Cleveland, OH 44115. (216) 687-4643.

Center For Applied Engineering. University of Missouri - Rolla, 206 Harris Hall, Rolla, MO 65401. (314) 4559.

Center of Design and Manufacturing Innovation. Lehigh University, H.S. Mohler Laboratory, Room 200, Bethleham, PA 18015. (215) 758-4114.

Center For Quality and Productivity Improvement. University of Wisconsin - Madison, 610 North Walnut Street, Madison, WI 53704. (608) 263-2520.

Engineering and Industrial Experiment Station. University of Florida, 300 Weil Hall, Gainesvill, FL 32611. (904) 392-0941.

SPECIFICATIONS AND STANDARDS

Automated Manufacturing. L.B. Garner, editor. American Society for Testing and Materials, 1916 Race Street, Philadelphia, PA 19103. (215) 299-5400. 1985. $38.00.

INDUSTRIAL MANAGEMENT

See also: CAM, INDUSTRIAL ENGINEERING, QUALITY CONTROL ENGINEERING

ABSTRACT SERVICES AND INDEXES

Applied Mechanics Reviews: An Assessment of World Literature in Engineering Sciences. 1948-present. American Society of Mechanical Engineers, 345 East 47th Street, New York, NY 10017. (212) 705-7703. Monthly. $360.00 per year.

Applied Science and Technology Index; A Cumulative Subject Index To English Language Periodicals in the Fields of Aeronautics and Space Science, Computer Technology, Chemistry, Construction Industry, Energy and Related Areas. H.W. Wilson Co., 950 University Avenue, Bronx, NY 10452. (800) 367-6770 or (212) 588-8400. FAX (718) 590-1617. From 1958 to present. Monthly. Inquire about cost and availability. Also available on CD-ROM and online.

Current Contents: Engineering, Technology, and Applied Sciences. Institute for Scientific Information, 3501 Market Street, Philadelphia, PA 19104. (215) 386-0100. FAX (215) 386-6362. 1970 to present. Weekly. $442.00 per year.

Engineering Index Monthly. Engineering Information, Inc., Castle Point on the Hudson, Hoboken, NJ 07030. (800) 221-1044. FAX (212) 832-1857. Monthly. $2300.00 per year. Also available online as COMPENDEX, and also on CD-ROM. Covers chemical engineering, computers, electrical engineering, civil engineering, metals and mining, industrial management, and mechanical engineering.

Index To Scientific and Technical Proceedings. Institute for Scientific Information, 3501 Market St., Philadelphia, PA 19104. (215) 386-0100. FAX (215) 386-6362. Monthly. $500.00 per year.

Mechanical Engineering Abstracts (formerly ISMEC), Cambridge Scientific Abstracts, 7200 Wisconsin Avenue, Bethesda, MD 20814-4823. (301) 961-6750. FAX (301) 961-6720. 1967 to present. Monthly. $895.00 per year. Summarizes world literature in mechanical engineering, production engineering, and engineering management. Also available online.

Manufacturing and Process Engineering Abstracts, Cambridge Scientific Abstracts, 7200 Wisconsin Avenue, Bethesda, MD 20814-4823. (301) 961-6750. FAX (301) 961-6720. 1993 to present. Monthly. $385.00 per year. Covers concurrent engineering, quality control, automated manufacturing, petroleum engineering, oil field operations and equipment, energy management, metallurgy and metallography, foundry practice.

Operations Research/management Science. Executive Sciences Institute, 1005 Mississippi Avenue, Davenport, IA 52803. (319) 324-4463. 1961 to present. Bi-monthly. $149.00.

Science Citation Index. Institute for Scientific Information, 3501 Market Street, Philadelphia, PA 19104. (215) 386-0100. FAX (215) 386-6362. Inquire about availability and cost. Also available on CD-ROM.

ASSOCIATIONS AND PROFESSIONAL SOCIETIES

American Association of Industrial Management. Stearns Bldg., 293 Bridge Street, Suite 324, Springfield, MA 01103. (413) 737-8766.

American Institute of Plant Engineers. 8180 Corporate Park Drive, Cincinnati, OH 45242. (513) 489-2473.

American Production and Inventory Control Society. 500 W. Annandale Road, Falls Church, VA 22046-4274. (703) 237-8344.

American Society For Engineering Management. PO Box 867, Annapolis, MD 21404-0867. (410) 263-7065. FAX (410) 263-7065.

American Society For Quality Control. 611 E Wisconsin Ave., PO Box 3005, Milwaukee, WI 53201-3005. (414) 272-8575 or (800) 952-6587.

American Society of Mechanical Engineers. 345 East 47th Street, New York, NY 10017-2398. (212) 705-7703.

Industrial Designers Society of America. 1142-E Walker Road, Great Falls, VA 22066. (703) 759-0100.

Institute of Industrial Engineers. 25 Technology Park/Atlanta, Norcross, GA 30092. (404) 449-0460.

Society of American Value Engineers. 60 Revere Drive, Suite 500, Northbrook, IL 60062. (708) 480-9080.

Society of Logistics Engineers. 8100 Professional Place, Suite 211, Hyattsville, MD 20785. (301) 459-8446.

Society of Manufacturing Engineers. One SME Drive, Box 930, Dearborn, MI 48128-1490. (313) 271-1500 or FAX (313) 271-2861.

DIRECTORIES AND BIOGRAPHICAL SOURCES

American Men & Women of Science, 1995-96. R.R. Bowker Staff, eds. 19th edition. 8 volumes. R.R. Bowker/Reed International Publishing Company, 121 Chanlon Road, New Providence, NJ 07974. (908) 464-6800 or (800) 521-8110. 1995. $850.00

Directory of Engineering Societies and Related Organizations. Gordon Davis. 13th edition. American Association of Engineering Societies, 1111 19th Street NW, Suite 608, Washington, DC 20036. (202) 296-2237 or (800) 658-8897. 1989. Inquire for price.

International Research Centers Directory. Gale Research, 835 Penobscot Building, Detroit, MI 48226-4094. (313) 961-2242. (800) 347-4253. FAX (313) 961-6083. 8th edition. 1995. $410.00.

Research Centers Directory. Gale Research, 835 Penobscot Building, Detroit, MI 48226-4094. (313) 961-2242. (800) 347-4253. FAX (313) 961-6083. $485.00.

Scientific and Technical Organizations and Agencies Directory. Gale Research, 835 Penobscot Building, Detroit, MI 48226-4094. (313) 961-2242. (800) 347-4253. FAX (313) 961-6083. 4th edition. 1996. $195.00.

Society of Logistics Engineers Membership Directory. Society of Logistics Engineers, 8100 Professional Place, Suite 211, Hyattsville, MD 20785. (301) 459-8446. Inquire for cost and availability.

Who's Who in Engineering. American Association of Engineering Societies, 1111 19th Street NW, Suite 608, Washington, DC 20036. (202) 296-2237 or (800) 658-8897. 8th edition. 1991. Inquire for price.

ENCYCLOPEDIAS AND DICTIONARIES

Industrial Engineering Terminology. Institute of Industrial Engineers staff. Revised edition. McGraw-Hill Publishing Company, 1221 Avenue of the Americas, New York, NY 10020. (800) 262-4729 or (212) 512-3825. 1992. $81.50.

Thesaurus of Scientific, Technical, and Engineering Terms. Hemisphere Publishing Corporation, 1900 Frost Road, Suite 101, Bristol, PA 19007-1598. (215) 785-5800 or (800) 821-8312. FAX (215) 785-5515. 1987. $173.00.

GENERAL WORKS

Analysis and Control of Production Systems. Elsayed A. Elsayed, Thomas O. Boucher. 2nd edition. Prentice Hall, 113 Sylvan Avenue, Route 9W, Englewood Cliffs, NJ 07632. (201) 592-2000 or (800) 922-0579. 1994. Inquire for cost and availability.

Catalysts For Change: Concepts and Principles For Enabling Innovation. William B. Rouse. John Wiley and Sons, Inc., 605 Third Avenue, New York, NY 10158. (800) 526-5368 or (212) 850-6000. 1993. $54.95.

Industrial Engineering and Management. Philip E. Hicks. 2nd edition. McGraw-Hill Publishing Company, 1221 Avenue of the Americas, New York, NY 10020. (800) 262-4729 or (212) 512-3825. 1993. Inquire for cost and availability.

Principles of Engineering Organization. Stephern Wearne. Thomas Telford Services Ltd., Thomas Telford House, 1 Heron Quay, London E14 4JD, England. Telephone 071-987-6999. FAX 071-538-4101. 1993. Inquire for cost and availability.

Production and Operations Management: Concepts, Models, & Behavior. Everett E. Adams Jr. & Ronald J. Ebert. 5th edition. Prentice Hall, 113 Sylvan Avenue, Route 9W, Englewood Cliffs, NJ 07632. (201) 592-2000 or (800) 922-0579. 1992. Inquire for cost and availability.

Production and Operations Management: Text & Cases. Ray Wild. 5th edition. Cassell Publishing, 215 Park Avenue S., 10th floor, New York, NY 10003. (212) 598-5717. FAX (212) 598-5740. 1994. $40.00.

HANDBOOKS AND MANUALS

Computer Integrated Manufacturing Handbook. V. David Hunt. Chapman and Hall, 29 West 35th Street, New York, NY 10001. (800) 842-3636. FAX (212) 563-2269. 1988. $57.50.

Handbook of Commercial and Industrial Facilities Management. William Wrenall and Lee Quartermain, eds. McGraw-Hill Publishing Company, 1221 Avenue of the Americas, New York, NY 10020. (800) 262-4729 or (212) 512-3825. 1994. Write for information.

Handbook of Industrial Engineering. Gavriel Salvendy. 2nd edition. John Wiley and Sons, Inc., 605 Third Avenue, New York, NY 10158. (800) 526-5368 or (212) 850-6000. 1991. $165.00.

Handbook of Innovation Management. Anton Cozijnsen and Willem Vrakking, eds. Blackwell Scientific Publications, 238 Main Street, Cambridge, MA 02142. (617) 876-7000 or (800) 759-6102. FAX (617) 876-7022. 1993. Write for information.

Maintenance Engineering Handbook. Lindley R. Higgins. 4th edition. McGraw-Hill Publishing Company, 1221 Avenue of the Americas, New York, NY 10020. (800) 262-4729 or (212) 512-3825. 1987. $99.50.

Production Handbook. John A. White Jr. 4th edition. John Wiley and Sons, Inc., 605 Third Avenue, New York, NY 10158. (800) 526-5368 or (212) 850-6000. 1987. $115.00.

Project and Cost Engineers' Handbook. Kenneth K. Humphreys and Lloyd M. English, eds. 3d rev. and enlarged ed. Marcel Dekker, Inc., 270 Madison Avenue, New York, NY 10016. (212) 696-9000. FAX (212) 685-4540. 1993. $65.00.

ONLINE DATABASES AND CD-ROMS

Compendex Plus. Engineering Information, Inc., 345 East 47th Street, New York, NY 10017. (212) 705-7600 or (800) 221-1044. Contains citations with abstracts to worldwide literature in engineering and technology, from 1970 to present. Available on online BRS, (800) 289-4277, DIALOG, (800) 334-2564, ORBIT (800) 456-7248, and STN International, FIZ Karlsruhe, P.O. Box 2465, W-7500, Karlsruhe 1, Germany, online services. Also available on CD-ROM. Inquire as to cost and availability.

CSA Engineering. Cambridge Scientific Abstracts, 7200 Wisconsin Avenue, Suite 601, Bethesda, MD 20814. (301) 961-6750 or (800) 843-7751. Contains citations and abstracts of international periodicals and research literature covering all fields of engineering and science and technology, including computer and information science, electronics, mechanical engineering, solid state materials, 1981 to present. Available on BRS, (800) 289-4277, online service. Inquire as to cost and availability.

Current Contents Search. Institute for Scientific Information, 3501 Market Street, Philadelphia, PA 19104. (215) 386-0100. FAX (215) 386-6362. Contains citations to articles listed in the table of contents of science and technology journals. Also articles in social sciences and life sciences journals. Available on BRS, (800) 289-4277, DIALOG, (800) 334-2564, online services. Inquire as to cost and availability.

Dissertation Abstracts. University Microfilms International, 300 North Zeeb Road, Ann Arbor, MI 48106. (800) 521-0600 or (313) 761-4700. Scope includes virtually all doctoral dissertations accepted at accredited American institutions from 1861 to present in 252 subject areas. Available on BRS, (800) 289-4277, DIALOG, (800) 334-2564, and OCLC EPIC, (800) 848-5878, online services. Also available on CD-ROM. Inquire as to cost and availability.

ISMEC: Mechanical Engineering Abstracts. Cambridge Scientific Abstracts, 7200 Wisconsin Avenue, Suite 601, Bethesda, MD 20814. (301) 961-6750 or (800) 843-7751. Contains citations to the literature in mechanical engineering, industrial and production engineering, energy, power, mechanics, devices and related areas, from 1973 to present. Available on the DIALOG, (800) 334-2564, online service. Inquire as to cost and availability.

NTIS Bibliographic Database. National Technical Information Service, 5285 Port Royal Road, Springfield, VA 22161. (703) 487-4929 or FAX (703) 321-8199. Broad coverage of government-sponsored science and technology research reports, 1964 to present. Available on BRS, (800) 289-4277, DIALOG, (800) 334-2564, ORBIT, (800) 456-7248, and STN International, FIZ Karlsruhe, P.O. Box 2465, W-7500, Karlsruhe 1, Germany, online services. Also available on CD-ROM. Inquire as to cost and availability.

SCISEARCH. Institute for Scientific Information, 3501 Market Street, Philadelphia, PA 19104. (800) 523-1850 or (215) 386-0100. Broad multidisciplinary title and author index to the international literature of science and technology, 1974 to present. Available on DIALOG, (800) 334-2564, and ORBIT, (800) 456-7248, online services. Also available on CD-ROM. Inquire as to cost and availability.

WILSONLINE. H.W. Wilson Company, 950 University Avenue, Bronx, NY 10452. (800) 367-6770 or (212) 588-8400. Makes available online versions of the printed H.W. Wilson indexes including Applied Science and Technology Index, Business Periodicals Index, General Science Index, and Readers' Guide to Periodical Literature. Period covered is generally 1983 to present. Available on BRS,(800) 289-4277, DIALOG,(800) 334-2564, and OCLC EPIC,(800) 848-5878, online services. Also available on CD-ROM. Inquire as to cost and availability.

PERIODICALS

Advanced Manufacturing Technology. Technical Insights Inc., 32 N Dean Street, Englewood, NJ 07631. (201) 568-4744. FAX (201) 568-8247. 1980 to present. Monthly. $575.00 per year.

American Industry. Publications for Industry, 21 Russell Woods Road, Great Neck, NY 11021. (516) 487-0990. FAX (516) 487-0809. 1946 to present. Monthly. $25.00 per year.

Assembly. Hitchcock Publishing, 191 S. Gary Avenue, Carol Stream, IL 60188. (708) 462-4641. FAX (708) 462-2205. 1958 to present. Monthly. $60.00 per year.

The Engineering Economist. Institute of Industrial Engineers, 25 Technology Park-Atlanta, Norcross, GA 30092. (404) 449-0460. FAX (404) 263-8532. 1955 to present. Quarterly. $24.00 per year.

IEEE Transactions On Industry Applications. IEEE (Institute of Electrical and Electronic Engineers), 445 Hoes Lane, Piscataway, NJ 08854. (908) 562-5545 or (800) 678-IEEE. 1965 to present. Bi-monthly. $240.00 per year.

IEEE Transactions On Engineering Management. IEEE (Institute of Electrical and Electronic Engineers), 445 Hoes Lane, Piscataway, NJ 08854. (908) 562-5545 or (800) 678-IEEE. 1954 to present. Quarterly. $135.00 per year.

IIE Transactions: Industrial Engineering Research and Development. Institute of Industrial Engineers, 25 Technology Park-Atlanta, Norcross, GA 30092. (404) 449-0460. FAX (404) 263-8532. 1969 to present. Quarterly. $95.00 per year for non-members.

Industrial Management Review. Institute of Industrial Engineers, 25 Technology Park-Atlanta, Norcross, GA 30092. (404) 449-0460. FAX (404) 263-8532. 1952 to present. Bi-monthly. $32.00 per year.

Industry Week. Penton Publishing, 110 Superior Ave., Cleveland, OH 44114-2543. 1882 to present. 23 times a year. $60.00 per year.

Industrial Management and Data Systems. MCBUniversity Press, 62 Toller Lane, Bradford, W. Yorks. BD8 9BY, England. Telephone 0274-499821. FAX 0274-547143. 11 times a year. $1849.95 per year.

Journal of Engineering For Industry. American Society of Mechanical Engineers, 345 E. 47th Street, New York, NY 10017. (212) 705-7722. 1970 to present. Quarterly. $100.00 per year.

Journal of Operations Management. Elsevier Science Inc., Box 882, Madison Square Station, New York, NY 10159. (212) 989-5800. 1980 to present. 4 times a year. $157.00 per year.

Journal of Quality Technology. American Society for Quality Control, ASQC Quality Press, 611 E Wisconsin Ave., PO Box 3005, Milwaukee, WI 53201-3005. (414) 272-8575 or (800) 952-6587. 1944 to present. Quarterly. $30.00 per year.

Manufacturing Engineering. Society of Manufacturing Engineers, One SME Drive, Box 930, Dearborn, MI 48128-1490. (313) 271-1500 or FAX (313) 271-2861. 1932 to present. Monthly. $60.00 per year.

Materials Engineering. Penton Publishing, 110 Superior Ave., Cleveland, OH 44114-2543. 1929 to present. Monthly. $50.00 per year.

Plant Engineering. Cahners Publishing Company (Des Plains), 1350 East Touhy Avenue, Des Plaines, IL 60017-5080. (708) 635-8800. FAX (708) 635-9950. 1947 to present. 19 times a year. $69.95 per year.

Production. Gardner Publications, 6600 Clough Pike, Cincinnati, OH 45244-4090. (513) 231-8020. FAX (513) 231-2818. 1934 to present. Monthly. $48.00 per year.

Quality. Hitchcock Publishing, 191 S. Gary Avenue, Carol Stream, IL 60188. (708) 462-4641. FAX (708) 462-2205. 1962 to present. Monthly. $65.00 per year.

Quality Assurance. Academic Press, Inc., Journals Division, 1250 Sixth Avenue, San Diego, CA 92101-4311. (619) 231-0926. FAX (619) 699-6715. 1991 to present. Quarterly. $167.00 per year.

Technovation. Elsevier Science [journals], 660 White Plains Rd., Tarrytown, NY 10159-5153. (919) 524-9200. FAX (919) 333-2444. 1981 to present. 10 times a year. $440.00 per year.

RESEARCH CENTERS AND INSTITUTES

Advanced Manufacturing Center. Cleveland State University, 1751 E. 23rd Street, Cleveland, OH 44114. (216) 687-4643. FAX (216) 687-2999.

Center For Applied Engineering. University of Missouri—Rolla, 206 Harris Hall, Rolla, MO 65401. (314) 341-4559.

Center For Manufacturing Systems Engineering. Lehigh University, H.S. Mohler Laboratory 200, Bethlehem, PA 18015. (215) 758-4667. FAX (215) 694-0542.

Center For Quality and Productivity Improvement. University of Wisconsin—Madison, 610 N. Walnut Street, Madison, WI 53705. (608) 263-2520. (608) 263-1425.

Engineering and Industrial Experiment Station. University of Florida, 300 Weil Hall, Gainesville, FL 32611-2083. (904) 392-6000. (904) 392-9673.

Ucla Center For Manufacturing and Automation Research. School of Engineering and Applied Science, 3531 Boelte Hall, Los Angeles, CA 90024-1597. (213) 825-2212.

INDUSTRIAL ROBOTS

See: ROBOTICS

INERTIAL GUIDANCE SYSTEMS

See: GUIDANCE SYSTEMS

INERTIAL NAVIGATION

See: NAVIGATION

INERTIAL WELDING

See: WELDING

INFRARED ASTRONOMY

See also: ASTRONOMY, ASTROPHYSICS, BIG BANG, COSMO-CHEMISTRY, COSMOLOGY, DARK MATTER, GALAXIES, INTER-STELLAR MEDIUM, MILKY WAY, NEBULAE, STARS, UNIVERSE

ABSTRACT SERVICES AND INDEXES

Applied Science and Technology Index; A Cumulative Subject Index To English Language Periodicals in the Fields of Aeronautics and Space Science, Computer Technology, Chemistry, Construction Industry, Energy and Related Areas. H.W. Wilson Co., 950 University Avenue, Bronx, NY 10452. (212) 588-8400. (800) 367-6770. FAX (718) 590-1617. From 1958 to present. Monthly. Inquire about cost and availability. Also available on CD-ROM and online. ISSN: 0003-6986.

Astronomy and Astrophysics Abstracts. Springer-Verlag New York, 175 Fifth Avenue, New York, NY 10010. (212) 460-1500. FAX (212) 473-6272. Published for the Astronomisches Rechen-Institut. Comprehensive coverage of all aspects of astronomy, astrophysics and related fields. 1969 to present. Two parts per year. Annual. ISSN: 0067-0022.

General Science Index. H.W. Wilson Company, 950 University Avenue, Bronx, NY 10452. (212) 588-8400. (800) 367-6770. FAX (718) 590-1617. From 1978 to present. Ten issues per year; quarterly and annual cumulations. Service basis. Available on CD-ROM and online. Inquire regarding cost and availability. ISSN: 0162-1963.

Government Reports Announcements and Index. U. S. National Technical Information Service (NTIS), 5285 Port Royal Road, Springfield, VA 22161. (703) 487-4650. FAX (703) 321-8547. From 1968 to present. Annual. $630.00 per year. Also available online as *NTIS Bibliographic Database* and on CD- ROM.

International Aerospace Abstracts. American Institute of Aeronautics and Astronautics. Technical Information Service, 555 West 57th Street, New York 10019. (212) 247- 6500. FAX (212) 582-4961. 1961 to present. Semi-monthly with annual index. $1295.00 per year. Also available online and on CD-ROM. ISSN: 0020-5842.

Meteorological and Geoastrophysical Abstracts. American Meteorological Society, c/o Inforonics, Inc., 550 Newtown Road, Littleton, MA 01460. (508) 486-8976. FAX (508) 486- 0027. Covers literature in environmental sciences, meteorology, astrophysics, hydrology, glaciology, and physical oceanography.

From 1950 to present. Monthly. $950.00 per year. Also available on CD-ROM and online. Inquire regarding cost and availability. ISSN: 0026-1130.

NTIS Alerts: Astronomy & Astrophysics. U. S. National Technical Information Service. 5285 Port Royal Road, Springfield, VA 22161. (703) 487-4650. FAX (703) 321-8547. Weekly. $140.00 per year.

Pascal 48: Environnement Cosmique Terrestre, Astronomie Et Geologie Extraterrestre. Centre National de la Recherche Scientifique, Institut de Information Scientifique et Technique, 2 aliee du Parc de Brabois, 54514 Vandoeuvre Les Nancy Cedex, France. TEL 83- 50-46-00. FAX: 83-50-46- 50. 1985 to present. Ten issues per year. 770 F. Also available on CD-ROM and online.

Physics Abstracts. INSPEC. Section A, Science Abstracts. Institute of Electrical Engineers, London, United Kingdom. Available from: INSPEC/IEEE - Institute of Electrical and Electronic Engineers, Box 1331, Hoes Lane, Piscataway, NJ 08855-1331. (908) 562-5549. 1898 to present. 24 issues per year. $2835.00 per year. ISSN: 0036-8091. Also available online and on CD-ROM.

Physics Briefs (Physikalische Berichte). Information Center for Energy, Physics, Mathematics; German Physical Society. V C H Publishers, Inc., 220 East 23rd Street, New York, NY 10010-4606. (212) 683-8333. 1845 to present. 24 issues per year. $2390.00 per year. Also available online. ISSN: 0179-7434.

Science Citation Index. SCI. Institute for Scientific Information, 3501 Market Street, Philadelphia, PA 19104. (215) 386-0100. (800) 523-1850. FAX (215) 386-2991. 1961 to present. Six issues per year, plus annual cumulation. $11650.00 per year. Also available online and on CD-ROM. Inquire about price and availability. ISSN: 0036-827X.

STAR. SCIENTIFIC and TECHNICAL AEROSPACE REPORTS. U.S. National Aeronautics and Space Administration. Distributed by U. S. Superintendent of Documents, Washington, DC 20402. 1963 to present. Semi-monthly, with semiannual and annual indexes. $114.00 per year. ISSN: 0036-8741. Also available online and on CD-ROM.

ANNUAL REVIEWS AND YEARBOOKS

The Astronomical Almanac. Superintendent of Documents, U. S. Government Printing office, Washington, DC 10402. (202) 783-3238. 1981 to present. Superceeds Astronomical Ephemeris and the American Ephermis and Nautical Almanac. Annual. $44.95. ISSN: 0737-6421.

Annual Review of Astronomy and Astrophysics; Original Reviews of Critical Literature and Current Developments in Astronomy and Astrophysics. Annual Reviews, Inc., 4139 El Camino Way, Palo Alto, CA 94306-0139. (415) 493-4400. (800) 523-8635. FAX (415) 855-9815. 1963 to date. Annual. $60.00. ISSN: 0066-4146.

Highlights of Astronomy. I. Appenzeller, editor. Kluwer Academic Publishers, Box 358, Accord Station, Hingham, MA 02018-0358. (617) 871-6600. International Astronomical Union Highlights series. ISBN: 07923-3553-8. Triennal. 1995. Price varies.

ASSOCIATIONS AND PROFESSIONAL SOCIETIES

Amateur Astronomers Association. 1010 Park Avenue, New York, NY 10028. (212) 535-2922.

American Association of Variable Star Observers. 23 Birch Street, Cambridge, MA 02138. (617) 354-0484. FAX: (617)354-0665.

American Astronomical Society. 2000 Florida Avenue NW, Suite 400, Washington, DC 20009. (202) 328-2010. FAX:(202) 234-2560.

American Institute of Physics. One Physics Ellipse, College Park, MD 20740-3843. (301) 209-3100.

Association of University For Research in Astronomy. Suite 701, 1625 Massachusetts Avenue, NW, Washington, DC 20036. (202) 483-2101. FAX (202) 483-2106.

Astronomical League. 2112 Kingfisher Lane East, Rolling Meadows, IL 60008. (708) 398-0562.

Astronomical Society of the Pacific. 390 Ashton Avenue, San Francisco, CA 94112. (415) 337-1100. FAX: (415) 337-5205.

International Astronomical Union. 98 bis Boulevard Arago, F-75014 Paris, France. Tel 1-43-25-83-58. FAX 1-43-25-26-16.

Royal Astronomical Society of Canada. 136 Dupont Street, Toronto. ON N5R 1V2 Canada. (416) 924-7973. FAX (416) 968-6687.

BIBLIOGRAPHIES

Astronomy and Astrophysics: A Source Guide. Gordon Press, P.O. Box 459, Bowling Green Station, New York, NY 10004. (718) 624-8419. 1991. $250.00. .

A Bibliography of Astronomy, 1970-1979. R. A. Sea and S. S. Martin. Libraries Unlimited, Inc., Littleton, CO 80160. 1982.

Catalog of the Naval Observatory Library. Naval Observatory Staff. G. K. Hall & Company, 866 Third Avenue, New York, NY 10022. (212)702-6789. (800) 257-5755. 6 volumes. 1977. $655.00

Science Books and Films. American Association for the Advancement of Science, 1333 H Street NW, Washington, DC 20005. (202) 326-6454. 1965 to present. Nine issues per year. $40.00 per year. ISSN: 0098-342X

Scientific and Technical Books and Serials in Print, 1995. R.R. Bowker Inc., 121 Chanlon Road, New Providence, NJ 07974. (908) 464-6800. (800) 521-8110. 4 volumes. 1994. $299.95. Also available on compact disc and online.

A Source Book On Astronomy and Astrophysics, 1900 - 1975. Kenneth Lang and Owen Gingerich, editors. Harvard University Press, 79 Garden Street, Cambridge MA. (617) 495-2600. 1980. $41.95.

DIRECTORIES AND BIOGRAPHICAL SOURCES

American Men and Women of Science: Physical and Biological Sciences. R. R. Bowker Inc., 121 Chanlon Road, New Providence, NJ 07974. (908) 464-6800. (800) 521-8110. 20th edition. 8 volumes. 1996. $850.00.

The Astronomers. Donald Goldsmith. St. Martin's Press, Inc., 175 Fifth Avenue, New York, NY 10010. (212) 674-5151. (800) 221-7945. 1993. $14.95 in paper.

Astronomical Centers of the World. Kevin Krisciunas. Cambridge University Press, 40 West 20th Street, New York,

NY 10011-4211. (212) 924-3900. (800) 872-7423. 1988. $34.95

The Biographical Dictionary of Scientists: Astronomers. D. Abbott, editor. Peter Bedrick Books, Inc., 2112 Broadway, Room 318, New York, NY 10023. (212) 496-0751. 1984.

Directory of Physics and Astronomy Staff Members. American Institute of Physics, One Physics Ellipse, College Park, MD 20740-3843. (301) 209-3100. Annual. $45.00.

Earth and Astronomical Research Centers. Stockton Press, 345 Park Avenue South, New York, NY 10010. 4th edition. 1995. $515.00. ISBN: 1-56169-0967.

Graduate Programs in Physics, Astronomy and Related Fields. 1995-1996. American Institute of Physics, One Physics Ellipse, College Park, MD 20740-3843. (301) 209-3100. 1978 to present. Annual. $45.00.

Research Centers Directory. Gale Research Inc., 835 Penobscot Building, Detroit, MI 48226-4094. (313) 961-2242. (800) 347-4253. 20th edition. 1995. $485.00.

ENCYCLOPEDIAS AND DICTIONARIES

Astronomy From A To Z: A Dictionary of Celestial Objects and Ideas. Charles A. Schweighauser. Illinois Issues, Sangamon State University, K-10, Springfield, IL 62794-9243. (217) 786-9243. 1991. $14.95 in paper.

Concise Dictionary of Astronomy. Jacqueline Mitton. Oxford University Press, Inc., 200 Madison Avenue, New York, NY 10016. (212) 725-6000. (800) 334-4249. 1992. $24.95.

Facts On File Dictionary of Astronomy. Valerie Illingworth.

Facts-on-File, Inc., 460 Park Avenue South, New York, NY 10016-7382. (212) 683-2244. (800) 322-8755. Fax (800) 683-3633. 1994. $27.95.

Encyclopedia of Astronomy and Astrophysics. Stephen Maran, editor. Van Nostrand Reinhold, 115 Fifth Avenue, New York, NY 10003. (212) 254-3232. (800) 842-3636. 1992. $129.95.

Landolt-boernstein Numerical Data and Functional Relationships in Science and Technology: Astronomy, Astrophysics and Space Research; A. O. Madelung and H. H. Boigt, editors. Springer Verlag New York, Inc., 175 Fifth Avenue, New York, NY 10010. (212) 460-1500. (800) 777- 4643. 1993. $550.00.

Oxford Illustrated Encyclopedia. VOLUME 8, the UNIVERSE. Archie Roy, editor. Oxford University Press, Inc., 200 Madison

Avenue, New York, NY 10016. (212) 725-6000. (800) 334-4249. 1993. $45.00.

McGraw-Hill Encyclopedia of Science and Technology. McGraw-Hill Book Company, Inc., 1221 Avenue of the Americas, New York, NY 10020. (212) 512-2000. (800) 262-4729. 7th edition. 20 volume set. 1992. $1900.00.

Stars, Galaxies, Cosmos. William R. Corliss. Sourcebook Project, P.O. Box 107, Glen Arm, MD 21057. (410) 668-6047. 1987. $17.95.

GENERAL WORKS

Astrophysics With Infrared Arrays. R. Elston, editor. Astronomical Society of the Pacific, 390 Ashton Avenue, San Francisco, CA 94112. (415) 337-1100. 1991. $25.00.

Cycles of Fire: Stars, Galaxies and the Wonder of Deep Space. William K. Hartmann. Workman Publishing Company, Inc., 708 Broadway, New York, NY 10003. (212) 254-5900. (800) 722-7202. 1988. $15.95 in paper.

Farthest Things in the Universe. Jay M. Pasachoff. Cambridge University Press, 40 West 20th Street, New York, NY 10011-4211. (212) 924-3900. (800) 872-7423. 1995. $29.95.

The First Three Minutes: A Modern View of the Origin of the Universe. Steven Weinberg. Basic Books, 10 East 53rd Street, New York, NY 10022. (212) 207-7057. (800) 242-7737. Reprint edition. 1993. $12.00.

Galactic and Extragalactic Background Radiation. Stuart Bowyer and Christoph Leinert, editors. Kluwer Academic Publishers, 101 Philip Drive, Assinippi Park, Norwell, MA 02061. (617) 871-6600. 1989. 1990. $65.00.

Infrared Astronomy. A. Mampaso, et al, editors. Cambridge University Press, 40 West 20th Street, New York, NY 10011-4211. (212) 924-3900. (800) 872-7423. 1994. $64.95.

Infrared Astronomy With Iso. T. Encrenaz and M. F. Kesler, editors. Nova Science Publishers, Inc., 6080 Jericho Turnpike, Suite 207, Commack, NY 11725-2808. (516) 499- 3103. 1992. $143.00.

The Infrared Spectral Region of Stars. Carlos Jaschek and Y. andrillat, editors. Cambridge University Press, 40 West 20th Street, New York, NY 10011-4211. (212) 924-3900. (800) 872-7423. 1991. $74.05.

The Left Hand of Creation: The Origin and Evolution of the Expanding Universe. John D. Barrow and Joseph Silk. Oxford University Press, Inc., 200 Madison Avenue, New York, NY 10016. (212) 725-6000. (800) 334-4249. 1994. $23.00.

Observational Astrophysics. Robert C. Smith. Cambridge University Press, 40 West 20th Street, New York, NY 10011- 4211. (212) 924-3900. (800) 872-7423. 1995. $69.95.

Physical Processes in Comets, Stars and Active Galaxies. W. Hildebrandt, et al editors. Springer-Verlag New York, Inc., 175 Fifth Avenue, New York, NY 10010. (212) 460-1500. (800) 777-4643. 1987. $47.00.

The Physics of Stars. A. C. PHillips. John Wiley & Sons, Inc., 605 Third Avenue, New York, NY 10158-0012. (212) 850-6000. (800) 225-5945. 1994. $32.95 in paper.

Star Formation, Galaxies and the Interstellar Medium. J. Franco, et al, editors. Cambridge University Press, 40 West 20th Street, New York, NY 10011-4211. (212) 924-3900. (800) 872-7423. 1993. $59.95.

Stars and Atoms: From the Big Bang To the Solar System. Stuart Clark. Oxford University Press, Inc., 200 Madison Avenue, New York, NY 10016. (212) 725-6000. (800) 334-4249. 1995. $35.00.

Universe of Galaxies. Paul W. Hodge, editor. W. H. Freeman & Co., 41 Madison Avenue, East 26th Avenue, 35th Floor, New York, NY 10010. (212) 576-0400. Readings from Scientific American. 1995.

HANDBOOKS AND MANUALS

Burnham's Celestial Handbook: An Observer's Guide To the Universe Beyond the Solar System. Robert Burnham, Jr. Dover Publications, Inc., 180 Varick Street, New York, NY 10014. (212) 255-3755. (800) 223-3130. 3 volumes. 1978. $41.85 for the set.

Cambridge Guide To the Constellations. Michael E. Bakich. Cambridge University Press, 40 West 20th Street, New York, NY 10011-4211. (212) 924-3900. (800) 872-7423. 1995. $69.95.

Encyclopedia of Astronomy and Astrophysics. Stephen Maran, editor. Van Nostrand Reinhold, 115 Fifth Avenue, New York, NY 10003. (212) 254-3232. (800) 842-3636. 1992. $129.95.

Greenwich Guide To Stars, Galaxies and Nebulae. Stuart Malin. Cambridge University Press, 40 West 20th Street, New York, NY 10011-4211. (212) 924-3900. (800) 872-7423.1990. $10.95.

Handbook of Astronomy, Astrophysics and Geophysics; Volume 2: Galaxies and Cosmology. V. M. Canuto and B. G. Elmegreen. Gordon & Breach Science Publishers, Inc., P.O. Box 200029, Riverfront Plaza Station, Newark, NJ 07102-0301. (201) 643-7500. 1988. $194.00.

Handbook of Space Astronomy and Astrophsyics. Martin V. Zumbeck. Cambridge University Press, 40 West 20th Street, New York, NY 10011-4211. (212) 924-3900. (800) 872-7423. 2nd edition. 1990. $79.95.

Landolt-borenstein Numerical Data and Functional Relationships in Science and Technology: Astronomy, Astrophysics and Space Research. GROUP VI. Springer-Verlag New York, Inc., 175 Fifth Avenue, New York, NY 10010. (212) 460-1500. (800) 777-4643. volumes priced individually.

McGraw-Hill Encyclopedia of Astronomy. Sybil P. Parker and Jay M. Pasachoff, editors. McGraw-Hill Publishing Company, Inc., 1221 Avenue of the Americas, New York, NY 10020. (212) 512-2000. (800) 262-4729. 2nd edition. 1993. $75.00

National Audobon Society Pocket Guide: Galaxies and Other Deep-sky Objects. Alfred A. Knopf, Inc. 201 E. 50th Street, New York, NY 10022. (800) 733-3000. 1995. $7.95.

Nearby Galaxies Catalog. R. Brent Tully and J. Richard Fisher. Cambridge University Press, 40 West 20th Street, New York,

INFRARED ASTRONOMY

Ency. of Physical Sciences and Engineering Info. Sources

NY 10011-4211. (212) 924-3900. (800) 872-7423. 1988. $59.95.

Observer's Guide To Astronomy. Patrick Martinez, editor. Cambridge University Press, 40 West 20th Street, New York, NY 10011-4211. (212) 924-3900. (800) 872-7423. 2 volumes. 1994. $159.90.

ONLINE DATABASES AND CD-ROMS

CA Search. Chemical Abstracts Service, P.O. Box 3012, Columbus, OH 43210-0012. (614) 447-3600. (800) 848-6533. FAX (614) 447-3709. Very comprehensive guide to worldwide chemical literature and related fields, 1972 to present. Available on BRS,(800) 289-4277, DIALOG, (800) 334-2564, ORBIT (800) 456-7248, and STN International, FIZ Karlsruhe, P.O. Box 2465, W-7500, Karlsruhe 1, Germany, online services. Inquire as to cost and availability.

Current Contents Search. Institute for Scientific Information, 3501 Market Street, Philadelphia, PA 19104. (215) 386-0100. FAX (215) 386-6362. Contains citations to articles listed in the table of contents of science and technology journals. Also articles in social sciences and life sciences journals. Available on BRS, (800) 289-4277, DIALOG, (800) 334-2564, online services. Inquire as to cost and availability.

Dissertation Abstracts Online. University Microfilms International, 300 North Zeeb Road, Ann Arbor, MI 48106. (800) 521-0600 or (313) 761-4700. Scope includes virtually all doctoral dissertations accepted at accredited American institutions from 1861 to present in 252 subject areas. Available on BRS, (800) 289-4277, DIALOG, (800) 334-2564,and OCLC EPIC, (800) 848-5878, online services. Also available on CD-ROM. Inquire as to cost and availability.

INSPEC. Institution of Electrical Engineers, Michael Faraday House, Six Hills Way, Stevenage, Herts. SG1 2AY, England. Telephone: 0438 313311 or FAX 0438 742840. Contains citations to the worldwide literature of physics, electronics and electrical engineering, computer technology, and related fields. Available on BRS, (800) 289-4277, DIALOG, (800) 334-2564, ORBIT, (800) 456-7248, and STN International, FIZ Karlsruhe, P.O. Box 2465, W-7500, Karlsruhe 1, Germany, online services. Inquire as to cost and availability.

International Aerospace Abstracts. American Institute of Aeronautics and Astronautics, 370 L'Enfant Promenade, S.W., Washington, DC 20024. (202) 646-7400. Contains referencesand abstracts of journal and monograph literature relating to aerospace science and technology, from 1963 to present. Available through the NASA/RECON system of the National Aeronautics and Space Administration only.

NTIS Bibliographic Database. National Technical Information Service, 5285 Port Royal Road, Springfield, VA 22161. (703) 487-4929 or FAX (703) 321-8199. Broad coverage of government-sponsored science and technology research reports, 1964 to present. Available on BRS,(800) 289-4277, DIALOG, (800) 334-2564, ORBIT, (800) 456-7248, and STN International, FIZ Karlsruhe, P.O. Box 2465, W-7500, Karlsruhe 1, Germany, online services. Also available on CD-ROM. Inquire as to cost and availability.

Physics Briefs. American Institute of Physics, 335 East 45th Street, New York, NY 10017. (212) 661-9260 or FAX (212) 661-2036. Contains citations with abstracts of the literature of physics and related fields, 1979 to present. Available on the STN

International, FIZ Karlsruhe, P.O. Box 2465, W-7500, Karlsruhe 1, Germany, online service. Inquire as to cost and availability.

Scientific and Technical Books and Serials in Print. R.R. Bowker Inc., 121 Chanlon Road, New Providence, NJ 07974. (800) 521-8110 or (908) 464-6800. List of currently published books and serials in the physical and biological sciences, engineering and technology, with subject, author and titles indexes. Available on ORBIT, (800) 456-7248, online service. Inquire as to cost and availability.

SCISEARCH. Institute for Scientific Information, 3501 Market Street, Philadelphia, PA 19104. (800) 523-1850 or (215) 386-0100. Broad multidisciplinary title and author index to the international literature of science and technology, 1974 to present. Available on DIALOG, (800) 334-2564, and ORBIT, (800) 456-7248, online services. Also available on CD-ROM. Inquire as to cost and availability.

WILSONLINE. H.W. Wilson Company, 950 University Avenue, Bronx, NY 10452. (800) 367-6770 or (212) 588-8400. Makes available online versions of the printed H.W. Wilson indexes including Applied Science and Technology Index, Business Periodicals Index, General Science Index, and Readers' Guide to Periodical Literature. Period covered is generally 1983 to present. Available on BRS, (800) 289-4277, DIALOG, (800) 334-2564, and OCLC EPIC, (800) 848-5878, online services.Also available on CD-ROM. Inquire as to cost and availability.

OTHER SOURCES

Atlas of the andromeda Galaxy. Paul W. Hodge. University of Washington Press, P. O. Box 50096. Seattle, WA 98145- 5096. (206) 543-4050. (800) 441-4115. 1981. $75.00.

Atlas of Deep-sky Spendors. Hans Vehrenberg. Sky Publishing Corp., 49 Bay State Road, Cambridge, MA 02138. (617) 864-7360. (800) 253-0245. 4th edition. 1983. $24.95.

Atlas of the Ultraviolet Sky. Richard C. Henry, Wayne B. Landsman, et al. Johns Hopkins University Press, 2715 North Charles Street, Baltimore, MD 21218-4319. (410) 516-6900. (800) 537-5487. 1988. $70.00.

Atlas of the Universe. Patrick Moore. Rand McNally & Company. 8255 North Central Park, Skokie, IL 60076-2970. (708) 673-9100. 1994. $29.95.

Cambridge Atlas of Astronomy. Jean Audouze, et al. Cambridge University Press, 40 West 20th Street, New York,

NY 10011-4211. (212) 924-3900. (800) 872-7423. 3rd edition. 1994. $75.00.

Cambridge Guide To the Constellations. Michael E. Bakich. Cambridge University Press, 40 West 20th Street, New York, NY 10011-4211. (212) 924-3900. (800) 872-7423. 1995. $19.95 in paper.

Carnegie Atlas of Galaxies. Allan Sandage and John Bedke. Carnegie Institution of Washington, 1530 P Street NW, Washington, DC 20005. (202) 387-6411. Two volumes. 1994. $92.00.

Color Atlas of Galaxies. James Wray. Cambridge University Press, 40 West 20th Street, New York, NY 10011-4211. (212) 924-3900. (800) 872-7423. 1988. $94.95.

Observer's Sky Atlas. E. Karkoschka. Springer-Verlag New York, Inc., 175 Fifth Avenue, New York, NY 10010. (212) 460-1500. (800) 777-4643. 1995. $15.95.

Observing Handbook and Catalog of Deep-sky Objects. Christian B. Luginbuhl and Brian A. Skiff. Cambridge University Press, 40 West 20th Street, New York, NY 10011- 4211. (212) 924-3900. (800) 872-7423. 1990. $49.95.

Photometric Atlas of Northern Bright Galaxies. Kodaira, Keichi, et al editors. Columbia University Press, 562 West 113th Street, New York, NY 10025. (212) 666-1000. (800) 944-8648. 1990. $180.00.

Third Reference Catalogue of Bright Galaxies. Gerard H. DE Vaucouleurs et al, editors. Springer-Verlag New York, Inc., 175 Fifth Avenue, New York, NY 10010. (212) 460-1500. (800) 777-4643. Three volumes. 1991. $198.00 set.

The Sky: A User's Guide. David H. Levy. Cambridge University Press, 40 West 20th Street, New York, NY 10011- 4211. (212) 924-3900. (800) 872-7423. 1993. $14.95 in paper.

PERIODICALS

Astronomical Journal. American Institute of Physics, One Physics Ellipse, College Park, MD 20740-3843. (301) 209-3000. Published for the American Astronomical Society. 1849 to present. Monthly. $280.00 per year. ISSN: 0004- 6256.

Astronomical Society of the Pacific. PUBLICATIONS. Astronomical Society of the Pacific, 390 Ashton Avenue, San Francisco, CA 94112. (415) 337-1100. FAX (415) 337-5205.1889 to present. Monthly. $175.00 per year. ISSN: 0004- 6280.

Astronomy. Kalmbach Publishing Company, Box 1612, Waukesha, WI 53187-1612. (414) 796-3476. FAX (414) 796-1142. 1973

to present. Monthly. $27.00 per year. ISSN: 0091-6358.

Astronomy and Astrophsyics. Springer-Verlag New York, Inc., 44 Hartz Way, Secaucus, NJ 07096-2491. (201) 348-4033. FAX(201) 348-4505. 1969 to date. 36 issues per year in 12 volumes. $1733.00 per year. ISSN: 0004-6361.

Astronomy Reports. American Institute of Physics, One Physics Ellipse, College Park, MD 20740-3843. (301) 209- 3000. English translation of Astronomicheksii Zhurnal. Formerly Soviet Astronomy AJ. 1957 to present. Bimonthly. $1160.00 per year. ISSN: 1063-7729.

Astrophysical Journal; An International Review of Astronomy and Astronomical Physics. University of Chicago Press, Journals Division, 5720 South Woodlawn Avenue, Chicago, IL 60637. (312) 753-3347. FAX (312) 753-0811. 1895 to date. three issues per month. $740.00 per year. ISSN: 0004-637X.

Astrophpysics and Space Science; An International Journal of Cosmic Physics. Kluwer Academic Publishers, Box 358, Accord Station, Hingham, MA 02018-0358. (617) 871-6600. FAX (617) 871-6528. 1968 to present. 24 issues per year. $2544.00 per year. ISSN: 0004-640X.

Celestial Mechanics and Dynamical Astronomy; An International Journal of Space Dynamics. Kluwer Academic Publishers, Box 358, Accord Station, Hingham, MA 02018-0358. (617)

871-6600. FAX (617) 871-6528. 1969 to present. Monthly. $703.50 per year. ISSN: 0923-2958.

Experimental Astronomy; An International Journal On Astronomical Instrumentation and Data Analysis. Kluwer Academic Publishers, Box 358, Accord Station, Hingham, MA 02018-0358. (617) 871-6600. FAX (617) 871-6528. 1989 to present. Quarterly. $161.00 per year. ISSN: 0922-6435.

Icarus; International Journal of Solar System Studies. Academic Press, Inc., Journal Division, 525 B Street, Suite 1900, San Diego, CA 92101-4495. (619) 230-1840. FAX (619) 699-6800. 1962 to present. Monthly. $1080.00 per year. ISSN: 0019-1035.

Landolt-boernstein, Zahlenwerte Und Funktionen Aus Naturwissenschaften Und Technik. GROUP 6: ASTRONOMY. Springer-Verlag New York, Inc., 175 Fifth Avenue, New York, NY 10010. (212) 460-1500. FAX (212) 473-6272. 1965. Irregular. Price varies. ISSN: 0075-7896.

Mercury. Astronomical Society of the Pacific. 390 Ashton Avenue, San Francisco, CA 94112. (415) 337-1100. FAX: (415) 337-5205. 1972 to present. Bimonthly. $175.00 per year. ISSN: 0047-6773.

Monthly Notices of the Royal Astronomical Society. Blackwell Scientific Publication LT., Osney Mead, Oxford OX2 OEL, England. TEL 0865-240201. FAX 0865-721205. 1827 to present. Fortnightly. $1899.00 per year. ISSN: 0035-8711.

Observatory. Space and Astrophysics Division, Rutherford Appleton Laboratory, Chilton, Didcot, Oxon OX11 OQX England. FAX 0235-445848. 1877 to present. Bimonthly. $42.00 per year.

Sky and Telescope. Sky Publishing Corporation, Box 9111, Belmont, MA 02178. (617) 864-7360. FAX (617) 864-6117.1941 to present. Monthly. $27.00 per year. ISSN: 0037-6604.

Space Science Reviews. Kluwer Academic Publishers, Box 358, Accord Station, Hingham, MA 02018-0358. (617) 871-6600. FAX (617) 871-6528. 1962 to present. Sixteen issues per year. $940.00 per year. ISSN: 0038-6308.

Vistas in Astronomy; An International Review Journal. Albert C. Beer, editor. Elsevier Science Publishing Company, Inc., 660 White Plains Road, Terrytown, NY 10591- 5153. (914) 524-9200. FAX: (914) 333-2444. 1958 to present. Quarterly. $415.00 per year. ISSN: 0083-6656.

Transactions of the International Astronomical Union. Proceedings of the General Assembly. Kluwer Academic Publishers, Box 358, Accord Station, Hingham, MA 02018- 0358. (617) 871-6600. Issues in two parts: Part A, Reports, and Part B, Proceedings. Triennial. XXI - 1994. Price varies. ISSN: 0080-1372.

Webb Society Quarterly Journal. Webb Society, 194 Foundry Lane, Freemantle, Southampton, Hampshire SO1 3EQ, England. 1969 to present. Quarterly. $19.50 per year. ISSN: 0043- 1680.

RESEARCH CENTERS AND INSTITUTES

Arecibo Observatory. Cornell University, Box 995, Arecibo, PR 00613. (809) 878-2612. FAX (809) 878-1861.

INFRARED ASTRONOMY

Ency. of Physical Sciences and Engineering Info. Sources

Astronomy Program. University of Maryland, 1207 Computer & Space Sciences Building, College Park, MD 20742. (301) 405- 3001. FAX (301) 314-9067.

Joint Institute For Laboratory Astrophyics. University of Colorado, Box 440, Boulder, CO 80309-0440. (303) 492-7780. FAX (303) 492-5235.

Lowell Observatory. 1400 West Mars Hill Road. Flagstaff, AZ 86001, (602) 774-3358. FAX (602) 774-6296.

McDonald Observatory, University of Texas. P.O. Box 1337, Fort Davis, TX 79734. (915) 426-3263.

National Astronomy and Ionosphere Center. Cornell University, Space Sciences Building, Ithaca, NY 14853. (607) 255-3735. (607) 255-8803.

National Radio Astronomy Observatory. Edgemont Road, Charlottesville, VA 22903. (804) 296-0211.

Palomar Observatory. California Institute of Technology. 105-24, Pasadena CA 91125. (818) 356-4033.

Space Telescope Science Institute. 3700 San Martin Drive, Baltimore, MD 21218. (301) 338-4700.

Steward Observatory. University of Arizona. Tuscon, AZ 85721. (602) 621-6524. FAX (602) 621-1532.

INFRARED PHOTOGRAPHY

See also: AERIAL PHOTOGRAPHY, PHOTOCHEMISTRY, PHOTO-GRAMMETRY, PHOTOGRAPHIC FILM, PHOTOGRAPHY

ABSTRACT SERVICES AND INDEXES

Applied Science and Technology Index; A Cumulative Subject Index To English Language Periodicals in the Fields of Aeronautics and Space Science, Computer Technology, Chemistry, Construction Industry, Energy and Related Areas. H.W. Wilson Co., 950 University Avenue, Bronx, NY 10452. (800) 367-6770 or (212) 588-8400. FAX (718) 590-1617. From 1958 to present. Monthly. Inquire about cost and availability. Also available on CD-ROM and online.

Chemical Abstracts - Applied Chemistry and Chemical Engineering Sections. Chemical Abstracts Service, 2540 Olentangy River Road, Box 3012, Columbus, OH 43210. (800) 848-6538 or (614) 447-3600, FAX (614) 447-3713. Bi-weekly. $1410.00 per year.

Engineering Index Monthly. Engineering Information, Inc., Castle Point on the Hudson, Hoboken, NJ 07030. (800) 221-1044. FAX (212) 832-1857. Monthly. $2300.00 per year. Also available online as COMPENDEX, and also on CD-ROM. Covers chemical engineering, computers, electrical engineering, civil engineering, metals and mining, industrial management, and mechanical engineering.

Imaging Abstracts (formerly Photographic Abstracts), Pergamon Press Inc., Maxwell House, Fairview Park, Elmsford, NY 10523. (914) 592-7700. Fax (914) 592-3625. Bimonthly. Inquire for price.

Physics Abstracts. Institute of Electrical Engineers, Michael Faraday House, Six Hill Way, Stevenage, Herts, SG1 2AY, England. Distributed by IEEE, 445 Hoes Lane, Piscataway, NJ 08854. (908) 562-5549. 1898 to present. Monthly. $2700.00 per year. Also available online as INSPEC.

SCIENCE CITATION INDEX. Institute for Scientific Information, 3501 Market Street, Philadelphia, PA 19104. (215) 386-0100. FAX (215) 386-6362. Inquire about availability and cost. Also available on CD-ROM.

Scientific and Technical Aerospace Reports (star). National Aeronautics and Space Administration. NASA Center for Aerospace Information, Box 8757, BWI Airport, Baltimore, MD 21240. (301) 621-0153. Monthly. Inquire about availability and cost. ALso available through the NASA online retrieval service (RECON), and through the Aerospace Database through DIALOG.

ASSOCIATIONS AND PROFESSIONAL SOCIETIES

American Society For Photogrammetry and Remote Sensing, 5410 Grosvenor Lane, Suite 210, Bethesda, MD 20814. (301) 493-0290. FAX (301) 493-0208.

The International Society For Optical Engineering (spie), Po Box 10, 1000-20th Street, Bellingham, WA 98227-0010.

Optical Society of America, 2010 Massachusetts Avenue Nw, Washington, DC 20036. (202) 223-8130.

The Society For Imaging Science and Technology, 7003 Kilworth Lane, Springfield, VA 22151. (703) 642-9090.

Society of Photographic Scientists and Engineers. 7003 Kilworth Lane, Springfield, VA 22151. (703) 642-9090.

DIRECTORIES AND BIOGRAPHICAL SOURCES

Directory of Engineering Societies and Related Organizations. Gordon Davis. 13th edition. American Association of Engineering Societies, 1111 19th Street NW, Suite 608, Washington, DC 20036. (202) 296-2237 or (800) 658-8897. 1989. Inquire for price.

International Research Centers Directory. Gale Research, 835 Penobscot Building, Detroit, MI 48226-4094. (313) 961-2242. (800) 347-4253. FAX (313) 961-6083. 8th edition. 1995. $410.00

Research Centers Directory. Gale Research, 835 Penobscot Building, Detroit, MI 48226-4094. (313) 961-2242. (800) 347-4253. FAX (313) 961-6083. $485.00.

Scientific and Technical Organizations and Agencies Directory. Gale Research, 835 Penobscot Building, Detroit, MI 48226-4094. (313) 961-2242. (800) 347-4253. FAX (313) 961-6083. 4th edition. 1996. $195.00.

Who's Who in Engineering. American Association of Engineering Societies, 1111 19th Street NW, Suite 608, Washington, DC 20036. (202) 296-2237 or (800) 658-8897. 8th edition. 1991. Inquire for price.

ENCYCLOPEDIAS AND DICTIONARIES

McGraw-Hill Encyclopedia of Science and Technology. Sybil P. Parker, ed. 7th edition. 20 volumes. McGraw-Hill Publishing Company, 1221 Avenue of the Americas, New York, NY 10020. (800) 262-4729 or (212) 512-3825. 1992. $1900.00

Thesaurus of Photographic Science and Engineering. Society of Photographic Scientists and Engineers Staff. Books on Demand, 300 N. Zeeb Road, Ann Arbor, MI 48106-1346. (313) 761-4700 or (800) 521-0600. 1968. $34.30.

GENERAL WORKS

Applied Infrared Photography. Eastman Kodak Company. Saunders PhotoGraphic Inc., 21 Jet View Drive, Rochester, NY 14624. (716) 328-7200 or (800) 234-3686. 1981. Inquire for cost and availability.

The Art of Infrared Photography. Joseph Paduano. Revised edition. Amherst Media, 418 Homecrest Drive, Amherst, NY 14226. (716) 874-4450 or (800) 622-3278. FAX (716) 874-4508. 1990. $17.95.

An Introduction To Infrared Image Acquisition and Classification Systems. L.F. Pau & M.Y. El Nahas. John Wiley and Sons, Inc., 605 Third Avenue, New York, NY 10158. (800) 526-5368 or (212) 850-6000. 1983. $124.50.

Photography By Infrared: Its Principles and Applications. H. Lou Gibson. 3rd edition. John Wiley and Sons, Inc., 605 Third Avenue, New York, NY 10158. (800) 526-5368 or (212) 850-6000. 1978. Inquire for cost and availability.

HANDBOOKS AND MANUALS

Handbook of Photography. Ronald P. Lovell, et al. Delmar Publishers, Division of Thomson Educational Publishing, Inc., 3 Columbia Circle, Box 15015, Albany, NY 12205. (518) 464-3500 or (800) 347-7707. 1993. $27.95.

The Infrared Handbook. W. L. Wolfe & J. Zissis. SPIE, PO Box 10, 1000-20th Street, Bellingham, WA 98227-0010. 1985. $75.00.

Kodak Professional Photoguide. Eastman Kodak Company staff. Saunders PhotoGraphic Inc., 21 Jet View Drive, Rochester, NY 14624. (716) 328-7200 or (800) 234-3686. 1993. $19.95.

ONLINE DATABASES AND CD-ROMS

Compendex Plus. Engineering Information, Inc., 345 East 47th Street, New York, NY 10017. (212) 705-7600 or (800) 221-1044. Contains citations with abstracts to worldwide literature in engineering and technology, from 1970 to present. Available on online BRS,(800) 289-4277, DIALOG, (800) 334-2564, ORBIT (800) 456-7248, and STN International, FIZ Karlsruhe, P.O. Box 2465, W-7500, Karlsruhe 1, Germany, online services. Also available on CD-ROM. Inquire as to cost and availability.

CSA Engineering. Cambridge Scientific Abstracts, 7200 Wisconsin Avenue, Suite 601, Bethesda, MD 20814. (301) 961-6750 or (800) 843-7751. Contains citations and abstracts of international periodicals and research literature covering all fields of engineering and science and technology,including computer

and information science, electronics, mechanical engineering, solid state materials, 1981 to present. Available on BRS,(800) 289-4277, online service.Inquire as to cost and availability.

Current Contents Search. Institute for Scientific Information, 3501 Market Street, Philadelphia, PA 19104. (215) 386-0100. FAX (215) 386-6362. Contains citations to articles listed in the table of contents of science and technology journals. Also articles in social sciences and life sciences journals. Available on BRS,(800) 289-4277, DIALOG,(800) 334-2564, online services. Inquire as to cost and availability.

Dissertation Abstracts. University Microfilms International, 300 North Zeeb Road, Ann Arbor, MI 48106. (800) 521-0600 or (313) 761-4700. Scope includes virtually all doctoral dissertations accepted at accredited American institutions from 1861 to present in 252 subject areas. Available on BRS,(800) 289-4277, DIALOG,(800) 334-2564, and OCLC EPIC,(800) 848-5878, online services. Also available on CD-ROM. Inquire as to cost and availability.

NTIS Bibliographic Database. National Technical Information Service, 5285 Port Royal Road, Springfield, VA 22161. (703) 487-4929 or FAX (703) 321-8199. Broad coverage of government-sponsored science and technology research reports, 1964 to present. Available on BRS,(800) 289-4277, DIALOG, (800) 334-2564, ORBIT, (800) 456-7248, and STN International, FIZ Karlsruhe, P.O. Box 2465, W-7500, Karlsruhe 1, Germany, online services. Also available on CD-ROM. Inquire as to cost and availability.

SCISEARCH. Institute for Scientific Information, 3501 Market Street, Philadelphia, PA 19104. (800) 523-1850 or (215) 386-0100. Broad multidisciplinary title and author index to the international literature of science and technology, 1974 to present. Available on DIALOG,(800) 334-2564, and ORBIT,(800) 456-7248, online services. Also available on CD-ROM. Inquire as to cost and availability.

WILSONLINE. H.W. Wilson Company, 950 University Avenue, Bronx, NY 10452. (800) 367-6770 or (212) 588-8400. Makes available online versions of the printed H.W. Wilson indexes including Applied Science and Technology Index, Business Periodicals Index, General Science Index, and Readers' Guide to Periodical Literature. Period covered is generally 1983 to present. Available on BRS,(800) 289-4277, DIALOG,(800) 334-2564, and OCLC EPIC,(800) 848-5878, online services. Also available on CD-ROM. Inquire as to cost and availability.

PERIODICALS

British Journal of Photography. Bouverie Publishing Company Ltd., 147-151 Temple Chambers, Temple Avenue, London EC47 0DT, England. Telephone 071-583-3030. FAX 071-583-4068. 1854 to present. Weekly. Inquire for price and availability.

Industrial Photography. PTN Publishing Corporation, 445 Broad Hollow Road, Melville, NY 11747-4722. (516) 845-2700. FAX (516) 845-7109. 1952 to present. Monthly. $60.00 per year.

Journal of Photographic Science. The Barn, Whitehall, Near Middle Marwood, Barnstaple, N. Devon EX31 4EQ, England. Telephone 0271-72482. FAX 0271-72482. 1952 to present. Bimonthly. Inquire for price and availability.

Photogrammetric Engineering and Remote Sensing. American Society for Photogrammetry and Remote Sensing, 5410 Grosvenor Lane, Suite 210, Bethesda, MD 20814. (301) 493-

0290. FAX (301) 493-0208. 1934-present. Monthly. $160.00 per year.

Scientific and Applied Photography and Cinematography. Gordon and Breach Science Publishers, 820 Town Center Drive, Langhorne, PA 19047. (215) 750-2642. FAX (215) 750-6343. 6 times a year. Inquire for price.

RESEARCH CENTERS AND INSTITUTES

Image Permanence Institute. Rochester Institute of Technology, RIT City Center, 50 W. Main Street, Rochester, NY 14614. (716) 475-5199. FAX (716) 475-5097.

University of Idaho, Idaho Remote Sensing Research Unit. Moscow, ID 83843. (208) 885-7209.

University of North Dakota, Institute For Remote Sensing. Geography Department, Grand Forks, ND 58202-8274. (701) 777-42426.

INFRARED SPECTROSCOPY

See: SPECTROSCOPY

INFRARED STARS

See: STARS

INORGANIC CHEMISTRY

See also: ANALYTICAL CHEMISTRY, CHEMISTRY, ORGANIC CHEMISTRY, PHYSICAL CHEMISTRY

ABSTRACT SERVICES AND INDEXES

applied science and technology index; a cumulative subject index to english language periodicals in the fields of

Aeronautics and Space Science, Computer Technology, Chemistry, Construction Industry, Energy and Related Areas. H. W. Wilson Co., 950 University Avenue, Bronx, NY 10452- 9978. (212) 588-8400. (800) 367-6770. FAX (718) 590-1617. From 1958 to present. Monthly. Inquire about cost and availability. Also available online (BRS and WILSONLINE) and on CD-ROM. ISSN: 0003-6986.

Chemical Abstracts. Chemical Abstracts Service. 2540 Olentangy River Road, Box 3012, Columbus, OH 43210-0012. (614) 447-3600. FAX (614) 447-3713. 1907 to present. Weekly. $16,800.00 per year. Available online and on CD- ROM. CA is also available in five section groupings. Inquire regarding cost and availability.

Current Contents: Physical, Chemical and Earth Sciences. Institute for Scientific Information, 3501 Market Street, Philadelphia, PA 19104. (215) 386-0100. FAX (215) 386-2291. From 1961 to present. Weekly. $442.00 per year. Also available on

CD-ROM and online. Inquire regarding cost and availability. ISSN: 0163-2574.

Engineering Index Monthly. Engineering Information, Inc., Castle Point on the Hudson, Hoboken, NJ 07030. (201) 216-8500. (800) 221-1044. FAX(201) 216-8532. Monthly. $2300.00 per year for monthly issues. $1980 per year for annual cumulation. Available on CD-ROM and online as COMPENDEX. ISSN: 0742-1974; 0360-8557.

General Science Index. H.W. Wilson Company, 950 University Avenue, Bronx, NY 10452. (212) 588-8400. (800) 367-6770. FAX (718) 590-1617. From 1978 to present. Ten issues per year; quarterly and annual cumulations. Service basis. Available on CD-ROM and online. Inquire regarding cost and availability. ISSN: 0162-1963.

Physics Abstracts. INSPEC. Section A, Science Abstracts. Institution of Electrical Engineers (IEE), London, United Kingdom. Available from: INSPEC/IEEE -Institute of Electrical and Electronic Engineers, Box 1331, 445 Hoes Lane, Piscataway, NJ 08855-1331. (908) 562-5549. 1898 to present. 24 issues per year. $2835.00 per year. ISSN: 0036-8091. Also available online and on CD-ROM.

Science Citation Index. SCI. Institute for Scientific Information, 3501 Market Street, Philadelphia, PA 19104. (215) 386-0100. (800) 523-1850. FAX (215) 386-2991. 1961 to present. Six issues per year, plus annual cumulation. $11650.00 per year. Also available online and on CD-ROM. Inquire about price and availability. ISSN: 0036-827X.

ANNUAL REVIEWS AND YEARBOOKS

Advanceds in Inorganic and Radiochemistry. Academic Press, Inc., 1250 Sixth Avenue, San Diego, CA 92101. (619) 231-0926. FAX (619) 699-6715. 1959 to present. Irregular. Price varies, inquire.

ASSOCIATIONS AND PROFESSIONAL SOCIETIES

American Chemical Society, 1155 16th Street, Nw, Washington, DC 20036. (202) 872-4600.

American Institute of Chemical Engineers. 345 East 47th Street, New York, NY 10017. (212) 705-7338.

American Institute of Chemists. 501 Wythe Street, Alexandria, VA 22314-1917. (703) 836-2090. FAX (703) 836- 2091.

Association of Consulting Chemists and Chemical Engineers, 50 East 41st Street, Suite 92, New York, NY 10017. (212) 684-6255.

Association of official Analytical Chemists. 2200 Wilson Blvd, Suite 400, Arlington, VA 22001. (703) 522-3032.

Electrochemical Society. 10 South Main Street, Pennington, NJ 08534. (609) 737-1902.

BIBLIOGRAPHIES

Chemical Information. H. R. Collier, editor. Springer- Verlag New York, Inc., 175 Fifth Avenue, New York, NY 10010. (212) 460-1500. (800) 777-4643. 1990. $91.00.

Chemical Information Sources. Gary Wiggins. McGraw-Hill Publishing Company, 1221 Avenue of the Americas, New York, NY 10020. (800) 262-4729 or (212) 512-3825. 1991. $42.50.

Information Sources in Chemistry. R.T. Bottle and J.F.B Rowland, editors. R.R. Bowker Inc., 121 Chanlon Road, New Providence, NJ 07974. (800) 521-8110 or (908) 464-6800. Fourth edition. 1993. $75.00.

Scientific and Technical Books and Serials in Print; An Index To Literature in Science and Technology. R.R. Bowker Inc., 121 Chanlon Road, New Providence, NJ 07974. (908) 464-6800. (800) 521-8110. FAX (908) 665-3502. 1972 to present. Annual. 4 volumes. 1994. $299.95. Also available on compact disc and online. ISSN: 0000-054X.

DIRECTORIES AND BIOGRAPHICAL SOURCES

American Institute of Chemists - Professional Directory. American Institute of Chemists, 501 Wythe Street, Alexandria, VA 22314-1917. (703) 836-2090. FAX (703) 836-2091. Annual. $65.00.

American Men and Women of Science. Physical and Biological Sciences. R.R. Bowker Inc., 121 Chanlon Road, New Providence, NJ 07974. (908) 464-6800. (800) 521-8110. 20th edition. 8 volumes. 1996. $850.00

Chemical Sources International. Mike Desing and Kurt Gandenberger, editors. Chemical Sources International, PO Box 1824, Clemson, SC 29633. (803) 646-7840. 1994. $285.00.

Chemical Week — Buyers Guide Issue. Chemical Week Associates, 888 Seventh Avenue, New York, NY 10106. (212) 621-4900. FAX (212) 621-4949. Annual, October. $50.00.

Directory of Chemistry Software 1992. Wendy Warr, Peter Willett, Geoff Downs. American Chemical Society, 1155 16th Street, NW, Washington, DC 20036. (202) 872-4600. 1992. $35.95.

Research Centers Directory. Gale Research Company, Inc.,

835 Penobscot Building, Detroit, MI 48226-4094. (313) 961-

2242. (800) 347-4253. 20th edition. 1995. $485.00.

ISSN: 0080-1518.

ENCYCLOPEDIAS AND DICTIONARIES

Academic Press Dictionary of Science and Technology. Christopher Morris, editor. Academic Press, Inc., 1250 Sixth Avenue, San Diego, CA 92101. (619) 231-0926. FAX (619) 699-6715. 1991. $115.00.

Concise Encyclopedia Chemistry. Translated and revised by Mary Eagleson. Walter de Gruyter, Inc., 200 Saw Mill River Road, Hawthorne, New York, 10532. (914) 747-0110 or Fax (914) 747-1326. 1994. $69.95.

Dictionary of Inorganic Compounds. Jane E. Macintyre, editor. Chapman & Hall, 1 Penn Plaza, New York, NY 10119. (212) 564-1060. 1992. $39.95.

Encyclopedia of Applied Physics. VCH Publishers, Inc., 303 Northwest 12th Avenue, Deerfield Beach, FL 33442. (800) 367-8249. 1991-. Twenty volumes. $6000.00.

Encyclopedia of Chemical Processing and Design. McKetta. Dekker, Inc., 270 Madison Avenue, New York, NY 10016. (212) 696-9000. (800) 228-1160. 1976 - . $175.00 per volume.

Grant & Hawks Chemical Dictionary. R. A. Grant. McGraw-Hill Publishing Company, Inc., 1221 Avenue of the Americas, New York, NY 10020. (212) 512-2000. (800) 262-4729. 5th edition. 1987. $64.50.

Hawley's Condensed Chemical Dictionary. Richard J. Lewis. Van Nostrand Reinhold, 115 Fifth Avenue, New York, NY 10003. (212) 254-3232. (800) 842-3636. 12th edition. 1993. $69.95.

Kirk-Othmer Encyclopedia of Chemical Technology. John Wiley and Sons, Inc., 605 Third Avenue, New York, NY 10158. (800) 526-5368 or (212) 850-6000. Fourth edition. 1991-. Twenty-seven volumes. $5400.00.

McGraw-Hill Encyclopedia of Chemistry. Sybil P.Parker, editor. McGraw-Hill Book, Incorporated, 1221 Avenue of the Americas, New York, NY 10020. (212) 997-3675. Second edition. 1993. $95.50.

McGraw-Hill Encyclopedia of Science and Technology. McGraw-Hill Book, Incorporated, 1221 Avenue of the Americas, New York, NY 10020. (212) 997-3675. Seventeenth edition. Twenty volumes. 1992. $1900.00.

GENERAL WORKS

Basic Inorganic Chemistry. Albert Cotton, et al. John Wiley & Sons, Inc., 605 Third Avenue, New York, NY 10158- 0012. (212) 850-6000. (800) 225-5945. 3rd edition. 1994. $72.00.

The Chemical Bond: Structure and Dynamics. Ahmed Zewail. Academic Press, Inc., 1250 Sixth Avenue, San Diego, CA 92101-4311. (619) 231-0926. FAX (619) 699-6715. 1992. $49.95

Concepts and Models of Inorganic Chemistry. B. E. Douglas, et al. John Wiley & Sons, Inc., 605 Third Avenue, New York, NY 10158-0012. (212) 850-6000. (800) 225-5945. 3rd edition. 1994. $59.95.

The Development of Chemical Principles. Cooper H. Langford and Ralph A. Beebe. Dover Publications, Inc., 180 Varick Street, New York, NY 10014. (212) 255-3755. (800) 223- 3130. 1995. $9.95.

Exploring Chemical Elements and their Compounds. David L. Heiserman. TAB Books, P.O. Box 40, Blue Summit, PA 17294-0850. (717) 794-2191 or (800) 822-8138. 1992. $17.95.

Inorganic Chemical Nomenclature: Principles and Practice. B. Peter Brock, et al, editors. American Chemical Society, 1155 16th Street, NW, Washington, DC 20036. (202) 872-4600. (800) 333-9511. FAX (614) 447-3671. 1990. $59.95.

INORGANIC CHEMISTRY

Ency. of Physical Sciences and Engineering Info. Sources

HANDBOOKS AND MANUALS

Chemical Formulary. H. Bennett, editor. Chemical Publishing Company, Inc., 80 Eighth Avenue, New York, NY 10011. (212) 255-1950. Volumes 1-30. $70.00 per volume.

Chemometrics: Chemical and Sensory Data. David R. Burgard and James T. Kuznicki. CRC Publishers, Inc., 2000 Corporate Blvd., N.W., Boca Raton, FL 33431. (407) 994-0555 or (800) 333-8300. 1990. $120.00.

Chemical Reference Handbook. Gordon Press, PO Box 459, Bowling Green Station, New York, NY 10004. (718) 624-8419. 1995. $260.00.

CRC Handbook of Chemistry and Physics. David R. Kide, editor. CRC Press Inc., 2000 Corporate Blvd., NW, Boca Raton, FL 33431. (407) 994-0555. (800) 333-8300. 77th edition. 1996. $99.95.

Guide To Basic Chemical Compounds. D.R. Lide, Jr. CRC Publishers, Inc., 2000 Corporate Blvd., N.W., Boca Raton, FL 33431. (407) 994-0555. (800) 333-8300. 1993. $120.00.

Inorganic Chemicals Handbook. John J. McKetta. Marcel Dekker, Inc., 270 Madison Avenue, New York, NY 10016. (212) 696-9000. (800) 228-1160. 2 volumes. 1993. $400.00.

Laboratory Handbook of Materials, Equipment, and Techniques. Gary S. Coyne. Prentice Hall (Division of Simon and Schuster), 15 Columbus Circle, New York, NY 10023. (212) 373-8500 or (800) 922-0579. 1992. $45.00.

Lange's Handbook of Chemistry. John A. Dean, editor. McGraw-Hill Publishing Company, 1221 Avenue of the Americas, New York, NY 10020. (800) 262-4729 or (212) 512-3825. Fourteenth edition. 1992. $79.50.

Riegel's Handbook of Industrial Chemistry. James A. Kent, editor. Van Nostrand Reinhold, 115 Fifth Avenue, New York, NY 10003. (212) 254-3232 or (800) 926-2665. Ninth edition. 1992. $114.95.

ONLINE DATABASES AND CD-ROMS

Analytical Abstracts Online. Royal Society of Chemistry, Information Services, Thomas Graham House, Science Park, Milton Road, Cambridge, CB4 4WF, England. Contains citations, mostly with abstracts, of the worldwide literature on analytical chemistry, from 1980 to present. Available on DIALOG, (800) 334-2564, and STN International, FIZ Karlsruhe, P.O. Box 2465, W-7500, Karlsruhe 1, Germany, online services. Inquire as to cost and availability.

CA Search. Chemical Abstracts Service, P.O. Box 3012, Columbus, OH 43210-0012. (614) 447-3600. (800) 848-6533. FAX (614) 447-3709. Very comprehensive guide to worldwide chemical literature and related fields, 1972 to present. Available on BRS,(800) 289-4277, DIALOG, (800) 334-2564, ORBIT (800) 456-7248, and STN International, FIZ Karlsruhe, P.O. Box 2465, W-7500, Karlsruhe 1, Germany, online services. Inquire as to cost and availability.

Chemical Journals of the American Chemical Society. American Chemical Society, 1155 16th Street, N.W., Washington, DC 20036. (202) 872-4381 or (800) 424-6747. Contains complete text of approximately 90,000 articles from 22 primary journals published by the American Chemical Society, from mostly 1982 to present. Available on STN International, FIZ Karlsruhe, P.O. Box 2465, W-7500, Karlsruhe 1, Germany, online service. Inquire as to cost and availability.

Compendex Plus. Engineering Information, Inc., 345 East 47th Street, New York, NY 10017. (212) 705-7600 or (800) 221-1044. Contains citations with abstracts to worldwide literature in engineering and technology, from 1970 to present. Available on online BRS,(800) 289-4277, DIALOG, (800) 334-2564, ORBIT (800) 456-7248, and STN International, FIZ Karlsruhe, P.O. Box 2465, W-7500, Karlsruhe 1, Germany, online services. Also available on CD-ROM. Inquire as to cost and availability.

Current Contents Search. Institute for Scientific Information, 3501 Market Street, Philadelphia, PA 19104. (215) 386-0100. FAX (215) 386-6362. Contains citations to articles listed in the table of contents of science and technology journals. Also articles in social sciences and life sciences journals. Available on BRS,(800) 289-4277, DIALOG,(800) 334-2564, online services. Inquire as to cost and availability.

Dissertation Abstracts. University Microfilms International, 300 North Zeeb Road, Ann Arbor, MI 48106. (800) 521-0600 or (313) 761-4700. Scope includes virtually all doctoral dissertations accepted at accredited American institutions from 1861 to present in 252 subject areas. Available on BRS,(800) 289-4277, DIALOG,(800) 334-2564, and OCLC EPIC,(800) 848-5878, online services. Also available on CD-ROM. Inquire as to cost and availability.

Gmelin Database. Gmelin-Institut fur Anorganische Chemie und Grenzgebiete, Varrentrapperstrasse, 40-42, Carl-Bosch-Haus, D-6000, Frankfurt am Main 90, Germany. Contains structural and factual data relating to inorganic and organometallic chemistry. Provides data from the Gmelin Handbook of Inorganic and Organometallic Chemistry. Covers the period 1817 to 1975; 1988-89. Available on STN International, FIZ Karlsruhe, P.O. Box 2465, W-7500, Karlsruhe 1, Germany, online service. Inquire as to cost and availability.

Kirk-Othmer Encyclopedia of Chemical Technology. John Wiley and Sons, Inc., 605 Third Avenue, New York, NY 10158. (800) 526-5368 or (212) 850-6000. Contains the complete text of all chapters in the 27 volume fourth edition of the KIRK-Othmer ENCYCLOPEDIA of CHEMICAL TECHNOLOGY. 1991. Available on BRS,(800) 289-4277, DIALOG,(800) 334-2564, online services. Inquire as to cost and availability.

NTIS Bibliographic Database. National Technical Information Service, 5285 Port Royal Road, Springfield, VA 22161. (703) 487-4929 or FAX (703) 321-8199. Broad coverage of government-sponsored science and technology research reports, 1964 to present. Available on BRS,(800) 289-4277, DIALOG, (800) 334-2564, ORBIT, (800) 456-7248, and STN International, FIZ Karlsruhe, P.O. Box 2465, W-7500, Karlsruhe 1, Germany, online services. Also available on CD-ROM. Inquire as to cost and availability.

SCISEARCH. Institute for Scientific Information, 3501 Market Street, Philadelphia, PA 19104. (800) 523-1850 or (215) 386-0100. Broad multidisciplinary title and author index to the international literature of science and technology, 1974 to present. Available on DIALOG,(800) 334-2564, and ORBIT,(800) 456-7248, online services. Also available on CD-ROM. Inquire as to cost and availability.

WILSONLINE. H.W. Wilson Company, 950 University Avenue, Bronx, NY 10452. (800) 367-6770 or (212) 588-8400. Makes

Ency. of Physical Sciences and Engineering Info. Sources

INORGANIC CHEMISTRY

available online versions of the printed H.W. Wilson indexes including Applied Science and Technology Index, Business Periodicals Index, General Science Index, and Readers' Guide to Periodical Literature. Period covered is generally 1983 to present. Available on BRS,(800) 289-4277, DIALOG,(800) 334-2564, and OCLC EPIC,(800) 848-5878, online services. Also available on CD-ROM. Inquire as to cost and availability.

PERIODICALS

Angewandte Chemie. VCH Publishers, Inc., 220 East 23rd Street, New York, NY 10010-4606. (212) 683-8333. (800) 367-8249. 1888 to present. 22 issues per year. $840.00 per year. ISSN: 0044-8249.

Chemical Physics. Elsevier Science Publishing Co., Inc., Box 882, Madison Square Station, New York, NY 10159. (212) 989-5800. FAX (212) 633-3990. 1973 to present. 33 issues per year in 11 volumes. $2759.00 per year. ISSN: 0301-0104.

Chemical Reviews. American Chemical Society, Box 3337, Columbus, OH 43210. (614) 447-3776. 1924 to present. 8 issues per year. $346.00 per year. ISSN: 0009-2665.

Chemical Week; Includes Annual Buyers Guide. Chemical Week Associates, 888 Seventh Avenue, New York, NY 10106. (212) 621-4900. FAX (212) 621-4949. 1914 to present. Weekly. $99.00 per year. ISSN: 0009-272X.

Chemtech. American Chemical Society. Box 3337, Columbus, OH 43210. (614) 447-3776. 1970 to present. Monthly. $370.00 per year. ISSN: 0009-2703.

Faraday Discussions . Royal Society of Chemistry, Thomas Graham House, Science Park, Milton, Road, Cambridge CB4 4WF, England. TEL 0223-420066. FAX 0223-423623. 1946 to present. Semi-annual. $259.00. ISSN: 0301-7249.

Fresenius' Journal of Analytical Chemistry. Springer-Verlag, 44 Hartz Way, Seacus, NJ 07096-2491. (201) 348-4033. 1862 to present. 36 issues per year. $1556.00 per year. ISSN: 0937-0633.

Inorganic Chemistry. American Chemical Society, Box 3337, Columbus, OH 43210. (614) 447-3776. 1962 to present. Semi-monthly. $500.00 per year. ISSN: 0538- 8066. *Journal of the American Chemical Society.* American Chemical Society, Box 3337, Columbus, OH 43210. (614) 447-3776. 1879 to present. Biweekly. $1055.00 per year. ISSN: 0002-7863.

Polyhedron. Elsevier Science Publishing Company, Inc., 660 White Plains Road, Tarrytown, NY 10591-5153. (914-524-9200. FAX (914) 524-9200. 1982 to present. 24 issues per year. $2175.00 per year. ISSN: 0277-5387.

RESEARCH CENTERS AND INSTITUTES

Center For Chemical Characterization and Analysis. Texas A & M Univeristy, College Station, TX 77843-3144. (409) 845-2341. FAX (409) 845-1655.

Center For Process Analytical Chemistry. University of Washington, BG-10, Seattle, WA 98195. (206) 545-2326. FAX (206) 543-6506

Chemical Laboratories. Harvard University, Oxford Street, Cambridge, MA 02138. (617) 495-4283. FAX (617) 496-5618.

Theoretical Chemistry Institute. University of Wisconsin, Madison. 1101 University Avenue, Madison, WI 53706. (608) 262-1511.

University/industry Chemical Research Center. Mississippi State University, Department of Chemistry, P.O. Drawer CH, Mississippi State, MS 39762. (601) 325-3584.

INSTRUMENTATION

See also: METROLOGY

ABSTRACT SERVICES AND INDEXES

Applied Mechanics Reviews: An Assessment of World Literature in Engineering Sciences. 1948-present. American Society of Mechanical Engineers, 345 East 47th Street, New York, NY 10017. (212) 705-7703. Monthly. $360.00 per year.

Applied Science and Technology Index; A Cumulative Subject Index To English Language Periodicals in the Fields of Aeronautics and Space Science, Computer Technology, Chemistry, Construction Industry, Energy and Related Areas. H.W. Wilson Co., 950 University Avenue, Bronx, NY 10452. (800) 367-6770 or (212) 588-8400. FAX (718) 590-1617. From 1958 to present. Monthly. Inquire about cost and availability. Also available on CD-ROM and online.

Electrical and Electronics Abstracts. Institute of Electrical Engineers, Michael Faraday House, Six Hill Way, Stevenage, Herts, SG1 2AY, England. Distributed by IEEE, 445 Hoes Lane, Piscataway, NJ 08854. (908) 562-5549. 1898 to present. Monthly. $2200.00 per year. Also available on CD-ROM and online as INSPEC.

Engineering Index Monthly. Engineering Information, Inc., Castle Point on the Hudson, Hoboken, NJ 07030. (800) 221-1044. FAX (212) 832-1857. Monthly. $2300.00 per year. Also available online as COMPENDEX, and also on CD-ROM. Covers chemical engineering, computers, electrical engineering, civil engineering, metals and mining, industrial management, and mechanical engineering.

Index To Scientific and Technical Proceedings. Institute for Scientific Information, 3501 Market St., Philadelphia, PA 19104. (215) 386-0100. FAX (215) 386-6362. Monthly. $500.00 per year.

Index To Scientific Reviews. Institute for Scientific Information, 3501 Market St., Philadelphia, PA 19104. (215) 386-0100. FAX (215) 386-6362. Semi-annual.

Key Abstracts—electronic Instrumentation. IEEE Service Center, 445 Hoes Lane, Piscataway, NJ 08854. (908) 562-5549. 1976 to present. Monthly. $160.00 per year.

Key Abstracts—measurements in Physics. IEEE Service Center, 445 Hoes Lane, Piscataway, NJ 08854. (908) 562-5549. 1976 to present. Monthly. $160.00 per year.

Physics Abstracts. Institute of Electrical Engineers, Michael Faraday House, Six Hill Way, Stevenage, Herts, SG1 2AY, England. Distributed by IEEE, 445 Hoes Lane, Piscataway, NJ 08854. (908) 562-5549. 1898 to present. Monthly. $2700.00 per year. Also available online as INSPEC.

PHYSICS BRIEFS. American Institute of Physics, 335 East 45th Street, New York, NY 10017 (212) 661-9404 or FAX (516)349-9704. 1980 to present. Six times per year. $2600.00 per year.

Science Citation Index. Institute for Scientific Information, 3501 Market Street, Philadelphia, PA 19104. (215) 386-0100. FAX (215) 386-6362. Inquire about availability and cost. Also available on CD-ROM.

Standard Actions. American National Standards Institute, 11 W. 42nd Street, 13th Floor, New York, NY 10036. (212) 642-4900. FAX (212) 302-1286. 1970 to present. Bi-weekly. $100.00 per year.

ASSOCIATIONS AND PROFESSIONAL SOCIETIES

American Institute of Physics. 335 East 45th Street, New York, NY 10017. (212) 661-9404 or (800) 445-6638.

American National Standards Institute. 11 W. 42nd Street, 13th Floor, New York, NY 10036. (212) 642-4900. FAX (212) 302-1286.

American Society For Testing and Materials. 1916 Race Street, Philadelphia, PA 19103. (215) 299-5585.

Analytical Instrument Association. 225 Reinekers Lane, Suite 265, Alexandria, VA 22314-2875. (703) 836-1360. FAX (703) 836-6644.

Edison Electric Institute. 701 Pennsylvania Avenue NW, Washington, DC 20004-2696. (202) 508-5000.

Institute of Electrical and Electronics Engineers. 345 E. 47th Street, New York, NY 10017. (908) 981-0060. FAX (908) 981-9667.

Instrument Society of America. 67 Alexander Drive, Box 12277, Research Triangle Park, NC 27709. (919) 549-8411. FAX (919) 549-8288.

Standards Engineering Society. 1706 Darst Avenue, Dayton, OH 45403-3104. (513) 258-1955.

U.S. National Institute of STandARDS and TECHNOLOGY. U.S. Department of Commerce, Gaithersburg, MD 20899. (301) 975-3069.

DIRECTORIES AND BIOGRAPHICAL SOURCES

American Men & Women of Science, 1995-96. R.R. Bowker Staff, eds. 19th edition. 8 volumes. R.R. Bowker/Reed International Publishing Company, 121 Chanlon Road, New Providence, NJ 07974. (908) 464-6800 or (800) 521-8110. 1995. $850.00

Directory of Engineering Societies and Related Organizations. Gordon Davis. 13th edition. American Association of Engineering Societies, 1111 19th Street NW, Suite 608, Washington,

DC 20036. (202) 296-2237 or (800) 658-8897. 1989. Inquire for price.

International Research Centers Directory. Gale Research, 835 Penobscot Building, Detroit, MI 48226-4094. (313) 961-2242. (800) 347-4253. FAX (313) 961-6083. 8th edition. 1995. $410.00

Isa Directory of Instrumentation. Instrument Society of America, 67 Alexander Drive, Box 12277, Research Triangle Park, NC 27709. (919) 549-8411. FAX (919) 549-8288. 1979 to present. Annual. $100.00.

Research Centers Directory. Gale Research, 835 Penobscot Building, Detroit, MI 48226-4094. (313) 961-2242. (800) 347-4253. FAX (313) 961-6083. $485.00.

Scientific and Technical Organizations and Agencies Directory. Gale Research, 835 Penobscot Building, Detroit, MI 48226-4094. (313) 961-2242. (800) 347-4253. FAX (313) 961-6083. 4th edition. 1996. $195.00.

Who's Who in Engineering. American Association of Engineering Societies, 1111 19th Street NW, Suite 608, Washington, DC 20036. (202) 296-2237 or (800) 658-8897. 8th edition. 1991. Inquire for price.

ENCYCLOPEDIAS AND DICTIONARIES

Concise Encyclopedia of Measurement and Instrumentation. L. Finkelstein and K.T.V. Grattan. Pergamon Press Inc., Maxwell House, Fairview Park, Elmsford, NY 10523. (914) 592-7700. Fax (914) 592-3625. 1993. Inquire for cost.

McGraw-Hill Dictionary of Scientific and Technical Terms. Sybil P. Parker, ed. 5th edition. McGraw-Hill Publishing Company, 1221 Avenue of the Americas, New York, NY 10020. (800) 262-4729 or (212) 512-3825. 1993. $110.50.

McGraw-Hill Encyclopedia of Engineering. Sybil P. Parker, ed. 2nd edition. McGraw-Hill Publishing Company, 1221 Avenue of the Americas, New York, NY 10020. (800) 262-4729 or (212) 512-3825. 1993. $95.50.

McGraw-Hill Encyclopedia of Science and Technology. Sybil P. Parker, ed. 7th edition. 20 volumes. McGraw-Hill Publishing Company, 1221 Avenue of the Americas, New York, NY 10020. (800) 262-4729 or (212) 512-3825. 1992. $1900.00

Thesaurus of Scientific, Technical, and Engineering Terms. Hemisphere Publishing Corporation, 1900 Frost Road, Suite 101, Bristol, PA 19007-1598. (215) 785-5800 or (800) 821-8312. FAX (215) 785-5515. 1987. $173.00.

GENERAL WORKS

Instrumentation and Control: Fundamentals and Applications. Chester L. Nachtigal, editor. John Wiley and Sons, Inc., 605 Third Avenue, New York, NY 10158. (800) 526-5368 or (212) 850-6000. 1990. $130.00.

Instrumentation For Engineering Measurements. James W. Dally, et al. John Wiley and Sons, Inc., 605 Third Avenue, New York, NY 10158. (800) 526-5368 or (212) 850-6000. 1993. Inquire for cost and availability.

Ency. of Physical Sciences and Engineering Info. Sources

INSTRUMENTATION

Instruments. Osiris/University of Chicago Press, 5801 Ellis Avenue, 4th Floor, Chicago, IL 60637. (312) 702-7700. FAX (312) 702-9756. 1994. $39.00.

Mechanical Measurements. Thomas G. Beckwith, et al. 5th edition. Addison-Wesley Publishing Company, 1 Jacob Way, Reading, MA 01867. (617) 944-3700 or (800) 447-2226. FAX 617-942-1117. 1993. $62.50.

Principles of Electronic Instrumentation. A. James Dufenderfer & Brian E. Holton. Saunders College Publishing/ the Curtis Center, Independence Square W., Philadelphia, PA 19106. (215) 238-7800. 1994. $61.25.

Principles of Measurement and Instrumentation. Alan S. Morris. Prentice Hall, 113 Sylvan Avenue, Route 9W, Englewood Cliffs, NJ 07632. (201) 592-2000 or (800) 922-0579. 1993. $62.00.

HANDBOOKS AND MANUALS

Handbook of Measurement Science. P.H. Sydenham, editor. 3 volumes. John Wiley and Sons, Inc., 605 Third Avenue, New York, NY 10158. (800) 526-5368 or (212) 850-6000. 1982 (vol.1), 1983 (vol.2), 1992 (vol.3). $270.00 each for vols.1-2, $199.95 (vol.3).

ONLINE DATABASES AND CD-ROMS

Compendex Plus. Engineering Information, Inc., 345 East 47th Street, New York, NY 10017. (212) 705-7600 or (800) 221-1044. Contains citations with abstracts to worldwide literature in engineering and technology, from 1970 to present. Available on online BRS,(800) 289-4277, DIALOG, (800) 334-2564, ORBIT (800) 456-7248, and STN International, FIZ Karlsruhe, P.O. Box 2465, W-7500, Karlsruhe 1, Germany, online services. Also available on CD-ROM. Inquire as to cost and availability.

CSA Engineering. Cambridge Scientific Abstracts, 7200 Wisconsin Avenue, Suite 601, Bethesda, MD 20814. (301) 961-6750 or (800) 843-7751. Contains citations and abstracts of international periodicals and research literature covering all fields of engineering and science and technology,including computer and information science, electronics, mechanical engineering, solid state materials, 1981 to present. Available on BRS,(800) 289-4277, online service.Inquire as to cost and availability.

Current Contents Search. Institute for Scientific Information, 3501 Market Street, Philadelphia, PA 19104. (215) 386-0100. FAX (215) 386-6362. Contains citations to articles listed in the table of contents of science and technology journals. Also articles in social sciences and life sciences journals. Available on BRS,(800) 289-4277, DIALOG,(800) 334-2564, online services. Inquire as to cost and availability.

Dissertation Abstracts. University Microfilms International, 300 North Zeeb Road, Ann Arbor, MI 48106. (800) 521-0600 or (313) 761-4700. Scope includes virtually all doctoral dissertations accepted at accredited American institutions from 1861 to present in 252 subject areas. Available on BRS,(800) 289-4277, DIALOG,(800) 334-2564, and OCLC EPIC,(800) 848-5878, online services. Also available on CD-ROM. Inquire as to cost and availability.

ISMEC: Mechanical Engineering Abstracts. Cambridge Scientific Abstracts, 7200 Wisconsin Avenue, Suite 601, Bethesda, MD 20814. (301) 961-6750 or (800) 843-7751. Contains citations to the literature in mechanical engineering, industrial and production engineering, energy, power, mechanics, devices and related areas, from 1973 to present. Available on the DIALOG,(800) 334-2564, online service. Inquire as to cost and availability.

NTIS Bibliographic Database. National Technical Information Service, 5285 Port Royal Road, Springfield, VA 22161. (703) 487-4929 or FAX (703) 321-8199. Broad coverage of government-sponsored science and technology research reports, 1964 to present. Available on BRS,(800) 289-4277, DIALOG, (800) 334-2564, ORBIT, (800) 456-7248, and STN International, FIZ Karlsruhe, P.O. Box 2465, W-7500, Karlsruhe 1, Germany, online services. Also available on CD-ROM. Inquire as to cost and availability.

SCISEARCH. Institute for Scientific Information, 3501 Market Street, Philadelphia, PA 19104. (800) 523-1850 or (215) 386-0100. Broad multidisciplinary title and author index to the international literature of science and technology, 1974 to present. Available on DIALOG,(800) 334-2564, and ORBIT,(800) 456-7248, online services. Also available on CD-ROM. Inquire as to cost and availability.

WILSONLINE. H.W. Wilson Company, 950 University Avenue, Bronx, NY 10452. (800) 367-6770 or (212) 588-8400. Makes available online versions of the printed H.W. Wilson indexes including Applied Science and Technology Index, Business Periodicals Index, General Science Index, and Readers' Guide to Periodical Literature. Period covered is generally 1983 to present. Available on BRS,(800) 289-4277, DIALOG,(800) 334-2564, and OCLC EPIC,(800) 848-5878, online services. Also available on CD-ROM. Inquire as to cost and availability.

PERIODICALS

Astm Standardization News. American Society for Testing and Materials (ASTM), 1916 Race Street, Philadelphia, PA 19103. (215) 299-5585. FAX (215) 977-9679. 1973 to present. Monthly. $18.00 per year.

Control Engineering. Cahners Publishing Company (Des Plains), 1350 East Touhy Avenue, Des Plaines, IL 60017-5080. (708) 635-8800. FAX (708) 635-9950. 1954 to present. Monthly. $79.95 per year.

IEEE Transactions On Instrumentation and Measurement. IEEE (Institute of Electrical and Electronic Engineers), 345 E. 47th Street, New York, NY 10017-2394. 1952 to present. Bi-monthly. $165.00 per year for non-members.

Industrial and Scientific Instruments. Wilmington Publishing, Wilmington House, Church Hill, Dartford, Kent UA2 7EF, England. Telephone 0332-277788. FAX 0332-276476. 1960 to present. Monthly. $139.00 per year.

Instrumentation and Automation News. Chilton Co., One Chilton Way, Radnor PA 19089. (215) 964-4028. 1953 to present. Monthly. $35.00 per year.

Instrumentation and Control Systems. Chilton Co., One Chilton Way, Radnor PA 19089. (215) 964-4028. 1928 to present. Monthly. $60.00 per year.

Intech. Instrument Society of America, 67 Alexander Drive, Box 12277, Research Triangle Park, NC 27709. (919) 549-8411. FAX (919) 549-8288. 1954 to present. Monthly. $65.00 per year.

INSTRUMENTATION

Ency. of Physical Sciences and Engineering Info. Sources

Measurement Science and Technology. IOP Publishing, Institute of Physics. Distributed by American Institute of Physics, Subscriber Services, 500 Sunnyside Boulevard, Woodbury, NY 11797-2999. 1968 to present. Monthly. $759.00 per year.

Measurement and Control. Institute of Measurement and Control, 87 Gower Street, London WC1E 6AA, England. Telephone 44-71-387-4949. FAX 44-71-388-8431. 1968 to present. Monthly. Inquire for cost.

Measurements and Control. Measurements and Data Corporation, 2994 W. Liberty Avenue, Pittsburgh, PA 15216. (412) 343-4666. 1967 to present. Bi-monthly. $22.00 per year.

Measurements and Control News. Measurements and Data Corporation, 2994 W. Liberty Avenue, Pittsburgh, PA 15216. (412) 343-4666. 6 times a year. Inquire for cost.

Metrologia. Bureau International des Poids et Mesures, Pavillon de Breteuil, 92312 Sevres Decex, France. FAX 45-34-20-21. 1965 to present. 6 times a year. $298.00 per year.

National Institute of Standards and Technology Journal of Research. U.S. National Institute of Standards and Technology, U.S. Department of Commerce, Gaithersburg, MD 20899. (301) 975-3069. 1928 to present. Bi-monthly. $27.00 per year.

Review of Scientific Instruments. American Institute of Physics, One Physics Ellipse, College Park, MD 20740. (301) 209-3000. 1930 to present. Monthly. $800.00 per year.

Standards Engineering. Standards Engineering Society, 1706 Darst Avenue, Dayton, OH 45403-3104. (513) 258-1955. Bi-monthly. $40.00 per year.

RESEARCH CENTERS AND INSTITUTES

Instrumentation and Control Laboratory, Princeton University, Dept. of MAE, Engineering Quadrangle, Princeton, NJ 08544. (609) 452-5154.

Major Analytical Instrumentation Center, University of Florida. College of Engineering, Rhines Hall, Gainesville, FL 32611. (904) 392-6985.

University of California—davis, Facility For Advanced Instrumentation. Davis, CA 95616. (916) 752-0284.

INTEGRAL CALCULUS

See: CALCULUS

INTEGRATED CIRCUITS

See also: ELECTRONIC CIRCUITS and COMPONENTS, ELECTRONICS, ELECTRONICS ENGINEERING, MICROELECTRICS, VLSI..

ABSTRACT SERVICES AND INDEXES

Applied Science and Technology Index; A Cumulative Subject Index To English Language Periodicals in the Fields of Aero-

nautics and Space Science, Computer Technology, Chemistry, Construction Industry, Energy and Related Areas.

H.W. Wilson Co., 950 University Avenue, Bronx, NY 10452. (212) 588-8400. (800) 367-6770. FAX (718) 590-1617. From 1958 to present. Monthly. Inquire about cost and availability. Also available on CD-ROM and online. ISSN: 0003-6986.

Current Contents: Engineering, Technology and Applied Sciences. Institute for Scientific Information, 3501 Market Street, Philadelphia, PA 19104. (215) 386-0100. FAX (215) 386-6362. 1970 to present. Weekly. $442.00 per year. Also available on CD-ROM and online. Inquire regarding cost and availability. ISSN: 0095-7917.

Current Papers in Electrical and Electronics Engineering. Institution of Electrical Engineers (IEE), London. Distributed by INSPEC/IEEE, Box 1331, 445 Hoes Lane, Piscataway, NJ 08855-1331. (908) 562-5549. 1969 to present. Monthly. $345.00 per year. ISSN: 0011-3778.

Electrical and Electronics Abstracts. Institution of Electrical Engineers (IEE), London. Available from INSPEC/IEEE - Institute of Electrical and Electronic Engineers, Box 1331, Hoes Lane, Piscataway, NJ 08855-1331. (908) 562-5549. 1898 to present. Monthly. $2200.00 per year. Also available on CD-ROM and online as INSPEC.

Engineering Index Monthly; Indexes and Abstracts the World's Engineering and Technical Literature. Engineering Information, Inc., Castle Point on the Hudson, Hoboken, NJ 07030. (201) 216-8500. (800) 221-1044. FAX (201) 216-8532. Monthly. $2300.00 per year. Available online as COMPENDEX and also on CD-ROM. ISSN: 0742-1974.

General Science Index. H.W. Wilson Company, 950 University Avenue, Bronx, NY 10452. (212) 588-8400. (800) 367-6770. FAX (718) 590-1617. From 1978 to present. Ten issues per year; quarterly and annual cumulations. Service basis. Available on CD-ROM and online. Inquire regarding cost and availability. ISSN: 0162-1963.

Government Reports Announcements and Index. U. S. National Technical Information Service (NTIS), 5285 Port Royal Road, Springfield, VA 22161. (703) 487-4650. FAX (703) 321-8547. From 1968 to present. Annual. $630.00 per year. Also available online as NTIS Bibliographic Database and on CD-ROM.

Index To IEEE Publications. IEEE Service Center, 445 Hoes Lane, Piscataway, NJ 08855-1331. (908) 981-1393. (800) 678-IEEE. FAX (908) 981-9667. 1973 to present. Annual. ISSN: 0099-1368.

Physics Abstracts. INSPEC. Section A, Science Abstracts. Institution of Electrical Engineers (IEE), London. Available from: INSPEC/IEEE - Institute of Electrical and Electronic Engineers, Box 1331, Hoes Lane, Piscataway, NJ 08855-1331. (908) 562-5549. 1898 to present. 24 issues per year. $2835.00 per year. Also available online and on CD-ROM. ISSN: 0036-8091.

Physics Briefs (Physikalische Berichte). Information Center for Energy, Physics, Mathematics; German Physical Society. V C H Publishers, Inc., 220 East 23rd Street, New York, NY 10010-4606. (212) 683-8333. 1845 to present. 24 issues per year. $2390.00 per year. Also available online. ISSN: 0179-7434.

Science Citation Index. SCI. Institute for Scientific Information, 3501 Market Street, Philadelphia, PA 19104. (215) 386-0100.

(800) 523-1850. FAX (215) 386-2991. 1961 to present. Six issues per year, plus annual cumulation. $11650.00 per year. Also available online and on CD-ROM.Inquire about price and availability. ISSN: 0036-827X.

Solid State and Superconductivity Abstracts: Covers theory, Production and Application of Solid State Materials. Cambridge Scientific Abstracts, 7200 Wisconsin Avenue, Bethesda, MD 20824. (301) 961-6750. FAX (301) 961-6720. 1957 to present. Bimonthly. $1320.00 per year. Also available online. ISSN: 0896-5900.

STAR. SCIENTIFIC and TECHNICAL AEROSPACE REPORTS. U.S. National Aeronautics and Space Administration, Scientific and Technical Information Facility, Box 8757, Baltimore-Washington International Airport, MD 21240. (301) 621-0153. Distributed by U. S. Superintendent of Documents, Washington, DC 20402. From 1963 to present. Semimonthly, with semiannual and annual indexes. $114.00 per year. Also available online and on CD-ROM. ISSN: 0036-8741.

ANNUAL REVIEWS AND YEARBOOKS

Advances in Electronics and Electron Physics. Academic Press, Inc., 6277 Sea Harbor Drive, Orlando, FL 32887. (800) 321-5068. From 1948 to present. Irregular. Price varies, inquire. ISSN: 0065-2539.

Critical Reviews in Solid State and Materials Sciences. CRC Press, Inc., 2000 Corporate Boulevard, NW, Boca Raton, FL

33431. (407) 994-0555. (800) 272-7737. FAX (407) 998-9784. 1970 to present. Bimonthly. $265.00 per year. ISSN: 1040-8436.

ASSOCIATIONS AND PROFESSIONAL SOCIETIES

American Association of Engineering Societies, 1111 19th Street, Suite 608, Washington, DC 20036-3603. (202) 296-2237. FAX (202) 296-1151.

American Electronics Association. 5201 Great America Way, Suite 520, P.O. 52990, Santa Clara, CA 95056. (408) 987-4200. FAX (408) 970-8565.

American Institute of Physics. One Physics Ellipse, College Park, MD 20740-3843. (301) 209-3100.

Edison Electric Institute. 701 Pennsylvania Avenue NW, Washington, DC 20004-2696. (202) 508-5000, 5454. FAX (202) 508-5360.

Electric Power Research Institute. 3412 Hillview Avenue, Palo Alto, CA 94304. (415) 855-2000. FAX (415) 855-2954.

Electronics Industries Association. 2500 Wilson Boulevard, Arlington, VA 22201. (703) 907-7500. FAX (202) 457-4985.

IEEE Circuits and Systems Society. c/o Institute of Electrical and Electronic Engineers. 345 East 47th Street. New York, NY 10017. (212) 705-7900. FAX (212)705-4929.

IEEE Solid-state Circuits Council. c/o Institute of Electrical and Electronic Engineers. 345 East 47th Street. New York, NY 10017. (212) 705-7900. FAX (212)705-4929.

Institute For Interconnecting and Packaging Electronic Electronic Circuits. 2215 Sanders Road, Northbrook, IL 60062-6135. (708) 509-9700. (708) 509-9798.

Institute of Electrical and Electronics Engineers. 345 East 47th Street. New York, NY 10017. (212) 705-7900. FAX (212) 705-4929.

INTERNATIONAL SOCIETY FOR HYBRID MICROELECTRONICS. 1850 Centennial Park Drive, Suite 105, Reston, VA 22091. (703) 758-1060. FAX (703) 758-1066.

National Electrical Manufacturers Association. 1300 North 17th Street, Suite 1847, Rosslyn VA 22209. (703) 841-3200. FAX (703) 841-3300.

DIRECTORIES AND BIOGRAPHICAL SOURCES

American Men and Women of Science: Physical and Biological Sciences. R. R. Bowker Inc., 121 Chanlon Road, New Providence, NJ 07974. (908) 464-6800. (800) 521-8110. 20th edition. 8 volumes. 1996. $850.00.

Directory of Engineering College Research and Graduate Study. American Society for Engineering Education, Ohio State University, 2070 Neil Avenue, Columbus, OH 43210-1226. Annual. $20.00.

Directory of Engineering Societies and Related Organizations. American Association of Engineering Societies, 1111 19th Street, Suite 608, Washington, DC 20036-3603. (202) 296-2237. Semi-annual. $150.00.

Directory of Engineers in Private Practice. National Society of Professional Engineers, 1420 King Street, Alexandria, VA 22314-2750. (703) 684-2835. Biennial. $30.00.

E E M - Electronic Engineer's Master. Hearst Business Communications, Inc., 645 Stewart Avenue, Garden City NY 11530. (516) 227-1300. ISSN: 0732-9016.

Electrical and Electronic Trades Directory: The Blue Book. Institution of Electrical Engineers, Michael Faraday House, Six Hills Way, Stevenage, Herts. SG1 2AY, England. Telephone: 0438 313311 or FAX 0438 742840. 1995. $140.00

Engineering Research Centers: Incorporating Electronics Research Centers. Stockton Press, 345 Park Avenue, New York, NY 10010. (212) 689-9200. (800) 221-2123. 4th edition. 1995. $515.00.

Ic Master. Hearst Business Communications, Inc., 645 Stewart Avenue, Garden City, NY 11530. (516) 227-1300. 1975 to present. Annual. $170.00 per year. ISSN: 0894-6809.

IEEE Membership Directory. Institute of Electrical and Electronics Engineers, IEEE Service Center, 445 Hoes Lane, Piscataway, NY 08854. (908) 981-1393. (800) 678-IEEE. FAX (908) 981-9667. 2 volumes. Annual. $190.00.

International Directory of Abbreviations and Acroynms of Electronics, Electrical Engineering, Computer Technology and Information Processing. Peter Wennrich. K. G. Saur, 121 Chanlon Road, New Providence, NJ 07974. (908) 464-6800. (800) 521-8110 2 volumes. 1992. $230.00.

INTEGRAL CALCULUS

Ency. of Physical Sciences and Engineering Info. Sources

International Engineering Directory. American Consulting Engineers Council, 1015 15th Street, N.W. Suite 802, Washington, DC 20005-2670. (202) 347-7474. Annual. $10.00.

Research Centers Directory. Gale Research Company Inc., 835 Penobscot Building, Detroit, MI 48226-4094. (313) 961-2242. (800) 347-4253. 20th edition. 1995. $485.00. ISSN: 0080- 1518.

Who's Who in Electric Transmission and Distribution. Utility Data Institute, 1200 G Street NY, Suite 250, Washington, DC 20005-3802. (202) 942-8788. (800) 486-3660. 1995. $150.00.

Who's Who in Engineering. Gordon Davis, editor. American Association of Engineering Societies. 1111 19th Street, NY, Suite 608, Washington, DC 20036. (202) 296-2237. (800) 658-8897. 9th edition. 1995. $220.00.

Who's Who in Technology. Gale Research, 835 Penobscot Building, Detroit, MI 48226-4094. (313 961-1242. (800) 521-4253. 7th edition. 1995. $195.00. ISN 0-8103-7467-6.

ENCYCLOPEDIAS AND DICTIONARIES

Chambers Science and Technology Dictionary. Peter M.B. Walker, editor. Cambridge University Press, 40 West 20th Street, New York, NY 10011-4211. (212) 924-3900. 1988. $39.95.

Encyclopedia of Applied Physics. George Trigg, editor. VCH Publications, Inc., 220 East 23rd Street, Suite 909, New York, NY 10010-4606. (212) 683-8333. (800) 422-8824. 20 volume set. 1991 - . $5990.00.

Encyclopedia of Integrated Circuits. Walter A. Buchsbaum.

Prentice Hall , 113 Sylvan Avenue, Route 9W, Englewood Cliffs, NJ 07632. (201) 592-2000. (800) 922-0579. 1987. $39.95.

IEEE Standard Dictionary of Electrical and Electronics Terms. Christopher J. Booth, editor. IEEE Service Center, 445 Hoes Lane, Piscataway, NJ 08855-1331. (908) 981-1393. (800) 678-IEEE. FAX (908) 981-9667. IEEE Standard 100- 1992. 5th edition. 1993. $150.00.

Illustrated Dictionary of Electronics. Stan Gibilsco. TAB Books, P.O. Box 40, Blue Summit, PA 17294-0850. (717) 794- 2191. (800) 233-1128. 7th edition. 1994. $34.95.

International Encyclopedia of Integrated Circuits. Arthur A. Seidman. TAB Books, P.O. Box 40, Blue Summit, PA 17294-0850. (717) 794-2191. (800) 233-1128. 1991. $75.00.

McGraw-Hill Electronics Dictionary. J. Markus. McGraw-HillBook Company, Inc., 1221 Avenue of the Americas, New York, NY 10020. (212) 997-3675. 5th edition. 1994. $49.95.

McGraw-Hill Encyclopedia of Engineering. Sybil P. Parker, editor. McGraw-Hill Book Company, Inc. 1221 Avenue of the Americas, New York, NY 10020. (212) 997-3675. Second edition. 1993. $95.50. ISBN: 0-07-051392-9.

McGraw-Hill Encyclopedia of Science and Technology. McGraw-Hill Book, Incorporated, 1221 Avenue of the Americas, New York, NY 10020. (212) 997-3675. (800) 262-4729. Seventh edition. Twenty volumes. 1992. $1900.00.

GENERAL WORKS

Design and Technology of Integrated Circuits. Donald DeCogan. John Wiley & Sons, Inc., 605 Third Avenue, New York, NY 10158-0012. (212) 850-6000. (800) 225-5945. 1990. $75.00.

Diode Lasers and Photonic Integrated Circuits. David J. Comer. John Wiley & Sons, Inc., 605 Third Avenue, New York, NY 10158-0012. (212) 850-6000. (800) 225-5945. 1995. $64.95.

Electric Circuits: Principles, Applications and Computer Analysis. David A. Bell. Prentice Hall , 113 Sylvan Avenue, Route 9W, Englewood Cliffs, NJ 07632. (201) 592- 2000. (800) 922-0579. 1993. $72.00.

Exploring Electronic Devices. Mark E. Hazen. SCP: Third World Literature Publishing House, P.O. Box 482, Lithonia, GA 30058-0482. (404) 785-7725. 1991. $50.75.

Integrated Circuits and Microprocessors. Robin C. Holland.

Elsevier Science Publishing Company, Inc., 655 Avenue of the Americas, New York, NY 10010. (212) 989-5800. FAX (914) 333-2444. 1986. $30.00 in paper.

Introduction To Fields and Circuits. Gordon Lancaster. Oxford University Press, Inc., 200 Madison Avenue, New York, NY 10016. (212) 725-6000. (800) 334-4249. 1992 $39.95

Introduction To Vlsi Techology. T. E. Price. Prentice Hall, 113 Sylvan Avenue, Route 9W, Englewood Cliffs, NJ 07632. (201) 592-2000. (800) 922-0579. 1994. $60.00.

Mastering Ic Circuits. Joseph J. Carr. TAB Books, P.O. Box 40, Blue Summit, PA 17294-0850. (717) 794-2191. (800) 233-1128. 1991. $32.95.

Microelectronic Circuits. Adel S. Sedra and Kenneth C. Smith. Harcourt Brace College Pubs., 6277 Sea Harbor Drive, Orlando, FL 32887. (407) 345-2526. (800) 787-8717. 2nd edition. 1987. $65.25.

Operational Amplifiers, Integrated and Hybrid Circuits. George B. Rutkowski. Prentice Hall, 113 Sylvan Avenue, Route 9W, Englewood Cliffs, NJ 07632. (201) 592-2000. (800) 922-0579. 1993. $87.95.

Transistors: Fundamentals For the Integrated-circuit Engineer. R. M. Warner Jr., and B. L. Grung. Krieger Publishing Company, P.O. Box 9542, Melbourne, FL 32902-9542.(407) 724-9542. 1990. $72.50.

HANDBOOKS AND MANUALS

Active Electronic Component Handbook. Charles A. Harper and Harold C. Jones. McGraw-Hill Book Company, 1221 Avenue of the Americas, New York, NY 10020. (212) 512-2000. (800) 262-4729. 2nd edition. 1996. $79.50.

Electronics Handbook. Jerry C. Whitaker. CRC Press, Inc., 2000 Corporate Boulevard, NW, Boca Raton, FL 33431. (407) 994-0555. (800) 272-7737. 1996. $120.00.

Electrical Engineering Handbook. Richard C. Dorf, editor. CRC Press, Inc., 2000 Corporate Boulevard, NW, Boca Raton, FL 33431. (407) 994-0555. (800) 272-7737. 1993. $99.95.

Handbook of Electrical and Electronic Technology. Curtis Johnson. Prentice Hall , 113 Sylvan Avenue, Route 9W, Englewood Cliffs, NJ 07632. (201) 592-2000. (800) 922- 0579. 1996. $88.00.

Handbook of Quality Integrated Circuit Manufacturing. Robert Zurich. Academic Press, Inc., 6277 Sea Harbor Drive, Orlando, FL. (800) 321-5068. 1991. $59.95.

Handbook of Vlsi Chip Design and Expert Systems. A. F. Schwartz. Academic Press, Inc., 6277 Sea Harbor Drive, Orlando, FL. (800) 321-5068.

Integrated Circuits Handbook. Rector Press, Ltd., 130 Rattlesnake, Leverett, MA 01054-9726. (413) 548-9708.(800) 247-3473. 3 volumes. 1995. $295.00 in paper.

Standard Handbook For Electrical Engineers. Donald Fink. McGraw-Hill Publishing Company, 1221 Avenue of the Americas, New York, NY 10020. (212) 512-2000. (800) 262- 4729. 13th edition. 1996. $110.00.

ONLINE DATABASES AND CD-ROMS

CA Search. Chemical Abstracts Service, P.O. Box 3012, Columbus, OH 43210-0012. (614) 447-3600. (800) 848-6533. FAX (614) 447-3709. Very comprehensive guide to worldwide chemical literature and related fields, 1972 to present. Available on BRS,(800) 289-4277, DIALOG, (800) 334-2564, ORBIT (800) 456-7248, and STN International, FIZ Karlsruhe, P.O. Box 2465, W-7500, Karlsruhe 1, Germany, online services. Inquire as to cost and availability.

Compendex Plus. Engineering Information, Inc., 345 East 47th Street, New York, NY 10017. (212) 705-7600 or (800) 221-1044. Contains citations with abstracts to worldwide literature in engineering and technology, from 1970 to present. Available on online BRS,(800) 289-4277, DIALOG, (800) 334-2564, ORBIT (800) 456-7248, and STN International, FIZ Karlsruhe, P.O. Box 2465, W-7500, Karlsruhe 1, Germany, online services. Also available on CD-ROM. Inquire as to cost and availability.

Current Contents Search. Institute for Scientific Information, 3501 Market Street, Philadelphia, PA 19104. (215) 386-0100. FAX (215) 386-6362. Contains citations to articles listed in the table of contents of science and technology journals. Also articles in social sciences and life sciences journals. Available on BRS,(800) 289-4277, DIALOG,(800) 334-2564, online services. Inquire as to cost and availability.

Dissertation Abstracts Online. University Microfilms International, 300 North Zeeb Road, Ann Arbor, MI 48106. (800) 521-0600 or (313) 761-4700. Scope includes virtually all doctoral dissertations accepted at accredited American institutions from 1861 to present in 252 subject areas. Available on BRS, (800) 289-4277, DIALOG, (800) 334-2564, and OCLC EPIC,(800) 848-5878, online services. Also available on CD-ROM. Inquire as to cost and availability.

INSPEC. Institution of Electrical Engineers, Michael Faraday House, Six Hills Way, Stevenage, Herts. SG1 2AY, England. Telephone: 0438 313311 or FAX 0438 742840. Contains citations to the worldwide literature of physics, electronics and electrical engineering, computer technology, and related fields. Available on BRS, (800) 289-4277, DIALOG, (800) 334-2564, ORBIT, (800) 456-7248, and STN International, FIZ Karlsruhe, P.O. Box 2465, W-7500, Karlsruhe 1, Germany, online services. Inquire as to cost and availability.

NTIS Bibliographic Database. National Technical Information Service, 5285 Port Royal Road, Springfield, VA 22161. (703) 487-4929 or FAX (703) 321-8199. Broad coverage of government-sponsored science and technology research reports, 1964 to present. Available on BRS,(800) 289-4277, DIALOG, (800) 334-2564, ORBIT, (800) 456-7248, and STN International, FIZ Karlsruhe, P.O. Box 2465, W-7500, Karlsruhe 1, Germany, online services. Also available on CD-ROM. Inquire as to cost and availability.

Physics Briefs. American Institute of Physics, 335 East 45th Street, New York, NY 10017. (212) 661-9260 or FAX (212) 661-2036. Contains citations with abstracts of the literature of physics and related fields, 1979 to present. Available on the STN International, FIZ Karlsruhe, P.O. Box 2465, W-7500, Karlsruhe 1, Germany, online service. Inquire as to cost and availability.

SCISEARCH. Institute for Scientific Information, 3501 Market Street, Philadelphia, PA 19104. (800) 523-1850 or (215) 386-0100. Broad multidisciplinary title and author index to the international literature of science and technology, 1974 to present. Available on DIALOG,(800) 334-2564, and ORBIT,(800) 456-7248, online services. Also available on CD-ROM. Inquire as to cost and availability.

WILSONLINE. H.W. Wilson Company, 950 University Avenue, Bronx, NY 10452. (800) 367-6770 or (212) 588-8400. Makes available online versions of the printed H.W. Wilson indexes including Applied Science and Technology Index, Business Periodicals Index, General Science Index, and Readers' Guide to Periodical Literature. Period covered is generally 1983 to present. Available on BRS,(800) 289-4277, DIALOG,(800) 334-2564, and OCLC EPIC,(800) 848-5878, online services. Also available on CD-ROM. Inquire as to cost and availability.

PERIODICALS

Circuits, Systems, and Signal Processing. Birkhauser, 675 Massachusetts Avenue, Cambridge, MA 02139-3309. FAX (201) 348-4505. 1982 to present. Quarterly. $184.00 per year. ISSN: 0278-081X.

Electrical World. McGraw-Hill, Inc., Box 513 Hightstown, NJ 08520. (212) 512-3288. 1874 to present. Monthly. $55.00 per year. ISSN: 0013-4457.

Electronic Design. Penton Publishing, San Jose Gateway, Suite 354. 2025 Gateway Place, San Jose, CA 95110. (408) 441-0550. 1952 to present. Fortnightly. $95.00. ISSN: 0013-4872.

Electronics. Penton Publishing, San Jose Gateway, Suite 354. 2025 Gateway Place, San Jose, CA 95110. (408) 441-0550. 1930 to present. Semi-weekly. $98.00 per year. ISSN: 0883-4989.

IEEE Circuits and Devices Magazine. Institute of Electrical and Electronics Engineers, Inc., Box 1331, 445 Hoes Lane, Piscataway, NJ 08855-1331. (908) 981-0060. 1985 to present. Bi-monthly. $120.00 per year. ISSN: 8755-3996.

IEEE Journal of Solid State Circuits. Institute of Electrical and Electronics Engineers, Inc., Box 1331, 445 Hoes Lane, Piscataway, NJ 08855-1331. (908) 981-0060. 1966 to present. Bi-monthly. $275.00 per year. ISSN: 0018- 9200.

IEEE Spectrum. Institute of Electrical and Electronics Engineers, Inc., Box 1331, 445 Hoes Lane, Piscataway, NJ 08855-1331.

INTEGRAL CALCULUS

Ency. of Physical Sciences and Engineering Info. Sources

(908) 981-0060. 1964 to present. Monthly. $157.00 per year. ISSN: 0018-9235.

IEEE Transactions On Circuits and Systems. Part 1: Fundamental theory and Applications. Institute of

Electrical and Electronics Engineers, Inc., Box 1331, 445 Hoes Lane, Piscataway, NJ 08855-1331. (908) 981-0060. 1952to present. Monthly. $241.00 per year. ISSN: 1057-7122.

IEEE Transactions On Electron Devices. Institute of Electrical and Electronics Engineers, Inc., Box 1331, 445 Hoes Lane, Piscataway, NJ 08855-1331. (908) 981-0060. 1952 to present. Monthly. $395.00 per year. ISSN: 0018-9393.

IEEE Transactions On Signal Processing. Institute of Electrical and Electronics Engineers, Inc., Box 1331, 445 Hoes Lane, Piscataway, NJ 08855-1331. (908) 981-0060.

1951 to present. Monthly. $400.00. ISSN: 1053-587X.

Institute of Electrical and Electronics Engineers. Proceedings. Institute of Electrical and Electronics Engineers, Inc., Box 1331, 445 Hoes Lane, Piscataway, NJ 08855-1331. (908) 981-0060. 1913 to present. Monthly. $275.00. ISSN: 0018-9219.

Semiconductor International. Cahners Publishing Co., 44 Cook Street, Denver, CO 80206. (708) 635-8000. FAX (708) 390-2770. (800) 662-7776. 1978 to present. 13 issues per year. $84.95 per year. ISSN: 0163-3767.

Solid State Electronics. Elsevier Science, 660 White Plains Road, Tarrytown, NY 10591-5153. (914) 524-9200. FAX (914) 333-2444. 1960 to present. Monthly. $1025.00 per year. ISSN: 0038-1101.

RESEARCH CENTERS AND INSTITUTES

Alabama Microelectronics Science and Technology Center. Auburn University, 200 Broun Hall, Auburn University, AL 36849. (205) 844-1871. (205) 844-1809.

Center For Integrated Sensors and Circuits. University of Michigan, 1246 EECS Building, Ann Arbor, MI 48109-2122. (313) 764-3346. FAX (313) 747-1781.

Center For Integrated Circuit Packaging. Lehigh University, Sherman Fairchild Laboratory, Building 161. Bethlehem, PA 18015. (215) 758-4409. FAX (215) 758-4561.

Electrical Engineering Research Laboratories. Purdue University. Electrical Engineering Building, West Lafayette, in 47907. (317) 494-3536. FAX (317) 494-6440.

Electronics Research Laboratory. University of California, Berkeley, 253 Cory Hall, Berkeley, CA 94720. (415) 642-2301.

Engineering Research Center. University of Maryland, Clark School of Engineering, Potomac Building, Room 2104. (301) 405-3906. FAX (301_ 403-4105.

INTERNAL COMBUSTION ENGINE

See: ENGINES

INTERSTELLAR MATTER

See also: ASTRONOMY, BIG BANG, COSMOCHEMISTRY, COSMOLOGY, DARK MATTER, MILKY WAY, NEBULAE, STARS, UNIVERSE

ABSTRACT SERVICES AND INDEXES

*Applied Science and Technology Index; A Cumulative Subject Index To English Language Periodicals in the Fields ofaeronautics and Space Science, Computer Technology,chemistry, Construction Industry, Energy and Related Areas.*H.W. Wilson Co., 950 University Avenue, Bronx, NY 10452.(212) 588-8400. (800) 367-6770. FAX (718) 590-1617. From1958 to present. Monthly. Inquire about cost andavailability. Also available on CD-ROM and online. ISSN: 0003-6986.

Astronomy and Astrophysics Abstracts. Springer-Verlag New York, 175 Fifth Avenue, New York, NY 10010. (212) 460-1500.FAX (212) 473-6272. Published for the AstronomischesRechen-Institut. Comprehensive coverage of all aspects ofastronomy, astrophysics and related fields. 1969 to present. Two parts per year. Annual. ISSN: 0067-0022.

Chemical Abstracts. Chemical Abstracts Service, 2540Olentangy River Road, Box 3012, Columbus, OH 43210-0012.(614) 447-3600. (800) 848-6538. FAX (614) 447-3713.From 1907 to present. Weekly. $16800.00. Available onlineand on CD-ROM. CA is also available in five section groupings. Inquire regarding cost and availability.

Chemical Abstracts — Physical, Inorganic and Analytical Chemistry. Chemical Abstracts Service, 2540 Olentangy RiverRoad, Box 3012, Columbus, OH 43210-0012. (614) 447-3600.(800) 848-6538. FAX (614) 447-3713. Bi-weekly. $1870.00per year. Also available on CD-ROM and online. Inquireregarding cost and availability.

General Science Index. H.W. Wilson Company, 950 University Avenue, Bronx, NY 10452. (212) 588-8400. (800) 367-6770.FAX (718) 590-1617. From 1978 to present. Ten issues peryear; quarterly and annual cumulations. Service basis.Available on CD-ROM and online. Inquire regarding cost and availability. ISSN: 0162-1963.

Government Reports Announcements and Index. U. S. National Technical Information Service (NTIS), 5285 Port Royal Road,Springfield, VA 22161. (703) 487-4650. FAX (703) 321-8547.From 1968 to present. Annual. $630.00 per year. Also available online as *NTIS Bibliographic Database* and on CD-ROM.

International Aerospace Abstracts. American Institute of Aeronautics and Astronautics. Technical InformationService, 555 West 57th Street, New York 10019. (212) 247-6500. FAX (212) 582-4961. 1961 to present. Semi-monthly with annual index. $1295.00 per year. Also available online and on CD-ROM. ISSN: 0020-5842.

Meteorological and Geoastrophysical Abstracts. American Meteorological Society, c/o Inforonics, Inc., 550 NewtownRoad, Littleton, MA 01460. (508) 486-8976. FAX (508) 486-0027. Covers literature in environmental sciences,meteorology, astrophysics, hydrology, glaciology, and physical oceanography. From 1950 to present. Monthly.$950.00 per year. Also available on CD-ROM and online.Inquire regarding cost and availability. ISSN: 0026-1130.

NTIS Alerts: Astronomy & Astrophysics. U. S. National Technical Information Service. 5285 Port Royal Road,Springfield, VA 22161. (703) 487-4650. FAX (703) 321-8547.Weekly. $140.00 per year.

Pascal 48: Environnement Cosmique Terrestre, Astronomie Et Geologie Extraterrestre. Centre National de la RechercheScientifique, Institut de Information Scientifique et Technique, 2 aliee du Parc de Brabois, 54514 Vandoeuvre LesNancy Cedex, France. TEL 83-50-46-00. FAX: 83-50-46-50. 1985 to present. Ten issuesper year. 770 F. Also available on CD-ROM and online.

Physics Abstracts. INSPEC. Section A, Science Abstracts.Institute of Electrical Engineers, London, United Kingdom. Available from: INSPEC/IEEE - Institute of Electrical and Electronic Engineers, Box 1331, Hoes Lane, Piscataway, NJ 08855-1331. (908) 562-5549. 1898 to present. 24 issues per year. $2835.00 per year. ISSN: 0036-8091. Also available online and on CD-ROM.

Physics Briefs (Physikalische Berichte). InformationCenter for Energy, Physics, Mathematics; German Physical Society. V C H Publishers, Inc., 220 East 23rd Street, New York, NY 10010-4606. (212) 683-8333. 1845 to present. 24 issues per year. $2390.00 per year. Also available online.ISSN: 0179-7434.

Science Citation Index. SCI. Institute for ScientificInformation, 3501 Market Street, Philadelphia, PA 19104. (215) 386-0100. (800) 523-1850. FAX (215) 386-2991. 1961 to present. Six issues per year, plus annual cumulation. $11650.00 per year. Also available online and on CD-ROM. Inquire about price and availability. ISSN: 0036-827X.

Star. Scientific and Technical Aerospace Reports. U.S. National Aeronautics and Space Administration. Distributed by U. S. Superintendent of Documents, Washington, DC 20402. 1963 to present. Semi-monthly, with semiannual and annual indexes. $114.00 per year. ISSN: 0036-8741. Also available online and on CD-ROM.

ANNUAL REVIEWS AND YEARBOOKS

The Astronomical Almanac. Superintendent of Documents, U. S. Government Printing office, Washington, DC 10402. (202) 783-3238. 1981 to present. Superceeds Astronomical Ephemeris and the American Ephermis and Nautical Almanac. Annual. $44.95. ISSN: 0737-6421.

Annual Review of Astronomy and Astrophysics; Original Reviews of Critical Literature and Current Developments in Astronomy and Astrophysics. Annual Reviews, Inc., 4139 El Camino Way, Palo Alto, CA 94306-0139. (415) 493-4400. (800) 523-8635. FAX (415) 855-9815. 1963 to date. Annual. $60.00. ISSN: 0066-4146.

Highlights of Astronomy. I. Appenzeller, editor. Kluwer Academic Publishers, Box 358, Accord Station, Hingham, MA 02018-0358. (617) 871-6600. International Astronomical Union Highlights series. ISBN: 07923-3553-8. Triennal.1995. Price varies.

ASSOCIATIONS AND PROFESSIONAL SOCIETIES

Amateur Astronomers Association. 1010 Park Avenue, New York, NY 10028. (212) 535-2922.

American Association of Variable Star Observers. 23 Birch Street, Cambridge, MA 02138. (617) 354-0484. FAX: (617)354-0665.

American Astronomical Society. 2000 Florida Avenue NW, Suite 400, Washington, DC 20009. (202) 328-2010. FAX:(202) 234-2560.

American Institute of Physics. One Physics Ellipse, College Park, MD 20740-3843. (301) 209-3100.

Association of University For Research in Astronomy. Suite 701, 1625 Massachusetts Avenue, NW, Washington, DC 20036. (202) 483-2101. FAX (202) 483-2106.

Astronomical League. 2112 Kingfisher Lane East, Rolling Meadows, IL 60008. (708) 398-0562.

Astronomical Society of the Pacific. 390 Ashton Avenue, San Francisco, CA 94112. (415) 337-1100. FAX: (415) 337-5205.

International Astronomical Union. 98 bis Boulevard Arago, F-75014 Paris, France. Tel 1-43-25-83-58. FAX 1-43-25-26-16.

Joint Institute For Laboratory Astrophyics. University of Colorado, Box 440, Boulder, CO 80309-0440. (303) 492-7780. FAX (303) 492-5235.

National Council For Geocosmic Research. PO Box 1220, Dunkirk, MD 20754. (301) 855-2747. (800) PTOLEMY. FAX (410) 257-2824.

Royal Astronomical Society of Canada. 136 Dupont Street, Toronto. ON N5R 1V2 Canada. (416) 924-7973. FAX (416) 968-6687.

BIBLIOGRAPHIES

Astronomy and Astrophysics: A Source Guide. Gordon Press, P.O. Box 459, Bowling Green Station, New York, NY 10004. (718) 624-8419. 1991. $250.00. .

A Bibliography of Astronomy, 1970-1979. R. A. Sea and S. S. Martin. Libraries Unlimited, Inc., Littleton, CO 80160. 1982.

Catalog of the Naval Observatory Library. Naval Observatory Staff. G. K. Hall & Company, 866 Third Avenue, New York, NY 10022. (212)702-6789. (800) 257-5755. 6 volumes. 1977. $655.00

Science Books and Films. American Association for the Advancement of Science, 1333 H Street NW, Washington, DC 20005. (202) 326-6454. 1965 to present. Nine issues per year. $40.00 per year. ISSN: 0098-342X

Scientific and Technical Books and Serials in Print, 1995. R.R. Bowker Inc., 121 Chanlon Road, New Providence, NJ 07974. (908) 464-6800. (800) 521-8110. 4 volumes. 1994. $299.95. Also available on compact disc and online.

A Source Book On Astronomy and Astrophysics, 1900 - 1975. Kenneth Lang and Owen Gingerich, editors. Harvard University Press, 79 Garden Street, Cambridge MA. (617) 495-2600. 1980. $41.95.

INTERSTELLAR MATTER

Ency. of Physical Sciences and Engineering Info. Sources

DIRECTORIES AND BIOGRAPHICAL SOURCES

American Men and Women of Science: Physical and Biological Sciences. R. R. Bowker Inc., 121 Chanlon Road, New Providence, NJ 07974. (908) 464-6800. (800) 521-8110. 20th edition. 8 volumes. 1996. $850.00.

The Astronomers. Donald Goldsmith. St. Martin's Press, Inc., 175 Fifth Avenue, New York, NY 10010. (212) 674-5151. (800) 221-7945. 1993. $14.95 in paper.

Astronomical Centers of the World. Kevin Krisciunas. Cambridge University Press, 40 West 20th Street, New York, NY 10011-4211. (212) 924-3900. (800) 872-7423. 1988. $34.95

The Biographical Dictionary of Scientists: Astronomers. D. Abbott, editor. Peter Bedrick Books, Inc., 2112 Broadway, Room 318, New York, NY 10023. (212) 496-0751. 1984.

Directory of Physics and Astronomy Staff Members. American Institute of Physics, One Physics Ellipse, College Park, MD 20740-3843. (301) 209-3100. Annual. $45.00.

Earth and Astronomical Research Centers. Stockton Press, 345 Park Avenue South, New York, NY 10010. 4th edition. 1995. $515.00. ISBN: 1-56169-0967.

Graduate Programs in Physics, Astronomy and Related Fields. 1995-1996. American Institute of Physics, One Physics Ellipse, College Park, MD 20740-3843. (301) 209-3100. 1978 to present. Annual. $45.00.

Research Centers Directory. Gale Research Inc., 835 Penobscot Building, Detroit, MI 48226-4094. (313) 961-2242. (800) 347-4253. 20th edition. 1995. $485.00.

ENCYCLOPEDIAS AND DICTIONARIES

Astronomy From A To Z: A Dictionary of Celestial Objects and Ideas. Charles A. Schweighauser. Illinois Issues, Sangamon State University, K-10, Springfield, IL 62794-9243. (217) 786-9243. 1991. $14.95 in paper.

Concise Dictionary of Astronomy. Jacqueline Mitton. Oxford University Press, Inc., 200 Madison Avenue, New York, NY 10016. (212) 725-6000. (800) 334-4249. 1992. $24.95.

Facts On File Dictionary of Astronomy. Valerie Illingworth. Facts-on-File, Inc., 460 Park Avenue South, New York, NY 10016-7382. (212) 683-2244. (800) 322-8755. Fax (800) 683-3633. 1994. $27.95.

Encyclopedia of Astronomy and Astrophysics. Stephen Maran, editor. Van Nostrand Reinhold, 115 Fifth Avenue, New York, NY 10003. (212) 254-3232. (800) 842-3636. 1992. $129.95.

Landolt-boernstein Numerical Data and Functional Relationships in Science and Technology: Astronomy, Astrophysics and Space Research; A. O. Madelung and H. H. Boigt, editors. Springer Verlag New York, Inc., 175 Fifth Avenue, New York, NY 10010. (212) 460-1500. (800) 777- 4643. 1993. $550.00.

Oxford Illustrated Encyclopedia. Volume 8, the Universe. Archie Roy, editor. Oxford University Press, Inc., 200 Madison Avenue, New York, NY 10016. (212) 725-6000. (800) 334-4249. 1993. $45.00.

McGraw-Hill Encyclopedia of Science and Technology. McGraw-Hill Book Company, Inc., 1221 Avenue of the Americas, New York, NY 10020. (212) 512-2000. (800) 262-4729. 7th edition. 20 volume set. 1992. $1900.00.

Stars, Galaxies, Cosmos. William R. Corliss. Sourcebook Project, P.O. Box 107, Glen Arm, MD 21057. (410) 668-6047. 1987. $17.95.

GENERAL WORKS

Astronomer's Universe: Stars, Galaxies and Cosmos. Herbert Friedman. W. W. Norton & Company, 500 Fifth Avenue, New York, NY 10110. (212) 354-5500. (800) 223-2584. 1990. $24.95.

Brief History of Time: From the Big Bang To Black Holes. Stephen W. Hawking. Bantam Books, Inc., 1540 Broadway, New York, NY 10036-4094. (212) 354-6500. (800) 223-6934.1990. $13.95 in paper.

Black Holes and Baby Universes and Other Essays. Stephen W. Hawking. Bantam Books, Inc., 1540 Broadway, New York, NY 10036-4094. (212) 354-6500. (800) 223-6934. 1993.$21.95.

Cosmic Questions, Galactic Halos, Cold Dark Matter and the End of Time. Richard Morris. John Wiley & Sons, Inc., 605 Third Avenue, New York, NY 10158-0012. (212) 850-6000.(800) 225-5945. 1993. $24.95.

Cycles of Fire: Stars, Galaxies and the Wonder of Deep Space. William K. Hartmann. Workman Publishing Company, Inc., 708 Broadway, New York, NY 10003. (212) 254-5900. (800) 722-7202. 1988. $15.95 in paper.

The Discovery of Our Galaxy. Charles A. Whitney. Iowa State University Press, 2121 South State Avenue, Ames, IA 50014-8300. (515) 292-0140. (800) 862-6657. Reprint edition. 1988. $13.95 in paper.

The Early Universe: Facts and Fiction. Gerhard Borner. Springer-Verlag New York, Inc., 175 Fifth Avenue, New York, NY 10010. (212) 460-1500. (800) 777-4643. 1995. $69.00.

The Environment and Evolution of Galaxies. J. Michael Shull, editor. Kluwer Academic Publishers, 101 Philip Drive, Assinippi Park, Norwell, MA 02061. (617) 871-6600. 1993. $41.00.

Evolution of Interstellar Matter and Dynamics of Galaxies. J. Palous et al, editors. Cambridge University Press, 40 West 20th Street, New York, NY 10011-4211. (212) 924-3900. (800) 872-7423. 1992. $69.95.

Farthest Things in the Universe. Jay M. Pasachoff. Cambridge University Press, 40 West 20th Street, New York,

NY 10011-4211. (212) 924-3900. (800) 872-7423. 1995. $29.95.

The First Three Minutes: A Modern View of the Origin of the Universe. Steven Weinberg. Basic Books, 10 East 53rd Street, New York, NY 10022. (212) 207-7057. (800) 242-7737. Reprint edition. 1993. $12.00.

Galaxies and the Universe. David J. Eicher, editor. Kalmbach Publishing Company, P.O. Box 1612, Waukesha, WI 53187.

Ency. of Physical Sciences and Engineering Info. Sources

INTERSTELLAR MATTER

(414) 796-0126. (800) 446-5489. Readings from Deep Sky Magazine. 1992. $14.95.

Universe. William J. Kaufmann. W. H. Freeman & Co., 41 Madison Avenue, East 26th Avenue, 35th Floor, New York, NY 10010. (212) 576-0400. 4th edition. 1995. $54.95.

The Left Hand of Creation: The Origin and Evolution of the Expanding Universe. John D. Barrow and Joseph Silk. Oxford University Press, Inc., 200 Madison Avenue, New York, NY 10016. (212) 725-6000. (800) 334-4249. 1994. $23.00.

Man Discovers the Galaxies. Richard Berendzen. Columbia University Press, 562 West 113th Street, New York, NY 10025. (212) 666-1000. (800) 944-8648. 1984. $17.50 in paper.

The Nearest Active Galaxies. J. E. Beckman, editor. Kluwer Academic Publishers, 101 Philip Drive, Assinippi Park, Norwell, MA 02061. (617) 871-6600. 1993. $87.50.

The New Cosmos. A. Unsold and B. Baschek. Springer-Verlag New York, Inc., 175 Fifth Avenue, New York, NY 10010. (212) 460-1500. (800) 777-4643 . 4th revised edition. 1991. $59.00.

Observational Astrophysics. Robert C. Smith. Cambridge University Press, 40 West 20th Street, New York, NY 10011- 4211. (212) 924-3900. (800) 872-7423. 1995. $69.95.

The Origin and Evolution of Galaxies. B. J. Jones and J. E. Jones. Kluwer Academic Publishers, 101 Philip Drive, Assinippi Park, Norwell, MA 02061. (617) 871-6600. 1982. $52.50 in paper.

The Outer Galaxy. Leo Blitz and F. J. Lockman, editors. Springer-Verlag New York, Inc., 175 Fifth Avenue, New York, NY 10010. (212) 460-1500. (800) 777-4643. 1988. $39.00.

Physical Processes in Comets, Stars and Active Galaxies. W. Hildebrandt, et al editors. Springer-Verlag New York, Inc., 175 Fifth Avenue, New York, NY 10010. (212) 460-1500. (800) 777-4643. 1987. $47.00.

The Physics of Stars. A. C. PHillips. John Wiley & Sons, Inc., 605 Third Avenue, New York, NY 10158-0012. (212) 850- 6000. (800) 225-5945. 1994. $32.95 in paper.

Star Formation, Galaxies and the Interstellar Medium. J. Franco, et al, editors. Cambridge University Press, 40 West 20th Street, New York, NY 10011-4211. (212) 924-3900.(800) 872-7423. 1993. $59.95.

Stars and Atoms: From the Big Bang To the Solar System. Stuart Clark. Oxford University Press, Inc., 200 Madison Avenue, New York, NY 10016. (212) 725-6000. (800) 334-4249. 1995. $35.00.

Stars and Galaxies: Astronomy's Guide To Exploring the Cosmos. Astronomy Magazine Staff. Kalmbach Publishing Company, P.O. Box 1612, Waukesha, WI 53187. (414) 796-0126. (800) 446-5489. 1992. $29.95 in paper.

Structure and Evolution of Galaxies. Roger J. Tayler. Cambridge University Press, 40 West 20th Street, New York, NY 10011- 4211. (212) 924-3900. (800) 872-7423. 1993. $49.95. $22.95 in paper.

Universe of Galaxies. Paul W. Hodge, editor. W. H. Freeman & Co., 41 Madison Avenue, East 26th Avenue, 35th Floor, New

York, NY 10010. (212) 576-0400. Readings from Scientific American. 1995.

Visual Astronomy of the Deep Sky. Roger N. Clark. Cambridge University Press, 40 West 20th Street, New York, NY 10011-4211. (212) 924-3900. (800) 872-7423. 1991. $39.95.

The World of Galaxies. H. G. Corwin and L. Bettinelli, editors. Springer-Verlag New York, Inc., 175 Fifth Avenue, New York, NY 10010. (212) 460-1500. (800) 777-4643. 1989. $79.00

HANDBOOKS AND MANUALS

Burnham's Celestial Handbook: An Observer's Guide To the Universe Beyond the Solar System. Robert Burnham, Jr. Dover Publications, Inc., 180 Varick Street, New York, NY 10014. (212) 255-3755. (800) 223-3130. 3 volumes. 1978. $41.85 for the set.

Cambridge Guide To the Constellations. Michael E. Bakich. Cambridge University Press, 40 West 20th Street, New York, NY 10011-4211. (212) 924-3900. (800) 872-7423. 1995. $69.95.

Encyclopedia of Astronomy and Astrophysics. Stephen Maran, editor. Van Nostrand Reinhold, 115 Fifth Avenue, New York, NY 10003. (212) 254-3232. (800) 842-3636. 1992. $129.95.

Greenwich Guide To Stars, Galaxies and Nebulae. Stuart Malin. Cambridge University Press, 40 West 20th Street, New York, NY 10011-4211. (212) 924-3900. (800) 872-7423.1990. $10.95.

Handbook of Astronomy, Astrophysics and Geophysics; Volume 2: Galaxies and Cosmology. V. M. Canuto and B. G. Elmegreen. Gordon & Breach Science Publishers, Inc., P.O. Box 200029, Riverfront Plaza Station, Newark, NJ 07102-0301.(201) 643-7500. 1988. $194.00.

Handbook of Space Astronomy and Astrophsyics. Martin V. Zumbeck. Cambridge University Press, 40 West 20th Street, New York, NY 10011-4211. (212) 924-3900. (800) 872-7423.2nd edition. 1990. $79.95.

Landolt-borenstein Numerical Data and Functional Relationships in Science and Technology: Astronomy, Astrophysics and Space Research. GROUP VI. Springer- Verlag New York, Inc., 175 Fifth Avenue, New York, NY 10010. (212) 460-1500. (800) 777-4643. volumes priced individually.

McGraw-Hill Encyclopedia of Astronomy. Sybil P. Parker and Jay M. Pasachoff, editors. McGraw-Hill Publishing Company, Inc., 1221 Avenue of the Americas, New York, NY 10020. (212) 512-2000. (800) 262-4729. 2nd edition. 1993. $75.00

National Audobon Society Pocket Guide: Galaxies and Other Deep-sky Objects. Alfred A. Knopf, Inc. 201 E. 50th Street, New York, NY 10022. (800) 733-3000. 1995. $7.95.

Nearby Galaxies Catalog. R. Brent Tully and J. Richard Fisher. Cambridge University Press, 40 West 20th Street, New York, NY 10011-4211. (212) 924-3900. (800) 872-7423.1988. $59.95.

Observer's Guide To Astronomy. Patrick Martinez, editor. Cambridge University Press, 40 West 20th Street, New York, NY 10011-4211. (212) 924-3900. (800) 872-7423. 2 volumes. 1994. $159.90.

INTERSTELLAR MATTER

Ency. of Physical Sciences and Engineering Info. Sources

ONLINE DATABASES AND CD-ROMS

CA Search. Chemical Abstracts Service, P.O. Box 3012, Columbus, OH 43210-0012. (614) 447-3600. (800) 848-6533. FAX (614) 447-3709. Very comprehensive guide to worldwide chemical literature and related fields, 1972 to present. Available on BRS,(800) 289-4277, DIALOG, (800) 334-2564, ORBIT (800) 456-7248, and STN International, FIZ Karlsruhe, P.O. Box 2465, W-7500, Karlsruhe 1, Germany, online services. Inquire as to cost and availability.

Current Contents Search. Institute for Scientific Information, 3501 Market Street, Philadelphia, PA 19104. (215) 386-0100. FAX (215) 386-6362. Contains citations to articles listed in the table of contents of science and technology journals. Also articles in social sciences and life sciences journals. Available on BRS, (800) 289-4277, DIALOG, (800) 334-2564, online services. Inquire as to cost and availability.

Dissertation Abstracts Online. University Microfilms International, 300 North Zeeb Road, Ann Arbor, MI 48106. (800) 521-0600 or (313) 761-4700. Scope includes virtually all doctoral dissertations accepted at accredited American institutions from 1861 to present in 252 subject areas. Available on BRS, (800) 289-4277, DIALOG, (800) 334-2564, and OCLC EPIC, (800) 848-5878, online services. Also available on CD-ROM. Inquire as to cost and availability.

INSPEC. Institution of Electrical Engineers, Michael Faraday House, Six Hills Way, Stevenage, Herts. SG1 2AY, England. Telephone: 0438 313311 or FAX 0438 742840. Contains citations to the worldwide literature of physics, electronics and electrical engineering, computer technology, and related fields. Available on BRS, (800) 289-4277, DIALOG, (800) 334-2564, ORBIT, (800) 456-7248, and STN International, FIZ Karlsruhe, P.O. Box 2465, W-7500, Karlsruhe 1, Germany, online services. Inquire as to cost and availability.

International Aerospace Abstracts. American Institute of Aeronautics and Astronautics, 370 L'Enfant Promenade, S.W., Washington, DC 20024. (202) 646-7400. Contains referencesand abstracts of journal and monograph literature relating to aerospace science and technology, from 1963 to present. Available through the NASA/RECON system of the National Aeronautics and Space Administration only.

NTIS Bibliographic Database. National Technical Information Service, 5285 Port Royal Road, Springfield, VA 22161. (703) 487-4929 or FAX (703) 321-8199. Broad coverage of government-sponsored science and technology research reports, 1964 to present. Available on BRS,(800) 289-4277, DIALOG, (800) 334-2564, ORBIT, (800) 456-7248, and STN International, FIZ Karlsruhe, P.O. Box 2465, W-7500, Karlsruhe 1, Germany, online services. Also available on CD-ROM. Inquire as to cost and availability.

Physics Briefs. American Institute of Physics, 335 East 45th Street, New York, NY 10017. (212) 661-9260 or FAX (212) 661-2036. Contains citations with abstracts of the literature of physics and related fields, 1979 to present. Available on the STN International, FIZ Karlsruhe, P.O. Box 2465, W-7500, Karlsruhe 1, Germany, online service. Inquire as to cost and availability.

Scientific and Technical Books and Serials in Print. R.R. Bowker Inc., 121 Chanlon Road, New Providence, NJ 07974. (800) 521-8110 or (908) 464-6800. List of currently published books and serials in the physical and biological sciences, engineering and technology, with subject, author and titles indexes. Available

on ORBIT, (800) 456-7248, online service. Inquire as to cost and availability.

SCISEARCH. Institute for Scientific Information, 3501 Market Street, Philadelphia, PA 19104. (800) 523-1850 or (215) 386-0100. Broad multidisciplinary title and author index to the international literature of science and technology, 1974 to present. Available on DIALOG, (800) 334-2564, and ORBIT, (800) 456-7248, online services. Also available on CD-ROM. Inquire as to cost and availability.

WILSONLINE. H.W. Wilson Company, 950 University Avenue, Bronx, NY 10452. (800) 367-6770 or (212) 588-8400. Makes available online versions of the printed H.W. Wilson indexes including Applied Science and Technology Index, Business Periodicals Index, General Science Index, and Readers' Guide to Periodical Literature. Period covered is generally 1983 to present. Available on BRS, (800) 289-4277, DIALOG, (800) 334-2564, and OCLC EPIC, (800) 848-5878, online services. Also available on CD-ROM. Inquire as to cost and availability.

OTHER SOURCES

Atlas of the Andromeda Galaxy. Paul W. Hodge. University of Washington Press, P. O. Box 50096. Seattle, WA 98145- 5096. (206) 543-4050. (800) 441-4115. 1981. $75.00.

Atlas of Deep-sky Spendors. Hans Vehrenberg. Sky Publishing Corp., 49 Bay State Road, Cambridge, MA 02138. (617) 864-7360. (800) 253-0245. 4th edition. 1983.$24.95.

Atlas of the Ultraviolet Sky. Richard C. Henry, Wayne B. Landsman, et al. Johns Hopkins University Press, 2715 North Charles Street, Baltimore, MD 21218-4319. (410) 516-6900. (800) 537-5487. 1988. $70.00.

Atlas of the Universe. Patrick Moore. Rand McNally & Company. 8255 North Central Park, Skokie, IL 60076-2970. (708) 673-9100. 1994. $29.95.

Cambridge Atlas of Astronomy. Jean Audouze, et al. Cambridge University Press, 40 West 20th Street, New York, NY 10011-4211. (212) 924-3900. (800) 872-7423. 3rd edition. 1994. $75.00.

Cambridge Guide To the Constellations. Michael E. Bakich. Cambridge University Press, 40 West 20th Street, New York, NY 10011-4211. (212) 924-3900. (800) 872-7423. 1995. $19.95 in paper.

Carnegie Atlas of Galaxies. Allan Sandage and John Bedke. Carnegie Institution of Washington, 1530 P Street NW, Washington, DC 20005. (202) 387-6411. Two volumes. 1994.$92.00.

Color Atlas of Galaxies. James Wray. Cambridge University Press, 40 West 20th Street, New York, NY 10011-4211. (212) 924-3900. (800) 872-7423. 1988. $94.95.

Nearby Galaxy Atlas. R. Brent Tully and J. Richard Fisher. Cambridge University Press, 40 West 20th Street, New York, NY 10011-4211. (212) 924-3900. (800) 872-7423., 1987. $74.95.

Observer's Sky Atlas. E. Karkoschka. Springer-Verlag New York, Inc., 175 Fifth Avenue, New York, NY 10010. (212) 460-1500. (800) 777-4643. 1995. $15.95.

Observing Handbook and Catalog of Deep-sky Objects. Christian B. Luginbuhl and Brian A. Skiff. Cambridge University Press, 40 West 20th Street, New York, NY 10011-4211. (212) 924-3900. (800) 872-7423. 1990. $49.95.

Photometric Atlas of Northern Bright Galaxies. Kodaira, Keichi, et al editors. Columbia University Press, 562 West 113th Street, New York, NY 10025. (212) 666-1000. (800) 944-8648. 1990. $180.00.

Third Reference Catalogue of Bright Galaxies. Gerard H. DE Vaucouleurs et al, editors. Springer-Verlag New York, Inc., 175 Fifth Avenue, New York, NY 10010. (212) 460-1500. (800) 777-4643. Three volumes. 1991. $198.00 set.

The Sky: A User's Guide. David H. Levy. Cambridge University Press, 40 West 20th Street, New York, NY 10011- 4211. (212) 924-3900. (800) 872-7423. 1993. $14.95 in paper.

PERIODICALS

Astronomical Journal. American Institute of Physics, One Physics Ellipse, College Park, MD 20740-3843. (301) 209-3000. Published for the American Astronomical Society. 1849 to present. Monthly. $280.00 per year. ISSN: 0004- 6256.

Astronomical Society of the Pacific. PUBLICATIONS. Astronomical Society of the Pacific, 390 Ashton Avenue, San Francisco, CA 94112. (415) 337-1100. FAX (415) 337-5205.1889 to present. Monthly. $175.00 per year. ISSN: 0004- 6280.

Astronomy. Kalmbach Publishing Company, Box 1612, Waukesha, WI 53187-1612. (414) 796-8776. FAX (414) 796-1142. 1973 to present. Monthly. $27.00 per year. ISSN: 0091-6358.

Astronomy and Astrophsyics. Springer-Verlag New York, Inc., 44 Hartz Way, Secaucus, NJ 07096-2491. (201) 348-4033. FAX (201) 348-4505. 1969 to date. 36 issues per year in 12 volumes. $1733.00 per year. ISSN: 0004-6361.

Astronomy Reports. American Institute of Physics, One Physics Ellipse, College Park, MD 20740-3843. (301) 209-3000. English translation of Astronomicheksii Zhurnal. Formerly Soviet Astronomy AJ. 1957 to present. Bimonthly. $1160.00 per year. ISSN: 1063-7729.

Astrophysical Journal; An International Review of Astronomy and Astronomical Physics. University of Chicago Press, Journals Division, 5720 South Woodlawn Avenue, Chicago, IL 60637. (312) 753-3347. FAX (312) 753-0811. 1895 to date.three issues per month. $740.00 per year. ISSN: 0004-637X.

Astrophpysics and Space Science; An International Journal of Cosmic Physics. Kluwer Academic Publishers, Box 358, Accord Station, Hingham, MA 02018-0358. (617) 871-6600. FAX (617) 871-6528. 1968 to present. 24 issues per year. $2544.00 per year. ISSN: 0004-640X.

Celestial Mechanics and Dynamical Astronomy; An International Journal of Space Dynamics. Kluwer Academic Publishers, Box 358, Accord Station, Hingham, MA 02018-0358. (617) 871-6600. FAX (617) 871-6528. 1969 to present. Monthly. $703.50 per year. ISSN: 0923-2958.

Experimental Astronomy; An International Journal On Astronomical Instrumentation and Data Analysis. Kluwer Academic

Publishers, Box 358, Accord Station, Hingham, MA 02018-0358. (617) 871-6600. FAX (617) 871-6528. 1989 to present. Quarterly. $161.00 per year. ISSN: 0922-6435.

Icarus; International Journal of Solar System Studies. Academic Press, Inc., Journal Division, 525 B Street, Suite 1900, San Diego, CA 92101-4495. (619) 230-1840. FAX (619) 699-6800. 1962 to present. Monthly. $1080.00 per year. ISSN: 0019-1035.

Landolt-boernstein, Zahlenwerte Und Funktionen Aus Naturwissenschaften Und Technik. GROUP 6: ASTRONOMY. Springer-Verlag New York, Inc., 175 Fifth Avenue, New York, NY 10010. (212) 460-1500. FAX (212) 473-6272. 1965. Irregular. Price varies. ISSN: 0075-7896.

Mercury. Astronomical Society of the Pacific. 390 Ashton Avenue, San Francisco, CA 94112. (415) 337-1100. FAX: (415) 337-5205. 1972 to present. Bimonthly. $175.00 per year. ISSN: 0047-6773.

Monthly Notices of the Royal Astronomical Society. Blackwell Scientific Publication LT., Osney Mead, Oxford OX2 OEL, England. TEL 0865-240201. FAX 0865-721205. 1827 to present. Fortnightly. $1899.00 per year. ISSN: 0035-8711.

Observatory. Space and Astrophysics Division, Rutherford Appleton Laboratory, Chilton, Didcot, Oxon OX11 OQX England. FAX 0235-445848. 1877 to present. Bimonthly. $42.00 per year.

Sky and Telescope. Sky Publishing Corporation, Box 9111, Belmont, MA 02178. (617) 864-7360. FAX (617) 864-6117.1941 to present. Monthly. $27.00 per year. ISSN: 0037-6604.

Space Science Reviews. Kluwer Academic Publishers, Box 358, Accord Station, Hingham, MA 02018-0358. (617) 871-6600. FAX (617) 871-6528. 1962 to present. Sixteen issues per year. $940.00 per year. ISSN: 0038-6308.

Vistas in Astronomy; An International Review Journal. Albert C. Beer, editor. Elsevier Science Publishing Company, Inc., 660 White Plains Road, Terrytown, NY 10591- 5153. (914) 524-9200. FAX: (914) 333-2444. 1958 to present. Quarterly. $415.00 per year. ISSN: 0083-6656.

Transactions of the International Astronomical Union. Proceedings of the General Assembly. Kluwer Academic Publishers, Box 358, Accord Station, Hingham, MA 02018-0358. (617) 871-6600. Issues in two parts: Part A, Reports, and Part B, Proceedings. Triennial. XXI - 1994. Price varies. ISSN: 0080-1372.

Webb Society Quarterly Journal. Webb Society, 194 Foundry Lane, Freemantle, Southampton, Hampshire SO1 3EQ, England. 1969 to present. Quarterly. $19.50 per year. ISSN: 0043- 1680.

RESEARCH CENTERS AND INSTITUTES

Astronomy Program. University of Maryland, 1207 Computer & Space Sciences Building, College Park, MD 20742. (301) 405- 3001. FAX (301) 314-9067.

Carnegie Institution of Washington Observatories, 813 Santa Barbara Street, Pasadena, CA 91101. (818) 577-1122. FAX (818) 759-8136.

INTERSTELLAR MATTER

Ency. of Physical Sciences and Engineering Info. Sources

Center For Astronomy and Space Astrophysics. University of Colorado - Boulder, Campus Box 391, Boulder, CO 80309. (303) 492-4050. (FAX) 492-7178.

Center For Space Research. Massachusetts Institute of Technology, 77 Massachusetts Avenue, Camridge MA 02139. (617) 253-7501.

David Dunlap Observatory. University of Toronto. Box 360, Richmond Hill ON, Canada L4C 4Y6. (416) 884-9562. FAX (416) 978-3921.

Lick Observatory. University of California, Santa Cruz, CA 95064. (408) 459-2991. FAX (408) 426-3115.

Lowell Observatory. 1400 West Mars Hill Road. Flagstaff, AZ 86001, (602) 774-3358. FAX (602) 774-6296.

National Astronomy and Ionosphere Center. Cornell University, Space Sciences Building, Ithaca, NY 14853. (607) 255-3735. (607) 255-8803.

National Optical Astronomy Observatories. 950 North Cherry Avenue, Tucson, AZ 85719. (602) 325-9282. FAX (602) 325-9360.

National Radio Astronomy Observatory. Edgemont Road, Charlottesville, VA 22903. (804) 296-0211.

Smithsonian Center For Astrophysics. Harvard University, 60 Garden Street, Cambridge, MA 02138. (617) 495-7461. FAX (617) 495-7326.

Space Telescope Science Institute. 3700 San Martin Drive, Baltimore, MD 21218. (301) 338-4700.

Theoretical Astrophysics Program. University of Arizona, Planetary Sciences Department, Tuscon, AZ 85721. (602) 621-6963.

IODINE

See also: BROMINE, CHLORINE, ELEMENTS, FLUORINE, HALIDES

ABSTRACT SERVICES AND INDEXES

Applied Science and Technology Index; a cumulative subject index to English language periodicals in the fields of aeronautics and space science, computer technology, chemistry, construction industry, energy and related areas. H.W. Wilson Co., 950 University Avenue, Bronx, NY 10452. (212) 588-8400. (800) 367-6770. FAX (718) 590-1617. From 1958 to present. Monthly. Inquire about cost and availability. Also available on CD-ROM and online. ISSN: 0003-6986.

Chemical Abstracts. Chemical Abstracts Service. 2540 Olentangy River Road, Box 3012, Columbus, OH 43210-0012. (614) 447-3600. FAX (614) 447-3713. 1907 to present. Weekly. $16,800.00 per year. Available online and on CD-ROM. CA is also available in five section groupings. Inquire regarding cost and availability.

Current Contents: Physical, Chemical and Earth Sciences. Institute for Scientific Information, 3501 Market Street, Philadel-

phia, PA 19104. (215) 386-0100. FAX (215) 386-6362. Weekly. $360.00 per year.

General Science Index. H.W. Wilson Company, 950 University Avenue, Bronx, NY 10452. (212) 588-8400. (800) 367-6770. FAX (718) 590-1617. From 1978 to present. Ten issues per year; quarterly and annual cumulations. Service basis. Available on CD-ROM and online. Inquire regarding cost and availability. ISSN: 0162-1963.

Physics Abstracts. INSPEC. Section A, Science Abstracts. Institution of Electrical Engineers (IEE), London. Available from: INSPEC/IEEE - Institute of Electrical and Electronic Engineers, Box 1331, Hoes Lane, Piscataway, NJ 08855-1331. (908) 562-5549. 1898 to present. 24 issues per year. $2835.00 per year. Also available online and on CD-ROM. ISSN: 0036-8091.

Science Citation Index. SCI. Institute for Scientific Information, 3501 Market Street, Philadelphia, PA 19104. (215) 386-0100. (800) 523-1850. FAX (215) 386-2991. 1961 to present. Six issues per year, plus annual cumulation. $11650.00 per year. Also available online and on CD-ROM. Inquire about price and availability. ISSN: 0036-827X.

ASSOCIATIONS AND PROFESSIONAL SOCIETIES

American Chemical Society. 1155 16th Street, NW, Washington, DC 20036. (202) 872-4600.

Association of Consulting Chemists and Chemical Engineers, 295 Madison Avenue, 27th Floor, New York, NY 10017. (212) 983-3160. FAX (212) 983-3161.

Chemical Manufacturers Association. 2501 M Street, N.W., Washington, DC 20037. (202) 887-1100. FAX (202) 887-1237.

Chemical Specialties Manufacturers Association, 1913 Eye Street, Nw, Washington, Cd 20006, (202) 872-8110. FAX (202) 872-8114.

BIBLIOGRAPHIES

Chemical Information Sources. Gary Wiggins. McGraw-Hill Publishing Company, 1221 Avenue of the Americas, New York, NY 10020. (800) 262-4729 or (212) 512-3825. 1991. $42.50.

Information Sources in Chemistry. R.T. Bottle and J.F.B Rowland, editors. R.R. Bowker Inc., 121 Chanlon Road, New Providence, NJ 07974. (800) 521-8110 or (908) 464-6800. Fourth edition. 1993. $75.00.

Handbooks and Tables in Science and Technology. Russell H. Powell, editor. Oryx Press, 4041 North Central, Suite 700, Phoenix, AZ 85012-3330. (602) 265-2651 or (800) 279-6799. Third edition. 1994. $65.00.

Scientific and Technical Books and Serials in Print; An Index To Literature in Science and Technology. R.R. Bowker Inc., 121 Chanlon Road, New Providence, NJ 07974. (908) 464-6800. (800) 521-8110. FAX (908) 665-3502. 1972 to present. Annual. 4 volumes. 1994. $299.95. Also available on compact disc and online. ISSN: 0000-054X.

Ency. of Physical Sciences and Engineering Info. Sources

IODINE

DIRECTORIES AND BIOGRAPHICAL SOURCES

American Men and Women of Science. Physical and Biological Sciences. R.R. Bowker Company, 121 Chanlon Road, New Providence, NJ 07974. (908) 464-6800. (800) 521-8110. 20th edition. 8 volumes. 1996. $850.00.

Chemical Week — Buyers Guide Issue. Chemical Week Associates, 888 Seventh Avenue, New York, NY 10106. (212) 621-4900. FAX (212) 621-4949. Annual, October. $50.00.

Consulting Services: Chemists and Chemical Engineers. Association of Consulting Chemists and Chemical Engineers, 50 East 41st Street, Suite 92, New York, NY 10017. (212)

684-6255. Biennial. $60.00.

Directory of Chemistry Software 1992. Wendy Warr, Peter Willett, Geoff Downs. American Chemical Society, 1155 16th Street, NW, Washington, DC 20036. (202) 872-4600. 1992. $35.95.

Research Centers Directory. Gale Research Company Inc., 835 Penobscot Building, Detroit, MI 48226-4094. (313) 961-2242. (800) 347-4253. 20th edition. 1995. $485.00. ISSN: 0080-1518.

ENCYCLOPEDIAS AND DICTIONARIES

Academic Press Dictionary of Science and Technology. Christopher Morris, editor. Academic Press, Inc., 1250 Sixth Avenue, San Diego, CA 92101. (619) 231-0926. FAX (619) 699-6715. 1991. $115.00.

Dictionary of Named Processes in Chemical Technology. Alan E. Comyns. Oxford University Press, Inc., 200 Madison Avenue, New York, NY 10016. (212) 725-6000. (800) 334-4249. 1994. $75.00.

Consulting Services: Chemists and Chemical Engineers. Association of Consulting Chemists and Chemical Engineers, 50 East 41st Street, Suite 92, New York, NY 10017. (212)

684-6255. Biennial. $60.00.

Encyclopedia of Chemical Processing and Design. McKetta

Marcel Dekker, Inc., 270 Madison Avenue, New York, NY 10016. (212) 696-9000. (800) 228-1160. 1976 - . $175.00 per volume.

Kirk-Othmer Encyclopedia of Chemical Technology. John Wiley and Sons, Inc., 605 Third Avenue, New York, NY 10158. (800) 526-5368 or (212) 850-6000. Fourth edition. 1991 - . Twenty-seven volumes. $5400.00.

McGraw-Hill Encyclopedia of Science and Technology. McGraw-Hill Book, Incorporated, 1221 Avenue of the Americas, New York, NY 10020. (212) 997-3675. (800) 262-4729. Seventh edition. Twenty volumes. 1992. $1900.00.

Ullman's Encyclopedia of Industrial Chemistry. VCH Publications, Inc., 220 East 23rd Street, Suite 909, New York, NY 10010-4606. (212) 683-8333. (800) 422-8824. 5th edition. 1984 - . Price varies per volume.

GENERAL WORKS

The Chemical Bond: Structure and Dynamics. Ahmed Zewail. Academic Press, Inc., 6277 Sea Harbor Drive, Orlando, FL. (800) 321-5068. 1992. $49.95.

Chemistry of Halides, Pseudohalides and Azides, Part 1 and 2. Chemistry of Functional Groups series, supplement no. D. Saul Patai and Zvi Rappaport. Books on Demand, 300 North Zeeb Road, Ann Arbor, MI 48106-1346. (313) 761-4700. (800) 521-0600. 1983. $340.00 set.

Chemistry of the Elements. N. N. Greenwood and A. Earnshaw. Pergamon Press, Inc., Maxwell House, Fairview Park, Elmsford, NY 10523. (914) 592-7700. Fax (914) 592-3625. 1984. $143.00.

The Elements. John Emsley. Oxford University Press, Inc., 200 Madison Avenue, New York, NY 10016. (212) 725-6000. (800) 334-4249. Second edition. 1991. $49,95.

Exploring Chemical Elements and their Compounds. David L.Heiserman. TAB Books, P.O. Box 40, Blue Summit, PA 17294-0850. (717 794-2191. (800) 233-1128. 1992. $17.95.

Halogen Chemistry. V. Gutman, editor. Academic Press, Incorporated, 6277 Sea Harbor Drive, Orlando, FL 32821. (800) 321-5068. Three volumes. 1967. Inquire.

Iodine. Gordon Press Publications, P.O. Box 459, Bowling Green Station, New York, NY 10004. (718) 624-8419.. Metals and Minerals Series. 1993. $146.00

Organic Chemistry of Polycoordinated Iodine. A.Vargolis. VCH Publications, Inc., 220 East 23rd Street, Suite 909, New York, NY 10010-4606. (212) 683-8333. (800) 422-8824. . 1992. $115.00

Story of Iodine Deficiency: An International Challenge in Nutrition. Basil A. Hertzel. Oxford University Press, . 1989. $35.00.

Toxicology of Halogenated Hydrocarbons; Health and Ecological Effects. M. A. Khan and R. H. Stanton, editors. Pergamon Press, Inc., Maxwell House, Fairview Park, Elmsford, NY 10523. (914) 592-7700. Fax (914) 592-3625. 1981. $72.50.

HANDBOOKS AND MANUALS

Chemical Formulary. H. Bennett, editor. Chemical Publishing, Co., Inc. 80 Eighth Avenue, New York, NY 10011. (212 255-1950. Volumes 1 - 30. $60.00 per volume.

CRC Handbook of Chemistry and Physics. David R. Kide, editor. CRC Press Inc., 2000 Corporate Blvd., NW, Boca Raton, FL 33431. (407) 994-0555. (800) 333-8300. 77th edition. 1996. $99.95.

Guide To Basic Chemical Compounds. D.R. Lide, Jr. CRC Publishers, Inc., 2000 Corporate Blvd., N.W., Boca Raton, FL 33431. (407) 994-0555 or (800) 333-8300. 1993. $120.00.

Lange's Handbook of Chemistry. John A. Dean, editor. McGraw-Hill Publishing Company, 1221 Avenue of the Americas, New York, NY 10020. (800) 262-4729 or (212) 512-3825. Fourteenth editon. 1992. $79.50.

IODINE

Ency. of Physical Sciences and Engineering Info. Sources

Riegel's Handbook of Industrial Chemistry. James A. Kent, editor. Van Nostrand Reinhold, 115 Fifth Avenue, New York, NY 10003. (212) 254-3232. (800) 926-2665. Ninth edition. 1992. $114.95.

ONLINE DATABASES AND CD-ROMS

CA Search. Chemical Abstracts Service, P.O. Box 3012, Columbus, OH 43210-0012. (614) 447-3600. (800) 848-6533. FAX (614) 447-3709. Very comprehensive guide to worldwide chemical literature and related fields, 1972 to present. Available on BRS,(800) 289-4277, DIALOG, (800) 334-2564, ORBIT (800) 456-7248, and STN International, FIZ Karlsruhe, P.O. Box 2465, W-7500, Karlsruhe 1, Germany, online services. Inquire as to cost and availability.

Chemical Journals of the American Chemical Society. American Chemical Society, 1155 16th Street, N.W., Washington, DC 20036. (202) 872-4381 or (800) 424-6747. Contains complete text of approximately 90,000 articles from 22 primary journals published by the American Chemical Society, from mostly 1982 to present. Available on STN International, FIZ Karlsruhe, P.O. Box 2465, W-7500, Karlsruhe 1, Germany, online service. Inquire as to cost and availability.

Current Contents Search. Institute for Scientific Information, 3501 Market Street, Philadelphia, PA 19104. (215) 386-0100. FAX (215) 386-6362. Contains citations to articles listed in the table of contents of science and technology journals. Also articles in social sciences and life sciences journals. Available on BRS,(800) 289-4277, DIALOG,(800) 334-2564, online services. Inquire as to cost and availability.

Dissertation Abstracts. University Microfilms International, 300 North Zeeb Road, Ann Arbor, MI 48106. (800) 521-0600 or (313) 761-4700. Scope includes virtually all doctoral dissertations accepted at accredited American institutions from 1861 to present in 252 subject areas. Available on BRS,(800) 289-4277, DIALOG,(800) 334-2564, and OCLC EPIC,(800) 848-5878, online services. Also available on CD-ROM. Inquire as to cost and availability.

Gmelin Database. Gmelin-Institut fur Anorganische Chemie und Grenzgebiete, Varrentrapperstrasse, 40-42, Carl-Bosch-Haus, D-6000, Frankfurt am Main 90, Germany. Contains structural and factual data relating to inorganic and organometallic chemistry. Provides data from the Gmelin Handbook of Inorganic and Organometallic Chemistry. Covers the period 1817 to 1975; 1988-89. Available on STN International, FIZ Karlsruhe, P.O. Box 2465, W-7500, Karlsruhe 1, Germany, online service. Inquire as to cost and availability.

Kirk-Othmer Encyclopedia of Chemical Technology. John Wiley and Sons, Inc., 605 Third Avenue, New York, NY 10158. (800) 526-5368 or (212) 850-6000. Contains the complete text of all chapters in the 27 volume fourth edition of the Kirk-othmer Encyclopedia of Chemical Technology. 1991. Available on BRS,(800) 289-4277, DIALOG,(800) 334-2564, online services. Inquire as to cost and availability.

NTIS Bibliographic Database. National Technical Information Service, 5285 Port Royal Road, Springfield, VA 22161. (703) 487-4929 or FAX (703) 321-8199. Broad coverage of government-sponsored science and technology research reports, 1964 to present. Available on BRS,(800) 289-4277, DIALOG, (800) 334-2564, ORBIT, (800) 456-7248, and STN International, FIZ Karlsruhe, P.O. Box 2465, W-7500, Karlsruhe 1, Germany,

online services. Also available on CD-ROM. Inquire as to cost and availability.

SCISEARCH. Institute for Scientific Information, 3501 Market Street, Philadelphia, PA 19104. (800) 523-1850 or (215) 386-0100. Broad multidisciplinary title and author index to the international literature of science and technology, 1974 to present. Available on DIALOG,(800) 334-2564, and ORBIT,(800) 456-7248, online services. Also available on CD-ROM. Inquire as to cost and availability.

WILSONLINE. H.W. Wilson Company, 950 University Avenue, Bronx, NY 10452. (800) 367-6770 or (212) 588-8400. Makes available online versions of the printed H.W. Wilson indexes including Applied Science and Technology Index, Business Periodicals Index, General Science Index, and Readers' Guide to Periodical Literature. Period covered is generally 1983 to present. Available on BRS,(800) 289-4277, DIALOG,(800) 334-2564, and OCLC EPIC,(800) 848-5878, online services. Also available on CD-ROM. Inquire as to cost and availability.

PERIODICALS

Analytical Chemistry. American Chemical Society, Box 3337, Columbus, OH 43210. (614) 447-3776. 1929 to present. Semimonthly. $415.00 per year. ISSN: 0003-2700.

Angewandte Chemie. VCH Publishers, Inc., 220 East 23rd Street, New York, NY 10010-4606. (212) 683-8333. (800) 367-8249. 1888 to present. 22 issues per year. $840.00 per year. ISSN: 0044-8249.

Chemical Reviews. American Chemical Society, Box 3337, Columbus, OH 43210. (614) 447-3776. 1924 to present. 8 issues per year. $346.00 per year. ISSN: 0009-2665.

Chemical Week; Includes Annual Buyers Guide. Chemical Week Associates, 888 Seventh Avenue, New York, NY 10106. (212) 621-4900. FAX (212) 621-4949. 1914 to present. Weekly. $99.00 per year. ISSN: 0009-272X.

Chemtech. American Chemical Society. Box 3337, Columbus, OH 43210. (614) 447-3776. 1970 to present. Monthly. $370.00 per year. ISSN: 0009-2703.

Inorganic Chemistry. American Chemical Society, Box 3337, Columbus, OH 43210. (614) 447-3776. 1962 to present. Semimonthly. $500.00 per year. ISSN: 0020-1669.

Journal of the American Chemical Society. American Chemical Society, Box 3337, Columbus, OH 43210. (614) 447-3776. 1879 to present. Biweekly. $1055.00 per year. ISSN: 0002- 7863.

Polyhedron. Elsevier Science Publishing Company, Inc., 660 White Plains Road, Tarrytown, NY 10591-5153. (914-524-9200. FAX (914) 524-9200. 1982 to present. 24 issues per year. $2175.00 per year. ISSN: 0277-5387.

RESEARCH CENTERS AND INSTITUTES

Chemical Laboratories. Harvard Universiy. Oxford Street, Cambridge, MA 02138. (617) 495-4283. FAX (617) 496-5618.

Chemistry Laboratories. Rensselaer Polytechnic Institute. Cogswell Laboratory, Troy, NY 12180-3590. (518) 276-8981.

Research Program in Chemistry and Biochemistry. Southern Illinois University at Carbondale, Carbondale, IL 62901. (618) 453-5721.

Theoretical Chemistry Institute. University of Wisconsin, Madison, 1101 University Avenue, Madison, WI 53706. (608) 262-1511.

University/industry Chemical Research Center. Mississippi State University, Department of Chemistry, P.O. Drawer CH, Mississippi State, MS 39762. (601) 325-3584.

ION ACCELERATOR

See: PARTICLE ACCELERATORS

ION LASER

See: LASERS

IONIZATION

See also: CHEMICAL BONDS, CHEMISTRY, PHYSICAL CHEMISTRY

ABSTRACT SERVICES AND INDEXES

Analytical Abstracts. Royal Society of Chemistry, Information Services, Thomas Graham House, Science Park, Milton Road, Cambridge, CB4 4WF, England. Contains citations, mostly with abstracts, of the worldwide literature on analytical chemistry, from 1954 to present. Monthly. $636.00 per year. Also available online.

Applied Science and Technology Index; A Cumulative Subject Index To English Language Periodicals in the Fields of Aeronautics and Space Science, Computer Technology, Chemistry, Construction Industry, Energy and Related Areas. H. W. Wilson Co., 950 University Avenue, Bronx, NY 10452- 9978. (212) 588-8400. (800) 367-6770. FAX (718) 590-1617. From 1958 to present. Monthly. Inquire about cost and availability. Also available online (BRS and WILSONLINE) and on CD-ROM. ISSN: 0003-6986.

Chemical Abstracts. Chemical Abstracts Service. 2540 Olentangy River Road, Box 3012, Columbus, OH 43210-0012. (614) 447-3600. FAX (614) 447-3713. 1907 to present. Weekly. $16,800.00 per year. Available online and on CD- ROM. CA is also available in five section groupings. Inquire regarding cost and availability.

Current Contents: Physical, Chemical and Earth Sciences. Institute for Scientific Information, 3501 Market Street, Philadelphia, PA 19104. (215) 386-0100. FAX (215) 386-2291. <boFrom 1961 to present. Weekly. $442.00 per year. Also available on CD-ROM and online. Inquire regarding cost and availability. ISSN: 0163-2574.

Engineering Index Monthly. Engineering Information, Inc., Castle Point on the Hudson, Hoboken, NJ 07030. (201) 216-8500. (800) 221-1044. FAX(201) 216-8532. Monthly. $2300.00 per year for monthly issues. $1980 per year for annual cumulation. Available on CD-ROM and online as COMPENDEX. ISSN: 0742-1974; 0360-8557.

General Science Index. H.W. Wilson Company, 950 University Avenue, Bronx, NY 10452. (212) 588-8400. (800) 367-6770. FAX (718) 590-1617. From 1978 to present. Ten issues per year; quarterly and annual cumulations. Service basis. Available on CD-ROM and online. Inquire regarding cost and availability. ISSN: 0162-1963.

Physics Abstracts. INSPEC. Section A, Science Abstracts. Institution of Electrical Engineers (IEE), London, United Kingdom. Available from: INSPEC/IEEE -Institute of Electrical and Electronic Engineers, Box 1331, 445 Hoes Lane, Piscataway, NJ 08855-1331. (908) 562-5549. 1898 to present. 24 issues per year. $2835.00 per year. ISSN: 0036-8091. Also available online and on CD-ROM.

Science Citation Index. SCI. Institute for Scientific Information, 3501 Market Street, Philadelphia, PA 19104. (215) 386-0100. (800) 523-1850. FAX (215) 386-2991. 1961 to present. Six issues per year, plus annual cumulation. $11650.00 per year. Also available online and on CD-ROM. Inquire about price and availability. ISSN: 0036-827X.

ANNUAL REVIEWS AND YEARBOOKS

Annual Review of Physical Chemistry. Annual Reviews, Inc., 4139 El Camino Way, Palo Alto, CA 94306-0897. (415) 493-4400. Fax (415) 855-9815. 1951 to present. Annual. $55.00 per year. ISSN: 0066-426X

ASSOCIATIONS AND PROFESSIONAL SOCIETIES

American Chemical Society, 1155 16th Street, Nw, Washington, DC 20036. (202) 872-4600.

American Institute of Chemists. 501 Wythe Street, Alexandria, VA 22314-1917. (703) 836-2090. FAX (703) 836-2091.

Association of Consulting Chemists and Chemical Engineers, 50 East 41st Street, Suite 92, New York, NY 10017. (212) 684-6255.

Chemical Manufacturers Association. 2501 M Street, N.W., Washington, DC 20037. (202) 887-1182

Electrochemical Society. 10 South Main Street, Pennington, NJ 08534. (609) 737-1902.

BIBLIOGRAPHIES

Chemical Information. H. R. Collier, editor. Springer- Verlag New York, Inc., 175 Fifth Avenue, New York, NY 10010. (212) 460-1500. (800) 777-4643. 1990. $91.00.

Chemical Information Sources. Gary Wiggins. McGraw-Hill Publishing Company, 1221 Avenue of the Americas, New York, NY 10020. (800) 262-4729 or (212) 512-3825. 1991. $42.50.

Information Sources in Chemistry. R.T. Bottle and J.F.B Rowland, editors. R.R. Bowker Inc., 121 Chanlon Road, New Providence, NJ 07974. (800) 521-8110 or (908) 464-6800. Fourth edition. 1993. $75.00.

IONIZATION

Ency. of Physical Sciences and Engineering Info. Sources

Scientific and Technical Books and Serials in Print; An Index To Literature in Science and Technology. R.R. Bowker Inc., 121 Chanlon Road, New Providence, NJ 07974. (908) 464-6800. (800) 521-8110. FAX (908) 665-3502. 1972 to present. Annual. 4 volumes. 1994. $299.95. Also available on compact disc and online. ISSN: 0000-054X.

DIRECTORIES AND BIOGRAPHICAL SOURCES

American Institute of Chemists - Professional Directory. American Institute of Chemists, 501 Wythe Street, Alexandria, VA 22314-1917. (703) 836-2090. FAX (703) 836-2091. Annual. $65.00.

American Men and Women of Science. Physical and Biological Sciences. R.R. Bowker Inc., 121 Chanlon Road, New Providence, NJ 07974. (908) 464-6800. (800) 521-8110. 20th edition. 8 volumes. 1996. $850.00

Chemical Sources International. Mike Desing and Kurt Gandenberger, editors. Chemical Sources International, PO Box 1824, Clemson, SC 29633. (803) 646-7840. 1994.$285.00.

Chemical Week — Buyers Guide Issue. Chemical Week Associates, 888 Seventh Avenue, New York, NY 10106. (212) 621-4900. FAX (212) 621-4949. Annual, October. $50.00.

Directory of Chemistry Software 1992. Wendy Warr, Peter Willett, Geoff Downs. American Chemical Society, 1155 16th Street, NW, Washington, DC 20036. (202) 872-4600. 1992. $35.95.

Research Centers Directory. Gale Research Company,Inc., 835 Penobscot Building, Detroit, MI 48226-4094. (313) 961- 2242. (800) 347-4253. 20th edition. 1995. $485.00. ISSN: 0080-1518.

ENCYCLOPEDIAS AND DICTIONARIES

Academic Press Dictionary of Science and Technology. Christopher Morris, editor. Academic Press, Inc., 1250 Sixth Avenue, San Diego, CA 92101. (619) 231-0926. FAX (619) 699-6715. 1991. $115.00.

Concise Encyclopedia Chemistry. Translated and revised by Mary Eagleson. Walter de Gruyter, Inc., 200 Saw Mill River Road, Hawthorne, New York, 10532. (914) 747-0110 or Fax (914) 747-1326. 1994. $69.95.

Dictionary of Chemical Names and Synomyms. Philip A. Howard and Michael W. Neil. Lewis Pub.CRC Press, Inc., 2000 Corporate Boulevard, NW, Boca Raton, FL 33431. (407) 994- 0555. (800) 272-7737. 1991. $149.95.

Dictionary of Physical Chemistry. Satish Anand and Raj Kumar. South Asia Publications, PO Box 502 Collumbus, MO 65205. (314) 474-0116. 1990. $23.50.

Encyclopedic Dictionary of Science. Candida Hunt and Monica Byles, editors. Facts-on-File, Inc., 460 Park Avenue South, New York, NY 10016-7382. (800) 322-8755. Fax (800) 683-3633. 1988. $32.95.

Encyclopedia of Applied Physics. VCH Publishers, Inc., 303 Northwest 12th Avenue, Deerfield Beach, FL 33442. (800) 367-8249. 1991-. Twenty volumes. $6000.00.

Encyclopedia of Chemical Processing and Design. McKetta. Dekker, Inc., 270 Madison Avenue, New York, NY 10016. (212) 696-9000. (800) 228-1160. 1976 - . $175.00 per volume.

Grant & Hawks Chemical Dictionary. R. A. Grant. McGraw-Hill Publishing Company, Inc., 1221 Avenue of the Americas, New York, NY 10020. (212) 512-2000. (800) 262-4729. 5th edition. 1987. $64.50.

Hawley's Condensed Chemical Dictionary. Richard J. Lewis. Van Nostrand Reinhold, 115 Fifth Avenue, New York, NY 10003. (212) 254-3232. (800) 842-3636. 12th edition. 1993. $69.95.

Industrial Chemical thesaurus. Michael Ash and Irene Ash. VCH Publishers, Inc., 220 East 23rd Street, New York, NY 10010-4606. (800) 367-8249. 1992. $295.00.

Kirk-Othmer Encyclopedia of Chemical Technology. John Wiley and Sons, Inc., 605 Third Avenue, New York, NY 10158. (800) 526-5368 or (212) 850-6000. Fourth edition. 1991-. Twenty-seven volumes. $5400.00.

McGraw-Hill Encyclopedia of Chemistry. Sybil P.Parker, editor. McGraw-Hill Book, Incorporated, 1221 Avenue of the Americas, New York, NY 10020. (212) 997-3675. Second edition. 1993. $95.50.

McGraw-Hill Encyclopedia of Science and Technology. McGraw-Hill Book, Incorporated, 1221 Avenue of the Americas, New York, NY 10020. (212) 997-3675. Seventeenth edition. Twenty volumes. 1992. $1900.00.

GENERAL WORKS

Atoms, Radiation and Radiation Protection. James E. Turner. 2nd editor. John Wiley & Sons, Inc., 605 Third Avenue, New York, NY 10158-0012. (212) 850-6000. (800) 225-5945. 1995. $69.95.

The Chemical Bond: Structure and Dynamics. Ahmed Zewail. Academic Press, Inc., 1250 Sixth Avenue, San Diego, CA 92101-4311. (619) 231-0926. FAX (619) 699-6715. 1992. $49.95

Determination of Ionization Constants: A Laboratory Manual. Adrien Albert and E. P. Serjeant. Chapman & Hall, 1 Penn Plaza, New York, NY 10119. (212) 564-1060. 1984. $47.50.

Dynamics of the Liquid State. Umberto Balucani and Marco Zoppi. Oxford University Press, Inc., 200 Madison Avenue, New York, NY 10016. (212) 725-6000. (800) 334-4249. 1995. $95.00.

Ions, Electrodes and Membranes. Jiri Koryta. John Wiley & Sons, Inc., 605 Third Avenue, New York, NY 10158-0012. (212) 850-6000. (800) 225-5945. 2nd edition. 1991. $106.00.

Low Energy Ion Surface Interactions. J. Wayne Rabalais, editor. John Wiley & Sons, Inc., 605 Third Avenue, New York, NY 10158-0012. (212) 850-6000. (800) 225-5945. 1994. $139.00.

Physics and Chemistry of Aqueous Solutions. M. C. Bellisent-funel and G. W. Neilson. Kluwer Academic Publishers, 101 Philip Drive, Assinippi Park, Norwell, MA 02061. (617) 871-6600. 1987. $154.50.

Physics of Highly Charged Ions. Martin Stocki and Patrick Richard, editors. American Institute of Physics, One Physics Ellipse, College Park, MD 20740-3843. (301) 209- 3100. 1993. $175.00.

Qualitative Analysis and Ionic Equilibrium. Schenk. Houghton Mifflin Co., 222 Berkeley Street, Boston, MA 02116. (617) 351-5000. (800) 225-3362. 1990. $17.00 in paper.

Radiation and Radioactivity On Earth and Beyond. Ivan Draganic, et al, editors. CRC Press Inc., 2000 Corporate Boulevard, NW, Boca Raton, FL 33431. (407) 994-0555. (800) 272-7737. 1989. $39.95.

Solid State Ionics. Tetsuichi Kudo and Kazuo Fueki. VCH Publications, Inc., 220 East 23rd Street, Suite 909, New York, NY 10010-4606. (212) 683-8333. (800) 422-8824.. 1990. $135.00.

HANDBOOKS AND MANUALS

Chemical Formulary. H. Bennett, editor. Chemical Publishing Company, Inc., 80 Eighth Avenue, New York, NY 10011. (212) 255-1950. Volumes 1-30. $70.00 per volume.

Chemometrics: Chemical and Sensory Data. David R. Burgard and James T. Kuznicki. CRC Publishers, Inc., 2000 Corporate Blvd., N.W., Boca Raton, FL 33431. (407) 994-0555 or (800) 333-8300. 1990. $120.00.

Chemical Reference Handbook. Gordon Press, PO Box 459, Bowling Green Station, New York, NY 10004. (718) 624-8419. 1995. $260.00.

CRC Handbook of Chemistry and Physics. David R. Kide, editor. CRC Press Inc., 2000 Corporate Blvd., NW, Boca Raton, FL 33431. (407) 994-0555. (800) 333-8300. 77th edition. 1996. $99.95.

Guide To Basic Chemical Compounds. D.R. Lide, Jr. CRC Publishers, Inc., 2000 Corporate Blvd., N.W., Boca Raton, FL 33431. (407) 994-0555. (800) 333-8300. 1993. $120.00.

Improving Safety in the Chemical Laboratory: A Practical Guide. Jay A. Young, editor. John Wiley and Sons, Inc., 605 Third Avenue, New York, NY 10158. (800) 526-5368 or (212) 850-6000. Second edition. 1991. $75.00.

Laboratory Handbook of Materials, Equipment, and Techniques. Gary S. Coyne. Prentice Hall (Division of Simon and Schuster), 15 Columbus Circle, New York, NY 10023. (212) 373-8500 or (800) 922-0579. 1992. $45.00.

Lange's Handbook of Chemistry. John A. Dean, editor. McGraw-Hill Publishing Company, 1221 Avenue of the Americas, New York, NY 10020. (800) 262-4729 or (212) 512-3825. Fourteenth edition. 1992. $79.50.

Riegel's Handbook of Industrial Chemistry. James A. Kent, editor. Van Nostrand Reinhold, 115 Fifth Avenue, New York, NY 10003. (212) 254-3232 or (800) 926-2665. Ninth edition. 1992. $114.95.

ONLINE DATABASES AND CD-ROMS

Analytical Abstracts Online. Royal Society of Chemistry, Information Services, Thomas Graham House, Science Park, Milton Road, Cambridge, CB4 4WF, England. Contains citations, mostly with abstracts, of the worldwide literature on analytical chemistry, from 1980 to present. Available on DIALOG, (800) 334-2564, and STN International, FIZ Karlsruhe, P.O. Box 2465, W-7500, Karlsruhe 1, Germany, online services. Inquire as to cost and availability.

CA Search. Chemical Abstracts Service, P.O. Box 3012, Columbus, OH 43210-0012. (614) 447-3600. (800) 848-6533. FAX (614) 447-3709. Very comprehensive guide to worldwide chemical literature and related fields, 1972 to present. Available on BRS,(800) 289-4277, DIALOG, (800) 334-2564, ORBIT (800) 456-7248, and STN International, FIZ Karlsruhe, P.O. Box 2465, W-7500, Karlsruhe 1, Germany, online services. Inquire as to cost and availability.

Chemical Journals of the American Chemical Society. American Chemical Society, 1155 16th Street, N.W., Washington, DC 20036. (202) 872-4381 or (800) 424-6747. Contains complete text of approximately 90,000 articles from 22 primary journals published by the American Chemical Society, from mostly 1982 to present. Available on STN International, FIZ Karlsruhe, P.O. Box 2465, W-7500, Karlsruhe 1, Germany, online service. Inquire as to cost and availability.

Compendex Plus. Engineering Information, Inc., 345 East 47th Street, New York, NY 10017. (212) 705-7600 or (800) 221-1044. Contains citations with abstracts to worldwide literature in engineering and technology, from 1970 to present. Available on online BRS,(800) 289-4277, DIALOG, (800) 334-2564, ORBIT (800) 456-7248, and STN International, FIZ Karlsruhe, P.O. Box 2465, W-7500, Karlsruhe 1, Germany, online services. Also available on CD-ROM. Inquire as to cost and availability.

Current Contents Search. Institute for Scientific Information, 3501 Market Street, Philadelphia, PA 19104. (215) 386-0100. FAX (215) 386-6362. Contains citations to articles listed in the table of contents of science and technology journals. Also articles in social sciences and life sciences journals. Available on BRS,(800) 289-4277, DIALOG,(800) 334-2564, online services. Inquire as to cost and availability.

Dissertation Abstracts. University Microfilms International, 300 North Zeeb Road, Ann Arbor, MI 48106. (800) 521-0600 or (313) 761-4700. Scope includes virtually all doctoral dissertations accepted at accredited American institutions from 1861 to present in 252 subject areas. Available on BRS,(800) 289-4277, DIALOG,(800) 334-2564, and OCLC EPIC,(800) 848-5878, online services. Also available on CD-ROM. Inquire as to cost and availability.

Gmelin Database. Gmelin-Institut fur Anorganische Chemie und Grenzgebiete, Varrentrapperstrasse, 40-42, Carl-Bosch-Haus, D-6000, Frankfurt am Main 90, Germany. Contains structural and factual data relating to inorganic and organometallic chemistry. Provides data from the Gmelin Handbook of Inorganic and Organometallic Chemistry. Covers the period 1817 to 1975; 1988-89. Available on STN International, FIZ Karlsruhe, P.O. Box 2465, W-7500, Karlsruhe 1, Germany, online service. Inquire as to cost and availability.

Kirk-Othmer Encyclopedia of Chemical Technology. John Wiley and Sons, Inc., 605 Third Avenue, New York, NY 10158. (800) 526-5368 or (212) 850-6000. Contains the complete text of all chapters in the 27 volume fourth edition of the Kirk-othmer Encyclopedia of Chemical Technology. 1991. Available on BRS,(800) 289-4277, DIALOG,(800) 334-2564, online services. Inquire as to cost and availability.

IONIZATION

Ency. of Physical Sciences and Engineering Info. Sources

NTIS Bibliographic Database. National Technical Information Service, 5285 Port Royal Road, Springfield, VA 22161. (703) 487-4929 or FAX (703) 321-8199. Broad coverage of government-sponsored science and technology research reports, 1964 to present. Available on BRS,(800) 289-4277, DIALOG, (800) 334-2564, ORBIT, (800) 456-7248, and STN International, FIZ Karlsruhe, P.O. Box 2465, W-7500, Karlsruhe 1, Germany, online services. Also available on CD-ROM. Inquire as to cost and availability.

SCISEARCH. Institute for Scientific Information, 3501 Market Street, Philadelphia, PA 19104. (800) 523-1850 or (215) 386-0100. Broad multidisciplinary title and author index to the international literature of science and technology, 1974 to present. Available on DIALOG,(800) 334-2564, and ORBIT,(800) 456-7248, online services. Also available on CD-ROM. Inquire as to cost and availability.

WILSONLINE. H.W. Wilson Company, 950 University Avenue, Bronx, NY 10452. (800) 367-6770 or (212) 588-8400. Makes available online versions of the printed H.W. Wilson indexes including Applied Science and Technology Index, Business Periodicals Index, General Science Index, and Readers' Guide to Periodical Literature. Period covered is generally 1983 to present. Available on BRS,(800) 289-4277, DIALOG,(800) 334-2564, and OCLC EPIC,(800) 848-5878, online services. Also available on CD-ROM. Inquire as to cost and availability.

PERIODICALS

American Laboratory. American Laboratory, 30 Controls Drive, Box 870, Shelton, CT 06484-6111. (203) 926-9300. FAX (203) 926-9310. 1968 to present. Monthly. $210.00 per year. ISSN: 0044-7749.

Analytical Chemistry. American Chemical Society, Box 3337, Columbus, OH 43210. (614) 447-3776. 1929 to present. Semi-monthly. $415.00 per year. ISSN: 0003-2700.

Analytical Methods and Instrumentation. John Wiley and Sons, Inc., 605 Third Avenue, New York, NY 10158. (800) 526-5368 or (212) 850-6000. Bimonthly. $250.00 per year. ISSN: 1063-5246.

Angewandte Chemie. VCH Publishers, Inc., 220 East 23rd Street, New York, NY 10010-4606. (212) 683-8333. (800) 367-8249. 1888 to present. 22 issues per year. $840.00 per year. ISSN: 0044-8249.

Chemical Physics. Elsevier Science Publishing Co., Inc., Box 882, Madison Square Station, New York, NY 10159. (212) 989-5800. FAX (212) 633-3990. 1973 to present. 33 issues per year in 11 volumes. $2759.00 per year. ISSN: 0301-0104.

Chemical Physics Letters. Elsevier Science Inc., Publishing Co., Inc., Box 882, Madison Square Station, New York, NY 10159. (212) 989-5800. (212) 633-3990. 1967 to present. 102 issues per year in 17 volumes. $5063.00 per year. ISSN: 0009-2614.

Chemical Reviews. American Chemical Society, Box 3337, Columbus, OH 43210. (614) 447-3776. 1924 to present. 8 issues per year. $346.00 per year. ISSN: 0009-2665.

Chemical Week; Includes Annual Buyers Guide. Chemical Week Associates, 888 Seventh Avenue, New York, NY 10106. (212) 621-4900. FAX (212) 621-4949. 1914 to present. Weekly. $99.00 per year. ISSN: 0009-272X.

Chemtech. American Chemical Society. Box 3337, Columbus, OH 43210. (614) 447-3776. 1970 to present. Monthly. $370.00 per year. ISSN: 0009-2703.

Inorganic Chemistry. American Chemical Society, Box 3337, Columbus, OH 43210. (614) 447-3776. 1962 to present. Semi-monthly. $500.00 per year. ISSN: 0020-1669.

International Journal of Chemical Kinetics. John Wiley and Sons, Inc., 605 Third Avenue, New York, NY 10158. (800) 526-5368 or (212) 850-6000. 1968 to present. Monthly. $730.00 per year. ISSN: 0538- 8066.

Journal of the American Chemical Society. American Chemical Society, Box 3337, Columbus, OH 43210. (614) 447-3776. 1879 to present. Biweekly. $1055.00 per year. ISSN: 0002-7863.

Journal of Organic Chemistry. American Chemical Society, Box 3337, Columbus, OH 43210. (614) 447-3776. (800) 333- 9511. FAX (614) 447-3671. 1936 to present. Bi-weekly. $785.00 per year. ISSN: 0022-3263.

Journal of Physical Chemistry. American Chemical Society, Box 3337, Columbus, OH 43210. (614) 447-3776. (800) 333- 9511. FAX (614) 447-3671. 1896 to present. Weekly. $1140.00 per year to non-members. ISSN: 0894-3230.

RESEARCH CENTERS AND INSTITUTES

Barnett Institute of Chemical Analysis and Materials Science. Northeastern University, 360 Huntington Avenue, Boston, MA 02115. (617) 437-2864.

Chemical Laboratories. Harvard University, Oxford Street, Cambridge, MA 02138. (617) 495-4283. FAX (617) 496-5618.

Chemistry Laboratories. Rensselaer Polytechnic Institute, Cogswell Laboratory, Troy, NY 12180-3590. (518) 276-8981.

Laboratory For Research On the Structure of Matter. University of Pennsylvania, 3231 Walnut Street, Philadelphia, PA 19104. (215) 898-8571.

Research Program in Chemistry and Biochemistry. Southern Illinois University at Carbondale, Carbondale, IL 62901. (618) 453-5721.

Theoretical Chemistry Institute. University of Wisconsin, Madison. 1101 University Avenue, Madison, WI 53706. (608) 262-1511.

IONOSPHERE

See: METEOROLOGY, GEOPHYSICS

IONS

See: METEOROLOGY, GEOPHYSICS

IONS

See: IONIZATION

IRON

See: CAST IRON; STEEL AND STEEL MAKING

ISOMERS

See: ORGANIC CHEMISTRY

ISOTOPES

See: IONS

J

J PARTICLE

See: PARTICLE PHYSICS

JET PROPULSION

See also: AERODYNAMICS, AERONAUTICS, AERONAUTICAL ENGINEERING, AIRCRAFT ENGINES, AEROSPACE ENGINEERING

ABSTRACT SERVICES AND INDEXES

Aeronautical Engineering. Scientific and Technical Branch, National Aeronautics and Space Administration. National Technical Information Service (NTIS), 5285 Port Royal Road, Springfield, VA 22161. (703) 487-4650. FAX (703) 321-8547. Monthly. A selection of annotated references to unclassified reports and journal articles that were introduced into the NASA scientific and technical information system and announced in STAR and IAA.

Alloys Index. American Society for Metals, Metals Park, OH 44073. (216) 338-5151. FAX (216) 338-4634. 1974 to present. Monthly. $380.00. Also available on CD-ROM and online via METADEX, STN International, and DIALOG.

Applied Mechanics Reviews: An Assessment of World Literature in Engineering Sciences. From 1948 to present. American Society of Mechanical Engineers, 345 East 47th Street, New York, NY 10017. (212) 705-7703. Monthly. $360.00 per year.

Applied Science and Technology Index. A cumulative subject index to English language periodicals in the fields of Aeronautics and Space Science, Computer Technology, Chemistry, Construction Industry, Energy and related areas. H.W. Wilson Co., 950 University Avenue, Bronx, NY 10452. (800) 367-6770 or (212) 588-8400. FAX (718) 590-1617. From 1958 to present. Monthly. Inquire about cost and availability. Also available on CD-ROM.

Current Contents: Engineering, Technology, and Applied Sciences. Institute for Scientific Information, 3501 Market Street, Philadelphia, PA 19104. (215) 386-0100. FAX (215) 386-6362. Weekly. $360.00 per year.

Engineering Index Monthly. Engineering Information, Inc., Castle Point on the Hudson, Hoboken, NJ 07030. (800) 221-

1044. FAX (212) 832-1857. Monthly. $2200.00 per year. Also available online as COMPENDEX, and also on CD-ROM.

International Aerospace Abstracts. Technical Information Service, American Institute of Aeronautics and Astronautics, Inc., 555 West 57th St., New York, NY 10019. (212) 247-6500. FAX (212) 582-4861. Semi-monthly. $1295.00 per year.

Metals Abstracts and Metals Abstracts Index. American Society for Metals, Metals Park, OH 44073. (216) 338-5151. 1968 to present. Monthly. Abstracts are $1500.00 per year and Index is $460.00 per year.

Science Citation Index. Institute for Scientific Information, 3501 Market Street, Philadelphia, PA 19104. (215) 386-0100. FAX (215) 386-6362. Inquire about availability and cost. Also available on CD-ROM.

Scientific and Technical Aerospace Reports (STAR). National Aeronautics and Space Administration. NASA Center for Aerospace Information, Box 8757, BWI Airport, Baltimore, MD 21240. (301) 621-0153. Monthly. Inquire about availability and cost. ALso available through the NASA online retrieval service (RECON), and through the Aerospace Database through DIALOG.

ASSOCIATIONS AND PROFESSIONAL SOCIETIES

American Institute of Aeronautics and Astronautics, 370 L'Enfant Promenade SW, Washington, DC 20024-2518.

Jet Pioneers Association of the United States of America. c/o G.V. Henderson, 82 Bartholomew Street, Peabody, MA 01960. (508) 531-6559.

Society of Automotive Engineers. SAE, Inc., 400 Commonwealth Dr., Warrendale, PA 15096. (412) 776-4841.

BIBLIOGRAPHIES

Aeronautical Engineering: A Continuing Bibliography With Indexes. National Aeronautics and Space Administration, Center for Aerospace Information, Box 8757, BWI Airport, Baltimore, MD 21240. (301) 621-0153. Inquire for cost and availability.

JET PROPULSION

Ency. of Physical Sciences and Engineering Info. Sources

Scientific and Technical Books and Serials in Print. R.R. Bowker Inc., 121 Chanlon Road, New Providence, NJ 07974. (800) 521-8110 or (908) 464-6800. 3 volumes, annually. $285.00. Also available on CD-ROM.

DIRECTORIES AND BIOGRAPHICAL SOURCES

Research Centers Directory. Gale Research, 835 Penobscot Building, Detroit, MI 48226-4094. (313) 961-2242. (800) 347-4253. FAX (313) 961-6083. $485.00.

Who's Who in Engineering. American Association of Engineering Societies, 1111 19th St. NW, Suite 608, Washington, DC 20036. 7th ed. 1988. $200.00.

World Aviation Directory. McGraw-Hill, Aviation Week Group (NY), 1221 Avenue of the Americas, New York, NY 10020. (215) 237-4112. FAX (215) 586-3232. Semiannual. $250.00.

ENCYCLOPEDIAS AND DICTIONARIES

Dictionary of Aeronautical Terms. Dale Crane, comp. Aviation Supplies and Academics, Inc., 7005 132nd Place SE, Newcastle, WA 98059. (206) 235-1500 or (800) 426-8338. FAX (206) 235-0128. 1991. $15.95.

Encyclopedia of Physical Science and Technology. Robert A. Meyers, ed. Academic Press Inc., 6277 Sea Harbor Drive, Orlando, FL 32887. (800) 321-5068. 1992. $2100.00.

Jane's Aerospace Dictionary. B. Guston. 3d ed. Jane's Information Group, 1340 Braddock Place, Suite 300, Alexandria, VA 22314-1651. (703) 683-3700. FAX (703) 836-0029. 1991. $45.00.

McGraw-Hill Encyclopedia of Science and Technology, Sybil P. Parker, ed. 7th ed. 20 vols. McGraw-Hill Publishing Company, 1221 Avenue of the Americas, New York, NY 10020. (800) 262-4729 or (212) 512-3825. $1900.00

GENERAL WORKS

High-Speed Flight Propulsion Systems. S.N.B. Murthy and E.T. Curran. American Institute of Aeronautics and Astronautics, 370 L'Enfant Promenade, S.W., Washington, DC 20024. (202) 646-7400. 1991. $86.95.

The Jet Engine. 4th edition. Rolls-Royce Plc, Derby, England. 1986. Inquire for cost and availability.

Jet Engines and Jet Aircraft: A Source Guide. Gordon Press Publishers, PO Box 459, Bowling Green Station, New York, NY 10004. (718) 624-8419. 1991. $75.00.

Mechanics and thermodynamics of Propulsion. Philip G. Hill and Carl R. Peterson. 2d ed. Addison-Wesley Publishing Company, 1 Jacob Way, Reading, MA 01867. (617) 944-3700 or (800) 447-2226. FAX 617-942-1117. 1992. $72.25.

Modern Research Topics in Aerospace Propulsion. G. Angelino, L. de Luca, and W.A. Sirignano, eds. Springer-Verlag, 175 Fifth Avenue, New York, NY 10010. (212) 460-1500 or (800) 777-4643. FAX (212) 473-6272. 1991. $79.00.

ONLINE DATABASES AND CD-ROMS

Aerospace Database. American Institute of Aeronautics and Astronautics, 370 L'Enfant Promenage, S.W., Washington, DC 20024. (202) 646-7400. Worldwide published literature on research and development in aerospace and related areas, with abstracts. Covers 1962 to present. Available on DIALOG, (800) 334-2564, online service. Also available on CD-ROM. Inquire as to cost and availability.

Compendex Plus. Engineering Information, Inc., 345 East 47th Street, New York, NY 10017. (212) 705-7600 or (800) 221-1044. Contains citations with abstracts to worldwide literature in engineering and technology, from 1970 to present. Available on online BRS,(800) 289-4277, DIALOG, (800) 334-2564, ORBIT (800) 456-7248, and STN International, FIZ Karlsruhe, P.O. Box 2465, W-7500, Karlsruhe 1, Germany, online services. Also available on CD-ROM. Inquire as to cost and availability.

CSA Engineering. Cambridge Scientific Abstracts, 7200 Wisconsin Avenue, Suite 601, Bethesda, MD 20814. (301) 961-6750 or (800) 843-7751. Contains citations and abstracts of international periodicals and research literature covering all fields of engineering and science and technology, including computer and information science, electronics, mechanical engineering, solid state materials, 1981 to present. Available on BRS,(800) 289-4277, online service.Inquire as to cost and availability.

Current Contents Search. Institute for Scientific Information, 3501 Market Street, Philadelphia, PA 19104. (215) 386-0100. FAX (215) 386-6362. Contains citations to articles listed in the table of contents of science and technology journals. Also articles in social sciences and life sciences journals. Available on BRS,(800) 289-4277, DIALOG,(800) 334-2564, online services. Inquire as to cost and availability.

Dissertation Abstracts. University Microfilms International, 300 North Zeeb Road, Ann Arbor, MI 48106. (800) 521-0600 or (313) 761-4700. Scope includes virtually all doctoral dissertations accepted at accredited American institutions from 1861 to present in 252 subject areas. Available on BRS,(800) 289-4277, DIALOG,(800) 334-2564, and OCLC EPIC,(800) 848-5878, online services. Also available on CD-ROM. Inquire as to cost and availability.

International Aerospace Abstracts. American Institute of Aeronautics and Astronautics, 370 L'Enfant Promenade, S.W., Washington, DC 20024. (202) 646-7400. Contains references and abstracts of journal and monograph literature relating to aerospace science and technology, from 1963 to present. Available through the NASA/RECON system of the National Aeronautics and Space Administration only.

NASA Database. American Institute of Aeronautics and Astronautics, 370 L'Enfant Promenade, S.W., Washington, DC 20024. (202) 646-7400. Citations and abstracts of aeronautics and astronautics literature, 1962 to present. Also contains citations from STAR, Scientific and Technical Aerospace Reports, and International Aerospace Abstracts. Available through NASA/RECON online service. Inquire as to cost and availability.

NTIS Bibliographic Database. National Technical Information Service, 5285 Port Royal Road, Springfield, VA 22161. (703) 487-4929 or FAX (703) 321-8199. Broad coverage of government-sponsored science and technology research reports, 1964 to present. Available on BRS,(800) 289-4277, DIALOG, (800) 334-2564, ORBIT, (800) 456-7248, and STN International, FIZ Karlsruhe, P.O. Box 2465, W-7500, Karlsruhe 1, Germany,

online services. Also available on CD-ROM. Inquire as to cost and availability.

Scientific and Technical Books and Serials in Print. R.R. Bowker Inc., 121 Chanlon Road, New Providence, NJ 07974. (800) 521-8110 or (908) 464-6800. List of currently published books and serials in the physical and biological sciences, engineering and technology, with subject, author and titles indexes. Available on ORBIT,(800) 456-7248, online service.Inquire as to cost and availability.

SCISEARCH. Institute for Scientific Information, 3501 Market Street, Philadelphia, PA 19104. (800) 523-1850 or (215) 386-0100. Broad multidisciplinary title and author index to the international literature of science and technology, 1974 to present. Available on DIALOG,(800) 334-2564, and ORBIT,(800) 456-7248, online services. Also available on CD-ROM. Inquire as to cost and availability.

WILSONLINE. H.W. Wilson Company, 950 University Avenue, Bronx, NY 10452. (800) 367-6770 or (212) 588-8400. Makes available online versions of the printed H.W. Wilson indexes including Applied Science and Technology Index, Business Periodicals Index, General Science Index, and Readers' Guide to Periodical Literature. Period covered is generally 1983 to present. Available on BRS,(800) 289-4277, DIALOG,(800) 334-2564, and OCLC EPIC,(800) 848-5878, online services. Also available on CD-ROM. Inquire as to cost and availability.

PERIODICALS

Aeronautical Journal. The Royal Aeronautical Society, 4 Hamilton Place, London W1V 0BQ, England. 1897 to present. Monthly except June and August. Inquire for price information in U.S.

Aerospace America. American Institute of Aeronautics and Astronautics, 370 L'Enfant Promenade SW, Washington, DC 20024-2518. (202) 646-7471. 1932 to present. Monthly. $70 annually (non-members).

Aerospace Engineering. Society of Automotive Engineers, 400 Commonwealth Dr., Warrendale, PA 15096. (412) 776-4841. 1981 to present. Monthly. $48.00 per year.

Aerospace Propulsion. McGraw-Hill Inc., Aviation Week Group, 1200 G Street NW, Suite 200, Washington, DC 20005. (202) 383-2350. 1963 to present. 25 times a year. $550.00 a year.

AIAA Journal. American Institute of Aeronautics and Astronautics, 370 L'Enfant Promenade SW, Washington, DC 20024-2518. (202) 646-7471. 1963 to present. Monthly. $52.00 per year (members) and $435.00 per year (non-members).

Aircraft Engineering and Aerospace Technology. BunHill Publications Ltd., 127 Stanstead Rd., London SE23 1JE, England. 1929 to present. Monthly. $102.00 per year (U.S.).

Aviation Mechanics Bulletin. Flight Safety Foundation, Inc., 2200 Wilson Blvd., Suite 500, Arlington, VA 22201-3306. (703) 522-8300. FAX (703) 525-6047. 1953 to present. Bimonthly. $15.00 per year (non-members).

Aviation Week and Space Technology. 1221 Avenue of the Americas, New York, NY 10020. (212) 512-2999 or (800) 525-5003. 1916 to present. Weekly. $82.00 per year.

Canadian Aeronautics and Space Journal. Canadian Aeronautics and Space Institute, Suite 818, the National Building, 130 Slater St., Ottawa, Canada K1P 6E2. 1955 to present. Monthly. $75.00 per year (U.S.).

IEEE Transactions On Aerospace and Electronic Systems. Institute of Electrical and Electronics Engineers, 445 Hoes Lane, PO Box 1331, Piscataway, NJ 08855-1331. (908) 981-0060. 1965 to present. Quarterly. $10.00 per issue (members), $20.00 per issue (non-members).

International Journal of Turbo and Jet Engines. Freund Publishing House, Ltd., PO Box 35010, Tel Aviv 63150, Israel. 972-2-5662925. FAX 972-3-5605335. 1983 to present. Quarterly. $190.00 per year.

Journal of Aircraft. American Institute of Aeronautics and Astronautics, 370 L'Enfant Promenade SW, Washington, DC 20024-2518. 1963 to present. Bimonthly. $42.00 (members) and $230.00 (non-members).

Journal of Guidance, Control and Dynamics. American Institute of Aeronautics and Astronautics, 370 L'Enfant Promenade SW, Washington, DC 20024-2518. 1978 to present. Bimonthly. $42.00 (members) and $250.00 (non-members).

Journal of Propulsion and Power. American Institute of Aeronautics and Astronautics, 370 L'Enfant Promenade SW, Washington, DC 20024-2518. 1985 to present. Bimonthly. $38.00 (members) and $260.00 (non-members).

Propellants, Explosives, Pyrotechnics. VCH Publishers, Inc., 220 E. 23rd St., New York, NY 10010-4606. 6 times a year. $415.00 per year.

RESEARCH CENTERS AND INSTITUTES

Flight Research Laboratory, University of Kansas, Raymond Nichols Hall, Lawrence, KS 66045. (913) 864-3043. FAX (913) 864-7789.

Joint Institute For Advancement of Flight Sciences, George Washington University, Langley Research Center, Mail Stop 269, Hampton, VA 23665. (804) 864-1982. FAX (804) 864-5894.

Machinery and Engine Technology Laboratory, Institute For Mechanical Engineering, Bldg. M-7, Montreal Rd., Ottawa, Ontario, Canada K1A 0R6. (613) 993-2425. (613) 957-3281.

Pennsylvania State University, Propulsion Engineering Research Center, 106 Research Bldg. E, Bigler Rd., University Park, PA 16802. (814) 863-6272. FAX (814) 865-3389.

Purdue University, Thermal Sciences and Propulsion Center, Lafayette, IN 47907. (317) 494-1503.

JET STREAM

See also: METEOROLOGY, WEATHER

ABSTRACT SERVICES AND INDEXES

Applied Science and Technology Index. A cumulative subject index to English language periodicals in the fields of Aeronautics and Space Science, Computer Technology, Chemistry, Construction Industry, Energy and related areas. H.W. Wilson Co., 950 University Avenue, Bronx, NY 10452-9978. (212) 588-8400. (800) 367-6770. FAX (718) 590-1617. From 1958 to present. Monthly. Inquire about cost and availability. Also available online (BRS and WILSONLINE) and on CD-ROM. ISSN: 0003-6986.

General Science Index. H.W. Wilson Co., 950 University Avenue, Bronx, NY 10452. (212) 588-8400. (800) 367-6770. FAX (718) 590-1617. From 1978 to present. Ten issues per year; quarterly and annual cumulations. Service basis. Available on CD-ROM and online. Inquire regarding cost and availability. ISSN: 0162-1963.

Government Reports Announcements and Index. National Technical Information Service (NTIS), 5285 Port Royal Road, Springfield, VA 22161. (703) 487-4650. FAX (703) 321-8547. From 1968 to present. Annual. $630.00 per year. Also available online as *NTIS Bibliographic Database* and on CD-ROM. ISSN:

Meteorological and Geoastrophysical Abstracts. American Meteorological Society, c/o Inforonics, Inc., 550 Newtown Road, Box 458, Littleton, MA 01460. (508) 486-8976. FAX (508) 486-0027. Covers literature in environmental sciences, meteorology, astrophysics, hydrology, glaciology, and physical oceanography. 1950 to present. Monthly. $950.00 per year. ISSN: 0026-1130. Also available online (DIALOG) and on CD-ROM.

Oceanic Abstracts. Cambridge Scientific Abstracts, 7200 Wisconsin Avenue, Bethesda, MD 20814. (301) 961-6750. Fax (301) 961-6720. Bimonthly. $995.00 per year.

Science Citation Index (SCI). Institute for Scientific Information, 3501 Market Street, Philadelphia, PA 19104. (215) 386-0100. (800) 523-1850. FAX (215) 386-2991. 1961 to present. Six issues per year, plus annual cumulation. $11650.00 per year. Also available online and on CD-ROM. Inquire about price and availability. ISSN: 0036-827X.

ANNUAL REVIEWS AND YEARBOOKS

Annual Review and Earth and Planetary Sciences. Annual Reviews, Inc., 4139 El Camino Way, Palo Alto, CA 94306-0897. (415) 493-4400. Fax (415) 855-9815. 1973 to present. Annual. $65.00 per year.

Developments in Atmospheric Science Series. Elsevier Science Publishing Company, Inc., 655 Avenue of the Americas, New York, NY 10010. (212) 989-5800. Irregular. Inquire about price and availability.

ASSOCIATIONS AND PROFESSIONAL SOCIETIES

American Geophysical Union, 2000 Florida Avenue, N.W., Washington, DC 20009. (202) 462-6900.

American Meteorological Society, 45 Beacon Street, Boston, MA 02108-3693. (617) 227-2425. FAX (617) 742-8718.

International Association of Meteorology and Atmospheric Physics. c/o UCAR, P.O. Box 3000, Boulder CO 80307-3000. (303) 497-1673.

National Environmental Satellite Data, and Information Service. 2069 Federal Building 4 Room 2060, Washington, DC 20233. (301) 763-71990. FAX (301) 763-4011.

National Weather Association, 6704 Wolke Court, Montgomery, AL 36116-2134. (334) 213-0388. FAX (334) 213-0388.

University Corporation For Atmospheric Research. P.O. Box 3000, 1850 Table Mesa Drive, Boulder, CO 80307-3000. (303) 497-1673. FAX (303) 497-1654.

BIBLIOGRAPHIES

Information Sources in the Earth Sciences. David N. Wood, Joan E. Hardy, and Anthony P. Harvey, editors. Bowker-Saur/K.G. Saur. Distributed by R.R. Bowker, 121 Chanlon Road, New Providence, NJ 07974. (800) 521-8110 or (908) 464-6800. Second edition. 1989. $85.00.

Scientific and Technical Books and Serials in Print: An Index to Literature in Science and Technology. R.R. Bowker Co., 205 E. 42nd Street, New York, NY 10017. (800) 521-8110 or (212) 916-1600. 1992. $250.00.

DIRECTORIES AND BIOGRAPHICAL SOURCES

American Men and Women of Science: Physical and Biological Sciences. Eighteenth edition. R.R. Bowker Company, 245 West 17th Street, New York, NY 10011. (800) 521-8810 or (212) 916- 1600. $750.00.

Meteorological Services of the World. World Meteorological Organization, 45 Beacon Street, Boston, MA 02108. (617) 227-2425. Annual. $70.00.

National Weather Service offices and Stations. National Oceanic and Atmospheric Administration, Department of Commerce, Silver Spring, MD 20910. (301) 427-7698. Annual. Free.

Research Centers Directory. Gale Research, 835 Penobscot Building, Detroit, MI 48226-4094. (800) 347-4253 or (313) 961-2242. 18th edition. 1995. $485.00.

ENCYCLOPEDIAS AND DICTIONARIES

Concise Oxford Dictionary of the Earth Sciences. Ailsa Allaby and Michael Allaby, editors. Oxford University Press, Inc., 200 Madison Avenue, New York, NY 10016. (800) 334-4249 or (212) 679-7300. 1990. $42.95.

Encyclopedia of Climatology. John E. Oliver and Rhodes W. Fairbridge, editors. Chapman & Hall, 1 Penn Plaza, New York, NY 10119. (212) 564-1060. 1986. $115.00.

Handbook in Applied Meteorology. David D. Houghton. John Wiley & Sons, Inc. 605 Third Avenue, New York, NY 10158-0012. (212) 850-6000. (800) 225-5945. 1985. $168.00.

McGraw-Hill Encyclopedia of Ocean and Atmospheric Sciences. Sybil P. Parker, editor. McGraw-Hill Publishing Company, 1221 Avenue of the Americas, New York, NY 10020. (212) 512-2000. (800) 262-4729. 1979. $79.95.

McGraw-Hill Encyclopedia of Science and Technology. McGraw-Hill Publishing Company, 1221 Avenue of the Americas, New York, NY 10020. (800) 262-4729 or (212) 512-2000. Seventh edition. 1992. $1900.00.

Magill's Survey of Science: Earth Science Series. Salem Press, Inc., P.O. Box 1097, Englewood Cliffs, NJ 07632. (800) 221-1592 or (201) 871-3700. Five volumes. 1990. $400.00 for the set.

GENERAL WORKS

The Atmosphere: An Introduction to Meteorology. Frederick K. Lutgens and Edward J. Tarbuck. Prentice Hall, 15 Columbus Circle, New York, NY 10023. (212) 373-8500. (800) 223-2348. Sixth edition. 1994. $49.00.

Atmospheric Data Analysis. Roger Daly. Cambridge University Press, 40 West 20th Street, New York, NY 10011-4211. (212) 924-3900. 1991. $79.50.

Fundamentals of Atmospheric Dynamics and Thermodynamics. C. Riegel. World Scientific Publishing Company, Inc., 1060 Main Street, River Edge, NJ 07661. (800) 227-7562 or (201) 487-9655. 1992. $59.00.

The Jet Stream. H. Riehl, et el. American Meteorological Society, 42 Beacon Street, Boston, MA 02108-3693. (617) 227-2425. 1954. $17.00 in paper.

Jet-Stream Meteorology. Elmar R. Reiter. Books on Demand, 300 North Zeeb Street, Ann Arbor, MI 48106-1346. (313) 761-4700. (800) 521-0600. $150.80 in paper.

Jet Streams: How Do they Affect Our Weather? Elmar R. Reiter. Greenwood Press, 88 Post Road, West, Box 5007, Westport, CT 06881. (800) 225-5800. 1979. $35.00.

Mesoscale Meteorology and Forecasting. Peter S. Ray, editor. American Meteorological Society, 45 Beacon Street, Boston, MA 02108-3693. (617) 227-2425. FAX (617) 742-8718. 1986. $66.25.

Meteorology: The Atmosphere and the Science of Weather. Joseph M. Moran. Macmillan Publishing, 866 3rd Avenue, New York, NY 10024. (212) 689-9140. Fourth edition. 1994. $49.95.

Severe and Unusual Weather. Joe R. Eagleman. Trimedia Publishing Company, 12008 West 87th Street, Suite 117, Lenexa, KS 66215. (913) 599-0505. Second edition. 1990. $41.95.

Weather Cycles: Real or Imaginary? William James Burroughs. Cambridge University Press, 40 West 20th Street, New York, NY 10011-4211. (212) 924-3900. 1992. $39.95.

HANDBOOKS AND MANUALS

Author's Guide to the Journals of the American Meteorological Society. American Meteorological Society, 45 Beacon Street, Boston, MA 02108-3693. (617) 227-2425. FAX (617) 742-8718. 1983. $15.00 in paper.

Handbook in Applied Meteorology. David D. Houghton, editor. John Wiley and Sons, Inc., 605 Third Avenue, New York, NY 10158. (800) 526-5368 or (212) 850-6000. 1985. $168.00.

ONLINE DATABASES AND CD-ROMS

Climate Assessment Database. National Weather Service, National Meteorological Center, 5200 Auth Road, Suite 101, Camp Springs, MD 20233. (301) 763-8016. Contains daily, weekly and monthly summaries of North American and world climatological data. Also provides five to ten weather forecasts, and 30 to 90 day weather outlook. Subscription required. Inquire as to cost and availability.

Current Contents Search. Institute for Scientific Information, 3501 Market Street, Philadelphia, PA 19104. (215) 386-0100. FAX (215) 386-6362. Contains citations to articles listed in the table of contents of science and technology journals. Also articles in social sciences and life sciences journals. Available on BRS,(800) 289-4277, DIALOG,(800) 334-2564, online services. Inquire as to cost and availability.

International Aerospace Abstracts. American Institute of Aeronautics and Astronautics, 370 L'Enfant Promenade, S.W., Washington, DC 20024. (202) 646-7400. Contains references and abstracts of journal and monograph literature relating to aerospace science and technology, from 1963 to present. Available through the NASA/RECON system of the National Aeronautics and Space Administration only.

Meteorological and Geoastrophysical Abstracts. American Meteorological Society, 45 Beacon Street, Boston, MA 02108-3693. (617) 227-2425. FAX (617) 742-8718. Contains citations and abstracts to the worldwide literature on significant research in meteorology and geoastrophysics. Related topics include physical oceanography, hydrology, environmental sciences and glaciology. Covers the period 1972 to present. Available on DIALOG,(800) 334-2564, online service. Inquire as to cost and availability.

NTIS Bibliographic Database. National Technical Information Service, 5285 Port Royal Road, Springfield, VA 22161. (703) 487-4929 or FAX (703) 321-8199. Broad coverage of government-sponsored science and technology research reports, 1964 to present. Available on BRS,(800) 289-4277, DIALOG, (800) 334-2564, ORBIT, (800) 456-7248, and STN International, FIZ Karlsruhe, P.O. Box 2465, W-7500, Karlsruhe 1, Germany, online services. Also available on CD-ROM. Inquire as to cost and availability.

SCISEARCH. Institute for Scientific Information, 3501 Market Street, Philadelphia, PA 19104. (800) 523-1850 or (215) 386-0100. Broad multidisciplinary title and author index to the international literature of science and technology, 1974 to present. Available on DIALOG and ORBIT online services. Also available on CD-ROM. Inquire as to cost and availability.

WILSONLINE. H.W. Wilson Company, 950 University Avenue, Bronx, NY 10452. (800) 367-6770 or (212) 588-8400. Makes available online versions of the printed H.W. Wilson indexes including Applied Science and Technology Index, Business Periodicals Index, General Science Index, and Readers' Guide to Periodical Literature. Period covered is generally 1983 to present. Available on BRS,(800) 289-4277, DIALOG,(800) 334-2564, and OCLC EPIC,(800) 848-5878, online services. Also available on CD-ROM. Inquire as to cost and availability.

World Weatherdisc. WeatherDisc Associates, Inc., 4584 N.E. 89th Street, Seattle, WA 98115. (206) 524-4314. FAX (206) 543-0308. Meteorological data on CD-ROM which describes the climate of the earth today and for the past few hundred years. First edition 1989. Approximately $295.00 per year with biannual updates.

PERIODICALS

American Meteorological Society Bulletin. American Meteorological Society, 45 Beacon Street, Boston, MA 02108-3693. (617) 227-2425. FAX (617) 742-8718. Monthly. $60.00 per year.

American Weather Observer. Association of American Weather Observers, 401 Whitney Boulevard, Box 455, Belvedere, IL 61008. (815) 544-5665. FAX (815) 544-6334. Monthly. $21.00 per year.

Climatic Change: Atmosphere-Ocean. Canadian Meteorological and Oceanographic Society, P.O. Box 334, Newmarket, Ontario, L3Y 4X7, Canada. (416) 898-1040. FAX (416) 898-7937. Quarterly. $30.00 per year.

Dynamics of Atmospheres and Oceans. Elsevier Science Publishing Company, Inc., 655 Avenue of the Americas, New York, NY 10010. (212) 989-5800. Six times per year. $205.00 per year.

Earth. Kalmbach Publishing Company, P.O. Box 1612, Waukesha, WI 53187. (414) 796-0126 or (800) 558-1544. 1991 to present. Bimonthly. $14.95 per year.

JGR: Journal of Geophysical Research: Atmosphere. American Geophysical Union, 2000 Florida Avenue, N.W., Washington, DC 20009. (202) 462-6903. Monthly. $90.00 per year to members.

Journal of Applied Meteorology. American Meteorological Society, 45 Beacon Street, Boston, MA 02108-3693. (617) 227-2425. FAX (617) 742-8718. Monthly. $165.00 per year.

Journal of the Atmospheric Sciences. American Meteorological Society, 45 Beacon Street, Boston, MA 02108-3693. (617) 227-2425. FAX (617) 742-8718. Semi-monthly. $320.00 per year.

Monthly Weather Review. American Meteorological Society, 45 Beacon Street, Boston, MA 02108-3693. (617) 227-2425. FAX (617) 742-8718. Monthly. $205.00 per year.

National Weather Digest. National Weather Association, 4400 Stamp Road, Room 404, Temple Hills, MD 20748. (301) 899-3784.

Royal Meteorological Society Quarterly Journal. Royal Meteorological Society, 104 Oxford Road, Reading, Berks RG1 7LJ, England. Six times per year. $250.00 per year.

Weather. Royal Meteorological Society, 104 Oxford Road, Reading, Berks RG1 7LJ, England. Monthly. $44.00 per year.

Weatherwise. Heldref Publications, 1319 Eighteenth Street, N.W., Washington, DC 20036-1802. (202) 296-6267. FAX (202) 296-5149. Bi-monthly. $28.00 per year.

RESEARCH CENTERS AND INSTITUTES

Cooperative Institute For Mesoscale Meteorological Studies. University of Oklahoma, 401 East Boyd, Norman, OK 73019. (405) 325-3041.

National Center For Atmospheric Research. P.O. Box 3000, Boulder, CO 80307. (303) 496-1000. FAX (303) 497-1654.

National Severe Storms Forecast Center. 601 East 12th Street, Kansas City, MO 64106. (816) 374-5922.

JEWEL BEARINGS

See: BEARINGS AND BALL BEARINGS

JOINTS

See also: BUILDING MATERIALS, CIVIL ENGINEERING, CONSTRUCTION ENGINEERING, STRESS AND STRAIN, STRUCTURAL ENGINEERING

ABSTRACT SERVICES AND INDEXES

Applied Mechanics Reviews: An Assessment of World Literature in Engineering Sciences. 1948-present. American Society of Mechanical Engineers, 345 East 47th Street, New York, NY 10017. (212) 705-7703. Monthly. $360.00 per year.

Applied Science and Technology Index. A cumulative subject index to English language periodicals in the fields of Aeronautics and Space Science, Computer Technology, Chemistry, Construction Industry, Energy and related areas. H.W. Wilson Co., 950 University Avenue, Bronx, NY 10452. (800) 367-6770 or (212) 588-8400. FAX (718) 590-1617. From 1958 to present. Monthly. Inquire about cost and availability. Also available on CD-ROM and online.

ASCE Combined Annual Combined Index. American Society of Civil Engineers, 345 East 47th Street, New York, NY 10017-2398. (212) 705-7520 or (800) 548-2723. Annual. $48.00.

ASCE Publications Information. 1966-present. American Society of Civil Engineers, 345 East 47th Street, New York, NY 10017-2398. (212) 705-7520 or (800) 548-2723. Bi-monthly. $160.00 per year for non-members.

Civil and Structural Engineering Abstracts, Cambridge Scientific Abstracts, 7200 Wisconsin Avenue, Bethesda, MD 20814-4823. (301) 961-6750. FAX (301) 961-6720. 1993 to present. Monthly. $385.00 per year. Topics covered include structural design, construction equipment and methods, civil defense and military engineering, surveying, highway engineering, maritime and port structures, materials, land reclamation, and soil mechanics.

Current Contents: Engineering, Technology, and Applied Sciences. Institute for Scientific Information, 3501 Market Street, Philadelphia, PA 19104. (215) 386-0100. FAX (215) 386-6362. 1970 to present. Weekly. $442.00 per year.

Engineering Index Monthly. Engineering Information, Inc., Castle Point on the Hudson, Hoboken, NJ 07030. (800) 221-1044. FAX (212) 832-1857. Monthly. $2300.00 per year. Also available online as COMPENDEX, and also on CD-ROM. Covers chemical engineering, computers, electrical engineering, civil engineering, metals and mining, industrial management, and mechanical engineering.

Mechanical Engineering Abstracts (formerly *ISMEC*), Cambridge Scientific Abstracts, 7200 Wisconsin Avenue, Bethesda, MD 20814-4823. (301) 961-6750. FAX (301) 961-6720. 1967 to present. Monthly. $895.00 per year. Summarizes world literature in mechanical engineering, production engineering, and engineering management. Also available online.

ASSOCIATIONS AND PROFESSIONAL SOCIETIES

American Society of Civil Engineers. 345 East 47th Street, New York, NY 10017-2398. (212) 705-7520 or (800) 548-2723.

American Society of Mechanical Engineers. 345 East 47th Street, New York, NY 10017-2398. (212) 705-7722.

Expansion Joint Manufacturers Association. 25 N. Broadway, Tarrytown, NY 10591-3201. (202) 842-0292. FAX (212) 842-4840.

Research Council On Structural Connections. Sagent & Lyndy Engineers, 55 E. Monroe Street, Chicago, IL 60603. (312) 269-2424.

Society For Experimental Mechanics. 7 School Street, Bethel, CT 06801. (203) 790-6373.

DIRECTORIES AND BIOGRAPHICAL SOURCES

Research Centers Directory. Gale Research, 835 Penobscot Building, Detroit, MI 48226-4094. (313) 961-2242. (800) 347-4253. FAX (313) 961-6083. $485.00.

Scientific and Technical Organizations and Agencies Directory. Gale Research, 835 Penobscot Building, Detroit, MI 48226-4094. (313) 961-2242. (800) 347-4253. FAX (313) 961-6083. 4th edition. 1996. $195.00.

Who's Who in Engineering. American Association of Engineering Societies, 1111 19th Street NW, Suite 608, Washington, DC 20036. (202) 296-2237 or (800) 658-8897. 8th edition. 1991. Inquire for price.

ENCYCLOPEDIAS AND DICTIONARIES

McGraw-Hill Encyclopedia of Engineering. Sybil P. Parker, ed. 2nd edition. McGraw-Hill Publishing Company, 1221 Avenue of the Americas, New York, NY 10020. (800) 262-4729 or (212) 512-3825. 1993. $95.50.

GENERAL WORKS

Couplings and Joints: Design, Selection, and Application. Jon R. Mancuso. Marcel Dekker, Inc., 270 Madison Avenue, New York, NY 10016. (212) 696-9000. FAX (212) 685-4540. 1986. $125.00.

Design of Mechanical Joints. Alexander Blake. Marcel Dekker, Inc., 270 Madison Avenue, New York, NY 10016. (212) 696-9000. FAX (212) 685-4540. 1985. Inquire for cost and availability.

Joining of Advanced Materials. Robert W. Messler Jr. Butterworth-Heinemann, 313 Washington Street, Newton, MA 02158. (617) 928-2500 or (800) 366-2665. FAX (617) 928-2620. 1993. $125.00.

Structural Connections: Stability and Strength. R. Narayanan, editor. Elsevier Science B.V., PO Box 945, Madison Square Station, New York, NY 10159-0945. (212) 633-3650. FAX (212) 633-3680. 1989. $117.00.

HANDBOOKS AND MANUALS

Standard Handbook of Fastening and Joining. Robert O. Parmley, editor-in-chief. 2nd edition. McGraw-Hill Publishing Company, 1221 Avenue of the Americas, New York, NY 10020. (800) 262-4729 or (212) 512-3825. 1989. $79.50.

ONLINE DATABASES AND CD-ROMS

Compendex Plus. Engineering Information, Inc., 345 East 47th Street, New York, NY 10017. (212) 705-7600 or (800) 221-1044. Contains citations with abstracts to worldwide literature in engineering and technology, from 1970 to present. Available on online BRS,(800) 289-4277, DIALOG, (800) 334-2564, ORBIT (800) 456-7248, and STN International, FIZ Karlsruhe, P.O. Box 2465, W-7500, Karlsruhe 1, Germany, online services. Also available on CD-ROM. Inquire as to cost and availability.

CSA Engineering. Cambridge Scientific Abstracts, 7200 Wisconsin Avenue, Suite 601, Bethesda, MD 20814. (301) 961-6750 or (800) 843-7751. Contains citations and abstracts of international periodicals and research literature covering all fields of engineering and science and technology,including computer and information science, electronics, mechanical engineering, solid state materials, 1981 to present. Available on BRS,(800) 289-4277, online service.Inquire as to cost and availability.

ISMEC: Mechanical Engineering Abstracts. Cambridge Scientific Abstracts, 7200 Wisconsin Avenue, Suite 601, Bethesda, MD 20814. (301) 961-6750 or (800) 843-7751. Contains citations to the literature in mechanical engineering, industrial and production engineering, energy, power, mechanics, devices and related areas, from 1973 to present. Available on the DIALOG,(800) 334-2564, online service. Inquire as to cost and availability.

NTIS Bibliographic Database. National Technical Information Service, 5285 Port Royal Road, Springfield, VA 22161. (703) 487-4929 or FAX (703) 321-8199. Broad coverage of government-sponsored science and technology research reports, 1964 to present. Available on BRS,(800) 289-4277, DIALOG, (800) 334-2564, ORBIT, (800) 456-7248, and STN International, FIZ Karlsruhe, P.O. Box 2465, W-7500, Karlsruhe 1, Germany, online services. Also available on CD-ROM. Inquire as to cost and availability.

SCISEARCH. Institute for Scientific Information, 3501 Market Street, Philadelphia, PA 19104. (800) 523-1850 or (215) 386-0100. Broad multidisciplinary title and author index to the international literature of science and technology, 1974 to present. Available on DIALOG,(800) 334-2564, and ORBIT,(800) 456-

7248, online services. Also available on CD-ROM. Inquire as to cost and availability.

WILSONLINE. H.W. Wilson Company, 950 University Avenue, Bronx, NY 10452. (800) 367-6770 or (212) 588-8400. Makes available online versions of the printed H.W. Wilson indexes including Applied Science and Technology Index, Business Periodicals Index, General Science Index, and Readers' Guide to Periodical Literature. Period covered is generally 1983 to present. Available on BRS,(800) 289-4277, DIALOG,(800) 334-2564, and OCLC EPIC,(800) 848-5878, online services. Also available on CD-ROM. Inquire as to cost and availability.

PERIODICALS

Engineering Fracture Mechanics. Elsevier Science, 660 White Plains Rd., Tarrytown, NY 10159-5153. (919) 524-9200. FAX (919) 333-2444. 1968 to present. 18 times a year. $1945.00 per year.

Fatigue and Fracture of Engineering Materials and Structures. Structural Integrity Research Institute, University of Sheffield, Sheffield S1 3JD, England. Telephone 0742-825239. FAX 0742-753671. 1979 to present. Monthly. Inquire for cost.

Journal of Engineering Mechanics. American Society of Civil Engineers, Engineering Mechanics Division, 345 East 47th Street, New York, NY 10017-2398. (212) 705-7520 or (800) 548-2723. 1956 to present. Monthly. $284.00 per year.

Journal of Structural Engineering. American Society of Civil Engineers, 345 East 47th Street, New York, NY 10017-2398. (212) 705-7520 or (800) 548-2723. 1956 to present. Monthly. $300.00 per year for non-members.

Mechanical Engineering. American Society of Mechanical Engineers, 345 E. 47th Street, New York, NY 10017. (212) 705-7722. 1906 to present. Monthly. $45.00 for non-members.

Structural Engineer. Structural Engineers Trading Organization Ltd., 11 Upper Belgrave Street, London SW1X 8BH, England. Telephone 071-235-4535. FAX 071-235-4294. 1922 to present. 24 times a year. Inquire for cost.

Structural Engineering International. International Association for Bridge and Structural Engineering, ETH-Hoenggerberg, CH-8093 Zurich, Switzerland. Telephone 01-3772647. FAX 01-3712131. Quarterly. Inquire for cost and availability.

Structural Engineering Reviews. Elsevier Science [journals], 660 White Plains Rd., Tarrytown, NY 10159-5153. (919) 524-9200. FAX (919) 333-2444. 1988 to present. 4 times a year. $190.00 per year.

JUPITER

See also: ASTROGEOLOGY, ASTEROIDS, ASTRONOMY, COMETS, METEORITES, PLANETARY SCIENCE, SATELLITES (NATURAL), SEISMOLOGY, SOLAR SYSTEM, VOLCANOLOGY

ABSTRACT SERVICES AND INDEXES

Applied Science and Technology Index. A cumulative subject index to English language periodicals in the fields of Aeronau-

tics and Space Science, Computer Technology, Chemistry, Construction Industry, Energy and related areas. H. W. Wilson Co., 950 University Avenue, Bronx, NY 10452-9978. (212) 588-8400. (800) 367-6770. FAX (718) 590-1617. From 1958 to present. Monthly. Inquire about cost and availability. Also available online (BRS and WILSONLINE) and on CD-ROM. ISSN: 0003-6986.

Astronomy and Astrophysics Abstracts. Springer-Verlag New York, 175 Fifth Avenue, New York, NY 10010. (212) 460-1500. FAX (212) 473-6272. Published for the Astronomisches Rechen-Institut. Comprehensive coverage of all aspects of astronomy, astrophysics and related fields. 1969 to present. Two parts per year. Annual. ISSN: 0067-0022.

General Science Index. H.W. Wilson Company, 950 University Avenue, Bronx, NY 10452. (212) 588-8400. (800) 367-6770. FAX (718) 590-1617. From 1978 to present. Ten issues per year; quarterly and annual cumulations. Service basis. Available on CD-ROM and online. Inquire regarding cost and availability. ISSN: 0162-1963.

Meteorological and Geoastrophysical Abstracts. American Meteorological Society, c/o Inforonics, Inc., 550 Newtown Road, Box 458, Littleton, MA 01460. (508) 486-8976. FAX (508) 486-0027. Covers literature in environmental sciences, meteorology, astrophsyics, hydrology, glaciology and physical oceanography. 1950 to present. Monthly. $950.00 per year. Also available online (DIALOG) and on CD-ROM. ISSN: 0026-1130.

NTIS Alerts: Astronomy & Astrophysics. U. S. National Technical Information Service. 5285 Port Royal Road, Springfield, VA 22161. (703) 487-4650. FAX (703) 321-8547. Weekly. $140.00 per year.

Science Citation Index (SCI). Institute for Scientific Information, 3501 Market Street, Philadelphia, PA 19104. (215) 386-0100. (800) 523-1850. FAX (215) 386-2991. 1961 to present. Six issues per year, plus annual cumulation. $11650.00 per year. Also available online and on CD-ROM. Inquire about price and availability. ISSN: 0036-827X.

STAR (Scientific and Technical Aerospace Reports). U.S. National Aeronautics and Space Administration. Distributed by U. S. Superintendent of Documents, Washington, DC 20402. 1963 to present. Semi-monthly, with semiannual and annual indexes. $114.00 per year. ISSN: 0036-8741. Also available online and on CD-ROM.

ANNUAL REVIEWS AND YEARBOOKS

The Astronomical Almanac. Superintendent of Documents, U. S. Government Printing office, Washington, DC 10402. (202) 783-3238. 1981 to present. Supercedes Astronomical Ephemeris and the American Ephermis and Nautical Almanac. Annual. $44.95. ISSN: 0737-6421.

Annual Review of Astronomy and Astrophysics. Original reviews of critical literature and current developments in astronomy and astrophysics. Annual Reviews, Inc., 4139 El Camino Way, Palo Alto, CA 94303-0139. (415) 493-4400. (800) 523-8635. Fax (415) 855-9815. 1963 to date. Annual. $60.00. ISSN: 0066-4146.

Annual Review of Earth and Planetary Sciences. Annual Reviews, Inc., 4139 El Camino Way, Palo Alto, CA 94303-0139. (415) 493-4400. (800) 523-8635. Fax (415) 855-9815. 1973 to date. Annual. $62.00. ISSN: 0084-6597.

Proceedings of the Lunar and Planetary Science Conference. Lunar and Planetary Institute, 3600 Bay Area Boulevard, Houston, TX 77058. (713) 486-2143. 1970 to date. Annual. ISSN: 0270-9511.

ASSOCIATIONS AND PROFESSIONAL SOCIETIES

American Astronomical Society. 2000 Florida Avenue NW, Suite 400, Washington, DC 20009. (202) 328-2010. FAX: (202) 234-2560.

American Geophysical Union. 2000 Florida Avenue NW, Washington, DC 20009. (202) 462-6900. (800) 966-AGU1. FAX (202) 328-0566.

American Institute of Physics. One Physics Ellipse, College Park, MD 20740-3843. (301) 209-3100.

Association of Universities For Research in Astronomy, Inc. (AURA). Suite 701, 1625 Massachusetts Avenue, NW, Washington, DC 20036. (202) 483-2101. FAX (202) 483-2106.

Astronomical Society of the Pacific. 390 Ashton Avenue, San Francisco, CA 94112. (415) 337-1100. FAX: (415) 337-5205.

Lunar and Planetary Institute. 3600 Bay Area Boulevard, Houston, TX 77058. (713) 486-2143.

Planetary Society. 65 North Catalina Avenue, Pasadena, CA 91106. (818) 793-5100. (800) WOW-MARS. FAX (818) 793-5528.

BIBLIOGRAPHIES

A Bibliography of Astronomy: 1970-1979. R. A. Sea and S. S. Martin. Libraries Unlimited, Inc., Littleton, CO 80160. 1982. $37.50.

Science Books & Films. American Association for the Advancement of Science, 1333 H Street NW, Washington, DC 20005. (202) 326-6434. Reviews of print, film and software materials in all sciences.. 1965 to present. Nine issues per year. $40.00 per year. ISSN: 0098-342X.

Scientific and Technical Books and Serials in Print: An Index to Literature in Science and Technology. R. R. Bowker. 121 Chanlon Road, New Providence, NJ 07974. (908) 464-6800. FAX (908) 665-3502. (800) 521-8110. 1972 to present. Annual. $299.95 per year. Also available on CD-ROM. ISSN: 0000-054X.

DIRECTORIES AND BIOGRAPHICAL SOURCES

American Astronomical Society: Membership Directory. American Astronomical Society. 2000 Florida Avenue NW, Suite 400, Washington, DC 20009. (202) 328-2010. FAX (202) 234-2560 Annual. Included in membership dues. ISSN: 1061-9038.

American Men and Women of Science: Physical and Biological Sciences. R. R. Bowker Inc., 121 Chanlon Road, New Providence, NJ 07974. (908) 464-6800. (800) 521-8110. 20th edition. 8 volumes. 1996. $850.00.

The Astronomers. Donald Goldsmith. St. Martin's Press, Inc., 175 Fifth Avenue, New York, NY 10010. (212) 674-5151. (800) 221-7945. 1993. $14.95 in paper.

Astronomical Centers of the World. Kevin Krisciunas. Cambridge University Press, 40 West 20th Street, New York, NY 10011-4211. (212) 924-3900. (800) 872-7423. 1988. $34.95

Directory of Physics and Astronomy Staff. American Institute of Physics, One Physics Ellipse, College Park, MD 20740-3843. (301) 209-3100. 1975/76 to present. Annual. $60.00. ISSN: 0361-2228.

Earth and Astronomical Research Centers. Stockton Press, 345 Park Avenue South, New York, NY 10010. 4th edition. 1995. $515.00. ISBN: 1-56169-0967.

Graduate Programs in Physics, Astronomy and Related Fields. American Institute of Physics, One Physics Ellipse, College Park, MD 20740-3843. (301) 209-3100. 1978 to present. Annual. $45.00. ISSN: 0147-1821.

Research Centers Directory. Gale Research, 835 Penobscot Building, Detroit, MI 48226-4094. (313) 961-2242. (800) 347-4253. 20th edition. 1995. $485.00.

ENCYCLOPEDIAS AND DICTIONARIES

Concise Dictionary of Astronomy. Jacqueline Mitton. Oxford University Press, Inc., 200 Madison Avenue, New York, NY 10016. (212) 725-6000. (800) 334-4249. 1992. $24.95.

Encyclopedia of Astronomy and Astrophysics. Stephen Maran, editor. Van Nostrand Reinhold, 115 Fifth Avenue, New York, NY 10003. (212) 254-3232. (800) 842-3636. 1992. $129.95.

McGraw-Hill Encyclopedia of Science and Technology. McGraw-Hill Book Company, Inc., 1221 Avenue of the Americas, New York, NY 10020. (212) 512-2000. (800) 262-4729. 7th edition. 20 volume set. 1992. $1900.00.

Moons of the Solar System: An Illustrated Encyclopedia. John Stewart. McFarland & Company, Inc., Box 611, Jefferson, MC 28640. (910) 246-4460. (800) 253-2187. 1991. $49.95.

New Guide to the Planets. Patrick Moore. Trans-Atlantic Publications, Inc., 311 Bainbridge Street, Philadelphia, PA 19147. (215) 925-5083. 1993. $37.50.

Stars and Planets: The Sierra Club Guide to Sky Watching and Direction Finding. W. S. Kals. The Sierra Press, 4988 Gold Leaf Drive, Mariposa, CA 95338. (209) 966-5071. (800) 745-2631. $15.00.

GENERAL WORKS

Evolution of the Earth and Planets. E. Takahashi, et al, editors. American Geophysical Union, 2000 Florida Avenue, N.W., Washington, DC 20009. (202) 462-6903. (800) 966-2481. 1993. $28.00.

Exploring the Planets. W. Kenneth Hamblin and Eric H. Christiansen. Macmillan Publishing Company, Inc,. 200 Old Tappan Road, Old Tappan, NJ 07675. (800) 233-2336. 2nd edition. 1995. $62.00.

Galileo: Exploration of Jupiter's System. C. Yeates, et al. NASA Special Publication. U. S. Government Printing office, Washington D. C. 20402-9325. (202) 783-3238. 1085.

The Geology of Multi-Ring Impact Basins: The Moon and Other Planets. P. D. Spudis. Cambridge University Press, 40 West 20th Street, New York, NY 10011-4211. (212) 924-3900. (800) 872-7423. Planetary Sciences Series. 1993. $59.95.

The Giant Planet Jupiter. John H. Rogers. Cambridge University Press, 40 West 20th Street, New York, NY 10011-4211. (212) 924-3900. (800) 872-7423. 1995.

Jupiter: The Giant Planet. Reta Beebe. Smithsonian. Institution Press, 470 L'Enfant Plaza, Suite 7100, Washington, DC 20560. (202) 287-3738. (800) 782-4612. $29.95.

Moons and Planets. William K. Hartmann. Wadsworth Publishing Co., 10 Davis Drive, Belmont, CA 94002. (415) 595-2350. (800) 354-9706. 3rd edition. 1993. $49.95.

Planetary Landscapes. Ronald Greeley. Chapman & Hall, 1 Penn Plaza, New York, NY 10119. (212) 564-1060. Div. of Routledge. 2nd edition. 1994. $49.95.

Planetary Rings. Richard Greenberg, editor. University of Arizona Press, 1230 North Park Avenue, Number 102, Tucson, AZ 85719. (520) 621-1441. (800) 426-3797. 1984. $65.00.

Rings: Discoveries From Galileo to Voyager. James C. Elliot and Richard Kerr. MIT Press, 55 Hayward Street, Cambridge, MA 02142. (617) 253-8569. Reprint edition. 1987. $9.95 in paper.

Satellites of Jupiter. David Morrison, editor. Books on Demand, 300 North Zeeb Road, Ann Arbor, MI 48106-1346. (313) 761-4700. (800) 521-0600. Reprint edition. $180.00.

Space Technology and Planetary Astronomy. Joseph N. Tatarewicz. Indiana University Pres, 601 North Morton Street, Bloomington, in 47404-3797. (812) 855-6804. (800) 842-6796. 1990. $29.95.

Voyage to Jupiter: The Voyager Mission. Gordon Press. Publications, P.O. Box 459, Bowling Green Station, New York, NY 10004. (718) 624-8419. 1994. $260.00.

HANDBOOKS AND MANUALS

Astrophysical Data: Planets and Stars. Kenneth R. Lanb. Springer-Verlag New York, Inc., 175 Fifth Avenue, New York, NY 10010. (212) 460-1500. (800) 777-4643. 1993. $59.00.

Field Guide to the Stars and Planets. Jay M. Pasachoff and Donald H. Menzel. Houghton Mifflin Co., 222 Berkeley Street, Boston, MA 02116. (617) 351-5000. (800) 225-3362. Revised edition. 1992. $24.95.

New Guide to the Planets. Patrick Moore. Trans-Atlantic Publications, Inc., 311 Bainbridge Street, Philadelphia, PA 19147. (215) 925-5083. 1993. $37.50.

Planet Observer's Handbook. Fred W. Price. Cambridge University Press, 40 West 20th Street, New York, NY 10011-4211. (212) 924-3900. (800) 872-7423. 1994. $34.95.

Planets and their Moons. Aububon Society Staff. Alfred A. Knopf, Inc. 201 E. 50th Street, New York, NY 10022. (800) 733-3000. 1995. $7.99.

ONLINE DATABASES AND CD-ROMS

CA Search. Chemical Abstracts Service, P.O. Box 3012, Columbus, OH 43210-0012. (614) 447-3600. (800) 848-6533. FAX (614) 447-3709. Very comprehensive guide to worldwide chemical literature and related fields, 1972 to present. Available on BRS,(800) 289-4277, DIALOG, (800) 334-2564, ORBIT (800) 456-7248, and STN International, FIZ Karlsruhe, P.O. Box 2465, W-7500, Karlsruhe 1, Germany, online services. Inquire as to cost and availability.

Dissertation Abstracts. University Microfilms International, 300 North Zeeb Road, Ann Arbor, MI 48106. (800) 521-0600 or (313) 761-4700. Scope includes virtually all doctoral dissertations accepted at accredited American institutions from 1861 to present in 252 subject areas. Available on BRS, (800) 289-4277, DIALOG, (800) 334-2564, and OCLC EPIC, (800) 848-5878, online services. Also available on CD-ROM. Inquire as to cost and availability.

INSPEC. Institution of Electrical Engineers, Michael Faraday House, Six Hills Way, Stevenage, Herts. SG1 2AY, England. Telephone: 0438 313311 or FAX 0438 742840. Contains citations to the worldwide literature of physics, electronics and electrical engineering, computer technology, and related fields. Available on BRS, (800) 289-4277, DIALOG, (800) 334-2564, ORBIT, (800) 456-7248, and STN International, FIZ Karlsruhe, P.O. Box 2465, W-7500, Karlsruhe 1, Germany, online services. Inquire as to cost and availability.

NTIS Bibliographic Database. National Technical Information Service, 5285 Port Royal Road, Springfield, VA 22161. (703) 487-4929 or FAX (703) 321-8199. Broad coverage of government-sponsored science and technology research reports, 1964 to present. Available on BRS,(800) 289-4277, DIALOG, (800) 334-2564, ORBIT, (800) 456-7248, and STN International, FIZ Karlsruhe, P.O. Box 2465, W-7500, Karlsruhe 1, Germany, online services. Also available on CD-ROM. Inquire as to cost and availability.

SCISEARCH. Institute for Scientific Information, 3501 Market Street, Philadelphia, PA 19104. (800) 523-1850 or (215) 386-0100. Broad multidisciplinary title and author index to the international literature of science and technology, 1974 to present. Available on DIALOG, (800) 334-2564, and ORBIT, (800) 456-7248, online services. Also available on CD-ROM. Inquire as to cost and availability.

WILSONLINE. H.W. Wilson Company, 950 University Avenue, Bronx, NY 10452. (800) 367-6770 or (212) 588-8400. Makes available online versions of the printed H.W. Wilson indexes including Applied Science and Technology Index, Business Periodicals Index, General Science Index, and Readers' Guide to Periodical Literature. Period covered is generally 1983 to present. Available on BRS, (800) 289-4277, DIALOG, (800) 334-2564, and OCLC EPIC, (800) 848-5878, online services. Also available on CD-ROM. Inquire as to cost and availability.

OTHER SOURCES

Atlas of the Planets. Vincent DeCallatay and Audouin Dollfus. Books on Demand, 300 North Zeeb Road, Ann Arbor, MI 48106-1346. (313) 761-4700. (800) 521-0600. $45.60.

Atlas of the Solar System. B. Yenne. Simon & Schuster, Inc. 1230 Avenue of the Americas, New York, NY 10020. (212) 698-7000. (800) 223-2348. 1987. $12.98.

Cambridge Atlas of Astronomy. Jean Audouze, et al. Cambridge University Press, 40 West 20th Street, New York, NY 10011-4211. (212) 924-3900. (800) 872-7423. 3rd edition. 1994. $75.00.

The Sky: A User's Guide. David H. Levy. Cambridge University Press, 40 West 20th Street, New York, NY 10011-4211. (212) 924-3900. (800) 872-7423. 1993. $14.95.

The View From Space: Photographic Exploration of the Planets. Merton Davies and Bruce C. Murray. Columbia University Press, 562 West 113th Street, New York, NY 10025. (212) 666-1000. (800) 944-8648. 1973. $17.50.

PERIODICALS

Association of Lunar and Planetary Observers Journal. Association of Lunar and Planetary Observers. Box 16131, San Francisco, CA 94116. (415) 566-5786. FAX (415) 731-8242.

Astronomical Journal. American Institute of Physics, One Physics Ellipse, College Park, MD 20740-3843. (301) 209-3000. Published for the American Astronomical Society. 1849 to present. Monthly. $280.00 per year. ISSN: 0004-6256.

Astronomical Society of the Pacific Publications. Astronomical Society of the Pacific, 390 Ashton Avenue, San Francisco, CA 94112. (415) 337-1100. FAX (415) 337-5205. 1889 to present. Monthly. $175.00 per year. ISSN: 0004-6280.

Astronomy. Kalmbach Publishing Company, Box 1612, Waukesha, WI 53187-1612. (414) 796-8776. FAX (414) 796-1142. 1973 to present. Monthly. $27.00 per year. ISSN: 0091-6358.

Astronomy and Astrophysics. Springer-Verlag New York, Inc., 44 Hartz Way, Secaucus, NJ 07096-2491. (201) 348-4033. FAX (201) 348-4505. 1969 to date. 36 issues per year in 12 volumes. $1733.00 per year. ISSN: 0004-6361.

Earth, Moon and Planets: An International Journal of Comparative Planetology. Kluwer Academic Publishers, Box 358, Accord Station, Hingham, MA 02018-0358. (617) 871-6600. FAX (617) 871-6528. 1969 to present. Monthly. $840.00 per year. ISSN: 0167-9295.

Geochemica et Cosmochimica Acta. Elsevier Science. 660 White Plains Road, Tarrytown, NY 10591-5153. (914) 524-9200. FAX (914) 333-2444. 1950 to the present. Biweekly. $895.00 per year. ISSN: 0016-7037.

Icarus: International Journal of Solar System Studies. Academic Press, Inc., Journal Division, 525 B Street, Suite 1900, San Diego, CA 92101-4495. (619) 230-1840. FAX (619) 699-6800. 1962 to the present. Monthly. $1080.00. ISSN: 0019-1035.

J G R: Journal of Geophysical Research: Planets. American Geophysical Union, 2000 Florida Avenue, NW, Washington, CD 20009. (202) 462-6900. FAX (202) 328-0566. 1991 to present. Monthly. $597.00 per year. ISSN: 0148-0227.

Lunar and Planetary Information Bulletin. Lunar and Planetary Institute. 3600 Bay Area Boulevard, Houston, TX 77058-1113.

(713) 486-2175. FAX (713) 486-2125. 1970 to present. Quarterly. Free. Also available online.

Mercury. Astronomical Society of the Pacific. 390 Ashton Avenue, San Francisco, CA 94112. (415) 337-1100. FAX: (415) 337-5205. 1972 to present. Bimonthly. $175.00 per year. ISSN: 0047-6773.

Meteoritics. Meteoritical Society, Department of Chemistry, University of Arkansas, Fayetteville, AR 72701. (501) 575-7625. FAX (501) 575-7778. 1953 to present. Bimonthly. $210.00 per year. ISSN: 0026-1114.

Planetary and Space Science. Elsevier Science Publishing Company, Inc., 660 White Plains Road, Terrytown, NY 10591-5153. (914) 524-9200. FAX: (914) 333-2444. 1959 to present. Monthly. $1355.00 per year. ISSN: 0032-0633.

Planetary Report. Planetary Society. 65 North Catalina, Pasadena, CA 91106-2301. (818) 793-5100. FAX (818) 793-5528. 1980 to present. $25.00 per year. ISSN: 0736-3680.

Solar System Research. Plenum Publishing Corp., Consultants Bureau, 233 Spring Street, New York, NY 10013-1578. (212) 620-8468. FAX (212) 463-0742. English translation of *Astronomicheskii Vestnik.* 1967 to present. $1080.00 per year. ISSN: 0038-0946.

Sky and Telescope. Sky Publishing Corporation, Box 9111, Belmont, MA 02178. (617) 864-7360. FAX (617) 864-6117. 1941 to present. Monthly. $27.00 per year. ISSN: 0037-6604.

RESEARCH CENTERS AND INSTITUTES

Center for Radiophysics and Space Research. Cornell University, Space Sciences Building, Ithaca NY 14853. (607) 255-4341. FAX (607) 255-9888.

Center for Space Science and Astrophysics. Stanford University. 325 Durand Building, Stanford, CA 04305. (415) 723-3582.

Center for Space Sciences. University of Texas at Dallas, P. O. Box 830688, MS-F022, Richardson, TX 75083-0688. (214) 690-2851. FAX (214) 690-2848.

Institute for Astronomy. University of Hawaii at Manoa, 2680 Woodlawn Drive, Honolulu, HI 96822. (808) 956-8312. FAX (808) 988-2790.

Institute of Geophysics and Planetary Physics. University of California, Riverside, CA 92521. (714) 787-4503. FAX (714) 787-4529.

Laboratory for Atmospheric and Space Physics. University of Colorado-Boulder, Boulder, CO 80309-0392. (303) 492-7677. (303) 492-6946.

Laboratory for Planetary Geology. Arizona State University, Department of Geology, Tempe, AZ 85281. (601) 965-7029.

Laboratory for Planetary Studies. Cornell University, Space Sciences Building, Ithaca, NY 14853. (607) 255-4971.

Lunar and Planetary Institute. 3303 NASA Road One, Houston, TX 77058-4399. (713) 486-2139. FAX 713-496-2162.

JUPITER

Ency. of Physical Sciences and Engineering Info. Sources

Lunar and Planetary Laboratory. University of Arizona, Tucson, AZ 85721. (602) 621-6963. FAX (602) 621-4933.

McDonnell Center for the Space Sciences. Washington University, Box 1105, One Brookings Drive, St. Louis MO 63130-4899. (314) 889-6255. FAX (314) 889-6219.

Smithsonian Astrophysical Observatory. 60 Garden Street, Cambridge, MA 02138. (617) 495-7461. FAX (617) 495-7105.

K

KARST

See also: CARBONATES, CAVES AND CAVING, GEOLOGY, SEDIMENTARY ROCK

ABSTRACT SERVICES AND INDEXES

Bibliography and Index of Geology. American Geological Institute, 4220 King Street, Alexandria, VA 22302. (703) 379-2480. Fax (703) 379-7563. Monthly. $1295.00 per year. Also available online as GEOREF.

Current Contents: Physical, Chemical and Earth Sciences. Institute for Scientific Information, 3501 Market Street, Philadelphia, PA 19104. (215) 386-0100. FAX (215) 386-6362. Weekly. $360.00 per year.

Geological Abstracts. Elsevier - Geo Abstracts, Regency House, 34 Duke Street, Norwich NR3 3AP, England. Monthly. $760.00 per year. Also available online as GEOBASE.

Geological Society of America. Abstracts with Programs. Geological Society of America. 3300 Penrose Place, P.O. Box 9140, Boulder, CO 80301-9140. (303) 447-2020. Abstracts and programs of the annual conference. Annual. $69.00.

ASSOCIATIONS AND PROFESSIONAL SOCIETIES

American Geological Institute. 4220 King Street, Alexandria, VA 22302. (703) 379-2480. Fax (703) 379-7563.

British Cave Research Association. c/0 Bryan Ellis, 20 Woodland Avenue, Westonzoyland, Bridgwater, Somerset TA7 0LQ, England.

Cave Research Foundation. 4074 W Redwing Street, Tucson, AZ 85741. (602) 744-2243.

Geological Society of America. 3300 Penrose Place, PO Box 9140, Denver, CO 80301-9140. (303) 447-2020.

International Association of Hydrogeology. National Rivers Authority, Sapphire E 550, St.sBrook Road, Solihull, W. Midlands B91 1QT, England. Telephone 121-7112324. FAX 121-7112794.

National Caves Association. Rt.9, Box 106, McMinnville, TV 37110. (615) 668-3925. FAX (615) 668-3988.

National Speleological Society Inc., 2813 Cave Avenue, Huntsville, AL 35810-4431. (205) 852-1300. FAX (205) 851-9241.

DIRECTORIES AND BIOGRAPHICAL SOURCES

American Men & Women of Science, 1995-96. R.R. Bowker Staff, eds. 19th edition. 8 volumes. R.R. Bowker/Reed International Publishing Company, 121 Chanlon Road, New Providence, NJ 07974. (908) 464-6800 or (800) 521-8110. 1995. $850.00.

Caves and Caverns: National Caves Association Directory. National Caves Association, Rt.9, Box 106, McMinnville, TV 37110. (615) 668-3925. FAX (615) 668-3988. Annual. 1973 to date. Free.

Geological Society of America Membership Directory. 3300 Penrose Place, PO Box 9140, Denver, CO 80301-9140. (303) 447-2020. Annual. Inquire for cost and availability.

International Research Centers Directory. Gale Research, 835 Penobscot Building, Detroit, MI 48226-4094. (313) 961-2242. (800) 347-4253. FAX (313) 961-6083. 8th edition. 1995. $410.00.

Research Centers Directory. Gale Research, 835 Penobscot Building, Detroit, MI 48226-4094. (313) 961-2242. (800) 347-4253. FAX (313) 961-6083. $485.00.

Scientific and Technical Organizations and Agencies Directory. Gale Research, 835 Penobscot Building, Detroit, MI 48226-4094. (313) 961-2242. (800) 347-4253. FAX (313) 961-6083. 4th edition. 1996. $195.00.

ENCYCLOPEDIAS AND DICTIONARIES

The Encyclopedia of the Solid Earth Sciences. Philip Kearey, editor-in-chief. Blackwell Scientific Publications, 238 Main Street, Cambridge, MA 02142. (617) 876-7000 or (800) 759-6102. FAX (617) 876-7022. 1993. $60.00.

KARST

Ency. of Physical Sciences and Engineering Info. Sources

Encyclopedia of Physical Science and Technology. Robert A. Meyers, ed. 18 volumes. Academic Press Inc., 6277 Sea Harbor Drive, Orlando, FL 32887. (800) 321-5068. 1992. $2100.00.

Glossary of Geology. Robert Bates, et al. Reprint of third edition. American Geological Institute, 4220 King Street, Alexandria, VA 22302-1507. (703) 379-2480. FAX (301) 379-7563. 1990. $225.00.

GENERAL WORKS

Geomorphology and Geoecology—Karst. J. Nicod, et al. Lubrecht & Cramer Ltd., 38 Country Route 48, Forestburgh, NY 12777-6400. (914) 794-8539. FAX (914) 791-7575. 1992. $56.70.

Karst Geomorphology and Hydrology. Derek C. Ford and Paul W. Williams. Routledge, Chapman and Hall, Inc., 29 West 35th Street, New York, NY 10001-2291. (212) 244-3336 or FAX (212) 563-2269. 1989. $140.00 (hardback) or $49.95 (paperback).

Karst Hydrology. O. Bonacci. Springer-Verlag, 175 Fifth Avenue, New York, NY 10010. (212) 460-1500 or (800) 777-4643. FAX (212) 473-6272. 1987. $97.00.

HANDBOOKS AND MANUALS

Handbook of Hydrology. David R. Maidment, editor in chief. McGraw-Hill Publishing Company, 1221 Avenue of the Americas, New York, NY 10020. (800) 262-4729 or (212) 512-3825. 1993. $115.00.

ONLINE DATABASES AND CD-ROMS

Earth Sciences. U.S. Geological Survey, 12201 Sunrise Valley Drive, Reston, VA 22092-9998. (703) 648-4460. CD-ROM of earth science databases including the U.S. Geological Survey Library, Earth Science Data Directory, and GEOINDEX, citations to published geological maps. $350.00 per year with quarterly updates.

Geoarchive. Geosystems, P.O. Box 1024, Westminster, London, England, SW1 P 2JL. Citations to literature on geoscience, 1969 to present. Inquire as to online cost and availability.

Geobase. Elsevier - Geo Abstracts, Regency House, 34 Duke Street, Norwich NR3 3AP, England. Contains citations to the worldwide earth science literature from 1980 to date. Available on DIALOG, ORBIT online services. Inquire as to cost and availability.

GEOREF: Bibliography and Index of Geology. American Geological Institute, 4220 King Street, Alexandria, VA 22302. (703) 379-2480. Fax (703) 379-7563. Monthly. Inquire for price and availability.

PERIODICALS

Cave Geology. National Speleological Society, Section of Cave Geology and Geography, 542 Glenn Roard, State College, PA 16803. (814) 237-3187. Irregular. Price varies.

Cave Science: Transactions of the British Cave Research Association. British Cave Research Association, c/0 Bryan Ellis, 20 Woodland Avenue, Westonzoyland, Bridgwater, Somerset TA7 0LQ, England. 1974 to date. Irregular. Inquire for price in U.S.

Caves and Caving. British Cave Research Association, c/0 Bryan Ellis, 20 Woodland Avenue, Westonzoyland, Bridgwater, Somerset TA7 0LQ, England. 1973 to date. Quarterly. Inquire for price in U.S.

Geological Society of America Bulletin. 3300 Penrose Place, PO Box 9140, Denver, CO 80301-9140. (303) 447-2020. 1888 to present. Monthly. $205.00 per year.

Geology. Geological Society of America, 3300 Penrose Place, PO Box 9140, Denver, CO 80301-9140. (303) 447-2020. 1973 to present. Monthly. $170.00 per year for non-members.

Sedimentary Geology. Elsevier Science Inc., Box 882, Madison Square Station, New York, NY 10159. (212) 989-5800. FAX (212) 633-3990. 1967 to present. 24 times a year. $1187.00.

Sedimentology. Blackwell Scientific Publishing, Inc., 3 Cambridge Court, Suite 208, Cambridge, MA 02142. (617) 225-0401. 1952 to present. Six times a year. $434.00.

RESEARCH CENTERS AND INSTITUTES

Western Kentucky University Center For Cave and Karst Studies. Department of Geography and Geology, Bowling Green, KY 42101. (502) 745-5989.

KINEMATICS

See: KINETICS

KINETICS

See also: CHAOS, DYNAMICS, MECHANICS, PHYSICAL CHEMISTRY

ABSTRACT SERVICES AND INDEXES

Applied Mechanics Reviews: An Assessment of World Literature in Engineering Sciences. American Society of Mechanical Engineers, 345 East 47th Street, New York, NY 10017. (212) 705-7703. 1948 to present. Monthly. $360.00 per year. ISSN: 0003-6900.

Applied Science and Technology Index; A Cumulative Subject Index To English Language Periodicals in the Fields ofaeronautics and Space Science, Computer Technology,chemistry, Construction Industry, Energy and Related Areas. H.W. Wilson Co., 950 University Avenue, Bronx, NY 10452. (212) 588-8400. (800) 367-6770. FAX (718) 590-1617. From 1958 to present. Monthly. Inquire about cost andavailability. Also available on CD-ROM and online. ISSN:0003-6986.

Chemical Abstracts. Chemical Abstracts Service, 2540Olentangy River Road, Box 3012, Columbus, OH 43210-0012.(614) 447-3600. (800) 848-6538. FAX (614) 447-3713.From 1907 to present. Weekly. $16800.00 per year. Alsoavailable on CD-ROM and online. Inquire regarding cost and availability.

Current Contents: Physical, Chemical and Earth Sciences. Institute for Scientific Information, 3501 Market Street,Philadelphia, PA 19104. (215) 386-0100. FAX (215) 386-2291.From 1961 to present. Weekly. $442.00 per year. Alsoavailable on CD-ROM and online. Inquire regarding cost andavailability. ISSN: 0163-2574.

Current Papers in Physics. Institute of Electrical Engineers, Michael Faraday House, Six Hill Way,Stevenage,Herts, SG1 2AY, England. Distributed by INSPEC/IEEE, Box1331, 445 Hoes Lane, Piscataway, NJ 08855-1331. (908) 562-5549. 1966 to present. Fortnightly. $410.00 per year. ISSN: 0011-3786.

Engineering Index Monthly. Engineering Information, Inc.,Castle Point on the Hudson, Hoboken, NJ 07030. (201) 216-8500. FAX (201) 216-8532. Monthly. $2200.00 per year.Also available online as COMPENDEX and on CD-ROM. ISSN: 0742-1974.

General Science Index. H.W. Wilson Company, 950 University Avenue, Bronx, NY 10452. (212) 588-8400. (800) 367-6770.FAX (718) 590-1617. From 1978 to present. Ten issues peryear; quarterly and annual cumulations. Service basis. Available on CD-ROM and online. Inquire regarding cost and availability. ISSN: 0162-1963.

Government Reports Announcements and Index. U. S. National Technical Information Service (NTIS), 5285 Port Royal Road, Springfield, VA 22161. (703) 487-4650. FAX (703) 321-8547. From 1968 to present. Annual. $630.00 per year. Also available online as NTIS Bibliographic Database and on CD- ROM. ISSN:

*ISMEC: Mechanical Engineering Abstracts.*Cambridge Scientific Abstracts, 7200 Wisconsin Avenue, Suite 601, Bethesda, MD 20814. (301) 961-6750 or (800) 843-7751. Contains citations to the literature in mechanical engineering, industrial and production engineering, energy, power, mechanics, devices and related areas, from 1973 to present. Available on the DIA-LOG,(800) 334-2564, online service. Inquire as to cost and availability.

NTIS Alerts: Physics. U. S. NationalTechnicalInformation Service. 5285 Port Royal Road, Springfield, VA22161. (703) 487-4650. FAX (703) 321-8547. Weekly. $140.00 per year.

Physics Abstracts. INSPEC. Section A, Science Abstracts. Institution of Electrical Engineers (IEE). Available from INSPEC/IEEE - Institute of Electrical and Electronic Engineers,, Box 1331, 445 Hoes Lane, Piscaway NJ 08855-1331. (908) 562-5549. 1898 to present. 24 issues per year. $2835.00 per year. Also available on CD-ROM and online.ISSN: 0036-8091.

Physics Briefs (Physikalische Berichte). Information Centerfor Energy, Physics, Mathematics; German Physical Society. V C H Publishers, Inc., 220 East 23rd Street, New York, NY 10010-4606. (212) 683-8333. 1845 to present. 24 issues per year. $2390.00 per year. Also available online. ISSN: 0179-7434.

Science Citation Index. SCI. Institute for ScientificInformation, 3501 Market Street, Philadelphia, PA 19104. (215) 386-0100. (800) 523-1850. FAX (215) 386-2991. 1961to present. Six issues per year, plus annual cumulation. $11650.00 per year. Also available online and on CD-ROM.Inquire about price and availability. ISSN: 0036-827X.

ANNUAL REVIEWS AND YEARBOOKS

Advances in Applied Mechanics. Academic Press, Inc., 6277 Sea Harbor Drive, Orlando, FL. (800) 321-5068. 1948 to present. Irregular. ISSN: 0065-2156.

Advances in Physics. Taylor & Francis, Ltd., Rankine Road, Basingstoke, Hants RG24 8PR, England. 0256-840366. FAX 0256-47938. 1952 to present. Bimonthly. $511.00 per year. ISSN: 0001-8732.

Topics in Applied Physics. Springer-Verlag New York, Inc., 175 Fifth Avenue, New York, NY 10010. (212) 460-1500. (800) 777-4643. FAX (212) 473-6272. ISSN: 0303-4216.

ASSOCIATIONS AND PROFESSIONAL SOCIETIES

American Chemical Society, 1155 16th Street NW, Washington DC 20036. (202) 872-4600. (800) 424-6747.

American Institute of Physics. One Physics Ellipse, College Park, MD 20740-3843. (301) 209-3100.

American Physical Society. One Physics Ellipse, College Park, MD 20740-3843. (301) 209-3200. FAX (301) 209-0865.

American Society of Civil Engineers. 1015 15th Street NW, Suite 600, Washington, DC 20005. (202) 789-2200.

American Society of Mechanical Engineers, 345 East 47th Street, New York, NY 10017. (212) 705-7722. FAX (212) 705-7739.

DIRECTORIES AND BIOGRAPHICAL SOURCES

American Men and Women of Science: Physical and Biological Sciences. R. R. Bowker Inc., 121 Chanlon Road, New Providence, NJ 07974. (908) 464-6800. (800) 521-8110. 20th edition. 8 volumes. 1996. $850.00.

American Physical Society. Membership Directory Bulletin Issue. American Physical Society, One Physics Ellipse, College Park, MD 20740-3843. (301) 209-3200. FAX (301) 209-0865. Biennial. $50.00.

Directory of Physics and Astronomy Staff Members. American Institute of Physics, One Physics Ellipse, College Park, MD 20740-3843. (301) 209-3100. Annual. $45.00.

Graduate Programs in Physics, Astronomy and Related Fields. 1995-1996. American Institute of Physics, One Physics Ellipse, College Park, MD 20740-3843. (301) 209-3100. Annual. $45.00. ISSN: 0147-1821.

International Directory of Engineering Societies and Related Organizations. American Society for Engineering Education, 1818 N Street NW, Suite 600, Washington, DC 20036. (202) 331-3526. 15th edition. 1996. $185.00.

Research Centers Directory. Gale Research Inc., 835 Penobscot Building, Detroit, MI 48226-4094. (313) 961-2242. (800) 347-4253. 20th edition. 1995. $485.00. ISSN: 0080- 1518.

Scientific and Technical Organizations and Agencies Directory. Gale Research, 835 Penobscot Building, Detroit, MI 48226-

KINETICS

Ency. of Physical Sciences and Engineering Info. Sources

4094. (313) 961-2242. (800) 347-4253. 4th edition. 1996. $195.00.

Who's Who in Engineering. Gordon Davis, editor. American Society for Engineering Education, 1111 19th Street, NW, Suite 608, Washington, DC 20036. 9th edition. 1995. $220.00.

ENCYCLOPEDIAS AND DICTIONARIES

A Concise Dictionary of Physics. Oxford University Press, Inc., 200 Madison Avenue, New York, NY 10016. (212) 725- 6000. (800) 334-4249. 1990. $10.95.

Dictionary of the Physical Sciences: Terms, Formulas, Data. Cesare Emiliani. Oxford University Press, Inc., 200 Madison Avenue, New York, NY 10016. (212) 725-6000. (800) 334-4249. 1989. $19.95.

Encyclopedia of Applied Physics. George Trigg, editor. VCH Publications, Inc., 220 East 23rd Street, Suite 909, New York, NY 10010-4606. (212) 683-8333. (800) 422-8824. 20 volume set. 1991-. $5990.00.

Encyclopedia of Physics. Robert M. Besancon. Chapman & Hall, 1 Penn Plaza, New York, NY 10119. (212) 564-1060. 3rd edition. 1990. $54.95 in paper.

Encyclopedia of Physical Science and Technology. Academic Press, Inc., 6277 Sea Harbor Drive, Orlando, FL. (800) 321- 5068. 2nd edition. 18 volume set. 1992. $2625.00.

McGraw-Hill Encyclopedia of Science and Technology. McGraw-Hill Book Company, Inc., 1221 Avenue of the Americas, New York, NY 10020. (212) 512-2000. (800) 262-4729. 7th edition. 20 volume set. 1992. $1900.00.

Penguin Dictionary of Physics. Valerie Illingworth, editor. Viking Penguin, 375 Hudson Street, New York, NY 10014-3657. (212) 366-2000. (800) 331-4624 2nd edition. 1991.

GENERAL WORKS

Applied Nonlinear Dynamics: a Primer on Stability, Chaos and Fractals. Leon Glass and Daniel Kaplan. Springer-Verlag New York, Inc., 175 Fifth Avenue, New York, NY 10010. (212) 460- 1500. (800) 777-4643. FAX (212) 473-6272. 1995. $44.95.

Chaos, Dynamics, and Fractals: An Algoritmic Approach To Deterministic Chaos. Joseph L. McCauley. Cambridge University Press, 40 West 20th Street, New York, NY 10011- 4211. (212) 924-3900. (800) 872-7423. 1993. $94.95.

Introduction To the Kinetic theory of Gases and Magnetoplasm. L. C. Woods. Oxford University Press, Inc., 200 Madison Avenue, New York, NY 10016. (212) 725-6000. (800) 334-4249. 1993. $42.50.

Statistical Mechanics of Nonequilibrium Liquids. D. J. Evans. Academic Press, Inc., 6277 Sea Harbor Drive, Orlando, FL. (800) 321-5068. 1990. $88.00.

Physics. David Haliday and Robert Resnick. John Wiley & Sons, Inc., 605 Third Avenue, New York, NY 10158-0012. (212) 850- 6000. (800) 225-5945. 2 volumes. 4th edition. 1992.

Quantum Kinematics and Dynamics. Julian Schwinger. Addison-Wesley Publishing Co., Inc., 1 Jacob Way, Reading, MA 01867. (617) 944-3700. (800) 447-2226. 1991. $29.75.

Statistical Mechanics and Thermodynamics. Claude Garrod. Oxford University Press, Inc., 200 Madison Avenue, New York, NY 10016. (212) 725-6000. (800) 334-4249. 1995. $45.00.

HANDBOOKS AND MANUALS

Chemical and Physical Data. Arthur M. James and M. P. Lord, Van Nostrand Reinhold, 115 Fifth Avenue, New York, NY 10003. (212) 254-3232. (800) 842-3636. 1992. $59.95.

CRC Handbook of Chemistry and Physics. David R. Kide, editor. CRC Press, Inc., 2000 Corporate Boulevard, NW, Boca Raton, FL 33431. (407) 994-0555. (800) 272-7737. 77th edition. 1996. $99.95.

Handbook of Physical Quantities. Igor S. Grigoriev and Evgenil Z. Melikhov, editors. CRC Press, Inc., 2000 Corporate Boulevard, NW, Boca Raton, FL 33431. (407) 994- 0555. (800) 272-7737. 1995. $99.00.

Physics Problem Solver. Research and Education Association Staff. Research & Education Association, 61 Ethel Road, West Piscataway, NJ 08854. (908) 819-8880. Revisededition. 1994. $23.95 in paper.

Tables of Physical and Chemical Constants and Some Mathematical Functions. G. W. Kaye and T. M. Laby. Halsted Press 605 Third Avenue, New York, 10158-0012. (212) 850- 6400. 15th edition. 1986. $74.95.

ONLINE DATABASES AND CD-ROMS

CA Search. Chemical Abstracts Service, P.O. Box 3012, Columbus, OH 43210-0012. (614) 447-3600. (800) 848-6533. FAX (614) 447-3709. Very comprehensive guide to worldwide chemical literature and related fields, 1972 to present. Available on BRS,(800) 289-4277, DIALOG, (800) 334-2564, ORBIT (800) 456-7248, and STN International, FIZ Karlsruhe, P.O. Box 2465, W-7500, Karlsruhe 1, Germany, online services. Inquire as to cost and availability.

Current Contents Search. Institute for Scientific Information, 3501 Market Street, Philadelphia, PA 19104. (215) 386-0100. FAX (215) 386-6362. Contains citations to articles listed in the table of contents of science and technology journals. Also articles in social sciences and life sciences journals. Available on BRS, (800) 289-4277, DIALOG, (800) 334-2564, online services. Inquire as to cost and availability.

Compendex Plus. Engineering Information, Inc., 345 East 47th Street, New York, NY 10017. (212) 705-7600 or (800) 221- 1044. Contains citations with abstracts to worldwide literature in engineering and technology, from 1970 to present. Available on online BRS,(800) 289-4277, DIALOG, (800) 334-2564, ORBIT (800) 456-7248, and STN International, FIZ Karlsruhe, P.O. Box 2465, W-7500, Karlsruhe 1, Germany, online services. Also available on CD-ROM. Inquire as to cost and availability.

Dissertation Abstracts Online. University Microfilms International, 300 North Zeeb Road, Ann Arbor, MI 48106. (800) 521- 0600 or (313) 761-4700. Scope includes virtually all doctoral dissertations accepted at accredited American institutions from

1861 to present in 252 subject areas. Available on BRS, (800) 289-4277, DIALOG, (800) 334-2564, and OCLC EPIC, (800) 848-5878, online services. Also available on CD-ROM. Inquire as to cost and availability.

INSPEC. Institution of Electrical Engineers, Michael Faraday House, Six Hills Way, Stevenage, Herts. SG1 2AY, England. Telephone: 0438 313311 or FAX 0438 742840. Contains citations to the worldwide literature of physics, electronics and electrical engineering, computer technology, and related fields. Available on BRS, (800) 289-4277, DIALOG, (800) 334-2564, ORBIT, (800) 456-7248, and STN International, FIZ Karlsruhe, P.O. Box 2465, W-7500, Karlsruhe 1, Germany, online services. Inquire as to cost and availability.

NTIS Bibliographic Database. National Technical Information Service, 5285 Port Royal Road, Springfield, VA 22161. (703) 487-4929 or FAX (703) 321-8199. Broad coverage of government-sponsored science and technology research reports, 1964 to present. Available on BRS,(800) 289-4277, DIALOG, (800) 334-2564, ORBIT, (800) 456-7248, and STN International, FIZ Karlsruhe, P.O. Box 2465, W-7500, Karlsruhe 1, Germany, online services. Also available on CD-ROM. Inquire as to cost and availability.

Physics Briefs. American Institute of Physics, 335 East 45th Street, New York, NY 10017. (212) 661-9260 or FAX (212) 661-2036. Contains citations with abstracts of the literature of physics and related fields, 1979 to present. Available on the STN International, FIZ Karlsruhe, P.O. Box 2465, W-7500, Karlsruhe 1, Germany, online service. Inquire as to cost and availability.

SCISEARCH. Institute for Scientific Information, 3501 Market Street, Philadelphia, PA 19104. (800) 523-1850 or (215) 386-0100. Broad multidisciplinary title and author index to the international literature of science and technology, 1974 to present. Available on DIALOG, (800) 334-2564, and ORBIT, (800) 456-7248, online services. Also available on CD-ROM. Inquire as to cost and availability.

WILSONLINE. H.W. Wilson Company, 950 University Avenue, Bronx, NY 10452. (800) 367-6770 or (212) 588-8400. Makes available online versions of the printed H.W. Wilson indexes including Applied Science and Technology Index, Business Periodicals Index, General Science Index, and Readers' Guide to Periodical Literature. Period covered is generally 1983 to present. Available on BRS, (800) 289-4277, DIALOG, (800) 334-2564, and OCLC EPIC, (800) 848-5878, online services. Also available on CD-ROM. Inquire as to cost and availability.

PERIODICALS

American Journal of Physics. American Association of Physics Teachers, 5112 Berywn Road, College Park, MD 20740. (301) 345-4200. 1933 to present. Monthly. $215.00 per year. ISSN: 0002-9505.

Archive For Rational Mechanics and Analysis. Springer-Verlag New York, 44 Hartz Way, Secacus, NJ 07096-2491. (201) 348-4033. 1957 to present. 16 issues per year in 4 volumes. $1238.00. ISSN: 0003-9527.

International Journal of Chemical Kinetics. John Wiley & Sons, Inc., 605 Third Avenue, New York, NY 10158-0012. (212) 850-6000. 1968 to present. Monthly. $730.00 per year. ISSN: 0538-8066.

International Journal of Solids and Structures. Elsevier Science, 660 White Plains Road, Tarrytown, NY 10591-5153. (914) 524-9200. FAX (914) 333-2444. 1965 to present. 24 issues per year. $2035.00. ISSN: 0020-7683.

Journal of Applied Mechanics. American Society of Mechanical Engineers, 345 East 47th Street, New York, NY 10017. (212) 705-7722. FAX (212) 705-7739. 1935 to present. Quarterly. $120.00. ISSN: 0021-8936.

Journal of Engineering Mechanics. American Society of Civil Engineers, 345 East 45th Street, New York, NY 10017. (212) 682-7341. 1956 to present. Monthly. $284.00. ISSN: 0733-9399.

Journal of Mechanics and Physics of Solids. Elsevier Science, 660 White Plains Road, Tarrytown, NY 10591-5153. (914) 524-9200. FAX (914) 333-2444. 1952 to present. Monthly. $1335.00. ISSN: 0022-5096.

Quarterly Journal of Mechanics and Applied Mathematics. Oxford University Press, Inc., 2001 Evans Road, Cary, NC 27513. (919) 677-0977. 1948 to present. Quarterly. $214.00 per year. ISSN: 0033-5614.   <subhead2>Research Centers and Institutes

Center for the Advancement of Computational Mechanics. Georgia Institute of Technology, Atlanta, GA 30332. (404) 894-2758.

Laboratory For Experimental Mechanics Research. State University of New York at Stony Brook, Stony Brook, NY 11794-2300. (516) 632-8311. FAX (516) 632-8720.

Solid Mechanics Laboratory. University of Michigan, Department of Mechanical Engineering and Applied Mechanics, 2246A G. G. Brown Building, Ann Arbor, MI 48109. (313) 763-0684.

L

LAND RECLAMATION

ABSTRACT SERVICES AND INDEXES

Civil and Structural Engineering Abstracts, Cambridge Scientific Abstracts, 7200 Wisconsin Avenue, Bethesda, MD 20814-4823. (301) 961-6750. FAX (301) 961-6720. 1993 to present. Monthly. $385.00 per year. Topics covered include structural design, construction equipment and methods, civil defense and military engineering, surveying, highway engineering, maritime and port structures, materials, land reclamation, and soil mechanics.

Current Contents: Engineering, Technology, and Applied Sciences. Institute for Scientific Information, 3501 Market Street, Philadelphia, PA 19104. (215) 386-0100. FAX (215) 386-6362. 1970 to present. Weekly. $442.00 per year.

Current Contents: Physical, Chemical and Earth Sciences. Institute for Scientific Information, 3501 Market Street, Philadelphia, PA 19104. (215) 386-0100. FAX (215) 386-6362. Weekly. $360.00 per year.

Environmental Abstracts. R.R. Bowker Inc., 121 Chanlon Road, New Providence, NJ 07974. (800) 521-8110 or (908) 464-6800. Contains citations, most with abstracts, to literature relating to environmental issues and problems, from 1971 to present. Inquire as to cost and availability.

Environmental Engineering Abstracts, Cambridge Scientific Abstracts, 7200 Wisconsin Avenue, Bethesda, MD 20814-4823. (301) 961-6750. FAX (301) 961-6720. Monthly. Covers hazardous materials, environmental impact and protection, treatment of sewage and industrial wastes, hydroelectric power, tidal and wind power, artic and tropical engineering.

ASSOCIATIONS AND PROFESSIONAL SOCIETIES

International Association For Land Reclamation and Improvement (iari). Postbus 45, NL-6700 AA Wageningen, Netherlands. Telephone 8370-90144. FAX 8370-17187.

National Association of State Land Reclamationists. 459B Carlisle Drive, Herndon, VA 22070. (703) 709-8654. FAX (703) 709-8655.

DIRECTORIES AND BIOGRAPHICAL SOURCES

American Men & Women of Science, 1995-96. R.R. Bowker Staff, eds. 19th edition. 8 volumes. R.R. Bowker/Reed International Publishing Company, 121 Chanlon Road, New Providence, NJ 07974. (908) 464-6800 or (800) 521-8110. 1995. $850.00

Conservation Directory. National Wildlife Federation, 1400 16th Street NW, Washington, DC 20036-2266. (202) 797-6800. FAX (703) 442-7332. 1955 to present. Annual. $20.00.

Directory of Engineering Societies and Related Organizations. Gordon Davis. 13th edition. American Association of Engineering Societies, 1111 19th Street NW, Suite 608, Washington, DC 20036. (202) 296-2237 or (800) 658-8897. 1989. Inquire for price.

International Research Centers Directory. Gale Research, 835 Penobscot Building, Detroit, MI 48226-4094. (313) 961-2242. (800) 877-4253. FAX (313) 961-6083. 8th edition. 1995. $410.00.

Research Centers Directory. Gale Research, 835 Penobscot Building, Detroit, MI 48226-4094. (313) 961-2242. (800) 877-4253. FAX (313) 961-6083. $485.00.

Scientific and Technical Organizations and Agencies Directory. Gale Research, 835 Penobscot Building, Detroit, MI 48226-4094. (313) 961-2242. (800) 877-4253. FAX (313) 961-6083. 4th edition. 1996. $195.00.

Who's Who in Engineering. American Association of Engineering Societies, 1111 19th Street NW, Suite 608, Washington, DC 20036. (202) 296-2237 or (800) 658-8897. 8th edition. 1991. Inquire for price.

ENCYCLOPEDIAS AND DICTIONARIES

Dictionary of Civil Engineering. V.N. Vazirani. South Asia Books, Box 502, Columbia, MO 65205. (314) 474-0116. FAX (314) 474-8124. 1992. $20.00.

The Encyclopedia of the Solid Earth Sciences. Philip Kearey, editor-in-chief. Blackwell Scientific Publications, 238 Main Street, Cambridge, MA 02142. (617) 876-7000 or (800) 759-6102. FAX (617) 876-7022. 1993. $60.00.

LAND RECLAMATION

Ency. of Physical Sciences and Engineering Info. Sources

Encyclopedia of Physical Science and Technology. Robert A. Meyers, ed. 18 volumes. Academic Press Inc., 6277 Sea Harbor Drive, Orlando, FL 32887. (800) 321-5068. 1992. $2100.00

GENERAL WORKS

Contaminated Land: Assessment and Redevelopment. Richard A. Failey and Amanda J. Scrivens. State Mutual Book & Periodical Service Ltd., 521 Fifth Ave., 17th floor, New York, NY 10175. (718) 261-1704. FAX (516) 537-0412. 1994. $150.00.

Contaminated Land: Reclamation, Redevelopment and Re-use in the United States and the European Union. Peter B. Meyer, et al. Ashgate Publishing Co., Old Post Rd., Brookfield, VT 05036. (802) 276-3162. FAX (802) 276-3837. 1996. $63.95.

Environmental Restoration: Science and Strategies For Restoring the Earth. Island Press, 1718 Connecticut Avenue, N.W., Suite 300, Washington, DC 20009. (202) 232-7933. FAX (202) 234-1328. 1990. $22.00 (hardback) or $13.00 (paperback).

Land Reclamation: Advances in Research and Technology. American Society of Agricultural Engineers, 2950 Niles Road, St. Joseph, MI 49085-9659. (616) 429-0300. FAX (616) 429-3852. 1993. $53.00.

Reclamation of Land: An End To Dereliction? M.C. Davies. Elsevier Science Publishing Company, Inc., 655 Avenue of the Americas, New York, NY 10010. (212) 989-5800. 1991. $102.00.

Reclaiming Wasted Lands For Our Future. J.T. Mensah and Larry Soule. Dorrance Publishing Company Inc., 643 Smithfield Street, Pittsburgh, PA 15222. (800) 788-7654. FAX (412) 288-1786. 1993. $14.95.

HANDBOOKS AND MANUALS

Practical Handbook of Disturbed Land Revegetation. Frank F. Munshower. Lewis Publishers, 121 S. Main St., PO Box 519, Chelsea, MI 48118. (313) 475-8619 or (800) 272-7737. 1993. $75.00.

ONLINE DATABASES AND CD-ROMS

Dissertation Abstracts. University Microfilms International, 300 North Zeeb Road, Ann Arbor, MI 48106. (800) 521-0600 or (313) 761-4700. Scope includes virtually all doctoral dissertations accepted at accredited American institutions from 1861 to present in 252 subject areas. Available on BRS,(800) 289-4277, DIALOG,(800) 334-2564, and OCLC EPIC,(800) 848-5878, online services. Also available on CD-ROM. Inquire as to cost and availability.

Earth Sciences. U.S. Geological Survey, 12201 Sunrise Valley Drive, Reston, VA 22092-9998. (703) 648-4460. CD-ROM of earth science databases including the U.S. Geological Survey Library, Earth Science Data Directory, and GEOINDEX, citations to published geological maps. $350.00 per year with quarterly updates.

Geoarchive. Geosystems, P.O. Box 1024, Westminster, London, England, SW1 P 2JL. Citations to literature on geoscience, 1969 to present. Inquire as to online cost and availability.

Geobase. Elsevier - Geo Abstracts, Regency House, 34 Duke Street, Norwich NR3 3AP, England. Contains citations to the worldwide earth science literature from 1980 to date. Available on DIALOG, ORBIT online services. Inquire as to cost and availability.

NTIS Bibliographic Database. National Technical Information Service, 5285 Port Royal Road, Springfield, VA 22161. (703) 487-4929 or FAX (703) 321-8199. Broad coverage of government-sponsored science and technology research reports, 1964 to present. Available on BRS,(800) 289-4277, DIALOG, (800) 334-2564, ORBIT, (800) 456-7248, and STN International, FIZ Karlsruhe, P.O. Box 2465, W-7500, Karlsruhe 1, Germany, online services. Also available on CD-ROM. Inquire as to cost and availability.

SCISEARCH. Institute for Scientific Information, 3501 Market Street, Philadelphia, PA 19104. (800) 523-1850 or (215) 386-0100. Broad multidisciplinary title and author index to the international literature of science and technology, 1974 to present. Available on DIALOG,(800) 334-2564, and ORBIT,(800) 456-7248, online services. Also available on CD-ROM. Inquire as to cost and availability.

PERIODICALS

Land Degradation and Rehabilitation. J. Wiley & Sons, Journals, 605 Third Avenue, New York, NY 10158-0012. (212) 692-6000. 1989 to present. Quarterly. $165.00.

Landscape and Urban Planning. Elsevier Science Inc., Box 882, Madison Square Station, New York, NY 10159. (212) 989-5800. FAX (212) 633-3990. 1974 to present. Nine times a year. $584.00.

Naslr Newsletter. National Association of State Land Reclamationists, 459B Carlisle Drive, Herndon, VA 22070. (703) 709-8654. FAX (703) 709-8655. Quarterly. Inquire for price and availability.

Restoration and Management Notes. University of Wisconsin Press, Journals Division, 114 N. Murray Street, Madison, WI 53715. (608) 262-4952. FAX (608) 262-7560. 1981 to present. Semi-annual. $18.00.

RESEARCH CENTERS AND INSTITUTES

North Dakota State University, Land Reclamation Research Center. MDSU-Reclamation, PO Box 459, Mandan, ND 58554. (701) 663-6445. FAX (701) 667-1811.

Utah State University, Institute For Land Rehabilitation. College of Natural Resources, Logan, UT 84322-5230. (801) 750-2547.

LANDSAT

See: SATELLITES (ARTIFICIAL)

LASER DISKS

See: OPTICAL DISKS

LASERS

See also: FIBER OPTICS, HOLOGRAPHY, OPTICAL COMMUNICATIONS, OPTICAL DISKS, OPTICS, OPTOELECTRONICS, PHYSICS, SPECTROSCOPY

ABSTRACT SERVICES AND INDEXES

Applied Science and Technology Index; a cumulative subject index to English language periodicals in the fields of aeronautics and space science, computer technology, chemistry, construction industry, energy and related areas. H.W. Wilson Co., 950 University Avenue, Bronx, NY 10452. (212) 588-8400. (800) 367-6770. FAX (718) 590-1617. From 1958 to present. Monthly. Inquire about cost and availability. Also available on CD-ROM and online. ISSN: 0003-6986.

Chemical Abstracts. Chemical Abstracts Service, 2540 Olentangy River Road, Box 3012, Columbus, OH 43210-0012. (614) 447-3600. (800) 848-6538. FAX (614) 447-3713. From 1907 to present. Weekly. $16800.00 per year. Also available on CD-ROM and online. Inquire regarding cost and availability.

Current Contents: Engineering, Technology and Applied Sciences. Institute for Scientific Information, 3501 Market Street, Philadelphia, PA 19104. (215) 386-0100. FAX (215) 386-2291. From 1961 to present. Weekly. $442.00 per year. Also available on CD-ROM and online. Inquire regarding cost and availability. ISSN: 0095-7917.

Current Papers in Physics. Institute of Electrical Engineers, Michael Faraday House, Six Hill Way, Stevenage, Herts, SG1 2AY, England. Distributed by INSPEC/IEEE, Box 1331, 445 Hoes Lane, Piscataway, NJ 08855-1331. (908) 562-5549. 1966 to present. Fortnightly. $410.00 per year. ISSN: 0011-3786.

Electrical and Electronics Abstracts. Science Abstracts Section B. Institute of Electrical Engineers, Michael Faraday House, Six Hill Way, Stevenage, Herts, SG1 2AY, England. Distributed by IEEE, 445 Hoes Lane, Piscataway, NJ 08854. (908) 562-5549. 1898 to present. Monthly. $2200.00 per year. Also available on CD-ROM and online as INSPEC. ISSN: 0036-8105.

Engineering Index Monthly. Engineering Information, Inc., Castle Point on the Hudson, Hoboken, NJ 07030. (201) 216-8500. FAX (201) 216-8532. Monthly. $2200.00 per year. Also available online as COMPENDEX and on CD-ROM. ISSN: 0742-1974.

General Science Index. H.W. Wilson Company, 950 University Avenue, Bronx, NY 10452. (212) 588-8400. (800) 367-6770. FAX (718) 590-1617. From 1978 to present. Ten issues per year; quarterly and annual cumulations. Service basis. Available on CD-ROM and online. Inquire regarding cost and availability. ISSN: 0162-1963.

Government Reports Announcements and Index. U. S. National Technical Information Service (NTIS), 5285 Port Royal Road, Springfield, VA 22161. (703) 487-4650. FAX (703) 321-8547. From 1968 to present. Annual. $630.00 per year. Also available online as *NTIS Bibliographic Database* and on CD-ROM.

NTIS Alerts: Electrotechnology. U. S. National Technical Information Service. 5285 Port Royal Road, Springfield, VA 22161. (703) 487-4650. FAX (703) 321-8547. Weekly. $140.00 per year.

Physics Abstracts. INSPEC. Section A, Science Abstracts. Institution of Electrical Engineers (IEE). Available from INSPEC/IEEE - Institute of Electrical and Electronic Engineers,, Box 1331, 445 Hoes Lane, Piscaway NJ 08855-1331. (908) 562-5549. 1898 to present. 24 issues per year. $2835.00 per year. Also available on CD-ROM and online. ISSN: 0036-8091.

Physics Briefs (Physikalische Berichte). Information Center for Energy, Physics, Mathematics; German Physical Society. V C H Publishers, Inc., 220 East 23rd Street, New York, NY 10010-4606. (212) 683-8333. 1845 to present. 24 issues per year. $2390.00 per year. Also available online. ISSN: 0179-7434.

Science Citation Index. SCI. Institute for Scientific Information, 3501 Market Street, Philadelphia, PA 19104. (215) 386-0100. (800) 523-1850. FAX (215) 386-2991. 1961 to present. Six issues per year, plus annual cumulation. $11650.00 per year. Also available online and on CD-ROM. Inquire about price and availability. ISSN: 0036-827X.

ANNUAL REVIEWS AND YEARBOOKS

Advanced Solid State Lasers. Optical Society of America, 2010 Massachusetts Avenue NW, Washington, DC 20036. (202) 223-8130. FAX (202) 223-1096. Irregular. $82.00.

Advances in Electronics and Electron Physics. Academic Press, Inc., 6277 Sea Harbor Drive, Orlando, FL. (800) 321-5068. From 1948 to present. Irregular. ISSN: 0065-2539.

Progress in Optics. Elsevier Science Inc., 660 White Plains Road, Tarrytown, NY 10591-5153. (914) 524-9200. FAX (914) 333-2444. 1961 to present. Irregular. Price varies. ISSN: 0079-6638.

Springer Series in Optical Sciences. Springer-Verlag New York, Inc., 175 Fifth Avenue, New York, NY 10010. (212) 460-1500. (800) 777-4643. FAX (212) 473-6272. 1976 to present. Irregular. Price varies.

ASSOCIATIONS AND PROFESSIONAL SOCIETIES

American Institute of Physics. One Physics Ellipse, College Park, MD 20740-3843. (301) 209-3100.

IEEE Lasers and Electro-optics Society. c/o IEEE, 445 Hoes Lane, PO Box 1331, Piscataway, NJ 08855. (908) 562-3892. FAX (908) 562-8434.

Institute of Electrical and Electronics Engineers. 345 East 47th Street, New York, NY 10017. (212) 705-7900. FAX (212) 705-4929.

Laser Institute of America. 12424 Research Parkway, Suite 130, Orlando, FL 32926. (407) 380-1553. FAX (407) 380-5588.

Optical Society of America. 2010 Massachusetts Avenue NW, Washington, DC 20036. (202) 223-8130. FAX (202) 223-1096.

Society For Imaging Science and Technology. 7003 Kilworth Lane, Springfield, VA 22151. (703) 642-9090. FAX (703) 642-9094.

Spie - the International Society For Optical Engineering. PO Box 10, 1000 20th Street, Bellingham, WA 98227. (206) 676-3290. FAX (206) 647-1445.

LASERS

Ency. of Physical Sciences and Engineering Info. Sources

BIBLIOGRAPHIES

Science Books and Films. American Association for the Advancement of Science, 1333 H Street NW, Washington, DC 20005. (202) 326-6454. Reviews of print, film and software materials in all sciences. 1965 to present. Nine issues per year. $40.00 per year. ISSN: 0098-342X

Scientific and Technical Books and Serials in Print, 1995; An Index To Literature in Science and Technology. R.R. Bowker Inc., 121 Chanlon Road, New Providence, NJ 07974. (908) 464-6800. (800) 521-8110. 4 volumes. 1972 to present. Annual. $299.95. Also available on compact disc and online. ISSN: 0000-054X.

DIRECTORIES AND BIOGRAPHICAL SOURCES

American Men and Women of Science: Physical and Biological Sciences. R. R. Bowker Inc., 121 Chanlon Road, New Providence, NJ 07974. (908) 464-6800. (800) 521-8110. 20th edition. 8 volumes. 1996. $850.00.

American Physical Society. Membership Directory Bulletin Issue. American Physical Society, One Physics Ellipse, College Park, MD 20740-3843. (301) 209-3200. FAX (301) 209-0865. Biennial.

IEEE Membership Directory. Institute of Electrical and Electronics Engineers. 345 East 47th Street, New York, NY 10017. (212) 705-7900. FAX (212) 705-4929.

International Directory of Engineering Societies and Related Organizations. American Society for Engineering Education, 1818 N Street NW, Suite 600, Washington, DC 20036. (202) 331-3526. 15th edition. 1996. $185.00.

Laser Focus World Buyers Guide. PennWell Publishing Co., 1421 Sheridan Road, Tulsa, OK 74112. (918) 831-9421. Annual. $75.00.

Laser Pioneers. Jeff Hecht. Academic Press, Inc., 6277 Sea Harbor Drive, Orlando, FL. (800) 321-5068. 2nd edition. 1991. $49.95.

Lasers and Optoelectronics Buying Guide. Gordon Publications. P.O. Box 650, Morris Plains, NJ 07950. (908) 292-5100. Annual. $95.00.

Optical Society of America. Membership Directory. Optical Society of America. 2010 Massachusetts Avenue NW, Washington, DC 20036-1023. (202) 223-8130. (800) 762-6960. Biennial. $12.00. $95.00.

Optics Education. SPIE - the International Society for Optical Engineering, PO Box 10, 1000 20th Street, Bellingham, WA 98227. (206) 676-3290. FAX (206) 647-1445. Annual. No charge.

Research Centers Directory. Gale Research Inc., 835 Penobscot Building, Detroit, MI 48226-4094. (313) 961-2242. (800) 877-4253. 20th edition. 1995. $485.00. ISSN: 0080-1518.

Scientific and Technical Organizations and Agencies Directory. Gale Research, 835 Penobscot Building, Detroit, MI 48226-4094. (313) 961-2242. (800) 877-4253. 4th edition. 1996. $195.00.

Who's Who in Engineering. Gordon Davis, editor. American Society for Engineering Education, 1111 19th Street, NW, Suite 608, Washington, DC 20036. 9th edition. 1995. $220.00.

Who's Who in Technology: Biographies and Index. Amy L. Unterburger, editor. Gale Research, 835 Penobscot Building, Detroit, MI 48226-4094. (313) 961-2242. (800) 877-4253. 7th edition. 1995. $195.00.

ENCYCLOPEDIAS AND DICTIONARIES

Encyclopedia of Applied Physics. George Trigg, editor. VCH Publications, Inc., 220 East 23rd Street, Suite 909, New York, NY 10010-4606. (212) 683-8333. (800) 422-8824. 20 volume set. 1991-. $5990.00.

Encyclopedia of Lasers and Optical Technology. Robert A. Meyers, editor. Academic Press, Inc., 6277 Sea Harbor Drive, Orlando, FL. (800) 321-5068. 1991. $92.00.

Encyclopedia of Physical Science and Technology. Academic Press, Inc., 6277 Sea Harbor Drive, Orlando, FL. (800) 321-5068. 2nd edition. 18 volume set. 1992. $2625.00.

Fiber Optics Standard Dictionary. Martin H. Weik. 2nd edition. Van Nostrand Reinhold, 115 Fifth Avenue, New York, NY 10003. (212) 254-3232. (800) 842-3636. 2nd edition. 1989. $49.95.

Glossary of Optical Terminology. Thomas K. Farrell. Butterworth-Heinemann, 313 Washington Avenue, Newton, MA 02158. (617) 928-2500. (800) 366-2665. 1985. $24.95.

McGraw-Hill Encyclopedia of Science and Technology. McGraw-Hill Book Company, Inc., 1221 Avenue of the Americas, New York, NY 10020. (212) 512-2000. (800) 262-4729. 7th edition. 20 volume set. 1992. $1900.00.

GENERAL WORKS

Coherent Optics. W. Lauterborn, et al. Springer-Verlag New York, Inc., 175 Fifth Avenue, New York, NY 10010. (212) 460-1500. (800) 777-4643. FAX (212) 473-6272. 1995. $49.00.

Fundamentals of Laser Optics. K. Iga. Plenum Publishing Corp., 233 Spring Street, New York, NY. (212) 620-8000. (800) 2221-9369. FAX (212) 463-0742. 1994. $49.50.

Integrated Optoelectronics: Waveguide Optics, Photonics, Semiconductors. Karl J. Ebling. Springer-Verlag New York, Inc., 175 Fifth Avenue, New York, NY 10010. (212) 460-1500. (800) 777-4643. FAX (212) 473-6272. 1993. $64.50.

Introduction To Laser Physics. K. Shimoda. Springer-Verlag New York, Inc., 175 Fifth Avenue, New York, NY 10010. (212) 460-1500. (800) 777-4643. FAX (212) 473-6272. 2nd edition. 1991. $49.00 in paper.

Lasers and Holography. V. V. Rampal and P. C. Mehta. World Scientific Publishing Company, Inc., 1060 Main Street, River Edge, NJ 07661. (201) 487-9655. (800) 227-7562. 1993. $98.00.

Lasers and Holography: An Introduction To Coherent Optics. Winston E. Kock. Revised edition. Dover Publications, Inc., 180 Varick Street, New York, NY 10014. (212) 255-3755. (800) 223-3130. 1981. $4.95.

Lasers and Optical Engineering. P. K. Das. Springer-Verlag New York, Inc., 175 Fifth Avenue, New York, NY 10010. (212) 460-1500. (800) 777-4643. FAX (212) 473-6272. 1990.$65.00.

Lasers: Invention To Application. National Academy of Engineering Staff. National Academy Press, 2101 Constitution Avenue NW, Washington, DC 20055. (202) 334-3313. (800) 624-6242. 1987. $14.95.

Lasers: Theory and Practice. John Hawkes and Ian Latimer. Prentice-Hall, 113 Sylvan Avenue, Route 9W, Englewood Cliffs, NJ 07632. (201) 592-2000. (800) 922-0579 1993. 1994. $39.00.

Nonlinear Fiber Optics. Govind P. Agrawal, Academic Press, Inc., 6277 Sea Harbor Drive, Orlando, FL. (800) 321-5068. 2nd edition. 1995. $59.95.

Optics and Lasers: Including Fibers and Optical Waveguides. M. Young. Springer-Verlag New York, Inc., 175 Fifth Avenue, New York, NY 10010. (212) 460-1500. (800) 777-4643. FAX (212) 473-6272. 4th revised edition. 1993. $43.50.

Principles of Optics, Optoelectronics and Photonics. Alan Billings. Prentice-Hall, 113 Sylvan Avenue, Route 9W, Englewood Cliffs, NJ 07632. (201) 592-2000. (800) 922- 0579. 1994. $40.00 in paper.

Semiconductor Laser Physics. W. W. Chow, et al. Springer-Verlag New York, Inc., 175 Fifth Avenue, New York, NY 10010. (212) 460-1500. (800) 777-4643. FAX (212) 473-6272 1994. $49.95.

Studies in Classical and Quantum Nonlinear Optics. Ole Keller, editor. Nova Science Publishers, Inc., 6080 Jericho Turnpike, Suite 207, Commack, NY 11725-2808. (516) 499- 3103. 1994. $98.00.

Tunable Lasers. L. F. Mollenauer, et al, editors. Springer-Verlag New York, Inc., 175 Fifth Avenue, New York, NY 10010. (212) 460-1500. (800) 777-4643. FAX (212) 473- 6272. 2nd edition. 1992. $69.00 in paper.

HANDBOOKS AND MANUALS

Handbook of Nonlinear Optical Crystals. V. G. Dmitriev, et al. Springer-Verlag New York, Inc., 175 Fifth Avenue, New York, NY 10010. (212) 460-1500. (800) 777-4643. FAX (212) 473-6272. Revised edition. 1991. $69.00.

Handbook of Optics. Optical Society of America. McGraw- Hill Publishing Company, Inc., 1221 Avenue of the Americas, New York, NY 10020. (212) 512-2000. (800) 262- 4729. 2 volumes. 2nd edition. 1993. $99.50 each.

Industrial Laser Handbook. M. Levitt and D. Belforte, editors. Springer-Verlag New York, Inc., 175 Fifth Avenue, New York, NY 10010. (212) 460-1500. (800) 777-4643. FAX (212) 473-6272. 1992. $163.00.

Laser Experimenter's Handbook. Delton T. Horn. TAB Books, P.O. Box 40, Blue Summit, PA 17294-0850. (717) 794-2191. (800) 233-1128. 2nd edition. 1988. $21.95.

Laser Handbook. F. T. Areechi, E. D. Schultz-Dubois, et al, editors. Elsevier Science Inc., 660 White Plains Road, Tarrytown, NY 10591-5153. (914) 524-9200. FAX (914) 333- 2444. 6 volumes. 1973 -

ONLINE DATABASES AND CD-ROMS

CA Search. Chemical Abstracts Service, P.O. Box 3012, Columbus, OH 43210-0012. (614) 447-3600. (800) 848-6533. FAX (614) 447-3709. Very comprehensive guide to worldwide chemical literature and related fields, 1972 to present. Available on BRS,(800) 289-4277, DIALOG, (800) 334-2564, ORBIT (800) 456-7248, and STN International, FIZ Karlsruhe, P.O. Box 2465, W-7500, Karlsruhe 1, Germany, online services. Inquire as to cost and availability.

Compendex Plus. Engineering Information, Inc., 345 East 47th Street, New York, NY 10017. (212) 705-7600 or (800) 221-1044. Contains citations with abstracts to worldwide literature in engineering and technology, from 1970 to present. Available on online BRS,(800) 289-4277, DIALOG, (800) 334-2564, ORBIT (800) 456-7248, and STN International, FIZ Karlsruhe, P.O. Box 2465, W-7500, Karlsruhe 1, Germany, online services. Also available on CD-ROM. Inquire as to cost and availability.

Current Contents Search. Institute for Scientific Information, 3501 Market Street, Philadelphia, PA 19104. (215) 386-0100. FAX (215) 386-6362. Contains citations to articles listed in the table of contents of science and technology journals. Also articles in social sciences and life sciences journals. Available on BRS, (800) 289-4277, DIALOG, (800) 334-2564, online services. Inquire as to cost and availability.

Dissertation Abstracts Online. University Microfilms International, 300 North Zeeb Road, Ann Arbor, MI 48106. (800) 521-0600 or (313) 761-4700. Scope includes virtually all doctoral dissertations accepted at accredited American institutions from 1861 to present in 252 subject areas. Available on BRS, (800) 289-4277, DIALOG, (800) 334-2564, and OCLC EPIC, (800) 848-5878, online services. Also available on CD-ROM. Inquire as to cost and availability.

INSPEC. Institution of Electrical Engineers, Michael Faraday House, Six Hills Way, Stevenage, Herts. SG1 2AY, England. Telephone: 0438 313311 or FAX 0438 742840. Contains citations to the worldwide literature of physics, electronics and electrical engineering, computer technology, and related fields. Available on BRS, (800) 289-4277, DIALOG, (800) 334-2564, ORBIT, (800) 456-7248, and STN International, FIZ Karlsruhe, P.O. Box 2465, W-7500, Karlsruhe 1, Germany, online services. Inquire as to cost and availability.

NTIS Bibliographic Database. National Technical Information Service, 5285 Port Royal Road, Springfield, VA 22161. (703) 487-4929 or FAX (703) 321-8199. Broad coverage of government-sponsored science and technology research reports, 1964 to present. Available on BRS,(800) 289-4277, DIALOG, (800) 334-2564, ORBIT, (800) 456-7248, and STN International, FIZ Karlsruhe, P.O. Box 2465, W-7500, Karlsruhe 1, Germany, online services. Also available on CD-ROM. Inquire as to cost and availability.

Physics Briefs. American Institute of Physics, 335 East 45th Street, New York, NY 10017. (212) 661-9260 or FAX (212) 661-2036. Contains citations with abstracts of the literature of physics and related fields, 1979 to present. Available on the STN International, FIZ Karlsruhe, P.O. Box 2465, W-7500, Karlsruhe 1, Germany, online service. Inquire as to cost and availability.

Scientific and Technical Books and Serials in Print. R.R. Bowker Inc., 121 Chanlon Road, New Providence, NJ 07974. (800) 521-8110 or (908) 464-6800. List of currently published books and serials in the physical and biological sciences, engineering and technology, with subject, author and titles indexes. Available on ORBIT, (800) 456-7248, online service. Inquire as to cost and availability.

SCISEARCH. Institute for Scientific Information, 3501 Market Street, Philadelphia, PA 19104. (800) 523-1850 or (215) 386-0100. Broad multidisciplinary title and author index to the international literature of science and technology, 1974 to present. Available on DIALOG, (800) 334-2564, and ORBIT, (800) 456-7248, online services. Also available on CD-ROM. Inquire as to cost and availability.

WILSONLINE. H.W. Wilson Company, 950 University Avenue, Bronx, NY 10452. (800) 367-6770 or (212) 588-8400. Makes available online versions of the printed H.W. Wilson indexes including Applied Science and Technology Index, Business Periodicals Index, General Science Index, and Readers' Guide to Periodical Literature. Period covered is generally 1983 to present. Available on BRS, (800) 289-4277, DIALOG, (800) 334-2564, and OCLC EPIC, (800) 848-5878, online services. Also available on CD-ROM. Inquire as to cost and availability.

PERIODICALS

Applied Optics. Optical Society of America. 2010 Massachusetts Avenue NW, Washington, DC 20036-1023. (202) 223-8130. 1962 to present. 36 issues per year. $1090.00 per year. ISSN: 0003-6935.

Applied Physics B: Photophysics and Laser Chemistry. Springer-Verlag New York, Inc., 44 Hartz Way, Seacucus, NJ 07096-2491. (201) 348-4033. FAX (201) 348-4505. 1981 to present. Monthly. $1030.00 per year. ISSN: 0721-7269.

Fiber and Integrated Optics. Taylor and Francis, 1900 Frost Road, Suite 101. Bristol, PA 19007. (215) 785-5800. FAX (215) 785-5515. 1977 to present. Quarterly. $180.00 per year. ISSN: 0146-8030.

IEEE Journal of Quantum Electronics. IEEE Lasers and Electro-optics Society/IEEE, 445 Hoes Lane, PO Box 1331, Piscataway, NJ 08855. (908) 981-0060. FAX (908) 981-9667. 1965 to present. Monthly. $535.00 per year. ISSN: 0018- 9197.

International Journal of Nonlinear Optical Physics. World Scientific Publishing Co., 1060 Main Street, Suite 18, River Edge, NJ 07661. (800) 227-7562. 1992 to present. Quarterly. $348.00 per year. ISSN: 0218-1991.

Journal of Applied Physics. American Institute of Physics. One Physics Ellipse, College Park, MD 20740-3843. (301) 209-3100. 1931 to present. Semi-monthly. $1655.00 per year. ISSN: 0021-8979.

Journal of Laser Applications. Laser Institute of America, 12424 Research Parkway, Suite 130, Orlando, FL 32926. (407) 380-1553. FAX (407) 380-5588.

Journal of Lightwave Technology. Institute of Electrical and Electronics Engineers, 445 Hoes Lane, PO Box 1331, Piscataway, NJ 08855. (908) 562-3892. FAX (908) 562-8434. 1983 to present. Monthly. $425.00 per year. ISSN: 0733- 8724.

Journal of the Optical Society of America. Part A: Optics and Image Science. Optical Society of America, Inc., 2010 Massachusetts Avenue, NW, Washington, DC 20036-1023. (202)223-8130. 1917 to present. Monthly. $610.00 per year. ISSN: 0740-3232.

Journal of the Optical Society of America. Part B: Optical Physics. Optical Society of America, Inc., 2010 Massachusetts Avenue, NW, Washington, DC 20036-1023. (202) 223-8130. 1917 to present. Monthly. $610.00 per year. ISSN: 0740-3224.

Laser Focus World. PennWell Publishing Co., 10 Tara Boulevard, Nashua, NH 03062-2891. (603) 891-0123. FAX(603) 891-0574. 1964 to present. Monthly. $104.00 per year. ISSN: 1043-8092.

Optical Engineering. SPIE - International Society for Optical Engineering, Box 10, 1000 20th Street, Bellingham, WA 98227-0010. (206) 676-3290. FAX (206) 647-1445. 1962 to present. Monthly. $170.00 per year. ISSN: 0091-3286.

Optics and Photonics News. Optical Society of America, Inc., 2010 Massachusetts Avenue, NW, Washington, DC 20036-1023. (202) 223-8130. 1975 to present. Monthly. $99.00 per year. ISSN: 1047-6938.

Optics and Lasers in Enginering. Elsevier Science, 660 White Plains Road, Tarrytown, NY 10591-5153. (914) 545- 9200. FAX (914) 333-2444. 1980 to present. Ten issues per year. $500.00 per year. ISSN: 0143-8166.

Optics Communications. Elsevier Science, 660 White Plains Road, Tarrytown, NY 10591-5153. (914) 545-9200. FAX (914) 333-2444. 1969 to present. 54 issues per year. $2121.00. per year. ISSN: 0030-4018.

Optics Letters. Optical Society of America, Inc., 2010 Massachusetts Avenue, NW, Washington, DC 20036-1023. (202) 223-8130. 1977 to present. Semi-Monthly. $625.00 per year. ISSN: 0146-9592.

Optics Materials. Elsevier Science Inc., Box 882, Madison Square Station, New York, NY 10159. (212) 989-5800. FAX (212) 633-4990. 1991 to present. Quarterly. $206.00 per year. ISSN: 0925-3467.

Spie - International Society For Optical Engineering. Proceedings. SPIE - International Society for Optical Engineering, Box 10, 1000 20th Street, Bellingham, WA 98227-0010. (206) 676-3290. FAX (206) 647-1445. 1963 to present. Approximately 50 numbers per year. Individually priced.

RESEARCH CENTERS AND INSTITUTES

Center for Applied Optics. University of Alabama in Huntsville, Research Institute Building, Huntsville, AL 25899. (205) 895-6102.

Center For Laser Science and Engineering. University of Iowa, Iowa City, IA 52242. (319) 335-4580. FAX (319) 335-2951.

Center For Optics, Lasers and Holography. New York Institute of Technology, 100 Glen Cove, NY 11542. (516) 686-7863.

Center For Research in Electro-optics and Lasers.University of Central Florida, 12424 Research Parkway, Orlando, FL 32826. (407) 658-6800. FAX (407) 658-6880.

Laboratory For Laser Energetics. University of Rochester, 250 East River Road, Rochester, NY 14623-1299. (716) 2752-5101.

Laser Research Center. Massachusetts Institute of Technology, 77 Massachusetts Avenue, Cambridge, MA 02139. (617) 253-7700.

LEAD

See also: METALLURGY, METALS AND METALWORKING

ABSTRACT SERVICES AND INDEXES

Applied Science and Technology Index; A Cumulative Subject Index To English Language Periodicals in the Fields of Aeronautics and Space Science, Computer Technology, Chemistry, Construction Industry, Energy and Related Areas. H.W. Wilson Co., 950 University Avenue, Bronx, NY 10452. (800) 367-6770 or (212) 588-8400. FAX (718) 590-1617. From 1958 to present. Monthly. Inquire about cost and availability. Also available on CD-ROM and online.

Current Contents: Engineering, Technology, and Applied Sciences. Institute for Scientific Information, 3501 Market Street, Philadelphia, PA 19104. (215) 386-0100. FAX (215) 386-6362. 1970 to present. Weekly. $442.00 per year.

Current Contents: Physical, Chemical and Earth Sciences. Institute for Scientific Information, 3501 Market Street, Philadelphia, PA 19104. (215) 386-0100. FAX (215) 386-6362. Weekly. $360.00 per year.

Engineering Index Monthly. Engineering Information, Inc., Castle Point on the Hudson, Hoboken, NJ 07030. (800) 221-1044. FAX (212) 832-1857. Monthly. $2300.00 per year. Also available online as COMPENDEX, and also on CD-ROM. Covers chemical engineering, computers, electrical engineering, civil engineering, metals and mining, industrial management, and mechanical engineering.

Environmental Abstracts. R.R. Bowker Inc., 121 Chanlon Road, New Providence, NJ 07974. (800) 521-8110 or (908) 464-6800. Contains citations, most with abstracts, to literature relating to environmental issues and problems, from 1971 to present. Inquire as to cost and availability.

Environmental Engineering Abstracts, Cambridge Scientific Abstracts, 7200 Wisconsin Avenue, Bethesda, MD 20814-4823. (301) 961-6750. FAX (301) 961-6720. Monthly. Covers hazardous materials, environmental impact and protection, treatment of sewage and industrial wastes, hydroelectric power, tidal and wind power, artic and tropical engineering.

Index To Scientific and Technical Proceedings. Institute for Scientific Information, 3501 Market St., Philadelphia, PA 19104. (215) 386-0100. FAX (215) 386-6362. Monthly. $500.00 per year.

Index To Scientific Reviews. Institute for Scientific Information, 3501 Market St., Philadelphia, PA 19104. (215) 386-0100. FAX (215) 386-6362. Semi-annual.

Leadscan (formerly Lead Abstracts), Lead Development Association, 42 Weymouth Street, London W1x 3lq, England. 1958 to present. Quarterly. Inquire for cost.

Metals Abstracts and Metals Abstracts Index. American Society for Metals, Metals Park, OH 44073. (216) 338-5151. 1968 to present. Monthly. Abstracts are $1500.00 per year and Index is $460.00 per year.

Science Citation Index. Institute for Scientific Information, 3501 Market Street, Philadelphia, PA 19104. (215) 386-0100. FAX (215) 386-6362. Inquire about availability and cost. Also available on CD-ROM.

ASSOCIATIONS AND PROFESSIONAL SOCIETIES

American Society For Metals. Materials Park, OH 44073. (216) 338-5151 or FAX (216) 338-4634.

International Lead and Zinc Study Group. 58 St.Jame's Street, London SW1A 1LD, England. Telephone 071-499-9373.

International Lead-zinc Research Organization. 2525 Meridian Parkway, PO Box 12036, Research Triangle Park, NC 27709. (919) 361-4647.

Lead Development Association. 42 Weymouth Street, London W1X 3LQ, England.

DIRECTORIES AND BIOGRAPHICAL SOURCES

Directory of Engineering Societies and Related Organizations. Gordon Davis. 13th edition. American Association of Engineering Societies, 1111 19th Street NW, Suite 608, Washington, DC 20036. (202) 296-2237 or (800) 658-8897. 1989. Inquire for price.

International Research Centers Directory. Gale Research, 835 Penobscot Building, Detroit, MI 48226-4094. (313) 961-2242. (800) 877-4253. FAX (313) 961-6083. 8th edition. 1995. $410.00

Lead Poisoning Prevention: Directory of State Contacts 1993. Doug Farquhar & Linda South, compilers. National Conference of State Legislatures, 1560 Broadway, No. 700, Denver, CO 80202-5140. (303) 830-2200. Inquire for cost and availability.

1992 Census of Mineral Industries, Preliminary Report, Industry Series: Copper Ores, Lead and Zinc Ores, Gold Ores, and Silver Ores. U.S. Department of Commerce, Bureau of the Census. Superintendent of Documents, U.S. Government Printing office, Box 371954, Pittsburgh, PA 15250-7954. (202) 783-3238. FAX (202) 512-2233. 1994. Inquire for cost and availability.

Metal Finishing Guidebook & Directory. Elsevier Science Publishing Company, Inc., 655 Avenue of the Americas, New York, NY 10010. (212) 989-5800. Annual. Supplement to *Metal Finishing* periodical.

Research Centers Directory. Gale Research, 835 Penobscot Building, Detroit, MI 48226-4094. (313) 961-2242. (800) 877-4253. FAX (313) 961-6083. $485.00.

Scientific and Technical Organizations and Agencies Directory. Gale Research, 835 Penobscot Building, Detroit, MI 48226-4094. (313) 961-2242. (800) 877-4253. FAX (313) 961-6083. 4th edition. 1996. $195.00.

Who's Who in Engineering. American Association of Engineering Societies, 1111 19th Street NW, Suite 608, Washington,

LEAD

Ency. of Physical Sciences and Engineering Info. Sources

DC 20036. (202) 296-2237 or (800) 658-8897. 8th edition. 1991. Inquire for price.

ENCYCLOPEDIAS AND DICTIONARIES

Encyclopedia of Physical Science and Technology. Robert A. Meyers, ed. 18 volumes. Academic Press Inc., 6277 Sea Harbor Drive, Orlando, FL 32887. (800) 321-5068. 1992. $2100.00.

McGraw-Hill Dictionary of Scientific and Technical Terms. Sybil P. Parker, ed. 5th edition. McGraw-Hill Publishing Company, 1221 Avenue of the Americas, New York, NY 10020. (800) 262-4729 or (212) 512-3825. 1993. $110.50.

GENERAL WORKS

The Availability of Primary Lead and Zinc in the Market Economy Countries, A Minerals Availability Appraisal. Carl A. DiFrancesco, et al. U.S. Department of the Interior, Bureau of Mines. Superintendent of Documents, U.S. Government Printing office, Box 371954, Pittsburgh, PA 15250-7954. (202) 783-3238. FAX (202) 512-2233. 1993. Inquire for cost and availability.

Lead. Gordon Press Publishers, PO Box 459, Bowling Green Station, New York, NY 10004. (718) 624-8419. 1993. $250.95.

Lead: A Guidebook To Hazard Detection, Remediation, and Control. Paul N. Cheremisinoff & Nicholas P. Cheremisinoff. Prentice Hall, 113 Sylvan Avenue, Route 9W, Englewood Cliffs, NJ 07632. (201) 592-2000 or (800) 922-0579. 1993. $50.00.

Lead and Its Alloys. D.R. Blaskett & D. Boxall. E. Horwood Publishing, distributed by Prentice 113 Sylvan Avenue, Route 9W, Englewood Cliffs, NJ 07632. (201) 592-2000 or (800) 922-0579. 1990. Inquire for cost and availability.

Lead in Food. Torsten Berg. Council of Europe Press/ Manhattan Publishing Company, PO Box 650, Croton-on-Hudson, NY 10520. (914) 271-5194. 1994. Inquire for cost and availability.

HANDBOOKS AND MANUALS

Smithell's Metals Reference Book. E.A. Brandes & G.B. Brook. 7th edition. Butterworth-Heinemann, 313 Washington Street, Newton, MA 02158. (617) 928-2500 or (800) 366-2665. FAX (617) 928-2620. 1992. $250.00.

ONLINE DATABASES AND CD-ROMS

Compendex Plus. Engineering Information, Inc., 345 East 47th Street, New York, NY 10017. (212) 705-7600 or (800) 221-1044. Contains citations with abstracts to worldwide literature in engineering and technology, from 1970 to present. Available on online BRS,(800) 289-4277, DIALOG, (800) 334-2564, ORBIT (800) 456-7248, and STN International, FIZ Karlsruhe, P.O. Box 2465, W-7500, Karlsruhe 1, Germany, online services. Also available on CD-ROM. Inquire as to cost and availability.

CSA Engineering. Cambridge Scientific Abstracts, 7200 Wisconsin Avenue, Suite 601, Bethesda, MD 20814. (301) 961-6750 or (800) 843-7751. Contains citations and abstracts of international periodicals and research literature covering all fields of engineering and science and technology, including computer

and information science, electronics, mechanical engineering, solid state materials, 1981 to present. Available on BRS,(800) 289-4277, online service.Inquire as to cost and availability.

Current Contents Search. Institute for Scientific Information, 3501 Market Street, Philadelphia, PA 19104. (215) 386-0100. FAX (215) 386-6362. Contains citations to articles listed in the table of contents of science and technology journals. Also articles in social sciences and life sciences journals. Available on BRS,(800) 289-4277, DIALOG,(800) 334-2564, online services. Inquire as to cost and availability.

Dissertation Abstracts. University Microfilms International, 300 North Zeeb Road, Ann Arbor, MI 48106. (800) 521-0600 or (313) 761-4700. Scope includes virtually all doctoral dissertations accepted at accredited American institutions from 1861 to present in 252 subject areas. Available on BRS,(800) 289-4277, DIALOG,(800) 334-2564, and OCLC EPIC,(800) 848-5878, online services. Also available on CD-ROM. Inquire as to cost and availability.

Metadex. Jointly produced by ASM International and the Institute of Materials. Contains more than 925,000 records from the international literature on metals and alloys, concerning properties, processes, materials classes, applications, and metallurgical systems. Updated monthly. Available from ORBIT-QUESTEL (703) 442-0900.

NTIS Bibliographic Database. National Technical Information Service, 5285 Port Royal Road, Springfield, VA 22161. (703) 487-4929 or FAX (703) 321-8199. Broad coverage of government-sponsored science and technology research reports, 1964 to present. Available on BRS,(800) 289-4277, DIALOG, (800) 334-2564, ORBIT, (800) 456-7248, and STN International, FIZ Karlsruhe, P.O. Box 2465, W-7500, Karlsruhe 1, Germany, online services. Also available on CD-ROM. Inquire as to cost and availability.

SCISEARCH. Institute for Scientific Information, 3501 Market Street, Philadelphia, PA 19104. (800) 523-1850 or (215) 386-0100. Broad multidisciplinary title and author index to the international literature of science and technology, 1974 to present. Available on DIALOG,(800) 334-2564, and ORBIT,(800) 456-7248, online services. Also available on CD-ROM. Inquire as to cost and availability.

WILSONLINE. H.W. Wilson Company, 950 University Avenue, Bronx, NY 10452. (800) 367-6770 or (212) 588-8400. Makes available online versions of the printed H.W. Wilson indexes including Applied Science and Technology Index, Business Periodicals Index, General Science Index, and Readers' Guide to Periodical Literature. Period covered is generally 1983 to present. Available on BRS,(800) 289-4277, DIALOG,(800) 334-2564, and OCLC EPIC,(800) 848-5878, online services. Also available on CD-ROM. Inquire as to cost and availability.

PERIODICALS

Engineering & Mining Journal. Maclean Hunter Publishing Company, 29 N. Wacker Dr., Chicago, IL 60606. (312) 726-2802. FAX (312) 726-2574. Monthly. $60.00 per year.

Inorganic Chemistry. American Chemical Society, 1155 16th Street NW< Washington, DC 20036. (800) 333-9511. FAX (914) 447-3671. 1962 to present. Semi-monthly. $949.00 per year.

Institution of Mining and Metallurgy. Transactions. Section A: Mining Industry. Institution of Mining and Metallurgy, 44 Port-

Ency. of Physical Sciences and Engineering Info. Sources

LIGHTNING

land Pl., London W1N 4BR, England. 071-580-3802. FAX 071-436-5388. Three times a year. $126.00 a year.

Iron Age. Hitchcock Publishing, 191 S. Gary Avenue, Carol Stream, IL 60188. (708) 462-4641. FAX (708) 462-2205. 1867-1993. Monthly. $55.00 per year.

Lead and Zinc Statistics. International Lead and Zinc Study Group, 58 St.Jame's Street, London SW1A 1LD, England. Telephone 071-499-9373. 1961 to present. Monthly. $75.00 per year.

Main Group Metal Chemistry. Freund Publishing House Ltd., Suite 500, Chelsea House, 150 Regent Street, London W1R 5FA, England. 1972 to present. Monthly. $360.00 per year.

Metal Bulletin Monthly. Metal Bulletin Inc., 220 Fifth Avenue, 19th Floor, New York, NY 10001-7781. (800) 638-2525. FAX (212) 213-6273. 1972 to present. Monthly. Inquire for cost.

Metal Finishing. Elsevier Science Publishing Company, Inc., 655 Avenue of the Americas, New York, NY 10010. (212) 989-5800. 1903 to present. 12 times a year. $40.00 per year.

Mining Engineer. Institution of Mining Engineers, Danum House, South Parade, Doncaster DN1 2DY, England. Phone 0302-320486. FAX 0302-340554. Monthly. Inquire for price.

Mining Engineering. Society for Mining, Metallurgy, and Exploration, Inc., 8307 Shaffer Parkway, PO Box 625002, Littleton, CO 80162-5002. (303) 973-9550. FAX (303) 973-3845. Monthly. $100.00 per year.

R&d Focus. International Lead Zinc Research Organization Inc., 2525 Meridian Parkway, PO Box 12036, Research Triangle Park, North Carolina 27709. (919) 361-4647. 1986 to present. 3 times a year. Free.

RESEARCH CENTERS AND INSTITUTES

International Lead Zinc Research Organization Inc., 2525 Meridian Parkway, PO Box 12036, Research Triangle Park, North Carolina 27709. (919) 361-4647.

SPECIFICATIONS AND STANDARDS

*Annual Book of Astm Standards, Volume 02.*04. American Society for Testing and Materials (ASTM), 1916 Race Street, Philadelphia, PA 19103. (215) 299-5585. Annual. Inquire for cost and availability.

LENSES

See: OPTICS

LEPTONS

See: PARTICLE PHYSICS

LIGHTING

See: ILLUMINATION

LIGHTNING

See also: METEORLOGY, RAIN, THUNDERSTORMS

ABSTRACT SERVICES AND INDEXES

Applied Science and Technology Index; A Cumulative Subject Index To English Language Periodicals in the Fields of Aeronautics and Space Science, Computer Technology, Chemistry, Construction Industry, Energy and Related Areas. H.W. Wilson Co., 950 University Avenue, Bronx, NY 10452-9978. (212) 588-8400 (800) 367-6770. FAX (718) 590-1617. From 1958 to present. Monthly. Inquire about cost and availability. Also available online (BRS and WILSONLINE) and on CD-ROM. ISSN: 0003-6986.

Current Contents: Physical and Chemical Science. Institute For Scientific Information, 3501 Market Street, Philadelphia, PA 19104. (215) 386-0100. *(800) 523-1850.* FAX (215) 386-2291. 1961 to present. Weekly. $442.00 per year. Also available online (BRS, DIALOG) and on CD-ROM. Inquire regarding cost and availability. ISSN: 0163-2574.

General Science Index. H.W. Wilson Co., 950 University Avenue, Bronx, NY 10452. (212) 588-8400. (800) 367-6770. FAX (718) 590-1617. From 1978 to present. Ten issues per year; quarterly and annual cumulations. Service basis. Available on CD-ROM and online. Inquire regarding cost and availability. ISSN: 0162-1963.

Meteorological and Geoastrophysical Abstracts. American Meteorological Society, c/o Inforonics, Inc., 550 Newtown Road, Box 458, Littleton, MA 01460. (508) 486-8976. FAX (508) 486-0027. Covers literature in environmental sciences, meteorology, astrophysics, hydrology, glaciology, and physical oceanography. 1950 to present. Monthly. $950.00 per year. ISSN: 0026-1130. Also available online (DIALOG) and on CD-ROM.

Physics Abstracts. Institute of Electrical Engineers, London, United Kingdom. Available from: Institute of Electrical and Electronic Engineers (IEEE), 345 East 47th Street, New York, NY 10017. (212) 705-7900.

Science Citation Index. SCI. Institute for Scientific Information, 3501 Market Street, Philadelphia, PA 19104. (215) 386-0100. (800) 523-1850. FAX (215) 386-2991. 1961 to present. Six issues per year, plus annual cumulation. $11650.00 per year. Also available online and on CD-ROM. Inquire about price and availability. ISSN: 0036-827X.

Selected Water Resources Abstracts. U.S. Geological Survey. Distributed by National Technical Information Service (NTIS), 5285 Port Royal Road, Springfield, VA 22161. (703) 487-4650. Monthly. $115.00 per year.

ANNUAL REVIEWS AND YEARBOOKS

Developments in Atmospheric Science Series. Elsevier Science Publishing Company, Inc., 655 Avenue of the Americas, New York, NY 10010. (212) 989-5800. Irregular. Inquire about price and availability.

LIGHTNING

Ency. of Physical Sciences and Engineering Info. Sources

American Meteorological Society. Meteorological Monographs. 45 Beacon Street, Boston, MA 02108-3693. (617) 227-2425. FAX (617) 742-8718. Irregular. Price varies.

ASSOCIATIONS AND PROFESSIONAL SOCIETIES

American Association of State Climatologists. c/o Myron Moinau, University of Idaho, Agriculture Engineering Department, State Climate Service, Moscow, ID 834844-2040.

American Meteorological Society, 45 Beacon Street, Boston, MA 02108-3693. (617) 227-2425. FAX (617) 742-8718.

Association of American Weather Observers, 401 Whitney Boulevard, Box 455, Belvedere, IL 61008. (815) 544-5665. FAX (815) 544-6334.

International Association of Severe Weather Specialists. c/o Warren Faidley, PO Box 31808, Tuscon, AZ 85751. (520) 751-9964. FAX (520) 751-1185.

Lightning Protection Institute. 3365 North Arlington Heights Road, Suite J, Arlington Heights, IL 60004. (800) 488-6864.

National Environmental Satellite Data, and Information Service. 2069 Federal Building 4 Room 2060, Washington, CD 20233. (301) 763-71990. FAX (301) 763-4011.

National Weather Association, 6704 Wolke Court, Montgomery, AL 36116-2134. (334) 213-0388. FAX (334) 213-0388.

University Corporation For Atmospheric Research. P.O. Box 3000, Boulder, CO 80307-3000. (303) 497-1650. FAX (303) 497-1654.

BIBLIOGRAPHIES

Information Sources in the Earth Sciences. David N. Wood, Joan E. Hardy, and Anthony P. Harvey, editors. Bowker-Saur/K.G. Saur. Distributed by R.R. Bowker, 121 Chanlon Road, New Providence, NJ 07974. (800) 521-8110 or (908) 464-6800. Second edition. 1989. $85.00.

Lightining Protection of Buildings: Sources of Information. Building Services Information Association. State Mutual Book Service, Ltd. 521 Fifth Avenue, New York, NY 10175. (212) 682-5844. 1979. $60.00.

DIRECTORIES AND BIOGRAPHICAL SOURCES

American Men and Women of Science: Physical and Biological Sciences. Eighteenth edition. R.R. Bowker Company, 245 West 17th Street, New York, NY 10011. (800) 521-8810 or (212) 916-1600. $750.00.

American Meteorological Society. PRofESSIONAL DIRECTORY. American Meteorological Society. 45 Beacon Street, Boston, MA 02108. (617) 227-2425. Included in each issue of the Bulletin of the Society.

Meteorological Services of the World. American Meteorological Society, 45 Beacon Street, Boston, MA 02108- 3693. (617) 227-2425. FAX (617) 742-8718. Annual. $70.00. National Weather Service Offices and Stations. National Oceanic and Atmospheric Administration, Deparatment of Commerce, Silver Spring, MD 20910. (301) 427-7698. Annual. Free.

Research Centers Directory. Gale Research, 835 Penobscot Building, Detroit, MI 48226-4094. (313) 961-2242. (800) 877-4253. 20th edition. 1995. $485.00. ISSN: 0080-1518.

ENCYCLOPEDIAS AND DICTIONARIES

Climates of the States. Bair. Gale Research, 835 Penobscot Building, Detroit, MI 48226-4094. (313) 961-2242. (800) 877-4253. 5th edition. 1995. $255.00.

Concise Oxford Dictionary of the Earth Sciences. Ailsa Allaby and Michael Allaby, editors. Oxford University Press, Inc., 200 Madison Avenue, New York, NY 10016. (800) 334-4249 or (212) 679-7300. 1990. $42.95.

Encyclopedia of Climatology. Rhodes W. Fairbridge, and John E. Oliver. Chapman & Hall, 1 Penn Plaza, New York, NY 10119. (212) 564-1060. 1986. $115.00.

Handbook of Applied Meteorology. David D. Houghton, editor. John Wiley & Sons, Inc. 605 Third Avenue, New York, NY 10158-0012. (212) 850-6000. (800) 225-5945. 1985.

McGraw-Hill Encyclopedia of Ocean and Atmospheric Sciences. Sybil P. Parker, editor. McGraw-Hill Publishing Company, 1221 Avenue of the Americas, New York, NY 10020. (800) 262-4729 or (212) 512-3825. 1979. $79.95.

McGraw-Hill Encyclopedia of Science and Technology. McGraw-Hill Publishing Company, 1221 Avenue of the Americas, New York, NY 10020. (800) 262-4729 or (212) 512-3825. Seventh edition. 1992. $1900.00.

GENERAL WORKS

All about Lightning. Martin A. Uman. Dover Publications, Inc., 180 Varick Street, New York, NY 10014. (212) 255-3755. (800) 223-3130.. 1966. $5.95.

Ball Lightning and Bead Lightning: Extreme Forms of Atmospheric Electricity. James D. Barry. Plenum Publishing Corp., 233 Spring Street, New York, NY. (212) 620-8000. (800) 221-9369. FAX (212) 463-0742. 1980. $75.00.

Lightining, Auroras, Noctural Lights and Related Luminous Phenomena. William R. Corliss. Sourcebook Project, P.O. Box 107, Glen Arm, MD 21057. (410) 668 6047. 1982. $34.95.

Lightning and Its Spectrum: An Atlas of Photographs. Leon E. Salanave. University of Arizona Press, 1230 North Park Avenue, Number 102, Tucson, AZ 85719. (520) 621-1441. (800) 426-3797. 1980. $30.00.

The Lightning Discharge. Martin A. Uman. Academic Press, Inc., 6277 Sea Harbor Drive, Orlando, FL. (800) 321-5068. 1987. $72.00.

Lightning Protection. J. Lawrence Marshall. Books on Demand, 300 North Zeeb Road, Ann Arbor, MI 48106-1346. (313) 761-4700. (800) 521-0600. $58.00 in paper.

Lightning Protection For People and Property. Marvin M. Frydenland. Van Nostrand Reinhold, 115 Fifth Avenue, New York, NY 10003. (212) 254-3232. (800) 842-3636. 1993. $54.95

Severe and Unusual Weather. Joe R. Eagleman. Trimedia Publishing Company, 12008 West 87th Street, Suite 117, Lenexa, KS 66215. (913) 599-0505. Second edition. 1990. $41.95.

Thunderstorm Morphology and Dynamics. Edwin Kessler. University of Okalahoma Press.1005 Asp Avenue, Norman, OK 73019-0445. (405) 325-5111. 2nd revised edition. 1992. $34.95.

HANDBOOKS AND MANUAL

Author's Guide To the Journals of the American Meteorological Society. American Meteorological Society, 45 Beacon Street, Boston, MA 02108-3693. (617) 227-2425. FAX (617) 742-8718. 1983. $15.00 in paper.

Handbook in Applied Meteorology. David D. Houghton, editor. John Wiley and Sons, Inc., 605 Third Avenue, New York, NY 10158. (800) 526-5368 or (212) 850-6000. 1985. $114.00.

Lightning Protection Code. National Fire Protection Association, Batterymarch Park, Quincy, MA 02269. (617) 770-3000. 1992. $20.25 in appear.

ONLINE DATABASES AND CD-ROMS

Climate Assessment Database. National Weather Service, National Meteorological Center, 5200 Auth Road, Suite 101, Camp Springs, MD 20233. (301) 763-8016. Contains daily, weekly and monthly summaries of North American and world climatological data. Also provides five to ten weather forecasts, and 30 to 90 day weather outlook. Subscription required. Inquire as to cost and availability.

CA Search. Chemical Abstracts Service, P.O. Box 3012, Columbus, OH 43210-0012. (614) 447-3600. (800) 848-6533. FAX (614) 447-3709. Very comprehensive guide to worldwide chemical literature and related fields, 1972 to present. Available on BRS,(800) 289-4277, DIALOG, (800) 334-2564, ORBIT (800) 456-7248, and STN International, FIZ Karlsruhe, P.O. Box 2465, W-7500, Karlsruhe 1, Germany, online services. Inquire as to cost and availability.

Compendex Plus. Engineering Information, Inc., 345 East 47th Street, New York, NY 10017. (212) 705-7600 or (800) 221-1044. Contains citations with abstracts to worldwide literature in engineering and technology, from 1970 to present. Available on online BRS,(800) 289-4277, DIALOG, (800) 334-2564, ORBIT (800) 456-7248, and STN International, FIZ Karlsruhe, P.O. Box 2465, W-7500, Karlsruhe 1, Germany, online services. Also available on CD-ROM. Inquire as to cost and availability.

Current Contents Search. Institute for Scientific Information, 3501 Market Street, Philadelphia, PA 19104. (215) 386-0100. FAX (215) 386-6362. Contains citations to articles listed in the table of contents of science and technology journals. Also articles in social sciences and life sciences journals. Available on BRS,(800) 289-4277, DIALOG,(800) 334-2564, online services. Inquire as to cost and availability.

Dissertation Abstracts. University Microfilms International, 300 North Zeeb Road, Ann Arbor, MI 48106. (800) 521-0600 or (313) 761-4700. Scope includes virtually all doctoral dissertations accepted at accredited American institutions from 1861 to present in 252 subject areas. Available on BRS,(800) 289-4277, DIALOG,(800) 334-2564, and OCLC EPIC,(800) 848-5878, online services. Also available on CD-ROM. Inquire as to cost and availability.

INSPEC. Institution of Electrical Engineers, Michael Faraday House, Six Hills Way, Stevenage, Herts. SG1 2AY, England. Telephone: 0438 313311 or FAX 0438 742840. Contains citations to the worldwide literature of physics, electronics and electrical engineering, computer technology, and related fields. Available on BRS,(800) 289-4277, DIALOG, (800) 334-2564, ORBIT, (800) 456-7248, and STN International, FIZ Karlsruhe, P.O. Box 2465, W-7500, Karlsruhe 1, Germany, online services. Inquire as to cost and availability.

International Aerospace Abstracts. American Institute of Aeronautics and Astronautics, 370 L'Enfant Promenade, S.W., Washington, DC 20024. (202) 646-7400. Contains references and abstracts of journal and monograph literature relating to aerospace science and technology, from 1963 to present. Available through the NASA/RECON system of the National Aeronautics and Space Administration only.

Meteorological and Geoastrophysical Abstracts. American Meteorological Society, 45 Beacon Street, Boston, MA 02108-3693. (617) 227-2425. FAX (617) 742-8718. Contains citations and abstracts to the worldwide literature on significant research in meteorology and geoastrophysics. Related topics include physical oceanography, hydrology, environmental sciences and glaciology. Covers the period 1972 to present. Available on DIALOG,(800) 334-2564, online service. Inquire as to cost and availability.

NTIS Bibliographic Database. National Technical Information Service, 5285 Port Royal Road, Springfield, VA 22161. (703) 487-4929 or FAX (703) 321-8199. Broad coverage of government-sponsored science and technology research reports, 1964 to present. Available on BRS,(800) 289-4277, DIALOG, (800) 334-2564, ORBIT, (800) 456-7248, and STN International, FIZ Karlsruhe, P.O. Box 2465, W-7500, Karlsruhe 1, Germany, online services. Also available on CD-ROM. Inquire as to cost and availability.

Physics Briefs. American Institute of Physics, 335 East 45th Street, New York, NY 10017. (212) 661-9260 or FAX (212) 661-2036. Contains citations with abstracts of the literature of physics and related fields, 1979 to present. Available on the STN International, FIZ Karlsruhe, P.O. Box 2465, W-7500, Karlsruhe 1, Germany, online service. Inquire as to cost and availability.

SCISEARCH. Institute for Scientific Information, 3501 Market Street, Philadelphia, PA 19104. (800) 523-1850 or (215) 386-0100. Broad multidisciplinary title and author index to the international literature of science and technology, 1974 to present. Available on DIALOG and ORBIT online services. Also available on CD-ROM. Inquire as to cost and availability.

WILSONLINE. H.W. Wilson Company, 950 University Avenue, Bronx, NY 10452. (800) 367-6770 or (212) 588-8400. Makes available online versions of the printed H.W. Wilson indexes including Applied Science and Technology Index, Business Periodicals Index, General Science Index, and Readers' Guide to Periodical Literature. Period covered is generally 1983 to present. Available on BRS,(800) 289-4277, DIALOG,(800) 334-2564, and OCLC EPIC,(800) 848-5878, online services. Also available on CD-ROM. Inquire as to cost and availability.

World Climate Disc. Chadwyck-Healey Inc., 1101 King Street, Alexandria, VA 22314. (703) 683-4890. FAX (703) 683-7589. Weather and climate data from approximately 5,000 weather stations worldwide, covering the years 1854 to 1990 on CD-ROM. First edition 1992. Approximately $1200.00 per year with annual updates.

World Weatherdisc. WeatherDisc Associates, Inc., 4584 N.E. 89th Street, Seattle, WA 98115. (206) 524-4314. FAX (206) 543-0308. Meteorological data on CD-ROM which describes the climate of the earth today and for the past few hundred years. First edition 1989. Approximately $295.00 per year with biannual updates.

PERIODICALS

American Meteorological Society. Bulletin. American Meteorological Society, 45 Beacon Street, Boston, MA 02108-3693. (617) 227-2425. FAX (617) 742-8718. Monthly. $60.00 per year.

American Meteorological Society. Meteorological Monographs. American Meteorological Society, 45 Beacon Street, Boston, MA 02108-3693. (617) 227-2425. FAX (617) 742-8718. Irregular. Price varies.

American Weather Observer. Association of American Weather Observers, 401 Whitney Boulevard, Box 455, Belvedere, IL 61008. (815) 544-5665. FAX (815) 544-6334. Monthly. $21.00 per year.

Atmosphere - Ocean. Canadian Meteorological and Oceanographic Society, P.O. Box 334, Newmarket, Ontario, L3Y 4X7, Canada. (416) 898-1040. FAX (416) 898-7937. Quarterly. $30.00 per year.

Journal of Applied Meteorology. American Meteorological Society, 45 Beacon Street, Boston, MA 02108-3693. (617) 227-2425. FAX (617) 742-8718. Monthly. $165.00 per year.

Journal of the Atmospheric Sciences. American Meteorological Society, 45 Beacon Street, Boston, MA 02108-3693. (617) 227-2425. FAX (617) 742-8718. Semi-monthly. $320.00 per year.

Monthly Weather Review. American Meteorological Society, 45 Beacon Street, Boston, MA 02108-3693. (617) 227-2425. FAX (617) 742-8718. Monthly. $205.00 per year.

National Weather Digest. National Weather Association, 4400 Stamp Road, Room 404, Temple Hills, MD 20748. (301) 899-3784.

Weather. Royal Meteorological Society, 104 Oxford Road, Reading, Berks RG1 7LJ, England. Monthly. $44.00 per year.

Weatherwise. Heldref Publications, 1319 Eighteenth Street, N.W., Washington, DC 20036-1802. (202) 296-6267. FAX (202) 296-5149. Bi-monthly. $28.00 per year.

RESEARCH CENTERS AND INSTITUTES

Desert Research Institute - Air Motions Laboratory. University of Nevada at Reno, P.O. Box 60220, Reno, NV 89506. (702) 677-3201.

Institute of Atmospheric Physics. University of Arizona, Tucson, AZ 85721. (602) 621-6831.

Institute of Atmospheric Sciences. South Dakota School of Mines and Technology, 501 East St. Joseph Street, Rapid City, SD 57701-3995. (605) 394-2291.

Irving Langmiuir Laboratory For Atmospheric Research. New Mexico Institute of Mining and Technology. Socorro, NM 87801. (505) 835-5423.

National Center For Atmospheric Research. P.O. Box 3000, Boulder, CO 80307. (303) 496-1000.

National Severe Storms Forecast Center. 601 East 12th Street, Kansas City, MO 64106. (816) 374-5922.

LINEAR ACCELERATORS

See: PARTICLE ACCELERATORS

LINEAR ALGEBRA

See: ALGEBRA

LINEAR PROGRAMMING

See: MATHEMATICS

LIQUID CRYSTALS

See also: CRYSTALLOGRAPHY, OPTICS

ABSTRACT SERVICES AND INDEXES

Applied Science and Technology Index; A Cumulative Subject Index To English Language Periodicals in the Fields of Aeronautics and Space Science, Computer Technology, Chemistry, Construction Industry, Energy and Related Areas. H.W. Wilson Co., 950 University Avenue, Bronx, NY 10452. (800) 367-6770 or (212) 588-8400. FAX (718) 590-1617. From 1958 to present. Monthly. Inquire about cost and availability. Also available on CD-ROM and online.

Current Contents: Engineering, Technology, and Applied Sciences. Institute for Scientific Information, 3501 Market Street, Philadelphia, PA 19104. (215) 386-0100. FAX (215) 386-6362. 1970 to present. Weekly. $442.00 per year.

Index To Scientific and Technical Proceedings. Institute for Scientific Information, 3501 Market St., Philadelphia, PA 19104. (215) 386-0100. FAX (215) 386-6362. Monthly. $500.00 per year.

Index To Scientific Reviews. Institute for Scientific Information, 3501 Market St., Philadelphia, PA 19104. (215) 386-0100. FAX (215) 386-6362. Semi-annual.

Science Citation Index. Institute for Scientific Information, 3501 Market Street, Philadelphia, PA 19104. (215) 386-0100. FAX

(215) 386-6362. Inquire about availability and cost. Also available on CD-ROM.

ASSOCIATIONS AND PROFESSIONAL SOCIETIES

The International Society For Optical Engineering (spie), Po Box 10, 1000-20th Street, Bellingham, WA 98227-0010. (206) 676-3290.

Optical Society of America, 2010 Massachusetts Avenue Nw, Washington, DC 20036. (202) 223-8130.

DIRECTORIES AND BIOGRAPHICAL SOURCES

Directory of Engineering Societies and Related Organizations. Gordon Davis. 13th edition. American Association of Engineering Societies, 1111 19th Street NW, Suite 608, Washington, DC 20036. (202) 296-2237 or (800) 658-8897. 1989. Inquire for price.

International Research Centers Directory. Gale Research, 835 Penobscot Building, Detroit, MI 48226-4094. (313) 961-2242. (800) 877-4253. FAX (313) 961-6083. 8th edition. 1995. $410.00.

Research Centers Directory. Gale Research, 835 Penobscot Building, Detroit, MI 48226-4094. (313) 961-2242. (800) 877-4253. FAX (313) 961-6083. $485.00.

Scientific and Technical Organizations and Agencies Directory. Gale Research, 835 Penobscot Building, Detroit, MI 48226-4094. (313) 961-2242. (800) 877-4253. FAX (313) 961-6083. 4th edition. 1996. $195.00.

Who's Who in Engineering. American Association of Engineering Societies, 1111 19th Street NW, Suite 608, Washington, DC 20036. (202) 296-2237 or (800) 658-8897. 8th edition. 1991. Inquire for price.

ENCYCLOPEDIAS AND DICTIONARIES

Encyclopedia of Physical Science and Technology. Robert A. Meyers, ed. 18 volumes. Academic Press Inc., 6277 Sea Harbor Drive, Orlando, FL 32887. (800) 321-5068. 1992. $2100.00

McGraw-Hill Dictionary of Scientific and Technical Terms. Sybil P. Parker, ed. 5th edition. McGraw-Hill Publishing Company, 1221 Avenue of the Americas, New York, NY 10020. (800) 262-4729 or (212) 512-3825. 1993. $110.50.

McGraw-Hill Encyclopedia of Engineering. Sybil P. Parker, ed. 2nd edition. McGraw-Hill Publishing Company, 1221 Avenue of the Americas, New York, NY 10020. (800) 262-4729 or (212) 512-3825. 1993. $95.50.

McGraw-Hill Encyclopedia of Science and Technology. Sybil P. Parker, ed. 7th edition. 20 volumes. McGraw-Hill Publishing Company, 1221 Avenue of the Americas, New York, NY 10020. (800) 262-4729 or (212) 512-3825. 1992. $1900.00

Thesaurus of Scientific, Technical, and Engineering Terms. Hemisphere Publishing Corporation, 1900 Frost Road, Suite 101, Bristol, PA 19007-1598. (215) 785-5800 or (800) 821-8312. FAX (215) 785-5515. 1987. $173.00.

GENERAL WORKS

Liquid Crystals. S. Chandrasekhar. 2nd edition. Cambridge University Press, 40 West 20th Street, New York, NY 10011-4211. (212) 924-3900 or (800) 872-7423. FAX (914) 937-4712. 1992. $39.95.

Liquid Crystals: Applications and Uses. Birendra Bahadur, editor. World Scientific Publishing Company, Inc., 1060 Main Street, Unit B, River Edge, NJ 07661. (800) 227-7562 or (201) 487-9655. 1990. $109.00.

Liquid Crystals: Nature's Delicate Phase of Matter. Peter J. Collings. Princeton University Press, 41 William Street, Princeton, NJ 08540. (609) 258-4900 or (800) 777-4726. FAX (609) 895-1081. 1990. $39.50 (hardback), $10.95 (paperback).

Liquid Crystals: Physical Properties and Nonlinear Optical Phenomena. I.C. Khoo. John Wiley and Sons, Inc., 605 Third Avenue, New York, NY 10158. (800) 526-5368 or (212) 850-6000. 1995. Inquire for cost and availability.

Modern Topics in Liquid Crystals: From Neutron Scattering To Ferroelectricity. Agnes Buka, editor. World Scientific Publishing Company, Inc., 1060 Main Street, Unit B, River Edge, NJ 07661. (800) 227-7562 or (201) 487-9655. 1993. $86.00.

The Physics of Liquid Crystals. P.G. de Gennes and J. Prost. Oxford University Press, Inc., 198 Madison Avenue, New York, NY 10016-4314. (212) 726-6000. FAX (212) 726-6446. 1993. $90.00.

Rheology of Liquid Crystals. P.K. Khabibullaev, et al. Allerton Press Inc., 150 Fifth Avenue, New York, NY 10011. 212) 924-3950. FAX (212) 463-9684. 1994. $95.00.

Theory and Applications of Liquid Crystals. J.L. Ericksen and D. Kinderlehrer, editors. Springer-Verlag, 175 Fifth Avenue, New York, NY 10010. (212) 460-1500 or (800) 777-4643. FAX (212) 473-6272. 1987. $52.00.

ONLINE DATABASES AND CD-ROMS

Compendex Plus. Engineering Information, Inc., 345 East 47th Street, New York, NY 10017. (212) 705-7600 or (800) 221-1044. Contains citations with abstracts to worldwide literature in engineering and technology, from 1970 to present. Available on online BRS,(800) 289-4277, DIALOG, (800) 334-2564, ORBIT (800) 456-7248, and STN International, FIZ Karlsruhe, P.O. Box 2465, W-7500, Karlsruhe 1, Germany, online services. Also available on CD-ROM. Inquire as to cost and availability.

CSA Engineering. Cambridge Scientific Abstracts, 7200 Wisconsin Avenue, Suite 601, Bethesda, MD 20814. (301) 961-6750 or (800) 843-7751. Contains citations and abstracts of international periodicals and research literature covering all fields of engineering and science and technology,including computer and information science, electronics, mechanical engineering, solid state materials, 1981 to present. Available on BRS,(800) 289-4277, online service.Inquire as to cost and availability.

Current Contents Search. Institute for Scientific Information, 3501 Market Street, Philadelphia, PA 19104. (215) 386-0100. FAX (215) 386-6362. Contains citations to articles listed in the table of contents of science and technology journals. Also articles in social sciences and life sciences journals. Available on

LIQUID CRYSTALS

Ency. of Physical Sciences and Engineering Info. Sources

BRS,(800) 289-4277, DIALOG,(800) 334-2564, online services. Inquire as to cost and availability.

Dissertation Abstracts. University Microfilms International, 300 North Zeeb Road, Ann Arbor, MI 48106. (800) 521-0600 or (313) 761-4700. Scope includes virtually all doctoral dissertations accepted at accredited American institutions from 1861 to present in 252 subject areas. Available on BRS,(800) 289-4277, DIALOG,(800) 334-2564, and OCLC EPIC,(800) 848-5878, online services. Also available on CD-ROM. Inquire as to cost and availability.

NTIS Bibliographic Database. National Technical Information Service, 5285 Port Royal Road, Springfield, VA 22161. (703) 487-4929 or FAX (703) 321-8199. Broad coverage of government-sponsored science and technology research reports, 1964 to present. Available on BRS,(800) 289-4277, DIALOG, (800) 334-2564, ORBIT, (800) 456-7248, and STN International, FIZ Karlsruhe, P.O. Box 2465, W-7500, Karlsruhe 1, Germany, online services. Also available on CD-ROM. Inquire as to cost and availability.

SCISEARCH. Institute for Scientific Information, 3501 Market Street, Philadelphia, PA 19104. (800) 523-1850 or (215) 386-0100. Broad multidisciplinary title and author index to the international literature of science and technology, 1974 to present. Available on DIALOG,(800) 334-2564, and ORBIT,(800) 456-7248, online services. Also available on CD-ROM. Inquire as to cost and availability.

WILSONLINE. H.W. Wilson Company, 950 University Avenue, Bronx, NY 10452. (800) 367-6770 or (212) 588-8400. Makes available online versions of the printed H.W. Wilson indexes including Applied Science and Technology Index, Business Periodicals Index, General Science Index, and Readers' Guide to Periodical Literature. Period covered is generally 1983 to present. Available on BRS,(800) 289-4277, DIALOG,(800) 334-2564, and OCLC EPIC,(800) 848-5878, online services. Also available on CD-ROM. Inquire as to cost and availability.

PERIODICALS

Liquid Crystals. Taylor & Francis, 1900 Frost Road, Suite 101, Bristol, PA 19007. (215) 785-5800. FAX (215) 785-5515. 1986 to present. Monthly. $1375.00 per year.

RESEARCH CENTERS AND INSTITUTES

Center For Advanced Liquid Crystalline Optical Materials. Liquid Crystal Institute, Kent State University, Kent, OH 44242. (216) 672-2654. FAX (216) 672-2796.

Liquid Crystalline Polymer Research Center. University of Connecticut, Institute of Materials Science, 97 N. Eagleville Road, Storrs, CT 06269-3136. (203) 486-3582.

LORAN

See: NAVIGATION

LOW-TEMPERATURE PHYSICS

See: CRYOGENICS

LUBRICATION

See also: AUTOMOTIVE ENGINEERING, MACHINERY

ABSTRACT SERVICES AND INDEXES

A.P.I. ABSTRACTS. American Petroleum Institute, Central Abstracting and Information Services, 275 Seventh Avenue, New York, NY 10001. (212) 366-4040 or FAX (212) 366-4298. Contains citations with abstracts to literature relating to the petroleum refining and the petrochemical industries, from 1964 to present.

Applied Mechanics Reviews: An Assessment of World Literature in Engineering Sciences. 1948-present. American Society of Mechanical Engineers, 345 East 47th Street, New York, NY 10017. (212) 705-7703. Monthly. $360.00 per year.

Applied Science and Technology Index; A Cumulative Subject Index To English Language Periodicals in the Fields of Aeronautics and Space Science, Computer Technology, Chemistry, Construction Industry, Energy and Related Areas. H.W. Wilson Co., 950 University Avenue, Bronx, NY 10452. (800) 367-6770 or (212) 588-8400. FAX (718) 590-1617. From 1958 to present. Monthly. Inquire about cost and availability. Also available on CD-ROM and online.

Chemical Abstracts - Applied Chemistry and Chemical Engineering Sections. Chemical Abstracts Service, 2540 Olentangy River Road, Box 3012, Columbus, OH 43210. (800) 848-6538 or (614) 447-3600, FAX (614) 447-3713. Bi-weekly. $1410.00 per year.

Corrosion Abstracts; Abstracts of the World's Literature On Corrosion and Corrosion Mitigation. National Association of Corrosion Engineers (NACE), Box 218340, Houston, TX 77218. (713) 492-0535 or FAX (713) 492-8254. 1962 to present. Bi-monthly. $195.00 per year. Also available on CD-ROM.

Current Contents: Engineering, Technology, and Applied Sciences. Institute for Scientific Information, 3501 Market Street, Philadelphia, PA 19104. (215) 386-0100. FAX (215) 386-6362. 1970 to present. Weekly. $442.00 per year.

Engineering Index Monthly. Engineering Information, Inc., Castle Point on the Hudson, Hoboken, NJ 07030. (800) 221-1044. FAX (212) 832-1857. Monthly. $2300.00 per year. Also available online as COMPENDEX, and also on CD-ROM. Covers chemical engineering, computers, electrical engineering, civil engineering, metals and mining, industrial management, and mechanical engineering.

Mechanical Engineering Abstracts (formerly ISMEC), Cambridge Scientific Abstracts, 7200 Wisconsin Avenue, Bethesda, MD 20814-4823. (301) 961-6750. FAX (301) 961-6720. 1967 to present. Monthly. $895.00 per year. Summarizes world literature in mechanical engineering, production engineering, and engineering management. Also available online.

Materials Science and Engineering Abstracts, Cambridge Scientific Abstracts, 7200 Wisconsin Avenue, Bethesda, MD 20814-

4823. (301) 961-6750. FAX (301) 961-6720. 1993 to present. Monthly. $385.00 per year. Focuses on mechanical and physical properties of materials and commercial or industrial applications for materials, methods for strength testing, effects of vibration and other stresses, corrosion and protective coatings, storage and handling, ceramics, composites, metals, wood, plastics, and polymers.

Petroleum Abstracts. 600 South College, Tulsa, OK 74104. (918) 631-2296 or (800) 247-8678. Contains citations with abstracts to the worldwide literature and patents on the exploration, development, and production of petroleum resources, from 1965 to present.

Science Citation Index. Institute for Scientific Information, 3501 Market Street, Philadelphia, PA 19104. (215) 386-0100. FAX (215) 386-6362. Inquire about availability and cost. Also available on CD-ROM.

Tribology and Corrosion Abstracts. Elsevier Science Publishing Company, Inc., 655 Avenue of the Americas, New York, NY 10010. (212) 989-5800. Covers literature and research in friction, wear, lubrication and lubricants, and corrosion problems and protection, 1984 to present. Thirteen times per year. $450.00 per year.

Tribos - Tribology Abstracts. STI Limited, 4 Kings Meadow, Ferry Hinksey Road, Oxford, OX2 0DU, England. Distributed in the United States by Air Science Company, Box 143, Corning NY 14830. (607) 962-5591. Contains citations of literature and research in friction, wear, lubrication and lubricants, and corrosion problems and protection, from 1967 to present. Monthly. $375.00 per year.

ASSOCIATIONS AND PROFESSIONAL SOCIETIES

American Society For Metals. Materials Park, OH 44073. (216) 338-5151 or FAX (216) 338-4634.

American Society of Mechanical Engineers. 345 East 47th Street, New York, NY 10017-2398. (212) 705-7703.

Independent Lubricant Manufacturers Association. 651 S. Washington Street, Alexandria, VA 22314. (703) 684-5574. FAX (703) 836-8503.

Metallurgical Society of the Aime (american Institute of Mining, Metallurgical, and Petroleum Engineers). 345 E.47th Street, 14th Floor, New York, NY 10017. (212) 705-7695.

National Association of Corrosion Engineers. Box 218340, Houston, TX 77218. (713) 492-0535. FAX (713) 492-8254.

National Lubricating Grease Institute. 4635 Wyandotte Street, Kansas City, MO 64112. (816) 931-9480. FAX (816) 753-5026.

Society of Automotive Engineers International. 400 Commonwealth Drive, Warrendale, PA 15096-0001.

Society of Tribologists and Lubrication Engineers. 840 Busse Highway, Park Ridge, IL 60068-2376. (708) 825-5536. FAX (708) 825-1456.

DIRECTORIES AND BIOGRAPHICAL SOURCES

Directory of Engineering Societies and Related Organizations. Gordon Davis. 13th edition. American Association of Engineering Societies, 1111 19th Street NW, Suite 608, Washington, DC 20036. (202) 296-2237 or (800) 658-8897. 1989. Inquire for price.

International Research Centers Directory. Gale Research, 835 Penobscot Building, Detroit, MI 48226-4094. (313) 961-2242. (800) 877-4253. FAX (313) 961-6083. 8th edition. 1995. $410.00

Research Centers Directory. Gale Research, 835 Penobscot Building, Detroit, MI 48226-4094. (313) 961-2242. (800) 877-4253. FAX (313) 961-6083. $485.00.

Scientific and Technical Organizations and Agencies Directory. Gale Research, 835 Penobscot Building, Detroit, MI 48226-4094. (313) 961-2242. (800) 877-4253. FAX (313) 961-6083. 4th edition. 1996. $195.00.

Who's Who in Engineering. American Association of Engineering Societies, 1111 19th Street NW, Suite 608, Washington, DC 20036. (202) 296-2237 or (800) 658-8897. 8th edition. 1991. Inquire for price.

ENCYCLOPEDIAS AND DICTIONARIES

Dictionary of Corrosion and Corrosion Control. Helmut Gross, editor in chief. Elsevier Science Publishing Company, Inc., 655 Avenue of the Americas, New York, NY 10010. (212) 989-5800. 1986. $133.50.

McGraw-Hill Dictionary of Scientific and Technical Terms. Sybil P. Parker, ed. 5th edition. McGraw-Hill Publishing Company, 1221 Avenue of the Americas, New York, NY 10020. (800) 262-4729 or (212) 512-3825. 1993. $110.50.

McGraw-Hill Encyclopedia of Engineering. Sybil P. Parker, ed. 2nd edition. McGraw-Hill Publishing Company, 1221 Avenue of the Americas, New York, NY 10020. (800) 262-4729 or (212) 512-3825. 1993. $95.50.

McGraw-Hill Encyclopedia of Science and Technology. Sybil P. Parker, ed. 7th edition. 20 volumes. McGraw-Hill Publishing Company, 1221 Avenue of the Americas, New York, NY 10020. (800) 262-4729 or (212) 512-3825. 1992. $1900.00

Thesaurus of Scientific, Technical, and Engineering Terms. Hemisphere Publishing Corporation, 1900 Frost Road, Suite 101, Bristol, PA 19007-1598. (215) 785-5800 or (800) 821-8312. FAX (215) 785-5515. 1987. $173.00.

GENERAL WORKS

Chemistry and Technology of Lubricants. R.M. Mortier and S.T. Orszulik, eds. VCH Publishers, Inc., 303 Northwest 12th Avenue, Deerfield Beach, FL 33442. (800) 422-8824. 1992. Inquire for cost and availability.

Engine Tribology. SAE International, 400 Commonwealth Drive, Warrendale, PA 15096-0001. (412) 776-4970. FAX (412) 776-0790. 1992. $39.00.

LUBRICATION

Ency. of Physical Sciences and Engineering Info. Sources

Lubricants and their Applications. Robert W. Miller. McGraw-Hill Publishing Company, 1221 Avenue of the Americas, New York, NY 10020. (800) 262-4729 or (212) 512-3825. 1993. $37.00.

Tribology in the Usa and the Former Soviet Union: Studies and Applications. Vladimir A. Belyi, et al., editors. Allerton Press Inc., 150 Fifth Avenue, New York, NY 10011.(212) 924-3950. FAX (212) 463-9684. 1995. $47.00.

HANDBOOKS AND MANUALS

ASM Handbook Volume 18: Friction, Lubrication, and Wear Technology. ASM International Handbook Committee. ASM International, Materials Information, Materials Park, OH 44073-0002. (216) 338-5151 or (800) 336-5152. FAX (216) 338-4634. 1992. $147.00.

CRC Handbook of Lubrication and Tribology. E.R. Booser, editor. 3 volumes. CRC Press, 2000 Corporate Blvd., N.W., Boca Raton, FL 33431. (407) 994-0555 or (800) 272-7737. FAX (407) 994-0949. 1983. Inquire for cost and availability.

Lubrication: A Tribology Handbook. M.J. Neale. Society of Automotive Engineers, 400 Commonwealth Dr., Warrendale, PA 15096. (412) 776-4841. 1993. $39.00.

ONLINE DATABASES AND CD-ROMS

Compendex Plus. Engineering Information, Inc., 345 East 47th Street, New York, NY 10017. (212) 705-7600 or (800) 221-1044. Contains citations with abstracts to worldwide literature in engineering and technology, from 1970 to present. Available on online BRS,(800) 289-4277, DIALOG, (800) 334-2564, ORBIT (800) 456-7248, and STN International, FIZ Karlsruhe, P.O. Box 2465, W-7500, Karlsruhe 1, Germany, online services. Also available on CD-ROM. Inquire as to cost and availability.

CSA Engineering. Cambridge Scientific Abstracts, 7200 Wisconsin Avenue, Suite 601, Bethesda, MD 20814. (301) 961-6750 or (800) 843-7751. Contains citations and abstracts of international periodicals and research literature covering all fields of engineering and science and technology,including computer and information science, electronics, mechanical engineering, solid state materials, 1981 to present. Available on BRS,(800) 289-4277, online service.Inquire as to cost and availability.

Current Contents Search. Institute for Scientific Information, 3501 Market Street, Philadelphia, PA 19104. (215) 386-0100. FAX (215) 386-6362. Contains citations to articles listed in the table of contents of science and technology journals. Also articles in social sciences and life sciences journals. Available on BRS,(800) 289-4277, DIALOG,(800) 334-2564, online services. Inquire as to cost and availability.

Dissertation Abstracts. University Microfilms International, 300 North Zeeb Road, Ann Arbor, MI 48106. (800) 521-0600 or (313) 761-4700. Scope includes virtually all doctoral dissertations accepted at accredited American institutions from 1861 to present in 252 subject areas. Available on BRS,(800) 289-4277, DIALOG,(800) 334-2564, and OCLC EPIC,(800) 848-5878, online services. Also available on CD-ROM. Inquire as to cost and availability.

ISMEC: Mechanical Engineering Abstracts. Cambridge Scientific Abstracts, 7200 Wisconsin Avenue, Suite 601, Bethesda, MD 20814. (301) 961-6750 or (800) 843-7751. Contains cita-

tions to the literature in mechanical engineering, industrial and production engineering, energy, power, mechanics, devices and related areas, from 1973 to present. Available on the DIA-LOG,(800) 334-2564, online service. Inquire as to cost and availability.

NTIS Bibliographic Database. National Technical Information Service, 5285 Port Royal Road, Springfield, VA 22161. (703) 487-4929 or FAX (703) 321-8199. Broad coverage of government-sponsored science and technology research reports, 1964 to present. Available on BRS,(800) 289-4277, DIALOG, (800) 334-2564, ORBIT, (800) 456-7248, and STN International, FIZ Karlsruhe, P.O. Box 2465, W-7500, Karlsruhe 1, Germany, online services. Also available on CD-ROM. Inquire as to cost and availability.

SAE Global Mobility Database. Society of Automative Engineers (SAE), Electronic Publishing Division, 400 Commonwealth Drive, Warrendale, PA 15098. (412) 776-4841. Contains citations with abstracts to technical papers on automotive and aerospace technology and vehicular-related industries that have been presented at SAE conferences. Covers 1906 to present. Available on ORBIT,(800) 456-7248,online service. Inquire as to cost and availability.

SCISEARCH. Institute for Scientific Information, 3501 Market Street, Philadelphia, PA 19104. (800) 523-1850 or (215) 386-0100. Broad multidisciplinary title and author index to the international literature of science and technology, 1974 to present. Available on DIALOG,(800) 334-2564, and ORBIT,(800) 456-7248, online services. Also available on CD-ROM. Inquire as to cost and availability.

WILSONLINE. H.W. Wilson Company, 950 University Avenue, Bronx, NY 10452. (800) 367-6770 or (212) 588-8400. Makes available online versions of the printed H.W. Wilson indexes including Applied Science and Technology Index, Business Periodicals Index, General Science Index, and Readers' Guide to Periodical Literature. Period covered is generally 1983 to present. Available on BRS,(800) 289-4277, DIALOG,(800) 334-2564, and OCLC EPIC,(800) 848-5878, online services. Also available on CD-ROM. Inquire as to cost and availability.

PERIODICALS

Industrial Lubrication and Tribology. MCBUniversity Press, 62 Toller Lane, Bradford, W. Yorks. BD8 9BY, England. Telephone 0274-499821. FAX 0274-547143. 1948 to present. Bi-monthly. $649.95 per year.

Japanese Journal of Tribology. Allerton Press Incorporated, 150 Fifth Avenue, New York, NY 10011. (212) 924-3950. FAX (212) 463-9684. 12 times a year. $590.00 per year.

Journal of Friction and Wear. Allerton Press Incorporated, 150 Fifth Avenue, New York, NY 10011. (212) 924-3950. FAX (212) 463-9684. 6 times a year. $845.00 per year.

Journal of Tribology. American Society of Mechanical Engineers, 345 East 47th Street, New York, NY 10017-2398. (212) 705-7703. 1967 to present. Quarterly. $120.00 per year for non-members.

Lubrication Engineering. Society of Tribologists and Lubrication Engineers, 840 Busse Highway, Park Ridge, IL 60068-2376. (708) 825-5536. FAX (708) 825-1456. 1945 to present. Monthly. $61.00 per year.

Lubrication Science. Leaf Coppin Publishing Company, PO Box 111, Deal, Kent CT14 6SX, England. Telephone 0304-360241. 1988 to present. Quarterly. $170.00 per year.

N.L.G.I. Spokesman. National Lubricating Grease Institute, 4635 Wyandotte Street, Kansas City, MO 64112. (816) 931-9480. FAX (816) 753-5026. 1937 to present. Monthly. $20.00 per year.

The Oil Daily's Lubricants World. Oil Daily Company, 1401 New York Avenue NW, No. 500, Washington, DC 20005. (800) 621-0050. FAX (202) 783-5918. 1991 to present. Monthly. $150.00 per year.

Tribology International. Butterworth-Heinemann, Linacre House, Jordan Hill, Oxford OX2 8DP, England. Telephone 0865-310366. FAX 0865-310898. 1968 to present. Bi-monthly. Inquire for cost.

Tribology Series. Elsevier Science Inc., Box 882, Madison Square Station, New York, NY 10159. (212) 989-5800. 1978 to present. Irregular. Inquire for cost and availability.

Wear. Elsevier Science Inc., Box 882, Madison Square Station, New York, NY 10159. (212) 989-5800. 1958 to present. 20 times a year. $2585.00.

RESEARCH CENTERS AND INSTITUTES

Coordinating Research Council Inc. 219 Perimeter Center Parkway, Atlanta, GA 30346. (404) 396-3400.

Fluid Power Research Center, Oklahoma State University. 1724 W. Tyler, Stillwater, OK 74078. (405) 744-7375.

Institute For Wear Control and Tribology. Rensselaer Polytechnic Institute, Mechanical Engineering Department, Jonsson Engineering Center, Troy, NY 12181. (518) 276-6992. FAX (518) 276-8788.

Northwestern University Center For Engineering Tribology. Technical Institute, Evanston, IL 60208. (708) 491-3112. FAX (708) 491-4133.

SPECIFICATIONS AND STANDARDS

*Annual Book of ASTM Standards, Volumes 05.*01-05.03. American Society for Testing and Materials (ASTM), 1916 Race Street, Philadelphia, PA 19103. (215) 299-5585. Annual. Inquire for cost and availability.

LUNAR GEOLOGY

See: ASTROGEOLOGY, PLANETARY SCIENCES

M

MACHINE DESIGN

See also: CAD (COMPUTER-AIDED DESIGN), MACHINERY, MECHANICAL ENGINEERING

ABSTRACT SERVICES AND INDEXES

Applied Mechanics Reviews: An Assessment of World Literature in Engineering Sciences. 1948-Present. American Society of Mechanical Engineers, 345 East 47th Street, New York, NY 10017. (212) 705-7703. Monthly. $360.00 per year.

Applied Science and Technology Index; A Cumulative Subject Index To English Language Periodicals in the Fields of Aeronautics and Space Science, Computer Technology, Chemistry, Construction Industry, Energy and Related Areas. H.W. Wilson Co., 950 University Avenue, Bronx, NY 10452. (800) 367-6770 or (212) 588-8400. FAX (718) 590-1617. From 1958 to present. Monthly. Inquire about cost and availability. Also available on CD-ROM and online.

CAD/CAM Abstracts. R.R. Bowker, Bowker A&I Publishing, 121 Chanlon Road, New Providence, NJ 07974. (908) 771-7714. Fax (908) 771-7725. 1984 to present. Monthly. $495.00 per year.

Computer Abstracts, MCB University Press Ltd., PO Box 10812, Birmingham, AL 35201-0812. (800) 633-4931. FAX (205) 995-1588. Monthly. Covers computer theory, data, hardware, systems, networks, human-computer interaction, artificial intelligence, as well as applications of computers in aerospace, business, CAD/CAM, cartography, civil engineering, electronics and electrical engineering, industrial engineering, mechanical engineering, medicine, structural engineering, etc.

Current Contents: Engineering, Technology, and Applied Sciences. Institute for Scientific Information, 3501 Market Street, Philadelphia, PA 19104. (215) 386-0100. FAX (215) 386-6362. 1970 to present. Weekly. $442.00 per year.

Engineering Index Monthly. Engineering Information, Inc., Castle Point on the Hudson, Hoboken, NJ 07030. (800) 221-1044. FAX (212) 832-1857. Monthly. $2300.00 per year. Also available online as COMPENDEX, and also on CD-ROM. Covers chemical engineering, computers, electrical engineering, civil engineering, metals and mining, industrial management, and mechanical engineering.

Mechanical Engineering Abstracts (formerly ISMEC), Cambridge Scientific Abstracts. 7200 Wisconsin Avenue, Bethesda, MD 20814-4823. (301) 961-6750. FAX (301) 961-6720. 1967 to present. Monthly. $895.00 per year. Summarizes world literature in mechanical engineering, production engineering, and engineering management. Also available online.

Physics Abstracts. Institute of Electrical Engineers, Michael Faraday House, Six Hill Way, Stevenage, Herts, SG1 2AY, England. Distributed by IEEE, 445 Hoes Lane, Piscataway, NJ 08854. (908) 562-5549. 1898 to present. Monthly. $2700.00 per year. Also available online as INSPEC.

Science Citation Index. Institute for Scientific Information, 3501 Market Street, Philadelphia, PA 19104. (215) 386-0100. FAX (215) 386-6362. Inquire about availability and cost. Also available on CD-ROM.

ASSOCIATIONS AND PROFESSIONAL SOCIETIES

American Society of Mechanical Engineers. 345 E. 47th Street, New York, NY 10017. (212) 705-7722.

Association for Computing Machinery. 1515 Broadway, 17th Floor, New York, NY 10036-5701. (212) 869-7440. FAX (212) 944-1318.

Institute of Electrical and Electronic Engineers. 345 E. 47th Street, New York, NY 10017-2394.

National Conference On Fluid Power. 3333 N. Mayfair Road, Suite 111, Milwaukee, WI 53226. (414) 778-3368.

Vibration Institute. 6262 S. Kingery Highway, Suite 212, Willowbrook, IL 60514. (708) 654-2254.

BIBLIOGRAPHIES

Scientific and Technical Books in Print. R.R. Bowker Inc., 121 Chanlon Road, New Providence, NJ 07974. (800) 521-8110 or (908) 464-6800. List of currently published books and serials in the physical and biological sciences, engineering and technology, with subject, author and titles indexes. Inquire as to cost and availability. Also available on ORBIT online service, (800) 456-7248.

MACHINE DESIGN

Ency. of Physical Sciences and Engineering Info. Sources

DIRECTORIES AND BIOGRAPHICAL SOURCES

CAD/CAM International Software Directory. TAB Books, P.O. Box 40, Blue Ridge Summit, PA 17294-0850. (717) 794-2191 or (800) 233-1128. FAX (717) 794-2080. 1991. Inquire about cost and availability.

Computer-Aided Design (CAD) Directory. TAB Books, P.O. Box 40, Blue Ridge Summit, PA 17294-0850. (717) 794-2191 or (800) 233-1128. FAX (717) 794-2080. 1991. Inquire for cost and availability.

Directory of Engineering Societies and Related Organizations. Gordon Davis. 13th edition. American Association of Engineering Societies, 1111 19th Street NW, Suite 608, Washington, DC 20036. (202) 296-2237 or (800) 658-8897. 1989. Inquire for price.

Machinery Buyers Guide. Findlay Publications Inc. Ltd., Franks Hall, Franks Lane, Horton Kirby, Kent DA4 9LL, England. 1926 to present. Annual. Inquire for cost and availability.

Research Centers Directory. Gale Research, 835 Penobscot Building, Detroit, MI 48226-4094. (313) 961-2242. (800) 347-4253. FAX (313) 961-6083. $485.00.

Scientific and Technical Organizations and Agencies Directory. Gale Research, 835 Penobscot Building, Detroit, MI 48226-4094. (313) 961-2242. (800) 347-4253. FAX (313) 961-6083. 4th edition. 1996. $195.00.

Who's Who in Engineering. American Association of Engineering Societies, 1111 19th Street NW, Suite 608, Washington, DC 20036. (202) 296-2237 or (800) 658-8897. 8th edition. 1991. Inquire for price.

ENCYCLOPEDIAS AND DICTIONARIES

CAD/CAM Dictionary. Preston & Crawford. Marcel Dekker, Inc., 270 Madison Avenue, New York, NY 10016. (212) 696-9000. FAX (212) 685-4540. 1985. $89.75.

CAD/CAM Glossary. CIM Systems, 2425 N. Central, No. 432, Richardson, TX 75080-3549. (214) 437-5171. $19.25.

McGraw-Hill Encyclopedia of Engineering. Sybil P. Parker, ed. 2nd edition. McGraw-Hill Publishing Company, 1221 Avenue of the Americas, New York, NY 10020. (800) 262-4729 or (212) 512-3825. 1993. $95.50.

McGraw-Hill Encyclopedia of Science and Technology. Sybil P. Parker, ed. 7th edition. 20 volumes. McGraw-Hill Publishing Company, 1221 Avenue of the Americas, New York, NY 10020. (800) 262-4729 or (212) 512-3825. 1992. $1900.00

Thesaurus of Scientific, Technical, and Engineering Terms. Hemisphere Publishing Corporation, 1900 Frost Road, Suite 101, Bristol, PA 19007-1598. (215) 785-5800 or (800) 821-8312. FAX (215) 785-5515. 1987. $173.00.

GENERAL WORKS

Computer-Assisted Mechanical Design. J. Ed Akin. Prentice Hall, 113 Sylvan Avenue, Route 9W, Englewood Cliffs, NJ 07632. (201) 592-2000 or (800) 922-0579. 1990. $54.00.

Fundamentals of Machine Component Design. Robert C. Juvinall. 2nd edition. John Wiley and Sons, Inc., 605 Third Avenue, New York, NY 10158. (800) 526-5368 or (212) 850-6000. 1991. Inquire for cost and availability.

Machine Design. Anthony Esposito, J. Robert Thrower. 2nd edition. Delmar Publishers, Division of Thomson Educational Publishing, Inc., 3 Columbia Circle, Box 15015, Albany, NY 12205. (518) 464-3500 or (800) 347-7707. 1991. $44.95.

Machine Design for Mobile and Industrial Applications. Gary W. Krutz, et al. SAE International, 400 Commonwealth Drive, Warrendale, PA 15096-0001. (412) 776-4970. FAX (412) 776-0790. 1994. $69.00.

The Mechanical Design Process. David G. Ullman. McGraw-Hill Publishing Company, 1221 Avenue of the Americas, New York, NY 10020. (800) 262-4729 or (212) 512-3825. 1992. Inquire for cost and availability.

Precision Machine Design. Alexander H. Slocum. Prentice Hall, 113 Sylvan Avenue, Route 9W, Englewood Cliffs, NJ 07632. (201) 592-2000 or (800) 922-0579. 1991. $60.00.

HANDBOOKS AND MANUALS

Handbook of Mechanics, Materials, and Structures. Alexander Blake, ed. John Wiley and Sons, Inc., 605 Third Avenue, New York, NY 10158. (800) 526-5368 or (212) 850-6000. 1985. $120.00.

Marks' Standard Handbook for Mechanical Engineers. E.A. Avallone and T. Baumeister III. 9th edition. McGraw-Hill Publishing Company, 1221 Avenue of the Americas, New York, NY 10020. (800) 262-4729 or (212) 512-3825. 1987. $133.00.

Machine Design Fundamentals: A Mechanical Designers' Workbook. Joseph E. Shigley & Charles R. Mischke, editors-in-chief. McGraw-Hill Publishing Company, 1221 Avenue of the Americas, New York, NY 10020. (800) 262-4729 or (212) 512-3825. 1989. $31.95.

Machine Design Data Handbook. K. Lingaiah. McGraw-Hill Publishing Company, 1221 Avenue of the Americas, New York, NY 10020. (800) 262-4729 or (212) 512-3825. 1994. $89.50.

Machinery's Handbook. Erik Oberg, et al., editors. 24th edition. Industrial Press Inc., 200 Madison Avenue, New York, NY 10016. (212) 889-6330. FAX (212) 545-8327. 1992. $75.00.

Mechanical Design and Systems Handbook. Harold A. Rothbart, editor-in-chief. McGraw-Hill Publishing Company, 1221 Avenue of the Americas, New York, NY 10020. (800) 262-4729 or (212) 512-3825. 1985. Inquire for cost and availability.

Mechanical Engineer's Data Handbook. James Carvill. CRC Publishers, Inc., 2000 Corporate Blvd., N.W., Boca Raton, FL 33431. (407) 994-0555 or (800) 272-7737. FAX (407) 994-0949. 1993. $69.95.

Mechanical Engineer's Reference Book. Edward H. Smith, editor. 12th edition. SAE International, 400 Commonwealth Drive, Warrendale, PA 15096-0001. (412) 776-4970. FAX (412) 776-0790. 1994. $155.00.

Standard Handbook of Machine Design. Joseph E. Shigley & Charles R. Mischke, editors-in-chief. McGraw-Hill Publishing

Company, 1221 Avenue of the Americas, New York, NY 10020. (800) 262-4729 or (212) 512-3825. 1986. $124.50.

ONLINE DATABASES AND CD-ROMS

Compendex Plus. Engineering Information, Inc., 345 East 47th Street, New York, NY 10017. (212) 705-7600 or (800) 221-1044. Contains citations with abstracts to worldwide literature in engineering and technology, from 1970 to present. Available on online BRS,(800) 289-4277, DIALOG, (800) 334-2564, ORBIT (800) 456-7248, and STN International, FIZ Karlsruhe, P.O. Box 2465, W-7500, Karlsruhe 1, Germany, online services. Also available on CD-ROM. Inquire as to cost and availability.

CSA Engineering. Cambridge Scientific Abstracts, 7200 Wisconsin Avenue, Suite 601, Bethesda, MD 20814. (301) 961-6750 or (800) 843-7751. Contains citations and abstracts of international periodicals and research literature covering all fields of engineering and science and technology, including computer and information science, electronics, mechanical engineering, solid state materials, 1981 to present. Available on BRS,(800) 289-4277, online service. Inquire as to cost and availability.

Current Contents Search. Institute for Scientific Information, 3501 Market Street, Philadelphia, PA 19104. (215) 386-0100. FAX (215) 386-6362. Contains citations to articles listed in the table of contents of science and technology journals. Also articles in social sciences and life sciences journals. Available on BRS,(800) 289-4277; DIALOG (800) 334-2564; online services. Inquire as to cost and availability.

Dissertation Abstracts. UMI, 300 North Zeeb Road, Ann Arbor, MI 48106. (800) 521-0600 or (313) 761-4700. Scope includes virtually all doctoral dissertations accepted at accredited American institutions from 1861 to present in 252 subject areas. Available on BRS,(800) 289-4277; DIALOG,(800) 334-2564; and OCLC EPIC,(800) 848-5878; online services. Also available on CD-ROM. Inquire as to cost and availability.

ISMEC: Mechanical Engineering Abstracts. Cambridge Scientific Abstracts, 7200 Wisconsin Avenue, Suite 601, Bethesda, MD 20814. (301) 961-6750 or (800) 843-7751. Contains citations to the literature in mechanical engineering, industrial and production engineering, energy, power, mechanics, devices and related areas, from 1973 to present. Available on the DIALOG,(800) 334-2564, online service. Inquire as to cost and availability.

NTIS Bibliographic Database. National Technical Information Service, 5285 Port Royal Road, Springfield, VA 22161. (703) 487-4929 or FAX (703) 321-8199. Broad coverage of government-sponsored science and technology research reports, 1964 to present. Available on BRS,(800) 289-4277, DIALOG, (800) 334-2564, ORBIT, (800) 456-7248, and STN International, FIZ Karlsruhe, P.O. Box 2465, W-7500, Karlsruhe 1, Germany, online services. Also available on CD-ROM. Inquire as to cost and availability.

Scisearch. Institute for Scientific Information, 3501 Market Street, Philadelphia, PA 19104. (800) 523-1850 or (215) 386-0100. Broad multidisciplinary title and author index to the international literature of science and technology, 1974 to present. Available on DIALOG,(800) 334-2564, and ORBIT,(800) 456-7248, online services. Also available on CD-ROM. Inquire as to cost and availability.

Wilsonline. H.W. Wilson Company, 950 University Avenue, Bronx, NY 10452. (800) 367-6770 or (212) 588-8400. Makes available online versions of the printed H.W. Wilson indexes, including Applied Science and Technology Index, Business Periodicals Index, General Science Index, and Readers' Guide to Periodical Literature. Period covered is generally 1983 to present. Available on BRS,(800) 289-4277, DIALOG,(800) 334-2564, and OCLC EPIC,(800) 848-5878, online services. Also available on CD-ROM. Inquire as to cost and availability.

PERIODICALS

International Communications in Heat and Mass Transfer. Elsevier Science [journals], 660 White Plains Rd., Tarrytown, NY 10159-5153. (919) 524-9200. FAX (919) 333-2444. 1974 to present. 6 times a year. $430.00 per year.

International Journal for Numerical Methods in Engineering. J. Wiley & Sons Ltd., Journals, Baffins Lane, Chichester, Sussex PO19 1UD, England. Telephone 0243-779777. FAX 0243-775878. 1964 to present. 24 times a year. $1795.00 per year.

International Journal of Heat and Mass Transfer. Elsevier Science [journals], 660 White Plains Rd., Tarrytown, NY 10159-5153. (919) 524-9200. FAX (919) 333-2444. 1960 to present. 18 times a year. $2070.00 per year.

Journal of Applied Mechanics. American Society of Mechanical Engineers, 345 E. 47th Street, New York, NY 10017. (212) 705-7722. 1935 to present. Quarterly. $120.00.

Journal of Engineering for Industry. American Society of Mechanical Engineers, 345 E. 47th Street, New York, NY 10017. (212) 705-7722. 1970 to present. Quarterly. $100.00 per year.

Journal of Fluid Control. Delbridge Publishing Company, Box 160817, Cupertino, CA 95016. (408) 446-3131. 1967 to present. Quarterly. $145.00 per volume.

Journal of Heat Transfer. American Society of Mechanical Engineers, 345 E. 47th Street, New York, NY 10017. (212) 705-7722. 1970 to present. Quarterly. $155.00 per year.

Journal of Mechanical Design. American Society of Mechanical Engineers, 345 E. 47th Street, New York, NY 10017. (212) 705-7722. 1978 to present. Quarterly. $100.00 per year.

Journal of Pressure Vessel Technology. American Society of Mechanical Engineers, 345 E. 47th Street, New York, NY 10017. (212) 705-7722. 1974 to present. Quarterly. $100.00 per year.

Lubrication Engineering. Society of Tribologist and Lubrication Engineers, 840 Busse Highway, Park Ridge, IL 60068-2376. (708) 825-5536. FAX (708) 825-1456. 1945 to present. Monthly. $61.00 per year.

Machine Design. Penton Publishing, 110 Superior Ave., Cleveland, OH 44114-2543. 1929 to present. 6 times a year. $180.00 per year.

Mechanical Engineering. American Society of Mechanical Engineers, 345 E. 47th Street, New York, NY 10017. (212) 705-7722. 1906 to present. Monthly. $45.00 per year.

RESEARCH CENTERS AND INSTITUTES

Laboratory for Experimental Mechanics Research. SUNY, Stony Brook, NY 11794-2300. (516) 632-8311. FAX (516) 632-8720.

MACHINE DESIGN

Ency. of Physical Sciences and Engineering Info. Sources

Mechanical Engineering Design Laboratory. University of Florida, 237 Mechanical Engineering Bldg., Gainesville, FL 32611. (904) 392-0827. FAX (904) 392-1071.

Mechanical Engineering Research Laboratory. Kansas State University, Durland Hall, Manhattan, KS 66506. (913) 532-5610.

MACHINERY

See also: CAD (COMPUTER-AIDED DESIGN), MACHINE DESIGN, MECHANICAL ENGINEERING

ABSTRACT SERVICES AND INDEXES

Applied Mechanics Reviews: An Assessment of World Literature in Engineering Sciences. 1948-present. American Society of Mechanical Engineers, 345 East 47th Street, New York, NY 10017. (212) 705-7703. Monthly. $360.00 per year.

Applied Science and Technology Index; A Cumulative Subject Index To English Language Periodicals in the Fields of Aeronautics and Space Science, Computer Technology, Chemistry, Construction Industry, Energy and Related Areas. H.W. Wilson Co., 950 University Avenue, Bronx, NY 10452. (800) 367-6770 or (212) 588-8400. FAX (718) 590-1617. From 1958 to present. Monthly. Inquire about cost and availability. Also available on CD-ROM and online.

CAD/CAM Abstracts. R.R. Bowker, Bowker A&I Publishing, 121 Chanlon Road, New Providence, NJ 07974. (908) 771-7714. Fax (908) 771-7725. 1984 to present. Monthly. $495.00 per year.

Computer Abstracts, MCB University Press Ltd., PO Box 10812, Birmingham, AL 35201-0812. (800) 633-4931. FAX (205) 995-1588. Monthly. Covers computer theory, data, hardware, systems, networks, human-computer interaction, artificial intelligence, as well as applications of computers in aerospace, business, CAD/CAM, cartography, civil engineering, electronics and electrical engineering, industrial engineering, mechanical engineering, medicine, structural engineering, etc.

Current Contents: Engineering, Technology, and Applied Sciences. Institute for Scientific Information, 3501 Market Street, Philadelphia, PA 19104. (215) 386-0100. FAX (215) 386-6362. 1970 to present. Weekly. $442.00 per year.

Engineering Index Monthly. Engineering Information, Inc., Castle Point on the Hudson, Hoboken, NJ 07030. (800) 221-1044. FAX (212) 832-1857. Monthly. $2300.00 per year. Also available online as COMPENDEX, and also on CD-ROM. Covers chemical engineering, computers, electrical engineering, civil engineering, metals and mining, industrial management, and mechanical engineering.

Mechanical Engineering Abstracts (formerly ISMEC). Cambridge Scientific Abstracts, 7200 Wisconsin Avenue, Bethesda, MD 20814-4823. (301) 961-6750. FAX (301) 961-6720. 1967 to present. Monthly. $895.00 per year. Summarizes world literature in mechanical engineering, production engineering, and engineering management. Also available online.

Science Citation Index. Institute for Scientific Information, 3501 Market Street, Philadelphia, PA 19104. (215) 386-0100. FAX

(215) 386-6362. Inquire about availability and cost. Also available on CD-ROM.

ASSOCIATIONS AND PROFESSIONAL SOCIETIES

American Society of Mechanical Engineers. 345 E. 47th Street, New York, NY 10017. (212) 705-7722.

Association For Computing Machinery. 1515 Broadway, 17th Floor, New York, NY 10036-5701. (212) 869-7440. FAX (212) 944-1318.

Institute of Electrical and Electronic Engineers. 345 E. 47th Street, New York, NY 10017-2394.

Vibration Institute. 6262 S. Kingery Highway, Suite 212, Willowbrook, IL 60514. (708) 654-2254.

BIBLIOGRAPHIES

Scientific and Technical Books in Print. R.R. Bowker Inc., 121 Chanlon Road, New Providence, NJ 07974. (800) 521-8110 or (908) 464-6800. List of currently published books and serials in the physical and biological sciences, engineering and technology, with subject, author and titles indexes. Inquire as to cost and availability. Also available on ORBIT online service, (800) 456-7248.

DIRECTORIES AND BIOGRAPHICAL SOURCES

CAD/CAM International Software Directory. TAB Books, P.O. Box 40, Blue Ridge Summit, PA 17294-0850. (717) 794-2191 or (800) 233-1128. FAX (717) 794-2080. 1991. Inquire for cost and availability.

Computer-Aided Design (CAD) Directory. TAB Books, P.O. Box 40, Blue Ridge Summit, PA 17294-0850. (717) 794-2191 or (800) 233-1128. FAX (717) 794-2080. 1991. Inquire for cost and availability.

Directory of Engineering Societies and Related Organizations. Gordon Davis. 13th edition. American Association of Engineering Societies, 1111 19th Street NW, Suite 608, Washington, DC 20036. (202) 296-2237 or (800) 658-8897. 1989. Inquire for price.

Machinery Buyers Guide. Findlay Publishers Inc. Ltd., Franks Hall, Franks Lane, Horton Kirby, Kent DA4 9LL, England. 1926 to present. Annual. Inquire for cost and availability.

Research Centers Directory. Gale Research, 835 Penobscot Building, Detroit, MI 48226-4094. (313) 961-2242. (800) 347-4253. FAX (313) 691-6083. $485.00.

Scientific and Technical Organizations and Agencies Directory. Gale Research, 835 Penobscot Building, Detroit, MI 48226-4094. (313) 961-2242. (800) 347-4253. FAX (313) 691-6083. 4th edition. 1996. $195.00.

Who's Who in Engineering. American Association of Engineering Societies, 1111 19th Street NW, Suite 608, Washington, DC 20036. (202) 296-2237 or (800) 658-8897. 8th edition. 1991. Inquire for price.

ENCYCLOPEDIAS AND DICTIONARIES

CAD/CAM Dictionary. Preston & Crawford. Marcel Dekker, Inc., 270 Madison Avenue, New York, NY 10016. (212) 696-9000. FAX (212) 685-4540. 1985. $89.75.

CAD/CAM Glossary. CIM Systems, 2425 N. Central, No. 432, Richardson, TX 75080-3549. (214) 437-5171. $19.25.

McGraw-Hill Encyclopedia of Engineering. Sybil P. Parker, ed. 2nd edition. McGraw-Hill Publishing Company, 1221 Avenue of the Americas, New York, NY 10020. (800) 262-4729 or (212) 512-3825. 1993. $95.50.

McGraw-Hill Encyclopedia of Science and Technology. Sybil P. Parker, ed. 7th edition. 20 volumes. McGraw-Hill Publishing Company, 1221 Avenue of the Americas, New York, NY 10020. (800) 262-4729 or (212) 512-3825. 1992. $1900.00

Thesaurus of Scientific, Technical, and Engineering Terms. Hemisphere Publishing Corporation, 1900 Frost Road, Suite 101, Bristol, PA 19007-1598. (215) 785-5800 or (800) 821-8312. FAX (215) 785-5515.FAX (215) 785-5515. 1987. $173.00.

GENERAL WORKS

Computer-Assisted Mechanical Design. J. Ed Akin. Prentice Hall, 113 Sylvan Avenue, Route 9W, Englewood Cliffs, NJ 07632. (201) 592-2000 or (800) 922-0579. 1990. $54.00.

Fundamentals of Machine Component Design. Robert C. Juvinall & Kurt M. Marshek. John Wiley and Sons, Inc., 605 Third Avenue, New York, NY 10158. (800) 526-5368 or (212) 850-6000. 1991. Inquire for cost and availability.

Machine Design. Anthony Esposito & J.R. Thrower. Delmar Publishers, Division of Thomson Educational Publishing, Inc., 3 Columbia Circle, Box 15015, Albany, NY 12205. (518) 464-3500 or (800) 347-7707. 1991. $44.95.

Machine Design Fundamentals. Joseph E. Shigley & Charles R. Mischke. McGraw-Hill Publishing Company, 1221 Avenue of the Americas, New York, NY 10020. (800) 262-4729 or (212) 512-3825. 1989. $31.95.

HANDBOOKS AND MANUALS

Handbook of Mechanics, Materials, and Structures. Alexander Blake, editor. John Wiley and Sons, Inc., 605 Third Avenue, New York, NY 10158. (800) 526-5368 or (212) 850-6000. 1985. $120.00.

Machine Design Data Handbook. K. Lingaiah. McGraw-Hill Publishing Company, 1221 Avenue of the Americas, New York, NY 10020. (800) 262-4729 or (212) 512-3825. 1994. $89.50.

Machinery's Handbook. Erik Oberg, et al., editors. 24th edition. Industrial Press Inc., 200 Madison Avenue, New York, NY 10016. (212) 889-6330. FAX (212) 545-8327. 1992. $75.00.

Marks' Standard Handbook For Mechanical Engineers. E.A. Avallone and T. Baumeister III. 9th edition. McGraw-Hill Publishing Company, 1221 Avenue of the Americas, New York, NY 10020. (800) 262-4729 or (212) 512-3825. 1987. $133.00.

Mechanical Engineer's Data Handbook. J. Carvill. CRC Publishers, Inc., 2000 Corporate Blvd., N.W., Boca Raton, FL 33431. (407) 994-0555 or (800) 272-7737. FAX (407) 994-0949. 1993. $69.95.

Mechanical Engineer's Handbook. Joseph E. Shigley & Charles R. Mischke. McGraw-Hill Publishing Company, 1221 Avenue of the Americas, New York, NY 10020. (800) 262-4729 or (212) 512-3825. 1986. $124.50.

Pressure Vessel Design Handbook: Illustrated Procedures For Solving Every Major Pressure Vessel Design Problem. Dennis R. Moss. Gulf Publishing Company, P.O. Box 2608, Houston, TX. (713) 529-4301 or (800) 231-6275. 1987. $56.50.

Standard Handbook of Machine Design. Joseph E. Shigley & Charles R. Mischke, editors-in-chief. McGraw-Hill Publishing Company, 1221 Avenue of the Americas, New York, NY 10020. (800) 262-4729 or (212) 512-3825. 1986. $124.50.

ONLINE DATABASES AND CD-ROMS

Compendex Plus. Engineering Information, Inc., 345 East 47th Street, New York, NY 10017. (212) 705-7600 or (800) 221-1044. Contains citations with abstracts to worldwide literature in engineering and technology, from 1970 to present. Available on online BRS,(800) 289-4277, DIALOG, (800) 334-2564, ORBIT (800) 456-7248, and STN International, FIZ Karlsruhe, P.O. Box 2465, W-7500, Karlsruhe 1, Germany, online services. Also available on CD-ROM. Inquire as to cost and availability.

CSA Engineering. Cambridge Scientific Abstracts, 7200 Wisconsin Avenue, Suite 601, Bethesda, MD 20814. (301) 961-6750 or (800) 843-7751. Contains citations and abstracts of international periodicals and research literature covering all fields of engineering and science and technology, including computer and information science, electronics, mechanical engineering, solid state materials, 1981 to present. Available on BRS,(800) 289-4277, online service. Inquire as to cost and availability.

Current Contents Search. Institute for Scientific Information, 3501 Market Street, Philadelphia, PA 19104. (215) 386-0100. FAX (215) 386-6362. Contains citations to articles listed in the table of contents of science and technology journals. Also articles in social sciences and life sciences journals. Available on BRS,(800) 289-4277, DIALOG,(800) 334-2564, online services. Inquire as to cost and availability.

Dissertation Abstracts. University Microfilms International, 300 North Zeeb Road, Ann Arbor, MI 48106. (800) 521-0600 or (313) 761-4700. Scope includes virtually all doctoral dissertations accepted at accredited American institutions from 1861 to present in 252 subject areas. Available on BRS,(800) 289-4277, DIALOG,(800) 334-2564, and OCLC EPIC,(800) 848-5878, online services. Also available on CD-ROM. Inquire as to cost and availability.

ISMEC: Mechanical Engineering Abstracts. Cambridge Scientific Abstracts, 7200 Wisconsin Avenue, Suite 601, Bethesda, MD 20814. (301) 961-6750 or (800) 843-7751. Contains citations to the literature in mechanical engineering, industrial and production engineering, energy, power, mechanics, devices and related areas, from 1973 to present. Available on the DIALOG,(800) 334-2564, online service. Inquire as to cost and availability.

NTIS Bibliographic Database. National Technical Information Service, 5285 Port Royal Road, Springfield, VA 22161. (703)

MACHINERY

Ency. of Physical Sciences and Engineering Info. Sources

487-4929 or FAX (703) 321-8199. Broad coverage of government-sponsored science and technology research reports, 1964 to present. Available on BRS,(800) 289-4277, DIALOG, (800) 334-2564, ORBIT, (800) 456-7248, and STN International, FIZ Karlsruhe, P.O. Box 2465, W-7500, Karlsruhe 1, Germany, online services. Also available on CD-ROM. Inquire as to cost and availability.

Scisearch. Institute for Scientific Information, 3501 Market Street, Philadelphia, PA 19104. (800) 523-1850 or (215) 386-0100. Broad multidisciplinary title and author index to the international literature of science and technology, 1974 to present. Available on DIALOG,(800) 334-2564, and ORBIT,(800) 456-7248, online services. Also available on CD-ROM. Inquire as to cost and availability.

Wilsonline. H.W. Wilson Company, 950 University Avenue, Bronx, NY 10452. (800) 367-6770 or (212) 588-8400. Makes available online versions of the printed H.W. Wilson indexes, including Applied Science and Technology Index, Business Periodicals Index, General Science Index, and Readers' Guide to Periodical Literature. Period covered is generally 1983 to present. Available on BRS,(800) 289-4277, DIALOG,(800) 334-2564, and OCLC EPIC,(800) 848-5878, online services. Also available on CD-ROM. Inquire as to cost and availability.

PERIODICALS

International Communications in Heat and Mass Transfer. Elsevier Science [journals], 660 White Plains Rd., Tarrytown, NY 10159-5153. (919) 524-9200. FAX (919) 333-2444. 1974 to present. 6 times a year. $430.00 per year.

International Journal For Numerical Methods in Engineering. J. Wiley & Sons Ltd., Journals, Baffins Lane, Chichester, Sussex PO19 1UD, England. Telephone 0243-779777. FAX 0243-775878. 1964 to present. 24 times a year. $1795.00 per year.

International Journal of Heat and Mass Transfer. Elsevier Science [journals], 660 White Plains Rd., Tarrytown, NY 10159-5153. (919) 524-9200. FAX (919) 333-2444. 1960 to present. 18 times a year. $2070.00 per year.

International Journal of Machine Tools & Manufacture. Elsevier Science [journals], 660 White Plains Rd., Tarrytown, NY 10159-5153. (919) 524-9200. FAX (919) 333-2444. 1961 to present. 8 times a year. $710.00 per year.

International Journal of Multiphase Flow. Elsevier Science [journals], 660 White Plains Rd., Tarrytown, NY 10159-5153. (919) 524-9200. FAX (919) 333-2444. 1974 to present. 7 times a year. $900.00 per year.

Journal of Applied Mechanics. American Society of Mechanical Engineers, 345 E. 47th Street, New York, NY 10017. (212) 705-7722. 1935 to present. Quarterly. $120.00.

Journal of Engineering For Industry. American Society of Mechanical Engineers, 345 E. 47th Street, New York, NY 10017. (212) 705-7722. 1970 to present. Quarterly. $100.00 per year.

Journal of Fluid Control. Delbridge Publishing Company, Box 160817, Cupertino, CA 95016. (408) 446-3131. 1967 to present. Quarterly. $145.00 per volume.

Journal of Heat Transfer. American Society of Mechanical Engineers, 345 E. 47th Street, New York, NY 10017. (212) 705-7722. 1970 to present. Quarterly. $155.00 per year.

Journal of Mechanical Design. American Society of Mechanical Engineers, 345 E. 47th Street, New York, NY 10017. (212) 705-7722. 1978 to present. Quarterly. $100.00 per year.

Journal of Pressure Vessel Technology. American Society of Mechanical Engineers, 345 E. 47th Street, New York, NY 10017. (212) 705-7722. 1974 to present. Quarterly. $100.00 per year.

Lubrication Engineering. Society of Tribologist and Lubrication Engineers, 840 Busse Highway, Park Ridge, IL 60068-2376. (708) 825-5536. FAX (708) 825-1456. 1945 to present. Monthly. $61.00 per year.

Machine Design. Penton Publishing, 110 Superior Ave., Cleveland, OH 44114-2543. 1929 to present. 6 times a year. $180.00 per year.

Mechanical Engineering. American Society of Mechanical Engineers, 345 E. 47th Street, New York, NY 10017. (212) 705-7722. 1906 to present. Monthly. $45.00 per year.

RESEARCH CENTERS AND INSTITUTES

Laboratory For Experimental Mechanics Research. SUNY, Stony Brook, NY 11794-2300. (516) 632-8311. FAX (516) 632-8720.

Mechanical Engineering Design Laboratory. University of Florida, 237 Mechanical Engineering Bldg., Gainesville, FL 32611. (904) 392-0827. FAX (904) 392-1071.

Mechanical Engineering Research Laboratory. Kansas State University, Durland Hall, Manhattan, KS 66506. (913) 532-5610.

MACHINING

See also: CAD (COMPUTER-AIDED DESIGN), MACHINE DESIGN, MACHINERY, MECHANICAL ENGINEERING

ABSTRACT SERVICES AND INDEXES

Applied Mechanics Reviews: An Assessment of World Literature in Engineering Sciences. 1948-present. American Society of Mechanical Engineers, 345 East 47th Street, New York, NY 10017. (212) 705-7703. Monthly. $360.00 per year.

Applied Science and Technology Index; A Cumulative Subject Index To English Language Periodicals in the Fields of Aeronautics and Space Science, Computer Technology, Chemistry, Construction Industry, Energy and Related Areas. H.W. Wilson Co., 950 University Avenue, Bronx, NY 10452. (800) 367-6770 or (212) 588-8400. FAX (718) 590-1617. From 1958 to present. Monthly. Inquire about cost and availability. Also available on CD-ROM and online.

CAD/CAM Abstracts. R.R. Bowker, Bowker A&I Publishing, 121 Chanlon Road, New Providence, NJ 07974. (908) 771-7714. Fax (908) 771-7725. 1984 to present. Monthly. $495.00 per year.

Computer Abstracts. MCB University Press Ltd., PO Box 10812, Birmingham, AL 35201-0812. (800) 633-4931. FAX (205) 995-1588. Monthly. Covers computer theory, data, hardware, systems, networks, human-computer interaction, artificial intelligence, as well as applications of computers in aerospace, business, CAD/CAM, cartography, civil engineering, electronics and electrical engineering, industrial engineering, mechanical engineering, medicine, structural engineering, etc.

Current Contents: Engineering, Technology, and Applied Sciences. Institute for Scientific Information, 3501 Market Street, Philadelphia, PA 19104. (215) 386-0100. FAX (215) 386-6362. 1970 to present. Weekly. $442.00 per year.

Engineering Index Monthly. Engineering Information, Inc., Castle Point on the Hudson, Hoboken, NJ 07030. (800) 221-1044. FAX (212) 832-1857. Monthly. $2300.00 per year. Also available online as COMPENDEX, and also on CD-ROM. Covers chemical engineering, computers, electrical engineering, civil engineering, metals and mining, industrial management, and mechanical engineering.

Mechanical Engineering Abstracts (formerly ISMEC), Cambridge Scientific Abstracts, 7200 Wisconsin Avenue, Bethesda, MD 20814-4823. (301) 961-6750. FAX (301) 961-6720. 1967 to present. Monthly. $895.00 per year. Summarizes world literature in mechanical engineering, production engineering, and engineering management. Also available online.

Science Citation Index. Institute for Scientific Information, 3501 Market Street, Philadelphia, PA 19104. (215) 386-0100. FAX (215) 386-6362. Inquire about availability and cost. Also available on CD-ROM.

ASSOCIATIONS AND PROFESSIONAL SOCIETIES

American Society of Mechanical Engineers. 345 E. 47th Street, New York, NY 10017. (212) 705-7722.

American Society For Metals. Materials Park, OH 44073-0002. (216) 338-5151 or (800) 336-5152. FAX (216) 338-4634.

Machining Technology Association of SME. PO Box 930, One SME Drive, Dearborn, MI 48121-0930. (313) 271-1500. FAX (313) 271-2861.

BIBLIOGRAPHIES

Scientific and Technical Books in Print. R.R. Bowker Inc., 121 Chanlon Road, New Providence, NJ 07974. (800) 521-8110 or (908) 464-6800. List of currently published books and serials in the physical and biological sciences, engineering and technology, with subject, author and titles indexes. Inquire as to cost and availability. Also available on ORBIT online service, (800) 456-7248.

DIRECTORIES AND BIOGRAPHICAL SOURCES

Dun's Industrial Guide: The Metalworking Directory, 1993-94. Dun & Bradstreet Information Services Staff. 3 volumes. Dun & Bradstreet Information Services, 899 Eaton Avenue, Bethlehem, PA 18025. (610) 882-7000 or (800) 526-0651. FAX (610) 882-7269. 1993. $795.00

Machinery Buyers Guide. Findlay Publications Inc. Ltd., Franks Hall, Franks Lane, Horton Kirby, Kent DA4 9LL, England. 1926 to present. Annual. Inquire for cost and availability.

Who's Who in Engineering. American Association of Engineering Societies, 1111 19th Street NW, Suite 608, Washington, DC 20036. (202) 296-2237 or (800) 658-8897. 8th edition. 1991. Inquire for price.

GENERAL WORKS

CNC Machining Technology. Graham T. Smith. 4 volumes. Springer-Verlag, 175 Fifth Avenue, New York, NY 10010. (212) 460-1500 or (800) 777-4643. FAX (212) 473-6272. 1993. $29.00 per volume.

Metal Machining and Forming Technology. Joseph P. Vidosic. Books on Demand, Division of University Microfilms International, 300 North Zeeb Road, Ann Arbor, MI 48106-1346. (313) 761-4700 or (800) 521-0600. $142.30.

Modern Machining Technology. Richard Baril. Delmar Publishers, Division of Thomson Educational Publishing, Inc., 3 Columbia Circle, Box 15015, Albany, NY 12205. (518) 464-3500 or (800) 347-7707. 1987. $45.95.

Precision Machining Technology & Machine Development and Improvement. M. Jouaneh & S.S. Rangwala, editors. American Society of Mechanical Engineers, 345 East 47th Street, New York, NY 10017. (212) 705-7703. 1992. $57.50.

HANDBOOKS AND MANUALS

ASM Handbook Volume 16: Machining. ASM International Handbook Committee. ASM International, Materials Information, Materials Park, OH 44073-0002. (216) 338-5151 or (800) 336-5152. FAX (216) 338-4634. 1989. $147.00.

Handbook of Mechanics, Materials, and Structures. Alexander Blake, editor. John Wiley and Sons, Inc., 605 Third Avenue, New York, NY 10158. (800) 526-5368 or (212) 850-6000. 1985. $120.00.

Marks' Standard Handbook For Mechanical Engineers. E.A. Avallone and T. Baumeister III. 9th edition. McGraw-Hill Publishing Company, 1221 Avenue of the Americas, New York, NY 10020. (800) 262-4729 or (212) 512-3825. 1987. $133.00.

Machinery's Handbook. Erik Oberg, et al., editors. 24th edition. Industrial Press Inc., 200 Madison Avenue, New York, NY 10016. (212) 889-6330. FAX (212) 545-8327. 1992. $75.00.

McGraw-Hill Machining and Metalworking Handbook. Ronald A. Walsh. McGraw-Hill Publishing Company, 1221 Avenue of the Americas, New York, NY 10020. (800) 262-4729 or (212) 512-3825. 1994. $69.50.

Mechanical Design and Systems Handbook. Harold A. Rothbart, editor-in-chief. McGraw-Hill Publishing Company, 1221 Avenue of the Americas, New York, NY 10020. (800) 262-4729 or (212) 512-3825. 1985. Inquire for cost and availability.

Mechanical Engineer's Reference Book. Edward H. Smith, editor. 12th edition. SAE International, 400 Commonwealth Drive, Warrendale, PA 15096-0001. (412) 776-4970. FAX (412) 776-0790. 1994. $155.00.

Standard Handbook of Machine Design. Joseph E. Shigley & Charles R. Mischke, editors-in-chief. McGraw-Hill Publishing Company, 1221 Avenue of the Americas, New York, NY 10020. (800) 262-4729 or (212) 512-3825. 1986. $124.50.

Tool and Manufacturing Engineer's Handbook, Volume 1: Machining. C. Wick & T. Drozda, editors. 4th edition. Society of Manufacturing Engineers, One SME Drive, Box 930, Dearborn, MI 48128-1490. (313) 271-1500 or FAX (313) 271-2861. 1983. $120.00.

ONLINE DATABASES AND CD-ROMS

Compendex Plus. Engineering Information, Inc., 345 East 47th Street, New York, NY 10017. (212) 705-7600 or (800) 221-1044. Contains citations with abstracts to worldwide literature in engineering and technology, from 1970 to present. Available on online BRS,(800) 289-4277, DIALOG, (800) 334-2564, ORBIT (800) 456-7248, and STN International, FIZ Karlsruhe, P.O. Box 2465, W-7500, Karlsruhe 1, Germany, online services. Also available on CD-ROM. Inquire as to cost and availability.

CSA Engineering. Cambridge Scientific Abstracts, 7200 Wisconsin Avenue, Suite 601, Bethesda, MD 20814. (301) 961-6750 or (800) 843-7751. Contains citations and abstracts of international periodicals and research literature covering all fields of engineering and science and technology, including computer and information science, electronics, mechanical engineering, solid state materials, 1981 to present. Available on BRS,(800) 289-4277, online service. Inquire as to cost and availability.

Current Contents Search. Institute for Scientific Information, 3501 Market Street, Philadelphia, PA 19104. (215) 386-0100. FAX (215) 386-6362. Contains citations to articles listed in the table of contents of science and technology journals. Also articles in social sciences and life sciences journals. Available on BRS,(800) 289-4277, DIALOG,(800) 334-2564, online services. Inquire as to cost and availability.

ISMEC: Mechanical Engineering Abstracts. Cambridge Scientific Abstracts, 7200 Wisconsin Avenue, Suite 601, Bethesda, MD 20814. (301) 961-6750 or (800) 843-7751. Contains citations to the literature in mechanical engineering, industrial and production engineering, energy, power, mechanics, devices and related areas, from 1973 to present. Available on the DIALOG,(800) 334-2564, online service. Inquire as to cost and availability.

NTIS Bibliographic Database. National Technical Information Service, 5285 Port Royal Road, Springfield, VA 22161. (703) 487-4929 or FAX (703) 321-8199. Broad coverage of government-sponsored science and technology research reports, 1964 to present. Available on BRS,(800) 289-4277, DIALOG, (800) 334-2564, ORBIT, (800) 456-7248, and STN International, FIZ Karlsruhe, P.O. Box 2465, W-7500, Karlsruhe 1, Germany, online services. Also available on CD-ROM. Inquire as to cost and availability.

Wilsonline. H.W. Wilson Company, 950 University Avenue, Bronx, NY 10452. (800) 367-6770 or (212) 588-8400. Makes available online versions of the printed H.W. Wilson indexes, including Applied Science and Technology Index, Business Periodicals Index, General Science Index, and Readers' Guide to Periodical Literature. Period covered is generally 1983 to present. Available on BRS,(800) 289-4277, DIALOG,(800) 334-2564, and OCLC EPIC,(800) 848-5878, online services. Also available on CD-ROM. Inquire as to cost and availability.

PERIODICALS

Abrasive Engineering Society Magazine. Abrasive Engineering Society, Meadowlark Technical Services, 108 Elliot Drive, Butler, PA 16001. (412) 282-6210. 1963 to present. Quarterly. $40.00 per year.

American Machinist. Penton Publishing, 110 Superior Ave., Cleveland, OH 44114-2543. (216) 696-7000. FAX (216) 696-0177. 1877 to present. Monthly. $65.00 per year.

Automatic Machining. Screw Machine Publishing Company Inc., 100 Seneca Avenue, Rochester, NY 14621. (716) 338-1522. 1939 to present. Monthly. $40.00 per year.

Cutting Tool Engineering. CTE Publications Inc., 400 Skokie Blvd., Suite 395, Northbrook, IL 60062-7403. (708) 441-7520. FAX (708) 441-8740. 1955 to present. Quarterly. $30.00 per year.

Finer Points. Industrial Diamond Association of America, 355 Lexington Avenue, New York, NY 10017. (212) 661-4261. FAX (212) 370-9047. 4 times a year. Inquire for cost.

Industrial Diamond Review. DE Beers Industrial Diamond Division Pty. Ltd., Charters, SunningHill, Ascot, Berks. SL5 9PX, England. FAX 0344-28188. 1940 to present. Bi-monthly. Inquire for cost and availability.

International Journal of Machine Tools & Manufacture. Elsevier Science [journals], 660 White Plains Rd., Tarrytown, NY 10159-5153. (919) 524-9200. FAX (919) 333-2444. 1961 to present. 8 times a year. $710.00 per year.

Journal of Applied Mechanics. American Society of Mechanical Engineers, 345 E. 47th Street, New York, NY 10017. (212) 705-7722. 1935 to present. Quarterly. $120.00.

Journal of Engineering For Industry. American Society of Mechanical Engineers, 345 E. 47th Street, New York, NY 10017. (212) 705-7722. 1970 to present. Quarterly. $100.00 per year.

Machine Design. Penton Publishing, 110 Superior Ave., Cleveland, OH 44114-2543. 1929 to present. 6 times a year. $180.00 per year.

Machining Technology. Society of Mechanical Engineers, One SME Drive, Box 930, Dearborn, MI 48121-0930. (313) 271-1500. FAX (313) 271=2861. Quarterly. $45.00 per year.

Metal Working Production. Morgan-Grampian Ltd., Morgan-Grampian House, 30 Calderwood Street, London SE18 6QH, England. Telephone 081-855-7777. FAX 081-854-7476. 1900 to present. Monthly. Inquire for cost.

Modern Machine Shop. Gardner Publications, 6600 Clough Pike, Cincinnati, OH 45244-4090. (513) 231-8020. FAX (513) 231-2818. 1928 to present. Monthly. $42.00 per year.

Precision Toolmaker. International Business & Technical Magazines, Ltd., Queensway House, 2 Queensway, RedHill, Surrey RH1 1QS England. Telephone 0737-768611. FAX 0737-761685. 1983 to present. Bi-monthly. $160.00 per year.

Utensil. ETM s.r.l., Via Principe Eugenio 3, 20155 Milan, Italy. Telephone 02-48010095. FAX 02-48010011. 1978 to present. 9 times a year. Inquire for cost and availability.

Ency. of Physical Sciences and Engineering Info. Sources

MAGNESIUM

RESEARCH CENTERS AND INSTITUTES

Mechanical Engineering Design Laboratory. University of Florida, 237 Mechanical Engineering Bldg., Gainesville, FL 32611. (904) 392-0827. FAX (904) 392-1071.

Mechanical Engineering Research Laboratory. Kansas State University, Durland Hall, Manhattan, KS 66506. (913) 532-5610.

MAGNESIUM

See also: ALLOYS, MATERIALS SCIENCE

ABSTRACT SERVICES AND INDEXES

Alloys Index. American Society for Metals, Metals Park, OH 44073. (216) 338-5151. FAX (216) 338-4634. 1974 to present. Monthly. $380.00. Also available on CD-ROM and online via METADEX, STN International, and DIALOG.

Aluminum Industry Abstracts. Aluminum Association, Materials Park, OH 44073. (216) 338-5151. FAX (216) 338-4634. 1968 to present. Monthly. $525.00 per year.

Applied Science and Technology Index; A Cumulative Subject Index To English Language Periodicals in the Fields of Aeronautics and Space Science, Computer Technology, Chemistry, Construction Industry, Energy and Related Areas. H.W. Wilson Co., 950 University Avenue, Bronx, NY 10452. (800) 367-6770 or (212) 588-8400. FAX (718) 590-1617. From 1958 to present. Monthly. Inquire about cost and availability. Also available on CD-ROM and online.

Current Contents: Engineering, Technology, and Applied Sciences. Institute for Scientific Information, 3501 Market Street, Philadelphia, PA 19104. (215) 386-0100. FAX (215) 386-6362. 1970 to present. Weekly. $442.00 per year.

Engineering Index Monthly. Engineering Information, Inc., Castle Point on the Hudson, Hoboken, NJ 07030. (800) 221-1044. FAX (212) 832-1857. Monthly. $2300.00 per year. Also available online as COMPENDEX, and also on CD-ROM. Covers chemical engineering, computers, electrical engineering, civil engineering, metals and mining, industrial management, and mechanical engineering.

Metals Abstracts and Metals Abstracts Index. American Society for Metals, Metals Park, OH 44073. (216) 338-5151. 1968 to present. Monthly. Abstracts are $1500.00 per year and Index is $460.00 per year.

ANNUAL REVIEWS AND YEARBOOKS

Annual Review of Materials Science. Annual Reviews Inc., 4139 El Camino Way, PO Box 10139, Palo Alto, CA 94303-0139. (800) 523-8635 or (415) 493-4400. FAX (415) 855-9815. Annual. $75.00 (1994 edition).

Non-Ferrous Metal Data. American Bureau of Metal Statistics, Inc., Box 1405 Plaza Station, 400 Plaza Drive, Secaucus, NJ 07094-0405. (201) 863-6900. FAX (201) 863-6050. 1920 to present. Annual. $350.00 per year.

ASSOCIATIONS AND PROFESSIONAL SOCIETIES

American Society For Metals. Materials Park, OH 44073-0002. (216) 338-5151 or (800) 336-5152. FAX (216) 338-4634.

International Magnesium Association. 1303 Vincent Place No.1, McLean, VA 22101-2615. (703) 442-8888. FAX (703) 821-1824.

Metallurgical Society of the AIME (American Institute of Mining, Metallurgical, and Petroleum Engineers). 345 E.47th Street, 14th Floor, New York, NY 10017. (212) 705-7695.

BIBLIOGRAPHIES

Scientific and Technical Books in Print. R.R. Bowker Inc., 121 Chanlon Road, New Providence, NJ 07974. (800) 521-8110 or (908) 464-6800. List of currently published books and serials in the physical and biological sciences, engineering and technology, with subject, author and titles indexes. Also available on ORBIT, (800) 456-7248, online service. Inquire as to cost and availability.

DIRECTORIES AND BIOGRAPHICAL SOURCES

Dun's Industrial Guide: The Metalworking Directory, 1993-94. Dun & Bradstreet Information Services Staff. 3 volumes. Dun & Bradstreet Information Services, 899 Eaton Avenue, Bethlehem, PA 18025. (610) 882-7000 or (800) 526-0651. FAX (610) 882-7269. 1993. $795.00.

International Magnesium Association—Buyer's Guide. International Magnesium Association, 1303 Vincent Place No.1, McLean, VA 22101-2615. (703) 442-8888. FAX (703) 821-1824. Inquire for cost and availability.

International Research Centers Directory. Gale Research, 835 Penobscot Building, Detroit, MI 48226-4094. (313) 961-2242. (800) 347-4253. FAX (313) 961-6083. 8th edition. 1995. $410.00

Research Centers Directory. Gale Research, 835 Penobscot Building, Detroit, MI 48226-4094. (313) 961-2242. (800) 347-4253. FAX (313) 961-6083. $485.00.

Scientific and Technical Organizations and Agencies Directory. Gale Research, 835 Penobscot Building, Detroit, MI 48226-4094. (313) 961-2242. (800) 347-4253. FAX (313) 961-6083. 4th edition. 1996. $195.00.

Who's Who in Engineering. American Association of Engineering Societies, 1111 19th Street NW, Suite 608, Washington, DC 20036. (202) 296-2237 or (800) 658-8897. 8th edition. 1991. Inquire for price.

Who's Who in Technology. Gale Research, 835 Penobscot Building, Detroit, MI 48226-4094. (313) 961-2242. (800) 347-4253. FAX (313) 961-6083. 1995. $195.00.

ENCYCLOPEDIAS AND DICTIONARIES

Encyclopedia of Physical Science and Technology. Robert A. Meyers, ed. 18 volumes. Academic Press Inc., 6277 Sea Harbor Drive, Orlando, FL 32887. (800) 321-5068. 1992. $2100.00.

McGraw-Hill Dictionary of Scientific and Technical Terms. Sybil P. Parker, ed. 5th edition. McGraw-Hill Publishing Company, 1221 Avenue of the Americas, New York, NY 10020. (800) 262-4729 or (212) 512-3825. 1993. $110.50.

McGraw-Hill Encyclopedia of Engineering. Sybil P. Parker, ed. 2nd edition. McGraw-Hill Publishing Company, 1221 Avenue of the Americas, New York, NY 10020. (800) 262-4729 or (212) 512-3825. 1993. $95.50.

McGraw-Hill Encyclopedia of Science and Technology. Sybil P. Parker, ed. 7th edition. 20 volumes. McGraw-Hill Publishing Company, 1221 Avenue of the Americas, New York, NY 10020. (800) 262-4729 or (212) 512-3825. 1992. $1900.00

Thesaurus of Scientific, Technical, and Engineering Terms. Hemisphere Publishing Corporation, 1900 Frost Road, Suite 101, Bristol, PA 19007-1598. (215) 785-5800 or (800) 821-8312. FAX (215) 785-5515. 1987. $173.00.

GENERAL WORKS

Magnesium and Magnesium Compounds. Gordon Press Publishers, PO Box 459, Bowling Green Station, New York, NY 10004. (718) 624-8419. 1993. $250.95.

Magnesium in Vehicle Design. Society of Automotive Engineers, 400 Commonwealth Drive, Warrendale, PA 15096-0001. (412) 776-4970. FAX (412) 776-0790. 1995. $34.00.

Magnesium Products Design. Robert S. Busk. Marcel Dekker, Inc., 270 Madison Avenue, New York, NY 10016. (212) 696-9000. FAX (212) 685-4540. 1987. $150.00.

HANDBOOKS AND MANUALS

ASM Handbook Volume 2: Properties and Selection: Nonferrous Alloys and Special-Purpose Materials. ASM International Handbook Committee. ASM International, Materials Information, Materials Park, OH 44073-0002. (216) 338-5151 or (800) 336-5152. FAX (216) 338-4634. 1991. $147.00.

ASM Handbook Volume 3: Alloy Phase Diagrams. ASM International Handbook Committee. ASM International, Materials Information, Materials Park, OH 44073-0002. (216) 338-5151 or (800) 336-5152. FAX (216) 338-4634. 1992. $147.00.

ASM Metals Reference Book. 3rd edition. ASM International, Materials Information, Materials Park, OH 44073-0002. (216) 338-5151 or (800) 336-5152. FAX (216) 338-4634. 1993. $111.00.

CRC Handbook of Chemistry and Physics. David R. Lide, editor. 75th edition. CRC Press, 2000 Corporate Blvd., N.W., Boca Raton, FL 33431. (407) 994-0555 or (800) 272-7737. FAX (407) 994-0949. $99.50.

Metals Handbook. ASM International, Materials Information, Materials Park, OH 44073-0002. (216) 338-5151 or (800) 336-5152. FAX (216) 338-4634. $154.00.

Smithell's Metals Reference Book. E.A. Brandes & G.B. Brook. 7th edition. Butterworth-Heinemann, 313 Washington Street, Newton, MA 02158. (617) 928-2500 or (800) 366-2665. FAX (617) 928-2620. 1992. $250.00.

ONLINE DATABASES AND CD-ROMS

Compendex Plus. Engineering Information, Inc., 345 East 47th Street, New York, NY 10017. (212) 705-7600 or (800) 221-1044. Contains citations with abstracts to worldwide literature in engineering and technology, from 1970 to present. Available on online BRS,(800) 289-4277, DIALOG, (800) 334-2564, ORBIT (800) 456-7248, and STN International, FIZ Karlsruhe, P.O. Box 2465, W-7500, Karlsruhe 1, Germany, online services. Also available on CD-ROM. Inquire as to cost and availability.

CSA Engineering. Cambridge Scientific Abstracts, 7200 Wisconsin Avenue, Suite 601, Bethesda, MD 20814. (301) 961-6750 or (800) 843-7751. Contains citations and abstracts of international periodicals and research literature covering all fields of engineering and science and technology, including computer and information science, electronics, mechanical engineering, solid state materials, 1981 to present. Available on BRS,(800) 289-4277, online service. Inquire as to cost and availability.

Current Contents Search. Institute for Scientific Information, 3501 Market Street, Philadelphia, PA 19104. (215) 386-0100. FAX (215) 386-6362. Contains citations to articles listed in the table of contents of science and technology journals. Also articles in social sciences and life sciences journals. Available on BRS,(800) 289-4277, DIALOG,(800) 334-2564, online services. Inquire as to cost and availability.

Metadex. Jointly produced by ASM International and the Institute of Materials. Contains more than 925,000 records from the international literature on metals and alloys, concerning properties, processes, materials classes, applications, and metallurgical systems. Updated monthly. Available from ORBIT-QUESTEL (703) 442-0900.

NASA Database. American Institute of Aeronautics and Astronautics, 370 L'Enfant Promenade, S.W., Washington, DC 20024. (202) 646-7400. Citations and abstracts of aeronautics and astronautics literature, 1962 to present. Also contains citations from STAR, Science and Technical Aerospace Reports, and International Aerospace Abstracts available through NASA/RECON online service. Inquire as to cost and availability.

NTIS Bibliographic Database. National Technical Information Service, 5285 Port Royal Road, Springfield, VA 22161. (703) 487-4929 or FAX (703) 321-8199. Broad coverage of government-sponsored science and technology research reports, 1964 to present. Available on BRS,(800) 289-4277, DIALOG, (800) 334-2564, ORBIT, (800) 456-7248, and STN International, FIZ Karlsruhe, P.O. Box 2465, W-7500, Karlsruhe 1, Germany, online services. Also available on CD-ROM. Inquire as to cost and availability.

Wilsonline. H.W. Wilson Company, 950 University Avenue, Bronx, NY 10452. (800) 367-6770 or (212) 588-8400. Makes available online versions of the printed H.W. Wilson indexes, including Applied Science and Technology Index, Business Periodicals Index, General Science Index, and Readers' Guide to Periodical Literature. Period covered is generally 1983 to present. Available on BRS,(800) 289-4277, DIALOG,(800) 334-2564, and OCLC EPIC,(800) 848-5878, online services. Also available on CD-ROM. Inquire as to cost and availability.

PERIODICALS

Alloy Digest. Alloy Digest Inc., 27 Canfield Street, Orange, NJ 07050. (201) 677-9161. 1952 to present. Monthly. $140.00 per year.

International Magnesium Association, World Magnesium Conference Proceedings. 1303 Vincent Place No.1, McLean, VA 22101-2615. (703) 442-8888. FAX (703) 821-1824. 1943 to present. Annual. Inquire for cost and availability.

JOM (Journal of Metals). Minerals, Metals and Materials Society, 420 Commonwealth Drive, Warrendale, PA 15086. (412) 776-9080. 1949 to present. Monthly. $50.00 per year.

*Journal of Phase Equilibria (*formerly *Bulletin of Alloy Phase Diagrams).* ASM International, Materials Park, OH 44073-0002. (216) 338-5151. FAX (216) 338-4634. 6 times a year. $658.00 per year for non-members.

Light Metal Age. Fellom Publishing, 170 S. Spruce Avenue, Suite 120, San Francisco, CA 94080. (415) 588-8832. FAX (415) 588-0901. 1943 to present. Bi-monthly. $35.00 per year.

Magnesium Monthly Review. 106 Spring Forest Road, Greenville, SC 29615-2241. (803) 244-5718. 1971 to present. Monthly. $45.00 per year.

Magnesium Newsletter. International Magnesium Association, 1303 Vincent Place No.1, McLean, VA 22101-2615. (703) 442-8888. FAX (703) 821-1824. 1944 to present. 9 times a year. $90.00 per year.

Magnesium Research. John Libbey & Company Ltd., 13 Smithsyard, Summerly Street, London SW18 4HR, England. Telephone 081-947-2777. FAX 061-947-2664. 1988 to present. 4 times a year. Inquire for cost.

Metallurgical Transactions A: Physical Metallurgy and Materials Science. ASM International, Materials Park, OH 44073. (216) 338-5151 or FAX (216) 338-4634. 1970 to present. Monthly. $520.00 per year.

Metals Week. McGraw-Hill Publishing Company, 1221 Avenue of the Americas, New York, NY 10020. (800) 262-4729 or (212) 512-3825. 1930 to present. Weekly. $770.00 per year.

Mineral Industry Surveys, Magnesium. U.S. Bureau of Mines, Production and Distribution, Cochrans Mill Rd., Box 18070, Pittsburgh, PA 15236. (412) 892-4411. Quarterly. Inquire for price and availability.

RESEARCH CENTERS AND INSTITUTES

Gannon University Engineering Research Institute. University Square, Erie, PA 16541. (814) 871-7619.

SPECIFICATIONS AND STANDARDS

*Annual Book of ASTM Standards, Volume 02.*02. American Society for Testing and Materials (ASTM), 1916 Race Street, Philadelphia, PA 19103. (215) 299-5585. Annual. Inquire for cost and availability.

MAGNETIC RESONANCE

See: NUCLEAR MAGNETIC RESONANCE

MAGNETISM

See: ELECTROMAGNETISM

MAGNETOHYDRODYNAMICS

See also: ELECTROMAGNETISM, PHYSICS, PLASMA PHYSICS, ROCKETS (AERONAUTICS)

ABSTRACT SERVICES AND INDEXES

Applied Mechanics Review: An Assessment of World Literature in Engineering Sciences. American Society of Mechanical Engineers, 345 East 47th Street, New York, NY 10017. (212) 705-7703. 1948 to present. Monthly. $360.00 ISSN: 0003-6900.

Applied Science and Technology Index; A Cumulative Subject Index To English Language Periodicals in the Fields of Aeronautics and Space Science, Computer Technology, Chemistry, Construction Industry, Energy and Related Areas. H.W. Wilson Co., 950 University Avenue, Bronx, NY 10452. (212) 588-8400. (800) 367-6770. FAX (718) 590-1617. From 1958 to present. Monthly. Inquire about cost and availability. Also available on CD-ROM and online. ISSN: 0003-6986.

Current Contents: Engineering, Technology and Applied Sciences. Institute for Scientific Information, 3501 Market Street, Philadelphia, PA 19104. (215) 386-0100. FAX (215) 386-6362. 1970 to present. Weekly. $442.00 per year. Also available on CD-ROM and online. Inquire regarding cost and availability. ISSN: 0095-7917.

Current Papers in Electrical and Electronics Engineering. Institution of Electrical Engineers (IEE), London. Distributed by INSPEC/IEEE, Box 1331, 445 Hoes Lane, Piscataway, NJ 08855-1331. (908) 562-5549. 1969 to present. Monthly. $345.00 per year. ISSN: 0011-3778.

Electrical and Electronics Abstracts. Institution of Electrical Engineers (IEE), London. Available from INSPEC/IEEE-Institute of Electrical and Electronic Engineers, Box 1331, Hoes Lane, Piscataway, NJ 08855-1331. (908) 562-5549. 1898 to present. Monthly. $2200.00 per year. Also available on CD-ROM and online as INSPEC. ISSN: 0036-8105.

Engineering Index Monthly. Indexes and abstracts of the world's engineering and technical literature. Engineering Information, Inc., Castle Point on the Hudson, Hoboken, NJ 07030. (201) 216-8500. (800) 221-1044. FAX (201) 216-8532. Monthly. $2300.00 per year. Available online as COMPENDEX and also on CD-ROM. ISSN: 0742-1974.

General Science Index. H.W. Wilson Company, 950 University Avenue, Bronx, NY 10452. (212) 588-8400. (800) 367-6770. FAX (718) 590-1617. From 1978 to present. Ten issues per year; quarterly and annual cumulations. Service basis. Available on CD-ROM and online. Inquire regarding cost and availability. ISSN: 0162-1963.

Index To IEEE Publications. IEEE Service Center, 445 Hoes Lane, Piscataway, NJ 08855-1331. (908) 981-1393. (800) 678-IEEE. FAX (908) 981-9667. 1973 to present. Annual. ISSN: 0099-1368.

Physics Abstracts. INSPEC. Section A, Science Abstracts. Institution of Electrical Engineers (IEE), London. Available from:

MAGNETOHYDRODYNAMICS

Ency. of Physical Sciences and Engineering Info. Sources

INSPEC/IEEE-Institute of Electrical and Electronic Engineers, Box 1331, Hoes Lane, Piscataway, NJ 08855-1331. (908) 562-5549. 1898 to present. 24 issues per year. $2835.00 per year. Also available online and on CD-ROM. ISSN: 0036-8091.

Physics Briefs (Physikalische Berichte). Information Center for Energy, Physics, Mathematics; German Physical Society. VCH Publishers, Inc., 220 East 23rd Street, New York, NY 10010-4606. (212) 683-8333. 1845 to present. 24 issues per year. $2390.00 per year. Also available online. ISSN: 0179-7434.

Science Citation Index. SCI. Institute for Scientific Information, 3501 Market Street, Philadelphia, PA 19104. (215) 386-0100. (800) 523-1850. FAX (215) 386-2991. 1961 to present. Six issues per year, plus annual cumulation. $11650.00 per year. Also available online and on CD-ROM. Inquire about price and availability. ISSN: 0036-827X.

ASSOCIATIONS AND PROFESSIONAL SOCIETIES

American Institute of Physics. One Physics Ellipse, College Park, MD 20740-3843. (301) 209-3100. FAX (415) 855-2954.

Electromagnetic Energy Association. 1255 23rd Street NW Suite 850, Washington, DC 20017, (202) 452-1070.

IEEE (Institute of Electrical and Electronic Engineers). 345 East 47th Street. New York, NY 10017. (212) 705-7900.FAX (212) 705-4929.

IEEE Power Engineering Society. c/o Institute of Electrical and Electronic Engineers. 345 East 47th Street, New York, NY 10017. (212) 705-7900. FAX (212) 705-4929.

DIRECTORIES AND BIOGRAPHICAL SOURCES

American Men and Women of Science: Physical and Biological Sciences. R. R. Bowker Inc., 121 Chanlon Road, New Providence, NJ 07974. (908) 464-6800. (800) 521-8110. 20th edition. 8 volumes. 1996. $850.00.

Directory of Engineering Societies and Related Organizations. American Association of Engineering Societies, 1111 19th Street, Suite 608, Washington, DC 20036-3603. (202) 296-2237. Semi-annual. $150.00.

Engineering Research Centers: Incorporating Electronics Research Centers. Stockton Press, 345 Park Avenue, New York, NY 10010. (212) 689-9200. (800) 221-2123. 4th edition. 1995. $515.00.

International Engineering Directory. American Consulting Engineers Council, 1015 15th Street, N.W. Suite 802, Washington, DC 20005-2670. (202) 347-7474. Annual. $10.00.

Research Centers Directory. Gale Research Company Inc., 835 Penobscot Building, Detroit, MI 48226-4094. (313) 961-2242. (800) 347-4253. 20th edition. 1995. $485.00. ISSN: 0080-1518.

Scientific and Technical Organizations and Agencies Directory. Gale Research, 835 Penobscot Building, Detroit, MI 48226-4094. (313) 961-2242. (800) 347-4253. 4th edition. 1996. $195.00.

Who's Who in Electric Transmission and Distribution. Utility Data Institute, 1200 G Street NY, Suite 250, Washington, DC 20005-3802. (202) 942-8788. (800) 486-3660. 1995. $150.00.

Who's Who in Engineering. Gordon Davis, editor. American Association of Engineering Societies. 1111 19th Street, NY, Suite 608, Washington, DC 20036. (202) 296-2237. (800) 658-8897. 9th edition. 1995. $220.00.

Who's Who in Technology. Kimberley McGrath, editor. Gale Research Company, 835 Penobscot Building, Detroit, MI 48226-4094. (313 961-2242. (800) 521-4253. 7th edition. 1995. $195.00. ISSN 0-8103-7467-6.

ENCYCLOPEDIAS AND DICTIONARIES

Chambers Science and Technology Dictionary. Peter M.B. Walker, editor. Cambridge University Press, 40 West 20th Street, New York, NY 10011-4211. (212) 924-3900. 1988. $39.95.

Encyclopedia of Applied Physics. George Trigg, editor. VCH Publications, Inc., 220 East 23rd Street, Suite 909, New York, NY 10010-4606. (212) 683-8333. (800) 422-8824. 20 volume set. 1991-. $5990.00.

McGraw-Hill Encyclopedia of Science and Technology. McGraw-Hill Book, Incorporated, 1221 Avenue of the Americas, New York, NY 10020. (212) 997-3675. (800) 262-4729. Seventh edition. 20 volumes. 1992. $1900.00.

GENERAL WORKS

Basic Electrical Power and Machines. David Bradley. Chapman & Hall, 1 Penn Plaza, New York, NY 10119. (212) 564-1060.1994. $30.95.

Essentials of Electromagnetism. David E. Dugdale. American Institute of Physics. One Physics Ellipse, College Park, MD 20740-3843. (301) 209-3100. 1993. $55.00.

Fundamentals of Magnetohydrodynamics. R.V. Polovin and V.P. Demutskii. Plenum Publishing Corp., 233 Spring Street, New York, NY. (212) 620-8000. (800) 2221-9369. FAX (212) 463-0742. 1990. $105.00.

Interplanetary Magnetohydrodynamics. Leonard F. Burlaga. Oxford University Press, Inc., 200 Madison Avenue, New York, NY 10016. (212) 725-6000. (800) 334-4249. 1995. $70.00.

Nonlinear Magnetohydrodynamics. Dieter Kiskamp. Cambridge University Press, 40 West 20th Street, New York, NY 10011-4211. (212) 924-3900. (800) 872-7423. 1993. $84.95.

Plasma Physics: An Introduction To the Theory of Astrophysical, Geophysical and Laboratory Plasmas. Peter A. Sturrock. Cambridge University Press, 40 West 20th Street, New York, NY 10011-4211. (212) 924-3900. (800) 872-7423. 1994. $24.95 in paper.

Magnetohydrodynamic Energy Conversion. Richard J. Rosa. Hemisphere Publishing Corp., 1900 Frost Road, Suite 101, Bristol PA 19007. (215) 785-5800. (800) 821-8312. Revised edition. 1987. $68.00.

Magneto-hydro-dynamic Electrical Power Generation. Hugo K. Messerie. John Wiley & Sons, Inc., 605 Third Avenue, New

York, NY 10158-0012. (212) 850-6000. (800) 225-5945. 1994. $34.95 in paper.

HANDBOOKS AND MANUALS

Standard Handbook for Electrical Engineers. Donald Fink. McGraw-Hill Publishing Company, 1221 Avenue of the Americas, New York, NY 10020. (212) 512-2000. (800) 262-4729. 13th edition. 1996. $110.00.

ONLINE DATABASES AND CD-ROMS

CA Search. Chemical Abstracts Service, P.O. Box 3012, Columbus, OH 43210-0012. (614) 447-3600. (800) 848-6533. FAX (614) 447-3709. Very comprehensive guide to worldwide chemical literature and related fields, 1972 to present. Available on BRS,(800) 289-4277, DIALOG, (800) 334-2564, ORBIT (800) 456-7248, and STN International, FIZ Karlsruhe, P.O. Box 2465, W-7500, Karlsruhe 1, Germany, online services. Inquire as to cost and availability.

Compendex Plus. Engineering Information, Inc., 345 East 47th Street, New York, NY 10017. (212) 705-7600 or (800) 221-1044. Contains citations with abstracts to worldwide literature in engineering and technology, from 1970 to present. Available on online BRS,(800) 289-4277, DIALOG, (800) 334-2564, ORBIT (800) 456-7248, and STN International, FIZ Karlsruhe, P.O. Box 2465, W-7500, Karlsruhe 1, Germany, online services. Also available on CD-ROM. Inquire as to cost and availability.

Current Contents Search. Institute for Scientific Information, 3501 Market Street, Philadelphia, PA 19104. (215) 386-0100. FAX (215) 386-6362. Contains citations to articles listed in the table of contents of science and technology journals. Also articles in social sciences and life sciences journals. Available on BRS,(800) 289-4277, DIALOG,(800) 334-2564, online services. Inquire as to cost and availability.

Dissertation Abstracts Online. University Microfilms International, 300 North Zeeb Road, Ann Arbor, MI 48106. (800) 521-0600 or (313) 761-4700. Scope includes virtually all doctoral dissertations accepted at accredited American institutions from 1861 to present in 252 subject areas. Available on BRS, (800) 289-4277, DIALOG, (800) 334-2564, and OCLC EPIC,(800) 848-5878, online services. Also available on CD-ROM. Inquire as to cost and availability.

INSPEC. Institution of Electrical Engineers, Michael Faraday House, Six Hills Way, Stevenage, Herts. SG1 2AY, England. Telephone: 0438 313311 or FAX 0438 742840. Contains citations to the worldwide literature of physics, electronics and electrical engineering, computer technology, and related fields. Available on BRS, (800) 289-4277, DIALOG, (800) 334-2564, ORBIT, (800) 456-7248, and STN International, FIZ Karlsruhe, P.O. Box 2465, W-7500, Karlsruhe 1, Germany, online services. Inquire as to cost and availability.

NTIS Bibliographic Database. National Technical Information Service, 5285 Port Royal Road, Springfield, VA 22161. (703) 487-4929 or FAX (703) 321-8199. Broad coverage of government-sponsored science and technology research reports, 1964 to present. Available on BRS,(800) 289-4277, DIALOG, (800) 334-2564, ORBIT, (800) 456-7248, and STN International, FIZ Karlsruhe, P.O. Box 2465, W-7500, Karlsruhe 1, Germany, online services. Also available on CD-ROM. Inquire as to cost and availability.

Physics Briefs. American Institute of Physics, 335 East 45th Street, New York, NY 10017. (212) 661-9260 or FAX (212) 661-2036. Contains citations with abstracts of the literature of physics and related fields, 1979 to present. Available on the STN International, FIZ Karlsruhe, P.O. Box 2465, W-7500, Karlsruhe 1, Germany, online service. Inquire as to cost and availability.

Scisearch. Institute for Scientific Information, 3501 Market Street, Philadelphia, PA 19104. (800) 523-1850 or (215) 386-0100. Broad multidisciplinary title and author index to the international literature of science and technology, 1974 to present. Available on DIALOG,(800) 334-2564, and ORBIT,(800) 456-7248, online services. Also available on CD-ROM. Inquire as to cost and availability.

Wilsonline. H.W. Wilson Company, 950 University Avenue, Bronx, NY 10452. (800) 367-6770 or (212) 588-8400. Makes available online versions of the printed H.W. Wilson indexes, including Applied Science and Technology Index, Business Periodicals Index, General Science Index, and Readers' Guide to Periodical Literature. Period covered is generally 1983 to present. Available on BRS,(800) 289-4277, DIALOG,(800) 334-2564, and OCLC EPIC,(800) 848-5878, online services. Also available on CD-ROM. Inquire as to cost and availability.

PERIODICALS

Electric Machines and Power Systems. Taylor & Francis, 1900 Frost Road, Suite 101, Bristol, PA 19007-1598. (800) 821-8312, FAX (215) 785-5515. 1976 to present. Monthly. $355.00 per year. ISSN: 0731-356X.

IEEE Transactions On Electromagnetic Compatibility. Institute of Electrical and Electronics Engineers, Inc., Box 1331, 445 Hoes Lane, Piscataway, NJ 08855-1331. (908) 981- 0060. 1959 to present. Quarterly. $93.00 per year. ISSN: 0018-9375.

IEEE Transactions On Magentics. Institute of Electrical and Electronics Engineers, Inc., Box 1331, 445 Hoes Lane, Piscataway, NJ 08855-1331. (908) 981-0060. 1965 to present. Bi-monthly. $263.00 per year. ISSN: 0018-9464.

Journal of Magnetism and Magnetic Materials. Elsevier Science Inc., Publishing Co., Inc., Box 882, Madison Square Station, New York, NY 10159. (212) 989-5800. FAX (212) 633-3990. 1976 to present. 36 issues per year. $3347.00 per year. ISSN: 0304-8853.

Journal of Magnetohydrodynamics and Plasma Research. Nova Science Publisher, Inc., 6080 Jericho Turnpike, Suite 207, Commack, NY 11725-2808. (516) 499-3103. 1988 to present. Quarterly. $285.00 per year. ISSN: 0891-9801.

Magnetohydrodynamics. English translation of Magnitnaya Gidrodinmaika. Plenum. Consultants Bureau, 223 Spring Street, New York, NY 10013-1578. (212) 620-8468. 1965 to present. Quarterly. $1045.00 per year. ISSN: 0024-998X.

RESEARCH CENTERS AND INSTITUTES

Aerospace Sciences Laboratory. Purdue University. Hanger #3, Purdue Airport, West Lafayette, in 47906. (317) 494-3340.

Diagnostic Instrumentation and Analysis Center. Mississippi State University, College of Engineering, PO Drawer MM, Mississippi State, MS 39762. (601) 325-2105. FAX (601) 325- 8465.

MAGNETOHYDRODYNAMICS

Ency. of Physical Sciences and Engineering Info. Sources

High Temperature Gas Dynamics Laboratory. Stanford University, Mechanical Engineering Department, Stanford, CA 04305. (415) 723-1745. FAX (415) 723-1748.

MANNED SPACE FLIGHT

See also: AEROSPACE ENGINEERING, ASTRONAUTICS, ROCKETS, SPACECRAFT

ABSTRACT SERVICES AND INDEXES

Aeronautical Engineering. Scientific and Technical Branch, National Aeronautics and Space Administration. National Technical Information Service (NTIS), 5285 Port Royal Road, Springfield, VA 22161. (703) 487-4650. FAX (703) 321-8547. Monthly. A selection of annotated references to unclassified reports and journal articles that were introduced into the NASA scientific and technical information system and announced in STAR and IAA.

Applied Science and Technology Index; A Cumulative Subject Index To English Language Periodicals in the Fields of Aeronautics and Space Science, Computer Technology, Chemistry, Construction Industry, Energy and Related Areas. H.W. Wilson Co., 950 University Avenue, Bronx, NY 10452. (800) 367-6770 or (212) 588-8400. FAX (718) 590-1617. From 1958 to present. Monthly. Inquire about cost and availability. Also available on CD-ROM.

Astronomy and Astrophysics Abstracts. Springer-Verlag, 175 Fifth Avenue, New York, NY 10010. (212) 460-1500. Irregular. Price varies, inquire.

Current Contents: Engineering, Technology and Applied Sciences. Institute for Scientific Information, 3501 Market Street, Philadelphia, PA 19104. (215) 386-0100. FAX (215) 386-6362. Weekly. $360.00 per year.

Engineering Index Monthly. Engineering Information, Inc., Castle Point on the Hudson, Hoboken, NJ 07030. (800) 221-1044. FAX (212) 832-1857. Monthly. $2200.00 per year. Also available online as COMPENDEX, and also on CD-ROM.

Index To Scientific and Technical Proceedings. Institute for Scientific Information, 3501 Market St., Philadelphia, PA 19104. (215) 386-0100. FAX (215) 386-6362. Monthly. $500.00 per year.

Index To Scientific Reviews. Institute for Scientific Information, 3501 Market St., Philadelphia, PA 19104. (215) 386-0100. FAX (215) 386-6362. Semi-annual.

International Aerospace Abstracts. Technical Information Service, American Institute of Aeronautics and Astronautics, Inc., 555 West 57th St., New York, NY 10019. (212) 247-6500. FAX (212) 582-4861. Semi-monthly. $1295.00 per year.

NASA Patent Abstracts Bibliography. National Aeronautics and Space Administration, Center for Aerospace Information, Box 8757, BWI Airport, Baltimore, MD 21240. (301) 621-0153. Semi-annual. Inquire for price information.

Numerical/Chronological/Author Index. An index to articles and papers published by or for the American Astronautical Society, including: Journal of the Astronautical Sciences, Astronautical

Sciences Review, Advances in the Astronautical Sciences, Science and Technology, AAS history series, and AAS microfiche series. Horace Jacobs and Robert H. Jacobs. Univelt, Inc. PO Box 28130, San Diego, CA 92198. (619) 746-4005. FAX (619) 746-3139. Inquire for price and availability.

Science Citation Index. Institute for Scientific Information, 3501 Market Street, Philadelphia, PA 19104. (215) 386-0100. FAX (215) 386-6362. Inquire about availability and cost. Also available on CD-ROM.

Scientific and Technical Aerospace Reports (STAR). National Aeronautics and Space Administration. NASA Center for Aerospace Information, Box 8757, BWI Airport, Baltimore, MD 21240. (301) 621-0153. Monthly. Inquire about availability and cost. Also available through the NASA online retrieval service (RECON), and through the Aerospace Database through DIALOG.

ASSOCIATIONS AND PROFESSIONAL SOCIETIES

Aerospace Industries Association of America. 1250 Eye Street NW, Washington, DC 20005. (202) 371-8400.

American Astronautical Society. 6352 Rolling Mill Place, Suite 102, Springfield, VA 22152.

American Institute of Aeronautics and Astronautics. 370 L'Enfant Promenade SW, Washington, DC 20024. (202) 646-7400.

British Interplanetary Society, 27/29 South Lambeth Road, London Sw8 1sz, England. (071) 735-3160. FAX (071) 820-1504.

Canadian Aeronautics and Space Institute, 222 Somerset Street West, Suite 601, Ottawa, Ontario, Canada K2p 0j1.

National Space Society, 922 Pennsylvania Avenue SE, Washington, DC 20003-2140. (202) 543-1900. FAX (202) 546-4189.

DIRECTORIES AND BIOGRAPHICAL SOURCES

American Astronautical Society Directory. American Astronautical Society, 6352 Rolling Mill Place, Suite 102, Springfield, VA 22152. Inquire for cost and availability.

American Institute of Aeronautics and Astronautics Roster. American Institute of Aeronautics and Astronautics. 370 L'Enfant Promenade SW, Washington, DC 20024. (202) 646-7400. Inquire for cost and availability.

Jane's Space Directory. Jane's Information Group, 1340 Braddock Place, Suite 300, Alexandria, VA 22314-1651. (703) 683-3700. FAX (703) 836-0029. Annual. $245.00. Also available on CD-ROM.

Research Centers Directory. Gale Research, 835 Penobscot Building, Detroit, MI 48226-4094. (313) 961-2242. (800) 347-4253. FAX (313) 961-6083. $485.00.

Scientific and Technical Organizations and Agencies Directory. Gale Research, 835 Penobscot Building, Detroit, MI 48226-4094. (313) 961-2242. (800) 347-4253. FAX (313) 961-6083. 4th ed. 1995. $195.00.

Who's Who in Engineering. 7th ed. American Association of Engineering Societies, 1111 19th St. NW, Suite 608, Washington, DC 20036. (800)658-8897. 1988. $200.00.

GENERAL WORKS

Almanac of Soviet Manned Space Flight. Dennis Newkirk. Gulf Publishing Company, P.O. Box 2608, Houston, TX (713) 529-4301 or (800) 231-6275. 1990. $29.95.

Astronauts and Cosmonauts Biographical and Statistical Data Report. Congressional Research Service. Superintendent of Documents, U.S. Government Printing Office, Box 371954, Pittsburgh, PA 15250-7954. (202) 783-3238. FAX (202) 512-2233. 1994. Inquire for price and availability.

Astronautics and Aeronautics, 1979-1984: A Chronology. Bette R. Janson and Eleanor H. Ritchie. National Aeronautics and Space Administration, Scientific and Technical Information Division. 1990. Order from: Superintendent of Documents, U.S. Government Printing Office, Box 371954, Pittsburgh, PA 15250-7954. (202) 783-3238. FAX (202) 512-2233. Inquire for price and availability.

Choosing the Right Stuff: The Psychological Selection of Astronauts and Cosmonauts. Patricia A. Santry. Greenwood Publishing Group, 88 Post Road W., Box 5007, Westport, CT 06881. (203) 226-3571 or (800) 225-5800. FAX (203) 222-1502. 1994. Write for price and availability information.

Introduction To Space Flight. Francis J. Hale. Prentice Hall, 113 Sylvan Avenue, Route 9W, Englewood Cliffs, NJ 07632. (201) 592-2000 or (800) 922-0579. 1994. $39.50 (hardback), $32.50 (paperback).

Leaving the Cradle: Human Exploration of Space in the 21st Century. Thomas O. Paine. Univelt, Inc. PO Box 28130, San Diego, CA 92198. (619) 746-4005. FAX (619) 746-3139. 1991. $70.00 (hardbound) and $55.00 (paperback).

Race To the Moon: America's Duel With the Soviets. William B. Breuer. Greenwood Publishing Group, 88 Post Road W., Box 5007, Westport, CT 06881. (203) 226-3571 or (800) 225-5800. FAX (203) 222-1502. 1993. $24.95.

Space Almanac. Anthony R. Curtis, ed. Gulf Publishing Company, P.O. Box 2608, Houston, TX (713) 529-4301 or (800) 231-6275. 1992. $36.95 (hardbound) and $24.95 (paperback).

Space Exploration: A Reference Handbook. Mrinal Bali. ABC-CLIO Information Services, PO Box 1911, Santa Barbara, CA 93116-1911. (800) 368-6868. FAX (805) 685-9685. 1990. $39.50.

Space Mission Analysis and Design. Wiley J. Larson and James R. Wertz, eds. 2d ed. Kluwer Academic Publishers, P.O. Box 358, Accord Station, Hingham, MA 02018-0358. (617) 871-6000. FAX (617) 871-6528. 1992. $99.00.

Thirty Years of Progress in Space. L.G. Napolitano. Pergamon Press, Maxwell House, Fairview Park, Elmsford, NY 10523. (914) 592-7700. Fax (914) 592-3625. 1988. $150.00.

U.S. Space Gear: Outfitting the Astronaut. Lillian D. Kozloski. Smithsonian Institution Press, 470 L'Enfant Plaza, Suite 7100, Washington, DC 20560. Write for price and availability information.

ONLINE DATABASES AND CD-ROMS

Aerospace Database. American Institute of Aeronautics and Astronautics, 370 L'Enfant Promenage, S.W., Washington, DC 20024. (202) 646-7400. Worldwide published literature on research and development in aerospace and related areas, with abstracts. Covers 1962 to present. Available on DIALOG, (800) 334-2564, online service. Also available on CD-ROM. Inquire as to cost and availability.

Compendex Plus. Engineering Information, Inc., 345 East 47th Street, New York, NY 10017. (212) 705-7600 or (800) 221-1044. Contains citations with abstracts to worldwide literature in engineering and technology, from 1970 to present. Available on online BRS,(800) 289-4277, DIALOG, (800) 334-2564, ORBIT (800) 456-7248, and STN International, FIZ Karlsruhe, P.O. Box 2465, W-7500, Karlsruhe 1, Germany, online services. Also available on CD-ROM. Inquire as to cost and availability.

CSA Engineering. Cambridge Scientific Abstracts, 7200 Wisconsin Avenue, Suite 601, Bethesda, MD 20814. (301) 961-6750 or (800) 843-7751. Contains citations and abstracts of international periodicals and research literature covering all fields of engineering and science and technology, including computer and information science, electronics, mechanical engineering, solid state materials, 1981 to present. Available on BRS,(800) 289-4277, online service. Inquire as to cost and availability.

Current Contents Search. Institute for Scientific Information, 3501 Market Street, Philadelphia, PA 19104. (215) 386-0100. FAX (215) 386-6362. Contains citations to articles listed in the table of contents of science and technology journals. Also articles in social sciences and life sciences journals. Available on BRS,(800) 289-4277, DIALOG,(800) 334-2564, online services. Inquire as to cost and availability.

Dissertation Abstracts. University Microfilms International, 300 North Zeeb Road, Ann Arbor, MI 48106. (800) 521-0600 or (313) 761-4700. Scope includes virtually all doctoral dissertations accepted at accredited American institutions from 1861 to present in 252 subject areas. Available on BRS,(800) 289-4277, DIALOG,(800) 334-2564, and OCLC EPIC,(800) 848-5878, online services. Also available on CD-ROM. Inquire as to cost and availability.

International Aerospace Abstracts. American Institute of Aeronautics and Astronautics, 370 L'Enfant Promenade, S.W., Washington, DC 20024. (202) 646-7400. Contains references and abstracts of journal and monograph literature relating to aerospace science and technology, from 1963 to present. Available through the NASA/RECON system of the National Aeronautics and Space Administration only.

NASA Database. American Institute of Aeronautics and Astronautics, 370 L'Enfant Promenade, S.W., Washington, DC 20024. (202) 646-7400. Citations and abstracts of aeronautics and astronautics literature, 1962 to present. Also contains citations from STAR, Scientific and Technical Aerospace Reports, and International Aerospace Abstracts. Available through NASA/RECON online service. Inquire as to cost and availability.

NTIS Bibliographic Database. National Technical Information Service, 5285 Port Royal Road, Springfield, VA 22161. (703) 487-4929 or FAX (703) 321-8199. Broad coverage of government-sponsored science and technology research reports, 1964 to present. Available on BRS,(800) 289-4277, DIALOG, (800) 334-2564, ORBIT, (800) 456-7248, and STN International, FIZ Karlsruhe, P.O. Box 2465, W-7500, Karlsruhe 1, Germany,

MANNED SPACE FLIGHT

Ency. of Physical Sciences and Engineering Info. Sources

online services. Also available on CD-ROM. Inquire as to cost and availability.

Scientific and Technical Books and Serials in Print. R.R. Bowker Inc., 121 Chanlon Road, New Providence, NJ 07974. (800) 521-8110 or (908) 464-6800. List of currently published books and serials in the physical and biological sciences, engineering and technology, with subject, author and titles indexes. Available on ORBIT,(800) 456-7248, online service. Inquire as to cost and availability.

Scisearch. Institute for Scientific Information, 3501 Market Street, Philadelphia, PA 19104. (800) 523-1850 or (215) 386-0100. Broad multidisciplinary title and author index to the international literature of science and technology, 1974 to present. Available on DIALOG,(800) 334-2564, and ORBIT,(800) 456-7248, online services. Also available on CD-ROM. Inquire as to cost and availability.

Wilsonline. H.W. Wilson Company, 950 University Avenue, Bronx, NY 10452. (800) 367-6770 or (212) 588-8400. Makes available online versions of the printed H.W. Wilson indexes, including Applied Science and Technology Index, Business Periodicals Index, General Science Index, and Readers' Guide to Periodical Literature. Period covered is generally 1983 to present. Available on BRS,(800) 289-4277, DIALOG,(800) 334-2564, and OCLC EPIC,(800) 848-5878, online services. Also available on CD-ROM. Inquire as to cost and availability.

PERIODICALS

Acta Astronautica. Pergamon Press, 660 White Plains Rd., Tarrytown, NY 10591-5153. 14 times a year. Inquire for price and availability.

Ad Astra. National Space Society, 922 Pennsylvania Avenue SE, Washington, DC 20003-2140. (202) 543-1900. FAX (202) 546-4189. 1989 to present. 6 times a year. $35.00 a year.

Advances in Astronautical Sciences. American Astronautical Society, 6352 Rolling Mill Place, Suite 102, Springfield, VA 22152. (703) 866-0020. FAX (703) 866-3526. Quarterly. Write or call for price information.

Advances in Space Research. Pergamon Press, 660 White Plains Rd., Tarrytown, NY 10591-5153. 12 times a year. Inquire for price and availability.

Aeronautical Journal. Royal Aeronautical Society, PO Box 139, Tonbridge, Kent, TN9 1EW, England. 0732-770823. FAX 0723-361708. 1897 to present. Ten times a year. Write for price and availability.

Aerospace. Royal Aeronautical Society, PO Box 139, Tonbridge, Kent, TN9 1EW, England. 0732-770823. FAX 0723-361708. Monthly. Write for price and availability.

Aerospace America. American Institute of Aeronautics and Astronautics, 370 L'Enfant Promenade SW, Washington, DC 20024-2518. (202) 646-7471. 1932 to present. Monthly. $70 per year (non-members).

Aerospace Engineering. Society of Automotive Engineers, SAE, Inc., 400 Commonwealth Dr., Warrendale, PA 15096. (412) 776-4841. 1981 to present. Monthly. $48 a year.

AIAA Journal. American Institute of Aeronautics and Astronautics, 370 L'Enfant Promenade SW, Washington, DC 20024-2518. (202) 646-7471. 1963 to present. Monthly. $52.00 per year (members) and $435.00 per year (non-members).

Aviation & Aerospace. Baxter Publishing Company, 310 Dupont Street, Toronto, Ontario M5R 1V9 Canada. (416) 968-7252. FAX (416) 968-2377. Bi-monthly. $42.00 per year.

Aviation Week and Space Technology. 1221 Avenue of the Americas, New York, NY 10020. (212) 512-2999 or (800) 525-5003. 1916 to present. Weekly. $82.00 per year.

ESA Bulletin. European Space Agency, Publications Division, Keplerlaan 1, 22000 AG Nordwijk, Netherlands. Phone 31-1719-86555. FAX 31-1719-85433. Quarterly. Free.

ESA Journal. European Space Agency, Publications Division, Keplerlaan 1, 22000 AG Nordwijk, Netherlands. Phone 31-1719-86555. FAX 31-1719-85433. Quarterly. Free.

Interavia/Aerospace World. Aerospace Media Publishing, Swissair Centre, 31 Route de l'Aeroport, POB Box 437, 1215 Geneva 15, Switzerland. (022) 788 27 88. FAX (022) 788 27 26. Monthly. $128.00 per year (U.S.).

Journal of the Astronautical Sciences. American Astronautical Society, 6352 Rolling Mill Place, Suite 102, Springfield, VA 22152. Quarterly. $35 per year.

Journal of the British Interplanetary Society (JBIS). British Interplanetary Society, 27/29 South Lambeth Road, London SW8 1SZ, England. (071) 735-3160. FAX (071) 820-1504. Monthly. $370.00 per year.

Journal of Spacecraft and Rockets. American Institute of Aeronautics and Astronautics, the Aerospace Center, 370 L'Enfant Promenade SW, Washington, DC 20024. (202) 646-7400. FAX (202) 646-7508. 1964 to present. Bi-monthly. $185.00 per year.

Progress in Aerospace Sciences. Pergamon Press, 660 White Plains Rd., Tarrytown, NY 10591-5153. 1961 to present. Quarterly. $320.00 per year.

Space Science Reviews. Kluwer Academic Publishers, P.O. Box 358, Accord Station, Hingham, MA 02018-0358. (617) 871-6000. Monthly except March, June, September, December. $258.50 per volume.

Spaceflight. British Interplanetary Society, 27/29 South Lambeth Road, London SW8 1SZ, England. (071) 735-3160. FAX (071) 820-1504. Monthly. $69 annual membership includes subscription.

RESEARCH CENTERS AND INSTITUTES

Aerospace Research Applications Center, Indianapolis Center For Advanced Research, 611 N. Capitol Ave., Indianapolis, in 46204. (317) 262-5003. FAX (317) 262-5044.

Aerospace Research Center, 1250 Eye St., NW, Washington, DC 20005. (202) 371-8400. FAX (202) 371-8470.

Center For Aerospace Technology, Weber State College, Ogden, UT 84408-1805. (801) 626-7272.

Center For Space Research, Massachusetts Institute of Technology, 77 Massachusetts Ave., Cambridge, MA 02139. (617) 253-7501.

Charles Stark Draper Laboratory Inc., 555 Technology Square, Cambridge, MA 02139. (617) 258-1000.

Florida A&M University, Space Life Sciences Training Program, 106 Honor House, Tallahassee, FL 32307. (904) 599-3636.

Goddard Space Flight Center, Greenbelt, MD 20771. (301) 286-7351.

Houston Advanced Research Center, Space Technology and Research Center, 4802 Research Forest Dr., the Woodlands, TX 77381. (713) 363-7922. FAX (713) 363-7914.

Man-Vehicle Laboratory, Massachusetts Institute of Technology, Rm. 37-219, Cambridge, MA 02139. (617) 253-7805.

George C. Marshall Space Flight Center, Huntsville, AL 35812. (205) 544-0024.

Ohio Aerospace Institute, 2001 Aerospace Parkway, Brook Park, OH 44142. (216) 891-2100. FAX (216) 891-2140.

Space Science and Engineering Center, University of Wisconsin—Madison, 1225 W. Dayton St., Madison, WI 53706. (608) 262-0544.

MAPPING

See: CARTOGRAPHY

MARINE ENGINEERING

See also: MECHANICAL ENGINEERING, NAVAL ARCHITECTURE

ABSTRACTS AND INDEXES

Applied Mechanics Reviews: An Assessment of World Literature in Engineering Sciences. 1948-present. American Society of Mechanical Engineers, 345 East 47th Street, New York, NY 10017. (212) 705-7703. Monthly. $360.00 per year.

Applied Science and Technology Index; A Cumulative Subject Index To English Language Periodicals in the Fields of Aeronautics and Space Science, Computer Technology, Chemistry, Construction Industry, Energy and Related Areas. H.W. Wilson Co., 950 University Avenue, Bronx, NY 10452. (800) 367-6770 or (212) 588-8400. FAX (718) 590-1617. From 1958 to present. Monthly. Inquire about cost and availability. Also available on CD-ROM and online.

CAD/CAM Abstracts. R.R. Bowker, Bowker A&I Publishing, 121 Chanlon Road, New Providence, NJ 07974. (908) 771-7714. Fax (908) 771-7725. 1984 to present. Monthly. $495.00 per year.

Civil and Structural Engineering Abstracts, Cambridge Scientific Abstracts, 7200 Wisconsin Avenue, Bethesda, MD 20814-4823. (301) 961-6750. FAX (301) 961-6720. 1993 to present. Monthly. $385.00 per year. Topics covered include structural

design, construction equipment and methods, civil defense and military engineering, surveying, highway engineering, maritime and port structures, materials, land reclamation, and soil mechanics.

Computer Abstracts, MCB University Press Ltd., PO Box 10812, Birmingham, AL 35201-0812. (800) 633-4931. FAX (205) 995-1588. Monthly. Covers computer theory, data, hardware, systems, networks, human-computer interaction, artificial intelligence, as well as applications of computers in aerospace, business, CAD/CAM, cartography, civil engineering, electronics and electrical engineering, industrial engineering, mechanical engineering, medicine, structural engineering, etc.

Current Contents: Engineering, Technology, and Applied Sciences. Institute for Scientific Information, 3501 Market Street, Philadelphia, PA 19104. (215) 386-0100. FAX (215) 386-6362. 1970 to present. Weekly. $442.00 per year.

Engineering Index Monthly. Engineering Information, Inc., Castle Point on the Hudson, Hoboken, NJ 07030. (800) 221-1044. FAX (212) 832-1857. Monthly. $2300.00 per year. Also available online as COMPENDEX, and also on CD-ROM. Covers chemical engineering, computers, electrical engineering, civil engineering, metals and mining, industrial management, and mechanical engineering.

Mechanical Engineering Abstracts (formerly ISMEC), Cambridge Scientific Abstracts, 7200 Wisconsin Avenue, Bethesda, MD 20814-4823. (301) 961-6750. FAX (301) 961-6720. 1967 to present. Monthly. $895.00 per year. Summarizes world literature in mechanical engineering, production engineering, and engineering management. Also available online.

Naval Engineers' Journal Cumulative Index, 1889-1979. American Society of Naval Engineers, 1452 Duke Street, Alexandria, VA 22314. (703) 836-6727. FAX (703) 836-7491. Inquire for price and availability.

Physics Abstracts. Institute of Electrical Engineers, Michael Faraday House, Six Hill Way, Stevenage, Herts, SG1 2AY, England. Distributed by IEEE, 445 Hoes Lane, Piscataway, NJ 08854. (908) 562-5549. 1898 to present. Monthly. $2700.00 per year. Also available online as INSPEC.

Science Citation Index. Institute for Scientific Information, 3501 Market Street, Philadelphia, PA 19104. (215) 386-0100. FAX (215) 386-6362. Inquire about availability and cost. Also available on CD-ROM.

ASSOCIATIONS AND PROFESSIONAL SOCIETIES

American Society of Mechanical Engineers. 345 East 47th Street, New York, NY 10017-2398. (212) 705-7703.

American Bureau of Shipping. Two World Trade Center, 106th Floor, New York, NY 10048. (212) 839-5000. FAX (212) 839-5130.

American Society of Naval Engineers. 1452 Duke Street, Alexandria, VA 22314. (703) 836-6727. FAX (703) 836-7491.

Institute of Marine Engineers. 76 Mark Lane, London EC3R 7JN, England. Telephone 71-4818493. FAX 71-4818854.

Marine Technology Society. 1828 L Street NW, No. 906, Washington, DC 20036-5104. (202) 775-5966. FAX (202) 429-9417.

MARINE ENGINEERING

Ency. of Physical Sciences and Engineering Info. Sources

Society of Marine Port Engineers. PO Box 466, Avenel, NJ 07001-2402. (908) 381-7673. FAX (908) 381-2046.

Society of Naval Architects and Marine Engineers. 601 Pavonia Avenue, Suite 400, Jersey City, NJ 07306. (201) 798-4800. FAX (201) 798-4975.

Vibration Institute. 6262 S. Kingery Highway, Suite 212, Willowbrook, IL 60514. (708) 654-2254.

DIRECTORIES AND BIOGRAPHICAL SOURCES

American Society of Naval Engineers Membership Directory. 1452 Duke Street, Alexandria, VA 22314. (703) 836-6727. FAX (703) 836-7491. Biennial. Inquire for cost and availability.

Directory of Engineering Societies and Related Organizations. Gordon Davis. 13th edition. American Association of Engineering Societies, 1111 19th Street NW, Suite 608, Washington, DC 20036. (202) 296-2237 or (800) 658-8897. 1989. Inquire for price.

Marine Engineers Review Directory of Marine Diesel Engines. Learned Information Inc., 143 Old Marlton Pike, Medford, NJ 08055-8750. (607) 654-6266. FAX (607) 654-4309. 1988 to present. Annual. Inquire for price.

Marine Equipment Catalog. Maritime Activity Reports Inc., 118 E. 25th Street, New York, NY 10010. (212) 477-6700. 1984 to present. Annual. $65.00.

Research Centers Directory. Gale Research, 835 Penobscot Building, Detroit, MI 48226-4094. (313) 961-2242. (800) 347-4253. FAX (313) 961-6083. $485.00.

Scientific and Technical Organizations and Agencies Directory. Gale Research, 835 Penobscot Building, Detroit, MI 48226-4094. (313) 961-2242. (800) 347-4253. FAX (313) 961-6083. 4th edition. 1996. $195.00.

Who's Who in Engineering. American Association of Engineering Societies, 1111 19th Street NW, Suite 608, Washington, DC 20036. (202) 296-2237 or (800) 658-8897. 8th edition. 1991. Inquire for price.

ENCYCLOPEDIAS AND DICTIONARIES

Marine Technology Reference Book. Nina Morgan, editor. Butterworth-Heinemann, 313 Washington Street, Newton, MA 02158. (617) 928-2500 or (800) 366-2665. FAX (617) 928-2620. 1990. $225.00.

McGraw-Hill Encyclopedia of Engineering. Sybil P. Parker, ed. 2nd edition. McGraw-Hill Publishing Company, 1221 Avenue of the Americas, New York, NY 10020. (800) 262-4729 or (212) 512-3825. 1993. $95.50.

GENERAL WORKS

Composite Materials in Maritime Structures. R.A. Schenoi & J.F. Wellicome, editors. 2 volumes. Cambridge University Press, 40 West 20th Street, New York, NY 10011-4211. (212) 924-3900 or (800) 872-7423. FAX (914) 937-4712. 1993. $69.95 per volume.

Design of Marine Facilities. J. Gaythwaite. Chapman and Hall, 29 West 35th Street, New York, NY 10001. (800) 842-3636. FAX (212) 563-2269. 1990. $85.00.

Marine Structures Engineering. Gregory Tsinker. Chapman & Hall, 115 Fifth Avenue, New York, NY 10211-0906. 1995. $65.95.

A Half Century of Maritime Technology, 1943-1993. Harry Benford, editor. Society of Naval Architects and Marine Engineers. 601 Pavonia Avenue, Suite 400, Jersey City, NJ 07306. (201) 798-4800. FAX (201) 798-4975. 1993. $45.00.

Introduction To Marine Engineering. D.A. Taylor. 2nd edition. Butterworth-Heinemann, 313 Washington Street, Newton, MA 02158. (617) 928-2500 or (800) 366-2665. FAX (617) 928-2620. 1990. $36.95.

Marine Engineering. Roy L. Harrington, ed. Revised edition. Society of Naval Architects and Marine Engineers, 601 Pavonia Avenue, Suite 400, Jersey City, NJ 07306. (201) 798-4800. FAX (201) 798-4975. 1992. $190.00 for non-members.

Marine Engineering: Design and Operation of Ships and offshore Structures. T.K.S. Murthy & C.A. Brebbia, editors. Computational Mechanics Publications, 25 Bridge Street, Billerica, MA 01821. (508) 667-5841. FAX (508) 667-7582. 1994. $134.00.

HANDBOOKS AND MANUALS

Handbook of Mechanics, Materials, and Structures. Alexander Blake, editor. John Wiley and Sons, Inc., 605 Third Avenue, New York, NY 10158. (800) 526-5368 or (212) 850-6000. 1985. $120.00.

Mechanical Design and Systems Handbook. Harold A. Rothbart, editor-in-chief. McGraw-Hill Publishing Company, 1221 Avenue of the Americas, New York, NY 10020. (800) 262-4729 or (212) 512-3825. 1985. Inquire for cost and availability.

Marine Engine Room Blue Book. William D. Eglinton. 4th edition. Cornell Maritime Press Inc., PO Box 456, Centreville, MD 21617. (410) 758-1075 or (800) 638-7641. FAX (410) 758-6849. 1993. $25.00.

Marks' Standard Handbook For Mechanical Engineers. E.A. Avallone and T. Baumeister III. 9th edition. McGraw-Hill Publishing Company, 1221 Avenue of the Americas, New York, NY 10020. (800) 262-4729 or (212) 512-3825. 1987. $133.00.

Pressure Vessel Design Handbook: Illustrated Procedures For Solving Every Major Pressure Vessel Design Problem. Dennis R. Moss. Gulf Publishing Company, P.O. Box 2608, Houston, TX. (713) 529-4301 or (800) 231-6275. 1987. $56.50.

Pressure Vessel Handbook. E.F. Megyesy. 9th edition. Pressure Vessel Handbook Publishing Inc., PO Box 35365, Tulsa, OK 74153-0365. (918) 742-9637. 1992. $108.95.

Standard Handbook of Machine Design. Joseph E. Shigley & Charles R. Mischke, editors-in-chief. McGraw-Hill Publishing Company, 1221 Avenue of the Americas, New York, NY 10020. (800) 262-4729 or (212) 512-3825. 1986. $124.50.

ONLINE DATABASES AND CD-ROMS

Compendex Plus. Engineering Information, Inc., 345 East 47th Street, New York, NY 10017. (212) 705-7600 or (800) 221-1044. Contains citations with abstracts to worldwide literature in engineering and technology, from 1970 to present. Available on online BRS,(800) 289-4277, DIALOG, (800) 334-2564, ORBIT (800) 456-7248, and STN International, FIZ Karlsruhe, P.O. Box 2465, W-7500, Karlsruhe 1, Germany, online services. Also available on CD-ROM. Inquire as to cost and availability.

CSA Engineering. Cambridge Scientific Abstracts, 7200 Wisconsin Avenue, Suite 601, Bethesda, MD 20814. (301) 961-6750 or (800) 843-7751. Contains citations and abstracts of international periodicals and research literature covering all fields of engineering and science and technology, including computer and information science, electronics, mechanical engineering, solid state materials, 1981 to present. Available on BRS,(800) 289-4277, online service. Inquire as to cost and availability.

ISMEC: Mechanical Engineering Abstracts. Cambridge Scientific Abstracts, 7200 Wisconsin Avenue, Suite 601, Bethesda, MD 20814. (301) 961-6750 or (800) 843-7751. Contains citations to the literature in mechanical engineering, industrial and production engineering, energy, power, mechanics, devices and related areas, from 1973 to present. Available on the DIALOG,(800) 334-2564, online service. Inquire as to cost and availability.

NTIS Bibliographic Database. National Technical Information Service, 5285 Port Royal Road, Springfield, VA 22161. (703) 487-4929 or FAX (703) 321-8199. Broad coverage of government-sponsored science and technology research reports, 1964 to present. Available on BRS,(800) 289-4277, DIALOG, (800) 334-2564, ORBIT, (800) 456-7248, and STN International, FIZ Karlsruhe, P.O. Box 2465, W-7500, Karlsruhe 1, Germany, online services. Also available on CD-ROM. Inquire as to cost and availability.

Scisearch. Institute for Scientific Information, 3501 Market Street, Philadelphia, PA 19104. (800) 523-1850 or (215) 386-0100. Broad multidisciplinary title and author index to the international literature of science and technology, 1974 to present. Available on DIALOG,(800) 334-2564, and ORBIT,(800) 456-7248, online services. Also available on CD-ROM. Inquire as to cost and availability.

Tris (Transportation Research Information). National Academy of Science, 2101 Constitution Avenue, N.W., Washington, DC 20418. (202) 334-3313 or (800) 624-6242. Citations with abstracts of literature on transportation, including air, highway, rail, maritime and other modes from 1968 to present. Available on DIALOG,(800) 334-2564, online service. Inquire as to cost and availability.

Wilsonline. H.W. Wilson Company, 950 University Avenue, Bronx, NY 10452. (800) 367-6770 or (212) 588-8400. Makes available online versions of the printed H.W. Wilson indexes, including Applied Science and Technology Index, Business Periodicals Index, General Science Index, and Readers' Guide to Periodical Literature. Period covered is generally 1983 to present. Available on BRS,(800) 289-4277, DIALOG,(800) 334-2564, and OCLC EPIC,(800) 848-5878, online services. Also available on CD-ROM. Inquire as to cost and availability.

PERIODICALS

Institute of Marine Engineers Transactions. Learned Information Inc., 143 Old Marlton Pike, Medford, NJ 08055-8750. (607) 654-6266. FAX (607) 654-4307. 1889 to present. Six times a year. Inquire for price.

Journal of Applied Mechanics. American Society of Mechanical Engineers, 345 E. 47th Street, New York, NY 10017. (212) 705-7722. 1935 to present. Quarterly. $120.00.

Journal of Engineering For Industry. American Society of Mechanical Engineers, 345 E. 47th Street, New York, NY 10017. (212) 705-7722. 1970 to present. Quarterly. $100.00 per year.

Journal of Mechanical Design. American Society of Mechanical Engineers, 345 E. 47th Street, New York, NY 10017. (212) 705-7722. 1978 to present. Quarterly. $100.00 per year.

Journal of Pressure Vessel Technology. American Society of Mechanical Engineers, 345 E. 47th Street, New York, NY 10017. (212) 705-7722. 1974 to present. Quarterly. $100.00 per year.

Journal of Ship Production. Society of Naval Architects and Marine Engineers, 601 Pavonia Avenue, Suite 400, Jersey City, NJ 07306. (201) 798-4800. FAX (201) 798-4975. 1985 to present. Quarterly. $65.00 per year.

Journal of Ship Research. Society of Naval Architects and Marine Engineers, 601 Pavonia Avenue, Suite 400, Jersey City, NJ 07306. (201) 798-4800. FAX (201) 798-4975. 1957 to present. Quarterly. $80.00 per year.

Machine Design. Penton Publishing, 110 Superior Ave., Cleveland, OH 44114-2543. 1929 to present. 6 times a year. $180.00 per year.

Marine Engineers Review. Learned Information Inc., 143 Old Marlton Pike, Medford, NJ 08055-8750. (607) 654-6266. FAX (607) 654-4307. 1967 to present. Monthly. Inquire for price.

Marine Technology. Society of Naval Architects and Marine Engineers, 601 Pavonia Avenue, Suite 400, Jersey City, NJ 07306. (201) 798-4800. FAX (201) 798-4975. 1964 to present. Quarterly. $70.00 per year for non-members.

Marine Technology Society Journal. Marine Technology Society, 1828 L Street NW, No. 906, Washington, DC 20036-5104. (202) 775-5966. FAX (202) 429-9417. Quarterly. $70.00 per year.

Mechanical Engineering. American Society of Mechanical Engineers, 345 E. 47th Street, New York, NY 10017. (212) 705-7722. 1906 to present. Monthly. $45.00 per year.

Naval Engineers Journal. American Society of Naval Engineers, 1452 Duke Street, Alexandria, VA 22314. (703) 836-6727. FAX (703) 836-7491. 1889 to present. Bi-monthly. $65.00 per year.

RESEARCH CENTERS AND INSTITUTES

Institute For Marine Dynamics. National Research Council of Canada, PO Box 12093, Postal Station "A," St. John's, NF Canada A1B 3T5. (709) 772-2469. FAX (709) 772-2462.

MARINE ENGINEERING

Ency. of Physical Sciences and Engineering Info. Sources

Maritime Research Department. Webb Institute of Naval Architecture, Crescent Beach Road, Glen Cove, NY 11542. (516) 671-2356. FAX (516) 674-9838.

Ship Hydrodynamics Laboratory. University of Michigan, 126 W. Engineering Bldg., Ann Arbor, MI 48109. (313) 764-9432.

MARINE NAVIGATION

See: NAVIGATION

MARINE POLLUTION

See also: OCEANOGRAPHY, POLLUTION, WATER POLLUTION

ABSTRACT SERVICES AND INDEXES

CA Selects: Environmental Pollution. Chemical Abstracts Service, 2540 Olentangy River Road, Box 3012, Columbus, OH 43210. (800) 848-6538 or (614) 447-3600, FAX (614) 447-3713. Semi-weekly. $210.00 per year.

CA Selects: Pollution Monitoring. Chemical Abstracts Service, 2540 Olentangy River Road, Box 3012, Columbus, OH 43210. (800) 848-6538 or (614) 447-3600, FAX (614) 447-3713. Semi-weekly. $210.00 per year.

Current Contents: Physical, Chemical and Earth Sciences. Institute for Scientific Information, 3501 Market Street, Philadelphia, PA 19104. (215) 386-0100. FAX (215) 386-6362. Weekly. $360.00 per year.

Environmental Abstracts. R.R. Bowker Inc., 121 Chanlon Road, New Providence, NJ 07974. (800) 521-8110 or (908) 464-6800. Contains citations, most with abstracts, to literature relating to environmental issues and problems, from 1971 to present. Inquire as to cost and availability.

Environmental Engineering Abstracts, Cambridge Scientific Abstracts, 7200 Wisconsin Avenue, Bethesda, MD 20814-4823. (301) 961-6750. FAX (301) 961-6720. Monthly. Covers hazardous materials, environmental impact and protection, treatment of sewage and industrial wastes, hydroelectric power, tidal and wind power, arctic and tropical engineering.

Hydro-Abstracts (formerly Water Resources Abstracts). HydroScience Press, 2527 Jackson Street, N.E., Minneapolis, MN 55418. (612) 781-9081. Monthly. $150.00 per year.

NTIS Alerts: Environmental Pollution and Control. National Technical Information Service (NTIS), 5285 Port Royal Road, Springfield, VA 22161. (703) 487-4650. Weekly. $150.00 per year.

Pollution Abstracts. Cambridge Scientific Abstracts, 7200 Wisconsin Avenue, Bethesda, MD 20814-4823. (301) 961-6750. FAX (301) 961-6720. Inquire for cost and availability.

Science Citation Index. Institute for Scientific Information, 3501 Market Street, Philadelphia, PA 19104. (215) 386-0100. FAX (215) 386-6362. Inquire about availability and cost. Also available on CD-ROM.

Selected Water Resources Abstracts. U.S. Geological Survey, Water Resources Scientific Information Center, 425 National Center, Reston, VA 22092. (703) 648-6820. Monthly. $115.00 per year. Also available on CD-ROM.

ASSOCIATIONS AND PROFESSIONAL SOCIETIES

Environmental Defense Fund. 257 Park Avenue South, New York, NY 10010. (212) 505-2100. FAX (212) 505-2375.

Environmental Industry Council. 529 14th Street NW, Suite 655, Washington, DC 20045. (202) 737-3018.

International Center For the Solution of Environmental Problems. 535 Lovett Blvd., Houston, TX 77006. (713) 527-8711. FAX (713) 527-8025.

National Association of Environmental Professionals. 5165 MacArthur Blvd. NW, PO Box 9400, Washington, DC 20016-3315. (202) 966-1500. FAX (202) 966-1977.

BIBLIOGRAPHIES

Environmental Periodicals Bibliography. International Academy at Santa Barbara, Environmental Studies Institute, 800 Garden Street, Suite D, Santa Barbara, CA 93101-1552. (805) 965-5010. 1972 to present. Bi-monthly. Price varies.

DIRECTORIES AND BIOGRAPHICAL SOURCES

Government Research Directory. 9th edition. Gale Research, 835 Penobscot Building, Detroit, MI 48226-4094. (313) 961-2242. (800) 347-4253. 1996. $465.00.

International Research Centers Directory. Gale Research, 835 Penobscot Building, Detroit, MI 48226-4094. (313) 961-2242. (800) 347-4253. 8th edition. 1995. $410.00

Research Centers Directory. Gale Research, 835 Penobscot Building, Detroit, MI 48226-4094. (313) 961-2242. (800) 347-4253. $470.00

Scientific and Technical Organizations and Agencies Directory. Gale Research, 835 Penobscot Building, Detroit, MI 48226-4094. (313) 961-2242. (800) 347-4253. 4th edition. 1996. $195.00.

Who's Who in Technology. Gale Research, 835 Penobscot Building, Detroit, MI 48226-4094. (313) 961-2242. (800) 347-4253. 1995. $195.00.

ENCYCLOPEDIAS AND DICTIONARIES

Encyclopedia of Physical Science and Technology. Robert A. Meyers, ed. 18 volumes. Academic Press, Inc., 1250 Sixth Avenue, San Diego, CA 92101-4311. (619) 231-0926. FAX (619) 699-6715. 1992. $2100.00

McGraw-Hill Encyclopedia of Science and Technology. Sybil P. Parker, ed. 7th edition. 20 volumes. McGraw-Hill Publishing Company, 1221 Avenue of the Americas, New York, NY 10020. (800) 262-4729 or (212) 512-3825. 1992. $1900.00

GENERAL WORKS

Environmental Hazards: Marine Pollution. Martha Gorman. ABC-CLIO Information Services, PO Box 1911, Santa Barbara, CA 93116-1911. (800) 368-6868. FAX (805) 685-9685. 1993. $39.50.

Ocean Environmental Management. Ernst G. Frankel. Prentice Hall, 113 Sylvan Avenue, Route 9W, Englewood Cliffs, NJ 07632. (201) 592-2000 or (800) 922-0579. 1995. $65.00.

Ocean Pollution: Effects On Living Resources and Humans. Carl J. Sindermann. CRC Press, 2000 Corporate Blvd., N.W., Boca Raton, FL 33431. (407) 994-0555 or (800) 272-7737. FAX (407) 994-0949. 1996. Inquire for price.

HANDBOOKS AND MANUALS

Environmental Science and Technology Handbook. Porter C. Knowles, et al. Government Institutes Inc., 4 Research Place, Suite 200, Rockville, MD 20850. (301) 921-2355. 1994. $75.00.

Practical Handbook of Estuarine and Marine Pollution. Michael J. Kennish. CRC Press, 2000 Corporate Blvd., N.W., Boca Raton, FL 33431. (407) 994-0555 or (800) 272-7737. FAX (407) 994-0949. 1996. Inquire for price.

ONLINE DATABASES AND CD-ROMS

CA Search. Chemical Abstracts Service, P.O. Box 3012, Columbus, OH 43210-0012. (614) 447-3600. (800) 848-6533. FAX (614) 447-3709. Very comprehensive guide to worldwide chemical literature and related fields, 1972 to present. Available on BRS,(800) 289-4277, DIALOG, (800) 334-2564, ORBIT (800) 456-7248, and STN International, FIZ Karlsruhe, P.O. Box 2465, W-7500, Karlsruhe 1, Germany, online services. Inquire as to cost and availability.

NTIS Bibliographic Database. National Technical Information Service, 5285 Port Royal Road, Springfield, VA 22161. (703) 487-4929 or FAX (703) 321-8199. Broad coverage of government-sponsored science and technology research reports, 1964 to present. Available on BRS,(800) 289-4277, DIALOG, (800) 334-2564, ORBIT, (800) 456-7248, and STN International, FIZ Karlsruhe, P.O. Box 2465, W-7500, Karlsruhe 1, Germany, online services. Also available on CD-ROM. Inquire as to cost and availability.

Waternet. American Water Works Association, Technical Library. Available on DIALOG online services. Citations to literature on water quality, water utility management, water analysis, water pollution, and related areas, 1971 to present. Available on DIALOG,(800) 334-2564, online service. Inquire as to cost and availability.

PERIODICALS

Air & Water Pollution Control. Bureau of National Affairs, 1231 25th Street NW, Washington, DC 20037. (202) 452-4200. FAX (202) 822-8092. 1986 to present. Bi-weekly. $301.00 per year.

Atmospheric Environment. Elsevier Science [journals], 660 White Plains Rd., Tarrytown, NY 10159-5153. (919) 524-9200. FAX (919) 333-2444. 1967 to present. 22 times a year. $1995.00 per year.

Bulletin of Environmental Contamination and Toxicology. Springer-Verlag, 175 Fifth Avenue, New York, NY 10010. (212) 460-1500 or (800) 777-4643. FAX (212) 473-6272. 1966 to present. Monthly. $399.00 per year.

Environmental Engineering. Mechanical Engineering Publications Ltd., Northgate Avenue, Bury St. Edmunds, Suffolk IP32 6BW, England. Telephone 0284-763277. FAX 0284-704006. 1972 to present. Quarterly. Inquire for price.

Environmental Monitoring and Assessment. Kluwer Academic Publishers, P.O. Box 358, Accord Station, Hingham, MA 02018-0358. (617) 871-6000. FAX (617) 871-6528. 1981 to present. 15 times a year. $765.00 per year.

Environmental Pollution. Elsevier Science [journals], 660 White Plains Rd., Tarrytown, NY 10159-5153. (919) 524-9200. FAX (919) 333-2444. 1970 to present. 12 times a year. $1250.00 per year.

Environmental Science and Technology. American Chemical Society, 1155 16th Street, N.W., Washington, DC 20036. (202) 872-4381 or (800) 333-9511. FAX (614) 447-3671. 1967 to present. Monthly. $89.00 per year for non-members.

EPA Journal. Superintendent of Documents, U.S. Government Printing Office, Box 371954, Pittsburgh, PA 15250-7954. (202) 783-3238. FAX (202) 512-2233. 1975 to present. Bi-monthly. $10.00 per year.

Marine Pollution Bulletin. Elsevier Science [journals], 660 White Plains Rd., Tarrytown, NY 10159-5153. (919) 524-9200. FAX (919) 333-2444. 1970 to present. 24 times a year. $430.00 per year.

Pollution Engineering. Cahners Publishing Company (Des Plains), 1350 East Touhy Avenue, Des Plaines, IL 60017-5080. (800) 323-4958. FAX (708) 390-2779. 1969 to present. 13 times a year. $70.00 per year.

Water, Air and Soil Pollution. Kluwer Academic Publishers, P.O. Box 358, Accord Station, Hingham, MA 02018-0358. (617) 871-6000. FAX (617) 871-6528. 1971 to present. 28 times a year. $1519.00.

RESEARCH CENTERS AND INSTITUTES

Nova University Oceanographic Center. 8000 N. Ocean Drive, Dania, FL 33004. (305) 920-1909.

Oceanic Society. 218 D Street SE, Washington, DC 20003. (202) 544-2600. FAX (202) 543-4710.

University of Hawaii At Manoa, Hawaii Undersea Research Laboratory. 1000 Pope Road, MSB 303, Honolulu, HI 96822. (808) 956-6335. FAX (808) 956-2136.

University of North Carolina At Wilmington, Noaa National Undersea Research Center. 7205 Wrightsville Avenue, Wilmington, NC 28403. (919) 256-5133.

MARKOV PROCESSES

Ency. of Physical Sciences and Engineering Info. Sources

MARKOV PROCESSES

See also: MATHEMATICS, PROBABILITY, STATISTICS, STOCHASTIC PROCESSES

ABSTRACT SERVICES AND INDEXES

Applied Science and Technology Index; A Cumulative Subject Index To English Language Periodicals in the Fields of Aeronautics and Space Science, Computer Technology, Chemistry, Construction Industry, Energy and Related Areas. H. W. Wilson and Company, 950 University Avenue, Bronx, NY 10452. (212) 588-8400. (800) 367-6770. FAX (718) 590-1617. 1958 to present. Monthly. Available on-line from BRS and Wilsonline. Also available on CD-ROM. Inquire as to cost and availability. ISSN: 0003-6986.

Compactmath-Compact Mathematics Library. Cumulative CD-ROM edition of Zentralblatt fuer Mathematik-Mathematics Abstracts. Springer-Verlag New York, Inc. 44 Hartz Way. Secaucus. NJ. 07096-2491. (201) 348-4033. FAX (201) 348-4505. 1993 to present. Annual. Available only on CD-ROM. Inquire as to cost and availability. ISSN: 0938-3174.

Current Index To Statistics: Applications-Methods-Theory. American Statistical Association, 1429 Duke Street, Alexandria, VA 22314-3402. (703) 684-1221. FAX (703) 684-2037. 1976 to present. Annual. $50.00 per year. Available on-line from BRS, Dialog, European Space Agency. ISSN: 0364- 1228.

Current Mathematical Publications. American Mathematical Society. P.O. Box 1571, Annex Station, Providence, RI 02901-9930. (800) 556-7774 or (401) 455-4000. FAX (401) 331-3842. 1969 to present. Seventeen times per year. $377.00 per year. Available online from BRX, DIALOG, European Space Agency. Also available on CD-ROM. Inquire as to cost and availability. ISSN: 0361-4794.

Index To Mathematical Tables. Alan Fletcher. Addison- Wesley Publishing Co., Inc., 1 Jacob Way, Reading, MA 01867. (617) 944-3700. (800) 447-2226. 2nd edition. 1962. 2 volumes.

Mathematical Reviews: A Review Journal Covering the World Literature of Mathematical Research. American Mathematical Society, P.O. Box 1571, Annex Station, Providence, RI 02901-9930. (800) 556-7774 or (401) 455-4000. 1940 to present. Monthly. $4594.00 per year. Also available via network (MATHSCINET), online and on CD-ROM. Inquire regarding cost and availability. ISSN: 0025-5629.

NTIS Alerts: Mathematical Sciences. U.S. National Technical Information Service, 5285 Port Royal Road, Springfield, VA 22161. (703) 487-4650. FAX (703) 321-8547. Weekly. $140.00.

Science Citation Index. Institute for Scientific Information. 3501 Market Street, Philadelphia, PA 19104. (215) 386-0100. (800) 523-1850. FAX (215) 386-2991. 1961 to present. Six issues per year, plus annual cumulation. $11650.00 per year. Also available online and on CD-ROM. Inquire regarding cost and availability. ISSN: 0036-827X.

Statistical Theory and Method Abstracts. International Statistical Institute. Prinses Beatrixiaan 428, Postbus 950, 2270 AZ Voorburg, Netherlands. 31-70-3375737. FAX 31-70-3860025. 1959 to present. Quarterly. $191.82.

ANNUAL REVIEWS AND YEARBOOKS

Advances in Probability and Related Topics. Marcel Dekker, Inc., 270 Madison Avenue, New York, NY 10016. (212) 696-9000. FAX (212) 685-4540. 1971 to present. Irregular.

Advances in Applied Mathematics. Academic Press, Inc., 6277 Sea Harbor Drive, Orlando, FL 32821. (800) 321-5068. Irregular. Price varies.

Ergebnisse Der Mathematik Und Ihrer Grenzgebiete. Springer-Verlag New York, Inc. 44 Hartz Way. Secaucus. NJ. 07096-2491. (201) 348-4033. FAX (201) 348-4505. Irregular. Price varies.

Frontiers in Pure and Applied Probability. Books International, Inc., Box 605 Herndon, VA 22070. (703) 435-7064. FAC (703) 689-0660. 1993 to present. Irregular. Price varies.

Lecture Notes in Mathematics. Springer-Verlag New York, Inc. 44 Hartz Way. Secaucus. NJ. 07096-2491. (201) 348-4033. FAX (201) 348-4505. Irregular. Price varies.

Probability and Mathematical Statistics. Academic Press, Inc., 6277 Sea Harbor Drive, Orlando, FL (800) 321-5068. 1967 to present. Irregular. Price varies.

Probability: Pure and Applied. Marcel Dekker, Inc., 270 Madison Avenue, New York, NY 10016. (212) 696-9000. FAX (212) 685-4540. (800) 228-1160. 1984 to present. Irregular. Price varies.

Statistics: Textbooks and Monographs Series. Marcel Dekker, Inc., 270 Madison Avenue, New York, NY 10016. (212) 696-9000. FAX (212) 685-4540. 1972 to present. Irregular. Price varies.

ASSOCIATIONS AND PROFESSIONAL SOCIETIES

American Mathematical Society. P.O. Box 6248, Providence, RI 02940. (401) 455-4000. FAX (401) 331-3842.

American Statistical Association. 1429 Duke Street, Alexandria, VA 22314-3402. (703) 684-1221. FAX (703) 684-2037.

Association For Women in Mathematics. 4114 Computer and Space Science Building, University of Maryland, College Park, MD 20742-2461. (301) 405-7892.

Biometric Society, Eastern North American Region. c/o Randy Price, 11250 Roger Bacon Drive, Suite 8, Reston, VA 22090-5202. (703) 525-1191. FAX (703) 276-8196.

Caucus For Women in Statistics. c/o Cyntha Struthers, St. Jerome's College, Waterloo, ON, Canada N2L 3G3. (519) 888-4801. FAX (519) 746-6530.

Econometric Society. Northwestern University, Department of Economics, Evanston, IL 60208-2600. (708) 491-3615.

Institute of Mathematical Statistics. 3401 Investment Boulevard, Suite 7, Hayward CA 94545-3819. (510) 783-8141. FAX (510) 783-4131.

International Association For Statistical Computing. Prinses Beatrixiaan 428, Postbus 950, NL-2270 AZ Voorburg, Netherlands. PH: 70 3375737. FX: 70 3860025.

International Biometric Society. 1429 Duke Street, Suite 401, Alexandria, VA 22314. (703) 836-8311.

International Statistical Institute. Prinses Beatrixiaan 428, Postbus 950, NL-2270 AZ Voorburg, Netherlands, PH 70-3375737. FX 70 3860025.

Mathematical Association of America. 1529 18th Street, NW, Washington, DC 20036. (202) 387-5200. FAX (202) 265-2384.

National Institute of Statistical Sciences. PO Box 14162, Research Triangle Park, NC 27709. (919) 541-6255. FAX (919) 541-7102.

Operations Research Society of America. 1314 Guilford Avenue, Baltimore, MD. 21202. (410) 528-4146. FAX (410) 528-8556.

Royal Statistical Society. 12 Errol Street, London EC1Y 8LX, England. PH 171 7235882. FX 171 7061710.

Society For Industrial and Applied Mathematics. 3600 University City Science Center, Philadelphia, PA 19104-2688. (215) 382-9800. FAX (215) 386-7999.

BIBLIOGRAPHIES

Bibliography of Statistical Literature. Maurice Kendall and Alison Doig. Ayer Company Pubs., Inc., Lower Mill Road, North Stratford, NH 03590. (603) 922-5105. (800) 282-5413. 1981. 3 volumes: Pre-1940, 1940 to 1949, 1950 to 1958. $82.95, set.

Chronological Annotated Bibliography of Order Statistics, Volume Vi. H. Leon Harter. American Sciences Press, 20 Cross Road, Syracuse, NY 13224-2144. 1991. $115.00.

Guide To Statistical Methods and To the Pertinent Literature. L. Sachs. Springer-Verlag New York, Inc., 175 Fifth Avenue, New York, NY 10010. (212) 460-1500. (800) 777-4643. 1986. $39.00 in paper.

How To Find Out About Statistics. Gillian Burrington. Franklin Book Co., Inc., Elkins Park, PA 19117. (215) 635-5252. 1972. reprint edition. $74.00.

Mathematical Journals: An Annotated Guide. Diana F. Liang. Scarecrow Press. Distributed by University Press of America, 4720 Boston Way, Lanham, MD 20706. (301) 459-3366. (800) 462-6420. $29.50. ISBN: 0-8108-2585-6.

DIRECTORIES AND BIOGRAPHICAL SOURCES

Assistantships and Graduate Fellowships in the Mathematical Sciences. American Mathematical Society, P.O. Box 6248, Providence, RI 02940. Annual. (401) 455-4000. (800) 321- 4267. Annual. $19.00. ISSN: 1040-7650.

Combined Membership List. American Mathematical Society, P.O. Box 6248, Providence, RI 02940. (401) 455-4000. (800) 321-4267. Annual. $50.00. ISSN: 0569-6461.

Directory of Statisticians. American Statistical Association, 1429 Duke Street, Alexandria, VA 22314-3402. (703) 684-1221. FAX (703) 684-2037. Periodic.

Research Centers Directory. Gale Research Company, 835 Penobscot Building, Detroit, MI 48226-4094. (313) 961- 2242. (800) 347-4253. 20th edition. 1995. Annual. $485.00. ISSN: 0080-1518.

Mathematical Science Professional Directory. American Mathematical Society, P.O. Box 6248, Providence, RI 02940. (401) 455-4000. (800) 321-4267. Annual. $45.00. ISBN: 0-8218-0173-2.

World Directory of Mathematicians. G.D. Mostow, editor. International Mathematical Union, Helsinki. 1958 to present. 10th revised edition. 1995. $45.00. Available from: American Mathematical Society, P.O. Box 6248, Providence, RI 02940. (401) 455-4000. ISSN: 0512-2740.

ENCYCLOPEDIAS AND DICTIONARIES

Dictionary of Gaming, Modeling & Simulation. G. Ian Gibbs. Books on Demand, 300 North Zeeb Road, Ann Arbor, MI 48106-1346. (313) 761-4700. (800) 521-0600. $49.60 in paper.

Dictionary-Outline of Basic Statistics. John Freund and Frank J. Williams. Dover Publications, Inc., 180 Varick Street, New York, NY 10014. (212) 255-3755. (800) 223-3130. Reprint edition. $5.95 in paper.

Dictionary of Statistical Terms. F.H. Marriott. Halsted Press, 605 Third Avenue, New York, 10158-0012. (212) 850-6400. 5th edition. 1990. $76.95 in paper.

Dictionary of Statistics. Michael G. Mulhall. Gordon Press Publications, P.O. Box 459, Bowling Green Station, New York, NY 10004. (718) 624-8419. 1972. $300.00.

Encyclopedia of Statistical Sciences. Samuel Kotz, Norman L. Johnson, and Campbell B. Read, editors. John Wiley & Sons, Inc., 605 Third Avenue, New York, NY 10158-0012. (212) 850-6000. (800) 225-5945. 1988. 9 volumes. ISBN: 0-471-05544-1. $1440.00 set.

Encyclopedic Dictionary of Mathematics. Kiyosi Ito, editor. MIT Press, 55 Hayward Street, Cambridge, MA 02142. (617) 625-8569. 2nd edition, reprint. 1993. 2 volumes. ISBN: 0- 262-59020-4. $70.00.

HarperCollins Dictionary of Statistics. Roger Porkess. HarperCollins Publ. 10 East 53rd Street, New York, NY 10022-5299. (800) 331-3761. 1991. $14.00 in paper.

International Encyclopedia of Statistics, 2. William H. Kruskal and Judith M. Tanur, editors. The Free Press, 1230 Avenue of the Americas, New York, NY 10020. (212) 698-7000. (800) 223-2348. 1978. $90.00.

Mathematics Dictionary. Robert C. James, editor. Van Nostrand Reinhold, 115 Fifth Avenue, New York, NY 10003. (212) 254-3232. (800)-842-3636. 5th edition. 1992. $42.95. ISBN: 0-442-00741-8.

Russian-English, Russian-English Dictionary On Probability, Statistics and Combinatorics. K. A. Borokov. Society for Industrial and Applied Mathematics, 3600 University City Sci-

MARKOV PROCESSES

Ency. of Physical Sciences and Engineering Info. Sources

ence Center, Philadelphia, PA 19104-2688. (215) 382-9800. FAX (215) 386-7999. (800) 447-7426. 1993. $47.50 in paper.

Topical Dictionary of Statistics: Advanced Industrial Terminology. Gary L. Teitjen. Chapman & Hall, 1 Penn Plaza, New York, NY 10119. (212) 564-1060. 1985. $24.50.

GENERAL WORKS

Applied Probability and Stochastic Processes in Engineering and the Physical Sciences. Michael K. Ochi. John Wiley & Sons, Inc., 605 Third Avenue, New York, NY 10158-0012. (212) 850-6000. (800) 225-5945. 1990. $114.00.

The Art of Probability For Scientists and Engineers. Richard W. Hamming. Addison-Wesley Publishing Co., Inc., 1 Jacob Way, Reading, MA 01867. (617) 944-3700. (800) 447-2226. 1991. $45.95.

Chance and Chaos. David Ruelle. Princeton University Press, 41 William Street, Princeton, NJ 08540. (609) 258-4900. (800) 777-4726. 1993. $35.00.

Complex Stochastic Systems and Engineering. D.M. Titterington, editor. Oxford University Press, Inc., 200 Madison Avenue, New York, NY 10016. (212) 725-6000. (800) 334-4249. 1995. $80.00.

Denumerable Markov Chains. J.G. Kemeny, et al. Springer-Verlag New York, Inc., 175 Fifth Avenue, New York, NY 10010. (212) 460-1500. (800) 777-4643. 2nd edition. 1991. $49.00.

Electrical Engineering Probability. Richard Williams. West Publishing Co., 610 Opperman Drive, P. O. Box 64526, Saint Paul, MN 55164-0526. (612) 687-8000. (800) 328-9424. 1991. $69.25.

Excursions of Markov Processes. R.M. Blumenthal. Birkhauser Boston, 675 Massachusetts Avenue, Cambridge, MA 02139. (617) 876-2333. (800) 777-4643. 1991. $64.50.

From Markov Chains to Non-Equilibrium Particle Systems. M.F. Chen. World Scientific Publishing Company, Inc., 1060 Main Street, River Edge, NJ 07661. (201) 487-9655. (800) 227-7562. 1992. $74.00.

General Theory of Markov Processes. Michael Sharpe. Academic Press, Inc., 6277 Sea Harbor Drive, Orlando, FL 32821. (800) 321-5068. 1988. $80.00.

Geometry of Random Motion. R. Durrett and M. Pinsky. American Mathematical Society, P.O. Box 6248, Providence, RI 02940. (401) 455-4000. (800) 321-4267. 1988. $41.00.

Hidden Markov Models: Estimation and Control. Robert J. Elliot, et al. Springer-Verlag New York, Inc., 175 Fifth Avenue, New York, NY 10010. (212) 460-1500. (800) 777-4643. 1994. $49.00.

Introduction To the Numerical Solution of Markov Processes. William J. Stewart. Princeton University Press, 41 William Street, Princeton, NJ 08540. (609) 258-4900. (800) 777- 4726. 1994. $49.50.

Markov Chains and Stochastic Stability. S.P. Meyn and R. L. Tweedie. Springer-Verlag New York, Inc., 175 Fifth Avenue,

New York, NY 10010. (212) 460-1500. (800) 777-4643. 1994. $99.00.

Markov Decision Processes: Discrete Stochastic Dynamic Programming. Martin L. Puterman. John Wiley & Sons, Inc., 605 Third Avenue, New York, NY 10158-0012. (212) 850-6000. 1994. $89.95.

Markov Processes: An Introduction For Physical Scientists. Daniel T. Gillespie. Academic Press, Inc., 6277 Sea Harbor Drive, Orlando, FL 32821. (800) 321-5068. 1991. $54.95.

Markov Random Fields: Theory and Applications. Rama Chellappa and Anil Jain, editors. Academic Press, Inc., 6277 Sea Harbor Drive, Orlando, FL 32821. (800) 321-5068. 1993. $75.00.

Nonlinear Dynamic and Stochastic Mechanics. Wolfgang H. Kliemann. CRC Press Inc., 2000 Corporate Boulevard, NW, Coca Raton, FL 33431. (407) 994-0555. (800) 272-7737. 1994. $79.95.

Probability and Random Processes. Geoffery Grimmett and David Stirzaker. Oxford University Press, Inc., 200 Madison Avenue, New York, NY 10016. (212) 725-6000. (800) 334-4249. 2nd edition. 1992. $45.95.

Random Vibration of Mechanical and Structural Systems. T. T. Soong and Mircea Grigoriu. Prentice-Hall, 15 Columbus Circle, New York, NY 10023. (212) 373-8500. (800) 223-2348. 1992. $82.00

Regular and Chaotic Dynamics. A. J. Lichtenberg and M. A. Lichtenberg. Springer-Verlag New York, Inc., 175 Fifth Avenue, New York, NY 10010. (212) 460-1500. (800) 777-4643. 1992. $59.95.

HANDBOOKS AND MANUALS

Biometrika Tables For Statistics. E. S. Pearson and H. O. Hartley, editors. Lubrecht & Cramer, Ltd., 38 County Route 48, Forestburgh, NY 12777-6400. (914) 794-8539. volume 1, 3rd edition. volume 2, reprint edition. 1976. $75.00 each.

Computational Handbook of Statistics. James L. Bruning, and B. L. Kintz. HarperCollins Publ. 10 East 53rd Street, New York, NY 10022-5299. (800) 331-376. 3rd edition. 1990. $38.00 in paper.

Handbook of Applicable Mathematics. Walter Ledermann, editor. John Wiley & Sons, Inc., 605 Third Avenue, New York, NY 10158-0012. (212) 850-6000. (800) 225-5945. 1985. $1496.00 6 volumes. ISBN: 0-471-90804-5.

Handbook of Applied Mathematics: Selected Results and Methods. Carl E. Pearsen, editor. Chapman & Hall, 1 Penn Plaza, New York NY 10119. (212) 564-1060. 1990. $46.95. ISBN: 0-442-00521-0.

Handbook of Mathematical Functions With Formulas, Graphs, and Mathematical Tables. Milton Abramowitz and Irene A. Stegun, editors. U.S. National Bureau of Standards. U.S. Government Printing Office, Washington, D.C. 10th edition. 1972. $43.00. ISBN: 0-16-000202-8.

Handbook of Mathematics. I. N. Bronshtein and K. A. Semedyayev. Van Nostrand Reinhold, 115 Fifth Avenue, New

York, NY 10003. (800)-842-3636. 20th edition. $59.95. ISBN: 0-442-21171-6.

Handbook of Stochastic Methods. C. W. Gardiner. Springer-Verlag New York, Inc., 175 Fifth Avenue, New York, NY 10010. (212) 460-1500. (800) 777-4643. 1994. $59.00 in paper.

Handbook of Tables For Probability and Statistics. William H. Beyer, editor. CRC Press, 2000 Corporate Boulevard, Boca Raton, FL 33431. (800) 272-7737. 2nd edition. 1968. $92.95. ISBN: 0-8493-0692-2.

Standard Probability and Statistics Tables and Formulas. William H. Beyer. CRC Press, Inc., 2000 Corporate Boulevard, NW, Coca Raton, FL 33431. (407) 994-0555. (800) 272-7737. 1990. $34.95.

Standard Mathematical Tables and Formulas. William H. Beyer, editor. CRC Press, 2000 Corporate Boulevard, Boca Raton, FL 33431. (800) 272-7737. 29th edition. 1991. $41.95. ISBN: 084-930- 6299.

Stochastic Models. D. P. Heyman and M. J. Sobel. Elsevier Science Publishing Company, Inc., 655 Avenue of the Americas, New York, NY 10010. (212) 989-5800. 1990. $115.00.

Tables of Integrals, Series and Products: Corrected and Enlarged Edition. I. S. Gradshteyn and I. M. Ryzhik. Academic Press, Inc., 6277 Sea Harbor Drive, Orlando, FL 32821. (800) 321-5068. 5th edition. 1993. $61.00. ISBN: 0-12-294755-X.

ONLINE DATABASES AND CD-ROMS

CA SEARCH. Chemical Abstracts Service, P.O. Box 3012, Columbus, OH 43210-0012. (614) 447-3600. (800) 848-6533. FAX (614) 447-3709. Very comprehensive guide to worldwide chemical literature and related fields, 1972 to present. Available on BRS,(800) 289-4277, DIALOG, (800) 334-2564, ORBIT (800) 456-7248, and STN International, FIZ Karlsruhe, P.O. Box 2465, W-7500, Karlsruhe 1, Germany, online services. Inquire as to cost and availability.

Compendex Plus. Engineering Information, Inc., 345 East 47th Street, New York, NY 10017. (212) 705-7600 or (800) 221-1044. Contains citations with abstracts to worldwide literature in engineering and technology, from 1970 to present. Available on online BRS,(800) 289-4277, DIALOG, (800) 334-2564, ORBIT (800) 456-7248, and STN International, FIZ Karlsruhe, P.O. Box 2465, W-7500, Karlsruhe 1, Germany, online services. Also available on CD-ROM. Inquire as to cost and availability.

Current Contents Search. Institute for Scientific Information, 3501 Market Street, Philadelphia, PA 19104. (215) 386-0100. FAX (215) 386-6362. Contains citations to articles listed in the table of contents of science and technology journals. Also articles in social sciences and life sciences journals. Available on BRS,(800) 289-4277, DIALOG,(800) 334-2564, online services. Inquire as to cost and availability.

Dissertation Abstracts. University Microfilms International, 300 North Zeeb Road, Ann Arbor, MI 48106. (800) 521-0600 or (313) 761-4700. Scope includes virtually all doctoral dissertations accepted at accredited American institutions from 1861 to present in 252 subject areas. Available on BRS,(800) 289-4277, DIALOG,(800) 334-2564, and OCLC EPIC,(800) 848-5878, online services. Also available on CD-ROM. Inquire as to cost and availability.

INSPEC. Institution of Electrical Engineers, Michael Faraday House, Six Hills Way, Stevenage, Herts. SG1 2AY, England. Telephone: 0438 313311 or FAX 0438 742840. Contains citations to the worldwide literature of physics, electronics and electrical engineering, computer technology, and related fields. Available on BRS,(800) 289-4277, DIALOG, (800) 334-2564, ORBIT, (800) 456-7248, and STN International, FIZ Karlsruhe, P.O. Box 2465, W-7500, Karlsruhe 1, Germany, online services. Inquire as to cost and availability.

Mathsci and Mathscinet. American Mathematical Society, P.O. Box 6248, Providence, RI 02940. (800) 321-4667 or (401) 455-4000 or FAX (401) 331-3842. Scope includes pure and applied mathematics and related areas of physics, statistics, engineering, computer science, and operations research literature since 1959. Available on DIALOG,(800) 334-2564, online service. Also available on CD-ROM. Inquire as to cost and availability.

NTIS Bibliographic Database. National Technical Information Service, 5285 Port Royal Road, Springfield, VA 22161. (703) 487-4929 or FAX (703) 321-8199. Broad coverage of government-sponsored science and technology research reports, 1964 to present. Available on BRS,(800) 289-4277, DIALOG, (800) 334-2564, ORBIT, (800) 456-7248, and STN International, FIZ Karlsruhe, P.O. Box 2465, W-7500, Karlsruhe 1, Germany, online services. Also available on CD-ROM. Inquire as to cost and availability.

Scisearch. Institute for Scientific Information, 3501 Market Street, Philadelphia, PA 19104. (800) 523-1850 or (215) 386-0100. Broad multidisciplinary title and author index to the international literature of science and technology, 1974 to present. Available on DIALOG,(800) 334-2564, and ORBIT,(800) 456-7248, online services. Also available on CD-ROM. Inquire as to cost and availability.

Wilsonline. H.W. Wilson Company, 950 University Avenue, Bronx, NY 10452. (800) 367-6770 or (212) 588-8400. Makes available online versions of the printed H.W. Wilson indexes, including Applied Science and Technology Index, Business Periodicals Index, General Science Index, and Readers' Guide to Periodical Literature. Period covered is generally 1983 to present. Available on BRS,(800) 289-4277, DIALOG,(800) 334-2564, and OCLC EPIC,(800) 848-5878, online services. Also available on CD-ROM. Inquire as to cost and availability.

PERIODICALS

Advances in Applied Probability. Applied Probability Trust, School of Mathematics, University of Sheffield, Sheffield S3 7RH, England. TEL 44-742-824269. FAX 44-742-729782. 1969 to present. Quarterly. $56.00 per year. ISSN: 0001-8678.

American Statistician. American Statistical Association. 1429 Duke Street, Alexandria, VA 22314-3402. (703) 684-1221. FAX (703) 684-2037. 1947 to present. Quarterly. $55.00 per year. ISSN: 0003-1305.

Annals of Probability. Institute of Mathematical Statistics. 3401 Investment Boulevard, Suite 7, Hayward, CA 94545. (510) 783-8141. 1973 to present. Quarterly. $130.00 per year. ISSN: 0090-5364.

Canadian Journal of Statistics/Revue Canadienne de Statistique. Statistical Society of Canada, Room 612 Dunton Tower, Carleton University, Ottawa, ON K1S 5B6, Canada. (613) 788-2167. 1973 to present. Quarterly. $98.00 per year. ISSN: 0319-5724.

Chance. Springer-Verlag, 175 Fifth Avenue, New York, NY 10010. (212) 460-1612. FAX (212) 473-6272. 1988 to present. Quarterly. $69.00 per year. ISSN: 0933-2480.

Communications in Statistics, Part A: Theory and Methods. Marcel Dekker Journals, 270 Madison Avenue, New York, NY, 10016. (212) 696-9000. FAX (212) 685-4540. 1973 to present. Twelve issues per year. $1295.00 per year. ISSN: 0361-0926.

Institut Henri Poincare. Annales. Probabilities et Statistics. Gauthier-Villars, 15 rue Gossin, 92543 Montrouge Cedex, France. TEL 33-1-40-92-65-00. FAX 33-1-40-92-65-97. 1930 to present. Quarterly. 1230F. ISSN: 0246-0203.

JASA. Journal of the American Statistical Association. American Statistical Association, 1429 Duke Street, Alexandria, VA 22314-3402. (703) 684-1221. FAX (703) 684-2037. 1888 to present. Quarterly. $210.00 per year. ISSN: 0162-1259.

Journal of Applied Probability. Applied Probability Trust, University of Sheffield, Sheffield S3 7RH, England. TEL 44-742-824269. FAX 44-742-729782. 1964 to present. Quarterly. $168.00 per year. ISSN: 0021-9002.

Metrika; International Journal For Theoretical and Applied Statistics. Physica-Verlag GmbH und Co. Distributed by Springer-Verlag New York, Inc., 175 Fifth Avenue, New York, NY 10010. (212) 460-1500. 1953 to present. Bimonthly. DM360. ISSN: 0026-1335.

Probability theory and Related Fields. Springer-Verlag New York, Inc., 175 Fifth Avenue, New York, NY 10010. (212) 460-1500. (800) 777-4643. 1962 to present. Monthly. $1553.00. ISSN: 0178-8051.

Random Structures and Algorithms. John Wiley & Sons, Inc., 605 Third Avenue, New York, NY 10158-0012. (212) 850-6000. FAX (212) 850-6088. 1990 to present. Quarterly. $185.00 per year. ISSN: 1042-9832.

Risk Analysis. Society for Risk Analysis. Plenum Publishing Corp. 233 Spring Street, New York, NY 10013-1578. (212) 620-8000. FAX (212) 463-0742. 1980 to present. Bimonthly. $320.00 per year. ISSN: 0272-4332.

Royal Statistical Society. Journal. Series A-C. Basil Blackwell Ltd., 108 Cowley Road, Oxford OX4, 1JF, Eng. TEL 0865-791100. FAX 0865-791347. Series A: Statistics in Society; 1838 to present. Series B: Methodological; 1934 to present. Series C: Applied Statistics, 1952 to present. Each series, three issues per year; $73.00 per year each. ISSN: 0035-9238, 9246, 9254.

SIAM Journal On Applied Mathematics. Society for Industrial and Applied Mathematics, 3600 University City Science Center, Philadelphia, PA 19104-2688. (215) 382-9800. FAX (215) 386-7999. 1953 to present. Bimonthly. $242.00 per year. ISSN: 0036-1399.

Siam Journal On Control and Optimization. Society for Industrial and Applied Mathematics, 3600 University City Science Center, Philadelphia, PA 19104-2688. (215) 382-9800. FAX (215) 386-7999. 1963 to present. Bimonthly. $287.00 per year. ISSN: 0363-0129.

Siam Review. Society for Industrial and Applied Mathematics, 3600 University City Science Center, Philadelphia, PA 19104-2688. (215) 382-9800. 1959 to present. Quarterly. $148.00 per year. ISSN: 0036-1445.

Stochastic Analysis and Applications. Marcel Dekker Journals, 270 Madison Avenue, New York, NY, 10016. (212) 696-9000. FAX (212) 685-4540. 1983 to present. Five issues per year. $535.00 per year. ISSN: 0736-2992.

Stochastic Processes and their Applications. Elsevier Science, Inc., Box 882, Madison Square Station, New York, NY 10159. (212) 989-5800. FAX (212) 633-3990. 1973 to present. Monthly. $1171.00 per year. ISSN: 0304-4149.

Technometrics. American Statistical Association, 1429 Duke Street, Alexandria, VA 22314-3402. (703) 684-1221. FAX(703) 684-2037. 1959 to present. Quarterly. $50.00. ISSN: 0040-1706.

RESEARCH CENTERS AND INSTITUTES

Center for Pure and Applied Mathematics. University of California, Berkeley, 977 Evans Hall, Berkeley, CA 94720. (415) 642-0116. FAX (415) 642-6726.

Courant Institute of Mathematical Sciences. New York University, 251 Mercer Street, New York, NY 10012. (212) 998-3000.

Institute For Mathematics and Its Applications. University of Minnesota, 514 Vincent Hall, 206 Church Street, S.E., Minneapolis, MN 55455. (612) 624-6066. FAX (612) 626-7370.

Institute of Applied Mathematics. University of Missouri- Rolla, Rolla, MO 65401. (341) 341-4641.

Laboratory For Research in Statistics and Probability. Carleton University, Dunton Tower, Room 611, Ottawa, ON, Canada K1S 5B6. (613) 788-2167.

Mathematical Sciences Institute. Cornell University, 201 Caldwell Hall, Ithaca, NY 14853. (607) 255-8005.

Mathematical Statistics Program. University of Maryland, 1107 Mathematics Building, College Park, MD 20742. (301) 405-5061. FAX (301) 314-0827.

Statistical Laboratory. Kansas State University. Dickens Hall, Manhattan, KS 66506-0802. (913) 532-6882. (913) 532-7736.

MARS

See also: ASTROGEOLOGY, ASTEROIDS, ASTRONOMY, COMETS, METEORITES, NAMES of INDIVIDUAL PLANETS, PLANETARY SCIENCES, SATELLITES (NATURAL), SEISMOLOGY, SOLAR SYSTEM, TEKTITES, VOLCANOLOGY

ABSTRACT SERVICES AND INDEXES

Meteorological and Geoastrophysical Abstracts. Applied Science and Technology Index; a cumulative subject index to English language periodicals in the fields of aeronautics and space science, computer technology, chemistry, construction industry, energy and related areas. H. W. Wilson Co., 950 University Avenue, Bronx, NY 10452- 9978. (212) 588-8400. (800) 367-6770. FAX (718) 590-1617.From 1958 to present. Monthly. Inquire about cost and availability. Also available online (BRS and WILSONLINE) and on CD-ROM. ISSN: 0003-6986.

Astronomy and Astrophysics Abstracts. Springer-Verlag New York, 175 Fifth Avenue, New York, NY 10010. (212) 460-1500. FAX (212) 473-6272. Published for the Astronomisches Rechen-Institut. Comprehensive coverage of all aspects of astronomy, astrophysics and related fields. 1969 to present. Two parts per year. Annual. ISSN: 0067-0022.

Bibliography and Index of Geology. American Geological Institute, 4220 King Street, Alexandria, VA 22302-1507. (703) 379-2480. FAX (703) 379-7563. From 1969 to present. Monthly. $1295.00 per year. ISSN: 0098-2784. Also available as GEOREF online (CISTI, DIALOG, Orbit, STN) and on CD-ROM. Inquire about price and availability.

Chemical Abstracts. Chemical abstracts Service. 2540 Oleantangy River Road, Box 3012, Columbus, OH 43210-0012. (614) 447-3600. FAX (614) 447-3713. 1907 to present. Weekly. $16,800.00 per year. Available online and on CD-ROM. CA is also available in five section groupings. Inquire regarding cost and availability.

Current Contents: Physical, Chemical, and Earth Sciences. Institute for Scientific Information, 3501 Market Street, Philadelphia, PA 19104. (215) 386-0100. (800) 523-1850. FAX (215) 386-2291. 1961 to present. Weekly. $442.00 per year. Also available online (BRS, DIALOG) and on CD-ROM. Inquire about price and availability.

General Science Index. H.W. Wilson Company, 950 University Avenue, Bronx, NY 10452. (212) 588-8400. (800) 367-6770. FAX (718) 590-1617. From 1978 to present. Ten issues per year; quarterly and annual cumulations. Service basis. Available on CD-ROM and online. Inquire regarding cost and availability. ISSN: 0162-1963.

Meteorological and Geoastrophysical Abstracts. American Meteorological Society, c/o Inforonics, Inc., 550 Newtown Road, Box 458, Littleton, MA 01460. (508) 486-8976. FAX (508) 486-0027. Covers literature in environmental sciences, meteorology, astrophysics, hydrology, glaciology and physical oceanography. 1950 to present. Monthly. $950.00 per year. Also available online (DIALOG) and on CD-ROM. ISSN: 0026-1130.

NTIS Alerts: Astronomy & Astrophysics. U.S. National Technical Information Service. 5285 Port Royal Road, Springfield, VA 22161. (703) 487-4650. FAX (703) 321-8547. Weekly. $140.00 per year.

Physics Abstracts. INSPEC. Section A, Science Abstracts. Institute of Electrical Engineers, London, United Kingdom. Available from: INSPEC/IEEE-Institute of Electrical and Electronic Engineers, Box 1331, Hoes Lane, Piscataway, NJ 08855-1331. (908) 562-5549. 1898 to present. 24 issues per year. $2835.00 per year. ISSN: 0036-8091. Also available online and on CD-ROM.

Science Citation Index. SCI. Institute for Scientific Information, 3501 Market Street, Philadelphia, PA 19104. (215) 386-0100. (800) 523-1850. FAX (215) 386-2291. 1961 to present. Six issues per year, plus annual cumulation. $11650.00 per year. Also available online and on CD-ROM. Inquire about price and availability. ISSN: 0036-827X.

STAR. Scientific and Technical Aerospace Reports. U.S. National Aeronautics and Space Administration. Distributed by U. S. Superintendent of Documents, Washington, DC 20402. 1963 to present. Semi-monthly, with semiannual and annual indexes. $114.00 per year. ISSN: 0036-8741. Also available online and on CD-ROM.

ANNUAL REVIEWS AND YEARBOOKS

The Astronomical Almanac; Data For Astronomy, Space Sciences, Geodesy, Surveying, Navigation and Other Applications. Superintendent of Documents, U.S. Government Printing Office, Washington, DC 10402. (202) 783-3238.1981 to present. Supersedes Astronomical Ephemeris and the American Ephermeris and Nautical Almanac. Annual. $44.95. ISSN: 0737-6421.

Annual Review of Astronomy and Astrophysics. Original reviews of critical literature and current developments in astronomy and astrophysics. Annual Reviews, Inc., 4139 El Camino Way, Palo Alto, CA 94303-0139. (415) 493-4400. (800) 523-8635. Fax (415) 855-9815. 1963 to date. Annual. $60.00. ISSN: 0066-4146.

Annual Review of Earth and Planetary Sciences. Annual Reviews, Inc., 4139 El Camino Way, Palo Alto, CA 94303-0139. (415) 493-4400. (800) 523-8635. Fax (415) 855-9815. 1973 to date. Annual. $62.00. ISSN: 0084-6597.

Proceedings of the Lunar and Planetary Science Conference. Lunar and Planetary Institute, 3600 Bay Area Boulevard, Houston, TX 77058. (713) 486-2143. 1970 to date. Annual.

ASSOCIATIONS AND PROFESSIONAL SOCIETIES

American Astronomical Society. 2000 Florida Avenue NW, Suite 400, Washington, DC 20009. (202) 328-2010. FAX (202) 234-2560.

American Geophysical Union. 2000 Florida Avenue NW, Washington, DC 20009. (202) 462-6900. (800) 966-AGU1. FAX (202) 328-0566.

American Institute of Physics. One Physics Ellipse, College Park, MD 20740-3843. (301) 209-3100.

Astronomical League. 2112 Kingfisher Lane East, Rolling Meadows, IL 60008. (708) 398-0562.

Association of Universities For Research in Astronomy, Inc. (AURA). Suite 701, 1625 Massachusetts Avenue, NW, Washington, DC 20036. (202) 483-2101. FAX (202) 483-2106.

Astronomical Society of the Pacific. 390 Ashton Avenue, San Francisco, CA 94112. (415) 337-1100. FAX: (415) 337-5205.

Geological Society of America. 3300 Penrose Place, PO Box 9140. Boulder, CO 80301-9140. (303) 447-2020. FAX (303) 447-1133.

IEEE Geoscience and Remote Sensing Society. c/o Institute of Electrical and Electronics Engineers. 345 East 47th Street, New York, NY 10017. (212) 705-7900. FAX (212) 705-4929.

Meteoritical Society. University of Massachusetts, 125 Marston Hall, Amherst, MA 01003. (413) 545-0300. (413) 545-0724.

Planetary Society. 65 North Catalina Avenue, Pasadena, CA 91106. (818) 793-5100. (800) WOW-MARS. FAX (818) 793-5528.

MARS

Ency. of Physical Sciences and Engineering Info. Sources

BIBLIOGRAPHIES

A Bibliography of Astronomy, 1970-1979. R. A. Sea and S. S. Martin. Libraries Unlimited, Inc., Littleton, CO 80160. 1982. $37.50.

Scientific and Technical Books and Serials in Print; An Index To Literature in Science and Technology. R. R. Bowker. 121 Chanlon Road, New Providence, NJ 07974. (908) 464-6800. FAX (908) 665-3502. (800) 521-8110. 1972 to present. Annual. $299.95 per year. Also available on CD-ROM. ISSN: 0000-054X.

DIRECTORIES AND BIOGRAPHICAL SOURCES

American Astronomical Society. Membership Directory. 2000 Florida Avenue, NW, Suite 300. Washington, DC 20009. (202) 328-2010. FAX (202) 234-2560. 1980 to present. Monthly. Membership only.

American Men and Women of Science: Physical and Biological Sciences. R. R. Bowker Inc., 121 Chanlon Road, New Providence, NJ 07974. (908) 464-6800. (800) 521-8110. 20th edition. 8 volumes. 1996. $850.00.

The Astronomers. Donald Goldsmith. St. Martin's Press, Inc., 175 Fifth Avenue, New York, NY 10010. (212) 674-5151. (800) 221-7945. 1993. $14.95 in paper.

Astronomical Centers of the World. Kevin Krisciunas. Cambridge University Press, 40 West 20th Street, New York, NY 10011-4211. (212) 924-3900. (800) 872-7423. 1988. $34.95

Directory of Physics and Astronomy Staff. American Institute of Physics, One Physics Ellipse, College Park, MD 20740-3843. (301) 209-3100. 1975/76 to present. Annual. $60.00. ISSN: 0361-2228.

Earth and Astronomical Research Centers. Stockton Press, 345 Park Avenue South, New York, NY 10010. 4th edition. 1995. $515.00. ISBN: 1-56169-0967.

ENCYCLOPEDIAS AND DICTIONARIES

Concise Dictionary of Astronomy. Jacqueline Mitton. Oxford University Press, Inc., 200 Madison Avenue, New York, NY 10016. (212) 725-6000. (800) 334-4249. 1992. $24.95.

Encyclopedia of Astronomy and Astrophysics. Stephen Maran, editor. Van Nostrand Reinhold, 115 Fifth Avenue, New York, NY 10003. (212) 254-3232. (800) 842-3636. 1992. $129.95.

McGraw-Hill Encyclopedia of Science and Technology. McGraw-Hill Book Company, Inc., 1221 Avenue of the Americas, New York, NY 10020. (212) 512-2000. (800) 262-4729. 7th edition. 20 volume set. 1992. $1900.00.

Moons of the Solar System: An Illustrated Encyclopedia. John Stewart. McFarland & Company, Inc., Box 611, Jefferson, MC 28640. (910) 246-4460. (800) 253-2187. 1991. $49.95.

New Guide To the Planets. Patrick Moore. Trans-Atlantic Publications, Inc., 311 Bainbridge Street, Philadelphia, PA 19147. (215) 925-5083. 1993. $37.50.

Stars and Planets: The Sierra Club Guide To Sky Watching and Direction Finding. W. S. Kals. The Sierra Press, 4988 Gold Leaf Drive, Mariposa, CA 95338. (209) 966-5071. (800) 745-2631. $15.00.

GENERAL WORKS

Birth of the Earth. David E. Fisher. Columbia University Press, 562 West 113th Street, New York, NY 10025. (212) 666-1000. (800) 944-8648. 1988. $16.50 in paper.

The Case For Mars Ii. Christopher P. McKay. Univelt Inc., P.O. Box 28130, San Diego, CA 92198-0198. (619) 746-4005. AAS Science & Technology Series, volume 65. 1985. $30.00.

The Channels of Mars. Victor R. Baker. University of Texas Press, P.O. Box 7819, Austin, TX 78713-7819. (512) 471-7233. (800) 252-3206. 1982. $55.00.

Chemistry and Physics of the Terrestrial Planets. Surendra K. Saxwna. Springer-Verlag New York, Inc., 175 Fifth Avenue, New York, NY 10010. (212) 460-1500. (800) 777-4643. 1986. $109.00.

Evolution of the Earth and Planets. E. Takahashi, et al, editors. American Geophysical Union, 2000 Florida Avenue, N.W., Washington, DC 20009. (202) 462-6903. (800) 966-2481. 1993. $28.00.

Exploration of Venus & Mars Atmospheres. G. M. Keating, editor. Pergamon Press, Inc., Maxwell House, Fairview Park, Elmsford, NY 10523. (914) 592-7700. Fax (914) 592-3625. 1995. 94.00.

Exploring the Moon and Mars: Choices For the Nation. Diane Publishing Co., 600 Upland Avenue, Upland, PA 19015. (610) 499-7415. 1992. $29.95 in paper.

Exploring the Planets. W. Kenneth Hamblin and Eric H. Christiansen. Macmillan Publishing Company, Inc,. 200 Old Tappan Road, Old Tappan, NJ 07675. (800) 233-2336. 2nd edition. 1995. $62.00.

Geology of Mars. Thomas A. Mutch, et al. Princeton University Press, 41 William Street, Princeton, NJ 08540. (609) 258-4900. (800) 777-4726. 1976. $90.00.

The Geology of Multi-Ring Impact Basins: The Moon and Other Planets. P. D. Spudis. Cambridge University Press, 40 West 20th Street, New York, NY 10011-4211. (212) 924-3900. (800) 872-7423. Planetary Sciences Series. 1993. $59.95.

Geology of the Terrestrial Planets. Michael H. Carr. U.S. Government Printing Office, Superintendent of Documents, Washington, DC 20402-9325. (202) 783-3238. 1985. $16.00

An Introduction To Cosmochemistry. Charles R. Cowley. Cambridge University Press, 40 West 20th Street, New York, NY 10011-4211. (212) 924-3900. (800) 872-7423. 1995. $29.95 in paper.

Life on Mars. Patrick Moore and Francis Jackson. W. W. Norton & Company, 500 Fifth Avenue, New York, NY 10110. (212) 354-5500. (800) 223-2584. 1966. $4.50.

Mars and the Development of Life. H. Hanson. Prentice-Hall, 113 Sylvan Avenue, Route 9W, Englewood Cliffs, NJ 07632. (201) 592-2000. (800) 922-0579. 1991. $69.95.

Mars Beckons: The Mysterious, the Challenges, the Expectation of Our Next Great Adventure in Space. John N. Wilford. Alfred A. Knopf, Inc. 201 E. 50th Street, New York, NY 10022. (800) 733-3000. 1990. $24.95.

Mars: Past, Present & Future. E.B. Prichard, editor. AIAA, 370 L'Efant Promenade SW, Washington, DC 20024-2518. (202) 646-7400. (800) 682-2422. 1992. $69.95.

Meteorites and the Origin of Planets. John A. Wood. Books on Demand, 300 North Zeeb Road, Ann Arbor, MI 48106-1346. (313) 761-4700. (800) 521-0600. Reprint edition. $35.00 in paper.

Moons and Planets. William K. Hartmann. Wadsworth Publishing Co., 10 Davis Drive, Belmont, CA 94002. (415) 595-2350. (800) 354-9706. 3rd edition. 1993. $49.95.

Music of the Heavens: Kepler's Harmonic Astronomy. Bruce Stephenson. Princeton University Press. 41 William Street, Princeton, NJ 08540. (609) 258-4900. (800) 777-4726. 1994. $39.50.

The Near Planets. Time-Life Books Editors. Time-Life, Inc., 777 Duke Street, Alexandria, VA 22314. (703) 8388-7000. (800) 621-7026. Voyage through the Universe Series. 1992.

On Mars: Exploration of the Red Planet, 1958-1978. E and L. Ezell. NASA Special Publication SP-4212. U.S. Government Printing Office, Superintendent of Documents, Washington, DC 20402-9325. (202) 783-3238. 1984.

Photochemistry of the Atmosphere of Mars and Venus. V. A. Krasnopolosky. Springer-Verlag New York, Inc., 175 Fifth Avenue, New York, NY 10010. (212) 460-1500. (800) 777-4643. 1986.

Planetary Landscapes. Ronald Greeley. Chapman & Hall, 1 Penn Plaza, New York, NY 10119. (212) 564-1060. Div. of Routledge, 2nd edition. 1994. $49.95.

The Planetary System. Tobias Owen and David Morrison. Addison-Wesley Publishing Co., Inc., 1 Jacob Way, Reading, MA 01867. (617) 944-3700. (800) 447-2226. 1988. $49.50.

Planets and Their Atmospheres. John S. Lewis and Ronald G. Primm. Academic Press Inc., 6277 Sea Harbor Drive, Orlando, FL. (800) 321-5068. 1983. $65.00 in paper.

Resources of Near-Earth Space. John S. Lewis, et al. University of Arizona Press, 1230 North Park Avenue, Number 102, Tucson, AZ 85719. (520) 621-1441. (800) 426-3797. 1993. $75.00.

Return To the Red Planet. Eric Burgess. Columbia University Press, 562 West 113th Street, New York, NY 10025. (212) 666-1000. (800) 944-8648. 1993. $34.95.

Rings: Discoveries From Galileo To Voyager. James C. Elliot and Richard Kerr. MIT Press, 55 Hayward Street, Cambridge, MA 02142. (617) 253-8569. Reprint edition. 1987. $9.95 in paper.

Solar and Planetary Dynamos. M.R. Proctor, et al, editors. Cambridge University Press, 40 West 20th Street, New York, NY 10011-4211. (212) 924-3900. (800) 872-7423. Publications of the Newton Institute. 1994. $49.95.

Space Technology and Planetary Astronomy. Joseph N. Tatarewicz. Indiana University Pres, 601 North Morton Street, Bloomington, in 47404-3797. (812) 855-6804. (800) 842-6796. 1990. $29.95.

The Story of the Red Planet. P. Cattermole. Chapman & Hall, 1 Penn Plaza, New York, NY 10119. (212) 564-1060. Div. of Routledge. 1992. $35.00.

Terraforming: Engineering Planetary Environments. Martyn J. Fogg. Society of Automotive Engineers (SAE), 400 Commonwealth Drive, Warrendale, PA 15096. (412) 776-4841. 1995. $49.00.

Venus and Mars: Atmospheres, Ionospheres, and Solar Wind Interactions. J.G. Luhman, et al editors. American Geophysical Union, 2000 Florida Avenue NW, Washington, DC 20009. (202) 462-6900. (800) 966-2481. 1992. $59.00.

Worlds Apart: A Textbook in Planetary Sciences. Guy Consolmagno. Prentice-Hall, 113 Sylvan Avenue, Route 9W, Englewood Cliffs, NJ 07632. (201) 592-2000. (800) 922-0579. 1994. $62.00.

HANDBOOKS AND MANUALS

Apparent Places of Fundamental Stars. Astronomisches Rechen-Institut. Verlag G. Braun, Karl-Friedrich-Stresse, 76122 Karlsruhe, Germany. Annual. ISSN: 0174-254X.

Astrophysical Data: Planets and Stars. Kenneth R. Lanb. Springer-Verlag New York, Inc., 175 Fifth Avenue, New York, NY 10010. (212) 460-1500. (800) 777-4643. 1993. $59.00.

Field Guide to the Stars and Planets. Jay M. Pasachoff and Donald H. Menzel. Houghton Mifflin Co., 222 Berkeley Street, Boston, MA 02116. (617) 351-5000. (800) 225-3362. Revised edition. 1992. $24.95.

The Mars Reference Atmospheres. A. J. Kliore, editor. Elsevier Science Publishing Company, Inc., 655 Avenue of the Americas, New York, NY 10010. (212) 989-5800. Volume 2, Part 2. 1982. $43.00.

New Guide To the Planets. Patrick Moore. Trans-Atlantic Publications, Inc., 311 Bainbridge Street, Philadelphia, PA 19147. (215) 925-5083. 1993. $37.50.

Planet Observer's Handbook. Fred W. Price. Cambridge University Press, 40 West 20th Street, New York, NY 10011-4211. (212) 924-3900. (800) 872-7423. 1994. $34.95.

Planets and their Moons. Audubon Society Staff. Alfred A. Knopf, Inc. 201 E. 50th Street, New York, NY 10022. (800) 733-3000. 1995. $7.99.

ONLINE DATABASES AND CD-ROMS

CA SEARCH. Chemical Abstracts Service, P.O. Box 3012, Columbus, OH 43210-0012. (614) 447-3600. (800) 848-6533. FAX (614) 447-3709. Very comprehensive guide to worldwide chemical literature and related fields, 1972 to present. Available on BRS,(800) 289-4277, DIALOG, (800) 334-2564, ORBIT (800) 456-7248, and STN International, FIZ Karlsruhe, P.O. Box 2465, W-7500, Karlsruhe 1, Germany, online services. Inquire as to cost and availability.

Dissertation Abstracts. University Microfilms International, 300 North Zeeb Road, Ann Arbor, MI 48106. (800) 521-0600 or (313) 761-4700. Scope includes virtually all doctoral dissertations accepted at accredited American institutions from 1861 to present in 252 subject areas. Available on BRS, (800) 289-4277, DIALOG, (800) 334-2564, and OCLC EPIC, (800) 848-5878, online services. Also available on CD-ROM. Inquire as to cost and availability.

GEOREF. American Geological Institute, 4220 King Street, Alexandria, VA 22302. (800) 336-4764 or (703) 379-2480. Geology and geosciences literature, 1785 to present for North America. Available on DIALOG,(800) 334-2564, ORBIT (800) 456-7248, online services. Also available on CD-ROM. Inquire as to cost and availability.

INSPEC. Institution of Electrical Engineers, Michael Faraday House, Six Hills Way, Stevenage, Herts. SG1 2AY, England. Telephone: 0438 313311 or FAX 0438 742840. Contains citations to the worldwide literature of physics, electronics and electrical engineering, computer technology, and related fields. Available on BRS, (800) 289-4277, DIALOG, (800) 334-2564, ORBIT, (800) 456-7248, and STN International, FIZ Karlsruhe, P.O. Box 2465, W-7500, Karlsruhe 1, Germany, online services. Inquire as to cost and availability.

NTIS Bibliographic Database. National Technical Information Service, 5285 Port Royal Road, Springfield, VA 22161. (703) 487-4929 or FAX (703) 321-8199. Broad coverage of government-sponsored science and technology research reports, 1964 to present. Available on BRS,(800) 289-4277, DIALOG, (800) 334-2564, ORBIT, (800) 456-7248, and STN International, FIZ Karlsruhe, P.O. Box 2465, W-7500, Karlsruhe 1, Germany, online services. Also available on CD-ROM. Inquire as to cost and availability.

Scisearch. Institute for Scientific Information, 3501 Market Street, Philadelphia, PA 19104. (800) 523-1850 or (215) 386-0100. Broad multidisciplinary title and author index to the international literature of science and technology, 1974 to present. Available on DIALOG, (800) 334-2564, and ORBIT, (800) 456-7248, online services. Also available on CD-ROM. Inquire as to cost and availability.

Wilsonline. H.W. Wilson Company, 950 University Avenue, Bronx, NY 10452. (800) 367-6770 or (212) 588-8400. Makes available online versions of the printed H.W. Wilson indexes, including Applied Science and Technology Index, Business Periodicals Index, General Science Index, and Readers' Guide to Periodical Literature. Period covered is generally 1983 to present. Available on BRS, (800) 289-4277, DIALOG, (800) 334-2564, and OCLC EPIC, (800) 848-5878, online services. Also available on CD-ROM. Inquire as to cost and availability.

OTHER SOURCES

Atlas of the Planets. Vincent DeCallatay and Audouin Dollfus. Books on Demand, 300 North Zeeb Road, Ann Arbor, MI 48106-1346. (313) 761-4700. (800) 521-0600. $45.60.

Atlas of the Solar System. B. Yenne. Simon & Schuster, Inc. 1230 Avenue of the Americas, New York, NY 10020. (212) 698-7000. (800) 223-2348. 1987. $12.98.

Martian Enigmas: A Photographic Album of the Investigation of the Face And Other Unusual Objects On the Surface of Mars. Mark Cariotto. North Atlantic Books, P.O. Box 12327, Berkeley, CA 94712. (510) 559-8277. 1991. $29.95.

Planetary, Lunar and Solar Positions 601 B.C. TO A.D. One at Five-Day and Ten-Day Intervals. Bryant Tuckerman. American Philosophical Society, 104 South 5th Street, Philadelphia, PA 19106-3387. (215) 440-3400. FAX (215) 440-3436. Memoirs Series, volume 56. 1979. $20.00.

Planetary, Lunar and Solar Positions, A.D. 2 TO A.D. 1649 At Five-Day and Ten-Day Intervals. Bryant Tuckerman. American Philosophical Society, 104 South 5th Street, Philadelphia, PA 19106-3387. (215) 440-3400. FAX (215)440-3436. Memoirs Series, volume 59. 1964. $30.00.

Planetary, Lunar and Solar Positions, 1650-1805. Owen Gingerich and Barbara L. Welther. American Philosophical Society, 104 South 5th Street, Philadelphia, PA 19106-3387. (215) 440-3400. FAX (215)440-3436. Memoirs Series, volume 59S. 1983. $20.00.

The Sky: A User's Guide. David H. Levy. Cambridge University Press, 40 West 20th Street, New York, NY 10011-4211. (212) 924-3900. (800) 872-7423. 1993. $14.95.

The View from Space: Photographic Exploration of the Planets. Merton Davies and Bruce C. Murray. Columbia University Press, 562 West 113th Street, New York, NY 10025. (212) 666-1000. (800) 944-8648. 1973. $17.50.

PERIODICALS

Association of Lunar and Planetary Observers. Journal. Association of Lunar and Planetary Observers, Box 16131, San Francisco, CA 94116. (415) 566-5786. FAX (415) 731-8242. 1947 to present. Quarterly. $14.00 per year. ISSN: 0039-2502.

Astronomical Journal. American Institute of Physics, One Physics Ellipse, College Park, MD 20740-3843. (301) 209-3000. Published for the American Astronomical Society. 1849 to present. Monthly. $280.00 per year. ISSN: 0004-6256.

Astronomical Society of the Pacific. Publications. Astronomical Society of the Pacific, 390 Ashton Avenue, San Francisco, CA 94112. (415) 337-1100. FAX (415) 337-5205. 1889 to present. Monthly. $175.00 per year. ISSN: 0004-6280.

Astronomy. Kalmbach Publishing Company, Box 1612, Waukesha, WI 53187-1612. (414) 796-8776. FAX (414) 796-1142. 1973 to present. Monthly. $27.00 per year. ISSN: 0091-6358.

Astronomy and Astrophysics. Springer-Verlag New York, Inc., 44 Hartz Way, Secaucus, NJ 07096-2491. (201) 348-4033. FAX (201) 348-4505. 1969 to date. 36 issues per year in 12 volumes. $1733.00 per year. ISSN: 0004-6361.

Ciel Et Terre. Societe Royale Belge d'Astronomie, de Meteorologie et de Physique du Globe, 3 Avenue Circulaire, Uccle, Brussels, Belgium. FAX 32-2-374-9822. 1880 to present. Bimonthly. 1600 BEF per year. ISSN: 0009-6709.

Earth, Moon and Planets; An International Journal of Comparative Planetology. Kluwer Academic Publishers, Box 358, Accord Station, Hingham, MA 02018-0358. (617) 871-6600. FAX (617) 871-6528. 1969 to present. Monthly. $840.00 per year. ISSN: 0167-9295.

Geochemica Et Cosmochimica Acta. Elsevier Science. 660 White Plains Road, Tarrytown, NY 10591-5153. (914) 524-9200. FAX (914) 333-2444. 1950 to the present. Biweekly. $895.00 per year. ISSN: 0016-7037.

Icarus; International Journal of Solar System Studies. American Astronomical Society. Academic Press, Inc., Journal Division, 525 B Street, Suite 1900, San Diego, CA 92101-4495. (619) 230-1840. FAX (619) 699-6800. (800) 543-9534. 1962 to present. Monthly. $1080.00. ISSN: 0019-1035.

JGR: Journal of Geophysical Research: Planets. American Geophysical Union, 2000 Florida Avenue, NW, Washington, CD 20009. (202) 462-6900. FAX (202) 328-0566. 1991 to present. Monthly. $597.00 per year. ISSN: 0148-0227.

Lunar and Planetary Information Bulletin. Lunar and Planetary Institute. 3600 Bay Area Boulevard, Houston, TX 77058-1113. (713) 486-2175. FAX (713) 486-2125. 1970 to present. Quarterly. Free. Also available online.

Mercury. Astronomical Society of the Pacific. 390 Ashton Avenue, San Francisco, CA 94112. (415) 337-1100. FAX: (415) 337-5205. 1972 to present. Bimonthly. $175.00 per year. ISSN: 0047-6773.

Meteoritics. Meteoritical Society, Department of Chemistry, University of Arkansas, Fayetteville, AR 72701. (501) 575-7625. FAX (501) 575-7778. 1953 to present. Bimonthly. $210.00 per year. ISSN: 0026-1114.

Observatory. c/o Dr. D.J. Stickland Space and Astrophysics Division, Rutherford Appleton Laboratory, Chilton, Didcot, Oxon OX11 OQX England. FAX 0235-445848. 1877 to present. Bimonthly. $42.00 per year. ISSN: 0029-7704.

Planetary and Space Science. Elsevier Science Publishing Company, Inc., 660 White Plains Road, Terrytown, NY 10591-5153. (914) 524-9200. FAX: (914) 333-2444. 1959 to present. Monthly. $1355.00 per year. ISSN: 0032-0633.

Planetary Report. Planetary Society. 65 North Catalina, Pasadena, CA 91106-2301. (818) 793-5100. FAX (818) 793-5528. 1980 to present. $25.00 per year. ISSN: 0736-3680.

Solar System Research. (English translation of Astronomicheskii Vestink) Plenum Publishing Corp., Consultants Bureau, 233 Spring Street, New York, NY 10013-1578. (212) 620-8468. FAX (212) 463-0742. 1967 to present. $1080.00. ISSN: 0038-0946.

Sky and Telescope. Sky Publishing Corporation, Box 9111, Belmont, MA 02178. (617) 864-7360. FAX (617) 864-6117. 1941 to present. Monthly. $27.00 per year. ISSN: 0037-6604.

RESEARCH CENTERS AND INSTITUTES

Center For Radiophysics and Space Research. Cornell University, Space Sciences Building, Ithaca NY 14853. (607) 255-4341. FAX (607) 255-9888.

Center For Space Science and Astrophysics. Stanford University. 325 Durand Building, Stanford, CA 04305. (415) 723-3582.

Center For Space Sciences. University of Texas at Dallas, P. O. Box 830688, MS-F022, Richardson, TX 75083-0688. (214) 690-2851. FAX (214) 690-2848.

Earth and Planetary Remote Sensing Laboratory. Washington University, Department of Earth and Planetary Sciences. Campus Box 1169, 1 Brookings Drive, St. Louis, MO 63130. (314) 889-5679. FAX (314) 889-5799.

Institute For Astronomy. University of Hawaii at Manoa, 2680 Woodlawn Drive, Honolulu, HI 96822. (808) 956-8312. FAX (808) 988-2790.

Institute For Terrestrial and Planetary Atmospheres. State University of New York at Stony Brook, Stony Brook, NY 11794-2300. (516) 632-6170. FAX (516) 632-6251.

Institute of Geophysics and Planetary Physics. University of California, Riverside, CA 92521. (714) 787-4503. FAX (714) 787-4529.

Laboratory For Atmospheric and Space Physics. University of Colorado-Boulder, Boulder, CO 80309-0392. (303) 492-7677. (303) 492-6946.

Laboratory For Planetary Geology. Arizona State University, Department of Geology, Tempe, AZ 85281. (601) 965-7029.

Laboratory For Planetary Studies. Cornell University, 302 Space Sciences Building, Ithaca, NY 14853. (607) 255-4971.

Lunar and Planetary Institute. 3303 NASA Road One, Houston, TX 77058-4399. (713) 486-2139. FAX 713-496-2162.

Lunar and Planetary Laboratory. University of Arizona, Tucson, AZ 85721. (602) 621-6963. FAX (602) 621-4933.

McDonnell Center For the Space Sciences. Washington University, Box 1105, One Brookings Drive, St. Louis MO 63130-4899. (314) 889-6255. FAX (314) 889-6219.

MASER

See: LASERS

MASKING

See: SEMICONDUCTORS

MASS SPECTROMETRY

Ency. of Physical Sciences and Engineering Info. Sources

MASS SPECTROMETRY

See also: ANALYTICAL CHEMISTRY, MASS SPECTROMETRY, NUCLEAR MAGNETIC RESONANCE, SPECTROSCOPY

ABSTRACT SERVICES AND INDEXES

Applied Science and Technology Index; A Cumulative Subject Index to English Language Periodicals in the Fields of Aeronautics and Space Science, Computer Technology, Chemistry, Construction Industry, Energy and Related Areas. H.W. Wilson Co., 950 University Avenue, Bronx, NY 10452. (212) 588-8400. (800) 367-6770. FAX (718) 590-1617. From 1958 to present. Monthly. Inquire about cost and availability. Also available on CD-ROM and online. ISSN: 0003-6986.

Chemical Abstracts. Chemical Abstracts Service. 2540 Olentangy River Road, Box 3012, Columbus, OH 43210-0012. (614) 447-3600. FAX (614) 447-3713. 1907 to present. Weekly. $16,800.00 per year. Available online and on CD-ROM. CA is also available in five section groupings. Inquire regarding cost and availability.

Current Contents: Physical, Chemical and Earth Sciences. Institute for Scientific Information, 3501 Market Street, Philadelphia, PA 19104. (215) 386-0100. FAX (215) 386-2291. From 1961 to present. Weekly. $442.00 per year. Also available on CD-ROM and online. Inquire regarding cost and availability. ISSN: 0163-2574.

Engineering Index Monthly; Indexes and Abstracts the World's Engineering and Technical Literature. Engineering Information, Inc., Castle Point on the Hudson, Hoboken, NJ 07030. (201) 216-8500. (800) 221-1044. FAX (201) 216-8532. Monthly. $2300.00 per year. Available online as COMPENDEX and also on CD-ROM. ISSN: 0742-1974.

General Science Index. H.W. Wilson Company, 950 University Avenue, Bronx, NY 10452. (212) 588-8400. (800) 367-6770. FAX (718) 590-1617. From 1978 to present. Ten issues per year; quarterly and annual cumulations. Service basis. Available on CD-ROM and online. Inquire regarding cost and availability. ISSN: 0162-1963.

Mass Spectrometry Bulletin. Royal Society of Chemistry. Available from: Turpin Distribution Services, Ltd., Blackhorse Road, Letchworth, Herts SG6 1HN, England. 1966 to present. Monthly. $910.00 per year. ISSN: 0025-4738.

Physics Abstracts. INSPEC. Section A, Science Abstracts. Institution of Electrical Engineers (IEE), London. Available from: INSPEC/IEEE-Institute of Electrical and Electronic Engineers, Box 1331, Hoes Lane, Piscataway, NJ 08855-1331. (908) 562-5549. 1898 to present. 24 issues per year. $2835.00 per year. Also available online and on CD-ROM. ISSN: 0036-8091.

Physics Briefs (Physikalische Berichte). Information Center for Energy, Physics, Mathematics; German Physical Society. VCH Publishers, Inc., 220 East 23rd Street, New York, NY 10010-4606. (212) 683-8333. 1845 to present. 24 issues per year. $2390.00 per year. Also available online. ISSN: 0179-7434.

Science Citation Index. SCI. Institute for Scientific Information, 3501 Market Street, Philadelphia, PA 19104.(215) 386-0100. (800) 523-1850. FAX (215) 386-2991. 1961 to present. Six issues per year, plus annual cumulation. $11650.00 per year. Also available online and on CD-ROM. Inquire about price and availability. ISSN: 0036-827X.

ASSOCIATIONS AND PROFESSIONAL SOCIETIES

American Chemical Society. 1155 16th Street NW, Washington, DC 20036. (202) 872-4600.

American Society For Mass Spectroscopy. 1201 Don Diego, Santa Fe, NM 87505. (505) 989-4517.

Federation of Analytical Chemistry and Spectroscopy Societies. c/o Jo Ann Brown, 201-B Broadway Street, Frederick, MD 21701. (301) 846-4797. FAX (301) 694-6860.

Society For Applied Spectroscopy. 201B Broadway Street, Frederick, MD 21701. (301) 694-8122. FAX (301) 694-6860.

DIRECTORIES AND BIOGRAPHICAL SOURCES

American Men and Women of Science: Physical and Biological Sciences. R. R. Bowker Inc., 121 Chanlon Road, New Providence, NJ 07974. (908) 464-6800. (800) 521-8110. 20th edition. 8 volumes. 1996. $850.00.

Engineering Research Centers: Incorporating Electronics Research Centers. Stockton Press, 345 Park Avenue, New York, NY 10010. (212) 689-9200. (800) 221-2123. 4th edition. 1995. $515.00.

International Engineering Directory. American Consulting Engineers Council, 1015 15th Street, N.W. Suite 802, Washington, DC 20005-2670. (202) 347-7474. Annual. $10.00.

Research Centers Directory. Gale Research Company Inc., 835 Penobscot Building, Detroit, MI 48226-4094. (313) 961-2242. (800) 347-4253. 20th edition. 1995. $485.00. ISSN: 0080-1518.

Scientific and Technical Organizations and Agencies Directory. Gale Research, 835 Penobscot Building, Detroit, MI 48226-4094.(313) 961-2242. (800) 347-4253. 4th edition. 1996. $195.00.

Society For Applied Spectroscopy Membership Directory. Society for Applied Spectroscopy, 201B Broadway Street, Frederick, MD 21701. (301) 694-8122. FAX (301) 694-6860. Irregular.

ENCYCLOPEDIAS AND DICTIONARIES

Chambers Nuclear Energy and Radiation Dictionary. Peter M.B. Walker, editor. Cambridge University Press, 40 West 20th Street, New York, NY 10011-4211. (212) 924-3900. 1992. $40.00.

Encyclopedia of Applied Physics. George Trigg, editor. VCH Publications, Inc., 220 East 23rd Street, Suite 909, New York, NY 10010-4606. (212) 683-8333. (800) 422-8824. 20 volume set. 1991- . $5990.00.

McGraw-Hill Encyclopedia of Science and Technology. McGraw-Hill Book, Incorporated, 1221 Avenue of the Americas, New York, NY 10020. (212) 997-3675. (800) 262-4729. Seventh edition. Twenty volumes. 1992. $1900.00.

GENERAL WORKS

Introduction To Mass Spectrometry. Hilson C. Hill. Books on Demand, 300 North Zeeb Road, Ann Arbor, MI 48106-1346. (313) 761-4700. (800) 521-0600. Revised edition. $36.50.

Lasers and Mass Spectrometry. David M. Lubman, editor. Oxford University Press, Inc., 200 Madison Avenue, New York, NY 10016. (212) 725-6000. (800) 334-4249. 1990. $75.00.

Mass Spectrometry. H.E. Duckworth, et al. Cambridge University Press, 40 West 20th Street, New York, NY 10011-4211. (212) 924-3900. (800) 872-7423. 1990. $34.95.

Practical Applications of Ion Trap Mass Spectrometry. Raymond E. March and John F. Todd. CRC Press, Inc., 2000 Corporate Boulevard, NW, Boca Raton, FL 33431. (407) 994-0555. (800) 272-7737. 1995. $89.95.

Practical Organic Mass Spectrometry. J. R. Chapman. John Wiley & Sons, Inc., 605 Third Avenue, New York, NY 10158-0012. (212) 850-6000. (800)225-5945. 2nd edition. 1993. $54.95.

HANDBOOKS AND MANUALS

CRC Handbook of Chemistry and Physics. David R. Kide, editor. CRC Press Inc., 2000 Corporate Blvd., NW, Boca Raton, FL 33431. (407) 994-0555. (800) 333-8300. 77th edition. 1996. $99.95.

Handbook of Mass Spectra of Environmental Contaminants. Ronald A. Hites. Lewis Pubs. Inc., 2000 Corporate Boulevard, NW, Boca Raton, FL 33431. (407) 994-0555. (800) 272-7737. 2nd edition. 1992. $89.95.

Handbook of Nuclear Spectrometry. Juhnai Kantele. Academic Press, Inc., 6277 Sea Harbor Drive, Orlando, FL. (800) 321-5068. 1995. $79.95.

Practical Handbook of Spectroscopy. James W. Robinson. CRC Press, Inc., 2000 Corporate Boulevard, NW, Boca Raton, FL 33431. (407) 994-0555. (800) 272-7737. 1991. $96.50.

ONLINE DATABASES AND CD-ROMS

CA Search. Chemical Abstracts Service, P.O. Box 3012, Columbus, OH 43210-0012. (614) 447-3600. (800) 848-6533. FAX (614) 447-3709. Very comprehensive guide to worldwide chemical literature and related fields, 1972 to present. Available on BRS,(800) 289-4277, DIALOG, (800) 334-2564, ORBIT (800) 456-7248, and STN International, FIZ Karlsruhe, P.O. Box 2465, W-7500, Karlsruhe 1, Germany, online services. Inquire as to cost and availability.

Current Contents Search. Institute for Scientific Information, 3501 Market Street, Philadelphia, PA 19104. (215) 386-0100. FAX (215) 386-6362. Contains citations to articles listed in the table of contents of science and technology journals. Also articles in social sciences and life sciences journals. Available on BRS,(800) 289-4277, DIALOG,(800) 334-2564, online services. Inquire as to cost and availability.

Dissertation Abstracts Online. University Microfilms International, 300 North Zeeb Road, Ann Arbor, MI 48106. (800) 521-0600 or (313) 761-4700. Scope includes virtually all doctoral dissertations accepted at accredited American institutions from 1861 to present in 252 subject areas. Available on BRS, (800) 289-4277, DIALOG, (800) 334-2564, and OCLC EPIC,(800) 848-5878, online services. Also available on CD-ROM. Inquire as to cost and availability.

INSPEC. Institution of Electrical Engineers, Michael Faraday House, Six Hills Way, Stevenage, Herts. SG1 2AY, England. Telephone: 0438 313311 or FAX 0438 742840. Contains citations to the worldwide literature of physics, electronics and electrical engineering, computer technology, and related fields. Available on BRS, (800) 289-4277, DIALOG, (800) 334-2564, ORBIT, (800) 456-7248, and STN International, FIZ Karlsruhe, P.O. Box 2465, W-7500, Karlsruhe 1, Germany, online services. Inquire as to cost and availability.

NTIS Bibliographic Database. National Technical Information Service, 5285 Port Royal Road, Springfield, VA 22161. (703) 487-4929 or FAX (703) 321-8199. Broad coverage of government-sponsored science and technology research reports, 1964 to present. Available on BRS,(800) 289-4277, DIALOG, (800) 334-2564, ORBIT, (800) 456-7248, and STN International, FIZ Karlsruhe, P.O. Box 2465, W-7500, Karlsruhe 1, Germany, online services. Also available on CD-ROM. Inquire as to cost and availability.

Physics Briefs. American Institute of Physics, 335 East 45th Street, New York, NY 10017. (212) 661-9260 or FAX (212) 661-2036. Contains citations with abstracts of the literature of physics and related fields, 1979 to present. Available on the STN International, FIZ Karlsruhe, P.O. Box 2465, W-7500, Karlsruhe 1, Germany, online service. Inquire as to cost and availability.

Scisearch. Institute for Scientific Information, 3501 Market Street, Philadelphia, PA 19104. (800) 523-1850 or (215) 386-0100. Broad multidisciplinary title and author index to the international literature of science and technology, 1974 to present. Available on DIALOG,(800) 334-2564, and ORBIT,(800) 456-7248, online services. Also available on CD-ROM. Inquire as to cost and availability.

Wilsonline. H.W. Wilson Company, 950 University Avenue, Bronx, NY 10452. (800) 367-6770 or (212) 588-8400. Makes available online versions of the printed H.W. Wilson indexes, including Applied Science and Technology Index, Business Periodicals Index, General Science Index, and Readers' Guide to Periodical Literature. Period covered is generally 1983 to present. Available on BRS,(800) 289-4277, DIALOG,(800) 334-2564, and OCLC EPIC,(800) 848-5878, online services. Also available on CD-ROM. Inquire as to cost and availability.

PERIODICALS

Applied Spectroscopy. Society for Applied Spectroscopy, Thomas Johnson Drive, Suite 2, Frederick, MD 21702. (301) 694-8122. 1946 to present. Monthly. $180.00 per year. ISSN: 0003-7028.

Applied Spectroscopy Reviews. Marcel Dekker Journals, 270 Madison Avenue, New York, NY 10016. (212) 696-9000. FAX (212) 685-4540. 1964 to present. Quarterly. $395.00 per year. ISSN: 0570-4928.

International Journal of Mass Spectrometry and Ion Processes. Elsevier Science Inc., Publishing Co., Inc., Box 882, Madison Square Station, New York, NY 10159. (212) 989-5800. (212) 633-3990. 1968 to present. 36 issues per year in 12 volumes. $2504.00 per year. ISSN: 0168-1176.

MASS SPECTROMETRY

Ency. of Physical Sciences and Engineering Info. Sources

Journal of Electron Spectroscopy and Related Phenomena. Elsevier Science Inc., Publishing Co., Inc., Box 882, Madison Square Station, New York, NY 10159. (212) 989-5800. (212) 633-3990. 1972 to present. 20 issues per year. $1030.00 per year. ISSN: 0368-2048.

Journal of Quantitative Spectroscopy and Radiative Transfer. Elsevier Science Inc., Publishing Co., Inc., Box 882, Madison Square Station, New York, NY 10159. (212) 989-5800. (212) 633-3990. 1961 to present. Monthly. $1815.00 per year. ISSN: 0022-4073.

Mass Spectrometry. Royal Society of Chemistry. Available from: Turpin Distribution Services, Ltd., Blackhorse Road, Letchworth, Herts SG6 1HN, England. 1971 to present. Biennial. price varies. ISSN: 0305-9987.

Mass Spectrometry Reviews. John Wiley and Sons, Inc., 605 Third Avenue, New York, NY 10158. (800) 526-5368 or (212) 850-6000. 1982 to present. Bi-monthly. $330.00 per year. ISSN: 0277-7037.

Spectroscopy Letters. Marcel Dekker Journals, 270 Madison Avenue, New York, NY 10016. (212) 696-9000. FAX (212) 685-4540. 1985 to present. 9 issues per year. $59.00 per year. ISSN: 0887-6703.

RESEARCH CENTERS AND INSTITUTES

Chromatography-Mass Spectrometry Facility. University of Missouri-Columbia, Room 4, Agriculture Building, Columbia, MO 65211. (314) 882-2608.

Mass Spectrometry Laboratory. University of Illinois, School of Chemical Sciences, 505 South Matthews, Urbana, IL 61801. (217) 333-2545.

Midwest Center For Mass Spectroscopy. University of Nebraska-Lincoln, Room 18, Hamilton Hall, Department of Chemistry, Lincoln NE 68588. (402) 472-3507. FAX (402) 472-3109.

NIH/MSU Mass Spectrometry Facility. Michigan State University, Department of Biochemistry, East Lansing, MI 48824. (517) 353-0612. FAX (517) 353-9334.

Nuclear Structure Research Laboratory. University of Rochester, Rochester, NY 14627. (716) 275-4934. FAX (716) 275-8527.

MATERIALS HANDLING

See also: INDUSTRIAL ENGINEERING

ABSTRACT SERVICES AND INDEXES

Applied Mechanics Reviews: An Assessment of World Literature in Engineering Sciences. 1948-present. American Society of Mechanical Engineers, 345 East 47th Street, New York, NY 10017. (212) 705-7703. Monthly. $360.00 per year.

Applied Science and Technology Index; A Cumulative Subject Index To English Language Periodicals in the Fields of Aeronautics and Space Science, Computer Technology, Chemistry, Construction Industry, Energy and Related Areas. H.W. Wilson Co., 950 University Avenue, Bronx, NY 10452. (800) 367-

6770 or (212) 588-8400. FAX (718) 590-1617. From 1958 to present. Monthly. Inquire about cost and availability. Also available on CD-ROM and online.

Current Contents: Engineering, Technology, and Applied Sciences. Institute for Scientific Information, 3501 Market Street, Philadelphia, PA 19104. (215) 386-0100. FAX (215) 386-6362. 1970 to present. Weekly. $442.00 per year.

Engineering Index Monthly. Engineering Information, Inc., Castle Point on the Hudson, Hoboken, NJ 07030. (800) 221-1044. FAX (212) 832-1857. Monthly. $2300.00 per year. Also available online as COMPENDEX, and also on CD-ROM. Covers chemical engineering, computers, electrical engineering, civil engineering, metals and mining, industrial management, and mechanical engineering.

International Powder & Bulk Solids Abstracts. Childwall University Press Ltd., Box 78, London NW11 0PG, England. Telephone 01-455-0011. 1984 to present. Three times a year. $168.00 per year.

Mechanical Engineering Abstracts (formerly ISMEC), Cambridge Scientific Abstracts, 7200 Wisconsin Avenue, Bethesda, MD 20814-4823. (301) 961-6750. FAX (301) 961-6720. 1967 to present. Monthly. $895.00 per year. Summarizes world literature in mechanical engineering, production engineering, and engineering management. Also available online.

Materials Science and Engineering Abstracts, Cambridge Scientific Abstracts, 7200 Wisconsin Avenue, Bethesda, MD 20814-4823. (301) 961-6750. FAX (301) 961-6720. 1993 to present. Monthly. $385.00 per year. Focuses on mechanical and physical properties of materials and commercial or industrial applications for materials, methods for strength testing, effects of vibration and other stresses, corrosion and protective coatings, storage and handling, ceramics, composites, metals, wood, plastics, and polymers.

Science Citation Index. Institute for Scientific Information, 3501 Market Street, Philadelphia, PA 19104. (215) 386-0100. FAX (215) 386-6362. Inquire about availability and cost. Also available on CD-ROM.

ASSOCIATIONS AND PROFESSIONAL SOCIETIES

American Society of Mechanical Engineers. 345 E. 47th Street, New York, NY 10017. (212) 705-7722.

Industrial Truck Association. 1750 K Street NW, Washington, DC 20006. (202) 296-9880.

Material Handling & Management Society. 8720 Red Oak Blvd., Suite 201, Charlotte, NC 28217. (704) 522-8644.

Material Handling Equipment Distributors Association. 201 Route 45, Vernon Hills, IL 60061. (708) 680-3500.

Material Handling Institute. 8720 Red Oak Blvd., Suite 201, Charlotte, NC 28217. (704) 522-8644.

DIRECTORIES AND BIOGRAPHICAL SOURCES

Material Handling Engineering Handbook and Directory. Penton Publishing, 110 Superior Ave., Cleveland, OH 44114-2543. (800) 321-7003. Biennial. $35.00 per year.

Materials Handling Buyers Guide. Turret Group Plc., Turret House, 171 High Street, Richmansworth, Herts WD3 1SN, England. Telephone 0923-777000. FAX 0923-771297. 1963 to present. Annual. Inquire for cost and availability.

Modern Materials Handling Casebook and Directory Issue. Cahners Publishing Company, 248 W. 17th St., New York, NY 10011. (212) 645-0067. FAX (212) 463-6560. Annual supplement to Modern Materials Handling periodical (see below).

Directory of Engineering Societies and Related Organizations. Gordon Davis. 13th edition. American Association of Engineering Societies, 1111 19th Street NW, Suite 608, Washington, DC 20036. (202) 296-2237 or (800) 658-8897. 1989. Inquire for price.

International Research Centers Directory. Gale Research, 835 Penobscot Building, Detroit, MI 48226-4094. (313) 961-2242. (800) 347-4253. FAX (313) 961-6083. 8th edition. 1995. $410.00

Research Centers Directory. Gale Research, 835 Penobscot Building, Detroit, MI 48226-4094. (313) 961-2242. (800) 347-4253. FAX (313) 961-6083. $485.00.

Scientific and Technical Organizations and Agencies Directory. Gale Research, 835 Penobscot Building, Detroit, MI 48226-4094. (313) 961-2242. (800) 347-4253. FAX (313) 961-6083. 4th edition. 1996. $195.00.

Who's Who in Engineering. American Association of Engineering Societies, 1111 19th Street NW, Suite 608, Washington, DC 20036. (202) 296-2237 or (800) 658-8897. 8th edition. 1991. Inquire for price.

ENCYCLOPEDIAS AND DICTIONARIES

The Encyclopedia of Operations Research and Management Science. Saul I. Gass & Carl M. Harris, editors. Kluwer Academic Publishers, PO Box 358, Accord Station, Hingham, MA 02018-0358. (617) 871-6600. FAX (617) 871-6528. 1996. $295.00.

GENERAL WORKS

A Guide To Manual Materials Handling. Anil Mital, et al. Taylor & Francis, 1900 Frost Road, Suite 101, Bristol, PA 19007. (215) 785-5800. FAX (215) 785-5515. 1993. $25.00.

Material Flow Systems in Manufacturing. J.M.A. Tanchoco, editor. Chapman and Hall, 29 West 35th Street, New York, NY 10001. (800) 842-3636. FAX (212) 563-2269. 1994. Inquire for cost and availability.

Materials Handling. Robert M. Eastman. Marcel Dekker, Inc., 270 Madison Avenue, New York, NY 10016. (212) 696-9000. FAX (212) 685-4540. 1987. $125.00.

Materials Handling: Principles and Practice. T.H. Allegri Sr. Krieger Publishing Company, P.O. Box 9542, Melbourne, FL 32902-9542. (407) 724-9542. FAX (407) 951-3671. 1992 (reprint of 1984 edition). $66.00.

Modern Materials Handling. J. Chowdhury, editor. McGraw-Hill Publishing Company, 1221 Avenue of the Americas, New York, NY 10020. (800) 262-4729 or (212) 512-3825. 1989. Inquire for cost and availability.

HANDBOOKS AND MANUALS

Bulk Materials Handling Handbook. Jacob Fruchtman. Van Nostrand Reinhold, 115 Fifth Avenue, New York, NY 10003. (212) 254-3232 or FAX (212) 254-9499. 1988. $89.95.

Materials Handling Handbook. Raymond A. Kulwiec, editor. John Wiley and Sons, Inc., 605 Third Avenue, New York, NY 10158. (800) 526-5368 or (212) 850-6000. 1985. $146.00.

Modern Materials Handling Planning Guidebook. Cahners Publishing Company, 248 W. 17th St., New York, NY 10011. (212) 645-0067. FAX (212) 463-6560. Annual supplement to Modern Materials Handling (see below).

ONLINE DATABASES AND CD-ROMS

Compendex Plus. Engineering Information, Inc., 345 East 47th Street, New York, NY 10017. (212) 705-7600 or (800) 221-1044. Contains citations with abstracts to worldwide literature in engineering and technology, from 1970 to present. Available on online BRS,(800) 289-4277, DIALOG, (800) 334-2564, ORBIT (800) 456-7248, and STN International, FIZ Karlsruhe, P.O. Box 2465, W-7500, Karlsruhe 1, Germany, online services. Also available on CD-ROM. Inquire as to cost and availability.

CSA Engineering. Cambridge Scientific Abstracts, 7200 Wisconsin Avenue, Suite 601, Bethesda, MD 20814. (301) 961-6750 or (800) 843-7751. Contains citations and abstracts of international periodicals and research literature covering all fields of engineering and science and technology, including computer and information science, electronics, mechanical engineering, solid state materials, 1981 to present. Available on BRS,(800) 289-4277, online service. Inquire as to cost and availability.

Current Contents Search. Institute for Scientific Information, 3501 Market Street, Philadelphia, PA 19104. (215) 386-0100. FAX (215) 386-6362. Contains citations to articles listed in the table of contents of science and technology journals. Also articles in social sciences and life sciences journals. Available on BRS,(800) 289-4277, DIALOG,(800) 334-2564, online services. Inquire as to cost and availability.

ISMEC: Mechanical Engineering Abstracts. Cambridge Scientific Abstracts, 7200 Wisconsin Avenue, Suite 601, Bethesda, MD 20814. (301) 961-6750 or (800) 843-7751. Contains citations to the literature in mechanical engineering, industrial and production engineering, energy, power, mechanics, devices and related areas, from 1973 to present. Available on the DIALOG,(800) 334-2564, online service. Inquire as to cost and availability.

NTIS Bibliographic Database. National Technical Information Service, 5285 Port Royal Road, Springfield, VA 22161. (703) 487-4929 or FAX (703) 321-8199. Broad coverage of government-sponsored science and technology research reports, 1964 to present. Available on BRS,(800) 289-4277, DIALOG, (800) 334-2564, ORBIT, (800) 456-7248, and STN International, FIZ Karlsruhe, P.O. Box 2465, W-7500, Karlsruhe 1, Germany, online services. Also available on CD-ROM. Inquire as to cost and availability.

Scisearch. Institute for Scientific Information, 3501 Market Street, Philadelphia, PA 19104. (800) 523-1850 or (215) 386-

0100. Broad multidisciplinary title and author index to the international literature of science and technology, 1974 to present. Available on DIALOG, (800) 334-2564, and ORBIT, (800) 456-7248, online services. Also available on CD-ROM. Inquire as to cost and availability.

TRIS (Transportation Research Information). National ACADEMY of Science, 2101 Constitution Avenue, N.W., Washington, DC 20418. (202) 334-3313 or (800) 624-6242. Citations with abstracts of literature on transportation, including air, highway, rail, maritime and other modes from 1968 to present. Available on DIALOG, (800) 334-2564, online service. Inquire as to cost and availability.

Wilsonline. H.W. Wilson Company, 950 University Avenue, Bronx, NY 10452. (800) 367-6770 or (212) 588-8400. Makes available online versions of the printed H.W. Wilson indexes, including Applied Science and Technology Index, Business Periodicals Index, General Science Index, and Readers' Guide to Periodical Literature. Period covered is generally 1983 to present. Available on BRS, (800) 289-4277, DIALOG, (800) 334-2564, and OCLC EPIC, (800) 848-5878, online services. Also available on CD-ROM. Inquire as to cost and availability.

PERIODICALS

Bulk Solids Handling. Trans-Tech Publications, Hardstr. 13, PO Box 100, CH-4714, Aedermannsdorf, Switzerland. FAX 062-741058. 1981 to present. Quarterly. Inquire for cost.

Material Handling Engineering. Penton Publishing, 110 Superior Ave., Cleveland, OH 44114-2543. 1945 to present. Monthly. $45.00 per year.

Materials Handling and Distribution. Intermedia Group Pty. Ltd., PO Box 606, Roselle, New South Wales 2039, Australia. Telephone 61-2-818-4738. FAX 61-2-818-4738. 1970 to present. Bi-monthly. Inquire for cost.

Materials Handling News. Nexus Business Communications Ltd., Warwick House, Azalea Drive, Swanley, Kent BR8 8HY, England. Telephone 0322-660070. FAX 0322-667633. 1955 to present. Monthly. $185.00 per year.

Modern Materials Handling. Cahners Publishing Company, 248 W. 17th St., New York, NY 10011. (212) 645-0067. FAX (212) 463-6560. 1946 to present. 14 times a year. $80.00 per year.

Powder and Bulk Engineering. CSC Publishing Inc., 1300 E. 66th Street, Minneapolis, MN 55423. (612) 866-2242. FAX (612) 866-1939. 1987 to present. Monthly. $60.00 per year.

Powder Handling & Processing. Trans-Tech Publications, Hardstr. 13, PO Box 100, CH-4714, Aedermannsdorf, Switzerland. FAX 062-741058. 1989 to present. Quarterly. Inquire for cost.

RESEARCH CENTERS AND INSTITUTES

Center For Manufacturing Systems Engineering. Lehigh University, H.S. Mohler Laboratory 200, Bethlehem, PA 18015. (215) 758-4667. FAX (215) 694-0542.

Web Handling Research Center, Oklahoma State University. 111 Engineering N., Stillwater, OK 74078-0535. (405) 774-5140.

MATERIALS SCIENCE

See also: ALLOYS, FERROALLOYS, METALLURGICAL ENGINEERING, METALLURGY, METALS AND METALWORKING, NONDESTRUCTIVE TESTING

ABSTRACT SERVICES AND INDEXES

Aluminum Industry Abstracts. Aluminum Association, Materials Park, OH 44073. (216) 338-5151. FAX (216) 338-4634. 1968 to present. Monthly. $525.00 per year.

Applied Mechanics Reviews: An Assessment of World Literature in Engineering Sciences. 1948-present. American Society of Mechanical Engineers, 345 East 47th Street, New York, NY 10017. (212) 705-7703. Monthly. $360.00 per year.

Applied Science and Technology Index; A Cumulative Subject Index To English Language Periodicals in the Fields of Aeronautics and Space Science, Computer Technology, Chemistry, Construction Industry, Energy and Related Areas. H.W. Wilson Co., 950 University Avenue, Bronx, NY 10452. (800) 367-6770 or (212) 588-8400. FAX (718) 590-1617. From 1958 to present. Monthly. Inquire about cost and availability. Also available on CD-ROM and online.

C2C Abstracts: Japan—Materials Science. Scan C2C, 1001 Pennsylvania Ave., NW, No. 1300, Washington, DC 20024-2505. (800) 525-3865. FAX (202) 863-3855. 1990 to present. Monthly. $200.00 per year. Contains abstracts of articles from Japanese scientific, business and technical journals on aggregates, coatings, composites, fibers and whiskers, powder technology, processing, refractories, and wood products. Also available on CD-ROM and online.

CA Selects: Ceramic Materials (journals). Chemical Abstracts Service, 2450 Olentagy River Road, Box 3012, Columbus, OH 43210-0012. (614) 447-3600. FAX (614) 447-3713. 1988 to present. Semi-weekly. $210.00 per year.

Ceramic Abstracts. American Ceramic Society, Inc., 757 Brooksedge Plaza Drive, Westerville, OH 43081-2821. (614) 890-6136. Bi-monthly. 1922 to present. $120.00 per year.

Chemical Abstracts-Applied Chemistry and Chemical Engineering Sections. Chemical Abstracts Service, 2540 Olentangy River Road, Box 3012, Columbus, OH 43210. (800) 848-6538 or (614) 447-3600, FAX (614) 447-3713. Bi-weekly. $1410.00 per year.

Corrosion Abstracts; Abstracts of the World's Literature On Corrosion and Corrosion Mitigation. National Association of Corrosion Engineers (NACE), Box 218340, Houston, TX 77218. (713) 492-0535 or FAX (713) 492-8254. 1962 to present. Bi-monthly. $195.00 per year. Also available on CD-ROM.

Current Contents: Engineering, Technology, and Applied Sciences. Institute for Scientific Information, 3501 Market Street, Philadelphia, PA 19104. (215) 386-0100. FAX (215) 386-6362. 1970 to present. Weekly. $442.00 per year.

Engineered Materials Abstracts. ASM (American Society of Metals) International, Materials Information, Materials Park, OH 44073. (216) 338-5151 or FAX (216) 338-4634. Covers literature on technical developments in polymer, ceramic, and composite materials and engineering. 1986 to present. Monthly. $1200.00 per year. Also available on CD-ROM and online. Covers published materials on technical developments in polymer, ceramic, and composite materials in an engineering environment.

Engineering Index Monthly. Engineering Information, Inc., Castle Point on the Hudson, Hoboken, NJ 07030. (800) 221-1044. FAX (212) 832-1857. Monthly. $2300.00 per year. Also available online as COMPENDEX, and also on CD-ROM. Covers chemical engineering, computers, electrical engineering, civil engineering, metals and mining, industrial management, and mechanical engineering.

International Aerospace Abstracts. Technical Information Service, American Institute of Aeronautics and Astronautics, Inc., 555 West 57th St., New York, NY 10019. (212) 247-6500. FAX (212) 582-4861. Semi-monthly. $1295.00 per year.

Mechanical Engineering Abstracts (formerly ISMEC), Cambridge Scientific Abstracts, 7200 Wisconsin Avenue, Bethesda, MD 20814-4823. (301) 961-6750. FAX (301) 961-6720. 1967 to present. Monthly. $895.00 per year. Summarizes world literature in mechanical engineering, production engineering, and engineering management. Also available online.

Materials Science and Engineering Abstracts, Cambridge Scientific Abstracts, 7200 Wisconsin Avenue, Bethesda, MD 20814-4823. (301) 961-6750. FAX (301) 961-6720. 1993 to present. Monthly. $385.00 per year. Focuses on mechanical and physical properties of materials and commercial or industrial applications for materials, methods for strength testing, effects of vibration and other stresses, corrosion and protective coatings, storage and handling, ceramics, composites, metals, wood, plastics, and polymers.

Materials Science Citation Index. Institute for Scientific Information, 3501 Market Street, Philadelphia, PA 19104. (215) 386-0100. FAX (215) 386-0100. Bi-monthly. $975.00 per year. Also available on CD-ROM.

Metals Abstracts and Metals Abstracts Index. American Society for Metals, Metals Park, OH 44073. (216) 338-5151. 1968 to present. Monthly. Abstracts are $1500.00 per year and Index is $460.00 per year.

Physics Abstracts. Institute of Electrical Engineers, Michael Faraday House, Six Hill Way, Stevenage, Herts, SG1 2AY, England. Distributed by IEEE, 445 Hoes Lane, Piscataway, NJ 08854. (908) 562-5549. 1898 to present. Monthly. $2700.00 per year. Also available online as INSPEC.

Science Citation Index. Institute for Scientific Information, 3501 Market Street, Philadelphia, PA 19104. (215) 386-0100. FAX (215) 386-6362. Inquire about availability and cost. Also available on CD-ROM.

World Ceramic Abstracts. Pergamon Press, Maxwell House, Fairview Park, Elmsford, NY 10523. (914) 592-7700. Monthly. 1958 to present. $310.00 per year.

ANNUAL REVIEWS AND INDEXES

Annual Review of Materials Science. Annual Reviews, Inc., 4139 Camino Way, P.O. Box 10139, Palo Alto, CA 94303-0897. (800) 523-8635 or (415) 493-4400. FAX (415) 855-9815. 1971 to present. Annual. $75.00 per year.

ASSOCIATIONS AND PROFESSIONAL SOCIETIES

American Ceramic Society. 735 Ceramic Place, Westerville, OH 43081. (614) 890-4700. FAX (614) 899-6109.

American Powder Metallurgy Institute. 105 College Road E., Princeton, NJ 08540. (609) 452-7700. FAX (609) 987-8523.

American Society For Metals. Metals Park, OH 44073. (216) 338-5151.

American Society For Nondestructive Testing. 1711 Arlingate Lane, PO Box 28518, Columbus, OH 43228-0518. (614) 274-6003.

American Society For Testing and Materials (ASTM). 1916 Race Street, Philadelphia, PA 19103. (215) 299-5585. FAX (215) 977-9679.

American Society of Mechanical Engineers. 345 E. 47th Street, New York, NY 10017. (212) 705-7722.

Materials Properties Council. 345 E. 47th Street, New York, NY 10017. (212) 705-7693.

Materials Research Society, 9800 Mcknight Road, Pittsburg, PA 15237. (412) 367-3003. FAX (412) 367-4373.

Metallurgical Society of the AIME (American Institute of Mining, Metallurgical, and Petroleum Engineers). 345 E.47th Street, 14th Floor, New York, NY 10017. (212) 705-7695.

National Association of Corrosion Engineers. Box 218340, Houston, TX 77218. (713) 492-0535 or FAX (713) 492-8254.

National Materials Advisory Board. 2101 Constitution Avenue NW, Washington, DC 20418. (202) 334-3505.

Society For the Advancement of Material and Process Engineering. PO Box 2459, Covina, CA 91722. (818) 331-0616.

DIRECTORIES AND BIOGRAPHICAL SOURCES

Directory of Engineering Societies and Related Organizations. Gordon Davis. 13th edition. American Association of Engineering Societies, 1111 19th Street NW, Suite 608, Washington, DC 20036. (202) 296-2237 or (800) 658-8897. 1989. Inquire for price.

International Research Centers Directory. Gale Research, 835 Penobscot Building, Detroit, MI 48226-4094. (313) 961-2242. (800) 347-4253. FAX (313) 961-6083. 8th edition. 1995. $410.00

Research Centers Directory. Gale Research, 835 Penobscot Building, Detroit, MI 48226-4094. (313) 961-2242. (800) 347-4253. FAX (313) 961-6083. $485.00.

Scientific and Technical Organizations and Agencies Directory. Gale Research, 835 Penobscot Building, Detroit, MI 48226-4094. (313) 961-2242. (800) 347-4253. FAX (313) 961-6083. 4th edition. 1996. $195.00.

Who's Who in Engineering. American Association of Engineering Societies, 1111 19th Street NW, Suite 608, Washington,

DC 20036. (202) 296-2237 or (800) 658-8897. 8th edition. 1991. Inquire for price.

ENCYCLOPEDIAS AND DICTIONARIES

Encyclopedia of Physical Science and Technology. Robert A. Meyers, ed. 18 volumes. Academic Press Inc., 6277 Sea Harbor Drive, Orlando, FL 32887. (800) 321-5068. 1992. $2100.00

McGraw-Hill Dictionary of Scientific and Technical Terms. Sybil P. Parker, ed. 5th edition. McGraw-Hill Publishing Company, 1221 Avenue of the Americas, New York, NY 10020. (800) 262-4729 or (212) 512-3825. 1993. $110.50.

McGraw-Hill Encyclopedia of Science and Technology. Sybil P. Parker, ed. 7th edition. 20 volumes. McGraw-Hill Publishing Company, 1221 Avenue of the Americas, New York, NY 10020. (800) 262-4729 or (212) 512-3825. 1992. $1900.00

Thesaurus of Scientific, Technical, and Engineering Terms. Hemisphere Publishing Corporation, 1900 Frost Road, Suite 101, Bristol, PA 19007-1598. (215) 785-5800 or (800) 821-8312. FAX (215) 785-5515. 1987. $173.00.

GENERAL WORKS

Engineering Materials: Properties and Selection. Kenneth G. Budinski. 4th edition. Prentice Hall, 113 Sylvan Avenue, Route 9W, Englewood Cliffs, NJ 07632. (201) 592-2000 or (800) 922-0579. 1992. $51.00.

Engineering Materials Technology: Structures, Processing, Properties & Selection. James A. Jacobs, Thomas F. Kilduff. 2nd edition. Butterworths Publishers, 80 Montvale Avenue, Stoneham, MA 02180. (617) 438-8464 or (800) 366-2665. 1994. $31.50.

Introduction To Engineering Materials. Vernon John. 3rd edition. Industrial Press, Inc. 200 Madison Avenue, New York, NY 10016. (212) 889-6330. FAX (212) 545-8327. 1992. $49.95.

Introduction To Engineering Materials: Behavior, Properties & Selection. G.T. Murray. Marcel Dekker, Inc., 270 Madison Avenue, New York, NY 10016. (212) 696-9000. FAX (212) 685-4540. 1993. $150.00.

Introduction To Materials Engineering. Frank L. Bouquet. 4th edition. Systems Company, PO Box 339, Carlsborg, WA 98324. (206) 452-4987. 1994. $69.00.

Introduction To Materials Science For Engineers. James F. Shackelford. 4th edition. Macmillan Publishing, 200 Old Tappan Road, Old Tappan, NJ 07675. (800) 223-2336. FAX (800) 445-6991. 1995. Inquire for price and availability.

Materials Science and Engineering. Giles F. Carter & Donald E. Paul. ASM (American Society of Metals) International, Materials Information, Materials Park, OH 44073. (216) 338-5151 or FAX (216) 338-4634. 1991. $82.00.

Materials Science and Engineering: An Introduction. William D. Callister, Jr. 3rd edition. John Wiley and Sons, Inc., 605 Third Avenue, New York, NY 10158. (800) 526-5368 or (212) 850-6000. 1993. Inquire for price and availability.

Materials Science and Technology: A Comprehensive Treatment. R.W. Cahn, P. Haasen, E.J. Kramer, editors. Multiple volumes. VCH Publishers, Inc., 220 E. 23rd Street, Suite 909, New York, NY 10010-4606. (800) 422-8824. 1991 to present. Inquire for specific volume prices and availability.

The Properties of Engineering Materials. Raymond A. Higgins. 2nd edition. Industrial Press, Inc. 200 Madison Avenue, New York, NY 10016. (212) 889-6330. FAX (212) 545-8327. 1994. $34.95.

HANDBOOKS AND MANUALS

Materials Handbook: An Encyclopedia For Managers, Technical Professionals, Purchasing and Production Managers, Technicians, Supervisors and Foremen. G.S. Brady & H.R. Clauser. 14th edition. McGraw-Hill Publishing Company, 1221 Avenue of the Americas, New York, NY 10020. (800) 262-4729 or (212) 512-3825. 1991. Inquire for cost and availability.

Materials Science & Engineering Handbook. James F. Shackelford. 2nd edition. CRC Press, 2000 Corporate Blvd., N.W., Boca Raton, FL 33431. (407) 994-0555 or (800) 272-7737. FAX (407) 994-0949. 1994. Inquire for price and availability.

Physical Properties of Materials For Engineers. Daniel D. Pollock. 2nd edition. CRC Press, 2000 Corporate Blvd., N.W., Boca Raton, FL 33431. (407) 994-0555 or (800) 272-7737. FAX (407) 994-0949. 1993. $69.95.

ONLINE DATABASES AND CD-ROMS

Compendex Plus. Engineering Information, Inc., 345 East 47th Street, New York, NY 10017. (212) 705-7600 or (800) 221-1044. Contains citations with abstracts to worldwide literature in engineering and technology, from 1970 to present. Available on online BRS,(800) 289-4277, DIALOG, (800) 334-2564, ORBIT (800) 456-7248, and STN International, FIZ Karlsruhe, P.O. Box 2465, W-7500, Karlsruhe 1, Germany, online services. Also available on CD-ROM. Inquire as to cost and availability.

CSA Engineering. Cambridge Scientific Abstracts, 7200 Wisconsin Avenue, Suite 601, Bethesda, MD 20814. (301) 961-6750 or (800) 843-7751. Contains citations and abstracts of international periodicals and research literature covering all fields of engineering and science and technology, including computer and information science, electronics, mechanical engineering, solid state materials, 1981 to present. Available on BRS,(800) 289-4277, online service. Inquire as to cost and availability.

Current Contents Search. Institute for Scientific Information, 3501 Market Street, Philadelphia, PA 19104. (215) 386-0100. FAX (215) 386-6362. Contains citations to articles listed in the table of contents of science and technology journals. Also articles in social sciences and life sciences journals. Available on BRS,(800) 289-4277, DIALOG,(800) 334-2564, online services. Inquire as to cost and availability.

Dissertation Abstracts. University Microfilms International, 300 North Zeeb Road, Ann Arbor, MI 48106. (800) 521-0600 or (313) 761-4700. Scope includes virtually all doctoral dissertations accepted at accredited American institutions from 1861 to present in 252 subject areas. Available on BRS,(800) 289-4277, DIALOG,(800) 334-2564, and OCLC EPIC,(800) 848-5878, online services. Also available on CD-ROM. Inquire as to cost and availability.

ISMEC: Mechanical Engineering Abstracts. Cambridge Scientific Abstracts, 7200 Wisconsin Avenue, Suite 601, Bethesda, MD 20814. (301) 961-6750 or (800) 843-7751. Contains citations to the literature in mechanical engineering, industrial and production engineering, energy, power, mechanics, devices and related areas, from 1973 to present. Available on the DIALOG,(800) 334-2564, online service. Inquire as to cost and availability.

Metadex. Jointly produced by ASM International and the Institute of Materials. Contains more than 925,000 records from the international literature on metals and alloys, concerning properties, processes, materials classes, applications, and metallurgical systems. Updated monthly. Available from ORBIT-QUESTEL (703) 442-0900.

NTIS Bibliographic Database. National Technical Information Service, 5285 Port Royal Road, Springfield, VA 22161. (703) 487-4929 or FAX (703) 321-8199. Broad coverage of government-sponsored science and technology research reports, 1964 to present. Available on BRS,(800) 289-4277, DIALOG, (800) 334-2564, ORBIT, (800) 456-7248, and STN International, FIZ Karlsruhe, P.O. Box 2465, W-7500, Karlsruhe 1, Germany, online services. Also available on CD-ROM. Inquire as to cost and availability.

Scisearch. Institute for Scientific Information, 3501 Market Street, Philadelphia, PA 19104. (800) 523-1850 or (215) 386-0100. Broad multidisciplinary title and author index to the international literature of science and technology, 1974 to present. Available on DIALOG,(800) 334-2564, and ORBIT,(800) 456-7248, online services. Also available on CD-ROM. Inquire as to cost and availability.

Wilsonline. H.W. Wilson Company, 950 University Avenue, Bronx, NY 10452. (800) 367-6770 or (212) 588-8400. Makes available online versions of the printed H.W. Wilson indexes, including Applied Science and Technology Index, Business Periodicals Index, General Science Index, and Readers' Guide to Periodical Literature. Period covered is generally 1983 to present. Available on BRS,(800) 289-4277, DIALOG,(800) 334-2564, and OCLC EPIC,(800) 848-5878, online services. Also available on CD-ROM. Inquire as to cost and availability.

PERIODICALS

Journal of Engineering For Industry. American Society of Mechanical Engineers, 345 E. 47th Street, New York, NY 10017. (212) 705-7722. 1970 to present. Quarterly. $100.00 per year.

Journal of Materials Science. Chapman & Hall, Journals Promotion Department, One Penn Plaza, 41st floor, New York, NY 10019. (212) 564-1060. FAX (212) 564-1505. 1966 to present. 58 times a year. $3170.00 per year (subscription also includes Journals of Materials Science Letters, Journal of Materials Science: Materials in Medicine, and Journal of Materials: Materials in Electronics).

Journal of Materials Research. Materials Research Society, 9800 McKnight Road, Pittsburg, PA 15237. (412) 367-3003. FAX (412) 367-4373. 1986 to present. Monthly. $440.00 per year.

JOM (Journal of Metals). Minerals, Metals and Materials Society, 420 Commonwealth Drive, Warrendale, PA 15086. (412) 776-9080. 1949 to present. Monthly. $50.00 per year.

Journal of Nondestructive Evaluation. Plenum Publishing Corporation, 233 Spring Street, New York, NY 10013. (212) 620-8000. FAX (212) 463-0742. 1980 to present. Quarterly. $230.00 per year.

Journal of Testing and Evaluation. American Society for Testing and Materials (ASTM), 1916 Race Street, Philadelphia, PA 19103. (215) 299-5585. FAX (215) 977-9679. 1966 to present. Bi-monthly. $103.00 per year for non-members.

Materials Engineering. Penton Publishing, 110 Superior Ave., Cleveland, OH 44114-2543. 1929 to present. Monthly. $50.00 per year.

Materials Evaluation. American Society for Nondestructive Testing, 1711 Arlingate Lane, Box 28518, Columbus, OH 43228-0158. (614) 274-6003. FAX (614) 274-6899. 1942 to present. Monthly. $85.00 per year.

Materials Performance. National Association of Corrosion Engineers, P.O. Box 218340, 1440 South Creek, Houston, TX 77218. (713) 492-0535. 1962 to present. Monthly. $65.00 per year.

Materials Science and Engineering A: Structural Materials: Properties, Microstructures, and Processing. Elsevier Science Inc., Box 882, Madison Square Station, New York, NY 10159. (212) 989-5800. 1967 to present. 30 times a year. $3367.00 per year.

Materials Science and Engineering R: Reports. Elsevier Science Inc., Box 882, Madison Square Station, New York, NY 10159. (212) 989-5800. 1986 to present. 16 times a year. $449.00 a year.

Materials Science and Technology. Institute of Materials, 1 Carlton House Terrace, London SW1Y 5DB, England. Telephone 071-839-4071. 1985 to present. Monthly. $37.00 per year.

Materials Transactions, JIM. Japan Institute of Metals, Nihon Kinzoku Gakkai, Aoba Aramaki, Aoba-Ku, Sendai 980, Japan. Telephone 022-223-3685. FAX 8-22-223-6312. 1960 to present. Monthly. Inquire for cost.

Progress in Materials Science. Elsevier Science [journals], 660 White Plains Rd., Tarrytown, NY 10159-5153. (919) 524-9200. FAX (919) 333-2444. 1949 to present. 5 times a year. $485.00 per year.

SAMPE Journal. Society for the Advancement of Material and Process Engineering, Covina, CA 91724. (818) 331-0610. FAX (818) 332-8929. 1965 to present. Bi-monthly. $65.00 per year.

RESEARCH CENTERS AND INSTITUTES

Center For Materials Science and Engineering. Massachusetts Institute of Technology, Bldg. 13, Room 2090, 77 Massachusetts Avenue, Cambridge, MA 02139. (617) 253-6701. FAX (617) 258-6883.

Center For Research in Materials Science and Engineering. University of Texas at Austin, ETC 9.104, Austin, TX 78712. (512) 471-1504.

Center For Materials Research. Ohio State University, 4108 Smith Laboratory, 174 W. 18th Avenue, Columbus, OH 43210. (614) 292-1133. FAX (614) 292-7557.

Materials Research Laboratory. University of Illinois, Urbana, IL 61801. (217) 333-1370.

Materials Science Center. Cornell University, Clark Hall of Science, Ithaca, NY 14853-2501. (607) 255-4272. FAX (607) 255-6428.

MATHEMATICAL PHYSICS

See Also: ELASTICITY, ELECTRICITY, HYDRODYNAMICS, OPTICS, PHYSICS, THERMODYNAMCIS

ABSTRACT SERVICES AND INDEXES

Applied Science and Technology Index; A Cumulative Subject Index To English Language Periodicals in the Fields of aeronautics and Space Science, Computer Technology, chemistry, Construction Industry, Energy and Related Areas. H.W. Wilson Co., 950 University Avenue, Bronx, NY 10452.(212) 588-8400. (800) 367-6770. FAX (718) 590-1617. From 1958 to present. Monthly. Inquire about cost and availability. Also available on CD-ROM and online. ISSN: 0003-6986.

Current Contents: Physical, Chemical and Earth Sciences. Institute for Scientific Information, 3501 Market Street, Philadelphia, PA 19104. (215) 386-0100. FAX (215) 386-2291. From 1961 to present. Weekly. $442.00 per year. Also available on CD-ROM and online. Inquire regarding cost andavailability. ISSN: 0163-2574.

Current Papers in Physics. Institute of Electrical Engineers, Michael Faraday House, Six Hill Way, Stevenage, Herts, SG1 2AY, England. Distributed by INSPEC/IEEE, Box 1331, 445 Hoes Lane, Piscataway, NJ 08855-1331. (908) 562-5549. 1966 to present. Fortnightly. $410.00 per year. ISSN: 0011-3786.

General Science Index. H.W. Wilson Company, 950 University Avenue, Bronx, NY 10452. (212) 588-8400. (800) 367-6770. FAX (718) 590-1617. From 1978 to present. Ten issues per year; quarterly and annual cumulations. Service basis. Available on CD-ROM and online. Inquire regarding cost and availability. ISSN: 0162-1963.

Mathematical Reviews: A Review Journal Covering the world Literature of Mathematical Research. American Mathematical Society, P.O. Box 1571, Annex Station, Providence, RI 02901-9930. (401) 455-4000. (800) 556-7774. 1940 to present. Monthly. $4594.00 per year. Also available via network (MATHSCINET), online and on CD-ROM. Inquire regarding cost and availability. ISSN: 0025-5629.

NTIS Alerts: Physics. U.S. National Technical Information Service. 5285 Port Royal Road, Springfield, VA22161. (703) 487-4650. FAX (703) 321-8547. Weekly. $140.00 per year.

Pascal 10: Mechanique, Acoustique Et Transport DE Chaleur. Centre National de la Recherche Scientifique, Institue del'Information Scientifique et Technique, 2 aliee du Parc deBrabois, 54514 Vandoeuvre-Les Nancy Cedex, France. TEC 83-50-46-00. FAX 83-50-46-50. 1984 to present. Ten issues per year. 1530 F per year. Also available on CD-ROM and online. ISSN: 1136-5107.

Physics Abstracts. INSPEC. Section A, Science Abstracts. Institution of Electrical Engineers (IEE). Available from INSPEC/

IEEE-Institute of Electrical and Electronic Engineers, Box 1331, 445 Hoes Lane, Piscataway, NJ 08855-1331.(908) 562-5549. 1898 to present. 24 issues per year. $2835.00 per year. Also available on CD-ROM and online. ISSN: 0036-8091.

Physics Briefs (Physikalische Berichte). Information Center for Energy, Physics, Mathematics; German Physical Society. VCH Publishers, Inc., 220 East 23rd Street, New York, NY 10010-4606. (212) 683-8333. 1845 to present. 24 issues per year. $2390.00 per year. Also available online. ISSN: 0179-7434.

Science Citation Index. SCI. Institute for Scientific Information, 3501 Market Street, Philadelphia, PA 19104.(215) 386-0100. (800) 523-1850. FAX (215) 386-2991. 1961 to present. Six issues per year, plus annual cumulation. $11650.00 per year. Also available online and on CD-ROM. Inquire about price and availability. ISSN: 0036-827X.

ANNUAL REVIEWS AND YEARBOOKS

Advances in Physics. Taylor & Francis, Ltd., Rankine Road, Basingstoke, Hants RG24 8PR, England. 0256-840366. FAX 0256-47938. 1952 to present. Bimonthly. $511.00 per year. ISSN: 0001-8732.

Boundary Value Problems of Mathematical Physics. AM Mathematical Society, P.O. Box 6248, Providence, RI 02940. (401) 455-4000. FAX (401) 331-3842. Proceedings of the Steklov Institute of Mathematics, 1967 to present. Irregular. Priced individually. ISSN: 0081-5438.

ASSOCIATIONS AND PROFESSIONAL SOCIETIES

American Institute of Physics. One Physics Ellipse, College Park, MD 20740-3843. (301) 209-3100.

American Mathematical Society. P.O. Box 6248, Providence, RI 02940. (401) 455-4000. FAX (401) 331-3842.

American Physical Society. One Physics Ellipse, College Park, MD 20740-3843. (301) 209-3200. FAX (301) 209-0865.

Universities Research Association, 1111 19th Street NW, Suite 400, Washington, DC 20036. (202) 293-1382.

BIBLIOGRAPHIES

Guide To the Archival Collections in the Niels Bohr Libary At the American Institute of Physics. Niels Bohr Library Staff. American Institute of Physics, One Physics Ellipse, College Park, MD 20740-3843. (301) 209-3100. 1995. $140.00

DIRECTORIES AND BIOGRAPHICAL SOURCES

American Men and Women of Science: Physical and Biological Sciences. R.R. Bowker Inc., 121 Chanlon Road, New Providence, NJ 07974. (908) 464-6800. (800) 521-8110. 20th edition. 8 volumes. 1996. $850.00.

American Physical Society. Membership Directory Bulletin Issue. American Physical Society, One Physics Ellipse, College Park, MD 20740-3843. (301) 209-3200. FAX (301) 209-0865. Biennial. $50.00.

The Biographical Dictionary of Scientists: Physicists. D. Abbott, editor. Peter Bedrick Books, Inc., 2112 Broadway, Room 318, New York, NY 10023. (212) 496-0751. 1984.

Directory of Physics and Astronomy Staff Members. American Institute of Physics, One Physics Ellipse, College Park, MD 20740-3843. (301) 209-3100. Annual. $45.00.

Graduate Programs in Physics, Astronomy and Related Fields. 1995-1996. American Institute of Physics, One Physics Ellipse, College Park, MD 20740-3843. (301) 209-3100. Annual. $45.00.

Legends in Their Own Time: A Century of American Physical Scientists. Anthony Serafini. Plenum Publishing Corp., 233 Spring Street, New York, NY. (212) 620-8000. (800)2221-9369. FAX (212) 463-0742. 1993. $27.50.

The Nobel Prize Winners. Frank N. Magill, editor. Salem Press, Inc., P.O. Box 1097, Englewood Cliffs, NJ 07632. (201) 871-3700. (800) 221-1592. 3 volumes. 1989. $210.00.

Research Centers Directory. Gale Research Inc., 835 Penobscot Building, Detroit, MI 48226-4094. (313) 961-2242. (800) 347-4253. 20th edition. 1995. $485.00.

Scientific and Technical Organizations and Agencies Directory. Gale Research, 835 Penobscot Building, Detroit, MI 48226-4094. (313) 961-2242. (800) 347-4253.

World Guide To Scientific Associations and Learned Societies. K. G. Saur Inc.,

ENCYCLOPEDIAS AND DICTIONARIES

A Concise Dictionary of Physics. Oxford University Press, Inc., 200 Madison Avenue, New York, NY 10016. (212) 725-6000. (800) 334-4249. 1990. $10.95.

Dictionary of Effects and Phenomena in Physics. Joachim Schubert. VCH Publications, Inc., 220 East 23rd Street, Suite 909, New York, NY 10010-4606. (212) 683-8333. (800) 422-8824. 1987. $35.00.

Dictionary of the Physical Sciences: Terms, Formulas, Data. Cesare Emiliani. Oxford University Press, Inc., 200 Madison Avenue, New York, NY 10016. (212) 725-6000. (800) 334-4249. 1989. $19.95.

Encyclopedia of Applied Physics. George Trigg, editor. VCH Publications, Inc., 220 East 23rd Street, Suite 909, New York, NY 10010-4606. (212) 683-8333. (800) 422-8824. 20 volume set. 1991-. $5990.00.

Encyclopedia of Physics. Robert M. Besancon. Chapman & Hall, 1 Penn Plaza, New York, NY 10119. (212) 564-1060. 3rd edition. 1990. $54.95 in paper.

Equations of Mathematical Physics. A. N. Tikhonov. Dover Publications, Inc., 180 Varick Street, New York, NY 10014. (212) 255-3755. (800) 223-3130. 1990. $16.95 in paper.

Facts On File Dictionary of Physics. John Daintith, editor. Facts-on-File, Inc., 460 Park Avenue South, New York, NY 10016-7382. (212) 683-2244. (800) 322-8755. Fax (800) 683-3633. Revised edition. 1990. $12.95.

Encyclopedia of Physical Science and Technology. Academic Press, Inc., 6277 Sea Harbor Drive, Orlando, FL. (800) 321-5068. 2nd edition. 18 volume set. 1992. $2625.00.

McGraw-Hill Encyclopedia of Science and Technology. McGraw-Hill Book Company, Inc., 1221 Avenue of the Americas, New York, NY 10020. (212) 512-2000. (800) 262-4729. 7th edition. 20 volume set. 1992. $1900.00.

GENERAL WORKS

Applied Functional Analysis: Applications of Mathematical Physics. Eberhard Zeidler. Springer-Verlag New York, Inc., 175 Fifth Avenue, New York, NY 10010. (212) 460-1500. (800) 777-4643. FAX (212) 473-6272. 1995. $65.00.

Computation in Modern Physics. William R. Gibbs. World Scientific Publishing Company, Inc., 1060 Main Street, River Edge, NJ 07661. (201) 487-9655. (800) 227-7562. 1994. $55.00.

Course in Mathematical Physics. Vols. 1 & 2: Classical Dynamical Systems. W. Thirring and E. M. Harrell. Springer-Verlag New York, Inc., 175 Fifth Avenue, New York, NY 10010. (212) 460-1500. (800) 777-4643. FAX (212) 473-6272. 2nd edition. 1991. $74.95 in paper.

Differential Geometry and Mathematical Physics. John Beem and Krishan L. Duggal, editors. American Mathematical Society, P.O. Box 6248, Providence, RI 02940. (401) 455-4000. (800) 321-4267. 1994. $41.00.

From Newton To Mandelbrot: A Primer in theoretical Physics. Deitrich Stauffer and H. Eugene Stanley. Springer-Verlag New York, Inc., 175 Fifth Avenue, New York, NY 10010. (212) 460-1500. (800) 777-4643. FAX (212) 473-6272. 1991. $19.95 in paper.

Introduction To Methods and Concepts of Mathematical Physics. Wa Wong Chun. Oxford University Press, Inc., 200 Madison Avenue, New York, NY 10016. (212) 725-6000. (800) 334-4249. 1991. $49.95.

Landmark Experiments in Twentieth Century Physics. George L. Trigg. Dover Publications, Inc., 180 Varick Street, New York, NY 10014. (212) 255-3755. (800) 223-3130. 1995. $9.95 in paper.

Mathematical Methods For Physicists. George B. Arkfen and Hans-Jurgen Weber, editors. Academic Press Inc., 6277 Sea Harbor Drive, Orlando, FL. (800) 321-5068. 4th edition. 1993.

Methods of Mathematical Physics. Richard Courant and David Hilbert. John Wiley & Sons, Inc., 605 Third Avenue, New York, NY 10158-0012. (212) 850-6000. (800) 225-5945. 1989. Two volumes. $61.95 each in paper.

Physics. David Haliday and Robert Resnick. John Wiley & Sons, Inc., 605 Third Avenue, New York, NY 10158-0012. (212) 850-6000. (800) 225-5945. 2 volumes. 4th edition. 1992.

Practical Quantum Mechanics. S. Fluegge. Springer-Verlag New York, Inc., 175 Fifth Avenue, New York, NY 10010. (212) 460-1500. (800) 777-4643. FAX (212) 473-6272. 1994. $49.95 in paper.

The Six Core theories of Modern Physics. Charles P. Stevens. MIT Press, 55 Hayward Street, Cambridge, MA 02142. (617)

253-8569. 1995. $30.00.

HANDBOOKS AND MANUALS

Chemical and Physical Data. Arthur M. James and M. P. Lord, Van Nostrand Reinhold, 115 Fifth Avenue, New York, NY 10003. (212) 254-3232. (800) 842-3636. 1992. $59.95.

CRC Handbook of Chemistry and Physics. David R. Kide, editor. CRC Press, Inc., 2000 Corporate Boulevard, NW, Boca Raton, FL 33431. (407) 994-0555. (800) 272-7737. 77th edition. 1996. $99.95.

Fundamental Formulas of Physics. Donald H. Menzel. Dover Publications, Inc., 180 Varick Street, New York, NY 10014. (212) 255-3755. (800) 223-3130. 2nd edition. 1960. 2 volumes. 19.90.

Handbook of Physical Quantities. Igor S. Grigoriev and Evgenil Z. Melikhov, editors. CRC Press, Inc., 2000 Corporate Boulevard, NW, Boca Raton, FL 33431. (407) 994-0555. (800) 272-7737. 1995. $99.00.

McGraw-Hill Encyclopedia of Physics. Sybil P. Parker, editor. McGraw-Hill Publishing Company, Inc., 1221 Avenue of the Americas, New York, NY 10020. (212) 512-2000.(800) 262-4729. 2nd edition. 1993. $95.50

A Physicist's Desk Reference. Herbert L. Anderson, editor. American Institute of Physics, One Physics Ellipse, College Park, MD 20740-3843. (301) 209-3100. 2nd edition. 1989. $70.00.

Physics Problem Solver. Research and Education Association Staff. Research & Education Association, 61 Ethel Road, West Piscataway, NJ 08854. (908) 819-8880. Revised edition. 1994. $23.95 in paper.

Tables of Physical and Chemical Constants and Some Mathematical Functions. G. W. Kaye and T. M. Laby. Halsted Press 605 Third Avenue, New York, 10158-0012. (212) 850-6400. 15th edition. 1986. $74.95.

ONLINE DATABASES AND CD-ROMS

CA Search. Chemical Abstracts Service, P.O. Box 3012, Columbus, OH 43210-0012. (614) 447-3600. (800) 848-6533. FAX (614) 447-3709. Very comprehensive guide to worldwide chemical literature and related fields, 1972 to present. Available on BRS,(800) 289-4277, DIALOG, (800) 334-2564, ORBIT (800) 456-7248, and STN International, FIZ Karlsruhe, P.O. Box 2465, W-7500, Karlsruhe 1, Germany, online services. Inquire as to cost and availability.

Current Contents Search. Institute for Scientific Information, 3501 Market Street, Philadelphia, PA 19104. (215) 386-0100. FAX (215) 386-6362. Contains citations to articles listed in the table of contents of science and technology journals. Also articles in social sciences and life sciences journals. Available on BRS, (800) 289-4277, DIALOG, (800) 334-2564, online services. Inquire as to cost and availability.

Dissertation Abstracts Online. University Microfilms International, 300 North Zeeb Road, Ann Arbor, MI 48106. (800) 521-0600 or (313) 761-4700. Scope includes virtually all doctoral dissertations accepted at accredited American institutions from 1861 to present in 252 subject areas. Available on BRS, (800)

289-4277, DIALOG, (800) 334-2564, and OCLC EPIC, (800) 848-5878, online services. Also available on CD-ROM. Inquire as to cost and availability.

INSPEC. Institution of Electrical Engineers, Michael Faraday House, Six Hills Way, Stevenage, Herts. SG1 2AY, England. Telephone: 0438 313311 or FAX 0438 742840. Contains citations to the worldwide literature of physics, electronics and electrical engineering, computer technology, and related fields. Available on BRS, (800) 289-4277, DIALOG, (800) 334-2564, ORBIT, (800) 456-7248, and STN International, FIZ Karlsruhe, P.O. Box 2465, W-7500, Karlsruhe 1, Germany, online services. Inquire as to cost and availability.

Mathsci. American Mathematical Society, P.O. Box 6248, Providence, RI 02940. (800) 321-4667 or (401) 455-4000 or FAX (401) 331-3842. Scope includes pure and applied mathematics and related areas of physics, statistics, engineering, computer science, and operations research literature since 1959. Available on DIALOG,(800) 334-2564, online service. Also available on CD-ROM. Inquire as to cost and availability.

NTIS Bibliographic Database. National Technical Information Service, 5285 Port Royal Road, Springfield, VA 22161. (703) 487-4929 or FAX (703) 321-8199. Broad coverage of government-sponsored science and technology research reports, 1964 to present. Available on BRS,(800) 289-4277, DIALOG, (800) 334-2564, ORBIT, (800) 456-7248, and STN International, FIZ Karlsruhe, P.O. Box 2465, W-7500, Karlsruhe 1, Germany, online services. Also available on CD-ROM. Inquire as to cost and availability.

Physics Briefs. American Institute of Physics, 335 East 45th Street, New York, NY 10017. (212) 661-9260 or FAX (212) 661-2036. Contains citations with abstracts of the literature of physics and related fields, 1979 to present. Available on the STN International, FIZ Karlsruhe, P.O. Box 2465, W-7500, Karlsruhe 1, Germany, online service. Inquire as to cost and availability.

Scientific and Technical Books and Serials in Print. R.R. Bowker Inc., 121 Chanlon Road, New Providence, NJ 07974. (800) 521-8110 or (908) 464-6800. List of currently published books and serials in the physical and biological sciences, engineering and technology, with subject, author and titles indexes. Available on ORBIT, (800) 456-7248, online service. Inquire as to cost and availability.

Scisearch. Institute for Scientific Information, 3501 Market Street, Philadelphia, PA 19104. (800) 523-1850 or (215) 386-0100. Broad multidisciplinary title and author index to the international literature of science and technology, 1974 to present. Available on DIALOG, (800) 334-2564, and ORBIT, (800) 456-7248, online services. Also available on CD-ROM. Inquire as to cost and availability.

Wilsonline. H.W. Wilson Company, 950 University Avenue, Bronx, NY 10452. (800) 367-6770 or (212) 588-8400. Makes available online versions of the printed H.W. Wilson indexes, including Applied Science and Technology Index, Business Periodicals Index, General Science Index, and Readers' Guide to Periodical Literature. Period covered is generally 1983 to present. Available on BRS, (800) 289-4277, DIALOG, (800) 334-2564, and OCLC EPIC, (800) 848-5878, online services. Also available on CD-ROM. Inquire as to cost and availability.

PERIODICALS

American Journal of Physics. American Association of Physics Teachers, 5112 Berywn Road, College Park, MD 20740. (301) 345-4200. 1933 to present. Monthly. $215.00 per year. ISSN: 0002-9505.

Chaos; An Interdisciplinary Journal of Nonlinear Science. American Institute of Physics, One Physics Ellipse, College Park, MD 20740-3843. (301) 209-3000. 1991 to present. Quarterly. $225.00 per year. ISSN: 1054-1500.

Contemporary Physics. Taylor & Francis, Ltd, Rankine Road, Basingstoke, Hants RG24 8PR, England. TEL 0256-840366. FAX 0256-479438. 1959 to present. Bimonthly. $318.00 per year. ISSN: 0010-7514.

European Journal of Physics. IOP Publishing. U.S. subscriptions to: American Institute of Physics, 500 Sunnyside Boulevard, Woodbury, NY 11797-2999. (516) 575-2270. 1980 to present. Bimonthly. $339.00 per year. ISSN: 0143-0807.

International Journal of Modern Physics C: Physics and Computers. World Scientific Publishing Co., 1060 Main Street, Suite 18, River Edge, NJ 07661. (800) 227-7562. 1990 to present. Bimonthly. $350.00 per year. ISSN: 0129-1831.

Journal of Computational Physics. Academic Press, Inc., Box 620000, Orlando, FL 32891-8340. (800) 543-9534. 1966 to present. Monthly. $1296.00 per year. ISSN: 0021-9991.

Journal of Mathematical Physics. American Institute of Physics, One Physics Ellipse, College Park, MD 20740-3843. (301) 209-3000. 1960 to present. Monthly. $1215.00 per year. ISSN: 0022-2488.

Journal of Physics A: Mathematical and General. IOP Publishing. U. S. orders to: American Institute of Physics, 500 Sunnyside Boulevard, Woodbury, NY 11797-2999. (516) 576-2200. 1968 to present. Twenty-four issues per year. $2487.00 per year. ISSN: 0305-4470.

Letters in Mathematical Physics. Kluwer Academic Publishers, Box 358 Accord Station, Hingham, MA 02018-0358. (617) 871-6600. FAX (617) 871-6528. 1975 to present. Monthly. $549.00 per year. ISSN: 0377-9017.

Physical Review A (general Physics). American Institute of Physics, One Physics Ellipse, College Park, MD 20740-3843. (301) 209-3000. 1970 to present. Monthly. $990.00 per year. ISSN: 1050-2947.

Physics Today. American Institute of Physics, One Physics Ellipse, College Park, MD 20740-3843. (301) 209-3000. 1948 to present. Monthly. $140.00 per year. ISSN: 0031-9228.

Reports On Progress in Physics. IOP Publishing. U.S. orders to: American Institute of Physics, 500 Sunnyside Boulevard, Woodbury, NY 11797-2999. (516) 576-2200. 1934 to present. Monthly. $1214.00 per year. ISSN: 0034-4885.

Reviews of Modern Physics. American Institute of Physics, One Physics Ellipse, College Park, MD 20740-3843. (301) 209-3000. 1929 to present. Quarterly. $300.00 per year. ISSN: 0034-6861.

Royal Society of London, Philosophical Transactions. Series A, Physical Sciences and Engineering. Royal Society of London, 6

Carlton House Terrace, London SW1Y 5AG, England. TEL 071-839-5561. FAX 071-976-1837. 1665 to present. Monthly. L485 per year. ISSN: 0962-8428.

RESEARCH CENTERS AND INSTITUTES

Center For theoretical Physics. Massachusetts Institute of Technology, Massachusetts Avenue, Cambridge, MA 02139. (671) 253-4852. FAX (617) 258-6923.

Courant Institute of Mathematical Sciences. New York University, 251 Mercer Street, New York, NY 10012. (212) 998-3000.

Institute For Physical Science and Technology. University of Maryland, 4203 Computer and Space Sciences Building, College Park, 20742. (301) 405-4874. FAX (301) 314-9363.

Institute For theoretical Physics. State University of New York at Stony Brook, NY 11794. (516) 689-6000.

Institute of theoretical Science. University of Oregon, Eugene, OR 97403. (503) 346-5204.

Mathematical Sciences Institute. Cornell University, 201 Caldwell Hall, Ithaca, NY 14853. (607) 255-8005.

MATHEMATICS

See also: ALGEBRA, CALCULUS, GEOMETRY, STATISTICS

ABSTRACT SERVICES AND INDEXES

Applied Mechanics Reviews: An Assessment of World Literature in Engineering Sciences. American Society of Mechanical Engineers, 345 East 47th Street, New York, NY 10017. (212) 705-7703. 1948 to present. Monthly. $360.00 per year. ISSN: 0003-6900.

Applied Science and Technology Index; A Cumulative Subject Index To English Language Periodicals in the Fields of Aeronautics and Space Science, Computer Technology, Chemistry, Construction Industry, Energy and Related Areas. H.W. Wilson Company, 950 University Avenue, Bronx, NY 10452-9978. (212) 588-8400. (800) 367-6770. FAX (718) 590-1617. 1958 to present. Monthly. Available on-line from BRS and Wilsonline. Also available on CD-ROM. Inquire about cost and availability. ISSN: 0003-6986.

Compactmath-Compact Mathematics Library. Cumulative CD-ROM edition of Zentralblatt fuer Mathematik-Mathematics Abstracts. Springer-Verlag New York, Inc. 44 Hartz Way. Secaucus. NJ. 07096-2491. (201) 348-4033. FAX (201) 348-4505. 1993 to present. Annual. Available only on CD-ROM. Inquire regarding cost and availability. ISSN: 0938-3174.

Compumath Citation Index. Institute for Scientific Information, 3501 Market Street, Philadelphia, PA 19104. (800) 523-1850 or (215)-386-0100. FAX (215) 386-2991. Three times per year. $1955.00. Also available online and on CD-ROM. Inquire regarding cost and availability. ISSN: 0730-6199.

Current Index To Statistics; Applications-Methods-Theory. American Statistical Association, 1429 Duke Street, Alexandria, VA 22314-3402. (703) 684-1221. FAX (703) 684-2037. 1976

MATHEMATICS

Ency. of Physical Sciences and Engineering Info. Sources

to present. Annual. $50.00 per year. Available on-line from BRS, Dialog, European Space Agency. ISSN: 0364-1228.

Current Mathematical Publications. American Mathematical Society. P.O. Box 1571, Annex Station, Providence, RI 02901-9930. (800) 556-7774 or (401) 455-4000. FAX (401) 331-3842. 1969 to present. Seventeen times per year. $377.00 per year. Available online from BRS, DIALOG, European Space Agency. Also available on CD-ROM. Inquire regarding cost and availability. ISSN: 0361-4794.

General Science Index. H.W. Wilson Company, 950 University Avenue, Bronx, NY 10452. (212) 588-8400. (800) 367-6770. FAX (718) 590-1617. From 1978 to present. Ten issues per year; quarterly and annual cumulations. Service basis. Available on CD-ROM and online. Inquire regarding cost and availability. ISSN: 0162-1963.

Index To Mathematical Tables. Alan Fletcher. Addison-Wesley Publishing Co., Inc., 1 Jacob Way, Reading, MA 01867. (617) 944-3700. (800) 447-2226. 2nd edition. 1962. 2 volumes.

Mathematical Reviews: A Review Journal Covering the World Literature of Mathematical Research. American Mathematical Society, P.O. Box 1571, Annex Station, Providence, RI 02901-9930. (800) 556-7774 or (401) 455-4000. 1940 to present. Monthly. $4594.00 per year. Also available online and on CD-ROM. Inquire regarding cost and availability. ISSN: 0025-5629.

NTIS Alerts: Mathematical Sciences. U.S. National Technical Information Service, 5285 Port Royal Road, Springfield, VA 22161. (703) 487-4650. FAX (703) 321-8547. Weekly. $140.00.

Science Citation Index. Institute for Scientific Information. 3501 Market Street, Philadelphia, PA 19104. (215) 386-0100. (800) 523-1850. FAX (215) 386-2991. 1961 to present. Six issues per year, plus annual cumulation. $11650.00 per year. Also available online and on CD-ROM. Inquire about price and availability. ISSN: 0036-827X.

Statistical theory and Method Abstracts. International Statistical Institute. Prinses Beatrixiaan 428, Postbus 950, 2270 AZ Voorburg, Netherlands. 31-70-3375737. FAX 31-70-3860025. 1959 to present. Quarterly. $191.82.

Zentralblatt Fuer Mathematik Und Ihre Grenzebiete/Mathematics Abstracts. Heidelberger Akademie der Wissenschaften/Springer-Verlag New York, Inc., 44 Hartz Way, Secaucus, NJ 07096-2491. (201) 348-4033. FAX (201) 348-4505. 1931 to present. 30 issues per year. DM 8340. ISSN: 0044-4235. Also available online and on CD-ROM. Inquire regarding cost and availability.

ANNUAL REVIEWS AND YEARBOOKS

Advances in Applied Mathematics. Academic Press, Inc., 6277 Sea Harbor Drive, Orlando, FL 32821-8340. (800) 543-9534. 1980 to present. Irregular, Price varies; inquire.

Applied Mathematical Sciences. Springer-Verlag, 175 Fifth Avenue, New York, NY 10010. (212) 460-2500. FAX (212) 473-6272. 1972 to present. Irregular. ISSN: 0066-5452.

CBMS-NSF Regional Conference Series in Applied Mathematics. Society for Industrial and Applied Mathematics, 3600 University City Science Center, Philadelphia, PA 19104-2688. (215)

382-9800. FAX (215) 386-7999. 1971 to present. Irregular. ISSN: 0163-9439.

Ergebnisse Der Mathematik Und Ihrer Grenzgebiete. Neue folge. Springer-Verlag New York, Inc. 175 Fifth Avenue, New York, NY 10010. (212) 460-1500. FAX (212) 473-6272. 1953 to present. Irregular. Price varies; inquire. ISSN: 0071-1136.

Lecture Notes in Mathematics. Springer-Verlag New York, Inc. 44 Hartz Way. Secaucus. NJ. 07096-2491. (201) 348-4033. FAX (201) 348-4505. Irregular. Price varies; inquire.

ASSOCIATIONS AND PROFESSIONAL SOCIETIES

American Mathematical Society. P.O. Box 6248, Providence, RI 02940. (401) 455-4000. FAX (401) 331-3842.

American Statistical Association. 1429 Duke Street, Alexandria, VA 22314-3402. (703) 684-1221. FAX (703) 684-2037.

Association For Symbolic Logic. University of Illinois, Department of Mathematics, 1409 West Green Street, Urbana IL 61801. (217) 244-7902. FAX (217) 333-9576.

Association For Women in Mathematics. 4114 Computer and Space Science Building, University of Maryland, College Park, MD 20742-2461. (301) 405-7892.

Mathematical Association of America. 1529 18th Street, NW, Washington, DC 20036. (202) 387-5200. FAX (202) 265-2384.

National Council of Teachers of Mathematics. 1906 Association Drive, Reston, VA 22091. (703) 620-9840. FAX (703) 475-2970.

Operations Research Society of America. 1314 Guilford Avenue, Baltimore, MD. 21202. (410) 528-4146. FAX (410)528-8556.

Society For Industrial and Applied Mathematics. 3600 University City Science Center, Philadelphia, PA 19104-2688. (215) 382-9800. FAX (215) 386-7999.

Special Interest Group On Numerical Mathematics (signum). c/o ACM. 1515 Broadway, New York, NY 10036-5701. (212) 869-7440. FAX (212) 944-1318.

Special Interest Group For Symbolic and Algebraic Manipulation (SIGSAM). c/o ACM. 1515 Broadway, New York, NY 10036-5701. (212) 869-7440. FAX (212) 944-1318.

BIBLIOGRAPHIES

Books You Can Count On: Linking Mathematics & Literature. Rachel Griffiths and Margaret Clyne. Heinemann Publ. 361 Hanover Street, Portsmouth, NH 03801-3912. (603) 431-7894. (800) 541-2086. 1991. $15.00. ISBN: 0-435-08322-8.

Bibliography of Mathematical Works Printed in America Through 1850. Louis C. Karpinski. Supplements 1 & 2, Bernard I. Cohen, editor. Ayer Company Publ. Inc. Lower Mill Road, North Stratford, NH 03590. (603) 922-5105. (800) 282-5413. Reprint edition. 1980. $66.95.

International Mathematical Olympiads 1978-1985. Murray S. Klamkin. National Council of Teachers of Mathematics. 1906 Association Drive, Reston, VA 22091. (703) 620-9840. (800) 235-7566. 1987. $11.75. ISBN: 0-88385-631-X.

Mathematical Book Review Index, 1800-1940. Louise S. Grinstein. Garland Publ. Fifth Avenue, Suite 2500, New York, NY 10022-8101. (212) 751-7447. 1992. $72.00. ISBN: 0-8240-4114-3.

Mathematical Journals: An Annotated Guide. Diana F. Liang. Scarecrow Press. Distributed by University Press of America, 4720 Boston Way, Lanham, MD 20706. (301) 459-3366. (800) 462-6420. $29.50. ISBN: 0-8108-2585-6.

DIRECTORIES AND BIOGRAPHICAL SOURCES

Assistantship and Graduate Fellowships in Mathematical Sciences. American Mathematical Society, P.O. Box 6248, Providence, RI 02940. (401) 455-4000. (800) 321-4267. Annual. $19.00. ISSN: 1040-7650.

Combined Membership List. American Mathematical Society, P.O. Box 6248, Providence, RI 02940. (401) 455-4000. (800) 321-4267. Annual. $50.00. ISSN: 0569-6461.

Mathematical Science Professional Directory. American Mathematical Society, P.O. Box 6248, Providence, RI 02940. (401) 455-4000. (800) 321-4267. Annual. $45.00. ISBN: 0-8218-0173-2.

Research Centers Directory. Gale Research Company, 835 Penobscot Building, Detroit, MI 48226-4094. (313) 961-2242. (800) 347-4253. 20th edition. 1995. Annual. $485.00. ISSN: 0080-1518.

World Directory of Mathematicians. G.D. Mostow, editor. International Mathematical Union, Helsinki. 1958 to present. 9th revised edition. 1990. $40.00. Available from: American Mathematical Society, P.O. Box 6248, Providence, RI 02940. (401) 455-4000. ISSN: 0512-2740.

ENCYCLOPEDIAS AND DICTIONARIES

Beyond Numeracy: An Uncommon Dictionary of Mathematics. John A. Paulos. Alfred A. Knopf Inc., 201 East 50th Street, New York, NY 10022. (800) 733-3000. 1991. $21.50. ISBN: 0-394-58640-9.

Concise Oxford Dictionary of Mathematics. Christopher Clapham. Oxford University Press, 200 Madison Avenue, New York, NY 10016. (212) 725-6000. (800) 334-4249. 1990. $10.95. ISBN: 0-19-286103-4.

Encyclopedia of Mathematics. 1987. 10 volumes. $147.00. ISBN: 0-685-74040-4, 1-55608-010-7.

Encyclopedia of Mathematics and Its Applications. Addison-Wesley Publishing Co., Inc., 1 Jacob Way, Reading, MA 01867. (617) 944-3700. (800) 447-2226. 1976 to present. Irregular. Volumes are individually priced.

Encyclopedic Dictionary of Mathematics. Kiyosi Ito, editor. MIT Press, 55 Hayward Street, Cambridge, MA 02142. (617) 625-8569. 2nd edition, reprinted. 1993. 2 volumes. $70.00. ISBN: 0-262-59020-4.

HarperCollins Dictionary of Mathematics. E. J. Borowski and J. J. Borwein. HarperCollins Publ. 10 East 53rd Street, New York, NY 10022-5299. (800) 331-3761. 1990. $10.95. ISBN: 0-19-286103-4.

Mathematics Dictionary. Robert C. James, editor. Van Nostrand Reinhold, 115 Fifth Avenue, New York, NY 10003. (800)-842-3636. 5th edition. 1992. $42.95. ISBN: 0-442-00741-8.

Penguin Dictionary of Mathematics. J. Daintith and R. D. Nelson. Viking Penguin, 375 Hudson Street, New York, NY 10014-3657. (800) 331-4624. 1989. $12.00. ISBN: 0-14-051119-9.

The Words of Mathematics: An Etymological Dictionary of Mathematical Terms Used in English. Steven Schwartzman. Mathematical Association of America, 18th Street NW, Washington, DC 20036. (800) 331-1622. 1994. $29.50. ISBN: 0-88385-551.

GENERAL WORKS

Applied Mathematics; Getting Started, Getting It Done. William T. Shaw and Jason Tigg. Addison-Wesley Publishing Company, Inc., 1 Jacob Way, Reading, MA 01867. (617) 944-3700. (800) 447-2226. 1994. $31.25. ISBN: 0-201-54217-X.

Fractals Everywhere. Michael F. Barnsley. Academic Press Inc., 6277 Sea Harbor Drive, Orlando, FL 32887. (800) 321-5068. 1988. $49.00. ISBN: 0-12-079062-9.

Fundamentals of Mathematics. H. Behnke et al, editors. MIT Press, MIT Press, 55 Howard Street, Cambridge, MA 02142. (617) 625-8569. Volume 1, Foundations of Mathematics: Real Number Systems and Algebra; Volume 2, Geometry; Volume 3, Analysis. 1984; reprint edition. $75.00.

The Magic of Numbers. Eric T. Bell. Dover Publications Inc., 180 Varick Street, New York, NY 10014. (212) 255-3755. (800) 223-3130. reprint edition. $8.95.

Mathematica: A System For Doing Mathematics By Computer. Stephen Wolfram. Addison-Wesley Publishing Company, Inc., 1 Jacob Way, Reading, MA 01867. (617) 944-3700. (800) 447-2226. 2nd edition. 1991. ISBN: 0-201-51502-4

Mathematics: Queen and Servant of Science. Eric T. Bell. Mathematical Association of America, 18th Street NW, Washington, DC 20036. (202) 387-5200. (800) 331-1622. 1988. $16.50; reprint edition. ISBN: 0-88385-447-3.

Mathematics For the Millions: How To Master the Magic of Numbers. Lancelot Hogben. W. W. Norton & Company, Inc., 500 Fifth Avenue, New York, NY 10010. (212) 345-5500.(800) 223-2584. 1993. $14.95. ISBN: 0-393-31071-X.

The Maple Handbook. Darren Redfern. Springer Verlag New York, Inc., 175 Fifth Avenue, New York, NY 10010. (212) 460-1500. (800) 777-4643. 2nd edition. 1995. $29.00. 0-387-94331-5.

Merging of Disciplines: New Directions in Pure, Applied and Computational Mathematics. R. Ewing et al, editors. Springer-Verlag New York, Inc., 175 Fifth Avenue, New York, NY 10010. (212) 460-1500. (800) 777-4643. 1986. $39.00. ISBN: 0-387-06414-3.

MATHEMATICS

Ency. of Physical Sciences and Engineering Info. Sources

Numerical Analysis Problem Solver. Research & Education Association, 61 Ethel Road, West, Piscataway, NJ 08854. (908) 819-8880. 1994; revised edition. $29.95. ISBN: 0-87891-549-4.

Poetry of the Universe: A Mathematical Exploration of the Cosmos. Robert Osserman. Doubleday & Co., 1540 Broadway, New York, NY 10036-4094. (212) 345-6500. (800) 223-6834. 1995. $18.95. ISBN: 0-385-47340-0.

Sourcebook in Mathematics. David E. Smith. Dover Publications Inc., 180 Varick Street, New York, NY 10014. (212) 255-3755. (800) 223-3130. 1984; reprint edition. $13.95. ISBN: 0-486-64690-4.

Statistics For Research. S. Dowdy and S. Wearden. John Wiley & Sons, Inc. 605 Third Avenue, New York, NY 10158. (800) 526-5368. 2nd edition. 1991. $69.95. ISBN: 0-474-85703-3.

The Visual Mind: Art and Mathematics. Michele Emmer. MIT Press, 55 Hayward Street, Cambridge, MA 02142. (617) 625-8569. 1993. $45.00. ISBN: 0-262-05048-X.

Writing Mathematics Well. Leonard Gillman. Mathematical Association of America, 18th Street NW, Washington, DC 20036. (202) 387-5200. (800) 331-1622. 1987. $8.00. ISBN: 0-88385-443-0.

The World Treasury of Physics, Astronomy, and Mathematics. Timothy Ferris. Little, Brown & Co., 1271 Avenue of the Americas, New York, NY 10020. (212) 522-8700. (800) 343-9204. 1993. $19.95. ISBN: 0-316-28133-6.

HANDBOOKS AND MANUALS

Handbook of Applicable Mathematics. Walter Ledermann, editor. Wiley 1 Wiley Drive, Somerset, NJ 08875-1272. 1985. $1496.00 6 volumes. ISBN: 0-471-90804-5.

Handbook of Applied Mathematics; Selected Results and Methods. Carl E. Pearsen, editor. Chapman & Hall, 1 Penn Plaza, New York NY 10119. (212) 564-1060. 1990. $46.95. ISBN: 0-442-00521-0.

Handbook of Mathematical Functions with Formulas, Graphs, and Mathematical Tables. Milton Abramowitz and Irene A. Stegun, editors. U.S. National Bureau of Standards. U.S. Government Printing Office, Washington, D.C. 10th edition. 1972. $43.00. ISBN: 0-16-000202-8.

Handbook of Mathematics. I. N. Bronshtein and K. A. Semedyayev. Van Nostrand Reinhold, 115 Fifth Avenue, New York, NY 10003. (800)842-3636. 20th edition. $59.95. ISBN: 0-442-21171-6.

Handbook of Mathematical Sciences. William H. Beyer, editor. CRC Press, 2000 Corporate Boulevard, Boca Raton, FL 33431. (800) 272-7737. 6th edition. 1987. $91.95. ISBN: 0-8493-0656-6.

Handbook of Tables For Probability and Statistics. William H. Beyer, editor. CRC Press, 2000 Corporate Boulevard, Boca Raton, FL 33431. (800) 272-7737. 2nd edition. 1968.$92.95. ISBN: 0-8493-0692-2.

Handbook of Writing For the Mathematical Sciences. Society for Industrial and Applied Mathematics, 3600 University City

Science Center, Philadelphia, PA 19104-2688. (215)382-9800. FAX (215) 386-7999. 1993. $21.50. ISBN: 0-89871-314-5.

Standard Mathematical Tables and Formulas. William H. Beyer, editor. CRC Press, 2000 Corporate Boulevard, Boca Raton, FL 33431. (800) 272-7737. 29th edition. 1991. $41.95. ISBN: 084-930-6299.

Tables of Integrals, Series and Products: Corrected and Enlarged Edition. I. S. Gradshteyn and I. M. Ryzhik. Academic Press, Inc., 6277 Sea Harbor Drive, Orlando, FL 32821. (800) 321-5068. 5th edition. 1993. $61.00. ISBN: 0-12-294755-X.

ONLINE DATABASES AND CD-ROMS

CA Search. Chemical Abstracts Service, P.O. Box 3012, Columbus, OH 43210-0012. (614) 447-3600. (800) 848-6533. FAX (614) 447-3709. Very comprehensive guide to worldwide chemical literature and related fields, 1972 to present. Available on BRS,(800) 289-4277, DIALOG, (800) 334-2564, ORBIT (800) 456-7248, and STN International, FIZ Karlsruhe, P.O. Box 2465, W-7500, Karlsruhe 1, Germany, online services. Inquire as to cost and availability.

Compendex Plus. Engineering Information, Inc., 345 East 47th Street, New York, NY 10017. (212) 705-7600 or (800) 221-1044. Contains citations with abstracts to worldwide literature in engineering and technology, from 1970 to present. Available on online BRS,(800) 289-4277, DIALOG, (800) 334-2564, ORBIT (800) 456-7248, and STN International, FIZ Karlsruhe, P.O. Box 2465, W-7500, Karlsruhe 1, Germany, online services. Also available on CD-ROM. Inquire as to cost and availability.

Current Contents Search. Institute for Scientific Information, 3501 Market Street, Philadelphia, PA 19104. (215) 386-0100. FAX (215) 386-6362. Contains citations to articles listed in the table of contents of science and technology journals. Also articles in social sciences and life sciences journals. Available on BRS,(800) 289-4277, DIALOG,(800) 334-2564, online services. Inquire as to cost and availability.

Dissertation Abstracts. University Microfilms International, 300 North Zeeb Road, Ann Arbor, MI 48106. (800) 521-0600 or (313) 761-4700. Scope includes virtually all doctoral dissertations accepted at accredited American institutions from 1861 to present in 252 subject areas. Available on BRS,(800) 289-4277, DIALOG,(800) 334-2564, and OCLC EPIC,(800) 848-5878, online services. Also available on CD-ROM. Inquire as to cost and availability.

INSPEC. Institution of Electrical Engineers, Michael Faraday House, Six Hills Way, Stevenage, Herts. SG1 2AY, England. Telephone: 0438 313311 or FAX 0438 742840. Contains citations to the worldwide literature of physics, electronics and electrical engineering, computer technology, and related fields. Available on BRS,(800) 289-4277, DIALOG, (800) 334-2564, ORBIT, (800) 456-7248, and STN International, FIZ Karlsruhe, P.O. Box 2465, W-7500, Karlsruhe 1, Germany, online services. Inquire as to cost and availability.

Mathsci and Mathscinet. American Mathematical Society, P.O. Box 6248, Providence, RI 02940. (800) 321-4667 or (401) 455-4000 or FAX (401) 331-3842. Scope includes pure and applied mathematics and related areas of physics, statistics, engineering, computer science, and operations research literature since 1959. Available on DIALOG,(800) 334-2564, online service. Also available on CD-ROM and MATHSCINET, on the Internet. Inquire as to cost and availability.

NTIS Bibliographic Database. National Technical Information Service, 5285 Port Royal Road, Springfield, VA 22161. (703) 487-4929 or FAX (703) 321-8199. Broad coverage of government-sponsored science and technology research reports, 1964 to present. Available on BRS,(800) 289-4277, DIALOG, (800) 334-2564, ORBIT, (800) 456-7248, and STN International, FIZ Karlsruhe, P.O. Box 2465, W-7500, Karlsruhe 1, Germany, online services. Also available on CD-ROM. Inquire as to cost and availability.

Scientific and Technical Books and Serials in Print. R.R. Bowker Inc., 121 Chanlon Road, New Providence, NJ 07974. (800) 521-8110 or (908) 464-6800. List of currently published books and serials in the physical and biological sciences, engineering and technology, with subject, author and titles indexes. Available on ORBIT,(800) 456-7248, online service. Inquire as to cost and availability.

Scisearch. Institute for Scientific Information, 3501 Market Street, Philadelphia, PA 19104. (800) 523-1850 or (215) 386-0100. Broad multidisciplinary title and author index to the international literature of science and technology, 1974 to present. Available on DIALOG,(800) 334-2564, and ORBIT,(800) 456-7248, online services. Also available on CD-ROM. Inquire as to cost and availability.

Wilsonline. H.W. Wilson Company, 950 University Avenue, Bronx, NY 10452. (800) 367-6770 or (212) 588-8400. Makes available online versions of the printed H.W. Wilson indexes, including Applied Science and Technology Index, Business Periodicals Index, General Science Index, and Readers' Guide to Periodical Literature. Period covered is generally 1983 to present. Available on BRS,(800) 289-4277, DIALOG,(800) 334-2564, and OCLC EPIC,(800) 848-5878, online services. Also available on CD-ROM. Inquire as to cost and availability.

PERIODICALS

American Journal of Mathematics. Johns Hopkins University Press, 2715 North Charles Street, Baltimore, MD 21218. (410) 516-6987. FAX (410) 516-6968. 1878 to present. Bimonthly. $162.00 per year. ISSN: 0002-9327.

American Mathematical Monthly. Mathematical Association of America. 1529 Eighteenth Street, NW, Washington, DC 20036. (202) 387-5200. 1894 to present. Ten issues per year. $160.00. ISSN: 0002-9890.

Canadian Applied Mathematical Quarterly. Canadian Mathematical Society. Distributed by Rocky Mountain Mathematics Consortium, Department of Mathematics, Arizona State University, Tempe AZ 85287-1904. 1992 to present. Quarterly. $175.00 per year. ISSN: 1073-1849.

College Mathematics Journal. Mathematical Association of America. 1529 18th Street NW, Washington D.C. 20036. (202) 387-5200. 1970 to present. Five issues per year. $98.00. ISSN: 0746-8342.

Communications On Pure and Applied Mathematics. Courant Institute of Mathematical Sciences. Distributed by John Wiley & Sons, Inc. 605 Third Avenue, New York, NY 10158-0012. (212) 850-6000. FAX (212) 850-6088. 1939 to present. $730.00 per year. ISSN: 0010-3640.

IMA Journal of Applied Mathematics. Oxford University Press, 2001 Evans Road, Cary, NC 27513. (919) 677-0977. 1981 to present. Bimonthly. $410.00 per year. ISSN: 0272-4960.

Industrial Mathematics. Industrial Mathematics Society, Box 159 Roseville, MI 48066. (313) 771-0403. 1950 to present. Semiannual. $15.00 per year. ISSN: 0019-8528.

Journal of Recreational Mathematics. Baywood Publishing Co. Inc., 26 Austin Avenue, Box 337, Amityville, NY 11701. (516) 691-1770. 1968 to present. Quarterly. $78.00. ISSN: 0022-412X.

Mathematical Intelligencer. Springer-Verlag, 175 Fifth Avenue, New York, NY 10010. (212) 460-1500. FAX (212) 473-6272. (800) 777-4643. 1978 to present. Quarterly. $39.00 per year. ISSN: -343-6993.

Mathematics Magazine. Mathematical Association of America. 1529 Eighteenth Street, NW, Washington, DC 20036. (202) 387-5200. 1926 to present. Bimonthly. $80.00 per year. ISSN: 0025-570X.

Quarterly Journal of Mechanics and Applied Mathematics. Oxford University Press, 2001 Evans Road, Cary, NC 27513. (919) 677-0977. 1948 to present. Quarterly. $214.00 per year. ISSN: 0033-5614.

Risk Analysis. Society for Risk Analysis. Plenum Publishing Corp. 233 Spring Street, New York, NY 10013-1578. (212) 620-8000. FAX (212) 463-0742. 1980 to present. Bimonthly. $320.00 per year. ISSN: 0272-4332.

Siam Journal On Applied Mathematics. Society for Industrial and Applied Mathematics, 3600 University City Science Center, Philadelphia, PA 19104-2688. (215) 382-9800. FAX (215) 386-7999. 1953 to present. Bimonthly. $242.00 per year. ISSN: 0036-1399.

Siam Journal On Discrete Mathematics. Society for Industrial and Applied Mathematics, 3600 University City Science Center, Philadelphia, PA 19104-2688. (215) 382-9800. FAX (215) 386-7999. 1980 to present. Quarterly. $220.00 per year. ISSN: 0895-4801.

Siam Journal On Mathematical Analysis. Society for Industrial and Applied Mathematics, 3600 University City Science Center, Philadelphia, PA 19104-2688. (215) 382-9800. FAX (215) 386-7999. 1970 to present. Bimonthly. $352.00 per year. ISSN: 0036-1410.

Siam Journal On Numerical Analysis. Society for Industrial and Applied Mathematics, 3600 University City Science Center, Philadelphia, PA 19104-2688. (215) 382-9800. FAX (215) 386-7999. Bimonthly. $255.00 per year. ISSN: 0036-1429

Siam Review. Society for Industrial and Applied Mathematics, 3600 University City Science Center, Philadelphia, PA 19104-2688. (215) 382-9800. 1959 to present. Quarterly. $148.00 per year. ISSN: 0036-1445.

RESEARCH CENTERS AND INSTITUTES

Center For Applied Mathematics. Cornell University. 305 Sage Hall, Ithaca, NY 14853-6201. (607) 255-4335. FAX(607) 255-4336.

Center For Applied Mathematics (cam). University of Georgia. Tucker Hall, Athens, GA 30602. (404) 542-3491.

Center For Mathematical Science Research. Rutgers University, New Brunswick, NJ 08903. (201) 932-3117. FAX (201) 932-

MATHEMATICS

Ency. of Physical Sciences and Engineering Info. Sources

5530.

Center For Mathematical Sciences. University of Wisconsin-Madison, 610 Walnut Street, Madison, WI 53705. (608) 263-2696. FAX (608) 263-2841.

Center For Pure and Applied Mathematics. University of California, Berkeley, 977 Evans Hall, Berkeley, CA 94720. (415) 642-0116. FAX (415) 642-6726.

Courant Institute of Mathematical Sciences. New York University. 251 Mercer Street, New York, NY 10012. (212) 998-3000.

Institute For Computer Applications in Science and Engineering (ICASE). Universities Space Research Association. Mail Stop 132C, NASA, Langley Research Center, Hampton, VA 23665. (804) 865-2513. FAX (804 864-6134.

Institute of Applied Mathematics (IAM). University of British Columbia. 222-1984 Mathematics Road, Vancouver, BC Canada V6T 1Y4. (604) 228-4584. FAX (604) 228-6074.

Lefschetz Center For Dynamical Systems. Brown University. 182 George Street, Providence, RI 02912. (401) 863-2358.

MECHANICAL ENGINEERING

ABSTRACT SERVICES AND INDEXES

Applied Mechanics Reviews: An Assessment of World Literature in Engineering Sciences. 1948-present. American Society of Mechanical Engineers, 345 East 47th Street, New York, NY 10017. (212) 705-7703. Monthly. $360.00 per year.

Applied Science and Technology Index; A Cumulative Subject Index To English Language Periodicals in the Fields of Aeronautics and Space Science, Computer Technology, Chemistry, Construction Industry, Energy and Related Areas. H.W. Wilson Co., 950 University Avenue, Bronx, NY 10452. (800) 367-6770 or (212) 588-8400. FAX (718) 590-1617. From 1958 to present. Monthly. Inquire about cost and availability. Also available on CD-ROM and online.

Computer Abstracts, MCB University Press Ltd., PO Box 10812, Birmingham, AL 35201-0812. (800) 633-4931. FAX (205) 995-1588. Monthly. Covers computer theory, data, hardware, systems, networks, human-computer interaction, artificial intelligence, as well as applications of computers in aerospace, business, CAD/CAM, cartography, civil engineering, electronics and electrical engineering, industrial engineering, mechanical engineering, medicine, structural engineering, etc.

Current Contents: Engineering, Technology, and Applied Sciences. Institute for Scientific Information, 3501 Market Street, Philadelphia, PA 19104. (215) 386-0100. FAX (215) 386-6362. 1970 to present. Weekly. $442.00 per year.

Engineering Index Monthly. Engineering Information, Inc., Castle Point on the Hudson, Hoboken, NJ 07030. (800) 221-1044. FAX (212) 832-1857. Monthly. $2300.00 per year. Also available online as COMPENDEX, and also on CD-ROM. Covers chemical engineering, computers, electrical engineering, civil

engineering, metals and mining, industrial management, and mechanical engineering.

Index To Scientific and Technical Proceedings. Institute for Scientific Information, 3501 Market St., Philadelphia, PA 19104. (215) 386-0100. FAX (215) 386-6362. Monthly. $500.00 per year.

Mechanical Engineering Abstracts (formerly ISMEC), Cambridge Scientific Abstracts, 7200 Wisconsin Avenue, Bethesda, MD 20814-4823. (301) 961-6750. FAX (301) 961-6720. 1967 to present. Monthly. $895.00 per year. Summarizes world literature in mechanical engineering, production engineering, and engineering management. Also available online.

Science Citation Index. Institute for Scientific Information, 3501 Market Street, Philadelphia, PA 19104. (215) 386-0100. FAX (215) 386-6362. Inquire about availability and cost. Also available on CD-ROM.

ASSOCIATIONS AND PROFESSIONAL SOCIETIES

American Society of Mechanical Engineers. 345 E. 47th Street, New York, NY 10017. (212) 705-7722.

Vibration Institute. 6262 S. Kingery Highway, Suite 212, Willowbrook, IL 60514. (708) 654-2254.

BIBLIOGRAPHIES

Scientific and Technical Books and Serials in Print. R.R. Bowker Inc., 121 Chanlon Road, New Providence, NJ 07974. (800) 521-8110 or (908) 464-6800. List of currently published books and serials in the physical and biological sciences, engineering and technology, with subject, author and titles indexes. Inquire as to cost and availability. Also available on ORBIT online service, (800) 456-7248.

DIRECTORIES AND BIOGRAPHICAL SOURCES

Directory of Engineering Societies and Related Organizations. Gordon Davis. 13th edition. American Association of Engineering Societies, 1111 19th Street NW, Suite 608, Washington, DC 20036. (202) 296-2237 or (800) 658-8897. 1989. Inquire for price.

International Research Centers Directory. Gale Research, 835 Penobscot Building, Detroit, MI 48226-4094. (313) 961-2242. (800) 347-4253. FAX (313) 961-6083. 8th edition. 1995. $410.00

Research Centers Directory. Gale Research, 835 Penobscot Building, Detroit, MI 48226-4094. (313) 961-2242. (800) 347-4253. FAX (313) 961-6083. $485.00.

Scientific and Technical Organizations and Agencies Directory. Gale Research, 835 Penobscot Building, Detroit, MI 48226-4094. (313) 961-2242. (800) 347-4253. FAX (313) 961-6083. 4th edition. 1996. $195.00.

Who's Who in Engineering. American Association of Engineering Societies, 1111 19th Street NW, Suite 608, Washington, DC 20036. (202) 296-2237 or (800) 658-8897. 8th edition. 1991. Inquire for price.

ENCYCLOPEDIAS AND DICTIONARIES

Dictionary of Mechanical Engineering. G.H. Nayler. 3rd edition. Butterworth-Heinemann, 313 Washington Street, Newton, MA 02158. (617) 928-2500 or (800) 366-2665. FAX (617) 928-2620. 1985. $84.95.

Encyclopedia of Fluid Mechanics. 10 volumes plus 3 supplements. Gulf Publishing Company, P.O. Box 2608, Houston, TX. (713) 529-4301 or (800) 231-6275. 1986-90 (vols. 1-10), 1993-94 (supplements). $195.00 each (vols. 1-10), $139.00 each (supplements).

McGraw-Hill Encyclopedia of Engineering. Sybil P. Parker, ed. 2nd edition. McGraw-Hill Publishing Company, 1221 Avenue of the Americas, New York, NY 10020. (800) 262-4729 or (212) 512-3825. 1993. $95.50.

Thesaurus of Scientific, Technical, and Engineering Terms. Hemisphere Publishing Corporation, 1900 Frost Road, Suite 101, Bristol, PA 19007-1598. (215) 785-5800 or (800) 821-8312. FAX (215) 785-5515. 1987. $173.00.

GENERAL WORKS

Basic Mechanics of Machines. Samuel Doughty. John Wiley and Sons, Inc., 605 Third Avenue, New York, NY 10158. (800) 526-5368 or (212) 850-6000. 1988. Inquire for cost and availability.

Engineering Mechanics. D.K. Anand & P.F. Cunniff. Prentice Hall, 113 Sylvan Avenue, Route 9W, Englewood Cliffs, NJ 07632. (201) 592-2000 or (800) 922-0579. 1984. Inquire for cost and availability.

Engineering Mechanics in Civil Engineering. A.P. Boresi & K.P. Chong, editors. 2 volumes. American Society of Civil Engineers, 345 East 47th Street, New York, NY 10017-2398. (212) 705-7520 or (800) 548-2723. 1984. $134.00.

Form, Structure, and Mechanism. M. French. Springer-Verlag, 175 Fifth Avenue, New York, NY 10010. (212) 460-1500 or (800) 777-4643. FAX (212) 473-6272. 1992. $49.95.

Mechanical Engineering Design. J.E. Shigley & C.R. Mischke. 5th revised edition. McGraw-Hill Publishing Company, 1221 Avenue of the Americas, New York, NY 10020. (800) 262-4729 or (212) 512-3825. 1989. Inquire for cost and availability.

Theory of Machines and Mechanisms. J.E. Shigley & J.J. Uicker. 2nd edition. McGraw-Hill Publishing Company, 1221 Avenue of the Americas, New York, NY 10020. (800) 262-4729 or (212) 512-3825. 1994. $68.13.

HANDBOOKS AND MANUALS

Handbook of Mechanical Engineering. W. Beitz & K.H. Kuttner, editors. Springer-Verlag, 175 Fifth Avenue, New York, NY 10010. (212) 460-1500 or (800) 777-4643. FAX (212) 473-6272. 1994. Inquire for cost and availability.

Handbook of Mechanics, Materials, and Structures. Alexander Blake, editor. John Wiley and Sons, Inc., 605 Third Avenue, New York, NY 10158. (800) 526-5368 or (212) 850-6000. 1985. $120.00.

Machine Design Fundamentals: A Mechanical Designers' Workbook. Joseph E. Shigley & Charles R. Mischke, editors-in-chief. McGraw-Hill Publishing Company, 1221 Avenue of the Americas, New York, NY 10020. (800) 262-4729 or (212) 512-3825. 1989. $31.95.

Marks' Standard Handbook For Mechanical Engineers. E.A. Avallone and T. Baumeister III. 9th edition. McGraw-Hill Publishing Company, 1221 Avenue of the Americas, New York, NY 10020. (800) 262-4729 or (212) 512-3825. 1987. $133.00.

Mechanical Design and Systems Handbook. Harold A. Rothbart, editor-in-chief. McGraw-Hill Publishing Company, 1221 Avenue of the Americas, New York, NY 10020. (800) 262-4729 or (212) 512-3825. 1985. Inquire for cost and availability.

Mechanical Engineer's Data Handbook. James Carvill. CRC Press, 2000 Corporate Blvd., N.W., Boca Raton, FL 33431. (407) 994-0555 or (800) 272-7737. FAX (407) 994-0949. 1993. $69.95.

Mechanical Engineer's Reference Book. Edward H. Smith, editor. 12th edition. SAE International, 400 Commonwealth Drive, Warrendale, PA 15096-0001. (412) 776-4970. FAX (412) 776-0790. 1994. $155.00.

Mechanical Engineering Reference Manual. Michael R. Lindeburg. 9th edition. Professional Publications, 1250 Fifth Avenue, Belmont, CA 94002. (415) 593-9119 or (800) 426-1178. 1994. $48.95.

Standard Handbook of Machine Design. Joseph E. Shigley & Charles R. Mischke, editors-in-chief. McGraw-Hill Publishing Company, 1221 Avenue of the Americas, New York, NY 10020. (800) 262-4729 or (212) 512-3825. 1986. $124.50.

ONLINE DATABASES AND CD-ROMS

Compendex Plus. Engineering Information, Inc., 345 East 47th Street, New York, NY 10017. (212) 705-7600 or (800) 221-1044. Contains citations with abstracts to worldwide literature in engineering and technology, from 1970 to present. Available on online BRS,(800) 289-4277, DIALOG, (800) 334-2564, ORBIT (800) 456-7248, and STN International, FIZ Karlsruhe, P.O. Box 2465, W-7500, Karlsruhe 1, Germany, online services. Also available on CD-ROM. Inquire as to cost and availability.

CSA Engineering. Cambridge Scientific Abstracts, 7200 Wisconsin Avenue, Suite 601, Bethesda, MD 20814. (301) 961-6750 or (800) 843-7751. Contains citations and abstracts of international periodicals and research literature covering all fields of engineering and science and technology, including computer and information science, electronics, mechanical engineering, solid state materials, 1981 to present. Available on BRS,(800) 289-4277, online service. Inquire as to cost and availability.

Current Contents Search. Institute for Scientific Information, 3501 Market Street, Philadelphia, PA 19104. (215) 386-0100. FAX (215) 386-6362. Contains citations to articles listed in the table of contents of science and technology journals. Also articles in social sciences and life sciences journals. Available on BRS,(800) 289-4277, DIALOG,(800) 334-2564, online services. Inquire as to cost and availability.

Dissertation Abstracts. University Microfilms International, 300 North Zeeb Road, Ann Arbor, MI 48106. (800) 521-0600 or (313) 761-4700. Scope includes virtually all doctoral disserta-

MECHANICAL ENGINEERING

Ency. of Physical Sciences and Engineering Info. Sources

tions accepted at accredited American institutions from 1861 to present in 252 subject areas. Available on BRS,(800) 289-4277, DIALOG,(800) 334-2564, and OCLC EPIC,(800) 848-5878, online services. Also available on CD-ROM. Inquire as to cost and availability.

ISMEC: Mechanical Engineering Abstracts. Cambridge Scientific Abstracts, 7200 Wisconsin Avenue, Suite 601, Bethesda, MD 20814. (301) 961-6750 or (800) 843-7751. Contains citations to the literature in mechanical engineering, industrial and production engineering, energy, power, mechanics, devices and related areas, from 1973 to present. Available on the DIALOG,(800) 334-2564, online service. Inquire as to cost and availability.

NTIS Bibliographic Database. National Technical Information Service, 5285 Port Royal Road, Springfield, VA 22161. (703) 487-4929 or FAX (703) 321-8199. Broad coverage of government-sponsored science and technology research reports, 1964 to present. Available on BRS,(800) 289-4277, DIALOG, (800) 334-2564, ORBIT, (800) 456-7248, and STN International, FIZ Karlsruhe, P.O. Box 2465, W-7500, Karlsruhe 1, Germany, online services. Also available on CD-ROM. Inquire as to cost and availability.

Scisearch. Institute for Scientific Information, 3501 Market Street, Philadelphia, PA 19104. (800) 523-1850 or (215) 386-0100. Broad multidisciplinary title and author index to the international literature of science and technology, 1974 to present. Available on DIALOG,(800) 334-2564, and ORBIT,(800) 456-7248, online services. Also available on CD-ROM. Inquire as to cost and availability.

Wilsonline. H.W. Wilson Company, 950 University Avenue, Bronx, NY 10452. (800) 367-6770 or (212) 588-8400. Makes available online versions of the printed H.W. Wilson indexes, including Applied Science and Technology Index, Business Periodicals Index, General Science Index, and Readers' Guide to Periodical Literature. Period covered is generally 1983 to present. Available on BRS,(800) 289-4277, DIALOG,(800) 334-2564, and OCLC EPIC,(800) 848-5878, online services. Also available on CD-ROM. Inquire as to cost and availability.

PERIODICALS

International Communications in Heat and Mass Transfer. Elsevier Science [journals], 660 White Plains Rd., Tarrytown, NY 10159-5153. (919) 524-9200. FAX (919) 333-2444. 1974 to present. 6 times a year. $430.00 per year.

International Journal For Numerical Methods in Engineering. J. Wiley & Sons Ltd., Journals, Baffins Lane, Chichester, Sussex PO19 1UD, England. Telephone 0243-779777. FAX 0243-775878. 1964 to present. 24 times a year. $1795.00 per year.

International Journal of Heat and Mass Transfer. Elsevier Science [journals], 660 White Plains Rd., Tarrytown, NY 10159-5153. (919) 524-9200. FAX (919) 333-2444. 1960 to present. 18 times a year. $2070.00 per year.

International Journal of Multiphase Flow. Elsevier Science [journals], 660 White Plains Rd., Tarrytown, NY 10159-5153. (919) 524-9200. FAX (919) 333-2444. 1974 to present. 7 times a year. $900.00 per year.

Journal of Applied Mechanics. American Society of Mechanical Engineers, 345 E. 47th Street, New York, NY 10017. (212) 705-7722. 1935 to present. Quarterly. $120.00.

Journal of Engineering For Industry. American Society of Mechanical Engineers, 345 E. 47th Street, New York, NY 10017. (212) 705-7722. 1970 to present. Quarterly. $100.00 per year.

Journal of Fluid Control. Delbridge Publishing Company, Box 160817, Cupertino, CA 95016. (408) 446-3131. 1967 to present. Quarterly. $145.00 per volume.

Journal of Heat Transfer. American Society of Mechanical Engineers, 345 E. 47th Street, New York, NY 10017. (212) 705-7722. 1970 to present. Quarterly. $155.00 per year.

Journal of Mechanical Design. American Society of Mechanical Engineers, 345 E. 47th Street, New York, NY 10017. (212) 705-7722. 1978 to present. Quarterly. $100.00 per year.

Journal of Pressure Vessel Technology. American Society of Mechanical Engineers, 345 E. 47th Street, New York, NY 10017. (212) 705-7722. 1974 to present. Quarterly. $100.00 per year.

Lubrication Engineering. Society of Tribologist and Lubrication Engineers, 840 Busse Highway, Park Ridge, IL 60068-2376. (708) 825-5536. FAX (708) 825-1456. 1945 to present. Monthly. $61.00 per year.

Mechanical Engineering. American Society of Mechanical Engineers, 345 E. 47th Street, New York, NY 10017. (212) 705-7722. 1906 to present. Monthly. $45.00 for non-members.

RESEARCH CENTERS AND INSTITUTES

Laboratory For Experimental Mechanics Research. SUNY, Stony Brook, NY 11794-2300. (516) 632-8311. FAX (516) 632-8720.

Mechanical Engineering Design Laboratory. University of Florida, 237 Mechanical Engineering Bldg., Gainesville, FL 32611. (904) 392-0827. FAX (904) 392-1071.

Mechanical Engineering Laboratories. Stevens Institute of Technology, Hoboken, NJ 07030. (201) 420-5591. FAX (201) 420-6978.

Mechanical Engineering Research Laboratory. Kansas State University, Durland Hall, Manhattan, KS 66506. (913) 532-5610.

MECHANICS

See also: FLUID MECHANICS, KINEMATICS, MECHANICAL ENGINEERING, PHYSICS, QUANTUM MECHANICS, STATISTICAL MECHANICS

ABSTRACT SERVICES AND INDEXES

Applied Mechanics Reviews: An Assessment of World Literature in Engineering Sciences. American Society of Mechanical Engineers, 345 East 47th Street, New York, NY 10017. (212) 705-7703. 1948 to present. Monthly. $360.00 per year. ISSN: 0003-6900.

Applied Science and Technology Index; A Cumulative Subject Index To English Language Periodicals in the Fields of Aeronautics and Space Science, Computer Technology, Chemistry, Construction Industry, Energy and Related Areas. H.W. Wilson Co., 950 University Avenue, Bronx, NY 10452. (212) 588-8400. (800) 367-6770. FAX (718) 590-1617. From 1958 to present. Monthly. Inquire about cost and availability. Also available on CD-ROM and online. ISSN: 0003-6986.

Chemical Abstracts. Chemical Abstracts Service, 2540 Olentangy River Road, Box 3012, Columbus, OH 43210-0012. (614) 447-3600. (800) 848-6538. FAX (614) 447-3713. From 1907 to present. Weekly. $16800.00 per year. Also available on CD-ROM and online. Inquire regarding cost and availability. ISSN:

Current Contents: Physical, Chemical and Earth Sciences. Institute for Scientific Information, 3501 Market Street, Philadelphia, PA 19104. (215) 386-0100. FAX (215) 386-2291. From 1961 to present. Weekly. $442.00 per year. Also available on CD-ROM and online. Inquire regarding cost and availability. ISSN: 0163-2574.

Current Papers in Physics. Institute of Electrical Engineers, Michael Faraday House, Six Hill Way, Stevenage, Herts, SG1 2AY, England. Distributed by INSPEC/IEEE, Box 1331, 445 Hoes Lane, Piscataway, NJ 08855-1331. (908) 562-5549. 1966 to present. Fortnightly. $410.00 per year. ISSN: 0011-3786.

Engineering Index Monthly. Engineering Information, Inc., Castle Point on the Hudson, Hoboken, NJ 07030. (201) 216-8500. FAX (201) 216-8532. Monthly. $2200.00 per year. Also available online as COMPENDEX and on CD-ROM. ISSN: 0742-1974.

General Science Index. H.W. Wilson Company, 950 University Avenue, Bronx, NY 10452. (212) 588-8400. (800) 367-6770. FAX (718) 590-1617. From 1978 to present. Ten issues per year; quarterly and annual cumulations. Service basis. Available on CD-ROM and online. Inquire regarding cost and availability. ISSN: 0162-1963.

Government Reports Announcements and Index. U.S. National Technical Information Service (NTIS), 5285 Port Royal Road, Springfield, VA 22161. (703) 487-4650. FAX (703) 321-8547. From 1968 to present. Annual. $630.00 per year. Also available online as NTIS Bibliographic Database and on CD-ROM. ISSN:

ISMEC Bulletin (Information Service in Mechanical Engineering). Cambridge Scientific Abstracts, 5161 River Road, Bethesda, MD 20816. (301) 951-1400. 1973 to present. Monthly. $750.00 per year.

Mathematical Reviews: A Review Journal Covering the World Literature of Mathematical Research. American Mathematical Society, P.O. Box 1571, Annex Station, Providence, RI 02901-9930. (800) 556-7774 or (401) 455-4000. 1940 to present. Monthly. $4594.00 per year. Also available via network (MATHSCINET), online and on CD-ROM. Inquire regardingcost and availability. ISSN: 0025-5629.

NTIS Alerts: Physics. U.S. National Technical Information Service. 5285 Port Royal Road, Springfield, VA 22161. (703) 487-4650. FAX (703) 321-8547. Weekly. $140.00 per year.

Pascal 10: Mechanique, Acoustique Et Transport DE Chaleur. Centre National de la Recherche Scientifique, Institue de l'Information Scientifique et Technique, 2 aliee du Parc de Brabois, 54514 Vandoeuvre-Les Nancy Cedex, France. TEC 83-50-46-00. FAX 83-50-46-50. 1984 to present. Ten issues per year. 1530 F per year. Also available on CD-ROM and online. ISSN: 1136-5107.

Physics Abstracts. INSPEC. Section A, Science Abstracts. Institution of Electrical Engineers (IEE). Available from INSPEC/IEEE-Institute of Electrical and Electronic Engineers,, Box 1331, 445 Hoes Lane, Piscataway NJ 08855-1331. (908) 562-5549. 1898 to present. 24 issues per year. $2835.00 per year. Also available on CD-ROM and online. ISSN: 0036-8091.

Physics Briefs (Physikalische Berichte). Information Center for Energy, Physics, Mathematics; German Physical Society. VCH Publishers, Inc., 220 East 23rd Street, New York, NY 10010-4606. (212) 683-8333. 1845 to present. 24 issues per year. $2390.00 per year. Also available online. ISSN: 0179-7434.

Science Citation Index. SCI. Institute for Scientific Information, 3501 Market Street, Philadelphia, PA 19104. (215) 386-0100. (800) 523-1850. FAX (215) 386-2991. 1961 to present. Six issues per year, plus annual cumulation. $11650.00 per year. Also available online and on CD-ROM. Inquire about price and availability. ISSN: 0036-827X.

ANNUAL REVIEWS AND YEARBOOKS

Advances in Applied Mechanics. Academic Press, Inc., 6277 Sea Harbor Drive, Orlando, FL. (800) 321-5068. 1948 to present. Irregular. ISSN: 0065-2156.

Advances in Atomic, Molecular and Optical Physics. Academic Press, Inc., 6277 Sea Harbor Drive, Orlando, FL. (800) 321-5068. 1965 to present. Irregular. ISSN: 1049-250X.

Advances in Electronics and Electron Physics. Academic Press, Inc., 6277 Sea Harbor Drive, Orlando, FL. (800) 321-5068. From 1948 to present. Irregular. ISSN: 0065-2539.

Advances in Heat Transfer. Academic Press, Inc., 6277 Sea Harbor Drive, Orlando, FL. (800) 321-5068. 1964 to present. Irregular. ISSN: 0065-2717.

Advances in Physics. Taylor & Francis, Ltd., Rankine Road, Basingstoke, Hants RG24 8PR, England. 0256-840366. FAX 0256-47938. 1952 to present. Bimonthly. $511.00 per year. ISSN: 0001-8732.

Annual Review of Fluid Mechanics. Annual Reviews, Inc., 4139 El Camino Way, Palo Alto, CA 94303-0139. (415) 493-4400. (800) 523-8635. Fax (415) 855-9815. 1969 to present. Annual. $47.00 per year. ISSN: 0066-4189.

Solid State Physics: Advances in Research and Applications. Academic Press, Inc., 6277 Sea Harbor Drive, Orlando, FL. (800) 321-5068. 1955 to present. Irregular. ISSN: 0081-1947.

Topics in Applied Physics. Springer-Verlag New York, Inc., 175 Fifth Avenue, New York, NY 10010. (212) 460-1500. (800) 777-4643. FAX (212) 473-6272. ISSN: 0303-4216.

ASSOCIATIONS AND PROFESSIONAL SOCIETIES

American Academy of Mechanics. Virginia Tech, ESM Department, Blacksburg, VA 24061-0219. (703) 231-6841. FAX (703) 231-4574.

American Chemical Society. 1155 16TH Street NW, Washington DC 20036. (202) 872-4600. (800) 424-6747.

MECHANICS

Ency. of Physical Sciences and Engineering Info. Sources

American Institute of Physics. One Physics Ellipse, College Park, MD 20740-3843. (301) 209-3100.

American Physical Society. One Physics Ellipse, College Park, MD 20740-3843. (301) 209-3200. FAX (301) 209-0865.

American Society of Mechanical Engineers.345 East 47th Street, New York, NY 10017. (212) 705-7722. FAX (212) 705-7739.

Universities Research Association, 1111 19th Street NW, Suite 400, Washington, DC 20036. (202) 293-1382.

BIBLIOGRAPHIES

Guide To the Archival Collections in the Niels Bohr Library At the American Institute of Physics. Niels Bohr Library Staff. American Institute of Physics, One Physics Ellipse, College Park, MD 20740-3843. (301) 209-3100. 1995. $140.00

Science Books and Films. American Association for the Advancement of Science, 1333 H Street NW, Washington, DC 20005. (202) 326-6454. Reviews of print, film and software materials in all sciences. 1965 to present. Nine issues per year. $40.00 per year. ISSN: 0098-342X

Scientific and Technical Books and Serials in Print, 1995. An index to literature in science and technology. R.R. Bowker Inc., 121 Chanlon Road, New Providence, NJ 07974. (908) 464-6800. (800) 521-8110. 4 volumes. Annual. 1972 to present. $299.95. Also available on compact disc and online. ISSN: 0000-054X

DIRECTORIES AND BIOGRAPHICAL SOURCES

American Men and Women of Science: Physical and Biological Sciences. R. R. Bowker Inc., 121 Chanlon Road, New Providence, NJ 07974. (908) 464-6800. (800) 521-8110. 20th edition. 8 volumes. 1996. $850.00.

American Physical Society. Membership Directory Bulletin Issue.American Physical Society, One Physics Ellipse, College Park, MD 20740-3843. (301) 209-3200. FAX (301) 209-0865. Biennial. $50.00.

The Biographical Dictionary of Scientists: Physicists. D. Abbott, editor. Peter Bedrick Books, Inc., 2112 Broadway, Room 318, New York, NY 10023. (212) 496-0751. 1984.

Directory of Physics and Astronomy Staff Members. American Institute of Physics, One Physics Ellipse, College Park, MD 20740-3843. (301) 209-3100. Annual. $45.00.

Graduate Programs in Physics, Astronomy and Related Fields. 1995-1996. American Institute of Physics, One Physics Ellipse, College Park, MD 20740-3843. (301) 209-3100. Annual. $45.00. ISSN: 0147-1821.

International Directory of Engineering Societies and Related Organizations. American Society for Engineering Education, 1818 N Street NW, Suite 600, Washington, DC 20036. (202) 331-3526. 15th edition. 1996. $185.00.

Legends in Their Own Time: A Century of American Physical Scientists. Anthony Serafini. Plenum Publishing Corp., 233 Spring Street, New York, NY. (212) 620-8000. (800) 2221-9369. FAX (212) 463-0742. 1993. $27.50.

The Nobel Prize Winners. Frank N. Magill, editor. Salem Press, Inc., P.O. Box 1097, Englewood Cliffs, NJ 07632. (201) 871-3700. (800) 221-1592. 3 volumes. 1989. $210.00.

Research Centers Directory. Gale Research Inc., 835 Penobscot Building, Detroit, MI 48226-4094. (313) 961-2242. (800) 347-4253. 20th edition. 1995. $485.00. ISSN: 0080-1518.

Scientific and Technical Organizations and Agencies Directory. Gale Research, 835 Penobscot Building, Detroit, MI 48226-4094. (313) 961-2242. (800) 347-4253. 4th edition. 1996. $195.00.

Who's Who in Engineering. Gordon Davis, editor. American Society for Engineering Education, 1111 19th Street, NW, Suite 608, Washington, DC 20036. 9th edition. 1995. $220.00.

ENCYCLOPEDIAS AND DICTIONARIES

A Concise Dictionary of Physics. Oxford University Press, Inc., 200 Madison Avenue, New York, NY 10016. (212) 725-6000. (800) 334-4249. 1990. $10.95.

Dictionary of Effects and Phenomena in Physics. Joachim Schubert. VCH Publications, Inc., 220 East 23rd Street, Suite 909, New York, NY 10010-4606. (212) 683-8333. (800) 422-8824. 1987. $35.00.

Dictionary of the Physical Sciences: Terms, Formulas, Data. Cesare Emiliani. Oxford University Press, Inc., 200 Madison Avenue, New York, NY 10016. (212) 725-6000. (800) 334-4249. 1989. $19.95.

Encyclopedia of Applied Physics. George Trigg, editor. VCH Publications, Inc., 220 East 23rd Street, Suite 909, New York, NY 10010-4606. (212) 683-8333. (800) 422-8824. 20 volume set. 1991-. $5990.00.

Encyclopedia of Physics. Robert M. Besancon. Chapman & Hall, 1 Penn Plaza, New York, NY 10119. (212) 564-1060. 3rd edition. 1990. $54.95 in paper.

Facts On File Dictionary of Physics. John Daintith, editor. Facts-on-File, Inc., 460 Park Avenue South, New York, NY 10016-7382. (212) 683-2244. (800) 322-8755. Fax (800) 683-3633. Revised edition. 1990. $12.95.

Encyclopedia of Physical Science and Technology. Academic Press, Inc., 6277 Sea Harbor Drive, Orlando, FL. (800) 321-5068. 2nd edition. 18 volume set. 1992. $2625.00.

McGraw-Hill Encyclopedia of Science and Technology. McGraw-Hill Book Company, Inc., 1221 Avenue of the Americas, New York, NY 10020. (212) 512-2000. (800) 262-4729. 7th edition. 20 volume set. 1992. $1900.00.

Penguin Dictionary of Physics. Valerie Illingworth, editor. Viking Penguin, 375 Hudson Street, New York, NY 10014-3657. (212) 366-2000. (800) 331-4624. 2nd edition. 1991.

GENERAL WORKS

Albert Einstein, Three Complete Books: Relativity, Ideas and Opinions, and Out of My Later Years. Albert Einstein. Random House Inc., 201 East 50th Street, New York, NY 10022. (212) 751-2600. (800) 733-3000. 1991. $19.50. 1993. $7.99 in paper.

Big Science: The Growth of Large-Scale Research. Peter Galison and Bruce Hevly, editors. Stanford University Press, Stanford, CA 04305-2235. (413) 723-1593. 1992. $47.50.

Classical General Relativity. Subrahmanyan Chandrasekhar and Morton D. Hull, editors. Oxford University Press, Inc., 200 Madison Avenue, New York, NY 10016. (212) 725-6000. (800) 334-4249. 1993. $45.00.

Cosmic Catastrophes. C. P. Chapman and D. Morrison, editors. Plenum Publishing Corp., 233 Spring Street, New York, NY. (212) 620-8000. (800) 2221-9369. 1989. $22.95.

Dialogues Concerning Two New Sciences. Galilei Galileo. Prometheus Books, 89 John Glenn Drive, Amherst, NY 14228-2197. (716) 691-0133. (800) 421-0351. 1991, $9.95 in paper.

Doing Physics: How Physicists Take Hold of the World. Martin Krieger. Indiana University Press, 601 North Morton Street, Bloomington, in 47404-3797. (812) 855-6804. (800) 842-6796. 1992. $29.95.

Frontiers of Fundamental Physics. M. Barone and F. Selleri, editors. Plenum Publishing Corp., 233 Spring Street, New York, NY. (212) 620-8000. (800) 2221-9369. 1994. $145.00.

Fractals in Physics: Essays in Honor of Benoit B. Mandelbrot. A. Aharony and Jens Feder, editors. Elsevier Science Publishing Company, Inc., 655 Avenue of the Americas, New York, NY 10010. (212) 989-5800. FAX (914) 333-2444. 1990. $45.25 in paper.

Fundamental Particles and Forces. Richard A. Carrigan, Jr., and W. Peter Trower. W. H. Freeman & Co., 41 Madison Avenue, East 26th Avenue, 35th Floor, New York, NY 10010. (212) 576-0400. 1995. $13.95.

Gravitation. Charles W. Misner, et al. W. H. Freeman & Co., 41 Madison Avenue, East 26th Avenue, 35th Floor, New York, NY 10010. (212) 576-0400. 1995.

Landmark Experiments in Twentieth Century Physics. George L. Trigg. Dover Publications, Inc., 180 Varick Street, New York, NY 10014. (212) 255-3755. (800) 223-3130. 1995. $9.95 in paper.

Mathematics For Physicists. Robert L. Zimmerman and Frederick I. Olness. Addison-Wesley Publishing Co., Inc., 1 Jacob Way, Reading, MA 01867. (617) 944-3700. (800) 447-2226. 1995. $43.25 in paper.

Matter and Energy: Physics in Everyday Life. John Clark, editor. Oxford University Press, Inc., 200 Madison Avenue, New York, NY 10016. (212) 725-6000. (800) 334-4249. 1995. $35.00.

Mind, Matter, and Quantum Mechanics. Henry P. Stapp. Springer-Verlag New York, Inc., 175 Fifth Avenue, New York, NY 10010. (212) 460-1500. (800) 777-4643. FAX (212) 473-6272. 1994. $34.50.

Physics. David Haliday and Robert Resnick. John Wiley & Sons, Inc., 605 Third Avenue, New York, NY 10158-0012. (212) 850-6000. (800) 225-5945. 2 volumes. 4th edition. 1992.

Physics For Poets. Robert H. March. McGraw Publishing Company, Inc., 1221 Avenue of the Americas, New York, NY 10020. (212) 512-2000. (800) 262-4729. 3rd edition. 1992.

Rainbows, Snowflakes, and Quarks: Physics and the World Around Us. Hans C. Von Baeyer. Random House, Inc., 201 East 50th Street, New York, NY 10022. (212) 751-2600. (800) 733-3000. 1991. $19.50. 1993. $12.00 in paper.

Statistical Physics and the Atomic Theory of Matter, From Boyle and Newton To Landau and Onsager. Stephen G. Brush. Princeton University Press, 41 William Street, Princeton, NJ 08540. (609) 258-4900. (800) 777-4726. 1983. $72.50.

Visit To A Small Universe. Virginia Trimble. American Institute of Physics, One Physics Ellipse, College Park, MD 20740-3843. (301) 209-3100. 1992. $24.95.

HANDBOOKS AND MANUALS

Chemical and Physical Data. Arthur M. James and M. P. Lord, Van Nostrand Reinhold, 115 Fifth Avenue, New York, NY 10003. (212) 254-3232. (800) 842-3636. 1992. $59.95.

CRC Handbook of Chemistry and Physics. David R. Kide, editor. CRC Press, Inc., 2000 Corporate Boulevard, NW, Boca Raton, FL 33431. (407) 994-0555. (800) 272-7737. 77th edition. 1996. $99.95.

Handbook of Physical Quantities. Igor S. Grigoriev and Evgenil Z. Melikhov, editors. CRC Press, Inc., 2000 Corporate Boulevard, NW, Boca Raton, FL 33431. (407) 994-0555. (800) 272-7737. 1995. $99.00.

Handbook of Space Astronomy and Astrophysics. Martin V. Zumbeck. Cambridge University Press, 40 West 20th Street, New York, NY 10011-4211. (212) 924-3900. (800) 872-7423. 2nd edition. 1990. $79.95.

Landolt-Borenstein Numerical Data and Functional Relationships in Science and Technology: Nuclear Particle and Physics. Group I. Springer-Verlag New York, Inc., 175 Fifth Avenue, New York, NY 10010. (212) 460-1500. (800) 777-4643. Volumes priced individually.

McGraw-Hill Encyclopedia of Physics. Sybil P. Parker, editor. McGraw-Hill Publishing Company, Inc., 1221 Avenue of the Americas, New York, NY 10020. (212) 512-2000. (800) 262-4729. 2nd edition. 1993. $95.50

Newnes Engineering and Physical Science Pocketbook. J. O. Bird and P. J. Chivers. Butterworth-Heinmann, 313 Washington Avenue, Newton, MA 02158. (617) 928-2500. (800) 366-2665. 1993. $49.95.

A Physicist's Desk Reference. Herbert L. Anderson, editor. American Institute of Physics, One Physics Ellipse, College Park, MD 20740-3843. (301) 209-3100. 2nd edition. 1989. $70.00.

The Physics of Sports. Angelo Armenti, Jr., editor. American Institute of Physics, One Physics Ellipse, College Park, MD 20740-3843. (301) 209-3100. 1992. $35.00.

Physics Problem Solver. Research and Education Association Staff. Research & Education Association, 61 Ethel Road, West Piscataway, NJ 08854. (908) 819-8880. Revised edition. 1994. $23.95 in paper.

MECHANICS

Ency. of Physical Sciences and Engineering Info. Sources

Tables of Physical and Chemical Constants and Some Mathematical Functions. G. W. Kaye and T. M. Laby. Halsted Press 605 Third Avenue, New York, 10158-0012. (212) 850-6400. 15th edition. 1986. $74.95.

ONLINE DATABASES AND CD-ROMS

CA Search. Chemical Abstracts Service, P.O. Box 3012, Columbus, OH 43210-0012. (614) 447-3600. (800) 848-6533. FAX (614) 447-3709. Very comprehensive guide to worldwide chemical literature and related fields, 1972 to present. Available on BRS,(800) 289-4277, DIALOG, (800) 334-2564, ORBIT (800) 456-7248, and STN International, FIZ Karlsruhe, P.O. Box 2465, W-7500, Karlsruhe 1, Germany, online services. Inquire as to cost and availability.

Current Contents Search. Institute for Scientific Information, 3501 Market Street, Philadelphia, PA 19104. (215) 386-0100. FAX (215) 386-6362. Contains citations to articles listed in the table of contents of science and technology journals. Also articles in social sciences and life sciences journals. Available on BRS, (800) 289-4277, DIALOG, (800) 334-2564, online services. Inquire as to cost and availability.

Compendex Plus. Engineering Information, Inc., 345 East 47th Street, New York, NY 10017. (212) 705-7600 or (800) 221-1044. Contains citations with abstracts to worldwide literature in engineering and technology, from 1970 to present. Available on online BRS,(800) 289-4277, DIALOG, (800) 334-2564, ORBIT (800) 456-7248, and STN International, FIZ Karlsruhe, P.O. Box 2465, W-7500, Karlsruhe 1, Germany, online services. Also available on CD-ROM. Inquire as to cost and availability.

Dissertation Abstracts Online. University Microfilms International, 300 North Zeeb Road, Ann Arbor, MI 48106. (800) 521-0600 or (313) 761-4700. Scope includes virtually all doctoral dissertations accepted at accredited American institutions from 1861 to present in 252 subject areas. Available on BRS, (800) 289-4277, DIALOG, (800) 334-2564, and OCLC EPIC, (800) 848-5878, online services. Also available on CD-ROM. Inquire as to cost and availability.

INSPEC. Institution of Electrical Engineers, Michael Faraday House, Six Hills Way, Stevenage, Herts. SG1 2AY, England. Telephone: 0438 313311 or FAX 0438 742840. Contains citations to the worldwide literature of physics, electronics and electrical engineering, computer technology, and related fields. Available on BRS, (800) 289-4277, DIALOG, (800) 334-2564, ORBIT, (800) 456-7248, and STN International, FIZ Karlsruhe, P.O. Box 2465, W-7500, Karlsruhe 1, Germany, online services. Inquire as to cost and availability.

Mathsci. American Mathematical Society, P.O. Box 6248, Providence, RI 02940. (800) 321-4667 or (401) 455-4000 or FAX (401) 331-3842. Scope includes pure and applied mathematics and related areas of physics, statistics, engineering, computer science, and operations research literature since 1959. Available on DIALOG,(800) 334-2564, online service. Also available on CD-ROM. Inquire as to cost and availability.

NTIS Bibliographic Database. National Technical Information Service, 5285 Port Royal Road, Springfield, VA 22161. (703) 487-4929 or FAX (703) 321-8199. Broad coverage of government-sponsored science and technology research reports, 1964 to present. Available on BRS,(800) 289-4277, DIALOG, (800) 334-2564, ORBIT, (800) 456-7248, and STN International, FIZ Karlsruhe, P.O. Box 2465, W-7500, Karlsruhe 1, Germany, online services. Also available on CD-ROM. Inquire as to cost and availability.

Physics Briefs. American Institute of Physics, 335 East 45th Street, New York, NY 10017. (212) 661-9260 or FAX (212) 661-2036. Contains citations with abstracts of the literature of physics and related fields, 1979 to present. Available on the STN International, FIZ Karlsruhe, P.O. Box 2465, W-7500, Karlsruhe 1, Germany, online service. Inquire as to cost and availability.

Scisearch. Institute for Scientific Information, 3501 Market Street, Philadelphia, PA 19104. (800) 523-1850 or (215) 386-0100. Broad multidisciplinary title and author index to the international literature of science and technology, 1974 to present. Available on DIALOG, (800) 334-2564, and ORBIT, (800) 456-7248, online services. Also available on CD-ROM. Inquire as to cost and availability.

Wilsonline. H.W. Wilson Company, 950 University Avenue, Bronx, NY 10452. (800) 367-6770 or (212) 588-8400. Makes available online versions of the printed H.W. Wilson indexes, including Applied Science and Technology Index, Business Periodicals Index, General Science Index, and Readers' Guide to Periodical Literature. Period covered is generally 1983 to present. Available on BRS, (800) 289-4277, DIALOG, (800) 334-2564, and OCLC EPIC, (800) 848-5878, online services. Also available on CD-ROM. Inquire as to cost and availability.

PERIODICALS

American Journal of Physics. American Association of Physics Teachers, 5112 Berywn Road, College Park, MD 20740. (301) 345-4200. 1933 to present. Monthly. $215.00 per year. ISSN: 0002-9505.

Applied Physics Letters. American Institute of Physics, One Physics Ellipse, College Park, MD 20740-3843. (301) 209-3000. 1962 to present. Weekly. $1060.00 per year. ISSN: 0003-6951.

Bioimaging. IOP Publishing. U.S. subscriptions to: American Institute of Physics, 500 Sunnyside Boulevard, Woodbury, NY 11797-2999. (516) 575-2270. 1993 to present. Quarterly. $236.00 per year. ISSN: 0966-9051.

Canadian Journal of Physics. National Research Council of Canada. Research Journals, Ottawa K1A OR6, Canada. (613) 993-9084. FAX (613) 952-7656. 1929 to present. Monthly. $285.00 per year. ISSN: 0008-4204.

Chaos; An Interdisciplinary Journal of Nonlinear Science. American Institute of Physics, One Physics Ellipse, College Park, MD 20740-3843. (301) 209-3000. 1991 to present. Quarterly. $225.00 per year. ISSN: 1054-1500.

Contemporary Physics. Taylor & Francis, Ltd, Rankine Road, Basingstoke, Hants RG24 8PR, England. TEL 0256-840366. FAX 0256-479438. 1959 to present. Bimonthly. $318.00 per year. ISSN: 0010-7514.

Distributed Systems Engineering. IOP Publishing. U.S. subscriptions to: American Institute of Physics, 500 Sunnyside Boulevard, Woodbury, NY 11797-2999. (516) 575-2270. 1993 to present. Quarterly. $305.00 per year. ISSN: 0967-1846.

European Journal of Physics. IOP Publishing. U.S. subscriptions to: American Institute of Physics, 500 Sunnyside Boulevard, Woodbury, NY 11797-2999. (516) 575-2270. 1980 to present. Bimonthly. $339.00 per year. ISSN: 0143-0807.

Franklin Institute. Journal. Elsevier Science, 660 White Plains Road, Tarrytown, NY 10591-5153. (914) 524-9200. FAX (914) 333-2444. 1826 to present. Six issues per year. $740.00 per year. ISSN: 0016-0032.

General Relativity and Gravitation. Plenum Publishing Corp., 244 Spring Street, New York, NY 10013-1578. (212) 620-8000. FAX (212) 463-0742. 1970 to present. Monthly. $625.00.

Institut Henri Poincare. Annales. Section A: Theoretical Physics. Gauthier-Villars, 15 rue Gossin, 92543 Montrouge Cedex, France. TEL 33-140-92-65-00. FAX 331-40-92-65-97. 1930 to present. Eight issues per year. 2000F per year. ISSN: 0020-2339.

International Journal of Modern Physics A. World Scientific Publishing Co., 1060 Main Street, Suite 18, River Edge, NJ 07661. (800) 227-7562. 1986 to present. 32 issues per year. $1710.00 per year. ISSN: 0217-751X.

Journal of Applied Physics. American Institute of Physics, One Physics Ellipse, College Park, MD 20740-3843. (301) 209-3000. 1931 to present. Semimonthly. $1655.00 per year. ISSN: 0021-8979.

Journal of Physics. A: Mathematical and General. IOP Publishing. U.S. orders to: American Institute of Physics, 500 Sunnyside Boulevard, Woodbury, NY 11797-2999. (516) 576-2200. 1968 to present. Twenty-four issues per year. $2487.00 per year. ISSN: 0305-4470.

Landolt-Boernstein, Zahlenwerte Und Funktionen Aus Naturwissenschaften Und Technik. Group 3: Crystal Physics. Springer-Verlag New York, Inc., 175 Fifth Avenue, New York, NY 10010. (212) 460-1500. FAX (212) 473-6272. 1965. Irregular. Price varies. ISSN: 0075-7896.

Letters in Mathematical Physics. Kluwer Academic Publishers, Box 358 Accord Station, Hingham, MA 02018-0358. (617) 871-6600. FAX (617) 871-6528. 1975 to present. Monthly. $549.00 per year. ISSN: 0377-9017.

Nanotechnology. IOP Publishing. U. S. orders to: American Institute of Physics, 500 Sunnyside Boulevard, Woodbury, NY 11797-2999. (516) 576-2200. 1990 to present. Quarterly. $358.00 per year. ISSN: 0957-4484.

Philosophical Magazine. Parts A & B. Taylor & Francis Ltd., Rankine Road, Basingstoke, Hants RG24 8PR, England. TEL 0256-840366. FAX 0256-479438. 1798 to present. Monthly. $1168.00 per year. ISSN: 0031-8086.

Physical Review A (general Physics). American Institute of Physics, One Physics Ellipse, College Park, MD 20740-3843. (301) 209-3000. 1970 to present. Monthly. $990.00 per year. ISSN: 1050-2947.

Physics Reports-A Review Section of Physics Letters. Elsevier Science Publishing Company, Inc., 655 Avenue of the Americas, New York, NY 10010. (212) 989-5800. FAX (212) 633-3990. 1971 to present. 72 issues per year in 12 volumes. $2342.00 per year. ISSN: 0370-1573.

Physics Today. American Institute of Physics, One Physics Ellipse, College Park, MD 20740-3843. (301) 209-3000. 1948 to present. Monthly. $140.00 per year. ISSN: 0031-9228.

Reports On Progress in Physics. IOP Publishing. U.S. orders to: American Institute of Physics, 500 Sunnyside Boulevard, Woodbury, NY 11797-2999. (516) 576-2200. 1934 to present. Monthly. $1214.00 per year. ISSN: 0034-4885.

Reviews of Modern Physics. American Institute of Physics, One Physics Ellipse, College Park, MD 20740-3843. (301) 209-3000. 1929 to present. Quarterly. $300.00 per year. ISSN: 0034-6861.

Royal Society of London, Philosophical Transactions. Series A, Physical Sciences and Engineering. Royal Society of London, 6 Carlton House Terrace, London SW1Y 5AG, England. TEL 071-839-5561. FAX 071-976-1837. 1665 to present. Monthly. L485 per year. ISSN: 0962-8428.

RESEARCH CENTERS AND INSTITUTES

Center for the Advancement of Computational Mechanics. Georgia Institute of Technology, Atlanta, GA 30332. (404) 894-2758.

Laboratory For Experimental Mechanics Research. State University of New York at Stony Brook, Stony Brook, NY 11794-2300. (516) 632-8311. FAX (516) 632-8720.

Solid Mechanics Laboratory. University of Michigan, Department of Mechanical Engineering and Applied Mechanics, 2246A G. G. Brown Building, Ann Arbor, MI 48109. (313) 763-0684.

MERCURY

ABSTRACT SERVICES AND INDEXES

Current Contents: Engineering, Technology, and Applied Sciences. Institute for Scientific Information, 3501 Market Street, Philadelphia, PA 19104. (215) 386-0100. FAX (215) 386-6362. 1970 to present. Weekly. $442.00 per year.

Current Contents: Physical, Chemical and Earth Sciences. Institute for Scientific Information, 3501 Market Street, Philadelphia, PA 19104. (215) 386-0100. FAX (215) 386-6362. Weekly. $360.00 per year.

Engineering Index Monthly. Engineering Information, Inc., Castle Point on the Hudson, Hoboken, NJ 07030. (800) 221-1044. FAX (212) 832-1857. Monthly. $2300.00 per year. Also available online as COMPENDEX, and also on CD-ROM. Covers chemical engineering, computers, electrical engineering, civil engineering, metals and mining, industrial management, and mechanical engineering.

Environmental Abstracts. R.R. Bowker Inc., 121 Chanlon Road, New Providence, NJ 07974. (800) 521-8110 or (908) 464-6800. Contains citations, most with abstracts, to literature relating to environmental issues and problems, from 1971 to present. Inquire as to cost and availability.

Environmental Engineering Abstracts, Cambridge Scientific Abstracts, 7200 Wisconsin Avenue, Bethesda, MD 20814-4823.

(301) 961-6750. FAX (301) 961-6720. Monthly. Covers hazardous materials, environmental impact and protection, treatment of sewage and industrial wastes, hydroelectric power, tidal and wind power, arctic and tropical engineering.

Index To Scientific and Technical Proceedings. Institute for Scientific Information, 3501 Market St., Philadelphia, PA 19104. (215) 386-0100. FAX (215) 386-6362. Monthly. $500.00 per year.

Index To Scientific Reviews. Institute for Scientific Information, 3501 Market St., Philadelphia, PA 19104. (215) 386-0100. FAX (215) 386-6362. Semi-annual.

Materials Science and Engineering Abstracts, Cambridge Scientific Abstracts, 7200 Wisconsin Avenue, Bethesda, MD 20814-4823. (301) 961-6750. FAX (301) 961-6720. 1993 to present. Monthly. $385.00 per year. Focuses on mechanical and physical properties of materials and commercial or industrial applications for materials, methods for strength testing, effects of vibration and other stresses, corrosion and protective coatings, storage and handling, ceramics, composites, metals, wood, plastics, and polymers.

Materials Science Citation Index. Institute for Scientific Information, 3501 Market Street, Philadelphia, PA 19104. (215) 386-0100. FAX (215) 386-0100. Bi-monthly. $975.00 per year. Also available on CD-ROM.

Metals Abstracts and Metals Abstracts Index. American Society for Metals, Metals Park, OH 44073. (216) 338-5151. 1968 to present. Monthly. Abstracts are $1500.00 per year and Index is $460.00 per year.

Science Citation Index. Institute for Scientific Information, 3501 Market Street, Philadelphia, PA 19104. (215) 386-0100. FAX (215) 386-6362. Inquire about availability and cost. Also available on CD-ROM.

ASSOCIATIONS AND PROFESSIONAL SOCIETIES

American Society For Metals. ASM International, Materials Information, Materials Park, OH 44073-0002. (216) 338-5151 or (800) 336-5152. FAX (216) 338-4634.

DIRECTORIES AND BIOGRAPHICAL SOURCES

Directory of Engineering Societies and Related Organizations. Gordon Davis. 13th edition. American Association of Engineering Societies, 1111 19th Street NW, Suite 608, Washington, DC 20036. (202) 296-2237 or (800) 658-8897. 1989. Inquire for price.

International Research Centers Directory. Gale Research, 835 Penobscot Building, Detroit, MI 48226-4094. (313) 961-2242. (800) 347-4253. FAX (313) 961-6083. 8th edition. 1995. $410.00

Research Centers Directory. Gale Research, 835 Penobscot Building, Detroit, MI 48226-4094. (313) 961-2242. (800) 347-4253. FAX (313) 961-6083. $485.00.

Scientific and Technical Organizations and Agencies Directory. Gale Research, 835 Penobscot Building, Detroit, MI 48226-4094. (313) 961-2242. (800) 347-4253. FAX (313) 961-6083. 4th edition. 1996. $195.00.

Who's Who in Engineering. American Association of Engineering Societies, 1111 19th Street NW, Suite 608, Washington, DC 20036. (202) 296-2237 or (800) 658-8897. 8th edition. 1991. Inquire for price.

ENCYCLOPEDIAS AND DICTIONARIES

Encyclopedia of Physical Science and Technology. Robert A. Meyers, ed. 18 volumes. Academic Press Inc., 6277 Sea Harbor Drive, Orlando, FL 32887. (800) 321-5068. 1992. $2100.00.

McGraw-Hill Dictionary of Scientific and Technical Terms. Sybil P. Parker, ed. 5th edition. McGraw-Hill Publishing Company, 1221 Avenue of the Americas, New York, NY 10020. (800) 262-4729 or (212) 512-3825. 1993. $110.50.

McGraw-Hill Encyclopedia of Engineering. Sybil P. Parker, ed. 2nd edition. McGraw-Hill Publishing Company, 1221 Avenue of the Americas, New York, NY 10020. (800) 262-4729 or (212) 512-3825. 1993. $95.50.

McGraw-Hill Encyclopedia of Science and Technology. Sybil P. Parker, ed. 7th edition. 20 volumes. McGraw-Hill Publishing Company, 1221 Avenue of the Americas, New York, NY 10020. (800) 262-4729 or (212) 512-3825. 1992. $1900.00

Thesaurus of Scientific, Technical, and Engineering Terms. Hemisphere Publishing Corporation, 1900 Frost Road, Suite 101, Bristol, PA 19007-1598. (215) 785-5800 or (800) 821-8312. FAX (215) 785-5515. 1987. $173.00.

GENERAL WORKS

Advances in Mercury Toxicology. T. Suzuki, et al., editors. Plenum Publishing Corporation, 233 Spring Street, New York, NY 10013. (212) 620-8000. FAX (212) 463-0742. 1991. $115.00.

Copper, Cilver, Gold and Zinc, Cadmium, Mercury, Oxides and Hydroxides. T.P. Dirkse, editor. Pergamon Press Inc., Maxwell House, Fairview Park, Elmsford, NY 10523. (914) 592-7700. Fax (914) 592-3625. 1986. Inquire for cost and availability.

Hazardous Metals in the Environment. M. Stoeppler, editor. Elsevier Science Publishing Company, Inc., 655 Avenue of the Americas, New York, NY 10010. (212) 989-5800. 1992. Inquire for cost and availability.

The Heavy Elements: Chemistry, Environmental Impact, and Health Effects. Jack E. Fergusson. Elsevier Science Publishing Company, Inc., 655 Avenue of the Americas, New York, NY 10010. (212) 989-5800. 1990. $54.00.

HANDBOOKS AND MANUALS

ASM Metals Reference Book. 3rd edition. ASM International, Materials Information, Materials Park, OH 44073-0002. (216) 338-5151 or (800) 336-5152. FAX (216) 338-4634. 1993. $111.00.

CRC Handbook of Chemistry and Physics. David R. Lide, editor. 75th edition. CRC Press, 2000 Corporate Blvd., N.W., Boca Raton, FL 33431. (407) 994-0555 or (800) 272-7737. FAX (407) 994-0949. $99.50.

Metals Handbook. ASM International, Materials Information, Materials Park, OH 44073-0002. (216) 338-5151 or (800) 336-5152. FAX (216) 338-4634. $154.00.

Smithell's Metals Reference Book. E.A. Brandes & G.B. Brook. 7th edition. Butterworth-Heinemann, 313 Washington Street, Newton, MA 02158. (617) 928-2500 or (800) 366-2665. FAX (617) 928-2620. 1992. $250.00.

ONLINE DATABASES AND CD-ROMS

Compendex Plus. Engineering Information, Inc., 345 East 47th Street, New York, NY 10017. (212) 705-7600 or (800) 221-1044. Contains citations with abstracts to worldwide literature in engineering and technology, from 1970 to present. Available on online BRS,(800) 289-4277, DIALOG, (800) 334-2564, ORBIT (800) 456-7248, and STN International, FIZ Karlsruhe, P.O. Box 2465, W-7500, Karlsruhe 1, Germany, online services. Also available on CD-ROM. Inquire as to cost and availability.

CSA Engineering. Cambridge Scientific Abstracts, 7200 Wisconsin Avenue, Suite 601, Bethesda, MD 20814. (301) 961-6750 or (800) 843-7751. Contains citations and abstracts of international periodicals and research literature covering all fields of engineering and science and technology, including computer and information science, electronics, mechanical engineering, solid state materials, 1981 to present. Available on BRS,(800) 289-4277, online service. Inquire as to cost and availability.

Current Contents Search. Institute for Scientific Information, 3501 Market Street, Philadelphia, PA 19104. (215) 386-0100. FAX (215) 386-6362. Contains citations to articles listed in the table of contents of science and technology journals. Also articles in social sciences and life sciences journals. Available on BRS,(800) 289-4277, DIALOG,(800) 334-2564, online services. Inquire as to cost and availability.

Dissertation Abstracts. University Microfilms International, 300 North Zeeb Road, Ann Arbor, MI 48106. (800) 521-0600 or (313) 761-4700. Scope includes virtually all doctoral dissertations accepted at accredited American institutions from 1861 to present in 252 subject areas. Available on BRS,(800) 289-4277, DIALOG,(800) 334-2564, and OCLC EPIC,(800) 848-5878, online services. Also available on CD-ROM. Inquire as to cost and availability.

Metadex. Jointly produced by ASM International and the Institute of Materials. Contains more than 925,000 records from the international literature on metals and alloys, concerning properties, processes, materials classes, applications, and metallurgical systems. Updated monthly. Available from ORBIT-QUESTEL (703) 442-0900.

NTIS Bibliographic Database. National Technical Information Service, 5285 Port Royal Road, Springfield, VA 22161. (703) 487-4929 or FAX (703) 321-8199. Broad coverage of government-sponsored science and technology research reports, 1964 to present. Available on BRS,(800) 289-4277, DIALOG, (800) 334-2564, ORBIT, (800) 456-7248, and STN International, FIZ Karlsruhe, P.O. Box 2465, W-7500, Karlsruhe 1, Germany, online services. Also available on CD-ROM. Inquire as to cost and availability.

Scisearch. Institute for Scientific Information, 3501 Market Street, Philadelphia, PA 19104. (800) 523-1850 or (215) 386-0100. Broad multidisciplinary title and author index to the international literature of science and technology, 1974 to present. Available on DIALOG,(800) 334-2564, and ORBIT,(800) 456-7248, online services. Also available on CD-ROM. Inquire as to cost and availability.

Wilsonline. H.W. Wilson Company, 950 University Avenue, Bronx, NY 10452. (800) 367-6770 or (212) 588-8400. Makes available online versions of the printed H.W. Wilson indexes, including Applied Science and Technology Index, Business Periodicals Index, General Science Index, and Readers' Guide to Periodical Literature. Period covered is generally 1983 to present. Available on BRS,(800) 289-4277, DIALOG,(800) 334-2564, and OCLC EPIC,(800) 848-5878, online services. Also available on CD-ROM. Inquire as to cost and availability.

PERIODICALS

Inorganic Chemistry. American Chemical Society, 1155 16th Street NW. Washington, DC 20036. (800) 333-9511. FAX (914) 447-3671. 1962 to present. Semi-monthly. $949.00 per year.

JOM (Journal of Metals). Minerals, Metals and Materials Society, 420 Commonwealth Drive, Warrendale, PA 15086. (412) 776-9080. 1949 to present. Monthly. $50.00 per year.

Main Group Metal Chemistry. Freund Publishing House Ltd., Suite 500, Chelsea House, 150 Regent Street, London W1R 5FA, England. 1972 to present. Monthly. $360.00 per year.

Metal Bulletin Monthly. Metal Bulletin Inc., 220 Fifth Avenue, 19th Floor, New York, NY 10001-7781. (800) 638-2525. FAX (212) 213-6273. 1972 to present. Monthly. Inquire for cost.

Mineral Industry Surveys, Mercury. U.S. Bureau of Mines, Production and Distribution, Cochrans Mill Rd., Box 18070, Pittsburgh, PA 15236. (412) 892-4411. Quarterly and annually. Inquire for price and availability.

Powder Metallurgy and Metal Ceramics. Plenum Publishing Corporation, 233 Spring Street, New York, NY 10013. (212) 620-8000. FAX (212) 463-0742. 1962 to present. Monthly. $1215.00 per year.

MERCURY

See also: ASTROGEOLOGY, SATELLITES (NATURAL), SOLAR SYSTEM

ABSTRACT SERVICES AND INDEXES

Applied Science and Technology Index; A Cumulative Subject Index to English Language Periodicals in the Fields of Aeronautics and Space Science, Computer Technology, Chemistry, Construction Industry, Energy and Related Areas. H. W. Wilson Co., 950 University Avenue, Bronx, NY 10452- 9978. (212) 588-8400. (800) 367-6770. FAX (718) 590-1617. From 1958 to present. Monthly. Inquire about cost and availability. Also available online (BRS and Wilsonline) and on CD-ROM. ISSN: 0003-6986.

Astronomy and Astrophysics Abstracts. Springer-Verlag New York, 175 Fifth Avenue, New York, NY 10010. (212) 460-1500. FAX (212) 473-6272. Published for the Astronomisches Rechen-Institut. Comprehensive coverage of all aspects of astronomy, astrophysics and related fields. 1969 to present. Two parts per year. Annual. ISSN: 0067-0022.

MERCURY

Ency. of Physical Sciences and Engineering Info. Sources

Bibliography and Index of Geology. American Geological Institute, 4220 King Street, Alexandria, VA 22302-1507. (703) 379-2480. FAX (703) 379-7563. From 1969 to present. Monthly. $1295.00 per year. ISSN: 0098-2784. Also available as GEOREF online (CISTI, DIALOG, Orbit, STN) and on CD-ROM. Inquire about price and availability.

Chemical Abstracts. Chemical abstracts Service. 2540 Oleantangy River Road, Box 3012, Columbus, OH 43210-0012. (614) 447-3600. FAX (614) 447-3713. 1907 to present. Weekly. $16,800.00 per year. Available online and on CD-ROM. CA is also available in five section groupings. Inquire regarding cost and availability.

Current Contents: Physical, Chemical, and Earth Sciences. Institute for Scientific Information, 3501 Market Street, Philadelphia, PA 19104. (215) 386-0100. (800) 523-1850. FAX (215) 386-2291. 1961 to present. Weekly. $442.00 per year. Also available online (BRS, DIALOG) and on CD-ROM. Inquire about price and availability.

General Science Index. H.W. Wilson Company, 950 University Avenue, Bronx, NY 10452. (212) 588-8400. (800) 367-6770. FAX (718) 590-1617. From 1978 to present. Ten issues per year; quarterly and annual cumulations. Service basis. Available on CD-ROM and online. Inquire regarding cost and availability. ISSN: 0162-1963.

Meteorological and Geoastrophysical Abstracts. American Meteorological Society, c/o Inforonics, Inc., 550 Newtown Road, Box 458, Littleton, MA 01460. (508) 486-8976. FAX (508) 486-0027. Covers literature in environmental sciences, meteorology, astrophysics, hydrology, glaciology and physical oceanography. 1950 to present. Monthly. $950.00 per year. Also available online (DIALOG) and on CD-ROM. ISSN: 0026-1130.

NTIS Alerts: Astronomy & Astrophysics. U.S. National Technical Information Service. 5285 Port Royal Road, Springfield, VA 22161. (703) 487-4650. FAX (703) 321-8547. Weekly. $140.00 per year.

Physics Abstracts. INSPEC. Section A, Science Abstracts. Institute of Electrical Engineers, London, United Kingdom. Available from: INSPEC/IEEE-Institute of Electrical and Electronic Engineers, Box 1331, Hoes Lane, Piscataway, NJ 08855-1331. (908) 562-5549. 1898 to present. 24 issues per year. $2835.00 per year. ISSN: 0036-8091. Also available online and on CD-ROM.

Science Citation Index. SCI. Institute for Scientific Information, 3501 Market Street, Philadelphia, PA 19104. (215) 386-0100. (800) 523-1850. FAX (215) 386-2291. 1961 to present. Six issues per year, plus annual cumulation. $11650.00 per year. Also available online and on CD-ROM. Inquire about price and availability. ISSN: 0036-827X.

STAR. Scientific and Technical Aerospace Reports. U.S. National Aeronautics and Space Administration. Distributed by U.S. Superintendent of Documents, Washington, DC 20402. 1963 to present. Semi-monthly, with semiannual and annual indexes. $114.00 per year. ISSN: 0036-8741. Also available online and on CD-ROM.

ANNUAL REVIEWS AND YEARBOOKS

The Astronomical Almanac. Superintendent of Documents, U.S. Government Printing Office, Washington, DC 10402. (202) 783-

3238. 1981 to present. Supersedes Astronomical Ephemeris. Annual. $44.95. ISSN: 0737-6421.

Annual Review of Astronomy and Astrophysics. Original reviews of critical literature and current developments in astronomy and astrophysics. Annual Reviews, Inc., 4139 El Camino Way, Palo Alto, CA 94303-0139. (415) 493-4400. (800)523-8635. Fax (415) 855-9815. 1963 to date. Annual. $60.00. ISSN: 0066-4146.

Annual Review of Earth and Planetary Sciences. Annual Reviews, Inc., 4139 El Camino Way, Palo Alto, CA 94303-0139. (415) 493-4400. (800) 523-8635. Fax (415) 855-9815. 1973 to date. Annual. $62.00. ISSN: 0084-6597.

Proceedings of the Lunar and Planetary Science Conference. Lunar and Planetary Institute, 3600 Bay Area Boulevard, Houston, TX 77058. (713) 486-2143. 1970 to date. Annual. ISSN: 0270-9511.

ASSOCIATIONS AND PROFESSIONAL SOCIETIES

American Astronomical Society. 2000 Florida Avenue NW, Suite 400, Washington, DC 20009. (202) 328-2010. FAX (202) 234-2560.

American Geophysical Union. 2000 Florida Avenue NW, Washington, DC 20009. (202) 462-6900. (800) 966-AGU1. FAX (202) 328-0566.

American Institute of Physics. 1 Physics Ellipse, College Park, MD 20740-3843. (301) 209-3100.

Association of Universities For Research in Astronomy, Inc. (AURA). Suite 701, 1625 Massachusetts Avenue, NW, Washington, DC 20036. (202) 483-2101. FAX (202) 483-2106.

Astronomical Society of the Pacific. 390 Ashton Avenue, San Francisco, CA 94112. (415) 337-1100. FAX: (415) 337-5205.

Geological Society of America. 3300 Penrose Place, PO Box 9140. Boulder, CO 80301-9140. (303) 447-2020. FAX (303) 447-1133.

Planetary Society. 65 North Catalina Avenue, Pasadena, CA 91106. (818) 793-5100. (800) WOW-MARS. FAX (818) 793-5528.

DIRECTORIES AND BIOGRAPHICAL SOURCES

American Astronomical Society. Membership Directory. American Astronomical Society. 2000 Florida Avenue NW, Suite 400, Washington, DC 20009. (202) 328-2010. FAX (202)234-2560 Annual. Included in membership dues. ISSN: 1061-9038.

American Men and Women of Science: Physical and Biological Sciences. R. R. Bowker Inc., 121 Chanlon Road, New Providence, NJ 07974. (908) 464-6800. (800) 521-8110. 20th edition. 8 volumes. 1996. $850.00.

The Astronomers. Donald Goldsmith. St. Martin's Press, Inc., 175 Fifth Avenue, New York, NY 10010. (212) 674-5151. (800) 221-7945. 1993. $14.95 in paper.

Astronomical Centers of the World. Kevin Krisciunas. Cambridge University Press, 40 West 20th Street, New York, NY 10011-4211. (212) 924-3900. (800) 872-7423. 1988. $34.95

Directory of Physics and Astronomy Staff Members. American Institute of Physics, One Physics Ellipse, College Park, MD 20740-3843. (301) 209-3100. 1975/76 to present. Annual. $60.00. ISSN: 0361-2228.

Earth and Astronomical Research Centers. Stockton Press, 345 Park Avenue South, New York, NY 10010. 4th edition. 1995. $515.00. ISBN: 1-56169-0967.

ENCYCLOPEDIAS AND DICTIONARIES

Concise Dictionary of Astronomy. Jacqueline Mitton. Oxford University Press, Inc., 200 Madison Avenue, New York, NY 10016. (212) 725-6000. (800) 334-4249. 1992. $24.95.

Encyclopedia of Astronomy and Astrophysics. Stephen Maran, editor. Van Nostrand Reinhold, 115 Fifth Avenue, New York, NY 10003. (212) 254-3232. (800) 842-3636. 1992. $129.95.

McGraw-Hill Encyclopedia of Science and Technology. McGraw-Hill Book Company, Inc., 1221 Avenue of the Americas, New York, NY 10020. (212) 512-2000. (800) 262-4729. 7th edition. 20 volume set. 1992. $1900.00.

Moons of the Solar System: An Illustrated Encyclopedia. John Stewart. McFarland & Company, Inc., Box 611, Jefferson, MC 28640. (910) 246-4460. (800) 253-2187. 1991. $49.95.

New Guide To the Planets. Patrick Moore. Trans-Atlantic Publications, Inc., 311 Bainbridge Street, Philadelphia, PA 19147. (215) 925-5083. 1993. $37.50.

Stars and Planets: The Sierra Club Guide To Sky Watching and Direction Finding. W. S. Kals. The Sierra Press, 4988 Gold Leaf Drive, Mariposa, CA 95338. (209) 966-5071. (800) 745-2631. $15.00.

GENERAL WORKS

Evolution of the Earth and Planets. E. Takahashi, et al, editors. American Geophysical Union, 2000 Florida Avenue, N.W., Washington, DC 20009. (202) 462-6903. (800) 966-2481. 1993. $28.00.

Mercury. Faith Vilas, et al, editors. University of Arizona Press, 1230 North Park Avenue, Number 102, Tucson, AZ 85719. (520) 621-1441. (800) 426-3797. 1988. $50.00.

Mercury: The Elusive Planet. Robert B. Strom. Smithsonian Institution Press, 470 L'Enfant Plaza, Suite 7100, Washington, DC 20560. (202) 287-3738. (800) 782-4612. 1987. $29.95.

Mercury's Perihelion From Le Verrier To Einstein. N. T. Roseveare. Oxford University Press, 40 West 20th Street, New York, NY 10011-4211. (212) 924-3900. (800) 872-7423. 1982. $47.50.

Nearest the Sun: The Planet Mercury. Isaac Asimov. Gareth Stevens, Inc., River Center Building, 1555 North River Center Drive, Milwaukee, WI 53212. (414) 225-0333. (800) 341-3569. 1995.

Resources of Near-Earth Space. John S. Lewis, et al. University of Arizona Press, 1230 North Park Avenue, Number 102, Tucson, AZ 85719. (520) 621-1441. (800) 426-3797. 1993. $75.00.

Worlds Apart: A Textbook in Planetary Sciences. Guy Consolmagno. Prentice-Hall, 113 Sylvan Avenue, Route 9W, Englewood Cliffs, NJ 07632. (201) 592-2000. (800) 922-0579. 1994. $62.00.

HANDBOOKS AND MANUALS

Astrophysical Data: Planets and Stars. Kenneth R. Lanb. Springer-Verlag New York, Inc., 175 Fifth Avenue, New York, NY 10010. (212) 460-1500. (800) 777-4643. 1993. $59.00.

Field Guide to the Stars and Planets. Jay M. Pasachoff and Donald H. Menzel. Houghton Mifflin Co., 222 Berkeley Street, Boston, MA 02116. (617) 351-5000. (800) 225-3362. Revised edition. 1992. $24.95.

New Guide To the Planets. Patrick Moore. Trans-Atlantic Publications, Inc., 311 Bainbridge Street, Philadelphia, PA 19147. (215) 925-5083. 1993. $37.50.

Planet Observer's Handbook. Fred W. Price. Cambridge University Press, 40 West 20th Street, New York, NY 10011-4211. (212) 924-3900. (800) 872-7423. 1994. $34.95.

Planets and Their Moons. Audubon Society Staff. Alfred A. Knopf, Inc. 201 E. 50th Street, New York, NY 10022. (800) 733-3000. 1995. $7.99.

ONLINE DATABASES AND CD-ROMS

CA Search. Chemical Abstracts Service, P.O. Box 3012, Columbus, OH 43210-0012. (614) 447-3600. (800) 848-6533. FAX (614) 447-3709. Very comprehensive guide to worldwide chemical literature and related fields, 1972 to present. Available on BRS, (800) 289-4277, DIALOG, (800) 334-2564, ORBIT (800) 456-7248, and STN International, FIZ Karlsruhe, P.O. Box 2465, W-7500, Karlsruhe 1, Germany, online services. Inquire as to cost and availability.

Dissertation Abstracts. University Microfilms International, 300 North Zeeb Road, Ann Arbor, MI 48106. (800) 521-0600 or (313) 761-4700. Scope includes virtually all doctoral dissertations accepted at accredited American institutions from 1861 to present in 252 subject areas. Available on BRS, (800) 289-4277, DIALOG, (800) 334-2564, and OCLC EPIC, (800) 848-5878, online services. Also available on CD-ROM. Inquire as to cost and availability.

GEOREF. American Geological Institute, 4220 King Street, Alexandria, VA 22302. (800) 336-4764 or (703) 379-2480. Geology and geosciences literature, 1785 to present for North America. Available on DIALOG, (800) 334-2564, ORBIT (800) 456-7248, online services. Also available on CD-ROM. Inquire as to cost and availability.

INSPEC. Institution of Electrical Engineers, Michael Faraday House, Six Hills Way, Stevenage, Herts. SG1 2AY, England. Telephone: 0438 313311 or FAX 0438 742840. Contains citations to the worldwide literature of physics, electronics and electrical engineering, computer technology, and related fields. Available on BRS, (800) 289-4277, DIALOG, (800) 334-2564,

ORBIT, (800) 456-7248, and STN International, FIZ Karlsruhe, P.O. Box 2465, W-7500, Karlsruhe 1, Germany, online services. Inquire as to cost and availability.

NTIS Bibliographic Database. National Technical Information Service, 5285 Port Royal Road, Springfield, VA 22161. (703) 487-4929 or FAX (703) 321-8199. Broad coverage of government-sponsored science and technology research reports, 1964 to present. Available on BRS,(800) 289-4277, DIALOG, (800) 334-2564, ORBIT, (800) 456-7248, and STN International, FIZ Karlsruhe, P.O. Box 2465, W-7500, Karlsruhe 1, Germany, online services. Also available on CD-ROM. Inquire as to cost and availability.

Scisearch. Institute for Scientific Information, 3501 Market Street, Philadelphia, PA 19104. (800) 523-1850 or (215) 386-0100. Broad multidisciplinary title and author index to the international literature of science and technology, 1974 to present. Available on DIALOG, (800) 334-2564, and ORBIT, (800) 456-7248, online services. Also available on CD-ROM. Inquire as to cost and availability.

Wilsonline. H.W. Wilson Company, 950 University Avenue, Bronx, NY 10452. (800) 367-6770 or (212) 588-8400. Makes available online versions of the printed H.W. Wilson indexes, including Applied Science and Technology Index, Business Periodicals Index, General Science Index, and Readers' Guide to Periodical Literature. Period covered is generally 1983 to present. Available on BRS, (800) 289-4277, DIALOG, (800) 334-2564, and OCLC EPIC, (800) 848-5878, online services. Also available on CD-ROM. Inquire as to cost and availability.

OTHER SOURCES

Atlas of the Planets. Vincent DeCallatay and Audouin Dollfus. Books on Demand, 300 North Zeeb Road, Ann Arbor, MI 48106-1346. (313) 761-4700. (800) 521-0600. $45.60.

Atlas of the Solar System. B. Yenne. Simon & Schuster, Inc. 1230 Avenue of the Americas, New York, NY 10020. (212) 698-7000. (800) 223-2348. 1987. $12.98.

Planetary, Lunar and Solar Positions 601 B.C. to A.D. One at Five-Day and Ten-Day Intervals. Bryant Tuckerman. American Philosophical Society, 104 South 5th Street, Philadelphia, PA 19106-3387. (215) 440-3400. FAX (215) 440-3436. Memoirs Series, volume 56. 1979. $20.00.

Planetary, Lunar and Solar Positions, A.D. 2 to A.D. 1649 at Five-Day and Ten-Day Intervals. Bryant Tuckerman. American Philosophical Society, 104 South 5th Street, Philadelphia, PA 19106-3387. (215) 440-3400. FAX (215)440-3436. Memoirs Series, volume 59. 1964. $30.00.

Planetary, Lunar and Solar Positions, 1650-1805. Owen Gingerich and Barbara L. Welther. American Philosophical Society, 104 South 5th Street, Philadelphia, PA 19106-3387. (215) 440-3400. FAX (215)440-3436. Memoirs Series, volume 59S. 1983. $20.00.

The Sky: A User's Guide. David H. Levy. Cambridge University Press, 40 West 20th Street, New York, NY 10011-4211. (212) 924-3900. (800) 872-7423. 1993. $14.95.

The View from Space: Photographic Exploration of the Planets. Merton Davies and Bruce C. Murray. Columbia University Press, 562 West 113th Street, New York, NY 10025. (212) 666-1000. (800) 944-8648. 1973. $17.50.

PERIODICALS

Astronomical Journal. American Institute of Physics, One Physics Ellipse, College Park, MD 20740-3843. (301) 209-3000. Published for the American Astronomical Society. 1849 to present. Monthly. $280.00 per year. ISSN: 0004-6256.

Astronomical Society of the Pacific. Publications. Astronomical Society of the Pacific, 390 Ashton Avenue, San Francisco, CA 94112. (415) 337-1100. FAX (415) 337-5205. 1889 to present. Monthly. $175.00 per year. ISSN: 0004-6280.

Astronomy. Kalmbach Publishing Company, Box 1612, Waukesha, WI 53187-1612. (414) 796-8776. FAX (414) 796-1142. 1973 to present. Monthly. $27.00 per year. ISSN: 0091-6358.

Astronomy and Astrophysics. Springer-Verlag New York, Inc., 44 Hartz Way, Secaucus, NJ 07096-2491. (201) 348-4033. FAX(201) 348-4505. 1969 to date. 36 issues per year in 12 volumes. $1733.00 per year. ISSN: 0004-6361.

Earth, Moon and Planets; An International Journal of Comparative Planetology. Kluwer Academic Publishers, Box 358, Accord Station, Hingham, MA 02018-0358. (617) 871-6600. FAX (617) 871-6528. 1969 to present. Monthly. $840.00 per year. ISSN: 0167-9295.

Geochemica Et Cosmochimica Acta. Elsevier Science. 660 White Plains Road, Tarrytown, NY 10591-5153. (914) 524-9200. FAX (914) 333-2444. 1950 to the present. Biweekly. $895.00 per year. ISSN: 0016-7037.

Icarus; International Journal of Solar System Studies. Academic Press, Inc., Journal Division, 525 B Street, Suite 1900, San Diego, CA 92101-4495. (619) 230-1840. FAX (619) 699-6800. 1962 to the present. Monthly. $1080.00. ISSN: 0019-1035.

JGR: Journal of Geophysical Research: Planets. American Geophysical Union, 2000 Florida Avenue, NW, Washington, CD 20009. (202) 462-6900. FAX (202) 328-0566. 1991 to present. Monthly. $597.00 per year. ISSN: 0148-0227.

Lunar and Planetary Information Bulletin. Lunar and Planetary Institute. 3600 Bay Area Boulevard, Houston, TX 77058-1113. (713) 486-2175. FAX (713) 486-2125. 1970 to present. Quarterly. Free. Also available online.

Mercury. Astronomical Society of the Pacific. 390 Ashton Avenue, San Francisco, CA 94112. (415) 337-1100. FAX: (415) 337-5205. 1972 to present. Bimonthly. $175.00 per year. ISSN: 0047-6773.

Planetary and Space Science. Elsevier Science Publishing Company, Inc., 660 White Plains Road, Terrytown, NY 10591-5153. (914) 524-9200. FAX: (914) 333-2444. 1959 to present. Monthly. $1355.00 per year. ISSN: 0032-0633.

Planetary Report. Planetary Society. 65 North Catalina, Pasadena, CA 91106-2301. (818) 793-5100. FAX (818) 793-5528. 1980 to present. $25.00 per year. ISSN: 0736-3680.

Sky and Telescope. Sky Publishing Corporation, Box 9111, Belmont, MA 02178. (617) 864-7360. FAX (617) 864-6117.

1941 to present. Monthly. $27.00 per year. ISSN: 0037-6604.

RESEARCH CENTERS AND INSTITUTES

Center For Space Science and Astrophysics. Stanford University. 325 Durand Building, Stanford, CA 04305. (415) 723-3582.

Institute For Astronomy. University of Hawaii at Manoa, 2680 Woodlawn Drive, Honolulu, HI 96822. (808) 956-8312. FAX (808) 988-2790.

Institute of Geophysics and Planetary Physics. University of California, Riverside, CA 92521. (714) 787-4503. FAX (714) 787-4529.

Laboratory For Planetary Studies. Cornell University, Space Sciences Building, Ithaca, NY 14853. (607) 255-4971.

Lunar and Planetary Institute. 3303 NASA Road One, Houston, TX 77058-4399. (713) 486-2139. FAX 713-496-2162.

Smithsonian Astrophysical Observatory. 60 Garden Street, Cambridge, MA 02138. (617) 495-7461. FAX (617) 495-7105.

MESON

See: PARTICLE PHYSICS

METALLURGICAL ENGINEERING

See also: ALLOYS, FERROALLOYS, MACHINING, METALS and METAL WORKING, METALLURGY

ABSTRACT SERVICES AND INDEXES

Alloys Index. American Society for Metals, Metals Park, OH 44073. (216) 338-5151. FAX (216) 338-4634. 1974 to present. Monthly. $380.00. Also available on CD-ROM and online via METADEX, STN International, and DIALOG.

Aluminum Industry Abstracts. Aluminum Association, Materials Park, OH 44073. (216) 338-5151. FAX (216) 338-4634. 1968 to present. Monthly. $525.00 per year.

Applied Mechanics Reviews: An Assessment of World Literature in Engineering Sciences. 1948-present. American Society of Mechanical Engineers, 345 East 47th Street, New York, NY 10017. (212) 705-7703. Monthly. $360.00 per year.

Applied Science and Technology Index; A Cumulative Subject Index To English Language Periodicals in the Fields of Aeronautics and Space Science, Computer Technology, Chemistry, Construction Industry, Energy and Related Areas. H.W. Wilson Co., 950 University Avenue, Bronx, NY 10452. (800) 367-6770 or (212) 588-8400. FAX (718) 590-1617. From 1958 to present. Monthly. Inquire about cost and availability. Also available on CD-ROM and online.

Corrosion Abstracts; Abstracts of the World's Literature On Corrosion and Corrosion Mitigation. National Association of Corrosion Engineers (NACE), Box 218340, Houston, TX 77218.

(713) 492-0535 or FAX (713) 492-8254. 1962 to present. Bi-monthly. $195.00 per year. Also available on CD-ROM.

Current Contents: Engineering, Technology, and Applied Sciences. Institute for Scientific Information, 3501 Market Street, Philadelphia, PA 19104. (215) 386-0100. FAX (215) 386-6362. 1970 to present. Weekly. $442.00 per year.

Engineering Index Monthly. Engineering Information, Inc., Castle Point on the Hudson, Hoboken, NJ 07030. (800) 221-1044. FAX (212) 832-1857. Monthly. $2300.00 per year. Also available online as COMPENDEX, and also on CD-ROM. Covers chemical engineering, computers, electrical engineering, civil engineering, metals and mining, industrial management, and mechanical engineering.

I.M.M. Abstracts and Index. Institution of Mining and Metallurgy, 44 Portland Pl., London W1N 4BR, England. 071-580-3802. FAX 071-436-5388. Bi-monthly. $364 for non-members.

Leadscan (formerly Lead Abstracts), Lead Development Association, 42 Weymouth Street, London W1x 3lq, England. 1958 to present. Quarterly. Inquire for cost.

Manufacturing and Process Engineering Abstracts, Cambridge Scientific Abstracts, 7200 Wisconsin Avenue, Bethesda, MD 20814-4823. (301) 961-6750. FAX (301) 961-6720. 1993 to present. Monthly. $385.00 per year. Covers concurrent engineering, quality control, automated manufacturing, petroleum engineering, oil field operations and equipment, energy management, metallurgy and metallography, foundry practice.

Materials Science and Engineering Abstracts, Cambridge Scientific Abstracts, 7200 Wisconsin Avenue, Bethesda, MD 20814-4823. (301) 961-6750. FAX (301) 961-6720. 1993 to present. Monthly. $385.00 per year. Focuses on mechanical and physical properties of materials and commercial or industrial applications for materials, methods for strength testing, effects of vibration and other stresses, corrosion and protective coatings, storage and handling, ceramics, composites, metals, wood, plastics, and polymers.

Materials Science Citation Index. Institute for Scientific Information, 3501 Market Street, Philadelphia, PA 19104. (215) 386-0100. FAX (215) 386-0100. Bi-monthly. $975.00 per year. Also available on CD-ROM.

Metals Abstracts and Metals Abstracts Index. American Society for Metals, Metals Park, OH 44073. (216) 338-5151. 1968 to present. Monthly. Abstracts are $1925.00 per year and Index is $460.00 per year.

Physics Abstracts. Institute of Electrical Engineers, Michael Faraday House, Six Hill Way, Stevenage, Herts, SG1 2AY, England. Distributed by IEEE, 445 Hoes Lane, Piscataway, NJ 08854. (908) 562-5549. 1898 to present. Monthly. $2700.00 per year. Also available online as INSPEC.

Science Citation Index. Institute for Scientific Information, 3501 Market Street, Philadelphia, PA 19104. (215) 386-0100. FAX (215) 386-6362. Inquire about availability and cost. Also available on CD-ROM.

ASSOCIATIONS AND PROFESSIONAL SOCIETIES

Aluminum Association. 900 19th Street NW, Suite 300, Washington, DC 20006. (202) 862-5100. FAX (202) 862-5164.

American Powder Metallurgy Institute. 105 College Road E., Princeton, NJ 08540. (609) 452-7700. FAX (609) 987-8523.

American Society For Metals. Materials Park, OH 44073. (216) 338-5151 or FAX (216) 338-4634.

American Society For Testing and Materials (ASTM). 1916 Race Street, Philadelphia, PA 19103. (215) 299-5585.

American Society of Mechanical Engineers. 345 East 47th Street, New York, NY 10017-2398. (212) 705-7703.

Association of Iron and Steel Engineers. Three Gateway Center, Suite 2350, Pittsburgh, PA 15222. (412) 281-6323.

Iron and Steel Society. 410 Commonwealth Drive, Warrendale, PA 15086. (412) 776-9460. FAX (412) 776-0430.

Metallurgical Society of the AIME (American Institute of Mining, Metallurgical, and Petroleum Engineers). 345 E.47th Street, 14th Floor, New York, NY 10017. (212) 705-7695.

Minerals, Metals and Materials Society. 420 Commonwealth Drive, Warrendale, PA 15086. (412) 776-9080.

National Association of Corrosion Engineers. P.O. Box 218340, 1440 South Creek, Houston, TX 77218. (713) 492-0535.

BIBLIOGRAPHIES

The History of Metal Mining and Metallurgy: An Annotated Bibliography. Peter M. Molloy. Garland Publishing Inc., 717 Fifth Avenue, Suite 2500, New York, NY 10022-8102. (212) 751-7447. FAX (212) 308-9399. 1986. $65.00.

DIRECTORIES AND BIOGRAPHICAL SOURCES

Directory of Engineering Societies and Related Organizations. Gordon Davis. 13th edition. American Association of Engineering Societies, 1111 19th Street NW, Suite 608, Washington, DC 20036. (202) 296-2237 or (800) 658-8897. 1989. Inquire for price.

Dun's Industrial Guide: The Metalworking Directory, 1993-94. Dun & Bradstreet Information Services Staff. 3 volumes. Dun & Bradstreet Information Services, 899 Eaton Avenue, Bethlehem, PA 18025. (610) 882-7000 or (800) 526-0651. FAX (610) 882-7269. 1993. $795.00.

International Research Centers Directory. Gale Research, 835 Penobscot Building, Detroit, MI 48226-4094. (313) 961-2242. (800) 347-4253. FAX (313) 961-6083. 8th edition. 1995. $410.00

Machinery Buyers Guide. Findlay Publications Inc. Ltd., Franks Hall, Franks Lane, Horton Kirby, Kent DA4 9LL, England. 1926 to present. Annual. Inquire for cost and availability.

Metallurgical Society of AIME—Membership List. 345 E.47th Street, 14th Floor, New York, NY 10017. (212) 705-7695. Inquire for cost and availability.

Research Centers Directory. Gale Research, 835 Penobscot Building, Detroit, MI 48226-4094. (313) 961-2242. (800) 347-4253. FAX (313) 961-6083. $485.00.

Scientific and Technical Organizations and Agencies Directory. Gale Research, 835 Penobscot Building, Detroit, MI 48226-4094. (313) 961-2242. (800) 347-4253. FAX (313) 961-6083. 4th edition. 1996. $195.00.

Who's Who in Engineering. American Association of Engineering Societies, 1111 19th Street NW, Suite 608, Washington, DC 20036. (202) 296-2237 or (800) 658-8897. 8th edition. 1991. Inquire for price.

ENCYCLOPEDIAS AND DICTIONARIES

ASM Materials Engineering Dictionary. J.R. Davis, editor. ASM International, Materials Information, Materials Park, OH 44073-0002. (216) 338-5151 or (800) 336-5152. FAX (216) 338-4634. 1992. $101.00.

Dictionary of Physical Metallurgy in Five Languages. R. Freiwillig, et al. Elsevier Science Publishing Company, Inc., 655 Avenue of the Americas, New York, NY 10010. (212) 989-5800. 1987. $113.00.

Dictionary of Metallurgy and Foundry Technology. Karl Stolzel, editor. Elsevier Science Publishing Company, Inc., 655 Avenue of the Americas, New York, NY 10010. (212) 989-5800. 1987. $250.00.

McGraw-Hill Dictionary of Scientific and Technical Terms. Sybil P. Parker, ed. 5th edition. McGraw-Hill Publishing Company, 1221 Avenue of the Americas, New York, NY 10020. (800) 262-4729 or (212) 512-3825. 1993. $110.50.

McGraw-Hill Encyclopedia of Engineering. Sybil P. Parker, ed. 2nd edition. McGraw-Hill Publishing Company, 1221 Avenue of the Americas, New York, NY 10020. (800) 262-4729 or (212) 512-3825. 1993. $95.50.

McGraw-Hill Encyclopedia of Science and Technology. Sybil P. Parker, ed. 7th edition. 20 volumes. McGraw-Hill Publishing Company, 1221 Avenue of the Americas, New York, NY 10020. (800) 262-4729 or (212) 512-3825. 1992. $1900.00

Thesaurus of Scientific, Technical, and Engineering Terms. Hemisphere Publishing Corporation, 1900 Frost Road, Suite 101, Bristol, PA 19007-1598. (215) 785-5800 or (800) 821-8312. FAX (215) 785-5515. 1987. $173.00.

GENERAL WORKS

Fundamentals of Metallurgical Processes. Lucien Coudurier, et al. Pergamon/ Franklin Book Company, 7804 Montgomery Avenue, Elkins Park, PA 19117. (215) 635-5252. FAX (215) 635-6155. 1985. $174.00 (hardback), $44.00 (paperback).

The Role of Technology in Iron and Steel Development. Organization for Economic Cooperation and Development, 2001 L St. NW, Suite 650, Washington, DC 20036. (202) 785-6323 or (800) 456-6323. FAX (202) 785-0350. 1989. $21.00.

The Technology of Metallurgy. William K. Dalton. Maxwell Macmillan International, 866 Third Avenue, 21st Floor, New York, NY 10022. (212) 702-2000 or (800) 257-5755. 1994. Inquire for cost and availability.

HANDBOOKS AND MANUALS

ASM Metals Reference Book. 3rd edition. ASM International, Materials Information, Materials Park, OH 44073-0002. (216) 338-5151 or (800) 336-5152. FAX (216) 338-4634. 1993. $111.00.

CRC Handbook of Chemistry and Physics. David R. Lide, editor. 75th edition. CRC Press, 2000 Corporate Blvd., N.W., Boca Raton, FL 33431. (407) 994-0555 or (800) 272-7737. FAX (407) 994-0949. $99.50.

Metals Handbook. ASM International, Materials Information, Materials Park, OH 44073-0002. (216) 338-5151 or (800) 336-5152. FAX (216) 338-4634. $154.00.

Smithell's Metals Reference Book. E.A. Brandes & G.B. Brook. 7th edition. Butterworth-Heinemann, 313 Washington Street, Newton, MA 02158. 1992. $250.00.

ONLINE DATABASES AND CD-ROMS

Compendex Plus. Engineering Information, Inc., 345 East 47th Street, New York, NY 10017. (212) 705-7600 or (800) 221-1044. Contains citations with abstracts to worldwide literature in engineering and technology, from 1970 to present. Available on online BRS,(800) 289-4277, DIALOG, (800) 334-2564, ORBIT (800) 456-7248, and STN International, FIZ Karlsruhe, P.O. Box 2465, W-7500, Karlsruhe 1, Germany, online services. Also available on CD-ROM. Inquire as to cost and availability.

CSA Engineering. Cambridge Scientific Abstracts, 7200 Wisconsin Avenue, Suite 601, Bethesda, MD 20814. (301) 961-6750 or (800) 843-7751. Contains citations and abstracts of international periodicals and research literature covering all fields of engineering and science and technology, including computer and information science, electronics, mechanical engineering, solid state materials, 1981 to present. Available on BRS,(800) 289-4277, online service. Inquire as to cost and availability.

Current Contents Search. Institute for Scientific Information, 3501 Market Street, Philadelphia, PA 19104. (215) 386-0100. FAX (215) 386-6362. Contains citations to articles listed in the table of contents of science and technology journals. Also articles in social sciences and life sciences journals. Available on BRS,(800) 289-4277, DIALOG,(800) 334-2564, online services. Inquire as to cost and availability.

Dissertation Abstracts. University Microfilms International, 300 North Zeeb Road, Ann Arbor, MI 48106. (800) 521-0600 or (313) 761-4700. Scope includes virtually all doctoral dissertations accepted at accredited American institutions from 1861 to present in 252 subject areas. Available on BRS,(800) 289-4277, DIALOG,(800) 334-2564, and OCLC EPIC,(800) 848-5878, online services. Also available on CD-ROM. Inquire as to cost and availability.

ISMEC: Mechanical Engineering Abstracts. Cambridge Scientific Abstracts, 7200 Wisconsin Avenue, Suite 601, Bethesda, MD 20814. (301) 961-6750 or (800) 843-7751. Contains citations to the literature in mechanical engineering, industrial and production engineering, energy, power, mechanics, devices and related areas, from 1973 to present. Available on the DIALOG,(800) 334-2564, online service. Inquire as to cost and availability.

Metadex. Jointly produced by ASM International and the Institute of Materials. Contains more than 925,000 records from the international literature on metals and alloys, concerning properties, processes, materials classes, applications, and metallurgical systems. Updated monthly. Available from ORBIT-QUESTEL (703) 442-0900.

NTIS Bibliographic Database. National Technical Information Service, 5285 Port Royal Road, Springfield, VA 22161. (703) 487-4929 or FAX (703) 321-8199. Broad coverage of government-sponsored science and technology research reports, 1964 to present. Available on BRS,(800) 289-4277, DIALOG, (800) 334-2564, ORBIT, (800) 456-7248, and STN International, FIZ Karlsruhe, P.O. Box 2465, W-7500, Karlsruhe 1, Germany, online services. Also available on CD-ROM. Inquire as to cost and availability.

Scisearch. Institute for Scientific Information, 3501 Market Street, Philadelphia, PA 19104. (800) 523-1850 or (215) 386-0100. Broad multidisciplinary title and author index to the international literature of science and technology, 1974 to present. Available on DIALOG,(800) 334-2564, and ORBIT,(800) 456-7248, online services. Also available on CD-ROM. Inquire as to cost and availability.

Wilsonline. H.W. Wilson Company, 950 University Avenue, Bronx, NY 10452. (800) 367-6770 or (212) 588-8400. Makes available online versions of the printed H.W. Wilson indexes, including Applied Science and Technology Index, Business Periodicals Index, General Science Index, and Readers' Guide to Periodical Literature. Period covered is generally 1983 to present. Available on BRS,(800) 289-4277, DIALOG,(800) 334-2564, and OCLC EPIC,(800) 848-5878, online services. Also available on CD-ROM. Inquire as to cost and availability.

PERIODICALS

Chemical Engineering. McGraw-Hill Publishing Company, 1221 Avenue of the Americas, New York, NY 10020. (800) 262-4729 or (212) 512-3825. 1902 to present. Monthly. $35.50 per year.

International Journal of Powder Metallurgy. American Powder Metallurgy Institute, 105 College Road E., Princeton, NJ 08540. (609) 452-7700. FAX (609) 987-8523. 1965 to present. Quarterly. $70.00 per year.

Iron and Steel Engineer. Association of Iron and Steel Engineers, Three Gateway Center, Suite 2350, Pittsburgh, PA 15222. (412) 281-6323. 1924 to present. Monthly. $50.00 per year.

Iron and Steel Society. 410 Commonwealth Drive, Warrendale, PA 15086. (412) 776-9460. FAX (412) 776-0430.

Iron & Steelmaker. Iron and Steel Society, 410 Commonwealth Drive, Warrendale, PA 15086. (412) 776-9460. FAX (412) 776-0430. 1974 to present. Monthly. $50.00 per year.

Ironmaking and Steelmaking. Institute of Materials, 1 Carlton House Terrace, London SW1Y 5DB, England. Telephone 071-839-4071. FAX 071-839-2078. 1974 to present. Bi-monthly. $420.00 per year for non-members.

JOM (Journal of Metals). Minerals, Metals and Materials Society, 420 Commonwealth Drive, Warrendale, PA 15086. (412) 776-9080. 1949 to present. Monthly. $50.00 per year.

METALLURGICAL ENGINEERING

Ency. of Physical Sciences and Engineering Info. Sources

Journal of Alloys and Compounds. Elsevier Science [journals], 660 White Plains Rd., Tarrytown, NY 10159-5153. (919) 524-9200. FAX (919) 333-2444. 1959 to present. 30 times a year. $3980.00 per year.

Journal of Applied Mechanics. American Society of Mechanical Engineers, 345 E. 47th Street, New York, NY 10017. (212) 705-7722. 1935 to present. Quarterly. $120.00.

Journal of Engineering For Industry. American Society of Mechanical Engineers, 345 E. 47th Street, New York, NY 10017. (212) 705-7722. 1970 to present. Quarterly. $100.00 per year.

Metallurgia: The Journal of Metals Technology, Metal Forming and thermal Processing. FMJ International Publications, Ltd., Queensway House, 2 Queensway, RedHill, Surrey RH1 1QS England. Telephone 0737-768611. FAX 0737-761685. 1929 to present. Monthly. $168.50 per year.

Metallurgical Transactions A: Physical Metallurgy and Materials Science. ASM (American Society for Metals) International, Materials Information, Materials Park, OH 44073. (216) 338-5151 or FAX (216) 338-4634. 1970 to present. Monthly. $520.00 per year for non-members.

Metallurgical Transactions B: Process Metallurgy. ASM (American Society for Metals) International, Materials Information, Materials Park, OH 44073. (216) 338-5151 or FAX (216) 338-4634. 1975 to present. Bi-monthly. $375.00 per year for non-members.

Mining Engineering. Society for Mining, Metallurgy, and Exploration, Inc., 8307 Shaffer Parkway, PO Box 625002, Littleton, CO 80162-5002. (303) 973-9550. FAX (303) 973-3845. Monthly. $100.00 per year.

Scripta Metallurgia Et Materialia. Elsevier Science [journals], 660 White Plains Rd., Tarrytown, NY 10159-5153. (919) 524-9200. FAX (919) 333-2444. 1967 to present. 24 times a year. $570.00 per year.

RESEARCH CENTERS AND INSTITUTES

Cooperative Program in Metallurgy, Pennsylvania State University. 208A Steidle Bldg., University Park, PA 16802. (814) 865-5446.

University of Connecticut Institute of Materials Science. U-136, 97 N. Eagleville Road, Storrs, CT 06268. (203) 486-4623. FAX (203) 486-4745.

University of Florida Department of Materials Science and Engineering. Gainesville, FL 32611. (904) 392-1454. FAX (904) 392-6359.

METALLURGY

See also: ALLOYS, FERROALLOYS, MACHINING, METALLURGICAL ENGINEERING, METALS and METAL WORKING, WELDING

ABSTRACT SERVICES AND INDEXES

Alloys Index. American Society for Metals, Metals Park, OH 44073. (216) 338-5151. FAX (216) 338-4634. 1974 to present.

Monthly. $380.00. Also available on CD-ROM and online via METADEX, STN International, and DIALOG.

Aluminum Industry Abstracts. Aluminum Association, Materials Park, OH 44073. (216) 338-5151. FAX (216) 338-4634. 1968 to present. Monthly. $525.00 per year.

Applied Mechanics Reviews: An Assessment of World Literature in Engineering Sciences. 1948-present. American Society of Mechanical Engineers, 345 East 47th Street, New York, NY 10017. (212) 705-7703. Monthly. $360.00 per year.

Applied Science and Technology Index; A Cumulative Subject Index To English Language Periodicals in the Fields of Aeronautics and Space Science, Computer Technology, Chemistry, Construction Industry, Energy and Related Areas. H.W. Wilson Co., 950 University Avenue, Bronx, NY 10452. (800) 367-6770 or (212) 588-8400. FAX (718) 590-1617. From 1958 to present. Monthly. Inquire about cost and availability. Also available on CD-ROM and online.

Corrosion Abstracts; Abstracts of the World's Literature On Corrosion and Corrosion Mitigation. National Association of Corrosion Engineers (NACE), Box 218340, Houston, TX 77218. (713) 492-0535 or FAX (713) 492-8254. 1962 to present. Bi-monthly. $195.00 per year. Also available on CD-ROM.

Current Contents: Engineering, Technology, and Applied Sciences. Institute for Scientific Information, 3501 Market Street, Philadelphia, PA 19104. (215) 386-0100. FAX (215) 386-6362. 1970 to present. Weekly. $442.00 per year.

Engineering Index Monthly. Engineering Information, Inc., Castle Point on the Hudson, Hoboken, NJ 07030. (800) 221-1044. FAX (212) 832-1857. Monthly. $2300.00 per year. Also available online as COMPENDEX, and also on CD-ROM. Covers chemical engineering, computers, electrical engineering, civil engineering, metals and mining, industrial management, and mechanical engineering.

I.M.M. Abstracts and Index. Institution of Mining and Metallurgy, 44 Portland Pl., London W1N 4BR, England. 071-580-3802. FAX 071-436-5388. Bi-monthly. $364 for non-members.

Leadscan (formerly Lead Abstracts), Lead Development Association, 42 Weymouth Street, London W1x 3lq, England. 1958 to present. Quarterly. Inquire for cost.

Manufacturing and Process Engineering Abstracts, Cambridge Scientific Abstracts, 7200 Wisconsin Avenue, Bethesda, MD 20814-4823. (301) 961-6750. FAX (301) 961-6720. 1993 to present. Monthly. $385.00 per year. Covers concurrent engineering, quality control, automated manufacturing, petroleum engineering, oil field operations and equipment, energy management, metallurgy and metallography, foundry practice.

Materials Science and Engineering Abstracts, Cambridge Scientific Abstracts, 7200 Wisconsin Avenue, Bethesda, MD 20814-4823. (301) 961-6750. FAX (301) 961-6720. 1993 to present. Monthly. $385.00 per year. Focuses on mechanical and physical properties of materials and commercial or industrial applications for materials, methods for strength testing, effects of vibration and other stresses, corrosion and protective coatings, storage and handling, ceramics, composites, metals, wood, plastics, and polymers.

Materials Science Citation Index. Institute for Scientific Information, 3501 Market Street, Philadelphia, PA 19104. (215) 386-

0100. FAX (215) 386-6362. Bi-monthly. $975.00 per year. Also available on CD-ROM.

Metals Abstracts and Metals Abstracts Index. American Society for Metals, Metals Park, OH 44073. (216) 338-5151. 1968 to present. Monthly. Abstracts are $1925.00 per year and Index is $460.00 per year.

Physics Abstracts. Institute of Electrical Engineers, Michael Faraday House, Six Hill Way, Stevenage, Herts, SG1 2AY, England. Distributed by IEEE, 445 Hoes Lane, Piscataway, NJ 08854. (908) 562-5549. 1898 to present. Monthly. $2700.00 per year. Also available online as INSPEC.

Science Citation Index. Institute for Scientific Information, 3501 Market Street, Philadelphia, PA 19104. (215) 386-0100. FAX (215) 386-6362. Inquire about availability and cost. Also available on CD-ROM.

ASSOCIATIONS AND PROFESSIONAL SOCIETIES

Aluminum Association. 900 19th Street NW, Suite 300, Washington, DC 20006. (202) 862-5100. FAX (202) 862-5164.

American Powder Metallurgy Institute. 105 College Road E., Princeton, NJ 08540. (609) 452-7700. FAX (609) 987-8523.

American Society For Metals. Materials Park, OH 44073. (216) 338-5151 or FAX (216) 338-4634.

American Society For Testing and Materials (ASTM). 1916 Race Street, Philadelphia, PA 19103. (215) 299-5585.

American Society of Mechanical Engineers. 345 East 47th Street, New York, NY 10017-2398. (212) 705-7703.

Association of Iron and Steel Engineers. Three Gateway Center, Suite 2350, Pittsburgh, PA 15222. (412) 281-6323.

Iron and Steel Society. 410 Commonwealth Drive, Warrendale, PA 15086. (412) 776-9460. FAX (412) 776-0430.

Metallurgical Society of the AIME (American Institute of Mining, Metallurgical, and Petroleum Engineers). 345 E.47th Street, 14th Floor, New York, NY 10017. (212) 705-7695.

Minerals, Metals and Materials Society. 420 Commonwealth Drive, Warrendale, PA 15086. (412) 776-9080.

National Association of Corrosion Engineers. P.O. Box 218340, 1440 South Creek, Houston, TX 77218. (713) 492-0535.

BIBLIOGRAPHIES

The History of Metal Mining and Metallurgy: An Annotated Bibliography. Peter M. Molloy. Garland Publishing Inc., 717 Fifth Avenue, Suite 2500, New York, NY 10022-8102. (212) 751-7447. FAX (212) 308-9399. 1986. $65.00.

DIRECTORIES AND BIOGRAPHICAL SOURCES

Directory of Engineering Societies and Related Organizations. Gordon Davis. 13th edition. American Association of Engineering Societies, 1111 19th Street NW, Suite 608, Washington,

DC 20036. (202) 296-2237 or (800) 658-8897. 1989. Inquire for price.

Dun's Industrial Guide: The Metalworking Directory, 1993-94. Dun & Bradstreet Information Services Staff. 3 volumes. Dun & Bradstreet Information Services, 899 Eaton Avenue, Bethlehem, PA 18025. (610) 882-7000 or (800) 526-0651. FAX (610) 882-7269. 1993. $795.00

International Research Centers Directory. Gale Research, 835 Penobscot Building, Detroit, MI 48226-4094. (313) 961-2242. (800) 347-4253. FAX (313) 961-6083. 8th edition. 1995. $410.00.

Machinery Buyers Guide. Findlay Publications Inc. Ltd., Franks Hall, Franks Lane, Horton Kirby, Kent DA4 9LL, England. 1926 to present. Annual. Inquire for cost and availability.

Metallurgical Society of AIME—Membership List. 345 E.47th Street, 14th Floor, New York, NY 10017. (212) 705-7695. Inquire for cost and availability.

Research Centers Directory. Gale Research, 835 Penobscot Building, Detroit, MI 48226-4094. (313) 961-2242. (800) 347-4253. FAX (313) 961-6083. $485.00.

Scientific and Technical Organizations and Agencies Directory. Gale Research, 835 Penobscot Building, Detroit, MI 48226-4094. (313) 961-2242. (800) 347-4253. FAX (313) 961-6083. 4th edition. 1996. $195.00.

Who's Who in Engineering. American Association of Engineering Societies, 1111 19th Street NW, Suite 608, Washington, DC 20036. (202) 296-2237 or (800) 658-8897. 8th edition. 1991. Inquire for price.

ENCYCLOPEDIAS AND DICTIONARIES

ASM Materials Engineering Dictionary. J.R. Davis, editor. ASM International, Materials Information, Materials Park, OH 44073-0002. (216) 338-5151 or (800) 336-5152. FAX (216) 338-4634. 1992. $101.00.

Dictionary of Physical Metallurgy in Five Languages. R. Freiwillig, et al. Elsevier Science Publishing Company, Inc., 655 Avenue of the Americas, New York, NY 10010. (212) 989-5800. 1987. $113.00.

Dictionary of Metallurgy and Foundry Technology. Karl Stolzel, editor. Elsevier Science Publishing Company, Inc., 655 Avenue of the Americas, New York, NY 10010. (212) 989-5800. 1987. $250.00.

McGraw-Hill Dictionary of Scientific and Technical Terms. Sybil P. Parker, ed. 5th edition. McGraw-Hill Publishing Company, 1221 Avenue of the Americas, New York, NY 10020. (800) 262-4729 or (212) 512-3825. 1993. $110.50.

McGraw-Hill Encyclopedia of Engineering. Sybil P. Parker, ed. 2nd edition. McGraw-Hill Publishing Company, 1221 Avenue of the Americas, New York, NY 10020. (800) 262-4729 or (212) 512-3825. 1993. $95.50.

McGraw-Hill Encyclopedia of Science and Technology. Sybil P. Parker, ed. 7th edition. 20 volumes. McGraw-Hill Publishing Company, 1221 Avenue of the Americas, New York, NY 10020. (800) 262-4729 or (212) 512-3825. 1992. $1900.00

Thesaurus of Scientific, Technical, and Engineering Terms. Hemisphere Publishing Corporation, 1900 Frost Road, Suite 101, Bristol, PA 19007-1598. (215) 785-5800 or (800) 821-8312. FAX (215) 785-5515. 1987. $173.00.

GENERAL WORKS

The Extraction and Refining of Metals. Colin Bodsworth. CRC Press, 2000 Corporate Blvd., N.W., Boca Raton, FL 33431. (407) 994-0555 or (800) 272-7737. FAX (407) 994-0949. 1994. Inquire for cost.

Principles of Extractive Metallurgy. Ahindra Ghosh & Hem Shanker Ray. 2nd edition. John Wiley and Sons, Inc., 605 Third Avenue, New York, NY 10158. (800) 526-5368 or (212) 850-6000. 1991. $43.95.

Principles of Metal Refining. T. Abel Engh, et al. Oxford University Press, Inc., 198 Madison Avenue, New York, NY 10016-4314. (212) 726-6000. FAX (212) 726-6446. 1992. $98.00.

The Technology of Metallurgy. William K. Dalton. Maxwell Macmillan International, 866 Third Avenue, 21st Floor, New York, NY 10022. (212) 702-2000 or (800) 257-5755. 1994. Inquire for cost and availability.

HANDBOOKS AND MANUALS

ASM Metals Reference Book. 3rd edition. ASM International, Materials Information, Materials Park, OH 44073-0002. (216) 338-5151 or (800) 336-5152. FAX (216) 338-4634. 1993. $111.00.

CRC Handbook of Chemistry and Physics. David R. Lide, editor. 75th edition. CRC Press, 2000 Corporate Blvd., N.W., Boca Raton, FL 33431. (407) 994-0555 or (800) 272-7737. FAX (407) 994-0949. $99.50.

Metals Handbook. ASM International, Materials Information, Materials Park, OH 44073-0002. (216) 338-5151 or (800) 336-5152. FAX (216) 338-4634. $154.00.

Smithell's Metals Reference Book. E.A. Brandes & G.B. Brook. 7th edition. Butterworth-Heinemann, 313 Washington Street, Newton, MA 02158. (617) 928-2500 or (800) 366-2665. FAX (617) 928-2620. 1992. $250.00.

Powder Metallurgy Design Manual. Metal Powder Industries Federation, 105 College Road E., Princeton, NJ 08540. (609) 452-7700. 1989. $75.00.

Woldman's Engineering Alloys. J. Frick, editor. 8th edition. ASM International, Materials Information, Materials Park, OH 44073-0002. (216) 338-5151 or (800) 336-5152. FAX (216) 338-4634. 1994. $175.00.

ONLINE DATABASES AND CD-ROMS

Compendex Plus. Engineering Information, Inc., 345 East 47th Street, New York, NY 10017. (212) 705-7600 or (800) 221-1044. Contains citations with abstracts to worldwide literature in engineering and technology, from 1970 to present. Available on online BRS,(800) 289-4277, DIALOG, (800) 334-2564, ORBIT (800) 456-7248, and STN International, FIZ Karlsruhe,

P.O. Box 2465, W-7500, Karlsruhe 1, Germany, online services. Also available on CD-ROM. Inquire as to cost and availability.

CSA Engineering. Cambridge Scientific Abstracts, 7200 Wisconsin Avenue, Suite 601, Bethesda, MD 20814. (301) 961-6750 or (800) 843-7751. Contains citations and abstracts of international periodicals and research literature covering all fields of engineering and science and technology, including computer and information science, electronics, mechanical engineering, solid state materials, 1981 to present. Available on BRS,(800) 289-4277, online service. Inquire as to cost and availability.

Current Contents Search. Institute for Scientific Information, 3501 Market Street, Philadelphia, PA 19104. (215) 386-0100. FAX (215) 386-6362. Contains citations to articles listed in the table of contents of science and technology journals. Also articles in social sciences and life sciences journals. Available on BRS,(800) 289-4277, DIALOG,(800) 334-2564, online services. Inquire as to cost and availability.

Dissertation Abstracts. University Microfilms International, 300 North Zeeb Road, Ann Arbor, MI 48106. (800) 521-0600 or (313) 761-4700. Scope includes virtually all doctoral dissertations accepted at accredited American institutions from 1861 to present in 252 subject areas. Available on BRS,(800) 289-4277, DIALOG,(800) 334-2564, and OCLC EPIC,(800) 848-5878, online services. Also available on CD-ROM. Inquire as to cost and availability.

ISMEC: Mechanical Engineering Abstracts. Cambridge Scientific Abstracts, 7200 Wisconsin Avenue, Suite 601, Bethesda, MD 20814. (301) 961-6750 or (800) 843-7751. Contains citations to the literature in mechanical engineering, industrial and production engineering, energy, power, mechanics, devices and related areas, from 1973 to present. Available on the DIALOG,(800) 334-2564, online service. Inquire as to cost and availability.

Metadex. Jointly produced by ASM International and the Institute of Materials. Contains more than 925,000 records from the international literature on metals and alloys, concerning properties, processes, materials classes, applications, and metallurgical systems. Updated monthly. Available from ORBIT-QUESTEL (703) 442-0900.

NTIS Bibliographic Database. National Technical Information Service, 5285 Port Royal Road, Springfield, VA 22161. (703) 487-4929 or FAX (703) 321-8199. Broad coverage of government-sponsored science and technology research reports, 1964 to present. Available on BRS,(800) 289-4277, DIALOG, (800) 334-2564, ORBIT, (800) 456-7248, and STN International, FIZ Karlsruhe, P.O. Box 2465, W-7500, Karlsruhe 1, Germany, online services. Also available on CD-ROM. Inquire as to cost and availability.

Scisearch. Institute for Scientific Information, 3501 Market Street, Philadelphia, PA 19104. (800) 523-1850 or (215) 386-0100. Broad multidisciplinary title and author index to the international literature of science and technology, 1974 to present. Available on DIALOG,(800) 334-2564, and ORBIT,(800) 456-7248, online services. Also available on CD-ROM. Inquire as to cost and availability.

Wilsonline. H.W. Wilson Company, 950 University Avenue, Bronx, NY 10452. (800) 367-6770 or (212) 588-8400. Makes available online versions of the printed H.W. Wilson indexes, including Applied Science and Technology Index, Business Periodicals Index, General Science Index, and Readers' Guide to Periodical Literature. Period covered is generally 1983 to present. Available on BRS,(800) 289-4277, DIALOG,(800) 334-2564, and OCLC EPIC,(800) 848-5878, online services. Also available on CD-ROM. Inquire as to cost and availability.

PERIODICALS

Alloy Digest. Alloy Digest Inc., 27 Canfield Street, Orange, NJ 07050. (201) 677-9161. 1952 to present. Monthly. $140.00 per year.

CIM Bulletin. Canadian Institute of Mining, Metallurgy, and Petroleum, Xerox Tower, 3400 de Maisonneuve Blvd. W., Suite 1210, Montreal PQH3Z 3B8, Canada. (514) 939-2710. FAX (514) 939-2714. 1898 to present. Monthly. $105.00 per year.

International Journal of Powder Metallurgy. American Powder Metallurgy Institute, 105 College Road E., Princeton, NJ 08540. (609) 452-7700. FAX (609) 987-8523. 1965 to present. Quarterly. $70.00 per year.

Iron and Steel Engineer. Association of Iron and Steel Engineers, Three Gateway Center, Suite 2350, Pittsburgh, PA 15222. (412) 281-6323. 1924 to present. Monthly. $50.00 per year.

Iron & Steelmaker. Iron and Steel Society, 410 Commonwealth Drive, Warrendale, PA 15086. (412) 776-9460. FAX (412) 776-0430. 1974 to present. Monthly. $50.00 per year.

Ironmaking and Steelmaking. Institute of Materials, 1 Carlton House Terrace, London SW1Y 5DB, England. Telephone 071-839-4071. FAX 071-839-2078. 1974 to present. Bi-monthly. $420.00 per year for non-members.

JOM (Journal of Metals). Minerals, Metals and Materials Society, 420 Commonwealth Drive, Warrendale, PA 15086. (412) 776-9080. 1949 to present. Monthly. $50.00 per year.

Journal of Alloys and Compounds. Elsevier Science [journals], 660 White Plains Rd., Tarrytown, NY 10159-5153. (919) 524-9200. FAX (919) 333-2444. 1959 to present. 30 times a year. $3980.00 per year.

Journal of Applied Mechanics. American Society of Mechanical Engineers, 345 E. 47th Street, New York, NY 10017. (212) 705-7722. 1935 to present. Quarterly. $120.00.

Journal of Engineering For Industry. American Society of Mechanical Engineers, 345 E. 47th Street, New York, NY 10017. (212) 705-7722. 1970 to present. Quarterly. $100.00 per year.

Light Metal Age. Fellom Publishing, 170 S. Spruce Avenue, Suite 120, San Francisco, CA 94080. (415) 588-8832. FAX (415) 588-0901. 1943 to present. Bi-monthly. $35.00 per year.

Materials Science and Technology. Institute of Materials, 1 Carlton House Terrace, London SW1Y 5DB, England. Telephone 071-839-4071. FAX 071-839-2078. 1985 to present. Monthly. $837.00 per year.

Materials Transactions, JIM. Japan Institute of Metals, Nihon Kinzoku Gakkai, Aoba Aramaki, Aoba-Ku, Sendai 980, Japan. Telephone 022-223-3685. FAX 8-22-223-6312. 1960 to present. Monthly. Inquire for cost.

Metallurgia: The Journal of Metals Technology, Metal Forming and thermal Processing. FMJ International Publications,

Ltd., Queensway House, 2 Queensway, RedHill, Surrey RH1 1QS England. Telephone 0737-768611. FAX 0737-761685. 1929 to present. Monthly. $168.50 per year.

Metallurgical Transactions A: Physical Metallurgy and Materials Science. ASM (American Society for Metals) International, Materials Information, Materials Park, OH 44073. (216) 338-5151 or FAX (216) 338-4634. 1970 to present. Monthly. $520.00 per year for non-members.

Metallurgical Transactions B: Process Metallurgy. ASM (American Society for Metals) International, Materials Information, Materials Park, OH 44073. (216) 338-5151 or FAX (216) 338-4634. 1975 to present. Bi-monthly. $375.00 per year for non-members.

Metals Week. McGraw-Hill Publishing Company, 1221 Avenue of the Americas, New York, NY 10020. (800) 262-4729 or (212) 512-3825. 1930 to present. Weekly. $770.00 per year.

Scripta Metallurgia Et Materialia. Elsevier Science [journals], 660 White Plains Rd., Tarrytown, NY 10159-5153. (919) 524-9200. FAX (919) 333-2444. 1967 to present. 24 times a year. $570.00 per year.

RESEARCH CENTERS AND INSTITUTES

Cooperative Program in Metallurgy, Pennsylvania State University. 208A Steidle Bldg., University Park, PA 16802. (814) 865-5446.

University of Connecticut Institute of Materials Science. U-136, 97 N. Eagleville Road, Storrs, CT 06268. (203) 486-4623. FAX (203) 486-4745.

University of Florida Department of Materials Science and Engineering. Gainesville, FL 32611. (904) 392-1454. FAX (904) 392-6359.

METALS AND METALWORKING

See also: ALLOYS, FERROALLOYS, MACHINING, METALLURGICAL ENGINEERING, METALLURGY, WELDING

ABSTRACT SERVICES AND INDEXES

Alloys Index. American Society for Metals, Metals Park, OH 44073. (216) 338-5151. FAX (216) 338-4634. 1974 to present. Monthly. $380.00. Also available on CD-ROM and online via METADEX, STN International, and DIALOG.

Aluminum Industry Abstracts. Aluminum Association, Materials Park, OH 44073. (216) 338-5151. FAX (216) 338-4634. 1968 to present. Monthly. $525.00 per year.

Applied Mechanics Reviews: An Assessment of World Literature in Engineering Sciences. 1948-present. American Society of Mechanical Engineers, 345 East 47th Street, New York, NY 10017. (212) 705-7703. Monthly. $360.00 per year.

Applied Science and Technology Index; A Cumulative Subject Index To English Language Periodicals in the Fields of Aeronautics and Space Science, Computer Technology, Chemistry, Construction Industry, Energy and Related Areas. H.W. Wilson Co., 950 University Avenue, Bronx, NY 10452. (800) 367-6770 or (212) 588-8400. FAX (718) 590-1617. From 1958 to present. Monthly. Inquire about cost and availability. Also available on CD-ROM and online.

Corrosion Abstracts; Abstracts of the World's Literature On Corrosion and Corrosion Mitigation. National Association of Corrosion Engineers (NACE), Box 218340, Houston, TX 77218. (713) 492-0535 or FAX (713) 492-8254. 1962 to present. Bi-monthly. $195.00 per year. Also available on CD-ROM.

Current Contents: Engineering, Technology, and Applied Sciences. Institute for Scientific Information, 3501 Market Street, Philadelphia, PA 19104. (215) 386-0100. FAX (215) 386-6362. 1970 to present. Weekly. $442.00 per year.

Engineering Index Monthly. Engineering Information, Inc., Castle Point on the Hudson, Hoboken, NJ 07030. (800) 221-1044. FAX (212) 832-1857. Monthly. $2300.00 per year. Also available online as COMPENDEX, and also on CD-ROM. Covers chemical engineering, computers, electrical engineering, civil engineering, metals and mining, industrial management, and mechanical engineering.

I.M.M. Abstracts and Index. Institution of Mining and Metallurgy, 44 Portland Pl., London W1N 4BR, England. 071-580-3802. FAX 071-436-5388. Bi-monthly. $364 for non-members.

Leadscan (formerly Lead Abstracts), Lead Development Association, 42 Weymouth Street, London W1x 3lq, England. 1958 to present. Quarterly. Inquire for cost.

Manufacturing and Process Engineering Abstracts, Cambridge Scientific Abstracts, 7200 Wisconsin Avenue, Bethesda, MD 20814-4823. (301) 961-6750. FAX (301) 961-6720. 1993 to present. Monthly. $385.00 per year. Covers concurrent engineering, quality control, automated manufacturing, petroleum engineering, oil field operations and equipment, energy management, metallurgy and metallography, foundry practice.

Materials Science and Engineering Abstracts, Cambridge Scientific Abstracts, 7200 Wisconsin Avenue, Bethesda, MD 20814-4823. (301) 961-6750. FAX (301) 961-6720. 1993 to present. Monthly. $385.00 per year. Focuses on mechanical and physical properties of materials and commercial or industrial applications for materials, methods for strength testing, effects of vibration and other stresses, corrosion and protective coatings, storage and handling, ceramics, composites, metals, wood, plastics, and polymers.

Materials Science Citation Index. Institute for Scientific Information, 3501 Market Street, Philadelphia, PA 19104. (215) 386-0100. FAX (215) 386-0100. Bi-monthly. $975.00 per year. Also available on CD-ROM.

Metals Abstracts and Metals Abstracts Index. American Society for Metals, Metals Park, OH 44073. (216) 338-5151. 1968 to present. Monthly. Abstracts are $1925.00 per year and Index is $460.00 per year.

Metalworking Digest Literature Review. Gordon Publications Inc., 301 Gibraltar Drive, Box 350, Morris Plains, NJ 07950-0650. (201) 292-5100. FAX (201) 898-9281. Annual. Included with subscription to *Metalworking Digest* periodical (see below).

Physics Abstracts. Institute of Electrical Engineers, Michael Faraday House, Six Hill Way, Stevenage, Herts, SG1 2AY,

England. Distributed by IEEE, 445 Hoes Lane, Piscataway, NJ 08854. (908) 562-5549. 1898 to present. Monthly. $2700.00 per year. Also available online as INSPEC.

Science Citation Index. Institute for Scientific Information, 3501 Market Street, Philadelphia, PA 19104. (215) 386-0100. FAX (215) 386-6362. Inquire about availability and cost. Also available on CD-ROM.

ASSOCIATIONS AND PROFESSIONAL SOCIETIES

Alliance of Metalworking Industries. 27027 Chardon Road, Richmond Heights, OH 44143-1113. (216) 585-8800. FAX (216) 585-3126.

Aluminum Association. 900 19th Street NW, Suite 300, Washington, DC 20006. (202) 862-5100. FAX (202) 862-5164.

American Powder Metallurgy Institute. 105 College Road E., Princeton, NJ 08540. (609) 452-7700. FAX (609) 987-8523.

American Society For Metals. Materials Park, OH 44073. (216) 338-5151 or FAX (216) 338-4634.

American Society For Testing and Materials (ASTM). 1916 Race Street, Philadelphia, PA 19103. (215) 299-5585.

American Society of Mechanical Engineers. 345 East 47th Street, New York, NY 10017-2398. (212) 705-7703.

Association of Iron and Steel Engineers. Three Gateway Center, Suite 2350, Pittsburgh, PA 15222. (412) 281-6323.

Iron and Steel Society. 410 Commonwealth Drive, Warrendale, PA 15086. (412) 776-9460. FAX (412) 776-0430.

Metallurgical Society of the AIME (American Institute of Mining, Metallurgical, and Petroleum Engineers). 345 E.47th Street, 14th Floor, New York, NY 10017. (212) 705-7695.

Minerals, Metals and Materials Society. 420 Commonwealth Drive, Warrendale, PA 15086. (412) 776-9080.

National Association of Corrosion Engineers. P.O. Box 218340, 1440 South Creek, Houston, TX 77218. (713) 492-0535.

DIRECTORIES AND BIOGRAPHICAL SOURCES

Directory of Engineering Societies and Related Organizations. Gordon Davis. 13th edition. American Association of Engineering Societies, 1111 19th Street NW, Suite 608, Washington, DC 20036. (202) 296-2237 or (800) 658-8897. 1989. Inquire for price.

Dun's Industrial Guide: The Metalworking Directory, 1993-94. Dun & Bradstreet Information Services Staff. 3 volumes. Dun & Bradstreet Information Services, 899 Eaton Avenue, Bethlehem, PA 18025. (610) 882-7000 or (800) 526-0651. FAX (610) 882-7269. 1993. $795.00

International Research Centers Directory. Gale Research, 835 Penobscot Building, Detroit, MI 48226-4094. (313) 961-2242. (800) 347-4253. FAX (313) 961-6083. 8th edition. 1995. $410.00

Machinery Buyers Guide. Findlay Publications Inc. Ltd., Franks Hall, Franks Lane, Horton Kirby, Kent DA4 9LL, England. 1926 to present. Annual. Inquire for cost and availability.

Metallurgical Society of AIME—Membership List. 345 E.47th Street, 14th Floor, New York, NY 10017. (212) 705-7695. Inquire for cost and availability.

Research Centers Directory. Gale Research, 835 Penobscot Building, Detroit, MI 48226-4094. (313) 961-2242. (800) 347-4253. FAX (313) 961-6083. $485.00.

Scientific and Technical Organizations and Agencies Directory. Gale Research, 835 Penobscot Building, Detroit, MI 48226-4094. (313) 961-2242. (800) 347-4253. FAX (313) 961-6083. 4th edition. 1996. $195.00.

Who's Who in Engineering. American Association of Engineering Societies, 1111 19th Street NW, Suite 608, Washington, DC 20036. (202) 296-2237 or (800) 658-8897. 8th edition. 1991. Inquire for price.

ENCYCLOPEDIAS AND DICTIONARIES

ASM Materials Engineering Dictionary. J.R. Davis, editor. ASM International, Materials Information, Materials Park, OH 44073-0002. (216) 338-5151 or (800) 336-5152. FAX (216) 338-4634. 1992. $101.00.

Dictionary of Metallurgy and Foundry Technology. Karl Stolzel, editor. Elsevier Science Publishing Company, Inc., 655 Avenue of the Americas, New York, NY 10010. (212) 989-5800. 1987. $250.00.

McGraw-Hill Dictionary of Scientific and Technical Terms. Sybil P. Parker, ed. 5th edition. McGraw-Hill Publishing Company, 1221 Avenue of the Americas, New York, NY 10020. (800) 262-4729 or (212) 512-3825. 1993. $110.50.

McGraw-Hill Encyclopedia of Engineering. Sybil P. Parker, ed. 2nd edition. McGraw-Hill Publishing Company, 1221 Avenue of the Americas, New York, NY 10020. (800) 262-4729 or (212) 512-3825. 1993. $95.50.

McGraw-Hill Encyclopedia of Science and Technology. Sybil P. Parker, ed. 7th edition. 20 volumes. McGraw-Hill Publishing Company, 1221 Avenue of the Americas, New York, NY 10020. (800) 262-4729 or (212) 512-3825. 1992. $1900.00

Thesaurus of Scientific, Technical, and Engineering Terms. Hemisphere Publishing Corporation, 1900 Frost Road, Suite 101, Bristol, PA 19007-1598. (215) 785-5800 or (800) 821-8312. FAX (215) 785-5515. 1987. $173.00.

GENERAL WORKS

Metal Machining and Forming Technology. Joseph P. Vidosic. Books on Demand, Division of University Microfilms International, 300 North Zeeb Road, Ann Arbor, MI 48106-1346. (313) 761-4700 or (800) 521-0600. $142.30.

Metalworking Science and Engineering. Edward M. Mielnik. McGraw-Hill Publishing Company, 1221 Avenue of the Americas, New York, NY 10020. (800) 262-4729 or (212) 512-3825. 1991. Inquire for cost and availability.

Metalworking Technology. Richard L. Little. McGraw-Hill Publishing Company, 1221 Avenue of the Americas, New York, NY 10020. (800) 262-4729 or (212) 512-3825. 1976. $43.95.

HANDBOOKS AND MANUALS

ASM Specialty Handbook: Tool Materials. J.R. Davis, editor. ASM International Handbook Committee. ASM International, Materials Information, Materials Park, OH 44073-0002. (216) 338-5151 or (800) 336-5152. FAX (216) 338-4634. 1995. $159.00.

McGraw-Hill Machining and Metalworking Handbook. Ronald A. Welsh. McGraw-Hill Publishing Company, 1221 Avenue of the Americas, New York, NY 10020. (800) 262-4729 or (212) 512-3825. 1994. $69.50.

Metals Handbook. ASM International, Materials Information, Materials Park, OH 44073-0002. (216) 338-5151 or (800) 336-5152. FAX (216) 338-4634. $154.00.

The Metal Worker's Benchtop Reference. TAB Books, P.O. Box 40, Blue Ridge Summit, PA 17294-0850. (717) 794-2191 or (800) 233-1128. FAX (717) 794-2080. 1986. $25.95.

Metalworking: A Manual of Techniques. Mike George. Trafalgar Square, PO Box 257, North Pomfret, VT 05053. (800) 423-4525. FAX (802) 457-1913. 1991. $39.95.

Smithell's Metals Reference Book. E.A. Brandes & G.B. Brook. 7th edition. Butterworth-Heinemann, 313 Washington Street, Newton, MA 02158. (617) 928-2500 or (800) 366-2665. FAX (617) 928-2620. 1992. $250.00.

Woldman's Engineering Alloys. J. Frick, editor. 8th edition. ASM International, Materials Information, Materials Park, OH 44073-0002. (216) 338-5151 or (800) 336-5152. FAX (216) 338-4634. 1994. $175.00.

ONLINE DATABASES AND CD-ROMS

Compendex Plus. Engineering Information, Inc., 345 East 47th Street, New York, NY 10017. (212) 705-7600 or (800) 221-1044. Contains citations with abstracts to worldwide literature in engineering and technology, from 1970 to present. Available on online BRS,(800) 289-4277, DIALOG, (800) 334-2564, ORBIT (800) 456-7248, and STN International, FIZ Karlsruhe, P.O. Box 2465, W-7500, Karlsruhe 1, Germany, online services. Also available on CD-ROM. Inquire as to cost and availability.

ISMEC: Mechanical Engineering Abstracts. Cambridge Scientific Abstracts, 7200 Wisconsin Avenue, Suite 601, Bethesda, MD 20814. (301) 961-6750 or (800) 843-7751. Contains citations to the literature in mechanical engineering, industrial and production engineering, energy, power, mechanics, devices and related areas, from 1973 to present. Available on the DIALOG,(800) 334-2564, online service. Inquire as to cost and availability.

Metadex. Jointly produced by ASM International and the Institute of Materials. Contains more than 925,000 records from the international literature on metals and alloys, concerning properties, processes, materials classes, applications, and metallurgical systems. Updated monthly. Available from ORBIT-QUESTEL (703) 442-0900.

NTIS Bibliographic Database. National Technical Information Service, 5285 Port Royal Road, Springfield, VA 22161. (703) 487-4929 or FAX (703) 321-8199. Broad coverage of government-sponsored science and technology research reports, 1964 to present. Available on BRS,(800) 289-4277, DIALOG, (800) 334-2564, ORBIT, (800) 456-7248, and STN International, FIZ Karlsruhe, P.O. Box 2465, W-7500, Karlsruhe 1, Germany, online services. Also available on CD-ROM. Inquire as to cost and availability.

Scisearch. Institute for Scientific Information, 3501 Market Street, Philadelphia, PA 19104. (800) 523-1850 or (215) 386-0100. Broad multidisciplinary title and author index to the international literature of science and technology, 1974 to present. Available on DIALOG,(800) 334-2564, and ORBIT,(800) 456-7248, online services. Also available on CD-ROM. Inquire as to cost and availability.

Wilsonline. H.W. Wilson Company, 950 University Avenue, Bronx, NY 10452. (800) 367-6770 or (212) 588-8400. Makes available online versions of the printed H.W. Wilson indexes, including Applied Science and Technology Index, Business Periodicals Index, General Science Index, and Readers' Guide to Periodical Literature. Period covered is generally 1983 to present. Available on BRS,(800) 289-4277, DIALOG,(800) 334-2564, and OCLC EPIC,(800) 848-5878, online services. Also available on CD-ROM. Inquire as to cost and availability.

PERIODICALS

Iron and Steel Engineer. Association of Iron and Steel Engineers, Three Gateway Center, Suite 2350, Pittsburgh, PA 15222. (412) 281-6323. 1924 to present. Monthly. $50.00 per year.

Iron and Steel Society. 410 Commonwealth Drive, Warrendale, PA 15086. (412) 776-9460. FAX (412) 776-0430.

Iron & Steelmaker. Iron and Steel Society, 410 Commonwealth Drive, Warrendale, PA 15086. (412) 776-9460. FAX (412) 776-0430. 1974 to present. Monthly. $50.00 per year.

Ironmaking and Steelmaking. Institute of Materials, 1 Carlton House Terrace, London SW1Y 5DB, England. Telephone 071-839-4071. FAX 071-839-2078. 1974 to present. Bi-monthly. $420.00 per year for non-members.

JOM (Journal of Metals). Minerals, Metals and Materials Society, 420 Commonwealth Drive, Warrendale, PA 15086. (412) 776-9080. 1949 to present. Monthly. $50.00 per year.

Journal of Alloys and Compounds. Elsevier Science [journals], 660 White Plains Rd., Tarrytown, NY 10159-5153. (919) 524-9200. FAX (919) 333-2444. 1959 to present. 30 times a year. $3980.00 per year.

Journal of Applied Mechanics. American Society of Mechanical Engineers, 345 E. 47th Street, New York, NY 10017. (212) 705-7722. 1935 to present. Quarterly. $120.00.

Journal of Engineering For Industry. American Society of Mechanical Engineers, 345 E. 47th Street, New York, NY 10017. (212) 705-7722. 1970 to present. Quarterly. $100.00 per year.

Metals Week. McGraw-Hill Publishing Company, 1221 Avenue of the Americas, New York, NY 10020. (800) 262-4729 or (212) 512-3825. 1930 to present. Weekly. $770.00 per year.

Metalworking Digest. Gordon Publications Inc., 301 Gibraltar Drive, Box 350, Morris Plains, NJ 07950-0650. (201) 292-5100. FAX (201) 898-9281. 1968 to present. 12 times a year. $48.00 per year.

Metalworking Interfaces. MK Infotech Company, 98 Schiller Lane, Lake Zurich, IL 60047. 1976 to present. Bi-monthly. $55.00 per year.

Metalworking Production. Morgan-Grampian Publishers Ltd., Morgan-Grampian House, 30 Calderwood Street, London SE18 6QH, England. Telephone 081-855-7777. FAX 081-854-7476. 1900 to present. Monthly. Inquire for cost.

Metalworking Production & Purchasing. Action Communications Inc., 135 Spy Court, Markham, Ontario OL3R 5H6, Canada. (905) 477-3222. FAX (905) 477-4320. 1974 to present. 6 times a year. Inquire for cost.

RESEARCH CENTERS AND INSTITUTES

Cooperative Program in Metallurgy, Pennsylvania State University. 208A Steidle Bldg., University Park, PA 16802. (814) 865-5446.

University of Connecticut Institute of Materials Science. U-136, 97 N. Eagleville Road, Storrs, CT 06268. (203) 486-4623. FAX (203) 486-4745.

University of Florida Department of Materials Science and Engineering. Gainesville, FL 32611. (904) 392-1454. FAX (904) 392-6359.

METAMORPHIC ROCKS

See also: CRUST, GEOCHEMISTRY, GEOPHYSICS, IGNEOUS ROCKS, MINERALOGY, PETROLOGY, PHYSICAL GEOLOGY, SEDIMENTARY ROCKS, SEDIMENTOLOGY, VOLCANOLOGY

ABSTRACT SERVICES AND INDEXES

Bibliography and Index of Geology. American Geological Institute, 4220 King Street, Alexandria, VA 22302. (703) 379-2480. Fax (703) 379-7563. Monthly. $1295.00 per year. Also available online as GEOREF.

Bibliography of Economic Geology. Geosystems, Box 40, Didcot Oxon Ox11 9BX, England. Bimonthly. $150.00 per year. Also available online as GeoArchive.

Chemical Abstracts-Applied Chemistry and Chemical Engineering Sections. Chemical Abstracts Service, 2540 Olentangy River Road, Box 3012, Columbus, OH 43210. (800) 848-6538 or (614) 447-3600, FAX (614) 447-3713. Bi-weekly. $1410.00 per year. Also available online as CA (Chemical Abstracts).

Deep-Sea Research. Part A: Oceanographic Research Papers. Deep-Sea Research. Part B: Oceanographic Literature Review. Pergamon Press, Inc., Maxwell House, Fairview Park, Elmsford, NY 10523. (914) 592-7700. Fax (914) 592-3625. Twelve times per year. $1370 per year for Parts A and B.

Geological Abstracts. Elsevier-Geo Abstracts, Regency House, 34 Duke Street, Norwich NR3 3AP, England. Monthly. $760.00 per year. Also available online as GEOBASE.

Geological Society of America. Abstracts with Programs. Geological Society of America. 3300 Penrose Place, P.O. Box 9140, Boulder, CO 80301-9140. (303) 447-2020. Abstracts and programs of the annual conference. Annual. $69.00.

Geophysics Abstracts. Pergamon Press, Inc., Maxwell House, Fairview Park, Elmsford, NY 10523. (914) 592-7700. Fax (914) 592-3625. Twelve times per year. $565.00 per year. Also available in microform.

Mineralogical Abstracts. Mineralogical Society and the Mineralogical Society of America, 41 Queen's Gate, London, SW7 5HR, England. Quarterly. $235.00 per year.

Oceanic Abstracts. Cambridge Scientific Abstracts, 7200 Wisconsin Avenue, Bethesda, MD 20814. (301) 961-6750. Fax (301) 961-6720. Bimonthly. $995.00 per year.

Science Citation Index. Institute for Scientific Information, 3501 Market Street, Philadelphia, PA 19104. (800) 523-1850 or (215) 386-0100. Inquire about price and availability.

General Science Index. H.W. Wilson Co., 950 University Avenue, Bronx, NY 10452. (800) 367-6770 or (212) 588-8400. Inquire about price and availability.

ANNUAL REVIEWS AND YEARBOOKS

Annual Review and Earth and Planetary Sciences. Annual Reviews, Inc., 4139 El Camino Way, Palo Alto, CA 94306-0897. (415) 493-4400. Fax (415) 855-9815. Annual. $55.00 per year.

Minerals Yearbook. Bureau of Mines, U.S. Department of the Interior. Available from U.S. Government Printing Office, Washington, DC 20402. (202) 783-3238. Annual. Three volumes. $45.00.

ASSOCIATIONS AND PROFESSIONAL SOCIETIES

American Geological Institute. 4220 King Street, Alexandria, VA 22302. (703) 379-2480. Fax (703) 379-7563.

American Geophysical Union, 2000 Florida Avenue, N.W., Washington, DC 20009. (202) 462-6900.

American Institute of Professional Geologists. 7828 Vance Drive, Suite 103, Arvada, CO 80003. (303) 431-0831.

Association of Engineering Geologists. 323 Boston Post Road, Suite 2D, Sudbury, MA 01776. (508) 443-4639.

Geochemical Society. Department of Terrestrial Magnetism, 5241 Broad Branch Road, N.W., Washington, DC 20015. (202) 686-4370.

Geological Society of America. 3300 Penrose Place, P.O. Box 9140, Boulder, CO 80301-9140. (303) 447-2020.

Society of Independent Professional Earth Scientists. 4925 Greenville Avenue, Suite 170, Dallas, TX 75206. (214) 363-1780.

BIBLIOGRAPHIES

Geologic Reference Sources: A Subject and Regional Bibliography of Publications and Maps in the Geological Sciences. Dederick C. Ward et al. Scarecrow Press, Inc., 52 Liberty Street, Box 4167, Metuchen, NJ 08840. (800) 537-7107 or (201) 548-8600. 1981. $49.50.

Information Sources in the Earth Sciences. David N. Wood, Joan E. Hardy, and Anthony P. Harvey, editors. Bowker-Saur/K.G. Saur. Distributed by R.R. Bowker, 121 Chanlon Road, New Providence, NJ 07974. (800) 521-8110 or (908) 464-6800. Second edition. 1989. $85.00.

DIRECTORIES AND BIOGRAPHICAL SOURCES

American Men and Women of Science: Physical and Biological Sciences. Eighteenth edition. R.R. Bowker Company, 245 West 17th Street, New York, NY 10011. (800) 521-8810 or (212) 916-1600. $750.00.

American Institute of Professional Geologists. Membership Directory. American Institute of Professional Geologists, 7828 Vance Drive, Suite 103, Arvada, CO 80003. (303) 431-0831. Annual. $18.00.

Directory of Geoscience Departments: United States and Canada. American Geological Institute. 4220 King Street, Alexandria, VA 22302-1507. (703) 379-2480. Fax (703) 379-7563. Twenty-six edition. 1987. $22.00.

Geological Society of America. Membership Directory. Geological Society of America, 3300 Penrose Place, Boulder, CO 80301. (303) 447-2020. Annual. Available to members only.

Research Centers Directory. Gale Research, 835 Penobscot Building, Detroit, MI 48226-4094. (800) 347-4253 or (313) 961-2242. 17th edition, 1992. $400.00.

ENCYCLOPEDIAS AND DICTIONARIES

Concise Oxford Dictionary of the Earth Sciences. Ailsa Allaby and Michael Allaby, editors. Oxford University Press, Inc., 200 Madison Avenue, New York, NY 10016. (800) 334-4249 or (212) 679-7300. 1990. $42.95.

Dictionary of Geological Terms. American Geological Institute. Doubleday and Company, Inc., 245 Park Avenue, New York, NY 10017. (800) 645-6156 or (212) 953-4561. Third edition. 1984. $12.00 in paper.

Dictionary of Petrology. S.I. Tomkeieff. John Wiley and Sons, Inc., 605 Third Avenue, New York, NY 10158. (800) 526-5368 or (212) 850-6000. 1983. $180.00.

Encyclopedia of Igneous and Metamorphic Petrology. D.R. Bowes. Van Nostrand Reinhold, 115 Fifth Avenue, New York, NY 10003. (212) 254-3232 or (800) 926-2665. 1989. $125.95.

Glossary of Geology. Robert L. Bates and Julia A. Jackson. American Geological Institute, 4220 King Street, Alexandria, VA 22032. (703) 379-2480. Third edition. 1987. $75.00.

METAMORPHIC ROCKS

Ency. of Physical Sciences and Engineering Info. Sources

GENERAL WORKS

Atlas of Metamorphic Rocks and Their Textures. Bruce W. Yardley, et al. Halsted Press (Division of John Wiley and Sons, Inc.), 605 Third Avenue, New York, NY 10158. (800) 526-5368 or (212) 850-6418. 1990. $49.95.

Introduction To Metamorphic Petrology. Bruce W. Yardley. Halsted Press (Division of John Wiley and Sons, Inc.), 605 Third Avenue, New York, NY 10158. (800) 526-5368 or (212) 850-6418. 1989. $39.95.

Introduction To Metamorphic Textures and Microstructures. A.J. Barker. Chapman and Hall, Inc., 29 West 35th Street, New York, NY 10001-2291. (212) 244-3336. 1990. $69.95.

Petrogenesis of Metamorphic Rocks. K. Bucher and M. Frey. 6th edition. Springer-Verlag, 175 Fifth Avenue, New York, NY 10010. (212) 460-1500 or (800) 777-4643. FAX (212) 473-6272. 1994. $42.95.

Petrology of Igneous and Metamorphic Rocks. Donald W. Hyndman. McGraw-Hill Publishing Company, 1221 Avenue of the Americas, New York, NY 10020. (800) 262-4729 or (212) 512-3825. Second edition. 1985. $49.95.

HANDBOOKS AND MANUALS

Agi Data Sheets For Geology in the Field, Laboratory, and office. J.T. Dutro, R.V. Dietrich, and R.M. Foose, editors. American Geological Institute, 4220 King Street, Alexandria, VA 223020-1507 (800) 336-4764 or (703) 379-2480. 1989. $24.95.

Field Description of Metamorphic Rocks. Norman Fry. John Wiley and Sons, Inc., 605 Third Avenue, New York, NY 10158. (800) 526-5368 or (212) 850-6418. 1991. $24.95.

Field Mapping For Geology Students. F. Ahmed and D.C. Almond. Unwin Hyman, Inc., 10 East 53rd Street, New York, NY 10022. (212) 207-7626 or (800) 242-7737. 1983. $18.95.

Geology in the Field. R.R. Compton. John Wiley and Sons, Inc., 605 Third Avenue, New York, NY 10158. (800) 526-5368 or (212) 850-6418. 1985. $37.95.

A Manual of Geology For Civil Engineers. John A. Pitts. World Scientific Publishing Company, Inc., 1060 Main Street, Unit B, River Edge, NJ 07661. (800) 227-7562 or (201) 487-9655. 1985. $38.00.

The Mapping of Geological Structures. K.R. McClay. John Wiley and Sons, Inc., 605 Third Avenue, New York, NY 10158. (800) 526-5368 or (212) 850-6000. 1988. $18.95.

ONLINE DATABASES AND CD-ROMS

Earth Sciences. U.S. Geological Survey, 12201 Sunrise Valley Drive, Reston, VA 22092-9998. (703) 648-4460. CD-ROM of earth science databases, including the U.S. Geological Survey Library, Earth Science Data Directory, and GEOINDEX, citations to published geological maps. $350.00 per year with quarterly updates.

Geoarchive. Geosystems, P.O. Box 1024, Westminster, London, England, SW1 P 2JL. Citations to literature on geoscience, 1969 to present. Inquire as to online cost and availability.

Geobase. Elsevier-Geo Abstracts, Regency House, 34 Duke Street, Norwich NR3 3AP, England. Contains citations to the worldwide earth science literature from 1980 to date. Available on DIALOG, ORBIT online services. Inquire as to cost and availability.

Geomechanics Abstracts. University of London, Imperial College of Science and Technology, Rock Mechanics Information Service, Royal School of Mines, Prince Consort Road, London SW7 2BP, England. Telephone: 071-589 5111 or FAX 071-589 6806. Scope includes worldwide literature on engineering geology, hydrogeology, mining, rock mechanics, soil mechanics, and tunneling, 1977 to present. Available on the ORBIT online service. Inquire as to cost and availability.

GEOREF. American Geological Institute, 4220 King Street, Alexandria, VA 22302. (800) 336-4764 or (703) 379-2480. Geology and geosciences literature, 1785 to present for North America. Available on DIALOG, ORBIT online services. Also available on CD-ROM. Inquire as to cost and availability.

NTIS Bibliographic Database. National Technical Information Service, 5285 Port Royal Road, Springfield, VA 22161. (703) 487-4630. Broad coverage of government-sponsored research reports, 1964 to present. Available on BRS, DIALOG, ORBIT and STN online services. Also available on CD-ROM. Inquire as to cost and availability.

Scisearch. Institute for Scientific Information, 3501 Market Street, Philadelphia, PA 19104. (800) 523-1850 or (215) 386-0100. Broad multidisciplinary title and author index to the international literature of science and technology, 1974 to present. Available on DIALOG and ORBIT online services. Also available on CD-ROM. Inquire as to cost and availability.

PERIODICALS

American Journal of Science. Kline Geology Laboratory, Yale University, Box 6666, Yale Station, New Haven, CT 06511-8130. (203) 432-3131. Ten times per year. $40.00 per year.

Canadian Journal of Earth Sciences. National Research Council of Canada, Research Journals, Ottawa, Ont. K1A OR6, Canada. (613) 993-9084. Monthly. $112.00 per year.

Earth and Planetary Science Letters. Elsevier Science Publishing Company, Inc., 655 Avenue of the Americas, New York, NY 10010. (212) 989-5800. Twenty times per year. $945.00 per year.

Geochimica Et Cosmochimica Acta. Pergamon Journals, Inc., Maxwell House, Fairview Park, Elmsford, NY 10523. (914) 592-7700. Fax (914) 592-3625. Monthly. $580.00 per year.

Geoforum. Pergamon Press, Maxwell House, Fairview Park, Elmsford, NY 10523. (914) 592-7700. Fax (914) 592-3625. Quarterly. $305.00 per year.

Geological Magazine. Cambridge University Press, 40 West 20th Street, New York, NY 10011-4211. (212) 924-3900. Bimonthly. $205.00 per year.

Geological Society. Journal. Geological Society of London. Geological Society Publishing House, Unit 7, Brassmill Enterprise Centre, Brassmill Lane, Bath, Avon BA1 3JN, England. 6 times per year. $432.00.

Geological Society of America Bulletin. P.O. Box 9140, 3300 Penrose Place, Boulder, CO 80301. (303) 447-2020. Fax (303) 447-1133. Monthly. $150.00.

Geology. Geological Society of America. P.O. Box 9140, 3300 Penrose Place, Boulder, CO 80301. (303) 447-2020. Fax (303) 447-1133. Monthly. $120.00 per year.

Geophysics. Society of Exploration Geophysicists, P.O. Box 702740, Tulsa, OK 74170. (918) 493-3516. Monthly. $175.00 per year.

Journal of Geology. University of Chicago Press, 5720 Woodlawn Avenue, Chicago, IL 60637. (312) 753-3347. Bimonthly. $38.00 per year.

Journal of Geophysical Research. American Geophysical Union, 2000 Florida Avenue, N.W., Washington, DC 20009. (202) 462-6903. Monthly. $2170.00 per year.

Journal of Petrology. Oxford University Press, Inc., 200 Madison Avenue, New York, NY 10016. (800) 334-4249 or (212) 679-7300. Bimonthly. $240.00.

Journal of Structural Geology. Pergamon Journals, Inc., Maxwell House, Fairview Park, Elmsford, NY 10523. (914) 592-7700. Ten times per year. $375.00 per year.

Journal of Volcanology and Geothermal Research. Elsevier Science Publishing Company, Inc., 655 Avenue of the Americas, New York, NY 10010. (212) 989-5800. Monthly. $500.00 per year.

Lithos; An International Journal of Mineralogy, Petrology, and Geochemistry. Elsevier Science Publishing Company, Inc., 655 Avenue of the Americas, New York, NY 10010. (212) 989-5800. Quarterly. $275.00 per year.

RESEARCH CENTERS AND INSTITUTES

Carnegie Institution of Washington. Geophysical Laboratory. 5251 Broad Branch Road, N.W., Washington, DC 20015. (202) 686-2410. Fax (202) 686-2419.

Geological Society of America. P.O. Box 9140, 3300 Penrose Place, Boulder, CO 80301. (303) 447-2020.

U.S. Geological Survey, Geologic Division, National Center, 12201 Sunrise Valley Drive, Reston, VA 22092. (703) 648-6600. The major geological research agency of the federal government conducting research in most areas of pure and applied research in the geosciences.

METEORITES

See: METEORS

METEOROLOGY

See also: CLIMATE, CLOUDS, DROUGHT, FOG, GREENHOUSE EFFECT, HAIL, HURRICANES, ICE, JET STREAM, LIGHTNING, OZONE, RAIN, SNOW, THUNDERSTORMS, TORNADO, WEATHER MODIFICATION

ABSTRACT SERVICES AND INDEXES

Applied Science and Technology Index; A Cumulative Subject Index To English Language Periodicals in the Fields of Aeronautics and Space Science, Computer Technology, Chemistry, Construction Industry, Energy and Related Areas. H.W. Wilson Co., 950 University Avenue, Bronx, NY 10452-9978. (212) 588-8400. (800) 367-6770. FAX (718) 590-1617. From 1958 to present. Monthly. Inquire about cost and availability. Also available online (BRS and WILSONLINE) and on CD-ROM. ISSN: 0003-6986.

Bibliography and Index of Geology. American Geological Institute, 4220 King Street, Alexandria, VA 22302-1507. (703) 379-2480. FAX (703) 379-7563. From 1969 to present. Monthly. $1295.00 per year. Also available as GEOREF online (CISTI, DIALOG, Orbit, STN) and on CD-ROM. Inquire about price and availability. ISSN: 0098-2784.

Current Contents: Physical and Chemical Science. Institute for Scientific Information, 3501 Market Street, Philadelphia, PA 19104. (215) 386-0100. (800) 523-1850. FAX (215) 386-2291. 1961 to present. Weekly. $442.00 per year. Also available online (BRS, DIALOG) and on CD-ROM. Inquire regarding cost and availability. ISSN: 0163-2574.

Deep-Sea Research. Part A: Oceanographic Research Papers. Deep-Sea Research. Part B: Oceanographic Literature Review. Pergamon Press, Inc., Maxwell House, Fairview Park, Elmsford, NY 10523. (914) 592-7700. Fax (914) 592-3625. 1953 to present. Twelve times per year. $2000.00 per year for Parts A and B. Oceanographic Literature Review also available on CD-ROM. Inquire about price and availability.

General Science Index. H.W. Wilson Co., 950 University Avenue, Bronx, NY 10452. (212) 588-8400. (800) 367-6770. FAX (718) 590-1617. From 1978 to present. Ten issues per year; quarterly and annual cumulations. Service basis. Available on CD-ROM and online. Inquire regarding cost and availability. ISSN: 0162-1963.

Government Reports Announcements and Index. National Technical Information Service (NTIS), 5285 Port Royal Road, Springfield, VA 22161. (703) 487-4650. FAX (703) 321-8547. From 1968 to present. Annual. $630.00 per year. Also available online as NTIS Bibliographic Database and on CD-ROM. ISSN:

Meteorological and Geoastrophysical Abstracts. American Meteorological Society, c/o Inforonics, Inc., 550 Newtown Road, Box 458, Littleton, MA 01460. (508) 486-8976. FAX (508) 486-0027. Covers literature in environmental sciences, meteorology, astrophysics, hydrology, glaciology, and physical oceanography. 1950 to present. Monthly. $950.00 per year. ISSN: 0026-1130. Also available online (DIALOG) and on CD-ROM.

Oceanic Abstracts. Cambridge Scientific Abstracts, 7200 Wisconsin Avenue, Bethesda, MD 20814. (301) 961-6750. Fax (301) 961-6720. Bimonthly. $995.00 per year.

Physics Abstracts. INSPEC. Section A, Science Abstracts. Institution of Electrical Engineers (IEE), London, United Kingdom. Available from: INSPEC/IEEE-Institute of Electrical and Electronic Engineers, Box 1331, 445 Hoes Lane, Piscataway, NJ 08855-1331. (908) 562-5549. 1898 to present. 24 issues per

METEOROLOGY

Ency. of Physical Sciences and Engineering Info. Sources

year. $2835.00 per year. ISSN: 0036-8091. Also available online and on CD-ROM.

Science Citation Index. SCI. Institute for Scientific Information, 3501 Market Street, Philadelphia, PA 19104. (215) 386-0100. (800) 523-1850. FAX (215) 386-2991. 1961 to present. Six issues per year, plus annual cumulation. $11650.00 per year. Also available online and on CD-ROM. Inquire about price and availability. ISSN: 0036-827X.

Selected Water Resources Abstracts. U.S. Geological Survey. Distributed by National Technical Information Service (NTIS), 5285 Port Royal Road, Springfield, VA 22161. (703) 487-4650. Monthly. $115.00 per year.

ANNUAL REVIEWS AND YEARBOOKS

American Meteorological Society. Meteorological Monographs. American Meteorological Society, 45 Beacon Street, Boston, MA 02108-3693. (617) 227-2425. FAX (617) 742-8718. Irregular. Price varies.

Annual Review of Earth and Planetary Sciences. Annual Reviews, Inc., 4139 El Camino Way, Palo Alto, CA 94306-0897. (415) 493-4400. Fax (415) 855-9815. Annual. $55.00 per year.

Developments in Atmospheric Science Series. Elsevier Science Publishing Company, Inc., 655 Avenue of the Americas, New York, NY 10010. (212) 989-5800. Irregular. Inquire about price and availability.

ASSOCIATIONS AND PROFESSIONAL SOCIETIES

American Meteorological Society, 45 Beacon Street, Boston, MA 02108-3693. (617) 227-2425. FAX (617) 742-8718. Irregular. Price varies.

Association of American Weather Observers, 401 Whitney Boulevard, Box 455, Belvedere, IL 61008. (815) 544-5665. FAX (815) 544-6334.

Canadian Meteorological and Oceanographic Society, P.O. Box 334, Newmarket, Ontario, L3Y 4X7, Canada. (416) 898-1040. FAX (416) 898-7937.

National Weather Association, 6704 Wolke Court, Montgomery, Al. (334) 213-0388.

University Corporation For Atmospheric Research. PO Box 3000 1850 Table Mesa Drive, Boulder, CO 80307. (303) 497-1000.

Weather Modification Association. PO Box 8116, Fresno CA 93747. (209) 291-8466.

BIBLIOGRAPHIES

Information Sources in the Earth Sciences. David N. Wood, Joan E. Hardy, and Anthony P. Harvey, editors. Bowker-Saur/K.G. Saur. Distributed by R.R. Bowker, 121 Chanlon Road, New Providence, NJ 07974. (800) 521-8110 or (908) 464-6800. Second edition. 1989. $85.00.

International Bibliography of Meteorology: From the Beginning of Printing to 1889. Oliver L. Fassig. Diane Publishing

Co., 600 Upland Avenue, Upland, PA 19015. (610) 499-7415. Reprint edition. 1994. $195.00

Scientific and Technical Books and Serials in Print; An Index To Literature in Science and Technology. R.R. Bowker Co., 205 E. 42nd Street, New York, NY 10017. (800) 521-8110 or (212) 916-1600. 1992. $250.00.

DIRECTORIES AND BIOGRAPHICAL SOURCES

American Men and Women of Science: Physical and Biological Sciences. Eighteenth edition. R.R. Bowker Company, 245 West 17th Street, New York, NY 10011. (800) 521-8810 or (212) 916-1600. $750.00.

American Meteorological Society. Professional Directory. American Meteorological Society. 45 Beacon Street, Boston, MA 02108. (617) 227-2425. Included in each issue of the Bulletin of the Society.

Meteorological Services of the World. American Meteorological Society, 45 Beacon Street, Boston, MA 02108-3693. (617) 227-2425.FAX (617) 742-8718. Annual. $70.00.

National Weather Service offices and Stations. National Oceanic and Atmospheric Administration, Department of Commerce, Silver Spring, MD 20910. (301) 427-7698. Annual. Free

Research Centers Directory. Gale Research, 835 Penobscot Building, Detroit, MI 48226-4094. (313) 961-2242. (800) 347-4253. 20th edition. 1995. $485.00. ISSN: 0080-1518.

ENCYCLOPEDIAS AND DICTIONARIES

Concise Oxford Dictionary of the Earth Sciences. Ailsa Allaby and Michael Allaby, editors. Oxford University Press, Inc., 200 Madison Avenue, New York, NY 10016. (212) 725-6000. (800) 334-4249. 1990. $42.95.

Dictionary of Technical Information: Meteorology. C. Rudolf and S. Dezone. State Mutual Book Service, Ltd. 521 Fifth Avenue, New York, NY 10175. (212) 682-5844. 1986. $120.00.

McGraw-Hill Encyclopedia of Ocean and Atmospheric Sciences. Sybil P. Parker, editor. McGraw-Hill Publishing Company, 1221 Avenue of the Americas, New York, NY 10020. (212) 512-2000. (800) 262-4729. 1979. $79.95.

McGraw-Hill Encyclopedia of Science and Technology. McGraw-Hill Publishing Company, 1221 Avenue of the Americas, New York, NY 10020. (800) 262-4729 or (212) 512-2000. Seventh edition. 1992. $1900.00.

Magill's Survey of Science: Earth Science Series. Salem Press, Inc., P.O. Box 1097, Englewood Cliffs, NJ 07632. (201) 871-3700. (800) 221-1592. Five volumes. 1990. $400.00 for the set.

Smithsonian Meteorological Tables. Books on Demand, 300 North Zeeb Road, Ann Arbor, MI 48106-1346. (313) 761-4700. (800) 521-0600. Reprint edition. $153.90.

Weather Satellite Handbook. American Radio Relay League, 225 Main Street, Newington, CT 06111. (203) 666-1541. 1994. $20.00.

GENERAL WORKS

The Atmosphere: An Introduction to Meteorology. Frederick K. Lutgens and Edward J. Tarbuck. Prentice Hall, 15 Columbus Circle, New York, NY 10023. (212) 373-8500. (800) 223-2348. 5th edition. 1991. $49.00.

Calculating the Weather: Meteorology in the 20th Century. Frederik Nebeker. Academic Press, Inc., 6277 Sea Harbor Drive, Orlando, FL (800) 321-5068. 1995. $59.95.

Climate in Human Perspective. A Tribute to Helmut A. Landsberg. F. Baer, et al editors. Kluwer Academic Publishers, 101 Philip Drive, Assinippi Park, Norwell, MA 02061. (617) 871-6600. 1991. $61.50.

Coastal Meteorology. Hsu Shih-Ang, editor. Academic Press, Inc., 6277 Sea Harbor Drive, Orlando, FL (800) 321-5068. 1988. $71.00.

Dynamics of Atmospheric Motion. John A. Dutton. Dover Publications, Inc., 180 Varick Street, New York, NY 10014. (212) 255-3755. (800) 223-3130. 1995. $18.95.

Fundamentals of Atmospheric Dynamics and Thermodynamics. C. Riegel. World Scientific Publishing Company, Inc., 1060 Main Street, River Edge, NJ 07661. (201) 487-9655. (800) 227-7562. 1992. $33.00 in paper.

History of American Weather. David M. Ludlum. American Meteorological Society. 45 Beacon Street, Boston, MA 02108-3693. (617) 227-2425. FAX (617) 742-8718. 1970. Reprint edition. 4 vol. $55.00.

Mesoscale Meteorology and Forecasting. Peter S. Ray, editor. American Meteorological Society, 45 Beacon Street, Boston, MA 02108-3693. (617) 227-2425. FAX (617) 742-8718. 1986. $50.00.

Meteorology: The Atmosphere and the Science of Weather. Joseph M. Moran. Macmillan Publishing Company, Inc,. 200 Old Tappan Road, Old Tappan, NJ 07675. (800) 233-2336. 4th edition. 1994. $49.95.

Satellite Meteorology: An Introduction. Stanley Q. Kidder and Thomas H. Von der Haar. Academic Press, Inc., 6277 Sea Harbor Drive, Orlando, FL (800) 321-5068. 1992. 1995. $59.95.

Severe and Unusual Weather. Joe R. Eagleman. Trimedia Publishing Company, 12008 West 87th Street, Suite 117, Lenexa, KS 66215. (913) 599-0505. Second edition. 1990. $41.95.

Storm & Cloud Dynamics. William R. Cotton. Academic Press, Inc., 6277 Sea Harbor Drive, Orlando, FL (800) 321-5068. 1992. $69.95.

Weather Companion: An Album of Meteorological History, Science, Legend, and Folklore. Gary Lockhart. John Wiley and Sons, Inc., 605 Third Avenue, New York, NY 10158-0012. (212) 850-6000. (800) 225-5945. 1988. $12.95.

Weather Cycles: Real OR Imaginary? William James Burroughs. Cambridge University Press, 40 West 20th Street, New York, NY 10011-4211. (212) 924-3900. 1992. $39.95.

Wonders of the Sky: Observing Rainbows, Comets, Eclipses, the Stars, and Other Phenomena. Dover Publications, Inc., 180 Varick Street, New York, NY 10014. (212) 255-3755. (800) 223-3130 1983. $7.95.

HANDBOOKS AND MANUALS

Author's Guide To the Journals of the American Meteorological Society. American Meteorological Society, 45 Beacon Street, Boston, MA 02108-3693. (617) 227-2425. FAX (617) 742-8718. 1983. $15.00 in paper.

Basic Meteorology Lab Manual. Thomas A. Leavy. Allegheny Press, P.O. Box 220, Elgin, PA 19413. (814) 664-8504. Second edition. 1969. $7.95.

Handbook in Applied Meteorology. David D. Houghton, editor. John Wiley and Sons, Inc., 605 Third Avenue, New York, NY 10158-0012. (212) 850-6000. (800) 225-5945. 1985. $168.00.

Handbook of Agricultural Meteorology. John F. Griffin. Oxford University Press, Inc., 200 Madison Avenue, New York, NY 10016. (212) 726-6000. 1994. $85.00.

Handbook of Applied Meteorology. David D. Houghton, editor. John Wiley & Sons, Inc., 605 Third Avenue, New York, NY 10158-0012. (212) 850-6000. (800) 225-5945. 1985. $168.00.

ONLINE DATABASES AND CD-ROMS

Climate Assessment Database. National Weather Service, National Meteorological Center, 5200 Auth Road, Suite 101, Camp Springs, MD 20233. (301) 763-8016. Contains daily, weekly and monthly summaries of North American and world climatological data. Also provides five to ten Weather forecasts, and 30 to 90 day Weather outlook. Subscription required. Inquire as to cost and availability.

CA Search. Chemical Abstracts Service, P.O. Box 3012, Columbus, OH 43210-0012. (614) 447-3600. (800) 848-6533. FAX (614) 447-3709. Very comprehensive guide to worldwide chemical literature and related fields, 1972 to present. Available on BRS,(800) 289-4277, DIALOG, (800) 334-2564, ORBIT (800) 456-7248, and STN International, FIZ Karlsruhe, P.O. Box 2465, W-7500, Karlsruhe 1, Germany, online services. Inquire as to cost and availability.

Current Contents Search. Institute for Scientific Information, 3501 Market Street, Philadelphia, PA 19104. (215) 386-0100. FAX (215) 386-6362. Contains citations to articles listed in the table of contents of science and technology journals. Also articles in social sciences and life sciences journals. Available on BRS,(800) 289-4277, DIALOG,(800) 334-2564, online services. Inquire as to cost and availability.

Dissertation Abstracts. University Microfilms International, 300 North Zeeb Road, Ann Arbor, MI 48106. (800) 521-0600 or (313) 761-4700. Scope includes virtually all doctoral dissertations accepted at accredited American institutions from 1861 to present in 252 subject areas. Available on BRS,(800) 289-4277, DIALOG,(800) 334-2564, and OCLC EPIC,(800) 848-5878, online services. Also available on CD-ROM. Inquire as to cost and availability.

INSPEC. Institution of Electrical Engineers, Michael Faraday House, Six Hills Way, Stevenage, Herts. SG1 2AY, England. Telephone: 0438 313311 or FAX 0438 742840. Contains citations to the worldwide literature of physics, electronics and elec-

METEOROLOGY

Ency. of Physical Sciences and Engineering Info. Sources

trical engineering, computer technology, and related fields. Available on BRS,(800) 289-4277, DIALOG, (800) 334-2564, ORBIT, (800) 456-7248, and STN International, FIZ Karlsruhe, P.O. Box 2465, W-7500, Karlsruhe 1, Germany, online services. Inquire as to cost and availability.

Meteorological and Geoastrophysical Abstracts. American Meteorological Society, 45 Beacon Street, Boston, MA 02108-3693. (617) 227-2425. FAX (617) 742-8718. Contains citations and abstracts to the worldwide literature on significant research in meteorology and geoastrophysics. Related topics include physical oceanography, hydrology, environmental sciences and glaciology. Covers the period 1972 to present. Available on DIALOG,(800) 334-2564, online service. Inquire as to cost and availability.

NTIS Bibliographic Database. National Technical Information Service, 5285 Port Royal Road, Springfield, VA 22161. (703) 487-4929 or FAX (703) 321-8199. Broad coverage of government-sponsored science and technology research reports, 1964 to present. Available on BRS,(800) 289-4277, DIALOG, (800) 334-2564, ORBIT, (800) 456-7248, and STN International, FIZ Karlsruhe, P.O. Box 2465, W-7500, Karlsruhe 1, Germany, online services. Also available on CD-ROM. Inquire as to cost and availability.

Scisearch. Institute for Scientific Information, 3501 Market Street, Philadelphia, PA 19104. (800) 523-1850 or (215) 386-0100. Broad multidisciplinary title and author index to the international literature of science and technology, 1974 to present. Available on DIALOG and ORBIT online services. Also available on CD-ROM. Inquire as to cost and availability.

Wilsonline. H.W. Wilson Company, 950 University Avenue, Bronx, NY 10452. (800) 367-6770 or (212) 588-8400. Makes available online versions of the printed H.W. Wilson indexes, including Applied Science and Technology Index, Business Periodicals Index, General Science Index, and Readers' Guide to Periodical Literature. Period covered is generally 1983 to present. Available on BRS,(800) 289-4277, DIALOG,(800) 334-2564, and OCLC EPIC,(800) 848-5878, online services. Also available on CD-ROM. Inquire as to cost and availability.

World Climate Disc. Chadwyck-Healey Inc., 1101 King Street, Alexandria, VA 22314. (703) 683-4890. FAX (703) 683-7589. Weather and climate data from approximately 5,000 Weather stations worldwide, covering the years 1854 to 1990 on CD-ROM. First edition 1992. Approximately $1200.00 per year with annual updates.

World Weatherdisc. WeatherDisc Associates, Inc., 4584 N.E. 89th Street, Seattle, WA 98115. (206) 524-4314. FAX (206) 543-0308. Meteorological data on CD-ROM which describes the climate of the earth today and for the past few hundred years. First edition 1989. Approximately $295.00 per year with biannual updates.

PERIODICALS

Agricultural and Forest Meteorology. Elsevier Science Publishing Company, Inc., 655 Avenue of the Americas, New York, NY 10010. (212) 989-5800. Twenty times per year. $750.00 per year.

American Meteorological Society. BULLETIN. American Meteorological Society, 45 Beacon Street, Boston, MA 02108-3693. (617) 227-2425. FAX (617) 742-8718. Monthly. $60.00 per year.

Atmosphere-Ocean. Canadian Meteorological and Oceanographic Society, P.O. Box 334, Newmarket, Ontario, L3Y 4X7, Canada. (416) 898-1040. FAX (416) 898-7937. Quarterly. $30.00 per year.

Boundary-Layer Meteorology: An International Journal of Physical and Biological Processes in the Atmospheric Boundary Layer. Kluwer Academic Publishers, P.O. Box 358, Accord Station, Hingham, MA 02018-0358. (617) 871-6600. Sixteen per year. $785.00 per year.

Dynamics of Atmospheres and Oceans. Elsevier Science Publishing Company, Inc., 655 Avenue of the Americas, New York, NY 10010. (212) 989-5800. Six times per year. $205.00 per year.

JGR: Journal of Geophysical Research: Atmosphere. American Geophysical Union, 2000 Florida Avenue, N.W., Washington, DC 20009. (202) 462-6903. Monthly. $90.00 per year to members.

JGR: Journal of Geophysical Research: Oceans. American Geophysical Union, 2000 Florida Avenue, N.W., Washington, DC 20009. (202) 462-6903. (800) 966-2481. Monthly. $1545.00 per year.

Journal of Applied Meteorology. American Meteorological Society, 45 Beacon Street, Boston, MA 02108-3693. (617) 227-2425. FAX (617) 742-8718. Monthly. $165.00 per year.

Journal of Atmospheric Sciences. American Meteorological Society, 45 Beacon Street, Boston, MA 02108-3693. (617) 227-2425. FAX (617) 742-8718. Semi-monthly. $320.00 per year.

Monthly Weather Review. American Meteorological Society, 45 Beacon Street, Boston, MA 02108-3693. (617) 227-2425. FAX (617) 742-8718. Monthly. $205.00 per year.

National Weather Digest. National Weather Association, 6704 Wolke Court, Montgomery, AL. (334) 213-0388.

Royal Meteorological Society. Quarterly Journal. Royal Meteorological Society, 104 Oxford Road, Reading, Berks RG1 7LJ, England. Six times per year. $250.00 per year.

Weather. Royal Meteorological Society, 104 Oxford Road, Reading, Berks RG1 7LJ, England. Monthly. $44.00 per year.

Weatherwise. Heldref Publications, 1319 Eighteenth Street, N.W., Washington, DC 20036-1802. (202) 296-6267. FAX (202) 296-5149. Bi-monthly. $28.00 per year.

RESEARCH CENTERS AND INSTITUTES

Atlantic Oceanographic and Meteorological Laboratory. 4301 Rickenbacker Causeway, Miami, FL 33149. (305) 361-4300.

Center For Atmospheric theory and Analysis. University of Colorado-Boulder, Campus Box 391, Boulder CO 80309-0391. (303) 492-6487. FAX (303) 492-0642.

Cooperative Institute For Mesoscale Meteorological Studies. University of Oklahoma, 401 East Boyd, Norman, OK 73019. (405) 325-3041.

Joint Institute For Marine and Atmospheric Research. University of Hawaii at Manoa, 1000 Pope Road, MSB 312, Honolulu, HI 96822. (808) 956-8083. FAX (808) 956-4104.

National Center For Atmospheric Research. P.O. Box 3000, Boulder, CO 80307. (303) 496-1000. FAX (303) 497-1654.

National Hurricane Center. 1320 South Dixie Highway, Coral Gables, FL 33146. (305) 666-4612.

National Meteorological Center. World Weather Building, Room 101, 5200 Auth Road, Camp Springs, MD 20746. (301) 763-8016.

National Weather Service. 1325 East-West Highway, Silver Spring, MD 20910. (301) 427-7689.

University Corporation for Atmospheric Research. P.O. Box 3000, Boulder, CO 80307-3000. Consortium of 58 universities with doctoral programs in atmospheric and/or oceanic sciences. (303) 497-1650. FAX (303) 497-1654.

METEORS

See also: ASTEROIDS, COMETS, SATELLITES (NATURAL) PLANETARY SCIENCES, SOLAR SYSTEM

ABSTRACTS AND INDEXES

Applied Science and Technology Index; A Cumulative Subject Index To English Language Periodicals in the Fields of Aeronautics and Space Science, Computer Technology, Chemistry, Construction Industry, Energy and Related Areas. H. W. Wilson Co., 950 University Avenue, Bronx, NY 10452-9978. (212) 588-8400. (800) 367-6770. FAX (718) 590-1617. From 1958 to present. Monthly. Inquire about cost and availability. Also available online (BRS and WILSONLINE) and on CD-ROM. ISSN: 0003-6986.

Astronomy and Astrophysics Abstracts. Springer-Verlag New York, 175 Fifth Avenue, New York, NY 10010. (212) 460-1500. FAX (212) 473-6272. Published for the Astronomisches Rechen-Institut. Comprehensive coverage of all aspects of astronomy, astrophysics and related fields. 1969 to present. Two parts per year. Annual. ISSN: 0067-0022.

Bibliography and Index of Geology. American Geological Institute, 4220 King Street, Alexandria, VA 22302-1507. (703) 379-2480. FAX (703) 379-7563. From 1969 to present. Monthly. $1295.00 per year. ISSN: 0098-2784. Also available as GEOREF online (CISTI, DIALOG, Orbit, STN) and on CD-ROM. Inquire about price and availability.

Meteorological and Geoastrophysical Abstracts. American Meteorological Society, c/o Inforonics, Inc., 550 Newtown Road, Box 458, Littleton, MA 01460. (508) 486-8976. FAX (508) 486-0027. Covers literature in environmental sciences, meteorology, astrophysics, hydrology, glaciology and physical oceanography. 1950 to present. Monthly. $950.00 per year. ISSN: 0026-1130. Also available online (DIALOG) and on CD-ROM.

Physics Abstracts. INSPEC. Section A, Science Abstracts. Institute of Electrical Engineers, London, United Kingdom. Available from: INSPEC/IEEE-Institute of Electrical and Electronic Engineers, Box 1331, Hoes Lane, Piscataway, NJ 08855-1331.

(908) 562-5549. 1898 to present. 24 issues per year. $2835.00 per year. ISSN: 0036-8091. Also available online and on CD-ROM.

Science Citation Index. SCI. Institute for Scientific Information, 3501 Market Street, Philadelphia, PA 19104. (215) 386-0100. (800) 523-1850. FAX (215) 386-2991. 1961 to present. Six issues per year, plus annual cumulation. $11650.00 per year. Also available online and on CD-ROM. Inquire about price and availability. ISSN: 0036-827X.

STAR. Scientific and Technical Aerospace Reports. U.S. National Aeronautics and Space Administration. Distributed by U.S. Superintendent of Documents, Washington, DC 20402. 1963 to present. Semi-monthly, with semiannual and annual indexes. $114.00 per year. ISSN: 0036-8741. Also available online and on CD-ROM.

ANNUAL REVIEWS AND YEARBOOKS

Annual Review of Astronomy and Astrophysics. Original reviews of critical literature and current developments in astronomy and astrophysics. Annual Reviews, Inc., 4139 El Camino Way, Palo Alto, CA 94303-0139. (415) 493-4400. (800) 523-8635. Fax (415) 855-9815. 1963 to date. Annual. $60.00. ISSN: 0066-4146.

Annual Review of Earth and Planetary Sciences. Annual Reviews, Inc., 4139 El Camino Way, Palo Alto, CA 94303-0139. (415) 493-4400. (800) 523-8635. Fax (415) 855-9815. 1973 to date. Annual. $62.00. ISSN: 0084-6597.

ASSOCIATIONS AND PROFESSIONAL SOCIETIES

Amateur Astronomers Association. 1010 Park Avenue, New York, NY 10028. (212) 535-2922.

American Astronomical Society. 2000 Florida Avenue NW, Suite 400, Washington, DC 20009. (202) 328-2010. FAX (202) 234-2560.

American Meteor Society. SUNY-Geneseo, 1 College Circle, Department of Physics and Astronomy, Geneseo, NY 14454. (716) 245-5282.

Astronomical Society of the Pacific. 390 Ashton Avenue, San Francisco, CA 94112. (415) 337-1100. FAX: (415) 337-5205.

Geological Society of America. 3300 Penrose Place, PO Box 9140. Boulder, CO 80301-9140. (303) 447-2020. FAX (303) 447-1133.

Meteoritical Society. University of Massachusetts, 125 Marston Hall, Amherst, MA 01003. (413) 545-0300. (413) 545-0724.

Planetary Society. 65 North Catalina Avenue, Pasadena, CA 91106. (818) 793-5100. (800) WOW-MARS. FAX (818) 793-5528.

DIRECTORIES AND BIOGRAPHICAL SOURCES

American Astronomical Society. Membership Directory. American Astronomical Society. 2000 Florida Avenue NW, Suite 400, Washington, DC 20009. (202) 328-2010. FAX (202)234-2560. Annual. Included in membership dues. ISSN: 1061-9038.

American Men and Women of Science. Physical and Biological Sciences. R.R. Bowker Inc., 121 Chanlon Road, New Providence, NJ 07974. (908) 464-6800. (800) 521-8110. 20th edition. 8 volumes. 1996. $850.00.

Directory of Physics and Astronomy Staff. American Institute of Physics, One Physics Ellipse, College Park, MD 20740-3843. (301) 209-3100. 1975/76 to present. Annual. $60.00. ISSN: 0361-2228.

Earth and Astronomical Research Centers. Stockton Press, 345 Park Avenue South, New York, NY 10010. 4th edition. 1995. $515.00. ISBN: 1-56169-0967.

ENCYCLOPEDIAS AND DICTIONARIES

Asteroid Name Encyclopedia. Jacob Schwartz. Llewellyn Publications, P.O. Box 64383, Saint Paul, MN 55164-0383. (612) 291-1970. (800) 843-6666. 1995. $19.95 in paper.

Dictionary of Minor Planet Names. Lutz D. Schmadel. Springer-Verlag New York, Inc., 175 Fifth Avenue, New York, NY 10010. (212) 460-1500. (800) 777-4643. 2nd edition, revised and enlarged. 1993. $69.00.

Encyclopedia of Astronomy and Astrophysics. **Stephen Maran,** editor. Van Nostrand Reinhold, 115 Fifth Avenue, New York, NY 10003. (212) 254-3232. (800) 842-3636. 1992. $129.95.

Stars and Planets: The Sierra Club Guide To Sky Watching and Direction Finding. W. S. Kals. The Sierra Press, 4988 Gold Leaf Drive, Mariposa, CA 95338. (209) 966-5071. (800) 745-2631. $15.00.

GENERAL WORKS

Asteroids. Tom Gehrels, et al, editors. Books on Demand, 300 North Zeeb Road, Ann Arbor, MI 48106-1346. (313) 761-4700. (800) 521-0600. $180.00.

Asteroids Ii. Richard P. Binzel, et al, editors. University of Arizona Press, 1230 North Park Avenue, Number 102, Tucson, AZ 85719. (520) 621-1441. (800) 426-3797. 1990. $65.00.

Comets, Asteroids, and Meteorites. Time-Life Books Editors. Time-Life, Inc., 777 Duke Street, Alexandria, VA 22314. (703) 8388-7000. (800) 621-7026. Voyage through the Universe Series. Revised edition. 1992.

Dynamics and Evolution of Minor Bodies With Galactic and Geological Implications. S. V. Clube, et al, editors. Kluwer Academic Publishers, 101 Philip Drive, Assinippi Park, Norwell, MA 02061. (617) 871-6600. 1992. $149.00.

The Geology of Multi-Ring Impact Basins: The Moon and Other Planets. P. D. Spudis. Cambridge University Press, 40 West 20th Street, New York, NY 10011-4211. (212) 924-3900. (800) 872-7423. Planetary Sciences Series. 1993. $59.95.

Hazards Due To Comets and Asteroids. Tom Gehrels, editor. University of Arizona Press, 1230 North Park Avenue, Number 102, Tucson, AZ 85719. (520) 621-1441. (800) 426-3797. 1995. $75.00.

Meteorites: An Introduction. Fritz Heide. Springer-Verlag New York, Inc., 175 Fifth Avenue, New York, NY 10010. (212) 460-1500. (800) 777-4643. 1995. $24.00 in paper.

Meteors. Neil Bone. Sky Publishing Corp. 49 Bay State Road, Cambridge, MA 02138. (617) 864-7360. (800) 263-0245. 1993. $18.95 in paper.

Mrs. O'Leary's Comet. Mel Waskin. Academy Chicago Pubs., Ltd., 361 West Erie Street, Chicago, IL 60610-3125. (312) 751-7300. (800) 248-7323. 1985. $10.00.

Observing Comets, Asteroids, Meteors and the Zodiacal Light. Stephen J. Edberg and David H. Levy. Cambridge University Press, 40 West 20th Street, New York, NY 10011-4211. (212) 924-3900. (800) 872-7423. Practical Astronomy Handbooks Series. 1994. $29.95.

Orbital Debris: A Space Environmental Problem. U.S. Government Printing Office, Superintendent of Documents, Washington, DC 20402-9325. (202) 783-3238. 1990. $3.00 in paper.

Physics of Comets and Meteors. V. P. Konopleva. Coronet Books, 311 Bainbridge Street, Philadelphia PA 19147. (215) 925-5083. 1966. $28.50.

Rocks From Space: Meteorites and Meteorite Hunters. O. Richard Norton. Mountain Press, P.O. Box 2399, Missoula, MT 59806. (406) 728-1900. (800) 234-5308. 1994. $20.00 in paper.

Rogue Asteroids and Doomsday Comets. Duncan Steel. John Wiley & Sons, Inc., 605 Third Avenue, New York, NY 10158-0012. (212) 850-6000. (800) 225-5945. 1995. $24.95.

HANDBOOKS AND MANUALS

Field Guide To the Stars and Planets. Jay M. Pasachoff and Donald H. Menzel. Houghton Mifflin Co., 222 Berkeley Street, Boston, MA 02116. (617) 351-5000. (800) 225-3362. Revised edition. 1992. $24.95.

Handbook For Visual Meteor Observations. Paul Roggemans, editor. Sky Publishing Corp. 49 Bay State Road, Cambridge, MA 02138. (617) 864-7360. (800) 263-0245. 2nd edition. 1989. $18.95.

ONLINE DATABASES AND CD-ROMS

CA Search. Chemical Abstracts Service, P.O. Box 3012, Columbus, OH 43210-0012. (614) 447-3600. (800) 848-6533. FAX (614) 447-3709. Very comprehensive guide to worldwide chemical literature and related fields, 1972 to present. Available on BRS,(800) 289-4277, DIALOG, (800) 334-2564, ORBIT (800) 456-7248, and STN International, FIZ Karlsruhe, P.O. Box 2465, W-7500, Karlsruhe 1, Germany, online services. Inquire as to cost and availability.

Dissertation Abstracts. University Microfilms International, 300 North Zeeb Road, Ann Arbor, MI 48106. (800) 521-0600 or (313) 761-4700. Scope includes virtually all doctoral dissertations accepted at accredited American institutions from 1861 to present in 252 subject areas. Available on BRS, (800) 289-4277, DIALOG, (800) 334-2564, and OCLC EPIC, (800) 848-5878, online services. Also available on CD-ROM. Inquire as to cost and availability.

Georef. American Geological Institute, 4220 King Street, Alexandria, VA 22302. (800) 336-4764 or (703) 379-2480. Geology and geosciences literature, 1785 to present for North America. Available on DIALOG,(800) 334-2564, ORBIT (800) 456-7248, online services. Also available on CD-ROM. Inquire as to cost and availability.

INSPEC. Institution of Electrical Engineers, Michael Faraday House, Six Hills Way, Stevenage, Herts. SG1 2AY, England. Telephone: 0438 313311 or FAX 0438 742840. Contains citations to the worldwide literature of physics, electronics and electrical engineering, computer technology, and related fields. Available on BRS, (800) 289-4277, DIALOG, (800) 334-2564, ORBIT, (800) 456-7248, and STN International, FIZ Karlsruhe, P.O. Box 2465, W-7500, Karlsruhe 1, Germany, online services. Inquire as to cost and availability.

NTIS Bibliographic Database. National Technical Information Service, 5285 Port Royal Road, Springfield, VA 22161. (703) 487-4929 or FAX (703) 321-8199. Broad coverage of government-sponsored science and technology research reports, 1964 to present. Available on BRS,(800) 289-4277, DIALOG, (800) 334-2564, ORBIT, (800) 456-7248, and STN International, FIZ Karlsruhe, P.O. Box 2465, W-7500, Karlsruhe 1, Germany, online services. Also available on CD-ROM. Inquire as to cost and availability.

Scisearch. Institute for Scientific Information, 3501 Market Street, Philadelphia, PA 19104. (800) 523-1850 or (215) 386-0100. Broad multidisciplinary title and author index to the international literature of science and technology, 1974 to present. Available on DIALOG, (800) 334-2564, and ORBIT, (800) 456-7248, online services. Also available on CD-ROM. Inquire as to cost and availability.

Wilsonline. H.W. Wilson Company, 950 University Avenue, Bronx, NY 10452. (800) 367-6770 or (212) 588-8400. Makes available online versions of the printed H.W. Wilson indexes, including Applied Science and Technology Index, Business Periodicals Index, General Science Index, and Readers' Guide to Periodical Literature. Period covered is generally 1983 to present. Available on BRS, (800) 289-4277, DIALOG, (800) 334-2564, and OCLC EPIC, (800) 848-5878, online services. Also available on CD-ROM. Inquire as to cost and availability.

OTHER SOURCES

Cambridge Atlas of Astronomy. Cambridge University Press, 40 West 20th Street, New York, NY 10011-4211. (212) 924-3900. (800) 872-7423. 3rd edition. 1994. $74.00.

The Sky: A User's Guide. David H. Levy. Cambridge University Press, 40 West 20th Street, New York, NY 10011-4211. (212) 924-3900. (800) 872-7423. 1993. $14.95.

PERIODICALS

Astronomical Society of the Pacific. Publications. Astronomical Society of the Pacific, 390 Ashton Avenue, San Francisco, CA 94112. (415) 337-1100. FAX (415) 337-5205.1889 to present. Monthly. $175.00 per year. ISSN: 0004-6280.

Astronomy. Kalmbach Publishing Company, Box 1612, Waukesha, WI 53187-1612. (414) 796-8776. FAX (414) 796-1142. 1973 to present. Monthly. $27.00 per year. ISSN: 0091-6358.

Earth, Moon and Planets; An International Journal of Comparative Planetology. Kluwer Academic Publishers, Box 358, Accord Station, Hingham, MA 02018-0358. (617) 871-6600. FAX (617) 871-6528. 1969 to present. Monthly. $840.00 per year. ISSN: 0167-9295.

Geochemica Et Cosmochemica Acta. Elsevier Science. 660 White Plains Road, Tarrytown, NY 10591-5153. (914) 524-9200. FAX (914) 333-2444. 1950 to the present. Biweekly. $895.00 per year. ISSN: 0016-7037.

Icarus; International Journal of Solar System Studies. Academic Press, Inc., Journal Division, 525 B Street, Suite 1900, San Diego, CA 92101-4495. (619) 230-1840. FAX (619) 699-6800. 1962 to the present. Monthly. $1080.00. ISSN: 0019-1035.

JGR: Journal of Geophysical Research: Planets. American Geophysical Union, 2000 Florida Avenue, NW, Washington, CD 20009. (202) 462-6900. FAX (202) 328-0566. 1991 to present. Monthly. $597.00 per year. ISSN: 0148-0227.

Mercury. Astronomical Society of the Pacific. 390 Ashton Avenue, San Francisco, CA 94112. (415) 337-1100. FAX: (415) 337-5205. 1972 to present. Bimonthly. $175.00 per year. ISSN: 0047-6773.

Meteoritics. Meteoritical Society, Department of Chemistry, University of Arkansas, Fayetteville, AR 72701. (501) 575-7625. FAX (501) 575-7778. 1953 to present. Bimonthly. $210.00 per year. ISSN: 0026-1114.

Observatory. c/o Dr. D. J. Stickland, Space and Astrophysics Division, Rutherford Appleton Laboratory, Chilton, Didcot, Oxon OX11 OQX England. FAX 0235-445848. 1877 to present. Bimonthly. $42.00 per year. ISSN: 0029-7704.

Sky and Telescope. Sky Publishing Corporation, Box 9111, Belmont, MA 02178. (617) 864-7360. FAX (617) 864-6117. 1941 to present. Monthly. $27.00 per year. ISSN: 0037- 6604.

Special Publication. University of New Mexico. Institute of meteoritics, Albuquerque, NM 87131. (505) 277-2747. FAX (505) 277-3577. 1970 to present. Irregular (2-4 per year), $10.00 per issue or on exchange basis. ISSN: 0085-3968.

RESEARCH CENTERS AND INSTITUTES

Center For Meteorite Studies. Arizona State University, Tempe, AZ 85287-2504. (602) 965-6511.

Herzberg Institute of Astrophysics. National Research Council of Canada. 100 Sussex Drive, Ottawa, ON, Canada K1A 0R6. (613) 990-0907. FAX (613) 952-6602.

Institute For Astronomy. University of Hawaii at Manoa, 2680 Woodlawn Drive, Honolulu, HI 96822. (808) 956-8312. FAX (808) 988-2790.

Lowell Observatory. 1400 West Mars Road, Flagstaff, AZ 86001. (602) 774-3358. FAX (602) 774-6296.

Lunar and Planetary Laboratory. University of Arizona, Tucson, AZ 85721. (602) 621-6963. FAX (602) 621-4933.

McDonnell Center For the Space Sciences. Washington University, Box 1105, One Brookings Drive, St. Louis MO 63130-4899. (314) 889-6255. FAX (314) 889-6219.

National Astronomy and Ionosphere Center. Cornell University, Space Sciences Building, Ithaca, NY 14853. (697) 255-3735. FAX (607) 255-8803.

METROLOGY

See also: GEODESY, INSTRUMENTATION, SURVEYING

ABSTRACT SERVICES AND INDEXES

Applied Mechanics Reviews: An Assessment of World Literature in Engineering Sciences. 1948-present. American Society of Mechanical Engineers, 345 East 47th Street, New York, NY 10017. (212) 705-7703. Monthly. $360.00 per year.

Applied Science and Technology Index; A Cumulative Subject Index To English Language Periodicals in the Fields of Aeronautics and Space Science, Computer Technology, Chemistry, Construction Industry, Energy and Related Areas. H.W. Wilson Co., 950 University Avenue, Bronx, NY 10452. (800) 367-6770 or (212) 588-8400. FAX (718) 590-1617. From 1958 to present. Monthly. Inquire about cost and availability. Also available on CD-ROM and online.

Electrical and Electronics Abstracts. Institute of Electrical Engineers, Michael Faraday House, Six Hill Way, Stevenage, Herts, SG1 2AY, England. Distributed by IEEE, 445 Hoes Lane, Piscataway, NJ 08854. (908) 562-5549. 1898 to present. Monthly. $2200.00 per year. Also available on CD-ROM and online as INSPEC.

Engineering Index Monthly. Engineering Information, Inc., Castle Point on the Hudson, Hoboken, NJ 07030. (800) 221-1044. FAX (212) 832-1857. Monthly. $2300.00 per year. Also available online as COMPENDEX, and also on CD-ROM. Covers chemical engineering, computers, electrical engineering, civil engineering, metals and mining, industrial management, and mechanical engineering.

Index To Scientific and Technical Proceedings. Institute for Scientific Information, 3501 Market St., Philadelphia, PA 19104. (215) 386-0100. FAX (215) 386-6362. Monthly. $500.00 per year.

Index to Scientific Reviews. Institute for Scientific Information, 3501 Market St., Philadelphia, PA 19104. (215) 386-0100. FAX (215) 386-6362. Semi-annual.

Key Abstracts—Electronic Instrumentation. IEEE Service Center, 445 Hoes Lane, Piscataway, NJ 08854. (908) 562-5549. 1976 to present. Monthly. $160.00 per year.

Key Abstracts—Measurements in Physics. IEEE Service Center, 445 Hoes Lane, Piscataway, NJ 08854. (908) 562-5549. 1976 to present. Monthly. $160.00 per year.

Physics Abstracts. Institute of Electrical Engineers, Michael Faraday House, Six Hill Way, Stevenage, Herts, SG1 2AY, England. Distributed by IEEE, 445 Hoes Lane, Piscataway, NJ 08854. (908) 562-5549. 1898 to present. Monthly. $2700.00 per year. Also available online as INSPEC.

Physics Briefs. American Institute of Physics, 335 East 45th Street, New York, NY 10017 (212) 661-9404 or FAX (516)349-9704. 1980 to present. Six times per year. $2600.00 per year.

Science Citation Index. Institute for Scientific Information, 3501 Market Street, Philadelphia, PA 19104. (215) 386-0100. FAX (215) 386-6362. Inquire about availability and cost. Also available on CD-ROM.

Standard Actions. American National Standards Institute, 11 W. 42nd Street, 13th Floor, New York, NY 10036. (212) 642-4900. FAX (212) 302-1286. 1970 to present. Bi-weekly. $100.00 per year.

ASSOCIATIONS AND PROFESSIONAL SOCIETIES

American Institute of Physics. 335 East 45th Street, New York, NY 10017. (212) 661-9404 or (800) 445-6638.

American Society For Testing and Materials. 1916 Race Street, Philadelphia, PA 19103. (215) 299-5585.

Analytical Instrument Association. 225 Reinekers Lane, Suite 265, Alexandria, VA 22314-2875. (703) 836-1360. FAX (703) 836-6644.

Edison Electric Institute. 701 Pennsylvania Avenue NW, Washington, DC 20004-2696. (202) 508-5000.

Institute of Electrical and Electronics Engineers. 345 E. 47th Street, New York, NY 10017. (908) 981-0060. FAX (908) 981-9667.

Instrument Society of America. PO Box 12277, 67 Alexander Drive, Research Triangle Park, NC 27709. (919) 549-8411.

DIRECTORIES AND BIOGRAPHICAL SOURCES

American Men & Women of Science, 1995-96. R.R. Bowker Staff, eds. 19th edition. 8 volumes. R.R. Bowker/Reed International Publishing Company, 121 Chanlon Road, New Providence, NJ 07974. (908) 464-6800 or (800) 521-8110. 1995. $850.00

Directory of Engineering Societies and Related Organizations. Gordon Davis. 13th edition. American Association of Engineering Societies, 1111 19th Street NW, Suite 608, Washington, DC 20036. (202) 296-2237 or (800) 658-8897. 1989. Inquire for price.

International Research Centers Directory. Gale Research, 835 Penobscot Building, Detroit, MI 48226-4094. (313) 961-2242. (800) 347-4253. FAX (313) 961-6083. 8th edition. 1995. $410.00

Research Centers Directory. Gale Research, 835 Penobscot Building, Detroit, MI 48226-4094. (313) 961-2242. (800) 347-4253. FAX (313) 961-6083. $485.00.

Scientific and Technical Organizations and Agencies Directory. Gale Research, 835 Penobscot Building, Detroit, MI 48226-4094. (313) 961-2242. (800) 347-4253. FAX (313) 961-6083. 4th edition. 1996. $195.00.

Ency. of Physical Sciences and Engineering Info. Sources

METROLOGY

Who's Who in Engineering. American Association of Engineering Societies, 1111 19th Street NW, Suite 608, Washington, DC 20036. (202) 296-2237 or (800) 658-8897. 8th edition. 1991. Inquire for price.

ENCYCLOPEDIAS AND DICTIONARIES

Concise Encyclopedia of Measurement and Instrumentation. L. Finkelstein and K.T.V. Grattan. Pergamon Press Inc., Maxwell House, Fairview Park, Elmsford, NY 10523. (914) 592-7700. Fax (914) 592-3625. 1993. Inquire for cost.

McGraw-Hill Dictionary of Scientific and Technical Terms. Sybil P. Parker, ed. 5th edition. McGraw-Hill Publishing Company, 1221 Avenue of the Americas, New York, NY 10020. (800) 262-4729 or (212) 512-3825. 1993. $110.50.

McGraw-Hill Encyclopedia of Engineering. Sybil P. Parker, ed. 2nd edition. McGraw-Hill Publishing Company, 1221 Avenue of the Americas, New York, NY 10020. (800) 262-4729 or (212) 512-3825. 1993. $95.50.

McGraw-Hill Encyclopedia of Science and Technology. Sybil P. Parker, ed. 7th edition. 20 volumes. McGraw-Hill Publishing Company, 1221 Avenue of the Americas, New York, NY 10020. (800) 262-4729 or (212) 512-3825. 1992. $1900.00

Thesaurus of Scientific, Technical, and Engineering Terms. Hemisphere Publishing Corporation, 1900 Frost Road, Suite 101, Bristol, PA 19007-1598. (215) 785-5800 or (800) 821-8312. FAX (215) 785-5515. 1987. $173.00.

GENERAL WORKS

The Art of Measurement: Metrology in Fundamental and Applied Physics. Bernhard Kramer. VCH Publishers, Inc., 303 Northwest 12th Avenue, Deerfield Beach, FL 33442. (800) 422-8824. 1988. $85.00.

Materials Metrology and Standards For Structural Performance. B.F. Dyson, et al., editors. Chapman and Hall, 29 West 35th Street, New York, NY 10001. (800) 842-3636. FAX (212) 563-2269. 1995. Inquire for cost and availability.

Metrology At the Frontiers of Physics and Technology. L. Crovini & T.J. Quinn, editors. Elsevier Science B.V., PO Box 945, Madison Square Station, New York, NY 10159-0945. (212) 633-3650. FAX (212) 633-3680. 1992. Inquire for cost and availability.

Optical Methods in Engineering Metrology. D.C. Williams. Routledge, Chapman and Hall, Inc., 29 West 35th Street, New York, NY 10001-2291. (212) 244-3336 or FAX (212) 563-2269. 1993. Inquire for cost and availability.

The Origins of Metrology. C.J. Scarre, editor. McDonald Institute for Archaeological Research/ Cambridge University Press, 40 West 20th Street, New York, NY 10011-4211. (212) 924-3900 or (800) 872-7423. FAX (914) 937-4712. 1992. Inquire for cost and availability.

HANDBOOKS AND MANUALS

Handbook of Dimensional Measurement. F.T. Farago & M.A. Curtis, editors. 3rd edition. Industrial Press, Inc. 200 Madison

Avenue, New York, NY 10016. (212) 889-6330. FAX (212) 545-8327. 1994. $55.95.

Handbook of Measurement Science. P.H. Sydenham, editor. 3 volumes. John Wiley and Sons, Inc., 605 Third Avenue, New York, NY 10158. (800) 526-5368 or (212) 850-6000. 1982 (vol.1), 1983 (vol.2), 1992 (vol.3). $270.00 each for vols.1-2, $199.95 (vol.3).

ONLINE DATABASES AND CD-ROMS

Compendex Plus. Engineering Information, Inc., 345 East 47th Street, New York, NY 10017. (212) 705-7600 or (800) 221-1044. Contains citations with abstracts to worldwide literature in engineering and technology, from 1970 to present. Available on online BRS,(800) 289-4277, DIALOG, (800) 334-2564, ORBIT (800) 456-7248, and STN International, FIZ Karlsruhe, P.O. Box 2465, W-7500, Karlsruhe 1, Germany, online services. Also available on CD-ROM. Inquire as to cost and availability.

CSA Engineering. Cambridge Scientific Abstracts, 7200 Wisconsin Avenue, Suite 601, Bethesda, MD 20814. (301) 961-6750 or (800) 843-7751. Contains citations and abstracts of international periodicals and research literature covering all fields of engineering and science and technology, including computer and information science, electronics, mechanical engineering, solid state materials, 1981 to present. Available on BRS,(800) 289-4277, online service. Inquire as to cost and availability.

Current Contents Search. Institute for Scientific Information, 3501 Market Street, Philadelphia, PA 19104. (215) 386-0100. FAX (215) 386-6362. Contains citations to articles listed in the table of contents of science and technology journals. Also articles in social sciences and life sciences journals. Available on BRS,(800) 289-4277, DIALOG,(800) 334-2564, online services. Inquire as to cost and availability.

Dissertation Abstracts. University Microfilms International, 300 North Zeeb Road, Ann Arbor, MI 48106. (800) 521-0600 or (313) 761-4700. Scope includes virtually all doctoral dissertations accepted at accredited American institutions from 1861 to present in 252 subject areas. Available on BRS,(800) 289-4277, DIALOG,(800) 334-2564, and OCLC EPIC,(800) 848-5878, online services. Also available on CD-ROM. Inquire as to cost and availability.

ISMEC: Mechanical Engineering Abstracts. Cambridge Scientific Abstracts, 7200 Wisconsin Avenue, Suite 601, Bethesda, MD 20814. (301) 961-6750 or (800) 843-7751. Contains citations to the literature in mechanical engineering, industrial and production engineering, energy, power, mechanics, devices and related areas, from 1973 to present. Available on the DIALOG,(800) 334-2564, online service. Inquire as to cost and availability.

NTIS Bibliographic Database. National Technical Information Service, 5285 Port Royal Road, Springfield, VA 22161. (703) 487-4929 or FAX (703) 321-8199. Broad coverage of government-sponsored science and technology research reports, 1964 to present. Available on BRS,(800) 289-4277, DIALOG, (800) 334-2564, ORBIT, (800) 456-7248, and STN International, FIZ Karlsruhe, P.O. Box 2465, W-7500, Karlsruhe 1, Germany, online services. Also available on CD-ROM. Inquire as to cost and availability.

Scisearch. Institute for Scientific Information, 3501 Market Street, Philadelphia, PA 19104. (800) 523-1850 or (215) 386-0100. Broad multidisciplinary title and author index to the in-

METROLOGY

Ency. of Physical Sciences and Engineering Info. Sources

ternational literature of science and technology, 1974 to present. Available on DIALOG,(800) 334-2564, and ORBIT,(800) 456-7248, online services. Also available on CD-ROM. Inquire as to cost and availability.

Wilsonline. H.W. Wilson Company, 950 University Avenue, Bronx, NY 10452. (800) 367-6770 or (212) 588-8400. Makes available online versions of the printed H.W. Wilson indexes, including Applied Science and Technology Index, Business Periodicals Index, General Science Index, and Readers' Guide to Periodical Literature. Period covered is generally 1983 to present. Available on BRS,(800) 289-4277, DIALOG,(800) 334-2564, and OCLC EPIC,(800) 848-5878, online services. Also available on CD-ROM. Inquire as to cost and availability.

PERIODICALS

ASTM Standardization News. American Society for Testing and Materials (ASTM), 1916 Race Street, Philadelphia, PA 19103. (215) 299-5585. FAX (215) 977-9679. 1973 to present. Monthly. $18.00 per year.

IEEE Transactions On Instrumentation and Measurement. IEEE (Institute of Electrical and Electronic Engineers), 345 E. 47th Street, New York, NY 10017-2394. 1952 to present. Bi-monthly. $165.00 per year for non-members.

Metrologia. Bureau International des Poids et Mesures, Pavillon de Breteuil, 92312 Sevres Decex, France. FAX 45-34-20-21. 1965 to present. 6 times a year. $298.00 per year.

National Institute of Standards and Technology Journal of Research. U.S. National Institute of Standards and Technology, U.S. Department of Commerce, Gaithersburg, MD 20899. (301) 975-3069. 1928 to present. Bi-monthly. $27.00 per year.

Standards Engineering. Standards Engineering Society, 1706 Darst Avenue, Dayton, OH 45403-3104. (513) 258-1955. Bi-monthly. $40.00 per year.

RESEARCH CENTERS AND INSTITUTES

Instrumentation and Control Laboratory, Princeton University, Dept. of MAE, Engineering Quadrangle, Princeton, NJ 08544. (609) 452-5154.

Major Analytical Instrumentation Center, University of Florida. College of Engineering, Rhines Hall, Gainesville, FL 32611. (904) 392-6985.

University of California—Davis, Facility For Advanced Instrumentation. Davis, CA 95616. (916) 752-0284.

SPECIFICATIONS AND STANDARDS

Annual Book of ASTM Standards, Volume 14.02. American Society for Testing and Materials (ASTM), 1916 Race Street, Philadelphia, PA 19103. (215) 299-5585. Annual. Inquire for cost and availability.

MICROCIRCUITRY

See: MICROELECTRONICS

MICROELECTRONICS

See also: ELECTRICAL ENGINEERING, ELECTRICITY, ELECTRONIC CIRCUITS and COMPONENTS, ELECTRONICS, ELECTRONICS ENGINEERING, MICROPROCESSORS

ABSTRACT SERVICES AND INDEXES

Applied Science and Technology Index; A Cumulative Subject Index To English Language Periodicals in the Fields of Aeronautics and Space Science, Computer Technology, Chemistry, Construction Industry, Energy and Related Areas. H.W. Wilson Co., 950 University Avenue, Bronx, NY 10452. (212) 588-8400. (800) 367-6770. FAX (718) 590-1617. From 1958 to present. Monthly. Inquire about cost and availability. Also available on CD-ROM and online. ISSN: 0003-6986.

Computer and Control Abstracts. Science Abstracts, section C Institution of Electrical Engineers (IEE), London. Available from INSPEC/IEEE-Institute of Electrical and Electronic Engineers, Box 1331, Hoes Lane, Piscataway, NJ 08855-1331. (908) 562-5549. 1966 to present. Monthly. $1455.00. ISSN: 0036-8113. Also available on CD-ROM and online as INSPEC.

Current Contents: Engineering, Technology and Applied Sciences. Institute for Scientific Information, 3501 Market Street, Philadelphia, PA 19104. (215) 386-0100. FAX (215)386-6362. 1970 to present. Weekly. $442.00 per year. Also available on CD-ROM and online. Inquire regarding cost and availability. ISSN: 0095-7917.

Current Papers in Electrical and Electronics Engineering. Institution of Electrical Engineers (IEE), London. Distributed by INSPEC/IEEE, Box 1331, 445 Hoes Lane, Piscataway, NJ 08855-1331. (908) 562-5549. 1969 to present. Monthly. $345.00 per year. ISSN: 0011-3778.

Electrical and Electronics Abstracts. Institution of Electrical Engineers (IEE), London. Available from INSPEC/IEEE-Institute of Electrical and Electronic Engineers, Box 1331, Hoes Lane, Piscataway, NJ 08855-1331.(908) 562-5549. 1898 to present. Monthly. $2200.00 per year. Also available on CD-ROM and online as INSPEC.

Engineering Index Monthly; Indexes and Abstracts the World's Engineering and Technical Literature. Engineering Information, Inc., Castle Point on the Hudson, Hoboken, NJ 07030. (201) 216-8500. (800) 221-1044. FAX (201) 216-8532. Monthly. $2300.00 per year. Available online as COMPENDEX and also on CD-ROM. ISSN: 0742-1974.

General Science Index. H.W. Wilson Company, 950 University Avenue, Bronx, NY 10452. (212) 588-8400. (800) 367-6770. FAX (718) 590-1617. From 1978 to present. Ten issues per year; quarterly and annual cumulations. Service basis. Available on CD-ROM and online. Inquire regarding cost and availability. ISSN: 0162-1963.

Government Reports Announcements and Index. U.S. National Technical Information Service (NTIS), 5285 Port Royal Road, Springfield, VA 22161. (703) 487-4650. FAX (703) 321-8547. From 1968 to present. Annual. $630.00 per year. Also available online as NTIS Bibliographic Database and on CD-ROM.

Index To IEEE Publications. IEEE Service Center, 445 Hoes Lane, Piscataway, NJ 08855-1331. (908) 981-1393. (800) 678-IEEE. FAX (908) 981-9667. 1973 to present. Annual. ISSN: 0099-1368.

Physics Abstracts. INSPEC. Section A, Science Abstracts. Institution of Electrical Engineers (IEE), London. Available from: INSPEC/IEEE-Institute of Electrical and Electronic Engineers, Box 1331, Hoes Lane, Piscataway, NJ 08855-1331. (908) 562-5549. 1898 to present. 24 issues per year. $2835.00 per year. Also available online and on CD-ROM. ISSN: 0036-8091.

Physics Briefs (Physikalische Berichte). Information Center for Energy, Physics, Mathematics; German Physical Society. VCH Publishers, Inc., 220 East 23rd Street, New York, NY 10010-4606. (212) 683-8333. 1845 to present. 24 issues per year. $2390.00 per year. Also available online. ISSN: 0179-7434.

Science Citation Index. SCI. Institute for Scientific Information, 3501 Market Street, Philadelphia, PA 19104. (215) 386-0100. (800) 523-1850. FAX (215) 386-2991. 1961 to present. Six issues per year, plus annual cumulation. $11650.00 per year. Also available online and on CD-ROM. Inquire about price and availability. ISSN: 0036-827X.

Solid State and Superconductivity Abstracts: Covers theory, Production and Application of Solid State Materials. Cambridge Scientific Abstracts, 7200 Wisconsin Avenue, Bethesda, MD 20824. (301) 961-6750. FAX (301) 961-6720. 1957 to present. Bimonthly. $1320.00 per year. Also available online. ISSN: 0896-5900.

ASSOCIATIONS AND PROFESSIONAL SOCIETIES

American Association of Engineering Societies, 1111 19th Street, Suite 608, Washington, DC 20036-3603. (202) 296-2237. FAX (202) 296-1151.

American Electronics Association. 5201 Great America Way, Suite 520, P.O. 52990, Santa Clara, CA 95056. (408) 987-4200. FAX (408) 970-8565.

American Institute of Physics. One Physics Ellipse, College Park, MD 20740-3843. (301) 209-3100.

Electronics Industries Association. 2500 Wilson Boulevard, Arlington, VA 22201. (703) 907-7500. FAX (202) 457-4985.

IEEE Circuits and Systems Society. c/o Institute of Electrical and Electronic Engineers. 345 East 47th Street. New York, NY 10017. (212) 705-7900. FAX (212) 705-4929.

Institute of Electrical and Electronics Engineers. 345 East 47th Street. New York, NY 10017. (212) 705-7900. FAX (212) 705-4929.

International Society for Hybrid Microelectronics. 1850 Centennial Park Drive, Suite 105, Reston, VA 22091. (703) 758-1060. FAX (703) 758-1066.

National Electrical Manufacturers Association. 1300 North 17th Street, Suite 1847, Rosslyn VA 22209. (703) 841-3200. FAX (703) 841-3300.

DIRECTORIES AND BIOGRAPHICAL SOURCES

American Electronics Association Directory. American Electronics Association, 5201 Great America Way, Suite 520, P.O. 52990, Santa Clara, CA 95056. (408) 987-4200. FAX (408) 970-8565. Annual.

EEM-Electronic Engineer's Master. Hearst Business Communications, Inc., 645 Stewart Avenue, Garden City NY 11530. (516) 227-1300. ISSN: 0732-9016.

Electrical and Electronic Trades Directory: The Blue Book. Institution of Electrical Engineers, Michael Faraday House, Six Hills Way, Stevenage, Herts. SG1 2AY, England. Telephone: 0438 313311 or FAX 0438 742840. 1995. $140.00

EEM-Electronic Engineer's Master. Hearst Business Communications, Inc., 645 Stewart Avenue, Garden City NY 11530. (516) 227-1300. ISSN: 0732-9016.

Engineering Research Centers: Incorporating Electronics Research Centers. Stockton Press, 345 Park Avenue, New York, NY 10010. (212) 689-9200. (800) 221-2123. 4th edition. 1995. $515.00.

IC Master. Hearst Business Communications, Inc., 645 Stewart Avenue, Garden City, NY 11530. (516) 227-1300. 1975 to present. Annual. $170.00 per year. ISSN: 0894-6809.

IEEE Membership Directory. Institute of Electrical and Electronics Engineers, IEEE Service Center, 445 Hoes Lane, Piscataway, NY 08854. (908) 981-1393. (800) 678-IEEE. FAX(908) 981-9667. 2 volumes. Annual. $190.00. ISSN:

International Directory of Abbreviations and Acronyms of Electronics, Electrical Engineering, Computer Technology and Information Processing. Peter Wennrich. K. G. Saur, 121 Chanlon Road, New Providence, NJ 07974. (908) 464-6800. (800) 521-8110 2 volumes. 1992. $230.00.

International Engineering Directory. American Consulting Engineers Council, 1015 15th Street, N.W. Suite 802, Washington, DC 20005-2670. (202) 347-7474. Annual. $10.00.

Research Centers Directory. Gale Research Company Inc., 835 Penobscot Building, Detroit, MI 48226-4094. (313) 961-2242. (800) 347-4253. 20th edition. 1995. $485.00. ISSN: 0080-1518.

Who's Who in Engineering. Gordon Davis, editor. American Association of Engineering Societies. 1111 19th Street, NY, Suite 608, Washington, DC 20036. (202) 296-2237. (800) 658-8897. 9th edition. 1995. $220.00.

Who's Who in Technology; Biographies and Index. Gale Research Company, 835 Penobscot Building, Detroit, MI 48226-4094. (313 961-2242. (800)521-4253. 7th edition. 1995. $195.00. ISBN 0-8103-7467-6.

ENCYCLOPEDIAS AND DICTIONARIES

Encyclopedia of Applied Physics. George Trigg, editor. VCH Publications, Inc., 220 East 23rd Street, Suite 909, New York, NY 10010-4606. (212) 683-8333. (800) 422-8824. 20 volume set. 1991 -. $5990.00.

IEEE Standard Dictionary of Electrical and Electronics Terms. Christopher J. Booth, editor. IEEE Service Center, 445 Hoes Lane, Piscataway, NJ 08855-1331. (908) 981-1393. (800) 678-IEEE. FAX (908) 981-9667. IEEE Standard 100-1992. 5th edition. 1993. $150.00.

Illustrated Dictionary of Electronics. Stan Gibilsco. TAB Books, P.O. Box 40, Blue Summit, PA 17294-0850. (717) 794-2191. (800) 233-1128. 7th edition. 1994. $34.95.

International Encyclopedia of Integrated Circuits. Arthur A. Seidman. TAB Books, P.O. Box 40, Blue Summit, PA 17294-0850. (717) 794-2191. (800) 233-1128. 1991. $75.00.

McGraw-Hill Electronics Dictionary. J. Markus. McGraw-Hill Book Company, Inc., 1221 Avenue of the Americas, New York, NY 10020. (212) 997-3675. 5th edition. 1994. $49.95.

McGraw-Hill Encyclopedia of Science and Technology. McGraw-Hill Book, Incorporated, 1221 Avenue of the Americas, New York, NY 10020. (212) 997-3675. (800) 262-4729. Seventh edition. Twenty volumes. 1992. $1900.00.

GENERAL WORKS

Analysis of Microelectronic Materials and Devices. M. Grasserbauer and H. W. Werner, editors. John Wiley & Sons, Inc., 605 Third Avenue, New York, NY 10158-0012. (212) 850-6000. (800) 225-5945. 1991. $275.00.

Computer Controlled Systems: Theory and Applications. George AL Perdikaris. Kluwer Academic Publishers, 101 Philip Drive, Assinippi Park, Norwell, MA 02061. (617) 871-6600. 1991. $142.00

Conquest of the Microchip. Hans Quiesser. Harvard University Press, 79 Garden Street, Cambridge MA. (617) 495-2600. 1990. $9.95 in paper.

Digital and Microprocessor Engineering. S. J. Cahill and I. McGrum. Routledge Chapman & Hall, Inc., 29 West 35th Street, New York, NY 10001-3291. (212) 244-3336. 2nd edition. 1993.

Digital Protection: Protective Relaying From Electromechanical To Microprocessor. L.P. Singh. John Wiley & Sons, Inc., 605 Third Avenue, New York, NY 10158-0012. (212) 850-6000. (800) 225-5945. 1995. $29.95.

Hybrid Assemblies and Multichip Modules. Fred W. Kear. Marcel Dekker, Inc., 270 Madison Avenue, New York, NY 10016. (212) 696-9000. (800) 228-1160. 1992. $99.75.

Laser Technology in Microelectronics. V.M. Koleshko. World Scientific Publishing Company, Inc., 1060 Main Street, River Edge, NJ 07661. (201) 487-9655. (800) 227-7562. 1995. $86.00.

Microelectronic Circuits. Adel S. Sedra and Kenneth C. Smith. Harcourt Brace College Pubs., 6277 Sea Harbor Drive, Orlando, FL 32887. (407) 345-2526. (800) 787-8717. 2nd edition. 1987. $65.25.

Microelectronic Materials. C. Grovenor. IOP Publishing, Public Ledger Building, Suite 1035, Independence Square, Philadelphia, PA 19106. (215) 627-0880. (800) 358-4677. 1989. $43.00 in paper.

Microprocessor Technology. Stuart Anderson. Butterworth-Heinemann, 313 Washington Street, Newton, MA 02158. (618) 928-2500. (800) 366-2665. 1994. $14.95

Microprocessors: Fundamental Concepts and Applications. Victor E. Gibson. McGraw-Hill Book Company, 1221 Avenue of the Americas, New York, NY 10020. (212) 512-2000. (800) 262-4729. 1994. $42.95.

Microsensors. R.S. Muller, et al, editors. Institute of Electrical and Electronic Engineers (IEEE), 345 East 47th Street, New York, NY 10017. (212) 705-7900. 1991. $59.95.

Physics of Submicron Devices. David L. Ferry and R. O. Grondin. Plenum Publishing Corp., 233 Spring Street, New York, NY. (212) 620-8000. (800) 2221-9369. FAX (212) 463-0742. 1991. $85.00.

Polymers For Microelectronics: Resists and Dielectrics. Larry F. Thompson. American Chemical Society, 1155 16th Street, NW, Washington, DC 20036. (202) 872-4600. (800) 333-9511. FAX (614) 447-3671. 1993. $119.95.

HANDBOOKS AND MANUALS

Active Electronic Component Handbook. Charles A. Harper and Harold C. Jones. McGraw-Hill Book Company, 1221 Avenue of the Americas, New York, NY 10020. (212) 512-2000. (800) 262-4729. 2nd edition. 1996. $79.50.

Electronics Handbook. Jerry C. Whitaker. CRC Press, Inc., 2000 Corporate Boulevard, NW, Boca Raton, FL 33431. (407) 994-0555. (800) 272-7737. 1996. $120.00.

Embedded Systems Handbook. William Barden Jr. TAB Books, P.O. Box 40, Blue Summit, PA 17294-0850. (717) 794-2191. (800) 233-1128. 1991. $54.95.

A Guide To RISC Microprocessors. Michael Slater, editor. Academic Press, Inc., 6277 Sea Harbor Drive, Orlando, FL. (800) 321-5068. 1992. $49.95.

Handbook of Electrical and Electronic Technology. Curtis Johnson. Prentice Hall , 113 Sylvan Avenue, Route 9W, Englewood Cliffs, NJ 07632. (201) 592-2000. (800) 922-0579. 1996. $88.00.

Handbook of Quality Integrated Circuit Manufacturing. Robert Zurich. Academic Press, Inc., 6277 Sea Harbor Drive, Orlando, FL. (800) 321-5068. 1991. $59.95.

Handbook of Vlsi Chip Design and Expert Systems. A.F. Schwartz. Academic Press, Inc., 6277 Sea Harbor Drive, Orlando, FL. (800) 321-5068.

Integrated Circuits Handbook. Rector Press, Ltd., 130 Rattlesnake, Leverett, MA 01054-9726. (413) 548-9708.(800) 247-3473. 3 volumes. 1995. $295.00 in paper.

Microprocessor Support Chip Sourcebook. Alan Clements. McGraw-Hill Publishing Company, 1221 Avenue of the Americas, New York, NY 10020. (212) 512-2000. (800) 262-4729. 1992. $90.00.

Standard Handbook For Electrical Engineers. Donald Fink. McGraw-Hill Publishing Company, 1221 Avenue of the Americas, New York, NY 10020. (212) 512-2000. (800) 262-4729. 13th edition. 1996. $110.00.

ONLINE DATABASES AND CD-ROMS

CA SEARCH. Chemical Abstracts Service, P.O. Box 3012, Columbus, OH 43210-0012. (614) 447-3600. (800) 848-6533. FAX (614) 447-3709. Very comprehensive guide to worldwide chemical literature and related fields, 1972 to present. Available on

BRS,(800) 289-4277, DIALOG, (800) 334-2564, ORBIT (800) 456-7248, and STN International, FIZ Karlsruhe, P.O. Box 2465, W-7500, Karlsruhe 1, Germany, online services. Inquire as to cost and availability.

Compendex Plus. Engineering Information, Inc., 345 East 47th Street, New York, NY 10017. (212) 705-7600 or (800) 221-1044. Contains citations with abstracts to worldwide literature in engineering and technology, from 1970 to present. Available on online BRS,(800) 289-4277, DIALOG, (800) 334-2564, ORBIT (800) 456-7248, and STN International, FIZ Karlsruhe, P.O. Box 2465, W-7500, Karlsruhe 1, Germany, online services. Also available on CD-ROM. Inquire as to cost and availability.

Current Contents Search. Institute for Scientific Information, 3501 Market Street, Philadelphia, PA 19104. (215) 386-0100. FAX (215) 386-6362. Contains citations to articles listed in the table of contents of science and technology journals. Also articles in social sciences and life sciences journals. Available on BRS,(800) 289-4277, DIALOG,(800) 334-2564, online services. Inquire as to cost and availability.

Dissertation Abstracts Online. University Microfilms International, 300 North Zeeb Road, Ann Arbor, MI 48106. (800) 521-0600 or (313) 761-4700. Scope includes virtually all doctoral dissertations accepted at accredited American institutions from 1861 to present in 252 subject areas. Available on BRS, (800) 289-4277, DIALOG, (800) 334-2564, and OCLC EPIC,(800) 848-5878, online services. Also available on CD-ROM. Inquire as to cost and availability.

INSPEC. Institution of Electrical Engineers, Michael Faraday House, Six Hills Way, Stevenage, Herts. SG1 2AY, England. Telephone: 0438 313311 or FAX 0438 742840. Contains citations to the worldwide literature of physics, electronics and electrical engineering, computer technology, and related fields. Available on BRS, (800) 289-4277, DIALOG, (800) 334-2564, ORBIT, (800) 456-7248, and STN International, FIZ Karlsruhe, P.O. Box 2465, W-7500, Karlsruhe 1, Germany, online services. Inquire as to cost and availability.

NTIS Bibliographic Database. National Technical Information Service, 5285 Port Royal Road, Springfield, VA 22161. (703) 487-4929 or FAX (703) 321-8199. Broad coverage of government-sponsored science and technology research reports, 1964 to present. Available on BRS,(800) 289-4277, DIALOG, (800) 334-2564, ORBIT, (800) 456-7248, and STN International, FIZ Karlsruhe, P.O. Box 2465, W-7500, Karlsruhe 1, Germany, online services. Also available on CD-ROM. Inquire as to cost and availability.

Physics Briefs. American Institute of Physics, 335 East 45th Street, New York, NY 10017. (212) 661-9260 or FAX (212) 661-2036. Contains citations with abstracts of the literature of physics and related fields, 1979 to present. Available on the STN International, FIZ Karlsruhe, P.O. Box 2465, W-7500, Karlsruhe 1, Germany, online service. Inquire as to cost and availability.

Scisearch. Institute for Scientific Information, 3501 Market Street, Philadelphia, PA 19104. (800) 523-1850 or (215) 386-0100. Broad multidisciplinary title and author index to the international literature of science and technology, 1974 to present. Available on DIALOG,(800) 334-2564, and ORBIT,(800) 456-7248, online services. Also available on CD-ROM. Inquire as to cost and availability.

Wilsonline. H.W. Wilson Company, 950 University Avenue, Bronx, NY 10452. (800) 367-6770 or (212) 588-8400. Makes available online versions of the printed H.W. Wilson indexes,

including Applied Science and Technology Index, Business Periodicals Index, General Science Index, and Readers' Guide to Periodical Literature. Period covered is generally 1983 to present. Available on BRS,(800) 289-4277, DIALOG,(800) 334-2564, and OCLC EPIC,(800) 848-5878, online services. Also available on CD-ROM. Inquire as to cost and availability.

PERIODICALS

Circuits, Systems, and Signal Processing. Birkhauser, 675 Massachusetts Avenue, Cambridge, MA 02139-3309. FAX (201) 348-4505. 1982 to present. Quarterly. $184.00 per year. ISSN: 0278-081X.

Electrical World. McGraw-Hill, Inc., Box 513 Hightstown, NJ 08520. (212) 512-3288. 1874 to present. Monthly. $55.00 per year. ISSN: 0013-4457.

Electronic Design. Penton Publishing, San Jose Gateway, Suite 354. 2025 Gateway Place, San Jose, CA 95110. (408) 441-0550. 1952 to present. Fortnightly. $95.00. ISSN: 0013-4872.

Electronics. Penton Publishing, San Jose Gateway, Suite 354. 2025 Gateway Place, San Jose, CA 95110. (408) 441-0550. 1930 to present. Semi-weekly. $98.00 per year. ISSN: 0883-4989.

Electronics Now. Gernsback Publications, Inc., 500-B Bi-County Boulevard, Farmingdale, NY 11735. (516) 293-3000. FAX (516) 293-3115. 1929 to present. Monthly. $19.95. ISSN: 0033-7862.

IEEE Circuits and Devices Magazine. Institute of Electrical and Electronics Engineers, Inc., Box 1331, 445 Hoes Lane, Piscataway, NJ 08855-1331. (908) 981-0060. 1985 to present. Bi-monthly. $120.00 per year. ISSN: 8755-3996.

IEEE Journal of Solid State Circuits. Institute of Electrical and Electronics Engineers, Inc., Box 1331, 445 Hoes Lane, Piscataway, NJ 08855-1331. (908) 981-0060. 1966 to present. Bi-monthly. $275.00 per year. ISSN: 0018-9200.

IEEE Spectrum. Institute of Electrical and Electronics Engineers, Inc., Box 1331, 445 Hoes Lane, Piscataway, NJ 08855-1331. (908) 981-0060. 1964 to present. Monthly. $157.00 per year. ISSN: 0018-9235.

IEEE Transactions On Circuits and Systems. Part 1: Fundamental theory and Applications. Institute of Electrical and Electronics Engineers, Inc., Box 1331, 445 Hoes Lane, Piscataway, NJ 08855-1331. (908) 981-0060. 1952 to present. Monthly. $241.00 per year. ISSN: 1057-7122.

IEEE Transactions On Electron Devices. Institute of Electrical and Electronics Engineers, Inc., Box 1331, 445 Hoes Lane, Piscataway, NJ 08855-1331. (908) 981-0060. 1952 to present. Monthly. $395.00 per year. ISSN: 0018-9393.

Institute of Electrical and Electronics Engineers. PROCEEDINGS. Institute of Electrical and Electronics Engineers, Inc., Box 1331, 445 Hoes Lane, Piscataway, NJ 08855-1331. (908) 981-0060. 1913 to present. Monthly. $275.00. ISSN: 0018-9219.

Microelectronic Engineering: An Interdisciplinary Journal of Semiconductor Manufacturing Technology. Elsevier Science, 660 White Plains Road, Tarrytown, NY 10591-5153. (914) 524-9200. FAX (914) 333-2444. 1983 to present. 16 issues per year. $845.00 per year. ISSN: 0167-9317.

Microelectronics and Reliability. Elsevier Science, 660 White Plains Road, Tarrytown, NY 10591-5153. (914) 524-9200. FAX (914) 333-2444. 1962 to present. 15 issues per year. $1145.00 per year. ISSN: 0026-2714.

Microelectronics Journal. Elsevier Science, 660 White Plains Road, Tarrytown, NY 10591-5153. (914) 524-9200. FAX (914) 333-2444. 1967 to present. 8 issues per year. $455.00 per year. ISSN: 0026-2692.

Semiconductor International. Cahners Publishing Co., 44 Cook Street, Denver, CO 80206. (708) 635-8000. FAX (708) 390-2770. (800) 662-7776. 1978 to present. 13 issues per year. $84.95 per year. ISSN: 0163-3767.

RESEARCH CENTERS AND INSTITUTES

Alabama Microelectronics Science and Technology Center. Auburn University, 200 Broun Hall, Auburn University, AL 36849. (205) 844-1871. (205) 844-1809.

Center For Integrated Sensors and Circuits. University of Michigan, 1246 EECS Building, Ann Arbor, MI 48109-2122. (313) 764-3346. FAX (313) 747-1781.

Center For Integrated Systems. Stanford University, School of Engineering, Stanford, CA 94305-4028. (415) 723-9038.

Electrical Engineering Research Laboratories. Purdue University. Electrical Engineering Building, West Lafayette, in 47907. (317) 494-3536. FAX (317) 494-6440.

Electronics Research Laboratory. University of California, Berkeley, 253 Cory Hall, Berkeley, CA 94720. (415) 642-2301.

Engineering Research Center. University of Maryland, Clark School of Engineering, Potomac Building, Room 2104. (301) 405-3906. FAX (301) 403-4105.

Laboratory For Electromagnetic and Electronic Systems. Massachusetts Institute of Technology, 77 Massachusetts Avenue, Cambridge, MA 02139. (617) 253) 4631.

Solid State Electronics Laboratory. North Carolina State University, 432 Daniels Hall, Raleigh, NC 27695. (919) 737-2336.

MICROPROCESSORS

See also: ELECTRICAL ENGINEERING, ELECTRICITY, ELECTRONIC CIRCUITS and COMPONENTS, ELECTRONICS, ELECTRONICS ENGINEERING, MICROELECTRONICS

ABSTRACT SERVICES AND INDEXES

Applied Science and Technology Index; A Cumulative Subject Index To English Language Periodicals in the Fields of Aeronautics and Space Science, Computer Technology, Chemistry, Construction Industry, Energy and Related Areas. H.W. Wilson Co., 950 University Avenue, Bronx, NY 10452. (212) 588-8400. (800) 367-6770. FAX (718) 590-1617. From 1958 to present. Monthly. Inquire about cost and availability. Also available on CD-ROM and online. ISSN: 0003-6986.

Computer and Control Abstracts. Science Abstracts, section C Institution of Electrical Engineers (IEE), London. Available from INSPEC/IEEE-Institute of Electrical and Electronic Engineers, Box 1331, Hoes Lane, Piscataway, NJ 08855-1331. (908) 562-5549. 1966 to present. Monthly. $1455.00. ISSN: 0036-8113. Also available on CD-ROM and online as INSPEC.

Current Contents: Engineering, Technology and Applied Sciences. Institute for Scientific Information, 3501 Market Street, Philadelphia, PA 19104. (215) 386-0100. FAX (215) 386-6362. 1970 to present. Weekly. $442.00 per year. Also available on CD-ROM and online. Inquire regarding cost and availability. ISSN: 0095-7917.

Current Papers in Electrical and Electronics Engineering. Institution of Electrical Engineers (IEE), London. Distributed by INSPEC/IEEE, Box 1331, 445 Hoes Lane, Piscataway, NJ 08855-1331. (908) 562-5549. 1969 to present. Monthly. $345.00 per year. ISSN: 0011-3778.

Electrical and Electronics Abstracts. Institution of Electrical Engineers (IEE), London. Available from INSPEC/IEEE-Institute of Electrical and Electronic Engineers, Box 1331, Hoes Lane, Piscataway, NJ 08855-1331. (908) 562-5549. 1898 to present. Monthly. $2200.00 per year. Also available on CD-ROM and online as INSPEC.

Engineering Index Monthly; Indexes and Abstracts the World's Engineering and Technical Literature. Engineering Information, Inc., Castle Point on the Hudson, Hoboken, NJ 07030. (201) 216-8500. (800) 221-1044. FAX (201) 216-8532. Monthly. $2300.00 per year. Available online as COMPENDEX and also on CD-ROM. ISSN: 0742-1974.

General Science Index. H.W. Wilson Company, 950 University Avenue, Bronx, NY 10452. (212) 588-8400. (800) 367-6770. FAX (718) 590-1617. From 1978 to present. Ten issues per year; quarterly and annual cumulations. Service basis. Available on CD-ROM and online. Inquire regarding cost and availability. ISSN: 0162-1963.

Government Reports Announcements and Index. U.S. National Technical Information Service (NTIS), 5285 Port Royal Road, Springfield, VA 22161. (703) 487-4650. FAX (703) 321-8547. From 1968 to present. Annual. $630.00 per year. Also available online as NTIS Bibliographic Database and on CD-ROM.

Index To IEEE Publications. IEEE Service Center, 445 Hoes Lane, Piscataway, NJ 08855-1331. (908) 981-1393. (800) 678-IEEE. FAX (908) 981-9667. 1973 to present. Annual. ISSN: 0099-1368.

Physics Abstracts. INSPEC. Section A, Science Abstracts. Institution of Electrical Engineers (IEE), London. Available from: INSPEC/IEEE-Institute of Electrical and Electronic Engineers, Box 1331, Hoes Lane, Piscataway, NJ 08855-1331. (908) 562-5549. 1898 to present. 24 issues per year. $2835.00 per year. Also available online and on CD-ROM. ISSN: 0036-8091.

Physics Briefs (Physikalische Berichte). Information Center for Energy, Physics, Mathematics; German Physical Society. V C H Publishers, Inc., 220 East 23rd Street, New York, NY 10010-4606. (212) 683-8333. 1845 to present. 24 issues per year. $2390.00 per year. Also available online. ISSN: 0179-7434.

Science Citation Index. SCI. Institute for Scientific Information, 3501 Market Street, Philadelphia, PA 19104. (215) 386-0100. (800) 523-1850. FAX (215) 386-2991. 1961 to present. Six issues per year, plus annual cumulation. $11650.00 per year. Also available online and on CD-ROM. Inquire about price and availability. ISSN: 0036-827X.

Solid State and Superconductivity Abstracts: Covers theory, Production and Application of Solid State Materials. Cambridge Scientific Abstracts, 7200 Wisconsin Avenue, Bethesda, MD 20824. (301) 961-6750. FAX (301) 961-6720. 1957 to present. Bimonthly. $1320.00 per year. Also available online. ISSN: 0896-5900.

STAR. Scientific and Technical Aerospace Reports. U.S. National Aeronautics and Space Administration, Scientific and Technical Information Facility, Box 8757, Baltimore-Washington International Airport, MD 21240. (301) 621-0153. Distributed by U. S. Superintendent of Documents, Washington, DC 20402. From 1963 to present. Semimonthly, with semiannual and annual indexes. $114.00 per year. Also available online and on CD-ROM. ISSN: 0036-8741.

ANNUAL REVIEWS AND YEARBOOKS

Advances in Electronics and Electron Physics. Academic Press, Inc., 6277 Sea Harbor Drive, Orlando, FL 32887. (800) 321-5068. From 1948 to present. Irregular. Price varies, inquire. ISSN: 0065-2539.

Critical Reviews in Solid State and Materials Sciences. CRC Press, Inc., 2000 Corporate Boulevard, NW, Boca Raton, FL 33431. (407) 994-0555. (800) 272-7737. FAX (407) 998-9784. 1970 to present. Bimonthly. $265.00 per year. ISSN: 1040-8436.

ASSOCIATIONS AND PROFESSIONAL SOCIETIES

American Association of Engineering Societies, 1111 19th Street, Suite 608, Washington, DC 20036-3603. (202) 296-2237. FAX (202) 296-1151.

American Electronics Association. 5201 Great America Way, Suite 520, P.O. 52990, Santa Clara, CA 95056. (408) 987-4200. FAX (408) 970-8565.

American Institute of Physics. One Physics Ellipse, College Park, MD 20740-3843. (301) 209-3100.

Electronics Industries Association. 2500 Wilson Boulevard, Arlington, VA 22201. (703) 907-7500. FAX (202) 457-4985.

IEEE Circuits and Systems Society. c/o Institute of Electrical and Electronic Engineers. 345 East 47th Street. New York, NY 10017. (212) 705-7900. FAX (212) 705-4929.

Institute of Electrical and Electronics Engineers. 345 East 47th Street. New York, NY 10017. (212) 705-7900. FAX (212) 705-4929.

International Society for Hybrid Microelectronics. 1850 Centennial Park Drive, Suite 105, Reston, VA 22091. (703) 758-1060. FAX (703) 758-1066.

National Electrical Manufacturers Association. 1300 North 17th Street, Suite 1847, Rosslyn VA 22209. (703) 841-3200. FAX (703) 841-3300.

DIRECTORIES AND BIOGRAPHICAL SOURCES

American Electronics Association Directory. American Electronics Association, 5201 Great America Way, Suite 520, P.O. 52990, Santa Clara, CA 95056. (408) 987-4200. FAX (408) 970-8565. Annual.

EEM-Electronic Engineer's Master. Hearst Business Communications, Inc., 645 Stewart Avenue, Garden City NY 11530. (516) 227-1300. ISSN: 0732-9016.

Electrical and Electronic Trades Directory: The Blue Book. Institution of Electrical Engineers, Michael Faraday House, Six Hills Way, Stevenage, Herts. SG1 2AY, England. Telephone: 0438 313311 or FAX 0438 742840. 1995. $140.00

EEM-Electronic Engineer's Master. Hearst Business Communications, Inc., 645 Stewart Avenue, Garden City NY 11530. (516) 227-1300. ISSN: 0732-9016.

Engineering Research Centers: Incorporating Electronics Research Centers. Stockton Press, 345 Park Avenue, New York, NY 10010. (212) 689-9200. (800) 221-2123. 4th edition. 1995. $515.00.

IC Master. Hearst Business Communications, Inc., 645 Stewart Avenue, Garden City, NY 11530. (516) 227-1300. 1975 to present. Annual. $170.00 per year. ISSN: 0894-6809.

IEEE Membership Directory. Institute of Electrical and Electronics Engineers, IEEE Service Center, 445 Hoes Lane, Piscataway, NY 08854. (908) 981-1393. (800) 678-IEEE. FAX (908) 981-9667. 2 volumes. Annual. $190.00. ISSN:

International Directory of Abbreviations and Acronyms of Electronics, Electrical Engineering, Computer Technology and Information Processing. Peter Wennrich. K. G. Saur, 121 Chanlon Road, New Providence, NJ 07974. (908) 464-6800. (800) 521-8110 2 volumes. 1992. $230.00.

International Engineering Directory. American Consulting Engineers Council, 1015 15th Street, N.W. Suite 802, Washington, DC 20005-2670. (202) 347-7474. Annual. $10.00.

Research Centers Directory. Gale Research Company Inc., 835 Penobscot Building, Detroit, MI 48226-4094. (313) 961-2242.(800) 347-4253. 20th edition. 1995. $485.00. ISSN: 0080-1518.

Who's Who in Electric Transmission and Distribution. Utility Data Institute, 1200 G Street NY, Suite 250, Washington, DC 20005-3802. (202) 942-8788. (800) 486-3660. 1995. $150.00.

Who's Who in Engineering. Gordon Davis, editor. American Association of Engineering Societies. 1111 19th Street, NY, Suite 608, Washington, DC 20036. (202) 296-2237. (800) 658-8897. 9th edition. 1995. $220.00.

Who's Who in Technology; Biographies and Index. Gale Research Company, 835 Penobscot Building, Detroit, MI 48226-4094. (313 961-2242. (800) 521-4253. 7th edition. 1995. $195.00. ISBN 0-8103-7467-6.

MICROPROCESSORS

Ency. of Physical Sciences and Engineering Info. Sources

ENCYCLOPEDIAS AND DICTIONARIES

Encyclopedia of Applied Physics. George Trigg, editor. VCH Publications, Inc., 220 East 23rd Street, Suite 909, New York, NY 10010-4606. (212) 683-8333. (800) 422-8824. 20 volume set. 1991-. $5990.00.

IEEE Standard Dictionary of Electrical and Electronics Terms. Christopher J. Booth, editor. IEEE Service Center, 445 Hoes Lane, Piscataway, NJ 08855-1331. (908) 981-1393. (800) 678-IEEE. FAX (908) 981-9667. IEEE Standard 100-1992. 5th edition. 1993. $150.00.

Illustrated Dictionary of Electronics. Stan Gibilsco. TAB Books, P.O. Box 40, Blue Summit, PA 17294-0850. (717) 794-2191. (800) 233-1128. 7th edition. 1994. $34.95.

International Encyclopedia of Integrated Circuits. Arthur A. Seidman. TAB Books, P.O. Box 40, Blue Summit, PA 17294-0850. (717) 794-2191. (800) 233-1128. 1991. $75.00.

McGraw-Hill Electronics Dictionary. J. Markus. McGraw-Hill Book Company, Inc., 1221 Avenue of the Americas, New York, NY 10020. (212) 997-3675. 5th edition. 1994. $49.95.

McGraw-Hill Encyclopedia of Science and Technology. McGraw-Hill Book, Incorporated, 1221 Avenue of the Americas, New York, NY 10020. (212) 997-3675. (800) 262-4729. Seventh edition. Twenty volumes. 1992. $1900.00.

GENERAL WORKS

Advanced Microprocessors. Daniel Tabak. McGraw-Hill Book Company, 1221 Avenue of the Americas, New York, NY 10020. (212) 512-2000. (800) 262-4729. 2nd edition. 1991. $50.00.

Computer Controlled Systems: Theory and Applications. George AL Perdikaris. Kluwer Academic Publishers, 101 Philip Drive, Assinippi Park, Norwell, MA 02061. (617) 871-6600. 1991. $142.00

Computers and Microprocessors. A.C. Downton. Chapman & Hall, 1 Penn Plaza, New York, NY 10119. (212) 564-1060. 1994. $26.95 in paper.

Design and Technology of Integrated Circuits. Donald DeCogan. John Wiley & Sons, Inc., 605 Third Avenue, New York, NY 10158-0012. (212) 850-6000. (800) 225-5945. 1990. $75.00.

Digital and Microprocessor Engineering. S. J. CaHill and I. McGrum. Routledge, Chapman & Hall, Inc., 29 West 35th Street, New York, NY 10001-3291. (212) 244-3336. 2nd edition. 1993.

Digital Protection: Protective Relaying From Electromechanical To Microprocessor. L.P. Singh. John Wiley & Sons, Inc., 605 Third Avenue, New York, NY 10158-0012. (212) 850-6000. (800) 225-5945. 1995. $29.95.

Integrated Circuits and Microprocessors. Robin C. Holland. Elsevier Science Publishing Company, Inc., 655 Avenue of the Americas, New York, NY 10010. (212) 989-5800. FAX (914) 333-2444. 1986. $30.00 in paper.

Linear Optimization and Extensions: Theory and Algorithms. Shu-Cheng Fang and Sarat C. Puthenpura. Prentice-Hall, 113

Sylvan Avenue, Route 9W, Englewood Cliffs, NJ 07632. (201) 592-2000. (800) 922-0579. 1993. $69.00.

Mastering IC Circuits. Joseph J. Carr. TAB Books, P.O. Box 40, Blue Summit, PA 17294-0850. (717) 794-2191. (800) 233-1128. 1991. $32.95.

Microelectronic Circuits. Adel S. Sedra and Kenneth C. Smith. Harcourt Brace College Pubs., 6277 Sea Harbor Drive, Orlando, FL 32887. (407) 345-2526. (800) 787-8717. 2nd edition. 1987. $65.25.

Microprocessor Architectures and Systems: Risc, Cisc and Dsp Processors. Steve Heath. Butterworth-Heinemann, 313 Washington Street, Newton, MA 02158. (618) 928-2500. (800) 366-2665. 1993. $25.95 in paper.

Microprocessor Programming and Applications For Scientists and Engineers. R. R. Smardzewski. Elsevier Science Publishing Company, Inc., 655 Avenue of the Americas, New York, NY 10010. (212) 989-5800. FAX (914) 333-2444. 1984. $99.00.

Microprocessor System Design Fundamentals. Kenneth J. Breeding. Prentice-Hall, 113 Sylvan Avenue, Route 9W, Englewood Cliffs, NJ 07632. (201) 592-2000. (800) 922-0579. 1994. $65.00.

Microprocessor Technology. Stuart Anderson. Butterworth-Heinemann, 313 Washington Street, Newton, MA 02158. (618) 928-2500. (800) 366-2665. 1994. $14.95

Microprocessors: Fundamental Concepts and Applications. Victor E. Gibson. McGraw-Hill Book Company, 1221 Avenue of the Americas, New York, NY 10020. (212) 512-2000. (800) 262-4729. 1994. $42.95.

Schaum's Microprocessor Fundamentals. Roger L. Tokheim. McGraw-Hill Book Company, 1221 Avenue of the Americas, New York, NY 10020. (212) 512-2000. (800) 262-4729. 2nd edition. 1990. $12.95.

Transistors: Fundamentals For the Integrated-Circuit Engineer. R. M. Warner Jr., and B. L. Grung. Krieger Publishing Company, P.O. Box 9542, Melbourne, FL 32902-9542. (407) 724-9542. 1990. $72.50.

HANDBOOKS AND MANUALS

Active Electronic Component Handbook. Charles A. Harper and Harold C. Jones. McGraw-Hill Book Company, 1221 Avenue of the Americas, New York, NY 10020. (212) 512-2000. (800) 262-4729. 2nd edition. 1996. $79.50.

Electronics Handbook. Jerry C. Whitaker. CRC Press, Inc., 2000 Corporate Boulevard, NW, Boca Raton, FL 33431. (407) 994-0555. (800) 272-7737. 1996. $120.00.

A Guide To Risc Microprocessors. Michael Slater, editor. Academic Press, Inc., 6277 Sea Harbor Drive, Orlando, FL. (800) 321-5068. 1992. $49.95.

Handbook of Electrical and Electronic Technology. Curtis Johnson. Prentice Hall , 113 Sylvan Avenue, Route 9W, Englewood Cliffs, NJ 07632. (201) 592-2000. (800) 922-0579. 1996. $88.00.

Handbook of Quality Integrated Circuit Manufacturing. Robert Zurich. Academic Press, Inc., 6277 Sea Harbor Drive, Orlando, FL. (800) 321-5068. 1991. $59.95.

Handbook of Vlsi Chip Design and Expert Systems. A. F. Schwartz. Academic Press, Inc., 6277 Sea Harbor Drive, Orlando, FL. (800) 321-5068.

Integrated Circuits Handbook. Rector Press, Ltd., 130 Rattlesnake, Leverett, MA 01054-9726. (413) 548-9708. (800) 247-3473. 3 volumes. 1995. $295.00 in paper.

Microprocessor Support Chip Sourcebook. Alan Clements. McGraw-Hill Publishing Company, 1221 Avenue of the Americas, New York, NY 10020. (212) 512-2000. (800) 262-4729. 1992. $90.00.

Standard Handbook For Electrical Engineers. Donald Fink. McGraw-Hill Publishing Company, 1221 Avenue of the Americas, New York, NY 10020. (212) 512-2000. (800) 262- 4729. 13th edition. 1996. $110.00.

ONLINE DATABASES AND CD-ROMS

CA SEARCH. Chemical Abstracts Service, P.O. Box 3012, Columbus, OH 43210-0012. (614) 447-3600. (800) 848-6533. FAX (614) 447-3709. Very comprehensive guide to worldwide chemical literature and related fields, 1972 to present. Available on BRS,(800) 289-4277, DIALOG, (800) 334-2564, ORBIT (800) 456-7248, and STN International, FIZ Karlsruhe, P.O. Box 2465, W-7500, Karlsruhe 1, Germany, online services. Inquire as to cost and availability.

Compendex Plus. Engineering Information, Inc., 345 East 47th Street, New York, NY 10017. (212) 705-7600 or (800) 221-1044. Contains citations with abstracts to worldwide literature in engineering and technology, from 1970 to present. Available on online BRS,(800) 289-4277, DIALOG, (800) 334-2564, ORBIT (800) 456-7248, and STN International, FIZ Karlsruhe, P.O. Box 2465, W-7500, Karlsruhe 1, Germany, online services. Also available on CD-ROM. Inquire as to cost and availability.

Current Contents Search. Institute for Scientific Information, 3501 Market Street, Philadelphia, PA 19104. (215) 386-0100. FAX (215) 386-6362. Contains citations to articles listed in the table of contents of science and technology journals. Also articles in social sciences and life sciences journals. Available on BRS,(800) 289-4277, DIALOG,(800) 334-2564, online services. Inquire as to cost and availability.

Dissertation Abstracts Online. University Microfilms International, 300 North Zeeb Road, Ann Arbor, MI 48106. (800) 521-0600 or (313) 761-4700. Scope includes virtually all doctoral dissertations accepted at accredited American institutions from 1861 to present in 252 subject areas. Available on BRS, (800) 289-4277, DIALOG, (800) 334-2564, and OCLC EPIC,(800) 848-5878, online services. Also available on CD-ROM. Inquire as to cost and availability.

INSPEC. Institution of Electrical Engineers, Michael Faraday House, Six Hills Way, Stevenage, Herts. SG1 2AY, England. Telephone: 0438 313311 or FAX 0438 742840. Contains citations to the worldwide literature of physics, electronics and electrical engineering, computer technology, and related fields. Available on BRS, (800) 289-4277, DIALOG, (800) 334-2564, ORBIT, (800) 456-7248, and STN International, FIZ Karlsruhe, P.O. Box 2465, W-7500, Karlsruhe 1, Germany, online services. Inquire as to cost and availability.

NTIS Bibliographic Database. National Technical Information Service, 5285 Port Royal Road, Springfield, VA 22161. (703) 487-4929 or FAX (703) 321-8199. Broad coverage of government-sponsored science and technology research reports, 1964 to present. Available on BRS,(800) 289-4277, DIALOG, (800) 334-2564, ORBIT, (800) 456-7248, and STN International, FIZ Karlsruhe, P.O. Box 2465, W-7500, Karlsruhe 1, Germany, online services. Also available on CD-ROM. Inquire as to cost and availability.

Physics Briefs. American Institute of Physics, 335 East 45th Street, New York, NY 10017. (212) 661-9260 or FAX (212) 661-2036. Contains citations with abstracts of the literature of physics and related fields, 1979 to present. Available on the STN International, FIZ Karlsruhe, P.O. Box 2465, W-7500, Karlsruhe 1, Germany, online service. Inquire as to cost and availability.

Scisearch. Institute for Scientific Information, 3501 Market Street, Philadelphia, PA 19104. (800) 523-1850 or (215) 386-0100. Broad multidisciplinary title and author index to the international literature of science and technology, 1974 to present. Available on DIALOG,(800) 334-2564, and ORBIT,(800) 456-7248, online services. Also available on CD-ROM. Inquire as to cost and availability.

Wilsonline. H.W. Wilson Company, 950 University Avenue, Bronx, NY 10452. (800) 367-6770 or (212) 588-8400. Makes available online versions of the printed H.W. Wilson indexes, including Applied Science and Technology Index, Business Periodicals Index, General Science Index, and Readers' Guide to Periodical Literature. Period covered is generally 1983 to present. Available on BRS,(800) 289-4277, DIALOG,(800) 334-2564, and OCLC EPIC,(800) 848-5878, online services. Also available on CD-ROM. Inquire as to cost and availability.

PERIODICALS

Circuits, Systems, and Signal Processing. Birkhauser, 675 Massachusetts Avenue, Cambridge, MA 02139-3309. FAX (201) 348-4505. 1982 to present. Quarterly. $184.00 per year. ISSN: 0278-081X.

Electrical World. McGraw-Hill, Inc., Box 513 Hightstown, NJ 08520. (212) 512-3288. 1874 to present. Monthly. $55.00 per year. ISSN: 0013-4457.

Electronic Design. Penton Publishing, San Jose Gateway, Suite 354. 2025 Gateway Place, San Jose, CA 95110. (408) 441-0550. 1952 to present. Fortnightly. $95.00. ISSN: 0013-4872.

Electronics. Penton Publishing, San Jose Gateway, Suite 354. 2025 Gateway Place, San Jose, CA 95110. (408) 441-0550. 1930 to present. Semi-weekly. $98.00 per year. ISSN: 0883-4989.

Electronics Now. Gernsback Publications, Inc., 500-B Bi-County Boulevard, Farmingdale, NY 11735. (516) 293-3000. FAX (516) 293-3115. 1929 to present. Monthly. $19.95. ISSN: 0033-7862.

IEEE Circuits and Devices Magazine. Institute of Electrical and Electronics Engineers, Inc., Box 1331, 445 Hoes Lane, Piscataway, NJ 08855-1331. (908) 981-0060. 1985 to present. Bi-monthly. $120.00 per year. ISSN: 8755-3996.

IEEE Journal of Solid State Circuits. Institute of Electrical and Electronics Engineers, Inc., Box 1331, 445 Hoes Lane,

Piscataway, NJ 08855-1331. (908) 981-0060. 1966 to present. Bi-monthly. $275.00 per year. ISSN: 0018-9200.

IEEE Spectrum. Institute of Electrical and Electronics Engineers, Inc., Box 1331, 445 Hoes Lane, Piscataway, NJ 08855-1331. (908) 981-0060. 1964 to present. Monthly. $157.00 per year. ISSN: 0018-9235.

IEEE Transactions On Circuits and Systems. Part 1: Fundamental theory and Applications. Institute of Electrical and Electronics Engineers, Inc., Box 1331, 445 Hoes Lane, Piscataway, NJ 08855-1331. (908) 981-0060. 1952 to present. Monthly. $241.00 per year. ISSN: 1057-7122.

IEEE Transactions On Electron Devices. Institute of Electrical and Electronics Engineers, Inc., Box 1331, 445 Hoes Lane, Piscataway, NJ 08855-1331. (908) 981-0060. 1952 to present. Monthly. $395.00 per year. ISSN: 0018-9393.

Institute of Electrical and Electronics Engineers. PROCEEDINGS. Institute of Electrical and Electronics Engineers, Inc., Box 1331, 445 Hoes Lane, Piscataway, NJ 08855-1331. (908) 981-0060. 1913 to present. Monthly. $275.00. ISSN: 0018-9219.

Microelectronic Engineering: An Interdisciplinary Journal of Semiconductor Manufacturing Technology. Elsevier Science, 660 White Plains Road, Tarrytown, NY 10591-5153. (914) 524-9200. FAX (914) 333-2444. 1983 to present. 16 issues per year. $845.00 per year. ISSN: 0167-9317.

Microelectronics and Reliability. Elsevier Science, 660 White Plains Road, Tarrytown, NY 10591-5153. (914) 524-9200. FAX (914) 333-2444. 1962 to present. 15 issues per year. $1145.00 per year. ISSN: 0026-2714.

Microelectronics Journal. Elsevier Science, 660 White Plains Road, Tarrytown, NY 10591-5153. (914) 524-9200. FAX (914) 333-2444. 1967 to present. 8 issues per year. $455.00 per year. ISSN: 0026-2692.

Semiconductor International. Cahners Publishing Co., 44 Cook Street, Denver, CO 80206. (708) 635-8000. FAX (708) 390-2770. (800) 662-7776. 1978 to present. 13 issues per year. $84.95 per year. ISSN: 0163-3767.

RESEARCH CENTERS AND INSTITUTES

Alabama Microelectronics Science and Technology Center. Auburn University, 200 Broun Hall, Auburn University, AL 36849. (205) 844-1871. (205) 844-1809.

Center For Integrated Sensors and Circuits. University of Michigan, 1246 EECS Building, Ann Arbor, MI 48109-2122. (313) 764-3346. FAX (313) 747-1781.

Center For Integrated Systems. Stanford University, School of Engineering, Stanford, CA 94305-4028. (415) 723-9038.

Electroscience Laboratory, Ohio State University, 1320 Kinnear Road, Columbus, Ohio 43212. (614) 292-7981, FAX (614) 292-7297.

Engineering Research Center. University of Maryland, Clark School of Engineering, Potomac Building, Room 2104. (301) 405-3906. FAX (301) 403-4105.

Laboratory For Electromagnetic and Electronic Systems. Massachusetts Institute of Technology, 77 Massachusetts Avenue, Cambridge, MA 02139. (617) 253) 4631.

Solid State Electronics Laboratory North Carolina State University, 432 Daniels Hall, Raleigh, NC 27695. (919) 737-2336.

MICROSCOPY

See also: ELECTRON MICROSCOPY, OPTICS

ABSTRACT SERVICES AND INDEXES

Applied Science and Technology Index; A Cumulative Subject Index to English Language Periodicals in the Fields of Aeronautics and Space Science, Computer Technology, Chemistry, Construction Industry, Energy and Related Areas. H.W. Wilson Co., 950 University Avenue, Bronx, NY 10452. (212) 588-8400. (800) 367-6770. FAX (718) 590-1617. From 1958 to present. Monthly. Inquire about cost and availability. Also available on CD-ROM and online. ISSN: 0003-6986.

Bibliography and Index of Geology. American Geological Institute, 4220 King Street, Alexandria, VA 22302-1507. (703) 379-2480. FAX (703) 379-7563. From 1969 to present. Monthly. $1295.00 per year. Also available as GEOREF online (CISTI, DIALOG, Orbit, STN) and on CD-ROM. Inquire about price and availability. ISSN: 0098-2784.

Chemical Abstracts. Chemical Abstracts Service, 2540 Olentangy River Road, Box 3012, Columbus, OH 43210-0012. (614) 447-3600. (800) 848-6538. FAX (614) 447-3713. From 1907 to present. Weekly. $16800.00 per year. Also available on CD-ROM and online. Inquire regarding cost and availability. ISSN:

Current Contents: Engineering, Technology and Applied Sciences. Institute for Scientific Information, 3501 Market Street, Philadelphia, PA 19104. (215) 386-0100. FAX (215) 386-2291. From 1961 to present. Weekly. $442.00 per year. Also available on CD-ROM and online. Inquire regarding cost and availability. ISSN: 0095-7917.

Engineering Index Monthly. Engineering Information, Inc., Castle Point on the Hudson, Hoboken, NJ 07030. (201) 216-8500. FAX (201) 216-8532. Monthly. $2200.00 per year. Also available online as COMPENDEX and on CD-ROM. ISSN: 0742-1974.

General Science Index. H.W. Wilson Company, 950 University Avenue, Bronx, NY 10452. (212) 588-8400. (800) 367-6770. FAX (718) 590-1617. From 1978 to present. Ten issues per year; quarterly and annual cumulations. Service basis. Available on CD-ROM and online. Inquire regarding cost and availability. ISSN: 0162-1963.

NTIS Alerts: Physics. U.S. National Technical Information Service. 5285 Port Royal Road, Springfield, VA 22161. (703) 487-4650. FAX (703) 321-8547. Weekly. $140.00 per year.

Physics Abstracts. INSPEC. Section A, Science Abstracts. Institution of Electrical Engineers (IEE). Available from INSPEC/IEEE-Institute of Electrical and Electronic Engineers,, Box 1331, 445 Hoes Lane, Piscataway, NJ 08855-1331. (908) 562-5549.

1898 to present. 24 issues per year. $2835.00 per year. Also available on CD-ROM and online. ISSN: 0036-8091.

Physics Briefs (Physikalische Berichte). Information Center for Energy, Physics, Mathematics; German Physical Society. V C H Publishers, Inc., 220 East 23rd Street, New York, NY 10010-4606. (212) 683-8333. 1845 to present. 24 issues per year. $2390.00 per year. Also available online. ISSN: 0179-7434.

Science Citation Index. SCI. Institute for Scientific Information, 3501 Market Street, Philadelphia, PA 19104. (215) 386-0100. (800) 523-1850. FAX (215) 386-2991. 1961 to present. Six issues per year, plus annual cumulation. $11650.00 per year. Also available online and on CD-ROM. Inquire about price and availability. ISSN: 0036-827X.

ANNUAL REVIEWS AND YEARBOOKS

Advances in Optical and Electron Microscopy. Academic Press, Inc., 6277 Sea Harbor Drive, Orlando, FL. (800) 321-5068. 1966 to present. Irregular. ISSN: 0065-3012

ASSOCIATIONS AND PROFESSIONAL SOCIETIES

American Institute of Physics. One Physics Ellipse, College Park, MD 20740-3843. (301) 209-3100.

IEEE Lasers and Electro-Optics Society. c/o IEEE, 445 Hoes Lane, PO Box 1331, Piscataway, NJ 08855. (908) 562-3892. FAX (908) 562-8434.

Microbeam Analysis Society. c/o VCH Publishers, Inc., 303 NW 12th Avenue, Deerfield Beach, FL 33442-1788. (800) 367-8249. FAX (305) 428-8201.

Microscopy Society of America. PO Box MSA, Woods Hole, MA 02543. (508) 540-7630. FAX (508) 548-9053.

Optical Society of America. 2010 Massachusetts Avenue NW, Washington, DC 20036. (202) 223-8130. FAX (202) 223-1096.

Spie-the International For Optical Engineering. PO Box 10, Bellingham, WA 98227. (206) 676-3290. FAX (206) 647-1445.

DIRECTORIES AND BIOGRAPHICAL SOURCES

American Men and Women of Science: Physical and Biological Sciences. R. R. Bowker Inc., 121 Chanlon Road, New Providence, NJ 07974. (908) 464-6800. (800) 521-8110. 20th edition. 8 volumes. 1996. $850.00.

International Directory of Engineering Societies and Related Organizations. American Society for Engineering Education, 1818 N Street NW, Suite 600, Washington, DC 20036. (202) 331-3526. 15th edition. 1996. $185.00.

Research Centers Directory. Gale Research Inc., 835 Penobscot Building, Detroit, MI 48226-4094. (313) 961-2242. (800) 347-4253. 20th edition. 1995. $485.00. ISSN: 0080-1518.

Scientific and Technical Organizations and Agencies Directory. Gale Research, 835 Penobscot Building, Detroit, MI 48226-4094. (313) 961-2242. (800) 347-4253. 4th edition. 1996. $195.00.

Who's Who in Engineering. Gordon Davis, editor. American Society for Engineering Education, 1111 19th Street, NW, Suite 608, Washington, DC 20036. 9th edition. 1995. $220.00.

Who's Who in Technology: Biographies and Index. Gale Research, 835 Penobscot Building, Detroit, MI 48226-4094. (313) 961-2242. (800) 347-4253. 7th edition. 1995. $195.00.

ENCYCLOPEDIAS AND DICTIONARIES

Encyclopedia of Applied Physics. George Trigg, editor. VCH Publications, Inc., 220 East 23rd Street, Suite 909, New York, NY 10010-4606. (212) 683-8333. (800) 422-8824. 20 volume set. 1991-. $5990.00.

Encyclopedia of Lasers and Optical Technology. Robert A. Meyers, editor. Academic Press, Inc., 6277 Sea Harbor Drive, Orlando, FL. (800) 321-5068. 1991. $92.00.

Glossary of Optical Terminology. Thomas K. Farrell. Butterworth-Heinemann, 313 Washington Avenue, Newton, MA 02158. (617) 928-2500. (800) 366-2665. 1985. $24.95.

McGraw-Hill Encyclopedia of Science and Technology. McGraw-Hill Book Company, Inc., 1221 Avenue of the Americas, New York, NY 10020. (212) 512-2000. (800) 262-4729. 7th edition. 20 volume set. 1992. $1900.00.

RMS Dictionary of Light Microscopy. Royal Microscopy Society, Nomenclature Committee Staff. Oxford University Press, Inc., 200 Madison Avenue, New York, NY 10016. (212) 725-6000. (800) 334-4249. 1989. $32.00.

GENERAL WORKS

Acousto-Otics and Acoustic Microscopy. S. M. Gracewski and T. Kundu, editors. American Society of Mechanical Engineers, 345 East 47th Street, New York, NY 10017. (212) 705-7722. (800) 843-2763. 1992. $52.00.

Applied Optics and Optical Design. A. E. Conrady. Dover Publications, Inc., 180 Varick Street, New York, NY 10014. (212) 255-3755. (800) 223-3130. 2 volumes. Reprint edition. 1992. $10.95.

Applied Optics and Optical Engineering; A Comprehensive Treatise. Rudolf Kingslake, et al, editors. Academic Press 9 volumes. 1965- . Volumes individually priced.

The Great Age of the Microscope: The Collection of the Royal Microscopical Society Through 150 Years. Gerard Turner. IOP Publishing, Public Ledger Building, Suite 1035, Independence Square, Philadelphia, PA 19106. (215) 627-0880. (800) 358-4677. 1989. $103.00.

High-Resolution Transmission Electron Microscopy and Associated Technologies. Peter Buseck, et al. Oxford University Press, Inc., 200 Madison Avenue, New York, NY 10016. (212) 725-6000. (800) 334-4249. 1989. $75.00.

Introduction To Classical and Modern Optics. Jurgen R. Meyer-Arendt. Prentice-Hall, 113 Sylvan Avenue, Route 9W, Englewood Cliffs, NJ 07632. (201) 592-2000. (800) 922-0579. 4th edition. 1994. $70.00.

MICROSCOPY

Ency. of Physical Sciences and Engineering Info. Sources

Introduction To Microscopy By Means of Light, Electrons, X-Rays OR Acoustics. Theodore G. Rochow and Paul A. Tucker. Plenum Publishing Corp., 233 Spring Street, New York, NY. (212) 620-8000. (800) 2221-9369. FAX (212) 463-0742. 1994. $49.50.

Light and Electron Microscopy. Elizabeth M. Slayter and Henry S. Slayter. Cambridge University Press, 40 West 20th Street, New York, NY 10011-4211. (212) 924-3900. (800)872-7423. 1992. $29.95 in paper.

Microscopy and Photomicrography. Robert F. Smith. Telford Press, P.O. Box 287, West Caldwell, NJ 07006. (201) 228-7744. 1989. $45.00.

Optical Microscopy: Emerging Methods and Applications. Brian Herman and John J. Lemasers. Academic Press, Inc., 6277 Sea Harbor Drive, Orlando, FL. (800) 321-5068. 1992. $79.95.

Practical Electron Microscopy: A Beginner's Illustrated Guide. Elaine Hunter. Cambridge University Press, 40 West 20th Street, New York, NY 10011-4211. (212) 924-3900. (800) 872-7423. 1993. 34.95 in paper.

HANDBOOKS AND MANUALS

CRC Handbook of Chemistry and Physics. David R. Kide, editor. CRC Press, Inc., 2000 Corporate Boulevard, NW, Boca Raton, FL 33431. (407) 994-0555. (800) 272-7737. 77th edition. 1996. $99.95.

Handbook of Optics. Optical Society of America. McGraw-Hill Publishing Company, Inc., 1221 Avenue of the Americas, New York, NY 10020. (212) 512-2000. (800) 262-4729. 2 volumes. 2nd edition. 1993. $99.50 each.

Light Microscopy: Essential Data. C. Rubbi. John Wiley & Sons, Inc., 605 Third Avenue, New York, NY 10158-0012. (212) 850-6000. (800) 225-5945. 1994. $19.95 in paper.

ONLINE DATABASES AND CD-ROMS

CA Search. Chemical Abstracts Service, P.O. Box 3012, Columbus, OH 43210-0012. (614) 447-3600. (800) 848-6533. FAX (614) 447-3709. Very comprehensive guide to worldwide chemical literature and related fields, 1972 to present. Available on BRS,(800) 289-4277, DIALOG, (800) 334-2564, ORBIT (800) 456-7248, and STN International, FIZ Karlsruhe, P.O. Box 2465, W-7500, Karlsruhe 1, Germany, online services. Inquire as to cost and availability.

Compendex Plus. Engineering Information, Inc., 345 East 47th Street, New York, NY 10017. (212) 705-7600 or (800) 221-1044. Contains citations with abstracts to worldwide literature in engineering and technology, from 1970 to present. Available on online BRS,(800) 289-4277, DIALOG, (800) 334-2564, ORBIT (800) 456-7248, and STN International, FIZ Karlsruhe, P.O. Box 2465, W-7500, Karlsruhe 1, Germany, online services. Also available on CD-ROM. Inquire as to cost and availability.

Current Contents Search. Institute for Scientific Information, 3501 Market Street, Philadelphia, PA 19104. (215) 386-0100. FAX (215) 386-6362. Contains citations to articles listed in the table of contents of science and technology journals. Also articles in social sciences and life sciences journals. Available on

BRS, (800) 289-4277, DIALOG, (800) 334-2564, online services. Inquire as to cost and availability.

Dissertation Abstracts Online. University Microfilms International, 300 North Zeeb Road, Ann Arbor, MI 48106. (800) 521-0600 or (313) 761-4700. Scope includes virtually all doctoral dissertations accepted at accredited American institutions from 1861 to present in 252 subject areas. Available on BRS, (800) 289-4277, DIALOG, (800) 334-2564, and OCLC EPIC, (800) 848-5878, online services. Also available on CD-ROM. Inquire as to cost and availability.

GEOREF. American Geological Institute, 4220 King Street, Alexandria, VA 22302. (800) 336-4764 or (703) 379-2480. Geology and geosciences literature, 1785 to present for North America. Available on DIALOG,(800) 334-2564, ORBIT (800) 456-7248, online services. Also available on CD-ROM. Inquire as to cost and availability.

INSPEC. Institution of Electrical Engineers, Michael Faraday House, Six Hills Way, Stevenage, Herts. SG1 2AY, England. Telephone: 0438 313311 or FAX 0438 742840. Contains citations to the worldwide literature of physics, electronics and electrical engineering, computer technology, and related fields. Available on BRS, (800) 289-4277, DIALOG, (800) 334-2564, ORBIT, (800) 456-7248, and STN International, FIZ Karlsruhe, P.O. Box 2465, W-7500, Karlsruhe 1, Germany, online services. Inquire as to cost and availability.

NTIS Bibliographic Database. National Technical Information Service, 5285 Port Royal Road, Springfield, VA 22161. (703) 487-4929 or FAX (703) 321-8199. Broad coverage of government-sponsored science and technology research reports, 1964 to present. Available on BRS,(800) 289-4277, DIALOG, (800) 334-2564, ORBIT, (800) 456-7248, and STN International, FIZ Karlsruhe, P.O. Box 2465, W-7500, Karlsruhe 1, Germany, online services. Also available on CD-ROM. Inquire as to cost and availability.

Physics Briefs. American Institute of Physics, 335 East 45th Street, New York, NY 10017. (212) 661-9260 or FAX (212) 661-2036. Contains citations with abstracts of the literature of physics and related fields, 1979 to present. Available on the STN International, FIZ Karlsruhe, P.O. Box 2465, W-7500, Karlsruhe 1, Germany, online service. Inquire as to cost and availability.

Scisearch. Institute for Scientific Information, 3501 Market Street, Philadelphia, PA 19104. (800) 523-1850 or (215) 386-0100. Broad multidisciplinary title and author index to the international literature of science and technology, 1974 to present. Available on DIALOG, (800) 334-2564, and ORBIT, (800) 456-7248, online services. Also available on CD-ROM. Inquire as to cost and availability.

Wilsonline. H.W. Wilson Company, 950 University Avenue, Bronx, NY 10452. (800) 367-6770 or (212) 588-8400. Makes available online versions of the printed H.W. Wilson indexes, including Applied Science and Technology Index, Business Periodicals Index, General Science Index, and Readers' Guide to Periodical Literature. Period covered is generally 1983 to present. Available on BRS, (800) 289-4277, DIALOG, (800) 334-2564, and OCLC EPIC, (800) 848-5878, online services. Also available on CD-ROM. Inquire as to cost and availability.

PERIODICALS

Applied Optics. Optical Society of America. 2010 Massachusetts Avenue NW, Washington, DC 20036-1023. (202) 223-

8130. 1962 to present. 36 issues per year. $1090.00 per year. ISSN: 0003-6935.

Journal of Microscopy. Blackwell Scientific Publications, Inc., 52 Beacon Street, Boston, MA 02108. (800) 325-4177. 1878 to present. Monthly. $552.00 per year. ISSN: 0022-2720.

Journal of Microscopy Research and Technique. John Wiley & Sons, Inc., Journal, 605 Third Avenue, New York, NY 10158. (212) 850-6000. FAX (212) 850-6088. 1984 to present. Sixteen issues per year. $898.00 per year. ISSN: 1059-910X.

Optical Engineering. SPIE-International Society for Optical Engineering, Box 10, 1000 20th Street, Bellingham, WA 98227-0010. (206) 676-3290. FAX (206) 647-1445. 1962 to present. Monthly. $170.00 per year. ISSN: 0091-3286.

Scanning Electron Microscopy. Scanning Microscopy International, Inc., Box 66507, AMF O'Hare, Chicago, IL 60666. (708) 529-6677. FAX (708) 980-6698. 1968 to present. Quarterly. $164.00 per year. ISSN: 0891-7035.

SPIE-International Society for Optical Engineering. Proceedings. Box 10, 1000 20th Street, Bellingham, WA 98227-0010. (206) 676-3290. FAX (206) 647-1445. 1963 to present. Approximately 50 numbers per year. Individually priced.

RESEARCH CENTERS AND INSTITUTES

Center for Applied Optics. University of Alabama in Huntsville, Research Institute Building, Huntsville, AL 25899. (205) 895-6102.

Center For Applied Optics. University of Texas at Dallas, P.O. Box 830688, Richardson, TX 75803-0688. (214) 690-2868. FAX (214) 690-2848.

Center For Electron Microscopy. University of Illinois, 74 Bevier Hall, 905 S. Goodwin Avenue, Urbana, IL 61801. (217) 333-2108. FAX (217) 333-1656.

Center For Imaging Science. Rochester Institute of Technology, One Lomb Memorial Drive, Rochester, NY 14623. (716) 475-2774.

Institute For Physical Science and Technology. University of Maryland, 4203 Computer and Space Sciences Building, College Park, MD 20742. (301) 454-4874. FAX (301) 314-9363.

Institute of Optics. University of Rochester, Wilmot Building, Rochester, NY 14627. (716) 275-5248.

National Center For Electron Microscopy. Lawrence Berkeley Laboratory, 1 Cyclotron Road, Berkeley, CA 94720. (415) 486-5006. FAX (415) 486-5888.

Optical Sciences Center. University of Arizona, Tucson, AZ 85721. (602) 621-6997.

MICROWAVE SPECTROSCOPY

See: SPECTROSCOPY

MICROWAVES

See also: ELECTRICAL ENGINEERING, ELECTROMAGNETISM

ABSTRACT SERVICES AND INDEXES

Applied Science and Technology Index; A Cumulative Subject Index To English Language Periodicals in the Fields of Aeronautics and Space Science, Computer Technology, Chemistry, Construction Industry, Energy and Related Areas. H.W. Wilson Co., 950 University Avenue, Bronx, NY 10452. (212) 588-8400. (800) 367-6770. FAX (718) 590-1617. From 1958 to present. Monthly. Inquire about cost and availability. Also available on CD-ROM and online. ISSN: 0003-6986.

Current Contents: Engineering, Technology and Applied Sciences. Institute for Scientific Information, 3501 Market Street, Philadelphia, PA 19104. (215) 386-0100. FAX (215) 386-6362. 1970 to present. Weekly. $442.00 per year. Also available on CD-ROM and online. Inquire regarding cost and availability. ISSN: 0095-7917.

Current Papers in Electrical and Electronics Engineering. Institution of Electrical Engineers (IEE), London. Distributed by INSPEC/IEEE, Box 1331, 445 Hoes Lane, Piscataway, NJ 08855-1331. (908) 562-5549. 1969 to present. Monthly. $345.00 per year. ISSN: 0011-3778.

Electrical and Electronics Abstracts. Institution of Electrical Engineers (IEE), London. Available from INSPEC/IEEE-Institute of Electrical and Electronic Engineers, Box 1331, Hoes Lane, Piscataway, NJ 08855-1331. (908) 562-5549. 1898 to present. Monthly. $2200.00 per year. Also available on CD-ROM and online as INSPEC.

Engineering Index Monthly; Indexes and Abstracts the World's Engineering and Technical Literature. Engineering Information, Inc., Castle Point on the Hudson, Hoboken, NJ 07030. (201) 216-8500. (800) 221-1044. FAX (201) 216-8532. Monthly. $2300.00 per year. Available online as COMPENDEX and also on CD-ROM. ISSN: 0742-1974.

General Science Index. H.W. Wilson Company, 950 University Avenue, Bronx, NY 10452. (212) 588-8400. (800) 367-6770. FAX (718) 590-1617. From 1978 to present. Ten issues per year; quarterly and annual cumulations. Service basis. Available on CD-ROM and online. Inquire regarding cost and availability. ISSN: 0162-1963.

Index To IEEE Publications. IEEE Service Center, 445 Hoes Lane, Piscataway, NJ 08855-1331. (908) 981-1393. (800) 678-IEEE. FAX (908) 981-9667. 1973 to present. Annual. ISSN: 0099-1368.

Physics Abstracts. INSPEC. Section A, Science Abstracts. Institution of Electrical Engineers (IEE), London. Available from: INSPEC/IEEE-Institute of Electrical and Electronic Engineers, Box 1331, Hoes Lane, Piscataway, NJ 08855-1331. (908) 562-5549. 1898 to present. 24 issues per year. $2835.00 per year. Also available online and on CD-ROM. ISSN: 0036-8091.

Physics Briefs (Physikalische Berichte). Information Center for Energy, Physics, Mathematics; German Physical Society. V C H Publishers, Inc., 220 East 23rd Street, New York, NY 10010-4606. (212) 683-8333. 1845 to present. 24 issues per year. $2390.00 per year. Also available online. ISSN: 0179-7434.

Science Citation Index. SCI. Institute for Scientific Information, 3501 Market Street, Philadelphia, PA 19104. (215) 386-0100. (800) 523-1850. FAX (215) 386-2991. 1961 to present. Six issues per year, plus annual cumulation. $11650.00 per year. Also available online and on CD-ROM. Inquire about price and availability. ISSN: 0036-827X.

ASSOCIATIONS AND PROFESSIONAL SOCIETIES

American Institute of Physics. One Physics Ellipse, College Park, MD 20740-3843. (301) 209-3100.

Edison Electric Institute. 701 Pennsylvania Avenue NW, Washington, DC 20004-2696. (202) 508-5000, 5454. FAX (202) 508-5360.

Electric Power Research Institute. 3412 Hillview Avenue, Palo Alto, CA 94304. (415) 855-2000. FAX (415) 855-2954.

IEEE (Institute of Electrical and Electronic Engineers). 345 East 47th Street. New York, NY 10017. (212) 705-7900. FAX (212) 705-4929.

IEEE Microwave theory and Techniques. c/o Institute of Electrical and Electronics Engineers, 345 East 47th Street, New York, NY 10017. (212) 705-7900. FAX (212) 705-4929.

International Microwave Power Institute. 10210 Leatherleaf Court, Manassas, VA 22111. (703) 257-1415. FAX (703) 257-0213.

BIBLIOGRAPHIES

Science Books and Films. American Association for the Advancement of Science, 1333 H Street NW, Washington, DC 20005. (202) 326-6454. 1965 to present. Nine issues per year. $40.00 per year. ISSN: 0098-342X.

Scientific and Technical Books and Serials in Print; An Index To Literature in Science and Technology. R.R. Bowker Inc., 121 Chanlon Road, New Providence, NJ 07974. (908) 464-6800. (800) 521-8110. FAX (908) 665-3502. 1972 to present. Annual. 4 volumes. 1994. $299.95. Also available on compact disc and online. ISSN: 0000-054X.

DIRECTORIES AND BIOGRAPHICAL SOURCES

American Men and Women of Science: Physical and Biological Sciences. R. R. Bowker Inc., 121 Chanlon Road, New Providence, NJ 07974. (908) 464-6800. (800) 521-8110. 20th edition. 8 volumes. 1996. $850.00.

Directory of Engineering Societies and Related Organizations. American Association of Engineering Societies, 1111 19th Street, Suite 608, Washington, DC 20036-3603. (202) 296-2237. Semi-annual. $150.00.

EEM-Electronic Engineer's Master. Hearst Business Communications, Inc., 645 Stewart Avenue, Garden City NY 11530. (516) 227-1300. ISSN: 0732-9016.

Engineering Research Centers: Incorporating Electronics Research Centers. Stockton Press, 345 Park Avenue, New York, NY 10010. (212) 689-9200. (800) 221-2123. 4th edition. 1995. $515.00.

IEEE Membership Directory. Institute of Electrical and Electronics Engineers, IEEE Service Center, 445 Hoes Lane, Piscataway, NY 08854. (908) 981-1393. (800) 678-IEEE. FAX (908) 981-9667. 2 volumes. Annual. $190.00. ISSN:

International Directory of Abbreviations and Acronyms of Electronics, Electrical Engineering, Computer Technology and Information Processing. Peter Wennrich. K. G. Saur, 121 Chanlon Road, New Providence, NJ 07974. (908) 464-6800. (800) 521-8110 2 volumes. 1992. $230.00.

Microwaves and RF. Directory Issue. Penton Publishing, 611 Route 46 West, Hasbrouck Heights, NJ 07604. (201) 393-6289. Annual. $50.00.

Research Centers Directory. Gale Research Company Inc., 835 Penobscot Building, Detroit, MI 48226-4094. (313) 961-2242. (800) 347-4253. 20th edition. 1995. $485.00. ISSN: 0080-1518.

Who's Who in Engineering. Gordon Davis, editor. American Association of Engineering Societies. 1111 19th Street, NY, Suite 608, Washington, DC 20036. (202) 296-2237. (800) 658-8897. 9th edition. 1995. $220.00.

Who's Who in Technology. Gale Research Company, 835 Penobscot Building, Detroit, MI 48226-4094. (313 961-2242. (800)521-4253. 7th edition. 1995. $195.00. ISBN 0-8103-7467-6.

ENCYCLOPEDIAS AND DICTIONARIES

Chambers Science and Technology Dictionary. Peter M.B. Walker, editor. Cambridge University Press, 40 West 20th Street, New York, NY 10011-4211. (212) 924-3900. 1988. $39.95.

Encyclopedia of Applied Physics. George Trigg, editor. VCH Publications, Inc., 220 East 23rd Street, Suite 909, New York, NY 10010-4606. (212) 683-8333. (800) 422-8824. 20 volume set. 1991-. $5990.00.

IEEE Standard Dictionary of Electrical and Electronics Terms. Christopher J. Booth, editor. IEEE Service Center, 445 Hoes Lane, Piscataway, NJ 08855-1331. (908) 981-1393. (800) 678-IEEE. FAX (908) 981-9667. IEEE Standard 100-1992. 5th edition. 1993. $150.00.

Illustrated Dictionary of Electronics. Stan Gibilsco. TAB Books, P.O. Box 40, Blue Summit, PA 17294-0850. (717) 794-2191. (800) 233-1128. 7th edition. 1994. $34.95.

McGraw-Hill Electronics Dictionary. J. Markus. McGraw-HillBook Company, Inc., 1221 Avenue of the Americas, New York, NY 10020. (212) 997-3675. 5th edition. 1994. $49.95.

McGraw-Hill Encyclopedia of Science and Technology. McGraw-Hill Book, Incorporated, 1221 Avenue of the Americas, New York, NY 10020. (212) 997-3675. (800) 262-4729. Seventh edition. Twenty volumes. 1992. $1900.00.

GENERAL WORKS

Applications of High-Power Microwaves. Andrei V. Gaponov and Victor L. Granatstein. Artech House, 685 Canton Street, Norwood, MA 02062. (617) 769-9750. (800) 225-9977. 1994. $88.00.

Classical Theory of Electricity and Magnetism. A. K. Roychawdjuri. Oxford University Press, Inc., 200 Madison Avenue, New York, NY 10016. (212) 725-6000. (800) 334-4249. 1991. $35.00.

Computational Methods For Electromagnetics and Microwaves. Richard C. Booton. John Wiley & Sons, Inc., 605 Third Avenue, New York, NY 10158-0012. (212) 850-6000. (800) 225-5945. 1992. $59.95.

Foundations For Microwave Engineering. Robert E. Collin.

McGraw-Hill Publishing Company, Inc., 1221 Avenue of the Americas, New York, NY 10020. (212) 512-2000. (800) 262-4729. 2nd edition. 1992. $55.00.

An Introduction To Applied Electromagnetism. Christos Christopoulos. John Wiley & Sons, Inc., 605 Third Avenue, New York, NY 10158-0012. (212) 850-6000. (800) 225-5945. 1990. $59.95.

Microwaves: An Introduction To Microwave theory and Techniques. A. J. Baden-Fuller. Elsevier Science Publishing Company, Inc., 655 Avenue of the Americas, New York, NY 10010. (212) 989-5800. FAX (914) 333-2444. 1990. $76.00.

Solid State Microwave Power Oscillator Design. Eric L. Holtzman and Ralston S. Robertson. Artech House, 685 Canton Street, Norwood, MA 02062. (617) 769-9750. (800) 225-9977. 1992. $88.00

Understanding Microwaves. Allan W. Scott. John Wiley & Sons, Inc., 605 Third Avenue, New York, NY 10158-0012. (212) 850-6000. (800) 225-5945. 1993. $95.00.

HANDBOOKS AND MANUALS

Active Electronic Component Handbook. Charles A. Harper and Harold C. Jones. McGraw-Hill Book Company, 1221 Avenue of the Americas, New York, NY 10020. (212) 512-2000. (800) 262-4729. 2nd edition. 1996. $79.50.

Electrical Engineering Handbook. Richard C. Dorf, editor. CRC Press, Inc., 2000 Corporate Boulevard, NW, Boca Raton, FL 33431. (407) 994-0555. (800) 272-7737. 1993. $99.95.

Electrical Engineer's Handbook. Donald Christiansen, editor. McGraw Book Company Inc., 1221 Avenue of the Americas, New York, NY 10020. (212) 512-2000. (800) 262- 4729. 4th edition. 1996. $110.00.

Electronics Handbook. Jerry C. Whitaker. CRC Press, Inc., 2000 Corporate Boulevard, NW, Boca Raton, FL 33431. (407) 994-0555. (800) 272-7737. 1996. $120.00.

Engineering Tables and Data. A.M. Howatson, P.G. Lund and J.D. Todd. Van Nostrand Reinhold, 115 Fifth Avenue, New York, NY 10003. (212) 254-3232. (800) 926-2665. Second Edition. 1991. $39.95.

Handbook of Electrical and Electronic Technology. Curtis Johnson. Prentice Hall , 113 Sylvan Avenue, Route 9W, Englewood Cliffs, NJ 07632. (201) 592-2000. (800) 922-0579. 1996. $88.00.

Handbook of Microwave and Optical Components. Kai Chang. John Wiley & Sons, Inc., 605 Third Avenue, New York, NY 10158-0012. (212) 850-6000. (800) 225-5945. 4 volumes.1989. $324.00 set.

Handbook of Microwave Technology. T. Koryu Ishii, editor. Academic Press, Inc., 6277 Sea Harbor Drive, Orlando, FL. (800) 321-5068. 2 volumes. 1995. $300.00 set.

Industrial Electronics and Systems Handbook. Irwin, CRC Press, Inc., 2000 Corporate Boulevard, NW, Boca Raton, FL 33431. (407) 994-0555. (800) 272-7737. 1996. $129.95.

Newnes Electrical Pocket Book. E. A. Reeves, editor. Butterworth-Heinemann, 313 Washington Street, Newton, MA 02158. (618) 928-2500. (800) 366-2665. 21th edition. 1992. $32.95 in paper.

Standard Handbook For Electrical Engineers. Donald Fink. McGraw-Hill Publishing Company, 1221 Avenue of the Americas, New York, NY 10020. (212) 512-2000. (800) 262-4729. 13th edition. 1996. $110.00.

ONLINE DATABASES AND CD-ROMS

CA Search. Chemical Abstracts Service, P.O. Box 3012, Columbus, OH 43210-0012. (614) 447-3600. (800) 848-6533. FAX (614) 447-3709. Very comprehensive guide to worldwide chemical literature and related fields, 1972 to present. Available on BRS,(800) 289-4277, DIALOG, (800) 334-2564, ORBIT (800) 456-7248, and STN International, FIZ Karlsruhe, P.O. Box 2465, W-7500, Karlsruhe 1, Germany, online services. Inquire as to cost and availability.

Compendex Plus. Engineering Information, Inc., 345 East 47th Street, New York, NY 10017. (212) 705-7600 or (800) 221-1044. Contains citations with abstracts to worldwide literature in engineering and technology, from 1970 to present. Available on online BRS,(800) 289-4277, DIALOG, (800) 334-2564, ORBIT (800) 456-7248, and STN International, FIZ Karlsruhe, P.O. Box 2465, W-7500, Karlsruhe 1, Germany, online services. Also available on CD-ROM. Inquire as to cost and availability.

Current Contents Search. Institute for Scientific Information, 3501 Market Street, Philadelphia, PA 19104. (215) 386-0100. FAX (215) 386-6362. Contains citations to articles listed in the table of contents of science and technology journals. Also articles in social sciences and life sciences journals. Available on BRS,(800) 289-4277, DIALOG,(800) 334-2564, online services. Inquire as to cost and availability.

Dissertation Abstracts Online. University Microfilms International, 300 North Zeeb Road, Ann Arbor, MI 48106. (800) 521-0600 or (313) 761-4700. Scope includes virtually all doctoral dissertations accepted at accredited American institutions from 1861 to present in 252 subject areas. Available on BRS, (800) 289-4277, DIALOG, (800) 334-2564, and OCLC EPIC,(800) 848-5878, online services. Also available on CD-ROM. Inquire as to cost and availability.

INSPEC. Institution of Electrical Engineers, Michael Faraday House, Six Hills Way, Stevenage, Herts. SG1 2AY, England. Telephone: 0438 313311 or FAX 0438 742840. Contains citations to the worldwide literature of physics, electronics and electrical engineering, computer technology, and related fields. Available on BRS, (800) 289-4277, DIALOG, (800) 334-2564, ORBIT, (800) 456-7248, and STN International, FIZ Karlsruhe,

MICROWAVES

Ency. of Physical Sciences and Engineering Info. Sources

P.O. Box 2465, W-7500, Karlsruhe 1, Germany, online services. Inquire as to cost and availability.

NTIS Bibliographic Database. National Technical Information Service, 5285 Port Royal Road, Springfield, VA 22161. (703) 487-4929 or FAX (703) 321-8199. Broad coverage of government-sponsored science and technology research reports, 1964 to present. Available on BRS,(800) 289-4277, DIALOG, (800) 334-2564, ORBIT, (800) 456-7248, and STN International, FIZ Karlsruhe, P.O. Box 2465, W-7500, Karlsruhe 1, Germany, online services. Also available on CD-ROM. Inquire as to cost and availability.

Physics Briefs. American Institute of Physics, 335 East 45th Street, New York, NY 10017. (212) 661-9260 or FAX (212) 661-2036. Contains citations with abstracts of the literature of physics and related fields, 1979 to present. Available on the STN International, FIZ Karlsruhe, P.O. Box 2465, W-7500, Karlsruhe 1, Germany, online service. Inquire as to cost and availability.

Scientific and Technical Books and Serials in Print. R.R. Bowker Inc., 121 Chanlon Road, New Providence, NJ 07974. (800) 521-8110 or (908) 464-6800. List of currently published books and serials in the physical and biological sciences, engineering and technology, with subject, author and titles indexes. Available on ORBIT,(800) 456-7248, online service. Inquire as to cost and availability.

Scisearch. Institute for Scientific Information, 3501 Market Street, Philadelphia, PA 19104. (800) 523-1850 or (215) 386-0100. Broad multidisciplinary title and author index to the international literature of science and technology, 1974 to present. Available on DIALOG,(800) 334-2564, and ORBIT,(800) 456-7248, online services. Also available on CD-ROM. Inquire as to cost and availability.

Wilsonline. H.W. Wilson Company, 950 University Avenue, Bronx, NY 10452. (800) 367-6770 or (212) 588-8400. Makes available online versions of the printed H.W. Wilson indexes, including Applied Science and Technology Index, Business Periodicals Index, General Science Index, and Readers' Guide to Periodical Literature. Period covered is generally 1983 to present. Available on BRS,(800) 289-4277, DIALOG,(800) 334-2564, and OCLC EPIC,(800) 848-5878, online services. Also available on CD-ROM. Inquire as to cost and availability.

PERIODICALS

Electric Light and Power. PennWell Publishing Co. 1421 South Sheridan Road, Box 1260, Tulsa OK 74101. PH: (918)835-3161. FAX (918) 831-9497. 1922 to present. Monthly. $42.00 per year.

Electric Machines and Power Systems. Taylor & Francis, 1900 Frost Road, Suite 101, Bristol, PA 19007-1598. (800) 821-8312, FAX (215) 785-5515. 1976 to present. Monthly. $355.00 per year. ISSN: 0731-356X.

Electrical World. McGraw-Hill, Inc., Box 513 Hightstown, NJ 08520. (212) 512-3288. 1874 to present. Monthly. $55.00 per year. ISSN: 0013-4457.

Electronic Design. Penton Publishing, San Jose Gateway, Suite 354. 2025 Gateway Place, San Jose, CA 95110. (408) 441-0550. 1952 to present. Fortnightly. $95.00. ISSN: 0013-4872.

Electronics. Penton Publishing, San Jose Gateway, Suite 354. 2025 Gateway Place, San Jose, CA 95110. (408) 441-0550. 1930 to present. Semi-weekly. $98.00 per year. ISSN: 0883-4989.

IEEE Microwave and Guided Wave Letters. Institute of Electrical and Electronics Engineers, Inc., Box 1331, 445 Hoes Lane, Piscataway, NJ 08855-1331. (908) 981-0060. 1991 to present. Monthly. $150.00 per year. ISSN: 0149-645X.

IEEE Power Engineering Review. Institute of Electrical and Electronics Engineers, Inc., Box 1331, 445 Hoes Lane, Piscataway, NJ 08855-1331. (908) 981-0060. 1981 to present. Monthly. $115.00 per year. ISSN: 0272-1724.

IEEE Spectrum. Institute of Electrical and Electronics Engineers, Inc., Box 1331, 445 Hoes Lane, Piscataway, NJ 08855-1331. (908) 981-0060. 1964 to present. Monthly. $157.00. ISSN: 0018-9235.

IEEE Transactions On Magentics. Institute of Electrical and Electronics Engineers, Inc., Box 1331, 445 Hoes Lane, Piscataway, NJ 08855-1331. (908) 981-0060. 1965 to present. Bi-monthly. $263.00 per year. ISSN: 0018-9464.

IEEE Transactions On Microwave theory and Techniques. Institute of Electrical and Electronics Engineers, Inc., Box 1331, 445 Hoes Lane, Piscataway, NJ 08855-1331. (908) 981-0060. 1953 to present. Monthly. $285.00 per year. ISSN: 0018-9480.

Institute of Electrical and Electronics Engineers. PROCEEDINGS. Institute of Electrical and Electronics Engineers, Inc., Box 1331, 445 Hoes Lane, Piscataway, NJ 08855-1331. (908) 981-0060. 1913 to present. Monthly. $275.00. ISSN: 0018-9219.

IEE Proceedings. Part H: Microwaves, Antennas and Propagation. Institution of Electrical Engineers. Available from: Institute of Electrical and Electronic Engineers (IEEE) Inc., Box 1331, 445 Hoes Lane, Piscataway, NJ 08855-1331. (908) 981-0060. 1980 to present. Bi-monthly. ISSN: 0950-107X.

International Journal of Microwave and Millimeter-Wave Computer Aided Engineering. John Wiley and Sons, Inc., 605 Third Avenue, New York, NY 10158. (800) 526-5368 or (212) 850-6000. 1991 to present. Quarterly. $220.00 per year. ISSN: 1050-1827.

Journal of Microwave Power and Electromagnetic Energy. International Microwave Power Institute, 10210 Leatherleaf Court, Manassas, VA 22111. (703) 257-1415. FAX (703) 257-0213. 1966 to present. Quarterly. $125.00 per year. ISSN: 0832-7823.

Microwave Journal. Horizon House, Inc., 685 Canton Street, Dedham, MA 02062. (617) 769-9750. 1958 to present. Monthly. $67.00 per year. ISSN: 0192-6225.

Microwaves and RF. Penton Publishing Co., Circulation Department, 1100 Superior Avenue, Cleveland, OH 44114. (201) 393-6060. FAX (201) 393-6297. From 1962 to present. Monthly. $60.00. ISSN: 0745-2992.

RESEARCH CENTERS AND INSTITUTES

Edward L. Ginzton Laboratory. Stanford University, Via Palou, Stanford, CA 94305. (415) 723-0200.

Microwave Device and Physical Electronics Laboratory. University of Utah, Electrical Engineering Department, MEB 3280,

Salt Lake City, UT 84112. (801) 581-7634.

Microwave Microelectronics Laboratory. Texas A&M University, Electrical Engineering Department, College Station, TX 77843. (409) 845-5285. FAX (409) 845-5285.

Weber Research Institute. Polytechnic University. Route 110, Farmingdale, NY 11735. (516) 755-4250, FAX (516) 755-4404.

MILKY WAY

See also: ASTRONOMY, BIG BANG, COSMOCHEMISTRY, COSMOLOGY, DARK MATTER, GALAXIES, INTERSTELLAR MEDIUM, NEBULAE, STARS, UNIVERSE

ABSTRACT SERVICES AND INDEXES

Applied Science and Technology Index; A Cumulative Subject Index To English Language Periodicals in the Fields of Aeronautics and Space Science, Computer Technology, Chemistry, Construction Industry, Energy and Related Areas. H.W. Wilson Co., 950 University Avenue, Bronx, NY 10452. (212) 588-8400. (800) 367-6770. FAX (718) 590-1617. From 1958 to present. Monthly. Inquire about cost and availability. Also available on CD-ROM and online. ISSN: 0003-6986.

Astronomy and Astrophysics Abstracts. Springer-Verlag New York, 175 Fifth Avenue, New York, NY 10010. (212) 460-1500. FAX (212) 473-6272. Published for the Astronomisches Rechen-Institut. Comprehensive coverage of all aspects of astronomy, astrophysics and related fields. 1969 to present. Two parts per year. Annual. ISSN: 0067-0022.

Chemical Abstracts. Chemical Abstracts Service, 2540 Olentangy River Road, Box 3012, Columbus, OH 43210-0012. (614) 447-3600. (800) 848-6538. FAX (614) 447-3713. From 1907 to present. Weekly. $16800.00. Available online and on CD-ROM. CA is also available in five section groupings. Inquire regarding cost and availability. ISSN:

Chemical Abstracts—Physical, Inorganic and Analytical Chemistry. Chemical Abstracts Service, 2540 Olentangy River Road, Box 3012, Columbus, OH 43210-0012. (614) 447-3600. (800) 848-6538. FAX (614) 447-3713. Bi-weekly. $1870.00 per year. Also available on CD-ROM and online. Inquire regarding cost and availability.

General Science Index. H.W. Wilson Company, 950 University Avenue, Bronx, NY 10452. (212) 588-8400. (800) 367-6770. FAX (718) 590-1617. From 1978 to present. Ten issues per year; quarterly and annual cumulations. Service basis. Available on CD-ROM and online. Inquire regarding cost and availability. ISSN: 0162-1963.

Government Reports Announcements and Index. U.S. National Technical Information Service (NTIS), 5285 Port Royal Road, Springfield, VA 22161. (703) 487-4650. FAX (703) 321-8547. From 1968 to present. Annual. $630.00 per year. Also available online as NTIS Bibliographic Database and on CD-ROM. ISSN:

International Aerospace Abstracts. American Institute of Aeronautics and Astronautics. Technical Information Service, 555 West 57th Street, New York 10019. (212) 247-6500. FAX (212) 582-4961. 1961 to present. Semi-monthly with annual index.

$1295.00 per year. Also available online and on CD-ROM. ISSN: 0020-5842.

Meteorological and Geoastrophysical Abstracts. American Meteorological Society, c/o Inforonics, Inc., 550 Newtown Road, Littleton, MA 01460. (508) 486-8976. FAX (508) 486-0027. Covers literature in environmental sciences, meteorology, astrophysics, hydrology, glaciology, and physical oceanography. From 1950 to present. Monthly. $950.00 per year. Also available on CD-ROM and online. Inquire regarding cost and availability. ISSN: 0026-1130.

NTIS Alerts: Astronomy & Astrophysics. U. S. National Technical Information Service. 5285 Port Royal Road, Springfield, VA 22161. (703) 487-4650. FAX (703) 321-8547. Weekly. $140.00 per year.

Pascal 48: Environment Cosmique Terrestre, Astronomie Et Geologie Extraterrestre. Centre National de la Recherche Scientifique, Institut de Information Scientifique et Technique, 2 aliee du Parc de Brabois, 54514 Vandoeuvre Les Nancy Cedex, France. TEL 83-50-46-00. FAX: 83-50-46-50. 1985 to present. Ten issues per year. 770 F. Also available on CD-ROM and online.

Physics Abstracts. INSPEC. Section A, Science Abstracts. Institute of Electrical Engineers, London, United Kingdom. Available from: INSPEC/IEEE-Institute of Electrical and Electronic Engineers, Box 1331, Hoes Lane, Piscataway, NJ 08855-1331. (908) 562-5549. 1898 to present. 24 issues per year. $2835.00 per year. ISSN: 0036-8091. Also available online and on CD-ROM.

Physics Briefs (Physikalische Berichte). Information Center for Energy, Physics, Mathematics; German Physical Society. V C H Publishers, Inc., 220 East 23rd Street, New York, NY 10010-4606. (212) 683-8333. 1845 to present. 24 issues per year. $2390.00 per year. Also available online. ISSN: 0179-7434.

Science Citation Index. SCI. Institute for Scientific Information, 3501 Market Street, Philadelphia, PA 19104. (215) 386-0100. (800) 523-1850. FAX (215) 386-2991. 1961 to present. Six issues per year, plus annual cumulation. $11650.00 per year. Also available online and on CD-ROM. Inquire about price and availability. ISSN: 0036-827X.

STAR. Scientific and Technical Aerospace Reports. U.S. National Aeronautics and Space Administration. Distributed by U.S. Superintendent of Documents, Washington, DC 20402. 1963 to present. Semi-monthly, with semiannual and annual indexes. $114.00 per year. ISSN: 0036-8741. Also available online and on CD-ROM.

ANNUAL REVIEWS AND YEARBOOKS

The Astronomical Almanac. Superintendent of Documents, U. S. Government Printing Office, Washington, DC 10402. (202) 783-3238. 1981 to present. Supersedes Astronomical Ephemeris and the American Ephermis and Nautical Almanac. Annual. $44.95. ISSN: 0737-6421.

Annual Review of Astronomy and Astrophysics; Original Reviews of Critical Literature and Current Developments in Astronomy and Astrophysics. Annual Reviews, Inc., 4139 El Camino Way, Palo Alto, CA 94306-0139. (415) 493-4400.(800) 523-8635.FAX (415) 855-9815. 1963 to date. Annual. $60.00. ISSN: 0066-4146.

MILKY WAY

Ency. of Physical Sciences and Engineering Info. Sources

Highlights of Astronomy. I. Appenzeller, editor. Kluwer Academic Publishers, Box 358, Accord Station, Hingham, MA 02018-0358. (617) 871-6600. International Astronomical Union Highlights series. ISBN: 07923-3553-8. Triennal. 1995. Price varies.

ASSOCIATIONS AND PROFESSIONAL SOCIETIES

Amateur Astronomers Association. 1010 Park Avenue, New York, NY 10028. (212) 535-2922.

American Association of Variable Star Observers. 23 Birch Street, Cambridge, MA 02138. (617) 354-0484. FAX: (617) 354-0665.

American Astronomical Society. 2000 Florida Avenue NW, Suite 400, Washington, DC 20009. (202) 328-2010. FAX: (202) 234-2560.

American Institute of Physics. One Physics Ellipse, College Park, MD 20740-3843. (301) 209-3100.

Association of University For Research in Astronomy. Suite 701, 1625 Massachusetts Avenue, NW, Washington, DC 20036. (202) 483-2101. FAX (202) 483-2106.

Astronomical League. 2112 Kingfisher Lane East, Rolling Meadows, IL 60008. (708) 398-0562.

Astronomical Society of the Pacific. 390 Ashton Avenue, San Francisco, CA 94112. (415) 337-1100. FAX: (415) 337-5205.

International Astronomical Union. 98 bis Boulevard Arago, F-75014 Paris, France. Tel 1-43-25-83-58. FAX 1-43-25-26-16.

Joint Institute For Laboratory Astrophysics. University of Colorado, Box 440, Boulder, CO 80309-0440. (303) 492-7780. FAX (303) 492-5235.

National Council For Geocosmic Research. PO Box 1220, Dunkirk, MD 20754. (301) 855-2747. (800) PTOLEMY. FAX (410) 257-2824.

Royal Astronomical Society of Canada. 136 Dupont Street, Toronto. ON N5R 1V2 Canada. (416) 924-7973. FAX (416) 968-6687.

BIBLIOGRAPHIES

Astronomy and Astrophysics: A Source Guide. Gordon Press, P.O. Box 459, Bowling Green Station, New York, NY 10004. (718) 624-8419. 1991. $250.00.

A Bibliography of Astronomy, 1970-1979. R. A. Sea and S. S. Martin. Libraries Unlimited, Inc., Littleton, CO 80160. 1982.

Catalog of the Naval Observatory Library. Naval Observatory Staff. G. K. Hall & Company, 866 Third Avenue, New York, NY 10022. (212)702-6789. (800) 257-5755. 6 volumes. 1977. $655.00

Science Books and Films. American Association for the Advancement of Science, 1333 H Street NW, Washington, DC 20005. (202) 326-6454. 1965 to present. Nine issues per year. $40.00 per year. ISSN: 0098-342X

Scientific and Technical Books and Serials in Print, 1995. R.R. Bowker Inc., 121 Chanlon Road, New Providence, NJ 07974. (908) 464-6800. (800) 521-8110. 4 volumes. 1994. $299.95. Also available on compact disc and online.

A Source Book On Astronomy and Astrophysics, 1900-1975. Kenneth Lang and Owen Gingerich, editors. Harvard University Press, 79 Garden Street, Cambridge MA. (617) 495-2600. 1980. $41.95.

DIRECTORIES AND BIOGRAPHICAL SOURCES

American Men and Women of Science: Physical and Biological Sciences. R. R. Bowker Inc., 121 Chanlon Road, New Providence, NJ 07974. (908) 464-6800. (800) 521-8110. 20th edition. 8 volumes. 1996. $850.00.

The Astronomers. Donald Goldsmith. St. Martin's Press, Inc., 175 Fifth Avenue, New York, NY 10010. (212) 674-5151. (800) 221-7945. 1993. $14.95 in paper.

Astronomical Centers of the World. Kevin Krisciunas. Cambridge University Press, 40 West 20th Street, New York, NY 10011-4211. (212) 924-3900. (800) 872-7423. 1988. $34.95

The Biographical Dictionary of Scientists: Astronomers. D. Abbott, editor. Peter Bedrick Books, Inc., 2112 Broadway, Room 318, New York, NY 10023. (212) 496-0751. 1984.

Directory of Physics and Astronomy Staff Members. American Institute of Physics, One Physics Ellipse, College Park, MD 20740-3843. (301) 209-3100. Annual. $45.00.

Earth and Astronomical Research Centers. Stockton Press, 345 Park Avenue South, New York, NY 10010. 4th edition. 1995. $515.00. ISBN: 1-56169-0967.

Graduate Programs in Physics, Astronomy and Related Fields. 1995-1996. American Institute of Physics, One Physics Ellipse, College Park, MD 20740-3843. (301) 209-3100. 1978 to present. Annual. $45.00.

Research Centers Directory. Gale Research Inc., 835 Penobscot Building, Detroit, MI 48226-4094. (313) 961-2242. (800) 347-4253. 20th edition. 1995. $485.00.

ENCYCLOPEDIAS AND DICTIONARIES

Astronomy From A To Z: A Dictionary of Celestial Objects and Ideas. Charles A. Schweighauser. Illinois Issues, Sangamon State University, K-10, Springfield, IL 62794-9243. (217) 786-9243. 1991. $14.95 in paper.

Concise Dictionary of Astronomy. Jacqueline Mitton. Oxford University Press, Inc., 200 Madison Avenue, New York, NY 10016. (212) 725-6000. (800) 334-4249. 1992. $24.95.

Facts On File Dictionary of Astronomy. Valerie Illingworth. Facts-on-File, Inc., 460 Park Avenue South, New York, NY 10016-7382. (212) 683-2244. (800) 322-8755. Fax (800) 683-3633. 1994. $27.95.

Encyclopedia of Astronomy and Astrophysics. Stephen Maran, editor. Van Nostrand Reinhold, 115 Fifth Avenue, New York, NY 10003. (212) 254-3232. (800) 842-3636. 1992. $129.95.

Oxford Illustrated Encyclopedia. Volume 8, The Universe. Archie Roy, editor. Oxford University Press, Inc., 200 Madison Avenue, New York, NY 10016. (212) 725-6000. (800) 334-4249. 1993. $45.00.

McGraw-Hill Encyclopedia of Science and Technology. McGraw-Hill Book Company, Inc., 1221 Avenue of the Americas, New York, NY 10020. (212) 512-2000. (800) 262-4729. 7th edition. 20 volume set. 1992. $1900.00.

Stars, Galaxies, Cosmos. William R. Corliss. Sourcebook Project, P.O. Box 107, Glen Arm, MD 21057. (410) 668-6047. 1987. $17.95.

GENERAL WORKS

The Center, Bulge and Disk of the Milky Way. Leo Blitz, editor. Kluwer Academic Publishers, 101 Philip Drive, Assinippi Park, Norwell, MA 02061. (617) 871-6600. 1992. $59.50.

Cosmic Questions, Galactic Halos, Cold Dark Matter and the End of Time. Richard Morris. John Wiley & Sons, Inc., 605 Third Avenue, New York, NY 10158-0012. (212) 850-6000.(800) 225-5945. 1993. $24.95.

The Discovery of Our Galaxy. Charles A. Whitney. Iowa State University Press, 2121 South State Avenue, Ames, IA 50014-8300. (515) 292-0140. (800) 862-6657. Reprint edition. 1988. $13.95 in paper.

The Early Universe: Facts and Fiction. Gerhard Borner. Springer-Verlag New York, Inc., 175 Fifth Avenue, New York, NY 10010. (212) 460-1500. (800) 777-4643. 1995. $69.00.

The Environment and Evolution of Galaxies. J. Michael Shull, editor. Kluwer Academic Publishers, 101 Philip Drive, Assinippi Park, Norwell, MA 02061. (617) 871-6600. 1993. $41.00.

Evolution of Interstellar Matter and Dynamics of Galaxies. J. Palous et al, editors. Cambridge University Press, 40 West 20th Street, New York, NY 10011-4211. (212) 924-3900. (800) 872-7423. 1992. $69.95.

Galaxies and the Universe. David J. Eicher, editor. Kalmbach Publishing Company, P.O. Box 1612, Waukesha, WI 53187. (414) 796-0126. (800) 446-5489. Readings from Deep Sky Magazine. 1992. $14.95.

The Galaxy and the Solar System. Roman Smoluchowski, et al, editors. University of Arizona Press, 1230 North Park Avenue, Number 102, Tucson, AZ 85719. (520) 621-1441. (800) 426-3797. 1987. $35.00.

Galaxy Evolution: The Milky Way Perspective. S.R. Majewski, editor. Astronomical Society of the Pacific, 390 Ashton Avenue, San Francisco, CA 94112. (415) 337-1100. 1993. $40.00.

The Guide To the Galaxy. Nigel Henbest and Heather Cooper. Cambridge University Press, 40 West 20th Street, New York, NY 10011-4211. (212) 924-3900. (800) 872-7423. 1994. $24.95 in paper.

The Left Hand of Creation: The Origin and Evolution of the Expanding Universe. John D. Barrow and Joseph Silk. Oxford University Press, Inc., 200 Madison Avenue, New York, NY 10016. (212) 725-6000. (800) 334-4249. 1994. $23.00.

The Milky Way. Bart J. Bok and Priscilla R. Bok. Harvard University Press, 79 Garden Street, Cambridge MA. (617) 495-2600. 1981. $40.00.

The Milky Galaxy and Statistical Cosmology, 1890-1924. Erich R. Paul. Cambridge University Press, 40 West 20th Street, New York, NY 10011-4211. (212) 924-3900. (800) 872-7423. 1993. $44.95.

The Milky Way As A Galaxy. Ivan King, et al. University Science Books, 55D Gate Five Road, Sausalito, CA 94965. (415) 332-5390. 1990. $44.50.

Minnesota Lectures On the Structure and Dynamics of the Milky Way. R. M. Humphreys, editor. Astronomical Society of the Pacific, 390 Ashton Avenue, San Francisco, CA 94112. (415) 337-1100. 1993. $40.00.

The Nearest Active Galaxies. J.E. Beckman, editor. Kluwer Academic Publishers, 101 Philip Drive, Assinippi Park, Norwell, MA 02061. (617) 871-6600. 1993. $87.50.

Star Formation, Galaxies and the Interstellar Medium. J. Franco, et al, editors. Cambridge University Press, 40 West 20th Street, New York, NY 10011-4211. (212) 924-3900.(800) 872-7423. 1993. $59.95.

Stars and Atoms: From the Big Bang To the Solar System. Stuart Clark. Oxford University Press, Inc., 200 Madison Avenue, New York, NY 10016. (212) 725-6000. (800) 334-4249. 1995. $35.00.

Stars and Galaxies: Astronomy's Guide To Exploring the Cosmos. Astronomy Magazine Staff. Kalmbach Publishing Company, P.O. Box 1612, Waukesha, WI 53187. (414) 796-0126. (800) 446-5489. 1992. $29.95 in paper.

Structure and Evolution of Galaxies. Roger J. Tayler. Cambridge University Press, 40 West 20th Street, New York, NY 10011-4211. (212) 924-3900. (800) 872-7423. 1993. $49.95. $22.95 in paper.

Universe of Galaxies. Paul W. Hodge, editor. W. H. Freeman & Co., 41 Madison Avenue, East 26th Avenue, 35th Floor, New York, NY 10010. (212) 576-0400. Readings from Scientific American. 1995.

Visual Astronomy of the Deep Sky. Roger N. Clark. Cambridge University Press, 40 West 20th Street, New York, NY 10011-4211. (212) 924-3900. (800) 872-7423. 1991. $39.95.

The World of Galaxies. H. G. Corwin and L. Bettinelli, editors. Springer-Verlag New York, Inc., 175 Fifth Avenue, New York, NY 10010. (212) 460-1500. (800) 777-4643. 1989. $79.00

HANDBOOKS AND MANUALS

Burnham's Celestial Handbook: An Observer's Guide To the Universe Beyond the Solar System. Robert Burnham, Jr. Dover Publications, Inc., 180 Varick Street, New York, NY 10014. (212) 255-3755. (800) 223-3130. 3 volumes. 1978. $41.85 for the set.

Cambridge Guide To the Constellations. Michael E. Bakich. Cambridge University Press, 40 West 20th Street, New York, NY 10011-4211. (212) 924-3900. (800) 872-7423. 1995. $69.95.

MILKY WAY

Ency. of Physical Sciences and Engineering Info. Sources

Encyclopedia of Astronomy and Astrophysics. Stephen Maran, editor. Van Nostrand Reinhold, 115 Fifth Avenue, New York, NY 10003. (212) 254-3232. (800) 842-3636. 1992. $129.95.

Greenwich Guide To Stars, Galaxies and Nebulae. Stuart Malin. Cambridge University Press, 40 West 20th Street, New York, NY 10011-4211. (212) 924-3900. (800) 872-7423. 1990. $10.95.

Handbook of Astronomy, Astrophysics and Geophysics; Volume 2: Galaxies and Cosmology. V. M. Canuto and B. G. Elmegreen. Gordon & Breach Science Publishers, Inc., P.O. Box 200029, Riverfront Plaza Station, Newark, NJ 07102-0301.(201) 643-7500. 1988. $194.00.

Handbook of Space Astronomy and Astrophysics. Martin V. Zumbeck. Cambridge University Press, 40 West 20th Street, New York, NY 10011-4211. (212) 924-3900. (800) 872-7423. 2nd edition. 1990. $79.95.

Landolt-Borenstein Numerical Data and Functional Relationships in Science and Technology: Astronomy, Astrophysics and Space Research. Group VI. Springer-Verlag New York, Inc., 175 Fifth Avenue, New York, NY 10010. (212) 460-1500. (800) 777-4643. volumes priced individually.

McGraw-Hill Encyclopedia of Astronomy. Sybil P. Parker and Jay M. Pasachoff, editors. McGraw-Hill Publishing Company, Inc., 1221 Avenue of the Americas, New York, NY 10020. (212) 512-2000. (800) 262-4729. 2nd edition. 1993. $75.00

National Audubon Society Pocket Guide: Galaxies and Other Deep-sky Objects. Alfred A. Knopf, Inc. 201 E. 50th Street, New York, NY 10022. (800) 733-3000. 1995. $7.95.

Nearby Galaxies Catalog. R. Brent Tully and J. Richard Fisher. Cambridge University Press, 40 West 20th Street, New York, NY 10011-4211. (212) 924-3900. (800) 872-7423.1988. $59.95.

Observer's Guide To Astronomy. Patrick Martinez, editor. Cambridge University Press, 40 West 20th Street, New York, NY 10011-4211. (212) 924-3900. (800) 872-7423. 2 volumes. 1994. $159.90.

ONLINE DATABASES AND CD-ROMS

CA Search. Chemical Abstracts Service, P.O. Box 3012, Columbus, OH 43210-0012. (614) 447-3600. (800) 848-6533. FAX (614) 447-3709. Very comprehensive guide to worldwide chemical literature and related fields, 1972 to present. Available on BRS,(800) 289-4277, DIALOG, (800) 334-2564, ORBIT (800) 456-7248, and STN International, FIZ Karlsruhe, P.O. Box 2465, W-7500, Karlsruhe 1, Germany, online services. Inquire as to cost and availability.

Current Contents Search. Institute for Scientific Information, 3501 Market Street, Philadelphia, PA 19104. (215) 386-0100. FAX (215) 386-6362. Contains citations to articles listed in the table of contents of science and technology journals. Also articles in social sciences and life sciences journals. Available on BRS, (800) 289-4277, DIALOG, (800) 334-2564, online services. Inquire as to cost and availability.

Dissertation Abstracts Online. University Microfilms International, 300 North Zeeb Road, Ann Arbor, MI 48106. (800) 521-0600 or (313) 761-4700. Scope includes virtually all doctoral dissertations accepted at accredited American institutions from 1861 to present in 252 subject areas. Available on BRS, (800)

289-4277, DIALOG, (800) 334-2564, and OCLC EPIC, (800) 848-5878, online services. Also available on CD-ROM. Inquire as to cost and availability.

INSPEC. Institution of Electrical Engineers, Michael Faraday House, Six Hills Way, Stevenage, Herts. SG1 2AY, England. Telephone: 0438 313311 or FAX 0438 742840. Contains citations to the worldwide literature of physics, electronics and electrical engineering, computer technology, and related fields. Available on BRS, (800) 289-4277, DIALOG, (800) 334-2564, ORBIT, (800) 456-7248, and STN International, FIZ Karlsruhe, P.O. Box 2465, W-7500, Karlsruhe 1, Germany, online services. Inquire as to cost and availability.

International Aerospace Abstracts. American Institute of Aeronautics and Astronautics, 370 L'Enfant Promenade, S.W., Washington, DC 20024. (202) 646-7400. Contains references and abstracts of journal and monograph literature relating to aerospace science and technology, from 1963 to present. Available through the NASA/RECON system of the National Aeronautics and Space Administration only.

NTIS Bibliographic Database. National Technical Information Service, 5285 Port Royal Road, Springfield, VA 22161. (703) 487-4929 or FAX (703) 321-8199. Broad coverage of government-sponsored science and technology research reports, 1964 to present. Available on BRS,(800) 289-4277, DIALOG, (800) 334-2564, ORBIT, (800) 456-7248, and STN International, FIZ Karlsruhe, P.O. Box 2465, W-7500, Karlsruhe 1, Germany, online services. Also available on CD-ROM. Inquire as to cost and availability.

Physics Briefs. American Institute of Physics, 335 East 45th Street, New York, NY 10017. (212) 661-9260 or FAX (212) 661-2036. Contains citations with abstracts of the literature of physics and related fields, 1979 to present. Available on the STN International, FIZ Karlsruhe, P.O. Box 2465, W-7500, Karlsruhe 1, Germany, online service. Inquire as to cost and availability.

Scientific and Technical Books and Serials in Print. R.R. Bowker Inc., 121 Chanlon Road, New Providence, NJ 07974. (800) 521-8110 or (908) 464-6800. List of currently published books and serials in the physical and biological sciences, engineering and technology, with subject, author and titles indexes. Available on ORBIT, (800) 456-7248, online service. Inquire as to cost and availability.

Scisearch. Institute for Scientific Information, 3501 Market Street, Philadelphia, PA 19104. (800) 523-1850 or (215) 386-0100. Broad multidisciplinary title and author index to the international literature of science and technology, 1974 to present. Available on DIALOG, (800) 334-2564, and ORBIT, (800) 456-7248, online services. Also available on CD-ROM. Inquire as to cost and availability.

Wilsonline. H.W. Wilson Company, 950 University Avenue, Bronx, NY 10452. (800) 367-6770 or (212) 588-8400. Makes available online versions of the printed H.W. Wilson indexes, including Applied Science and Technology Index, Business Periodicals Index, General Science Index, and Readers' Guide to Periodical Literature. Period covered is generally 1983 to present. Available on BRS, (800) 289-4277, DIALOG, (800) 334-2564, and OCLC EPIC, (800) 848-5878, online services. Also available on CD-ROM. Inquire as to cost and availability.

OTHER SOURCES

Atlas of Deep-Sky Splendors. Hans Vehrenberg. Sky Publishing Corp., 49 Bay State Road, Cambridge, MA 02138. (617) 864-7360. (800) 253-0245. 4th edition. 1983. $24.95.

Atlas of the Universe. Patrick Moore. Rand McNally & Company. 8255 North Central Park, Skokie, IL 60076-2970. (708) 673-9100. 1994. $29.95.

Cambridge Atlas of Astronomy. Jean Audouze, et al. Cambridge University Press, 40 West 20th Street, New York, NY 10011-4211. (212) 924-3900. (800) 872-7423. 3rd edition. 1994. $75.00.

Cambridge Guide To the Constellations. Michael E. Bakich. Cambridge University Press, 40 West 20th Street, New York, NY 10011-4211. (212) 924-3900. (800) 872-7423. 1995. $19.95 in paper.

Carnegie Atlas of Galaxies. Allan Sandage and John Bedke. Carnegie Institution of Washington, 1530 P Street NW, Washington, DC 20005. (202) 387-6411. Two volumes. 1994.$92.00.

Color Atlas of Galaxies. James Wray. Cambridge University Press, 40 West 20th Street, New York, NY 10011-4211. (212) 924-3900. (800) 872-7423. 1988. $94.95.

Nearby Galaxy Atlas. R. Brent Tully and J. Richard Fisher. Cambridge University Press, 40 West 20th Street, New York, NY 10011-4211. (212) 924-3900. (800) 872-7423. 1987. $74.95.

Observer's Sky Atlas. E. Karkoschka. Springer-Verlag New York, Inc., 175 Fifth Avenue, New York, NY 10010. (212) 460-1500. (800) 777-4643. 1995. $15.95.

Observing Handbook and Catalog of Deep-Sky Objects. Christian B. Luginbuhl and Brian A. Skiff. Cambridge University Press, 40 West 20th Street, New York, NY 10011-4211. (212) 924-3900. (800) 872-7423. 1990. $49.95.

Third Reference Catalogue of Bright Galaxies. Gerard H. DE Vaucouleurs et al, editors. Springer-Verlag New York, Inc., 175 Fifth Avenue, New York, NY 10010. (212) 460-1500. (800) 777-4643. Three volumes. 1991. $198.00 set.

The Sky: A User's Guide. David H. Levy. Cambridge University Press, 40 West 20th Street, New York, NY 10011-4211. (212) 924-3900. (800) 872-7423. 1993. $14.95 in paper.

PERIODICALS

Astronomical Journal. American Institute of Physics, One Physics Ellipse, College Park, MD 20740-3843. (301) 209-3000. Published for the American Astronomical Society. 1849 to present. Monthly. $280.00 per year. ISSN: 0004-6256.

Astronomical Society of the Pacific. Publications. Astronomical Society of the Pacific, 390 Ashton Avenue, San Francisco, CA 94112. (415) 337-1100. FAX (415) 337-5205. 1889 to present. Monthly. $175.00 per year. ISSN: 0004-6280.

Astronomy. Kalmbach Publishing Company, Box 1612, Waukesha, WI 53187-1612. (414) 796-8776. FAX (414) 796-1142. 1973 to present. Monthly. $27.00 per year. ISSN: 0091-6358.

Astronomy and Astrophysics. Springer-Verlag New York, Inc., 44 Hartz Way, Secaucus, NJ 07096-2491. (201) 348-4033. FAX (201) 348-4505. 1969 to date. 36 issues per year in 12 volumes. $1733.00 per year. ISSN: 0004-6361.

Astronomy Reports. American Institute of Physics, One Physics Ellipse, College Park, MD 20740-3843. (301) 209-3000. English translation of Astronomicheksii Zhurnal. Formerly Soviet Astronomy AJ. 1957 to present. Bimonthly. $1160.00 per year. ISSN: 1063-7729.

Astrophysical Journal; An International Review of Astronomy and Astronomical Physics. University of Chicago Press, Journals Division, 5720 South Woodlawn Avenue, Chicago, IL 60637. (312) 753-3347. FAX (312) 753-0811. 1895 to date. Three issues per month. $740.00 per year. ISSN: 0004-637X.

Astrophpysics and Space Science; An International Journal of Cosmic Physics. Kluwer Academic Publishers, Box 358, Accord Station, Hingham, MA 02018-0358. (617) 871-6600. FAX (617) 871-6528. 1968 to present. 24 issues per year. $2544.00 per year. ISSN: 0004-640X.

Celestial Mechanics and Dynamical Astronomy; An International Journal of Space Dynamics. Kluwer Academic Publishers, Box 358, Accord Station, Hingham, MA 02018-0358. (617) 871-6600. FAX (617) 871-6528. 1969 to present. Monthly. $703.50 per year. ISSN: 0923-2958.

Experimental Astronomy; An International Journal On Astronomical Instrumentation and Data Analysis. Kluwer Academic Publishers, Box 358, Accord Station, Hingham, MA 02018-0358. (617) 871-6600. FAX (617) 871-6528. 1989 to present. Quarterly. $161.00 per year. ISSN: 0922-6435.

Icarus; International Journal of Solar System Studies. Academic Press, Inc., Journal Division, 525 B Street, Suite 1900, San Diego, CA 92101-4495. (619) 230-1840. FAX (619) 699-6800. 1962 to present. Monthly. $1080.00 per year. ISSN: 0019-1035.

Landolt-Boernstein, Zahlenwerte Und Funktionen Aus Naturwissenschaften Und Technik. Group 6: Astronomy. Springer-Verlag New York, Inc., 175 Fifth Avenue, New York, NY 10010. (212) 460-1500. FAX (212) 473-6272. 1965. Irregular. Price varies. ISSN: 0075-7896.

Mercury. Astronomical Society of the Pacific. 390 Ashton Avenue, San Francisco, CA 94112. (415) 337-1100. FAX: (415) 337-5205. 1972 to present. Bimonthly. $175.00 per year. ISSN: 0047-6773.

Monthly Notices of the Royal Astronomical Society. Blackwell Scientific Publication LT., Osney Mead, Oxford OX2 OEL, England. TEL 0865-240201. FAX 0865-721205. 1827 to present. Fortnightly. $1899.00 per year. ISSN: 0035-8711.

Observatory. Space and Astrophysics Division, Rutherford Appleton Laboratory, Chilton, Didcot, Oxon OX11 OQX England. FAX 0235-445848. 1877 to present. Bimonthly. $42.00 per year.

Sky and Telescope. Sky Publishing Corporation, Box 9111, Belmont, MA 02178. (617) 864-7360. FAX (617) 864-6117. 1941 to present. Monthly. $27.00 per year. ISSN: 0037-6604.

Space Science Reviews. Kluwer Academic Publishers, Box 358, Accord Station, Hingham, MA 02018-0358. (617) 871-6600.

MILKY WAY

Ency. of Physical Sciences and Engineering Info. Sources

FAX (617) 871-6528. 1962 to present. Sixteen issues per year. $940.00 per year. ISSN: 0038-6308.

Vistas in Astronomy; An International Review Journal. Albert C. Beer, editor. Elsevier Science Publishing Company, Inc., 660 White Plains Road, Terrytown, NY 10591-5153. (914) 524-9200. FAX: (914) 333-2444. 1958 to present. Quarterly. $415.00 per year. ISSN: 0083-6656.

Transactions of the International Astronomical Union. Proceedings of the General Assembly. Kluwer Academic Publishers, Box 358, Accord Station, Hingham, MA 02018-0358. (617) 871-6600. Issues in two parts: Part A, Reports, and Part B, Proceedings. Triennial. XXI-1994. Price varies. ISSN: 0080-1372.

Webb Society Quarterly Journal. Webb Society, 194 Foundry Lane, Freemantle, Southampton, Hampshire SO1 3EQ, England. 1969 to present. Quarterly. $19.50 per year. ISSN: 0043-1680.

RESEARCH CENTERS AND INSTITUTES

Astronomy Program. University of Maryland, 1207 Computer & Space Sciences Building, College Park, MD 20742. (301) 405-3001. FAX (301) 314-9067.

Carnegie Institution of Washington Observatories, 813 Santa Barbara Street, Pasadena, CA 91101. (818) 577-1122. FAX (818) 759-8136.

Center For Astronomy and Space Astrophysics. University of Colorado-Boulder, Campus Box 391, Boulder, CO 80309. (303) 492-4050. (FAX) 492-7178.

Center For Space Research. Massachusetts Institute of Technology, 77 Massachusetts Avenue, Cambridge MA 02139. (617) 253-7501.

David Dunlap Observatory. University of Toronto. Box 360, Richmond Hill ON, Canada L4C 4Y6. (416) 884-9562. FAX (416) 978-3921.

Lick Observatory. University of California, Santa Cruz, CA 95064. (408) 459-2991. FAX (408) 426-3115.

Lowell Observatory. 1400 West Mars Hill Road. Flagstaff, AZ 86001, (602) 774-3358. FAX (602) 774-6296.

National Astronomy and Ionosphere Center. Cornell University, Space Sciences Building, Ithaca, NY 14853. (607) 255-3735. (607) 255-8803.

National Optical Astronomy Observatories. 950 North Cherry Avenue, Tucson, AZ 85719. (602) 325-9282. FAX (602) 325-9360.

National Radio Astronomy Observatory. Edgemont Road, Charlottesville, VA 22903. (804) 296-0211.

Smithsonian Center For Astrophysics. Harvard University, 60 Garden Street, Cambridge, MA 02138. (617) 495-7461. FAX (617) 495-7326.

Space Astronomy Laboratory. 1810 NW 6th Street, Gainsville, FL 32609-3530. (904) 371-4778. FAX (904) 372-5042.

Space Telescope Science Institute. 3700 San Martin Drive, Baltimore, MD 21218. (301) 338-4700.

Theoretical Astrophysics Program. University of Arizona, Planetary Sciences Department, Tucson, AZ 85721. (602) 621-6963.

MINERAL PROSPECTING AND EXPLORATION

See also: GEOLOGY, GEOPHYSICS, GEOCHEMISTRY, MINERALOGY, REMOTE SENSING

ABSTRACT SERVICES AND INDEXES

Bibliography and Index of Geology. American Geological Institute, 4220 King Street, Alexandria, VA 22302. (703) 379-2480. Fax (703) 379-7563. Monthly. $1295.00 per year. Also available online as GEOREF.

Geodex Retrieval System For Geotechnical Abstracts. Geodex Retrieval Systems, 669 Broadway, Box 279, Sonoma, CA 95476. (707) 939-8476. FAX (707) 996-8734. 1970 to present. Three times a year. $645.00 per year includes Geotechnical Abstracts, as Geodex Systems-GRS.

Geological Abstracts. Elsevier-Geo Abstracts, Regency House, 34 Duke Street, Norwich NR3 3AP, England. Monthly. $760.00 per year. Also available online as GEOBASE.

Geological Society of America. Abstracts with Programs. Geological Society of America. 3300 Penrose Place, P.O. Box 9140, Boulder, CO 80301-9140. (303) 447-2020. Abstracts and programs of the annual conference. Annual. $69.00.

Mineralogical Abstracts. Mineralogical Society and the Mineralogical Society of America, 41 Queen's Gate, London, SW7 5HR, England. Quarterly. $235.00 per year.

Petroleum Abstracts. 600 South College, Tulsa, OK 74104. (918) 631-2296 or (800) 247-8678. Contains citations with abstracts to the worldwide literature and patents on the exploration, development, and production of petroleum resources, from 1965 to present.

Remote Sensing of Earth Resources. University of New Mexico Earth Data Analysis Center, 2500 Yale Blvd., Suite 100, Albuquerque, NM 87131-6031. (505) 277-3622. FAX (505) 277-3614. 1974 to present (1991 to present under current title). Quarterly. $175.00 per year.

Science Citation Index. Institute for Scientific Information, 3501 Market Street, Philadelphia, PA 19104. (215) 386-0100. FAX (215) 386-6362. Inquire about availability and cost. Also available on CD-ROM.

ANNUAL REVIEWS AND YEARBOOKS

Annual Review of Earth and Planetary Sciences. Annual Reviews Inc., 4139 El Camino Way, PO Box 10139, Palo Alto, CA 94303-0139. (800) 523-8635. FAX (415) 855-9815. Annual. $62.00 (1995 edition).

U.S. Department of the Interior, Bureau of Mines, Minerals Yearbook. Superintendent of Documents, U.S. Government Printing Office, Box 371954, Pittsburgh, PA 15250-7954. (202)

783-3238. FAX (202) 512-2233. 1932 to present. Annual. Price varies.

ASSOCIATIONS AND PROFESSIONAL SOCIETIES

American Association of Petroleum Geologists. Box 979, Tulsa, OK 74101. (918) 584-2555. FAX (918) 584-0469.

American Geological Institute. 4220 King Street, Alexandria, VA 22302. (703) 379-2480. Fax (703) 379-7563.

American Institute of Professional Geologists. 7828 Vance Drive, Suite 103, Arvada, CO 80003. (303) 431-0831. FAX (303) 431-1332.

Association of Engineering Geologists. 323 Boston Post Road, Suite 2D, Sudbury, MA 01776. (508) 443-4639.

Geological Society of America. 3300 Penrose Place, PO Box 9140, Denver, CO 80301-9140. (303) 447-2020. FAX (303) 447-1133.

Mineralogical Society of America, 1130 17th St. NW, Suite 330, Washington, DC 20036. (202) 775-4344. FAX (202) 775-0018.

The Minerals, Metals & Materials Society, 410 Commonwealth Drive, Warrendale, PA 15086.

Mining and Metallurgical Society of America, 275 Madison Ave., Room 2301, New York, NY 10016. (212) 684-4150.

SEPM (formerly Society For Economic Paleontologists and Mineralogists). 1731 E 71st Street, Tulsa, OK, 74136-5108. FAX (918) 493-2093.)

Society For Mining, Metallurgy and Exploration, Inc., 8307 Shaffer Parkway, PO Box 625002, Littleton, CO 80162-5002. (303) 973-9550. FAX (303) 973-3845.

Society of Economic Geologists. 5808 S. Rapp Street, Suite 209, Littleton, CO 80120-1942. (303) 797-0332. FAX (303) 797-0417.

Society of Exploration Geophysicists. PO Box 702740, Tulsa, OK 74170-2740. (918) 493-3516. FAX (918) 493-2074.

DIRECTORIES AND BIOGRAPHICAL SOURCES

American Men & Women of Science, 1995-96. R.R. Bowker Staff, eds. 19th edition. 8 volumes. R.R. Bowker/Reed International Publishing Company, 121 Chanlon Road, New Providence, NJ 07974. (908) 464-6800 or (800) 521-8110. 1995. $850.00

American Institute of Professional Geologists Membership Directory. 7828 Vance Drive, Suite 103, Arvada, CO 80003. (303) 431-0831. FAX (303) 431-1332. Annual. Inquire for price and availability.

Association of Engineering Geologists Directory. 323 Boston Post Road, Suite 2D, Sudbury, MA 01776. (508) 443-4639. Annual. Inquire for price and availability.

Geological Society of America Membership Directory. 3300 Penrose Place, PO Box 9140, Denver, CO 80301-9140. (303) 447-2020. Annual. Inquire for cost and availability.

International Research Centers Directory. Gale Research, 835 Penobscot Building, Detroit, MI 48226-4094. (313) 961-2242. (800) 347-4253. FAX (313) 961-6083. 8th edition. 1995. $410.00.

Research Centers Directory. Gale Research, 835 Penobscot Building, Detroit, MI 48226-4094. (313) 961-2242. (800) 347-4253. FAX (313) 961-6083. $485.00.

Who's Who in Technology. Gale Research, 835 Penobscot Building, Detroit, MI 48226-4094. (313) 961-2242. (800) 347-4253. FAX (313) 961-6083. 1995. $195.00.

ENCYCLOPEDIAS AND DICTIONARIES

Concise Encyclopedia of Mineral Resources. Donald D. Carr & Norman Herz, eds. MIT Press, 55 Hayward Street, Cambridge, MA 02142. (617) 253-2884 or (800) 356-0343. FAX (617) 253-1709. 1989. $145.00.

The Encyclopedia of the Solid Earth Sciences. Philip Kearey, editor-in-chief. Blackwell Scientific Publications, 238 Main Street, Cambridge, MA 02142. (617) 876-7000 or (800) 759-6102. FAX (617) 876-7022. 1993. $60.00.

Glossary of Geology. Robert Bates, et al. Reprint of third edition. American Geological Institute, 4220 King Street, Alexandria, VA 22302-1507. (703) 379-2480. FAX (301) 379-7563. 1990. $225.00.

GENERAL WORKS

Introduction To Geophysical Exploration. P. Kearey, M. Brooks. 2nd edition. Blackwell Scientific Publications, 238 Main Street, Cambridge, MA 02142. (617) 876-7000 or (800) 759-6102. FAX (617) 876-7022. 1991. $115.00 (hardback) or $49.95 (paperback).

Marine Mineral Exploration. H. Kunzendorf, editor. Elsevier Science Publishing Company, Inc., 655 Avenue of the Americas, New York, NY 10010. (212) 989-5800. 1986. $89.75.

Mineral Exploration Decisions, A Guide To Economic Analysis and Modeling. Deverle P. Harris. John Wiley and Sons, Inc., 605 Third Avenue, New York, NY 10158. (800) 526-5368 or (212) 850-6000. 1990. $99.95.

Prospecting and Exploration of Mineral Deposits. H. Kuzvart and M. Bohmer. Second revised edition. Elsevier Science Publishing Company, Inc., 655 Avenue of the Americas, New York, NY 10010. (212) 989-5800. 1986. $146.25.

Remote Sensing & Geographic Information Systems: Geological Mapping, Mineral Exploration & Mining. Christopher Legg. Routledge, Chapman and Hall, Inc., 29 West 35th Street, New York, NY 10001-2291. (212) 244-3336 or FAX (212) 563-2269. 1993. $59.95.

World Mineral Exploration: Trends & Economic Issues. John E. Tilton, et al., editors. Resources for the Future, Inc., 1616 P Street NW, Room 627, Washington, DC 20036. (202) 328-5086. FAX (202) 949-3460. 1988. $75.00.

MINERAL PROSPECTING AND EXPLORATION

Ency. of Physical Sciences and Engineering Info. Sources

HANDBOOKS AND MANUALS

Mineral Prospecting Manual. J. B. Chaussier and J. Morer. Elsevier Science Publishing Company, Inc., 655 Avenue of the Americas, New York, NY 10010. (212) 989-5800. 1987. $48.75.

ONLINE DATABASES AND CD-ROMS

Current Contents Search. Institute for Scientific Information, 3501 Market Street, Philadelphia, PA 19104. (215) 386-0100. FAX (215) 386-6362. Contains citations to articles listed in the table of contents of science and technology journals. Also articles in social sciences and life sciences journals. Available on BRS,(800) 289-4277, DIALOG,(800) 334-2564, online services. Inquire as to cost and availability.

GEOREF: Bibliography and Index of Geology. American Geological Institute, 4220 King Street, Alexandria, VA 22302. (703) 379-2480. Fax (703) 379-7563. Monthly. Inquire for price and availability.

NTIS Bibliographic Database. National Technical Information Service, 5285 Port Royal Road, Springfield, VA 22161. (703) 487-4929 or FAX (703) 321-8199. Broad coverage of government-sponsored science and technology research reports, 1964 to present. Available on BRS,(800) 289-4277, DIALOG, (800) 334-2564, ORBIT, (800) 456-7248, and STN International, FIZ Karlsruhe, P.O. Box 2465, W-7500, Karlsruhe 1, Germany, online services. Also available on CD-ROM. Inquire as to cost and availability.

Scisearch. Institute for Scientific Information, 3501 Market Street, Philadelphia, PA 19104. (800) 523-1850 or (215) 386-0100. Broad multidisciplinary title and author index to the international literature of science and technology, 1974 to present. Available on DIALOG,(800) 334-2564, and ORBIT,(800) 456-7248, online services. Also available on CD-ROM. Inquire as to cost and availability.

PERIODICALS

American Mineralogist. Mineralogical Society of America, 1130 17th St. NW, Suite 330, Washington, DC 20036. (202) 775-4344. FAX (202) 775-0018. Bi-monthly. $225.00 per year.

Colorado School of Mines Quarterly. Colorado School of Mines Press, 1500 Illinois, Golden, CO 80401-1887. (303) 273-3000. FAX (303) 273-3310. Quarterly. $50.00 per year.

Engineering & Mining Journal. Maclean Hunter Publishing Company, 29 N. Wacker Dr., Chicago, IL 60606. (312) 726-2802. FAX (312) 726-2574. Monthly. $60.00 per year.

Exploration and Mining Geology. Elsevier Science [journals], 660 White Plains Rd., Tarrytown, NY 10159-5153. (919) 524-9200. FAX (919) 333-2444. 1992 to present. Four times a year. $170.00 per year.

Exploration Geophysics. Australian Society of Exploration Geophysicists, PO Box 354, Hawthorn, VIC 3122. Telephone 61-3-818-1272. FAX 61-3-818-1286. 1970 to present. Quarterly. Inquire for price.

Geophysical Prospecting. Blackwell Scientific Publishing, Inc., 3 Cambridge Court, Suite 208, Cambridge, MA 02142. (617) 225-0401. 1953 to present. Eight times a year. $369.00.

Journal of Geochemical Exploration. Elsevier Science Inc., Box 882, Madison Square Station, New York, NY 10159. (212) 989-5800. FAX (212) 633-3990. 1972 to present. Nine times a year. $642.00.

RESEARCH CENTERS AND INSTITUTES

Earth Science Laboratory. University of Utah Research Institute, 391 Chipeta Way, Suite C, Salt Lake City, UT 84108. (801) 524-3422. FAX (801) 524-3453.

Mineral Exploration Research Institute. C.P. 6079, Station A., Montreal, PQ, Canada H3C 3A7. (514) 340-4991. FAX (514) 340-4191.

New Mexico Bureau of Mines and Mineral Resources. Institute of Mining & Technology, Campus Station, Socorro, NM 87801. (505) 835-5420. FAX (505) 835-6333.

Oklahoma Geological Survey. Energy Center, 100 E. Boyd, Room N-131, Norman, OK 73019-0628. (405) 325-3031. FAX (405) 325-3180.

Southern Illinois University At Carbondale, Illinois Mining and Mineral Resources Research Institute. Coal Extraction and Utilization Research Center, Carbondale, IL 62901. (618) 536-6637. FAX (618) 453-7455.

Texas A&M University Center For Engineering Geosciences. College Station, TX 77843. (409) 845-3224. FAX (409) 845-6162.

University of Kentucky Institute For Mining and Minerals Research. 233 Mining and Mineral Resources Bldg., Lexington, KY 40506. (606) 257-8636. FAX (606) 258-1049.

University of Utah, Utah Mining and Mineral Resources Research Institute. 113 Energy and Mineral Research office Bldg., Salt Lake City, UT 84112. (801) 581-8006. FAX (801) 581-8119.

MINERALOGY

See also: CRUST, GEOCHEMISTRY, GEOCHRONOLOGY, GEOPHYSICS, IGNEOUS ROCKS, METAMORPHIC ROCKS, MINING ENGINEERING, PETROLOGY, PHYSICAL GEOLOGY, SEDIMENTARY ROCKS, SEISMOLOGY, VOLCANOLOGY

ABSTRACT SERVICES AND INDEXES

Bibliography and Index of Geology. American Geological Institute, 4220 King Street, Alexandria, VA 22302. (703) 379-2480. Fax (703) 379-7563. Monthly. $1295.00 per year. Also available online as GEOREF.

Bibliography of Economic Geology. Geosystems, Box 40, Didcot Oxon Ox11 9BX, England. Bimonthly. $150.00 per year. Also available online as GeoArchive.

Chemical Abstracts-Applied Chemistry and Chemical Engineering Sections. Chemical Abstracts Service, 2540 Olentangy River Road, Box 3012, Columbus, OH 43210. (800) 848-6538 or (614) 447-3600, FAX (614) 447-3713. Bi-weekly. $1410.00 per year. Also available online as CA (Chemical Abstracts).

Deep-Sea Research. Part A: Oceanographic Research Papers. Deep-Sea Research. Part B: Oceanographic Literature Review. Pergamon Press, Inc., Maxwell House, Fairview Park, Elmsford, NY 10523. (914) 592-7700. Fax (914) 592-3625. Twelve times per year. $1370 per year for Parts A and B.

Geological Abstracts. Elsevier-Geo Abstracts, Regency House, 34 Duke Street, Norwich NR3 3AP, England. Monthly. $760.00 per year. Also available online as GEOBASE.

Geological Society of America. Abstracts with Programs. Geological Society of America. 3300 Penrose Place, P.O. Box 9140, Boulder, CO 80301-9140. (303) 447-2020. Abstracts and programs of the annual conference. Annual. $69.00.

Mineralogical Abstracts. Mineralogical Society and the Mineralogical Society of America, 41 Queen's Gate, London, SW7 5HR, England. Quarterly. $235.00 per year.

Oceanic Abstracts. Cambridge Scientific Abstracts, 7200 Wisconsin Avenue, Bethesda, MD 20814. (301) 961-6750. Fax (301) 961-6720. Bimonthly. $995.00 per year.

Science Citation Index. Institute for Scientific Information, 3501 Market Street, Philadelphia, PA 19104. (800) 523-1850 or (215) 386-0100. Inquire about price and availability.

General Science Index. H.W. Wilson Co., 950 University Avenue, Bronx, NY 10452. (800) 367-6770 or (212) 588-8400. Inquire about price and availability.

ANNUAL REVIEWS AND YEARBOOKS

Annual Review and Earth and Planetary Sciences. Annual Reviews, Inc., 4139 El Camino Way, Palo Alto, CA 94306-0897. (415) 493-4400. Fax (415) 855-9815. Annual. $55.00 per year.

Minerals Yearbook. Bureau of Mines, U.S. Department of the Interior. Available from U.S. Government Printing Office, Washington, DC 20402. (202) 783-3238. Annual. Three volumes. $45.00.

ASSOCIATIONS AND PROFESSIONAL SOCIETIES

American Association of Petroleum Geologists. P.O. Box 979, Tulsa, OK 74101. (918) 584-2555.

American Geological Institute. 4220 King Street, Alexandria, VA 22302. (703) 379-2480. Fax (703) 379-7563.

Geochemical Society. Department of Terrestrial Magnetism, 5241 Broad Branch Road, N.W., Washington, DC 20015. (202) 686-4370.

Geological Society of America. 3300 Penrose Place, P.O. Box 9140, Boulder, CO 80301-9140. (303) 447-2020.

Mineralogy Society of America. 1130 17th Street, Suite 330, N.W., Washington, DC 20036-4604. (202) 735-4344.

Society of Economic Geologists. P.O. Box 571, Golden, CO 80402. (303) 277-1716.

Society of Economic Paleontologists and Mineralogists. P.O. Box 4756, Tulsa, OK 74159-0756. (918) 743-9765.

Society of Independent Professional Earth Scientists. 4925 Greenville Avenue, Suite 170, Dallas, TX 75206. (214) 363-1780.

BIBLIOGRAPHIES

Geologic Reference Sources: A Subject and Regional Bibliography of Publications and Maps in the Geological Sciences. Dederick C. Ward et al. Scarecrow Press, Inc., 52 Liberty Street, Box 4167, Metuchen, NJ 08840. (800) 537-7107 or (201) 548-8600. 1981. $49.50.

Information Sources in the Earth Sciences. David N. Wood, Joan E. Hardy, and Anthony P. Harvey, editors. Bowker-Saur/K.G. Saur. Distributed by R.R. Bowker, 121 Chanlon Road, New Providence, NJ 07974. (800) 521-8110 or (908) 464-6800. Second edition. 1989. $85.00.

World Databases in Geography and Geology. C.J. Armstrong, editor. Bowker-Saur, 121 Chanlon Road, New Providence, NJ 07974. (800) 521-8110. FAX (908) 665-6707. 1995. $200.00.

DIRECTORIES AND BIOGRAPHICAL SOURCES

American Institute of Professional Geologists. Membership Directory. American Institute of Professional Geologists, 7828 Vance Drive, Suite 103, Arvada, CO 80003. (303) 431-0831. Annual. $18.00.

Directory of Geoscience Departments: United States and Canada. American Geological Institute. 4220 King Street, Alexandria, VA 22302-1507. (703) 379-2480. Fax (703) 379-7563. Twenty-six edition. 1987. $22.00.

Geological Society of America. Membership Directory. Geological Society of America, 3300 Penrose Place, Boulder, CO 80301. (303) 447-2020. Annual. Available to members only.

Research Centers Directory. Gale Research, 835 Penobscot Building, Detroit, MI 48226-4094. (800) 347-4253 or (313) 961-2242. 17th edition, 1992. $400.00.

ENCYCLOPEDIAS AND DICTIONARIES

Concise Encyclopedia of Mineral Resources. Donald D. Carr and Norman Herz, editors. MIT Press, 55 Hayward Street, Cambridge, MA 02142. (617) 253-2884 or (800) 356-0343. 1989. $145.00.

Dictionary of Gemmology. P.G. Read. Butterworths Publishers, 80 Montvale Avenue, Stoneham, MA 02180. (617) 438-8464 or (800) 366-2665. 1988. $49.95.

Elsevier's Dictionary of Mining and Mineralogy. A.F. Dorian, compiler. Elsevier Science Publishing Company, Inc., 655 Avenue of the Americas, New York, NY 10010. (212) 989-5800. 1993. $187.50.

Encyclopedia of Minerals. Willard L. Roberts, Thomas J. Campbell, and George R. Rapp, editors. Van Nostrand Reinhold, 115 Fifth Avenue, New York, NY 10003. (212) 254-3232 or (800) 926-2665. Second edition. 1990. $100.00.

Facts On File Dictionary of Geology and Geophysics. Dorothy Farris and Donald R. Coates, editors. Facts on File, Inc., 460

MINERALOGY

Ency. of Physical Sciences and Engineering Info. Sources

Park Avenue South, New York, NY 10016-7382. (800) 322-8755. Fax (800) 683-3633. 1989. $24.95.

Glossary of Geology. Robert L. Bates and Julia A. Jackson. American Geological Institute, 4220 King Street, Alexandria, VA 22032. (703) 379-2480. Third edition. 1987. $75.00.

GENERAL WORKS

Earth Materials and Earth Processes: An Introduction. Lynn Fichter, et al. Macmillan Publishing, 866 3rd Avenue, New York, NY 10024. (212) 689-9140. Third edition. 1991. $19.50.

Gems, Granites, and Gravels: Knowing and Using Rocks and Minerals. R.V. Dietrich and Brian J. Skinner. Cambridge University Press, 40 West 20th Street, New York, NY 10011-4211. (212) 924-3900. 1990. $24.95.

Introduction To Mineral Sciences. Andrew Putnis. Cambridge University Press, 40 West 20th Street, New York, NY 10011-4211. (212) 924-3900 or (800) 872-7423. FAX (914) 937-4712. 1993. $39.95.

Introduction To Minerals and Rocks. Joseph C. Cepeda. Macmillan Publishing, 200 Old Tappan Road, Old Tappan, NJ 07675. (800) 223-2336. FAX (800) 445-6991. 1993. $32.00.

Mineralogy: Concepts and Principles. Tibor Zoltai & James H. Stout. 2nd edition. Macmillan Publishing, 200 Old Tappan Road, Old Tappan, NJ 07675. (800) 223-2336. FAX (800) 445-6991. 1996. Inquire for price.

Rocks, Minerals and Fossils of the World. Chris Pellant. Little, Brown and Company, 34 Beacon Street, Boston, MA 02108. (617) 227-0730 or (800) 343-9204. 1990. $17.95.

HANDBOOKS AND MANUALS

Agi Data Sheets For Geology in the Field, Laboratory, and office. J.T. Dutro, R.V. Dietrich, and R.M. Foose, editors. American Geological Institute, 4220 King Street, Alexandria, VA 223020-1507 (800) 336-4764 or (703) 379-2480. 1989. $24.95.

CRC Practical Handbook of Physical Properties of Rocks and Minerals. Robert S. Carmichael. CRC Publishers, Inc., 2000 Corporate Blvd., N.W., Boca Raton, FL 33431. (407) 994-0555 or (800) 333-8300. 1988. $54.95.

Gemology. Cornelius S. Hurlbut and Robert C. Kammerling. Wiley-Interscience, a division of John Wiley and Sons, Inc., 605 Third Avenue, New York, NY 10158. (800) 526-5368 or (212) 850-6000. Second edition. 1991. $58.00.

Handbook of Mineralogy. John W. Anthony, editor. Mineral Data Publishing, 710 W. Bangalor Drive, Oro Valley, AZ 85737. (520) 297-4862. FAX (520) 297-6330. Inquire for price and availability.

Manual of Mineralogy. Cornelis Klein & Cornelis Hurlbut. 21st revised edition. John Wiley and Sons, Inc., 605 Third Avenue, New York, NY 10158. (800) 526-5368 or (212) 850-6000. 1994. $69.50.

The Mineral Index. A. Clark. Chapman and Hall, Inc., 29 West 35th Street, New York, NY 10001-2291. (212) 244-3336. Third edition. 1992. $95.00.

ONLINE DATABASES AND CD-ROMS

Earth Sciences. U.S. Geological Survey, 12201 Sunrise Valley Drive, Reston, VA 22092-9998. (703) 648-4460. CD-ROM of earth science databases, including the U.S. Geological Survey Library, Earth Science Data Directory, and GEOINDEX, citations to published geological maps. $350.00 per year with quarterly updates.

Geoarchive. Geosystems, P.O. Box 1024, Westminster, London, England, SW1 P 2JL. Citations to literature on geoscience, 1969 to present. Inquire as to online cost and availability.

Geobase. Elsevier-Geo Abstracts, Regency House, 34 Duke Street, Norwich NR3 3AP, England. Contains citations to the worldwide earth science literature from 1980 to date. Available on DIALOG, ORBIT online services. Inquire as to cost and availability.

GEOREF. American Geological Institute, 4220 King Street, Alexandria, VA 22302. (800) 336-4764 or (703) 379-2480. Geology and geosciences literature, 1785 to present for North America. Available on DIALOG, ORBIT online services. Also available on CD-ROM. Inquire as to cost and availability.

NTIS Bibliographic Database. National Technical Information Service, 5285 Port Royal Road, Springfield, VA 22161. (703) 487-4630. Broad coverage of government-sponsored research reports, 1964 to present. Available on BRS, DIALOG, ORBIT and STN online services. Also available on CD-ROM. Inquire as to cost and availability.

Scisearch. Institute for Scientific Information, 3501 Market Street, Philadelphia, PA 19104. (800) 523-1850 or (215) 386-0100. Broad multidisciplinary title and author index to the international literature of science and technology, 1974 to present. Available on DIALOG and ORBIT online services. Also available on CD-ROM. Inquire as to cost and availability.

PERIODICALS

American Mineralogist. Mineralogical Society of America, 1130 17th Street, Suite 330, N.W., Washington, DC 20036-4604. (202) 735-4344. Bimonthly. $150.00 per year.

Canadian Journal of Earth Sciences. National Research Council of Canada, Research Journals, Ottawa, Ont. K1A OR6, Canada. (613) 993-9084. Monthly. $112.00 per year.

Earth and Planetary Science Letters. Elsevier Science Publishing Company, Inc., 655 Avenue of the Americas, New York, NY 10010. (212) 989-5800. Twenty times per year. $945.00 per year.

Economic Geology and the Bulletin of the Society of Economic Geologists. Society of Economic Geologists, Economic Geology Publishing Company, 101 Vowell Hall, University of Texas at El Paso, El Paso, TX 79968. 8 times per year. $51.00.

Geochimica Et Cosmochimica Acta. Pergamon Journals, Inc., Maxwell House, Fairview Park, Elmsford, NY 10523. (914) 592-7700. Fax (914) 592-3625. Monthly. $580.00 per year.

Geological Magazine. Cambridge University Press, 40 West 20th Street, New York, NY 10011-4211. (212) 924-3900. Bimonthly. $205.00 per year.

Geological Society. Journal. Geological Society of London. Geological Society Publishing House, Unit 7, Brassmill Enterprise Centre, Brassmill Lane, Bath, Avon BA1 3JN, England. 6 times per year. $432.00.

Geological Society of America Bulletin. P.O. Box 9140, 3300 Penrose Place, Boulder, CO 80301. (303) 447-2020. Fax (303) 447-1133. Monthly. $150.00.

Geology. Geological Society of America. P.O. Box 9140, 3300 Penrose Place, Boulder, CO 80301. (303) 447-2020. Fax (303) 447-1133. Monthly. $120.00 per year.

Journal of Volcanology and Geothermal Research. Elsevier Science Publishing Company, Inc., 655 Avenue of the Americas, New York, NY 10010. (212) 989-5800. Monthly. $500.00 per year.

Lithos; An International Journal of Mineralogy, Petrology, and Geochemistry. Elsevier Science Publishing Company, Inc., 655 Avenue of the Americas, New York, NY 10010. (212) 989-5800. Quarterly. $275.00 per year.

Mineralogy and Petrology. Springer-Verlag, Moelkerbastei 5, Postfach 367, A-1011 Vienna, Austria. Irregular. $360.00 per year.

Mineralogy Magazine. Mineralogical Society, 41 Queen's Gate, London, SW7 5HR, England. Quarterly. $205.00 per year.

Physics of the Earth and Planetary Interiors. Elsevier Science Publishing Company, Inc., 655 Avenue of the Americas, New York, NY 10010. (212) 989-5800. Twenty-four times per year. $900.00 per year.

RESEARCH CENTERS AND INSTITUTES

Carnegie Institution of Washington. Geophysical Laboratory. 5251 Broad Branch Road, NW, Washington, DC 20015. (202) 686-2410 or FAX (202) 686-2419.

National Science Foundation. Directorate for Geosciences. Division of Earth Sciences. Experimental and theoretical Geochemistry Program. 1800 G Street, NW, Washington, DC 20550. (202) 357-7498.

U.S. Geological Survey, Geologic Division, National Center, 12201 Sunrise Valley Drive, Reston, VA 22092. (703) 648-6600. The major geological research agency of the federal government conducting research in most areas of pure and applied research in the geosciences.

MINING ENGINEERING

See also: COAL MINING, GEOTECHNICAL ENGINEERING, GOLD and GOLD MINING, PETROLEUM ENGINEERING, QUARRIES and QUARRYING, ROCK MECHANICS

ABSTRACT SERVICES AND INDEXES

Applied Mechanics Reviews: An Assessment of World Literature in Engineering Sciences. American Society of Mechanical Engineers, 345 East 47th Street, New York, NY 10017. (212) 705-7703. 1948 to present. Monthly. $360.00 per year.

Applied Science and Technology Index; A Cumulative Subject Index To English Language Periodicals in the Fields of Aeronautics and Space Science, Computer Technology, Chemistry, Construction Industry, Energy and Related Areas. H.W. Wilson Co., 950 University Avenue, Bronx, NY 10452. (800) 367-6770 or (212) 588-8400. FAX (718) 590-1617. From 1958 to present. Monthly. Inquire about cost and availability. Also available on CD-ROM.

Bibliography and Index of Geology. American Geological Institute, 4220 King Street, Alexandria, VA 22302. (703) 379-2480. Fax (703) 379-7563. Monthly. $1295.00 per year. Also available online as GEOREF.

Civil and Structural Engineering Abstracts, Cambridge Scientific Abstracts, 7200 Wisconsin Avenue, Bethesda, MD 20814-4823. (301) 961-6750. FAX (301) 961-6720. Monthly. Topics covered include structural design, construction equipment and methods, civil defense and military engineering, surveying, highway engineering, maritime and port structures, materials, land reclamation, and soil mechanics.

Engineering Index Monthly. Engineering Information, Inc., Castle Point on the Hudson, Hoboken, NJ 07030. (800) 221-1044. FAX (212) 832-1857. Monthly. $2200.00 per year. Also available online as Compendex, and also on CD-ROM.

Geological Society of America. Abstracts with Programs. Geological Society of America. 3300 Penrose Place, P.O. Box 9140, Boulder, CO 80301-9140. (303) 447-2020. Abstracts and programs of the annual conference. Annual. $69.00.

Geotechnical Abstracts. Deutsche Gesellschaft fure Erd- und Grundbau EV, Hohenzollernstrasse 52, 4300 Essen 1, Germany. Also sponsored by International Society for Soil Mechanics and Foundation Engineering. Covers soil mechanics, foundation engineering, and engineering geology. 1970 to present. Monthly. $200.00 per year.

International Journal of Rock Mechanics and Mining Sciences and Geomechanics Abstracts. Elsevier Science, 660 White Plains Rd., Tarrytown, NY 10591-5153. (914) 524-9200. FAX (914) 333-2444. Six times a year. $845 per year.

I.M.M. Abstracts and Index. Institution of Mining and Metallurgy, 44 Portland Pl., London W1N 4BR, England. 071-580-3802. FAX 071-436-5388. Bi-monthly. $364 for non-members.

Mechanical Engineering Abstracts (formerly ISMEC), Cambridge Scientific Abstracts, 7200 Wisconsin Avenue, Bethesda, MD 20814-4823. (301) 961-6750. FAX (301) 961-6720. Monthly. Summarizes world literature in mechanical engineering, production engineering, and engineering management.

Mineralogical Abstracts. Mineralogical Society and the Mineralogical Society of America, 41 Queen's Gate, London, SW7 5HR, England. Quarterly. $235.00 per year.

ASSOCIATIONS AND PROFESSIONAL SOCIETIES

American Institute of Mining, Metallurgical & Petroleum Engineers, 345 East 47th Street, New York, NY 10017.

American Mining Congress, 1920 N St., NW, Suite 300, Washington, DC 20036-1662. (202) 861-2800. FAX (202) 861-7535.

American Society of Civil Engineers, 345 East 47th Street, New York, NY 10017.

Mineralogical Society of America, 1130 17th St. NW, Suite 330, Washington, DC 20036. (202) 775-4344. FAX (202) 775-0018.

The Minerals, Metals & Materials Society, 410 Commonwealth Drive, Warrendale, PA 15086.

Mining and Metallurgical Society of America, 275 Madison Ave., Room 2301, New York, NY 10016. (212) 684-4150.

Society For Mining, Metallurgy and Exploration, Inc., 8307 Shaffer Parkway, PO Box 625002, Littleton, CO 80162-5002. (303) 973-9550. FAX (303) 973-3845.

Society of Mining Engineers of AIME, Caller Number D, Littleton, CO 80127. (303) 973-9550.

DIRECTORIES AND BIOGRAPHICAL SOURCES

American Mines Handbook. Southam Magazine Group, 1450 Don Mills Rd., Don Mills, ON M3B 2X7, Canada. (416) 442-2004. FAX (416) 442-2261. Annual. Inquire for price information.

Directory of Engineering Societies and Related Organizations. Gordon Davis. 13th edition. American Association of Engineering Societies, 1111 19th Street NW, Suite 608, Washington, DC 20036. (202) 296-2237 or (800) 658-8897. 1989. Inquire for price.

E & MJ International Directory of Mining. Maclean Hunter Publishing Company, Mining Information Services, 29 N. Wacker Dr., Chicago, IL 60606. (312) 726-2802. FAX (312) 726-2574. Annual. $120.00.

Financial Times International Year Books: Mining. St. James Press, 425 N. Michigan Ave., Chicago, IL 60611. Annual. Inquire for price.

Industrial Minerals Directory—World Guide To Producers and Processors. Metal Bulletin, Inc., 220 Fifth Ave., New York, NY 10001-7781. (212) 213-6202. FAX (212) 213-1870. Irregular. Inquire for price and availability.

International Research Centers Directory. Gale Research, 835 Penobscot Building, Detroit, MI 48226-4094. (313) 961-2242. (800) 347-4253. FAX (313) 961-6083. 8th edition. 1995. $410.00.

Mining Directory: Mines & Mining Equipment and Service Companies Worldwide (year). Don Nelson Publications Ltd., PPO Box 193, Barnet, Herts EN4 8LP, England. Telephone 081-368-5534. FAX 081-368-7010. Annual. Inquire for price.

Research Centers Directory. Gale Research, 835 Penobscot Building, Detroit, MI 48226-4094. (313) 961-2242. (800) 347-4253. FAX (313) 961-6083. $485.00.

Who's Who in Engineering. American Association of Engineering Societies, 1111 19th Street NW, Suite 608, Washington, DC 20036. (202) 296-2237 or (800) 658-8897. 8th edition. 1991. Inquire for price.

ENCYCLOPEDIAS AND DICTIONARIES

Concise Encyclopedia of Mineral Resources. Donald D. Carr & Norman Herz, eds. MIT Press, 55 Hayward Street, Cambridge, MA 02142. (617) 253-2884 or (800) 356-0343. FAX (617) 253-1709. 1989. $145.00.

The Contractors' Dictionary of Equipment, Tools, and Techniques: For Civil Engineering, Construction, Forestry, Open-Pit Mining, and Public Works. L.F. Webster. John Wiley and Sons, Inc., 605 Third Avenue, New York, NY 10158. (800) 526-5368 or (212) 850-6000. 1995. Inquire for price.

Dictionary of Mining & Processing. Helmut Schmidt. Collets/State Mutual Book & Periodical Service Ltd., 521 Fifth Ave., 17th floor, New York, NY 10175. (212) 682-5844. FAX (516) 537-0412. 1980. $99.00.

The Encyclopedia of the Solid Earth Sciences. Philip Kearey, editor-in-chief. Blackwell Scientific Publications, 238 Main Street, Cambridge, MA 02142. (617) 876-7000 or (800) 759-6102. FAX (617) 876-7022. 1993. $60.00.

McGraw-Hill Encyclopedia of Engineering. Sybil P. Parker, ed. 2nd edition. McGraw-Hill Publishing Company, 1221 Avenue of the Americas, New York, NY 10020. (800) 262-4729 or (212) 512-3825. 1993. $95.50.

McGraw-Hill Encyclopedia of Science and Technology. Sybil P. Parker, ed. 7th edition. 20 volumes. McGraw-Hill Publishing Company, 1221 Avenue of the Americas, New York, NY 10020. (800) 262-4729 or (212) 512-3825. 1992. $1900.00

Thesaurus of Scientific, Technical, and Engineering Terms. Hemisphere Publishing Corporation, 1900 Frost Road, Suite 101, Bristol, PA 19007-1598. (215) 785-5800 or (800) 821-8312. FAX (215) 785-5515. 1987. $173.00.

GENERAL WORKS

Automation in Mining, Mineral and Metal Processing 1992. Gu Yan and Chen Zhen-Yu. Pergamon Press Inc., Maxwell House, Fairview Park, Elmsford, NY 10523. (914) 592-7700. Fax (914) 592-3625. 1993. $120.00.

Introductory Mining Engineering. Howard L. Hartman. John Wiley and Sons, Inc., 605 Third Avenue, New York, NY 10158. (800) 526-5368 or (212) 850-6000. 1987. Inquire for price.

Logistics of Underground Coal Mining. James J. Hanslovan & Richard G. Visovsky. Noyes Publications, 120 Mill Road, Park Ridge, NJ 07656. (201) 391-8484. FAX (201) 391-6833. 1984. $32.00.

Rock Mechanics: For Underground Mining. B.H.G. Brady and E. T. Brown. 2nd ed. Routledge, Chapman and Hall, Inc., 29 West 35th Street, New York, NY 10001-2291. (212) 244-3336 or FAX (212) 563-2269. 1993. Inquire for price.

HANDBOOKS AND MANUALS

Practical Handbook For Underground Rock Mechanics. T.R. Stacey. Trans Tech/ LPS Distribution Center, 52 LaBombard Rd., Lebanon, NH 03766. (603) 448-0037. FAX (603) 448-2576. 1986. $38.00.

Ency. of Physical Sciences and Engineering Info. Sources

MINING ENGINEERING

Underground Mining Methods Handbook. William J. Hustrulid. Society for Mining, Metallurgy and Exploration, Inc., 8307 Shaffer Parkway, PO Box 625002, Littleton, CO 80162-5002. (303) 973-9550. FAX (303) 973-3845. 1982. $120.00.

ONLINE DATABASES AND CD-ROMS

Compendex Plus. Engineering Information, Inc., 345 East 47th Street, New York, NY 10017. (212) 705-7600 or (800) 221-1044. Contains citations with abstracts to worldwide literature in engineering and technology, from 1970 to present. Available on online BRS,(800) 289-4277, DIALOG, (800) 334-2564, ORBIT (800) 456-7248, and STN International, FIZ Karlsruhe, P.O. Box 2465, W-7500, Karlsruhe 1, Germany, online services. Also available on CD-ROM. Inquire as to cost and availability.

CSA Engineering. Cambridge Scientific Abstracts, 7200 Wisconsin Avenue, Suite 601, Bethesda, MD 20814. (301) 961-6750 or (800) 843-7751. Contains citations and abstracts of international periodicals and research literature covering all fields of engineering and science and technology, including computer and information science, electronics, mechanical engineering, solid state materials, 1981 to present. Available on BRS,(800) 289-4277, online service. Inquire as to cost and availability.

Current Contents Search. Institute for Scientific Information, 3501 Market Street, Philadelphia, PA 19104. (215) 386-0100. FAX (215) 386-6362. Contains citations to articles listed in the table of contents of science and technology journals. Also articles in social sciences and life sciences journals. Available on BRS,(800) 289-4277, DIALOG,(800) 334-2564, online services. Inquire as to cost and availability.

Dissertation Abstracts. University Microfilms International, 300 North Zeeb Road, Ann Arbor, MI 48106. (800) 521-0600 or (313) 761-4700. Scope includes virtually all doctoral dissertations accepted at accredited American institutions from 1861 to present in 252 subject areas. Available on BRS,(800) 289-4277, DIALOG,(800) 334-2564, and OCLC EPIC,(800) 848-5878, online services. Also available on CD-ROM. Inquire as to cost and availability.

GEOREF: Bibliography and Index of Geology. American Geological Institute, 4220 King Street, Alexandria, VA 22302. (703) 379-2480. Fax (703) 379-7563. Monthly. Inquire for price and availability.

ISMEC: Mechanical Engineering Abstracts. Cambridge Scientific Abstracts, 7200 Wisconsin Avenue, Suite 601, Bethesda, MD 20814. (301) 961-6750 or (800) 843-7751. Contains citations to the literature in mechanical engineering, industrial and production engineering, energy, power, mechanics, devices and related areas, from 1973 to present. Available on the DIALOG,(800) 334-2564, online service. Inquire as to cost and availability.

NTIS Bibliographic Database. National Technical Information Service, 5285 Port Royal Road, Springfield, VA 22161. (703) 487-4929 or FAX (703) 321-8199. Broad coverage of government-sponsored science and technology research reports, 1964 to present. Available on BRS,(800) 289-4277, DIALOG, (800) 334-2564, ORBIT, (800) 456-7248, and STN International, FIZ Karlsruhe, P.O. Box 2465, W-7500, Karlsruhe 1, Germany, online services. Also available on CD-ROM. Inquire as to cost and availability.

Scisearch. Institute for Scientific Information, 3501 Market Street, Philadelphia, PA 19104. (800) 523-1850 or (215) 386-0100. Broad multidisciplinary title and author index to the international literature of science and technology, 1974 to present. Available on DIALOG,(800) 334-2564, and ORBIT,(800) 456-7248, online services. Also available on CD-ROM. Inquire as to cost and availability.

Wilsonline. H.W. Wilson Company, 950 University Avenue, Bronx, NY 10452. (800) 367-6770 or (212) 588-8400. Makes available online versions of the printed H.W. Wilson indexes, including Applied Science and Technology Index, Business Periodicals Index, General Science Index, and Readers' Guide to Periodical Literature. Period covered is generally 1983 to present. Available on BRS,(800) 289-4277, DIALOG,(800) 334-2564, and OCLC EPIC,(800) 848-5878, online services. Also available on CD-ROM. Inquire as to cost and availability.

PERIODICALS

AMC Journal. American Mining Congress, 1920 N St., NW, Suite 300, Washington, DC 20036-1662. (202) 861-2800. FAX (202) 861-7535. Monthly. $36 per year.

Advances in Mining Science and Technology. Elsevier Science Inc., Box 882, Madison Square Station, New York, NY 10159. (212) 989-5800. Irregular. Inquire for price.

American Mineralogist. Mineralogical Society of America, 1130 17th St. NW, Suite 330, Washington, DC 20036. (202) 775-4344. FAX (202) 775-0018. Bi-monthly. $225.00 per year.

Canadian Mining Journal. Southam Magazine Group, 1450 Don Mills Rd., Don Mills, ON M3B 2X7, Canada. (416) 442-2004. FAX (416) 442-2261. Bi-monthly. $46.00 per year.

CIM Bulletin. Canadian Institute of Mining, Metallurgy, and Petroleum, Xerox Tower, 3400 de Maisonneuve Blvd. W., Suite 1210, Montreal PQH3Z 3B8, Canada. (514) 939-2710. FAX (514) 939-2714. 1898 to present. Monthly. $105.00 per year.

Colorado School of Mines Quarterly. Colorado School of Mines Press, 1500 Illinois, Golden, CO 80401-1887. (303) 273-3000. FAX (303) 273-3310. Quarterly. $50.00 per year.

Engineering & Mining Journal. Maclean Hunter Publishing Company, 29 N. Wacker Dr., Chicago, IL 60606. (312) 726-2802. FAX (312) 726-2574. Monthly. $60.00 per year.

Institution of Mining and Metallurgy. Transactions. Section A: Mining Industry. Institution of Mining and Metallurgy, 44 Portland Pl., London W1N 4BR, England. 071-580-3802. FAX 071-436-5388. Three times a year. $126.00 a year.

Mine and Quarry. IML Group, Blair House, High St., Tonbridge, Kent TN (1BQ, England). Telephone 0732-359990. FAX 0732-770049. Monthly. Inquire for price.

Mine & Quarry Trader. Allied Publications, 7355 N. Woodland, Box 603, Indianapolis, in 46206-0603. (317) 297-5500. FAX (317) 299-1356. Monthly. $18.00 per year.

Mining Engineer. Institution of Mining Engineers, Danum House, South Parade, Doncaster DN1 2DY, England. Phone 0302-320486. FAX 0302-340554. Monthly. Inquire for price.

Mining Engineering. Society for Mining, Metallurgy, and Exploration, Inc., 8307 Shaffer Parkway, PO Box 625002, Littleton,

MINING ENGINEERING

Ency. of Physical Sciences and Engineering Info. Sources

CO 80162-5002. (303) 973-9550. FAX (303) 973-3845. Monthly. $100.00 per year.

Pit and Quarry. Advanstar Communications Inc., 7500 Old Oak Blvd., Cleveland, OH 44130. (216) 826-2839. FAX (216) 891-2726. 1916 to present. Monthly. $35.00 per year.

World Mining Equipment. Metal Bulletin, Inc., 220 Fifth Ave., New York, NY 10001-7781. (212) 213-6202. FAX (212) 213-1870. 11 times a year. Inquire for price.

RESEARCH CENTERS AND INSTITUTES

Excavation Engineering & Earth Mechanics Institute. Colorado School of Mines, Golden, CO 80401. (303) 273-3419.

Michigan Technological University, Mining & Mineral Resources Center. Department of Metallurgical & Materials Engineering, 1400 Townsend Drive, Houghton, MI 49931. (906) 487-2630.

Mining & Excavation Research Institute. 1825 K St NW, Suite 218, Washington, DC 20006-1202. (202) 785-3756. FAX (202) 429-9417.

New Mexico Bureau of Mines and Mineral Resources. Institute of Mining & Technology, Campus Station, Socorro, NM 87801. (505) 835-5420. FAX (505) 835-6333.

South Dakota School of Mines and Technology, Mining and Mineral Resources Research Institute. 501 E St. Josephs St., Rapid City, SD 57701-3995. (605) 394-1971.

Southern Illinois University At Carbondale, Illinois Mining and Mineral Resources Research Institute. Coal Extraction and Utilization Research Center, Carbondale, IL 62901. (618) 536-6637. FAX (618) 453-7455.

University of Kentucky Institute For Mining and Minerals Research. 233 Mining and Mineral Resources Bldg., Lexington, KY 40506. (606) 257-8636. FAX (606) 258-1049.

University of Utah, Utah Mining and Mineral Resources Research Institute. 113 Energy and Mineral Research office Bldg., Salt Lake City, UT 84112. (801) 581-8006. FAX (801) 581-8119.

MIRROR OPTICS

See: OPTICS

MISSILES

See: GUIDED MISSILES

MODEMS

See: COMPUTER COMMUNICATIONS

MOLECULAR PHYSICS

See also: CHEMICAL BONDING, KINETICS, PHYSICAL CHEMISTRY, PHYSICS

ABSTRACT SERVICES AND INDEXES

Applied Science and Technology Index; A Cumulative Subject Index to English Language Periodicals in the Fields of Aeronautics and Space Science, Computer Technology, Chemistry, Construction Industry, Energy and Related Areas. H. W. Wilson Co., 950 University Avenue, Bronx, NY 10452-9978. (212) 588-8400. (800) 367-6770. FAX (718) 590-1617. From 1958 to present. Monthly. Inquire about cost and availability. Also available online (BRS and Wilsonline) and on CD-ROM. ISSN: 0003-6986.

Chemical Abstracts. Chemical Abstracts Service. 2540 Olentangy River Road, Box 3012, Columbus, OH 43210-0012. (614) 447-3600. FAX (614) 447-3713. 1907 to present. Weekly. $16,800.00 per year. Available online and on CD-ROM. CA is also available in five section groupings. Inquire regarding cost and availability.

Current Contents: Physical, Chemical and Earth Sciences. Institute for Scientific Information, 3501 Market Street, Philadelphia, PA 19104. (215) 386-0100. FAX (215) 386-2291. From 1961 to present. Weekly. $442.00 per year. Also available on CD-ROM and online. Inquire regarding cost and availability. ISSN: 0163-2574.

General Science Index. H.W. Wilson Company, 950 University Avenue, Bronx, NY 10452. (212) 588-8400. (800) 367-6770. FAX (718) 590-1617. From 1978 to present. Ten issues per year; quarterly and annual cumulations. Service basis. Available on CD-ROM and online. Inquire regarding cost and availability. ISSN: 0162-1963.

Physics Abstracts. INSPEC. Section A, Science Abstracts. Institution of Electrical Engineers (IEE), London, United Kingdom. Available from: INSPEC/IEEE-Institute of Electrical and Electronic Engineers, Box 1331, 445 Hoes Lane, Piscataway, NJ 08855-1331. (908) 562-5549. 1898 to present. 24 issues per year. $2835.00 per year. ISSN: 0036-8091. Also available online and on CD-ROM.

Physics Briefs (Physikalische Berichte). Information Center for Energy, Physics, Mathematics; German Physical Society. VCH Publishers, Inc., 220 East 23rd Street, New York, NY 10010-4606. (212) 683-8333. 1845 to present. 24 issues per year. $2390.00 per year. Also available online. ISSN: 0179-7434.

Science Citation Index. SCI. Institute for Scientific Information, 3501 Market Street, Philadelphia, PA 19104. (215) 386-0100. (800) 523-1850. FAX (215) 386-2991. 1961 to present. Six issues per year, plus annual cumulation. $11650.00 per year. Also available online and on CD-ROM. Inquire about price and availability. ISSN: 0036-827X.

ASSOCIATIONS AND PROFESSIONAL SOCIETIES

American Chemical Society, 1155 16th Street, NW, Washington, DC 20036. (202) 872-4600.

American Institute of Physics. One Physics Ellipse, College Park, MD 20740-3843. (301) 209-3100.

American Physical Society. One Physics Ellipse, College Park, MD 20740-3843. (301) 209-3200. FAX (301) 209-0865.

BIBLIOGRAPHIES

Scientific and Technical Books and Serials in Print; An Index to Literature in Science and Technology. R.R. Bowker Inc., 121 Chanlon Road, New Providence, NJ 07974. (908) 464-6800. (800) 521-8110. FAX (908) 665-3502. 1972 to present. Annual. 4 volumes. 1994. $299.95. Also available on compact disc and online. ISSN: 0000-054X.

DIRECTORIES AND BIOGRAPHICAL SOURCES

American Men and Women of Science: Physical and Biological Sciences. R.R. Bowker Inc., 121 Chanlon Road, New Providence, NJ 07974. (908) 464-6800. (800) 521-8110. 20th edition. 8 volumes. 1996. $850.00

American Physical Society. Membership Directory Bulletin Issue. American Physical Society, One Physics Ellipse, College Park, MD 20740-3843. (301) 209-3200. FAX (301) 209-0865. Biennial. $50.00.

Directory of Physics and Astronomy Staff Members. American Institute of Physics, One Physics Ellipse, College Park, MD 20740-3843. (301) 209-3100. Annual. $45.00.

Research Centers Directory. Gale Research Company, Inc., 835 Penobscot Building, Detroit, MI 48226-4094. (313) 961-2242. (800) 347-4253. 20th edition. 1995. $485.00. ISSN: 0080-1518.

Scientific and Technical Organizations and Agencies Directory. Gale Research, 835 Penobscot Building, Detroit, MI 48226-4094. (313) 961-2242. (800) 347-4253. 4th edition. 1996. $195.00.

ENCYCLOPEDIAS AND DICTIONARIES

Academic Press Dictionary of Science and Technology. Christopher Morris, editor. Academic Press, Inc., 1250 Sixth Avenue, San Diego, CA 92101. (619) 231-0926. FAX (619) 699-6715. 1991. $115.00.

Dictionary of the Physical Sciences: Terms, Formulas, Data. Cesare Emiliani. Oxford University Press, Inc., 200 Madison Avenue, New York, NY 10016. (212) 725-6000. (800) 334-4249. 1989. $19.95.

Encyclopedia of Applied Physics. George Trigg, editor. VCH Publications, Inc., 220 East 23rd Street, Suite 909, New York, NY 10010-4606. (212) 683-8333. (800) 422-8824. 20 volume set. 1991-. $5990.00.

Encyclopedia of Physics. Robert M. Besancon. Chapman & Hall, 1 Penn Plaza, New York, NY 10119. (212) 564-1060. 3rd edition. 1990. $54.95 in paper.

Encyclopedia of Physical Science and Technology. Academic Press, Inc., 6277 Sea Harbor Drive, Orlando, FL. (800) 321-5068. 2nd edition. 18 volume set. 1992. $2625.00.

McGraw-Hill Encyclopedia of Science and Technology. McGraw-Hill Book, Incorporated, 1221 Avenue of the Ameri-

cas, New York, NY 10020. (212) 997-3675. Seventeenth edition. Twenty volumes. 1992. $1900.00.

GENERAL WORKS

Clusters of Atoms and Molecules: Theory, Experiment and Clusters of Atoms. Hellmut Haberland, editor. Springer-Verlag, New York, Inc., 175 Fifth Avenue, New York, NY 10010. (212) 460-1500. (800) 777-4643. FAX (212) 473- 6272. 1995. $89.00.

Dynamics of the Liquid State. Umberto Balucani and Marco Zoppi. Oxford University Press, Inc., 200 Madison Avenue, New York, NY 10016. (212) 725-6000. (800) 334-4249. 1995. $95.00.

Electronic Structure of Materials. Adrian Sutton. Oxford University Press, Inc., 200 Madison Avenue, New York, NY 10016. (212) 725-6000. (800) 334-4249. 1993. $26.50 in paper.

Experimental Methods in the Physical Sciences: Atomic, Molecular, and Optical Physics. F. B. Dunning, et al, editors. Academic Press, Inc., 6277 Sea Harbor Drive, Orlando, FL. (800) 321-5068. 1995. $99.00.

Gases, Liquids and Solids: and Other States of Matter. D. Tabor. Cambridge University Press, 40 West 20th Street, New York, NY 10011-4211. (212) 924-3900. 1992. $80.00.

Intermolecular and Surface Forces. Jacob Israelmachvilli.

Academic Press, Inc., 6277 Sea Harbor Drive, Orlando, FL. (800) 321-5068. 2nd edition. 1991. $53.00.

Introduction To the Theory of Atomic and Molecular Collisions. J. N. Murrell and S. D. Bosanac. John Wiley & Sons, Inc., 605 Third Avenue, New York, NY 10158-0012. (212) 850-6000. (800) 225-5945. 2nd edition. 1989. $79.95.

Molecular Physics. T. Buyana. World Scientific Publishing Company, Inc., 1060 Main Street, River Edge, NJ 07661. (201) 487-9655. (800) 227-7562. 1995. $55.00.

Molecular Physics and Elements of Quantum Chemistry. H. Haken and H. C. Wolf. Springer-Verlag New York, Inc., 175 Fifth Avenue, New York, NY 10010. (212) 460-1500. (800) 777-4643. FAX (212) 473-6272. 1995. $49.50.

Molecular Solid State Physics. G.G. Hall. Springer-Verlag New York, Inc., 175 Fifth Avenue, New York, NY 10010. (212) 460-1500. (800) 777-4643. FAX (212) 473-6272. 1991. $39.95.

Multiple Bonds Between Metal Atoms. F. Albert Cotton and Richard A. Walton. Oxford University Press, Inc., 200 Madison Avenue, New York, NY 10016. (212) 725-6000. (800) 334-4249. 2nd edition. 1993. $95.00.

HANDBOOKS AND MANUALS

Atomic, Molecular, and Optical Physics Handbook. Gordon W. Drake, editor. American Institute of Physics, One Physics Ellipse, College Park, MD 20740-3843. (301) 209-3100. 1996. $130.00.

CRC Handbook of Chemistry and Physics. David R. Kide, editor. CRC Press Inc., 2000 Corporate Blvd., NW, Boca Raton,

MOLECULAR PHYSICS

Ency. of Physical Sciences and Engineering Info. Sources

FL 33431. (407) 994-0555. (800) 333-8300. 77th edition. 1996. $99.95.

Tables of Physical and Chemical Constants and Some Mathematical Functions. G. W. Kaye and T. M. Laby. Halsted Press 605 Third Avenue, New York, 10158-0012. (212) 850-6400. 15th edition. 1986. $74.95.

ONLINE DATABASES AND CD-ROMS

CA Search. Chemical Abstracts Service, P.O. Box 3012, Columbus, OH 43210-0012. (614) 447-3600. (800) 848-6533. FAX (614) 447-3709. Very comprehensive guide to worldwide chemical literature and related fields, 1972 to present. Available on BRS,(800) 289-4277, DIALOG, (800) 334-2564, ORBIT (800) 456-7248, and STN International, FIZ Karlsruhe, P.O. Box 2465, W-7500, Karlsruhe 1, Germany, online services. Inquire as to cost and availability.

Current Contents Search. Institute for Scientific Information, 3501 Market Street, Philadelphia, PA 19104. (215) 386-0100. FAX (215) 386-6362. Contains citations to articles listed in the table of contents of science and technology journals. Also articles in social sciences and life sciences journals. Available on BRS,(800) 289-4277, DIALOG,(800) 334-2564, online services. Inquire as to cost and availability.

Dissertation Abstracts. University Microfilms International, 300 North Zeeb Road, Ann Arbor, MI 48106. (800) 521-0600 or (313) 761-4700. Scope includes virtually all doctoral dissertations accepted at accredited American institutions from 1861 to present in 252 subject areas. Available on BRS,(800) 289-4277, DIALOG,(800) 334-2564, and OCLC EPIC,(800) 848-5878, online services. Also available on CD-ROM. Inquire as to cost and availability.

Kirk-Othmer Encyclopedia of Chemical Technology. John Wiley and Sons, Inc., 605 Third Avenue, New York, NY 10158. (800) 526-5368 or (212) 850-6000. Contains the complete text of all chapters in the 27 volume fourth edition of the *Kirk-Othmer Encyclopedia of Chemical Technology.* 1991. Available on BRS,(800) 289-4277, DIALOG,(800) 334-2564, online services. Inquire as to cost and availability.

NTIS Bibliographic Database. National Technical Information Service, 5285 Port Royal Road, Springfield, VA 22161. (703) 487-4929 or FAX (703) 321-8199. Broad coverage of government-sponsored science and technology research reports, 1964 to present. Available on BRS,(800) 289-4277, DIALOG, (800) 334-2564, ORBIT, (800) 456-7248, and STN International, FIZ Karlsruhe, P.O. Box 2465, W-7500, Karlsruhe 1, Germany, online services. Also available on CD-ROM. Inquire as to cost and availability.

Scisearch. Institute for Scientific Information, 3501 Market Street, Philadelphia, PA 19104. (800) 523-1850 or (215) 386-0100. Broad multidisciplinary title and author index to the international literature of science and technology, 1974 to present. Available on DIALOG,(800) 334-2564, and ORBIT,(800) 456-7248, online services. Also available on CD-ROM. Inquire as to cost and availability.

Wilsonline. H.W. Wilson Company, 950 University Avenue, Bronx, NY 10452. (800) 367-6770 or (212) 588-8400. Makes available online versions of the printed H.W. Wilson indexes, including Applied Science and Technology Index, Business Periodicals Index, General Science Index, and Readers' Guide to Periodical Literature. Period covered is generally 1983 to present. Available on BRS,(800) 289-4277, DIALOG,(800) 334-2564, and OCLC EPIC,(800) 848-5878, online services. Also available on CD-ROM. Inquire as to cost and availability.

PERIODICALS

American Journal of Physics. American Association of Physics Teachers, 5112 Berywn Road, College Park, MD 20740. (301) 345-4200. 1933 to present. Monthly. $215.00 per year. ISSN: 0002-9505.

Chemical Physics. Elsevier Science Publishing Co., Inc., Box 882, Madison Square Station, New York, NY 10159. (212) 989-5800. FAX (212) 633-3990. 1973 to present. 33 issues per year in 11 volumes. $2759.00 per year. ISSN: 0301-0104.

Chemical Physics Letters. Elsevier Science Inc., Publishing Co., Inc., Box 882, Madison Square Station, New York, NY 10159. (212) 989-5800. (212) 633-3990. 1967 to present. 102 issues per year in 17 volumes. $5063.00 per year. ISSN: 0009-2614.

International Journal of Chemical Kinetics. John Wiley and Sons, Inc., 605 Third Avenue, New York, NY 10158. (800) 526-5368 or (212) 850-6000. 1968 to present. Monthly. $730.00 per year. ISSN: 0538-8066.

Journal of Physics B: Atomic and Molecular Physics. IOP Publishing. Orders to: American Institute of Physics, 500 Sunnyside Boulevard, Woodbury, NY 11797-2999. (516) 576-2200. 1968 to present. 24 issues per year. $2240.00 per year. ISSN: 0953-4075.

Molecular Physics. Taylor & Francis Ltd., Rankine Road, Basingstoke, Hants RG24 8PR, England. TEL 0256-840366. FAX 0256-479438. 1958 to present. 18 issues per year. $1667.00 per year. ISSN: 0026-8976.

Physical Review B (condensed Matter). American Institute of Physics, One Physics Ellipse, College Park, MD 20740-3843. (301)209-3000. 1970 to present. 48 issues per year. $3180.00 per year. ISSN: 0163-1829.

Reviews of Modern Physics. American Institute of Physics, One Physics Ellipse, College Park, MD 20740-3843. (301) 209-3000. 1929 to present. Quarterly. $300.00 per year. ISSN: 0034-6861.

RESEARCH CENTERS AND INSTITUTES

Atomic Physics and Chemical Physics Laboratory. University of Nevada-Reno, Department of Physics, Reno, NV 89557. (702) 784-4920.

Center For Chemical Physics. University of Florida, Williamson Hall, Gainsville, FL 32611. (904) 392-6977. FAX (904) 392-8722.

Institute For Physical Science and Technology. University of Maryland, College Park, MD 20742. (301) 405-4874. FAX (301) 405-9363.

Institute For Theoretical Atomic and Molecular Physics. Harvard University, Oxford Street, Cambridge, MA 02138. (617) 495-9524.

Laboratory For Research On the Structure of Matter. University of Pennsylvania, 3231 Walnut Street, Philadelphia, PA 19104. (215) 898-8571.

MONOPOLE

See: ELECTROMAGNETISM

MONORAIL TECHNOLOGY

See: RAILROAD ENGINEERING

MONTE CARLO METHOD

See: STATISTICS

MOON

See also: ASTROGEOLOGY, METEORITES, PLANETARY SCIENCES, SATELLITES (NATURAL), SOLAR SYSTEM,

ABSTRACT SERVICES AND INDEXES

Applied Science and Technology Index; A Cumulative Subject Index to English Language Periodicals in the Fields of Aeronautics and Space Science, Computer Technology, Chemistry, Construction Industry, Energy and Related Areas. H. W. Wilson Co., 950 University Avenue, Bronx, NY 10452- 9978. (212) 588-8400. (800) 367-6770. FAX (718) 590-1617. From 1958 to present. Monthly. Inquire about cost and availability. Also available online (BRS and Wilsonline) and on CD-ROM. ISSN: 0003-6986.

Astronomy and Astrophysics Abstracts. Springer-Verlag New York, 175 Fifth Avenue, New York, NY 10010. (212) 460-1500. FAX (212) 473-6272. Published for the Astronomisches Rechen-Institut. Comprehensive coverage of all aspects of astronomy, astrophysics and related fields. 1969 to present. Two parts per year. Annual. ISSN: 0067-0022.

Bibliography and Index of Geology. American Geological Institute, 4220 King Street, Alexandria, VA 22302-1507. (703) 379-2480. FAX (703) 379-7563. From 1969 to present. Monthly. $1295.00 per year. ISSN: 0098-2784. Also available as GEOREF online (CISTI, DIALOG, Orbit, STN) and on CD-ROM. Inquire about price and availability.

Chemical Abstracts. Chemical abstracts Service. 2540 Oleantangy River Road, Box 3012, Columbus, OH 43210-0012. (614) 447-3600. FAX (614) 447-3713. 1907 to present. Weekly. $16,800.00 per year. Available online and on CD-ROM. CA is also available in five section groupings. Inquire regarding cost and availability.

Current Contents: Physical, Chemical, and Earth Sciences. Institute for Scientific Information, 3501 Market Street, Philadelphia, PA 19104. (215) 386-0100. (800) 523-1850. FAX (215) 386-2291. 1961 to present. Weekly. $442.00 per year. Also avail-

able online (BRS, DIALOG) and on CD-ROM. Inquire about price and availability.

General Science Index. H.W. Wilson Company, 950 University Avenue, Bronx, NY 10452. (212) 588-8400. (800) 367-6770. FAX (718) 590-1617. From 1978 to present. Ten issues per year; quarterly and annual cumulations. Service basis. Available on CD-ROM and online. Inquire regarding cost and availability. ISSN: 0162-1963.

Meteorological and Geoastrophysical Abstracts. Physics Abstracts. Inspec. Section A, Science Abstracts. Institute of Electrical Engineers, London, United Kingdom. Available from: INSPEC/IEEE-Institute of Electrical and Electronic Engineers, Box 1331, Hoes Lane, Piscataway, NJ 08855-1331. (908) 562-5549. 1898 to present. 24 issues per year. $2835.00 per year. ISSN: 0036-8091. Also available online and on CD-ROM.

NTIS Alerts: Astronomy & Astrophysics. U.S. National Technical Information Service. 5285 Port Royal Road, Springfield, VA 22161. (703) 487-4650. FAX (703) 321-8547. Weekly. $140.00 per year.

Science Citation Index. SCI. Institute for Scientific Information, 3501 Market Street, Philadelphia, PA 19104. (215) 386-0100. (800) 523-1850. FAX (215) 386-2991. 1961 to present. Six issues per year, plus annual cumulation. $11650.00 per year. Also available online and on CD-ROM. Inquire about price and availability. ISSN: 0036-827X.

STAR. Scientific and Technical Aerospace Reports. U.S. National Aeronautics and Space Administration. Distributed by U.S. Superintendent of Documents, Washington, DC 20402. 1963 to present. Semi-monthly, with semiannual and annual indexes. $114.00 per year. ISSN: 0036-8741. Also available online and on CD-ROM.

ANNUAL REVIEWS AND YEARBOOKS

The Astronomical Almanac. Superintendent of Documents, U. S. Government Printing Office, Washington, DC 10402. (202) 783-3238. 1981 to present. Supercedes Astronomical Ephemeris. Annual. $44.95. ISSN: 0737-6421.

Annual Review of Astronomy and Astrophysics. Original reviews of critical literature and current developments in astronomy and astrophysics. Annual Reviews, Inc., 4139 El Camino Way, Palo Alto, CA 94303-0139. (415) 493-4400. (800) 523-8635. Fax (415) 855-9815. 1963 to date. Annual. $60.00. ISSN: 0066-4146.

Annual Review of Earth and Planetary Sciences. Annual Reviews, Inc., 4139 El Camino Way, Palo Alto, CA 94303-0139. (415) 493-4400. (800) 523-8635. Fax (415) 855-9815. 1973 to date. Annual. $62.00. ISSN: 0084-6597.

Proceedings of the Lunar and Planetary Science Conference. Lunar and Planetary Institute, 3600 Bay Area Boulevard, Houston, TX 77058. (713) 486-2143. 1970 to date. Annual. ISSN: 0270-9511.

ASSOCIATIONS AND PROFESSIONAL SOCIETIES

American Astronomical Society. 2000 Florida Avenue NW, Suite 400, Washington, DC 20009. (202) 328-2010. FAX (202)234-2560.

American Geophysical Union. 2000 Florida Avenue NW, Washington, DC 20009. (202) 462-6900. (800) 966-AGU1. FAX (202) 328-0566.

American Institute of Physics. 1 Physics Ellipse, College Park, MD 20740-3843. (301) 209-3100.

Association of Universities For Research in Astronomy, Inc. (AURA). Suite 701, 1625 Massachusetts Avenue, NW, Washington, DC 20036. (202) 483-2101. FAX (202) 483-2106.

Astronomical Society of the Pacific. 390 Ashton Avenue, San Francisco, CA 94112. (415) 337-1100. FAX: (415) 337-5205.

Geological Society of America. 3300 Penrose Place, PO Box 9140. Boulder, CO 80301-9140. (303) 447-2020. FAX (303) 447-1133.

IEEE Geoscience and Remote Sensing Society. c/o Institute of Electrical and Electronics Engineers. 345 East 47th Street, New York, NY 10017. (212) 705-7900. FAX (212) 705-4929.

Meteoritical Society. University of Massachusetts, 125 Marston Hall, Amherst, MA 01003. (413) 545-0300. (413) 545-0724.

Planetary Society. 65 North Catalina Avenue, Pasadena, CA 91106. (818) 793-5100. (800) WOW-MARS. FAX (818) 793-5528.

BIBLIOGRAPHIES

A Bibliography of Astronomy, 1970-1979. R. A. Sea and S. S. Martin. Libraries Unlimited, Inc., Littleton, CO 80160. 1982. $37.50.

Lunar Sourcebook: A User's Guide To the Moon. Grant Heiken, et al, editors. Cambridge University Press, 40 West 20th Street, New York, NY 10011-4211. (212) 924-3900. (800)872-7423. 1991. $69.95.

DIRECTORIES AND BIOGRAPHICAL SOURCES

American Astronomical Society. Membership Directory. 2000 Florida Avenue, NW, Suite 300. Washington, DC 20009. (202) 328-2010. FAX (202) 234-2560. 1980 to present. Monthly. Membership only.

American Men and Women of Science: Physical and Biological Sciences. R. R. Bowker Inc., 121 Chanlon Road, New Providence, NJ 07974. (908) 464-6800. (800) 521-8110. 20th edition. 8 volumes. 1996. $850.00.

The Astronomers. Donald Goldsmith. St. Martin's Press, Inc., 175 Fifth Avenue, New York, NY 10010. (212) 674-5151. (800) 221-7945. 1993. $14.95 in paper.

Astronomical Centers of the World. Kevin Krisciunas. Cambridge University Press, 40 West 20th Street, New York, NY 10011-4211. (212) 924-3900. (800) 872-7423. 1988. $34.95

Directory of Physics and Astronomy Staff Members. American Institute of Physics, One Physics Ellipse, College Park, MD 20740-3843. (301) 209-3100. 1975/76 to present. Annual. $60.00. ISSN: 0361-2228.

Earth and Astronomical Research Centers. Stockton Press, 345 Park Avenue South, New York, NY 10010. 4th edition. 1995. $515.00. ISBN: 1-56169-0967.

ENCYCLOPEDIAS AND DICTIONARIES

Concise Dictionary of Astronomy. Jacqueline Mitton. Oxford University Press, Inc., 200 Madison Avenue, New York, NY 10016. (212) 725-6000. (800) 334-4249. 1992. $24.95.

Encyclopedia of Astronomy and Astrophysics. Stephen Maran, editor. Van Nostrand Reinhold, 115 Fifth Avenue, New York, NY 10003. (212) 254-3232. (800) 842-3636. 1992. $129.95.

McGraw-Hill Encyclopedia of Science and Technology. McGraw-Hill Book Company, Inc., 1221 Avenue of the Americas, New York, NY 10020. (212) 512-2000. (800) 262-4729. 7th edition. 20 volume set. 1992. $1900.00.

Moons of the Solar System: An Illustrated Encyclopedia. John Stewart. McFarland & Company, Inc., Box 611, Jefferson, MC 28640. (910) 246-4460. (800) 253-2187. 1991. $49.95.

New Guide To the Planets. Patrick Moore. Trans-Atlantic Publications, Inc., 311 Bainbridge Street, Philadelphia, PA 19147. (215) 925-5083. 1993. $37.50.

Stars and Planets: The Sierra Club Guide To Sky Watching and Direction Finding. W. S. Kals. The Sierra Press, 4988 Gold Leaf Drive, Mariposa, CA 95338. (209) 966-5071. (800) 745-2631. $15.00.

GENERAL WORKS

The Book of the Moon; A Lunar Introduction To Astronomy, Geology, Space, Physics and Space Travel. Thomas A. Hockey. Prentice-Hall, 113 Sylvan Avenue, Route 9W, Englewood Cliffs, NJ 07632. (201) 592-2000. (800) 922-0579. 1986. $19.95.

Evolution of the Earth and Planets. E. Takahashi, et al, editors. American Geophysical Union, 2000 Florida Avenue, N.W., Washington, DC 20009. (202) 462-6903. (800) 966-2481. 1993. $28.00.

Exploring the Moon Through Binoculars and Small Telescopes. Ernst H. Cherrington. Publications, Inc., 180 Varick Street, New York, NY 10014. (212) 255-3755. (800) 223- 3130.

Exploring the Planets. W. Kenneth Hamblin and Eric H. Christiansen. Macmillan Publishing Company, Inc,. 200 Old Tappan Road, Old Tappan, NJ 07675. (800) 233-2336. 2nd edition. 1995. $62.00.

The Geology of Multi-Ring Impact Basins: The Moon and Other Planets. P. D. Spudis. Cambridge University Press, 40 West 20th Street, New York, NY 10011-4211. (212) 924-3900. (800) 872-7423. Planetary Sciences Series. 1993. $59.95.

July Twentieth, Nineteen Sixty-Nine: The Day Men Landed On the Moon. William H. Whitten. Eagle Press, 5959 Pleasant Farm Drive, Beaufort, SC 29902. (803) 522-1594. $25.00 in paper.

The Moon in the Post-Apollo Era. Zdenek Kopal. Kluwer Academic Publishers, 101 Philip Drive, Assinippi Park, Norwell, MA 02061. (617) 871-6600. 1974. $80.00.

Moons and Planets. William K. Hartmann. Wadsworth Publishing Co., 10 Davis Drive, Belmont, CA 94002. (415) 595-2350. (800) 354-9706. 3rd edition. 1993. $49.95.

The Motion of the Moon. Alan Cook. IOP Publishing, Public Ledger Building, Suite 1035, Independence Square, Philadelphia, PA 19106. (215) 627-0880. (800) 358-4677. 1988. $106.00.

Origin of the Moon. William K. Hartmann, et al editors. Lunar and Planetary Institute, 3600 Bay Area Boulevard, Houston, TX 77058. (713) 486-2143. 1986. $25.00.

Outpost On Apollo's Moon. Eric Burgess. Columbia University Press, 562 West 113th Street, New York, NY 10025. (212) 666-1000. (800) 944-8648. 1993. $34.95.

A Portfolio of Lunar Drawings. Harold Hill. Cambridge University Press, 40 West 20th Street, New York, NY 10011-4211. (212) 924-3900. (800) 872-7423. 1991. $49.95.

Proceedings of Lunar and Planetary Science, Vol. 22. Graham Ryder and Virgil L. Sharpton, editors. Lunar and Planetary Institute, 3600 Bay Area Boulevard, Houston, TX 77058. (713) 486-2143. 1992. $50.00.

To A Rocky Moon: A Geologic History of Lunar Exploration. Don E. Wilhelms. University of Arizona Press, 1230 North Park Avenue, Number 102, Tucson, AZ 85719. (520) 621-1441. (800) 426-3797. Reprint edition. 1991. $19.95 in paper.

HANDBOOKS AND MANUALS

Astrophysical Data: Planets and Stars. Kenneth R. Lanb. Springer-Verlag New York, Inc., 175 Fifth Avenue, New York, NY 10010. (212) 460-1500. (800) 777-4643. 1993. $59.00.

Field Guide to the Stars and Planets. Jay M. Pasachoff and Donald H. Menzel. Houghton Mifflin Co., 222 Berkeley Street, Boston, MA 02116. (617) 351-5000. (800) 225-3362. Revised edition. 1992. $24.95.

The Moon: An Observing Guide For Backyard Telescopes. Michael Kitt. Kalmbach Publishing Company, P.O. Box 1612, Waukesha, WI 53187. (414) 796-0126. (800) 446-5489. 1992. $11.95.

The Moon Observer's Handbook. Fred W. Price. Cambridge University Press, 40 West 20th Street, New York, NY 10011-4211. (212) 924-3900. (800) 872-7423. 1989. $39.95.

New Guide To the Planets. Patrick Moore. Trans-Atlantic Publications, Inc., 311 Bainbridge Street, Philadelphia, PA 19147. (215) 925-5083. 1993. $37.50.

Planet Observer's Handbook. Fred W. Price. Cambridge University Press, 40 West 20th Street, New York, NY 10011-4211. (212) 924-3900. (800) 872-7423. 1994. $34.95.

Planets and their Moons. Audubon Society Staff. Alfred A. Knopf, Inc. 201 E. 50th Street, New York, NY 10022. (800) 733-3000. 1995. $7.99.

ONLINE DATABASES AND CD-ROMS

CA Search. Chemical Abstracts Service, P.O. Box 3012, Columbus, OH 43210-0012. (614) 447-3600. (800) 848-6533. FAX (614) 447-3709. Very comprehensive guide to worldwide chemical literature and related fields, 1972 to present. Available on BRS,(800) 289-4277, DIALOG, (800) 334-2564, ORBIT (800) 456-7248, and STN International, FIZ Karlsruhe, P.O. Box 2465, W-7500, Karlsruhe 1, Germany, online services. Inquire as to cost and availability.

Dissertation Abstracts. University Microfilms International, 300 North Zeeb Road, Ann Arbor, MI 48106. (800) 521-0600 or (313) 761-4700. Scope includes virtually all doctoral dissertations accepted at accredited American institutions from 1861 to present in 252 subject areas. Available on BRS, (800) 289-4277, DIALOG, (800) 334-2564, and OCLC EPIC, (800) 848-5878, online services. Also available on CD-ROM. Inquire as to cost and availability.

GEOREF. American Geological Institute, 4220 King Street, Alexandria, VA 22302. (800) 336-4764 or (703) 379-2480. Geology and geosciences literature, 1785 to present for North America. Available on DIALOG,(800) 334-2564, ORBIT (800) 456-7248, online services. Also available on CD-ROM. Inquire as to cost and availability.

INSPEC. Institution of Electrical Engineers, Michael Faraday House, Six Hills Way, Stevenage, Herts. SG1 2AY, England. Telephone: 0438 313311 or FAX 0438 742840. Contains citations to the worldwide literature of physics, electronics and electrical engineering, computer technology, and related fields. Available on BRS, (800) 289-4277, DIALOG, (800) 334-2564, ORBIT, (800) 456-7248, and STN International, FIZ Karlsruhe, P.O. Box 2465, W-7500, Karlsruhe 1, Germany, online services. Inquire as to cost and availability.

NTIS Bibliographic Database. National Technical Information Service, 5285 Port Royal Road, Springfield, VA 22161. (703) 487-4929 or FAX (703) 321-8199. Broad coverage of government-sponsored science and technology research reports, 1964 to present. Available on BRS,(800) 289-4277, DIALOG, (800) 334-2564, ORBIT, (800) 456-7248, and STN International, FIZ Karlsruhe, P.O. Box 2465, W-7500, Karlsruhe 1, Germany, online services. Also available on CD-ROM. Inquire as to cost and availability.

Scisearch. Institute for Scientific Information, 3501 Market Street, Philadelphia, PA 19104. (800) 523-1850 or (215) 386-0100. Broad multidisciplinary title and author index to the international literature of science and technology, 1974 to present. Available on DIALOG, (800) 334-2564, and ORBIT, (800) 456-7248, online services. Also available on CD-ROM. Inquire as to cost and availability.

Wilsonline. H.W. Wilson Company, 950 University Avenue, Bronx, NY 10452. (800) 367-6770 or (212) 588-8400. Makes available online versions of the printed H.W. Wilson indexes, including Applied Science and Technology Index, Business Periodicals Index, General Science Index, and Readers' Guide to Periodical Literature. Period covered is generally 1983 to present. Available on BRS, (800) 289-4277, DIALOG, (800) 334-2564, and OCLC EPIC, (800) 848-5878, online services. Also available on CD-ROM. Inquire as to cost and availability.

MOON

Ency. of Physical Sciences and Engineering Info. Sources

OTHER SOURCES

Atlas of the Moon. Antonio Ruki. Kalmbach Publishing Company, P.O. Box 1612, Waukesha, WI 53187. (414) 796-0126. (800) 446-5489. 1992. $29.95.

Atlas of the Planets. Vincent DeCallatay and Audouin Dollfus. Books on Demand, 300 North Zeeb Road, Ann Arbor, MI 48106-1346. (313) 761-4700. (800) 521-0600. $45.60.

Atlas of the Solar System. B. Yenne. Simon & Schuster, Inc. 1230 Avenue of the Americas, New York, NY 10020. (212) 698-7000. (800) 223-2348. 1987. $12.98.

Lunar Tables and Programs 4000 B.C. TO A.D. 8000. Chaproni-Touze and Chapront. Willmann-Bell, Inc., P.O. Box 35025, Richmond, VA 23235. (804) 320-7016. 1991. $19.95.

New and Full Moons 1001 B.C. to A.D. 1651. Herman H. Goldstine. American Philosophical Society, 104 South 5th Street, Philadelphia, PA 19106-3387. (215) 440-3400. FAX (215) 440-3436. Memoirs Series, volume 94. 1973. $12.00.

Planetary, Lunar and Solar Positions 601 B.C. to A.D. 1 at Five-Day and Ten-Day Intervals. Bryant Tuckerman. American Philosophical Society, 104 South 5th Street, Philadelphia, PA 19106-3387. (215) 440-3400. FAX (215) 440-3436. Memoirs Series, volume 56. 1979. $20.00.

Planetary, Lunar and Solar Positions, A.D. 2 to A.D. 1649 at Five-Day and Ten-Day Intervals. Bryant Tuckerman. American Philosophical Society, 104 South 5th Street, Philadelphia, PA 19106-3387. (215) 440-3400. FAX (215)440-3436. Memoirs Series, volume 59. 1964. $30.00.

Planetary, Lunar and Solar Positions, 1650-1805. Owen Gingerich and Barbara L. Welther. American Philosophical Society, 104 South 5th Street, Philadelphia, PA 19106-3387. (215) 440-3400. FAX (215)440-3436. Memoirs Series, volume 59S. 1983. $20.00.

The Sky: A User's Guide. David H. Levy. Cambridge University Press, 40 West 20th Street, New York, NY 10011-4211. (212) 924-3900. (800) 872-7423. 1993. $14.95.

Tables of the Motion of the Moon. Ernst Brown and Henry B. Hedrick. Elliot's Books, P.O. 6, Northfield, CT 06472. (203) 484-2184. 3 volumes. 1920. $750.00

PERIODICALS

Astronomical Journal. American Institute of Physics, One Physics Ellipse, College Park, MD 20740-3843. (301) 209-3000. Published for the American Astronomical Society. 1849 to present. Monthly. $280.00 per year. ISSN: 0004-6256.

Astronomical Society of the Pacific. Publications. Astronomical Society of the Pacific, 390 Ashton Avenue, San Francisco, CA 94112. (415) 337-1100. FAX (415) 337-5205. 1889 to present. Monthly. $175.00 per year. ISSN: 0004-6280.

Astronomy. Kalmbach Publishing Company, Box 1612, Waukesha, WI 53187-1612. (414) 796-8776. FAX (414) 796-1142. 1973 to present. Monthly. $27.00 per year. ISSN: 0091-6358.

Astronomy and Astrophysics. Springer-Verlag New York, Inc., 44 Hartz Way, Secaucus, NJ 07096-2491. (201) 348-4033. FAX (201) 348-4505. 1969 to date. 36 issues per year in 12 volumes. $1733.00 per year. ISSN: 0004-6361.

Earth, Moon and Planets; An International Journal of Comparative Planetology. Kluwer Academic Publishers, Box 358, Accord Station, Hingham, MA 02018-0358. (617) 871-6600. FAX (617) 871-6528. 1969 to present. Monthly. $840.00 per year. ISSN: 0167-9295.

Geochemica Et Cosmochimica Acta. Elsevier Science. 660 White Plains Road, Tarrytown, NY 10591-5153. (914) 524-9200. FAX (914) 333-2444. 1950 to the present. Biweekly. $895.00 per year. ISSN: 0016-7037.

ICARUS; International Journal of Solar System Studies. Academic Press, Inc., Journal Division, 525 B Street, Suite 1900, San Diego, CA 92101-4495. (619) 230-1840. FAX (619) 699-6800. 1962 to the present. Monthly. $1080.00. ISSN: 0019-1035.

JGR: Journal of Geophysical Research: Planets. American Geophysical Union, 2000 Florida Avenue, NW, Washington, CD 20009. (202) 462-6900. FAX (202) 328-0566. 1991 to present. Monthly. $597.00 per year. ISSN: 0148-0227.

Lunar and Planetary Information Bulletin. Lunar and Planetary Institute. 3600 Bay Area Boulevard, Houston, TX 77058-1113. (713) 486-2175. FAX (713) 486-2125. 1970 to present. Quarterly. Free. Also available online.

Mercury. Astronomical Society of the Pacific. 390 Ashton Avenue, San Francisco, CA 94112. (415) 337-1100. FAX: (415) 337-5205. 1972 to present. Bimonthly. $175.00 per year. ISSN: 0047-6773.

Meteoritics. Meteoritical Society, Department of Chemistry, University of Arkansas, Fayetteville, AR 72701. (501) 575-7625. FAX (501) 575-7778. 1953 to present. Bimonthly. $210.00 per year. ISSN: 0026-1114.

Observatory. c/o Dr. D.J. Stickland Space and Astrophysics Division, Rutherford Appleton Laboratory, Chilton, Didcot, Oxon OX11 OQX England. FAX 0235-445848. 1877 to present. Bimonthly. $42.00 per year. ISSN: 0029-7704.

Planetary and Space Science. Elsevier Science Publishing Company, Inc., 660 White Plains Road, Terrytown, NY 10591-5153. (914) 524-9200. FAX: (914) 333-2444. 1959 to present. Monthly. $1355.00 per year. ISSN: 0032-0633.

Planetary Report. Planetary Society. 65 North Catalina, Pasadena, CA 91106-2301. (818) 793-5100. FAX (818) 793-5528. 1980 to present. $25.00 per year. ISSN: 0736-3680.

Sky and Telescope. Sky Publishing Corporation, Box 9111, Belmont, MA 02178. (617) 864-7360. FAX (617) 864-6117. 1941 to present. Monthly. $27.00 per year. ISSN: 0037-6604.

RESEARCH CENTERS AND INSTITUTES

Center For Radiophysics and Space Research. Cornell University, Space Sciences Building, Ithaca NY 14853. (607) 255-4341. FAX (607) 255-9888.

Center For Space Sciences. University of Texas at Dallas, P. O. Box 830688, MS-F022, Richardson, TX 75083-0688. (214) 690-

2851. FAX (214) 690-2848.

Earth and Planetary Remote Sensing Laboratory. Washington University, Department of Earth and Planetary Sciences. Campus Box 1169, 1 Brookings Drive, St. Louis, MO 63130. (314) 889-5679. FAX (314) 889-5799.

Institute For Astronomy. University of Hawaii at Manoa, 2680 Woodlawn Drive, Honolulu, HI 96822. (808) 956-8312. FAX (808) 988-2790.

Institute of Geophysics and Planetary Physics. University of California, Riverside, CA 92521. (714) 787-4503. FAX (714) 787-4529.

Laboratory For Planetary Geology. Arizona State University, Department of Geology, Tempe, AZ 85281. (601) 965-7029.

Laboratory For Planetary Studies. Cornell University, Space Sciences Building, Ithaca, NY 14853. (607) 255-4971.

Lunar and Planetary Institute. 3303 NASA Road One, Houston, TX 77058-4399. (713) 486-2139. FAX 713-496-2162.

Lunar and Planetary Laboratory. University of Arizona, Tucson, AZ 85721. (602) 621-6963. FAX (602) 621-4933.

MOTORS

See also: DIESEL ENGINES, MACHINERY, TURBINES

ABSTRACT SERVICES AND INDEXES

Applied Mechanics Reviews: An Assessment of World Literature in Engineering Sciences. 1948-present. American Society of Mechanical Engineers, 345 East 47th Street, New York, NY 10017. (212) 705-7703. Monthly. $360.00 per year.

Applied Science and Technology Index; A Cumulative Subject Index To English Language Periodicals in the Fields of Aeronautics and Space Science, Computer Technology, Chemistry, Construction Industry, Energy and Related Areas. H.W. Wilson Co., 950 University Avenue, Bronx, NY 10452. (800) 367-6770 or (212) 588-8400. FAX (718) 590-1857. From 1958 to present. Monthly. Inquire about cost and availability. Also available on CD-ROM and online.

CAD/CAM Abstracts. R.R. Bowker, Bowker A&I Publishing, 121 Chanlon Road, New Providence, NJ 07974. (908) 771-7714. Fax (908) 771-7725. 1984 to present. Monthly. $495.00 per year.

Current Contents: Engineering, Technology, and Applied Sciences. Institute for Scientific Information, 3501 Market Street, Philadelphia, PA 19104. (215) 386-0100. FAX (215) 386-6362. 1970 to present. Weekly. $442.00 per year.

Electrical and Electronics Abstracts. Institute of Electrical Engineers, Michael Faraday House, Six Hill Way, Stevenage, Herts, SG1 2AY, England. Distributed by IEEE, 445 Hoes Lane, Piscataway, NJ 08854. (908) 562-5549. 1898 to present. Monthly. $2200.00 per year. Also available on CD-ROM and online as INSPEC.

Engineering Index Monthly. Engineering Information, Inc., Castle Point on the Hudson, Hoboken, NJ 07030. (800) 221-1044. FAX (212) 832-1857. Monthly. $2300.00 per year. Also available online as COMPENDEX, and also on CD-ROM. Covers chemical engineering, computers, electrical engineering, civil engineering, metals and mining, industrial management, and mechanical engineering.

Mechanical Engineering Abstracts (formerly ISMEC), Cambridge Scientific Abstracts, 7200 Wisconsin Avenue, Bethesda, MD 20814-4823. (301) 961-6750. FAX (301) 961-6720. 1967 to present. Monthly. $895.00 per year. Summarizes world literature in mechanical engineering, production engineering, and engineering management. Also available online.

Science Citation Index. Institute for Scientific Information, 3501 Market Street, Philadelphia, PA 19104. (215) 386-0100. FAX (215) 386-6362. Inquire about availability and cost. Also available on CD-ROM.

ASSOCIATIONS AND PROFESSIONAL SOCIETIES

American Society of Mechanical Engineers. 345 E. 47th Street, New York, NY 10017. (212) 705-7722.

Institute of Electrical and Electronic Engineers. 345 E. 47th Street, New York, NY 10017-2394.

Society of Automotive Engineers. 400 Commonwealth Drive, Warrendale, PA 15096. (412) 776-4841.

Vibration Institute. 6262 S. Kingery Highway, Suite 212, Willowbrook, IL 60514. (708) 654-2254.

BIBLIOGRAPHIES

Scientific and Technical Books in Print. R.R. Bowker Inc., 121 Chanlon Road, New Providence, NJ 07974. (800) 521-8110 or (908) 464-6800. List of currently published books and serials in the physical and biological sciences, engineering and technology, with subject, author and titles indexes. Inquire as to cost and availability. Also available on ORBIT online service, (800) 456-7248.

DIRECTORIES AND BIOGRAPHICAL SOURCES

Directory of Engineering Societies and Related Organizations. Gordon Davis. 13th edition. American Association of Engineering Societies, 1111 19th Street NW, Suite 608, Washington, DC 20036. (202) 296-2237 or (800) 658-8897. 1989. Inquire for price.

Machinery Buyers Guide. Findlay Publishers Inc. Ltd., Franks Hall, Franks Lane, Horton Kirby, Kent DA4 9LL, England. 1926 to present. Annual. Inquire for cost and availability.

Research Centers Directory. Gale Research, 835 Penobscot Building, Detroit, MI 48226-4094. (313) 961-2242. (800) 347-4253. FAX (313) 961-6083. $485.00.

Scientific and Technical Organizations and Agencies Directory. Gale Research, 835 Penobscot Building, Detroit, MI 48226-4094. (313) 961-2242. (800) 347-4253. FAX (313) 961-6083. 4th edition. 1996. $195.00.

MOTORS

Ency. of Physical Sciences and Engineering Info. Sources

Who's Who in Engineering. American Association of Engineering Societies, 1111 19th Street NW, Suite 608, Washington, DC 20036. (202) 296-2237 or (800) 658-8897. 8th edition. 1991. Inquire for price.

Who's Who in Technology. Gale Research, 835 Penobscot Building, Detroit, MI 48226-4094. (313) 961-2242. (800) 347-4253. FAX (313) 961-6083. 1995. $195.00.

ENCYCLOPEDIAS AND DICTIONARIES

Encyclopedia of Physical Science and Technology. Robert A. Meyers, ed. 18 volumes. Academic Press Inc., 6277 Sea Harbor Drive, Orlando, FL 32887. (800) 321-5068. 1992. $2100.00.

McGraw-Hill Dictionary of Scientific and Technical Terms. Sybil P. Parker, ed. 5th edition. McGraw-Hill Publishing Company, 1221 Avenue of the Americas, New York, NY 10020. (800) 262-4729 or (212) 512-3825. 1993. $110.50.

McGraw-Hill Encyclopedia of Engineering. Sybil P. Parker, ed. 2nd edition. McGraw-Hill Publishing Company, 1221 Avenue of the Americas, New York, NY 10020. (800) 262-4729 or (212) 512-3825. 1993. $95.50.

McGraw-Hill Encyclopedia of Science and Technology. Sybil P. Parker, ed. 7th edition. 20 volumes. McGraw-Hill Publishing Company, 1221 Avenue of the Americas, New York, NY 10020. (800) 262-4729 or (212) 512-3825. 1992. $1900.00

Thesaurus of Scientific, Technical, and Engineering Terms. Hemisphere Publishing Corporation, 1900 Frost Road, Suite 101, Bristol, PA 19007-1598. (215) 785-5800 or (800) 821-8312. FAX (215) 785-5515. 1987. $173.00.

GENERAL WORKS

Design of Small Electrical Machines. E.S. Hamdi. John Wiley and Sons, Inc., 605 Third Avenue, New York, NY 10158. (800) 526-5368 or (212) 850-6000. 1994. Inquire for cost and availability.

Electric Machines—Dynamics and Control. S.A. NASAR. CRC Press, 2000 Corporate Blvd., N.W., Boca Raton, FL 33431. (407) 994-0555 or (800) 272-7737. FAX (407) 994-0949. 1993. Inquire for cost and availability.

Electric Motors and Their Controls. Takahashi Kenjo. Oxford University Press, Inc., 198 Madison Avenue, New York, NY 10016-4314. (212) 726-6000. FAX (212) 726-6446. 1991. $24.95.

Energy-Efficient Motors and Their Applications. Howard E. Jordan. 2nd edition. Plenum Publishing Corporation, 233 Spring Street, New York, NY 10013. (212) 620-8000. FAX (212) 463-0742. 1994. Inquire for cost and availability.

HANDBOOKS AND MANUALS

Electric/Electronic Motor Handbook. Martin Clifford. Prentice Hall, 113 Sylvan Avenue, Route 9W, Englewood Cliffs, NJ 07632. (201) 592-2000 or (800) 922-0579. 1990. $38.00.

Handbook of Mechanics, Materials, and Structures. Alexander Blake, editor. John Wiley and Sons, Inc., 605 Third Avenue,

New York, NY 10158. (800) 526-5368 or (212) 850-6000. 1985. $120.00.

Marks' Standard Handbook For Mechanical Engineers. E.A. Avallone and T. Baumeister III. 9th edition. McGraw-Hill Publishing Company, 1221 Avenue of the Americas, New York, NY 10020. (800) 262-4729 or (212) 512-3825. 1987. $133.00.

Mechanical Design and Systems Handbook. Harold A. Rothbart, editor-in-chief. McGraw-Hill Publishing Company, 1221 Avenue of the Americas, New York, NY 10020. (800) 262-4729 or (212) 512-3825. 1985. Inquire for cost and availability.

Mechanical Engineer's Data Handbook. James Carvill. CRC Press, 2000 Corporate Blvd., N.W., Boca Raton, FL 33431. (407) 994-0555 or (800) 272-7737. FAX (407) 994-0949. 1993. $69.95.

Standard Handbook of Machine Design. Joseph E. Shigley & Charles R. Mischke, editors-in-chief. McGraw-Hill Publishing Company, 1221 Avenue of the Americas, New York, NY 10020. (800) 262-4729 or (212) 512-3825. 1986. $124.50.

ONLINE DATABASES AND CD-ROMS

Compendex Plus. Engineering Information, Inc., 345 East 47th Street, New York, NY 10017. (212) 705-7600 or (800) 221-1044. Contains citations with abstracts to worldwide literature in engineering and technology, from 1970 to present. Available on online BRS,(800) 289-4277, DIALOG, (800) 334-2564, ORBIT (800) 456-7248, and STN International, FIZ Karlsruhe, P.O. Box 2465, W-7500, Karlsruhe 1, Germany, online services. Also available on CD-ROM. Inquire as to cost and availability.

CSA Engineering. Cambridge Scientific Abstracts, 7200 Wisconsin Avenue, Suite 601, Bethesda, MD 20814. (301) 961-6750 or (800) 843-7751. Contains citations and abstracts of international periodicals and research literature covering all fields of engineering and science and technology, including computer and information science, electronics, mechanical engineering, solid state materials, 1981 to present. Available on BRS,(800) 289-4277, online service. Inquire as to cost and availability.

Current Contents Search. Institute for Scientific Information, 3501 Market Street, Philadelphia, PA 19104. (215) 386-0100. FAX (215) 386-6362. Contains citations to articles listed in the table of contents of science and technology journals. Also articles in social sciences and life sciences journals. Available on BRS,(800) 289-4277, DIALOG,(800) 334-2564, online services. Inquire as to cost and availability.

NTIS Bibliographic Database. National Technical Information Service, 5285 Port Royal Road, Springfield, VA 22161. (703) 487-4929 or FAX (703) 321-8199. Broad coverage of government-sponsored science and technology research reports, 1964 to present. Available on BRS,(800) 289-4277, DIALOG, (800) 334-2564, ORBIT, (800) 456-7248, and STN International, FIZ Karlsruhe, P.O. Box 2465, W-7500, Karlsruhe 1, Germany, online services. Also available on CD-ROM. Inquire as to cost and availability.

SAE Global Mobility Database. Society of Automotive Engineers (SAE), Electronic Publishing Division, 400 Commonwealth Drive, Warrendale, PA 15098. (412) 776-4841. Contains citations with abstracts to technical papers on automotive and aerospace technology and vehicular-related industries that have been presented at SAE conferences. Covers 1906 to present.

Available on ORBIT,(800) 456-7248, online service. Inquire as to cost and availability.

TRIS (Transportation Research Information). National Academy of Science, 2101 Constitution Avenue, N.W., Washington, DC 20418. (202) 334-3313 or (800) 624-6242. Citations with abstracts of literature on transportation, including air, highway, rail, maritime and other modes from 1968 to present. Available on DIALOG,(800) 334-2564, online service. Inquire as to cost and availability.

Wilsonline. H.W. Wilson Company, 950 University Avenue, Bronx, NY 10452. (800) 367-6770 or (212) 588-8400. Makes available online versions of the printed H.W. Wilson indexes, including Applied Science and Technology Index, Business Periodicals Index, General Science Index, and Readers' Guide to Periodical Literature. Period covered is generally 1983 to present. Available on BRS,(800) 289-4277, DIALOG,(800) 334-2564, and OCLC EPIC,(800) 848-5878, online services. Also available on CD-ROM. Inquire as to cost and availability.

PERIODICALS

Automotive Engineering Magazine. SAE International, 400 Commonwealth Drive, Warrendale, PA 15096-0001. (412) 776-4841. 1905 to present. Monthly. $72.00.

Institution of Diesel and Gas Turbine Engineers Transactions. Mechanical Engineering Publications Ltd., Northgate Avenue, Bury St. Edmunds, Suffolk IP32 6BW, England. Telephone 0284-763277. FAX 0284-704006. 6 times a year. Inquire for cost.

International Communications in Heat and Mass Transfer. Elsevier Science [journals], 660 White Plains Rd., Tarrytown, NY 10159-5153. (919) 524-9200. FAX (919) 333-2444. 1974 to present. 6 times a year. $430.00 per year.

International Journal For Numerical Methods in Engineering. J. Wiley & Sons Ltd., Journals, Baffins Lane, Chichester, Sussex PO19 1UD, England. Telephone 0243-779777. FAX 0243-775878. 1964 to present. 24 times a year. $1795.00 per year.

International Journal of Heat and Mass Transfer. Elsevier Science [journals], 660 White Plains Rd., Tarrytown, NY 10159-5153. (919) 524-9200. FAX (919) 333-2444. 1960 to present. 18 times a year. $2070.00 per year.

International Journal of Multiphase Flow. Elsevier Science [journals], 660 White Plains Rd., Tarrytown, NY 10159-5153. (919) 524-9200. FAX (919) 333-2444. 1974 to present. 7 times a year. $900.00 per year.

Journal of Applied Mechanics. American Society of Mechanical Engineers, 345 E. 47th Street, New York, NY 10017. (212) 705-7722. 1935 to present. Quarterly. $120.00.

Journal of Engineering For Industry. American Society of Mechanical Engineers, 345 E. 47th Street, New York, NY 10017. (212) 705-7722. 1970 to present. Quarterly. $100.00.

Journal of Fluid Control. Delbridge Publishing Company, Box 160817, Cupertino, CA 95016. (408) 446-3131. 1967 to present. Quarterly. $145.00 per volume.

Journal of Heat Transfer. American Society of Mechanical Engineers, 345 E. 47th Street, New York, NY 10017. (212) 705-7722. 1970 to present. Quarterly. $155.00 per year.

Journal of Mechanical Design. American Society of Mechanical Engineers, 345 E. 47th Street, New York, NY 10017. (212) 705-7722. 1978 to present. Quarterly. $100.00 per year.

Journal of Pressure Vessel Technology. American Society of Mechanical Engineers, 345 E. 47th Street, New York, NY 10017. (212) 705-7722. 1974 to present. Quarterly. $100.00 per year.

Lubrication Engineering. Society of Tribologist and Lubrication Engineers, 840 Busse Highway, Park Ridge, IL 60068-2376. (708) 825-5536. FAX (708) 825-1456. 1945 to present. Monthly. $61.00 per year.

Machine Design. Penton Publishing, 110 Superior Ave., Cleveland, OH 44114-2543. 1929 to present. 6 times a year. $180.00 per year.

Mechanical Engineering. American Society of Mechanical Engineers, 345 E. 47th Street, New York, NY 10017. (212) 705-7722. 1906 to present. Monthly. $45.00 per year.

RESEARCH CENTERS AND INSTITUTES

Laboratory For Experimental Mechanics Research. SUNY, Stony Brook, NY 11794-2300. (516) 632-8311. FAX (516) 632-8720.

Mechanical Engineering Design Laboratory. University of Florida, 237 Mechanical Engineering Bldg., Gainesville, FL 32611. (904) 392-0827. FAX (904) 392-1071.

Mechanical Engineering Laboratories. Stevens Institute of Technology, Hoboken, NJ 07030. (201) 420-5591. FAX (201) 420-6978.

Mechanical Engineering Research Laboratory. Kansas State University, Durland Hall, Manhattan, KS 66506. (913) 532-5610.

MOUNTAINS AND MOUNTAIN BUILDING

See also: GEOLOGY

ABSTRACT SERVICES AND INDEXES

Bibliography and Index of Geology. American Geological Institute, 4220 King Street, Alexandria, VA 22302. (703) 379-2480. Fax (703) 379-7563. Monthly. $1295.00 per year. Also available online as GEOREF.

Current Contents: Physical, Chemical and Earth Sciences. Institute for Scientific Information, 3501 Market Street, Philadelphia, PA 19104. (215) 386-0100. FAX (215) 386-6362. Weekly. $360.00 per year.

Geological Abstracts. Elsevier-Geo Abstracts, Regency House, 34 Duke Street, Norwich NR3 3AP, England. Monthly. $760.00 per year. Also available online as GEOBASE.

Geological Society of America. Abstracts with Programs. Geological Society of America. 3300 Penrose Place, P.O. Box 9140,

MOUNTAINS AND MOUNTAIN BUILDING

Ency. of Physical Sciences and Engineering Info. Sources

Boulder, CO 80301-9140. (303) 447-2020. Abstracts and programs of the annual conference. Annual. $69.00.

Science Citation Index. Institute for Scientific Information, 3501 Market Street, Philadelphia, PA 19104. (215) 386-0100. FAX (215) 386-6362. Inquire about availability and cost. Also available on CD-ROM.

ANNUAL REVIEWS AND YEARBOOKS

Annual Review of Earth and Planetary Sciences. Annual Reviews Inc., 4139 El Camino Way, PO Box 10139, Palo Alto, CA 94303-0139. (800) 523-8635. FAX (415) 855-9815. Annual. $62.00 (1995 edition).

ASSOCIATIONS AND PROFESSIONAL SOCIETIES

American Geological Institute. 4220 King Street, Alexandria, VA 22302. (703) 379-2480. Fax (703) 379-7563.

American Geophysical Union. 2000 Florida Avenue, N.W., Washington, DC 20009. (202) 462-6900. FAX (202) 328-0566.

Geological Society of America. 3300 Penrose Place, PO Box 9140, Denver, CO 80301-9140. (303) 447-2020. FAX (303) 447-1133.

DIRECTORIES AND BIOGRAPHICAL SOURCES

American Men & Women of Science, 1995-96. R.R. Bowker Staff, eds. 19th edition. 8 volumes. R.R. Bowker/Reed International Publishing Company, 121 Chanlon Road, New Providence, NJ 07974. (908) 464-6800 or (800) 521-8110. 1995. $850.00

Directory of Engineering Societies and Related Organizations. Gordon Davis. 13th edition. American Association of Engineering Societies, 1111 19th Street NW, Suite 608, Washington, DC 20036. (202) 296-2237 or (800) 658-8897. 1989. Inquire for price.

Geological Society of America Membership Directory. 3300 Penrose Place, PO Box 9140, Denver, CO 80301-9140. (303) 447-2020. Annual. Inquire for cost and availability.

International Research Centers Directory. Gale Research, 835 Penobscot Building, Detroit, MI 48226-4094. (313) 961-2242. (800) 347-4253. FAX (313) 961-6083. 8th edition. 1995. $410.00

Research Centers Directory. Gale Research, 835 Penobscot Building, Detroit, MI 48226-4094. (313) 961-2242. (800) 347-4253. FAX (313) 961-6083. $485.00.

Scientific and Technical Organizations and Agencies Directory. Gale Research, 835 Penobscot Building, Detroit, MI 48226-4094. (313) 961-2242. (800) 347-4253. FAX (313) 961-6083. 4th edition. 1996. $195.00.

ENCYCLOPEDIAS AND DICTIONARIES

The Encyclopedia of the Solid Earth Sciences. Philip Kearey, editor-in-chief. Blackwell Scientific Publications, 238 Main

Street, Cambridge, MA 02142. (617) 876-7000 or (800) 759-6102. FAX (617) 876-7022. 1993. $60.00.

Encyclopedia of Physical Science and Technology. Robert A. Meyers, ed. 18 volumes. Academic Press Inc., 6277 Sea Harbor Drive, Orlando, FL 32887. (800) 321-5068. 1992. $2100.00.

Glossary of Geology. Robert Bates, et al. Reprint of third edition. American Geological Institute, 4220 King Street, Alexandria, VA 22302-1507. (703) 379-2480. FAX (301) 379-7563. 1990. $225.00.

GENERAL WORKS

An Anatomy of Mountain Ranges. Jean-Paul Schaer and John Rodgers, editors. Princeton University Press, 41 William Street, Princeton, NJ 08540. (609) 258-4900 or (800) 777-4726. FAX (609) 895-1081. 1987. $52.50.

The Earth and Its Mountains. Raymond A. Lyttleton. Reprint edition. Books on Demand, Division of University Microfilms International, 300 North Zeeb Road, Ann Arbor, MI 48106-1346. (313) 761-4700 or (800) 521-0600. $66.20. Inquire for availability.

Mountain Building Processes. K.J. Hsu, editor. Academic Press Inc., 6277 Sea Harbor Drive, Orlando, FL 32887. (800) 321-5068. 1983. Out of print.

Principles of Terrane Analysis: New Applications For Global Tectonics. David G. Howell. Second edition (formerly *Tectonics of Suspect Terranes: Mountain Building and Continental Growth,* 1989). Chapman and Hall, 29 West 35th Street, New York, NY 10001. (800) 842-3636. FAX (212) 563-2269. 1995. Inquire for price.

ONLINE DATABASES AND CD-ROMS

Current Contents Search. Institute for Scientific Information, 3501 Market Street, Philadelphia, PA 19104. (215) 386-0100. FAX (215) 386-6362. Contains citations to articles listed in the table of contents of science and technology journals. Also articles in social sciences and life sciences journals. Available on BRS,(800) 289-4277, DIALOG,(800) 334-2564, online services. Inquire as to cost and availability.

Dissertation Abstracts. University Microfilms International, 300 North Zeeb Road, Ann Arbor, MI 48106. (800) 521-0600 or (313) 761-4700. Scope includes virtually all doctoral dissertations accepted at accredited American institutions from 1861 to present in 252 subject areas. Available on BRS,(800) 289-4277, DIALOG,(800) 334-2564, and OCLC EPIC,(800) 848-5878, online services. Also available on CD-ROM. Inquire as to cost and availability.

Earth Sciences. U.S. Geological Survey, 12201 Sunrise Valley Drive, Reston, VA 22092-9998. (703) 648-4460. CD-ROM of earth science databases, including the U.S. Geological Survey Library, Earth Science Data Directory, and GEOINDEX, citations to published geological maps. $350.00 per year with quarterly updates.

Geoarchive. Geosystems, P.O. Box 1024, Westminster, London, England, SW1 P 2JL. Citations to literature on geoscience, 1969 to present. Inquire as to online cost and availability.

Geobase. Elsevier-Geo Abstracts, Regency House, 34 Duke Street, Norwich NR3 3AP, England. Contains citations to the worldwide earth science literature from 1980 to date. Available on DIALOG, ORBIT online services. Inquire as to cost and availability.

GEOREF: Bibliography and Index of Geology. American Geological Institute, 4220 King Street, Alexandria, VA 22302. (703) 379-2480. Fax (703) 379-7563. Monthly. Inquire for price and availability.

Scisearch. Institute for Scientific Information, 3501 Market Street, Philadelphia, PA 19104. (800) 523-1850 or (215) 386-0100. Broad multidisciplinary title and author index to the international literature of science and technology, 1974 to present. Available on DIALOG,(800) 334-2564, and ORBIT,(800) 456-7248, online services. Also available on CD-ROM. Inquire as to cost and availability.

PERIODICALS

American Journal of Science. Box 6666, Yale Station, New Haven, CT 06511-8130. (203) 432-3131. FAX (203) 432-5668. 1818 to present. Monthly except July and August. $40.00 per year.

Geological Magazine. Cambridge University Press, Journals Department, 40 West 20th Street, New York, NY 10011-4211. (212) 924-3900. 1864 to present. Bi-monthly. $263.00 per year.

Geological Society Journal. Geological Society Publishing House, Unit 7, Brassmill Enterprise Centre, Brassmill Lane, Bath BA1 3JN, England. Telephone 0225-445046. FAX 0225-442836. 1845 to present. Six times a year. $524.00.

Geological Society of America Bulletin. 3300 Penrose Place, PO Box 9140, Denver, CO 80301-9140. (303) 447-2020. 1888 to present. Monthly. $205.00 per year.

Geology. Geological Society of America, 3300 Penrose Place, PO Box 9140, Denver, CO 80301-9140. (303) 447-2020. 1973 to present. Monthly. $170.00 per year for non-members.

Journal of Geology. University of Chicago Press, Journals Division, 5720 S. Woodlawn Avenue, Chicago, IL 60637. (312) 753-3347. FAX (312) 753-0811. 1893 to present. Bi-monthly. $45.00 per year.

RESEARCH CENTERS AND INSTITUTES

Columbia University, Lamont-Doherty Geological Observatory. Palisades, NY 10964. (914) 359-2900. FAX (914) 365-2312.

Cornell University Institute For the Study of Continents. 3122 Snee Hall, Ithaca, NY 14853-1504. (607) 255-3474.

Texas A&M University, Center For Tectonophysics. College Station, TX 77843-3113. (409) 845-3251. FAX (409) 845-6780.

Texas A&M University Geodynamics Research Center. College Station, TX 77843-3364. (409) 845-8477. fax (409) 845-3138.

MULTIVARIATE ANALYSIS

See: MATHEMATICS

N

NATURAL GAS

See also: PETROLEUM GEOLOGY, PETROLEUM ENGINEERING, PIPELINE TECHNOLOGY

ABSTRACT SERVICES AND INDEXES

Applied Science and Technology Index; A Cumulative Subject Index To English Language Periodicals in the Fields of Aeronautics and Space Science, Computer Technology, Chemistry, Construction Industry, Energy and Related Areas. H.W. Wilson Co., 950 University Avenue, Bronx, NY 10452. (800) 367-6770 or (212) 588-8400. FAX (718) 590-1617. From 1958 to present. Monthly. Inquire about cost and availability. Also available on CD-ROM and online.

Bibliography and Index of Geology. American Geological Institute, 4220 King Street, Alexandria, VA 22302. (703) 379-2480. Fax (703) 379-7563. Monthly. $1295.00 per year. Also available online as GEOREF.

Current Contents: Engineering, Technology, and Applied Sciences. Institute for Scientific Information, 3501 Market Street, Philadelphia, PA 19104. (215) 386-0100. FAX (215) 386-6362. 1970 to present. Weekly. $442.00 per year.

Energy Research Abstracts. 1976-present. U.S. Department of Energy, office of Scientific and Technical Information, Box 62, Oak Ridge, TN 37831. (615) 574-0733. Subscriptions: Superintendent of Documents, U.S. Government Printing office, Box 371954, Pittsburgh, PA 15250-7954. (202) 783-3238. FAX (202) 512-2233. Monthly. $164.00 per year. Abstracts all the scientific and technical reports, journal articles, conference proceedings, patents, theses, and monographs sponsored by the U.S. Energy Research and Development Administration.

Engineering Index Monthly. Engineering Information, Inc., Castle Point on the Hudson, Hoboken, NJ 07030. (800) 221-1044. FAX (212) 832-1857. Monthly. $2300.00 per year. Also available online as COMPENDEX, and also on CD-ROM. Covers chemical engineering, computers, electrical engineering, civil engineering, metals and mining, industrial management, and mechanical engineering.

Fuel and Energy Abstracts. Butterworth-Heinemann, Linacre House, Jordan Hill, Ocford OX2 8DP, England. Telephone 0865-310366. FAX 0865-310898. Summary of world literature on all scientific, technical, commercial, and environmental aspects of fuel and energy. 1960 to present. Six times a year. Inquire for price in U.S.

Geological Abstracts. Elsevier - Geo Abstracts, Regency House, 34 Duke Street, Norwich NR3 3AP, England. Monthly. $760.00 per year. Also available online as GEOBASE.

Geological Society of America. ABSTRACTS WITH PROGRAMS. Geological Society of America. 3300 Penrose Place, P.O. Box 9140, Boulder, CO 80301-9140. (303) 447-2020. Abstracts and programs of the annual conference. Annual. $69.00.

Offshore Engineering Abstracts. STI Limited, 4 Kings Meadow, Ferry Hinksey Road, Oxford, OX2 ODU, England. Distributed in the United States by Air Science Company, Box 143, Corning NY 14830. (607) 962-5591. Contains citations on planning, design, and construction of offshore platforms and pipelines, from 1985 to present. Monthly. $400.00 per year.

Petroleum Abstracts. 600 South College, Tulsa, OK 74104. (918) 631-2296 or (800) 247-8678. Contains citations with abstracts to the worldwide literature and patents on the exploration, development, and production of petroleum resources, from 1965 to present. <subhead2>Annual Reviews and Yearbooks

International Oil and Gas Exploration Developments, 1991. Diane Publishing Company, 600 Upland Avenue, Upland, PA 19015. (610) 499-7415. FAX (610) 499-7429. 1991. $49.95.

Natural Gas Annual. Gordon Press Publishers, PO Box 459, Bowling Green Station, New York, NY 10004. (718) 624-8419. 1995. $250.00.

Natural Gas Annual. U.S. Department of Energy, Energy Information Administration, National Energy Information Center, El-231, Room 1F-048, Forrestal Bldg., 1000 Independence Avenue SW, Washington, DC 20585. (202) 586-8800. 1990 to present. Annual. $13.00.

Natural Gas Annual, 1992. Diane Publishing Company, 600 Upland Avenue, Upland, PA 19015. (610) 499-7415. FAX (610) 499-7429. 1994. $95.00.

Natural Gas Yearbook. Executive Enterprises Publications Company, 22 W. 21st Street, New York, NY 10010-6904. (212) 645-7880. FAX (212) 645-1160. 1988 to present. Annual. $95.00.

NATURAL GAS

Ency. of Physical Sciences and Engineering Info. Sources

The 1994 Natural Gas Yearbook. Robert E. Willett, editor. John Wiley and Sons, Inc., 605 Third Avenue, New York, NY 10158. (800) 526-5368 or (212) 850-6000. 1994. $95.00.

Oil & Gas Information, 1993. OECD staff. Organization for Economic Cooperation and Development, 2001 L St. NW, Suite 650, Washington, DC 20036. (202) 785-6323 or (800) 456-6323. FAX (202) 785-0350. 1994. $125.00.

ASSOCIATIONS AND PROFESSIONAL SOCIETIES

American Association of Petroleum Geologists. Box 979, Tulsa, OK 74101. (918) 584-2555. FAX (918) 584-0469.

American Gas Association. 1515 Wilson Blvd., Arlington, VA 22209. (703) 841-8400. FAX (703) 841-8406.

American Geological Institute. 4220 King Street, Alexandria, VA 22302. (703) 379-2480. Fax (703) 379-7563.

American Geophysical Union. 2000 Florida Avenue, N.W., Washington, DC 20009. (202) 462-6900. FAX (202) 328-0566.

Gas Processors' Association. 6526 E 60th Street, Tulsa, OK 74145. (918) 493-4872. FAX (918) 493-3875.

Gas Research Institute. 8600 W Bryn Mawr Avenue, Chicago, IL 60631-3562. (312) 399-8100. FAX (312) 399-8170.

Institute of Gas Technology. 3424 S. State Street, Chicago, IL 60616-3896. (312) 949-3650. FAX (312) 949-3776.

Institution of Gas Engineers. 17 Grosvenor Crescent, London SW1X 7ES, England.

Society of Exploration Geophysicists. PO Box 70240, Tulsa, OK 74170-2640. (918) 493-3516. FAX (918) 493-2074.

DIRECTORIES AND BIOGRAPHICAL SOURCES

American Men & Women of Science, 1995-96. R.R. Bowker Staff, eds. 19th edition. 8 volumes. R.R. Bowker/Reed International Publishing Company, 121 Chanlon Road, New Providence, NJ 07974. (908) 464-6800 or (800) 521-8110. 1995. $850.00.

Gas Industry Directory. Benn Business Information Services, Riverbank House, Angel Lane, Tonbridge, Kent TN9 1SE. Telephone 0732-362666. FAX 0732-767301. 1896 to present. Annual. Inquire for price in U.S.

Gas Pipeline Personnel Directory, 1993. Oil Pipeline Research Institute (OPRI), PO Box 106, Boulder, CO 80306. (303) 545-5459. FAX (303) 444-1286. 1992. $250.00.

International Research Centers Directory. Gale Research, 835 Penobscot Building, Detroit, MI 48226-4094. (313) 961-2242. (800) 347-4253. FAX (313) 961-6083. 8th edition. 1995. $410.00

Natural Gas Companies in the U.S.: A Detailed Profile. Diane Publishing Company, 600 Upland Avenue, Upland, PA 19015. (610) 499-7415. FAX (610) 499-7429. 1994. $60.00.

Natural Gas Industry Directory. PennWell Books, 1421 S. Sheridan Road, Box 1260, Tulsa, OK 74112. (918) 835-3161 or (800) 752-9764. FAX (918) 831-9555. 1991 to present. Annual. $150.00.

Oil & Gas Directory. Geophysical Directory Inc., 2200 Welch Avenue, Box 130508, Houston, TX 77219. (713) 529-8789. FAX (713) 529-3646. 1970 to present. Annual. $60.00.

Research Centers Directory. Gale Research, 835 Penobscot Building, Detroit, MI 48226-4094. (313) 961-2242. (800) 347-4253. FAX (313) 961-6083. $485.00.

Scientific and Technical Organizations and Agencies Directory. Gale Research, 835 Penobscot Building, Detroit, MI 48226-4094. (313) 961-2242. (800) 347-4253. FAX (313) 961-6083. 4th edition. 1996. $195.00.

Who's Who in Engineering. American Association of Engineering Societies, 1111 19th Street NW, Suite 608, Washington, DC 20036. (202) 296-2237 or (800) 658-8897. 8th edition. 1991. Inquire for price.

Who's Who in Natural Gas. Stalsby/Wilson Press, PO Box 19976, Houston, TX 77224. (713) 496-1734. FAX (713) 531-7229. $125.00. Inquire for availability.

ENCYCLOPEDIAS AND DICTIONARIES

The Natural Gas Pipeline Encyclopedia, 1993. Donald Liebson. Revised edition. Oil Pipeline Research Institute (OPRI), PO Box 106, Boulder, CO 80306. (303) 545-5459. FAX (303) 444-1286. $2000.00. Inquire for availability.

The Natural Gas Pipeline Sourcebook, 1993. Donald Liebson. Revised edition. Oil Pipeline Research Institute (OPRI), PO Box 106, Boulder, CO 80306. (303) 545-5459. FAX (303) 444-1286. $1000.00. Inquire for availability.

Oil & Gas Dictionary. Paul Stevens. Nichols Publishing Company, PO Box 6036, East Brunswick, NJ 08816-6036. (908) 297-2862. FAX (908) 940-0549. 1988. $78.50.

GENERAL WORKS

The Economics of Natural Gas: Pricing, Planning & Policy. DeAnne Julius and Afsaneh Mashayekhi. Oxford University Press, Inc., 198 Madison Avenue, New York, NY 10016-4314. (212) 726-6000. FAX (212) 726-6446. 1990. $49.95.

Largest U.S. Oil & Gas Fields. Diane Publishing Company, 600 Upland Avenue, Upland, PA 19015. (610) 499-7415. FAX (610) 499-7429. 1994. $45.00.

Natural Gas: Basic Science and Technology. A. Melvin. IOP Publishing, Public Ledger Building, Suite 1035, Independence Square, Philadelphia, PA 19106. (800) 358-4677. FAX (215) 627-0879. 1987. $87.00.

Natural Gas Engineering: Production and Storage. Donald Katz and Robert Lee. McGraw-Hill Publishing Company, 1221 Avenue of the Americas, New York, NY 10020. (800) 262-4729 or (212) 512-3825. 1990. Inquire for price.

Ency. of Physical Sciences and Engineering Info. Sources

NATURAL GAS

Natural Gas, Issues & Trends. Gordon Press Publishers, PO Box 459, Bowling Green Station, New York, NY 10004. (718) 624-8419. 1995. $250.00.

Natural Gas Statistics Sourcebook. PennWell Books, 1421 S. Sheridan Road, Box 1260, Tulsa, OK 74112. (918) 835-3161 or (800) 752-9764. FAX (918) 831-9555. 1994. $185.00.

Oil & Gas Pipeline Fundamentals. John L. Kennedy. Second edition. PennWell Books, 1421 S. Sheridan Road, Box 1260, Tulsa, OK 74112. (918) 835-3161 or (800) 752-9764. FAX (918) 831-9555. 1993. $59.95.

Practical Natural Gas Engineering. R.V. Smith. Second edition. PennWell Books, 1421 S. Sheridan Road, Box 1260, Tulsa, OK 74112. (918) 835-3161 or (800) 752-9764. FAX (918) 831-9555. 1990. $25.00.

HANDBOOKS AND MANUALS

Petroleum Geology Handbook. N. A. Eremenko and George V. Chilingarian, editors. Optimization Software, Inc., 1100 Glendon Avenue, Suite 1447, Los Angeles, CA 90024. (310) 208-5674. 1991. $200.00.

ONLINE DATABASES AND CD-ROMS

CA Search. Chemical Abstracts Service, P.O. Box 3012, Columbus, OH 43210-0012. (614) 447-3600. (800) 848-6533. FAX (614) 447-3709. Very comprehensive guide to worldwide chemical literature and related fields, 1972 to present. Available on BRS,(800) 289-4277, DIALOG, (800) 334-2564, ORBIT (800) 456-7248, and STN International, FIZ Karlsruhe, P.O. Box 2465, W-7500, Karlsruhe 1, Germany, online services. Inquire as to cost and availability.

Compendex Plus. Engineering Information, Inc., 345 East 47th Street, New York, NY 10017. (212) 705-7600 or (800) 221-1044. Contains citations with abstracts to worldwide literature in engineering and technology, from 1970 to present. Available on online BRS,(800) 289-4277, DIALOG, (800) 334-2564, ORBIT (800) 456-7248, and STN International, FIZ Karlsruhe, P.O. Box 2465, W-7500, Karlsruhe 1, Germany, online services. Also available on CD-ROM. Inquire as to cost and availability.

Earth Sciences. U.S. Geological Survey, 12201 Sunrise Valley Drive, Reston, VA 22092-9998. (703) 648-4460. CD-ROM of earth science databases including the U.S. Geological Survey Library, Earth Science Data Directory, and GEOINDEX, citations to published geological maps. $350.00 per year with quarterly updates.

Geoarchive. Geosystems, P.O. Box 1024, Westminster, London, England, SW1 P 2JL. Citations to literature on geoscience, 1969 to present. Inquire as to online cost and availability.

Geobase. Elsevier - Geo Abstracts, Regency House, 34 Duke Street, Norwich NR3 3AP, England. Contains citations to the worldwide earth science literature from 1980 to date. Available on DIALOG, ORBIT online services. Inquire as to cost and availability.

GEOREF: Bibliography and Index of Geology. American Geological Institute, 4220 King Street, Alexandria, VA 22302. (703) 379-2480. Fax (703) 379-7563. Monthly. Inquire for price and availability.

NTIS Bibliographic Database. National Technical Information Service, 5285 Port Royal Road, Springfield, VA 22161. (703) 487-4929 or FAX (703) 321-8199. Broad coverage of government-sponsored science and technology research reports, 1964 to present. Available on BRS,(800) 289-4277, DIALOG, (800) 334-2564, ORBIT, (800) 456-7248, and STN International, FIZ Karlsruhe, P.O. Box 2465, W-7500, Karlsruhe 1, Germany, online services. Also available on CD-ROM. Inquire as to cost and availability.

SCISEARCH. Institute for Scientific Information, 3501 Market Street, Philadelphia, PA 19104. (800) 523-1850 or (215) 386-0100. Broad multidisciplinary title and author index to the international literature of science and technology, 1974 to present. Available on DIALOG,(800) 334-2564, and ORBIT,(800) 456-7248, online services. Also available on CD-ROM. Inquire as to cost and availability.

WILSONLINE. H.W. Wilson Company, 950 University Avenue, Bronx, NY 10452. (800) 367-6770 or (212) 588-8400. Makes available online versions of the printed H.W. Wilson indexes including Applied Science and Technology Index, Business Periodicals Index, General Science Index, and Readers' Guide to Periodical Literature. Period covered is generally 1983 to present. Available on BRS,(800) 289-4277, DIALOG,(800) 334-2564, and OCLC EPIC,(800) 848-5878, online services. Also available on CD-ROM. Inquire as to cost and availability.

PERIODICALS

A A P G Bulletin. American Association of Petroleum Geologists, Box 979, Tulsa, OK 74101. (918) 584-2555. FAX (918) 584-0469. 1917 to present. Monthly. $135.00.

A A P G Explorer. American Association of Petroleum Geologists, Box 979, Tulsa, OK 74101. (918) 584-2555. FAX (918) 584-0469. 1979 to present. Monthly. $45.00.

A A P G Studies in Geology Series. American Association of Petroleum Geologists, Box 979, Tulsa, OK 74101. (918) 584-2555. FAX (918) 584-0469. Irregular. Price varies.

American Gas. American Gas Association, 1515 Wilson Blvd., Arlington, VA 22209. (703) 841-8400. FAX (703) 841-8406. 1919 to present. 11 times a year. $39.00.

American Oil and Gas Reporter. National Publishers Group Inc., Box 343, Derby, KS 67037-0343. FAX (316) 788-7568. 1958 to present. 12 times a year. $28.00.

Gas Engineering & Management. Institution of Gas Engineers. 17 Grosvenor Crescent, London SW1X 7ES, England. 1961 to present. Ten times a year. Inquire for price in U.S.

Gas World International. Oakfield House, Perrymount Road, Haywards Heath, W. Sussex RH16 3DH, England. Telephone 0732-364422. 1884 to present. Monthly. Inquire for price in U.S.

G R I D (gas Research Institute Digest). Gas Research Institute, 8600 W Bryn Mawr Avenue, Chicago, IL 60631-3562. (312) 399-8100. FAX (312) 399-8170. 1976 to present. Quarterly. Free.

Journal of Petroleum Geology. Scientific Press Ltd., PO Box 21, Beaconsfield, Bucks HP9 1NS, England. Telephone 0494-

NATURAL GAS

Ency. of Physical Sciences and Engineering Info. Sources

675139. FAX 0494-670155. 1978 to present. Quarterly. $264.00 per year.

Natural Gas. Oakfield House, Perrymount Road, Haywards Heath, W. Sussex RH16 3DH, England. Telephone 0732-364422. 1928 to present. Bi-monthly. Inquire for price in U.S.

Natural Gas. Executive Enterprises Publications Company, 22 W. 21st Street, New York, NY 10010-6904. (212) 645-7880. FAX (212) 645-1160. 1949 to present. Monthly. $295.00.

Natural Gas Monthly. U.S. Department of Energy, Energy Information Administration, National Energy Information Center, El-231, Room 1F-048, Forrestal Bldg., 1000 Independence Avenue SW, Washington, DC 20585. (202) 586-8800. Monthly. $73.00.

Oil & Gas Journal. PennWell Publishing Company, Box 1260, Tulsa, OK 74101. (918) 935-3161. 1902 to present. Weekly. $95.00 per year.

Petroleum Geology: A Digest of Russian Literature On Petroleum Geology. Box 171, McLean, VA 22101. 1959 to present. Monthly. $40.00 per year.

RESEARCH CENTERS AND INSTITUTES

Gas Research Institute. 8600 W Bryn Mawr Avenue, Chicago, IL 60631-3562. (312) 399-8100. FAX (312) 399-8170.

Institute of Gas Technology. 3424 S. State Street, Chicago, IL 60616. (312) 567-3650. FAX (312) 567-5209.

University of Oklahoma Energy Center. Norman, OK 73019. (405) 325-3821. FAX (405) 325-3180.

NAVAL ARCHITECTURE

See also: MARINE ENGINEERING

ABTRACT SERVICES AND INDEXES

Applied Mechanics Reviews: An Assessment of World Literature in Engineering Sciences. 1948-present. American Society of Mechanical Engineers, 345 East 47th Street, New York, NY 10017. (212) 705-7703. Monthly. $360.00 per year.

Applied Science and Technology Index; A Cumulative Subject Index To English Language Periodicals in the Fields of Aeronautics and Space Science, Computer Technology, Chemistry, Construction Industry, Energy and Related Areas. H.W. Wilson Co., 950 University Avenue, Bronx, NY 10452. (800) 367-6770 or (212) 588-8400. FAX (718) 590-1617. From 1958 to present. Monthly. Inquire about cost and availability. Also available on CD-ROM and online.

CAD/cam Abstracts. R.R. Bowker, Bowker A&I Publishing, 121 Chanlon Road, New Providence, NJ 07974. (908) 771-7714. Fax (908) 771-7725. 1984 to present. Monthly. $495.00 per year.

Civil and Structural Engineering Abstracts, Cambridge Scientific Abstracts, 7200 Wisconsin Avenue, Bethesda, MD 20814-4823. (301) 961-6750. FAX (301) 961-6720. 1993 to present.

Monthly. $385.00 per year. Topics covered include structural design, construction equipment and methods, civil defense and military engineering, surveying, highway engineering, maritime and port structures, materials, land reclamation, and soil mechanics.

Computer Abstracts, MCBUniversity Press Ltd., PO Box 10812, Birmingham, AL 35201-0812. (800) 633-4931. FAX (205) 995-1588. Monthly. Covers computer theory, data, hardware, systems, networks, human-computer interaction, artificial intelligence, as well as applications of computers in aerospace, business, CAD/CAM, cartography, civil engineering, electronics and electrical engineering, industrial engineering, mechanical engineering, medicine, structural engineering, etc.

Current Contents: Engineering, Technology, and Applied Sciences. Institute for Scientific Information, 3501 Market Street, Philadelphia, PA 19104. (215) 386-0100. FAX (215) 386-6362. 1970 to present. Weekly. $442.00 per year.

Engineering Index Monthly. Engineering Information, Inc., Castle Point on the Hudson, Hoboken, NJ 07030. (800) 221-1044. FAX (212) 832-1857. Monthly. $2300.00 per year. Also available online as COMPENDEX, and also on CD-ROM. Covers chemical engineering, computers, electrical engineering, civil engineering, metals and mining, industrial management, and mechanical engineering.

Mechanical Engineering Abstracts (formerly ISMEC), Cambridge Scientific Abstracts, 7200 Wisconsin Avenue, Bethesda, MD 20814-4823. (301) 961-6750. FAX (301) 961-6720. 1967 to present. Monthly. $895.00 per year. Summarizes world literature in mechanical engineering, production engineering, and engineering management. Also available online.

Naval Engineers' Journal Cumulative Index, 1889-1979. American Society of Naval Engineers, 1452 Duke Street, Alexandria, VA 22314. (703) 836-6727. FAX (703) 836-7491. Inquire for price and availability.

Oceanic Abstracts. Cambridge Scientific Abstracts, 7200 Wisconsin Avenue, Bethesda, MD 20814. (301) 961-6750. Fax (301) 961-6720. Bimonthly. $995.00 per year.

Science Citation Index. Institue for Scientific Information, 3501 Market Street, Philadelphia, PA 19104. (215) 386-0100. FAX (215) 386-6362. Inquire about availability and cost. Also available on CD-ROM.

ASSOCIATIONS AND PROFESSIONAL SOCIETIES

American Bureau of Shipping. Two World Trade Center, 106th Floor, New York, NY 10048. (212) 839-5000. FAX (212) 839-5130.

American Society of Mechanical Engineers. 345 E. 47th Street, New York, NY 10017. (212) 705-7722.

American Society of Naval Engineers. 1452 Duke Street, Alexandria, VA 22314. (703) 836-6727. FAX (703) 836-7491.

Institute of Marine Engineers. 76 Mark Lane, London EC3R 7JN, England. Telephone 71-4818493. FAX 71-4818854.

Marine Technology Society. 1828 L Street NW, No. 906, Washington, DC 20036-5104. (202) 775-5966. FAX (202) 429-9417.

Royal Society of Naval Architects. 10 Upper Belgrave Street, London SW1 X8BQ, England. Telephone 071-235-4622. FAX 071-245-6959.

Society of Naval Architects and Marine Engineers. 601 Pavonia Avenue, Suite 400, Jersey City, NJ 07306. (201) 798-4800. FAX (201) 798-4975.

DIRECTORIES AND BIOGRAPHICAL SOURCES

American Society of Naval Engineers—membership Directory. American Society of Naval Engineers, 1452 Duke Street, Alexandria, VA 22314. (703) 836-6727. FAX (703) 836-7491. Inquire for price and availability.

Directory of Engineering Societies and Related Organizations. Gordon Davis. 13th edition. American Association of Engineering Societies, 1111 19th Street NW, Suite 608, Washington, DC 20036. (202) 296-2237 or (800) 658-8897. 1989. Inquire for price.

Marine Equipment Catalog. Maritime Activity Reports, Inc., 118 E 25th Street, New York, NY 10010. (212) 477-6700. 1984 to present. Annual. $65.00.

Research Centers Directory. Gale Research, 835 Penobscot Building, Detroit, MI 48226-4094. (313) 961-2242. (800) 347-4253. FAX (313) 961-6083. $485.00.

Scientific and Technical Organizations and Agencies Directory. Gale Research, 835 Penobscot Building, Detroit, MI 48226-4094. (313) 961-2242. (800) 347-4253. FAX (313) 961-6083. 4th edition. 1996. $195.00.

Who's Who in Engineering. American Association of Engineering Societies, 1111 19th Street NW, Suite 608, Washington, DC 20036. (202) 296-2237 or (800) 658-8897. 8th edition. 1991. Inquire for price.

GENERAL WORKS

Basic Ship theory. K.J. Rawson & E.C. Tupper. 4th edition. Halsted Press (Division of John Wiley and Sons, Inc.), 605 Third Avenue, New York, NY 10158. (800) 526-5368 or (212) 850-6418. 1994. Inquire for cost and availability.

Cfd and CAD in Ship Design. G. Van Oortmerssen. Elsevier Science B.V., PO Box 945, Madison Square Station, New York, NY 10159-0945. (212) 633-3650. FAX (212) 633-3680. 1990. $130.00.

Management of Marine Design. Stian Erichsen. Butterworth-Heinemann, 313 Washington Street, Newton, MA 02158. (617) 928-2500 or (800) 366-2665. FAX (617) 928-2620. 1989. $110.00.

Marine Engineering: Design and Operation of Ships and offshore Structures. T.K.S. Murthy & C.A. Brebbia, editors. Computational Mechanics Publications, 25 Bridge Street, Billerica, MA 01821. (508) 667-5841. FAX (508) 667-7582. 1994. $134.00.

Naval Architecture For Non-naval Architects. Harry Benford. Society of Naval Architects and Marine Engineers, 601 Pavonia Avenue, Suite 400, Jersey City, NJ 07306. (201) 798-4800. FAX (201) 798-4975. 1991. $35.00.

Principles of Naval Architecture. Edward V. Lewis, editor. Society of Naval Architects and Marine Engineers, 601 Pavonia Avenue, Suite 400, Jersey City, NJ 07306. (201) 798-4800. FAX (201) 798-4975. 1988. 3-volume set, $240.00.

Ship Construction. 3rd edition. Butterworth-Heinemann, 313 Washington Street, Newton, MA 02158. (617) 928-2500 or (800) 366-2665. FAX (617) 928-2620. 1988. $34.95.

Ship Design For Efficiency and Economy. Herbert Schneekluth. Butterworth-Heinemann, 313 Washington Street, Newton, MA 02158. (617) 928-2500 or (800) 366-2665. FAX (617) 928-2620. 1987. $115.00.

Ships, Submarines, and the Sea. P.J. Gates and N.M. Lynn. Brassey's, 1313 Dolley Madison Blvd., Suite 101, McLean, VA 22102-3926. (703) 442-4535 or (800) 775-2518. FAX (703) 442-9848. 1990. $40.00 (hardback), $25.00 (paperback).

HANDBOOKS AND MANUALS

Marine Engine Room Blue Book. William D. Eglinton. 4th edition. Cornell Maritime Press Inc., PO Box 456, Centreville, MD 21617. (410) 758-1075 or (800) 638-7641. FAX (410) 758-6849. 1993. $25.00.

Marks' Standard Handbook For Mechanical Engineers. E.A. Avallone and T. Baumeister III. 9th edition. McGraw-Hill Publishing Company, 1221 Avenue of the Americas, New York, NY 10020. (800) 262-4729 or (212) 512-3825. 1987. $133.00.

Mechanical Engineer's Data Handbook. James Carvill. CRC Press, 2000 Corporate Blvd., N.W., Boca Raton, FL 33431. (407) 994-0555 or (800) 272-7737. FAX (407) 994-0949. 1993. $69.95.

ONLINE DATABASES AND CD-ROMS

Compendex Plus. Engineering Information, Inc., 345 East 47th Street, New York, NY 10017. (212) 705-7600 or (800) 221-1044. Contains citations with abstracts to worldwide literature in engineering and technology, from 1970 to present. Available on online BRS,(800) 289-4277, DIALOG, (800) 334-2564, ORBIT (800) 456-7248, and STN International, FIZ Karlsruhe, P.O. Box 2465, W-7500, Karlsruhe 1, Germany, online services. Also available on CD-ROM. Inquire as to cost and availability.

CSA Engineering. Cambridge Scientific Abstracts, 7200 Wisconsin Avenue, Suite 601, Bethesda, MD 20814. (301) 961-6750 or (800) 843-7751. Contains citations and abstracts of international periodicals and research literature covering all fields of engineering and science and technology,including computer and information science, electronics, mechanical engineering, solid state materials, 1981 to present. Available on BRS,(800) 289-4277, online service.Inquire as to cost and availability.

Current Contents Search. Institute for Scientific Information, 3501 Market Street, Philadelphia, PA 19104. (215) 386-0100. FAX (215) 386-6362. Contains citations to articles listed in the table of contents of science and technology journals. Also articles in social sciences and life sciences journals. Available on BRS,(800) 289-4277, DIALOG,(800) 334-2564, online services. Inquire as to cost and availability.

ISMEC: Mechanical Engineering Abstracts. Cambridge Scientific Abstracts, 7200 Wisconsin Avenue, Suite 601, Bethesda,

NAVAL ARCHITECTURE

Ency. of Physical Sciences and Engineering Info. Sources

MD 20814. (301) 961-6750 or (800) 843-7751. Contains citations to the literature in mechanical engineering, industrial and production engineering, energy, power, mechanics, devices and related areas, from 1973 to present. Available on the DIALOG, (800) 334-2564, online service. Inquire as to cost and availability.

NTIS Bibliographic Database. National Technical Information Service, 5285 Port Royal Road, Springfield, VA 22161. (703) 487-4929 or FAX (703) 321-8199. Broad coverage of government-sponsored science and technology research reports, 1964 to present. Available on BRS, (800) 289-4277, DIALOG, (800) 334-2564, ORBIT, (800) 456-7248, and STN International, FIZ Karlsruhe, P.O. Box 2465, W-7500, Karlsruhe 1, Germany, online services. Also available on CD-ROM. Inquire as to cost and availability.

SCISEARCH. Institute for Scientific Information, 3501 Market Street, Philadelphia, PA 19104. (800) 523-1850 or (215) 386-0100. Broad multidisciplinary title and author index to the international literature of science and technology, 1974 to present. Available on DIALOG, (800) 334-2564, and ORBIT, (800) 456-7248, online services. Also available on CD-ROM. Inquire as to cost and availability.

Tris (transportation Research Information). National Academy of Science, 2101 Constitution Avenue, N.W., Washington, DC 20418. (202) 334-3313 or (800) 624-6242. Citations with abstracts of literature on transportation, including air, highway, rail, maritime and other modes from 1968 to present. Available on DIALOG, (800) 334-2564, online service. Inquire as to cost and availability.

WILSONLINE. H.W. Wilson Company, 950 University Avenue, Bronx, NY 10452. (800) 367-6770 or (212) 588-8400. Makes available online versions of the printed H.W. Wilson indexes including Applied Science and Technology Index, Business Periodicals Index, General Science Index, and Readers' Guide to Periodical Literature. Period covered is generally 1983 to present. Available on BRS, (800) 289-4277, DIALOG, (800) 334-2564, and OCLC EPIC, (800) 848-5878, online services. Also available on CD-ROM. Inquire as to cost and availability.

PERIODICALS

Fairplay International Shipping Weekly (includes Quarterly Newbuildings). Fairplay Publications Ltd., Box 354, Germantown, NY 12526. (518) 537-6682. FAX (518) 5379-6667. 1883 to present. Weekly. $265.00 per year.

Marine Log. Simmons-Boardman Publishing Corporation, 345 Hudson St., New York, NY 10014. (212) 620-7200. 1878 to present. Monthly. $35.00 per year.

Journal of Ship Production. Society of Naval Architects and Marine Engineers, 601 Pavonia Avenue, Suite 400, Jersey City, NJ 07306. (201) 798-4800. FAX (201) 798-4975. 1985 to present. Quarterly. $65.00 per year.

Journal of Ship Research. Society of Naval Architects and Marine Engineers, 601 Pavonia Avenue, Suite 400, Jersey City, NJ 07306. (201) 798-4800. FAX (201) 798-4975. 1957 to present. Quarterly. $80.00 per year.

Marine Engineers Review. Learned Information Inc., 143 Old Marlton Pike, Medford, NJ 08055-8750. (607) 654-6266. FAX (607) 654-4307. 1967 to present. Monthly. Inquire for price.

Marine Technology. Society of Naval Architects and Marine Engineers, 601 Pavonia Avenue, Suite 400, Jersey City, NJ 07306. (201) 798-4800. FAX (201) 798-4975. 1964 to present. Quarterly. $70.00 per year for non-members.

Marine Technology Society Journal. Marine Technology Society, 1828 L Street NW, No. 906, Washington, DC 20036-5104. (202) 775-5966. FAX (202) 429-9417. Quarterly. $70.00 per year.

Motor Ship. Reed Business Publishing Group, Quadrant House, the Quadrant, Sutton, Surrey SM2 5AS, England. Telephone 081-652-3369. FAX 081-652-8180. 1920 to present. Monthly. $127.00 per year.

Naval Architect. Royal Society of Naval Architects, 10 Upper Belgrave Street, London SW1 X8BQ, England. Telephone 071-235-4622. FAX 071-245-6959. 1971 to present. Ten times a year. Inquire for price.

Naval Engineers Journal. American Society of Naval Engineers, 1452 Duke Street, Alexandria, VA 22314. (703) 836-6727. FAX (703) 836-7491. 1889 to present. Bi-monthly. $65.00 per year.

Shipping World & Shipbuilder. Marine Publications International Ltd., 4 Hubbard Road, Houndmills, Basingstoke, Hampshire RG21 2UH, England. Telephone 0256-840444. FAX 0256-817347. 1883 to present. Ten times a year. Inquire for price.

Transactions of the Society of Naval Architects and Marine Engineers. Society of Naval Architects and Marine Engineers, 601 Pavonia Avenue, Suite 400, Jersey City, NJ 07306. (201) 798-4800. FAX (201) 798-4975. 1893 to present. Annual. $67.50 per year.

RESEARCH CENTERS AND INSTITUTES

Institute For Marine Dynamics. National Research Council of Canada, PO Box 12093, Postal Station "A," St. John's, NF Canada A1B 3T5. (709) 772-2469. FAX (709) 772-2462.

Maritime Research Department. Webb Institute of Naval Architecture, Crescent Beach Road, Glen Cove, NY 11542. (516) 671-2356. FAX (516) 674-9838.

Ship Hydrodynamics Laboratory. University of Michigan, 126 W. Engineering Bldg., Ann Arbor, MI 48109. (313) 764-9432.

NAVIGATION

See also: AVIONICS, BUOYS, COMPASSES, GUIDANCE SYSTEMS, GYROSCOPES, RADAR

ABSTRACT SERVICES AND INDEXES

Applied Science and Technology Index; A Cumulative Subject Index To English Language Periodicals in the Fields of Aeronautics and Space Science, Computer Technology, Chemistry, Construction Industry, Energy and Related Areas. H.W. Wilson Co., 950 University Avenue, Bronx, NY 10452. (800) 367-6770 or (212) 588-8400. FAX (718) 590-1617. From 1958 to present. Monthly. Inquire about cost and availability. Also available on CD-ROM and online.

Computer Abstracts, MCBUniversity Press Ltd., PO Box 10812, Birmingham, AL 35201-0812. (800) 633-4931. FAX (205) 995-1588. Monthly. Covers computer theory, data, hardware, systems, networks, human-computer interaction, artificial intelligence, as well as applications of computers in aerospace, business, CAD/CAM, cartography, civil engineering, electronics and electrical engineering, industrial engineering, mechanical engineering, medicine, structural engineering, etc.

Electronics and Communications Abstracts Journal, Cambridge Scientific Abstracts, 7200 Wisconsin Avenue, Bethesda, MD 20814-4823. (301) 961-6750. FAX (301) 961-6720. Monthly. $975.00 per year.

Engineering Index Monthly. Engineering Information, Inc., Castle Point on the Hudson, Hoboken, NJ 07030. (800) 221-1044. FAX (212) 832-1857. Monthly. $2300.00 per year. Also available online as COMPENDEX, and also on CD-ROM. Covers chemical engineering, computers, electrical engineering, civil engineering, metals and mining, industrial management, and mechanical engineering.

International Aerospace Abstracts. Technical Information Service, American Institute of Aeronautics and Astronautics, Inc., 555 West 57th St., New York, NY 10019. (212) 247-6500. FAX (212) 582-4861. Semi-monthly. $1295.00 per year.

Scientific and Technical Aerospace Reports (star). National Aeronautics and Space Administration. NASA Center for Aerospace Information, Box 8757, BWI Airport, Baltimore, MD 21240. (301) 621-0153. Monthly. Inquire about availability and cost. ALso available through the NASA online retrieval service (RECON), and through the Aerospace Database through DIALOG.

ASSOCIATIONS AND PROFESSIONAL SOCIETIES

Aeronautical Navigator Association. 640 Brumbaugh Drive, New Carlisle, OH 45344-2523. (513) 849-6082.

Institute of Navigation. 1800 Diagonal Road, Suite 480, Alexandria, VA 22314-2840. (703) 683-7101. FAX (703) 683-7105.

International Navigation Association, Po Box 2324, Arlington, VA 22202-0324. (604) 276-4626.

Permanent International Association of Navigation Congresses, United States Section. CECW-PK, 20 Massachusetts Avenue NW, Washington, DC 20314-1000. (202) 504-4312.

Wild Goose Association. Operations, 150 S. Plains Road, the Plains, OH 45780. (614) 797-2081.

DIRECTORIES AND BIOGRAPHICAL SOURCES

Directory of Engineering Societies and Related Organizations. Gordon Davis. 13th edition. American Association of Engineering Societies, 1111 19th Street NW, Suite 608, Washington, DC 20036. (202) 296-2237 or (800) 658-8897. 1989. Inquire for price.

International Research Centers Directory. Gale Research, 835 Penobscot Building, Detroit, MI 48226-4094. (313) 961-2242. (800) 347-4253. FAX (313) 961-6083. 8th edition. 1995. $410.00

Research Centers Directory. Gale Research, 835 Penobscot Building, Detroit, MI 48226-4094. (313) 961-2242. (800) 347-4253. FAX (313) 961-6083. $485.00.

Scientific and Technical Organizations and Agencies Directory. Gale Research, 835 Penobscot Building, Detroit, MI 48226-4094. (313) 961-2242. (800) 347-4253. FAX (313) 961-6083. 4th edition. 1996. $195.00.

Who's Who in Engineering. American Association of Engineering Societies, 1111 19th Street NW, Suite 608, Washington, DC 20036. (202) 296-2237 or (800) 658-8897. 8th edition. 1991. Inquire for price.

ENCYCLOPEDIAS AND DICTIONARIES

The Norton Encyclopedic Dictionary of Navigation. David F. Tver. W.W. Norton & Co., Inc., 500 Fifth Ave, New York, NY 11010. (800) 223-2584. FAX (800) 233-2588. 1987. Inquire for cost and availability.

GENERAL WORKS

Basic and Intermediate Celestial Navigation. W. Bruce Paulk. Hearst Marine Books/ William Morrow and Company, Inc., 1350 Avenue of the Americas, New York, NY 10019. (212) 261-6500 or (800) 843-9389. FAX (212) 779-0965. 1989. $16.95.

Dutton's Navigation and Piloting. Elbert S. Maloney. 14th edition. U.S. Naval Institute Press, Preble Hall, 118 Maryland Avenue, Annapolis, MD 21402-5035. (410) 268-6110 or (800) 233-8764. 1985. $45.95.

Electronic Aids To Navigation: Position Fixing. L. Tetley & David Calcutt. 2nd edition. Edward Arnold/ Routledge, Chapman and Hall, Inc., 29 West 35th Street, New York, NY 10001-2291. (212) 244-3336 or FAX (212) 563-2269. 1992. $59.95.

Electronic Aids To Navigation: Radar and Arpa. Roger Lownsborough & David Calcutt. Edward Arnold/ Routledge, Chapman and Hall, Inc., 29 West 35th Street, New York, NY 10001-2291. (212) 244-3336 or FAX (212) 563-2269. 1993. Inquire for price and availability.

Global Positioning System: An Overview. Yehuda Bock & Norman Leppard, editors. Springer-Verlag, 175 Fifth Avenue, New York, NY 10010. (212) 460-1500 or (800) 777-4643. FAX (212) 473-6272. 1990. $64.00.

Global Positioning System Applications. Richard K. Miller & Terri C. Walker. SEAI Technical Publications, 5880 Live Oak Parkway, Suite 270, Norcross, GA 30093. (404) 416) 0006. $285.00.

Navigation: Land, Sea, Air, and Space. Myron Katon, editor. IEEE (Institute of Electrical and Electronic Engineers), 445 Hoes Lane, Piscataway, NJ 08854. (908) 562-5545 or (800) 678-IEEE. 1989. $64.00.

Navigational & Surveying Instruments: Industry and Trade Summary. Sundar A. Shetty. Diane Publishing Company, 600 Upland Avenue, Upland, PA 19015. (215) 499-7415. 1994. $40.00.

Radar and Electronic Navigation. G. J. Sonnenberg. 6th edition. Butterworths Publishers, 80 Montvale Avenue, Stoneham, MA 02180. (617) 438-8464 or (800) 366-2665. 1988. $75.00.

NAVIGATION

Ency. of Physical Sciences and Engineering Info. Sources

HANDBOOKS AND MANUALS

The Navigation Control Manual. A.G. Bole, W.O. Dineley, C.E. Nicholls. 2nd edition. Butterworths Publishers, 80 Montvale Avenue, Stoneham, MA 02180. (617) 438-8464 or (800) 366-2665. 1992. $59.95.

ONLINE DATABASES AND CD-ROMS

Aerospace Database. American Institute of Aeronautics and Astronautics, 370 L'Enfant Promenade, S.W., Washington, DC 20024. (202) 646-7400. Worldwide published literature on research and development in aerospace and related areas, with abstracts. Covers 1962 to present. Available on DIALOG, (800) 334-2564, online service. Also available on CD-ROM. Inquire as to cost and availability.

Compendex Plus. Engineering Information, Inc., 345 East 47th Street, New York, NY 10017. (212) 705-7600 or (800) 221-1044. Contains citations with abstracts to worldwide literature in engineering and technology, from 1970 to present. Available on online BRS,(800) 289-4277, DIALOG, (800) 334-2564, ORBIT (800) 456-7248, and STN International, FIZ Karlsruhe, P.O. Box 2465, W-7500, Karlsruhe 1, Germany, online services. Also available on CD-ROM. Inquire as to cost and availability.

CSA Engineering. Cambridge Scientific Abstracts, 7200 Wisconsin Avenue, Suite 601, Bethesda, MD 20814. (301) 961-6750 or (800) 843-7751. Contains citations and abstracts of international periodicals and research literature covering all fields of engineering and science and technology,including computer and information science, electronics, mechanical engineering, solid state materials, 1981 to present. Available on BRS,(800) 289-4277, online service.Inquire as to cost and availability.

Current Contents Search. Institute for Scientific Information, 3501 Market Street, Philadelphia, PA 19104. (215) 386-0100. FAX (215) 386-6362. Contains citations to articles listed in the table of contents of science and technology journals. Also articles in social sciences and life sciences journals. Available on BRS,(800) 289-4277, DIALOG,(800) 334-2564, online services. Inquire as to cost and availability.

International Aerospace Abstracts. American Institute of Aeronautics and Astronautics, 370 L'Enfant Promenade, S.W., Washington, DC 20024. (202) 646-7400. Contains references and abstracts of journal and monograph literature relating to aerospace science and technology, from 1963 to present. Available through the NASA/RECON system of the National Aeronautics and Space Administration only.

NASA Database. American Institute of Aeronautics and Astronautics, 370 L'Enfant Promenade, S.W., Washington, DC 20024. (202) 646-7400. Citations and abstracts of aeronautics and astronautics literature, 1962 to present. Also Contains Citations from Star, Scientific and Technical Aerospace Reports, and International Aerospace Abstracts. Available through NASA/RECON online service. Inquire as to cost and availability.

NTIS Bibliographic Database. National Technical Information Service, 5285 Port Royal Road, Springfield, VA 22161. (703) 487-4929 or FAX (703) 321-8199. Broad coverage of government-sponsored science and technology research reports, 1964 to present. Available on BRS,(800) 289-4277, DIALOG, (800) 334-2564, ORBIT, (800) 456-7248, and STN International, FIZ Karlsruhe, P.O. Box 2465, W-7500, Karlsruhe 1, Germany, online services. Also available on CD-ROM. Inquire as to cost and availability.

SAE Global Mobility Database. Society of Automotive Engineers (SAE), Electronic Publishing Division, 400 Commonwealth Drive, Warrendale, PA 15098. (412) 776-4841. Contains citations with abstracts to technical papers on automotive and aerospace technology and vehicular-related industries that have been presented at SAE conferences. Covers 1906 to present. Available on ORBIT,(800) 456-7248,online service. Inquire as to cost and availability.

Tris (transportation Research Information). National Academy of Science, 2101 Constitution Avenue, N.W., Washington, DC 20418. (202) 334-3313 or (800) 624-6242. Citations with abstracts of literature on transportation, including air, highway, rail, maritime and other modes from 1968 to present. Available on DIALOG,(800) 334-2564, online service. Inquire as to cost and availability.

WILSONLINE. H.W. Wilson Company, 950 University Avenue, Bronx, NY 10452. (800) 367-6770 or (212) 588-8400. Makes available online versions of the printed H.W. Wilson indexes including Applied Science and Technology Index, Business Periodicals Index, General Science Index, and Readers' Guide to Periodical Literature. Period covered is generally 1983 to present. Available on BRS,(800) 289-4277, DIALOG,(800) 334-2564, and OCLC EPIC,(800) 848-5878, online services. Also available on CD-ROM. Inquire as to cost and availability.

PERIODICALS

Aerospace America. American Institute of Aeronautics and Astronautics, 370 L'Enfant Promenade SW, Washington, DC 20024-2518. (202) 646-7471. 1932 to present. Monthly. $70 per year (non-members).

Aerospace Engineering. SAE, Inc., 400 Commonwealth Dr., Warrendale, PA 15096. (412) 776-4841. 1981 to present. Monthly. $48.00 per year.

AIAA Journal. American Institute of Aeronautics and Astronautics, 370 L'Enfant Promenade SW, Washington, DC 20024-2518. (202) 646-7471. 1963 to present. Monthly. $52.00 per year (members) and $435.00 per year (non-members).

Aviation Week and Space Technology. 1221 Avenue of the Americas, New York, NY 10020. (212) 512-2999 or (800) 525-5003. 1916 to present. Weekly. $82.00 per year.

Institute of Navigation, Proceedings of Annual Meeting. Institute of Navigation, 1800 Diagonal Road, Suite 480, Alexandria, VA 22314-2840. Annual. $80.00 per year for non-members.

Journal of Navigation. Cambridge University Press, Journals Department, 40 West 20th Street, New York, NY 10011-4211. (212) 924-3900. 1947 to present. Three times a year. $165.00 per year.

Journal of Ship Research. Society of Naval Architects and Marine Engineers, 601 Pavonia Avenue, Suite 400, Jersey City, NJ 07306. (201) 798-4800. FAX (201) 798-4975. 1957 to present. Quarterly. Inquire for price.

Journal of Spacecraft and Rockets. American Institute of Aeronautics and Astronautics, the Aerospace Center, 370 L'Enfant Promenade SW, Washington, DC 20024. (202) 646-7400. FAX (202) 646-7508. 1964 to present. Bi-monthly. $185.00 per year.

RESEARCH CENTERS AND INSTITUTES

Aviation Safety Institute, 6797 N. High St., Suite 316, PO Box 304, Worthington, OH 43085. (614) 885-4242. FAX (614) 885-5891.

Avionics Engineering Center. Ohio University, 239 Stocker Center, Athens, OH 45701. (614) 593-1534. FAX (614) 593-1604.

Charles Stark Draper Laboratory Inc., 555 Technology Square, Cambridge, MA 02139. (617) 258-1000.

NEBULAE

See also: ASTRONOMY, BIG BANG, COSMOCHEMISTRY, COSMOLOGY, DARK MATTER, GALAXIES, INTERSTELLAR MEDIUM, MILKY WAY, STARS

ABSTRACT SERVICES AND INDEXES

Applied Science and Technology Index; A Cumulative Subject Index To English Language Periodicals in the Fields of Aeronautics and Space Science, Computer Technology, Chemistry, Construction Industry, Energy and Related Areas. H.W. Wilson Co., 950 University Avenue, Bronx, NY 10452. (212) 588-8400. (800) 367-6770. FAX (718) 590-1617. From 1958 to present. Monthly. Inquire about cost and availability. Also available on CD-ROM and online. ISSN: 0003-6986.

Astronomy and Astrophysics Abstracts. Springer-Verlag New York, 175 Fifth Avenue, New York, NY 10010. (212) 460-1500. FAX (212) 473-6272. Published for the Astronomisches Rechen-Institut. Comprehensive coverage of all aspects of astronomy, astrophysics and related fields. 1969 to present. Two parts per year. Annual. ISSN: 0067-0022.

Chemical Abstracts. Chemical Abstracts Service, 2540 Olentangy River Road, Box 3012, Columbus, OH 43210-0012. (614) 447-3600. (800) 848-6538. FAX (614) 447-3713. From 1907 to present. Weekly. $16800.00. Available online and on CD-ROM. CA is also available in five section groupings. Inquire regarding cost and availability.

General Science Index. H.W. Wilson Company, 950 University Avenue, Bronx, NY 10452. (212) 588-8400. (800) 367-6770. FAX (718) 590-1617. From 1978 to present. Ten issues per year; quarterly and annual cumulations. Service basis. Available on CD-ROM and online. Inquire regarding cost and availability. ISSN: 0162-1963.

Government Reports Announcements and Index. U. S. National Technical Information Service (NTIS), 5285 Port Royal Road, Springfield, VA 22161. (703) 487-4650. FAX (703) 321-8547. From 1968 to present. Annual. $630.00 per year. Also available online as *NTIS Bibliographic Database* and on CD- ROM. ISSN:

Meteorological and Geoastrophysical Abstracts. American Meteorological Society, c/o Inforonics, Inc., 550 Newtown Road, Littleton, MA 01460. (508) 486-8976. FAX (508) 486- 0027. Covers literature in environmental sciences, meteorology, astrophysics, hydrology, glaciology, and physical oceanography. From 1950 to present. Monthly. $950.00 per year. Also available on CD-ROM and online. Inquire regarding cost and availability. ISSN: 0026-1130.

NTIS Alerts: Astronomy & Astrophysics. U. S. National Technical Information Service. 5285 Port Royal Road, Springfield, VA 22161. (703) 487-4650. FAX (703) 321-8547. Weekly. $140.00 per year.

Pascal 48: Environment Cosmique Terrestre, Astronomie Et Geologie Extraterrestre. Centre National de la Recherche Scientifique, Institut de Information Scientifique et Technique, 2 aliee du Parc de Brabois, 54514 Vandoeuvre Les Nancy Cedex, France. TEL 83- 50-46-00. FAX: 83-50-46- 50. 1985 to present. Ten issues per year. 770 F. Also available on CD-ROM and online.

Physics Abstracts. INSPEC. Section A, Science Abstracts. Institute of Electrical Engineers, London, United Kingdom. Available from: INSPEC/IEEE - Institute of Electrical and Electronic Engineers, Box 1331, Hoes Lane, Piscataway, NJ 08855-1331. (908) 562-5549. 1898 to present. 24 issues per year. $2835.00 per year. ISSN: 0036-8091. Also available online and on CD-ROM.

Science Citation Index. SCI. Institute for Scientific Information, 3501 Market Street, Philadelphia, PA 19104. (215) 386-0100. (800) 523-1850. FAX (215) 386-2991. 1961 to present. Six issues per year, plus annual cumulation. $11650.00 per year. Also available online and on CD-ROM. Inquire about price and availability. ISSN: 0036-827X.

STAR. Scientific and Technical Aerospace Reports. U.S. National Aeronautics and Space Administration. Distributed by U. S. Superintendent of Documents, Washington, DC 20402. 1963 to present. Semi-monthly, with semiannual and annual indexes. $114.00 per year. ISSN: 0036-8741. Also available online and on CD-ROM.

ANNUAL REVIEWS AND YEARBOOKS

The Astronomical Almanac. Superintendent of Documents, U. S. Government Printing office, Washington, DC 10402. (202) 783-3238. 1981 to present. Supercedes Astronomical Ephemeris and the American Ephermis and Nautical Almanac. Annual. $44.95. ISSN: 0737-6421.

Annual Review of Astronomy and Astrophysics; Original Reviews of Critical Literature and Current Developments in Astronomy and Astrophysics. Annual Reviews, Inc., 4139 El Camino Way, Palo Alto, CA 94306-0139. (415) 493-4400. (800) 523-8635. FAX (415) 855-9815. 1963 to date. Annual. $60.00. ISSN: 0066-4146.

Highlights of Astronomy. I. Appenzeller, editor. Kluwer Academic Publishers, Box 358, Accord Station, Hingham, MA 02018-0358. (617) 871-6600. International Astronomical Union Highlights series. ISBN: 07923-3553-8. Triennal. 1995. Price varies.

ASSOCIATIONS AND PROFESSIONAL SOCIETIES

Amateur Astronomers Association. 1010 Park Avenue, New York, NY 10028. (212) 535-2922.

American Association of Variable Star Observers. 23 Birch Street, Cambridge, MA 02138. (617) 354-0484. FAX: (617) 354-0665.

NEBULAE

Ency. of Physical Sciences and Engineering Info. Sources

American Astronomical Society. 2000 Florida Avenue NW, Suite 400, Washington, DC 20009. (202) 328-2010. FAX: (202) 234-2560.

American Institute of Physics. One Physics Ellipse, College Park, MD 20740-3843. (301) 209-3100.

Association of University For Research in Astronomy. Suite 701, 1625 Massachusetts Avenue, NW, Washington, DC 20036. (202) 483-2101. FAX (202) 483-2106.

Astronomical League. 2112 Kingfisher Lane East, Rolling Meadows, IL 60008. (708) 398-0562.

Astronomical Society of the Pacific. 390 Ashton Avenue, San Francisco, CA 94112. (415) 337-1100. FAX: (415) 337-5205.

Joint Institute For Laboratory Astrophyics. University of Colorado, Box 440, Boulder, CO 80309-0440. (303) 492-7780. FAX (303) 492-5235.

Royal Astronomical Society of Canada. 136 Dupont Street, Toronto. ON N5R 1V2 Canada. (416) 924-7973. FAX (416) 968-6687.

BIBLIOGRAPHIES

Astronomy and Astrophysics: A Source Guide. Gordon Press, P.O. Box 459, Bowling Green Station, New York, NY 10004. (718) 624-8419. 1991. $250.00. .

A Bibliography of Astronomy, 1970-1979. R. A. Sea and S. S. Martin. Libraries Unlimited, Inc., Littleton, CO 80160. 1982.

Catalog of the Naval Observatory Library. Naval Observatory Staff. G. K. Hall & Company, 866 Third Avenue, New York, NY 10022. (212)702-6789. (800) 257-5755. 6 volumes. 1977. $655.00

Science Books and Films. American Association for the Advancement of Science, 1333 H Street NW, Washington, DC 20005. (202) 326-6454. 1965 to present. Nine issues per year. $40.00 per year. ISSN: 0098-342X

Scientific and Technical Books and Serials in Print, 1995. R.R. Bowker Inc., 121 Chanlon Road, New Providence, NJ 07974. (908) 464-6800. (800) 521-8110. 4 volumes. 1994. $299.95. Also available on compact disc and online.

A Source Book On Astronomy and Astrophysics, 1900 - 1975. Kenneth Lang and Owen Gingerich, editors. Harvard University Press, 79 Garden Street, Cambridge MA. (617) 495-2600. 1980. $41.95.

DIRECTORIES AND BIOGRAPHICAL SOURCES

American Astronomical Society. Membership Directory. American Astronomical Society. 2000 Florida Avenue NW, Suite 400, Washington, DC 20009. (202) 328-2010. FAX (202)234-2560 Annual. Included in membership dues. ISSN: 1061- 9038.

American Men and Women of Science: Physical and Biological Sciences. R. R. Bowker Inc., 121 Chanlon Road, New Providence, NJ 07974. (908) 464-6800. (800) 521-8110. 20th edition. 8 volumes. 1996. $850.00.

The Astronomers. Donald Goldsmith. St. Martin's Press, Inc., 175 Fifth Avenue, New York, NY 10010. (212) 674-5151. (800) 221-7945. 1993. $14.95 in paper.

Astronomical Centers of the World. Kevin Krisciunas. Cambridge University Press, 40 West 20th Street, New York, NY 10011-4211. (212) 924-3900. (800) 872-7423. 1988. $34.95

The Biographical Dictionary of Scientists: Astronomers. D. Abbott, editor. Peter Bedrick Books, Inc., 2112 Broadway, Room 318, New York, NY 10023. (212) 496-0751. 1984.

Directory of Physics and Astronomy Staff Members. American Institute of Physics, One Physics Ellipse, College Park, MD 20740-3843. (301) 209-3100. Annual. $45.00.

Earth and Astronomical Research Centers. Stockton Press, 345 Park Avenue South, New York, NY 10010. 4th edition. 1995. $515.00. ISBN: 1-56169-0967.

ENCYCLOPEDIAS AND DICTIONARIES

Astronomy From A To Z: A Dictionary of Celestial Objects and Ideas. Charles A. Schweighauser. Illinois Issues, Sangamon State University, K-10, Springfield, IL 62794-9243. (217) 786-9243. 1991. $14.95 in paper.

Concise Dictionary of Astronomy. Jacqueline Mitton. Oxford University Press, Inc., 200 Madison Avenue, New York, NY 10016. (212) 725-6000. (800) 334-4249. 1992. $24.95.

Facts On File Dictionary of Astronomy. Valerie Illingworth. Facts-on-File, Inc., 460 Park Avenue South, New York, NY 10016-7382. (212) 683-2244. (800) 322-8755. Fax (800) 683-3633. 1994. $27.95.

Encyclopedia of Astronomy and Astrophysics. Stephen Maran, editor. Van Nostrand Reinhold, 115 Fifth Avenue, New York, NY 10003. (212) 254-3232. (800) 842-3636. 1992. $129.95.

Oxford Illustrated Encyclopedia. Volume 8, the Universe. Archie Roy, editor. Oxford University Press, Inc., 200 Madison Avenue, New York, NY 10016. (212) 725-6000. (800) 334-4249. 1993. $45.00.

McGraw-Hill Encyclopedia of Science and Technology. McGraw-Hill Book Company, Inc., 1221 Avenue of the Americas, New York, NY 10020. (212) 512-2000. (800) 262-4729. 7th edition. 20 volume set. 1992. $1900.00.

Stars, Galaxies, Cosmos. William R. Corliss. Sourcebook Project, P.O. Box 107, Glen Arm, MD 21057. (410) 668-6047. 1987. $17.95.

GENERAL WORKS

Astronomer's Universe: Stars, Galaxies and Cosmos. Herbert Friedman. W. W. Norton & Company, 500 Fifth Avenue, New York, NY 10110. (212) 354-5500. (800) 223-2584. 1990. $24.95.

Astrophysics of Gaseous Nebulae and Active Galactic Nuclei. Donald E. Osterbrock. University Science Books, 55D Gate Five Road, Sausalito, CA 94965. (415) 332-5390. 1989. $36.00.

Atoms, Stars and Nebulae. Lawrence H. Aller. Books on Demand, 300 North Zeeb Road, Ann Arbor, MI 48106-1346. (313) 761-4700. (800) 521-0600. Revised edition. $90.80.

Cycles of Fire: Stars, Galaxies and the Wonder of Deep SPACE. William K. Hartmann. Workman Publishing Company, Inc., 708 Broadway, New York, NY 10003. (212) 254-5900. (800) 722-7202. 1988. $15.95 in paper.

Evolution of Interstellar Matter and Dynamics of Galaxies. J. Palous et al, editors. Cambridge University Press, 40 West 20th Street, New York, NY 10011-4211. (212) 924-3900. (800) 872-7423. 1992. $69.95.

Farthest Things in the Universe. Jay M. Pasachoff. Cambridge University Press, 40 West 20th Street, New York,

NY 10011-4211. (212) 924-3900. (800) 872-7423. 1995. $29.95.

Messier's Nebulae and Star Clusters. Kenneth G. Jones. Cambridge University Press, 40 West 20th Street, New York, NY 10011-4211. (212) 924-3900. (800) 872-7423. 1991. $59.95.

Nebulae and Interstellar Matter. Barbara Middlehurst and Lawrence H. Aller, editors. University of Chicago Press, 5801 Ellis Avenue, Chicago, IL 60637. (312) 702-7700. 1968. $60.00.

Observational Astrophysics. Robert C. Smith. Cambridge University Press, 40 West 20th Street, New York, NY 10011- 4211. (212) 924-3900. (800) 872-7423. 1995. $69.95.

Planetary Nebulae. Stuart R. Pottasch. Kluwer Academic Publishers, 101 Philip Drive, Assinippi Park, Norwell, MA 02061. (617) 871-6600. 1983. $107.50.

The Realm of the Nebulae. Edwin Hubble. Yale University Press, 302 Temple Street, New Haven, CT 06511. (203) 432- 0960. Reprint edition. 1982. $14.00 in paper.

Star Formation, Galaxies and the Interstellar Medium. J. Franco, et al, editors. Cambridge University Press, 40 West 20th Street, New York, NY 10011-4211. (212) 924-3900. (800) 872-7423. 1993. $59.95.

Stars, Nebulae and the Interstellar Medium. C. R. Kitchin. IOP Publishing, Public Ledger Building, Suite 1035, Independence Square, Philadelphia, PA 19106. (215) 627- 0880. (800) 358-4677. 1987. $43.00 in paper.

Structure and Evolution of Galaxies. Roger J. Tayler. Cambridge University Press, 40 West 20th Street, New York, NY 10011-4211. (212) 924-3900. (800) 872-7423. 1993. $49.95. $22.95 in paper.

Visual Astronomy of the Deep Sky. Roger N. Clark. Cambridge University Press, 40 West 20th Street, New York, NY 10011-4211. (212) 924-3900. (800) 872-7423. 1991. $39.95.

HANDBOOKS AND MANUALS

Burnham's Celestial Handbook: An Observer's Guide To the Universe Beyond the Solar System. Robert Burnham, Jr. Dover Publications, Inc., 180 Varick Street, New York, NY 10014. (212) 255-3755. (800) 223-3130. 3 volumes. 1978. $41.85 for the set.

Cambridge Guide To the Constellations. Michael E. Bakich. Cambridge University Press, 40 West 20th Street, New York, NY 10011-4211. (212) 924-3900. (800) 872-7423. 1995. $69.95.

Encyclopedia of Astronomy and Astrophysics. Stephen Maran, editor. Van Nostrand Reinhold, 115 Fifth Avenue, New York, NY 10003. (212) 254-3232. (800) 842-3636. 1992. $129.95.

Greenwich Guide To Stars, Galaxies and Nebulae. Stuart Malin. Cambridge University Press, 40 West 20th Street, New York, NY 10011-4211. (212) 924-3900. (800) 872-7423.1990. $10.95.

Handbook of Astronomy, Astrophysics and Geophysics; Volume 2: Galaxies and Cosmology. V. M. Canuto and B. G. Elmegreen. Gordon & Breach Science Publishers, Inc., P.O. Box 200029, Riverfront Plaza Station, Newark, NJ 07102-0301. (201) 643-7500. 1988. $194.00.

Handbook of Space Astronomy and Astrophsyics. Martin V. Zumbeck. Cambridge University Press, 40 West 20th Street, New York, NY 10011-4211. (212) 924-3900. (800) 872-7423. 2nd edition. 1990. $79.95.

Landolt-borenstein Numerical Data and Functional Relationships in Science and Technology: Astronomy, Astrophysics and Space Research. Group VI. Springer-Verlag New York, Inc., 175 Fifth Avenue, New York, NY 10010. (212) 460-1500. (800) 777-4643. volumes priced individually.

McGraw-Hill Encyclopedia of Astronomy. Sybil P. Parker and Jay M. Pasachoff, editors. McGraw-Hill Publishing Company, Inc., 1221 Avenue of the Americas, New York, NY 10020. (212) 512-2000. (800) 262-4729. 2nd edition. 1993. $75.00

National Audobon Society Pocket Guide: Galaxies and Other Deep-sky Objects. Alfred A. Knopf, Inc. 201 E. 50th Street, New York, NY 10022. (800) 733-3000. 1995. $7.95.

Observer's Guide To Astronomy. Patrick Martinez, editor. Cambridge University Press, 40 West 20th Street, New York, NY 10011-4211. (212) 924-3900. (800) 872-7423. 2 volumes. 1994. $159.90.

ONLINE DATABASES AND CD-ROMS

CA Search. Chemical Abstracts Service, P.O. Box 3012, Columbus, OH 43210-0012. (614) 447-3600. (800) 848-6533. FAX (614) 447-3709. Very comprehensive guide to worldwide chemical literature and related fields, 1972 to present. Available on BRS,(800) 289-4277, DIALOG, (800) 334-2564, ORBIT (800) 456-7248, and STN International, FIZ Karlsruhe, P.O. Box 2465, W-7500, Karlsruhe 1, Germany, online services. Inquire as to cost and availability.

Current Contents Search. Institute for Scientific Information, 3501 Market Street, Philadelphia, PA 19104. (215) 386-0100. FAX (215) 386-6362. Contains citations to articles listed in the table of contents of science and technology journals. Also articles in social sciences and life sciences journals. Available on BRS, (800) 289-4277, DIALOG, (800) 334-2564, online services. Inquire as to cost and availability.

Dissertation Abstracts Online. University Microfilms International, 300 North Zeeb Road, Ann Arbor, MI 48106. (800) 521-0600 or (313) 761-4700. Scope includes virtually all doctoral dissertations accepted at accredited American institutions from 1861 to present in 252 subject areas. Available on BRS, (800)

NEBULAE

Ency. of Physical Sciences and Engineering Info. Sources

289-4277, DIALOG, (800) 334-2564,and OCLC EPIC, (800) 848-5878, online services. Also available on CD-ROM. Inquire as to cost and availability.

INSPEC. Institution of Electrical Engineers, Michael Faraday House, Six Hills Way, Stevenage, Herts. SG1 2AY, England. Telephone: 0438 313311 or FAX 0438 742840. Contains citations to the worldwide literature of physics, electronics and electrical engineering, computer technology, and related fields. Available on BRS, (800) 289-4277, DIALOG, (800) 334-2564, ORBIT, (800) 456-7248, and STN International, FIZ Karlsruhe, P.O. Box 2465, W-7500, Karlsruhe 1, Germany, online services. Inquire as to cost and availability.

International Aerospace Abstracts. American Institute of Aeronautics and Astronautics, 370 L'Enfant Promenade, S.W., Washington, DC 20024. (202) 646-7400. Contains referencesand abstracts of journal and monograph literature relating to aerospace science and technology, from 1963 to present. Available through the NASA/RECON system of the National Aeronautics and Space Administration only.

NTIS Bibliographic Database. National Technical Information Service, 5285 Port Royal Road, Springfield, VA 22161. (703) 487-4929 or FAX (703) 321-8199. Broad coverage of government-sponsored science and technology research reports, 1964 to present. Available on BRS,(800) 289-4277, DIALOG, (800) 334-2564, ORBIT, (800) 456-7248, and STN International, FIZ Karlsruhe, P.O. Box 2465, W-7500, Karlsruhe 1, Germany, online services. Also available on CD-ROM. Inquire as to cost and availability.

Physics Briefs. American Institute of Physics, 335 East 45th Street, New York, NY 10017. (212) 661-9260 or FAX (212) 661-2036. Contains citations with abstracts of the literature of physics and related fields, 1979 to present. Available on the STN International, FIZ Karlsruhe, P.O. Box 2465, W-7500, Karlsruhe 1, Germany, online service. Inquire as to cost and availability.

SCISEARCH. Institute for Scientific Information, 3501 Market Street, Philadelphia, PA 19104. (800) 523-1850 or (215) 386-0100. Broad multidisciplinary title and author index to the international literature of science and technology, 1974 to present. Available on DIALOG, (800) 334-2564, and ORBIT, (800) 456-7248, online services. Also available on CD-ROM. Inquire as to cost and availability.

WILSONLINE. H.W. Wilson Company, 950 University Avenue, Bronx, NY 10452. (800) 367-6770 or (212) 588-8400. Makes available online versions of the printed H.W. Wilson indexes including Applied Science and Technology Index, Business Periodicals Index, General Science Index, and Readers' Guide to Periodical Literature. Period covered is generally 1983 to present. Available on BRS, (800) 289-4277, DIALOG, (800) 334-2564, and OCLC EPIC, (800) 848-5878, online services.Also available on CD-ROM. Inquire as to cost and availability.

OTHER SOURCES

Atlas of Deep-sky Spendors. Hans Vehrenberg. Sky Publishing Corp., 49 Bay State Road, Cambridge, MA 02138. (617) 864-7360. (800) 253-0245. 4th edition. 1983. $24.95.

Atlas of the Ultraviolet Sky. Richard C. Henry, Wayne B. Landsman, et al. Johns Hopkins University Press, 2715 North Charles Street, Baltimore, MD 21218-4319. (410) 516-6900. (800) 537-5487. 1988. $70.00.

Atlas of the Universe. Patrick Moore. Rand McNally & Company. 8255 North Central Park, Skokie, IL 60076-2970. (708) 673-9100. 1994. $29.95.

Cambridge Atlas of Astronomy. Jean Audouze, et al. Cambridge University Press, 40 West 20th Street, New York, NY 10011-4211. (212) 924-3900. (800) 872-7423. 3rd edition. 1994. $75.00.

Cambridge Guide To the Constellations. Michael E. Bakich. Cambridge University Press, 40 West 20th Street, New York, NY 10011-4211. (212) 924-3900. (800) 872-7423. 1995. $19.95 in paper. Carnegie Atlas of Galaxies. Allan Sandage and John Bedke. Carnegie Institution of Washington, 1530 P Street NW, Washington, DC 20005. (202) 387-6411. Two volumes. 1994. $92.00.

Color Atlas of Galaxies. James Wray. Cambridge University Press, 40 West 20th Street, New York, NY 10011-4211. (212) 924-3900. (800) 872-7423. 1988. $94.95.

NGC 2000.0: The Complete New General Catalog (Ngc) and Index Catalogue of Nebulae and Star Clusters. Roger W. Sinnott. Cambridge University Press, 40 West 20th Street, New York,

NY 10011-4211. (212) 924-3900. (800) 872-7423. 1989. $21.95 in paper.

Observer's Sky Atlas. E. Karkoschka. Springer-Verlag New York, Inc., 175 Fifth Avenue, New York, NY 10010. (212) 460-1500. (800) 777-4643. 1995. $15.95.

Observing Handbook and Catalog of Deep-sky Objects. Christian B. Luginbuhl and Brian A. Skiff. Cambridge University Press, 40 West 20th Street, New York, NY 10011- 4211. (212) 924-3900. (800) 872-7423. 1990. $49.95.

The Sky: A User's Guide. David H. Levy. Cambridge University Press, 40 West 20th Street, New York, NY 10011- 4211. (212) 924-3900. (800) 872-7423. 1993. $14.95 in paper.

PERIODICALS

Astronomical Journal. American Institute of Physics, One Physics Ellipse, College Park, MD 20740-3843. (301) 209-3000. Published for the American Astronomical Society. 1849 to present. Monthly. $280.00 per year. ISSN: 0004- 6256.

Astronomical Society of the Pacific. PUBLICATIONS. Astronomical Society of the Pacific, 390 Ashton Avenue, San Francisco, CA 94112. (415) 337-1100. FAX (415) 337-5205.1889 to present. Monthly. $175.00 per year. ISSN: 0004- 6280.

Astronomy. Kalmbach Publishing Company, Box 1612, Waukesha, WI 53187-1612. (414) 796-8776. FAX (414) 796-1142. 1973 to present. Monthly. $27.00 per year. ISSN: 0091-6358.

Astronomy and Astrophsyics. Springer-Verlag New York, Inc., 44 Hartz Way, Secaucus, NJ 07096-2491. (201) 348-4033. FAX(201) 348-4505. 1969 to date. 36 issues per year in 12 volumes. $1733.00 per year. ISSN: 0004-6361.

Astronomy Reports. American Institute of Physics, One Physics Ellipse, College Park, MD 20740-3843. (301) 209-3000. English translation of Astronomicheksii Zhurnal. Formerly Soviet

Astronomy AJ. 1957 to present. Bimonthly. $1160.00 per year. ISSN: 1063-7729.

Astrophysical Journal; An International Review of Astronomy and Astronomical Physics. University of Chicago Press, Journals Division, 5720 South Woodlawn Avenue, Chicago, IL 60637. (312) 753-3347. FAX (312) 753-0811. 1895 to date. three issues per month. $740.00 per year. ISSN: 0004-637X.

Astrophysics and Space Science; An International Journal of Cosmic Physics. Kluwer Academic Publishers, Box 358, Accord Station, Hingham, MA 02018-0358. (617) 871-6600. FAX (617) 871-6528. 1968 to present. 24 issues per year. $2544.00 per year. ISSN: 0004-640X.

Celestial Mechanics and Dynamical Astronomy; An International Journal of Space Dynamics. Kluwer Academic Publishers, Box 358, Accord Station, Hingham, MA 02018-0358. (617) 871-6600. FAX (617) 871-6528. 1969 to present. Monthly. $703.50 per year. ISSN: 0923-2958.

Experimental Astronomy; An International Journal On Astronomical Instrumentation and Data Analysis. Kluwer Academic Publishers, Box 358, Accord Station, Hingham, MA 02018-0358. (617) 871-6600. FAX (617) 871-6528. 1989 to present. Quarterly. $161.00 per year. ISSN: 0922-6435.

Mercury. Astronomical Society of the Pacific. 390 Ashton Avenue, San Francisco, CA 94112. (415) 337-1100. FAX: (415) 337-5205. 1972 to present. Bimonthly. $175.00 per year. ISSN: 0047-6773.

Monthly Notices of the Royal Astronomical Society. Blackwell Scientific Publication LT., Osney Mead, Oxford OX2 OEL, England. TEL 0865-240201. FAX 0865-721205. 1827 to present. Fortnightly. $1899.00 per year. ISSN: 0035-8711.

Observatory. Space and Astrophysics Division, Rutherford Appleton Laboratory, Chilton, Didcot, Oxon OX11 OQX England. FAX 0235-445848. 1877 to present. Bimonthly. $42.00 per year.

Sky and Telescope. Sky Publishing Corporation, Box 9111, Belmont, MA 02178. (617) 864-7360. FAX (617) 864-6117.1941 to present. Monthly. $27.00 per year. ISSN: 0037-6604.

Space Science Reviews. Kluwer Academic Publishers, Box 358, Accord Station, Hingham, MA 02018-0358. (617) 871-6600. FAX (617) 871-6528. 1962 to present. Sixteen issues per year. $940.00 per year. ISSN: 0038-6308.

Vistas in Astronomy; An International Review Journal. Albert C. Beer, editor. Elsevier Science Publishing Company, Inc., 660 White Plains Road, Terrytown, NY 10591- 5153. (914) 524-9200. FAX: (914) 333-2444. 1958 to present. Quarterly. $415.00 per year. ISSN: 0083-6656.

Transactions of the International Astronomical Union. Proceedings of the General Assembly. Kluwer Academic Publishers, Box 358, Accord Station, Hingham, MA 02018- 0358. (617) 871-6600. Issues in two parts: Part A, Reports, and Part B, Proceedings. Triennial. XXI - 1994. Price varies. ISSN: 0080-1372.

Webb Society Quarterly Journal. Webb Society, 194 Foundry Lane, Freemantle, Southampton, Hampshire SO1 3EQ, England. 1969 to present. Quarterly. $19.50 per year. ISSN: 0043- 1680.

RESEARCH CENTERS AND INSTITUTES

David Dunlap Observatory. University of Toronto. Box 360, Richmond Hill ON, Canada L4C 4Y6. (416) 884-9562. FAX (416) 978-3921.

McDonald Observatory. University of Texas at Austin. P.O. Box 1337, Fort Davis, TX 79734. (915) 426-3263.

National Radio Astronomy Observatory. Edgemont Road, Charlottesville, VA 22903. (804) 296-0211.

Palomar Observatory. California Institute of Technology, 105-24, Pasadena, CA 91125. (818) 356-4033.

Smithsonian Center For Astrophysics. Harvard University, 60 Garden Street, Cambridge, MA 02138. (617) 495-7461. FAX (617) 495-7326.

Space Telescope Science Institute. 3700 San Martin Drive, Baltimore, MD 21218. (301) 338-4700.

Washburn Observatory. University of Wisconsin - Madison, Madison, WI 53706. (608) 262-3072. FAX (608) 263-0361.

NEPTUNE

See also: ASTRONOMY, PLANETARY SCIENCES, SOLAR SYSTEM,

ABSTRACT SERVICES AND INDEXES

Applied Science and Technology Index; a cumulative subject index to English language periodicals in the fields of aeronautics and space science, computer technology, chemistry, construction industry, energy and related areas. H. W. Wilson Co., 950 University Avenue, Bronx, NY 10452- 9978. (212) 588-8400. (800) 367-6770. FAX (718) 590-1617.From 1958 to present. Monthly. Inquire about cost and availability. Also available online (BRS and WILSONLINE) and on CD-ROM. ISSN: 0003-6986.

Astronomy and Astrophysics Abstracts. Springer-Verlag New York, 175 Fifth Avenue, New York, NY 10010. (212) 460-1500. FAX (212) 473-6272. Published for the Astronomisches Rechen-Institut. Comprehensive coverage of all aspects of astronomy, astrophysics and related fields. 1969 to present. Two parts per year. Annual. ISSN: 0067-0022.

Bibliography and Index of Geology. American Geological Institute, 4220 King Street, Alexandria, VA 22302-1507. (703) 379-2480. FAX (703) 379-7563. From 1969 to present. Monthly. $1295.00 per year. ISSN: 0098-2784. Also available as GEOREF online (CISTI, DIALOG, Orbit, STN) and on CD-ROM. Inquire about price and availability.

Chemical Abstracts. Chemical abstracts Service. 2540 Oleantangy River Road, Box 3012, Columbus, OS 43210-0012. (614) 447-3600. FAX (614) 447-3713. 1907 to present. Weekly. $16,800.00 per year. Available online and on CD- ROM. CA is also available in five section groupings. Inquire regarding cost and availability.

Current Contents: Physical, Chemical, and Earth Sciences. Institute for Scientific Information, 3501 Market Street, Philadel-

Ency. of Physical Sciences and Engineering Info. Sources

phia, PA 19104. (215) 386-0100. (800) 523-1850. FAX (215) 386-2291. 1961 to present. Weekly. $442.00 per year. Also available online (BRS, DIALOG) and on CD- ROM. Inquire about price and availability.

General Science Index. H.W. Wilson Company, 950 University Avenue, Bronx, NY 10452. (212) 588-8400. (800) 367-6770. FAX (718) 590-1617. From 1978 to present. Ten issues per year; quarterly and annual cumulations. Service basis. Available on CD-ROM and online. Inquire regarding cost and availability. ISSN: 0162-1963.

Meteorological and Geoastrophysical Abstracts. American Meteorological Society, c/o Inforonics, Inc., 550 Newtown Road, Box 458, Littleton, MA 01460. (508) 486-8976. FAX (508) 486-0027. Covers literature in environmental sciences, meteorology, astrophsyics, hydrology, glaciology and physical oceanography. 1950 to present. Monthly. $950.00 per year. Also available online (DIALOG) and on CD-ROM. ISSN: 0026-1130.

NTIS Alerts: Astronomy & Astrophysics. U. S. National Technical Information Service. 5285 Port Royal Road, Springfield, VA 22161. (703) 487-4650. FAX (703) 321-8547. Weekly. $140.00 per year.

Physics Abstracts. INSPEC. Section A, Science Abstracts. Institute of Electrical Engineers, London, United Kingdom. Available from: INSPEC/IEEE - Institute of Electrical and Electronic Engineers, Box 1331, Hoes Lane, Piscataway, NJ 08855-1331. (908) 562-5549. 1898 to present. 24 issues per year. $2835.00 per year. ISSN: 0036-8091. Also available online and on CD-ROM.

Science Citation Index. SCI. Institute for Scientific Information, 3501 Market Street, Philadelphia, PA 19104. (215) 386-0100. (800) 523-1850. FAX (215) 386-2991. 1961 to present. Six issues per year, plus annual cumulation. $11650.00 per year. Also available online and on CD-ROM.Inquire about price and availability. ISSN: 0036-827X.

STAR. Scientific and Technical Aerospace Reports. U.S. National Aeronautics and Space Administration. Distributed by U. S. Superintendent of Documents, Washington, DC 20402. 1963 to present. Semi-monthly, with semiannual and annual indexes. $114.00 per year. ISSN: 0036-8741. Also available online and on CD-ROM.

ANNUAL REVIEWS AND YEARBOOKS

The Astronomical Almanac. Superintendent of Documents, U. S. Government Printing office, Washington, DC 10402. (202) 783-3238. 1981 to present. Supercedes Astronomical Ephemeris. Annual. $44.95. ISSN: 0737-6421.

Annual Review of Astronomy and Astrophysics. Original reviews of critical literature and current developments in astronomy and astrophysics. Annual Reviews, Inc., 4139 El Camino Way, Palo Alto, CA 94303-0139. (415) 493-4400. (800)523-8635. Fax (415) 855-9815. 1963 to date. Annual. $60.00. ISSN: 0066-4146.

Annual Review of Earth and Planetary Sciences. Annual Reviews, Inc., 4139 El Camino Way, Palo Alto, CA 94303-0139. (415) 493-4400. (800) 523-8635. Fax (415) 855-9815. 1973 to date. Annual. $62.00. ISSN: 0084-6597.

ASSOCIATIONS AND PROFESSIONAL SOCIETIES

American Astronomical Society. 2000 Florida Avenue NW, Suite 400, Washington, DC 20009. (202) 328-2010. FAX\;(202) 234-2560.

American Geophysical Union. 2000 Florida Avenue NW, Washington, DC 20009. (202) 462-6900. (800) 966-AGU1. FAX (202) 328-0566.

American Institute of Physics. 1 Physics Ellipse, College Park, MD 20740-3843. (301) 209-3100.

Association of Universities For Research in Astronomy, Inc. (AURA). Suite 701, 1625 Massachusetts Avenue, NW, Washington, DC 20036. (202) 483-2101. FAX (202) 483-2106.

Astronomical Society of the Pacific. 390 Ashton Avenue, San Francisco, CA 94112. (415) 337-1100. FAX: (415) 337-5205.

Planetary Society. 65 North Catalina Avenue, Pasadena, CA 91106. (818) 793-5100. (800) WOW-MARS. FAX (818) 793-5528.

DIRECTORIES AND BIOGRAPHICAL SOURCES

American Astronomical Society. Membership Directory.

American Astronomical Society. 2000 Florida Avenue NW, Suite 400, Washington, DC 20009. (202) 328-2010. FAX (202)234-2560 Annual. Included in membership dues. ISSN: 1061- 9038.

american men and women of science: physical and biological

Sciences. R. R. Bowker Inc., 121 Chanlon Road, New Providence, NJ 07974. (908) 464-6800. (800) 521-8110. 20th edition. 8 volumes. 1996. $850.00.

The Astronomers. Donald Goldsmith. St. Martin's Press, Inc., 175 Fifth Avenue, New York, NY 10010. (212) 674-5151. (800) 221-7945. 1993. $14.95 in paper.

Astronomical Centers of the World. Kevin Krisciunas. Cambridge University Press, 40 West 20th Street, New York, NY 10011-4211. (212) 924-3900. (800) 872-7423. 1988. $34.95

Directory of Physics and Astronomy Staff Members. American Institute of Physics, One Physics Ellipse, College Park, MD 20740-3843. (301) 209-3100. 1975/76 to present. Annual. $60.00. ISSN: 0361-2228.

Earth and Astronomical Research Centers. Stockton Press, 345 Park Avenue South, New York, NY 10010. 4th edition. 1995. $515.00. ISBN: 1-56169-0967.

ENCYCLOPEDIAS AND DICTIONARIES

Concise Dictionary of Astronomy. Jacqueline Mitton. Oxford University Press, Inc., 200 Madison Avenue, New York, NY 10016. (212) 725-6000. (800) 334-4249. 1992. $24.95.

Encyclopedia of Astronomy and Astrophysics. Stephen Maran, editor. Van Nostrand Reinhold, 115 Fifth Avenue, New York, NY 10003. (212) 254-3232. (800) 842-3636. 1992. $129.95.

McGraw-Hill Encyclopedia of Science and Technology. McGraw-Hill Book Company, Inc., 1221 Avenue of the Americas, New York, NY 10020. (212) 512-2000. (800) 262-4729. 7th edition. 20 volume set. 1992. $1900.00.

Moons of the Solar System: An Illustrated Encyclopedia. John Stewart. McFarland & Company, Inc., Box 611, Jefferson, MC 28640. (910) 246-4460. (800) 253-2187. 1991. $49.95.

New Guide To the Planets. Patrick Moore. Trans-Atlantic Publications, Inc., 311 Bainbridge Street, Philadelphia, PA 19147. (215) 925-5083. 1993. $37.50.

Stars and Planets: The Sierra Club Guide To Sky Watching and Direction Finding. W. S. Kals. The Sierra Press, 4988 Gold Leaf Drive, Mariposa, CA 95338. (209) 966-5071. (800) 745-2631. $15.00.

GENERAL WORKS

Far Encounter. Eric Burgess. Columbia University Press, 562 West 113th Street, New York, NY 10025. (212) 666-1000. (800) 944-8648. 1988. $40.00.

Moons and Planets. William K. Hartmann. Wadsworth Publishing Co., 10 Davis Drive, Belmont, CA 94002. (415) 595-2350. (800) 354-9706. 3rd edition. 1993. $49.95.

Neptune and Triton. Dale P. Cruikshank, editor. University of Arizona Press, 1230 North Park Avenue, Number 102, Tucson, AZ 85719. (520) 621-1441. (800) 426-3797. 1996. $110.00.

The Planet Neptune: An Historical Survey Before Voyager. Patrick Moore. John Wiley & Sons, Inc., 605 Third Avenue, New York, NY 10158-0012. (212) 850-6000. (800) 225-5945.2995. $49.95.

Space Technology and Planetary Astronomy. Joseph N. Tatarewicz. Indiana University Pres, 601 North Morton Street, Bloomington, in 47404-3797. (812) 855-6804. (800) 842-6796. 1990. $29.95.

Uranus and Neptune: The Distant Giants. Eric Burgess. Columbia University Press, 562 West 113th Street, New York, NY 10025. (212) 666-1000. (800) 944-8648. 1988. $40.00.

Voyager Neptune Travel Guide. Gordon Press Publications, P.O. Box 459, Bowling Green Station, New York, NY 10004. (718) 624-8419. 1994. $260.75.

Worlds Apart: A Textbook in Planetary Sciences. Guy Consolmagno. Prentice-Hall, 113 Sylvan Avenue, Route 9W, Englewood Cliffs, NJ 07632. (201) 592-2000. (800) 922-0579. 1994. $62.00.

HANDBOOKS AND MANUALS

Astrophysical Data: Planets and Stars. Kenneth R. Lanb. Springer-Verlag New York, Inc., 175 Fifth Avenue, New York, NY 10010. (212) 460-1500. (800) 777-4643. 1993. $59.00.

Field Guide to the Stars and Planets. Jay M. Pasachoff and Donald H. Menzel. Houghton Mifflin Co., 222 Berkeley Street, Boston, MA 02116. (617) 351-5000. (800) 225-3362. Revised edition. 1992. $24.95.

New Guide To the Planets. Patrick Moore. Trans-Atlantic Publications, Inc., 311 Bainbridge Street, Philadelphia, PA 19147. (215) 925-5083. 1993. $37.50.

Planet Observer's Handbook. Fred W. Price. Cambridge University Press, 40 West 20th Street, New York, NY 10011-4211. (212) 924-3900. (800) 872-7423. 1994. $34.95.

ONLINE DATABASES AND CD-ROMS

CA Search. Chemical Abstracts Service, P.O. Box 3012, Columbus, OH 43210-0012. (614) 447-3600. (800) 848-6533. FAX (614) 447-3709. Very comprehensive guide to worldwide chemical literature and related fields, 1972 to present. Available on BRS,(800) 289-4277, DIALOG, (800) 334-2564, ORBIT (800) 456-7248, and STN International, FIZ Karlsruhe, P.O. Box 2465, W-7500, Karlsruhe 1, Germany, online services. Inquire as to cost and availability.

Dissertation Abstracts. University Microfilms International, 300 North Zeeb Road, Ann Arbor, MI 48106. (800) 521-0600 or (313) 761-4700. Scope includes virtually all doctoral dissertations accepted at accredited American institutions from 1861 to present in 252 subject areas. Available on BRS, (800) 289-4277, DIALOG, (800) 334-2564, and OCLC EPIC, (800) 848-5878, online services. Also available on CD-ROM. Inquire as to cost and availability.

GEOREF. American Geological Institute, 4220 King Street, Alexandria, VA 22302. (800) 336-4764 or (703) 379-2480. Geology and geosciences literature, 1785 to present for North America. Available on DIALOG,(800) 334-2564, ORBIT (800) 456-7248, online services. Also available on CD-ROM. Inquire as to cost and availability.

INSPEC. Institution of Electrical Engineers, Michael Faraday House, Six Hills Way, Stevenage, Herts. SG1 2AY, England. Telephone: 0438 313311 or FAX 0438 742840. Contains citations to the worldwide literature of physics, electronics and electrical engineering, computer technology, and related fields. Available on BRS, (800) 289-4277, DIALOG, (800) 334-2564, ORBIT, (800) 456-7248, and STN International, FIZ Karlsruhe, P.O. Box 2465, W-7500, Karlsruhe 1, Germany, online services. Inquire as to cost and availability.

NTIS Bibliographic Database. National Technical Information Service, 5285 Port Royal Road, Springfield, VA 22161. (703) 487-4929 or FAX (703) 321-8199. Broad coverage of government-sponsored science and technology research reports, 1964 to present. Available on BRS,(800) 289-4277, DIALOG, (800) 334-2564, ORBIT, (800) 456-7248, and STN International, FIZ Karlsruhe, P.O. Box 2465, W-7500, Karlsruhe 1, Germany, online services. Also available on CD-ROM. Inquire as to cost and availability.

SCISEARCH. Institute for Scientific Information, 3501 Market Street, Philadelphia, PA 19104. (800) 523-1850 or (215) 386-0100. Broad multidisciplinary title and author index to the international literature of science and technology, 1974 to present. Available on DIALOG, (800) 334-2564, and ORBIT, (800) 456-7248, online services. Also available on CD-ROM. Inquire as to cost and availability.

WILSONLINE. H.W. Wilson Company, 950 University Avenue, Bronx, NY 10452. (800) 367-6770 or (212) 588-8400. Makes available online versions of the printed H.W. Wilson indexes including Applied Science and Technology Index, Business Periodicals Index, General Science Index, and Readers' Guide

NEPTUNE

Ency. of Physical Sciences and Engineering Info. Sources

to Periodical Literature. Period covered is generally 1983 to present. Available on BRS, (800) 289-4277, DIALOG, (800) 334-2564, and OCLC EPIC, (800) 848-5878, online services. Also available on CD-ROM. Inquire as to cost and availability.

OTHER SOURCES

Atlas of the Solar System. B. Yenne. Simon & Schuster, Inc. 1230 Avenue of the Americas, New York, NY 10020. (212) 698-7000. (800) 223-2348. 1987. $12.98.

Atlas of Neptune. Garry E. Hunt and Patrick Moore. Cambridge University Press, 40 West 20th Street, New York, NY 10011-4211. (212) 924-3900. (800) 872-7423. 1994. $27.95.

The Sky: A User's Guide. David H. Levy. Cambridge University Press, 40 West 20th Street, New York, NY 10011- 4211. (212) 924-3900. (800) 872-7423. 1993. $14.95.

The View from Space: Photographic Exploration of the Planets. Merton Davies and Bruce C. Murray. Columbia University Press, 562 West 113th Street, New York, NY 10025. (212) 666-1000. (800) 944-8648. 1973. $17.50.

PERIODICALS

Association of Lunar and Planetary Observers. JOURNAL. Association of Lunar and Planetary Observers, Box 16131, San Francisco, CA 94116. (415) 566-5786. FAX (415) 731-8242. 1947 to present. Quarterly. $14.00 per year. ISSN: 0039- 2502.

Astronomical Journal. American Institute of Physics, One Physics Ellipse, College Park, MD 20740-3843. (301) 209- 3000. Published for the American Astronomical Society. 1849 to present. Monthly. $280.00 per year. ISSN: 0004- 6256.

Astronomical Society of the Pacific. PUBLICATIONS. Astronomical Society of the Pacific, 390 Ashton Avenue, San Francisco, CA 94112. (415) 337-1100. FAX (415) 337-5205. 1889 to present. Monthly. $175.00 per year. ISSN: 0004- 6280.

Astronomy. Kalmbach Publishing Company, Box 1612, Waukesha, WI 53187-1612. (414) 796-8776. FAX (414) 796-1142. 1973 to present. Monthly. $27.00 per year. ISSN: 0091- 6358.

Astronomy and Astrophysics. Springer-Verlag New York, Inc., 44 Hartz Way, Secaucus, NJ 07096-2491. (201) 348-4033. FAX (201) 348-4505. 1969 to date. 36 issues per year in 12 volumes. $1733.00 per year. ISSN: 0004-6361.

Earth, Moon and Planets; An International Journal of Comparative Planetology. Kluwer Academic Publishers, Box 358, Accord Station, Hingham, MA 02018-0358. (617) 871-6600. FAX (617) 871-6528. 1969 to present. Monthly. $840.00 per year. ISSN: 0167-9295.

Geochemica Et Cosmochimica Acta. Elsevier Science. 660 White Plains Road, Tarrytown, NY 10591-5153. (914) 524-9200. FAX (914) 333-2444. 1950 to the present. Biweekly. $895.00 per year. ISSN: 0016-7037.

Icarus; International Journal of Solar System Studies. Academic Press, Inc., Journal Division, 525 B Street, Suite 1900,

San Diego, CA 92101-4495. (619) 230-1840. FAX (619) 699-6800. 1962 to the present. Monthly. $1080.00. ISSN: 0019-1035.

J G R: Journal of Geophysical Research: Planets. American Geophysical Union, 2000 Florida Avenue, NW, Washington, CD 20009. (202) 462-6900. FAX (202) 328-0566. 1991 to present. Monthly. $597.00 per year. ISSN: 0148-0227.

Lunar and Planetary Information Bulletin. Lunar and Planetary Institute. 3600 Bay Area Boulevard, Houston, TX 77058-1113. (713) 486-2175. FAX (713) 486-2125. 1970 to present. Quarterly. Free. Also available online.

Mercury. Astronomical Society of the Pacific. 390 Ashton Avenue, San Francisco, CA 94112. (415) 337-1100. FAX: (415) 337-5205. 1972 to present. Bimonthly. $175.00 per year. ISSN: 0047-6773.

Meteoritics. Meteoritical Society, Department of Chemistry, University of Arkansas, Fayetteville, AR 72701. (501) 575-7625. FAX (501) 575-7778. 1953 to present. Bimonthly. $210.00 per year. ISSN: 0026-1114.

Planetary and Space Science. Elsevier Science Publishing Company, Inc., 660 White Plains Road, Terrytown, NY 10591- 5153. (914) 524-9200. FAX: (914) 333-2444. 1959 to present. Monthly. $1355.00 per year. ISSN: 0032-0633.

Planetary Report. Planetary Society. 65 North Catalina, Pasadena, CA 91106-2301. (818) 793-5100. FAX (818) 793- 5528. 1980 to present. $25.00 per year. ISSN: 0736-3680.

Sky and Telescope. Sky Publishing Corporation, Box 9111, Belmont, MA 02178. (617) 864-7360. FAX (617) 864-6117. 1941 to present. Monthly. $27.00 per year. ISSN: 0037- 6604.

RESEARCH CENTERS AND INSTITUTES.

Center For Radiophysics and Space Research. Cornell University, Space Sciences Building, Ithaca NY 14853. (607) 255-4341. FAX (607) 255-9888.

Center For Space Science and Astrophysics. Stanford University. 325 Durand Building, Stanford, CA 04305. (415) 723-3582.

Laboratory For Planetary Geology. Arizona State University, Department of Geology, Tempe, AZ 85281. (601) 965-7029.

Laboratory For Planetary Studies. Cornell University, Space Sciences Building, Ithaca, NY 14853. (607) 255-4971.

Lunar and Planetary Institute. 3303 NASA Road One, Houston, TX 77058-4399. (713) 486-2139. FAX 713-496-2162.

Lunar and Planetary Laboratory. University of Arizona, Tucson, AZ 85721. (602) 621-6963. FAX (602) 621-4933.

Smithsonian Astrophysical Observatory. 60 Garden Street, Cambridge, MA 02138. (617) 495-7461. FAX (617) 495-7105.

NEUTRINOS

See: PARTICLE PHYSICS

NEUTRON OPTICS

See: OPTICS

NEUTRON SPECTROSCOPY

See: SPECTROSCOPY

NEUTRON STARS

See also: ASTRONOMY, BIG BANG, COSMOLOGY, DARK MATTER, GALAXIES, INTERSTELLAR MEDIUM, MILKY WAY, PULSARS, QUASARS, RADIO ASTRONOMY, STARS, UNIVERSE

ABSTRACT SERVICES AND INDEXES

Applied Science and Technology Index; A Cumulative Subject Index To English Language Periodicals in the Fields of Aeronautics and Space Science, Computer Technology, Chemistry, Construction Industry, Energy and Related Areas. H.W. Wilson Co., 950 University Avenue, Bronx, NY 10452. (212) 588-8400. (800) 367-6770. FAX (718) 590-1617. From 1958 to present. Monthly. Inquire about cost and availability. Also available on CD-ROM and online. ISSN: 0003-6986.

Astronomy and Astrophysics Abstracts. Springer-Verlag New York, 175 Fifth Avenue, New York, NY 10010. (212) 460-1500. FAX (212) 473-6272. Published for the Astronomisches Rechen-Institut. Comprehensive coverage of all aspects of astronomy, astrophysics and related fields. 1969 to present. Two parts per year. Annual. ISSN: 0067-0022.

General Science Index. H.W. Wilson Company, 950 University Avenue, Bronx, NY 10452. (212) 588-8400. (800) 367-6770. FAX (718) 590-1617. From 1978 to present. Ten issues per year; quarterly and annual cumulations. Service basis. Available on CD-ROM and online. Inquire regarding cost and availability. ISSN: 0162-1963.

Government Reports Announcements and Index. U. S. National Technical Information Service (NTIS), 5285 Port Royal Road, Springfield, VA 22161. (703) 487-4650. FAX (703) 321-8547. From 1968 to present. Annual. $630.00 per year. Also available online as *NTIS Bibliographic Database* and on CD-ROM.

Meteorological and Geoastrophysical Abstracts. American Meteorological Society, c/o Inforonics, Inc., 550 Newtown Road, Littleton, MA 01460. (508) 486-8976. FAX (508) 486-0027. Covers literature in environmental sciences, meteorology, astrophysics, hydrology, glaciology, and physical oceanography. From 1950 to present. Monthly. $950.00 per year. Also available on CD-ROM and online. Inquire regarding cost and availability. ISSN: 0026-1130.

NTIS Alerts: Astronomy & Astrophysics. U. S. National Technical Information Service. 5285 Port Royal Road, Springfield, VA 22161. (703) 487-4650. FAX (703) 321-8547. Weekly. $140.00 per year.

Pascal 48: Environnement Cosmique Terrestre, Astronomie Et Geologie Extraterrestre. Centre National de la Recherche Scientifique, Institut de Information Scientifique et Technique, 2 aliee du Parc de Brabois, 54514 Vandoeuvre Les Nancy Cedex, France. TEL 83- 50-46-00. FAX: 83-50-46-50. 1985 to present.

Ten issues per year. 770 F. Also available on CD-ROM and online.

Physics Abstracts. INSPEC. Section A, Science Abstracts. Institute of Electrical Engineers, London, United Kingdom. Available from: INSPEC/IEEE - Institute of Electrical and Electronic Engineers, Box 1331, Hoes Lane, Piscataway, NJ 08855-1331. (908) 562-5549. 1898 to present. 24 issues per year. $2835.00 per year. ISSN: 0036-8091. Alsoavailable online and on CD-ROM.

Physics Briefs (Physikalische Berichte). Information Center for Energy, Physics, Mathematics; German Physical Society. V C H Publishers, Inc., 220 East 23rd Street, New York, NY 10010-4606. (212) 683-8333. 1845 to present. 24 issues per year. $2390.00 per year. Also available online. ISSN: 0179-7434.

Science Citation Index. SCI. Institute for Scientific Information, 3501 Market Street, Philadelphia, PA 19104. (215) 386-0100. (800) 523-1850. FAX (215) 386-2991. 1961 to present. Six issues per year, plus annual cumulation. $11650.00 per year. Also available online and on CD-ROM.Inquire about price and availability. ISSN: 0036-827X.

STAR. Scientific and Technical Aerospace Reports. U.S. National Aeronautics and Space Administration. Distributed by U. S. Superintendent of Documents, Washington, DC 20402. 1963 to present. Semi-monthly, with semiannual and annual indexes. $114.00 per year. ISSN: 0036-8741. Also available online and on CD-ROM.

ANNUAL REVIEWS AND YEARBOOKS

The Astronomical Almanac. Superintendent of Documents, U. S. Government Printing office, Washington, DC 10402. (202) 783-3238. 1981 to present. Supercedes Astronomical Ephemeris and the American Ephermis and Nautical Almanac. Annual. $44.95. ISSN: 0737-6421.

Annual Review of Astronomy and Astrophysics; Original Reviews of Critical Literature and Current Developments in Astronomy and Astrophysics. Annual Reviews, Inc., 4139 El Camino Way, Palo Alto, CA 94306-0139. (415) 493-4400. (800) 523-8635. FAX (415) 855-9815. 1963 to date. Annual. $60.00. ISSN: 0066-4146.

Highlights of Astronomy. I. Appenzeller, editor. Kluwer Academic Publishers, Box 358, Accord Station, Hingham, MA 02018-0358. (617) 871-6600. International Astronomical Union Highlights series. ISBN: 07923-3553-8. Triennal. 1995. Price varies.

ASSOCIATIONS AND PROFESSIONAL SOCIETIES

Amateur Astronomers Association. 1010 Park Avenue, New York, NY 10028. (212) 535-2922.

American Association of Variable Star Observers. 23 Birch Street, Cambridge, MA 02138. (617) 354-0484. FAX: (617) 354-0665.

American Astronomical Society. 2000 Florida Avenue NW, Suite 400, Washington, DC 20009. (202) 328-2010. FAX: (202) 234-2560.

NEUTRON STARS

Ency. of Physical Sciences and Engineering Info. Sources

American Institute of Physics. One Physics Ellipse, College Park, MD 20740-3843. (301) 209-3100.

Association of University For Research in Astronomy. Suite 701, 1625 Massachusetts Avenue, NW, Washington, DC 20036. (202) 483-2101. FAX (202) 483-2106.

Astronomical League. 2112 Kingfisher Lane East, Rolling Meadows, IL 60008. (708) 398-0562.

Astronomical Society of the Pacific. 390 Ashton Avenue, San Francisco, CA 94112. (415) 337-1100. FAX: (415) 337-5205.

International Astronomical Union. 98 bis Boulevard Arago, F-75014 Paris, France. Tel 1-43-25-83-58. FAX 1-43-25-26-16.

Royal Astronomical Society of Canada. 136 Dupont Street, Toronto. ON N5R 1V2 Canada. (416) 924-7973. FAX (416) 968-6687.

BIBLIOGRAPHIES

Astronomy and Astrophysics: A Source Guide. Gordon Press, P.O. Box 459, Bowling Green Station, New York, NY 10004. (718) 624-8419. 1991. $250.00. .

A Bibliography of Astronomy, 1970-1979. R. A. Sea and S. S. Martin. Libraries Unlimited, Inc., Littleton, CO 80160. 1982.

Catalog of the Naval Observatory Library. Naval Observatory Staff. G. K. Hall & Company, 866 Third Avenue, New York, NY 10022. (212)702-6789. (800) 257-5755. 6 volumes. 1977. $655.00

Science Books and Films. American Association for the Advancement of Science, 1333 H Street NW, Washington, DC 20005. (202) 326-6454. 1965 to present. Nine issues per year. $40.00 per year. ISSN: 0098-342X

Scientific and Technical Books and Serials in Print, 1995. R.R. Bowker Inc., 121 Chanlon Road, New Providence, NJ 07974. (908) 464-6800. (800) 521-8110. 4 volumes. 1994. $299.95. Also available on compact disc and online.

A Source Book On Astronomy and Astrophysics, 1900 - 1975. Kenneth Lang and Owen Gingerich, editors. Harvard University Press, 79 Garden Street, Cambridge MA. (617) 495-2600. 1980. $41.95.

DIRECTORIES AND BIOGRAPHICAL SOURCES

American Men and Women of Science: Physical and Biological Sciences. R. R. Bowker Inc., 121 Chanlon Road, New Providence, NJ 07974. (908) 464-6800. (800) 521-8110. 20th edition. 8 volumes. 1996. $850.00.

The Astronomers. Donald Goldsmith. St. Martin's Press, Inc., 175 Fifth Avenue, New York, NY 10010. (212) 674-5151. (800) 221-7945. 1993. $14.95 in paper.

Astronomical Centers of the World. Kevin Krisciunas. Cambridge University Press, 40 West 20th Street, New York, NY 10011-4211. (212) 924-3900. (800) 872-7423. 1988. $34.95

The Biographical Dictionary of Scientists: Astronomers. D. Abbott, editor. Peter Bedrick Books, Inc., 2112 Broadway, Room 318, New York, NY 10023. (212) 496-0751. 1984.

Directory of Physics and Astronomy Staff Members. American Institute of Physics, One Physics Ellipse, College Park, MD 20740-3843. (301) 209-3100. Annual. $45.00.

Earth and Astronomical Research Centers. Stockton Press, 345 Park Avenue South, New York, NY 10010. 4th edition. 1995. $515.00. ISBN: 1-56169-0967.

Graduate Programs in Physics, Astronomy and Related Fields. 1995-1996. American Institute of Physics, One Physics Ellipse, College Park, MD 20740-3843. (301) 209-3100. 1978 to present. Annual. $45.00.

Research Centers Directory. Gale Research Inc., 835 Penobscot Building, Detroit, MI 48226-4094. (313) 961-2242. (800) 347-4253. 20th edition. 1995. $485.00.

ENCYCLOPEDIAS AND DICTIONARIES

Astronomy From A To Z: A Dictionary of Celestial Objects and Ideas. Charles A. Schweighauser. Illinois Issues, Sangamon State University, K-10, Springfield, IL 62794-9243. (217) 786-9243. 1991. $14.95 in paper.

Concise Dictionary of Astronomy. Jacqueline Mitton. Oxford University Press, Inc., 200 Madison Avenue, New York, NY 10016. (212) 725-6000. (800) 334-4249. 1992. $24.95.

Facts On File Dictionary of Astronomy. Valerie Illingworth.

Facts-on-File, Inc., 460 Park Avenue South, New York, NY 10016-7382. (212) 683-2244. (800) 322-8755. Fax (800) 683-3633. 1994. $27.95.

Encyclopedia of Astronomy and Astrophysics. Stephen Maran, editor. Van Nostrand Reinhold, 115 Fifth Avenue, New York, NY 10003. (212) 254-3232. (800) 842-3636. 1992. $129.95.

Landolt-boernstein Numerical Data and Functional Relationships in Science and Technology: Astronomy, Astrophysics and Space Research; A. O. Madelung and H. H. Boigt, editors. Springer Verlag New York, Inc., 175 Fifth Avenue, New York, NY 10010. (212) 460-1500. (800) 777- 4643. 1993. $550.00.

Oxford Illustrated Encyclopedia. VOLUME 8, the UNIVERSE. Archie Roy, editor. Oxford University Press, Inc., 200 Madison Avenue, New York, NY 10016. (212) 725-6000. (800)334-4249. 1993. $45.00.

McGraw-Hill Encyclopedia of Science and Technology. McGraw-Hill Book Company, Inc., 1221 Avenue of the Americas, New York, NY 10020. (212) 512-2000. (800) 262-4729. 7th edition. 20 volume set. 1992. $1900.00.

Stars, Galaxies, Cosmos. William R. Corliss. Sourcebook Project, P.O. Box 107, Glen Arm, MD 21057. (410) 668-6047. 1987. $17.95.

Ency. of Physical Sciences and Engineering Info. Sources

NEUTRON STARS

GENERAL WORKS

Asteroids To Quasars. Phyliss Lugger, editor. Cambridge University Press, 40 West 20th Street, New York, NY 10011- 4211. (212) 924-3900. (800) 872-7423. 1991. $69.95.

ASTROPHYSICS of NEUTRON STARS. M. Lipunov. Springer-Verlag New York, Inc., 175 Fifth Avenue, New York, NY 10010. (212) 460-1500. (800) 777-4643. FAX (212) 473-6272. 1992. $98.00.

Black Holes, White Dwarfs, and Neutron Stars: The Physics of Compact Objects. Stuart L. Shapiro and Saul A. Teukolsky. John Wiley & Sons, Inc., 605 Third Avenue, New York, NY 10158-0012. (212) 850-6000. (800) 225-5945. 1983. $79.95.

Cosmic Questions, Galactic Halos, Cold Dark Matter and the End of Time. Richard Morris. John Wiley & Sons, Inc., 605 Third Avenue, New York, NY 10158-0012. (212) 850-6000. (800) 225-5945. 1993. $24.95.

Cycles of Fire: Stars, Galaxies and the Wonder of Deep Space. William K. Hartmann. Workman Publishing Company, Inc., 708 Broadway, New York, NY 10003. (212) 254-5900. (800) 722-7202. 1988. $15.95 in paper.

Farthest Things in the Universe. Jay M. Pasachoff. Cambridge University Press, 40 West 20th Street, New York, NY 10011-4211. (212) 924-3900. (800) 872-7423.

1995. $29.95.

The First Three Minutes: A Modern View of the Origin of the Universe. Steven Weinberg. Basic Books, 10 East 53rd Street, New York, NY 10022. (212) 207-7057. (800) 242- 7737. Reprint edition. 1993. $12.00.

High-energy Radiation From Magnetized Neutron Stars. Peter Meszaros. University of Chicago Press, 5801 Ellis Avenue, Chicago, IL 60637. (312) 702-7700. 1992. $39.95 in paper.

The Left Hand of Creation: The Origin and Evolution of the Expanding Universe. John D. Barrow and Joseph Silk. Oxford University Press, Inc., 200 Madison Avenue, New York, NY 10016. (212) 725-6000. (800) 334-4249. 1994. $23.00.

Neutron Stars: Theory and Observation. Joseph Ventura and David Pines, editors. Kluwer Academic Publishers, 101 Philip Drive, Assinippi Park, Norwell, MA 02061. (617) 871- 6600. 1991. $174.00.

New Perspectives On Stellar Pulsation and Pulsating Variable Stars. James M. Nemec and Jaymie M. Matthews. Cambridge U. Press, 40 West 20th Street, New York, NY 10011- 4211. (212) 924-3900. 1993. $59.95.

Observational Astrophysics. Robert C. Smith. Cambridge University Press, 40 West 20th Street, New York, NY 10011- 4211. (212) 924-3900. (800) 872-7423. 1995. $69.95.

Origin and Evolution of Neutron Stars. D. J. Helfand and J. H. Huang. Kluwer Academic Publishers, 101 Philip Drive, Assinippi Park, Norwell, MA 02061. (617) 871- 6600. 1987. $66.50 in paper.

Physical Processes in Comets, Stars and Active Galaxies. W. Hildebrandt, et al editors. Springer-Verlag New York, Inc., 175 Fifth Avenue, New York, NY 10010. (212) 460-1500. (800) 777-4643. 1987. $47.00.

Physics of Neutron Stars. A. M. Kaminker. Nova Science Publishers, Inc., 6080 Jericho Turnpike, Suite 207, Commack, NY 11725-2808. (516) 499-3103. 1994. $97.00.

The Physics of Stars. A. C. PHillips. John Wiley & Sons, Inc., 605 Third Avenue, New York, NY 10158-0012. (212) 850-6000. (800) 225-5945. 1994. $32.95 in paper.

Planets Around Pulsars. J. A. PHillips, et al, editors. Astronomical Society of the Pacific, 390 Ashton Avenue, San Francisco, CA 94112. (415) 337-1100. 1993. $40.00.

Pulsar Astronomy. andrew Lyne and Francis Graham-Smith. Cambridge University Press, 40 West 20th Street, New York,

NY 10011-4211. (212) 924-3900. 1990. $79.95.

Pulsars As Physics Laboratories. R. D. B landord, et al, editors. Oxford University Press, Inc., 200 Madison Avenue, New York, NY 10016. (212) 725-6000. (800) 334-4249. 1994. $52.50 in paper.

Star Formation, Galaxies and the Interstellar Medium. j. Franco, et al, editors. Cambridge University Press, 40 West 20th Street, New York, NY 10011-4211. (212) 924-3900. (800) 872-7423. 1993. $59.95.

Stars and Atoms: From the Big Bang To the Solar System. Stuart Clark. Oxford University Press, Inc., 200 Madison Avenue, New York, NY 10016. (212) 725-6000. (800) 334-4249. 1995. $35.00.

Structure and Evolution of Neutron Stars. R. Tamagaki, et al. Addison-Wesley Publishing Co., Inc., 1 Jacob Way, Reading, MA 01867. (617) 944-3700. (800) 447-2226. . 1992. $49.95.

Universe of Galaxies. Paul W. Hodge, editor. W. H. Freeman & Co., 41 Madison Avenue, East 26th Avenue, 35th Floor, New York, NY 10010. (212) 576-0400. Readings from Scientific American. 1995.

Visual Astronomy of the Deep Sky. Roger N. Clark. Cambridge University Press, 40 West 20th Street, New York, NY 10011-4211. (212) 924-3900. (800) 872-7423. 1991. $39.95.

X-ray Binaries and Recycled Pulsars. E. P. Van den Heuvel and S. A. Rappaport, editors. Kluwer Academic Publishers, 101 Philip Drive, Assinippi Park, Norwell, MA 02061. (617) 871-6600. 1992. $222.00.

HANDBOOKS AND MANUALS

Burnham's Celestial Handbook: An Observer's Guide To the Universe Beyond the Solar System. Robert Burnham, Jr. Dover Publications, Inc., 180 Varick Street, New York, NY 10014. (212) 255-3755. (800) 223-3130. 3 volumes. 1978. $41.85 for the set.

Cambridge Guide To the Constellations. Michael E. Bakich. Cambridge University Press, 40 West 20th Street, New York, NY 10011-4211. (212) 924-3900. (800) 872-7423. 1995. $69.95.

NEUTRON STARS

Ency. of Physical Sciences and Engineering Info. Sources

Encyclopedia of Astronomy and Astrophysics. Stephen Maran, editor. Van Nostrand Reinhold, 115 Fifth Avenue, New York, NY 10003. (212) 254-3232. (800) 842-3636. 1992. $129.95.

Greenwich Guide To Stars, Galaxies and Nebulae. Stuart Malin. Cambridge University Press, 40 West 20th Street, New York, NY 10011-4211. (212) 924-3900. (800) 872-7423.1990. $10.95.

Handbook of Astronomy, Astrophysics and Geophysics; Volume 2: Galaxies and Cosmology. V. M. Canuto and B. G. Elmegreen. Gordon & Breach Science Publishers, Inc., P.O. Box 200029, Riverfront Plaza Station, Newark, NJ 07102-0301. (201) 643-7500. 1988. $194.00.

Handbook of Space Astronomy and Astrophsyics. Martin V. Zumbeck. Cambridge University Press, 40 West 20th Street, New York, NY 10011-4211. (212) 924-3900. (800) 872-7423. 2nd edition. 1990. $79.95.

Landolt-borenstein Numerical Data and Functional Relationships in Science and Technology: Astronomy, Astrophysics and Space Research. GROUP VI. Springer-Verlag New York, Inc., 175 Fifth Avenue, New York, NY 10010. (212) 460-1500. (800) 777-4643. volumes priced individually.

McGraw-Hill Encyclopedia of Astronomy. Sybil P. Parker and Jay M. Pasachoff, editors. McGraw-Hill Publishing Company, Inc., 1221 Avenue of the Americas, New York, NY 10020. (212) 512-2000. (800) 262-4729. 2nd edition. 1993. $75.00

National Audobon Society Pocket Guide: Galaxies and Other Deep-sky Objects. Alfred A. Knopf, Inc. 201 E. 50th Street, New York, NY 10022. (800) 733-3000. 1995. $7.95.

Nearby Galaxies Catalog. R. Brent Tully and J. Richard Fisher. Cambridge University Press, 40 West 20th Street, New York, NY 10011-4211. (212) 924-3900. (800) 872-7423. 1988. $59.95.

Observer's Guide To Astronomy. Patrick Martinez, editor. Cambridge University Press, 40 West 20th Street, New York, NY 10011-4211. (212) 924-3900. (800) 872-7423. 2 volumes. 1994. $159.90.

ONLINE DATABASES AND CD-ROMS

CA Search. Chemical Abstracts Service, P.O. Box 3012, Columbus, OH 43210-0012. (614) 447-3600. (800) 848-6533. FAX (614) 447-3709. Very comprehensive guide to worldwide chemical literature and related fields, 1972 to present. Available on BRS,(800) 289-4277, DIALOG, (800) 334-2564, ORBIT (800) 456-7248, and STN International, FIZ Karlsruhe, P.O. Box 2465, W-7500, Karlsruhe 1, Germany, online services. Inquire as to cost and availability.

Current Contents Search. Institute for Scientific Information, 3501 Market Street, Philadelphia, PA 19104. (215) 386-0100. FAX (215) 386-6362. Contains citations to articles listed in the table of contents of science and technology journals. Also articles in social sciences and life sciences journals. Available on BRS, (800) 289-4277, DIALOG, (800) 334-2564, online services. Inquire as to cost and availability.

Dissertation Abstracts Online. University Microfilms International, 300 North Zeeb Road, Ann Arbor, MI 48106. (800) 521-0600 or (313) 761-4700. Scope includes virtually all doctoral dissertations accepted at accredited American institutions from 1861 to present in 252 subject areas. Available on BRS, (800)

289-4277, DIALOG, (800) 334-2564,and OCLC EPIC, (800) 848-5878, online services. Also available on CD-ROM. Inquire as to cost and availability.

INSPEC. Institution of Electrical Engineers, Michael Faraday House, Six Hills Way, Stevenage, Herts. SG1 2AY, England. Telephone: 0438 313311 or FAX 0438 742840. Contains citations to the worldwide literature of physics, electronics and electrical engineering, computer technology, and related fields. Available on BRS, (800) 289-4277, DIALOG, (800) 334-2564, ORBIT, (800) 456-7248, and STN International, FIZ Karlsruhe, P.O. Box 2465, W-7500, Karlsruhe 1, Germany, online services. Inquire as to cost and availability.

International Aerospace Abstracts. American Institute of Aeronautics and Astronautics, 370 L'Enfant Promenade, S.W., Washington, DC 20024. (202) 646-7400. Contains referencesand abstracts of journal and monograph literature relating to aerospace science and technology, from 1963 to present. Available through the NASA/RECON system of the National Aeronautics and Space Administration only.

NTIS Bibliographic Database. National Technical Information Service, 5285 Port Royal Road, Springfield, VA 22161. (703) 487-4929 or FAX (703) 321-8199. Broad coverage of government-sponsored science and technology research reports, 1964 to present. Available on BRS,(800) 289-4277, DIALOG, (800) 334-2564, ORBIT, (800) 456-7248, and STN International, FIZ Karlsruhe, P.O. Box 2465, W-7500, Karlsruhe 1, Germany, online services. Also available on CD-ROM. Inquire as to cost and availability.

Physics Briefs. American Institute of Physics, 335 East 45th Street, New York, NY 10017. (212) 661-9260 or FAX (212) 661-2036. Contains citations with abstracts of the literature of physics and related fields, 1979 to present. Available on the STN International, FIZ Karlsruhe, P.O. Box 2465, W-7500, Karlsruhe 1, Germany, online service. Inquire as to cost and availability.

Scientific and Technical Books and Serials in Print. R.R. Bowker Inc., 121 Chanlon Road, New Providence, NJ 07974. (800) 521-8110 or (908) 464-6800. List of currently published books and serials in the physical and biological sciences, engineering and technology, with subject, author and titles indexes. Available on ORBIT, (800) 456-7248, online service. Inquire as to cost and availability.

SCISEARCH. Institute for Scientific Information, 3501 Market Street, Philadelphia, PA 19104. (800) 523-1850 or (215) 386-0100. Broad multidisciplinary title and author index to the international literature of science and technology, 1974 to present. Available on DIALOG, (800) 334-2564, and ORBIT, (800) 456-7248, online services. Also available on CD-ROM. Inquire as to cost and availability.

WILSONLINE. H.W. Wilson Company, 950 University Avenue, Bronx, NY 10452. (800) 367-6770 or (212) 588-8400. Makes available online versions of the printed H.W. Wilson indexes including Applied Science and Technology Index, Business Periodicals Index, General Science Index, and Readers' Guide to Periodical Literature. Period covered is generally 1983 to present. Available on BRS, (800) 289-4277, DIALOG, (800) 334-2564, and OCLC EPIC, (800) 848-5878, online services.Also available on CD-ROM. Inquire as to cost and availability. subhead2> Other Sources

Atlas of Deep-sky Spendors. Hans Vehrenberg. Sky Publishing Corp., 49 Bay State Road, Cambridge, MA 02138. (617) 864-7360. (800) 253-0245. 4th edition. 1983. $24.95.

Atlas of the Ultraviolet Sky. Richard C. Henry, Wayne B. Landsman, et al. Johns Hopkins University Press, 2715 North Charles Street, Baltimore, MD 21218-4319. (410) 516- 6900. (800) 537-5487. 1988. $70.00.

Atlas of the Universe. Patrick Moore. Rand McNally & Company. 8255 North Central Park, Skokie, IL 60076-2970. (708) 673-9100. 1994. $29.95.

Cambridge Atlas of Astronomy. Jean Audouze, et al. Cambridge University Press, 40 West 20th Street, New York,

NY 10011-4211. (212) 924-3900. (800) 872-7423. 3rd edition. 1994. $75.00.

Carnegie Atlas of Galaxies. Allan Sandage and John Bedke. Carnegie Institution of Washington, 1530 P Street NW, Washington, DC 20005. (202) 387-6411. Two volumes. 1994. $92.00.

Observer's Sky Atlas. E. Karkoschka. Springer-Verlag New York, Inc., 175 Fifth Avenue, New York, NY 10010. (212) 460-1500. (800) 777-4643. 1995. $15.95.

Observing Handbook and Catalog of Deep-sky Objects. Christian B. Luginbuhl and Brian A. Skiff. Cambridge University Press, 40 West 20th Street, New York, NY 10011- 4211. (212) 924-3900. (800) 872-7423. 1990. $49.95.

Photometric Atlas of Northern Bright Galaxies. Kodaira, Keichi, et al editors. Columbia University Press, 562 West 113th Street, New York, NY 10025. (212) 666-1000. (800) 944-8648. 1990. $180.00.

Third Reference Catalogue of Bright Galaxies. Gerard H. DE Vaucouleurs et al, editors. Springer-Verlag New York, Inc., 175 Fifth Avenue, New York, NY 10010. (212) 460-1500. (800) 777-4643. Three volumes. 1991. $198.00 set.

PERIODICALS

Astronomical Journal. American Institute of Physics, One Physics Ellipse, College Park, MD 20740-3843. (301) 209-3000. Published for the American Astronomical Society. 1849 to present. Monthly. $280.00 per year. ISSN: 0004- 6256.

Astronomical Society of the Pacific. PUBLICATIONS. Astronomical Society of the Pacific, 390 Ashton Avenue, San Francisco, CA 94112. (415) 337-1100. FAX (415) 337-5205.1889 to present. Monthly. $175.00 per year. ISSN: 0004- 6280.

Astronomy. Kalmbach Publishing Company, Box 1612, Waukesha, WI 53187-1612. (414) 796-8776. FAX (414) 796-1142. 1973

to present. Monthly. $27.00 per year. ISSN: 0091-6358.

Astronomy and Astrophsyics. Springer-Verlag New York, Inc., 44 Hartz Way, Secaucus, NJ 07096-2491. (201) 348-4033. FAX(201) 348-4505. 1969 to date. 36 issues per year in 12 volumes. $1733.00 per year. ISSN: 0004-6361.

Astronomy Reports. American Institute of Physics, One Physics Ellipse, College Park, MD 20740-3843. (301) 209- 3000. English translation of Astronomicheksii Zhurnal. Formerly Soviet Astronomy AJ. 1957 to present. Bimonthly. $1160.00 per year. ISSN: 1063-7729.

Astrophysical Journal; An International Review of Astronomy and Astronomical Physics. University of Chicago Press, Journals Division, 5720 South Woodlawn Avenue, Chicago, IL 60637. (312) 753-3347. FAX (312) 753-0811. 1895 to date. three issues per month. $740.00 per year. ISSN: 0004-637X.

Astrophpysics and Space Science; An International Journal of Cosmic Physics. Kluwer Academic Publishers, Box 358, Accord Station, Hingham, MA 02018-0358. (617) 871-6600. FAX (617) 871-6528. 1968 to present. 24 issues per year. $2544.00 per year. ISSN: 0004-640X.

Celestial Mechanics and Dynamical Astronomy; An International Journal of Space Dynamics. Kluwer Academic Publishers, Box 358, Accord Station, Hingham, MA 02018-0358. (617) 871-6600. FAX (617) 871-6528. 1969 to present. Monthly. $703.50 per year. ISSN: 0923-2958.

Experimental Astronomy; An International Journal On Astronomical Instrumentation and Data Analysis. Kluwer Academic Publishers, Box 358, Accord Station, Hingham, MA 02018-0358. (617) 871-6600. FAX (617) 871-6528. 1989 to present. Quarterly. $161.00 per year. ISSN: 0922-6435.

Icarus; International Journal of Solar System Studies. Academic Press, Inc., Journal Division, 525 B Street, Suite 1900, San Diego, CA 92101-4495. (619) 230-1840. FAX (619) 699-6800. 1962 to present. Monthly. $1080.00 per year. ISSN: 0019-1035.

Landolt-boernstein, Zahlenwerte Und Funktionen Aus Naturwissenschaften Und Technik. GROUP 6: ASTRONOMY. Springer-Verlag New York, Inc., 175 Fifth Avenue, New York, NY 10010. (212) 460-1500. FAX (212) 473-6272. 1965. Irregular. Price varies. ISSN: 0075-7896.

Mercury. Astronomical Society of the Pacific. 390 Ashton Avenue, San Francisco, CA 94112. (415) 337-1100. FAX: (415) 337-5205. 1972 to present. Bimonthly. $175.00 per year. ISSN: 0047-6773.

Monthly Notices of the Royal Astronomical Society. Blackwell Scientific Publication LT., Osney Mead, Oxford OX2 OEL, England. TEL 0865-240201. FAX 0865-721205. 1827 to present. Fortnightly. $1899.00 per year. ISSN: 0035-8711.

Observatory. Space and Astrophysics Division, Rutherford Appleton Laboratory, Chilton, Didcot, Oxon OX11 OQX England. FAX 0235-445848. 1877 to present. Bimonthly. $42.00 per year.

Sky and Telescope. Sky Publishing Corporation, Box 9111, Belmont, MA 02178. (617) 864-7360. FAX (617) 864-6117.1941 to present. Monthly. $27.00 per year. ISSN: 0037-6604.

Space Science Reviews. Kluwer Academic Publishers, Box 358, Accord Station, Hingham, MA 02018-0358. (617) 871-6600. FAX (617) 871-6528. 1962 to present. Sixteen issues per year. $940.00 per year. ISSN: 0038-6308.

Vistas in Astronomy; An International Review Journal. Albert C. Beer, editor. Elsevier Science Publishing Company, Inc., 660 White Plains Road, Terrytown, NY 10591- 5153. (914) 524-9200. FAX: (914) 333-2444. 1958 to present. Quarterly. $415.00 per year. ISSN: 0083-6656.

NEUTRON STARS

Ency. of Physical Sciences and Engineering Info. Sources

Transactions of the International Astronomical Union. Proceedings of the General Assembly. Kluwer Academic Publishers, Box 358, Accord Station, Hingham, MA 02018- 0358. (617) 871-6600. Issues in two parts: Part A, Reports, and Part B, Proceedings. Triennial. XXI - 1994. Price varies. ISSN: 0080-1372.

Webb Society Quarterly Journal. Webb Society, 194 Foundry Lane, Freemantle, Southampton, Hampshire SO1 3EQ, England. 1969 to present. Quarterly. $19.50 per year. ISSN: 0043- 1680.

RESEARCH CENTERS AND INSTITUTES

ARECIBO Observatory. Cornell University, Box 995, Arecibo, PR 00613. (809) 878-2612. FAX (809) 878-1861.

Astronomy Program. University of Maryland, 1207 Computer & Space Sciences Building, College Park, MD 20742. (301) 405- 3001. FAX (301) 314-9067.

Fred Lawrence Whipple Observatory. P. O. Box 97, Amado, AZ 85645. (602) 670-6741. FAX (602) 670-6779.

Joint Institute For Laboratory Astrophyics. University of Colorado, Box 440, Boulder, CO 80309-0440. (303) 492-7780. FAX (303) 492-5235.

Lick Observatory. University of California, Santa Cruz, CA 95064. (408) 459-2991. FAX (408) 426-3115.

Lowell Observatory. 1400 West Mars Hill Road. Flagstaff, AZ 86001, (602) 774-3358. FAX (602) 774-6296.

McDonald Observatory, University of Texas. P.O. Box 1337, Fort Davis, TX 79734. (915) 426-3263.

National Astronomy and Ionosphere Center. Cornell University, Space Sciences Building, Ithaca, NY 14853. (607) 255-3735. (607) 255-8803.

National Radio Astronomy Observatory. Edgemont Road, Charlottesville, VA 22903. (804) 296-0211.

Palomar Observatory. California Institute of Technology. 105-24, Pasadena CA 91125. (818) 356-4033.

Radio Observatory. University of Florida, 211 Space Sciences Research Building, Gainsville, FL 32611. (904) 392-2066.

Space Telescope Science Institute. 3700 San Martin Drive, Baltimore, MD 21218. (301) 338-4700.

Steward Observatory. University of Arizona. Tuscon, AZ 85721. (602) 621-6524. FAX (602) 621-1532.

NOISE CONTROL

See also: ACOUSTICS, VIBRATION

ABSTRACT SERVICES AND INDEXES

Acoustics Abstracts. Multi-Science Publishing Company, Limited, 107 High Street, Brentwood, Essex, CM14 4RX, England. 1967 to present. Monthly. $600.00 per year.

Applied Science and Technology Index; A Cumulative Subject Index To English Language Periodicals in the Fields of Aeronautics and Space Science, Computer Technology, Chemistry, Construction Industry, Energy and Related Areas. H.W. Wilson Co., 950 University Avenue, Bronx, NY 10452. (800) 367-6770 or (212) 588-8400. FAX (718) 590-1617. From 1958 to present. Monthly. Inquire about cost and availability. Also available on CD-ROM and online.

Current Contents: Engineering, Technology, and Applied Sciences. Institute for Scientific Information, 3501 Market Street, Philadelphia, PA 19104. (215) 386-0100. FAX (215) 386-6362. 1970 to present. Weekly. $442.00 per year.

Engineering Index Monthly. Engineering Information, Inc., Castle Point on the Hudson, Hoboken, NJ 07030. (800) 221-1044. FAX (212) 832-1857. Monthly. $2300.00 per year. Also available online as COMPENDEX, and also on CD-ROM. Covers chemical engineering, computers, electrical engineering, civil engineering, metals and mining, industrial management, and mechanical engineering.

Environmental Abstracts. R.R. Bowker Inc., 121 Chanlon Road, New Providence, NJ 07974. (800) 521-8110 or (908) 464-6800. Contains citations, most with abstracts, to literature relating to environmental issues and problems, from 1971 to present. Inquire as to cost and availability.

Pollution Abstracts, Cambridge Scientific Abstracts, 7200 Wisconsin Avenue, Bethesda, MD 20814-4823. (301) 961-6750. FAX (301) 961-6720. Inquire about price.

ASSOCIATIONS AND PROFESSIONAL SOCIETIES

Acoustical Society of America. 500 Sunnyside Blvd., Woodbury, NY 11797. (516) 576-2360. FAX (516) 349-7669.

Institute of Noise Control Engineering. PO Box 3206, Arlington Br., Poughkeepsie, NY 12603. (914) 462-4006. FAX (914) 473-9325.

National Association of Noise Control officials. c/o Edward J. DiPolvere, 53 Cubberly Road, Trenton, NJ 08690. (609) 586-2684.

Noise (National Organization To Insure A Sound-controlled Environment). 1620 Eye Street NW, Suite 300, Washington, DC 20005. (202) 682-3901.

DIRECTORIES AND BIOGRAPHICAL SOURCES

Directory of Engineering Societies and Related Organizations. Gordon Davis. 13th edition. American Association of Engineering Societies, 1111 19th Street NW, Suite 608, Washington, DC 20036. (202) 296-2237 or (800) 658-8897. 1989. Inquire for price.

International Research Centers Directory. Gale Research, 835 Penobscot Building, Detroit, MI 48226-4094. (313) 961-2242.

Ency. of Physical Sciences and Engineering Info. Sources

NOISE CONTROL

(800) 347-4253. FAX (313) 961-6083. 8th edition. 1995. $410.00.

The 1991 International Directory of Noise & Vibration Manufacturers & Suppliers. S. Barrett. Elsevier Science Inc., 660 White Plains Road, Tarrytown, NY 10591-5153. (914) 524-9200. FAX (914) 333-2444. 1991. $54.00.

Research Centers Directory. Gale Research, 835 Penobscot Building, Detroit, MI 48226-4094. (313) 961-2242. (800) 347-4253. FAX (313) 961-6083. $485.00.

Scientific and Technical Organizations and Agencies Directory. Gale Research, 835 Penobscot Building, Detroit, MI 48226-4094. (313) 961-2242. (800) 347-4253. FAX (313) 961-6083. 4th edition. 1996. $195.00.

Who's Who in Engineering. American Association of Engineering Societies, 1111 19th Street NW, Suite 608, Washington, DC 20036. (202) 296-2237 or (800) 658-8897. 8th edition. 1991. Inquire for price.

GENERAL WORKS

Active Control of Noise and Vibration. C. Hansen & S.D. Snyder. Chapman & Hall, 115 Fifth Avenue, New York, NY 10211-0906. 1996. $275.00.

Engineering Noise Control. D.A. Bies & C.H. Hansen. Chapman & Hall, 115 Fifth Avenue, New York, NY 10211-0906. 1996. $49.95.

Fundamentals of Noise Control Engineering. R.K. Miller & A. Thumann. 2nd edition. Fairmont Press, 700 Indian Trail, Lilburn, GA 30247. (404) 925-9388. 1989. $58.00.

Industrial Noise Control. Bell & Bell, eds. Marcel Dekker, Inc., 270 Madison Avenue, New York, NY 10016. (212) 696-9000. FAX (212) 685-4540. Inquire for cost and availability.

Industrial Noise Control. Paul N. Cheremisinoff. Prentice Hall, 113 Sylvan Avenue, Route 9W, Englewood Cliffs, NJ 07632. (201) 592-2000 or (800) 922-0579. 1993. $45.00.

Noise and Vibration Control Engineering. Leo L. Beranek and Istvan L. Ver, eds. John Wiley and Sons, Inc., 605 Third Avenue, New York, NY 10158. (800) 526-5368 or (212) 850-6000. 1992. $89.95.

Noise Control in Buildings: A Guide For Architects and Engineers. Cyril M. Harris, ed. McGraw-Hill Publishing Company, 1221 Avenue of the Americas, New York, NY 10020. (800) 262-4729 or (212) 512-3825. 1994. $45.00.

Noise Control Management. Howard K. Pelton. Van Nostrand Reinhold, 115 Fifth Avenue, New York, NY 10003. (212) 254-3232 or FAX (212) 254-9499. 1993. Write for information.

HANDBOOKS AND MANUALS

Handbook of Architecture and Noise Control: A Manual For Architects & Engineers. Michael Rettinger. TAB Books, P.O. Box 40, Blue Ridge Summit, PA 17294-0850. (717) 794-2191 or (800) 233-1128. FAX (717) 794-2080. 1988. $32.95.

Noise Control Manual. David Harris. Chapman and Hall, 115 Fifth Avenue, New York, NY 10211-0906. 1992. $64.95.

ONLINE DATABASES AND CD-ROMS

Compendex Plus. Engineering Information, Inc., 345 East 47th Street, New York, NY 10017. (212) 705-7600 or (800) 221-1044. Contains citations with abstracts to worldwide literature in engineering and technology, from 1970 to present. Available on online BRS,(800) 289-4277, DIALOG, (800) 334-2564, ORBIT (800) 456-7248, and STN International, FIZ Karlsruhe, P.O. Box 2465, W-7500, Karlsruhe 1, Germany, online services. Also available on CD-ROM. Inquire as to cost and availability.

CSA Engineering. Cambridge Scientific Abstracts, 7200 Wisconsin Avenue, Suite 601, Bethesda, MD 20814. (301) 961-6750 or (800) 843-7751. Contains citations and abstracts of international periodicals and research literature covering all fields of engineering and science and technology,including computer and information science, electronics, mechanical engineering, solid state materials, 1981 to present. Available on BRS,(800) 289-4277, online service.Inquire as to cost and availability.

Current Contents Search. Institute for Scientific Information, 3501 Market Street, Philadelphia, PA 19104. (215) 386-0100. FAX (215) 386-6362. Contains citations to articles listed in the table of contents of science and technology journals. Also articles in social sciences and life sciences journals. Available on BRS,(800) 289-4277, DIALOG,(800) 334-2564, online services. Inquire as to cost and availability.

Dissertation Abstracts. University Microfilms International, 300 North Zeeb Road, Ann Arbor, MI 48106. (800) 521-0600 or (313) 761-4700. Scope includes virtually all doctoral dissertations accepted at accredited American institutions from 1861 to present in 252 subject areas. Available on BRS,(800) 289-4277, DIALOG,(800) 334-2564, and OCLC EPIC,(800) 848-5878, online services. Also available on CD-ROM. Inquire as to cost and availability.

ISMEC: Mechanical Engineering Abstracts. Cambridge Scientific Abstracts, 7200 Wisconsin Avenue, Suite 601, Bethesda, MD 20814. (301) 961-6750 or (800) 843-7751. Contains citations to the literature in mechanical engineering, industrial and production engineering, energy, power, mechanics, devices and related areas, from 1973 to present. Available on the DIALOG,(800) 334-2564, online service. Inquire as to cost and availability.

NTIS Bibliographic Database. National Technical Information Service, 5285 Port Royal Road, Springfield, VA 22161. (703) 487-4929 or FAX (703) 321-8199. Broad coverage of government-sponsored science and technology research reports, 1964 to present. Available on BRS,(800) 289-4277, DIALOG, (800) 334-2564, ORBIT, (800) 456-7248, and STN International, FIZ Karlsruhe, P.O. Box 2465, W-7500, Karlsruhe 1, Germany, online services. Also available on CD-ROM. Inquire as to cost and availability.

WILSONLINE. H.W. Wilson Company, 950 University Avenue, Bronx, NY 10452. (800) 367-6770 or (212) 588-8400. Makes available online versions of the printed H.W. Wilson indexes including Applied Science and Technology Index, Business Periodicals Index, General Science Index, and Readers' Guide to Periodical Literature. Period covered is generally 1983 to present. Available on BRS,(800) 289-4277, DIALOG,(800) 334-2564, and OCLC EPIC,(800) 848-5878, online services. Also available on CD-ROM. Inquire as to cost and availability.

NOISE CONTROL

Ency. of Physical Sciences and Engineering Info. Sources

PERIODICALS

Noise Control Engineering Journal. Institute of Noise Control Engineering, Department of Mechanical Engineering, Auburn University, Auburn, AL 36849-5341. (205) 844-3306. FAX (205) 844-3307. 1973 to present. Bi-monthly. $60.00 per year.

Sound and Vibration. Acoustical Publications Inc., Box 40416, Bay Village, OH 44140. (216) 835-0101. FAX (216) 835-0303. 1967 to present. Monthly. Inquire for cost.

RESEARCH CENTERS AND INSTITUTES

Acoustics and Vibration Laboratory. University of Hartford, College of Engineering, West Hartford, CT 06117. (203) 243-4614.

Center For Sound Control and Vibration. North Carolina State University, Campus Box 7910, Raleigh, NC 27695. (919) 737-3024. FAX (919) 737-7968.

Noise Control Laboratory. Pennsylvania State University, 157 Hammond Bldg., University Park, PA 16802. (814) 865-2761. FAX (814) 863-7222.

Institute of Noise Control Engineering. PO Box 3206, Arlington Br., Poughkeepsie, NY 12603. (914) 462-4006. FAX (914) 473-9325.

Institute of Noise Control Engineering. Department of Mechanical Engineering, Auburn University, Auburn, AL 36849-5341. (205) 844-3306. FAX (205) 844-3307.

SPECIFICATIONS AND STANDARDS

*Annual Book of Astm Standards, Volume 04.*06. American Society for Testing and Materials (ASTM), 1916 Race Street, Philadelphia, PA 19103. (215) 299-5585. Annual. Inquire for cost and availability.

NONDESTRUCTIVE TESTING

See also: ALLOYS, FERROALLOYS, MATERIALS SCIENCE, METALLURGICAL ENGINEERING, METALLURGY, METALS AND METAL WORKING

ABSTRACT SERVICES AND INDEXES

Applied Mechanics Reviews: An Assessment of World Literature in Engineering Sciences. 1948-present. American Society of Mechanical Engineers, 345 East 47th Street, New York, NY 10017. (212) 705-7703. Monthly. $360.00 per year.

Applied Science and Technology Index; A Cumulative Subject Index To English Language Periodicals in the Fields of Aeronautics and Space Science, Computer Technology, Chemistry, Construction Industry, Energy and Related Areas. H.W. Wilson Co., 950 University Avenue, Bronx, NY 10452. (800) 367-6770 or (212) 588-8400. FAX (718) 590-1617. From 1958 to present. Monthly. Inquire about cost and availability. Also available on CD-ROM and online.

Ceramic Abstracts. American Ceramic Society, Inc., 757 Brooksedge Plaza Drive, Westerville, OH 43081-2821. (614) 890-6136. Bi-monthly. 1922 to present. $120.00 per year.

Current Contents: Engineering, Technology, and Applied Sciences. Institute for Scientific Information, 3501 Market Street, Philadelphia, PA 19104. (215) 386-0100. FAX (215) 386-6362. 1970 to present. Weekly. $442.00 per year.

Engineering Index Monthly. Engineering Information, Inc., Castle Point on the Hudson, Hoboken, NJ 07030. (800) 221-1044. FAX (212) 832-1857. Monthly. $2300.00 per year. Also available online as COMPENDEX, and also on CD-ROM. Covers chemical engineering, computers, electrical engineering, civil engineering, metals and mining, industrial management, and mechanical engineering.

International Aerospace Abstracts. Technical Information Service, American Institute of Aeronautics and Astronautics, Inc., 555 West 57th St., New York, NY 10019. (212) 247-6500. FAX (212) 582-4861. Semi-monthly. $1295.00 per year.

Materials Science and Engineering Abstracts, Cambridge Scientific Abstracts, 7200 Wisconsin Avenue, Bethesda, MD 20814-4823. (301) 961-6750. FAX (301) 961-6720. 1993 to present. Monthly. $385.00 per year. Focuses on mechanical and physical properties of materials and commercial or industrial applications for materials, methods for strength testing, effects of vibration and other stresses, corrosion and protective coatings, storage and handling, ceramics, composites, metals, wood, plastics, and polymers.

Mechanical Engineering Abstracts (formerly ISMEC), Cambridge Scientific Abstracts, 7200 Wisconsin Avenue, Bethesda, MD 20814-4823. (301) 961-6750. FAX (301) 961-6720. 1967 to present. Monthly. $895.00 per year. Summarizes world literature in mechanical engineering, production engineering, and engineering management. Also available online.

Metals Abstracts and Metals Abstracts Index. American Society for Metals, Metals Park, OH 44073. (216) 338-5151. 1968 to present. Monthly. Abstracts are $1500.00 per year and Index is $460.00 per year.

Physics Abstracts. Institute of Electrical Engineers, Michael Faraday House, Six Hill Way, Stevenage, Herts, SG1 2AY, England. Distributed by IEEE, 445 Hoes Lane, Piscataway, NJ 08854. (908) 562-5549. 1898 to present. Monthly. $2700.00 per year. Also available online as INSPEC.

Science Citation Index. Institute for Scientific Information, 3501 Market Street, Philadelphia, PA 19104. (215) 386-0100. FAX (215) 386-6362. Inquire about availability and cost. Also available on CD-ROM.

ASSOCIATIONS AND PROFESSIONAL SOCIETIES

American Ceramic Society. 757 Brooksedge Plaza Drive, Westerville, OH 43081-2821. (614) 890-6136.

American Powder Metallurgy Institute. 105 College Road E., Princeton, NJ 08540. (609) 452-7700. FAX (609) 987-8523.

American Society For Metals. Materials Park, OH 44073. (216) 338-5151 or FAX (216) 338-4634.

American Society For Nondestructive Testing. 1711 Arlingate Lane, Box 28518, Columbus, OH 43228-0158. (614) 274-6003. FAX (614) 274-6899.

American Society For Testing and Materials. 1916 Race Street, Philadelphia, PA 19103. (215) 299-5585.

American Society of Mechanical Engineers. 345 East 47th Street, New York, NY 10017-2398. (212) 705-7703.

Materials Properties Council. 345 East 47th Street, New York, NY 10017-2398. (212) 705-7693.

Materials Research Society. 9800 McKnight Road, Pittsburgh, PA 15237. (412) 367-3003.

National Association of Corrosion Engineers. P.O. Box 218340, 1440 South Creek, Houston, TX 77218. (713) 492-0535.

National Materials Advisory Board. 2101 Constitution Avenue NW, Washington, DC 20418. (202) 334-3505.

Society For the Advancement of Material and Process Engineering (SAMPE). PO Box 2459, Covina, CA 91722. (818) 331-0616.

DIRECTORIES AND BIOGRAPHICAL SOURCES

Directory of Engineering Societies and Related Organizations. Gordon Davis. 13th edition. American Association of Engineering Societies, 1111 19th Street NW, Suite 608, Washington, DC 20036. (202) 296-2237 or (800) 658-8897. 1989. Inquire for price.

Guide To Engineering Materials Producers. J.C. Bittence. ASM International, Materials Information, Materials Park, OH 44073-0002. (216) 338-5151 or (800) 336-5152. FAX (216) 338-4634. 1993. $141.00

International Research Centers Directory. Gale Research, 835 Penobscot Building, Detroit, MI 48226-4094. (313) 961-2242. (800) 347-4253. FAX (313) 961-6083. 8th edition. 1995. $410.00

Research Centers Directory. Gale Research, 835 Penobscot Building, Detroit, MI 48226-4094. (313) 961-2242. (800) 347-4253. FAX (313) 961-6083. $485.00.

Scientific and Technical Organizations and Agencies Directory. Gale Research, 835 Penobscot Building, Detroit, MI 48226-4094. (313) 961-2242. (800) 347-4253. FAX (313) 961-6083. 4th edition. 1996. $195.00.

Who's Who in Engineering. American Association of Engineering Societies, 1111 19th Street NW, Suite 608, Washington, DC 20036. (202) 296-2237 or (800) 658-8897. 8th edition. 1991. Inquire for price.

ENCYCLOPEDIAS AND DICTIONARIES

Asm Engineered Materials Reference Book. Michael L. Bauccio, editor. 2nd edition. ASM International, Materials Information, Materials Park, OH 44073-0002. (216) 338-5151 or (800) 336-5152. FAX (216) 338-4634. 1993. $111.00.

Asm Materials Engineering Dictionary. J.R. Davis, editor. ASM International, Materials Information, Materials Park, OH 44073-0002. (216) 338-5151 or (800) 336-5152. FAX (216) 338-4634. 1992. $101.00.

Encyclopedia of Physical Science and Technology. Robert A. Meyers, ed. 18 volumes. Academic Press Inc., 6277 Sea Harbor Drive, Orlando, FL 32887. (800) 321-5068. 1992. $2100.00.

McGraw-Hill Dictionary of Scientific and Technical Terms. Sybil P. Parker, ed. 5th edition. McGraw-Hill Publishing Company, 1221 Avenue of the Americas, New York, NY 10020. (800) 262-4729 or (212) 512-3825. 1993. $110.50.

McGraw-Hill Encyclopedia of Engineering. Sybil P. Parker, ed. 2nd edition. McGraw-Hill Publishing Company, 1221 Avenue of the Americas, New York, NY 10020. (800) 262-4729 or (212) 512-3825. 1993. $95.50.

McGraw-Hill Encyclopedia of Science and Technology. Sybil P. Parker, ed. 7th edition. 20 volumes. McGraw-Hill Publishing Company, 1221 Avenue of the Americas, New York, NY 10020. (800) 262-4729 or (212) 512-3825. 1992. $1900.00

Thesaurus of Scientific, Technical, and Engineering Terms. Hemisphere Publishing Corporation, 1900 Frost Road, Suite 101, ristol, PA 19007-1598. (215) 785-5800 or (800) 821-8312. FAX (215) 785-5515. 1987. $173.00.

GENERAL WORKS

Nondestructive Testing. Louis Cartz. ASM International, Materials Information, Materials Park, OH 44073-0002. (216) 338-5151 or (800) 336-5152. FAX (216) 338-4634. 1995. $98.00.

Non-destructive Testing. R. Halmshaw. 2d ed. Routledge, Chapman and Hall, Inc., 29 West 35th Street, New York, NY 10001-2291. (212) 244-3336 or FAX (212) 563-2269. 1991. $39.95.

Nondestructive Testing Techniques. Don E. Bray and Don McBride, eds. John Wiley and Sons, Inc., 605 Third Avenue, New York, NY 10158. (800) 526-5368 or (212) 850-6000. 1992. $89.95.

HANDBOOKS AND MANUALS

Asm Handbook Volume 17: Nondestructive Evaluation and Quality Control. ASM International Handbook Committee. ASM International, Materials Information, Materials Park, OH 44073-0002. (216) 338-5151 or (800) 336-5152. FAX (216) 338-4634. 1989. $147.00.

Asm Specialty Handbook: Tool Materials. J.R. Davis, editor. ASM International Handbook Committee. ASM International, Materials Information, Materials Park, OH 44073-0002. (216) 338-5151 or (800) 336-5152. FAX (216) 338-4634. 1995. $159.00.

Materials Handbook: An Encyclopedia For Managers, Technical Professionals, Purchasing and Production Managers, Technicians, Supervisors and Foremen. G.S. Brady & H.R. Clauser. 14th edition. McGraw-Hill Publishing Company, 1221 Avenue of the Americas, New York, NY 10020. (800) 262-4729 or (212) 512-3825. 1991. Inquire for cost and availability.

NONDESTRUCTIVE TESTING

Ency. of Physical Sciences and Engineering Info. Sources

Materials Science & Engineering Handbook. James F. Shackelford. 2nd edition. CRC Press, 2000 Corporate Blvd., N.W., Boca Raton, FL 33431. (407) 994-0555 or (800) 272-7737. FAX (407) 994-0949. 1994. Inquire for price and availability.

ONLINE DATABASES AND CD-ROMS

Compendex Plus. Engineering Information, Inc., 345 East 47th Street, New York, NY 10017. (212) 705-7600 or (800) 221-1044. Contains citations with abstracts to worldwide literature in engineering and technology, from 1970 to present. Available on online BRS,(800) 289-4277, DIALOG, (800) 334-2564, ORBIT (800) 456-7248, and STN International, FIZ Karlsruhe, P.O. Box 2465, W-7500, Karlsruhe 1, Germany, online services. Also available on CD-ROM. Inquire as to cost and availability.

CSA Engineering. Cambridge Scientific Abstracts, 7200 Wisconsin Avenue, Suite 601, Bethesda, MD 20814. (301) 961-6750 or (800) 843-7751. Contains citations and abstracts of international periodicals and research literature covering all fields of engineering and science and technology,including computer and information science, electronics, mechanical engineering, solid state materials, 1981 to present. Available on BRS,(800) 289-4277, online service.Inquire as to cost and availability.

Current Contents Search. Institute for Scientific Information, 3501 Market Street, Philadelphia, PA 19104. (215) 386-0100. FAX (215) 386-6362. Contains citations to articles listed in the table of contents of science and technology journals. Also articles in social sciences and life sciences journals. Available on BRS,(800) 289-4277, DIALOG,(800) 334-2564, online services. Inquire as to cost and availability.

ISMEC: Mechanical Engineering Abstracts. Cambridge Scientific Abstracts, 7200 Wisconsin Avenue, Suite 601, Bethesda, MD 20814. (301) 961-6750 or (800) 843-7751. Contains citations to the literature in mechanical engineering, industrial and production engineering, energy, power, mechanics, devices and related areas, from 1973 to present. Available on the DIALOG,(800) 334-2564, online service. Inquire as to cost and availability.

NTIS Bibliographic Database. National Technical Information Service, 5285 Port Royal Road, Springfield, VA 22161. (703) 487-4929 or FAX (703) 321-8199. Broad coverage of government-sponsored science and technology research reports, 1964 to present. Available on BRS,(800) 289-4277, DIALOG, (800) 334-2564, ORBIT, (800) 456-7248, and STN International, FIZ Karlsruhe, P.O. Box 2465, W-7500, Karlsruhe 1, Germany, online services. Also available on CD-ROM. Inquire as to cost and availability.

SCISEARCH. Institute for Scientific Information, 3501 Market Street, Philadelphia, PA 19104. (800) 523-1850 or (215) 386-0100. Broad multidisciplinary title and author index to the international literature of science and technology, 1974 to present. Available on DIALOG,(800) 334-2564, and ORBIT,(800) 456-7248, online services. Also available on CD-ROM. Inquire as to cost and availability.

WILSONLINE. H.W. Wilson Company, 950 University Avenue, Bronx, NY 10452. (800) 367-6770 or (212) 588-8400. Makes available online versions of the printed H.W. Wilson indexes including Applied Science and Technology Index, Business Periodicals Index, General Science Index, and Readers' Guide to Periodical Literature. Period covered is generally 1983 to present. Available on BRS,(800) 289-4277, DIALOG,(800) 334-

2564, and OCLC EPIC,(800) 848-5878, online services. Also available on CD-ROM. Inquire as to cost and availability.

PERIODICALS

International Advances in Nondestructive Testing. Gordon and Breach Science Publishers, 820 Town Center Drive, Langhorne, PA 19047. (215) 750-2642. FAX (215) 750-6343. 1969 to present. Irregular. Inquire for cost and availability of latest volume.

Journal of Engineering For Industry. American Society of Mechanical Engineers, 345 E. 47th Street, New York, NY 10017. (212) 705-7722. 1970 to present. Quarterly. $100.00 per year.

Journal of Materials Research. Materials Research Society, 9800 McKnight Road, Pittsburg, PA 15237. (412) 367-3003. FAX (412) 367-4373. 1986 to present. Monthly. $440.00 per year.

Journal of Nondestructive Evaluation. Plenum Publishing Corporation, 233 Spring Street, New York, NY 10013. (212) 620-8000. FAX (212) 463-0742. 1980 to present. Quarterly. $230.00 per year.

Journal of Testing and Evaluation. American Society for Testing and Materials (ASTM), 1916 Race Street, Philadelphia, PA 19103. (215) 299-5585. FAX (215) 977-9679. 1966 to present. Bi-monthly. $103.00 per year for non-members.

Materials Engineering. Penton Publishing, 110 Superior Ave., Cleveland, OH 44114-2543. 1929 to present. Monthly. $50.00 per year.

Materials Evaluation. American Society for Nondestructive Testing, 1711 Arlingate Lane, Box 28518, Columbus, OH 43228-0158. (614) 274-6003. FAX (614) 274-6899. 1942 to present. Monthly. $85.00 per year.

Materials Performance. National Association of Corrosion Engineers, P.O. Box 218340, 1440 South Creek, Houston, TX 77218. (713) 492-0535. 1962 to present. Monthly. $65.00 per year.

Materials Science and Technology. Institute of Materials, 1 Carlton House Terrace, London SW1Y 5DB, England. Telephone 071-839-4071. 1985 to present. Monthly. $37.00 per year.

Materials Transactions, Jim. Japan Institute of Metals, Nihon Kinzoku Gakkai, Aoba Aramaki, Aoba-Ku, Sendai 980, Japan. Telephone 022-223-3685. FAX 8-22-223-6312. 1960 to present. Monthly. Inquire for cost.

SAMPE Journal. Society for the Advancement of Material and Process Engineering, Covina, CA 91724. (818) 331-0610. FAX (818) 332-8929. 1965 to present. Bi-monthly. $65.00 per year.

Russian Journal of Nondestructive Testing. Plenum Publishing Corporation, Consultants Bureau, 233 Spring Street, New York, NY 10013-8468. (212) 620-8000. FAX (212) 463-0742. 1964 to present. Monthly. $1345.00 per year.

RESEARCH CENTERS AND INSTITUTES

Center For Materials Research. Ohio State University, 4108 Smith Laboratory, 174 W. 18th Avenue, Columbus, OH 43210. (614) 292-1133. FAX (614) 292-7557.

Center For Materials Science and Engineering. Massachusetts Institute of Technology, Bldg. 13, Room 2090, 77 Massachusetts Avenue, Cambridge, MA 02139. (617) 253-6701. FAX (617) 258-6883.

Center For Nondestructive Evaluation. Iowa State University, 115A Applied Sciences Complex II, 1915 Scholl Road, Ames, IA 50011. (515) 294-8152. FAX (515) 294-7771.

Center For Nondestructive Evaluation. The Johns Hopkins University, 102 Maryland Hall, 3400 N. Charles Street, Baltimore, MD 21218. (410) 338-6115. FAX (410) 338-5293.

Center For Research in Materials Science and Engineering. University of Texas at Austin, ETC 9.104, Austin, TX 78712. (512) 471-1504.

Materials Research Laboratory. University of Illinois, Urbana, IL 61801. (217) 333-1370.

Materials Science Center. Cornell University, Clark Hall of Science, Ithaca, NY 14853-2501. (607) 255-4272. FAX (607) 255-6428.

Nondestructive Testing & Evaluation Programs. Pennsylvania State University, 159 Materials Research Laboratory, University Park, PA 16802. (814) 863-2843.

SPECIFICATIONS AND STANDARDS

*Annual Book of Astm Standards, Volume 03.*03. American Society for Testing and Materials (ASTM), 1916 Race Street, Philadelphia, PA 19103. (215) 299-5585. Annual. Inquire for cost and availability.

NUCLEAR CHEMISTRY

See also: NUCLEAR ENERGY, NUCLEAR REACTORS

ABSTRACT SERVICES AND INDEXES

Applied Science and Technology Index; a cumulative subject index to English language periodicals in the fields of aeronautics and space science, computer technology, chemistry, construction industry, energy and related areas. H. W. Wilson Co., 950 University Avenue, Bronx, NY 10452- 9978. (212) 588-8400. (800) 367-6770. FAX (718) 590-1617.From 1958 to present. Monthly. Inquire about cost and availability. Also available online (BRS and WILSONLINE) and on CD-ROM. ISSN: 0003-6986.

Chemical Abstracts. Chemical Abstracts Service. 2540 Olentangy River Road, Box 3012, Columbus, OH 43210-0012. (614) 447-3600. FAX (614) 447-3713. 1907 to present. Weekly. $16,800.00 per year. Available online and on CD-ROM. CA is also available in five section groupings. Inquire regarding cost and availability.

Current Contents: engineering, technology and applied *Sciences*. Institute for Scientific Information, 3501 Market Street, Philadelphia, PA 19104. (215) 386-0100. FAX (215) 386-2991. From 1970 to present. Weekly. $442.00 per year.Also available online. Inquire regarding cost and availability. ISSN: 0095-7917.

Engineering Index Monthly. Engineering Information, Inc., Castle Point on the Hudson, Hoboken, NJ 07030. (201) 216-8500. (800) 221-1044. FAX (201) 216-8532. Monthly. $2300.00 per year for monthly issues. $1980 per year for annual cumulation. Available on CD-ROM and online as COMPENDEX. ISSN: 0742-1974; 0360-8557.

General Science Index. H.W. Wilson Company, 950 University Avenue, Bronx, NY 10452. (212) 588-8400. (800) 367-6770. FAX (718) 590-1617. From 1978 to present. Ten issues per year; quarterly and annual cumulations. Service basis. Available on CD-ROM and online. Inquire regarding cost and availability. ISSN: 0162-1963.

Physics Abstracts. INSPEC. Section A, Science Abstracts. Institution of Electrical Engineers (IEE), London, United Kingdom. Available from: INSPEC/IEEE - Institute of Electrical and Electronic Engineers, Box 1331, 445 Hoes Lane, Piscataway, NJ 08855-1331. (908) 562-5549. 1898 to present. 24 issues per year. $2835.00 per year. ISSN: 0036-8091. Also available online and on CD-ROM.

Science Citation Index. SCI. Institute for Scientific Information, 3501 Market Street, Philadelphia, PA 19104. (215) 386-0100. (800) 523-1850. FAX (215) 386-2991. 1961 to present. Six issues per year, plus annual cumulation. $11650.00 per year. Also available online and on CD-ROM. Inquire about price and availability. ISSN: 0036-827X.

ANNUAL REVIEWS AND YEARBOOKS

Advances in Chemical Engineering. Academic Press, Inc., 6277 Sea Harbor Drive, Orlando, FL 32887. (800) 321-5068. 1956 to present. Irregular. ISSN: 0065-2377.

Advances in Nuclear Science and Technology. Plenum Publishing Corp., 233 Spring Street, New York, NY. (212) 620-8000. (800) 2221-9369. FAX (212) 463-0742. 1977 to present. Irregular.

ASSOCIATIONS AND PROFESSIONAL SOCIETIES

American Chemical Society, 1155 16th Street, NW, Washington, DC 20036. (202) 872-4600.

American Institute of Chemical Engineers. 345 East 47th Street, New York, NY 10017. (212) 705-7338.

American Institute of Physics. One Physics Ellipse, College Park, MD 20740-3843. (301) 209-3100.

American Nuclear Society. 555 North Kensington Avenue, LA Grange Park, IL 60525. (708) 352-6611. (800) NUC-NEWS.

Fusion Power Associates. 2 Professional Drive, Suite 248, Gaithersburg, MD 20879. (301) 258-0545. FAX (301) 975-9869.

Institute of Nuclear Materials Management. 60 Revere Drive, Northbrook, IL 60062. (847) 480-9573. FAX (947) 480-9282.

BIBLIOGRAPHIES

Chemical Information Sources. Gary Wiggins. McGraw-Hill Publishing Company, 1221 Avenue of the Americas, New York, NY 10020. (212) 512-2000. (800) 262-4729. 1991. $42.50.

NUCLEAR CHEMISTRY

Ency. of Physical Sciences and Engineering Info. Sources

Handbooks and Tables in Science and Technology. Russell H. Powell, editor. Oryx Press, 4041 North Central, Suite 700, Phoenix, AZ 85012-3330. (602) 265-2651 or (800) 279-6799. Third edition. 1994. $65.00.

Information Sources in Chemistry. R.T. Bottle and J.F.B Rowland, editors. R.R. Bowker Inc., 121 Chanlon Road, New Providence, NJ 07974. (800) 521-8110 or (908) 464-6800. Fourth edition. 1993. $75.00.

Information Sources in Engineering. Peter Hicks, editor.

Bowker-Saur, 121 Chanlon Road, New Providence, NJ 07974. (800) 521-8110 or (908) 464-6800. 3rd edition

1996. $125.00.

Scientific and Technical Books and Serials in Print; An Index To Literature in Science and Technology. R.R. Bowker Inc., 121 Chanlon Road, New Providence, NJ 07974. (908) 464-6800. (800) 521-8110. FAX (908) 665-3502. 1972 to present. Annual. 4 volumes. 1994. $299.95. Also available on compact disc and online. ISSN: 0000-054X.

DIRECTORIES AND BIOGRAPHICAL SOURCES

American Institute of Chemical Engineers. Directory. American Institute of Chemical Engineers. 345 East 47th Street, New York, NY 10017-2396. (212) 705-7663. Annual.

Energy and Nuclear Science International Who's Who. Gale Research, 835 Penobscot Building, Detroit, MI 48226- 4094. (313) 961-2242. (800) 347-4253. Triennial. 1993. $325.00.

Energy Information Centers Directory. U. S. Council for Energy Awareness, 1776 I Street, NW, Suite 400, Washington, DC 20006. (202) 293-0770. Triennial. No charge.

Engineering Research Centers: Incorporating Electronics Research Centers. Stockton Press, 345 Park Avenue, New York, NY 10010. (212) 689-9200. (800) 221-2123. 4th edition. 1995. $515.00.

International Directory of Nuclear Utilities. Nuexco, 1515 Arapahoe Street, Suite 900, Denver CO 80202. (303) 899- 4500. Annual. $220.00 per year

Nuclear News - Directory Issues. American Nuclear Society, 555 North Kensington Avenue, LA Grange Park, IL 60525. (708) 352-6611. (800) NUC-NEWS. March issue contains Buyer's Guide; $65.00. World List of Nuclear Power Plants. Semi-annual in February and August. $15.00 per issue.

Research Centers Directory. Gale Research, 835 Penobscot Building, Detroit, MI 48226-4094. (313) 961-2242. (800) 347-4253. 20th edition. 1995. $485.00. ISSN: 0080-1518.

Who's Who in Engineering. Gordon Davis, editor. American Association of Engineering Societies. 1111 19th Street, NY, Suite 608, Washington, DC 20036. (202) 296-2237. (800) 658-8897. 9th edition. 1995. $220.00.

World Energy and Nuclear Directory. Gale Research Company, Inc., 835 Penobscot Building, Detroit, MI 48226-4094. (313) 961-2242. (800) 347-4253. Biennial, even years. $450.00.

ENCYCLOPEDIAS AND DICTIONARIES

Dictionary of Named Processes in Chemical Technology. Alan E. Comyns. Oxford University Press, Inc., 200 Madison Avenue, New York, NY 10016. (212) 725-6000. (800) 334-4249. 1994. $75.00.

Dictionary of Nuclear Technology. Pro/Am Music Resources, Inc., 63 Prospect Street, White Plains, New York, 10606. (914) 948-7436. 1986. $245.00.

Encyclopedia of Applied Physics. VCH Publishers, Inc., 303 Northwest 12th Avenue, Deerfield Beach, FL 33442. (800) 367-8249. 1991-. Twenty volumes. $6000.00.

Encyclopedia of Chemical Processing and Design. McKetta Marcel Dekker, Inc., 270 Madison Avenue, New York, NY 10016. (212) 696-9000. (800) 228-1160. 1976 - . $175.00 per volume. ISSN:

IEEE Nuclear Science Standards Collection. IEEE Standards office, Box 1331, 445 Hoes Lane, Piscataway, NJ 08855-1331. (908) 562-3822. (800) 678-4333. 1991. $90.00 in paper.

Kirk-Othmer Encyclopedia of Chemical Technology. John Wiley and Sons, Inc., 605 Third Avenue, New York, NY 10158. (800) 526-5368 or (212) 850-6000. Fourth edition. 1991 - . Twenty-seven volumes. $5400.00.

McGraw-Hill Encyclopedia of Science and Technology. McGraw-Hill Book, Incorporated, 1221 Avenue of the Americas, New York, NY 10020. (212) 997-3675. (800) 262-4729. Seventh edition. Twenty volumes. 1992. $1900.00.

Ullman's Encyclopedia of Industrial Chemistry. VCH Publications, Inc., 220 East 23rd Street, Suite 909, New York, NY 10010-4606. (212) 683-8333. (800) 422-8824. 5th edition. 1984 - . Price varies per volume. ISSN:

GENERAL WORKS

Energy Levels in Atoms and Molecules. W. Graham Richards. Oxford University Press, Inc., 200 Madison Avenue, New York, NY 10016. (212) 725-6000. (800) 334-4249. 1995. $9.95 in paper.

Nuclear Chemistry. James D. Navratil. Prentice-Hall, 113 Sylvan Avenue, Route 9W, Englewood Cliffs, NJ 07632. (201) 592-2000. (800) 922-0579. 1993.

Nuclear Engineering: An Introduction. K. Almenas and R. Lee. Springer-Verlag New York, Inc., 175 Fifth Avenue, New York, NY 10010. (212) 460-1500. (800) 777-4643. FAX (212) 473-6272. 1992. $89.50.

Radiochemistry and Nuclear Chemistry: Nuclear Chemistry, theory and Appplications. Gregory R. Choppin, et al. Butterworth-Heinemann, 313 Washington Street, Newton, MA 02158. (618-928-2500. (800) 366-2665. 2nd edition. 1995. $64.95.

Radiochemistry and Nuclear Methods of Analysis. William D. Ehmann, and Diane Vance. John Wiley & Sons, Inc., 605 Third Avenue, New York, NY 10158-0012. (212) 850-6000. (800) 225-5945. 1991. $135.00

HANDBOOKS AND MANUALS

Chemical Engineering Reference Manual. Randall N. Robinson. Professional Publications, Inc., 1250 Fifth Avenue, Belmont, CA 94002. (415) 593-0110. (800) 426-1178. 1988. $49.95.

CRC Handbook of Chemistry and Physics. David R. Kide, editor. CRC Press Inc., 2000 Corporate Blvd., NW, Boca Raton, FL 33431. (407) 994-0555. (800) 333-8300. 77th edition. 1996. $99.95.

Handbook of Hot Atom Chemistry. J. P. Adloff, et al, editors. VCH Publications, Inc., 220 East 23rd Street, Suite 909, New York, NY 10010-4606. (212) 683-8333. (800) 422-8824. 1992. $225.00.

Information Center On Nuclear Standards. American Nuclear Society, 555 North Kensington Avenue, LA Grange Park, IL 60525. (708) 352-6611. Standards for all aspects of the design, construction, operation and maintenance of nuclear power plants.

Perry's Chemical Engineers' Handbook. Robert H. Perry and Donald W. Green, editors. McGraw-Hill Publishing Company, Inc., 1221 Avenue of the Americas, New York, NY 10020. (212) 512-2000. (800) 262-4729. 6th edition. 1996. $129.50.

ONLINE DATABASES AND CD-ROMS

CA Search. Chemical Abstracts Service, P.O. Box 3012, Columbus, OH 43210-0012. (614) 447-3600. (800) 848-6533. FAX (614) 447-3709. Very comprehensive guide to worldwide chemical literature and related fields. 1972 to present. Available on BRS,(800) 289-4277, DIALOG, (800) 334-2564, ORBIT (800) 456-7248, and STN International, FIZ Karlsruhe, P.O. Box 2465, W-7500, Karlsruhe 1, Germany, online services. Inquire as to cost and availability.

Compendex Plus. Engineering Information, Inc., 345 East 47th Street, New York, NY 10017. (212) 705-7600 or (800) 221-1044. Contains citations with abstracts to worldwide literature in engineering and technology, from 1970 to present. Available on online BRS,(800) 289-4277, DIALOG, (800) 334-2564, ORBIT (800) 456-7248, and STN International, FIZ Karlsruhe, P.O. Box 2465, W-7500, Karlsruhe 1, Germany, online services. Also available on CD-ROM. Inquire as to cost and availability.

Dissertation Abstracts. University Microfilms International, 300 North Zeeb Road, Ann Arbor, MI 48106. (800) 521-0600 or (313) 761-4700. Scope includes virtually all doctoral dissertations accepted at accredited American institutions from 1861 to present in 252 subject areas. Available on BRS, (800) 289-4277, DIALOG, (800) 334-2564, and OCLC EPIC, (800) 848-5878, online services. Also available on CD-ROM. Inquire as to cost and availability.

INSPEC. Institution of Electrical Engineers, Michael Faraday House, Six Hills Way, Stevenage, Herts. SG1 2AY, England. Telephone: 0438 313311 or FAX 0438 742840. Contains citations to the worldwide literature of physics, electronics and electrical engineering, computer technology, and related fields. Available on BRS, (800) 289-4277, DIALOG, (800) 334-2564, ORBIT, (800) 456-7248, and STN International, FIZ Karlsruhe, P.O. Box 2465, W-7500, Karlsruhe 1, Germany, online services. Inquire as to cost and availability.

Kirk-Othmer Encyclopedia of Chemical Technology. John Wiley and Sons, Inc., 605 Third Avenue, New York, NY 10158. (800) 526-5368 or (212) 850-6000. Contains the complete text of all chapters in the 27 volume fourth edition of the Kirk-othmer Encyclopedia of Chemical Technology. 1991. Available on BRS,(800) 289-4277, DIALOG,(800) 334-2564, online services. Inquire as to cost and availability.

NTIS Bibliographic Database. National Technical Information Service, 5285 Port Royal Road, Springfield, VA 22161. (703) 487-4929 or FAX (703) 321-8199. Broad coverage of government-sponsored science and technology research reports, 1964 to present. Available on BRS,(800) 289-4277, DIALOG, (800) 334-2564, ORBIT, (800) 456-7248, and STN International, FIZ Karlsruhe, P.O. Box 2465, W-7500, Karlsruhe 1, Germany, online services. Also available on CD-ROM. Inquire as to cost and availability.

Nuclear Science Abstracts. U.S. Department of Energy, office of Scientific and Technical Energy, P.O. Box 62, Oak Ridge, TN 37831. (615) 576-1189. Contains citations with abstracts to literature in all fields of nuclear science and energy. 1948 to 1976. Available on DIALOG,(800) 334-2564, online service. Inquire as to cost and availability.

SCISEARCH. Institute for Scientific Information, 3501 Market Street, Philadelphia, PA 19104. (800) 523-1850 or (215) 386-0100. Broad multidisciplinary title and author index to the international literature of science and technology, 1974 to present. Available on DIALOG, (800) 334-2564, and ORBIT, (800) 456-7248, online services. Also available on CD-ROM. Inquire as to cost and availability.

WILSONLINE. H.W. Wilson Company, 950 University Avenue, Bronx, NY 10452. (800) 367-6770 or (212) 588-8400. Makes available online versions of the printed H.W. Wilson indexes including Applied Science and Technology Index, Business Periodicals Index, General Science Index, and Readers' Guide to Periodical Literature. Period covered is generally 1983 to present. Available on BRS, (800) 289-4277, DIALOG, (800) 334-2564, and OCLC EPIC, (800) 848-5878, online services. Also available on CD-ROM. Inquire as to cost and availability.

PERIODICALS

Aiche Journal. American Institute of Chemical Engineers. 345 East 47th Street, New York, NY 10017. (212) 705-7338. 1955 to present. Monthly. $295.00 per year.

American Nuclear Society Transactions. American Nuclear Society, 555 North Kensington Avenue, LA Grange Park, IL 60525. (708) 352-6611. 1958 to present. 3 issues per year. $340.00 per year. ISSN: 0003-018X

Annals of Nuclear Energy. Elsevier Science Publishing Company, Inc., 660 White Plains Road, Tarrytown, NY 10591-5133. (914) 524-9200. 1974 to present. Monthly. $815.00 per year. ISSN: 0306-4549.

Bulletin of the Atomic Scientists. Educational Foundation for Nuclear Science, 6042 South Kimbark Avenue, Chicago, IL 60637. (312) 702-2555. 1945 to present. 10 issues per year. $30.00 per year. ISSN: 0096-3402.

Fusion Technology. American Nuclear Society, 555 North Kensington Avenue, LA Grange Park, IL 60525. (708) 352-6611. 1981 to present. 8 issues per year. $$485.00 per year. ISSN: 0748-1896.

Journal of Fusion Energy. Plenum Publishing Corporation, 233 Spring Street, New York, NY 10013- 1578. (212) 620- 8000. 1981 to present. Quarterly. $315.00 per year. 0164-0313.

Nuclear Engineer. Institution of Nuclear Engineers, 1 Penerley Road, London, SE6 2LQ, England. 1959 to present. Bi-monthly. L77 per year. ISSN: 0262-5091.

Nuclear Engineering and Design. Elsevier Science Publishing Company, Inc., 660 White Plains Road,

Tarrytown, NY 10591-5133. (914) 524-9200. 1965 to present. 24 issues per year. $2748.00 per year.

Nuclear Engineering International. Reed Business Pulbishing Group, Quadrant House, the Quadrant, Sutton, Surrey, SM2 5AS, Englnad. 1956 to present. Monthly. $236.00 per year. ISSN: 0029-5507.

Nuclear Science and Engineering. American Nuclear Society, 555 North Kensington Avenue, LA Grange Park, IL 60525. (708) 352-6611. 1956 to present. Monthly. $405.00 per year. ISSN: 0029-5639.

Nuclear Technology. American Nuclear Society,

555 North Kensington Avenue, LA Grange Park, IL 60525. (708) 352-6611. 1965 to present. Monthly. $440.00 per year. ISSN: 0029-5450.

RESEARCH CENTERS AND INSTITUTES

Center for Applied Energy Research. University of Kentucky, 3572 Iron Works Pike, Lexington, KY (606) 257-0305. FAX (606) 257-0220.

Center For Energy Studies. University of Texas at Austin, 10100 Burnet Road, Building 133, Austin, TX 78758. (512) 471-4496. FAX (512) 471-0781.

Fusion Research Center. University of Texas at Austin. RLM 11.222, Austin, TX 78712. ((512) 471-3926. FAX (512) 471-8865.

Institute of Plasma and Fusion Research. University of California, Los Angeles, 44-139 Engineering IV 405 Hilgard Avenue, Los Angeles, CA 90024-1597. (213) 825-1613. FAX (213) 206-4832.

Nuclear Physics Laboratory. University of Washington, Seattle, WA 98195. (206) 543-4080.

Oak Ridge National Laboratory. U. S. Department of Energy, P. O. Box 2008, Oak Ridge, TN 37831-6255. (615) 576-2900.FAX (615) 576-2912

NUCLEAR ENERGY

See also: ENERGY, FISSION, FUSION, NATURAL GAS, NUCLEAR ENERGY, NUCLEAR ENGINEERING, NUCLEAR PHYSICS, PLUTONIUM

ABSTRACT SERVICES AND INDEXES

Applied Science and Technology Index; a cumulative subject index to English language periodicals in the fields of aeronautics and space science, computer technology, chemistry, construction industry, energy and related areas. H. W. Wilson Co., 950 University Avenue, Bronx, NY 10452- 9978. (212) 588-8400. (800) 367-6770. FAX (718) 590-1617. From 1958 to present. Monthly. Inquire about cost and availability. Also available online (BRS and WILSONLINE) and on CD-ROM. ISSN: 0003-6986.

Chemical Abstracts. Chemical Abstracts Service. 2540 Olentangy River Road, Box 3012, Columbus, OH 43210-0012. (614) 447-3600. FAX (614) 447-3713. 1907 to present. Weekly. $16,800.00 per year. Available online and on CD- ROM. CA is also available in five section groupings. Inquire regarding cost and availability.

Current Contents: Engineering, Technology and Applied Sciences. Institute for Scientific Information, 3501 Market Street, Philadelphia, PA 19104. (215) 386-0100. FAX (215) 386-2991. From 1970 to present. Weekly. $442.00 per year. Also available online. Inquire regarding cost and availability. ISSN: 0095-7917.

Energy Research Abstracts. U. S. Department of Energy, office of Scientific and Technical Information, Box 62, Oak Ridge, TN 37831. (615) 674-0733. Subscriptions to: U. S. Government Printing office, Box 371954, Pittsburgh, PA 15250-7954. (202) 783-3238. 1976 to present. Monthly. $164.00 per year. ISSN: 0160-3604.

Engineering Index Monthly. Engineering Information, Inc., Castle Point on the Hudson, Hoboken, NJ 07030. (201) 216-8500. (800) 221-1044. FAX (201) 216-8532. Monthly. $2300.00 per year for monthly issues. $1980 per year for annual cumulation. Available on CD-ROM and online as COMPENDEX. ISSN: 0742-1974; 0360-8557.

General Science Index. H.W. Wilson Company, 950 University Avenue, Bronx, NY 10452. (212) 588-8400. (800) 367-6770. FAX (718) 590-1617. From 1978 to present. Ten issues per year; quarterly and annual cumulations. Service basis. Available on CD-ROM and online. Inquire regarding cost and availability. ISSN: 0162-1963.

N T I S Alerts: Energy. U.S. National Technical Information Service, 5285 Port Royal Road, Springfield, VA 22161. (703) 487-4929. FAX (703) 321-8547. Weekly. $160.00 per year.

Physics Abstracts. INSPEC. Section A, Science Abstracts. Institution of Electrical Engineers (IEE), London, United Kingdom. Available from: INSPEC/IEEE - Institute of Electrical and Electronic Engineers, Box 1331, 445 Hoes Lane, Piscataway, NJ 08855-1331. (908) 562-5549. 1898 to present. 24 issues per year. $2835.00 per year. ISSN: 0036-8091. Also available online and on CD-ROM.

Science Citation Index. SCI. Institute for Scientific Information, 3501 Market Street, Philadelphia, PA 19104. (215) 386-0100. (800) 523-1850. FAX (215) 386-2991. 1961 to present. Six issues per year, plus annual cumulation. $11650.00 per year. Also available online and on CD-ROM. Inquire about price and availability. ISSN: 0036-827X.

ANNUAL REVIEWS AND YEARBOOKS

Advances in Energy Systems and Technology. Academic Press,Inc., 6277 Sea Harbor Drive, Orlando, FL 32887. (800) 321- 5068. 1979 to present. Irregular. Inquire.

Advances in Nuclear Science and Technology. Plenum Publishing Corp., 233 Spring Street, New York, NY. (212) 620-8000. (800) 2221-9369. FAX (212) 463-0742. 1977 to present. Irregular.

ASSOCIATIONS AND PROFESSIONAL SOCIETIES

American Institute of Chemical Engineers. 345 East 47th Street, New York, NY 10017. (212) 705-7338.

American Institute of Physics. One Physics Ellipse, College Park, MD 20740-3843. (301) 209-3100.

American Nuclear Society. 555 North Kensington Avenue, LA Grange, IL 60525. (312) 352-6611. fax (708) 352-0499.

Fusion Power Associates. 2 Professional Drive, Suite 248, Gaithersburg, MD 20879. (301) 258-0545. FAX (301) 975-9869.

IEEE Nuclear and Plasma Sciences Society. c/o Instiute of Electrical and Electronics Engineers, 345 East 47th Street, New York, NY 10017. (212) 705-7900. FAX (212) 705-4929.

Institute of Nuclear Materials Management. 60 Revere Drive, Northbrook, IL 60062. (847) 480-9573. FAX (947) 480-9282.

BIBLIOGRAPHIES

Information Sources in Engineering. Peter Hicks, editor. Bowker-Saur, 121 Chanlon Road, New Providence, NJ 07974. (800) 521-8110 or (908) 464-6800. 3rd edition. 1996. $125.00.

Scientific and Technical Books and Serials in Print; An Index To Literature in Science and Technology. R.R. Bowker Inc., 121 Chanlon Road, New Providence, NJ 07974. (908) 464-6800. (800) 521-8110. FAX (908) 665-3502. 1972 to present. Annual. 4 volumes. 1994. $299.95. Also available on compact disc and online. ISSN: 0000-054X.

DIRECTORIES AND BIOGRAPHICAL SOURCES

Energy and Nuclear Science International WHO'S WHO. Gale Research, 835 Penobscot Building, Detroit, MI 48226- 4094. (313) 961-2242. (800) 347-4253. Triennial. 1993. $325.00.

Energy Information Centers Directory. U. S. Council for Energy Awareness, 1776 I Street, NW, Suite 400, Washington, DC 20006. (202) 293-0770. Triennial. No charge.

Engineering Research Centers: Incorporating Electronics Research Centers. Stockton Press, 345 Park Avenue, New York,

NY 10010. (212) 689-9200. (800) 221-2123. 4th edition. 1995. $515.00.

Research Centers Directory. Gale Research, 835 Penobscot Building, Detroit, MI 48226-4094. (313) 961-2242. (800) 347-4253. 20th edition. 1995. $485.00. ISSN: 0080-1518.

Who's Who in Engineering. Gordon Davis, editor. American Association of Engineering Societies. 1111 19th Street, NY, Suite 608, Washington, DC 20036. (202) 296-2237. (800) 658-8897. 9th edition. 1995. $220.00.

World Energy and Nuclear Directory. Gale Research Company,Inc., 835 Penobscot Building, Detroit, MI 48226-4094. (313) 961-2242. (800) 347-4253. Biennial, even years. $450.00.

ENCYCLOPEDIAS AND DICTIONARIES

Chamber's Nuclear Energy and Radiation Dictionary. Peter Walker, editor. LKC, 95 Madison AVenue, New York, NY 10018. (800) 497-1657. 1992. $ 40.00.

Dictionary of Nuclear Power and Waste Management With Abbreviations and Acronyms. Foo-Sun Lau. John Wiley & Sons, Inc., 605 Third Avenue, New York, NY 10158-0012. (212) 850-6000. (800) 225-5945. 1987. $385.00.

Encyclopedia of Applied Physics. VCH Publishers, Inc., 303 Northwest 12th Avenue, Deerfield Beach, FL 33442. (800) 367-8249. 1991-. Twenty volumes. $6000.00.

Kirk-Othmer Encyclopedia of Chemical Technology. John Wiley and Sons, Inc., 605 Third Avenue, New York, NY 10158. (800) 526-5368 or (212) 850-6000. Fourth edition. 1991 - . Twenty-seven volumes. $5400.00.

McGraw-Hill Encyclopedia of Science and Technology. McGraw-Hill Book, Incorporated, 1221 Avenue of the Americas, New York, NY 10020. (212) 997-3675. (800) 262-4729. Seventh edition. Twenty volumes. 1992. $1900.00.

Ullman's Encyclopedia of Industrial Chemistry. VCH Publications, Inc., 220 East 23rd Street, Suite 909, New York, NY 10010-4606. (212) 683-8333. (800) 422-8824. 5th edition. 1984 - . Price varies per volume. ISSN:

GENERAL WORKS

Energy Conversion: Systems, Flow Physics and Engineering. Reiner Decher. Oxford University Press, Inc., 200 Madison Avenue, New York, NY 10016. (212) 725-6000. (800) 334-4249. 1994. $69.95.

The Future of Nuclear Power. G. Greenhaigh. Kluwer Academic Publishers, 101 Philip Drive, Assinippi Park, Norwell, MA 02061. (617) 871-6600. 1989. $102.50.

Nuclear Energy: An Introduction To the Concepts, Systems and Applications of Nuclear Processes. Raymond L. Murry .

Elsevier Science Publishing Company, Inc., 655 Avenue of the Americas, New York, NY 10010. (212) 989-5800. FAX (914) 333-2444. 4th edition 1993. $104.00.

NUCLEAR ENERGY

Ency. of Physical Sciences and Engineering Info. Sources

Nuclear Energy and the Public. J. Van Der Plight. Blackwell Publications, 238 Main Street, Cambridge, MA 02142. (800) 216-2522. 1993. $31.95.

Nuclear Energy Simplified. Frank L. Bouquet. Systems Company, Inc., P.O.Box 339, Carlborg, WA 98324. (360)1. 2nd edition. 1994. $45.00 in paper

Nuclear Power At the Crossroads: Challenges and Prospects For the Twenty-first Century. Thomas C. Lowinger, editor. International Research Center for Energy and Economic Development, 909 14th Street, Suite 201, Boulder CO 80302. (303) 492-7667. 1994. $24.00.

Nuclear Engineering: Theory and Technology of Commercial Nuclear Power. Ronald A. Knief. Hemisphere Publishing Corp., 1900 Frost Road, Suite 101, Bristol PA 19007. (215) 785-5800. (800) 821-8312. 2nd edition. 1992. $156.00.

Nuclear Fission Reactors. I. R. Cameron. Plenum Publishing Corp., 233 Spring Street, New York, NY. (212) 620-8000. (800) 2221-9369. FAX (212) 463-0742.. 1992. $75.00.

Nuclear Physics. Enrico Fermi. University of Chicago Press, 5801 Ellis Avenue, Chicago, IL 60637. (312) 702- 7700. Revised edition, reprinted. 1974. $17.95 in paper.

Nucear Power from Underseas to Outer Space. John W. Simpson. American Nuclear Society, 555 North Kensington Avenue, LA Grange, IL 60525. (312) 352-6611. fax (708) 352-0499. 1994. $50.00

Nuclear Power, Nuclear Fuel Cycle and Waste Management: Status and Trends 1992. UNIPUB, 4511-F Assembly Drive, Lanham, MD 20706-4391. (301) 459-7666. (800) 274-4888.1992. $19.00.

HANDBOOKS AND MANUALS

Annual Book of Astm Standards. Volume 12.01, 12.02: Nuclear energy. American Society for Testing and Materials, 1916 Race Street, Philadelphia, PA 19103. (215) 299-5419. 1996. $92.00 each.

Energy Engineering Handbook. Fairmont Press Staff. Prentice-Hall, 113 Sylvan Avenue, Route 9W, Englewood Cliffs, NJ 07632. (201) 592-2000. (800) 922-0579. 3rd edition. 1995. $75.00.

Fuel Cells: A Handbook. J. M. Hirschenfoffer. Business/Technology Information Services, P. O. Box 574, Orinda, CA 94563. (510) 299-1829. (800) 808-8811. 1996. $115.00.

Perry's Chemical Engineers' Handbook. Robert H. Perry and Donald W. Green, editors. McGraw-Hill Publishing Company, Inc., 1221 Avenue of the Americas, New York, NY 10020. (212) 512-2000. (800) 262-4729. 6th edition. 1996. $129.50.

ONLINE DATABASES AND CD-ROMS

CA Search. Chemical Abstracts Service, P.O. Box 3012, Columbus, OH 43210-0012. (614) 447-3600. (800) 848-6533. FAX (614) 447-3709. Very comprehensive guide to worldwide chemical literature and related fields. 1972 to present. Available on BRS,(800) 289-4277, DIALOG, (800) 334-2564, ORBIT (800) 456-7248, and STN International, FIZ Karlsruhe, P.O. Box 2465,

W-7500, Karlsruhe 1, Germany, online services. Inquire as to cost and availability.

Compendex Plus. Engineering Information, Inc., 345 East 47th Street, New York, NY 10017. (212) 705-7600 or (800) 221-1044. Contains citations with abstracts to worldwide literature in engineering and technology, from 1970 to present. Available on online BRS,(800) 289-4277, DIALOG, (800) 334-2564, ORBIT (800) 456-7248, and STN International, FIZ Karlsruhe, P.O. Box 2465, W-7500, Karlsruhe 1, Germany, online services. Also available on CD-ROM. Inquire as to cost and availability.

Dissertation Abstracts. University Microfilms International, 300 North Zeeb Road, Ann Arbor, MI 48106. (800) 521-0600 or (313) 761-4700. Scope includes virtually all doctoral dissertations accepted at accredited American institutions from 1861 to present in 252 subject areas. Available on BRS, (800) 289-4277, DIALOG, (800) 334-2564, and OCLC EPIC, (800) 848-5878, online services. Also available on CD-ROM. Inquire as to cost and availability.

INSPEC. Institution of Electrical Engineers, Michael Faraday House, Six Hills Way, Stevenage, Herts. SG1 2AY, England. Telephone: 0438 313311 or FAX 0438 742840. Contains citations to the worldwide literature of physics, electronics and electrical engineering, computer technology, and related fields. Available on BRS, (800) 289-4277, DIALOG, (800) 334-2564, ORBIT, (800) 456-7248, and STN International, FIZ Karlsruhe, P.O. Box 2465, W-7500, Karlsruhe 1, Germany, online services. Inquire as to cost and availability.

Kirk-Othmer Encyclopedia of Chemical Technology. John Wiley and Sons, Inc., 605 Third Avenue, New York, NY 10158. (800) 526-5368 or (212) 850-6000. Contains the complete text of all chapters in the 27 volume fourth edition of the Kirk-othmer Encyclopedia of Chemical Technology. 1991. Available on BRS,(800) 289-4277, DIALOG,(800) 334-2564, online services. Inquire as to cost and availability.

NTIS Bibliographic Database. National Technical Information Service, 5285 Port Royal Road, Springfield, VA 22161. (703) 487-4929 or FAX (703) 321-8199. Broad coverage of government-sponsored science and technology research reports, 1964 to present. Available on BRS,(800) 289-4277, DIALOG, (800) 334-2564, ORBIT, (800) 456-7248, and STN International, FIZ Karlsruhe, P.O. Box 2465, W-7500, Karlsruhe 1, Germany, online services. Also available on CD-ROM. Inquire as to cost and availability.

Physics Briefs. American Institute of Physics, 335 East 45th Street, New York, NY 10017. (212) 661-9260 or FAX (212) 661-2036. Contains citations with abstracts of the literature of physics and related fields, 1979 to present. Available on the STN International, FIZ Karlsruhe, P.O. Box 2465, W-7500, Karlsruhe 1, Germany, online service. Inquire as to cost and availability.

SCISEARCH. Institute for Scientific Information, 3501 Market Street, Philadelphia, PA 19104. (800) 523-1850 or (215) 386-0100. Broad multidisciplinary title and author index to the international literature of science and technology, 1974 to present. Available on DIALOG, (800) 334-2564, and ORBIT, (800) 456-7248, online services. Also available on CD-ROM. Inquire as to cost and availability.

WILSONLINE. H.W. Wilson Company, 950 University Avenue, Bronx, NY 10452. (800) 367-6770 or (212) 588-8400. Makes available online versions of the printed H.W. Wilson indexes including Applied Science and Technology Index, Business Periodicals Index, General Science Index, and Readers' Guide

to Periodical Literature. Period covered is generally 1983 to present. Available on BRS, (800) 289-4277, DIALOG, (800) 334-2564, and OCLC EPIC, (800) 848-5878, online services. Also available on CD-ROM. Inquire as to cost and availability.

Who's Who in Technology. Gale Research Inc., 835 Penobscot Building, Detroit, MI 48226-4094. (313) 961-2242. (800) 347-4253. Contains biographical information of contemporary American scieNTISts and engineers. Available on DIA-LOG,(800) 334-2564, online service. Inquire as to cost and availability.

PERIODICALS

American Nuclear Society Transactions. American Nuclear Society, 555 North Kensington Avenue, LA Grange Park, IL 60525. (708) 352-6611. 1958 to present. 3 issues per year. $340.00 per year. ISSN: 0003-018X

Annals of Nuclear Energy. Elsevier Science Publishing Company, Inc., 660 White Plains Road, Tarrytown, NY 10591- 5133. (914) 524-9200. 1974 to present. Monthly. $815.00 per year. ISSN: 0306-4549.

Bulletin of the Atomic Scientists. Educational Foundation for Nuclear Science, 6042 South Kimbark Avenue, Chicago, IL 60637. (312) 702-2555. 1945 to present. 10 issues per year. $30.00 per year. ISSN: 0096-3402.

Fusion Technology. American Nuclear Society, 555 North Kensington Avenue, LA Grange Park, IL 60525. (708) 352-6611. 1981 to present. 8 issues per year. $$485.00 per year. ISSN: 0748-1896.

Journal of Fusion Energy. Plenum Publishing Corporation, 233 Spring Street, New York, NY 10013- 1578. (212) 620- 8000. 1981 to present. Quarterly. $315.00 per year.

0164-0313.

Nuclear Engineer. Institution of Nuclear Engineers, 1 Penerley Road, London, SE6 2LQ, England. 1959 to present. Bi-monthly. L77 per year. ISSN: 0262-5091.

Nuclear Engineering and Design. Elsevier Science Publishing Company, Inc., 660 White Plains Road,

Tarrytown, NY 10591-5133. (914) 524-9200. 1965 to present. 24 issues per year. $2748.00 per year.

Nuclear Engineering International. Reed Business Pulbishing Group, Quadrant House, the Quadrant, Sutton, Surrey, SM2 5AS, Englnad. 1956 to present. Monthly. $236.00 per year. ISSN: 0029-5507.

Nuclear Science and Engineering. American Nuclear Society, 555 North Kensington Avenue, LA Grange Park, IL 60525. (708) 352-6611. 1956 to present. Monthly. $405.00 per year. ISSN: 0029-5639.

Nuclear Technology. American Nuclear Society, 555 North Kensington Avenue, LA Grange Park, IL 60525. (708) 352-6611. 1965 to present. Monthly. $440.00 per year. ISSN: 0029-5450.

RESEARCH CENTERS AND INSTITUTES

Department of Nuclear Engineering and Engineering Physics. University of Wisconsin - Madison, 1500 Johnson Drive, Madison, WI 53706. (608) 263-1648. FAX (608) 262-6707

Ford Nuclear Reactor. Phoenix Memorial Laboratory. University of Michigan, 2301 Bonisteel Boulevard, Ann Arbor MI 48109-2100. (313) 764-6223. FAX (313) 764-4540.

Frank H. Neely Nuclear Research Center. Georgia Institute of Technology, 900 Atantic Drive NW, Atlanta, GA 30332. (404) 894-3600.

Fusion Research Center. University of Texas at Austin. RLM 11.222, Austin, TX 78712. ((512) 471-3926. FAX (512) 471-8865.

Institute of Plasma and Fusion Research. University of California, Los Angeles, 44-139 Engineering IV 405 Hilgard Avenue, Los Angeles, CA 90024-1597. (213) 825-1613. FAX (213) 206-4832.

Nuclear Physics Laboratory. University of Washington, Seattle, WA 98195. (206) 543-4080.

Radiation Science and Engineering Center. Pennsylvania State University, University Park, PA 16802. (814) 865- 6351. FAX (814) 863-4840.

Stanford Synchrotron Radiation Laboratory. Stanford University, SLAC Bin #69, P.O. Box 4349, Stanford, CA 943050210. (415) 926-3153. FAX (415) 926-4100.

NUCLEAR ENGINEERING

See also: NUCLEAR CHEMISTRY, NUCLEAR ENERGY, NUCLEAR REACTORS

ABSTRACT SERVICES AND INDEXES

Applied Mechanics Reviews: An assessment of world literature in engineering sciences. American Society of Mechanical Engineers, 345 East 47th Street, New York, NY 10017. (212) 705-7703. 1948 to present. Monthly. $360.00 per year. ISSN: 0003-6900.

Applied Science and Technology Index; A Cumulative Subject Index To English Language Periodicals in the Fields of Aeronautics and Space Science, Computer Technology, Chemistry, Construction Industry, Energy and Related Areas. H. W. Wilson Co., 950 University Avenue, Bronx, NY 10452- 9978. (212) 588-8400. (800) 367-6770. FAX (718) 590-1617. From 1958 to present. Monthly. Inquire about cost and availability. Also available online (BRS and WILSONLINE) and on CD-ROM. ISSN: 0003-6986.

Chemical Abstracts. Chemical Abstracts Service. 2540 Olentangy River Road, Box 3012, Columbus, OH 43210-0012. (614) 447-3600. FAX (614) 447-3713. 1907 to present. Weekly. $16,800.00 per year. Available online and on CD- ROM. CA is also available in five section groupings. Inquire regarding cost and availability.

Current Contents: Engineering, Technology and Applied Sciences. Institute for Scientific Information, 3501 Market Street, Philadelphia, PA 19104. (215) 386-0100. FAX (215) 386-2991. From 1970 to present. Weekly. $442.00 per year. Also available online. Inquire regarding cost and availability. ISSN: 0095-7917.

Engineering Index Monthly. Engineering Information, Inc., Castle Point on the Hudson, Hoboken, NJ 07030. (201) 216-8500. (800) 221-1044. FAX (201) 216-8532. Monthly. $2300.00 per year for monthly issues. $1980 per year for annual cumulation. Available on CD-ROM and online as COMPENDEX. ISSN: 0742-1974; 0360-8557.

General Science Index. H.W. Wilson Company, 950 University Avenue, Bronx, NY 10452. (212) 588-8400. (800) 367-6770. FAX (718) 590-1617. From 1978 to present. Ten issues per year; quarterly and annual cumulations. Service basis. Available on CD-ROM and online. Inquire regarding cost and availability. ISSN: 0162-1963.

ISMEC: Mechanical Engineering Abstracts. Cambridge Scientific Abstracts, 7200 Wisconsin Avenue, Suite 601, Bethesda, MD 20814. (301) 961-6750 or (800) 843-7751. Contains citations to the literature in mechanical engineering, industrial and production engineering, energy, power, mechanics, devices and related areas, from 1973 to present. Available on the DIALOG, (800) 334-2564, online service. Inquire as to cost and availability.

N T I S Alerts: Energy. U.S. National Technical Information Service, 5285 Port Royal Road, Springfield, VA 22161. (703) 487-4929. FAX (703) 321-8547. Weekly. $160.00 per year.

Physics Abstracts. INSPEC. Section A, Science Abstracts. Institution of Electrical Engineers (IEE), London, United Kingdom. Available from: INSPEC/IEEE - Institute of Electrical and Electronic Engineers, Box 1331, 445 Hoes Lane, Piscataway, NJ 08855-1331. (908) 562-5549. 1898 to present. 24 issues per year. $2835.00 per year. ISSN: 0036-8091. Also available online and on CD-ROM.

Process and Chemical Engineering/chemical Engineering Abstracts. Royal Society of Chemistry, Information Services, Thomas Graham House, Science Park, Milton Road, Cambridge, CB4 4WF, England. Contains citations, mostly with abstracts, of the worldwide literature on chemical engineering; from 1982 to present. Monthly. $610.00 per year. Also available online. ISSN: 0960-5045.

Science Citation Index. SCI. Institute for Scientific Information, 3501 Market Street, Philadelphia, PA 19104. (215) 386-0100. (800) 523-1850. FAX (215) 386-2991. 1961 to present. Six issues per year, plus annual cumulation. $11650.00 per year. Also available online and on CD-ROM. Inquire about price and availability. ISSN: 0036-827X.

ANNUAL REVIEWS AND YEARBOOKS

Advances in Chemical Engineering. Academic Press, Inc., 6277 Sea Harbor Drive, Orlando, FL 32887. (800) 321-5068. 1956 to present. Irregular. ISSN: 0065-2377.

Advances in Nuclear Science and Technology. Plenum Publishing Corp., 233 Spring Street, New York, NY. (212) 620-8000. (800) 2221-9369. FAX (212) 463-0742. 1977 to present. Irregular.

ASSOCIATIONS AND PROFESSIONAL SOCIETIES

American Institute of Chemical Engineers. 345 East 47th Street, New York, NY 10017. (212) 705-7338.

American Nuclear Society. 555 North Kensington Avenue, LA Grange Park, IL 60525. (708) 352-6611. (800) NUC-NEWS.

Fusion Power Associates. 2 Professional Drive, Suite 248, Gaithersburg, MD 20879. (301) 258-0545. FAX (301) 975-9869.

Institute of Nuclear Materials Management. 60 Revere Drive, Northbrook, IL 60062. (847) 480-9573. FAX (947) 480-9282.

IEEE Nuclear and Plasma Sciences Society. c/o Institute of Electrical and Electronics Engineers, 345 East 47th Street, New York, NY 10017. (212) 705-7900. FAX 212) 705- 4929.

BIBLIOGRAPHIES

Handbooks and Tables in Science and Technology. Russell H. Powell, editor. Oryx Press, 4041 North Central, Suite 700, Phoenix, AZ 85012-3330. (602) 265-2651 or (800) 279-6799. Third edition. 1994. $65.00.

Information Sources in Engineering. Peter Hicks, editor. Bowker-Saur, 121 Chanlon Road, New Providence, NJ 07974. (800) 521-8110 or (908) 464-6800. 3rd edition. 1996. $125.00.

Scientific and Technical Books and Serials in Print; An Index To Literature in Science and Technology. R.R. Bowker Inc., 121 Chanlon Road, New Providence, NJ 07974. (908) 464-6800. (800) 521-8110. FAX (908) 665-3502. 1972 to present. Annual. 4 volumes. 1994. $299.95. Also available on compact disc and online. ISSN: 0000-054X.

DIRECTORIES AND BIOGRAPHICAL SOURCES

Energy and Nuclear Science International Who's Who. Gale Research, 835 Penobscot Building, Detroit, MI 48226- 4094. (313) 961-2242. (800) 347-4253. Triennial. 1993. $325.00.

Energy Information Centers Directory. U. S. Council for Energy Awareness, 1776 I Street, NW, Suite 400, Washington, DC 20006. (202) 293-0770. Triennial. No charge.

Engineering Research Centers: Incorporating Electronics Research Centers. Stockton Press, 345 Park Avenue, New York, NY 10010. (212) 689-9200. (800) 221-2123. 4th edition. 1995. $515.00.

International Directory of Nuclear Utilities. Nuexco, 1515 Arapahoe Street, Suite 900, Denver CO 80202. (303) 899- 4500. Annual. $220.00 per year

Nuclear News - Directory Issues. American Nuclear Society, 555 North Kensington Avenue, LA Grange Park, IL 60525. (708) 352-6611. (800) NUC-NEWS. March issue contains Buyer's Guide; $65.00. World List of Nuclear Power Plants. Semi-annual in February and August. $15.00 per issue.

Nuclear Reactors Built, Being Built OR Planned in the United States. National Technical Information Service, Springfield, VA 22161. (703) 487-4780. 1989. $17.00.

Research Centers Directory. Gale Research, 835 Penobscot Building, Detroit, MI 48226-4094. (313) 961-2242. (800) 347-4253. 20th edition. 1995. $485.00. ISSN: 0080-1518.

Who's Who in Engineering. Gordon Davis, editor. American Association of Engineering Societies. 1111 19th Street, NY, Suite 608, Washington, DC 20036. (202) 296-2237. (800) 658-8897. 9th edition. 1995. $220.00.

World Energy and Nuclear Directory. Gale Research, 835 Penobscot Building, Detroit, MI 48226-4094. (313)961-2242. (800) 347-4253. Biennial, even years. $450.00.

ENCYCLOPEDIAS AND DICTIONARIES

Dictionary of Named Processes in Chemical Technology. Alan E. Comyns. Oxford University Press, Inc., 200 Madison Avenue, New York, NY 10016. (212) 725-6000. (800) 334-4249. 1994. $75.00.

Dictionary of Nuclear Technology. Pro/Am Music Resources, Inc., 63 Prospect Street, White Plains, New York, 10606. (914) 948-7436. 1986. $245.00.

Encyclopedia of Applied Physics. VCH Publishers, Inc., 303 Northwest 12th Avenue, Deerfield Beach, FL 33442. (800) 367-8249. 1991-. Twenty volumes. $6000.00.

Encyclopedia of Chemical Processing and Design. McKetta Marcel Dekker, Inc., 270 Madison Avenue, New York, NY 10016. (212) 696-9000. (800) 228-1160. 1976 - . $175.00 per volume. ISSN:

IEEE Nuclear Science Standards Collection. IEEE Standards office, Box 1331, 445 Hoes Lane, Piscataway, NJ 08855-1331. (908) 562-3822. (800) 678-4333. 1991. $90.00 in paper.

Illustrated Chemistry Laboratory Terminology. Gerbert W. Ockerman. CRC Press, Inc., 2000 Corporate Boulevard, NW, Boca Raton, FL 33431. (407) 994-0555. (800) 272-7737. 1991. $32.95.

Kirk-Othmer Encyclopedia of Chemical Technology. John Wiley and Sons, Inc., 605 Third Avenue, New York, NY 10158. (800) 526-5368 or (212) 850-6000. Fourth edition. 1991 - . Twenty-seven volumes. $5400.00.

McGraw-Hill Encyclopedia of Science and Technology. McGraw-Hill Book, Incorporated, 1221 Avenue of the Americas, New York, NY 10020. (212) 997-3675. (800) 262-4729. Seventh edition. Twenty volumes. 1992. $1900.00.

Ullman's Encyclopedia of Industrial Chemistry. VCH Publications, Inc., 220 East 23rd Street, Suite 909, New York, NY 10010-4606. (212) 683-8333. (800) 422-8824. 5th edition. 1984 - . Price varies per volume.

GENERAL WORKS

Basic Nuclear Engineering. K. Sri Ram. South Asia Books, PO Box 502, Columbia, MO 65205. (314) 474-0116. 1990. $32.00.

Chain Raction: Expert Debate and Public Participation in American Commercial Nuclear Power, 1945 - 1975. Brian Balough. Cambridge University Press, 40 West 20th Street, New York, NY 10011-4211. (212) 924-3900. (800) 872-7423. 1993. $17.95 in paper.

Chemical Reactor Analysis and Design. G. F. Froment and Kenneth B. Bischoff. John Wiley & Sons, Inc., 605 Third Avenue, New York, NY 10158-0012. (212) 850-6000. (800) 225-5945. 2nd edition. 1990.

Electronics For Nuclear Instrumentation. Hai H. Chiang. Hemisphere Publishing Corp., 1900 Frost Road, Suite 101, Bristol PA 19007. (215) 785-5800. (800) 821-8312. 1985. $63.50.

Fifieth Anniversary of Nuclear Fission. L. V. Drapchinsky, editor. Nova Science Publishers, Inc., 6080 Jericho Turnpike, Suite 207, Commack, NY 11725-2808. (516) 499- 3103. 1994. $145.00.

Introduction To Nuclear Engineering. John R. Lamarch. Addison-Wesley Publishing Co., Inc., 1 Jacob Way, Reading, MA 01867. (617) 944-3700. (800) 447-2226. 2nd edition. 1983. $67.75.

Legacy of Chernobyl. Zhores A. Medvedev. W. W. Norton & Company, 500 Fifth Avenue, New York, NY 10110. (212) 354-5500. (800) 223-2584. 1992. $10.95 in paper.

Nuclear Choices: A Citizen's Guide To Nuclear Technology. Richard Wolfson. MIT Press, 55 Hayward Street, Cambridge, MA 02142. (617) 253-8569. Revised edition. 1993. $19.95 in paper.

Nuclear Engineering: An Introduction. K. Almenas and R. Lee. Springer-Verlag New York, Inc., 175 Fifth Avenue, New York, NY 10010. (212) 460-1500. (800) 777-4643. FAX (212) 473-6272. 1992. $89.50.

Nuclear Engineering: Theory and Technology of Commercial Nuclear Power. Ronald A. Knief. Hemisphere Publishing Corp., 1900 Frost Road, Suite 101, Bristol PA 19007. (215) 785-5800. (800) 821-8312. 2nd edition. 1992. $69.50 in paper.

Nuclear Power For Beginners. C. Croall and S. Sempler. State Mutual Book Service, Ltd. 521 Fifth Avenue, New York, NY 10175. (212) 682-5844. 1990. $35.00.

nuclear techniques for analytical and industrial *Applications.* G. Vourvopoulos and T. Paradellis. World Scientific Publishing Company, Inc., 1060 Main Street, River Edge, NJ 07661. (201) 487-9655. (800) 227-7562. 1993. $78.00. *Principles of Nuclear Science and Engineering.* A. A. Harms. John Wiley & Sons, Inc., 605 Third Avenue, New York, NY 10158-0012. (212) 850-6000. (800) 225-5945. 1987.$138.00.

HANDBOOKS AND MANUALS

CRC Handbook of Nuclear Reactors Calculations. CRC Press, Inc., 2000 Corporate Blvd., N.W., Boca Raton, FL 33431. (407) 994-0555 or (800) 333-8300. 3 volumes. 1986.

A Guide To Nuclear Power Technology: A Resource For Decision Making. Frank J. Rahn, et al. Krieger Publishing Company, P.O. Box 9542, Melbourne, FL 32902-9542. (407) 724-9542. Reprint editon. 1992. $145.00.

Information Center On Nuclear Standards. American Nuclear Society, 555 North Kensington Avenue, LA Grange Park, IL

NUCLEAR ENGINEERING

Ency. of Physical Sciences and Engineering Info. Sources

60525. (708) 352-6611. Standards for all aspects of the design, construction, operation and maintenance of nuclear power plants.

Nuclear Engineering Databases, Standards and Numerical Analysis. Jack Jedruch. Van Nostrand Reinhold, 115 Fifth Avenue, New York, NY 10003. (212) 254-3232. (800) 842- 3636. 1985.

Nuclear Engineering Sourcebook: A Compilation of Documents For Nuclear Equipment Qualification. IEEE Standards office, Box 1331, 445 Hoes Lane, Piscataway, NJ 08855-1331. (908) 562-3822. (800) 678-4333. 1992.

Perry's Chemical Engineers' Handbook. Robert H. Perry and Donald W. Green, editors. McGraw-Hill Publishing Company, Inc., 1221 Avenue of the Americas, New York, NY 10020. (212) 512-2000. (800) 262-4729. 6th edition. 1996. $129.50.

Summary Description of Design Criteria, Codes, Standards, and Regulatory Provisions Typically Used For the Civil and Structural Design of Nuclear Fuel Cycle Facilities. American Society of Civil Engineers. 345 East 47th Street, New York, NY 10017. (212) 705-7496. (800) 548-2723. 1988. $17.00

ONLINE DATABASES AND CD-ROMS

CA Search. Chemical Abstracts Service, P.O. Box 3012, Columbus, OH 43210-0012. (614) 447-3600. (800) 848-6533. FAX (614) 447-3709. Very comprehensive guide to worldwide chemical literature and related fields. 1972 to present. Available on BRS,(800) 289-4277, DIALOG, (800) 334-2564, ORBIT (800) 456-7248, and STN International, FIZ Karlsruhe, P.O. Box 2465, W-7500, Karlsruhe 1, Germany, online services. Inquire as to cost and availability.

Compendex Plus. Engineering Information, Inc., 345 East 47th Street, New York, NY 10017. (212) 705-7600 or (800) 221-1044. Contains citations with abstracts to worldwide literature in engineering and technology, from 1970 to present. Available on online BRS,(800) 289-4277, DIALOG, (800) 334-2564, ORBIT (800) 456-7248, and STN International,FIZ Karlsruhe, P.O. Box 2465, W-7500, Karlsruhe 1, Germany, online services. Also available on CD-ROM. Inquire as to cost and availability.

Current Contents Search. Institute for Scientific Information, 3501 Market Street, Philadelphia, PA 19104. (215) 386-0100. FAX (215) 386-6362. Contains citations to articles listed in the table of contents of science and technology journals. Also articles in social sciences and life sciences journals. Available on BRS,(800) 289-4277, DIALOG,(800) 334-2564, online services. Inquire as to cost and availability.

Dissertation Abstracts. University Microfilms International, 300 North Zeeb Road, Ann Arbor, MI 48106. (800) 521-0600 or (313) 761-4700. Scope includes virtually all doctoral dissertations accepted at accredited American institutions from 1861 to present in 252 subject areas. Available on BRS, (800) 289-4277, DIALOG, (800) 334-2564, and OCLC EPIC, (800) 848-5878, online services. Also available on CD-ROM. Inquire as to cost and availability.

INSPEC. Institution of Electrical Engineers, Michael Faraday House, Six Hills Way, Stevenage, Herts. SG1 2AY, England. Telephone: 0438 313311 or FAX 0438 742840. Contains citations to the worldwide literature of physics, electronics and electrical engineering, computer technology, and related fields. Available on BRS, (800) 289-4277, DIALOG, (800) 334-2564, ORBIT, (800) 456-7248, and STN International, FIZ Karlsruhe,

P.O. Box 2465, W-7500, Karlsruhe 1, Germany, online services. Inquire as to cost and availability.

NTIS Bibliographic Database. National Technical Information Service, 5285 Port Royal Road, Springfield, VA 22161. (703) 487-4929 or FAX (703) 321-8199. Broad coverage of government-sponsored science and technology research reports, 1964 to present. Available on BRS,(800) 289-4277, DIALOG, (800) 334-2564, ORBIT, (800) 456-7248, and STN International, FIZ Karlsruhe, P.O. Box 2465, W-7500, Karlsruhe 1, Germany, online services. Also available on CD-ROM. Inquire as to cost and availability.

*ISMEC: Mechanical Engineering Abstracts.*Cambridge Scientific Abstracts, 7200 Wisconsin Avenue, Suite 601, Bethesda, MD 20814. (301) 961-6750 or (800) 843-7751. Contains citations to the literature in mechanical engineering, industrial and production engineering, energy, power, mechanics, devices and related areas, from 1973 to present. Available on the DIALOG,(800) 334-2564, online service. Inquire as to cost and availability.

Kirk-Othmer Encyclopedia of Chemical Technology. John Wiley and Sons, Inc., 605 Third Avenue, New York, NY 10158. (800) 526-5368 or (212) 850-6000. Contains the complete text of all chapters in the 27 volume fourth edition of the KIRK-Othmer Encyclopedia of Chemical Technology. 1991. Available on BRS,(800) 289-4277, DIALOG,(800) 334-2564, online services. Inquire as to cost and availability.

NTIS Bibliographic Database. National Technical Information Service, 5285 Port Royal Road, Springfield, VA 22161. (703) 487-4929 or FAX (703) 321-8199. Broad coverage of government-sponsored science and technology research reports, 1964 to present. Available on BRS,(800) 289-4277, DIALOG, (800) 334-2564, ORBIT, (800) 456-7248, and STN International, FIZ Karlsruhe, P.O. Box 2465, W-7500, Karlsruhe 1, Germany, online services. Also available on CD-ROM. Inquire as to cost and availability.

Nuclear Science Abstracts. U.S. Department of Energy, office of Scientific and Technical Energy, P.O. Box 62, Oak Ridge, TN 37831. (615) 576-1189. Contains citations with abstracts to literature in all fields of nuclear science and energy. 1948 to 1976. Available on DIALOG,(800) 334-2564, online service. Inquire as to cost and availability.

Physics Briefs. American Institute of Physics, 335 East 45th Street, New York, NY 10017. (212) 661-9260 or FAX (212) 661-2036. Contains citations with abstracts of the literature of physics and related fields, 1979 to present. Available on the STN International, FIZ Karlsruhe, P.O. Box 2465, W-7500, Karlsruhe 1, Germany, online service. Inquire as to cost and availability.

SCISEARCH. Institute for Scientific Information, 3501 Market Street, Philadelphia, PA 19104. (800) 523-1850 or (215) 386-0100. Broad multidisciplinary title and author index to the international literature of science and technology, 1974 to present. Available on DIALOG, (800) 334-2564, and ORBIT, (800) 456-7248, online services. Also available on CD-ROM. Inquire as to cost and availability.

Scientific and Technical Books and Serials in Print. R.R. Bowker Inc., 121 Chanlon Road, New Providence, NJ 07974. (800) 521-8110 or (908) 464-6800. List of currently published books and serials in the physical and biological sciences, engineering and technology, with subject, author and titles indexes. Available on ORBIT, (800) 456-7248, online service. Inquire as to cost and availability.

WILSONLINE. H.W. Wilson Company, 950 University Avenue, Bronx, NY 10452. (800) 367-6770 or (212) 588-8400. Makes available online versions of the printed H.W. Wilson indexes including Applied Science and Technology Index, Business Periodicals Index, General Science Index, and Readers' Guide to Periodical Literature. Period covered is generally 1983 to present. Available on BRS, (800) 289-4277, DIALOG, (800) 334-2564, and OCLC EPIC, (800) 848-5878, online services. Also available on CD-ROM. Inquire as to cost and availability.

Who's Who in Technology. Gale Research Inc., 835 Penobscot Building, Detroit, MI 48226-4094. (313) 961-2242. (800) 347-4253. Contains biographical information of contemporary American scieNTISts and engineers. Available on DIALOG,(800) 334-2564, online service. Inquire as to cost and availability.

PERIODICALS

Aiche Journal. American Institute of Chemical Engineers. 345 East 47th Street, New York, NY 10017. (212) 705-7338.

1955 to present. Monthly. $295.00 per year.

American Nuclear Society Transactions. American Nuclear Society, 555 North Kensington Avenue, LA Grange Park, IL 60525. (708) 352-6611. 1958 to present. 3 issues per year. $340.00 per year. ISSN: 0003-018X

Annals of Nuclear Energy. Elsevier Science Publishing Company, Inc., 660 White Plains Road, Tarrytown, NY 10591- 5133. (914) 524-9200. 1974 to present. Monthly. $815.00 per year. ISSN: 0306-4549.

Bulletin of the Atomic Scientists. Educational Foundation for Nuclear Science, 6042 South Kimbark Avenue, Chicago, IL 60637. (312) 702-2555. 1945 to present. 10 issues per year. $30.00 per year. ISSN: 0096-3402.

Fusion Technology. American Nuclear Society, 555 North Kensington Avenue, LA Grange Park, IL 60525. (708) 352-6611. 1981 to present. 8 issues per year. $$485.00 per year. ISSN: 0748-1896.

Journal of Fusion Energy. Plenum Publishing Corporation, 233 Spring Street, New York, NY 10013- 1578. (212) 620- 8000. 1981 to present. Quarterly. $315.00 per year. 0164-0313.

Nuclear Engineer. Institution of Nuclear Engineers, 1 Penerley Road, London, SE6 2LQ, England. 1959 to present. Bi-monthly. L77 per year. ISSN: 0262-5091.

Nuclear Engineering and Design. Elsevier Science Publishing Company, Inc., 660 White Plains Road, Tarrytown, NY 10591-5133. (914) 524-9200. 1965 to present. 24 issues per year. $2748.00 per year.

Nuclear Engineering International. Reed Business Pulbishing Group, Quadrant House, the Quadrant, Sutton, Surrey, SM2 5AS, Englnad. 1956 to present. Monthly. $236.00 per year. ISSN: 0029-5507.

Nuclear Science and Engineering. American Nuclear Society, 555 North Kensington Avenue, LA Grange Park, IL 60525. (708) 352-6611. 1956 to present. Monthly. $405.00 per year. ISSN: 0029-5639.

Nuclear Technology. American Nuclear Society, 555 North Kensington Avenue, LA Grange Park, IL 60525. (708) 352-6611. 1965 to present. Monthly. $440.00 per year. ISSN: 0029-5450.

RESEARCH CENTERS AND INSTITUTES

Department of Nuclear Engineering and Engineering Physics. University of Wisconsin - Madison, 1500 Johnson Drive, Madison, WI 53706. (608) 263-1648. FAX (608) 262-6707.

Frank H. Neely Nuclear Research Center. Georgia Institute of Technology, 900 Atantic Drive NW, Atlanta, GA 30332. (404) 894-3600.

Fusion Research Center. University of Texas at Austin. RLM 11.222, Austin, TX 78712. ((512) 471-3926. FAX (512) 471-8865.

Institute of Nuclear Science and Engineering. Oregon State University Radiation Center- A100, 35th and Jefferson Streets, Corvallis, OR 97331-5903. (503) 737-2341. FAX (503) 737-0480.

Institute of Plasma and Fusion Research. University of California, Los Angeles, 44-139 Engineering IV 405 Hilgard Avenue, Los Angeles, CA 90024-1597. (213) 825-1613. FAX (213) 206-4832.

Radiation Science and Engineering Center. Pennsylvania State University, University park, PA 16802. (814) 865- 6351. FAX (814) 863-4840.

Whiteshell Laboratories. Research Company, Atomic Energy of Canada Limited, Pinawa, MB, Canada ROE ILO. (204) 753-2311. FAX (204) 753-8404.

NUCLEAR FUELS

See: NUCLEAR ENERGY

O

OBSERVATORIES

See also: ASTRONOMY, ASTROPHYSICS, GALAXIES, OPTICS, RADIO TELESCOPES, SOLAR SYSTEM, TELESCOPES

ABSTRACT SERVICES AND INDEXES

Applied Science and Technology Index: A Cumulative Subject Index to English Language Periodicals in the Fields of Aeronautics and Space Science, Computer Technology, Chemistry, Construction Industry, Energy and Related Areas. H. W. Wilson Co., 950 University Avenue, Bronx, NY 10452-9978. (212) 588-8400. (800) 367-6770. FAX (718) 590-1617.From 1958 to present. Monthly. Inquire about cost andavailability. Also available online (BRS and WILSONLINE)and on CD-ROM. ISSN: 0003-6986.

Astronomy and Astrophysics Abstracts. Springer-Verlag New York, 175 Fifth Avenue, New York, NY 10010. (212) 460-1500.FAX (212) 473-6272. Published for the AstronomischesRechen-Institut. Comprehensive coverage of all aspects of astronomy, astrophysics and related fields. 1969 to present. Two parts per year. Annual. ISSN: 0067-0022.

Meteorological and Geoastrophysical Abstracts. American Meteorological Society, c/o Inforonics, Inc., 550 NewtownRoad, Box 458, Littleton, MA 01460. (508) 486-8976. FAX(508) 486-0027. Covers literature in environmentalsciences, meteorology, astrophsyics, hydrology, glaciology and physical oceanography. 1950 to present. Monthly. $950.00 per year. Also available online (DIALOG) and on CD-ROM. ISSN: 0026-1130.

Physics Abstracts. INSPEC. Section A, Science Abstracts. Institute of Electrical Engineers, London, United Kingdom. Available from: INSPEC/IEEE-Institute of Electrical and Electronic Engineers, Box 1331, Hoes Lane, Piscataway, NJ08855-1331. (908) 562-5549. 1898 to present. 24 issues per year. $2835.00 per year. ISSN: 0036-8091. Also available online and on CD-ROM.

Science Citation Index. SCI. Institute for Scientific Information, 3501 Market Street, Philadelphia, PA 19104.(215) 386-0100. (800) 523-1850. FAX (215) 386-2991. 1961 to present. Six issues per year, plus annual cumulation.$11650.00 per year. Also available online and on CD-ROM. Inquire about price and availability. ISSN: 0036-827X.

STAR. Scientific and Technical Aerospace Reports. U.S. National Aeronautics and Space Administration. Distributed by U. S.

Superintendent of Documents, Washington, DC 20402. 1963 to present. Semi-monthly, with semiannual and annual indexes. $114.00 per year. ISSN: 0036-8741. Also available online and on CD-ROM.

ANNUAL REVIEWS AND YEARBOOKS

The Astronomical Almanac. Superintendent of Documents, U. S. Government Printing office, Washington, DC 10402. (202) 783-3238. 1981 to present. Superceeds Astronomical Ephemeris and the American Ephermis and Nautical Almanac. Annual. $44.95. ISSN: 0737-6421.

Annual Review of Astronomy and Astrophysics. Original reviews of critical literature and current developments in astronomy and astrophysics. Annual Reviews, Inc., 4139 El Camino Way, Palo Alto, CA 94303-0139. (415) 493-4400. (800) 523-8635. Fax (415) 855-9815. 1963 to date. Annual. $60.00. ISSN: 0066-4146.

ASSOCIATIONS AND PROFESSIONAL SOCIETIES

American Astronomical Society. 2000 Florida Avenue NW, Suite 400, Washington, DC 20009. (202) 328-2010. FAX (202) 234-2560.

American Institute of Physics. One Physics Ellipse, College Park, MD 20740-3843. (301) 209-3100.

American Association of Variable Star Observers. 23 Birch Street, Cambridge, MA 02138. (617) 354-0484. FAX: (617) 354-0665.

Association of Universities For Research in Astronomy, Inc. (AURA). Suite 701, 1625 Massachusetts Avenue, NW, Washington, DC 20036. (202) 483-2101. FAX (202) 483-2106.

Astronomical League. 2112 Kingfisher Lane East, Rolling Meadows, IL 60008. (708) 398-0562.

Astronomical Society of the Pacific. 390 Ashton Avenue, San Francisco, CA 94112. (415) 337-1100. FAX: (415) 337-5205.

Institute of Electrical and Electronics Engineers (IEEE). 345 East 47th Street, New York, NY 10017. (212) 705-7900.

OBSERVATORIES

Ency. of Physical Sciences and Engineering Info. Sources

Optical Society of America. 1816 Jefferson Place, N.W., Washington, DC 20036. (202) 223-8130.

Society of Photo-optical Instrumentation Engineers; the International Society of Optical Engineering. P.O. Box 10, 1022 19th Street, Bellingham, WA 98227. (206) 676-3290.

ATLASES

Cambridge Atlas of Astronomy. Cambridge University Press, 40 West 20th Street, New York, NY 10011-4211. (212) 924-3900. (800) 872-7423. 3rd edition. 1994. $74.00.

The Sky: A User's Guide. David H. Levy. Cambridge University Press, 40 West 20th Street, New York, NY 10011-4211. (212) 924-3900. (800) 872-7423. 1993. $14.95.

BIBLIOGRAPHIES

A Bibliography of Astronomy: 1970-1979. R. A. Sea and S. S. Martin. Libraries Unlimited, Inc., Littleton, CO 80160. 1982. $37.50.

Science Books & Films. American Association for the Advancement of Science, 1333 H Street NW, Washington, DC 20005. (202) 326-6434. Reviews of print, film and software materials in all sciences. 1965 to present. Nine issues per year. $40.00 per year. ISSN: 0098-342X.

Scientific and Technical Books and Serials in Print; An Index To Literature in Science and Technology. R. R. Bowker. 121 Chanlon Road, New Providence, NJ 07974. (908) 464-6800. FAX (908) 665-3502. (800) 521-8110. 1972 to present. Annual. $299.95 per year. Also available on CD-ROM. ISSN: 0000-054X.

DIRECTORIES AND BIOGRAPHICAL SOURCES

American Astronomical Society Membership Directory. American Astronomical Society. 2000 Florida Avenue NW, Suite 400, Washington, DC 20009. (202) 328-2010. FAX (202) 234-2560 Annual. Included in membership dues. ISSN: 1061- 9038.

American Men and Women of Science: Physical and Biological Sciences. R. R. Bowker Inc., 121 Chanlon Road, New Providence, NJ 07974. (908) 464-6800. (800) 521-8110. 20th edition. 8 volumes. 1996. $850.00.

The Astronomers. Donald Goldsmith. St. Martin's Press, Inc., 175 Fifth Avenue, New York, NY 10010. (212) 674-5151. (800) 221-7945. 1993. $14.95 in paper.

Astronomical Centers of the World. Kevin Krisciunas. Cambridge University Press, 40 West 20th Street, New York, NY 10011-4211. (212) 924-3900. (800) 872-7423. 1988. $34.95

Directory of Physics and Astronomy Staff. American Institute of Physics, One Physics Ellipse, College Park, MD 20740-3843. (301) 209-3100. 1975/76 to present. Annual. $60.00. ISSN: 0361-2228.

Earth and Astronomical Research Centers. Stockton Press, 345 Park Avenue South, New York, NY 10010. 4th edition. 1995. $515.00. ISBN: 1-56169-0967.

Graduate Programs in Physics, Astronomy and Related Fields. American Institute of Physics, One Physics Ellipse, College Park, MD 20740-3843. (301) 209-3100. 1978 to present. Annual. $45.00. ISSN: 0147-1821.

Research Centers Directory. Gale Research, 835 Penobscot Building, Detroit, MI 48226-4094. (313) 961-2242. (800) 347-4253. 20th edition. 1995. $485.00.

ENCYCLOPEDIAS AND DICTIONARIES

Concise Dictionary of Astronomy. Jacqueline Mitton. Oxford University Press, Inc., 200 Madison Avenue, New York, NY 10016. (212) 725-6000. (800) 334-4249. 1992. $24.95.

Encyclopedia of Astronomy and Astrophysics. Stephen Maran, editor. Van Nostrand Reinhold, 115 Fifth Avenue, New York, NY 10003. (212) 254-3232. (800) 842-3636. 1992. $129.95.

McGraw-Hill Encyclopedia of Science and Technology. McGraw-Hill Book Company, Inc., 1221 Avenue of the Americas, New York, NY 10020. (212) 512-2000. (800) 262-4729. 7th edition. 20 volume set. 1992. $1900.00.

GENERAL WORKS

Astronomical Observations: An Optical Perspective. Gordon Walker. Cambridge University Press, 40 West 20th Street, New York, NY 10011-4211. (212) 924-3900. (800) 872-7423.1987. $37.95 in paper.

Big and Bright: A History of the McDonald Observatory. David S. Evans and J. Derral Mulholland. University of Texas Press, P. O. Box 7819, Austin, TX 78713-7819. (512) 471-7233. (800) 252-3206. 1986. $19.95.

The Cosmic Inquirers: Modern Telescopes and their Makers. Wallace Tucker and Karen Tucker. Harvard University Press, 79 Garden Street, Cambridge MA. (617) 495-2600. Reprint edition. 1989. $10.95 in paper.

The Explorers of Mars Hill: A Centennial History of Lowell Observatory. William L. Putnam. Phoenix Publishing, Main Street, Sugar Hill, NJ 03585. (603) 823-8531. 1994. $30.00.

Galactic and Extragalactic Radio Astronomy. G. K. Verschuur and K. Kellermann. Springer-Verlag New York, 175 Fifth Avenue, New York, NY 10010. (212) 460-1500. (800) 777- 4643. 1991. $55.00 in paper.

History of Radio Astronomy and the National Radio Astronomy Observatory: Evolution Toward Big Science. Benjamin K. Malphrus. Kreiger Publishing Company, P.O. Box 9542, Melbourne, FL 32902-9542. (407) 724-9542. 1995. $wfi

National Radio Astronomy Observatory Very Large Array Sky Survey. J. J. Condon, et al. Rector Press, Ltd., 130 Rattlesnake, Leverett, MA 01054-9726. (413) 548-9708. (800) 247-3473. 1994. $45.00 in paper.

Optical Astronomy From the Earth and Moon. Diane M. Pyper and Ronald Angione, editors. Astronomical Society of the Pacific, 390 Ashton Avenue, San Francisco, CA 94112. (415) 337-1100. 1994. $40.00.

Pauper and Prince: Richey, Hale and Big American Telescopes. Donald E. Osterbrock. University of Arizona Press, 1230 North Park Avenue, Number 102, Tucson, AZ 85719. (520) 621- 1441. (800) 426-3797. 1993. $45.00.

Realm of the Long Eyes: A Brief History of Kitt Peak National Observatory. James E. Kloeppel. 2nd edition. Univelt Inc., P.O. Box 28130, San Diego, CA 92198-0198. (619) 746-4005. 1991. $15.00 in paper.

Robotic Telescopes in the 1990s. A. Filipenko. Astronomical Society of the Pacific, 390 Ashton Avenue, San Francisco, CA 94112. (415) 337-1100. 1992. $40.00.

Seeing the Deep Sky: Telescopic Astronomy Projects Beyond the Solar System. Fred Schaaf. John Wiley & Sons, Inc., 605 Third Avenue, New York, NY 10158-0012. (212) 850-6000. (800) 225-5945. 1992. $14.95 in paper.

Space Technology and Planetary Astronomy. Joseph N. Tatarewicz. Indiana University Pres, 601 North Morton Street, Bloomington, in 47404-3797. (812) 855-6804. (800) 842-6796. 1990. $29.95.

Stairway to the Stars: The Story of the World's Largest Observatory. B. Parker. Plenum Publishing Corp., 233 Spring Street, New York, NY. (212) 620-8000. (800) 2221- 9369. FAX (212) 463-0742. 1994. $27.95.

Unusual Telescopes. Peter L. Manly. Cambridge University Press, 40 West 20th Street, New York, NY 10011-4211. (212) 924-3900. (800) 872-7423. 1992. $39.95.

Visual Astronomy of the Deep Sky. Roger N. Clark. Cambridge University Press, 40 West 20th Street, New York, NY 10011-4211. (212) 924-3900. (800) 872-7423. 1991. $39.95.

Voice of the Universe: Building the Jodrell Bank Telescope. Barnard Lovell. Greenwood Press, 1600 Larkin Street, No. 104, San Francisco, CA 94109. (415) 928-4142. 1987. $14.95 in paper.

HANDBOOKS AND MANUALS

Astrophysical Data: Planets and Stars. Kenneth R. Lanb. Springer-Verlag New York, Inc., 175 Fifth Avenue, New York, NY 10010. (212) 460-1500. (800) 777-4643. 1993. $59.00.

Planet Observer's Handbook. Fred W. Price. Cambridge University Press, 40 West 20th Street, New York, NY 10011- 4211. (212) 924-3900. (800) 872-7423. 1994. $34.95.

ONLINE DATABASES AND CD-ROMS

Dissertation Abstracts. University Microfilms International, 300 North Zeeb Road, Ann Arbor, MI 48106. (800) 521-0600 or (313) 761-4700. Scope includes virtually all doctoral dissertations accepted at accredited American institutions from 1861 to present in 252 subject areas. Available on BRS, (800) 289-4277, DIALOG, (800) 334-2564, and OCLC EPIC, (800) 848-5878, online services. Also available on CD-ROM. Inquire as to cost and availability.

INSPEC. Institution of Electrical Engineers, Michael Faraday House, Six Hills Way, Stevenage, Herts. SG1 2AY, England. Telephone: 0438 313311 or FAX 0438 742840. Contains citations to the worldwide literature of physics, electronics and electrical engineering, computer technology, and related fields. Available on BRS, (800) 289-4277, DIALOG, (800) 334-2564, ORBIT, (800) 456-7248, and STN International, FIZ Karlsruhe, P.O. Box 2465, W-7500, Karlsruhe 1, Germany, online services. Inquire as to cost and availability.

NTIS Bibliographic Database. National Technical Information Service, 5285 Port Royal Road, Springfield, VA 22161. (703) 487-4929 or FAX (703) 321-8199. Broad coverage of government-sponsored science and technology research reports, 1964 to present. Available on BRS,(800) 289-4277, DIALOG, (800) 334-2564, ORBIT, (800) 456-7248, and STN International, FIZ Karlsruhe, P.O. Box 2465, W-7500, Karlsruhe 1, Germany, online services. Also available on CD-ROM. Inquire as to cost and availability.

SCISEARCH. Institute for Scientific Information, 3501 Market Street, Philadelphia, PA 19104. (800) 523-1850 or (215) 386-0100. Broad multidisciplinary title and author index to the international literature of science and technology, 1974 to present. Available on DIALOG, (800) 334-2564, and ORBIT, (800) 456-7248, online services. Also available on CD-ROM. Inquire as to cost and availability.

WILSONLINE. H.W. Wilson Company, 950 University Avenue, Bronx, NY 10452. (800) 367-6770 or (212) 588-8400. Makes available online versions of the printed H.W. Wilson indexes including *Applied Science and Technology Index, Business Periodicals Index, General Science Index,* and *Readers' Guide to Periodical Literature.* Period covered is generally 1983 to present. Available on BRS, (800) 289-4277, DIALOG, (800) 334-2564, and OCLC EPIC, (800) 848-5878, online services.Also available on CD-ROM. Inquire as to cost and availability.

PERIODICALS

Astronomical Journal. American Institute of Physics, One Physics Ellipse, College Park, MD 20740-3843. (301) 209-3000. Published for the American Astronomical Society. 1849 to present. Monthly. $280.00 per year. ISSN: 0004- 6256.

Astronomical Society of the Pacific Publications. Astronomical Society of the Pacific, 390 Ashton Avenue, San Francisco, CA 94112. (415) 337-1100. FAX (415) 337-5205.1889 to present. Monthly. $175.00 per year. ISSN: 0004- 6280.

Astronomy. Kalmbach Publishing Company, Box 1612, Waukesha, WI 53187-1612. (414) 796-8776. FAX (414) 796-1142. 1973 to present. Monthly. $27.00 per year. ISSN: 0091-6358.

Astronomy and Astrophysics. Springer-Verlag New York, Inc., 44 Hartz Way, Secaucus, NJ 07096-2491. (201) 348-4033. FAX (201) 348-4505. 1969 to date. 36 issues per year in 12 volumes. $1733.00 per year. ISSN: 0004-6361.

Astrophysical Journal. University of Chicago Press, 5720 South Woodlawn Avenue, Chicago, IL 60637. (312) 753-3347. FAX (312) 753-0811. 1895 to date. 3 issues per monthly. $740.00 per year. ISSN: 0004-637X.

Astrophysics and Space Science. Kluwer Academic Publishers, Box 358, Accord Station, Hingham MA 02018-0358. (617) 871-

6600. FAX (617) 871-6528. 1968 to present. 24 issues per year. $2544.00 per year. ISSN: 0004-640X.

Celestial Mechanics: An International Journal of Space Dynamics. Kluwer Academic Publishers, Box 358, Accord Station, Hingham MA 02018-0358. (617) 871-6600. FAX (617) 871-6528. 1969 to date. 12 issues per year. $703.50 per year. ISSN: 0923-2958.

Icarus: International Journal of Solar System Studies. Academic Press, Inc., Journal Division, 525 B Street, Suite 1900, San Diego, CA 92101-4495. (619) 230-1840. FAX (619) 699-6800. 1962 to the present. Monthly. $1080.00. ISSN: 0019-1035.

Mercury. Astronomical Society of the Pacific. 390 Ashton Avenue, San Francisco, CA 94112. (415) 337-1100. FAX: (415) 337-5205. 1972 to present. Bimonthly. $175.00 per year. ISSN: 0047-6773.

Monthly Notices of the Royal Astronomical Society. Blackwell Scientific Publications, Inc., Osney Mead, Oxford OX2 QEL England. TEL 0865-240201. FAX 0865-721205. 1827 to present. Fortnightly. $1899.00 per year. ISSN: 0035- 8711.

Observatory. c/o Dr. D. J. Strickland, Space and Astrophysics Division, Rutherford Appleton Laboaratory, Chlton, Didcot, Oxon, OX11 OQX, England. FAX 0235-445848.1877 to present. Bimonthly. $42.00. ISSN: 0029-7704.

Planetary and Space Science. Elsevier Science Publishing Company, Inc., 660 White Plains Road, Terrytown, NY 10591- 5153. (914) 524-9200. FAX: (914) 333-2444. 1959 topresent. Monthly. $1355.00 per year. ISSN: 0032-0633.

Sky and Telescope. Sky Publishing Corporation, Box 9111, Belmont, MA 02178. (617) 864-7360. FAX (617) 864-6117.1941 to present. Monthly. $27.00 per year. ISSN: 0037-6604.

Vistas in Astronomy. Elsevier Science Publishing Company, Inc., 660 White Plains Road, Terrytown, NY 10591- 5153. (914) 524-9200. FAX: (914) 333-2444. 1958 to date. Quarterly. $415.00 per year. ISSN: 0083-6656.

RESEARCH CENTERS AND INSTITUTES

Arecobo Observatory. Cornell University, Box 995. Arecibo, PR 00613. (809) 878-2612. FAX (809) 878-1861.

Haystack Observatory. Massachusetts Institute of Technology. Westford, MA 01886. (508) 692-4765. FAX (617) 981-0590.

Herzberg Institute of Astrophsyics. National Research Council of Canada, 100 Sussex Drive, Ottawa, ON Canada K1A 0R6. (613) 990-0907. FAX (613) 952-6602.

National Astronomy and Ionosphere Center. Space Sciences Building, Ithaca, NY 14853. (607) 255-3735. FAX (607) 255-8803.

National Radio Astronomy Observatory. Edgemont Road, Charlottesville, VA 22903. (804) 296-0211.

Owens Valley Radio Observatory. California Institute of Technology. Pasadena, CA 91125. (818) 795-6811.

Radio Astronomy Laboratory. University of California, Berkeley, 601 Campbell Hall, Berkeley, CA 94720. (415) 642-5275. FAX (415) 642-3411.

Radio Astronomy Observatory. University of Michigan, 810 Dennison Building, Ann Arbor, MI 48109. (313) 426-8441.

Smithsonian Astrophysical Observatory. 60 Garden Street, Cambridge, MA 02138. (617) 495-7461. FAX (617) 495-7105.

OCEAN CURRENTS

See: OCEANOGRAPHY

OCEAN ENGINEERING

See: MARINE ENGINEERING, OCEANOGRAPHY

OCEAN TRENCHES

See: OCEANOGRAPHY

OCEANOGRAPHY

See also: COASTS, CONTINENTAL MARGIN, GEOLOGY, METEO-ROLOGY, OCEAN ENGINEERING, TSUNAMIS

ABSTRACT SERVICES AND INDEXES

Applied Science and Technology Index: A Cumulative Subject Index to English Language Periodicals in the Fields of Aeronautics and Space Science, Computer Technology, Chemistry, Construction Industry, Energy and Related Areas. H.W. Wilson Co., 950 University Avenue, Bronx, NY 10452. (800) 367-6770 or (212) 588-8400. FAX (718) 590-1617. From 1958 to present. Monthly. Inquire about cost and availability. Also available on CD-ROM and online.

Aquatic Sciences and Fisheries Abstracts, Part 2: Ocean Technology, Policy and Non-living Resources. Cambridge Scientific Abstracts, 7200 Wisconsin Avenue, Bethesda, MD 20814. (301) 961-6750. Fax (301) 961-6720. 1969 to present. Monthly. $685.00.

Bibliography and Index of Geology. American Geological Institute, 4220 King Street, Alexandria, VA 22302. (703) 379-2480. Fax (703) 379-7563. Monthly. $1295.00 per year. Also available online as GEOREF.

Current Contents: Physical, Chemical and Earth Sciences. Institute for Scientific Information, 3501 Market Street, Philadelphia, PA 19104. (215) 386-0100. FAX (215) 386-6362. Weekly. $360.00 per year.

Deep-Sea Research. Part A: *Oceanographic Research Papers.* Part B: *Oceanographic Literature Review.* Pergamon Press, Inc.,

Maxwell House, Fairview Park, Elmsford, NY 10523. (914) 592-7700. Fax (914) 592-3625. Twelve times per year. $1370 per year for Parts A and B.

Index to Scientific Reviews. Institute for Scientific Information, 3501 Market St., Philadelphia, PA 19104. (215) 386-0100. FAX (215) 386-6362. Semi-annual.

Oceanic Abstracts. Cambridge Scientific Abstracts, 7200 Wisconsin Avenue, Bethesda, MD 20814. (301) 961-6750. Fax (301) 961-6720. Bimonthly. $995.00 per year.

Physics Abstracts. Institute of Electrical Engineers, Michael Faraday House, Six Hill Way, Stevenage, Herts, SG1 2AY, England. Distributed by IEEE, 445 Hoes Lane, Piscataway, NJ 08854. (908) 562-5549. 1898 to present. Monthly. $2700.00 per year. Also available online as INSPEC.

Science Citation Index. Institute for Scientific Information, 3501 Market Street, Philadelphia, PA 19104. (215) 386-0100. FAX (215) 386-6362. Inquire about availability and cost. Also available on CD-ROM.

ANNUAL REVIEWS AND YEARBOOKS

Annual Review of Earth and Planetary Sciences. Annual Reviews Inc., 4139 El Camino Way, P.O. Box 10139, Palo Alto, CA 94303-0139. (800) 523-8635 or (415) 493-4400. FAX (415) 855-9815. Annual. $62.00 (1995 edition).

Ocean Yearbook. Elizabeth M. Borgese, et al, editors. University of Chicago Press, Journals Division, 5720 S. Woodlawn Avenue, Chicago, IL 60637. (312) 753-3347. FAX (312) 753-0811. 1995. $77.00.

ASSOCIATIONS AND PROFESSIONAL SOCIETIES

American Geophysical Union. 2000 Florida Avenue NW, Washington, DC 20009-1277. (202) 462-6900.

American Meteorological Society. 45 Beacon Street, Boston, MA 02108-3693. (617) 227-2425. FAX (617) 742-8718.

American Society of Limnology and Oceanography. c/o Virginia Institute of Marine Science, College of William and Mary, Route 1208, Gloucester Point, VA 23062. (804) 642-7242. FAX (804) 357-0422.

Marine Technology Society. 1828 L Street NW, No. 906, Washington, DC 20036-5104. (202) 775-5966. FAX (202) 429-9417.

The Oceanography Society. 4052 Timber Ridge Drive, Virginia Beach, VA 23455-7017. (804) 464-0131. FAX (804) 464-1759.

DIRECTORIES AND BIOGRAPHICAL SOURCES

American Men & Women of Science, 1995-96. R.R. Bowker Staff, eds. 19th edition. 8 volumes. R.R. Bowker/Reed International Publishing Company, 121 Chanlon Road, New Providence, NJ 07974. (908) 464-6800 or (800) 521-8110. 1995. $850.00.

International Research Centers Directory. Gale Research, 835 Penobscot Building, Detroit, MI 48226-4094. (313) 961-2242.

(800) 347-4253. FAX (313) 961-6083. 8th edition. 1995. $410.00

Research Centers Directory. Gale Research, 835 Penobscot Building, Detroit, MI 48226-4094. (313) 961-2242. (800) 347-4253. FAX (313) 961-6083. $485.00.

Scientific and Technical Organizations and Agencies Directory. Gale Research, 835 Penobscot Building, Detroit, MI 48226-4094. (313) 961-2242. (800) 347-4253. FAX (313) 961-6083. 4th edition. 1996. $195.00.

ENCYCLOPEDIAS AND DICTIONARIES

Encyclopedia of Physical Science and Technology. Robert A. Meyers, ed. 18 volumes. Academic Press Inc., 6277 Sea Harbor Drive, Orlando, FL 32887. (800) 321-5068. 1992. $2100.00.

Encyclopedia of Underwater Investigations. Robert Teather. Best Publishing Company, PO Box 30100, Flagstaff, AZ 86003-0100. (800) 468-1055. FAX (520) 526-0370. 1994. $67.50.

McGraw-Hill Dictionary of Scientific and Technical Terms. Sybil P. Parker, ed. 5th edition. McGraw-Hill Publishing Company, 1221 Avenue of the Americas, New York, NY 10020. (800) 262-4729 or (212) 512-3825. 1993. $110.50.

McGraw-Hill Encyclopedia of Science and Technology. Sybil P. Parker, ed. 7th edition. 20 volumes. McGraw-Hill Publishing Company, 1221 Avenue of the Americas, New York, NY 10020. (800) 262-4729 or (212) 512-3825. 1992. $1900.00

Thesaurus of Scientific, Technical, and Engineering Terms. Hemisphere Publishing Corporation, 1900 Frost Road, Suite 101, Bristol, PA 19007-1598. (215) 785-5800 or (800) 821-8312. FAX (215) 785-5515. 1987. $173.00.

GENERAL WORKS

Introduction to Oceanography. David A. Ross. Harpercollins Publishers, Inc., 10 E. 53rd Street, New York, NY 10022-5299. (800) 331-3761. FAX (212) 207-7145. 1995. $41.95.

Oceanography: An Illustrated Text. C.P. Summerhayes & S.A. Thorpe, editors. John Wiley and Sons, Inc., 605 Third Avenue, New York, NY 10158. (800) 526-5368 or (212) 850-6000. 1996. $74.95.

Oceanography: An Introduction. Dale E. Ingmanson, editor. 5th edition. Wadsworth Publishing Company, 10 Davis Drive, Belmont, CA 94002. (415) 595-2350 or FAX (606) 525-0978. 1995. $44.95.

Oceanography: An Introduction to the Marine Science. Chamberlain. Mosby-Year Book Inc., 11830 Westline Industrial Drive, St. Louis, MO 63146. (800) 426-4545. FAX (314) 432-1380. 1994. $48.50.

Oceanography: An Introduction to the Planet Oceanus. Paul Pinet. West Publishing Company, 610 Opperman Drive, PO Box 64526, St. Paul, MN 55164-0526. (800) 328-9424. 1992. $61.00.

Principles of Oceanography. M. Grant Gross. Macmillan Publishing, 200 Old Tappan Road, Old Tappan, NJ 07675. (800) 223-2336. FAX (800) 445-6991. 1994. $22.00.

OCEANOGRAPHY

Ency. of Physical Sciences and Engineering Info. Sources

Regional Oceanography: An Introduction. Matthias Tomczak and J. Stuart Godfrey. Elsevier Science Publishing Company, Inc., 655 Avenue of the Americas, New York, NY 10010. (212) 989-5800. 1994. $120.00 (hardbound) and $37.50 (paperbound).

HANDBOOKS AND MANUALS

CRC Handbook of Chemistry and Physics. David R. Lide, editor. 75th edition. CRC Press, 2000 Corporate Blvd., N.W., Boca Raton, FL 33431. (407) 994-0555 or (800) 272-7737. FAX (407) 994-0949. $99.50.

ONLINE DATABASES AND CD-ROMS

Current Contents Search. Institute for Scientific Information, 3501 Market Street, Philadelphia, PA 19104. (215) 386-0100. FAX (215) 386-6362. Contains citations to articles listed in the table of contents of science and technology journals. Also articles in social sciences and life sciences journals. Available on BRS,(800) 289-4277, DIALOG,(800) 334-2564, online services. Inquire as to cost and availability.

Dissertation Abstracts. University Microfilms International, 300 North Zeeb Road, Ann Arbor, MI 48106. (800) 521-0600 or (313) 761-4700. Scope includes virtually all doctoral dissertations accepted at accredited American institutions from 1861 to present in 252 subject areas. Available on BRS,(800) 289-4277, DIALOG,(800) 334-2564, and OCLC EPIC,(800) 848-5878, online services. Also available on CD-ROM. Inquire as to cost and availability.

NTIS Bibliographic Database. National Technical Information Service, 5285 Port Royal Road, Springfield, VA 22161. (703) 487-4929 or FAX (703) 321-8199. Broad coverage of government-sponsored science and technology research reports, 1964 to present. Available on BRS,(800) 289-4277, DIALOG, (800) 334-2564, ORBIT, (800) 456-7248, and STN International, FIZ Karlsruhe, P.O. Box 2465, W-7500, Karlsruhe 1, Germany, online services. Also available on CD-ROM. Inquire as to cost and availability.

SCISEARCH. Institute for Scientific Information, 3501 Market Street, Philadelphia, PA 19104. (800) 523-1850 or (215) 386-0100. Broad multidisciplinary title and author index to the international literature of science and technology, 1974 to present. Available on DIALOG,(800) 334-2564, and ORBIT (800) 456-7248, online services. Also available on CD-ROM. Inquire as to cost and availability.

PERIODICALS

Bulletin of Marine Science. Rosenstiel School of Marine and Atmospheric Sciences, 4600 Rickenbacker Causeway, Miami, FL 33149. (305) 361-4190. 1950 to present. Bi-monthly. $70.00.

Coastal Management. Taylor & Francis, 1900 Frost Road, Suite 101, Bristol, PA 19007. (215) 785-5800. FAX (215) 785-5515. 1973 to present. Quarterly. $140.00.

Continental Shelf Research. Elsevier Science [journals], 660 White Plains Rd., Tarrytown, NY 10159-5153. (919) 524-9200. FAX (919) 333-2444. 1982 to present. 15 times a year. $760.00.

Deep-Search Research Part 1: Oceanographic Research Papers. Elsevier Science [journals], 660 White Plains Rd., Tarrytown, NY 10159-5153. (919) 524-9200. FAX (919) 333-

2444. 1953 to present. Twelve times a year. $1565.00 per year (includes Part 2).

Deep-Search Research Part 2: Topical Studies in Oceanography. Elsevier Science [journals], 660 White Plains Rd., Tarrytown, NY 10159-5153. (919) 524-9200. FAX (919) 333-2444. 1993 to present. Six times a year. $1565.00 per year (includes Part 1).

Dynamics of Atmospheres and Oceans. Elsevier Science Inc., Box 882, Madison Square Station, New York, NY 10159. (212) 989-5800. FAX (212) 633-3990. 1977 to present. Eight times a year. $461.00.

Estuarine Coastal and Shelf Science. Academic Press Ltd./ Harcourt Brace & Company, Ltd., Foots Cray High Street, Sidcup, Kent DA14 5HP, England. Telephone 44-81-300-3322. FAX 44-81-309-0807. 1973 to present. Monthly. Inquire for cost.

Journal of Atmospheric and Oceanic Technology. American Meteorological Society, 45 Beacon Street, Boston, MA 02108-3693. (617) 227-2425. FAX (617) 742-8718. 1984 to present. Bi-monthly. $135.00.

Journal of Oceanography. Terra Scientific Publishing Company, 302 Jiyugaoka Komatsu Bldg., Midorigaoka 2-24-17, Meguro, Tokyo 152, Japan. Telephone 03-3718-4403. FAX 81-3-3843-5015. 1941 to present. Quarterly. Inquire for price.

Journal of Physical Oceanography. American Meteorological Society, 45 Beacon Street, Boston, MA 02108-3693. (617) 227-2425. FAX (617) 742-8718. 1971 to present. Monthly. $235 per year.

Journal of Waterway, Port, Coastal, and Ocean Engineering. American Society of Civil Engineers, Waterway, Port, Coastal, and Ocean Division, 345 East 47th Street, New York, NY 10017-2398. (212) 705-7520 or (800) 548-2723. 1956 to present. Bi-monthly. $100.00 per year.

Limnology and Oceanography. American Society of Limnology and Oceanography, School of Oceanography, WB-10, University of Washington, Seattle, WA 98195. FAX (206) 543-8655. 1956 to present. Eight times a year. $160.00.

Marine Geology. Elsevier Science Inc., Box 882, Madison Square Station, New York, NY 10159. (212) 989-5800. FAX (212) 633-3990. 1964 to present. 28 times a year. $1385.00.

Marine Technology Society Journal. Marine Technology Society, 1828 L Street NW, No. 906, Washington, DC 20036-5104. (202) 775-5966. FAX (202) 429-9417. Quarterly. $70.00 per year.

Ocean and Coastal Management. Elsevier Science [journals], 660 White Plains Rd., Tarrytown, NY 10159-5153. (919) 524-9200. FAX (919) 333-2444. 1973 to present. Twelve times a year. $565.00.

Ocean Challenge. Challenger Society for Marine Science, Deacon Laboratory, Institute of Oceanographic Sciences, Wormley, Godalming, Surrey GU8 5UB, England. FAX 051-653-6269. 1990 to present. Four times a year. Inquire for price.

Ocean Engineering. Elsevier Science [journals], 660 White Plains Rd., Tarrytown, NY 10159-5153. (919) 524-9200. FAX

(919) 333-2444. 1968 to present. 8 times a year. $595.00 per year.

Oceanus. Wood Hole Oceanographic Institution, Woods Hole, MA 02543. (508) 548-1400. FAX (508) 457-2172. 1952 to present. Semi-annual. Inquire for price.

Progress in Oceanography. Elsevier Science [journals], 660 White Plains Rd., Tarrytown, NY 10159-5153. (919) 524-9200. FAX (919) 333-2444. 1963 to present. Eight times a year. $705.00.

Sea Frontiers. 400 SE Second Avenue, 4th Floor, Knight Centre, Miami, FL 33131. (305) 375-9498. 1954 to present. Bi-monthly. $24.00.

RESEARCH CENTERS AND INSTITUTES

Harbor Branch Oceanographic Institution Inc. 5600 Old Dixie Highway, Ft. Pierce, FL 34946. (407) 465-2400. FAX (407) 468-0757.

Nova University Oceanographic Center. 800 N. Ocean Drive, Dania, FL 33004. (305) 920-1909.

Skidaway Institute of Oceanography. P.O. Box 13687, McWhorter Drive, Skidaway Island, Savannah, GA 31416. (912) 356-2453. FAX (912) 356-2751.

Woods Hole Oceanographic Institution. Woods Hole, MA 02543. (508) 548-1400. FAX (508) 457-2172.

OFFSHORE ENGINEERING

See: MARINE ENGINEERING

OPTICAL COMMUNICATIONS

See also: FIBER OPTICS, HOLOGRAPHY, LASERS, OPTICAL DISKS, OPTICS, OPTOELECTONICS, PHYSICS, SPECTROSCOPY

ABSTRACT SERVICES AND INDEXES

Applied Science and Technology Index. A cumulative subject index to English language periodicals in the fields of aeronautics and space science, computer technology, chemistry, construction industry, energy and related areas. H.W. Wilson Co., 950 University Avenue, Bronx, NY 10452. (212) 588-8400. (800) 367-6770. FAX (718) 590-1617. From 1958 to present. Monthly. Inquire about cost and availability. Also available on CD-ROM and online. ISSN: 0003-6986.

Chemical Abstracts. Chemical Abstracts Service, 2540 Olentangy River Road, Box 3012, Columbus, OH 43210-0012. (614) 447-3600. (800) 848-6538. FAX (614) 447-3713. From 1907 to present. Weekly. $16800.00 per year. Also available on CD-ROM and online. Inquire regarding cost and availability. ISSN:

Current Contents: Engineering, Technology and Applied Sciences. Institute for Scientific Information, 3501 Market Street,

Philadelphia, PA 19104. (215) 386-0100. FAX (215) 386-2291. From 1961 to present. Weekly. $442.00 per year. Also available on CD-ROM and online. Inquire regarding cost and availability. ISSN: 0095-7917.

Current Papers in Physics. Institute of Electrical Engineers, Michael Faraday House, Six Hill Way, Stevenage, Herts, SG1 2AY, England. Distributed by INSPEC/IEEE, Box 1331, 445 Hoes Lane, Piscataway, NJ 08855-1331. (908) 562-5549. 1966 to present. Fortnightly. $410.00 per year. ISSN: 0011-3786.

Engineering Index Monthly. Engineering Information, Inc., Castle Point on the Hudson, Hoboken, NJ 07030. (201) 216-8500. FAX (201) 216-8532. Monthly. $2200.00 per year. Also available online as COMPENDEX and on CD-ROM. ISSN:0742-1974.

General Science Index. H.W. Wilson Company, 950 University Avenue, Bronx, NY 10452. (212) 588-8400. (800) 367-6770. FAX (718) 590-1617. From 1978 to present. Ten issues per year; quarterly and annual cumulations. Service basis. Available on CD-ROM and online. Inquire regarding cost and availability. ISSN: 0162-1963.

Government Reports Announcements and Index. U. S. National Technical Information Service (NTIS), 5285 Port Royal Road, Springfield, VA 22161. (703) 487-4650. FAX (703) 321-8547. From 1968 to present. Annual. $630.00 per Year. Also Available Online as *NTIS Bibliographic Database* and on CD-ROM.

NTIS Alerts: Physics. U.S. National Technical Information Service. 5285 Port Royal Road, Springfield, VA 22161. (703) 487-4650. FAX (703) 321-8547. Weekly. $140.00 per year.

Pascal 10: Mechanique, Acoustique et Transport de Chaleur. Centre National de la Recherche Scientifique, Institue de l'Information Scientifique et Technique, 2 aliee du Parc de Brabois, 54514 Vandoeuvre-Les Nancy Cedex, France. TEC 83-50-46-00. FAX 83-50-46-50. 1984 to present. Ten issues per year. 1530 F per year. Also available on CD-ROM and online. ISSN: 1136-5107.

Physics Abstracts. INSPEC. Section A, Science Abstracts. Institution of Electrical Engineers (IEE). Available from INSPEC/IEEE-Institute of Electrical and Electronic Engineers,, Box 1331, 445 Hoes Lane, Piscaway NJ 08855-1331. (908) 562-5549. 1898 to present. 24 issues per year. $2835.00 per year. Also available on CD-ROM and online. ISSN: 0036-8091.

Physics Briefs (Physikalische Berichte). Information Center for Energy, Physics, Mathematics; German Physical Society. VCH Publishers, Inc., 220 East 23rd Street, New York, NY 10010-4606. (212) 683-8333. 1845 to present. 24 issues per year. $2390.00 per year. Also available online. ISSN: 0179-7434.

Science Citation Index (SCI). Institute for Scientific Information, 3501 Market Street, Philadelphia, PA 19104. (215) 386-0100. (800) 523-1850. FAX (215) 386-2991. 1961 to present. Six issues per year, plus annual cumulation. $11650.00 per year. Also available online and on CD-ROM. Inquire about price and availability. ISSN: 0036-827X.

ANNUAL REVIEWS AND YEARBOOKS

Advances in Electronics and Electron Physics. Academic Press, Inc., 6277 Sea Harbor Drive, Orlando, FL. (800) 321-5068. From 1948 to present. Irregular. ISSN: 0065-2539.

Advances in Solid State Technology. Kluwer Academic Publishers, 101 Philip Drive, Assinippi Park, Norwell, MA 02061. (617) 871-6600. 1985 to present. Irregular. Price varies.

Critical Reviews in Solid State and Materials Sciences. CRC Press, Inc., 2000 Corporate Boulevard, NW, Boca Raton, FL 33431. (407) 994-0555. (800) 272-7737. FAX (407) 998-9784. 1970 to present. Bimonthly. $265.00 per year. ISSN: 1040-8436.

Progress in Optics. Elsevier Science Inc., 660 White Plains Road, Tarrytown, NY 10591-5153. (914) 524-9200. FAX (914) 333-2444. 1961 to present. Irregular. Price varies. ISSN: 0079-6638.

Progress in Quantum Electronics. Elsevier Science Inc., 660 White Plains Road, Tarrytown, NY 10591-5153. (914) 524-9200. FAX (914) 333-2444. 1969 to present. 5 issues per year. $395.00 per year. ISSN: 0079-6727.

Solid State Physics: Advances in Research and Applications. Academic Press, Inc., 6277 Sea Harbor Drive, Orlando, FL. (800) 321-5068. 1955 to present. Irregular. ISSN: 0081-1947.

ASSOCIATIONS AND PROFESSIONAL SOCIETIES

American Institute of Physics. One Physics Ellipse, College Park, MD 20740-3843. (301) 209-3100.

IEEE Lasers and Electro-optics Society. c/o IEEE, 445 Hoes Lane, PO Box 1331, Piscataway, NJ 08855. (908) 562-3892. FAX (908) 562-8434.

Laser Institute of America. 12424 Research Parkway, Suite 130, Orlando, FL 32926. (407) 380-1553. FAX (407) 380-5588.

Optical Society of America. 2010 Massachusetts Avenue NW, Washington, DC 20036. (202) 223-8130. FAX (202) 223-1096.

Society For Imagining Science and Technology. 7003 Kilworth Lane, Springfield, VA 22151. (703) 642-9090. FAX (703) 642-9094.

SPIE: The International Society for Optical Engineering. P.O. Box 10, 1000 20th Street, Bellingham, WA 98227. (206) 676-3290. FAX (206) 647-1445.

BIBLIOGRAPHIES

Science Books and Films. American Association for the Advancement of Science, 1333 H Street N.W., Washington, DC 20005. (202) 326-6454. Reviews of print, film and software materials in all sciences. 1965 to present. Nine issues per year. $40.00 per year. ISSN: 0098-342X.

Scientific and Technical Books and Serials in Print: An Index to Literature in Science and Technology. R.R. Bowker Inc., 121 Chanlon Road, New Providence, NJ 07974. (908) 464-6800. (800) 521-8110. 4 volumes. 1972 to present. Annual. $299.95. Also available on compact disc and online. ISSN: 0000-054X.

DIRECTORIES AND BIOGRAPHICAL SOURCES

American Men and Women of Science: Physical and Biological Sciences. R. R. Bowker Inc., 121 Chanlon Road, New Providence, NJ 07974. (908) 464-6800. (800) 521-8110. 20th edition. 8 volumes. 1996. $850.00.

American Physical Society. Membership Directory Bulletin Issue. American Physical Society, One Physics Ellipse, College Park, MD 20740-3843. (301) 209-3200. FAX (301) 209-0865. Biennial.

International Directory of Engineering Societies and Related Organizations. American Society for Engineering Education, 1818 N Street NW, Suite 600, Washington, DC 20036. (202) 331-3526. 15th edition. 1996. $185.00.

Graduate Programs in Physics, Astronomy and Related Fields. 1995-1996. American Institute of Physics, One Physics Ellipse, College Park, MD 20740-3843. (301) 209-3100. Annual. $45.00. ISSN: 0147-1821.

Optics Education. SPIE-The International Society for Optical Engineering, PO Box 10, 1000 20th Street, Bellingham, WA 98227. (206) 676-3290. FAX (206) 647-1445. Annual. No charge.

Research Centers Directory. Gale Research, 835 Penobscot Building, Detroit, MI 48226-4094. (313) 961-2242. (800) 347-4253. 20th edition. 1995. $485.00. ISSN: 0080-1518.

Scientific and Technical Organizations and Agencies Directory. Gale Research, 835 Penobscot Building, Detroit, MI 48226-4094. (313) 961-2242. (800) 347-4253. 4th edition. 1996. $195.00.

Who's Who in Engineering. Gordon Davis, editor. American Society for Engineering Education, 1111 19th Street, NW, Suite 608, Washington, DC 20036. 9th edition. 1995. $220.00.

Who's Who in Technology. Gale Research, 835 Penobscot Building, Detroit, MI 48226-4094. (313) 961-2242. (800) 347-4253. 7th edition. 1995. $195.00.

ENCYCLOPEDIAS AND DICTIONARIES

Dictionary of Optical Communications Technology. J. P. Rehahn and N. Schafer. VCH Publications, Inc., 220 East 23rd Street, Suite 909, New York, NY 10010-4606. (212) 683-8333. (800) 422-8824. 1988. $65.00.

Dictionary of the Physical Sciences: Terms, Formulas, Data. Cesare Emiliani. Oxford University Press, Inc., 200 Madison Avenue, New York, NY 10016. (212) 725-6000. (800) 334-4249. 1989. $19.95.

Encyclopedia of Applied Physics. George Trigg, editor. VCH Publications, Inc., 220 East 23rd Street, Suite 909, New York, NY 10010-4606. (212) 683-8333. (800) 422-8824. 20 volume set. 1991-. $5990.00.

Encyclopedia of Physics. Robert M. Besancon. Chapman & Hall, 1 Penn Plaza, New York, NY 10119. (212) 564-1060. 3rd edition. 1990. $54.95 in paper.

Encyclopedial of Lasers and Optical Technology. Robert A. Meyers, editor. Academic Press, Inc., 6277 Sea Harbor Drive, Orlando, FL. (800) 321-5068. 1991. $92.00.

Encyclopedia of Physical Science and Technology. Academic Press, Inc., 6277 Sea Harbor Drive, Orlando, FL. (800) 321-5068. 2nd edition. 18 volume set. 1992. $2625.00.

Fiber Optics Standard Dictionary. Martin H. Weik. 2nd edition. Van Nostrand Reinhold, 115 Fifth Avenue, New York, NY 10003. (212) 254-3232. (800) 842-3636. 2nd edition. 1989. $49.95.

Glossary of Optical Terminology. Thomas K. Farrell. Butterworth-Heinemann, 313 Washington Avenue, Newton, MA 02158. (617) 928-2500. (800) 366-2665. 1985. $24.95.

McGraw-Hill Encyclopedia of Science and Technology. McGraw-Hill Book Company, Inc., 1221 Avenue of the Americas, New York, NY 10020. (212) 512-2000. (800) 262-4729. 7th edition. 20 volume set. 1992. $1900.00.

GENERAL WORKS

Applied Optics and Optical Engineering: A Comprehensive Treatise. Rudolf Kingslake, et al, editors. Academic Press, 9 volumes. 1965- . Volumes individually priced.

Coherent Optical Communications Systems. Silvello, Bett, et al. John Wiley & Sons, Inc., 605 Third Avenue, New York, NY 10158-0012. (212) 850-6000. (800) 225-5945. 1995. $74.95.

Fiber Optic Communications. Lynne D. Green. CRC Press, Inc., 2000 Corporate Boulevard, NW, Boca Raton, FL 33431. (407) 994-0555. (800) 272-7737. 1992. $54.95.

Fundamentals of Multiaccess Optical Fiber Networks. Denis Mestdagh. Prentice-Hall, 113 Sylvan Avenue, Route 9W, Englewood Cliffs, NJ 07632. (201) 592-2000. 1994. $77.00.

Fundamentals of Optical Fibers. John A. Buck. John Wiley & Sons, Inc., 605 Third Avenue, New York, NY 10158-0012. (212) 850-6000. (800) 225-5945. 1995. $59.95.

Microwaves and Optical Transmission. A. David Olver. John Wiley & Sons, Inc., 605 Third Avenue, New York, NY 10158-0012. (212) 850-6000. (800) 225-5945. 1992. $98.00.

Optical Communications Systems. John Gowar. Prentice-Hall, 113 Sylvan Avenue, Route 9W, Englewood Cliffs, NJ 07632. (201) 592-2000. (800) 922-0579 1993. $64.00.

Optical Fiber Amplifiers: Design and Sytstem Applications. Anders Bjarklev. Artech House, 685 Canton Street, Norwood, MA 02062. (617) 769-9750. (800) 225-9977. 1993. $88.00.

Optical Fibre Sensor Technology. Ken Grattan and Beverley Meggitt. Prentice-Hall, 113 Sylvan Avenue, Route 9W, Englewood Cliffs, NJ 07632. (201) 592-2000. (800) 922-0579 1993. 1994. $33.33.

Optics and Lasers, Including Fibers and Optical Waveguides. M. Young. Springer-Verlag New York, Inc., 175 Fifth Avenue, New York, NY 10010. (212) 460-1500. (800) 777-4643. FAX (212) 473-6272. 4th revised edition. 1993. $34.50.

Principles of Optical Fiber Communications. Van Etten and Van Der Platts. Prentice-Hall, 113 Sylvan Avenue, Route 9W, Englewood Cliffs, NJ 07632. (201) 592-2000. (800) 922- 0579. 1991. $56.00.

Principles of Optics, Optoelectronics and Photonics. Alan Billings. Prentice-Hall, 113 Sylvan Avenue, Route 9W, Englewood

Cliffs, NJ 07632. (201) 592-2000. (800) 922- 0579. 1994. $40.00 in paper.

Technologies of Light. E. U. Kotte, et al, editors. Springer-Verlag New York, Inc., 175 Fifth Avenue, New York, NY 10010. (212) 460-1500. (800) 777-4643. FAX (212) 473- 6272. 1988. $66.00.

Understanding Telecommunications and Lightwave Systems. John G. Nellist. Institute of Electrical and Electronic Engineers (IEEE), 345 East 47th Street, New York, NY 10017. (212) 705-7900. 2nd edition. 1995.

HANDBOOKS AND MANUALS

CRC Handbook of Chemistry and Physics. David R. Kide, editor. CRC Press, Inc., 2000 Corporate Boulevard, NW, Boca Raton, FL 33431. (407) 994-0555. (800) 272-7737. 77th edition. 1996. $99.95.

Design Handbook for Optical Fiber Systems. R. P. DePaula and E. Udd. Information Gatekeepers, 214 Harvard Avenue, Boston, MA 02134. (617) 232-3111. (800) 323-1088. $75.00.

Fiber Optic Communication Design Handbook. Robert H. Hoss. Prentice-Hall, 113 Sylvan Avenue, Route 9W, Englewood Cliffs, NJ 07632. (201) 592-2000. (800) 922-0579. 1990. $66.00.

Fiber Optic Communications Handbook. TAB Books, P.O. Box 40, Blue Summit, PA 17294-0850. (717) 794-2191. (800) 233-1128. 2nd edition. 1990. $89.50.

Fiber Optic Metrology and Standards. O. D. Soares. SPIE-International Society for Optical Engineering, 1000 20th Street, Bellingham, WA 98227. (206) 676-3290. FAX (206) 647-1445. 1991. $68.00.

Handbook of Optics. Optical Society of America. McGraw- Hill Publishing Company, Inc., 1221 Avenue of the Americas, New York, NY 10020. (212) 512-2000. (800) 262- 4729. 2 volumes. 2nd edition. 1993. $99.50 each.

Handbook of Physical Quantities. Igor S. Grigoriev and Evgenil Z. Melikhov, editors. CRC Press, Inc., 2000 Corporate Boulevard, NW, Boca Raton, FL 33431. (407) 994-0555. (800) 272-7737. 1995. $99.00.

Handbook of Tables of Functions for Applied Optics. Leo Levi, editor. Franklin Book Co., Inc., Elkins Park, PA 19117. (215) 635-5252. CRC Press Reprints. 1974. $152.95.

McGraw-Hill Encyclopedia of Physics. Sybil P. Parker, editor. McGraw-Hill Publishing Company, Inc., 1221 Avenue of the Americas, New York, NY 10020. (212) 512-2000. (800) 262-4729. 2nd edition. 1993. $95.50

Newnes Engineering and Physical Science Pocketbook. J. O. Bird and P. J. Chivers. Butterworth-Heinmann, 313 Washington Avenue, Newton, MA 02158. (617) 928-2500. (800) 366-2665. 1993. $49.95.

A Physicist's Desk Reference. Herbert L. Anderson, editor. American Institute of Physics, One Physics Ellipse, College Park, MD 20740-3843. (301) 209-3100. 2nd edition. 1989. $70.00.

Optics Problem Solver. Research and Education Association Staff. Research & Education Association, 61 Ethel Road, West

Piscataway, NJ 08854. (908) 819-8880. Revised edition. 1994. $29.95 in paper.

ONLINE DATABASES AND CD-ROMS

CA Search. Chemical Abstracts Service, P.O. Box 3012, Columbus, OH 43210-0012. (614) 447-3600. (800) 848-6533. FAX (614) 447-3709. Very comprehensive guide to worldwide chemical literature and related fields, 1972 to present. Available on BRS,(800) 289-4277, DIALOG, (800) 334-2564, ORBIT (800) 456-7248, and STN International, FIZ Karlsruhe, P.O. Box 2465, W-7500, Karlsruhe 1, Germany, online services. Inquire as to cost and availability.

Compendex Plus. Engineering Information, Inc., 345 East 47th Street, New York, NY 10017. (212) 705-7600 or (800) 221-1044. Contains citations with abstracts to worldwide literature in engineering and technology, from 1970 to present. Available on online BRS,(800) 289-4277, DIALOG, (800) 334-2564, ORBIT (800) 456-7248, and STN International, FIZ Karlsruhe, P.O. Box 2465, W-7500, Karlsruhe 1, Germany, online services. Also available on CD-ROM. Inquire as to cost and availability.

Current Contents Search. Institute for Scientific Information, 3501 Market Street, Philadelphia, PA 19104. (215) 386-0100. FAX (215) 386-6362. Contains citations to articles listed in the table of contents of science and technology journals. Also articles in social sciences and life sciences journals. Available on BRS, (800) 289-4277, DIALOG, (800) 334-2564, online services. Inquire as to cost and availability.

Dissertation Abstracts Online. University Microfilms International, 300 North Zeeb Road, Ann Arbor, MI 48106. (800) 521-0600 or (313) 761-4700. Scope includes virtually all doctoral dissertations accepted at accredited American institutions from 1861 to present in 252 subject areas. Available on BRS, (800) 289-4277, DIALOG, (800) 334-2564,and OCLC EPIC, (800) 848-5878, online services. Also available on CD-ROM. Inquire as to cost and availability.

INSPEC. Institution of Electrical Engineers, Michael Faraday House, Six Hills Way, Stevenage, Herts. SG1 2AY, England. Telephone: 0438 313311 or FAX 0438 742840. Contains citations to the worldwide literature of physics, electronics and electrical engineering, computer technology, and related fields. Available on BRS, (800) 289-4277, DIALOG, (800) 334-2564, ORBIT, (800) 456-7248, and STN International, FIZ Karlsruhe, P.O. Box 2465, W-7500, Karlsruhe 1, Germany, online services. Inquire as to cost and availability.

NTIS Bibliographic Database. National Technical Information Service, 5285 Port Royal Road, Springfield, VA 22161. (703) 487-4929 or FAX (703) 321-8199. Broad coverage of government-sponsored science and technology research reports, 1964 to present. Available on BRS,(800) 289-4277, DIALOG, (800) 334-2564, ORBIT, (800) 456-7248, and STN International, FIZ Karlsruhe, P.O. Box 2465, W-7500, Karlsruhe 1, Germany, online services. Also available on CD-ROM. Inquire as to cost and availability.

Physics Briefs. American Institute of Physics, 335 East 45th Street, New York, NY 10017. (212) 661-9260 or FAX (212) 661-2036. Contains citations with abstracts of the literature of physics and related fields, 1979 to present. Available on the STN International, FIZ Karlsruhe, P.O. Box 2465, W-7500, Karlsruhe 1, Germany, online service. Inquire as to cost and availability.

Scientific and Technical Books and Serials in Print. R.R. Bowker Inc., 121 Chanlon Road, New Providence, NJ 07974. (800) 521-8110 or (908) 464-6800. List of currently published books and serials in the physical and biological sciences, engineering and technology, with subject, author and titles indexes. Available on ORBIT, (800) 456-7248, online service. Inquire as to cost and availability.

SCISEARCH. Institute for Scientific Information, 3501 Market Street, Philadelphia, PA 19104. (800) 523-1850 or (215) 386-0100. Broad multidisciplinary title and author index to the international literature of science and technology, 1974 to present. Available on DIALOG, (800) 334-2564, and ORBIT, (800) 456-7248, online services. Also available on CD-ROM. Inquire as to cost and availability.

WILSONLINE. H.W. Wilson Company, 950 University Avenue, Bronx, NY 10452. (800) 367-6770 or (212) 588-8400. Makes available online versions of the printed H.W. Wilson indexes including *Applied Science and Technology Index, Business Periodicals Index, General Science Index,* and *Readers' Guide to Periodical Literature.* Period covered is generally 1983 to present. Available on BRS, (800) 289-4277, DIALOG, (800) 334-2564, and OCLC EPIC, (800) 848-5878, online services.Also available on CD-ROM. Inquire as to cost and availability.

PERIODICALS

Applied Optics. Optical Society of America. 2010 Massachusetts Avenue NW, Washington, DC 20036-1023. (202)223-8130. 1962 to present. 36 issues per year. $1090.00 per year. ISSN: 0003-6935.

Applied Spectroscopy. Society for Applied Spectroscopy, 198 Thomas Johnson Drive, Suite 2, Frederick, MD 21702. (301) 694-8122. FAX (301) 694-6860. 1946 to present. Monthly. $180.00 per year. ISSN: 0003-7028.

Fiber and Integrated Optics. Taylor and Francis, 1900 Frost Road, Suite 101. Bristol, PA 19007. (215) 785-5800. FAX (215) 785-5515. 1977 to present. Quarterly. $180.00 per year. ISSN: 0146-8030.

International Journal of Nonlinear Optical Physics. World Scientific Publishing Co., 1060 Main Street, Suite 18, River Edge, NJ 07661. (800) 227-7562. 1992 to present. Quarterly. $348.00 per year. ISSN: 0218-1991.

Journal of Lightwave Technology. Institute of Electrical and Electronics Engineers, 445 Hoes Lane, PO Box 1331, Piscataway, NJ 08855. (908) 562-3892. FAX (908) 562-8434. 1983 to present. Monthly. $425.00 per year. ISSN: 0733- 8724.

IEEE Journal of Quantum Electronics. IEEE Lasers and Electro-optics Society/IEEE, 445 Hoes Lane, PO Box 1331, Piscataway, NJ 08855. (908) 981-0060. FAX (908) 981-9667. 1965 to present. Monthly. $535.00 per year. ISSN: 0018-9197.

Journal of the Optical Society of America. Part A: Optics and Image Science. Optical Society of America, Inc., 2010 Massachusetts Avenue, NW, Washington, DC 20036-1023. (202)223-8130. 1917 to present. Monthly. $610.00 per year. ISSN: 0740-3232.

Journal of the Optical Society of America. Part B: Optical Physics. Optical Society of America, Inc., 2010 Massachusetts Avenue, NW, Washington, DC 20036-1023. (202) 223-8130. 1917 to present. Monthly. $610.00 per year. ISSN: 0740-3224.

Laser Focus World. PennWell Publishing Co., 10 Tara Boulevard, Nashua, NH 03062-2891. (603) 891-0123. FAX (603) 891-0574. 1964 to present. Monthly. $104.00 per year. ISSN: 1043-8092.

Optical Engineering. SPIE-International Society for Optical Engineering, Box 10, 1000 20th Street, Bellingham, WA 98227-0010. (206) 676-3290. FAX (206) 647-1445. 1962 to present. Monthly. $170.00 per year. ISSN: 0091-3286.

Optics and Photonics News. Optical Society of America, Inc., 2010 Massachusetts Avenue, NW, Washington, DC 20036-1023. (202) 223-8130. 1975 to present. Monthly. $99.00 per year. ISSN: 1047-6938.

Optics Communications. Elsevier Science Inc., Box 882, Madison Square Station, New York, NY 10159. (212) 989-5800. FAX (212) 633-4990. 1969 to present. 54 issues per year. $2121.00. per year. ISSN: 0030-4018.

Optics Letters. Optical Society of America, Inc., 2010 Massachusetts Avenue, NW, Washington, DC 20036-1023. (202) 223-8130. 1977 to present. Semi-Monthly. $625.00 per year. ISSN: 0146-9592.

Optics Materials. Elsevier Science Inc., Box 882, Madison Square Station, New York, NY 10159. (212) 989-5800. FAX (212) 633-4990. 1991 to present. Quarterly. $206.00 per year. ISSN: 0925-3467.

Spie-International Society for Optical Engineering. Proceedings. SPIE-International Society for Optical Engineering, Box 10, 1000 20th Street, Bellingham, WA 98227-0010. (206) 676-3290. FAX (206) 647-1445. 1963 to present. Approximately 50 numbers per year. Individually priced.

RESEARCH CENTERS AND INSTITUTES

Institute of Optics. University of Rochester, Wilmot Building, Rochester, NY 14627. (716) 275-5248.

Center for Advanced Technology in Telecommunications. Polytechnic University, 333 Jay Street, Brooklyn, NY 11201. (718) 260-3050.

Center for Applied Optics. University of Alabama in Huntsville, Research Institute Building, Huntsville, AL 25899. (205) 895-6102.

Center for Applied Optics. University of Texas at Dallas, P.O. Box 830688, Richardson, TX 75803-0688. (214) 690-2868. FAX (214) 690-2848.

Center for Electronic and Electro-optic Materials. State University of New York at Buffalo, 217 C Bonner Hall, Buffalo, NY 14260. (716) 636-3109. FAX (716) 636-2528.

Center for Imaging Science. Rochester Institute of Technology, One Lomb Memorial Drive, Rochester, NY 14623. (716) 475-2774.

Center for Research in Electro-optics and Lasers. University of Central Florida, 12424 Research Parkway, Orlando, FL 32826. (407) 658-6800. FAX (407) 658-6880.

Institute of Optics. University of Rochester, Wilmot Building, Rochester, NY 14627. (716) 275-5248.

OPTICAL DISKS

See also: LASERS, OPTICAL DISKS, OPTICS, TELEVISION, VIDEO TECHNOLOGY

ABSTRACT SERVICES AND INDEXES

Applied Science and Technology Index. A cumulative subject index to English language periodicals in the fields of aeronautics and space science, computer technology, chemistry, construction industry, energy and related areas. H.W. Wilson Co., 950 University Avenue, Bronx, NY 10452. (212) 588-8400. (800) 367-6770. FAX (718) 590-1617. From 1958 to present. Monthly. Inquire about cost and availability. Also available on CD-ROM and online. ISSN: 0003-6986.

Current Contents: Engineering, Technology and Applied Sciences. Institute for Scientific Information, 3501 Market Street, Philadelphia, PA 19104. (215) 386-0100. FAX (215) 386-2291. From 1961 to present. Weekly. $442.00 per year. Also available on CD-ROM and online. Inquire regarding cost and availability. ISSN: 0095-7917.

Electrical and Electronics Abstracts. Science Abstracts Section B. Institute of Electrical Engineers, Michael Faraday House, Six Hill Way, Stevenage, Herts, SG1 2AY, England. Distributed by IEEE, 445 Hoes Lane, Piscataway, NJ 08854. (908) 562-5549. 1898 to present. Monthly. $2200.00 per year. Also available on CD-ROM and online as INSPEC. ISSN: 0036-8105.

Engineering Index Monthly. Engineering Information, Inc., Castle Point on the Hudson, Hoboken, NJ 07030. (201) 216-8500. FAX (201) 216-8532. Monthly. $2200.00 per year. Also available online as COMPENDEX and on CD-ROM. ISSN: 0742-1974.

General Science Index. H.W. Wilson Company, 950 University Avenue, Bronx, NY 10452. (212) 588-8400. (800) 367-6770. FAX (718) 590-1617. From 1978 to present. Ten issues per year; quarterly and annual cumulations. Service basis. Available on CD-ROM and online. Inquire regarding cost and availability. ISSN: 0162-1963.

NTIS Alerts: Electrotechnology. U. S. National Technical Information Service. 5285 Port Royal Road, Springfield, VA 22161. (703) 487-4650. FAX (703) 321-8547. Weekly. $140.00 per year.

Physics Abstracts. INSPEC. Section A, Science Abstracts. Institution of Electrical Engineers (IEE). Available from INSPEC/IEEE-Institute of Electrical and Electronic Engineers, Box 1331, 445 Hoes Lane, Piscaway NJ 08855-1331. (908) 562-5549. 1898 to present. 24 issues per year. $2835.00 per year. Also available on CD-ROM and online. ISSN: 0036-8091.

Physics Briefs (Physikalische Berichte). Information Center for Energy, Physics, Mathematics; German Physical Society. VCH Publishers, Inc., 220 East 23rd Street, New York, NY 10010-

OPTICAL DISKS

Ency. of Physical Sciences and Engineering Info. Sources

4606. (212) 683-8333. 1845 to present. 24 issues per year. $2390.00 per year. Also available online. ISSN: 0179-7434.

Science Citation Index (SCI). Institute for Scientific Information, 3501 Market Street, Philadelphia, PA 19104. (215) 386-0100. (800) 523-1850. FAX (215) 386-2991. 1961 to present. Six issues per year, plus annual cumulation. $11650.00 per year. Also available online and on CD-ROM. Inquire about price and availability. ISSN: 0036-827X.

ANNUAL REVIEWS AND YEARBOOKS

Progress in Optics. Elsevier Science Inc., 660 White Plains Road, Tarrytown, NY 10591-5153. (914) 524-9200. FAX (914) 333-2444. 1961 to present. Irregular. Price varies. ISSN: 0079-6638.

Springer Series in Optical Sciences. Springer-Verlag New York, Inc., 175 Fifth Avenue, New York, NY 10010. (212) 460-1500. (800) 777-4643. FAX (212) 473-6272. 1976 to present. Irregular. Price varies.

ASSOCIATIONS AND PROFESSIONAL SOCIETIES

American Electronics Association. 5201 Great America Way, Suite 520, P.O. 52990, Santa Clara, CA 95056. (408) 987-4200. FAX (408) 970-8565.

American Institute of Physics. One Physics Ellipse, College Park, MD 20740-3843. (301) 209-3100.

Electronics Industries Association. 2500 Wilson Boulevard, Arlington, VA 22201. (703) 907-7500. FAX (202) 457-4985.

IEEE Lasers and Electro-optics Society. c/o IEEE, 445 Hoes Lane, PO Box 1331, Piscataway, NJ 08855. (908) 562-3892. FAX (908) 562-8434.

Institute of Electrical and Electronics Engineers. 345 East 47th Street, New York, NY 10017. (212) 705-7900. FAX (212) 705-4929.

Laser Institute of America. 12424 Research Parkway, Suite 130, Orlando, FL 32926. (407) 380-1553. FAX (407) 380-5588.

Optical Society of America. 2010 Massachusetts Avenue NW, Washington, DC 20036. (202) 223-8130. FAX (202) 223-1096.

Society for Imaging Science and Technology. 7003 Kilworth Lane, Springfield, VA 22151. (703) 642-9090. FAX (703)642-9094.

Spie-The International Society for Optical Engineering. P.O. Box 10, 1000 20th Street, Bellingham, WA 98227. (206) 676-3290. FAX (206) 647-1445.

BIBLIOGRAPHIES

Science Books and Films. American Association for the Advancement of Science, 1333 H Street NW, Washington, DC 20005. (202) 326-6454. Reviews of print, film and software materials in all sciences. 1965 to present. Nine issues per year. $40.00 per year. ISSN: 0098-342X.

Scientific and Technical Books and Serials in Print: An Index to Literature in Science and Technology. R.R. Bowker Inc., 121 Chanlon Road, New Providence, NJ 07974. (908) 464-6800. (800) 521-8110. 4 volumes. 1972 to present. Annual. $299.95. Also available on compact disc and online. ISSN: 0000-054X.

DIRECTORIES AND BIOGRAPHICAL SOURCES

American Men and Women of Science: Physical and Biological Sciences. R. R. Bowker Inc., 121 Chanlon Road, New Providence, NJ 07974. (908) 464-6800. (800) 521-8110. 20th edition. 8 volumes. 1996. $850.00.

CD-ROMs in Print; 1994. An International Guide to CD-ROM, CD-1, CDTV and Electronic Book Products. Mecklermedia, 20 Ketchum Street, Westport, CT 0680. (203) 226-6967. 1994. $95.00 in paper.

IEEE Membership Directory. Institute of Electrical and Electronics Engineers. IEEE Service Center, 445 Hoes Lane, Piscataway, NY 08854. (908) 981-1393. (800) 678-IEEE. FAX (908) 981-9667. 2 volumes. Annual. $190.00.

International Directory of Engineering Societies and Related Organizations. American Society for Engineering Education, 1818 N Street NW, Suite 600, Washington, DC 20036. (202) 331-3526. 15th edition. 1996. $185.00.

Laser Focus World Buyers Guide. PennWell Publishing Co., 1421 Sheridan Road, Tulsa, OK 74112. (918) 831-9421. Annual. $75.00.

Laser Pioneers. Jeff Hecht. Academic Press, Inc., 6277 Sea Harbor Drive, Orlando, FL. (800) 321-5068. 2nd edition. 1991. $49.95.

Optical Society of America. Membership Directory. Optical Society of America. 2010 Massachusetts Avenue NW, Washington, DC 20036-1023. (202) 223-8130. (800) 762-6960. Biennial. $12.00. $95.00.

Optics Education. SPIE-The International Society for Optical Engineering, PO Box 10, 1000 20th Street, Bellingham, WA 98227. (206) 676-3290. FAX (206) 647-1445. Annual. No charge.

Research Centers Directory. Gale Research, 835 Penobscot Building, Detroit, MI 48226-4094. (313) 961-2242. (800) 347-4253. 20th edition. 1995. $485.00. ISSN: 0080-1518.

Scientific and Technical Organizations and Agencies Directory. Gale Research, 835 Penobscot Building, Detroit, MI 48226-4094. (313) 961-2242. (800) 347-4253. 4th edition. 1996. $195.00.

Who's Who in Engineering. Gordon Davis, editor. American Society for Engineering Education, 1111 19th Street, NW, Suite 608, Washington, DC 20036. (202) 296-2237. (800) 658-8897. 9th edition. 1995. $220.00.

Who's Who in Technology. Gale Research, 835 Penobscot Building, Detroit, MI 48226-4094. (313) 961-2242. (800)347-4253. 7th edition. 1995. $195.00.

Ency. of Physical Sciences and Engineering Info. Sources

OPTICAL DISKS

ENCYCLOPEDIAS AND DICTIONARIES

Encyclopedia of Lasers and Optical Technology. Robert A. Meyers, editor. Academic Press, Inc., 6277 Sea Harbor Drive, Orlando, FL. (800) 321-5068. 1991. $92.00.

Glossary of Optical Terminology. Thomas K. Farrell. Butterworth-Heinemann, 313 Washington Avenue, Newton, MA 02158. (617) 928-2500. (800) 366-2665. 1985. $24.95.

IEEE Standard Dictionary of Electrical and Electronics Terms. Christopher J. Booth, editor. IEEE Service Center, 445 Hoes Lane, Piscataway, NJ 08855-1331. (908) 981-1393. (800) 678-IEEE. FAX (908) 981-9667. IEEE Standard 100-1992. 5th edition. 1993. $150.00.

McGraw-Hill Encyclopedia of Science and Technology. McGraw-Hill Book Company, Inc., 1221 Avenue of the Americas, New York, NY 10020. (212) 512-2000. (800) 262-4729. 7th edition. 20 volume set. 1992. $1900.00.

Videodisc and Related Technologies: A Glossary of Terms. R. L. Miller and John H. Sayers. Future Systems, 5929 Lee Highway, Arlington, VA 22207. (703) 241-1799. 1986. $7.95.

GENERAL WORKS

CD-ROM Mastering For Information and Image Managment. Tom Thiel, et al. Association for Information and Image Management, 1100 Wayne Avenue, Silver Spring, MD 20910. (301) 587-8202. 1992. $33.00 in paper.

Compact Discs and Computers: Converging Technologies. J. Megarry. Routledge, Chapman & Hall, Inc., 29 West 35th Street, New York, NY 10001-3291. (212) 244-3336. 1994. $23.00 in paper.

Digital Audio and Compact Disc Technology. Luc Baert, et al. Butterworth-Heinemann, 313 Washington Street, Newton MA 02158. (617) 928-2500. (800) 366-2665. 3rd edition. 1995. $37.95 in paper.

Digital Image Processing. W. K. Pratt. John Wiley & Sons, Inc., 605 Third Avenue, New York, NY 10158-0012. (212) 850-6000. (800) 225-5945. 2nd edition. 1991. $98.00.

Fundamentals of Laser Optics. K. Iga. Plenum Publishing Corp., 233 Spring Street, New York, NY. (212) 620-8000. (800) 2221-9369. FAX (212) 463-0742. 1994. $49.50.

Lasers: Invention to Application. National Academy of Engineering Staff. National Academy Press, 2101 Constitution Avenue NW, Washington, DC 20055. (202) 334-3313. (800) 624-6242. 1987. $14.95.

Optical Discs. Jonathan L. Mayo. TAB Books, P.O. Box 40, Blue Summit, PA 17294-0850. (717) 794-2191. (800) 233-1128. 1990. $24.95.

Optical Disk Technology: A Source Guide. Gordon Press Publications, P.O. Box 459, Bowling Green Station, New York, NY 10004. (718) 624-8419. 1991. $79.00.

Troubleshooting and Repairing Compact Disc Players. Homer L. Davidson. TAB Books, P.O. Box 40, Blue Summit, PA 17294-0850. (717) 794-2191. (800) 233-1128. 1989. $26.95.

Videodisc and Optical Memory Technologies. Jordan Isailovic. Prentice-Hall, 113 Sylvan Avenue, Route 9W, Englewood Cliffs, NJ 07632. (201) 592-2000. (800) 922-0579. 1985. $68.00.

HANDBOOKS AND MANUALS

Handbook of Nonlinear Optical Crystals. V. G. Dmitriev, et al. Springer-Verlag New York, Inc., 175 Fifth Avenue, New York, NY 10010. (212) 460-1500. (800) 777-4643. FAX (212) 473-6272. Revised edition. 1991. $69.00.

Handbook of Optical Memory Systems. C. Peter Waegmann. Optical Disk Institute, P. O. Box 289, Newtonville, MA 02160. (617) 964-3923. 1990. $195.00.

Handbook of Optics. Optical Society of America. McGraw-Hill Publishing Company, Inc., 1221 Avenue of the Americas, New York, NY 10020. (212) 512-2000. (800) 262-4729. 2 volumes. 2nd edition. 1993. $99.50 each.

Industrial Laser Handbook. M. Levitt and D. Belforte, editors. Springer-Verlag New York, Inc., 175 Fifth Avenue, New York, NY 10010. (212) 460-1500. (800) 777-4643. FAX (212) 473-6272. 1992. $163.00.

Laser Experimenter's Handbook. Delton T. Horn. TAB Books, P.O. Box 40, Blue Summit, PA 17294-0850. (717) 794-2191. (800) 233-1128. 2nd edition. 1988. $21.95.

Laser Handbook. F. T. Areechi, E. D. Schultz-Dubois, et al, editors. Elsevier Science Inc., 660 White Plains Road, Tarrytown, NY 10591-5153. (914) 524-9200. FAX (914) 333-2444. 6 volumes. 1973-present.

ONLINE DATABASES AND CD-ROMS

CA Search. Chemical Abstracts Service, P.O. Box 3012, Columbus, OH 43210-0012. (614) 447-3600. (800) 848-6533. FAX (614) 447-3709. Very comprehensive guide to worldwide chemical literature and related fields, 1972 to present. Available on BRS, (800) 289-4277, DIALOG, (800) 334-2564, ORBIT (800) 456-7248, and STN International, FIZ Karlsruhe, P.O. Box 2465, W-7500, Karlsruhe 1, Germany, online services. Inquire as to cost and availability.

Compendex Plus. Engineering Information, Inc., 345 East 47th Street, New York, NY 10017. (212) 705-7600 or (800) 221-1044. Contains citations with abstracts to worldwide literature in engineering and technology, from 1970 to present. Available on online BRS, (800) 289-4277, DIALOG, (800) 334-2564, ORBIT (800) 456-7248, and STN International, FIZ Karlsruhe, P.O. Box 2465, W-7500, Karlsruhe 1, Germany, online services. Also available on CD-ROM. Inquire as to cost and availability.

Current Contents Search. Institute for Scientific Information, 3501 Market Street, Philadelphia, PA 19104. (215) 386-0100. FAX (215) 386-6362. Contains citations to articles listed in the table of contents of science and technology journals. Also articles in social sciences and life sciences journals. Available on BRS, (800) 289-4277, DIALOG, (800) 334-2564, online services. Inquire as to cost and availability.

Dissertation Abstracts Online. University Microfilms International, 300 North Zeeb Road, Ann Arbor, MI 48106. (800) 521-0600 or (313) 761-4700. Scope includes virtually all doctoral dissertations accepted at accredited American institutions from

1861 to present in 252 subject areas. Available on BRS, (800) 289-4277, DIALOG, (800) 334-2564, and OCLC EPIC, (800) 848-5878, online services. Also available on CD-ROM. Inquire as to cost and availability.

INSPEC. Institution of Electrical Engineers, Michael Faraday House, Six Hills Way, Stevenage, Herts. SG1 2AY, England. Telephone: 0438 313311 or FAX 0438 742840. Contains citations to the worldwide literature of physics, electronics and electrical engineering, computer technology, and related fields. Available on BRS, (800) 289-4277, DIALOG, (800) 334-2564, ORBIT, (800) 456-7248, and STN International, FIZ Karlsruhe, P.O. Box 2465, W-7500, Karlsruhe 1, Germany, online services. Inquire as to cost and availability.

NTIS Bibliographic Database. National Technical Information Service, 5285 Port Royal Road, Springfield, VA 22161. (703) 487-4929 or FAX (703) 321-8199. Broad coverage of government-sponsored science and technology research reports, 1964 to present. Available on BRS, (800) 289-4277, DIALOG, (800) 334-2564, ORBIT, (800) 456-7248, and STN International, FIZ Karlsruhe, P.O. Box 2465, W-7500, Karlsruhe 1, Germany, online services. Also available on CD-ROM. Inquire as to cost and availability.

Physics Briefs. American Institute of Physics, 335 East 45th Street, New York, NY 10017. (212) 661-9260 or FAX (212) 661-2036. Contains citations with abstracts of the literature of physics and related fields, 1979 to present. Available on the STN International, FIZ Karlsruhe, P.O. Box 2465, W-7500, Karlsruhe 1, Germany, online service. Inquire as to cost and availability.

Scientific and Technical Books and Serials in Print. R.R. Bowker Inc., 121 Chanlon Road, New Providence, NJ 07974. (800) 521-8110 or (908) 464-6800. List of currently published books and serials in the physical and biological sciences, engineering and technology, with subject, author and titles indexes. Available on ORBIT, (800) 456-7248, online service. Inquire as to cost and availability.

SCISEARCH. Institute for Scientific Information, 3501 Market Street, Philadelphia, PA 19104. (800) 523-1850 or (215) 386-0100. Broad multidisciplinary title and author index to the international literature of science and technology, 1974 to present. Available on DIALOG, (800) 334-2564, and ORBIT, (800) 456-7248, online services. Also available on CD-ROM. Inquire as to cost and availability.

WILSONLINE. H.W. Wilson Company, 950 University Avenue, Bronx, NY 10452. (800) 367-6770 or (212) 588-8400. Makes available online versions of the printed H.W. Wilson indexes including *Applied Science and Technology Index, Business Periodicals Index, General Science Index,* and *Readers' Guide to Periodical Literature.* Period covered is generally 1983 to present. Available on BRS, (800) 289-4277, DIALOG, (800) 334-2564, and OCLC EPIC, (800) 848-5878, online services. Also available on CD-ROM. Inquire as to cost and availability.

PERIODICALS

Applied Optics. Optical Society of America. 2010 Massachusetts Avenue NW, Washington, DC 20036-1023. (202)223-8130. 1962 to present. 36 issues per year. $1090.00 per year. ISSN: 0003-6935.

Applied Physics B: Photophysics and Laser Chemistry. Springer-Verlag New York, Inc., 44 Hartz Way, Seacucus, NJ 07096-

2491. (201) 348-4033. FAX (201) 348-4505. 1981 to present. Monthly. $1030.00 per year. ISSN: 0721-7269.

Electronic Design. Penton Publishing, San Jose Gateway, Suite 354. 2025 Gateway Place, San Jose, CA 95110. (408) 441-0550. 1952 to present. Fortnightly. $95.00. ISSN: 0013-4872.

Electronics. Penton Publishing, San Jose Gateway, Suite 354. 2025 Gateway Place, San Jose, CA 95110. (408) 441-0550. 1930 to present. Semi-weekly. $98.00 per year. ISSN: 0883-4989.

Fiber and Integrated Optics. Taylor and Francis, 1900 Frost Road, Suite 101. Bristol, PA 19007. (215) 785-5800. FAX (215) 785-5515. 1977 to present. Quarterly. $180.00 per year. ISSN: 0146-8030.

IEEE Journal of Quantum Electronics. IEEE Lasers and Electro-optics Society/IEEE, 445 Hoes Lane, PO Box 1331, Piscataway, NJ 08855. (908) 981-0060. FAX (908) 981-9667. 1965 to present. Monthly. $535.00 per year. ISSN: 0018-9197.

International Journal of Nonlinear Optical Physics. World Scientific Publishing Co., 1060 Main Street, Suite 18, River Edge, NJ 07661. (800) 227-7562. 1992 to present. Quarterly. $348.00 per year. ISSN: 0218-1991.

Journal of Applied Physics. American Institute of Physics. One Physics Ellipse, College Park, MD 20740-3843. (301) 209-3100. 1931 to present. Semi-monthly. $1655.00 per year. ISSN: 0021-8979.

Journal of Laser Applications. Laser Institute of America, 12424 Research Parkway, Suite 130, Orlando, FL 32926. (407) 380-1553. FAX (407) 380-5588.

Journal of Lightwave Technology. Institute of Electrical and Electronics Engineers, 445 Hoes Lane, PO Box 1331, Piscataway, NJ 08855. (908) 562-3892. FAX (908) 562-8434. 1983 to present. Monthly. $425.00 per year. ISSN: 0733-8724.

Journal of the Optical Society of America. Part A: Optics and Image Science. Optical Society of America, Inc., 2010 Massachusetts Avenue, NW, Washington, DC 20036-1023. (202) 223-8130. 1917 to present. Monthly. $610.00 per year. ISSN: 0740-3232.

Journal of the Optical Society of America. Part B: Optical Physics. Optical Society of America, Inc., 2010 Massachusetts Avenue, NW, Washington, DC 20036-1023. (202) 223-8130. 1917 to present. Monthly. $610.00 per year. ISSN: 0740-3224.

Laser Focus World. PennWell Publishing Co., 10 Tara Boulevard, Nashua, NH 03062-2891. (603) 891-0123. FAX (603) 891-0574. 1964 to present. Monthly. $104.00 per year. ISSN: 1043-8092.

Optical Engineering. SPIE-International Society for Optical Engineering, Box 10, 1000 20th Street, Bellingham,

WA 98227-0010. (206) 676-3290. FAX (206) 647-1445. 1962 to present. Monthly. $170.00 per year. ISSN: 0091-3286.

Optics and Photonics News. Optical Society of America, Inc., 2010 Massachusetts Avenue, NW, Washington, DC 20036-1023. (202) 223-8130. 1975 to present. Monthly. $99.00 per year. ISSN: 1047-6938.

Optics and Lasers in Engineering. Elsevier Science, 660 White Plains Road, Tarrytown, NY 10591-5153. (914) 545- 9200. FAX (914) 333-2444. 1980 to present. Ten issues per year. $500.00 per year. ISSN: 0143-8166.

Optics Communications. Elsevier Science, 660 White Plains Road, Tarrytown, NY 10591-5153. (914) 545-9200. FAX (914) 333-2444. 1969 to present. 54 issues per year. $2121.00. per year. ISSN: 0030-4018.

Optics Letters. Optical Society of America, Inc., 2010 Massachusetts Avenue, NW, Washington, DC 20036-1023. (202) 223-8130. 1977 to present. Semi-Monthly. $625.00 per year. ISSN: 0146-9592.

SPIE-International Society For Optical Engineering. Proceedings. SPIE-International Society for Optical Engineering, Box 10, 1000 20th Street, Bellingham, WA 98227-0010. (206) 676-3290. FAX (206) 647-1445. 1963 to present. Approximately 50 numbers per year. Individually priced.

RESEARCH CENTERS AND INSTITUTES

American Video Institute. Rochester Institute of Technology, One Lomb Memorial Drive, Building 7, P.O. Box 9887, Rochester, NY 14623-0887. (716) 475-6625. FAX (716) 475-5804.

Center for Applied Optics. University of Alabama in Huntsville, Research Institute Building, Huntsville, AL 25899. (205) 895-6102.

Communications and Information Processing Group. Rensselaer Polytechnic Institute, Electrical Computer and Systems Engineering Department, Troy, NY 12180-3590. (518) 266-6486.

Institute of Optics. University of Rochester, Wilmot Building, Rochester, NY 14627. (716) 275-5248.

OPTICAL FIBERS

See: Fiber Optics

OPTICS

See Also: FIBER OPTICS, HOLOGRAPHY, OPTICAL COMMUNICATIONS, OPTICAL DISKS, OPTOELECTRONICSi, PHYSICS, SPECTROSCOPY

ABSRACT SERVICES AND INDEXES

Applied Science and Technology Index: A Cumulative Subject Index to English Language Periodicals in the Fields of Aeronautics and Space Science, Computer Technology, Chemistry, Construction Industry, Energy and Related Areas. H.W. Wilson Co., 950 University Avenue, Bronx, NY 10452. (212) 588-8400. (800) 367-6770. FAX (718) 590-1617. From 1958 to present. Monthly. Inquire about cost and availability. Also available on CD-ROM and online. ISSN: 0003-6986.

Chemical Abstracts. Chemical Abstracts Service, 2540 Olentangy River Road, Box 3012, Columbus, OH 43210-0012.

(614) 447-3600. (800) 848-6538. FAX (614) 447-3713.From 1907 to present. Weekly. $16800.00 per year. Also available on CD-ROM and online. Inquire regarding cost and availability.

Current Contents: Engineering, Technology and Applied Sciences. Institute for Scientific Information, 3501 Market Street, Philadelphia, PA 19104. (215) 386-0100. FAX (215)386-2291. From 1961 to present. Weekly. $442.00 per year. Also available on CD-ROM and online. Inquire regarding cost and availability. ISSN: 0095-7917.

Current Papers in Physics. Institute of Electrical Engineers, Michael Faraday House, Six Hill Way, Stevenage, Herts, SG1 2AY, England. Distributed by INSPEC/IEEE, Box 1331, 445 Hoes Lane, Piscataway, NJ 08855-1331. (908) 562- 5549. 1966 to present. Fortnightly. $410.00 per year. ISSN: 0011-3786.

Engineering Index Monthly. Engineering Information, Inc., Castle Point on the Hudson, Hoboken, NJ 07030. (201) 216-8500. FAX (201) 216-8532. Monthly. $2200.00 per year. Also available online as COMPENDEX and on CD-ROM. ISSN: 0742-1974.

General Science Index. H.W. Wilson Company, 950 University Avenue, Bronx, NY 10452. (212) 588-8400. (800) 367-6770. FAX (718) 590-1617. From 1978 to present. Ten issues per year; quarterly and annual cumulations. Service basis. Available on CD-ROM and online. Inquire regarding cost and availability. ISSN: 0162-1963.

Government Reports Announcements and Index. U. S. National Technical Information Service (NTIS), 5285 Port Royal Road, Springfield, VA 22161. (703) 487-4650. FAX (703) 321-8547. From 1968 to present. Annual. $630.00 per year. Also available online as *NTIS Bibliographic Database* and on CD-ROM.

Journal of Current Laser Abstracts. PennWell Publishing Co., 10 Tara Boulevard, Nashua, NH 03062-2801. (603) 891- 9177. FAX (603) 891-0539. From 1964 to present. Monthly. $450.00 per year. ISSN: 0022-0264.

NTIS Alerts: Physics. U. S. National Technical Information Service. 5285 Port Royal Road, Springfield, VA 22161. (703) 487-4650. FAX (703) 321-8547. Weekly. $140.00 per year.

Pascal 10: Mechanique, Acoustique et Transport dE Chaleur. Centre National de la Recherche Scientifique, Institue de l'Information Scientifique et Technique, 2 aliee du Parc de Brabois, 54514 Vandoeuvre-Les Nancy Cedex, France. TEC 83-50-46-00. FAX 83-50-46-50. 1984 to present. Ten issues per year. 1530 F per year. Also available on CD-ROM and online. ISSN: 1136-5107.

Physics Abstracts. INSPEC. Section A, Science Abstracts. Institution of Electrical Engineers (IEE). Available from INSPEC/IEEE-Institute of Electrical and Electronic Engineers,, Box 1331, 445 Hoes Lane, Piscaway NJ 08855-1331. (908) 562-5549. 1898 to present. 24 issues per year. $2835.00 per year. Also available on CD-ROM and online. ISSN: 0036-8091.

Physics Briefs (Physikalische Berichte). Information Center for Energy, Physics, Mathematics; German Physical Society. V C H Publishers, Inc., 220 East 23rd Street, New York, NY 10010-4606. (212) 683-8333. 1845 to present. 24 issues per year. $2390.00 per year. Also available online. ISSN: 0179-7434.

Science Citation Index (SCI). Institute for Scientific Information, 3501 Market Street, Philadelphia, PA 19104. (215) 386-

OPTICS

Ency. of Physical Sciences and Engineering Info. Sources

0100. (800) 523-1850. FAX (215) 386-2991. 1961 to present. Six issues per year, plus annual cumulation. $11650.00 per year. Also available online and on CD-ROM. Inquire about price and availability. ISSN: 0036-827X.

Scientific and Technical Aerospace Reports (STAR). U.S. National Aeronautics and Space Administration. Distributed by U. S. Superintendent of Documents, Washington, DC 20402. 1963 to present. Semi-monthly, with semiannual and annual indexes. $114.00 per year. ISSN: 0036-8741. Also available online and on CD-ROM.

ANNUAL REVIEWS AND YEARBOOKS

Advances in Atomic, Molecular and Optical Physics. Academic Press, Inc., 6277 Sea Harbor Drive, Orlando, FL. (800) 321-5068. 1965 to present. Irregular. ISSN: 1049-250X.

Advances in Electronics and Electron Physics. Academic Press, Inc., 6277 Sea Harbor Drive, Orlando, FL. (800) 321-5068. From 1948 to present. Irregular. ISSN: 0065-2539.

Advances in Physics. Taylor & Francis, Ltd., Rankine Road, Basingstoke, Hants RG24 8PR, England. 0256-840366. FAX 0256-47938. 1952 to present. Bimonthly. $511.00 per year. ISSN: 0001-8732.

Advances in Solid State Technology. Kluwer Academic Publishers, 101 Philip Drive, Assinippi Park, Norwell, MA 02061. (617) 871-6600. 1985 to present. Irregular. Price varies.

Critical Reviews in Solid State and Materials Sciences. CRC Press, Inc., 2000 Corporate Boulevard, NW, Boca Raton, FL 33431. (407) 994-0555. (800) 272-7737. FAX (407) 998-9784. 1970 to present. Bimonthly. $265.00 per year. ISSN: 1040-8436.

Progress in Optics. Elsevier Science Inc., 660 White Plains Road, Tarrytown, NY 10591-5153. (914) 524-9200. FAX (914) 333-2444. 1961 to present. Irregular. Price varies. ISSN: 0079-6638.

Solid State Physics: Advances in Research and Applications. Academic Press, Inc., 6277 Sea Harbor Drive, Orlando, FL. (800) 321-5068. 1955 to present. Irregular. ISSN: 0081-1947.

Topics in Applied Physics. Springer-Verlag New York, Inc., 175 Fifth Avenue, New York, NY 10010. (212) 460-1500. 1(800) 777-4643. FAX (212) 473-6272. ISSN: 0303-4216.

ASSOCIATIONS AND PROFESSIONAL SOCIETIES

American Astronomical Society. 2000 Florida Avenue NW, Suite 400, Washington, DC 20009. (202) 328-2010. FAX: (202) 234-2560.

American Institute of Physics. One Physics Ellipse, College Park, MD 20740-3843. (301) 209-3100.

American Precision Optics Manufacturers Association. University of Rochester, CPU Box 276386, Rochester, NY 14627-6386. (716) 275-2753.

IEEE Lasers and Electro-optics Society. c/o IEEE, 445 Hoes Lane, Piscataway, NJ 08855. (908) 562-3892. FAX (908)981-9667.

Optical Society of America. 2010 Massachusetts Avenue NW, Washington, DC 20036. (202) 223-8130. FAX (202) 223-1096.

Spie-The International For Optical Engineering. PO Box 10, Bellingham, WA 98227. (206) 676-3290. FAX (206) 647-1445.

BIBLIOGRAPHIES

Science Books and Films. American Association for the Advancement of Science, 1333 H Street NW, Washington, DC 20005. (202) 326-6454. Reviews of print, film and software materials in all sciences. 1965 to present. Nine issues per year. $40.00 per year. ISSN: 0098-342X

Scientific and Technical Books and Serials in Print, 1995; An Index To Literature in Science and Technology. R.R. Bowker Inc., 121 Chanlon Road, New Providence, NJ 07974. (908) 464-6800. (800) 521-8110. 4 volumes. 1972 to present. Annual. $299.95. Also available on compact disc and online. ISSN: 0000-054X.

DIRECTORIES AND BIOGRAPHICAL SOURCES

AMERICAN MEN and WOMEN of SCIENCE: Physical and Biological Sciences. R. R. Bowker Inc., 121 Chanlon Road, New Providence, NJ 07974. (908) 464-6800. (800) 521-8110. 20th edition. 8 volumes. 1996. $850.00.

American Physical Society. MEMBERSHIP DIRECTORY BULLETIN ISSUE. American Physical Society, One Physics Ellipse, College Park, MD 20740-3843. (301) 209-3200. FAX (301) 209-0865. Biennial. $50.00.

International Directory of Engineering Societies and Related Organizations. American Society for Engineering Education, 1111 19th Street, NW, Suite 608, Washington, DC 20036. 15th edition. 1996. $185.00.

Graduate Programs in Physics, Astronomy and Related Fields. 1995-1996. American Institute of Physics, One Physics Ellipse, College Park, MD 20740-3843. (301) 209-3100. Annual. $45.00. ISSN: 0147-1821.

Optics Education. SPIE International P.O. Box 10, Bellingham, WA 98227. (206) 676-3290. FAX (206) 647-1445. Annual.

Research Centers Directory. Gale Research, 835 Penobscot Building, Detroit, MI 48226-4094. (313) 961-2242. (800) 347-4253. 20th edition. 1995. $485.00. ISSN: 0080-1518.

Scientific and Technical Organizations and Agencies Directory. Gale Research, 835 Penobscot Building, Detroit, MI 48226-4094. (313) 961-2242. (800) 347-4253. 4th edition. 1996. $195.00.

Who's Who in Engineering. Gordon Davis, editor. American Society for Engineering Education, 1111 19th Street, NW, Suite 608, Washington, DC 20036. 9th edition. 1995. $220.00.

Who's Who in Technology. Gale Research, 835 Penobscot Building, Detroit, MI 48226-4094. (313) 961-2242. (800) 347-4253. 7th edition. 1995. $195.00.

ENCYCLOPEDIAS AND DICTIONARIES

Dictionary of the Physical Sciences: Terms, Formulas, Data. Cesare Emiliani. Oxford University Press, Inc., 200 Madison Avenue, New York, NY 10016. (212) 725-6000. (800) 334-4249.1989. $19.95.

Encyclopedia of Applied Physics. George Trigg, editor. VCH Publications, Inc., 220 East 23rd Street, Suite 909, New York, NY 10010-4606. (212) 683-8333. (800) 422-8824. 20 volume set. 1991-present. $5990.00.

Encyclopedia of Physics. Robert M. Besancon. Chapman & Hall, 1 Penn Plaza, New York, NY 10119. (212) 564-1060. 3rd edition. 1990. $54.95 in paper.

Encyclopedial of Lasers and Optical Technology. Robert A. Meyers, editor. Academic Press, Inc., 6277 Sea Harbor Drive, Orlando, FL. (800) 321-5068. 1991. $92.00.

Encyclopedia of Physical Science and Technology. Academic Press, Inc., 6277 Sea Harbor Drive, Orlando, FL. (800) 321-5068. 2nd edition. 18 volume set. 1992. $2625.00.

Glossary of Optical Terminology. Thomas K. Farrell. Butterworth-Heinemann, 313 Washington Avenue, Newton, MA 02158. (617) 928-2500. (800) 366-2665. 1985. $24.95.

McGraw-Hill Encyclopedia of Science and Technology. McGraw-Hill Book Company, Inc., 1221 Avenue of the Americas, New York, NY 10020. (212) 512-2000. (800) 262-4729. 7th edition. 20 volume set. 1992. $1900.00.

GENERAL WORKS

All Was Light: An Introduction to Newton's Opticks. A. Rupert Hall. Oxford University Press Inc., 200 Madison Avenue, New York, NY 10016. (212) 725-6000. (800) 334-4249. 1993. $52.50.

Applied Optics and Optical Design. A. E. Conrady. Dover Publications, Inc., 180 Varick Street, New York, NY 10014. (212) 255-3755. (800) 223-3130. 2 volumes. Reprint edition. 1992. $10.95.

Applied Optics and Optical Engineering: A Comprehensive Treatise. Rudolf Kingslake, et al, editors. Academic Press 9 volumes. 1965-present . Volumes individually priced.

Coherent Optics. W. Lauterborn, et al. Springer-Verlag New York, Inc., 175 Fifth Avenue, New York, NY 10010. (212) 460-1500. (800) 777-4643. FAX (212) 473-6272. 1995. $49.00.

Color and Light in Nature. David K. Lynch and William Livingston. Cambridge University Press, 40 West 20th Street, New York, NY 10011-4211. (212) 924-3900. (800) 872-7423. 1995. $69.95.

Integrated Optics: Devices and Applications. Joseph T. Boyd, editor. Institute of Electrical and Electronic Engineers (IEEE), 345 East 47th Street, New York, NY 10017. (212) 705-7900. 1991. $59.95.

Introduction to Classical and Modern Optics. Jurgen R. Meyer-Arendt. Prentice-Hall, 113 Sylvan Avenue, Route 9W,

Englewood Cliffs, NJ 07632. (201) 592-2000. (800) 922-0579. 4th editon. 1994. $70.00.

Lasers and Optical Engineering. P.K. Das. Springer-Verlag New York, Inc., 175 Fifth Avenue, New York, NY 10010. (212) 460-1500. (800) 777-4643. FAX (212) 473-6272. 1990. $65.00.

Nonlinear Optics and Optical Physics. Iam-Choon Khoo, et al, editors. World Scientific Publishing Company, Inc., 1060 Main Street, River Edge, NJ 07661. (201) 487-9655. (800) 227-7562. 1994. $38.00 in paper.

Optics and Lasers, Including Fibers and Optical Waveguides. M. Young. Springer-Verlag New York, Inc., 175 Fifth Avenue, New York, NY 10010. (212) 460-1500. (800) 777-4643. FAX (212) 473-6272. 4th revised edition. 1993. $34.50.

Principles of Optics, Electromagnetic theory of Propagation, Interference and Diffraction of Light. Max Born and E. Wolf. Elsevier Science Publishing Company, Inc., 655 Avenue of the Americas, New York, NY 10010. (212) 989-5800. FAX (914) 333-2444. 6th edition. 1980. $59.00 in paper.

Principles of Optics, Optoelectronics and Photonics. Alan Billings. Prentice-Hall, 113 Sylvan Avenue, Route 9W, Englewood Cliffs, NJ 07632. (201) 592-2000. (800) 922- 0579. 1994. $40.00 in paper.

HANDBOOKS AND MANUALS

CRC Handbook of Chemistry and Physics. David R. Kide, editor. CRC Press, Inc., 2000 Corporate Boulevard, NW, Boca Raton, FL 33431. (407) 994-0555. (800) 272-7737. 77th edition. 1996. $99.95.

Handbook of Optics. Optical Society of America. McGraw- Hill Publishing Company, Inc., 1221 Avenue of the Americas, New York, NY 10020. (212) 512-2000. (800) 262-4729. 2 volumes. 2nd edition. 1993. $99.50 each.

Handbook of Physical Quantities. Igor S. Grigoriev and Evgenil Z. Melikhov, editors. CRC Press, Inc., 2000 Corporate Boulevard, NW, Boca Raton, FL 33431. (407) 994-0555. (800) 272-7737. 1995. $99.00.

Handbook of Tables of Functions for Applied Optics. Leo Levi, editor. Franklin Book Co., Inc., Elkins Park, PA 19117. (215) 635-5252. CRC Press Reprints. 1974. $152.95.

McGraw-Hill Encyclopedia of Physics. Sybil P. Parker, editor. McGraw-Hill Publishing Company, Inc., 1221 Avenue of the Americas, New York, NY 10020. (212) 512-2000. (800) 262-4729. 2nd edition. 1993. $95.50

Newnes Engineering and Physical Science Pocketbook. J. O. Bird and P. J. Chivers. Butterworth-Heinmann, 313 Washington Avenue, Newton, MA 02158. (617) 928-2500. (800) 366-2665. 1993. $49.95.

A Physicist's Desk Reference. Herbert L. anderson, editor. American Institute of Physics, One Physics Ellipse, College Park, MD 20740-3843. (301) 209-3100. 2nd edition. 1989. $70.00.

Optics Problem Solver. Research and Education Association Staff. Research & Education Association, 61 Ethel Road, West Piscataway, NJ 08854. (908) 819-8880. Revised edition. 1994. $29.95 in paper.

OPTICS

Ency. of Physical Sciences and Engineering Info. Sources

ONLINE DATABASES AND CD-ROMS

CA Search. Chemical Abstracts Service, P.O. Box 3012, Columbus, OH 43210-0012. (614) 447-3600. (800) 848-6533. FAX (614) 447-3709. Very comprehensive guide to worldwide chemical literature and related fields, 1972 to present. Available on BRS,(800) 289-4277, DIALOG, (800) 334-2564, ORBIT (800) 456-7248, and STN International, FIZ Karlsruhe, P.O. Box 2465, W-7500, Karlsruhe 1, Germany, online services. Inquire as to cost and availability.

Compendex Plus. Engineering Information, Inc., 345 East 47th Street, New York, NY 10017. (212) 705-7600 or (800) 221-1044. Contains citations with abstracts to worldwide literature in engineering and technology, from 1970 to present. Available on online BRS,(800) 289-4277, DIALOG, (800) 334-2564, ORBIT (800) 456-7248, and STN International, FIZ Karlsruhe, P.O. Box 2465, W-7500, Karlsruhe 1, Germany, online services. Also available on CD-ROM. Inquire as to cost and availability.

Current Contents Search. Institute for Scientific Information, 3501 Market Street, Philadelphia, PA 19104. (215) 386-0100. FAX (215) 386-6362. Contains citations to articles listed in the table of contents of science and technology journals. Also articles in social sciences and life sciences journals. Available on BRS, (800) 289-4277, DIALOG, (800) 334-2564, online services. Inquire as to cost and availability.

Dissertation Abstracts Online. University Microfilms International, 300 North Zeeb Road, Ann Arbor, MI 48106. (800) 521-0600 or (313) 761-4700. Scope includes virtually all doctoral dissertations accepted at accredited American institutions from 1861 to present in 252 subject areas. Available on BRS, (800) 289-4277, DIALOG, (800) 334-2564,and OCLC EPIC, (800) 848-5878, online services. Also available on CD-ROM. Inquire as to cost and availability.

INSPEC. Institution of Electrical Engineers, Michael Faraday House, Six Hills Way, Stevenage, Herts. SG1 2AY, England. Telephone: 0438 313311 or FAX 0438 742840. Contains citations to the worldwide literature of physics, electronics and electrical engineering, computer technology, and related fields. Available on BRS,(800) 289-4277, DIALOG, (800) 334-2564, ORBIT, (800) 456-7248, and STN International, FIZ Karlsruhe, P.O. Box 2465, W-7500, Karlsruhe 1, Germany, online services. Inquire as to cost and availability.

NTIS Bibliographic Database. National Technical Information Service, 5285 Port Royal Road, Springfield, VA 22161. (703) 487-4929 or FAX (703) 321-8199. Broad coverage of government-sponsored science and technology research reports, 1964 to present. Available on BRS,(800) 289-4277, DIALOG, (800) 334-2564, ORBIT, (800) 456-7248, and STN International, FIZ Karlsruhe, P.O. Box 2465, W-7500, Karlsruhe 1, Germany, online services. Also available on CD-ROM. Inquire as to cost and availability.

Physics Briefs. American Institute of Physics, 335 East 45th Street, New York, NY 10017. (212) 661-9260 or FAX (212) 661-2036. Contains citations with abstracts of the literature of physics and related fields, 1979 to present. Available on the STN International, FIZ Karlsruhe, P.O. Box 2465, W-7500, Karlsruhe 1, Germany, online service. Inquire as to cost and availability.

Scientific and Technical Books and Serials in Print. R.R. Bowker Inc., 121 Chanlon Road, New Providence, NJ 07974. (800) 521-8110 or (908) 464-6800. List of currently published books and serials in the physical and biological sciences, engineering and technology, with subject, author and titles indexes. Available

on ORBIT, (800) 456-7248, online service. Inquire as to cost and availability.

SCISEARCH. Institute for Scientific Information, 3501 Market Street, Philadelphia, PA 19104. (800) 523-1850 or (215) 386-0100. Broad multidisciplinary title and author index to the international literature of science and technology, 1974 to present. Available on DIALOG, (800) 334-2564, and ORBIT, (800) 456-7248, online services. Also available on CD-ROM. Inquire as to cost and availability.

WILSONLINE. H.W. Wilson Company, 950 University Avenue, Bronx, NY 10452. (800) 367-6770 or (212) 588-8400.Makes available online versions of the printed H.W. Wilson indexes including *Applied Science and Technology Index, Business Periodicals Index, General Science Index,* and *Readers' Guide to Periodical Literature.* Period covered is generally 1983 to present. Available on BRS, (800) 289-4277, DIALOG, (800) 334-2564, and OCLC EPIC, (800) 848-5878, online services.Also available on CD-ROM. Inquire as to cost and availability.

PERIODICALS

Applied Optics. Optical Society of America. 2010 Massachusetts Avenue NW, Washington, DC 20036-1023. (202)223-8130. 1962 to present. 36 issues per year. $1090.00 per year. ISSN: 0003-6935.

Applied Spectroscopy. Society for Applied Spectroscopy, 198 Thomas Johnson Drive, Suite 2, Frederick, MD 21702. (301) 694-8122. FAX (301) 694-6860. 1946 to present. Monthly. $180.00 per year. ISSN: 0003-7028.

Fiber and Integrated Optics. Taylor and Francis, 1900 Frost Road, Suite 101. Bristol, PA 19007. (215) 785-5800. FAX (215) 785-5515. 1977 to present. Quarterly. $180.00 per year. ISSN: 0146-8030.

International Journal of Nonlinear Optical Physics. World Scientific Publishing Co., 1060 Main Street, Suite 18, River Edge, NJ 07661. (800) 227-7562. 1992 to present. Quarterly. $348.00 per year. ISSN: 0218-1991.

Journal of Lightwave Technology. Institute of Electrical and Electronics Engineers, 445 Hoes Lane, PO Box 1331, Piscataway, NJ 08855. (908) 562-3892. FAX (908) 562-8434. 1983 to present. Monthly. $425.00 per year. ISSN: 0733- 8724.

IEEE Journal of Quantum Electronics. IEEE Lasers and Electro-optics Society/IEEE, 445 Hoes Lane, PO Box 1331, Piscataway, NJ 08855. (908) 981-0060. FAX (908) 981-9667. 1965 to present. Monthly. $535.00 per year. ISSN: 0018- 9197.

Journal of the Optical Society of America. Part A: Optics and Image Science. Optical Society of America, Inc., 2010 Massachusetts Avenue, NW, Washington, DC 20036-1023. (202) 223-8130. 1917 to present. Monthly. $610.00 per year. ISSN: 0740-3232.

Journal of the Optical Society of America. Part B: Optical Physics. Optical Society of America, Inc., 2010 Massachusetts Avenue, NW, Washington, DC 20036-1023. (202) 223-8130. 1917 to present. Monthly. $610.00 per year. ISSN: 0740-3224.

Laser Focus World. PennWell Publishing Co., 10 Tara Boulevard, Nashua, NH 03062-2891. (603) 891-0123. FAX (603) 891-

0574. 1964 to present. Monthly. $104.00 per year. ISSN: 1043-8092.

Optical Engineering. SPIE-International Society for Optical Engineering, Box 10, 1000 20th Street, Bellingham, WA 98227-0010. (206) 676-3290. FAX (206) 647-1445. 1962 to present. Monthly. $170.00 per year. ISSN: 0091-3286.

Optics and Photonics News. Optical Society of America, Inc., 2010 Massachusetts Avenue, NW, Washington, DC 20036-1023. (202) 223-8130. 1975 to present. Monthly. $99.00 per year. ISSN: 1047-6938.

Optics Communications. Elsevier Science Inc., Box 882, Madison Square Station, New York, NY 10159. (212) 989-5800. FAX (212) 633-4990. 1969 to present. 54 issues per year. $2121.00. per year. ISSN: 0030-4018.

Optics Letters. Optical Society of America, Inc., 2010 Massachusetts Avenue, NW, Washington, DC 20036-1023. (202) 223-8130. 1977 to present. Semi-Monthly. $625.00 per year. ISSN: 0146-9592.

Optics Materials. Elsevier Science Inc., Box 882, Madison Square Station, New York, NY 10159. (212) 989-5800. FAX (212) 633-4990. 1991 to present. Quarterly. $206.00 per year. ISSN: 0925-3467.

Spie-International Society for Optical Engineering. Proceedings. SPIE-International Society for Optical Engineering, Box 10, 1000 20th Street, Bellingham, WA 98227-0010. (206) 676-3290. FAX (206) 647-1445. 1963 to present. Approximately 50 numbers per year. Individually priced.

RESEARCH CENTERS AND INSTITUTES

Center for Applied Optics. University of Alabama in Huntsville, Research Institute Building, Huntsville, AL 25899. (205) 895-6102.

Center for Applied Optics. University of Texas at Dallas, P.O. Box 830688, Richardson, TX 75803-0688. (214) 690-2868. FAX (214) 690-2848.

Center for Imaging Science. Rochester Institute of Technology, One Lomb Memorial Drive, Rochester, NY 14623. (716) 475-2774.

Center for Research in Electro-optics and Lasers. University of Central Florida, 12424 Research Parkway, Orlando, FL 32826. (407) 658-6800. FAX (407) 658-6880.

Institute for Physical Science and Technology. University of Maryland, 4203 Computer and Space Sciences Building, College Park, MD 20742. (301) 454-4874. FAX (301) 314-9363.

Institute of Optics. University of Rochester, Wilmot Building, Rochester, NY 14627. (716) 275-5248.

Optical Sciences Center. University of Arizona, Tucson, AZ 85721. (602) 621-6997.

OPTOELECTRONICS

See also: ELECTRONICS, ELECTRONICS ENGINEERING, FIBER OPTICS, HOLOGRAPHY, LASERS, OPTICAL COMMUNICATION, OPTICAL DISKS, OPTICS

ABSTRACT SERVICES AND INDEXES

Applied Science and Technology Index. A cumulative subject index to English language periodicals in the fields of aeronautics and space science, computer technology, chemistry, construction industry, energy and related areas. H.W. Wilson Co., 950 University Avenue, Bronx, NY 10452. (212) 588-8400. (800) 367-6770. FAX (718) 590-1617. From 1958 to present. Monthly. Inquire about cost and availability. Also available on CD-ROM and online. ISSN: 0003-6986.

Chemical Abstracts. Chemical Abstracts Service, 2540 Olentangy River Road, Box 3012, Columbus, OH 43210-0012. (614) 447-3600. (800) 848-6538. FAX (614) 447-3713. From 1907 to present. Weekly. $16800.00 per year. Also available on CD-ROM and online. Inquire regarding cost and availability.

Current Contents: Engineering, Technology and Applied Sciences. Institute for Scientific Information, 3501 Market Street, Philadelphia, PA 19104. (215) 386-0100. FAX (215) 386-2291. From 1961 to present. Weekly. $442.00 per year. Also available on CD-ROM and online. Inquire regarding cost and availability. ISSN: 0095-7917.

Current Papers in Physics. Institute of Electrical Engineers, Michael Faraday House, Six Hill Way, Stevenage, Herts, SG1 2AY, England. Distributed by INSPEC/IEEE, Box 1331, 445 Hoes Lane, Piscataway, NJ 08855-1331. (908) 562-5549. 1966 to present. Fortnightly. $410.00 per year. ISSN: 0011-3786.

Electrical and Electronics Abstracts. Institution of Electrical Engineers (IEE), London. Available from INSPEC/IEEE - Institute of Electrical and Electronic Engineers, Box 1331, Hoes Lane, Piscataway, NJ 08855-1331. (908) 562-5549. 1898 to present. Monthly. $2200.00 per year. Also available on CD-ROM and online as INSPEC. ISSN: 0036-8105.

Engineering Index Monthly. Engineering Information, Inc., Castle Point on the Hudson, Hoboken, NJ 07030. (201) 216-8500. FAX (201) 216-8532. Monthly. $2200.00 per year. Also available online as COMPENDEX and on CD-ROM. ISSN: 0742-1974.

General Science Index. H.W. Wilson Company, 950 University Avenue, Bronx, NY 10452. (212) 588-8400. (800) 367-6770. FAX (718) 590-1617. From 1978 to present. Ten issues per year; quarterly and annual cumulations. Service basis. Available on CD-ROM and online. Inquire regarding cost and availability. ISSN: 0162-1963.

Government Reports Announcements and Index. U. S. National Technical Information Service (NTIS), 5285 Port Royal Road, Springfield, VA 22161. (703) 487-4650. FAX (703) 321-8547. From 1968 to present. Annual. $630.00 per year. Also available online as *NTIS Bibliographic Database* and on CD-ROM.

NTIS Alerts: Electrotechnology. U. S. National Technical Information Service. 5285 Port Royal Road, Springfield, VA 22161. (703) 487-4650. FAX (703) 321-8547. Weekly. $140.00 per year.

OPTOELECTRONICS

Ency. of Physical Sciences and Engineering Info. Sources

Physics Abstracts. INSPEC. Section A, Science Abstracts. Institution of Electrical Engineers (IEE). Available from INSPEC/IEEE-Institute of Electrical and Electronic Engineers,, Box 1331, 445 Hoes Lane, Piscaway NJ 08855-1331. (908) 562-5549. 1898 to present. 24 issues per year. $2835.00 per year. Also available on CD-ROM and online. ISSN: 0036-8091.

Physics Briefs (Physikalische Berichte). Information Center for Energy, Physics, Mathematics; German Physical Society. VCH Publishers, Inc., 220 East 23rd Street, New York, NY 10010-4606. (212) 683-8333. 1845 to present. 24 issues per year. $2390.00 per year. Also available online. ISSN: 0179-7434.

Science Citation Index (SCI). Institute for Scientific Information, 3501 Market Street, Philadelphia, PA 19104. (215) 386-0100. (800) 523-1850. FAX (215) 386-2991. 1961 to present. Six issues per year, plus annual cumulation. $11650.00 per year. Also available online and on CD-ROM. Inquire about price and availability. ISSN: 0036-827X.

ANNUAL REVIEWS AND YEARBOOKS

Advances in Atomic, Molecular and Optical Physics. Academic Press, Inc., 6277 Sea Harbor Drive, Orlando, FL. (800) 321-5068. 1965 to present. Irregular. ISSN: 1049-250X.

Advances in Electronics and Electron Physics. Academic Press, Inc., 6277 Sea Harbor Drive, Orlando, FL. (800) 321-5068. From 1948 to present. Irregular. ISSN: 0065-2539.

Advances in Physics. Taylor & Francis, Ltd., Rankine Road, Basingstoke, Hants RG24 8PR, England. 0256-840366. FAX 0256-47938. 1952 to present. Bimonthly. $511.00 per year. ISSN: 0001-8732.

Advances in Solid State Technology. Kluwer Academic Publishers, 101 Philip Drive, Assinippi Park, Norwell, MA 02061. (617) 871-6600. 1985 to present. Irregular. Price varies.

Critical Reviews in Solid State and Materials Sciences. CRC Press, Inc., 2000 Corporate Boulevard, NW, Boca Raton, FL 33431. (407) 994-0555. (800) 272-7737. FAX (407) 998-9784. 1970 to present. Bimonthly. $265.00 per year. ISSN: 1040-8436.

Progress in Optics. Elsevier Science Inc., 660 White Plains Road, Tarrytown, NY 10591-5153. (914) 524-9200. FAX (914) 333-2444. 1961 to present. Irregular. Price varies. ISSN: 0079-6638.

Progress in Quantum Electronics. Elsevier Science Inc., 660 White Plains Road, Tarrytown, NY 10591-5153. (914) 524-9200. FAX (914) 333-2444. 1969 to present. 5 issues per year. $395.00 per year. ISSN: 0079-6727.

Solid State Physics: Advances in Research and Applications. Academic Press, Inc., 6277 Sea Harbor Drive, Orlando, FL. (800) 321-5068. 1955 to present. Irregular. ISSN: 0081-1947.

ASSOCIATIONS AND PROFESSIONAL SOCIETIES

American Institute of Physics. One Physics Ellipse, College Park, MD 20740-3843. (301) 209-3100.

IEEE Lasers and Electro-optics Society. c/o IEEE, 445 Hoes Lane, PO Box 1331, Piscataway, NJ 08855. (908) 562-3892. FAX (908) 562-8434.

Institute of Electrical and Electronics Engineers. 345 East 47th Street, New York, NY 10017. (212) 705-7900. FAX (212) 705-4929.

Laser Institute of America. 12424 Research Parkway, Suite 130, Orlando, FL 32926. (407) 380-1553. FAX (407) 380-5588.

Optical Society of America. 2010 Massachusetts Avenue NW, Washington, DC 20036. (202) 223-8130. FAX (202) 223-1096.

Society for Imagining Science and Technology. 7003 Kilworth Lane, Springfield, VA 22151. (703) 642-9090. FAX (703) 642-9094.

Spie-The International Society for Optical Engineering. P.O. Box 10, 1000 20th Street, Bellingham, WA 98227. (206) 676-3290. FAX (206) 647-1445.

BIBLIOGRAPHIES

Science Books and Films. American Association for the Advancement of Science, 1333 H Street NW, Washington, DC 20005. (202) 326-6454. Reviews of print, film and software materials in all sciences. 1965 to present. Nine issues per year. $40.00 per year. ISSN: 0098-342X

Scientific and Technical Books and Serials in Print: An Index to Literature in Science and Technology. 1995. R.R. Bowker Inc., 121 Chanlon Road, New Providence, NJ 07974. (908) 464-6800. (800) 521-8110. 4 volumes. 1972 to present. Annual. $299.95. Also available on compact disc and online. ISSN: 0000-054X.

DIRECTORIES AND BIOGRAPHICAL SOURCES

American Men and Women of Science: Physical and Biological Sciences. R. R. Bowker Inc., 121 Chanlon Road, New Providence, NJ 07974. (908) 464-6800. (800) 521-8110. 20th edition. 8 volumes. 1996. $850.00.

IEEE Membership Directory. Institute of Electrical and Electronics Engineers. IEEE Service Center, 445 Hoes Lane, Piscataway, NY 08854. (908) 981-1393. (800) 678-IEEE. FAX (908) 981-9667. 2 volumes. Annual. $190.00. ISSN:

International Directory of Engineering Societies and Related Organizations. American Society for Engineering Education, 1818 N Street NW, Suite 600, Washington, DC 20036. (202) 331-3526. 15th edition. 1996. $185.00.

Graduate Programs in Physics, Astronomy and Related Fields. 1995-1996. American Institute of Physics, One Physics Ellipse, College Park, MD 20740-3843. (301) 209-3100. Annual. $45.00. ISSN: 0147-1821.

Optics Education. SPIE-The International Society for Optical Engineering, PO Box 10, 1000 20th Street, Bellingham, WA 98227. (206) 676-3290. FAX (206) 647-1445. Annual. No charge.

Research Centers Directory. Gale Research, 835 Penobscot Building, Detroit, MI 48226-4094. (313) 961-2242. (800) 347-4253. 20th edition. 1995. $485.00. ISSN: 0080-1518.

Scientific and Technical Organizations and Agencies Directory. Gale Research, 835 Penobscot Building, Detroit, MI 48226-

4094. (313) 961-2242. (800) 347-4253. 4th edition. 1996. $195.00.

Who's Who in Engineering. Gordon Davis, editor. American Society for Engineering Education, 1111 19th Street, NW, Suite 608, Washington, DC 20036. 9th edition. 1995. $220.00.

Who's Who in Technology. Gale Research, 835 Penobscot Building, Detroit, MI 48226-4094. (313) 961-2242. (800) 347-4253. 7th edition. 1995. $195.00.

ENCYCLOPEDIAS AND DICTIONARIES

Dictionary of Optical Communications Technology. J. P. Rehahn and N. Schafer. VCH Publications, Inc., 220 East 23rd Street, Suite 909, New York, NY 10010-4606. (212) 683-8333. (800) 422-8824. 1988. $65.00.

Encyclopedia of Applied Physics. George Trigg, editor. VCH Publications, Inc., 220 East 23rd Street, Suite 909, New York, NY 10010-4606. (212) 683-8333. (800) 422-8824. 20 volume set. 1991-present. $5990.00.

Encyclopedia of Physics. Robert M. Besancon. Chapman & Hall, 1 Penn Plaza, New York, NY 10119. (212) 564-1060. 3rd edition. 1990. $54.95 in paper.

Encyclopedia of Lasers and Optical Technology. Robert A. Meyers, editor. Academic Press, Inc., 6277 Sea Harbor Drive, Orlando, FL. (800) 321-5068. 1991. $92.00.

Encyclopedia of Physical Science and Technology. Academic Press, Inc., 6277 Sea Harbor Drive, Orlando, FL. (800) 321-5068. 2nd edition. 18 volume set. 1992. $2625.00.

Fiber Optics Standard Dictionary. Martin H. Weik. 2nd edition. Van Nostrand Reinhold, 115 Fifth Avenue, New York, NY 10003. (212) 254-3232. (800) 842-3636. 2nd edition. 1989. $49.95.

Glossary of Optical Terminology. Thomas K. Farrell. Butterworth-Heinemann, 313 Washington Avenue, Newton, MA 02158. (617) 928-2500. (800) 366-2665. 1985. $24.95.

McGraw-Hill Encyclopedia of Science and Technology. McGraw-Hill Book Company, Inc., 1221 Avenue of the Americas, New York, NY 10020. (212) 512-2000. (800) 262-4729. 7th edition. 20 volume set. 1992. $1900.00.

GENERAL WORKS

Applied Optics and Optical Engineering: A Comprehensive Treatise. Rudolf Kingslake, et al, editors. Academic Press, Inc., 6277 Sea Harbor Drive, Orlando, FL. (800) 321-5068. 9 volumes. 1965-present . Volumes individually priced.

Coherent Optics. W. Lauterborn, et al. Springer-Verlag New York, Inc., 175 Fifth Avenue, New York, NY 10010. (212) 460-1500. (800) 777-4643. FAX (212) 473-6272. 1995. $49.00.

Electro-optical Systems Performance Modelling. Gary Waldman and John Wootton. Artech House, 685 Canton Street, Norwood, MA 02062. (617) 769-9750. (800) 225-9977. 1992. $85.00.

Integrated Optoelectronics. Mario Dagenais, et al, editors. Academic Press Inc., 6277 Sea Harbor Drive, Orlando, FL. (800) 321-5068. 1994. $84.95.

Integrated Optoelectronics: Waveguide Optics, Photonics, Semiconductors. Karl J. Ebling. Springer-Verlag New York, Inc., 175 Fifth Avenue, New York, NY 10010. (212) 460- 1500. (800) 777-4643. FAX (212) 473-6272. 1993. $64.50.

Lasers and Optical Engineering. P. K. Das. Springer-Verlag New York, Inc., 175 Fifth Avenue, New York, NY 10010. (212) 460-1500. (800) 777-4643. FAX (212) 473-6272. 1990. $65.00.

Microwaves and Optical Transmission. A. David Olver. John Wiley & Sons, Inc., 605 Third Avenue, New York, NY 10158-0012. (212) 850-6000. (800) 225-5945. 1992. $98.00.

Nonlinear Optics and Optical Physics. Iam-Choon Khoo, et al, editors. World Scientific Publishing Company, Inc., 1060 Main Street, River Edge, NJ 07661. (201) 487-9655. (800) 227-7562. 1994. $38.00 in paper.

Optical Network Theory. Yitzhak Weissman. Artech House, 685 Canton Street, Norwood, MA 02062. (617) 769-9750. (800) 225-9977. 1992. $78.00.

Optoelectronics and Lightwave Technology. J. E. Midwinter and Y. L. Gou. John Wiley & Sons, Inc., 605 Third Avenue, New York, NY 10158-0012. (212) 850-6000. (800) 225-5945. 1992. $89.95.

Principles of Optics, Electromagnetic theory of Propagation, Interference and Diffraction of Light. Max Born and E. Wolf. Elsevier Science Publishing Company, Inc., 655 Avenue of the Americas, New York, NY 10010. (212) 989-5800. FAX (914) 333-2444. 6th edition. 1980. $59.00 in paper.

Principles of Optics, Optoelectronics and Photonics. Alan Billings. Prentice-Hall, 113 Sylvan Avenue, Route 9W, Englewood Cliffs, NJ 07632. (201) 592-2000. (800) 922- 0579. 1994. $40.00 in paper.

Selected Papers On Fundamentals of Quantum Optics. Girsh S. Agarwal, editor. SPIE-International Society for Optical Engineering, 1000 20th Street, Bellingham, WA 98227. (206) 676-3290. FAX (206) 647-1994.

Semiconductor Optoelectronics. Pallab Bhattacharya. Prentice-Hall, 113 Sylvan Avenue, Route 9W, Englewood Cliffs, NJ 07632. (201) 592-2000. (800) 922-0579 1993. 1993. $75.00.

Understanding Telecommunications and Lightwave Systems. John G. Nellist. 1995. 2nd edition. Institute of Electrical and Electronic Engineers (IEEE), 345 East 47th Street, New York, NY 10017. (212) 705-7900.

HANDBOOKS AND MANUALS

Fiber Optic Communications Handbook. TAB Books, P.O. Box 40, Blue Summit, PA 17294-0850. (717) 794-2191. (800) 233-1128. 2nd edition. 1990. $89.50.

Fiber Optic Metrology and Standards. O. D. Soares. SPIE-International Society for Optical Engineering, 1000 20th Street, Bellingham, WA 98227. (206) 676-3290. FAX (206) 647-1445. 1991. $68.00.

Handbook of Optics. Optical Society of America. McGraw-Hill Publishing Company, Inc., 1221 Avenue of the Americas, New York, NY 10020. (212) 512-2000. (800) 262-4729. 2 volumes. 2nd edition. 1993. $99.50 each.

Handbook of Tables of Functions for Applied Optics. Leo Levi, editor. Franklin Book Co., Inc., Elkins Park, PA 19117. (215) 635-5252. CRC Press Reprints. 1974. $152.95.

Optoelectronics and Image Sensing Data Book. Texas Instruments Inc., 1991. $14.95.

ONLINE DATABASES AND CD-ROMS

CA Search. Chemical Abstracts Service, P.O. Box 3012, Columbus, OH 43210-0012. (614) 447-3600. (800) 848-6533. FAX (614) 447-3709. Very comprehensive guide to worldwide chemical literature and related fields, 1972 to present. Available on BRS,(800) 289-4277, DIALOG, (800) 334-2564, ORBIT (800) 456-7248, and STN International, FIZ Karlsruhe, P.O. Box 2465, W-7500, Karlsruhe 1, Germany, online services. Inquire as to cost and availability.

Compendex Plus. Engineering Information, Inc., 345 East 47th Street, New York, NY 10017. (212) 705-7600 or (800) 221-1044. Contains citations with abstracts to worldwide literature in engineering and technology, from 1970 to present. Available on online BRS,(800) 289-4277, DIALOG, (800) 334-2564, ORBIT (800) 456-7248, and STN International, FIZ Karlsruhe, P.O. Box 2465, W-7500, Karlsruhe 1, Germany, online services. Also available on CD-ROM. Inquire as to cost and availability.

Current Contents Search. Institute for Scientific Information, 3501 Market Street, Philadelphia, PA 19104. (215) 386-0100. FAX (215) 386-6362. Contains citations to articles listed in the table of contents of science and technology journals. Also articles in social sciences and life sciences journals. Available on BRS, (800) 289-4277, DIALOG, (800) 334-2564, online services. Inquire as to cost and availability.

Dissertation Abstracts Online. University Microfilms International, 300 North Zeeb Road, Ann Arbor, MI 48106. (800) 521-0600 or (313) 761-4700. Scope includes virtually all doctoral dissertations accepted at accredited American institutions from 1861 to present in 252 subject areas. Available on BRS, (800) 289-4277, DIALOG, (800) 334-2564,and OCLC EPIC, (800) 848-5878, online services. Also available on CD-ROM. Inquire as to cost and availability.

INSPEC. Institution of Electrical Engineers, Michael Faraday House, Six Hills Way, Stevenage, Herts. SG1 2AY, England. Telephone: 0438 313311 or FAX 0438 742840. Contains citations to the worldwide literature of physics, electronics and electrical engineering, computer technology, and related fields. Available on BRS, (800) 289-4277, DIALOG, (800) 334-2564, ORBIT, (800) 456-7248, and STN International, FIZ Karlsruhe, P.O. Box 2465, W-7500, Karlsruhe 1, Germany, online services. Inquire as to cost and availability.

NTIS Bibliographic Database. National Technical Information Service, 5285 Port Royal Road, Springfield, VA 22161. (703) 487-4929 or FAX (703) 321-8199. Broad coverage of government-sponsored science and technology research reports, 1964 to present. Available on BRS,(800) 289-4277, DIALOG, (800) 334-2564, ORBIT, (800) 456-7248, and STN International, FIZ Karlsruhe, P.O. Box 2465, W-7500, Karlsruhe 1, Germany, online services. Also available on CD-ROM. Inquire as to cost and availability.

Physics Briefs. American Institute of Physics, 335 East 45th Street, New York, NY 10017. (212) 661-9260 or FAX (212) 661-2036. Contains citations with abstracts of the literature of physics and related fields, 1979 to present. Available on the STN International, FIZ Karlsruhe, P.O. Box 2465, W-7500, Karlsruhe 1, Germany, online service. Inquire as to cost and availability.

Scientific and Technical Books and Serials in Print. R.R. Bowker Inc., 121 Chanlon Road, New Providence, NJ 07974. (800) 521-8110 or (908) 464-6800. List of currently published books and serials in the physical and biological sciences, engineering and technology, with subject, author and titles indexes. Available on ORBIT, (800) 456-7248, online service. Inquire as to cost and availability.

SCISEARCH. Institute for Scientific Information, 3501 Market Street, Philadelphia, PA 19104. (800) 523-1850 or (215) 386-0100. Broad multidisciplinary title and author index to the international literature of science and technology, 1974 to present. Available on DIALOG, (800) 334-2564, and ORBIT, (800) 456-7248, online services. Also available on CD-ROM. Inquire as to cost and availability.

WILSONLINE. H.W. Wilson Company, 950 University Avenue, Bronx, NY 10452. (800) 367-6770 or (212) 588-8400. Makes available online versions of the printed H.W. Wilson indexes including Applied Science and Technology Index, Business Periodicals Index, General Science Index, and Readers' Guide to Periodical Literature. Period covered is generally 1983 to present. Available on BRS, (800) 289-4277, DIALOG, (800) 334-2564, and OCLC EPIC, (800) 848-5878, online services.Also available on CD-ROM. Inquire as to cost and availability.

PERIODICALS

Applied Optics. Optical Society of America. 2010 Massachusetts Avenue NW, Washington, DC 20036-1023. (202)223-8130. 1962 to present. 36 issues per year. $1090.00 per year. ISSN: 0003-6935.

Applied Spectroscopy. Society for Applied Spectroscopy, 198 Thomas Johnson Drive, Suite 2, Frederick, MD 21702. (301) 694-8122. FAX (301) 694-6860. 1946 to present. Monthly. $180.00 per year. ISSN: 0003-7028.

Fiber and Integrated Optics. Taylor and Francis, 1900 Frost Road, Suite 101. Bristol, PA 19007. (215) 785-5800. FAX (215) 785-5515. 1977 to present. Quarterly. $180.00 per year. ISSN: 0146-8030.

IEEE Journal of Quantum Electronics. IEEE Lasers and Electro-optics Society/IEEE, 445 Hoes Lane, PO Box 1331, Piscataway, NJ 08855. (908) 981-0060. FAX (908) 981-9667. 1965 to present. Monthly. $535.00 per year. ISSN: 0018-9197.

International Journal of Nonlinear Optical Physics. World Scientific Publishing Co., 1060 Main Street, Suite 18, River Edge, NJ 07661. (800) 227-7562. 1992 to present. Quarterly. $348.00 per year. ISSN: 0218-1991.

Journal of Lightwave Technology. Institute of Electrical and Electronics Engineers, 445 Hoes Lane, PO Box 1331, Piscataway, NJ 08855. (908) 562-3892. FAX (908) 562-8434. 1983 to present. Monthly. $425.00 per year. ISSN: 0733-8724.

Journal of the Optical Society of America. Part A: Optics and Image Science. Optical Society of America, Inc., 2010 Massachusetts Avenue, NW, Washington, DC 20036-1023. (202) 223-8130. 1917 to present. Monthly. $610.00 per year. ISSN: 0740-3232.

Journal of the Optical Society of America. Part B: Optical Physics. Optical Society of America, Inc., 2010 Massachusetts Avenue, NW, Washington, DC 20036-1023. (202) 223-8130. 1917 to present. Monthly. $610.00 per year. ISSN: 0740-3224.

Laser Focus World. PennWell Publishing Co., 10 Tara Boulevard, Nashua, NH 03062-2891. (603) 891-0123. FAX (603) 891-0574. 1964 to present. Monthly. $104.00 per year. ISSN: 1043-8092.

Optical Engineering. SPIE-International Society for Optical Engineering, Box 10, 1000 20th Street, Bellingham, WA 98227-0010. (206) 676-3290. FAX (206) 647-1445. 1962 to present. Monthly. $170.00 per year. ISSN: 0091-3286.

Optics and Photonics News. Optical Society of America, Inc., 2010 Massachusetts Avenue, NW, Washington, DC 20036-1023. (202) 223-8130. 1975 to present. Monthly. $99.00 per year. ISSN: 1047-6938.

Optics Communications. Elsevier Science Inc., Box 882, Madison Square Station, New York, NY 10159. (212) 989-5800. FAX (212) 633-4990. 1969 to present. 54 issues per year. $2121.00. per year. ISSN: 0030-4018.

Optics Letters. Optical Society of America, Inc., 2010 Massachusetts Avenue, NW, Washington, DC 20036-1023. (202) 223-8130. 1977 to present. Semi-Monthly. $625.00 per year. ISSN: 0146-9592.

Optics Materials. Elsevier Science Inc., Box 882, Madison Square Station, New York, NY 10159. (212) 989-5800. FAX (212) 633-4990. 1991 to present. Quarterly. $206.00 per year. ISSN: 0925-3467.

Spie-International Society for Optical Engineering. Proceedings. SPIE-International Society for Optical Engineering, Box 10, 1000 20th Street, Bellingham, WA 98227-0010. (206) 676-3290. FAX (206) 647-1445. 1963 to present. Approximately 50 numbers per year. Individually priced.

RESEARCH CENTERS AND INSTITUTES

Center for Advanced Technology in Telecommunications. Polytechnic University, 333 Jay Street, Brooklyn, NY 11201. (718) 260-3050.

Center for Applied Optics. University of Alabama in Huntsville, Research Institute Building, Huntsville, AL 25899. (205) 895-6102.

Center for Applied Optics. University of Texas at Dallas, P.O. Box 830688, Richardson, TX 75803-0688. (214) 690-2868. FAX (214) 690-2848.

Center for Compound Semiconductor Microelectronics. University of Illinois, 127 Microelectronics Laboratory, 208 North Wright Street, Urbana IL 61801. (217) 333-3097. FAX (217) 244-6375.

Center for Electronic and Electro-optic Materials. State University of New York at Buffalo, 217 C Bonner Hall, Buffalo, NY 14260. (716) 636-3109. FAX (716) 636-2528.

Center for High Technology Materials. University of New Mexico, Room 125, Electrical and Computer Engineering Building, Albuquerque, NM 87131. (505) 277-3317. FAX (505) 277-6433.

Institute of Optics. University of Rochester, Wilmot Building, Rochester, NY 14627. (716) 275-5248.

ORGANIC CHEMISTRY

See also: CARBON, CHEMISTRY, CHEMICAL ENGINEERING, HETEROCYCLIC CHEMISTRY, PLASTICS

ABSTRACT SERVICES AND INDEXES

Applied Science and Technology Index: A Cumulative Subject Index to English Language Periodicals in the Fields of Aeronautics and Space Science, Computer Technology, Chemistry, Construction Industry, Energy and Related Areas. H.W. Wilson Co., 950 University Avenue, Bronx, NY 10452-9978. (212) 588-8400. (800) 367-6770. FAX (718) 590-1617. From 1958 to present. Monthly. Inquire about cost and availability. Also available online (BRS and WILSONLINE) and on CD-ROM. ISSN: 0003-6986.

Chemical Abstracts. Chemical Abstracts Service. 2540 Oleantangy River Road, Box 3012, Columbus, OS 43210-0012. (614) 447-3600. FAX (614) 447-3713. 1907 to present. Weekly. $16,800.00 per year. Available online and on CD-ROM. CA is also available in five section groupings. Inquire regarding cost and availability.

Current Contents: Physical, Chemical and Earth Sciences. Institute for Scientific Information, 3501 Market Street, Philadelphia, PA 19104. (215) 386-0100. (800) 523-1850. FAX (215) 386-2291. 1961 to present. Weekly. $442.00 per year. Also available online (BRS, DIALOG) and on CD-ROM. Inquire regarding cost and availability. ISSN: 0163-2574.

General Science Index. H.W. Wilson Company, 950 University Avenue, Bronx, NY 10452. (212) 588-8400. (800) 367-6770. FAX (718) 590-1617. From 1978 to present. Ten issues per year; quarterly and annual cumulations. Service basis. Available on CD-ROM and online. Inquire regarding cost and availability. ISSN: 0162-1963.

Physics Abstracts. INSPEC. Section A, Science Abstracts. Institution of Electrical Engineers (IEE), London. Available from: INSPEC/IEEE - Institute of Electrical and Electronic Engineers, Box 1331, Hoes Lane, Piscataway, NJ 08855-1331. (908) 562-5549. 1898 to present. 24 issues per year. $2835.00 per year. Also available online and on CD-ROM. ISSN: 0036-8091.

Science Citation Index (SCI). Institute for Scientific Information, 3501 Market Street, Philadelphia, PA 19104. (215) 386-0100. (800) 523-1850. FAX (215) 386-2991. 1961 to present. Six issues per year, plus annual cumulation. $11650.00 per year. Also available online and on CD-ROM. Inquire about price and availability. ISSN: 0036-827X.

ORGANIC CHEMISTRY

Ency. of Physical Sciences and Engineering Info. Sources

ANNUAL REVIEWS AND YEARBOOKS

Advances in Heterocyclic Chemistry. Alan R. Kitritzky, editor. Academic Press.Inc., 525 B Street, Suite 1900, San Diego, CA 92101-4495. (619) 231-0926. FAX (619) 699-6715. 1963 to present. Irregular. price varies. ISSN: 0065-2725.

Advances in Physical Organic Chemistry. Academic Press, Inc., 525 B Street, Suite 1900, San Diego, CA 92101-4495. (619) 231-0926. FAX (619) 699-6715. 1963 to present. Irregular. Price varies. ISSN: 0065-3160.

Studies in Organic Chemistry. Elsevier Science Publishers., Box 882, Madison Square Station, New York, NY 10159. (212) 989-5800. 1973 to present. Irregular. Price varies. ISSN: 0165-3253.

ASSOCIATIONS AND PROFESSIONAL SOCIETIES

American Carbon Society. c/o Carbone of America, 215 Stackpole street, St. Marys, PS 15857. (814) 781-8410. FAX (814) 781-8570.

American Chemical Society. 1155 16th Street, NW, Washington, DC 20036. (202) 872-4600.

American Institute of Chemical Engineers. 345 East 47th Street, New York, NY 10017-2396. (212) 705-7663.

American Oil Chemists' Society. PO Box 3489, Champaign, IL 61826-3489. (217) 359-2344. FAX (217) 351-8091.

Association of Official Analytical Chemists. 2200 Wilson Boulevard, Suite 400, Arlington, VA 22001. (703) 522-3032.

Chemical Manufacturers Association. 2501 M Street, N.W., Washington, DC 20037. (202) 887-1182

BIBLIOGRAPHIES

Chemical Information Sources. Gary Wiggins. McGraw-Hill Publishing Company, 1221 Avenue of the Americas, New York, NY 10020. (800) 262-4729 or (212) 512-3825. 1991. $42.50.

Information Sources in Chemistry. R.T. Bottle and J.F.B Rowland, editors. R.R. Bowker Inc., 121 Chanlon Road, New Providence, NJ 07974. (800) 521-8110 or (908) 464-6800. Fourth edition. 1993. $75.00.

Handbooks and Tables in Science and Technology. Russell H. Powell, editor. Oryx Press, 4041 North Central, Suite 700, Phoenix, AZ 85012-3330. (602) 265-2651 or (800) 279-6799. Third edition. 1994. $65.00. *Organic Reaction Mechanisms: An Annual Survey of Literature.* A. C. Knipe and W. E. Watts, editors. Wiley. vol. 92, 1992. price varies.

Scientific and Technical Books and Serials in Print: An Index to Literature in Science and Technology. R.R. Bowker Inc., 121 Chanlon Road, New Providence, NJ 07974. (908) 464-6800. (800) 521-8110. FAX (908) 665-3502. 1972 to present. Annual. 4 volumes. 1994. $299.95. Also available on compact disc and online. ISSN: 0000-054X.

DIRECTORIES AND BIOGRAPHICAL SOURCES

American Men and Women of Science. Physical and Biological Sciences. R.R. Bowker Company, 121 Chanlon Road, New Providence, NJ 07974. (908) 464-6800. (800) 521-8110. 20th edition. 8 volumes. 1996. $850.00.

Directory of Chemistry Software 1992. Wendy Warr, Peter Willett, Geoff Downs. American Chemical Society, 1155 16th Street, NW, Washington, DC 20036. (202) 872-4600. 1992. $35.95.

Consulting Services: Chemists and Chemical Engineers. Association of Consulting Chemists and Chemical Engineers, 50 East 41st Street, Suite 92, New York, NY 10017. (212)

684-6255. Biennial. $60.00.

Research Centers Directory. Gale Research, 835 Penobscot Building, Detroit, MI 48226-4094. (313) 961-2242.(800) 347-4253. 20th edition. 1995. $485.00. ISSN: 0080-1518.

ENCYCLOPEDIAS AND DICTIONARIES

Academic Press Dictionary of Science and Technology. Christopher Morris, editor. Academic Press, Inc., 1250 Sixth Avenue, San Diego, CA 92101. (619) 231-0926. FAX (619) 699-6715. 1991. $115.00.

Concise Encyclopedia Chemistry. Translated and revised by Mary Eagleson. Walter de Gruyter, Inc., 200 Saw Mill River Road, Hawthorne, New York, 10532. (914) 747-0110 or Fax (914) 747-1326. 1994. $69.95.

Dictionary of Organic Compounds. John Buckingham, et al, editors. Chapman & Hall, 1 Penn Plaza, New York, NY 10119. (212) 564-1060. Fifth edition. 7 volumes. $2550.00 set.

Encyclopedic Dictionary of Science. Candida Hunt and Monica Byles, editors. Facts-on-File, Inc., 460 Park Avenue South, New York, NY 10016-7382. (800) 322-8755. Fax (800) 683-3633. 1988. $32.95.

Encyclopedia of Applied Physics. VCH Publishers, Inc., 303 Northwest 12th Avenue, Deerfield Beach, FL 33442. (800) 367-8249. 1991-present. Twenty volumes. $6000.00.

Industrial Chemical Thesaurus. Michael Ash and Irene Ash. VCH Publishers, Inc., 220 East 23rd Street, New York, NY 10010-4606. (800) 367-8249. 1992. $295.00.

Kirk-Othmer Encyclopedia of Chemical Technology. John Wiley and Sons, Inc., 605 Third Avenue, New York, NY 10158. (800) 526-5368 or (212) 850-6000. Fourth edition. 1991-present. Twenty-seven volumes. $5400.00.

McGraw-Hill Encyclopedia of Chemistry. Sybil P.Parker, editor. McGraw-Hill Book, Incorporated, 1221 Avenue of the Americas, New York, NY 10020. (212) 997-3675. Second edition. 1993. $95.50.

McGraw-Hill Encyclopedia of Science and Technology. McGraw-Hill Book, Incorporated, 1221 Avenue of the Americas, New York, NY 10020. (212) 997-3675. Seventeenth edition. Twenty volumes. 1992. $1900.00.

GENERAL WORKS

Advanced Organic Chemistry. Frank A. Carey and R. J. Sundberg. Plenum Publishing Corp., 233 Spring Street, New York, NY. (212) 620-8000. (800) 221-9369. FAX (212) 463-0742. 3rd edition. 1990. Part A: Structure and Mechanisms, $32.50; Part B: Reactions and synthesis, $34.50 in paper.

Advanced Organic Chemistry. Jerry March. John Wiley & Sons, Inc., 605 Third Avenue, New York, NY 10158-0012. (212) 850-6000. (800) 225-5945. 4th edition. 1992. $59.95.

The Art of Problem Solving in Organic Chemistry. Miguel E. Alonso. John Wiley & Sons, Inc., 605 Third Avenue, New York, NY 10158-0012. (212) 850-6000. (800) 225-5945. 1987. $64.95.

The Chemical Bond: Structure and Dynamics. Ahmed Zewail. Academic Press, Inc., 1250 Sixth Avenue, San Diego, CA 92101-4311. (619) 231-0926. FAX (619) 699-6715. 1992. $49.95.

Comprehensive Organic Chemistry. Derek H. Barton and W. D. Ollis, editors. Elsevier Science Publishing Company, Inc., 655 Avenue of the Americas, New York, NY 10010. (212) 989-5800. FAX (914) 333-2444. 6 volume set. 1979. $800.00 in paper.

Contemporary Heterocyclic Chemistry: Synthesis, Reactions and Applications. G. R. Newkome and W. W. Paudler. John Wiley & Sons, Inc., 605 Third Avenue, New York, NY 10158-0012. (212) 850-6000. (800) 225-5945. 1982. $99.95.

Desk Reference for Organic Chemists. Michael B. East and David J. Ager. Krieger Publishing Company, P.O. Box 9542, Melbourne, FL 32902-9542. (407) 724-9542. 1995. $64.50.

Introduction to Organic and Biological Chemistry. Stuart J. Baum and John W. Hill. Macmillan Publishing Company, Inc,. 200 Old Tappan Road, Old Tappan, NJ 07675. (800) 233-2336. 1993. wfi

Organic Acids in Geologic Processes. E. D. Pittman and M. D. Lewan. Springer-Verlag New York, Inc., 175 Fifth Avenue, New York, NY 10010. (212) 460-1500. (800) 777-4643. FAX (212) 473-6272. 1994. $89.00.

Organic Chemistry: Its Language and Its State of the Art. Volkan Kisakurek, editor. VCH Publications, Inc., 220 East 23rd Street, Suite 909, New York, NY 10010-4606. (212) 683- 8333. (800) 422-8824. 1993. $117.00.

Organic Reactions: Simplicity and Logic. Pierre Laszlo. John Wiley & Sons, Inc., 605 Third Avenue, New York, NY 10158-0012. (212) 850-6000. (800) 225-5945. 1994.$74.95.

Physical Basis of Organic Chemistry. Howard Maskill. Oxford University Press, Inc., 200 Madison Avenue, New York, NY 10016. (212) 725-6000. (800) 334-4249. 1986. $35.00.

Reaction Mechanisms in Environmental Organic Chemistry. Richard A. Larson and Erick J. Weber. Lewis Pubs., CRC Press, Inc., 2000 Corporate Boulevard, NW, Boca Raton, FL 33431. (407) 994-0555. (800) 272-7737. 1994. $69.95.

HANDBOOKS AND MANUALS

Chemical Formulary. H. Bennett, editor. Chemical Publishing, Co., Inc. 80 Eighth Avenue, New York, NY 10011. (212) 255-1950. Volumes 1-30. $60.00 per volume.

Chemometrics: Chemical and Sensory Data. David R. Burgard and James T. Kuznicki. CRC Publishers, Inc., 2000 Corporate Blvd., N.W., Boca Raton, FL 33431. (407) 994-0555 or (800) 333-8300. 1990. $120.00.

Comprehensive Organic Synthesis: Selectivity, Strategy, and Efficiency in Modern Organic Chemical. Barry M. Trost and others, editors. Pergamon Press, Maxwell House, Fairview Park, Elmsford, NY 10523. (914) 592-7700. Fax (914) 592-3625. 1991. Nine volumes. $3900.00.

CRC Handbook of Chemistry and Physics. David R. Kide, editor. CRC Press Inc., 2000 Corporate Blvd., NW, Boca Raton, FL 33431. (407) 994-0555. (800) 333-8300. 77th edition. 1996. $99.95.

CRC Handbook of Data on Organic Compounds. R. C. Weast and M. J. Astle, editors. CRC Press Inc., 2000 Corporate Blvd., NW, Boca Raton, FL 33431. (407) 994-0555. (800) 333-8300. 1985. 2 volumes.

Guide to Basic Chemical Compounds. D.R. Lide, Jr. CRC Publishers, Inc., 2000 Corporate Blvd., N.W., Boca Raton, FL 33431. (407) 994-0555 or (800) 333-8300. 1993. $120.00.

Guide to Iupac Nomenclature of Organic Compounds. J. C. Richer, et al, editors. CRC Press Inc., 2000 Corporate Blvd., NW, Boca Raton, FL 33431. (407) 994-0555. (800) 333-8300. 1993. $39.95.

Handbook of Data and Organic Coumpounds. Robert C. Weast. CRC Press Inc., 2000 Corporate Blvd., NW, Boca Raton, FL 33431. (407) 994-0555. (800) 333-8300. 2nd edition. 1988. $2655.00.

Handbook of Organic Chemistry. Hans Beyer and Wolfgang Walter. Prentice Hall General Reference & Travel, 15 Columbus Circle, New York, NY 10023. (212) 373-8500. (800) 223-2348. 1995. $96.00.

Handbook of Tables for Organic Compound Identification. Zvi Rappoport. CRC Press Inc., 2000 Corporate Blvd., NW, Boca Raton, FL 33431. (407) 994-0555. (800) 333-8300. 1966. $76.95.

Laboratory Handbook of Materials, Equipment, and Techniques. Gary S. Coyne. Prentice Hall (Division of Simon and Schuster), 15 Columbus Circle, New York, NY 10023. (212) 373-8500 or (800) 922-0579. 1992. $45.00.

Lange's Handbook of Chemistry. John A. Dean, editor. McGraw-Hill Publishing Company, 1221 Avenue of the Americas, New York, NY 10020. (800) 262-4729 or (212) 512-3825. Fourteenth edition. 1992. $79.50.

Riegel's Handbook of Industrial Chemistry. James A. Kent, editor. Van Nostrand Reinhold, 115 Fifth Avenue, New York, NY 10003. (212) 254-3232 or (800) 926-2665. Ninth edition. 1992. $114.95.

ORGANIC CHEMISTRY

Ency. of Physical Sciences and Engineering Info. Sources

Tables of Spectral Data For Structure Determination of Organic Compounds. E. Pretsch, et al. Springer-Verlag New York, Inc., 175 Fifth Avenue, New York, NY 10010. (212) 460-1500. (800) 777-4643. FAX (212) 473-6272. 1994. $49.95 in paper.

ONLINE DATABASES AND CD-ROMS

Beilstein Online. Beilstein-Institut fur Literatur der Organischen Chemie, Varrentrapperstrasse, 40-42, D-6000, Frankfurt am Main 90, Germany. Contains data on carbon compounds from the Beilstein Handbook of Organic Chemistry. Structural and factual data for more than 3.4 million heterocyclic, isocyclic, and acyclic compounds included. Covers the period 1830 to present. Available on DIALOG, (800) 334-2564, ORBIT (800) 456-7248, and STN International, FIZ Karlsruhe, P.O. Box 2465, W-7500, Karlsruhe 1, Germany, online services. Inquire as to cost and availability.

CA Search. Chemical Abstracts Service, P.O. Box 3012, Columbus, OH 43210-0012. (614) 447-3600. (800) 848-6533. FAX (614) 447-3709. Very comprehensive guide to worldwide chemical literature and related fields, 1972 to present. Available on BRS,(800) 289-4277, DIALOG, (800) 334-2564, ORBIT (800) 456-7248, and STN International, FIZ Karlsruhe, P.O. Box 2465, W-7500, Karlsruhe 1, Germany, online services. Inquire as to cost and availability.

Chemical Journals of the American Chemical Society. American Chemical Society, 1155 16th Street, N.W., Washington, DC 20036. (202) 872-4381 or (800) 424-6747. Contains complete text of approximately 90,000 articles from 22 primary journals published by the American Chemical Society, from mostly 1982 to present. Available on STN International, FIZ Karlsruhe, P.O. Box 2465, W-7500, Karlsruhe 1, Germany, online service. Inquire as to cost and availability.

Compendex Plus. Engineering Information, Inc., 345 East 47th Street, New York, NY 10017. (212) 705-7600 or (800) 221-1044. Contains citations with abstracts to worldwide literature in engineering and technology, from 1970 to present. Available on online BRS,(800) 289-4277, DIALOG, (800) 334-2564, ORBIT (800) 456-7248, and STN International, FIZ Karlsruhe, P.O. Box 2465, W-7500, Karlsruhe 1, Germany, online services. Also available on CD-ROM. Inquire as to cost and availability.

Current Contents Search. Institute for Scientific Information, 3501 Market Street, Philadelphia, PA 19104. (215) 386-0100. FAX (215) 386-6362. Contains citations to articles listed in the table of contents of science and technology journals. Also articles in social sciences and life sciences journals. Available on BRS,(800) 289-4277, DIALOG,(800) 334-2564, online services. Inquire as to cost and availability.

Dissertation Abstracts. University Microfilms International, 300 North Zeeb Road, Ann Arbor, MI 48106. (800) 521-0600 or (313) 761-4700. Scope includes virtually all doctoral dissertations accepted at accredited American institutions from 1861 to present in 252 subject areas. Available on BRS,(800) 289-4277, DIALOG,(800) 334-2564, and OCLC EPIC,(800) 848-5878, online services. Also available on CD-ROM. Inquire as to cost and availability.

Kirk-Othmer Encyclopedia of Chemical Technology. John Wiley and Sons, Inc., 605 Third Avenue, New York, NY 10158. (800) 526-5368 or (212) 850-6000. Contains the complete text of all chapters in the 27 volume fourth edition of the KIRK-Othmer ENCYCLOPEDIA of CHEMICAL TECHNOLOGY. 1991.

Available on BRS,(800) 289-4277, DIALOG,(800) 334-2564, online services. Inquire as to cost and availability.

NTIS Bibliographic Database. National Technical Information Service, 5285 Port Royal Road, Springfield, VA 22161. (703) 487-4929 or FAX (703) 321-8199. Broad coverage of government-sponsored science and technology research reports, 1964 to present. Available on BRS,(800) 289-4277, DIALOG, (800) 334-2564, ORBIT, (800) 456-7248, and STN International, FIZ Karlsruhe, P.O. Box 2465, W-7500, Karlsruhe 1, Germany, online services. Also available on CD-ROM. Inquire as to cost and availability.

SCISEARCH. Institute for Scientific Information, 3501 Market Street, Philadelphia, PA 19104. (800) 523-1850 or (215) 386-0100. Broad multidisciplinary title and author index to the international literature of science and technology, 1974 to present. Available on DIALOG,(800) 334-2564, and ORBIT,(800) 456-7248, online services. Also available on CD-ROM. Inquire as to cost and availability.

WILSONLINE. H.W. Wilson Company, 950 University Avenue, Bronx, NY 10452. (800) 367-6770 or (212) 588-8400. Makes available online versions of the printed H.W. Wilson indexes including *Applied Science and Technology Index, Business Periodicals Index, General Science Index,* and *Readers' Guide to Periodical Literature.* Period covered is generally 1983 to present. Available on BRS,(800) 289-4277, DIALOG,(800) 334-2564, and OCLC EPIC,(800) 848-5878, online services. Also available on CD-ROM. Inquire as to cost and availability.

PERIODICALS

American Oil Chemists' Society. Journal. American Oil Chemists' Society, 1608 Broadmoor Drive, Box 3489, Champaign, IL 61821-0489. (217) 359-2344. 1917 to present. Monthly. $185.00 per year. ISSN: 0003-021X.

Aiche Journal. American Institute of Chemical Engineers, 345 47th Street, New York, NY 10017-2396. (212) 705-7338. FAX (212) 752-3294. 1955 to present. Monthly. $395.00 per year. ISSN: 0001-1541.

Carbon. Elsevier Science Publishing Company, Inc., 660 White Plains Road, Tarrytown, NY 10591-5153. (914)524-9200. FAX (914) 524-9200. 1963 to present. Bi-monthly. $1045.00 ISSN: 0008-6223.

Chemical Week. Includes Annual Buyers Guide. Chemical Week Associates, 888 Seventh Avenue, New York, NY 10106. (212) 621-4900. FAX (212) 621-4949. 1914 to present. Weekly. $99.00 per year. ISSN: 0009-272X.

Chemtech. American Chemical Society, Box 3337, Columbus, OH 43210. (614) 447-3776. (800) 333-9511. FAX (614) 447-3671. 1970 to present. Monthly. $79.00 per year. ISSN: 0009-2703.

Hydrocarbon Processing. Gulf Publishing Company, P.O. Box 2608, Houston, TX 77001. (713) 520-4433. 1922 to present. Monthly. $22.00. ISSN: 0887-0284

Journal of Heterocyclic Chemistry. American Chemical Society, Box 3337, Columbus, OH 43210. (614) 447-3776. (800) 333-9511. FAX (614) 447-3671. 1964 to present. Bi-monthly. $420.00 per year. ISSN: 0022-152X.

Journal of Organic Chemistry. American Chemical Society, Box 3337, Columbus, OH 43210. (614) 447-3776. (800) 333- 9511. FAX (614) 447-3671. 1936 to present. Bi-weekly. $785.00 per year. ISSN: 0022-3263.

Journal of Polymer Science. Part A: Polymer Chemistry. John Wiley and Sons, Inc., 605 Third Avenue, New York, NY 10158. (800) 526-5368 or (212) 850-6000. 1962 to present. 13 issues per year. $2540.00 per year. ISSN: 9887-624X.

Organometallics. American Chemical Society, Box 3337, Columbus, OH 43210. (614) 447-3776. (800) 333-9511. FAX(614) 447-3671. 1982 to present. Monthly. $850.00 per year. ISSN: 0276-7333.

Tetrahedron. Elsevier Science Publishing Company, Inc., 660 White Plains Road, Tarrytown, NY 10591-5153. (914)524-9200.FAX (914) 524-9200. 1957 to present. Sixty times per year. $5365.00 per year. ISSN: 0040-4020.

Tetrahedron Letters. Elsevier Science Publishing Company, Inc., 660 White Plains Road, Tarrytown, NY 10591-5153. (914)524-9200. FAX (914) 524-9200. 1959 to present. Weekly. $4745.00 per year. ISSN: 0040-4039.

RESEARCH CENTERS AND INSTITUTES

Center for Catalytic Science and Technology. University of Delaware, Departments of Chemical Engineering and Chemistry, Newark, DE 19716. (302) 451-8056. FAX (302) 451-1048.

Chemical Laboratories. Harvard Universiy. Oxford Street, Cambridge, MA 02138. (617) 495-4283. FAX (617) 496-5618.

Chemistry Laboratories. Rensselaer Polytechnic Institute. Cogswell Laboratory, Troy, NY 12180-3590. (518) 276-8981.

Chemical Research Laboratory. Brown University, Providence, RI 02912. (401) 863-2256.

University/Industry Chemical Research Center. Mississippi State University, Department of Chemistry, P.O. Drawer CH, Mississippi State, MS 39762. (601) 325-3584.

ORGANOMETALLIC COMPOUNDS

See: ORGANIC CHEMISTRY

OSCILLOSCOPE

See: INSTRUMENTATION

OXYGEN

See also: ELECTROCHEMISTRY, ELEMENTS, HYDROGEN, NAMES of INDIVIDUAL ELEMENTS.

ABSTRACT SERVICES AND INDEXES

Analytical Abstracts. Royal Society of Chemistry, Information Services, Thomas Graham House, Science Park, Milton Road, Cambridge, CB4 4WF, England. Contains citations, mostly with abstracts, of the worldwide literature on analytical chemistry, from 1954 to present. Monthly. $636.00 per year. Also available online.

Applied Science and Technology Index: A Cumulative Subject Index to English Language Periodicals in the Fields of Aeronautics and Space Science, Computer Technology, Chemistry, Construction Industry, Energy and Related Areas. H. W. Wilson Co., 950 University Avenue, Bronx, NY 10452- 9978. (212) 588-8400. (800) 367-6770. FAX (718) 590-1617.From 1958 to present. Monthly. Inquire about cost and availability. Also available online (BRS and WILSONLINE) and on CD-ROM. ISSN: 0003-6986.

Chemical Abstracts. Chemical Abstracts Service. 2540 Olentangy River Road, Box 3012, Columbus, OH 43210-0012. (614) 447-3600. FAX (614) 447-3713. 1907 to present. Weekly. $16,800.00 per year. Available online and on CD- ROM. CA is also available in five section groupings. Inquire regarding cost and availability.

Current Contents: Physical, Chemical and Earth Sciences. Institute for Scientific Information, 3501 Market Street, Philadelphia, PA 19104. (215) 386-0100. FAX (215) 386-2291. From 1961 to present. Weekly. $442.00 per year. Also available on CD-ROM and online. Inquire regarding cost and availability. ISSN: 0163-2574.

Engineering Index Monthly. Engineering Information, Inc., Castle Point on the Hudson, Hoboken, NJ 07030. (800) 221-1044. FAX (201) 216-8532.Monthly. $2200.00 per year. Also available online as COMPENDEX, and also on CD-ROM.

General Science Index. H.W. Wilson Company, 950 University Avenue, Bronx, NY 10452. (212) 588-8400. (800) 367-6770. FAX (718) 590-1617. From 1978 to present. Ten issues per year; quarterly and annual cumulations. Service basis. Available on CD-ROM and online. Inquire regarding cost and availability. ISSN: 0162-1963.

Physics Abstracts. INSPEC. Section A, Science Abstracts. Institution of Electrical Engineers (IEE), London, United Kingdom. Available from: INSPEC/IEEE -Institute of Electrical and Electronic Engineers, Box 1331, 445 Hoes Lane, Piscataway, NJ 08855-1331. (908) 562-5549. 1898 to present. 24 issues per year. $2835.00 per year. ISSN: 0036-8091. Also available online and on CD-ROM.

Science Citation Index (SCI). Institute for Scientific Information, 3501 Market Street, Philadelphia, PA 19104. (215) 386-0100. (800) 523-1850. FAX (215) 386-2991. 1961 to present. Six issues per year, plus annual cumulation. $11650.00 per year. Also available online and on CD-ROM. Inquire about price and availability. ISSN: 0036-827X.

ASSOCIATIONS AND PROFESSIONAL SOCIETIES

American Chemical Society, 1155 16th Street, Nw, Washington, DC 20036. (202) 872-4600.

American Institute of Chemical Engineers. 345 East 47th Street, New York, NY 10017-2396. (212) 705-7663. (202) 752-3294.

OXYGEN

Ency. of Physical Sciences and Engineering Info. Sources

American Institute of Chemists. 7315 Wisconson Avenue, Bethesda, MD 20814. (301) 652-2447. FAX (301) 657-3549.

American Institute of Physics, 335 East 45th Street, New York, NY 10017 (212) 661-9404 OR Fax (516)349-9704.

Association of Consulting Chemists and Chemical Engineers, 50 East 41st Street, Suite 92, New York, NY 10017. (212) 684-6255.

Chemical Manufacturers Association. 2501 M Street, N.W., Washington, DC 20037. (202) 887-1100. FAX (202) 202-1237.

International Oxygen Manufacturers Association. PO Box 16248, Cleveland, OH 44116-0248. (216) 228-2166. FAX (216) 228-5810.

BIBLIOGRAPHIES

Chemical Information. H. R. Collier, editor. Springer- Verlag New York, Inc., 175 Fifth Avenue, New York, NY 10010. (212) 460-1500. (800) 777-4643. 1990. $91.00.

Chemical Information Sources. Gary Wiggins. McGraw-Hill Publishing Company, 1221 Avenue of the Americas, New York, NY 10020. (212) 512-2000. (800) 262-4729. 1991. $42.50.

Information Sources in Chemistry. R.T. Bottle and J.F.B Rowland, editors. R.R. Bowker Inc., 121 Chanlon Road, New Providence, NJ 07974. (908) 464-6800. (800) 521-8110. Fourth edition. 1993. $75.00.

Scientific and Technical Books and Serials in Print: An Index to Literature in Science and Technology. R.R. Bowker Inc., 121 Chanlon Road, New Providence, NJ 07974. (908) 464-6800. (800) 521-8110. FAX (908) 665-3502. 1972 to present. Annual. 4 volumes. 1994. $299.95. Also available on compact disc and online. ISSN: 0000-054X.

DIRECTORIES AND BIOGRAPHICAL SOURCES

American Institute of Chemists: Professional Directory. American Institute of Chemists, 501 Wythe Street, Alexandria, VA 22314-1917. (703) 836-2090. FAX (703) 836-2091. Annual. $65.00.

American Men and Women of Science: Physical and Biological Sciences. R.R. Bowker Inc., 121 Chanlon Road, New Providence, NJ 07974. (908) 464-6800. (800) 521-8110. 20thedition. 8 volumes. 1996. $850.00

Chemical Week: Buyers Guide Issue. Chemical Week Associates, 888 Seventh Avenue, New York, NY 10106. (212) 621-4900. FAX (212) 621-4949. Annual, October. $50.00.

Chemical Sources International. Mike Desing and Kurt Gandenberger, editors. Chemical Sources International, Inc., P.O. Box 1824, Clemson, SC 29633. (803) 646-7840. 1994. $285.00.

Directory of Chemistry Software 1992. Wendy Warr, Peter Willett, Geoff Downs. American Chemical Society, 1155 16th Street, NW, Washington, DC 20036. (202) 872-4600. 1992. $35.95.

Hydrogen Energy International Directory. Rector Press, Ltd., 130 Rattlesnake, Leverett, MA 01054-9726. (413) 548-9708. (800) 247-3473. 1994. $395.00.

Research Centers Directory. Gale Research, 835 Penobscot Building, Detroit, MI 48226-4094. (313) 961-2242. (800) 347-4253. 20th edition. 1995. $485.00. ISSN: 0080-1518.

ENCYCLOPEDIAS AND DICTIONARIES

Academic Press Dictionary of Science and Technology. Christopher Morris, editor. Academic Press, Inc., 6277 Sea Harbor Drive, Orlando, FL. (800) 321-5068. 1991. $115.00.

Concise Encyclopedia Chemistry. Translated and revised by Mary Eagleson. Walter de Gruyter, Inc., 200 Saw Mill River Road, Hawthorne, New York, 10532. (914) 747-0110. Fax (914) 747-1326. 1994. $69.95.

Concise Encyclopedia of Chemical Technology. John Wiley & Sons, Inc., 605 Third Avenue, New York, NY 10158-0012. (212) 850-6000. (800) 225-5945. 2nd edition. 1989. $99.95.

Dictionary of Chemical Names and Synonyms. Philip A. Howard and Michael W. Neil. Lewis Pub./ CRC Press, Inc., 2000 Corporate Boulevard, NW, Boca Raton, FL 33431. (407) 994- 0555. (800) 272-7737. 1991. $149.95.

Encyclopedia of Applied Physics. VCH Publications, Inc., 220 East 23rd Street, Suite 909, New York, NY 10010-4606. (212) 683-8333. (800) 422-8824. 1991-. Twenty volumes. $6000.00.

Gardner's Chemical Synonyms and Trade Names. Michael Ash and Irene Ash. Ashgate Publishing Co., Old Post Road, Brookfield, VT 05036. (802) 276-3162 10th edition. 1994. $195.00.

Grant & Hawks Chemical Dictionary. R. A. Grant. McGraw-Hill Publishing Company, 1221 Avenue of the Americas, New York, NY 10020. (212) 512-2000. (800) 262-4729. 5th edition. 1987. $64.50.

Hawley's Condensed Chemical Dictionary. Richard J. Lewis. Van Nostrand Reinhold, 115 Fifth Avenue, New York, NY 10003. (212) 254-3232. (800) 842-3636. 12th edition. 1993. $69.95.

Kirk-Othmer Encyclopedia of Chemical Technology. John Wiley and Sons, Inc., 605 Third Avenue, New York, NY 10158-0012. (212) 850-6000. (800) 225-5945. Fourth edition. 1991-present. Twenty-seven volumes. $5400.00.

McGraw-Hill Encyclopedia of Chemistry. Sybil P.Parker, editor. McGraw-Hill Book, Incorporated, 1221 Avenue of the Americas, New York, NY 10020. (212) 512-2000. (800) 262-4729. Second edition. 1993. $95.50.

McGraw-Hill Encyclopedia of Science and Technology. McGraw-Hill Book, Incorporated, 1221 Avenue of the Americas, New York, NY 10020. (212) 512-2000. (800) 262-4729. Seventh edition. Twenty volumes. 1992. $1900.00.

Mcketta's Encyclopedia of Chemical Processes and Design. Marcel Dekker, Inc., 270 Madison Avenue, New York, NY 10016. (212) 696-9000. (800) 228-1160., 1979-present . $175.00 per volume.

GENERAL WORKS

Advances in Oxygenated Processes. Alfons L. Baumstark. Jai Press, 55 Old Post Road, Number 2, P. O. Box 1678, Greenwich, CT 06836. (203) 661-7602. 1988. $73.25.

Bulk Oxygen Systems At Consumer Sites. National Fire Protection Association, Batterymarch Park, Quincy, MA 02269. (617) 770-3000. 1990. $16.75 in paper.

Determining the Molar Volume of Oxygen. John N. Bedenbaugh, et al. Chemical Education Resources, Inc., P. O. Box 357, Palmrya, PA 17078. (717) 838-3165. 1990. $1.25.

Electrochemical Oxygen Technology. Kim Kinoshita. John Wiley & Sons, Inc., 605 Third Avenue, New York, NY 10158- 0012. (212) 850-6000. (800) 225-5945. 1992. $195.00.

Oxygen and Ozone: Gas Solubilities. Battino. Franklin Book Co., Inc., Elkins Park, PA 19117. (215) 635-5252. 1981. $216.00.

Oxygen Chemistry. Donald T. Sawyer. Oxford University Press, Inc., 200 Madison Avenue, New York, NY 10016. (212) 725-6000. (800) 334-4249. 1991. $35.00.

Oxyacetylene Welding. Ronald J. Baird. Goodheart-Wilcox, Co. 123 West Taft Drive, South Holland, IL 60473-2089. (708) 333-7200. (800) 323-0440. 1991. $14.40 in paper.

Preparing and Studying Oxygen and Some of Its Compounds. H. A. Neidig, editor. Chemical Education Resources, Inc., P. O. Box 357, Palmrya, PA 17078. (717) 838-3165. 1992. $1.25.

Thermodynamic Properties of Oxygen. V. V. Sychev. Hemisphere Publishing Corp., 1900 Frost Road, Suite 101, Bristol PA 19007. (215) 785-5800. (800) 821-8312. 1987. $167.00.

HANDBOOKS AND MANUALS

Chemical Formulary. H. Bennett, editor. Chemical Publishing Company, Inc., 80 Eighth Avenue, New York, NY 10011. (212) 255-1950. Volumes 1-30. $70.00 per volume.

Chemical Reference Handbook. Gordon Press Publications, P.O. Box 459, Bowling Green Station, New York, NY 10004. (718) 624-8419. 1995. $260.00.

CRC Handbook of Chemistry and Physics. David R. Kide, editor. CRC Press Inc., 2000 Corporate Blvd., NW, Boca Raton, FL 33431. (407) 994-0555. (800) 333-8300. 77th edition. 1996. $99.95.

CRC Handbook of Methods for Oxygen Radical Research. Robert A. Greenwald. Franklin Book Co., Inc., Elkins Park, PA 19117. (215) 635-5252. 1985. $195.00.

Guide to Basic Chemical Compounds. D.R. Lide, Jr. CRC Publishers, Inc., 2000 Corporate Blvd., N.W., Boca Raton, FL 33431. (407) 994-0555 or (800) 333-8300. 1993. $120.00.

Handbook of Properties of Condensed Phases of Hydrogen and Oxygen. B. I. Verkin. Hemisphere Publishing Corp., 1900 Frost Road, Suite 101, Bristol PA 19007. (215) 785-5800. (800) 821-8312. 1991. $116.00.

Lange's Handbook of Chemistry. John A. Dean, editor. McGraw-Hill Publishing Company, 1221 Avenue of the Americas, New York, NY 10020. (800) 262-4729 or (212) 512-3825. Fourteenth edition. 1992. $79.50.

Riegel's Handbook of Industrial Chemistry. James A. Kent, editor. Van Nostrand Reinhold, 115 Fifth Avenue, New York, NY 10003. (212) 254-3232 or (800) 926-2665. Ninth edition. 1992. $114.95.

ONLINE DATABASES AND CD-ROMS

Analytical Abstracts Online. Royal Society of Chemistry, Information Services, Thomas Graham House, Science Park, Milton Road, Cambridge, CB4 4WF, England. Contains citations, mostly with abstracts, of the worldwide literature on analytical chemistry, from 1980 to present. Available on DIALOG, (800) 334-2564, and STN International, FIZ Karlsruhe, P.O. Box 2465, W-7500, Karlsruhe 1, Germany, online services. Inquire as to cost and availability.

Beilstein Online. Beilstein-Institut fur Literatur der Organischen Chemie, Varrentrapperstrasse, 40-42, D-6000, Frankfurt am Main 90, Germany. Contains data on carbon compounds from the Beilstein Handbook of Organic Chemistry. Structural and factual data for more than 3.4 million heterocyclic, isocyclic, and acyclic compounds included. Covers the period 1830 to present. Available on DIALOG, (800) 334-2564, ORBIT (800) 456-7248, and STN International, FIZ Karlsruhe, P.O. Box 2465, W-7500, Karlsruhe 1, Germany, online services. Inquire as to cost and availability.

CA Search. Chemical Abstracts Service, P.O. Box 3012, Columbus, OH 43210-0012. (614) 447-3600. (800) 848-6533. FAX (614) 447-3709. Very comprehensive guide to worldwide chemical literature and related fields, 1972 to present. Available on BRS,(800) 289-4277, DIALOG, (800) 334-2564, ORBIT (800) 456-7248, and STN International, FIZ Karlsruhe, P.O. Box 2465, W-7500, Karlsruhe 1, Germany, online services. Inquire as to cost and availability.

Chemical Journals of the American Chemical Society. American Chemical Society, 1155 16th Street, N.W., Washington, DC 20036. (202) 872-4381 or (800) 424-6747. Contains complete text of approximately 90,000 articles from 22 primary journals published by the American Chemical Society, from mostly 1982 to present. Available on STN International, FIZ Karlsruhe, P.O. Box 2465, W-7500, Karlsruhe 1, Germany, online service. Inquire as to cost and availability.

Compendex Plus. Engineering Information, Inc., 345 East 47th Street, New York, NY 10017. (212) 705-7600 or (800) 221-1044. Contains citations with abstracts to worldwide literature in engineering and technology, from 1970 to present. Available on online BRS,(800) 289-4277, DIALOG, (800) 334-2564, ORBIT (800) 456-7248, and STN International, FIZ Karlsruhe, P.O. Box 2465, W-7500, Karlsruhe 1, Germany, online services. Also available on CD-ROM. Inquire as to cost and availability.

Current Contents Search. Institute for Scientific Information, 3501 Market Street, Philadelphia, PA 19104. (215) 386-0100. FAX (215) 386-6362. Contains citations to articles listed in the table of contents of science and technology journals. Also articles in social sciences and life sciences journals. Available on BRS,(800) 289-4277, DIALOG,(800) 334-2564, online services. Inquire as to cost and availability.

OXYGEN

Ency. of Physical Sciences and Engineering Info. Sources

Dissertation Abstracts. University Microfilms International, 300 North Zeeb Road, Ann Arbor, MI 48106. (800) 521-0600 or (313) 761-4700. Scope includes virtually all doctoral dissertations accepted at accredited American institutions from 1861 to present in 252 subject areas. Available on BRS,(800) 289-4277, DIALOG,(800) 334-2564, and OCLC EPIC,(800) 848-5878, online services. Also available on CD-ROM. Inquire as to cost and availability.

Gmelin Database. Gmelin-Institut fur Anorganische Chemie und Grenzgebiete, Varrentrapperstrasse, 40-42, Carl-Bosch-Haus, D-6000, Frankfurt am Main 90, Germany. Contains structural and factual data relating to inorganic and organometallic chemistry. Provides data from the Gmelin Handbook of Inorganic and Organometallic Chemistry. Covers the period 1817 to 1975; 1988-89. Available on STN International, FIZ Karlsruhe, P.O. Box 2465, W-7500, Karlsruhe 1, Germany, online service. Inquire as to cost and availability.

Kirk-Othmer Encyclopedia of Chemical Technology. John Wiley and Sons, Inc., 605 Third Avenue, New York, NY 10158. (800) 526-5368 or (212) 850-6000. Contains the complete text of all chapters in the 27 volume fourth edition of the KIRK-Othmer *Encyclopedia of Chemical Technology.* 1991. Available on BRS,(800) 289-4277, DIALOG,(800) 334-2564, online services. Inquire as to cost and availability.

Metals Datafile. Materials Information, ASM International, Materials Park, OH 44073. (216) 338-5151.Contains designation and specification numbers for ferrous and nonferrous metals and alloys. Covers the period 1982 to present. Available on ORBIT,(800) 456-7248, and STN International, FIZ Karlsruhe, P.O. Box 2465, W-7500, Karlsruhe 1, Germany, online services. Inquire as to cost and availability.

NTIS Bibliographic Database. National Technical Information Service, 5285 Port Royal Road, Springfield, VA 22161. (703) 487-4929 or FAX (703) 321-8199. Broad coverage of government-sponsored science and technology research reports, 1964 to present. Available on BRS,(800) 289-4277, DIALOG, (800) 334-2564, ORBIT, (800) 456-7248, and STN International, FIZ Karlsruhe, P.O. Box 2465, W-7500, Karlsruhe 1, Germany, online services. Also available on CD-ROM. Inquire as to cost and availability.

Physics Briefs. American Institute of Physics, 335 East 45th Street, New York, NY 10017. (212) 661-9260 or FAX (212) 661-2036. Contains citations with abstracts of the literature of physics and related fields, 1979 to present. Available on the STN International, FIZ Karlsruhe, P.O. Box 2465, W-7500, Karlsruhe 1, Germany, online service. Inquire as to cost and availability.

SCISEARCH. Institute for Scientific Information, 3501 Market Street, Philadelphia, PA 19104. (800) 523-1850 or (215) 386-0100. Broad multidisciplinary title and author index to the international literature of science and technology, 1974 to present. Available on DIALOG,(800) 334-2564, and ORBIT,(800) 456-7248, online services. Also available on CD-ROM. Inquire as to cost and availability.

WILSONLINE. H.W. Wilson Company, 950 University Avenue, Bronx, NY 10452. (800) 367-6770 or (212) 588-8400. Makes available online versions of the printed H.W. Wilson indexes including Applied Science and Technology Index, Business Periodicals Index, General Science Index, and Readers' Guide to Periodical Literature. Period covered is generally 1983 to present. Available on BRS,(800) 289-4277, DIALOG,(800) 334-2564, and OCLC EPIC,(800) 848-5878, online services. Also available on CD-ROM. Inquire as to cost and availability.

PERIODICALS

Analytical Chemistry. American Chemical Society, Box 3337, Columbus, OH 43210. (614) 447-3776. 1929 to present. Semimonthly. $415.00 per year. ISSN: 0003-2700.

Angewandte Chemie. VCH Publishers, Inc., 220 East 23rd Street, New York, NY 10010-4606. (212) 683-8333. (800) 367-8249. 1888 to present. 22 issues per year. $840.00 per year. ISSN: 0044-8249.

Chemical Reviews. American Chemical Society, Box 3337, Columbus, OH 43210. (614) 447-3776. 1924 to present. 8 issues per year. $346.00 per year. ISSN: 0009-2665.

Chemical Week. Includes Annual Buyers Guide. Chemical Week Associates, 888 Seventh Avenue, New York, NY 10106. (212) 621-4900. FAX (212) 621-4949. 1914 to present. Weekly. $99.00 per year. ISSN: 0009-272X.

Chemtech. American Chemical Society. Box 3337, Columbus, OH 43210. (614) 447-3776. 1970 to present. Monthly. $370.00 per year. ISSN: 0009-2703.

Inorganic Chemistry. American Chemical Society, Box 3337, Columbus, OH 43210. (614) 447-3776. 1962 to present. Semimonthly. $500.00 per year. ISSN: 0020-1669.

Journal of the American Chemical Society. American Chemical Society, Box 3337, Columbus, OH 43210. (614) 447-3776. 1879 to present. Biweekly. $1055.00 per year. ISSN: 0002-7863.

Polyhedron. Elsevier Science Publishing Company, Inc., 660 White Plains Road, Tarrytown, NY 10591-5153. (914)524-9200. FAX (914) 524-9200. 1982 to present. 24 issues per year. $2175.00 per year. ISSN: 0277-5387.

RESEARCH CENTERS AND INSTITUTES

Chemical Laboratories. Harvard University, Oxford Street, Cambridge, MA 02138. (617) 495-4283. FAX (617) 496-5618.

Chemistry Laboratories. Rensselaer Polytechnic Institute, Cogswell Laboratory, Troy, NY 12180-3590. (518) 276-8981.

Lawrence Berkeley Laboratory, Chemical Sciences Division. One Cyclotron Road, Building 66, Berkeley, CA 94720. (510) 486-6062. FAX (510) 486-4995.

Theoretical Chemistry Institute. University of Wisconsin, Madison. 1101 University Avenue, Madison, WI 53706. (608) 262-1511.

University/Industry Chemical Research Center. Mississippi State University, Department of Chemistry, P.O. Drawer CH, Mississippi State, MS 39762. (601) 325-3584.

OZONE

See also: AIR POLLUTION, ATMOSPHERE, GREENHOUSE EFFECT, METEOROLOGY

Ency. of Physical Sciences and Engineering Info. Sources

OZONE

ABSTRACT SERVICES AND INDEXES

Applied Science and Technology Index: A Cumulative Subject Index to English Language Periodicals in the Fields of Aeronautics and Space Science, Computer Technology, Chemistry, Construction Industry, Energy and Related Areas. H.W. Wilson Co., 950 University Avenue, Bronx, NY 10452-9978. (212) 588-8400. (800) 367-6770. FAX (718) 590-1617. From 1958 to present. Monthly. Inquire about cost and availability. Also available online (BRS and WILSONLINE) and on CD-ROM. ISSN: 0003-6986.

Chemical Abstracts. Chemical Abstracts Service. 2540 Oleantangy River Road, Box 3012, Columbus, OS 43210-0012. (614) 447-3600. FAX (614) 447-3713. 1907 to present. Weekly. $16,800.00 per year. Available online and on CD-ROM. CA is also available in five section groupings. Inquire regarding cost and availability.

General Science Index. H.W. Wilson Co., 950 University Avenue, Bronx, NY 10452. (212) 588-8400. (800) 367-6770. FAX (718) 590-1617. From 1978 to present. Ten issues per year; quarterly and annual cumulations. Service basis. Available on CD-ROM and online. Inquire regarding cost and availability. ISSN: 0162-1963.

Government Reports Announcements and Index. National Technical Information Service (NTIS), 5285 Port Royal Road, Springfield, VA 22161. (703) 487-4650. FAX (703) 321-8547. From 1968 to present. Annual. $630.00 per year. Also available online as *NTIS Bibliographic Database* and on CD-ROM.

Meteorological and Geoastrophysical Abstracts. American Meteorological Society, c/o Inforonics, Inc., 550 Newtown Road, Box 458, Littleton, MA 01460. (508) 486-8976. FAX (508) 486-0027. Covers literature in environmental sciences, meteorology, astrophysics, hydrology, glaciology, and physical oceanography. 1950 to present. Monthly. $950.00 per year. ISSN: 0026-1130. Also available online (DIALOG) and on CD-ROM.

Oceanic Abstracts. Cambridge Scientific Abstracts, 7200 Wisconsin Avenue, Bethesda, MD 20814. (301) 961-6750. Fax (301) 961-6720. Bimonthly. $995.00 per year.

Science Citation Index (SCI). Institute for Scientific Information, 3501 Market Street, Philadelphia, PA 19104. (215) 386-0100. (800) 523-1850. FAX (215) 386-2991. 1961 to present. Six issues per year, plus annual cumulation. $11650.00 per year. Also available online and on CD-ROM. Inquire about price and availability. ISSN: 0036-827X.

Selected Water Resources Abstracts. U.S. Geological Survey. Distributed by National Technical Information Service (NTIS), 5285 Port Royal Road, Springfield, VA 22161. (703) 487-4650. Monthly. $115.00 per year.

ANNUAL REVIEWS AND YEARBOOKS

American Meteorological Society. Meteorological Monographs. American Meteorological Society, 45 Beacon Street, Boston, MA 02108-3693. (617) 227-2425. FAX (617) 742-8718. Irregular. Price varies.

Annual Review of Earth and Planetary Sciences. Annual Reviews, Inc., 4139 El Camino Way, Palo Alto, CA 94306-0897. (415) 493-4400. Fax (415) 855-9815. Annual. $55.00 per year.

Developments in Atmospheric Science Series. Elsevier Science Publishing Company, Inc., 655 Avenue of the Americas, New York, NY 10010. (212) 989-5800. Irregular. Inquire about price and availability.

Ocean Yearbook. Elisabeth M. Borgese, et al, editors. University of Chicago Press, 5801 Ellis Avenue, Chicago, IL 60637. (312) 702-7700. 1995. $77.00.

ASSOCIATIONS AND PROFESSIONAL SOCIETIES

American Association of Radon Scientists and Technologists. P.O. Box 70, Park Ridge, NJ 07656. (201) 391-6445. FAX (201) 391-0670.

American Association of State Climatologists. c/o Myron Moinau, University of Idaho, Agriculture Engineering Department, State Climate Service, Moscow, ID 834844-2040.

American Meteorological Society, 45 Beacon Street, Boston, MA 02108-3693. (617) 227-2425. FAX (617) 742-8718. Irregular. Price varies.

Greenhouse Crisis Foundation. 1130 17th Street NW, Suite 630, Washington, DC 20036. (202) 466-2823.

International Association of Meteorology and Atmospheric Physics. UCAR, P.O. Box 3000, Boulder CO 80307-3000.

(303) 497-1673.

International Ozone Association. c/o Pan American Group, 31 Strawberry Hill Avenue, Stamford, CT 06902. (203) 348-3542. FAX (203) 967-4845.

National Weather Association, 6704 Wolke Court, Montgomery, AL 36116-2134. (334) 213-0388. FAX (334) 213-0388.

University Corporation for Atmospheric Research. P.O. Box 3000, 1850 Table Mesa Drive, Boulder, CO 80307-3000. (303) 497-1673. FAX (303) 497-1654.

BIBLIOGRAPHIES

Bibliography On Ozone: Its Biological Effects and Technical Applications. Harald Rosenthal and J. Scott Wilson. Diane Publishing Co., 600 Upland Avenue, Upland, PA 19015. (610) 499-7415. 1993. $95.00

The Greenhouse Effect: A Bibliography. Joan Nordquist. Reference and Research Services, 511 Lincoln Street, Santa Cruz, CA 95060. (408) 426-4479. 1990. $15.00.

Information Sources in the Earth Sciences. David N. Wood, Joan E. Hardy, and Anthony P. Harvey, editors. Bowker-Saur/K.G. Saur. Distributed by R.R. Bowker, 121 Chanlon Road, New Providence, NJ 07974. (800) 521-8110 or (908) 464-6800. Second edition. 1989. $85.00.

Scientific and Technical Books and Serials in Print: An Index to Literature in Science and Technology. R.R. Bowker Co., 205 E. 42nd Street, New York, NY 10017. (800) 521-8110 or (212) 916-1600. 1992. $250.00.

OZONE

Ency. of Physical Sciences and Engineering Info. Sources

DIRECTORIES AND BIOGRAPHICAL SOURCES

American Men and Women of Science: Physical and Biological Sciences. Eighteenth edition. R.R. Bowker Company, 245 West 17th Street, New York, NY 10011. (800) 521-8810 or (212) 916-1600. $750.00.

American Meteorological Society. Professional Directory. American Meteorological Society. 45 Beacon Street, Boston, MA 02108. (617) 227-2425. Included in each issue of the Bulletin of the Society.

Meteorological Services of the World. American Meteorological Society, 45 Beacon Street, Boston, MA 02108-3693. (617) 227-2425. FAX (617) 742-8718. Annual. $70.00.

National Weather Service Offices and Stations. National Oceanic and Atmospheric Administration, Deparatment of Commerce, Silver Spring, MD 20910. (301) 427-7698. Annual. Free

Research Centers Directory. Gale Research, 835 Penobscot Building, Detroit, MI 48226-4094. (313) 961-2242. (800) 347-4253. 20th edition. 1995. $485.00. ISSN: 0080-1518.

Who's Who in the World of Ozone. International Ozone Association. c/o Pan American Group, 31 Strawberry Hill Avenue, Stamford, CT 06902. (203) 348-3542. annual. $65.00.

ENCYCLOPEDIAS AND DICTIONARIES

Concise Oxford Dictionary of the Earth Sciences. Ailsa Allaby and Michael Allaby, editors. Oxford University Press, Inc., 200 Madison Avenue, New York, NY 10016. (800) 334-4249 or (212) 679-7300. 1990. $42.95.

McGraw-Hill Encyclopedia of Science and Technology. McGraw-Hill Publishing Company, 1221 Avenue of the Americas, New York, NY 10020. (800) 262-4729 or (212) 512-3825. Seventh edition. 1992. $1900.00.

GENERAL WORKS

Global Alert: The Ozone Pollution Crisis. Jack Fishman and Robert Kalish. Plenum Publishing Corp., 233 Spring Street, New York, NY. (212) 620-8000. (800) 2221-9369. FAX (212) 463-0742. 1990. $24.50.

The Changing Atmosphere: A Global Challenge. John Flour. Yale University Press, 302 Temple Street, New Haven, CT 06511. (203) 432-0940. 1992. $9.00 in paper.

Global and Regional Changes in Atmospheric Composition. Erno Meszaros. Lewis Pubs., CRC Press, Inc., 2000 Corporate Boulevard, NW, Boca Raton, FL 33431. (407) 994-0555. (800) 272-7737. 1993. $59.95.

Halocarbons and the Stratospheric Ozone Layer: A Source Guide. Gordon Press Publications, P.O. Box 459, Bowling Green Station, New York, NY 10004. (718) 624-8419. 1991. $75.00.

The Ozone Holes. Ellen Thiro. ABC-CLIO, Inc., P. O. Box 1911, Santa Barbara, CA 93116-1911. (805) 968-1911. (800) 368-6868. 1994. $39.50.

The Urban Climate. Helmut Landsberg. Academic Press, Inc., 6277 Sea Harbor Drive, Orlando, FL. (800) 321-5068. 1981. $65.00

HANDBOOKS AND MANUALS

Author's Guide to the Journals of the American Meteorological Society. American Meteorological Society, 45 Beacon Street, Boston, MA 02108-3693. (617) 227-2425. FAX (617) 742-8718. 1983. $15.00 in paper.

ASTM Standards On Enviromental Sampling. American Society for Testing and Materials, 1916 Race Street, Philadelphia, PA 19103. (215) 299-5419. 1995. $.

Handbook in Applied Meteorology. David D. Houghton, editor. John Wiley and Sons, Inc., 605 Third Avenue, New York, NY 10158. (800) 526-5368 or (212) 850-6000. 1985. $168.00.

Handbook of Agricultural Meteorology. John F. Griffin. Oxford University Press, Inc., 200 Madison Avenue, New York, NY 10016. (212) 725-6000. (800) 334-4249. 1994. $85.00.

Handbook of Atmospherics, Volumes I & II. Hans Volland, editor, Franklin Book Co., Inc., Elkins Park, PA 19117. (215) 635-5252. 1982. $299.90.

Handbook of Ozone Technology and Application II. Pan American Group, 31 Strawberry Hill Avenue, Stamford, CT 06902. (203) 348-3542. $69.95 for set.

Ozone Systems Design Mamual. Pan American Group, 31 Strawberry Hill Avenue, Stamford, CT 06902. (203) 348-3542. 1990. $49.00

ONLINE DATABASES AND CD-ROMS

Climate Assessment Database. National Weather Service, National Meteorological Center, 5200 Auth Road, Suite 101, Camp Springs, MD 20233. (301) 763-8016. Contains daily, weekly and monthly summaries of North American and world climatological data. Also provides five to ten weather forecasts, and 30 to 90 day weather outlook. Subscription required. Inquire as to cost and availability.

CA Search. Chemical Abstracts Service, P.O. Box 3012, Columbus, OH 43210-0012. (614) 447-3600. (800) 848-6533. FAX (614) 447-3709. Very comprehensive guide to worldwide chemical literature and related fields, 1972 to present. Available on BRS,(800) 289-4277, DIALOG, (800) 334-2564, ORBIT (800) 456-7248, and STN International, FIZ Karlsruhe, P.O. Box 2465, W-7500, Karlsruhe 1, Germany, online services. Inquire as to cost and availability.

Current Contents Search. Institute for Scientific Information, 3501 Market Street, Philadelphia, PA 19104. (215) 386-0100. FAX (215) 386-6362. Contains citations to articles listed in the table of contents of science and technology journals. Also articles in social sciences and life sciences journals. Available on BRS,(800) 289-4277, DIALOG,(800) 334-2564, online services. Inquire as to cost and availability.

Dissertation Abstracts. University Microfilms International, 300 North Zeeb Road, Ann Arbor, MI 48106. (800) 521-0600 or (313) 761-4700. Scope includes virtually all doctoral dissertations accepted at accredited American institutions from 1861 to

present in 252 subject areas. Available on BRS,(800) 289-4277, DIALOG,(800) 334-2564, and OCLC EPIC,(800) 848-5878, online services. Also available on CD-ROM. Inquire as to cost and availability.

Meteorological and Geoastrophysical Abstracts. American Meteorological Society, 45 Beacon Street, Boston, MA 02108-3693. (617) 227-2425. FAX (617) 742-8718. Contains citations and abstracts to the worldwide literature on significant research in meteorology and geoastrophysics. Related topics include physical oceanography, hydrology, environmental sciences and glaciology. Covers the period 1972 to present. Available on DIALOG,(800) 334-2564, online service. Inquire as to cost and availability.

NTIS Bibliographic Database. National Technical Information Service, 5285 Port Royal Road, Springfield, VA 22161. (703) 487-4929 or FAX (703) 321-8199. Broad coverage of government-sponsored science and technology research reports, 1964 to present. Available on BRS,(800) 289-4277, DIALOG, (800) 334-2564, ORBIT, (800) 456-7248, and STN International, FIZ Karlsruhe, P.O. Box 2465, W-7500, Karlsruhe 1, Germany, online services. Also available on CD-ROM. Inquire as to cost and availability.

Physics Briefs. American Institute of Physics, 335 East 45th Street, New York, NY 10017. (212) 661-9260 or FAX (212) 661-2036. Contains citations with abstracts of the literature of physics and related fields, 1979 to present. Available on the STN International, FIZ Karlsruhe, P.O. Box 2465, W-7500, Karlsruhe 1, Germany, online service. Inquire as to cost and availability.

SCISEARCH. Institute for Scientific Information, 3501 Market Street, Philadelphia, PA 19104. (800) 523-1850 or (215) 386-0100. Broad multidisciplinary title and author index to the international literature of science and technology, 1974 to present. Available on DIALOG and ORBIT online services. Also available on CD-ROM. Inquire as to cost and availability.

WILSONLINE. H.W. Wilson Company, 950 University Avenue, Bronx, NY 10452. (800) 367-6770 or (212) 588-8400. Makes available online versions of the printed H.W. Wilson indexes including *Applied Science and Technology Index, Business Periodicals Index, General Science Index,* and *Readers' Guide to Periodical Literature.* Period covered is generally 1983 to present. Available on BRS,(800) 289-4277, DIALOG,(800) 334-2564, and OCLC EPIC,(800) 848-5878, online services. Also available on CD-ROM. Inquire as to cost and availability.

World Climate Disc. Chadwyck-Healey Inc., 1101 King Street, Alexandria, VA 22314. (703) 683-4890. FAX (703) 683-7589. Weather and climate data from approximately 5,000 weather stations worldwide, covering the years 1854 to 1990 on CD-ROM. First edition 1992. Approximately $1200.00 per year with annual updates.

World Weatherdisc. WeatherDisc Associates, Inc., 4584 N.E. 89th Street, Seattle, WA 98115. (206) 524-4314. FAX (206) 543-0308. Meteorological data on CD-ROM which describes the climate of the earth today and for the past few hundred years. First edition 1989. Approximately $295.00 per year with biannual updates.

PERIODICALS

Ozone Science and Engineering. International Ozone Association. c/o Pan American Group, 31 Strawberry Hill Avenue, Stamford, CT 06902. (203) 348-3542. $225.00.

Agricultural and Forest Meteorology. Elsevier Science Publishing Company, Inc., 655 Avenue of the Americas, New York, NY 10010. (212) 989-5800. Twenty times per year. $750.00 per year.

American Meteorological Society. Bulletin. American Meteorological Society, 45 Beacon Street, Boston, MA 02108-3693. (617) 227-2425. FAX (617) 742-8718. Monthly. $60.00 per year.

Atmosphere-Ocean. Canadian Meteorological and Oceanographic Society, P.O. Box 334, Newmarket, Ontario, L3Y 4X7, Canada. (416) 898-1040. FAX (416) 898-7937. Quarterly. $30.00 per year.

Boundary-layer Meteorology: An International Journal of Physical and Biological Processes in the Atmospheric Boundary Layer. Kluwer Academic Publishers, P.O. Box 358, Accord Station, Hingham, MA 02018-0358. (617) 871-6000. Sixteen per year. $785.00 per year.

Climate Change. Kluwer Academic Publishers, P.O. Box 358, Accord Station, Hingham, MA 02018-0358. (617) 871-6000. Six times per year. $327.00 per year.

Dynamics of Atmospheres and Oceans. Elsevier Science Publishing Company, Inc., 655 Avenue of the Americas, New York, NY 10010. (212) 989-5800. Six times per year. $205.00 per year.

International Journal of Climatology. Royal Meteorological Society, distributed by John Wiley and Sons, Inc., 605 Third Avenue, New York, NY 10158. (800) 526-5368 or (212) 850-6000. Eight times per year. $425.00 per year.

JGR: Journal of Geophysical Research: Atmosphere. American Geophysical Union, 2000 Florida Avenue, N.W., Washington, DC 20009. (202) 462-6903. Monthly. $90.00 per year to members.

JGR: Journal of Geophysical Research: Oceans. American Geophysical Union, 2000 Florida Avenue, N.W., Washington, DC 20009. (202) 462-6903. Monthly. $1545.00 per year.

Journal of Applied Meteorology. American Meteorological Society, 45 Beacon Street, Boston, MA 02108-3693. (617) 227-2425. FAX (617) 742-8718. Monthly. $165.00 per year.

Journal of Climate. American Meteorological Society, 45 Beacon Street, Boston, MA 02108-3693. (617) 227-2425. FAX (617) 742-8718. Monthly. $175.00 per year.

Journal of the Atmospheric Sciences. American Meteorological Society, 45 Beacon Street, Boston, MA 02108-3693. (617) 227-2425. FAX (617) 742-8718. Semi-monthly. $320.00 per year.

RESEARCH CENTERS AND INSTITUTES

Center for Global Change. University of Maryland, Executive Building, Suite 401, 7100 Baltimore Avenue, College park, MD 20740. (301) 403-4165. FAX (301) 454-0954.

Interdisciplinary Center for Aeronomy and Other Atmospheric Sciences. University of Florida, 311 Space Sciences Research Building, Gainsville, FL 32611. (904) 2001.

OZONE

Ency. of Physical Sciences and Engineering Info. Sources

National Center for Atmospheric Research. P.O. Box 3000, Boulder, CO 80307. (303) 497-1000. FAX (303) 497-1654.

Alaska Sar (synthetic Aperature Radar) Facility. University of Alaska Fairbanks, Geophysical Institute, Fairbanks, AK 99775-0800. (907) 474-7954.

Atlantic Oceanographic and Meteorological Laboratory. 4301 Rickenbacker Causeway, Miami, FL 33149. (305) 361-4300.

Atmospheric Sciences Research Center. State University of New York, 100 Fuller Road. Albany NY 12205. (518) 442-3820. FAX (518) 442-3867.

Center for Atmospheric Theory and Analysis. University of Colorado-Boulder, Campus Box 391, Boulder CO 80309-0391. (303) 492-6487. FAX (303) 492-0642.

Center for Climate Research. Columbia University, Lamont-Doherty Geological Observatory. Palisades, NY 10964. (914) 359-2900.

Centre for Climate and Global Change Research. McGill University, 805 Sherbrooke Street, West, Montreal, PQ Canada H3A 2K6. (514) 398-3759. (514) 398-6115.

Centre for Climatic Chemistry. Institute of Ocean Sciences. Canadian Department of Fisheries and Oceans, P.O. Box 6000, 9860 West Saanich Road, Sidney, BC, Canada V9L 4B2. (604) 356-6329.

Climatic Research Institute. Oregon State University. Strand Hall, Room 326, Corvallis, OR 97331-2209. (503) 737-5705.

Climate Analysis Center. National Weather Service, 5200 Auth Road, Camp Springs, MD 20746. (301) 763-8167.

Cooperative Institute for Mesoscale Meteorological Studies. University of Oklahoma, 401 East Boyd, Norman, OK 73019. (405) 325-3041.

Cooperative Institute for Meteorological Satellite Studies. Space Science & Engineering Center, University of Wisconsin-Madison, 1225 West Dayton Street, Madison, WI 53706. (608) 263-7435.

Cooperative Institute for Research in the Atmosphere. Colorado State University, FootHills Campus. Fort Collins, CO 80523. (303) 491-8448.

Florida State Climate Center. Florida State University. 431 Love Building, Tallahassee, FL 32306. (904) 644-3417.

Geophysical Fluid Dynamics Laboratory. Princeton University, P.O. Box 308, Princeton, NJ 08542. (609) 452-6500.

Geophysical Institute. University of Alaska Fairbanks, C.T. Elvey Building, Fairbanks, AK 99770-0800. (907) 474-7282. FAX (907) 474-7290.

Goddard Space Flight Center. Laboratory for Atmospheres, Mail Code 610, Greenbelt, MD 20771. (301) 286-5002.

Hurricane Research Division. Atlantic Oceanographic and Meteorological Laboratory. 4301 Rickenbacker Causeway, Miami, FL 33149. (305) 361-4400.

Institute of Atmospheric Sciences. South Dakota School of Mines and Technology. 501 East Saint Joseph Street, Rapid City, SD 57701-3995. (605) 394-2291. FAX (605) 394-6131.

Institute of Snow Research. Michigan Technological University. Houghton, MI 49931. (906) 487-2750. FAX (906) 487-2202.

Joint Institute for Marine and Atmospheric Research. University of Hawaii at Manoa, 1000 Pope Road, MSB 312, Honolulu, HI 96822. (808) 956-8083. FAX (808) 956-4104.

National Center for Atmospheric Research. P.O. Box 3000, Boulder, CO 80307. (303) 496-1000. FAX (303) 497-1654.

National Hurricane Center. 1320 South Dixie Highway, Coral Gables, FL 33146. (305) 666-4612.

National Meteorological Center. World Weather Building, Room 101, 5200 Auth Road, Camp Springs, MD 20746. (301) 763-8016.

National Severe Storms Forecast Center. 601 East 12th Street, Kansas City, MO 64106. (816) 374-5922.

National Severe Storms Laboratory. National Oceanic and Atmospheric Administration Environment Research Laboratories, 1313 Halley Circle, Norman, OK 73069. (405) 360-3620. FAX (405) 366-0472.

National Weather Service. 1325 East-West Highway, Silver Spring, MD 20910. (301) 427-7689.

Program in Atmospheric and Ocenic Sciences. Princeton University. P.O. CN 710, Sayre Hall, Forrestal Campus, Princeton, NJ 08544-0710. (609) 258-6571. FAX (609) 258-2850.

Radar Systems and Remote Sensing Laboratory. University of Kansas. 2291 Irving Hill Road, Lawrence, Kansas. 66045-2969. (913) 864-4835. FAX (913) 864-7789.

Radar Weather Observatory. McGill University, Macdonald College P.O. Box 241, Saint Anne de Bellevue, PQ Canada, H9X 1C0. (514) 398-7733. FAX (514) 398-7755.

University Corporation for Atmospheric Research. P.O. Box 3000, Boulder, CO 80307-3000. Consortium of 58 universities with doctoral programs in atmospheric and/or oceanic sciences. (303) 497-1650. FAX (303) 497-1654.

P

PAINTS AND PIGMENTS

See also: DYES AND DYEING, LEAD, ORGANIC CHEMISTRY, SURFACE CHEMISTRY

ABSTRACT SERVICES AND INDEXES

Applied Science and Technology Index. A cumulative subject index to English language periodicals in the fields of aeronautics and space science, computer technology, chemistry, construction industry, energy and related areas. H. W. Wilson Co., 950 University Avenue, Bronx, NY 10452-9978. (212) 588-8400. (800) 367-6770. FAX (718) 590-1617. From 1958 to present. Monthly. Inquire about cost and availability. Also available online (BRS and WILSONLINE) and on CD-ROM. ISSN: 0003-6986.

Chemical Abstracts. Chemical Abstracts Service. 2540 Olentangy River Road, Box 3012, Columbus, OH 43210-0012. (614) 447-3600. FAX (614) 447-3713. 1907 to present. Weekly. $16,800.00 per year. Available online and on CD- ROM. CA is also available in five section groupings. Inquire regarding cost and availability.

Current Contents: Physical, Chemical and Earth Sciences. Institute for Scientific Information, 3501 Market Street, Philadelphia, PA 19104. (215) 386-0100. (800) 523-1850. FAX (215) 386-2291. 1961 to present. Weekly. $442.00 per year. Also available online (BRS, DIALOG) and on CD-ROM. Inquire regarding cost and availability. ISSN: 0163-2574.

General Science Index. H.W. Wilson Company, 950 University Avenue, Bronx, NY 10452. (212) 588-8400. (800) 367-6770. FAX (718) 590-1617. From 1978 to present. Ten issues per year; quarterly and annual cumulations. Service basis. Available on CD-ROM and online. Inquire regarding cost and availability. ISSN: 0162-1963.

Physics Abstracts. INSPEC. Section A, Science Abstracts. Institution of Electrical Engineers (IEE), London, United Kingdom. Available from: INSPEC/IEEE-Institute of Electrical and Electronic Engineers, Box 1331, 445 Hoes Lane, Piscataway, NJ 08855-1331. (908) 562-5549. 1898 to present. 24 issues per year. $2835.00 per year. ISSN: 0036-8091. Also available online and on CD-ROM.

Process and Chemical Engineering/Chemical Engineering Abstracts. Royal Society of Chemistry, Information Services, Thomas Graham House, Science Park, Milton Road, Cambridge, CB4 4WF, England. Contains citations, mostly with abstracts, of the worldwide literature on chemical engineering; from 1982 to present. Monthly. $610.00 per year. Also available online. ISSN: 0960-5045.

Science Citation Index (SCI). Institute for Scientific Information, 3501 Market Street, Philadelphia, PA 19104. (215) 386-0100. (800) 523-1850. FAX (215) 386-2991. 1961 to present. Six issues per year, plus annual cumulation. $11650.00 per year. Also available online and on CD-ROM. Inquire about price and availability. ISSN: 0036-827X.

Theoretical Chemical Engineering. Royal Society of Chemistry, Information Services, Thomas Graham House, Science Park, Milton Road, Cambridge, CB4 4WF, England. Covers theoretical chemical engineering, including theory and laboratory experimentation. 1964 to present. Monthly. $235.00. ISSN: 0960-5053.

ANNUAL REVIEWS AND YEARBOOKS

Advances in Photochemistry. John Wiley and Sons, Inc., 605 Third Avenue, New York, NY 10158. (800) 526-5368 or (212) 850-6000. 1963 to present. Irregular, price varies.

Advances in Physical Organic Chemistry. Academic Press, Inc., 1250 Sixth Avenue, San Diego, CA 92101-4311. (619) 231-0926. FAX (619) 699-6715. 1963 to present. Irregular, price varies.

Annual Review of Physical Chemistry. Annual Reviews, Inc., 4139 El Camino Way, Palo Alto, CA 94306-0897. (415) 493-4400. Fax (415) 855-9815. 1951 to present. Annual. $55.00 per year.

ASSOCIATIONS AND PROFESSIONAL SOCIETIES

American Chemical Society, 1155 16th Street, NW, Washington.

American Electroplaters and Surface Finishers Society, Inc., 12644 Research Parkway, Orlando, FL 32826-3298. (407) 281-6441. FAX (407) 281-6446.

American Institute of Chemical Engineers. 345 East 47th Street, New York, NY 10017. (212) 705-7338.

PAINTS AND PIGMENTS

Ency. of Physical Sciences and Engineering Info. Sources

Association of Consulting Chemists and Chemical Engineers, 50 East 41st Street, Suite 92, New York, NY 10017. (212) 684-6255.

Chemical Manufacturers Association. 2501 M Street, N.W., Washington, DC 20037. (202) 887-1182

Color Pigments Manufacturers Association. P.O. Box 20839, Alexandria VA 22320. (703) 684-4044. FAX (703) 684-1795.

Federation of Societies for Coating Technology. 429 Norristown Road, Blue Bell, PA 19422. (215) 940-0777. FAX (215) 940-0292.

National Paint and Coating Association. 1500 Rhode Island Avenue NW, Washington, DC 20005. (202) 462-6272.

Steel Structures Painting Council. 40 24th Street, Pittsburgh, PA 15222-4643. (412) 291-2331. FAX (412) 281-9992.

BIBLIOGRAPHIES

Chemical Information Sources. Gary Wiggins. McGraw-Hill Publishing Company, 1221 Avenue of the Americas, New York, NY 10020. (800) 262-4729 or (212) 512-3825. 1991. $42.50.

Scientific and Technical Books and Serials in Print: An Index to Literature in Science and Technology. R.R. Bowker Inc., 121 Chanlon Road, New Providence, NJ 07974. (908) 464-6800. (800) 521-8110. FAX (908) 665-3502. 1972 to present. Annual. 4 volumes. 1994. $299.95. Also available on compact disc and online. ISSN: 0000-054X.

DIRECTORIES AND BIOGRAPHICAL SOURCES

American Institute of Chemical Engineers. Directory. American Institute of Chemical Engineers. 345 East 47th Street, New York, NY 10017-2396. (212) 705-7663. Annual.

American Men and Women of Science: Physical and Biological Sciences. R.R. Bowker Inc., 121 Chanlon Road, New Providence, NJ 07974. (908) 464-6800. (800) 521-8110. 20th edition. 8 volumes. 1996. $850.00

Chemical Week: Buyers Guide.

Consulting Services: Chemists and Chemical Engineers. Association of Consulting Chemists and Chemical Engineers, 40 East 45th Street, New York, NY 10017. (212) 983-3160. FAX (212) 983-3161. Biennial. $60.00.

Research Centers Directory. Gale Research Company, 835 Penobscot Building, Detroit, MI 48226-4094. (313) 961-2242. (800) 347-4253. 20th edition. 1995. $485.00. ISSN: 0080-1518.

ENCYCLOPEDIAS AND DICTIONARIES

Dictionary of Analytical Reagents. R. Lobinski, editor. Routledge, Chapman and Hall, Inc., 29 West 35th Street, New York, NY 10001-2291. (212) 244-3336. 1993. $1000.00.

Encyclopedia of Chemical Processing and Design. McKetta Marcel Dekker, Inc., 270 Madison Avenue, New York, NY

10016. (212) 696-9000. (800) 228-1160. 1976-present . $175.00 per volume.

Industrial Chemical Thesaurus. Michael Ash and Irene Ash. VCH Publishers, Inc., 220 East 23rd Street, New York, NY 10010-4606. (800) 367-8249. 1992. $295.00.

Kirk-Othmer Encyclopedia of Chemical Technology. John Wiley and Sons, Inc., 605 Third Avenue, New York, NY 10158. (800) 526-5368 or (212) 850-6000. Fourth edition. 1991-present. Twenty-seven volumes. $5400.00.

McGraw-Hill Encyclopedia of Chemistry. Sybil P.Parker, editor. McGraw-Hill Book, Incorporated, 1221 Avenue of the Americas, New York, NY 10020. (212) 997-3675. Second edition. 1993. $95.50.

Paint Manufacturing Industry: Guides to Pollution Prevention. Diane Publishing Co., 600 Upland Avenue, Upland, PA 19015. (610) 499-7415. 1993. $45.00.

Painting Materials: A Short Encyclopedia. Rutherford J. Gettens and George L. Stoug. Dover Publications, Inc., 180 Varick Street, New York, NY 10014. (212) 255-3755. (800) 223-3130. l965. $6.95.

Ullman's Encyclopedia of Industrial Chemistry. VCH Publications, Inc., 220 East 23rd Street, Suite 909, New York, NY 10010-4606. (212) 683-8333. (800) 422-8824. 5th edition. 1984-present . Price varies per volume.

GENERAL WORKS

ASTM Standards on Lead-based Paint Assesment in Buildings. ASTM Committee E-6 on Performance of Buildings. American Society for Testing and Materials, 1916 Race Street, Philadelphia, PA 19103. (215) 299-5419. 1994. $69.00.

Color Chemistry: Syntheses, Properties and Applicaitons of Organic Dyes and Pigments. Heinrich Zollinger. VCH Publishers, Inc., 220 East 23rd Street, Suite 909, New York, NY 10010-4606. (212) 683-8333. (800) 422-8824. 1991. $150.00.

Gilding and Lacquering. Rodrigo Titian. H. Holt & Co., 115 West 18th Street, New York, NY 10011. (212) 886-9200. (800) 488-5233. 1994. $15.95 in paper.

Industrial Inorganic Pigments. Gunter Buxbaum, editor. VCH Publishers, Inc., 220 East 23rd Street, Suite 909, New York, NY 10010-4606. (212) 683-8333. (800) 422-8824. 1993. $92.00.

Industrial Organic Pigments: Production, Properties, Applications. Willy Herbst and Kaus Hunger. VCH Publishers, Inc., 220 East 23rd Street, Suite 909, New York, NY 10010-4606. (212) 683-8333. (800) 422-8824. 1993. $195.00.

Liquid Paint Finishing Defects. Robert D. Grear. SME, PO Box 930, 1 SME Drive, Dearborn, MI 48121. (313) 271-1500. (800) 733-4763. 1991. $18.00 in paper.

Outlines of Paint Technology. W. M. Morgan. Wiley. John Wiley & Sons, Inc., 605 Third Avenue, New York, NY 10158-0012. (212) 850-6000. (800) 225-5945. 3rd edition. 1990. $139.00.

Paint and Surface Coatings. Ron Lambourne. Prentice-Hall, 113 Sylvan Avenue, Route 9W, Englewood Cliffs, NJ 07632. (201) 592-2000. (800) 922-0579. 1993. $205.00.

Paints, Coatings and Solvents. Dieter Stoye, editor. Publishers, Inc., 220 East 23rd Street, Suite 909, New York, NY 10010-4606. (212) 683-8333. (800) 422-8824. 1993. $125.00.

Paints and Protective Coatings. Gordon Press Publications, P.O. Box 459, Bowling Green Station, New York, NY 10004. (718)624-8419. 1991. $73.00.

Principles of Paint Formulation. R. Woodbridge, editor. Routledge, Chapman & Hall, Inc., 29 West 35th Street, New York, NY 10001-3291. (212) 244-3336. 1991. $170.00.

HANDBOOKS AND MANUALS

CRC Handbook of Chemistry and Physics. David R. Kide, editor. CRC Press Inc., 2000 Corporate Blvd., NW, Boca Raton, FL 33431. (407) 994-0555. (800) 333-8300. 77th edition. 1996. $99.95.

Handbook of Paint Raw Materials. Ernest W. Flich. Noyes, 120 Mill Road, Park Ridge, NJ 07656. (201) 391-8484. 1989. $98.00.

Lange's Handbook of Chemistry. John A. Dean, editor. McGraw-Hill Publishing Company, 1221 Avenue of the Americas, New York, NY 10020. (800) 262-4729 or (212) 512-3825. Fourteenth edition. 1992. $79.50.

Lead-Based Paint Handbook. Jan W. Gooch. Plenum Publishing Corp., 233 Spring Street, New York, NY. (212) 620-8000. (800) 2221-9369. FAX (212) 463-0742. 1993. $59.50.

Paint Handbook. Guy E. Weismantel, editor. McGraw-Hill Publishing Company, Inc., 1221 Avenue of the Americas, New York, NY 10020. (212) 512-2000. (800) 262-4729. 1980. $74.50.

Paint Manufacturing Industry: Guides to Pollution Prevention. Diane Publishing Co., 600 Upland Avenue, Upland, PA 19015. (610) 499-7415. 1993. $45.00 in paper.

Pigment Handbook. Peter A. Lewis. John Wiley & Sons, Inc., 605 Third Avenue, New York, NY 10158-0012. (212) 850-6000. (800) 225-5945. 3 volumes. 1988. $608.00 set.

Riegel's Handbook of Industrial Chemistry. James A. Kent, editor. Van Nostrand Reinhold, 115 Fifth Avenue, New York, NY 10003. (212) 254-3232 or (800) 926-2665. Ninth edition. 1992. $114.95.

Steel Structures Painting Manual. SSPC-Steel Structures Painting Council, 40 24th Street, Pittsburgh, PA 15222-4643. (412) 291-2331. FAX (412) 281-9992. 2 volumes. 2nd edition. $150.00 set.

ONLINE DATABASES AND CD-ROMS

Analytical Abstracts Online. Royal Society of Chemistry, Information Services, Thomas Graham House, Science Park, Milton Road, Cambridge, CB4 4WF, England. Contains citations, mostly with abstracts, of the worldwide literature on analytical chemistry, from 1980 to present. Available on DIALOG and STN online services. Inquire as to cost and availability.

CA Search. Chemical Abstracts Service, P.O. Box 3012, Columbus, OH 43210-0012. (614) 447-3600. (800) 848-6533. FAX (614) 447-3709. Guide to worldwide chemical literature and related fields, 1972 to present. Available on BRS, DIALOG, ORBIT and STN online services. Inquire as to cost and availability.

Chemical Journals of the American Chemical Society. American Chemical Society, 1155 16th Street, N.W., Washington, DC 20036. (202) 872-4381 or (800) 424-6747. Contains complete text of approximately 90,000 articles from 22 primary journals published by the American Chemical Society, from mostly 1982 to present. Available on STN online service. Inquire as to cost and availability.

Dissertation Abstracts. University Microfilms International, 300 North Zeeb Road, Ann Arbor, MI 48106. (800) 521-0600 or (313) 761-4700. Scope includes virtually all doctoral dissertations accepted at accredited American institutions from 1861 to present in 252 subject areas. Available on BRS, DIALOG, and OCLC EPIC online services. Also available on CD-ROM. Inquire as to cost and availability.

INSPEC. Institution of Electrical Engineers, Michael Faraday House, Six Hills Way, Stevenage, Herts. SG1 2AY, England. Telephone: 0438 313311 or FAX 0438 742840. Contains citations to the worldwide literature of physics, electronics and electrical engineering, computer technology, and related fields. Available on BRS, DIALOG, ORBIT and STN online services. Inquire as to cost and availability.

Kirk-Othmer Encyclopedia of Chemical Technology. John Wiley and Sons, Inc., 605 Third Avenue, New York, NY 10158. (800) 526-5368 or (212) 850-6000. Contains the complete text of all chapters in the 27 volume fourth edition of the KIRK-Othmer *Encyclopedia of Chemical Technology.* 1991. Available on BRS, DIALOG, online services. Inquire as to cost and availability.

NTIS. National Technical Information Service, 5285 Port Royal Road, Springfield, VA 22161. (703) 487-4929 or FAX (703) 321-8199. Broad coverage of government-sponsored research reports, 1964 to present. Available on BRS, DIALOG, ORBIT and STN online services. Also available on CD-ROM. Inquire as to cost and availability.

SCISEARCH. Institute for Scientific Information, 3501 Market Street, Philadelphia, PA 19104. (800) 523-1850 or (215) 386-0100. Broad multidisciplinary title and author index to the international literature of science and technology, 1974 to present. Available on DIALOG and ORBIT online services. Also available on CD-ROM. Inquire as to cost and availability.

WILSONLINE. H.W. Wilson Company, 950 University Avenue, Bronx, NY 10452. (800) 367-6770 or (212) 588-8400. Makes available online versions of the printed H.W. Wilson indexes including *Applied Science and Technology Index, Business Periodicals Index,* and *Readers' Guide to Periodical Literature.* Period covered is generally 1983 to present. Also available on CD-ROM. Inquire as to cost and availability.

Who's Who in Technology. Gale Research, 835 Penobscot Building, Detroit, MI 48226-4094. (313) 961-2242. (800) 347-4253. Contains biographical information of contemporary American scieNTISts and engineers. Available on DIALOG online service. Inquire as to cost and availability.

PAINTS AND PIGMENTS

Ency. of Physical Sciences and Engineering Info. Sources

PERIODICALS

Chemical and Engineering News. American Chemical Society, 1155 16th Street, NW, Washington, DC 20036. (202) 872-4600. 1923 to present. Weekly. $115.00 per year. ISSN: 0009-2347.

Chemical Engineering Progress. American Institute of Chemical Engineers, 345 East 47th Street, New York, NY 10017. (212) 705-7338. FAX (212) 752-3294. 1947 to present. Monthly. $75.00 per year. ISSN: 0360-7275.

Chemical Week. Includes Annual Buyers Guide. Chemical Week Associates, 888 Seventh Avenue, New York, NY 10106. (212) 621-4900. FAX (212) 621-4949. 1914 to present. Weekly. $99.00 per year. ISSN: 0009-272X.

Chemtech. American Chemical Society. Box 3337, Columbus, OH 43210. (614) 447-3776. 1970 to present. Monthly. $370.00 per year. ISSN: 0009-2703.

Industrial and Engineering Chemistry Research. American Chemical Society, Box 3337, Columbus, OH 43210. (614) 447-3776. (800) 333-9511. FAX (614) 447-3671. 1962 to present. Monthly. $567.00 per year. ISSN: 0888-5885.

Journal of Coatings Technology. Federation of Societies for Coating Technology, 429 Norristown Road, Blue Bell, PA 19422. (215) 940-0777. FAX (215) 940-0292. Monthly. Included in membership dues. ISSN: 0361-8773.

Journal of Dispersion Science and Technology. Marcel Dekker, Inc., 270 Madison Avenue, New York, NY 10016. (212) 696-9000. FAX (212) 685-4540. Subscriptions to: Box 5017 Monticello, NY 12701. 1980 to present. Six times per year. $445.00 per year. ISSN: 0193-2691.

Journal of Physical Chemistry. American Chemical Society, Box 3337, Columbus, OH 43210. (614) 447-3776. (800) 227-5558. 1896 to present. Weekly. $1140.00 per year. ISSN: 0022-3654.

Metal Finishing. Elsevier Science Inc., 655 Avenue of the Americas, New York, NY 10010. (212) 989-5800. FAX (212) 633-3990. 1903 to present. Monthly, plus 2 supplements. $40.00 per year. ISSN: 0026-0576.

Modern Paint and Coatings. Argus Inc., 6151 Powers Ferry Road, NW, Atlanta, GA 30339-2941. (404) 955-2500. FAX (404) 955-0400. 1910 to present. Monthly. $49.00. ISSN: 0098-7786.

Plating and Surface Finishing. American Electroplaters and Surface Finishers Society, Inc., 12644 Research Parkway, Orlando, FL 32826-3298. (407) 281-6441. FAX (407) 281-6446. 1909 to present. Monthly. $35.00. ISSN: 0369-3164.

RESEARCH CENTERS AND INSTITUTES

Paint Research Associates. 430 West Forest, Ypsilanti, MI 48197. (313) 483-3401. FAX (313) 487-8755.

Science and Technology Laboratory Group. Gerogia Institute of Technology, Georgia Tech Research Institute, Atlanta, GA 30332. (404) 894-3530. (404) 894-6285.

Steel Structures Painting Council. 4400 Fifth Avenue, Pittsburgh, PA 15213-2683. (412) 268-3327. FAX (412) 268-7048.

PAPER CHEMISTRY

See also: ADHESIVES, CHEMICAL ENGINEERING, CHEMISTRY, ORGANIC CHEMISTRY, PLASTICS

ABSTRACT SERVICES AND INDEXES

Applied Science and Technology Index: A Cumulative Subject Index to English Language Periodicals in the Fields of Aeronautics and Space Science, Computer Technology, Chemistry, Construction Industry, Energy and Related Areas. H. W. Wilson Co., 950 University Avenue, Bronx, NY 10452- 9978. (212) 588-8400. (800) 367-6770. FAX (718) 590-1617.From 1958 to present. Monthly. Inquire about cost and availability. Also available online (BRS and WILSONLINE) and on CD-ROM. ISSN: 0003-6986.

Chemical Abstracts. Chemical Abstracts Service. 2540 Olentangy River Road, Box 3012, Columbus, OH 43210-0012. 614) 447-3600. FAX (614) 447-3713. 1907 to present. Weekly. $16,800.00 per year. Available online and on CD- ROM. CA is also available in five section groupings. Inquire Regarding Cost and Availability.

Chemical Engineering Abstracts. Royal Society of Chemistry. Information Services, Thomas Graham House, Science Park, Milton Road, Cambridge, Cb4 4wf, England. Contains citations, Mostly With Abstracts, of the Worldwide Literature On Chemical Engineering, From 1982 to Present. Monthly. $450.00 per year. Also Available Online.

Current Contents: Engineering, Technology and Applied Sciences. Institute for Scientific Information, 3501 Market Street, Philadelphia, PA 19104. (215) 386-0100. FAX (215) 386-2991. From 1970 to present. Weekly. $442.00 per year. Also available online. Inquire regarding cost and availability. ISSN: 0095-7917.

Engineering Index Monthly. Engineering Information, Inc., Castle Point on the Hudson, Hoboken, NJ 07030. (201) 216-8500. (800) 221-1044. FAX (201) 216-8532. Monthly. $2300.00 per year for monthly issues. $1980 per year for annual cumulation. Available on CD-ROM and online as COMPENDEX. ISSN: 0742-1974; 0360-8557.

General Science Index. H.W. Wilson Company, 950 University Avenue, Bronx, NY 10452. (212) 588-8400. (800) 367-6770. FAX (718) 590-1617. From 1978 to present. Ten issues per year; quarterly and annual cumulations. Service basis. Available on CD-ROM and online. Inquire regarding cost and availability. ISSN: 0162-1963.

Institute of Paper Science and Technology Abstract Bulletin. Institute of Paper Science and Technology, 500 10th Street, NW, Atlanta, GA 30318. (404) 853-9500. 1930 to present.Monthly. Price varies. ISSN: 1047-2088.

Paper and Board Abstracts. PIRA International, Randalls Road, Leatherhead, Surrey KT22 7RU, England. TEL 0372-376161. FAX 0372-360104. 1965 to present. Monthly. $907.00 per year. Issn: 0307-0778.

Process and Chemical Engineering/Chemical Engineering Abstracts. Royal Society of Chemistry, Information Services, Thomas Graham House, Science Park, Milton Road, Cambridge, Cb4 4wf, England. Contains citations, mostly with abstracts, of the worldwide literature on chemical engineering; from 1982 to

Present. Monthly. $610.00 per year. Also available online. Issn: 0960-5045.

Science Citation Index (SCI). Institute for Scientific Information, 3501 Market Street, Philadelphia, PA 19104. (215) 386-0100. (800) 523-1850. FAX (215) 386-2991. 1961 to Present. Six issues per year, plus annual Cumulation. $11650.00 per year. Also available online and On CD-ROM Inquire about price and availability. ISSN: 0036-827X.

Theoretical Chemical Engineering. Royal Society of Chemistry, Information Services, Thomas Graham House, Science Park, Milton Road, Cambridge, Cb4 4wf, England. Covers theoretical chemical engineering, including theory and laboratory Experimentation. 1964 to present. Monthly. $235.00. ISSN: 0960-5053.

ASSOCIATIONS AND PROFESSIONAL SOCIETIES

American Chemical Society, 1155 16th Street, NW, Washington, DC 20036. (202) 872-4600.

American Institute of Chemical Engineers. 345 East 47th Street, New York, NY 10017. (212) 705-7338.

American Forest and Paper Association. 1111 19th Street NW, Suite 800, Washington, DC 20036. (202) 463-2700. FAX (202) 463-2785.

Association of Consulting Chemists and Chemical Engineers, 50 East 41st Street, Suite 92, New York, NY 10017. (212) 684-6255.

Chemical Manufacturers Association. 2501 M Street, N.W., Washington, DC 20037. (202) 887-1182.

Institute of Paper Science and Technology, 500 10th Street, Atlanta, GA 30318. (404) 853-9500.

Paper Industry Management Association. 1699 Wall Street, Suite 212, Mount Prospect, IL 60056-5782. (708) 956-0250. (708) 956-0520.

TAPPI: Technical Association of the Pulp and Paper Industry. Technology Park, Post office Box 105113, Atlanta, GA 30348. (770) 446-1400. FAX (770) 446-6947.

BIBLIOGRAPHIES

Chemical Information Sources. Gary Wiggins. McGraw-Hill Publishing Company, 1221 Avenue of the Americas, New York, NY 10020. (800) 262-4729 or (212) 512-3825. 1991. $42.50.

Scientific and Technical Books and Serials in Print: An Index to Literature in Science and Technology. R.R. Bowker Inc., 121 Chanlon Road, New Providence, NJ 07974. (908) 464-6800. (800) 521-8110. FAX (908) 665-3502. 1972 to present. Annual. 4 volumes. 1994. $299.95. Also available on compact disc and online. ISSN: 0000-054X.

DIRECTORIES AND BIOGRAPHICAL SOURCES

American Institute of Chemical Engineers. Directory. American Institute of Chemical Engineers. 345 East 47th Street, New York, NY 10017-2396. (212) 705-7663. Annual.

American Men and Women of Science. Physical and Biological Sciences. R.R. Bowker Inc., 121 Chanlon Road, New Providence, NJ 07974. (908) 464-6800. (800) 521-8110. 20th edition. 8 volumes. 1996. $850.00

Chemical Engineering Faculties. American Institute of Chemical Engineers. 345 East 47th Street, New York, NY 10017-2396. (212) 705-7663. Annual. $75.00.

Pulp and Paper-Directory Issues. Miller Freeman Publications, 600 Harrison, Street, San Francisco, CA 94107. (415) 905-2200. Annual: Buyers Guide issued November; $25.00. North American Factbook issued December; $275.00.

Research Centers Directory. Gale Research Company, 835 Penobscot Building, Detroit, MI 48226-4094. (313) 961-2242. (800) 347-4253. 20th edition. 1995. $485.00. ISSN: 0080-1518.

TAPPI Membership Directory and Company Guide. Technical Association of the Pulp and Paper Industry, Technology Park, Post office Box 105113, Atlanta, GA 30348. (404) 446-1400. (800) 332-8686. Annual. $115.00.

Who's Who in Engineering. Gordon Davis, editor. American Association of Engineering Societies. 1111 19th Street, NY, Suite 608, Washington, DC 20036. (202) 296-2237. (800) 658-8897. 9th edition. 1995. $220.00.

ENCYCLOPEDIAS AND DICTIONARIES

Encyclopedia of Chemical Processing and Design. McKetta Marcel Dekker, Inc., 270 Madison Avenue, New York, NY 10016. (212) 696-9000. (800) 228-1160. 1976-present. $175.00 per volume. ISSN:

Hawley's Condensed Chemical Dictionary. Van Nostrand Reinhold, 115 Fifth Avenue, New York, NY 10003. (212) 254-3232. (800) 842-3636. 13th edition. 1996. $99.95.

Industrial Chemical Thesaurus. Michael Ash and Irene Ash. VCH Publishers, Inc., 220 East 23rd Street, New York, NY 10010-4606. (800) 367-8249. 1992. $295.00.

Kirk-Othmer Encyclopedia of Chemical Technology. John Wiley and Sons, Inc., 605 Third Avenue, New York, NY 10158. (800) 526-5368 or (212) 850-6000. Fourth edition. 1991-present. Twenty-seven volumes. $5400.00.

McGraw-Hill Encyclopedia of Science and Technology. McGraw-Hill Book, Incorporated, 1221 Avenue of the Americas, New York, NY 10020. (212) 997-3675. Seventeenth editon. 1992. $1900.00.

Pulp and Paper Dictionary. Kenneth L. Patrick, editor. Miller Freeman, Inc., 600 Harrison Street, San Francisco, CA 94107. (415) 905-2470. 1993. $65.00.

Ullman's Encyclopedia of Industrial Chemistry. VCH Publications, Inc., 220 East 23rd Street, Suite 909, New York, NY 10010-4606. (212) 683-8333. (800) 422-8824. 5th edition. 1984-present. Price varies per volume.

GENERAL WORKS

The Manufacture of Pulp and Paper: Science and Concepts. TAPPI-Technical Association of the Pulp and Paper Industry,

15 Technology Parkway, Atlanta, GA 30092. (404) 435-1400. (800) 332-8686. 1988. $6.00.

IEEE Annual Pulp and Paper Industry Conference. Institute of Electrical and Electronic Engineers (IEEE), 345 East 47th Street, New York, NY 10017. (212) 705-7900. priced individually, inquire.

Paper Basics: Forestry, Manufacture, Selection, Purchasing, Recycling. David Saltman. Krieger Publishing Company, P.O. Box 9542, Melbourne, FL 32902-9542. (407) 724-9542. Reprint Edition. 1992. $27.50.

Paper: An Engineered Stochastic Structure. M. Deng and C. T. Dodson. TAPPI-Technical Association of the Pulp and Paper Industry, 15 Technology Parkway, Atlanta, GA 30092. (404) 435-1400. (800) 332-8686. 1994. $69.00.

Permanence of Paper For Publications and Documents in Libraries and Archives. NISO. Transaction Publications, Rutgers University, New Brusnwick, NJ 08903. (9088) 445- 2280. 1993. $30.00 in paper.

Pulp and Paper: Chemistry and Chemical Technolgy. James F. Casey, editor. John Wiley & Sons, Inc., 605 Third Avenue, New York, NY 10158-0012. (212) 850-6000. (800) 225-5945. 3rd edition. 1982. 4 volumes, 568.00 for set.

Recycling Paper From Fiber to Finished Product. Matthew Coleman, editor. TAPPI-Technical Association of the Pulp and Paper Industry, 15 Technology Parkway, Atlanta, GA 30092. (404) 435-1400. (800) 332-8686. 2 volumes. 1990. $148.00.

Surface Phenenomena and Additives in Water-based Coatings and Printing Technology. M. K. Sharma, editor. Plenum Publishing Corp., 233 Spring Street, New York, NY. (212) 620-8000. (800) 2221-9369. FAX (212) 463-0742. 1992. $95.00.

Technology of Wood Bonding: Principles in Practice. Alan A. Marra. Chapman & Hall, 1 Penn Plaza, New York, NY 10119. (212) 564-1060. 1992. $59.95.

HANDBOOKS AND MANUALS

Chemical Formulary. H. Bennett, editor. Chemical Publishing Co., Inc., 80 Eighth Avenue, New York, NY 10011. (212) 255-1950. Volumes 1-30. $70.00 per volume.

CRC Handbook of Chemistry and Physics. David R. Kide, editor. CRC Press Inc., 2000 Corporate Blvd., NW, Boca Raton, FL 33431. (407) 994-0555. (800) 333-8300. 77th edition. 1996. $99.95.

Guide to Basic Chemical Compounds. D.R. Lide, Jr. CRC Publishers, Inc., 2000 Corporate Blvd., N.W., Boca Raton, FL 33431. (407) 994-0555 or (800) 333-8300. 1993. $120.00.

Handbook of Adhesion. D. E. Packham, editor. Halsted Press, 605 Third Avenue, New York, 10158-0012. (212) 850-6400. 1993. $250.00.

Handbook of Physical and Mechanical Testing of Paper and Paperboard. Mark. Marcel Dekker, Inc., 270 Madison Avenue, New York, NY 10016. (212) 696-9000. (800) 228-1160. 1983. Volume 1, $190.00.

Handbook of Pressure-Sensitive Adhesive Technology. Chapman & Hall, 1 Penn Plaza, New York, NY 10119. (212) 564-1060. 2nd edition. 1989. $105.00.

Lange's Handbook of Chemistry. John A. Dean, editor. McGraw-Hill Publishing Company, 1221 Avenue of the Americas, New York, NY 10020. (800) 262-4729 or (212) 512-3825. Fourteenth editon. 1992. $79.50.

Perry's Chemical Engineers' Handbook. Robert H. Perry and Donald W. Green, editors. McGraw-Hill Publishing Company, Inc., 1221 Avenue of the Americas, New York, NY 10020. (212) 512-2000. (800) 262-4729. 6th edition. 1996. $129.50.

ONLINE DATABASES AND CD-ROMS

CA Search. Chemical Abstracts Service, P.O. Box 3012, Columbus, OH 43210-0012. (614) 447-3600. (800) 848-6533. FAX (614) 447-3709. Very comprehensive guide to worldwide chemical literature and related fields, 1972 to present. Available on BRS,(800) 289-4277, DIALOG, (800) 334-2564, ORBIT (800) 456-7248, and STN International, FIZ Karlsruhe, P.O. Box 2465, W-7500, Karlsruhe 1, Germany, online services. Inquire as to cost and availability.

Chemical Journals of the American Chemical Society. American Chemical Society, 1155 16th Street, N.W., Washington, DC 20036. (202) 872-4381 or (800) 424-6747. Contains complete text of approximately 90,000 articles from 22 primary journals published by the American Chemical Society, from mostly 1982 to present. Available on STN International, FIZ Karlsruhe, P.O. Box 2465, W-7500, Karlsruhe 1, Germany, online service. Inquire as to cost and availability.

Dissertation Abstracts. University Microfilms International, 300 North Zeeb Road, Ann Arbor, MI 48106. (800) 521-0600 or (313) 761-4700. Scope includes virtually all doctoral dissertations accepted at accredited American institutions from 1861 to present in 252 subject areas. Available on BRS, (800) 289-4277, DIALOG, (800) 334-2564, and OCLC EPIC, (800) 848-5878, online services. Also available on CD-ROM. Inquire as to cost and availability.

NTIS Bibliographic Database. National Technical Information Service, 5285 Port Royal Road, Springfield, VA 22161. (703) 487-4929 or FAX (703) 321-8199. Broad coverage of government-sponsored science and technology research reports, 1964 to present. Available on BRS,(800) 289-4277, DIALOG, (800) 334-2564, ORBIT, (800) 456-7248, and STN International, FIZ Karlsruhe, P.O. Box 2465, W-7500, Karlsruhe 1, Germany, online services. Also available on CD-ROM. Inquire as to cost and availability.

SCISEARCH. Institute for Scientific Information, 3501 Market Street, Philadelphia, PA 19104. (800) 523-1850 or (215) 386-0100. Broad multidisciplinary title and author index to the international literature of science and technology, 1974 to present. Available on DIALOG, (800) 334-2564, and ORBIT, (800) 456-7248, online services. Also available on CD-ROM. Inquire as to cost and availability.

WILSONLINE. H.W. Wilson Company, 950 University Avenue, Bronx, NY 10452. (800) 367-6770 or (212) 588-8400. Makes available online versions of the printed H.W. Wilson indexes including Applied Science and Technology Index, Business Periodicals Index, General Science Index, and Readers' Guide to Periodical Literature. Period covered is generally 1983 to present. Available on BRS, (800) 289-4277, DIALOG, (800)

334-2564, and OCLC EPIC, (800) 848-5878, online services. Also available on CD-ROM. Inquire as to cost and availability.

PERIODICALS

ADHESIVES AGE. Argus Inc., 6151 Powers Perry Road, NW, Atlanta, GA 30339-2941. (404) 955-2500. FAX (404) 955-0400, 1957 to present. Monthly, plus annual directory. $52.00 per year. ISSN: 0001-821X.

Chemical Week. Includes Annual Buyers Guide. Chemical Week Associates, 888 Seventh Avenue, New York, NY 10106. (212) 621-4900. FAX (212) 621-4949. 1914 to present. Weekly. $99.00 per year. ISSN: 0009-272X.

Chemtech. American Chemical Society. Box 3337, Columbus, OH 43210. (614) 447-3776. 1970 to present. Monthly. $370.00 per year. ISSN: 0009-2703.

Journal of Pulp and Paper Science. Canadian Pulp and Paper Association, Sun Life Building, 1155 Metcalfe Street, Montreal, PQ h3B 2X9, Canada. (514) 866-6621. FAX (514) 866-3035. 1975 to present. Bimonthly. $65.00. ISSN: 0826-6220.

Pulp and Paper. Miller Freeman, Inc., 600 Harrison Street, San Francisco, CA 94107. (415) 905-2200. FAX (415) 905-2232. 1927 to present. Monthly. $100.00 per year. ISSN: 0033-4081

Tappi Journal. Technical Association of the Pulp and Paper Industry, 15 Technology Parkway, Atlanta, GA 30092. (404) 446-1400. (800) 332-8686. FAX (404) 446-6947. 1949 to present. Monthly. Subscription included in membership. ISSN: 0734-1415.

RESEARCH CENTERS AND INSTITUTES

Institute of Paper Science and Technology, 500 10th Street, Atlanta, GA 30318. (404) 853-9500.

Pulp and Paper Laboratory. North Carolina State University, Box 8005, Raleigh, NC 27965. (919) 737-3181.

Pulp and Paper Research Institute of Canada. 570 Jt. John Boulevard, Pointe Claire, PQ Canada, H9R 3J9. (514) 630-4100.

Tappi-Technical Association of the Pulp and Paper Industry. Technology Park, Post office Box 105113, Atlanta, GA 30348. (404) 446-1400. (800) 332-8686.

Wisconsin Center for Structural and Materials Testing, University of Wisconsin at Madison, 1415 Johnson Drive, Madison, WI 53706. (608) 262-3205.

PARALLEL COMPUTERS

See also: ARTIFICIAL INTELLIGENCE, COMPUTER OPERATING SYSTEMS, COMPUTER PROGRAMMING, COMPUTER PROGRAMMING LANGUAGES, COMPUTERS, SOFTWARE ENGINEERING, SYSTEMS ENGINEERING

ABSTRACT SERVICES AND INDEXES

ACM Guide to Computing Literature. Association for Computing Machinery, Association for Computing Machinery, 1515 Broadway, 17th Floor, New York, NY 10036-5701. (212) 869-7440. FAX (212) 944-1318. 1964 to present. Monthly. $190.00 per year for non-members.

Applied Science and Technology Index: A Cumulative Subject Index to English Language Periodicals in the Fields of Aeronautics and Space Science, Computer Technology, Chemistry, Construction Industry, Energy and Related Areas. H.W. Wilson Co., 950 University Avenue, Bronx, NY 10452. (800) 367-6770 or (212) 588-8400. FAX (718) 590-1617. From 1958 to present. Monthly. Inquire about cost and availability. Also available on CD-ROM and online.

Computer Abstracts, MCBUniversity Press Ltd., PO Box 10812, Birmingham, AL 35201-0812. (800) 633-4931. FAX (205) 995-1588. Monthly. Covers computer theory, data, hardware, systems, networks, human-computer interaction, artificial intelligence, as well as applications of computers in aerospace, business, CAD/CAM, cartography, civil engineering, electronics and electrical engineering, industrial engineering, mechanical engineering, medicine, structural engineering, etc. Also available on CD-ROM.

Computer & Control Abstracts (inspec). INSPEC/IEEE, Box 1331, 445 Hoes Lane, Piscataway, NJ 08855-1331. (908) 562-5549. Abstracts organized by subjects of international technological information. Monthly. $1455.00 per year. Also available on CD-ROM and online via BRS Online Products, DIALOG Information Services, Orbit Search Service, and STN International.

Computer and Information Systems Abstracts Journal. Cambridge Scientific Abstracts, 7200 Wisconsin Avenue, Bethesda, MD 20814. (301) 961-6750. Fax (301) 961-6720. 1962 to present. Monthly. $1265.00 per year. Also available online via STN International.

Computer Literature Index. Applied Computer Research Inc., Box 82266, Phoenix, AZ 85071-2266. (800) 234-2227. 1971 to present. Quarterly plus annual cumulation. $198.50 per year. Bibliography of books, articles, and reports.

Computing Journal Abstracts. Techgnosis Ltd., Blade House, Battersea Road, Stockport, Cheshire 3AE, England. Telephone 061-442-2639. FAX 061-443-1162. 1969 to present. Monthly. Inquire for price.

Computing Reviews. Association for Computing Machinery, 1515 Broadway, 17th Floor, New York, NY 10036-5701. (212) 869-7440. FAX (212) 944-1318. 1960 to present. Monthly. $130.00 per year for non-members. Also available online via DIALOG Information Services.

Current Contents: Engineering, Technology, and Applied Sciences. Institute for Scientific Information, 3501 Market Street, Philadelphia, PA 19104. (215) 386-0100. FAX (215) 386-6362. 1970 to present. Weekly. $442.00 per year.

Engineering Index Monthly. Engineering Information, Inc., Castle Point on the Hudson, Hoboken, NJ 07030. (800) 221-1044. FAX (212) 832-1857. Monthly. $2300.00 per year. Also available online as COMPENDEX, and also on CD-ROM. Covers chemical engineering, computers, electrical engineering, civil engineering, metals and mining, industrial management, and mechanical engineering.

Index to Scientific and Technical Proceedings. Institute for Scientific Information, 3501 Market St., Philadelphia, PA 19104. (215) 386-0100. FAX (215) 386-6362. Monthly. $500.00 per year.

Index to Scientific Reviews. Institute for Scientific Information, 3501 Market St., Philadelphia, PA 19104. (215) 386-0100. FAX (215) 386-6362. Semi-annual.

Physics Abstracts. Institute of Electrical Engineers, Michael Faraday House, Six Hill Way, Stevenage, Herts, SG1 2AY, England. Distributed by IEEE, 445 Hoes Lane, Piscataway, NJ 08854. (908) 562-5549. 1898 to present. Monthly. $2700.00 per year. Also available online as INSPEC.

Science Citation Index. Institute for Scientific Information, 3501 Market Street, Philadelphia, PA 19104. (215) 386-0100. FAX (215) 386-6362. Inquire about availability and cost. Also available on CD-ROM.

ANNUAL REVIEWS AND YEARBOOKS

ACM Monograph Series. Association for Computing Machinery (ACM), 11 West 42nd Street, New York, NY 10036. (212) 869-7440. Fax (212) 869-0481. Irregular. Price varies, inquire.

Advances in Computers. Academic Press Inc., 6277 Sea Harbor Drive, Orlando, FL 32887. (800) 321-5068. Irregular. Price varies, inquire.

Advances in Computing Research. JAI Press, Inc., 55 Old Post Road, Number 2, Box 1678, Greenwich, CT 06830. (203) 661-7602. FAX (203) 661-0792. Annual. $68.50 per year.

Computer Science and Scientific Computing. Academic Press Inc., 6277 Sea Harbor Drive, Orlando, FL 32887. (800) 321-5068. Irregular. Price varies, inquire.

ASSOCIATIONS AND PROFESSIONAL SOCIETIES

American Federation of Information Processing Societies, 1899 Preston White Drive, Reston, VA 22091-5435. (703) 620-8937.

Association For Computing Machinery. 1515 Broadway, 17th Floor, New York, NY 10036-5701. (212) 869-7440. FAX (212) 944-1318.

Computer and Automation Systems Association of SME. 1 SME Drive, P.O. Box 930, Dearborn, MI 48121-0930. (313) 271-1500.

IEEE Computer Society. 1730 Massachusetts Avenue NW, Washington, DC 20036-1992. (202) 371-0101. FAX (202) 728-9614.

Society for Computer Simulation International, 4838 Ronson Court Lane, San Diego, CA 92111-1810. (619) 277-3888.

Special Interest Group On Software Engineering. c/o ACM, 1515 Broadway, New York, NY 10036. (212) 626-0611. FAX (212) 302-5826.

BIBLIOGRAPHIES

ACM Guide to Computing Literature. Association for Computing Machinery, 1515 Broadway, 17th Floor, New York, NY 10036-5701. (212) 869-7440. FAX (212) 944-1318. 1964 to present. Annual. $190.00 for non-members. Also available online via DIALOG.

DIRECTORIES AND BIOGRAPHICAL SOURCES

American Men & Women of Science, 1995-96. R.R. Bowker Staff, eds. 19th edition. 8 volumes. R.R. Bowker/Reed International Publishing Company, 121 Chanlon Road, New Providence, NJ 07974. (908) 464-6800 or (800) 521-8110. 1995. $850.00

Asis Handbook and Directory, American Society For Information Science, 8720 Georgia Avenue, Suite 501, Silver Spring, MD 20910-3602. (301) 495-0900. Annual. $100.00.

Computing Information Directory. Hildebrandt Inc., P.O. Box 576, Pullman, WA 99114. 1981 to present. Annual. $199.95 per year.

Directory of Engineering Societies and Related Organizations. Gordon Davis. 13th edition. American Association of Engineering Societies, 1111 19th Street NW, Suite 608, Washington, DC 20036. (202) 296-2237 or (800) 658-8897. 1989. Inquire for price.

International Research Centers Directory. Gale Research, 835 Penobscot Building, Detroit, MI 48226-4094. (313) 961-2242. (800) 347-4253. FAX (313) 961-6083. 8th edition. 1995. $410.00

Research Centers Directory. Gale Research, 835 Penobscot Building, Detroit, MI 48226-4094. (313) 961-2242. (800) 347-4253. FAX (313) 961-6083. $485.00.

Scientific and Technical Organizations and Agencies Directory. Gale Research, 835 Penobscot Building, Detroit, MI 48226-4094. (313) 961-2242. (800) 347-4253. FAX (313) 961-6083. 4th edition. 1996. $195.00.

Who's Who in Technology. Gale Research, 835 Penobscot Building, Detroit, MI 48226-4094. (313) 961-2242. (800) 347-4253. FAX (313) 961-6083. 1995. $195.00.

ENCYCLOPEDIAS AND DICTIONARIES

Computer Dictionary. Donald D. Spencer. 4th edition. Camelot Publishing Company, 709 SW 80th Blvd., Gainesville, FL 32607-1537. (904) 331-0952. 1993. $24.95.

Dictionary of Computing. S.M.H. Collin. 2nd edition. Peter Collin Publishing, 8 the Causeway, Teddington, TW11 0HE, England. FAX 0181-943-3386. 1994. Inquire for cost.

Encyclopedia of Computer Science and Engineering. Anthony Ralston & Edwin D. Reilly Jr., editors. 3rd revised edition. Van Nostrand Reinhold, 115 Fifth Avenue, New York, NY 10003. (212) 254-3232 or FAX (212) 254-9499. 1993. $125.00.

The Encyclopedia of Operations Research and Management Science. Saul I. Gass & Carl M. Harris, editors. Kluwer Academic Publishers, PO Box 358, Accord Station, Hingham, MA

02018-0358. (617) 871-6600. FAX (617) 871-6528. 1996. $295.00.

Microsoft Press Computer Dictionary: The Comprehensive Standard for Business, School, Library, and Home. 2nd edition. Microsoft Press, One Microsoft Way, Redmond, WA 98052-6399. (206) 936-0055. FAX (206) 823-8101. 1994. $19.95.

GENERAL WORKS

Introduction to Parallel Computing. Ted G. Lewis. Prentice Hall, 113 Sylvan Avenue, Route 9W, Englewood Cliffs, NJ 07632. (201) 592-2000 or (800) 922-0579. 1992. $48.00.

Introduction to Parallel Computing: Design and Analysis of Parallel Algorithms. Vipin Kumar. Benjamin-Cummings Publishing Company, 390 Bridge Parkway, Redwood City, CA 94065. (800) 950-2665. FAX (415) 594-4409. 1994. $55.95.

Parallel Computing: An Introduction. Edward L. Lafferty. Noyes Publications, 120 Mill Road, Park Ridge, NJ 07656. (201) 391-8484. FAX (201) 391-6833. 1993. $45.00.

Parallel Computing: Principles and Practice. T.J. Fountain. Cambridge University Press, 40 West 20th Street, New York, NY 10011-4211. (212) 924-3900 or (800) 872-7423. FAX (914) 937-4712. 1995. $39.95.

Parallel and Vector Computing: A Practical Introduction. Ernst L. Leiss. McGraw-Hill Publishing Company, 1221 Avenue of the Americas, New York, NY 10020. (800) 262-4729 or (212) 512-3825. 1995. $55.00.

Past, Present, Parallel: A Survey of Available Parallel Computer Systems. Arthur Trew and Greg Wilson. Springer-Verlag, 175 Fifth Avenue, New York, NY 10010. (212) 460-1500 or (800) 777-4643. FAX (212) 473-6272. 1991. $69.00.

ONLINE DATABASES AND CD-ROMS

Compendex Plus. Engineering Information, Inc., 345 East 47th Street, New York, NY 10017. (212) 705-7600 or (800) 221-1044. Contains citations with abstracts to worldwide literature in engineering and technology, from 1970 to present. Available on online BRS,(800) 289-4277, DIALOG, (800) 334-2564, ORBIT (800) 456-7248, and STN International, FIZ Karlsruhe, P.O. Box 2465, W-7500, Karlsruhe 1, Germany, online services. Also available on CD-ROM. Inquire as to cost and availability.

Compuscience. FIZ Karlsruhe, D-7514, Eggenstein-Leopoldshafen 2, Germany. Contains citations with abstracts to European and North American literature on computer science, 1972 to present. Available on STN International, FIZ Karlsruhe, P.O. Box 2465, W-7500, Karlsruhe 1, Germany, online service. Inquire as to cost and availability.

Computer and Information Systems Abstracts. Cambridge Scientific Abstracts, 7200 Wisconsin Avenue, Suite 601, Bethesda, MD 20814. (301) 961-6750 or (800) 843-7751. Contains citations to worldwide literature in theoretical and applied computer science and related areas, from 1981 to present. Inquire as to cost and availability.

Computer and Mathematics Search. Institute for Scientific Information, 3501 Market Street, Philadelphia, PA 19104. (215) 386-0100. FAX (215) 386-6362. Covers worldwide literature

in computer science and mathematics, from 1980 to present. Available on BRS,(800) 289-4277, online service. Inquire as to cost and availability.

Computer Database. Information Access Company, 362 Lakeside Drive, Foster City, CA 94404. (415) 378-5000 or (800) 227-8431. Contains citations with abstracts to literature from trade journals covering the computer,telecommunications,and electronics industries. Available on the BRS, (800) 289-4277, Compuserve Information Service,(800) 848-8990, and DIALOG,(800) 334-2564, online services. Inquire as to cost and availability.

CSA Engineering. Cambridge Scientific Abstracts, 7200 Wisconsin Avenue, Suite 601, Bethesda, MD 20814. (301) 961-6750 or (800) 843-7751. Contains citations and abstracts of international periodicals and research literature covering all fields of engineering and science and technology,including computer and information science, electronics, mechanical engineering, solid state materials, 1981 to present. Available on BRS,(800) 289-4277, online service.Inquire as to cost and availability.

Current Contents Search. Institute for Scientific Information, 3501 Market Street, Philadelphia, PA 19104. (215) 386-0100. FAX (215) 386-6362. Contains citations to articles listed in the table of contents of science and technology journals. Also articles in social sciences and life sciences journals. Available on BRS,(800) 289-4277, DIALOG,(800) 334-2564, online services. Inquire as to cost and availability.

Dissertation Abstracts. University Microfilms International, 300 North Zeeb Road, Ann Arbor, MI 48106. (800) 521-0600 or (313) 761-4700. Scope includes virtually all doctoral dissertations accepted at accredited American institutions from 1861 to present in 252 subject areas. Available on BRS,(800) 289-4277, DIALOG,(800) 334-2564, and OCLC EPIC,(800) 848-5878, online services. Also available on CD-ROM. Inquire as to cost and availability.

INSPEC. Institution of Electrical Engineers, Michael Faraday House, Six Hills Way, Stevenage, Herts. SG1 2AY, England. Telephone: 0438 313311 or FAX 0438 742840. Contains citations to the worldwide literature of physics, electronics and electrical engineering, computer technology, and related fields. Available on BRS,(800) 289-4277, DIALOG, (800) 334-2564, ORBIT, (800) 456-7248, and STN International, FIZ Karlsruhe, P.O. Box 2465, W-7500, Karlsruhe 1, Germany, online services. Inquire as to cost and availability.

NTIS Bibliographic Database. National Technical Information Service, 5285 Port Royal Road, Springfield, VA 22161. (703) 487-4929 or FAX (703) 321-8199. Broad coverage of government-sponsored science and technology research reports, 1964 to present. Available on BRS,(800) 289-4277, DIALOG, (800) 334-2564, ORBIT, (800) 456-7248, and STN International, FIZ Karlsruhe, P.O. Box 2465, W-7500, Karlsruhe 1, Germany, online services. Also available on CD-ROM. Inquire as to cost and availability.

SCISEARCH. Institute for Scientific Information, 3501 Market Street, Philadelphia, PA 19104. (800) 523-1850 or (215) 386-0100. Broad multidisciplinary title and author index to the international literature of science and technology, 1974 to present. Available on DIALOG,(800) 334-2564, and ORBIT,(800) 456-7248, online services. Also available on CD-ROM. Inquire as to cost and availability.

WILSONLINE. H.W. Wilson Company, 950 University Avenue, Bronx, NY 10452. (800) 367-6770 or (212) 588-8400. Makes

PARALLEL COMPUTERS

Ency. of Physical Sciences and Engineering Info. Sources

available online versions of the printed H.W. Wilson indexes including *Applied Science and Technology Index, Business Periodicals Index, General Science Index,* and *Readers' Guide to Periodical Literature.* Period covered is generally 1983 to present. Available on BRS,(800) 289-4277, DIALOG,(800) 334-2564, and OCLC EPIC,(800) 848-5878, online services. Also available on CD-ROM. Inquire as to cost and availability.

PERIODICALS

Communications of the ACM. Association for Computing Machinery, 11 West 42nd Street, New York, NY 10036. (212) 869-7440. Fax (212) 869-0481. 1958 to present. Monthly. $105.00 per year.

IEEE Transactions on Computers. IEEE Computer Society, 10662 Vaqueros Circle, Box 3014, Los Alamitos, CA 90720. (714) 821-8380. 1952 to present. Monthly. $335.00 per year.

Journal of Parallel and Distributed Computing. Academic Press, Inc., 1250 Sixth Avenue, San Diego, CA 92101. (619) 231-0926. FAX (619) 699-6715. 1984 to present. Monthly. $325.00 per year.

Parallel Processing Letters. World Scientific Publishing Company, Inc., 1060 Main Street, Unit B, River Edge, NJ 07661. (800) 227-7562 or (201) 487-9655. 1991 to present. Quarterly. $70.00 per year.

Software Engineering. Auerbach Publishers, 1 Penn Plaza, New York, NY 10119-0118. 1989 to present. Bimonthly. $145.00 per year.

RESEARCH CENTERS AND INSTITUTES

Center for Research on Parallel Computation, Computer and Information Technology Institute. Rice University, PO Box 1892, Houston, TX 77251. (713) 527-6077. FAX (713) 285-5136.

Ohio State University, Computer and Information Science Research Center. 2036 Neil Avenue Mall, Columbus, OH 43210. (614) 292-6374. FAX (614) 292-9021.

University of Virginia, Institute for Parallel Computing. Thornton Hall, Charlottesville, VA 22903. (804) 924-1043.

PARTICLE ACCELERATORS

See also: FUSION, PARTICLE PHYSICS, PHYSICS

ABSTRACT SERVICES AND INDEXES

Applied Science and Technology Index. A cumulative subject index to English language periodicals in the fields of aeronautics and space science, computer technology, chemistry, construction industry, energy and related areas. H.W. Wilson Co., 950 University Avenue, Bronx, NY 10452. (212) 588-8400. (800) 367-6770. FAX (718) 590-1617. From 1958 to present. Monthly. Inquire about cost and availability. Also available on CD-ROM and online. ISSN: 0003-6986.

Chemical Abstracts. Chemical Abstracts Service, 2540 Olentangy River Road, Box 3012, Columbus, OH 43210-0012. (614) 447-3600. (800) 848-6538. FAX (614) 447-3713. From 1907 to present. Weekly. $16800.00 per year. Also available on CD-ROM and online. Inquire regarding cost and availability.

Current Contents: Physical, Chemical and Earth Sciences. Institute for Scientific Information, 3501 Market Street, Philadelphia, PA 19104. (215) 386-0100. FAX (215) 386-2291. From 1961 to present. Weekly. $442.00 per year. Also available on CD-ROM and online. Inquire regarding cost and availability. ISSN: 0163-2574.

Current Papers in Physics. Institute of Electrical Engineers, Michael Faraday House, Six Hill Way, Stevenage, Herts, SG1 2AY, England. Distributed by INSPEC/IEEE, Box 1331, 445 Hoes Lane, Piscataway, NJ 08855-1331. (908) 562-5549. 1966 to present. Fortnightly. $410.00 per year. ISSN: 0011-3786.

General Science Index. H.W. Wilson Company, 950 University Avenue, Bronx, NY 10452. (212) 588-8400. (800) 367-6770. FAX (718) 590-1617. From 1978 to present. Ten issues per year; quarterly and annual cumulations. Service basis. Available on CD-ROM and online. Inquire regarding cost and availability. ISSN: 0162-1963.

Government Reports Announcements and Index. U. S. National Technical Information Service (NTIS), 5285 Port Royal Road, Springfield, VA 22161. (703) 487-4650. FAX (703) 321-8547. From 1968 to present. Annual. $630.00 per year. Also available online as NTIS Bibliographic Database and on CD-ROM. ISSN:

NTIS Alerts: Physics. U. S. National Technical Information Service. 5285 Port Royal Road, Springfield, VA 22161. (703) 487-4650. FAX (703) 321-8547. Weekly. $140.00 per year.

Physics Abstracts. INSPEC. Section A, Science Abstracts. Institution of Electrical Engineers (IEE). Available from INSPEC/IEEE-Institute of Electrical and Electronic Engineers,, Box 1331, 445 Hoes Lane, Piscaway NJ 08855-1331.(908) 562-5549. 1898 to present. 24 issues per year. $2835.00 per year. Also available on CD-ROM and online. ISSN: 0036-8091.

Physics Briefs (Physikalische Berichte). Information Center for Energy, Physics, Mathematics; German Physical Society. V C H Publishers, Inc., 220 East 23rd Street, New York, NY 10010-4606. (212) 683-8333. 1845 to present. 24 issues per year. $2390.00 per year. Also available online. ISSN: 0179-7434.

Science Citation Index (SCI). Institute for Scientific Information, 3501 Market Street, Philadelphia, PA 19104. (215) 386-0100. (800) 523-1850. FAX (215) 386-2991. 1961 to present. Six issues per year, plus annual cumulation. $11650.00 per year. Also available online and on CD-ROM. Inquire about price and availability. ISSN: 0036-827X.

ANNUAL REVIEWS AND YEARBOOKS

Annual Review of Nuclear and Particle Science. Annual Reviews, Inc., 4139 El Camino Way, Palo Alto, CA 94303-0139. (415) 493-4400. (800) 523-8635. Fax (415) 855-9815. 1952 to present. Annual. $62.00. ISSN: 0163-8998.

Progress in Particle and Nuclear Physics. Elsevier Science Publishing Company, Inc., 655 Avenue of the Americas, New York, NY 10010. (212) 989-5800. FAX (914) 333-2444. 1977 to present. Two issues per year. $660.00 per year. ISSN: 0146-6410.

ASSOCIATIONS AND PROFESSIONAL SOCIETIES

American Institute of Physics. One Physics Ellipse, College Park, MD 20740-3843. (301) 209-3100.

American Physical Society. One Physics Ellipse, College Park, MD 20740-3843. (301) 209-3200. FAX (301) 209-0865.

Universities Research Association, 1111 19th Street Nw, Suite 400, Washington, DC 20036. (202) 293-1382.

DIRECTORIES AND BIOGRAPHICAL SOURCES

American Men and Women of Science: Physical and Biological Sciences. R. R. Bowker Inc., 121 Chanlon Road, New Providence, NJ 07974. (908) 464-6800. (800) 521-8110. 20th edition. 8 volumes. 1996. $850.00.

American Physical Society. Membership Directory Bulletin Issue. American Physical Society, One Physics Ellipse, College Park, MD 20740-3843. (301) 209-3200. FAX (301) 209-0865. Biennial. $ 50.00.

Directory of Physics and Astronomy Staff Members. American Institute of Physics, One Physics Ellipse, College Park, MD 20740-3843. (301) 209-3100. Annual. $45.00.

Graduate Programs in Physics, Astronomy and Related Fields. 1995-1996. American Institute of Physics, One Physics Ellipse, College Park, MD 20740-3843. (301) 209-3100. Annual. $45.00. ISSN: 0147-1821.

Research Centers Directory. Gale Research, 835 Penobscot Building, Detroit, MI 48226-4094. (313) 961-2242. (800) 347-4253. 20th edition. 1995. $485.00. ISSN: 0080-1518.

Scientific and Technical Organizations and Agencies Directory. Gale Research, 835 Penobscot Building, Detroit, MI 48226-4094. (313) 961-2242. (800) 347-4253. 4th edition. 1996. $195.00.

World Guide to Scientific Associations and Learned Societies. K. G. Saur Inc., 121 Chanlon Road, New Providence, NJ 07974. (908) 464-6800. (800) 521-8110. 5th 1990. $225.00.

ENCYCLOPEDIAS AND DICTIONARIES

A Concise Dictionary of Physics. Oxford University Press, Inc., 200 Madison Avenue, New York, NY 10016. (212) 725-6000. (800) 334-4249. 1990. $10.95.

Dictionary of Effects and Phenomena in Physics. Joachim Schubert. VCH Publications, Inc., 220 East 23rd Street, Suite 909, New York, NY 10010-4606. (212) 683-8333. (800) 422-8824. 1987. $35.00.

Dictionary of the Physical Sciences: Terms, Formulas, Data. Cesare Emiliani. Oxford University Press, Inc., 200 Madison Avenue, New York, NY 10016. (212) 725-6000. (800) 334-4249. 1989. $19.95.

Encyclopedia of Applied Physics. George Trigg, editor. VCH Publications, Inc., 220 East 23rd Street, Suite 909, New York, NY 10010-4606. (212) 683-8333. (800) 422-8824. 20 volume set. 1991-present. $5990.00.

Encyclopedia of Physics. Robert M. Besancon. Chapman & Hall, 1 Penn Plaza, New York, NY 10119. (212) 564-1060. 3rd edition. 1990. $54.95 in paper.

Facts on File Dictionary of Physics. John Daintith, editor. Facts-on-File, Inc., 460 Park Avenue South, New York, NY 10016-7382. (212) 683-2244. (800) 322-8755. FAX (800) 683-3633. Revised edition. 1990. $12.95.

Encyclopedia of Physical Science and Technology. Academic Press, Inc., 6277 Sea Harbor Drive, Orlando, FL. (800) 321-5068. 2nd edition. 18 volume set. 1992. $2625.00.

McGraw-Hill Encyclopedia of Science and Technology. McGraw-Hill Book Company, Inc., 1221 Avenue of the Americas, New York, NY 10020. (212) 512-2000. (800) 262-4729. 7th edition. 20 volume set. 1992. $1900.00.

Penguin Dictionary of Physics. Valerie Illingworth, editor. Viking Penguin, 375 Hudson Street, New York, NY 10014-3657. (212) 366-2000. (800) 331-4624 2nd edition. 1991.

GENERAL WORKS

The Development of Colliders. Cladio Pellegrini and andrew M. Sessler, editors. American Institute of Physics, One Physics Ellipse, College Park, MD 20740-3843. (301) 209-3100. 1995.

Introduction to the Physics of High Energy Accelerators. D. A. Edwards and M. H. Syphers. John Wiley & Sons, Inc., 605 Third Avenue, New York, NY 10158-0012. (212) 850-6000. (800) 225-5945. 1992. $69.95.

The Physics of Intense Beams and Storage Rings. Demitri Patrikov and Nikolai Dikansky. American Institute of Physics, One Physics Ellipse, College Park, MD 20740-3843. (301) 209-3100. 1994. $70.00.

Particle Accelerator Physics: Basic Principles and Linear Beam Dynamics. Helmut Wiedemann. Springer-Verlag New York, Inc., 175 Fifth Avenue, New York, NY 10010. (212) 460-1500. (800) 777-4643. FAX (212) 473-6272. 1993. $62.00.

Principles of Charged Particle Accelerators. Stanley Humphries. John Wiley & Sons, Inc., 605 Third Avenue, New York, NY 10158-0012. (212) 850-6000. (800) 225-5945. 1986. $140.00.

Theory and Design of Charged Particle Beams. Martin Reiser. John Wiley & Sons, Inc., 605 Third Avenue, New York, NY 10158-0012. (212) 850-6000. (800) 225-5945. 1994. $74.95.

HANDBOOKS AND MANUALS

Handbook of Physical Quantities. Igor S. Grigoriev and Evgenil Z. Melikhov, editors. CRC Press, Inc., 2000 Corporate Boulevard, NW, Boca Raton, FL 33431. (407) 994-0555. (800) 272-7737. 1995. $99.00.

Landolt-borenstein Numerical Data and Functional Relationships in Science and Technology: Nuclear Particle and Physics. GROUP I. Springer-Verlag New York, Inc., 175 Fifth Av-

PARTICLE ACCELERATORS

Ency. of Physical Sciences and Engineering Info. Sources

enue, New York, NY 10010. (212) 460-1500. (800) 777-4643. Volumes priced individually.

McGraw-Hill Encyclopedia of Physics. Sybil P. Parker, editor. McGraw-Hill Publishing Company, Inc., 1221 Avenue of the Americas, New York, NY 10020. (212) 512-2000. (800) 262-4729. 2nd edition. 1993. $95.50

A Physicist's Desk Reference. Herbert L. Anderson, editor. American Institute of Physics, One Physics Ellipse, College Park, MD 20740-3843. (301) 209-3100. 2nd edition. 1989. $70.00.

Physics Problem Solver. Research and Education Association Staff. Research & Education Association, 61 Ethel Road, West Piscataway, NJ 08854. (908) 819-8880. Revised edition. 1994. $23.95 in paper.

Tables of Physical and Chemical Constants and Some Mathematical Functions. G. W. Kaye and T. M. Laby. Halsted Press 605 Third Avenue, New York, 10158-0012. (212) 850- 6400. 15th edition. 1986. $74.95.

ONLINE DATABASES AND CD-ROMS

CA Search. Chemical Abstracts Service, P.O. Box 3012, Columbus, OH 43210-0012. (614) 447-3600. (800) 848-6533. FAX (614) 447-3709. Very comprehensive guide to worldwide chemical literature and related fields, 1972 to present. Available on BRS,(800) 289-4277, DIALOG, (800) 334-2564, ORBIT (800) 456-7248, and STN International, FIZ Karlsruhe, P.O. Box 2465, W-7500, Karlsruhe 1, Germany, online services. Inquire as to cost and availability.

Current Contents Search. Institute for Scientific Information, 3501 Market Street, Philadelphia, PA 19104. (215) 386-0100. FAX (215) 386-6362. Contains citations to articles listed in the table of contents of science and technology journals. Also articles in social sciences and life sciences journals. Available on BRS, (800) 289-4277, DIALOG, (800) 334-2564, online services. Inquire as to cost and availability.

Dissertation Abstracts Online. University Microfilms International, 300 North Zeeb Road, Ann Arbor, MI 48106. (800) 521-0600 or (313) 761-4700. Scope includes virtually all doctoral dissertations accepted at accredited American institutions from 1861 to present in 252 subject areas. Available on BRS, 289-4277, DIALOG, (800) 334-2564,and OCLC EPIC, (800) 848-5878, online services. Also available on CD-ROM. Inquire as to cost and availability.

INSPEC. Institution of Electrical Engineers, Michael Faraday House, Six Hills Way, Stevenage, Herts. SG1 2AY, England. Telephone: 0438 313311 or FAX 0438 742840. Contains citations to the worldwide literature of physics, electronics and electrical engineering, computer technology, and related fields. Available on BRS, (800) 289-4277, DIALOG, (800) 334-2564, ORBIT, (800) 456-7248, and STN International, FIZ Karlsruhe, P.O. Box 2465, W-7500, Karlsruhe 1, Germany, online services. Inquire as to cost and availability.

Mathsci. American Mathematical Society, P.O. Box 6248, Providence, RI 02940. (800) 321-4667 or (401) 455-4000 or FAX (401) 331-3842. Scope includes pure and applied mathematics and related areas of physics, statistics, engineering, computer science, and operations research literature since 1959. Available on DIALOG,(800) 334-2564,online service. Also available on CD-ROM. Inquire as to cost and availability.

NTIS Bibliographic Database. National Technical Information Service, 5285 Port Royal Road, Springfield, VA 22161. (703) 487-4929 or FAX (703) 321-8199. Broad coverage of government-sponsored science and technology research reports, 1964 to present. Available on BRS,(800) 289-4277, DIALOG, (800) 334-2564, ORBIT, (800) 456-7248, and STN International, FIZ Karlsruhe, P.O. Box 2465, W-7500, Karlsruhe 1, Germany, online services. Also available on CD-ROM. Inquire as to cost and availability.

Physics Briefs. American Institute of Physics, 335 East 45th Street, New York, NY 10017. (212) 661-9260 or FAX (212) 661-2036. Contains citations with abstracts of the literature of physics and related fields, 1979 to present. Available on the STN International, FIZ Karlsruhe, P.O. Box 2465, W-7500, Karlsruhe 1, Germany, online service. Inquire as to cost and availability.

SCISEARCH. Institute for Scientific Information, 3501 Market Street, Philadelphia, PA 19104. (800) 523-1850 or (215) 386-0100. Broad multidisciplinary title and author index to the international literature of science and technology, 1974 to present. Available on DIALOG, (800) 334-2564, and ORBIT, (800) 456-7248, online services. Also available on CD-ROM. Inquire as to cost and availability.

WILSONLINE. H.W. Wilson Company, 950 University Avenue, Bronx, NY 10452. (800) 367-6770 or (212) 588-8400. Makes available online versions of the printed H.W. Wilson indexes including Applied Science and Technology Index, Business Periodicals Index, General Science Index, and Readers' Guide to Periodical Literature. Period covered is generally 1983 to present. Available on BRS, (800) 289-4277, DIALOG, (800) 334-2564, and OCLC EPIC, (800) 848-5878, online services. Also available on CD-ROM. Inquire as to cost and availability.

PERIODICALS

American Journal of Physics. American Association of Physics Teachers, 5112 Berywn Road, College Park, MD 20740. (301) 345-4200. 1933 to present. Monthly. $215.00 per year. ISSN: 0002-9505.

Canadian Journal of Physics. National Research Council of Canada. Research Journals, Ottawa K1A OR6, Canada. (613) 993-9084. FAX (613) 952-7656. 1929 to present. Monthly. $285.00 per year. ISSN: 0008-4204.

International Journal of Modern Physics E: Report On Nuclear Physics. World Scientific Publishing Co., 1060 Main Street, Suite 18, River Edge, NJ 07661. (800) 227-7562. 1992 to present. Quarterly. $195.00 per year. ISSN: 0218-3013.

Journal of Physics G: Nuclear and Particle Physics. IOP Publishing. Orders to: American Institute of Physics, 500 Sunnyside Boulevard, Woodbury, NY 11797-2999. (516) 576-2200. 1975 to present. Monthly. $1426.00 per year. ISSN: 0954-3899.

Particle Accelerators. Gordon and Breach Science Publishers, Inc., 820 Town Center Drive, Langhorne, PA 19047. (215) 750-2642. 1969 to present. Monthly. 212 ECU. ISSN: 0031-2460.

Physical Review D (particles and Fields). American Institute of Physics, One Physics Ellipse, College Park, MD 20740-3843. (301) 209-3000. 1970 to present. 24 issues per year. $1280.00 per year. ISSN: 0556-2821.

Physics Letters B: Nuclear, Elementary Particle and High-Energy Physics. Elsevier Science Publishing Company, Inc., 655 Avenue of the Americas, New York, NY 10010. (212) 989-5800. FAX (212) 633-3990. 1962 to present. 96 issues per year in 24 volumes. $4683.00 per year. ISSN: 0370-2693.

Physics Today. American Institute of Physics, One Physics Ellipse, College Park, MD 20740-3843. (301) 209-3000. 1948 to present. Monthly. $140.00 per year. ISSN: 0031-9228.

RESEARCH CENTERS AND INSTITUTES

Argonne National Laboratory. 9700 South Cass Avenue, Argonne, IL 60439-4832. (708) 972-3872.

Fermi National Accelerator Laboratory. P.O. Box 500 Batavia, IL 60510. (708) 840-3351. FAX (708) 840-4343.

High Energy Physics and Elementary Particle Research Program. University of Pennsylvania, 33rd and Walnut Streets, Philadelphia. PA 19104. (215) 898-5960. FAX (215) 898-8512.

Lawrence Berkeley Laboratory, Physics Division. Building 50, Room 256, 1 Cyclotron Road, Berkeley, CA 04720. (415) 486-5421. FAX (415) 486-4553.

PARTICLE PHYSICS

See also: CONDENSED MATTER, COSMIC RAYS, DARK MATTER, NUCLEAR PHYSICS, PARTICLE ACCELERATORS, PHYSICS, QUANTUM FIELD THEORY, QUANTUM MECHANICS, QUARKS

ABSTRACT SERVICES AND INDEXES

Applied Science and Technology Index. A cumulative subject index to English language periodicals in the fields of aeronautics and space science, computer technology, chemistry, construction industry, energy and related areas. H.W. Wilson Co., 950 University Avenue, Bronx, NY 10452. (212) 588-8400. (800) 367-6770. FAX (718) 590-1617. From 1958 to present. Monthly. Inquire about cost and availability. Also available on CD-ROM and online. ISSN: 0003-6986.

Chemical Abstracts. Chemical Abstracts Service, 2540 Olentangy River Road, Box 3012, Columbus, OH 43210-0012. (614) 447-3600. (800) 848-6538. FAX (614) 447-3713. From 1907 to present. Weekly. $16800.00 per year. Also available on CD-ROM and online. Inquire regarding cost and availability.

Current Contents: Physical, Chemical and Earth Sciences. Institute for Scientific Information, 3501 Market Street, Philadelphia, PA 19104. (215) 386-0100. FAX (215) 386-2291. From 1961 to present. Weekly. $442.00 per year. Also available on CD-ROM and online. Inquire regarding cost and availability. ISSN: 0163-2574.

Current Papers in Physics. Institute of Electrical Engineers, Michael Faraday House, Six Hill Way, Stevenage, Herts, SG1 2AY, England. Distributed by INSPEC/IEEE, Box 1331, 445 Hoes Lane, Piscataway, NJ 08855-1331. (908) 562-5549. 1966 to present. Fortnightly. $410.00 per year. ISSN: 0011-3786.

General Science Index. H.W. Wilson Company, 950 University Avenue, Bronx, NY 10452. (212) 588-8400. (800) 367-6770.

FAX (718) 590-1617. From 1978 to present. Ten issues per year; quarterly and annual cumulations. Service basis. Available on CD-ROM and online. Inquire regarding cost and availability. ISSN: 0162-1963.

Government Reports Announcements and Index. U. S. National Technical Information Service (NTIS), 5285 Port Royal Road, Springfield, VA 22161. (703) 487-4650. FAX (703) 321-8547. From 1968 to present. Annual. $630.00 per year. Also available online as NTIS Bibliographic Database and on CD-ROM.

Mathematical Reviews: A Review Journal Covering the World Literature of Mathematical Research. American Mathematical Society, P.O. Box 1571, Annex Station, Providence, RI 02901-9930. (800) 556-7774 or (401) 455-4000. 1940 to present. Monthly. $4594.00 per year. Also available via network (MATHSCINET), online and on CD-ROM. Inquire regarding cost and availability. ISSN: 0025-5629.

NTIS Alerts: Physics. U. S. National Technical Information Service. 5285 Port Royal Road, Springfield, VA 22161. (703) 487-4650. FAX (703) 321-8547. Weekly. $140.00 per year.

Physics Abstracts. INSPEC. Section A, Science Abstracts. Institution of Electrical Engineers (IEE). Available from INSPEC/IEEE-Institute of Electrical and Electronic Engineers, Box 1331, 445 Hoes Lane, Piscaway NJ 08855-1331. (908) 562-5549. 1898 to present. 24 issues per year. $2835.00 per year. Also available on CD-ROM and online. ISSN: 0036-8091.

Physics Briefs (Physikalische Berichte). Information Center for Energy, Physics, Mathematics; German Physical Society. VCH Publishers, Inc., 220 East 23rd Street, New York, NY 10010-4606. (212) 683-8333. 1845 to present. 24 issues per year. $2390.00 per year. Also available online. ISSN: 0179-7434.

Science Citation Index (SCI). Institute for Scientific Information, 3501 Market Street, Philadelphia, PA 19104. (215) 386-0100. (800) 523-1850. FAX (215) 386-2991. 1961 to present. Six issues per year, plus annual cumulation. $11650.00 per year. Also available online and on CD-ROM. Inquire about price and availability. ISSN: 0036-827X.

ANNUAL REVIEWS AND YEARBOOKS

Advances in Nuclear Physics. Plenum Publishing Corp., 233 Spring Street, New York, NY. (212) 620-8000. (800) 2221-9369. FAX (212) 463-0742. 1968 to present. Irregular. Price varies. ISSN: 0065-2970.

Annual Review of Nuclear and Particle Science. Annual Reviews, Inc., 4139 El Camino Way, Palo Alto, CA 94303-0139. (415) 493-4400. (800) 523-8635. Fax (415) 855-9815. 1952 to present. Annual. $62.00.

Progress in Particle and Nuclear Physics. Elsevier Science Publishing Company, Inc., 655 Avenue of the Americas, New York, NY 10010. (212) 989-5800. FAX (914) 333-2444. 1977 to present. Two issues per year. $660.00 per year. ISSN: 0146-6410.

ASSOCIATIONS AND PROFESSIONAL SOCIETIES

American Institute of Physics. One Physics Ellipse, College Park, MD 20740-3843. (301) 209-3100.

American Physical Society. One Physics Ellipse, College Park, MD 20740-3843. (301) 209-3200. FAX (301) 209-0865.

Universities Research Association, 1111 19th Street Nw, Suite 400, Washington, DC 20036. (202) 293-1382.

BIBLIOGRAPHIES

Guide to the Archival Collections in the Niels Bohr Library at the American Institute of Physics. Niels Bohr Library Staff. American Institute of Physics, One Physics Ellipse, College Park, MD 20740-3843. (301) 209-3100. 1995. $140.00

Science Books and Films. American Association for the Advancement of Science, 1333 H Street NW, Washington, DC 20005. (202) 326-6454. Reviews of print, film and software materials in all sciences. 1965 to present. Nine issues per year. $40.00 per year. ISSN: 0098-342X.

Scientific and Technical Books and Serials in Print: An Index to Literature in Science and Technology. R.R. Bowker Inc., 121 Chanlon Road, New Providence, NJ 07974. (908) 464-6800. (800) 521-8110. 4 volumes. Annual. 1972 to Present. $299.95. Also available on compact disc and online. ISSN: 0000-054X

DIRECTORIES AND BIOGRAPHICAL SOURCES

American Men and Women of Science: Physical and Biological Sciences. R. R. Bowker Inc., 121 Chanlon Road, New Providence, NJ 07974. (908) 464-6800. (800) 521-8110. 20th edition. 8 volumes. 1996. $850.00.

American Physical Society. Membership Directory Bulletin Issue. American Physical Society, One Physics Ellipse, College Park, MD 20740-3843. (301) 209-3200. FAX (301) 209-0865. Biennial. $ 50.00.

The Biographical Dictionary of Scientists: Physicists. D. Abbott, editor. Peter Bedrick Books, Inc., 2112 Broadway, Room 318, New York, NY 10023. (212) 496-0751. 1984.

Directory of Physics and Astronomy Staff Members. American Institute of Physics, One Physics Ellipse, College Park, MD 20740-3843. (301) 209-3100. Annual. $45.00.

Graduate Programs in Physics, Astronomy and Related Fields. 1995-1996. American Institute of Physics, One Physics Ellipse, College Park, MD 20740-3843. (301) 209-3100. Annual. $45.00. ISSN: 0147-1821.

Legends in Their Own Time: A Century of American Physical Scientists. Anthony Serafini. Plenum Publishing Corp., 233 Spring Street, New York, NY. (212) 620-8000. (800) 2221-9369. FAX (212) 463-0742. 1993. $27.50.

The Nobel Prize Winners. Frank N. Magill, editor. Salem Press, Inc., P.O. Box 1097, Englewood Cliffs, NJ 07632. (201) 871-3700. (800) 221-1592. 3 volumes. 1989. $210.00.

Research Centers Directory. Gale Research, 835 Penobscot Building, Detroit, MI 48226-4094. (313) 961-2242. (800) 347-4253. 20th edition. 1995. $485.00. ISSN: 0080- 1518.

Scientific and Technical Organizations and Agencies Directory. Gale Research, 835 Penobscot Building, Detroit, MI 48226-

4094. (313) 961-2242. (800) 347-4253. 4th edition. 1996. $195.00.

World Guide to Scientific Associations and Learned Societies. K. G. Saur Inc., 121 Chanlon Road, New Providence, NJ 07974. (908) 464-6800. (800) 521-8110. 5th 1990. $225.00.

ENCYCLOPEDIAS AND DICTIONARIES

A Concise Dictionary of Physics. Oxford University Press, Inc., 200 Madison Avenue, New York, NY 10016. (212) 725-6000. (800) 334-4249. 1990. $10.95.

Dictionary of Effects and Phenomena in Physics. Joachim Schubert. VCH Publications, Inc., 220 East 23rd Street, Suite 909, New York, NY 10010-4606. (212) 683-8333. (800) 422-8824. 1987. $35.00.

Dictionary of the Physical Sciences: Terms, Formulas, Data. Cesare Emiliani. Oxford University Press, Inc., 200 Madison Avenue, New York, NY 10016. (212) 725-6000. (800) 334-4249. 1989. $19.95.

Encyclopedia of Applied Physics. George Trigg, editor. VCH Publications, Inc., 220 East 23rd Street, Suite 909, New York, NY 10010-4606. (212) 683-8333. (800) 422-8824. 20 volume set. 1991-. $5990.00.

Encyclopedia of Physics. Robert M. Besancon. Chapman & Hall, 1 Penn Plaza, New York, NY 10119. (212) 564-1060. 3rd edition. 1990. $54.95 in paper.

Facts on File Dictionary of Physics. John Daintith, editor. Facts-on-File, Inc., 460 Park Avenue South, New York, NY 10016-7382. (212) 683-2244. (800) 322-8755. Fax (800) 683-3633. Revised edition. 1990. $12.95.

Encyclopedia of Physical Science and Technology. Academic Press, Inc., 6277 Sea Harbor Drive, Orlando, FL. (800) 321-5068. 2nd edition. 18 volume set. 1992. $2625.00.

Landolt-boernstein Numerical Data and Functional Relationships in Science and Technology: Nuclear and Particle Physics. K. H. Hellwege editor. Springer Verlag New York, Inc., 175 Fifth Avenue, New York, NY 10010. (212) 460-1500. (800) 777-4643. FAX (212) 473-6272. 1993. Volumes priced separately.

McGraw-Hill Encyclopedia of Science and Technology. McGraw-Hill Book Company, Inc., 1221 Avenue of the Americas, New York, NY 10020. (212) 512-2000. (800) 262-4729. 7th edition. 20 volume set. 1992. $1900.00.

Penguin Dictionary of Physics. Valerie Illingworth, editor. Viking Penguin, 375 Hudson Street, New York, NY 10014-3657. (212) 366-2000. (800) 331-4624 2nd edition. 1991.

GENERAL WORKS

Classical Dynamics of Particles and Systems. Jerry B. Marion and Stephen T. Thornton. SCP: Third World Literature Publishing House, P.O. Box 482, Lithonia, GA 30058-0482. (404) 785-7725. 3rd edition. 1988. $58.75.

The Experimental Foundations of Particle Physics. Robert N. Cahn and Gerson Goldhaber. Cambridge University Press, 40

West 20th Street, New York, NY 10011-4211. (212) 924-3900. (800) 872-7423. 1992. $32.95 in paper.

Fundamental Particles and Forces. Richard A. Carrigan, Jr., and W. Peter Trower. W. H. Freeman & Co., 41 Madison Avenue, East 26th Avenue, 35th Floor, New York, NY 10010. (212) 576-0400. 1995. $13.95.

The Fundamental Particles and their Interactions. William B. Rolnick. Addison-Wesley Publishing Co., Inc., 1 Jacob Way, Reading, MA 01867. (617) 944-3700. (800) 447-2226. 1994. $60.25.

The Ideas of Particle Physics: An Introduction For Scientists. G. D. Coughlan and J. E. Dodd. Cambridge University Press, 40 West 20th Street, New York, NY 10011- 4211. (212) 924-3900. (800) 872-7423. 1991. $34.95 in paper.

Introduction to Gauge Field Theory. D. Bailin and A. Love. IOP Publishing, Public Ledger Building, Suite 1035, Independence Square, Philadelphia, PA 19106. (215) 627- 0880. (800) 358-4677. 1993. $49.00 in paper.

Introduction to the Physics of High Energy Accelerators. D. A. Edwards and M. H. Syphers. John Wiley & Sons, Inc., 605 Third Avenue, New York, NY 10158-0012. (212) 850-6000. (800) 225-5945. 1992. $69.95.

Nuclear and Particle Physics. W. E. Burcham and M. Jobes. John Wiley & Sons, Inc., 605 Third Avenue, New York, NY 10158-0012. (212) 850-6000. (800) 225-5945. 1995.$59.95 in paper.

The Particle Garden: Our Universe as Understood by Particle Physicists. Gordon Kane. Addison-Wesley Publishing Co., Inc., 1 Jacob Way, Reading, MA 01867. (617) 944-3700. (800) 447-2226. 1995. $21.15.

Particle Physics. B. R. Martin and G. Shaw, John Wiley & Sons, Inc., 605 Third Avenue, New York, NY 10158-0012. (212) 850-6000. (800) 225-5945. 1992. $49.95 in paper.

Particle Physics in the Cosmos. Richard A. Carrigan and W. Peter Trower, editors. W. H. Freeman & Co., 41 Madison Avenue, East 26th Avenue, 35th Floor, New York, NY 10010. (212) 576-0400. 1995. $13.95.

HANDBOOKS AND MANUALS

Chemical and Physical Data. Arthur M. James and M. P. Lord, Van Nostrand Reinhold, 115 Fifth Avenue, New York, NY 10003. (212) 254-3232. (800) 842-3636. 1992. $59.95.

CRC Handbook of Chemistry and Physics. David R. Kide, editor. CRC Press, Inc., 2000 Corporate Boulevard, NW, Boca Raton, FL 33431. (407) 994-0555. (800) 272-7737. 77th edition. 1996. $99.95.

Handbook of Physical Quantities. Igor S. Grigoriev and Evgenil Z. Melikhov, editors. CRC Press, Inc., 2000 Corporate Boulevard, NW, Boca Raton, FL 33431. (407) 994- 0555. (800) 272-7737. 1995. $99.00.

Handbook of Space Astronomy and Astrophsyics. Martin V. Zumbeck. Cambridge University Press, 40 West 20th Street, New York, NY 10011-4211. (212) 924-3900. (800) 872-7423. 2nd edition. 1990. $79.95.

Landolt-borenstein Numerical Data and Functional Relationships in Science and Technology: Nuclear Particle and Physics. GROUP I. Springer-Verlag New York, Inc., 175 Fifth Avenue, New York, NY 10010. (212) 460-1500. (800) 777-4643. Volumes priced individually.

McGraw-Hill Encyclopedia of Physics. Sybil P. Parker, editor. McGraw-Hill Publishing Company, Inc., 1221 Avenue of the Americas, New York, NY 10020. (212) 512-2000. (800) 262-4729. 2nd edition. 1993. $95.50

Newnes Engineering and Physical Science Pocketbook. J. O. Bird and P. J. Chivers. Butterworth-Heinmann, 313 Washington Avenue, Newton, MA 02158. (617) 928-2500. (800) 366-2665. 1993. $49.95.

A Physicist's Desk Reference. Herbert L. anderson, editor. American Institute of Physics, One Physics Ellipse, College Park, MD 20740-3843. (301) 209-3100. 2nd edition. 1989. $70.00.

Physics Problem Solver. Research and Education Association Staff. Research & Education Association, 61 Ethel Road, West Piscataway, NJ 08854. (908) 819-8880. Revised edition. 1994. $23.95 in paper.

Tables of Physical and Chemical Constants and Some Mathematical Functions. G. W. Kaye and T. M. Laby. Halsted Press 605 Third Avenue, New York, 10158-0012. (212) 850-6400. 15th edition. 1986. $74.95.

ONLINE DATABASES AND CD-ROMS

CA Search. Chemical Abstracts Service, P.O. Box 3012, Columbus, OH 43210-0012. (614) 447-3600. (800) 848-6533. FAX (614) 447-3709. Very comprehensive guide to worldwide chemical literature and related fields, 1972 to present. Available on BRS,(800) 289-4277, DIALOG, (800) 334-2564, ORBIT (800) 456-7248, and STN International, FIZ Karlsruhe, P.O. Box 2465, W-7500, Karlsruhe 1, Germany, online services. Inquire as to cost and availability.

Current Contents Search. Institute for Scientific Information, 3501 Market Street, Philadelphia, PA 19104. (215) 386-0100. FAX (215) 386-6362. Contains citations to articles listed in the table of contents of science and technology journals. Also articles in social sciences and life sciences journals. Available on BRS, (800) 289-4277, DIALOG, (800) 334-2564, online services. Inquire as to cost and availability.

Dissertation Abstracts Online. University Microfilms International, 300 North Zeeb Road, Ann Arbor, MI 48106. (800) 521-0600 or (313) 761-4700. Scope includes virtually all doctoral dissertations accepted at accredited American institutions from 1861 to present in 252 subject areas. Available on BRS, (800) 289-4277, DIALOG, (800) 334-2564,and OCLC EPIC, (800) 848-5878, online services. Also available on CD-ROM. Inquire as to cost and availability.

INSPEC. Institution of Electrical Engineers, Michael

Faraday House, Six Hills Way, Stevenage, Herts. SG1 2AY, England. Telephone: 0438 313311 or FAX 0438 742840. Contains citations to the worldwide literature of physics, electronics and electrical engineering, computer technology, and related fields. Available on BRS, (800) 289-4277, DIALOG, (800) 334-2564, ORBIT, (800) 456-7248, and STN International, FIZ

Karlsruhe, P.O. Box 2465, W-7500, Karlsruhe 1, Germany, online services. Inquire as to cost and availability.

MATHSCI. American Mathematical Society, P.O. Box 6248, Providence, RI 02940. (800) 321-4667 or (401) 455-4000 or FAX (401) 331-3842. Scope includes pure and applied mathematics and related areas of physics, statistics, engineering, computer science, and operations research literature since 1959. Available on DIALOG,(800) 334-2564,online service. Also available on CD-ROM. Inquire as to cost and availability.

NTIS Bibliographic Database. National Technical Information Service, 5285 Port Royal Road, Springfield, VA 22161. (703) 487-4929 or FAX (703) 321-8199. Broad coverage of government-sponsored science and technology research reports, 1964 to present. Available on BRS,(800) 289-4277, DIALOG, (800) 334-2564, ORBIT, (800) 456-7248, and STN International, FIZ Karlsruhe, P.O. Box 2465, W-7500, Karlsruhe 1, Germany, online services. Also available on CD-ROM. Inquire as to cost and availability.

Physics Briefs. American Institute of Physics, 335 East 45th Street, New York, NY 10017. (212) 661-9260 or FAX (212) 661-2036. Contains citations with abstracts of the literature of physics and related fields, 1979 to present. Available on the STN International, FIZ Karlsruhe, P.O. Box 2465, W-7500, Karlsruhe 1, Germany, online service. Inquire as to cost and availability.

Scientific and Technical Books and Serials in Print. R.R. Bowker Inc., 121 Chanlon Road, New Providence, NJ 07974. (800) 521-8110 or (908) 464-6800. List of currently published books and serials in the physical and biological sciences, engineering and technology, with subject, author and titles indexes. Available on ORBIT, (800) 456-7248, online service. Inquire as to cost and availability.

SCISEARCH. Institute for Scientific Information, 3501 Market Street, Philadelphia, PA 19104. (800) 523-1850 or (215) 386-0100. Broad multidisciplinary title and author index to the international literature of science and technology, 1974 to present. Available on DIALOG, (800) 334-2564, and ORBIT, (800) 456-7248, online services. Also available on CD-ROM. Inquire as to cost and availability.

WILSONLINE. H.W. Wilson Company, 950 University Avenue, Bronx, NY 10452. (800) 367-6770 or (212) 588-8400. Makes available online versions of the printed H.W. Wilson indexes including *Applied Science and Technology Index, Business Periodicals Index, General Science Index,* and *Readers' Guide to Periodical Literature.* Period covered is generally 1983 to present. Available on BRS, (800) 289-4277, DIALOG, (800) 334-2564, and OCLC EPIC, (800) 848-5878, online services.Also available on CD-ROM. Inquire as to cost and availability.

PERIODICALS

American Journal of Physics. American Association of Physics Teachers, 5112 Berywn Road, College Park, MD 20740. (301) 345-4200. 1933 to present. Monthly. $215.00 per year. ISSN: 0002-9505.

Canadian Journal of Physics. National Research Council of Canada. Research Journals, Ottawa K1A OR6, Canada. (613) 993-9084. FAX (613) 952-7656. 1929 to present. Monthly. $285.00 per year. ISSN: 0008-4204.

Chaos: An Interdisciplinary Journal of Nonlinear Science. American Institute of Physics, One Physics Ellipse, College Park, MD 20740-3843. (301) 209-3000. 1991 to present. Quarterly. $225.00 per year. ISSN: 1054-1500.

Contemporary Physics. Taylor & Francis, Ltd, Rankine Road, Basingstoke, Hants RG24 8PR, England. TEL 0256-840366. FAX 0256-479438. 1959 to present. Bimonthly. $318.00 per year. ISSN: 0010-7514.

International Journal of Modern Physics E: Report On Nuclear Physics. World Scientific Publishing Co., 1060 Main Street, Suite 18, River Edge, NJ 07661. (800) 227-7562. 1992 to present. Quarterly. $195.00 per year. ISSN: 0218-3013.

Journal of Physics G: Nuclear and Particle Physics. I O P Publishing. Orders to: American Institute of Physics, 500 Sunnyside Boulevard, Woodbury, NY 11797-2999. (516) 576-2200. 1975 to present. Monthly. $1426.00 per year. ISSN: 0954-3899.

Physical Review D (Particles and Fields). American Institute of Physics, One Physics Ellipse, College Park, MD 20740-3843. (301) 209-3000. 1970 to present. 24 issues per year. $1280.00 per year. ISSN: 0556-2821.

Physics Letters B: Nuclear, Elementary Particle and High-Energy Physics. Elsevier Science Publishing Company, Inc., 655 Avenue of the Americas, New York, NY 10010. (212) 989- 5800. FAX (212) 633-3990. 1962 to present. 96 issues per year in 24 volumes. $4683.00 per year. ISSN: 0370-2693.

Physics Reports: A Review Section of Physics Letters. Elsevier Science Publishing Company, Inc., 655 Avenue of the Americas, New York, NY 10010. (212) 989-5800. FAX (212) 633-3990. 1971 to present. 72 issues per year in 12 volumes. $2342.00 per year. ISSN: 0370-1573.

Physics Today. American Institute of Physics, One Physics Ellipse, College Park, MD 20740-3843. (301) 209-3000. 1948 to present. Monthly. $140.00 per year. ISSN: 0031- 9228.

Reports On Progress in Physics. I O P Publishing. U. S. orders to: American Institute of Physics, 500 Sunnyside Boulevard, Woodbury, NY 11797-2999. (516) 576-2200. 1934 to present. Monthly. $1214.00 per year. ISSN: 0034-4885.

Reviews of Modern Physics. American Institute of Physics, One Physics Ellipse, College Park, MD 20740-3843. (301) 209-3000. 1929 to present. Quarterly. $300.00 per year. ISSN: 0034-6861.

RESEARCH CENTERS AND INSTITUTES

Argonne National Laboratory. 9700 South Cass Avenue, Argonne, IL 60439-4832. (708) 972-3872.

Fermi National Accelerator Laboratory. P.O. Box 500 Batavia, IL 60510. (708) 840-3351. FAX (708) 840-4343.

High Energy Particle Physics Group. Florida State University. Keen Building, Tallahassee, FL 32306. (904) 644-1492. (904) 644-6735.

High Energy Physics and Elementary Particle Research Program. University of Pennsylvania, 33rd and Walnut Streets, Philadelphia. PA 19104. (215) 898-5960. FAX (215) 898-8512.

Lawrence Berkeley Laboratory, Physics Division. Building 50, Room 256, 1 Cyclotron Road, Berkeley, CA 04720. (415) 486-5421. FAX (415) 486-4553.

PATTERN RECOGNITION

See: COMPUTER VISION

PASCAL

See: COMPUTER PROGRAMMING LANGUAGES

PETROLEUM ENGINEERING

See also: PETROLEUM GEOLOGY, PIPELINE TECHNOLOGY

ABSTRACT SERVICES AND INDEXES

Bibliography and Index of Geology. American Geological Institute, 4220 King Street, Alexandria, VA 22302. (703) 379-2480. Fax (703) 379-7563. Monthly. $1295.00 per year. Also available online as GeoRef.

Chemical Abstracts: Applied Chemistry and Chemical Engineering Sections. Chemical Abstracts Service, 2540 Olentangy River Road, Box 3012, Columbus, OH 43210. (800) 848-6538 or (614) 447-3600, FAX (614) 447-3713. Bi-weekly. $1410.00 per year.

Comprehensive Index of the Publications of the American Association of Petroleum Geologists, 1986-1990. American Association of Petroleum Geologists, P.O. Box 979, Tulsa, OK 74101-0979. (918) 584-2555 or (800) 364-2274. FAX (918) 584-0469. 1991. $43.00.

Current Contents: Engineering, Technology, and Applied Sciences. Institute for Scientific Information, 3501 Market Street, Philadelphia, PA 19104. (215) 386-0100. FAX (215) 386-6362. 1970 to present. Weekly. $442.00 per year.

Deep-Sea Research. Part A: Oceanographic Research Papers. Part B: Oceanographic Literature Review. Pergamon Press, Inc., Maxwell House, Fairview Park, Elmsford, NY 10523. (914) 592-7700. Fax (914) 592-3625. Twelve times per year. $1370 per year for Parts A and B.

Energy Research Abstracts. 1976-present. U.S. Department of Energy, office of Scientific and Technical Information, Box 62, Oak Ridge, TN 37831. (615) 574-0733. Subscriptions: Superintendent of Documents, U.S. Government Printing office, Box 371954, Pittsburgh, PA 15250-7954. (202) 783-3238. FAX (202) 512-2233. Monthly. $164.00 per year. Abstracts all the scientific and technical reports, journal articles, conference proceedings, patents, theses, and monographs sponsored by the U.S. Energy Research and Development Administration.

Fuel and Energy Abstracts. Butterworth-Heinemann, Linacre House, Jordan Hill, Ocford OX2 8DP, England. Telephone 0865-310366. FAX 0865-310898. Summary of world literature on all scientific, technical, commercial, and environmental aspects of fuel and energy. 1960 to present. Six times a year. Inquire for price in U.S.

Geological Abstracts. Elsevier-Geo Abstracts, Regency House, 34 Duke Street, Norwich NR3 3AP, England. Monthly. $760.00 per year. Also available online as GEOBASE.

Geological Society of America. Abstracts with Programs. Geological Society of America. 3300 Penrose Place, P.O. Box 9140, Boulder, CO 80301-9140. (303) 447-2020. Abstracts and programs of the annual conference. Annual. $69.00.

International Petroleum Abstracts. John Wiley & Sons Ltd., Journals, Baffins Lane, Chichester, Sussex, PO19 1UD, England. Telephone 0243-779777. FAX 0243-775878. 1973 to present. Quarterly. $825.00 per year. Review of oil and gas literature including oil and gas exploration and development, petroleum refining and products, economics and offshore technology.

Literature Abstracts (formerly *API Abstracts—Literature*). American Petroleum Institute, Central Abstracting & Information Services, 275 Seventh Avenue, New York, NY 10001-6708. (212) 366-4040. FAX (212) 366-4298. 1954 to present. Weekly. Price varies.

Literature Abstracts: Catalysts and Catalysis. American Petroleum Institute, Central Abstracting and Information Services, 275 Seventh Avenue, New York, NY 10001. (212) 366-4040 or FAX (212) 366-4298. Contains citations with abstracts to literature relating to catalysts used in petroleum refining and the petrochemical industries, from 1985 to present. Weekly. $160.00 per year. Also available online.

Literature Abstracts—Petroleum Refining and Petrochemicals. American Petroleum Institute, Central Abstracting & Information Services, 275 Seventh Avenue, New York, NY 10001-6708. (212) 366-4040. FAX (212) 366-4298. 1954 to present. Weekly. Price varies.

Manufacturing and Process Engineering Abstracts, Cambridge Scientific Abstracts, 7200 Wisconsin Avenue, Bethesda, MD 20814-4823. (301) 961-6750. FAX (301) 961-6720. 1993 to present. Monthly. $385.00 per year. Covers concurrent engineering, quality control, automated manufacturing, petroleum engineering, oil field operations and equipment, energy management, metallurgy and metallography, foundry practice.

Oceanic Abstracts. Cambridge Scientific Abstracts, 7200 Wisconsin Avenue, Bethesda, MD 20814. (301) 961-6750. Fax (301) 961-6720. Bimonthly. $995.00 per year.

Petroleum Abstracts. 600 South College, Tulsa, OK 74104. (918) 631-2296 or (800) 247-8678. Contains citations with abstracts to the worldwide literature and patents on the exploration, development, and production of petroleum resources, from 1965 to present.

ASSOCIATIONS AND PROFESSIONAL SOCIETIES

American Institute of Mining, Metallurgical and Petroleum Engineers. 345 East 47th Street, New York, NY 10017-2398. (212) 705-7695.

American Petroleum Institute. 1220 L Street NW, Washington, DC 20005-8029. (202) 682-8000. FAX (202) 682-8115.

Institute of Petroleum. 61 New Cavendish Street, London W1M 8AR, England. Telephone 071-636-1004. FAX 071-255-1472.

Society of Petroleum Engineers. P.O. Box 833836, Richardson, TX 75083-3836. (214) 952-9393. FAX (214) 952-9435.

DIRECTORIES AND BIOGRAPHICAL SOURCES

Directory of Engineering Societies and Related Organizations. Gordon Davis. 13th edition. American Association of Engineering Societies, 1111 19th Street NW, Suite 608, Washington, DC 20036. (202) 296-2237 or (800) 658-8897. 1989. Inquire for price.

International Petroleum Industry. Midwest Register Inc., 1140 E. 4th Street, Tulsa, OK 74120. (918) 582-2000. FAX (918) 587-9349. Annual. $75.00.

International Research Centers Directory. Gale Research, 835 Penobscot Building, Detroit, MI 48226-4094. (313) 961-2242. (800) 347-4253. FAX (313) 961-6083. 8th edition. 1995. $410.00

Oil & Gas Directory. Geophysical Directory Inc., 2200 Welch Avenue, Box 130508, Houston, TX 77219. (713) 529-8789. FAX (713) 529-3646. 1970 to present. Annual. $60.00.

Research Centers Directory. Gale Research, 835 Penobscot Building, Detroit, MI 48226-4094. (313) 961-2242. (800) 347-4253. FAX (313) 961-6083. $485.00.

Southeast Petroleum Industry. Midwest Register Inc., 1140 E. 4th Street, Tulsa, OK 74120. (918) 582-2000. FAX (918) 587-9349. 1945 to present. Annual. $40.00.

Texas Petroleum Industry. Midwest Register Inc., 1140 E. 4th Street, Tulsa, OK 74120. (918) 582-2000. FAX (918) 587-9349. 1945 to present. Annual. $40.00.

USA Oil Industry Directory. PennWell Publishing Company, Box 1260, Tulsa, OK 74101. (918) 935-3161. 1962 to present. Annual. $125.00.

USA Oilfield Service, Supply and Manufacturers Directory. PennWell Publishing Company, Box 1260, Tulsa, OK 74101. (918) 935-3161. 1983 to present. Annual. $115.00.

Western Petroleum Industry. Midwest Register Inc., 1140 E. 4th Street, Tulsa, OK 74120. (918) 582-2000. FAX (918) 587-9349. 1945 to present. Annual. $40.00.

Who's Who in Engineering. American Association of Engineering Societies, 1111 19th Street NW, Suite 608, Washington, DC 20036. (202) 296-2237 or (800) 658-8897. 8th edition. 1991. Inquire for price.

Who's Who in Technology. Gale Research, 835 Penobscot Building, Detroit, MI 48226-4094. (313) 961-2242. (800) 347-4253. FAX (313) 961-6083. 1995. $195.00.

Worldwide Petrochemical Directory. PennWell Books, 1421 S. Sheridan Road, Box 1260, Tulsa, OK 74112. (918) 835-3161 or (800) 752-9764. FAX (918) 831-9555. 1962 to present. Annual. $140.00.

Worldwide Refining and Gas Processing Directory. PennWell Books, 1421 S. Sheridan Road, Box 1260, Tulsa, OK 74112. (918) 835-3161 or (800) 752-9764. FAX (918) 831-9555. 1942 to present. Annual. $140.00.

ENCYCLOPEDIAS AND DICTIONARIES

Basic Petroleum Data Book. American Petroleum Institute, Publications Section, 1220 L Street NW, Washington, DC 20005. (202) 682-8378. 1974 to present. Three times a year. $150.00 per year for non-members.

Encyclopedia of Physical Science and Technology. Robert A. Meyers, ed. 18 volumes. Academic Press Inc., 6277 Sea Harbor Drive, Orlando, FL 32887. (800) 321-5068. 1992. $2100.00

The Encyclopedia of the Solid Earth Sciences. Philip Kearey, editor-in-chief. Blackwell Scientific Publications, 238 Main Street, Cambridge, MA 02142. (617) 876-7000 or (800) 759-6102. FAX (617) 876-7022. 1993. $60.00.

Illustrated Glossary of Petroleum Geochemistry. Jennifer A. Miles. Reprint edition. Oxford University Press, Inc., 198 Madison Avenue, New York, NY 10016-4314. (212) 726-6000. FAX (212) 726-6446. 1994. $21.95.

The Illustrated Petroleum Reference Dictionary. Robert D. Langenkamp. PennWell Books, 1421 S. Sheridan Road, Box 1260, Tulsa, OK 74112. (918) 835-3161 or (800) 752-9764. FAX (918) 831-9555. 1994. Inquire for cost and availability.

International Petroleum Encyclopedia. PennWell Books, 1421 S. Sheridan Road, Box 1260, Tulsa, OK 74112. (918) 835-3161 or (800) 752-9764. FAX (918) 831-9555. 1968 to present. Annual. $95.00.

GENERAL WORKS

Basics of Reservoir Engineering. A. Leblond, et al. Gulf Publishing Company, P.O. Box 2608, Houston, TX. (713) 529-4301 or (800) 231-6275. 1993. $75.00.

Fundmentals and Applications in the Petroleum Industry. Laurier A. Schramm, editor. American Chemical Society, 1155 16th Street, N.W., Washington, DC 20036. (202) 872-4381 or (800) 333-9511. FAX (614) 447-3671. 1994. $129.95.

Modern Petroleum: A Basic Primer of the Industry. Bill D. Berger and Kenneth E. anderson. 3rd edition. PennWell Books, 1421 S. Sheridan Road, Box 1260, Tulsa, OK 74112. (918) 835-3161 or (800) 752-9764. FAX (918) 831-9555. 1992. $59.95.

Petroleum Economics and Engineering. Abdel-Aal, et al., editors. 2nd edition. Marcel Dekker, Inc., 270 Madison Avenue, New York, NY 10016. (212) 696-9000. FAX (212) 685-4540. 1992. $185.00.

Practical Petroleum Engineer. W. Lyons. 2 volumes. 6th edition. Gulf Publishing Company, P.O. Box 2608, Houston, TX. (713) 529-4301 or (800) 231-6275. 1996. Inquire for cost and availability.

Principles of Petroleum Reservoir Engineering. Gian Luigi Chierici. Two volumes. Springer-Verlag, 175 Fifth Avenue, New York, NY 10010. (212) 460-1500 or (800) 777-4643. FAX (212) 473-6272. 1994. $99.00 (Vol.1) and $109.00 (Vol. 2).

HANDBOOKS AND MANUALS

Petroleum Engineering Handbook for the Practicing Engineer. M. A. Mian. 2 volumes. PennWell Books, 1421 S. Sheridan Road, Box 1260, Tulsa, OK 74112. (918) 835-3161 or (800) 752-9764. FAX (918) 831-9555. 1992. $125.95 per volume.

ONLINE DATABASES AND CD-ROMS

CSA Engineering. Cambridge Scientific Abstracts, 7200 Wisconsin Avenue, Suite 601, Bethesda, MD 20814. (301) 961-6750 or (800) 843-7751. Contains citations and abstracts of international periodicals and research literature covering all fields of engineering and science and technology,including computer and information science, electronics, mechanical engineering, solid state materials, 1981 to present. Available on BRS,(800) 289-4277, online service.Inquire as to cost and availability.

Current Contents Search. Institute for Scientific Information, 3501 Market Street, Philadelphia, PA 19104. (215) 386-0100. FAX (215) 386-6362. Contains citations to articles listed in the table of contents of science and technology journals. Also articles in social sciences and life sciences journals. Available on BRS,(800) 289-4277, DIALOG,(800) 334-2564, online services. Inquire as to cost and availability.

Dissertation Abstracts. University Microfilms International, 300 North Zeeb Road, Ann Arbor, MI 48106. (800) 521-0600 or (313) 761-4700. Scope includes virtually all doctoral dissertations accepted at accredited American institutions from 1861 to present in 252 subject areas. Available on BRS,(800) 289-4277, DIALOG,(800) 334-2564, and OCLC EPIC,(800) 848-5878, online services. Also available on CD-ROM. Inquire as to cost and availability.

Georef: Bibliography and Index of Geology. American Geological Institute, 4220 King Street, Alexandria, VA 22302. (703) 379-2480. Fax (703) 379-7563. Monthly. Inquire for price and availability.

NTIS Bibliographic Database. National Technical Information Service, 5285 Port Royal Road, Springfield, VA 22161. (703) 487-4929 or FAX (703) 321-8199. Broad coverage of government-sponsored science and technology research reports, 1964 to present. Available on BRS,(800) 289-4277, DIALOG, (800) 334-2564, ORBIT, (800) 456-7248, and STN International, FIZ Karlsruhe, P.O. Box 2465, W-7500, Karlsruhe 1, Germany, online services. Also available on CD-ROM. Inquire as to cost and availability.

SCISEARCH. Institute for Scientific Information, 3501 Market Street, Philadelphia, PA 19104. (800) 523-1850 or (215) 386-0100. Broad multidisciplinary title and author index to the international literature of science and technology, 1974 to present. Available on DIALOG,(800) 334-2564, and ORBIT,(800) 456-7248, online services. Also available on CD-ROM. Inquire as to cost and availability.

PERIODICALS

Developments in Petroleum Engineering. Elsevier Science, 660 White Plains Road, Tarrytown, NY 10591-5153. (914) 524-9200. 1985 to present. Irregular. Price varies.

JPT: Journal of Petroleum Technology. Society of Petroleum Engineers, Inc., Box 833836, Richardson, TX 75083-3836. (214)

952-9393. 1949 to present. Monthly. $45.00 per year for non-members.

Journal of Petroleum Science and Engineering. Elsevier Science Inc., Box 882, Madison Square Station, New York, NY 10159. (212) 989-5800. FAX (212) 633-3990. 1987 to present. Quarterly. $396.00 per year.

Offshore, Incorporating the Oilman. PennWell Publishing Company, Box 1260, Tulsa, OK 74101. (918) 935-3161. 1954 to present. Twelve times a year. $55.00 per year.

Oil & Gas Journal. PennWell Publishing Company, Box 1260, Tulsa, OK 74101. (918) 935-3161. 1902 to present. Weekly. $95.00 per year.

Petroleum Review. Institute of Petroleum, 61 New Cavendish Street, London W1M 8AR, England. Telephone 071-636-1004. FAX 071-255-1472. 1947 to present. Monthly. Inquire for price in U.S.

Society of Petroleum Engineers: Transactions. Society of Petroleum Engineers, Inc., Box 833836, Richardson, TX 75083-3836. (214) 952-9393. 1925 to present. Annual. $90.00 for non-members.

SPE Reservoir Engineering. Society of Petroleum Engineers, Inc., PO Box 833836, Richardson, TX 75083-3836. (214) 952-9393. 1986 to present. Quarterly. $60.00 for non-members.

World Oil (formerly Oil Weekly). Gulf Publishing Company, Box 2608, Houston, TC 77252-2608. (713) 529-4301. FAX (713) 520-4433. 1916 to present. Monthly. $24.00 per year.

RESEARCH CENTERS AND INSTITUTES

American Petroleum Institute. 1220 L Street NW, Washington, DC 20005. (202) 682-8000. FAX (202) 682-8030.

National Institute for Petroleum and Energy Research. PO Box 2128, 220 NW Virginia Avenue, Bartlesville, OK 74005. (918) 336-2400. FAX (918) 337-4365.

New Mexico Institute of Mining and Technology, New Mexico Petroleum Recovery Research Center. Kelly Bldg., Socorro, NM 87801. (505) 835-5142. FAX (505) 835-6031.

Texas A&M University, Crisman Institute For Petroleum Reservoir Management. Department of Petroleum Engineering, College Station, TX 77843. (409) 845-2241. FAX (409) 845-1307.

University of Alaska Fairbanks, Petroleum Development Laboratory. 437 Duckering Bldg., Fairbanks, AK 99775-1260. (907) 474-7743.

University of Oklahoma Energy Center. Norman, OK 73019. (405) 325-3821. FAX (405) 325-3180.

University of Texas at Austin, Center for Petroleum and Geosystems Engineering. Austin, TX 78712. (512) 471-7234.

University of Wyoming, Enhanced Oil Recovery Institute. Box 3036 University Station, Laramie, WY 82071. (307) 766-4933.

PETROLEUM GEOLOGY

Ency. of Physical Sciences and Engineering Info. Sources

PETROLEUM GEOLOGY

See also: GEOLOGY, GEOPHYSICS, GEOCHEMISTRY, PETROLEUM ENGINEERING

ABSTRACT SERVICES AND INDEXES

Bibliography and Index of Geology. American Geological Institute, 4220 King Street, Alexandria, VA 22302. (703) 379-2480. Fax (703) 379-7563. Monthly. $1295.00 per year. Also available online as GeoRef.

Comprehensive Index of the Publications of the American Association of Petroleum Geologists, 1986-1990. American Association of Petroleum Geologists, PO Box 979, Tulsa, OK 74101-0979. (918) 584-2555 or (800) 364-2274. FAX (918) 584-0469. 1991. $43.00.

Deep-Sea Research. Part A: Oceanographic Research Papers. Part B: Oceanographic Literature Review. Pergamon Press, Inc., Maxwell House, Fairview Park, Elmsford, NY 10523. (914) 592-7700. Fax (914) 592-3625. Twelve times per year. $1370 per year for Parts A and B.

Fuel and Energy Abstracts. Butterworth-Heinemann, Linacre House, Jordan Hill, Ocford OX2 8DP, England. Telephone 0865-310366. FAX 0865-310898. Summary of world literature on all scientific, technical, commercial, and environmental aspects of fuel and energy. 1960 to present. Six times a year. Inquire for price in U.S.

Geological Abstracts. Elsevier-Geo Abstracts, Regency House, 34 Duke Street, Norwich NR3 3AP, England. Monthly. $760.00 per year. Also available online as GEOBASE.

Geological Society of America. Abstracts with Programs. Geological Society of America. 3300 Penrose Place, P.O. Box 9140, Boulder, CO 80301-9140. (303) 447-2020. Abstracts and programs of the annual conference. Annual. $69.00.

International Petroleum Abstracts. John Wiley & Sons Ltd., Journals, Baffins Lane, Chichester, Sussex, PO19 1UD, England. Telephone 0243-779777. FAX 0243-775878. 1973 to present. Quarterly. $825.00 per year. Review of oil and gas literature including oil and gas exploration and development, petroleum refining and products, economics and offshore technology.

Literature Abstracts. American Petroleum Institute, Central Abstracting & Information Services, 275 Seventh Avenue, New York, NY 10001-6708. (212) 366-4040. FAX (212) 366-4298. 1954 to present. Weekly. Price varies.

Mineralogical Abstracts. Mineralogical Society and the Mineralogical Society of America, 41 Queen's Gate, London, SW7 5HR, England. Quarterly. $235.00 per year.

Oceanic Abstracts. Cambridge Scientific Abstracts, 7200 Wisconsin Avenue, Bethesda, MD 20814. (301) 961-6750. Fax (301) 961-6720. Bimonthly. $995.00 per year.

Petroleum Abstracts. 600 South College, Tulsa, OK 74104. (918) 631-2296 or (800) 247-8678. Contains citations with abstracts to the worldwide literature and patents on the exploration, development, and production of petroleum resources, from 1965 to present. 50 times a year. Inquire for price and availability.

ASSOCIATIONS AND PROFESSIONAL SOCIETIES

American Association of Petroleum Geologists. Box 979, Tulsa, OK 74101. (918) 584-2555. FAX (918) 584-0469.

Institute of Petroleum. 61 New Cavendish Street, London W1M 8AR, England. Telephone 071-636-1004. FAX 071-255-1472.

International Oil Scouts Association. Box 272949, Houston, TX 77277-2949.

Society of Petroleum Engineers. PO Box 833836, Richardson, TX 75083-3836. (214) 952-9393.

DIRECTORIES AND BIOGRAPHICAL SOURCES

Directory of Engineering Societies and Related Organizations. Gordon Davis. 13th edition. American Association of Engineering Societies, 1111 19th Street NW, Suite 608, Washington, DC 20036. (202) 296-2237 or (800) 658-8897. 1989. Inquire for price.

International Oil Scouts Association Directory. Box 272949, Houston, TX 77277-2949. Inquire for price and availability.

International Petroleum Industry. Midwest Register Inc., 1140 E. 4th Street, Tulsa, OK 74120. (918) 582-2000. FAX (918) 587-9349. Annual. $75.00.

International Research Centers Directory. Gale Research, 835 Penobscot Building, Detroit, MI 48226-4094. (313) 961-2242. (800) 347-4253. FAX (313) 961-6083. 8th edition. 1995. $410.00

Oil & Gas Directory. Geophysical Directory Inc., 2200 Welch Avenue, Box 130508, Houston, TX 77219. (713) 529-8789. FAX (713) 529-3646. 1970 to present. Annual. $60.00.

Research Centers Directory. Gale Research, 835 Penobscot Building, Detroit, MI 48226-4094. (313) 961-2242. (800) 347-4253. FAX (313) 961-6083. $485.00.

USA Oil Industry Directory. PennWell Publishing Company, Box 1260, Tulsa, OK 74101. (918) 935-3161. 1962 to present. Annual. $125.00.

Who's Who in Engineering. American Association of Engineering Societies, 1111 19th Street NW, Suite 608, Washington, DC 20036. (202) 296-2237 or (800) 658-8897. 8th edition. 1991. Inquire for price.

ENCYCLOPEDIAS AND DICTIONARIES

Basic Petroleum Data Book. American Petroleum Institute, Publications Section, 1220 L Street NW, Washington, DC 20005. (202) 682-8378. 1974 to present. Three times a year. $150.00 per year for non-members.

Encyclopedia of Physical Science and Technology. Robert A. Meyers, ed. 18 volumes. Academic Press Inc., 6277 Sea Harbor Drive, Orlando, FL 32887. (800) 321-5068. 1992. $2100.00.

The Encyclopedia of the Solid Earth Sciences. Philip Kearey, editor-in-chief. Blackwell Scientific Publications, 238 Main

Street, Cambridge, MA 02142. (617) 876-7000 or (800) 759-6102. FAX (617) 876-7022. 1993. $60.00.

The Illustrated Petroleum Reference Dictionary. Robert D. Langenkamp. PennWell Books, 1421 S. Sheridan Road, Box 1260, Tulsa, OK 74112. (918) 835-3161 or (800) 752-9764. FAX (918) 831-9555. 1994. Inquire for cost and availability.

International Petroleum Encyclopedia. PennWell Books, 1421 S. Sheridan Road, Box 1260, Tulsa, OK 74112. (918) 835-3161 or (800) 752-9764. FAX (918) 831-9555. 1968 to present. Annual. $95.00.

GENERAL WORKS

Basic Petroleum Geology. Peter K. Link. 2nd revised edition. Oil & Gas Consultants International, Inc., 4554 S. Harvard, Tulsa, OK 74135-2980. (800) 821-5933. 1990. $50.00.

The Business of Petroleum Geology. Richard Steinmetz, editor. American Association of Petroleum Geologists, PO Box 979, Tulsa, OK 74101-0979. (918) 584-2555 or (800) 364-2274. FAX (918) 584-0469. 1992. $26.00.

Geology of Petroleum. Arville I. Levorsen. 2nd edition. W.H. Freeman and Company, 41 Madison Avenue, East 26th, 35th Floor, New York, NY 10010. (212) 576-9400. FAX (212) 689-2383. 1995. Inquire for cost and availability.

Petroleum Formation and Occurrence. B. P. Tissot and Dietrich H. Welte. Revised and enlarged edition. Springer-Verlag, 175 Fifth Avenue, New York, NY 10010. (212) 460-1500 or (800) 777-4643. FAX (212) 473-6272. 1992. $113.00.

HANDBOOKS AND MANUALS

Petroleum Geology Handbook. N. A. Eremenko and George V. Chilingarian, editors. Optimization Software, Inc., 1100 Glendon Avenue, Suite 1447, Los Angeles, CA 90024. (310) 208-5674. 1991. $200.00.

ONLINE DATABASES AND CD-ROMS

Georef: Bibliography and Index of Geology. American Geological Institute, 4220 King Street, Alexandria, VA 22302. (703) 379-2480. Fax (703) 379-7563. Monthly. Inquire for price and availability.

NTIS Bibliographic Database. National Technical Information Service, 5285 Port Royal Road, Springfield, VA 22161. (703) 487-4929 or FAX (703) 321-8199. Broad coverage of government-sponsored science and technology research reports, 1964 to present. Available on BRS,(800) 289-4277, DIALOG, (800) 334-2564, ORBIT, (800) 456-7248, and STN International, FIZ Karlsruhe, P.O. Box 2465, W-7500, Karlsruhe 1, Germany, online services. Also available on CD-ROM. Inquire as to cost and availability.

SCISEARCH. Institute for Scientific Information, 3501 Market Street, Philadelphia, PA 19104. (800) 523-1850 or (215) 386-0100. Broad multidisciplinary title and author index to the international literature of science and technology, 1974 to present. Available on DIALOG,(800) 334-2564, and ORBIT,(800) 456-7248, online services. Also available on CD-ROM. Inquire as to cost and availability.

PERIODICALS

AAPG Bulletin. American Association of Petroleum Geologists, Box 979, Tulsa, OK 74101. (918) 584-2555. FAX (918) 584-0469. 1917 to present. Monthly. $135.00.

AAPG Explorer. American Association of Petroleum Geologists, Box 979, Tulsa, OK 74101. (918) 584-2555. FAX (918) 584-0469. 1979 to present. Monthly. $45.00.

AAPG Studies in Geology Series. American Association of Petroleum Geologists, Box 979, Tulsa, OK 74101. (918) 584-2555. FAX (918) 584-0469. Irregular. Price varies.

Journal of Petroleum Geology. Scientific Press Ltd., PO Box 21, Beaconsfield, Bucks HP9 1NS, England. Telephone 0494-675139. FAX 0494-670155. 1978 to present. Quarterly. $264.00 per year.

Journal of Petroleum Science and Engineering. Elsevier Science Inc., Box 882, Madison Square Station, New York, NY 10159. (212) 989-5800. FAX (212) 633-3990. 1987 to present. Quarterly. $396.00 per year.

Offshore, Incorporating the Oilman. PennWell Publishing Company, Box 1260, Tulsa, OK 74101. (918) 935-3161. 1954 to present. Twelve times a year. $55.00 per year.

Oil & Gas Journal. PennWell Publishing Company, Box 1260, Tulsa, OK 74101. (918) 935-3161. 1902 to present. Weekly. $95.00 per year.

Oilfield Review. Elsevier Science Inc., Box 882, Madison Square Station, New York, NY 10159. (212) 989-5800. FAX (212) 633-3990. 1950 to present. Four times a year. $195.00 per year.

Petroleum Frontiers. Petroleum Information Corporation, Box 2612, Denver, CO 80201-2612. Quarterly. $200.00 per year.

Petroleum Geology: A Digest of Russian Literature On Petroleum Geology. Box 171, McLean, VA 22101. 1959 to present. Monthly. $40.00 per year.

Petroleum Review. Institute of Petroleum, 61 New Cavendish Street, London W1M 8AR, England. Telephone 071-636-1004. FAX 071-255-1472. 1947 to present. Monthly. Inquire for price in U.S.

SPE Formation Evaluation. Society of Petroleum Engineers, Inc., PO Box 833836, Richardson, TX 75083-3836. (214) 952-9393. 1986 to present. Quarterly. $60.00 for non-members.

World Oil (formerly Oil Weekly). Gulf Publishing Company, Box 2608, Houston, TC 77252-2608. (713) 529-4301. FAX (713) 520-4433. 1916 to present. Monthly. $24.00 per year.

RESEARCH CENTERS AND INSTITUTES

Texas A&M University Center for Petroleum Reservoir Geology. College of Geosciences, College Station, TX 77843-3115. (409) 845-2460. FAX (409) 845-6162.

PETROLOGY

See also: CRUST, CRYSTALLOGRAPHY, GEOCHEMISTRY, GEOLOGY, METAMORPHIC ROCKS, MINERALOGY, PHYSICAL GEOLOGY, SEDIMENTARY ROCKS, SEISMOLOGY, VOLCANOLOGY

ABSTRACT SERVICES AND INDEXES

Bibliography and Index of Geology. American Geological Institute, 4220 King Street, Alexandria, VA 22302. (703) 379-2480. Fax (703) 379-7563. Monthly. $1295.00 per year. Also available online as GeoRef.

Bibliography of Economic Geology. Geosystems, Box 40, Didcot Oxon Ox11 9BX, England. Bimonthly. $150.00 per year. Also available online as GeoArchive.

Chemical Abstracts. Applied Chemistry and Chemical Engineering Sections. Chemical Abstracts Service, 2540 Olentangy River Road, Box 3012, Columbus, OH 43210. (800) 848-6538 or (614) 447-3600, FAX (614) 447-3713. Bi-weekly. $1410.00 per year. Also available online as CA (Chemical Abstracts).

Deep-Sea Research. Part A: Oceanographic Research Papers. Part B: Oceanographic Literature Review. Pergamon Press, Inc., Maxwell House, Fairview Park, Elmsford, NY 10523. (914) 592-7700. Fax (914) 592-3625. Twelve times per year. $1370 per year for Parts A and B.

Geological Abstracts. Elsevier-Geo Abstracts, Regency House, 34 Duke Street, Norwich NR3 3AP, England. Monthly. $760.00 per year. Also available online as GEOBASE.

Geological Society of America. Abstracts with Programs. Geological Society of America. 3300 Penrose Place, P.O. Box 9140, Boulder, CO 80301-9140. (303) 447-2020. Abstracts and programs of the annual conference. Annual. $69.00.

Mineralogical Abstracts. Mineralogical Society and the Mineralogical Society of America, 41 Queen's Gate, London, SW7 5HR, England. Quarterly. $235.00 per year.

Oceanic Abstracts. Cambridge Scientific Abstracts, 7200 Wisconsin Avenue, Bethesda, MD 20814. (301) 961-6750. Fax (301) 961-6720. Bimonthly. $995.00 per year.

Science Citation Index. Institute for Scientific Information, 3501 Market Street, Philadelphia, PA 19104. (800) 523-1850 or (215) 386-0100. Inquire about price and availability.

ANNUAL REVIEWS AND YEARBOOKS

Annual Review and Earth and Planetary Sciences. Annual Reviews, Inc., 4139 El Camino Way, Palo Alto, CA 94306-0897. (415) 493-4400. Fax (415) 855-9815. Annual. $55.00 per year.

Minerals Yearbook. Bureau of Mines, U.S. Department of the Interior. Available from U.S. Government Printing office, Washington, DC 20402. (202) 783-3238. Annual. Three volumes. $45.00.

ASSOCIATIONS AND PROFESSIONAL SOCIETIES

American Geological Institute. 4220 King Street, Alexandria, VA 22302. (703) 379-2480. Fax (703) 379-7563.

American Institute of Professional Geologists. 7828 Vance Drive, Suite 103, Arvada, CO 80003. (303) 431-0831.

Association of Engineering Geologists. 323 Boston Post Road, Suite 2D, Sudbury, MA 01776. (508) 443-4639.

Geochemical Society. Department of Terrestrial Magnetism, 5241 Broad Branch Road, N.W., Washington, DC 20015. (202) 686-4370.

Geological Society of America. 3300 Penrose Place, P.O. Box 9140, Boulder, CO 80301-9140. (303) 447-2020.

Society of Economic Paleontologists and Mineralogists. P.O. Box 4756, Tulsa, OK 74159-0756. (918) 743-9765.

Society of Independent Professional Earth Scientists. 4925 Greenville Avenue, Suite 170, Dallas, TX 75206. (214) 363-1780.

BIBLIOGRAPHIES

Geologic Reference Sources: A Subject and Regional Bibliography of Publications and Maps in the Geological Sciences. Dederick C. Ward et al. Scarecrow Press, Inc., 52 Liberty Street, Box 4167, Metuchen, NJ 08840. (800) 537-7107 or (201) 548-8600. 1981. $49.50.

Information Sources in the Earth Sciences. David N. Wood, Joan E. Hardy, and Anthony P. Harvey, editors. Bowker-Saur/K.G. Saur. Distributed by R.R. Bowker, 121 Chanlon Road, New Providence, NJ 07974. (800) 521-8110 or (908) 464-6800. Second edition. 1989. $85.00.

DIRECTORIES AND BIOGRAPHICAL SOURCES

American Institute of Professional Geologists. Membership Directory. American Institute of Professional Geologists, 7828 Vance Drive, Suite 103, Arvada, CO 80003. (303) 431-0831. Annual. $18.00.

Directory of Geoscience Departments: United States and Canada. American Geological Institute. 4220 King Street, Alexandria, VA 22302-1507. (703) 379-2480. Fax (703) 379-7563. Twenty-six edition. 1987. $22.00.

Geological Society of America. Membership Directory. Geological Society of America, 3300 Penrose Place, Boulder, CO 80301. (303) 447-2020. Annual. Available to members only.

Research Centers Directory. Gale Research, 835 Penobscot Building, Detroit, MI 48226-4094. (800) 347-4253 or (313) 961-2242. 17th edition, 1992. $400.00.

ENCYCLOPEDIAS AND DICTIONARIES

Concise Oxford Dictionary of the Earth Sciences. Ailsa Allaby and Michael Allaby, editors. Oxford University Press, Inc., 200 Madison Avenue, New York, NY 10016. (800) 334-4249 or (212) 679-7300. 1990. $42.95.

Dictionary of Geological Terms. American Geological Institute. Doubleday and Company, Inc., 245 Park Avenue, New York, NY 10017. (800) 645-6156 or (212) 953-4561. Third edition. 1984. $12.00 in paper.

Dictionary of Petrology. S.I. Tomkeieff. John Wiley and Sons, Inc., 605 Third Avenue, New York, NY 10158. (800) 526-5368 or (212) 850-6000. 1983. $180.00.

Encyclopedia of Igneous and Metamorphic Petrology. D.R. Bowes. Van Nostrand Reinhold, 115 Fifth Avenue, New York, NY 10003. (212) 254-3232 or (800) 926-2665. 1989. $125.95.

Glossary of Geology. Robert L. Bates and Julia A. Jackson. American Geological Institute, 4220 King Street, Alexandria, VA 22032. (703) 379-2480. Third edition. 1987. $75.00.

GENERAL WORKS

Basic Analytical Petrology. Paul C. Ragland. Oxford University Press, Inc., 200 Madison Avenue, New York, NY 10016. (800) 334-4249 or (212) 679-7300. 1989. $45.00.

Gems, Granites, and Gravels: Knowing and Using Rocks and Minerals. Richard Vincent Dietrich and Brian J. Skinner. Cambridge University Press, 40 West 20th Street, New York, NY 10011-4211. (212) 924-3900. 1990. $24.95.

Igneous Petrology. Anthony Hall. 2d edition. Addison-Wesley Publishing Company, 1 Jacob Way, Reading, MA 01867. (617) 944-3700 or (800) 447-2226. FAX 617-942-1117. 1996. $89.75.

Igneous Petrology. Alexander R. McBirney. Jones and Bartlett Publishers, Inc., 20 Park Plaza, Boston, MA 02116. (800) 832-0034 or (617) 482-3900. Second edition. 1991. $50.00.

Introduction to Rock Physics. Yves Gueguen & Victor Palciauskas. Princeton University Press, 41 William Street, Princeton, NJ 08540. (609) 258-4900 or (800) 777-4726. FAX (609) 895-1081. 1994. $49.50.

Origins of Igneous Rocks. Paul C. Hess. Harvard University Press, 79 Garden Street, Cambridge, MA 02138. (617) 495-2600. 1989. $65.00.

Petrology: Igneous, Sedimentary & Metamorphic. Ernest G. Ehlers, et al. W.H. Freeman and Company, 41 Madison Avenue, East 26th, 35th Floor, New York, NY 10010. (212) 576-9400. FAX (212) 689-2383. 1995. Inquire for price.

Petrology: The Study of Igneous, Sedimentary & Metamorphic Rocks. Loren A. Raymond. William C. Brown Publications, 2460 Kerper Blvd., Dubuque, IA 52001. (800) 338-5578. 1994. Inquire for price.

Sedimentary Petrology. Harvey Blatt. 2nd edition. W.H. Freeman and Company, 41 Madison Avenue, East 26th, 35th Floor, New York, NY 10010. (212) 576-9400. FAX (212) 689-2383. 1995. Inquire for price.

HANDBOOKS AND MANUALS

Agi Data Sheets for Geology in the Field, Laboratory, and Office. J.T. Dutro, R.V. Dietrich, and R.M. Foose, editors. American Geological Institute, 4220 King Street, Alexandria, VA 223020-1507 (800) 336-4764 or (703) 379-2480. 1989. $24.95.

Field Mapping for Geology Students. F. Ahmed and D.C. Almond. Unwin Hyman, Inc., 10 East 53rd Street, New York, NY 10022. (212) 207-7626 or (800) 242-7737. 1983. $18.95.

Geology in the Field. R.R. Compton. John Wiley and Sons, Inc., 605 Third Avenue, New York, NY 10158. (800) 526-5368 or (212) 850-6418. 1985. $37.95.

Manual of Mineralogy. Cornelis Klein & Cornelis Hurlbut. 21st revised edition. John Wiley and Sons, Inc., 605 Third Avenue, New York, NY 10158. (800) 526-5368 or (212) 850-6000. 1994. $69.50.

The Mapping of Geological Structures. K.R. McClay. John Wiley and Sons, Inc., 605 Third Avenue, New York, NY 10158. (800) 526-5368 or (212) 850-6000. 1988. $18.95.

The Mineral Index. A. Clark, editor. Chapman and Hall, Inc., 29 West 35th Street, New York, NY 10001-2291. (212) 244-3336. Third edition. 1992. $95.00.

ONLINE DATABASES AND CD-ROMS

Earth Sciences. U.S. Geological Survey, 12201 Sunrise Valley Drive, Reston, VA 22092-9998. (703) 648-4460. CD-ROM of earth science databases including the U.S. Geological Survey Library, Earth Science Data Directory, and GEOINDEX, citations to published geological maps. $350.00 per year with quarterly updates.

Geoarchive. Geosystems, P.O. Box 1024, Westminster, London, England, SW1 P 2JL. Citations to literature on geoscience, 1969 to present. Inquire as to online cost and availability.

Geobase. Elsevier-Geo Abstracts, Regency House, 34 Duke Street, Norwich NR3 3AP, England. Contains citations to the worldwide earth science literature from 1980 to date. Available on DIALOG, ORBIT online services. Inquire as to cost and availability.

Geomechanics Abstracts. University of London, Imperial College of Science and Technology, Rock Mechanics Information Service, Royal School of Mines, Prince Consort Road, London SW7 2BP, England. Telephone: 071-589 5111 or FAX 071-589 6806. Scope includes worldwide literature on engineering geology, hydrogeology, mining, rock mechanics, soil mechanics, and tunnelling, 1977 to present. Available on the ORBIT online service. Inquire as to cost and availability.

Georef. American Geological Institute, 4220 King Street, Alexandria, VA 22302. (800) 336-4764 or (703) 379-2480. Geology and geosciences literature, 1785 to present for North America. Available on DIALOG, ORBIT online services. Also available on CD-ROM. Inquire as to cost and availability.

NTIS Bibliographic Database. National Technical Information Service, 5285 Port Royal Road, Springfield, VA 22161. (703) 487-4630. Broad coverage of government-sponsored research reports, 1964 to present. Available on BRS, DIALOG, ORBIT and STN online services. Also available on CD-ROM. Inquire as to cost and availability.

SCISEARCH. Institute for Scientific Information, 3501 Market Street, Philadelphia, PA 19104. (800) 523-1850 or (215) 386-0100. Broad multidisciplinary title and author index to the international literature of science and technology, 1974 to present. Available on DIALOG and ORBIT online services. Also available on CD-ROM. Inquire as to cost and availability.

PETROLOGY

Ency. of Physical Sciences and Engineering Info. Sources

PERIODICALS

American Journal of Science. Kline Geology Laboratory, Yale University, Box 6666, Yale Station, New Haven, CT 06511-8130. (203) 432-3131. Ten times per year. $40.00 per year.

Canadian Journal of Earth Sciences. National Research Council of Canada, Research Journals, Ottawa, Ont. K1A OR6, Canada. (613) 993-9084. Monthly. $112.00 per year.

Earth and Planetary Science Letters. Elsevier Science Publishing Company, Inc., 655 Avenue of the Americas, New York, NY 10010. (212) 989-5800. Twenty times per year. $945.00 per year.

Geochimica et Cosmochimica Acta. Pergamon Journals, Inc., Maxwell House, Fairview Park, Elmsford, NY 10523. (914) 592-7700. Fax (914) 592-3625. Monthly. $580.00 per year.

Geoforum. Pergamon Press, Maxwell House, Fairview Park, Elmsford, NY 10523. (914) 592-7700. Fax (914) 592-3625. Quarterly. $305.00 per year.

Geological Magazine. Cambridge University Press, 40 West 20th Street, New York, NY 10011-4211. (212) 924-3900. Bimonthly. $205.00 per year.

Geological Society. Journal. Geological Society of London. Geological Society Publishing House, Unit 7, Brassmill Enterprise Centre, Brassmill Lane, Bath, Avon BA1 3JN, England. 6 times per year. $432.00.

Geological Society of America Bulletin. P.O. Box 9140, 3300 Penrose Place, Boulder, CO 80301. (303) 447-2020. Fax (303) 447-1133. Monthly. $150.00.

Geology. Geological Society of America. P.O. Box 9140, 3300 Penrose Place, Boulder, CO 80301. (303) 447-2020. Fax (303) 447-1133. Monthly. $120.00 per year.

Journal of Geology. University of Chicago Press, 5720 Woodlawn Avenue, Chicago, IL 60637. (312) 753-3347. Bimonthly. $38.00 per year.

Journal of Petrology. Oxford University Press, Inc., 200 Madison Avenue, New York, NY 10016. (800) 334-4249 or (212) 679-7300. Bimonthly. $240.00.

Journal of Structural Geology. Pergamon Journals, Inc., Maxwell House, Fairview Park, Elmsford, NY 10523. (914) 592-7700. Ten times per year. $375.00 per year.

Journal of Volcanology and Geothermal Research. Elsevier Science Publishing Company, Inc., 655 Avenue of the Americas, New York, NY 10010. (212) 989-5800. Monthly. $500.00 per year.

Lithos: An International Journal of Mineralogy, Petrology, and Geochemistry. Elsevier Science Publishing Company, Inc., 655 Avenue of the Americas, New York, NY 10010. (212) 989-5800. Quarterly. $275.00 per year.

Physics of the Earth and Planetary Interiors. Elsevier Science Publishing Company, Inc., 655 Avenue of the Americas, New York, NY 10010. (212) 989-5800. Twenty-four times per year. $900.00 per year.

Sedimentary Geology: International Journal of Pure and Applied Sedimentology. Elsevier Science Publishing Company, Inc., 655 Avenue of the Americas, New York, NY 10010. (212) 989-5800. Twenty times per year. $800.00 per year.

Sedimentology. Blackwell Scientific Publishing, Inc., 3 Cambridge Court, Suite 208, Cambridge, MA 02142. (617) 225-0401. Six times per year. $400.00 per year.

Tectonics. American Geophysical Union, 2000 Florida Avenue, N.W., Washington, DC 20009. (202) 462-6900. Bimonthly. $230.00 per year to individuals.

Tectonophysics: An International Journal of Geotectonics and the Geology and Physics of the Interior of the Earth. Elsevier Science Publishing Company, Inc., 655 Avenue of the Americas, New York, NY 10010. (212) 989-5800. Fifty-six times per year. $1275.00 per year.

RESEARCH CENTERS AND INSTITUTES

Carnegie Institution of Washington. Geophysical Laboratory. 5251 Broad Branch Road, N.W., Washington, DC 20015. (202) 686-2410. Fax (202) 686-2419.

Geological Society of America. P.O. Box 9140, 3300 Penrose Place, Boulder, CO 80301. (303) 447-2020.

Harvard University. Hoffman Laboratory of Experimental Geology. Department of Earth and Planetary Sciences, 20 Oxford Street, Cambridge, MA 02138. (617) 495-2351.

U.S. Geological Survey, Geologic Division, National Center, 12201 Sunrise Valley Drive, Reston, VA 22092. (703) 648-6600. The major geological research agency of the federal government conducting research in most areas of pure and applied research in the geosciences.

PHONOGRAPHS

See: AUDIO ENGINEERING

PHOTOCHEMISTRY

See also: COLOR, LASERS, PHOTOGRAPHIC FILM PHOTOGRAPHY, PHYSICAL CHEMISTRY

ABSTRACT SERVICES AND INDEXES

Applied Science and Technology Index: A Cumulative Subject Index to English Language Periodicals in the Fields of Aeronautics and Space Science, Computer Technology, Chemistry, Construction Industry, Energy and Related Areas.

*H.*W. Wilson Co., 950 University Avenue, Bronx, NY 10452. (212) 588-8400. (800) 367-6770. FAX (718) 590-1617. From 1958 to present. Monthly. Inquire about cost and availability. Also available on CD-ROM and online. ISSN: 0003-6986.

Chemical Abstracts. Chemical Abstracts Service. 2540 Oleantangy River Road, Box 3012, Columbus, OS 43210-0012.

(614) 447-3600. FAX (614) 447-3713. 1907 to present. Weekly. $16,800.00 per year. Available online and on CD-ROM. CA is also available in five section groupings. Inquire regarding cost and availability.

Current Contents: Engineering, Technology and Applied Sciences. Institute for Scientific Information, 3501 Market Street, Philadelphia, PA 19104. (215) 386-0100. FAX (215) 386-6362. 1970 to present. Weekly. $442.00 per year. Also available on CD-ROM and online. Inquire regarding cost and availability. ISSN: 0095-7917.

Engineering Index Monthly: Indexes and Abstracts the World's Engineering and Technical Literature. Engineering Information, Inc., Castle Point on the Hudson, Hoboken, NJ 07030. (201) 216-8500. (800) 221-1044. FAX (201) 216-8532. Monthly. $2300.00 per year. Available online as COMPENDEX and also on CD-ROM. ISSN: 0742-1974.

General Science Index. H.W. Wilson Company, 950 University Avenue, Bronx, NY 10452. (212) 588-8400. (800) 367-6770. FAX (718) 590-1617. From 1978 to present. Ten issues per year; quarterly and annual cumulations. Service basis. Available on CD-ROM and online. Inquire regarding cost and availability. ISSN: 0162-1963.

Imaging Abstracts. Royal Photographic Society of Great Britain, Imaging Science and Technology Group. Order from: Elsevier Science, 660 White Plains Road, Tarrytown, NY 10591-5153. (914-524-9200. FAX (914) 524-9200. 1921 to present. Bi-monthly. $460.00. ISSN: 0896-100X.

Index to IEEE Publications. IEEE Service Center, 445 Hoes Lane, Piscataway, NJ 08855-1331. (908) 981-1393. (800) 678-IEEE. FAX (908) 981-9667. 1973 to present. Annual. ISSN: 0099-1368.

Physics Abstracts. INSPEC. Section A, Science Abstracts. Institution of Electrical Engineers (IEE), London. Available from: INSPEC/IEEE-Institute of Electrical and Electronic Engineers, Box 1331, Hoes Lane, Piscataway, NJ 08855-1331. (908) 562-5549. 1898 to present. 24 issues per year. $2835.00 per year. Also available online and on CD-ROM. ISSN: 0036-8091.

Physics Briefs (Physikalische Berichte). Information Center for Energy, Physics, Mathematics; German Physical Society. V C H Publishers, Inc., 220 East 23rd Street, New York, NY 10010-4606. (212) 683-8333. 1845 to present. 24 issues per year. $2390.00 per year. Also available online. ISSN: 0179-7434.

Science Citation Index (SCI). Institute for Scientific Information, 3501 Market Street, Philadelphia, PA 19104. (215) 386-0100. (800) 523-1850. FAX (215) 386-2991. 1961 to present. Six issues per year, plus annual cumulation. $11650.00 per year. Also available online and on CD-ROM. Inquire about price and availability. ISSN: 0036-827X.

ANNUAL REVIEWS AND YEARBOOKS

Advances in Photochemistry. John Wiley and Sons, Inc., 605 Third Avenue, New York, NY 10158. (800) 526-5368 or (212) 850-6000. 1963 to present. Irregular. $145.00.

ASSOCIATIONS AND PROFESSIONAL SOCIETIES

American Chemical Society. 1155 16th Street, NW, Washington DC 20036. (202) 872-4600.

American Institute of Chemical Engineers. 345 East 47th Street, New York, NY 10017. (212) 705-7338.

Optical Society of America, 2010 Massachusetts Avenue, N.W., Washington, DC 20036-1023. (202) 223-8130. FAX (202) 223-1096.

Society for Imaging Science and Technology. 7003 Kilworth Lane, Springfield, VA 22151. (703) 642-9090. FAX (703) 642-9094.

Spie-The International Society for Optical Engineering. PO Box 10, 1000 20th Street, Bellingham, WA 98227. (206) 676-3290.

DIRECTORIES AND BIOGRAPHICAL SOURCES

Directory of Engineering Societies and Related Organizations. American Association of Engineering Societies, 1111 19th Street, Suite 608, Washington, DC 20036-3603. (202) 296-2237. Semi-annual. $150.00.

Engineering Research Centers: Incorporating Electronics Research Centers. Stockton Press, 345 Park Avenue, New York, NY 10010. (212) 689-9200. (800) 221-2123. 4th edition. 1995. $515.00.

Research Centers Directory. Gale Research Company, 835 Penobscot Building, Detroit, MI 48226-4094. (313) 961-2242. (800) 347-4253. 20th edition. 1995. $485.00. ISSN: 0080-1518.

Who's Who in Engineering. Gordon Davis, editor. American Association of Engineering Societies. 1111 19th Street, NY, Suite 608, Washington, DC 20036. (202) 296-2237. (800) 658-8897. 9th edition. 1995. $220.00.

ENCYCLOPEDIAS AND DICTIONARIES

Encyclopedia of Applied Physics. VCH Publishers, Inc., 303 Northwest 12th Avenue, Deerfield Beach, FL 33442. (800) 367-8249. 1991- . Twenty volumes. $6000.00.

Encyclopedia of Chemical Processing and Design. McKetta Marcel Dekker, Inc., 270 Madison Avenue, New York, NY 10016. (212) 696-9000. (800) 228-1160. 1976- . $175.00 per volume. ISSN:

Kirk-Othmer Encyclopedia of Chemical Technology. John Wiley and Sons, Inc., 605 Third Avenue, New York, NY 10158. (800) 526-5368 or (212) 850-6000. Fourth edition. 1991- . Twenty-seven volumes. $5400.00.

Thesaurus of Photographic Science and Engineering. Society of Photographic Scientists and Engineers. Books on Demand, 300 North Zeeb Road, Ann Arbor, MI 48106-1346. (313) 761-4700. (800) 521-0600. $39.00.

PHOTOCHEMISTRY

Ency. of Physical Sciences and Engineering Info. Sources

GENERAL WORKS

Chemistry and Light. Paul Suppan. CRC Press, Inc., 2000 Corporate Boulevard, NW, Boca Raton, FL 33431. (407) 994- 0555. (800) 272-7737. 1994. $45.00.

Current Trends in Polymer Photochemistry. Norman S. Allen, et al, editors. Chapman & Hall, 1 Penn Plaza, New York, NY 10119. (212) 564-1060. 1995.

Elements of Inorganic Photochemistry. G. J. Ferraudi. John Wiley & Sons, Inc., 605 Third Avenue, New York, NY 10158-0012. (212) 850-6000. (800) 225-5945. 1988. $89.95.

Organic Photochemistry and Electronic Structure. Josef Michl and Vlasta Bonacic-Koutecky. John Wiley & Sons, Inc., 605 Third Avenue, New York, NY 10158-0012. (212) 850-6000. (800) 225-5945. 1990. $94.95.

Photochemical Processing of Electronic Materials. J. W. Boyd and R. B. Jackman, editors. Academic Press, Inc., 6277 Sea Harbor Drive, Orlando, FL. (800) 321-5068. 1991. $176.00.

Photochemical Technology. andre M. Braun. John Wiley & Sons, Inc., 605 Third Avenue, New York, NY 10158-0012. (212) 850-6000. (800) 225-5945. 1991. $325.00.

The Physics and Technology of Xerographic Processes. Edgar M. Williams. Krieger Publishing Company, P.O. Box 9542, Melbourne, FL 32902-9542. (407) 724-9542. . 1993. Revised edition. $64.50.

Principles and Applications of Photochemistry. Richard P. Wayne. Oxford University Press, Inc., 200 Madison Avenue, New York, NY 10016. (212) 725-6000. (800) 334-4249. 1988.$24.95 in paper.

Principles and Practice of Photochemical Machining and Photoetching. D. M. Allen. IOP Publishing, Public Ledger Building, Suite 1035, Independence Square, Philadelphia, PA 19106. (215) 627-0880. (800) 358-4677. 1986. $59.00.

HANDBOOKS AND MANUALS

Handbook of Organic Photochemistry and Photobiology. William M. Horspool. CRC Press Inc., 2000 Corporate Boulevard, NW, Boca Raton, FL 33431. (407) 994-0555. (800) 272-7737. 1994. wfi.

Handbook of Photochemistry. Steven L. Murov et al. Marcel Dekker, Inc., 270 Madison Avenue, New York, NY 10016. (212) 696-9000. (800) 228-1160. 2nd edition revised. 1993. $140.00.

Handbook of Photographic Science and Engineering. Society of Photographic Scientists and Engineers, 7003 Kilworth Lane, Springfield, VA 22151. (703) 642-9090. FAX (703)642-9094. Inquire.

Kodal Professional Photoguide. Eastman Kodak Company Staff. Saunders PhotoGraphic Inc., 21 Jet View Avenue, Rochester, NY 14624. (716) 328-7800. 1995. $29.95 in paper.

ONLINE DATABASES AND CD-ROMS

CA SEARCH. Chemical Abstracts Service, P.O. Box 3012, Columbus, OH 43210-0012. (614) 447-3600. (800) 848-6533. FAX (614) 447-3709. Very comprehensive guide to worldwide chemical literature and related fields, 1972 to present. Available on BRS,(800) 289-4277, DIALOG, (800) 334-2564, ORBIT (800) 456-7248, and STN International, FIZ Karlsruhe, P.O. Box 2465, W-7500, Karlsruhe 1, Germany, online services. Inquire as to cost and availability.

Compendex Plus. Engineering Information, Inc., 345 East 47th Street, New York, NY 10017. (212) 705-7600 or (800) 221-1044. Contains citations with abstracts to worldwide literature in engineering and technology, from 1970 to present. Available on online BRS,(800) 289-4277, DIALOG, (800) 334-2564, ORBIT (800) 456-7248, and STN International, FIZ Karlsruhe, P.O. Box 2465, W-7500, Karlsruhe 1, Germany, online services. Also available on CD-ROM. Inquire as to cost and availability.

INSPEC. Institution of Electrical Engineers, Michael Faraday House, Six Hills Way, Stevenage, Herts. SG1 2AY, England. Telephone: 0438 313311 or FAX 0438 742840. Contains citations to the worldwide literature of physics, electronics and electrical engineering, computer technology, and related fields. Available on BRS, (800) 289-4277, DIALOG, (800) 334-2564, ORBIT, (800) 456-7248, and STN International, FIZ Karlsruhe, P.O. Box 2465, W-7500, Karlsruhe 1, Germany, online services. Inquire as to cost and availability.

NTIS Bibliographic Database. National Technical Information Service, 5285 Port Royal Road, Springfield, VA 22161. (703) 487-4929 or FAX (703) 321-8199. Broad coverage of government-sponsored science and technology research reports, 1964 to present. Available on BRS,(800) 289-4277, DIALOG, (800) 334-2564, ORBIT, (800) 456-7248, and STN International, FIZ Karlsruhe, P.O. Box 2465, W-7500, Karlsruhe 1, Germany, online services. Also available on CD-ROM. Inquire as to cost and availability.

SCISEARCH. Institute for Scientific Information, 3501 Market Street, Philadelphia, PA 19104. (800) 523-1850 or (215) 386-0100. Broad multidisciplinary title and author index to the international literature of science and technology, 1974 to present. Available on DIALOG,(800) 334-2564, and ORBIT,(800) 456-7248, online services. Also available on CD-ROM. Inquire as to cost and availability.

WILSONLINE. H.W. Wilson Company, 950 University Avenue, Bronx, NY 10452. (800) 367-6770 or (212) 588-8400. Makes available online versions of the printed H.W. Wilson indexes including *Applied Science and Technology Index, Business Periodicals Index, General Science Index,* and *Readers' Guide to Periodical Literature.* Period covered is generally 1983 to present. Available on BRS, (800) 289-4277, DIALOG, (800) 334-2564, and OCLC EPIC, (800) 848-5878, online services.Also available on CD-ROM. Inquire as to cost and availability.

PERIODICALS

British Journal of Photography. Bouverie Publishing Co. Ltd., 147-151 Temple Chambers, Temple Avenue, London EC47 ODT, England. 1854 to present. Weekly. L45.50 per year. ISSN: 0007-1196

Journal of Imaging Science and Technology. Society for Imaging Science and Technology, 7003 Kilworth Lane, Springfield,

VA 22151. (703) 642-9090. FAX (703) 642-9094. 1950 to present. Bi-monthly. $120.00 per year. ISSN: 1062- 3701.

Journal of Photochemistry and Photobiology. Elsevier Science Publishing Company, Inc., 655 Avenue of the Americas, New York, NY 10010. (212) 989-5800. 1972 to present. 24 issues per year. $1986.00 per year. ISSN: 1010-6030.

Journal of Photographic Science. Royal Photographic Society of Great Britain, 7 Ladbroke Walk, London, W11, England. 1953 to present. Bi-monthly. L99 per year. ISSN: 0022-3638.

Kodak Tech Bits. Eastman Kodak Company, 343 State Street, Rochester, NY 14650. (716) 724-4000. 1963-1993.

RESEARCH CENTERS AND INSTITUTES

Center for Photochemical Sciences. Bowling Green State University, Bowling Green, OH 43403. (419) 372-2033. FAX (419) 372-9808.

Image Permanence Instiute. Rochester Institute of Technology, RIT City Center, 50 West Main Street, Rochester, NY 13614. (716) 475-5199. FAX (716) 475-5097.

Lawrence Berkeley Laboratory. Materials and Chemicals Sciences Division. Building 66, 1 Cyclotron Road, Berkeley, CA 94720. (415) 486-6062. FAX (415) 486-4995.

Laser Research Center. Massachusetts Institute of Technology, 77 Massachusetts Avenue, Cambridge, MA 02139. (617) 253-7700. FAX (617) 253-4513.

Photochemistry Unit. University of Western Ontario, London, ON Canada N6A 5B7. (519) 679-2111.

Radiation Chemistry Data Center. University of Notre Dame, Notre Dame in 46556. (219) 239-7502.

PHOTOELECTRICITY

See also: ELECTRICITY, ENERGY, PHYSICAL CHEMISTRY, PHYSICS

ABSTRACT SERVICES AND INDEXES

Applied Science and Technology Index: A Cumulative Subject Index to English Language Periodicals in the Fields of Aeronautics and Space Science, Computer Technology, Chemistry, Construction Industry, Energy and Related Areas. H. W. Wilson Co., 950 University Avenue, Bronx, NY 10452- 9978. (212) 588-8400. (800) 367-6770. FAX (718) 590-1617.From 1958 to present. Monthly. Inquire about cost and availability. Also available online (BRS and WILSONLINE) and on CD-ROM. ISSN: 0003-6986.

Chemical Abstracts. Chemical Abstracts Service. 2540 Olentangy River Road, Box 3012, Columbus, OH 43210-0012. (614) 447-3600. FAX (614) 447-3713. 1907 to present. Weekly. $16,800.00 per year. Available online and on CD- ROM. CA is also available in five section groupings. Inquire Regarding Cost and Availability.

Current Contents: Engineering, Technology and Applied Sciences. Institute for Scientific Information, 3501 Market Street, Philadelphia, PA 19104. (215) 386-0100. FAX (215) 386-2991. From 1970 to present. Weekly. $442.00 per year. Also available online. Inquire regarding cost and availability. ISSN: 0095-7917.

Electrical and Electronics Abstracts. Institution of Electrical Engineers (IEE), London. Available from INSPEC/IEEE- Institute of Electrical and Electronic Engineers, Box 1331, Hoes Lane, Piscataway, NJ 08855-1331. (908) 562-5549. 1898 to present. Monthly. $2200.00 per year. Also available on CD-ROM and online as INSPEC.

Energy Research Abstracts. U. S. Department of Energy, Office of Scientific and Technical Information, Box 62, Oak Ridge, TN 37831. (615) 674-0733. Subscriptions to: U. S. Government Printing office, Box 371954, Pittsburgh, PA 15250-7954. (202) 783-3238. 1976 to present. Monthly. $164.00 per year. ISSN: 0160-3604.

Engineering Index Monthly. Engineering Information, Inc., Castle Point on the Hudson, Hoboken, NJ 07030. (201) 216-8500. (800) 221-1044. FAX (201) 216-8532. Monthly. $2300.00 per year for monthly issues. $1980 per year for annual cumulation. Available on CD-ROM and online as COMPENDEX. ISSN: 0742-1974; 0360-8557.

General Science Index. H.W. Wilson Company, 950 University Avenue, Bronx, NY 10452. (212) 588-8400. (800) 367-6770. FAX (718) 590-1617. From 1978 to present. Ten issues per year; quarterly and annual cumulations. Service basis. Available on CD-ROM and online. Inquire regarding cost and availability. ISSN: 0162-1963.

Physics Abstracts. INSPEC. Section A, Science Abstracts. Institution of electrical engineers (IEE), London, United Kingdom. Available from: INSPEC/IEEE-Institute of Electrical and Electronic Engineers, Box 1331, 445 Hoes Lane, Piscataway, NJ 08855-1331. (908) 562-5549. 1898 to present. 24 Issues Per Year. $2835.00 per year. ISSN: 0036-8091. Also available online and on CD-ROM.

Science Citation Index (SCI). Institute for Scientific Information, 3501 Market Street, Philadelphia, PA 19104. (215) 386-0100. (800) 523-1850. FAX (215) 386-2991.

1961 To Present. Six issues per year, plus annual cumulation. $11650.00 per year. Also available online and on CD-ROM. Inquire about price and availability. ISSN: 0036-827X.

ANNUAL REVIEWS AND YEARBOOKS

Advances in Chemical Engineering. Academic Press, Inc., 6277 Sea Harbor Drive, Orlando, FL 32887. (800) 321-5068. 1956 To Present. Irregular. ISSN: 0065-2377.

Advances in Energy Systems and Technology. Academic Press, Inc., 6277 Sea Harbor Drive, Orlando, FL 32887. (800) 321-5068. 1979 to present. Irregular. Inquire.

ASSOCIATIONS AND PROFESSIONAL SOCIETIES

American Chemical Society, 1155 16th Street, NW, Washington, DC 20036. (202) 872-4600.

PHOTOELECTRICITY

Ency. of Physical Sciences and Engineering Info. Sources

American Institute of Chemical Engineers. 345 East 47th Street, New York, NY 10017. (212) 705-7338.

American Institute of Physics. One Physics Ellipse, College Park, MD 20740-3843. (301) 209-3100.

Association of Energy Engineers, 4025 Pleasantdale Road, Suite 340, Atlanta, GA 30340. (404) 447-5083.

IEEE (Institute of Electrical and Electronics Engineers) 345 East 47th Street, New York, NY 10017. (212) 705-7900. FAX (212) 705-4929.

Optical Society of America, 2010 Massachusetts Avenue, N.W., Washington, DC 20036-1023. (202) 223-8130. FAX (202) 223-1096.

Spie-The International Society for Optical Engineering. P.O. Box 10, 1000 20th Street, Bellingham, WA 98227. (206) 676-3290.

BIBLIOGRAPHIES

Information Sources in Engineering. Peter Hicks, editor. Bowker-Saur, 121 Chanlon Road, New Providence, NJ 07974. (800) 521-8110 or (908) 464-6800. 3rd edition. 1996. $125.00.

Scientific and Technical Books and Serials in Print: An Index to Literature in Science and Technology. R.R. Bowker Inc., 121 Chanlon Road, New Providence, NJ 07974. (908) 464-6800. (800) 521-8110. FAX (908) 665-3502. 1972 to present. Annual. 4 volumes. 1994. $299.95. Also available on compact disc and online. ISSN: 0000-054X.

DIRECTORIES AND BIOGRAPHICAL SOURCES

American Institute of Chemical Engineers. Directory. American Institute of Chemical Engineers. 345 East 47th Street, New York, NY 10017-2396. (212) 705-7663. Annual.

American Men and Women of Science: Physical and Biological Sciences. Fifteenth edition. R.R. Bowker Inc., 121 Chanlon Road, New Providence, NJ 07974. (908) 464-6800. (800) 521-8110. 20th edition. 8 volumes. 1996. $850.00

Energy Information Centers Directory. U. S. Council for Energy Awareness, 1776 I Street, NW, Suite 400, Washington, DC 20006. (202) 293-0770. Triennial. No charge.

Engineering Research Centers: Incorporating Electronics Research Centers. Stockton Press, 345 Park Avenue, New York, NY 10010. (212) 689-9200. (800) 221-2123. 4th edition. 1995. $515.00.

Research Centers Directory. Gale Research, 835 Penobscot Building, Detroit, MI 48226-4094. (313) 961-2242. (800) 347-4253. 20th edition. 1995. $485.00. ISSN: 0080-1518.

Who's Who in Engineering. Gordon Davis, editor. American Association of Engineering Societies. 1111 19th Street, NW, Suite 608, Washington, DC 20036. (202) 296-2237. (800) 658-8897. 9th edition. 1995. $220.00.

Who's Who in Technology. Gale Research, 835 Penobscot Building, Detroit, MI 48226-4094. (313) 961-2242. (800) 347-4253. 7th edition. 1995. $195.00. ISBN 0-8103-7467-6.

World Energy and Nuclear Directory. Gale Research, 835 Penobscot Building, Detroit, MI 48226-4094. (313) 961-2242. (800) 347-4253. Biennial, even years. $450.00.

ENCYCLOPEDIAS AND DICTIONARIES

Dictionary of Named Processes in Chemical Technology. Alan E. Comyns. Oxford University Press, Inc., 200 Madison Avenue, New York, NY 10016. (212) 725-6000. (800) 334-4249. 1994. $75.00.

Encyclopedia of Applied Physics. VCH Publishers, Inc., 303 Northwest 12th Avenue, Deerfield Beach, FL 33442. (800) 367-8249. 1991-. Twenty volumes. $6000.00.

Encyclopedia of Chemical Processing and Design. McKetta Marcel Dekker, Inc., 270 Madison Avenue, New York, NY 10016. (212) 696-9000. (800) 228-1160. 1976-. $175.00 per volume. ISSN:

Kirk-Othmer Encyclopedia of Chemical Technology. John Wiley and Sons, Inc., 605 Third Avenue, New York, NY 10158. (800) 526-5368 or (212) 850-6000. Fourth edition. 1991-. Twenty-seven volumes. $5400.00.

McGraw-Hill Encyclopedia of Science and Technology. McGraw-Hill Book, Incorporated, 1221 Avenue of the Americas, New York, NY 10020. (212) 997-3675. (800) 262-4729. Seventh edition. Twenty volumes. 1992. $1900.00.

Ullman's Encyclopedia of Industrial Chemistry. VCH Publications, Inc., 220 East 23rd Street, Suite 909, New York, NY 10010-4606. (212) 683-8333. (800) 422-8824. 5th edition. 1984-. Price varies per volume.

GENERAL WORKS

Applications of Photovoltaics. R. Hill and N. M. Pearsall, editors. IOP Publishing, Public Ledger Building, Suite 1035, Independence Square, Philadelphia, PA 19106. (215) 627-0880. (800) 358-4677. 1989. $49.00.

Environmental Electrochemistry. Krishnam Rajeshwar and Jorge Ibanez, editors. Academic Press, Inc., 6277 Sea Harbor Drive, Orlando, FL. (800) 321-5068. 1996. $95.00.

Harnessing Solar Power: The Photovoltaics Challenge. Ken Zweibel. Plenum Publishing Corp., 233 Spring Street, New York, NY. (212) 620-8000. (800) 2221-9369. FAX (212) 463-0742. 1990. $24.95.

Photoelectrochemistry and Photovoltaics of Layered Semiconductors. A. Aruchamy, editor. Kluwer Academic Publishers, 101 Philip Drive, Assinippi Park, Norwell, MA 02061. (617) 871-6600. 1992. $132.50.

Solar Electricity. Thomas Markvart. John Wiley & Sons, Inc., 605 Third Avenue, New York, NY 10158-0012. (212) 850-6000. (800) 225-5945. 1994. $39.95.

Solar Electricity: A Practical Guide to Small Photovoltaic Systems. Simon Roberts. Prentice-Hall, 113 Sylvan Avenue, Route 9W, Englewood Cliffs, NJ 07632. (201) 592-2000. (800) 922-0579. 1991. $35.00.

HANDBOOKS AND MANUALS

Chemical Engineering Reference Manual. Randall N. Robinson. Professional Publications, Inc., 1250 Fifth Avenue, Belmont, CA 94002. (415) 593-0110. (800) 426-1178. 1988. $49.95.

Perry's Chemical Engineers' Handbook. Robert H. Perry and Donald W. Green, editors. McGraw-Hill Publishing Company, Inc., 1221 Avenue of the Americas, New York, NY 10020. (212) 512-2000. (800) 262-4729. 6th edition. 1996. $129.50.

Photovoltaic Engineering Handbook. F. Lasnier and T. Ang. IOP Publishing, Public Ledger Building, Suite 1035, Independence Square, Philadelphia, PA 19106. (215) 627- 0880. (800) 358-4677. $209.00. 1990.

Photovoltaics Technical Information Guide. Gordon Press Publications, P.O. Box 459, Bowling Green Station, New York, NY 10004. (718) 624-8419. 1992. $88.95.

ONLINE DATABASES AND CD-ROMS

CA Search. Chemical Abstracts Service, P.O. Box 3012, Columbus, OH 43210-0012. (614) 447-3600. (800) 848-6533. FAX (614) 447-3709. Very comprehensive guide to worldwide chemical literature and related fields. 1972 to present. Available on BRS,(800) 289-4277, DIALOG, (800) 334-2564, ORBIT (800) 456-7248, and STN International, FIZ Karlsruhe, P.O. Box 2465, W-7500, Karlsruhe 1, Germany, online services. Inquire as to cost and availability.

Compendex Plus. Engineering Information, Inc., 345 East 47th Street, New York, NY 10017. (212) 705-7600 or (800) 221-1044. Contains citations with abstracts to worldwide literature in engineering and technology, from 1970 to present. Available on online BRS,(800) 289-4277, DIALOG, (800) 334-2564, ORBIT (800) 456-7248, and STN International, FIZ Karlsruhe, P.O. Box 2465, W-7500, Karlsruhe 1, Germany, online services. Also available on CD-ROM. Inquire as to cost and availability.

Current Contents Search. Institute for Scientific Information, 3501 Market Street, Philadelphia, PA 19104. (215) 386-0100. FAX (215) 386-6362. Contains citations to articles listed in the table of contents of science and technology journals. Also articles in social sciences and life sciences journals. Available on BRS,(800) 289-4277, DIALOG,(800) 334-2564, online services. Inquire as to cost and availability.

Dissertation Abstracts. University Microfilms International, 300 North Zeeb Road, Ann Arbor, MI 48106. (800) 521-0600 or (313) 761-4700. Scope includes virtually all doctoral dissertations accepted at accredited American institutions from 1861 to present in 252 subject areas. Available on BRS, (800) 289-4277, DIALOG, (800) 334-2564, and OCLC EPIC, (800) 848-5878, online services. Also available on CD-ROM. Inquire as to cost and availability.

INSPEC. Institution of Electrical Engineers, Michael Faraday House, Six Hills Way, Stevenage, Herts. SG1 2AY, England. Telephone: 0438 313311 or FAX 0438 742840. Contains citations to the worldwide literature of physics, electronics and electrical engineering, computer technology, and related fields. Available on BRS, (800) 289-4277, DIALOG, (800) 334-2564, ORBIT, (800) 456-7248, and STN International, FIZ Karlsruhe, P.O. Box 2465, W-7500, Karlsruhe 1, Germany, online services. Inquire as to cost and availability.

Kirk-Othmer Encyclopedia of Chemical Technology. John Wiley and Sons, Inc., 605 Third Avenue, New York, NY 10158. (800) 526-5368 or (212) 850-6000. Contains the complete text of all chapters in the 27 volume fourth edition of the KIRK-Othmer *Encyclopedia of Chemical Technology.* 1991. Available on BRS,(800) 289-4277, DIALOG,(800) 334-2564, online services. Inquire as to cost and availability.

NTIS Bibliographic Database. National Technical Information Service, 5285 Port Royal Road, Springfield, VA 22161. (703) 487-4929 or FAX (703) 321-8199. Broad coverage of government-sponsored science and technology research reports, 1964 to present. Available on BRS,(800) 289-4277, DIALOG, (800) 334-2564, ORBIT, (800) 456-7248, and STN International, FIZ Karlsruhe, P.O. Box 2465, W-7500, Karlsruhe 1, Germany, online services. Also available on CD-ROM. Inquire as to cost and availability.

SCISEARCH. Institute for Scientific Information, 3501 Market Street, Philadelphia, PA 19104. (800) 523-1850 or (215) 386-0100. Broad multidisciplinary title and author index to the international literature of science and technology, 1974 to present. Available on DIALOG, (800) 334-2564, and ORBIT, (800) 456-7248, online services. Also available on CD-ROM. Inquire as to cost and availability.

WILSONLINE. H.W. Wilson Company, 950 University Avenue, Bronx, NY 10452. (800) 367-6770 or (212) 588-8400. Makes available online versions of the printed H.W. Wilson indexes including Applied Science and Technology Index, Business Periodicals Index, General Science Index, and Readers' Guide to Periodical Literature. Period covered is generally 1983 to present. Available on BRS, (800) 289-4277, DIALOG, (800) 334-2564, and OCLC EPIC, (800) 848-5878, online services. Also available on CD-ROM. Inquire as to cost and availability.

PERIODICALS

Aiche Journal. American Institute of Chemical Engineers. 345 East 47th Street, New York, NY 10017. (212) 705-7338. 1955 To Present. Monthly. $295.00 per year.

Energy Engineering. Association of Energy Engineers. Available from Fairmont Press, 700 Indian Trail, Lilburn, GA 30324. (404) 925-9388. 1904 to present. Bimonthly. $99.00 per year. ISSN: 0199-8595.

Energy Sources. Taylor and Francis, 1900 Forst Road, Stuie 101, Bristol, PA 19007-1598. (215) 785-5800. 1973 to present. Quarterly. $225.00. ISSN: 0090-8312.

International Journal of Solar Energy. Harwood Academic Publishers, 820 Town Center Drive, Langhorne, PA 19047. (215) 750-2642. 1982 to present. 8 issues per year in 2 volumes. 127 ECU per volume. ISSN: 0142-5919.

Journal of Energy Resources Technology. American Society of Mechanical Engineers, 345 East 47th Street, New York, NY 10017. (212) 705-7703. 1979 to present. Quarterly. $100.00 per year. ISSN: 0195-0738.

Solar Energy. Elsevier Science, 660 White Plains Road, Tarrytown, NY 10591-5153. (914) 524-9200. FAX (914) 333-2444. 1957 to present. Monthly. $600.00 per year. ISSN: 0038-092X

PHOTOELECTRICITY

Ency. of Physical Sciences and Engineering Info. Sources

Solar Energy Materials and Solar Cells. Elsevier Science, 660 White Plains Road, Tarrytown, NY 10591-5153. (914) 524-9200. FAX (914) 333-2444. 1979 to present. 16 issues per year. $781.00 per year. ISSN: 0927-0248.

RESEARCH CENTERS AND INSTITUTES

Center for Applied Energy Research. University of Kentucky, 3572 Iron Works Pike, Lexington, KY (606) 257-0305. FAX (606) 257-0220.

Center for Energy Studies. University of Texas at Austin, 10100 Burnet Road, Building 133, Austin, TX 78758. (512) 471-4496. FAX (512) 471-0781.

Energy Research Laboratories. Canada Centre for Mineral and Energy Technology, 555 Booth Street, Ottawa, ON Canada KIA 0G1. (613) 996-8201. FAX (613) 995-9584.

Laboratory for Electrochemical Research. University of Texas at Austin, Austin TX 78712. (512) 471-3761.

PHOTOGRAMMETRY

See also: AERIAL PHOTOGRAPHY, CARTOGRAPHY, REMOTE SENSING, SURVEYING

ABSTRACTS AND INDEXES

Applied Science and Technology Index: A Cumulative Subject Index to English Language Periodicals in the Fields of Aeronautics and Space Science, Computer Technology, Chemistry, Construction Industry, Energy and Related Areas. H.W. Wilson Co., 950 University Avenue, Bronx, NY 10452. (800) 367-6770 or (212) 588-8400. FAX (718) 590-1617. From 1958 to present. Monthly. Inquire about cost and availability. Also available on CD-ROM.

Bibliography and Index of Geology. American Geological Institute, 4220 King Street, Alexandria, VA 22302. (703) 379-2480. Fax (703) 379-7563. Monthly. $1295.00 per year. Also available online as GeoRef.

Civil and Structural Engineering Abstracts. Cambridge Scientific Abstracts, 7200 Wisconsin Avenue, Bethesda, MD 20814-4823. (301) 961-6750. FAX (301) 961-6720. Monthly. Topics covered include structural design, construction equipment and methods, civil defense and military engineering, surveying, highway engineering, maritime and port structures, materials, land reclamation, and soil mechanics.

Current Contents: Engineering, Technology, and Applied Sciences. Institute for Scientific Information, 3501 Market Street, Philadelphia, PA 19104. (215) 386-0100. FAX (215) 386-6362. 1970 to present. Weekly. $442.00 per year.

Engineering Index Monthly. Engineering Information, Inc., Castle Point on the Hudson, Hoboken, NJ 07030. (800) 221-1044. FAX (212) 832-1857. Monthly. $2200.00 per year. Also available online as COMPENDEX, and also on CD-ROM.

Geological Abstracts. Elsevier-Geo Abstracts, Regency House, 34 Duke Street, Norwich NR3 3AP, England. Monthly. $760.00 per year. Also available online as GEOBASE.

Geological Society of America. Abstracts with Programs. Geological Society of America. 3300 Penrose Place, P.O. Box 9140, Boulder, CO 80301-9140. (303) 447-2020. Abstracts and programs of the annual conference. Annual. $69.00.

Imaging Abstracts (formerly *Photographic Abstracts*), Pergamon Press Inc., Maxwell House, Fairview Park, Elmsford, NY 10523. (914) 592-7700. Fax (914) 592-3625. Bimonthly.

Meteorological and Geoastrophysical Abstracts. American Meteorological Society, c/o Inforonics Inc., 550 Newtown Road, Littleton, MA 01460. (508) 486-8976. FAX (508) 486-0027. 1950 to present. Monthly. $950.00.

Physics Abstracts. Institute of Electrical Engineers, Michael Faraday House, Six Hill Way, Stevenage, Herts, SG1 2AY, England. Distributed by IEEE, 445 Hoes Lane, Piscataway, NJ 08854. (908) 562-5549. 1898 to present. Monthly. $2700.00 per year. Also available online as INSPEC.

Remote Sensing of Earth Resources. University of New Mexico Earth Data Analysis Center, 2500 Yale Blvd., Suite 100, Albuquerque, NM 87131-6031. (505) 277-3622. FAX (505) 277-3614. 1974 to present (1991 to present under current title). Quarterly. $175.00 per year.

Science Citation Index. Institute for Scientific Information, 3501 Market Street, Philadelphia, PA 19104. (215) 386-0100. FAX (215) 386-6362. Inquire about availability and cost. Also available on CD-ROM.

ASSOCIATIONS AND PROFESSIONAL SOCIETIES

American Society for Photogrammetry and Remote Sensing, 5410 Grosvenor Lane, Suite 210, Bethesda, MD 20814. (301) 493-0290. FAX (301) 493-0208.

The International Society for Optical Engineering (SPIE), P.O. Box 10, 1000-20th Street, Bellingham, WA 98227-0010.

Optical Society of America, 2010 Massachusetts Avenue NW, Washington, DC 20036. (202) 223-8130.

The Society for Imaging Science and Technology, 7003 Kilworth Lane, Springfield, VA 22151. (703) 642-9090.

Society of Photographic Scientists and Engineers. 7003 Kilworth Lane, Springfield, VA 22151. (703) 642-9090.

BIBLIOGRAPHIES

Scientific and Technical Books and Serials in Print. R.R. Bowker Inc., 121 Chanlon Road, New Providence, NJ 07974. (800) 521-8110 or (908) 464-6800. List of currently published books and serials in the physical and biological sciences, engineering and technology, with subject, author and titles indexes. Inquire as to cost and availability. Also available on ORBIT online service, (800) 456-7248.

A Selective Bibliography on Imagery Reconnaissance and Related Matters. Robert A. McDonald. 4th ed. Defense Intelligence College, Washington, DC. Order from Superintendent of Documents, U.S. Government Printing office, Box 371954, Pittsburgh, PA 15250-7954. (202) 783-3238. FAX (202) 512-2233. 1993. Inquire for price and availability.

DIRECTORIES AND BIOGRAPHICAL SOURCES

Directory of Engineering Societies and Related Organizations. Gordon Davis. 13th edition. American Association of Engineering Societies, 1111 19th Street NW, Suite 608, Washington, DC 20036. (202) 296-2237 or (800) 658-8897. 1989. Inquire for price.

International Research Centers Directory. Gale Research, 835 Penobscot Building, Detroit, MI 48226-4094. (313) 961-2242. (800) 347-4253. FAX (313) 961-6083. 8th edition. 1995. $410.00

Research Centers Directory. Gale Research, 835 Penobscot Building, Detroit, MI 48226-4094. (313) 961-2242. (800) 347-4253. FAX (313) 961-6083. $485.00.

Scientific and Technical Organizations and Agencies Directory. Gale Research, 835 Penobscot Building, Detroit, MI 48226-4094. (313) 961-2242. (800) 347-4253. FAX (313) 961-6083. 4th edition. 1996. $195.00.

Who's Who in Engineering. American Association of Engineering Societies, 1111 19th Street NW, Suite 608, Washington, DC 20036. (202) 296-2237 or (800) 658-8897. 8th edition. 1991. Inquire for price.

ENCYCLOPEDIAS AND DICTIONARIES

McGraw-Hill Dictionary of Scientific and Technical Terms. Sybil P. Parker, ed. 5th edition. McGraw-Hill Publishing Company, 1221 Avenue of the Americas, New York, NY 10020. (800) 262-4729 or (212) 512-3825. 1993. $110.50.

McGraw-Hill Encyclopedia of Science and Technology. Sybil P. Parker, ed. 7th edition. 20 volumes. McGraw-Hill Publishing Company, 1221 Avenue of the Americas, New York, NY 10020. (800) 262-4729 or (212) 512-3825. 1992. $1900.00

Thesaurus of Scientific, Technical, and Engineering Terms. Hemisphere Publishing Corporation, 1900 Frost Road, Suite 101, Bristol, PA 19007-1598. (215) 785-5800 or (800) 821-8312. FAX (215) 785-5515. 1987. $173.00.

GENERAL WORKS

Fundamentals of Remote Sensing and Airphoto Interpretation. Thomas Eugene Avery and Graydon Lennis Berlin. 5th ed. Macmillan Publishing, 200 Old Tappan Road, Old Tappan, NJ 07675. (800) 223-2336. FAX (800) 445-6991. 1992. Inquire for price.

Introduction to Environmental Remote Sensing. E.C. Barrett and L.F. Curtis. 3d ed. Routledge, Chapman and Hall, Inc., 29 West 35th Street, New York, NY 10001-2291. (212) 244-3336 or FAX (212) 563-2269. 1992. $39.95.

A Practical Guide to Aerial Photography: With an Introduction to Surveying. John A. Ciciarelli. Van Nostrand Reinhold, 115 Fifth Avenue, New York, NY 10003. (212) 254-3232. FAX (212) 254-9499. 1991. $49.95.

HANDBOOKS AND MANUALS

Manual of Photogrammetry. Chester C. Slama, ed. 4th edition. American Society for Photogrammetry and Remote Sensing, 5410 Grosvenor Lane, Suite 210, Bethesda, MD 20814. (301) 493-0290. FAX (301) 493-0208. 1980. Inquire for price and availability.

ONLINE DATABASES AND CD-ROMS

Compendex Plus. Engineering Information, Inc., 345 East 47th Street, New York, NY 10017. (212) 705-7600 or (800) 221-1044. Contains citations with abstracts to worldwide literature in engineering and technology, from 1970 to present. Available on online BRS, (800) 289-4277, DIALOG, (800) 334-2564, ORBIT (800) 456-7248, and STN International, FIZ Karlsruhe, P.O. Box 2465, W-7500, Karlsruhe 1, Germany, online services. Also available on CD-ROM. Inquire as to cost and availability.

CSA Engineering. Cambridge Scientific Abstracts, 7200 Wisconsin Avenue, Suite 601, Bethesda, MD 20814. (301) 961-6750 or (800) 843-7751. Contains citations and abstracts of international periodicals and research literature covering all fields of engineering and science and technology, including computer and information science, electronics, mechanical engineering, solid state materials, 1981 to present. Available on BRS, (800) 289-4277, online service. Inquire as to cost and availability.

Current Contents Search. Institute for Scientific Information, 3501 Market Street, Philadelphia, PA 19104. (215) 386-0100. FAX (215) 386-6362. Contains citations to articles listed in the table of contents of science and technology journals. Also articles in social sciences and life sciences journals. Available on BRS, (800) 289-4277, DIALOG, (800) 334-2564, online services. Inquire as to cost and availability.

Dissertation Abstracts. University Microfilms International, 300 North Zeeb Road, Ann Arbor, MI 48106. (800) 521-0600 or (313) 761-4700. Scope includes virtually all doctoral dissertations accepted at accredited American institutions from 1861 to present in 252 subject areas. Available on BRS, (800) 289-4277, DIALOG, (800) 334-2564, and OCLC EPIC, (800) 848-5878, online services. Also available on CD-ROM. Inquire as to cost and availability.

Georef: Bibliography and Index of Geology. American Geological Institute, 4220 King Street, Alexandria, VA 22302. (703) 379-2480. Fax (703) 379-7563. Monthly. Inquire for price and availability.

NTIS Bibliographic Database. National Technical Information Service, 5285 Port Royal Road, Springfield, VA 22161. (703) 487-4929 or FAX (703) 321-8199. Broad coverage of government-sponsored science and technology research reports, 1964 to present. Available on BRS, (800) 289-4277, DIALOG, (800) 334-2564, ORBIT, (800) 456-7248, and STN International, FIZ Karlsruhe, P.O. Box 2465, W-7500, Karlsruhe 1, Germany, online services. Also available on CD-ROM. Inquire as to cost and availability.

SCISEARCH. Institute for Scientific Information, 3501 Market Street, Philadelphia, PA 19104. (800) 523-1850 or (215) 386-0100. Broad multidisciplinary title and author index to the international literature of science and technology, 1974 to present. Available on DIALOG, (800) 334-2564, and ORBIT, (800) 456-7248, online services. Also available on CD-ROM. Inquire as to cost and availability.

PHOTOGRAMMETRY

Ency. of Physical Sciences and Engineering Info. Sources

WILSONLINE. H.W. Wilson Company, 950 University Avenue, Bronx, NY 10452. (800) 367-6770 or (212) 588-8400. Makes available online versions of the printed H.W. Wilson indexes including *Applied Science and Technology Index, Business Periodicals Index, General Science Index,* and *Readers' Guide to Periodical Literature.* Period covered is generally 1983 to present. Available on BRS,(800) 289-4277, DIALOG,(800) 334-2564, and OCLC EPIC,(800) 848-5878, online services. Also available on CD-ROM. Inquire as to cost and availability.

PERIODICALS

British Journal of Photography. Bouverie Publishing Company Ltd., 147-151 Temple Chambers, Temple Avenue, London EC47 0DT, England. Telephone 071-583-3030. FAX 071-583-4068. 1854 to present. Weekly. Inquire for price and availability.

Industrial Photography. PTN Publishing Corporation, 445 Broad Hollow Road, Melville, NY 11747-4722. (516) 845-2700. FAX (516) 845-7109. 1952 to present. Monthly. $60.00 per year.

Isprs Journal of Photogrammetry and Remote Sensing. Elsevier Science [journals], 660 White Plains Rd., Tarrytown, NY 10159-5153. (919) 524-9200. FAX (919) 333-2444. 1949 to present. Bi-monthly. $225.00 per year.

Journal of Imaging Science and Technology. Society for Imaging Science and Technology, 7003 Kilworth Lane, Springfield, VA 22151. (703) 642-9090. FAX (703) 642-9094. 1950 to present. Bi-monthly. $120.00 per year.

Journal of Photographic Science. The Barn, Whitehall, Near Middle Marwood, Barnstaple, N. Devon EX31 4EQ, England. Telephone 0271-72482. FAX 0271-72482. 1952 to present. Bi-monthly. Inquire for price and availability.

Mapping Sciences and Remote Sensing. V.H. Winston & Son Inc., c/o Bellwether Publishing Ltd., 8640 Guilford Road, Suite 200, Columbia, MD 21046. (410) 290-3870. FAX (410) 290-8726. 1962 to present, 4 times a year. $279.00 per year.

Photogrammetric Engineering and Remote Sensing. American Society for Photogrammetry and Remote Sensing, 5410 Grosvenor Lane, Suite 210, Bethesda, MD 20814. (301) 493-0290. FAX (301) 493-0208. 1934-present. Monthly. $160.00 per year.

Photogrammetric Record. Photogrammetric Society, Dept. of Photogrammetry & Surveying, University College London, Gower St., London WC1E 6BT England. Telephone 071-387-7050. FAX 071-380-0453. 1953-present. Semi-annual. $88.00 per year.

Remote Sensing of Environment. Elsevier Science Publishing Company, Inc., 655 Avenue of the Americas, New York, NY 10010. (212) 989-5800. 1969 to present. 12 times a year. $995.00 per year.

Remote Sensing Reviews. Harwood Academic, c/o Gordon & Breach Science Publishers, P.O. Box 786, Cooper Station, New York, NY 10276. (212) 206-8900. Quarterly. Inquire for cost.

Scientific and Applied Photography and Cinematography. Gordon and Breach Science Publishers, 820 Town Center Drive, Langhorne, PA 19047. (215) 750-2642. FAX (215) 750-6343. Six times a year. Inquire for price.

RESEARCH CENTERS AND INSTITUTES

Boston University Center for Remote Sensing and Cartography. 725 Commonwealth Avenue, Boston, MA 02215. (617) 353-5081. FAX (617) 353-3200.

Michigan State University Center for Remote Sensing. 302 Berkey Hall, East Lansing, MI 48824. (517) 353-7195. FAX (517) 353-1821.

University of Delaware Center for Remote Sensing. College of Marine Studies, Newark, DE 19711. (302) 451-2336.

University of Idaho, Idaho Remote Sensing Research Unit. Moscow, ID 83843. (208) 885-7209.

University of Minnesota Remote Sensing Laboratory. 1530 N. Cleveland Avenue, St. Paul, MN 55108. (612) 624-7764. FAX (612) 625-5212.

University of North Dakota, Institute for Remote Sensing. Geography Department, Grand Forks, ND 58202-8274. (701) 777-4246.

PHOTOGRAPHIC FILM

See also: AERIAL PHOTOGRAPHY, CAMERAS, COLOR, PHOTOCHEMISTRY, PHOTOGRAMMETRY, PHOTOGRAPHY

ABSTRACT SERVICES AND INDEXES

Applied Science and Technology Index: A Cumulative Subject Index to English Language Periodicals in the Fields of Aeronautics and Space Science, Computer Technology, Chemistry, Construction Industry, Energy and Related Areas. H.W. Wilson Co., 950 University Avenue, Bronx, NY 10452. (800) 367-6770 or (212) 588-8400. FAX (718) 590-1617. From 1958 to present. Monthly. Inquire about cost and availability. Also available on CD-ROM and online.

Chemical Abstracts: Applied Chemistry and Chemical Engineering Sections. Chemical Abstracts Service, 2540 Olentangy River Road, Box 3012, Columbus, OH 43210. (800) 848-6538 or (614) 447-3600, FAX (614) 447-3713. Bi-weekly. $1410.00 per year.

Current Contents: Engineering, Technology, and Applied Sciences. Institute for Scientific Information, 3501 Market Street, Philadelphia, PA 19104. (215) 386-0100. FAX (215) 386-6362. 1970 to present. Weekly. $442.00 per year.

Engineering Index Monthly. Engineering Information, Inc., Castle Point on the Hudson, Hoboken, NJ 07030. (800) 221-1044. FAX (212) 832-1857. Monthly. $2300.00 per year. Also available online as COMPENDEX, and also on CD-ROM. Covers chemical engineering, computers, electrical engineering, civil engineering, metals and mining, industrial management, and mechanical engineering.

Imaging Abstracts (formerly *Photographic Abstracts*), Pergamon Press Inc., Maxwell House, Fairview Park, Elmsford, NY 10523. (914) 592-7700. Fax (914) 592-3625. Bimonthly. Inquire for price.

Physics Abstracts. Institute of Electrical Engineers, Michael Faraday House, Six Hill Way, Stevenage, Herts, SG1 2AY, England. Distributed by IEEE, 445 Hoes Lane, Piscataway, NJ 08854. (908) 562-5549. 1898 to present. Monthly. $2700.00 per year. Also available online as INSPEC.

Science Citation Index. Institute for Scientific Information, 3501 Market Street, Philadelphia, PA 19104. (215) 386-0100. FAX (215) 386-6362. Inquire about availability and cost. Also available on CD-ROM.

ASSOCIATIONS AND PROFESSIONAL SOCIETIES

American Society of Photographers. Box 3191, Spartanburg, SC 29304-3191. (803) 582-3115. FAX (803) 582-4179.

The International Society for Optical Engineering (SPIE), P.O. Box 10, 1000-20th Street, Bellingham, WA 98227-0010.

Optical Society of America, 2010 Massachusetts Avenue Nw, Washington, DC 20036. (202) 223-8130.

Photographic Society of America. 3000 United Founders Blvd., Suite 103, Oklahoma City, OK 73112. (405) 843-1437.

The Society For Imaging Science and Technology, 7003 Kilworth Lane, Springfield, VA 22151. (703) 642-9090.

Society of Motion Picture and Television Engineers, 595 W. Hartsdale Avenue, White Plains, NY 10607.

Society of Photographic Scientists and Engineers. 7003 Kilworth Lane, Springfield, VA 22151. (703) 642-9090.

DIRECTORIES AND BIOGRAPHICAL SOURCES

Directory of Engineering Societies and Related Organizations. Gordon Davis. 13th edition. American Association of Engineering Societies, 1111 19th Street NW, Suite 608, Washington, DC 20036. (202) 296-2237 or (800) 658-8897. 1989. Inquire for price.

International Research Centers Directory. Gale Research, 835 Penobscot Building, Detroit, MI 48226-4094. (313) 961-2242. (800) 347-4253. FAX (313) 961-6083. 8th edition. 1995. $410.00

PTN Master Buying Guide and Directory. PTN Publishing Corporation, 445 Broad Hollow Road, Suite 21, Melville, NY 11747-4722. (516) 845-2700. FAX (516) 845-7109. 1937 to present. 1937 to present. Annual. Included with subscription to PTN (*Photographic Trade News*) (see below).

Research Centers Directory. Gale Research, 835 Penobscot Building, Detroit, MI 48226-4094. (313) 961-2242. (800) 347-4253. FAX (313) 961-6083. $485.00.

Scientific and Technical Organizations and Agencies Directory. Gale Research, 835 Penobscot Building, Detroit, MI 48226-4094. (313) 961-2242. (800) 347-4253. FAX (313) 961-6083. 4th edition. 1996. $195.00.

Who's Who in Engineering. American Association of Engineering Societies, 1111 19th Street NW, Suite 608, Washington,

DC 20036. (202) 296-2237 or (800) 658-8897. 8th edition. 1991. Inquire for price.

ENCYCLOPEDIAS AND DICTIONARIES

The Focal Encyclopedia of Photography. Leslie Stroebel & Richard D. Zakia, editors. 3rd edition. Butterworth-Heinemann, 313 Washington Street, Newton, MA 02158. (617) 928-2500 or (800) 366-2665. 1993. $125.00.

McGraw-Hill Dictionary of Scientific and Technical Terms. Sybil P. Parker, ed. 5th edition. McGraw-Hill Publishing Company, 1221 Avenue of the Americas, New York, NY 10020. (800) 262-4729 or (212) 512-3825. 1993. $110.50.

McGraw-Hill Encyclopedia of Science and Technology. Sybil P. Parker, ed. 7th edition. 20 volumes. McGraw-Hill Publishing Company, 1221 Avenue of the Americas, New York, NY 10020. (800) 262-4729 or (212) 512-3825. 1992. $1900.00

Thesaurus of Photographic Science and Engineering. Society of Photographic Scientists and Engineers Staff. Books on Demand, 300 N. Zeeb Road, Ann Arbor, MI 48106-1346. (313) 761-4700 or (800) 521-0600. 1968. $34.30.

Thesaurus of Scientific, Technical, and Engineering Terms. Hemisphere Publishing Corporation, 1900 Frost Road, Suite 101, Bristol, PA 19007-1598. (215) 785-5800 or (800) 821-8312. FAX (215) 785-5515. 1987. $173.00.

GENERAL WORKS

Science and Technology of Photography. Karlheinz Keller, editor. VCH Publishers, Inc., 303 Northwest 12th Avenue, Deerfield Beach, FL 33442. (800) 422-8824. 1993. Inquire for cost and availability.

The Story of Kodak. Douglas Collins. H.N. Abrams, 100 Fifth Avenue, New York, NY 10011. (212) 206-7715 or (800) 345-1359. $60.00.

The 35mm Film Source Book. Marc Levey. Focal Press, 80 Montvale Avenue, Stoneham, MA 02180. 1992. Inquire for cost and availability.

HANDBOOKS AND MANUALS

Handbook of Photography. Ronald P. Lovell, et al. Delmar Publishers, Division of Thomson Educational Publishing, Inc., 3 Columbia Circle, Box 15015, Albany, NY 12205. (518) 464-3500 or (800) 347-7707. 1993. $27.95.

The New 35mm Handbook: A Complete Course From Basic Techniques To Professional Applications. Michael Freeman. Revised edition. Running Press, 125 S. 22nd Street, Philadelphia, PA 19103-4394. (215) 567-5080 or (800) 345-5359. 1993. $24.95.

The New 35mm Photographer's Handbook. Julian Calder & John Garrett. 2nd revised edition. Crown Publishing Group, 201 E. 50th St., New York, NY 10022. (800) 726-0600 or (800) 733-3000. 1986. Inquire for cost and availability.

PHOTOGRAPHIC FILM

Ency. of Physical Sciences and Engineering Info. Sources

ONLINE DATABASES AND CD-ROMS

Compendex Plus. Engineering Information, Inc., 345 East 47th Street, New York, NY 10017. (212) 705-7600 or (800) 221-1044. Contains citations with abstracts to worldwide literature in engineering and technology, from 1970 to present. Available on online BRS,(800) 289-4277, DIALOG, (800) 334-2564, ORBIT (800) 456-7248, and STN International, FIZ Karlsruhe, P.O. Box 2465, W-7500, Karlsruhe 1, Germany, online services. Also available on CD-ROM. Inquire as to cost and availability.

CSA Engineering. Cambridge Scientific Abstracts, 7200 Wisconsin Avenue, Suite 601, Bethesda, MD 20814. (301) 961-6750 or (800) 843-7751. Contains citations and abstracts of international periodicals and research literature covering all fields of engineering and science and technology,including computer and information science, electronics, mechanical engineering, solid state materials, 1981 to present. Available on BRS,(800) 289-4277, online service.Inquire as to cost and availability.

Current Contents Search. Institute for Scientific Information, 3501 Market Street, Philadelphia, PA 19104. (215) 386-0100. FAX (215) 386-6362. Contains citations to articles listed in the table of contents of science and technology journals. Also articles in social sciences and life sciences journals. Available on BRS,(800) 289-4277, DIALOG,(800) 334-2564, online services. Inquire as to cost and availability.

NTIS Bibliographic Database. National Technical Information Service, 5285 Port Royal Road, Springfield, VA 22161. (703) 487-4929 or FAX (703) 321-8199. Broad coverage of government-sponsored science and technology research reports, 1964 to present. Available on BRS,(800) 289-4277, DIALOG, (800) 334-2564, ORBIT, (800) 456-7248, and STN International, FIZ Karlsruhe, P.O. Box 2465, W-7500, Karlsruhe 1, Germany, online services. Also available on CD-ROM. Inquire as to cost and availability.

SCISEARCH. Institute for Scientific Information, 3501 Market Street, Philadelphia, PA 19104. (800) 523-1850 or (215) 386-0100. Broad multidisciplinary title and author index to the international literature of science and technology, 1974 to present. Available on DIALOG,(800) 334-2564, and ORBIT,(800) 456-7248, online services. Also available on CD-ROM. Inquire as to cost and availability.

WILSONLINE. H.W. Wilson Company, 950 University Avenue, Bronx, NY 10452. (800) 367-6770 or (212) 588-8400. Makes available online versions of the printed H.W. Wilson indexes including *Applied Science and Technology Index, Business Periodicals Index, General Science Index,* and *Readers' Guide to Periodical Literature.* Period covered is generally 1983 to present. Available on BRS,(800) 289-4277, DIALOG,(800) 334-2564, and OCLC EPIC,(800) 848-5878, online services. Also available on CD-ROM. Inquire as to cost and availability.

PERIODICALS

British Journal of Photography. Bouverie Publishing Company Ltd., 147-151 Temple Chambers, Temple Avenue, London EC47 0DT, England. Telephone 071-583-3030. FAX 071-583-4068. 1854 to present. Weekly. Inquire for price and availability.

Industrial Photography. PTN Publishing Corporation, 445 Broad Hollow Road, Melville, NY 11747-4722. (516) 845-2700. FAX (516) 845-7109. 1952 to present. Monthly. $60.00 per year.

Journal of Imaging Science and Technology. Society for Imaging Science and Technology, 7003 Kilworth Lane, Springfield, VA 22151. (703) 642-9090. FAX (703) 642-9094. 1950 to present. Bi-monthly. $120.00 per year.

Journal of Photographic Science. The Barn, Whitehall, Near Middle Marwood, Barnstaple, N. Devon EX31 4EQ, England. Telephone 0271-72482. FAX 0271-72482. 1952 to present. Bi-monthly. Inquire for price and availability.

PSA Journal. Photographic Society of America, 3000 United Founders Blvd., No.103, Oklahoma City, OK 73112-3940. (405) 843-1437. FAX (405) 843-1438. 1934 to present. Monthly. Included with $35.00 annual membership.

PTN (Photographic Trade News). PTN Publishing Corporation, 445 Broad Hollow Road, Suite 21, Melville, NY 11747-4722. (516) 845-2700. FAX (516) 845-7109. 1937 to present. 24 times a year. $25.00 per year.

Scientific and Applied Photography and Cinematography. Gordon and Breach Science Publishers, 820 Town Center Drive, Langhorne, PA 19047. (215) 750-2642. FAX (215) 750-6343. 6 times a year. Inquire for price.

RESEARCH CENTERS AND INSTITUTES

George Eastman House, International Museum of Photography and Film. 900 East Avenue, Rochester, NY 14607. (716) 271-3361. FAX (716) 271-3970.

PHOTOGRAPHY

See also: AERIAL PHOTOGRAPHY, CAMERAS, COLOR, PHOTOCHEMISTRY, PHOTOGRAMMETRY, PHOTOGRAPHIC FILM, UNDERWATER PHOTOGRAPHY

ABSTRACT SERVICES AND INDEXES

Applied Science and Technology Index: A Cumulative Subject Index to English Language Periodicals in the Fields of Aeronautics and Space Science, Computer Technology, Chemistry, Construction Industry, Energy and Related Areas. H.W. Wilson Co., 950 University Avenue, Bronx, NY 10452. (800) 367-6770 or (212) 588-8400. FAX (718) 590-1617. From 1958 to present. Monthly. Inquire about cost and availability. Also available on CD-ROM and online.

Chemical Abstracts: Applied Chemistry and Chemical Engineering Sections. Chemical Abstracts Service, 2540 Olentangy River Road, Box 3012, Columbus, OH 43210. (800) 848-6538 or (614) 447-3600, FAX (614) 447-3713. Bi-weekly. $1410.00 per year.

Current Contents: Engineering, Technology, and Applied Sciences. Institute for Scientific Information, 3501 Market Street, Philadelphia, PA 19104. (215) 386-0100. FAX (215) 386-6362. 1970 to present. Weekly. $442.00 per year.

Engineering Index Monthly. Engineering Information, Inc., Castle Point on the Hudson, Hoboken, NJ 07030. (800) 221-1044. FAX (212) 832-1857. Monthly. $2300.00 per year. Also available online as COMPENDEX, and also on CD-ROM. Covers chemical engineering, computers, electrical engineering, civil engineering, metals and mining, industrial management, and mechanical engineering.

Imaging Abstracts (formerly *Photographic Abstracts*), Pergamon Press Inc., Maxwell House, Fairview Park, Elmsford, NY 10523. (914) 592-7700. Fax (914) 592-3625. Bimonthly. Inquire for price.

Index to Scientific and Technical Proceedings. Institute for Scientific Information, 3501 Market St., Philadelphia, PA 19104. (215) 386-0100. FAX (215) 386-6362. Monthly. $500.00 per year.

Index to Scientific Reviews. Institute for Scientific Information, 3501 Market St., Philadelphia, PA 19104. (215) 386-0100. FAX (215) 386-6362. Semi-annual.

Physics Abstracts. Institute of Electrical Engineers, Michael Faraday House, Six Hill Way, Stevenage, Herts, SG1 2AY, England. Distributed by IEEE, 445 Hoes Lane, Piscataway, NJ 08854. (908) 562-5549. 1898 to present. Monthly. $2700.00 per year. Also available online as INSPEC.

Science Citation Index. Institute for Scientific Information, 3501 Market Street, Philadelphia, PA 19104. (215) 386-0100. FAX (215) 386-6362. Inquire about availability and cost. Also available on CD-ROM.

ASSOCIATIONS AND PROFESSIONAL SOCIETIES

American Society For Photogrammetry and Remote Sensing, 5410 Grosvenor Lane, Suite 210, Bethesda, MD 20814. (301) 493-0290. FAX (301) 493-0208.

American Society of Photographers. Box 3191, Spartanburg, SC 29304-3191. (803) 582-3115. FAX (803) 582-4179.

The International Society for Optical Engineering (SPIE), P.O. Box 10, 1000-20th Street, Bellingham, WA 98227-0010.

Optical Society of America, 2010 Massachusetts Avenue NW, Washington, DC 20036. (202) 223-8130.

Photographic Society of America. 3000 United Founders Blvd., No.103, Oklahoma City, OK 73112-3940. (405) 843-1437. FAX (405) 843-1438.

The Society for Imaging Science and Technology, 7003 Kilworth Lane, Springfield, VA 22151. (703) 642-9090.

Society of Motion Picture and Television Engineers, 595 W. Hartsdale Avenue, White Plains, NY 10607.

Society of Photographic Scientists and Engineers. 7003 Kilworth Lane, Springfield, VA 22151. (703) 642-9090.

DIRECTORIES AND BIOGRAPHICAL SOURCES

Directory of Engineering Societies and Related Organizations. Gordon Davis. 13th edition. American Association of Engineering Societies, 1111 19th Street NW, Suite 608, Washington, DC 20036. (202) 296-2237 or (800) 658-8897. 1989. Inquire for price.

International Research Centers Directory. Gale Research, 835 Penobscot Building, Detroit, MI 48226-4094. (313) 961-2242. (800) 347-4253. FAX (313) 961-6083. 8th edition. 1995. $410.00.

PTN Master Buying Guide and Directory. PTN Publishing Corporation, 445 Broad Hollow Road, Suite 21, Melville, NY 11747-4722. (516) 845-2700. FAX (516) 845-7109. 1937 to present. 1937 to present. Annual. Included with subscription to *PTN* (*Photographic Trade News*) (see below).

Research Centers Directory. Gale Research, 835 Penobscot Building, Detroit, MI 48226-4094. (313) 961-2242. (800) 347-4253. FAX (313) 961-6083. $485.00.

Scientific and Technical Organizations and Agencies Directory. Gale Research, 835 Penobscot Building, Detroit, MI 48226-4094. (313) 961-2242. (800) 347-4253. FAX (313) 961-6083. 4th edition. 1996. $195.00.

Who's Who in Engineering. American Association of Engineering Societies, 1111 19th Street NW, Suite 608, Washington, DC 20036. (202) 296-2237 or (800) 658-8897. 8th edition. 1991. Inquire for price.

ENCYCLOPEDIAS AND DICTIONARIES

Encyclopedia of Physical Science and Technology. Robert A. Meyers, ed. 18 volumes. Academic Press Inc., 6277 Sea Harbor Drive, Orlando, FL 32887. (800) 321-5068. 1992. $2100.00.

The Focal Encyclopedia of Photography. Leslie Stroebel & Richard D. Zakia, editors. 3rd edition. Butterworth-Heinemann, 313 Washington Street, Newton, MA 02158. (617) 928-2500 or (800) 366-2665. 1993. $125.00.

McGraw-Hill Dictionary of Scientific and Technical Terms. Sybil P. Parker, ed. 5th edition. McGraw-Hill Publishing Company, 1221 Avenue of the Americas, New York, NY 10020. (800) 262-4729 or (212) 512-3825. 1993. $110.50.

McGraw-Hill Encyclopedia of Science and Technology. Sybil P. Parker, ed. 7th edition. 20 volumes. McGraw-Hill Publishing Company, 1221 Avenue of the Americas, New York, NY 10020. (800) 262-4729 or (212) 512-3825. 1992. $1900.00

Thesaurus of Photographic Science and Engineering. Society of Photographic Scientists and Engineers Staff. Books on Demand, 300 N. Zeeb Road, Ann Arbor, MI 48106-1346. (313) 761-4700 or (800) 521-0600. 1968. $34.30.

Thesaurus of Scientific, Technical, and Engineering Terms. Hemisphere Publishing Corporation, 1900 Frost Road, Suite 101, Bristol, PA 19007-1598. (215) 785-5800 or (800) 821-8312. FAX (215) 785-5515. 1987. $173.00.

GENERAL WORKS

Applied Photographic Optics: Lenses and Optical Systems for Photography, Film, Video, and Electronic Imaging. Sidney F. Ray. 2nd edition. Focal Press, 80 Montvale Avenue, Stoneham, MA 02180. 1994. $99.95.

Science and Technology of Photography. Karlheinz Keller, editor. VCH Publishers, Inc., 303 Northwest 12th Avenue, Deerfield

PHOTOGRAPHY

Ency. of Physical Sciences and Engineering Info. Sources

Beach, FL 33442. (800) 422-8824. 1993. Inquire for cost and availability.

The Story of Kodak. Douglas Collins. H.N. Abrams, 100 Fifth Avenue, New York, NY 10011. (212) 206-7715 or (800) 345-1359. $60.00.

HANDBOOKS AND MANUALS

Handbook of Photography. Ronald P. Lovell, et al. Delmar Publishers, Division of Thomson Educational Publishing, Inc., 3 Columbia Circle, Box 15015, Albany, NY 12205. (518) 464-3500 or (800) 347-7707. 1993. $27.95.

The New 35mm Handbook: A Complete Course From Basic Techniques To Professional Applications. Michael Freeman. Revised edition. Running Press, 125 S. 22nd Street, Philadelphia, PA 19103-4394. (215) 567-5080 or (800) 345-5359. 1993. $24.95.

The New 35mm Photographer's Handbook. Julian Calder & John Garrett. 2nd revised edition. Crown Publishing Group, 201 E. 50th St., New York, NY 10022. (800) 726-0600 or (800) 733-3000. 1986. Inquire for cost and availability.

ONLINE DATABASES AND CD-ROMS

Compendex Plus. Engineering Information, Inc., 345 East 47th Street, New York, NY 10017. (212) 705-7600 or (800) 221-1044. Contains citations with abstracts to worldwide literature in engineering and technology, from 1970 to present. Available on online BRS,(800) 289-4277, DIALOG, (800) 334-2564, ORBIT (800) 456-7248, and STN International, FIZ Karlsruhe, P.O. Box 2465, W-7500, Karlsruhe 1, Germany, online services. Also available on CD-ROM. Inquire as to cost and availability.

CSA Engineering. Cambridge Scientific Abstracts, 7200 Wisconsin Avenue, Suite 601, Bethesda, MD 20814. (301) 961-6750 or (800) 843-7751. Contains citations and abstracts of international periodicals and research literature covering all fields of engineering and science and technology,including computer and information science, electronics, mechanical engineering, solid state materials, 1981 to present. Available on BRS,(800) 289-4277, online service.Inquire as to cost and availability.

Current Contents Search. Institute for Scientific Information, 3501 Market Street, Philadelphia, PA 19104. (215) 386-0100. FAX (215) 386-6362. Contains citations to articles listed in the table of contents of science and technology journals. Also articles in social sciences and life sciences journals. Available on BRS,(800) 289-4277, DIALOG,(800) 334-2564, online services. Inquire as to cost and availability.

NTIS Bibliographic Database. National Technical Information Service, 5285 Port Royal Road, Springfield, VA 22161. (703) 487-4929 or FAX (703) 321-8199. Broad coverage of government-sponsored science and technology research reports, 1964 to present. Available on BRS,(800) 289-4277, DIALOG, (800) 334-2564, ORBIT, (800) 456-7248, and STN International, FIZ Karlsruhe, P.O. Box 2465, W-7500, Karlsruhe 1, Germany, online services. Also available on CD-ROM. Inquire as to cost and availability.

SCISEARCH. Institute for Scientific Information, 3501 Market Street, Philadelphia, PA 19104. (800) 523-1850 or (215) 386-0100. Broad multidisciplinary title and author index to the international literature of science and technology, 1974 to present.

Available on DIALOG,(800) 334-2564, and ORBIT,(800) 456-7248, online services. Also available on CD-ROM. Inquire as to cost and availability.

WILSONLINE. H.W. Wilson Company, 950 University Avenue, Bronx, NY 10452. (800) 367-6770 or (212) 588-8400. Makes available online versions of the printed H.W. Wilson indexes including *Applied Science and Technology Index, Business Periodicals Index, General Science Index,* and *Readers' Guide to Periodical Literature.* Period covered is generally 1983 to present. Available on BRS,(800) 289-4277, DIALOG,(800) 334-2564, and OCLC EPIC,(800) 848-5878, online services. Also available on CD-ROM. Inquire as to cost and availability.

PERIODICALS

British Journal of Photography. Bouverie Publishing Company Ltd., 147-151 Temple Chambers, Temple Avenue, London EC47 0DT, England. Telephone 071-583-3030. FAX 071-583-4068. 1854 to present. Weekly. Inquire for price and availability.

Image. George Eastman House, International Museum of Photography and FIlm, 900 East Avenue, Rochester, NY 14607. (716) 271-3361. FAX (716) 271-3970. 1952 to present. Twice a year. $40.00 membership.

Industrial Photography. PTN Publishing Corporation, 445 Broad Hollow Road, Melville, NY 11747-4722. (516) 845-2700. FAX (516) 845-7109. 1952 to present. Monthly. $60.00 per year.

Journal of Imaging Science and Technology. Society for Imaging Science and Technology, 7003 Kilworth Lane, Springfield, VA 22151. (703) 642-9090. FAX (703) 642-9094. 1950 to present. Bi-monthly. $120.00 per year.

Journal of Photographic Science. The Barn, Whitehall, Near Middle Marwood, Barnstaple, N. Devon EX31 4EQ, England. Telephone 0271-72482. FAX 0271-72482. 1952 to present. Bi-monthly. Inquire for price and availability.

Photogrammetric Engineering and Remote Sensing. American Society for Photogrammetry and Remote Sensing, 5410 Grosvenor Lane, Suite 210, Bethesda, MD 20814. (301) 493-0290. FAX (301) 493-0208. 1934-present. Monthly. $160.00 per year.

PSA Journal. Photographic Society of America, 3000 United Founders Blvd., No.103, Oklahoma City, OK 73112-3940. (405) 843-1437. FAX (405) 843-1438. 1934 to present. Monthly. Included with $35.00 annual membership.

PTN (Photographic Trade News). PTN Publishing Corporation, 445 Broad Hollow Road, Suite 21, Melville, NY 11747-4722. (516) 845-2700. FAX (516) 845-7109. 1937 to present. 24 times a year. $25.00 per year.

Scientific and Applied Photography and Cinematography. Gordon and Breach Science Publishers, 820 Town Center Drive, Langhorne, PA 19047. (215) 750-2642. FAX (215) 750-6343. 6 times a year. Inquire for price.

RESEARCH CENTERS AND INSTITUTES

George Eastman House, International Museum of Photography and Film. 900 East Avenue, Rochester, NY 14607. (716) 271-3361. FAX (716) 271-3970.

PHOTOMETRY

See: ASTRONOMY

PHYSICAL CHEMISTRY

See also: CATALYSIS, CHEMICAL BONDS, CHEMISTRY, COL-
LOIDS, CRYSTALLOGRAPHY, ELECTROCHEMISTRY, PHOTO-
CHEMISTRY, POLYMERS, QUANTUM CHEMISTRY

ABSTRACT SERVICES AND INDEXES

*Applied Science and Technology Index: A Cumulative Subject
Index to English Language Periodicals in the Fields of Aeronautics and Space Science, Computer Technology, Chemistry,
Construction Industry, Energy and Related Areas.* H. W. Wilson Co., 950 University Avenue, Bronx, NY 10452-9978. (212)
588-8400. (800) 367-6770. FAX (718) 590-1617. From 1958
to present. Monthly. Inquire about cost and availability. Also
available online (BRS and WILSONLINE) and on CD-ROM.
ISSN: 0003-6986.

Chemical Abstracts. Chemical Abstracts Service. 2540
Olentangy River Road, Box 3012, Columbus, OH 43210-0012.
(614) 447-3600. FAX (614) 447-3713. 1907 to present. Weekly.
$16,800.00 per year. Available online and on CD- ROM. CA is
also available in five section groupings. Inquire Regarding Cost
and availability.

Current Contents: Physical, Chemical and Earth Sciences. Institute For Scientific Information, 3501 Market Street, Philadelphia, PA 19104. (215) 386-0100. FAX (215) 386-2291. From
1961 to Present. Weekly. $442.00 per year. Also Available on
CD-ROM and online. Inquire regarding cost and availability.
ISSN: 0163-2574.

Engineering Index Monthly. Engineering Information, Inc.,
Castle Point on the Hudson, Hoboken, NJ 07030. (201) 216-
8500. (800) 221-1044. FAX(201) 216-8532. Monthly. $2300.00
per year for monthly issues. $1980 per year for annual cumula-
tion. Available on CD-ROM and online as COMPENDEX. ISSN:
0742-1974; 0360-8557.

General Science Index. H.W. Wilson Company, 950 University
Avenue, Bronx, NY 10452. (212) 588-8400. (800) 367-6770.
Fax (718) 590-1617. From 1978 to present. Ten issues per Year;
Quarterly and Annual Cumulations. Service basis. Available on
CD-ROM and online. Inquire regarding cost and availability.
ISSN: 0162-1963.

Physics Abstracts. INSPEC. Section A, Science Abstracts. Institution of Electrical Engineers (IEE), London, United Kingdom. Available from: INSPEC/IEEE-Institute of Electrical and
Electronic Engineers, Box 1331, 445 Hoes Lane, Piscataway,
NJ 08855-1331. (908) 562-5549. 1898 to present. 24 Issues
Per Year. $2835.00 per year. ISSN: 0036-8091. Also available
online and on CD-ROM.

Science Citation Index (SCI). Institute for Scientific Information, 3501 Market Street, Philadelphia, PA 19104. (215) 386-
0100. (800) 523-1850. FAX (215) 386-2991. 1961 To Present.
Six issues per year, plus annual cumulation. $11650.00 per year.
Also available online and on CD-ROM. Inquire about price and
availability. ISSN: 0036-827X.

ANNUAL REVIEWS AND YEARBOOKS

Annual Review of Physical Chemistry. Annual Reviews, Inc.,
4139 El Camino Way, Palo Alto, CA 94306-0897. (415) 493-
4400. Fax (415) 855-9815. 1951 to present. Annual. $55.00 per
year.

ASSOCIATIONS AND PROFESSIONAL SOCIETIES

American Chemical Society, 1155 16th Street, Nw, Washington, DC 20036. (202) 872-4600.

American Institute of Chemists. 501 Wythe Street, Alexandria,
VA 22314-1917. (703) 836-2090. FAX (703) 836- 2091.

Association of Consulting Chemists and Chemical Engineers,
50 East 41st Street, Suite 92, New York, NY 10017. (212) 684-
6255.

Chemical Manufacturers Association. 2501 M Street, N.W.,
Washington, DC 20037. (202) 887-1182

Division of Physical Chemistry (a Division of the American
Chemical Society). Iowa State University, Department of Chemistry, Wilhelm Hall, Room 303, Ames, IA 50011-4111. (515)
294-4111.

Electrochemical Society. 10 South Main Street, Pennington, NJ
08534. (609) 737-1902.

BIBLIOGRAPHIES

Chemical Information. H. R. Collier, editor. Springer- Verlag
New York, Inc., 175 Fifth Avenue, New York, NY 10010. (212)
460-1500. (800) 777-4643. 1990. $91.00.

Chemical Information Sources. Gary Wiggins. McGraw-Hill
Publishing Company, 1221 Avenue of the Americas, New York,
NY 10020. (800) 262-4729 or (212) 512-3825. 1991. $42.50.

Information Sources in Chemistry. R.T. Bottle and J.F.B
Rowland, editors. R.R. Bowker Inc., 121 Chanlon Road, New
Providence, NJ 07974. (800) 521-8110 or (908) 464-6800.
Fourth edition. 1993. $75.00.

*Scientific and Technical Books and Serials in Print: An Index to
Literature in Science and Technology.* R.R. Bowker Inc., 121
Chanlon Road, New Providence, NJ 07974. (908) 464-6800.
(800) 521-8110. FAX (908) 665-3502. 1972 to present. Annual.
4 volumes. 1994. $299.95. Also available on compact disc and
online. ISSN: 0000-054X.

DIRECTORIES AND BIOGRAPHICAL SOURCES

American Institute of Chemists: Professional Directory. American Institute of Chemists, 501 Wythe Street, Alexandria, VA
22314-1917. (703) 836-2090. FAX (703) 836-2091. Annual.
$65.00.

*American Men and Women of Science: Physical and Biological
Sciences.* R.R. Bowker Inc., 121 Chanlon Road, New Providence, NJ 07974. (908) 464-6800. (800) 521-8110. 20th edition. 8 volumes. 1996. $850.00

PHYSICAL CHEMISTRY

Ency. of Physical Sciences and Engineering Info. Sources

Chemical Sources International. Mike Desing and Kurt Gandenberger, editors. Chemical Sources International, P.O. Box 1824, Clemson, SC 29633. (803) 646-7840. 1994.$285.00.

Chemical Week: Buyers Guide Issue. Chemical Week Associates, 888 Seventh Avenue, New York, NY 10106. (212) 621-4900. FAX (212) 621-4949. Annual, October. $50.00.

Directory of Chemistry Software 1992. Wendy Warr, Peter Willett, Geoff Downs. American Chemical Society, 1155 16th Street, NW, Washington, DC 20036. (202) 872-4600. 1992. $35.95.

Research Centers Directory. Gale Research Company,835 Penobscot Building, Detroit, MI 48226-4094. (313) 961-2242. (800) 347-4253. 20th edition. 1995. $485.00. ISSN: 0080-1518.

ENCYCLOPEDIAS AND DICTIONARIES

Academic Press Dictionary of Science and Technology. Christopher Morris, editor. Academic Press, Inc., 1250 Sixth Avenue, San Diego, CA 92101. (619) 231-0926. FAX (619) 699-6715. 1991. $115.00.

Concise Encyclopedia Chemistry. Translated and revised by Mary Eagleson. Walter de Gruyter, Inc., 200 Saw Mill River Road, Hawthorne, New York, 10532. (914) 747-0110 or Fax (914) 747-1326. 1994. $69.95.

Dictionary of Chemical Names and Synomyms. Philip A. Howard and Michael W. Neil. Lewis Pub./CRC Press, Inc., 2000 Corporate Boulevard, NW, Boca Raton, FL 33431. (407) 994-0555. (800) 272-7737. 1991. $149.95.

Dictionary of Physical Chemistry. Satish Anand and Raj Kumar. South Asia Publications, PO Box 502 Collumbus, MO 65205. (314) 474-0116. 1990. $23.50.

Encyclopedic Dictionary of Science. Candida Hunt and Monica Byles, editors. Facts-on-File, Inc., 460 Park Avenue South, New York, NY 10016-7382. (800) 322-8755. Fax (800) 683-3633. 1988. $32.95.

Encyclopedia of Applied Physics. VCH Publishers, Inc., 303 Northwest 12th Avenue, Deerfield Beach, FL 33442. (800) 367-8249. 1991-. Twenty volumes. $6000.00.

Encyclopedia of Chemical Processing and Design. McKetta. Dekker, Inc., 270 Madison Avenue, New York, NY 10016. (212) 696-9000. (800) 228-1160. 1976 - . $175.00 per volume.

Grant & Hawks Chemical Dictionary. R. A. Grant. McGraw-Hill Publishing Company, Inc., 1221 Avenue of the Americas, New York, NY 10020. (212) 512-2000. (800) 262-4729. 5th edition. 1987. $64.50.

Hawley's Condensed Chemical Dictionary. Richard J. Lewis. Van Nostrand Reinhold, 115 Fifth Avenue, New York, NY 10003. (212) 254-3232. (800) 842-3636. 12th edition. 1993. $69.95.

Industrial Chemical Thesaurus. Michael Ash and Irene Ash. VCH Publishers, Inc., 220 East 23rd Street, New York, NY 10010-4606. (800) 367-8249. 1992. $295.00.

Kirk-Othmer Encyclopedia of Chemical Technology. John Wiley and Sons, Inc., 605 Third Avenue, New York, NY 10158. (800) 526-5368 or (212) 850-6000. Fourth edition. 1991-. Twenty-seven volumes. $5400.00.

McGraw-Hill Encyclopedia of Chemistry. Sybil P.Parker, editor. McGraw-Hill Book, Incorporated, 1221 Avenue of the Americas, New York, NY 10020. (212) 997-3675. Second edition. 1993. $95.50.

McGraw-Hill Encyclopedia of Science and Technology. McGraw-Hill Book, Incorporated, 1221 Avenue of the Americas, New York, NY 10020. (212) 997-3675. Seventeenth edition. Twenty volumes. 1992. $1900.00.

GENERAL WORKS

Basic Physical Chemistry for the Physical Sciences. Peter V. Hobbs. Cambridge University Press,1995. $49.95.

The Chemical Bond: Structure and Dynamics. Ahmed Zewail. Academic Press, Inc., 1250 Sixth Avenue, San Diego, CA 92101-4311. (619) 231-0926. FAX (619) 699-6715. 1992. $49.95

Dynamics of the Liquid State. Umberto Balucani and Marco Zoppi. Oxford University Press, Inc., 200 Madison Avenue, New York, NY 10016. (212) 725-6000. (800) 334-4249. 1995. $95.00.

Electronic Structure of Materials. Adrian Sutton. Oxford University Press, Inc., 200 Madison Avenue, New York, NY 10016. (212) 725-6000. (800) 334-4249. 1993. $26.50 in paper.

The Elements of Physical Chemistry. P. W. Atkins. W. H. Freeman, & Co., 41 Madison Avenue, East 26th Avenue, 35th Floor, New York, NY 10010. (212) 576-0400. 1995. $42.95.

Physical Chemistry Problem Solver. Research and Education Association, 61 Ethel Road, West Piscataway, NJ 08854. (908) 819-8880. Revised edition. 1994. $29.95 in paper.

Physical Methods For Chemists. Russell S. Drago. SCP: Third World Literature Publishing House, P.O. Box 482, Lithonia, GA 30058-0482. (404) 785-7725. 1992. $63.00.

Physical Organic Chemistry. Neil S. Isaacs. Halsted Press, 605 Third Avenue, New York, 10158-0012. (212) 850-6400. 2nd edition. 1995. $59.95 in paper.

Physical Properties of Polymers. James E. Mark, et al. American Chemical Society, 1155 16th Street, NW, Washington, DC 20036. (202) 872-4600. (800) 333-9511. FAX (614) 447- 3671. 2nd edition. 1993. $59.95.

Physics of Highly Charged Ions. Martin Stocki and Patrick Richard, editors. American Institute of Physics, One Physics Ellipse, College Park, MD 20740-3843. (301) 209-3100. 1993. $175.00.

Quantum States and Processes. George H. Duffey. Prentice-Hall, 113 Sylvan Avenue, Route 9W, Englewood Cliffs, NJ 07632. (201) 592-2000. (800) 922-0579. 1991. $94.00.

HANDBOOKS AND MANUALS

Chemical Formulary. H. Bennett, editor. Chemical Publishing Company, Inc., 80 Eighth Avenue, New York, NY 10011. (212) 255-1950. Volumes 1-30. $70.00 per volume.

Chemometrics: Chemical and Sensory Data. David R. Burgard and James T. Kuznicki. CRC Publishers, Inc., 2000 Corporate Blvd., N.W., Boca Raton, FL 33431. (407) 994-0555 or (800) 333-8300. 1990. $120.00.

Chemical Reference Handbook. Gordon Press, PO Box 459, Bowling Green Station, New York, NY 10004. (718) 624-8419. 1995. $260.00.

CRC Handbook of Chemistry and Physics. David R. Kide, editor. CRC Press Inc., 2000 Corporate Blvd., NW, Boca Raton, FL 33431. (407) 994-0555. (800) 333-8300. 77th edition. 1996. $99.95.

Guide to Basic Chemical Compounds. D.R. Lide, Jr. CRC Publishers, Inc., 2000 Corporate Blvd., N.W., Boca Raton, FL 33431. (407) 994-0555. (800) 333-8300. 1993. $120.00.

Improving Safety in the Chemical Laboratory: A Practical Guide. Jay A. Young, editor. John Wiley and Sons, Inc., 605 Third Avenue, New York, NY 10158. (800) 526-5368 or (212) 850-6000. Second edition. 1991. $75.00.

Laboratory Handbook of Materials, Equipment, and Techniques. Gary S. Coyne. Prentice Hall (Division of Simon and Schuster), 15 Columbus Circle, New York, NY 10023. (212) 373-8500 or (800) 922-0579. 1992. $45.00.

Lange's Handbook of Chemistry. John A. Dean, editor. McGraw-Hill Publishing Company, 1221 Avenue of the Americas, New York, NY 10020. (800) 262-4729 or (212) 512-3825. Fourteenth edition. 1992. $79.50.

Riegel's Handbook of Industrial Chemistry. James A. Kent, editor. Van Nostrand Reinhold, 115 Fifth Avenue, New York, NY 10003. (212) 254-3232 or (800) 926-2665. Ninth edition. 1992. $114.95.

ONLINE DATABASES AND CD-ROMS

Analytical Abstracts Online. Royal Society of Chemistry, Information Services, Thomas Graham House, Science Park, Milton Road, Cambridge, CB4 4WF, England. Contains citations, mostly with abstracts, of the worldwide literature on analytical chemistry, from 1980 to present. Available on DIALOG, (800) 334-2564, and STN International, FIZ Karlsruhe, P.O. Box 2465, W-7500, Karlsruhe 1, Germany, online services. Inquire as to cost and availability.

CA Search. Chemical Abstracts Service, P.O. Box 3012, Columbus, OH 43210-0012. (614) 447-3600. (800) 848-6533. FAX (614) 447-3709. Very comprehensive guide to worldwide chemical literature and related fields, 1972 to present. Available on BRS,(800) 289-4277, DIALOG, (800) 334-2564, ORBIT (800) 456-7248, and STN International, FIZ Karlsruhe, P.O. Box 2465, W-7500, Karlsruhe 1, Germany, online services. Inquire as to cost and availability.

Chemical Journals of the American Chemical Society. American Chemical Society, 1155 16th Street, N.W., Washington, DC 20036. (202) 872-4381 or (800) 424-6747. Contains complete text of approximately 90,000 articles from 22 primary journals published by the American Chemical Society, from mostly 1982 to present. Available on STN International, FIZ Karlsruhe, P.O. Box 2465, W-7500, Karlsruhe 1, Germany, online service. Inquire as to cost and availability.

Compendex Plus. Engineering Information, Inc., 345 East 47th Street, New York, NY 10017. (212) 705-7600 or (800) 221-1044. Contains citations with abstracts to worldwide literature in engineering and technology, from 1970 to present. Available on online BRS,(800) 289-4277, DIALOG, (800) 334-2564, ORBIT (800) 456-7248, and STN International, FIZ Karlsruhe, P.O. Box 2465, W-7500, Karlsruhe 1, Germany, online services. Also available on CD-ROM. Inquire as to cost and availability.

Current Contents Search. Institute for Scientific Information, 3501 Market Street, Philadelphia, PA 19104. (215) 386-0100. FAX (215) 386-6362. Contains citations to articles listed in the table of contents of science and technology journals. Also articles in social sciences and life sciences journals. Available on BRS,(800) 289-4277, DIALOG,(800) 334-2564, online services. Inquire as to cost and availability.

Dissertation Abstracts. University Microfilms International, 300 North Zeeb Road, Ann Arbor, MI 48106. (800) 521-0600 or (313) 761-4700. Scope includes virtually all doctoral dissertations accepted at accredited American institutions from 1861 to present in 252 subject areas. Available on BRS,(800) 289-4277, DIALOG,(800) 334-2564, and OCLC EPIC,(800) 848-5878, online services. Also available on CD-ROM. Inquire as to cost and availability.

Gmelin Database. Gmelin-Institut fur Anorganische Chemie und Grenzgebiete, Varrentrapperstrasse, 40-42, Carl-Bosch-Haus, D-6000, Frankfurt am Main 90, Germany. Contains structural and factual data relating to inorganic and organometallic chemistry. Provides data from the Gmelin Handbook of Inorganic and Organometallic Chemistry. Covers the period 1817 to 1975; 1988-89. Available on STN International, FIZ Karlsruhe, P.O. Box 2465, W-7500, Karlsruhe 1, Germany, online service. Inquire as to cost and availability.

Kirk-Othmer Encyclopedia of Chemical Technology. John Wiley and Sons, Inc., 605 Third Avenue, New York, NY 10158. (800) 526-5368 or (212) 850-6000. Contains the complete text of all chapters in the 27 volume fourth edition of the *Kirk-Othmer Encyclopedia of Chemical Technology.* 1991. Available on BRS,(800) 289-4277, DIALOG,(800) 334-2564, online services. Inquire as to cost and availability.

NTIS Bibliographic Database. National Technical Information Service, 5285 Port Royal Road, Springfield, VA 22161. (703) 487-4929 or FAX (703) 321-8199. Broad coverage of government-sponsored science and technology research reports, 1964 to present. Available on BRS,(800) 289-4277, DIALOG, (800) 334-2564, ORBIT, (800) 456-7248, and STN International, FIZ Karlsruhe, P.O. Box 2465, W-7500, Karlsruhe 1, Germany, online services. Also available on CD-ROM. Inquire as to cost and availability.

SCISEARCH. Institute for Scientific Information, 3501 Market Street, Philadelphia, PA 19104. (800) 523-1850 or (215) 386-0100. Broad multidisciplinary title and author index to the international literature of science and technology, 1974 to present. Available on DIALOG,(800) 334-2564, and ORBIT,(800) 456-7248, online services. Also available on CD-ROM. Inquire as to cost and availability.

WILSONLINE. H.W. Wilson Company, 950 University Avenue, Bronx, NY 10452. (800) 367-6770 or (212) 588-8400. Makes available online versions of the printed H.W. Wilson indexes including *Applied Science and Technology Index, Business Periodicals Index, General Science Index,* and *Readers' Guide to Periodical Literature.* Period covered is generally 1983 to present. Available on BRS,(800) 289-4277, DIALOG,(800) 334-2564, and OCLC EPIC,(800) 848-5878, online services. Also available on CD-ROM. Inquire as to cost and availability.

PERIODICALS

American Laboratory. American Laboratory, 30 Controls Drive, Box 870, Shelton, CT 06484-6111. (203) 926-9300. FAX (203) 926-9310. 1968 to present. Monthly. $210.00 per year. ISSN: 0044-7749.

Analytical Chemistry. American Chemical Society, Box 3337, Columbus, OH 43210. (614) 447-3776. 1929 to present. Semimonthly. $415.00 per year. ISSN: 0003-2700.

Analytical Methods and Instrumentation. John Wiley and Sons, Inc., 605 Third Avenue, New York, NY 10158. (800) 526-5368 or (212) 850-6000. Bimonthly. $250.00 per year. ISSN: 1063-5246.

Angewandte Chemie. VCH Publishers, Inc., 220 East 23rd Street, New York, NY 10010-4606. (212) 683-8333. (800) 367-8249. 1888 to present. 22 issues per year. $840.00 ISSN: 0044-8249.

Chemical Physics. Elsevier Science Publishing Co., Inc., Box 882, Madison Square Station, New York, NY 10159. (212) 989-5800. FAX (212) 633-3990. 1973 to present. 33 issues per year in 11 volumes. $2759.00 per year. ISSN: 0301-0104.

Chemical Physics Letters. Elsevier Science Inc., Publishing Co., Inc., Box 882, Madison Square Station, New York, NY 10159. (212) 989-5800. (212) 633-3990. 1967 to present. 102 issues per year in 17 volumes. $5063.00 per year. ISSN: 0009-2614.

Chemical Reviews. American Chemical Society, Box 3337, Columbus, OH 43210. (614) 447-3776. 1924 to present. 8 Issues Per Year. $346.00 per year. ISSN: 0009-2665.

Chemical Week. Includes Annual Buyers Guide. Chemical Week Associates, 888 Seventh Avenue, New York, NY 10106. (212) 621-4900. FAX (212) 621-4949. 1914 to present. Weekly. $99.00 per year. ISSN: 0009-272X.

Chemtech. American Chemical Society. Box 3337, Columbus, OH 43210. (614) 447-3776. 1970 to present. Monthly. $370.00 per year. ISSN: 0009-2703.

Inorganic Chemistry. American Chemical Society, Box 3337, Columbus, OH 43210. (614) 447-3776. 1962 to present. Semimonthly. $500.00 per year. ISSN: 0020-1669.

International Journal of Chemical Kinetics. John Wiley and Sons, Inc., 605 Third Avenue, New York, NY 10158. (800) 526-5368 or (212) 850-6000. 1968 to present. Monthly. $730.00 per year. ISSN: 0538-8066.

Journal of Catalysis. Academic Press, Inc., Box 620000, Orlando, FL 32891-8340. 1962 to present. Monthly. $1296.00 per year. ISSN: 0021-9517.

Journal of Physical Chemistry. American Chemical Society, Box 3337, Columbus, OH 43210. (614) 447-3776. (800) 333-511. FAX (614) 447-3671. 1896 to present. Weekly.$1140.00 per year to non-members. ISSN: 0894-3230.

Journal of Solid State Chemistry. Academic Press, Inc., Box 620000, Orlando, FL 32891-8340. 1969 to present. Monthly. $1248.00 per year. ISSN: 0022-4596.

Journal of the American Chemical Society. American Chemical Society, Box 3337, Columbus, OH 43210. (614) 447-3776. 1879 topresent. Biweekly. $1055.00 per year. ISSN: 0002-7863.

RESEARCH CENTERS AND INSTITUTES

Chemical Laboratories. Harvard University, Oxford Street, Cambridge, MA 02138. (617) 495-4283. FAX (617) 496-5618.

Institute for Theoretical Chemistry. University of Texas at Austin, Department of Chemistry, WEL 3.237, Austin, TX 78712. (512) 471-3114.

Laboratory for Research on the Structure of Matter. University of Pennsylvania, 3231 Walnut Street, Philadelphia, PA 19104. (215) 898-8571.

Materials Science Center. Cornell University, Clark Hall of Science, Ithaca, NY 14853-2501. (607) 255-4272. FAX (607) 255-6428.

Research Program in Chemistry and Biochemistry. Southern Illinois University at Carbondale, Carbondale, IL 62901. 618) 453-5721.

PHYSICAL GEOLOGY

See also: CRUST, GEOMORPHOLOGY, GEOPHYSICS, IGNEOUS ROCKS, METAMORPHIC ROCKS, PETROLOGY, SEDIMENTARY ROCKS, SEISMOLOGY, STRUCTURAL GEOLOGY, VOLCANOLOGY

ABSTRACT SERVICES AND INDEXES

Bibliography and Index of Geology. American Geological Institute, 4220 King Street, Alexandria, VA 22302. (703) 379-2480. Fax (703) 379-7563. Monthly. $1295.00 per year. Also available online as GeoRef.

Deep-sea Research. Part A: Oceanographic Research Papers. Part B: Oceanographic Literature Review. Pergamon Press, Inc., Maxwell House, Fairview Park, Elmsford, NY 10523. (914) 592-7700. Fax (914) 592-3625. Twelve times per year. $1370 per year for parts A and B.

Geological Abstracts. Elsevier-Geo Abstracts, Regency House, 34 Duke Street, Norwich NR3 3AP, England. Monthly. $760.00 per year. Also available online as GEOBASE.

Geological Society of America. Abstracts with Programs. Geological Society of America. 3300 Penrose Place, P.O. Box 9140, Boulder, CO 80301-9140. (303) 447-2020. Abstracts and programs of the annual conference. Annual. $69.00.

Geophysics Abstracts. Pergamon Press, Inc., Maxwell House, Fairview Park, Elmsford, NY 10523. (914) 592-7700. Fax (914) 592-3625. Twelve times per year. $565.00 per year. Also available in microform.

Oceanic Abstracts. Cambridge Scientific Abstracts, 7200 Wisconsin Avenue, Bethesda, MD 20814. (301) 961-6750. Fax (301) 961-6720. Bimonthly. $995.00 per year.

Science Citation Index. Institute for Scientific Information, 3501 Market Street, Philadelphia, PA 19104. (800) 523-1850 or (215) 386-0100. Inquire about price and availability.

General Science Index. H.W. Wilson Co., 950 University Avenue, Bronx, NY 10452. (800) 367-6770 or (212) 588-8400. Inquire about price and availability.

ANNUAL REVIEWS AND YEARBOOKS

Advances in Geophysics. Academic Press, Inc., 1250 Sixth Avenue, San Diego, CA 92101. (619) 231-0926. FAX (619) 699-6715. Irregular. Inquire about price and availability.

Annual Review and Earth and Planetary Sciences. Annual Reviews, Inc., 4139 El Camino Way, Palo Alto, CA 94306-0897. (415) 493-4400. Fax (415) 855-9815. Annual. $55.00 per year.

ASSOCIATIONS AND PROFESSIONAL SOCIETIES

American Geological Institute. 4220 King Street, Alexandria, VA 22302. (703) 379-2480. Fax (703) 379-7563.

American Geophysical Union, 2000 Florida Avenue, N.W., Washington, DC 20009. (202) 462-6900.

American Institute of Professional Geologists. 7828 Vance Drive, Suite 103, Arvada, CO 80003. (303) 431-0831.

Association of Engineering Geologists. 323 Boston Post Road, Suite 2D, Sudbury, MA 01776. (508) 443-4639.

Geological Society of America. 3300 Penrose Place, P.O. Box 9140, Boulder, CO 80301-9140. (303) 447-2020.

Society of Exploration Geophysicists. P.O. Box 702740, Tulsa, OK 74170. (918) 493-3516.

Society of Independent Professional Earth Scientists. 4925 Greenville Avenue, Suite 170, Dallas, TX 75206. (214) 363-1780.

BIBLIOGRAPHIES

Geologic Reference Sources: A Subject and Regional Bibliography of Publications and Maps in the Geological Sciences. Dederick C. Ward et al. Scarecrow Press, Inc., 52 Liberty Street, Box 4167, Metuchen, NJ 08840. (800) 537-7107 or (201) 548-8600. 1981. $49.50.

Information Sources in the Earth Sciences. David N. Wood, Joan E. Hardy, and Anthony P. Harvey, editors. Bowker-Saur/K.G. Saur. Distributed by R.R. Bowker, 121 Chanlon Road, New Providence, NJ 07974. (800) 521-8110 or (908) 464-6800. Second edition. 1989. $85.00.

Remote Sensing of Earth Resources: A Quarterly Bibliography. University of New Mexico, Technology Application Center, 2808 Central Avenue, S.E., Albuquerque, NM 87131-6031. (505) 277-3622. Fax (505) 277-3614. $150.00 per year.

DIRECTORIES AND BIOGRAPHICAL SOURCES

American Institute of Professional Geologists. Membership Directory. American Institute of Professional Geologists, 7828 Vance Drive, Suite 103, Arvada, CO 80003. (303) 431-0831. Annual. $18.00.

Association of Engineering Geologists. Membership directory. Association of Engineering Geologists, 323 Boston Post Road, Suite 2D, Sudbury, MA 01776. (508) 443-4639. Annual. $18.00.

Directory of Geoscience Departments: United States and Canada. American Geological Institute. 4220 King Street, Alexandria, VA 22302-1507. (703) 379-2480. Fax (703) 379-7563. Twenty-six edition. 1987. $22.00.

Geological Society of America. Membership Directory. Geological Society of America, 3300 Penrose Place, Boulder, CO 80301. (303) 447-2020. Annual. Available to members only.

ENCYCLOPEDIAS AND DICTIONARIES

Concise Oxford Dictionary of the Earth Sciences. Ailsa Allaby and Michael Allaby, editors. Oxford University Press, Inc., 200 Madison Avenue, New York, NY 10016. (800) 334-4249 or (212) 679-7300. 1990. $42.95.

Dictionary of Geological Terms. American Geological Institute. Doubleday and Company, Inc., 245 Park Avenue, New York, NY 10017. (800) 645-6156 or (212) 953-4561. Third edition. 1984. $12.00 in paper.

Facts on File Dictionary of Geology and Geophysics. Dorothy Farris and Donald R. Coates, editors. Facts on File, Inc., 460 Park Avenue South, New York, NY 10016-7382. (800) 322-8755. Fax (800) 683-3633. 1989. $24.95.

Glossary of Geology. Robert L. Bates and Julia A. Jackson. American Geological Institute, 4220 King Street, Alexandria, VA 22032. (703) 379-2480. Third edition. 1987. $75.00.

Magill's Survey of Science: Earth Science Series. Salem Press, Inc., P.O. Box 1097, Englewood Cliffs, NJ 07632. (800) 221-1592 or (201) 871-3700. Five volumes. 1990. $400.00 for the set.

GENERAL WORKS

Dynamic Earth: An Introduction to Physical Geology. B.J. Skinner and Stephen C. Porter. John Wiley and Sons, Inc., 605 Third Avenue, New York, NY 10158. (800) 526-5368 or (212) 850-6000. 1990. $51.95.

Earth Materials and Earth Processes: An Introduction. Lynn Fichter, et al. Macmillan Publishing, 866 3rd Avenue, New York, NY 10024. (212) 689-9140. Third edition. 1991. $19.50.

Evolution of the Earth. Robert H. Dott, Jr. and Roger L. Batten. McGraw-Hill Publishing Company, 1221 Avenue of the Americas, New York, NY 10020. (800) 262-4729 or (212) 512-3825. Fourth edition. 1988. $41.00.

Geology: The Science of A Changing Earth. Ira S. Allison et al. McGraw-Hill Publishing Company, 1221 Avenue of the Americas, New York, NY 10020. (800) 262-4729 or (212) 512-3825. Sixth edition. 1974. $18.95 in paper.

Mechanics in Structural Geology. Brian Bayly. Springer-Verlag, 175 Fifth Avenue, New York, NY 10010. (212) 460-1500. 1992. $39.00

HANDBOOKS AND MANUALS

Agi Data Sheets for Geology in the Field, Laboratory, and Office. J.T. Dutro, R.V. Dietrich, and R.M. Foose, editors. American Geological Institute, 4220 King Street, Alexandria, VA 23020-1507 (800) 336-4764 or (703) 379-2480. 1989. $24.95.

Field Mapping For Geology Students. F. Ahmed and D.C. Almond. Unwin Hyman, Inc., 10 East 53rd Street, New York, NY 10022. (212) 207-7626 or (800) 242-7737. 1983. $18.95.

Geology in the Field. R.R. Compton. John Wiley and Sons, Inc., 605 Third Avenue, New York, NY 10158. (800) 526-5368 or (212) 850-6418. 1985. $37.95.

A Manual of Geology For Civil Engineers. John A. Pitts. World Scientific Publishing Company, Inc., 1060 Main Street, Unit B, River Edge, NJ 07661. (800) 227-7562 or (201) 487-9655. 1985. $38.00.

The Mapping of Geological Structures. K.R. McClay. John Wiley and Sons, Inc., 605 Third Avenue, New York, NY 10158. (800) 526-5368 or (212) 850-6000. 1988. $18.95.

ONLINE DATABASES AND CD-ROMS

Earth Sciences. U.S. Geological Survey, 12201 Sunrise Valley Drive, Reston, VA 22092-9998. (703) 648-4460. CD-ROM of earth science databases including the U.S. Geological Survey Library, Earth Science Data Directory, and GEOINDEX, citations to published geological maps. $350.00 per year with quarterly updates.

Geoarchive. Geosystems, P.O. Box 1024, Westminster, London, England, SW1 P 2JL. Citations to literature on geoscience, 1969 to present. Inquire as to online cost and availability.

Geobase. Elsevier-Geo Abstracts, Regency House, 34 Duke Street, Norwich NR3 3AP, England. Contains citations to the worldwide earth science literature from 1980 to date. Available on DIALOG, ORBIT online services. Inquire as to cost and availability.

Geomechanics Abstracts. University of London, Imperial College of Science and Technology, Rock Mechanics Information Service, Royal School of Mines, Prince Consort Road, London SW7 2BP, England. Telephone: 071-589 5111 or FAX 071-589 6806. Scope includes worldwide literature on engineering geology, hydrogeology, mining, rock mechanics, soil mechanics, and tunnelling, 1977 to present. Available on the ORBIT online service. Inquire as to cost and availability.

Georef. American Geological Institute, 4220 King Street, Alexandria, VA 22302. (800) 336-4764 or (703) 379-2480. Geology

and geosciences literature, 1785 to present for North America. Available on DIALOG, ORBIT online services. Also available on CD-ROM. Inquire as to cost and availability.

NTIS Bibliographic Database. National Technical Information Service, 5285 Port Royal Road, Springfield, VA 22161. (703) 487-4630. Broad coverage of government- sponsored research reports, 1964 to present. Available on BRS, DIALOG, ORBIT and STN online services. Also available on CD-ROM. Inquire as to cost and availability.

SCISEARCH. Institute for Scientific Information, 3501 Market Street, Philadelphia, PA 19104. (800) 523-1850 or (215) 386-0100. Broad multidisciplinary title and author index to the international literature of science and technology, 1974 to present. Available on DIALOG and ORBIT online services. Also available on CD-ROM. Inquire as to cost and availability.

PERIODICALS

AAPG Bulletin. American Association of Petroleum Geologists, P.O. Box 979, Tulsa, OK 7401. (918) 584-2555. Monthly. $120.00 per year.

American Journal of Science. Kline Geology Laboratory, Yale University, Box 6666, Yale Station, New Haven, CT 06511-8130. (203) 432-3131. Ten times per year. $40.00 per year.

Canadian Journal of Earth Sciences. National Research Council of Canada, Research Journals, Ottawa, Ont. K1A OR6, Canada. (613) 993-9084. Monthly. $112.00 per year.

Earth and Planetary Science Letters. Elsevier Science Publishing Company, Inc., 655 Avenue of the Americas, New York, NY 10010. (212) 989-5800. Twenty times per year. $945.00 per year.

Geological Magazine. Cambridge University Press, 40 West 20th Street, New York, NY 10011-4211. (212) 924-3900. Bimonthly. $205.00 per year.

Geological Society. Journal. Geological Society of London. Geological Society Publishing House, Unit 7, Brassmill Enterprise Centre, Brassmill Lane, Bath, Avon BA1 3JN, England. 6 times per year. $432.00.

Geological Society of America Bulletin. P.O. Box 9140, 3300 Penrose Place, Boulder, CO 80301. (303) 447-2020. Fax (303) 447-1133. Monthly. $150.00.

Geology. Geological Society of America. P.O. Box 9140, 3300 Penrose Place, Boulder, CO 80301. (303) 447-2020. Fax (303) 447-1133. Monthly. $120.00 per year.

Geophysics. Society of Exploration Geophysicists, P.O. Box 702740, Tulsa, OK 74170. (918) 493-3516. Monthly. $175.00 per year.

Geotimes. American Geological Institute, 4220 King Street, Alexandria, VA 22032. (703) 379-2480. Monthly. $24.95 per year.

Journal of Geology. University of Chicago Press, 5720 Woodlawn Avenue, Chicago, IL 60637. (312) 753-3347.

Bimonthly. $38.00 per year.

Journal of Structural Geology. Pergamon Journals, Inc., Maxwell House, Fairview Park, Elmsford, NY 10523. (914) 592-7700. Ten times per year. $375.00 per year.

Journal of Volcanology and Geothermal Research. Elsevier Science Publishing Company, Inc., 655 Avenue of the Americas, New York, NY 10010. (212) 989-5800. Monthly. $500.00 per year.

Lithos: An International Journal of Mineralogy, Petrology, and Geochemistry. Elsevier Science Publishing Company, Inc., 655 Avenue of the Americas, New York, NY 10010. (212) 989-5800. Quarterly. $275.00 per year.

Mountain Geologist. Rocky Mountain Association of Geologists, 4201 West 51st Avenue, Denver, CO 80212-2902. Quarterly. $15.00 per year.

Physics of the Earth and Planetary Interiors. Elsevier Science Publishing Company, Inc., 655 Avenue of the Americas, New York, NY 10010. (212) 989-5800. Twenty-four times per year. $900.00 per year.

Rock Mechanics and Rock Engineering. Springer-Verlag New York, Inc., 175 Fifth Avenue, New York, NY 10010. (800) 526-7254 or (212) 460-1500. Quarterly. $122.00 per year.

Sedimentary Geology. Elsevier Science Publishing Company, Inc., 655 Avenue of the Americas, New York, NY 10010. (212) 989-5800. Twenty times per year. $901.00 per year.

Sedimentology. Blackwell Scientific Publishing, Inc., 3 Cambridge Court, Suite 208, Cambridge, MA 02142. (617) 225-0401. Six times per year. $370.00 per year.

Tectonics. American Geophysical Union, 2000 Florida Avenue, N.W., Washington, DC 20009. (202) 462-6900. Bimonthly. $230.00 per year to individuals.

Tectonophysics: An International Journal of Geotectonics and the Geology and Physics of the Interior of the Earth. Elsevier Science Publishing Company, Inc., 655 Avenue of the Americas, New York, NY 10010. (212) 989-5800. Fifty-six times per year. $1275.00 per year.

RESEARCH CENTERS AND INSTITUTES

Cornell University. Institute for the Study of Continents. 3122 Snee Hall, Ithaca, NY 14853-1504. (607) 255-3474.

Louisiana State University. Basin Research Institute. Baton Rouge, LA 70803-4101. (504) 388-8328. Fax (504) 388-5328.

University of California, Santa Barbara. Institute for Crustal Studies. Santa Barbara, CA 93105. (805) 961-8231.

University of Alaska Fairbanks. Geophysical Institute. C.T. Elvey Building, Fairbanks, AK 99775-0800. (907) 474-7282. Fax (907) 474-7290.

PHYSICS

See also: ASTRONOMY, COSMOLOGY, MATTER, MECHANICS, TIME, UNIVERSE

ABSTRACT SERVICES AND INDEXES

Acoustics Abstracts. Parts A & B. Multi-SciencePublishing Co., Ltd, Box 176, Avenel, NJ 07001. 1967 to present. Monthly. L265 per year. ISSN: 0001-4974.

Applied Science and Technology Index: A Cumulative Subject Index to English Language Periodicals in the Fields of Aeronautics and Space Science, Computer Technology, Chemistry, Construction Industry, Energy and Related Areas. H.W. Wilson Co., 950 University Avenue, Bronx, NY 10452. (212) 588-8400. (800) 367-6770. FAX (718) 590-1617. From 1958 to present. Monthly. Inquire about cost and availability. Also available on CD-ROM and online. ISSN: 0003-6986.

Chemical Abstracts. Chemical Abstracts Service, 2540 Olentangy River Road, Box 3012, Columbus, OH 43210-0012. (614) 447-3600. (800) 848-6538. FAX (614) 447-3713. From 1907 to present. Weekly. $16800.00 per year. Also available on CD-ROM and online. Inquire regarding cost and availability. ISSN:

Current Contents: Physical, Chemical and Earth Sciences. Institute for Scientific Information, 3501 Market Street, Philadelphia, PA 19104. (215) 386-0100. FAX (215) 386-2291. From 1961 to present. Weekly. $442.00 per year. Also available on CD-ROM and online. Inquire regarding cost and availability. ISSN: 0163-2574.

Current Papers in Physics. Institute of Electrical Engineers, Michael Faraday House, Six Hill Way, Stevenage, Herts, SG1 2AY, England. Distributed by INSPEC/IEEE, Box 1331, 445 Hoes Lane, Piscataway, NJ 08855-1331. (908) 562-5549. 1966 to present. Fortnightly. $410.00 per year. ISSN: 0011-3786.

General Science Index. H.W. Wilson Company, 950 University Avenue, Bronx, NY 10452. (212) 588-8400. (800) 367-6770. FAX (718) 590-1617. From 1978 to present. Ten issues peryear; quarterly and annual cumulations. Service basis. Available on CD-ROM and online. Inquire regarding cost and availability. ISSN: 0162-1963.

Government Reports Announcements and Index. U. S. National Technical Information Service (NTIS), 5285 Port Royal Road, Springfield, VA 22161. (703) 487-4650. FAX (703) 321-8547. From 1968 to present. Annual. $630.00 per year. Also available online as NTIS Bibliographic Database and on CD-ROM.

Journal of Current Laser Abstracts. PennWell Publishing Co., 10 Tara Boulevard, Nashua, NH 03062-2801. (603) 891-9177. FAX (603) 891-0539. From 1964 to present. Monthly.$450.00 per year. ISSN: 0022-0264.

Mathematical Reviews: A Review Journal Covering the World Literature of Mathematical Research. American Mathematical Society, P.O. Box 1571, Annex Station, Providence, RI 02901-9930. (800) 556-7774 or (401) 455-4000. 1940 to present. Monthly. $4594.00 per year. Also available via network (MATHSCINET), online and on CD-ROM. Inquire regarding cost and availability. ISSN: 0025-5629.

Meteorological and Geoastrophysical Abstracts. American Meteorological Society, c/o Inforonics, Inc., 550 Newtown Road, Littleton, MA 01460. (508) 486-8976. FAX (508) 486-0027. Covers literature in environmental sciences, meteorology, astrophysics, hydrology, glaciology, and physical oceanography. From 1950 to present. Monthly. $950.00 per year. Also avail-

PHYSICS

Ency. of Physical Sciences and Engineering Info. Sources

able on CD-ROM and online. Inquire regarding cost and availability. ISSN: 0026-1130.

NTIS Alerts: Physics. U. S. National Technical Information Service. 5285 Port Royal Road, Springfield, VA 22161. (703) 487-4650. FAX (703) 321-8547. Weekly. $140.00 per year.

Pascal 10: Mechanique, Acoustique et Transport de Chaleur. Centre National de la Recherche Scientifique, Institue de L'information Scientifique et Technique, 2 Aliee Du Parc de Brabois, 54514 Vandoeuvre-les Nancy Cedex, France. TEC 83-50-46-00. FAX 83-50-46-50. 1984 to present. Ten issues per year. Also available on CD-ROM and online. ISSN: 1136-5107.

Physics Abstracts. INSPEC. Section A, Science Abstracts. Institution of Electrical Engineers (IEE). Available from INSPEC/IEEE-Institute of Electrical and Electronic Engineers,, Box 1331, 445 Hoes Lane, Piscaway NJ 08855-1331. (908) 562-5549. 1898 to present. 24 issues per year. $2835.00 per year. Also available on CD-ROM and online. ISSN: 0036-8091.

Physics Briefs (Physikalische Berichte). Information Center for Energy, Physics, Mathematics; German Physical Society. V C H Publishers, Inc., 220 East 23rd Street, New York, NY 10010-4606. (212) 683-8333. 1845 to present. 24 issues per year. $2390.00 per year. Also available online. ISSN: 0179-7434.

Science Citation Index (SCI). Institute for Scientific Information, 3501 Market Street, Philadelphia, PA 19104. (215) 386-0100. (800) 523-1850. FAX (215) 386-2991. 1961 to present. Six issues per year, plus annual cumulation. $11650.00 per year. Also available online and on CD-ROM. Inquire about price and availability. ISSN: 0036-827X.

Solid State and Superconductivity Abstracts: Covers theory, Production and Application of Solid State Materials. Cambridge Scientific Abstracts, 7200 Wisconsin Avenue, Bethesda, MD 20824. (301) 961-6750. FAX (301) 961-6720. 1957 to present. Bimonthly. $1320.00 per year. Also available online. ISSN: 0896-5900.

STAR (Scientific and Technical Aerospace Reports). U.S. National Aeronautics and Space Administration. Distributed by U. S. Superintendent of Documents, Washington, DC 20402. 1963 to present. Semi-monthly, with semiannual and annual indexes. $114.00 per year. ISSN: 0036-8741. Also available online and on CD-ROM.

ANNUAL REVIEWS AND YEARBOOKS

Advances in Applied Mechanics. Academic Press, Inc., 6277 Sea Harbor Drive, Orlando, FL. (800) 321-5068. 1948 to present. Irregular. ISSN: 0065-2156.

Advances in Atomic, Molecular and Optical Physics. Academic Press, Inc., 6277 Sea Harbor Drive, Orlando, FL. (800) 321-5068. 1965 to present. Irregular. ISSN: 1049-250X.

Advances in Electronics and Electron Physics. Academic Press, Inc., 6277 Sea Harbor Drive, Orlando, FL. (800) 321-5068. From 1948 to present. Irregular. ISSN: 0065-2539.

Advances in Chemical Physics. John Wiley & Sons, Inc., 605 Third Avenue, New York, NY 10158-0012. (212) 850-6000. FAX (212) 850-6099. 1958 to present. Irregular. Price varies. ISSN: 0065-2385.

Advances in Heat Transfer. Academic Press, Inc., 6277 Sea Harbor Drive, Orlando, FL. (800) 321-5068. 1964 to present. Irregular. ISSN: 0065-2717.

Advances in Nuclear Physics. Plenum Publishing Corp., 233 Spring Street, New York, NY. (212) 620-8000. (800) 2221-9369. FAX (212) 463-0742. 1968 to present. Irregular. Price varies. ISSN: 0065-2970.

Advances in Physics. Taylor & Francis, Ltd., Rankine Road, Basingstoke, Hants RG24 8PR, England. 0256-840366. FAX 0256-47938. 1952 to present. Bimonthly. $511.00 per year. ISSN: 0001-8732.

Advances in Solid State Technology. Kluwer Academic Publishers, 101 Philip Drive, Assinippi Park, Norwell, MA 02061. (617) 871-6600. 1985 to present. Irregular. Price varies.

Annual Review of Fluid Mechanics. Annual Reviews, Inc., 4139 El Camino Way, Palo Alto, CA 94303-0139. (415) 493-4400. (800) 523-8635. Fax (415) 855-9815. 1969 to present. Annual. $47.00 per year. ISSN: 0066-4189.

Annual Review of Nuclear and Particle Science. Annual Reviews, Inc., 4139 El Camino Way, Palo Alto, CA 94303-0139. (415) 493-4400. (800) 523-8635. Fax (415) 855-9815. 1952 to present. Annual. $62.00.

Progress in Optics. Elsevier Science Inc., 660 White Plains Road, Tarrytown, NY 10591-5153. (914) 524-9200. FAX (914) 333-2444. 1961 to present. Irregular. Price varies. ISSN: 0079-6638.

Solid State Physics: Advances in Research and Applications. Academic Press, Inc., 6277 Sea Harbor Drive, Orlando, FL. (800) 321-5068. 1955 to present. Irregular. ISSN: 0081-1947.

Topics in Applied Physics. Springer-Verlag New York, Inc., 175 Fifth Avenue, New York, NY 10010. (212) 460-1500. (800) 777-4643. FAX (212) 473-6272. ISSN: 0303-4216.

ASSOCIATIONS AND PROFESSIONAL SOCIETIES

American Astronomical Society. 2000 Florida Avenue NW, Suite 400, Washington, DC 20009. (202) 328-2010. FAX: (202) 234-2560.

American Geophysical Union. 2000 Florida Avenue, NW, Washington, DC 20009. (202) 462-6900. (800) 966-AGU1. FAX(202) 328-0566.

American Institute of Physics. One Physics Ellipse, College Park, MD 20740-3843. (301) 209-3100.

American Physical Society. One Physics Ellipse, College Park, MD 20740-3843. (301) 209-3200. FAX (301) 209-0865.

Universities Research Association, 1111 19th Street Nw, Suite 400, Washington, DC 20036. (202) 293-1382.

BIBLIOGRAPHIES

Guide to the Archival Collectins in the Niels Bohr Libary at the American Institute of Physics. Niels Bohr Library Staff. American Institute of Physics, One Physics Ellipse, College Park, MD 20740-3843. (301) 209-3100. 1995. $140.00

Science Books and Films. American Association for the Advancement of Science, 1333 H Street NW, Washington, DC 20005. (202) 326-6454. Reviews of print, film and software materials in all sciences. 1965 to present. Nine issues per year. $40.00 per year. ISSN: 0098-342X

Scientific and Technical Books and Serials in Print: An Index to Literature in Science and Technology. R.R. Bowker Inc., 121 Chanlon Road, New Providence, NJ 07974. (908) 464-6800. (800) 521-8110. 4 volumes. Annual. 1972 to present. $299.95. Also available on compact disc and online. ISSN: 0000-054X

DIRECTORIES AND BIOGRAPHICAL SOURCES

American Association of Physics Teachers. Directory of members. American Association of Physics Teachers, 5112 Berywn Road, College Park, MD 20740. (301) 345-4200.

American Men and Women of Science: Physical and Biological Sciences. R. R. Bowker Inc., 121 Chanlon Road, New Providence, NJ 07974. (908) 464-6800. (800) 521-8110. 20th edition. 8 volumes. 1996. $850.00.

American Physical Society. Membership Directory Bulletin Issue. American Physical Society, One Physics Ellipse, College Park, MD 20740-3843. (301) 209-3200. FAX (301) 209- 0865. Biennial. $50.00.

The Biographical Dictionary of Scientists: Physicists. D. Abbott, editor. Peter Bedrick Books, Inc., 2112 Broadway, Room 318, New York, NY 10023. (212) 496-0751. 1984.

Directory of Physics and Astronomy Staff Members. American Institute of Physics, One Physics Ellipse, College Park, MD 20740-3843. (301) 209-3100. Annual. $45.00.

Graduate Programs in Physics, Astronomy and Related Fields. 1995-1996. American Institute of Physics, One Physics Ellipse, College Park, MD 20740-3843. (301) 209-3100. Annual. $45.00. ISSN: 0147-1821.

Legends in Their Own Time: A Century of American Physical Scientists. Anthony Serafini. Plenum Publishing Corp., 233 Spring Street, New York, NY. (212) 620-8000. (800) 2221-9369. FAX (212) 463-0742.. 1993. $27.50.

The Nobel Prize Winners. Frank N. Magill, editor. Salem Press, Inc., P.O. Box 1097, Englewood Cliffs, NJ 07632. (201) 871-3700. (800) 221-1592. 3 volumes. 1989. $210.00.

Research Centers Directory. Gale Research, 835 Penobscot Building, Detroit, MI 48226-4094. (313) 961-2242. (800) 347-4253. 20th edition. 1995. $485.00. ISSN: 0080-1518.

Scientific and Technical Organizations and Agencies Directory. Gale Research,835 Penobscot Building, Detroit, MI 48226-4094. (313) 961-2242. (800) 347-4253. 4th edition. 1996. $195.00.

World Guide to Scientific Associations and Learned Societies. K. G. Saur Inc., 121 Chanlon Road, New Providence, NJ 07974. (908) 464-6800. (800) 521-8110. 5th edition. 1990. $225.00.

ENCYCLOPEDIAS AND DICTIONARIES

A Concise Dictionary of Physics. Oxford University Press, Inc., 200 Madison Avenue, New York, NY 10016. (212) 725-6000. (800) 334-4249. 1990. $10.95.

Dictionary of Effects and Phenomena in Physics. Joachim Schubert. VCH Publications, Inc., 220 East 23rd Street, Suite 909, New York, NY 10010-4606. (212) 683-8333. (800) 422-8824. 1987. $35.00.

Dictionary of the Physical Sciences: Terms, Formulas, Data. Cesare Emiliani. Oxford University Press, Inc., 200 Madison Avenue, New York, NY 10016. (212) 725-6000. (800) 334-4249. 1989. $19.95.

Encyclopedia of Applied Physics. George Trigg, editor. VCH Publications, Inc., 220 East 23rd Street, Suite 909, New York, NY 10010-4606. (212) 683-8333. (800) 422-8824. 20 volume set. 1991-. $5990.00.

Encyclopedia of Physics. Robert M. Besancon. Chapman & Hall, 1 Penn Plaza, New York, NY 10119. (212) 564-1060. 3rd edition. 1990. $54.95 in paper.

Facts on File Dictionary of Physics. John Daintith, editor. Facts-on-File, Inc., 460 Park Avenue South, New York, NY 10016-7382. (212) 683-2244. (800) 322-8755. Fax (800) 683-3633. Revised edition. 1990. $12.95.

Encyclopedia of Physical Science and Technology. Academic Press, Inc., 6277 Sea Harbor Drive, Orlando, FL. (800) 321-5068. 2nd edition. 18 volume set. 1992. $2625.00.

Landolt-Boernstein Numerical Data and Functional Relationships in Science and Technology: Nuclear and Particle Physics. K. H. Hellwege editor. Springer Verlag New York, Inc., 175 Fifth Avenue, New York, NY 10010. (212) 460-1500. (800) 777-4643. FAX (212) 473-6272. 1993. Volumes priced separately.

McGraw-Hill Encyclopedia of Science and Technology. McGraw-Hill Book Company, Inc., 1221 Avenue of the Americas, New York, NY 10020. (212) 512-2000. (800) 262-4729. 7th edition. 20 volume set. 1992. $1900.00.

Penguin Dictionary of Physics. Valerie Illingworth, editor. Viking Penguin, 375 Hudson Street, New York, NY 10014-3657. (212) 366-2000. (800) 331-4624. 2nd edition, 1991.

GENERAL WORKS

Albert Einstein, Three Complete Books: Relativity, Ideas and Opinions, and Out of My Later Years. Albert Einstein. Random House Inc., 201 East 50th Street, New York, NY 10022. (212) 751-2600. (800) 733-3000. 1991. $19.50. 1993. $7.99 in paper.

Big Science: The Growth of Large-Scale Research. PeterGalison and Bruce Hevly, editors. Stanford University Press, Stanford, CA 04305-2235. (413) 723-1593. 1992. $47.50.

Classical General Relativity. Subrahmanyan Chandrasekhar and Morton D. Hull, editors. Oxford University Press, Inc., 200 Madison Avenue, New York, NY 10016. (212) 725-6000. (800) 334-4249. 1993. $45.00.

PHYSICS

Ency. of Physical Sciences and Engineering Info. Sources

Cosmic Catastrophies. C. P. Chapman and D. Morrison, editors. Plenum Publishing Corp., 233 Spring Street, New York, NY. (212) 620-8000. (800) 2221-9369. 1989. $22.95.

Dialogues Concerning Two New Sciences. Galilei Galileo. Prometheus Books, 89 John Glenn Drive, Amherst, NY 14228-2197. (716) 691-0133. (800) 421-0351. 1991, $9.95 in paper.

Doing Physics: How Physicists Take Hold of the World. Martin Krieger. Indiana University Press, 601 North Morton Street, Bloomington, IN 47404-3797. (812) 855-6804. (800) 842-6796. 1992. $29.95.

Frontiers of Fundamental Physics. M. Barone and F. Selleri, editors. Plenum Publishing Corp., 233 Spring Street, New York, NY. (212) 620-8000. (800) 2221-9369. 1994. $145.00.

Fractals in Physics: Essays in Honor of Benoit B. Mandelbrot. A. Aharony and Jens Feder, editors. Elsevier Science Publishing Company, Inc., 655 Avenue of the Americas, New York, NY 10010. (212) 989-5800. FAX (914) 333-2444. 1990. $45.25 in paper.

Fundamental Particles and Forces. Richard A. Carrigan, Jr., and W. Peter Trower. W. H. Freeman & Co., 41 Madison Avenue, East 26th Avenue, 35th Floor, New York, NY 10010. (212) 576-0400. 1995. $13.95.

Gravitation. Charles W. Misner, et al. W. H. Freeman & Co., 41 Madison Avenue, East 26th Avenue, 35th Floor, New York, NY 10010. (212) 576-0400. 1995.

Landmark Experiments in Twentieth Century Physics. George L. Trigg. Dover Publications, Inc., 180 Varick Street, New York, NY 10014. (212) 255-3755. (800) 223- 3130. 1995. $9.95 in paper.

Mathematics for Physicists. Robert L. Zimmerman and Frederick I. Olness. Addison-Wesley Publishing Co., Inc., 1 Jacob Way, Reading, MA 01867. (617) 944-3700. (800) 447-2226. 1995. $43.25 in paper.

Matter and Energy: Physics in Everyday Life. John Clark, editor. Oxford University Press, Inc., 200 Madison Avenue, New York, NY 10016. (212) 725-6000. (800) 334-4249. 1995. $35.00.

Mind, Matter, and Quantum Mechanics. Henry P. Stapp. Springer-Verlag New York, Inc., 175 Fifth Avenue, New York, NY 10010. (212) 460-1500. (800) 777-4643. FAX (212) 473-6272. 1994. $34.50.

Physics. David Haliday and Robert Resnick. John Wiley & Sons, Inc., 605 Third Avenue, New York, NY 10158-0012. (212) 850-6000. (800) 225-5945. 2 volumes. 4th edition. 1992.

Physics for Poets. Robert H. March. McGraw Publishing Company, Inc., 1221 Avenue of the Americas, New York, NY 10020. (212) 512-2000. (800) 262-4729. 3rd edition. 1992.

Rainbows, Snowflakes, and Quarks: Physics and the World Around Us. Hans C. Von Baeyer. Random House, Inc., 201 East 50th Street, New York, NY 10022. (212) 751-2600. (800) 733-3000. 1991. $19.50. 1993. $12.00 in paper.

Statistical Physics and the Atomic theory of Matter, From Boyle and Newton to Landau and Onsager. Stephen G. Brush.

Princeton University Press, 41 William Street, Princeton, NJ 08540. (609) 258-4900. (800) 777-4726. 1983. $72.50.

Visit to a Small Universe. Virginia Trimble. American Institute of Physics, One Physics Ellipse, College Park, MD 20740-3843. (301) 209-3100. 1992. $24.95.

HANDBOOKS AND MANUALS

Chemical and Physical Data. Arthur M. James and M. P. Lord, Van Nostrand Reinhold, 115 Fifth Avenue, New York, NY 10003. (212) 254-3232. (800) 842-3636. 1992. $59.95.

CRC Handbook of Chemistry and Physics. David R. Kide, editor. CRC Press, Inc., 2000 Corporate Boulevard, NW, Boca Raton, FL 33431. (407) 994-0555. (800) 272-7737. 77th edition. 1996. $99.95.

Handbook of Physical Quantities. Igor S. Grigoriev and Evgenil Z. Melikhov, editors. CRC Press, Inc., 2000 Corporate Boulevard, NW, Boca Raton, FL 33431. (407) 994-0555. (800) 272-7737. 1995. $99.00.

Handbook of Space Astronomy and Astrophsyics. Martin V. Zumbeck. Cambridge University Press, 40 West 20th Street, New York, NY 10011-4211. (212) 924-3900. (800) 872-7423. 2nd edition. 1990. $79.95.

Landolt-Borenstein Numerical Data and Functional Relationships in Science and Technology: Nuclear Particle and Physics. Group I. Springer-Verlag New York, Inc., 175 Fifth Avenue, New York, NY 10010. (212) 460-1500. (800) 777-4643. volumes priced individually.

McGraw-Hill Encyclopedia of Physics. Sybil P. Parker, editor. McGraw-Hill Publishing Company, Inc., 1221 Avenue of the Americas, New York, NY 10020. (212) 512-2000. (800) 262-4729. 2nd edition. 1993. $95.50

Newnes Engineering and Physical Science Pocketbook. J. O. Bird and P. J. Chivers. Butterworth-Heinemann, 313 Washington Avenue, Newton, MA 02158. (617) 928-2500. (800) 366-2665. 1993. $49.95.

A Physicist's Desk Reference. Herbert L. anderson, editor. American Institute of Physics, One Physics Ellipse, College Park, MD 20740-3843. (301) 209-3100. 2nd edition. 1989. $70.00.

The Physics of Sports. Angelo Armenti, Jr., editor. American Institute of Physics, One Physics Ellipse, College Park, MD 20740-3843. (301) 209-3100. 1992. $35.00.

Physics Problem Solver. Research and Education Association Staff. Research & Education Association, 61 Ethel Road, West Piscataway, NJ 08854. (908) 819-8880. Revised edition. 1994. $23.95 in paper.

Tables of Physical and Chemical Constants and Some Mathematical Functions. G. W. Kaye and T. M. Laby. Halsted Press 605 Third Avenue, New York, 10158-0012. (212) 850-6400. 15th edition. 1986. $74.95.

ONLINE DATABASES AND CD-ROMS

CA Search. Chemical Abstracts Service, P.O. Box 3012, Columbus, OH 43210-0012. (614) 447-3600. (800) 848-6533. FAX

(614) 447-3709. Very comprehensive guide to worldwide chemical literature and related fields, 1972 to present. Available on BRS,(800) 289-4277, DIALOG, (800) 334-2564, ORBIT (800) 456-7248, and STN International, FIZ Karlsruhe, P.O. Box 2465, W-7500, Karlsruhe 1, Germany, online services. Inquire as to cost and availability.

Current Contents Search. Institute for Scientific Information, 3501 Market Street, Philadelphia, PA 19104. (215) 386-0100. FAX (215) 386-6362. Contains citations to articles listed in the table of contents of science and technology journals. Also articles in social sciences and life sciences journals. Available on BRS, (800) 289-4277, DIALOG, (800) 334-2564, online services. Inquire as to cost and availability.

Dissertation Abstracts Online. University Microfilms International, 300 North Zeeb Road, Ann Arbor, MI 48106. (800) 521-0600 or (313) 761-4700. Scope includes virtually all doctoral dissertations accepted at accredited American institutions from 1861 to present in 252 subject areas. Available on BRS, (800) 289-4277, DIALOG, (800) 334-2564, and OCLC EPIC, (800) 848-5878, online services. Also available on CD-ROM. Inquire as to cost and availability.

INSPEC. Institution of Electrical Engineers, Michael Faraday House, Six Hills Way, Stevenage, Herts. SG1 2AY, England. Telephone: 0438 313311 or FAX 0438 742840. Contains citations to the worldwide literature of physics, electronics and electrical engineering, computer technology, and related fields. Available on BRS, (800) 289-4277, DIALOG, (800) 334-2564, ORBIT, (800) 456-7248, and STN International, FIZ Karlsruhe, P.O. Box 2465, W-7500, Karlsruhe 1, Germany, online services. Inquire as to cost and availability.

International Aerospace Abstracts. American Institute of Aeronautics and Astronautics, 370 L'Enfant Promenade, S.W., Washington, DC 20024. (202) 646-7400. Contains references and abstracts of journal and monograph literature relating to aerospace science and technology, from 1963 to present. Available through the NASA/RECON system of the National Aeronautics and Space Administration only.

Mathsci. American Mathematical Society, P.O. Box 6248, Providence, RI 02940. (800) 321-4667 or (401) 455-4000 or FAX (401) 331-3842. Scope includes pure and applied mathematics and related areas of physics, statistics, engineering, computer science, and operations research literature since 1959. Available on DIALOG,(800) 334-2564, online service. Also available on CD-ROM. Inquire as to cost and availability.

NTIS Bibliographic Database. National Technical Information Service, 5285 Port Royal Road, Springfield, VA 22161. (703) 487-4929 or FAX (703) 321-8199. Broad coverage of government-sponsored science and technology research reports, 1964 to present. Available on BRS,(800) 289-4277, DIALOG, (800) 334-2564, ORBIT, (800) 456-7248, and STN International, FIZ Karlsruhe, P.O. Box 2465, W-7500, Karlsruhe 1, Germany, online services. Also available on CD-ROM. Inquire as to cost and availability.

Physics Briefs. American Institute of Physics, 335 East 45th Street, New York, NY 10017. (212) 661-9260 or FAX (212) 661-2036. Contains citations with abstracts of the literature of physics and related fields, 1979 to present. Available on the STN International, FIZ Karlsruhe, P.O. Box 2465, W-7500, Karlsruhe 1, Germany, online service. Inquire as to cost and availability.

Scientific and Technical Books and Serials in Print. R.R. Bowker Inc., 121 Chanlon Road, New Providence, NJ 07974. (800) 521-

8110 or (908) 464-6800. List of currently published books and serials in the physical and biological sciences, engineering and technology, with subject, author and titles indexes. Available on ORBIT, (800) 456-7248, online service. Inquire as to cost and availability.

SCISEARCH. Institute for Scientific Information, 3501 Market Street, Philadelphia, PA 19104. (800) 523-1850 or (215) 386-0100. Broad multidisciplinary title and author index to the international literature of science and technology, 1974 to present. Available on DIALOG, (800) 334-2564, and ORBIT, (800) 456-7248, online services. Also available on CD-ROM. Inquire as to cost and availability.

WILSONLINE. H.W. Wilson Company, 950 University Avenue, Bronx, NY 10452. (800) 367-6770 or (212) 588-8400. Makes available online versions of the printed H.W. Wilson indexes including *Applied Science and Technology Index, Business Periodicals Index, General Science Index,* and *Readers' Guide to Periodical Literature.* Period covered is generally 1983 to present. Available on BRS, (800) 289-4277, DIALOG, (800) 334-2564, and OCLC EPIC, (800) 848-5878, online services. Also available on CD-ROM. Inquire as to cost and availability.

PERIODICALS

American Journal of Physics. American Association of Physics Teachers, 5112 Berywn Road, College Park, MD 20740. (301) 345-4200. 1933 to present. Monthly. $215.00 per year. ISSN: 0002-9505.

Applied Physics Letters. American Institute of Physics, One Physics Ellipse, College Park, MD 20740-3843. (301) 209-3000. 1962 to present. Weekly. $1060.00 per year. ISSN: 0003-6951.

Bioimaging. I O P Publishing. U. S. subscriptions to: American Institute of Physics, 500 Sunnyside Boulevard, Woodbury, NY 11797-2999. (516) 575-2270. 1993 to present. Quarterly. $236.00 per year. ISSN: 0966-9051.

Canadian Journal of Physics. National Research Council of Canada. Research Journals, Ottawa K1A OR6, Canada. (613) 993-9084. FAX (613) 952-7656. 1929 to present. Monthly. $285.00 per year. ISSN: 0008-4204.

Chaos; An Interdisciplinary Journal of Nonlinear Science. American Institute of Physics, One Physics Ellipse, College Park, MD 20740-3843. (301) 209-3000. 1991 to present. Quarterly. $225.00 per year. ISSN: 1054-1500.

Contemporary Physics. Taylor & Francis, Ltd, Rankine Road, Basingstoke, Hants RG24 8PR, England. TEL 0256-840366. FAX 0256-479438. 1959 to present. Bimonthly. $318.00 per year. ISSN: 0010-7514.

Critical Reviews in Solid State and Materials Sciences. CRC Press, Inc., 2000 Corporate Boulevard, NW, Boca Raton, FL 33431. (407) 994-0555. (800) 272-7737. FAX (407) 998-9784. 1970 to present. Bimonthly. $265.00 per year. ISSN: 1040-8436.

Distributed Systems Engineering. I O P Publishing. U. S. subscriptions to: American Institute of Physics, 500 Sunnyside Boulevard, Woodbury, NY 11797-2999. (516) 575-2270. 1993 to present. Quarterly. $305.00 per year. ISSN: 0967-1846.

European Journal of Physics. I O P Publishing. U. S. subscriptions to: American Institute of Physics, 500 Sunnyside Boulevard, Woodbury, NY 11797-2999. (516) 575- 2270. 1980 to present. Bimonthly. $339.00 per year. ISSN: 0143-0807.

Franklin Institute. JOURNAL. Elsevier Science, 660 White Plains Road, Tarrytown, NY 10591-5153. (914) 524-9200. FAX (914) 333-2444. 1826 to present. Six issues per year. $740.00 per year. ISSN: 0016-0032.

General Relativity and Gravitation. Plenum Publishing Corp., 244 Spring Street, New York, NY 10013-1578. (212) 620-8000. FAX (212) 463-0742. 1970 to present. Monthly. $625.00.

Institut Henri Poincare. Annales. Section A: Theoretical Physics. Gauthier-Villars, 15 rue Gossin, 92543 Montrouge Cedex, France. TEL 33-140-92-65-00. FAX 331-40-92-65-97. 1930 to present. Eight issues per year. 2000F per year. ISSN: 0020-2339.

International Journal of Modern Physics. World Scientific Publishing Co., 1060 Main Street, Suite 18, River Edge, NJ 07661. (800) 227-7562. 1986 to present. 32 issues per year. $1710.00 per year. ISSN: 0217-751X.

Journal of Applied Physics. American Institute of Physics, One Physics Ellipse, College Park, MD 20740-3843. (301) 209-3000. 1931 to present. Semimonthly. $1655.00 per year. ISSN: 0021-8979.

Journal of Physics. IOP Publishing. U. S. orders to: American Institute of Physics, 500 Sunnyside Boulevard, Woodbury, NY 11797-2999. (516) 576-2200. 1968 to present. Twenty-four issues per year. $2487.00 per year. ISSN: 0305-4470.

Landolt-Boernstein, Zahlenwerte und Funktionen Aus Naturwissenschaften und Technik. Group 3: Crystal Physics. Springer-Verlag New York, Inc., 175 Fifth Avenue, New York, NY 10010. (212) 460-1500. FAX (212) 473-6272. 1965. Irregular. Price varies. ISSN: 0075-7896.

Letters in Mathematical Physics. Kluwer Academic Publishers, Box 358 Accord Station, Hingham, MA 02018-0358. (617) 871-6600. FAX (617) 871-6528. 1975 to present. Monthly. $549.00 per year. ISSN: 0377-9017.

Nanotechnology. IOP Publishing. U. S. orders to: American Institute of Physics, 500 Sunnyside Boulevard, Woodbury, NY 11797-2999. (516) 576-2200. 1990 to present. Quarterly. $358.00 per year. ISSN: 0957-4484.

Philosophical Magazine. Parts A & B. Taylor & Francis Ltd., Rankine Road, Basingstoke, Hants RG24 8PR, England. TEL 0256-840366. FAX 0256-479438. 1798 to present. Monthly. $1168.00 per year. ISSN: 0031-8086.

Physical Review. American Institute of Physics, One Physics Ellipse, College Park, MD 20740-3843. (301) 209-3000. 1970 to present. Monthly. $990.00 per year. ISSN: 1050-2947.

Physics and Chemistry of the Earth. Elsevier Science Inc., 660 White Plains Road, Tarrytown, NY 10591-5153. (914) 524-9200. FAX (914) 333-2444. 1956 to present. Bi- monthly. $550.00 per year. ISSN: 0079-1946.

Physics Reports: A Review Section of Physics Letters. Elsevier Science Publishing Company, Inc., 655 Avenue of the Americas, New York, NY 10010. (212) 989-5800. FAX (212) 633-3990. 1971 to present. 72 issues per year in 12 volumes. $2342.00 per year. ISSN: 0370-1573.

Physics Today. American Institute of Physics, One Physics Ellipse, College Park, MD 20740-3843. (301) 209-3000.1948 to present. Monthly. $140.00 per year. ISSN: 0031-9228.

Progress in Particle and Nuclear Physics. Elsevier Science Publishing Company, Inc., 655 Avenue of the Americas, New York, NY 10010. (212) 989-5800. FAX (914) 333-2444. 1977 to present. Two issues per year. $660.00 per year. ISSN: 0146-6410.

Progress in Quantum Electronics. Elsevier Science Inc., 660 White Plains Road, Tarrytown, NY 10591-5153. (914) 524-9200. FAX (914) 333-2444. 1969 to present. 5 issues per year. $395.00 per year. ISSN: 0079-6727.

Reports On Progress in Physics. IOP Publishing. U. S. orders to: American Institute of Physics, 500 Sunnyside Boulevard, Woodbury, NY 11797-2999. (516) 576-2200. 1934 to present. Monthly. $1214.00 per year. ISSN: 0034-4885.

Reviews of Modern Physics. American Institute of Physics, One Physics Ellipse, College Park, MD 20740-3843. (301) 209-3000. 1929 to present. Quarterly. $300.00 per year. ISSN: 0034-6861.

Royal Society of London, Philosophical Transactions. Series A, Physical Sciences and Engineering. Royal Society of London, 6 Carlton House Terrace, London SW1Y 5AG, England. TEL 071-839-5561. FAX 071-976-1837. 1665 to present. Monthly. L485 per year. ISSN: 0962-8428.

PIEZOELECTRICITY

See also: CRYSTALLOGRAPHY, ELECTRICITY, SOLID STATE
PHYSICS

ABSTRACT SERVICES AND INDEXES

Applied Science and Technology Index: A Cumulative Subject Index to English Language Periodicals in the Fields of Aeronautics and Space Science, Computer Technology, Chemistry, Construction Industry, Energy and Related Areas. H.W. Wilson Co., 950 University Avenue, Bronx, NY 10452. (212) 588-8400. (800) 367-6770. FAX (718) 590-1617. From 1958 to present. Monthly. Inquire about cost and availability. Also available on CD-ROM and online. ISSN: 0003-6986.

Chemical Abstracts. Chemical Abstracts Service, 2540 Olentangy River Road, Box 3012, Columbus, OH 43210-0012. (614) 447-3600. (800) 848-6538. FAX (614) 447-3713. From 1907 to present. Weekly. $16800.00 per year. Also available on CD-ROM and online. Inquire regarding cost and availability. ISSN:

Current Contents: Engineering, Technology and Applied Sciences. Institute for Scientific Information, 3501 Market Street, Philadelphia, PA 19104. (215) 386-0100. FAX (215) 386-6362. 1970 to present. Weekly. $442.00 per year. Also available on CD-ROM and online. Inquire regarding cost and availability. ISSN: 0095-7917.

Current Papers in Electrical and Electronics Engineering. Institution of Electrical Engineers (IEE), London. Distributed by INSPEC/IEEE, Box 1331, 445 Hoes Lane, Piscataway, NJ 08855-1331. (908) 562-5549. 1969 to present. Monthly. $345.00 per year. ISSN: 0011-3778.

Electrical and Electronics Abstracts. Institution of Electrical Engineers (IEE), London. Available from INSPEC/IEEE- Institute of Electrical and Electronic Engineers, Box 1331, Hoes Lane, Piscataway, NJ 08855-1331. (908) 562-5549. 1898 to present. Monthly. $2200.00 per year. Also available on CD-ROM and online as INSPEC.

Engineering Index Monthly. Indexes and abstracts the world's engineering and technical literature. Engineering Information, Inc., Castle Point on the Hudson, Hoboken, NJ 07030. (201) 216-8500. (800) 221-1044. FAX (201) 216-8532. Monthly. $2300.00 per year. Available online as COMPENDEX and also on CD-ROM. ISSN: 0742-1974.

General Science Index. H.W. Wilson Company, 950 University Avenue, Bronx, NY 10452. (212) 588-8400. (800) 367-6770. FAX (718) 590-1617. From 1978 to present. Ten issues per year; quarterly and annual cumulations. Service basis. Available on CD-ROM and online. Inquire regarding cost and availability. ISSN: 0162-1963.

Physics Abstracts. INSPEC. Section A, Science Abstracts. Institution of Electrical Engineers (IEE). Available from INSPEC/IEEE-Institute of Electrical and Electronic Engineers,, Box 1331, 445 Hoes Lane, Piscaway NJ 08855-1331. (908) 562-5549. 1898 to present. 24 issues per year. $2835.00 per year. Also available on CD-ROM and online. ISSN: 0036-8091.

Physics Briefs (Physikalische Berichte). Information Center for Energy, Physics, Mathematics; German Physical Society. VCH Publishers, Inc., 220 East 23rd Street, New York, NY 10010-4606. (212) 683-8333. 1845 to present. 24 issues per year. $2390.00 per year. Also available online. ISSN: 0179-7434.

Science Citation Index (SCI). Institute for Scientific Information, 3501 Market Street, Philadelphia, PA 19104. (215) 386-0100. (800) 523-1850. FAX (215) 386-2991. 1961 to present. Six issues per year, plus annual cumulation. $11650.00 per year. Also available online and on CD-ROM. Inquire about price and availability. ISSN: 0036-827X.

Solid State and Superconductivity Abstracts. Covers theory, production and application of solid state materials. Cambridge Scientific Abstracts, 7200 Wisconsin Avenue, Bethesda, MD 20824. (301) 961-6750. FAX (301) 961-6720. 1957 to present. Bimonthly. $1320.00 per year. Also available online. ISSN: 0896-5900.

ANNUAL REVIEWS AND YEARBOOKS

Advances in Electronics and Electron Physics. Academic Press, Inc., 6277 Sea Harbor Drive, Orlando, FL 32887. (800) 321-5068. From 1948 to present. Irregular. Price varies, inquire. ISSN: 0065-2539.

Critical Reviews in Solid State and Materials Sciences. CRC Press, Inc., 2000 Corporate Boulevard, NW, Boca Raton, FL 33431. (407) 994-0555. (800) 272-7737. FAX (407) 998-9784. 1970 to present. Bimonthly. $265.00 per year. ISSN: 1040-8436.

Progress in Crystal Growth and Characterization. Elsevier Science Inc., 660 White Plains Road, Tarrytown, NY 10591-5153. (914) 524-9200. FAX (914) 333-2444. 1978 to present. 8 issues per year. $1005.00 per year. ISSN: 0960-8974.

Progress in Quantum Electronics. Elsevier Science Inc., 660 White Plains Road, Tarrytown, NY 10591-5153. (914) 524-9200.

FAX (914) 333-2444. 1969 to present. 5 issues per year. $395.00 per year. ISSN: 0079-6727.

Solid State Physics: Advances in Research and Applications. Academic Press, Inc., 6277 Sea Harbor Drive, Orlando, FL. (800) 321-5068. 1955 to present. Irregular. ISSN: 0081-1947.

ASSOCIATIONS AND PROFESSIONAL SOCIETIES

American Institute of Physics. One Physics Ellipse, College Park, MD 20740-3843. (301) 209-3100.

IEEE (Institute of Electrical and Electronic Engineers). 345 East 47th Street. New York, NY 10017. (212) 705-7900. FAX (212) 705-4929.

IEEE Ultrasonics, Ferroelectrics and Frequency Control Society. c/o Gary K. Montress, Raytheon Research Division, 131 Spring Street, Lexington, MA 02173. (617) 860-3053. FAX(617) 860-3194.

BIBLIOGRAPHIES

Science Books and Films. American Association for the Advancement of Science, 1333 H Street NW, Washington, DC 20005. (202) 326-6454. Reviews of print, film and software materials in all sciences. 1965 to present. Nine issues per year. $40.00 per year. ISSN: 0098-342X

Scientific and Technical Books and Serials in Print: An Index to Literature in Science and Technology. R.R. Bowker Inc., 121 Chanlon Road, New Providence, NJ 07974. (908) 464-6800. (800) 521-8110. 4 volumes. Annual. 1972to present. $299.95. Also available on compact disc and online. ISSN: 0000-054X

Sourcebook of Pyroelectricity. Sidney B. Lang. Gordon & Breach Science Publishers, Inc., P.O.Box 200029, Riverfront Plaza Station, Newark, NJ 07102-0301. (201) 643-7500.1974. $303.00.

DIRECTORIES AND BIOGRAPHICAL SOURCES

American Men and Women of Science: Physical and Biological Sciences. R. R. Bowker Inc., 121 Chanlon Road, New Providence, NJ 07974. (908) 464-6800. (800) 521-8110. 20th edition. 8 volumes. 1996. $850.00.

Engineering Research Centers: Incorporating Electronics Research Centers. Stockton Press, 345 Park Avenue, New York, NY 10010. (212) 689-9200. (800) 221-2123. 4th edition. 1995. $515.00.

IEEE Membership Directory. Institute of Electrical and Electronics Engineers, IEEE Service Center, 445 Hoes Lane, Piscataway, NY 08854. (908) 981-1393. (800) 678-IEEE. FAX (908) 981-9667. 2 volumes. Annual. $190.00. ISSN:

Research Centers Directory. Gale Research, 835 Penobscot Building, Detroit, MI 48226-4094. (313) 961-2242. (800) 347-4253. 20th edition. 1995. $485.00. ISSN: 0080- 1518.

Scientific and Technical Organizations and Agencies Directory. Gale Research, 835 Penobscot Building, Detroit, MI 48226-4094. (313) 961-2242. (800) 347-4253. 4th edition. 1996. $195.00.

PIEZOELECTRICITY

Ency. of Physical Sciences and Engineering Info. Sources

Who's Who in Engineering. Gordon Davis, editor. American Association of Engineering Societies. 1111 19th Street, NY, Suite 608, Washington, DC 20036. (202) 296-2237. (800) 658-8897. 9th edition. 1995. $220.00.

Who's Who in Technology. Gale Research, 835 Penobscot Building, Detroit, MI 48226-4094. (313) 961-2242. (800) 347-4253. 7th edition. 1995. $195.00. ISBN 0-8103-7467-6.

ENCYCLOPEDIAS AND DICTIONARIES

Concise Encyclopedia of Solid State Physics. Rita G. Lerner and George L. Trigg, editors. Addison-Wesley Publishing Co., Inc., 1 Jacob Way, Reading, MA 01867. (617) 944-3700. (800) 447-2226. 1983. $40.95 in paper.

Encyclopedia of Applied Physics. George Trigg, editor. VCH Publications, Inc., 220 East 23rd Street, Suite 909, New York, NY 10010-4606. (212) 683-8333. (800) 422-8824. 20 volume set. 1991-. $5990.00.

Encyclopedia of Physical Science and Technology. Academic Press, Inc., 6277 Sea Harbor Drive, Orlando, FL. (800) 321-5068. 2nd edition. 18 volume set. 1992. $2625.00.

IEEE Standard Dictionary of Electrical and Electronics Terms. Christopher J. Booth, editor. IEEE Service Center, 445 Hoes Lane, Piscataway, NJ 08855-1331. (908) 981-1393. (800) 678-IEEE. FAX (908) 981-9667. IEEE Standard 100-1992. 5th edition. 1993. $150.00.

McGraw-Hill Electronics Dictionary. J. Markus. McGraw-HillBook Company, Inc., 1221 Avenue of the Americas, New York, NY 10020. (212) 997-3675. 5th edition. 1994. $49.95.

McGraw-Hill Encyclopedia of Science and Technology. McGraw-Hill Book Company, Inc., 1221 Avenue of the Americas, New York, NY 10020. (212) 512-2000. (800) 262-4729. 7th edition. 20 volume set. 1992. $1900.00.

GENERAL WORKS

Electronic Properties of Materials: an Introduction for Engineers. Rolf E. Hummel. Springer-Verlag New York, Inc., 175 Fifth Avenue, New York, NY 10010. (212) 460-1500. (800) 777-4643. FAX (212) 473-6272. 2nd edition. 1994. $49.00.

Fundamentals of Piezoelectricity. T. Ikeda. Oxford University Press, Inc., 200 Madison Avenue, New York, NY 10016. (212) 725-6000. (800) 334-4249. 1994. $82.00. 1990. $98.00.

An Introduction to Ultrasonic Motors. T. Sashida and T. Kenjo. Oxford University Press, Inc., 200 Madison Avenue, New York, NY 10016. (212) 725-6000. (800) 334-4249. 1994. $82.00.

Piezoelectric Resonators and Their Applications. J. Zelenka. Elsevier Science Publishing Company, Inc., 655 Avenue of the Americas, New York, NY 10010. (212) 989-5800. FAX (914) 333-2444. 1986. $116.00.

Piezoelectricity. C. Rosen, et al, editors. American Institute of Physics. One Physics Ellipse, College Park, MD 20740-3843. (301) 209-3100. 1992. $125.00.

Solid State Physics for Engineering and Materials Science. John P. McKelvey. Krieger Publishing Company, P.O. Box 9542, Melbourne, FL 32902-9542. (407) 724-9542. 1993. $49.50.

Theory of Piezoelectric Plates and Shells. Rogacheva. CRC Press, Inc., 2000 Corporate Boulevard, NW, Boca Raton, FL 33431. (407) 994-0555. (800) 272-7737. 1993.

HANDBOOKS AND MANUALS

Chemical and Physical Data. Arthur M. James and M. P. Lord, Van Nostrand Reinhold, 115 Fifth Avenue, New York, NY 10003. (212) 254-3232. (800) 842-3636. 1992. $59.95.

Compendium of Thermophysical Property Measurements Methods. Kosta D. Maglic, et al, editors. Plenum Publishing Corp., 233 Spring Street, New York, NY. (212) 620-8000. (800)221-9369. FAX (212) 463-0742. 1991. $165.00.

CRC Handbook of Chemistry and Physics. David R. Kide, editor. CRC Press, Inc., 2000 Corporate Boulevard, NW, Boca Raton, FL 33431. (407) 994-0555. (800) 272-7737. 77th edition. 1996. $99.95.

Newnes Engineering and Physical Science Pocketbook. J. O. Bird and P. J. Chivers. Butterworth-Heinmann, 313 Washington Avenue, Newton, MA 02158. (617) 928-2500. (800) 366-2665. 1993. $49.95.

Tables of Physical and Chemical Constants and Some Mathematical Functions. G. W. Kaye and T. M. Laby. Halsted Press 605 Third Avenue, New York, 10158-0012. (212) 850-6400. 15th edition. 1986. $74.95.

ONLINE DATABASES AND CD-ROMS

CA Search. Chemical Abstracts Service, P.O. Box 3012, Columbus, OH 43210-0012. (614) 447-3600. (800) 848-6533. FAX (614) 447-3709. Very comprehensive guide to worldwide chemical literature and related fields, 1972 to present. Available on BRS, (800) 289-4277, DIALOG, (800) 334-2564, ORBIT (800) 456-7248, and STN International, FIZ Karlsruhe, P.O. Box 2465, W-7500, Karlsruhe 1, Germany, online services. Inquire as to cost and availability.

Current Contents Search. Institute for Scientific Information, 3501 Market Street, Philadelphia, PA 19104. (215) 386-0100. FAX (215) 386-6362. Contains citations to articles listed in the table of contents of science and technology journals. Also articles in social sciences and life sciences journals. Available on BRS, (800) 289-4277, DIALOG, (800) 334-2564, online services. Inquire as to cost and availability.

Dissertation Abstracts Online. University Microfilms International, 300 North Zeeb Road, Ann Arbor, MI 48106. (800) 521-0600 or (313) 761-4700. Scope includes virtually all doctoral dissertations accepted at accredited American institutions from 1861 to present in 252 subject areas. Available on BRS, (800) 289-4277, DIALOG, (800) 334-2564, and OCLC EPIC, (800) 848-5878, online services. Also available on CD-ROM. Inquire as to cost and availability.

INSPEC. Institution of Electrical Engineers, Michael Faraday House, Six Hills Way, Stevenage, Herts. SG1 2AY, England. Telephone: 0438 313311 or FAX 0438 742840. Contains citations to the worldwide literature of physics, electronics and electrical engineering, computer technology, and related fields. Available on BRS, (800) 289-4277, DIALOG, (800) 334-2564, ORBIT, (800) 456-7248, and STN International, FIZ Karlsruhe, P.O. Box 2465, W-7500, Karlsruhe 1, Germany, online services. Inquire as to cost and availability.

NTIS Bibliographic Database. National Technical Information Service, 5285 Port Royal Road, Springfield, VA 22161. (703) 487-4929 or FAX (703) 321-8199. Broad coverage of government-sponsored science and technology research reports, 1964 to present. Available on BRS,(800) 289-4277, DIALOG, (800) 334-2564, ORBIT, (800) 456-7248, and STN International, FIZ Karlsruhe, P.O. Box 2465, W-7500, Karlsruhe 1, Germany, online services. Also available on CD-ROM. Inquire as to cost and availability.

Physics Briefs. American Institute of Physics, 335 East 45th Street, New York, NY 10017. (212) 661-9260 or FAX (212) 661-2036. Contains citations with abstracts of the literature of physics and related fields, 1979 to present. Available on the STN International, FIZ Karlsruhe, P.O. Box 2465, W-7500, Karlsruhe 1, Germany, online service. Inquire as to cost and availability.

Scientific and Technical Books and Serials in Print. R.R. Bowker Inc., 121 Chanlon Road, New Providence, NJ 07974. (800) 521-8110 or (908) 464-6800. List of currently published books and serials in the physical and biological sciences, engineering and technology, with subject, author and titles indexes. Available on ORBIT, (800) 456-7248, online service. Inquire as to cost and availability.

SCISEARCH. Institute for Scientific Information, 3501 Market Street, Philadelphia, PA 19104. (800) 523-1850 or (215) 386-0100. Broad multidisciplinary title and author index to the international literature of science and technology, 1974 to present. Available on DIALOG, (800) 334-2564, and ORBIT, (800) 456-7248, online services. Also available on CD-ROM. Inquire as to cost and availability.

WILSONLINE. H.W. Wilson Company, 950 University Avenue, Bronx, NY 10452. (800) 367-6770 or (212) 588-8400. Makes available online versions of the printed H.W. Wilson indexes including *Applied Science and Technology Index, Business Periodicals Index, General Science Index,* and *Readers' Guide to Periodical Literature.* Period covered is generally 1983 to present. Available on BRS, (800) 289-4277, DIALOG, (800) 334-2564, and OCLC EPIC, (800) 848-5878, online services. Also available on CD-ROM. Inquire as to cost and availability.

PERIODICALS

Ferroelectrics Letters. Gordon & Breach Science Publishers, 820 Town Center Drive, Langhorne, PA 19047. Monthly, in two volumes. 116 ECU per volume, per year. ISSN: 0731-5171.

IEEE Transactions on Ultrasonics, Ferroelectrics and Frequency Control. Institute of Electrical and Electronics Engineers, Inc., Box 1331, 445 Hoes Lane, Piscataway, NJ 08855-1331. (908) 981-0060. 1954 to present. Bi-monthly. $230.00 per year. ISSN: 0885-3010.

Journal of Physics B: Atomic and Molecular Physics. IOP Publishing. U. S. orders to: American Institute of Physics, 500 Sunnyside Boulevard, Woodbury, NY 11797-2999. (516) 576-2200. 1968 to present. 24 issues per year. $2240.00 per year. ISSN: 0953-4075.

Journal of Physics: Condensed Matter. IOP Publishing. Orders to: American Institute of Physics, 500 Sunnyside Boulevard, Woodbury, NY 11797-2999. (516) 576-2200. 1968 to present. 50 issues per year. $4229.00 per year. ISSN: 0953-8984.

Journal of Physics and Chemistry of Solids. Elsevier Science, 660 White Plains Road, Tarrytown, NY 10591-5151. (914) 524-9200. FAX (914) 333-2444. 1956 to present. Monthly. $1670.00 per year. ISSN: 0022-3697.

RESEARCH CENTERS AND INSTITUTES

Dielectrics Research Group. University of Quebec at Trois-Rivieres, Case Postale 500, Boulevard des Forges, Trois-Rivieres, PQ Canada G9A 5H7. (819) 376-5108. FAX (819) 376-5012.

Materials Research Laboratory. Pennsylvania State University, 202 MRL, University Park, PA 16802. (814) 863-0180. FAX (814) 865-2326.

Microelectronics Research Center. Iowa State University, 1925 Scholl Road, Ames, IA 50011. (515) 294-7732. FAX (515) 294-9584.

Piezoelectricity Research Laboratory. York University, 4700 Keele Street, Yorth York, ON, Canada M3J 1P3. (416) 736) 5250.

University of Lowell Research Foundation. 450 Aiken Street, Lowell, MA 01854. (508) 458-2508. FAX (508) 458-6586.

PIPELINE TECHNOLOGY

See also: PETROLEUM ENGINEERING

ABSTRACTS AND INDEXES

API Abstracts. American Petroleum Institute, Central Abstracting and Information Services, 275 Seventh Avenue, New York, NY 10001. (212) 366-4040 or FAX (212) 366-4298. Contains citations with abstracts to literature relating to the petroleum refining and the petrochemical industries, from 1964 to present.

Applied Science and Technology Index: A Cumulative Subject Index to English Language Periodicals in the Fields of Aeronautics and Space Science, Computer Technology, Chemistry, Construction Industry, Energy and Related Areas. H.W. Wilson Co., 950 University Avenue, Bronx, NY 10452. (800) 367-6770 or (212) 588-8400. FAX (718) 590-1617. From 1958 to present. Monthly. Inquire about cost and availability. Also available on CD-ROM and online.

ASCE Combined Annual Combined Index. American Society of Civil Engineers, 345 East 47th Street, New York, NY 10017-2398. (212) 705-7520 or (800) 548-2723. Annual. $48.00.

ASCE Publications Information. 1966-present. American Society of Civil Engineers, 345 East 47th Street, New York, NY 10017-2398. (212) 705-7520 or (800) 548-2723. Bi-monthly. $160.00 per year for non-members.

Civil and Structural Engineering Abstracts. Cambridge Scientific Abstracts, 7200 Wisconsin Avenue, Bethesda, MD 20814-

PIPELINE TECHNOLOGY

Ency. of Physical Sciences and Engineering Info. Sources

4823. (301) 961-6750. FAX (301) 961-6720. 1993 to present. Monthly. $385.00 per year. Topics covered include structural design, construction equipment and methods, civil defense and military engineering, surveying, highway engineering, maritime and port structures, materials, land reclamation, and soil mechanics.

Current Contents: Engineering, Technology, and Applied Sciences. Institute for Scientific Information, 3501 Market Street, Philadelphia, PA 19104. (215) 386-0100. FAX (215) 386-6362. 1970 to present. Weekly. $442.00 per year.

Energy Research Abstracts. 1976-present. U.S. Department of Energy, office of Scientific and Technical Information, Box 62, Oak Ridge, TN 37831. (615) 574-0733. Subscriptions: Superintendent of Documents, U.S. Government Printing office, Box 371954, Pittsburgh, PA 15250-7954. (202) 783-3238. FAX (202) 512-2233. Monthly. $164.00 per year. Abstracts all the scientific and technical reports, journal articles, conference proceedings, patents, theses, and monographs sponsored by the U.S. Energy Research and Development Administration.

Engineering Index Monthly. Engineering Information, Inc., Castle Point on the Hudson, Hoboken, NJ 07030. (800) 221-1044. FAX (212) 832-1857. Monthly. $2300.00 per year. Also available online as COMPENDEX, and also on CD-ROM. Covers chemical engineering, computers, electrical engineering, civil engineering, metals and mining, industrial management, and mechanical engineering.

Fluid Abstracts: Civil Engineering. Elsevier Science [journals], 660 White Plains Rd., Tarrytown, NY 10159-5153. (919) 524-9200. FAX (919) 333-2444. 1991 to present. Monthly plus annual cumulation. $645.00 per year. Also available online as FLUIDEX. Covers civil engineering applications of fluid mechanics, hydraulics, flow metering and measuring, offshore engineering, environmental hydraulics, and related aspects of wind energy, the atmosphere, and aerodynamics.

Fuel and Energy Abstracts. Butterworth-Heinemann, Linacre House, Jordan Hill, Oxford, England. Telephone 0865-310366. FAX 0865-310898. 1960 to present. Bimonthly. Inquire for price. A summary of world literature on all technical, scientific, commercial and environmental aspects of fuel and energy.

Offshore Engineering Abstracts. STI Limited, 4 Kings Meadow, Ferry Hinksey Road, Oxford, OX2 0DU, England. Distributed in the United States by Air Science Company, Box 143, Corning NY 14830. (607) 962-5591. Contains citations on planning, design, and construction of offshore platforms and pipelines, from 1985 to present. Monthly. $400.00 per year.

Petroleum Abstracts. 600 South College, Tulsa, OK 74104. (918) 631-2296 or (800) 247-8678. Contains citations with abstracts to the worldwide literature and patents on the exploration, development, and production of petroleum resources, from 1965 to present.

ASSOCIATIONS AND PROFESSIONAL SOCIETIES

American Institute of Chemical Engineers. 345 East 47th Street, New York, NY 10017-2398. (212) 705-7338.

American Institute of Mining, Metallurgical and Petroleum Engineers. 345 East 47th Street, New York, NY 10017-2398. (212) 705-7695.

American Petroleum Institute. 1220 L Street NW, Washington, DC 20005-8029. (202) 682-8000. FAX (202) 682-8115.

American Society of Civil Engineers. 345 East 47th Street, New York, NY 10017-2398. (212) 705-7520 or (800) 548-2723.

American Society of Mechanical Engineers. 345 East 47th Street, New York, NY 10017-2398. (212) 705-7722.

Association of Oil Pipe Lines. 1725 K Street NW, Suite 1205, Washington, DC 20006. (202) 331-8228. FAX (202) 822-1964.

Pipe Fabrication Institute. Box 173, 612 Lenore Avenue, Springdale, PA 15144-1518. (412) 274-4722. FAX (412) 274-4722.

Pipe Line Contractors Association. 1700 Pacific Avenue, Suite 4100, Dallas, TX 75201. (214) 969-2700.

Society of Petroleum Engineers. P.O. Box 833836, Richardson, TX 75083-3836. (214) 952-9393.

Society of Piping Engineers and Designers. One Main Street, Suite N-723, Houston, TX 77002. (713) 221-8090. FAX (713) 221-8064.

TPF: The Tube & Pipe Fabricators Association International. 833 Featherstone Road, Rockford, IL 61107-6203. (815) 399-8700. FAX (815) 399-7279.

World Federation of Pipe Line Contractors Associations. 1700 Pacific Avenue, Suite 4100, Dallas, TX 75201. (214) 969-2700.

DIRECTORIES AND BIOGRAPHICAL SOURCES

Directory of Engineering Societies and Related Organizations. Gordon Davis. 13th edition. American Association of Engineering Societies, 1111 19th Street NW, Suite 608, Washington, DC 20036. (202) 296-2237 or (800) 658-8897. 1989. Inquire for price.

Gas Pipeline Personnel Directory, 1993. Oil Pipeline Research Institute (OPRI), PO Box 106, Boulder, CO 80306. (303) 545-5459. FAX (303) 444-1286. 1992. $250.00.

International Research Centers Directory. Gale Research, 835 Penobscot Building, Detroit, MI 48226-4094. (313) 961-2242. (800) 347-4253. FAX (313) 961-6083. 8th edition. 1995. $410.00

Pipeline & Gas Journal Annual Directory of Pipelines and Equipment. Oildom Publishing Co. of Texas, Inc., Box 219368, Houston, TX 77218-9368. (713) 558-6930. FAX (713) 558-7029. Annual. $75.00.

Pipeline & Utilities Construction. Annual Directory. Oildom Publishing Co. of Texas, Inc., Box 219368, Houston, TX 77218-9368. (713) 558-6930. FAX (713) 558-7029. Annual. $75.00.

Pipeline Industry. Midwest Register Inc., 1120 E. 4th St., Tulsa, OK 74120. (918) 582-2000. FAX (918) 582-9349. Annual. $50.00.

Research Centers Directory. Gale Research, 835 Penobscot Building, Detroit, MI 48226-4094. (313) 961-2242. (800) 347-4253. FAX (313) 961-6083. $485.00.

Scientific and Technical Organizations and Agencies Directory. Gale Research, 835 Penobscot Building, Detroit, MI 48226-4094. (313) 961-2242. (800) 347-4253. FAX (313) 961-6083. 4th edition. 1996. $195.00.

Who's Who in Engineering. American Association of Engineering Societies, 1111 19th Street NW, Suite 608, Washington, DC 20036. (202) 296-2237 or (800) 658-8897. 8th edition. 1991. Inquire for cost and availability.

World Pipelines and International Directory of Pipeline Organizations and Associations. Gulf Publishing Company, P.O. Box 2608, Houston, TX. (713) 529-4301 or (800) 231-6275. 1983. Inquire for cost and availability.

ENCYCLOPEDIAS AND DICTIONARIES

The Natural Gas Pipeline Encyclopedia, 1993. Donald Liebson. Revised edition. Oil Pipeline Research Institute (OPRI), PO Box 106, Boulder, CO 80306. (303) 545-5459. FAX (303) 444-1286. $2000.00. Inquire for availability.

The Natural Gas Pipeline Sourcebook, 1993. Donald Liebson. Revised edition. Oil Pipeline Research Institute (OPRI), PO Box 106, Boulder, CO 80306. (303) 545-5459. FAX (303) 444-1286. $1000.00. Inquire for availability.

GENERAL WORKS

Advances in Offshore Oil & Gas Pipeline Technology. R.F. de la Mare, ed. Gulf Publishing Company, P.O. Box 2608, Houston, TX. (713) 529-4301 or (800) 231-6275. 1985. Inquire for cost and availability.

Fundamentals of Pipeline Engineering. Jacques Vincent-Genod. Gulf Publishing Company, P.O. Box 2608, Houston, TX. (713) 529-4301 or (800) 231-6275. 1984. $45.00.

Hydraulics for Pipeliners. C.B. Lester. Second edition. Gulf Publishing Company, P.O. Box 2608, Houston, TX. (713) 529-4301 or (800) 231-6275. 1994. Inquire for price.

Offshore Pipeline Design Elements. John B. Herbich. Marcel Dekker/ Books on Demand, Division of University Microfilms International, 300 North Zeeb Road, Ann Arbor, MI 48106-1346. (313) 761-4700 or (800) 521-0600. 1981. $67.80.

Oil and Gas Pipeline Fundamentals. John L. Kennedy. 2d ed. PennWell Books, 1421 S. Sheridan Road, Box 1260, Tulsa, OK 74112. (918) 835-3161 or (800) 752-9764. FAX (918) 831-9555. 1993. $59.95.

HANDBOOKS AND MANUALS

Manual of Practical Pipeline Construction. B. Schurr. Gulf Publishing Company/ Books on Demand, Division of University Microfilms International, 300 North Zeeb Road, Ann Arbor, MI 48106-1346. (313) 761-4700 or (800) 521-0600. 1982. $45.40.

ONLINE DATABASES AND CD-ROMS

Compendex Plus. Engineering Information, Inc., 345 East 47th Street, New York, NY 10017. (212) 705-7600 or (800) 221-

1044. Contains citations with abstracts to worldwide literature in engineering and technology, from 1970 to present. Available on online BRS, (800) 289-4277, DIALOG, (800) 334-2564, ORBIT (800) 456-7248, and STN International, FIZ Karlsruhe, P.O. Box 2465, W-7500, Karlsruhe 1, Germany, online services. Also available on CD-ROM. Inquire as to cost and availability.

CSA Engineering. Cambridge Scientific Abstracts, 7200 Wisconsin Avenue, Suite 601, Bethesda, MD 20814. (301) 961-6750 or (800) 843-7751. Contains citations and abstracts of international periodicals and research literature covering all fields of engineering and science and technology, including computer and information science, electronics, mechanical engineering, solid state materials, 1981 to present. Available on BRS, (800) 289-4277, online service. Inquire as to cost and availability.

Current Contents Search. Institute for Scientific Information, 3501 Market Street, Philadelphia, PA 19104. (215) 386-0100. FAX (215) 386-6362. Contains citations to articles listed in the table of contents of science and technology journals. Also articles in social sciences and life sciences journals. Available on BRS, (800) 289-4277, DIALOG, (800) 334-2564, online services. Inquire as to cost and availability.

NTIS Bibliographic Database. National Technical Information Service, 5285 Port Royal Road, Springfield, VA 22161. (703) 487-4929 or FAX (703) 321-8199. Broad coverage of government-sponsored science and technology research reports, 1964 to present. Available on BRS, (800) 289-4277, DIALOG, (800) 334-2564, ORBIT, (800) 456-7248, and STN International, FIZ Karlsruhe, P.O. Box 2465, W-7500, Karlsruhe 1, Germany, online services. Also available on CD-ROM. Inquire as to cost and availability.

SCISEARCH. Institute for Scientific Information, 3501 Market Street, Philadelphia, PA 19104. (800) 523-1850 or (215) 386-0100. Broad multidisciplinary title and author index to the international literature of science and technology, 1974 to present. Available on DIALOG, (800) 334-2564, and ORBIT, (800) 456-7248, online services. Also available on CD-ROM. Inquire as to cost and availability.

WILSONLINE. H.W. Wilson Company, 950 University Avenue, Bronx, NY 10452. (800) 367-6770 or (212) 588-8400. Makes available online versions of the printed H.W. Wilson indexes including *Applied Science and Technology Index, Business Periodicals Index, General Science Index,* and *Readers' Guide to Periodical Literature.* Period covered is generally 1983 to present. Available on BRS, (800) 289-4277, DIALOG, (800) 334-2564, and OCLC EPIC, (800) 848-5878, online services. Also available on CD-ROM. Inquire as to cost and availability.

PERIODICALS

International Journal of Pressure Vessels and Piping. Elsevier Science [journals], 660 White Plains Rd., Tarrytown, NY 10159-5153. (919) 524-9200. FAX (919) 333-2444. 1972 to present. Monthly. $1505.00 per year.

Pipe Line Industry. Gulf Publishing Company, P.O. Box 2608, Houston, TX. (713) 529-4301 or (800) 231-6275. 1954-present. Monthly. $22.00 per year.

Pipeline & Gas Journal. Oildom Publishing Co. of Texas, Inc., Box 219368, Houston, TX 77218-9368. (713) 558-6930. FAX (713) 558-7029. 1859-present. Monthly. $22.00 per year.

PIPELINE TECHNOLOGY

Ency. of Physical Sciences and Engineering Info. Sources

Pipeline & Utilities Contruction. Oildom Publishing Co. of Texas, Inc., Box 219368, Houston, TX 77218-9368. (713) 558-6930. FAX (713) 558-7029. 1945-present. 12 times a year. $50.00 per year.

Pipeline Digest. Hart Publications Inc./PHillips Publishing International, Inc., 1900 Grant St., Suite 400, Box 1917, Denver, CO 80201. (303) 837-1917. FAX (303) 837-8585. 1963-present.

Pipes and Pipelines International. Scientific Surveys Ltd., P.O. Box 21, Beaconsfield, Bucks, HP9 1NS, England. Telephone 0494-675139. FAX 0494-670155. 1956-present. Bi-monthly. $139.00 per year.

Tube & Pipe Fabricating Technology Update. TPF: The Tube and Pipe Fabricators Association International, 833 Featherstone Road, Rockford, IL 61107-6301. (815) 399-8700. FAX (815) 399-7279. Quarterly. Inquire for cost and availability.

TPQ: The Tube and Pipe Quarterly. Croydon Group Ltd., 833 Featherstone Road, Rockford, IL 61107-6301. (815) 399-8700. FAX (815) 399-7279. 1990 to present. Quarterly. $15.00 per year.

Tube International. Publex International Ltd., 110 Station Road E., Oxted, Surrey, RH8 0QA, England. Telephone 0883-717755. FAX 0883-714554. 1982 to present. Bi-monthly. Free.

RESEARCH CENTERS AND INSTITUTES

Hydraulic Transients Unit, University of Michigan. 2340 G.G. Brown Bldg., Ann Arbor, MI 48109. (313) 764-8495. FAX (313) 764-4292.

SPECIFICATIONS AND STANDARDS

Annual Book of ASTM Standards. American Society for Testing and Materials (ASTM), 1916 Race Street, Philadelphia, PA 19103. (215) 299-5585. Annual. Inquire for cost and availability.

PLANETARY SCIENCES

See also: ASTROGEOLOGY, ASTEROIDS, ASTRONOMY, ASTRO-PHYSICS, COMETS, METEORS, NAMES OF INDIVIDUAL PLANETS, SATELLITES (NATURAL), SEISMOLOGY, SOLAR SYSTEM, TEKTITES, VOLCANOLOGY

ABSTRACT SERVICES AND INDEXES

Applied Science and Technology Index: A Cumulative Subject Index to English Language Periodicals in the Fields of Aeronautics and Space Science, Computer Technology, Chemistry, Construction Industry, Energy and Related Areas. H. W. Wilson Co., 950 University Avenue, Bronx, NY 10452- 9978. (212) 588-8400. (800) 367-6770. FAX (718) 590-1617. From 1958 to present. Monthly. Inquire about cost and availability. Also available online (BRS and WILSONLINE) and on CD-ROM. ISSN: 0003-6986.

Astronomy and Astrophysics Abstracts. Springer-Verlag New York, 175 Fifth Avenue, New York, NY 10010. (212) 460-1500. FAX (212) 473-6272. Published for the Astronomisches Rechen-Institut. Comprehensive coverage of all aspects of astronomy, astrophysics and related fields. 1969 to present. Two parts per year. Annual. ISSN: 0067-0022.

Bibliography and Index of Geology. American Geological Institute, 4220 King Street, Alexandria, VA 22302-1507. (703) 379-2480. FAX (703) 379-7563. From 1969 to present. Monthly. $1295.00 per year. ISSN: 0098-2784. Also available as GEOREF online (CISTI, DIALOG, Orbit, STN) and on CD-ROM. Inquire about price and availability.

Chemical Abstracts. Chemical abstracts Service. 2540 Oleantangy River Road, Box 3012, Columbus, OS 43210-0012. (614) 447-3600. FAX (614) 447-3713. 1907 to present. Weekly. $16,800.00 per year. Available online and on CD- ROM. CA is also available in five section groupings. Inquire regarding cost and availability.

Current Contents: Physical, Chemical, and Earth Sciences. Institute for Scientific Information, 3501 Market Street, Philadelphia, PA 19104. (215) 386-0100. (800) 523-1850. FAX (215) 386-2291. 1961 to present. Weekly. $442.00 per year. Also available online (BRS, DIALOG) and on CD-ROM. Inquire about price and availability.

General Science Index. H.W. Wilson Company, 950 University Avenue, Bronx, NY 10452. (212) 588-8400. (800) 367-6770. FAX (718) 590-1617. From 1978 to present. Ten issues per year; quarterly and annual cumulations. Service basis. Available on CD-ROM and online. Inquire regarding cost and availability. ISSN: 0162-1963.

Meteorological and Geoastrophysical Abstracts. American Meteorological Society, c/o Inforonics, Inc., 550 Newtown Road, Box 458, Littleton, MA 01460. (508) 486-8976. FAX (508) 486-0027. Covers literature in environmental sciences, meteorology, astrophsyics, hydrology, glaciology and physical oceanography. 1950 to present. Monthly. $950.00 per year. Also available online (DIALOG) and on CD-ROM. ISSN: 0026-1130.

NTIS Alerts: Astronomy & Astrophysics. U. S. National Technical Information Service. 5285 Port Royal Road, Springfield, VA 22161. (703) 487-4650. FAX (703) 321-8547. Weekly. $140.00 per year.

Pascale 49: Meteorologie, Glaciologie, Physique Des Oceans. Centre National de la Recherche Scientifique, Institut de Information Scientifique et Technique, 2 aliee du Parc de Brabois, 54514 Vandoeuvre Les Nancy Cedex, France. TEL 83- 50-46-00. FAX: 83-50-46-50. 1985 to present. Ten issues per year. 770 F. Also available on CD-ROM and online. ISSN: 1164-5997.

Physics Abstracts. INSPEC. Section A, Science Abstracts. Institute of Electrical Engineers, London, United Kingdom. Available from: INSPEC/IEEE-Institute of Electrical and Electronic Engineers, Box 1331, Hoes Lane, Piscataway, NJ 08855-1331. (908) 562-5549. 1898 to present. 24 issues per year. $2835.00 per year. ISSN: 0036-8091. Also available online and on CD-ROM.

Science Citation Index (SCI). Institute for Scientific Information, 3501 Market Street, Philadelphia, PA 19104. (215) 386-0100. (800) 523-1850. FAX (215) 386-2991. 1961 to present. Six issues per year, plus annual cumulation. $11650.00 per year. Also available online and on CD-ROM. Inquire about price and availability. ISSN: 0036-827X.

STAR. Scientific and Technical Aerospace Reports. U.S. National Aeronautics and Space Administration. Distributed by U. S. Superintendent of Documents, Washington, DC 20402. 1963 to present. Semi-monthly, with semiannual and annual indexes. $114.00 per year. ISSN: 0036-8741. Also available online and on CD-ROM.

ANNUAL REVIEWS AND YEARBOOKS

The Astronomical Almanac. Superintendent of Documents, U. S. Government Printing office, Washington, DC 10402. (202) 783-3238. 1981 to present. Superceeds Astronomical Ephemeris and the American Ephermis and Nautical Almanac. Annual. $44.95. ISSN: 0737-6421.

Annual Review of Astronomy and Astrophysics. Original reviews of critical literature and current developments in astronomy and astrophysics. Annual Reviews, Inc., 4139 El Camino Way, Palo Alto, CA 94303-0139. (415) 493-4400. (800) 523-8635. Fax (415) 855-9815. 1963 to date. Annual. $60.00. ISSN: 0066-4146.

Annual Review of Earth and Planetary Sciences. Annual Reviews, Inc., 4139 El Camino Way, Palo Alto, CA 94303-0139. (415) 493-4400. (800) 523-8635. Fax (415) 855-9815. 1973 to date. Annual. $62.00. ISSN: 0084-6597.

Proceedings of the Lunar and Planetary Science Conference. Lunar and Planetary Institute, 3600 Bay Area Boulevard, Houston, TX 77058. (713) 486-2143. 1970 to date. Annual. ISSN: 0270-9511.

ASSOCIATIONS AND PROFESSIONAL SOCIETIES

American Astronomical Society. 2000 Florida Avenue NW, Suite 400, Washington, DC 20009. (202) 328-2010. FAX (202)234-2560.

American Geophysical Union. 2000 Florida Avenue NW, Washington, DC 20009. (202) 462-6900. (800) 966-AGU1. FAX (202) 328-0566.

American Institute of Physics. One Physics Ellipse, College Park, MD 20740-3843. (301) 209-3100.

American Meteor Society. SUNY-Geneseo, 1 College Circle, Department of Physics and Astronomy, Geneseo, NY 14454. (716) 245-5282.

Association of Universities For Research in Astronomy, Inc. (AURA). Suite 701, 1625 Massachusetts Avenue, NW, Washington, DC 20036. (202) 483-2101. FAX (202) 483-2106.

Astronomical Society of the Pacific. 390 Ashton Avenue, San Francisco, CA 94112. (415) 337-1100. FAX: (415) 337-5205.

Geological Society of America. 3300 Penrose Place, PO Box 9140. Boulder, CO 80301-9140. (303) 447-2020. FAX (303) 447-1133.

IEEE Geoscience and Remote Sensing Society. c/o Institute of Electrical and Electronics Engineers. 345 East 47th Street, New York, NY 10017. (212) 705-7900. FAX (212) 705-4929.

Lunar and Planetary Institute. 3600 Bay Area Boulevard, Houston, TX 77058. (713) 486-2143.

Meteoritical Society. University of Massachusetts, 125 Marston Hall, Amherst, MA 01003. (413) 545-0300. (413) 545-0724.

Planetary Society. 65 North Catalina Avenue, Pasadena, CA 91106. (818) 793-5100. (800) WOW-MARS. FAX (818) 793-5528.

BIBLIOGRAPHIES

A Bibliography of Astronomy, 1970-1979. R. A. Sea and S. S. Martin. Libraries Unlimited, Inc., Littleton, CO 80160. 1982. $37.50.

Science Books & Films. American Association for the Advancement of Science, 1333 H Street NW, Washington, DC 20005. (202) 326-6434. Reviews of print, film and software materials in all sciences.. 1965 to present. Nine issues per year. $40.00 per year. ISSN: 0098-342X.

Scientific and Technical Books and Serials in Print: An Index to Literature in Science and Technology. R. R. Bowker. 121 Chanlon Road, New Providence, NJ 07974. (908) 464-6800. FAX (908) 665-3502. (800) 521-8110. 1972 to present. Annual. $299.95 per year. Also available on CD-ROM. ISSN: 0000-054X.

DIRECTORIES AND BIOGRAPHICAL SOURCES

American Astronomical Society. Membership Directory. American Astronomical Society. 2000 Florida Avenue NW, Suite 400, Washington, DC 20009. (202) 328-2010. FAX (202) 234-2560 Annual. Included in membership dues. ISSN: 1061-9038.

American Men and Women of Science: Physical and Biological Sciences. R. R. Bowker Inc., 121 Chanlon Road, New Providence, NJ 07974. (908) 464-6800. (800) 521-8110. 20th edition. 8 volumes. 1996. $850.00.

The Astronomers. Donald Goldsmith. St. Martin's Press, Inc., 175 Fifth Avenue, New York, NY 10010. (212) 674-5151. (800) 221-7945. 1993. $14.95 in paper.

Astronomical Centers of the World. Kevin Krisciunas. Cambridge University Press, 40 West 20th Street, New York, NY 10011-4211. (212) 924-3900. (800) 872-7423. 1988. $34.95

Directory of Physics and Astronomy Staff Members. American Institute of Physics, One Physics Ellipse, College Park, MD 20740-3843. (301) 209-3100. 1975/76 to present. Annual. $60.00. ISSN: 0361-2228.

Earth and Astronomical Research Centers. Stockton Press, 345 Park Avenue South, New York, NY 10010. 4th edition. 1995. $515.00. ISBN: 1-56169-0967.

Graduate Programs in Physics, Astronomy and Related Fields. American Institute of Physics, One Physics Ellipse, College Park, MD 20740-3843. (301) 209-3100. 1978 to present. Annual. $45.00. ISSN: 0147-1821.

Research Centers Directory. Gale Research, 835 Penobscot Building, Detroit, MI 48226-4094. (313) 961-2242. (800) 347-4253. 20th edition. 1995. $485.00.

PLANETARY SCIENCES

Ency. of Physical Sciences and Engineering Info. Sources

ENCYCLOPEDIAS AND DICTIONARIES

Concise Dictionary of Astronomy. Jacqueline Mitton. Oxford University Press, Inc., 200 Madison Avenue, New York, NY 10016. (212) 725-6000. (800) 334-4249. 1992. $24.95.

Encyclopedia of Astronomy and Astrophysics. Stephen Maran, editor. Van Nostrand Reinhold, 115 Fifth Avenue, New York, NY 10003. (212) 254-3232. (800) 842-3636. 1992. $129.95.

McGraw-Hill Encyclopedia of Science and Technology. McGraw-Hill Book Company, Inc., 1221 Avenue of the Americas, New York, NY 10020. (212) 512-2000. (800) 262-4729. 7th edition. 20 volume set. 1992. $1900.00.

Moons of the Solar System: An Illustrated Encyclopedia. John Stewart. McFarland & Company, Inc., Box 611, Jefferson, MC 28640. (910) 246-4460. (800) 253-2187. 1991. $49.95.

New Guide to the Planets. Patrick Moore. Trans-Atlantic Publications, Inc., 311 Bainbridge Street, Philadelphia, PA 19147. (215) 925-5083. 1993. $37.50.

Stars and Planets: The Sierra Club Guide to Sky Watching and Direction Finding. W. S. Kals. The Sierra Press, 4988 Gold Leaf Drive, Mariposa, CA 95338. (209) 966-5071. (800) 745-2631. $15.00.

GENERAL WORKS

Birth of the Earth. David E. Fisher. Columbia University Press, 562 West 113th Street, New York, NY 10025. (212) 666-1000. (800) 944-8648. 1988. $16.50 in paper.

Chemistry and Physics of the Terrestrial Planets. Surendra K. Saxwna. Springer-Verlag New York, Inc., 175 Fifth Avenue, New York, NY 10010. (212) 460-1500. (800) 777-4643. 1986. $109.00.

Clyde Tombaugh: Discoverer of Planet Pluto. David H. Levy. University of Arizona Press, 1230 North Park Avenue, Number 102, Tucson, AZ 85719. (520) 621-1441. (800) 426-3797. Reprint edition. 1992. $15.95 in paper.

Evolution of the Earth and Planets. E. Takahashi, et al, editors. American Geophysical Union, 2000 Florida Avenue, N.W., Washington, DC 20009. (202) 462-6903. (800) 966-2481. 1993. $28.00.

Exploration of Venus & Mars Atmospheres. G. M. Keating, editor. Pergamon Press, Inc., Maxwell House, Fairview Park, Elmsford, NY 10523. (914) 592-7700. Fax (914) 592-3625. 1995. 94.00.

Exploring the Planets. W. Kenneth Hamblin and Eric H. Christiansen. Macmillan Publishing Company, Inc,. 200 Old Tappan Road, Old Tappan, NJ 07675. (800) 233-2336. 2nd edition. 1995. $62.00.

The Geology of Multi-Ring Impact Basins: The Moon and Other Planets. P. D. Spudis. Cambridge University Press, 40 West 20th Street, New York, NY 10011-4211. (212) 924-3900. (800) 872-7423. Planetary Sciences Series. 1993. $59.95.

Geology of the Terrestrial Planets. Michael H. Carr. U. S. Government Printing office, Superintendent of Documents, Washington, DC 20402-9325. (202) 783-3238. 1985. $16.00

A History of Arabic Astronomy: Planetary Theories During the Golden Age of Islam. George Saliba. New York University Press, 70 Washington Square South, New York, NY 10012. (212) 998-2575. (800) 996-6987. 1994. $45.00.

An Introduction to Cosmochemistry. Charles R. Cowley. Cambridge University Press, 40 West 20th Street, New York, NY 10011-4211. (212) 924-3900. (800) 872-7423. 1995. $29.95 in paper.

Mercury: The Elusive Planet. Robert B. Strom. Smithsonian Institution Press, 470 L'Enfant Plaza, Suite 7100, Washington, DC 20560. (202) 287-3738. (800) 782-4612. 1987. $29.95.

Meteorites: An Introduction. Fritz Heide. Springer-Verlag New York, Inc., 175 Fifth Avenue, New York, NY 10010. (212) 460-1500. (800) 777-4643. 1995. $24.00 in paper.

Meteorites and the Origin of Planets. John A. Wood. Books on Demand, 300 North Zeeb Road, Ann Arbor, MI 48106-1346. (313) 761-4700. (800) 521-0600. Reprint edition. $35.00 in paper.

Moons and Planets. William K. Hartmann. Wadsworth Publishing Co., 10 Davis Drive, Belmont, CA 94002. (415) 595-2350. (800) 354-9706. 3rd edition. 1993. $49.95.

Music of the Heavens: Kepler's Harmonic Astronomy. Bruce Stephenson. Princeton University Press. 41 William Street, Princeton, NJ 08540. (609) 258-4900. (800) 777-4726. 1994. $39.50.

The Near Planets. Time-Life Books Editors. Time-Life, Inc., 777 Duke Street, Alexandria, VA 22314. (703) 8388- 7000. (800) 621-7026. Voyage through the Universe Series. 1992. $ wfi.

Planetary Nebulae. Hynes . Willmann-Bell, Inc., P.O. Box 35025, Richmond, VA 23235. (804)-320-7016. 1991. $24.95.

Planetary Landscapes. Ronald Greeley. Chapman & Hall, 1 Penn Plaza, New York, NY 10119. (212) 564-1060. Div. of Routledge, 2nd edition. 1994. $49.95.

The Planetary System. Tobias Owen and David Morrison. Addison-Wesley Publishing Co., Inc., 1 Jacob Way, Reading, MA 01867. (617) 944-3700. (800) 447-2226. 1988. $49.50.

Planets and their Atmospheres. John S. Lewis and Ronald G. Primm. Academic Press Inc., 6277 Sea Harbor Drive, Orlando, FL. (800) 321-5068. 1983. $65.00 in paper.

Proceedings of Lunar and Planetary Science, Vol.22. Graham Ryder and Virgil L. Sharpton, editors. Lunar and Planetary Institute, 3600 Bay Area Boulevard, Houston, TX 77058. (713) 486-2143. 1992. $50.00.

Resources of Near-Earth Space. John S. Lewis, et al. University of Arizona Press, 1230 North Park Avenue, Number 102, Tucson, AZ 85719. (520) 621-1441. (800) 426-3797. 1993. $75.00.

Rings: Discoveries From Galileo to Voyager. James C. Elliot and Richard Kerr. MIT Press, 55 Hayward Street, Cambridge,

MA 02142. (617) 253-8569. Reprint edition. 1987. $9.95 in paper.

Solar and Planetary Dynamos. M. R. Proctor, et al, editors. Cambridge University Press, 40 West 20th Street, New York, NY 10011-4211. (212) 924-3900. (800) 872-7423. Publications of the Newton Institue. 1994. $49.95.

Space Technology and Planetary Astronomy. Joseph N. Tatarewicz. Indiana University Pres, 601 North Morton Street, Bloomington, in 47404-3797. (812) 855-6804. (800) 842-6796. 1990. $29.95.

Story of the Earth. Peter Cattermole and Patrick Moore. Cambridge University Press, 40 West 20th Street, New York, NY 10011-4211. (212) 924-3900. (800) 872-7423. 1986. $5.95 in paper.

Terraforming: Engineering Planetary Environments. Martyn J. Fogg. Society of Automotive Engineers (SAE), 400 Commonwealth Drive, Warrendale, PA 15096. (412) 776-4841. 1995. $49.00.

Uranus and Neptune: The Distant Giants. Eric Burgess. Columbia University Press, 562 West 113th Street, New York, NY 10025. (212) 666-1000. (800) 944-8648. 1988. $40.00.

Venus: An Errant Twin. Eric Burgess. Columbia University Press, 562 West 113th Street, New York, NY 10025. (212) 666-1000. (800) 944-8648. 1985. $43.50.

Venus: A New Geology. Peter Cattermole. Johns Hopkins University Press, 2715 North Charles Street, Baltimore, MD 21218-4319. (410) 516-6900. (800) 537-5487. 1994. $49.95.

Worlds Apart: A Textbook in Planetary Sciences. Guy Consolmagno. Prentice-Hall, 113 Sylvan Avenue, Route 9W, Englewood Cliffs, NJ 07632. (201) 592-2000. (800) 922-0579. 1994. $62.00.

HANDBOOKS AND MANUALS

Astrophysical Data: Planets and Stars. Kenneth R. Lanb. Springer-Verlag New York, Inc., 175 Fifth Avenue, New York, NY 10010. (212) 460-1500. (800) 777-4643. 1993. $59.00.

Field Guide to the Stars and Planets. Jay M. Pasachoff and Donald H. Menzel. Houghton Mifflin Co., 222 Berkeley Street, Boston, MA 02116. (617) 351-5000. (800) 225-3362. Revised edition. 1992. $24.95.

New Guide to the Planets. Patrick Moore. Trans-Atlantic Publications, Inc., 311 Bainbridge Street, Philadelphia, PA 19147. (215) 925-5083. 1993. $37.50.

Planet Observer's Handbook. Fred W. Price. Cambridge University Press, 40 West 20th Street, New York, NY 10011-4211. (212) 924-3900. (800) 872-7423. 1994. $34.95.

Planets and Their Moons. Aububon Society Staff. Alfred A. Knopf, Inc. 201 E. 50th Street, New York, NY 10022. (800) 733-3000. 1995. $7.99.

Tables of Planetary Phenomena. Neil F. Michelsen. ACS Publications, P.O. Box 3447, San Diego, CA 92163-4487. (619-492-9919. (800) 888-9983. San Diego, CA. 2nd revised edition. 1993. $24.95.

ONLINE DATABASES AND CD-ROMS

CA Search. Chemical Abstracts Service, P.O. Box 3012, Columbus, OH 43210-0012. (614) 447-3600. (800) 848-6533. FAX (614) 447-3709. Very comprehensive guide to worldwide chemical literature and related fields, 1972 to present. Available on BRS,(800) 289-4277, DIALOG, (800) 334-2564, ORBIT (800) 456-7248, and STN International, FIZ Karlsruhe, P.O. Box 2465, W-7500, Karlsruhe 1, Germany, online services. Inquire as to cost and availability.

Dissertation Abstracts. University Microfilms International, 300 North Zeeb Road, Ann Arbor, MI 48106. (800) 521-0600 or (313) 761-4700. Scope includes virtually all doctoral dissertations accepted at accredited American institutions from 1861 to present in 252 subject areas. Available on BRS, (800) 289-4277, DIALOG, (800) 334-2564, and OCLC EPIC, (800) 848-5878, online services. Also available on CD-ROM. Inquire as to cost and availability.

Georef. American Geological Institute, 4220 King Street, Alexandria, VA 22302. (800) 336-4764 or (703) 379-2480. Geology and geosciences literature, 1785 to present for North America. Available on DIALOG,(800) 334-2564, ORBIT (800) 456-7248, online services. Also available on CD-ROM. Inquire as to cost and availability.

INSPEC. Institution of Electrical Engineers, Michael Faraday House, Six Hills Way, Stevenage, Herts. SG1 2AY, England. Telephone: 0438 313311 or FAX 0438 742840. Contains citations to the worldwide literature of physics, electronics and electrical engineering, computer technology, and related fields. Available on BRS, (800) 289-4277, DIALOG, (800) 334-2564, ORBIT, (800) 456-7248, and STN International, FIZ Karlsruhe, P.O. Box 2465, W-7500, Karlsruhe 1, Germany, online services. Inquire as to cost and availability.

NTIS Bibliographic Database. National Technical Information Service, 5285 Port Royal Road, Springfield, VA 22161. (703) 487-4929 or FAX (703) 321-8199. Broad coverage of government-sponsored science and technology research reports, 1964 to present. Available on BRS,(800) 289-4277, DIALOG, (800) 334-2564, ORBIT, (800) 456-7248, and STN International, FIZ Karlsruhe, P.O. Box 2465, W-7500, Karlsruhe 1, Germany, online services. Also available on CD-ROM. Inquire as to cost and availability.

SCISEARCH. Institute for Scientific Information, 3501 Market Street, Philadelphia, PA 19104. (800) 523-1850 or (215) 386-0100. Broad multidisciplinary title and author index to the international literature of science and technology, 1974 to present. Available on DIALOG, (800) 334-2564, and ORBIT, (800) 456-7248, online services. Also available on CD-ROM. Inquire as to cost and availability.

WILSONLINE. H.W. Wilson Company, 950 University Avenue, Bronx, NY 10452. (800) 367-6770 or (212) 588-8400. Makes available online versions of the printed H.W. Wilson indexes including *Applied Science and Technology Index, Business Periodicals Index, General Science Index,* and *Readers' Guide to Periodical Literature.* Period covered is generally 1983 to present. Available on BRS, (800) 289-4277, DIALOG, (800) 334-2564, and OCLC EPIC, (800) 848-5878, online services.Also available on CD-ROM. Inquire as to cost and availability.

PLANETARY SCIENCES

Ency. of Physical Sciences and Engineering Info. Sources

PERIODICALS

Association of Lunar and Planetary Observers. Journal. Association of Lunar and Planetary Observers, Box 16131, San Francisco, CA 94116. (415) 566-5786. FAX (415) 731-8242. 1947 to present. Quarterly. $14.00 per year. ISSN: 0039- 2502.

Astronomical Journal. American Institute of Physics, One Physics Ellipse, College Park, MD 20740-3843. (301) 209-3000. Published for the American Astronomical Society. 1849 to present. Monthly. $280.00 per year. ISSN: 0004-6256.

Astronomical Society of the Pacific. Publications. Astronomical Society of the Pacific, 390 Ashton Avenue, San Francisco, CA 94112. (415) 337-1100. FAX (415) 337-5205.1889 to present. Monthly. $175.00 per year. ISSN: 0004-6280.

Astronomy. Kalmbach Publishing Company, Box 1612, Waukesha, WI 53187-1612. (414) 796-8776. FAX (414) 796-1142. 1973 to present. Monthly. $27.00 per year. ISSN: 0091-6358.

Astronomy and Astrophysics. Springer-Verlag New York, Inc., 44 Hartz Way, Secaucus, NJ 07096-2491. (201) 348-4033. FAX(201) 348-4505. 1969 to date. 36 issues per year in 12 volumes. $1733.00 per year. ISSN: 0004-6361.

Celestial Mechanics and Dynamical Astronomy: An International Journal of Space Dynamics. Kluwer Academic Publishers, Box 358, Accord Station, Hingham, MA 02018-0358. (617) 871-6600. FAX (617) 871-6528. 1969 to present.Monthly. $703.50 per year. ISSN: 0923-2958.

Ciel et Terre. Societe Royale Belge d'Astronomie, de Meteorologie et de Physique du Globe, 3 avenue Circulaire, Uccle, Brusels, Belgium. FAX 32-2-374-9822. 1880 topresent. Bimonthly. 1600 BEF per year. ISSN: 0009-6709.

Earth, Moon and Planets: An International Journal of Comparative Planetology. Kluwer Academic Publishers, Box 358, Accord Station, Hingham, MA 02018-0358. (617) 871-6600. FAX (617) 871-6528. 1969 to present. Monthly. $840.00 per year. ISSN: 0167-9295.

Geochemica et Cosmochimica Acta. Elsevier Science. 660 White Plains Road, Tarrytown, NY 10591-5153. (914) 524-9200. FAX (914) 333-2444. 1950 to the present. Biweekly. $895.00 per year. ISSN: 0016-7037.

Icarus: International Journal of Solar System Studies. Academic Press, Inc., Journal Division, 525 B Street, Suite 1900, San Diego, CA 92101-4495. (619) 230-1840. FAX (619) 699-6800. 1962 to the present. Monthly. $1080.00. ISSN: 0019-1035.

JGR: Journal of Geophysical Research: Planets. American Geophysical Union, 2000 Florida Avenue, NW, Washington, CD 20009. (202) 462-6900. FAX (202) 328-0566. 1991 to present. Monthly. $597.00 per year. ISSN: 0148-0227.

Lunar and Planetary Information Bulletin. Lunar and Planetary Institute. 3600 Bay Area Boulevard, Houston, TX 77058-1113. (713) 486-2175. FAX (713) 486-2125. 1970 to present. Quarterly. Free. Also available online.

Mercury. Astronomical Society of the Pacific. 390 Ashton Avenue, San Francisco, CA 94112. (415) 337-1100. FAX: (415)

337-5205. 1972 to present. Bimonthly. $175.00 per year. ISSN: 0047-6773.

Meteoritics. Meteoritical Society, Department of Chemistry, University of Arkansas, Fayetteville, AR 72701. (501) 575-7625. FAX (501) 575-7778. 1953 to present. Bimonthly. $210.00 per year. ISSN: 0026-1114.

Observatory. c/o Dr. D.J. Stickland Space and Astrophysics Division, Rutherford Appleton Laboratory, Chilton, Didcot, Oxon OX11 OQX England. FAX 0235-445848. 1877 to present. Bimonthly. $42.00 per year. ISSN: 0029-7704.

Planetary and Space Science. Elsevier Science Publishing Company, Inc., 660 White Plains Road, Terrytown, NY 10591-5153. (914) 524-9200. FAX: (914) 333-2444. 1959 to present. Monthly. $1355.00 per year. ISSN: 0032-0633.

Planetary Report. Planetary Society. 65 North Catalina, Pasadena, CA 91106-2301. (818) 793-5100. FAX (818) 793-5528. 1980 to present. $25.00 per year. ISSN: 0736-3680.

Sky and Telescope. Sky Publishing Corporation, Box 9111, Belmont, MA 02178. (617) 864-7360. FAX (617) 864-6117.1941 to present. Monthly. $27.00 per year. ISSN: 0037-6604.

RESEARCH CENTERS AND INSTITUTES.

Center for Radiophyscs and Space Research. Cornell University, Space Sciences Building, Ithaca NY 14853. (607) 255-4341. FAX (607) 255-9888.

Center for Space Science and Astrophysics. Stanford University. 325 Durand Building, Stanford, CA 04305. (415) 723-3582.

Center for Space Sciences. University of Texas at Dallas, P.O.Box 830688, MS-F022, Richardson, TX 75083-0688. (214) 690-2851. FAX (214) 690-2848.

Earth and Planetary Remote Sensing Laboratory. Washington University, Department of Earth and Planetary Sciences. Campus Box 1169, 1 Brookings Drive, St. Louis, MO 63130. (314) 889-5679. FAX (314) 889-5799.

Institute For Astronomy. University of Hawaii at Manoa, 2680 Woodlawn Drive, Honolulu, HI 96822. (808) 956-8312. FAX (808) 988-2790.

Institute for Terrestrial and Planetary Atmospheres. State University of New York at Stony Brook, Stony Brook, NY 11794-2300. (516) 632-6170. FAX (516) 632-6251.

Institute of Geophysics and Planetary Physics. University of California, Riverside, CA 92521. (714) 787-4503. FAX (714) 787-4529.

Laboratory For Atmospheric and Space Physics. University of Colorado-Boulder, Boulder, CO 80309-0392. (303) 492-7677. (303) 492-6946.

Laboratory for Planetary Geology. Arizona State University, Department of Geology, Tempe, AZ 85281. (601) 965-7029.

Laboratory for Planetary Studies. Cornell University, Space Sciences Building, Ithaca, NY 14853. (607) 255-4971.

Lunar and Planetary Institute. 3303 NASA Road One, Houston, TX 77058-4399. (713) 486-2139. FAX 713-496-2162.

Lunar and Planetary Laboratory. University of Arizona, Tucson, AZ 85721. (602) 621-6963. FAX (602) 621-4933.

McDonnell Center for the Space Sciences. Washington University, Box 1105, One Brookings Drive, St. Louis MO 63130-4899. (314) 889-6255. FAX (314) 889-6219.

Smithsonian Astrophysical Observatory. 60 Garden Street, Cambridge, MA 02138. (617) 495-7461. FAX (617) 495-7105.

PLASMA PHYSICS

See also: ASTROPHYSICS, FUSION, PARTICLE ACCELERATORS, PHYSICS

ABSTRACT SERVICES AND INDEXES

Applied Science and Technology Index: A Cumulative Subject Index to English Language Periodicals in the Fields of Aeronautics and Space Science, Computer Technology, Chemistry, Construction Industry, Energy and Related Areas. H.W. Wilson Co., 950 University Avenue, Bronx, NY 10452. (212) 588-8400. (800) 367-6770. FAX (718) 590-1617. From 1958 to present. Monthly. Inquire about cost and availability. Also available on CD-ROM and online. ISSN: 0003-6986.

Chemical Abstracts. Chemical Abstracts Service, 2540 Olentangy River Road, Box 3012, Columbus, OH 43210-0012. (614) 447-3600. (800) 848-6538. FAX (614) 447-3713.From 1907 to present. Weekly. $16800.00 per year. Also available on CD-ROM and online. Inquire regarding cost and availability. ISSN:

Current Contents: Physical, Chemical and Earth Sciences. Institute for Scientific Information, 3501 Market Street, Philadelphia, PA 19104. (215) 386-0100. FAX (215) 386-2291. From 1961 to present. Weekly. $442.00 per year. Also available on CD-ROM and online. Inquire regarding cost and availability. ISSN: 0163-2574.

Current Papers in Physics. Institute of Electrical Engineers, Michael Faraday House, Six Hill Way, Stevenage, Herts, SG1 2AY, England. Distributed by INSPEC/IEEE, Box 1331, 445 Hoes Lane, Piscataway, NJ 08855-1331. (908) 562-5549. 1966 to present. Fortnightly. $410.00 per year. ISSN: 0011-3786.

General Science Index. H.W. Wilson Company, 950 University Avenue, Bronx, NY 10452. (212) 588-8400. (800) 367-6770. FAX (718) 590-1617. From 1978 to present. Ten issues per year; quarterly and annual cumulations. Service basis. Available on CD-ROM and online. Inquire regarding cost and availability. ISSN: 0162-1963.

Government Reports Announcements and Index. U. S. National Technical Information Service (NTIS), 5285 Port Royal Road, Springfield, VA 22161. (703) 487-4650. FAX (703) 321-8547. From 1968 to present. Annual. $630.00 per year. Also available online as NTIS Bibliographic Database and on CD-ROM.

NTIS Alerts: Physics. U. S. National Technical Information Service. 5285 Port Royal Road, Springfield, VA 22161. (703) 487-4650. FAX (703) 321-8547. Weekly. $140.00 per year.

Physics Abstracts. INSPEC. Section A, Science Abstracts. Institution of Electrical Engineers (IEE). Available from INSPEC/IEEE-Institute of Electrical and Electronic Engineers, Box 1331, 445 Hoes Lane, Piscaway NJ 08855-1331. (908) 562-5549. 1898 to present. 24 issues per year. $2835.00 per year. Also available on CD-ROM and online. ISSN: 0036-8091.

Physics Briefs (Physikalische Berichte). Information Center for Energy, Physics, Mathematics; German Physical Society. VCH Publishers, Inc., 220 East 23rd Street, New York, NY 10010-4606. (212) 683-8333. 1845 to present. 24 issues per year. $2390.00 per year. Also available online. ISSN: 0179-7434.

Science Citation Index (SCI). Institute for Scientific Information, 3501 Market Street, Philadelphia, PA 19104. (215) 386-0100. (800) 523-1850. FAX (215) 386-2991. 1961 to present. Six issues per year, plus annual cumulation. $11650.00 per year. Also available online and on CD-ROM. Inquire about price and availability. ISSN: 0036-827X.

Solid State and Superconductivity Abstracts: Covers Theory, Production and Application of Solid State Materials. Cambridge Scientific Abstracts, 7200 Wisconsin Avenue, Bethesda, MD 20824. (301) 961-6750. FAX (301) 961-6720.1957 to present. Bimonthly. $1320.00 per year. Also available online. ISSN: 0896-5900.

ANNUAL REVIEWS AND YEARBOOKS

Advances in Physics. Taylor & Francis, Ltd., Rankine Road, Basingstoke, Hants RG24 8PR, England. 0256-840366. FAX 0256-47938. 1952 to present. Bimonthly. $511.00 per year. ISSN: 0001-8732.

Annual Review of Fluid Mechanics. Annual Reviews, Inc., 4139 El Camino Way, Palo Alto, CA 94303-0139. (415) 493- 4400. (800) 523-8635. Fax (415) 855-9815. 1969 to present.Annual. $47.00 per year. ISSN: 0066-4189.

Annual Review of Nuclear and Particle Science. Annual Reviews, Inc., 4139 El Camino Way, Palo Alto, CA 94303-0139. (415) 493-4400. (800) 523-8635. Fax (415) 855-9815. 1952 to present. Annual. $62.00.

Reviews of Plasma Physics. Plenum Publishing Corp., 233 Spring Street, New York, NY. (212) 620-8000. (800) 2221-9369. FAX (212) 463-0742. 1965 to present. Irregular. Price varies. ISSN: 0080-2050.

ASSOCIATIONS AND PROFESSIONAL SOCIETIES

American Institute of Physics. One Physics Ellipse, College Park, MD 20740-3843. (301) 209-3100.

American Nuclear Society. 555 North Kensington Avenue, LA Grange Park, IL 60525. (312) 352-6611. FAX (708) 352-0499.

American Physical Society. One Physics Ellipse, College Park, MD 20740-3843. (301) 209-3200. FAX (301) 209-0865.

IEEE Nuclear and Plasma Sciences Society. c/o Institute of Electrical and Electronics Engineers, 345 East 47th Street, New York, NY 10017. (212) 705-7900. FAX 212) 705-4929.

PLASMA PHYSICS

Ency. of Physical Sciences and Engineering Info. Sources

DIRECTORIES AND BIOGRAPHICAL SOURCES

American Men and Women of Science: Physical and Biological Sciences. R. R. Bowker Inc., 121 Chanlon Road, New Providence, NJ 07974. (908) 464-6800. (800) 521-8110. 20th edition. 8 volumes. 1996. $850.00.

American Physical Society. Membership Directory Bulletin Issue. American Physical Society, One Physics Ellipse, College Park, MD 20740-3843. (301) 209-3200. FAX (301) 209-0865. Biennial. $50.00.

The Biographical Dictionary of Scientists: Physicists. D. Abbott, editor. Peter Bedrick Books, Inc., 2112 Broadway, Room 318, New York, NY 10023. (212) 496-0751. 1984.

Directory of Physics and Astronomy Staff Members. American Institute of Physics, One Physics Ellipse, College Park, MD 20740-3843. (301) 209-3100. Annual. $45.00.

Graduate Programs in Physics, Astronomy and Related Fields. 1995-1996. American Institute of Physics, One Physics Ellipse, College Park, MD 20740-3843. (301) 209-3100. Annual. $45.00. ISSN: 0147-1821.

Research Centers Directory. Gale Research, 835 Penobscot Building, Detroit, MI 48226-4094. (313) 961-2242. (800) 347-4253. 20th edition. 1995. $485.00. ISSN: 0080- 1518.

ENCYCLOPEDIAS AND DICTIONARIES

A Concise Dictionary of Physics. Oxford University Press, Inc., 200 Madison Avenue, New York, NY 10016. (212) 725-6000. (800) 334-4249. 1990. $10.95.

Dictionary of Effects and Phenomena in Physics. Joachim Schubert. VCH Publications, Inc., 220 East 23rd Street, Suite 909, New York, NY 10010-4606. (212) 683-8333. (800) 422-8824. 1987. $35.00.

Dictionary of the Physical Sciences: Terms, Formulas, Data. Cesare Emiliani. Oxford University Press, Inc., 200 Madison Avenue, New York, NY 10016. (212) 725-6000. (800) 334-4249.1989. $19.95.

Encyclopedia of Applied Physics. George Trigg, editor. VCH Publications, Inc., 220 East 23rd Street, Suite 909, New York, NY 10010-4606. (212) 683-8333. (800) 422-8824. 20 volume set. 1991-. $5990.00.

Encyclopedia of Physics. Robert M. Besancon. Chapman & Hall, 1 Penn Plaza, New York, NY 10119. (212) 564-1060. 3rd edition. 1990. $54.95 in paper.

Facts on File Dictionary of Physics. John Daintith, editor. Facts-on-File, Inc., 460 Park Avenue South, New York, NY 10016-7382. (212) 683-2244. (800) 322-8755. Fax (800) 683-3633. Revised edition. 1990. $12.95.

Encyclopedia of Physical Science and Technology. Academic Press, Inc., 6277 Sea Harbor Drive, Orlando, FL. (800) 321-5068. 2nd edition. 18 volume set. 1992. $2625.00.

McGraw-Hill Encyclopedia of Science and Technology. McGraw-Hill Book Company, Inc., 1221 Avenue of the Ameri-

cas, New York, NY 10020. (212) 512-2000. (800) 262-4729. 7th edition. 20 volume set. 1992. $1900.00.

Penguin Dictionary of Physics. Valerie Illingworth, editor. Viking Penguin, 375 Hudson Street, New York, NY 10014-3657. (212) 366-2000. (800) 331-4624 2nd edition. 1991.

GENERAL WORKS

Dusty and Dirty Plasmas: Noise and Chaos in Space and in the Laboratory. H. Kikuchi, editor. Plenum Publishing Corp., 233 Spring Street, New York, NY. (212) 620-8000. (800) 2221-9369. FAX (212) 463-0742. 1995. $139.00.

Physicist's ABCs on Plasma. L. A. Artsimovich. State Mutual Book Service, Ltd. 521 Fifth Avenue, New York, NY 10175. (212) 682-5844. 1985. $14.00 in paper.

Plasma Physics: An Introduction to the theory of Astrophysical, Geophysical and Laboratory Plasmas. Peter A. Sturrock. Cambridge University Press, 40 West 20th Street, New York, NY 10011-4211. (212) 924-3900. (800) 872-7423. 1994. $64.95.

Plamsa Technology: Fundamentals and Applications. M. Capitelli and C. Gorse, editors. Plenum Publishing Corp., 233 Spring Street, New York, NY. (212) 620-8000. (800) 2221-9369. FAX (212) 463-0742. 1992. $65.00.

Principles of Plasma Mechanics. B. Chakraborty. Halsted Press, 605 Third Avenue, New York, 10158-0012. (212) 850-6400. 1991. $ 74.95.

Radio Frequency Heating of Plasmas. R. A. Cairns. IOP Publishing, Public Ledger Building, Suite 1035, Independence Square, Philadelphia, PA 19106. (215) 627-0880. (800) 358-4677 1991. $82.00.

Theory of Tokamak Plasmas. R. B. White. Elsevier Science Publishing Company, Inc., 655 Avenue of the Americas, New York, NY 10010. (212) 989-5800. FAX (914) 333-2444. 1989. $49.95 in paper.

HANDBOOKS AND MANUALS

Basic Data of Plasma Physics. Sanborn Brown. American Institute of Physics, One Physics Ellipse, College Park, MD 20740-3843. (301) 209-3100. 1994. $35.00.

Formulary for Plasma Physics. Andre Anders. VCH Publications, Inc., 220 East 23rd Street, Suite 909, New York, NY 10010-4606. (212) 683-8333. (800) 422-8824. 1991. $55.00.

Handbook of Plasma Processing Technology. Stephen M. Rossnagel. Noyes, 120 Mill Road, Park Ridge, NJ 07656. (201) 391-8484. 1990. $86.00.

Handbook of Thermodynamics and Optical Properties of Plasma of Metals and Dielectrics. Y.A. Protasov, editor. Hemisphere Publishing Corp., 1900 Frost Road, Suite 101, Bristol PA 19007. (215) 785-5800. (800) 821-8312. 1991. $99.50.

ONLINE DATABASES AND CD-ROMS

CA Search. Chemical Abstracts Service, P.O. Box 3012, Columbus, OH 43210-0012. (614) 447-3600. (800) 848-6533. FAX

(614) 447-3709. Very comprehensive guide to worldwide chemical literature and related fields, 1972 to present. Available on BRS,(800) 289-4277, DIALOG, (800) 334-2564, ORBIT (800) 456-7248, and STN International, FIZ Karlsruhe, P.O. Box 2465, W-7500, Karlsruhe 1, Germany, online services. Inquire as to cost and availability.

Current Contents Search. Institute for Scientific Information, 3501 Market Street, Philadelphia, PA 19104. (215) 386-0100. FAX (215) 386-6362. Contains citations to articles listed in the table of contents of science and technology journals. Also articles in social sciences and life sciences journals. Available on BRS, (800) 289-4277, DIALOG, (800) 334-2564, online services. Inquire as to cost and availability.

Dissertation Abstracts Online. University Microfilms International, 300 North Zeeb Road, Ann Arbor, MI 48106. (800) 521-0600 or (313) 761-4700. Scope includes virtually all doctoral dissertations accepted at accredited American institutions from 1861 to present in 252 subject areas. Available on BRS, (800) 289-4277, DIALOG, (800) 334-2564, and OCLC EPIC, (800) 848-5878, online services. Also available on CD-ROM. Inquire as to cost and availability.

INSPEC. Institution of Electrical Engineers, Michael Faraday House, Six Hills Way, Stevenage, Herts. SG1 2AY, England. Telephone: 0438 313311 or FAX 0438 742840. Contains citations to the worldwide literature of physics, electronics and electrical engineering, computer technology, and related fields. Available on BRS, (800) 289-4277, DIALOG, (800) 334-2564, ORBIT, (800) 456-7248, and STN International, FIZ Karlsruhe, P.O. Box 2465, W-7500, Karlsruhe 1, Germany, online services. Inquire as to cost and availability.

Mathsci. American Mathematical Society, P.O. Box 6248, Providence, RI 02940. (800) 321-4667 or (401) 455-4000 or FAX (401) 331-3842. Scope includes pure and applied mathematics and related areas of physics, statistics, engineering, computer science, and operations research literature since 1959. Available on DIALOG,(800) 334-2564,online service. Also available on CD-ROM. Inquire as to cost and availability.

NTIS Bibliographic Database. National Technical Information Service, 5285 Port Royal Road, Springfield, VA 22161. (703) 487-4929 or FAX (703) 321-8199. Broad coverage of government-sponsored science and technology research reports, 1964 to present. Available on BRS,(800) 289-4277, DIALOG, (800) 334-2564, ORBIT, (800) 456-7248, and STN International, FIZ Karlsruhe, P.O. Box 2465, W-7500, Karlsruhe 1, Germany, online services. Also available on CD-ROM. Inquire as to cost and availability.

Physics Briefs. American Institute of Physics, 335 East 45th Street, New York, NY 10017. (212) 661-9260 or FAX (212) 661-2036. Contains citations with abstracts of the literature of physics and related fields, 1979 to present. Available on the STN International, FIZ Karlsruhe, P.O. Box 2465, W-7500, Karlsruhe 1, Germany, online service. Inquire as to cost and availability.

SCISEARCH. Institute for Scientific Information, 3501 Market Street, Philadelphia, PA 19104. (800) 523-1850 or (215) 386-0100. Broad multidisciplinary title and author index to the international literature of science and technology, 1974 to present. Available on DIALOG, (800) 334-2564, and ORBIT, (800) 456-7248, online services. Also available on CD-ROM. Inquire as to cost and availability.

WILSONLINE. H.W. Wilson Company, 950 University Avenue, Bronx, NY 10452. (800) 367-6770 or (212) 588-8400. Makes available online versions of the printed H.W. Wilson indexes including *Applied Science and Technology Index, Business Periodicals Index, General Science Index,* and *Readers' Guide to Periodical Literature.* Period covered is generally 1983 to present. Available on BRS, (800) 289-4277, DIALOG, (800) 334-2564, and OCLC EPIC, (800) 848-5878, online services.Also available on CD-ROM. Inquire as to cost and availability.

PERIODICALS

American Journal of Physics. American Association of Physics Teachers, 5112 Berywn Road, College Park, MD 20740. (301) 345-4200. 1933 to present. Monthly. $215.00 per year. ISSN: 0002-9505.

Contemporary Physics. Taylor & Francis, Ltd, Rankine Road, Basingstoke, Hants RG24 8PR, England. TEL 0256-840366. FAX 0256-479438. 1959 to present. Bimonthly. $318.00 per year. ISSN: 0010-7514.

Fusion Technology. American Nuclear Society, 555 North Kensington Avenue, LA Grange Park, IL 60525. (708) 352-6611. FAX (708) 352-0499. 1981 to present. 8 issues per year.

*$485.*00 per year. ISSN: 0748-1896.

Journal of Fusion Energy. Plenum Publishing Corporation, 233 Spring Street, New York, NY 10013. (212) 620-8000. FAX (212) 463-0742. 1981 to present. Quarterly. $315.00 per year. ISSN: 0164-0313.

International Journal of Thermophysics. Plenum Publishing Corporation, 233 Spring Street, New York, NY 10013. (212) 620-8000. FAX (212) 463-0742. 1980 to present. Bi-monthly. $535.00 per year. ISSN: 0195-928X

Physical Review Letters. American Institute of Physics, One Physics Ellipse, College Park, MD 20740-3843. (301) 209-3000. 1958 to present. Monthly. $1370.00. ISSN: 0031-9007.

Physics Letters B: Nuclear, Elementary Particle and High-Energy Physics. Elsevier Science Publishing Company, Inc., 655 Avenue of the Americas, New York, NY 10010. (212) 989- 5800. FAX (212) 633-3990. 1962 to present. 96 issues per year in 24 volumes. $4683.00 per year. ISSN: 0370-2693.

Physics of Plasmas. American Institute of Physics, One Physics Ellipse, College Park, MD 20740-3843. (301) 209-3000.1958 to present. Monthly. $1255.00 per year. ISSN: 1070- 664X.

Physics Reports: A Review Section of Physics Letters. Elsevier Science Publishing Company, Inc., 655 Avenue of the Americas, New York, NY 10010. (212) 989-5800. FAX (212) 633-3990. 1971 to present. 72 issues per year in 12 volumes. $2342.00 per year. ISSN: 0370-1573.

Physics Today. American Institute of Physics, One Physics Ellipse, College Park, MD 20740-3843. (301) 209-3000.1948 to present. Monthly. $140.00 per year. ISSN: 0031-9228.

Plasma Physics and Controlled Fusion. Elsevier Science Publishing Company, Inc., 655 Avenue of the Americas, New York, NY 10010. (212) 989-5800. FAX (212) 633-3990. 1959 to present. Monthly. $1226.00. ISSN: 0741-3335.

PLASMA PHYSICS

Ency. of Physical Sciences and Engineering Info. Sources

Plasma Physics Reports. American Institute of Physics, One Physics Ellipse, College Park, MD 20740-3843. (301) 209-3000. Translation of Fizika Plazmy. 1975 to present. Monthly. $1520.00 per year. ISSN: 1063-780X.

Plasma Chemistry and Plasma Processing. Plenum Publishing Corporation, 233 Spring Street, New York, NY 10013. (212) 620-8000. FAX (212) 463-0742. 1981 to present Quarterly. $295.00. ISSN: 0272-4324.

Plasma Sources Science and Technology. IOP Publishing. Orders to: American Institute of Physics, 500 Sunnyside Boulevard, Woodbury, NY 11797-2999. (516) 576-2200. 1992 to present. Quarterly. $272.00 per year. ISSN: 0963-0252.

Reports On Progress in Physics. IOP Publishing. U. S. orders to: American Institute of Physics, 500 Sunnyside Boulevard, Woodbury, NY 11797-2999. (516) 576-2200. 1934 to present. Monthly. $1214.00 per year. ISSN: 0034-4885.

RESEARCH CENTERS AND INSTITUTES

Fusion Technology Institute. University of Wisconsin- Madison, Department of Physics, 1150 University Avenue, Madison, WI 53706. (608) 263-2308. (608) 263-4499.

Laboratory for Plasma Research. University of Maryland, Energy Research Building, College Park, MD 20742. (301) 405-5039. (301) 314-9437.

Plasma Physics Laboratory. Columbia University, Applied Physics Department, 500 West 120th Street, New York NY 10027. (212) 854-4457. FAX (212) 854-8257.

Plasma Physics Laboratory. Princeton University, James Forrestal Research Campus, P.O. Box 451, Princeton, NJ 08543. (609) 243-2000. FAX (609) 243-2749.

PLASTICS

See also: ADHESIVES, CERAMICS, CHEMISTRY, COMPOSITES, MATERIALS SCIENCE, ORGANIC CHEMISTRY, PLATICS, THERMOPLASTICS

ABSTRACT SERVICES AND INDEXES

Applied Science and Technology Index: A Cumulative Subject Index to English Language Periodicals in the Fields of Aeronautics and Space Science, Computer Technology, Chemistry, Construction Industry, Energy and Related Areas. H.W. Wilson Co., 950 University Avenue, Bronx, NY 10452-9978. (212) 588-8400. (800) 367-6770. FAX (718) 590-1617. From 1958 to present. Monthly. Inquire about cost and availability. Also available online (BRS and WILSONLINE) and on CD-ROM. ISSN: 0003-6986.

Chemical Abstracts. Chemical Abstracts Service. 2540 Olentangy River Road, Box 3012, Columbus, OH 43210-0012. (614) 447-3600. FAX (614) 447-3713. 1907 to present. Weekly. $16,800.00 per year. Available online and on CD- ROM. CA is also available in five section groupings. Inquire Regarding Cost and Availability..

Current Contents: Physical, Chemical and Earth Sciences. Institute for Scientific Information, 3501 Market Street, Philadelphia, PA 19104. (215) 386-0100. (800) 523-1850. FAX (215) 386-2291. 1961 to present. Weekly. $442.00 per year. Also available online (BRS, DIALOG) and on CD-ROM. Inquire regarding cost and availability. ISSN: 0163-2574.

Engineered Materials Abstracts. ASM (American Society of Metals) International, Materials Information, Materials Park, OH 44073. (216) 338-5151 or FAX (216) 338-4634. Covers literature on technical developments in polymer, ceramic, and composite materials and engineering. 1986 to present. Monthly. $1175.00 per year. Also available on CD-ROM.

Engineering Index Monthly. Engineering Information, Inc., Castle Point on the Hudson, Hoboken, NJ 07030. (201) 216-8500. (800) 221-1044. FAX (201) 216-8532. Monthly. $2300.00 per year for monthly issues. $1980 per year for annual cumulation. Available on CD-ROM and online as COMPENDEX. ISSN: 0742-1974; 0360-8557.

General Science Index. H.W. Wilson Company, 950 University Avenue, Bronx, NY 10452. (212) 588-8400. (800) 367-6770. FAX (718) 590-1617. From 1978 to present. Ten issues per year; quarterly and annual cumulations. Service basis. Available on CD-ROM and online. Inquire regarding cost and availability. ISSN: 0162-1963.

Process and Chemical Engineering/Chemical Engineering Abstracts. Royal Society of Chemistry, Information Services, Thomas Graham House, Science Park, Milton Road, Cambridge, Cb4 4wf, England. Contains citations, mostly with abstracts, of the worldwide literature on chemical engineering; from 1982 to present. Monthly. $610.00 per year. Also available online. ISSN: 0960-5045.

Science Citation Index (SCI). Institute for Scientific Information, 3501 Market Street, Philadelphia, PA 19104. (215) 386-0100. (800) 523-1850. FAX (215) 386-2991. 1961 to present. Six issues per year, plus annual cumulation. $11650.00 per year. Also available online and on CD-ROM. Inquire about price and availability. ISSN: 0036-827X.

Theoretical Chemical Engineering. Royal Society of Chemistry, Information Services, Thomas Graham House, Science Park, Milton Road, Cambridge, Cb4 4wf, England. Covers theoretical chemical engineering, including theory and laboratory experimentation. 1964 to present. Monthly. $235.00. ISSN: 0960-5053.

ANNUAL REVIEWS AND YEARBOOKS

Advances in Chemical Engineering. Academic Press, Inc., 6277 Sea Harbor Drive, Orlando, FL 32887. (800) 321-5068. 1956 to present. Irregular. ISSN: 0065-2377.

Progress in Colloid and Polymer Science Series. Springer-Verlag, 175 Fifth Avenue, New York, NY 10010. (212) 460-1500. Irregular, price varies.

ASSOCIATIONS AND PROFESSIONAL SOCIETIES

American Chemical Society, 1155 16th Street, NW, Washington, DC 20036. (202) 872-4600.

American Institute of Chemical Engineers. 345 East 47th Street, New York, NY 10017-2396. (212) 705-7663.

Association of Consulting Chemists and Chemical Engineers, 50 East 41st Street, Suite 92, New York, NY 10017. (212) 684-6255.

Chemical Manufacturers Association. 2501 M Street, N.W., Washington, DC 20037. (202) 887-1182

Plastics Institute of America. 277 Fairfield Road, Suite 100, Fairfield, NJ 07004-1932. (201) 808-5950. FAX (201) 808-5953.

Society of Plastics Engineers, Inc., 14 Fairfield Drive, Brookfield, CT 06805-0403. (203) 775-0471. FAX (203) 775-8490.

Society of the Plastics Industry. 1275 K Street NW, Suite 400, Washington, DC 20005. (202) 371-5200. FAX (202) 371-1022.

BIBLIOGRAPHIES

Information Sources in Polymers and Plastics. R. T. Adkins. Bowker-Saur, 121 Chanlon Road, New Providence, NJ 07974. (908) 464-6800. (800) 521-8110. 1989. $75.00

Organic Reaction Mechanisms: An Annual Survey of Literature. A. C. Knipe and W. E. Watts, editors. John Wiley & Sons, Inc., 605 Third Avenue, New York, NY 10158-0012. (212) 850-6000. (800) 225-5945. vol. 92, 1992. price varies.

Scientific and Technical Books and Serials in Print: An Index to Literature in Science and Technology. R.R. Bowker Inc., 121 Chanlon Road, New Providence, NJ 07974. (908) 464-6800. (800) 521-8110. FAX (908) 665-3502. 1972 to present. Annual. 4 volumes. 1994. $299.95. Also available on compact disc and online. ISSN: 0000-054X.

DIRECTORIES AND BIOGRAPHICAL SOURCES

American Institute of Chemical Engineers. Directory. American Institute of Chemical Engineers. 345 East 47th Street, New York, NY 10017-2396. (212) 705-7663. Annual.

American Men and Women of Science. Physical and Biological Sciences. R.R. Bowker Company, 121 Chanlon Road, New Providence, NJ 07974. (908) 464-6800. (800) 521-8110. 20th edition. 8 volumes. 1996. $850.00

Chemical Engineering Faculties. American Institute of Chemical Engineers. 345 East 47th Street, New York, NY 10017-2396. (212) 705-7663. Annual. $75.00.

Consulting Services: Chemists and Chemical Engineers. Association of Consulting Chemists and Chemical Engineers, 40 East 45th Street, New York, NY 10017. (212) 983-3160. FAX (212) 983-3161. Biennial. $60.00.

Engineering Research Centers: Incorporating Electronics Research Centers. Stockton Press, 345 Park Avenue, New York, NY 10010. (212) 689-9200. (800) 221-2123. 4th edition. 1995. $515.00.

Modern Plastics Encyclopedia Directory of Trade Names. McGraw-Hill, 1221 Avenue of the Americas, New York, NY 10020. (212) 512-6266. Annual. $10.00.

Research Centers Directory. Gale Research, 835 Penobscot Building, Detroit, MI 48226-4094. (313) 961-2242. (800) 347-4253. 20th edition. 1995. $485.00. ISSN: 0080-1518.

Society of the Plastics Industry—Membership Directory and Buyers' Guide. 1275 K Street NW, Suite 400, Washington, DC 20005. (202) 371-5200. FAX (202) 371-1022. Annual. $180.00.

Who's Who in Engineering. Gordon Davis, editor. American Association of Engineering Societies. 1111 19th Street, NY, Suite 608, Washington, DC 20036. (202) 296-2237. (800) 658-8897. 9th edition. 1995. $220.00.

ENCYCLOPEDIAS AND DICTIONARIES

Atlas of Polymer and Plastics Analysis. Dieter O. Hummel. *VCH Publications, Inc.,* 220 East 23rd Street, Suite 909, New York, NY 10010-4606. (212) 683-8333. (800) 422-8824. Volume 1, Parts A and B. 3rd edition. 1991. $525.00.

Concise Encyclopedia of Polymer Science. Herman F. Mark. John Wiley & Sons, Inc., 605 Third Avenue, New York, NY 10158-0012. (212) 850-6000. (800) 225-5945. 1990. *$199.00*

Dictionary of Plastics Technology. R. J. Heath and A. W. Birley. Chapman & Hall, 1 Penn Plaza, New York, NY 10119. (212) 564-1060. 1992. $59.95.

Encyclopedia of Chemical Processing and Design. McKetta Marcel Dekker, Inc., 270 Madison Avenue, New York, NY 10016. (212) 696-9000. (800) 228-1160. 1976 - . $175.00 per volume.

Encyclopedia of Polymer Science and Engineering. Herman F. Mark et al, editors. John Wiley & Sons, Inc., 605 Third Avenue, New York, NY 10158-0012. (212) 850-6000. (800) 225-5945. 2nd edition. 1985 -. 22 volumes. $4788.00

Handbook of Polymer Science and Technology. Nicholas P. Cheremisinoff. Marcel Dekker, Inc., 270 Madison Avenue, New York, NY 10016. (212) 696-9000. (800) 228-1160. 4 volumes. 1987. $235.00 each.

Hawley's Condensed Chemical Dictionary. Van Nostrand Reinhold, 115 Fifth Avenue, New York, NY 10003. (212) 254-3232. (800) 842-3636. 13th edition. 1996. $99.95.

Illustrated Chemistry Laboratory Terminology. Gerbert W. Ockerman. CRC Press, Inc., 2000 Corporate Boulevard, NW, Boca Raton, FL 33431. (407) 994-0555. (800) 272-7737. 1991. $32.95.

Kirk-Othmer Encyclopedia of Chemical Technology. John Wiley and Sons, Inc., 605 Third Avenue, New York, NY 10158. (800) 526-5368 or (212) 850-6000. Fourth edition. 1991 - . Twenty-seven volumes. $5400.00.

Modern Plastics: Encyclopedia Issue. McGraw-Hill 1221 Avenue of the Americas, New York, NY 10020. (212) 512-6255. Annual; issued in October. $49.95. ISSN: 0085-3518.

Polymer Handbook: Polymerization and Depolymerization. J. Brandrup and E. H. Immergut. John Wiley & Sons, Inc., 605 Third Avenue, New York, NY 10158-0012. (212) 850-6000.(800) 225-5945. Volume 1. 3rd edition. 1989. $245.00.

Ullman's Encyclopedia of Industrial Chemistry. VCH Publications, Inc., 220 East 23rd Street, Suite 909, New York, NY 10010-4606. (212) 683-8333. (800) 422-8824. 5th edition. 1984 - . Price varies per volume.

GENERAL WORKS

IEEE Rubber and Plastics Industry Technical Conference. Institute of Electrical and Electronic Engineers (IEEE), 345 East 47th Street, New York, NY 10017. (212) 705-7900. Annual.

Introduction to Plastics. Hans-Georg Elias. VCH Publications, Inc., 220 East 23rd Street, Suite 909, New York, NY 10010-4606. (212) 683-8333. (800) 422-8824. 1993. $50.00.

Plastics Engineering. R. J. Crawford. Elsevier Science Publishing Company, Inc., 655 Avenue of the Americas, New York, NY 10010. (212) 989-5800. FAX (914) 333-2444 2nd edition. 1987. $44.00 in paper.

Plastics for Electronic Materials. William M. Alvino. McGraw-Hill Publishing Company, Inc., 1221 Avenue of the Americas, New York, NY 10020. (212) 512-2000. (800) 262- 4729. 1994. $50.00.

Plastics: Microstructures Properties and Applications. N. J. Mills. Halsted Press, 605 Third Avenue, New York, 10158- 0012. (212) 850-6400. 2nd edition. 1994. $52.95.

Plastics Recycling: Products and Processes. Raymond J. Ehrig. Hanser-Gardener Publications, 6600 Clough Pike, Cincinnati, OH 45244-4090. (513)527-8977. (800) 950-8977.1992. $62.50.

Polymer Chemistry: An Introduction. Raymond B. Seymour and Charles E. Carraher Jr. Marcel Dekker, Inc., 270 Madison Avenue, New York, NY 10016. (212) 696-9000. (800) 228-1160. 3rd edition. 1992. $59.75.

Pvc Plastics: Properties, Processing and Applications. W. V. Titow. Elsevier Science Publishing Company, Inc., 655 Avenue of the Americas, New York, NY 10010. (212) 989-5800. FAX (914) 333-2444 1991. $230.00

HANDBOOKS AND MANUALS

Chemical Formulary. H. Bennett, editor. Chemical Publishing, Co., Inc. 80 Eighth Avenue, New York, NY 10011. (212) 255-1950. Volumes 1-30. $60.00 per volume.

CRC Handbook of Chemistry and Physics. David R. Kide, editor. CRC Press Inc., 2000 Corporate Blvd., NW, Boca Raton, FL 33431. (407) 994-0555. (800) 333-8300. 77th edition. 1996. $99.95.

Flammability Handbook for Plastics. Carlos J Hilado, Editor. Technomic Publishing Co., 851 New Holland Avenue, Box 3535, Lancaster PA 17604. (800) 233-9936. 4th edition. 1990. $65.00.

Handbook of Organic Chemistry. Hans Beyer and Wolfgang Walter. Prentice-Hall General Referal Reference & Travel, 15

Columbus Circle, New York, NY 10023. (212) 373-8500. (800) 223-2348. 1995. $96.00.

Handbook of Plastics Materials and Technology. Irvin I. Rubin. John Wiley and Sons, Inc., 605 Third Avenue, New York, NY 10158. (800) 526-5368 or (212) 850-6000 1990. $185.00.

Improving Safety in the Chemical Laboratory: A Practical Guide. Jay A. Young, editor. John Wiley and Sons, Inc., 605 Third Avenue, New York, NY 10158. (800) 526-5368 or (212) 850-6000. Second edition. 1991. $75.00.

Handbook of Plastics, Elastomers and Composites. Charles A. Harper. McGraw-Hill Publishing Company, Inc., 1221 Avenue of the Americas, New York, NY 10020. (212) 512-2000. (800) 262-4729. 1992. $89.50.

Plastics Engineering Handbook of the Society of the Plastics Industry. Michael L. Berins, editor. Chapman & Hall 5th edition. 1995. $104.00.

Plastics Handbook. Modern Plastics Magazine Staff. McGraw-Hill Publishing Company, Inc., 1221 Avenue of the Americas, New York, NY 10020. (212) 512-2000. (800) 262-4729. 1995. $55.50.

Plastics Technology Handbook. Manas Chanda and Salil K. Roy. 2nd edition. Marcel Dekker, Inc., 270 Madison Avenue, New York, NY 10016. (212) 696-9000. (800) 228-1160. 1992. $215.00.

Riegel's Handbook of Industrial Chemistry. James A. Kent, editor. Van Nostrand Reinhold, 115 Fifth Avenue, New York, NY 10003. (212) 254-3232 or (800) 926-2665. Ninth edition. 1992. $114.95.

ONLINE DATABASES AND CD-ROMS

Beilstein Online. Beilstein-Institut fur Literatur der Organischen Chemie, Varrentrapperstrasse, 40-42, D-6000, Frankfurt am Main 90, Germany. Contains data on carbon compounds from the Beilstein Handbook of Organic Chemistry. Structural and factual data for more than 3.4 million heterocyclic, isocyclic, and acyclic compounds included. Covers the period 1830 to present. Available on DIALOG, (800) 334-2564, ORBIT (800) 456-7248, and STN International, FIZ Karlsruhe, P.O. Box 2465, W-7500, Karlsruhe 1, Germany, online services. Inquire as to cost and availability.

CA Search. Chemical Abstracts Service, P.O. Box 3012, Columbus, OH 43210-0012. (614) 447-3600. (800) 848-6533. FAX (614) 447-3709. Very comprehensive guide to worldwide chemical literature and related fields, 1972 to present. Available on BRS,(800) 289-4277, DIALOG, (800) 334-2564, ORBIT (800) 456-7248, and STN International, FIZ Karlsruhe, P.O. Box 2465, W-7500, Karlsruhe 1, Germany, online services. Inquire as to cost and availability.

Chemical Journals of the American Chemical Society. American Chemical Society, 1155 16th Street, N.W., Washington, DC 20036. (202) 872-4381 or (800) 424-6747. Contains complete text of approximately 90,000 articles from 22 primary journals published by the American Chemical Society, from mostly 1982 to present. Available on STN International, FIZ Karlsruhe, P.O. Box 2465, W-7500, Karlsruhe 1, Germany, online service. Inquire as to cost and availability.

Compendex Plus. Engineering Information, Inc., 345 East 47th Street, New York, NY 10017. (212) 705-7600 or (800) 221-1044. Contains citations with abstracts to worldwide literature in engineering and technology, from 1970 to present. Available on online BRS,(800) 289-4277, DIALOG, (800) 334-2564, ORBIT (800) 456-7248, and STN International, FIZ Karlsruhe, P.O. Box 2465, W-7500, Karlsruhe 1, Germany, online services. Also available on CD-ROM. Inquire as to cost and availability.

Current Contents Search. Institute for Scientific Information, 3501 Market Street, Philadelphia, PA 19104. (215) 386-0100. FAX (215) 386-6362. Contains citations to articles listed in the table of contents of science and technology journals. Also articles in social sciences and life sciences journals. Available on BRS,(800) 289-4277, DIALOG,(800) 334-2564, online services. Inquire as to cost and availability.

Dissertation Abstracts. University Microfilms International, 300 North Zeeb Road, Ann Arbor, MI 48106. (800) 521-0600 or (313) 761-4700. Scope includes virtually all doctoral dissertations accepted at accredited American institutions from 1861 to present in 252 subject areas. Available on BRS,(800) 289-4277, DIALOG,(800) 334-2564, and OCLC EPIC,(800) 848-5878, online services. Also available on CD-ROM. Inquire as to cost and availability.

Engineered Materials Abstracts. Materials Information, ASM International, Materials Park, OH 44073. (216) 338-5151. Contains citations with abstracts of worldwide literature in the development, processing, and production of ceramic, composite, and polymeric materials for engineering uses.Available on DIALOG, (800) 334-2564, ORBIT (800) 456-7248, and STN International, FIZ Karlsruhe, P.O. Box 2465, W-7500, Karlsruhe 1, Germany, online services. Inquire as to cost and availability.

Kirk-Othmer Encyclopedia of Chemical Technology. John Wiley and Sons, Inc., 605 Third Avenue, New York, NY 10158. (800) 526-5368 or (212) 850-6000. Contains the complete text of all chapters in the 27 volume fourth edition of the *KIRK-Othmer Encyclopedia of Chemical Technology.* 1991. Available on BRS,(800) 289-4277, DIALOG,(800) 334-2564, online services. Inquire as to cost and availability.

NTIS Bibliographic Database. National Technical Information Service, 5285 Port Royal Road, Springfield, VA 22161. (703) 487-4929 or FAX (703) 321-8199. Broad coverage of government-sponsored science and technology research reports, 1964 to present. Available on BRS,(800) 289-4277, DIALOG, (800) 334-2564, ORBIT, (800) 456-7248, and STN International, FIZ Karlsruhe, P.O. Box 2465, W-7500, Karlsruhe 1, Germany, online services. Also available on CD-ROM. Inquire as to cost and availability.

Scientific and Technical Books and Serials in Print. R.R. Bowker Inc., 121 Chanlon Road, New Providence, NJ 07974. (800) 521-8110 or (908) 464-6800. List of currently published books and serials in the physical and biological sciences, engineering and technology, with subject, author and titles indexes. Available on ORBIT,(800) 456-7248, online service.Inquire as to cost and availability.

SCISEARCH. Institute for Scientific Information, 3501 Market Street, Philadelphia, PA 19104. (800) 523-1850 or (215) 386-0100. Broad multidisciplinary title and author index to the international literature of science and technology, 1974 to present. Available on DIALOG,(800) 334-2564, and ORBIT,(800) 456-7248, online services. Also available on CD-ROM. Inquire as to cost and availability.

Who's Who in Technology. Gale Research, 835 Penobscot Building, Detroit, MI 48226-4094. (313) 961-2242. (800) 347-4253. Contains biographical information of contemporary American scientists and engineers. Available on DIALOG,(800) 334-2564, online service. Inquire as to cost and availability.

WILSONLINE. H.W. Wilson Company, 950 University Avenue, Bronx, NY 10452. (800) 367-6770 or (212) 588-8400. Makes available online versions of the printed H.W. Wilson indexes including *Applied Science and Technology Index, Business Periodicals Index, General Science Index,* and *Readers' Guide to Periodical Literature.* Period covered is generally 1983 to present. Available on BRS,(800) 289-4277, DIALOG,(800) 334-2564, and OCLC EPIC,(800) 848-5878, online services. Also available on CD-ROM. Inquire as to cost and availability.

PERIODICALS

Advances in Polymer Technology. John Wiley and Sons, Inc., 605 Third Avenue, New York, NY 10158. (800) 526-5368 or (212) 850-6000. 1977 to present.

Aiche Journal. American Institute of Chemical Engineers, 345 47th Street, New York, NY 10017-2396. (212) 705-7663. 1955 to present. Monthly. $295.00 per year.

Canadian Plastics. Southam Magazine Group, 1450 Don Mills Road, Don Mills, ON M3B 2X7. Canada. (416) 445-6641. FAX (416) 442-2213. 1943 to present. 8 issues per year. $55.00 per year. ISSN: 0008-4778.

Chemical Week: Includes Annual Buyers Guide. Chemical Week Associates, 888 Seventh Avenue, New York, NY 10106. (212) 621-4900. FAX (212) 621-4949. 1914 to present. Weekly. $99.00 per year. ISSN: 0009-272X.

Chemtech. American Chemical Society. Box 3337, Columbus, OH 43210. (614) 447-3776. 1970 to present. Monthly. $370.00 per year. ISSN: 0009-2703.

High Performance Plastics. Elsevier Science Publishing Company, Inc., 655 Avenue of the Americas, New York, NY 10010. (212) 989-5800. 1983 to present. Monthly. $390.00 per year. ISSN: 0264-7753.

Industrial and Engineering Chemistry Research. American Chemical Society, Box 3337, Columbus, OH 43210. (614) 447-3776. (800) 333-9511. FAX (614) 447-3671. 1962 to present. Monthly. $567.00 per year. ISSN: 0888-5885.

Journal of Cellular Plastics. Technomic Publishing Company, Inc., 851 New Holland Avenue, Box 3535, Lancaster, PA 17604. (717) 291-5609. FAX (717) 295-4538. 1965 to present. Bi-monthly. $175.00 per year. ISSN: 0021-955X

Journal of Elastomers and Plastics. Technomic Publishing Company, Inc., 851 New Holland Avenue, Box 3535, Lancaster, PA 17604. (717) 291-5609. FAX (717) 295-4538. 1969 to present. Quarterly. $205.00 per year. ISSN: 0095-2443.

Journal of Materials Processing and Manufacturing. Technomic Publishing Company, Inc., 851 New Holland Avenue, Box 3535, Lancaster, PA 17604. (717) 291-5609. 1992 to present. Quarterly. $195.00 per year.

Journal of Plastic Film and Sheeting. Technomic Publishing Company, Inc., 851 New Holland Avenue, Box 3535, Lancaster,

PLASTICS

Ency. of Physical Sciences and Engineering Info. Sources

PA 17604. (717) 291-5609. FAX (717) 295-4538. 1985 to present. Quarterly. $265.00. per year. ISSN: 8756-0879.

Journal of Polymer Science. Part A: Polymer Chemistry. John Wiley and Sons, Inc., 605 Third Avenue, New York, NY 10158. (212) 850-6088. FAX (212) 850-6088. 1962 to present. Subscription includes part B: Polymer Physics, and Symposia Proceedings. 13 issues per year. $2540.00 per year. ISSN: 0887-624X; 0887-6266; 0360-8905.

Modern Plastics. McGraw-Hill, Inc., 1221 Avenue of the Americas, New York, NY 10020. (212) 512-6267. FAX (212) 512-6111. 1925 to present. Monthly. $41.75 per year. ISSN: 0026-8275.

Plastics Compounding: For Resin Producers, Formulators and Compounders. Advanstar Communications, Inc., 7500 Old Oak Boulevard, Cleveland, OH 44130. (216) 826-2839. FAX (216) 891-2726. 1977 to present. Bi-monthly. $40.00 per year.ISSN: 0148-9119.

Plastics Design Forum. Advanstar Communications, Inc., 7500 Old Oak Boulevard, Cleveland, OH 44130. (216) 826-2839. FAX (216) 891-2726. 1976 to present. 7 issues per year. $40.00 per year. ISSN: 0362-9376.

Plastics Engineering. Society of Plastics Engineers, Inc., 14 Fairfield Drive, Brookfield, Center, CT 06805-0403. (203) 775-0471. FAX (203) 775-8490. 1945 to present. Monthly. $50.00 per year. ISSN: 0091-9578.

Plastics Technology. Bill Communications Inc., 355 Park Avenue South, New York, NY 10011-1706. (212) 592-6570. 1955 to present. 13 issues per year. $65.00 per year. ISSN: 0032-1257.

Polymer Composites. Society of Plastics Engineers, Inc., 14 Fairfield Drive, Brookfield, CT 06805-0403. (203) 775-0471. FAX (203) 775-8490. 1980 to presesnt. Bi-monthly. $200.00 per year. ISSN: 0272-8397.

Processing of Advanced Materials. Routledge, Chapman and Hall, Inc., 29 West 35th Street, New York, NY 10001-2291. (212) 244-3336. 1991 to present. Quarterly. $145.00 per year.

RESEARCH CENTERS AND INSTITUTES

Center for Plastics Recycling Research. Rutgers University, Building 3529, Busch Campus, PO Box 1179, Piscataway, NJ 08855-1179. (809) 932-3683. FAX (98) 932-5636.

Institute of Polymer Engineering. University of Akron, Akron, OH 44325-0301. (216) 972-6865. (216) 258-2339.

Plastics Institute of America. 277 Fairfield Road, Suite 100, Fairfield, NJ 07004-1932. (201) 808-5950. FAX (201) 808-5953.

Polymer Processing Institute. Stevens Institute of Technology, Castle Point Station, Hoboken, NJ 07030.

(201) 420-5819.

Polymer Research Center. University of Cincinnati, Mail Location 172, Cincinnati, OH 45221. (513) 465-2453.

Polymer Research Laboratory. University of Michigan, 2045 Dow Building, Ann Arbor, MI 48109. (313) 763-2240. FAX (313) 763-4788.

PLATE TECTONICS

See also: CONTINENTAL DRIFT, CRUST, GEOPHYSICS, PHYSICAL GEOLOGY, STRUCTURAL GEOLOGY, SEISMOLOGY

ABSTRACT SERVICES AND INDEXES

Bibliography and Index of Geology. American Geological Institute, 4220 King Street, Alexandria, VA 22302. (703) 379-2480. Fax (703) 379-7563. Monthly. $1295.00 per year. Also available online as GeoRef.

Deep-Sea Research. Part A: Oceanographic Research Papers. Part B: Oceanographic Literature Review. Pergamon Press, Inc., Maxwell House, Fairview Park, Elmsford, NY 10523. (914) 592-7700. Fax (914) 592-3625. Twelve times per year. $1370 per year for Parts A and B.

Geological Abstracts. Elsevier-Geo Abstracts, Regency House, 34 Duke Street, Norwich NR3 3AP, England. Monthly. $760.00 per year. Also available online as GEOBASE.

Geological Society of America. Abstracts with Programs. Geological Society of America. 3300 Penrose Place, P.O. Box 9140, Boulder, CO 80301-9140. (303) 447-2020. Abstracts and programs of the annual conference. Annual. $69.00.

Geophysics Abstracts. Pergamon Press, Inc., Maxwell House, Fairview Park, Elmsford, NY 10523. (914) 592-7700. Fax (914) 592-3625. Twelve times per year. $565.00 per year. Also available in microform.

Oceanic Abstracts. Cambridge Scientific Abstracts, 7200 Wisconsin Avenue, Bethesda, MD 20814. (301) 961-6750. Fax (301) 961-6720. Bimonthly. $995.00 per year.

ANNUAL REVIEWS AND YEARBOOKS

Advances in Geophysics. Academic Press, Inc., 1250 Sixth Avenue, San Diego, CA 92101. (619) 231-0926. FAX (619) 699-6715. Irregular. Inquire about price and availability.

Annual Review and Earth and Planetary Sciences. Annual Reviews, Inc., 4139 El Camino Way, Palo Alto, CA 94306-0897. (415) 493-4400. Fax (415) 855-9815. Annual. $55.00 per year.

ASSOCIATIONS AND PROFESSIONAL SOCIETIES

American Geophysical Union, 2000 Florida Avenue, N.W., Washington, DC 20009. (202) 462-6900.

Geological Society of America. 3300 Penrose Place, P.O. Box 9140, Boulder, CO 80301-9140. (303) 447-2020.

Society of Exploration Geophysicists. P.O. Box 702740, Tulsa, OK 74170. (918) 493-3516.

BIBLIOGRAPHIES

Geologic Reference Sources: A Subject and Regional Bibliography of Publications and Maps in the Geological Sciences. Dederick C. Ward et al. Scarecrow Press, Inc., 52 Liberty Street, Box 4167, Metuchen, NJ 08840. (800) 537-7107 or (201) 548-8600. 1981. $49.50.

Information Sources in the Earth Sciences. David N. Wood, Joan E. Hardy, and Anthony P. Harvey, editors. Bowker-Saur/K.G. Saur. Distributed by R.R. Bowker, 121 Chanlon Road, New Providence, NJ 07974. (800) 521-8110 or (908) 464-6800. Second edition. 1989. $85.00.

DIRECTORIES AND BIOGRAPHICAL SOURCES

Research Centers Directory. Gale Research, 835 Penobscot Building, Detroit, MI 48226-4094. (800) 347-4253 or (313) 961-2242. 17th edition, 1992. $400.00.

ENCYCLOPEDIAS AND DICTIONARIES

Dictionary of Geological Terms. American Geological Institute. Doubleday and Company, Inc., 245 Park Avenue, New York, NY 10017. (800) 645-6156 or (212) 953-4561. Third edition. 1984. $12.00 in paper.

Encyclopedia of Structural Geology and Plate Tectonics. (Encyclopedia of the Earth Sciences Series). Van Nostrand Reinhold, 115 Fifth Avenue, New York, NY 10003. (212) 254-3232 or (800) 926-2665. 1987. $114.95.

Glossary of Geology. Robert L. Bates and Julia A. Jackson. American Geological Institute, 4220 King Street, Alexandria, VA 22032. (703) 379-2480. Third edition. 1987. $75.00.

GENERAL WORKS

Behavior of the Earth: Continental and Seafloor Mobility. Claude Allegre. Harvard University Press, 79 Garden Street, Cambridge, MA 02138. (617) 495-2600. 1988. $39.95

Continents in Motion: The New Earth Debate. Walter Sullivan. American Institute of Physics, 335 East 45th Street, New York, NY 10017. (212) 661-9404 or (800) 445-6638. 1991. $50.00.

Earth's Changing Surface. M.J.Selby. Oxford University Press, Inc., 200 Madison Avenue, New York, NY 10016. (800) 334-4249 or (212) 679-7300. 1985. $35.00 in paper.

Evolution of the Earth. Robert H. Dott, Jr. and Roger L. Batten. McGraw-Hill Publishing Company, 1221 Avenue of the Americas, New York, NY 10020. (800) 262-4729 or (212) 512-3825. Fourth edition. 1988. $41.00.

Exposed Cross-Sections of the Continental Crust. Matthew H. Salisbury and David M. Fountain, editors. Kluwer Academic Publishers, P.O. Box 358, Hingham, MA 02018-0358. (617) 871-6000. 1990. $185.00.

Plate Tectonics and Crustal Evolution. Kent C. Condie. Pergamon Press, Maxwell House, Fairview Park, Elmsford, NY 10523. (914) 592-7700. Fax (914) 592-3625. Third edition. 1989. $55.00.

Seismology and Plate Tectonics. David Gubbins. Cambridge University Press, 40 West 20th Street, New York, NY 10011-4211. (212) 924-3900. 1990. $59.50.

Tectonic Evolution of North America. D.J. Blundell and A.D. Gibbs, editors. Oxford University Press, Inc., 200 Madison Avenue, New York, NY 10016. (800) 334-4249 or (212) 679-7300. 1991. $125.00.

HANDBOOKS AND MANUALS

Field Mapping for Geology Students. F. Ahmed and D.C. Almond. Unwin Hyman, Inc., 10 East 53rd Street, New York, NY 10022. (212) 207-7626 or (800) 242-7737. 1983. $18.95.

Geology in the Field. R.R. Compton. John Wiley and Sons, Inc., 605 Third Avenue, New York, NY 10158. (800) 526-5368 or (212) 850-6418. 1985. $37.95.

The Mapping of Geological Structures. K.R. McClay. John Wiley and Sons, Inc., 605 Third Avenue, New York, NY 10158. (800) 526-5368 or (212) 850-6000. 1988. $18.95.

ONLINE DATABASES AND CD-ROMS

Earth Sciences. U.S. Geological Survey, 12201 Sunrise Valley Drive, Reston, VA 22092-9998. (703) 648-4460. CD-ROM of earth science databases, including the U.S. Geological Survey Library, Earth Science Data Directory, and GEOINDEX, citations to published geological maps. $350.00 per year with quarterly updates.

Geoarchive. Geosystems, P.O. Box 1024, Westminster, London, England, SW1 P 2JL. Citations to literature on geoscience, 1969 to present. Inquire as to online cost and availability.

Geobase. Elsevier-Geo Abstracts, Regency House, 34 Duke Street, Norwich NR3 3AP, England. Contains citations to the worldwide earth science literature from 1980 to date. Available on DIALOG, ORBIT online services. Inquire as to cost and availability.

Geomechanics Abstracts. University of London, Imperial College of Science and Technology, Rock Mechanics Information Service, Royal School of Mines, Prince Consort Road, London SW7 2BP, England. Telephone: 071-589 5111 or FAX 071-589 6806. Scope includes worldwide literature on engineering geology, hydrogeology, mining, rock mechanics, soil mechanics, and tunnelling, 1977 to present. Available on the ORBIT online service. Inquire as to cost and availability.

Georef. American Geological Institute, 4220 King Street, Alexandria, VA 22302. (800) 336-4764 or (703) 379-2480. Geology and geosciences literature, 1785 to present for North America. Available on DIALOG, ORBIT online services. Also available on CD-ROM. Inquire as to cost and availability.

PERIODICALS

American Journal of Science. Kline Geology Laboratory, Yale University, Box 6666, Yale Station, New Haven, CT 06511-8130. (203) 432-3131. Ten times per year. $40.00 per year.

Continental Shelf Research. Pergamon Press, Maxwell House, Fairview Park, Elmsford, NY 10523. (914) 592-7700. Fax (914) 592-3625. Monthly. $420.00.

Earth and Planetary Science Letters. Elsevier Science Publishing Company, Inc., 655 Avenue of the Americas, New York, NY 10010. (212) 989-5800. Twenty times per year. $945.00 per year.

Geological Society. Journal. Geological Society of London. Geological Society Publishing House, Unit 7, Brassmill Enterprise Centre, Brassmill Lane, Bath, Avon BA1 3JN, England. 6 times per year. $432.00.

Geological Society of America Bulletin. P.O. Box 9140, 3300 Penrose Place, Boulder, CO 80301. (303) 447-2020. Fax (303) 447-1133. Monthly. $150.00.

Geology. Geological Society of America. P.O. Box 9140, 3300 Penrose Place, Boulder, CO 80301. (303) 447-2020. Fax (303) 447-1133. Monthly. $120.00 per year.

Geophysics. Society of Exploration Geophysicists, P.O. Box 702740, Tulsa, OK 74170. (918) 493-3516. Monthly. $175.00 per year.

Journal of Geology. University of Chicago Press, 5720 Woodlawn Avenue, Chicago, IL 60637. (312) 753-3347. Bimonthly. $38.00 per year.

Journal of Geophysical Research. American Geophysical Union, 2000 Florida Avenue, N.W., Washington, DC 20009. (202) 462-6903. Monthly. $2170.00 per year.

Journal of Structural Geology. Pergamon Journals, Inc., Maxwell House, Fairview Park, Elmsford, NY 10523. (914) 592-7700. Ten times per year. $375.00 per year.

Marine Geology. Elsevier Science Publishing Company, Inc., 655 Avenue of the Americas, New York, NY 10010. (212) 989-5800. Twenty-fours times per year. $995.00 per year.

Marine Geophysical Researches. Kluwer Academic Publishers, P.O. Box 358, Hingham, MA 02018-0358. (617) 871-6000. Quarterly. $237.50 per year.

Reviews of Geophysics. American Geophysical Union, 2000 Florida Avenue, N.W., Washington, DC 20009. (202) 462-6900. Quarterly. $300.00 per year.

Tectonics. American Geophysical Union, 2000 Florida Avenue, N.W., Washington, DC 20009. (202) 462-6900. Bimonthly. $230.00 per year to individuals.

Tectonophysics: An International Journal of Geotectonics and the Geology and Physics of the Interior of the Earth. Elsevier Science Publishing Company, Inc., 655 Avenue of the Americas, New York, NY 10010. (212) 989-5800. Fifty-six times per year. $1275.00 per year.

RESEARCH CENTERS AND INSTITUTES

Ocean Drilling Program. Joint Oceanographic Institutions, Inc., 1755 Massachusetts Avenue, NW, Suite 800, Washington, DC 20036. (202) 232-3900.

U.S. Geological Survey, Geologic Division, National Center, 12201 Sunrise Valley Drive, Reston, VA 22092. (703) 648-6600. The major geological research agency of the federal government conducting research in most areas of pure and applied research in the geosciences.

University of Alaska Fairbanks. Geophysical Institute. C.T. Elvey Building, Fairbanks, AK 99775-0800. (907) 474-7282. Fax (907) 474-7290.

University of California, San Diego. Scripps Institution of Oceanography, Geological Research Division. San Diego, CA 92093-0220. (619) 534-1829. FAX (619) 534-0784.

PLUTO

See also: NATURAL SATELLITES, PLANETARY SCIENCES, SOLAR SYSTEM

ABSTRACT SERVICES AND INDEXES

Applied Science and Technology Index: A Cumulative Subject Index to English Language Periodicals in the Fields of Aeronautics and Space Science, Computer Technology, Chemistry, Construction Industry, Energy and Related Areas. H. W. Wilson Co., 950 University Avenue, Bronx, NY 10452- 9978. (212) 588-8400. (800) 367-6770. FAX (718) 590-1617.From 1958 to present. Monthly. Inquire about cost and availability. Also available online (BRS and WILSONLINE) and on CD-ROM. ISSN: 0003-6986.

Astronomy and Astrophysics Abstracts. Springer-Verlag New York, 175 Fifth Avenue, New York, NY 10010. (212) 460-1500. FAX (212) 473-6272. Published for the Astronomisches Rechen-Institut. Comprehensive coverage of all aspects of astronomy, astrophysics and related fields. 1969 to present. Two parts per year. Annual. ISSN: 0067-0022.

General Science Index. H.W. Wilson Company, 950 University Avenue, Bronx, NY 10452. (212) 588-8400. (800) 367- 6770. FAX (718) 590-1617. From 1978 to present. Ten issues per year; quarterly and annual cumulations. Service basis. Available on CD-ROM and online. Inquireregarding cost and availability. ISSN: 0162-1963.

Meteorological and Geoastrophysical Abstracts. American Meteorological Society, c/o Inforonics, Inc., 550 Newtown Road, Box 458, Littleton, MA 01460. (508) 486-8976. FAX (508) 486-0027. Covers literature in environmental sciences, meteorology, astrophsyics, hydrology, glaciology and physical oceanography. 1950 to present. Monthly. $950.00 per year. Also available online (DIALOG) and on CD-ROM. ISSN: 0026-1130.

NTIS Alerts: Astronomy & Astrophysics. U. S. National Technical Information Service. 5285 Port Royal Road, Springfield, VA 22161. (703) 487-4650. FAX (703) 321-8547. Weekly. $140.00 per year.

Physics Abstracts. INSPEC. Section A, Science Abstracts. Institute of Electrical Engineers, London, United Kingdom. Available from: INSPEC/IEEE-Institute of Electrical and Electronic Engineers, Box 1331, Hoes Lane, Piscataway, NJ 08855-1331. (908) 562-5549. 1898 to present. 24 issues per year. $2835.00 per year. ISSN: 0036-8091. Also available online and on CD-ROM.

Science Citation Index (SCI). Institute for Scientific Information, 3501 Market Street, Philadelphia, PA 19104. (215) 386-0100. (800) 523-1850. FAX (215) 386-2991. 1961 to present. Six issues per year, plus annual cumulation. $11650.00 per year. Also available online and on CD-ROM. Inquire about price and availability. ISSN: 0036-827X.

STAR. Scientific and Technical Aerospace Reports. U.S. National Aeronautics and Space Administration. Distributed by U. S. Superintendent of Documents, Washington, DC 20402. 1963 to present. Semi-monthly, with semiannual and annual indexes. $114.00 per year. ISSN: 0036-8741. Also available online and on CD-ROM.

ANNUAL REVIEWS AND YEARBOOKS

The Astronomical Almanac. Superintendent of Documents, U. S. Government Printing office, Washington, DC 10402. (202) 783-3238. 1981 to present. Superceeds Astronomical Ephemeris. Annual. $44.95. ISSN: 0737-6421.

Annual Review of Astronomy and Astrophysics. Original reviews of critical literature and current developments in astronomy and astrophysics. Annual Reviews, Inc., 4139 El Camino Way, Palo Alto, CA 94303-0139. (415) 493-4400. (800)523-8635. Fax (415) 855-9815. 1963 to date. Annual. $60.00. ISSN: 0066-4146.

Annual Review of Earth and Planetary Sciences. Annual Reviews, Inc., 4139 El Camino Way, Palo Alto, CA 94303-0139. (415) 493-4400. (800) 523-8635. Fax (415) 855-9815. 1973 to date. $62.00. ISSN: 0084-6597.

ASSOCIATIONS AND PROFESSIONAL SOCIETIES

American Astronomical Society. 2000 Florida Avenue NW, Suite 400, Washington, DC 20009. (202) 328-2010. FAX(202) 234-2560.

American Geophysical Union. 2000 Florida Avenue NW, Washington, DC 20009. (202) 462-6900. (800) 966-AGU1. FAX (202) 328-0566.

American Institute of Physics. 1 Physics Ellipse, College Park, MD 20740-3843. (301) 209-3100.

Association of Universities for Research in Astronomy, Inc. (AURA). Suite 701, 1625 Massachusetts Avenue, NW, Washington, DC 20036. (202) 483-2101. FAX (202) 483-2106.

Astronomical Society of the Pacific. 390 Ashton Avenue, San Francisco, CA 94112. (415) 337-1100. FAX: (415) 337-5205.

Planetary Society. 65 North Catalina Avenue, Pasadena, CA 91106. (818) 793-5100. (800) WOW-MARS. FAX (818) 793-5528.

ATLASES

Atlas of the Solar System. B. Yenne. Simon & Schuster, Inc. 1230 Avenue of the Americas, New York, NY 10020. (212) 698-7000. (800) 223-2348. 1987. $12.98.

The Sky: A User's Guide. David H. Levy. Cambridge University Press, 40 West 20th Street, New York, NY 10011- 4211. (212) 924-3900. (800) 872-7423. 1993. $14.95.

The View From Space: Photographic Exploration of the Planets. Merton Davies and Bruce C. Murray. Columbia University Press, 562 West 113th Street, New York, NY 10025. (212) 666-1000. (800) 944-8648. 1973. $17.50.

BIBLIOGRAPHIES

A Bibliography of Astronomy, 1970-1979. R. A. Sea and S. S. Martin. Libraries Unlimited, Inc., Littleton, CO 80160. 1982. $37.50.

DIRECTORIES AND BIOGRAPHICAL SOURCES

American Astronomical Society. Membership Directory. American Astronomical Society. 2000 Florida Avenue NW, Suite 400, Washington, DC 20009. (202) 328-2010. FAX (202) 234-2560 Annual. Included in membership dues. ISSN: 1061- 9038.

American Men and Women of Science: Physical and Biological Sciences. R. R. Bowker Inc., 121 Chanlon Road, New Providence, NJ 07974. (908) 464-6800. (800) 521-8110. 20th edition. 8 volumes. 1996. $850.00.

The Astronomers. Donald Goldsmith. St. Martin's Press, Inc., 175 Fifth Avenue, New York, NY 10010. (212) 674-5151. (800) 221-7945. 1993. $14.95 in paper.

Directory of Physics and Astronomy Staff Members. American Institute of Physics, One Physics Ellipse, College Park, MD 20740-3843. (301) 209-3100. 1975/76 to present. Annual. $60.00. ISSN: 0361-2228.

Earth and Astronomical Research Centers. Stockton Press, 345 Park Avenue South, New York, NY 10010. 4th edition. 1995. $515.00. ISBN: 1-56169-0967.

ENCYCLOPEDIAS AND DICTIONARIES

Concise Dictionary of Astronomy. Jacqueline Mitton. Oxford University Press, Inc., 200 Madison Avenue, New York, NY 10016. (212) 725-6000. (800) 334-4249. 1992. $24.95.

Encyclopedia of Astronomy and Astrophysics. Stephen Maran, editor. Van Nostrand Reinhold, 115 Fifth Avenue, New York, NY 10003. (212) 254-3232. (800) 842-3636. 1992. $129.95.

McGraw-Hill Encyclopedia of Science and Technology. McGraw-Hill Book Company, Inc., 1221 Avenue of the Americas, New York, NY 10020. (212) 512-2000. (800) 262-4729. 7th edition. 20 volume set. 1992. $1900.00.

Moons of the Solar System: An Illustrated Encyclopedia. John Stewart. McFarland & Company, Inc., Box 611, Jefferson, MC 28640. (910) 246-4460. (800) 253-2187. 1991. $49.95.

New Guide to the Planets. Patrick Moore. Trans-Atlantic Publications, Inc., 311 Bainbridge Street, Philadelphia, PA 19147. (215) 925-5083. 1993. $37.50.

Stars and Planets: The Sierra Club Guide to Sky Watching and Direction Finding. W. S. Kals. The Sierra Press, 4988 Gold Leaf Drive, Mariposa, CA 95338. (209) 966-5071. (800) 745-2631. $15.00.

PLUTO

Ency. of Physical Sciences and Engineering Info. Sources

GENERAL WORKS

Clyde Tombaugh: Discoverer of Planet Pluto. David H. Levy. University of Arizona Press, 1230 North Park Avenue, Number 102, Tucson, AZ 85719. (520) 621-1441. (800) 426-3797. Reprint edition. 1992. $15.95 in paper.

Evolution of the Earth and Planets. E. Takahashi, et al, editors. American Geophysical Union, 2000 Florida Avenue, N.W., Washington, DC 20009. (202) 462-6903. (800) 966- 2481. 1993. $28.00.

Exploring the Planets. W. Kenneth Hamblin and Eric H. Christiansen. Macmillan Publishing Company, Inc,. 200 Old Tappan Road, Old Tappan, NJ 07675. (800) 233-2336. 2nd edition. 1995. $62.00.

Far Encounter. Eric Burgess. Columbia University Press, 562 West 113th Street, New York, NY 10025. (212) 666-1000. (800) 944-8648. 1988. $40.00.

Moons and Planets. William K. Hartmann. Wadsworth Publishing Co., 10 Davis Drive, Belmont, CA 94002. (415) 595-2350. (800) 354-9706. 3rd edition. 1993. $49.95.

Planets X and Pluto. William G. Hoyt. Books on Demand 1980. 300 North Zeeb Road, Ann Arbor, MI 48106. (800) 521-0600 or (313) 761-4700. $90.10.

Planetary Landscapes. Ronald Greeley. Chapman & Hall, 1 Penn Plaza, New York, NY 10119. (212) 564-1060. Div. of Routledge, 2nd edition. 1994. $49.95.

HANDBOOKS AND MANUALS

Astrophysical Data: Planets and Stars. Kenneth R. Lanb. Springer-Verlag New York, Inc., 175 Fifth Avenue, New York, NY 10010. (212) 460-1500. (800) 777-4643. 1993. $59.00.

Field Guide to the Stars and Planets. Jay M. Pasachoff and Donald H. Menzel. Houghton Mifflin Co., 222 Berkeley Street, Boston, MA 02116. (617) 351-5000. (800) 225-3362. Revised edition. 1992. $24.95.

New Guide to the Planets. Patrick Moore. Trans-Atlantic Publications, Inc., 311 Bainbridge Street, Philadelphia, PA 19147. (215) 925-5083. 1993. $37.50.

Planet Observer's Handbook. Fred W. Price. Cambridge University Press, 40 West 20th Street, New York, NY 10011-4211. (212) 924-3900. (800) 872-7423. 1994. $34.95.

ONLINE DATABASES AND CD-ROMS

Dissertation Abstracts. University Microfilms International, 300 North Zeeb Road, Ann Arbor, MI 48106. (800) 521-0600 or (313) 761-4700. Scope includes virtually all doctoral dissertations accepted at accredited American institutions from 1861 to present in 252 subject areas. Available on BRS, (800) 289-4277, DIALOG, (800) 334-2564, and OCLC EPIC, (800) 848-5878, online services. Also available on CD-ROM. Inquire as to cost and availability.

Georef. American Geological Institute, 4220 King Street, Alexandria, VA 22302. (800) 336-4764 or (703) 379-2480. Geology and geosciences literature, 1785 to present for North America.

Available on DIALOG,(800) 334-2564, ORBIT (800) 456-7248, online services. Also available on CD-ROM. Inquire as to cost and availability.

INSPEC. Institution of Electrical Engineers, Michael Faraday House, Six Hills Way, Stevenage, Herts. SG1 2AY, England. Telephone: 0438 313311 or FAX 0438 742840. Contains citations to the worldwide literature of physics, electronics and electrical engineering, computer technology, and related fields. Available on BRS, (800) 289-4277, DIALOG, (800) 334-2564, ORBIT, (800) 456-7248, and STN International, FIZ Karlsruhe, P.O. Box 2465, W-7500, Karlsruhe 1, Germany, online services. Inquire as to cost and availability.

NTIS Bibliographic Database. National Technical Information Service, 5285 Port Royal Road, Springfield, VA 22161. (703) 487-4929 or FAX (703) 321-8199. Broad coverage of government-sponsored science and technology research reports, 1964 to present. Available on BRS,(800) 289-4277, DIALOG, (800) 334-2564, ORBIT, (800) 456-7248, and STN International, FIZ Karlsruhe, P.O. Box 2465, W-7500, Karlsruhe 1, Germany, online services. Also available on CD-ROM. Inquire as to cost and availability.

SCISEARCH. Institute for Scientific Information, 3501 Market Street, Philadelphia, PA 19104. (800) 523-1850 or (215) 386-0100. Broad multidisciplinary title and author index to the international literature of science and technology, 1974 to present. Available on DIALOG, (800) 334-2564, and ORBIT, (800) 456-7248, online services. Also available on CD-ROM. Inquire as to cost and availability.

WILSONLINE. H.W. Wilson Company, 950 University Avenue, Bronx, NY 10452. (800) 367-6770 or (212) 588-8400. Makes available online versions of the printed H.W. Wilson indexes including *Applied Science and Technology Index, Business Periodicals Index, General Science Index,* and *Readers' Guide to Periodical Literature.* Period covered is generally 1983 to present. Available on BRS, (800) 289-4277, DIALOG, (800) 334-2564, and OCLC EPIC, (800) 848-5878, online services.Also available on CD-ROM. Inquire as to cost and availability.

PERIODICALS

Astronomical Journal. American Institute of Physics, One Physics Ellipse, College Park, MD 20740-3843. (301) 209-3000. Published for the American Astronomical Society. 1849 to present. Monthly. $280.00 per year. ISSN: 0004-6256.

Astronomical Society of the Pacific. Publications. Astronomical Society of the Pacific, 390 Ashton Avenue, San Francisco, CA 94112. (415) 337-1100. FAX (415) 337-5205.1889 to present. Monthly. $175.00 per year. ISSN: 0004-6280.

Astronomy. Kalmbach Publishing Company, Box 1612, Waukesha, WI 53187-1612. (414) 796-8776. FAX (414) 796-1142. 1973 to present. Monthly. $27.00 per year. ISSN: 0091-6358.

Astronomy and Astrophysics. Springer-Verlag New York, Inc., 44 Hartz Way, Secaucus, NJ 07096-2491. (201) 348-4033. FAX (201) 348-4505. 1969 to date. 36 issues per year in 12 volumes. $1733.00 per year. ISSN: 0004-6361.

Earth, Moon and Planets: An International Journal of Comparative Planetology. Kluwer Academic Publishers, Box 358, Accord Station, Hingham, MA 02018-0358. (617) 871- 6600.

FAX (617) 871-6528. 1969 to present. Monthly. $840.00 per year. ISSN: 0167-9295.

Geochemica et Cosmochimica Acta. Elsevier Science. 660 White Plains Road, Tarrytown, NY 10591-5153. (914) 524- 9200. FAX (914) 333-2444. 1950 to the present. Biweekly. $895.00 per year. ISSN: 0016-7037.

Icarus: International Journal of Solar System Studies. Academic Press, Inc., Journal Division, 525 B Street, Suite 1900, San Diego, CA 92101-4495. (619) 230-1840. FAX (619) 699-6800. 1962 to the present. Monthly. $1080.00. ISSN: 0019-1035.

JGR: Journal of Geophysical Research: Planets. American Geophysical Union, 2000 Florida Avenue, NW, Washington, CD 20009. (202) 462-6900. FAX (202) 328-0566. 1991 to present. Monthly. $597.00 per year. ISSN: 0148-0227.

Lunar and Planetary Information Bulletin. Lunar and Planetary Institute. 3600 Bay Area Boulevard, Houston, TX 77058-1113. (713) 486-2175. FAX (713) 486-2125. 1970 to present. Quarterly. Free. Also available online.

Mercury. Astronomical Society of the Pacific. 390 Ashton Avenue, San Francisco, CA 94112. (415) 337-1100. FAX: (415) 337-5205. 1972 to present. Bimonthly. $175.00 per year. ISSN: 0047-6773.

Planetary and Space Science. Elsevier Science Publishing Company, Inc., 660 White Plains Road, Terrytown, NY 10591- 5153. (914) 524-9200. FAX: (914) 333-2444. 1959 topresent. Monthly. $1355.00 per year. ISSN: 0032-0633.

Planetary Report. Planetary Society. 65 North Catalina, Pasadena, CA 91106-2301. (818) 793-5100. FAX (818) 793- 5528. 1980 to present. $25.00 per year. ISSN: 0736-3680.

Sky and Telescope. Sky Publishing Corporation, Box 9111, Belmont, MA 02178. (617) 864-7360. FAX (617) 864-6117.1941 to present. Monthly. $27.00 per year. ISSN: 0037-6604.

RESEARCH CENTERS AND INSTITUTES.

Center for Space Sciences. University of Texas at Dallas, P.O. Box 830688, MS-F022, Richardson, TX 75083-0688. (214) 690-2851. FAX (214) 690-2848.

Institute for Astronomy. University of Hawaii at Manoa, 2680 Woodlawn Drive, Honolulu, HI 96822. (808) 956-8312. FAX (808) 988-2790.

Institute for Terrestrial and Planetary Atmospheres. State University of New York at Stony Brook, Stony Brook, NY 11794-2300. (516) 632-6170. FAX (516) 632-6251.

Laboratory for Planetary Studies. Cornell University, Space Sciences Building, Ithaca, NY 14853. (607) 255-4971.

Lunar and Planetary Institute. 3303 NASA Road One, Houston, TX 77058-4399. (713) 486-2139. FAX 713-496-2162.

Lunar and Planetary Laboratory. University of Arizona, Tucson, AZ 85721. (602) 621-6963. FAX (602) 621-4933.

McDonnell Center for the Space Sciences. Washington University, Box 1105, One Brookings Drive, St. Louis MO 63130-4899. (314) 889-6255. FAX (314) 889-6219.

PLUTONIUM

See also: ELEMENTS (CHEMICAL), NUCLEAR CHEMISTRY, NUCLEAR ENGINEERING, NUCLEAR FUELS, NUCLEAR REACTORS

ABSTRACT SERVICES AND INDEXES

Applied Science and Technology Index: A Cumulative Subject Index to English Language Periodicals in the Fields of Aeronautics and Space Science, Computer Technology, Chemistry, Construction Industry, Energy and Related Areas. H. W. Wilson Co., 950 University Avenue, Bronx, NY 10452- 9978. (212) 588-8400. (800) 367-6770. FAX (718) 590-1617. From 1958 to present. Monthly. Inquire about cost and availability. Also available online (BRS and WILSONLINE) and on CD-ROM. ISSN: 0003-6986.

Chemical Abstracts. Chemical Abstracts Service. 2540 Olentangy River Road, Box 3012, Columbus, OH 43210-0012. (614) 447-3600. FAX (614) 447-3713. 1907 to present. Weekly. $16,800.00 per year. Available online and on CD- ROM. CA is also available in five section groupings. Inquire Regarding Cost and Availability.

Current Contents: Engineering, Technology and Applied Sciences. Institute for Scientific Information, 3501 Market Street, Philadelphia, PA 19104. (215) 386-0100. FAX (215) 386-2991. From 1970 to present. Weekly. $442.00 per year. Also available online. Inquire regarding cost and availability. ISSN: 0095-7917.

Energy Research Abstracts. U. S. Department of Energy, office of Scientific and Technical Information, Box 62, Oak Ridge, TN 37831. (615) 674-0733. Subscriptions to: U. S. Government Printing office, Box 371954, Pittsburgh, PA 15250-7954. (202) 783-3238. 1976 to present. Monthly. $164.00 per year. ISSN: 0160-3604.

Engineering Index Monthly. Engineering Information, Inc., Castle Point on the Hudson, Hoboken, NJ 07030. (201) 216-8500. (800) 221-1044. FAX (201) 216-8532. Monthly. $2300.00 per year for monthly issues. $1980 per year for annual cumulation. Available on CD-ROM and online as COMPENDEX. ISSN: 0742-1974; 0360-8557.

General Science Index. H.W. Wilson Company, 950 University Avenue, Bronx, NY 10452. (212) 588-8400. (800) 367-6770. FAX (718) 590-1617. From 1978 to present. Ten issues per year; quarterly and annual cumulations. Service basis. Available on CD-ROM and online. Inquire regarding cost and availability. ISSN: 0162-1963.

NTIS Alerts: Energy. U.S. National Technical Information Service, 5285 Port Royal Road, Springfield, VA 22161. (703) 487-4929. FAX (703) 321-8547. Weekly. $160.00 per year.

Physics Abstracts. INSPEC. Section A, Science Abstracts. Institution of Electrical Engineers (IEE), London, United Kingdom. Available from: INSPEC/IEEE-Institute of Electrical and Electronic Engineers, Box 1331, 445 Hoes Lane, Piscataway,

NJ 08855-1331. (908) 562-5549. 1898 to present. 24 issues per year. $2835.00 per year. ISSN: 0036-8091. Also available online and on CD-ROM.

Physics Briefs (Physikalische Berichte). Information center for energy, physics, mathematics; german physical society. VCH Publishers, Inc., 220 East 23rd Street, New York, NY 10010-4606. (212) 683-8333. 1845 to present. 24 issues per year. $2390.00 per year. Also available online. ISSN: 0179-7434.

Science Citation Index (SCI). Institute for Scientific Information, 3501 Market Street, Philadelphia, PA 19104. (215) 386-0100. (800) 523-1850. FAX (215) 386-2991. 1961 to Present. Six issues per year, plus annual cumulation. $11650.00 per year. Also available online and on CD-ROM. Inquire about price and availability. ISSN: 0036-827x.

ANNUAL REVIEWS AND YEARBOOKS

Advamces in Nuclear Science and Technology. Plenum Publishing Corp., 233 Spring Street, New York, NY. (212) 620-8000. (800) 2221-9369. FAX (212) 463-0742. 1977 to present. Irregular.

Annual Review of Energy and the Environment. Annual Reviews, Inc., 4139 El Camino Way, Palo Alto, CA 94303-0139. (415) 943-4400. (800) 523-8635. 1976 to present. Annual. $71.00. ISSN: 1056-3466.

ASSOCIATIONS AND PROFESSIONAL SOCIETIES

American Chemical Society, 1155 16th Street, NW, Washington, DC 20036. (202) 872-4600.

American Institute of Chemical Engineers. 345 East 47th Street, New York, NY 10017. (212) 705-7338.

American Institute of Physics. One Physics Ellipse, College Park, MD 20740-3843. (301) 209-3100.

American Nuclear Society. 555 North Kensington Avenue, La Grange Park, IL 60525. (312) 352-6611. (800) NUC-NEWS. FAX (708) 352-0499.

Association of Energy Engineers, 402225 Pleasantdale Road, Suite 340, Atlanta, GA 30340. (404) 447-5083.

Institute of Nuclear Materials Management. 60 Revere Drive, Northbrook, IL 60062. (847) 480-9573. FAX (947) 480-9282.

IEEE Nuclear and Plasma Sciences Society. c/o Institute of Electrical and Electronics Engineers, 345 East 47th Street, New York, NY 10017. (212) 705-7900. FAX (212) 705-4929.

BIBLIOGRAPHIES

Information Sources in Engineering. Peter Hicks, editor. Bowker-Saur, 121 Chanlon Road, New Providence, NJ 07974. (800) 521-8110 or (908) 464-6800. 3rd edition. 1996. $125.00.

Scientific and Technical Books and Serials in Print: An Index To Literature in Science and Technology. R.R. Bowker Inc., 121 Chanlon Road, New Providence, NJ 07974. (908) 464-6800. (800) 521-8110. FAX (908) 665-3502. 1972 to present. Annual.

4 volumes. 1994. $299.95. Also available on compact disc and online. ISSN: 0000-054X.

DIRECTORIES AND BIOGRAPHICAL SOURCES

Energy and Nuclear Science International Who's Who. Gale Research, 835 Penobscot Building, Detroit, MI 48226-4094. (313) 961-2242. (800) 347-4253. Triennial. 1993. $325.00.

Energy Information Centers Directory. U. S. Council for Energy Awareness, 1776 I Street, NW, Suite 400, Washington, DC 20006. (202) 293-0770. Triennial. No charge.

Engineering Research Centers: Incorporating Electronics Research Centers. Stockton Press, 345 Park Avenue, New York, NY 10010. (212) 689-9200. (800) 221-2123. 4th edition. 1995. $515.00.

Research Centers Directory. Gale Research, 835 Penobscot Building, Detroit, MI 48226-4094. (313) 961-2242. (800) 347-4253. 20th edition. 1995. $485.00. ISSN: 0080-1518.

Who's Who in Engineering. Gordon Davis, editor. American Association of Engineering Societies. 1111 19th Street, NY, Suite 608, Washington, DC 20036. (202) 296-2237. (800) 658-8897. 9th edition. 1995. $220.00.

World Energy and Nuclear Directory. Gale Research, 835 Penobscot Building, Detroit, MI 48226-4094. (313)961-2242. (800) 347-4253. Biennial, even years. $450.00.

World Inventory of Plutonium and Highly Enriched Uranium. David Albright, et al. OUP, 1993. $42.00.

ENCYCLOPEDIAS AND DICTIONARIES

Encyclopedia of Applied Physics. VCH Publishers, Inc., 303 Northwest 12th Avenue, Deerfield Beach, FL 33442. (800) 367-8249. 1991-. Twenty volumes. $6000.00.

Encyclopedia of Chemical Processing and Design. McKetta Marcel Dekker, Inc., 270 Madison Avenue, New York, NY 10016. (212) 696-9000. (800) 228-1160. 1976 - . $175.00 per volume. ISSN:

Industrial Chemical Thesaurus. Michael Ash and Irene Ash. VCH Publishers, Inc., 220 East 23rd Street, New York, NY 10010-4606. (800) 367-8249. 1992. $295.00.

Kirk-Othmer Encyclopedia of Chemical Technology. John Wiley and Sons, Inc., 605 Third Avenue, New York, NY 10158. (800) 526-5368 or (212) 850-6000. Fourth edition. 1991- . Twenty-seven volumes. $5400.00.

McGraw-Hill Encyclopedia of Science and Technology. McGraw-Hill Book, Incorporated, 1221 Avenue of the Americas, New York, NY 10020. (212) 997-3675. (800) 262-4729. Seventh edition. Twenty volumes. 1992. $1900.00.

Ullman's Encyclopedia of Industrial Chemistry. VCH Publications, Inc., 220 East 23rd Street, Suite 909, New York, NY 10010-4606. (212) 683-8333. (800) 422-8824. 5th edition. 1984- . Price varies per volume.

GENERAL WORKS

Crystal Chemistry of Simple Compounds of Uranium, Thorium, Plutonium Neptunium. Evgeniis S. Makarov. Books on Demand, 300 North Zeeb Road, Ann Arbor, MI 48106-1346. (313) 761-4700. (800) 521-0600. $43.00.

Nuclear Engineering: Theory and Technology of Commercial Nuclear Power. Ronald A. Knief. Hemisphere Publishing Corp., 1900 Frost Road, Suite 101, Bristol PA 19007. (215) 785-5800. (800) 821-8312. 2nd edition. 1992. $156.00.

Nuclear Power, Nuclear Fuel Cycle and Waste Management: Status and Trends 1992. UNIPUB, 4511-F Assembly Drive, Lanham, MD 20706-4391. (301) 459-7666. (800) 274-4888. 1992. $19.00.

Physics of Plutonium Recycling. OECD/NEA Staff. OECD, 2001 L Street NW, Suite 650, Washington, DC 20036. (202) 785-6323. (800) 456-6323. volumes 1-4. 1995. $140.00.

Plutonium Chemistry. William T. Carnall, editor. American Chemical Society, 1155 16th Street, Nw, Washington, DC 20036. (202) 872-4600. (800) 333-9511. FAX (614) 447-3671. 1983. $60.95.

The Plutonium Story: The Journals of Professor Glenn T. Seaborg, 1939-1946. Roland L. Kathren, et al. Battelle Press, 505 King Avenue, Columbus, OH 43201-2693. (614) 424-6393. (800) 451-3542. 1994. $49.95.

HANDBOOKS AND MANUALS

Energy Engineering Handbook. Fairmont Press Staff. Prentice-Hall, 113 Sylvan Avenue, Route 9W, Englewood Cliffs, NJ 07632. (201) 592-2000. (800) 922-0579. 3rd edition. 1995. $75.00.

Guide to Nuclear Power Technology: A Resource for Decision Making. John Wiley & Sons, Inc., 605 Third Avenue, New York, NY 10158-0012. (212) 850-6000. (800) 225-5945. 1984.

Perry's Chemical Engineers' Handbook. Robert H. Perry and Donald W. Green, editors. McGraw-Hill Publishing Company, Inc., 1221 Avenue of the Americas, New York, NY 10020. (212) 512-2000. (800) 262-4729. 6th edition. 1996. $129.50.

Plutonium Handbook. O. J. Wick. American Nuclear Society, 555 North Kensington Avenue, LaGrange Park, IL 60525. (312) 352-6611. (800) NUC-NEWS. FAX (708) 352-0499. 2 volumes. Revised edition. 1980. $98.00

ONLINE DATABASES AND CD-ROMS

CA SEARCH. Chemical Abstracts Service, P.O. Box 3012, Columbus, OH 43210-0012. (614) 447-3600. (800) 848-6533. FAX (614) 447-3709. Very comprehensive guide to worldwide chemical literature and related fields. 1972 to present. Available on BRS,(800) 289-4277, DIALOG, (800) 334-2564, ORBIT (800) 456-7248, and STN International, FIZ Karlsruhe, P.O. Box 2465, W-7500, Karlsruhe 1, Germany, online services. Inquire as to cost and availability.

Compendex Plus. Engineering Information, Inc., 345 East 47th Street, New York, NY 10017. (212) 705-7600 or (800) 221-1044. Contains citations with abstracts to worldwide literature in engineering and technology, from 1970 to present. Available on online BRS,(800) 289-4277, DIALOG, (800) 334-2564, ORBIT (800) 456-7248, and STN International, FIZ Karlsruhe, P.O. Box 2465, W-7500, Karlsruhe 1, Germany, online services. Also available on CD-ROM. Inquire as to cost and availability.

Current Contents Search. Institute for Scientific Information, 3501 Market Street, Philadelphia, PA 19104. (215) 386-0100. FAX (215) 386-6362. Contains citations to articles listed in the table of contents of science and technology journals. Also articles in social sciences and life sciences journals. Available on BRS,(800) 289-4277, DIALOG,(800) 334-2564, online services. Inquire as to cost and availability.

Dissertation Abstracts. University Microfilms International, 300 North Zeeb Road, Ann Arbor, MI 48106. (800) 521-0600 or (313) 761-4700. Scope includes virtually all doctoral dissertations accepted at accredited American institutions from 1861 to present in 252 subject areas. Available on BRS, (800) 289-4277, DIALOG, (800) 334-2564, and OCLC EPIC, (800) 848-5878, online services. Also available on CD-ROM. Inquire as to cost and availability.

INSPEC. Institution of Electrical Engineers, Michael Faraday House, Six Hills Way, Stevenage, Herts. SG1 2AY, England. Telephone: 0438 313311 or FAX 0438 742840. Contains citations to the worldwide literature of physics, electronics and electrical engineering, computer technology, and related fields. Available on BRS, (800) 289-4277, DIALOG, (800) 334-2564, ORBIT, (800) 456-7248, and STN International, FIZ Karlsruhe, P.O. Box 2465, W-7500, Karlsruhe 1, Germany, online services. Inquire as to cost and availability.

Kirk-Othmer Encyclopedia of Chemical Technology. John Wiley and Sons, Inc., 605 Third Avenue, New York, NY 10158. (800) 526-5368 or (212) 850-6000. Contains the complete text of all chapters in the 27 volume fourth edition of the *Kirk-Othmer Encyclopedia of Chemical Technology.* 1991. Available on BRS,(800) 289-4277, DIALOG,(800) 334-2564, online services. Inquire as to cost and availability.

NTIS Bibliographic Database. National Technical Information Service, 5285 Port Royal Road, Springfield, VA 22161. (703) 487-4929 or FAX (703) 321-8199. Broad coverage of government-sponsored science and technology research reports, 1964 to present. Available on BRS,(800) 289-4277, DIALOG, (800) 334-2564, ORBIT, (800) 456-7248, and STN International, FIZ Karlsruhe, P.O. Box 2465, W-7500, Karlsruhe 1, Germany, online services. Also available on CD-ROM. Inquire as to cost and availability.

Physics Briefs. American Institute of Physics, 335 East 45th Street, New York, NY 10017. (212) 661-9260 or FAX (212) 661-2036. Contains citations with abstracts of the literature of physics and related fields, 1979 to present. Available on the STN International, FIZ Karlsruhe, P.O. Box 2465, W-7500, Karlsruhe 1, Germany, online service. Inquire as to cost and availability.

SCISEARCH. Institute for Scientific Information, 3501 Market Street, Philadelphia, PA 19104. (800) 523-1850 or (215) 386-0100. Broad multidisciplinary title and author index to the international literature of science and technology, 1974 to present. Available on DIALOG, (800) 334-2564, and ORBIT, (800) 456-7248, online services. Also available on CD-ROM. Inquire as to cost and availability.

WILSONLINE. H.W. Wilson Company, 950 University Avenue, Bronx, NY 10452. (800) 367-6770 or (212) 588-8400. Makes available online versions of the printed H.W. Wilson indexes

PLUTONIUM

Ency. of Physical Sciences and Engineering Info. Sources

including *Applied Science and Technology Index, Business Periodicals Index, General Science Index,* and *Readers' Guide to Periodical Literature.* Period covered is generally 1983 to present. Available on BRS, (800) 289-4277, DIALOG, (800) 334-2564, and OCLC EPIC, (800) 848-5878, online services. Also available on CD-ROM. Inquire as to cost and availability.

PERIODICALS

American Nuclear Society Transactions. American Nuclear Society, 555 North Kensington Avenue, LA Grange Park, IL 60525. (708) 352-6611. 1958 to present. 3 issues per year. $340.00 per year. ISSN: 0003-018X

Annals of Nuclear Energy. Elsevier Science Publishing Company, Inc., 660 White Plains Road, Tarrytown, NY 10591- 5133. (914) 524-9200. 1974 to present. Monthly. $815.00 per year. ISSN: 0306-4549.

Bulletin of the Atomic Scientists. Educational Foundation for Nuclear Science, 6042 South Kimbark Avenue, Chicago, IL 60637. (312) 702-2555. 1945 to present. 10 issues per year. $30.00 per year. ISSN: 0096-3402.

Nuclear Engineer. Institution of Nuclear Engineers, 1 Penerley Road, London, SE6 2LQ, England. 1959 to present. Bi-monthly. L77 per year. ISSN: 0262-5091.

Nuclear Engineering and Design. Elsevier Science Publishing Company, Inc., 660 White Plains Road, Tarrytown, NY 10591- 5133. (914) 524-9200. 1965 to present. 24 issues per year. $2748.00 per year.

Nuclear Engineering International. Reed Business Pulbishing Group, Quadrant House, the Quadrant, Sutton, Surrey, SM2 5AS, England. 1956 to present. Monthly. $236.00 per year. ISSN: 0029-5507.

Nuclear Science and Engineering. American Nuclear Society, 555 North Kensington Avenue, LA Grange Park, IL 60525. (708) 352-6611. 1956 to present. Monthly. $405.00 per year. ISSN: 0029-5639.

Nuclear Technology. American Nuclear Society, 555 North Kensington Avenue, LA Grange Park, IL 60525. (708) 352-6611. 1965 to present. Monthly. $440.00 per year. ISSN: 0029-5450.

RESEARCH CENTERS AND INSTITUTES

Department of Nuclear Engineering and Engineering Physics. University of Wisconsin-Madison, 1500 Johnson Drive, Madison, WI 53706. (608) 263-1648. FAX (608) 262-6707.

Fusion Research Center. University of Texas at Austin. RLM 11.222, Austin, TX 78712. ((512) 471-3926. FAX (512) 471-8865.

Frank H. Neely Nuclear Research Center. Georgia Institute of Technology, 900 Atlantic Drive NW, Atlanta, GA 30332. (404) 894-3600.

Institute of Nuclear Science and Engineering. Oregon State University Radiation Center-A100, 35th and Jefferson Streets, Corvallis, OR 97331-5903. (503) 737-2341. Fax (503) 737-0480.

Institute of Plasma and Fusion Research. University of California, Los Angeles, 44-139 Engineering IV 405 Hilgard Avenue, Los Angeles, CA 90024-1597. (213) 825-1613. FAX (213) 206-4832.

Radiation Science and Engineering Center. Pennsylvania State University, University Park, PA 16802. (814) 865-6351. FAX (814) 863-4840.

Whiteshell Laboratories. Research Company, Atomic Energy of Canada Limited, Pinawa, MB, Canada ROE ILO. (204) 753-2311. Fax (204) 753-8404.

POLLUTION

See also: AIR POLLUTION, ENVIRONMENTAL ENGINEERING, GROUND WATER POLLUTION, WATER POLLUTION

ABSTRACT SERVICES AND INDEXES

Air Pollution Titles. Pennsylvania State University, Environmental Resources Research Lab, Center for Air Environmental Studies. 226 Fenske Lab, University Park, PA 16802. (814) 865-1415. FAX (814) 862-1696. 1965 to present. Bi-monthly. $120.00 per year.

CA Selects: Environmental Pollution. Chemical Abstracts Service, 2540 Olentangy River Road, Box 3012, Columbus, OH 43210. (800) 848-6538 or (614) 447-3600, FAX (614) 447-3713. Semi-weekly. $210.00 per year.

CA Selects: Pollution Monitoring. Chemical Abstracts Service, 2540 Olentangy River Road, Box 3012, Columbus, OH 43210. (800) 848-6538 or (614) 447-3600, FAX (614) 447-3713. Semi-weekly. $210.00 per year.

Current Contents: Engineering, Technology, and Applied Sciences. Institute for Scientific Information, 3501 Market Street, Philadelphia, PA 19104. (215) 386-0100. FAX (215) 386-2991. 1970 to present. Weekly. $442.00 per year.

Current Contents: Physical, Chemical and Earth Sciences. Institute for Scientific Information, 3501 Market Street, Philadelphia, PA 19104. (215) 386-0100. FAX (215) 386-6362. Weekly. $360.00 per year.

Engineering Index Monthly. Engineering Information, Inc., Castle Point on the Hudson, Hoboken, NJ 07030. (800) 221-1044. FAX (201) 216-8532. Monthly. $2300.00 per year. Also available online as COMPENDEX, and also on CD-ROM. Covers chemical engineering, computers, electrical engineering, civil engineering, metals and mining, industrial management, and mechanical engineering.

Environmental Abstracts. R.R. Bowker Inc., 121 Chanlon Road, New Providence, NJ 07974. (800) 521-8110 or (908) 464-6800. Contains citations, most with abstracts, to literature relating to environmental issues and problems, from 1971 to present. Inquire as to cost and availability.

Environmental Engineering Abstracts, Cambridge Scientific Abstracts, 7200 Wisconsin Avenue, Bethesda, MD 20814-4823. (301) 961-6750. FAX (301) 961-6720. Monthly. Covers hazardous materials, environmental impact and protection, treat-

Ency. of Physical Sciences and Engineering Info. Sources

POLLUTION

ment of sewage and industrial wastes, hydroelectric power, tidal and wind power, arctic and tropical engineering.

Hydro-Abstracts (formerly *Water Resources Abstracts*). HydroScience Press, 2527 Jackson Street, N.E., Minneapolis, MN 55418. (612) 781-9081. Monthly. $150.00 per year.

NTIS Alerts: Environmental Pollution and Control. National Techical Information Service (NTIS), 5285 Port Royal Road, Springfield, VA 22161. (703) 487-4650. Weekly. $150.00 per year.

Pollution Abstracts, Cambridge Scientific Abstracts, 7200 Wisconsin Avenue, Bethesda, MD 20814-4823. (301) 961-6750. FAX (301) 961-6720. Inquire for cost and availability.

Science Citation Index. Institute for Scientific Information, 3501 Market Street, Philadelphia, PA 19104. (215) 386-0100. FAX (215) 386-6362. Inquire about availability and cost. Also available on CD-ROM.

Selected Water Resources Abstracts. U.S. Geological Survey, Water Resources Scientific Information Center, 425 National Center, Reston, VA 22092. (703) 648-6820. Monthly. $115.00 per year. Also available on CD-ROM.

ASSOCIATIONS AND PROFESSIONALS SOCIETIES

Acid Rain Foundation. 1410 Varsity Drive, Raleigh, NC 27606-2010. (919) 828-9443. FAX (919) 515-3593.

Air & Waste Management Association. 1 Gateway Circle, Third floor, Pittsburgh, PA 15222. (800) 270-3444. FAX (412) 232-3450.

Association of State and Interstate Water Pollution Control Adminstrators. 750 First Street SE, Suite 910, Washington, DC 20002. (202) 898-0905.

Environmental Defense Fund. 257 Park Avenue South, New York, NY 10010. (212) 505-2100. FAX (212) 505-2375.

Environmental Industry Council. 529 14th Street NW, Suite 655, Washington, DC 20045. (202) 737-3018.

International Center For the Solution of Environmental Problems. 535 Lovett Blvd., Houston, TX 77006. (713) 527-8711. FAX (713) 527-8025.

National Association of Environmental Professionals. 5165 MacArthur Blvd. NW, P.O. Box 9400, Washington, DC 20016-3315. (202) 966-1500. FAX (202) 966-1977.

BIBLIOGRAPHIES

Environmental Periodicals Bibliography. International Academy at Santa Barbara, Environmental Studies Institute, 800 Garden Street, Suite D, Santa Barbara, CA 93101-1552. (805) 965-5010. 1972 to present. Bi-monthly. Price varies.

DIRECTORIES AND BIOGRAPHICAL SOURCES

Government Research Directory. 9th edition. Gale Research, 835 Penobscot Building, Detroit, MI 48226-4094. (313) 961-2242. (800) 347-4253. 1996. $465.00.

International Research Centers Directory. Gale Research, 835 Penobscot Building, Detroit, MI 48226-4094. (313) 961-2242. (800) 347-4253. 8th edition. 1995. $410.00

Research Centers Directory. Gale Research, 835 Penobscot Building, Detroit, MI 48226-4094. (313) 961-2242. (800) 347-4253. $470.00

Scientific and Technical Organizations and Agencies Directory. Gale Research, 835 Penobscot Building, Detroit, MI 48226-4094. (313) 961-2242. (800) 347-4253. 4th edition. 1996. $195.00.

Who's Who in Technology. Gale Research, 835 Penobscot Building, Detroit, MI 48226-4094. (313) 961-2242. (800) 347-4253. 1995. $195.00.

ENCYCLOPEDIAS AND DICTIONARIES

Encyclopedia of Environmental Control Technology, Volume 2: Air Pollution Technology. Paul N. Cheremisinoff, editor. Gulf Publishing Company, P.O. Box 2608, Houston, TX 77252-2608. (713) 529-4301 or (800) 231-6275. FAX (713) 525-9647. 1989. $155.00

Encyclopedia of Physical Science and Technology. Robert A. Meyers, ed. 18 volumes. Academic Press, Inc., 1250 Sixth Avenue, San Diego, CA 92101-4311. (619) 231-0926. FAX (619) 699-6715. 1992. $2100.00

McGraw-Hill Encyclopedia of Science and Technology. Sybil P. Parker, ed. 7th edition. 20 volumes. McGraw-Hill Publishing Company, 1221 Avenue of the Americas, New York, NY 10020. (800) 262-4729 or (212) 512-3825. 1992. $1900.00

GENERAL WORKS

Air Pollution Control Engineering. Noel DE Nevers. McGraw-Hill Publishing Company, 1221 Avenue of the Americas, New York, NY 10020. (800) 262-4729 or (212) 512-3825. 1995. Inquire for price.

Chemistry of Water & Water Pollution. Jan Dojilido & Gerald Best. Prentice Hall, 113 Sylvan Avenue, Route 9W, Englewood Cliffs, NJ 07632. (201) 592-2000 or (800) 922-0579. 1994. $73.00.

Earth Under Siege: Air Pollution & Global Change. Richard P. Turco. Oxford University Press, Inc., 198 Madison Avenue, New York, NY 10016-4314. (212) 726-6000. FAX (212) 726-6446. 1995. $50.00.

Fundamentals of Air Pollution. Richard W. Boubel, et al. 3d edition. Academic Press Inc., 6277 Sea Harbor Drive, Orlando, FL 32887. (800) 321-5068. 1994. $69.95.

Geochemistry, Groundwater and Pollution. C.A. Appelo and D. Postma. Ashgate Publishing Co., Old Post Rd., Brookfield, VT 05036. (802) 276-3162. FAX (802) 276-3837. 1993. $45.00.

POLLUTION

Ency. of Physical Sciences and Engineering Info. Sources

HANDBOOKS AND MANUALS

The Air Pollution Control Manual. Air & Waste Management Association staff. Van Nostrand Reinhold, 115 Fifth Avenue, New York, NY 10003. (212) 254-3232. FAX (212) 254-9499. 1992. $129.95.

CRC Handbook of Chemistry and Physics. David R. Lide, editor. 75th edition. CRC Press/Times Mirror Company, 2000 Corporate Blvd. NW, Boca Raton, FL 33431. (407) 994-0555 or (800) 272-7737. $99.50.

Environmental Science and Technology Handbook. Porter C. Knowles, et al. Government Institutes Inc., 4 Research Place, Suite 200, Rockville, MD 20850. (301) 921-2355. 1994. $75.00.

Handbook of Air Pollution Control Engineering. John C. Mycock. CRC Press/Times Mirror Company, 2000 Corporate Blvd. NW, Boca Raton, FL 33431. (407) 994-0555 or (800) 272-7737. 1995. $79.95.

Water Pollution Handbook of Environmental Chemistry Series, (Volume 5, Part A). O. Hutzinger, editor. Springer-Verlag, 175 Fifth Avenue, New York, NY 10010. (212) 460-1500 or (800) 777-4643. FAX (212) 473-6272. 1991. $128.00.

ONLINE DATABASES AND CD-ROMS

CA Search. Chemical Abstracts Service, P.O. Box 3012, Columbus, OH 43210-0012. (614) 447-3600. (800) 848-6533. FAX (614) 447-3709. Very comprehensive guide to worldwide chemical literature and related fields, 1972 to present. Available on BRS,(800) 289-4277, DIALOG, (800) 334-2564, ORBIT (800) 456-7248, and STN International, FIZ Karlsruhe, P.O. Box 2465, W-7500, Karlsruhe 1, Germany, online services. Inquire as to cost and availability.

NTIS Bibliographic Database. National Technical Information Service, 5285 Port Royal Road, Springfield, VA 22161. (703) 487-4929 or FAX (703) 321-8199. Broad coverage of government-sponsored science and technology research reports, 1964 to present. Available on BRS,(800) 289-4277, DIALOG, (800) 334-2564, ORBIT, (800) 456-7248, and STN International, FIZ Karlsruhe, P.O. Box 2465, W-7500, Karlsruhe 1, Germany, online services. Also available on CD-ROM. Inquire as to cost and availability.

Waternet. American Water Works Association, Technical Library. Available on DIALOG online services. Citations to literature on water quality, water utility management, water analysis, water pollution, and related areas, 1971 to present. Available on DIALOG,(800) 334-2564, online service. Inquire as to cost and availability.

PERIODICALS

Air & Water Pollution Control. Bureau of National Affairs, 1231 25th Street NW, Washington, DC 20037. (202) 452-4200. FAX (202) 822-8092. 1986 to present. Bi-weekly. $301.00 per year.

American Water Works Association Journal. 6666 W. Quincy Avenue, Denver, CO 80235. (303) 794-7711. FAX (303) 794-7310. 1914 to present. Monthly. $85.00 to libraries and government agencies only.

Atmospheric Environment. Elsevier Science [journals], 660 White Plains Rd., Tarrytown, NY 10159-5153. (919) 524-9200. FAX (919) 333-2444. 1967 to present. 22 times a year. $1995.00 per year.

Atmospheric Pollution & Abatement News. Business Communications Company, Inc., 25 Van Zant St., Norwalk, CT 06855. (203) 853-4266. FAX (203) 853-0348. Monthly. $295.00 per year.

Bulletin of Environmental Contamination and Toxicology. Springer-Verlag, 175 Fifth Avenue, New York, NY 10010. (212) 460-1500 or (800) 777-4643. FAX (212) 473-6272. 1966 to present. Monthly. $399.00 per year.

Environmental Engineering. Mechanical Engineering Publications Ltd., Northgate Avenue, Bury St. Edmunds, Suffolk IP32 6BW, England. Telephone 0284-763277. FAX 0284-704006. 1972 to present. Quarterly. Inquire for price.

Environmental Monitoring and Assessment. Kluwer Academic Publishers, P.O. Box 358, Accord Station, Hingham, MA 02018-0358. (617) 871-6000. FAX (617) 871-6528. 1981 to present. 15 times a year. $765.00 per year.

Environmental Pollution. Elsevier Science [journals], 660 White Plains Rd., Tarrytown, NY 10159-5153. (919) 524-9200. FAX (919) 333-2444. 1970 to present. 12 times a year. $1250.00 per year.

Environmental Science and Technology. American Chemical Society, 1155 16th Street, N.W., Washington, DC 20036. (202) 872-4381 or (800) 333-9511. FAX (614) 447-3671. 1967 to present. Monthly. $89.00 per year for non-members.

EPA Journal. Superintendent of Documents, U.S. Government Printing office, Box 371954, Pittsburgh, PA 15250-7954. (202) 783-3238. FAX (202) 512-2233. 1975 to present. Bi-monthly. $10.00 per year.

Ground Water. Ground Water Publishing Company, National Well Water Association, 6375 Riverside Drive, Dublin, OH 43017. (614) 761-1711. FAX (614) 761-3446. 1963 to present. Bi-monthly. $90.00 per year.

Journal of Contaminant Hydrology. Elsevier Science B.V., PO Box 945, Madison Square Station, New York, NY 10160-0757. (212) 633-3650. FAX (212) 633-3680. 1986 to present. 12 times a year. $553.00.

Journal of Environmental Engineering. American Society of Civil Engineers, Environmental Engineering Division, 345 East 47th Street, New York, NY 10017-2398. (212) 705-7520 or (800) 548-2723. FAX (212) 705-7712. 1956 to present. Bi-monthly. $136.00 per year for non-members.

Marine Pollution Bulletin. Elsevier Science [journals], 660 White Plains Rd., Tarrytown, NY 10159-5153. (919) 524-9200. FAX (919) 333-2444. 1970 to present. 24 times a year. $430.00 per year.

Pollution Engineering. Cahners Publishing Company (Des Plains), 1350 East Touhy Avenue, Des Plaines, IL 60017-5080. (800) 323-4958. FAX (708) 390-2779. 1969 to prsent. 13 times a year. $70.00 per year.

Techlink: Air Pollution. National Technical Information Service (NTIS), 5285 Port Royal Road, Springfield, VA 22161. (703) 487-4650. FAX (703) 321-8547. 1993 to present. Monthly. $200.00 per year.

Water, Air and Soil Pollution. Kluwer Academic Publishers, P.O. Box 358, Accord Station, Hingham, MA 02018-0358. (617) 871-6000. FAX (617) 871-6528. 1971 to present. 28 times a year. $1519.00.

RESEARCH CENTERS AND INSTITUTES

Center For Environmental Research. IIT Research Institute, 10 W. 35th Street, Chicago, IL 60616. (312) 567-4250. FAX (312) 567-4577.

Lehigh University, Environmental Studies Center. Chandler-Ullman Building #17, Bethlehem, PA 18015. (215) 758-3670.

Resources For the Future, Inc., Quality of the Environment Division. 1616 P Street NW, Washington, DC 20036. (202) 328-5000. FAX (202) 265-8069.

World Resources Institute. 1709 New York Avenue NW, Suite 700, Washington, DC 20006. (202) 638-6300. FAX (202) 638-0036.

POLYMERS

See also: ADHESIVES, CERAMICS, CHEMISTRY, COMPOSITES, MATERIALS SCIENCE, ORGANIC CHEMISTRY, PLATICS, THER-MOPLASTICS

ABSTRACT SERVICES AND INDEXES

Analytical Abstracts. Royal Society of Chemistry, Information Services, Thomas Graham House, Science Park, Milton Road, Cambridge, CB4 4WF, England. Contains citations, mostly with abstracts, of the worldwide literature on analytical chemistry, from 1954 to present. Monthly. $636.00 per year. Also available online.

API Abstracts. American Petroleum Institute, Central Abstracting and Information Services, 275 Seventh Avenue, New York, NY 10001. (212) 366-4040 or FAX (212) 366-4298. Contains citations with abstracts to literature relating to the petroleum refining and the petrochemical industries, from 1964 to present.

Applied Science and Technology Index: A Cumulative Subject Index to English Language Periodicals in the Fields of Aeronautics and Space Science, Computer Technology, Chemistry, Construction Industry, Energy and Related Areas. H.W. Wilson Co., 950 University Avenue, Bronx, NY 10452-9978. (212) 588-8400. (800) 367-6770. FAX (718) 590-1617. From 1958 to present. Monthly. Inquire about cost and availability. Also available online (BRS and WILSONLINE) and on CD-ROM. ISSN: 0003-6986.

Chemical Abstracts. Chemical Abstracts Service. 2540 Olentangy River Road, Box 3012, Columbus, OH 43210-0012. (614) 447-3600. FAX (614) 447-3713. 1907 to present. Weekly. $16,800.00 per year. Available online and on CD-ROM. CA is also available in five section groupings. Inquire Regarding Cost and Availability.

Chemical Engineering Abstracts. Royal Society of Chemistry, Information Services, Thomas Graham House, Science Park, Milton Road, Cambridge, Cb4 4wf, England. Contains citations, mostly with abstracts, of the worldwide literature on chemical engineering, from 1982 to present. Monthly. $450.00 per year. Also available online.

Current Contents: Physical, Chemical and Earth Sciences. Institute for Scientific Information, 3501 Market Street, Philadelphia, PA 19104. (215) 386-0100. (800) 523-1850. FAX (215) 386-2291. 1961 to present. Weekly. $442.00 per year. Also available online (BRS, DIALOG) and on CD-ROM. Inquire regarding cost and availability. ISSN: 0163-2574.

Engineered Materials Abstracts. ASM (American Society of Metals) International, Materials Information, Materials Park, OH 44073. (216) 338-5151 or FAX (216) 338-4634. Covers literature on technical developments in polymer, ceramic, and composite materials and engineering. 1986 to present. Monthly. $1175.00 per year. Also available on CD-ROM.

Engineering Index Monthly. Engineering Information, Inc., Castle Point on the Hudson, Hoboken, NJ 07030. (201) 216-8500. (800) 221-1044. FAX (201) 216-8532. Monthly. $2300.00 per year for monthly issues. $1980 per year for annual cumulation. Available on CD-ROM and online as COMPENDEX. ISSN: 0742-1974; 0360-8557.

General Science Index. H.W. Wilson Company, 950 University Avenue, Bronx, NY 10452. (212) 588-8400. (800) 367-6770. FAX (718) 590-1617. From 1978 to present. Ten issues per year; quarterly and annual cumulations. Service basis. Available on CD-ROM and online. Inquire regarding cost and availability. ISSN: 0162-1963.

Petroleum Abstracts. 600 South College, Tulsa, OK 74104. (918) 631-2296 or (800) 247-8678. Contains citations with abstracts to the worldwide literature and patents on the exploration, development, and production of petroleum resources, from 1965 to present.

Polymer Blends, Alloys, and Interpenetrating Polymer Network Abstracts. Technomic Publishing Company, Inc., 851 New Holland Avenue, Box 3535, Lancaster, PA 17604. (717) 291-5609. Covers literature in polymer science and product development, from 1987 to present. Monthly. $130.00 per year.

Polymer Contents: International Current Awareness Publication On Polymer Science and Engineering. Elsevier Science Publishing Company, Inc., 655 Avenue of the Americas, New York, NY 10010. (212) 989-5800. 1984 to present. Monthly. $350.00 per year.

Process and Chemical Engineering/Chemical Engineering Abstracts. Royal Society of Chemistry, Information Services, Thomas Graham House, Science Park, Milton Road, Cambridge, Cb4 4wf, England. Contains citations, mostly with abstracts, of the worldwide literature on chemical engineering; from 1982 to present. Monthly. $610.00 per year. Also available online. ISSN: 0960-5045.

Science Citation Index (SCI). Institute for Scientific Information, 3501 Market Street, Philadelphia, PA 19104. (215) 386-0100. (800) 523-1850. FAX (215) 386-2991. 1961 to present. Six issues per year, plus annual cumulation. $11650.00 per year. Also available online and on CD-ROM. Inquire about price and availability. ISSN: 0036-827X.

POLYMERS

Ency. of Physical Sciences and Engineering Info. Sources

Theoretical Chemical Engineering. Royal Society of Chemistry, Information Services, Thomas Graham House, Science Park, Milton Road, Cambridge, Cb4 4wf, England. Covers theoretical chemical engineering, including theory and laboratory Experimentation. 1964 to present. Monthly. $235.00. ISSN: 0960-5053.

ANNUAL REVIEWS AND YEARBOOKS

Advances in Catalysis. Academic Press, Inc., 1250 Sixth Avenue, San Diego, CA 92101-4311. (619) 231-0926. FAX (619) 699-6715. 1948 to present. Irregular. Price varies.

Advances in Chemical Engineering. Academic Press, Inc., 6277 Sea Harbor Drive, Orlando, FL 32887. (800) 321-5068. 1956 to present. Irregular. ISSN: 0065-2377.

Progress in Colloid and Polymer Science Series. Springer-Verlag, 175 Fifth Avenue, New York, NY 10010. (212) 460-1500. Irregular, price varies.

ASSOCIATIONS AND PROFESSIONAL SOCIETIES

American Chemical Society, 1155 16th Street, NW, Washington, DC 20036. (202) 872-4600.

American Institute of Chemical Engineers. 345 East 47th Street, New York, NY 10017-2396. (212) 705-7663.

Association of Consulting Chemists and Chemical Engineers, 50 East 41st Street, Suite 92, New York, NY 10017. (212) 684-6255.

Chemical Manufacturers Association. 2501 M Street, N.W., Washington, DC 20037. (202) 887-1182

Plastics Institute of America. 277 Fairfield Road, Suite 100, Fairfield, NJ 07004-1932. (201) 808-5950. FAX (201) 808-5953.

Society of Plastics Engineers, Inc., 14 Fairfield Drive, Brookfield, Center, CT 06805-0403. (203) 775-0471. FAX (203) 775-8490.

Society of the Plastics Industry. 1275 K Street NW, Suite 400, Washington, DC 20005. (202) 371-5200. FAX (202) 371-1022.

BIBLIOGRAPHIES

Information Sources in Polymers and Plastics. R. T. Adkins. Bowker-Saur, 121 Chanlon Road, New Providence, NJ 07974. (908) 464-6800. (800) 521-8110. 1989. $75.00

Organic Reaction Mechanisms: An Annual Survey of Literature. A. C. Knipe and W. E. Watts, editors. John Wiley & Sons, Inc., 605 Third Avenue, New York, NY 10158-0012. (212) 850-6000. (800) 225-5945. vol. 92, 1992. price varies.

Scientific and Technical Books and Serials in Print: An Index to Literature in Science and Technology. R.R. Bowker Inc., 121 Chanlon Road, New Providence, NJ 07974. (908) 464-6800. (800) 521-8110. FAX (908) 665-3502. 1972 to present. Annual. 4 volumes. 1994. $299.95. Also available on compact disc and online. ISSN: 0000-054X.

DIRECTORIES AND BIOGRAPHICAL SOURCES

American Institute of Chemical Engineers. Directory. American Institute of Chemical Engineers. 345 East 47th Street, New York, NY 10017-2396. (212) 705-7663. Annual.

American Men and Women of Science: Physical and Biological Sciences. R.R. Bowker Company, 121 Chanlon Road, New Providence, NJ 07974. (908) 464-6800. (800) 521-8110. 20th edition. 8 volumes. 1996. $850.00

Chemical Engineering Faculties. American Institute of Chemical Engineers. 345 East 47th Street, New York, NY 10017-2396. (212) 705-7663. Annual. $75.00.

Consulting Services: Chemists and Chemical Engineers. Association of Consulting Chemists and Chemical Engineers, 40 East 45th Street, New York, NY 10017. (212) 983-3160. FAX (212) 983-3161. Biennial. $60.00.

Engineering Research Centers: Incorporating Electronics Research Centers. Stockton Press, 345 Park Avenue, New York, NY 10010. (212) 689-9200. (800) 221-2123. 4th edition. 1995. $515.00.

Modern Plastics Encyclopedia Directory of Trade Names. McGraw-Hill, 1221 Avenue of the Americas, New York, NY 10020. (212) 512-6266. Annual. $10.00.

Research Centers Directory. Gale Research, 835 Penobscot Building, Detroit, MI 48226-4094. (313) 961-2242. (800) 347-4253. 20th edition. 1995. $485.00. ISSN: 0080-1518.

Society of the Plastics Industry-Membership Directory and Buyers' Guide. 1275 K Street NW, Suite 400, Washington, DC 20005. (202) 371-5200. FAX (202) 371-1022. Annual. $180.00.

Who's Who in Engineering. Gordon Davis, editor. American Association of Engineering Societies. 1111 19th Street, NY, Suite 608, Washington, DC 20036. (202) 296-2237. (800) 658-8897. 9th edition. 1995. $220.00.

ENCYCLOPEDIAS AND DICTIONARIES

Atlas of Polymer and Plastics Analysis. Dieter O. Hummel. VCH Publications, Inc., 220 East 23rd Street, Suite 909, New York, NY 10010-4606. (212) 683-8333. (800) 422-8824. Volume 1, Parts A and B. 3rd edition. 1991. $525.00.

Concise Encyclopedia of Polymer Science. Herman F. Mark. John Wiley & Sons, Inc., 605 Third Avenue, New York, NY 10158-0012. (212) 850-6000. (800) 225-5945. 1990.$199.00

Encyclopedia of Chemical Processing and Design. McKetta Marcel Dekker, Inc., 270 Madison Avenue, New York, NY 10016. (212) 696-9000. (800) 228-1160. 1976- . $175.00 per volume. ISSN:

Encyclopedia of Polymer Science and Engineering. Herman F. Mark et al, editors. John Wiley & Sons, Inc., 605 Third Avenue, New York, NY 10158-0012. (212) 850-6000. (800) 225-5945. 2nd edition. 1985-. 22 volumes. $4788.00

Handbook of Polymer Science and Technology. Nicholas P. Cheremisinoff. Marcel Dekker, Inc., 270 Madison Avenue, New

York, NY 10016. (212) 696-9000. (800) 228-1160. 4 volumes. 1987. $235.00 each.

Illustrated Chemistry Laboratory Terminology. Gerbert W. Ockerman. CRC Press, Inc., 2000 Corporate Boulevard, NW, Boca Raton, FL 33431. (407) 994-0555. (800) 272-7737. 1991. $32.95.

Kirk-Othmer Encyclopedia of Chemical Technology. John Wiley and Sons, Inc., 605 Third Avenue, New York, NY 10158. (800) 526-5368 or (212) 850-6000. Fourth edition. 1991 - . Twenty-seven volumes. $5400.00.

Modern Plastics-Encyclopedia Issue. McGraw-Hill 1221 Avenue of the Americas, New York, NY 10020. (212) 512-6255. Annual; issued in October. $49.95. ISSN: 0085-3518.

Polymer Handbook: Polymerization and Depolymerization. J. Brandrup and E. H. Immergut. John Wiley & Sons, Inc., 605 Third Avenue, New York, NY 10158-0012. (212) 850-6000. (800) 225-5945. Volume 1. 3rd edition. 1989. $245.00.

Ullman's Encyclopedia of Industrial Chemistry. VCH Publications, Inc., 220 East 23rd Street, Suite 909, New York, NY 10010-4606. (212) 683-8333. (800) 422-8824. 5th edition. 1984 - . Price varies per volume. ISSN:

GENERAL WORKS

Characterization of Polymers, Surfaces, Interfaces, Thin Films. Ho-Ming Tong, et al. Butterworth-Heinemann, 313 Washington Street, Newton, MA 02158. (618-928-2500. (800) 366-2665. 1994. $59.95.

Colloidal Polymer Particles. J. W. Goodwin and R. Buscall. Academic Press, Inc., 6277 Sea Harbor Drive, Orlando, FL. (800) 321-5068. 1995. $90.00.

Desk Reference for Organic Chemists. Michael B. East and David J. Ager. Krieger Publishing Company, P.O. Box 9542, Melbourne, FL 32902-9542. (407) 724-9542. 1995. $64.50.

Fundamental Principles of Polymeric Materials. Stephen L. Rosen. John Wiley & Sons, Inc., 605 Third Avenue, New York, NY 10158-0012. (212) 850-6000. (800) 225-5945. 2nd edition. 1993. $64.00.

Introduction to the Mechanical Properties of Solid Polymers. I. M. Ward, and D. W. Hadley. John Wiley & Sons, Inc., 605 Third Avenue, New York, NY 10158-0012. (212) 850-6000. (800) 225-5945. 1993. $44.95

Mechanical Properties of Polymers and Composites. L. Nielsen and Landei, editors. Marcel Dekker, Inc., 270 Madison Avenue, New York, NY 10016. (212) 696-9000. (800) 228-1160. 1993. $150.00.

Organic Reactions: Simplicity and Logic. Pierre Laszlo. John Wiley & Sons, Inc., 605 Third Avenue, New York, NY 10158-0012. (212) 850-6000. (800) 225-5945. 1994. *$74.95.*

Polymer Chemistry: An Introduction. Raymond B. Seymour and Charles E. Carraher Jr. Marcel Dekker, Inc., 270 Madison Avenue, New York, NY 10016. (212) 696-9000. (800) 228-1160. 3rd edition. 1992. $59.75.

Polymer Surfaces from Physics to Technology. Fabio Garbassi et al. Elsevier Science Publishing Company, Inc., 655 Avenue of the Americas, New York, NY 10010. (212) 989-5800. FAX (914) 333-2444. 1994. $95.00.

Polymeric Materials: Chemistry for the Future. Joseph Alper and Gordon L. Nelson. American Chemical Society, 1155 16th Street, NW, Washington, DC 20036. (202) 872-4600. (800) 333-9511. FAX (614) 447-3671. 1989. $24.95.

Polymers for Electronic Applications. Juey H. Lai, editor. CRC Press, Inc., 2000 Corporate Boulevard, NW, Boca Raton, FL 33431. (407) 994-0555. (800) 272-7737. 1989. $205.00.

Polymers from Biobased Materials. Helena L. Chum. Franklin Book Co., Inc., Elkins Park, PA 19117. (215) 635-5252. 1991. $45.00.

The Structure and Properties of Polymeric Materials. D. W. Clegg and A. A. Collyer. Ashgate Publishing Co., Old Post Road, Brookfield, VT 05036. (802) 276-3162. 1993. $70.00.

Thermal and Electrical Conductivity of Polymer Materials. Y.K. Godovsky and V.P. Privalko, editors. Springer-Verlag New York, Inc., 175 Fifth Avenue, New York, NY 10010. (212) 460-1500. (800) 777-4643. FAX (212) 473-6272. 1995. $86.00.

HANDBOOKS AND MANUALS

Chemometrics: Chemical and Sensory Data. David R. Burgard and James T. Kuznicki. CRC Publishers, Inc., 2000 Corporate Blvd., N.W., Boca Raton, FL 33431. (407) 994-0555 or (800) 333-8300. 1990. $120.00.

Comprehensive Organic Synthesis: Selectivity, Strategy, and Efficiency in Modern Organic Chemical. Barry M. Trost and others, editors. Pergamon Press, Maxwell House, Fairview Park, Elmsford, NY 10523. (914) 592-7700. Fax (914) 592-3625. 1991. Nine volumes. $3900.00.

CRC Handbook of Chemistry and Physics. David R. Kide, editor. CRC Press Inc., 2000 Corporate Blvd., NW, Boca Raton, FL 33431. (407) 994-0555. (800) 333-8300. 77th edition. 1996. $99.95.

Guide to Basic Chemical Compounds. D.R. Lide, Jr. CRC Publishers, Inc., 2000 Corporate Blvd., N.W., Boca Raton, FL 33431. (407) 994-0555 or (800) 333-8300. 1993. $120.00.

Handbook of Organic Chemistry. Hans Beyer and Wolfgang Walter. Prentice-Hall General Referal Reference & Travel, 15 Columbus Circle, New York, NY 10023. (212) 373-8500. (800) 223-2348. 1995. $96.00.

Handbook of Plastics, Elastomers and Composites. Charles A. Harper. McGraw-Hill Publishing Company, Inc., 1221 Avenue of the Americas, New York, NY 10020. (212) 512-2000. (800) 262-4729. 1992. $89.50.

Handbooks and Tables in Science and Technology. Russell H. Powell, editor. Oryx Press, 4041 North Central, Suite 700, Phoenix, AZ 85012-3330. (602) 265-2651 or (800) 279-6799. Third edition. 1994. $65.00.

Improving Safety in the Chemical Laboratory: A Practical Guide. Jay A. Young, editor. John Wiley and Sons, Inc., 605 Third

Avenue, New York, NY 10158. (800) 526-5368 or (212) 850-6000. Second edition. 1991. $75.00.

Laboratory Handbook of Materials, Equipment, and Techniques. Gary S. Coyne. Prentice Hall (Division of Simon and Schuster), 15 Columbus Circle, New York, NY 10023. (212) 373-8500 or (800) 922-0579. 1992. $45.00.

Lange's Handbook of Chemistry. John A. Dean, editor. McGraw-Hill Publishing Company, 1221 Avenue of the Americas, New York, NY 10020. (800) 262-4729 or (212) 512-3825. Fourteenth edition. 1992. $79.50.

Plastics Engineering Handbook of the Society of the Plastics Industry. Michael L. Berins, editor. Chapman & Hall 5th edition. 1995. $104.00.

Riegel's Handbook of Industrial Chemistry. James A. Kent, editor. Van Nostrand Reinhold, 115 Fifth Avenue, New York, NY 10003. (212) 254-3232 or (800) 926-2665. Ninth edition. 1992. $114.95.

ONLINE DATABASES AND CD-ROMS

Beilstein Online. Beilstein-Institut fur Literatur der Organischen Chemie, Varrentrapperstrasse, 40-42, D-6000, Frankfurt am Main 90, Germany. Contains data on carbon compounds from the Beilstein Handbook of Organic Chemistry. Structural and factual data for more than 3.4 million heterocyclic, isocyclic, and acyclic compounds included. Covers the period 1830 to present. Available on DIALOG, (800) 334-2564, ORBIT (800) 456-7248, and STN International, FIZ Karlsruhe, P.O. Box 2465, W-7500, Karlsruhe 1, Germany, online services. Inquire as to cost and availability.

CA Search. Chemical Abstracts Service, P.O. Box 3012, Columbus, OH 43210-0012. (614) 447-3600. (800) 848-6533. FAX (614) 447-3709. Very comprehensive guide to worldwide chemical literature and related fields, 1972 to present. Available on BRS,(800) 289-4277, DIALOG, (800) 334-2564, ORBIT (800) 456-7248, and STN International, FIZ Karlsruhe, P.O. Box 2465, W-7500, Karlsruhe 1, Germany, online services. Inquire as to cost and availability.

Chemical Journals of the American Chemical Society. American Chemical Society, 1155 16th Street, N.W., Washington, DC 20036. (202) 872-4381 or (800) 424-6747. Contains complete text of approximately 90,000 articles from 22 primary journals published by the American Chemical Society, from mostly 1982 to present. Available on STN International, FIZ Karlsruhe, P.O. Box 2465, W-7500, Karlsruhe 1, Germany, online service. Inquire as to cost and availability.

Compendex Plus. Engineering Information, Inc., 345 East 47th Street, New York, NY 10017. (212) 705-7600 or (800) 221-1044. Contains citations with abstracts to worldwide literature in engineering and technology, from 1970 to present. Available on online BRS,(800) 289-4277, DIALOG, (800) 334-2564, ORBIT (800) 456-7248, and STN International, FIZ Karlsruhe, P.O. Box 2465, W-7500, Karlsruhe 1, Germany, online services. Also available on CD-ROM. Inquire as to cost and availability.

Current Contents Search. Institute for Scientific Information, 3501 Market Street, Philadelphia, PA 19104. (215) 386-0100. FAX (215) 386-6362. Contains citations to articles listed in the table of contents of science and technology journals. Also articles in social sciences and life sciences journals. Available on

BRS,(800) 289-4277, DIALOG,(800) 334-2564, online services. Inquire as to cost and availability.

Dissertation Abstracts. University Microfilms International, 300 North Zeeb Road, Ann Arbor, MI 48106. (800) 521-0600 or (313) 761-4700. Scope includes virtually all doctoral dissertations accepted at accredited American institutions from 1861 to present in 252 subject areas. Available on BRS,(800) 289-4277, DIALOG,(800) 334-2564, and OCLC EPIC,(800) 848-5878, online services. Also available on CD-ROM. Inquire as to cost and availability.

Engineered Materials Abstracts. Materials Information, ASM International, Materials Park, OH 44073. (216) 338-5151. Contains citations with abstracts of worldwide literature in the development, processing, and production of ceramic, composite, and polymeric materials for engineering uses. Available on DIALOG, (800) 334-2564, ORBIT (800) 456-7248, and STN International, FIZ Karlsruhe, P.O. Box 2465, W-7500, Karlsruhe 1, Germany, online services. Inquire as to cost and availability.

Kirk-Othmer Encyclopedia of Chemical Technology. John Wiley and Sons, Inc., 605 Third Avenue, New York, NY 10158. (800) 526-5368 or (212) 850-6000. Contains the complete text of all chapters in the 27 volume fourth edition of the *Kirk-othmer Encyclopedia of Chemical Technology.* 1991. Available on BRS,(800) 289-4277, DIALOG,(800) 334-2564, online services. Inquire as to cost and availability.

NTIS Bibliographic Database. National Technical Information Service, 5285 Port Royal Road, Springfield, VA 22161. (703) 487-4929 or FAX (703) 321-8199. Broad coverage of government-sponsored science and technology research reports, 1964 to present. Available on BRS,(800) 289-4277, DIALOG, (800) 334-2564, ORBIT, (800) 456-7248, and STN International, FIZ Karlsruhe, P.O. Box 2465, W-7500, Karlsruhe 1, Germany, online services. Also available on CD-ROM. Inquire as to cost and availability.

Scientific and Technical Books and Serials in Print. R.R. Bowker Inc., 121 Chanlon Road, New Providence, NJ 07974. (800) 521-8110 or (908) 464-6800. List of currently published books and serials in the physical and biological sciences, engineering and technology, with subject, author and titles indexes. Available on ORBIT,(800) 456-7248, online service. Inquire as to cost and availability.

SCISEARCH. Institute for Scientific Information, 3501 Market Street, Philadelphia, PA 19104. (800) 523-1850 or (215) 386-0100. Broad multidisciplinary title and author index to the international literature of science and technology, 1974 to present. Available on DIALOG,(800) 334-2564, and ORBIT,(800) 456-7248, online services. Also available on CD-ROM. Inquire as to cost and availability.

Who's Who in Technology. Gale Research, 835 Penobscot Building, Detroit, MI 48226-4094. (313) 961-2242. (800) 347-4253. Contains biographical information of contemporary American scieNTISts and engineers. Available on DIALOG,(800) 334-2564, online service. Inquire as to cost and availability.

WILSONLINE. H.W. Wilson Company, 950 University Avenue, Bronx, NY 10452. (800) 367-6770 or (212) 588-8400. Makes available online versions of the printed H.W. Wilson indexes including Applied Science and Technology Index, Business Periodicals Index, General Science Index, and Readers' Guide to Periodical Literature. Period covered is generally 1983 to present. Available on BRS,(800) 289-4277, DIALOG,(800) 334-

2564, and OCLC EPIC,(800) 848-5878, online services. Also available on CD-ROM. Inquire as to cost and availability.

PERIODICALS

Advances in Polymer Technology. John Wiley and Sons, Inc., 605 Third Avenue, New York, NY 10158. (800) 526-5368 or (212) 850-6000. 1977 to present.

Aiche Journal. American Institute of Chemical Engineers, 345 47th Street, New York, NY 10017-2396. (212) 705-7663. 1955 to present. Monthly. $295.00 per year.

Canadian Plastics. Southam Magazine Group, 1450 Don Mills Road, Don Mills, ON M3B 2X7. Canada. (416) 445-6641. FAX (416) 442-2213. 1943 to present. 8 issues per year. $55.00 per year. ISSN: 0008-4778.

Chemical Week. Includes Annual Buyers Guide. Chemical Week Associates, 888 Seventh Avenue, New York, NY 10106. (212) 621-4900. FAX (212) 621-4949. 1914 to present. Weekly. $99.00 per year. ISSN: 0009-272X.

Chemtech. American Chemical Society. Box 3337, Columbus, OH 43210. (614) 447-3776. 1970 to present. Monthly. $370.00 per year. ISSN: 0009-2703.

High Performance Plastics. Elsevier Science Publishing Company, Inc., 655 Avenue of the Americas, New York, NY 10010. (212) 989-5800. 1983 to present. Monthly. $390.00 per year. ISSN: 0264-7753.

Industrial and Engineering Chemistry Research. American Chemical Society, Box 3337, Columbus, OH 43210. (614) 447-3776. (800) 333-9511. FAX (614) 447-3671. 1962 to present. Monthly. $567.00 per year. ISSN: 0888-5885.

Journal of Cellular Plastics. Technomic Publishing Company, Inc., 851 New Holland Avenue, Box 3535, Lancaster, PA 17604. (717) 291-5609. FAX (717) 295-4538. 1965 to present. Bi-monthly. $175.00 per year. ISSN: 0021-955X

Journal of Elastomers and Plastics. Technomic Publishing Company, Inc., 851 New Holland Avenue, Box 3535, Lancaster, PA 17604. (717) 291-5609. FAX (717) 295-4538. 1969 to present. Quarterly. $205.00 per year. ISSN: 0095-2443.

Journal of Materials Processing and Manufacturing. Technomic Publishing Company, Inc., 851 New Holland Avenue, Box 3535, Lancaster, PA 17604. (717) 291-5609. 1992 to present. Quarterly. $195.00 per year.

Journal of Plastic Film and Sheeting. Technomic Publishing Company, Inc., 851 New Holland Avenue, Box 3535, Lancaster, PA 17604. (717) 291-5609. FAX (717) 295-4538. 1985 to present. Quarterly. $265.00. per year. ISSN: 8756-0879.

Journal of Polymer Science. Part A: Polymer Chemistry. John Wiley and Sons, Inc., 605 Third Avenue, New York, NY 10158. (212) 850-6088. FAX (212) 850-6088. 1962 to present. Subscription includes part B: Polymer Physics, and Symposia Proceedings. 13 issues per year. $2540.00 per year. ISSN: 0887-624X; 0887-6266; 0360-8905.

Modern Plastics. McGraw-Hill, Inc., 1221 Avenue of the Americas, New York, NY 10020. (212) 512-6267. FAX (212) 512-6111. 1925 to present. Monthly. $41.75 per year. ISSN: 0026-8275.

Plastics Compounding: For Resin Producers, Formulators and Compounders. Advanstar Communications, Inc., 7500 Old Oak Boulevard, Cleveland, OH 44130. (216) 826-2839. FAX (216) 891-2726. 1977 to present. Bi-monthly. $40.00 per year.ISSN: 0148-9119.

Plastics Design Forum. Advanstar Communications, Inc., 7500 Old Oak Boulevard, Cleveland, OH 44130. (216) 826-2839. FAX (216) 891-2726. 1976 to present. 7 issues per year. $40.00 per year. ISSN: 0362-9376.

Plastics Engineering. Society of Plastics Engineers, Inc., 14 Fairfield Drive, Brookfield, Center, CT 06805-0403. (203) 775-0471. FAX (203) 775-8490. 1945 to present. Monthly. $50.00 per year. ISSN: 0091-9578.

Plastics Technology. Bill Communications Inc., 355 Park Avenue South, New York, NY 10011-1706. (212) 592-6570. 1955 to present. 13 issues per year. $65.00 per year. ISSN: 0032-1257.

Polymer Composites. Society of Plastics Engineers, Inc., 14 Fairfield Drive, Brookfield, Center, CT 06805-0403. (203) 775-0471. FAX (203) 775-8490. 1980 to presesnt. Bi-monthly. $200.00 per year. ISSN: 0272-8397.

Processing of Advanced Materials. Routledge, Chapman and Hall, Inc., 29 West 35th Street, New York, NY 10001-2291. (212) 244-3336. 1991 to present. Quarterly. $145.00 per year.

RESEARCH CENTERS AND INSTITUTES

Center for Plastics Recycling Research. Rutgers University, Building 3529, Busch Campus, PO Box 1179, Piscataway, NJ 08855-1179. (809) 932-3683. FAX (98) 932-5636.

Institute of Polymer Engineering. University of Akron, Akron, OH 44325-0301. (216) 972-6865. (216) 258-2339.

Materials Research Laboratory. Washington University, Campus Box 1087, 1 Brookings Drive, St. Louis, MO 63130. (314) 889- 6012. FAX (314) 726-4434.

Plastics Institute of America. 277 Fairfield Road, Suite 100, Fairfield, NJ 07004-1932. (201) 808-5950. FAX (201) 808-5953.

Polymer Processing Institute. Stevens Institute of Technology, Castle Point Station, Hoboken, NJ 07030. (201) 420-5819.

Polymer Research Center. University of Cincinnati, Mail Location 172, Cincinnati, OH 45221. (513) 465-2453.

Polymer Research Laboratory. University of Michigan, 2045 Dow Building, Ann Arbor, MI 48109. (313) 763-2240. FAX (313) 763-4788.

Polymer Technologies, Inc., University of Detroit, 4001 West McNichols Road, Detroit, MI 48221-9987. (313) 927-1270. FAX (313) 927-1409.

PORTLAND CEMENT

Ency. of Physical Sciences and Engineering Info. Sources

PORTLAND CEMENT

See also: BUILDING MATERIALS, CEMENT, CONCRETE

ABSTRACT SERVICES AND INDEXES

Applied Science and Technology Index: A Cumulative Subject Index to English Language Periodicals in the Fields of Aeronautics and Space Science, Computer Technology, Chemistry, Construction Industry, Energy and Related Areas. H.W. Wilson Co., 950 University Avenue, Bronx, NY 10452. (800) 367-6770 or (212) 588-8400. FAX (718) 590-1617. From 1958 to present. Monthly. Inquire about cost and availability. Also available on CD-ROM and online.

Concrete Abstracts. Covers articles, books, and other publications that report developments in concrete design, construction and technology. American Concrete Institute, PO Box 19150, Detroit, MI 48219. (313) 532-2600. FAX (213) 538-0655. 1972 to present. Bi-monthly. Inquire for price.

Engineering Index Monthly. Engineering Information, Inc., Castle Point on the Hudson, Hoboken, NJ 07030. (800) 221-1044. FAX (212) 832-1857. Monthly. $2300.00 per year. Also available online as COMPENDEX, and also on CD-ROM. Covers chemical engineering, computers, electrical engineering, civil engineering, metals and mining, industrial management, and mechanical engineering.

ANNUAL REVIEWS AND YEARBOOKS

U.S. Department of the Interior, Bureau of Mines, Minerals Yearbook. Superintendent of Documents, U.S. Government Printing office, Box 371954, Pittsburgh, PA 15250-7954. (202) 783-3238. FAX (202) 512-2233. 1932 to present. Annual. Price varies.

ASSOCIATIONS AND PROFESSIONAL SOCIETIES

American Concrete Institute. P.O. Box 19150, Redford Station, Detroit, MI 48219. (313) 532-2600. FAX (213) 538-0655.

American Concrete Pavement Association. 3800 N. Wilke Road, Suite 490, Arlington Heights, IL 60004. (708) 394-5577. FAX (708) 394-5610.

American Portland Cement Alliance. 1212 New York Avenue NW, Suite 500, Washington, DC 20005. (202) 408-9494. FAX (202) 408-9392.

American Society for Concrete Construction. 1902 Techny Court, Northbrook, IL 60062. (312) 291-0270.

American Society of Civil Engineers. 345 East 47th Street, New York, NY 10017-2398. (212) 705-7520 or (800) 548-2723.

National Concrete Masonry Association. 2302 Horse Pen Road, Herndon, VA 22071-3406. (703) 713-1900. FAX (703) 713-1910.

National Precast Concrete Association. 10333 North Meridian Street, Suite 272, Indianapolis, in 46290-9500. (800) 366-7731. FAX (317) 571-0041.

National Ready Mixed Concrete Association. 900 Spring Street, Silver Spring, MD 20910. (301) 587-1400. FAX (301) 585-4219.

Portland Cement Association. 5420 Old Orchard Road, Skokie, IL 60077. (708) 966-6200. FAX (708) 966-8389.

Precast/Prestressed Concrete Institute. 175 West Jackson Blvd., Chicago, IL 60604. (312) 786-0300. FAX (312) 786-0353.

Reinforced Concrete Research Council. 205 N. Mathews Avenue, Urbana, IL 61801. (217) 333-7384.

DIRECTORIES AND BIOGRAPHICAL SOURCES

Aberdeen's Concrete Sourcebook. Aberdeen Group, 426 S. Westgate Street, Addison, IL 60101. (708) 543-0870. FAX (708) 543-3112. Monthly. $30.00 per year.

American Concrete Institute-Directory. American Concrete Institute, PO Box 19150, Detroit, MI 48219. (313) 532-2600. FAX (213) 538-0655. Every 2-3 years. Write or call for price information.

National Precast Concrete Association Directory of Members and Precast Products. National Precast Concrete Association, 10333 North Meridian Street, Suite 272, Indianapolis, IN 46290-9500. (800) 366-7731. FAX (317) 571-0041. Annual. $295.00.

Who's Who in Engineering. American Association of Engineering Societies, 1111 19th Street NW, Suite 608, Washington, DC 20036. (202) 296-2237 or (800) 658-8897. 8th edition. 1991. Inquire for price.

World Cement Directory. European Cement Association/ Cimeurope, Rue d'Arlon 55, B-1040 Brussels, Belgium. Telephone 32-2-2341011. FAX 32-2-2304720. 1958 to present. Quinquennial. $600.00.

ENCYCLOPEDIAS AND DICTIONARIES

Cement and Concrete Terminology. American Concrete Institute Staff. American Concrete Institute, Box 19150, Redford Station, Detroit, MI 48219. (313) 532-2600. FAX (313) 538-0655. 1990. $33.95.

Concise Encyclopedia of Building and Construction Materials. Fred Moavenzadeh. Pergamon/ Elsevier Science B.V., P.O. Box 945, Madison Square Station, New York, NY 10159-0945. (212) 633-3650. FAX (212) 633-3680. 1990. $210.00.

Encyclopedia of Physical Science and Technology. Robert A. Meyers, ed. 18 volumes. Academic Press Inc., 6277 Sea Harbor Drive, Orlando, FL 32887. (800) 321-5068. 1992. $2100.00.

McGraw-Hill Dictionary of Scientific and Technical Terms. Sybil P. Parker, ed. 5th edition. McGraw-Hill Publishing Company, 1221 Avenue of the Americas, New York, NY 10020. (800) 262-4729 or (212) 512-3825. 1993. $110.50.

McGraw-Hill Encyclopedia of Engineering. Sybil P. Parker, ed. 2nd edition. McGraw-Hill Publishing Company, 1221 Avenue of the Americas, New York, NY 10020. (800) 262-4729 or (212) 512-3825. 1993. $95.50.

McGraw-Hill Encyclopedia of Science and Technology. Sybil P. Parker, ed. 7th edition. 20 volumes. McGraw-Hill Publishing Company, 1221 Avenue of the Americas, New York, NY 10020. (800) 262-4729 or (212) 512-3825. 1992. $1900.00

GENERAL WORKS

Cement Chemistry. H.F.W. Taylor. Academic Press Inc., 6277 Sea Harbor Drive, Orlando, FL 32887. (800) 321-5068. 1990. $120.00.

Design and Control of Concrete Mixtures. Steven H. Kosmatka and William C. Panarese. 13th edition. Portland Cement Association, 5420 Old Orchard Road, Skokie, IL 60077-1083. (708) 966-6200. $33.00.

Fundamentals of Portland Cement Concrete: A Quantitative Approach. Sandor Popovics. John Wiley and Sons, Inc., 605 Third Avenue, New York, NY 10158. (800) 526-5368 or (212) 850-6000. 1982. Inquire for cost and availability.

The Portland Cement Association's Guide to Concrete Homebuilding Systems. Pieter A. VanderWerf, W. Keith Munsell. McGraw-Hill Publishing Company, 1221 Avenue of the Americas, New York, NY 10020. (800) 262-4729 or (212) 512-3825. 1995. Inquire for cost and availability.

Portland Cement: Composition, Production, and Use. G.C. Bye. Pergamon/ Elsevier Science B.V., PO Box 945, Madison Square Station, New York, NY 10159-0945. (212) 633-3650. FAX (212) 633-3680. 1983. Inquire for cost and availability.

HANDBOOKS AND MANUALS

ACI Manual of Concrete Practice. 5 volumes. American Concrete Institute, P.O. Box 19150, Redford Station, Detroit, MI 48219. (313) 532-2600. FAX (213) 538-0655. 1967 to present. Annual. $423.00 per set for non-members.

Concrete Construction Handbook. Joseph J. Waddell & Jpseh A. Dobrowolski, editors. 3rd edition. McGraw-Hill Publishing Company, 1221 Avenue of the Americas, New York, NY 10020. (800) 262-4729 or (212) 512-3825. 1993. $92.00.

Concrete Manual. Materials Engineering Branch, Research and Laboratory Services Division, Denver office, U.S. Department of the Interior, Bureau of Reclamation. 9th edition. Superintendent of Documents, U.S. Government Printing office, Box 371954, Pittsburgh, PA 15250-7954. (202) 783-3238. FAX (202) 512-2233. 1992. Inquire for cost and availability.

Handbook of Concrete Engineering. Mark Fintel, editor. 2nd edition. Van Nostrand Reinhold, 115 Fifth Avenue, New York, NY 10003. (212) 254-3232 or FAX (212) 254-9499. 1985. $115.00.

Portland Cement Chemical Composition Standards-Blending, Packaging, and Testing: Standard Reference Materials. R. Keith Kirby and Howard M. Kanare. NBS Special Publication 260-110. U.S. Department of Commerce, National Bureau of Standards/ Superintendent of Documents, U.S. Government Printing office, Box 371954, Pittsburgh, PA 15250-7954. (202) 783-3238. FAX (202) 512-2233.1988. Inquire for cost and availability.

Structural Design Guide to the ASCI Specifications for Buildings. Paul F. Rice, et al. 3rd edition. Books on Demand, 300 N. Zeeb Road, Ann Arbor, MI 48106-1346. (313) 761-4700 or (800) 521-0600. 1985. $92.00.

ONLINE DATABASES AND CD-ROMS

Compendex Plus. Engineering Information, Inc., 345 East 47th Street, New York, NY 10017. (212) 705-7600 or (800) 221-1044. Contains citations with abstracts to worldwide literature in engineering and technology, from 1970 to present. Available on online BRS,(800) 289-4277, DIALOG, (800) 334-2564, ORBIT (800) 456-7248, and STN International, FIZ Karlsruhe, P.O. Box 2465, W-7500, Karlsruhe 1, Germany, online services. Also available on CD-ROM. Inquire as to cost and availability.

NTIS Bibliographic Database. National Technical Information Service, 5285 Port Royal Road, Springfield, VA 22161. (703) 487-4929 or FAX (703) 321-8199. Broad coverage of government-sponsored science and technology research reports, 1964 to present. Available on BRS,(800) 289-4277, DIALOG, (800) 334-2564, ORBIT, (800) 456-7248, and STN International, FIZ Karlsruhe, P.O. Box 2465, W-7500, Karlsruhe 1, Germany, online services. Also available on CD-ROM. Inquire as to cost and availability.

WILSONLINE. H.W. Wilson Company, 950 University Avenue, Bronx, NY 10452. (800) 367-6770 or (212) 588-8400. Makes available online versions of the printed H.W. Wilson indexes including *Applied Science and Technology Index, Business Periodicals Index, General Science Index,* and *Readers' Guide to Periodical Literature.* Period covered is generally 1983 to present. Available on BRS,(800) 289-4277, DIALOG,(800) 334-2564, and OCLC EPIC,(800) 848-5878, online services. Also available on CD-ROM. Inquire as to cost and availability.

PERIODICALS

Aberdeen's Concrete Construction. Aberdeen Group, 426 S. Westgate Street, Addison, IL 60101. (708) 543-0870. FAX (708) 543-3112. 1956 to present. Monthly. $24.00 per year.

Aberdeen's Concrete Journal. Aberdeen Group, 426 S. Westgate Street, Addison, IL 60101. (708) 543-0870. FAX (708) 543-3112. 1983 to present. Monthly. Inquire for cost.

ACI Materials Journal. American Concrete Institute, P.O. Box 19150, Detroit, MI 48219. (313) 532-2600. FAX (213) 538-0655. Monthly. $92.00 per year.

ACI Structural Journal. American Concrete Institute, P.O. Box 19150, Detroit, MI 48219. (313) 532-2600. FAX (213) 538-0655. Bi-monthly. $92.00 per year.

Cement and Concrete Composites. Elsevier Science [journals], 660 White Plains Rd., Tarrytown, NY 10159-5153. (919) 524-9200. FAX (919) 333-2444. 1979 to present. 4 times a year. $280.00 per year.

Cement and Concrete Research. Elsevier Science [journals], 660 White Plains Rd., Tarrytown, NY 10159-5153. (919) 524-9200. FAX (919) 333-2444. 1971 to present. 8 times a year. $590.00 per year.

Cement, Concrete and Aggregates. American Society for Testing and Materials, 1916 Race Street, Philadelphia, PA 19103.

PORTLAND CEMENT

Ency. of Physical Sciences and Engineering Info. Sources

(215) 299-5585. 1979-present. Semi-annual. $48.00 per year for non-members.

Concrete. The Concrete Society, Framewood Road, Wexham, Slough, Berks. SL3 6PJ, England. Telephone 0753-662226. FAX 0753-662126. 1966 to present. Bi-monthly. $75.00 per year.

Concrete International. American Concrete Institute, P.O. Box 19150, Detroit, MI 48219. (313) 532-2600. FAX (213) 538-0655. 1979 to present. Monthly. $101.00 per year.

Journal of Ferrocement. International Ferrocement Information Center, Asian Institute of Technology, PO Box 2754, Bangkok 10501, Thailand. Telephone 5160110. FAX 66-2-5162126. 1976 to present. Quarterly. $96.00 per year.

Journal of the Precast/Prestressed Concrete Institute. Precast/Prestressed Concrete Institute, 175 West Jackson Blvd., Chicago, IL 60604. (312) 786-0300. FAX (312) 786-0353. Bi-monthly. $29.00 per year.

Magazine of Concrete Research. Thomas Telford Services Ltd., Thomas Telford House, 1 Heron Quay, London E14 4JD, England. Telephone 071-987-6999. FAX 071-538-4101. 1949 to present. Quarterly. Inquire for cost.

Rock Products. McLean Hunter Publishing Company, 29 N. Wacker Drive, Chicago, IL 60606. (312) 726-2802. FAX (312) 726-2574. 1897 to present. Monthly. $31.25 per year.

RESEARCH CENTERS

American Concrete Institute. P.O. Box 19150, Redford Station, Detroit, MI 48219. (313) 532-2600. FAX (213) 538-0655.

American Society for Testing and Materials. 1916 Race Street, Philadelphia, PA 19103. (215) 299-5585.

Concrete Materials Research Council. P.O. Box 19150, Detroit, MI 48219. (313) 532-2600.

Concrete Research Laboratory. University of Michigan, G.G. Brown Laboratory, Ann Arbor, MI 48109. (313) 764-9660.

Portland Cement Association. 5420 Old Orchard Road, Skokie, IL 60077. (708) 966-6200. FAX (708) 966-8389.

SPECIFICATIONS AND STANDARDS

Annual Book of ASTM Standards, Volume 04.01. American Society for Testing and Materials (ASTM), 1916 Race Street, Philadelphia, PA 19103. (215) 299-5585. Annual. Inquire for cost and availability.

POWDER METALLURGY

See also: METALLURGY

ABSTRACT SERVICES AND INDEXES

Alloys Index. American Society for Metals, Metals Park, OH 44073. (216) 338-5151. FAX (216) 338-4634. 1974 to present.

Monthly. $380.00. Also available on CD-ROM and online via METADEX, STN International, and DIALOG.

Aluminum Industry Abstracts. Aluminum Association, Materials Park, OH 44073. (216) 338-5151. FAX (216) 338-4634. 1968 to present. Monthly. $525.00 per year.

Applied Mechanics Reviews: An Assessment of World Literature in Engineering Sciences. 1948-present. American Society of Mechanical Engineers, 345 East 47th Street, New York, NY 10017. (212) 705-7703. Monthly. $360.00 per year.

Applied Science and Technology Index: A Cumulative Subject Index to English Language Periodicals in the Fields of Aeronautics and Space Science, Computer Technology, Chemistry, Construction Industry, Energy and Related Areas. H.W. Wilson Co., 950 University Avenue, Bronx, NY 10452. (800) 367-6770 or (212) 588-8400. FAX (718) 590-1617. From 1958 to present. Monthly. Inquire about cost and availability. Also available on CD-ROM and online.

Current Contents: Engineering, Technology, and Applied Sciences. Institute for Scientific Information, 3501 Market Street, Philadelphia, PA 19104. (215) 386-0100. FAX (215) 386-6362. 1970 to present. Weekly. $442.00 per year.

Engineering Index Monthly. Engineering Information, Inc., Castle Point on the Hudson, Hoboken, NJ 07030. (800) 221-1044. FAX (212) 832-1857. Monthly. $2300.00 per year. Also available online as COMPENDEX, and also on CD-ROM. Covers chemical engineering, computers, electrical engineering, civil engineering, metals and mining, industrial management, and mechanical engineering.

I.M.M.Abstracts and Index. Institution of Mining and Metallurgy, 44 Portland Pl., London W1N 4BR, England. 071-580-3802. FAX 071-436-5388. Bi-monthly. $364 for non-members.

Manufacturing and Process Engineering Abstracts, Cambridge Scientific Abstracts, 7200 Wisconsin Avenue, Bethesda, MD 20814-4823. (301) 961-6750. FAX (301) 961-6720. 1993 to present. Monthly. $385.00 per year. Covers concurrent engineering, quality control, automated manufacturing, petroleum engineering, oil field operations and equipment, energy management, metallurgy and metallography, foundry practice.

Materials Science and Engineering Abstracts, Cambridge Scientific Abstracts, 7200 Wisconsin Avenue, Bethesda, MD 20814-4823. (301) 961-6750. FAX (301) 961-6720. 1993 to present. Monthly. $385.00 per year. Focuses on mechanical and physical properties of materials and commercial or industrial applications for materials, methods for strength testing, effects of vibration and other stresses, corrosion and protective coatings, storage and handling, ceramics, composites, metals, wood, plastics, and polymers.

Materials Science Citation Index. Institute for Scientific Information, 3501 Market Street, Philadelphia, PA 19104. (215) 386-0100. FAX (215) 386-6362. Bi-monthly. $975.00 per year. Also available on CD-ROM.

Metals Abstracts and Metals Abstracts Index. American Society for Metals, Metals Park, OH 44073. (216) 338-5151. 1968 to present. Monthly. Abstracts are $1925.00 per year and Index is $460.00 per year.

Science Citation Index. Institute for Scientific Information, 3501 Market Street, Philadelphia, PA 19104. (215) 386-0100. FAX

(215) 386-6362. Inquire about availability and cost. Also available on CD-ROM.

ASSOCIATIONS AND PROFESSIONAL SOCIETIES

American Powder Metallurgy Institute. 105 College Road E., Princeton, NJ 08540. (609) 452-7700. (609) 987-8523.

American Society for Metals. Materials Park, OH 44073. (216) 338-5151 or FAX (216) 338-4634.

Metallurgical Society of the AIME (American Institute of Mining, Metallurgical, and Petroleum Engineers). 345 E.47th Street, 14th Floor, New York, NY 10017. (212) 705-7695.

DIRECTORIES AND BIOGRAPHICAL SOURCES

Directory of Engineering Societies and Related Organizations. Gordon Davis. 13th edition. American Association of Engineering Societies, 1111 19th Street NW, Suite 608, Washington, DC 20036. (202) 296-2237 or (800) 658-8897. 1989. Inquire for price.

International Research Centers Directory. Gale Research, 835 Penobscot Building, Detroit, MI 48226-4094. (313) 961-2242. (800) 347-4253. FAX (313) 961-6083. 8th edition. 1995. $410.00

Research Centers Directory. Gale Research, 835 Penobscot Building, Detroit, MI 48226-4094. (313) 961-2242. (800) 347-4253. FAX (313) 961-6083. $485.00.

Scientific and Technical Organizations and Agencies Directory. Gale Research, 835 Penobscot Building, Detroit, MI 48226-4094. (313) 961-2242. (800) 347-4253. FAX (313) 961-6083. 4th edition. 1996. $195.00.

Who's Who in Engineering. American Association of Engineering Societies, 1111 19th Street NW, Suite 608, Washington, DC 20036. (202) 296-2237 or (800) 658-8897. 8th edition. 1991. Inquire for price.

Who's Who in Powder Metallurgy. American Powder Metallurgy Institute, 105 College Road E., Princeton, NJ 08540. (609) 452-7700. (609) 987-8523. Inquire for cost and availability.

ENCYCLOPEDIAS AND DICTIONARIES

ASM Materials Engineering Dictionary. J.R. Davis, editor. ASM International, Materials Information, Materials Park, OH 44073-0002. (216) 338-5151 or (800) 336-5152. FAX (216) 338-4634. 1992. $101.00.

Dictionary of Metallurgy and Foundry Technology. Karl Stolzel, editor. Elsevier Science Publishing Company, Inc., 655 Avenue of the Americas, New York, NY 10010. (212) 989-5800. 1987. $250.00.

McGraw-Hill Dictionary of Scientific and Technical Terms. Sybil P. Parker, ed. 5th edition. McGraw-Hill Publishing Company, 1221 Avenue of the Americas, New York, NY 10020. (800) 262-4729 or (212) 512-3825. 1993. $110.50.

McGraw-Hill Encyclopedia of Engineering. Sybil P. Parker, ed. 2nd edition. McGraw-Hill Publishing Company, 1221 Avenue of the Americas, New York, NY 10020. (800) 262-4729 or (212) 512-3825. 1993. $95.50.

McGraw-Hill Encyclopedia of Science and Technology. Sybil P. Parker, ed. 7th edition. 20 volumes. McGraw-Hill Publishing Company, 1221 Avenue of the Americas, New York, NY 10020. (800) 262-4729 or (212) 512-3825. 1992. $1900.00

Thesaurus of Scientific, Technical, and Engineering Terms. Hemisphere Publishing Corporation, 1900 Frost Road, Suite 101, Bristol, PA 19007-1598. (215) 785-5800 or (800) 821-8312. FAX (215) 785-5515. 1987. $173.00.

GENERAL WORKS

Powder Metallurgy. BCC Publishing, 11740 Shetland Heights Circle, Anchorage, AK 99516. (907) 345-1124. 1991. $225.00.

Powder Metallurgy: An Overview. Ivor Jenkins and J.V. Wood. Ashgate Publishing Co., Old Post Rd., Brookfield, VT 05036. (802) 276-3162. FAX (802) 276-3837. 1991. $147.00.

Powder Metallurgy Science. Randall M. German. 2nd edition. Metal Powder Industries Federation, APMI International, 105 College Road East, Princeton, NJ 08540-6692. (609) 452-7700. FAX (609) 987-8523. 1994. $60.00.

Powder Metallurgy: The Process and Its Products. Gordon Dowson. A. Hilger Publishers. Distributed by the American Institute of Physics, AIP Press, 500 Sunnyside Blvd., Woodbury, NY 11797-2999. (800) 809-2247. FAX (516) 576-2223. 1990. $65.00.

HANDBOOKS AND MANUALS

Metals Handbook. ASM International, Materials Information, Materials Park, OH 44073-0002. (216) 338-5151 or (800) 336-5152. FAX (216) 338-4634. $154.00.

Powder Metallurgy Design Manual. Metal Powder Industries Federation, 105 College Road E., Princeton, NJ 08540. (609) 452-7700. 1989. $75.00.

ONLINE DATABASES AND CD-ROMS

Compendex Plus. Engineering Information, Inc., 345 East 47th Street, New York, NY 10017. (212) 705-7600 or (800) 221-1044. Contains citations with abstracts to worldwide literature in engineering and technology, from 1970 to present. Available on online BRS,(800) 289-4277, DIALOG, (800) 334-2564, ORBIT (800) 456-7248, and STN International, FIZ Karlsruhe, P.O. Box 2465, W-7500, Karlsruhe 1, Germany, online services. Also available on CD-ROM. Inquire as to cost and availability.

CSA Engineering. Cambridge Scientific Abstracts, 7200 Wisconsin Avenue, Suite 601, Bethesda, MD 20814. (301) 961-6750 or (800) 843-7751. Contains citations and abstracts of international periodicals and research literature covering all fields of engineering and science and technology,including computer and information science, electronics, mechanical engineering, solid state materials, 1981 to present. Available on BRS,(800) 289-4277, online service.Inquire as to cost and availability.

Current Contents Search. Institute for Scientific Information, 3501 Market Street, Philadelphia, PA 19104. (215) 386-0100.

POWER METALLURGY

Ency. of Physical Sciences and Engineering Info. Sources

FAX (215) 386-6362. Contains citations to articles listed in the table of contents of science and technology journals. Also articles in social sciences and life sciences journals. Available on BRS,(800) 289-4277, DIALOG,(800) 334-2564, online services. Inquire as to cost and availability.

Dissertation Abstracts. University Microfilms International, 300 North Zeeb Road, Ann Arbor, MI 48106. (800) 521-0600 or (313) 761-4700. Scope includes virtually all doctoral dissertations accepted at accredited American institutions from 1861 to present in 252 subject areas. Available on BRS,(800) 289-4277, DIALOG,(800) 334-2564, and OCLC EPIC,(800) 848-5878, online services. Also available on CD-ROM. Inquire as to cost and availability.

ISMEC: Mechanical Engineering Abstracts. Cambridge Scientific Abstracts, 7200 Wisconsin Avenue, Suite 601, Bethesda, MD 20814. (301) 961-6750 or (800) 843-7751. Contains citations to the literature in mechanical engineering, industrial and production engineering, energy, power, mechanics, devices and related areas, from 1973 to present. Available on the DIALOG,(800) 334-2564, online service. Inquire as to cost and availability.

Metadex. Jointly produced by ASM International and the Institute of Materials. Contains more than 925,000 records from the international literature on metals and alloys, concerning properties, processes, materials classes, applications, and metallurgical systems. Updated monthly. Available from ORBIT-QUESTEL (703) 442-0900.

NTIS Bibliographic Database. National Technical Information Service, 5285 Port Royal Road, Springfield, VA 22161. (703) 487-4929 or FAX (703) 321-8199. Broad coverage of government-sponsored science and technology research reports, 1964 to present. Available on BRS,(800) 289-4277, DIALOG, (800) 334-2564, ORBIT, (800) 456-7248, and STN International, FIZ Karlsruhe, P.O. Box 2465, W-7500, Karlsruhe 1, Germany, online services. Also available on CD-ROM. Inquire as to cost and availability.

SCISEARCH. Institute for Scientific Information, 3501 Market Street, Philadelphia, PA 19104. (800) 523-1850 or (215) 386-0100. Broad multidisciplinary title and author index to the international literature of science and technology, 1974 to present. Available on DIALOG,(800) 334-2564, and ORBIT,(800) 456-7248, online services. Also available on CD-ROM. Inquire as to cost and availability.

WILSONLINE. H.W. Wilson Company, 950 University Avenue, Bronx, NY 10452. (800) 367-6770 or (212) 588-8400. Makes available online versions of the printed H.W. Wilson indexes including *Applied Science and Technology Index, Business Periodicals Index, General Science Index,* and *Readers' Guide to Periodical Literature.* Period covered is generally 1983 to present. Available on BRS,(800) 289-4277, DIALOG,(800) 334-2564, and OCLC EPIC,(800) 848-5878, online services. Also available on CD-ROM. Inquire as to cost and availability.

PERIODICALS

International Journal of Powder Metallurgy. American Powder Metallurgy Institute, 105 College Road E., Princeton, NJ 08540. (609) 452-7700. (609) 987-8523. 1965 to present. Quarterly. $70.00 per year.

JOM (Journal of Metals). Minerals, Metals and Materials Society, 420 Commonwealth Drive, Warrendale, PA 15086. (412) 776-9080. 1949 to present. Monthly. $50.00 per year.

Light Metal Age. Fellom Publishing, 170 S. Spruce Avenue, Suite 120, San Francisco, CA 94080. (415) 588-8832. FAX (415) 588-0901. 1943 to present. Bi-monthly. $35.00 per year.

Metal Powder Report. Elsevier Science [journals], 660 White Plains Rd., Tarrytown, NY 10159-5153. (919) 524-9200. FAX (919) 333-2444. 1946 to present. 8 times a year. $160.00 per year.

Metals Week. McGraw-Hill Publishing Company, 1221 Avenue of the Americas, New York, NY 10020. (800) 262-4729 or (212) 512-3825. 1930 to present. Weekly. $770.00 per year.

P-M Technology Newsletter. American Powder Metallurgy Institute, 105 College Road E., Princeton, NJ 08540. (609) 452-7700. (609) 987-8523. 1960 to present. Monthly. $70.00 per year.

Powder Metallurgy and Metal Ceramics. Plenum Publishing Corporation, 233 Spring Street, New York, NY 10013. (212) 620-8000. FAX (212) 463-0742. 1962 to present. Monthly. $1215.00 per year.

Powder Metallurgy. Institute of Materials, 1 Carlton House Terrace, London SW1Y 5DB, England. Telephone 071-839-4071. FAX 071-839-2078. 1958 to present. Quarterly. $170.00 per year.

Powder Metallurgy Science & Technology. Powder Metallurgy Development Centre Pvt. Ltd., P-26 Laxminagar, Saidabad, Hyderabad 500 659, India. 1989 to present. Quarterly. Inquire for cost.

Powder Technology. Elsevier Science Inc., Box 882, Madison Square Station, New York, NY 10159. (212) 989-5800. 1967 to present. 12 times a year. $1088.00 per year.

Science of Sintering. Jugoslovenska Knjiga, Export Department, Trg Repulike 5-VIII, 11000 Belgrade, Yugoslavia. 1969 to present. 3 times a year. $83.00 per year.

Scripta Metallurgia et Materialia. Elsevier Science [journals], 660 White Plains Rd., Tarrytown, NY 10159-5153. (919) 524-9200. FAX (919) 333-2444. 1967 to present. 24 times a year. $570.00 per year.

RESEARCH CENTERS AND INSTITUTES

American Powder Metallurgy Institute. 105 College Road E., Princeton, NJ 08540. (609) 452-7700. (609) 987-8523.

Cooperative Program in Metallurgy, Pennsylvania State University. 208A Steidle Bldg., University Park, PA 16802. (814) 865-5446.

University of Connecticut Institute of Materials Science. U-136, 97 N. Eagleville Road, Storrs, CT 06268. (203) 486-4623. FAX (203) 486-4745.

University of Florida Department of Materials Science and Engineering. Gainesville, FL 32611. (904) 392-1454. FAX (904) 392-6359.

SPECIFICATIONS AND STANDARDS

Annual Book of ASTM Standards, Volume 02.05. American Society for Testing and Materials (ASTM), 1916 Race Street, Philadelphia, PA 19103. (215) 299-5585. Annual. Inquire for cost and availability.

PRECIPITATION

See: RAIN, SNOW

PRESSURE VESSELS

See also: ELECTRIC POWER ENGINEERING, HEAT EXCHANGERS, HEAT TRANSFER, MECHANICAL ENGINEERING, NUCLEAR ENGINEERING, THERMODYNAMICS

ABSTRACT SERVICES AND INDEXES

Applied Mechanics Reviews: An Assessment of World Literature in Engineering Sciences. 1948-present. American Society of Mechanical Engineers, 345 East 47th Street, New York, NY 10017. (212) 705-7703. Monthly. $360.00 per year.

Applied Science and Technology Index: A Cumulative Subject Index to English Language Periodicals in the Fields of Aeronautics and Space Science, Computer Technology, Chemistry, Construction Industry, Energy and Related Areas. H.W. Wilson Co., 950 University Avenue, Bronx, NY 10452. (800) 367-6770 or (212) 588-8400. FAX (718) 590-1617. From 1958 to present. Monthly. Inquire about cost and availability. Also available on CD-ROM and online.

Current Contents: Engineering, Technology, and Applied Sciences. Institute for Scientific Information, 3501 Market Street, Philadelphia, PA 19104. (215) 386-0100. FAX (215) 386-6362. 1970 to present. Weekly. $442.00 per year.

Energy Research Abstracts. 1976-present. U.S. Department of Energy, office of Scientific and Technical Information, Box 62, Oak Ridge, TN 37831. (615) 574-0733. Subscriptions: Superintendent of Documents, U.S. Government Printing office, Box 371954, Pittsburgh, PA 15250-7954. (202) 783-3238. FAX (202) 512-2233. Monthly. $164.00 per year. Abstracts all the scientific and technical reports, journal articles, conference proceedings, patents, theses, and monographs sponsored by the U.S. Energy Research and Development Administration.

Engineering Index Monthly. Engineering Information, Inc., Castle Point on the Hudson, Hoboken, NJ 07030. (800) 221-1044. FAX (212) 832-1857. Monthly. $2300.00 per year. Also available online as COMPENDEX, and also on CD-ROM. Covers chemical engineering, computers, electrical engineering, civil engineering, metals and mining, industrial management, and mechanical engineering.

International Building Services Abstracts. Building Service Research and Information Association, Old Bracknell Lane W, Bracknell, Berks. RG12 7AH, England. Telephone 0344-426511. FAX 0344-487575. 1966 to present. Bi-monthly. Inquire for price. Also available online. A survey of world literature on mechanical and technical services of buildings, including heat-ing, sanitation, ventilation, air conditioning, lighting, and communications.

Mechanical Engineering Abstracts (formerly ISMEC), Cambridge Scientific Abstracts, 7200 Wisconsin Avenue, Bethesda, MD 20814-4823. (301) 961-6750. FAX (301) 961-6720. 1967 to present. Monthly. $895.00 per year. Summarizes world literature in mechanical engineering, production engineering, and engineering management. Also available online.

Physics Abstracts. Institute of Electrical Engineers, Michael Faraday House, Six Hill Way, Stevenage, Herts, SG1 2AY, England. Distributed by IEEE, 445 Hoes Lane, Piscataway, NJ 08854. (908) 562-5549. 1898 to present. Monthly. $2700.00 per year. Also available online as INSPEC.

Previews of Heat and Mass Transfer. Elsevier Science [journals], 660 White Plains Rd., Tarrytown, NY 10159-5153. (919) 524-9200. FAX (919) 333-2444. 1974 to present. Six times a year. $265.00 per year. Contains citations with abstracts to the worldwide literature on heat and mass transfer from more than 100 journals.

Science Citation Index. Institute for Scientific Information, 3501 Market Street, Philadelphia, PA 19104. (215) 386-0100. FAX (215) 386-6362. Inquire about availability and cost. Also available on CD-ROM.

ASSOCIATIONS AND PROFESSIONAL SOCIETIES

American Boiler Manufacturer Association. 950 N. Glebe Road, Suite 160, Arlington, VA 22203-1824. (703) 522-7350. FAX (703) 522-2665.

American Institute of Chemical Engineers. 345 East 47th Street, New York, NY 10017-2396. (212) 705-7663.

American Society of Heating, Refrigerating & Air-Conditioning Engineers, 1791 Tullie Circle, N.E., Atlanta, GA 30329. (404) 636-8400. FAX (404) 321-5478.

American Society of Mechanical Engineers. 345 East 47th Street, New York, NY 10017-2398. (212) 705-7703.

National Board of Boiler & Pressure Vessel Inspectors. 1055 Crupper Avenue, Columbus, OH 43229. (614) 888-8320. FAX (614) 888-0750.

Pressure Vessels Manufacturers Association. 600 S. Federal Street, Suite 400, Chicago, IL 60605-1895. (312) 922-6222. FAX (312) 922-2734.

Uniform Boiler & Pressure Vessel Laws Society. 308 N. Evergreen Road, Suite 240, Louisville, KY 40243-1010. (502) 244-6029. FAX (502) 244-6030.

DIRECTORIES AND BIOGRAPHICAL SOURCES

American Men & Women of Science, 1995-96. R.R. Bowker Staff, eds. 19th edition. 8 volumes. R.R. Bowker/Reed International Publishing Company, 121 Chanlon Road, New Providence, NJ 07974. (908) 464-6800 or (800) 521-8110. 1995. $850.00

Directory of Engineering Societies and Related Organizations. Gordon Davis. 13th edition. American Association of Engineer-

PRESSURE VESSELS

Ency. of Physical Sciences and Engineering Info. Sources

ing Societies, 1111 19th Street NW, Suite 608, Washington, DC 20036. (202) 296-2237 or (800) 658-8897. 1989. Inquire for price.

Plant Engineering Product Supplier Guide. Cahners Publishing Company (Des Plains), 1350 East Touhy Avenue, Des Plaines, IL 60017-5080. (800) 323-4958. FAX (708) 390-2779. Included with subscription to *Plant Engineering* (see below).

Research Centers Directory. Gale Research, 835 Penobscot Building, Detroit, MI 48226-4094. (313) 961-2242. (800) 347-4253. FAX (313) 961-6083. $485.00.

Scientific and Technical Organizations and Agencies Directory. Gale Research, 835 Penobscot Building, Detroit, MI 48226-4094. (313) 961-2242. (800) 347-4253. FAX (313) 961-6083. 4th edition. 1996. $195.00.

Who's Who in Engineering. American Association of Engineering Societies, 1111 19th Street NW, Suite 608, Washington, DC 20036. (202) 296-2237 or (800) 658-8897. 8th edition. 1991. Inquire for price.

ENCYCLOPEDIAS AND DICTIONARIES

Encyclopedia of Physical Science and Technology. Robert A. Meyers, ed. 18 volumes. Academic Press Inc., 6277 Sea Harbor Drive, Orlando, FL 32887. (800) 321-5068. 1992. $2100.00.

McGraw-Hill Encyclopedia of Engineering. Sybil P. Parker, ed. 2nd edition. McGraw-Hill Publishing Company, 1221 Avenue of the Americas, New York, NY 10020. (800) 262-4729 or (212) 512-3825. 1993. $95.50.

McGraw-Hill Encyclopedia of Science and Technology. Sybil P. Parker, ed. 7th edition. 20 volumes. McGraw-Hill Publishing Company, 1221 Avenue of the Americas, New York, NY 10020. (800) 262-4729 or (212) 512-3825. 1992. $1900.00

Thesaurus of Scientific, Technical, and Engineering Terms. Hemisphere Publishing Corporation, 1900 Frost Road, Suite 101, Bristol, PA 19007-1598. (215) 785-5800 or (800) 821-8312. FAX (215) 785-5515. 1987. $173.00.

GENERAL WORKS

High Pressure Vessels. Donald M. Fryer & John F. Harvey. Chapman & Hall, 115 Fifth Avenue, New York, NY 10211-0906. 1996. $74.95.

Pressure Vessel Design: Concepts and Principles. J. Spence and A. S. Tooth, editors. E & FN Spon/ Chapman and Hall, 115 Fifth Avenue, New York, NY 10211-0906. 1994. $125.95.

Pressure Vessels: The ASME Code Simplified. Robert Chuse and Bryce E. Carson Sr. 7th edition. McGraw-Hill Publishing Company, 1221 Avenue of the Americas, New York, NY 10020. (800) 262-4729 or (212) 512-3825. 1993. $53.00.

Technology for the '90s: A Decade of Progress, the ASME Pressure Vessels and Piping Division. M.K. Au-Yang, principal editor. ASME, 345 East 47th Street, New York, NY 10017-2398. (212) 705-7703. 1993. $200.00.

Theory and Design of Pressure Vessels. John F. Harvey. 2d ed. Van Nostrand Reinhold, 115 Fifth Avenue, New York, NY 10003. (212) 254-3232 or FAX (212) 254-9499. 1991. $69.95.

HANDBOOKS AND MANUALS

Ashrae Handbook: Fundamentals. 4 volumes. American Society of Heating, Refrigeration and Air Conditioning Engineers, 1791 Tullie Circle NE, Atlanta, GA 30329. (404) 636-8400. FAX (404) 321-5478. 1922 to present. Annual. $119.00.

HVAC Engineer's Handbook. Fred Porges. 9th ed. Butterworths Publishers, 80 Montvale Avenue, Stoneham, MA 02180. (617) 438-8464 or (800) 366-2665. 1991. $70.00.

Handbook of Heat Transfer Applications. W.M. Rohsenow, et al., editors. McGraw-Hill Publishing Company, 1221 Avenue of the Americas, New York, NY 10020. (800) 262-4729 or (212) 512-3825. 1985. Inquire for cost and availability.

Heat Exchanger Sourcebook. J. W. Palen. Hemisphere Publishing Corporation, 1900 Frost Road, Suite 101, Bristol, PA 19007-1598. (215) 785-5800 or (800) 821-8312. FAX (215) 785-5515. 1986. $81.00.

Pressure Vessel Design Handbook: Illustrated Procedures For Solving Every Major Pressure Vessel Design Problem. Dennis R. Moss. Gulf Publishing Company, P.O. Box 2608, Houston, TX. (713) 529-4301 or (800) 231-6275. 1987. $56.50.

Pressure Vessel Handbook. Eugene F. Megyesy. 9th edition. Pressure Vessel Handbook Publishing Inc., PO Box 35365, Tulsa, OK 74153-0365. (918) 742-9637. 1992. $108.95.

Pressure Vessels. R. Chuse and B.E. Carson Sr. Seventh edition. McGraw-Hill Publishing Company, 1221 Avenue of the Americas, New York, NY 10020. (800) 262-4729 or (212) 512-3825. 1993. $70.00.

Steam Tables: Thermodynamic Properties of Water Including Vapor, Liquid & Solid Phases. Joseph H. Keenan, et al., editors. 2nd edition. Krieger Publishing Company, P.O. Box 9542, Melbourne, FL 32902-9542. (407) 724-9542. FAX (407) 951-3671. 1992 (reprint of 1978 edition). $64.50.

ONLINE DATABASES AND CD-ROMS

Compendex Plus. Engineering Information, Inc., 345 East 47th Street, New York, NY 10017. (212) 705-7600 or (800) 221-1044. Contains citations with abstracts to worldwide literature in engineering and technology, from 1970 to present. Available on online BRS,(800) 289-4277, DIALOG, (800) 334-2564, ORBIT (800) 456-7248, and STN International, FIZ Karlsruhe, P.O. Box 2465, W-7500, Karlsruhe 1, Germany, online services. Also available on CD-ROM. Inquire as to cost and availability.

CSA Engineering. Cambridge Scientific Abstracts, 7200 Wisconsin Avenue, Suite 601, Bethesda, MD 20814. (301) 961-6750 or (800) 843-7751. Contains citations and abstracts of international periodicals and research literature covering all fields of engineering and science and technology,including computer and information science, electronics, mechanical engineering, solid state materials, 1981 to present. Available on BRS,(800) 289-4277, online service.Inquire as to cost and availability.

Current Contents Search. Institute for Scientific Information, 3501 Market Street, Philadelphia, PA 19104. (215) 386-0100. FAX (215) 386-6362. Contains citations to articles listed in the table of contents of science and technology journals. Also articles in social sciences and life sciences journals. Available on BRS,(800) 289-4277, DIALOG,(800) 334-2564, online services. Inquire as to cost and availability.

Dissertation Abstracts. University Microfilms International, 300 North Zeeb Road, Ann Arbor, MI 48106. (800) 521-0600 or (313) 761-4700. Scope includes virtually all doctoral dissertations accepted at accredited American institutions from 1861 to present in 252 subject areas. Available on BRS,(800) 289-4277, DIALOG,(800) 334-2564, and OCLC EPIC,(800) 848-5878, online services. Also available on CD-ROM. Inquire as to cost and availability.

ISMEC: Mechanical Engineering Abstracts. Cambridge Scientific Abstracts, 7200 Wisconsin Avenue, Suite 601, Bethesda, MD 20814. (301) 961-6750 or (800) 843-7751. Contains citations to the literature in mechanical engineering, industrial and production engineering, energy, power, mechanics, devices and related areas, from 1973 to present. Available on the DIALOG,(800) 334-2564, online service. Inquire as to cost and availability.

NTIS Bibliographic Database. National Technical Information Service, 5285 Port Royal Road, Springfield, VA 22161. (703) 487-4929 or FAX (703) 321-8199. Broad coverage of government-sponsored science and technology research reports, 1964 to present. Available on BRS,(800) 289-4277, DIALOG, (800) 334-2564, ORBIT, (800) 456-7248, and STN International, FIZ Karlsruhe, P.O. Box 2465, W-7500, Karlsruhe 1, Germany, online services. Also available on CD-ROM. Inquire as to cost and availability.

SCISEARCH. Institute for Scientific Information, 3501 Market Street, Philadelphia, PA 19104. (800) 523-1850 or (215) 386-0100. Broad multidisciplinary title and author index to the international literature of science and technology, 1974 to present. Available on DIALOG,(800) 334-2564, and ORBIT,(800) 456-7248, online services. Also available on CD-ROM. Inquire as to cost and availability.

WILSONLINE. H.W. Wilson Company, 950 University Avenue, Bronx, NY 10452. (800) 367-6770 or (212) 588-8400. Makes available online versions of the printed H.W. Wilson indexes including *Applied Science and Technology Index, Business Periodicals Index, General Science Index,* and *Readers' Guide to Periodical Literature.* Period covered is generally 1983 to present. Available on BRS,(800) 289-4277, DIALOG,(800) 334-2564, and OCLC EPIC,(800) 848-5878, online services. Also available on CD-ROM. Inquire as to cost and availability.

PERIODICALS

Ashrae Journal: Heating, Refrigerating, Air Conditioning, and Ventilating. American Society of Heating, Refrigeration and Air Conditioning Engineers, 1791 Tullie Circle NE, Atlanta, GA 30329. (404) 636-8400. FAX (404) 321-5478. 1914 to present. Monthly. $49.00 per year for non-members.

Ashrae Transactions. American Society of Heating, Refrigeration and Air Conditioning Engineers, 1791 Tullie Circle NE, Atlanta, GA 30329. (404) 636-8400. FAX (404) 321-5478. 1895 to present. Semi-annual. $170.00 per volume.

Heat Transfer Engineering. Taylor and Francis, Inc., 79 Madison Avenue, Suite 1106, New York, NY 10016. (212) 725-1999 or (800) 821-8312. 1979 to present. Quarterly. $176.00 per year.

Heating, Piping, Air Conditioning. Penton Publishing, 110 Superior Ave., Cleveland, OH 44114-2543. 1929 to present. Monthly. $50.00 per year.

International Communications in Heat and Mass Transfer. Elsevier Science [journals], 660 White Plains Rd., Tarrytown, NY 10159-5153. (919) 524-9200. FAX (919) 333-2444. 1974 to present. 6 times a year. $430.00 per year.

International Journal of Pressure Vessels and Piping. Elsevier Science [journals], 660 White Plains Rd., Tarrytown, NY 10159-5153. (919) 524-9200. FAX (919) 333-2444. 1972 to present. 12 times a year. $1505.00 per year.

Journal of Heat Transfer. American Society of Mechanical Engineers, 345 E. 47th Street, New York, NY 10017. (212) 705-7722. 1970 to present. Quarterly. $155.00 per year.

Journal of Pressure Vessel Technology. American Society of Mechanical Engineers, 345 East 47th Street, New York, NY 10017-2398. (212) 705-7703. 1974 to present. Quarterly. $100.00 per year.

Numerical Heat Transfer A: Applications. Taylor and Francis, Inc., 79 Madison Avenue, Suite 1106, New York, NY 10016. (212) 725-1999 or (800) 821-8312. 1978 to present. 8 times a year. $530.00 per year.

Plant Engineering. Cahners Publishing Company (Des Plains), 1350 East Touhy Avenue, Des Plaines, IL 60017-5080. (800) 323-4958. FAX (708) 390-2779. 1947 to present. 19 times a year. $69.95 per year.

Power. McGraw-Hill Publishing Company, 1221 Avenue of the Americas, New York, NY 10020. (800) 262-4729 or (212) 512-3825. 1882 to present. Monthly. Inquire for cost.

RESEARCH CENTERS AND INSTITUTES

Heat Transfer Laboratory. Massachusetts Institute of Technology, 77 Massachusetts Avenue, Cambridge, MA 02139. (716) 253-2248.

Heat Transfer Research. University of Wisconsin-Milwaukee, P.O. Box 784, Milwaukee, WI 53201. (414) 963-5001. FAX (414) 229-6958.

Institute of Thermo-Fluid Engineering and Science. Lehigh University, Bldg. A, Mountain Top Campus, Bethlehem, PA 18015. (215) 758-4091. FAX (215) 758-5057.

PRESTRESSED CONCRETE

See: CONCRETE

PROBABILITY

Ency. of Physical Sciences and Engineering Info. Sources

PROBABILITY

See also: MARKOV PROCESSES, MATHEMATICS, STATISTICS,
STOCHASTIC PROCESSES

ABSTRACT SERVICES AND INDEXES

Applied Mechanics Reviews: An Assessment of World Literature in Engineering Sciences. American Society of Mechanical Engineers, 345 East 47th Street, New York, NY 10017. (212) 705-7703. 1948 to present. Monthly. $360.00 per year. ISSN: 0003-6900.

Applied Science and Technology Index: A Cumulative Subject Index to English Language Periodicals in the Fields of Aeronautics and Space Science, Computer Technology, Chemistry, Construction Industry, Energy and Related Areas.. H. W. Wilson and Company, 950 University Avenue, Bronx, NY 10452. (800) 367-6770. FAX (718) 590-1617. 1958 to present. Monthly. Available on-line from BRS and WILSONLINE. Also available on CD-ROM. Inquire as to cost and availability. ISSN: 0003-6986.

Compactmath-Compact Mathematics Library. Cumulative CD-ROM edition of Zentralblatt fuer Mathematik-Mathematics Abstracts. Springer-Verlag New York, Inc. 44 Hartz Way. Secaucus. NJ. 07096-2491. (201) 348-4033. FAX (201) 348-4505. 1993 to present. Annual. Available only on CD- ROM. Inquire as to cost and availability. ISSN: 0938-3174.

Current Index to Statistics: Applications-Methods-Theory. American Statistical Association, 1429 Duke Street, Alexandria, VA 22314-3402. (703) 684-1221. FAX (703) 684- 2037. 1976 to present. Annual. $50.00 per year. Available on-line from BRS, Dialog, European Space Agency. ISSN: 0364- 1228.

Current Mathematical Publications. American Mathematical Society. P.O. Box 1571, Annex Station, Providence, RI 02901-9930. (800) 556-7774 or (401) 455-4000. FAX (401) 331-3842. 1969 to present. Seventeen times per year. $377.00 per year. Available online from BRS, DIALOG, European Space Agency. Also available on CD-ROM. Inquire as to cost and availability. ISSN: 0361-4794.

General Science Index. H.W. Wilson Company, 950 University Avenue, Bronx, NY 10452. (212) 588-8400. (800) 367-6770. FAX (718) 590-1617. From 1978 to present. Ten issues per year; quarterly and annual cumulations. Service basis. Available on CD-ROM and online. Inquire regarding cost and availability. ISSN: 0162-1963.

Index to Mathematical Tables. Alan Fletcher. Addison- Wesley Publishing Co., Inc., 1 Jacob Way, Reading, MA 01867. (617) 944-3700. (800) 447-2226. 2nd edition. 1962. 2 volumes.

Mathematical Reviews: A Review Journal Covering the World Literature of Mathematical Research. American Mathematical Society, P.O. Box 1571, Annex Station, Providence, RI 02901-9930. (800) 556-7774 or (401) 455-4000. 1940 to present. Monthly. $4594.00 per year. Also available via network (MATHSCINET), online and on CD-ROM. Inquire regarding cost and availability. ISSN: 0025-5629.

NTIS Alerts: Mathematical Sciences. U.S. National Technical Information Service, 5285 Port Royal Road, Springfield, VA 22161. (703) 487-4650. FAX (703) 321-8547. Weekly. $140.00.

Science Citation Index. Institute for Scientific Information. 3501 Market Street, Philadelphia, PA 19104. (215) 386-0100. (800) 523-1850. FAX (215) 386-2991. 1961 to present. Six issues per year, plus annual cumulation. $11650.00 per year. Also available online and on CD-ROM. Inquire about price and availability. ISSN: 0036-827X.

Statistical Theory and Method Abstracts. International Statistical Institute. Prinses Beatrixiaan 428, Postbus 950, 2270 AZ Voorburg, Netherlands. 31-70-3375737. FAX 31-70-3860025. 1959 to present. Quarterly. $191.82.

Zentralblatt Fuer Mathematik und Ihre Grenzebiete/mathematics Abstracts. Heidelberger Akademie der Wissenschaften/ Springer-Verlag New York, Inc., 44 Hartz Way, Seacaucus, NJ 07096-2491. (201) 348-4033. FAX (201) 348-4505. 1931 to present. 30 issues per year. DM 8340. ISSN: 0044-4235. Also available online and on CD-ROM. Inquire regarding cost and availability.

ANNUAL REVIEWS AND YEARBOOKS

Advances in Probability and Related Topics. Marcel Dekker, Inc., 270 Madison Avenue, New York, NY 10016. (212) 696-9000. FAX (212) 685-4540. 1971 to present. 1984? Irregular.

Advances in Applied Mathematics. Academic Press, Inc., 6277 Sea Harbor Drive, Orlando, FL 32821. (800) 321-5068. Irregular, Price varies.

Ergebnisse der Mathematik und Ihrer Grenzgebiete. Springer-Verlag New York, Inc. 44 Hartz Way. Secaucus. NJ. 07096-2491. (201) 348-4033. FAX (201) 348-4505. Irregular.Price varies.

Frontiers in Pure and Applied Probability. Books International, Inc., Box 605 Herndon, VA 22070. (703) 435- 7064. FAC (703) 689-0660. 1993 to present. Irregular. Price varies.

Lecture Notes in Mathematics. Springer-Verlag New York, Inc. 44 Hartz Way. Secaucus. NJ. 07096-2491. (201)348-4033. FAX (201) 348-4505. Irregular. Price varies.

Probability and Mathematical Statistics. Academic Press, Inc., 6277 Sea Harbor Drive, Orlando, FL (800) 321-5068. 1967 To Present. Irregular. Price varies.

Probability: Pure and Applied. Marcel Dekker, Inc., 270 Madison Avenue, New York, NY 10016. (212) 696-9000. FAX (212) 685-4540. (800) 228-1160. 1984 to present. Irregular. Price varies.

Statistics: Textbooks and Monographs Series. Marcel Dekker, Inc., 270 Madison venue, New York, NY 10016. (212) 696-9000. FAX (212) 685-4540. 1972 to present. Irregular. Price varies.

ASSOCIATIONS AND PROFESSIONAL SOCIETIES

American Mathematical Society. P.O. Box 6248, Providence, RI 02940. (401) 455-4000. FAX (401) 331-3842.

American Statistical Association. 1429 Duke Street, Alexandria, VA 22314-3402. (703) 684-1221. FAX (703) 684-2037.

Association For Women in Mathematics. 4114 Computer and Space Science Building, University of Maryland, College Park, MD 20742-2461. (301) 405-7892.

Biometric Society, Eastern North American Region. c/o Randy Price, 11250 Roger Bacon Drive, Suite 8, Reston, VA 22090-5202. (703) 525-1191. FAX (703) 276-8196.

Caucus for Women in Statistics. c/o Cyntha Struthers, St. Jerome's College, Waterloo, ON, Canada N2L 3G3. (519) 888-4801. FAX (519) 746-6530.

Econometric Society. Northwestern University, Department of Economics, Evanston, IL 60208-2600. (708) 491-3615.

Institute of Mathematical Statistics. 3401 Investment Boulevard, Suite 7, Hayward CA 94545-3819. (510) 783-8141. FAX (510) 783-4131.

International Association for Statistical Computing. Prinses Beatrixiaan 428, Postbus 950, NL-2270 AZ Voorburg, Netherlands. Telephone: 70 3375737. FAX: 70 3860025.

International Biometric Society. 1429 Duke Street, Suite 401, Alexandria, VA 22314. (703) 836-8311.

International Statistical Institute. Prinses Beatrixiaan 428, Postbus 950, NL-2270 AZ Voorburg, Netherlands. Telephone: 70-3375737. FAX: 70 3860025.

Mathematical Association of America. 1529 18th Street, NW, Washington, DC 20036. (202) 387-5200. FAX (202) 265-2384.

National Institute of Statistical Sciences. P.O. Box 14162, Research Triangle Park, NC 27709. (919) 541-6255. FAX (919) 541-7102.

Operations Research Society of America. 1314 Guilford Avenue, Baltimore, MD 21202. (410) 528-4146. FAX (410)528-8556.

Royal Statistical Society. 12 Errol Street, London EC1Y 8LX, England. Telephone: 171 7235882. FAX 171 7061710.

Society for Industrial and Applied Mathematics. 3600 University City Science Center, Philadelphia, PA 19104-2688. (215) 382-9800. FAX (215) 386-7999.

BIBLIOGRAPHIES

Bibliography of Statistical Literature. Maurice Kendall and Alison Doig. Ayer Company Pubs., Inc., Lower Mill Road, North Stratford, NH 03590. (603) 922-5105. (800) 282-5413. 1981. 3 volumes: Pre-1940, 1940 to 1949, 1950 to 1958. $82.95, set.

Chronological Annotated Bibliography of Order Statistics, Volume VI. H. Leon Harter. American Sciences Press, 20 Cross Road, Syracuse, NY 13224-2144. 1991. $115.00.

Guide to Statistical Methods and to the Pertinent Literature. L. Sachs. Springer-Verlag New York, Inc., 175 Fifth Avenue, New York, NY 10010. (212) 460-1500. (800) 777-4643. 1986. $39.00 in paper.

How to Find Out About Statistics. Gillian Burrington. Franklin Book Co., Inc., Elkins Park, PA 19117. (215) 635- 5252. 1972. reprint edition. $74.00.

Mathematical Journals: An Annotated Guide. Diana F. Liang. Scarecrow Press. Distributed by University Press of America, 4720 Boston Way, Lanham, MD 20706. (301) 459-3366. (800) 462-6420. $29.50. ISBN: 0-8108-2585-6.

DIRECTORIES AND BIOGRAPHICAL SOURCES

Assistantships and Graduate Fellowships in the Mathematical Sciences. American Mathematical Society, P.O. Box 6248, Providence, RI 02940. (401) 455-4000. (800) 321-4267. Annual. $19.00. ISSN: 1040-7650.

Combined Membership List. American Mathematical Society, P.O. Box 6248, Providence, RI 02940. (401) 455-4000. (800) 321-4267. Annual. $50.00. ISSN: 0569-6461.

Directory of Statisticians. American Statistical Association, 1429 Duke Street, Alexandria, VA 22314-3402. (703) 684-1221. FAX (703) 684-2037. Periodic.

Research Centers Directory. Gale Research, 835 Penobscot Building, Detroit, MI 48226-4094. (313) 961- 2242. (800) 347-4253. 20th edition. 1995. Annual. $485.00. ISSN: 0080-1518.

Mathematical Science Professional Directory. American Mathematical Society, P.O. Box 6248, Providence, RI 02940. (401) 455-4000. (800) 321-4267. Annual. $45.00. ISBN: 0-8218-0173-2.

World Directory of Mathematicians. G.D. Mostow, editor. International Mathematical Union, Helsinki. 1958 to present. 10th revised edition. 1995. $45.00. Available from: American Mathematical Society, P.O. Box 6248, Providence, RI 02940. (401) 455-4000. ISSN: 0512-2740.

ENCYCLOPEDIAS AND DICTIONARIES

Dictionary of Gaming, Modelling & Simulation. G. Ian Gibbs. Books on Demand, 300 North Zeeb Road, Ann Arbor, MI 48106-1346. (313) 761-4700. (800) 521-0600. $49.60 in paper.

Dictionary-Outline of Basic Statistics. John Freund and Frank J. Williams. Dover Publications, Inc., 180 Varick Street, New York, NY 10014. (212) 255-3755. (800) 223-3130. Reprint edition. $5.95 in paper.

Dictionary of Statistical Terms. F. H. Marriott. Halsted Press, 605 Third Avenue, New York, 10158-0012. (212) 850-6400. 5th edition. 1990. $76.95 in paper.

Dictionary of Statistics. Michael G. Mulhall. Gordon Press Publications, P.O. Box 459, Bowling Green Station, New York, NY 10004. (718) 624-8419. 1972. $300.00.

Encyclopedia of Statistical Sciences. Samuel Kotz, Norman L. Johnson, and Campbell B. Read, editors. John Wiley & Sons, Inc., 605 Third Avenue, New York, NY 10158-0012. (212) 850-6000. (800) 225-5945. 1988. 9 volumes. ISBN: 0- 471-05544-1. $1440.00 set.

Encyclopedic Dictionary of Mathematics. Kiyosi Ito, editor. MIT Press, 55 Hayward Street, Cambridge, MA 02142. (617) 625-

PROBABILITY

Ency. of Physical Sciences and Engineering Info. Sources

8569. 2nd edition, reprint. 1993. 2 volumes. ISBN: 0- 262-59020-4. $70.00.

HarperCollins Dictionary of Statistics. Roger Porkess. HarperCollins Publ. 10 East 53rd Street, New York, NY 10022-5299. (800) 331-3761. 1991. $14.00 in paper.

International Encyclopedia of Statistics. William H. Kruskal and Judith M. Tanur, editors. The Free Press, 1230 Avenue of the Americas, New York, NY 10020. (212) 698-7000. (800) 223-2348. 1978. $90.00.

Mathematics Dictionary. Robert C. James, editor. Van Nostrand Reinhold, 115 Fifth Avenue, New York, NY 10003. (212) 254-3232. (800)-842-3636. 5th edition. 1992. $42.95. ISBN: 0-442-00741-8.

Russian-English Dictionary on Probability, Statistics and Combinatorics. K. A. Borokov. Society for Industrial and Applied Mathematics, 3600 University City Science Center, Philadelphia, PA 19104-2688. (215) 382-9800. FAX (215) 386-7999. (800) 447-7426. 1993. $47.50 in paper.

Topical Dictionary of Statistics: Advanced Industrial Terminology. Gary L. Teitjen. Chapman & Hall, 1 Penn Plaza, New York, NY 10119. (212) 564-1060. 1985. $24.50.

GENERAL WORKS

The Art of Probability for Scientists and Engineers. Richard W. Hamming. Addison-Wesley Publishing Co., Inc., 1 Jacob Way, Reading, MA 01867. (617) 944-3700. (800) 447-2226. 1991. $45.95.

Beating Murphy's Law: The Amazing Science of Risk. R. Berger. Dell Publishing Co., Inc., 1540 Broadway, New York, NY 10036-4094. (212) 354-6500. (800) 223-6934. 1994. $11.95.

Chance and Chaos. David Ruelle. Princeton University Press, 41 William Street, Princeton, NJ 08540. (609) 258-4900. (800) 777-4726. 1993. $35.00.

Classical Probability in the Enlightenment. Lorraine Daston. Princeton University Press, 41 William Street, Princeton, NJ 08540. (609) 258-4900. (800) 777-4726. 1988. $59.00

The Empire of Chance: How Probability Changed Science and Everyday Life. Gerd Gigerenzer et al. Cambridge University Press, 40 West 20th Street, New York, NY 10011-4211. (212) 924-3900. 1990. $19.95 in paper.

Essentials of Probability. Research and Education Association, 61 Ethel Road, West Piscataway, NJ 08854. (908) 819-8880. Revised edition. 1994. $5.95 in paper.

Fallacies in Dealing With Probabilities. E. C. Poulton. Cambridge University Press, 40 West 20th Street, New York, NY 10011-4211. (212) 924-3900. 1994. $59.95.

History of Inverse Probability from Thomas Bayes to Karl Pearson. A. I. Dale. Springer-Verlag New York, Inc., 175 Fifth Avenue, New York, NY 10010. (212) 460-1500. (800) 777-4643. 1991. $59.00.

Introduction to Probability and Statistics. Narayan C. Giri. Marcel Dekker, Inc., 270 Madison Avenue, New York, NY 10016. (212) 696-9000. (800) 228-1160. 1993. $65.00.

Introduction to Probability Theory and Its Applications. William Feller. John Wiley & Sons, Inc., 605 Third Avenue, New York, NY 10158-0012. (212) 850-6000. (800) 225-5945. Volume 1, third edition. 1968. Volume 2, second edition. 1971.

Mathematical Statistics and Data Analysis. John A. Rice. Wadsworth Publishing Co., 10 Davis Drive, Belmont, CA 94002. (415) 595-2350. (800) 354-9706. 1995. $63.95.

Probability and Random Processes. Geoffery Grimmett and David Stirzaker. Oxford University Press, Inc., 200 Madison Avenue, New York, NY 10016. (212) 725-6000. (800) 334-4249.2nd edition. 1992. $45.50.

Probability and Statistics: A No Nonsense Approach. V.T. Alexander and E. G. Phodia. CT Publishing Co., P.O. Box 992197. Redding, CA 96099-2107. (916) 378-0511. (800) 767-0511. 1994. $26.75 in paper.

Probability and Statistics for Engineering and the Sciences. Jay L. Devore. Wadsworth Publishing Co., 10 Davis Drive, Belmont, CA 94002. (415) 595-2350. (800) 354-9706. 4th edition. 1995. $67.95.

Probability and Statisics in Experimental Physics. Byron P. Roe. Springer-Verlag New York, Inc., 175 Fifth Avenue, New York, NY 10010. (212) 460-1500. (800) 777-4643. 1994. $39.00.

Probability Problem Solver. Vance Berger. Research and Education Association, 61 Ethel Road, West Piscataway, NJ 08854. (908) 819-8880. 1995. $23.95 in paper.

Probability, Statistics and Truth. Richard Von Mises. Dover Publications, Inc., 180 Varick Street, New York, NY 10014. (212) 255-3755. (800) 223-3130. 1981. reprint edition. $5.95 in paper.

HANDBOOKS AND MANUALS

Biometrika Tables For Statistics. E. S. Pearson and H. O. Hartley, editors. Lubrecht & Cramer, Ltd., 38 County Route 48, Forestburgh, NY 12777-6400. (914) 794-8539. volume 1, 3rd edition. volume 2, reprint edition. 1976. $75.00 each.

Computational Handbook of Statistics. James L. Bruning, and B. L. Kintz. HarpCollins Publ. 10 East 53rd Street, New York, NY 10022-5299. (800) 331-376. 3rd edition. 1990. $38.00 in paper.

Guide to Tables in Mathematical Statistics. Joseph A. Greenwood, and H. O. Hartley. Books on Demand, 300 North Zeeb Road, Ann Arbor, MI 48106-1346. (313) 761-4700. (800)521-0600 $180.00.

Handbook of Applicable Mathematics. Walter Ledermann, editor. John Wiley & Sons, Inc., 605 Third Avenue, New York, NY 10158-0012. (212) 850-6000. (800) 225-5945. 1985. $1496.00 6 volumes. ISBN: 0-471-90804-5.

Handbook of Applied Mathematics: Selected Results and Methods. Carl E. Pearsen, editor. Chapman & Hall, 1 Penn Plaza, New York NY 10119. (212) 564-1060. 1990. $46.95. ISBN: 0-442-00521-0.

Handbook of Mathematical Functions with Formulas, Graphs, and Mathmetical Tables. Milton Abramowitz and Irene A. Stegun, editors. U.S. National Bureau of Standards. U.S. Government Printing office, Washington, D.C. 10th edition. 1972. $43.00. ISBN: 0-16-000202-8.

Handbook of Mathematical Sciences. William H. Beyer, editor. CRC Press, 2000 Corporate Boulevard, Boca Raton, FL 33431. (800) 272-7737. 6th edition. 1987. $91.95. ISBN: 0-8493-0656-6.

Handbook of Mathematics. I. N. Bronshtein and K. A. Semedyayev. Van Nostrand Reinhold, 115 Fifth Avenue, New York, NY 10003. (800)-842-3636. 20th edition. $59.95. ISBN: 0-442-21171-6.

Handbook of Tables For Probability and Statistics. William H. Beyer, editor. CRC Press, 2000 Corporate Boulevard, Boca Raton, FL 33431. (800) 272-7737. 2nd edition. 1968. $92.95. ISBN: 0-8493-0692-2.

Standard Probability and Statistics Tables and Formulas. William H. Beyer. CRC Press, Inc., 2000 Corporate Boulevard, NW, Coca Raton, FL 33431. (407) 994-0555. (800) 272-7737. 1990. $34.95.

Standard Mathematical Tables and Formulas. William H. Beyer, editor. CRC Press, 2000 Corporate Boulevard, Boca Raton, FL 33431. (800) 272-7737. 29th edition. 1991. $41.95. ISBN: 084-930-6299.

Tables of Integrals, Series and Products: Corrected and Enlarged Edition. I. S. Gradshteyn and I. M. Ryzhik. Academic Press, Inc., 6277 Sea Harbor Drive, Orlando, FL 32821. (800) 321-5068. 5th edition. 1993. $61.00. ISBN: 0-12-294755-X.

ONLINE DATABASES AND CD-ROMS

CA SEARCH. Chemical Abstracts Service, P.O. Box 3012, Columbus, OH 43210-0012. (614) 447-3600. (800) 848-6533. FAX (614) 447-3709. Very comprehensive guide to worldwide chemical literature and related fields, 1972 to present. Available on BRS,(800) 289-4277, DIALOG, (800) 334-2564, ORBIT (800) 456-7248, and STN International, FIZ Karlsruhe, P.O. Box 2465, W-7500, Karlsruhe 1, Germany, online services. Inquire as to cost and availability.

Compendex Plus. Engineering Information, Inc., 345 East 47th Street, New York, NY 10017. (212) 705-7600 or (800) 221-1044. Contains citations with abstracts to worldwide literature in engineering and technology, from 1970 to present. Available on online BRS,(800) 289-4277, DIALOG, (800) 334-2564, ORBIT (800) 456-7248, and STN International, FIZ Karlsruhe, P.O. Box 2465, W-7500, Karlsruhe 1, Germany, online services. Also available on CD-ROM. Inquire as to cost and availability.

Current Contents Search. Institute for Scientific Information, 3501 Market Street, Philadelphia, PA 19104. (215) 386-0100. FAX (215) 386-6362. Contains citations to articles listed in the table of contents of science and technology journals. Also articles in social sciences and life sciences journals. Available on BRS,(800) 289-4277, DIALOG,(800) 334-2564, online services. Inquire as to cost and availability.

Dissertation Abstracts. University Microfilms International, 300 North Zeeb Road, Ann Arbor, MI 48106. (800) 521-0600 or (313) 761-4700. Scope includes virtually all doctoral disserta-tions accepted at accredited American institutions from 1861 to present in 252 subject areas. Available on BRS,(800) 289-4277, DIALOG,(800) 334-2564, and OCLC EPIC,(800) 848-5878, online services. Also available on CD-ROM. Inquire as to cost and availability.

INSPEC. Institution of Electrical Engineers, Michael Faraday House, Six Hills Way, Stevenage, Herts. SG1 2AY, England. Telephone: 0438 313311 or FAX 0438 742840. Contains citations to the worldwide literature of physics, electronics and electrical engineering, computer technology, and related fields. Available on BRS,(800) 289-4277, DIALOG, (800) 334-2564, ORBIT, (800) 456-7248, and STN International, FIZ Karlsruhe, P.O. Box 2465, W-7500, Karlsruhe 1, Germany, online services. Inquire as to cost and availability.

Mathsci and Mathscinet. American Mathematical Society, P.O. Box 6248, Providence, RI 02940. (800) 321-4667 or (401) 455-4000 or FAX (401) 331-3842. Scope includes pure and applied mathematics and related areas of physics, statistics, engineering, computer science, and operations research literature since 1959. Available on DIALOG,(800) 334-2564, online service. Also available on CD-ROM. Inquire as to cost and availability.

NTIS Bibliographic Database. National Technical Information Service, 5285 Port Royal Road, Springfield, VA 22161. (703) 487-4929 or FAX (703) 321-8199. Broad coverage of government-sponsored science and technology research reports, 1964 to present. Available on BRS,(800) 289-4277, DIALOG, (800) 334-2564, ORBIT, (800) 456-7248, and STN International, FIZ Karlsruhe, P.O. Box 2465, W-7500, Karlsruhe 1, Germany, online services. Also available on CD-ROM. Inquire as to cost and availability.

Scientific and Technical Books and Serials in Print. R.R. Bowker Inc., 121 Chanlon Road, New Providence, NJ 07974. (800) 521-8110 or (908) 464-6800. List of currently published books and serials in the physical and biological sciences, engineering and technology, with subject, author and titles indexes. Available on ORBIT,(800) 456-7248, online service.Inquire as to cost and availability.

SCISEARCH. Institute for Scientific Information, 3501 Market Street, Philadelphia, PA 19104. (800) 523-1850 or (215) 386-0100. Broad multidisciplinary title and author index to the international literature of science and technology, 1974 to present. Available on DIALOG,(800) 334-2564, and ORBIT,(800) 456-7248, online services. Also available on CD-ROM. Inquire as to cost and availability.

WILSONLINE. H.W. Wilson Company, 950 University Avenue, Bronx, NY 10452. (800) 367-6770 or (212) 588-8400. Makes available online versions of the printed H.W. Wilson indexes including *Applied Science and Technology Index, Business Periodicals Index, General Science Index,* and *Readers' Guide to Periodical Literature.* Period covered is generally 1983 to present. Available on BRS,(800) 289-4277, DIALOG,(800) 334-2564, and OCLC EPIC,(800) 848-5878, online services. Also available on CD-ROM. Inquire as to cost and availability.

PERIODICALS

Advances in Applied Probability. Applied Probability Trust, School of Mathematics, University of Sheffield, Sheffield S3 7RH, England. TEL 44-742-824269. FAX 44-742-729782. 1969 to present. Quarterly. $56.00 per year. ISSN: 0001-8678.

PROBABILITY

Ency. of Physical Sciences and Engineering Info. Sources

American Statistician. American Statistical Association. 1429 Duke Street, Alexandria, VA 22314-3402. (703) 684- 1221. FAX (703) 684-2037. 1947 to present. Quarterly. $55.00 per year. ISSN: 0003-1305.

Annals of Probability. Institute of Mathematical Statistics. 3401 Investment Boulevard, Suite 7, Hayward, CA 94545. (510) 783-8141. 1973 to present. Quarterly. $130.00 per year. ISSN: 0090-5364.

Biometrics. Biometric Society. 1429 Duke Street, Suite 401, Alexandria, VA 22314-3402. (703) 836-8311. 1945 to present. Quarterly. $90.00 per year. ISSN: 0006-341X.

Canadian Journal of Statistics/revue Canadienne DE Statistique. Statistical Society of Canada, Room 612 Dunton Tower, Carleton University, Ottawa, ON K1S 5B6, Canada. (613) 788-2167. 1973 to present. Quarterly. $98.00 per year. ISSN: 0319-5724.

Chance. Springer-Verlag, 175 Fifth Avenue, New York, NY 10010. (212) 460-1612. FAX (212) 473-6272. 1988 to present. Quarterly. $69.00 per year. ISSN: 0933-2480.

Communications in Statistics, Part A: Theory and Methods. Marcel Dekker Journals, 270 Madison Avenue, New York, NY, 10016. (212) 696-9000. FAX (212) 685-4540. 1973 to present. Twelve issues per year. $1295.00 per year. ISSN: 0361-0926.

Institut Henri Poincare. Annales. Probabilities et Statistics. Gauthier-Villars, 15 rue Gossin, 92543 Montrouge Cedex, France. Telephone: 33-1-40-92-65-00. FAX 33-1-40-92-65-97. 1930 to present. Quarterly. 1230F. ISSN: 0246-0203.

International Journal of Game theory. Physica-Verlag GmbH. Distributed by Springer-Verlag New York, Inc., 175 Fifth Avenue, New York, NY 10010. (212) 460-1500. 1971 to present. Quarterly. DM 390. ISSN: 0020-7276.

JASA. Journal of the American Statistical Association. American Statistical Association, 1429 Duke Street, Alexandria, VA 22314-3402. (703) 684-1221. FAX (703) 684-2037. 1888 to present. Quarterly. $210.00 per year. ISSN: 0162-1259.

Journal of Applied Probability. Applied Probability Trust, University of Sheffield, Sheffield S3 7RH, England. Telephone: 44-742-824269. FAX 44-742-729782. 1964 to present. Quarterly. $168.00 per year. ISSN: 0021-9002.

Journal of Theoretical Probability. Plenum Publishing Corp., 233 Spring Street, New Yorik, NY 10013-1578. (212) 620-8000. FAX (212) 463-0742. 1988 to present. Quarterly. $225.00. ISSN: 0894-9846.

Metrika: International Journal for Theoretical and Applied Statistics. Physica-Verlag GmbH und Co. Distributed by Springer-Verlag New York, Inc., 175 Fifth Avenue, New York, NY 10010. (212) 460-1500. 1953 to present. Bimonthly. DM360. ISSN: 0026-1335.

Probability Theory and Related Fields. Springer-Verlag New York, Inc., 175 Fifth Avenue, New York, NY 10010. (212) 460-1500. (800) 777-4643. 1962 to present. Monthly. $1553.00. ISSN: 0178-8051.

Random Structures and Algorithms. John Wiley & Sons, Inc., 605 Third AVenue, New York, NY 10158-0012. (212) 850-6000.

FAX (212) 850-6088. 1990 to present. Quarterly. $185.00 per year. ISSN: 1042-9832.

Risk Analysis. Society for Risk Analysis. Plenum Publishing Corp. 233 Spring Street, New York, NY 10013- 1578. (212) 620-8000. FAX (212) 463-0742. 1980 to present. Bimonthly. $320.00 per year. ISSN: 0272-4332.

Royal Statistical Society. Journal. Series A-C. Basil Blackwell Ltd., 108 Cowley Road, Oxford OX4, 1JF, Eng. Telephone: 0865-791100. FAX 0865-791347. Series A: Statistics in Society; 1838 to present. Series B: Methodological; 1934 to present. Series C: Applied Statistics; 1952 to present. Each series, three issues per year; $73.00 per year each. ISSN: 0035-9238, 9246, 9254.

Siam Journal on Applied Mathematics. Society for Industrial and Applied Mathematics, 3600 University City Science Center, Philadelphia, PA 19104-2688. (215) 382-9800. FAX (215) 386-7999. 1953 to present. Bimonthly. $242.00 per year. ISSN: 0036-1399.

Siam Journal on Control and Optimization. Society for Industrial and Applied Mathematics, 3600 University City Science Center, Philadelphia, PA 19104-2688. (215) 382-9800. FAX (215) 386-7999. 1963 to present. Bimonthly. $287.00 per year. ISSN: 0363-0129.

Siam Journal on Scientific and Statistical Computing. Society for Industrial and Applied Mathematics, 3600 University City Science Center, Philadelphia, PA 19104-2688. (215) 382-9800. FAX (215) 386-7999.

Siam Review. Society for Industrial and Applied Mathematics, 3600 University City Science Center, Philadelphia, PA 19104-2688. (215) 382-9800. 1959 to present. Quarterly. $148.00 per year. ISSN: 0036-1445.

Statistical Science. Institute of Mathematical Statistics. 3401 Investment Boulevard, Suite 7, Hayward, CA 94545. (510) 783-8141. 1986 to present. Quarterly. $75.00 per year. ISSN: 0883-4237.

Statistics and Computing. Chapman & Hall, One Penn Plaza, 41st Floor, New York, NY 10019. (212) 564-1060, FAX (212) 564-1505. Quarterly. $190.00 per year. ISSN: 0960-3174.

Theory of Probability and Its Applications. Society for Industrial and Applied Mathematics, 3600 University City Science Center, Philadelphia, PA 19104-2688. (215) 382- 9800. 1956 to present. Quarterly. $354.00 per year. ISSN: 0040-585X.

Theory of Probability and Mathematical Statistics. American Mathematical Society, Box 1571 Annex Station, Providence, RI 02901-9930. (401) 455-4000. 1970 to present. Two issues per year. $359.00 per year. ISSN: 0094-9000.

Technometrics. American Statistical Association, 1429 Duke Street, Alexandria, VA 22314-3402. (703) 684-1221. FAX(703) 684-2037. 1959 to present. Quarterly. $50.00. *Issn: 0040-1706.*

RESEARCH CENTERS AND INSTITUTES

Center for Pure and Applied Mathematics. University of California, Berkeley, 977 Evans Hall, Berkeley, CA 94720. (415) 642-0116. FAX (415) 642-6726.

Courant Institute of Mathematical Sciences. New York University, 251 Mercer Street, New York, NY 10012. (212) 998-3000.

Institute for Mathematics and Its Applications. University of Minnesota, 514 Vincent Hall, 206 Church Street, S.E., Minneapolis, MN 55455. (612) 624-6066. FAX (612) 626-7370.

Laboratory for Research in Statistics and Probability. Carleton University, Dunton Tower, Room 611, Ottawa, ON, Canada K1S 5B6. (613) 788-2167.

Mathematical Sciences Institute. Cornell University, 201 Caldwell Hall, Ithaca, NY 14853. (607) 255-8005.

Mathematical Statistics Program. University of Maryland, 1107 Mathematics Building, College Park, MD 20742. (301) 405 5061. FAX (301) 314-0827.

Operations Research Center. Massachusetts Institute of Technology, Cambridge, MA 02139. (617) 253-3601. FAX (617) 258-9214.

Statistical Laboratory. Iowa State University, 102 Snedecor Hall, Ames, IA 50011. (515) 294-3440. FAX (515) 294-4040.

Statistics Department. Rutgers University, Hill Center, New Brunswick, NJ 08903. (201) 932-2691. FAX (201) 932-3428.

PRODUCTION ENGINEERING

See: INDUSTRIAL ENGINEERING

PROGRAMMMING

See: COMPUTER PROGRAMMING

PROGRAMMING LANGUAGES

See: COMPUTER PROGRAMMING LANGUAGES

PROGRAMS

See: SOFTWARE

PROJECTIVE GEOMETRY

See: GEOMETRY

PROPELLANTS

See also: AERONAUTICAL ENGINEERING, AEROSPACE ENGINEERING, CHEMICAL ENGINEERING, EXPLOSIVES, FUELS, JET PROPULSION, PROPULSION SYSTEMS, ROCKETS

ABSTRACT SERVICES AND INDEXES

Applied Science and Technology Index: A Cumulative Subject Index to English Language Periodicals in the Fields of Aeronautics and Space Science, Computer Technology, Chemistry, Construction Industry, Energy and Related Areas. H.W. Wilson Co., 950 University Avenue, Bronx, NY 10452. (800) 367-6770 or (212) 588-8400. FAX (718) 590-1617. From 1958 to present. Monthly. Inquire about cost and availability. Also available on CD-ROM.

Chemical Abstracts-Applied Chemistry and Chemical Engineering Sections. Chemical Abstracts Service, 2540 Olentangy River Road, Box 3012, Columbus, OH 43210. (800) 848-6538 or (614) 447-3600, FAX (614) 447-3713. Bi-weekly. $1410.00 per year.

Chemical Engineering Abstracts. Royal Society of Chemistry, Information Services, Thomas Graham House, Science Park, Milton Road, Cambridge, CB4 4WF, England. Contains citations, mostly with abstracts, of the worldwide literature on chemical engineering, from 1982 to present. Monthly. $450.00 per year. Also available online.

Current Contents: Engineering, Technology and Applied Sciences. Institute for Scientific Information, 3501 Market Street, Philadelphia, PA 19104. (215) 386-0100. FAX (215) 386-6362. Weekly. $360.00 per year.

Engineering Index Monthly. Engineering Information, Inc., Castle Point on the Hudson, Hoboken, NJ 07030. (800) 221-1044. FAX (212) 832-1857. Monthly. $2200.00 per year. Also available online as COMPENDEX, and also on CD-ROM.

Index to Scientific and Technical Proceedings. Institute for Scientific Information, 3501 Market St., Philadelphia, PA 19104. (215) 386-0100. FAX (215) 386-6362. Monthly. $500.00 per year.

Index to Scientific Reviews. Institute for Scientific Information, 3501 Market St., Philadelphia, PA 19104. (215) 386-0100. FAX (215) 386-6362. Semi-annual.

International Aerospace Abstracts. Technical Information Service, American Institute of Aeronautics and Astronautics, Inc., 555 West 57th St., New York, NY 10019. (212) 247-6500. FAX (212) 582-4861. Semi-monthly. $1295.00 per year.

NASA Patent Abstracts Bibliography. National Aeronautics and Space Administration, Center for Aerospace Information, Box 8757, BWI Airport, Baltimore, MD 21240. (301) 621-0153. Semi-annual. Inquire for price information.

Numerical/chronological/Author Index. An index to articles and papers published by or for the American Astronautical Society, including: *Journal of the Astronautical Sciences, Astronautical Sciences Review, Advances in the Astronautical Sciences, Science and Technology,* AAS history series, and AAS microfiche series. Horace Jacobs and Robert H. Jacobs. Univelt, Inc. PO Box 28130, San Diego, CA 92198. (619) 746-4005. FAX (619) 746-3139. Inquire for price and availability.

Science Citation Index. Institute for Scientific Information, 3501 Market Street, Philadelphia, PA 19104. (215) 386-0100. FAX (215) 386-6362. Inquire about availability and cost. Also available on CD-ROM.

Scientific and Technical Aerospace Reports (STAR). National Aeronautics and Space Administration. NASA Center for Aero-

PROPELLANTS

Ency. of Physical Sciences and Engineering Info. Sources

space Information, Box 8757, BWI Airport, Baltimore, MD 21240. (301) 621-0153. Monthly. Inquire about availability and cost. Also available through the NASA online retrieval service (RECON), and through the Aerospace Database through DIALOG.

ASSOCIATIONS AND PROFESSIONAL SOCIETIES

American Astronautical Society. 6352 Rolling Mill Place, Suite 102, Springfield, VA 22152.

American Chemical Society. 1155 16th Street, N.W., Washington, DC 20036. (202) 872-4381 or (800) 424-6747.

American Institute of Aeronautics and Astronautics. 370 L'Enfant Promenade SW, Washington, DC 20024. (202) 646-7400.

British Interplanetary Society, 27/29 South Lambeth Road, London Sw8 1sz, England. (071) 735-3160. FAX (071) 820-1504.

Canadian Aeronautics and Space Institute, 222 Somerset Street West, Suite 601, Ottawa, Ontario, Canada K2p 0j1.

National Space Society, 922 Pennsylvania Avenue Se, Washington, DC 20003-2140. (202) 543-1900. FAX (202) 546-4189.

DIRECTORIES AND BIOGRAPHICAL SOURCES

American Astronomical Society Directory. American Astronautical Society, 6352 Rolling Mill Place, Suite 102, Springfield, VA 22152. Inquire for cost and availability.

American Institute of Aeronautics and Astronautics Roster. American Institute of Aeronautics and Astronautics. 370 L'Enfant Promenade SW, Washington, DC 20024. (202) 646-7400. Inquire for cost and availability.

Jane's Space Directory. Jane's Information Group, 1340 Braddock Place, Suite 300, Alexandria, VA 22314-1651. (703) 683-3700. FAX (703) 836-0029. Annual. $245.00. Also available on CD-ROM.

Research Centers Directory. Gale Research, 835 Penobscot Building, Detroit, MI 48226-4094. (313) 961-2242. (800) 347-4253. FAX (313) 961-6083. $485.00.

Scientific and Technical Organizations and Agencies Directory. Gale Research, 835 Penobscot Building, Detroit, MI 48226-4094. (313) 961-2242. (800) 347-4253. FAX (313) 961-6083. 4th ed. 1995. $195.00.

Who's Who in Engineering. 7th ed. American Association of Engineering Societies, 1111 19th St. NW, Suite 608, Washington, DC 20036. (800)-658-8897. 1988. $200.00.

GENERAL WORKS

Chemistry of Energetic Materials. George A. Olah & David R. Squire, editors. Academic Press Inc., 6277 Sea Harbor Drive, Orlando, FL 32887. (800) 321-5068. 1991. $79.00.

Explosives, Propellants, and Pyrotechnics. A. Baily and S.G. Murray. Brassey's, 1313 Dolley Madison Blvd., Suite 101,

McLean, VA 22102-3926. (703) 442-4535 or (800) 775-2518. FAX (703) 442-9848. 1989. $40.00 (hardback) and $25.00 paperback.

Fundamentals of Solid-Propellant Combustion. Kenneth K. Kuo & Martin Summerfield. American Institute of Aeronautics and Astronautics, 370 L'Enfant Promenade, S.W., Washington, DC 20024. (202) 646-7400. 1984. Inquire for cost and availability.

Modern Engineering for Design of Liquid-Propellant Rocket Engines. Dieter K. Huzel and David H. Huang. Revised, updated, and enlarged. American Institute of Aeronautics and Astronautics, 370 L'Enfant Promenade, S.W., Washington, DC 20024. (202) 646-7400. 1992. Write for price and availability information.

Solid Rocket Propulsion Theory. Alain Davenas, editor. Pergamon/Elsevier Science, 660 White Plains Rd., Tarrytown, NY 10159-5153. (919) 524-9200. FAX (919) 333-2444. 1993. $235.00.

ONLINE DATABASES AND CD-ROMS

Aerospace Database. American Institute of Aeronautics and Astronautics, 370 L'Enfant Promenade, S.W., Washington, DC 20024. (202) 646-7400. Worldwide published literature on research and development in aerospace and related areas, with abstracts. Covers 1962 to present. Available on DIALOG, (800) 334-2564, online service. Also available on CD-ROM. Inquire as to cost and availability.

Applied Science and Technology Index. Indexes Key English-language Journals in Applied Sciences, Including Aeronautics and Space Science, Chemistry, Computer Science, Electronics, Energy, Geology, Mathematics, Physics, Transportation, and Engineering.

CA Search. Chemical Abstracts Service, P.O. Box 3012, Columbus, OH 43210-0012. (614) 447-3600. (800) 848-6533. FAX (614) 447-3709. Very comprehensive guide to worldwide chemical literature and related fields, 1972 to present. Available on BRS,(800) 289-4277, DIALOG, (800) 334-2564, ORBIT (800) 456-7248, and STN International, FIZ Karlsruhe, P.O. Box 2465, W-7500, Karlsruhe 1, Germany, online services. Inquire as to cost and availability.

Compendex Plus. Engineering Information, Inc., 345 East 47th Street, New York, NY 10017. (212) 705-7600 or (800) 221-1044. Contains citations with abstracts to worldwide literature in engineering and technology, from 1970 to present. Available on online BRS,(800) 289-4277, DIALOG, (800) 334-2564, ORBIT (800) 456-7248, and STN International, FIZ Karlsruhe, P.O. Box 2465, W-7500, Karlsruhe 1, Germany, online services. Also available on CD-ROM. Inquire as to cost and availability.

CSA Engineering. Cambridge Scientific Abstracts, 7200 Wisconsin Avenue, Suite 601, Bethesda, MD 20814. (301) 961-6750 or (800) 843-7751. Contains citations and abstracts of international periodicals and research literature covering all fields of engineering and science and technology,including computer and information science, electronics, mechanical engineering, solid state materials, 1981 to present. Available on BRS,(800) 289-4277, online service.Inquire as to cost and availability.

Current Contents Search. Institute for Scientific Information, 3501 Market Street, Philadelphia, PA 19104. (215) 386-0100. FAX (215) 386-6362. Contains citations to articles listed in the table of contents of science and technology journals. Also ar-

ticles in social sciences and life sciences journals. Available on BRS,(800) 289-4277, DIALOG,(800) 334-2564, online services. Inquire as to cost and availability.

Dissertation Abstracts. University Microfilms International, 300 North Zeeb Road, Ann Arbor, MI 48106. (800) 521-0600 or (313) 761-4700. Scope includes virtually all doctoral dissertations accepted at accredited American institutions from 1861 to present in 252 subject areas. Available on BRS,(800) 289-4277, DIALOG,(800) 334-2564, and OCLC EPIC,(800) 848-5878, online services. Also available on CD-ROM. Inquire as to cost and availability.

International Aerospace Abstracts. American Institute of Aeronautics and Astronautics, 370 L'Enfant Promenade, S.W., Washington, DC 20024. (202) 646-7400. Contains references and abstracts of journal and monograph literature relating to aerospace science and technology, from 1963 to present. Available through the NASA/RECON system of the National Aeronautics and Space Administration only.

NASA Database. American Institute of Aeronautics and Astronautics, 370 L'Enfant Promenade, S.W., Washington, DC 20024. (202) 646-7400. Citations and abstracts of aeronautics and astronautics literature, 1962 to present. Also contains citations from *STAR, Scientific and Technical Aerospace Reports*, and *International Aerospace Abstracts*. Available through NASA/RECON online service. Inquire as to cost and availability.

NTIS Bibliographic Database. National Technical Information Service, 5285 Port Royal Road, Springfield, VA 22161. (703) 487-4929 or FAX (703) 321-8199. Broad coverage of government-sponsored science and technology research reports, 1964 to present. Available on BRS,(800) 289-4277, DIALOG, (800) 334-2564, ORBIT, (800) 456-7248, and STN International, FIZ Karlsruhe, P.O. Box 2465, W-7500, Karlsruhe 1, Germany, online services. Also available on CD-ROM. Inquire as to cost and availability.

Scientific and Technical Books and Serials in Print. R.R. Bowker Inc., 121 Chanlon Road, New Providence, NJ 07974. (800) 521-8110 or (908) 464-6800. List of currently published books and serials in the physical and biological sciences, engineering and technology, with subject, author and titles indexes. Available on ORBIT online service,(800) 456-7248. Inquire as to cost and availability.

SCISEARCH. Institute for Scientific Information, 3501 Market Street, Philadelphia, PA 19104. (800) 523-1850 or (215) 386-0100. Broad multidisciplinary title and author index to the international literature of science and technology, 1974 to present. Available on DIALOG,(800) 334-2564, and ORBIT,(800) 456-7248, online services. Also available on CD-ROM. Inquire as to cost and availability.

WILSONLINE. H.W. Wilson Company, 950 University Avenue, Bronx, NY 10452. (800) 367-6770 or (212) 588-8400. Makes available online versions of the printed H.W. Wilson indexes including *Applied Science and Technology Index, Business Periodicals Index, General Science Index,* and *Readers' Guide to Periodical Literature.* Period covered is generally 1983 to present. Available on BRS,(800) 289-4277, DIALOG,(800) 334-2564, and OCLC EPIC,(800) 848-5878, online services. Also available on CD-ROM. Inquire as to cost and availability.

PERIODICALS

Aerospace Engineering. Society of Automotive Engineers, 400 Commonwealth Dr., Warrendale, PA 15096. (412) 776-4841. 1981 to present. Monthly. $48.00 per year.

Aerospace Propulsion. McGraw-Hill Inc., Aviation Week Group, 1200 G Street NW, Suite 200, Washington, DC 20005. (202) 383-2350. 25 times a year. $550.00 a year.

Aviation Week and Space Technology. 1221 Avenue of the Americas, New York, NY 10020. (212) 512-2999 or (800) 525-5003. 1916 to present. Weekly. $82.00 per year.

Journal of Propulsion and Power. American Institute of Aeronautics and Astronautics, 370 L'Enfant Promenade SW, Washington, DC 20024-2518. 1985 to present. Bimonthly. $38.00 (members) and $260.00 (non-members).

Journal of Spacecraft and Rockets. American Institute of Aeronautics and Astronautics, 370 L'Enfant Promenade SW, Washington, DC 20024-2518. 1964 to present. Bimonthly. $33.00 (members) and $215.00 (non-members).

Progress in Aerospace Sciences. Pergamon Press, 660 White Plains Rd., Tarrytown, NY 10591-5153. 1961 to present. Quarterly. $320.00 per year.

Propellants, Explosives, Pyrotechnics. VCH Publishers, Inc., 220 E. 23rd St., New York, NY 10010-4606. 6 times a year. $415.00 per year.

RESEARCH CENTERS AND INSTITUTES

Center for Aerospace Technology, Weber State College, Ogden, UT 84408-1805. (801) 626-7272.

Charles Stark Draper Laboratory Inc., 555 Technology Square, Cambridge, MA 02139. (617) 258-1000.

Goddard Space Flight Center, Greenbelt, MD 20771. (301) 286-7351.

Houston Advanced Research Center, Space Technology and Research Center, 4802 Research Forest Dr., the Woodlands, TX 77381. (713) 363-7922. FAX (713) 363-7914.

Innovative Nuclear Space Power and Propulsion Institute, College of Engineering, University of Florida, Gainesville, FL 32611. (904) 392-1427. FAX (904) 392-8656.

George C. Marshall Space Flight Center, Huntsville, AL 35812. (205) 544-0024.

Ohio Aerospace Institute, 2001 Aerospace Parkway, Brook Park, OH 44142. (216) 891-2100. FAX (216) 891-2140.

Pennsylvania State University, Propulsion Engineering Research Center, 106 Research Bldg. E, Bigler Rd., University Park, PA 16802. (814) 863-6272. FAX (814) 865-3389.

University of Tennessee, Center for Advanced Space Propulsion, P.O. Box 1385, Tullahoma, TN 37388-8897. (615) 454-9294. FAX (615) 455-6167.

PROPELLANTS

Ency. of Physical Sciences and Engineering Info. Sources

Space Science and Engineering Center, University of Wisconsin-Madison, 1225 W. Dayton St., Madison, WI 53706. (608) 262-0544.

Thermal Sciences and Propulsion Center, Purdue University, Lafayette, IN 47907. (317) 494-1503.

PROPULSION SYSTEMS

See also: AERONAUTICAL ENGINEERING, AEROSPACE ENGINEERING, GUIDED MISSILES, JET PROPULSION, PROPELLANTS, ROCKETS

ABSTRACT SERVICES AND INDEXES

Applied Science and Technology Index: A Cumulative Subject Index to English Language Periodicals in the Fields of Aeronautics and Space Science, Computer Technology, Chemistry, Construction Industry, Energy and Related Areas. H.W. Wilson Co., 950 University Avenue, Bronx, NY 10452. (800) 367-6770 or (212) 588-8400. FAX (718) 590-1617. From 1958 to present. Monthly. Inquire about cost and availability. Also available on CD-ROM.

Astronomy and Astrophysics Abstracts. Springer-Verlag, 175 Fifth Avenue, New York, NY 10010. (212) 460-1500. Irregular. Price varies, inquire.

Current Contents: Engineering, Technology and Applied Sciences. Institute for Scientific Information, 3501 Market Street, Philadelphia, PA 19104. (215) 386-0100. FAX (215) 386-6362. Weekly. $360.00 per year.

Engineering Index Monthly. Engineering Information, Inc., Castle Point on the Hudson, Hoboken, NJ 07030. (800) 221-1044. FAX (212) 832-1857. Monthly. $2200.00 per year. Also available online as COMPENDEX, and also on CD-ROM.

Index to Scientific and Technical Proceedings. Institute for Scientific Information, 3501 Market St., Philadelphia, PA 19104. (215) 386-0100. FAX (215) 386-6362. Monthly. $500.00 per year.

Index to Scientific Reviews. Institute for Scientific Information, 3501 Market St., Philadelphia, PA 19104. (215) 386-0100. FAX (215) 386-6362. Semi-annual.

International Aerospace Abstracts. Technical Information Service, American Institute of Aeronautics and Astronautics, Inc., 555 West 57th St., New York, NY 10019. (212) 247-6500. FAX (212) 582-4861. Semi-monthly. $1295.00 per year.

NASA Patent Abstracts Bibliography. National Aeronautics and Space Administration, Center for Aerospace Information, Box 8757, BWI Airport, Baltimore, MD 21240. (301) 621-0153. Semi-annual. Inquire for price information.

Numerical/Chronological/Author Index. An index to articles and papers published by or for the American Astronautical Society, including: Journal of the Astronautical Sciences, Astronautical Sciences Review, Advances in the Astronautical Sciences, Science and Technology, AAS history series, and AAS microfiche series. Horace Jacobs and Robert H. Jacobs. Univelt, Inc. PO Box 28130, San Diego, CA 92198. (619) 746-4005. FAX (619) 746-3139. Inquire for price and availability.

Science Citation Index. Institute for Scientific Information, 3501 Market Street, Philadelphia, PA 19104. (215) 386-0100. FAX (215) 386-6362. Inquire about availability and cost. Also available on CD-ROM.

Scientific and Technical Aerospace Reports (star). National Aeronautics and Space Administration. NASA Center for Aerospace Information, Box 8757, BWI Airport, Baltimore, MD 21240. (301) 621-0153. Monthly. Inquire about availability and cost. Also available through the NASA online retrieval service (RECON), and through the Aerospace Database through DIALOG.

ASSOCIATIONS AND PROFESSIONAL SOCIETIES

American Astronautical Society. 6352 Rolling Mill Place, Suite 102, Springfield, VA 22152.

American Institute of Aeronautics and Astronautics, 370 L'enfant Promenade Sw, Washington, DC 20024-2518.

National Space Society, 922 Pennsylvania Avenue Se, Washington, DC 20003-2140. (202) 543-1900. FAX (202) 546-4189.

Society of Automotive Engineers. SAE, Inc., 400 Commonwealth Dr., Warrendale, PA 15096. (412) 776-4841.

DIRECTORIES AND BIOGRAPHICAL SOURCES

American Astronautical Society Directory. American Astronautical Society, 6352 Rolling Mill Place, Suite 102, Springfield, VA 22152. Inquire for cost and availability.

American Institute of Aeronautics and Astronautics Roster. American Institute of Aeronautics and Astronautics. 370 L'Enfant Promenade SW, Washington, DC 20024. (202) 646-7400. Inquire for cost and availability.

Research Centers Directory. Gale Research, 835 Penobscot Building, Detroit, MI 48226-4094. (313) 961-2242. (800) 347-4253. FAX (313) 961-6083. $485.00.

Scientific and Technical Organizations and Agencies Directory. Gale Research, 835 Penobscot Building, Detroit, MI 48226-4094. (313) 961-2242. (800) 347-4253. FAX (313) 961-6083. 4th ed. 1995. $195.00.

Who's Who in Engineering. 7th ed. American Association of Engineering Societies, 1111 19th St. NW, Suite 608, Washington, DC 20036. (800)-658-8897. 1988. $200.00.

GENERAL WORKS

High-Speed Flight Propulsion Systems. S.N.B. Murthy and E.T. Curran. American Institute of Aeronautics and Astronautics, 370 L'Enfant Promenade, S.W., Washington, DC 20024. (202) 646-7400. 1991. $86.95.

History of Liquid Rocket Engine Development in the United States 1955-1980. Stephen E. Doyle and R. Cargill Hall, eds. Univelt, Inc. P.O. Box 28130, San Diego, CA 92198. (619) 746-4005. FAX (619) 746-3139. 1992. $50.00 (hardback) and $35.00 (paperback).

Mechanics and Thermodynamics of Propulsion. Philip G. Hill and Carl R. Peterson. 2d ed. Addison-Wesley Publishing Company, 1 Jacob Way, Reading, MA 01867. (617) 944-3700 or (800) 447-2226. FAX 617-942-1117. 1992. $64.50.

Modern Engineering for Design of Liquid-Propellant Rocket Engines. Dieter K. Huzel and David H. Huang. Revised, updated, and enlarged. American Institute of Aeronautics and Astronautics, 370 L'Enfant Promenade, S.W., Washington, DC 20024. (202) 646-7400. 1992. Write for price and availability information.

Modern Research Topics in Aerospace Propulsion. G. Angelino, L. de Luca, and W.A. Sirignano, eds. Springer-Verlag, 175 Fifth Avenue, New York, NY 10010. (212) 460-1500 or (800) 777-4643. FAX (212) 473-6272. 1991. $69.00.

ONLINE DATABASES AND CD-ROMS

Aerospace Database. American Institute of Aeronautics and Astronautics, 370 L'Enfant Promenade, S.W., Washington, DC 20024. (202) 646-7400. Worldwide published literature on research and development in aerospace and related areas, with abstracts. Covers 1962 to present. Available on DIALOG, (800) 334-2564, online service. Also available on CD-ROM. Inquire as to cost and availability.

Applied Science and Technology Index. Indexes key English-language journals in applied sciences, including aeronautics and space science, chemistry, computer science, electronics, energy, geology, mathematics, physics, transportation, and engineering.

Compendex Plus. Engineering Information, Inc., 345 East 47th Street, New York, NY 10017. (212) 705-7600 or (800) 221-1044. Contains citations with abstracts to worldwide literature in engineering and technology, from 1970 to present. Available on online BRS,(800) 289-4277, DIALOG, (800) 334-2564, ORBIT (800) 456-7248, and STN International, FIZ Karlsruhe, P.O. Box 2465, W-7500, Karlsruhe 1, Germany, online services. Also available on CD-ROM. Inquire as to cost and availability.

CSA Engineering. Cambridge Scientific Abstracts, 7200 Wisconsin Avenue, Suite 601, Bethesda, MD 20814. (301) 961-6750 or (800) 843-7751. Contains citations and abstracts of international periodicals and research literature covering all fields of engineering and science and technology,including computer and information science, electronics, mechanical engineering, solid state materials, 1981 to present. Available on BRS,(800) 289-4277, online service.Inquire as to cost and availability.

Current Contents Search. Institute for Scientific Information, 3501 Market Street, Philadelphia, PA 19104. (215) 386-0100. FAX (215) 386-6362. Contains citations to articles listed in the table of contents of science and technology journals. Also articles in social sciences and life sciences journals. Available on BRS,(800) 289-4277, DIALOG,(800) 334-2564, online services. Inquire as to cost and availability.

Dissertation Abstracts. University Microfilms International, 300 North Zeeb Road, Ann Arbor, MI 48106. (800) 521-0600 or (313) 761-4700. Scope includes virtually all doctoral dissertations accepted at accredited American institutions from 1861 to present in 252 subject areas. Available on BRS,(800) 289-4277, DIALOG,(800) 334-2564, and OCLC EPIC,(800) 848-5878, online services. Also available on CD-ROM. Inquire as to cost and availability.

International Aerospace Abstracts. American Institute of Aeronautics and Astronautics, 370 L'Enfant Promenade, S.W., Washington, DC 20024. (202) 646-7400. Contains references and abstracts of journal and monograph literature relating to aerospace science and technology, from 1963 to present. Available through the NASA/RECON system of the National Aeronautics and Space Administration only.

NASA Database. American Institute of Aeronautics and Astronautics, 370 L'Enfant Promenade, S.W., Washington, DC 20024. (202) 646-7400. Citations and abstracts of aeronautics and astronautics literature, 1962 to present. Also contains citations from *STAR, Scientific and Technical Aerospace Reports,* and *International Aerospace Abstracts.* Available through NASA/RECON online service. Inquire as to cost and availability.

NTIS Bibliographic Database. National Technical Information Service, 5285 Port Royal Road, Springfield, VA 22161. (703) 487-4929 or FAX (703) 321-8199. Broad coverage of government-sponsored science and technology research reports, 1964 to present. Available on BRS,(800) 289-4277, DIALOG, (800) 334-2564, ORBIT, (800) 456-7248, and STN International, FIZ Karlsruhe, P.O. Box 2465, W-7500, Karlsruhe 1, Germany, online services. Also available on CD-ROM. Inquire as to cost and availability.

Scientific and Technical Books and Serials in Print. R.R. Bowker Inc., 121 Chanlon Road, New Providence, NJ 07974. (800) 521-8110 or (908) 464-6800. List of currently published books and serials in the physical and biological sciences, engineering and technology, with subject, author and titles indexes. Available on ORBIT,(800) 456-7248, online service.Inquire as to cost and availability.

SCISEARCH. Institute for Scientific Information, 3501 Market Street, Philadelphia, PA 19104. (800) 523-1850 or (215) 386-0100. Broad multidisciplinary title and author index to the international literature of science and technology, 1974 to present. Available on DIALOG,(800) 334-2564, and ORBIT,(800) 456-7248, online services. Also available on CD-ROM. Inquire as to cost and availability.

WILSONLINE. H.W. Wilson Company, 950 University Avenue, Bronx, NY 10452. (800) 367-6770 or (212) 588-8400. Makes available online versions of the printed H.W. Wilson indexes including *Applied Science and Technology Index, Business Periodicals Index, General Science Index,* and *Readers' Guide to Periodical Literature.* Period covered is generally 1983 to present. Available on BRS,(800) 289-4277, DIALOG,(800) 334-2564, and OCLC EPIC,(800) 848-5878, online services. Also available on CD-ROM. Inquire as to cost and availability.

PERIODICALS

Advances in Astronautical Sciences. American Astronautical Society, 6352 Rolling Mill Place, Suite 102, Springfield, VA 22152. (703) 866-0020. FAX (703) 866-3526. Quarterly. Write or call for price information.

Advances in Space Research. Pergamon Press, Pergamon Press, 660 White Plains Rd., Tarrytown, NY 10591-5153. 12 times a year. Inquire for price and availability.

Aeronautical Journal. Royal Aeronautical Society, PO Box 139, Tonbridge, Kent, TN9 1EW, England. 0732-770823. FAX 0723-361708. 1897 to present. Ten times a year. Write for price and availability.

PROPULSION SYSTEMS

Ency. of Physical Sciences and Engineering Info. Sources

Aerospace. Royal Aeronautical Society, PO Box 139, Tonbridge, Kent, TN9 1EW, England. 0732-770823. FAX 0723-361708. Monthly. Write for price and availability.

Aerospace America. American Institute of Aeronautics and Astronautics, 370 L'Enfant Promenade SW, Washington, DC 20024-2518. (202) 646-7471. 1932 to present. Monthly. $70 per year (non-members).

Aerospace Engineering. Society of Automotive Engineers. SAE, Inc., 400 Commonwealth Dr., Warrendale, PA 15096. (412) 776-4841. 1981 to present. Monthly. $48 a year.

Aviation & Aerospace. Baxter Publishing Company, 310 Dupont Street, Toronto, Ontario M5R 1V9 Canada. (416) 968-7252. FAX (416) 968-2377. Bi-monthly. $42.00 per year.

Aviation Week and Space Technology. 1221 Avenue of the Americas, New York, NY 10020. (212) 512-2999 or (800) 525-5003. 1916 to present. Weekly. $82.00 per year.

ESA Bulletin. European Space Agency, Publications Division, Keplerlaan 1, 22000 AG Nordwijk, Netherlands. Phone 31-1719-86555. FAX 31-1719-85433. Quarterly. Free.

ESA Journal. European Space Agency, Publications Division, Keplerlaan 1, 22000 AG Nordwijk, Netherlands. Phone 31-1719-86555. FAX 31-1719-85433. Quarterly. Free.

Interavia/Aerospace World. Aerospace Media Publishing, Swissair Centre, 31 Route de l'Aeroport, POB Box 437, 1215 Geneva 15, Switzerland. (022) 788 27 88. FAX (022) 788 27 26. Monthly. $128.00 per year (U.S.).

Journal of the Astronautical Sciences. American Astronautical Society, 6352 Rolling Mill Place, Suite 102, Springfield, VA 22152. Quarterly. $35 per year.

Journal of the British Interplanetary Society (JBIS). British Interplanetary Society, 27/29 South Lambeth Road, London SW8 1SZ, England. (071) 735-3160. FAX (071) 820-1504. Monthly. $370.00 per year.

Journal of Guidance, Control and Dynamics. American Institute of Aeronautics and Astronautics, 370 L'Enfant Promenade SW, Washington, DC 20024-2518. 1978 to present. Bimonthly. $42.00 (members) and $250.00 (non-members).

Journal of Propulsion and Power. American Institute of Aeronautics and Astronautics, 370 L'Enfant Promenade SW, Washington, DC 20024-2518. 1985 to present. Bimonthly. $38.00 (members) and $260.00 (non-members).

Journal of Spacecraft and Rockets. American Institute of Aeronautics and Astronautics, the Aerospace Center, 370 L'Enfant Promenade SW, Washington, DC 20024. (202) 646-7400. FAX (202) 646-7508. 1964 to present. Bi-monthly. $185.00 per year.

Progress in Aerospace Sciences. Pergamon Press, 660 White Plains Rd., Tarrytown, NY 10591-5153. 1961 to present. Quarterly. $320.00 per year.

Propellants, Explosives, Pyrotechnics. VCH Publishers, Inc., 220 E. 23rd St., New York, NY 10010-4606. 6 times a year. $415.00 per year.

RESEARCH CENTERS AND INSTITUTES

Aerospace Research Applications Center, Indianapolis Center for Advanced Research, 611 N. Capitol Ave., Indianapolis, in 46204. (317) 262-5003. FAX (317) 262-5044.

Aerospace Research Center, 1250 Eye St., NW, Washington, DC 20005. (202) 371-8400. FAX (202) 371-8470.

Canadian Space Agency, Space Science Program, C/o National Research Council of Canada, 100 Sussex Drive, Ottawa, Ontario, Canada K1a 0r6. (613) 990-0799. FAX (613) 952-0974.

Center for Aerospace Technology, Weber State College, Ogden, UT 84408-1805. (801) 626-7272.

Center For Space Research, Massachusetts Institute of Technology, 77 Massachusetts Ave., Cambridge, MA 02139. (617) 253-7501.

Charles Stark Draper Laboratory Inc., 555 Technology Square, Cambridge, MA 02139. (617) 258-1000.

Goddard Space Flight Center, Greenbelt, MD 20771. (301) 286-7351.

Houston Advanced Research Center, Space Technology and Research Center, 4802 Research Forest Dr., the Woodlands, TX 77381. (713) 363-7922. FAX (713) 363-7914.

Innovative Nuclear Space Power and Propulsion Institute, College of Engineering, University of Florida, Gainesville, FL 32611. (904) 392-1427. FAX (904) 392-8656.

George C. Marshall Space Flight Center, Huntsville, AL 35812. (205) 544-0024.

Machinery & Engine Technology Laboratory. Institute for Mechanical Engineering, Bldg. M-7, Montreal Road, Ottawa, ON, Canada K1A 0R6. (613) 993-2425. FAX (613) 957-3281.

Ohio Aerospace Institute, 2001 Aerospace Parkway, Brook Park, OH 44142. (216) 891-2100. FAX (216) 891-2140.

Pennsylvania State University, Propulsion Engineering Research Center, 106 Research Bldg. E, Bigler Rd., University Park, PA 16802. (814) 863-6272. FAX (814) 865-3389.

University of Tennessee, Center for Advanced Space Propulsion, Po Box 1385, Tullahoma, TN 37388-8897. (615) 454-9294. FAX (615) 455-6167.

Space Science and Engineering Center, University of Wisconsin—madison, 1225 W. Dayton St., Madison, WI 53706. (608) 262-0544.

Thermal Sciences and Propulsion Center, Purdue University, Lafayette, IN 47907. (317) 494-1503.

PROSPECTING

See: MINERAL PROSPECTING

Ency. of Physical Sciences and Engineering Info. Sources

PULSARS

PULSARS

See also: ASTRONOMY, BIG BANG, COSMOLOGY, DARK MATTER, GALAXIES, INTERSTELLAR MEDIUM, MILKY WAY, QUASARS, RADIO ASTRONOMY, STARS, UNIVERSE

ABSTRACT SERVICES AND INDEXES

Applied Science and Technology Index: A Cumulative Subject Index to English Language Periodicals in the Fields of Aeronautics and Space Science, Computer Technology, Chemistry, Construction Industry, Energy and Related Areas. H.W. Wilson Co., 950 University Avenue, Bronx, NY 10452. (212) 588-8400. (800) 367-6770. FAX (718) 590-1617. From 1958 to present. Monthly. Inquire about cost and availability. Also available on CD-ROM and online. ISSN: 0003-6986.

Astronomy and Astrophysics Abstracts. Springer-Verlag New York, 175 Fifth Avenue, New York, NY 10010. (212) 460-1500.FAX (212) 473-6272. Published for the Astronomisches Rechen-Institut. Comprehensive coverage of all aspects of astronomy, astrophysics and related fields. 1969 to present. Two parts per year. Annual. ISSN: 0067-0022.

General Science Index. H.W. Wilson Company, 950 University Avenue, Bronx, NY 10452. (212) 588-8400. (800) 367-6770. FAX (718) 590-1617. From 1978 to present. Ten issues per year; quarterly and annual cumulations. Service basis. Available on CD-ROM and online. Inquire regarding cost and availability. ISSN: 0162-1963.

Government Reports Announcements and Index. U. S. National Technical Information Service (NTIS), 5285 Port Royal Road, Springfield, VA 22161. (703) 487-4650. FAX (703) 321-8547. From 1968 to present. Annual. $630.00 per year. Also available online as NTIS Bibliographic Database and on CD-ROM. ISSN:

Meteorological and Geoastrophysical Abstracts. American Meteorological Society, c/o Inforonics, Inc., 550 Newtown Road, Littleton, MA 01460. (508) 486-8976. FAX (508) 486-0027. Covers literature in environmental sciences, meteorology, astrophysics, hydrology, glaciology, and physical oceanography. From 1950 to present. Monthly. $950.00 per year. Also available on CD-ROM and online. Inquire regarding cost and availability. ISSN: 0026-1130.

NTIS Alerts: Astronomy & Astrophysics. U. S. National Technical Information Service. 5285 Port Royal Road, Springfield, VA 22161. (703) 487-4650. FAX (703) 321-8547. Weekly. $140.00 per year.

Pascal 48: Environnement Cosmique Terrestre, Astronomie et Geologie Extraterrestre. Centre National de la Recherche Scientifique, Institut de Information Scientifique et Technique, 2 aliee du Parc de Brabois, 54514 Vandoeuvre Les Nancy Cedex, France. TEL 83-50-46-00. FAX: 83-50-46-50. 1985 to present. Ten issues per year. 770 F. Also available on CD-ROM and online.

Physics Abstracts. INSPEC. Section A, Science Abstracts. Institute of Electrical Engineers, London, United Kingdom. Available from: INSPEC/IEEE-Institute of Electrical and Electronic Engineers, Box 1331, Hoes Lane, Piscataway, NJ 08855-1331. (908) 562-5549. 1898 to present. 24 issues per year. $2835.00 per year. ISSN: 0036-8091. Also available online and on CD-ROM.

Physics Briefs (Physikalische Berichte). Information Center for Energy, Physics, Mathematics; German Physical Society. VCH Publishers, Inc., 220 East 23rd Street, New York, NY 10010-4606. (212) 683-8333. 1845 to present. 24 issues per year. $2390.00 per year. Also available online. ISSN: 0179-7434.

Science Citation Index (SCI). Institute for Scientific Information, 3501 Market Street, Philadelphia, PA 19104. (215) 386-0100. (800) 523-1850. FAX (215) 386-2991. 1961 to present. Six issues per year, plus annual cumulation. $11650.00 per year. Also available online and on CD-ROM. Inquire about price and availability. ISSN: 0036-827X.

STAR. Scientific and Technical Aerospace Reports. U.S. National Aeronautics and Space Administration. Distributed by U. S. Superintendent of Documents, Washington, DC 20402. 1963 to present. Semi-monthly, with semiannual and annual indexes. $114.00 per year. ISSN: 0036-8741. Also available online and on CD-ROM.

ANNUAL REVIEWS AND YEARBOOKS

The Astronomical Almanac. Superintendent of Documents, U. S. Government Printing office, Washington, DC 10402. (202) 783-3238. 1981 to present. Superceeds Astronomical Ephemeris and the American Ephermis and Nautical Almanac. Annual. $44.95. ISSN: 0737-6421.

Annual Review of Astronomy and Astrophysics: Original Reviews of Critical Literature and Current Developments in Astronomy and Astrophysics. Annual Reviews, Inc., 4139 El Camino Way, Palo Alto, CA 94306-0139. (415) 493-4400.(800) 523-8635. FAX (415) 855-9815. 1963 to date. Annual. $60.00. ISSN: 0066-4146.

Highlights of Astronomy. I. Appenzeller, editor. Kluwer Academic Publishers, Box 358, Accord Station, Hingham, MA 02018-0358. (617) 871-6600. International Astronomical Union Highlights series. ISBN: 07923-3553-8. Triennal. 1995. Price varies.

ASSOCIATIONS AND PROFESSIONAL SOCIETIES

Amateur Astronomers Association. 1010 Park Avenue, New York, NY 10028. (212) 535-2922.

American Association of Variable Star Observers. 23 Birch Street, Cambridge, MA 02138. (617) 354-0484. FAX: (617) 354-0665.

American Astronomical Society. 2000 Florida Avenue NW, Suite 400, Washington, DC 20009. (202) 328-2010. FAX: (202) 234-2560.

American Institute of Physics. One Physics Ellipse, College Park, MD 20740-3843. (301) 209-3100.

Association of University for Research in Astronomy. Suite 701, 1625 Massachusetts Avenue, NW, Washington, DC 20036. (202) 483-2101. FAX (202) 483-2106.

Astronomical League. 2112 Kingfisher Lane East, Rolling Meadows, IL 60008. (708) 398-0562.

Astronomical Society of the Pacific. 390 Ashton Avenue, San Francisco, CA 94112. (415) 337-1100. FAX: (415) 337-5205.

International Astronomical Union. 98 bis Boulevard Arago, F-75014 Paris, France. Telephone: 1-43-25-83-58. FAX 1-43-25-26-16.

Royal Astronomical Society of Canada. 136 Dupont Street, Toronto. ON N5R 1V2 Canada. (416) 924-7973. FAX (416) 968-6687.

ATLASES

Atlas of Deep-Sky Spendors. Hans Vehrenberg. Sky Publishing Corp., 49 Bay State Road, Cambridge, MA 02138. (617) 864-7360. (800) 253-0245. 4th edition. 1983. $24.95.

Atlas of the Ultraviolet Sky. Richard C. Henry, Wayne B. Landsman, et al. Johns Hopkins University Press, 2715 North Charles Street, Baltimore, MD 21218-4319. (410) 516-6900. (800) 537-5487. 1988. $70.00.

Atlas of the Universe. Patrick Moore. Rand McNally & Company. 8255 North Central Park, Skokie, IL 60076-2970. (708) 673-9100. 1994. $29.95.

Cambridge Atlas of Astronomy. Jean Audouze, et al. Cambridge University Press, 40 West 20th Street, New York,

NY 10011-4211. (212) 924-3900. (800) 872-7423. 3rd edition. 1994. $75.00.

Carnegie Atlas of Galaxies. Allan Sandage and John Bedke. Carnegie Institution of Washington, 1530 P Street NW, Washington, DC 20005. (202) 387-6411. Two volumes. 1994. $92.00.

Observer's Sky Atlas. E. Karkoschka. Springer-Verlag New York, Inc., 175 Fifth Avenue, New York, NY 10010. (212) 460-1500. (800) 777-4643. 1995. $15.95.

Observing Handbook and Catalog of Deep-Sky Objects. Christian B. Luginbuhl and Brian A. Skiff. Cambridge University Press, 40 West 20th Street, New York, NY 10011-4211. (212) 924-3900. (800) 872-7423. 1990. $49.95.

Photometric Atlas of Northern Bright Galaxies. Kodaira, Keichi, et al editors. Columbia University Press, 562 West 113th Street, New York, NY 10025. (212) 666-1000. (800) 944-8648. 1990. $180.00.

Third Reference Catalogue of Bright Galaxies. Gerard H. de Vaucouleurs et al, editors. Springer-Verlag New York, Inc., 175 Fifth Avenue, New York, NY 10010. (212) 460-1500. (800) 777-4643. Three volumes. 1991. $198.00 set.

BIBLIOGRAPHIES

Astronomy and Astrophysics: A Source Guide. Gordon Press, P.O. Box 459, Bowling Green Station, New York, NY 10004. (718) 624-8419. 1991. $250.00..

A Bibliography of Astronomy: 1970-1979. R. A. Sea and S. S. Martin. Libraries Unlimited, Inc., Littleton, CO 80160. 1982.

Catalog of the Naval Observatory Library. Naval Observatory Staff. G. K. Hall & Company, 866 Third Avenue, New York, NY 10022. (212)702-6789. (800) 257-5755. 6 volumes. 1977. $655.00

Science Books and Films. American Association for the Advancement of Science, 1333 H Street NW, Washington, DC 20005. (202) 326-6454. 1965 to present. Nine issues per year. $40.00 per year. ISSN: 0098-342X.

Scientific and Technical Books and Serials in Print, 1995. R.R. Bowker Inc., 121 Chanlon Road, New Providence, NJ 07974. (908) 464-6800. (800) 521-8110. 4 volumes. 1994. $299.95. Also available on compact disc and online.

A Source Book on Astronomy and Astrophysics: 1900-1975. Kenneth Lang and Owen Gingerich, editors. Harvard University Press, 79 Garden Street, Cambridge MA. (617) 495-2600. 1980. $41.95.

DIRECTORIES AND BIOGRAPHICAL SOURCES

American Men and Women of Science: Physical and Biological Sciences. R. R. Bowker Inc., 121 Chanlon Road, New Providence, NJ 07974. (908) 464-6800. (800) 521-8110. 20th edition. 8 volumes. 1996. $850.00.

The Astronomers. Donald Goldsmith. St. Martin's Press, Inc., 175 Fifth Avenue, New York, NY 10010. (212) 674-5151. (800) 221-7945. 1993. $14.95 in paper.

Astronomical Centers of the World. Kevin Krisciunas. Cambridge University Press, 40 West 20th Street, New York, NY 10011-4211. (212) 924-3900. (800) 872-7423. 1988. $34.95

The Biographical Dictionary of Scientists: Astronomers. D. Abbott, editor. Peter Bedrick Books, Inc., 2112 Broadway, Room 318, New York, NY 10023. (212) 496-0751. 1984.

Directory of Physics and Astronomy Staff Members. American Institute of Physics, One Physics Ellipse, College Park, MD 20740-3843. (301) 209-3100. Annual. $45.00.

Earth and Astronomical Research Centers. Stockton Press, 345 Park Avenue South, New York, NY 10010. 4th edition. 1995. $515.00. ISBN: 1-56169-0967.

Graduate Programs in Physics, Astronomy and Related Fields. 1995-1996. American Institute of Physics, One Physics Ellipse, College Park, MD 20740-3843. (301) 209-3100. 1978 to present. Annual. $45.00.

Research Centers Directory. Gale Research, 835 Penobscot Building, Detroit, MI 48226-4094. (313) 961-2242. (800) 347-4253. 20th edition. 1995. $485.00.

ENCYCLOPEDIAS AND DICTIONARIES

Astronomy From A To Z: A Dictionary of Celestial Objects and Ideas. Charles A. Schweighauser. Illinois Issues, Sangamon State University, K-10, Springfield, IL 62794-9243. (217) 786-9243. 1991. $14.95 in paper.

Concise Dictionary of Astronomy. Jacqueline Mitton. Oxford University Press, Inc., 200 Madison Avenue, New York, NY 10016. (212) 725-6000. (800) 334-4249. 1992. $24.95.

Facts on File Dictionary of Astronomy. Valerie Illingworth. Facts-on-File, Inc., 460 Park Avenue South, New York, NY 10016-7382. (212) 683-2244. (800) 322-8755. Fax (800) 683-3633. 1994. $27.95.

Encyclopedia of Astronomy and Astrophysics. Stephen Maran, editor. Van Nostrand Reinhold, 115 Fifth Avenue, New York, NY 10003. (212) 254-3232. (800) 842-3636. 1992. $129.95.

Landolt-Boernstein Numerical Data and Functional Relationships in Science and Technology: Astronomy, Astrophysics and Space Research; A. O. Madelung and H. H. Boigt, editors. Springer Verlag New York, Inc., 175 Fifth Avenue, New York, NY 10010. (212) 460-1500. (800) 777- 4643. 1993. $550.00.

Oxford Illustrated Encyclopedia. Volume 8, The Universe. Archie Roy, editor. Oxford University Press, Inc., 200 Madison Avenue, New York, NY 10016. (212) 725-6000. (800) 334-4249. 1993. $45.00.

McGraw-Hill Encyclopedia of Science and Technology. McGraw-Hill Book Company, Inc., 1221 Avenue of the Americas, New York, NY 10020. (212) 512-2000. (800) 262-4729. 7th edition. 20 volume set. 1992. $1900.00.

Stars, Galaxies, Cosmos. William R. Corliss. Sourcebook Project, P.O. Box 107, Glen Arm, MD 21057. (410) 668-6047. 1987. $17.95.

GENERAL WORKS

Asteroids to Quasars. Phyliss Lugger, editor. Cambridge University Press, 40 West 20th Street, New York, NY 10011-4211. (212) 924-3900. (800) 872-7423. 1991. $69.95.

Black Holes, White Dwarfs, and Neutron Stars: The Physics of Compact Objects. Stuart L. Shapiro and Saul A. Teukolsky. John Wiley & Sons, Inc., 605 Third Avenue, New York, NY 10158-0012. (212) 850-6000. (800) 225-5945. 1983. $ .00

Cosmic Questions, Galactic Halos, Cold Dark Matter and the End of Time. Richard Morris. John Wiley & Sons, Inc., 605 Third Avenue, New York, NY 10158-0012. (212) 850-6000. (800) 225-5945. 1993. $24.95.

Cycles of Fire: Stars, Galaxies and the Wonder of Deep Space. William K. Hartmann. Workman Publishing Company, Inc., 708 Broadway, New York, NY 10003. (212) 254-5900. (800) 722-7202. 1988. $15.95 in paper.

Farthest Things in the Universe. Jay M. Pasachoff. Cambridge University Press, 40 West 20th Street, New York, NY 10011-4211. (212) 924-3900. (800) 872-7423. 1995. $29.95.

The First Three Minutes: A Modern View of the Origin of the Universe. Steven Weinberg. Basic Books, 10 East 53rd Street, New York, NY 10022. (212) 207-7057. (800) 242-7737. Reprint edition. 1993. $12.00.

High-Energy Radiation From Magnetized Neutron Stars. Peter Meszaros. University of Chicago Press, 5801 Ellis Avenue, Chicago, IL 60637. (312) 702-7700. 1992. $39.95 in paper.

The LEFT Hand of CREATION: The Origin and Evolution of the Expanding Universe. John D. Barrow and Joseph Silk. Oxford University Press, Inc., 200 Madison Avenue, New York, NY 10016. (212) 725-6000. (800) 334-4249. 1994. $23.00.

New Perspectives on Stellar Pulsation and Pulsating Variable Stars. James M. Nemec and Jaymie M. Matthews. Cambridge U. Press, 40 West 20th Street, New York, NY 10011- 4211. (212) 924-3900. 1993. $59.95.

Observational Astrophysics. Robert C. Smith. Cambridge University Press, 40 West 20th Street, New York, NY 10011- 4211. (212) 924-3900. (800) 872-7423. 1995. $69.95.

Physical Processes in Comets, Stars and Active Galaxies. W. Hildebrandt, et al editors. Springer-Verlag New York, Inc., 175 Fifth Avenue, New York, NY 10010. (212) 460-1500. (800) 777-4643. 1987. $47.00.

The Physics of Stars. A. C. PHillips. John Wiley & Sons, Inc., 605 Third Avenue, New York, NY 10158-0012. (212) 850-6000. (800) 225-5945. 1994. $32.95 in paper.

Planets Around Pulsars. J. A. PHillips, et al, editors. Astronomical Society of the Pacific, 390 Ashton Avenue, San Francisco, CA 94112. (415) 337-1100. 1993. $40.00.

Pulsar Astronomy. andrew Lyne and Francis Graham-Smith. Cambridge University Press, 40 West 20th Street, New York, NY 10011-4211. (212) 924-3900. 1990. $79.95.

Pulsars. A. D. Kuzmin, editor. Nova Science Publishers, Inc., 6080 Jericho Turnpike, Suite 207, Commack, NY 11725-2808. (516) 499-3103. Lebedev Physics Institute; Proceedings, volume 196. 1992. $112.00 in paper.

Pulsars As Physics Laboratories. R. D. B landord, et al, editors. Oxford University Press, Inc., 200 Madison Avenue, New York, NY 10016. (212) 725-6000. (800) 334-4249. 1994.$52.50 in paper.

Star Formation, Galaxies and the Interstellar Medium. J. Franco, et al, editors. Cambridge University Press, 40 West 20th Street, New York, NY 10011-4211. (212) 924-3900. (800) 872-7423. 1993. $59.95.

Stars and Atoms: From the Big Bang to the Solar System. Stuart Clark. Oxford University Press, Inc., 200 Madison Avenue, New York, NY 10016. (212) 725-6000. (800) 334-4249. 1995. $35.00.

Universe of Galaxies. Paul W. Hodge, editor. W. H. Freeman & Co., 41 Madison Avenue, East 26th Avenue, 35th Floor, New York, NY 10010. (212) 576-0400. Readings from Scientific American. 1995.

Visual Astronomy of the Deep Sky. Roger N. Clark. Cambridge University Press, 40 West 20th Street, New York, NY 10011-4211. (212) 924-3900. (800) 872-7423. 1991. $39.95.

HANDBOOKS AND MANUALS

Burnham's Celestial Handbook: An Observer's Guide to the Universe Beyond the Solar System. Robert Burnham, Jr. Dover Publications, Inc., 180 Varick Street, New York, NY 10014. (212) 255-3755. (800) 223-3130. 3 volumes. 1978. $41.85 for the set.

Cambridge Guide to the Constellations. Michael E. Bakich. Cambridge University Press, 40 West 20th Street, New York, NY 10011-4211. (212) 924-3900. (800) 872-7423. 1995. $69.95.

Encyclopedia of Astronomy and Astrophysics. Stephen Maran, editor. Van Nostrand Reinhold, 115 Fifth Avenue, New York, NY 10003. (212) 254-3232. (800) 842-3636. 1992. $129.95.

PULSARS

Ency. of Physical Sciences and Engineering Info. Sources

Greenwich Guide to Stars, Galaxies and Nebulae. Stuart Malin. Cambridge University Press, 40 West 20th Street, New York, NY 10011-4211. (212) 924-3900. (800) 872-7423. 1990. $10.95.

Handbook of Astronomy, Astrophysics and Geophysics; Volume 2: Galaxies and Cosmology. V. M. Canuto and B. G. Elmegreen. Gordon & Breach Science Publishers, Inc., P.O. Box 200029, Riverfront Plaza Station, Newark, NJ 07102-0301. (201) 643-7500. 1988. $194.00.

Handbook of Space Astronomy and Astrophsyics. Martin V. Zumbeck. Cambridge University Press, 40 West 20th Street, New York, NY 10011-4211. (212) 924-3900. (800) 872-7423. 2nd edition. 1990. $79.95.

Landolt-Borenstein Numerical Data and Functional Relationships in Science and Technology: Astronomy, Astrophysics and Space Research. Group VI. Springer-Verlag New York, Inc., 175 Fifth Avenue, New York, NY 10010. (212) 460-1500. (800) 777-4643. Volumes priced individually.

McGraw-Hill Encyclopedia of Astronomy. Sybil P. Parker and Jay M. Pasachoff, editors. McGraw-Hill Publishing Company, Inc., 1221 Avenue of the Americas, New York, NY 10020. (212) 512-2000. (800) 262-4729. 2nd edition. 1993. $75.00

National Audobon Society Pocket Guide: Galaxies and Other Deep-sky Objects. Alfred A. Knopf, Inc. 201 E. 50th Street, New York, NY 10022. (800) 733-3000. 1995. $7.95.

Nearby Galaxies Catalog. R. Brent Tully and J. Richard Fisher. Cambridge University Press, 40 West 20th Street, New York, NY 10011-4211. (212) 924-3900. (800) 872-7423. 1988. $59.95.

Observer's Guide to Astronomy. Patrick Martinez, editor. Cambridge University Press, 40 West 20th Street, New York, NY 10011-4211. (212) 924-3900. (800) 872-7423. 2 volumes. 1994. $159.90.

ONLINE DATABASES AND CD-ROMS

CA Search. Chemical Abstracts Service, P.O. Box 3012, Columbus, OH 43210-0012. (614) 447-3600. (800) 848-6533. FAX (614) 447-3709. Very comprehensive guide to worldwide chemical literature and related fields, 1972 to present. Available on BRS,(800) 289-4277, DIALOG, (800) 334-2564, ORBIT (800) 456-7248, and STN International, FIZ Karlsruhe, P.O. Box 2465, W-7500, Karlsruhe 1, Germany, online services. Inquire as to cost and availability.

Current Contents Search. Institute for Scientific Information, 3501 Market Street, Philadelphia, PA 19104. (215) 386-0100. FAX (215) 386-6362. Contains citations to articles listed in the table of contents of science and technology journals. Also articles in social sciences and life sciences journals. Available on BRS, (800) 289-4277, DIALOG, (800) 334-2564, online services. Inquire as to cost and availability.

Dissertation Abstracts Online. University Microfilms International, 300 North Zeeb Road, Ann Arbor, MI 48106. (800) 521-0600 or (313) 761-4700. Scope includes virtually all doctoral dissertations accepted at accredited American institutions from 1861 to present in 252 subject areas. Available on BRS, (800) 289-4277, DIALOG, (800) 334-2564, and OCLC EPIC, (800) 848-5878, online services. Also available on CD-ROM. Inquire as to cost and availability.

INSPEC. Institution of Electrical Engineers, Michael Faraday House, Six Hills Way, Stevenage, Herts. SG1 2AY, England. Telephone: 0438 313311 or FAX 0438 742840. Contains citations to the worldwide literature of physics, electronics and electrical engineering, computer technology, and related fields. Available on BRS, (800) 289-4277, DIALOG, (800) 334-2564, ORBIT, (800) 456-7248, and STN International, FIZ Karlsruhe, P.O. Box 2465, W-7500, Karlsruhe 1, Germany, online services. Inquire as to cost and availability.

International Aerospace Abstracts. American Institute of Aeronautics and Astronautics, 370 L'Enfant Promenade, S.W., Washington, DC 20024. (202) 646-7400. Contains references and abstracts of journal and monograph literature relating to aerospace science and technology, from 1963 to present. Available through the NASA/RECON system of the National Aeronautics and Space Administration only.

NTIS Bibliographic Database. National Technical Information Service, 5285 Port Royal Road, Springfield, VA 22161. (703) 487-4929 or FAX (703) 321-8199. Broad coverage of government-sponsored science and technology research reports, 1964 to present. Available on BRS,(800) 289-4277, DIALOG, (800) 334-2564, ORBIT, (800) 456-7248, and STN International, FIZ Karlsruhe, P.O. Box 2465, W-7500, Karlsruhe 1, Germany, online services. Also available on CD-ROM. Inquire as to cost and availability.

Physics Briefs. American Institute of Physics, 335 East 45th Street, New York, NY 10017. (212) 661-9260 or FAX (212) 661-2036. Contains citations with abstracts of the literature of physics and related fields, 1979 to present. Available on the STN International, FIZ Karlsruhe, P.O. Box 2465, W-7500, Karlsruhe 1, Germany, online service. Inquire as to cost and availability.

Scientific and Technical Books and Serials in Print. R.R. Bowker Inc., 121 Chanlon Road, New Providence, NJ 07974. (800) 521-8110 or (908) 464-6800. List of currently published books and serials in the physical and biological sciences, engineering and technology, with subject, author and titles indexes. Available on ORBIT, (800) 456-7248, online service. Inquire as to cost and availability.

SCISEARCH. Institute for Scientific Information, 3501 Market Street, Philadelphia, PA 19104. (800) 523-1850 or (215) 386-0100. Broad multidisciplinary title and author index to the international literature of science and technology, 1974 to present. Available on DIALOG, (800) 334-2564, and ORBIT, (800) 456-7248, online services. Also available on CD-ROM. Inquire as to cost and availability.

WILSONLINE. H.W. Wilson Company, 950 University Avenue, Bronx, NY 10452. (800) 367-6770 or (212) 588-8400. Makes available online versions of the printed H.W. Wilson indexes including *Applied Science and Technology Index, Business Periodicals Index, General Science Index,* and *Readers' Guide to Periodical Literature.* Period covered is generally 1983 to present. Available on BRS, (800) 289-4277, DIALOG, (800) 334-2564, and OCLC EPIC, (800) 848-5878, online services.Also available on CD-ROM. Inquire as to cost and availability.

PERIODICALS

Astronomical Journal. American Institute of Physics, One Physics Ellipse, College Park, MD 20740-3843. (301) 209-3000. Published for the American Astronomical Society. 1849 to present. Monthly. $280.00 per year. ISSN: 0004-6256.

Astronomical Society of the Pacific. Publications. Astronomical Society of the Pacific, 390 Ashton Avenue, San Francisco, CA 94112. (415) 337-1100. FAX (415) 337-5205. 1889 to present. Monthly. $175.00 per year. ISSN: 0004- 6280.

Astronomy. Kalmbach Publishing Company, Box 1612, Waukesha, WI 53187-1612. (414) 796-8776. FAX (414) 796-1142. 1973 to present. Monthly. $27.00 per year. ISSN: 0091-6358.

Astronomy and Astrophsyics. Springer-Verlag New York, Inc., 44 Hartz Way, Secaucus, NJ 07096-2491. (201) 348-4033. FAX (201) 348-4505. 1969 to date. 36 issues per year in 12 volumes. $1733.00 per year. ISSN: 0004-6361.

Astronomy Reports. American Institute of Physics, One Physics Ellipse, College Park, MD 20740-3843. (301) 209- 3000. English translation of Astronomicheksii Zhurnal. Formerly *Soviet Astronomy AJ.* 1957 to present. Bimonthly. $1160.00 per year. ISSN: 1063-7729.

Astrophysical Journal: An International Review of Astronomy and Astronomical Physics. University of Chicago Press, Journals Division, 5720 South Woodlawn Avenue, Chicago, IL 60637. (312) 753-3347. FAX (312) 753-0811. 1895 to date. three issues per month. $740.00 per year. ISSN: 0004-637X.

Astrophpysics and Space Science: An International Journal of Cosmic Physics. Kluwer Academic Publishers, Box 358, Accord Station, Hingham, MA 02018-0358. (617) 871-6600. FAX (617) 871-6528. 1968 to present. 24 issues per year. $2544.00 per year. ISSN: 0004-640X.

Celestial Mechanics and Dynamical Astronomy; An International Journal of Space Dynamics. Kluwer Academic Publishers, Box 358, Accord Station, Hingham, MA 02018-0358. (617) 871-6600. FAX (617) 871-6528. 1969 to present.Monthly. $703.50 per year. ISSN: 0923-2958.

Experimental Astronomy: An International Journal On Astronomical Instrumentation and Data Analysis. Kluwer Academic Publishers, Box 358, Accord Station, Hingham, MA 02018-0358. (617) 871-6600. FAX (617) 871-6528. 1989 topresent. Quarterly. $161.00 per year. ISSN: 0922-6435.

Icarus: International Journal of Solar System Studies. Academic Press, Inc., Journal Division, 525 B Street, Suite 1900, San Diego, CA 92101-4495. (619) 230-1840. FAX (619) 699-6800. 1962 to present. Monthly. $1080.00 per year. ISSN: 0019-1035.

Landolt-Boernstein, Zahlenwerte und Funktionen Aus Naturwissenschaften und Technik. Group 6: Astronomy. Springer-Verlag New York, Inc., 175 Fifth Avenue, New York, NY 10010. (212) 460-1500. FAX (212) 473-6272. 1965. Irregular. Price varies. ISSN: 0075-7896.

Mercury. Astronomical Society of the Pacific. 390 Ashton Avenue, San Francisco, CA 94112. (415) 337-1100. FAX: (415) 337-5205. 1972 to present. Bimonthly. $175.00 per year. ISSN: 0047-6773.

Monthly Notices of the Royal Astronomical Society. Blackwell Scientific Publication LT., Osney Mead, Oxford OX2 OEL, England. Telephone: 0865-240201. FAX 0865-721205. 1827 to present. Fortnightly. $1899.00 per year. ISSN: 0035-8711.

Observatory. Space and Astrophysics Division, Rutherford Appleton Laboratory, Chilton, Didcot, Oxon OX11 OQX England. FAX 0235-445848. 1877 to present. Bimonthly. $42.00 per year.

Sky and Telescope. Sky Publishing Corporation, Box 9111, Belmont, MA 02178. (617) 864-7360. FAX (617) 864-6117. 1941 to present. Monthly. $27.00 per year. ISSN: 0037- 6604.

Space Science Reviews. Kluwer Academic Publishers, Box 358, Accord Station, Hingham, MA 02018-0358. (617) 871- 6600. FAX (617) 871-6528. 1962 to present. Sixteen issues per year. $940.00 per year. ISSN: 0038-6308.

Vistas in Astronomy: An International Review Journal. Albert C. Beer, editor. Elsevier Science Publishing Company, Inc., 660 White Plains Road, Terrytown, NY 10591- 5153. (914) 524-9200. FAX: (914) 333-2444. 1958 to present. Quarterly. $415.00 per year. ISSN: 0083-6656.

Transactions of the International Astronomical Union. Proceedings of the General Assembly. Kluwer Academic Publishers, Box 358, Accord Station, Hingham, MA 02018- 0358. (617) 871-6600. Issues in two parts: Part A, Reports, and Part B, Proceedings. Triennial. XXI-1994. Price varies. ISSN: 0080-1372.

Webb Society Quarterly Journal. Webb Society, 194 Foundry Lane, Freemantle, Southampton, Hampshire SO1 3EQ, England. 1969 to present. Quarterly. $19.50 per year. ISSN: 0043- 1680.

RESEARCH CENTERS AND INSTITUTES

Arecibo Observatory. Cornell University, Box 995, Arecibo, PR 00613. (809) 878-2612. FAX (809) 878-1861.

Astronomy Program. University of Maryland, 1207 Computer & Space Sciences Building, College Park, MD 20742. (301) 405- 3001. FAX (301) 314-9067.

Fred Lawrence Whipple Observatory. P.O. Box 97, Amado, AZ 85645. (602) 670-6741. FAX (602) 670-6779.

Joint Institute for Laboratory Astrophyics. University of Colorado, Box 440, Boulder, CO 80309-0440. (303) 492-7780. FAX (303) 492-5235.

Lick Observatory. University of California, Santa Cruz, CA 95064. (408) 459-2991. FAX (408) 426-3115.

Lowell Observatory. 1400 West Mars Hill Road. Flagstaff, AZ 86001, (602) 774-3358. FAX (602) 774-6296.

McDonald Observatory, University of Texas. P.O. Box 1337, Fort Davis, TX 79734. (915) 426-3263.

National Astronomy and Ionosphere Center. Cornell University, Space Sciences Building, Ithaca, NY 14853. (607) 255-3735. (607) 255-8803.

National Radio Astronomy Observatory. Edgemont Road, Charlottesville, VA 22903. (804) 296-0211.

Palomar Observatory. California Institute of Technology. 105-24, Pasadena CA 91125. (818) 356-4033.

PULSARS

Ency. of Physical Sciences and Engineering Info. Sources

Radio Observatory. University of Florida, 211 Space Sciences Research Building, Gainsville, FL 32611. (904) 392-2066.

Space Telescope Science Institute. 3700 San Martin Drive, Baltimore, MD 21218. (301) 338-4700.

Steward Observatory. University of Arizona. Tuscon, AZ 85721. (602) 621-6524. FAX (602) 621-1532.

PUMPING MACHINERY

See also: PETROLEUM ENGINEERING, PIPELINE TECHNOLOGY

ABSTRACTS AND INDEXES

API Abstracts. American Petroleum Institute, Central Abstracting and Information Services, 275 Seventh Avenue, New York, NY 10001. (212) 366-4040 or FAX (212) 366-4298. Contains citations with abstracts to literature relating to the petroleum refining and the petrochemical industries, from 1964 to present.

Applied Mechanics Reviews: An Assessment of World Literature in Engineering Sciences. 1948-present. American Society of Mechanical Engineers, 345 East 47th Street, New York, NY 10017. (212) 705-7703. Monthly. $360.00 per year.

Applied Science and Technology Index: A Cumulative Subject Index to English Language Periodicals in the Fields of Aeronautics and Space Science, Computer Technology, Chemistry, Construction Industry, Energy and Related Areas. H.W. Wilson Co., 950 University Avenue, Bronx, NY 10452. (800) 367-6770 or (212) 588-8400. FAX (718) 590-1617. From 1958 to present. Monthly. Inquire about cost and availability. Also available on CD-ROM and online.

Current Contents: Engineering, Technology, and Applied Sciences. Institute for Scientific Information, 3501 Market Street, Philadelphia, PA 19104. (215) 386-0100. FAX (215) 386-6362. 1970 to present. Weekly. $442.00 per year.

Engineering Index Monthly. Engineering Information, Inc., Castle Point on the Hudson, Hoboken, NJ 07030. (800) 221-1044. FAX (212) 832-1857. Monthly. $2300.00 per year. Also available online as COMPENDEX, and also on CD-ROM. Covers chemical engineering, computers, electrical engineering, civil engineering, metals and mining, industrial management, and mechanical engineering.

Fluid Abstracts: Civil Engineering. Elsevier Science [journals], 660 White Plains Rd., Tarrytown, NY 10159-5153. (919) 524-9200. FAX (919) 333-2444. 1991 to present. Monthly plus annual cumulation. $645.00 per year. Also available online as FLUIDEX. Covers civil engineering applications of fluid mechanics, hydraulics, flow metering and measuring, offshore engineering, environmental hydraulics, and related aspects of wind energy, the atmosphere, and aerodynamics.

Fluid Abstracts: Process Engineering. Elsevier Science [journals], 660 White Plains Rd., Tarrytown, NY 10159-5153. (919) 524-9200. FAX (919) 333-2444. 1991 to present. Monthly plus annual cumulation. $645.00 per year. Also available online as FLUIDEX.

International Building Services Abstracts. Building Service Research and Information Association, Old Bracknell Lane W,

Bracknell, Berks. RG12 7AH, England. Telephone 0344-426511. FAX 0344-487575. 1966 to present. Bi-monthly. Inquire for price. Also available online. A survey of world literature on mechanical and technical services of buildings, including heating, sanitation, ventilation, air conditioning, lighting, and communications.

Mechanical Engineering Abstracts (formerly *ISMEC*), Cambridge Scientific Abstracts, 7200 Wisconsin Avenue, Bethesda, MD 20814-4823. (301) 961-6750. FAX (301) 961-6720. 1967 to present. Monthly. $895.00 per year. Summarizes world literature in mechanical engineering, production engineering, and engineering management. Also available online.

Manufacturing and Process Engineering Abstracts, Cambridge Scientific Abstracts, 7200 Wisconsin Avenue, Bethesda, MD 20814-4823. (301) 961-6750. FAX (301) 961-6720. 1993 to present. Monthly. $385.00 per year. Covers concurrent engineering, quality control, automated manufacturing, petroleum engineering, oil field operations and equipment, energy management, metallurgy and metallography, foundry practice.

Oceanic Abstracts. Cambridge Scientific Abstracts, 7200 Wisconsin Avenue, Bethesda, MD 20814. (301) 961-6750. Fax (301) 961-6720. Bimonthly. $995.00 per year.

Offshore Engineering Abstracts. STI Limited, 4 Kings Meadow, Ferry Hinksey Road, Oxford, OX2 ODU, England. Distributed in the United States by Air Science Company, Box 143, Corning NY 14830. (607) 962-5591. Contains citations on planning, design, and construction of offshore platforms and pipelines, from 1985 to present. Monthly. $400.00 per year.

Petroleum Abstracts. 600 South College, Tulsa, OK 74104. (918) 631-2296 or (800) 247-8678. Contains citations with abstracts to the worldwide literature and patents on the exploration, development, and production of petroleum resources, from 1965 to present.

ASSOCIATIONS AND PROFESSIONAL SOCIETIES

American Society of Heating, Refrigeration and Air Conditioning Engineers. 1791 Tullie Circle NE, Atlanta, GA 30329. (404) 636-8400.

American Society of Mechanical Engineers. 345 East 47th Street, New York, NY 10017-2398. (212) 705-7703.

American Water Works Association. 6666 W. Quincy Avenue, Denver, CO 80235. (303) 794-7711.

Gasoline Pump Manufacturers Association. c/o Mark Joelson, Morgan, Lewis and Bochius, 1800 M Street NW, Suite 600N, Washington, DC 20036. (202) 467-7240.

Submersible Wastewater Pump Association. 600 S. Federal Street, Suite 400, Chicago, IL 60605. (312) 922-6222.

Sump and Sewage Pump Manufacturers Association. P.O. Box 647, Northbrook, IL 60065-0647. (708) 559-9233.

DIRECTORIES AND BIOGRAPHICAL SOURCES

Directory of Engineering Societies and Related Organizations. Gordon Davis. 13th edition. American Association of Engineering Societies, 1111 19th Street NW, Suite 608, Washington,

DC 20036. (202) 296-2237 or (800) 658-8897. 1989. Inquire for price.

Research Centers Directory. Gale Research, 835 Penobscot Building, Detroit, MI 48226-4094. (313) 961-2242. (800) 347-4253. FAX (313) 961-6083. $485.00.

Who's Who in Engineering. American Association of Engineering Societies, 1111 19th Street NW, Suite 608, Washington, DC 20036. (202) 296-2237 or (800) 658-8897. 8th edition. 1991. Inquire for price.

GENERAL WORKS

Capture Pumping Technology. Kimo M. Welch. Pergamon/Elsevier Science B.V., PO Box 945, Madison Square Station, New York, NY 10159-0945. (212) 633-3650. FAX (212) 633-3680. 1991. $81.00.

Centrifugal Pump Sourcebook. J.W. Dufour and W. E. Nelson. McGraw-Hill Publishing Company, 1221 Avenue of the Americas, New York, NY 10020. (800) 262-4729 or (212) 512-3825. 1992. $47.50.

Centrifugal Pumps. I.J. Karassik & J.T. McGuire. 2nd edition. Chapman & Hall, 115 Fifth Avenue, New York, NY 10211-0906. 1996. $89.95.

Pump Application Desk Book. Paul N. Garay. 2nd ed. Fairmont Press, 700 Indian Trail, Lilburn, GA 30247. (404) 925-9388. 1993. $69.00.

Pumps. Harry L. Stewart, rev. by Rex Miller. 5th ed. Macmillan Publishing, 200 Old Tappan Road, Old Tappan, NJ 07675. (800) 223-2336. FAX (800) 445-6991. 1991. $35.00.

Pumps and Pumping Operations. Nicolas P. Cheremisinoff. Prentice Hall, 113 Sylvan Avenue, Route 9W, Englewood Cliffs, NJ 07632. (201) 592-2000 or (800) 922-0579. 1992. $72.00.

Pump Technology: New Challenges, Where Next. B. Glanfield, editor. Springer-Verlag, 175 Fifth Avenue, New York, NY 10010. (212) 460-1500 or (800) 777-4643. FAX (212) 473-6272. 1989. $33.00.

HANDBOOKS AND MANUALS

Centrifugal Pump User's Guidebook. S. Yedidiah.Chapman & Hall, 115 Fifth Avenue, New York, NY 10211-0906. 1996. $59.95.

Machinery's Handbook. Erik Oberg, et al., editors. 24th edition. Industrial Press Inc., 200 Madison Avenue, New York, NY 10016. (212) 889-6330. FAX (212) 545-8327. 1992. $75.00.

Marks' Standard Handbook for Mechanical Engineers. E.A. Avallone and T. Baumeister III. 9th edition. McGraw-Hill Publishing Company, 1221 Avenue of the Americas, New York, NY 10020. (800) 262-4729 or (212) 512-3825. 1987. $133.00.

Pump Applications Desk Book. Paul N. Garay. Prentice Hall, 113 Sylvan Avenue, Route 9W, Englewood Cliffs, NJ 07632. (201) 592-2000 or (800) 922-0579. 1992. $60.00.

Pump Handbook. I.J. Karassik & W.C. Krutzch. 2nd edition. McGraw-Hill Publishing Company, 1221 Avenue of the Americas, New York, NY 10020. (800) 262-4729 or (212) 512-3825. 1986. $115.50.

ONLINE DATABASES AND CD-ROMS

Compendex Plus. Engineering Information, Inc., 345 East 47th Street, New York, NY 10017. (212) 705-7600 or (800) 221-1044. Contains citations with abstracts to worldwide literature in engineering and technology, from 1970 to present. Available on online BRS,(800) 289-4277, DIALOG, (800) 334-2564, ORBIT (800) 456-7248, and STN International, FIZ Karlsruhe, P.O. Box 2465, W-7500, Karlsruhe 1, Germany, online services. Also available on CD-ROM. Inquire as to cost and availability.

CSA Engineering. Cambridge Scientific Abstracts, 7200 Wisconsin Avenue, Suite 601, Bethesda, MD 20814. (301) 961-6750 or (800) 843-7751. Contains citations and abstracts of international periodicals and research literature covering all fields of engineering and science and technology,including computer and information science, electronics, mechanical engineering, solid state materials, 1981 to present. Available on BRS,(800) 289-4277, online service.Inquire as to cost and availability.

Current Contents Search. Institute for Scientific Information, 3501 Market Street, Philadelphia, PA 19104. (215) 386-0100. FAX (215) 386-6362. Contains citations to articles listed in the table of contents of science and technology journals. Also articles in social sciences and life sciences journals. Available on BRS,(800) 289-4277, DIALOG,(800) 334-2564, online services. Inquire as to cost and availability.

ISMEC: Mechanical Engineering Abstracts. Cambridge Scientific Abstracts, 7200 Wisconsin Avenue, Suite 601, Bethesda, MD 20814. (301) 961-6750 or (800) 843-7751. Contains citations to the literature in mechanical engineering, industrial and production engineering, energy, power, mechanics, devices and related areas, from 1973 to present. Available on the DIALOG,(800) 334-2564, online service. Inquire as to cost and availability.

NTIS Bibliographic Database. National Technical Information Service, 5285 Port Royal Road, Springfield, VA 22161. (703) 487-4929 or FAX (703) 321-8199. Broad coverage of government-sponsored science and technology research reports, 1964 to present. Available on BRS,(800) 289-4277, DIALOG, (800) 334-2564, ORBIT, (800) 456-7248, and STN International, FIZ Karlsruhe, P.O. Box 2465, W-7500, Karlsruhe 1, Germany, online services. Also available on CD-ROM. Inquire as to cost and availability.

SCISEARCH. Institute for Scientific Information, 3501 Market Street, Philadelphia, PA 19104. (800) 523-1850 or (215) 386-0100. Broad multidisciplinary title and author index to the international literature of science and technology, 1974 to present. Available on DIALOG,(800) 334-2564, and ORBIT,(800) 456-7248, online services. Also available on CD-ROM. Inquire as to cost and availability.

WILSONLINE. H.W. Wilson Company, 950 University Avenue, Bronx, NY 10452. (800) 367-6770 or (212) 588-8400. Makes available online versions of the printed H.W. Wilson indexes including *Applied Science and Technology Index, Business Periodicals Index, General Science Index,* and *Readers' Guide to Periodical Literature.* Period covered is generally 1983 to present. Available on BRS,(800) 289-4277, DIALOG,(800) 334-

PUMPING MACHINERY

Ency. of Physical Sciences and Engineering Info. Sources

2564, and OCLC EPIC,(800) 848-5878, online services. Also available on CD-ROM. Inquire as to cost and availability.

PERIODICALS

Industrial Pump and Valve. Industrial Pump Company, 20509 W. 7 Mile Road, Detroit, MI 48219. (313) 531-4771. 1984 to present. Bi-monthly. Inquire for cost.

PYROMETALLURGY

See: METALLURGY

Q

QUALITY CONTROL ENGINEERING

See also: CHEMICAL ENGINEERING, CONTROL SYSTEMS, DESIGN ENGINEERING, INDUSTRIAL ENGINEERING, MATERIALS HANDLING, MECHANICAL ENGINEERING, OPERATIONS RESEARCH, ROBOTICS

ABSTRACT SERVICES AND INDEXES

Applied Mechanics Reviews: An Assessment of World Literature in Engineering Sciences. 1948-present. American Society of Mechanical Engineers, 345 East 47th Street, New York, NY 10017. (212) 705-7703. Monthly. $360.00 per year.

Applied Science and Technology Index. A Cumulative Subject Index To English Language Periodicals in the Fields of Aeronautics and Space Science, Computer Technology, Chemistry, Construction Industry, Energy and Related Areas. H.W. Wilson Co., 950 University Avenue, Bronx, NY 10452. (800) 367-6770 or (212) 588-8400. FAX (313) 590-1617. From 1958 to present. Monthly. Inquire about cost and availability. Also available on CD-ROM and online.

Current Contents: Engineering, Technology, and Applied Sciences. Institute for Scientific Information, 3501 Market Street, Philadelphia, PA 19104. (215) 386-0100. FAX (215) 386-6362. 1970 to present. Weekly. $442.00 per year.

Engineering Index Monthly. Engineering Information, Inc., Castle Point on the Hudson, Hoboken, NJ 07030. (800) 221-1044. FAX (212) 832-1857. Monthly. $2300.00 per year. Also available online as COMPENDEX, and also on CD-ROM. Covers chemical engineering, computers, electrical engineering, civil engineering, metals and mining, industrial management, and mechanical engineering.

Index To Scientific and Technical Proceedings. Institute for Scientific Information, 3501 Market St., Philadelphia, PA 19104. (215) 386-0100. FAX (215) 386-6362. Monthly. $500.00 per year.

Mechanical Engineering Abstracts (formerly ISMEC), Cambridge Scientific Abstracts, 7200 Wisconsin Avenue, Bethesda, MD 20814-4823. (301) 961-6750. FAX (301) 961-6720. 1967 to present. Monthly. $895.00 per year. Summarizes world literature in mechanical engineering, production engineering, and engineering management. Also available online.

Manufacturing and Process Engineering Abstracts. Cambridge Scientific Abstracts, 7200 Wisconsin Avenue, Bethesda, MD 20814-4823. (301) 961-6750. FAX (301) 961-6720. 1993 to present. Monthly. $385.00 per year. Covers concurrent engineering, quality control, automated manufacturing, petroleum engineering, oil field operations and equipment, energy management, metallurgy and metallography, foundry practice.

Science Citation Index. Institute for Scientific Information, 3501 Market Street, Philadelphia, PA 19104. (215) 386-0100. FAX (215) 386-6362. Inquire about availability and cost. Also available on CD-ROM.

ASSOCIATIONS AND PROFESSIONAL SOCIETIES

American Production and Inventory Control Society. 500 W. Annandale Road, Falls Church, VA 22046-4274. (703) 237-8344.

American Society For Quality Control. 611 E Wisconsin Ave., PO Box 3005, Milwaukee, WI 53201-3005. (414) 272-8575 or (800) 952-6587.

American Society of Mechanical Engineers. 345 East 47th Street, New York, NY 10017-2398. (212) 705-7703.

Association For Quality and Participation. 801-B W. 8th Street, Suite 501, Cincinnati, OH 45203-0070. (513) 381-1959. FAX (513) 381-0070.

Industrial Designers Society of America. 1142-E Walker Road, Great Falls, VA 22066. (703) 759-0100.

Quality and Productivity Management Association. 300 N. Martingale Road, Suite 230, Schaumburg, IL 60173. (708) 619-2909. FAX (708) 619-3383.

Society of American Value Engineers. 60 Revere Drive, Suite 500, Northbrook, IL 60062. (708) 480-9080.

Society of Logistics Engineers. 8100 Professional Place, Suite 211, Hyattsville, MD 20785. (301) 459-8446.

Society of Manufacturing Engineers. One SME Drive, Box 930, Dearborn, MI 48128-1490. (313) 271-1500 or FAX (313) 271-2861.

DIRECTORIES AND BIOGRAPHICAL SOURCES

American Men & Women of Science. R.R. Bowker Staff, eds. 19th edition. 8 volumes. R.R. Bowker/Reed International Publishing Company, 121 Chanlon Road, New Providence, NJ 07974. (908) 464-6800 or (800) 521-8110. 1995. $850.00

Directory of Engineering Societies and Related Organizations. Gordon Davis. 13th edition. American Association of Engineering Societies, 1111 19th Street NW, Suite 608, Washington, DC 20036. (202) 296-2237 or (800) 658-8897. 1989. Inquire for price.

International Research Centers Directory. Gale Research, 835 Penobscot Building, Detroit, MI 48226-4094. (313) 961-2242. (800) 877-4253. FAX (313) 961-6083. 8th edition. 1995. $410.00

Plant Engineering Product Supplier Guide. Cahners Publishing Company (Des Plaines), 1350 East Touhy Avenue, Des Plaines, IL 60017-5080. (800) 323-4958. FAX (708) 390-2779. 1965 to present. Annual. $25.00.

Research Centers Directory. Gale Research, 835 Penobscot Building, Detroit, MI 48226-4094. (313) 961-2242. (800) 877-4253. FAX (313) 961-6083. 20th edition. $485.00.

Scientific and Technical Organizations and Agencies Directory. Gale Research, 835 Penobscot Building, Detroit, MI 48226-4094. (313) 961-2242. (800) 877-4253. FAX (313) 961-6083. 4th edition. 1996. $195.00.

Who's Who in Engineering. American Association of Engineering Societies, 1111 19th Street NW, Suite 608, Washington, DC 20036. (202) 296-2237 or (800) 658-8897. 8th edition. 1991. Inquire for price.

ENCYCLOPEDIAS AND DICTIONARIES

The Encyclopedia of Operations Research and Management Science. Saul I. Gass & Carl M. Harris, editors. Kluwer Academic Publishers, PO Box 358, Accord Station, Hingham, MA 02018-0358. (617) 871-6600. FAX (617) 871-6528. 1996. $295.00.

GENERAL WORKS

Process Quality Control. E.R. Ott and E.G. ScHilling. Second edition. McGraw-Hill Publishing Company, 1221 Avenue of the Americas, New York, NY 10020. (800) 262-4729 or (212) 512-3825. 1990. $54.00.

Quality Control. Dale H. Besterfield. 3rd edition. Prentice Hall, 113 Sylvan Avenue, Route 9W, Englewood Cliffs, NJ 07632. (201) 592-2000 or (800) 922-0579. 1990. $47.00.

Quality Planning and Analysis From Product Development Through Use. J.M. Juran and Frank M. Gryna. 3d ed. McGraw-Hill Publishing Company, 1221 Avenue of the Americas, New York, NY 10020. (800) 262-4729 or (212) 512-3825. 1993. $53.82.

Total Engineering Quality Management. Ronald J. Cottman. Marcel Dekker, Inc., 270 Madison Avenue, New York, NY 10016. (212) 696-9000. FAX (212) 685-4540. 1992. $45.00.

Total Quality Development. Don Clausing. American Society of Mechanical Engineers, 345 East 47th Street, New York, NY 10017-2398. (212) 705-7703. 1994. Inquire for cost and availability.

HANDBOOKS AND MANUALS

Handbook of Industrial Engineering. Gavriel Salvendy. 2nd edition. John Wiley and Sons, Inc., 605 Third Avenue, New York, NY 10158. (800) 526-5368 or (212) 850-6000. 1991. $165.00.

Juran's Quality Control Handbook. Joseph N. Juran Jr. & F. M. Gryna, editors. 4th edition. McGraw-Hill Publishing Company, 1221 Avenue of the Americas, New York, NY 10020. (800) 262-4729 or (212) 512-3825. 1988. $102.50.

Plant Engineer's Handbook of Formulas, Charts & Tables. Donald A. Moffatt. 3rd edition. Prentice Hall, 113 Sylvan Avenue, Route 9W, Englewood Cliffs, NJ 07632. (201) 592-2000 or (800) 922-0579.1991. $69.95.

Plant Engineer's Reference Book. Dennis A. Snow. Butterworth-Heinemann, 313 Washington Street, Newton, MA 02158. (617) 928-2500 or (800) 366-2665. FAX (617) 928-2620. 1991. $199.00.

Production Handbook. John A. White Jr. 4th edition. John Wiley and Sons, Inc., 605 Third Avenue, New York, NY 10158. (800) 526-5368 or (212) 850-6000. 1987. $115.00.

Quality Engineering Handbook. Thomas Pyzdek and Roger W. Beiger, eds. Marcel Dekker, Inc., 270 Madison Avenue, New York, NY 10016. (212) 696-9000. FAX (212) 685-4540. 1992. $99.75.

ONLINE DATABASES AND CD-ROMS

Compendex Plus. Engineering Information, Inc., 345 East 47th Street, New York, NY 10017. (212) 705-7600 or (800) 221-1044. Contains citations with abstracts to worldwide literature in engineering and technology, from 1970 to present. Available on online BRS,(800) 289-4277, DIALOG, (800) 334-2564, ORBIT (800) 456-7248, and STN International, FIZ Karlsruhe, P.O. Box 2465, W-7500, Karlsruhe 1, Germany, online services. Also available on CD-ROM. Inquire as to cost and availability.

CSA Engineering. Cambridge Scientific Abstracts, 7200 Wisconsin Avenue, Suite 601, Bethesda, MD 20814. (301) 961-6750 or (800) 843-7751. Contains citations and abstracts of international periodicals and research literature covering all fields of engineering and science and technology,including computer and information science, electronics, mechanical engineering, solid state materials, 1981 to present. Available on BRS,(800) 289-4277, online service.Inquire as to cost and availability.

Current Contents Search. Institute for Scientific Information, 3501 Market Street, Philadelphia, PA 19104. (215) 386-0100. FAX (215) 386-6362. Contains citations to articles listed in the table of contents of science and technology journals. Also articles in social sciences and life sciences journals. Available on BRS,(800) 289-4277, DIALOG,(800) 334-2564, online services. Inquire as to cost and availability.

Dissertation Abstracts. University Microfilms International, 300 North Zeeb Road, Ann Arbor, MI 48106. (800) 521-0600 or (313) 761-4700. Scope includes virtually all doctoral dissertations accepted at accredited American institutions from 1861 to

present in 252 subject areas. Available on BRS,(800) 289-4277, DIALOG,(800) 334-2564, and OCLC EPIC,(800) 848-5878, online services. Also available on CD-ROM. Inquire as to cost and availability.

ISMEC: Mechanical Engineering Abstracts. Cambridge Scientific Abstracts, 7200 Wisconsin Avenue, Suite 601, Bethesda, MD 20814. (301) 961-6750 or (800) 843-7751. Contains citations to the literature in mechanical engineering, industrial and production engineering, energy, power, mechanics, devices and related areas, from 1973 to present. Available on the DIALOG,(800) 334-2564, online service. Inquire as to cost and availability.

NTIS Bibliographic Database. National Technical Information Service, 5285 Port Royal Road, Springfield, VA 22161. (703) 487-4929 or FAX (703) 321-8199. Broad coverage of government-sponsored science and technology research reports, 1964 to present. Available on BRS,(800) 289-4277, DIALOG, (800) 334-2564, ORBIT, (800) 456-7248, and STN International, FIZ Karlsruhe, P.O. Box 2465, W-7500, Karlsruhe 1, Germany, online services. Also available on CD-ROM. Inquire as to cost and availability.

SCISEARCH. Institute for Scientific Information, 3501 Market Street, Philadelphia, PA 19104. (800) 523-1850 or (215) 386-0100. Broad multidisciplinary title and author index to the international literature of science and technology, 1974 to present. Available on DIALOG,(800) 334-2564, and ORBIT,(800) 456-7248, online services. Also available on CD-ROM. Inquire as to cost and availability.

WILSONLINE. H.W. Wilson Company, 950 University Avenue, Bronx, NY 10452. (800) 367-6770 or (212) 588-8400. Makes available online versions of the printed H.W. Wilson indexes including Applied Science and Technology Index, Business Periodicals Index, General Science Index, and Readers' Guide to Periodical Literature. Period covered is generally 1983 to present. Available on BRS,(800) 289-4277, DIALOG,(800) 334-2564, and OCLC EPIC,(800) 848-5878, online services. Also available on CD-ROM. Inquire as to cost and availability.

PERIODICALS

Advanced Manufacturing Technology. Technical Insights Inc., 32 N Dean Street, Englewood, NJ 07631. (201) 568-4744. FAX (201) 568-8247. 1980 to present. Monthly. $575.00 per year.

American Industry. Publications for Industry, 21 Russell Woods Road, Great Neck, NY 11021. (516) 487-0990. FAX (516) 487-0809. 1946 to present. Monthly. $25.00 per year.

Assembly Engineering. Hitchcock Publishing, 191 S. Gary Avenue, Carol Stream, IL 60188. (708) 462-4641. FAX (708) 462-2205. 1958 to present. Monthly. $60.00 per year.

IEEE Transactions On Industry Applications. IEEE, 345 E. 47th Street, New York, NY 10017. (908) 981-0060. FAX (908) 981-9667. 1965 to present. Bi-monthly. $240.00 per year.

IIE Transactions: Industrial Engineering Research and Development. Institute of Industrial Engineers, 25 Technology Park-Atlanta, Norcross, GA 30092. (404) 449-0460. FAX (404) 263-8532. 1969 to present. Quarterly. $95.00 per year for non-members.

Journal of Engineering For Industry. American Society of Mechanical Engineers, 345 E. 47th Street, New York, NY 10017. (212) 705-7722. 1970 to present. Quarterly. $100.00 per year.

Journal of Quality Technology. American Society for Quality Control, 611 E Wisconsin Ave., PO Box 3005, Milwaukee, WI 53201-3005. (414) 272-8575 or (800) 952-6587. 1944 to present. Quarterly. $30.00.

Journal of Operations Management. Elsevier Science Inc., Box 882, Madison Square Station, New York, NY 10159. (212) 989-5800. 1980 to present. 4 times a year. $157.00 per year.

Manufacturing Engineering. Society of Manufacturing Engineers, One SME Drive, Box 930, Dearborn, MI 48128-1490. (313) 271-1500 or FAX (313) 271-2861. 1932 to present. Monthly. $60.00 per year.

Material Handling Engineering. Penton Publishing, 110 Superior Ave., Cleveland, OH 44114-2543. 1945 to present. Monthly. $45.00 per year.

Materials Engineering. Penton Publishing, 110 Superior Ave., Cleveland, OH 44114-2543. 1929 to present. Monthly. $50.00 per year.

Production. Gardner Publications, 6600 Clough Pike, Cincinnati, OH 45244-4090. (513) 231-8020. FAX (513) 231-2818. 1934 to present. Monthly. $48.00 per year.

Production and Inventory Management. American Production and Inventory Control Society, 500 W. Annandale Road, Falls Church, VA 22046-4274. (703) 237-8344. 1959 to present. Quarterly. $110.00 per year.

Quality. Hitchcock Publishing, 191 S. Gary Avenue, Carol Stream, IL 60188. (708) 462-4641. FAX (708) 462-2205. 1962 to present. Monthly. Inquire for cost.

Quality Assurance. Academic Press Inc., Journals Division, 1250 Sixth Avenue, San Diego, CA 92101-4311. (619) 231-0926. FAX (619) 699-6715. 1991 to present. Quarterly. $167.00 per year.

Quality Progress. American Society for Quality Control, ASQC Quality Press, 611 E Wisconsin Ave., PO Box 3005, Milwaukee, WI 53201-3005. (414) 272-8575 or (800) 952-6587. 1944 to present. Monthly. $50.00 per year.

RESEARCH CENTERS AND INSTITUTES

Advanced Manufacturing Center. Cleveland State University, 1751 E. 23rd Street, Cleveland, OH 44114. (216) 687-4643. FAX (216) 687-2999.

Center For Applied Engineering. University of Missouri—Rolla, 206 Harris Hall, Rolla, MO 65401. (314) 341-4559.

Center For Manufacturing Systems Engineering. Lehigh University, H.S. Mohler Laboratory 200, Bethlehem, PA 18015. (215) 758-4667. FAX (215) 694-0542.

Center For Quality and Productivity Improvement. University of Wisconsin—Madison, 610 N. Walnut Street, Madison, WI 53705. (608) 263-2520. (608) 263-1425.

Engineering and Industrial Experiment Station. University of Florida, 300 Weil Hall, Gainesville, FL 32611-2083. (904) 392-6000. (904) 392-9673.

QUANTUM CHEMISTRY

See also: CATALYSIS, CHEMICAL BONDS, CHEMISTRY, ELECTROCHEMISTRY, PHYSICAL CHEMISTRY

ABSTRACT SERVICES AND INDEXES

Applied Science and Technology Index. A Cumulative Subject Index To English Language Periodicals in the Fields of Aeronautics and Space Science, Computer Technology, Chemistry, Construction Industry, Energy and Related Areas. H. W. Wilson Co., 950 University Avenue, Bronx, NY 10452-9978. (212) 588-8400. (800) 367-6770. FAX (718) 590-1617. From 1958 to present. Monthly. Inquire about cost and availability. Also available online (BRS and WILSONLINE) and on CD-ROM. ISSN: 0003-6986.

Chemical Abstracts. Chemical Abstracts Service. 2540 Olentangy River Road, Box 3012, Columbus, OH 43210-0012. (614) 447-3600. FAX (614) 447-3713. 1907 to present. Weekly. $16,800.00 per year. Available online and on CD-ROM. CA is also available in five section groupings. Inquire regarding cost and availability.

Current Contents: Physical, Chemical and Earth Sciences. Institute for Scientific Information, 3501 Market Street, Philadelphia, PA 19104. (215) 386-0100. FAX (215) 386-2291. From 1961 to present. Weekly. $442.00 per year. Also available on CD-ROM and online. Inquire regarding cost and availability. ISSN: 0163-2574.

Engineering Index Monthly. Engineering Information, Inc., Castle Point on the Hudson, Hoboken, NJ 07030. (201) 216-8500. (800) 221-1044. FAX(201) 216-8532. Monthly. $2300.00 per year for monthly issues. $1980 per year for annual cumulation. Available on CD-ROM and online as COMPENDEX. ISSN: 0742-1974; 0360-8557.

General Science Index. H.W. Wilson Company, 950 University Avenue, Bronx, NY 10452. (212) 588-8400. (800) 367-6770. FAX (718) 590-1617. From 1978 to present. Ten issues per year; quarterly and annual cumulations. Service basis. Available on CD-ROM and online. Inquire regarding cost and availability. ISSN: 0162-1963.

Physics Abstracts. INSPEC. Section A, Science Abstracts. Institution of Electrical Engineers (IEE), London, United Kingdom. Available from: INSPEC/IEEE -Institute of Electrical and Electronic Engineers, Box 1331, 445 Hoes Lane, Piscataway, NJ 08855-1331. (908) 562-5549. 1898 to present. 24 issues per year. $2835.00 per year. ISSN: 0036-8091. Also available online and on CD-ROM.

Science Citation Index. SCI. Institute for Scientific Information, 3501 Market Street, Philadelphia, PA 19104. (215) 386-0100. (800) 523-1850. FAX (215) 386-2991. 1961 to present. Six issues per year, plus annual cumulation. $11650.00 per year. Also available online and on CD-ROM. Inquire about price and availability. ISSN: 0036-827X.

ANNUAL REVIEWS AND YEARBOOKS

Annual Review of Physical Chemistry. Annual Reviews, Inc., 4139 El Camino Way, Palo Alto, CA 94306-0897. (415) 493-4400. Fax (415) 855-9815. 1951 to present. Annual. $55.00 per year.

ASSOCIATIONS AND PROFESSIONAL SOCIETIES

American Chemical Society. 1155 16th Street, Nw, Washington, DC 20036. (202) 872-4600.

American Institute of Chemists. 501 Wythe Street, Alexandria, VA 22314-1917. (703) 836-2090. FAX (703) 836-2091.

Association of Consulting Chemists and Chemical Engineers, 50 East 41st Street, Suite 92, New York, NY 10017. (212) 684-6255.

Division of Physical Chemistry (a Division of the American Chemical Society). Iowa State University, Department of Chemistry, Wilhelm Hall, Room 303, Ames, IA 50011-4111. (515) 294-4111.

Electrochemical Society. 10 South Main Street, Pennington, NJ 08534. (609) 737-1902.

BIBLIOGRAPHIES

Chemical Information. H. R. Collier, editor. Springer- Verlag New York, Inc., 175 Fifth Avenue, New York, NY 10010. (212) 460-1500. (800) 777-4643. 1990. $91.00.

Chemical Information Sources. Gary Wiggins. McGraw-Hill Publishing Company, 1221 Avenue of the Americas, New York, NY 10020. (800) 262-4729 or (212) 512-3825. 1991. $42.50.

Information Sources in Chemistry. R.T. Bottle and J.F.B Rowland, editors. R.R. Bowker Inc., 121 Chanlon Road, New Providence, NJ 07974. (800) 521-8110 or (908) 464-6800. Fourth edition. 1993. $75.00.

Scientific and Technical Books and Serials in Print. An Index To Literature in Science and Technology. R.R. Bowker Inc., 121 Chanlon Road, New Providence, NJ 07974. (908) 464-6800. (800) 521-8110. FAX (908) 665-3502. 1972 to present. Annual. 4 volumes. 1994. $299.95. Also available on compact disc and online. ISSN: 0000-054X.

DIRECTORIES AND BIOGRAPHICAL SOURCES

American Institute of Chemists - Professional Directory. American Institute of Chemists, 501 Wythe Street, Alexandria, VA 22314-1917. (703) 836-2090. FAX (703) 836-2091. Annual. $65.00.

American Men and Women of Science. Physical and Biological Sciences. R.R. Bowker Inc., 121 Chanlon Road, New Providence, NJ 07974. (908) 464-6800. (800) 521-8110. 20th edition. 8 volumes. 1996. $850.00

Chemical Sources International. Mike Desing and Kurt Gandenberger, editors. Chemical Sources International, PO Box 1824, Clemson, SC 29633. (803) 646-7840. 1994.$285.00.

Chemical Week — Buyers Guide Issue. Chemical Week Associates, 888 Seventh Avenue, New York, NY 10106. (212) 621-4900. FAX (212) 621-4949. Annual, October. $50.00.

Directory of Chemistry Software 1992. Wendy Warr, Peter Willett, Geoff Downs. American Chemical Society, 1155 16th Street, NW, Washington, DC 20036. (202) 872-4600. 1992. $35.95.

Research Centers Directory. Gale Research Company,Inc., 835 Penobscot Building, Detroit, MI 48226-4094. (313) 961-2242. (800) 877-4253. 20th edition. 1995. $485.00. ISSN: 0080-1518.

ENCYCLOPEDIAS AND DICTIONARIES

Academic Press Dictionary of Science and Technology. Christopher Morris, editor. Academic Press, Inc., 1250 Sixth Avenue, San Diego, CA 92101. (619) 231-0926. FAX (619) 699-6715. 1991. $115.00.

Concise Encyclopedia Chemistry. Translated and revised by Mary Eagleson. Walter de Gruyter, Inc., 200 Saw Mill River Road, Hawthorne, New York, 10532. (914) 747-0110 or Fax (914) 747-1326. 1994. $69.95.

Dictionary of Chemical Names and Synomyms. Philip A. Howard and Michael W. Neil. Lewis Pub.,/ CRC Press, Inc., 2000 Corporate Boulevard, NW, Boca Raton, FL 33431. (407) 994- 0555. (800) 272-7737. 1991. $149.95.

Dictionary of Physical Chemistry. Satish Anand and Raj Kumar. South Asia Publications, PO Box 502 Collumbus, MO 65205. (314) 474-0116. 1990. $23.50.

Encyclopedia of Applied Physics. VCH Publishers, Inc., 303 Northwest 12th Avenue, Deerfield Beach, FL 33442. (800) 367-8249. 1991-. Twenty volumes. $6000.00.

Encyclopedia of Chemical Processing and Design. McKetta. Dekker, Inc., 270 Madison Avenue, New York, NY 10016. (212) 696-9000. (800) 228-1160. 1976 - . $175.00 per volume. ISSN:

Grant & Hawks Chemical Dictionary. R. A. Grant. McGraw-Hill Publishing Company, Inc., 1221 Avenue of the Americas, New York, NY 10020. (212) 512-2000. (800) 262-4729. 5th edition. 1987. $64.50.

Hawley's Condensed Chemical Dictionary. Richard J. Lewis. Van Nostrand Reinhold, 115 Fifth Avenue, New York, NY 10003. (212) 254-3232. (800) 842-3636. 12th edition. 1993. $69.95.

Kirk-Othmer Encyclopedia of Chemical Technology. John Wiley and Sons, Inc., 605 Third Avenue, New York, NY 10158. (800) 526-5368 or (212) 850-6000. Fourth edition. 1991-. Twenty-seven volumes. $5400.00.

McGraw-Hill Encyclopedia of Chemistry. Sybil P.Parker, editor. McGraw-Hill Book, Incorporated, 1221 Avenue of the Americas, New York, NY 10020. (212) 997-3675. Second edition. 1993. $95.50.

McGraw-Hill Encyclopedia of Science and Technology. McGraw-Hill Book, Incorporated, 1221 Avenue of the Americas, New York, NY 10020. (212) 997-3675. Seventeenth edition. Twenty volumes. 1992. $1900.00.

GENERAL WORKS

Algebraic Methods in Quantum Chemistry and Physics. Francisco M. Fernandez and Eduardo A. Castro. CRC Press, Inc., 2000 Corporate Boulevard, NW, Boca Raton, FL 33431. (407) 994-0555. (800) 272-7737. 1995. $89.95.

Basic Principles and Techniques of Molecular Quantum Mechanics. R. E. Christoffersen. Springer-Verlag New York,Inc., 175 Fifth Avenue, New York, NY 10010. (212) 460-1500. (800) 777-4643. FAX (212) 473-6272. 1989. $89.00.

The Chemical Bond: Structure and Dynamics. Ahmed Zewail. Academic Press, Inc., 1250 Sixth Avenue, San Diego, CA 92101-4311. (619) 231-0926. FAX (619) 699-6715. 1992. $49.95

Electronic Structure of Materials. Adrian Sutton. Oxford University Press, Inc., 200 Madison Avenue, New York, NY 10016. (212) 725-6000. (800) 334-4249. 1993. $26.50 in paper.

The Elements of Physical Chemistry. P. W. Atkins. W. H. Freeman, & Co., 41 Madison Avenue, East 26th Avenue, 35th Floor, New York, NY 10010. (212) 576-0400. 1995. $42.95.

Introduction to Quantum Chemistry. Clifford E. Dykstra. Prentice-Hall, 113 Sylvan Avenue, Route 9W, Englewood Cliffs, NJ 07632. (201) 592-2000. (800) 922-0579. 1993. $27.60.

The Meaning of Quantum Chemistry: A Guide For Students of Chemistry and Physics. J. Baggot. Oxford University Press, Inc., 200 Madison Avenue, New York, NY 10016. (212) 725- 6000. (800) 334-4249. 1992. $24.95.

Physical Methods For Chemists. Russell S. Drago. SCP: Third World Literature Publishing House, P.O. Box 482, Lithonia, GA 30058-0482. (404) 785-7725. 1992. $63.00.

Problems and Solutions in Quantum Chemistry and Physics. Charles S. Johnson Jr,. and Leo G. Pedersen. Dover Publications, Inc., 180 Varick Street, New York, NY 10014. (212) 255-3755. (800) 223-3130. 1986. $11.95.

Quantum Chemistry of Organic Compounds: Mechanisms and Reactions. V. L. Minkin. Springer-Verlag New York, Inc., 175 Fifth Avenue, New York, NY 10010. (212) 460-1500. (800) 777-4643. FAX (212) 473-6272. 1990. $119.00.

Quantum Mechanics in Chemistry. George C. Schatz, and Mark A. Ratner. Prentice-Hall, 113 Sylvan Avenue, Route 9W, Englewood Cliffs, NJ 07632. (201) 592-2000. (800) 922- 0579. 1993. $74.00.

Quantum States and Processes. George H. Duffey. Prentice-Hall, 113 Sylvan Avenue, Route 9W, Englewood Cliffs, NJ 07632. (201) 592-2000. (800) 922-0579. 1991. $94.00.

HANDBOOKS AND MANUALS

Chemical Formulary. H. Bennett, editor. Chemical Publishing Company, Inc., 80 Eighth Avenue, New York, NY 10011. (212) 255-1950. Volumes 1-30. $70.00 per volume.

Chemometrics: Chemical and Sensory Data. David R. Burgard and James T. Kuznicki. CRC Publishers, Inc., 2000 Corporate Blvd., N.W., Boca Raton, FL 33431. (407) 994-0555 or (800) 333-8300. 1990. $120.00.

Chemical Reference Handbook. Gordon Press, PO Box 459, Bowling Green Station, New York, NY 10004. (718) 624-8419. 1995. $260.00.

CRC Handbook of Chemistry and Physics. David R. Kide, editor. CRC Press Inc., 2000 Corporate Blvd., NW, Boca Raton, FL 33431. (407) 994-0555. (800) 333-8300. 77th edition. 1996. $99.95.

Guide To Basic Chemical Compounds. D.R. Lide, Jr. CRC Publishers, Inc., 2000 Corporate Blvd., N.W., Boca Raton, FL 33431. (407) 994-0555. (800) 333-8300. 1993. $120.00.

Improving Safety in the Chemical Laboratory: A Practical Guide. Jay A. Young, editor. John Wiley and Sons, Inc., 605 Third Avenue, New York, NY 10158. (800) 526-5368 or (212) 850-6000. Second edition. 1991. $75.00.

Laboratory Handbook of Materials, Equipment, and Techniques. Gary S. Coyne. Prentice Hall (Division of Simon and Schuster), 15 Columbus Circle, New York, NY 10023. (212) 373-8500 or (800) 922-0579. 1992. $45.00.

Lange's Handbook of Chemistry. John A. Dean, editor. McGraw-Hill Publishing Company, 1221 Avenue of the Americas, New York, NY 10020. (800) 262-4729 or (212) 512-3825. Fourteenth edition. 1992. $79.50.

Riegel's Handbook of Industrial Chemistry. James A. Kent, editor. Van Nostrand Reinhold, 115 Fifth Avenue, New York, NY 10003. (212) 254-3232 or (800) 926-2665. Ninth edition. 1992. $114.95.

ONLINE DATABASES AND CD-ROMS

Analytical Abstracts Online. Royal Society of Chemistry, Information Services, Thomas Graham House, Science Park, Milton Road, Cambridge, CB4 4WF, England. Contains citations, mostly with abstracts, of the worldwide literature on analytical chemistry, from 1980 to present. Available on DIALOG, (800) 334-2564, and STN International, FIZ Karlsruhe, P.O. Box 2465, W-7500, Karlsruhe 1, Germany, online services. Inquire as to cost and availability.

CA Search. Chemical Abstracts Service, P.O. Box 3012, Columbus, OH 43210-0012. (614) 447-3600. (800) 848-6533. FAX (614) 447-3709. Very comprehensive guide to worldwide chemical literature and related fields, 1972 to present. Available on BRS,(800) 289-4277, DIALOG, (800) 334-2564, ORBIT (800) 456-7248, and STN International, FIZ Karlsruhe, P.O. Box 2465, W-7500, Karlsruhe 1, Germany, online services. Inquire as to cost and availability.

Chemical Journals of the American Chemical Society. American Chemical Society, 1155 16th Street, N.W., Washington, DC 20036. (202) 872-4381 or (800) 424-6747. Contains complete text of approximately 90,000 articles from 22 primary journals published by the American Chemical Society, from mostly 1982 to present. Available on STN International, FIZ Karlsruhe, P.O. Box 2465, W-7500, Karlsruhe 1, Germany, online service. Inquire as to cost and availability.

Compendex Plus. Engineering Information, Inc., 345 East 47th Street, New York, NY 10017. (212) 705-7600 or (800) 221-1044. Contains citations with abstracts to worldwide literature in engineering and technology, from 1970 to present. Available on online BRS,(800) 289-4277, DIALOG, (800) 334-2564, ORBIT (800) 456-7248, and STN International, FIZ Karlsruhe,

P.O. Box 2465, W-7500, Karlsruhe 1, Germany, online services. Also available on CD-ROM. Inquire as to cost and availability.

Current Contents Search. Institute for Scientific Information, 3501 Market Street, Philadelphia, PA 19104. (215) 386-0100. FAX (215) 386-6362. Contains citations to articles listed in the table of contents of science and technology journals. Also articles in social sciences and life sciences journals. Available on BRS,(800) 289-4277, DIALOG,(800) 334-2564, online services. Inquire as to cost and availability.

Dissertation Abstracts. University Microfilms International, 300 North Zeeb Road, Ann Arbor, MI 48106. (800) 521-0600 or (313) 761-4700. Scope includes virtually all doctoral dissertations accepted at accredited American institutions from 1861 to present in 252 subject areas. Available on BRS,(800) 289-4277, DIALOG,(800) 334-2564, and OCLC EPIC,(800) 848-5878, online services. Also available on CD-ROM. Inquire as to cost and availability.

Gmelin Database. Gmelin-Institut fur Anorganische Chemie und Grenzgebiete, Varrentrapperstrasse, 40-42, Carl-Bosch-Haus, D-6000, Frankfurt am Main 90, Germany. Contains structural and factual data relating to inorganic and organometallic chemistry. Provides data from the Gmelin Handbook of Inorganic and Organometallic Chemistry. Covers the period 1817 to 1975; 1988-89. Available on STN International, FIZ Karlsruhe, P.O. Box 2465, W-7500, Karlsruhe 1, Germany, online service. Inquire as to cost and availability.

Kirk-Othmer Encyclopedia of Chemical Technology. John Wiley and Sons, Inc., 605 Third Avenue, New York, NY 10158. (800) 526-5368 or (212) 850-6000. Contains the complete text of all chapters in the 27 volume fourth edition of the *Kirk-othmer Encyclopedia of Chemical Technology.* 1991. Available on BRS,(800) 289-4277, DIALOG,(800) 334-2564, online services. Inquire as to cost and availability.

NTIS Bibliographic Database. National Technical Information Service, 5285 Port Royal Road, Springfield, VA 22161. (703) 487-4929 or FAX (703) 321-8199. Broad coverage of government-sponsored science and technology research reports, 1964 to present. Available on BRS,(800) 289-4277, DIALOG, (800) 334-2564, ORBIT, (800) 456-7248, and STN International, FIZ Karlsruhe, P.O. Box 2465, W-7500, Karlsruhe 1, Germany, online services. Also available on CD-ROM. Inquire as to cost and availability.

SCISEARCH. Institute for Scientific Information, 3501 Market Street, Philadelphia, PA 19104. (800) 523-1850 or (215) 386-0100. Broad multidisciplinary title and author index to the international literature of science and technology, 1974 to present. Available on DIALOG,(800) 334-2564, and ORBIT,(800) 456-7248, online services. Also available on CD-ROM. Inquire as to cost and availability.

WILSONLINE. H.W. Wilson Company, 950 University Avenue, Bronx, NY 10452. (800) 367-6770 or (212) 588-8400. Makes available online versions of the printed H.W. Wilson indexes including Applied Science and Technology Index, Business Periodicals Index, General Science Index, and Readers' Guide to Periodical Literature. Period covered is generally 1983 to present. Available on BRS,(800) 289-4277, DIALOG,(800) 334-2564, and OCLC EPIC,(800) 848-5878, online services. Also available on CD-ROM. Inquire as to cost and availability.

PERIODICALS

Angewandte Chemie. VCH Publishers, Inc., 220 East 23rd Street, New York, NY 10010-4606. (212) 683-8333. (800) 367-8249. 1888 to present. 22 issues per year. $840.00 per year. ISSN: 0044-8249.

Chemical Physics. Elsevier Science Publishing Co., Inc., Box 882, Madison Square Station, New York, NY 10159. (212) 989-5800. FAX (212) 633-3990. 1973 to present. 33 issues per year in 11 volumes. $2759.00 per year. ISSN: 0301-0104.

Chemical Physics Letters. Elsevier Science Inc., Publishing Co., Inc., Box 882, Madison Square Station, New York, NY 10159. (212) 989-5800. (212) 633-3990. 1967 to present. 102 issues per year in 17 volumes. $5063.00 per year. ISSN: 0009-2614.

Chemical Reviews. American Chemical Society, Box 3337, Columbus, OH 43210. (614) 447-3776. 1924 to present.

8 issues per year. $346.00 per year. ISSN: 0009-2665.

Chemical Week; Includes Annual Buyers Guide. Chemical Week Associates, 888 Seventh Avenue, New York, NY 10106. (212) 621-4900. FAX (212) 621-4949. 1914 to present. Weekly. $99.00 per year. ISSN: 0009-272X.

Chemtech. American Chemical Society. Box 3337, Columbus, OH 43210. (614) 447-3776. 1970 to present. Monthly. $370.00 per year. ISSN: 0009-2703.

Inorganic Chemistry. American Chemical Society, Box 3337, Columbus, OH 43210. (614) 447-3776. 1962 to present. Semi-monthly. $500.00 per year. ISSN: 0020-1669.

International Journal of Chemical Kinetics. John Wiley and Sons,Inc., 605 Third Avenue, New York, NY 10158. (800) 526-5368 or (212) 850-6000. 1968 to present. Monthly. $730.00 per year. ISSN: 0538- 8066.

Journal of Catalysis. Academic Press, Inc., Box 620000, Orlando, FL 32891-8340. 1962 to present. Monthly. $1296.00 per year. ISSN: 0021- 9517.

Journal of Physical Chemistry. American Chemical Society, Box 3337, Columbus, OH 43210. (614) 447-3776. (800) 333- 9511. FAX (614) 447-3671. 1896 to present. Weekly. $1140.00 per year to non-members. ISSN: 0894-3230.

Journal of Solid State Chemistry. Academic Press, Inc., Box 620000, Orlando, FL 32891-8340. 1969 to present. Monthly. $1248.00 per year. ISSN: 0022-4596.

Journal of the American Chemical Society. American Chemical Society, Box 3337, Columbus, OH 43210. (614) 447-3776. 1879 topresent. Biweekly. $1055.00 per year. ISSN: 0002-7863.

RESEARCH CENTERS AND INSTITUTES

Chemical Laboratories. Harvard University, Oxford Street, Cambridge, MA 02138. (617) 495-4283. FAX (617) 496-5618.

Institute For theoretical Chemistry. University of Texas at Austin, Department of Chemistry, WEL 3.237, Austin, TX 78712. (512) 471-3114.

Laboratory For Research On the Structure of Matter. University of Pennsylvania, 3231 Walnut Street, Philadelphia, PA 19104. (215) 898-8571.

Materials Science Center. Cornell University, Clark Hall of Science, Ithaca, NY 14853-2501. (607) 255-4272. FAX (607) 255-6428.

Research Program in Chemistry and Biochemistry. Southern Illinois University at Carbondale, Carbondale, IL 62901. (618) 453-5721.

QUANTUM ELECTRODYNAMICS

See: QUANTUM MECHANICS

QUANTUM MECHANICS

See also: MECHANICS, PHYSICS, STATISTICAL MECHANICS

ABSTRACT SERVICES AND INDEXES

Applied Mechanics Reviews: An Assessment of World Literature in Engineering Sciences. American Society of Mechanical Engineers, 345 East 47th Street, New York, NY 10017. (212) 705-7703. 1948 to present. Monthly. $360.00 per year.ISSN: 0003-6900.

Applied Science and Technology Index; A Cumulative Subject Index To English Language Periodicals in the Fields of Aeronautics and Space Science, Computer Technology, Chemistry, Construction Industry, Energy and Related Areas. H.W. Wilson Co., 950 University Avenue, Bronx, NY 10452. (212) 588-8400. (800) 367-6770. FAX (718) 590-1617. From 1958 to present. Monthly. Inquire about cost and availability. Also available on CD-ROM and online. ISSN: 0003-6986.

Chemical Abstracts. Chemical Abstracts Service, 2540 Olentangy River Road, Box 3012, Columbus, OH 43210-0012. (614) 447-3600. (800) 848-6538. FAX (614) 447-3713. From 1907 to present. Weekly. $16800.00 per year. Also available on CD-ROM and online. Inquire regarding cost and availability.

Current Contents: Physical, Chemical and Earth Sciences. Institute for Scientific Information, 3501 Market Street, Philadelphia, PA 19104. (215) 386-0100. FAX (215) 386-2291. From 1961 to present. Weekly. $442.00 per year. Also available on CD-ROM and online. Inquire regarding cost and availability. ISSN: 0163-2574.

Current Papers in Physics. Institute of Electrical Engineers, Michael Faraday House, Six Hill Way, Stevenage, Herts, SG1 2AY, England. Distributed by INSPEC/IEEE, Box 1331, 445 Hoes Lane, Piscataway, NJ 08855-1331. (908) 562-5549. 1966 to present. Fortnightly. $410.00 per year. ISSN: 0011-3786.

Engineering Index Monthly. Engineering Information, Inc., Castle Point on the Hudson, Hoboken, NJ 07030. (201) 216-8500. FAX (201) 216-8532. Monthly. $2200.00 per year. Also available online as COMPENDEX and on CD-ROM. ISSN: 0742-1974.

General Science Index. H.W. Wilson Company, 950 University Avenue, Bronx, NY 10452. (212) 588-8400. (800) 367-6770. FAX (718) 590-1617. From 1978 to present. Ten issues per year; quarterly and annual cumulations. Service basis. Available on CD-ROM and online. Inquire regarding cost and availability. ISSN: 0162-1963.

Government Reports Announcements and Index. U. S. National Technical Information Service (NTIS), 5285 Port Royal Road, Springfield, VA 22161. (703) 487-4650. FAX (703) 321-8547. From 1968 to present. Annual. $630.00 per year. Also available online as *NTIS Bibliographic Database* and on CD- ROM.

ISMEC: Mechanical Engineering Abstracts. Cambridge Scientific Abstracts, 7200 Wisconsin Avenue, Suite 601, Bethesda, MD 20814. (301) 961-6750 or (800) 843-7751. Contains citations to the literature in mechanical engineering, industrial and production engineering, energy, power, mechanics, devices and related areas, from 1973 to present. Available on the DIA-LOG,(800) 334-2564, online service. Inquire as to cost and availability.

NTIS Alerts: Physics. U. S. National Technical Information Service. 5285 Port Royal Road, Springfield, VA 22161. (703) 487-4650. FAX (703) 321-8547. Weekly. $140.00 per year.

Physics Abstracts. INSPEC. Section A, Science Abstracts. Institution of Electrical Engineers (IEE). Available from INSPEC/IEEE - Institute of Electrical and Electronic Engineers,, Box 1331, 445 Hoes Lane, Piscaway NJ 08855-1331.(908) 562-5549. 1898 to present. 24 issues per year. $2835.00 per year. Also available on CD-ROM and online. ISSN: 0036-8091.

Physics Briefs (Physikalische Berichte). Information Center for Energy, Physics, Mathematics; German Physical Society. V C H Publishers, Inc., 220 East 23rd Street, New York, NY 10010-4606. (212) 683-8333. 1845 to present. 24 issues per year. $2390.00 per year. Also available online. ISSN: 0179-7434.

Science Citation Index. SCI. Institute for Scientific Information, 3501 Market Street, Philadelphia, PA 19104. (215) 386-0100. (800) 523-1850. FAX (215) 386-2991. 1961 to present. Six issues per year, plus annual cumulation. $11650.00 per year. Also available online and on CD-ROM. Inquire about price and availability. ISSN: 0036-827X.

ASSOCIATIONS AND PROFESSIONAL SOCIETIES

American Chemical Society, 1155 16th Street Nw, Washington DC 20036. (202) 872-4600. (800) 424-6747.

American Institute of Physics. One Physics Ellipse, College Park, MD 20740-3843. (301) 209-3100.

American Physical Society. One Physics Ellipse, College Park, MD 20740-3843. (301) 209-3200. FAX (301) 209-0865.

American Society of Mechanical Engineers, 345 East 47th Street, New York, NY 10017. (212) 705-7722. FAX (212) 705- 7739.

DIRECTORIES AND BIOGRAPHICAL SOURCES

American Men and Women of Science: Physical and Biological Sciences. R. R. Bowker Inc., 121 Chanlon Road, New Providence, NJ 07974. (908) 464-6800. (800) 521-8110. 20th edition. 8 volumes. 1996. $850.00.

American Physical Society. Membership Directory Bulletin Issue. American Physical Society, One Physics Ellipse, College Park, MD 20740-3843. (301) 209-3200. FAX (301) 209-0865. Biennial. $50.00.

The Biographical Dictionary of Scientists: Physicists. D. Abbott, editor. Peter Bedrick Books, Inc., 2112 Broadway, Room 318, New York, NY 10023. (212) 496-0751. 1984.

Directory of Physics and Astronomy Staff Members. American Institute of Physics, One Physics Ellipse, College Park, MD 20740-3843. (301) 209-3100. Annual. $45.00.

Graduate Programs in Physics, Astronomy and Related Fields. 1995-1996. American Institute of Physics, One Physics Ellipse, College Park, MD 20740-3843. (301) 209-3100. Annual. $45.00. ISSN: 0147-1821.

International Directory of Engineering Societies and Related Organizations. American Society for Engineering Education, 1818 N Street NW, Suite 600, Washington, DC 20036. (202) 331-3526. 15th edition. 1996. $185.00.

Research Centers Directory. Gale Research Inc., 835 Penobscot Building, Detroit, MI 48226-4094. (313) 961-2242. (800) 877-4253. 20th edition. 1995. $485.00. ISSN: 0080- 1518.

Scientific and Technical Organizations and Agencies Directory. Gale Research, 835 Penobscot Building, Detroit, MI 48226-4094. (313) 961-2242. (800) 877-4253. 4th edition. 1996. $195.00.

Who's Who in Engineering. Gordon Davis, editor. American Society for Engineering Education, 1111 19th Street, NW, Suite 608, Washington, DC 20036. 9th edition. 1995. $220.00.

ENCYCLOPEDIAS AND DICTIONARIES

A Concise Dictionary of Physics. Oxford University Press, Inc., 200 Madison Avenue, New York, NY 10016. (212) 725- 6000. (800) 334-4249. 1990. $10.95.

Dictionary of the Physical Sciences: Terms, Formulas, Data. Cesare Emiliani. Oxford University Press, Inc., 200 Madison Avenue, New York, NY 10016. (212) 725-6000. (800) 334-4249.1989. $19.95.

Encyclopedia of Applied Physics. George Trigg, editor. VCH Publications, Inc., 220 East 23rd Street, Suite 909, New York, NY 10010-4606. (212) 683-8333. (800) 422-8824. 20 volume set. 1991-. $5990.00.

Encyclopedia of Physics. Robert M. Besancon. Chapman & Hall, 1 Penn Plaza, New York, NY 10119. (212) 564-1060. 3rd edition. 1990. $54.95 in paper.

Encyclopedia of Physical Science and Technology. Academic Press, Inc., 6277 Sea Harbor Drive, Orlando, FL. (800) 321-5068. 2nd edition. 18 volume set. 1992. $2625.00.

McGraw-Hill Encyclopedia of Science and Technology. McGraw-Hill Book Company, Inc., 1221 Avenue of the Americas, New York, NY 10020. (212) 512-2000. (800) 262-4729. 7th edition. 20 volume set. 1992. $1900.00.

Penguin Dictionary of Physics. Valerie Illingworth, editor. Viking Penguin, 375 Hudson Street, New York, NY 10014-3657. (212) 366-2000. (800) 331-4624 2nd edition. 1991.

GENERAL WORKS

Classical Topology and Quantum States. A. P. Balachandran, et al. World Scientific Publishing Company, Inc., 1060 Main Street, River Edge, NJ 07661. (201) 487-9655. (800) 227-7562. 1991. $36.00 in paper.

The Ghost in the Atom. A discussion of the mysteries of quantum physics. P. C. Davies and J. Brown, editors. Cambridge University Press, 40 West 20th Street, New York, NY 10011-4211. (212) 924-3900. (800) 872-7423. 1993. $9.95 in paper.

Introduction To Quantum Mechanics. H. Smith. World Scientific Publishing Company, Inc., 1060 Main Street, River Edge, NJ 07661. (201) 487-9655. (800) 227-7562. 1991. $50.00 in paper.

Men Who Made A New Physics: Physicists and the Quantum theory. Barbara L. Cline. University of Chicago Press, 5801 Ellis Avenue, Chicago, IL 60637. (312) 702-7700. 1987. $12.95 in paper.

Mind, Matter and Quantum Mechanics. Henry P. Stapp. Springer-Verlag New York, Inc., 175 Fifth Avenue, New York, NY 10010. (212) 460-1500. (800) 777-4643. FAX (212) 473-6272. 1994. $34.50.

Quantum Mechanics and Experience. David Z. Albert. Harvard University Press, 79 Garden Street, Cambridge MA. (617) 495-2600. 1994. $14.95 in paper.

Quantum theory of the Solid State. Joseph Calloway. Academic Press, Inc., 6277 Sea Harbor Drive, Orlando, FL. (800) 321-5068. 1991. $69.95.

The Secret Life of Quanta. M. Y. Han. McGraw-Hill Publishing Company, Inc., 1221 Avenue of the Americas, New York, NY 10020. (212) 512-2000. (800) 262-4729. 1992. $17.95.

Symbols, Pictures and Quantum Reality: On the Foundations of the Physical Universe. W. Schommers. World Scientific Publishing Company, Inc., 1060 Main Street, River Edge, NJ 07661. (201) 487-9655. (800) 227-7562. 1995. $55.00

HANDBOOKS AND MANUALS

Chemical and Physical Data. Arthur M. James and M. P. Lord, Van Nostrand Reinhold, 115 Fifth Avenue, New York, NY 10003. (212) 254-3232. (800) 842-3636. 1992. $59.95.

CRC Handbook of Chemistry and Physics. David R. Kide, editor. CRC Press, Inc., 2000 Corporate Boulevard, NW, Boca Raton, FL 33431. (407) 994-0555. (800) 272-7737. 77th edition. 1996. $99.95.

Handbook of Physical Quantities. Igor S. Grigoriev and Evgenil Z. Melikhov, editors. CRC Press, Inc., 2000 Corporate Boulevard, NW, Boca Raton, FL 33431. (407) 994-0555. (800) 272-7737. 1995. $99.00.

Physics Problem Solver. Research and Education Association Staff. Research & Education Association, 61 Ethel Road, West Piscataway, NJ 08854. (908) 819-8880. Revised edition. 1994. $23.95 in paper.

Tables of Physical and Chemical Constants and Some Mathematical Functions. G. W. Kaye and T. M. Laby. Halsted Press 605 Third Avenue, New York, 10158-0012. (212) 850-6400. 15th edition. 1986. $74.95.

ONLINE DATABASES AND CD-ROMS

CA Search. Chemical Abstracts Service, P.O. Box 3012, Columbus, OH 43210-0012. (614) 447-3600. (800) 848-6533. FAX (614) 447-3709. Very comprehensive guide to worldwide chemical literature and related fields, 1972 to present. Available on BRS,(800) 289-4277, DIALOG, (800) 334-2564, ORBIT (800) 456-7248, and STN International, FIZ Karlsruhe, P.O. Box 2465, W-7500, Karlsruhe 1, Germany, online services. Inquire as to cost and availability.

Current Contents Search. Institute for Scientific Information, 3501 Market Street, Philadelphia, PA 19104. (215) 386-0100. FAX (215) 386-6362. Contains citations to articles listed in the table of contents of science and technology journals. Also articles in social sciences and life sciences journals. Available on BRS, (800) 289-4277, DIALOG, (800) 334-2564, online services. Inquire as to cost and availability.

Compendex Plus. Engineering Information, Inc., 345 East 47th Street, New York, NY 10017. (212) 705-7600 or (800) 221-1044. Contains citations with abstracts to worldwide literature in engineering and technology, from 1970 to present. Available on online BRS,(800) 289-4277, DIALOG, (800) 334-2564, ORBIT (800) 456-7248, and STN International, FIZ Karlsruhe, P.O. Box 2465, W-7500, Karlsruhe 1, Germany, online services. Also available on CD-ROM. Inquire as to cost and availability.

Dissertation Abstracts Online. University Microfilms International, 300 North Zeeb Road, Ann Arbor, MI 48106. (800) 521-0600 or (313) 761-4700. Scope includes virtually all doctoral dissertations accepted at accredited American institutions from 1861 to present in 252 subject areas. Available on BRS, (800) 289-4277, DIALOG, (800) 334-2564,and OCLC EPIC, (800) 848-5878, online services. Also available on CD-ROM. Inquire as to cost and availability.

INSPEC. Institution of Electrical Engineers, Michael Faraday House, Six Hills Way, Stevenage, Herts. SG1 2AY, England. Telephone: 0438 313311 or FAX 0438 742840. Contains citations to the worldwide literature of physics, electronics and electrical engineering, computer technology, and related fields. Available on BRS, (800) 289-4277, DIALOG, (800) 334-2564, ORBIT, (800) 456-7248, and STN International, FIZ Karlsruhe, P.O. Box 2465, W-7500, Karlsruhe 1, Germany, online services. Inquire as to cost and availability.

NTIS Bibliographic Database. National Technical Information Service, 5285 Port Royal Road, Springfield, VA 22161. (703) 487-4929 or FAX (703) 321-8199. Broad coverage of government-sponsored science and technology research reports, 1964 to present. Available on BRS,(800) 289-4277, DIALOG, (800) 334-2564, ORBIT, (800) 456-7248, and STN International, FIZ Karlsruhe, P.O. Box 2465, W-7500, Karlsruhe 1, Germany, online services. Also available on CD-ROM. Inquire as to cost and availability.

Physics Briefs. American Institute of Physics, 335 East 45th Street, New York, NY 10017. (212) 661-9260 or FAX (212) 661-2036. Contains citations with abstracts of the literature of physics and related fields, 1979 to present. Available on the STN International, FIZ Karlsruhe, P.O. Box 2465, W-7500, Karlsruhe 1, Germany, online service. Inquire as to cost and availability.

SCISEARCH. Institute for Scientific Information, 3501 Market Street, Philadelphia, PA 19104. (800) 523-1850 or (215) 386-0100. Broad multidisciplinary title and author index to the international literature of science and technology, 1974 to present. Available on DIALOG, (800) 334-2564, and ORBIT, (800) 456-7248, online services. Also available on CD-ROM. Inquire as to cost and availability.

WILSONLINE. H.W. Wilson Company, 950 University Avenue, Bronx, NY 10452. (800) 367-6770 or (212) 588-8400. Makes available online versions of the printed H.W. Wilson indexes including Applied Science and Technology Index, Business Periodicals Index, General Science Index, and Readers' Guide to Periodical Literature. Period covered is generally 1983 to present. Available on BRS, (800) 289-4277, DIALOG, (800) 334-2564, and OCLC EPIC, (800) 848-5878, online services. Also available on CD-ROM. Inquire as to cost and availability.

PERIODICALS

American Journal of Physics. American Association of Physics Teachers, 5112 Berywn Road, College Park, MD 20740. (301) 345-4200. 1933 to present. Monthly. $215.00 per year. ISSN: 0002-9505.

Archive For Rational Mechanics and Analysis. Springer-Verlag New York, 44 Hartz Way, Secaucus, NJ 07096-2491. (201) 348-4033. 1957 to present. 16 issues per year in 4 volumes. $1238.00. ISSN: 0003-9527.

International Journal of Solids and Structures. Elsevier Science, 660 White Plains Road, Tarrytown, NY 10591-5153. (914) 524-9200. FAX (914) 333-2444. 1965 to present. 24 issues per year. $2035.00. ISSN: 0020-7683.

Journal of Applied Mechanics. American Society of Mechanical Engineers, 345 East 47th Street, New York, NY 10017. (212) 705-7722. FAX (212) 705-7739. 1935 to present. Quarterly. $120.00. ISSN: 0021-8936.

Journal of Engineering Mechanics. American Society of Civil Engineers, 345 East 45th Street, New York, NY 10017. (212) 682-7341. 1956 to present. Monthly. $284.00. ISSN: 0733-9399.

Journal of Mechanics and Physics of Solids. Elsevier Science, 660 White Plains Road, Tarrytown, NY 10591-5153. (914) 524-9200. FAX (914) 333-2444. 1952 to present. Monthly. $1335.00. ISSN: 0022-5096.

Quarterly Journal of Mechanics and Applied Mathematics. Oxford University Press, Inc., 2001 Evans Road, Cary, NC 27513. (919) 677-0977. 1948 to present. Quarterly. $214.00 per year. ISSN: 0033-5614.

Journal of Non-newtonian Fluid Mechanics. Elsevier Science, 660 White Plains Road, Tarrytown, NY 10591-5153. (914) 524-9200. FAX (914) 333-2444. 1976 to present. 15 issues per year. $1205.00. ISSN:. 0377-0257.

Journal of Rheology. Society of Rheology. Published by the American Institute of Physics, 500 Sunnyside Boulevard, Woodbury, NY 11797. 1957 to present. Bimonthly. $425.00. ISSN: 0148-6055.

Physical Review Letters. American Institute of Physics, One Physics Ellipse, College Park, MD 20740-3843. (301) 209-3000. 1958 to resent. Weekly. $1370. ISSN: 0031-9007.

RESEARCH CENTERS AND INSTITUTES

Center For Advanced Studies. University of New Mexico, Department of Physics and Astronomy, Alburquerque, NM 87131. (505) 277-2726.

Enrico Fermi Institute. University of Chicago, 5640 South Ellis Avenue, Chicago, IL 60637. (312) 702-7823. FAX (312) 702-5863.

Joint Institute For Laboratory Astrophysics, Quantum Physics Division. National Institute of Standards and Technology, University of Colorado, Boulder, CO 80309-0440. (303) 497-3527.

Laboratory For Experimental Mechanics Research. State University of New York at Stony Brook, Stony Brook, NY 11794-2300. (516) 632-8311. FAX (516) 632-8720.

Quantum Theory Project. University of Flolrida, 362 Williamson Halla, Gainsville, FL 32611. (904) 392-1597. FAX (904) 392-9822.

QUANTUM THEORY

See: QUANTUM MECHANICS

QUARKS

See also: CONDENSED MATTER, COSMIC RAYS, DARK MATTER, NUCLEAR PHYSICS, PARTICLE ACCELERATORS, PHYSICS, QUANTUM MECHANICS, QUANTUM THEORY, QUARKS,

ABSTRACT SERVICES AND INDEXES

Applied Science and Technology Index; a cumulative subject index to English language periodicals in the fields of aeronautics and space science, computer technology, chemistry, construction industry, energy and related areas. H.W. Wilson Co., 950 University Avenue, Bronx, NY 10452. (212) 588-8400. (800) 367-6770. FAX (718) 590-1617. From 1958 to present. Monthly. Inquire about cost and availability. Also available on CD-ROM and online. ISSN: 0003-6986.

Chemical Abstracts. Chemical Abstracts Service, 2540 Olentangy River Road, Box 3012, Columbus, OH 43210-0012. (614) 447-3600. (800) 848-6538. FAX (614) 447-3713. From 1907 to present. Weekly. $16800.00 per year. Also available on CD-ROM and online. Inquire regarding cost and availability.

Current Contents: Physical, Chemical and Earth Sciences. Institute for Scientific Information, 3501 Market Street, Philadelphia, PA 19104. (215) 386-0100. FAX (215) 386-2291. From 1961 to present. Weekly. $442.00 per year. Also available on CD-ROM and online. Inquire regarding cost and availability. ISSN: 0163-2574.

Current Papers in Physics. Institute of Electrical Engineers, Michael Faraday House, Six Hill Way, Stevenage, Herts, SG1 2AY, England. Distributed by INSPEC/IEEE, Box 1331, 445 Hoes Lane, Piscataway, NJ 08855-1331. (908) 562-5549. 1966 to present. Fortnightly. $410.00 per year. ISSN: 0011-3786.

General Science Index. H.W. Wilson Company, 950 University Avenue, Bronx, NY 10452. (212) 588-8400. (800) 367-6770. FAX (718) 590-1617. From 1978 to present. Ten issues per year; quarterly and annual cumulations. Service basis. Available on CD-ROM and online. Inquire regarding cost and availability. ISSN: 0162-1963.

Government Reports Announcements and Index. U. S. National Technical Information Service (NTIS), 5285 Port Royal Road, Springfield, VA 22161. (703) 487-4650. FAX (703) 321-8547. From 1968 to present. Annual. $630.00 per year. Also available online as *NTIS Bibliographic Database* and on CD- ROM.

Mathematical Reviews: A Review Journal Covering the World Literature of Mathematical Research. American Mathematical Society, P.O. Box 1571, Annex Station, Providence, RI 02901-9930. (800) 556-7774 or (401) 455-4000. 1940 to present. Monthly. $4594.00 per year. Also available via network (MATHSCINET), online and on CD-ROM. Inquire regarding cost and availability. ISSN: 0025-5629.

NTIS Alerts: Physics. U. S. National Technical Information Service. 5285 Port Royal Road, Springfield, VA 22161. (703) 487-4650. FAX (703) 321-8547. Weekly. $140.00 per year.

Physics Abstracts. INSPEC. Section A, Science Abstracts. Institution of Electrical Engineers (IEE). Available from INSPEC/IEEE - Institute of Electrical and Electronic Engineers,, Box 1331, 445 Hoes Lane, Piscaway NJ 08855-1331. (908) 562-5549. 1898 to present. 24 issues per year. $2835.00 per year. Also available on CD-ROM and online. ISSN: 0036-8091.

Physics Briefs (Physikalische Berichte). Information Center for Energy, Physics, Mathematics; German Physical Society. V C H Publishers, Inc., 220 East 23rd Street, New York, NY 10010-4606. (212) 683-8333. 1845 to present. 24 issues per year. $2390.00 per year. Also available online. ISSN: 0179-7434.

Science Citation Index. SCI. Institute for Scientific Information, 3501 Market Street, Philadelphia, PA 19104. (215) 386-0100. (800) 523-1850. FAX (215) 386-2991. 1961 to present. Six issues per year, plus annual cumulation. $11650.00 per year. Also available online and on CD-ROM.Inquire about price and availability. ISSN: 0036-827X.

ANNUAL REVIEWS AND YEARBOOKS

Advances in Nuclear Physics. Plenum Publishing Corp., 233 Spring Street, New York, NY. (212) 620-8000. (800) 2221-9369. FAX (212) 463-0742. 1968 to present. Irregular. Price varies. ISSN: 0065-2970.

Annual Review of Nuclear and Particle Science. Annual Reviews, Inc., 4139 El Camino Way, Palo Alto, CA 94303-0139. (415) 493-4400. (800) 523-8635. Fax (415) 855-9815. 1952 to present. Annual. $62.00.

Progress in Particle and Nuclear Physics. Elsevier Science Publishing Company, Inc., 655 Avenue of the Americas, New York, NY 10010. (212) 989-5800. FAX (914) 333-2444. 1977 to present. Two issues per year. $660.00 per year. ISSN: 0146-6410.

ASSOCIATIONS AND PROFESSIONAL SOCIETIES

American Institute of Physics. One Physics Ellipse, College Park, MD 20740-3843. (301) 209-3100.

American Physical Society. One Physics Ellipse, College Park, MD 20740-3843. (301) 209-3200. FAX (301) 209-0865.

Universities Research Association, 1111 19th Street Nw, Suite 400, Washington, DC 20036. (202) 293-1382.

DIRECTORIES AND BIOGRAPHICAL SOURCES

American Men and Women of Science: Physical and Biological Sciences. R. R. Bowker Inc., 121 Chanlon Road, New Providence, NJ 07974. (908) 464-6800. (800) 521-8110. 20th edition. 8 volumes. 1996. $850.00.

American Physical Society. Membership Directory Bulletin Issue. American Physical Society, One Physics Ellipse, College Park, MD 20740-3843. (301) 209-3200. FAX (301) 209-0865. Biennial. $ 50.00.

The Biographical Dictionary of Scientists: Physicists. D. Abbott, editor. Peter Bedrick Books, Inc., 2112 Broadway, Room 318, New York, NY 10023. (212) 496-0751. 1984.

Directory of Physics and Astronomy Staff Members. American Institute of Physics, One Physics Ellipse, College Park, MD 20740-3843. (301) 209-3100. Annual. $45.00.

Graduate Programs in Physics, Astronomy and Related Fields. 1995-1996. American Institute of Physics, One Physics Ellipse, College Park, MD 20740-3843. (301) 209-3100. Annual. $45.00. ISSN: 0147-1821.

Legends in their Own Time: A Century of American Physical Scientists. Anthony Serafini. Plenum Publishing Corp., 233 Spring Street, New York, NY. (212) 620-8000. (800) 2221-9369. FAX (212) 463-0742.. 1993. $27.50.

The Nobel Prize Winners. Frank N. Magill, editor. Salem Press, Inc., P.O. Box 1097, Englewood Cliffs, NJ 07632. (201) 871-3700. (800) 221-1592. 3 volumes. 1989. $210.00.

Research Centers Directory. Gale Research Inc., 835 Penobscot Building, Detroit, MI 48226-4094. (313) 961-2242. (800) 877-4253. 20th edition. 1995. $485.00. ISSN: 0080- 1518.

Scientific and Technical Organizations and Agencies Directory. Gale Research, 835 Penobscot Building, Detroit, MI 48226-4094. (313) 961-2242. (800) 877-4253. 4th edition. 1996. $195.00.

World Guide To Scientific Associations and Learned Societies. K. G. Saur Inc., 121 Chanlon Road, New Providence, NJ 07974. (908) 464-6800. (800) 521-8110. 5th edition. 1990. $225.00.

ENCYCLOPEDIAS AND DICTIONARIES

A Concise Dictionary of Physics. Oxford University Press, Inc., 200 Madison Avenue, New York, NY 10016. (212) 725- 6000. (800) 334-4249. 1990. $10.95.

Dictionary of Effects and Phenomena in Physics. Joachim Schubert. VCH Publications, Inc., 220 East 23rd Street, Suite 909, New York, NY 10010-4606. (212) 683-8333. (800) 422-8824. 1987. $35.00.

Dictionary of the Physical Sciences: Terms, Formulas, Data. Cesare Emiliani. Oxford University Press, Inc., 200 Madison

Avenue, New York, NY 10016. (212) 725-6000. (800) 334-4249. 1989. $19.95.

Encyclopedia of Applied Physics. George Trigg, editor. VCH Publications, Inc., 220 East 23rd Street, Suite 909, New York, NY 10010-4606. (212) 683-8333. (800) 422-8824. 20 volume set. 1991-. $5990.00.

Encyclopedia of Physics. Robert M. Besancon. Chapman & Hall, 1 Penn Plaza, New York, NY 10119. (212) 564-1060. 3rd edition. 1990. $54.95 in paper.

Facts On File Dictionary of Physics. John Daintith, editor. Facts-on-File, Inc., 460 Park Avenue South, New York, NY 10016-7382. (212) 683-2244. (800) 322-8755. Fax (800) 683-3633. Revised edition. 1990. $12.95.

Encyclopedia of Physical Science and Technology. Academic Press, Inc., 6277 Sea Harbor Drive, Orlando, FL. (800) 321-5068. 2nd edition. 18 volume set. 1992. $2625.00.

Landolt-boernstein Numerical Data and Functional Relationships in Science and Technology: Nuclear and Particle Physics. K. H. Hellwege editor. Springer Verlag New York, Inc., 175 Fifth Avenue, New York, NY 10010. (212) 460-1500. (800) 777-4643. FAX (212) 473-6272. 1993. Volumes priced separately.

McGraw-Hill Encyclopedia of Science and Technology. McGraw-Hill Book Company, Inc., 1221 Avenue of the Americas, New York, NY 10020. (212) 512-2000. (800) 262-4729. 7th edition. 20 volume set. 1992. $1900.00.

Penguin Dictionary of Physics. Valerie Illingworth, editor. Viking Penguin, 375 Hudson Street, New York, NY 10014-3657. (212) 366-2000. (800) 331-4624 2nd edition. 1991.

GENERAL WORKS

The Cosmic Onion: Quarks and the Nature of the Universe. Frank Close. American Institute of Physics, One Physics Ellipse, College Park, MD 20740-3843. (301) 209-3100. Reprint edition. $34.00.

Classical Dynamics of Particles and Systems. Jerry B. Marion and Stephen T. Thornton. SCP: Third World Literature Publishing House, P.O. Box 482, Lithonia, GA 30058-0482. (404) 785-7725. 3rd edition. 1988. $58.75.

The Quark Structure of Matter. Maurice Jacob, editor. World Scientific Publishing Company, Inc., 1060 Main Street, River Edge, NJ 07661. (201) 487-9655. (800) 227-7562. 1992. $118.00.

Quarks: A Source Guide. Gordon Press Publications, P.O. Box 459, Bowling Green Station, New York, NY 10004. (718) 624-8419. 1991. $76.00.

Quarks: Frontiers in Elementary Particle Physics. Y. Nambu. World Scientific Publishing Company, Inc., 1060 Main Street, River Edge, NJ 07661. (201) 487-9655. (800) 227-7562. 1985. $14.00 in paper.

Quarks: Selected Reprints. O. W. Greenberg, editor. American Association of Physics Teachers, One Physics Ellipse, College Park MD 20740-3842. (301) 209-3300. 1986. $18.00 in paper.

Quarks, Leptons and Gauge Fields. K. S. Huang. World Scientific Publishing Company, Inc., 1060 Main Street, River Edge, NJ 07661. (201) 487-9655. (800) 227-7562. 2nd edition. 1992. $44.00 in paper.

Quarks, Symmetries and Strings. M. Kaku, et al editors. World Scientific Publishing Company, Inc., 1060 Main Street, River Edge, NJ 07661. (201) 487-9655. (800) 227-7562. 1991. $44.00.

The Quest for Quarks. Brian McCusker. Cambridge University Press, 40 West 20th Street, New York, NY 10011-4211. (212) 924-3900. (800) 872-7423. 1984. $18.95

The Experimental Foundations of Particle Physics. Robert N. Cahn and Gerson Goldhaber. Cambridge University Press, 40 West 20th Street, New York, NY 10011-4211. (212) 924-3900. (800) 872-7423. 1992. $32.95 in paper.

Fundamental Particles and Forces. Richard A. Carrigan, Jr., and W. Peter Trower. W. H. Freeman & Co., 41 Madison Avenue, East 26th Avenue, 35th Floor, New York, NY 10010. (212) 576-0400. 1995. $13.95.

The Fundamental Particles and their Interactions. William B. Rolnick. Addison-Wesley Publishing Co., Inc., 1 Jacob Way, Reading, MA 01867. (617) 944-3700. (800) 447-2226. 1994. $60.25.

The Ideas of Particle Physics: An Introduction For Scientists. G. D. Coughlan and J. E. Dodd. Cambridge University Press, 40 West 20th Street, New York, NY 10011- 4211. (212) 924-3900. (800) 872-7423. 1991. $34.95 in paper.

Introduction to Gauge Field theory. D. Bailin and A. Love. IOP Publishing, Public Ledger Building, Suite 1035, Independence Square, Philadelphia, PA 19106. (215) 627- 0880. (800) 358-4677. 1993. $49.00 in paper.

Introduction to the Physics of High Energy Accelerators. D. A. Edwards and M. H. Syphers. John Wiley & Sons, Inc., 605 Third Avenue, New York, NY 10158-0012. (212) 850-6000. (800) 225-5945. 1992. $69.95.

Nuclear and Particle Physics. W. E. Burcham and M. Jobes. John Wiley & Sons, Inc., 605 Third Avenue, New York, NY 10158-0012. (212) 850-6000. (800) 225-5945. 1995. $59.95 in paper.

The Particle Garden: Our Universe As Understood By Particle Physicists. Gordon Kane. Addison-Wesley Publishing Co., Inc., 1 Jacob Way, Reading, MA 01867. (617) 944-3700. (800) 447-2226. 1995. $21.15.

Particle Physics. B. R. Martin and G. Shaw, John Wiley & Sons, Inc., 605 Third Avenue, New York, NY 10158-0012. (212) 850-6000. (800) 225-5945. 1992. $49.95 in paper.

Particle Physics in the Cosmos. Richard A. Carrigan and W. Peter Trower, editors. W. H. Freeman & Co., 41 Madison Avenue, East 26th Avenue, 35th Floor, New York, NY 10010. (212) 576-0400. 1995. $13.95.

HANDBOOKS AND MANUALS

Chemical and Physical Data. Arthur M. James and M. P. Lord, Van Nostrand Reinhold, 115 Fifth Avenue, New York, NY 10003. (212) 254-3232. (800) 842-3636. 1992. $59.95.

Ency. of Physical Sciences and Engineering Info. Sources

QUARKS

CRC Handbook of Chemistry and Physics. David R. Kide, editor. CRC Press, Inc., 2000 Corporate Boulevard, NW, Boca Raton, FL 33431. (407) 994-0555. (800) 272-7737. 77th edition. 1996. $99.95.

Handbook of Physical Quantities. Igor S. Grigoriev and Evgenil Z. Melikhov, editors. CRC Press, Inc., 2000 Corporate Boulevard, NW, Boca Raton, FL 33431. (407) 994- 0555. (800) 272-7737. 1995. $99.00.

Landolt-borenstein Numerical Data and Functional Relationships in Science and Technology: Nuclear Particle and Physics. GROUP I. Springer- Verlag New York, Inc., 175 Fifth Avenue, New York, NY 10010. (212) 460-1500. (800) 777-4643. volumes priced individually.

McGraw-Hill Encyclopedia of Physics. Sybil P. Parker, editor. McGraw-Hill Publishing Company, Inc., 1221 Avenue of the Americas, New York, NY 10020. (212) 512-2000. (800) 262-4729. 2nd edition. 1993. $95.50

A Physicist's Desk Reference. Herbert L. anderson, editor. American Institute of Physics, One Physics Ellipse, College Park, MD 20740-3843. (301) 209-3100. 2nd edition. 1989. $70.00.

Physics Problem Solver. Research and Education Association Staff. Research & Education Association, 61 Ethel Road, West Piscataway, NJ 08854. (908) 819-8880. Revised edition. 1994. $23.95 in paper.

Tables of Physical and Chemical Constants and Some Mathematical Functions. G. W. Kaye and T. M. Laby. Halsted Press 605 Third Avenue, New York, 10158-0012. (212) 850- 6400. 15th edition. 1986. $74.95.

ONLINE DATABASES AND CD-ROMS

CA Search. Chemical Abstracts Service, P.O. Box 3012, Columbus, OH 43210-0012. (614) 447-3600. (800) 848-6533. FAX (614) 447-3709. Very comprehensive guide to worldwide chemical literature and related fields, 1972 to present. Available on BRS,(800) 289-4277, DIALOG, (800) 334-2564, ORBIT (800) 456-7248, and STN International, FIZ Karlsruhe, P.O. Box 2465, W-7500, Karlsruhe 1, Germany, online services. Inquire as to cost and availability.

Current Contents Search. Institute for Scientific Information, 3501 Market Street, Philadelphia, PA 19104. (215) 386-0100. FAX (215) 386-6362. Contains citations to articles listed in the table of contents of science and technology journals. Also articles in social sciences and life sciences journals. Available on BRS, (800) 289-4277, DIALOG, (800) 334-2564, online services. Inquire as to cost and availability.

Dissertation Abstracts Online. University Microfilms International, 300 North Zeeb Road, Ann Arbor, MI 48106. (800) 521-0600 or (313) 761-4700. Scope includes virtually all doctoral dissertations accepted at accredited American institutions from 1861 to present in 252 subject areas. Available on BRS, (800) 289-4277, DIALOG, (800) 334-2564,and OCLC EPIC, (800) 848-5878, online services. Also available on CD-ROM. Inquire as to cost and availability.

INSPEC. Institution of Electrical Engineers, Michael Faraday House, Six Hills Way, Stevenage, Herts. SG1 2AY, England. Telephone: 0438 313311 or FAX 0438 742840. Contains citations to the worldwide literature of physics, electronics and electrical engineering, computer technology, and related fields.

Available on BRS, (800) 289-4277, DIALOG, (800) 334-2564, ORBIT, (800) 456-7248, and STN International, FIZ Karlsruhe, P.O. Box 2465, W-7500, Karlsruhe 1, Germany, online services. Inquire as to cost and availability.

Mathsci. American Mathematical Society, P.O. Box 6248, Providence, RI 02940. (800) 321-4667 or (401) 455-4000 or FAX (401) 331-3842. Scope includes pure and applied mathematics and related areas of physics, statistics, engineering, computer science, and operations research literature since 1959. Available on DIALOG,(800) 334-2564, online service. Also available on CD-ROM. Inquire as to cost and availability.

NTIS Bibliographic Database. National Technical Information Service, 5285 Port Royal Road, Springfield, VA 22161. (703) 487-4929 or FAX (703) 321-8199. Broad coverage of government-sponsored science and technology research reports, 1964 to present. Available on BRS,(800) 289-4277, DIALOG, (800) 334-2564, ORBIT, (800) 456-7248, and STN International, FIZ Karlsruhe, P.O. Box 2465, W-7500, Karlsruhe 1, Germany, online services. Also available on CD-ROM. Inquire as to cost and availability.

Physics Briefs. American Institute of Physics, 335 East 45th Street, New York, NY 10017. (212) 661-9260 or FAX (212) 661-2036. Contains citations with abstracts of the literature of physics and related fields, 1979 to present. Available on the STN International, FIZ Karlsruhe, P.O. Box 2465, W-7500, Karlsruhe 1, Germany, online service. Inquire as to cost and availability.

Scientific and Technical Books and Serials in Print. R.R. Bowker Inc., 121 Chanlon Road, New Providence, NJ 07974. (800) 521-8110 or (908) 464-6800. List of currently published books and serials in the physical and biological sciences, engineering and technology, with subject, author and titles indexes. Available on ORBIT, (800) 456-7248,online service. Inquire as to cost and availability.

SCISEARCH. Institute for Scientific Information, 3501 Market Street, Philadelphia, PA 19104. (800) 523-1850 or (215) 386-0100. Broad multidisciplinary title and author index to the international literature of science and technology, 1974 to present. Available on DIALOG, (800) 334-2564, and ORBIT, (800) 456-7248, online services. Also available on CD-ROM. Inquire as to cost and availability.

WILSONLINE. H.W. Wilson Company, 950 University Avenue, Bronx, NY 10452. (800) 367-6770 or (212) 588-8400. Makes available online versions of the printed H.W. Wilson indexes including Applied Science and Technology Index, Business Periodicals Index, General Science Index, and Readers' Guide to Periodical Literature. Period covered is generally 1983 to present. Available on BRS, (800) 289-4277, DIALOG, (800) 334-2564, and OCLC EPIC, (800) 848-5878, online services.Also available on CD-ROM. Inquire as to cost and availability.

PERIODICALS

American Journal of Physics. American Association of Physics Teachers, 5112 Berywn Road, College Park, MD 20740. (301) 345-4200. 1933 to present. Monthly. $215.00 per year. ISSN: 0002-9505.

Applied Physics Letters. American Institute of Physics, One Physics Ellipse, College Park, MD 20740-3843. (301) 209- 3000. 1962 to present. Weekly. $1060.00 per year. ISSN: 0003-6951.

Canadian Journal of Physics. National Research Council of Canada. Research Journals, Ottawa K1A OR6, Canada. (613) 993-9084. FAX (613) 952-7656. 1929 to present. Monthly. $285.00 per year. ISSN: 0008-4204.

Chaos; An Interdisciplinary Journal of Nonlinear Science. American Institute of Physics, One Physics Ellipse,

College Park, MD 20740-3843. (301) 209-3000. 1991 to present. Quarterly. $225.00 per year. ISSN: 1054-1500.

Contemporary Physics. Taylor & Francis, Ltd, Rankine Road, Basingstoke, Hants RG24 8PR, England. TEL 0256-840366. FAX 0256-479438. 1959 to present. Bimonthly. $318.00 per year. ISSN: 0010-7514.

Journal of Physics G: Nuclear and Particle Physics. I O P Publishing. Orders to: American Institute of Physics, 500 Sunnyside Boulevard, Woodbury, NY 11797-2999. (516) 576-2200. 1975 to present. Monthly. $1426.00 per year. ISSN: 0954-3899.

Physical Review D (Particles and Fields). American Institute of Physics, One Physics Ellipse, College Park, MD 20740-3843. (301) 209-3000. 1970 to present. 24 issues per year. $1280.00 per year. ISSN: 0556-2821.

Physics Letters B: Nuclear, Elementary Particle and High-Energy Physics. Elsevier Science Publishing Company, Inc., 655 Avenue of the Americas, New York, NY 10010. (212) 989-5800. FAX (212) 633-3990. 1962 to present. 96 issues per year in 24 volumes. $4683.00 per year. ISSN: 0370-2693.

Physics Today. American Institute of Physics, One Physics Ellipse, College Park, MD 20740-3843. (301) 209-3000. 1948 to present. Monthly. $140.00 per year. ISSN: 0031- 9228.

RESEARCH CENTERS AND INSTITUTES

Argonne National Laboratory. 9700 South Cass Avenue, Argonne, IL 60439-4832. (708) 972-3872.

Fermi National Accelerator Laboratory. P.O. Box 500 Batavia, IL 60510. (708) 840-3351. FAX (708) 840-4343.

High Energy Physics and Elementary Particle Research Program. University of Pennsylvania, 33rd and Walnut Streets, Philadelphia. PA 19104. (215) 898-5960. FAX (215) 898-8512.

Lawrence Berkeley Laboratory, Physics Division. Building 50, Room 256, 1 Cyclotron Road, Berkeley, CA 04720. (415) 486-5421. FAX (415) 486-4553.

QUARRIES AND QUARRYING

See also: MINING ENGINEERING, ROCK MECHANICS

ABSTRACTS AND INDEXES

Applied Mechanics Reviews: An Assessment of World Literature in Engineering Sciences. American Society of Mechanical Engineers, 345 East 47th Street, New York, NY 10017. (212) 705-7703. 1948 to present. Monthly. $360.00 per year.

Applied Science and Technology Index; A Cumulative Subject Index To English Language Periodicals in the Fields of Aeronautics and Space Science, Computer Technology, Chemistry, Construction Industry, Energy and Related Areas. H.W. Wilson Co., 950 University Avenue, Bronx, NY 10452. (800) 367-6770 or (212) 588-8400. FAX (718) 590-1617. From 1958 to present. Monthly. Inquire about cost and availability. Also available on CD-ROM.

Bibliography and Index of Geology. American Geological Institute, 4220 King Street, Alexandria, VA 22302. (703) 379-2480. Fax (703) 379-7563. Monthly. $1295.00 per year. Also available online as GEOREF.

Civil and Structural Engineering Abstracts. Cambridge Scientific Abstracts, 7200 Wisconsin Avenue, Bethesda, MD 20814-4823. (301) 961-6750. FAX (301) 961-6720. Monthly. Topics covered include structural design, construction equipment and methods, civil defense and military engineering, surveying, highway engineering, maritime and port structures, materials, land reclamation, and soil mechanics.

Engineering Index Monthly. Engineering Information, Inc., Castle Point on the Hudson, Hoboken, NJ 07030. (800) 221-1044. FAX (212) 832-1857. Monthly. $2200.00 per year. Also available online as COMPENDEX, and also on CD-ROM.

Geological Society of America. Geological Society of America. 3300 Penrose Place, P.O. Box 9140, Boulder, CO 80301-9140. (303) 447-2020. Abstracts and programs of the annual conference. Annual. $69.00.

Geotechnical Abstracts. Deutsche Gesellschaft fure Erd- und Grundbau e.V., Hohenzollernstrasse 52, 4300 Essen 1, Germany. Also sponsored by International Society for Soil Mechanics and Foundation Engineering. Covers soil mechanics, foundation engineering, and engineering geology. 1970 to present. Monthly. $200.00 per year.

International Joural of Rock Mechanics and Mining Sciences and Geomechanics Abstracts. Elsevier Science, 660 White Plains Rd., Tarrytown, NY 10591-5153. (914) 524-9200. FAX (914) 333-2444. Six times a year. $845 per year.

I.M.M. Abstracts and Index. Institution of Mining and Metallurgy, 44 Portland Pl., London W1N 4BR, England. 071-580-3802. FAX 071-436-5388. Bi-monthly. $364 for non-members.

Mechanical Engineering Abstracts (formerly ISMEC), Cambridge Scientific Abstracts, 7200 Wisconsin Avenue, Bethesda, MD 20814-4823. (301) 961-6750. FAX (301) 961-6720. Monthly. Summarizes world literature in mechanical engineering, production engineering, and engineering management.

Mineralogical Abstracts. Mineralogical Society and the Mineralogical Society of America, 41 Queen's Gate, London, SW7 5HR, England. Quarterly. $235.00 per year.

ASSOCIATIONS AND PROFESSIONAL SOCIETIES

American Institute of Mining, Metallurgical & Petroleum Engineers, 345 East 47th Street, New York, NY 10017.

American Mining Congress, 1920 N St., NW, Suite 300, Washington, DC 20036-1662. (202) 861-2800. FAX (202) 861-7535.

American Society of Civil Engineers, 345 East 47th Street, New York, NY 10017-2398. (212) 705-7520 or (800) 548-2723.

Mineralogical Society of America, 1130 17th St. NW, Suite 330, Washington, DC 20036. (202) 775-4344. FAX (202) 775-0018.

The Minerals, Metals & Materials Society, 410 Commonwealth Drive, Warrendale, PA 15086.

Mining and Metallurgical Society of America, 275 Madison Ave., Room 2301, New York, NY 10016. (212) 684-4150.

Society of Mining Engineers of Aime. Caller Number D, Littleton, CO 80127. (303) 973-9550.

DIRECTORIES AND BIOGRAPHICAL SOURCES

Mining Directory: Mines & Mining Equipment and Service Companies Worldwide (year). Don Nelson Publications Ltd., PPO Box 193, Barnet, Herts EN4 8LP, England. Telephone 081-368-5534. FAX 081-368-7010. Annual. Inquire for price.

Research Centers Directory. Gale Research, 835 Penobscot Building, Detroit, MI 48226-4094. (313) 961-2242. (800) 877-4253. FAX (313) 961-6083. 20th edition. $485.00.

Scientific and Technical Organizations and Agencies Directory. Gale Research, 835 Penobscot Building, Detroit, MI 48226-4094. (313) 961-2242. (800) 877-4253. FAX (313) 961-6083. 4th edition. 1996. $195.00.

Who's Who in Mineral Engineering. Society of Mining Engineers, Caller Number D, Littleton, CO 80127. (303) 973-9550. Annual. Inquire for cost and availability.

Who's Who in Engineering. American Association of Engineering Societies, 1111 19th Street NW, Suite 608, Washington, DC 20036. (202) 296-2237 or (800) 658-8897. 8th edition. 1991. Inquire for price.

ENCYCLOPEDIAS AND DICTIONARIES

Concise Encyclopedia of Mineral Resources. Donald D. Carr & Norman Herz, eds. MIT Press, 55 Hayward Street, Cambridge, MA 02142. (617) 253-2884 or (800) 356-0343. FAX (617) 253-1709. 1989. $145.00.

Dictionary of Mining & Processing. Helmut Schmidt. Collets/State Mutual Book & Periodical Service Ltd., 521 Fifth Ave., 17th floor, New York, NY 10175. (718) 261-1704. FAX (516) 537-0412. 1980. $99.00.

GENERAL WORKS

Introductory Mining Engineering. Howard L. Hartman. John Wiley and Sons, Inc., 605 Third Avenue, New York, NY 10158. (800) 526-5368 or (212) 850-6000. 1987. Inquire for price.

Mining & Quarrying Trends in the Metals & Industrial Minerals Industries. Gordon Press Publishers, PO Box 459, Bowling Green Station, New York, NY 10004. (718) 624-8419. 1993. $263.95.

Surface Mining & Quarrying: Mechanization, Technology & Capacity. T. Shaw & Vladimir Pavlovic. Prentice Hall, 113

Sylvan Avenue, Route 9W, Englewood Cliffs, NJ 07632. (201) 592-2000 or (800) 922-0579. 1992. Inquire for price and availability.

HANDBOOKS AND MANUALS

The Contractors' Dictionary of Equipment, Tools, and Techniques: For Civil Engineering, Construction, Forestry, Open-pit Mining, and Public Works. L.F. Webster. John Wiley and Sons, Inc., 605 Third Avenue, New York, NY 10158. (800) 526-5368 or (212) 850-6000. 1995. Inquire for price.

Pit and Quarry Mining Reference Manual & Buyers Guide. Advanstar Communications Inc., 7500 Old Oak Blvd., Cleveland, OH 44130. (216) 826-2839. FAX (216) 891-2726. 1907 to present. Annual. $25.00.

ONLINE DATABASES AND CD-ROMS

Compendex Plus. Engineering Information, Inc., 345 East 47th Street, New York, NY 10017. (212) 705-7600 or (800) 221-1044. Contains citations with abstracts to worldwide literature in engineering and technology, from 1970 to present. Available on online BRS,(800) 289-4277, DIALOG, (800) 334-2564, ORBIT (800) 456-7248, and STN International, FIZ Karlsruhe, P.O. Box 2465, W-7500, Karlsruhe 1, Germany, online services. Also available on CD-ROM. Inquire as to cost and availability.

CSA Engineering. Cambridge Scientific Abstracts, 7200 Wisconsin Avenue, Suite 601, Bethesda, MD 20814. (301) 961-6750 or (800) 843-7751. Contains citations and abstracts of international periodicals and research literature covering all fields of engineering and science and technology,including computer and information science, electronics, mechanical engineering, solid state materials, 1981 to present. Available on BRS,(800) 289-4277, online service.Inquire as to cost and availability.

Current Contents Search. Institute for Scientific Information, 3501 Market Street, Philadelphia, PA 19104. (215) 386-0100. FAX (215) 386-6362. Contains citations to articles listed in the table of contents of science and technology journals. Also articles in social sciences and life sciences journals. Available on BRS,(800) 289-4277, DIALOG,(800) 334-2564, online services. Inquire as to cost and availability.

Dissertation Abstracts. University Microfilms International, 300 North Zeeb Road, Ann Arbor, MI 48106. (800) 521-0600 or (313) 761-4700. Scope includes virtually all doctoral dissertations accepted at accredited American institutions from 1861 to present in 252 subject areas. Available on BRS,(800) 289-4277, DIALOG,(800) 334-2564, and OCLC EPIC,(800) 848-5878, online services. Also available on CD-ROM. Inquire as to cost and availability.

GEOREF: Bibliography and Index of Geology. American Geological Institute, 4220 King Street, Alexandria, VA 22302. (703) 379-2480. Fax (703) 379-7563. Monthly. Inquire for price and availability.

NTIS Bibliographic Database. National Technical Information Service, 5285 Port Royal Road, Springfield, VA 22161. (703) 487-4929 or FAX (703) 321-8199. Broad coverage of government-sponsored science and technology research reports, 1964 to present. Available on BRS,(800) 289-4277, DIALOG, (800) 334-2564, ORBIT, (800) 456-7248, and STN International, FIZ Karlsruhe, P.O. Box 2465, W-7500, Karlsruhe 1, Germany,

QUARRIES AND QUARRYING

Ency. of Physical Sciences and Engineering Info. Sources

online services. Also available on CD-ROM. Inquire as to cost and availability.

SCISEARCH. Institute for Scientific Information, 3501 Market Street, Philadelphia, PA 19104. (800) 523-1850 or (215) 386-0100. Broad multidisciplinary title and author index to the international literature of science and technology, 1974 to present. Available on DIALOG,(800) 334-2564, and ORBIT,(800) 456-7248, online services. Also available on CD-ROM. Inquire as to cost and availability.

WILSONLINE. H.W. Wilson Company, 950 University Avenue, Bronx, NY 10452. (800) 367-6770 or (212) 588-8400. Makes available online versions of the printed H.W. Wilson indexes including Applied Science and Technology Index, Business Periodicals Index, General Science Index, and Readers' Guide to Periodical Literature. Period covered is generally 1983 to present. Available on BRS,(800) 289-4277, DIALOG,(800) 334-2564, and OCLC EPIC,(800) 848-5878, online services. Also available on CD-ROM. Inquire as to cost and availability.

PERIODICALS

Amc Journal. American Mining Congress, 1920 N St., NW, Suite 300, Washington, DC 20036-1662. (202) 861-2800. FAX (202) 861-7535. Monthly. $36 per year.

Advances in Mining Science and Technology. Elsevier Science Inc., Box 882, Madison Square Station, New York, NY 10159. (212) 989-5800. Irregular. Inquire for price.

Canadian Mining Journal. Southam Magazine Group, 1450 Don Mills Rd., Don Mills, ON M3B 2X7, Canada. (416) 442-2004. FAX (416) 442-2261. Bi-monthly. $46.00 per year.

CIM Bulletin. Canadian Institute of Mining, Metallurgy, and Petroleum, Xerox Tower, 3400 de Maisonneuve Blvd. W., Suite 1210, Montreal PQH3Z 3B8, Canada. (514) 939-2710. FAX (514) 939-2714. 1898 to present. Monthly. $105.00 per year.

Colorado School of Mines Quarterly. Colorado School of Mines Press, 1500 Illinois, Golden, CO 80401-1887. (303) 273-3000. FAX (303) 273-3310. Quarterly. $50.00 per year.

Engineering & Mining Journal. Maclean Hunter Publishing Company, 29 N. Wacker Dr., Chicago, IL 60606. (312) 726-2802. FAX (312) 726-2574. Monthly. $60.00 per year.

Mine and Quarry. IML Group, Blair House, High St., Tonbridge, Kent TN9 1BQ, England. Telephone 0732-359990. FAX 0732-770049. Monthly. Inquire for price.

Mine & Quarry Trader. Allied Publications, 7355 N. Woodland, Box 603, Indianapolis, in 46206-0603. (317) 297-5500. FAX (317) 299-1356. Monthly. $18.00 per year.

Mining Engineer. Institution of Mining Engineers, Danum House, South Parade, Doncaster DN1 2DY, England. Phone 0302-320486. FAX 0302-340554. Monthly. Inquire for price.

Mining Engineering. Society for Mining, Metallurgy, and Exploration, Inc., 8307 Shaffer Parkway, PO Box 625002, Littleton, CO 80162-5002. (303) 973-9550. FAX (303) 973-3845. Monthly. $100.00 per year.

Pit and Quarry. Advanstar Communications Inc., 7500 Old Oak Blvd., Cleveland, OH 44130. (216) 826-2839. FAX (216) 891-

2726. 1916 to present. Monthly. $35.00 per year.

Quarry Management. Quarry Managers Journal Ltd., 7 Regent Street, Nottingham NG1 5BY, England. Telephone 0602-411315. FAX 0602-484035. 1918 to present. Monthly. Inquire for price.

Rock Products. Maclean Hunter Publishing Company, 29 North Wacker Drive, Chicago, IL 60606-3298. (312) 726-2802. 1897 to present. Monthly. $31.25 per year.

World Mining Equipment. Metal Bulletin, Inc., 220 Fifth Ave., New York, NY 10001-7781. (212) 213-6202. FAX (212) 213-1870. 11 times a year. Inquire for price.

RESEARCH CENTERS AND INSTITUTES

Bureau of Mines. U.S. Department of the Interior, 2401 E Street, NW, Washington, DC 20241. (202) 634-1004.

Excavation Engineering & Earth Mechanics Institute. Colorado School of Mines, Golden, CO 80401. (303) 273-3419.

Mining & Excavation Research Institute. 1825 K St NW, Suite 218, Washington, DC 20006-1202. (202) 785-3756. FAX (202) 429-9417.

New Mexico Bureau of Mines and Mineral Resources. Institute of Mining & Technology, Campus Station, Socorro, NM 87801. (505) 835-5420. FAX (505) 835-6333.

South Dakota School of Mines and Technology, Mining and Mineral Resources Research Institute. 501 E St. Josephs St., Rapid City, SD 57701-3995. (605) 394-1971.

Southern Illinois University At Carbondale, Illinois Mining and Mineral Resources Research Institute. Coal Extraction and Utilization Research Center, Carbondale, IL 62901. (618) 536-6637. FAX (618) 453-7455.

University of Kentucky Institute For Mining and Minerals Research. 233 Mining and Mineral Resources Bldg., Lexington, KY 40506. (606) 257-8636. FAX (606) 258-1049.

University of Utah, Utah Mining and Mineral Resources Research Institute. 113 Energy and Mineral Research office Bldg., Salt Lake City, UT 84112. (801) 581-8006. FAX (801) 581-8119.

QUASARS

See also: ASTRONOMY, BIG BANG, COSMOCHEMISTRY, COSMOLOGY, DARK MATTER, GALAXIES, INTERSTELLAR MEDIUM, MILKY WAY, NEBULAE, STARS, UNIVERSE

ABSTRACT SERVICES AND INDEXES

Applied Science and Technology Index; A Cumulative Subject Index To English Language Periodicals in the Fields of Aeronautics and Space Science, Computer Technology, Chemistry, Construction Industry, Energy and Related Areas. H.W. Wilson Co., 950 University Avenue, Bronx, NY 10452. (212) 588-8400. (800) 367-6770. FAX (718) 590-1617. From 1958 to present.

Monthly. Inquire about cost and availability. Also available on CD-ROM and online. ISSN: 0003-6986.

Astronomy and Astrophysics Abstracts. Springer-Verlag New York, 175 Fifth Avenue, New York, NY 10010. (212) 460-1500.FAX (212) 473-6272. Published for the Astronomisches Rechen-Institut. Comprehensive coverage of all aspects of astronomy, astrophysics and related fields. 1969 to present. Two parts per year. Annual. ISSN: 0067-0022.

General Science Index. H.W. Wilson Company, 950 University Avenue, Bronx, NY 10452. (212) 588-8400. (800) 367-6770. FAX (718) 590-1617. From 1978 to present. Ten issues per year; quarterly and annual cumulations. Service basis. Available on CD-ROM and online. Inquire regarding cost and availability. ISSN: 0162-1963.

Government Reports Announcements and Index. U. S. National Technical Information Service (NTIS), 5285 Port Royal Road, Springfield, VA 22161. (703) 487-4650. FAX (703) 321-8547. From 1968 to present. Annual. $630.00 per year. Also available online as *NTIS Bibliographic Database* and on CD-ROM. ISSN:

International Aerospace Abstracts. American Institute of Aeronautics and Astronautics. Technical Information Service, 555 West 57th Street, New York 10019. (212) 247- 6500. FAX (212) 582-4961. 1961 to present. Semi-monthlywith annual index. $1295.00 per year. Also available online and on CD-ROM. ISSN: 0020-5842.

Meteorological and Geoastrophysical Abstracts. American Meteorological Society, c/o Inforonics, Inc., 550 Newtown Road, Littleton, MA 01460. (508) 486-8976. FAX (508) 486- 0027. Covers literature in environmental sciences, meteorology, astrophysics, hydrology, glaciology, and physical oceanography. From 1950 to present. Monthly. $950.00 per year. Also available on CD-ROM and online. Inquire regarding cost and availability. ISSN: 0026-1130.

NTIS Alerts: Astronomy & Astrophysics. U. S. National Technical Information Service. 5285 Port Royal Road, Springfield, VA 22161. (703) 487-4650. FAX (703) 321-8547.Weekly. $140.00 per year.

Pascal 48: Environnement Cosmique Terrestre, Astronomie Et Geologie Extraterrestre. Centre National de la Recherche Scientifique, Institut de Information Scientifique et Technique, 2 aliee du Parc de Brabois, 54514 Vandoeuvre Les Nancy Cedex, France. TEL 83- 50-46-00. FAX: 83-50-46- 50. 1985 to present. Ten issues per year. 770 F. Also available on CD-ROM and online.

Physics Abstracts. INSPEC. Section A, Science Abstracts. Institute of Electrical Engineers, London, United Kingdom. Available from: INSPEC/IEEE - Institute of Electrical and Electronic Engineers, Box 1331, Hoes Lane, Piscataway, NJ 08855-1331. (908) 562-5549. 1898 to present. 24 issuesper year. $2835.00 per year. ISSN: 0036-8091. Also available online and on CD-ROM.

Physics Briefs (Physikalische Berichte). Information Center for Energy, Physics, Mathematics; German Physical Society. V C H Publishers, Inc., 220 East 23rd Street, New York, NY 10010-4606. (212) 683-8333. 1845 to present. 24 issues per year. $2390.00 per year. Also available online. ISSN: 0179-7434.

Science Citation Index. SCI. Institute for Scientific Information, 3501 Market Street, Philadelphia, PA 19104. (215) 386-0100.

(800) 523-1850. FAX (215) 386-2991. 1961 to present. Six issues per year, plus annual cumulation. $11650.00 per year. Also available online and on CD-ROM. Inquire about price and availability. ISSN: 0036-827X.

STAR. Scientific and Technical Aerospace Reports. U.S. National Aeronautics and Space Administration. Distributed by U. S. Superintendent of Documents, Washington, DC 20402. 1963 to present. Semi-monthly, with semiannual and annual indexes. $114.00 per year. ISSN: 0036-8741. Also available online and on CD-ROM.

ANNUAL REVIEWS AND YEARBOOKS

The Astronomical Almanac. Superintendent of Documents, U. S. Government Printing office, Washington, DC 10402. (202) 783-3238. 1981 to present. Supercedes Astronomical Ephemeris and the American Ephermis and Nautical Almanac. Annual. $44.95. ISSN: 0737-6421.

Annual Review of Astronomy and Astrophysics; Original Reviews of Critical Literature and Current Developments in Astronomy and Astrophysics. Annual Reviews, Inc., 4139 El Camino Way, Palo Alto, CA 94306-0139. (415) 493-4400.(800) 523-8635. FAX (415) 855-9815. 1963 to date. Annual. $60.00. ISSN: 0066-4146.

Highlights of Astronomy. I. Appenzeller, editor. Kluwer Academic Publishers, Box 358, Accord Station, Hingham, MA 02018-0358. (617) 871-6600. International Astronomical Union Highlights series. ISBN: 07923-3553-8. Triennal.1995. Price varies.

ASSOCIATIONS AND PROFESSIONAL SOCIETIES

Amateur Astronomers Association. 1010 Park Avenue, New York, NY 10028. (212) 535-2922.

American Association of Variable Star Observers. 23 Birch Street, Cambridge, MA 02138. (617) 354-0484. FAX: (617)354-0665.

American Astronomical Society. 2000 Florida Avenue NW, Suite 400, Washington, DC 20009. (202) 328-2010. FAX:(202) 234-2560.

American Institute of Physics. One Physics Ellipse, College Park, MD 20740-3843. (301) 209-3100.

Association of University For Research in Astronomy. Suite 701, 1625 Massachusetts Avenue, NW, Washington, DC 20036. (202) 483-2101. FAX (202) 483-2106.

Astronomical League. 2112 Kingfisher Lane East, Rolling Meadows, IL 60008. (708) 398-0562.

Astronomical Society of the Pacific. 390 Ashton Avenue, San Francisco, CA 94112. (415) 337-1100. FAX: (415) 337-5205.

International Astronomical Union. 98 bis Boulevard Arago, F-75014 Paris, France. Tel 1-43-25-83-58. FAX 1-43-25-26-16.

Royal Astronomical Society of Canada. 136 Dupont Street, Toronto. ON N5R 1V2 Canada. (416) 924-7973. FAX (416) 968-6687.

QUASARS

Ency. of Physical Sciences and Engineering Info. Sources

BIBLIOGRAPHIES

Astronomy and Astrophysics: A Source Guide. Gordon Press, P.O. Box 459, Bowling Green Station, New York, NY 10004. (718) 624-8419. 1991. $250.00. .

A Bibliography of Astronomy, 1970-1979. R. A. Sea and S. S. Martin. Libraries Unlimited, Inc., Littleton, CO 80160. 1982.

Catalog of the Naval Observatory Library. Naval Observatory Staff. G. K. Hall & Company, 866 Third Avenue, New York, NY 10022. (212)702-6789. (800) 257-5755. 6 volumes. 1977. $655.00

Science Books and Films. American Association for the Advancement of Science, 1333 H Street NW, Washington, DC 20005. (202) 326-6454. 1965 to present. Nine issues per year. $40.00 per year. ISSN: 0098-342X

Scientific and Technical Books and Serials in Print, 1995. R.R. Bowker Inc., 121 Chanlon Road, New Providence, NJ 07974. (908) 464-6800. (800) 521-8110. 4 volumes. 1994. $299.95. Also available on compact disc and online.

A Source Book On Astronomy and Astrophysics, 1900 - 1975. Kenneth Lang and Owen Gingerich, editors. Harvard University Press, 79 Garden Street, Cambridge MA. (617) 495-2600. 1980. $41.95.

DIRECTORIES AND BIOGRAPHICAL SOURCES

American Men and Women of Science: Physical and Biological Sciences. R. R. Bowker Inc., 121 Chanlon Road, New Providence, NJ 07974. (908) 464-6800. (800) 521-8110. 20th edition. 8 volumes. 1996. $850.00.

The Astronomers. Donald Goldsmith. St. Martin's Press, Inc., 175 Fifth Avenue, New York, NY 10010. (212) 674-5151. (800) 221-7945. 1993. $14.95 in paper.

Astronomical Centers of the World. Kevin Krisciunas. Cambridge University Press, 40 West 20th Street, New York, NY 10011-4211. (212) 924-3900. (800) 872-7423. 1988. $34.95

The Biographical Dictionary of Scientists: Astronomers. D. Abbott, editor. Peter Bedrick Books, Inc., 2112 Broadway, Room 318, New York, NY 10023. (212) 496-0751. 1984.

Directory of Physics and Astronomy Staff Members. American Institute of Physics, One Physics Ellipse, College Park, MD 20740-3843. (301) 209-3100. Annual. $45.00.

Earth and Astronomical Research Centers. Stockton Press, 345 Park Avenue South, New York, NY 10010. 4th edition. 1995. $515.00. ISBN: 1-56169-0967.

Graduate Programs in Physics, Astronomy and Related Fields. 1995-1996. American Institute of Physics, One Physics Ellipse, College Park, MD 20740-3843. (301) 209-3100. 1978 to present. Annual. $45.00.

Research Centers Directory. Gale Research Inc., 835 Penobscot Building, Detroit, MI 48226-4094. (313) 961-2242. (800) 877-4253. 20th edition. 1995. $485.00.

ENCYCLOPEDIAS AND DICTIONARIES

Astronomy From A To Z: A Dictionary of Celestial Objects and Ideas. Charles A. Schweighauser. Illinois Issues, Sangamon State University, K-10, Springfield, IL 62794-9243. (217) 786-9243. 1991. $14.95 in paper.

Concise Dictionary of Astronomy. Jacqueline Mitton. Oxford University Press, Inc., 200 Madison Avenue, New York, NY 10016. (212) 725-6000. (800) 334-4249. 1992. $24.95.

Facts On File Dictionary of Astronomy. Valerie Illingworth. Facts On File, Inc., 460 Park Avenue South, New York, NY 10016-7382. (212) 683-2244. (800) 322-8755. Fax (800) 683-3633. 1994. $27.95.

Encyclopedia of Astronomy and Astrophysics. Stephen Maran, editor. Van Nostrand Reinhold, 115 Fifth Avenue, New York, NY 10003. (212) 254-3232. (800) 842-3636. 1992. $129.95.

Landolt-boernstein Numerical Data and Functional Relationships in Science and Technology: Astronomy, Astrophysics and Space Research; A. O. Madelung and H. H. Boigt, editors. Springer Verlag New York, Inc., 175 Fifth Avenue, New York, NY 10010. (212) 460-1500. (800) 777- 4643. 1993. $550.00.

Oxford Illustrated Encyclopedia. VOLUME 8, the UNIVERSE. Archie Roy, editor. Oxford University Press, Inc., 200 Madison Avenue, New York, NY 10016. (212) 725-6000. (800) 334-4249. 1993. $45.00.

McGraw-Hill Encyclopedia of Science and Technology. McGraw-Hill Book Company, Inc., 1221 Avenue of the Americas, New York, NY 10020. (212) 512-2000. (800) 262-4729. 7th edition. 20 volume set. 1992. $1900.00.

Stars, Galaxies, Cosmos. William R. Corliss. Sourcebook Project, P.O. Box 107, Glen Arm, MD 21057. (410) 668-6047. 1987. $17.95.

GENERAL WORKS

Asteroids To Quasars. Phyliss Lugger, editor. Cambridge University Press, 40 West 20th Street, New York, NY 10011- 4211. (212) 924-3900. (800) 872-7423. 1991. $69.95.

Cosmic Questions, Galactic Halos, Cold Dark Matter and the End of Time. Richard Morris. John Wiley & Sons, Inc., 605 Third Avenue, New York, NY 10158-0012. (212) 850-6000. (800) 225-5945. 1993. $24.95.

Cycles of Fire: Stars, Galaxies and the Wonder of Deep Space. William K. Hartmann. Workman Publishing Company, Inc., 708 Broadway, New York, NY 10003. (212) 254-5900. (800) 722-7202. 1988. $15.95 in paper.

Farthest Things in the Universe. Jay M. Pasachoff. Cambridge University Press, 40 West 20th Street, New York, NY 10011-4211. (212) 924-3900. (800) 872-7423. 1995. $29.95.

The First Three Minutes: A Modern View of the Origin of the Universe. Steven Weinberg. Basic Books, 10 East 53rd Street, New York, NY 10022. (212) 207-7057. (800) 242-7737. Reprint edition. 1993. $12.00.

From Quarks To Quasars: A Tour of the Universe. James Jespersen and Jane Fitz-Randolph. Simon & Schuster, Inc., 1230 Avenue of the Americas, New York, NY 10020. (212) 698- 7000. (800) 223-2348. 1987. $16.95.

Galaxies, Quasars and Cosmology. Fang Lizhi. World Scientific Publishing Company, Inc., 1060 Main Street, River Edge, NJ 07661. (201) 487-9655. (800) 227-7562. 1985. $28.00.

Learning About Quasars. Sherwood Harrington, et al. Astronomical Society of the Pacific, 390 Ashton Avenue, San Francisco, CA 94112. (415) 337-1100. Revised edition. 1990. $4.00.

The Left Hand of Creation: The Origin and Evolution of the Expanding Universe. John D. Barrow and Joseph Silk. Oxford University Press, Inc., 200 Madison Avenue, New York, NY 10016. (212) 725-6000. (800) 334-4249. 1994. $23.00.

Observational Astrophysics. Robert C. Smith. Cambridge University Press, 40 West 20th Street, New York, NY 10011- 4211. (212) 924-3900. (800) 872-7423. 1995. $69.95.

Physical Processes in Comets, Stars and Active Galaxies. W. Hildebrandt, et al editors. Springer-Verlag New York, Inc., 175 Fifth Avenue, New York, NY 10010. (212) 460-1500. (800) 777-4643. 1987. $47.00.

The Physics of Stars. A. C. PHillips. John Wiley & Sons, Inc., 605 Third Avenue, New York, NY 10158-0012. (212) 850-6000. (800) 225-5945. 1994. $32.95 in paper.

Quarks, Quasars and Quandries. Gordon Aubrecht, editor. American Association of Physics Teachers, One Physics Ellipse, College Park MD 20740-3842. (301) 209-3300.. 1987. $20.00 in paper.

Quasar Absorption Lines: Probing the Universe. C. Blades, et al, editors. Cambridge University Press, 40 West 20th Street, New York, NY 10011-4211. (212) 924-3900. 1988. $54.95.

Quasar Astronomy. Daniel W. Weedman. Cambridge University Press, 40 West 20th Street, New York, NY 10011-4211. (212) 924-3900. 1988. $24.95 in paper.

Space Distribution of Quasars. D. Crampton, editor. Astronomical Society of the Pacific, 390 Ashton Avenue, San Francisco, CA 94112. (415) 337-1100. . 1991. $25.00.

Star Formation, Galaxies and the Interstellar Medium. J. Franco, et al, editors. Cambridge University Press, 40 West 20th Street, New York, NY 10011-4211. (212) 924-3900. (800) 872-7423. 1993. $59.95.

Stars and Atoms: From the Big Bang To the Solar System. Stuart Clark. Oxford University Press, Inc., 200 Madison Avenue, New York, NY 10016. (212) 725-6000. (800) 334-4249. 1995. $35.00.

Universe of Galaxies. Paul W. Hodge, editor. W. H. Freeman & Co., 41 Madison Avenue, East 26th Avenue, 35th Floor, New York, NY 10010. (212) 576-0400. Readings from Scientific American. 1995.

Visual Astronomy of the Deep Sky. Roger N. Clark. Cambridge University Press, 40 West 20th Street, New York, NY 10011-4211. (212) 924-3900. (800) 872-7423. 1991. $39.95.

HANDBOOKS AND MANUALS

Burnham's Celestial Handbook: An Observer's Guide To the Universe Beyond the Solar System. Robert Burnham, Jr. Dover Publications, Inc., 180 Varick Street, New York, NY 10014. (212) 255-3755. (800) 223-3130. 3 volumes. 1978. $41.85 for the set.

Cambridge Guide To the Constellations. Michael E. Bakich. Cambridge University Press, 40 West 20th Street, New York, NY 10011-4211. (212) 924-3900. (800) 872-7423. 1995. $69.95.

Encyclopedia of Astronomy and Astrophysics. Stephen Maran, editor. Van Nostrand Reinhold, 115 Fifth Avenue, New York, NY 10003. (212) 254-3232. (800) 842-3636. 1992. $129.95.

Greenwich Guide To Stars, Galaxies and Nebulae. Stuart Malin. Cambridge University Press, 40 West 20th Street, New York, NY 10011-4211. (212) 924-3900. (800) 872-7423. 1990. $10.95.

Handbook of Astronomy, Astrophysics and Geophysics; Volume 2: Galaxies and Cosmology. V. M. Canuto and B. G. Elmegreen. Gordon & Breach Science Publishers, Inc., P.O. Box 200029, Riverfront Plaza Station, Newark, NJ 07102-0301. (201) 643- 7500. 1988. $194.00.

Handbook of Space Astronomy and Astrophsyics. Martin V. Zumbeck. Cambridge University Press, 40 West 20th Street, New York, NY 10011-4211. (212) 924-3900. (800) 872-7423. 2nd edition. 1990. $79.95.

Landolt-borenstein Numerical Data and Functional Relationships in Science and Technology: Astronomy, Astrophysics and Space Research. GROUP VI. Springer-Verlag New York, Inc., 175 Fifth Avenue, New York, NY 10010. (212) 460-1500. (800) 777-4643. volumes priced individually.

McGraw-Hill Encyclopedia of Astronomy. Sybil P. Parker and Jay M. Pasachoff, editors. McGraw-Hill Publishing Company, Inc., 1221 Avenue of the Americas, New York, NY 10020. (212) 512-2000. (800) 262-4729. 2nd edition. 1993. $75.00

National Audobon Society Pocket Guide: Galaxies and Other Deep-Sky Objects. Alfred A. Knopf, Inc. 201 E. 50th Street, New York, NY 10022. (800) 733-3000. 1995. $7.95.

Nearby Galaxies Catalog. R. Brent Tully and J. Richard Fisher. Cambridge University Press, 40 West 20th Street, New York, NY 10011-4211. (212) 924-3900. (800) 872-7423. 1988. $59.95.

Observer's Guide To Astronomy. Patrick Martinez, editor. Cambridge University Press, 40 West 20th Street, New York, NY 10011-4211. (212) 924-3900. (800) 872-7423. 2 volumes. 1994. $159.90.

ONLINE DATABASES AND CD-ROMS

CA Search. Chemical Abstracts Service, P.O. Box 3012, Columbus, OH 43210-0012. (614) 447-3600. (800) 848-6533. FAX (614) 447-3709. Very comprehensive guide to worldwide chemical literature and related fields, 1972 to present. Available on BRS,(800) 289-4277, DIALOG, (800) 334-2564, ORBIT (800) 456-7248, and STN International, FIZ Karlsruhe, P.O. Box 2465, W-7500, Karlsruhe 1, Germany, online services. Inquire as to cost and availability.

Current Contents Search. Institute for Scientific Information, 3501 Market Street, Philadelphia, PA 19104. (215) 386-0100. FAX (215) 386-6362. Contains citations to articles listed in the table of contents of science and technology journals. Also articles in social sciences and life sciences journals. Available on BRS, (800) 289-4277, DIALOG, (800) 334-2564, online services. Inquire as to cost and availability.

Dissertation Abstracts Online. University Microfilms International, 300 North Zeeb Road, Ann Arbor, MI 48106. (800) 521-0600 or (313) 761-4700. Scope includes virtually all doctoral dissertations accepted at accredited American institutions from 1861 to present in 252 subject areas. Available on BRS, (800) 289-4277, DIALOG, (800) 334-2564,and OCLC EPIC, (800) 848-5878, online services. Also available on CD-ROM. Inquire as to cost and availability.

INSPEC. Institution of Electrical Engineers, Michael Faraday House, Six Hills Way, Stevenage, Herts. SG1 2AY, England. Telephone: 0438 313311 or FAX 0438 742840. Contains citations to the worldwide literature of physics, electronics and electrical engineering, computer technology, and related fields. Available on BRS, (800) 289-4277, DIALOG, (800) 334-2564, ORBIT, (800) 456-7248, and STN International, FIZ Karlsruhe, P.O. Box 2465, W-7500, Karlsruhe 1, Germany, online services. Inquire as to cost and availability.

International Aerospace Abstracts. American Institute of Aeronautics and Astronautics, 370 L'Enfant Promenade, S.W., Washington, DC 20024. (202) 646-7400. Contains references and abstracts of journal and monograph literature relating to aerospace science and technology, from 1963 to present. Available through the NASA/RECON system of the National Aeronautics and Space Administration only.

NTIS Bibliographic Database. National Technical Information Service, 5285 Port Royal Road, Springfield, VA 22161. (703) 487-4929 or FAX (703) 321-8199. Broad coverage of government-sponsored science and technology research reports, 1964 to present. Available on BRS,(800) 289-4277, DIALOG, (800) 334-2564, ORBIT, (800) 456-7248, and STN International, FIZ Karlsruhe, P.O. Box 2465, W-7500, Karlsruhe 1, Germany, online services. Also available on CD-ROM. Inquire as to cost and availability.

Physics Briefs. American Institute of Physics, 335 East 45th Street, New York, NY 10017. (212) 661-9260 or FAX (212) 661-2036. Contains citations with abstracts of the literature of physics and related fields, 1979 to present. Available on the STN International, FIZ Karlsruhe, P.O. Box 2465, W-7500, Karlsruhe 1, Germany, online service. Inquire as to cost and availability.

Scientific and Technical Books and Serials in Print. R.R. Bowker Inc., 121 Chanlon Road, New Providence, NJ 07974. (800) 521-8110 or (908) 464-6800. List of currently published books and serials in the physical and biological sciences, engineering and technology, with subject, author and titles indexes. Available on ORBIT, (800) 456-7248,online service. Inquire as to cost and availability.

SCISEARCH. Institute for Scientific Information, 3501 Market Street, Philadelphia, PA 19104. (800) 523-1850 or (215) 386-0100. Broad multidisciplinary title and author index to the international literature of science and technology, 1974 to present. Available on DIALOG, (800) 334-2564, and ORBIT, (800) 456-7248, online services. Also available on CD-ROM. Inquire as to cost and availability.

WILSONLINE. H.W. Wilson Company, 950 University Avenue, Bronx, NY 10452. (800) 367-6770 or (212) 588-8400. Makes available online versions of the printed H.W. Wilson indexes including Applied Science and Technology Index, Business Periodicals Index, General Science Index, and Readers' Guide to Periodical Literature. Period covered is generally 1983 to present. Available on BRS, (800) 289-4277, DIALOG, (800) 334-2564, and OCLC EPIC, (800) 848-5878, online services.Also available on CD-ROM. Inquire as to cost and availability.

OTHER SOURCES

Atlas of the Andromeda Galaxy. Paul W. Hodge. University of Washington Press, P. O. Box 50096. Seattle, WA 98145- 5096. (206) 543-4050. (800) 441-4115. 1981. $75.00.

Atlas of Deep-sky Spendors. Hans Vehrenberg. Sky Publishing Corp., 49 Bay State Road, Cambridge, MA 02138. (617) 864-7360. (800) 253-0245. 4th edition. 1983. $24.95.

Atlas of the Ultraviolet Sky. Richard C. Henry, Wayne B. Landsman, et al. Johns Hopkins University Press, 2715 North Charles Street, Baltimore, MD 21218-4319. (410) 516-6900. (800) 537-5487. 1988. $70.00.

Atlas of the Universe. Patrick Moore. Rand McNally & Company. 8255 North Central Park, Skokie, IL 60076-2970. (708) 673-9100. 1994. $29.95.

Cambridge Atlas of Astronomy. Jean Audouze, et al. Cambridge University Press, 40 West 20th Street, New York, NY 10011-4211. (212) 924-3900. (800) 872-7423. 3rd edition. 1994. $75.00.

Cambridge Guide To the Constellations. Michael E. Bakich. Cambridge University Press, 40 West 20th Street, New York, NY 10011-4211. (212) 924-3900. (800) 872-7423. 1995. $19.95 in paper.

Carnegie Atlas of Galaxies. Allan Sandage and John Bedke. Carnegie Institution of Washington, 1530 P Street NW, Washington, DC 20005. (202) 387-6411. Two volumes. 1994. $92.00.

Color Atlas of Galaxies. James Wray. Cambridge University Press, 40 West 20th Street, New York, NY 10011-4211. (212) 924-3900. (800) 872-7423. 1988. $94.95.

Nearby Galaxy Atlas. R. Brent Tully and J. Richard Fisher. Cambridge University Press, 40 West 20th Street, New York, NY 10011-4211. (212) 924-3900. (800) 872-7423., 1987. $74.95.

Observer's Sky Atlas. E. Karkoschka. Springer-Verlag New York, Inc., 175 Fifth Avenue, New York, NY 10010. (212) 460-1500. (800) 777-4643. 1995. $15.95.

Observing Handbook and Catalog of Deep-sky Objects. Christian B. Luginbuhl and Brian A. Skiff. Cambridge University Press, 40 West 20th Street, New York, NY 10011- 4211. (212) 924-3900. (800) 872-7423. 1990. $49.95.

Photometric Atlas of Northern Bright Galaxies. Kodaira, Keichi, et al editors. Columbia University Press, 562 West 113th Street, New York, NY 10025. (212) 666-1000. (800) 944-8648. 1990. $180.00.

Third Reference Catalogue of Bright Galaxies. Gerard H. DE Vaucouleurs et al, editors. Springer-Verlag New York, Inc., 175

Fifth Avenue, New York, NY 10010. (212) 460-1500. (800) 777-4643. Three volumes. 1991. $198.00 set.

The Sky: A User's Guide. David H. Levy. Cambridge University Press, 40 West 20th Street, New York, NY 10011- 4211. (212) 924-3900. (800) 872-7423. 1993. $14.95 in paper.

PERIODICALS

Astronomical Journal. American Institute of Physics, One Physics Ellipse, College Park, MD 20740-3843. (301) 209- 3000. Published for the American Astronomical Society. 1849 to present. Monthly. $280.00 per year. ISSN: 0004- 6256.

Astronomical Society of the Pacific Publications. Astronomical Society of the Pacific, 390 Ashton Avenue, San Francisco, CA 94112. (415) 337-1100. FAX (415) 337-5205. 1889 to present. Monthly. $175.00 per year. ISSN: 0004- 6280.

Astronomy. Kalmbach Publishing Company, Box 1612, Waukesha, WI 53187-1612. (414) 796-8776. FAX (414) 796-1142. 1973

to present. Monthly. $27.00 per year. ISSN: 0091-6358.

Astronomy and Astrophsyics. Springer-Verlag New York, Inc., 44 Hartz Way, Secaucus, NJ 07096-2491. (201) 348-4033. FAX (201) 348-4505. 1969 to date. 36 issues per year in 12 volumes. $1733.00 per year. ISSN: 0004-6361.

Astronomy Reports. American Institute of Physics, One Physics Ellipse, College Park, MD 20740-3843. (301) 209- 3000. English translation of Astronomicheksii Zhurnal. Formerly Soviet Astronomy AJ. 1957 to present. Bimonthly. $1160.00 per year. ISSN: 1063-7729.

Astrophysical Journal; An International Review of Astronomy and Astronomical Physics. University of Chicago Press, Journals Division, 5720 South Woodlawn Avenue, Chicago, IL 60637. (312) 753-3347. FAX (312) 753-0811. 1895 to date. three issues per month. $740.00 per year. ISSN: 0004-637X.

Astrophpysics and Space Science; An International Journal of Cosmic Physics. Kluwer Academic Publishers, Box 358, Accord Station, Hingham, MA 02018-0358. (617) 871-6600. FAX (617) 871-6528. 1968 to present. 24 issues per year. $2544.00 per year. ISSN: 0004-640X.

Celestial Mechanics and Dynamical Astronomy; An International Journal of Space Dynamics. Kluwer Academic Publishers, Box 358, Accord Station, Hingham, MA 02018-0358. (617) 871-6600. FAX (617) 871-6528. 1969 to present.Monthly. $703.50 per year. ISSN: 0923-2958.

Experimental Astronomy; An International Journal On Astronomical Instrumentation and Data Analysis. Kluwer Academic Publishers, Box 358, Accord Station, Hingham, MA 02018-0358. (617) 871-6600. FAX (617) 871-6528. 1989 topresent. Quarterly. $161.00 per year. ISSN: 0922-6435.

Icarus; International Journal of Solar System Studies. Academic Press, Inc., Journal Division, 525 B Street, Suite 1900, San Diego, CA 92101-4495. (619) 230-1840. FAX (619) 699-6800. 1962 to present. Monthly. $1080.00 per year. ISSN: 0019-1035.

Landolt-boernstein, Zahlenwerte Und Funktionen Aus Naturwissenschaften Und Technik. GROUP 6: ASTRONOMY. Springer-Verlag New York, Inc., 175 Fifth Avenue, New York, NY 10010. (212) 460-1500. FAX (212) 473-6272. 1965. Irregular. Price varies. ISSN: 0075-7896.

Mercury. Astronomical Society of the Pacific. 390 Ashton Avenue, San Francisco, CA 94112. (415) 337-1100. FAX: (415) 337-5205. 1972 to present. Bimonthly. $175.00 per year. ISSN: 0047-6773.

Monthly Notices of the Royal Astronomical Society. Blackwell Scientific Publication LT., Osney Mead, Oxford OX2 OEL, England. TEL 0865-240201. FAX 0865-721205. 1827 to present. Fortnightly. $1899.00 per year. ISSN: 0035-8711.

Observatory. Space and Astrophysics Division, Rutherford Appleton Laboratory, Chilton, Didcot, Oxon OX11 OQX England. FAX 0235-445848. 1877 to present. Bimonthly. $42.00 per year.

Sky and Telescope. Sky Publishing Corporation, Box 9111, Belmont, MA 02178. (617) 864-7360. FAX (617) 864-6117. 1941 to present. Monthly. $27.00 per year. ISSN: 0037- 6604.

Space Science Reviews. Kluwer Academic Publishers, Box 358, Accord Station, Hingham, MA 02018-0358. (617) 871- 6600. FAX (617) 871-6528. 1962 to present. Sixteen issues per year. $940.00 per year. ISSN: 0038-6308.

Vistas in Astronomy; An International Review Journal. Albert C. Beer, editor. Elsevier Science Publishing Company, Inc., 660 White Plains Road, Terrytown, NY 10591- 5153. (914) 524-9200. FAX: (914) 333-2444. 1958 to present. Quarterly. $415.00 per year. ISSN: 0083-6656.

Transactions of the International Astronomical Union. Proceedings of the General Assembly. Kluwer Academic Publishers, Box 358, Accord Station, Hingham, MA 02018- 0358. (617) 871-6600. Issues in two parts: Part A, Reports, and Part B, Proceedings. Triennial. XXI - 1994. Price varies. ISSN: 0080-1372.

Webb Society Quarterly Journal. Webb Society, 194 Foundry Lane, Freemantle, Southampton, Hampshire SO1 3EQ, England. 1969 to present. Quarterly. $19.50 per year. ISSN: 0043- 1680.

RESEARCH CENTERS AND INSTITUTES

Arecibo Observatory. Cornell University, Box 995, Arecibo, PR 00613. (809) 878-2612. FAX (809) 878-1861.

Astronomy Program. University of Maryland, 1207 Computer & Space Sciences Building, College Park, MD 20742. (301) 405- 3001. FAX (301) 314-9067.

Joint Institute For Laboratory Astrophyics. University of Colorado, Box 440, Boulder, CO 80309-0440. (303) 492-7780. FAX (303) 492-5235.

Lick Observatory. University of California, Santa Cruz, CA 95064. (408) 459-2991. FAX (408) 426-3115.

Lowell Observatory. 1400 West Mars Hill Road. Flagstaff, AZ 86001, (602) 774-3358. FAX (602) 774-6296.

QUASARS

Ency. of Physical Sciences and Engineering Info. Sources

McDonald Observatory, University of Texas. P.O. Box 1337, Fort Davis, TX 79734. (915) 426-3263.

National Astronomy and Ionosphere Center. Cornell University, Space Sciences Building, Ithaca, NY 14853. (607) 255-3735. (607) 255-8803.

National Radio Astronomy Observatory. Edgemont Road, Charlottesville, VA 22903. (804) 296-0211.

Palomar Observatory. California Institute of Technology. 105-24, Pasadena CA 91125. (818) 356-4033.

Radio Observatory. University of Florida, 211 Space Sciences Research Building, Gainsville, FL 32611. (904) 392-2066.

Space Telescope Science Institute. 3700 San Martin Drive, Baltimore, MD 21218. (301) 338-4700.

Steward Observatory. University of Arizona. Tuscon, AZ 85721. (602) 621-6524. FAX (602) 621-1532.

R

RADAR

See also: ACOUSTICS, ANTENNAS, ELECTRICAL ENGINEERING, ELECTRONIC CIRCUITS AND COMPONENTS, ELECTRONICS, ELECTRONICS ENGINEERING, MICROWAVES, RADIO, SIGNAL PROCESSING, SONAR

ABSTRACT SERVICES AND INDEXES

Applied Science and Technology Index; A Cumulative Subject Index To English Language Periodicals in the Fields ofaeronautics and Space Science, Computer Technology,chemistry, Construction Industry, Energy and Related Areas.

H.W. Wilson Co., 950 University Avenue, Bronx, NY 10452.(212) 588-8400. (800) 367-6770. FAX (718) 590-1617. From 1958 to present. Monthly. Inquire about cost andavailability. Also available on CD-ROM and online. ISSN: 0003-6986.

Current Contents: Engineering, Technology and Applied Sciences. Institute for Scientific Information, 3501 MarketStreet, Philadelphia, PA 19104. (215) 386-0100. FAX (215)386-6362. 1970 to present. Weekly. $442.00 per year.Also available on CD-ROM and online. Inquire regarding costand availability. ISSN: 0095-7917.

Electrical and Electronics Abstracts. Institution ofElectrical Engineers (IEE), London. Available fromINSPEC/IEEE - Institute of Electrical and ElectronicEngineers, Box 1331, Hoes Lane, Piscataway, NJ 08855-1331. (908) 562-5549. 1898 to present. Monthly. $2200.00 peryear. Also available on CD-ROM and online as INSPEC.

Engineering Index Monthly; Indexes and Abstracts the World's Engineering and Technical Literature. Engineering Information, Inc., Castle Point on the Hudson, Hoboken, NJ 07030. (201) 216-8500. (800) 221-1044. FAX (201) 216-8532. Monthly. $2300.00 per year. Available online as COMPENDEX and also on CD-ROM. ISSN: 0742-1974.

General Science Index. H.W. Wilson Company, 950 University Avenue, Bronx, NY 10452. (212) 588-8400. (800) 367-6770. FAX (718) 590-1617. From 1978 to present. Ten issues per year; quarterly and annual cumulations. Service basis. Available on CD-ROM and online. Inquire regarding cost and availability. ISSN: 0162-1963.

Government Reports Announcements and Index. U. S. National Technical Information Service (NTIS), 5285 Port Royal Road, Springfield, VA 22161. (703) 487-4650. FAX (703) 321-8547. From 1968 to present. Annual. $630.00 per year. Also available online as *NTIS Bibliographic Database* and on CD-ROM.

Index To IEEE Publications. IEEE Service Center, 445 HoesLane, Piscataway, NJ 08855-1331. (908) 981-1393. (800)678-IEEE. FAX (908) 981-9667. 1973 to present. Annual.ISSN: 0099-1368.

Physics Abstracts. INSPEC. Section A, Science Abstracts. Institution of Electrical Engineers (IEE), London.Available from: INSPEC/IEEE - Institute of Electrical andElectronic Engineers, Box 1331, Hoes Lane, Piscataway, NJ08855-1331. (908) 562-5549. 1898 to present. 24 issuesper year. $2835.00 per year. Also available online and onCD-ROM. ISSN: 0036-8091.

Physics Briefs (Physikalische Berichte). Information Centerfor Energy, Physics, Mathematics; German Physical Society.V C H Publishers, Inc., 220 East 23rd Street, New York, NY10010-4606. (212) 683-8333. 1845 to present. 24 issuesper year. $2390.00 per year. Also available online. ISSN:0179-7434.

Science Citation Index. SCI. Institute for ScientificInformation, 3501 Market Street, Philadelphia, PA 19104.(215) 386-0100. (800) 523-1850. FAX (215) 386-2991. 1961to present. Six issues per year, plus annual cumulation. $11650.00 per year. Also available online and on CD-ROM.Inquire about price and availability. ISSN: 0036-827X.

STAR. Scientific and Technical Aerospace Reports. U.S. National Aeronautics and Space Administration, Scientific and Technical Information Facility, Box 8757, Baltimore-Washington International Airport, MD 21240. (301) 621-0153. Distributed by U. S. Superintendent of Documents, Washington, DC 20402. From 1963 to present. Semimonthly, with semiannual and annual indexes. $114.00 per year. Also available online and on CD-ROM. ISSN: 0036-8741.

ANNUAL REVIEWS AND YEARBOOKS

Advances in Electronics and Electron Physics. Academic Press, Inc., 6277 Sea Harbor Drive, Orlando, FL 32887. (800) 321-5068. From 1948 to present. Irregular. Price varies, inquire. ISSN: 0065-2539.

RADAR

Ency. of Physical Sciences and Engineering Info. Sources

ASSOCIATIONS AND PROFESSIONAL SOCIETIES

American Electronics Association. 5201 Great America Way, Suite 520, P.O. 52990, Santa Clara, CA 95056. (408) 987- 4200. FAX (408) 970-8565.

American Institute of Physics. One Physics Ellipse, College Park, MD 20740-3843. (301) 209-3100.

Electronics Industries Association. 2500 Wilson Boulevard, Arlington, VA 22201. (703) 907-7500. FAX (202) 457-4985. <bodylead>IEEE Antennas and Propagation Society. c/o Institute of Electrical and Electronic Engineers, 345 East 47th Street. New York, NY 10017. (212) 705-7900. FAX (212) 705-4929.

IEEE (Institute of Electrical and Electronic Engineers). 345 East 47th Street. New York, NY 10017. (212) 705-7900. FAX (212) 705-4929,

National Association of Radio and Telecommunications Engineers. PO Box 678, Medway, MA 02053. (508) 533-8333. FAX (508) 533-3815.

United States National Committee For the International Union of Radio Science. Board on Physics and Astronomy, Naitonal Research Council, 2101 Constitution Avenue NW, Washington, DC 20418. (202) 334-3520. FAX (202) 3343-2791.

DIRECTORIES AND BIOGRAPHICAL SOURCES

American Electronics Association Directory. American Electronics Association, 5201 Great America Way, Suite 520, P.O. 52990, Santa Clara, CA 95056. (408) 987-4200. FAX (408) 970-8565. Annual. $175.00.

American Men and Women of Science: Physical and Biological Sciences. R. R. Bowker Inc., 121 Chanlon Road, New Providence, NJ 07974. (908) 464-6800. (800) 521-8110.20th edition. 8 volumes. 1996. $850.00.

E E M - Electronic Engineer's Master. Hearst Business Communications, Inc., 645 Stewart Avenue, Garden City NY 11530. (516) 227-1300. Annual. $90.00 ISSN: 0732-9016.

Engineering Research Centers: Incorporating Electronics Research Centers. Stockton Press, 345 Park Avenue, New York, NY 10010. (212) 689-9200. (800) 221-2123. 4th edition. 1995. $515.00.

IEEE Membership Directory. Institute of Electrical and Electronics Engineers, IEEE Service Center, 445 Hoes Lane, Piscataway, NY 08854. (908) 981-1393. (800) 678-IEEE. FAX (908) 981-9667. 2 volumes. Annual. $190.00. ISSN:

International Engineering Directory. American Consulting Engineers Council, 1015 15th Street, N.W. Suite 802, Washington, DC 20005-2670. (202) 347-7474. Annual. $10.00.

Research Centers Directory. Gale Research Inc., 835 Penobscot Building, Detroit, MI 48226-4094. (313) 961-2242. (800) 877-4253. 20th edition. 1995. $485.00. ISSN: 0080- 1518.

Who's Who in Engineering. Gordon Davis, editor. American Association of Engineering Societies. 1111 19th Street, NY, Suite 608, Washington, DC 20036. (202) 296-2237. (800) 658-8897. 9th edition. 1995. $220.00.

Who's Who in Technology. Gale Research, 835 Penobscot Building, Detroit, MI 48226-4094. (313) 961-2242. (800) 521-4253. 7th edition. 1995. $195.00. ISBN: 0-8103-7467-6.

ENCYCLOPEDIAS AND DICTIONARIES

Chambers Science and Technology Dictionary. Peter M.B. Walker, editor. Cambridge University Press, 40 West 20th Street, New York, NY 10011-4211. (212) 924-3900. 1988. $39.95.

IEEE Standard Dictionary of Electrical and Electronics Terms. Christopher J. Booth, editor. IEEE Service Center, 445 Hoes Lane, Piscataway, NJ 08855-1331. (908) 981-1393. (800) 678-IEEE. FAX (908) 981-9667. IEEE Standard 100- 1992. 5th edition. 1993. $150.00.

Illustrated Dictionary of Electronics. Stan Gibilsco. TAB Books, P.O. Box 40, Blue Summit, PA 17294-0850. (717) 794- 2191. (800) 233-1128. 7th edition. 1994. $34.95.

McGraw-Hill Electronics Dictionary. J. Markus. McGraw-HillBook Company, Inc., 1221 Avenue of the Americas, New York, NY 10020. (212) 997-3675. 5th edition. 1994. $49.95.

McGraw-Hill Encyclopedia of Science and Technology. McGraw-Hill Book, Incorporated, 1221 Avenue of the Americas, New York, NY 10020. (212) 997-3675. (800) 262-4729. Seventh edition. Twenty volumes. 1992. $1900.00.

GENERAL WORKS

Adaptive Radar Detection and Estimation. Simon Haykin and Allan Steinhardt. John Wiley & Sons, Inc., 605 Third Avenue, New York, NY 10158-0012. (212) 850-6000. (800) 225-5945. 1992. $108.00. <bodylead>*Antenna-based Signal Processing Techniques For Radar Systems*. Alfonse Farina. Artech House, 685 Canton Street, Norwood, MA 02062. (617) 769-9750. (800) 225-9977. 1992. $88.00.

Antennas For Radar and Communications: A Polarimetric Approach. Harold Mott. John Wiley & Sons, Inc., 605 Third Avenue, New York, NY 10158-0012. (212) 850-6000. (800) 225-5945. 1992. $99.95.

Electromagnetic and Acoustic Scattering By Simple Shapes. J. J. Bowman, et al, editors. Hemisphere Publishing Corp., 1900 Frost Road, Suite 101, Bristol PA 19007. (215) 785- 5800. (800) 821-8312. 1988. $63.00 in paper.

Elements of Signal Detection and Estimation. Carl W. Helstrom. Prentice Hall General Reference & Travel, 15 Columbus Circle, New York, NY 10023. (212) 373-8500. (800) 223-2348. 1994. $63.00.

High Resolution Radar. Donald R. Wehner. Artech House, 685 Canton Street, Norwood, MA 02062. (617) 769-9750. (800) 225-9977. 2nd edition. 1994. $89.00.

Images in Weather Forecasting: A Practical Guide For Interpreting Satellite and Radar Imagery. M. H. Bader et al, editors. Cambridge University Press, 40 West 20th Street, New York, NY 10011-4211. (212) 924-3900. (800) 872-7423. 1994. $wfi.

Methods and Techniques of Radar Recognition. V. G. Nebabin. Artech House, 685 Canton Street, Norwood, MA 02062. (617) 769-9750. (800) 225-9977. 1994. $79.00.

Radar and Electronic Navigation. G. J. Sonnenberg. Butterworth-Heinemann, 313 Washington Street, Newton, MA 02158. (618-928-2500. (800) 366-2665. 6th edition. 1988. $75.00.

Radar and Sonar. Richard E. Blahut, et al. Springer-Verlag New York, Inc., 175 Fifth Avenue, New York, NY 10010. (212) 460-1500. (800) 777-4643. FAX (212) 473-6272. Parts 1 & 2. 1991. $74.00 set.

Radar in World War II. Henry E. Guerlac. American Institute of Physics. One Physics Ellipse, College Park, MD 20740-3843. (301) 209-3100. 1987. $110.00.

Radar Meteorology. Henri Sauvegeot. Artech House, 685 Canton Street, Norwood, MA 02062. (617) 769-9750. (800) 225-9977. 1992. $78.00.

Signals, Noise and Active Sensors, Radar, Sonar and Laser Radar. John R. Minkoff. John Wiley & Sons, Inc., 605 Third Avenue, New York, NY 10158-0012. (212) 850-6000. (800) 225-5945. 1992. $67.95.

Understanding Antennas for Radar Communication and Avionics. Gregory A. Robertshaw. Van Nostrand Reinhold, 115 Fifth Avenue, New York, NY 10003. (212) 254-3232. (800) 842-3636. 1987. $74.95.

Understanding Radar. Harry Cole. Blackwell Scientific Publications, Inc., 238 Main Street, Cambridge MA 02142. (617) 876-7000. (800) 759-6102. 2nd edition. 1993.$59.95.

HANDBOOKS AND MANUALS

Antenna Engineering Handbook. Richard C. Johnson. McGraw-Hill Book Company, 1221 Avenue of the Americas, New York, NY 10020. (212) 512-2000. (800) 262-4729. 3rd edition, 1993. $119.50.

Electrical Engineer's Handbook. Donald Christiansen, editor. McGraw Book Company, 1221 Avenue of the Americas, New York, NY 10020. (212) 512-2000. (800) 262-4729. 4th edition. 1996. $110.00

Engineering Tables and Data. A.M. Howatson, P.G. Lund and J.D. Todd. Van Nostrand Reinhold, 115 Fifth Avenue, New York, NY 10003. (212) 254-3232. (800) 926-2665. Second Edition. 1991. $39.95.

Jane's Radar and Electronic Warfare Systems. B. Blake, editor. Jane's Information Group, Ltd., 1340 Braddock Place, Suite 300, Alexandria, VA 22314-1651. (703) 683-3700. 1994. $255.00.

Radar and ARPA Manual. A. G. Bole and W. O. Dineley. Butterworth-Heinemann, 313 Washington Street, Newton, MA 02158. (618-928-2500. (800) 366-2665. 1993. $49.95 in paper.

Radar Evaluation Handbook. Anro Engineering Staff, Inc., editors. Artech House, 685 Canton Street, Norwood, MA 02062. (617) 769-9750. (800) 225-9977. 1991. $75.00.

Radar Handbook. Merrill Skolnik. McGraw-Hill Publishing Company, Inc., 1221 Avenue of the Americas, New York, NY 10020. (212) 512-2000. (800) 262-4729. 2nd edition. 1990. $94.50.

Radio and Radar Reference Data. Gordon Press Publications, P.O. Box 459, Bowling Green Station, New York, NY 10004. (718) 624-8419. 1995. $258.95.

Standard Handbook For Electrical Engineers. Donald Fink. McGraw-Hill Publishing Company, 1221 Avenue of the Americas, New York, NY 10020. (212) 512-2000. (800) 262- 4729. 13th edition. 1996. $110.00.

Standard Handbook of Engineering Calculations. Tyler G. Hicks. McGraw-Hill Publishing Company, 1221 Avenue of the Americas, New York, NY 10020. (800) 262-4729 or (212) 512-3825. 3rd edition. 1994. $99.50. ISBN: 0-07-028812-7.

ONLINE DATABASES AND CD-ROMS

COMPENDEX PLUS. Engineering Information, Inc., 345 East 47th Street, New York, NY 10017. (212) 705-7600 or (800) 221-1044. Contains citations with abstracts to worldwide literature in engineering and technology, from 1970 to present. Available on online BRS,(800) 289-4277, DIALOG, (800) 334-2564, ORBIT (800) 456-7248, and STN International, FIZ Karlsruhe, P.O. Box 2465, W-7500, Karlsruhe 1, Germany, online services. Also available on CD-ROM. Inquire as to cost and availability.

Current Contents Search. Institute for Scientific Information, 3501 Market Street, Philadelphia, PA 19104. (215) 386-0100. FAX (215) 386-6362. Contains citations to articles listed in the table of contents of science and technology journals. Also articles in social sciences and life sciences journals. Available on BRS,(800) 289-4277, DIALOG,(800) 334-2564, online services. Inquire as to cost and availability.

Dissertation Abstracts. University Microfilms International, 300 North Zeeb Road, Ann Arbor, MI 48106. (800) 521-0600 or (313) 761-4700. Scope includes virtually all doctoral dissertations accepted at accredited American institutions from 1861 to present in 252 subject areas. Available on BRS,(800) 289-4277, DIALOG,(800) 334-2564, and OCLC EPIC,(800) 848-5878, online services. Also available on CD-ROM. Inquire as to cost and availability.

INSPEC. Institution of Electrical Engineers, Michael Faraday House, Six Hills Way, Stevenage, Herts. SG1 2AY, England. Telephone: 0438 313311 or FAX 0438 742840. Contains citations to the worldwide literature of physics, electronics and electrical engineering, computer technology, and related fields. Available on BRS,(800) 289-4277, DIALOG, (800) 334-2564, ORBIT, (800) 456-7248, and STN International, FIZ Karlsruhe, P.O. Box 2465, W-7500, Karlsruhe 1, Germany, online services. Inquire as to cost and availability. The National Aeronautics and Space Administration only.

NTIS Bibliographic Database. National Technical Information Service, 5285 Port Royal Road, Springfield, VA 22161. (703) 487-4929 or FAX (703) 321-8199. Broad coverage of government-sponsored science and technology research reports, 1964 to present. Available on BRS,(800) 289-4277, DIALOG, (800) 334-2564, ORBIT, (800) 456-7248, and STN International, FIZ Karlsruhe, P.O. Box 2465, W-7500, Karlsruhe 1, Germany, online services. Also available on CD-ROM. Inquire as to cost and availability.

SCISEARCH. Institute for Scientific Information, 3501 Market Street, Philadelphia, PA 19104. (800) 523-1850 or (215) 386-0100. Broad multidisciplinary title and author index to the international literature of science and technology, 1974 to present. Available on DIALOG,(800) 334-2564, and ORBIT,(800) 456-

RADAR

Ency. of Physical Sciences and Engineering Info. Sources

7248, online services. Also available on CD-ROM. Inquire as to cost and availability.

WILSONLINE. H.W. Wilson Company, 950 University Avenue, Bronx, NY 10452. (800) 367-6770 or (212) 588-8400. Makes available online versions of the printed H.W. Wilson indexes including Applied Science and Technology Index, Business Periodicals Index, General Science Index, and Readers' Guide to Periodical Literature. Period covered is generally 1983 to present. Available on BRS,(800) 289-4277, DIALOG,(800) 334-2564, and OCLC EPIC,(800) 848-5878, online services. Also available on CD-ROM. Inquire as to cost and availability.

PERIODICALS

Electronic Design. Penton Publishing, San Jose Gateway, Suite 354. 2025 Gateway Place, San Jose, CA 95110. (408) 441-0550. 1952 to present. Fortnightly. $95.00. ISSN: 0013-4872.

Electronics. Penton Publishing, San Jose Gateway, Suite 354. 2025 Gateway Place, San Jose, CA 95110. (408) 441-0550. 1930 to present. Semi-weekly. $98.00 per year. ISSN: 0883-4989.

Electronics World and Wireless World. Reed Business Publishing, Ltd., Quadrant House, the Quadrant, Sutton, Surrey SM2 5AS, England. 1911 to present. Monthly. L35. ISSN: 0266-3244.

IEE Proceedings. Part H: Microwaves, Antennas and Propagation. Institution of Electrical Engineers. Available from: Institute of Electrical and Electronic Engineers (IEEE) Inc., Box 1331, 445 Hoes Lane, Piscataway, NJ 08855- 1331. (908) 981-0060. 1980 to present. Bi-monthly. ISSN:0950-107X.

IEEE - Signal Processing Magazine. Institute of Electrical and Electronics Engineers, Inc., Box 1331, 445 Hoes Lane, Piscataway, NJ 08855-1331. (908) 981-0060. 1984 to present. Quarterly. $70.00 per year. ISSN: 1053-5888.

IEEE Circuits and Devices Magazine. Institute of Electrical and Electronics Engineers, Inc., Box 1331, 445 Hoes Lane, Piscataway, NJ 08855-1331. (908) 981-0060. 1985 to present. Bi-monthly. $120.00 per year. ISSN: 8755-3996.

IEEE Transactions On Antennas and Propagation. Institute of Electrical and Electronics Engineers, Inc., Box 1331, 445 Hoes Lane, Piscataway, NJ 08855-1331. (908) 981-0060. 1952 to present. Monthly. $212.00 per year. ISSN: 0018-926X.

International Journal of Microwave and Millimeter-wave Computer Aided Engineering. John Wiley and Sons, Inc., 605 Third Avenue, New York, NY 10158. (800) 526-5368 or (212) 850-6000. 1991 to present. Quarterly. $220.00 per year. ISSN: 1050-1827.

RESEARCH CENTERS AND INSTITUTES

Applied Microelectronics Institute. 1127 Barrington Street, Halifax, NS, Canada B3H 2P8. (902) 421-1250. FAX (902) 429-9983.

Applied Research Laboratories. University of Texas at Austin, PO Box 8029, University Station, Austin, TX 78713-8029. (512) 835-3200. FAX (512) 835-3259.

Radar and Instrumentation Development Laboratory. Georgia Institute of Technology, Georgia Tech Research Institute, Atlanta, GA 30332. (404) 528-7741..

Radar Systems and Remote Sensing Laboratory. University of Kansas Center for Research, 2291 Irving Hill Road, Lawrence, KS 66045. (913) 864-4832.

RADIATION

See also: ATMOSPHERE, GAMMA RAYS, NUCLEAR ENERGY, NUCLEAR ENGINEERING, PLUTONIUM, RADIATION SHIELDING, X-RAYS

ABSTRACT SERVICES AND INDEXES

Applied Science and Technology Index; A Cumulative Subject Index To English Language Periodicals in the Fields of Aeronautics and Space Science, Computer Technology, Chemistry, Construction Industry, Energy and Related Areas. H. W. Wilson Co., 950 University Avenue, Bronx, NY 10452- 9978. (212) 588-8400. (800) 367-6770. FAX (718) 590-1617. From 1958 to present. Monthly. Inquire about cost and availability. Also available online (BRS and WILSONLINE) and on CD-ROM. ISSN: 0003-6986.

Chemical Abstracts. Chemical Abstracts Service. 2540 Olentangy River Road, Box 3012, Columbus, OH 43210-0012. (614) 447-3600. FAX (614) 447-3713. 1907 to present. Weekly. $16,800.00 per year. Available online and on CD-ROM. CA is also available in five section groupings. Inquire regarding cost and availability.

Current Contents: Engineering, Technology and Applied Sciences. Institute for Scientific Information, 3501 Market Street, Philadelphia, PA 19104. (215) 386-0100. FAX (215) 386-2991. From 1970 to present. Weekly. $442.00 per year. Also available online. Inquire regarding cost and availability. ISSN: 0095-7917.

Engineering Index Monthly. Engineering Information, Inc., Castle Point on the Hudson, Hoboken, NJ 07030. (201) 216-8500. (800) 221-1044. FAX (201) 216-8532. Monthly. $2300.00 per year for monthly issues. $1980 per year for annual cumulation. Available on CD-ROM and online as COMPENDEX. ISSN: 0742-1974; 0360-8557.

General Science Index. H.W. Wilson Company, 950 University Avenue, Bronx, NY 10452. (212) 588-8400. (800) 367-6770. FAX (718) 590-1617. From 1978 to present. Ten issues per year; quarterly and annual cumulations. Service basis. Available on CD-ROM and online. Inquire regarding cost and availability. ISSN: 0162-1963.

Physics Abstracts. INSPEC. Section A, Science Abstracts. Institution of Electrical Engineers (IEE), London, United Kingdom. Available from: INSPEC/IEEE - Institute of Electrical and Electronic Engineers, Box 1331, 445 Hoes Lane, Piscataway, NJ 08855-1331. (908) 562-5549. 1898 to present. 24 issues per year. $2835.00 per year. ISSN: 0036-8091. Also available online and on CD-ROM.

Physics Briefs (Physikalische Berichte). Information Center for Energy, Physics, Mathematics; German Physical Society. V C H Publishers, Inc., 220 East 23rd Street, New York, NY 10010-4606. (212) 683-8333. 1845 to present. 24 issues per year. $2390.00 per year. Also available online. ISSN: 0179-7434.

Science Citation Index. SCI. Institute for Scientific Information, 3501 Market Street, Philadelphia, PA 19104. (215) 386-0100. (800) 523-1850. FAX (215) 386-2991. 1961 to present. Six issues per year, plus annual cumulation. $11650.00 per year. Also available online and on CD-ROM. Inquire about price and availability. ISSN: 0036-827X.

ANNUAL REVIEWS AND YEARBOOKS

Advances in Nuclear Science and Technology. Plenum Publishing Corp., 233 Spring Street, New York, NY. (212) 620-8000. (800) 2221-9369. FAX (212) 463-0742. 1977 to present. Irregular.

ASSOCIATIONS AND PROFESSIONAL SOCIETIES

American Nuclear Society. 555 North Kensington Avenue, LA Grange Park, IL 60525. (708) 352-6611. (800) NUC-NEWS. FAX (708) 352-0499.

American Society For Nondestructive Testing. 2722 Arlingate Lane, PO Box 28518, Columbus OH 43228-0518. (614) 274-6003.

Fusion Power Associates. 2 Professional Drive, Suite 248, Gaithersburg, MD 20879. (301) 258-0545. FAX (301) 975-9869.

IEEE Nuclear and Plasma Sciences Society. c/o Institute of Electrical and Electronics Engineers, 345 East 47th Street, New York, NY 10017. (212) 705-7900. FAX 212) 705-4929.

Institute of Nuclear Materials Management. 60 Revere Drive, Northbrook, IL 60062. (847) 480-9573. FAX (947) 480-9282.

Radiation Research Society. 2021 Spring Road, Suite 600, Oak Brook Road, IL 60521. (708) 571-2881. (708) 571-7837.

BIBLIOGRAPHIES

Handbooks and Tables in Science and Technology. Russell H. Powell, editor. Oryx Press, 4041 North Central, Suite 700, Phoenix, AZ 85012-3330. (602) 265-2651 or (800) 279-6799. Third edition. 1994. $65.00.

Information Sources in Chemistry. R.T. Bottle and J.F.B Rowland, editors. R.R. Bowker Inc., 121 Chanlon Road, New Providence, NJ 07974. (800) 521-8110 or (908) 464-6800. Fourth edition. 1993. $75.00.

Information Sources in Engineering. Peter Hicks, editor. Bowker-Saur, 121 Chanlon Road, New Providence, NJ 07974. (800) 521-8110 or (908) 464-6800. 3rd edition. 1996. $125.00.

Scientific and Technical Books and Serials in Print; An Index To Literature in Science and Technology. R.R. Bowker Inc., 121 Chanlon Road, New Providence, NJ 07974. (908) 464-6800. (800) 521-8110. FAX (908) 665-3502. 1972 to present. Annual. 4 volumes. 1994. $299.95. Also available on compact disc and online. ISSN: 0000-054X.

DIRECTORIES AND BIOGRAPHICAL SOURCES

American Men and Women of Science. Physical and Biological Sciences. Fifteenth edition. R.R. Bowker Inc., 121 Chanlon

Road, New Providence, NJ 07974. (908) 464-6800. (800) 521-8110. 20th edition. 8 volumes. 1996. $850.00

Energy and Nuclear Science International Who's Who. Gale Research, 835 Penobscot Building, Detroit, MI 48226- 4094. (313) 961-2242. (800) 877-4253. Triennial. 1993. $325.00.

Energy Information Centers Directory. U. S. Council for Energy Awareness, 1776 I Street, NW, Suite 400, Washington, DC 20006. (202) 293-0770. Triennial. No charge.

Engineering Research Centers: Incorporating Electronics Research Centers. Stockton Press, 345 Park Avenue, New York, NY 10010. (212) 689-9200. (800) 221-2123. 4th edition. 1995. $515.00.

Fusion Facilities Directory. Fusion Power Associates, 2 Professional Drive, Suite 248, Gaithersburg, MD 20879. (301) 258-0545. Biennial, odd years. $20.00.

International Directory of Nuclear Utilities. Nuexco, 1515 Arapahoe Street, Suite 900, Denver CO 80202. (303) 899- 4500. Annual. $220.00 per year

Nuclear Fusion - World Survey of Activities in Controlled Fusion Research. UNIPUB, 4611-F Assembly Drive, Lnaham MD 20706. (800) 274-4447. Triennial. 1991. $95.00.

Nuclear News - Directory Issues. American Nuclear Society, 555 North Kensington Avenue, LA Grange Park, IL 60525. (708) 352-6611. (800) NUC-NEWS. March issue contains Buyer's Guide; $65.00. World List of Nuclear Power Plants. Semi-annual in February and August. $15.00 per issue.

Radiation Research Society Membership Directory. 1891 Preston White Drive, Reston, VA 22091. (703) 648-3780. Included with membership. Annual.

Research Centers Directory. Gale Research Company, Inc., 835 Penobscot Building, Detroit, MI 48226-4094. (313) 961-2242. (800) 877-4253. 20th edition. 1995. $485.00. ISSN: 0080-1518.

Who's Who in Engineering. Gordon Davis, editor. American Association of Engineering Societies. 1111 19th Street, NY, Suite 608, Washington, DC 20036. (202) 296-2237. (800) 658-8897. 9th edition. 1995. $220.00.

World Energy and Nuclear Directory. Gale Research Company, Inc., 835 Penobscot Building, Detroit, MI 48226-4094. (313) 961-2242. (800) 877-4253. Biennial, even years. $450.00.

ENCYCLOPEDIAS AND DICTIONARIES

Dictionary of Nuclear Technology. Pro/Am Music Resources, Inc., 63 Prospect Street, White Plains, New York, 10606. (914) 948-7436. 1986. $245.00.

Encyclopedia of Applied Physics. VCH Publishers, Inc., 303 Northwest 12th Avenue, Deerfield Beach, FL 33442. (800) 367-8249. 1991-. Twenty volumes. $6000.00.

IEEE Nuclear Science Standards Collection. IEEE Standards office, Box 1331, 445 Hoes Lane, Piscataway, NJ 08855-1331. (908) 562-3822. (800) 678-4333. 1991. $90.00 in paper.

RADIATION

Ency. of Physical Sciences and Engineering Info. Sources

Kirk-Othmer Encyclopedia of Chemical Technology. John Wiley and Sons, Inc., 605 Third Avenue, New York, NY 10158. (800) 526-5368 or (212) 850-6000. Fourth edition. 1991 - . Twenty-seven volumes. $5400.00.

McGraw-Hill Encyclopedia of Science and Technology. McGraw-Hill Book, Incorporated, 1221 Avenue of the Americas, New York, NY 10020. (212) 997-3675. (800) 262-4729. Seventh edition. Twenty volumes. 1992. $1900.00.

Ullman's Encyclopedia of Industrial Chemistry. VCH Publications, Inc., 220 East 23rd Street, Suite 909, New York, NY 10010-4606. (212) 683-8333. (800) 422-8824. 5th edition. 1984 - . Price varies per volume. ISSN:

GENERAL WORKS

Atmospheric Radiation: Theoretical Basis. Richard M. Goodwin, et al,. Oxford University Press, Inc., 200 Madison Avenue, New York, NY 10016. (212) 725-6000. (800) 334-4249.1989. $47.00.

Atoms, Radiation and Radiation Protection. James E. Turner. John Wiley & Sons, Inc., 605 Third Avenue, New York, NY 10158-0012. (212) 850-6000. (800) 225-5945. 2nd edition.1994. $69.95.

Coherent Radiation Generation and Particle Accelerators. A. Prokhorov, et al. American Institute of Physics, One Physics Ellipse, College Park, MD 20740-3843. (301) 209- 3100. 1992. $120.00.

Basic Radiation Protection Technology. Daniel A. Gollnick. Pacific Radiation Corporation, 2945 StoneHill Drive, Altadena CA 91001. (818) 798-8100. 1994. $75.00 in paper.

Foundations of Radiation Theory and Quantum electrodynamics. Asim O. Barat, editor. Plenum Publishing Corp., 233 Spring Street, New York, NY. (212) 620-8000. (800) 2221-9369. FAX (212) 463-0742. 1980. $65.00.

Introduction To Radiometry and Photometry. Ross McCluney. Artech House, 685 Canton Street, Norwood, MA 02062. (617) 769-9750. (800) 225-9977. 1994. $85.00.

Legacy of Chernobyl. Zhores A. Medvedev. W. W. Norton & Company, 500 Fifth Avenue, New York, NY 10110. (212) 354-5500. (800) 223-2584. 1992. $10.95 in paper.

Our Radiant World. David W. Lillie. TAB Books, P.O. Box 40, Blue Summit, PA 17294-0850. (717) 794-2191. (800) 233-1128. 1987. $13.95 in paper.

Nuclear Safety: After Three Mile Island and Chernybol. G. M. Ballard, Elsevier Science Publishing Company, Inc., 655 Avenue of the Americas, New York, NY 10010. (212) 989-5800. FAX (914) 333-2444. 1988. $106.25.

Radiation Exchange: An Introduction. J. H. Taylor. Academic Press, Inc., 6277 Sea Harbor Drive, Orlando, FL. (800) 321-5068. 1990. $43.00.

Radiation theory and the Quantum Revolution. Joseph Agassi. Birkhauser, Boston, 675 Massachusetts Avenue, Cambridge, MA 02139. (617) 876-2333. (800) 777-4643. 1993. $59.00.

HANDBOOKS AND MANUALS

Handbook of Radiation Measurement and Protection. Allen Brodsky, editor. Franklin Franklin Book Co., Inc., Elkins Park, PA 19117. (215) 635-5252. Reprint edition. 2 volumes. 1979-1982. $737.00.

Handbook of Radiation and Scattering of Waves. Adrianus T. DE Hoop, editor. Academic Press, Inc., 6277 Sea Harbor Drive, Orlando, FL. (800) 321-5068. 1995. $199.00.

Handbook of Radiation Effects. andrew Holmes-Siedle and Len Adams. Oxford University Press, Inc., 200 Madison Avenue, New York, NY 10016. (212) 725-6000. (800) 334-4249. 1993. $75.00.

IEEE Nuclear Power Standards Archives Collection. IEEE Standards office, Box 1331, 445 Hoes Lane, Piscataway, NJ 08855-1331. (908) 562-3822. (800) 678-4333. Revisededition. 1990. $165.00.

Information Center On Nuclear Standards. American Nuclear Society, 555 North Kensington Avenue, LA Grange Park, IL 60525. (708) 352-6611. Standards for all aspects of the design, construction, operation and maintenance of nuclear power plants.

ONLINE DATABASES AND CD-ROMS

CA Search. Chemical Abstracts Service, P.O. Box 3012, Columbus, OH 43210-0012. (614) 447-3600. (800) 848-6533. FAX (614) 447-3709. Very comprehensive guide to worldwide chemical literature and related fields. 1972 to present. Available on BRS,(800) 289-4277, DIALOG, (800) 334-2564, ORBIT (800) 456-7248, and STN International, FIZ Karlsruhe, P.O. Box 2465, W-7500, Karlsruhe 1, Germany, online services. Inquire as to cost and availability.

Compendex Plus. Engineering Information, Inc., 345 East 47th Street, New York, NY 10017. (212) 705-7600 or (800) 221-1044. Contains citations with abstracts to worldwide literature in engineering and technology, from 1970 to present. Available on online BRS,(800) 289-4277, DIALOG, (800) 334-2564, ORBIT (800) 456-7248, and STN International, FIZ Karlsruhe, P.O. Box 2465, W-7500, Karlsruhe 1, Germany, online services. Also available on CD-ROM. Inquire as to cost and availability.

Current Contents Search. Institute for Scientific Information, 3501 Market Street, Philadelphia, PA 19104. (215) 386-0100. FAX (215) 386-6362. Contains citations to articles listed in the table of contents of science and technology journals. Also articles in social sciences and life sciences journals. Available on BRS,(800) 289-4277, DIALOG,(800) 334-2564, online services. Inquire as to cost and availability.

Dissertation Abstracts. University Microfilms International, 300 North Zeeb Road, Ann Arbor, MI 48106. (800) 521-0600 or (313) 761-4700. Scope includes virtually all doctoral dissertations accepted at accredited American institutions from 1861 to present in 252 subject areas. Available on BRS, (800) 289-4277, DIALOG, (800) 334-2564, and OCLC EPIC, (800) 848-5878, online services. Also available on CD-ROM. Inquire as to cost and availability.

INSPEC. Institution of Electrical Engineers, Michael Faraday House, Six Hills Way, Stevenage, Herts. SG1 2AY, England. Telephone: 0438 313311 or FAX 0438 742840. Contains citations to the worldwide literature of physics, electronics and electrical engineering, computer technology, and related fields. Available on BRS, (800) 289-4277, DIALOG, (800) 334-2564, OR-BIT, (800) 456-7248, and STN International, FIZ Karlsruhe, P.O. Box 2465, W-7500, Karlsruhe 1, Germany, online services. Inquire as to cost and availability.

Kirk-Othmer Encyclopedia of Chemical Technology. John Wiley and Sons, Inc., 605 Third Avenue, New York, NY 10158. (800) 526-5368 or (212) 850-6000. Contains the complete text of all chapters in the 27 volume fourth edition of the KIRK-Othmer ENCYCLOPEDIA of CHEMICAL TECHNOLOGY. 1991. Available on BRS,(800) 289-4277, DIALOG,(800) 334-2564, online services. Inquire as to cost and availability.

NTIS Bibliographic Database. National Technical Information Service, 5285 Port Royal Road, Springfield, VA 22161. (703) 487-4929 or FAX (703) 321-8199. Broad coverage of government-sponsored science and technology research reports, 1964 to present. Available on BRS,(800) 289-4277, DIALOG, (800) 334-2564, ORBIT, (800) 456-7248, and STN International, FIZ Karlsruhe, P.O. Box 2465, W-7500, Karlsruhe 1, Germany, online services. Also available on CD-ROM. Inquire as to cost and availability.

Nuclear Science Abstracts. U.S. Department of Energy, office of Scientific and Technical Energy, P.O. Box 62, Oak Ridge, TN 37831. (615) 576-1189. Contains citations with abstracts to literature in all fields of nuclear science and energy. 1948 to 1976. Available on DIALOG,(800) 334-2564, online service. Inquire as to cost and availability.

Physics Briefs. American Institute of Physics, 335 East 45th Street, New York, NY 10017. (212) 661-9260 or FAX (212) 661-2036. Contains citations with abstracts of the literature of physics and related fields, 1979 to present. Available on the STN International, FIZ Karlsruhe, P.O. Box 2465, W-7500, Karlsruhe 1, Germany, online service. Inquire as to cost and availability.

SCISEARCH. Institute for Scientific Information, 3501 Market Street, Philadelphia, PA 19104. (800) 523-1850 or (215) 386-0100. Broad multidisciplinary title and author index to the international literature of science and technology, 1974 to present. Available on DIALOG, (800) 334-2564, and ORBIT, (800) 456-7248, online services. Also available on CD-ROM. Inquire as to cost and availability.

WILSONLINE. H.W. Wilson Company, 950 University Avenue, Bronx, NY 10452. (800) 367-6770 or (212) 588-8400. Makes available online versions of the printed H.W. Wilson indexes including Applied Science and Technology Index, Business Periodicals Index, General Science Index, and Readers' Guide to Periodical Literature. Period covered is generally 1983 to present. Available on BRS, (800) 289-4277, DIALOG, (800) 334-2564, and OCLC EPIC, (800) 848-5878, online services. Also available on CD-ROM. Inquire as to cost and availability.

PERIODICALS

American Nuclear Society Transactions. American Nuclear Society, 555 North Kensington Avenue, LA Grange Park, IL 60525. (708) 352-6611. 1958 to present. 3 issues per year. $340.00 per year. ISSN: 0003-018X

Bulletin of the Atomic Scientists. Educational Foundation for Nuclear Science, 6042 South Kimbark Avenue, Chicago, IL 60637. (312) 702-2555. 1945 to present. 10 issues per year. $30.00 per year. ISSN: 0096-3402.

Journal of Environmental Radioactivity. Elsevier Science Publishing Company, Inc., 660 White Plains Road, Tarrytown, NY 10591-5133. (914) 524-9200. 1984 to present. Monthly. $645.00 per year. ISSN: 0265-931X.

Nuclear Science and Engineering. American Nuclear Society, 555 North Kensington Avenue, LA Grange Park, IL 60525. (708) 352-6611. 1956 to present. Monthly. $405.00 per year. ISSN: 0029-5639.

Radiation Research. Kluge Carden Jennings, 853 West Main Street, Charlottesville, VA 22903. 1954 to present. Monthly. $560.00 per year. ISSN: 0033-7587.

RESEARCH CENTERS AND INSTITUTES

Department of Nuclear Engineering and Engineering Physics. University of Wisconsin - Madison, 1500 Johnson Drive, Madison, WI 53706. (608) 263-1648. FAX (608) 262-6707

Ford Nuclear Reactor. Phoenix Memorial Laboratory. University of Michigan, 2301 Bonisteel Boulevard, Ann Arbor MI 48109-2100. (313) 764-6223. FAX (313) 764-4540.

Frank H. Neely Nuclear Research Center. Georgia Institute of Technology, 900 Atantic Drive NW, Atlanta, GA 30332. (404) 894-3600.

Fusion Research Center. University of Texas at Austin. RLM 11.222, Austin, TX 78712. ((512) 471-3926. FAX (512) 471-8865.

Institute of Plasma and Fusion Research. University of California, Los Angeles, 44-139 Engineering IV 405 Hilgard Avenue, Los Angeles, CA 90024-1597. (213) 825-1613. FAX (213) 206-4832.

Nuclear Physics Laboratory. University of Washington, Seattle, WA 98195. (206) 543-4080.

Radiation Science and Engineering Center. Pennsylvania State University, University Park, PA 16802. (814) 865- 6351. FAX (814) 863-4840.

Stanford Synchrotron Radiation Laboratory. Stanford University, SLAC Bin #69, P.O. Box 4349, Stanford, CA 943050210. (415) 926-3153. FAX (415) 926-4100.

RADIATION SHIELDING

See also: ATMOSPHERE, GAMMA RAYS, NUCLEAR ENERGY, NUCLEAR ENGINEERING, RADIATION, X-RAYS

ABSTRACT SERVICES AND INDEXES

Applied Science and Technology Index; A Cumulative Subject Index To English Language Periodicals in the Fields of Aeronautics and Space Science, Computer Technology, Chemistry, Construction Industry, Energy and Related Areas. H. W. Wilson Co., 950 University Avenue, Bronx, NY 10452- 9978. (212) 588-8400. (800) 367-6770. FAX (718) 590-1617. From 1958 to present. Monthly. Inquire about cost and availability. Also available online (BRS and WILSONLINE) and on CD-ROM. ISSN: 0003-6986.

Chemical Abstracts. Chemical Abstracts Service. 2540 Olentangy River Road, Box 3012, Columbus, OH 43210-0012. (614) 447-3600. FAX (614) 447-3713. 1907 to present. Weekly. $16,800.00 per year. Available online and on CD- ROM. CA is also available in five section groupings. Inquire regarding cost and availability.

Current Contents: Engineering, Technology and Applied Sciences. Institute for Scientific Information, 3501 Market Street, Philadelphia, PA 19104. (215) 386-0100. FAX (215) 386-2991. From 1970 to present. Weekly. $442.00 per year. Also available online. Inquire regarding cost and availability. ISSN: 0095-7917.

Engineering Index Monthly. Engineering Information, Inc., Castle Point on the Hudson, Hoboken, NJ 07030. (201) 216-8500. (800) 221-1044. FAX (201) 216-8532. Monthly. $2300.00 per year for monthly issues. $1980 per year for annual cumulation. Available on CD-ROM and online as COMPENDEX. ISSN: 0742-1974; 0360-8557.

General Science Index. H.W. Wilson Company, 950 University Avenue, Bronx, NY 10452. (212) 588-8400. (800) 367-6770. FAX (718) 590-1617. From 1978 to present. Ten issues per year; quarterly and annual cumulations. Service basis. Available on CD-ROM and online. Inquire regarding cost and availability. ISSN: 0162-1963.

Physics Abstracts. INSPEC. Section A, Science Abstracts. Institution of Electrical Engineers (IEE), London, United Kingdom. Available from: INSPEC/IEEE - Institute of Electrical and Electronic Engineers, Box 1331, 445 Hoes Lane, Piscataway, NJ 08855-1331. (908) 562-5549. 1898 to present. 24 issues per year. $2835.00 per year. ISSN: 0036-8091. Also available online and on CD-ROM.

Physics Briefs (Physikalische Berichte). Information Center for Energy, Physics, Mathematics; German Physical Society. V C H Publishers, Inc., 220 East 23rd Street, New York, NY 10010-4606. (212) 683-8333. 1845 to present. 24 issues per year. $2390.00 per year. Also available online. ISSN: 0179-7434.

Science Citation Index. SCI. Institute for Scientific Information, 3501 Market Street, Philadelphia, PA 19104. (215) 386-0100. (800) 523-1850. FAX (215) 386-2991. 1961 to present. Six issues per year, plus annual cumulation. $11650.00 per year. Also available online and on CD-ROM. Inquire about price and availability. ISSN: 0036-827X.

ANNUAL REVIEWS AND YEARBOOKS

Advances in Nuclear Science and Technology. Plenum Publishing Corp., 233 Spring Street, New York, NY. (212) 620-8000. (800) 2221-9369. FAX (212) 463-0742. 1977 to present. Irregular.

ASSOCIATIONS AND PROFESSIONAL SOCIETIES

American Nuclear Society. 555 North Kensington Avenue, LA Grange Park, IL 60525. (708) 352-6611. (800) NUC-NEWS. FAX (708) 352-0499.

American Society For Nondestructive Testing. 2722 Arlingate Lane, PO Box 28518, Columbus OH 43228-0518. (614) 274-6003.

Fusion Power Associates. 2 Professional Drive, Suite 248, Gaithersburg, MD 20879. (301) 258-0545. FAX (301) 975-9869.

IEEE Nuclear and Plasma Sciences Society. c/o Institute of Electrical and Electronics Engineers, 345 East 47th Street, New York, NY 10017. (212) 705-7900. FAX 212) 705- 4929.

Institute of Nuclear Materials Management. 60 Revere Drive, Northbrook, IL 60062. (847) 480-9573. FAX (947) 480-9282.

National Council On Radiation Protection and Measurements. 7910 Woodmont Avenue, Suite 1016, Bethesda, MD 10814. (301)657-2652.

Radiation Research Society. 2021 Spring Road, Suite 600, Oak Brook Road, IL 60521. (708) 571-2881. (708) 571-7837.

BIBLIOGRAPHIES

Information Sources in Engineering. Peter Hicks, editor. Bowker-Saur, 121 Chanlon Road, New Providence, NJ 07974. (800) 521-8110 or (908) 464-6800. 3rd edition. 1996. $125.00.

Scientific and Technical Books and Serials in Print; An Index To Literature in Science and Technology. R.R. Bowker Inc., 121 Chanlon Road, New Providence, NJ 07974. (908) 464-6800. (800) 521-8110. FAX (908) 665-3502. 1972 to present. Annual. 4 volumes. 1994. $299.95. Also available on compact disc and online. ISSN: 0000-054X.

DIRECTORIES AND BIOGRAPHICAL SOURCES

American Men and Women of Science. Physical and Biological Sciences. Fifteenth edition. R.R. Bowker Inc., 121 Chanlon Road, New Providence, NJ 07974. (908) 464-6800. (800) 521-8110. 20th edition. 8 volumes. 1996. $850.00

Energy and Nuclear Science International Who's Who. Gale Research, 835 Penobscot Building, Detroit, MI 48226- 4094. (313) 961-2242. (800) 877-4253. Triennial. 1993. $325.00.

Energy Information Centers Directory. U. S. Council for Energy Awareness, 1776 I Street, NW, Suite 400, Washington, DC 20006. (202) 293-0770. Triennial. No charge.

Fusion Facilities Directory. Fusion Power Associates, 2 Professional Drive, Suite 248, Gaithersburg, MD 20879. (301) 258-0545. Biennial, odd years. $20.00.

International Directory of Nuclear Utilities. Nuexco, 1515 Arapahoe Street, Suite 900, Denver CO 80202. (303) 899- 4500. Annual. $220.00 per year

Nuclear Fusion - World Survey of Activities in Controlled Fusion Research. UNIPUB, 4611-F Assembly Drive, Lnaham MD 20706. (800) 274-4447. Triennial. 1991. $95.00.

Nuclear News - Directory Issues. American Nuclear Society, 555 North Kensington Avenue, LA Grange Park, IL 60525. (708) 352-6611. (800) NUC-NEWS. March issue contains Buyer's Guide; $65.00. World List of Nuclear Power Plants. Semi-annual in February and August. $15.00 per issue.

Radiation Research Society Membership Directory. 1891 Preston White Drive, Reston, VA 22091. (703) 648-3780. Included with membership. Annual.

Research Centers Directory. Gale Research Company, Inc., 835 Penobscot Building, Detroit, MI 48226-4094. (313) 961-2242. (800) 877-4253. 20th edition. 1995. $485.00. ISSN: 0080-1518.

Who's Who in Engineering. Gordon Davis, editor. American Association of Engineering Societies. 1111 19th Street, NY, Suite 608, Washington, DC 20036. (202) 296-2237. (800) 658-8897. 9th edition. 1995. $220.00.

World Energy and Nuclear Directory. Gale Research Company, Inc., 835 Penobscot Building, Detroit, MI 48226-4094. (313)961-2242. (800) 877-4253. Biennial, even years. $450.00.

World List of Nuclear Power Plants. Anerican Nuclear Society, 555 North Kensington Avenue, LA Grange Park, IL 60525. (708) 352-6611. (800) NUC-NEWS. Semi-annual in February and August issues of Nuclear News. $15.00 per copy.

ENCYCLOPEDIAS AND DICTIONARIES

Dictionary of Nuclear Technology. Pro/Am Music Resources, Inc., 63 Prospect Street, White Plains, New York, 10606. (914) 948-7436. 1986. $245.00.

Encyclopedia of Applied Physics. VCH Publishers, Inc., 303 Northwest 12th Avenue, Deerfield Beach, FL 33442. (800) 367-8249. 1991-. Twenty volumes. $6000.00.

IEEE Nuclear Science Standards Collection. IEEE Standards office, Box 1331, 445 Hoes Lane, Piscataway, NJ 08855-1331. (908) 562-3822. (800) 678-4333. 1991. $90.00 in paper.

Kirk-Othmer Encyclopedia of Chemical Technology. John Wiley and Sons, Inc., 605 Third Avenue, New York, NY 10158. (800) 526-5368 or (212) 850-6000. Fourth edition. 1991 - . Twenty-seven volumes. $5400.00.

McGraw-Hill Encyclopedia of Science and Technology. McGraw-Hill Book, Incorporated, 1221 Avenue of the Americas, New York, NY 10020. (212) 997-3675. (800) 262-4729. Seventh edition. Twenty volumes. 1992. $1900.00.

Ullman's Encyclopedia of Industrial Chemistry. VCH Publications, Inc., 220 East 23rd Street, Suite 909, New York, NY 10010-4606. (212) 683-8333. (800) 422-8824. 5th edition. 1984 Price varies per volume. ISSN:

GENERAL WORKS

Atoms, Radiation and Radiation Protection. James E. Turner. John Wiley & Sons, Inc., 605 Third Avenue, New York, NY 10158-0012. (212) 850-6000. (800) 225-5945. 2nd edition.1995. $69.95.

Basic Radiation Protection Technology. Daniel A. Gollnick. Pacific Radiation Corporation, 2945 StoneHill Drive, Altadena CA 91001. (818) 798-8100. 1994. $75.00 in paper.

Emergency Response Resources Guide for Nuclear Power Plant Emergencies. Gordon Press Publications, P.O. Box 459, Bowling Green Station, New York, NY 10004. (718) 624-8419. 1995. $250.75.

An Introduction To Radiation Protection. Alan D. Martin and S. A. Harbison. Chapman & Hall, 1 Penn Plaza, New York, NY 10119. (212) 564-1060. 1986. $18.95 in paper.

Legacy of Chernobyl. Zhores A. Medvedev. W. W. Norton & Company, 500 Fifth Avenue, New York, NY 10110. (212) 354-5500. (800) 223-2584. 1992. $10.95 in paper.

Nuclear Reactors - Physics, Design and Safety. A. Gandini, et al editors. World Scientific Publishing Company, Inc., 1060 Main Street, River Edge, NJ 07661. (201) 487-9655. (800) 227-7562. 1995. $168.00.

Our Radiant World. David W. Lillie. TAB Books, P.O. Box 40, Blue Summit, PA 17294-0850. (717) 794-2191. (800) 233- 1128. 1987. $13.95.

Radiation Detectors: Physical Principles and Applications. C. F. Delaney. Oxford University Press, Inc., 200 Madison Avenue, New York, NY 10016. (212) 725-6000. (800) 334-4249. 1992. $98.00.

Radiation Protection. William M. Hallenbeck. Lewis Pubs., CRC Press, Inc., 2000 Corporate Boulevard, NW, Boca Raton, FL 33431. (407) 994-0555. (800) 272-7737. 1994. $59.95.

Radiation Shielding. Kenneth J. Shultis and Richard E. Faw. Prentice-Hall, 113 Sylvan Avenue, Route 9W, Englewood Cliffs, NJ 07632. (201) 592-2000. (800) 922-0579. 1996. $85.00.

Radioactivity Measurements: Principles and Practice. W. B. Mann, et al, editors. Butterworth-Heinemann, 313 Washington Street, Newton, MA 02158. (618-928-2500. (800) 366-2665.1991. $30.00.

HANDBOOKS AND MANUALS

Handbook of Radiation Measurement and Protection. Allen Brodsky, editor. Franklin Book Co., Inc., Elkins Park, PA 19117. (215) 635-5252. Reprint edition. 2 volumes. 1979-1982. $737.00.

Handbook of Radiation and Scattering of Waves. Adrianus T. DE Hoop, editor. Academic Press, Inc., 6277 Sea Harbor Drive, Orlando, FL. (800) 321-5068. 1995. $199.00.

Handbook of Radiation Effects. andrew Holmes-Siedle and Len Adams. Oxford University Press, Inc., 200 Madison Avenue, New York, NY 10016. (212) 725-6000. (800) 334-4249. 1993. $75.00.

IEEE Nuclear Power Standards Archives Collection. IEEE Standards office, Box 1331, 445 Hoes Lane, Piscataway, NJ 08855-1331. (908) 562-3822. (800) 678-4333. Revisededition. 1990. $165.00.

Information Center On Nuclear Standards. American Nuclear Society, 555 North Kensington Avenue, LA Grange Park, IL 60525. (708) 352-6611. Standards for all aspects of the design, construction, operation and maintenance of nuclear power plants.

ONLINE DATABASES AND CD-ROMS

CA Search. Chemical Abstracts Service, P.O. Box 3012, Columbus, OH 43210-0012. (614) 447-3600. (800) 848-6533. FAX (614) 447-3709. Very comprehensive guide to worldwide chemical literature and related fields. 1972 to present. Available on BRS,(800) 289-4277, DIALOG, (800) 334-2564, ORBIT (800) 456-7248, and STN International, FIZ Karlsruhe, P.O. Box 2465,

RADIATION SHIELDING

Ency. of Physical Sciences and Engineering Info. Sources

W-7500, Karlsruhe 1, Germany, online services. Inquire as to cost and availability.

Compendex Plus. Engineering Information, Inc., 345 East 47th Street, New York, NY 10017. (212) 705-7600 or (800) 221-1044. Contains citations with abstracts to worldwide literature in engineering and technology, from 1970 to present. Available on online BRS,(800) 289-4277, DIALOG, (800) 334-2564, ORBIT (800) 456-7248, and STN International, FIZ Karlsruhe, P.O. Box 2465, W-7500, Karlsruhe 1, Germany, online services. Also available on CD-ROM. Inquire as to cost and availability.

Current Contents Search. Institute for Scientific Information, 3501 Market Street, Philadelphia, PA 19104. (215) 386-0100. FAX (215) 386-6362. Contains citations to articles listed in the table of contents of science and technology journals. Also articles in social sciences and life sciences journals. Available on BRS,(800) 289-4277, DIALOG,(800) 334-2564, online services. Inquire as to cost and availability.

Dissertation Abstracts. University Microfilms International, 300 North Zeeb Road, Ann Arbor, MI 48106. (800) 521-0600 or (313) 761-4700. Scope includes virtually all doctoral dissertations accepted at accredited American institutions from 1861 to present in 252 subject areas. Available on BRS, (800) 289-4277, DIALOG, (800) 334-2564, and OCLC EPIC, (800) 848-5878, online services. Also available on CD-ROM. Inquire as to cost and availability.

INSPEC. Institution of Electrical Engineers, Michael Faraday House, Six Hills Way, Stevenage, Herts. SG1 2AY, England. Telephone: 0438 313311 or FAX 0438 742840. Contains citations to the worldwide literature of physics, electronics and electrical engineering, computer technology, and related fields. Available on BRS, (800) 289-4277, DIALOG, (800) 334-2564, ORBIT, (800) 456-7248, and STN International, FIZ Karlsruhe, P.O. Box 2465, W-7500, Karlsruhe 1, Germany, online services. Inquire as to cost and availability.

Kirk-Othmer Encyclopedia of Chemical Technology. John Wiley and Sons, Inc., 605 Third Avenue, New York, NY 10158. (800) 526-5368 or (212) 850-6000. Contains the complete text of all chapters in the 27 volume fourth edition of the KIRK-Othmer ENCYCLOPEDIA of CHEMICAL TECHNOLOGY. 1991. Available on BRS,(800) 289-4277, DIALOG,(800) 334-2564, online services. Inquire as to cost and availability.

NTIS Bibliographic Database. National Technical Information Service, 5285 Port Royal Road, Springfield, VA 22161. (703) 487-4929 or FAX (703) 321-8199. Broad coverage of government-sponsored science and technology research reports, 1964 to present. Available on BRS,(800) 289-4277, DIALOG, (800) 334-2564, ORBIT, (800) 456-7248, and STN International, FIZ Karlsruhe, P.O. Box 2465, W-7500, Karlsruhe 1, Germany, online services. Also available on CD-ROM. Inquire as to cost and availability.

Nuclear Science Abstracts. U.S. Department of Energy, office of Scientific and Technical Energy, P.O. Box 62, Oak Ridge, TN 37831. (615) 576-1189. Contains citations with abstracts to literature in all fields of nuclear science and energy. 1948 to 1976. Available on DIALOG,(800) 334-2564, online service. Inquire as to cost and availability.

Physics Briefs. American Institute of Physics, 335 East 45th Street, New York, NY 10017. (212) 661-9260 or FAX (212) 661-2036. Contains citations with abstracts of the literature of physics and related fields, 1979 to present. Available on the STN International, FIZ Karlsruhe, P.O. Box 2465, W-7500,

Karlsruhe 1, Germany, online service. Inquire as to cost and availability.

SCISEARCH. Institute for Scientific Information, 3501 Market Street, Philadelphia, PA 19104. (800) 523-1850 or (215) 386-0100. Broad multidisciplinary title and author index to the international literature of science and technology, 1974 to present. Available on DIALOG, (800) 334-2564, and ORBIT, (800) 456-7248, online services. Also available on CD-ROM. Inquire as to cost and availability.

WILSONLINE. H.W. Wilson Company, 950 University Avenue, Bronx, NY 10452. (800) 367-6770 or (212) 588-8400. Makes available online versions of the printed H.W. Wilson indexes including Applied Science and Technology Index, Business Periodicals Index, General Science Index, and Readers' Guide to Periodical Literature. Period covered is generally 1983 to present. Available on BRS, (800) 289-4277, DIALOG, (800) 334-2564, and OCLC EPIC, (800) 848-5878, online services. Also available on CD-ROM. Inquire as to cost and availability.

PERIODICALS

American Nuclear Society Transactions. American Nuclear Society, 555 North Kensington Avenue, LA Grange Park, IL 60525. (708) 352-6611. 1958 to present. 3 issues per year. $340.00 per year. ISSN: 0003-018X

Annals of Nuclear Energy. Elsevier Science Publishing Company, Inc., 660 White Plains Road, Tarrytown, NY 10591- 5133. (914) 524-9200. 1974 to present. Monthly. $815.00 per year. ISSN: 0306-4549.

Bulletin of the Atomic Scientists. Educational Foundation for Nuclear Science, 6042 South Kimbark Avenue, Chicago, IL 60637. (312) 702-2555. 1945 to present. 10 issues per year. $30.00 per year. ISSN: 0096-3402.

Journal of Environmental Radioactivity. Elsevier Science Publishing Company, Inc., 660 White Plains Road, Tarrytown, NY 10591-5133. (914) 524-9200. 1984 to present. Monthly. $645.00 per year. ISSN: 0265-931X.

Nuclear Engineer. Institution of Nuclear Engineers, 1 Penerley Road, London, SE6 2LQ, England. 1959 to present. Bi-monthly. L77 per year. ISSN: 0262-5091.

Nuclear Engineering and Design. Elsevier Science Publishing Company, Inc., 660 White Plains Road, Tarrytown, NY 10591-5133. (914) 524-9200. 1965 to present. 24 issues per year. $2748.00 per year.

Nuclear Engineering International. Reed Business Pulbishing Group, Quadrant House, the Quadrant, Sutton, Surrey, SM2 5AS, Englnad. 1956 to present. Monthly. $236.00 per year. ISSN: 0029-5507.

Nuclear Science and Engineering. American Nuclear Society, 555 North Kensington Avenue, LA Grange Park, IL 60525. (708) 352-6611. 1956 to present. Monthly. $405.00 per year. ISSN: 0029-5639.

Nuclear Technology. American Nuclear Society, 555 North Kensington Avenue, LA Grange Park, IL 60525. (708) 352-6611. 1965 to present. Monthly. $440.00 per year. ISSN: 0029-5450.

Radiation Research. Kluge Carden Jennings, 853 West Main Street, Charlottesville, VA 22903. 1954 to present. Monthly. $560.00 per year. ISSN: 0033-7587.

RESEARCH CENTERS AND INSTITUTES

Department of Nuclear Engineering and Engineering Physics. University of Wisconsin - Madison, 1500 Johnson Drive, Madison, WI 53706. (608) 263-1648. FAX (608) 262-6707

Ford Nuclear Reactor. Phoenix Memorial Laboratory. University of Michigan, 2301 Bonisteel Boulevard, Ann Arbor MI 48109-2100. (313) 764-6223. FAX (313) 764-4540.

Frank H. Neely Nuclear Research Center. Georgia Institute of Technology, 900 Atantic Drive NW, Atlanta, GA 30332. (404) 894-3600.

Fusion Research Center. University of Texas at Austin. RLM 11.222, Austin, TX 78712. ((512) 471-3926. FAX (512) 471-8865.

Institute of Plasma and Fusion Research. University of California, Los Angeles, 44-139 Engineering IV 405 Hilgard Avenue, Los Angeles, CA 90024-1597. (213) 825-1613. FAX (213) 206-4832.

Nuclear Physics Laboratory. University of Washington, Seattle, WA 98195. (206) 543-4080.

Radiation Science and Engineering Center. Pennsylvania State University, University Park, PA 16802. (814) 865- 6351. FAX (814) 863-4840.

Stanford Synchrotron Radiation Laboratory. Stanford University, SLAC Bin #69, P.O. Box 4349, Stanford, CA 943050210. (415) 926-3153. FAX (415) 926-4100.

RADIO

See also: AMATEUR RADIO, AMPLIFIERS, ANTENNAS, ELECTRICAL ENGINEERING, ELECTRONIC CIRCUITS and COMPONENTS, ELECTRONICS ENGINEERING, RADAR, TELEVISION

ABSTRACT SERVICES AND INDEXES

Amateur Radio Technical Abstracts. Graham Thornton, editor. Thornton Publishing, 1502 Howard Street, New Iberia, LA 70560. (318) 364-2752. (800) 551-3076. 1992. $25.00.

Applied Science and Technology Index; A Cumulative Subject Index To English Language Periodicals in the Fields of Aeronautics and Space Science, Computer Technology, Chemistry, Construction Industry, Energy and Related Areas. H.W. Wilson Co., 950 University Avenue, Bronx, NY 10452. (212) 588-8400. (800) 367-6770. FAX (718) 590-1617. From 1958 to present. Monthly. Inquire about cost and availability. Also available on CD-ROM and online. ISSN: 0003-6986.

Current Contents: Engineering, Technology and Applied Sciences. Institute for Scientific Information, 3501 Market Street, Philadelphia, PA 19104. (215) 386-0100. FAX (215) 386-6362. 1970 to present. Weekly. $442.00 per year. Also available on

CD-ROM and online. Inquire regarding cost and availability. ISSN: 0095-7917.

Electrical and Electronics Abstracts. Institution of Electrical Engineers (IEE), London. Available from INSPEC/IEEE - Institute of Electrical and Electronic Engineers, Box 1331, Hoes Lane, Piscataway, NJ 08855-1331. (908) 562-5549. 1898 to present. Monthly. $2200.00 per year. Also available on CD-ROM and online as INSPEC.

Engineering Index Monthly; Indexes and Abstracts the World's Engineering and Technical Literature. Engineering Information, Inc., Castle Point on the Hudson, Hoboken, NJ 07030. (201) 216-8500. (800) 221-1044. FAX (201) 216-8532. Monthly. $2300.00 per year. Available online as COMPENDEX and also on CD-ROM. ISSN: 0742-1974.

General Science Index. H.W. Wilson Company, 950 University Avenue, Bronx, NY 10452. (212) 588-8400. (800) 367-6770. FAX (718) 590-1617. From 1978 to present. Ten issues per year; quarterly and annual cumulations. Service basis. Available on CD-ROM and online. Inquire regarding cost and availability. ISSN: 0162-1963.

Index To IEEE Publications. IEEE Service Center, 445 Hoes Lane, Piscataway, NJ 08855-1331. (908) 981-1393. (800) 678-IEEE. FAX (908) 981-9667. 1973 to present. Annual. ISSN: 0099-1368.

Physics Abstracts. INSPEC. Section A, Science Abstracts. Institution of Electrical Engineers (IEE), London. Available from: INSPEC/IEEE - Institute of Electrical and Electronic Engineers, Box 1331, Hoes Lane, Piscataway, NJ 08855-1331. (908) 562-5549. 1898 to present. 24 issues per year. $2835.00 per year. Also available online and on CD-ROM. ISSN: 0036-8091.

Physics Briefs (Physikalische Berichte). Information Center for Energy, Physics, Mathematics; German Physical Society. V C H Publishers, Inc., 220 East 23rd Street, New York, NY 10010-4606. (212) 683-8333. 1845 to present. 24 issues per year. $2390.00 per year. Also available online. ISSN: 0179-7434.

ASSOCIATIONS AND PROFESSIONAL SOCIETIES

American Electronics Association. 5201 Great America Parkway, Suite 520, PO Box 54990, Santa Clara, CA 95056. (408) 987-4200. FAX (408) 970-8565.

American Radio Relay League. 225 Main Street, Newington, CT 06111. (203) 666-1541. FAX (203) 665-7531.

Audio Engineering Society. 60 East 42nd Street, New York, NY 10017. (212) 661-8528.

IEEE (Institute of Electrical and Electronics Engineers). 345 East 47th Street, New York, NY 10017. (212) 705-7900. FAX (212 705-4929.

National Amateur Radio Association. PO Box 201407, Arlington, TX 76006. (817) 860-0978. FAX (817) 860-0979.

National Association of Radio and Telecommunications Engineers. PO Box 678, Medway, MA 02053. (508) 533-8333. FAX (508) 533-3815.

Radio Amateur Satellite Corporation. 850 Sligo Avenue, Suite 600, Silver Spring, MD 20910. (301) 589-6062. FAX (301) 608-

RADIO

Ency. of Physical Sciences and Engineering Info. Sources

3410.

United States National Committee for the International Union of Radio Science. Borad on Physics and Astronomy, Naitonal Research Council, 2101 Constitution Avenue, NW, Washington, DC 20416. (202) 334-3559.

DIRECTORIES AND BIOGRAPHICAL SOURCES

Broadcast Engineering - Equipment Reference Manual. Intertec Publishing Corporation, Box 12901, Overland Park, KS 66212. (913) 888-4664. Annual. $20.00.

E E M - Electronic Engineer's Master. Hearst Business Communications, Inc., 645 Stewart Avenue, Garden City NY 11530. (516) 227-1300. Annual. $90.00. ISSN: 0732-9016.

IEEE Membership Directory. Institute of Electrical and Electronics Engineers, IEEE Service Center, 445 Hoes Lane, Piscataway, NY 08854. (908) 981-1393. (800) 678-IEEE. FAX (908) 981-9667. 2 volumes. Annual. $190.00. ISSN:

Shortwave Directory. Larry Van Horn, editor. Grove Enterprises, PO Box 98, Brasstown, NC 28902 (800) 438-8155. 8th edition. 1993. $24.95.

Who's Who in Engineering. Gordon Davis, editor. American Association of Engineering Societies. 1111 19th Street, NY, Suite 608, Washington, DC 20036. (202) 296-2237. (800) 658-8897. 9th edition. 1995. $220.00.

ENCYCLOPEDIAS AND DICTIONARIES

Amateur Radio Encyclopedia. Stan Gibilsco. McGraw-Hill Publishing Company, Inc., 1221 Avenue of the Americas, New York, NY 10020. (212) 512-2000. (800) 262-4729. 1993. $49.95.

IEEE Standard Dictionary of Electrical and Electronics Terms. Christopher J. Booth, editor. IEEE Service Center, 445 Hoes Lane, Piscataway, NJ 08855-1331. (908) 981-1393. (800) 678-IEEE. FAX (908) 981-9667. IEEE Standard 100- 1992. 5th edition. 1993. $150.00.

Illustrated Dictionary of Electronics. Stan Gibilsco. TAB Books, P.O. Box 40, Blue Summit, PA 17294-0850. (717) 794- 2191. (800) 233-1128. 7th edition. 1994. $34.95.

McGraw-Hill Electronics Dictionary. John Markus and Neil Sclater. McGraw-Hill-Hill Publishing Company, Inc., 1221 Avenue of the Americas, New York, NY 10020. (212) 512-2000. (800) 262-4729. 5th edition. 1994. $49.95.

GENERAL WORKS

Amateur Radio Equipment Fundamentals. Albert D. Helfrick. Prentice-Hall, 113 Sylvan Avenue, Route 9W, Englewood Cliffs, NJ 07632. (201) 592-2000. (800) 922-0579. 1982. $40.00.

The Art and Science of Radio. Linda J. Busby and Donald L. Parker. Allyn & Bacon, Inc., 160 Gould Street, Needham Heights, MA 02194-2310. (617) 455-1200. 1984. $58.00.

Broadcast Transmission Engineering Practice. Wharton, et al. Butterworth-Heinemann, 313 Washington Street, Newton, MA 02158. (618-928-2500. (800) 366-2665. 1992. $125.00.

Electric and Electronic Engineering For Scientists and Engineers. K. A. Krishnamurthy. Halsted Press, 605 Third Avenue, New York, 10158-0012. (212) 850-6400. 1944. $48.95.

Introduction To Radio Frequency and Microwave Engineering. Robert K. Feeney and David R. Hertling. TAB Books, P.O. Box 40, Blue Summit, PA 17294-0850. (717) 794-2191. (800) 233-1128. 1995. $44.50.

Low Power Communications. Richard H. Arland. Tiare Publications, P.O. Box 493, Lake Geneva, WI 53147. (800) 420-0579 volume 1: Basic QRP. 1992. $14.95.

Meteor Burst Communications: Theory and Practice. Donald L. ScHilling. John Wiley & Sons, Inc., 605 Third Avenue, New York, NY 10158-0012. (212) 850-6000. (800) 225-5945. 1993. $115.00.

Modern Radio Production. Lewis R. O'Donnel, et al. Wadsworth Publishing Co., 10 Davis Drive, Belmont, CA 94002. (415) 595-2350. (800) 354-9706. 1993. $29.95.

Modern Radio Science 1993. H Matsumoto, editor. Oxford University Press, Inc., 200 Madison Avenue, New York, NY 10016. (212) 725-6000. (800) 334-4249. 1993. $37.50.

Packet Radio Operator's Manual. Glynn B. Rogers. CQ Communications, Inc., 76 North Broadway, Hicksville, NY 11801. (516) 681-2922. 1993. $15.95 in paper.

Pioneers of Wireless. Ellison Hawks. Ayer Company Pubs., Inc., Lower Mill Road, North Stratford, NH 03590. (603) 922-5105. (800) 282-5413. Reprint edition. 1974. $29.95.

Radio Frequency Transistors: Principles and Practical Applications. Norman E. Dye and Helge O. Granberg. Butterworth-Heinemann, 313 Washington Street, Newton, MA 02158. (618-928-2500. (800) 366-2665. 1993. $39.95.

Radio-frequency Transmission Systems Design and Operation. Jerry Whitaker. McGraw—Hill Publishing Company, Inc., 1221 Avenue of the Americas, New York, NY 10020. (212) 512- 2000. (800) 262-4729. 1991. $50.00.

Radio Systems Technology. D. C. Green. Halsted Press, 605 Third Avenue, New York, 10158-0012. (212) 850-6400. 1990. $78.95.

Shortwave Radio Listening for Beginners. Anita L. McGormic. TAB Books, P.O. Box 40, Blue Summit, PA 17294-0850. (717) 794-2191. (800) 233-1128. 1993. $18.95.

Single-sideband Systems and Circuits. William E. Sabin and Edgar O. Schoenike. McGraw-Hill Publishing Company, Inc., 1221 Avenue of the Americas, New York, NY 10020. (212) 512-2000. (800) 262-4729. 1995. $75.00.

Solid State Design For the Radio Amateur. American Radio Relay League, 225 Main Street, Newington, CT 06111. (203) 666-1541. FAX (203) 665-7531. 1986. $15.00 in paper.

VHG and UHF Antennas. R. B. Burberry. Institute of Electrical and Electronic Engineers (IEEE), 345 East 47th Street, New York, NY 10017. (212) 705-7900. 1992. $95.00.

HANDBOOKS AND MANUALS

AARL Handbook For Radio Amateurs. American Radio Relay League, 225 Main Street, Newington, CT 06111. (203) 666-1541. FAX (203) 665-7531. 78th edition. 1996. $30.00. ISSN: 0079-9440.

American Electricians Handbook. Terrell Croft and Wilford I. Summers. McGraw-Hill Book Company, 1221 Avenue of the Americas, New York, NY 10020. (212) 512-2000. (800) 262-4729. 12th edition. 1992. $77.50.

Antenna Engineering Handbook. Richard C. Johnson. McGraw-Hill Publishing Company, Inc., 1221 Avenue of the Americas, New York, NY 10020. (212) 512-2000. (800) 262-4729. 3rd edition. 1993. $119.50.

The Calculations and Measurement of Inductance and Capacity: A Handbook For Experimenting With Tesla Coils and Radio. Gordon Press Publications, P.O. Box 459, Bowling Green Station, New York, NY 10004. (718) 624-8419. 1991. $88.95.

Complete Shortwave Listener's Handbook. Hank Bennett. TAB Books, P.O. Box 40, Blue Summit, PA 17294-0850. (717) 794-2191. (800) 233-1128. 4th edition. 1993. $19.95 in paper.

Electronics Handbook. Jerry C. Whitaker. CRC Press, Inc., 2000 Corporate Boulevard, NW, Boca Raton, FL 33431. (407) 994-0555. (800) 272-7737. 1996. $120.00.

Interferance Handbook. Diane Publishing Co., 600 Upland Avenue, Upland, PA 19015. (610) 499-7415. 1994. $24.95.

Mobile Radio Servicing Handbook. Roger Belcher, et al. Butterworth-Heinemann, 313 Washington Street, Newton, MA 02158. (618-928-2500. (800) 366-2665. 1990. $62.95.

Newnes Radio Engineer's Pocket Book. John Davies. Butterworth-Heinemann, 313 Washington Street, Newton, MA 02158. (618-928-2500. (800) 366-2665. 1995. $22.95

Practical Antenna Handbook. Joseph J. Carr. McGraw-Hill Publishing Company, Inc., 1221 Avenue of the Americas, New York, NY 10020. (212) 512-2000. (800) 262-4729. 1990. $32.95.

Radio Amateur and Listener's Data Handbook. Steve Money. Butterworth-Heinemann, 313 Washington Street, Newton, MA 02158. (618-928-2500. (800) 366-2665. 1995. $39.95 in paper.

Radio Amateur's Digital Communications Handbook. Jonathan L. Mayo. TAB Books, P.O. Box 40, Blue Summit, PA 17294-0850. (717) 794-2191. (800) 233-1128. 1992. $22.95.

Radio and Radar Reference Data. Gordon Press Publications, P.O. Box 459, Bowling Green Station, New York, NY 10004. (718) 624-8419. 1991. $79.00.

Standard Handbook For Electrical Engineers. Donald Fink. McGraw-Hill Publishing Company, 1221 Avenue of the Americas, New York, NY 10020. (212) 512-2000. (800) 262-4729. 13th edition. 1996. $110.00.

Standard Radio Communications Manual, With Instrumentation and Testing Techniques. R. Harold Kinley. Prentice-Hall, 113 Sylvan Avenue, Route 9W, Englewood Cliffs, NJ 07632. (201) 592-2000. (800) 922-0579. 1988. $16.95.

ONLINE DATABASES AND CD-ROMS

COMPENDEX PLUS. Engineering Information, Inc., 345 East 47th Street, New York, NY 10017. (212) 705-7600 or (800) 221-1044. Contains citations with abstracts to worldwide literature in engineering and technology, from 1970 to present. Available on online BRS,(800) 289-4277, DIALOG, (800) 334-2564, ORBIT (800) 456-7248, and STN International, FIZ Karlsruhe, P.O. Box 2465, W-7500, Karlsruhe 1, Germany, online services. Also available on CD-ROM. Inquire as to cost and availability.

INSPEC. Institution of Electrical Engineers, Michael Faraday House, Six Hills Way, Stevenage, Herts. SG1 2AY, England. Telephone: 0438 313311 or FAX 0438 742840. Contains citations to the worldwide literature of physics, electronics and electrical engineering, computer technology, and related fields. Available on BRS, (800) 289-4277, DIALOG, (800) 334-2564, ORBIT, (800) 456-7248, and STN International, FIZ Karlsruhe, P.O. Box 2465, W-7500, Karlsruhe 1, Germany, online services. Inquire as to cost and availability.

NTIS Bibliographic Database. National Technical Information Service, 5285 Port Royal Road, Springfield, VA 22161. (703) 487-4929 or FAX (703) 321-8199. Broad coverage of government-sponsored science and technology research reports, 1964 to present. Available on BRS,(800) 289-4277, DIALOG, (800) 334-2564, ORBIT, (800) 456-7248, and STN International, FIZ Karlsruhe, P.O. Box 2465, W-7500, Karlsruhe 1, Germany, online services. Also available on CD-ROM. Inquire as to cost and availability

WILSONLINE. H.W. Wilson Company, 950 University Avenue, Bronx, NY 10452. (800) 367-6770 or (212) 588-8400. Makes available online versions of the printed H.W. Wilson indexes including Applied Science and Technology Index, Business Periodicals Index, General Science Index, and Readers' Guide to Periodical Literature. Period covered is generally 1983 to present. Available on BRS, (800) 289-4277, DIALOG, (800) 334-2564, and OCLC EPIC, (800) 848-5878, online services.Also available on CD-ROM. Inquire as to cost and availability.

PERIODICALS

Audio Engineering Society Journal. Audio Engineering Society, 60 East 42nd Street, New York, NY 10017. (212) 661-8528. 1953 to present. 10 issues per year. $125.00 per year. ISSN: 0004-7554.

Broadcaster Engineering. Intertec Publishing Corporation, Box 12801, Overland Park, KS 66212. (913) 888-4664. 1959 to present. Monthly.

CQ: The Radio Amateur's Journal. CQ Communications, Inc., 76 North Broadway, Hicksville, NY 11801-2953. (516) 681-2922. FAX (516) 681-2926. 1945 to present. Monthly. $24.95. ISSN: 0007-893X.

DB, the Sound Engineering Magazine. 203 Commack Road, Suite 1010, Commack, NY 1725. (516) 586-6530. 1967 to present. $15.00 per year. ISSN: 0011-7145. .

Electronic Design. Penton Publishing, San Jose Gateway, Suite 354. 2025 Gateway Place, San Jose, CA 95110. (408) 441-0550. 1952 to present. Fortnightly. $95.00. ISSN: 0013-4872.

Electronics. Penton Publishing, San Jose Gateway, Suite 354. 2025 Gateway Place, San Jose, CA 95110. (408) 441-0550. 1930 to present. Semi-weekly. $98.00 per year. ISSN: 0883-4989.

Electronics Now. Gernsback Publications, Inc., 500-B Bi-County Boulevard, Farmingdale, NY 11735. (516) 293-3000. FAX (516) 293-3115. 1929 to present. Monthly. $19.95. ISSN: 0033-7862.

Electronics World and Wireless World. Reed Business Publishing, Ltd., Quadrant House, the Quadrant, Sutton, Surrey SM2 5AS, England. 1911 to present. Monthly. L35. ISSN: 0266-3244.

IEEE Antennas and Propagation Magazine. Institute of Electrical and Electronics Engineers, Inc., 445 Hoes Lane, Box 1331, Piscataway, NJ 08855-1331. (908) 981-0060. FAX (908) 981-9667. Bi-monthly. $85.00. ISSN: 1045-9243.

IEEE Circuits and Devices Magazine. Institute of Electrical and Electronics Engineers, Inc., Box 1331, 445 Hoes Lane, Piscataway, NJ 08855-1331. (908) 981-0060. 1985 to present. Bi-monthly. $120.00 per year. ISSN: 8755-3996.

IEEE Journal of Solid State Circuits. Institute of Electrical and Electronics Engineers, Inc., Box 1331, 445 Hoes Lane, Piscataway, NJ 08855-1331. (908) 981-0060. 1966 to present. Bi-monthly. $275.00 per year. ISSN: 0018- 9200.

IEEE Transactions On Broadcasting. Institute of Electrical and Electronics Engineers, Inc., Box 1331, 445 Hoes Lane, Piscataway, NJ 08855-1331. (908) 981-0060. 1955 to present. Bimonthly.

Microwaves and RF. Penton Publishing Co., Circulation Department, 1100 Superior Avenue, Cleveland, OH 44114. (201) 393-6060. FAX (201) 393-6297. From 1962 to present. Monthly. $60.00. ISSN: 0745-2992.

Radio Communications Report. Titsch Publishing Company, 2500 Curtis Street, Suite 200, Denver, CO 80217. (303) 573-1433. 1981 to present. Monthly.

RESEARCH CENTERS AND INSTITUTES

Center For Advanced Technology in Telecommunications. Polytechnic University, 333 Jay Street, Brooklyn, NY 11202. (718) 643-5160.

Communications Research Laboratory. McMaster University, 1280 Main Street West, Hamilton, ON Canada L85 4K1. (416) 525-9140.

Electronics Research Laboratory. Montana State Universiy, Bozeman, MT 59717. (406 994-2505.

RADIO ASTRONOMY

See also: ASTRONOMY, ASTROPHYSICS, GALAXIES, SOLAR SYSTEM, TELESCOPES, OBSERVATORIES, RADIO TELESCOPES

ABSTRACT SERVICES AND INDEXES

Applied Science and Technology Index; A Cumulative Subject Index To English Language Periodicals in the Fields of Aeronautics and Space Science, Computer Technology, Chemistry, Construction Industry, Energy and Related Areas. H. W. Wilson Co., 950 University Avenue, Bronx, NY 10452- 9978. (212) 588-8400. (800) 367-6770. FAX (718) 590-1617.From 1958 to present. Monthly. Inquire about cost and availability. Also available online (BRS and WILSONLINE) and on CD-ROM. ISSN: 0003-6986.

Astronomy and Astrophysics Abstracts. Springer-Verlag New York, 175 Fifth Avenue, New York, NY 10010. (212) 460-1500. FAX (212) 473-6272. Published for the Astronomisches Rechen-Institut. Comprehensive coverage of all aspects of astronomy, astrophysics and related fields. 1969 to present. Two parts per year. Annual. ISSN: 0067-0022.

Physics Abstracts. INSPEC. Section A, Science Abstracts. Institute of Electrical Engineers, London, United Kingdom. Available from: INSPEC/IEEE - Institute of Electrical and Electronic Engineers, Box 1331, Hoes Lane, Piscataway, NJ 08855-1331. (908) 562-5549. 1898 to present. 24 issues per year. $2835.00 per year. ISSN: 0036-8091. Also available online and on CD-ROM.

Science Citation Index. SCI. Institute for Scientific Information, 3501 Market Street, Philadelphia, PA 19104. (215) 386-0100. (800) 523-1850. FAX (215) 386-2991. 1961 to present. Six issues per year, plus annual cumulation. $11650.00 per year. Also available online and on CD-ROM. Inquire about price and availability. ISSN: 0036-827X.

STAR. Scientific and Technical Aerospace Reports. U.S. National Aeronautics and Space Administration. Distributed by U. S. Superintendent of Documents, Washington, DC 20402. 1963 to present. Semi-monthly, with semiannual and annual indexes. $114.00 per year. ISSN: 0036-8741. Also available online and on CD-ROM.

ANNUAL REVIEWS AND YEARBOOKS

The Astronomical Almanac. Superintendent of Documents, U. S. Government Printing office, Washington, DC 10402. (202) 783-3238. 1981 to present. Superceeds Astronomical Ephemeris and the American Ephermis and Nautical Almanac. Annual. $44.95. ISSN: 0737-6421.

Annual Review of Astronomy and Astrophysics. Original reviews of critical literature and current developments in astronomy and astrophysics. Annual Reviews, Inc., 4139 El Camino Way, Palo Alto, CA 94303-0139. (415) 493-4400. (800)523-8635. Fax (415) 855-9815. 1963 to date. Annual. $60.00. ISSN: 0066-4146.

Annual Review of Earth and Planetary Sciences. Annual Reviews, Inc., 4139 El Camino Way, Palo Alto, CA 94303-0139. (415) 493-4400. (800) 523-8635. Fax (415) 855-9815. 1973 to date. Annual. $62.00. ISSN: 0084-6597.

ASSOCIATIONS AND PROFESSIONAL SOCIETIES

American Astronomical Society. 2000 Florida Avenue NW, Suite 400, Washington, DC 20009. (202) 328-2010.FAX(202) 234-2560.

American Institute of Physics. One Physics Ellipse, College Park, MD 20740-3843. (301) 209-3100.

American Association of Variable Star Observers. 23 Birch Street, Cambridge, MA 02138. (617) 354-0484. FAX: (617)354-0665.

Association of Universities For Research in Astronomy, Inc. (AURA). Suite 701, 1625 Massachusetts Avenue, NW, Washington, DC 20036. (202) 483-2101. FAX (202) 483-2106.

Astronomical League. 2112 Kingfisher Lane East, Rolling Meadows, IL 60008. (708) 398-0562.

Astronomical Society of the Pacific. 390 Ashton Avenue, San Francisco, CA 94112. (415) 337-1100. FAX: (415) 337-5205.

Lunar and Planetary Institute. 3600 Bay Area Boulevard, Houston, TX 77058. (713) 486-2143.

Meteoritical Society. University of Massachusetts, 125 Marston Hall, Amherst, MA 01003. (413) 545-0300. (413) 545-0724.

Planetary Society. 65 North Catalina Avenue, Pasadena, CA 91106. (818) 793-5100. (800) WOW-MARS. FAX (818) 793-5528.

BIBLIOGRAPHIES

A Bibliography of Astronomy, 1970 - 1979. R. A. Sea and S. S. Martin. Libraries Unlimited, Inc., Littleton, CO 80160. 1982. $37.50.

Science Books & Films. American Association for the Advancement of Science, 1333 H Street NW, Washington, DC 20005. (202) 326-6434. Reviews of print, film and software materials in all sciences.. 1965 to present. Nine issues per year. $40.00 per year. ISSN: 0098-342X.

Scientific and Technical Books and Serials in Print; An Index To Literature in Science and Technology. R. R. Bowker. 121 Chanlon Road, New Providence, NJ 07974. (908) 464-6800. FAX (908) 665-3502. (800) 521-8110. 1972 to present. Annual. $299.95 per year. Also available on CD-ROM. ISSN: 0000-054X.

DIRECTORIES AND BIOGRAPHICAL SOURCES

American Astronomical Society. Membership Directory. American Astronomical Society. 2000 Florida Avenue NW, Suite 400, Washington, DC 20009. (202) 328-2010. FAX (202)234-2560 Annual. Included in membership dues. ISSN: 1061- 9038.

American Men and Women of Science: Physical and Biological Sciences. R. R. Bowker Inc., 121 Chanlon Road, New Providence, NJ 07974. (908) 464-6800. (800) 521-8110. 20th edition. 8 volumes. 1996. $850.00.

The Astronomers. Donald Goldsmith. St. Martin's Press, Inc., 175 Fifth Avenue, New York, NY 10010. (212) 674-5151. (800) 221-7945. 1993. $14.95 in paper.

Astronomical Centers of the World. Kevin Krisciunas. Cambridge University Press, 40 West 20th Street, New York, NY 10011-4211. (212) 924-3900. (800) 872-7423. 1988. $34.95

Directory of Physics and Astronomy Staff. American Institute of Physics, One Physics Ellipse, College Park, MD 20740-3843. (301) 209-3100. 1975/76 to present. Annual. $60.00. ISSN: 0361-2228.

Earth and Astronomical Research Centers. Stockton Press, 345 Park Avenue South, New York, NY 10010. 4th edition. 1995. $515.00. ISBN: 1-56169-0967.

Graduate Programs in Physics, Astronomy and Related Fields. American Institute of Physics, One Physics Ellipse, College Park, MD 20740-3843. (301) 209-3100. 1978 to present. Annual. $45.00. ISSN: 0147-1821.

Research Centers Directory. Gale Research Inc., 835 Penobscot Building, Detroit, MI 48226-4094. (313) 961-2242. (800) 877-4253. 20th edition. 1995. $485.00.

ENCYCLOPEDIAS AND DICTIONARIES

Concise Dictionary of Astronomy. Jacqueline Mitton. Oxford University Press, Inc., 200 Madison Avenue, New York, NY 10016. (212) 725-6000. (800) 334-4249. 1992. $24.95.

Encyclopedia of Astronomy and Astrophysics. Stephen Maran, editor. Van Nostrand Reinhold, 115 Fifth Avenue, New York, NY 10003. (212) 254-3232. (800) 842-3636. 1992. $129.95.

McGraw-Hill Encyclopedia of Science and Technology. McGraw-Hill Book Company, Inc., 1221 Avenue of the Americas, New York, NY 10020. (212) 512-2000. (800) 262-4729. 7th edition. 20 volume set. 1992. $1900.00.

Moons of the Solar System: An Illustrated Encyclopedia. John Stewart. McFarland & Company, Inc., Box 611, Jefferson, MC 28640. (910) 246-4460. (800) 253-2187. 1991. $49.95.

GENERAL WORKS

Classics in Radio Astronomy. W. Sullivan. Kluwer Academic Publishers, 101 Philip Drive, Assinippi Park, Norwell, MA 02061. (617) 871-6600. 1982. $131.50.

Galactic and Extragalactic Radio Astronomy. G. K. Verschuur and K. Kellermann. Springer-Verlag New York, 175 Fifth Avenue, New York, NY 10010. (212) 460-1500. (800) 777- 4643. 1991. $55.00 in paper.

Instrumentation and Techniques For Radio Astronomy. P. F. Goldsmith. Institute of Electrical and Electronic Engineers (IEEE), 345 East 47th Street, New York, NY 10017. (212) 705-7900. 1988. $59.95.

Interferometry and Synthesis in Radio Astronomy. A. Richard Thompson, et al. Krieger Publishing Company, P.O. Box 9542, Melbourne, FL 32902-9542. (407) 724-9542. Reprint edition. 1994. $59.50

Interferometry in Radioastronomy and Radio Telescopes. R. Wohlleben, et al. Kluwer Academic Publishers, 101 Philip Drive, Assinippi Park, Norwell, MA 02061. (617) 871-6600. 1991. $92.00.

Extragalactic Radio Sources From Beams To Jets. Jacques Roland, et al, editors. Cambridge University Press, 40 West 20th

RADIO ASTRONOMY

Ency. of Physical Sciences and Engineering Info. Sources

Street, New York, NY 10011-4211. (212) 924-3900.(800) 872-7423.

History of Radio Astronomy and the National Radio Astronomy Observatory: Evolution Toward Big Science. Benjamin K. Malphrus. Kreiger Publishing Company, P.O. Box 9542, Melbourne, FL 32902-9542. (407) 724-9542. 1995. $wfi

National Radio Astronomy Observatory Very Large Array Sky Survey. J. J. Condon, et al. Rector Press, Ltd., 130 Rattlesnake, Leverett, MA 01054-9726. (413) 548-9708. (800) 247-3473. 1994. $45.00 in paper.

Radio Astronomy. J. D. Kraus. CYGNUS-QUASAR Books, P.O. Box 85, Powell, OH 43065-1854. (614) 548-7895. 1986. $wfi, in paper.

Review of Radio Science 1990-1992. W. Ross Stone, editor.

Oxford University Press, Inc., 200 Madison Avenue, New York, NY 10016. (212) 725-6000. (800) 334-4249. 1993. $98.00.

Space Technology and Planetary Astronomy. Joseph N. Tatarewicz. Indiana University Pres, 601 North Morton Street, Bloomington, in 47404-3797. (812) 855-6804. (800) 842-6796. 1990. $29.95.

Tools of Radio Astronomy. K. Rolhfs. Springer-Verlag, 175 Fifth Avenue, New York, NY 10010. (212) 460-1500. (800) 777-4643. 2nd edition. 1990. $45.00.

HANDBOOKS AND MANUALS

The Amateur Radio Astronomy Ovserver's Handbooks. John P. Shields, Crown Publishers, Inc., 1 Park Avenue, New York, NY 10016.

Astrophysical Data: Planets and Stars. Kenneth R. Lanb. Springer-Verlag New York, Inc., 175 Fifth Avenue, New York, NY 10010. (212) 460-1500. (800) 777-4643. 1993. $59.00.

ONLINE DATABASES AND CD-ROMS

CA Search. Chemical Abstracts Service, P.O. Box 3012, Columbus, OH 43210-0012. (614) 447-3600. (800) 848-6533. FAX (614) 447-3709. Very comprehensive guide to worldwide chemical literature and related fields, 1972 to present. Available on BRS,(800) 289-4277, DIALOG, (800) 334-2564, ORBIT (800) 456-7248, and STN International, FIZ Karlsruhe, P.O. Box 2465, W-7500, Karlsruhe 1, Germany, online services. Inquire as to cost and availability.

Dissertation Abstracts. University Microfilms International, 300 North Zeeb Road, Ann Arbor, MI 48106. (800) 521-0600 or (313) 761-4700. Scope includes virtually all doctoral dissertations accepted at accredited American institutions from 1861 to present in 252 subject areas. Available on BRS, (800) 289-4277, DIALOG, (800) 334-2564, and OCLC EPIC, (800) 848-5878, online services. Also available on CD-ROM. Inquire as to cost and availability.

INSPEC. Institution of Electrical Engineers, Michael Faraday House, Six Hills Way, Stevenage, Herts. SG1 2AY, England. Telephone: 0438 313311 or FAX 0438 742840. Contains citations to the worldwide literature of physics, electronics and electrical engineering, computer technology, and related fields.

Available on BRS, (800) 289-4277, DIALOG, (800) 334-2564, ORBIT, (800) 456-7248, and STN International, FIZ Karlsruhe, P.O. Box 2465, W-7500, Karlsruhe 1, Germany, online services. Inquire as to cost and availability.

NTIS Bibliographic Database. National Technical Information Service, 5285 Port Royal Road, Springfield, VA 22161. (703) 487-4929 or FAX (703) 321-8199. Broad coverage of government-sponsored science and technology research reports, 1964 to present. Available on BRS,(800) 289-4277, DIALOG, (800) 334-2564, ORBIT, (800) 456-7248, and STN International, FIZ Karlsruhe, P.O. Box 2465, W-7500, Karlsruhe 1, Germany, online services. Also available on CD-ROM. Inquire as to cost and availability.

SCISEARCH. Institute for Scientific Information, 3501 Market Street, Philadelphia, PA 19104. (800) 523-1850 or (215) 386-0100. Broad multidisciplinary title and author index to the international literature of science and technology, 1974 to present. Available on DIALOG, (800) 334-2564, and ORBIT, (800) 456-7248, online services. Also available on CD-ROM. Inquire as to cost and availability.

WILSONLINE. H.W. Wilson Company, 950 University Avenue, Bronx, NY 10452. (800) 367-6770 or (212) 588-8400. Makes available online versions of the printed H.W. Wilson indexes including Applied Science and Technology Index, Business Periodicals Index, General Science Index, and Readers' Guide to Periodical Literature. Period covered is generally 1983 to present. Available on BRS, (800) 289-4277, DIALOG, (800) 334-2564, and OCLC EPIC, (800) 848-5878, online services.Also available on CD-ROM. Inquire as to cost and availability.

OTHER SOURCES

Atlas of the Planets. Vincent DeCallatay and Audouin Dollfus. Books on Demand, 300 North Zeeb Road, Ann Arbor, MI 48106-1346. (313) 761-4700. (800) 521-0600. $45.60.

Atlas of the Solar System. B. Yenne. Simon & Schuster, Inc. 1230 Avenue of the Americas, New York, NY 10020. (212) 698-7000. (800) 223-2348. 1987. $12.98.

Cambridge Atlas of Astronomy. Cambridge University Press, 40 West 20th Street, New York, NY 10011-4211. (212) 924- 3900. (800) 872-7423. 3rd edition. 1994. $74.00.

The Sky: A User's Guide. David H. Levy. Cambridge University Press, 40 West 20th Street, New York, NY 10011- 4211. (212) 924-3900. (800) 872-7423. 1993. $14.95.

The View from Space: Photographic Exploration of the Planets. Merton Davies and Bruce C. Murray. Columbia University Press, 562 West 113th Street, New York, NY 10025. (212) 666-1000. (800) 944-8648. 1973. $17.50.

PERIODICALS

Astronomical Journal. American Institute of Physics, One Physics Ellipse, College Park, MD 20740-3843. (301) 209- 3000. Published for the American Astronomical Society. 1849 to present. Monthly. $280.00 per year. ISSN: 0004- 6256.

Astronomical Society of the Pacific Publications. Astronomical Society of the Pacific, 390 Ashton Avenue, San Francisco, CA 94112. (415) 337-1100. FAX (415) 337-5205.1889 to present. Monthly. $175.00 per year. ISSN: 0004- 6280.

Astronomy. Kalmbach Publishing Company, Box 1612, Waukesha, WI 53187-1612. (414) 796-8776. FAX (414) 796-1142. 1973 to present. Monthly. $27.00 per year. ISSN: 0091-6358.

Astronomy and Astrophysics. Springer-Verlag New York, Inc., 44 Hartz Way, Secaucus, NJ 07096-2491. (201) 348-4033. FAX(201) 348-4505. 1969 to date. 36 issues per year in 12 volumes. $1733.00 per year. ISSN: 0004-6361.

Astrophysical Journal. University of Chicago Press, 5720 South Woodlawn Avenue, Chicago, IL 60637. (312) 753-3347. FAX (312) 753-0811. 1895 to date. 3 issues per monthly. $740.00 per year. ISSN: 0004-637X.

Astrophysics and Space Science. Kluwer Academic Publishers, Box 358, Accord Station, Hingham MA 02018-0358. (617) 871-6600. FAX (617) 871-6528. 1968 to present. 24 issues per year. $2544.00 per year. ISSN: 0004-640X.

Celestial Mechanics; An International Journal of Space Dynamics. Kluwer Academic Publishers, Box 358, Accord Station, Hingham MA 02018-0358. (617) 871-6600. FAX (617) 871-6528. 1969 to date. 12 issues per year. $703.50 per year. ISSN: 0923-2958.

Icarus; International Journal of Solar System Studies. Academic Press, Inc., Journal Division, 525 B Street, Suite 1900, San Diego, CA 92101-4495. (619) 230-1840. FAX (619) 699-6800. 1962 to the present. Monthly. $1080.00. ISSN: 0019-1035.

IEEE Transactions On Antennas and Propagation. Institute of Electrical and Electronics Engineers, Inc., Box 1331, 445 Hoes Lane, Piscataway, NJ 08855-1331. (908) 981-0060. 1952 to present. Monthly. $212.00 per year. ISSN: 0018-926X.

Mercury. Astronomical Society of the Pacific. 390 Ashton Avenue, San Francisco, CA 94112. (415) 337-1100. FAX: (415) 337-5205. 1972 to present. Bimonthly. $175.00 per year. ISSN: 0047-6773.

Monthly Notices of the Royal Astronomical Society. Blackwell Scientific Publications, Inc., Osney Mead, Oxford OX2 QEL England. TEL 0865-240201. FAX 0865-721205. 1827 to present. Fortnightly. $1899.00 per year. ISSN: 0035-8711.

Observatory. c/o Dr. D. J. Strickland, Space and Astrophysics Division, Rutherford Appleton Laboaratory, Chlton, Didcot, Oxon, OX11 OQX, England. FAX 0235-445848. 1877 to present. Bimonthly. $42.00. ISSN: 0029-7704.

Planetary and Space Science. Elsevier Science Publishing Company, Inc., 660 White Plains Road, Terrytown, NY 10591-5153. (914) 524-9200. FAX: (914) 333-2444. 1959 to present. Monthly. $1355.00 per year. ISSN: 0032-0633.

Sky and Telescope. Sky Publishing Corporation, Box 9111, Belmont, MA 02178. (617) 864-7360. FAX (617) 864-6117. 1941 to present. Monthly. $27.00 per year. ISSN: 0037-6604.

Space Science Reviews. Kluwer Academic Publishers, Box 358, Accord Station, Hingham MA 02018-0358. (617) 871-6600. FAX (617) 871-6528. 1962 to present. 16 issues per year. $940 issues per year. ISSN: 0038-6308.

Vistas in Astronomy. Elsevier Science Publishing Company, Inc., 660 White Plains Road, Terrytown, NY 10591-5153. (914) 524-

9200. FAX: (914) 333-2444. 1958 to date. Quarterly. $415.00 per year. ISSN: 0083-6656.

RESEARCH CENTERS AND INSTITUTES.

Arecobo Observatory. Cornell University, Box 995. Arecibo, PR 00613. (809) 878-2612. FAX (809) 878-1861.

Haystack Observatory. Massachusetts institute of Technology. Westford, MA 01886. (508) 692-4765. FAX (617) 981-0590.

Herzberg Institute of Astrophsyics. National Research Council of Canada, 100 Sussex Drive, Ottawa, ON Canada K1A 0R6. (613) 990-0907. FAX (613) 952-6602.

National Astronomy and Ionosphere Center. Space Sciences Building, Ithaca, NY 14853. (607) 255-3735. FAX (607) 255-8803.

National Radio Astronomy Observatory. Edgemont Road, Charlottesville, VA 22903. (804) 296-0211.

Owens Valley Radio Observatory. California Institute of Technology. Pasadena, CA 91125. (818) 795-6811.

Radio Astronomy Laboratory. University of California, Berkeley, 601 Campbell Hall, Berkeley, CA 94720. (415) 642-5275. FAX (415) 642-3411.

Radio Astronomy Observatory. University of Michigan, 810 Dennison Building, Ann Arbor, MI 48109. (313) 426-8441.

Smithsonian Astrophysical Observatory. 60 Garden Street, Cambridge, MA 02138. (617) 495-7461. FAX (617) 495-7105.

RADIO TELESCOPES

See also: ASTRONOMY, ASTROPHYSICS, GALAXIES, SOLAR SYSTEM, TELESCOPES, OBSERVATORIES, RADIO ASTRONOMY

ABSTRACT SERVICES AND INDEXES

Applied Science and Technology Index; A Cumulative Subject Index To English Language Periodicals in the Fields of Aeronautics and Space Science, Computer Technology, Chemistry, Construction Industry, Energy and Related Areas. H. W. Wilson Co., 950 University Avenue, Bronx, NY 10452-9978. (212) 588-8400. (800) 367-6770. FAX (718) 590-1617. From 1958 to present. Monthly. Inquire about cost and availability. Also available online (BRS and WILSONLINE) and on CD-ROM. ISSN: 0003-6986.

Astronomy and Astrophysics Abstracts. Springer-Verlag New York, 175 Fifth Avenue, New York, NY 10010. (212) 460-1500. FAX (212) 473-6272. Published for the Astronomisches Rechen-Institut. Comprehensive coverage of all aspects of astronomy, astrophysics and related fields. 1969 to present. Two parts per year. Annual. ISSN: 0067-0022.

Physics Abstracts. INSPEC. Section A, Science Abstracts. Institute of Electrical Engineers, London, United Kingdom. Available from: INSPEC/IEEE - Institute of Electrical and Electronic Engineers, Box 1331, Hoes Lane, Piscataway, NJ 08855-1331. (908) 562-5549. 1898 to present. 24 issues per year. $2835.00

per year. ISSN: 0036-8091. Also available online and on CD-ROM.

Science Citation Index. SCI. Institute for Scientific Information, 3501 Market Street, Philadelphia, PA 19104. (215) 386-0100. (800) 523-1850. FAX (215) 386-2991. 1961 to present. Six issues per year, plus annual cumulation. $11650.00 per year. Also available online and on CD-ROM.Inquire about price and availability. ISSN: 0036-827X.

STAR. Scientific and Technical Aerospace Reports. U.S. National Aeronautics and Space Administration. Distributed by U. S. Superintendent of Documents, Washington, DC 20402. 1963 to present. Semi-monthly, with semiannual and annual indexes. $114.00 per year. ISSN: 0036-8741. Also available online and on CD-ROM.

ANNUAL REVIEWS AND YEARBOOKS

The Astronomical Almanac. Superintendent of Documents, U. S. Government Printing office, Washington, DC 10402. (202) 783-3238. 1981 to present. Superceeds Astronomical Ephemeris and the American Ephermis and Nautical Almanac. Annual. $44.95. ISSN: 0737-6421.

Annual Review of Astronomy and Astrophysics. Original reviews of critical literature and current developments in astronomy and astrophysics. Annual Reviews, Inc., 4139 El Camino Way, Palo Alto, CA 94303-0139. (415) 493-4400. (800)523-8635. Fax (415) 855-9815. 1963 to date. Annual. $60.00. ISSN: 0066-4146.

ASSOCIATIONS AND PROFESSIONAL SOCIETIES

American Astronomical Society. 2000 Florida Avenue NW, Suite 400, Washington, DC 20009. (202) 328-2010.FAX(202) 234-2560.

American Institute of Physics. One Physics Ellipse, College Park, MD 20740-3843. (301) 209-3100.

American Association of Variable Star Observers. 23 Birch Street, Cambridge, MA 02138. (617) 354-0484. FAX: (617) 354-0665.

Association of Universities For Research in Astronomy, Inc. (AURA). Suite 701, 1625 Massachusetts Avenue, NW, Washington, DC 20036. (202) 483-2101. FAX (202) 483-2106.

Astronomical League. 2112 Kingfisher Lane East, Rolling Meadows, IL 60008. (708) 398-0562.

Astronomical Society of the Pacific. 390 Ashton Avenue, San Francisco, CA 94112. (415) 337-1100. FAX: (415) 337-5205.

Planetary Society. 65 North Catalina Avenue, Pasadena, CA 91106. (818) 793-5100. (800) WOW-MARS. FAX (818) 793-5528.

BIBLIOGRAPHIES

A Bibliography of Astronomy, 1970 - 1979. R. A. Sea and S. S. Martin. Libraries Unlimited, Inc., Littleton, CO 80160. 1982. $37.50.

Science Books & Films. American Association for the Advancement of Science, 1333 H Street NW, Washington, DC 20005. (202) 326-6434. Reviews of print, film and software materials in all sciences.. 1965 to present. Nine issues per year. $40.00 per year. ISSN: 0098-342X.

Scientific and Technical Books and Serials in Print; An Index To Literature in Science and Technology. R. R. Bowker. 121 Chanlon Road, New Providence, NJ 07974. (908) 464-6800. FAX (908) 665-3502. (800) 521-8110. 1972 to present. Annual. $299.95 per year. Also available on CD-ROM. ISSN: 0000-054X.

DIRECTORIES AND BIOGRAPHICAL SOURCES

American Astronomical Society. Membership Directory. American Astronomical Society. 2000 Florida Avenue NW, Suite 400, Washington, DC 20009. (202) 328-2010. FAX (202)234-2560 Annual. Included in membership dues. ISSN: 1061- 9038.

American Men and Women of Science: Physical and Biological Sciences. R. R. Bowker Inc., 121 Chanlon Road, New Providence, NJ 07974. (908) 464-6800. (800) 521-8110. 20th edition. 8 volumes. 1996. $850.00.

The Astronomers. Donald Goldsmith. St. Martin's Press, Inc., 175 Fifth Avenue, New York, NY 10010. (212) 674-5151. (800) 221-7945. 1993. $14.95 in paper.

Astronomical Centers of the World. Kevin Krisciunas. Cambridge University Press, 40 West 20th Street, New York, NY 10011-4211. (212) 924-3900. (800) 872-7423. 1988. $34.95

Directory of Physics and Astronomy Staff. American Institute of Physics, One Physics Ellipse, College Park, MD 20740-3843. (301) 209-3100. 1975/76 to present. Annual. $60.00. ISSN: 0361-2228.

Earth and Astronomical Research Centers. Stockton Press, 345 Park Avenue South, New York, NY 10010. 4th edition. 1995. $515.00. ISBN: 1-56169-0967.

Graduate Programs in Physics, Astronomy and Related Fields. American Institute of Physics, One Physics Ellipse, College Park, MD 20740-3843. (301) 209-3100. 1978 to present. Annual. $45.00. ISSN: 0147-1821.

Research Centers Directory. Gale Research Inc., 835 Penobscot Building, Detroit, MI 48226-4094. (313) 961-2242. (800) 877-4253. 20th edition. 1995. $485.00.

ENCYCLOPEDIAS AND DICTIONARIES

Concise Dictionary of Astronomy. Jacqueline Mitton. Oxford University Press, Inc., 200 Madison Avenue, New York, NY 10016. (212) 725-6000. (800) 334-4249. 1992. $24.95.

Encyclopedia of Astronomy and Astrophysics. Stephen Maran, editor. Van Nostrand Reinhold, 115 Fifth Avenue, New York, NY 10003. (212) 254-3232. (800) 842-3636. 1992. $129.95.

McGraw-Hill Encyclopedia of Science and Technology. McGraw-Hill Book Company, Inc., 1221 Avenue of the Americas, New York, NY 10020. (212) 512-2000. (800) 262-4729. 7th edition. 20 volume set. 1992. $1900.00.

GENERAL WORKS

An Amateur Radio Telescope. George W. Swenson. Pachart Publishing House, 1130 San Lucas Circle, Tuscon, AZ 85704. (520) 297-6760. 1980. $9.95 in paper.

Beyond Southern Skies: Radio Astronomy and the Parkes Telescope. Peter Robertson. Cambridge University Press, 40 West 20th Street, New York, NY 10011-4211. (212) 924-3900. (800) 872-7423. 1992. $75.00.

Classics in Radio Astronomy. W. Sullivan. Kluwer Academic Publishers, 101 Philip Drive, Assinippi Park, Norwell, MA 02061. (617) 871-6600. 1982. $131.50.

Galactic and Extragalactic Radio Astronomy. G. K. Verschuur and K. Kellermann. Springer-Verlag New York, 175 Fifth Avenue, New York, NY 10010. (212) 460-1500. (800) 777- 4643. 1991. $55.00 in paper.

Instrumentation and Techniques For Radio Astronomy. P. F. Goldsmith. Institute of Electrical and Electronic Engineers (IEEE), 345 East 47th Street, New York, NY 10017. (212) 705-7900. 1988. $59.95.

Interferometry and Synthesis in Radio Astronomy. A. Richard Thompson, et al. Krieger Publishing Company, P.O. Box 9542, Melbourne, FL 32902-9542. (407) 724-9542. Reprint edition. 1994. $59.50

Interferometry in Radioastronomy and Radio Telescopes. R. Wohlleben, et al. Kluwer Academic Publishers, 101 Philip Drive, Assinippi Park, Norwell, MA 02061. (617) 871-6600. 1991. $92.00.

Extragalactic Radio Sources From Beams To Jets. Jacques Roland, et al, editors. Cambridge University Press, 40 West 20th Street, New York, NY 10011-4211. (212) 924-3900.(800) 872-7423.

History of Radio Astronomy and the National Radio Astronomy Observatory: Evolution Toward Big Science. Benjamin K. Malphrus. Kreiger Publishing Company, P.O. Box 9542, Melbourne, FL 32902-9542. (407) 724-9542. 1995. $wfi

National Radio Astronomy Observatory Very Large Array Sky Survey. J. J. Condon, et al. Rector Press, Ltd., 130 Rattlesnake, Leverett, MA 01054-9726. (413) 548-9708. (800) 247-3473. 1994. $45.00 in paper.

Space Technology and Planetary Astronomy. Joseph N. Tatarewicz. Indiana University Pres, 601 North Morton Street, Bloomington, in 47404-3797. (812) 855-6804. (800) 842-6796. 1990. $29.95.

Tools of Radio Astronomy. K. Rolhfs. Springer-Verlag, 175 Fifth Avenue, New York, NY 10010. (212) 460-1500. (800) 777-4643. 2nd edition. 1990. $45.00.

HANDBOOKS AND MANUALS

Astrophysical Data: Planets and Stars. Kenneth R. Lanb. Springer-Verlag New York, Inc., 175 Fifth Avenue, New York, NY 10010. (212) 460-1500. (800) 777-4643. 1993. $59.00.

Encyclopedia of Astronomy and Astrophysics. Stephen Maran, editor. Van Nostrand Reinhold, 115 Fifth Avenue, New York, NY 10003. (212) 254-3232. (800) 842-3636. 1992. $129.95.

Handbook of Space Astronomy and Astrophsyics. Martin V. Zumbeck. Cambridge University Press, 40 West 20th Street, New York, NY 10011-4211. (212) 924-3900. (800) 872-7423. 2nd edition. 1990. $79.95.

ONLINE DATABASES AND CD-ROMS

CA Search. Chemical Abstracts Service, P.O. Box 3012, Columbus, OH 43210-0012. (614) 447-3600. (800) 848-6533. FAX (614) 447-3709. Very comprehensive guide to worldwide chemical literature and related fields, 1972 to present. Available on BRS,(800) 289-4277, DIALOG, (800) 334-2564, ORBIT (800) 456-7248, and STN International, FIZ Karlsruhe, P.O. Box 2465, W-7500, Karlsruhe 1, Germany, online services. Inquire as to cost and availability.

Dissertation Abstracts. University Microfilms International, 300 North Zeeb Road, Ann Arbor, MI 48106. (800) 521-0600 or (313) 761-4700. Scope includes virtually all doctoral dissertations accepted at accredited American institutions from 1861 to present in 252 subject areas. Available on BRS, (800) 289-4277, DIALOG, (800) 334-2564, and OCLC EPIC, (800) 848-5878, online services. Also available on CD-ROM. Inquire as to cost and availability.

INSPEC. Institution of Electrical Engineers, Michael Faraday House, Six Hills Way, Stevenage, Herts. SG1 2AY, England. Telephone: 0438 313311 or FAX 0438 742840. Contains citations to the worldwide literature of physics, electronics and electrical engineering, computer technology, and related fields. Available on BRS, (800) 289-4277, DIALOG, (800) 334-2564, ORBIT, (800) 456-7248, and STN International, FIZ Karlsruhe, P.O. Box 2465, W-7500, Karlsruhe 1, Germany, online services. Inquire as to cost and availability.

NTIS Bibliographic Database. National Technical Information Service, 5285 Port Royal Road, Springfield, VA 22161. (703) 487-4929 or FAX (703) 321-8199. Broad coverage of government-sponsored science and technology research reports, 1964 to present. Available on BRS,(800) 289-4277, DIALOG, (800) 334-2564, ORBIT, (800) 456-7248, and STN International, FIZ Karlsruhe, P.O. Box 2465, W-7500, Karlsruhe 1, Germany, online services. Also available on CD-ROM. Inquire as to cost and availability.

SCISEARCH. Institute for Scientific Information, 3501 Market Street, Philadelphia, PA 19104. (800) 523-1850 or (215) 386-0100. Broad multidisciplinary title and author index to the international literature of science and technology, 1974 to present. Available on DIALOG, (800) 334-2564, and ORBIT, (800) 456-7248, online services. Also available on CD-ROM. Inquire as to cost and availability.

WILSONLINE. H.W. Wilson Company, 950 University Avenue, Bronx, NY 10452. (800) 367-6770 or (212) 588-8400. Makes available online versions of the printed H.W. Wilson indexes including Applied Science and Technology Index, Business Periodicals Index, General Science Index, and Readers' Guide to Periodical Literature. Period covered is generally 1983 to present. Available on BRS, (800) 289-4277, DIALOG, (800) 334-2564, and OCLC EPIC, (800) 848-5878, online services.Also available on CD-ROM. Inquire as to cost and availability.

RADIO TELESCOPES

Ency. of Physical Sciences and Engineering Info. Sources

OTHER SOURCES

Cambridge Atlas of Astronomy. Cambridge University Press, 40 West 20th Street, New York, NY 10011-4211. (212) 924-3900. (800) 872-7423. 3rd edition. 1994. $74.00.

The Sky: A User's Guide. David H. Levy. Cambridge University Press, 40 West 20th Street, New York, NY 10011- 4211. (212) 924-3900. (800) 872-7423. 1993. $14.95.

PERIODICALS

Astronomical Journal. American Institute of Physics, One Physics Ellipse, College Park, MD 20740-3843. (301) 209- 3000. Published for the American Astronomical Society. 1849 to present. Monthly. $280.00 per year. ISSN: 0004- 6256.

Astronomical Society of the Pacific Publications. Astronomical Society of the Pacific, 390 Ashton Avenue, San Francisco, CA 94112. (415) 337-1100. FAX (415) 337-5205. 1889 to present. Monthly. $175.00 per year. ISSN: 0004- 6280.

Astronomy. Kalmbach Publishing Company, Box 1612, Waukesha, WI 53187-1612. (414) 796-8776. FAX (414) 796-1142. 1973 to present. Monthly. $27.00 per year. ISSN: 0091-6358.

Astronomy and Astrophysics. Springer-Verlag New York, Inc., 44 Hartz Way, Secaucus, NJ 07096-2491. (201) 348-4033. FAX(201) 348-4505. 1969 to date. 36 issues per year in 12 volumes. $1733.00 per year. ISSN: 0004-6361.

Astrophysical Journal. University of Chicago Press, 5720 South Woodlawn Avenue, Chicago, IL 60637. (312) 753-3347. FAX (312) 753-0811. 1895 to date. 3 issues per monthly. $740.00 per year. ISSN: 0004-637X.

Astrophysics and Space Science. Kluwer Academic Publishers, Box 358, Accord Station, Hingham MA 02018-0358. (617) 871- 6600. FAX (617) 871-6528. 1968 to present. 24 issues per year. $2544.00 per year. ISSN: 0004-640X.

Celestial Mechanics; An International Journal of Space Dynamics. Kluwer Academic Publishers, Box 358, Accord Station, Hingham MA 02018-0358. (617) 871-6600. FAX (617) 871-6528. 1969 to date. 12 issues per year. $703.50 per year. ISSN: 0923-2958.

Icarus; International Journal of Solar System Studies. Academic Press, Inc., Journal Division, 525 B Street, Suite 1900, San Diego, CA 92101-4495. (619) 230-1840. FAX (619) 699-6800. 1962 to the present. Monthly. $1080.00. ISSN: 0019-1035.

IEEE Transactions On Antennas and Propagation. Institute of Electrical and Electronics Engineers, Inc., Box 1331, 445 Hoes Lane, Piscataway, NJ 08855-1331. (908) 981-0060. 1952 to present. Monthly. $212.00 per year. ISSN: 0018-926X.

Mercury. Astronomical Society of the Pacific. 390 Ashton Avenue, San Francisco, CA 94112. (415) 337-1100. FAX: (415) 337-5205. 1972 to present. Bimonthly. $175.00 per year. ISSN: 0047-6773.

Monthly Notices of the Royal Astronomical Society. Blackwell Scientific Publications, Inc., Osney Mead, Oxford OX2 QEL England. TEL 0865-240201. FAX 0865-721205. 1827 to present. Fortnightly. $1899.00 per year. ISSN: 0035- 8711.

Observatory. c/o Dr. D. J. Strickland, Space and Astrophysics Division, Rutherford Appleton Laboaratory, Chlton, Didcot, Oxon, OX11 OQX, England. FAX 0235-445848. 1877 to present. Bimonthly. $42.00. ISSN: 0029-7704.

Planetary and Space Science. Elsevier Science Publishing Company, Inc., 660 White Plains Road, Terrytown, NY 10591- 5153. (914) 524-9200. FAX: (914) 333-2444. 1959 to present. Monthly. $1355.00 per year. ISSN: 0032-0633.

Sky and Telescope. Sky Publishing Corporation, Box 9111, Belmont, MA 02178. (617) 864-7360. FAX (617) 864-6117. 1941 to present. Monthly. $27.00 per year. ISSN: 0037- 6604.

Space Science Reviews. Kluwer Academic Publishers, Box 358, Accord Station, Hingham MA 02018-0358. (617) 871-6600. FAX (617) 871-6528. 1962 to present. 16 issues per year. $940 issues per year. ISSN: 0038-6308.

Vistas in Astronomy. Elsevier Science Publishing Company, Inc., 660 White Plains Road, Terrytown, NY 10591- 5153. (914) 524-9200. FAX: (914) 333-2444. 1958 to date. Quarterly. $415.00 per year. ISSN: 0083-6656.

RESEARCH CENTERS AND INSTITUTES.

Arecobo Observatory. Cornell University, Box 995. Arecibo, PR 00613. (809) 878-2612. FAX (809) 878-1861.

Haystack Observatory. Massachusetts institute of Technology. Westford, MA 01886. (508) 692-4765. FAX (617) 981-0590.

Herzberg Institute of Astrophsyics. National Research Council of Canada, 100 Sussex Drive, Ottawa, ON Canada K1A 0R6. (613) 990-0907. FAX (613) 952-6602.

National Astronomy and Ionosphere Center. Space Sciences Building, Ithaca, NY 14853. (607) 255-3735. FAX (607) 255-8803.

National Radio Astronomy Observatory. Edgemont Road, Charlottesville, VA 22903. (804) 296-0211.

Owens Valley Radio Observatory. California Institute of Technology. Pasadena, CA 91125. (818) 795-6811.

Radio Astronomy Laboratory. University of California, Berkeley, 601 Campbell Hall, Berkeley, CA 94720. (415) 642-5275. FAX (415) 642-3411.

Radio Astronomy Observatory. University of Michigan, 810 Dennison Building, Ann Arbor, MI 48109. (313) 426-8441.

Smithsonian Astrophysical Observatory. 60 Garden Street, Cambridge, MA 02138. (617) 495-7461. FAX (617) 495-7105.

RADIOACTIVITY

See: RADIATION

RADIOCARBON DATING

See also: ANALYTICAL CHEMISTRY, CARBON, CHEMICAL ENGINEERING, CHEMISTRY, PHYSICAL CHEMISTRY, SPECTROS-COPY

ABSTRACT SERVICES AND INDEXES

Applied Science and Technology Index; A Cumulative Subject Index To English Language Periodicals in the Fields of Aeronautics and Space Science, Computer Technology, Chemistry, Construction Industry, Energy and Related Areas.

H.W. Wilson Co., 950 University Avenue, Bronx, NY 10452. (212) 588-8400. (800) 367-6770. FAX (718) 590-1617. From 1958 to present. Monthly. Inquire about cost and availability. Also available on CD-ROM and online. ISSN: 0003-6986.

Chemical Abstracts. Chemical Abstracts Service. 2540 Oleantangy River Road, Box 3012, Columbus, OS 43210-0012. (614) 447-3600. FAX (614) 447-3713. 1907 to present. Weekly. $16,800.00 per year. Available online and on CD- ROM. CA is also available in five section groupings. Inquire regarding cost and availability.

Current Contents: Physical, Chemical and Earth Sciences. Institute for Scientific Information, 3501 Market Street, Philadelphia, PA 19104. (215) 386-0100. FAX (215) 386-2291. From 1961 to present. Weekly. $442.00 per year. Also available on CD-ROM and online. Inquire regarding cost and availability. ISSN: 0163-2574.

General Science Index. H.W. Wilson Company, 950 University Avenue, Bronx, NY 10452. (212) 588-8400. (800) 367-6770. FAX (718) 590-1617. From 1978 to present. Ten issues per year; quarterly and annual cumulations. Service basis. Available on CD-ROM and online. Inquire regarding cost and availability. ISSN: 0162-1963.

Physics Abstracts. INSPEC. Section A, Science Abstracts. Institution of Electrical Engineers (IEE), London. Available from: INSPEC/IEEE - Institute of Electrical and Electronic Engineers, Box 1331, Hoes Lane, Piscataway, NJ 08855-1331. (908) 562-5549. 1898 to present. 24 issues per year. $2835.00 per year. Also available online and on CD-ROM. ISSN: 0036-8091.

Physics Briefs (Physikalische Berichte). Information Center for Energy, Physics, Mathematics; German Physical Society. V C H Publishers, Inc., 220 East 23rd Street, New York, NY 10010-4606. (212) 683-8333. 1845 to present. 24 issues per year. $2390.00 per year. Also available online. ISSN: 0179-7434.

Science Citation Index. SCI. Institute for Scientific Information, 3501 Market Street, Philadelphia, PA 19104. (215) 386-0100. (800) 523-1850. FAX (215) 386-2991. 1961 to present. Six issues per year, plus annual cumulation. $11650.00 per year. Also available online and on CD-ROM. Inquire about price and availability. ISSN: 0036-827X.

ASSOCIATIONS AND PROFESSIONAL SOCIETIES

American Carbon Society. c/o Carbonne of America, 215 Stackpole St. Marys, PA 15857. (814) 781-8410.

American Chemical Society, 1155 16th Street, Nw, Washington, DC 20036. (202) 872-4600.

American Institute of Chemical Engineers. 345 East 47th Street, New York, NY 10017. (212) 705-7338.

Association of Consulting Chemists and Chemical Engineers, 50 East 41st Street, Suite 92, New York, NY 10017. (212) 684-6255.

Association of official Analytical Chemists. 2200 Wilson Blvd, Suite 400, Arlington, VA 22001. (703) 522-3032.

BIBLIOGRAPHIES

Chemical Information Sources. Gary Wiggins. McGraw-Hill Publishing Company, 1221 Avenue of the Americas, New York, NY 10020. (800) 262-4729 or (212) 512-3825. 1991. $42.50.

Scientific and Technical Books and Serials in Print; An Index To Literature in Science and Technology. R.R. Bowker Co., 205 E. 42nd Street, New York, NY 10017. (800) 521-8110 or (212) 916-1600. 1992. $250.00.

DIRECTORIES AND BIOGRAPHICAL SOURCES

American Institute of Chemists - Professional Directory. American Institute of Chemists, 501 Wythe Street, Alexandria, VA 22314-1917. (703) 836-2090. FAX (703) 836-2091. Annual. $65.00.

American Men and Women of Science. Physical and Biological Sciences. Fifteenth edition. R.R. Bowker Company, 205 East 42nd Street, New York, NY 10017. (800) 521-8110 or (212) 916-1600.

Consulting Services: Chemists and Chemical Engineers. Association of Consulting Chemists and Chemical Engineers, 40 East 45th Street, New York, NY 10017. (212) 983-3160. FAX (212) 983-3161. Biennial. $60.00.

Directory of Chemistry Software 1992. Wendy Warr, Peter Willett, Geoff Downs. American Chemical Society, 1155 16th Street, NW, Washington, DC 20036. (202) 872-4600. 1992. $35.95.

Research Centers Directory. Gale Research Company, Detroit, MI 48226. Seventeenth edition, 1992. (800) 521-0707. $400.00.

ENCYCLOPEDIAS AND DICTIONARIES

Kirk-Othmer Encyclopedia of Chemical Technology. John Wiley and Sons, Inc., 605 Third Avenue, New York, NY 10158. (800) 526-5368 or (212) 850-6000. Fourth edition. 1991-. Twenty-seven volumes. $5400.00.

McGraw-Hill Encyclopedia of Science and Technology. McGraw-Hill Book, Incorporated, 1221 Avenue of the Americas, New York, NY 10020. (212) 997-3675. Seventeenth editon. 1992. $1900.00.

GENERAL WORKS

Atlas of Carbon-Thirteen NMB Data, Volume 1: Copounds 1-1003. E. Breitmaier, et al. Books on Demand, 300 North Zeeb Road, Ann Arbor, MI 48106-1346. (313) 761-4700. (800)521-0600. $73.60.

Radiocarbon After Four Decades: An Interdisciplinary Perspective. R. E. Taylor, et al, editors. Springer-Verlag New York, Inc., 175 Fifth Avenue, New York, NY 10010. (212) 460-1500. (800) 777-4643. FAX (212) 473-6272. 1992. $89.00.

Radiocarbon Dating. Sheridan Bowman. University of California Press, 2120 Berkeley Way, Berkeley, CA 94720. (510) 642-4247. (800) 777-4726. 1990. $12.00.

Radiocarbon Dating. Willard F. Libby. Books on Demand, 300 North Zeeb Road, Ann Arbor, MI 48016-1346. (313) 761-4700. (800) 521-0600. 2nd edition. $53.30.

Radiocarbon Dating. R. E. Taylor. Academic Press, Inc., 6277 Sea Harbor Drive, Orlando, FL. (800) 321-5068. 1987. $66.00.

HANDBOOKS AND MANUALS

Lange's Handbook of Chemistry. John A. Dean, editor. McGraw-Hill Publishing Company, 1221 Avenue of the Americas, New York, NY 10020. (800) 262-4729 or (212) 512-3825. Fourteenth editon. 1992. $79.50.

ONLINE DATABASES AND CD-ROMS

Analytical Abstracts Online. Royal Society of Chemistry, Information Services, Thomas Graham House, Science Park, Milton Road, Cambridge, CB4 4WF, England. Contains citations, mostly with abstracts, of the worldwide literature on analytical chemistry, from 1980 to present. Available on DIALOG and STN online services. Inquire as to cost and availability.

CA Search. Chemical Abstracts Service, P.O. Box 3012, Columbus, OH 43210-0012. (614) 447-3600. (800) 848-6533. FAX (614) 447-3709. Guide to worldwide chemical literature and related fields, 1972 to present. Available on BRS, DIALOG, ORBIT and STN online services. Inquire as to cost and availability.

Dissertation Abstracts. University Microfilms International, 300 North Zeeb Road, Ann Arbor, MI 48106. (800) 521-0600 or (313) 761-4700. Scope includes virtually all doctoral dissertations accepted at accredited American institutions from 1861 to present in 252 subject areas. Available on BRS, DIALOG, and OCLC EPIC online services. Also available on CD-ROM. Inquire as to cost and availability.

INSPEC. Institution of Electrical Engineers, Michael Faraday House, Six Hills Way, Stevenage, Herts. SG1 2AY, England. Telephone: 0438 313311 or FAX 0438 742840. Contains citations to the worldwide literature of physics, electronics and electrical engineering, computer technology, and related fields. Available on BRS, DIALOG, ORBIT and STN online services. Inquire as to cost and availability.

NTIS. National Technical Information Service, 5285 Port Royal Road, Springfield, VA 22161. (703) 487-4929 or FAX (703) 321-8199. Broad coverage of government-sponsored research reports, 1964 to present. Available on BRS, DIALOG, ORBIT and STN online services. Also available on CD-ROM. Inquire as to cost and availability.

SCISEARCH. Institute for Scientific Information, 3501 Market Street, Philadelphia, PA 19104. (800) 523-1850 or (215) 386-0100. Broad multidisciplinary title and author index to the international literature of science and technology, 1974 to present.

Available on DIALOG and ORBIT online services. Also available on CD-ROM. Inquire as to cost and availability.

WILSONLINE. H.W. Wilson Company, 950 University Avenue, Bronx, NY 10452. (800) 367-6770 or (212) 588-8400. Makes available online versions of the printed H.W. Wilson indexes including Applied Science and Technology Index, Business Periodicals Index, and Readers' Guide to Periodical Literature. Period covered is generally 1983 to present. Also available on CD-ROM. Inquire as to cost and availability.

PERIODICALS

Radiocarbon. University of Arizona, Department of Geosciences, 4717 East Fort Lowell Road, Tucson, AZ 85712. (602) 881-0857. 1959 to present. 3 issues per year. $105.00 per year. ISSN: 0033-8222.

RESEARCH CENTERS AND INSTITUTES

Laboratory of Isotope Geochemistry. University of Arizona, 136 Gould-Simpson Building, Tucson, AZ 85721. (602) 621-6014. FAX (602) 621-2672.

Radiocarbon Dating Laboratory. University of California, Riverside, Riverside CA 92521. (714) 787-5521.

Radiogenic-Isotopes Laboratory. Ohio State University, Mendenhall Laboratory, Department of Geological Sciences, Columbus, OH 43210. (614) 292-2721.

RADIOCHEMISTRY

See: NUCLEAR CHEMISTRY

RADON

ABSTRACT SERVICES AND INDEXES

Applied Science and Technology Index; A Cumulative Subject Index To English Language Periodicals in the Fields of Aeronautics and Space Science, Computer Technology, Chemistry, Construction Industry, Energy and Related Areas. H.W. Wilson Co., 950 University Avenue, Bronx, NY 10452. (800) 367-6770 or (212) 588-8400. From 1958 to present. Monthly. Inquire about cost and availability. Also available on CD-ROM and online.

Current Contents: Physical, Chemical and Earth Sciences. Institute for Scientific Information, 3501 Market Street, Philadelphia, PA 19104. (215) 386-0100. FAX (215) 386-6362. Weekly. $360.00 per year.

Environmental Abstracts. R.R. Bowker Inc., 121 Chanlon Road, New Providence, NJ 07974. (800) 521-8110 or (908) 464-6800. Contains citations, most with abstracts, to literature relating to environmental issues and problems, from 1971 to present. Inquire as to cost and availability.

Environmental Engineering Abstracts, Cambridge Scientific Abstracts, 7200 Wisconsin Avenue, Bethesda, MD 20814-4823.

(301) 961-6750. FAX (301) 961-6720. Monthly. Covers hazardous materials, environmental impact and protection, treatment of sewage and industrial wastes, hydroelectric power, tidal and wind power, arctic and tropical engineering.

Index To Scientific and Technical Proceedings. Institute for Scientific Information, 3501 Market St., Philadelphia, PA 19104. (215) 386-0100. FAX (215) 386-6362. Monthly. $500.00 per year.

International Building Services Abstracts. Building Service Research and Information Association, Old Bracknell Lane W, Bracknell, Berks. RG12 7AH, England. Telephone 0344-426511. FAX 0344-487575. 1966 to present. Bi-monthly. Inquire for price. Also available online. A survey of world literature on mechanical and technical services of buildings, including heating, sanitation, ventilation, air conditioning, lighting, and communications.

Pollution Abstracts, Cambridge Scientific Abstracts, 7200 Wisconsin Avenue, Bethesda, MD 20814-4823. (301) 961-6750. FAX (301) 961-6720. Inquire for cost and availability.

Science Citation Index. Institute for Scientific Information, 3501 Market Street, Philadelphia, PA 19104. (215) 386-0100. FAX (215) 386-6362. Inquire about availability and cost. Also available on CD-ROM.

ASSOCIATIONS AND PROFESSIONAL SOCIETIES

American Association of Radon Scientists and Technologists. PO Box 70, Park Ridge, NJ 07656. (201) 391-6445. FAX (201) 391-0670.

DIRECTORIES AND BIOGRAPHICAL SOURCES

Directory of Engineering Societies and Related Organizations. Gordon Davis. 13th edition. American Association of Engineering Societies, 1111 19th Street NW, Suite 608, Washington, DC 20036. (202) 296-2237 or (800) 658-8897. 1989. Inquire for price.

International Research Centers Directory. Gale Research, 835 Penobscot Building, Detroit, MI 48226-4094. (313) 961-2242. (800) 877-4253. 8th edition. 1995. $410.00

Research Centers Directory. Gale Research, 835 Penobscot Building, Detroit, MI 48226-4094. (313) 961-2242. (800) 877-4253. $470.00

Scientific and Technical Organizations and Agencies Directory. Gale Research, 835 Penobscot Building, Detroit, MI 48226-4094. (313) 961-2242. (800) 877-4253. 4th edition. 1996. $195.00.

ENCYCLOPEDIAS AND DICTIONARIES

McGraw-Hill Dictionary of Scientific and Technical Terms. Sybil P. Parker, ed. 5th edition. McGraw-Hill Publishing Company, 1221 Avenue of the Americas, New York, NY 10020. (800) 262-4729 or (212) 512-3825. 1993. $110.50.

McGraw-Hill Encyclopedia of Science and Technology. Sybil P. Parker, ed. 7th edition. 20 volumes. McGraw-Hill Publishing

Company, 1221 Avenue of the Americas, New York, NY 10020. (800) 262-4729 or (212) 512-3825. 1992. $1900.00

GENERAL WORKS

Element of Risk: The Politics of Radon. Leonard A. Cole. American Association for the Advancement of Science, 1333 H Street NW, Washington, DC 20005. (202) 326-6446. 1993. $29.95.

Health Risks of Exposure To Radon: Time For Reassessment? National Research Council. National Academy Press, Division of the National Academy of Science, 2101 Constitution Avenue, N.W., Lockbox 285, Washington, DC 20055. (800) 624-6242. FAX (202) 334-2793. 1994. $27.00.

Indoor Air Pollution: Radon. Jack G. Kay, et al. Lewis Publishers, 2000 Corporate Blvd. NW, Boca Raton, FL 33431. (407) 994-0555 or (800) 272-7737. 1991. $85.00.

The Physics of Radon. A. Worley. Taylor & Francis, 1900 Frost Road, Suite 101, Bristol, PA 19007. (215) 785-5800. FAX (215) 785-5515. 1993. $75.00.

Radon: Prevalence, Measurements, Health Risks, & Control. Niren L. Nagda, editor. American Society for Testing and Materials (ASTM), 1916 Race Street, Philadelphia, PA 19103. (215) 299-5585. FAX (215) 477-9679. 1994. $59.00.

HANDBOOKS AND MANUALS

CRC Handbook of Chemistry and Physics. David R. Lide, editor. 75th edition. CRC Press/Times Mirror Company, 2000 Corporate Blvd. NW, Boca Raton, FL 33431. (407) 994-0555 or (800) 272-7737. $99.50.

Environmental Science and Technology Handbook. Porter C. Knowles, et al. Government Institutes Inc., 4 Research Place, Suite 200, Rockville, MD 20850. (301) 921-2355. 1994. $75.00.

Handbook of Radon. Stephen J. Wozniak. Medical Physics Publishing, 4513 Vernon Blvd., Madison, W I 53705-4964. (800) 442-5778. 1992. $24.00.

Handbook of Radon in Buildings: Detection, Safety, and Control. Muller Associates Incorporated staff, et al., compilers. Hemisphere Publishing Corporation, 1900 Frost Road, Suite 101, Bristol, PA 19007-1598. (215) 785-5800 or (800) 821-8312. FAX (215) 785-5515. 1988. $67.00.

ONLINE DATABASES AND CD-ROMS

CA Search. Chemical Abstracts Service, P.O. Box 3012, Columbus, OH 43210-0012. (614) 447-3600. (800) 848-6533. FAX (614) 447-3709. Very comprehensive guide to worldwide chemical literature and related fields, 1972 to present. Available on BRS,(800) 289-4277, DIALOG, (800) 334-2564, ORBIT (800) 456-7248, and STN International, FIZ Karlsruhe, P.O. Box 2465, W-7500, Karlsruhe 1, Germany, online services. Inquire as to cost and availability.

Compendex Plus. Engineering Information, Inc., 345 East 47th Street, New York, NY 10017. (212) 705-7600 or (800) 221-1044. Contains citations with abstracts to worldwide literature in engineering and technology, from 1970 to present. Available on online BRS,(800) 289-4277, DIALOG, (800) 334-2564,

RADON

Ency. of Physical Sciences and Engineering Info. Sources

ORBIT (800) 456-7248, and STN International, FIZ Karlsruhe, P.O. Box 2465, W-7500, Karlsruhe 1, Germany, online services. Also available on CD-ROM. Inquire as to cost and availability.

Current Contents Search. Institute for Scientific Information, 3501 Market Street, Philadelphia, PA 19104. (215) 386-0100. FAX (215) 386-6362. Contains citations to articles listed in the table of contents of science and technology journals. Also articles in social sciences and life sciences journals. Available on BRS,(800) 289-4277, DIALOG,(800) 334-2564, online services. Inquire as to cost and availability.

Dissertation Abstracts. University Microfilms International, 300 North Zeeb Road, Ann Arbor, MI 48106. (800) 521-0600 or (313) 761-4700. Scope includes virtually all doctoral dissertations accepted at accredited American institutions from 1861 to present in 252 subject areas. Available on BRS,(800) 289-4277, DIALOG,(800) 334-2564, and OCLC EPIC,(800) 848-5878, online services. Also available on CD-ROM. Inquire as to cost and availability.

NTIS Bibliographic Database. National Technical Information Service, 5285 Port Royal Road, Springfield, VA 22161. (703) 487-4929 or FAX (703) 321-8199. Broad coverage of government-sponsored science and technology research reports, 1964 to present. Available on BRS,(800) 289-4277, DIALOG, (800) 334-2564, ORBIT, (800) 456-7248, and STN International, FIZ Karlsruhe, P.O. Box 2465, W-7500, Karlsruhe 1, Germany, online services. Also available on CD-ROM. Inquire as to cost and availability.

SCISEARCH. Institute for Scientific Information, 3501 Market Street, Philadelphia, PA 19104. (800) 523-1850 or (215) 386-0100. Broad multidisciplinary title and author index to the international literature of science and technology, 1974 to present. Available on DIALOG,(800) 334-2564, and ORBIT,(800) 456-7248, online services. Also available on CD-ROM. Inquire as to cost and availability.

PERIODICALS

Journal of Environmental Radioactivity. Elsevier Science [journals], 660 White Plains Rd., Tarrytown, NY 10159-5153. (919) 524-9200. FAX (919) 333-2444. 1984 to present. 12 times a year. $645.00 per year.

RESEARCH CENTERS AND INSTITUTES

Ball State University, Center for Energy Research Education Service. 2000 University Avenue, AB G18, Muncie, in 47306. (317) 285-1135.

New Mexico Institute of Mining and Technology, Atmospheric Radioactivity Lab. Socorro, NM 87801. (505) 835-5341. FAX (505) 835-5895.

New Mexico Institute of Mining and Technology, Research and Economic Development. Campus Station, Socorro, NM 87801. (505) 835-5646. FAX (505) 835-5630.

RAILWAY ENGINEERING

See also: TRANSPORTATION ENGINEERING

ABSTRACT SERVICES AND INDEXES

Applied Science and Technology Index; A Cumulative Subject Index To English Language Periodicals in the Fields of Aeronautics and Space Science, Computer Technology, Chemistry, Construction Industry, Energy and Related Areas. H.W. Wilson Co., 950 University Avenue, Bronx, NY 10452. (800) 367-6770 or (212) 588-8400. FAX (718) 590-1617. From 1958 to present. Monthly. Inquire about cost and availability. Also available on CD-ROM and online.

Civil and Structural Engineering Abstracts, Cambridge Scientific Abstracts, 7200 Wisconsin Avenue, Bethesda, MD 20814-4823. (301) 961-6750. FAX (301) 961-6720. 1993 to present. Monthly. $385.00 per year. Topics covered include structural design, construction equipment and methods, civil defense and military engineering, surveying, highway engineering, maritime and port structures, materials, land reclamation, and soil mechanics.

Current Contents: Engineering, Technology, and Applied Sciences. Institute for Scientific Information, 3501 Market Street, Philadelphia, PA 19104. (215) 386-0100. FAX (215) 386-6362. 1970 to present. Weekly. $442.00 per year.

Engineering Index Monthly. Engineering Information, Inc., Castle Point on the Hudson, Hoboken, NJ 07030. (800) 221-1044. FAX (212) 832-1857. Monthly. $2300.00 per year. Also available online as COMPENDEX, and also on CD-ROM. Covers chemical engineering, computers, electrical engineering, civil engineering, metals and mining, industrial management, and mechanical engineering.

Metals Abstracts and Metals Abstracts Index. American Society for Metals, Metals Park, OH 44073. (216) 338-5151. 1968 to present. Monthly. Abstracts are $1925.00 per year and Index is $460.00 per year.

ANNUAL REVIEWS AND YEARBOOKS

Jane's World Railways. 1950-present. Jane's Information Group, 1340 Braddock Place, Suite 300, Alexandria, VA 22314-1651. (703) 683-3700. FAX (703) 836-0029. Annual. $285.00 per year.

ASSOCIATIONS AND PROFESSIONAL SOCIETIES

American Railway Engineering Association. 50 F Street NW, Suite 7702, Washington, DC 20001. (202) 639-2190. FAX (202) 639-2183.

High Speed Rail/magnetic Levitation Association. 206 Valley Court, Suite 800, Pittsburgh, PA 15237. (412) 366-6887. FAX (412) 369-8698.

National Railroad Construction and Maintenance Association. 122 C Street NW, Suite 850, Washington, DC 20001. (202) 638-7790. FAX (202) 638-1045.

Railway Engineering Maintenance Suppliers Association. 210 N. Little Falls Street, Suite 100, Falls Church, VA 22046. (703) 241-8514. FAX (703) 241-8589.

Railway Progress Institute. 700 North Fairfax Street, Alexandria, VA 22314. (703) 836-2332. DIRECTORIES

American Railway Engineering Association Directory. American Railway Engineering Association, 50 F Street NW, Suite 7702, Washington, DC 20001. (202) 639-2190. FAX (202) 639-2183. Inquire for price and availability.

The Best in Railroad Construction. National Railroad Construction and Maintenance Association, 122 C Street NW, Suite 850, Washington, DC 20001. (202) 638-7790. FAX (202) 638-1045. Annual. Inquire about availability and price.

High Speed Rail Directory. High Speed Rail/Magnetic Levitation Association, 206 Valley Court, Suite 800, Pittsburgh, PA 15237. (412) 366-6887. FAX (412) 369-8698. Periodic. Inquire for availability and price.

Railway Directory. Reed Business Publishing Group, Quadrant House, the Quadrant, Sutton, Surrey SM2 5AS, England. Telephone 081-652-3369. FAX 081-652-8180. Annual. $140.00.

ENCYCLOPEDIAS AND DICTIONARIES

Railway Dictionary. Alan A. Jackson. Alan Sutton Publishing, 1 Washington Street, Dover, NH 03820. (603) 743-4266. FAX (603) 743-4267. 1992. $34.00.

GENERAL WORKS

High Speed Trains: Fast Tracks To the Future. John Whitelegg, et al, editors. Leading Edge Press & Publishing Ltd., the Old Chapel, Burtersett, Hawes, North Yorkshire DL8 3PB, England. Telephone (069) 667566. Inquire for cost and availability.

The North American Railroad: Its Origin, Evolution, and Geography. James E. Vance. Johns Hopkins University Press, 2715 North Charles St., Baltimore, MD 21218. (410) 516-6900 or (800) 537-5487. FAX (410) 516-6998. 1995. $39.95.

Railroad Engineering, Volume One. William W. Hay. John Wiley and Sons, Inc., 605 Third Avenue, New York, NY 10158. (800) 526-5368 or (212) 850-6000. 1982. $107.95.

Railroad Issues. Transporation Research Board. National ACADemy Press, Division of the National ACADemy of Science, 2101 Constitution Avenue, N.W., Lockbox 285, Washington, DC 20055. (800) 624-6242. FAX (202) 334-2793. 1992. $21.00.

Recent Issues in Rail Research. National Research Council. National ACADemy Press, Division of the National ACADemy of Science, 2101 Constitution Avenue, N.W., Lockbox 285, Washington, DC 20055. (800) 624-6242. FAX (202) 334-2793. 1993. Write for information.

HANDBOOKS AND MANUALS

Manual For Railway Engineering. American Railway Engineering Association, 50 F Street NW, Suite 7702, Washington, DC 20001. (202) 639-2190. FAX (202) 639-2183. Write or call for availability.

ONLINE DATABASES AND CD-ROMS

Compendex Plus. Engineering Information, Inc., 345 East 47th Street, New York, NY 10017. (212) 705-7600 or (800) 221-1044. Contains citations with abstracts to worldwide literature

in engineering and technology, from 1970 to present. Available on online BRS,(800) 289-4277, DIALOG, (800) 334-2564, ORBIT (800) 456-7248, and STN International, FIZ Karlsruhe, P.O. Box 2465, W-7500, Karlsruhe 1, Germany, online services. Also available on CD-ROM. Inquire as to cost and availability.

CSA Engineering. Cambridge Scientific Abstracts, 7200 Wisconsin Avenue, Suite 601, Bethesda, MD 20814. (301) 961-6750 or (800) 843-7751. Contains citations and abstracts of international periodicals and research literature covering all fields of engineering and science and technology,including computer and information science, electronics, mechanical engineering, solid state materials, 1981 to present. Available on BRS,(800) 289-4277, online service.Inquire as to cost and availability.

Current Contents Search. Institute for Scientific Information, 3501 Market Street, Philadelphia, PA 19104. (215) 386-0100. FAX (215) 386-6362. Contains citations to articles listed in the table of contents of science and technology journals. Also articles in social sciences and life sciences journals. Available on BRS,(800) 289-4277, DIALOG,(800) 334-2564, online services. Inquire as to cost and availability.

Dissertation Abstracts. University Microfilms International, 300 North Zeeb Road, Ann Arbor, MI 48106. (800) 521-0600 or (313) 761-4700. Scope includes virtually all doctoral dissertations accepted at accredited American institutions from 1861 to present in 252 subject areas. Available on BRS,(800) 289-4277, DIALOG,(800) 334-2564, and OCLC EPIC,(800) 848-5878, online services. Also available on CD-ROM. Inquire as to cost and availability.

ISMEC: Mechanical Engineering Abstracts. Cambridge Scientific Abstracts, 7200 Wisconsin Avenue, Suite 601, Bethesda, MD 20814. (301) 961-6750 or (800) 843-7751. Contains citations to the literature in mechanical engineering, industrial and production engineering, energy, power, mechanics, devices and related areas, from 1973 to present. Available on the DIALOG,(800) 334-2564, online service. Inquire as to cost and availability.

NTIS BIBLIOGRAPHIC DATABASE. National Technical Information Service, 5285 Port Royal Road, Springfield, VA 22161. (703) 487-4929 or FAX (703) 321-8199. Broad coverage of government-sponsored science and technology research reports, 1964 to present. Available on BRS,(800) 289-4277, DIALOG, (800) 334-2564, ORBIT, (800) 456-7248, and STN International, FIZ Karlsruhe, P.O. Box 2465, W-7500, Karlsruhe 1, Germany, online services. Also available on CD-ROM. Inquire as to cost and availability.

SCISEARCH. Institute for Scientific Information, 3501 Market Street, Philadelphia, PA 19104. (800) 523-1850 or (215) 386-0100. Broad multidisciplinary title and author index to the international literature of science and technology, 1974 to present. Available on DIALOG,(800) 334-2564, and ORBIT,(800) 456-7248, online services. Also available on CD-ROM. Inquire as to cost and availability.

Tris (transportation Research Information). National ACADemy of Science, 2101 Constitution Avenue, N.W., Washington, DC 20418. (202) 334-3313 or (800) 624-6242. Citations with abstracts of literature on transportation, including air, highway, rail, maritime and other modes from 1968 to present. Available on DIALOG,(800) 334-2564, online service. Inquire as to cost and availability.

WILSONLINE. H.W. Wilson Company, 950 University Avenue, Bronx, NY 10452. (800) 367-6770 or (212) 588-8400. Makes

RAILWAY ENGINEERING

Ency. of Physical Sciences and Engineering Info. Sources

available online versions of the printed H.W. Wilson indexes including Applied Science and Technology Index, Business Periodicals Index, General Science Index, and Readers' Guide to Periodical Literature. Period covered is generally 1983 to present. Available on BRS,(800) 289-4277, DIALOG,(800) 334-2564, and OCLC EPIC,(800) 848-5878, online services. Also available on CD-ROM. Inquire as to cost and availability.

PERIODICALS

A R E A Bulletin. 1900-present. American Railway Engineering Association, 50 F Street NW, Suite 7702, Washington, DC 20001. (202) 639-2190. FAX (202) 639-2183. Five times a year. $78.00.

American Railway Engineering Association Proceedings. American Railway Engineering Association, 50 F Street NW, Suite 7702, Washington, DC 20001. (202) 639-2190. FAX (202) 639-2183. Annual. $78.00.

Institution of Mechanical Engineers. Proceedings: Part F: Journal of Rail and Rapid Transit. Mechanical Engineering Publications Ltd., Northgate Ave., Bury St. Edmunds, Suffolk IP32 6BW, England. Telephone 0284-763227. FAX 0284-704006. Twice a year. $159.00 per year.

On Track. 1977-present. National Railroad Construction and Maintenance Association. 122 C Street NW, Suite 850, Washington, DC 20001. (202) 638-7790. FAX (202) 638-1045. 24 times a year. Free to members, others inquire for price and availability.

Railway Age. 1856-present. Simmons-Boardman Publishing Corporation, 345 Hudson St., New York, NY 10014. (212) 620-7200. Monthly. $35.00 per year.

RESEARCH CENTERS AND INSTITUTES

Carnegie Mellon University High-speed Ground Transportation Center. Mellon Institute, 4400 Fifth Avenue, Pittsburgh, PA 15213-2683. (412) 268-2960. FAX (412) 268-6852.

Carnegie Mellon University Rail Systems Center. Mellon Institute, 4400 Fifth Avenue, Pittsburgh, PA 15213-2683. (412) 268-2960. FAX (412) 268-6852.

Railroad Advancement Through Information and Law Foundation. 121 Wintergreen Street, Michigan City, in 46360. (219) 872-3920.

RAIN

See also: ACID RAIN, METEOROLOGY, WEATHER MODIFICATION

ABSTRACT SERVICES AND INDEXES

Applied Science and Technology Index; a cumulative subject index to English language periodicals in the fields of aeronautics and space science, computer technology, chemistry, construction industry, energy and related areas. H.W. Wilson Co., 950 University Avenue, Bronx, NY 10452. - 9978. (212) 588-8400. (800) 367-6770. FAX (718) 590-1617.From 1958 to present. Monthly. Inquire about cost and availability. Also avail-

able online (BRS and WILSONLINE) and on CD-ROM. ISSN: 0003-6986.

Deep-sea Research. Part A: Oceanographic Research Papers. Deep-sea Research. Part B: Oceanographic Literature Review. Pergamon Press, Inc., Maxwell House, Fairview Park, Elmsford, NY 10523. (914) 592-7700. Fax (914) 592-3625.

1953 to present. Twelve times per year. $2000.00 per year for Parts A and B. Oceanographic Literature Review also available on CD-ROM. Inquire about price and availability.

General Science Index. H.W. Wilson Co., 950 University Avenue, Bronx, NY 10452. (212) 588-8400. (800) 367-6770. FAX (718) 590-1617. From 1978 to present. Ten issues per year; quarterly and annual cumulations. Service basis. Available on CD-ROM and online. Inquire regarding cost and availability. ISSN: 0162-1963.

Government Reports Announcements and Index. National Technical Information Service (NTIS), 5285 Port Royal Road, Springfield, VA 22161. (703) 487-4650. FAX (703) 321-8547. From 1968 to present. Annual. $630.00 per year. Also available online as NTIS BIBLIOGRAPHIC DATABASE and on CD-ROM. ISSN:

METEOROLOGICAL and GEOASTROPHYSICAL ABSTRACTS. American Meteorological Society, c/o Inforonics, Inc., 550 Newtown Road, Box 458, Littleton, MA 01460. (508) 486-8976. FAX (508) 486-0027. Covers literature in environmental sciences, meteorology, astrophysics, hydrology, glaciology, and physical oceanography. 1950 to present. Monthly. $950.00 per year. ISSN: 0026-1130. Also available online (DIALOG) and on CD-ROM.

SCIENCE CITATION INDEX. SCI. Institute for Scientific Information, 3501 Market Street, Philadelphia, PA 19104. (215) 386-0100. (800) 523-1850. FAX (215) 386-2991. 1961 to present. Six issues per year, plus annual cumulation. $11650.00 per year. Also available online and on CD-ROM. Inquire about price and availability. ISSN: 0036-827X.

Selected Water Resources Abstracts. U.S. Geological Survey. Distributed by National Technical Information Service (NTIS), 5285 Port Royal Road, Springfield, VA 22161. (703) 487-4650. Monthly. $115.00 per year.

ANNUAL REVIEWS AND YEARBOOKS

Annual Review of Earth and Planetary Sciences. Annual Reviews, Inc., 4139 El Camino Way, Palo Alto, CA 94306-0897. (415) 493-4400. Fax (415) 855-9815. Annual. $55.00 per year.

Developments in Atmospheric Science Series. Elsevier Science Publishing Company, Inc., 655 Avenue of the Americas, New York, NY 10010. (212) 989-5800. Irregular. Inquire about price and availability.

ASSOCIATIONS AND PROFESSIONAL SOCIETIES

American Meteorological Society, 45 Beacon Street, Boston, MA 02108-3693. (617) 227-2425. FAX (617) 742-8718.

International Association of Meteorology and Atmospheric Physics. c/o UCAR, P.O. Box 3000, Boulder CO 80307-3000. (303) 497-1673.

Ency. of Physical Sciences and Engineering Info. Sources

RAIN

NATIONAL WEAtheR ASSOCIATION, 6704 Wolke Court, Montgomery, AL 36116-2134. (334) 213-0388. FAX (334) 213-0388.

University Corporation For Atmospheric Research. PO Box 3000, 1850 Table Mesa Drive, Boulder, CO 80307-3000. (303) 497-1673. FAX (303) 497-1654.

Weather Modification Association. PO Box 26926, Fresno, CA 93729-6926. (209) 434-3486.

BIBLIOGRAPHIES

Information Sources in the Earth Sciences. David N. Wood, Joan E. Hardy, and Anthony P. Harvey, editors. Bowker-Saur/K.G. Saur. Distributed by R.R. Bowker, 121 Chanlon Road, New Providence, NJ 07974. (800) 521-8110 or (908) 464-6800. Second edition. 1989. $85.00.

DIRECTORIES AND BIOGRAPHICAL SOURCES

American Men and Women of Science: Physical and Biological Sciences. Eighteenth edition. R.R. Bowker Company, 245 West 17th Street, New York, NY 10011. (800) 521-8810 or (212) 916-1600. $750.00.

American Meteorological Society. Professional Directory. American Meteorological Society. 45 Beacon Street, Boston, MA 02108. (617) 227-2425. Included in each issue of the Bulletin of the Society.

Meteorological Services of the World. American Meteorological Society, 45 Beacon Street, Boston, MA 02108- 3693. (617) 227-2425. FAX (617) 742-8718. Annual. $70.00.

National Weather Service offices and Stations. National Oceanic and Atmospheric Administration, Deparatment of Commerce, Silver Spring, MD 20910. (301) 427-7698. Annual. Free

Research Centers Directory. Gale Research, 835 Penobscot Building, Detroit, MI 48226-4094. (313) 961-2242. (800) 877-4253. 20th edition. 1995. $485.00. ISSN: 0080-1518.

ENCYCLOPEDIAS AND DICTIONARIES

Climates of the States. Bair. Gale Research, 835 Penobscot Building, Detroit, MI 48226-4094. (313) 961-2242. (800) 877-4253. 5th edition. 1995. $255.00.

Concise Oxford Dictionary of the Earth Sciences. Ailsa Allaby and Michael Allaby, editors. Oxford University Press, Inc., 200 Madison Avenue, New York, NY 10016. (800) 334-4249 or (212) 679-7300. 1990. $42.95.

Encyclopedia of Climatology. Rhodes W. Fairbridge, and John E. Oliver. Chapman & Hall, 1 Penn Plaza, New York, NY 10119. (212) 564-1060. 1986. $115.00.

Handbook of Applied Meteorology. David D. Houghton, editor. John Wiley & Sons, Inc. 605 Third Avenue, New York, NY 10158-0012. (212) 850-6000. (800) 225-5945. 1985.

Ocean Yearbook. Elisabeth M. Borgese, et al, editors. University of Chicago Press, 5801 Ellis Avenue, Chicago, IL 60637. (312) 702-7700. 1995. $77.00.

Water Encyclopedia. Frits van der Leeden, Fred L. Troise, and David K. Todd. Lewis Pubs., CRC Press, Inc., 2000 Corporate Boulevard, NW, Boca Raton, FL 33431. 2nd edition. 1990. $ 151.95

Weather Almanac. Gale Research, 835 Penobscot Building, Detroit, MI 48226-4094. (313) 961-2242. (800)877-4253. 7th edition. 1996. $130.00.

GENERAL WORKS

Design of Urban Runoff Quality Controls. Larry A. Roesner et all, editors. American Society of Civil Engineers. 345 East 47th Street, New York, NY 10017. (212) 705-7496. (800) 548-2723. 1989. $43.00.

Evaporation in the Uplands. Ian R. Calder. Wiley. John Wiley & Sons, Inc., 605 Third Avenue, New York, NY 10158-0012. (212) 850-6000. (800) 225-5945. 1990. $89.95.

Fundamentals of Atmospheric Dynamics and thermodynamics. C. Riegel. World Scientific Publishing Company, Inc., 1060 Main Street, River Edge, NJ 07661. (800) 227-7562 or (201) 487-9655. 1992. $30.00 in paper.

A History of the theories of Rain and Other Forms of Precipitation. William E. Middleton. Books on Demand, Division of University Microfilms International, 300 North Zeeb Road, Ann Arbor, MI 48106-1346. (313) 761-4700 or (800) 521-0600. 1966. $58.00.

Municipal Storm Water Management. Thomas N. Dobo and andrew J. Reese. Lewis Pubs, CRC Press, Inc., 2000 Corporate Boulevard, NW, Boca Raton, FL 33431. (407) 994-0555. (800) 272-7737. 1995. $79.95.

Sampling and Analysis of Rain. American Society for Testing and Materials (ASTM), 1916 Race Street, Philadelphia, PA 19103. (215) 299-5585. ASTM-STP 823. 1984. $18.00

Severe and Unusual Weather. Joe R. Eagleman. Trimedia Publishing Company, 12008 West 87th Street, Suite 117, Lenexa, KS 66215. (913) 599-0505. Second edition. 1990. $41.95.

Statistics For the Environment 2: Water-related Issues. Vic Barnet, and K. Peridun Turkman, editors. John Wiley & Sons, Inc., 605 Third Avenue, New York, NY 10158-0012. (212) 850-6000. (800) 225-5945. 1994. $55.95.

Urban Stor Mwater Hydrology: A Guide To Engineerng Calculations. Technomic, 851 New Holland Avenue, Box 3535, Lancaster, PA 17604. (717) 291-5609. (800) 233-9936. 1992. $65.00 HandBOOKS and MANUAL

Author's Guide To the Journals of the American Meteorological Society. American Meteorological Society, 45 Beacon Street, Boston, MA 02108-3693. (617) 227-2425. FAX (617) 742-8718. 1983. $15.00 in paper.

Handbook in Applied Meteorology. David D. Houghton, editor. John Wiley and Sons, Inc., 605 Third Avenue, New York, NY 10158. (800) 526-5368 or (212) 850-6000. 1985. $114.00.

ONLINE DATABASES AND CD-ROMS

RAIN

Ency. of Physical Sciences and Engineering Info. Sources

Climate Assessment Database. National Weather Service, National Meteorological Center, 5200 Auth Road, Suite 101, Camp Springs, MD 20233. (301) 763-8016. Contains daily, weekly and monthly summaries of North American and world climatological data. Also provides five to ten weather forecasts, and 30 to 90 day weather outlook. Subscription required. Inquire as to cost and availability.

Current Contents Search. Institute for Scientific Information, 3501 Market Street, Philadelphia, PA 19104. (215) 386-0100. FAX (215) 386-6362. Contains citations to articles listed in the table of contents of science and technology journals. Also articles in social sciences and life sciences journals. Available on BRS,(800) 289-4277, DIALOG,(800) 334-2564, online services. Inquire as to cost and availability.

Dissertation Abstracts. University Microfilms International, 300 North Zeeb Road, Ann Arbor, MI 48106. (800) 521-0600 or (313) 761-4700. Scope includes virtually all doctoral dissertations accepted at accredited American institutions from 1861 to present in 252 subject areas. Available on BRS,(800) 289-4277, DIALOG,(800) 334-2564, and OCLC EPIC,(800) 848-5878, online services. Also available on CD-ROM. Inquire as to cost and availability.

Meteorological and Geoastrophysical Abstracts. American Meteorological Society, 45 Beacon Street, Boston, MA 02108-3693. (617) 227-2425. FAX (617) 742-8718. Contains citations and abstracts to the worldwide literature on significant research in meteorology and geoastrophysics. Related topics include physical oceanography, hydrology, environmental sciences and glaciology. Covers the period 1972 to present. Available on DIALOG,(800) 334-2564, online service. Inquire as to cost and availability.

NTIS Bibliographic Database. National Technical Information Service, 5285 Port Royal Road, Springfield, VA 22161. (703) 487-4929 or FAX (703) 321-8199. Broad coverage of government-sponsored science and technology research reports, 1964 to present. Available on BRS,(800) 289-4277, DIALOG, (800) 334-2564, ORBIT, (800) 456-7248, and STN International, FIZ Karlsruhe, P.O. Box 2465, W-7500, Karlsruhe 1, Germany, online services. Also available on CD-ROM. Inquire as to cost and availability.

SCISEARCH. Institute for Scientific Information, 3501 Market Street, Philadelphia, PA 19104. (800) 523-1850 or (215) 386-0100. Broad multidisciplinary title and author index to the international literature of science and technology, 1974 to present. Available on DIALOG and ORBIT online services. Also available on CD-ROM. Inquire as to cost and availability.

WILSONLINE. H.W. Wilson Company, 950 University Avenue, Bronx, NY 10452. (800) 367-6770 or (212) 588-8400. Makes available online versions of the printed H.W. Wilson indexes including Applied Science and Technology Index, Business Periodicals Index, General Science Index, and Readers' Guide to Periodical Literature. Period covered is generally 1983 to present. Available on BRS,(800) 289-4277, DIALOG,(800) 334-2564, and OCLC EPIC,(800) 848-5878, online services. Also available on CD-ROM. Inquire as to cost and availability.

World Climate Disc. Chadwyck-Healey Inc., 1101 King Street, Alexandria, VA 22314. (703) 683-4890. FAX (703) 683-7589. Weather and climate data from approximately 5,000 weather stations worldwide, covering the years 1854 to 1990 on CD-ROM. First edition 1992. Approximately $1200.00 per year with annual updates.

World Weatherdisc. WeatherDisc Associates, Inc., 4584 N.E. 89th Street, Seattle, WA 98115. (206) 524-4314. FAX (206) 543-0308. Meteorological data on CD-ROM which describes the climate of the earth today and for the past few hundred years. First edition 1989. Approximately $295.00 per year with biannual updates.

PERIODICALS

AGRICULTURAL and FOREST METEOROLOGY. Elsevier Science Publishing Company, Inc., 655 Avenue of the Americas, New York, NY 10010. (212) 989-5800. Twenty times per year. $750.00 per year.

American Meteorological Society. BULLETIN. American Meteorological Society, 45 Beacon Street, Boston, MA 02108-3693. (617) 227-2425. FAX (617) 742-8718. Monthly. $60.00 per year.

American Meteorological Society. METEOROLOGICAL MONOGRAPHS. American Meteorological Society, 45 Beacon Street, Boston, MA 02108-3693. (617) 227-2425. FAX (617) 742-8718. Irregular. Price varies.

Atmosphere - Ocean. Canadian Meteorological and Oceanographic Society, P.O. Box 334, Newmarket, Ontario, L3Y 4X7, Canada. (416) 898-1040. FAX (416) 898-7937. Quarterly. $30.00 per year.

Dynamics of Atmospheres and Oceans. Elsevier Science Publishing Company, Inc., 655 Avenue of the Americas, New York, NY 10010. (212) 989-5800. Six times per year. $205.00 per year.

International Journal of Climatology. Royal Meteorological Society, distributed by John Wiley and Sons, Inc., 605 Third Avenue, New York, NY 10158. (800) 526-5368 or (212) 850-6000. Eight times per year. $425.00 per year.

Journal of Applied Meteorology. American Meteorological Society, 45 Beacon Street, Boston, MA 02108-3693. (617) 227-2425. FAX (617) 742-8718. Monthly. $165.00 per year.

Journal of the Atmospheric Sciences. American Meteorological Society, 45 Beacon Street, Boston, MA 02108-3693. (617) 227-2425. FAX (617) 742-8718. Semi-monthly. $320.00 per year.

Monthly Weather Review. American Meteorological Society, 45 Beacon Street, Boston, MA 02108-3693. (617) 227-2425. FAX (617) 742-8718. Monthly. $205.00 per year.

National Weather Digest. National Weather Association, 4400 Stamp Road, Room 404, Temple Hills, MD 20748. (301) 899-3784.

Weather. Royal Meteorological Society, 104 Oxford Road, Reading, Berks RG1 7LJ, England. Monthly. $44.00 per year.

Weatherwise. Heldref Publications, 1319 Eighteenth Street, N.W., Washington, DC 20036-1802. (202) 296-6267. FAX (202) 296-5149. Bi-monthly. $28.00 per year.

RESEARCH CENTERS AND INSTITUTES

Cooperative Institute for Mesoscale Meteorological Studies. University of Oklahoma, 401 East Boyd, Norman, OK 73019. (405) 325-3041.

Cooperative Institute For Research in the Atmosphere. Colorado State University. FootHills Campus, Fort Collins, CO 80523. (303) 491-8448.

Joint Institute For Marine and Atmospheric Research. University of Hawaii at Manoa, 1000 Pope Road, Honolulu, HI 96822. (808) 956-8083. FAX (808) 956-4104.

National Center For Atmospheric Research. P.O. Box 3000, Boulder, CO 80307. (303) 497-1000. FAX (303) 497-1654.

National Severe Storms Forecast Center. 601 East 12th Street, Kansas City, MO 64106. (816) 374-5922.

National Weather Service. 1325 East-West Highway, Silver Spring, MD 20910. (301) 427-7689.

Weather Research Center. 3710 Mt. Vernon, Houston, TX 77006. (713) 529-3076.

RAMAN SPECTROSCOPY

See: SPECTROSCOPY

RAMJETS

See: JET PROPULSION

REACTORS

See: NUCLEAR REACTORS

RECURSIVE PROGRAMMING

See: COMPUTER PROGRAMMING

RED DWARFS

See: STARS

RED GIANTS

See: STARS

RED SHIFT

See: ASTROPHYSICS

RECYCLING

ABSTRACT SERVICES AND INDEXES

Aluminum Industry Abstracts. Aluminum Association, Materials Park, OH 44073. (216) 338-5151. FAX (216) 338-4634. 1968 to present. Monthly. $525.00 per year.

Applied Science and Technology Index; A Cumulative Subject Index To English Language Periodicals in the Fields of Aeronautics and Space Science, Computer Technology, Chemistry, Construction Industry, Energy and Related Areas. H.W. Wilson Co., 950 University Avenue, Bronx, NY 10452. (800) 367-6770 or (212) 588-8400. FAX (718) 590-1617. From 1958 to present. Monthly. Inquire about cost and availability. Also available on CD-ROM and online.

Environmental Abstracts. R.R. Bowker Inc., 121 Chanlon Road, New Providence, NJ 07974. (800) 521-8110 or (908) 464-6800. Contains citations, most with abstracts, to literature relating to environmental issues and problems, from 1971 to present. Inquire as to cost and availability.

Environmental Engineering Abstracts, Cambridge Scientific Abstracts, 7200 Wisconsin Avenue, Bethesda, MD 20814-4823. (301) 961-6750. FAX (301) 961-6720. Monthly. Covers hazardous materials, environmental impact and protection, treatment of sewage and industrial wastes, hydroelectric power, tidal and wind power, artic and tropical engineering.

Science Citation Index. Institute for Scientific Information, 3501 Market Street, Philadelphia, PA 19104. (215) 386-0100. FAX (215) 386-6362. Inquire about availability and cost. Also available on CD-ROM.

ANNUAL REVIEWS AND YEARBOOKS

Materials Recovery and Recycling Yearbook. Government Advisory Associates, 177 E. 87th Street, Suite 506-A, New York, NY 10128. (212) 410-4165. FAX (212) 410-6607. 1990 to present. Biennial. $250.00 to non-profit and government agencies.

ASSOCIATIONS AND PROFESSIONAL SOCIETIES

Environmental Defense Fund. 257 Park Avenue S., New York, NY 10010. (212) 505-2100. FAX (212) 505-2375.

Institute of Scrap Recycling Industries. 1325 G Street NW, #1000, Washington, DC 20005-3104. (202) 737-1770. fax (202) 626-0900.

National Association For Plastic Container Recovery. 4828 Parkway Plaza Blvd., Suite 260, Charlotte, NC 28217.

National Oil Recyclers' Association. 12429 Cedar Road, Suite 26, Cleveland, OH 44106-3172. (216) 791-7316. FAX (216) 791-6047.

National Solid Waste Management Association. Suite 1000, 1730 Rhode Island Avenue NW, Washington, DC 20036. (202) 659-4613. FAX (202) 659-0925.

Plastics Recycling Foundation Inc. PO Box 189, Kennett Square, PA 19348.

RECYCLING

Ency. of Physical Sciences and Engineering Info. Sources

Steel Recycling Institute. Foster Plaza 10, 680 andersen Drive, Pittsburgh, PA 15220-2700. (412) 922-2772. FAX (412) 922-3213.

DIRECTORIES AND BIOGRAPHICAL SOURCES

American Recycling Market Directory "arm." Robert Boulanger. The American Recycling Market Inc., Box 577, Ogdensburg, NY 13669. (800) 267-0707.

Recycling Market and Resource Directory. Diane Publishing Company, 600 Upland Avenue, Upland, PA 19015. (610) 499-7415. FAX (610) 499-7429. 1993. $30.00.

Recycling Sourcebook. Thomas J. Cichonski and Karen Hill, editors. Gale Research, 835 Penobscot Building, Detroit, MI 48226-4094. (313) 961-2242. (800) 877-4253. FAX (313) 961-6083. Biennial. Inquire for price and latest edition.

ENCYCLOPEDIAS AND DICTIONARIES

Encyclopedia of Physical Science and Technology. Robert A. Meyers, ed. 18 volumes. Academic Press Inc., 6277 Sea Harbor Drive, Orlando, FL 32887. (800) 321-5068. 1992. $2100.00.

McGraw-Hill Dictionary of Scientific and Technical Terms. Sybil P. Parker, ed. 5th edition. McGraw-Hill Publishing Company, 1221 Avenue of the Americas, New York, NY 10020. (800) 262-4729 or (212) 512-3825. 1993. $110.50.

McGraw-Hill Encyclopedia of Science and Technology. Sybil P. Parker, ed. 7th edition. 20 volumes. McGraw-Hill Publishing Company, 1221 Avenue of the Americas, New York, NY 10020. (800) 262-4729 or (212) 512-3825. 1992. $1900.00

Thesaurus of Scientific, Technical, and Engineering Terms. Hemisphere Publishing Corporation, 1900 Frost Road, Suite 101, Bristol, PA 19007-1598. (215) 785-5800 or (800) 821-8312. FAX (215) 785-5515. 1987. $173.00.

GENERAL WORKS

Aluminum Recycling. Data Notes Staff. Revised edition. Prosperity & Profits Unlimited, PO Box 416, Denver, CO 80201. (303) 575-5676. 1992. $17.95.

The Blueprint For Plastics Recycling. Diane Publishing Company, 600 Upland Avenue, Upland, PA 19015. (610) 499-7415. FAX (610) 499-7429. 1993. $20.00.

Recycling Equipment and Technology For Municipal Solid Waste: Material Recovery Facilities. Joseph T. Swartzbaugh. Noyes Publications, 120 Mill Road, Park Ridge, NJ 07656. (201) 391-8484. FAX (201) 391-6833. 1993. $45.00.

Waste Minimization and Recycling. Paul Cheremisinoff and S. Baliko, editors. Gulf Publishing Company, P.O. Box 2608, Houston, TX. (713) 529-4301 or (800) 231-6275. 1992. $155.00.

Waste Oil Recycling and Resource Recovery. Richard K. Miller and Marcia Kupnow. Future Technology Surveys, 700 Indian Trial, Lilburn, GA 30247. (404) 717-0779. FAX (404) 381-9865. 1991. $200.00.

Waste-to-energy. Richard K. Miller and Marcia Kupnow. Future Technology Surveys, 700 Indian Trial, Lilburn, GA 30247. (404) 717-0779. FAX (404) 381-9865. 1994. $200.00.

HANDBOOKS AND MANUALS

NSWMA Recycling Handbook. National Solid Waste Management staff. Lewis Publishers, 121 S. Main St., PO Box 519, Chelsea, MI 48118. (313) 475-8619 or (800) 272-7737. 1995. $55.00.

A Practical Recycling Handbook. American Society of Civil Engineers, 345 East 47th Street, New York, NY 10017-2398. (212) 705-7520 or (800) 548-2723. 1994. $39.00.

Recycling Handbook. Herbert F. Lund. McGraw-Hill Publishing Company, 1221 Avenue of the Americas, New York, NY 10020. (800) 262-4729 or (212) 512-3825. 1992. $87.50.

Recycling in America: A Reference Handbook. Debi Kimball. ABC-CLIO Information Services, PO Box 1911, Santa Barbara, CA 93116-1911. (800) 368-6868. FAX (805) 685-9685. 1992. $39.50.

Waste Age-recycling Times' Recycling Handbook. John T. Aquino, compiler. Lewis Publishers, 121 S. Main St., PO Box 519, Chelsea, MI 48118. (313) 475-8619 or (800) 272-7737. 1995. $55.00.

ONLINE DATABASES AND CD-ROMS

Compendex Plus. Engineering Information, Inc., 345 East 47th Street, New York, NY 10017. (212) 705-7600 or (800) 221-1044. Contains citations with abstracts to worldwide literature in engineering and technology, from 1970 to present. Available on online BRS,(800) 289-4277, DIALOG, (800) 334-2564, ORBIT (800) 456-7248, and STN International, FIZ Karlsruhe, P.O. Box 2465, W-7500, Karlsruhe 1, Germany, online services. Also available on CD-ROM. Inquire as to cost and availability.

CSA Engineering. Cambridge Scientific Abstracts, 7200 Wisconsin Avenue, Suite 601, Bethesda, MD 20814. (301) 961-6750 or (800) 843-7751. Contains citations and abstracts of international periodicals and research literature covering all fields of engineering and science and technology, including computer and information science, electronics, mechanical engineering, solid state materials, 1981 to present. Available on BRS,(800) 289-4277, online service. Inquire as to cost and availability.

Current Contents Search. Institute for Scientific Information, 3501 Market Street, Philadelphia, PA 19104. (215) 386-0100. FAX (215) 386-6362. Contains citations to articles listed in the table of contents of science and technology journals. Also articles in social sciences and life sciences journals. Available on BRS,(800) 289-4277, DIALOG,(800) 334-2564, online services. Inquire as to cost and availability.

Dissertation Abstracts. University Microfilms International, 300 North Zeeb Road, Ann Arbor, MI 48106. (800) 521-0600 or (313) 761-4700. Scope includes virtually all doctoral dissertations accepted at accredited American institutions from 1861 to present in 252 subject areas. Available on BRS,(800) 289-4277, DIALOG,(800) 334-2564, and OCLC EPIC,(800) 848-5878, online services. Also available on CD-ROM. Inquire as to cost and availability.

Ency. of Physical Sciences and Engineering Info. Sources

REFRIGERATION

NTIS Bibliographic Database. National Technical Information Service, 5285 Port Royal Road, Springfield, VA 22161. (703) 487-4929 or FAX (703) 321-8199. Broad coverage of government-sponsored science and technology research reports, 1964 to present. Available on BRS,(800) 289-4277, DIALOG, (800) 334-2564, ORBIT, (800) 456-7248, and STN International, FIZ Karlsruhe, P.O. Box 2465, W-7500, Karlsruhe 1, Germany, online services. Also available on CD-ROM. Inquire as to cost and availability.

SCISEARCH. Institute for Scientific Information, 3501 Market Street, Philadelphia, PA 19104. (800) 523-1850 or (215) 386-0100. Broad multidisciplinary title and author index to the international literature of science and technology, 1974 to present. Available on DIALOG,(800) 334-2564, and ORBIT,(800) 456-7248, online services. Also available on CD-ROM. Inquire as to cost and availability.

WILSONLINE. H.W. Wilson Company, 950 University Avenue, Bronx, NY 10452. (800) 367-6770 or (212) 588-8400. Makes available online versions of the printed H.W. Wilson indexes including Applied Science and Technology Index, Business Periodicals Index, General Science Index, and Readers' Guide to Periodical Literature. Period covered is generally 1983 to present. Available on BRS,(800) 289-4277, DIALOG,(800) 334-2564, and OCLC EPIC,(800) 848-5878, online services. Also available on CD-ROM. Inquire as to cost and availability.

PERIODICALS

Biocycle. J G Press Inc., 419 State Avenue, Emmaus, PA 18049. (215) 967-4135. 1960 to present. Monthly. $63.00.

Cycle. Environmental Action Coalition, 625 Broadway, New York, NY 10012. (212) 677-1601. 1970 to present. Quarterly. $20.00.

Garbage. Old-House Journal Corporation, 2 Main Street, Gloucester, MA 01930-5726. (508) 283-3200. FAX (508) 283-5715. 1989 to present. Six times a year. $21.00.

Plastics Recycling Update. Resource Recycling Inc., Box 10540, Portland, OR 97210. (503) 227-1319 or (800) 227-1424. FAX (503) 227-6135. 1988 to present. Monthly. $49.00.

Resource Recycling. Resource Recycling Inc., Box 10540, Portland, OR 97210. (503) 227-1319 or (800) 227-1424. FAX (503) 227-6135. 1982 to present. Monthly. $42.00.

Resources, Conservation and Recycling. Elsevier Science Inc., Box 882, Madison Square Station, New York, NY 10159. (212) 989-5800. FAX (212) 633-3990. 1975 to present. Twelve times a year. $746.00.

Reuse—Recycle. Technomic Publishing Company, Inc., 851 New Holland Avenue, Box 3535, Lancaster, PA 17604. (717) 291-5609. 1971 to present. Monthly. $185.00.

Recycling—Reclamation Digest. ASM International, Materials Information, Materials Park, OH 44073-0002. (216) 338-5151 or (800) 336-5152. FAX (216) 338-4634. 1991 to present. Monthly. $155.00 for non-members.

Recycling Times. National Solid Waste Management Association, Suite 1000, 1730 Rhode Island Avenue NW, Washington, DC 20036. (202) 659-4613. FAX (202) 659-0925. 1989. 26 times a year. $95.00.

Recycling Update. Update Publicare Company, c/o Prosperity & Profits Unlimited, PO Box 416, Denver, CO 80201. (303) 575-5676. 1983 to present. Semi-annual. $6.00.

Recycling World. Environmental Defense Fund, 257 Park Avenue S., New York, NY 10010. (212) 505-2100. FAX (212) 505-2375. Irregular. Free.

Waste Age. National Solid Waste Management Association, Suite 1000, 1730 Rhode Island Avenue NW, Washington, DC 20036. (202) 659-4613. FAX (202) 659-0925. 1970 to present. Monthly. $45.00.

Waste Recovery Report. I C O N Information Concepts Inc., 211 S. 45th Street, Philadelphia, PA 19104. (215) 349-6500. FAX (215) 349-6502. 1976 to present. Monthly. $50.00.

RESEARCH CENTERS AND INSTITUTES

Center For Waste Minimization and Management. Department of Chemical Engineering, North Carolina State University, PO Box 7905, Raleigh, NC 27695-7905. (919) 737-2325. FAX (919) 737-3465.

Environmental Defense Fund. 257 Park Avenue, New York, NY 10010. (212) 505-2100. FAX (212) 505-2375.

Illinois Institute of Technology, Industrial Waste Elimination Research Center. 3200 S. State Street, Chicago, IL 60616. (312) 567-3535. FAX (312) 567-3548.

Louisiana State University, Institute For Recyclable Materials. 3304 CEBA, Baton Rouge, LA 70803-6401. (504) 388-8650. FAX (504) 388-5990.

Rutgers University Center For Plastics Recycling Research. Bldg. 3529, Busch Campus, Piscataway, NJ 08855-1179. (908) 932-3683. FAX (908) 932-5636.

University of Alabama, Alabama Waste Exchange (awe). Box 870203, Tuscaloosa, AL 35487-0203. (205) 348-5889. FAX (205) 348-8573.

REFRIGERATION

See also: AIR CONDITIONING, COOLANTS

ABSTRACT SERVICES AND INDEXES

Applied Science and Technology Index; A Cumulative Subject Index To English Language Periodicals in the Fields of Aeronautics and Space Science, Computer Technology, Chemistry, Construction Industry, Energy and Related Areas. H.W. Wilson Co., 950 University Avenue, Bronx, NY 10452. (800) 367-6770 or (212) 588-8400. FAX (718) 590-1617. From 1958 to present. Monthly. Inquire about cost and availability. Also available on CD-ROM and online.

Computer Abstracts, MCBUniversity Press Ltd., PO Box 10812, Birmingham, AL 35201-0812. (800) 633-4931. FAX (205) 995-1588. Monthly. Covers computer theory, data, hardware, systems, networks, human-computer interaction, artificial intelligence, as well as applications of computers in aerospace, business, CAD/CAM, cartography, civil engineering, electronics and

REFRIGERATION

Ency. of Physical Sciences and Engineering Info. Sources

electrical engineering, industrial engineering, mechanical engineering, medicine, structural engineering, etc.

Current Contents: Engineering, Technology, and Applied Sciences. Institute for Scientific Information, 3501 Market Street, Philadelphia, PA 19104. (215) 386-0100. FAX (215) 386-6362. 1970 to present. Weekly. $442.00 per year.

Engineering Index Monthly. Engineering Information, Inc., Castle Point on the Hudson, Hoboken, NJ 07030. (800) 221-1044. FAX (212) 832-1857. Monthly. $2300.00 per year. Also available online as COMPENDEX, and also on CD-ROM. Covers chemical engineering, computers, electrical engineering, civil engineering, metals and mining, industrial management, and mechanical engineering.

International Building Services Abstracts. Building Service Research and Information Association, Old Bracknell Lane W, Bracknell, Berks. RG12 7AH, England. Telephone 0344-426511. FAX 0344-487575. 1966 to present. Bi-monthly. Inquire for price. Also available online. A survey of world literature on mechanical and technical services of buildings, including heating, sanitation, ventilation, air conditioning, lighting, and communications.

ASSOCIATIONS AND PROFESSIONAL SOCIETIES

Air-conditioning and Refrigeration Institute. 1501 Wilson Blvd., Arlington, VA 22209. (703) 524-8800.

Air Conditioning Contractors of America. 1712 New Hampshire Avenue NW, Washington, DC 20009. (202) 483-9370. FAX (202) 234-4721.

American ACADemy of Environmental Engineers. 132 Holiday Ct., Suite 206, Annapolis, MD 21041.

American Society of Heating, Refrigerating & Air-conditioning Engineers. 1791 Tullie Circle, N.E., Atlanta, GA 30329. (404) 636-8400. FAX (404) 321-5478.

Institute of Heating & Air Conditioning Industries. 606 N Larchmont Blvd., Suite 4A, Los Angeles, CA 90004. (213) 467-1158. FAX (213) 461-2588.

Refrigerating Engineers and Technicians Association. 401 N. Michigan Avenue, Chicago, IL 60611-4267. (312) 644-6610. FAX (312) 321-6869.

DIRECTORIES AND BIOGRAPHICAL SOURCES

ACCA Quality Contractor's Catalog of Materials, Products and Services. Air Conditioning Contractors of America, 1712 New Hampshire Avenue NW, Washington, DC 20009. (202) 483-9370. FAX (202) 234-4721. 1968 to present. Annual. Inquire for price and availability.

Air Conditioning, Heating and Refrigerating News—directory Issue. Business News Publishing Company, 755 W. Big Beaver Road, Suite 1000, Troy, MI 48084. (313) 362-3700. FAX (313) 362-0317.

Heating, Plumbing and Air Conditioning Buyers' Guide. Cowgate Communications Inc., 1370 Don Mills Road, Suite 300, Don Mills, ON M3B 3N7, Canada. (416) 759-2500. FAX (416) 759-6979. 1923 to present. Annual. $30.00

Research Centers Directory. Gale Research, 835 Penobscot Building, Detroit, MI 48226-4094. (313) 961-2242. (800) 877-4253. FAX (313) 961-6083. $485.00.

Scientific and Technical Organizations and Agencies Directory. Gale Research, 835 Penobscot Building, Detroit, MI 48226-4094. (313) 961-2242. (800) 877-4253. FAX (313) 961-6083. 4th edition. 1996. $195.00.

Who's Who in Engineering. American Association of Engineering Societies, 1111 19th Street NW, Suite 608, Washington, DC 20036. (202) 296-2237 or (800) 658-8897. 8th edition. 1991. Inquire for price.

Who's Who in Technology. Gale Research, 835 Penobscot Building, Detroit, MI 48226-4094. (313) 961-2242. (800) 877-4253. FAX (313) 961-6083. 1995. $195.00.

ENCYCLOPEDIAS AND DICTIONARIES

Dictionary of Refrigerating and Air Conditioning. K.M. Booth, editor. Elsevier Science Publishing Company, Inc., 655 Avenue of the Americas, New York, NY 10010. (212) 989-5800. 1971. $64.00.

GENERAL WORKS

Industrial Refrigeration: Principles, Design and Applications. P.C. Koelet. Marcel Dekker, Inc., 270 Madison Avenue, New York, NY 10016. (212) 696-9000. FAX (212) 685-4540. 1992. $140.00.

Principles of Refrigeration. C. Thomas Olivo and R.W. March, eds. 3d ed. Delmar Publishers, Division of Thomson Educational Publishing, Inc., 3 Columbia Circle, Box 15015, Albany, NY 12205. (518) 464-3500 or (800) 347-7707. 1990. $29.95.

HANDBOOKS AND MANUALS

ASHRAE Handbook: Fundamentals. 4 volumes. American Society of Heating, Refrigeration and Air Conditioning Engineers, 1791 Tullie Circle NE, Atlanta, GA 30329. (404) 636-8400. FAX (404) 321-5478. 1922 to present. Annual. $119.00.

Handbook of Air Conditioning and Refrigeration. Shan K. Wang. McGraw-Hill Publishing Company, 1221 Avenue of the Americas, New York, NY 10020. (800) 262-4729 or (212) 512-3825. 1993. $96.50.

ONLINE DATABASES AND CD-ROMS

Compendex Plus. Engineering Information, Inc., 345 East 47th Street, New York, NY 10017. (212) 705-7600 or (800) 221-1044. Contains citations with abstracts to worldwide literature in engineering and technology, from 1970 to present. Available on online BRS, (800) 289-4277, DIALOG, (800) 334-2564, ORBIT (800) 456-7248, and STN International, FIZ Karlsruhe, P.O. Box 2465, W-7500, Karlsruhe 1, Germany, online services. Also available on CD-ROM. Inquire as to cost and availability.

CSA Engineering. Cambridge Scientific Abstracts, 7200 Wisconsin Avenue, Suite 601, Bethesda, MD 20814. (301) 961-6750 or (800) 843-7751. Contains citations and abstracts of international periodicals and research literature covering all fields of engineering and science and technology, including computer and information science, electronics, mechanical engineering, solid state materials, 1981 to present. Available on BRS,(800) 289-4277, online service.Inquire as to cost and availability.

Current Contents Search. Institute for Scientific Information, 3501 Market Street, Philadelphia, PA 19104. (215) 386-0100. FAX (215) 386-6362. Contains citations to articles listed in the table of contents of science and technology journals. Also articles in social sciences and life sciences journals. Available on BRS,(800) 289-4277, DIALOG,(800) 334-2564, online services. Inquire as to cost and availability.

Dissertation Abstracts. University Microfilms International, 300 North Zeeb Road, Ann Arbor, MI 48106. (800) 521-0600 or (313) 761-4700. Scope includes virtually all doctoral dissertations accepted at accredited American institutions from 1861 to present in 252 subject areas. Available on BRS,(800) 289-4277, DIALOG,(800) 334-2564, and OCLC EPIC,(800) 848-5878, online services. Also available on CD-ROM. Inquire as to cost and availability.

ISMEC: Mechanical Engineering Abstracts. Cambridge Scientific Abstracts, 7200 Wisconsin Avenue, Suite 601, Bethesda, MD 20814. (301) 961-6750 or (800) 843-7751. Contains citations to the literature in mechanical engineering, industrial and production engineering, energy, power, mechanics, devices and related areas, from 1973 to present. Available on the DIALOG,(800) 334-2564, online service. Inquire as to cost and availability.

NTIS Bibliographic Database. National Technical Information Service, 5285 Port Royal Road, Springfield, VA 22161. (703) 487-4929 or FAX (703) 321-8199. Broad coverage of government-sponsored science and technology research reports, 1964 to present. Available on BRS,(800) 289-4277, DIALOG, (800) 334-2564, ORBIT, (800) 456-7248, and STN International, FIZ Karlsruhe, P.O. Box 2465, W-7500, Karlsruhe 1, Germany, online services. Also available on CD-ROM. Inquire as to cost and availability.

SCISEARCH. Institute for Scientific Information, 3501 Market Street, Philadelphia, PA 19104. (800) 523-1850 or (215) 386-0100. Broad multidisciplinary title and author index to the international literature of science and technology, 1974 to present. Available on DIALOG,(800) 334-2564, and ORBIT,(800) 456-7248, online services. Also available on CD-ROM. Inquire as to cost and availability.

WILSONLINE. H.W. Wilson Company, 950 University Avenue, Bronx, NY 10452. (800) 367-6770 or (212) 588-8400. Makes available online versions of the printed H.W. Wilson indexes including Applied Science and Technology Index, Business Periodicals Index, General Science Index, and Readers' Guide to Periodical Literature. Period covered is generally 1983 to present. Available on BRS,(800) 289-4277, DIALOG,(800) 334-2564, and OCLC EPIC,(800) 848-5878, online services. Also available on CD-ROM. Inquire as to cost and availability.

PERIODICALS

Air Conditioning, Heating, and Refrigeration News. Business News Publishing Company, 755 W. Big Beaver Road, Suite 1000, Troy, MI 48084. (313) 362-3700. FAX (313) 362-0317. 1926 to present. Weekly. $72.00 per year. Includes the annual directory and statistical summary issues.

ASHRAE Journal: Heating, Refrigerating, Air Conditioning, and Ventilating. American Society of Heating, Refrigeration and Air

Conditioning Engineers, 1791 Tullie Circle NE, Atlanta, GA 30329. (404) 636-8400. FAX (404) 321-5478. 1914 to present. Monthly. $49.00 per year for non-members.

ASHRAE Transactions. American Society of Heating, Refrigeration and Air Conditioning Engineers, 1791 Tullie Circle NE, Atlanta, GA 30329. (404) 636-8400. FAX (404) 321-5478. 1895 to present. Semi-annual. $170.00 per volume.

RESEARCH CENTERS AND INSTITUTES

Kansas State University Mechanical Engineering Research Laboratory. Durland Hall, Manhattan, KS 66506. (913) 532-5610.

Lehigh University Institute of thermo-fluid Engineering and Science. Bldg. A, Mountain Top Campus, Bethlehem, PA 18015. (215) 758-4091. FAX (215) 758-5057.

Refrigeration Research Foundation. 7315 Wisconsin Avenue, Bethesda, MD 20814. (301) 652-5674. FAX (301) 652-7269.

Texas A&M University Energy Systems Laboratory. College Station, TX 77843. (409) 845-6402.

RELATIVITY

See: CELESTIAL MECHANICS, PHYSICS

RELAYS

See also: CIRCUIT BREAKERS, ELECTRICAL ENGINEERING, ELECTRICITY, ELECTROMAGENTISM, ELECTRONICS CIRCUITS and COMPONENTS, ELECTRONICS, ELECTRONICS ENGINEERING

ABSTRACT SERVICES AND INDEXES

Applied Science and Technology Index; A Cumulative Subject Index To English Language Periodicals in the Fields of Aeronautics and Space Science, Computer Technology, Chemistry, Construction Industry, Energy and Related Areas.

H.W. Wilson Co., 950 University Avenue, Bronx, NY 10452. (212) 588-8400. (800) 367-6770. FAX (718) 590-1617. From 1958 to present. Monthly. Inquire about cost and availability. Also available on CD-ROM and online. ISSN: 0003-6986.

Current Contents: Engineering, Technology and Applied Sciences. Institute for Scientific Information, 3501 Market Street, Philadelphia, PA 19104. (215) 386-0100. FAX (215)386-6362. 1970 to present. Weekly. $442.00 per year. Also available on CD-ROM and online. Inquire regarding cost and availability. ISSN: 0095-7917.

Current Papers in Electrical and Electronics Engineering. Institution of Electrical Engineers (IEE), London. Distributed by INSPEC/IEEE, Box 1331, 445 Hoes Lane, Piscataway, NJ 08855-1331. (908) 562-5549. 1969 to present. Monthly. $345.00 per year. ISSN: 0011-3778.

Electric Power Industry Abstracts. Edison Electric Institute, c/o Utility Data Institute, 2011 I Street, NW, Suite 700, Washing-

RELAYS

Ency. of Physical Sciences and Engineering Info. Sources

ton, DC 20006. 1975 to present. Bimonthly. Inquire as to cost and availability.

Electrical and Electronics Abstracts. Institution of Electrical Engineers (IEE), London. Available from INSPEC/IEEE - Institute of Electrical and Electronic Engineers, Box 1331, Hoes Lane, Piscataway, NJ 08855-1331.(908) 562-5549. 1898 to present. Monthly. $2200.00 per year. Also available on CD-ROM and online as INSPEC.

Engineering Index Monthly; Indexes and Abstracts the World's Engineering and Technical Literature. Engineering Information, Inc., Castle Point on the Hudson, Hoboken, NJ 07030. (201) 216-8500. (800) 221-1044. FAX (201) 216-8532. Monthly. $2300.00 per year. Available online as COMPENDEX and also on CD-ROM. ISSN: 0742-1974.

General Science Index. H.W. Wilson Company, 950 University Avenue, Bronx, NY 10452. (212) 588-8400. (800) 367-6770. FAX (718) 590-1617. From 1978 to present. Ten issues per year; quarterly and annual cumulations. Service basis. Available on CD-ROM and online. Inquire regarding cost and availability. ISSN: 0162-1963.

Index To IEEE Publications. IEEE Service Center, 445 Hoes Lane, Piscataway, NJ 08855-1331. (908) 981-1393. (800) 678-IEEE. FAX (908) 981-9667. 1973 to present. Annual. ISSN: 0099-1368.

Physics Abstracts. INSPEC. Section A, Science Abstracts. Institution of Electrical Engineers (IEE), London. Available from: INSPEC/IEEE - Institute of Electrical and Electronic Engineers, Box 1331, Hoes Lane, Piscataway, NJ 08855-1331. (908) 562-5549. 1898 to present. 24 issues per year. $2835.00 per year. Also available online and on CD-ROM. ISSN: 0036-8091.

Physics Briefs (Physikalische Berichte). Information Center for Energy, Physics, Mathematics; German Physical Society. V C H Publishers, Inc., 220 East 23rd Street, New York, NY 10010-4606. (212) 683-8333. 1845 to present. 24 issues per year. $2390.00 per year. Also available online. ISSN: 0179-7434.

Science Citation Index. SCI. Institute for Scientific Information, 3501 Market Street, Philadelphia, PA 19104. (215) 386-0100. (800) 523-1850. FAX (215) 386-2991. 1961 to present. Six issues per year, plus annual cumulation. $11650.00 per year. Also available online and on CD-ROM.Inquire about price and availability. ISSN: 0036-827X.

ASSOCIATIONS AND PROFESSIONAL SOCIETIES

American Electronics Association. 5201 Great America Way, Suite 520, P.O. 52990, Santa Clara, CA 95056. (408) 987- 4200. FAX (408) 970-8565.

Edison Electric Institute. 701 Pennsylvania Avenue NW, Washington, DC 20004-2696. (202) 508-5000, 5454. FAX (202) 508-5360.

Electric Power Research Institute. 3412 Hillview Avenue, Palo Alto, CA 94304. (415) 855-2000. FAX (415) 855-2954.

Electronics Industries Association. 2500 Wilson Boulevard, Arlington, VA 22201. (703) 907-7500. FAX (202) 457-4985.

IEEE (institute of Electrical and Electronic Engineers). 345 East 47th Street. New York, NY 10017. (212) 705-7900. FAX (212) 705-4929.

IEEE Power Engineering Society. c/o Institute of Electrical and Electronic Engineers. 345 East 47th Street, New York, NY 10017. (212) 705-7900. FAX (212) 705-4929.

National Association of Power Engineers. 1 Springfield Street, Chicopee, MA 01013. (413) 592-6273. FAX (413) 592-1998.

DIRECTORIES AND BIOGRAPHICAL SOURCES

American Men and Women of Science: Physical and Biological Sciences. R. R. Bowker Inc., 121 Chanlon Road, New Providence, NJ 07974. (908) 464-6800. (800) 521-8110. 20th edition. 8 volumes. 1996. $850.00.

Directory of Engineering Societies and Related Organizations. American Association of Engineering Societies, 1111 19th Street, Suite 608, Washington, DC 20036-3603. (202) 296-2237. Semi-annual. $150.00.

E E M - Electronic Engineer's Master. Hearst Business Communications, Inc., 645 Stewart Avenue, Garden City NY 11530. (516) 227-1300. ISSN: 0732-9016.

Electrical and Electronic Trades Directory: The Blue Book.

Institution of Electrical Engineers, Michael Faraday House, Six Hills Way, Stevenage, Herts. SG1 2AY, England. Telephone: 0438 313311 or FAX 0438 742840. 1995. $140.00

Electrical World Directory of Electrical Utilities. McGraw-Hill Book Company, Inc., 1221 Avenue of the Americas, New York, NY 10020. (212) 997-3675. Annual.

Engineering Research Centers: Incorporating Electronics Research Centers. Stockton Press, 345 Park Avenue, New York, NY 10010. (212) 689-9200. (800) 221-2123. 4th edition. 1995. $515.00.

IEEE Membership Directory. Institute of Electrical and Electronics Engineers, IEEE Service Center, 445 Hoes Lane, Piscataway, NY 08854. (908) 981-1393. (800) 678-IEEE. FAX (908) 981-9667. 2 volumes. Annual. $190.00. ISSN:

Research Centers Directory. Gale Research Company Inc., 835 Penobscot Building, Detroit, MI 48226-4094. (313) 961-2242. (800) 877-4253. 20th edition. 1995. $485.00. ISSN: 0080- 1518.

Who's Who in Electric Transmission and Distribution. Utility Data Institute, 1200 G Street NY, Suite 250, Washington, DC 20005-3802. (202) 942-8788. (800) 486-3660. 1995. $150.00.

Who's Who in Engineering. Gordon Davis, editor. American Association of Engineering Societies. 1111 19th Street, NY, Suite 608, Washington, DC 20036. (202) 296-2237. (800) 658-8897. 9th edition. 1995. $220.00.

Who's Who in Technology. Gale Research, 835 Penobscot Building, Detroit, MI 48226-4094. (313 961-2242. (800)521-4253. 7th edition. 1995. $195.00. ISBN 0-8103-7467-6.

Ency. of Physical Sciences and Engineering Info. Sources

RELAYS

ENCYCLOPEDIAS AND DICTIONARIES

Chambers Science and Technology Dictionary. Peter M.B. Walker, editor. Cambridge University Press, 40 West 20th Street, New York, NY 10011-4211. (212) 924-3900. 1988. $39.95.

IEEE Standard Dictionary of Electrical and Electronics Terms. Christopher J. Booth, editor. IEEE Service Center, 445 Hoes Lane, Piscataway, NJ 08855-1331. (908) 981-1393. (800) 678-IEEE. FAX (908) 981-9667. IEEE Standard 100- 1992. 5th edition. 1993. $150.00.

Illustrated Dictionary of Electronics. Stan Gibilsco. TAB Books, P.O. Box 40, Blue Summit, PA 17294-0850. (717) 794- 2191. (800) 233-1128. 7th edition. 1994. $34.95.

McGraw-Hill Electronics Dictionary. J. Markus. McGraw-Hill Book Company, Inc., 1221 Avenue of the Americas, New York, NY 10020. (212) 997-3675. 5th edition. 1994. $49.95.

McGraw-Hill Encyclopedia of Science and Technology. McGraw-Hill Book, Incorporated, 1221 Avenue of the Americas, New York, NY 10020. (212) 997-3675. (800) 262-4729. Seventh edition. Twenty volumes. 1992. $1900.00.

GENERAL WORKS

Analog and Switching Circuit Design Using Integrated and Discrete Devices. J. Watson. John Wiley & Sons, Inc., 605 Third Avenue, New York, NY 10158-0012. (212) 850-6000. (800) 225-5945. 1989. $143.00.

Analog Switches. Delton Y. Horn. McGraw-Hill Publishing Company, Inc., 1221 Avenue of the Americas, New York, NY 10020. (212) 512-2000. (800) 262-4729. 1990. $12.95 in paper.

Designing With Analog Switches. Steve Moore. Marcel Dekker, Inc., 270 Madison Avenue, New York, NY 10016. (212) 696-9000. (800) 228-1160. 1991. $115.00.

Electric and Electronic Engineering For Scientists and Engineers. K. A. Krishnamurthy. Halsted Press, 605 Third Avenue, New York, 10158-0012. (212) 850-6400. 1994. $48.95.

High-frequency Switching Power Supplies: Theory and Design. George C. Chryssis. McGraw-Hill Publishing Company, Inc., 1221 Avenue of the Americas, New York, NY 10020. (212) 512- 2000. (800) 262-4729. 1989. $48.00.

Logic Design and Switching theory. Saburo Muroga. Krieger Publishing Company, P.O. Box 9542, Melbourne, FL 32902-9542. (407) 724-9542. 1990. $67.50.

Modern Digital Design and Switching theory. Fabricius. CRC Press, Inc., 2000 Corporate Boulevard, NW, Boca Raton, FL

33431. (407) 994-0555. (800) 272-7737. 1992. $62.95.

Protective Relaying theory and Applications. Walter Elmore, editor. Marcel Dekker, Inc., 270 Madison Avenue, New York, NY 10016. (212) 696-9000. (800) 228-1160. 1994. $99.75.

UNINTERRUPTIBLE POWER SUPPLIES. J. Platts and J. D. St. Aubyn. Institution of Electrical Engineers; IEE/INSPEC, Box

1331, 445 Hoes Lane, Piscataway, NJ 08855-1331. (908) 981-0060. 1992. $70.00.

HANDBOOKS AND MANUALS

AC Power Systems Handbook. Jerry Whitaker. CRC Press, Inc., 2000 Corporate Boulevard, NW, Boca Raton, FL 33431. (407) 994-0555. (800) 272-7737. 1991. $55.00.

Active Electronic Component Handbook. Charles A. Harper and Harold C. Jones. McGraw-Hill Book Company, 1221 Avenue of the Americas, New York, NY 10020. (212) 512-2000. (800) 262-4729. 2nd edition. 1996. $79.50.

Electrical Engineering Handbook. Richard C. Dorf, editor. CRC Press, Inc., 2000 Corporate Boulevard, NW, Boca Raton, FL 33431. (407) 994-0555. (800) 272-7737. 1993. $99.95.

Electrical Engineer's Handbook. Donald Christiansen, editor. McGraw Book Company, 1221 Avenue of the Americas, New York, NY 10020. (212) 512-2000. (800) 262-4729. 4th edition. 1996. $110.00.

Electrical Power Systems Safety Handbook. John CADick. McGraw-Hill Book Company, 1221 Avenue of the Americas, New York, NY 10020. (212) 512-2000. 1994. $54.50.

Electronics Handbook. Jerry C. Whitaker. CRC Press, Inc., 2000 Corporate Boulevard, NW, Boca Raton, FL 33431. (407) 994-0555. (800) 272-7737. 1996. $120.00.

Handbook of Electrical and Electronic Technology. Curtis Johnson. Prentice Hall , 113 Sylvan Avenue, Route 9W, Englewood Cliffs, NJ 07632. (201) 592-2000. (800) 922- 0579. 1996. $88.00.

McGraw-Hill's Handbook of Electrical Construction Calculations. J. F. McPartland. McGraw-Hill Publishing Company, 1221 Avenue of the Americas, New York, NY 10020. (212) 512-2000. (800) 262-4729. 1996. $54.50.

McGraw-Hill's National Electrical Code Handbook. J. F. McPartland. McGraw-Hill Publishing Company, 1221 Avenue of the Americas, New York, NY 10020. (212) 512-2000. (800) 262-4729. 22nd edition. 1996. $65.00.

Standard Handbook For Electrical Engineers. Donald Fink. McGraw-Hill Publishing Company, 1221 Avenue of the Americas, New York, NY 10020. (212) 512-2000. (800) 262- 4729. 13th edition. 1996. $110.00.

Standard Handbook of Engineering Calculations. Tyler G.

Hicks. McGraw-Hill Publishing Company, 1221 Avenue of the Americas, New York, NY 10020. (212) 512-2000. (800) 262-4729. 3rd edition. 1994. $99.50. ISBN: 0-07-028812-7.

Switch Engineering Handbook. John R. Mason. McGraw-Hill Publishing Company, 1221 Avenue of the Americas, New York, NY 10020. (212) 512-2000. (800) 262-4729. 1992. $64.50.

ONLINE DATABASES AND CD-ROMS

CA Search. Chemical Abstracts Service, P.O. Box 3012, Columbus, OH 43210-0012. (614) 447-3600. (800) 848-6533. FAX (614) 447-3709. Very comprehensive guide to worldwide chemical literature and related fields, 1972 to present. Available on BRS,(800) 289-4277, DIALOG, (800) 334-2564, ORBIT (800) 456-7248, and STN International, FIZ Karlsruhe, P.O. Box 2465, W-7500, Karlsruhe 1, Germany, online services. Inquire as to cost and availability.

Compendex Plus. Engineering Information, Inc., 345 East 47th Street, New York, NY 10017. (212) 705-7600 or (800) 221-1044. Contains citations with abstracts to worldwide literature in engineering and technology, from 1970 to present. Available on online BRS,(800) 289-4277, DIALOG, (800) 334-2564, ORBIT (800) 456-7248, and STN International, FIZ Karlsruhe, P.O. Box 2465, W-7500, Karlsruhe 1, Germany, online services. Also available on CD-ROM. Inquire as to cost and availability.

Current Contents Search. Institute for Scientific Information, 3501 Market Street, Philadelphia, PA 19104. (215) 386-0100. FAX (215) 386-6362. Contains citations to articles listed in the table of contents of science and technology journals. Also articles in social sciences and life sciences journals. Available on BRS,(800) 289-4277, DIALOG,(800) 334-2564, online services. Inquire as to cost and availability.

Dissertation Abstracts Online. University Microfilms International, 300 North Zeeb Road, Ann Arbor, MI 48106. (800) 521-0600 or (313) 761-4700. Scope includes virtually all doctoral dissertations accepted at accredited American institutions from 1861 to present in 252 subject areas. Available on BRS, (800) 289-4277, DIALOG, (800) 334-2564, and OCLC EPIC,(800) 848-5878, online services. Also available on CD-ROM. Inquire as to cost and availability.

INSPEC. Institution of Electrical Engineers, Michael Faraday House, Six Hills Way, Stevenage, Herts. SG1 2AY, England. Telephone: 0438 313311 or FAX 0438 742840. Contains citations to the worldwide literature of physics, electronics and electrical engineering, computer technology, and related fields. Available on BRS, (800) 289-4277, DIALOG, (800) 334-2564, ORBIT, (800) 456-7248, and STN International, FIZ Karlsruhe, P.O. Box 2465, W-7500, Karlsruhe 1, Germany, online services. Inquire as to cost and availability.

NTIS Bibliographic Database. National Technical Information Service, 5285 Port Royal Road, Springfield, VA 22161. (703) 487-4929 or FAX (703) 321-8199. Broad coverage of government-sponsored science and technology research reports, 1964 to present. Available on BRS,(800) 289-4277, DIALOG, (800) 334-2564, ORBIT, (800) 456-7248, and STN International, FIZ Karlsruhe, P.O. Box 2465, W-7500, Karlsruhe 1, Germany, online services. Also available on CD-ROM. Inquire as to cost and availability.

Physics Briefs. American Institute of Physics, 335 East 45th Street, New York, NY 10017. (212) 661-9260 or FAX (212) 661-2036. Contains citations with abstracts of the literature of physics and related fields, 1979 to present. Available on the STN International, FIZ Karlsruhe, P.O. Box 2465, W-7500, Karlsruhe 1, Germany, online service. Inquire as to cost and availability.

Scientific and Technical Books and Serials in Print. R.R. Bowker Inc., 121 Chanlon Road, New Providence, NJ 07974. (800) 521-8110 or (908) 464-6800. List of currently published books and serials in the physical and biological sciences, engineering and technology, with subject, author and titles indexes. Available on ORBIT,(800) 456-7248, online service.Inquire as to cost and availability.

SCISEARCH. Institute for Scientific Information, 3501 Market Street, Philadelphia, PA 19104. (800) 523-1850 or (215) 386-0100. Broad multidisciplinary title and author index to the international literature of science and technology, 1974 to present. Available on DIALOG,(800) 334-2564, and ORBIT,(800) 456-7248, online services. Also available on CD-ROM. Inquire as to cost and availability.

WILSONLINE. H.W. Wilson Company, 950 University Avenue, Bronx, NY 10452. (800) 367-6770 or (212) 588-8400. Makes available online versions of the printed H.W. Wilson indexes including Applied Science and Technology Index, Business Periodicals Index, General Science Index, and Readers' Guide to Periodical Literature. Period covered is generally 1983 to present. Available on BRS,(800) 289-4277, DIALOG,(800) 334-2564, and OCLC EPIC,(800) 848-5878, online services. Also available on CD-ROM. Inquire as to cost and availability.

PERIODICALS

Electric Light and Power. PennWell Publishing Co. 1421 South Sheridan Road, Box 1260, Tulsa OK 74101. (835-3161. FAX (918) 831-9497. 1922 to present. Monthly. $42.00 per year.

Electric Machines and Power Systems. Taylor & Francis, 1900 Frost Road, Suite 101, Bristol, PA 19007-1598. (800) 821-8312, FAX (215) 785-5515. 1976 to present. Monthly. $355.00 per year. ISSN: 0731-356X.

Electrical World. McGraw-Hill, Inc., Box 513 Hightstown, NJ 08520. (212) 512-3288. 1874 to present. Monthly. $55.00 per year. ISSN: 0013-4457.

Electronic Design. Penton Publishing, San Jose Gateway, Suite 354. 2025 Gateway Place, San Jose, CA 95110. (408) 441-0550. 1952 to present. Fortnightly. $95.00. ISSN: 0013-4872.

Electronics. Penton Publishing, San Jose Gateway, Suite 354. 2025 Gateway Place, San Jose, CA 95110. (408) 441-0550. 1930 to present. Semi-weekly. $98.00 per year. ISSN: 0883-4989.

IEEE Power Engineering Review. Institute of Electrical and Electronics Engineers, Inc., Box 1331, 445 Hoes Lane, Piscataway, NJ 08855-1331. (908) 981-0060. 1981 to present. Monthly. $115.00 per year. ISSN: 0272-1724.

IEEE Spectrum. Institute of Electrical and Electronics Engineers, Inc., Box 1331, 445 Hoes Lane, Piscataway, NJ 08855-1331. (908) 981-0060. 1964 to present. Monthly. $157.00. ISSN: 0018-9235.

IEEE Transactions On Power Delivery. Institute of Electrical and Electronics Engineers, Inc., Box 1331, 445 Hoes Lane, Piscataway, NJ 08855-1331. (908) 981-0060. 1986 to present. Quarterly. $170.00 per year. ISSN: 0885-8977.

Institute of Electrical and Electronics Engineers. Proceedings. Institute of Electrical and Electronics Engineers, Inc., Box 1331, 445 Hoes Lane, Piscataway, NJ 08855-1331. (908) 981-0060. 1913 to present. Monthly. $275.00. ISSN: 0018-9219.

RESEARCH CENTERS AND INSTITUTES

Edison Electric Institute. 701 Pennsylvania Avenue, NW, Washington, DC 20004-2696. (202) 508-5000, 5454. FAX (202) 508-5360.

Electroscience Laboratory. Ohio State University, 1320 Kinnear Road, Columbus, Ohio 43212. (614) 292-7981, FAX (614) 292-7297.

Electrical Engineering Research Laboratories. Purdue University. Electrical Engineering Building, West Lafayette, in 47907. (317) 494-3536. FAX (317) 494-6440.

Electrical Power Research Institute. 3412 Hillview Avenue, PO Box 10412, Palo Alto, CA 94303. (415) 855-2000. FAX (415) 855-2954.

Laboratory For Electromagnetic and Electronic Systems. Massachusetts Institute of Technology, 77 Massachusetts Avenue, Cambridge, MA 02139. (617) 253-4631. FAX (617) 258-6774.

Weber Research Institute. Polytechnic University. Route 110, Farmingdale, NY 11735. (516) 755-4250, FAX (516) 755-4404.

RELIABILITY ENGINEERING
  See: QUALITY CONTROL ENGINEERING

REMOTE SENSING

See also: AERIAL PHOTOGRAPHY, CARTOGRAPHY, PHOTOGRAM-METRY, SURVEYING

ABSTRACT SERVICES AND INDEXES

Applied Science and Technology Index; A Cumulative Subject Index To English Language Periodicals in the Fields of Aeronautics and Space Science, Computer Technology, Chemistry, Construction Industry, Energy and Related Areas. H.W. Wilson Co., 950 University Avenue, Bronx, NY 10452. (800) 367-6770 or (212) 588-8400. FAX (718) 590-1617. From 1958 to present. Monthly. Inquire about cost and availability. Also available on CD-ROM.

Bibliography and Index of Geology. American Geological Institute, 4220 King Street, Alexandria, VA 22302. (703) 379-2480. Fax (703) 379-7563. Monthly. $1295.00 per year. Also available online as GeoRef.

Civil and Structural Engineering Abstracts, Cambridge Scientific Abstracts, 7200 Wisconsin Avenue, Bethesda, MD 20814-4823. (301) 961-6750. FAX (301) 961-6720. Monthly. Topics covered include structural design, construction equipment and methods, civil defense and military engineering, surveying, highway engineering, maritime and port structures, materials, land reclamation, and soil mechanics.

Engineering Index Monthly. Engineering Information, Inc., Castle Point on the Hudson, Hoboken, NJ 07030. (800) 221-1044. FAX (212) 832-1857. Monthly. $2200.00 per year. Also available online as COMPENDEX, and also on CD-ROM.

Geological Abstracts. Elsevier - Geo Abstracts, Regency House, 34 Duke Street, Norwich NR3 3AP, England. Monthly. $760.00 per year. Also available online as GEOBASE.

Imaging Abstracts (formerly *Photographic Abstracts*), Pergamon Press Inc., Maxwell House, Fairview Park, Elmsford, NY 10523. (914) 592-7700. Fax (914) 592-3625. Bimonthly.

Meteorological and Geoastrophysical Abstracts. American Meteorological Society, c/o Inforonics Inc., 550 Newtown Road, Littleton, MA 01460. (508) 486-8976. FAX (508) 486-0027. 1950 to present. Monthly. $950.00.

Physics Abstracts. Institute of Electrical Engineers, Michael Faraday House, Six Hill Way, Stevenage, Herts, SG1 2AY, England. Distributed by IEEE, 445 Hoes Lane, Piscataway, NJ 08854. (908) 562-5549. 1898 to present. Monthly. $2700.00 per year. Also available online as INSPEC.

Remote Sensing of Earth Resources. University of New Mexico Earth Data Analysis Center, 2500 Yale Blvd., Suite 100, Albuquerque, NM 87131-6031. (505) 277-3622. FAX (505) 277-3614. 1974 to present (1991 to present under current title). Quarterly. $175.00 per year.

Science Citation Index. Institue for Scientific Information, 3501 Market Street, Philadelphia, PA 19104. (215) 386-0100. FAX (215) 386-6362. Inquire about availability and cost. Also available on CD-ROM.

ASSOCIATIONS AND PROFESSIONAL SOCIETIES

American Society For Photogrammetry and Remote Sensing, 5410 Grosvenor Lane, Suite 210, Bethesda, MD 20814. (301) 493-0290. FAX (301) 493-0208.

The International Society For Optical Engineering (spie), Po Box 10, 1000-20th Street, Bellingham, WA 98227-0010.

Optical Society of America, 2010 Massachusetts Avenue Nw, Washington, DC 20036. (202) 223-8130.

The Society For Imaging Science and Technology, 7003 Kilworth Lane, Springfield, VA 22151. (703) 642-9090.

Society of Photographic Scientists and Engineers. 7003 Kilworth Lane, Springfield, VA 22151. (703) 642-9090.

BIBLIOGRAPHIES

Scientific and Technical Books and Serials in Print. R.R. Bowker Inc., 121 Chanlon Road, New Providence, NJ 07974. (800) 521-8110 or (908) 464-6800. List of currently published books and serials in the physical and biological sciences, engineering and technology, with subject, author and titles indexes. Inquire as to cost and availability. Also available on ORBIT online service, (800) 456-7248.

A Selective Bibliography On Imagery Reconnaissance and Related Matters. Robert A. McDonald. 4th ed. Defense Intelligence College, Washington, DC. Order from Superintendent of Documents, U.S. Government Printing office, Box 371954, Pittsburgh, PA 15250-7954. (202) 783-3238. FAX (202) 512-2233. 1993. Inquire for price and availability.

DIRECTORIES AND BIOGRAPHICAL SOURCES

Directory of Engineering Societies and Related Organizations. Gordon Davis. 13th edition. American Association of Engineering Societies, 1111 19th Street NW, Suite 608, Washington, DC 20036. (202) 296-2237 or (800) 658-8897. 1989. Inquire for price.

REMOTE SENSING

Ency. of Physical Sciences and Engineering Info. Sources

International Research Centers Directory. Gale Research, 835 Penobscot Building, Detroit, MI 48226-4094. (313) 961-2242. (800) 877-4253. FAX (313) 961-6083. 8th edition. 1995. $410.00

Research Centers Directory. Gale Research, 835 Penobscot Building, Detroit, MI 48226-4094. (313) 961-2242. (800) 877-4253. FAX (313) 961-6083. $485.00.

Scientific and Technical Organizations and Agencies Directory. Gale Research, 835 Penobscot Building, Detroit, MI 48226-4094. (313) 961-2242. (800) 877-4253. FAX (313) 961-6083. 4th edition. 1996. $195.00.

Who's Who in Engineering. American Association of Engineering Societies, 1111 19th Street NW, Suite 608, Washington, DC 20036. (202) 296-2237 or (800) 658-8897. 8th edition. 1991. Inquire for price.

ENCYCLOPEDIAS AND DICTIONARIES

The Encyclopedia of the Solid Earth Sciences. Philip Kearey, editor-in-chief. Blackwell Scientific Publications, 238 Main Street, Cambridge, MA 02142. (617) 876-7000 or (800) 759-6102. FAX (617) 876-7022. 1993. $60.00.

McGraw-Hill Dictionary of Scientific and Technical Terms. Sybil P. Parker, ed. 5th edition. McGraw-Hill Publishing Company, 1221 Avenue of the Americas, New York, NY 10020. (800) 262-4729 or (212) 512-3825. 1993. $110.50.

McGraw-Hill Encyclopedia of Science and Technology. Sybil P. Parker, ed. 7th edition. 20 volumes. McGraw-Hill Publishing Company, 1221 Avenue of the Americas, New York, NY 10020. (800) 262-4729 or (212) 512-3825. 1992. $1900.00

Thesaurus of Scientific, Technical, and Engineering Terms. Hemisphere Publishing Corporation, 1900 Frost Road, Suite 101, Bristol, PA 19007-1598. (215) 785-5800 or (800) 821-8312. FAX (215) 785-5515. 1987. $173.00.

GENERAL WORKS

Fundamentals of Remote Sensing and Airphoto Interpretation. Thomas Eugene Avery and Graydon Lennis Berlin. 5th ed. Macmillan Publishing, 200 Old Tappan Road, Old Tappan, NJ 07675. (800) 223-2336. FAX (800) 445-6991. 1992. Inquire for price.

Introduction To Environmental Remote Sensing. E.C. Barrett and L.F. Curtis. 3d ed. Routledge, Chapman and Hall, Inc., 29 West 35th Street, New York, NY 10001-2291. (212) 244-3336 or FAX (212) 563-2269. 1992. $39.95.

Introduction To Remote Sensing. Arthur P. Cracknell and Ladson Hayes. Taylor & Francis, 1900 Frost Road, Suite 101, Bristol, PA 19007. (215) 785-5800. FAX (215) 785-5515. 1991. $86.00 (hardback) and $36.00 (paper).

Remote Sensing: Digital Image Analysis. J.A. Richards. 2d ed. Springer-Verlag, 175 Fifth Avenue, New York, NY 10010. (212) 460-1500 or (800) 777-4643. FAX (212) 473-6272. 1993. $69.00.

HANDBOOKS AND MANUALS

Manual of Photogrammetry. Chester C. Slama, ed. 4th edition. American Society for Photogrammetry and Remote Sensing, 5410 Grosvenor Lane, Suite 210, Bethesda, MD 20814. (301) 493-0290. FAX (301) 493-0208. 1980. Inquire for price and availability.

Manual of Remote Sensing. Robert N. Colwell, ed. 2nd edition. American Society for Photogrammetry and Remote Sensing, 5410 Grosvenor Lane, Suite 210, Bethesda, MD 20814. (301) 493-0290. FAX (301) 493-0208. 1983. Inquire for price and availability.

ONLINE DATABASES AND CD-ROMS

Compendex Plus. Engineering Information, Inc., 345 East 47th Street, New York, NY 10017. (212) 705-7600 or (800) 221-1044. Contains citations with abstracts to worldwide literature in engineering and technology, from 1970 to present. Available on online BRS,(800) 289-4277, DIALOG, (800) 334-2564, ORBIT (800) 456-7248, and STN International, FIZ Karlsruhe, P.O. Box 2465, W-7500, Karlsruhe 1, Germany, online services. Also available on CD-ROM. Inquire as to cost and availability.

CSA Engineering. Cambridge Scientific Abstracts, 7200 Wisconsin Avenue, Suite 601, Bethesda, MD 20814. (301) 961-6750 or (800) 843-7751. Contains citations and abstracts of international periodicals and research literature covering all fields of engineering and science and technology, including computer and information science, electronics, mechanical engineering, solid state materials, 1981 to present. Available on BRS,(800) 289-4277, online service. Inquire as to cost and availability.

Current Contents Search. Institute for Scientific Information, 3501 Market Street, Philadelphia, PA 19104. (215) 386-0100. FAX (215) 386-6362. Contains citations to articles listed in the table of contents of science and technology journals. Also articles in social sciences and life sciences journals. Available on BRS,(800) 289-4277, DIALOG,(800) 334-2564, online services. Inquire as to cost and availability.

Dissertation Abstracts. University Microfilms International, 300 North Zeeb Road, Ann Arbor, MI 48106. (800) 521-0600 or (313) 761-4700. Scope includes virtually all doctoral dissertations accepted at accredited American institutions from 1861 to present in 252 subject areas. Available on BRS,(800) 289-4277, DIALOG,(800) 334-2564, and OCLC EPIC,(800) 848-5878, online services. Also available on CD-ROM. Inquire as to cost and availability.

Georef: Bibliography and Index of Geology. American Geological Institute, 4220 King Street, Alexandria, VA 22302. (703) 379-2480. Fax (703) 379-7563. Monthly. Inquire for price and availability.

NTIS Bibliographic Database. National Technical Information Service, 5285 Port Royal Road, Springfield, VA 22161. (703) 487-4929 or FAX (703) 321-8199. Broad coverage of government-sponsored science and technology research reports, 1964 to present. Available on BRS,(800) 289-4277, DIALOG, (800) 334-2564, ORBIT, (800) 456-7248, and STN International, FIZ Karlsruhe, P.O. Box 2465, W-7500, Karlsruhe 1, Germany, online services. Also available on CD-ROM. Inquire as to cost and availability.

SCISEARCH. Institute for Scientific Information, 3501 Market Street, Philadelphia, PA 19104. (800) 523-1850 or (215) 386-

0100. Broad multidisciplinary title and author index to the international literature of science and technology, 1974 to present. Available on DIALOG,(800) 334-2564, and ORBIT,(800) 456-7248, online services. Also available on CD-ROM. Inquire as to cost and availability.

WILSONLINE. H.W. Wilson Company, 950 University Avenue, Bronx, NY 10452. (800) 367-6770 or (212) 588-8400. Makes available online versions of the printed H.W. Wilson indexes including Applied Science and Technology Index, Business Periodicals Index, General Science Index, and Readers' Guide to Periodical Literature. Period covered is generally 1983 to present. Available on BRS,(800) 289-4277, DIALOG,(800) 334-2564, and OCLC EPIC,(800) 848-5878, online services. Also available on CD-ROM. Inquire as to cost and availability.

PERIODICALS

Canadian Journal of Remote Sensing. Canadian Aeronautics and Space Institute, 130 Slater, Suite 818, Ottawa ON K1P 6E2, Canada. (613) 234-0191. FAX (613) 234-9039. 1975 to present. Quarterly. Inquire for cost.

IEEE Transactions On Geoscience and Remote Sensing. IEEE, 345 E. 47th Street, New York, NY 10017. (908) 981-0060. FAX (908) 981-9667. 1963 to present. Bi-monthly. $188.00 per year for non-members.

International Journal of Remote Sensing. Taylor & Francis Ltd., Rankine Road, Hants RG24 8PR, England. Telephone 0256-840366. FAX 0256-479438. 1980 to present. 18 times a year. $1270.00 per year.

Isprs Journal of Photogrammetry and Remote Sensing. Elsevier Science [journals], 660 White Plains Rd., Tarrytown, NY 10159-5153. (919) 524-9200. FAX (919) 333-2444. 1949 to present. Bi-monthly. $225.00 per year.

Journal of Imaging Science and Technology. Society for Imaging Science and Technology, 7003 Kilworth Lane, Springfield, VA 22151. (703) 642-9090. FAX (703) 642-9094. 1950 to present. Bi-monthly. $120.00 per year.

Mapping Sciences and Remote Sensing. V.H. Winston & Son Inc., c/o Bellwether Publishing Ltd., 8640 Guilford Road, Suite 200, Columbia, MD 21046. (410) 290-3870. FAX (410) 290-8726. 1962 to present. 4 times a year. $279.00 per year.

Photogrammetric Engineering and Remote Sensing. American Society for Photogrammetry and Remote Sensing, 5410 Grosvenor Lane, Suite 210, Bethesda, MD 20814. (301) 493-0290. FAX (301) 493-0208. 1934-present. Monthly. $160.00 per year.

Photogrammetric Record. Photogrammetric Society, Dept. of Photogrammetry & Surveying, University College London, Gower St., London WC1E 6BT England. Telephone 071-387-7050. FAX 071-380-0453. 1953-present. Semi-annual. $88.00 per year.

Remote Sensing of Environment. Elsevier Science Publishing Company, Inc., 655 Avenue of the Americas, New York, NY 10010. (212) 989-5800. 1969 to present. 12 times a year. $995.00 per year.

Remote Sensing Reviews. Harwood ACADemic, c/o Gordon & Breach Science Publishers, PO Box 786, Cooper Station, New York, NY 10276. (212) 206-8900. Quarterly. Inquire for cost.

Scientific and Applied Photography and Cinematography. Gordon and Breach Science Publishers, 820 Town Center Drive, Langhorne, PA 19047. (215) 750-2642. FAX (215) 750-6343. 6 times a year. Inquire for price.

RESEARCH CENTERS AND INSTITUTES

Boston University Center For Remote Sensing and Cartography. 725 Commonwealth Avenue, Boston, MA 02215. (617) 353-5081. FAX (617) 353-3200.

Michigan State University Center For Remote Sensing. 302 Berkey Hall, East Lansing, MI 48824. (517) 353-7195. FAX (517) 353-1821.

University of Delaware Center For Remote Sensing. College of Marine Studies, Newark, DE 19711. (302) 451-2336.

University of Idaho, Idaho Remote Sensing Research Unit. Moscow, ID 83843. (208) 885-7209.

University of Minnesota Remote Sensing Laboratory. 1530 N. Cleveland Avenue, St. Paul, MN 55108. (612) 624-7764. FAX (612) 625-5212.

UNIVERSITY of NORTH DAKOTA, INSTITUTE FOR REMOTE SENSING. Geography Department, Grand Forks, ND 58202-8274. (701) 777-4246.

RESERVOIRS

See: DAMS

RESINS

See: PLASTICS

RESISTORS

See also: ELECTRICAL ENGINEERING, ELECTRICITY, ELECTRONIC CIRCUITS and COMPONENTS, ELECTRONICS, ELECTRONICS ENGINEERING, MICROELECCTRONCS

ABSTRACT SERVICES AND INDEXES

Applied Science and Technology Index; a cumulative subject index to English language periodicals in the fields of aeronautics and space science, computer technology, chemistry, construction industry, energy and related areas. H.W. Wilson Co., 950 University Avenue, Bronx, NY 10452. (212) 588-8400. (800) 367-6770. FAX (718) 590-1617. From 1958 to present. Monthly. Inquire about cost and availability. Also available on CD-ROM and online. ISSN: 0003-6986.

Chemical Abstracts. Chemical Abstracts Service, 2540 Olentangy River Road, Box 3012, Columbus, OH 43210-0012. (614) 447-3600. (800) 848-6538. FAX (614) 447-3713. From 1907 to present. Weekly. $16800.00 per year. Also available on CD-ROM and online. Inquire regarding cost and availability.

Current Contents: Engineering, Technology and Applied Sciences. Institute for Scientific Information, 3501 Market Street, Philadelphia, PA 19104. (215) 386-0100. FAX (215) 386-6362. 1970 to present. Weekly. $442.00 per year. Also available on CD-ROM and online. Inquire regarding cost and availability. ISSN: 0095-7917.

Current Papers in Electrical and Electronics Engineering. Institution of Electrical Engineers (IEE), London. Distributed by INSPEC/IEEE, Box 1331, 445 Hoes Lane, Piscataway, NJ 08855-1331. (908) 562-5549. 1969 to present. Monthly. $345.00 per year. ISSN: 0011-3778.

Electrical and Electronics Abstracts. Science Abstracts Section B. Institute of Electrical Engineers, Michael Faraday House, Six Hill Way, Stevenage, Herts, SG1 2AY, England. Distributed by IEEE, 445 Hoes Lane, Piscataway, NJ 08854. (908) 562-5549. 1898 to present. Monthly. $2200.00 per year. Also available on CD-ROM and online as INSPEC. ISSN: 0036-8105.

Engineering Index Monthly. Engineering Information, Inc., Castle Point on the Hudson, Hoboken, NJ 07030. (201) 216-8500. FAX (201) 216-8532. Monthly. $2200.00 per year. Also available online as COMPENDEX and on CD-ROM. ISSN:0742-1974.

General Science Index. H.W. Wilson Company, 950 University Avenue, Bronx, NY 10452. (212) 588-8400. (800) 367-6770. FAX (718) 590-1617. From 1978 to present. Ten issues per year; quarterly and annual cumulations. Service basis. Available on CD-ROM and online. Inquire regarding cost and availability. ISSN: 0162-1963.

Index To IEEE Publications. IEEE Service Center, 445 Hoes Lane, Piscataway, NJ 08855-1331. (908) 981-1393. (800)678-IEEE. FAX (908) 981-9667. 1973 to present. Annual.ISSN: 0099-1368.

Physics Abstracts. INSPEC. Section A, Science Abstracts. Institution of Electrical Engineers (IEE). Available from INSPEC/IEEE - Institute of Electrical and Electronic Engineers, Box 1331, 445 Hoes Lane, Piscaway NJ 08855-1331.(908) 562-5549. 1898 to present. 24 issues per year. $2835.00 per year. Also available on CD-ROM and online. ISSN: 0036-8091.

Physics Briefs (Physikalische Berichte). Information Center for Energy, Physics, Mathematics; German Physical Society. V C H Publishers, Inc., 220 East 23rd Street, New York, NY 10010-4606. (212) 683-8333. 1845 to present. 24 issues per year. $2390.00 per year. Also available online. ISSN: 0179-7434.

Science Citation Index. SCI. Institute for Scientific Information, 3501 Market Street, Philadelphia, PA 19104. (215) 386-0100. (800) 523-1850. FAX (215) 386-2991. 1961 to present. Six issues per year, plus annual cumulation. $11650.00 per year. Also available online and on CD-ROM. Inquire about price and availability. ISSN: 0036-827X.

Solid State and Superconductivity Abstracts: Covers theory, Production and Application of Solid State Materials. Cambridge Scientific Abstracts, 7200 Wisconsin Avenue, Bethesda, MD 20824. (301) 961-6750. FAX (301) 961-6720. 1957 to present. Bimonthly. $1320.00 per year. Also available online. ISSN: 0896-5900.

ANNUAL REVIEWS AND YEARBOOKS

Advances in Electronics and Electron Physics. Academic Press, Inc., 6277 Sea Harbor Drive, Orlando, FL 32887. (800) 321-5068. From 1948 to present. Irregular. Price varies, inquire. ISSN: 0065-2539.

ASSOCIATIONS AND PROFESSIONAL SOCIETIES

American Electronics Association. 5201 Great America Way, Suite 520, P.O. 52990, Santa Clara, CA 95056. (408) 987- 4200. FAX (408) 970-8565.

American Institute of Physics. One Physics Ellipse, College Park, MD 20740-3843. (301) 209-3100.

Edison Electric Institute. 701 Pennsylvania Avenue NW, Washington, DC 20004-2696. (202) 508-5000, 5454. FAX (202) 508-5360.

Electronics Industries Association. 2500 Wilson Boulevard, Arlington, VA 22201. (703) 907-7500. FAX (202) 457-4985.

IEEE (institute of Electrical and Electronic Engineers). 345 East 47th Street. New York, NY 10017. (212) 705-7900. FAX (212) 705-4929,

International Society For Hydrid Microelectronics. 1850 Centennial Park Drive, Suite 105, Reston, VA 22091. (703) 758-1060. FAX (703) 758-1066.

National Electrical Manufacturers Association. 1300 North 17th Street, Suite 1847, Rosslyn VA 22209. (703) 841-3200. FAX (703) 841-3300.

DIRECTORIES AND BIOGRAPHICAL SOURCES

American Electronics Association Directory. American Electronics Association, 5201 Great America Way, Suite 520, P.O. 52990, Santa Clara, CA 95056. (408) 987-4200. FAX (408) 970-8565. Annual. $175.00.

American Men and Women of Science: Physical and Biological Sciences. R. R. Bowker Inc., 121 Chanlon Road, New Providence, NJ 07974. (908) 464-6800. (800) 521-8110. 20th edition. 8 volumes. 1996. $850.00.

Directory of Engineering Societies and Related Organizations. American Association of Engineering Societies, 1111 19th Street, Suite 608, Washington, DC 20036-3603. (202) 296-2237. Semi-annual. $150.00.

E E M - Electronic Engineer's Master. Hearst Business Communications, Inc., 645 Stewart Avenue, Garden City NY 11530. (516) 227-1300. Annual. $90.00 ISSN: 0732-9016.

Engineering Research Centers: Incorporating Electronics Research Centers. Stockton Press, 345 Park Avenue, New York, NY 10010. (212) 689-9200. (800) 221-2123. 4th edition. 1995. $515.00.

IC Master. Hearst Business Communications, Inc., 645 Stewart Avenue, Garden City, NY 11530. (516) 227-1300. 1975 to present. Annual. $170.00 per year. ISSN: 0894- 6809.

IEEE Membership Directory. Institute of Electrical and Electronics Engineers, IEEE Service Center, 445 Hoes Lane, Picastaway, NY 08854. (908) 981-1393. (800) 678-IEEE. FAX (908) 981-9667. 2 volumes. Annual. $190.00. ISSN:

International Directory of Abbreviations and Acronyms of Electronics, Electrical Engineering, Computer Technology and Information Processing. Peter Wennrich. K. G. Saur, 121 Chanlon Road, New Providence, NJ 07974. (908) 464-6800. (800) 521-8110 2 volumes. 1992. $230.00.

International Engineering Directory. American Consulting Engineers Council, 1015 15th Street, N.W. Suite 802, Washington, DC 20005-2670. (202) 347-7474. Annual. $10.00.

Research Centers Directory. Gale Research Inc., 835 Penobscot Building, Detroit, MI 48226-4094. (313) 961-2242. (800) 877-4253. 20th edition. 1995. $485.00. ISSN: 0080- 1518.

Scientific and Technical Organizations and Agencies Directory. Gale Research, 835 Penobscot Building, Detroit, MI 48226-4094. (313) 961-2242. (800) 877-4253. 4th edition. 1996. $195.00.

Who's Who in Engineering. Gordon Davis, editor. American Society for Engineering Education, 1111 19th Street, NW, Suite 608, Washington, DC 20036. (202) 296-2237. (800) 658-8897. 9th edition. 1995. $220.00.

Who's Who in Technology. Gale Research, 835 Penobscot Building, Detroit, MI 48226-4094. (313) 961-2242. (800) 877-4253. 7th edition. 1995. $195.00.

ENCYCLOPEDIAS AND DICTIONARIES

Encyclopedia of Applied Physics. George Trigg, editor. VCH Publications, Inc., 220 East 23rd Street, Suite 909, New York, NY 10010-4606. (212) 683-8333. (800) 422-8824. 20 volume set. 1991-. $5990.00.

Encyclopedia of Integrated Circuits. Walter A. Buchsbaum.

Prentice Hall, 113 Sylvan Avenue, Route 9W, Englewood Cliffs, NJ 07632. (201) 592-2000. (800) 922-0579. 1987. $39.95.

IEEE Standard Dictionary of Electrical and Electronics Terms. Christopher J. Booth, editor. IEEE Service Center, 445 Hoes Lane, Piscataway, NJ 08855-1331. (908) 981-1393. (800) 678-IEEE. FAX (908) 981-9667. IEEE Standard 100- 1992. 5th edition. 1993. $150.00.

Illustrated Dictionary of Electronics. Stan Gibilsco. TAB Books, P.O. Box 40, Blue Summit, PA 17294-0850. (717) 794- 2191. (800) 233-1128. 7th edition. 1994. $34.95.

International Encyclopedia of Integrated Circuits. Arthur A. Seidman. TAB Books, P.O. Box 40, Blue Summit, PA 17294-0850. (717) 794-2191. (800) 233-1128. 1991. $75.00.

McGraw-Hill Electronics Dictionary. J. Markus. McGraw-HillBook Company, Inc., 1221 Avenue of the Americas, New York, NY 10020. (212) 997-3675. 5th edition. 1994. $49.95.

McGraw-Hill Encyclopedia of Science and Technology. McGraw-Hill Book Company, Inc., 1221 Avenue of the Ameri-cas, New York, NY 10020. (212) 512-2000. (800) 262-4729. 7th edition. 20 volume set. 1992. $1900.00.

GENERAL WORKS

Capacitive and Resistive Electronic Components. D. S. Campbell and J. A. Hayes. Gordon & Breach Science Publishers, Inc., P.O.Box 200029, Riverfront Plaza Station, Newark, NJ 07102-0301. (201) 643-7500. 1994. 110.00

Exploring Electronic Devices. Mark E. Hazen. SCP: Third World Literature Publishing House, P.O. Box 482, Lithonia, GA 30058-0482. (404) 785-7725. 1991. $50.75.

High-voltage Engineering and Testing. H. M. Ryan, editor. Institution of Electrical Engineers; IEE/INSPEC, Box 1331, 445 Hoes Lane, Piscataway, NJ 08855-1331. (908) 981-0060. 1994. $95.00.

High Voltage Insulation Technology: Textbook For Electgrical Engineers. Dieter Kind, and Hermann Karner. Ballen Booksellers, Int., 125 Ricefield Lane, Hauppauge, NY 11788. (516) 543-5600. (800) 645-5237. 1985. $48.00 in paper.

Microwave Transition Design. Jamal S. Izadian and Shahin M. Izadian. Artech House, 685 Canton Street, Norwood, MA 02062. (617) 769-9750. (800) 225-9977. 1988. $59.00.

Principles and Practice of Impedance. Rufus P. Turner and Stan Gibilisco. TAB Books, P.O. Box 40, Blue Summit, PA 17294-0850. (717) 794-2191. (800) 233-1128. 1987. $21.95.

HANDBOOKS AND MANUALS

Active Electronic Component Handbook. Charles A. Harper and Harold C. Jones. McGraw-Hill Book Company, 1221 Avenue of the Americas, New York, NY 10020. (212) 512-2000. (800) 262-4729. 2nd edition. 1996. $79.50.

Electrical Engineer's Handbook. Donald Christiansen, editor. McGraw Book Company, 1221 Avenue of the Americas, New York, NY 10020. (212) 512-2000. (800) 262-4729. 4th edition. 1996. $110.00.

Electronics Handbook. Jerry C. Whitaker. CRC Press, Inc., 2000 Corporate Boulevard, NW, Boca Raton, FL 33431. (407) 994-0555. (800) 272-7737. 1996. $120.00.

Handbook of Electrical and Electronic Technology. Curtis Johnson. Prentice Hall, 113 Sylvan Avenue, Route 9W, Englewood Cliffs, NJ 07632. (201) 592-2000. (800) 922- 0579. 1996. $88.00.

Industrial Electronics and Systems Handbook. Irwin, CRC Press, Inc., 2000 Corporate Boulevard, NW, Boca Raton, FL

33431. (407) 994-0555. (800) 272-7737. 1996. $129.95.

Standard Handbook for Electrical Engineers. Donald Fink. McGraw-Hill Publishing Company, 1221 Avenue of the Americas, New York, NY 10020. (212) 512-2000. (800) 262- 4729. 13th edition. 1996. $110.00.

RESISTORS

Ency. of Physical Sciences and Engineering Info. Sources

Standard Handbook of Engineering Calculations. Tyler G.

Hicks. McGraw-Hill Publishing Company, 1221 Avenue of the Americas, New York, NY 10020. (212) 512-2000. (800) 262-4729. 3rd edition. 1994. $99.50. ISBN: 0-07-028812-7.

ONLINE DATABASES

CA Search. Chemical Abstracts Service, P.O. Box 3012, Columbus, OH 43210-0012. (614) 447-3600. (800) 848-6533. FAX (614) 447-3709. Very comprehensive guide to worldwide chemical literature and related fields, 1972 to present. Available on BRS,(800) 289-4277, DIALOG, (800) 334-2564, ORBIT (800) 456-7248, and STN International, FIZ Karlsruhe, P.O. Box 2465, W-7500, Karlsruhe 1, Germany, online services. Inquire as to cost and availability.

Compendex Plus. Engineering Information, Inc., 345 East 47th Street, New York, NY 10017. (212) 705-7600 or (800) 221-1044. Contains citations with abstracts to worldwide literature in engineering and technology, from 1970 to present. Available on online BRS,(800) 289-4277, DIALOG, (800) 334-2564, ORBIT (800) 456-7248, and STN International, FIZ Karlsruhe, P.O. Box 2465, W-7500, Karlsruhe 1, Germany, online services. Also available on CD-ROM. Inquire as to cost and availability.

Dissertation Abstracts. University Microfilms International, 300 North Zeeb Road, Ann Arbor, MI 48106. (800) 521-0600 or (313) 761-4700. Scope includes virtually all doctoral dissertations accepted at accredited American institutions from 1861 to present in 252 subject areas. Available on BRS, (800) 289-4277, DIALOG, (800) 334-2564, and OCLC EPIC, (800) 848-5878, online services. Also available on CD-ROM. Inquire as to cost and availability.

INSPEC. Institution of Electrical Engineers, Michael

Faraday House, Six Hills Way, Stevenage, Herts. SG1 2AY, England. Telephone: 0438 313311 or FAX 0438 742840. Contains citations to the worldwide literature of physics, electronics and electrical engineering, computer technology, and related fields. Available on BRS, (800) 289-4277, DIALOG, (800) 334-2564, ORBIT, (800) 456-7248, and STN International, FIZ Karlsruhe, P.O. Box 2465, W-7500, Karlsruhe 1, Germany, online services. Inquire as to cost and availability.

NTIS Bibliographic Database. National Technical Information Service, 5285 Port Royal Road, Springfield, VA 22161. (703) 487-4929 or FAX (703) 321-8199. Broad coverage of government-sponsored science and technology research reports, 1964 to present. Available on BRS,(800) 289-4277, DIALOG, (800) 334-2564, ORBIT, (800) 456-7248, and STN International, FIZ Karlsruhe, P.O. Box 2465, W-7500, Karlsruhe 1, Germany, online services. Also available on CD-ROM. Inquire as to cost and availability.

SCISEARCH. Institute for Scientific Information, 3501 Market Street, Philadelphia, PA 19104. (800) 523-1850 or (215) 386-0100. Broad multidisciplinary title and author index to the international literature of science and technology, 1974 to present. Available on DIALOG, (800) 334-2564, and ORBIT, (800) 456-7248, online services. Also available on CD-ROM. Inquire as to cost and availability.

WILSONLINE. H.W. Wilson Company, 950 University Avenue, Bronx, NY 10452. (800) 367-6770 or (212) 588-8400. Makes available online versions of the printed H.W. Wilson indexes including Applied Science and Technology Index, Business

Periodicals Index, General Science Index, and Readers' Guide to Periodical Literature. Period covered is generally 1983 to present. Available on BRS, (800) 289-4277, DIALOG, (800) 334-2564, and OCLC EPIC, (800) 848-5878, online services.Also available on CD-ROM. Inquire as to cost and availability.

PERIODICALS

Electronic Design. Penton Publishing, San Jose Gateway, Suite 354. 2025 Gateway Place, San Jose, CA 95110. (408) 441-0550. 1952 to present. Biweekly. $95.00 per year. ISSN: 0013-4872.

Electronic Engineering Times. CMP Publications, Inc., 600 Community Drice, Manhasset, NY 11030. (516) 562-5000. FAX (516) 562-5325. 1972 to present. Weekly. $159.00 per year. ISSN: 0192-1541.

Electronics. Penton Publishing, San Jose Gateway, Suite 354. 2025 Gateway Place, San Jose, CA 95110. (408) 441- 0550. 1930 to present. Semi-weekly. $98.00. ISSN: 0883-4989.

IEEE Circuits and Devices Magazine. Institute of Electrical and Electronics Engineers, Inc., Box 1331, 445 Hoes Lane, Piscataway, NJ 08855-1331. (908) 981-0060. 1985 to present. Bi-monthly. $120.00 per year. ISSN: 8755- 3996.

IEEE Journal of Solid State Circuits. Institute of Electrical and Electronics Engineers, Inc., Box 1331, 445 Hoes Lane, Piscataway, NJ 08855-1331. (908) 981-0060. 1966 to present. Bi-monthly. $275.00 per year to non-members.

IEEE TRANSACTIONS ON ELECTRON DEVICES. Institute of

Electrical and Electronics Engineers, Inc., Box 1331, 445 Hoes Lane, Piscataway, NJ 08855-1331. (908) 981-0060. 1952 to present. Monthly. $395.00 per year. ISSN: 0018-9383.

Institute of Electrical and Electronics Engineers. PROCEEDINGS. Institute of Electrical and Electronics Engineers, Inc., Box 1331, 445 Hoes Lane, Piscataway, NJ 08855-1331. (908) 981-0060.

Microelectronic Engineering: An Interdisciplinary Journal of Semiconductor Manufacturing Technology. Elsevier Science, 660 White Plains Road, Tarrytown, NY 10591-5153. (914) 524-9200. FAX (914) 333-2444. 1983 to present. 16 issues per year. $845.00 per year. ISSN: 0167-9317.

Microelectronics and Reliability. Elsevier Science, 660 White Plains Road, Tarrytown, NY 10591-5153. (914) 524- 9200. FAX (914) 333-2444. 1962 to present. 15 issues per year. $1145.00 per year. ISSN: 0026-2714.

Microelectronics Journal. Elsevier Science, 660 White Plains Road, Tarrytown, NY 10591-5153. (914) 524-9200. FAX (914) 333-2444. 1967 to present. 8 issues per year. $455.00 per year. ISSN: 0026-2692.

Semiconductor International. Cahners Publishing Company. 1350 East Touhy Avenue, Box 5080, Des Plaines, IL 60017-6080. (708) 635-8800. FAX (708) 390-2770. 1978 to present. 13 issues per year. $84.95 per year.

Solid State Electronics. Elsevier Science, 660 White Plains Road, Tarrytown, NY 10591-5153. (914) 524-9200. FAX (914) 333-

2444. 1960 to present. Monthly. $1025.00 per year.

RESEARCH CENTERS AND INSTITUTES

Electrical Engineering Research Laboratories. Purdue University, Electrical Engineer4ing Building, West Lafayette, in 47907. (317) 494-3536. FAX (317) 494-6440.

Electronics Research Center. University of Texas at Austin, 132 Engineering Science Building, Austin, TX.

Electronics Research Laboratory. University of California, Berkeley, 243 Cory Hall, Berkeley CA 94720. (415) 642-2301.

Laboratory For Electromagetic and Electronic Systems. Massachusetts Institute of Technology, 77 Massachusetts Avenue, Cambridge, MA 02139

RHEOLOGY

See: FLUID MECHANICS

RINGS (PLANETAY)

See: PLANETARY SCIENCES

ROAD BUILDING

See: HIGHWAY ENGINEERING

ROBOTICS

See also: ARTIFICIAL INTELLIGENCE, CAM, COMPUTERS, CONTROL SYSTEMS, DESIGN ENGINEERING, INDUSTRIAL ENGINEERING, MATERIALS HANDLING, MECHANICAL ENGINEERING

ABSTRACT SERVICES AND INDEXES

Applied Mechanics Reviews: An Assessment of World Literature in Engineering Sciences. 1948-present. American Society of Mechanical Engineers, 345 East 47th Street, New York, NY 10017. (212) 705-7703. Monthly. $360.00 per year.

Applied Science and Technology Index; A Cumulative Subject Index To English Language Periodicals in the Fields of Aeronautics and Space Science, Computer Technology, Chemistry, Construction Industry, Energy and Related Areas. H.W. Wilson Co., 950 University Avenue, Bronx, NY 10452. (800) 367-6770 or (212) 588-8400. FAX (718) 590-1617. From 1958 to present. Monthly. Inquire about cost and availability. Also available on CD-ROM and online.

Current Contents: Engineering, Technology, and Applied Sciences. Institute for Scientific Information, 3501 Market Street, Philadelphia, PA 19104. (215) 386-0100. FAX (215) 386-6362. 1970 to present. Weekly. $442.00 per year.

Electrical and Electronics Abstracts. Institute of Electrical Engineers, Michael Faraday House, Six Hill Way, Stevenage, Herts, SG1 2AY, England. Distributed by IEEE, 445 Hoes Lane, Piscataway, NJ 08854. (908) 562-5549. 1898 to present. Monthly. $2200.00 per year. Also available on CD-ROM and online as INSPEC.

Engineering Index Monthly. Engineering Information, Inc., Castle Point on the Hudson, Hoboken, NJ 07030. (800) 221-1044. FAX (212) 832-1957. Monthly. $2300.00 per year. Also available online as COMPENDEX, and also on CD-ROM. Covers chemical engineering, computers, electrical engineering, civil engineering, metals and mining, industrial management, and mechanical engineering.

Index To Scientific and Technical Proceedings. Institute for Scientific Information, 3501 Market St., Philadelphia, PA 19104. (215) 386-0100. FAX (215) 386-6362. Monthly. $500.00 per year.

Key Abstracts: Robotics & Control. INSPEC/IEEE, Box 1331, 445 Hoes Lane, Piscataway, NJ 08855-1331. (908) 562-5549. 1975 to present. Monthly. $160.00 per year. Covers robots and their applications to materials handling, industrial production systems, and transportation systems.

Mechanical Engineering Abstracts (formerly ISMEC), Cambridge Scientific Abstracts, 7200 Wisconsin Avenue, Bethesda, MD 20814-4823. (301) 961-6750. FAX (301) 961-6720. 1967 to present. Monthly. $895.00 per year. Summarizes world literature in mechanical engineering, production engineering, and engineering management. Also available online.

Science Citation Index. Institute for Scientific Information, 3501 Market Street, Philadelphia, PA 19104. (215) 386-0100. FAX (215) 386-6362. Inquire about availability and cost. Also available on CD-ROM.

ASSOCIATIONS AND PROFESSIONAL SOCIETIES

American Society of Mechanical Engineers. 345 East 47th Street, New York, NY 10017-2398. (212) 705-7703.

IEEE Robotics and Automation Council. IEEE, 345 E. 47th Street, New York, NY 10017. (202) 705-7900.

Industrial Designers Society of America. 1142-E Walker Road, Great Falls, VA 22066. (703) 759-0100.

Institute of Industrial Engineers. 25 Technology Park/Atlanta, Norcross, GA 30092. (404) 449-0460.

Instrument Society of America. 67 Alexander Drive, Box 12277, Research Triangle Park, NC 27709. (919) 549-8411. FAX (919) 549-8288.

International Service Robot Association. 900 Victors Way, Ann Arbor, MI 48106. (313) 994-6088.

Robotic Industries Association. PO Box 3724, 900 Victors Way, Ann Arbor, MI 48106. (313) 994-6088.

Robotics International of the Society of Manufacturing Engineers. 1 SME Drive, PO Box 0930, Dearborn, MI 48121-0930. (313) 271-1500. FAX (313) 271-2861.

ROBOTICS

Ency. of Physical Sciences and Engineering Info. Sources

Society of Automotive Engineers International, 400 Commonwealth Drive, Warrendale, PA 15096-0001.

Society of Logistics Engineers. 8100 Professional Place, Suite 211, Hyattsville, MD 20785. (301) 459-8446.

Society of Manufacturing Engineers. One SME Drive, Box 930, Dearborn, MI 48121-0930. (313) 271-1500. FAX (313) 271-2861.

DIRECTORIES AND BIOGRAPHICAL SOURCES

American Men & Women of Science, 1995-96. R.R. Bowker Staff, eds. 19th edition. 8 volumes. R.R. Bowker/Reed International Publishing Company, 121 Chanlon Road, New Providence, NJ 07974. (908) 464-6800 or (800) 521-8110. 1995. $850.00

International Research Centers Directory. Gale Research, 835 Penobscot Building, Detroit, MI 48226-4094. (313) 961-2242. (800) 877-4253. FAX (313) 961-6083. 8th edition. 1995. $410.00.

Plant Engineering Product Supplier Guide. Cahners Publishing Company (Des Plains), 1350 East Touhy Avenue, Des Plaines, IL 60017-5080. (800) 323-4958. FAX (708) 390-2779. 1965 to present. Annual. $25.00 (also included with subscription to PLANT ENGINEERING magazine).

Research Centers Directory. Gale Research, 835 Penobscot Building, Detroit, MI 48226-4094. (313) 961-2242. (800) 877-4253. FAX (313) 961-6083. $485.00.

Robotics World Directory. Argus Inc., 6151 Powers Ferry Road NW, Atlanta, Georgia 30339-2941. (404) 955-2500 or (800) 241-9834. FAX (404) 955-0400. Included with subscription to ROBOTICS WORLD magazine (see below).

Scientific and Technical Organizations and Agencies Directory. Gale Research, 835 Penobscot Building, Detroit, MI 48226-4094. (313) 961-2242. (800) 877-4253. FAX (313) 961-6083. 4th edition. 1996. $195.00.

Society of Logistics Engineers Membership Directory. 8100 Professional Place, Suite 211, Hyattsville, MD 20785. (301) 459-8446. Inquire for cost and availability.

Who's Who in Engineering. American Association of Engineering Societies, 1111 19th Street NW, Suite 608, Washington, DC 20036. (202) 296-2237 or (800) 658-8897. 8th edition. 1991. Inquire for price.

Who's Who in Technology. Gale Research, 835 Penobscot Building, Detroit, MI 48226-4094. (313) 961-2242. (800) 877-4253. FAX (313) 961-6083. 1995. $195.00.

ENCYCLOPEDIAS AND DICTIONARIES

Encyclopedic Dictionary of Industrial Automation and Computer Control. David W. South. Prentice Hall, 113 Sylvan Avenue, Route 9W, Englewood Cliffs, NJ 07632. (201) 592-2000 or (800) 922-0579. 1994. Inquire for price and availability.

GENERAL WORKS

Introduction To Robotics. Arthur J. Critchlow. 2d rev. ed. Macmillan Publishing, 200 Old Tappan Road, Old Tappan, NJ 07675. (800) 223-2336. FAX (800) 445-6991. 1993. Inquire for cost.

Introduction To Robotics. PHillip McKerrow. Addison-Wesley Publishing Company, 1 Jacob Way, Reading, MA 01867. (617) 944-3700 or (800) 447-2226. FAX 617-942-1117. 1991. Inquire for cost and availability.

Space Robotics: Dynamics and Control. Yangsheng Xu and Takeo Kanade, eds. Kluwer Academic Publishers, P.O. Box 358, Accord Station, Hingham, MA 02018-0358. (617) 871-6000. FAX (617) 871-6528. 1993. Inquire for cost and availability.

HANDBOOKS AND MANUALS

Handbook of Industrial Engineering. Gavriel Salvendy. 2nd edition. John Wiley and Sons, Inc., 605 Third Avenue, New York, NY 10158. (800) 526-5368 or (212) 850-6000. 1991. $165.00.

ONLINE DATABASES AND CD-ROMS

Aerospace Database. American Institute of Aeronautics and Astronautics, 370 L'Enfant Promenade, S.W., Washington, DC 20024. (202) 646-7400. Worldwide published literature on research and development in aerospace and related areas, with abstracts. Covers 1962 to present. Available on DIALOG, (800) 334-2564, online service. Also available on CD-ROM. Inquire as to cost and availability.

Compendex Plus. Engineering Information, Inc., 345 East 47th Street, New York, NY 10017. (212) 705-7600 or (800) 221-1044. Contains citations with abstracts to worldwide literature in engineering and technology, from 1970 to present. Available on online BRS,(800) 289-4277, DIALOG, (800) 334-2564, ORBIT (800) 456-7248, and STN International, FIZ Karlsruhe, P.O. Box 2465, W-7500, Karlsruhe 1, Germany, online services. Also available on CD-ROM. Inquire as to cost and availability.

CSA Engineering. Cambridge Scientific Abstracts, 7200 Wisconsin Avenue, Suite 601, Bethesda, MD 20814. (301) 961-6750 or (800) 843-7751. Contains citations and abstracts of international periodicals and research literature covering all fields of engineering and science and technology,including computer and information science, electronics, mechanical engineering, solid state materials, 1981 to present. Available on BRS,(800) 289-4277, online service.Inquire as to cost and availability.

Current Contents Search. Institute for Scientific Information, 3501 Market Street, Philadelphia, PA 19104. (215) 386-0100. FAX (215) 386-6362. Contains citations to articles listed in the table of contents of science and technology journals. Also articles in social sciences and life sciences journals. Available on BRS,(800) 289-4277, DIALOG,(800) 334-2564, online services. Inquire as to cost and availability.

Dissertation Abstracts. University Microfilms International, 300 North Zeeb Road, Ann Arbor, MI 48106. (800) 521-0600 or (313) 761-4700. Scope includes virtually all doctoral dissertations accepted at accredited American institutions from 1861 to present in 252 subject areas. Available on BRS,(800) 289-4277, DIALOG,(800) 334-2564, and OCLC EPIC,(800) 848-5878,

online services. Also available on CD-ROM. Inquire as to cost and availability.

International Aerospace Abstracts. American Institute of Aeronautics and Astronautics, 370 L'Enfant Promenade, S.W., Washington, DC 20024. (202) 646-7400. Contains references and abstracts of journal and monograph literature relating to aerospace science and technology, from 1963 to present. Available through the NASA/RECON system of the National Aeronautics and Space Administration only.

ISMEC: Mechanical Engineering Abstracts. Cambridge Scientific Abstracts, 7200 Wisconsin Avenue, Suite 601, Bethesda, MD 20814. (301) 961-6750 or (800) 843-7751. Contains citations to the literature in mechanical engineering, industrial and production engineering, energy, power, mechanics, devices and related areas, from 1973 to present. Available on the DIA-LOG,(800) 334-2564, online service. Inquire as to cost and availability.

NASA Database. American Institute of Aeronautics and Astronautics, 370 L'Enfant Promenade, S.W., Washington, DC 20024. (202) 646-7400. Citations and abstracts of aeronautics and astronautics literature, 1962 to present. Also contains citations from *STAR, Scientific and Technical Aerospace Reports*, and *International Aerospace Abstracts*. Available through NASA/RECON online service. Inquire as to cost and availability.

NTIS Bibliographic Database. National Technical Information Service, 5285 Port Royal Road, Springfield, VA 22161. (703) 487-4929 or FAX (703) 321-8199. Broad coverage of government-sponsored science and technology research reports, 1964 to present. Available on BRS,(800) 289-4277, DIALOG, (800) 334-2564, ORBIT, (800) 456-7248, and STN International, FIZ Karlsruhe, P.O. Box 2465, W-7500, Karlsruhe 1, Germany, online services. Also available on CD-ROM. Inquire as to cost and availability.

SAE Global Mobility Database. Society of Automotive Engineers (SAE), Electronic Publishing Division, 400 Commonwealth Drive, Warrendale, PA 15098. (412) 776-4841. Contains citations with abstracts to technical papers on automotive and aerospace technology and vehicular-related industries that have been presented at SAE conferences. Covers 1906 to present. Available on ORBIT,(800) 456-7248,online service. Inquire as to cost and availability.

SCISEARCH. Institute for Scientific Information, 3501 Market Street, Philadelphia, PA 19104. (800) 523-1850 or (215) 386-0100. Broad multidisciplinary title and author index to the international literature of science and technology, 1974 to present. Available on DIALOG,(800) 334-2564, and ORBIT,(800) 456-7248, online services. Also available on CD-ROM. Inquire as to cost and availability.

WILSONLINE. H.W. Wilson Company, 950 University Avenue, Bronx, NY 10452. (800) 367-6770 or (212) 588-8400. Makes available online versions of the printed H.W. Wilson indexes including Applied Science and Technology Index, Business Periodicals Index, General Science Index, and Readers' Guide to Periodical Literature. Period covered is generally 1983 to present. Available on BRS,(800) 289-4277, DIALOG,(800) 334-2564, and OCLC EPIC,(800) 848-5878, online services. Also available on CD-ROM. Inquire as to cost and availability.

PERIODICALS

Advanced Manufacturing Technology. Technical Insights Inc., 32 N Dean Street, Englewood, NJ 07631. (201) 568-4744. FAX (201) 568-8247. 1980 to present. Monthly. $575.00 per year.

Advances in Automation and Robotics. JAI Press, 55 Old Post Road No.2, PO Box 1678, Greenwich, CT 06836. (203) 661-7602. 1985 to present. Irregular. $28.75.

American Industry. Publications for Industry, 21 Russell Woods Road, Great Neck, NY 11021. (516) 487-0990. FAX (516) 487-0809. 1946 to present. Monthly. $25.00 per year.

Assembly. Hitchcock Publishing, 191 S. Gary Avenue, Carol Stream, IL 60188. (708) 462-4641. FAX (708) 462-2205. 1958 to present. Monthly. $60.00 per year.

IEEE Robotics and Automation Magazine. IEEE (Institute of Electrical and Electronic Engineers), 445 Hoes Lane, Piscataway, NJ 08854. (908) 562-5545 or (800) 678-IEEE. 1994 to present. Quarterly. $50.00 per year.

IEEE Transactions On Industry Applications. IEEE (Institute of Electrical and Electronic Engineers), 445 Hoes Lane, Piscataway, NJ 08854. (908) 562-5545 or (800) 678-IEEE. 1965 to present. Bi-monthly. $240.00 per year.

IIE Transactions: Industrial Engineering Research and Development. Institute of Industrial Engineers, 25 Technology Park-Atlanta, Norcross, GA 30092. (404) 449-0460. FAX (404) 263-8532. 1969 to present. Quarterly. $95.00 per year.

Industrial Robot. MCBUniversity Press, 62 Toller Lane, Bradford, W. Yorks. BD8 9BY, England. Telephone 0274-499821. FAX 0274-547143. 1973 to present. 6 times a year. $749.95 per year.

International Journal of Robotics Research. MIT Press, 55 Hayward Street, Cambridge, MA 02142. (617) 253-2884 or (800) 356-0343. 1982 to present. Bi-monthly. $80.00 per year.

Journal of Engineering For Industry. American Society of Mechanical Engineers, 345 E. 47th Street, New York, NY 10017. (212) 705-7722. 1970 to present. Quarterly. $100.00 per year.

Journal of Intelligent and Robotics Systems. Kluwer Academic Publishers, P.O. Box 358, Accord Station, Hingham, MA 02018-0358. (617) 871-6000. 1988 to present. Nine times a year. $420.00 per year.

Journal of Robotic Systems. J. Wiley & Sons, Journals, 605 Third Avenue, New York, NY 10158-0012. (212) 692-6000.1984 to present. 8 times a year. $525.00 per year.

Manufacturing Engineering. Society of Manufacturing Engineers, One SME Drive, Box 930, Dearborn, MI 48128-1490. (313) 271-1500 or FAX (313) 271-2861. 1932 to present. Monthly. $60.00 per year.

Robotica. Cambridge University Press, Journals Department, 40 W 20th Street, New York, NY 10011-4211. (212) 924-3900. 1983 to present. Bi-monthly. $119.00.

Robotics and Autonomous Systems. Elsevier Science Inc., Box 882, Madison Square Station, New York, NY 10159. (212) 989-5800. 1985 to present. 8 times a year. $440.00 per year.

ROCK MECHANICS

Ency. of Physical Sciences and Engineering Info. Sources

Robotics and Computer-integrated Manufacturing. Elsevier Science [journals], 660 White Plains Rd., Tarrytown, NY 10159-5153. (919) 524-9200. FAX (919) 333-2444. 1984 to present. 4 times a year. $415.00 per year.

Robotics and Expert Systems. Instrument Society of America, 67 Alexander Drive, Box 12277, Research Triangle Park, NC 27709. (919) 549-8411. FAX (919) 549-8288. 1985 to present. Irregular. Price varies.

Robotics Today. Society of Manufacturing Engineers, One SME Drive, Box 930, Dearborn, MI 48121-0930. (313) 271-1500. FAX (313) 271-2861. 1979 to present. Quarterly. $60.00 per year for non-members.

Robotics World. Argus Inc., 6151 Powers Ferry Road NW, Atlanta, Georgia 30339-2941. (404) 955-2500. FAX (404) 955-0400. 1982 to present. 4 times a year. $42.00 per year.

RESEARCH CENTERS AND INSTITUTES

General Robotics and Active Sensory Processing Laboratory. University of Pennsylvania, Room 301c, GRASP Lab, 3401 Walnut Street, Philadelphia, PA 19104-6228. (215) 898-0370.

Intelligent Computer Systems Research Institute. University of Miami, PO Box 248235, Coral Gables, FL 33124. (305) 284-5195.

Robotics Research Laboratory. New York University, Courant Institute of Mathematical Sciences, 719 Broadway, 12th Floor, New York, NY 10003. (212) 998-3472.

Robotics and Automation Laboratories. Rensselaer Polytechnic Institute, School of Engineering, Troy, NY 12180. (518) 276-6076. FAX (518) 276-8715.

Robotics Institute. Carnegie Mellon University, Pittsburgh, PA 15213. (412) 268-3818. FAX (412) 268-5016.

SPECIFICATIONS AND STANDARDS

*Annual Book of ASTM Standards, Volume 14.*01. American Society for Testing and Materials (ASTM), 1916 Race Street, Philadelphia, PA 19103. (215) 299-5585. Annual. Inquire for cost and availability.

ROCK MECHANICS

See also: CIVIL ENGINEERING, COAL MINING, EARTHQUAKE ENGINEERING, GEOTECHNICAL ENGINEERING, GOLD and GOLD MINING, MINING ENGINEERING, QUARRYING, SOIL MECHANICS ABTRACTS and INDEXES

Abstract Journal in Earthquake Engineering. University of California, Berkeley, Earthquake Engineering Research Center, 1301 South 46th Street, Richmond, CA 94804-4698. (510) 231-9413 or FAX (510) 231-9471. 1972 to present. Annual. $80.00 per year.

Applied Science and Technology Index; A Cumulative Subject Index To English Language Periodicals in the Fields of Aeronautics and Space Science, Computer Technology, Chemistry, Construction Industry, Energy and Related Areas. H.W. Wilson Co., 950 University Avenue, Bronx, NY 10452. (800) 367-6770 or (212) 588-8400. FAX (718) 590-1617. From 1958 to present. Monthly. Inquire about cost and availability. Also available on CD-ROM and online.

ASCE Combined Annual Combined Index. American Society of Civil Engineers, 345 East 47th Street, New York, NY 10017-2398. (212) 705-7520 or (800) 548-2723. Annual. $48.00.

ASCE Publications Information. 1966-present. American Society of Civil Engineers, 345 East 47th Street, New York, NY 10017-2398. (212) 705-7520 or (800) 548-2723. Bi-monthly. $160.00 per year for non-members.

Bibliography and Index of Geology. American Geological Institute, 4220 King Street, Alexandria, VA 22302. (703) 379-2480. Fax (703) 379-7563. Monthly. $1295.00 per year. Also available online as GeoRef.

Civil and Structural Engineering Abstracts, Cambridge Scientific Abstracts, 7200 Wisconsin Avenue, Bethesda, MD 20814-4823. (301) 961-6750. FAX (301) 961-6720. 1993 to present. Monthly. $385.00 per year. Topics covered include structural design, construction equipment and methods, civil defense and military engineering, surveying, highway engineering, maritime and port structures, materials, land reclamation, and soil mechanics.

Current Contents: Engineering, Technology, and Applied Sciences. Institute for Scientific Information, 3501 Market Street, Philadelphia, PA 19104. (215) 386-0100. FAX (215) 386-6362. 1970 to present. Weekly. $442.00 per year.

Engineering Index Monthly. Engineering Information, Inc., Castle Point on the Hudson, Hoboken, NJ 07030. (800) 221-1044. FAX (212) 832-1857. Monthly. $2300.00 per year. Also available online as COMPENDEX, and also on CD-ROM. Covers chemical engineering, computers, electrical engineering, civil engineering, metals and mining, industrial management, and mechanical engineering.

Geodex Retrieval System For Geotechnical Abstracts. Geodex Retrieval Systems, 669 Broadway, Box 279, Sonoma, CA 95476. (707) 939-8476. FAX (707) 996-8734. 1970 to present. Three times a year. $645.00 per year includes Geotechnical Abstracts, as Geodex Systems-GRS.

Geological Abstracts. Elsevier - Geo Abstracts, Regency House, 34 Duke Street, Norwich NR3 3AP, England. Monthly. $760.00 per year. Also available online as GEOBASE.

Geological Society of America. ABSTRACTS WITH PROGRAMS. Geological Society of America. 3300 Penrose Place, P.O. Box 9140, Boulder, CO 80301-9140. (303) 447-2020. Abstracts and programs of the annual conference. Annual. $69.00.

Geotechnical Abstracts. Geodex Retrieval Systems, 669 Broadway, Box 279, Sonoma, CA 95476. (707) 939-8476. FAX (707) 996-8734. 1970 to present. Three times a year. $280.00 per year. Covers soil mechanics, foundation engineering, rock mechanics, and engineering geology.

I.M.M. Abstracts and Index. Institution of Mining and Metallurgy, 44 Portland Pl., London W1N 4BR, England. 071-580-3802. FAX 071-436-5388. Bi-monthly. $364 for non-members.

International Journal of Rock Mechanics and Mining Sciences and Geomechanics Abstracts. Elsevier Science, 660 White Plains

Rd., Tarrytown, NY 10591-5153. (914) 524-9200. FAX (914) 333-2444. Six times a year. $845 per year.

KWIC Index of Rock Mechanics Literature: Part 2, 1969-1976. J.P. Jenkins & E.T. Brown, eds. Pergamon Reprint/Franklin Book Company, Inc., 7804 Montgomery Ave., Elkins, PA 19117. (215) 635-5252. FAX (215) 635-6155. 1979. $298.00.

Science Citation Index. Institute for Scientific Information, 3501 Market Street, Philadelphia, PA 19104. (215) 386-0100. FAX (215) 386-6362. Inquire about availability and cost. Also available on CD-ROM.

ASSOCIATIONS AND PROFESSIONAL SOCIETIES

American Geological Institute. 4220 King Street, Alexandria, VA 22302. (703) 379-2480. Fax (703) 379-7563.

American Society of Civil Engineers. 345 East 47th Street, New York, NY 10017-2398. (212) 705-7520 or (800) 548-2723.

Association of Engineering Geologists. 323 Boston Post Road, Suite 2D, Sudbury, MA 01776. (508) 443-4639.

Geological Society of America. 3300 Penrose Place, PO Box 9140, Denver, CO 80301-9140. (303) 447-2020.

International Society For Rock Mechanics. c/o Laboratorio Nacional de Engenharia Civil, Avenida do Brasil 101, Lisbon Codex, Portugal. Telephone 1-8482131. FAX 1-897660.

DIRECTORIES AND BIOGRAPHICAL SOURCES

American Men & Women of Science, 1995-96. R.R. Bowker Staff, eds. 19th edition. 8 volumes. R.R. Bowker/Reed International Publishing Company, 121 Chanlon Road, New Providence, NJ 07974. (908) 464-6800 or (800) 521-8110. 1995. $850.00

Directory of Engineering Societies and Related Organizations. Gordon Davis. 13th edition. American Association of Engineering Societies, 1111 19th Street NW, Suite 608, Washington, DC 20036. (202) 296-2237 or (800) 658-8897. 1989. Inquire for price.

International Research Centers Directory. Gale Research, 835 Penobscot Building, Detroit, MI 48226-4094. (313) 961-2242. (800) 877-4253. FAX (313) 961-6083. 8th edition. 1995. $410.00

Research Centers Directory. Gale Research, 835 Penobscot Building, Detroit, MI 48226-4094. (313) 961-2242. (800) 877-4253. FAX (313) 961-6083. $485.00.

Scientific and Technical Organizations and Agencies Directory. Gale Research, 835 Penobscot Building, Detroit, MI 48226-4094. (313) 961-2242. (800) 877-4253. FAX (313) 961-6083. 4th edition. 1996. $195.00.

Who's Who in Engineering. American Association of Engineering Societies, 1111 19th Street NW, Suite 608, Washington, DC 20036. (202) 296-2237 or (800) 658-8897. 8th edition. 1991. Inquire for price.

ENCYCLOPEDIAS AND DICTIONARIES

The Encyclopedia of the Solid Earth Sciences. Philip Kearey, editor-in-chief. Blackwell Scientific Publications, 238 Main Street, Cambridge, MA 02142. (617) 876-7000 or (800) 759-6102. FAX (617) 876-7022. 1993. $60.00.

Encyclopedia of Physical Science and Technology. Robert A. Meyers, ed. 18 volumes. Academic Press Inc., 6277 Sea Harbor Drive, Orlando, FL 32887. (800) 321-5068. 1992. $2100.00.

Glossary of Rock Engineering. Harith S. Faruqi. French & European Publications, Inc., Rockefeller Center Promenade, 610 Fifth Ave., New York, NY 10020-2479. (212) 581-8810. FAX (212) 265-1094. 1980. $75.00.

Thesaurus of Rock & Soil Mechanics Terms. J.P. Jenkins & A.M. Smith, eds. Franklin Book Company, Inc., 7804 Montgomery Ave., Elkins, PA 19117. (215) 635-5252. FAX (215) 635-6155. 1984. $37.00.

GENERAL WORKS

Introduction To Rock Mechanics. Richard E. Goodman. 2d ed. John Wiley and Sons, Inc., 605 Third Avenue, New York, NY 10158. (800) 526-5368 or (212) 850-6000. 1989. Inquire for price and availability.

An Outline of Soil & Rock Mechanics. Pierre Habib. Cambridge University Press, 40 West 20th Street, New York, NY 10011-4211. (212) 924-3900 or (800) 872-7423. FAX (914) 937-4712. 1983. $49.95/$18.95 (paperback).

Rock & Soil Mechanics. Derski. Elsevier Science Publishing Company, Inc., 655 Avenue of the Americas, New York, NY 10010. (212) 989-5800. 1989. $174.50.

Rock Engineering Systems. J.A. Hudson. E. Horwood Publishing, distributed by Prentice Hall, 113 Sylvan Avenue, Route 9W, Englewood Cliffs, NJ 07632. (201) 592-2000 or (800) 922-0579. 1992. $53.95.

Rock Mechanics. Walter Wittke. Springer-Verlag, 175 Fifth Avenue, New York, NY 10010. (212) 460-1500 or (800) 777-4643. FAX (212) 473-6272. 1990. Inquire for price.

Rock Mechanics: For Underground Mining. B.H.G. Brady and E. T. Brown. 2nd ed. Routledge, Chapman and Hall, Inc., 29 West 35th Street, New York, NY 10001-2291. (212) 244-3336 or FAX (212) 563-2269. 1993. Inquire for price.

HANDBOOKS AND MANUALS

Handbook On Mechanical Properties of Rocks, Vol. II. V.S. Vutukuri, et al. LPS Distribution Center, 52 LaBombard Rd., Lebanon, NH 03766. (603) 448-0037. FAX (603) 448-2576. 1977. $65.00.

Practical Handbook For Underground Rock Mechanics. T.R. Stacey & C.H. Page. LPS Distribution Center, 52 LaBombard Rd., Lebanon, NH 03766. (603) 448-0037. FAX (603) 448-2576. 1986. $38.00.

ONLINE DATABASES AND CD-ROMS

Compendex Plus. Engineering Information, Inc., 345 East 47th Street, New York, NY 10017. (212) 705-7600 or (800) 221-1044. Contains citations with abstracts to worldwide literature in engineering and technology, from 1970 to present. Available on online BRS,(800) 289-4277, DIALOG, (800) 334-2564, ORBIT (800) 456-7248, and STN International, FIZ Karlsruhe, P.O. Box 2465, W-7500, Karlsruhe 1, Germany, online services. Also available on CD-ROM. Inquire as to cost and availability.

CSA Engineering. Cambridge Scientific Abstracts, 7200 Wisconsin Avenue, Suite 601, Bethesda, MD 20814. (301) 961-6750 or (800) 843-7751. Contains citations and abstracts of international periodicals and research literature covering all fields of engineering and science and technology,including computer and information science, electronics, mechanical engineering, solid state materials, 1981 to present. Available on BRS,(800) 289-4277, online service.Inquire as to cost and availability.

Current Contents Search. Institute for Scientific Information, 3501 Market Street, Philadelphia, PA 19104. (215) 386-0100. FAX (215) 386-6362. Contains citations to articles listed in the table of contents of science and technology journals. Also articles in social sciences and life sciences journals. Available on BRS,(800) 289-4277, DIALOG,(800) 334-2564, online services. Inquire as to cost and availability.

Dissertation Abstracts. University Microfilms International, 300 North Zeeb Road, Ann Arbor, MI 48106. (800) 521-0600 or (313) 761-4700. Scope includes virtually all doctoral dissertations accepted at accredited American institutions from 1861 to present in 252 subject areas. Available on BRS,(800) 289-4277, DIALOG,(800) 334-2564, and OCLC EPIC,(800) 848-5878, online services. Also available on CD-ROM. Inquire as to cost and availability.

Earth Sciences. U.S. Geological Survey, 12201 Sunrise Valley Drive, Reston, VA 22092-9998. (703) 648-4460. CD-ROM of earth science databases including the U.S. Geological Survey Library, Earth Science Data Directory, and GEOINDEX, citations to published geological maps. $350.00 per year with quarterly updates.

Geoarchive. Geosystems, P.O. Box 1024, Westminster, London, England, SW1 P 2JL. Citations to literature on geoscience, 1969 to present. Inquire as to online cost and availability.

Geobase. Elsevier - Geo Abstracts, Regency House, 34 Duke Street, Norwich NR3 3AP, England. Contains citations to the worldwide earth science literature from 1980 to date. Available on DIALOG, ORBIT online services. Inquire as to cost and availability.

Geomechanics Abstracts. University of London, Imperial College of Science and Technology, Rock Mechanics Information Service, Royal School of Mines, Prince Consort Road, London SW7 2BP, England. Telephone: 071-589 5111 or FAX 071-589 6806. Scope includes worldwide literature on engineering geology, hydrogeology, mining, rock mechanics, soil mechanics, and tunnelling, 1977 to present. Available on the ORBIT (800) 456-7248, online service. Inquire as to cost and availability.

Georef: Bibliography and Index of Geology. American Geological Institute, 4220 King Street, Alexandria, VA 22302. (703) 379-2480. Fax (703) 379-7563. Monthly. Inquire for price and availability.

NTIS Bibliographic Database. National Technical Information Service, 5285 Port Royal Road, Springfield, VA 22161. (703) 487-4929 or FAX (703) 321-8199. Broad coverage of government-sponsored science and technology research reports, 1964 to present. Available on BRS,(800) 289-4277, DIALOG, (800) 334-2564, ORBIT, (800) 456-7248, and STN International, FIZ Karlsruhe, P.O. Box 2465, W-7500, Karlsruhe 1, Germany, online services. Also available on CD-ROM. Inquire as to cost and availability.

SCISEARCH. Institute for Scientific Information, 3501 Market Street, Philadelphia, PA 19104. (800) 523-1850 or (215) 386-0100. Broad multidisciplinary title and author index to the international literature of science and technology, 1974 to present. Available on DIALOG,(800) 334-2564, and ORBIT,(800) 456-7248, online services. Also available on CD-ROM. Inquire as to cost and availability.

WILSONLINE. H.W. Wilson Company, 950 University Avenue, Bronx, NY 10452. (800) 367-6770 or (212) 588-8400. Makes available online versions of the printed H.W. Wilson indexes including Applied Science and Technology Index, Business Periodicals Index, General Science Index, and Readers' Guide to Periodical Literature. Period covered is generally 1983 to present. Available on BRS,(800) 289-4277, DIALOG,(800) 334-2564, and OCLC EPIC,(800) 848-5878, online services. Also available on CD-ROM. Inquire as to cost and availability.

PERIODICALS

Construction Products. Gordon Publications/ Cahners-Reed Elsevier, 301 Gilbraltar Drive, Box 650, Morris Plains, NJ 07950. (201) 252-5100. FAX (201) 539-3476. 1892 to present. 6 times a year. $40.00 per year.

Engineering Geology. Elsevier Science Inc., Box 882, Madison Square Station, New York, NY 10159. (212) 989-5800. 1965 to present. 12 times a year. $650.00 per year.

Journal of Construction Engineering and Management. American Society of Civil Engineers, Construction Division, 345 East 47th Street, New York, NY 10017-2398. (212) 705-7520 or (800) 548-2723. 1956 to present. Quarterly. $112.00 per year.

Journal of Geotechnical Engineering. American Society of Civil Engineers, Geotechnical Engineering Division, 345 East 47th Street, New York, NY 10017-2398. (212) 705-7520 or (800) 548-2723. 1956 to present. Monthly. $212.00 per year.

Journal of Irrigation and Drainage. American Society of Civil Engineers, Irrigation and Drainage Division, 345 East 47th Street, New York, NY 10017-2398. (212) 705-7520 or (800) 548-2723. 1956 to present. Bi-monthly. $136.00 per year.

Rock Mechanics and Rock Engineering. Springer Verlag/Journal Fulfillment Services, Box 2485, Secaucus, NJ 07096-2491. (800) 777-4643. FAX (201) 348-4505. 1929 to present. Four times a year. $187.00 per year.

Rock Mechanics Supplement. Springer-Verlag, 175 Fifth Avenue, New York, NY 10010. (212) 460-1500. FAX (212) 473-6272. 1970 to present. Irregular. Inquire for price.

RESEARCH CENTERS AND INSTITUTES

Center For Concrete and Geomaterials. Northwestern University, Technological Institute, Room 2474, 2145 Sheridan Road, Evanston, IL 60208. (708) 491-3858.

Geotechnical Engineering Center. University of Texas at Austin, Department of Civil Engineering, ECJ 9.227, Austin, TX 78712. (512) 471-4929.

Rock Mechanics and Ground Control Laboratory. University of West Virginia, Department of Mining Engineering, PO Box 6070, Morgantown, WV 26506. (304) 293-7680. FAX (304) 293-5708.

Underground Space Center. University of Minnesota, 790 Civil & Mineral Engineering Bldg., 500 Pillsbury Drive SE, Minneapolis, MN 55455. (612) 624-0066. FAX (612) 624-0293.

ROCKETS

See also: AEROSPACE ENGINEERING, GUIDED MISSILES, PROPELLANTS, PROPULSION SYSTEMS, SPACECRAFT

ABSTRACT SERVICES AND INDEXES

Applied Science and Technology Index; A Cumulative Subject Index To English Language Periodicals in the Fields of Aeronautics and Space Science, Computer Technology, Chemistry, Construction Industry, Energy and Related Areas. H.W. Wilson Co., 950 University Avenue, Bronx, NY 10452. (800) 367-6770 or (212) 588-8400. FAX (718) 590-1617. From 1958 to present. Monthly. Inquire about cost and availability. Also available on CD-ROM.

Current Contents: Engineering, Technology and Applied Sciences. Institute for Scientific Information, 3501 Market Street, Philadelphia, PA 19104. (215) 386-0100. FAX (215) 386-6362. Weekly. $360.00 per year.

Engineering Index Monthly. Engineering Information, Inc., Castle Point on the Hudson, Hoboken, NJ 07030. (800) 221-1044. FAX (212) 832-1857. Monthly. $2200.00 per year. Also available online as COMPENDEX, and also on CD-ROM.

Index To Scientific and Technical Proceedings. Institute for Scientific Information, 3501 Market St., Philadelphia, PA 19104. (215) 386-0100. FAX (215) 386-6362. Monthly. $500.00 per year.

Index To Scientific Reviews. Institute for Scientific Information, 3501 Market St., Philadelphia, PA 19104. (215) 386-0100. FAX (215) 386-6362. Semi-annual.

International Aerospace Abstracts. Technical Information Service, American Institute of Aeronautics and Astronautics, Inc., 555 West 57th St., New York, NY 10019. (212) 247-6500. FAX (212) 582-4861. Semi-monthly. $1295.00 per year.

NASA Patent Abstracts Bibliography. National Aeronautics and Space Administration, Center for Aerospace Information, Box 8757, BWI Airport, Baltimore, MD 21240. (301) 621-0153. Semi-annual. Inquire for price information.

Numerical/chronological/author Index. An index to articles and papers published by or for the American Astronautical Society, including: Journal of the Astronautical Sciences, Astronautical Sciences Review, Advances in the Astronautical Sciences, Science and Technology, AAS history series, and AAS microfiche series. Horace Jacobs and Robert H. Jacobs. Univelt, Inc. PO Box 28130, San Diego, CA 92198. (619) 746-4005. FAX (619) 746-3139. Inquire for price and availability.

Science Citation Index. Institute for Scientific Information, 3501 Market Street, Philadelphia, PA 19104. (215) 386-0100. FAX (215) 386-6362. Inquire about availability and cost. Also available on CD-ROM.

Scientific and Technical Aerospace Reports (star). National Aeronautics and Space Administration. NASA Center for Aerospace Information, Box 8757, BWI Airport, Baltimore, MD 21240. (301) 621-0153. Monthly. Inquire about availability and cost. Also available through the NASA online retrieval service (RECON), and through the Aerospace Database through DIALOG.

ASSOCIATIONS AND PROFESSIONAL SOCIETIES

Aerospace Industries Association of America. 1250 Eye Street NW, Washington, DC 20005. (202) 371-8400.

American Astronautical Society. 6352 Rolling Mill Place, Suite 102, Springfield, VA 22152.

American Institute of Aeronautics and Astronautics. 370 L'Enfant Promenade SW, Washington, DC 20024. (202) 646-7400.

British Interplanetary Society, 27/29 South Lambeth Road, London Sw8 1sz, England. (071) 735-3160. FAX (071) 820-1504.

Canadian Aeronautics and Space Institute, 222 Somerset Street West, Suite 601, Ottawa, Ontario, Canada K2p 0j1.

National Space Society, 922 Pennsylvania Avenue Se, Washington, DC 20003-2140. (202) 543-1900. FAX (202) 546-4189.

DIRECTORIES AND BIOGRAPHICAL SOURCES

American Astronautical Society Directory. American Astronautical Society, 6352 Rolling Mill Place, Suite 102, Springfield, VA 22152.

American Institute of Aeronautics and Astronautics Roster. American Institute of Aeronautics and Astronautics. 370 L'Enfant Promenade SW, Washington, DC 20024. (202) 646-7400.

Jane's Space Directory. Jane's Information Group, 1340 Braddock Place, Suite 300, Alexandria, VA 22314-1651. (703) 683-3700. FAX (703) 836-0029. Annual. $245.00. Also available on CD-ROM.

Research Centers Directory. Gale Research, 835 Penobscot Building, Detroit, MI 48226-4094. (313) 961-2242. (800) 877-4253. FAX (313) 961-6083. $485.00.

Scientific and Technical Organizations and Agencies Directory. Gale Research, 835 Penobscot Building, Detroit, MI 48226-

ROCKETS

Ency. of Physical Sciences and Engineering Info. Sources

4094. (313) 961-2242. (800) 877-4253. FAX (313) 961-6083. 4th ed. 1995. $195.00.

Who's Who in Engineering. 7th ed. American Association of Engineering Societies, 1111 19th St. NW, Suite 608, Washington, DC 20036. (800)-658-8897. 1988. $200.00.

GENERAL WORKS

Astronautics and Aeronautics, 1979-1984: A Chronology. Bette R. Janson and Eleanor H. Ritchie. National Aeronautics and Space Administration, Scientific and Technical Information Division. 1990. Order from: Supt. of Documents, Government Printing office, Box 371954, Pittsburgh, PA 15250-7954. (202) 783-3238. FAX (202) 512-2233. Inquire for price and availability.

Doodlebugs and Rockets: The Battle of the Flying Bombs. Bob Ogley. Froglets Publishing, Kent, England. 1992. Inquire for cost and availability.

Europe in Space. Guy Collins. St. Martin's Press, Inc., 175 Fifth Avenue, New York, NY 10010. (212) 674-5151 or (800) 221-7945. FAX (212) 420-9314. 1991. $59.95.

Introduction To Space Flight. Francis J. Hale. Prentice Hall, 113 Sylvan Avenue, Route 9W, Englewood Cliffs, NJ 07632. (201) 592-2000 or (800) 922-0579. 1994. $39.50 (hardback), $32.50 (paperback).

Prospects For Interstellar Travel. John H. Maudlin. Univelt Inc., PO Box 28130, San Diego, CA 92198. (619) 746-4005. FAX (619) 746-3139. 1992. $27.00.

Space Mission Analysis and Design. Wiley J. Larson and James R. Wertz, eds. 2d ed. Kluwer Academic Publishers, P.O. Box 358, Accord Station, Hingham, MA 02018-0358. (617) 871-6000. FAX (617) 871-6528. 1992. $99.00.

ONLINE DATABASES AND CD-ROMS

Aerospace Database. American Institute of Aeronautics and Astronautics, 370 L'Enfant Promenade, S.W., Washington, DC 20024. (202) 646-7400. Worldwide published literature on research and development in aerospace and related areas, with abstracts. Covers 1962 to present. Available on DIALOG, (800) 334-2564, online service. Also available on CD-ROM. Inquire as to cost and availability.

Compendex Plus. Engineering Information, Inc., 345 East 47th Street, New York, NY 10017. (212) 705-7600 or (800) 221-1044. Contains citations with abstracts to worldwide literature in engineering and technology, from 1970 to present. Available on online BRS,(800) 289-4277, DIALOG, (800) 334-2564, ORBIT (800) 456-7248, and STN International, FIZ Karlsruhe, P.O. Box 2465, W-7500, Karlsruhe 1, Germany, online services. Also available on CD-ROM. Inquire as to cost and availability.

CSA Engineering. Cambridge Scientific Abstracts, 7200 Wisconsin Avenue, Suite 601, Bethesda, MD 20814. (301) 961-6750 or (800) 843-7751. Contains citations and abstracts of international periodicals and research literature covering all fields of engineering and science and technology,including computer and information science, electronics, mechanical engineering, solid state materials, 1981 to present. Available on BRS,(800) 289-4277, online service.Inquire as to cost and availability.

Current Contents Search. Institute for Scientific Information, 3501 Market Street, Philadelphia, PA 19104. (215) 386-0100. FAX (215) 386-6362. Contains citations to articles listed in the table of contents of science and technology journals. Also articles in social sciences and life sciences journals. Available on BRS,(800) 289-4277, DIALOG,(800) 334-2564, online services. Inquire as to cost and availability.

Dissertation Abstracts. University Microfilms International, 300 North Zeeb Road, Ann Arbor, MI 48106. (800) 521-0600 or (313) 761-4700. Scope includes virtually all doctoral dissertations accepted at accredited American institutions from 1861 to present in 252 subject areas. Available on BRS,(800) 289-4277, DIALOG,(800) 334-2564, and OCLC EPIC,(800) 848-5878, online services. Also available on CD-ROM. Inquire as to cost and availability.

International Aerospace Abstracts. American Institute of Aeronautics and Astronautics, 370 L'Enfant Promenade, S.W., Washington, DC 20024. (202) 646-7400. Contains references and abstracts of journal and monograph literature relating to aerospace science and technology, from 1963 to present. Available through the NASA/RECON system of the National Aeronautics and Space Administration only.

NASA Database. American Institute of Aeronautics and Astronautics, 370 L'Enfant Promenade, S.W., Washington, DC 20024. (202) 646-7400. Citations and abstracts of aeronautics and astronautics literature, 1962 to present. Also contains citations from *STAR, Scientific and Technical Aerospace Reports*, and *International Aerospace Abstracts*. Available through NASA/RECON online service. Inquire as to cost and availability.

NTIS Bibliographic Database. National Technical Information Service, 5285 Port Royal Road, Springfield, VA 22161. (703) 487-4929 or FAX (703) 321-8199. Broad coverage of government-sponsored science and technology research reports, 1964 to present. Available on BRS,(800) 289-4277, DIALOG, (800) 334-2564, ORBIT, (800) 456-7248, and STN International, FIZ Karlsruhe, P.O. Box 2465, W-7500, Karlsruhe 1, Germany, online services. Also available on CD-ROM. Inquire as to cost and availability.

SCISEARCH. Institute for Scientific Information, 3501 Market Street, Philadelphia, PA 19104. (800) 523-1850 or (215) 386-0100. Broad multidisciplinary title and author index to the international literature of science and technology, 1974 to present. Available on DIALOG,(800) 334-2564, and ORBIT,(800) 456-7248, online services. Also available on CD-ROM. Inquire as to cost and availability.

WILSONLINE. H.W. Wilson Company, 950 University Avenue, Bronx, NY 10452. (800) 367-6770 or (212) 588-8400. Makes available online versions of the printed H.W. Wilson indexes including Applied Science and Technology Index, Business Periodicals Index, General Science Index, and Readers' Guide to Periodical Literature. Period covered is generally 1983 to present. Available on BRS,(800) 289-4277, DIALOG,(800) 334-2564, and OCLC EPIC,(800) 848-5878, online services. Also available on CD-ROM. Inquire as to cost and availability.

PERIODICALS

Acta Astronautica. Pergamon Press, Pergamon Press, 660 White Plains Rd., Tarrytown, NY 10591-5153. 14 times a year. Inquire for price and availability.

Ad Astra. National Space Society, 922 Pennsylvania Avenue SE, Washington, DC 20003-2140. (202) 543-1900. FAX (202) 546-4189. 1989 to present. 6 times a year. $35.00 a year.

Advances in Astronautical Sciences. American Astronautical Society, 6352 Rolling Mill Place, Suite 102, Springfield, VA 22152. (703) 866-0020. FAX (703) 866-3526. Quarterly. Write or call for price information.

Advances in Space Research. Pergamon Press, Pergamon Press, 660 White Plains Rd., Tarrytown, NY 10591-5153. 12 times a year. Inquire for price and availability.

Aeronautical Journal. Royal Aeronautical Society, PO Box 139, Tonbridge, Kent, TN9 1EW, England. 0732-770823. FAX 0723-361708. 1897 to present. Ten times a year. Write for price and availability.

Aerospace. Royal Aeronautical Society, PO Box 139, Tonbridge, Kent, TN9 1EW, England. 0732-770823. FAX 0723-361708. Monthly. Write for price and availability.

Aerospace America. American Institute of Aeronautics and Astronautics, 370 L'Enfant Promenade SW, Washington, DC 20024-2518. (202) 646-7471. 1932 to present. Monthly. $70 per year (non-members).

Aerospace Engineering. SOCIETY of AUTOMOTIVE ENGINEERS. SAE, Inc., 400 Commonwealth Dr., Warrendale, PA 15096. (412) 776-4841. 1981 to present. Monthly. $48 a year.

Aviation & Aerospace. Baxter Publishing Company, 310 Dupont Street, Toronto, Ontario M5R 1V9 Canada. (416) 968-7252. FAX (416) 968-2377. Bi-monthly. $42.00 per year.

Aviation Week and Space Technology. 1221 Avenue of the Americas, New York, NY 10020. (212) 512-2999 or (800) 525-5003. 1916 to present. Weekly. $82.00 per year.

ESA Bulletin. European Space Agency, Publications Division, Keplerlaan 1, 22000 AG Nordwijk, Netherlands. Phone 31-1719-86555. FAX 31-1719-85433. Quarterly. Free.

ESA Journal. European Space Agency, Publications Division, Keplerlaan 1, 22000 AG Nordwijk, Netherlands. Phone 31-1719-86555. FAX 31-1719-85433. Quarterly. Free.

Interavia/Aerospace World. Aerospace Media Publishing, Swissair Centre, 31 Route de l'Aeroport, POB Box 437, 1215 Geneva 15, Switzerland. (022) 788 27 88. FAX (022) 788 27 26. Monthly. $128.00 per year (U.S.).

Journal of the Astronautical Sciences. American Astronautical Society, 6352 Rolling Mill Place, Suite 102, Springfield, VA 22152. Quarterly. $35 per year.

Journal of the British Interplanetary Society (JBIS). British Interplanetary Society, 27/29 South Lambeth Road, London SW8 1SZ, England. (071) 735-3160. FAX (071) 820-1504. Monthly. $370.00 per year.

Journal of Guidance, Control and Dynamics. American Institute of Aeronautics and Astronautics, 370 L'Enfant Promenade SW, Washington, DC 20024-2518. Bimonthly. $42.00 (members) and $250.00 (non-members).

Journal of Propulsion and Power. American Institute of Aeronautics and Astronautics, 370 L'Enfant Promenade SW, Washington, DC 20024-2518. Bimonthly. $38.00 (members) and $260.00 (non-members).

Journal of Spacecraft and Rockets. American Institute of Aeronautics and Astronautics, the Aerospace Center, 370 L'Enfant Promenade SW, Washington, DC 20024. (202) 646-7400. FAX (202) 646-7508. 1964 to present. Bi-monthly. $185.00 per year.

Progress in Aerospace Sciences. Pergamon Press, 660 White Plains Rd., Tarrytown, NY 10591-5153. 1961 to present. Quarterly. $320.00 per year.

Propellants, Explosives, Pyrotechnics. VCH Publishers, Inc., 220 E. 23rd St., New York, NY 10010-4606. 6 times a year. $415.00 per year.

Space Power, Pergamon Press, 660 White Plains Rd., Tarrytown, NY 10591-5153.

Space Science Reviews. Kluwer Academic Publishers, P.O. Box 358, Accord Station, Hingham, MA 02018-0358. (617) 871-6000. Monthly except March, June, September, December. $258.50 per volume.

Space Technology: Industrial and Commercial Applications. Pergamon Press, 660 White Plains Rd., Tarrytown, NY 10591-5153. Bimonthly. $535.00 per year.

Spaceflight. British Interplanetary Society, 27/29 South Lambeth Road, London SW8 1SZ, England. (071) 735-3160. FAX (071) 820-1504. Monthly. $69 annual membership includes subscription.

RESEARCH CENTERS AND INSTITUTES

Aerospace Research Applications Center, Indianapolis Center For Advanced Research, 611 N. Capitol Ave., Indianapolis, in 46204. (317) 262-5003. FAX (317) 262-5044.

Aerospace Research Center, 1250 Eye St., NW, Washington, DC 20005. (202) 371-8400. FAX (202) 371-8470.

Auburn University Space Power Institute, 231 Leach Center, Auburn, AL 36849. (205) 844-5894.

Canadian Space Agency, Space Science Program, C/o National Research Council of Canada, 100 Sussex Drive, Ottawa, Ontario, Canada K1a 0r6. (613) 990-0799. FAX (613) 952-0974.

Center For Aerospace Technology, Weber State College, Ogden, UT 84408-1805. (801) 626-7272.

Center For Space Research, Massachusetts Institute of Technology, 77 Massachusetts Ave., Cambridge, MA 02139. (617) 253-7501.

Center For Control and Systems Research, University of Texas At Austin, Aerospace Engineering and Engineering Mechanics, Wrw 408, Austin, TX 78712-1085. (512) 471-4908. FAX (512) 471-3788.

Charles Stark Draper Laboratory Inc., 555 Technology Square, Cambridge, MA 02139. (617) 258-1000.

Goddard Space Flight Center, Greenbelt, MD 20771. (301) 286-7351.

Houston Advanced Research Center, Space Technology and Research Center, 4802 Research Forest Dr., the Woodlands, TX 77381. (713) 363-7922. FAX (713) 363-7914.

Innovative Nuclear Space Power and Propulsion Institute, College of Engineering, University of Florida, Gainesville, FL 32611. (904) 392-1427. FAX (904) 392-8656.

Instrumentation and Control Laboratory, Princeton University, Dept. of MAE, Engineering Quadrangle, Princeton, NJ 08544. (609) 452-5154.

Man-vehicle Laboratory, Massachusetts Institute of Technology, Rm. 37-219, Cambridge, MA 02139. (617) 253-7805.

George C. Marshall Space Flight Center, Huntsville, AL 35812. (205) 544-0024.

Ohio Aerospace Institute, 2001 Aerospace Parkway, Brook Park, OH 44142. (216) 891-2100. FAX (216) 891-2140.

Pennsylvania State University, Propulsion Engineering Research Center, 106 Research Bldg. E, Bigler Rd., University Park, PA 16802. (814) 863-6272. FAX (814) 865-3389.

University of Tennessee, Center For Advanced Space Propulsion, Po Box 1385, Tullahoma, TN 37388-8897. (615) 454-9294. FAX (615) 455-6167.

Space Science and Engineering Center, University of Wisconsin—Madison, 1225 W. Dayton St., Madison, WI 53706. (608) 262-0544.

Thermal Sciences and Propulsion Center, Purdue University, Lafayette, IN 47907. (317) 494-1503.

ROCKET ENGINES

See: ROCKETS

S

SAFETY ENGINEERING

See also: ERGONOMICS, FIRE PROTECTION, INDUSTRIAL ENGINEERING, NOISE CONTROL

ABSTRACT SERVICES AND INDEXES

American Industrial Hygiene Association Conference Abstracts. American Industrial Hygiene Association. 2700 Prosperity Ave., Suite 250, Fairfax, VA 22031-4307. (703) 849-8888. Annual. $20.00.

Applied Mechanics Reviews: An Assessment of World Literature in Engineering Sciences. American Society of Mechanical Engineers, 345 East 47th Street, New York, NY 10017. (212) 705-7703. 1948 to present. Monthly. $360.00 per year.

Applied Science and Technology Index; A Cumulative Subject Index To English Language Periodicals in the Fields of Aeronautics and Space Science, Computer Technology, Chemistry, Construction Industry, Energy and Related Areas. H.W. Wilson Co., 950 University Avenue, Bronx, NY 10452. (800) 367-6770 or (212) 588-8400. FAX (718) 590-1617. From 1958 to present. Monthly. Inquire about cost and availability. Also available on CD-ROM.

CA Selects—Chemical Hazards, Health & Safety. Chemical Abstracts Service, 2540 Olentangy River Road, Box 3012, Columbus, OH 43210. (800) 848-6538 or (614) 447-3600, FAX (614) 447-3713. Semi-weekly. $210.00 per year.

CA Selects—Occupational Exposure & Hazards. Chemical Abstracts Service, 2540 Olentangy River Road, Box 3012, Columbus, OH 43210. (800) 848-6538 or (614) 447-3600, FAX (614) 447-3713. Semi-weekly. $210.00 per year.

Chemical Abstracts - Applied Chemistry and Chemical Engineering Sections. Chemical Abstracts Service, 2540 Olentangy River Road, Box 3012, Columbus, OH 43210. (800) 848-6538 or (614) 447-3600, FAX (614) 447-3713. Bi-weekly. $1410.00 per year.

Current Contents: Physical, Chemical and Earth Sciences. Institute for Scientific Information, 3501 Market Street, Philadelphia, PA 19104. (215) 386-0100 or (800) 523-1850. FAX (215) 386-6362. Weekly. $360.00 per year.

Engineering Index Monthly. Engineering Information, Inc., Castle Point on the Hudson, Hoboken, NJ 07030. (800) 221-1044. FAX (212) 832-1857. Monthly. $2200.00 per year. Also available online as COMPENDEX, and also on CD-ROM.

Health and Safety Science Abstracts. Cambridge Scientific Abstracts, 7200 Wisconsin Avenue, Bethesda, MD 20814-4823. (301) 961-6750. FAX (301) 961-6720. Five issues yearly. Inquire for cost.

Industrial Hygiene Digest. Industrial Health Foundation Inc. 34 Penn Circle W, Pittsburgh, PA 15206. (412) 363-6600. Monthly. $150.00 per year.# #*Science Citation Index.* Institute for Scientific Information, 3501 Market Street, Philadelphia, PA 19104. (215) 386-0100 or (800) 523-1850. FAX (215) 386-6362. Inquire about availability and cost. Also available on CD-ROM.

ASSOCIATIONS AND PROFESSIONAL SOCIETIES

American Industrial Hygiene Association. 2700 Prosperity Ave., Suite 250, Fairfax, VA 22031-4307. (703) 849-8888.

American Society of Safety Engineers. 1800 E Oakton St., Des Plaines IL 60018-2187. (708) 692-4121. FAX (708) 296-3769.

Board of Certified Safety Professionals. 208 Burwash Avenue, Savoy, IL 61874. (217) 359-9263.

Industrial Accident Prevention Association. 250 Yonge St., 28th Fl., Toronto ON M5B 2N4 Canada. (416) 506-8888. FAX (416) 506-8880.

Industrial Health Foundation Inc. 34 Penn Circle W, Pittsburgh, PA 15206. (412) 363-6600.

National Safety Council. 1121 Spring Lake Dr., Itasca IL 60143. (708) 285-1121.

DIRECTORIES AND BIOGRAPHICAL SOURCES

American Men & Women of Science, 1995-96. R.R. Bowker Staff, eds. 19th edition. 8 volumes. R.R. Bowker/Reed International Publishing Company, 121 Chanlon Road, New Providence, NJ 07974. (908) 464-6800 or (800) 521-8110. FAX (908) 464-3553. 1995. $850.00

SAFETY ENGINEERING

Ency. of Physical Sciences and Engineering Info. Sources

Best's Safety Directory. A.M. Best Co., Ambest Rd., Oldwick, NJ 08858. (908) 439-2200. FAX (908) 439-3296. Annual. $45.00.

CSP Directory. Board of Certified Safety Professionals, 208 Burwash, Savoy, IL 61874. (217) 359-9263. Biennial. $30.00.

Directory of Engineering Societies and Related Organizations. Gordon Davis. 13th edition. American Association of Engineering Societies, 1111 19th Street NW, Suite 608, Washington, DC 20036. (202) 296-2237 or (800) 658-8897. 1989. Inquire for price.

International Directory of Occupational Safety and Health Institutions. International Labour office, 1828 L Street NW, Suite 801, Washington, DC 20036. (202) 653-7652. FAX (202) 376-5083. 1990. $28.00.

International Research Centers Directory. Gale Research, 835 Penobscot Building, Detroit, MI 48226-4094. (313) 961-2242 or (800) 877-4253. FAX (313) 961-6083. 8th edition. 1995. $410.00

Research Centers Directory. Gale Research, 835 Penobscot Building, Detroit, MI 48226-4094. (313) 961-2242. (800) 347-4253. FAX (313) 961-6083. $485.00

Scientific and Technical Organizations and Agencies Directory. Gale Research, 835 Penobscot Building, Detroit, MI 48226-4094. (313) 961-2242. (800) 877-4253. FAX (313) 961-6083. 4th edition. 1996. $195.00.

Who's Who in Engineering. American Association of Engineering Societies, 1111 19th Street NW, Suite 608, Washington, DC 20036. (202) 296-2237 or (800) 658-8897. 8th edition. 1991. Inquire for price.

ENCYCLOPEDIAS AND DICTIONARIES

McGraw-Hill Dictionary of Scientific and Technical Terms. Sybil P. Parker, ed. 5th edition. McGraw-Hill Publishing Company, 1221 Avenue of the Americas, New York, NY 10020. (800) 262-4729 or (212) 512-3825. 1993. $110.50.

Thesaurus of Scientific, Technical, and Engineering Terms. Hemisphere Publishing Corporation, 1900 Frost Road, Suite 101, Bristol, PA 19007-1598. (215) 785-5800 or (800) 821-8312. FAX (215) 785-5515. 1987. $173.00.

GENERAL WORKS

Engineering Design For Safety. Thomas A. Hunter. McGraw-Hill Publishing Company, 1221 Avenue of the Americas, New York, NY 10020. (800) 262-4729 or (212) 512-3825. 1992. $57.00.

Industrial Safety and Health in the Age of High Technology: For Technologists, Engineers, and Managers. David L. Goetsch. Maxwell Macmillan International, Macmillan Publishing, 866 3rd Avenue, New York, NY 10024. (212) 689-9140. 1993. Inquire for cost and availability.

Occupational and Environmental Safety Engineering and Management. H.R. Kavianian, C.A. Wentz. Van Nostrand Reinhold, 115 Fifth Avenue, New York, NY 10003. (212) 254-3232. FAX (212) 254-9499. 1990. $62.95.

Safety Engineering. James CoVan. John Wiley and Sons, Inc., 605 Third Avenue, New York, NY 10158. (800) 526-5368 or (212) 850-6000. 1995. $64.95.

Safety Engineering. Gilbert Marshall. American Society of Safety Engineers, 1800 E Oakton St., Des Plaines IL 60018-2187. (708) 692-4121. FAX (708) 296-9221. 1994. $29.95.@ @Handbooks and Manuals

Accident Prevention Manual For Business and Industry. 2 volumes. 10th edition. National Safety Council, 1121 Spring Lake Dr., Itasca IL 60143. (708) 285-1121. 1992. $99.95 per volume.

Handbook of Environmental Health and Safety: Principles and Practices. Herman Koren. 2d ed. Lewis Publishers, 121 S. Main St., PO Box 519, Chelsea, MI 48118. (313) 475-8619 or (800) 272-7737. 1991. $69.95.

OSHA Compliance and Management Handbook. Charleston C.K. Wang. Noyes Publications, 120 Mill Road, Park Ridge, NJ 07656. (201) 391-8484. FAX (201) 391-6833. 1993. $72.00.

Safety Manager's Guide 1993-94. Bureau of Business Practice/ Prentice Hall, 113 Sylvan Avenue, Route 9W, Englewood Cliffs, NJ 07632. (201) 592-2000 or (800) 922-0579. $59.95.

ONLINE DATABASES AND CD-ROMS

Compendex Plus. Engineering Information, Inc., 345 East 47th Street, New York, NY 10017. (212) 705-7600 or (800) 221-1044. Contains citations with abstracts to worldwide literature in engineering and technology, from 1970 to present. Available on online BRS,(800) 289-4277, DIALOG, (800) 334-2564, ORBIT (800) 456-7248, and STN International, FIZ Karlsruhe, P.O. Box 2465, W-7500, Karlsruhe 1, Germany, online services. Also available on CD-ROM. Inquire as to cost and availability.

CSA Engineering. Cambridge Scientific Abstracts, 7200 Wisconsin Avenue, Suite 601, Bethesda, MD 20814. (301) 961-6750 or (800) 843-7751. Contains citations and abstracts of international periodicals and research literature covering all fields of engineering and science and technology,including computer and information science, electronics, mechanical engineering, solid state materials, 1981 to present. Available on BRS,(800) 289-4277, online service.Inquire as to cost and availability.

Current Contents Search. Institute for Scientific Information, 3501 Market Street, Philadelphia, PA 19104. (215) 386-0100. FAX (215) 386-6362. Contains citations to articles listed in the table of contents of science and technology journals. Also articles in social sciences and life sciences journals. Available on BRS,(800) 289-4277, DIALOG,(800) 334-2564, online services. Inquire as to cost and availability.

Dissertation Abstracts. University Microfilms International, 300 North Zeeb Road, Ann Arbor, MI 48106. (800) 521-0600 or (313) 761-4700. Scope includes virtually all doctoral dissertations accepted at accredited American institutions from 1861 to present in 252 subject areas. Available on BRS,(800) 289-4277, DIALOG,(800) 334-2564, and OCLC EPIC,(800) 848-5878, online services. Also available on CD-ROM. Inquire as to cost and availability.

NTIS Bibliographic Database. National Technical Information Service, 5285 Port Royal Road, Springfield, VA 22161. (703) 487-4929 or FAX (703) 321-8199. Broad coverage of government-sponsored science and technology research reports, 1964 to present. Available on BRS,(800) 289-4277, DIALOG, (800) 334-2564, ORBIT, (800) 456-7248, and STN International, FIZ Karlsruhe, P.O. Box 2465, W-7500, Karlsruhe 1, Germany, online services. Also available on CD-ROM. Inquire as to cost and availability.

SCISEARCH. Institute for Scientific Information, 3501 Market Street, Philadelphia, PA 19104. (800) 523-1850 or (215) 386-0100. Broad multidisciplinary title and author index to the international literature of science and technology, 1974 to present. Available on DIALOG,(800) 334-2564, and ORBIT,(800) 456-7248, online services. Also available on CD-ROM. Inquire as to cost and availability.

WILSONLINE. H.W. Wilson Company, 950 University Avenue, Bronx, NY 10452. (800) 367-6770 or (212) 588-8400. Makes available online versions of the printed H.W. Wilson indexes including Applied Science and Technology Index, Business Periodicals Index, General Science Index, and Readers' Guide to Periodical Literature. Period covered is generally 1983 to present. Available on BRS,(800) 289-4277, DIALOG,(800) 334-2564, and OCLC EPIC,(800) 848-5878, online services. Also available on CD-ROM. Inquire as to cost and availability.

PERIODICALS

Accident Prevention. Industrial Accident Prevention Association, 250 Yonge St., 28th Fl., Toronto ON M5B 2N4 Canada. (416) 506-8888. FAX (416) 506-8880. 6 times a year. $24.00 per year.

American Industrial Hygiene Association Journal. American Industrial Hygiene Association, 2700 Prosperity Ave., Suite 250, Fairfax, VA 22031-4307. (703) 849-8888. Monthly. $100.00 per year.

BCSP Newsletter. Board of Certified Safety Professionals, 208 Burwash, Savoy, IL 61874. (217) 359-9263. 3 times a year. Free with membership.

Chemical Hazards in Industry. Royal Society of Chemistry, Thomas Graham House, Science Park, Milton Rd., Cambridge CB4 4WF, England. Telephone 0223-420066. FAX 0223-423623. Monthly. $620.00 per year. Also available online.

Chilton's Industrial Safety & Hygiene News. Chilton Co., One Chilton Way, Radnor PA 19089. (215) 964-4028. Monthly. $50.00 per year.

Hazard Prevention. Systems Safety Society Inc., Technology Trading Park, 5 Export Drive, Suite A, Sterling, VA 22170-4421. (703) 444-6520. 1965 to present. Quarterly. $45.00 per year.

Health & Safety At Work. Tolley Publishing Co. Ltd., Tolley House, 2 Addiscombe Rd., Croydon, Surrey CR9 5AX England. Telephone 081-686-9141. FAX 081-686-3155. Monthly. Write for price information.

Industrial Health. National Institute of Industrial Health, Rodosho Sangyo Igaku Sogo Kenkyujo, 21-1 Nagao 6-chome, Tamaku, Kawasaki-shi, Kanagawa-ken 214, Japan. 1963 to present. Quarterly. Inquire for cost and availability.

Journal of Safety Research. Elsevier Science [journals], 660 White Plains Rd., Tarrytown, NY 10159-5153. (919) 524-9200.

FAX (919) 333-2444. 1982 to present. 4 times a year. $305.00 per year.

Occupational Hazards. Penton Publishing, 1100 Superior Ave., Cleveland, OH 44114-2543. Monthly. $45.00 per year.

Occupational Health & Safety. Stevens Publishing Corp., 225 N New Rd., Waco, TX 76710. (817) 776-9000. FAX (817) 776-9018. Monthly. $41.00 per year.

Occupational Health & Safety News. Stevens Publishing Corp., 225 N New Rd., Waco, TX 76710. (817) 776-9000. FAX (817) 776-9018. 24 times per year. $269.00 per year.

Occupational Safety & Health Reporter. Bureau of National Affairs Inc., 1231 25th St. NW, Washington DC 20037. (202) 452-4200. FAX (202) 822-8092. Weekly. $992.00 per year.

Professional Safety. American Society of Safety Engineers, 1800 E Oakton St., Des Plaines IL 60018-2187. (708) 692-4121. FAX (708) 296-3769. Monthly. $50.00 per year.

Safety & Health. National Safety Council, Periodicals Dept., 1121 Spring Lake Dr., Itasca IL 60143. (708) 285-1121. Monthly. $56.00 for non-members.

Safety Science. Elsevier Science Inc., Box 282 Madison Square Station, New York, NY 10159. (212) 989-5800. FAX (212) 633-3990. 8 times a year. $379.00 per year.

RESEARCH CENTERS AND INSTITUTES

Institute of Safety and Systems Management, University of Southern California, SSM-MC 0021, University Park, CA 90089. (213) 740-2411. FAX (213) 740-5943.

Institute For Advanced Safety Studies. 5950 Touhy Avenue, Niles, IL 60648. (708) 647-1101. FAX (708) 647-2047.

Center For Ergonomics. University of Michigan, 1205 Beal Street, I.O.E. Bldg., Ann Arbor, MI 48109-2117. (313) 763-2443. FAX (313) 764-3451.

SATELLITES (ARTIFICIAL)

See also: AEROSPACE ENGINEERING, REMOTE SENSING, TELE-COMMUNICATIONS, TELEVISION

ABSTRACT SERVICES AND INDEXES

Applied Science and Technology Index; A Cumulative Subject Index To English Language Periodicals in the Fields of Aeronautics and Space Science, Computer Technology, Chemistry, Construction Industry, Energy and Related Areas. H.W. Wilson Co., 950 University Avenue, Bronx, NY 10452. (800) 367-6770 or (212) 588-8400. FAX (718) 590-1617. From 1958 to present. Monthly. Inquire about cost and availability. Also available on CD-ROM.

Astronomy and Astrophysics Abstracts. Springer-Verlag, 175 Fifth Avenue, New York, NY 10010. (212) 460-1500 or (800) 777-4643. Irregular. Price varies, inquire.

SATELLITES (ARTIFICIAL)

Ency. of Physical Sciences and Engineering Info. Sources

Current Contents: Engineering, Technology and Applied Sciences. Institute for Scientific Information, 3501 Market Street, Philadelphia, PA 19104. (215) 386-0100 or (800) 523-1850. FAX (215) 386-6362. Weekly. $360.00 per year.

Engineering Index Monthly. Engineering Information, Inc., Castle Point on the Hudson, Hoboken, NJ 07030. (800) 221-1044. FAX (212) 832-1857. Monthly. $2200.00 per year. Also available online as COMPENDEX, and also on CD-ROM.

Index To Scientific and Technical Proceedings. Institute for Scientific Information, 3501 Market St., Philadelphia, PA 19104. (215) 386-0100 or (800) 523-1850. FAX (215) 386-6362. Monthly. $500.00 per year.

Index To Scientific Reviews. Institute for Scientific Information, 3501 Market St., Philadelphia, PA 19104. (215) 386-0100 or (800) 523-1850. FAX (215) 386-6362. Semi-annual.

International Aerospace Abstracts. Technical Information Service, American Institute of Aeronautics and Astronautics, Inc., 555 West 57th St., New York, NY 10019. (212) 247-6500. FAX (212) 582-4861. Semi-monthly. $1295.00 per year.

NASA Patent Abstracts Bibliography. National Aeronautics and Space Administration, Center for Aerospace Information, Box 8757, BWI Airport, Baltimore, MD 21240-0757. (301) 621-0153. Semi-annual. Inquire for price information.

Numerical/chronological/author Index. An index to articles and papers published by or for the American Astronautical Society, including: Journal of the Astronautical Sciences, Astronautical Sciences Review, Advances in the Astronautical Sciences, Science and Technology, AAS history series, and AAS microfiche series. Horace Jacobs and Robert H. Jacobs. Univelt, Inc. PO Box 28130, San Diego, CA 92198. (619) 746-4005. Inquire for price and availability.

Science Citation Index. Institute for Scientific Information, 3501 Market Street, Philadelphia, PA 19104. (215) 386-0100 or (800) 523-1850. FAX (215) 386-6362. Inquire about availability and cost. Also available on CD-ROM. # #*Scientific and Technical Aerospace Reports (star).* National Aeronautics and Space Administration. NASA Center for Aerospace Information, Box 8757, BWI Airport, Baltimore, MD 21240-0757. (301) 621-0153. Monthly. Inquire about availability and cost. Also available through the NASA online retrieval service (RECON), and through the Aerospace Database through DIALOG.

ASSOCIATIONS AND PROFESSIONAL SOCIETIES

Aerospace Industries Association of America. 1250 Eye Street NW, Washington, DC 20005. (202) 371-8400.

American Astronautical Society. 6352 Rolling Mill Place, Suite 102, Springfield, VA 22152.

American Institute of Aeronautics and Astronautics. 370 L'Enfant Promenade SW, Washington, DC 20024. (202) 646-7400.

British Interplanetary Society, 27/29 South Lambeth Road, London Sw8 1sz, England. (071) 735-3160. FAX (071) 820-1504.

Canadian Aeronautics and Space Institute, 222 Somerset Street West, Suite 601, Ottawa, Ontario, Canada K2p 0j1.

International Association of Satellite Users and Suppliers. PO Box DD, McLean, VA 22101-3615. (703) 442-8888. FAX (703) 821-1824.

National Space Society, 922 Pennsylvania Avenue Se, Washington, DC 20003-2140. (202) 543-1900. FAX (202) 546-4189.

Society of Satellite Professionals International. 2200 Wilson Blvd. #102-248, Arlington, VA 22201. (703) 243-8948. FAX (703) 528-4084.

DIRECTORIES AND BIOGRAPHICAL SOURCES

American Astronautical Society Directory. American Astronautical Society, 6352 Rolling Mill Place, Suite 102, Springfield, VA 22152.

American Institute of Aeronautics and Astronautics Roster. American Institute of Aeronautics and Astronautics. 370 L'Enfant Promenade SW, Washington, DC 20024. (202) 646-7400.

Jane's Space Directory. Janes Info Group (order from:) Dept. DSM, 1340 Braddock Pl., Suite 300, Box 1436, Alexandria, VA 22314-1651. (703) 683-3700. FAX (703) 836-0029. Annual. $245.00. Also available on CD-ROM.

Research Centers Directory. Gale Research, 835 Penobscot Building, Detroit, MI 48226-4094. (313) 961-2242 or (800) 877-4253. FAX (313) 961-6083. $485.00.

Scientific and Technical Organizations and Agencies Directory. Gale Research, 835 Penobscot Building, Detroit, MI 48226-4094. (313) 961-2242. (800) 877-4253. FAX (313) 961-6083. 4th ed. 1995. $195.00.

Who's Who in Engineering. 7th ed. American Association of Engineering Societies, 1111 19th St. NW, Suite 608, Washington, DC 20036. (800)-658-8897. 1988. $200.00.

World Satellite Directory. PHillips Publishing Inc., 7811 Montrose Rd., Potomac, MD 20854. (301) 340-2100 or (800) 722-9000. FAX (301) 424-4297. Annual. $227.00 per year.

GENERAL WORKS

Global Positioning System: Theory and Practice. B. Hofmann-Wellenhof, et al. 2d ed. Springer-Verlag, 175 Fifth Avenue, New York, NY 10010. (212) 460-1500 or (800) 777-4643. FAX (212) 473-6272. 1993. $59.00.

Satellite Operations: Systems Approach To Design and Control. John T. Garner. E. Horwood Publishing, distributed by Prentice Hall, 113 Sylvan Avenue, Route 9W, Englewood Cliffs, NJ 07632. (201) 592-2000 or (800) 922-0579. 1990. Inquire for cost and availability.

Satellite Systems: Principles and Technologies. Bruno Pattan. Van Nostrand Reinhold, 115 Fifth Avenue, New York, NY 10003. (212) 254-3232. FAX (212) 254-9499. 1993. $72.95.

Satellite Technology and Its Applications. P.R.K. Chetty. 2nd edition. McGraw-Hill Publishing Company, 1221 Avenue of the Americas, New York, NY 10020. (800) 262-4729 or (212) 512-3825. 1991. $49.95.

World Satellite Almanac: The Global Guide To Satellite Transmission and Technology. Mark Long. 3d ed. MLE, Inc., Division of Mark Long Enterprises, 150 N. Federal Highway, Suite 230, Ft. Lauderdale, FL 33301. (305-767-4687). 1992. $149.95.

ONLINE DATABASES AND CD-ROMS

Aerospace Database. American Institute of Aeronautics and Astronautics, 370 L'Enfant Promenage, S.W., Washington, DC 20024. (202) 646-7400. Worldwide published literature on research and development in aerospace and related areas, with abstracts. Covers 1962 to present. Available on DIALOG, (800) 334-2564, online service. Also available on CD-ROM. Inquire as to cost and availability.

Compendex Plus. Engineering Information, Inc., 345 East 47th Street, New York, NY 10017. (212) 705-7600 or (800) 221-1044. Contains citations with abstracts to worldwide literature in engineering and technology, from 1970 to present. Available on online BRS,(800) 289-4277, DIALOG, (800) 334-2564, ORBIT (800) 456-7248, and STN International, FIZ Karlsruhe, P.O. Box 2465, W-7500, Karlsruhe 1, Germany, online services. Also available on CD-ROM. Inquire as to cost and availability.

CSA Engineering. Cambridge Scientific Abstracts, 7200 Wisconsin Avenue, Suite 601, Bethesda, MD 20814. (301) 961-6750 or (800) 843-7751. Contains citations and abstracts of international periodicals and research literature covering all fields of engineering and science and technology,including computer and information science, electronics, mechanical engineering, solid state materials, 1981 to present. Available on BRS,(800) 289-4277, online service.Inquire as to cost and availability.

Current Contents Search. Institute for Scientific Information, 3501 Market Street, Philadelphia, PA 19104. (215) 386-0100. FAX (215) 386-6362. Contains citations to articles listed in the table of contents of science and technology journals. Also articles in social sciences and life sciences journals. Available on BRS,(800) 289-4277, DIALOG,(800) 334-2564, online services. Inquire as to cost and availability.

Dissertation Abstracts. University Microfilms International, 300 North Zeeb Road, Ann Arbor, MI 48106. (800) 521-0600 or (313) 761-4700. Scope includes virtually all doctoral dissertations accepted at accredited American institutions from 1861 to present in 252 subject areas. Available on BRS,(800) 289-4277, DIALOG,(800) 334-2564, and OCLC EPIC,(800) 848-5878, online services. Also available on CD-ROM. Inquire as to cost and availability.

International Aerospace Abstracts. American Institute of Aeronautics and Astronautics, 370 L'Enfant Promenade, S.W., Washington, DC 20024. (202) 646-7400. Contains references and abstracts of journal and monograph literature relating to aerospace science and technology, from 1963 to present. Available through the NASA/RECON system of the National Aeronautics and Space Administration only.

NASA Database. American Institute of Aeronautics and Astronautics, 370 L'Enfant Promenade, S.W., Washington, DC 20024. (202) 646-7400. Citations and abstracts of aeronautics and astronautics literature, 1962 to present. Also contains citations from STAR, SCIENTIFIC and TECHNICAL AEROSPACE REPORTS, and INTERNATIONAL AEROSPACE ABSTRACTS. Available through NASA/RECON online service. Inquire as to cost and availability.

NTIS Bibliographic Database. National Technical Information Service, 5285 Port Royal Road, Springfield, VA 22161. (703) 487-4929 or FAX (703) 321-8199. Broad coverage of government-sponsored science and technology research reports, 1964 to present. Available on BRS,(800) 289-4277, DIALOG, (800) 334-2564, ORBIT, (800) 456-7248, and STN International, FIZ Karlsruhe, P.O. Box 2465, W-7500, Karlsruhe 1, Germany, online services. Also available on CD-ROM. Inquire as to cost and availability.# *#SCISEARCH.* Institute for Scientific Information, 3501 Market Street, Philadelphia, PA 19104. (800) 523-1850 or (215) 386-0100. Broad multidisciplinary title and author index to the international literature of science and technology, 1974 to present. Available on DIALOG,(800) 334-2564, and ORBIT,(800) 456-7248, online services. Also available on CD-ROM. Inquire as to cost and availability.

WILSONLINE. H.W. Wilson Company, 950 University Avenue, Bronx, NY 10452. (800) 367-6770 or (212) 588-8400. Makes available online versions of the printed H.W. Wilson indexes including Applied Science and Technology Index, Business Periodicals Index, General Science Index, and Readers' Guide to Periodical Literature. Period covered is generally 1983 to present. Available on BRS,(800) 289-4277, DIALOG,(800) 334-2564, and OCLC EPIC,(800) 848-5878, online services. Also available on CD-ROM. Inquire as to cost and availability.

PERIODICALS

Acta Astronautica. Pergamon Press, Pergamon Press, 660 White Plains Rd., Tarrytown, NY 10591-5153. 14 times a year. Inquire for price and availability.

Ad Astra. National Space Society, 922 Pennsylvania Avenue SE, Washington, DC 20003-2140. (202) 543-1900. FAX (202) 546-4189. 1989 to present. 6 times a year. $35.00 a year.

Advances in Astronautical Sciences. Univelt, Inc. PO Box 28130, San Diego, CA 92198. (619) 746-4005. Irregular. Price varies, inquire for price and availability.

Advances in Space Research. Pergamon Press, Pergamon Press, 660 White Plains Rd., Tarrytown, NY 10591-5153. 12 times a year. Inquire for price and availability.

Aeronautical Journal. Royal Aeronautical Society, PO Box 139, Tonbridge, Kent, TN9 1EW, England. 0732-770823. FAX 0723-361708. 1963 to present. Ten times a year. Write for price and availability.

Aerospace. Royal Aeronautical Society, PO Box 139, Tonbridge, Kent, TN9 1EW, England. 0732-770823. FAX 0723-361708. Monthly. Write for price and availability.

Aerospace America. American Institute of Aeronautics and Astronautics, 370 L'Enfant Promenade SW, Washington, DC 20024-2518. (202) 646-7471. 1932 to present. Monthly. $70 per year (non-members).

Aerospace Engineering. SAE International, Inc., 400 Commonwealth Dr., Warrendale, PA 15096. (412) 776-4841. 1981 to present. Monthly. $48 a year.

Aviation & Aerospace. Baxter Publishing Company, 310 Dupont Street, Toronto, Ontario M5R 1V9 Canada. (416) 968-7252. FAX (416) 968-2377. 1928 to present. Bi-monthly. $42.00 per year.

Aviation Week and Space Technology. 1221 Avenue of the Americas, New York, NY 10020. (212) 512-2999 or (800) 525-5003. 1916 to present. Weekly. $82.00 per year.

ESA Bulletin. European Space Agency, Publications Division, Keplerlaan 1, 22000 AG Nordwijk, Netherlands. Phone 31-1719-86555. FAX 31-1719-85433. Quarterly. Free.

ESA Journal. European Space Agency, Publications Division, Keplerlaan 1, 22000 AG Nordwijk, Netherlands. Phone 31-1719-86555. FAX 31-1719-85433. Quarterly. Free.

Journal of the British Interplanetary Society (JBIS). British Interplanetary Society, 27/29 South Lambeth Road, London SW8 1SZ, England. (071) 735-3160. FAX (071) 820-1504. Monthly. $370.00 per year.

Interavia/Aerospace World. Aerospace Media Publishing, Swissair Centre, 31 Route de l'Aeroport, POB Box 437, 1215 Geneva 15, Switzerland. (022) 788 27 88. FAX (022) 788 27 26. 1987 to present. Monthly. $128.00 per year (U.S.).

International Journal of Satellite Communications. John Wiley and Sons Ltd. Journals, Baffins Ln., Chichester, Sussex PO19 1UD, England. Phone 0243-77977. FAX 0243-775878. Bimonthly. $525.00 per year.

Journal of the Astronautical Sciences. American Astronautical Society, 6352 Rolling Mill Place, Suite 102, Springfield, VA 22152. Quarterly. $35 per year.

Journal of Guidance, Control and Dynamics. American Institute of Aeronautics and Astronautics, 370 L'Enfant Promenade SW, Washington, DC 20024-2518. 1978 to present. Bimonthly. $42.00 (members) and $250.00 (non-members).

Journal of Spacecraft and Rockets. American Institute of Aeronautics and Astronautics, the Aerospace Center, 370 L'Enfant Promenade SW, Washington, DC 20024. (202) 646-7400. FAX (202) 646-7508. 1964 to present. Bi-monthly. $185.00 per year.

Planetary Report. Planetary Society, 65 N. Catalina, Pasadena, CA 91106-2301. (818) 793-5100. FAX (818) 793-5528. Bimonthly. $25.00 per year.

Progress in Aerospace Sciences. Pergamon Press, 660 White Plains Rd., Tarrytown, NY 10591-5153. 1961 to present. Quarterly. $320.00 per year.

Satellite Communications. Cardiff Publishing Co., 6300 S. Syracuse Way, Ste. 650, Englewood, CO 80111. (303) 220-0600. Monthly. $29.00 per year.

Satellite News. PHillips Publishing Inc., 7811 Montrose Rd., Potomac, MD 20854. (800) 777-5006. Weekly. $797.00 per year.

Space Science Reviews. Kluwer Academic Publishers, P.O. Box 358, Accord Station, Hingham, MA 02018-0358. (617) 871-6000. Monthly except March, June, September, December. $258.50 per volume.

Space Technology: Industrial and Commercial Applications. Pergamon Press, 660 White Plains Rd., Tarrytown, NY 10591-5153. Bimonthly. $535.00 per year.

Spaceflight. British Interplanetary Society, 27/29 South Lambeth Road, London SW8 1SZ, England. (071) 735-3160. FAX (071)

820-1504. Monthly. $69 annual membership includes subscription.

RESEARCH CENTERS AND INSTITUTES

Aerospace Research Applications Center, Indianapolis Center For Advanced Research, 611 N. Capitol Ave., Indianapolis, in 46204. (317) 262-5003. FAX (317) 262-5044.

Aerospace Research Center, 1250 Eye St., NW, Washington, DC 20005. (202) 371-8400. FAX (202) 371-8470.

Auburn University Space Power Institute, 231 Leach Center, Auburn, AL 36849. (205) 844-5894.

Canadian Space Agency, Space Science Program, C/o National Research Council of Canada, 100 Sussex Drive, Ottawa, Ontario, Canada K1a 0r6. (613) 990-0799. FAX (613) 952-0974.

Center For Aerospace Technology, Weber State College, Ogden, UT 84408-1805. (801) 626-7272.

Center For Space Research, Massachusetts Institute of Technology, 77 Massachusetts Ave., Cambridge, MA 02139. (617) 253-7501.

Center For Control and Systems Research, University of Texas At Austin, Aerospace Engineering and Engineering Mechanics, Wrw 408, Austin, TX 78712-1085. (512) 471-4908. FAX (512) 471-3788.

Charles Stark Draper Laboratory Inc., 555 Technology Square, Cambridge, MA 02139. (617) 258-1000.

Goddard Space Flight Center, Greenbelt, MD 20771. (301) 286-7351.

Houston Advanced Research Center, Space Technology and Research Center, 4802 Research Forest Dr., the Woodlands, TX 77381. (713) 363-7922. FAX (713) 363-7914.

Innovative Nuclear Space Power and Propulsion Institute, College of Engineering, University of Florida, Gainesville, FL 32611. (904) 392-1427. FAX (904) 392-8656.

Institute For Space and Security Studies, 5115 Highway A1a, Melbourne Beach, FL 32951. (407) 952-0600.

Instrumentation and Control Laboratory, Princeton University, Dept. of MAE, Engineering Quadrangle, Princeton, NJ 08544. (609) 452-5154.

George C. MARSHALL SPACE FLIGHT CENTER, Huntsville, AL 35812. (205) 544-0024.

Ohio Aerospace Institute, 2001 Aerospace Parkway, Brook Park, OH 44142. (216) 891-2100. FAX (216) 891-2140.

Space Science and Engineering Center, University of Wisconsin—madison, 1225 W. Dayton St., Madison, WI 53706. (608) 262-0544.

SATELLITES (NATURAL)

See also: ARTIFICIAL SATELLITES, ASTROGEOLOGY, ASTEROIDS, ASTRONOMY, COMETS, METEORS, MOON, NAMES of INDIVIDUAL PLANETS, PLANETARY SCIENCES, SEISMOLOGY, SOLAR SYSTEM, SUn

ABSTRACT SERVICES AND INDEXES

Applied Science and Technology Index; A Cumulative Subject Index To English Language Periodicals in the Fields of Aeronautics and Space Science, Computer Technology, Chemistry, Construction Industry, Energy and Related Areas. H. W. Wilson Co., 950 University Avenue, Bronx, NY 10452- 9978. (212) 588-8400. (800) 367-6770. FAX (718) 590-1617.From 1958 to present. Monthly. Inquire about cost and availability. Also available online (BRS and WILSONLINE) and on CD-ROM. ISSN: 0003-6986.

Astronomy and Astrophysics Abstracts. Springer-Verlag New York, 175 Fifth Avenue, New York, NY 10010. (212) 460-1500.FAX (212) 473-6272. Published for the Astronomisches Rechen-Institut. Comprehensive coverage of all aspects of astronomy, astrophysics and related fields. 1969 to present. Two parts per year. Annual. ISSN: 0067-0022.

Bibliography and Index of Geology. American Geological Institute, 4220 King Street, Alexandria, VA 22302-1507. (703) 379-2480. FAX (703) 379-7563. From 1969 to present. Monthly. $1295.00 per year. ISSN: 0098-2784. Also available as GEOREF online (CISTI, DIALOG, Orbit, STN) and on CD-ROM. Inquire about price and availability.

Chemical Abstracts. Chemical abstracts Service. 2540Oleantangy River Road, Box 3012, Columbus, OS 43210-0012. (614) 447-3600. FAX (614) 447-3713. 1907 to present.Weekly. $16,800.00 per year. Available online and on CD- ROM. CA is also available in five section groupings. Inquire regarding cost and availability. ISSN:

Current Contents: Physical, Chemical, and Earth Sciences. Institute for Scientific Information, 3501 Market Street,Philadelphia, PA 19104. (215) 386-0100. (800) 523-1850. FAX (215) 386-2291. 1961 to present. Weekly. $442.00peryear. Also available online (BRS, DIALOG) and on CD- ROM. Inquire about price and availability.

General Science Index. H.W. Wilson Company, 950 UniversityAvenue, Bronx, NY 10452. (212) 588-8400. (800) 367-6770.FAX (718) 590-1617. From 1978 to present. Ten issues peryear; quarterly and annual cumulations. Service basis.Available on CD-ROM and online. Inquire regarding cost andavailability. ISSN: 0162-1963.

Meteorological and Geoastrophysical Abstracts. American Meteorological Society, c/o Inforonics, Inc., 550 NewtownRoad, Box 458, Littleton, MA 01460. (508) 486-8976. FAX(508) 486-0027. Covers literature in environmentalsciences, meteorology, astrophsyics, hydrology, glaciologyand physical oceanography. 1950 to present. Monthly. $950.00 per year. Also available online (DIALOG) and onCD-ROM. ISSN: 0026-1130.

NTIS Alerts: Astronomy & Astrophysics. U. S. National Technical Information Service. 5285 Port Royal Road,Springfield, VA 22161. (703) 487-4650. FAX (703) 321-8547.Weekly. $140.00 per year.

Physics Abstracts. INSPEC. Section A, Science Abstracts.Institute of Electrical Engineers, London, United Kingdom.Available from: INSPEC/IEEE - Institute of Electrical andElectronic Engineers, Box 1331, Hoes Lane, Piscataway, NJ08855-1331. (908) 562-5549. 1898 to present. 24 issuesper year. $2835.00 per year. ISSN: 0036-8091. Alsoavailable online and on CD-ROM.

Science Citation Index. SCI. Institute for ScientificInformation, 3501 Market Street, Philadelphia, PA 19104.(215) 386-0100. (800) 523-1850. FAX (215) 386-2991. 1961to present. Six issues per year, plus annual cumulation.$11650.00 per year. Also available online and on CD-ROM.Inquire about price and availability. ISSN: 0036-827X.

STAR. Scientific and Technical Aerospace Reports. U.S. National Aeronautics and Space Administration. Distributed by U. S. Superintendent of Documents, Washington, DC 20402. 1963 to present. Semi-monthly, with semiannual and annual indexes. $114.00 per year. ISSN: 0036-8741. Also available online and on CD-ROM.

ANNUAL REVIEWS AND YEARBOOKS

The Astronomical Almanac. Superintendent of Documents, U. S. Government Printing office, Washington, DC 10402. (202) 783-3238. 1981 to present. Superceeds Astronomical Ephemeris. Annual. $44.95. ISSN: 0737-6421.

Annual Review of Astronomy and Astrophysics. Original reviews of critical literature and current developments in astronomy and astrophysics. Annual Reviews, Inc., 4139 El Camino Way, Palo Alto, CA 94303-0139. (415) 493-4400. (800)523-8635. Fax (415) 855-9815. 1963 to date. Annual. $60.00. ISSN: 0066-4146.

Annual Review of Earth and Planetary Sciences. Annual Reviews, Inc., 4139 El Camino Way, Palo Alto, CA 94303-0139. (415) 493-4400. (800) 523-8635. Fax (415) 855-9815. 1973 to date. Annual. $62.00. ISSN: 0084-6597.

Proceedings of the Lunar and Planetary Science Conference. Lunar and Planetary Institute, 3600 Bay Area Boulevard, Houston, TX 77058. (713) 486-2143. 1970 to date. Annual. ISSN: 0270-9511.

World Satellite Annual. Mark Long. MLE Inc., division of Mark Long Enterprises, Inc., 150 North Federal Highway, Suite 230, Fort Lauderdale, FL 33301. (305) 767-4687. 1992. $50.95, in paper.

ASSOCIATIONS AND PROFESSIONAL SOCIETIES

American Astronomical Society. 2000 Florida Avenue NW, Suite 400, Washington, DC 20009. (202) 328-2010. FAX\; (202) 234-2560.

American Geophysical Union. 2000 Florida Avenue NW, Washington, DC 20009. (202) 462-6900. (800) 966-AGU1. FAX (202) 328-0566.

American Institute of Physics. 1 Physics Ellipse, College Park, MD 20740-3843. (301) 209-3100.

Association of Universities For Research in Astronomy, Inc. (AURA). Suite 701, 1625 Massachusetts Avenue, NW, Washington, DC 20036. (202) 483-2101. FAX (202) 483-2106.

Astronomical Society of the Pacific. 390 Ashton Avenue, San Francisco, CA 94112. (415) 337-1100. FAX: (415) 337-5205.

Geological Society of America. 3300 Penrose Place, PO Box 9140. Boulder, CO 80301-9140. (303) 447-2020. FAX (303) 447-1133.

IEEE Geoscience and Remote Sensing Society. c/o Institute of Electrical and Electronics Engineers. 345 East 47th Street, New York, NY 10017. (212) 705-7900. FAX (212) 705-4929.

Meteoritical Society. University of Massachusetts, 125 Marston Hall, Amherst, MA 01003. (413) 545-0300. (413) 545-0724.

Planetary Society. 65 North Catalina Avenue, Pasadena, CA 91106. (818) 793-5100. (800) WOW-MARS. FAX (818) 793-5528.

DIRECTORIES AND BIOGRAPHICAL SOURCES

American Astronomical Society Membership Directory. American Astronomical Society. 2000 Florida Avenue NW, Suite 400, Washington, DC 20009. (202) 328-2010. FAX (202) 234-2560 Annual. Included in membership dues. ISSN: 1061- 9038.

American Men and Women of Science: Physical and Biological Sciences. R. R. Bowker Inc., 121 Chanlon Road, New Providence, NJ 07974. (908) 464-6800. (800) 521-8110. 20th edition. 8 volumes. 1996. $850.00.

The Astronomers. Donald Goldsmith. St. Martin's Press, Inc., 175 Fifth Avenue, New York, NY 10010. (212) 674-5151. (800) 221-7945. 1993. $14.95 in paper.

Astronomical Centers of the World. Kevin Krisciunas. Cambridge University Press, 40 West 20th Street, New York, NY 10011-4211. (212) 924-3900. (800) 872-7423. 1988. $34.95

Directory of Physics and Astronomy Staff Members. American Institute of Physics, One Physics Ellipse, College Park, MD 20740-3843. (301) 209-3100. 1975/76 to present. Annual. $60.00. ISSN: 0361-2228.

Earth and Astronomical Research Centers. Stockton Press, 345 Park Avenue South, New York, NY 10010. 4th edition. 1995. $515.00. ISBN: 1-56169-0967.

Graduate Programs in Physics, Astronomy and Related Fields. American Institute of Physics, One Physics Ellipse, College Park, MD 20740-3843. (301) 209-3100. 1978 to present. Annual. $45.00. ISSN: 0147-1821. @ @Encyclopedias and Dictionaries

Concise Dictionary of Astronomy. Jacqueline Mitton. Oxford University Press, Inc., 200 Madison Avenue, New York, NY 10016. (212) 725-6000. (800) 334-4249. 1992. $24.95.

Encyclopedia of Astronomy and Astrophysics. Stephen Maran, editor. Van Nostrand Reinhold, 115 Fifth Avenue, New York, NY 10003. (212) 254-3232. (800) 842-3636. 1992. $129.95.

McGraw-Hill Encyclopedia of Science and Technology. McGraw-Hill Book Company, Inc., 1221 Avenue of the Americas, New York, NY 10020. (212) 512-2000. (800) 262-4729. 7th edition. 20 volume set. 1992. $1900.00.

Moons of the Solar System: An Illustrated Encyclopedia. John Stewart. McFarland & Company, Inc., Box 611, Jefferson, MC 28640. (910) 246-4460. (800) 253-2187. 1991. $49.95. # #*New Guide To the Planets.* Patrick Moore. Trans-Atlantic Publications, Inc., 311 Bainbridge Street, Philadelphia, PA 19147. (215) 925-5083. 1993. $37.50.

Stars and Planets: The Sierra Club Guide To Sky Watching and Direction Finding. W. S. Kals. The Sierra Press, 4988 Gold Leaf Drive, Mariposa, CA 95338. (209) 966-5071. (800) 745-2631. $15.00.

GENERAL WORKS

Birth of the Earth. David E. Fisher. Columbia University Press, 562 West 113th Street, New York, NY 10025. (212) 666-1000. (800) 944-8648. 1988. $16.50 in paper.

Evolution of the Earth and Planets. E. Takahashi, et al, editors. American Geophysical Union, 2000 Florida Avenue, N.W., Washington, DC 20009. (202) 462-6903. (800) 966- 2481. 1993. $28.00.

Exploration of Venus & Mars Atmospheres. G. M. Keating, editor. Pergamon Press, Inc., Maxwell House, Fairview Park, Elmsford, NY 10523. (914) 592-7700. Fax (914) 592-3625. 1995. 94.00.

Exploring the Planets. W. Kenneth Hamblin and Eric H. Christiansen. Macmillan Publishing Company, Inc,. 200 Old Tappan Road, Old Tappan, NJ 07675. (800) 233-2336. 2nd edition. 1995. $62.00.

The Geology of Multi-ring Impact Basins: The Moon and Other Planets. P. D. Spudis. Cambridge University Press, 40 West 20th Street, New York, NY 10011-4211. (212) 924-3900. (800) 872-7423. Planetary Sciences Series. 1993. $59.95.

Hazards Due To Comets and Asteroids. Tom Gehrels, editor. University of Arizona Press, 1230 North Park Avenue, Number 102, Tucson, AZ 85719. (520) 621-1441. (800) 426-3797. 1994. $75.00.

Mercury: The Elusive Planet. Robert B. Strom. Smithsonian Institution Press, 470 L'Enfant Plaza, Suite 7100, Washington, DC 20560. (202) 287-3738. (800) 782-4612. 1987. $29.95.

Meteorites: An Introduction. Fritz Heide. Springer-Verlag New York, Inc., 175 Fifth Avenue, New York, NY 10010. (212) 460-1500. (800) 777-4643. 1995. $24.00 in paper.

Mission To the Planets: The Illustrated Story of Man's Exploration of the Solar System. Patrick Moore. W. W. Norton & Company, 500 Fifth Avenue, New York, NY 10110. (212) 354-5500. (800) 223-2584. 1990. $24.95.

Moons and Planets. William K. Hartmann. Wadsworth Publishing Co., 10 Davis Drive, Belmont, CA 94002. (415) 595-2350. (800) 354-9706. 3rd edition. 1993. $49.95. # #*Morphology of the Rocky Members of the Solar System.* E. Uehupi and K. O. Emery. Springer-Verlag New York, Inc., 175 Fifth Avenue, New York, NY 10010. (212) 460-1500. (800)777-4643. 1994. $99.00.

Moons and Rings. Time-Life Books Editors. Time-Life, Inc., 777 Duke Street, Alexandria, VA 22314. (703) 8388- 7000. (800) 621-7026. Voyage through the Universe Series. 1991.

The New Solar System. J. Kelly Beatty and andrew Chaikin, editors. Cambridge University Press, 40 West 20th Street, New York, NY 10011-4211. (212) 924-3900. (800) 872-7423 .3rd edition. 1990. $44.95.

Physics and Chemistry of the Solar System. John S. Lewis, Jr., Academic Press, 6277 Sea Harbor Drive, Orlando, FL. (800) 321-5068. 1995. $69.95 in paper.

Planetary Landscapes. Ronald Greeley. Chapman & Hall, 1 Penn Plaza, New York, NY 10119. (212) 564-1060. Div. of Routledge,.. 2nd edition. 1994. $49.95.

The Planetary System. Tobias Owen and David Morrison. Addison-Wesley Publishing Co., Inc., 1 Jacob Way, Reading, MA 01867. (617) 944-3700. (800) 447-2226. 1988. $49.50.

Resources of Near-earth Space. John S. Lewis, et al. University of Arizona Press, 1230 North Park Avenue, Number 102, Tucson, AZ 85719. (520) 621-1441. (800) 426-3797. 1993. $75.00.

Rings: Discoveries From Galileo To Voyager. James C. Elliot and Richard Kerr. MIT Press, 55 Hayward Street, Cambridge, MA 02142. (617) 253-8569. Reprint edition. 1987. $9.95 in paper.

Rogue Astroids and Doomsday Comets. Duncan Stasiuk. John Wiley & Sons, Inc., 605 Third Avenue, New York, NY 10158-0012. (212) 850-6000. (800) 225-5945. 1995. $24.95.

Satellites of the Outer Planets: Worlds in their Own Right. David A. Rothery. Oxford University Press, Inc., 200 Madison Avenue, New York, NY 10016. (212) 725-6000. (800) 334-4249. 1992. $35.00 in paper.

Solar and Planetary Dynamos. M. R. Proctor, et al, editors. Cambridge University Press, 40 West 20th Street, New York, NY 10011-4211. (212) 924-3900. (800) 872-7423. Publications of the Newton Institue. 1994. $49.95.

Solar System Evolution: A New Perspective. Stuart R. Taylor. Cambridge University Press, 40 West 20th Street, New York, NY 10011-4211. (212) 924-3900. (800) 872-7423. 1989. 1992. $54.95.

Solar System: From the Renaissance To the Ninteenth Century. Part A. Curtis Wilson, editor. Cambridge University Press, 40 West 20th Street, New York, NY 10011-4211. (212) 924-3900. (800) 872-7423. 1989. $59.95.

Stardust To Planets: A Geological Tour of the Solar System. Harry Y. McSween, Jr., St. Martin's Press, Inc., 175 Fifth Avenue, New York, NY 10010. (212) 674-5151. (800) 221-7945. 1993. $22.95.

Wanderers in Space: Exploration and Discovery in the Solar System. Kenneth R. Lang. Cambridge University Press, 40 West 20th Street, New York, NY 10011-4211. (212) 924-3900. (800) 872-7423. 1991. $24.95 in paper.

The Wrong-way Comet and Other Mysteries of the Solar System. Barry Evans. McGraw-Hill Publishing Company, Inc., 1221 Avenue of the Americas, New York, NY 10020. (212) 512-2000. (800) 262-4729. 1992. $22.95 in paper.

HANDBOOKS AND MANUALS

Astrophysical Data: Planets and Stars. Kenneth R. Lanb. Springer-Verlag New York, Inc., 175 Fifth Avenue, New York, NY 10010. (212) 460-1500. (800) 777-4643. 1993. $59.00.

Burnham's Celestial Handbook: An Observer's Guide To the Universe Beyond the Solar System. Robert Burnham, Jr., and Herbert A. Luft. 1978. Dover Publications, Inc., 180 Varick Street, New York, NY 10014. (212) 255-3755. (800) 223-3130. $13.95.

Field Guide To the Stars and Planets. Jay M. Pasachoff and Donald H. Menzel. Houghton Mifflin Co., 222 Berkeley Street, Boston, MA 02116. (617) 351-5000. (800) 225-3362. Revised edition. 1992. $24.95.

New Guide To the Planets. Patrick Moore. Trans-Atlantic Publications, Inc., 311 Bainbridge Street, Philadelphia, PA 19147. (215) 925-5083. 1993. $37.50.

Planet Observer's Handbook. Fred W. Price. Cambridge University Press, 40 West 20th Street, New York, NY 10011-4211. (212) 924-3900. (800) 872-7423. 1994. $34.95.

Planets and their Moons. Aububon Society Staff. Alfred A. Knopf, Inc. 201 E. 50th Street, New York, NY 10022. (800) 733-3000. 1995. $7.99.

Space Satellite Handbook. Anthony R. Curtis. Gulf Publishing Co., P.O. Box 2608, Houston, TX 77252-2608. (713) 520-4444. (800) 231-6275. 3rd edition. $39.90.

ONLINE DATABASES AND CD-ROMS

CA Search. Chemical Abstracts Service, P.O. Box 3012, Columbus, OH 43210-0012. (614) 447-3600. (800) 848-6533. FAX (614) 447-3709. Very comprehensive guide to worldwide chemical literature and related fields, 1972 to present. Available on BRS,(800) 289-4277, DIALOG, (800) 334-2564, ORBIT (800) 456-7248, and STN International, FIZ Karlsruhe, P.O. Box 2465, W-7500, Karlsruhe 1, Germany, online services. Inquire as to cost and availability.

Dissertation Abstracts. University Microfilms International, 300 North Zeeb Road, Ann Arbor, MI 48106. (800) 521-0600 or (313) 761-4700. Scope includes virtually all doctoral dissertations accepted at accredited American institutions from 1861 to present in 252 subject areas. Available on BRS, (800) 289-4277, DIALOG, (800) 334-2564, and OCLC EPIC, (800) 848-5878, online services. Also available on CD-ROM. Inquire as to cost and availability.

GEOREF. American Geological Institute, 4220 King Street, Alexandria, VA 22302. (800) 336-4764 or (703) 379-2480. Geology and geosciences literature, 1785 to present for North America. Available on DIALOG,(800) 334-2564, ORBIT (800) 456-7248, online services. Also available on CD-ROM. Inquire as to cost and availability.

INSPEC. Institution of Electrical Engineers, Michael Faraday House, Six Hills Way, Stevenage, Herts. SG1 2AY, England. Telephone: 0438 313311 or FAX 0438 742840. Contains citations to the worldwide literature of physics, electronics and electrical engineering, computer technology, and related fields. Available on BRS, (800) 289-4277, DIALOG, (800) 334-2564, ORBIT, (800) 456-7248, and STN International, FIZ Karlsruhe,

P.O. Box 2465, W-7500, Karlsruhe 1, Germany, online services. Inquire as to cost and availability.

NTIS Bibliographic Database. National Technical Information Service, 5285 Port Royal Road, Springfield, VA 22161. (703) 487-4929 or FAX (703) 321-8199. Broad coverage of government-sponsored science and technology research reports, 1964 to present. Available on BRS,(800) 289-4277, DIALOG, (800) 334-2564, ORBIT, (800) 456-7248, and STN International, FIZ Karlsruhe, P.O. Box 2465, W-7500, Karlsruhe 1, Germany, online services. Also available on CD-ROM. Inquire as to cost and availability.

SCISEARCH. Institute for Scientific Information, 3501 Market Street, Philadelphia, PA 19104. (800) 523-1850 or (215) 386-0100. Broad multidisciplinary title and author index to the international literature of science and technology, 1974 to present. Available on DIALOG, (800) 334-2564, and ORBIT, (800) 456-7248, online services. Also available on CD-ROM. Inquire as to cost and availability.

WILSONLINE. H.W. Wilson Company, 950 University Avenue, Bronx, NY 10452. (800) 367-6770 or (212) 588-8400. Makes available online versions of the printed H.W. Wilson indexes including Applied Science and Technology Index, Business Periodicals Index, General Science Index, and Readers' Guide to Periodical Literature. Period covered is generally 1983 to present. Available on BRS, (800) 289-4277, DIALOG, (800) 334-2564, and OCLC EPIC, (800) 848-5878, online services.Also available on CD-ROM. Inquire as to cost and availability.

OTHER SOURCES

Atlas of the Planets. Vincent DeCallatay and Audouin Dollfus. Books on Demand, 300 North Zeeb Road, Ann Arbor, MI 48106-1346. (313) 761-4700. (800) 521-0600. $45.60.

Atlas of the Solar System. B. Yenne. Simon & Schuster, Inc. 1230 Avenue of the Americas, New York, NY 10020. (212) 698-7000. (800) 223-2348. 1987. $12.98.

Planetary, Lunar and Solar Positions Six Hundred-One B.C. to A.D. One at Five-Day and Ten-Day Intervals. Bryant Tuckerman. American Philosophical Society, 104 South 5th Street, Philadelphia, PA 19106-3387. (215) 440-3400. FAX (215) 440-3436. Memoirs Series, volume 56. 1979. $20.00.

Planetary, Lunar and Solar Positions, A.D.2 to A.D. 1649 at Five Day and Ten Day Intervals. Bryant Tuckerman. American Philosophical Society, 104 South 5th Street, Philadelphia, PA 19106-3387. (215) 440-3400. FAX (215)440-3436. Memoirs Series, volume 59. 1964. $30.00.

Planetary, Lunar and Solar Positions, 1650 - 1805. Owen Gingerich and Barbara L. Welther. American Philosophical Society, 104 South 5th Street, Philadelphia, PA 19106-3387. (215) 440-3400. FAX (215)440-3436. Memoirs Series, volume 59S. 1983. $20.00.

The Sky: A User's Guide. David H. Levy. Cambridge University Press, 40 West 20th Street, New York, NY 10011- 4211. (212) 924-3900. (800) 872-7423. 1993. $14.95.

The View From Space: Photographic Exploration of the Planets. Merton Davies and Bruce C. Murray. Columbia University Press, 562 West 113th Street, New York, NY 10025. (212) 666-1000. (800) 944-8648. 1973. $17.50.

PERIODICALS

Astronomical Journal. American Institute of Physics, One Physics Ellipse, College Park, MD 20740-3843. (301) 209- 3000. Published for the American Astronomical Society. 1849 to present. Monthly. $280.00 per year. ISSN: 0004- 6256.

Astronomical Society of the Pacific. PUBLICATIONS. Astronomical Society of the Pacific, 390 Ashton Avenue, San Francisco, CA 94112. (415) 337-1100. FAX (415) 337-5205. 1889 to present. Monthly. $175.00 per year. ISSN: 0004- 6280.

Astronomy. Kalmbach Publishing Company, Box 1612, Waukesha, WI 53187-1612. (414) 796-8776. FAX (414) 796-1142. 1973 to present. Monthly. $27.00 per year. ISSN: 0091-6358.

Astronomy and Astrophysics. Springer-Verlag New York, Inc., 44 Hartz Way, Secaucus, NJ 07096-2491. (201) 348-4033. FAX(201) 348-4505. 1969 to date. 36 issues per year in 12 volumes. $1733.00 per year. ISSN: 0004-6361.

Earth, Moon and Planets; An International Journal of Comparative Planetology. Kluwer Academic Publishers, Box 358, Accord Station, Hingham, MA 02018-0358. (617) 871-6600. FAX (617) 871-6528. 1969 to present. Monthly. $840.00 per year. ISSN: 0167-9295.

Geochemica Et Cosmochimica Acta. Elsevier Science. 660 White Plains Road, Tarrytown, NY 10591-5153. (914) 524-9200. FAX (914) 333-2444. 1950 to the present. Biweekly. $895.00 per year. ISSN: 0016-7037.

Icarus; International Journal of Solar System Studies. Academic Press, Inc., Journal Division, 525 B Street, Suite 1900, San Diego, CA 92101-4495. (619) 230-1840. FAX (619) 699-6800. 1962 to the present. Monthly. $1080.00. ISSN: 0019-1035.

J G R: Journal of Geophysical Research: Planets. American Geophysical Union, 2000 Florida Avenue, NW, Washington, CD 20009. (202) 462-6900. FAX (202) 328-0566. 1991 topresent. Monthly. $597.00 per year. ISSN: 0148-0227.

Lunar and Planetary Information Bulletin. Lunar and Planetary Institute. 3600 Bay Area Boulevard, Houston, TX 77058-1113. (713) 486-2175. FAX (713) 486-2125. 1970 topresent. Quarterly. Free. Also available online.

Mercury. Astronomical Society of the Pacific. 390 Ashton Avenue, San Francisco, CA 94112. (415) 337-1100. FAX: (415) 337-5205. 1972 to present. Bimonthly. $175.00 per year. ISSN: 0047-6773.

Meteoritics. Meteoritical Society, Department of Chemistry, University of Arkansas, Fayetteville, AR 72701. (501) 575-7625. FAX (501) 575-7778. 1953 to present. Bimonthly. $210.00 per year. ISSN: 0026-1114.

Planetary and Space Science. Elsevier Science Publishing Company, Inc., 660 White Plains Road, Terrytown, NY 10591- 5153. (914) 524-9200. FAX: (914) 333-2444. 1959 to present. Monthly. $1355.00 per year. ISSN: 0032-0633.

Ency. of Physical Sciences and Engineering Info. Sources

SATURN

Planetary Report. Planetary Society. 65 North Catalina, Pasadena, CA 91106-2301. (818) 793-5100. FAX (818) 793- 5528. 1980 to present. $25.00 per year. ISSN: 0736-3680.

Sky and Telescope. Sky Publishing Corporation, Box 9111, Belmont, MA 02178. (617) 864-7360. FAX (617) 864-6117. 1941 to present. Monthly. $27.00 per year. ISSN: 0037- 6604.

RESEARCH CENTERS AND INSTITUTES.

Center For Space Science and Astrophysics. Stanford University. 325 Durand Building, Stanford, CA 04305. (415) 723-3582.

Earth and Planetary Remote Sensing Laboratory. Washington University, Department of Earth and Planetary Sciences. Campus Box 1169, 1 Brookings Drive, St. Louis, MO 63130. (314) 889-5679. FAX (314) 889-5799.

Institute For Astronomy. University of Hawaii at Manoa, 2680 Woodlawn Drive, Honolulu, HI 96822. (808) 956-8312. FAX (808) 988-2790.

Institute of Geophysics and Planetary Physics. University of California, Riverside, CA 92521. (714) 787-4503. FAX (714) 787-4529.

Laboratory For Planetary Studies. Cornell University, Space Sciences Building, Ithaca, NY 14853. (607) 255-4971.

Lunar and Planetary Institute. 3303 NASA Road One, Houston, TX 77058-4399. (713) 486-2139. FAX 713-496-2162.

SATURN

See also: ASTROGEOLOGY, NATURAL SATELLITES, PLANETARY SCIENCES, SOLAR SYSTEM

ABSTRACT SERVICES AND INDEXES

Applied Science and Technology Index; a cumulative subject index to English language periodicals in the fields of aeronautics and space science, computer technology, chemistry, construction industry, energy and related areas. H. W. Wilson Co., 950 University Avenue, Bronx, NY 10452- 9978. (212) 588-8400. (800) 367-6770. FAX (718) 590-1617.From 1958 to present. Monthly. Inquire about cost and availability. Also available online (BRS and WILSONLINE) and on CD-ROM. ISSN: 0003-6986.

Astronomy and Astrophysics Abstracts. Springer-Verlag New York, 175 Fifth Avenue, New York, NY 10010. (212) 460-1500. FAX (212) 473-6272. Published for the Astronomisches Rechen-Institut. Comprehensive coverage of all aspects of astronomy, astrophysics and related fields. 1969 to present. Two parts per year. Annual. ISSN: 0067-0022.

Bibliography and Index of Geology. American Geological Institute, 4220 King Street, Alexandria, VA 22302-1507. (703) 379-2480. FAX (703) 379-7563. From 1969 to present. Monthly. $1295.00 per year. ISSN: 0098-2784. Also available as GEOREF online (CISTI, DIALOG, Orbit, STN) and on CD-ROM. Inquire about price and availability.

General Science Index. H.W. Wilson Company, 950 University Avenue, Bronx, NY 10452. (212) 588-8400. (800) 367-6770. FAX (718) 590-1617. From 1978 to present. Ten issues per year; quarterly and annual cumulations. Service basis. Available on CD-ROM and online. Inquire regarding cost and availability. ISSN: 0162-1963.

Meteorological and Geoastrophysical Abstracts. American Meteorological Society, c/o Inforonics, Inc., 550 Newtown Road, Box 458, Littleton, MA 01460. (508) 486-8976. FAX (508) 486-0027. Covers literature in environmental sciences, meteorology, astrophsyics, hydrology, glaciology and physical oceanography. 1950 to present. Monthly. $950.00 per year. Also available online (DIALOG) and on CD-ROM. ISSN: 0026-1130.

NTIS Alerts: Astronomy & Astrophysics. U. S. National Technical Information Service. 5285 Port Royal Road, Springfield, VA 22161. (703) 487-4650. FAX (703) 321-8547. Weekly. $140.00 per year.

Physics Abstracts. INSPEC. Section A, Science Abstracts. Institute of Electrical Engineers, London, United Kingdom. Available from: INSPEC/IEEE - Institute of Electrical and Electronic Engineers, Box 1331, Hoes Lane, Piscataway, NJ 08855-1331. (908) 562-5549. 1898 to present. 24 issues per year. $2835.00 per year. ISSN: 0036-8091. Alsoavailable online and on CD-ROM.

Science Citation Index. SCI. Institute for Scientific Information, 3501 Market Street, Philadelphia, PA 19104. (215) 386-0100. (800) 523-1850. FAX (215) 386-2991. 1961 to present. Six issues per year, plus annual cumulation. $11650.00 per year. Also available online and on CD-ROM.Inquire about price and availability. ISSN: 0036-827X.

STAR. Scientific and Technical Aerospace Reports. U.S. National Aeronautics and Space Administration. Distributed by U. S. Superintendent of Documents, Washington, DC 20402. 1963 to present. Semi-monthly, with semiannual and annual indexes. $114.00 per year. ISSN: 0036-8741. Also available online and on CD-ROM.

ANNUAL REVIEWS AND YEARBOOKS

The Astronomical Almanac. Superintendent of Documents, U. S. Government Printing office, Washington, DC 10402. (202) 783-3238. 1981 to present. Superceeds Astronomical Ephemeris. Annual. $44.95. ISSN: 0737-6421.

Annual Review of Astronomy and Astrophysics. Original reviews of critical literature and current developments in astronomy and astrophysics. Annual Reviews, Inc., 4139 El Camino Way, Palo Alto, CA 94303-0139. (415) 493-4400. (800)523-8635. Fax (415) 855-9815. 1963 to date. Annual. $60.00. ISSN: 0066-4146.

Annual Review of Earth and Planetary Sciences. Annual Reviews, Inc., 4139 El Camino Way, Palo Alto, CA 94303-0139. (415) 493-4400. (800) 523-8635. Fax (415) 855-9815. 1973 to date. Annual. $62.00. ISSN: 0084-6597.

ASSOCIATIONS AND PROFESSIONAL SOCIETIES

American Astronomical Society. 2000 Florida Avenue NW, Suite 400, Washington, DC 20009. (202) 328-2010. FAX\; (202) 234-2560.

SATURN

Ency. of Physical Sciences and Engineering Info. Sources

American Geophysical Union. 2000 Florida Avenue NW, Washington, DC 20009. (202) 462-6900. (800) 966-AGU1. FAX (202) 328-0566.

American Institute of Physics. 1 Physics Ellipse, College Park, MD 20740-3843. (301) 209-3100.

Association of Universities For Research in Astronomy, Inc. (AURA). Suite 701, 1625 Massachusetts Avenue, NW, Washington, DC 20036. (202) 483-2101. FAX (202) 483-2106.

Astronomical Society of the Pacific. 390 Ashton Avenue, San Francisco, CA 94112. (415) 337-1100. FAX: (415) 337-5205.

IEEE Geoscience and Remote Sensing Society. c/o Institute of Electrical and Electronics Engineers. 345 East 47th Street, New York, NY 10017. (212) 705-7900. FAX (212) 705- 4929.

Planetary Society. 65 North Catalina Avenue, Pasadena, CA 91106. (818) 793-5100. (800) WOW-MARS. FAX (818) 793-5528.

DIRECTORIES AND BIOGRAPHICAL SOURCES

American Astronomical Society. MEMBERSHIP DIRECTORY. American Astronomical Society. 2000 Florida Avenue NW, Suite 400, Washington, DC 20009. (202) 328-2010. FAX (202)234-2560 Annual. Included in membership dues. ISSN: 1061- 9038.

American Men and Women of Science: Physical and Biological Sciences. R. R. Bowker Inc., 121 Chanlon Road, New Providence, NJ 07974. (908) 464-6800. (800) 521-8110. 20th edition. 8 volumes. 1996. $850.00.

The Astronomers. Donald Goldsmith. St. Martin's Press, Inc., 175 Fifth Avenue, New York, NY 10010. (212) 674-5151. (800) 221-7945. 1993. $14.95 in paper.

Astronomical Centers of the World. Kevin Krisciunas. Cambridge University Press, 40 West 20th Street, New York, NY 10011-4211. (212) 924-3900. (800) 872-7423. 1988. $34.95

Directory of Physics and Astronomy Staff Members. American Institute of Physics, One Physics Ellipse, College Park, MD 20740-3843. (301) 209-3100. 1975/76 to present. Annual. $60.00. ISSN: 0361-2228.

Earth and Astronomical Research Centers. Stockton Press, 345 Park Avenue South, New York, NY 10010. 4th edition. 1995. $515.00. ISBN: 1-56169-0967.

ENCYCLOPEDIAS AND DICTIONARIES

Concise Dictionary of Astronomy. Jacqueline Mitton. Oxford University Press, Inc., 200 Madison Avenue, New York, NY 10016. (212) 725-6000. (800) 334-4249. 1992. $24.95.

Encyclopedia of Astronomy and Astrophysics. Stephen Maran, editor. Van Nostrand Reinhold, 115 Fifth Avenue, New York, NY 10003. (212) 254-3232. (800) 842-3636. 1992. $129.95.

McGraw-Hill Encyclopedia of Science and Technology. McGraw-Hill Book Company, Inc., 1221 Avenue of the Ameri-

cas, New York, NY 10020. (212) 512-2000. (800) 262-4729. 7th edition. 20 volume set. 1992. $1900.00.

Moons of the Solar System: An Illustrated Encyclopedia. John Stewart. McFarland & Company, Inc., Box 611, Jefferson, MC 28640. (910) 246-4460. (800) 253-2187. 1991. $49.95.

New Guide To the Planets. Patrick Moore. Trans-Atlantic Publications, Inc., 311 Bainbridge Street, Philadelphia, PA 19147. (215) 925-5083. 1993. $37.50.

Stars and Planets: The Sierra Club Guide To Sky Watching and Direction Finding. W. S. Kals. The Sierra Press, 4988 Gold Leaf Drive, Mariposa, CA 95338. (209) 966-5071. (800) 745-2631. $15.00.

GENERAL WORKS

Evolution of the Earth and Planets. E. Takahashi, et al, editors. American Geophysical Union, 2000 Florida Avenue, N.W., Washington, DC 20009. (202) 462-6903. (800) 966-2481. 1993. $28.00.

Exploring the Planets. W. Kenneth Hamblin and Eric H. Christiansen. Macmillan Publishing Company, Inc,. 200 Old Tappan Road, Old Tappan, NJ 07675. (800) 233-2336. 2nd edition. 1995. $62.00.

Far Encounter. Eric Burgess. Columbia University Press, 562 West 113th Street, New York, NY 10025. (212) 666-1000. (800) 944-8648. 1988. $40.00.

Maxwell On Saturn's Rings. James C. Maxwell. Stephen B. Brush, editor. MIT Press, 55 Hayward Street, Cambridge, MA 02142. (617) 253-8569. 1983. $37.50.

Moons and Planets. William K. Hartmann. Wadsworth Publishing Co., 10 Davis Drive, Belmont, CA 94002. (415) 595-2350. (800) 354-9706. 3rd edition. 1993. $49.95.

Planetary Landscapes. Ronald Greeley. Chapman & Hall, 1 Penn Plaza, New York, NY 10119. (212) 564-1060. Div. of Routledge,.. 2nd edition. 1994. $49.95.

Planetary Rings. Richard Greenberg, editor. University of Arizona Press, 1230 North Park Avenue, Number 102, Tucson, AZ 85719. (520) 621-1441. (800) 426-3797. 1984. $65.00.

The Planetary System. Tobias Owen and David Morrison. Addison-Wesley Publishing Co., Inc., 1 Jacob Way, Reading, MA 01867. (617) 944-3700. (800) 447-2226. 1988. $49.50.

Rings: Discoveries From Galileo To Voyager. James C. Elliot and Richard Kerr. MIT Press, 55 Hayward Street, Cambridge, MA 02142. (617) 253-8569. Reprint edition. 1987. $9.95 in paper.

Saturn. Tom Gehrels and Mildred S. Matthews, editors. University of Arizona Press, 1230 North Park Avenue, Number 102, Tucson, AZ 85719. (520) 621-1441. (800) 426-3797. 1984. $65.00.

Space Technology and Planetary Astronomy. Joseph N. Tatarewicz. Indiana University Pres, 601 North Morton Street, Bloomington, in 47404-3797. (812) 855-6804. (800) 842-6796. 1990. $29.95.

Voyage To Saturn: The Voyager Mission. Gordon Press.

*Gordon Press Publications, P.*O. Box 459, Bowling Green Station, New York, NY 10004. (718) 624-8419. 1994. $260.95.

Worlds Apart: A Textbook in Planetary Sciences. Guy Consolmagno. Prentice-Hall, 113 Sylvan Avenue, Route 9W, Englewood Cliffs, NJ 07632. (201) 592-2000. (800) 922- 0579. 1994. $62.00.

HANDBOOKS AND MANUALS

Astrophysical Data: Planets and Stars. Kenneth R. Lanb. Springer-Verlag New York, Inc., 175 Fifth Avenue, New York, NY 10010. (212) 460-1500. (800) 777-4643. 1993. $59.00.

Field Guide To the Stars and Planets. Jay M. Pasachoff and Donald H. Menzel. Houghton Mifflin Co., 222 Berkeley Street, Boston, MA 02116. (617) 351-5000. (800) 225-3362. Revised edition. 1992. $24.95.

New Guide To the Planets. Patrick Moore. Trans-Atlantic Publications, Inc., 311 Bainbridge Street, Philadelphia, PA 19147. (215) 925-5083. 1993. $37.50.

Planet Observer's Handbook. Fred W. Price. Cambridge University Press, 40 West 20th Street, New York, NY 10011- 4211. (212) 924-3900. (800) 872-7423. 1994. $34.95.

Planets and their Moons. Aububon Society Staff. Alfred A. Knopf, Inc. 201 E. 50th Street, New York, NY 10022. (800) 733-3000. 1995. $7.99.

ONLINE DATABASES AND CD-ROMS

Dissertation Abstracts. University Microfilms International, 300 North Zeeb Road, Ann Arbor, MI 48106. (800) 521-0600 or (313) 761-4700. Scope includes virtually all doctoral dissertations accepted at accredited American institutions from 1861 to present in 252 subject areas. Available on BRS, (800) 289-4277, DIALOG, (800) 334-2564, and OCLC EPIC, (800) 848-5878, online services. Also available on CD-ROM. Inquire as to cost and availability.

GEOREF. American Geological Institute, 4220 King Street, Alexandria, VA 22302. (800) 336-4764 or (703) 379-2480. Geology and geosciences literature, 1785 to present for North America. Available on DIALOG,(800) 334-2564, ORBIT (800) 456-7248, online services. Also available on CD-ROM. Inquire as to cost and availability.

INSPEC. Institution of Electrical Engineers, Michael Faraday House, Six Hills Way, Stevenage, Herts. SG1 2AY, England. Telephone: 0438 313311 or FAX 0438 742840. Contains citations to the worldwide literature of physics, electronics and electrical engineering, computer technology, and related fields. Available on BRS, (800) 289-4277, DIALOG, (800) 334-2564, ORBIT, (800) 456-7248, and STN International, FIZ Karlsruhe, P.O. Box 2465, W-7500, Karlsruhe 1, Germany, online services. Inquire as to cost and availability.

NTIS Bibliographic Database. National Technical Information Service, 5285 Port Royal Road, Springfield, VA 22161. (703) 487-4929 or FAX (703) 321-8199. Broad coverage of government-sponsored science and technology research reports, 1964 to present. Available on BRS,(800) 289-4277, DIALOG, (800)

334-2564, ORBIT, (800) 456-7248, and STN International, FIZ Karlsruhe, P.O. Box 2465, W-7500, Karlsruhe 1, Germany, online services. Also available on CD-ROM. Inquire as to cost and availability.

SCISEARCH. Institute for Scientific Information, 3501 Market Street, Philadelphia, PA 19104. (800) 523-1850 or (215) 386-0100. Broad multidisciplinary title and author index to the international literature of science and technology, 1974 to present. Available on DIALOG, (800) 334-2564, and ORBIT, (800) 456-7248, online services. Also available on CD-ROM. Inquire as to cost and availability.

WILSONLINE. H.W. Wilson Company, 950 University Avenue, Bronx, NY 10452. (800) 367-6770 or (212) 588-8400. Makes available online versions of the printed H.W. Wilson indexes including Applied Science and Technology Index, Business Periodicals Index, General Science Index, and Readers' Guide to Periodical Literature. Period covered is generally 1983 to present. Available on BRS, (800) 289-4277, DIALOG, (800) 334-2564, and OCLC EPIC, (800) 848-5878, online services.Also available on CD-ROM. Inquire as to cost and availability.

OTHER SOURCES

Atlas of the Solar System. B. Yenne. Simon & Schuster, Inc. 1230 Avenue of the Americas, New York, NY 10020. (212) 698-7000. (800) 223-2348. 1987. $12.98.

The Sky: A User's Guide. David H. Levy. Cambridge University Press, 40 West 20th Street, New York, NY 10011- 4211. (212) 924-3900. (800) 872-7423. 1993. $14.95.

The View From Space: Photographic Exploration of the Planets. Merton Davies and Bruce C. Murray. Columbia University Press, 562 West 113th Street, New York, NY 10025. (212) 666-1000. (800) 944-8648. 1973. $17.50.

PERIODICALS

Association of Lunar and Planetary Observers Journal. Association of Lunar and Planetary Observers, Box 16131, San Francisco, CA 94116. (415) 566-5786. FAX (415) 731-8242. 1947 to present. Quarterly. $14.00 per year. ISSN: 0039- 2502.

Astronomical Journal. American Institute of Physics, One Physics Ellipse, College Park, MD 20740-3843. (301) 209-3000. Published for the American Astronomical Society. 1849 to present. Monthly. $280.00 per year. ISSN: 0004- 6256.

Astronomical Society of the Pacific Publications. Astronomical Society of the Pacific, 390 Ashton Avenue, San Francisco, CA 94112. (415) 337-1100. FAX (415) 337-5205.1889 to present. Monthly. $175.00 per year. ISSN: 0004- 6280.

Astronomy. Kalmbach Publishing Company, Box 1612, Waukesha, WI 53187-1612. (414) 796-8776. FAX (414) 796-1142. 1973 to present. Monthly. $27.00 per year. ISSN: 0091-6358.

Earth, Moon and Planets; An International Journal of Comparative Planetology. Kluwer Academic Publishers, Box 358, Accord Station, Hingham, MA 02018-0358. (617) 871-6600. FAX (617) 871-6528. 1969 to present. Monthly. $840.00 per year. ISSN: 0167-9295.

SATURN

Ency. of Physical Sciences and Engineering Info. Sources

Geochemica Et Cosmochimica Acta. Elsevier Science. 660 White Plains Road, Tarrytown, NY 10591-5153. (914) 524-9200. FAX (914) 333-2444. 1950 to the present. Biweekly. $895.00 per year. ISSN: 0016-7037. # #*ICARUS; International Journal of Solar System Studies.* Academic Press, Inc., Journal Division, 525 B Street, Suite 1900, San Diego, CA 92101-4495. (619) 230-1840. FAX (619) 699-6800. 1962 to the present. Monthly. $1080.00. ISSN: 0019-1035.

J G R: Journal of Geophysical Research: Planets. American Geophysical Union, 2000 Florida Avenue, NW, Washington, CD 20009. (202) 462-6900. FAX (202) 328-0566. 1991 to present. Monthly. $597.00 per year. ISSN: 0148-0227.

Lunar and Planetary Information Bulletin. Lunar and Planetary Institute. 3600 Bay Area Boulevard, Houston, TX 77058-1113. (713) 486-2175. FAX (713) 486-2125. 1970 to present. Quarterly. Free. Also available online.

Mercury. Astronomical Society of the Pacific. 390 Ashton Avenue, San Francisco, CA 94112. (415) 337-1100. FAX: (415) 337-5205. 1972 to present. Bimonthly. $175.00 per year. ISSN: 0047-6773.

Observatory. c/o Dr. D. J. Stickland Space and Astrophysics Division, Rutherford Appleton Laboratory, Chilton, Didcot, Oxon OX11 OQX England. FAX 0235-445848. 1877 to present. Bimonthly. $42.00 per year. ISSN: 0029-7704.

Planetary and Space Science. Elsevier Science Publishing Company, Inc., 660 White Plains Road, Terrytown, NY 10591- 5153. (914) 524-9200. FAX: (914) 333-2444. 1959 to present. Monthly. $1355.00 per year. ISSN: 0032-0633.

Planetary Report. Planetary Society. 65 North Catalina, Pasadena, CA 91106-2301. (818) 793-5100. FAX (818) 793- 5528. 1980 to present. $25.00 per year. ISSN: 0736-3680.

Sky and Telescope. Sky Publishing Corporation, Box 9111, Belmont, MA 02178. (617) 864-7360. FAX (617) 864-6117.1941 to present. Monthly. $27.00 per year. ISSN: 0037-6604.

RESEARCH CENTERS AND INSTITUTES.

Center For Space Plasma and Aeronomic Research. University of Alabama in Huntsville, Huntsville, AL 35899. (205) 895-6258. (205) 805-6382.

Center For Space Science and Astrophysics. Stanford University. 325 Durand Building, Stanford, CA 04305. (415) 723-3582.

Center For Space Sciences. University of Texas at Dallas, P. O. Box 830688, MS_F022, Richardson, TX 75083-0688. (214) 690-2851. FAX (214) 690-2848.

Institute For Astronomy. University of Hawaii at Manoa, 2680 Woodlawn Drive, Honolulu, HI 96822. (808) 956-8312. FAX (808) 988-2790. # #Laboratory For Planetary Studies. Cornell University, Space Sciences Building, Ithaca, NY 14853. (607) 255-4971.

Lunar and Planetary Institute. 3303 NASA Road One, Houston, TX 77058-4399. (713) 486-2139. FAX 713-496-2162.

SCIENTIFIC INSTRUMENTS
See: INSTRUMENTATION, METROLOGY

SCINTILLATION
See: RADIATION

SCRUBBERS
See: AIR POLLUTION

SEA ICE
See: ICE

SEA WATER
See: OCEANOGRAPHY

SEDIMENTARY ROCKS

See also: CRUST, GEOCHEMISTRY, IGNEOUS ROCKS, METAMORPHIC ROCKS, MINERALOGY, PETROLEUM GEOLOGY, PETROLOGY, PHYSICAL GEOLOGY, SEDIMENTOLOGY

ABSTRACT SERVICES AND INDEXES

Bibliography and Index of Geology. American Geological Institute, 4220 King Street, Alexandria, VA 22302. (703) 379-2480. Fax (703) 379-7563. Monthly. $1295.00 per year. Also available online as GEOREF.

Bibliography of Economic Geology. Geosystems, Box 40, Didcot Oxon Ox11 9BX, England. Bimonthly. $150.00 per year. Also available online as GeoArchive.

Chemical Abstracts - Applied Chemistry and Chemical Engineering Sections. Chemical Abstracts Service, 2540 Olentangy River Road, Box 3012, Columbus, OH 43210. (800) 848-6538 or (614) 447-3600, FAX (614) 447-3713. Bi-weekly. $1410.00 per year. Also available online as CA (Chemical Abstracts).

Deep-Sea Research. Part A: Oceanographic Research Papers. Deep-Sea research. Part B: Oceanographic Literature Review. Pergamon Press, Inc., Maxwell House, Fairview Park, Elmsford, NY 10523. (914) 592-7700. Fax (914) 592-3625. Twelve times per year. $1370 per year for Parts A and B.

Geological Abstracts. Elsevier - Geo Abstracts, Regency House, 34 Duke Street, Norwich NR3 3AP, England. Monthly. $760.00 per year. Also available online as GEOBASE.

Mineralogical Abstracts. Mineralogical Society and the Mineralogical Society of America, 41 Queen's Gate, London, SW7 5HR, England. Quarterly. $235.00 per year.

Oceanic Abstracts. Cambridge Scientific Abstracts, 7200 Wisconsin Avenue, Bethesda, MD 20814. (301) 961-6750. Fax (301) 961-6720. Bimonthly. $995.00 per year.

Science Citation Index. Institute for Scientific Information, 3501 Market Street, Philadelphia, PA 19104. (800) 523-1850 or (215) 386-0100. Inquire about price and availability.

ANNUAL REVIEWS AND YEARBOOKS

Annual Review and Earth and Planetary Sciences. Annual Reviews, Inc., 4139 El Camino Way, Palo Alto, CA 94306-0897. (415) 493-4400. Fax (415) 855-9815. Annual. $55.00 per year.

Minerals Yearbook. Bureau of Mines, U.S. Department of the Interior. Available from U.S. Government Printing office, Washington, DC 20402. (202) 783-3238. Annual. Three volumes. $45.00.

ASSOCIATIONS AND PROFESSIONAL SOCIETIES

American Geological Institute. 4220 King Street, Alexandria, VA 22302. (703) 379-2480. Fax (703) 379-7563.

American Institute of Professional Geologists. 7828 Vance Drive, Suite 103, Arvada, CO 80003. (303) 431-0831.

Geochemical Society. Department of Terrestrial Magnetism, 5241 Broad Branch Road, N.W., Washington, DC 20015. (202) 686-4370.

Geological Society of America. 3300 Penrose Place, P.O. Box 9140, Boulder, CO 80301-9140. (303) 447-2020.

Society of Economic Paleontologists and Mineralogists. P.O. Box 4756, Tulsa, OK 74159-0756. (918) 743-9765.

Society of Independent Professional Earth Scientists. 4925 Greenville Avenue, Suite 170, Dallas, TX 75206. (214) 363-1780.

BIBLIOGRAPHIES

Geologic Reference Sources: A Subject and Regional Bibliography of Publications and Maps in the Geological Sciences. Dederick C. Ward et al. Scarecrow Press, Inc., 52 Liberty Street, Box 4167, Metuchen, NJ 08840. (800) 537-7107 or (201) 548-8600. 1981. $49.50.

Information Sources in the Earth Sciences. David N. Wood, Joan E. Hardy, and Anthony P. Harvey, editors. Bowker-Saur/K.G. Saur. Distributed by R.R. Bowker, 121 Chanlon Road, New Providence, NJ 07974. (800) 521-8110 or (908) 464-6800. Second edition. 1989. $85.00.

DIRECTORIES AND BIOGRAPHICAL SOURCES

American Institute of Professional Geologists. Membership Directory. American Institute of Professional Geologists, 7828 Vance Drive, Suite 103, Arvada, CO 80003. (303) 431-0831. Annual. $18.00.

Directory of Geoscience Departments: United States and Canada. American Geological Institute. 4220 King Street, Alexandria, VA 22302-1507. (703) 379-2480. Fax (703) 379-7563. Twenty-six edition. 1987. $22.00.

Geological Society of America. Membership Directory. Geological Society of America, 3300 Penrose Place, Boulder, CO 80301. (303) 447-2020. Annual. Available to members only.

Research Centers Directory. Gale Research, 835 Penobscot Building, Detroit, MI 48226-4094. (800) 877-4253 or (313) 961-2242. 20th edition, 1995. $485.00.

ENCYCLOPEDIAS AND DICTIONARIES

Concise Oxford Dictionary of the Earth Sciences. Ailsa Allaby and Michael Allaby, editors. Oxford University Press, Inc., 200 Madison Avenue, New York, NY 10016. (800) 334-4249 or (212) 679-7300. 1990. $42.95.

Dictionary of Geological Terms. American Geological Institute. Doubleday and Company, Inc., 245 Park Avenue, New York, NY 10017. (800) 645-6156 or (212) 953-4561. Third edition. 1984. $12.00 in paper.

Dictionary of Petrology. S.I. Tomkeieff. John Wiley and Sons, Inc., 605 Third Avenue, New York, NY 10158. (800) 526-5368 or (212) 850-6000. 1983. $180.00.

Encyclopedia Of Earth System Science. William A. Nierenberg, editor. Academic Press, Inc., 1250 Sixth Avenue, San Diego, CA 92101. (619) 231-0926. FAX (619) 699-6715. Four volumes. 1992. $950.00

Glossary of Geology. Robert L. Bates and Julia A. Jackson. American Geological Institute, 4220 King Street, Alexandria, VA 22032. (703) 379-2480. Third edition. 1987. $75.00.

GENERAL WORKS

Atlas of Sedimentary Rocks Under the Microscope. A.E. Adams. Halsted Press (Division of John Wiley and Sons, Inc.), 6057 Third Avenue, New York, NY 10158. (800) 526-5368 or (212) 850-6418. 1984. $32.95.

Origin of Sedimentary Rocks. H. Blatt. Prentice Hall (Division of Simon and Schuster), 15 Columbus Circle, New York, NY 10023. (212) 373-8500 or (800) 922-0579. Second edition. 1980.

Petrology of Sedimentary Rocks. Robert L. Folk. HempHill Publishing Company, 1400 Wathen Avenue, Austin, TX 78703. (512) 476-9422. Fourth edition. 1980. $17.95.

Petrology of the Sedimentary Rocks. J.T. Greensmith. Unwin Hyman, Inc., 10 East 53rd Street, New York, NY 10022. (212) 207-7626 or (800) 242-7737. Seventh edition. 1988. $24.95.

HANDBOOKS AND MANUALS

AGI Data Sheets For Geology in the Field, Laboratory, and Office. J.T. Dutro, R.V. Dietrich, and R.M. Foose, editors. American Geological Institute, 4220 King Street, Alexandria, VA 223020-1507 (800) 336-4764 or (703) 379-2480. 1989. $24.95.

Field Description of Sedimentary Rocks. Maurice E. Tucker. Halsted Press (Division of John Wiley and Sons, Inc.), 6057

SEDIMENTARY ROCKS

Ency. of Physical Sciences and Engineering Info. Sources

Third Avenue, New York, NY 10158. (800) 526-5368 or (212) 850-6418. 1982. $19.95.

Field Mapping For Geology Students. F. Ahmed and D.C. Almond. Unwin Hyman, Inc., 10 East 53rd Street, New York, NY 10022. (212) 207-7626 or (800) 242-7737. 1983. $18.95.

Geology in the Field. R.R. Compton. John Wiley and Sons, Inc., 605 Third Avenue, New York, NY 10158. (800) 526-5368 or (212) 850-6418. 1985. $37.95.

A Manual of Geology For Civil Engineers. John A. Pitts. World Scientific Publishing Company, Inc., 1060 Main Street, Unit B, River Edge, NJ 07661. (800) 227-7562 or (201) 487-9655. 1985. $38.00.

The Mapping of Geological Structures. K.R. McClay. John Wiley and Sons, Inc., 605 Third Avenue, New York, NY 10158. (800) 526-5368 or (212) 850-6000. 1988. $18.95.

ONLINE DATABASES AND CD-ROMS

Earth Sciences. U.S. Geological Survey, 12201 Sunrise Valley Drive, Reston, VA 22092-9998. (703) 648-4460. CD-ROM of earth science databases including the U.S. Geological Survey Library, Earth Science Data Directory, and GEOINDEX, citations to published geological maps. $350.00 per year with quarterly updates.

Geoarchive. Geosystems, P.O. Box 1024, Westminster, London, England, SW1 P 2JL. Citations to literature on geoscience, 1969 to present. Inquire as to online cost and availability.

Geobase. Elsevier - Geo Abstracts, Regency House, 34 Duke Street, Norwich NR3 3AP, England. Contains citations to the worldwide earth science literature from 1980 to date. Available on DIALOG, ORBIT online services. Inquire as to cost and availability.

Geomechanics Abstracts. University of London, Imperial College of Science and Technology, Rock Mechanics Information Service, Royal School of Mines, Prince Consort Road, London SW7 2BP, England. Telephone: 071-589 5111 or FAX 071-589 6806. Scope includes worldwide literature on engineering geology, hydrogeology, mining, rock mechanics, soil mechanics, and tunnelling, 1977 to present. Available on the ORBIT online service. Inquire as to cost and availability.

GEOREF. American Geological Institute, 4220 King Street, Alexandria, VA 22302. (800) 336-4764 or (703) 379-2480. Geology and geosciences literature, 1785 to present for North America. Available on DIALOG, ORBIT online services. Also available on CD-ROM. Inquire as to cost and availability.

NTIS Bibliographic Database. National Technical Information Service, 5285 Port Royal Road, Springfield, VA 22161. (703) 487-4630. Broad coverage of government- sponsored research reports, 1964 to present. Available on BRS, DIALOG, ORBIT and STN online services. Also available on CD-ROM. Inquire as to cost and availability.

SCISEARCH. Institute for Scientific Information, 3501 Market Street, Philadelphia, PA 19104. (800) 523-1850 or (215) 386-0100. Broad multidisciplinary title and author index to the international literature of science and technology, 1974 to present. Available on DIALOG and ORBIT online services. Also available on CD-ROM. Inquire as to cost and availability.

PERIODICALS

Canadian Journal of Earth Sciences. National Research Council of Canada, Research Journals, Ottawa, Ont. K1A OR6, Canada. (613) 993-9084. Monthly. $112.00 per year.

Geochimica Et Cosmochimica Acta. Pergamon Journals, Inc., Maxwell House, Fairview Park, Elmsford, NY 10523. (914) 592-7700. Fax (914) 592-3625. Monthly. $580.00 per year.

Geoforum. Pergamon Press, Maxwell House, Fairview Park, Elmsford, NY 10523. (914) 592-7700. Fax (914) 592-3625. Quarterly. $305.00 per year.

Geological Magazine. Cambridge University Press, 40 West 20th Street, New York, NY 10011-4211. (212) 924-3900. Bimonthly. $205.00 per year.

Geological Society. Journal. Geological Society of London. Geological Society Publishing House, Unit 7, Brassmill Enterprise Centre, Brassmill Lane, Bath, Avon BA1 3JN, England. 6 times per year. $432.00.

Geological Society of America Bulletin. P.O. Box 9140, 3300 Penrose Place, Boulder, CO 80301. (303) 447-2020. Fax (303) 447-1133. Monthly. $150.00.

Geology. Geological Society of America. P.O. Box 9140, 3300 Penrose Place, Boulder, CO 80301. (303) 447-2020. Fax (303) 447-1133. Monthly. $120.00 per year.

Journal of Geology. University of Chicago Press, 5720 Woodlawn Avenue, Chicago, IL 60637. (312) 753-3347.

Bimonthly. $38.00 per year.

Journal of Petrology. Oxford University Press, Inc., 200 Madison Avenue, New York, NY 10016. (800) 334-4249 or (212) 679-7300. Bimonthly. $240.00 per year.

Journal of Sedimentary Petrology. Society of Economic Paleontologists and Mineralogists. P.O. Box 4756, Tulsa, OK 74159-0756. (918) 743-9765. Bimonthly. $120.00 per year.

Journal of Structural Geology. Pergamon Journals, Inc., Maxwell House, Fairview Park, Elmsford, NY 10523. (914) 592-7700. Ten times per year. $375.00 per year.

Lithos; An International Journal of Mineralogy, Petrology, and Geochemistry. Elsevier Science Publishing Company, Inc., 655 Avenue of the Americas, New York, NY 10010. (212) 989-5800. Quarterly. $275.00 per year.

Sedimentary Geology: International Journal of Pure and Applied Sedimentology. Elsevier Science Publishing Company, Inc., 655 Avenue of the Americas, New York, NY 10010. (212) 989-5800. Twenty times per year. $800.00 per year.

Sedimentology. Blackwell Scientific Publishing, Inc., 3 Cambridge Court, Suite 208, Cambridge, MA 02142. (617) 225-0401. Six times per year. $400.00 per year.

RESEARCH CENTERS AND INSTITUTES

Geological Society of America. P.O. Box 9140, 3300 Penrose Place, Boulder, CO 80301. (303) 447-2020.

Louisiana State University. Basin Research Institute. Baton Rouge, LA 70803-4101. (504) 388-8328. Fax (504) 388-5328.

Texas A&m University. Earth Resources Institute. College Station, TX 77843. (409) 845-3651.

U.S. Geological Survey, Geologic Division, National Center, 12201 Sunrise Valley Drive, Reston, VA 22092. (703) 648-6600. The major geological research agency of the federal government conducting research in most areas of pure and applied research in the geosciences.

University of Miami. Comparative Sedimentology Laboratory. RSMAS-MGG, 4600 Rickenbacker Causeway, Miami, FL 33149. (305) 672-1840.

SEDIMENTOLOGY

See also: CRUST, GEOCHEMISTRY, GEOLOGY, IGNEOUS ROCKS, METAMORPHIC ROCKS, MINERALOGY, PETROLEUM GEOLOGY, PETROLOGY, PHYSICAL GEOLOGY, SEDIMENTARY ROCKS

ABSTRACT SERVICES AND INDEXES

Bibliography and Index of Geology. American Geological Institute, 4220 King Street, Alexandria, VA 22302. (703) 379-2480. Fax (703) 379-7563. Monthly. $1295.00 per year. Also available online as GEOREF.

Bibliography of Economic Geology. Geosystems, Box 40, Didcot Oxon Ox11 9BX, England. Bimonthly. $150.00 per year. Also available online as GeoArchive.

Chemical Abstracts - Applied Chemistry and Chemical Engineering Sections. Chemical Abstracts Service, 2540 Olentangy River Road, Box 3012, Columbus, OH 43210. (800) 848-6538 or (614) 447-3600, FAX (614) 447-3713. Bi-weekly. $1410.00 per year. Also available online as CA (Chemical Abstracts).

Deep-Sea Research. Part A: Oceanographic Research Papers. Deep-Sea Research. Part B: Oceanographic Literature Review. Pergamon Press, Inc., Maxwell House, Fairview Park, Elmsford, NY 10523. (914) 592-7700. Fax (914) 592-3625. Twelve times per year. $1370 per year for Parts A and B.

Geological Abstracts. Elsevier - Geo Abstracts, Regency House, 34 Duke Street, Norwich NR3 3AP, England. Monthly. $760.00 per year. Also available online as GEOBASE.

Mineralogical Abstracts. Mineralogical Society and the Mineralogical Society of America, 41 Queen's Gate, London, SW7 5HR, England. Quarterly. $235.00 per year.

Oceanic Abstracts. Cambridge Scientific Abstracts, 7200 Wisconsin Avenue, Bethesda, MD 20814. (301) 961-6750. Fax (301) 961-6720. Bimonthly. $995.00 per year.

Science Citation Index. Institute for Scientific Information, 3501 Market Street, Philadelphia, PA 19104. (800) 523-1850 or (215) 386-0100. Inquire about price and availability.

ANNUAL REVIEWS AND YEARBOOKS

Annual Review and Earth and Planetary Sciences. Annual Reviews, Inc., 4139 El Camino Way, Palo Alto, CA 94306-0897. (415) 493-4400. Fax (415) 855-9815. Annual. $55.00 per year.

Minerals Yearbook. Bureau of Mines, U.S. Department of the Interior. Available from U.S. Government Printing office, Washington, DC 20402. (202) 783-3238. Annual. Three volumes. $45.00.

ASSOCIATIONS AND PROFESSIONAL SOCIETIES# #AMERICAN GEOLOGICAL INSTITUTE. 4220 KING STREET, ALEXANDRIA, VA 22302. (703) 379-2480. FAX (703) 379-7563.

American Institute of Professional Geologists. 7828 Vance Drive, Suite 103, Arvada, CO 80003. (303) 431-0831.

Geochemical Society. Department of Terrestrial Magnetism, 5241 Broad Branch Road, N.W., Washington, DC 20015. (202) 686-4370.

Geological Society of America. 3300 Penrose Place, P.O. Box 9140, Boulder, CO 80301-9140. (303) 447-2020.

Society of Economic Paleontologists and Mineralogists. P.O. Box 4756, Tulsa, OK 74159-0756. (918) 743-9765.

Society of Independent Professional Earth Scientists. 4925 Greenville Avenue, Suite 170, Dallas, TX 75206. (214) 363-1780.

BIBLIOGRAPHIES

Geologic Reference Sources: A Subject and Regional Bibliography of Publications and Maps in the Geological Sciences. Dederick C. Ward et al. Scarecrow Press, Inc., 52 Liberty Street, Box 4167, Metuchen, NJ 08840. (800) 537-7107 or (201) 548-8600. 1981. $49.50.

Information Sources in the Earth Sciences. David N. Wood, Joan E. Hardy, and Anthony P. Harvey, editors. Bowker-Saur/K.G. Saur. Distributed by R.R. Bowker, 121 Chanlon Road, New Providence, NJ 07974. (800) 521-8110 or (908) 464-6800. Second edition. 1989. $85.00.

DIRECTORIES AND BIOGRAPHICAL SOURCES

American Institute of Professional Geologists. Membership Directory. American Institute of Professional Geologists, 7828 Vance Drive, Suite 103, Arvada, CO 80003. (303) 431-0831. Annual. $18.00.

Directory of Geoscience Departments: United States and Canada. American Geological Institute. 4220 King Street, Alexandria, VA 22302-1507. (703) 379-2480. Fax (703) 379-7563. Twenty-six edition. 1987. $22.00.

Geological Society of America. Membership Directory. Geological Society of America, 3300 Penrose Place, Boulder, CO 80301. (303) 447-2020. Annual. Available to members only.

SEDIMENTOLOGY

Ency. of Physical Sciences and Engineering Info. Sources

Research Centers Directory. Gale Research, 835 Penobscot Building, Detroit, MI 48226-4094. (800) 877-4253 or (313) 961-2242. 20th edition, 1995. $485.00.

Encyclopedias and Dictionaries# #*Concise Oxford Dictionary of the Earth Sciences.* Ailsa Allaby and Michael Allaby, editors. Oxford University Press, Inc., 200 Madison Avenue, New York, NY 10016. (800) 334-4249 or (212) 679-7300. 1990. $42.95.

Dictionary of Geological Terms. American Geological Institute. Doubleday and Company, Inc., 245 Park Avenue, New York, NY 10017. (800) 645-6156 or (212) 953-4561. Third edition. 1984. $12.00 in paper.

Dictionary of Petrology. S.I. Tomkeieff. John Wiley and Sons, Inc., 605 Third Avenue, New York, NY 10158. (800) 526-5368 or (212) 850-6000. 1983. $180.00.

Encyclopedia Of Earth System Science. William A. Nierenberg, editor. Academic Press, Inc., 1250 Sixth Avenue, San Diego, CA 92101. (619) 231-0926. FAX (619) 699-6715. Four volumes. 1992. $950.00

Glossary of Geology. Robert L. Bates and Julia A. Jackson. American Geological Institute, 4220 King Street, Alexandria, VA 22032. (703) 379-2480. Third edition. 1987. $75.00.

GENERAL WORKS

Applied Sedimentology. Richard C. Selley. Academic Press, Inc., 1250 Sixth Avenue, San Diego, CA 92101. (619) 231-0926. FAX (619) 699-6715. 1988. $88.00

Petrology of the Sedimentary Rocks. J.T. Greensmith. Unwin Hyman, Inc., 10 East 53rd Street, New York, NY 10022. (212) 207-7626 or (800) 242-7737. Seventh edition. 1988. $24.95.

Principles of Physical Sedimentology. John R. Allen. Unwin Hyman, Inc., 10 East 53rd Street, New York, NY 10022. (212) 207-7626 or (800) 242-7737. 1985. $65.00.

Physical Principles of Sedimentology. K.J. Hsu. Springer-Verlag, 175 Fifth Avenue, New York, NY 10010. (212) 460-1500. 1989. $29.00.

Sedimentology. H. Chamley. Springer-Verlag, 175 Fifth Avenue, New York, NY 10010. (212) 460-1500. 1990. $34.00.

Sedimentology: Process and Product. M.R. Leeder. Unwin Hyman, Inc., 10 East 53rd Street, New York, NY 10022. (212) 207-7626 or (800) 242-7737. 1982. $34.95.

HANDBOOKS AND MANUALS

AGI Data Sheets For Geology in the Field, Laboratory, and Office. J.T. Dutro, R.V. Dietrich, and R.M. Foose, editors. American Geological Institute, 4220 King Street, Alexandria, VA 223020-1507 (800) 336-4764 or (703) 379-2480. 1989. $24.95.# #*Field Description of Sedimentary Rocks.* Maurice E. Tucker. Halsted Press (Division of John Wiley and Sons, Inc.), 6057 Third Avenue, New York, NY 10158. (800) 526-5368 or (212) 850-6418. 1982. $19.95.

Field Mapping For Geology Students. F. Ahmed and D.C. Almond. Unwin Hyman, Inc., 10 East 53rd Street, New York, NY 10022. (212) 207-7626 or (800) 242-7737. 1983. $18.95.

Geology in the Field. R.R. Compton. John Wiley and Sons, Inc., 605 Third Avenue, New York, NY 10158. (800) 526-5368 or (212) 850-6418. 1985. $37.95.

ONLINE DATABASES AND CD-ROMS

Earth Sciences. U.S. Geological Survey, 12201 Sunrise Valley Drive, Reston, VA 22092-9998. (703) 648-4460. CD-ROM of earth science databases including the U.S. Geological Survey Library, Earth Science Data Directory, and GEOINDEX, citations to published geological maps. $350.00 per year with quarterly updates.

Geoarchive. Geosystems, P.O. Box 1024, Westminster, London, England, SW1 P 2JL. Citations to literature on geoscience, 1969 to present. Inquire as to online cost and availability.

Geobase. Elsevier - Geo Abstracts, Regency House, 34 Duke Street, Norwich NR3 3AP, England. Contains citations to the worldwide earth science literature from 1980 to date. Available on DIALOG, ORBIT online services. Inquire as to cost and availability.

Geomechanics Abstracts. University of London, Imperial College of Science and Technology, Rock Mechanics Information Service, Royal School of Mines, Prince Consort Road, London SW7 2BP, England. Telephone: 071-589 5111 or FAX 071-589 6806. Scope includes worldwide literature on engineering geology, hydrogeology, mining, rock mechanics, soil mechanics, and tunnelling, 1977 to present. Available on the ORBIT online service. Inquire as to cost and availability.

GEOREF. American Geological Institute, 4220 King Street, Alexandria, VA 22302. (800) 336-4764 or (703) 379-2480. Geology and geosciences literature, 1785 to present for North America. Available on DIALOG, ORBIT online services. Also available on CD-ROM. Inquire as to cost and availability.

NTIS Bibliographic Database. National Technical Information Service, 5285 Port Royal Road, Springfield, VA 22161. (703) 487-4630. Broad coverage of government- sponsored research reports, 1964 to present. Available on BRS, DIALOG, ORBIT and STN online services. Also available on CD-ROM. Inquire as to cost and availability.

SCISEARCH. Institute for Scientific Information, 3501 Market Street, Philadelphia, PA 19104. (800) 523-1850 or (215) 386-0100. Broad multidisciplinary title and author index to the international literature of science and technology, 1974 to present. Available on DIALOG and ORBIT online services. Also available on CD-ROM. Inquire as to cost and availability.

PERIODICALS

Canadian Journal of Earth Sciences. National Research Council of Canada, Research Journals, Ottawa, Ont. K1A OR6, Canada. (613) 993-9084. Monthly. $112.00 per year.

Geochimica Et Cosmochimica Acta. Pergamon Journals, Inc., Maxwell House, Fairview Park, Elmsford, NY 10523. (914) 592-7700. Fax (914) 592-3625. Monthly. $580.00 per year.

Geoforum. Pergamon Press, Maxwell House, Fairview Park, Elmsford, NY 10523. (914) 592-7700. Fax (914) 592-3625. Quarterly. $305.00 per year.

Geological Magazine. Cambridge University Press, 40 West 20th Street, New York, NY 10011-4211. (212) 924-3900. Bimonthly. $205.00 per year.

Geological Society. Journal. Geological Society of London. Geological Society Publishing House, Unit 7, Brassmill Enterprise Centre, Brassmill Lane, Bath, Avon BA1 3JN, England. 6 times per year. $432.00.

Geological Society of America Bulletin. P.O. Box 9140, 3300 Penrose Place, Boulder, CO 80301. (303) 447-2020. Fax (303) 447-1133. Monthly. $150.00.

Geology. Geological Society of America. P.O. Box 9140, 3300 Penrose Place, Boulder, CO 80301. (303) 447-2020. Fax (303) 447-1133. Monthly. $120.00 per year.

Journal of Geology. University of Chicago Press, 5720 Woodlawn Avenue, Chicago, IL 60637. (312) 753-3347. Bimonthly. $38.00 per year.

Journal of Petrology. Oxford University Press, Inc., 200 Madison Avenue, New York, NY 10016. (800) 334-4249 or (212) 679-7300. Bimonthly. $240.00 per year.

Journal of Sedimentary Petrology. Society of Economic Paleontologists and Mineralogists. P.O. Box 4756, Tulsa, OK 74159-0756. (918) 743-9765. Bimonthly. $120.00 per year.

Journal of Structural Geology. Pergamon Journals, Inc., Maxwell House, Fairview Park, Elmsford, NY 10523. (914) 592-7700. Ten times per year. $375.00 per year.

Lithos; An International Journal of Mineralogy, Petrology, and Geochemistry. Elsevier Science Publishing Company, Inc., 655 Avenue of the Americas, New York, NY 10010. (212) 989-5800. Quarterly. $275.00 per year.

Sedimentary Geology: International Journal of Pure and Applied Sedimentology. Elsevier Science Publishing Company, Inc., 655 Avenue of the Americas, New York, NY 10010. (212) 989-5800. Twenty times per year. $800.00 per year.

Sedimentology. Blackwell Scientific Publishing, Inc., 3 Cambridge Court, Suite 208, Cambridge, MA 02142. (617) 225-0401. Six times per year. $400.00 per year.

RESEARCH CENTERS AND INSTITUTES

Geological Society of America. P.O. Box 9140, 3300 Penrose Place, Boulder, CO 80301. (303) 447-2020.

Texas A&M University. Earth Resources Institute. College Station, TX 77843. (409) 845-3651.

University of Miami. Comparative Sedimentology Laboratory. RSMAS-MGG, 4600 Rickenbacker Causeway, Miami, FL 33149. (305) 672-1840.

University of South Carolina At Columbia. Earth Sciences and Resources Institute. Byrnes International Center, 901 Sumter Street, Columbia, SC 29208. (803) 777-6484.

SEISMOLOGY

See also: EARTHQUAKE ENGINEERING, EARTHQUAKES, GEOLOGY, GEOPHYSICS, MINERAL EXPLORATION, PHYSICAL GEOLOGY, PLATE TECTONICS, VOLCANOLOGY

ABSTRACT SERVICES AND INDEXES

Bibliography and Index of Geology. American Geological Institute, 4220 King Street, Alexandria, VA 22302. (703) 379-2480. Fax (703) 379-7563. Monthly. $1295.00 per year. Also available online as GEOREF.

Deep-Sea Research. Part A: Oceanographic Research Papers. Deep-Sea Research. Part B: Oceanographic Literature Review. Pergamon Press, Inc., Maxwell House, Fairview Park, Elmsford, NY 10523. (914) 592-7700. Fax (914) 592-3625. Twelve times per year. $1370.00 per year for Parts A and B.

Geological Abstracts. Elsevier - Geo Abstracts, Regency House, 34 Duke Street, Norwich NR3 3AP, England. Monthly. $760.00 per year. Also available online as GEOBASE.

Geological Society of America. Abstracts with Programs. Geological Society of America. 3300 Penrose Place, P.O. Box 9140, Boulder, CO 80301-9140. (303) 447-2020. Abstracts and programs of the annual conference. Annual. $69.00.

Geophysics Abstracts. Pergamon Press, Inc., Maxwell House, Fairview Park, Elmsford, NY 10523. (914) 592-7700. Fax (914) 592-3625. Twelve times per year. $565.00 per year. Also available in microform.

Mineralogical Abstracts. Mineralogical Society and the Mineralogical Society of America, 41 Queen's Gate, London, SW7 5HR, England. Quarterly. $235.00 per year.

Oceanic Abstracts. Cambridge Scientific Abstracts, 7200 Wisconsin Avenue, Bethesda, MD 20814. (301) 961-6750. Fax (301) 961-6720. Bimonthly. $995.00 per year.

Science Citation Index. Institute for Scientific Information, 3501 Market Street, Philadelphia, PA 19104. (800) 523-1850 or (215) 386-0100. Inquire about price and availability.

General Science Index. H.W. Wilson Co., 950 University Avenue, Bronx, NY 10452. (800) 367-6770 or (212) 588-8400. Inquire about price and availability.

ANNUAL REVIEWS AND YEARBOOKS

Advances in Geophysics. Academic Press, Inc., 1250 Sixth Avenue, San Diego, CA 92101. (619) 231-0926. FAX (619) 699-6715. Irregular. Inquire about price and availability.

Annual Review and Earth and Planetary Sciences. Annual Reviews, Inc., 4139 El Camino Way, Palo Alto, CA 94306-0897. (415) 493-4400. Fax (415) 855-9815. Annual. $55.00 per year.

ASSOCIATIONS AND PROFESSIONAL SOCIETIES

American Association of Petroleum Geologists. P.O. Box 979, Tulsa, OK 74101. (918) 584-2555.

SEISMOLOGY

Ency. of Physical Sciences and Engineering Info. Sources

American Geological Institute. 4220 King Street, Alexandria, VA 22302. (703) 379-2480. Fax (703) 379-7563.

American Geophysical Union, 2000 Florida Avenue, N.W., Washington, DC 20009. (202) 462-6900.

American Institute of Professional Geologists. 7828 Vance Drive, Suite 103, Arvada, CO 80003. (303) 431-0831.

Association of Engineering Geologists. 323 Boston Post Road, Suite 2D, Sudbury, MA 01776. (508) 443-4639.

Geological Society of America. 3300 Penrose Place, P.O. Box 9140, Boulder, CO 80301-9140. (303) 447-2020.

Geoscience Information Society. American Geological Institute. 4220 King Street, Alexandria, VA 22302. (703) 379-2480. Fax (703) 379-7563.

Seismological Society of America. 201 Plaza Professional Building, El Cerrito, CA 94530. (510) 525-5474 or Fax (510) 525-7204

Society of Economic Geologists. P.O. Box 571, Golden, CO 80402. (303) 277-1716.

Society of Exploration Geophysicists. P.O. Box 702740, Tulsa, OK 74170. (918) 493-3516.

Society of Independent Professional Earth Scientists. 4925 Greenville Avenue, Suite 170, Dallas, TX 75206. (214) 363-1780.

BIBLIOGRAPHIES

Earthquake Prediction. David A. Tyckoson, editor. Oryx Press, 4041 North Central, Suite 700, Phoenix, AZ 85012-3330. (602) 265-2651 OR (800) 279-6799. 1986. $18.75.

Geologic Reference Sources: A Subject and Regional Bibliography of Publications and Maps in the Geological Sciences. Dederick C. Ward et al. Scarecrow Press, Inc., 52 Liberty Street, Box 4167, Metuchen, NJ 08840. (800) 537-7107 or (201) 548-8600. 1981. $49.50.

Information Sources in the Earth Sciences. David N. Wood, Joan E. Hardy, and Anthony P. Harvey, editors. Bowker-Saur/K.G. Saur. Distributed by R.R. Bowker, 121 Chanlon Road, New Providence, NJ 07974. (800) 521-8110 or (908) 464-6800. Second edition. 1989. $85.00.

DICTIONARIES AND BIOGRAPHICAL SOURCES

American Institute of Professional Geologists. Membership Directory. American Institute of Professional Geologists, 7828 Vance Drive, Suite 103, Arvada, CO 80003. (303) 431-0831. Annual. $18.00.

Association of Engineering Geologists. Membership directory. Association of Engineering Geologists, 323 Boston Post Road, Suite 2D, Sudbury, MA 01776. (508) 443-4639. Annual. $18.00.

Directory of Geoscience Departments: United States and Canada. American Geological Institute. 4220 King Street, Al-

exandria, VA 22302-1507. (703) 379-2480. Fax (703) 379-7563. Twenty-six edition. 1987. $22.00.

Geological Society of America. Membership Directory. Geological Society of America, 3300 Penrose Place, Boulder, CO 80301. (303) 447-2020. Annual. Available to members only.# #*Research Centers Directory.* Gale Research, 835 Penobscot Building, Detroit, MI 48226-4094. (800) 877-4253 or (313) 961-2242. 20th edition, 1995. $485.00.

ENCYCLOPEDIAS AND DICTIONARIES

Concise Oxford Dictionary of the Earth Sciences. Ailsa Allaby and Michael Allaby, editors. Oxford University Press, Inc., 200 Madison Avenue, New York, NY 10016. (800) 334-4249 or (212) 679-7300. 1990. $42.95.

Dictionary of Geological Terms. American Geological Institute. Doubleday and Company, Inc., 245 Park Avenue, New York, NY 10017. (800) 645-6156 or (212) 953-4561. Third edition. 1984. $12.00 in paper.

Encyclopedia of Applied Physics. VCH Publishers, Inc., 220 East 23rd Street, New York, NY 10010. (800) 367-8249. Twenty volumes. 1991 - . $5950.00 for set.

Encyclopedia of Earth System Science. Academic Press, Inc., 1250 Sixth Street, San Diego, CA 92101. (619) 231-0926 or (800) 321-5068. Four volumes, 1992. $950.00.

Encyclopedia of Solid Earth Geophysics. David E. James, editor. Van Nostrand Reinhold, 115 Fifth Avenue, New York, NY 10003. (212) 254-3232 or (800) 926-2665. 1989. $124.95.

Glossary of Geology. Robert L. Bates and Julia A. Jackson. American Geological Institute, 4220 King Street, Alexandria, VA 22032. (703) 379-2480. Third edition. 1987. $75.00.

Magill's Survey of Science: Earth Science Series. Salem Press, Inc., P.O. Box 1097, Englewood Cliffs, NJ 07632. (800) 221-1592 or (201) 871-3700. Five volumes. 1990. $400.00 for the set.

GENERAL WORKS

An Introduction To Mining Seismology. S.J. Gibowicz & A. Kijko. Academic Press Inc., 6277 Sea Harbor Drive, Orlando, FL 32887. (800) 321-5068. 1994. $69.95.

An Introduction To Seismological Research: History and Development. Benjamin F. Howell, Jr. Cambridge University Press, 40 West 20th Street, New York, NY 10011-4211. (212) 924-3900. 1990. $44.95.

Modern Global Seismology. Thorne Lay & Terry C. Wallace. Academic Press Inc., 6277 Sea Harbor Drive, Orlando, FL 32887. (800) 321-5068. 1995. $54.95.

Practical Seismic Interpretation. R. Badley. Prentice Hall (Division of Simon and Schuster), 15 Columbus Circle, New York, NY 10023. (212) 373-8500 or (800) 922-0579. 1988. $54.00.

Seismology. Hugh A. Doyle. John Wiley and Sons, Inc., 605 Third Avenue, New York, NY 10158. (800) 526-5368 or (212) 850-6000. 1995. $46.95.

HANDBOOKS AND MANUALS

Field Mapping For Geology Students. F. Ahmed and D.C. Almond. Unwin Hyman, Inc., 10 East 53rd Street, New York, NY 10022. (212) 207-7626 or (800) 242-7737. 1983. $18.95.

Geology in the Field. R.R. Compton. John Wiley and Sons, Inc., 605 Third Avenue, New York, NY 10158. (800) 526-5368 or (212) 850-6418. 1985. $37.95.

A Manual of Geology For Civil Engineers. John A. Pitts. World Scientific Publishing Company, Inc., 1060 Main Street, Unit B, River Edge, NJ 07661. (800) 227-7562 or (201) 487-9655. 1985. $38.00.

The Mapping of Geological Structures. K.R. McClay. John Wiley and Sons, Inc., 605 Third Avenue, New York, NY 10158. (800) 526-5368 or (212) 850-6000. 1988. $18.95.

ONLINE DATABASES AND CD-ROMS

Earth Sciences. U.S. Geological Survey, 12201 Sunrise Valley Drive, Reston, VA 22092-9998. (703) 648-4460. CD-ROM of earth science databases including the U.S. Geological Survey Library, Earth Science Data Directory, and GEOINDEX, citations to published geological maps. $350.00 per year with quarterly updates.

Geoarchive. Geosystems, P.O. Box 1024, Westminster, London, England, SW1 P 2JL. Citations to literature on geoscience, 1969 to present. Inquire as to online cost and availability.

Geobase. Elsevier - Geo Abstracts, Regency House, 34 Duke Street, Norwich NR3 3AP, England. Contains citations to the worldwide earth science literature from 1980 to date. Available on DIALOG, ORBIT online services. Inquire as to cost and availability.

GEOREF. American Geological Institute, 4220 King Street, Alexandria, VA 22302. (800) 336-4764 or (703) 379-2480. Geology and geosciences literature, 1785 to present for North America. Available on DIALOG, ORBIT online services. Also available on CD-ROM. Inquire as to cost and availability.

NTIS Bibliographic Database. National Technical Information Service, 5285 Port Royal Road, Springfield, VA 22161. (703) 487-4630. Broad coverage of government- sponsored research reports, 1964 to present. Available on BRS, DIALOG, ORBIT and STN online services. Also available on CD-ROM. Inquire as to cost and availability.

SCISEARCH. Institute for Scientific Information, 3501 Market Street, Philadelphia, PA 19104. (800) 523-1850 or (215) 386-0100. Broad multidisciplinary title and author index to the international literature of science and technology, 1974 to present. Available on DIALOG and ORBIT online services. Also available on CD-ROM. Inquire as to cost and availability.

PERIODICALS

Canadian Journal of Earth Sciences. National Research Council of Canada, Research Journals, Ottawa, Ont. K1A OR6, Canada. (613) 993-9084. Monthly. $112.00 per year.

Earth and Planetary Science Letters. Elsevier Science Publishing Company, Inc., 655 Avenue of the Americas, New York, NY 10010. (212) 989-5800. Twenty times per year. $945.00 per year.

Geological Society of America Bulletin. P.O. Box 9140, 3300 Penrose Place, Boulder, CO 80301. (303) 447-2020. Fax (303) 447-1133. Monthly. $150.00.

Geology. Geological Society of America. P.O. Box 9140, 3300 Penrose Place, Boulder, CO 80301. (303) 447-2020. Fax (303) 447-1133. Monthly. $120.00 per year.

Geophysical Research Letters. American Geophysical Union, 2000 Florida Avenue, N.W., Washington, DC 20009. (202) 462-6900. Monthly. $480.00 per year.

Geophysics. Society of Exploration Geophysicists, P.O. Box 702740, Tulsa, OK 74170. (918) 493-3516. Monthly. $175.00 per year.

Geotimes. American Geological Institute, 4220 King Street, Alexandria, VA 22032. (703) 379-2480. Monthly. $24.95 per year.

JGR: Journal of Geophysical Research: Solid Earth. American Geophysical Union, 2000 Florida Avenue, N.W., Washington, DC 20009. (202) 462-6903. Monthly. $1385.00 per year.

Journal of Geology. University of Chicago Press, 5720 Woodlawn Avenue, Chicago, IL 60637. (312) 753-3347. Bimonthly. $38.00 per year.

Journal of Structural Geology. Pergamon Journals, Inc., Maxwell House, Fairview Park, Elmsford, NY 10523. (914) 592-7700. Ten times per year. $375.00 per year.

Journal of Volcanology and Geothermal Research. Elsevier Science Publishing Company, Inc., 655 Avenue of the Americas, New York, NY 10010. (212) 989-5800. Monthly. $500.00 per year.

Reviews of Geophysics. American Geophysical Union, 2000 Florida Avenue, N.W., Washington, DC 20009. (202) 462-6900. Quarterly. $220.00 per year.

Rock Mechanics and Rock Engineering. Springer-Verlag New York, Inc., 175 Fifth Avenue, New York, NY 10010. (800) 526-7254 or (212) 460-1500. Quarterly. $122.00 per year.

Seismological Research Letters. Seismological Society of America. 201 Plaza Professional Building, El Cerrito, CA 94530. (510) 525-5474 or Fax (510) 525-7204. Quarterly. $15.00.

Seismological Society of America Bulletin. Seismological Society of America. 201 Plaza Professional Building, El Cerrito, CA 94530. (510) 525-5474 or Fax (510) 525-7204. Bimonthly. $125.00 per year.

Tectonics. American Geophysical Union, 2000 Florida Avenue, N.W., Washington, DC 20009. (202) 462-6900. Bimonthly. $230.00 per year to individuals.

Tectonophysics; An International Journal of Geotectonics and the Geology and Physics of the Interior of the Earth. Elsevier Science Publishing Company, Inc., 655 Avenue of the Americas, New York, NY 10010. (212) 989-5800. Fifty-six times per year. $1275.00 per year.

SEISMOLOGY

Ency. of Physical Sciences and Engineering Info. Sources

RESEARCH CENTERS AND INSTITUTES

California Institue of Technology. Seismological Laboratory. Pasadena, CA 91125. (818) 356-6914.

Carnegie Institution of Washington. Geophysical Laboratory. 5251 Broad Branch Road, NW, Washington, DC 20015. (202) 686-2410 or FAX (202) 686-2419.

University of Alaska Fairbanks. Geophysical Institute. C.T. Elvey Building, Fairbanks, AK 99775-0800. (907) 474-7282. Fax (907) 474-7290.

University of California, Santa Barbara. Institute for Crustal Studies. Santa Barbara, CA 93105. (805) 961-8231.

SEMICONDUCTORS

See also: DIODES, ELECTRICAL ENGINEERING, ELECTRICITY, ELECTRONIC CIRCUITS and COMPONENTS, ELECTRONIC ENGINEERING

ABSTRACT SERVICES AND INDEXES

Applied Science and Technology Index; A Cumulative Subject Index To English Language Periodicals in the Fields of Aeronautics and Space Science, Computer Technology, Chemistry, Construction Industry, Energy and Related Areas. H.W. Wilson Co., 950 University Avenue, Bronx, NY 10452. (212) 588-8400. (800) 367-6770. FAX (718) 590-1617. From 1958 to present. Monthly. Inquire about cost and availability. Also available on CD-ROM and online. ISSN: 0003-6986.

Current Contents: Engineering, Technology and Applied Sciences. Institute for Scientific Information, 3501 Market Street, Philadelphia, PA 19104. (215) 386-0100. FAX (215) 386-6362. 1970 to present. Weekly. $442.00 per year. Also available on CD-ROM and online. Inquire regarding cost and availability. ISSN: 0095-7917.

Current Papers in Electrical and Electronics Engineering. Institution of Electrical Engineers (IEE), London. Distributed by INSPEC/IEEE, Box 1331, 445 Hoes Lane, Piscataway, NJ 08855-1331. (908) 562-5549. 1969 to present. Monthly. $345.00 per year. ISSN: 0011-3778.

Electrical and Electronics Abstracts. Institution of Electrical Engineers (IEE), London. Available from INSPEC/IEEE - Institute of Electrical and Electronic Engineers, Box 1331, Hoes Lane, Piscataway, NJ 08855-1331. (908) 562-5549. 1898 to present. Monthly. $2200.00 per year. Also available on CD-ROM and online as INSPEC.

Engineering Index Monthly; Indexes and Abstracts the World's Engineering and Technical Literature. Engineering Information, Inc., Castle Point on the Hudson, Hoboken, NJ 07030. (201) 216-8500. (800) 221-1044. FAX (201) 216-8532. Monthly. $2300.00 per year. Available online as COMPENDEX and also on CD-ROM. ISSN: 0742-1974.

General Science Index. H.W. Wilson Company, 950 University Avenue, Bronx, NY 10452. (212) 588-8400. (800) 367-6770. FAX (718) 590-1617. From 1978 to present. Ten issues per year; quarterly and annual cumulations. Service basis. Available on

CD-ROM and online. Inquire regarding cost and availability. ISSN: 0162-1963.

Index To IEEE Publications. IEEE Service Center, 445 Hoes Lane, Piscataway, NJ 08855-1331. (908) 981-1393. (800) 678-IEEE. FAX (908) 981-9667. 1973 to present. Annual. ISSN: 0099-1368.

Physics Abstracts. INSPEC. Section A, Science Abstracts. Institution of Electrical Engineers (IEE), London. Available from: INSPEC/IEEE - Institute of Electrical and Electronic Engineers, Box 1331, Hoes Lane, Piscataway, NJ 08855-1331. (908) 562-5549. 1898 to present. 24 issues per year. $2835.00 per year. Also available online and on CD-ROM. ISSN: 0036-8091.

Physics Briefs (Physikalische Berichte). Information Center for Energy, Physics, Mathematics; German Physical Society. V C H Publishers, Inc., 220 East 23rd Street, New York, NY 10010-4606. (212) 683-8333. 1845 to present. 24 issues per year. $2390.00 per year. Also available online. ISSN: 0179-7434.

Science Citation Index. SCI. Institute for Scientific Information, 3501 Market Street, Philadelphia, PA 19104. (215) 386-0100. (800) 523-1850. FAX (215) 386-2991. 1961 to present. Six issues per year, plus annual cumulation. $11650.00 per year. Also available online and on CD-ROM. Inquire about price and availability. ISSN: 0036-827X.

Solid State and Superconductivity Abstracts: Covers theory, Production and Application of Solid State Materials. Cambridge Scientific Abstracts, 7200 Wisconsin Avenue, Bethesda, MD 20824. (301) 961-6750. FAX (301) 961-6720. 1957 to present. Bimonthly. $1320.00 per year. Also available online. ISSN: 0896-5900.

ANNUAL REVIEWS AND YEARBOOKS

Advances in Electronics and Electron Physics. Academic Press, Inc., 6277 Sea Harbor Drive, Orlando, FL 32887. (800) 321-5068. From 1948 to present. Irregular. Price varies, inquire. ISSN: 0065-2539.

Critical Reviews in Solid State and Materials Sciences. CRC Press, Inc., 2000 Corporate Boulevard, NW, Boca Raton, FL

33431. (407) 994-0555. (800) 272-7737. FAX (407) 998-9784. 1970 to present. Bimonthly. $265.00 per year. ISSN: 1040-8436.

Semiconductors and Semimetals. Robert K. Willardson and Albert C. Beer, editors. Academic Press, Inc., 6277 Sea Harbor Drive, Orlando, FL. (800) 321-5068. 1966 to present. Irregular. Each volume has distinctive title. Price varies

ASSOCIATIONS AND PROFESSIONAL SOCIETIES

American Electronics Association. 5201 Great America Way, Suite 520, P.O. 52990, Santa Clara, CA 95056. (408) 987-4200. FAX (408) 970-8565.

American Institute of Physics. One Physics Ellipse, College Park, MD 20740-3843. (301) 209-3100.

Electronics Industries Association. 2500 Wilson Boulevard, Arlington, VA 22201. (703) 907-7500. FAX (202) 457-4985.

IEEE Circuits and Systems Society. c/o Institute of Electrical and Electronic Engineers. 345 East 47th Street. New York, NY 10017. (212) 705-7900. FAX (212) 705-4929.

IEEE Solid-state Circuits Council. c/o Institute of Electrical and Electronic Engineers. 345 East 47th Street. New York, NY 10017. (212) 705-7900. FAX (212) 705-4929.

Institute of Electrical and Electronics Engineers. 345 East 47th Street. New York, NY 10017. (212) 705-7900. FAX (212) 705-4929.

International Society For Hydrid Microelectronics. 1850 Centennial Park Drive, Suite 105, Reston, VA 22091. (703) 758-1060. FAX (703) 758-1066.

National Electrical Manufacturers Association. 1300 North 17th Street, Suite 1847, Rosslyn VA 22209. (703) 841-3200. FAX (703) 841-3300.

DIRECTORIES AND BIOGRAPHICAL SOURCES

American Electronics Association. DIRECTORY. 5201 Great America Way, Suite 520, P.O. 52990, Santa Clara, CA 95056. (408) 987-4200. FAX (408) 970-8565. Annual. $175.00.

E E M - Electronic Engineer's Master. Hearst Business Communications, Inc., 645 Stewart Avenue, Garden City NY 11530. (516) 227-1300. ISSN: 0732-9016.

Engineering Research Centers: Incorporating Electronics Research Centers. Stockton Press, 345 Park Avenue, New York, NY 10010. (212) 689-9200. (800) 221-2123. 4th edition. 1995. $515.00.

IC Master. Hearst Business Communications, Inc., 645 Stewart Avenue, Garden City, NY 11530. (516) 227-1300. 1975 to present. Annual. $170.00 per year. ISSN: 0894-6809.

IEEE Membership Directory. Institute of Electrical and Electronics Engineers, IEEE Service Center, 445 Hoes Lane, Piscataway, NY 08854. (908) 981-1393. (800) 678-IEEE. FAX (908) 981-9667. 2 volumes. Annual. $190.00. ISSN:

International Directory of Abbreviations and Acroynms of Electronics, Electrical Engineering, Computer Technology and Information Processing. Peter Wennrich. K. G. Saur, 121 Chanlon Road, New Providence, NJ 07974. (908) 464-6800. (800) 521-8110 2 volumes. 1992. $230.00.

Research Centers Directory. Gale Research, 835 Penobscot Building, Detroit, MI 48226-4094. (313) 961-2242. (800) 877-4253. 20th edition. 1995. $485.00. ISSN: 0080-1518.

Semiconductor Industry Association- Yearbook/Directory. Semiconductor Industry Association, 10201 Torre Avenue, # 275, Cupertino, CA 95014. (408) 973-0073. Biennial. $75.00.

Semiconductor International - Directory Issue. Cahners Publishing Co., 1350 Touhy Avenue, Des Plaines, IL 80018. (708) 635-8800. Annual, February issue. $50.00.

Who's Who in Engineering. Gordon Davis, editor. American Association of Engineering Societies. 1111 19th Street, NY, Suite 608, Washington, DC 20036. (202) 296-2237. (800) 658-8897. 9th edition. 1995. $220.00. # #*Who's Who in Technol-*

ogy. Gale Research, 835 Penobscot Building, Detroit, MI 48226-4094. (313) 961-2242. (800) 521-4253. 7th edition. 1995. $195.00. ISN 0-8103-7467-6.

ENCYCLOPEDIAS AND DICTIONARIES

Chambers Science and Technology Dictionary. Peter M.B. Walker, editor. Cambridge University Press, 40 West 20th Street, New York, NY 10011-4211. (212) 924-3900. 1988. $39.95.

Encyclopedia of Applied Physics. George Trigg, editor. VCH Publications, Inc., 220 East 23rd Street, Suite 909, New York, NY 10010-4606. (212) 683-8333. (800) 422-8824. 20 volume set. 1991 - . $5990.00.

Encyclopedia of Integrated Circuits. Walter A. Buchsbaum. Prentice Hall, 113 Sylvan Avenue, Route 9w, Englewood Cliffs, NJ 07632. (201) 592-2000. (800) 922-0579. 1987. $39.95.

IEEE Standard Dictionary of Electrical and Electronics Terms. Christopher J. Booth, editor. IEEE Service Center, 445 Hoes Lane, Piscataway, NJ 08855-1331. (908) 981-1393. (800) 678-IEEE. FAX (908) 981-9667. IEEE Standard 100- 1992. 5th edition. 1993. $150.00.

Illustrated Dictionary of Electronics. Stan Gibilsco. TAB Books, P.O. Box 40, Blue Summit, PA 17294-0850. (717) 794- 2191. (800) 233-1128. 7th edition. 1994. $34.95.

Illustrated Encyclopedic Dictionary of Electronic Circuits. John Douglas-Young. Prentice Hall, 113 Sylvan Avenue, Route 9W, Englewood Cliffs, NJ 07632. (201) 592-2000. (800) 922-0579. 1983. $32.95.

International Encyclopedia of Integrated Circuits. Arthur A. Seidman. TAB Books, P.O. Box 40, Blue Summit, PA 17294-0850. (717) 794-2191. (800) 233-1128. 1991. $75.00.

McGraw-Hill Electronics Dictionary. J. Markus. McGraw-HillBook Company, Inc., 1221 Avenue of the Americas, New York, NY 10020. (212) 997-3675. 5th edition. 1994. $49.95.

McGraw-Hill Encyclopedia of Engineering. Sybil P. Parker, editor. McGraw-Hill Book Company, Inc. 1221 Avenue of the Americas, New York, NY 10020. (212) 997-3675. Second edition. 1993. $95.50. ISBN: 0-07-051392-9.

McGraw-Hill Encyclopedia of Science and Technology. McGraw-Hill Book, Incorporated, 1221 Avenue of the Americas, New York, NY 10020. (212) 997-3675. (800) 262-4729. Seventh edition. Twenty volumes. 1992. $1900.00.

GENERAL WORKS

Design and Technology of Integrated Circuits. Donald DeCogan. John Wiley & Sons, Inc., 605 Third Avenue, New York, NY 10158-0012. (212) 850-6000. (800) 225-5945. 1990. $75.00.

Design of Solid-state Power Supplies. Eugene R. Hnatek. Van Nostrand Reinhold, 115 Fifth Avenue, New York, NY 10003. (212) 254-3232. (800) 842-3636. 1989. $74.95.

Electric Circuits: Principles, Applications and Computer Analysis. David A. Bell. Prentice Hall, 113 Sylvan Avenue, Route 9W, Englewood Cliffs, NJ 07632. (201) 592- 2000. (800) 922-0579. 5th edition. 1995. $72.00.

Electron Transport Phenomena in Semiconductors. B. M. Askerov. World Scientific Publishing Company, Inc., 1060 Main Street, River Edge, NJ 07661. (201) 487-9655. (800) 227-7562. 1994. $86.00.

Fundamentals of Semiconductor Processing Technologies. Badih El-Kareh. Kluwer Academic Publishers, 101 Philip Drive, Assinippi Park, Norwell, MA 02061. (617) 871-6600. 1994. $110.00.

Microwave Semiconductor Devices. Sigfried Yngvesson. Kluwer Academic Publishers, 101 Philip Drive, Assinippi Park, Norwell, MA 02061. (617) 871-6600.

Microwave Solid-state Circuits and Applications. Kai Chang. John Wiley & Sons, Inc., 605 Third Avenue, New York, NY 10158-0012. (212) 850-6000. (800) 225-5945. 1994. $74.95.

Photoelectric Properties of Semiconductors. Richard H. Bube. Cambridge University Press, 40 West 20th Street, New York, NY 10011-4211. (212) 924-3900. (800) 872-7423. 1992. $34.95 in paper.

Semiconductor Optoelectric Devices. Pallab Bhattacharya. Prentice-Hall, 113 Sylvan Avenue, Route 9w, Englewood Cliffs, NJ 07632. (201) 592-2000. (800) 922-0579. 1993. $75.00.

Semiconductors and Electronic Devices. Adir Bar-Lev. Prentice-Hall, 113 Sylvan Avenue, Route 9w, Englewood Cliffs, NJ 07632. (201) 592-2000. (800) 922-0579. 3rd edition. 1993. $62.00 in paper.

Testing Semiconducctor Memories: Theory and Practice. A. J. Van de Goor. John Wiley & Sons, Inc., 605 Third Avenue, New York, NY 10158-0012. (212) 850-6000. (800) 225-5945. 1991. $98.00.

HANDBOOKS AND MANUALS

Active Electronic Component Handbook. Charles A. Harper and Harold C. Jones. McGraw-Hill Book Company, 1221 Avenue of the Americas, New York, NY 10020. (212) 512-2000. (800) 262-4729. 2nd edition. 1996. $79.50.

Discrete Semiconductor Products Handbook. Rector Press, Ltd., 130 Rattlesnake, Leverett, MA 01054-9726. (413) 548-9708. (800) 247-3473. 1995. $125.00.

EDN Designer's Companion. Ian Hickman and Bill Travis, editors. Butterworth-Heinemann, 313 Washington Street, Newton, MA 02158. (618-928-2500. (800) 366-2665. 1994.$29.95.

Electronics Circuits Handbook. M. Tooley. Butterworth-Heinemann, 313 Washington Street, Newton, MA 02158. (618-928-2500. (800) 366-2665. 1993. $49.95.

Electronics Handbook. Jerry C. Whitaker. CRC Press, Inc., 2000 Corporate Boulevard, NW, Boca Raton, FL 33431. (407) 994-0555. (800) 272-7737. 1996. $120.00.

Handbook of Compound Semiconductors: Growth, Processing, Characterization and Devices. Paul H. Holloway and Gary E. McGuire. Noyes, 120 Mill Road, Park Ridge, NJ 07656. (201) 391-8484. 1996. $98.00.

Handbook of Electrical and Electronic Technology. Curtis Johnson. Prentice Hall , 113 Sylvan Avenue, Route 9W, Englewood Cliffs, NJ 07632. (201) 592-2000. (800) 922-0579. 1996. $88.00.

Handbook of Quality Integrated Circuit Manufacturing. Robert Zurich. Academic Press, Inc., 6277 Sea Harbor Drive, Orlando, FL. (800) 321-5068. 1991. $59.95.

Handbook of Semiconductor Lasers and Photonic Integrated Circuits. Chapman & Hall, 1 Penn Plaza, New York, NY 10119. (212) 564-1060. Y. Suematsu and A. R. Adams, editors. 1994. $163.95.

Handbook of VLSI Chip Design and Expert Systems. A. F. Schwartz. Academic Press, Inc., 6277 Sea Harbor Drive, Orlando, FL. (800) 321-5068.

Integrated Circuits Handbook. Rector Press, Ltd., 130 Rattlesnake, Leverett, MA 01054-9726. (413) 548-9708. (800) 247-3473. 3 volumes. 1995. $295.00 in paper.

Standard Handbook For Electrical Engineers. Donald Fink. McGraw-Hill Publishing Company, 1221 Avenue of the Americas, New York, NY 10020. (212) 512-2000. (800) 262- 4729. 13th edition. 1996. $110.00.

ONLINE DATABASES AND CD-ROMS

COMPENDEX PLUS. Engineering Information, Inc., 345 East 47th Street, New York, NY 10017. (212) 705-7600 or (800) 221-1044. Contains citations with abstracts to worldwide literature in engineering and technology, from 1970 to present. Available on online BRS,(800) 289-4277, DIALOG, (800) 334-2564, ORBIT (800) 456-7248, and STN International, FIZ Karlsruhe, P.O. Box 2465, W-7500, Karlsruhe 1, Germany, online services. Also available on CD-ROM. Inquire as to cost and availability.

Current Contents Search. Institute for Scientific Information, 3501 Market Street, Philadelphia, PA 19104. (215) 386-0100. FAX (215) 386-6362. Contains citations to articles listed in the table of contents of science and technology journals. Also articles in social sciences and life sciences journals. Available on BRS,(800) 289-4277, DIALOG,(800) 334-2564, online services. Inquire as to cost and availability.

Dissertation Abstracts Online. University Microfilms International, 300 North Zeeb Road, Ann Arbor, MI 48106. (800) 521-0600 or (313) 761-4700. Scope includes virtually all doctoral dissertations accepted at accredited American institutions from 1861 to present in 252 subject areas. Available on BRS, (800) 289-4277, DIALOG, (800) 334-2564, and OCLC EPIC,(800) 848-5878, online services. Also available on CD-ROM. Inquire as to cost and availability.

INSPEC. Institution of Electrical Engineers, Michael Faraday House, Six Hills Way, Stevenage, Herts. SG1 2AY, England. Telephone: 0438 313311 or FAX 0438 742840. Contains citations to the worldwide literature of physics, electronics and electrical engineering, computer technology, and related fields. Available on BRS, (800) 289-4277, DIALOG, (800) 334-2564, ORBIT, (800) 456-7248, and STN International, FIZ Karlsruhe, P.O. Box 2465, W-7500, Karlsruhe 1, Germany, online services. Inquire as to cost and availability.

NTIS Bibliographic Database. National Technical Information Service, 5285 Port Royal Road, Springfield, VA 22161. (703)

487-4929 or FAX (703) 321-8199. Broad coverage of government-sponsored science and technology research reports, 1964 to present. Available on BRS,(800) 289-4277, DIALOG, (800) 334-2564, ORBIT, (800) 456-7248, and STN International, FIZ Karlsruhe, P.O. Box 2465, W-7500, Karlsruhe 1, Germany, online services. Also available on CD-ROM. Inquire as to cost and availability.

Physics Briefs. American Institute of Physics, 335 East 45th Street, New York, NY 10017. (212) 661-9260 or FAX (212) 661-2036. Contains citations with abstracts of the literature of physics and related fields, 1979 to present. Available on the STN International, FIZ Karlsruhe, P.O. Box 2465, W-7500, Karlsruhe 1, Germany, online service. Inquire as to cost and availability.

SCISEARCH. Institute for Scientific Information, 3501 Market Street, Philadelphia, PA 19104. (800) 523-1850 or (215) 386-0100. Broad multidisciplinary title and author index to the international literature of science and technology, 1974 to present. Available on DIALOG,(800) 334-2564, and ORBIT,(800) 456-7248, online services. Also available on CD-ROM. Inquire as to cost and availability.

WILSONLINE. H.W. Wilson Company, 950 University Avenue, Bronx, NY 10452. (800) 367-6770 or (212) 588-8400. Makes available online versions of the printed H.W. Wilson indexes including Applied Science and Technology Index, Business Periodicals Index, General Science Index, and Readers' Guide to Periodical Literature. Period covered is generally 1983 to present. Available on BRS,(800) 289-4277, DIALOG,(800) 334-2564, and OCLC EPIC,(800) 848-5878, online services. Also available on CD-ROM. Inquire as to cost and availability.

PERIODICALS

Circuits, Systems, and Signal Processing. Birkhauser, 675 Massachusetts Avenue, Cambridge, MA 02139-3309. FAX (201) 348-4505. 1982 to present. Quarterly. $184.00 per year. ISSN: 0278-081X.

Electrical World. McGraw-Hill, Inc., Box 513 Hightstown, NJ 08520. (212) 512-3288. 1874 to present. Monthly. $55.00 per year. ISSN: 0013-4457.

Electronic Design. Penton Publishing, San Jose Gateway, Suite 354. 2025 Gateway Place, San Jose, CA 95110. (408) 441-0550. 1952 to present. Fortnightly. $95.00. ISSN: 0013-4872.

Electronic Engineering Times. CMP Publications, Inc., 600 Community Drice, Manhasset, NY 11030. (516) 562-5000. FAX (516) 562-5325. 1972 to present. Weekly. $159.00 per year. ISSN: 0192-1541.

Electronics. Penton Publishing, San Jose Gateway, Suite 354. 2025 Gateway Place, San Jose, CA 95110. (408) 441-0550. 1930 to present. Semi-weekly. $98.00 per year. ISSN: 0883-4989.

Electronics World and Wireless World. Reed Business Publishing, Ltd., Quadrant House, the Quadrant, Sutton, Surrey SM2 5AS, England. 1911 to present. Monthly. L35. ISSN: 0266-3244.

IEEE Circuits and Devices Magazine. Institute of Electrical and Electronics Engineers, Inc., Box 1331, 445 Hoes Lane, Piscataway, NJ 08855-1331. (908) 981-0060. 1985 to present. Bi-monthly. $120.00 per year. ISSN: 8755-3996.

IEEE Electron Device Letters. Institute of Electrical and Electronics Engineers, Inc., Box 1331, 445 Hoes Lane, Piscataway, NJ 08855-1331. (908) 981-0060. Monthly.

IEEE Journal of Solid State Circuits. Institute of Electrical and Electronics Engineers, Inc., Box 1331, 445 Hoes Lane, Piscataway, NJ 08855-1331. (908) 981-0060. 1966 to present. Bi-monthly. $275.00 per year. ISSN: 0018- 9200.

IEEE Spectrum. Institute of Electrical and Electronics Engineers, Inc., Box 1331, 445 Hoes Lane, Piscataway, NJ 08855-1331. (908) 981-0060. 1964 to present. Monthly. $157.00 per year. ISSN: 0018-9235.

IEEE Transactions On Circuits and Systems. Part 1: Fundamental theory and Applications. Institute of Electrical and Electronics Engineers, Inc., Box 1331, 445 Hoes Lane, Piscataway, NJ 08855-1331. (908) 981-0060. 1952 to present. Monthly. $241.00 per year. ISSN: 1057-7122.

IEEE Transactions On Electron Devices. Institute of Electrical and Electronics Engineers, Inc., Box 1331, 445 Hoes Lane, Piscataway, NJ 08855-1331. (908) 981-0060. 1952 to present. Monthly. $395.00 per year. ISSN: 0018-9393.

Institute of Electrical and Electronics Engineers. PROCEEDINGS. Institute of Electrical and Electronics Engineers, Inc., Box 1331, 445 Hoes Lane, Piscataway, NJ 08855-1331. (908) 981-0060. 1913 to present. Monthly. $275.00. ISSN: 0018-9219.

Journal of Physics: Condensed Matter. I O P Publishing. Orders to: American Institute of Physics, 500 Sunnyside Boulevard, Woodbury, NY 11797-2999. (516) 576-2200. 1968 to Present. 50 issues per year. $4229.00 per year. ISSN: 0953-8984.

Journal of Physics and Chemistry of Solids. Elsevier Science, 660 White Plains Road, Tarrytown, NY 10591-5151. (914) 524-9200. FAX (914) 333-2444. 1956 to present. Monthly. $1670.00 per year. ISSN: 0022-3697.

Physical Review B (condensed Matter). American Institute of Physics, 500 Sunnyside Boulevard, Woodbury, NY 11797-2999.(516) 576-2200. 1970 to present. 48 issues per year. $3180.00 per year. ISSN: 0163-1829.

Semiconductor International. Cahners Publishing Co., 44 Cook Street, Denver, CO 80206. (708) 635-8000. FAX (708) 390-2770. (800) 662-7776. 1978 to present. 13 issues per year. $84.95 per year. ISSN: 0163-3767.

Semiconductor Science and Technology. I O P Publishing. Orders to: American Institute of Physics, 500 Sunnyside Boulevard, Woodbury, NY 11808-2999. (516) 576-2200. 1986 to present. Monthly. $1238.00 per year. ISSN: 0268-1242.

Solid State Electronics. Elsevier Science, 660 White Plains Road, Tarrytown, NY 10591-5153. (914) 524-9200. FAX (914) 333-2444. 1960 to present. Monthly. $1025.00 per year. ISSN: 0038-1101.

Solid State Technology. PennWell Publishing Co., 10 Tara Boulevard, Nausha, NH 03062-2801. (603) 891-0123. FAX(609) 891-0597. 1958 to present. Monthly. $110.00 per year. ISSN: 0038-111X.

SEMICONDUCTORS

Ency. of Physical Sciences and Engineering Info. Sources

RESEARCH CENTERS AND INSTITUTES

Automation in Semiconductor Manufacturing Program. University of Michigan, Electrical Engineering and Computer Science Department, 1301 Beal Avenue, Room 1246, Ann Arbor MI 48109. (313) 764-3346. FAX (313) 747-1781.

Center For Semiconductor Device Reliabilty Research. Clemson University, Clemson, SC 29634-0915. (803) 656-5912.

Semiconductor Device Laboratory. University of Virginia, Thornton Hall, Department of Electrical Engineering, Charlottesville, VA 22903. (804) 924-7693. FAX (804) 924-8818.

Semiconductor Processing Facility. Univeristy of Arizona, Electrical and Computer Engineering Department, Tucson, AZ 85721. (602) 621-8237.

Solid State Electronics Laboratory. North Carolina State University, 432 Daniels Hall, Raleigh, NC 27695. (919) 737-2336. FAX (919) 737-3027.

SERVOMECHANISMS

See: ROBOTICS

SEWAGE TREATMENT

See also: ENVIRONMENTAL ENGINEERING, GROUND WATER POLLUTION, SLUDGE, WATER POLLUTION, WATER TREATMENT

ABSTRACT SERVICES AND INDEXES

Environmental Abstracts. Congressional Information Service, Reed Elsevier, 4520 East-West Highway, Bethesda, MD 20814-3389. (800) 638-8380. FAX (301) 654-4033. 1970 to present. Monthly. $1070.00 per year. Also available on-line and on CD-ROM.

Environmental Abstracts Annual. Congressional Information Service, Reed Elsevier, 4520 East-West Highway, Bethesda, MD 20814-3389. (800) 638-8380. FAX (301) 654-4033. 1970 to present. Annual. $495.00.

Environmental Engineering Abstracts. Cambridge Scientific Abstracts, 7200 Wisconsin Avenue, 6th Floor, Bethesda, MD 20814. (301) 961-6750. FAX (301) 961-6720. 1993 to present. Monthly. $385.00 per year.

ASSOCIATIONS AND PROFESSIONAL SOCIETIES

American Society of Sanitary Engineering. PO Box 40362, Bay Village, OH 44140. (216) 835-3040. FAX (216) 835-3488.

Association of Metropolitan Sewerage Agencies. 1000 Connecticut Avenue NW, Suite 410, Washington, DC 20036-5302. (202) 833-2672. FAX (202) 833-4657.

Municipal Waste Management Association. 16720 I Street NW, 4th floor, Washington, DC 20006. (202) 293-7330.

National Association of Sewer Service Companies. 101 Wymore Road, Suite 521, Altamonte, FL 32714. (407) 774-0304. FAX (407) 682-3111.

Water and Sewer Distributors of America. 1900 Arch Street, Philadelphia, PA 19103-1498. (215) 564-3483. FAX (215) 564-2175.

DIRECTORIES AND BIOGRAPHICAL SOURCES

Directory of Water-sewer and Related Industries Professionals. National Association of Regulatory Utility Commissioners, 1102 ICC Bldg., PO Box 684, Washington, DC 20044. (202) 898-2200.

ENCYCLOPEDIAS AND DICTIONARIES

Encyclopedia of Physical Science and Technology. Robert A. Meyers, ed. 18 volumes. Academic Press, Inc., 1250 Sixth Avenue, San Diego, CA 92101-4311. (619) 231-0926. FAX (619) 699-6715. 1992. $2100.00

McGraw-Hill Dictionary of Scientific and Technical Terms. Sybil P. Parker, ed. 5th edition. McGraw-Hill Publishing Company, 1221 Avenue of the Americas, New York, NY 10020. (800) 262-4729 or (212) 512-3825. 1993. $110.50.

GENERAL WORKS

Advanced Waste Treatment. Kenneth D. Kerri. 3d revised edition. California State University—Sacramento, office of Water Programs, 6000 J Street, Sacramento, CA 95819-6025. (916) 278-6142. 1995. $20.00.

Waste Treatment and Disposal. R.E. Hester and R.M. Harrison, editors. CRC Press, 2000 Corporate Blvd., N.W., Boca Raton, FL 33431. (407) 994-0555 or (800) 272-7737. FAX (407) 994-0949. 1995. $29.95.

Waste Treatment Plants. C.A. Sastry, et al., editors. John Wiley and Sons, Inc., 605 Third Avenue, New York, NY 10158. (800) 526-5368 or (212) 850-6000. 1996. $45.00.

Wastewater Treatment Plants: Planning, Design, and Operation. Syed R. Qasim. Technomic Publishing Company, Inc., 851 New Holland Avenue, Box 3535, Lancaster, PA 17604. (717) 291-5609 or (800) 233-9936. FAX (717) 295-4538. 1994. $58.00.

Wastewater Treatment Troubleshooting and Problem Solving. Glenn M. Tillman. Ann Arbor Press, PO Box 310, Chelsea, MI 48118. (800) 858-5299. 1996. $34.95.

HANDBOOKS AND MANUALS

Handbook of Water and Wastewater Treatment Technology. Paul N. Cheremisinoff. Marcel Dekker, Inc., 270 Madison Avenue, New York, NY 10016. (212) 696-9000. FAX (212) 685-4540. 1994. $195.00.

ONLINE DATABASES AND CD-ROMS

CA Search. Chemical Abstracts Service, P.O. Box 3012, Columbus, OH 43210-0012. (614) 447-3600. (800) 848-6533. FAX (614) 447-3709. Very comprehensive guide to worldwide chemical literature and related fields, 1972 to present. Available on BRS,(800) 289-4277, DIALOG, (800) 334-2564, ORBIT (800) 456-7248, and STN International, FIZ Karlsruhe, P.O. Box 2465, W-7500, Karlsruhe 1, Germany, online services. Inquire as to cost and availability.

Compendex Plus. Engineering Information, Inc., 345 East 47th Street, New York, NY 10017. (212) 705-7600 or (800) 221-1044. Contains citations with abstracts to worldwide literature in engineering and technology, from 1970 to present. Available on online BRS,(800) 289-4277, DIALOG, (800) 334-2564, ORBIT (800) 456-7248, and STN International, FIZ Karlsruhe, P.O. Box 2465, W-7500, Karlsruhe 1, Germany, online services. Also available on CD-ROM. Inquire as to cost and availability.

CSA Engineering. Cambridge Scientific Abstracts, 7200 Wisconsin Avenue, Suite 601, Bethesda, MD 20814. (301) 961-6750 or (800) 843-7751. Contains citations and abstracts of international periodicals and research literature covering all fields of engineering and science and technology,including computer and information science, electronics, mechanical engineering, solid state materials, 1981 to present. Available on BRS,(800) 289-4277, online service.Inquire as to cost and availability.

Current Contents Search. Institute for Scientific Information, 3501 Market Street, Philadelphia, PA 19104. (215) 386-0100. FAX (215) 386-6362. Contains citations to articles listed in the table of contents of science and technology journals. Also articles in social sciences and life sciences journals. Available on BRS,(800) 289-4277, DIALOG,(800) 334-2564, online services. Inquire as to cost and availability.

Dissertation Abstracts. University Microfilms International, 300 North Zeeb Road, Ann Arbor, MI 48106. (800) 521-0600 or (313) 761-4700. Scope includes virtually all doctoral dissertations accepted at accredited American institutions from 1861 to present in 252 subject areas. Available on BRS,(800) 289-4277, DIALOG,(800) 334-2564, and OCLC EPIC,(800) 848-5878, online services. Also available on CD-ROM. Inquire as to cost and availability.

NTIS Bibliographic Database. National Technical Information Service, 5285 Port Royal Road, Springfield, VA 22161. (703) 487-4929 or FAX (703) 321-8199. Broad coverage of government-sponsored science and technology research reports, 1964 to present. Available on BRS,(800) 289-4277, DIALOG, (800) 334-2564, ORBIT, (800) 456-7248, and STN International, FIZ Karlsruhe, P.O. Box 2465, W-7500, Karlsruhe 1, Germany, online services. Also available on CD-ROM. Inquire as to cost and availability.# #*SCISEARCH.* Institute for Scientific Information, 3501 Market Street, Philadelphia, PA 19104. (800) 523-1850 or (215) 386-0100. Broad multidisciplinary title and author index to the international literature of science and technology, 1974 to present. Available on DIALOG,(800) 334-2564, and ORBIT,(800) 456-7248, online services. Also available on CD-ROM. Inquire as to cost and availability.

PERIODICALS

Water and Waste Treatment. Faversham House Group Ltd., 111 St. Jame's Road, Croydon, Surrey CR9 2TH, England. 1950 to present. Monthly. Inquire for price.

Water Science and Technology. Elsevier Science [journals], 660 White Plains Rd., Tarrytown, NY 10159-5153. (919) 524-9200. FAX (919) 333-2444. 1972 to present. 24 times a year. $2310.00 per year.

World Water and Environmental Engineering. Thomas Telford Services Ltd., Thomas Telford House, 1 Heron Quay, London E14 4JD, England. Telephone 071-987-6999. FAX 071-538-4101. 1978 to present. Monthly. Inquire for cost.

RESEARCH CENTERS AND INSTITUTES

United States Environmental Protection Agency, Test and Evaluation Facility. 1600 Gest Street, Cincinnati, OH 45204. (513) 684-2621. FAX (513) 684-2628.

University of California At Berkeley, Sanitary Engineering and Environmental Health Research Lab. Building 112, 1031 S 46th, Richmond, CA 94804. (415) 231-9449. FAX (415) 231-9500.

University of Florida, Environmental Engineering Research Center. College of Engineering, 217 Black Hall, Gainesville, FL 32611. (904) 392-0841. FAX (904) 392-3026.

Virginia Polytechnic Institute, Environmental Engineering and Sciences Group. 330 Norris Hall, Department of Civil Engineering, Blacksburg, VA 24061. (703) 961-6635.

SHIP DESIGN

See: NAVAL ARCHITECTURE

SHIPBUILDING

See: NAVAL ARCHITECTURE

SIDE LOOKING RADAR

See: RADAR

SIGNAL PROCESSING

See also: ELECTRICAL ENGINEERING, ELECTRONIC CIRCUITS AND COMPONENTS, ELECTRONICS ENGINEERING

ABSTRACT SERVICES AND INDEXES

Applied Science and Technology Index; A Cumulative Subject Index To English Language Periodicals in the Fields of Aeronautics and Space Science, Computer Technology, Chemistry, Construction Industry, Energy and Related Areas. H.W. Wilson Co., 950 University Avenue, Bronx, NY 10452. (212) 588-8400. (800) 367-6770. FAX (718) 590-1617. From 1958 to present. Monthly. Inquire about cost and availability. Also available on CD-ROM and online. ISSN: 0003-6986.

SIGNAL PROCESSING

Ency. of Physical Sciences and Engineering Info. Sources

Current Contents: Engineering, Technology and Applied Sciences. Institute for Scientific Information, 3501 Market Street, Philadelphia, PA 19104. (215) 386-0100. FAX (215) 386-6362. 1970 to present. Weekly. $442.00 per year. Also available on CD-ROM and online. Inquire regarding cost and availability. ISSN: 0095-7917.

Current Papers in Electrical and Electronics Engineering. Institution of Electrical Engineers (IEE), London. Distributed by INSPEC/IEEE, Box 1331, 445 Hoes Lane, Piscataway, NJ 08855-1331. (908) 562-5549. 1969 to present. Monthly. $345.00 per year. ISSN: 0011-3778.

Electrical and Electronics Abstracts. Institution of Electrical Engineers (IEE), London. Available from INSPEC/IEEE - Institute of Electrical and Electronic Engineers, Box 1331, Hoes Lane, Piscataway, NJ 08855-1331. (908) 562-5549. 1898 to present. Monthly. $2200.00 per year. Also available on CD-ROM and online as INSPEC. ISSN: 0036-8105.

Engineering Index Monthly; Indexes and Abstracts the World's Engineering and Technical Literature. Engineering Information, Inc., Castle Point on the Hudson, Hoboken, NJ 07030. (201) 216-8500. (800) 221-1044. FAX (201) 216-8532. Monthly. $2300.00 per year. Available online as COMPENDEX and also on CD-ROM. ISSN: 0742-1974.

General Science Index. H.W. Wilson Company, 950 University Avenue, Bronx, NY 10452. (212) 588-8400. (800) 367-6770. FAX (718) 590-1617. From 1978 to present. Ten issues per year; quarterly and annual cumulations. Service basis. Available on CD-ROM and online. Inquire regarding cost and availability. ISSN: 0162-1963.

Government Reports Announcements and Index. U. S. National Technical Information Service (NTIS), 5285 Port Royal Road, Springfield, VA 22161. (703) 487-4650. FAX (703) 321-8547. From 1968 to present. Annual. $630.00 per year. Also available online as NTIS Bibliographic Database and on CD-ROM.

Index To IEEE Publications. IEEE Service Center, 445 Hoes Lane, Piscataway, NJ 08855-1331. (908) 981-1393. (800) 678-IEEE. FAX (908) 981-9667. 1973 to present. Annual. ISSN: 0099-1368.

Physics Abstracts. INSPEC. Section A, Science Abstracts. Institution of Electrical Engineers (IEE), London. Available from: INSPEC/IEEE - Institute of Electrical and Electronic Engineers, Box 1331, Hoes Lane, Piscataway, NJ 08855-1331. (908) 562-5549. 1898 to present. 24 issues per year. $2835.00 per year. Also available online and on CD-ROM. ISSN: 0036-8091.

Physics Briefs (Physikalische Berichte). Information Center for Energy, Physics, Mathematics; German Physical Society. V C H Publishers, Inc., 220 East 23rd Street, New York, NY 10010-4606. (212) 683-8333. 1845 to present. 24 issues per year. $2390.00 per year. Also available online. ISSN: 0179-7434.

Science Citation Index. SCI. Institute for Scientific Information, 3501 Market Street, Philadelphia, PA 19104. (215) 386-0100. (800) 523-1850. FAX (215) 386-2991. 1961 to present. Six issues per year, plus annual cumulation. $11650.00 per year. Also available online and on CD-ROM. Inquire about price and availability. ISSN: 0036-827X.

Solid State and Superconductivity Abstracts: Covers theory, Production and Application of Solid State Materials. Cambridge Scientific Abstracts, 7200 Wisconsin Avenue, Bethesda, MD

20824. (301) 961-6750. FAX (301) 961-6720. 1957 to present. Bimonthly. $1320.00 per year. Also available online. ISSN: 0896-5900.

ANNUAL REVIEWS AND YEARBOOKS

Advances in Electronics and Electron Physics. Academic Press, Inc., 6277 Sea Harbor Drive, Orlando, FL 32887. (800) 321-5068. From 1948 to present. Irregular. Price varies, inquire. ISSN: 0065-2539.

Critical Reviews in Solid State and Materials Sciences. CRC Press, Inc., 2000 Corporate Boulevard, NW, Boca Raton, FL 33431. (407) 994-0555. (800) 272-7737. FAX (407) 998-9784. 1970 to present. Bimonthly. $265.00 per year. ISSN: 1040-8436.

ASSOCIATIONS AND PROFESSIONAL SOCIETIES

American Association of Engineering Societies, 1111 19th Street, Suite 608, Washington, DC 20036-3603. (202) 296-2237. FAX (202) 296-1151.

American Electronics Association. 5201 Great America Way, Suite 520, P.O. 52990, Santa Clara, CA 95056. (408) 987-4200. FAX (408) 970-8565.

American Institute of Physics. One Physics Ellipse, College Park, MD 20740-3843. (301) 209-3100.

Association of Old Crows (electronic Warfare). 1000 North Payne Street, Alexandria, VA 22314. (703) 549-1600.

Edison Electric Institute. 701 Pennsylvania Avenue NW, Washington, DC 20004-2696. (202) 508-5000, 5454. FAX (202) 508-5360.

Electric Power Research Institute. 3412 Hillview Avenue, Palo Alto, CA 94304. (415) 855-2000. FAX (415) 855-2954.

Electronics Industries Association. 2500 Wilson Boulevard, Arlington, VA 22201. (703) 907-7500. FAX (202) 457-4985.

IEEE Circuits and Systems Society. c/o Institute of Electrical and Electronic Engineers. 345 East 47th Street. New York, NY 10017. (212) 705-7900. FAX (212) 705-4929.

IEEE Solid-state Circuits Council. c/o Institute of Electrical and Electronic Engineers. 345 East 47th Street. New York, NY 10017. (212) 705-7900. FAX (212) 705-4929.

Institute For Interconnecting and Packaging Electronic Electronic Circuits. 2215 Sanders Road, Northbrook, IL 60062-6135. (708) 509-9700. (708) 509-9798.

Institute of Electrical and Electronics Engineers. 345 East 47th Street. New York, NY 10017. (212) 705-7900. FAX (212) 705-4929.

International Society For Hydrid Microelectronics. 1850 Centennial Park Drive, Suite 105, Reston, VA 22091. (703) 758-1060. FAX (703) 758-1066.

National Electrical Manufacturers Association. 1300 North 17th Street, Suite 1847, Rosslyn VA 22209. (703) 841-3200. FAX (703) 841-3300.

DIRECTORIES AND BIOGRAPHICAL SOURCES

American Electronics Association Directory. 5201 Great America Way, Suite 520, P.O. 52990, Santa Clara, CA 95056. (408) 987-4200. FAX (408) 970-8565. Annual. $175.00.

American Men and Women of Science: Physical and Biological Sciences. R. R. Bowker Inc., 121 Chanlon Road, New Providence, NJ 07974. (908) 464-6800. (800) 521-8110. 20th edition. 8 volumes. 1996. $850.00.

Directory of Engineering Societies and Related Organizations. American Association of Engineering Societies, 1111 19th Street, Suite 608, Washington, DC 20036-3603. (202) 296-2237. Semi-annual. $150.00.

E E M - Electronic Engineer's Master. Hearst Business Communications, Inc., 645 Stewart Avenue, Garden City NY 11530. (516) 227-1300. ISSN: 0732-9016.

Electrical and Electronics Trades Directory: The Blue Book. Institution of Electrical Engineers. c.o IEEE Service Center, 445 Hoes Lane, Piscataway, NY 08854. (908) 981-1393. (800) 678-IEEE. FAX (908) 981-9667. Annual. 1995. $140.00

Engineering Research Centers: Incorporating Electronics Research Centers. Stockton Press, 345 Park Avenue, New York, NY 10010. (212) 689-9200. (800) 221-2123. 4th edition. 1995. $515.00.

Ic Master. Hearst Business Communications, Inc., 645 Stewart Avenue, Garden City, NY 11530. (516) 227-1300. 1975 to present. Annual. $170.00 per year. ISSN: 0894- 6809.

IEEE Membership Directory. Institute of Electrical and Electronics Engineers, IEEE Service Center, 445 Hoes Lane, Piscataway, NY 08854. (908) 981-1393. (800) 678-IEEE. FAX (908) 981-9667. 2 volumes. Annual. $190.00.

International Directory of Abbreviations and Acroynms of Electronics, Electrical Engineering, Computer Technology and Information Processing. Peter Wennrich. K. G. Saur, 121 Chanlon Road, New Providence, NJ 07974. (908) 464-6800. (800) 521-8110 2 volumes. 1992. $230.00.

International Engineering Directory. American Consulting Engineers Council, 1015 15th Street, N.W. Suite 802, Washington, DC 20005-2670. (202) 347-7474. Annual. $10.00.

Research Centers Directory. Gale Research Company Inc., 835 Penobscot Building, Detroit, MI 48226-4094. (313) 961-2242. (800) 877-4253. 20th edition. 1995. $485.00. ISSN: 0080- 1518.

Who's Who in Engineering. Gordon Davis, editor. American Association of Engineering Societies. 1111 19th Street, NY, Suite 608, Washington, DC 20036. (202) 296-2237. (800) 658-8897. 9th edition. 1995. $220.00.

Who's Who in Technology. Gale Research, 835 Penobscot Building, Detroit, MI 48226-4094. (313 961-2242. (800)521-4253. 7th edition. 1995. $195.00. ISN 0-8103-7467-6.

ENCYCLOPEDIAS AND DICTIONARIES

Chambers Science and Technology Dictionary. Peter M.B. Walker, editor. Cambridge University Press, 40 West 20th Street, New York, NY 10011-4211. (212) 924-3900. 1988. $39.95.

Encyclopedia of Applied Physics. George Trigg, editor. VCH Publications, Inc., 220 East 23rd Street, Suite 909, New York, NY 10010-4606. (212) 683-8333. (800) 422-8824. 20 volume set. 1991 -. $5990.00.

Encyclopedia of Integrated Circuits. Walter A. Buchsbaum. Prentice Hall, 113 Sylvan Avenue, Route 9w, Englewood Cliffs, NJ 07632. (201) 592-2000. (800) 922-0579. 1987. $39.95.

IEEE Standard Dictionary of Electrical and Electronics Terms. Christopher J. Booth, editor. IEEE Service Center, 445 Hoes Lane, Piscataway, NJ 08855-1331. (908) 981-1393. (800) 678-IEEE. FAX (908) 981-9667. IEEE Standard 100- 1992. 5th edition. 1993. $150.00.

Illustrated Dictionary of Electronics. Stan Gibilsco. TAB Books, P.O. Box 40, Blue Summit, PA 17294-0850. (717) 794- 2191. (800) 233-1128. 7th edition. 1994. $34.95.

Illustrated Encyclopedic Dictionary of Electronic Circuits. John Douglas-Young. Prentice Hall, 113 Sylvan Avenue, Route 9W, Englewood Cliffs, NJ 07632. (201) 592-2000. (800) 922-0579. 1983. $32.95.

International Encyclopedia of Integrated Circuits. Arthur A. Seidman. TAB Books, P.O. Box 40, Blue Summit, PA 17294-0850. (717) 794-2191. (800) 233-1128. 1991. $75.00.

McGraw-Hill Electronics Dictionary. J. Markus. McGraw-Hill Book Company, Inc., 1221 Avenue of the Americas, New York, NY 10020. (212) 997-3675. 5th edition. 1994. $49.95.

McGraw-Hill Encyclopedia of Engineering. Sybil P. Parker, editor. McGraw-Hill Book Company, Inc. 1221 Avenue of the Americas, New York, NY 10020. (212) 997-3675. Second edition. 1993. $95.50. ISBN: 0-07-051392-9.

McGraw-Hill Encyclopedia of Science and Technology. McGraw-Hill Book, Incorporated, 1221 Avenue of the Americas, New York, NY 10020. (212) 997-3675. (800) 262-4729. Seventh edition. Twenty volumes. 1992. $1900.00.

GENERAL WORKS

Design and Technology of Integrated Circuits. Donald DeCogan. John Wiley & Sons, Inc., 605 Third Avenue, New York, NY 10158-0012. (212) 850-6000. (800) 225-5945. 1990. $75.00.

Diode Lasers and Photonic Integrated Circuits. David J. Comer. John Wiley & Sons, Inc., 605 Third Avenue, New York, NY 10158-0012. (212) 850-6000. (800) 225-5945. 1995. $64.95.

Computer Hardware Diagnostics For Engineers. Ronald E. Howland. McGraw-Hill Publishing Company, Inc., 1221 Avenue of the Americas, New York, NY 10020. (212) 512-2000. (800) 262-4729. 1995. $55.00.

Electric Circuits: Principles, Applications and Computer Analysis. David A. Bell. Prentice Hall, 113 Sylvan Avenue, Route

SIGNAL PROCESSING

Ency. of Physical Sciences and Engineering Info. Sources

9W, Englewood Cliffs, NJ 07632. (201) 592- 2000. (800) 922-0579. 5th edition. 1995. $72.00.

Exploring Electronic Devices. Mark E. Hazen. SCP: Third World Literature Publishing House, P.O. Box 482, Lithonia, GA 30058-0482. (404) 785-7725. 1991. $50.75.

Integrated Circuits and Microprocessors. Robin C. Holland. Elsevier Science Publishing Company, Inc., 655 Avenue of the Americas, New York, NY 10010. (212) 989-5800. FAX (914) 333-2444. 1986. $30.00 in paper.

Introduction To Electronics Design. Fred H. Mitchell and P. H. Mitchell, Sr. Prentice Hall, 113 Sylvan Avenue, Route 9W, Englewood Cliffs, NJ 07632. (201) 592-2000. (800) 922- 0579. 2nd edition. 1991. $81.00

Introduction To Fields and Circuits. Gordon Lancaster. Oxford University Press, Inc., 200 Madison Avenue, New York, NY 10016. (212) 725-6000. (800) 334-4249. 1992 $39.95.

Introduction To VLSI Techology. T. E. Price. Prentice Hall, 113 Sylvan Avenue, Route 9W, Englewood Cliffs, NJ 07632. (201) 592-2000. (800) 922-0579. 1994. $60.00.

Mastering IC Circuits. Joseph J. Carr. TAB Books, P.O. Box 40, Blue Summit, PA 17294-0850. (717) 794-2191. (800) 233-1128. 1991. $32.95.

Operational Amplifiers, Integrated and Hybrid Circuits. George B. Rutkowski. Prentice Hall, 113 Sylvan Avenue, Route 9W, Englewood Cliffs, NJ 07632. (201) 592-2000. (800) 922-0579. 1993. $87.95.

Simplified Design of Linear Power Supplies. John D. Lenk. Butterworth-Heinemann, 313 Washington Street, Newton, MA 02158. (618-928-2500. (800) 366-2665. 1994. $29.95.

Transistors: Fundamentals For the Integrated-circuit Engineer. R. M. Warner Jr., and B. L. Grung. Krieger Publishing Company, P.O. Box 9542, Melbourne, FL 32902-9542. (407) 724-9542. 1990. $72.50.

HANDBOOKS AND MANUALS

Active Electronic Component Handbook. Charles A. Harper and Harold C. Jones. McGraw-Hill Book Company, 1221 Avenue of the Americas, New York, NY 10020. (212) 512-2000. (800) 262-4729. 2nd edition. 1996. $79.50.

EDN Designer's Companion. Ian Hickman and Bill Travis, editors. Butterworth-Heinemann, 313 Washington Street, Newton, MA 02158. (618-928-2500. (800) 366-2665. 1994. $29.95.

Electrical Engineering Handbook. Richard C. Dorf, editor. CRC Press, Inc., 2000 Corporate Boulevard, NW, Boca Raton, FL 33431. (407) 994-0555. (800) 272-7737. 1993. $99.95.

Electronics Circuits Handbook. M. Tooley. Butterworth-Heinemann, 313 Washington Street, Newton, MA 02158. (618) 928-2500. (800) 366-2665. 1993. $49.95.

Electronics Handbook. Jerry C. Whitaker. CRC Press, Inc., 2000 Corporate Boulevard, NW, Boca Raton, FL 33431. (407) 994-0555. (800) 272-7737. 1996. $120.00.

Handbook of Electrical and Electronic Technology. Curtis Johnson. Prentice Hall , 113 Sylvan Avenue, Route 9W, Englewood Cliffs, NJ 07632. (201) 592-2000. (800) 922- 0579. 1996. $88.00.

Handbook of Quality Integrated Circuit Manufacturing. Robert Zurich. Academic Press, Inc., 6277 Sea Harbor Drive, Orlando, FL. (800) 321-5068. 1991. $59.95.

Handbook of VLSI Chip Design and Expert Systems. A. F. Schwartz. Academic Press, Inc., 6277 Sea Harbor Drive, Orlando, FL. (800) 321-5068.

Integrated Circuits Handbook. Rector Press, Ltd., 130 Rattlesnake, Leverett, MA 01054-9726. (413) 548-9708.(800) 247-3473. 3 volumes. 1995. $295.00 in paper.

Standard Handbook For Electrical Engineers. Donald Fink. McGraw-Hill Publishing Company, 1221 Avenue of the Americas, New York, NY 10020. (212) 512-2000. (800) 262- 4729. 13th edition. 1996. $110.00.

ONLINE DATABASES AND CD-ROMS

COMPENDEX PLUS. Engineering Information, Inc., 345 East 47th Street, New York, NY 10017. (212) 705-7600 or (800) 221-1044. Contains citations with abstracts to worldwide literature in engineering and technology, from 1970 to present. Available on online BRS,(800) 289-4277, DIALOG, (800) 334-2564, ORBIT (800) 456-7248, and STN International, FIZ Karlsruhe, P.O. Box 2465, W-7500, Karlsruhe 1, Germany, online services. Also available on CD-ROM. Inquire as to cost and availability.

Current Contents Search. Institute for Scientific Information, 3501 Market Street, Philadelphia, PA 19104. (215) 386-0100. FAX (215) 386-6362. Contains citations to articles listed in the table of contents of science and technology journals. Also articles in social sciences and life sciences journals. Available on BRS,(800) 289-4277, DIALOG,(800) 334-2564, online services. Inquire as to cost and availability.

Dissertation Abstracts Online. University Microfilms International, 300 North Zeeb Road, Ann Arbor, MI 48106. (800) 521-0600 or (313) 761-4700. Scope includes virtually all doctoral dissertations accepted at accredited American institutions from 1861 to present in 252 subject areas. Available on BRS, (800) 289-4277, DIALOG, (800) 334-2564, and OCLC EPIC,(800) 848-5878, online services. Also available on CD-ROM. Inquire as to cost and availability.

INSPEC. Institution of Electrical Engineers, Michael Faraday House, Six Hills Way, Stevenage, Herts. SG1 2AY, England. Telephone: 0438 313311 or FAX 0438 742840. Contains citations to the worldwide literature of physics, electronics and electrical engineering, computer technology, and related fields. Available on BRS, (800) 289-4277, DIALOG, (800) 334-2564, ORBIT, (800) 456-7248, and STN International, FIZ Karlsruhe, P.O. Box 2465, W-7500, Karlsruhe 1, Germany, online services. Inquire as to cost and availability.

NTIS Bibliographic Database. National Technical Information Service, 5285 Port Royal Road, Springfield, VA 22161. (703) 487-4929 or FAX (703) 321-8199. Broad coverage of government-sponsored science and technology research reports, 1964 to present. Available on BRS,(800) 289-4277, DIALOG, (800) 334-2564, ORBIT, (800) 456-7248, and STN International, FIZ Karlsruhe, P.O. Box 2465, W-7500, Karlsruhe 1, Germany,

online services. Also available on CD-ROM. Inquire as to cost and availability.

Physics Briefs. American Institute of Physics, 335 East 45th Street, New York, NY 10017. (212) 661-9260 or FAX (212) 661-2036. Contains citations with abstracts of the literature of physics and related fields, 1979 to present. Available on the STN International, FIZ Karlsruhe, P.O. Box 2465, W-7500, Karlsruhe 1, Germany, online service. Inquire as to cost and availability.

SCISEARCH. Institute for Scientific Information, 3501 Market Street, Philadelphia, PA 19104. (800) 523-1850 or (215) 386-0100. Broad multidisciplinary title and author index to the international literature of science and technology, 1974 to present. Available on DIALOG, (800) 334-2564, and ORBIT, (800) 456-7248, online services. Also available on CD-ROM. Inquire as to cost and availability.

WILSONLINE. H.W. Wilson Company, 950 University Avenue, Bronx, NY 10452. (800) 367-6770 or (212) 588-8400. Makes available online versions of the printed H.W. Wilson indexes including Applied Science and Technology Index, Business Periodicals Index, General Science Index, and Readers' Guide to Periodical Literature. Period covered is generally 1983 to present. Available on BRS, (800) 289-4277, DIALOG, (800) 334-2564, and OCLC EPIC, (800) 848-5878, online services. Also available on CD-ROM. Inquire as to cost and availability.

PERIODICALS

Circuits, Systems, and Signal Processing. Birkhauser, 675 Massachusetts Avenue, Cambridge, MA 02139-3309. FAX (201) 348-4505. 1982 to present. Quarterly. $184.00 per year. ISSN: 0278-081X.

Electrical World. McGraw-Hill, Inc., Box 513 Hightstown, NJ 08520. (212) 512-3288. 1874 to present. Monthly. $55.00 per year. ISSN: 0013-4457.

Electronic Design. Penton Publishing, San Jose Gateway, Suite 354. 2025 Gateway Place, San Jose, CA 95110. (408) 441-0550. 1952 to present. Fortnightly. $95.00. ISSN: 0013-4872.

Electronic Engineering Times. CMP Publications, Inc., 600 Community Drice, Manhasset, NY 11030. (516) 562-5000. FAX (516) 562-5325. 1972 to present. Weekly. $159.00 per year. ISSN: 0192-1541.

Electronics. Penton Publishing, San Jose Gateway, Suite 354. 2025 Gateway Place, San Jose, CA 95110. (408) 441-0550. 1930 to present. Semi-weekly. $98.00 per year. ISSN: 0883-4989.

Electronics World and Wireless World. Reed Business Publishing, Ltd., Quadrant House, the Quadrant, Sutton, Surrey SM2 5AS, England. 1911 to present. Monthly. L35. ISSN: 0266-3244.

IEEE Circuits and Devices Magazine. Institute of Electrical and Electronics Engineers, Inc., Box 1331, 445 Hoes Lane, Piscataway, NJ 08855-1331. (908) 981-0060. 1985 to present. Bi-monthly. $120.00 per year. ISSN: 8755-3996.

IEEE Journal of Solid State Circuits. Institute of Electrical and Electronics Engineers, Inc., Box 1331, 445 Hoes Lane, Piscataway, NJ 08855-1331. (908) 981-0060. 1966 to present. Bi-monthly. $275.00 per year. ISSN: 0018- 9200.

IEEE - Signal Processing Magazine. Institute of Electrical and Electronics Engineers, Inc., Box 1331, 445 Hoes Lane, Piscataway, NJ 08855-1331. (908) 981-0060. 1984 to present. Quarterly. $70.00 per year. ISSN: 1053-5888.

IEEE Spectrum. Institute of Electrical and Electronics Engineers, Inc., Box 1331, 445 Hoes Lane, Piscataway, NJ 08855-1331. (908) 981-0060. 1964 to present. Monthly. $157.00 per year. ISSN: 0018-9235.

IEEE Journal of Solid State Circuits. Institute of Electrical and Electronics Engineers, Inc., Box 1331, 445 Hoes Lane, Piscataway, NJ 08855-1331. (908) 981-0060. 1966 to present. Bi-monthly. $275.00 per year. ISSN: 0018- 9200.

IEEE Transactions On Circuits and Systems. Part 1: Fundamental theory and Applications. Institute of Electrical and Electronics Engineers, Inc., Box 1331, 445 Hoes Lane, Piscataway, NJ 08855-1331. (908) 981-0060. 1952 to present. Monthly. $241.00 per year. ISSN: 1057-7122.

IEEE Transactions On Electron Devices. Institute of Electrical and Electronics Engineers, Inc., Box 1331, 445 Hoes Lane, Piscataway, NJ 08855-1331. (908) 981-0060. 1952 to present. Monthly. $395.00 per year. ISSN: 0018-9393.

IEEE Transactions On Signal Processing. Institute of Electrical and Electronics Engineers, Inc., Box 1331, 445 Hoes Lane, Piscataway, NJ 08855-1331. (908) 981-0060. 1951 To Present. Monthly. $400.00. ISSN: 1053-587X.

Institute of Electrical and Electronics Engineers. PROCEEDINGS. Institute of Electrical and Electronics Engineers, Inc., Box 1331, 445 Hoes Lane, Piscataway, NJ 08855-1331. (908) 981-0060. 1913 to present. Monthly. $275.00. ISSN: 0018-9219.

IEE Proceedings. Part H: Microwaves, Antennas and Propagation. Institution of Electrical Engineers. Available from: Institute of Electrical and Electronic Engineers (IEEE) Inc., Box 1331, 445 Hoes Lane, Piscataway, NJ 08855-1331. (908) 981-0060. 1980 to present. Bi-monthly. ISSN: 0950-107X.

Semiconductor International. Cahners Publishing Co., 44 Cook Street, Denver, CO 80206. (708) 635-8000. FAX (708) 390-2770. (800) 662-7776. 1978 to present. 13 issues per year. $84.95 per year. ISSN: 0163-3767.

Solid State Electronics. Elsevier Science, 660 White Plains Road, Tarrytown, NY 10591-5153. (914) 524-9200. FAX (914) 333-2444. 1960 to present. Monthly. $1025.00 per year. ISSN: 0038-1101.

RESEARCH CENTERS AND INSTITUTES

Center for Integrated Sensors and Circuits. University of Michigan, 1246 EECS Building, Ann Arbor, MI 48109-2122. (313) 764-3346. FAX (313) 747-1781.

Center For Integrated Systems. Stanford University, School of Engineering, Stanford, CA 94305-4028. (415) 723-9038.

Center For Research in Engineering and Applied Sciences. Arizona State University, Tempe, AZ 85287-5506. (602) 965- 1725. FAX (602) 965-8296.

Electrical Engineering Research Laboratories. Purdue University. Electrical Engineering Building, West Lafayette, in 47907.

SIGNAL PROCESSING

Ency. of Physical Sciences and Engineering Info. Sources

(317) 494-3536. FAX (317) 494-6440.

Electronics Research Laboratory. University of California, Berkeley, 253 Cory Hall, Berkeley, CA 94720. (415) 642-2301.

Electroscience Laboratory. Ohio State University, 1320 Kinnear Road, Columbus, Ohio 43212. (614) 292-7981, FAX (614) 292-7297.

Engineering Research Center. University of Maryland, Clark School of Engineering, Potomac Building, Room 2104. (301) 405-3906. FAX (301) 403-4105.

Solid State Electronics Laboratory. North Carolina State University, 432 Daniels Hall, Raleigh, NC 27695. (919) 737-2336. FAX (919) 737-3027.

Telecommunications and Information Systems Laboratory. University of Kansas, Nichols Hall, Lawrence, KS 66045. (913) 864-4832. Fax (913) 864-7789.

SILICON

See also: BORON, CHEMISTRY, ELECTRONIC CIRCUITS and COMPONENTS, GLASS, INORGANIC CHEMISTRY, MATERIALS SCIENCE, MICROELECTRONICS, ORGANIC CHEMISTRY

ABSTRACT SERVICES AND INDEXES

Applied Science and Technology Index; a cumulative subject index to english language periodicals in the fields of Aeronautics and Space Science, Computer Technology, Chemistry, Construction Industry, Energy and Related Areas. H. W. Wilson Co., 950 University Avenue, Bronx, NY 10452- 9978. (212) 588-8400. (800) 367-6770. FAX (718) 590-1617. From 1958 To Present. Monthly. Inquire about cost and availability. Also available online (BRS and WILSONLINE) and on CD-ROM. ISSN: 0003-6986.

Chemical Abstracts. Chemical Abstracts Service. 2540 Olentangy River Road, Box 3012, Columbus, OH 43210-0012. (614) 447-3600. FAX (614) 447-3713. 1907 to present. Weekly. $16,800.00 per year. Available online and on CD- Rom. CA is also available in five section groupings. Inquire Regarding Cost and Availability.

Current Contents: Physical, Chemical and Earth Sciences. Institute For Scientific Information, 3501 Market Street, Philadelphia, PA 19104. (215) 386-0100. FAX (215) 386-2291. From 1961 To Present. Weekly. $442.00 per year. Also Available On Cd-rom and Online. Inquire regarding cost and availability. ISSN: 0163-2574.

Engineering Index Monthly. Engineering Information, Inc., Castle Point on the Hudson, Hoboken, NJ 07030. (201) 216-8500. (800) 221-1044. FAX (201) 216-8532. Monthly. $2300.00 per year for monthly issues. $1980 per year for annual cumulation. Available on CD-ROM and online as COMPENDEX. ISSN: 0742-1974; 0360-8557.

General Science Index. H.W. Wilson Company, 950 University Avenue, Bronx, NY 10452. (212) 588-8400. (800) 367-6770. Fax (718) 590-1617. From 1978 to present. Ten issues per Year; Quarterly and Annual Cumulations. Service basis. Available On

Cd-rom and Online. Inquire regarding cost and availability. ISSN: 0162-1963.

Physics Abstracts. INSPEC. Section A, Science Abstracts. Institution of Electrical Engineers (IEE), london, united Kingdom. Available from: INSPEC/IEEE -Institute of Electrical and Electronic Engineers, Box 1331, 445 Hoes Lane, Piscataway, NJ 08855-1331. (908) 562-5549. 1898 to present. 24 Issues Per Year. $2835.00 per year. ISSN: 0036-8091. Also available online and on CD-ROM.

Science Citation Index. SCI. Institute for Scientific Information, 3501 Market Street, Philadelphia, PA 19104. (215) 386-0100. (800) 523-1850. FAX (215) 386-2991. 1961 To Present. Six issues per year, plus annual Cumulation. $11650.00 per year. Also available online and On CD-ROM. Inquire about price and availability. ISSN:0036-827X.

ASSOCIATIONS AND PROFESSIONAL SOCIETIES

American Chemical Society, 1155 16th Street, NW, Washington, DC 20036. (202) 872-4600.

American Institute of Physics, 335 East 45th Street, New York, NY 10017 (212) 661-9404 or Fax (516)349-9704.

American Society For Testing and Materials. 1916 Race Street, Philadelphia, PA 19103. (215) 299-5400.

BIBLIOGRAPHIES

Chemical Information. H. R. Collier, editor. Springer- Verlag New York, Inc., 175 Fifth Avenue, New York, NY 10010. (212) 460-1500. (800) 777-4643. 1990. $91.00.

Chemical Information Sources. Gary Wiggins. McGraw-Hill Publishing Company, 1221 Avenue of the Americas, New York, NY 10020. (800) 262-4729 or (212) 512-3825. 1991. $42.50.

Electrochemical Synthesis of Inorganic Compounds: A Bibliography. Zoltan Nagy, editor. Plenum Publishing Corp., 233 Spring Street, New York, NY. (212) 620-8000. (800) 2221-9369. 1985. $125.00

Information Sources in Chemistry. R.T. Bottle and J.F.B Rowland, editors. R.R. Bowker Inc., 121 Chanlon Road, New Providence, NJ 07974. (800) 521-8110 or (908) 464-6800. Fourth edition. 1993. $75.00.

Scientific and Technical Books and Serials in Print; An Index To Literature in Science and Technology. R.R. Bowker Inc., 121 Chanlon Road, New Providence, NJ 07974. (908) 464-6800. (800) 521-8110. FAX (908) 665-3502. 1972 to present. Annual. 4 volumes. 1994. $299.95. Also available on compact disc and online. ISSN: 0000-054X.

DIRECTORIES AND BIOGRAPHICAL SOURCES# #*AMERICAN INSTITUTE OF CHEMISTS - PROFESSIONAL DIRECTORY.* AMERICAN INSTITUTE OF CHEMISTS, 501 WYTHE STREET, ALEXANDRIA, VA 22314-1917. (703) 836-2090. FAX (703) 836-2091. ANNUAL. $65.00.

American Men and Women of Science. Physical and Biological Sciences. R.R. Bowker Inc., 121 Chanlon Road, New Provi-

Ency. of Physical Sciences and Engineering Info. Sources

SILICON

dence, NJ 07974. (908) 464-6800. (800) 521-8110. 20th edition. 8 volumes. 1996. $850.00

Chemical Sources International. Mike Desing and Kurt Gandenberger, editors. Chemical Sources International, PO Box 1824, Clemson, SC 29633. (803) 646-7840. 1994. $285.00.

Chemical Week — Buyers Guide Issue. Chemical Week Associates, 888 Seventh Avenue, New York, NY 10106. (212) 621-4900. FAX (212) 621-4949. Annual, October. $50.00.

Research Centers Directory. Gale Research, 835 Penobscot Building, Detroit, MI 48226-4094. (313) 961-2242. (800) 877-4253. 20th edition. 1995. $485.00. ISSN: 0080-1518.

ENCYCLOPEDIAS AND DICTIONARIES

Academic Press Dictionary of Science and Technology. Christopher Morris, editor. Academic Press, Inc., 1250 Sixth Avenue, San Diego, CA 92101. (619) 231-0926. FAX (619) 699-6715. 1991. $115.00.

Concise Encyclopedia Chemistry. Translated and revised by Mary Eagleson. Walter de Gruyter, Inc., 200 Saw Mill River Road, Hawthorne, New York, 10532. (914) 747-0110 or Fax (914) 747-1326. 1994. $69.95.

Dictionary of Inorganic Compounds. Jane E. Macintyre, editor. Chapman & Hall, 1 Penn Plaza, New York, NY 10119. (212) 564-1060. 1992. $39.95.

Encyclopedia of Applied Physics. VCH Publishers, Inc., 303 Northwest 12th Avenue, Deerfield Beach, FL 33442. (800) 367-8249. 1991-. Twenty volumes. $6000.00.

Encyclopedia of Chemical Processing and Design. McKetta. Dekker, Inc., 270 Madison Avenue, New York, NY 10016. (212) 696-9000. (800) 228-1160. 1976 - . $175.00 per volume.

Hawley's Condensed Chemical Dictionary. Richard J. Lewis. Van Nostrand Reinhold, 115 Fifth Avenue, New York, NY 10003. (212) 254-3232. (800) 842-3636. 12th edition. 1993. $69.95.

Kirk-Othmer Encyclopedia of Chemical Technology. John Wiley and Sons, Inc., 605 Third Avenue, New York, NY 10158. (800) 526-5368 or (212) 850-6000. Fourth edition. 1991-. Twenty-seven volumes. $5400.00.

McGraw-Hill Encyclopedia of Chemistry. Sybil P. Parker, editor. McGraw-Hill Book, Incorporated, 1221 Avenue of the Americas, New York, NY 10020. (212) 997-3675. Second edition. 1993. $95.50.

McGraw-Hill Encyclopedia of Science and Technology. McGraw-Hill Book, Incorporated, 1221 Avenue of the Americas, New York, NY 10020. (212) 997-3675. Seventeenth edition. Twenty volumes. 1992. $1900.00.

GENERAL WORKS

Advanced Silicon and Semiconducting Silicon Alloy-based Materials and Devices. John F. Nijs, editor. IOP Publishing, Public Ledger Building, Suite 1035, Independence Square, Philadelphia, PA 19106. (215) 627-0880. (800) 358- 4677. 1994. $160.00.

Amorphous Silicon Materials and Solar Cells. Byron L. Stafford. American Institute of Physics, One Physics Ellipse, College Park, MD 20740-3843. (301) 209-3100. 1991. $90.00.

Catalyzed Direct Reactions of Silicon. Kenrich M. Lewis and David G. Rethwisch. Elsevier Science Publishing Company, Inc., 655 Avenue of the Americas, New York, NY 10010. (212) 989-5800. FAX (914) 333-2444. 1993.

Novel Silicon Based Technologies. R. A. Levy, editor. John Wiley & Sons, Inc., 605 Third Avenue, New York, NY 10158-0012. (212) 850-6000. (800) 225-5945. 1991. $98.00.

Physical Chemistry of, in and on Silicon. G. F. Cerofolini and L. Meds. Springer-Verlag New York, Inc., 175 Fifth Avenue, New York, NY 10010. (212) 460-1500. (800) 777- 4643. FAX (212) 473-6272. 1989. $34.50.

Phosphorus and Silicon in Waters, Effluents and Sludges. UNIPUB, 4511-F Assembly Drive, Lanham, MD 20706-4391. (301) 459-7666. (800) 274-4888. 1992. $18.00 in paper.

Physics of Silicon Sensors. S. Middelhoek and Sl. Audet, editors. Academic Press Inc., 6277 Sea Harbor Drive, Orlando, FL. (800) 321-5068. 1989. $97.00.

Porous Silicon Science and Technology. J. C. Vail and J. Derrien, editors. Springer-Verlag New York, Inc., 175 Fifth Avenue, New York, NY 10010. (212) 460-1500. (800) 777- 4643. FAX (212) 473-6272. 1995. $89.00.

Semiconductor Silicon. G. C. Harbeke, et al, editors. Springer-Verlag New York, Inc., 175 Fifth Avenue, New York, NY 10010. (212) 460-1500. (800) 777-4643. FAX (212) 473- 6272. 1989. $74.00.

Silicon-Based Millimeter-Wave Devices: Series in Electronics and Photonics. J. F. Luy and P. Ruser. Springer-Verlag New York, Inc., 175 Fifth Avenue, New York, NY 10010. (212) 460-1500. (800) 777-4643. FAX (212) 473-6272. 1994. $69.00.

Silicon Geochemistry and Biogeochemistry. S. R. Aston. Academic Press, Inc., 6277 Sea Harbor Drive, Orlando, FL. (800) 321-5068. 1983. $103.00.

HANDBOOKS AND MANUALS

Chemical Formulary. H. Bennett, editor. Chemical Publishing Company, Inc., 80 Eighth Avenue, New York, NY 10011. (212) 255-1950. Volumes 1-30. $70.00 per volume.

Chemometrics: Chemical and Sensory Data. David R. Burgard and James T. Kuznicki. CRC Publishers, Inc., 2000 Corporate Blvd., N.W., Boca Raton, FL 33431. (407) 994-0555 or (800) 333-8300. 1990. $120.00.

Chemical Reference Handbook. Gordon Press, PO Box 459, Bowling Green Station, New York, NY 10004. (718) 624-8419. 1995. $260.00.

CRC Handbook of Chemistry and Physics. David R. Kide, editor. CRC Press Inc., 2000 Corporate Blvd., NW, Boca Raton, FL 33431. (407) 994-0555. (800) 333-8300. 77th edition. 1996. $99.95.

Gmelin Handbook of Inorganic and Organometallic Chemistry. Springer-Verlag New York, Inc., 175 Fifth Avenue, New York, NY 10010. (212) 460-1500. (800) 777-4643. Various volumes.

Guide To Basic Chemical Compounds. D.R. Lide, Jr. CRC Publishers, Inc., 2000 Corporate Blvd., N.W., Boca Raton, FL 33431. (407) 994-0555. (800) 333-8300. 1993. $120.00.

Inorganic Chemicals Handbook. John J. McKetta. Marcel Dekker, Inc., 270 Madison Avenue, New York, NY 10016. (212) 696-9000. (800) 228-1160. 2 volumes. 1993. $400.00.

Laboratory Handbook of Materials, Equipment, and Techniques. Gary S. Coyne. Prentice Hall (Division of Simon and Schuster), 15 Columbus Circle, New York, NY 10023. (212) 373-8500 or (800) 922-0579. 1992. $45.00.

Lange's Handbook of Chemistry. John A. Dean, editor. McGraw-Hill Publishing Company, 1221 Avenue of the Americas, New York, NY 10020. (800) 262-4729 or (212) 512-3825. Fourteenth edition. 1992. $79.50.

Riegel's Handbook of Industrial Chemistry. James A. Kent, editor. Van Nostrand Reinhold, 115 Fifth Avenue, New York, NY 10003. (212) 254-3232 or (800) 926-2665. Ninth edition. 1992. $114.95.

ONLINE DATABASES AND CD-ROMS

Analytical Abstracts Online. Royal Society of Chemistry, Information Services, Thomas Graham House, Science Park, Milton Road, Cambridge, CB4 4WF, England. Contains citations, mostly with abstracts, of the worldwide literature on analytical chemistry, from 1980 to present. Available on DIALOG, (800) 334-2564, and STN International, FIZ Karlsruhe, P.O. Box 2465, W-7500, Karlsruhe 1, Germany, online services. Inquire as to cost and availability.

CA Search. Chemical Abstracts Service, P.O. Box 3012, Columbus, OH 43210-0012. (614) 447-3600. (800) 848-6533. FAX (614) 447-3709. Very comprehensive guide to worldwide chemical literature and related fields, 1972 to present. Available on BRS,(800) 289-4277, DIALOG, (800) 334-2564, ORBIT (800) 456-7248, and STN International, FIZ Karlsruhe, P.O. Box 2465, W-7500, Karlsruhe 1, Germany, online services. Inquire as to cost and availability.

Chemical Journals of the American Chemical Society. American Chemical Society, 1155 16th Street, N.W., Washington, DC 20036. (202) 872-4381 or (800) 424-6747. Contains complete text of approximately 90,000 articles from 22 primary journals published by the American Chemical Society, from mostly 1982 to present. Available on STN International, FIZ Karlsruhe, P.O. Box 2465, W-7500, Karlsruhe 1, Germany, online service. Inquire as to cost and availability.

Compendex Plus. Engineering Information, Inc., 345 East 47th Street, New York, NY 10017. (212) 705-7600 or (800) 221-1044. Contains citations with abstracts to worldwide literature in engineering and technology, from 1970 to present. Available on online BRS,(800) 289-4277, DIALOG, (800) 334-2564, ORBIT (800) 456-7248, and STN International, FIZ Karlsruhe, P.O. Box 2465, W-7500, Karlsruhe 1, Germany, online services. Also available on CD-ROM. Inquire as to cost and availability.

Current Contents Search. Institute for Scientific Information, 3501 Market Street, Philadelphia, PA 19104. (215) 386-0100. FAX (215) 386-6362. Contains citations to articles listed in the table of contents of science and technology journals. Also articles in social sciences and life sciences journals. Available on BRS,(800) 289-4277, DIALOG,(800) 334-2564, online services. Inquire as to cost and availability.

Dissertation Abstracts. University Microfilms International, 300 North Zeeb Road, Ann Arbor, MI 48106. (800) 521-0600 or (313) 761-4700. Scope includes virtually all doctoral dissertations accepted at accredited American institutions from 1861 to present in 252 subject areas. Available on BRS,(800) 289-4277, DIALOG,(800) 334-2564, and OCLC EPIC,(800) 848-5878, online services. Also available on CD-ROM. Inquire as to cost and availability.

Gmelin Database. Gmelin-Institut fur Anorganische Chemie und Grenzgebiete, Varrentrapperstrasse, 40-42, Carl-Bosch-Haus, D-6000, Frankfurt am Main 90, Germany. Contains structural and factual data relating to inorganic and organometallic chemistry. Provides data from the Gmelin Handbook of Inorganic and Organometallic Chemistry. Covers the period 1817 to 1975; 1988-89. Available on STN International, FIZ Karlsruhe, P.O. Box 2465, W-7500, Karlsruhe 1, Germany, online service. Inquire as to cost and availability.

Kirk-Othmer Encyclopedia of Chemical Technology. John Wiley and Sons, Inc., 605 Third Avenue, New York, NY 10158. (800) 526-5368 or (212) 850-6000. Contains the complete text of all chapters in the 27 volume fourth edition of the KIRK-Othmer Encyclopedia of Chemical Technology. 1991. Available on BRS,(800) 289-4277, DIALOG,(800) 334-2564, online services. Inquire as to cost and availability.

NTIS Bibliographic Database. National Technical Information Service, 5285 Port Royal Road, Springfield, VA 22161. (703) 487-4929 or FAX (703) 321-8199. Broad coverage of government-sponsored science and technology research reports, 1964 to present. Available on BRS,(800) 289-4277, DIALOG, (800) 334-2564, ORBIT, (800) 456-7248, and STN International, FIZ Karlsruhe, P.O. Box 2465, W-7500, Karlsruhe 1, Germany, online services. Also available on CD-ROM. Inquire as to cost and availability.

SCISEARCH. Institute for Scientific Information, 3501 Market Street, Philadelphia, PA 19104. (800) 523-1850 or (215) 386-0100. Broad multidisciplinary title and author index to the international literature of science and technology, 1974 to present. Available on DIALOG,(800) 334-2564, and ORBIT,(800) 456-7248, online services. Also available on CD-ROM. Inquire as to cost and availability.

WILSONLINE. H.W. Wilson Company, 950 University Avenue, Bronx, NY 10452. (800) 367-6770 or (212) 588-8400. Makes available online versions of the printed H.W. Wilson indexes including Applied Science and Technology Index, Business Periodicals Index, General Science Index, and Readers' Guide to Periodical Literature. Period covered is generally 1983 to present. Available on BRS,(800) 289-4277, DIALOG,(800) 334-2564, and OCLC EPIC,(800) 848-5878, online services. Also available on CD-ROM. Inquire as to cost and availability.

PERIODICALS

Angewandte Chemie. VCH Publishers, Inc., 220 East 23rd Street, New York, NY 10010-4606. (212) 683-8333. (800) 367-8249.

1888 to present. 22 issues per year. $840.00 Per Year. ISSN: 0044-8249.

Industrial and Engineering Chemistry Research. American Chemical Society, Box 3337, Columbus, OH 43210. (614) 447-3776. (800) 333- 9511. FAX (614) 447-3671. 1962 to present. Monthly. $567.00 per year. ISSN: 0888-5885.

Inorganic Chemistry. American Chemical Society, Box 3337, Columbus, OH 43210. (614) 447-3776. 1962 to Present. Semimonthly. $500.00 per year. ISSN: 002O-1669.

Journal of the American Chemical Society. American Chemical Society, Box 3337, Columbus, OH 43210. (614) 447-3776. 1879 topresent. Biweekly. $1055.00 per year. ISSN: 0002-7863.

Semiconductor International. Cahners Publishing Co., 44 Cook Street, Denver, CO 80206. (708) 635-8000. FAX (708) 390-2770. (800) 662- 7776. 1978 to present. 13 issues per year. $84.95 per year. ISSN: 0163-3767.

Solid State Electronics. Elsevier Science, 660 White Plains Road, Tarrytown, NY 10591-5153. (914) 524-9200. FAX (914) 333-2444. 1960to present. Monthly. $1025.00 per year. ISSN: 0038-1101.

RESEARCH CENTERS AND INSTITUTES

Center For Chemical Characterization and Analysis. Texas A & M Univeristy, College Station, TX 77843-3144. (409) 845-2341. FAX (409) 845-1655.

Center For Process Analytical Chemistry. University of Washington, BG-10, Seattle, WA 98195. (206) 545-2326. FAX (206) 543-6506.

Chemical Laboratories. Harvard University, Oxford Street, Cambridge, MA 02138. (617) 495-4283. FAX (617) 496-5618.

Laboratory For Research On the Structure of Matter. University of Pennsylvania, 3231 Walnut Street, Philadelphia, PA 19104. (215) 898-8571.

Theoretical Chemistry Institute. University of Wisconsin, Madison. 1101 University Avenue, Madison, WI 53706. (608) 262-1511.

University/industry Chemical Research Center. Mississippi State University, Department of Chemistry, P.O. Drawer CH, Mississippi State, MS 39762. (601) 325-3584.

SILICON CHIPS

See: INTEGRATED CIRCUITS

SINTERING

See: POWDER METALLURGY

SLUDGE

See also: ENVIRONMENTAL ENGINEERING, GROUND WATER POLLUTION, SEWAGE TREATMENT, POLLUTION, WATER TREATMENT

ABSTRACT SERVICES AND INDEXES

Current Contents: Physical, Chemical and Earth Sciences. Institute for Scientific Information, 3501 Market Street, Philadelphia, PA 19104. (215) 386-0100. FAX (215) 386-6362. Weekly. $360.00 per year.

Engineering Index Monthly. Engineering Information, Inc., Castle Point on the Hudson, Hoboken, NJ 07030. (800) 221-1044. FAX (201) 216-8532. Monthly. $2300.00 per year. Also available online as COMPENDEX, and also on CD-ROM. Covers chemical engineering, computers, electrical engineering, civil engineering, metals and mining, industrial management, and mechanical engineering.

Environmental Abstracts. Congressional Information Service, Reed Elsevier, 4520 East-West Highway, Bethesda, MD 20814-3389. (800) 638-8380. FAX (301) 654-4033. 1970 to present. Monthly. $1070.00 per year. Also available on-line and on CD-ROM.

Environmental Abstracts Annual. Congressional Information Service, Reed Elsevier, 4520 East-West Highway, Bethesda, MD 20814-3389. (800) 638-8380. FAX (301) 654-4033. 1970 to present. Annual. $495.00.

Environmental Engineering Abstracts. Cambridge Scientific Abstracts, 7200 Wisconsin Avenue, 6th Floor, Bethesda, MD 20814. (301) 961-6750. FAX (301) 961-6720. 1993 to present. Monthly. $385.00 per year.

Pollution Abstracts, Cambridge Scientific Abstracts, 7200 Wisconsin Avenue, Bethesda, MD 20814-4823. (301) 961-6750. FAX (301) 961-6720. Inquire for cost and availability.

Science Citation Index. Institute for Scientific Information, 3501 Market Street, Philadelphia, PA 19104. (215) 386-0100. FAX (215) 386-6362. Inquire about availability and cost. Also available on CD-ROM.

Selected Water Resources Abstracts. U.S. Geological Survey, Water Resources Scientific Information Center, 425 National Center, Reston, VA 22092. (703) 648-6820. Monthly. $115.00 per year. Also available on CD-ROM.

ASSOCIATIONS AND PROFESSIONAL SOCIETIES

American Public Works Association. 106 West 11th Street, Suite 1800, Kansas City, MO 64015-1806. (816) 472-6100. FAX (816) 472-1610.

American Society of Civil Engineers. 345 East 47th Street, New York, NY 10017-2398. (212) 705-7520 or (800) 548-2723.

American Water Works Association. 6666 W. Quincy Avenue, Denver, CO 80235. (303) 794-7711. FAX (303) 794-7310.

Association of Ground Water Scientists and Engineers. A division of the National Water Well Association, 6375 Riverside Drive, Dublin, OH 43017. (614) 761-1711. FAX (614) 761-

SLUDGE

Ency. of Physical Sciences and Engineering Info. Sources

3446. # #Association of Metropolitan Sewerage Agencies. 1000 Connecticut Avenue NW, Suite 410, Washington, DC 20036-5302. (202) 833-2672. FAX (202) 833-4657.

Municipal Waste Management Association. 16720 I Street NW, 4th floor, Washington, DC 20006. (202) 293-7330.

National Water Resources Association. 3800 N. Fairfax Drive, Suite 4, Arlington, VA 22203-1703. (703) 524-1544. FAX (703) 524-1548.

National Water Well Association. 6375 Riverside Drive, Dublin, OH 43017. (614) 761-1711. FAX (614) 761-3446.

Water and Sewer Distributors of America. 1900 Arch Street, Philadelphia, PA 19103-1498. (215) 564-3483. FAX (215) 564-2175.

Water Quality Association. 4151 Naperville Road, Lisle, IL 60532. (708) 505-0160. FAX (708) 505-9637.

DIRECTORIES AND BIOGRAPHICAL SOURCES

Directory of Water-Sewer and Related Industries Professionals. National Association of Regulatory Utility Commissioners, 1102 ICC Bldg., PO Box 684, Washington, DC 20044. (202) 898-2200.

GENERAL WORKS

Municipal Sewage Sludge Management, Volume 4: Processing, Utilization and Disposal. Cecil Lue-Hing, et al., editors. Technomic Publishing Company, Inc., 851 New Holland Avenue, Box 3535, Lancaster, PA 17604. (717) 291-5609 or (800) 233-9936. FAX (717) 295-4538. 1992. $95.00.

Sludge: Handling and Disposal. American Water Works Association, 6666 W. Quincy Avenue, Denver, CO 80235. (303) 794-7711. FAX (303) 794-7310. 1989. $31.50.

Sludge: Management and Disposal. Paul N. Cheremisinoff. Prentice-Hall, 113 Sylvan Avenue, Route 9W, Englewood Cliffs, NJ 07632. (201) 592-2000 or (800) 922-0579. 1993. $56.00.

Sludge Treatment. W. Wesley Eckenfelder Jr. and Chakre J. Santhanam, editors. Reprint edition. Books on Demand, Division of University Microfilms International, 300 North Zeeb Road, Ann Arbor, MI 48106-1346. (313) 761-4700 or (800) 521-0600. 1994. $173.00.

Surface Disposal of Sewage Sludge. Diane Publishing Company, 600 Upland Avenue, Upland, PA 19015. (610) 499-7415. FAX (610) 499-7429. 1995. $30.00.

HANDBOOKS AND MANUALS# #*HANDBOOK OF WATER AND WASTEWATER TREATMENT TECHNOLOGY.* **PAUL N. CHEREMISINOFF. MARCEL DEKKER, INC., 270 MADISON AVENUE, NEW YORK, NY 10016. (212) 696-9000. FAX (212) 685-4540. 1994. $195.00.**

ONLINE DATABASES AND CD-ROMS

CA Search. Chemical Abstracts Service, P.O. Box 3012, Columbus, OH 43210-0012. (614) 447-3600. (800) 848-6533. FAX

(614) 447-3709. Very comprehensive guide to worldwide chemical literature and related fields, 1972 to present. Available on BRS,(800) 289-4277, DIALOG, (800) 334-2564, ORBIT (800) 456-7248, and STN International, FIZ Karlsruhe, P.O. Box 2465, W-7500, Karlsruhe 1, Germany, online services. Inquire as to cost and availability.

Compendex Plus. Engineering Information, Inc., 345 East 47th Street, New York, NY 10017. (212) 705-7600 or (800) 221-1044. Contains citations with abstracts to worldwide literature in engineering and technology, from 1970 to present. Available on online BRS,(800) 289-4277, DIALOG, (800) 334-2564, ORBIT (800) 456-7248, and STN International, FIZ Karlsruhe, P.O. Box 2465, W-7500, Karlsruhe 1, Germany, online services. Also available on CD-ROM. Inquire as to cost and availability.

Current Contents Search. Institute for Scientific Information, 3501 Market Street, Philadelphia, PA 19104. (215) 386-0100. FAX (215) 386-6362. Contains citations to articles listed in the table of contents of science and technology journals. Also articles in social sciences and life sciences journals. Available on BRS,(800) 289-4277, DIALOG,(800) 334-2564, online services. Inquire as to cost and availability.

NTIS Bibliographic Database. National Technical Information Service, 5285 Port Royal Road, Springfield, VA 22161. (703) 487-4929 or FAX (703) 321-8199. Broad coverage of government-sponsored science and technology research reports, 1964 to present. Available on BRS,(800) 289-4277, DIALOG, (800) 334-2564, ORBIT, (800) 456-7248, and STN International, FIZ Karlsruhe, P.O. Box 2465, W-7500, Karlsruhe 1, Germany, online services. Also available on CD-ROM. Inquire as to cost and availability.

SCISEARCH. Institute for Scientific Information, 3501 Market Street, Philadelphia, PA 19104. (800) 523-1850 or (215) 386-0100. Broad multidisciplinary title and author index to the international literature of science and technology, 1974 to present. Available on DIALOG,(800) 334-2564, and ORBIT,(800) 456-7248, online services. Also available on CD-ROM. Inquire as to cost and availability.

Waternet. American Water Works Association, Technical Library. Available on DIALOG online services. Citations to literature on water quality, water utility management, water analysis, water pollution, and related areas, 1971 to present. Available on DIALOG,(800) 334-2564, online service. Inquire as to cost and availability.

PERIODICALS

American Water Works Association Journal. 6666 W. Quincy Avenue, Denver, CO 80235. (303) 794-7711. FAX (303) 794-7310. 1914 to present. Monthly. $85.00 to libraries and government agencies only.

Biocycle. J G Press Inc., 419 State Avenue, Emmaus, PA 18049. (215) 967-4135. 1960 to present. Monthly. $63.00.

Bioresource Technology. Elsevier Science [journals], 660 White Plains Rd., Tarrytown, NY 10159-5153. (919) 524-9200. FAX (919) 333-2444. 1991 to present. 12 times a year. $940.00 per year.

Journal of Environmental Engineering. ASCE, Enviromental Engineering Division, 345 East 47th Street, New York, NY 10017-2398. (212) 705-7520 or (800) 548-2723. FAX (212) 705-

7712. 1956 to present. Bi-monthly. $136.00 per year for non-members.

Journal of Environmental Science and Health, Part A: Environmental Science and Engineering. Marcel Dekker, Inc., 270 Madison Avenue, New York, NY 10016. (212) 696-9000. FAX (212) 685-4540. 1968 to present. 10 times a year. $725.00.

Journal of Water Resources Planning and Management. American Society of Civil Engineers, Water Resources Planning and Management Division, 345 East 47th Street, New York, NY 10017-2398. (212) 705-7520 or (800) 548-2723. Bi-monthly. $112.00 per year.

Public Works. Public Works Journal Corporation, 200 S Broad Street, Ridgewood, NJ 07451. (201) 445-5800. FAX (201) 445-5170. 1896 to present. 13 times a year. $45.00 per year.

Sanitary Maintenance. Trade Press Publishing Corporation, 2100 W. Florist Avenue, Milwaukee, WI 53209. (414) 228-7701. FAX (414) 228-1135. 1943 to present. Monthly. $49.00 per year.

Sludge Newsletter. Business Publishers Inc., 951 Pershing Drive, Silver Spring, MD 20910-4464. (301) 587-6300. FAX (301) 585-9075. 1978 to present. Bi-weekly. $325.00 per year.

Water Engineering and Management. Scranton Gillette Communications, Inc., 380 East Northwest Highway, Des Plaines, IL 60016. (708) 298-6622. FAX (708) 390-0408. 1882 to present. Monthly. $25.00 per year.

Waste Management and Research. Harcourt Brace & Co., Ltd., Foots Cray High Street, Sidcup, Kent DA14 5HP, England. Telephone 44-81-300-3322. FAX 44-81-309-0807. 1983 to present. Bi-monthly. Inquire for price.

Water, Air and Soil Pollution. Kluwer Academic Publishers, P.O. Box 358, Accord Station, Hingham, MA 02018-0358. (617) 871-6000. FAX (617) 871-6528. 1971 to present. 28 times a year. $1519.00 per year.

Water and Waste Treatment. Faversham House Group Ltd., 111 St. Jame's Road, Croydon, Surrey CR9 2TH, England. 1950 to present. Monthly. Inquire for price.

Water Science and Technology. Elsevier Science [journals], 660 White Plains Rd., Tarrytown, NY 10159-5153. (919) 524-9200. FAX (919) 333-2444. 1972 to present. 24 times a year. $2310.00 per year.

World Wastes. Argus Inc., 6151 Powers Ferry Road NW, Atlanta, GA 30339-2941. (404) 955-2500. FAX (404) 955-0400. 1957 to present. Monthly. $48.00 per year.

RESEARCH CENTERS AND INSTITUTES

United States Environmental Protection Agency, Test and Evaluation Facility. 1600 Gest Street, Cincinnati, OH 45204. (513) 684-2621. FAX (513) 684-2628.

University of California At Berkeley, Sanitary Engineering and Environmental Health Research Lab. Building 112, 1031 S 46th, Richmond, CA 94804. (415) 231-9449. FAX (415) 231-9500.

University of Florida, Environmental Engineering Research Center. College of Engineering, 217 Black Hall, Gainesville, FL 32611. (904) 392-0841. FAX (904) 392-3026.

Virginia Polytechnic Institute, Environmental Engineering and Sciences Group. 330 Norris Hall, Department of Civil Engineering, Blacksburg, VA 24061. (703) 961-6635.

SMELTING

See: METALLURGY

SNOW

See also: CLOUDS, ICE, METEOROLOGY, PRECIPITATION, RAIN, WATER RESOURCES, WEATHER FORECASTING

ABSTRACT SERVICES AND INDEXES

Applied Science and Technology Index; A Cumulative Subject Index To English Language Periodicals in the Fields of Aeronautics and Space Science, Computer Technology, Chemistry, Construction Industry, Energy and Related Areas. H.W. Wilson Co., 950 University Avenue, Bronx, NY 10452. - 9978. (212) 588-8400. (800) 367-6770. FAX (718) 590-1617. From 1958 to present. Monthly. Inquire about cost and availability. Also available online (BRS and WILSONLINE) and on CD-ROM. ISSN: 0003-6986.

General Science Index. H.W. Wilson Co., 950 University Avenue, Bronx, NY 10452. (212) 588-8400. (800) 367-6770. FAX (718) 590-1617. From 1978 to present. Ten issues per year; quarterly and annual cumulations. Service basis. Available on CD-ROM and online. Inquire regarding cost and availability. ISSN: 0162-1963.

Government Reports Announcements and Index. National Technical Information Service (NTIS), 5285 Port Royal Road, Springfield, VA 22161. (703) 487-4650. FAX (703) 321-8547. From 1968 to present. Annual. $630.00 per year. Also available online as NTIS Bibliographic Database and on CD-ROM.

Meteorological and Geoastrophysical Abstracts. American Meteorological Society, c/o Inforonics, Inc., 550 Newtown Road, Box 458, Littleton, MA 01460. (508) 486-8976. FAX (508) 486-0027. Covers literature in environmental sciences, meteorology, astrophysics, hydrology, glaciology, and physical oceanography. 1950 to present. Monthly. $950.00 per year. ISSN: 0026-1130. Also available online (DIALOG) and on CD-ROM.

Science Citation Index. SCI. Institute for Scientific Information, 3501 Market Street, Philadelphia, PA 19104. (215) 386-0100. (800) 523-1850. FAX (215) 386-2991. 1961 to present. Six issues per year, plus annual cumulation. $11650.00 per year. Also available online and on CD-ROM. Inquire about price and availability. ISSN: 0036-827X.

Selected Water Resources Abstracts. U.S. Geological Survey. Distributed by National Technical Information Service (NTIS), 5285 Port Royal Road, Springfield, VA 22161. (703) 487-4650. Monthly. $115.00 per year.

SNOW

Ency. of Physical Sciences and Engineering Info. Sources

ANNUAL REVIEWS AND YEARBOOKS

Annual Review of Earth and Planetary Sciences. Annual Reviews, Inc., 4139 El Camino Way, Palo Alto, CA 94306-0897. (415) 493-4400. (800) 523-8635. FAX (415) 855-9815. Annual. $55.00 per year.

Developments in Atmospheric Science Series. Elsevier Science Publishing Company, Inc., 655 Avenue of the Americas, New York, NY 10010. (212) 989-5800. Irregular. Inquire about price and availability.

ASSOCIATIONS AND PROFESSIONAL SOCIETIES

American Meteorological Society, 45 Beacon Street, Boston, MA 02108-3693. (617) 227-2425. FAX (617) 742-8718.

Association of American Weather Observers, 401 Whitney Boulevard, Box 455, Belvedere, IL 61008. (815) 544-5665. FAX (815) 544-6334.

Canadian Meteorological and Oceanographic Society, P.O. Box 334, Newmarket, Ontario, L3Y 4X7, Canada. (416) 898-1040. FAX (416) 898-7937.

National Environmental Satellite Data, and Information Service. 2069 Federal Building 4 Room 2060, Washington, DC 20233. (301) 763-71990. FAX (301) 763-4011.

National Weather Association, 6704 Wolke Court, Montgomery, AL 36116-2134. (334) 213-0388. FAX (334) 213-0388.

University Corporation For Atmospheric Research. PO Box 3000, 1850 Table Mesa Drive, Boulder, CO 80307-3000. (303) 497-1673. FAX (303) 497-1654.

BIBLIOGRAPHIES

Information Sources in the Earth Sciences. David N. Wood, Joan E. Hardy, and Anthony P. Harvey, editors. Bowker-Saur/K.G. Saur. Distributed by R.R. Bowker, 121 Chanlon Road, New Providence, NJ 07974. (800) 521-8110 or (908) 464-6800. Second edition. 1989. $85.00.

DIRECTORIES AND BIOGRAPHICAL SOURCES

American Men and Women of Science: Physical and Biological Sciences. Eighteenth edition. R.R. Bowker Company, 245 West 17th Street, New York, NY 10011. (800) 521-8810 or (212) 916-1600. $750.00.

American Meteorological Society Professional Directory. American Meteorological Society. 45 Beacon Street, Boston, MA 02108. (617) 227-2425. Included in each issue of the Bulletin of the Society.

Meteorological Services of the World. American Meteorological Society, 45 Beacon Street, Boston, MA 02108- 3693. (617) 227-2425. FAX (617) 742-8718. Annual. $70.00.

National Weather Service Offices and Stations. National Oceanic and Atmospheric Administration, Department of Commerce, Silver Spring, MD 20910. (301) 427-7698. Annual. Free.

Research Centers Directory. Gale Research, 835 Penobscot Building, Detroit, MI 48226-4094. (313) 961-2242. (800) 877-4253. 20th edition. 1995. $485.00. ISSN: 0080-1518.

ENCYCLOPEDIAS AND DICTIONARIES

Climates of the States. Bair. Gale Research, 835 Penobscot Building, Detroit, MI 48226-4094. (313) 961-2242. (800) 877-4253. 5th edition. 1995. $255.00.

Concise Oxford Dictionary of the Earth Sciences. Ailsa Allaby and Michael Allaby, editors. Oxford University Press, Inc., 200 Madison Avenue, New York, NY 10016. (800) 334-4249 or (212) 679-7300. 1990. $42.95.

Encyclopedia of Climatology. Rhodes W. Fairbridge, and John E. Oliver. Chapman & Hall, 1 Penn Plaza, New York, NY 10119. (212) 564-1060. 1986. $115.00.

McGraw-Hill Encyclopedia of Ocean and Atmospheric Sciences. Sybil P. Parker, editor. McGraw-Hill Publishing Company, 1221 Avenue of the Americas, New York, NY 10020. (800) 262-4729 or (212) 512-3825. 1979. $79.95.

GENERAL WORKS

The Atmosphere: An Introduction To Meteorology. Frederick K. Lutgens and Edward J. Tarbuck. Prentice Hall (Division of Simon and Schuster), 15 Columbus Circle, New York, NY 10023. (212) 373-8500 or (800) 922-0579. Fifth edition. 1991. $49.00.

Cold Climate Hydrometeorology. D.S. Upadhyay. Halsted Press, 605 Third Avenue, New York, NY 10158-0012. (212) 850-6400. (800) 526-5368. 1995. $44.95.

Dynamics of Snow and Ice Masses. Samuel C. Colbeck. Academic Press, Inc., 6277 Sea Harbor Drive, Orlando FL 32887. (800) 3221-5068. 1980. $118.00.

A History of the theories of Rain and Other Forms of Precipitation. William E. Middleton. Books on Demand, Division of University Microfilms International, 300 North Zeeb Road, Ann Arbor, MI 48106-1346. (313) 761-4700 or (800) 521-0600. 1966. $58.00.

Mesoscale Meteorology and Forecasting. Peter S. Ray, editor. American Meteorological Society, 45 Beacon Street, Boston, MA 02108-3693. (617) 227-2425. FAX (617) 742-8718. 1986. $66.25.

Meteorology and Environmental Sciences. R. Guzzi and World Scientific Publishing Company, Inc., 1060 Main Street, River Edge, NJ 07661. (800) 227-7562 or (201) 487-9655. 1990. $120.00.

Meteorology: The Atmosphere and the Science of Weather. Joseph M. Moran. Macmillan Publishing, 866 3rd Avenue, New York, NY 10024. (212) 689-9140. Third edition. 1991. $49.95.

Severe and Unusual Weather. Joe R. Eagleman. Trimedia Publishing Company, 12008 West 87th Street, Suite 117, Lenexa, KS 66215. (913) 599-0505. Second edition. 1990. $41.95.

Snow Crystals. W.A. Bentley and W.J. Humphreys. Dover Publications, Inc., 180 Varick Street, New York, NY 10014. (212) 255-3755 or (800) 223-3130. 1931. $15.95 in paper.

Weather Companion: An Album of Meteorological History, Science, Legend, and Folklore. Gary Lockhart. John Wiley and Sons, Inc., 605 Third Avenue, New York, NY 10158. (800) 526-5368 or (212) 850-6000. 1988. $12.95.

HANDBOOKS AND MANUALS

Author's Guide To the Journals of the American Meteorological Society. American Meteorological Society, 45 Beacon Street, Boston, MA 02108-3693. (617) 227-2425. FAX (617) 742-8718. 1983. $15.00 in paper.

Basic Meteorology Lab Manual. Thomas A. Leavy. Allegheny Press, P.O. Box 220, Elgin, PA 19413. (814) 664-8504. Second edition. 1969. $7.95.

Guidelines For the Control of Blowing and Drifting Snow. Ronald D. Tabler. SHRP - Strategic Highway Research Program. 2101 Constitution Avenue NW, Washington, DC 20055. (202) 334-3313. (800) 624-6242. 1994. $20.00.

Handbook in Applied Meteorology. David D. Houghton, editor. John Wiley and Sons, Inc., 605 Third Avenue, New York, NY 10158. (800) 526-5368 or (212) 850-6000. 1985. $114.00.

Weather Almanac. Gale Research, 835 Penobscot Building, Detroit, MI 48226-4094. (313) 961-2242. (800) 877- 4253. 7th edition. 1996. $130.00.

Weather Satellite Handbook. American Radio Relay League, 225 Main Street, Newington CT -6111. (203) 666-1541. 1994. $20.00.

ONLINE DATABASES AND CD-ROMS

Climate Assessment Database. National Weather Service, National Meteorological Center, 5200 Auth Road, Suite 101, Camp Springs, MD 20233. (301) 763-8016. Contains daily, weekly and monthly summaries of North American and world climatological data. Also provides five to ten weather forecasts, and 30 to 90 day weather outlook. Subscription required. Inquire as to cost and availability.

Current Contents Search. Institute for Scientific Information, 3501 Market Street, Philadelphia, PA 19104. (215) 386-0100. FAX (215) 386-6362. Contains citations to articles listed in the table of contents of science and technology journals. Also articles in social sciences and life sciences journals. Available on BRS,(800) 289-4277, DIALOG,(800) 334-2564, online services. Inquire as to cost and availability.

Dissertation Abstracts. University Microfilms International, 300 North Zeeb Road, Ann Arbor, MI 48106. (800) 521-0600 or (313) 761-4700. Scope includes virtually all doctoral dissertations accepted at accredited American institutions from 1861 to present in 252 subject areas. Available on BRS,(800) 289-4277, DIALOG,(800) 334-2564, and OCLC EPIC,(800) 848-5878, online services. Also available on CD-ROM. Inquire as to cost and availability.

Meteorological and Geoastrophysical Abstracts. American Meteorological Society, 45 Beacon Street, Boston, MA 02108-

3693. (617) 227-2425. FAX (617) 742-8718. Contains citations and abstracts to the worldwide literature on significant research in meteorology and geoastrophysics. Related topics include physical oceanography, hydrology, environmental sciences and glaciology. Covers the period 1972 to present. Available on DIALOG,(800) 334-2564, online service. Inquire as to cost and availability.

NTIS Bibliographic Database. National Technical Information Service, 5285 Port Royal Road, Springfield, VA 22161. (703) 487-4929 or FAX (703) 321-8199. Broad coverage of government-sponsored science and technology research reports, 1964 to present. Available on BRS,(800) 289-4277, DIALOG, (800) 334-2564, ORBIT, (800) 456-7248, and STN International, FIZ Karlsruhe, P.O. Box 2465, W-7500, Karlsruhe 1, Germany, online services. Also available on CD-ROM. Inquire as to cost and availability.

SCISEARCH. Institute for Scientific Information, 3501 Market Street, Philadelphia, PA 19104. (800) 523-1850 or (215) 386-0100. Broad multidisciplinary title and author index to the international literature of science and technology, 1974 to present. Available on DIALOG and ORBIT online services. Also available on CD-ROM. Inquire as to cost and availability.

WILSONLINE. H.W. Wilson Company, 950 University Avenue, Bronx, NY 10452. (800) 367-6770 or (212) 588-8400. Makes available online versions of the printed H.W. Wilson indexes including Applied Science and Technology Index, Business Periodicals Index, General Science Index, and Readers' Guide to Periodical Literature. Period covered is generally 1983 to present. Available on BRS,(800) 289-4277, DIALOG,(800) 334-2564, and OCLC EPIC,(800) 848-5878, online services. Also available on CD-ROM. Inquire as to cost and availability.

World Climate Disc. Chadwyck-Healey Inc., 1101 King Street, Alexandria, VA 22314. (703) 683-4890. FAX (703) 683-7589. Weather and climate data from approximately 5,000 weather stations worldwide, covering the years 1854 to 1990 on CD-ROM. First edition 1992. Approximately $1200.00 per year with annual updates.

World Weatherdisc. WeatherDisc Associates, Inc., 4584 N.E. 89th Street, Seattle, WA 98115. (206) 524-4314. FAX (206) 543-0308. Meteorological data on CD-ROM which describes the climate of the earth today and for the past few hundred years. First edition 1989. Approximately $295.00 per year with biannual updates.

PERIODICALS

Agricultural and Forest Meteorology. Elsevier Science Publishing Company, Inc., 655 Avenue of the Americas, New York, NY 10010. (212) 989-5800. Twenty times per year. $750.00 per year. # #*American Meteorological Society Bulletin.* American Meteorological Society, 45 Beacon Street, Boston, MA 02108-3693. (617) 227-2425. FAX (617) 742-8718. Monthly. $60.00 per year.

American Meteorological Society Meteorological Monographs. American Meteorological Society, 45 Beacon Street, Boston, MA 02108-3693. (617) 227-2425. FAX (617) 742-8718. Irregular. Price varies.

American Weather Observer. Association of American Weather Observers, 401 Whitney Boulevard, Box 455, Belvedere, IL 61008. (815) 544-5665. FAX (815) 544-6334. Monthly. $21.00 per year.

SNOW

Ency. of Physical Sciences and Engineering Info. Sources

Atmosphere - Ocean. Canadian Meteorological and Oceanographic Society, P.O. Box 334, Newmarket, Ontario, L3Y 4X7, Canada. (416) 898-1040. FAX (416) 898-7937. Quarterly. $30.00 per year.

Boundary-layer Meteorology: An International Journal of Physical and Biological Processes in the Atmospheric Boundary Layer. Kluwer Academic Publishers, P.O. Box 358, Accord Station, Hingham, MA 02018-0358. (617) 871-6000. Sixteen per year. $785.00 per year.

Dynamics of Atmospheres and Oceans. Elsevier Science Publishing Company, Inc., 655 Avenue of the Americas, New York, NY 10010. (212) 989-5800. Six times per year. $205.00 per year.

International Journal of Climatology. Royal Meteorological Society, distributed by John Wiley and Sons, Inc., 605 Third Avenue, New York, NY 10158. (800) 526-5368 or (212) 850-6000. Eight times per year. $425.00 per year.

Journal of Applied Meteorology. American Meteorological Society, 45 Beacon Street, Boston, MA 02108-3693. (617) 227-2425. FAX (617) 742-8718. Monthly. $165.00 per year.

Journal of Climate. American Meteorological Society, 45 Beacon Street, Boston, MA 02108-3693. (617) 227-2425. FAX (617) 742-8718. Monthly. $175.00 per year.

Journal of Atmospheric Sciences. American Meteorological Society, 45 Beacon Street, Boston, MA 02108-3693. (617) 227-2425. FAX (617) 742-8718. Semi-monthly. $320.00 per year.

Monthly Weather Review. American Meteorological Society, 45 Beacon Street, Boston, MA 02108-3693. (617) 227-2425. FAX (617) 742-8718. Monthly. $205.00 per year.

National Weather Digest. National Weather Association, 4400 Stamp Road, Room 404, Temple Hills, MD 20748. (301) 899-3784.

Royal Meteorological Society Quarterly Journal. Royal Meteorological Society, 104 Oxford Road, Reading, Berks RG1 7LJ, England. Six times per year. $250.00 per year.

Weather. Royal Meteorological Society, 104 Oxford Road, Reading, Berks RG1 7LJ, England. Monthly. $44.00 per year.

Weatherwise. Heldref Publications, 1319 Eighteenth Street, N.W., Washington, DC 20036-1802. (202) 296-6267. FAX (202) 296-5149. Bi-monthly. $28.00 per year.

RESEARCH CENTERS AND INSTITUTES

Climate Analysis Center. National Weather Service, 5200 Auth Road, Camp Springs, MD 20746. (301) 763-8167.

Cooperative Institute For Mesoscale Meteorological Studies. University of Oklahoma, 401 East Boyd, Norman, OK 73019. (405) 325-3041.

Desert Research Institute, Atmospheric Water Resources Laboratory. P.O. Box 60220, Reno NV 89506. (702) 677-3123.

Institute of Snow Research. Michigan Technological University, Houghton, MI 49931. (906) 487-2750. FAX (906)487-2202.

National Center For Atmospheric Research. P.O. Box 3000, Boulder, CO 80307. (303) 496-1000. FAX (303) 497-1654.

National Snow and Ice Data Center. University of Colorado - Boulder, CB 449, Boulder, CO 80309-0449. (303) 492-5171. FAX (303) 492-2468.

SOFTWARE

See also: ARTIFICIAL INTELLIGENCE, COMPUTER PROGRAMMING, COMPUTER PROGRAMMING LANGUAGES, PARALLEL COMPUTERS, SOFTWARE ENGINEERING

ABSTRACT SERVICES AND INDEXES

ACM Guide To Computing Literature. Association for Computing Machinery, Association for Computing Machinery, 1515 Broadway, 17th Floor, New York, NY 10036-5701. (212) 869-7440. FAX (212) 944-1318. 1964 to present. Monthly. $190.00 per year for non-members.

Applied Science and Technology Index; A Cumulative Subject Index To English Language Periodicals in the Fields of Aeronautics and Space Science, Computer Technology, Chemistry, Construction Industry, Energy and Related Areas. H.W. Wilson Co., 950 University Avenue, Bronx, NY 10452. (800) 367-6770 or (212) 588-8400. FAX (718) 590-1617. From 1958 to present. Monthly. Inquire about cost and availability. Also available on CD-ROM and online.

Computer Abstracts, MCB University Press Ltd., PO Box 10812, Birmingham, AL 35201-0812. (800) 633-4931. FAX (205) 995-1588. Monthly. Covers computer theory, data, hardware, systems, networks, human-computer interaction, artificial intelligence, as well as applications of computers in aerospace, business, CAD/CAM, cartography, civil engineering, electronics and electrical engineering, industrial engineering, mechanical engineering, medicine, structural engineering, etc. Also available on CD-ROM.

Computer & Control Abstracts (inspec). INSPEC/IEEE, Box 1331, 445 Hoes Lane, Piscataway, NJ 08855-1331. (908) 562-5549. Abstracts organized by subjects of international technological information. Monthly. $1455.00 per year. Also available on CD-ROM and online via BRS Online Products, DIALOG Information Services, Orbit Search Service, and STN International.

Computer and Information Systems Abstracts Journal. Cambridge Scientific Abstracts, 7200 Wisconsin Avenue, Bethesda, MD 20814. (301) 961-6750. Fax (301) 961-6720. 1962 to present. Monthly. $1265.00 per year. Also available online via STN International.

Computer Literature Index. Applied Computer Research Inc., Box 82266, Phoenix, AZ 85071-2266. (800) 234-2227. 1971 to present. Quarterly plus annual cumulation. $198.50 per year. Bibliography of books, articles, and reports.

Computing Journal Abstracts. Techgnosis Ltd., Blade House, Battersea Road, Stockport, Cheshire 3AE, England. Telephone 061-442-2639. FAX 061-443-1162. 1969 to present. Monthly. Inquire for price.

Computing Reviews. Association for Computing Machinery, 1515 Broadway, 17th Floor, New York, NY 10036-5701. (212) 869-7440. FAX (212) 944-1318. 1960 to present. Monthly. $130.00 per year for non-members. Also available online via DIALOG Information Services.

Current Contents: Engineering, Technology, and Applied Sciences. Institute for Scientific Information, 3501 Market Street, Philadelphia, PA 19104. (215) 386-0100. FAX (215) 386-2991. 1970 to present. Weekly. $442.00 per year.

Engineering Index Monthly. Engineering Information, Inc., Castle Point on the Hudson, Hoboken, NJ 07030. (800) 221-1044. FAX (212) 832-1857. Monthly. $2300.00 per year. Also available online as COMPENDEX, and also on CD-ROM. Covers chemical engineering, computers, electrical engineering, civil engineering, metals and mining, industrial management, and mechanical engineering.

Index To Scientific and Technical Proceedings. Institute for Scientific Information, 3501 Market St., Philadelphia, PA 19104. (215) 386-0100. FAX (215) 386-6362. Monthly. $500.00 per year.

Index To Scientific Reviews. Institute for Scientific Information, 3501 Market St., Philadelphia, PA 19104. (215) 386-0100. FAX (215) 386-6362. Semi-annual.

Key Abstracts—Software Engineering. INSPEC/IEEE, IEEE (Institute of Electrical and Electronic Engineers), 445 Hoes Lane, Piscataway, NJ 08854. (908) 562-5545 or (800) 678-IEEE. 1987 to present. Monthly. $160.00 per year.

ASSOCIATIONS AND PROFESSIONAL SOCIETIES

American Federation of Information Processing Societies, 1899 Preston White Drive, Reston, VA 22091-5435. (703) 620-8937.

Association For Computing Machinery. 1515 Broadway, 17th Floor, New York, NY 10036-5701. (212) 869-7440. FAX (212) 944-1318.

Computer and Automation Systems Association of Sme. 1 SME Drive, PO Box 930, Dearborn, MI 48121-0930. (313) 271-1500.

Computing Research Association. 1875 Connecticut Avenue NW, Suite 718, Washington, DC 20009-5728. (202) 234-2111. FAX (202) 667-1066.

IEEE Computer Society. 1730 Massachusetts Avenue NW, Washington, DC 20036-1992. (202) 371-0101. FAX (202) 728-9614.

Society For Computer Simulation International, 4838 Ronson Court Lane, San Diego, CA 92111-1810. (619) 277-3888.

Software Management Association. PO Box 12004 #297, Vallejo, CA 94590. (707) 643-4423. FAX (707) 643-4423.

Software Publishers Association. 1730 M Street NW, Suite 700, Washington, DC 20036-4510. (202) 452-1600. FAX (202) 223-8756.

Special Interest Group On Programming Languages. c/o ACM, 1515 Broadway, New York, NY 10036. (212) 869-7440. FAX (212) 869-0481.

Special Interest Group On Software Engineering. c/o ACM, 1515 Broadway, New York, NY 10036. (212) 626-0611. FAX (212) 302-5826.

BIBLIOGRAPHIES

ACM Guide To Computing Literature. Association for Computing Machinery, 1515 Broadway, 17th Floor, New York, NY 10036-5701. (212) 869-7440. FAX (212) 944-1318. 1964 to present. Annual. $190.00 for non-members. Also available online via DIALOG.

DIRECTORIES AND BIOGRAPHICAL SOURCES

American Men & Women of Science, 1995-96. R.R. Bowker Staff, eds. 19th edition. 8 volumes. R.R. Bowker/Reed International Publishing Company, 121 Chanlon Road, New Providence, NJ 07974. (908) 464-6800 or (800) 521-8110. 1995. $850.00# *#ASIS Handbook and Directory*, American Society For Information Science, 8720 Georgia Avenue, Suite 501, Silver Spring, MD 20910-3602. (301) 495-0900. Annual. $100.00.

Computing Information Directory. Hildebrandt Inc., PO Box 576, Pullman, WA 99114. 1981 to present. Annual. $199.95 per year.

Directory of Engineering Societies and Related Organizations. Gordon Davis. 13th edition. American Association of Engineering Societies, 1111 19th Street NW, Suite 608, Washington, DC 20036. (202) 296-2237 or (800) 658-8897. 1989. Inquire for price.

International Research Centers Directory. Gale Research, 835 Penobscot Building, Detroit, MI 48226-4094. (313) 961-2242. (800) 877-4253. FAX (313) 961-6083. 8th edition. 1995. $410.00

Research Centers Directory. Gale Research, 835 Penobscot Building, Detroit, MI 48226-4094. (313) 961-2242. (800) 877-4253. FAX (313) 961-6083. $485.00.

Scientific and Technical Organizations and Agencies Directory. Gale Research, 835 Penobscot Building, Detroit, MI 48226-4094. (313) 961-2242. (800) 877-4253. FAX (313) 961-6083. 4th edition. 1996. $195.00.

The Softpub Yellow Pages: Directory of Suppliers To the Software Publishing Industry. SoftPub Group Inc., PO Box 356, Maple Valley, WA 98038. (800) 763-8782. FAX (206) 413-1098. Inquire for cost and availability.

Software Engineering Standards: An Annotated Index and Directory. S. Magee and L. Tripp. Global Engineering Documents, 15 Inverness Way East, Englewood, CO 80112-5704. (800) 624-3974. FAX (303) 397-7935. 1994. $79.50.

Who's Who in Computing. Canadian M I S Database Inc., 268 Lakeshore Road E., Suite 510, Mississauga, ON L5G 1H1, Canada. (905) 271-1601. FAX (905) 271-4522. 1987 to present. Annual. Inquire for price in U.S.

SOFTWARE

Ency. of Physical Sciences and Engineering Info. Sources

ENCYCLOPEDIAS AND DICTIONARIES

Computer Dictionary. Donald D. Spencer. 4th edition. Camelot Publishing Company, 709 SW 80th Blvd., Gainesville, FL 32607-1537. (904) 331-0952. 1993. $24.95.

Dictionary of Computing. S.M.H. Collin. 2nd edition. Peter Collin Publishing, 8 the Causeway, Teddington, TW11 0HE, England. FAX 0181-943-3386. 1994. Inquire for cost.

Encyclopedia of Computer Science and Engineering. Anthony Ralston & Edwin D. Reilly Jr., editors. 3rd revised edition. Van Nostrand Reinhold, 115 Fifth Avenue, New York, NY 10003. (212) 254-3232 FAX (212) 254-9499. 1993. $125.00.# #*Microsoft Press Computer Dictionary: The Comprehensive Standard For Business, School, Library, and Home.* 2nd edition. Microsoft Press, One Microsoft Way, Redmond, WA 98052-6399. (206) 936-0055. FAX (206) 823-8101. 1994. $19.95.

GENERAL WORKS

Art of Software Design. Isensee. Van Nostrand Reinhold, 115 Fifth Avenue, New York, NY 10003. (212) 254-3232 or (800) 926-2665. 1995. $42.95.

Elements of Friendly Software Design. Paul Heckel. Sybex, Inc. 2021 Challenger Drive, Alameda, CA 94501. (800) 227-2346. FAX (510) 523-2373. 1994. $17.99.

Software Architecture and Design: Principles, Models and Methods. Bernard I. Witt, et al. Van Nostrand Reinhold, 115 Fifth Avenue, New York, NY 10003. (212) 254-3232 or (800) 926-2665. 1994. $44.95.

Software Design. David Budgen. Addison-Wesley Publishing Company, 1 Jacob Way, Reading, MA 01867. (617) 944-3700 or (800) 447-2226. FAX 617-942-1117. 1994. $34.95.

Software Process Design. Jacqueline Holdsworth. McGraw-Hill Publishing Company, 1221 Avenue of the Americas, New York, NY 10020. (800) 262-4729 or (212) 512-3825. 1994. $35.00.

ONLINE DATABASES AND CD-ROMS

Compendex Plus. Engineering Information, Inc., 345 East 47th Street, New York, NY 10017. (212) 705-7600 or (800) 221-1044. Contains citations with abstracts to worldwide literature in engineering and technology, from 1970 to present. Available on online BRS,(800) 289-4277, DIALOG, (800) 334-2564, ORBIT (800) 456-7248, and STN International, FIZ Karlsruhe, P.O. Box 2465, W-7500, Karlsruhe 1, Germany, online services. Also available on CD-ROM. Inquire as to cost and availability.

Compuscience. FIZ Karlsruhe, D-7514, Eggenstein-Leopoldshafen 2, Germany. Contains citations with abstracts to European and North American literature on computer science, 1972 to present. Available on STN International, FIZ Karlsruhe, P.O. Box 2465, W-7500, Karlsruhe 1, Germany, online service. Inquire as to cost and availability.

Computer and Information Systems Abstracts. Cambridge Scientific Abstracts, 7200 Wisconsin Avenue, Suite 601, Bethesda, MD 20814. (301) 961-6750 or (800) 843-7751. Contains citations to worldwide literature in theoretical and applied computer science and related areas, from 1981 to present.Inquire as to cost and availability.

Computer and Mathematics Search. Institute for Scientific Information, 3501 Market Street, Philadelphia, PA 19104. (215) 386-0100. FAX (215) 386-6362. Covers worldwide literature in computer science and mathematics, from 1980 to present. Available on BRS,(800) 289-4277, online service. Inquire as to cost and availability.

Computer Database. Information Access Company, 362 Lakeside Drive, Foster City, CA 94404. (415) 378-5000 or (800) 227-8431. Contains citations with abstracts to literature from trade journals covering the computer,telecommunications,and electronics industries. Available on the BRS, (800) 289-4277, Compuserve Information Service,(800) 848-8990, and DIALOG,(800) 334-2564, online services. Inquire as to cost and availability.

Dissertation Abstracts. University Microfilms International, 300 North Zeeb Road, Ann Arbor, MI 48106. (800) 521-0600 or (313) 761-4700. Scope includes virtually all doctoral dissertations accepted at accredited American institutions from 1861 to present in 252 subject areas. Available on BRS,(800) 289-4277, DIALOG,(800) 334-2564, and OCLC EPIC,(800) 848-5878, online services. Also available on CD-ROM. Inquire as to cost and availability.

INSPEC. Institution of Electrical Engineers, Michael Faraday House, Six Hills Way, Stevenage, Herts. SG1 2AY, England. Telephone: 0438 313311 or FAX 0438 742840. Contains citations to the worldwide literature of physics, electronics and electrical engineering, computer technology, and related fields. Available on BRS,(800) 289-4277, DIALOG, (800) 334-2564, ORBIT, (800) 456-7248, and STN International, FIZ Karlsruhe, P.O. Box 2465, W-7500, Karlsruhe 1, Germany, online services. Inquire as to cost and availability.

Mathsci. American Mathematical Society, P.O. Box 6248, Providence, RI 02940. (800) 321-4667 or (401) 455-4000 or FAX (401) 331-3842. Scope includes pure and applied mathematics and related areas of physics, statistics, engineering, computer science, and operations research literature since 1959. Available on DIALOG,(800) 334-2564, online service. Also available on CD-ROM. Inquire as to cost and availability.

NTIS Bibliographic Database. National Technical Information Service, 5285 Port Royal Road, Springfield, VA 22161. (703) 487-4929 or FAX (703) 321-8199. Broad coverage of government-sponsored science and technology research reports, 1964 to present. Available on BRS,(800) 289-4277, DIALOG, (800) 334-2564, ORBIT, (800) 456-7248, and STN International, FIZ Karlsruhe, P.O. Box 2465, W-7500, Karlsruhe 1, Germany, online services. Also available on CD-ROM. Inquire as to cost and availability.

SCISEARCH. Institute for Scientific Information, 3501 Market Street, Philadelphia, PA 19104. (800) 523-1850 or (215) 386-0100. Broad multidisciplinary title and author index to the international literature of science and technology, 1974 to present. Available on DIALOG,(800) 334-2564, and ORBIT,(800) 456-7248, online services. Also available on CD-ROM. Inquire as to cost and availability.

WILSONLINE. H.W. Wilson Company, 950 University Avenue, Bronx, NY 10452. (800) 367-6770 or (212) 588-8400. Makes available online versions of the printed H.W. Wilson indexes including Applied Science and Technology Index, Business Periodicals Index, General Science Index, and Readers' Guide

to Periodical Literature. Period covered is generally 1983 to present. Available on BRS,(800) 289-4277, DIALOG,(800) 334-2564, and OCLC EPIC,(800) 848-5878, online services. Also available on CD-ROM. Inquire as to cost and availability.

PERIODICALS

ACM Transactions On Programming Languages and Systems. Association for Computing Machinery, 1515 Broadway, 17th Floor, New York, NY 10036-5701. (212) 869-7440. FAX (212) 944-1318. 1979 to present. Quarterly. $105.00 for non-members.

ACM Transactions On Software Engineering and Methodology. Association for Computing Machinery, 11 West 42nd Street, New York, NY 10036. (212) 869-7440. Fax (212) 869-0481. 1991 to present. Quarterly. $85.00 per year.

Byte. McGraw-Hill Inc., Byte Publications, One Phoenix Mill Lane, Peterborough, NH 03458. (603) 924-9281. FAX (603) 924-2550. 1975 to present. Monthly. $29.95 per year.

Communications of the A C M. Association for Computing Machinery, 1515 Broadway, 17th Floor, New York, NY 10036-5701. (212) 869-7440. FAX (212) 944-1318. 1958 to present. Monthly. $114.00 for non-members.

Computer Languages. Elsevier Science [journals], 660 White Plains Rd., Tarrytown, NY 10159-5153. (919) 524-9200. FAX (919) 333-2444. 1976 to present. Four times a year. $445.00 per year.

Dr. Dobb's Journal of Software Tools. M and T Publishing, Inc. 501 Galveston Drive, Redwood City, CA 94063. (415) 366-3600. 1976 to present. Monthly. $30.00 per year.

IEEE Software. IEEE Computer Society Press, 10662 Vaqueros Circle, Box 3014, Los Alamitos, CA 90720. (714) 821-8380. 1984 to present. Bi-monthly. $175.00 per year.

IEEE Transactions On Software Engineering. IEEE, 345 E. 47th Street, New York, NY 10017. (908) 981-0060. FAX (908) 981-9667. 1975 to present. Monthly. $400.00 per year for non-members.

Journal of Computer and Software Engineering. Ablex Publishing Corporation, 355 Chestnut Street, Norwood, NJ 07648. (201) 767-8450 or FAX (201) 767-6717. 1993 to present. Quarterly. $110.00 per year.

Journal of Functional Programming. Cambridge University Press, Journals Department, 40 West 20th Street, New York, NY 10011-4211. (212) 924-3900. 1991 to present. Quarterly. Inquire for price.

Journal of Logic Programming. Elsevier Science Publishing Company, Inc., 655 Avenue of the Americas, New York, NY 10010. (212) 989-5800. 1984 to present. Twelve times a year. $545.00 per year (institutions).

Journal of Object-oriented Programming. Sigs Publications Inc., 588 Broadway, Suite 604, New York, NY 10012-5408. (212) 274-0640. FAX (212) 274-0646. 1988 to present. Nine times a year. $59.00 per year.

Journal of Programming Languages. Chapman & Hall, Journals Promotion Department, One Penn Plaza, 41st Floor, New York, NY 10019. (212) 564-1060. FAX (212) 564-1505. 1992 to present. Four times a year. $210.00 per year.

Letters On Programming Languages and Systems. Association for Computing Machinery, 1515 Broadway, 17th Floor, New York, NY 10036-5701. (212) 869-7440. FAX (212) 944-1318. Irregular. $105.00 for non-members.

Programmer's Journal. Oakley Publishing Company, 150 North 4th Street, Springfield, OR 97477. (503) 747-0800. 1983 to present. Bimonthly. $20.00 per year.

Science of Computer Programming. Elsevier Science Inc., Box 882, Madison Square Station, New York, NY 10159. (212) 989-5800. FAX (212) 633-3990. 1981 to present. Six times a year. $411.00 per year.

Software Engineering. Auerbach Publishers, 1 Penn Plaza, New York, NY 10119-0118. 1989 to present. Bimonthly. $145.00 per year.

Software: Practice and Experience. John Wiley and Sons, Inc., 605 Third Avenue, New York, NY 10158. (800) 526-5368 or (212) 850-6000. 1971 to present. Monthly. $440.00 per year.

RESEARCH CENTERS AND INSTITUTES

George Mason University Center For Software and Systems Engineering. Room 203, Science and Technology Bldg., 4400 University Drive, Fairfax, VA 22030. (703) 323-3891.

Georgia Institute of Technology, Software Research Institute. College of Computing, Room 265, Atlanta, GA 30332-0280. (404) 894-3180.

Industry/university Cooperative Research Center For Software Engineering (serc). University of Florida, Computer and Information Sciences Department, Gainesville, FL 32611. (904) 392-1211. FAX (904) 392-1220.

University of Houston—clear Lake, Software Engineering Research Center. 2700 Bay Area Blvd., Houston, TX 77058. (713) 283-3830.

SOFTWARE ENGINEERING

See also: ARTIFICIAL INTELLIGENCE, COMPUTER PROGRAMMING, COMPUTER PROGRAMMING LANGUAGES, COMPUTER OPERATING SYSTEMS, PARALLEL COMPUTING, SOFTWARE, SYSTEMS ENGINEERING

ABSTRACT SERVICES AND INDEXES

ACM Guide To Computing Literature. Association for Computing Machinery, Association for Computing Machinery, 1515 Broadway, 17th Floor, New York, NY 10036-5701. (212) 869-7440. FAX (212) 944-1318. 1964 to present. Monthly. $190.00 per year for non-members.

Applied Science and Technology Index; A Cumulative Subject Index To English Language Periodicals in the Fields of Aeronautics and Space Science, Computer Technology, Chemistry, Construction Industry, Energy and Related Areas. H.W. Wilson Co., 950 University Avenue, Bronx, NY 10452. (800) 367-

SOFTWARE ENGINEERING

Ency. of Physical Sciences and Engineering Info. Sources

6770 or (212) 588-8400. FAX (718) 590-1617. From 1958 to present. Monthly. Inquire about cost and availability. Also available on CD-ROM and online.

Computer Abstracts, MCBUniversity Press Ltd., PO Box 10812, Birmingham, AL 35201-0812. (800) 633-4931. FAX (205) 995-1588. Monthly. Covers computer theory, data, hardware, systems, networks, human-computer interaction, artificial intelligence, as well as applications of computers in aerospace, business, CAD/CAM, cartography, civil engineering, electronics and electrical engineering, industrial engineering, mechanical engineering, medicine, structural engineering, etc. Also available on CD-ROM.

Computer & Control Abstracts (inspec). INSPEC/IEEE, Box 1331, 445 Hoes Lane, Piscataway, NJ 08855-1331. (908) 562-5549. Abstracts organized by subjects of international technological information. Monthly. $1455.00 per year. Also available on CD-ROM and online via BRS Online Products, DIALOG Information Services, Orbit Search Service, and STN International.

Computer and Information Systems Abstracts Journal. Cambridge Scientific Abstracts, 7200 Wisconsin Avenue, Bethesda, MD 20814. (301) 961-6750. Fax (301) 961-6720. 1962 to present. Monthly. $1265.00 per year. Also available online via STN International.

Computer Literature Index. Applied Computer Research Inc., Box 82266, Phoenix, AZ 85071-2266. (800) 234-2227. 1971 to present. Quarterly plus annual cumulation. $198.50 per year. Bibliography of books, articles, and reports.

Computing Journal Abstracts. Techgnosis Ltd., Blade House, Battersea Road, Stockport, Cheshire 3AE, England. Telephone 061-442-2639. FAX 061-443-1162. 1969 to present. Monthly. Inquire for price.

Computing Reviews. Association for Computing Machinery, 1515 Broadway, 17th Floor, New York, NY 10036-5701. (212) 869-7440. FAX (212) 944-1318. 1960 to present. Monthly. $130.00 per year for non-members. Also available online via DIALOG Information Services.

Current Contents: Engineering, Technology, and Applied Sciences. Institute for Scientific Information, 3501 Market Street, Philadelphia, PA 19104. (215) 386-0100 or FAX (215) 386-2991. (800) 523-1850. 1970 to present. Weekly. $442.00 per year.

Engineering Index Monthly. Engineering Information, Inc., Castle Point on the Hudson, Hoboken, NJ 07030. (800) 221-1044. FAX (212) 832-1857. Monthly. $2300.00 per year. Also available online as COMPENDEX, and also on CD-ROM. Covers chemical engineering, computers, electrical engineering, civil engineering, metals and mining, industrial management, and mechanical engineering.

Index To Scientific and Technical Proceedings. Institute for Scientific Information, 3501 Market St., Philadelphia, PA 19104. (215) 386-0100 or (800) 523-1850. FAX (215) 386-6362. Monthly. $500.00 per year.

Index To Scientific Reviews. Institute for Scientific Information, 3501 Market St., Philadelphia, PA 19104. (215) 386-0100 or (800) 523-1850. FAX (215) 386-6362. Semi-annual.

Key Abstracts—Software Engineering. INSPEC/IEEE, IEEE (Institute of Electrical and Electronic Engineers), 445 Hoes Lane, Piscataway, NJ 08854. (908) 562-5545 or (800) 678-IEEE. 1987 to present. Monthly. $160.00 per year.

ASSOCIATIONS AND PROFESSIONAL SOCIETIES

American Federation of Information Processing Societies, 1899 Preston White Drive, Reston, VA 22091-5435. (703) 620-8937.

Association For Computing Machinery. 1515 Broadway, 17th Floor, New York, NY 10036-5701. (212) 869-7440. FAX (212) 944-1318.

Computer and Automation Systems Association of Sme. 1 SME Drive, PO Box 930, Dearborn, MI 48121-0930. (313) 271-1500.

Computing Research Association. 1875 Connecticut Avenue NW, Suite 718, Washington, DC 20009-5728. (202) 234-2111. FAX (202) 667-1066.

IEEE Computer Society. 1730 Massachusetts Avenue NW, Washington, DC 20036-1992. (202) 371-0101. FAX (202) 728-9614.

Society For Computer Simulation International, 4838 Ronson Court Lane, San Diego, CA 92111-1810. (619) 277-3888.

Software Management Association. PO Box 12004 #297, Vallejo, CA 94590. (707) 643-4423. FAX (707) 643-4423.

Software Publishers Association. 1730 M Street NW, Suite 700, Washington, DC 20036-4510. (202) 452-1600. FAX (202) 223-8756.

Special Interest Group On Programming Languages. c/o ACM, 1515 Broadway, New York, NY 10036. (212) 869-7440. FAX (212) 869-0481.

Special Interest Group On Software Engineering. c/o ACM, 1515 Broadway, New York, NY 10036. (212) 626-0611. FAX (212) 302-5826.

BIBLIOGRAPHIES

ACM Guide To Computing Literature. Association for Computing Machinery, 1515 Broadway, 17th Floor, New York, NY 10036-5701. (212) 869-7440. FAX (212) 944-1318. 1964 to present. Annual. $190.00 for non-members. Also available online via DIALOG.

DIRECTORIES AND BIOGRAPHICAL SOURCES

American Men & Women of Science, 1995-96. R.R. Bowker Staff, eds. 19th edition. 8 volumes. R.R. Bowker/Reed International Publishing Company, 121 Chanlon Road, New Providence, NJ 07974. (908) 464-6800 or (800) 521-8110. FAX (908) 464-3553. 1995. $850.00

ASIS Handbook and Directory, American Society For Information Science, 8720 Georgia Avenue, Suite 501, Silver Spring, MD 20910-3602. (301) 495-0900. Annual. $100.00.

Computing Information Directory. Hildebrandt Inc., Box 576, Pullman, WA 99163-0576. 1981 to present. Annual. $199.95 per year.

Directory of Engineering Societies and Related Organizations. Gordon Davis. 13th edition. American Association of Engineering Societies, 1111 19th Street NW, Suite 608, Washington, DC 20036. (202) 296-2237 or (800) 658-8897. 1989. Inquire for price.

International Research Centers Directory. Gale Research, 835 Penobscot Building, Detroit, MI 48226-4094. (313) 961-2242. (800) 877-4253. FAX (313) 961-6083. 8th edition. 1995. $410.00

Research Centers Directory. Gale Research, 835 Penobscot Building, Detroit, MI 48226-4094. (313) 961-2242. (800) 877-4253. FAX (313) 961-6083. $485.00.# #*Scientific and Technical Organizations and Agencies Directory.* Gale Research, 835 Penobscot Building, Detroit, MI 48226-4094. (313) 961-2242. (800) 877-4253. FAX (313) 961-6083. 4th edition. 1996. $195.00.

The Softpub Yellow Pages: Directory of Suppliers To the Software Publishing Industry. SoftPub Group Inc., PO Box 356, Maple Valley, WA 98038. (800) 763-8782. FAX (206) 413-1098.

Software Engineering Standards: An Annotated Index and Directory. S. Magee and L. Tripp. Global Engineering Documents, 15 Inverness Way East, Englewood, CO 80112-5704. (800) 854-7179. FAX (303) 397-7935. 1994. $79.50.

Who's Who in Computing. Canadian M I S Database Inc., 268 Lakeshore Road E., Suite 510, Mississauga, ON L5G 1H1, Canada. (905) 271-1601. FAX (905) 271-4522. 1987 to present. Annual. Inquire for price in U.S.

ENCYCLOPEDIAS AND DICTIONARIES

Computer Dictionary. Donald D. Spencer. 4th edition. Camelot Publishing Company, 709 SW 80th Blvd., Gainesville, FL 32607-1537. (904) 331-0952. 1993. $24.95.

Dictionary of Computing. S.M.H. Collin. 2nd edition. Peter Collin Publishing, 8 the Causeway, Teddington, TW11 0HE, England. FAX 0181-943-3386. 1994. Inquire for cost.

Encyclopedia of Computer Science and Engineering. Anthony Ralston & Edwin D. Reilly Jr., editors. 3rd revised edition. Van Nostrand Reinhold, 115 Fifth Avenue, New York, NY 10003. (212) 254-3232. FAX (212) 254-9499. 1993. $125.00.

Microsoft Press Computer Dictionary: The Comprehensive Standard For Business, School, Library, and Home. 2nd edition. Microsoft Press, One Microsoft Way, Redmond, WA 98052-6399. (206) 936-0055. FAX (206) 823-8101. 1994. $19.95.

GENERAL WORKS

Computers: An Introduction to Hardware and Software Design. Larry L. Wear. McGraw-Hill Publishing Company, 1221 Avenue of the Americas, New York, NY 10020. (800) 262-4729 or (212) 512-3825. 1991. Inquire for cost and availability.

Fundamentals of Software Engineering. Carlo Ghezzi, et al. Prentice Hall, 113 Sylvan Avenue, Route 9W, Englewood Cliffs, NJ 07632. (201) 592-2000 or (800) 922-0579. 1991. $67.00.

Software Architecture and Design: Principles, Models and Methods. Bernard I. Witt, et al. Van Nostrand Reinhold, 115 Fifth Avenue, New York, NY 10003. (212) 254-3232. FAX (212) 254-9499. 1994. $44.95.

Software Engineering: Principles and Practice. J.C. Van Vliet. John Wiley and Sons, Inc., 605 Third Avenue, New York, NY 10158. (800) 526-5368 or (212) 850-6000. 1993. $41.95.

Systems, Software & Quality Engineering: Theory and Fundamentals. Arthur E. Ferdinand. Van Nostrand Reinhold, 115 Fifth Avenue, New York, NY 10003. (212) 254-3232. FAX (212) 254-9499. 1993. $59.95.

ONLINE DATABASES AND CD-ROMS

Compendex Plus. Engineering Information, Inc., 345 East 47th Street, New York, NY 10017. (212) 705-7600 or (800) 221-1044. Contains citations with abstracts to worldwide literature in engineering and technology, from 1970 to present. Available on online BRS,(800) 289-4277, DIALOG, (800) 334-2564, ORBIT (800) 456-7248, and STN International, FIZ Karlsruhe, P.O. Box 2465, W-7500, Karlsruhe 1, Germany, online services. Also available on CD-ROM. Inquire as to cost and availability.

Compuscience. FIZ Karlsruhe, D-7514, Eggenstein-Leopoldshafen 2, Germany. Contains citations with abstracts to European and North American literature on computer science, 1972 to present. Available on STN International, FIZ Karlsruhe, P.O. Box 2465, W-7500, Karlsruhe 1, Germany, online service. Inquire as to cost and availability.

Computer and Information Systems Abstracts. Cambridge Scientific Abstracts, 7200 Wisconsin Avenue, Suite 601, Bethesda, MD 20814. (301) 961-6750 or (800) 843-7751. Contains citations to worldwide literature in theoretical and applied computer science and related areas, from 1981 to present.Inquire as to cost and availability.

Computer and Mathematics Search. Institute for Scientific Information, 3501 Market Street, Philadelphia, PA 19104. (215) 386-0100. FAX (215) 386-6362. Covers worldwide literature in computer science and mathematics, from 1980 to present. Available on BRS,(800) 289-4277, online service. Inquire as to cost and availability.

Computer Database. Information Access Company, 362 Lakeside Drive, Foster City, CA 94404. (415) 378-5000 or (800) 227-8431. Contains citations with abstracts to literature from trade journals covering the computer,telecommunications,and electronics industries. Available on the BRS, (800) 289-4277, Compuserve Information Service,(800) 848-8990, and DIALOG,(800) 334-2564, online services. Inquire as to cost and availability.

Dissertation Abstracts. University Microfilms International, 300 North Zeeb Road, Ann Arbor, MI 48106. (800) 521-0600 or (313) 761-4700. Scope includes virtually all doctoral dissertations accepted at accredited American institutions from 1861 to present in 252 subject areas. Available on BRS,(800) 289-4277, DIALOG,(800) 334-2564, and OCLC EPIC,(800) 848-5878, online services. Also available on CD-ROM. Inquire as to cost and availability.

SOFTWARE ENGINEERING

Ency. of Physical Sciences and Engineering Info. Sources

INSPEC. Institution of Electrical Engineers, Michael Faraday House, Six Hills Way, Stevenage, Herts. SG1 2AY, England. Telephone: 0438 313311 or FAX 0438 742840. Contains citations to the worldwide literature of physics, electronics and electrical engineering, computer technology, and related fields. Available on BRS,(800) 289-4277, DIALOG, (800) 334-2564, ORBIT, (800) 456-7248, and STN International, FIZ Karlsruhe, P.O. Box 2465, W-7500, Karlsruhe 1, Germany, online services. Inquire as to cost and availability.

Mathsci. American Mathematical Society, P.O. Box 6248, Providence, RI 02940. (800) 321-4667 or (401) 455-4000 or FAX (401) 331-3842. Scope includes pure and applied mathematics and related areas of physics, statistics, engineering, computer science, and operations research literature since 1959. Available on DIALOG,(800) 334-2564, online service. Also available on CD-ROM. Inquire as to cost and availability.

NTIS Bibliographic Database. National Technical Information Service, 5285 Port Royal Road, Springfield, VA 22161. (703) 487-4929 or FAX (703) 321-8199. Broad coverage of government-sponsored science and technology research reports, 1964 to present. Available on BRS,(800) 289-4277, DIALOG, (800) 334-2564, ORBIT, (800) 456-7248, and STN International, FIZ Karlsruhe, P.O. Box 2465, W-7500, Karlsruhe 1, Germany, online services. Also available on CD-ROM. Inquire as to cost and availability.

SCISEARCH. Institute for Scientific Information, 3501 Market Street, Philadelphia, PA 19104. (800) 523-1850 or (215) 386-0100. Broad multidisciplinary title and author index to the international literature of science and technology, 1974 to present. Available on DIALOG,(800) 334-2564, and ORBIT,(800) 456-7248, online services. Also available on CD-ROM. Inquire as to cost and availability.

WILSONLINE. H.W. Wilson Company, 950 University Avenue, Bronx, NY 10452. (800) 367-6770 or (212) 588-8400. Makes available online versions of the printed H.W. Wilson indexes including Applied Science and Technology Index, Business Periodicals Index, General Science Index, and Readers' Guide to Periodical Literature. Period covered is generally 1983 to present. Available on BRS,(800) 289-4277, DIALOG,(800) 334-2564, and OCLC EPIC,(800) 848-5878, online services. Also available on CD-ROM. Inquire as to cost and availability.

PERIODICALS

ACM Transactions On Programming Languages and Systems. Association for Computing Machinery, 1515 Broadway, 17th Floor, New York, NY 10036-5701. (212) 869-7440. FAX (212) 944-1318. 1979 to present. Quarterly. $105.00 for non-members.# #*ACM Transactions On Software Engineering and Methodology.* Association for Computing Machinery, 11 West 42nd Street, New York, NY 10036. (212) 869-7440. Fax (212) 869-0481. 1991 to present. Quarterly. $85.00 per year.

Byte. McGraw-Hill Inc., Byte Publications, One Phoenix Mill Lane, Peterborough, NH 03458. (603) 924-9281. FAX (603) 924-2550. 1975 to present. Monthly. $29.95 per year.

Communications of the A C M. Association for Computing Machinery, 1515 Broadway, 17th Floor, New York, NY 10036-5701. (212) 869-7440. FAX (212) 944-1318. 1958 to present. Monthly. $114.00 for non-members.

Computer Languages. Elsevier Science [journals], 660 White Plains Rd., Tarrytown, NY 10159-5153. (919) 524-9200. FAX

(919) 333-2444. 1976 to present. Four times a year. $445.00 per year.

Dr. Dobb's Journal of Software Tools. M and T Publishing, Inc. 501 Galveston Drive, Redwood City, CA 94063. (415) 366-3600. 1976 to present. Monthly. $30.00 per year.

IEEE Software. IEEE Computer Society Press, 10662 Vaqueros Circle, Box 3014, Los Alamitos, CA 90720. (714) 821-8380. 1984 to present. Bi-monthly. $175.00 per year.

IEEE Transactions On Software Engineering. IEEE, 345 E. 47th Street, New York, NY 10017. (908) 981-0060. FAX (908) 981-9667. 1975 to present. Monthly. $400.00 per year for non-members.

Journal of Computer and Software Engineering. Ablex Publishing Corporation, 355 Chestnut Street, Norwood, NJ 07648. (201) 767-8450 or FAX (201) 767-6717. 1993 to present. Quarterly. $110.00 per year.

Journal of Functional Programming. Cambridge University Press, Journals Department, 40 West 20th Street, New York, NY 10011-4211. (212) 924-3900. 1991 to present. Quarterly. Inquire for price.

Journal of Logic Programming. Elsevier Science Publishing Company, Inc., 655 Avenue of the Americas, New York, NY 10010. (212) 989-5800. 1984 to present. Twelve times a year. $545.00 per year (institutions).

Journal of Object-oriented Programming. Sigs Publications Inc., 588 Broadway, Suite 604, New York, NY 10012-5408. (212) 274-0640. FAX (212) 274-0646. 1988 to present. Nine times a year. $59.00 per year.

Journal of Programming Languages. Chapman & Hall, Journals Promotion Department, One Penn Plaza, 41st Floor, New York, NY 10019. (212) 564-1060. FAX (212) 564-1505. 1992 to present. Four times a year. $210.00 per year.

Letters On Programming Languages and Systems. Association for Computing Machinery, 1515 Broadway, 17th Floor, New York, NY 10036-5701. (212) 869-7440. FAX (212) 944-1318. Irregular. $105.00 for non-members.

Programmer's Journal. Oakley Publishing Company, 150 North 4th Street, Springfield, OR 97477. (503) 747-0800. 1983 to present. Bimonthly. $20.00 per year.

Science of Computer Programming. Elsevier Science Inc., Box 882, Madison Square Station, New York, NY 10159. (212) 989-5800. FAX (212) 633-3990. 1981 to present. Six times a year. $411.00 per year.

Software Engineering. Auerbach Publishers, 1 Penn Plaza, New York, NY 10119-0118. 1989 to present. Bimonthly. $145.00 per year.

Software: Practice and Experience. John Wiley and Sons, Inc., 605 Third Avenue, New York, NY 10158. (800) 526-5368 or (212) 850-6000. 1971 to present. Monthly. $440.00 per year.

RESEARCH CENTERS AND INSTITUTES

Carnegie Mellon University, Software Engineering Institute. Pittsburgh, PA 15213. (412) 268-5800. FAX (412) 268-5758.

George Mason University Center For Software and Systems Engineering. Room 203, Science and Technology Bldg., 4400 University Drive, Fairfax, VA 22030. (703) 323-3891.

Industry/university Cooperative Research Center For Software Engineering (serc). University of Florida, Computer and Information Sciences Department, Gainesville, FL 32611. (904) 392-1211. FAX (904) 392-1220.

University of Houston—clear Lake, Software Engineering Research Center. 2700 Bay Area Blvd., Houston, TX 77058. (713) 283-3830.

SOIL CHEMISTRY

See also: AGRICULTURAL ENGINEERING, CIVIL ENGINEERING, EARTHQUAKE ENGINEERING, GEOTECHNICAL ENGINEERING, ROCK MECHANICS, SOIL MECHANICS, SOIL SCIENCE

ABSTRACT SERVICES AND INDEXES

Abstract Journal in Earthquake Engineering. University of California, Berkeley, Earthquake Engineering Research Center, 1301 South 46th Street, Richmond, CA 94804-4698. (510) 231-9413 or FAX (510) 231-9471. 1972 to present. Annual. $80.00 per year.

Applied Science and Technology Index; A Cumulative Subject Index To English Language Periodicals in the Fields of Aeronautics and Space Science, Computer Technology, Chemistry, Construction Industry, Energy and Related Areas. H.W. Wilson Co., 950 University Avenue, Bronx, NY 10452. (800) 367-6770 or (212) 588-8400. FAX (718) 590-1617. From 1958 to present. Monthly. Inquire about cost and availability. Also available on CD-ROM and online.

ASCE Combined Annual Combined Index. American Society of Civil Engineers, 345 East 47th Street, New York, NY 10017-2398. (212) 705-7520 or (800) 548-2723. Annual. $48.00.

ASCE Publications Information. 1966-present. American Society of Civil Engineers, 345 East 47th Street, New York, NY 10017-2398. (212) 705-7520 or (800) 548-2723. Bi-monthly. $160.00 per year for non-members.

Bibliography and Index of Geology. American Geological Institute, 4220 King Street, Alexandria, VA 22302. (703) 379-2480. Fax (703) 379-7563. Monthly. $1295.00 per year. Also available online as GEOREF.

Civil and Structural Engineering Abstracts, Cambridge Scientific Abstracts, 7200 Wisconsin Avenue, Bethesda, MD 20814-4823. (301) 961-6750. FAX (301) 961-6720. 1993 to present. Monthly. $385.00 per year. Topics covered include structural design, construction equipment and methods, civil defense and military engineering, surveying, highway engineering, maritime and port structures, materials, land reclamation, and soil mechanics.

Current Contents: Engineering, Technology, and Applied Sciences. Institute for Scientific Information, 3501 Market Street, Philadelphia, PA 19104. (215) 386-0100 or (800) 523-1850. FAX (215) 386-2991. 1970 to present. Weekly. $442.00 per year.

Engineering Index Monthly. Engineering Information, Inc., Castle Point on the Hudson, Hoboken, NJ 07030. (800) 221-1044. FAX (212) 832-1857. Monthly. $2300.00 per year. Also available online as COMPENDEX, and also on CD-ROM. Covers chemical engineering, computers, electrical engineering, civil engineering, metals and mining, industrial management, and mechanical engineering.

Geodex Retrieval System For Geotechnical Abstracts. Geodex Retrieval Systems, 669 Broadway, Box 279, Sonoma, CA 95476. (707) 939-8476. FAX (707) 996-8734. 1970 to present. Three times a year. $645.00 per year includes Geotechnical Abstracts, as Geodex Systems-GRS.

Geotechnical Abstracts. Geodex Retrieval Systems, 669 Broadway, Box 279, Sonoma, CA 95476. (707) 939-8476. FAX (707) 996-8734. 1970 to present. Three times a year. $280.00 per year. Covers soil mechanics, foundation engineering, rock mechanics, and engineering geology.# #*International Journal of Rock Mechanics and Mining Sciences and Geomechanics Abstracts.* Elsevier Science, 660 White Plains Rd., Tarrytown, NY 10591-5153. (914) 524-9200. FAX (914) 333-2444. Six times a year. $845 per year.

Science Citation Index. Institute for Scientific Information, 3501 Market Street, Philadelphia, PA 19104. (215) 386-0100 or (800) 523-1850. FAX (215) 386-6362. Inquire about availability and cost. Also available on CD-ROM.

ASSOCIATIONS AND PROFESSIONAL SOCIETIES

American Geological Institute. 4220 King Street, Alexandria, VA 22302. (703) 379-2480. Fax (703) 379-7563.

American Society of Agricultural Engineers. 2950 Niles Road, St. Joseph, MI 49085-9659. (616) 429-0300. FAX (616) 429-3852.

American Society of Agronomy. 677 S. Segoe Road, Madison, WI 53711. (608) 273-8080. FAX (608) 273-2021.

American Society of Civil Engineers. 345 East 47th Street, New York, NY 10017-2398. (212) 705-7520 or (800) 548-2723.

Association of Engineering Geologists. 323 Boston Post Road, Suite 2D, Sudbury, MA 01776. (508) 443-4639.

Geological Society of America. 3300 Penrose Place, PO Box 9140, Denver, CO 80301-9140. (303) 447-2020.

International Society For Soil Mechanics and Foundation Engineers. University Engineering Department, Trumpington Street, Cambridge CB2 1PZ, England. Telephone 1223-355020. FAX 1223-359675.

Soil Conservation Society of America. 7515 Northeast Ankeny Road, Ankeny, IA 50021. (515) 289-2331. FAX (515) 289-1227.

Soil Science Society of America. 677 S. Segoe Road, Madison, WI 53711. (608) 273-8080. FAX (608) 273-2021.

SOIL CHEMISTRY

Ency. of Physical Sciences and Engineering Info. Sources

DIRECTORIES AND BIOGRAPHICAL SOURCES

American Men & Women of Science, 1995-96. R.R. Bowker Staff, eds. 19th edition. 8 volumes. R.R. Bowker/Reed International Publishing Company, 121 Chanlon Road, New Providence, NJ 07974. (908) 464-6800 or (800) 521-8110. FAX (908) 464-3553. 1995. $850.00

Directory of Engineering Societies and Related Organizations. Gordon Davis. 13th edition. American Association of Engineering Societies, 1111 19th Street NW, Suite 608, Washington, DC 20036. (202) 296-2237 or (800) 658-8897. 1989. Inquire for price.

International Research Centers Directory. Gale Research, 835 Penobscot Building, Detroit, MI 48226-4094. (313) 961-2242. (800) 877-4253. FAX (313) 961-6083. 8th edition. 1995. $410.00

Research Centers Directory. Gale Research, 835 Penobscot Building, Detroit, MI 48226-4094. (313) 961-2242. (800) 877-4253. FAX (313) 961-6083. $485.00.

Scientific and Technical Organizations and Agencies Directory. Gale Research, 835 Penobscot Building, Detroit, MI 48226-4094. (313) 961-2242. (800) 877-4253. FAX (313) 961-6083. 4th edition. 1996. $195.00.

GENERAL WORKS

Environmental Soil Chemistry. Donald L. Sparks. Academic Press, Inc., 6277 Sea Harbor Drive, Orlando, FL 32887. 1995. $39.95.

Principles of Soil Chemistry. Kim Tan. Marcel Dekker, Inc., 270 Madison Avenue, New York, NY 10016. (212) 696-9000 or (800) 228-1160. FAX (212) 685-4540. 1992. $69.75.

Soil Chemistry and Its Applications. M.S. Cresser, et al. Cambridge University Press, 40 West 20th Street, New York, NY 10011-4211. (212) 924-3900 or (800) 872-7423. FAX (914) 937-4712. 1993. $49.95.

ONLINE DATABASES AND CD-ROMS

Compendex Plus. Engineering Information, Inc., 345 East 47th Street, New York, NY 10017. (212) 705-7600 or (800) 221-1044. Contains citations with abstracts to worldwide literature in engineering and technology, from 1970 to present. Available on online BRS,(800) 289-4277, DIALOG, (800) 334-2564, ORBIT (800) 456-7248, and STN International, FIZ Karlsruhe, P.O. Box 2465, W-7500, Karlsruhe 1, Germany, online services. Also available on CD-ROM. Inquire as to cost and availability.

Current Contents Search. Institute for Scientific Information, 3501 Market Street, Philadelphia, PA 19104. (215) 386-0100. FAX (215) 386-6362. Contains citations to articles listed in the table of contents of science and technology journals. Also articles in social sciences and life sciences journals. Available on BRS,(800) 289-4277, DIALOG,(800) 334-2564, online services. Inquire as to cost and availability.

Dissertation Abstracts. University Microfilms International, 300 North Zeeb Road, Ann Arbor, MI 48106. (800) 521-0600 or (313) 761-4700. Scope includes virtually all doctoral dissertations accepted at accredited American institutions from 1861 to

present in 252 subject areas. Available on BRS,(800) 289-4277, DIALOG,(800) 334-2564, and OCLC EPIC,(800) 848-5878, online services. Also available on CD-ROM. Inquire as to cost and availability.# #*GEOREF: Bibliography and Index of Geology.* American Geological Institute, 4220 King Street, Alexandria, VA 22302. (703) 379-2480. Fax (703) 379-7563. Monthly. Inquire for price and availability.

NTIS Bibliographic Database. National Technical Information Service, 5285 Port Royal Road, Springfield, VA 22161. (703) 487-4929 or FAX (703) 321-8199. Broad coverage of government-sponsored science and technology research reports, 1964 to present. Available on BRS,(800) 289-4277, DIALOG, (800) 334-2564, ORBIT, (800) 456-7248, and STN International, FIZ Karlsruhe, P.O. Box 2465, W-7500, Karlsruhe 1, Germany, online services. Also available on CD-ROM. Inquire as to cost and availability.

SCISEARCH. Institute for Scientific Information, 3501 Market Street, Philadelphia, PA 19104. (800) 523-1850 or (215) 386-0100. Broad multidisciplinary title and author index to the international literature of science and technology, 1974 to present. Available on DIALOG,(800) 334-2564, and ORBIT,(800) 456-7248, online services. Also available on CD-ROM. Inquire as to cost and availability.

WILSONLINE. H.W. Wilson Company, 950 University Avenue, Bronx, NY 10452. (800) 367-6770 or (212) 588-8400. Makes available online versions of the printed H.W. Wilson indexes including Applied Science and Technology Index, Business Periodicals Index, General Science Index, and Readers' Guide to Periodical Literature. Period covered is generally 1983 to present. Available on BRS,(800) 289-4277, DIALOG,(800) 334-2564, and OCLC EPIC,(800) 848-5878, online services. Also available on CD-ROM. Inquire as to cost and availability.

PERIODICALS

Agricultural Engineering. American Society of Agricultural Engineers, 2950 Niles Road, St. Joseph, MI 49085-9659. (616) 429-0300. FAX (616) 429-3852. 1920 to present. Bi-monthly. $42.50 for non-members.

Canadian Journal of Soil Science. Agricultural Institute of Canada, 151 Slater Street, Suit 907, Ottawa, Ontario K1P 5H4, Canada. Telephone (613) 232-9459. 1921 to present. Quarterly. Inquire for price in U.S.

Communications in Soil Science and Plant Analysis. Marcel Dekker, Inc., Journals, 270 Madison Avenue, New York, NY 10016. (212) 696-9000. FAX (212) 685-4540. 1970 to present. 20 times a year. $995.00.

Geoderma. Elsevier Science Inc., Box 882, Madison Square Station, New York, NY 10159. (212) 989-5800. FAX (212) 633-3990. 1967 to present. 20 times a year. $1003.00.

Journal of Irrigation and Drainage. American Society of Civil Engineers, Irrigation and Drainage Division, 345 East 47th Street, New York, NY 10017-2398. (212) 705-7520 or (800) 548-2723. 1956 to present. Bi-monthly. $136.00 per year.

Journal of Soil and Water Conservation. Soil and Water Conservation Society of America, 7515 Northeast Ankeny Road, Ankeny, IA 50021. (515) 289-2331. FAX (515) 289-1227. 1946 to present. Bi-monthly. $39.00.

Soil Science. Williams & Wilkins, 428 E Preston Street, Baltimore, MD 21202. (800) 527-5597. FAX (410) 528-4422. 1916 to present. Monthly. $128.00.

Soil Science Society of America Journal. Soil Science Society of America, 677 S. Segoe Road, Madison, WI 53711. (608) 273-8080. FAX (608) 273-2021. 1936 to present. Bi-monthly. $108.00.

RESEARCH CENTERS AND INSTITUTES

Soil and Environmental Chemistry Lab. Pennsylvania State University, 104 Research Unit A, University Park, PA 16802. (814) 865-1221. FAX (814) 865-3378.

Soil and Water Research. Louisiana State University, PO Box 25071, University Station, Baton Rouge, LA 70894-5071. (504) 387-2783.

University of California, Kearney Foundation of Soil Science. LAWR-127 Veihmeyer Hall, Davis, CA 95616. (916) 752-0683. FAX (916) 752-1552.

SOIL MECHANICS

See also: AGRICULTURAL ENGINEERING, CIVIL ENGINEERING, EARTHQUAKE ENGINEERING, GEOTECHNICAL ENGINEERING, MINING ENGINEERING, QUARRIES and QUARRYING, ROCK MECHANICS, SOIL CHEMISTRY, SOIL SCIENCE

ABSTRACT SERVICES AND INDEXES

Abstract Journal in Earthquake Engineering. University of California, Berkeley, Earthquake Engineering Research Center, 1301 South 46th Street, Richmond, CA 94804-4698. (510) 231-9413 or FAX (510) 231-9471. 1972 to present. Annual. $80.00 per year.

Applied Science and Technology Index; A Cumulative Subject Index To English Language Periodicals in the Fields of Aeronautics and Space Science, Computer Technology, Chemistry, Construction Industry, Energy and Related Areas. H.W. Wilson Co., 950 University Avenue, Bronx, NY 10452. (800) 367-6770 or (212) 588-8400. FAX (718) 590-1617. From 1958 to present. Monthly. Inquire about cost and availability. Also available on CD-ROM and online.

ASCE Combined Annual Combined Index. American Society of Civil Engineers, 345 East 47th Street, New York, NY 10017-2398. (212) 705-7520 or (800) 548-2723. Annual. $48.00.

ASCE Publications Information. 1966-present. American Society of Civil Engineers, 345 East 47th Street, New York, NY 10017-2398. (212) 705-7520 or (800) 548-2723. Bi-monthly. $160.00 per year for non-members.

Bibliography and Index of Geology. American Geological Institute, 4220 King Street, Alexandria, VA 22302. (703) 379-2480. Fax (703) 379-7563. Monthly. $1295.00 per year. Also available online as GEOREF.

Civil and Structural Engineering Abstracts, Cambridge Scientific Abstracts, 7200 Wisconsin Avenue, Bethesda, MD 20814-4823. (301) 961-6750. FAX (301) 961-6720. 1993 to present.

Monthly. $385.00 per year. Topics covered include structural design, construction equipment and methods, civil defense and military engineering, surveying, highway engineering, maritime and port structures, materials, land reclamation, and soil mechanics.

Current Contents: Engineering, Technology, and Applied Sciences. Institute for Scientific Information, 3501 Market Street, Philadelphia, PA 19104. (215) 386-0100 or (800) 523-1850. FAX (215) 386-2991. 1970 to present. Weekly. $442.00 per year.

Engineering Index Monthly. Engineering Information, Inc., Castle Point on the Hudson, Hoboken, NJ 07030. (800) 221-1044. FAX (212) 832-1857. Monthly. $2300.00 per year. Also available online as COMPENDEX, and also on CD-ROM. Covers chemical engineering, computers, electrical engineering, civil engineering, metals and mining, industrial management, and mechanical engineering.

Geodex Retrieval System For Geotechnical Abstracts. Geodex Retrieval Systems, 669 Broadway, Box 279, Sonoma, CA 95476. (707) 939-8476. FAX (707) 996-8734. 1970 to present. Three times a year. $645.00 per year includes Geotechnical Abstracts, as Geodex Systems-GRS.

Geotechnical Abstracts. Geodex Retrieval Systems, 669 Broadway, Box 279, Sonoma, CA 95476. (707) 939-8476. FAX (707) 996-8734. 1970 to present. Three times a year. $280.00 per year. Covers soil mechanics, foundation engineering, rock mechanics, and engineering geology.

International Journal of Rock Mechanics and Mining Sciences and Geomechanics Abstracts. Elsevier Science, 660 White Plains Rd., Tarrytown, NY 10591-5153. (914) 524-9200. FAX (914) 333-2444. Six times a year. $845 per year.

Science Citation Index. Institute for Scientific Information, 3501 Market Street, Philadelphia, PA 19104. (215) 386-0100 or (800) 523-1850. FAX (215) 386-6362. Inquire about availability and cost. Also available on CD-ROM.

ASSOCIATIONS AND PROFESSIONAL SOCIETIES

American Geological Institute. 4220 King Street, Alexandria, VA 22302. (703) 379-2480. Fax (703) 379-7563.

American Society of Agricultural Engineers. 2950 Niles Road, St. Joseph, MI 49085-9659. (616) 429-0300. FAX (616) 429-3852.

American Society of Civil Engineers. 345 East 47th Street, New York, NY 10017-2398. (212) 705-7520 or (800) 548-2723.

Association of Engineering Geologists. 323 Boston Post Road, Suite 2D, Sudbury, MA 01776. (508) 443-4639.

International Society For Soil Mechanics and Foundation Engineers. University Engineering Department, Trumpington Street, Cambridge CB2 1PZ, England. Telephone 1223-355020. FAX 1223-359675.

Geological Society of America. 3300 Penrose Place, PO Box 9140, Denver, CO 80301-9140. (303) 447-2020.

Soil Science Society of America. 677 S. Segoe Road, Madison, WI 53711. (608) 273-8080. FAX (608) 273-2021.

SOIL MECHANICS

Ency. of Physical Sciences and Engineering Info. Sources

DIRECTORIES AND BIOGRAPHICAL SOURCES

American Men & Women of Science, 1995-96. R.R. Bowker Staff, eds. 19th edition. 8 volumes. R.R. Bowker/Reed International Publishing Company, 121 Chanlon Road, New Providence, NJ 07974. (908) 464-6800 or (800) 521-8110. FAX (908) 464-3553. 1995. $850.00

Directory of Engineering Societies and Related Organizations. Gordon Davis. 13th edition. American Association of Engineering Societies, 1111 19th Street NW, Suite 608, Washington, DC 20036. (202) 296-2237 or (800) 658-8897. 1989. Inquire for price.

International Research Centers Directory. Gale Research, 835 Penobscot Building, Detroit, MI 48226-4094. (313) 961-2242. (800) 877-4253. FAX (313) 961-6083. 8th edition. 1995. $410.00

List of Members: International Society For Soil Mechanics and Foundation Engineering. Ashgate Publishing Co., Old Post Rd., Brookfield, VT 05036. (802) 276-3162. FAX (802) 276-3837. 1993. $70.00.

Research Centers Directory. Gale Research, 835 Penobscot Building, Detroit, MI 48226-4094. (313) 961-2242. (800) 877-4253. FAX (313) 961-6083. $485.00.

Scientific and Technical Organizations and Agencies Directory. Gale Research, 835 Penobscot Building, Detroit, MI 48226-4094. (313) 961-2242. (800) 877-4253. FAX (313) 961-6083. 4th edition. 1996. $195.00.

ENCYCLOPEDIAS AND DICTIONARIES

The Encyclopedia of the Solid Earth Sciences. Philip Kearey, editor-in-chief. Blackwell Scientific Publications, 3 Cambridge Court, Suite 208, Cambridge, MA 02142. (617) 225-0401. 1993. $60.00.

GENERAL WORKS

Basic Soil Mechanics. R. Whitlow. 3rd edition. Longman Publishing Group, the Longman Building, 10 Bank Street, White Plains, NY 10606-1951. (914) 993-5000 or (800) 266-8855. FAX (914) 997-8115. 1995. $49.95.

Essentials of Soil Mechanics and Foundations: Basic Geotechnics. David F. McCarthy. Fourth edition. Prentice Hall, 113 Sylvan Avenue, Route 9W, Englewood Cliffs, NJ 07632. (201) 592-2000 or (800) 922-0579. 1992. $69.00.

Fundamentals of Soil Behavior. James K. Mitchell. 2nd edition. John Wiley and Sons, Inc., 605 Third Avenue, New York, NY 10158. (800) 526-5368 or (212) 850-6000. 1993. $74.95.

Geotechnical Engineering: Soil Mechanics. John N. Cernica. John Wiley and Sons, Inc., 605 Third Avenue, New York, NY 10158. (800) 526-5368 or (212) 850-6000. 1994. $75.95.

An Introduction To the Mechanics of Soils and Foundations. John Atkinson. McGraw-Hill Publishing Company, 1221 Avenue of the Americas, New York, NY 10020. (800) 262-4729 or (212) 512-3825. 1995. Inquire for price and availability.

Soil Mechanics. William Powrie. Chapman & Hall, 115 Fifth Avenue, New York, NY 10211-0906. 1996. $34.00.

HANDBOOKS AND MANUALS

Foundation Engineering Handbook. H.Y. Fang. Chapman and Hall, 29 West 35th Street, New York, NY 10001. (800) 842-3636. FAX (212) 563-2269. 1990. $129.95.

Handbook of Soil Mechanics Volume 3: Soil Mechanics of Earthworks, Foundations, and Highway Engineering. Elsevier Science Inc., Box 945, Madison Square Station, New York, NY 10160-0757. (212) 633-3650. FAX (212) 633-3680. 1990. $123.00.

Handbook of Soil Mechanics Volume 4: Applications of Soil Mechanics in Practice. A. Kezdi and L. Rethati. Elsevier Science Inc., Box 945, Madison Square Station, New York, NY 10160-0757. (212) 633-3650. FAX (212) 633-3680. 1990. $118.00.

ONLINE DATABASES AND CD-ROMS

Compendex Plus. Engineering Information, Inc., 345 East 47th Street, New York, NY 10017. (212) 705-7600 or (800) 221-1044. Contains citations with abstracts to worldwide literature in engineering and technology, from 1970 to present. Available on online BRS,(800) 289-4277, DIALOG, (800) 334-2564, ORBIT (800) 456-7248, and STN International, FIZ Karlsruhe, P.O. Box 2465, W-7500, Karlsruhe 1, Germany, online services. Also available on CD-ROM. Inquire as to cost and availability.

Current Contents Search. Institute for Scientific Information, 3501 Market Street, Philadelphia, PA 19104. (215) 386-0100. FAX (215) 386-6362. Contains citations to articles listed in the table of contents of science and technology journals. Also articles in social sciences and life sciences journals. Available on BRS,(800) 289-4277, DIALOG,(800) 334-2564, online services. Inquire as to cost and availability.

Dissertation Abstracts. University Microfilms International, 300 North Zeeb Road, Ann Arbor, MI 48106. (800) 521-0600 or (313) 761-4700. Scope includes virtually all doctoral dissertations accepted at accredited American institutions from 1861 to present in 252 subject areas. Available on BRS,(800) 289-4277, DIALOG,(800) 334-2564, and OCLC EPIC,(800) 848-5878, online services. Also available on CD-ROM. Inquire as to cost and availability.

GEOREF: Bibliography and Index of Geology. American Geological Institute, 4220 King Street, Alexandria, VA 22302. (703) 379-2480. Fax (703) 379-7563. Monthly. Inquire for price and availability.

NTIS Bibliographic Database. National Technical Information Service, 5285 Port Royal Road, Springfield, VA 22161. (703) 487-4929 or FAX (703) 321-8199. Broad coverage of government-sponsored science and technology research reports, 1964 to present. Available on BRS,(800) 289-4277, DIALOG, (800) 334-2564, ORBIT, (800) 456-7248, and STN International, FIZ Karlsruhe, P.O. Box 2465, W-7500, Karlsruhe 1, Germany, online services. Also available on CD-ROM. Inquire as to cost and availability.

SCISEARCH. Institute for Scientific Information, 3501 Market Street, Philadelphia, PA 19104. (800) 523-1850 or (215) 386-0100. Broad multidisciplinary title and author index to the in-

Ency. of Physical Sciences and Engineering Info. Sources

SOIL SCIENCE

ternational literature of science and technology, 1974 to present. Available on DIALOG,(800) 334-2564, and ORBIT,(800) 456-7248, online services. Also available on CD-ROM. Inquire as to cost and availability.

WILSONLINE. H.W. Wilson Company, 950 University Avenue, Bronx, NY 10452. (800) 367-6770 or (212) 588-8400. Makes available online versions of the printed H.W. Wilson indexes including Applied Science and Technology Index, Business Periodicals Index, General Science Index, and Readers' Guide to Periodical Literature. Period covered is generally 1983 to present. Available on BRS,(800) 289-4277, DIALOG,(800) 334-2564, and OCLC EPIC,(800) 848-5878, online services. Also available on CD-ROM. Inquire as to cost and availability.@ @Periodicals

Engineering Geology. Elsevier Science Inc., Box 882, Madison Square Station, New York, NY 10159. (212) 989-5800. FAX (212) 633-3990. 1965 to present. Twelve times a year. $650.00.

Geotechnique International. Thomas Telford Services Ltd., Thomas Telford House, 1 Heron Quay, London E14 4JD, England. Telephone 071-987-6999. FAX 071-538-4101. 1948 to present. Quarterly. Inquire for price in U.S.A.

Journal of Geotechnical Engineering. American Society of Civil Engineers, Geotechnical Engineering Division, 345 East 47th Street, New York, NY 10017-2398. (212) 705-7520 or (800) 548-2723. 1956 to present. Monthly. $212.00 per year.

Journal of Irrigation and Drainage. American Society of Civil Engineers, Irrigation and Drainage Division, 345 East 47th Street, New York, NY 10017-2398. (212) 705-7520 or (800) 548-2723. 1956 to present. Bi-monthly. $136.00 per year.

Rock Mechanics and Rock Engineering. Springer Verlag/Journal Fulfillment Services, Box 2485, Secaucus, NJ 07096-2491. (800) 777-4643. FAX (201) 348-4505. 1929 to present. Four times a year. $187.00 per year.

Soil Dynamics and Earthquake Engineering. Elsevier Science [journals], 660 White Plains Rd., Tarrytown, NY 10159-5153. (919) 524-9200. FAX (919) 333-2444. 1981 to present. Six times a year. $515.00.

Soil Mechanics and Foundation Engineering. Plenum Publishing Corporation, Consultants Bureau, 233 Spring Street, New York, NY 10013-1578. (212) 620-8000. FAX (212) 463-0742. 1964 to present. Bi-monthly. $1075.00.

Soils and Foundations. Japanese Society of Soil Mechanics and Foundation Engineering, Doshitsu Kogakkai, 4F, Sugayama Bldg., Kanda Awaji-cho, 2-23, Chiyoda-ku, Tokyo 101, Japan. 1960 to present. Quarterly. $148.50.

Soil Technology. Elsevier Science Inc., Box 882, Madison Square Station, New York, NY 10159. (212) 989-5800. FAX (212) 633-3990. 1988 to present. Quarterly. $176.00.

RESEARCH CENTERS AND INSTITUTES

Mcgill University Geotechnical Research Centre. 817 Sherbrooke Street W., Montreal PQ, Canada H3A 2K6. (514) 398-6672. (514) 398-7361.

Rensselaer Polytechnic Institute, Soil Mechanics Laboratory. Civil Engineering Department, Troy, NY 12180-3590. (518)

276-6213. FAX (518) 276-4833.# #University of Michigan Soil Mechanics Laboratory. Department of Mechanical Engineering and Applied Mechanics, 2246A G.G. Brown Bldg., Ann Arbor, MI 48109-2125. (313) 763-0684.

SOIL SCIENCE

See also: AGRICULTURAL ENGINEERING, CIVIL ENGINEERING, EARTHQUAKE ENGINEERING, GEOTECHNICAL ENGINEERING, ROCK MECHANICS, SOIL CHEMISTRY, SOIL MECHANICS

ABSTRACT SERVICES AND INDEXES

Abstract Journal in Earthquake Engineering. University of California, Berkeley, Earthquake Engineering Research Center, 1301 South 46th Street, Richmond, CA 94804-4698. (510) 231-9413 or FAX (510) 231-9471. 1972 to present. Annual. $80.00 per year.

Applied Science and Technology Index; A Cumulative Subject Index To English Language Periodicals in the Fields of Aeronautics and Space Science, Computer Technology, Chemistry, Construction Industry, Energy and Related Areas. H.W. Wilson Co., 950 University Avenue, Bronx, NY 10452. (800) 367-6770 or (212) 588-8400. FAX (718) 590-1617. From 1958 to present. Monthly. Inquire about cost and availability. Also available on CD-ROM and online.

ASCE Combined Annual Combined Index. American Society of Civil Engineers, 345 East 47th Street, New York, NY 10017-2398. (212) 705-7520 or (800) 548-2723. Annual. $48.00.

ASCE Publications Information. 1966-present. American Society of Civil Engineers, 345 East 47th Street, New York, NY 10017-2398. (212) 705-7520 or (800) 548-2723. Bi-monthly. $160.00 per year for non-members.

Bibliography and Index of Geology. American Geological Institute, 4220 King Street, Alexandria, VA 22302. (703) 379-2480. Fax (703) 379-7563. Monthly. $1295.00 per year. Also available online as GEOREF.

Civil and Structural Engineering Abstracts, Cambridge Scientific Abstracts, 7200 Wisconsin Avenue, Bethesda, MD 20814-4823. (301) 961-6750. FAX (301) 961-6720. 1993 to present. Monthly. $385.00 per year. Topics covered include structural design, construction equipment and methods, civil defense and military engineering, surveying, highway engineering, maritime and port structures, materials, land reclamation, and soil mechanics.

Current Contents: Engineering, Technology, and Applied Sciences. Institute for Scientific Information, 3501 Market Street, Philadelphia, PA 19104. (215) 386-0100. FAX (215) 386-6362. 1970 to present. Weekly. $442.00 per year.

Engineering Index Monthly. Engineering Information, Inc., Castle Point on the Hudson, Hoboken, NJ 07030. (800) 221-1044. FAX (212) 832-1857. Monthly. $2300.00 per year. Also available online as COMPENDEX, and also on CD-ROM. Covers chemical engineering, computers, electrical engineering, civil engineering, metals and mining, industrial management, and mechanical engineering.

SOIL SCIENCE

Ency. of Physical Sciences and Engineering Info. Sources

Geodex Retrieval System For Geotechnical Abstracts. Geodex Retrieval Systems, 669 Broadway, Box 279, Sonoma, CA 95476. (707) 939-8476. FAX (707) 996-8734. 1970 to present. Three times a year. $645.00 per year includes Geotechnical Abstracts, as Geodex Systems-GRS.

Geotechnical Abstracts. Geodex Retrieval Systems, 669 Broadway, Box 279, Sonoma, CA 95476. (707) 939-8476. FAX (707) 996-8734. 1970 to present. Three times a year. $280.00 per year. Covers soil mechanics, foundation engineering, rock mechanics, and engineering geology.

International Journal of Rock Mechanics and Mining Sciences and Geomechanics Abstracts. Elsevier Science, 660 White Plains Rd., Tarrytown, NY 10591-5153. (914) 524-9200. FAX (914) 333-2444. Six times a year. $845 per year.

Science Citation Index. Institute for Scientific Information, 3501 Market Street, Philadelphia, PA 19104. (215) 386-0100. FAX (215) 386-6362. Inquire about availability and cost. Also available on CD-ROM.

ASSOCIATIONS AND PROFESSIONAL SOCIETIES

American Geological Institute. 4220 King Street, Alexandria, VA 22302. (703) 379-2480. Fax (703) 379-7563.

American Society of Agricultural Engineers. 2950 Niles Road, St. Joseph, MI 49085-9659. (616) 429-0300. FAX (616) 429-3852.

American Society of Agronomy. 677 S. Segoe Road, Madison, WI 53711. (608) 273-8080. FAX (608) 273-2021.

American Society of Civil Engineers. 345 East 47th Street, New York, NY 10017-2398. (212) 705-7520 or (800) 548-2723.

Association of Engineering Geologists. 323 Boston Post Road, Suite 2D, Sudbury, MA 01776. (508) 443-4639.

Geological Society of America. 3300 Penrose Place, PO Box 9140, Denver, CO 80301-9140. (303) 447-2020.

International Society For Soil Mechanics and Foundation Engineering. University Engineering Department, Trumpington Street, Cambridge CB2 1PZ, England. Telephone 1223-355020. FAX 1223-359675.

International Society of Soil Sciences. Universitat fur Bodenkultur, Institut fur Bodenforschung und Baugeologie, Gregor-Mendel-Str 33, A-1180 Vienna, Austria. Telephone 1-3106026. FAX 1-3106027.

Soil and Water Conservation Society of America. 7515 Northeast Ankeny Road, Ankeny, IA 50021. (515) 289-2331. FAX (515) 289-1227.

Soil Science Society of America. 677 S. Segoe Road, Madison, WI 53711. (608) 273-8080. FAX (608) 273-2021.

BIBLIOGRAPHIES

The Literature of Soil Science. Peter McDonald, editor. Cornell University Press, 512 E. State St., PO Box 250, Ithaca, NY 14851. (800) 666-2211. FAX (607) 277-2374. 1994. $67.50.

World Databases in Agriculture. C. J. Armstrong, editor. Bowker-Saur, 121 Chanlon Road, New Providence, NJ 07974. (800) 521-8110. FAX (908) 665-6707. 1995. $165.00.

DIRECTORIES AND BIOGRAPHICAL SOURCES

American Men & Women of Science, 1995-96. R.R. Bowker Staff, eds. 19th edition. 8 volumes. R.R. Bowker/Reed International Publishing Company, 121 Chanlon Road, New Providence, NJ 07974. (908) 464-6800 or (800) 521-8110. 1995. $850.00

Directory of Engineering Societies and Related Organizations. Gordon Davis. 13th edition. American Association of Engineering Societies, 1111 19th Street NW, Suite 608, Washington, DC 20036. (202) 296-2237 or (800) 658-8897. 1989. Inquire for price.

International Research Centers Directory. Gale Research, 835 Penobscot Building, Detroit, MI 48226-4094. (313) 961-2242. (800) 877-4253. FAX (313) 961-6083. 8th edition. 1995. $410.00

List of Members: International Society For Soil Mechanics and Foundation Engineering. Ashgate Publishing Co., Old Post Rd., Brookfield, VT 05036. (802) 276-3162. FAX (802) 276-3837. 1993. $70.00.

Research Centers Directory. Gale Research, 835 Penobscot Building, Detroit, MI 48226-4094. (313) 961-2242. (800) 877-4253. FAX (313) 961-6083. $485.00.

Scientific and Technical Organizations and Agencies Directory. Gale Research, 835 Penobscot Building, Detroit, MI 48226-4094. (313) 961-2242. (800) 877-4253. FAX (313) 961-6083. 4th edition. 1996. $195.00.

ENCYCLOPEDIAS AND DICTIONARIES

The Encyclopedia of the Solid Earth Sciences. Philip Kearey, editor-in-chief. Blackwell Scientific Publications, 238 Main Street, Cambridge, MA 02142. (617) 876-7000 or (800) 759-6102. FAX (617) 876-7022. 1993. $60.00.

GENERAL WORKS

Environmental Soil Science. Kim H. Tan. Marcel Dekker, Inc., 270 Madison Avenue, New York, NY 10016. (212) 696-9000. FAX (212) 685-4540. 1994. $65.00.

History of Soil Science: From Its Inception To the Present. I.A. Krupenikov. Ashgate Publishing Co., Old Post Rd., Brookfield, VT 05036. (802) 276-3162. FAX (802) 276-3837. 1993. $95.00.

An Introduction To the Mechanics of Soils and Foundations. McGraw-Hill Publishing Company, 1221 Avenue of the Americas, New York, NY 10020. (800) 262-4729 or (212) 512-3825. 1995. Inquire for price and availability.

Principles of Soil Dynamics. Braja M. Das. PWS Publishers, 20 Park Plaza, Boston, MA 02116. (617) 542-3377. FAX (617) 338-6134. 1993. $74.95.

Soil Science: Methods and Applications. D.L. Rowell. Halsted Press (Division of John Wiley and Sons, Inc.), 605 Third Avenue, New York, NY 10158. (800) 526-5368 or (212) 850-6418. 1994. $49.95.

Soil Science Simplified. Helmut Kohnke and D.P. Franzmeier. Fourth revised edition. Waveland Press Inc., PO Box 400, Prospect Heights, IL 60070. (708) 634-0081. FAX (708) 634-9501. 1995. $6.95.

Soils in Our Environment. Raymond W. Miller, et al. 7th revised edition. Prentice Hall, 113 Sylvan Avenue, Route 9W, Englewood Cliffs, NJ 07632. (201) 592-2000 or (800) 922-0579. 1995. $78.00.

HANDBOOKS AND MANUALS

Field Measurement of Soil Erosion and Runoff. N.W. Hudson. UNIPUB, 4611-F Assembly Drive, Lanham, MD 20706-4391. (800) 233-0504. FAX (301) 459-0056. 1994. $17.00.

ONLINE DATABASES AND CD-ROMS

Compendex Plus. Engineering Information, Inc., 345 East 47th Street, New York, NY 10017. (212) 705-7600 or (800) 221-1044. Contains citations with abstracts to worldwide literature in engineering and technology, from 1970 to present. Available on online BRS,(800) 289-4277, DIALOG, (800) 334-2564, ORBIT (800) 456-7248, and STN International, FIZ Karlsruhe, P.O. Box 2465, W-7500, Karlsruhe 1, Germany, online services. Also available on CD-ROM. Inquire as to cost and availability.

Current Contents Search. Institute for Scientific Information, 3501 Market Street, Philadelphia, PA 19104. (215) 386-0100. FAX (215) 386-6362. Contains citations to articles listed in the table of contents of science and technology journals. Also articles in social sciences and life sciences journals. Available on BRS,(800) 289-4277, DIALOG,(800) 334-2564, online services. Inquire as to cost and availability.

Dissertation Abstracts. University Microfilms International, 300 North Zeeb Road, Ann Arbor, MI 48106. (800) 521-0600 or (313) 761-4700. Scope includes virtually all doctoral dissertations accepted at accredited American institutions from 1861 to present in 252 subject areas. Available on BRS,(800) 289-4277, DIALOG,(800) 334-2564, and OCLC EPIC,(800) 848-5878, online services. Also available on CD-ROM. Inquire as to cost and availability.

GEOREF: Bibliography and Index of Geology. American Geological Institute, 4220 King Street, Alexandria, VA 22302. (703) 379-2480. Fax (703) 379-7563. Monthly. Inquire for price and availability.

NTIS Bibliographic Database. National Technical Information Service, 5285 Port Royal Road, Springfield, VA 22161. (703) 487-4929 or FAX (703) 321-8199. Broad coverage of government-sponsored science and technology research reports, 1964 to present. Available on BRS,(800) 289-4277, DIALOG, (800) 334-2564, ORBIT, (800) 456-7248, and STN International, FIZ Karlsruhe, P.O. Box 2465, W-7500, Karlsruhe 1, Germany, online services. Also available on CD-ROM. Inquire as to cost and availability.

SCISEARCH. Institute for Scientific Information, 3501 Market Street, Philadelphia, PA 19104. (800) 523-1850 or (215) 386-0100. Broad multidisciplinary title and author index to the international literature of science and technology, 1974 to present. Available on DIALOG,(800) 334-2564, and ORBIT,(800) 456-

7248, online services. Also available on CD-ROM. Inquire as to cost and availability.

WILSONLINE. H.W. Wilson Company, 950 University Avenue, Bronx, NY 10452. (800) 367-6770 or (212) 588-8400. Makes available online versions of the printed H.W. Wilson indexes including Applied Science and Technology Index, Business Periodicals Index, General Science Index, and Readers' Guide to Periodical Literature. Period covered is generally 1983 to present. Available on BRS,(800) 289-4277, DIALOG,(800) 334-2564, and OCLC EPIC,(800) 848-5878, online services. Also available on CD-ROM. Inquire as to cost and availability.

PERIODICALS

Advances in Soil Sciences. Springer-Verlag, 175 Fifth Avenue, New York, NY 10010. (212) 460-1500. FAX (212) 473-6272. 1985 to present. Irregular. Price varies.

Agricultural Engineering. American Society of Agricultural Engineers, 2950 Niles Road, St. Joseph, MI 49085-9659. (616) 429-0300. FAX (616) 429-3852. 1920 to present. Bi-monthly. $42.50 for non-members.

Canadian Journal of Soil Science. Agricultural Institute of Canada, 151 Slater Street, Suit 907, Ottawa, Ontario K1P 5H4, Canada. Telephone (613) 232-9459. 1921 to present. Quarterly. Inquire for price in U.S.

Communications in Soil Science and Plant Analysis. Marcel Dekker, Inc., Journals, 270 Madison Avenue, New York, NY 10016. (212) 696-9000. FAX (212) 685-4540. 1970 to present. 20 times a year. $995.00.

Engineering Geology. Elsevier Science Inc., Box 882, Madison Square Station, New York, NY 10159. (212) 989-5800. FAX (212) 633-3990. 1965 to present. Twelve times a year. $650.00.

Geoderma. Elsevier Science Inc., Box 882, Madison Square Station, New York, NY 10159. (212) 989-5800. FAX (212) 633-3990. 1967 to present. 20 times a year. $1003.00.

Journal of Irrigation and Drainage. American Society of Civil Engineers, Irrigation and Drainage Division, 345 East 47th Street, New York, NY 10017-2398. (212) 705-7520 or (800) 548-2723. 1956 to present. Bi-monthly. $136.00 per year.

Journal of Soil and Water Conservation. Soil and Water Conservation Society of America, 7515 Northeast Ankeny Road, Ankeny, IA 50021. (515) 289-2331. FAX (515) 289-1227. 1946 to present. Bi-monthly. $39.00.

Rock Mechanics and Rock Engineering. Springer Verlag/Journal Fulfillment Services, Box 2485, Secaucus, NJ 07096-2491. (800) 777-4643. FAX (201) 348-4505. 1929 to present. Four times a year. $187.00 per year.

Soil Dynamics and Earthquake Engineering. Elsevier Science [journals], 660 White Plains Rd., Tarrytown, NY 10159-5153. (919) 524-9200. FAX (919) 333-2444. 1981 to present. Six times a year. $515.00.

Soil Mechanics and Foundation Engineering. Plenum Publishing Corporation, Consultants Bureau, 233 Spring Street, New York, NY 10013-1578. (212) 620-8000. FAX (212) 463-0742. 1964 to present. Bi-monthly. $1075.00.

SOIL SCIENCE

Ency. of Physical Sciences and Engineering Info. Sources

Soil Science. Williams & Wilkins, 428 E Preston Street, Baltimore, MD 21202. (800) 527-5597. FAX (410) 528-4422. 1916 to present. Monthly. $128.00.

Soil Science Society of America Journal. Soil Science Society of America, 677 S. Segoe Road, Madison, WI 53711. (608) 273-8080. FAX (608) 273-2021. 1936 to present. Bi-monthly. $108.00.

Soils and Foundations. Japanese Society of Soil Mechanics and Foundation Engineering, Doshitsu Kogakkai, 4F, Sugayama Bldg., Kanda Awaji-cho, 2-23, Chiyoda-ku, Tokyo 101, Japan. 1960 to present. Quarterly. $148.50.

Soil Technology. Elsevier Science Inc., Box 882, Madison Square Station, New York, NY 10159. (212) 989-5800. FAX (212) 633-3990. 1988 to present. Quarterly. $176.00.

RESEARCH CENTERS AND INSTITUTES

Iowa State University, Geotechnical/civil Engineering Materials Research Laboratories. Materials Lab/Sprangler Geotechnical Lab, Ames, IA 50011. (515) 294-9470.

New Mexico State University, Soil, Water, and Plant Testing Laboratory. Agronomy and Horticulture, Department 3Q, PO Box 30003, Las Cruces, NM 88003-0003. (505) 646-4422. FAX (505) 646-6041.

Soil and Water Research. Louisiana State University, PO Box 25071, University Station, Baton Rouge, LA 70894-5071. (504) 387-2783.

University of California, Kearney Foundation of Soil Science. LAWR-127 Veihmeyer Hall, Davis, CA 95616. (916) 752-0683. FAX (916) 752-1552.

SOLAR ENERGY

See also: COAL, ENERGY, FUSION, GEOtheRMAL POWER, HYDRO-ELECTRIC POWER, HYDROGEN, NATURAL GAS, NUCLEAR ENERGY, PETROLEUM

ABSTRACT SERVICES AND INDEXES

Applied Science and Technology Index; A Cumulative Subject Index To English Language Periodicals in the Fields of Aeronautics and Space Science, Computer Technology, Chemistry, Construction Industry, Energy and Related Areas. H. W. Wilson Co., 950 University Avenue, Bronx, NY 10452-9978. (212) 588-8400. (800) 367-6770. FAX (718) 590-1617.From 1958 to present. Monthly. Inquire about cost and availability. Also available online (BRS and WILSONLINE) and on CD-ROM. ISSN: 0003-6986.

Chemical Abstracts. Chemical Abstracts Service. 2540 Olentangy River Road, Box 3012, Columbus, OH 43210-0012. (614) 447-3600. FAX (614) 447-3713. 1907 to present. Weekly. $16,800.00 per year. Available online and on CD-ROM. CA is also available in five section groupings. Inquire regarding cost and availability.

Current Contents: Engineering, Technology and Applied Sciences. Institute for Scientific Information, 3501 Market Street,

Philadelphia, PA 19104. (215) 386-0100. FAX (215) 386-2991. From 1970 to present. Weekly. $442.00 per year. Also available online. Inquire regarding cost and availability. ISSN: 0095-7917.

Energy Research Abstracts. U. S. Department of Energy, office of Scientific and Technical Information, Box 62, Oak Ridge, TN 37831. (615) 674-0733. Subscriptions to: U. S. Government Printing office, Box 371954, Pittsburgh, PA 15250-7954. (202) 783-3238. 1976 to present. Monthly. $164.00 per year. ISSN: 0160-3604.

Engineering Index Monthly. Engineering Information, Inc., Castle Point on the Hudson, Hoboken, NJ 07030. (201) 216-8500. (800) 221-1044. FAX (201) 216-8532. Monthly. $2300.00 per year for monthly issues. $1980 per year for annual cumulation. Available on CD-ROM and online as COMPENDEX. ISSN: 0742-1974; 0360-8557.

General Science Index. H.W. Wilson Company, 950 University Avenue, Bronx, NY 10452. (212) 588-8400. (800) 367-6770. FAX (718) 590-1617. From 1978 to present. Ten issues per year; quarterly and annual cumulations. Service basis. Available on CD-ROM and online. Inquire regarding cost and availability. ISSN: 0162-1963.

Index To IEEE Publications. IEEE Service Center, 445 Hoes Lane, Piscataway, NJ 08855-1331. (908) 981-1393. (800) 678-IEEE. FAX (908) 981-9667. 1973 to present. Annual. ISSN: 0099-1368.

*ISMEC: Mechanical Engineering Abstracts.*Cambridge Scientific Abstracts, 7200 Wisconsin Avenue, Suite 601, Bethesda, MD 20814. (301) 961-6750 or (800) 843-7751. Contains citations to the literature in mechanical engineering, industrial and production engineering, energy, power, mechanics, devices and related areas, from 1973 to present. Available on the DIALOG,(800) 334-2564, online service. Inquire as to cost and availability.

N T I S Alerts: Energy. U.S. National Technical Information Service, 5285 Port Royal Road, Springfield, VA 22161. (703) 487-4929. FAX (703) 321-8547. Weekly. $160.00 per year.

Physics Abstracts. INSPEC. Section A, Science Abstracts. Institution of Electrical Engineers (IEE), London, United Kingdom. Available from: INSPEC/IEEE - Institute of Electrical and Electronic Engineers, Box 1331, 445 Hoes Lane, Piscataway, NJ 08855-1331. (908) 562-5549. 1898 to present. 24 issues per year. $2835.00 per year. ISSN: 0036-8091. Also available online and on CD-ROM.

Physics Briefs (Physikalische Berichte). Information Center for Energy, Physics, Mathematics; German Physical Society. V C H Publishers, Inc., 220 East 23rd Street, New York, NY 10010-4606. (212) 683-8333. 1845 to present. 24 issues per year. $2390.00 per year. Also available online. ISSN: 0179-7434.

Process and Chemical Engineering/chemical Engineering Abstracts. Royal Society of Chemistry, Information Services, Thomas Graham House, Science Park, Milton Road, Cambridge, CB4 4WF, England. Contains citations, mostly with abstracts, of the worldwide literature on chemical engineering; from 1982 to present. Monthly. $610.00 per year. Also available online. ISSN: 0960-5045.

Science Citation Index. SCI. Institute for Scientific Information, 3501 Market Street, Philadelphia, PA 19104. (215) 386-

0100. (800) 523-1850. FAX (215) 386-2991. 1961 to present. Six issues per year, plus annual cumulation. $11650.00 per year. Also available online and on CD-ROM. Inquire about price and availability. ISSN: 0036-827X.

ANNUAL REVIEWS AND YEARBOOKS

Advances in Energy Systems and Technology. Academic Press, Inc., 6277 Sea Harbor Drive, Orlando, FL 32887. (800) 321-5068. 1979 to present. Irregular. Inquire.

Annual Review of Energy and the Environment. Annual Reviews, Inc., 4139 El Camino Way, Palo Alto, CA 94303-0139. (415) 943-4400. (800) 523-8635. 1976 to present. Annual. $71.00. ISSN: 1056-3466.

ASSOCIATIONS AND PROFESSIONAL SOCIETIES

American Solar Energy Association. c/o John Lillard, 124 South Street, Annapolis, MD 21401. (410) 268-1900.

American Solar Energy Society. 2400 Central Avenue, G-1, Boulder, CO 80301. (303) 443-3130. FAX (303) 443-3212.

Association of Energy Engineers, 402225 Pleasantdale Road, Suite 340, Atlanta, GA 30340. (770) 447-5083.

Edison Electric Institute, 701 Pennsylvania Avenue Nw, Washington, DC 20004-2696. (202) 508-5000.

IEEE (Institute of Electrical and Electronics Engineers) 345 East 47th Street, New York, NY 10017. (212) 705-7900. FAX (212) 705-4929.

Solar Energy Industries Association. 122 C Street NW, Washington, DC 20001. (202) 383-2600. FAX (202) 383-2620.

Solartherm. 1315 Apple Avenue, Silver Spring, MD 20910. (301) 587-8686. FAX (301) 587-8688.

BIBLIOGRAPHIES

Appropriate Technology Sourcebook. Ken Darrow and Mike Saxenian. Appropriate Technology Institute, W100 Engineering Research Center, Colorado State University, Fort Collins, CO 80523. (800) 648-8043. FAX (970) 491-2729.1993. $29.95. # #*Information Sources in Engineering.* Peter Hicks, editor. Bowker-Saur, 121 Chanlon Road, New Providence, NJ 07974. (800) 521-8110 or (908) 464-6800. 3rd edition. 1996. $125.00.

Scientific and Technical Books and Serials in Print; An Index To Literature in Science and Technology. R.R. Bowker Inc., 121 Chanlon Road, New Providence, NJ 07974. (908) 464-6800. (800) 521-8110. FAX (908) 665-3502. 1972 to present. Annual. 4 volumes. 1994. $299.95. Also available on compact disc and online. ISSN: 0000-054X.

Solar Living Sourcebook. John Schaeffer and Doug Pratt. Chelsea Green Publishing Co., PO Box 428, 205 Gates-Briggs Building, White River Junction, VT 05001. (802) 295-6300. (800) 639-4099. 1996. $30.00.

DIRECTORIES AND BIOGRAPHICAL SOURCES

American Men and Women of Science. Physical and Biological Sciences. Fifteenth edition. R.R. Bowker Inc., 121 Chanlon Road, New Providence, NJ 07974. (908) 464-6800. (800) 521-8110. 20th edition. 8 volumes. 1996. $850.00

Energy and Nuclear Science International Who's Who. Gale Research, 835 Penobscot Building, Detroit, MI 48226- 4094. (313) 961-2242. (800) 877-4253. Triennial. 1993. $325.00.

Energy Information Centers Directory. U. S. Council for Energy Awareness, 1776 I Street, NW, Suite 400, Washington, DC 20006. (202) 293-0770. Triennial. No charge.

Engineering Research Centers: Incorporating Electronics Research Centers. Stockton Press, 345 Park Avenue, New York, NY 10010. (212) 689-9200. (800) 221-2123. 4th edition. 1995. $515.00.

Research Centers Directory. Gale Research, 835 Penobscot Building, Detroit, MI 48226-4094. (313) 961-2242. (800) 877-4253. 20th edition. 1995. $485.00. ISSN: 0080-1518.

Who's Who in Engineering. Gordon Davis, editor. American Association of Engineering Societies. 1111 19th Street, Suite 608, Washington, DC 20036. (202) 296-2237. (800) 658-8897. 9th edition. 1995. $220.00.

Who's Who in Technology. Gale Research, 835 Penobscot Building, Detroit, MI 48226-4094. (313) 961-2242. (800) 521-4253. 7th edition. 1995. $195.00. ISBN: 0-8103-7467-6.

World Energy and Nuclear Directory. Gale Research, 835 Penobscot Building, Detroit, MI 48226-4094. (313)961-2242. (800) 877-4253. Biennial, even years. $450.00.

ENCYCLOPEDIAS AND DICTIONARIES

Encyclopedia of Applied Physics. VCH Publishers, Inc., 303 Northwest 12th Avenue, Deerfield Beach, FL 33442. (800) 367-8249. 1991-. Twenty volumes. $6000.00.

Encyclopedia of Chemical Processing and Design. McKetta Marcel Dekker, Inc., 270 Madison Avenue, New York, NY 10016. (212) 696-9000. (800) 228-1160. 1976 - . $175.00 per volume.

Kirk-Othmer Encyclopedia of Chemical Technology. John Wiley and Sons, Inc., 605 Third Avenue, New York, NY 10158. (800) 526-5368 or (212) 850-6000. Fourth edition. 1991 - . Twenty-seven volumes. $5400.00.

McGraw-Hill Encyclopedia of Science and Technology. McGraw-Hill Book, Incorporated, 1221 Avenue of the Americas, New York, NY 10020. (212) 997-3675. (800) 262-4729. Seventh edition. Twenty volumes. 1992. $1900.00.

Ullman's Encyclopedia of Industrial Chemistry. VCH Publications, Inc., 220 East 23rd Street, Suite 909, New York, NY 10010-4606. (212) 683-8333. (800) 422-8824. 5th edition. 1984 - . Price varies per volume.

SOLAR ENERGY

Ency. of Physical Sciences and Engineering Info. Sources

GENERAL WORKS

Active Solar Systems. George Lof, editor. MIT Press, 55 Hayward Street, Cambridge, MA 02142. (617) 253-8569. 1992. $80.00.

Electricity Supply: Efforts To Develop Solar and Wind Energy. Diane Publishing Co., 600 Upland Avenue, Upland, PA 19015. (610) 499-7415. $60.00 in paper.

Introduction To Solar Energy For Scinetists and Engineers. Sol Wieder. Krieger Publishing Company, P.O. Box 9542, Melbourne, FL 32902-9542. (407) 724-9542. 1992. $59.95.

Passive Solar Buildings. J. Douglas Blacomb. MIT Press, 55 Hayward Street, Cambridge, MA 02142. (617) 253-8569. 1992. $57.50.

Photochemical Conversion and Storage of Solar Energy. Mario Schiavello, editor. Kluwer Academic Publishers, 101 Philip Drive, Assinippi Park, Norwell, MA 02061. (617) 871-6600. 1991. $227.50.

The New Solar Home Book. Bruce anderson and Michael Riordan. Brick House Publishing Co., 4 Limbo Lane, PO Box 266, Amherst, NH 03031. (602) 672-5112. (800) 446-8642. 2nd edition. 1996. $20.00.

Solar Air Conditioning and Refrigeration. A. A. Sayigh and J. C. McVeigh, editors. Elsevier Science Publishing Company, Inc., 655 Avenue of the Americas, New York, NY 10010. (212) 989-5800. FAX (914) 333-2444. 1992. $130.00.

Solar Collectors, Energy Storage and Materials. Francis DeWinter. MIT Press, 55 Hayward Street, Cambridge, MA 02142. (617) 253-8569. 1991. $90.00.

Solar Energy Technology. B. Norton, Springer-Verlag New York, Inc., 175 Fifth Avenue, New York, NY 10010. (212) 460-1500. (800) 777-4643. FAX (212) 473-6272. 1991. $139.95.

Solar Power Satellites: The Emerging Energy Option. Peter Glaser, et al, editors. John Wiley & Sons, Inc., 605 Third Avenue, New York, NY 10158-0012. (212) 850-6000. (800) 225-5945. $98.00.

HANDBOOKS AND MANUALS

Energy Engineering Handbook. Fairmont Press Staff. Prentice-Hall, 113 Sylvan Avenue, Route 9W, Englewood Cliffs, NJ 07632. (201) 592-2000. (800) 922-0579. 3rd edition. 1995. $75.00.

Fuel Cells: A Handbook. J. M. Hirschenffoffer. Business/Technology Information Services, P. O. Box 574, Orinda, CA 94563. (510) 299-1829. (800) 808-8811. 1996. $115.00.

Perry's Chemical Engineers' Handbook. Robert H. Perry and Donald W. Green, editors. McGraw-Hill Publishing Company, Inc., 1221 Avenue of the Americas, New York, NY 10020. (212) 512-2000. (800) 262-4729. 6th edition. 1996. $129.50.

ONLINE DATABASES AND CD-ROMS

CA Search. Chemical Abstracts Service, P.O. Box 3012, Columbus, OH 43210-0012. (614) 447-3600. (800) 848-6533. FAX (614) 447-3709. Very comprehensive guide to worldwide chemical literature and related fields. 1972 to present. Available on BRS,(800) 289-4277, DIALOG, (800) 334-2564, ORBIT (800) 456-7248, and STN International, FIZ Karlsruhe, P.O. Box 2465, W-7500, Karlsruhe 1, Germany, online services. Inquire as to cost and availability.

Compendex Plus. Engineering Information, Inc., 345 East 47th Street, New York, NY 10017. (212) 705-7600 or (800) 221-1044. Contains citations with abstracts to worldwide literature in engineering and technology, from 1970 to present. Available on online BRS,(800) 289-4277, DIALOG, (800) 334-2564, ORBIT (800) 456-7248, and STN International, FIZ Karlsruhe, P.O. Box 2465, W-7500, Karlsruhe 1, Germany, online services. Also available on CD-ROM. Inquire as to cost and availability.

Current Contents Search. Institute for Scientific Information, 3501 Market Street, Philadelphia, PA 19104. (215) 386-0100. FAX (215) 386-6362. Contains citations to articles listed in the table of contents of science and technology journals. Also articles in social sciences and life sciences journals. Available on BRS,(800) 289-4277, DIALOG,(800) 334-2564, online services. Inquire as to cost and availability.

Dissertation Abstracts. University Microfilms International, 300 North Zeeb Road, Ann Arbor, MI 48106. (800) 521-0600 or (313) 761-4700. Scope includes virtually all doctoral dissertations accepted at accredited American institutions from 1861 to present in 252 subject areas. Available on BRS, (800) 289-4277, DIALOG, (800) 334-2564, and OCLC EPIC, (800) 848-5878, online services. Also available on CD-ROM. Inquire as to cost and availability.

INSPEC. Institution of Electrical Engineers, Michael Faraday House, Six Hills Way, Stevenage, Herts. SG1 2AY, England. Telephone: 0438 313311 or FAX 0438 742840. Contains citations to the worldwide literature of physics, electronics and electrical engineering, computer technology, and related fields. Available on BRS, (800) 289-4277, DIALOG, (800) 334-2564, ORBIT, (800) 456-7248, and STN International, FIZ Karlsruhe, P.O. Box 2465, W-7500, Karlsruhe 1, Germany, online services. Inquire as to cost and availability.

Kirk-Othmer Encyclopedia of Chemical Technology. John Wiley and Sons, Inc., 605 Third Avenue, New York, NY 10158. (800) 526-5368 or (212) 850-6000. Contains the complete text of all chapters in the 27 volume fourth edition of the KIRK-Othmer Encyclopedia of Chemical Technology. 1991. Available on BRS,(800) 289-4277, DIALOG,(800) 334-2564, online services. Inquire as to cost and availability.

NTIS Bibliographic Database. National Technical Information Service, 5285 Port Royal Road, Springfield, VA 22161. (703) 487-4929 or FAX (703) 321-8199. Broad coverage of government-sponsored science and technology research reports, 1964 to present. Available on BRS,(800) 289-4277, DIALOG, (800) 334-2564, ORBIT, (800) 456-7248, and STN International, FIZ Karlsruhe, P.O. Box 2465, W-7500, Karlsruhe 1, Germany, online services. Also available on CD-ROM. Inquire as to cost and availability.

Physics Briefs. American Institute of Physics, 335 East 45th Street, New York, NY 10017. (212) 661-9260 or FAX (212) 661-2036. Contains citations with abstracts of the literature of physics and related fields, 1979 to present. Available on the STN International, FIZ Karlsruhe, P.O. Box 2465, W-7500, Karlsruhe 1, Germany, online service. Inquire as to cost and availability.

SCISEARCH. Institute for Scientific Information, 3501 Market Street, Philadelphia, PA 19104. (800) 523-1850 or (215) 386-0100. Broad multidisciplinary title and author index to the international literature of science and technology, 1974 to present. Available on DIALOG, (800) 334-2564, and ORBIT, (800) 456-7248, online services. Also available on CD-ROM. Inquire as to cost and availability.# #*WILSONLINE.* H.W. Wilson Company, 950 University Avenue, Bronx, NY 10452. (800) 367-6770 or (212) 588-8400. Makes available online versions of the printed H.W. Wilson indexes including Applied Science and Technology Index, Business Periodicals Index, General Science Index, and Readers' Guide to Periodical Literature. Period covered is generally 1983 to present. Available on BRS, (800) 289-4277, DIALOG, (800) 334-2564, and OCLC EPIC, (800) 848-5878, online services. Also available on CD-ROM. Inquire as to cost and availability.

Who's Who in Technology. Gale Research, 835 Penobscot Building, Detroit, MI 48226-4094. (313) 961-2242. (800) 877-4253. Contains biographical information of contemporary American scientists and engineers. Available on DIALOG,(800) 334-2564, online service. Inquire as to cost and availability.

PERIODICALS

Energy and Fuels. American Chemical Society, Box 3337, Columbus, OH 43210. (614) 447-3774. 1987 to present. Bimonthly. $345.00 per year. ISSN: 0887-0624.

Energy Engineering. Association of Energy Engineers. Available from Fairmont Press, 700 Indian Trail, Lilburn, GA 30324. (404) 925-9388. 1904 to present. Bimonthly. $99.00 per year. ISSN: 0199-8595.

Energy Sources. Taylor and Francis, 1900 Forst Road, Stuie 101, Bristol, PA 19007-1598. (215) 785-5800. 1973 to present. Quarterly. $225.00. ISSN: 0090-8312.

International Journal of Solar Energy. Harwood Academic Publishers, 820 Town Center Drive, Langhorne, PA 19047. (215) 750-2642. 1982 to present. 8 issues per year. 254 ECU per year. ISSN: 0142-5919.

Journal of Energy Resources Technology. American Society of Mechanical Engineers, 345 East 47th Street, New York, NY 10017. (212) 705-7703. 1979 to present. Quarterly. $100.00 per year. ISSN: 0195-0738.

Journal of Solar Energy Engineering. American Society of Mechanical Engineers, 345 East 47th Street, New York, NY 10017. (212) 705-7703, 1980 to present. Quarterly. $100 per year. ISSN: 0199-6231.

Solar Energy. Elsevier Science Publishing Company, Inc., 660 White Plains Road, Tarrytown, NY 10591-5133. (914) 524-9200. 1957 to present. Monthly. $600.00 per year. ISSN: 0038-092X.

Solar Energy Materials and Solar Cells. Elsevier Science Publishing Company, Inc., 660 White Plains Road, Tarrytown, NY 10591-5133. (914) 524-9200. 1979 to present. 16 issues per year in 4 volumes. $781.00 per year. ISSN: 0927-0248.

Sunworld. International Solar Energy Society, c/o W. R. Read, 26 Railway Avenue, PO Box 124, Calufield East, Victoria, 3145, Australia. 1976 to present. Quarterly. $40.00 per year. ISSN: 0149-1938.

RESEARCH CENTERS AND INSTITUTES

Alabama Solar Energy Center. University of Alabama in Huntsville, Johnson Research Center, Huntsville, AL 35899. (205) 895-6707.

Florida Solar Energy Center. 300 State Road, 401, Cape Canaveral, FL 32920. (407) 782-0300. FAX (407) 783-2571.

Solar and Energy Research Facility. University of Arizona, College of Engineering and Mines, Tucson, AZ 85721. (602) 621-7496.

Solar Energy and Energy Conversion Laboratory. University of Florida, Department of Mechanical Engineering, Gainesville, FL 32611. (904) 392-0812. FAX (904) 392-1071.

Solar Energy Applications Laboratory. Colorado State University, College of Engineering, Fort Collins, CO 80523. (303) 491-8617. FAX (303) 491-8544.

Solar Energy Research Institute. 1617 Cole Boulevard, Golden CO 80401-3393. (303) 231-1000.

SOLAR PHYSICS

See: SUN

SOLAR SYSTEM

See also: ASTROGEOLOGY, ASTEROIDS, ASTRONOMY, COMETS, METEORS, MOON, NAMES of INDIVIDUAL PLANETS, NATURAL SATELLITES, PLANETARY SCIENCES, SEISMOLOGY, SUN, TEKTITES, VOLCANOLOGY

ABSTRACT SERVICES AND INDEXES

Applied Science and Technology Index; A Cumulative Subject Index To English Language Periodicals in the Fields of Aeronautics and Space Science, Computer Technology, Chemistry, Construction Industry, Energy and Related Areas. H. W. Wilson Co., 950 University Avenue, Bronx, NY 10452-9978. (212) 588-8400. (800) 367-6770. FAX (718) 590-1617.From 1958 to present. Monthly. Inquire about cost and availability. Also available online (BRS and WILSONLINE) and on CD-ROM. ISSN: 0003-6986.

Astronomy and Astrophysics Abstracts. Springer-Verlag New York, 175 Fifth Avenue, New York, NY 10010. (212) 460-1500. FAX (212) 473-6272. Published for the Astronomisches Rechen-Institut. Comprehensive coverage of all aspects of astronomy, astrophysics and related fields. 1969 to present. Two parts per year. Annual. ISSN: 0067-0022.

Bibliography and Index of Geology. American Geological Institute, 4220 King Street, Alexandria, VA 22302-1507. (703) 379-2480. FAX (703) 379-7563. From 1969 to present. Monthly. $1295.00 per year. ISSN: 0098-2784. Also available as GEOREF online (CISTI, DIALOG, Orbit, STN) and on CD-ROM. Inquire about price and availability.

SOLAR SYSTEM

Ency. of Physical Sciences and Engineering Info. Sources

Chemical Abstracts. Chemical abstracts Service. 2540 Oleantangy River Road, Box 3012, Columbus, OS 43210-0012. (614) 447-3600. FAX (614) 447-3713. 1907 to present. Weekly. $16,800.00 per year. Available online and on CD- ROM. CA is also available in five section groupings. Inquire regarding cost and availability.

Current Contents: Physical, Chemical, and Earth Sciences. Institute for Scientific Information, 3501 Market Street, Philadelphia, PA 19104. (215) 386-0100. (800) 523-1850. FAX (215) 386-2291. 1961 to present. Weekly. $442.00 per year. Also available online (BRS, DIALOG) and on CD- ROM. Inquire about price and availability.

General Science Index. H.W. Wilson Company, 950 University Avenue, Bronx, NY 10452. (212) 588-8400. (800) 367-6770. FAX (718) 590-1617. From 1978 to present. Ten issues per year; quarterly and annual cumulations. Service basis. Available on CD-ROM and online. Inquire regarding cost and availability. ISSN: 0162-1963.

Meteorological and Geoastrophysical Abstracts. American Meteorological Society, c/o Inforonics, Inc., 550 Newtown Road, Box 458, Littleton, MA 01460. (508) 486-8976. FAX (508) 486-0027. Covers literature in environmental sciences, meteorology, astrophsyics, hydrology, glaciology and physical oceanography. 1950 to present. Monthly. $950.00 per year. Also available online (DIALOG) and on CD-ROM. ISSN: 0026-1130.

NTIS Alerts: Astronomy & Astrophysics. U. S. National Technical Information Service. 5285 Port Royal Road, Springfield, VA 22161. (703) 487-4650. FAX (703) 321-8547. Weekly. $140.00 per year.

Physics Abstracts. INSPEC. Section A, Science Abstracts. Institute of Electrical Engineers, London, United Kingdom. Available from: INSPEC/IEEE - Institute of Electrical and Electronic Engineers, Box 1331, Hoes Lane, Piscataway, NJ 08855-1331. (908) 562-5549. 1898 to present. 24 issues per year. $2835.00 per year. ISSN: 0036-8091. Also available online and on CD-ROM.

Science Citation Index. SCI. Institute for Scientific Information, 3501 Market Street, Philadelphia, PA 19104. (215) 386-0100. (800) 523-1850. FAX (215) 386-2991. 1961 to present. Six issues per year, plus annual cumulation. $11650.00 per year. Also available online and on CD-ROM. Inquire about price and availability. ISSN: 0036-827X.

STAR. Scientific and Technical Aerospace Reports. U.S. National Aeronautics and Space Administration. Distributed by U. S. Superintendent of Documents, Washington, DC 20402. 1963 to present. Semi-monthly, with semiannual and annual indexes. $114.00 per year. ISSN: 0036-8741. Also available online and on CD-ROM.

ANNUAL REVIEWS AND YEARBOOKS

The Astronomical Almanac. Superintendent of Documents, U. S. Government Printing office, Washington, DC 10402. (202) 783-3238. 1981 to present. Supercedes Astronomical Ephemeris. Annual. $44.95. ISSN: 0737-6421.

Annual Review of Astronomy and Astrophysics. Original reviews of critical literature and current developments in astronomy and astrophysics. Annual Reviews, Inc., 4139 El Camino Way, Palo Alto, CA 94303-0139. (415) 493-4400. (800)523-8635. Fax (415) 855-9815. 1963 to date. Annual. $60.00. ISSN: 0066-4146.

Annual Review of Earth and Planetary Sciences. Annual Reviews, Inc., 4139 El Camino Way, Palo Alto, CA 94303-0139. (415) 493-4400. (800) 523-8635. Fax (415) 855-9815. 1973 to date. Annual. $62.00. ISSN: 0084-6597.

Proceedings of the Lunar and Planetary Science Conference. Lunar and Planetary Institute, 3600 Bay Area Boulevard, Houston, TX 77058. (713) 486-2143. 1970 to date. Annual. ISSN: 0270-9511.

ASSOCIATIONS AND PROFESSIONAL SOCIETIES

American Astronomical Society. 2000 Florida Avenue NW, Suite 400, Washington, DC 20009. (202) 328-2010. FAX (202) 234-2560.

American Geophysical Union. 2000 Florida Avenue NW, Washington, DC 20009. (202) 462-6900. (800) 966-AGU1. FAX (202) 328-0566.

American Institute of Physics. 1 Physics Ellipse, College Park, MD 20740-3843. (301) 209-3100.

Association of Universities For Research in Astronomy, Inc. (AURA). Suite 701, 1625 Massachusetts Avenue, NW, Washington, DC 20036. (202) 483-2101. FAX (202) 483-2106.

Astronomical Society of the Pacific. 390 Ashton Avenue, San Francisco, CA 94112. (415) 337-1100. FAX: (415) 337-5205.

Geological Society of America. 3300 Penrose Place, PO Box 9140. Boulder, CO 80301-9140. (303) 447-2020. FAX (303) 447-1133.

Meteoritical Society. University of Massachusetts, 125 Marston Hall, Amherst, MA 01003. (413) 545-0300. (413) 545-0724.

Planetary Society. 65 North Catalina Avenue, Pasadena, CA 91106. (818) 793-5100. (800) WOW-MARS. FAX (818) 793-5528.

BIBLIOGRAPHIES

A Bibliography of Astronomy, 1970 - 1979. R. A. Sea and S. S. Martin. Libraries Unlimited, Inc., Littleton, CO 80160. 1982. $37.50.

DIRECTORIES AND BIOGRAPHICAL SOURCES

American Astronomical Society Membership Directory. American Astronomical Society. 2000 Florida Avenue NW, Suite 400, Washington, DC 20009. (202) 328-2010. FAX (202) 234-2560 Annual. Included in membership dues. ISSN: 1061- 9038.

American Men and Women of Science: Physical and Biological Sciences. R. R. Bowker Inc., 121 Chanlon Road, New Providence, NJ 07974. (908) 464-6800. (800) 521-8110. 20th edition. 8 volumes. 1996. $850.00.

The Astronomers. Donald Goldsmith. St. Martin's Press, Inc., 175 Fifth Avenue, New York, NY 10010. (212) 674-5151. (800) 221-7945. 1993. $14.95 in paper.

Astronomical Centers of the World. Kevin Krisciunas. Cambridge University Press, 40 West 20th Street, New York, NY 10011-4211. (212) 924-3900. (800) 872-7423. 1988. $34.95

Directory of Physics and Astronomy Staff Members. American Institute of Physics, One Physics Ellipse, College Park, MD 20740-3843. (301) 209-3100. 1975/76 to present. Annual. $60.00. ISSN: 0361-2228.

Earth and Astronomical Research Centers. Stockton Press, 345 Park Avenue South, New York, NY 10010. 4th edition. 1995. $515.00. ISBN: 1-56169-0967.

ENCYCLOPEDIAS AND DICTIONARIES

Concise Dictionary of Astronomy. Jacqueline Mitton. Oxford University Press, Inc., 200 Madison Avenue, New York, NY 10016. (212) 725-6000. (800) 334-4249. 1992. $24.95.

Encyclopedia of Astronomy and Astrophysics. Stephen Maran, editor. Van Nostrand Reinhold, 115 Fifth Avenue, New York, NY 10003. (212) 254-3232. (800) 842-3636. 1992. $129.95.

McGraw-Hill Encyclopedia of Science and Technology. McGraw-Hill Book Company, Inc., 1221 Avenue of the Americas, New York, NY 10020. (212) 512-2000. (800) 262-4729. 7th edition. 20 volume set. 1992. $1900.00.

Moons of the Solar System: An Illustrated Encyclopedia. John Stewart. McFarland & Company, Inc., Box 611, Jefferson, MC 28640. (910) 246-4460. (800) 253-2187. 1991. $49.95.

New Guide To the Planets. Patrick Moore. Trans-Atlantic Publications, Inc., 311 Bainbridge Street, Philadelphia, PA 19147. (215) 925-5083. 1993. $37.50.

Stars and Planets: The Sierra Club Guide To Sky Watching and Direction Finding. W. S. Kals. The Sierra Press, 4988 Gold Leaf Drive, Mariposa, CA 95338. (209) 966-5071. (800) 745-2631. $15.00.

GENERAL WORKS

Birth of the Earth. David E. Fisher. Columbia University Press, 562 West 113th Street, New York, NY 10025. (212) 666-1000. (800) 944-8648. 1988. $16.50 in paper.

Chemistry and Physics of the Terrestrial Planets. Surendra K. Saxwna. Springer-Verlag New York, Inc., 175 Fifth Avenue, New York, NY 10010. (212) 460-1500. (800) 777- 4643. 1986. $109.00.

Evolution of the Earth and Planets. E. Takahashi, et al, editors. American Geophysical Union, 2000 Florida Avenue, N.W., Washington, DC 20009. (202) 462-6903. (800) 966- 2481. 1993. $28.00.

Exploration of Venus & Mars Atmospheres. G. M. Keating, editor. Pergamon Press, Inc., Maxwell House, Fairview Park, Elmsford, NY 10523. (914) 592-7700. Fax (914) 592-3625. 1995. 94.00.

Exploring the Planets. W. Kenneth Hamblin and Eric H. Christiansen. Macmillan Publishing Company, Inc,. 200 Old Tappan Road, Old Tappan, NJ 07675. (800) 233-2336. 2nd edition. 1995. $62.00.

Fundamental Astronomy and Solar System Dynamics. Raynor L. Duncombe, et al, editors. Kluwer Academic Publishers, 101 Philip Drive, Assinippi Park, Norwell, MA 02061. (617) 871-6600. 1986. $77.50.

The Geology of Multi-ring Impact Basins: The Moon and Other Planets. P. D. Spudis. Cambridge University Press, 40 West 20th Street, New York, NY 10011-4211. (212) 924-3900. (800) 872-7423. Planetary Sciences Series. 1993. $59.95.

Geology of the Terrestrial Planets. Michael H. Carr. U. S. Government Printing office, Superintendent of Documents, Washington, DC 20402-9325. (202) 783-3238. 1985. $16.00

History of Astronomy From Thales To Kepler. John L. Dryer. Dover Publications, Inc., 180 Varick Street, New York, NY 10014. (212) 255-3755. (800) 223-3130. $9.95.

An Introduction To Cosmochemistry. Charles R. Cowley. Cambridge University Press, 40 West 20th Street, New York, NY 10011-4211. (212) 924-3900. (800) 872-7423. 1995. $29.95 in paper.

Introduction To the Solar System. Donald Gelman. Whittier Publications, Inc., 121 Regent Drive, Lido Beach, NY 11561. (516) 432-8120. 1995. $33.00 in paper.

Mercury: The Elusive Planet. Robert B. Strom. Smithsonian Institution Press, 470 L'Enfant Plaza, Suite 7100, Washington, DC 20560. (202) 287-3738. (800) 782-4612. 1987. $29.95.

Meteorites: An Introduction. Fritz Heide. Springer-Verlag New York, Inc., 175 Fifth Avenue, New York, NY 10010. (212) 460-1500. (800) 777-4643. 1995. $24.00 in paper.

Meteorites and the Origin of Planets. John A. Wood. Books on Demand, 300 North Zeeb Road, Ann Arbor, MI 48106-1346. (313) 761-4700. (800) 521-0600. Reprint edition. $35.00 in paper.

Mission To the Planets: The Illustrated Story of Man's Exploration of the Solar System. Patrick Moore. W. W. Norton & Company, 500 Fifth Avenue, New York, NY 10110. (212) 354-5500. (800) 223-2584. 1990. $24.95.

Moons and Planets. William K. Hartmann. Wadsworth Publishing Co., 10 Davis Drive, Belmont, CA 94002. (415) 595-2350. (800) 354-9706. 3rd edition. 1993. $49.95.

Moons and Rings. Time-Life Books Editors. Time-Life, Inc., 777 Duke Street, Alexandria, VA 22314. (703) 8388- 7000. (800) 621-7026. Voyage through the Universe Series. 1991.

Music of the Heavens: Kepler's Harmonic Astronomy. Bruce Stephenson. Princeton University Press. 41 William Street, Princeton, NJ 08540. (609) 258-4900. (800) 777-4726. 1994. $39.50.

The Near Planets. Time-Life Books Editors. Time-Life, Inc., 777 Duke Street, Alexandria, VA 22314. (703) 8388- 7000. (800) 621-7026. Voyage through the Universe Series. 1992.

The New Solar System. J. Kelly Beatty and andrew Chaikin, editors. Cambridge University Press, 40 West 20th Street, New York, NY 10011-4211. (212) 924-3900. (800) 872-7423 . 3rd edition. 1990. $44.95.

SOLAR SYSTEM

Ency. of Physical Sciences and Engineering Info. Sources

Physics and Chemistry of the Solar System. John S. Lewis, Jr., Academic Press, 6277 Sea Harbor Drive, Orlando, FL. (800) 321-5068. 1995. $69.95 in paper.

Planetary Landscapes. Ronald Greeley. Chapman & Hall, 1 Penn Plaza, New York, NY 10119. (212) 564-1060. Div. of Routledge, 2nd edition. 1994. $49.95.

The Planetary System. Tobias Owen and David Morrison. Addison-Wesley Publishing Co., Inc., 1 Jacob Way, Reading, MA 01867. (617) 944-3700. (800) 447-2226. 1988. $49.50.

Planets and Their Atmospheres. John S. Lewis and Ronald G. Primm. Academic Press Inc., 6277 Sea Harbor Drive, Orlando, FL. (800) 321-5068. 1983. $65.00 in paper.

Resources of Near-earth Space. John S. Lewis, et al. University of Arizona Press, 1230 North Park Avenue, Number 102, Tucson, AZ 85719. (520) 621-1441. (800) 426-3797. 1993. $75.00.

Rings: Discoveries From Galileo To Voyager. James C. Elliot and Richard Kerr. MIT Press, 55 Hayward Street, Cambridge, MA 02142. (617) 253-8569. Reprint edition. 1987. $9.95 in paper.

Solar and Planetary Dynamos. M. R. Proctor, et al, editors. Cambridge University Press, 40 West 20th Street, New York, NY 10011-4211. (212) 924-3900. (800) 872-7423. Publications of the Newton Institue. 1994. $49.95.

Solar System Evolution: A New Perspective. Stuart R. Taylor. Cambridge University Press, 40 West 20th Street, New York, NY 10011-4211. (212) 924-3900. (800) 872-7423. 1989. 1992. $54.95.

Solar System: From the Renaissance To the Ninteenth Century. Part A. Curtis Wilson, editor. Cambridge University Press, 40 West 20th Street, New York, NY 10011-4211. (212) 924-3900. (800) 872-7423. 1989. $59.95.

Stardust To Planets: A Geological Tour of the Solar System. Harry Y. McSween, Jr., St. Martin's Press, Inc., 175 Fifth Avenue, New York, NY 10010. (212) 674-5151. (800) 221-7945. 1993. $22.95.

Story of the Earth. Peter Cattermole and Patrick Moore. Cambridge University Press, 40 West 20th Street, New York, NY 10011-4211. (212) 924-3900. (800) 872-7423. 1986. $5.95 in paper.

Uranus and Neptune: The Distant Giants. Eric Burgess. Columbia University Press, 562 West 113th Street, New York, NY 10025. (212) 666-1000. (800) 944-8648. 1988. $40.00.

Venus, A New Geology. Peter Cattermole. Johns Hopkins University Press, 2715 North Charles Street, Baltimore, MD 21218-4319. (410) 516-6900. (800) 537-5487. 1994. $49.95.

Wanderers in Space: Exploration and Discovery in the Solar System. Kenneth R. Lang. Cambridge University Press, 40 West 20th Street, New York, NY 10011-4211. (212) 924-3900. (800) 872-7423. 1991. $24.95 in paper.

Worlds Apart: A Textbook in Planetary Sciences. Guy Consolmagno. Prentice-Hall, 113 Sylvan Avenue, Route 9W, Englewood Cliffs, NJ 07632. (201) 592-2000. (800) 922-0579. 1994. $62.00.

The Wrong-way Comet and Other Mysteries of the Solar System. Barry Evans. McGraw-Hill Publishing Company, Inc., 1221 Avenue of the Americas, New York, NY 10020. (212) 512-2000. (800) 262-4729. 1992. $22.95 in paper.

HANDBOOKS AND MANUALS

Astrophysical Data: Planets and Stars. Kenneth R. Lanb. Springer-Verlag New York, Inc., 175 Fifth Avenue, New York, NY 10010. (212) 460-1500. (800) 777-4643. 1993. $59.00.

Burnham's Celestial Handbook: an Observer's Guide to the Universe Beyond the Solar System. Robert Burnham, Jr., and Herbert A. Luft. 1978. Dover Publications, Inc., 180 Varick Street, New York, NY 10014. (212) 255-3755. (800) 223-3130. $13.95.

Field Guide To the Stars and Planets. Jay M. Pasachoff and Donald H. Menzel. Houghton Mifflin Co., 222 Berkeley Street, Boston, MA 02116. (617) 351-5000. (800) 225-3362. Revised edition. 1992. $24.95.

New Guide To the Planets. Patrick Moore. Trans-Atlantic Publications, Inc., 311 Bainbridge Street, Philadelphia, PA 19147. (215) 925-5083. 1993. $37.50.

Planet Observer's Handbook. Fred W. Price. Cambridge University Press, 40 West 20th Street, New York, NY 10011-4211. (212) 924-3900. (800) 872-7423. 1994. $34.95.

Planets and their Moons. Aububon Society Staff. Alfred A. Knopf, Inc. 201 E. 50th Street, New York, NY 10022. (800) 733-3000. 1995. $7.99.

Solar System Photometry Handbook. Russell M. Genet, editor. Willmann-Bell, Inc., P.O. Box 35025, Richmond, VA 23235. (804) 320-7016. 1983. $17.95.

Tables of Planetary Phenomena. Neil F. Michelsen. ACS Publications, P.O. Box 3447, San Diego, CA 92163-4487. (619-492-9919. (800) 888-9983. 2nd revised edition. 1993. $24.95.

ONLINE DATABASES AND CD-ROMS

CA Search. Chemical Abstracts Service, P.O. Box 3012, Columbus, OH 43210-0012. (614) 447-3600. (800) 848-6533. FAX (614) 447-3709. Very comprehensive guide to worldwide chemical literature and related fields, 1972 to present. Available on BRS,(800) 289-4277, DIALOG, (800) 334-2564, ORBIT (800) 456-7248, and STN International, FIZ Karlsruhe, P.O. Box 2465, W-7500, Karlsruhe 1, Germany, online services. Inquire as to cost and availability.

Dissertation Abstracts. University Microfilms International, 300 North Zeeb Road, Ann Arbor, MI 48106. (800) 521-0600 or (313) 761-4700. Scope includes virtually all doctoral dissertations accepted at accredited American institutions from 1861 to present in 252 subject areas. Available on BRS, (800) 289-4277, DIALOG, (800) 334-2564, and OCLC EPIC, (800) 848-5878, online services. Also available on CD-ROM. Inquire as to cost and availability.

GEOREF. American Geological Institute, 4220 King Street, Alexandria, VA 22302. (800) 336-4764 or (703) 379-2480. Geology and geosciences literature, 1785 to present for North America. Available on DIALOG,(800) 334-2564, ORBIT (800)

456-7248, online services. Also available on CD-ROM. Inquire as to cost and availability.

INSPEC. Institution of Electrical Engineers, Michael Faraday House, Six Hills Way, Stevenage, Herts. SG1 2AY, England. Telephone: 0438 313311 or FAX 0438 742840. Contains citations to the worldwide literature of physics, electronics and electrical engineering, computer technology, and related fields. Available on BRS, (800) 289-4277, DIALOG, (800) 334-2564, ORBIT, (800) 456-7248, and STN International, FIZ Karlsruhe, P.O. Box 2465, W-7500, Karlsruhe 1, Germany, online services. Inquire as to cost and availability.

NTIS Bibliographic Database. National Technical Information Service, 5285 Port Royal Road, Springfield, VA 22161. (703) 487-4929 or FAX (703) 321-8199. Broad coverage of government-sponsored science and technology research reports, 1964 to present. Available on BRS,(800) 289-4277, DIALOG, (800) 334-2564, ORBIT, (800) 456-7248, and STN International, FIZ Karlsruhe, P.O. Box 2465, W-7500, Karlsruhe 1, Germany, online services. Also available on CD-ROM. Inquire as to cost and availability.

SCISEARCH. Institute for Scientific Information, 3501 Market Street, Philadelphia, PA 19104. (800) 523-1850 or (215) 386-0100. Broad multidisciplinary title and author index to the international literature of science and technology, 1974 to present. Available on DIALOG, (800) 334-2564, and ORBIT, (800) 456-7248, online services. Also available on CD-ROM. Inquire as to cost and availability.

WILSONLINE. H.W. Wilson Company, 950 University Avenue, Bronx, NY 10452. (800) 367-6770 or (212) 588-8400. Makes available online versions of the printed H.W. Wilson indexes including Applied Science and Technology Index, Business Periodicals Index, General Science Index, and Readers' Guide to Periodical Literature. Period covered is generally 1983 to present. Available on BRS, (800) 289-4277, DIALOG, (800) 334-2564, and OCLC EPIC, (800) 848-5878, online services.Also available on CD-ROM. Inquire as to cost and availability.

OTHER SOURCES

Atlas of the Planets. Vincent DeCallatay and Audouin Dollfus. Books on Demand, 300 North Zeeb Road, Ann Arbor, MI 48106-1346. (313) 761-4700. (800) 521-0600. $45.60.

Atlas of the Solar System. Patrick Moore. Random House, Inc., 201 East 50th Street, New York, NY 10022. (212) 751- 2600. (800) 733-3000. 1991. $19.50.

Atlas of the Solar System. B. Yenne. Simon & Schuster, Inc. 1230 Avenue of the Americas, New York, NY 10020. (212) 698-7000. (800) 223-2348. 1987. $12.98.

Planetary, Lunar and Solar Positions Six Hundred-One B.C. to A.D. One at Five-Day and Ten-Day Intervals. Bryant Tuckerman. American Philosophical Society, 104 South 5th Street, Philadelphia, PA 19106-3387. (215) 440-3400. FAX (215) 440-3436. Memoirs Series, volume 56. 1979. $20.00.

Planetary, Lunar and Solar Positions, A.D.2 to A.D. 1649 at Five Day and Ten Day Intervals. Bryant Tuckerman. American Philosophical Society, 104 South 5th Street, Philadelphia, PA 19106-3387. (215) 440-3400. FAX (215)440-3436. Memoirs Series, volume 59. 1964. $30.00.

Planetary, Lunar and Solar Positions, 1650 - 1805. Owen Gingerich and Barbara L. Welther. American Philosophical Society, 104 South 5th Street, Philadelphia, PA 19106-3387. (215) 440-3400. FAX (215)440-3436. Memoirs Series, volume59S. 1983. $20.00.

The Sky: A User's Guide. David H. Levy. Cambridge University Press, 40 West 20th Street, New York, NY 10011- 4211. (212) 924-3900. (800) 872-7423. 1993. $14.95.

Space Almanac. Anthony R. Curtis, editor. Gulf Publishing Co., P.O. Box 2608, Houston, TX 77252-2608. (713) 520-4444. (800) 231-6275. 2nd edition. 1992. $24.95 in paper.

The View From Space: Photographic Exploration of the Planets. Merton Davies and Bruce C. Murray. Columbia University Press, 562 West 113th Street, New York, NY 10025. (212) 666-1000. (800) 944-8648. 1973. $17.50.

PERIODICALS

Astronomical Journal. American Institute of Physics, One Physics Ellipse, College Park, MD 20740-3843. (301) 209-3000. Published for the American Astronomical Society. 1849 to present. Monthly. $280.00 per year. ISSN: 0004- 6256. # #*Astronomical Society of the Pacific Publications.* Astronomical Society of the Pacific, 390 Ashton Avenue, San Francisco, CA 94112. (415) 337-1100. FAX (415) 337-5205.1889 to present. Monthly. $175.00 per year. ISSN: 0004- 6280.

Astronomy. Kalmbach Publishing Company, Box 1612, Waukesha, WI 53187-1612. (414) 796-8776. FAX (414) 796-1142. 1973 to present. Monthly. $27.00 per year. ISSN: 0091-6358.

Astronomy and Astrophysics. Springer-Verlag New York, Inc., 44 Hartz Way, Secaucus, NJ 07096-2491. (201) 348-4033. FAX(201) 348-4505. 1969 to date. 36 issues per year in 12 volumes. $1733.00 per year. ISSN: 0004-6361.

Ciel Et Terre. Societe Royale Belge d'Astronomie, de Meteorologie et de Physique du Globe, 3 avenue Circulaire, Uccle, Brusels, Belgium. FAX 32-2-374-9822. 1880 to present. Bimonthly. 1600 BEF per year. ISSN: 0009-6709.

Earth, Moon and Planets; An International Journal of Comparative Planetology. Kluwer Academic Publishers, Box 358, Accord Station, Hingham, MA 02018-0358. (617) 871- 6600. FAX (617) 871-6528. 1969 to present. Monthly. $840.00 per year. ISSN: 0167-9295.

Geochemica Et Cosmochimica Acta. Elsevier Science. 660 White Plains Road, Tarrytown, NY 10591-5153. (914) 524- 9200. FAX (914) 333-2444. 1950 to the present. Biweekly. $895.00 per year. ISSN: 0016-7037.

Icarus; International Journal of Solar System Studies. Academic Press, Inc., Journal Division, 525 B Street, Suite 1900, San Diego, CA 92101-4495. (619) 230-1840. FAX (619) 699-6800. 1962 to the present. Monthly. $1080.00. ISSN: 0019-1035.

J G R: Journal of Geophysical Research: Planets. American Geophysical Union, 2000 Florida Avenue, NW, Washington, CD 20009. (202) 462-6900. FAX (202) 328-0566. 1991 to present. Monthly. $597.00 per year. ISSN: 0148-0227.

SOLAR SYSTEM

Ency. of Physical Sciences and Engineering Info. Sources

Lunar and Planetary Information Bulletin. Lunar and Planetary Institute. 3600 Bay Area Boulevard, Houston, TX 77058-1113. (713) 486-2175. FAX (713) 486-2125. 1970 to present. Quarterly. Free. Also available online.

Mercury. Astronomical Society of the Pacific. 390 Ashton Avenue, San Francisco, CA 94112. (415) 337-1100. FAX: (415) 337-5205. 1972 to present. Bimonthly. $175.00 per year. ISSN: 0047-6773.

Meteoritics. Meteoritical Society, Department of Chemistry, University of Arkansas, Fayetteville, AR 72701. (501) 575-7625. FAX (501) 575-7778. 1953 to present. Bimonthly. $210.00 per year. ISSN: 0026-1114. # #*Observatory.* c/o Dr. D. J. Stickland Space and Astrophysics Division, Rutherford Appleton Laboratory, Chilton, Didcot, Oxon OX11 OQX England. FAX 0235-445848. 1877 to present. Bimonthly. $42.00 per year. ISSN: 0029-7704.

Planetary and Space Science. Elsevier Science Publishing Company, Inc., 660 White Plains Road, Terrytown, NY 10591- 5153. (914) 524-9200. FAX: (914) 333-2444. 1959 topresent. Monthly. $1355.00 per year. ISSN: 0032-0633.

Planetary Report. Planetary Society. 65 North Catalina, Pasadena, CA 91106-2301. (818) 793-5100. FAX (818) 793- 5528. 1980 to present. $25.00 per year. ISSN: 0736-3680.

Sky and Telescope. Sky Publishing Corporation, Box 9111, Belmont, MA 02178. (617) 864-7360. FAX (617) 864-6117.1941 to present. Monthly. $27.00 per year. ISSN: 0037-6604.

RESEARCH CENTERS AND INSTITUTES.

Earth and Planetary Remote Sensing Laboratory. Washington University, Department of Earth and Planetary Sciences. Campus Box 1169, 1 Brookings Drive, St. Louis, MO 63130. (314) 889-5679. FAX (314) 889-5799.

Institute For Astronomy. University of Hawaii at Manoa, 2680 Woodlawn Drive, Honolulu, HI 96822. (808) 956-8312. FAX (808) 988-2790.

Institute For Terrestrial and Planetary Atmospheres. State University of New York at Stony Brook, Stony Brook, NY 11794-2300. (516) 632-6170. FAX (516) 632-6251.

Institute of Geophysics and Planetary Physics. University of California, Riverside, CA 92521. (714) 787-4503. FAX (714) 787-4529.

Laboratory For Planetary Geology. Arizona State University, Department of Geology, Tempe, AZ 85281. (601) 965-7029.

Laboratory For Planetary Studies. Cornell University, Space Sciences Building, Ithaca, NY 14853. (607) 255-4971.

Lunar and Planetary Institute. 3303 NASA Road One, Houston, TX 77058-4399. (713) 486-2139. FAX 713-496-2162.

Lunar and Planetary Laboratory. University of Arizona, Tucson, AZ 85721. (602) 621-6963. FAX (602) 621-4933.

Smithsonian Astrophysical Observatory. 60 Garden Street, Cambridge, MA 02138. (617) 495-7461. FAX (617) 495-7105.

SOLAR WIND

See also: SOLAR SYSTEM; SUN

ABSTRACT SERVICES AND INDEXES

Applied Science and Technology Index; A Cumulative Subject Index To English Language Periodicals in the Fields of Aeronautics and Space Science, Computer Technology, Chemistry, Construction Industry, Energy and Related Areas. H. W. Wilson Co., 950 University Avenue, Bronx, NY 10452- 9978. (212) 588-8400. (800) 367-6770. FAX (718) 590-1617.From 1958 to present. Monthly. Inquire about cost and availability. Also available online (BRS and WILSONLINE) and on CD-ROM. ISSN: 0003-6986.

Astronomy and Astrophysics Abstracts. Springer-Verlag New York, 175 Fifth Avenue, New York, NY 10010. (212) 460-1500. FAX (212) 473-6272. Published for the Astronomisches Rechen-Institut. Comprehensive coverage of all aspects of astronomy, astrophysics and related fields. 1969 to present. Two parts per year. Annual. ISSN: 0067-0022.

Current Contents: Physical, Chemical, and Earth Sciences. Institute for Scientific Information, 3501 Market Street, Philadelphia, PA 19104. (215) 386-0100. (800) 523-1850. FAX (215) 386-2291. 1961 to present. Weekly. $442.00 per year. Also available online (BRS, DIALOG) and on CD- ROM.

General Science Index. H.W. Wilson Company, 950 University Avenue, Bronx, NY 10452. (212) 588-8400. (800) 367- 6770. FAX (718) 590-1617. From 1978 to present. Ten issues per year; quarterly and annual cumulations. Service basis. Available on CD-ROM and online. Inquireregarding cost and availability. ISSN: 0162-1963.

Physics Abstracts. INSPEC. Section A, Science Abstracts. Institute of Electrical Engineers, London, United Kingdom. Available from: INSPEC/IEEE - Institute of Electrical and Electronic Engineers, Box 1331, Hoes Lane, Piscataway, NJ 08855-1331. (908) 562-5549. 1898 to present. 24 issues per year. $2835.00 per year. ISSN: 0036-8091. Also available online and on CD-ROM.

Science Citation Index. SCI. Institute for Scientific Information, 3501 Market Street, Philadelphia, PA 19104. (215) 386-0100. (800) 523-1850. FAX (215) 386-2991. 1961 to present. Six issues per year, plus annual cumulation. $11650.00 per year. Also available online and on CD-ROM. Inquire about price and availability. ISSN: 0036-827X.

STAR. Scientific and Technical Aerospace Reports. U.S. National Aeronautics and Space Administration. Distributed by U. S. Superintendent of Documents, Washington, DC 20402. 1963 to present. Semi-monthly, with semiannual and annual indexes. $114.00 per year. ISSN: 0036-8741. Also available online and on CD-ROM.

ANNUAL REVIEWS AND YEARBOOKS

The Astronomical Almanac. Superintendent of Documents, U. S. Government Printing office, Washington, DC 10402. (202) 783-3238. 1981 to present. Supercedes Astronomical Ephemeris and the American Ephermis and Nautical Almanac. Annual. $44.95. ISSN: 0737-6421.

Ency. of Physical Sciences and Engineering Info. Sources

SOLAR WIND

Annual Review of Astronomy and Astrophysics. Original reviews of critical literature and current developments in astronomy and astrophysics. Annual Reviews, Inc., 4139 El Camino Way, Palo Alto, CA 94303-0139. (415) 493-4400. (800) 523-8635. Fax (415) 855-9815. 1963 to date. Annual. $60.00. ISSN: 0066-4146.

ASSOCIATIONS AND PROFESSIONAL SOCIETIES

Amateur Astronomers Association. 1010 Park Avenue, New York, NY 10028. (212) 535-2922.

American Astronomical Society. 2000 Florida Avenue NW, Suite 400, Washington, DC 20009. (202) 328-2010. FAX\;(202) 234-2560.

American Association of Variable Star Observers. 23 Birch Street, Cambridge, MA 02138. (617) 354-0484. FAX: (617)354-0665.

American Geophysical Union. 2000 Florida Avenue NW, Washington, DC 20009. (202) 462-6900. (800) 966-AGU1. FAX (202) 328-0566.

American Institute of Physics. 1 Physics Ellipse, College Park, MD 20740-3843. (301) 209-3100.

Association of Universities For Research in Astronomy, Inc. (AURA). Suite 701, 1625 Massachusetts Avenue, NW, Washington, DC 20036. (202) 483-2101. FAX (202) 483-2106.

Astronomical League. 2112 Kingfisher Lane East, Rolling Meadows, IL 60008. (708) 398-0562.

Astronomical Society of the Pacific. 390 Ashton Avenue, San Francisco, CA 94112. (415) 337-1100. FAX: (415) 337-5205.

BIBLIOGRAPHIES

Astronomy and Astrophysics: A Source Guide. Gordon Press, P.O. Box 459, Bowling Green Station, New York, NY 10004. (718) 624-8419. 1991. $250.00.

A Bibliography of Astronomy, 1970 - 1979. R. A. Sea and S. S. Martin. Libraries Unlimited, Inc., Littleton, CO 80160. 1982. $37.50.

Science Books & Films. American Association for the Advancement of Science, 1333 H Street NW, Washington, DC 20005. (202) 326-6434. Reviews of print, film and software materials in all sciences. 1965 to present. Nine issues per year. $40.00 per year. ISSN: 0098-342X.

Scientific and Technical Books and Serials in Print; An Index To Literature in Science and Technology. R. R. Bowker. 121 Chanlon Road, New Providence, NJ 07974. (908) 464-6800. FAX (908) 665-3502. (800) 521-8110. 1972 to present. Annual. $299.95 per year. Also available on CD-ROM. ISSN: 0000-054X.

DIRECTORIES AND BIOGRAPHICAL SOURCES

American Astronomical Society. Membership Directory. American Astronomical Society. 2000 Florida Avenue NW, Suite 400, Washington, DC 20009. (202) 328-2010. FAX (202) 234-2560 Annual. Included in membership dues. ISSN: 1061- 9038.

American Men and Women of Science: Physical and Biological Sciences. R. R. Bowker Inc., 121 Chanlon Road, New Providence, NJ 07974. (908) 464-6800. (800) 521-8110. 20th edition. 8 volumes. 1996. $850.00.

The Astronomers. Donald Goldsmith. St. Martin's Press, Inc., 175 Fifth Avenue, New York, NY 10010. (212) 674-5151. (800) 221-7945. 1993. $14.95 in paper.

Astronomical Centers of the World. Kevin Krisciunas. Cambridge University Press, 40 West 20th Street, New York, NY 10011-4211. (212) 924-3900. (800) 872-7423. 1988. $34.95

Directory of Physics and Astronomy Staff. American Institute of Physics, One Physics Ellipse, College Park, MD 20740-3843. (301) 209-3100. 1975/76 to present. Annual. $60.00. ISSN: 0361-2228.

Earth and Astronomical Research Centers. Stockton Press, 345 Park Avenue South, New York, NY 10010. 4th edition. 1995. $515.00. ISBN: 1-56169-0967.

Graduate Programs in Physics, Astronomy and Related Fields. American Institute of Physics, One Physics Ellipse, College Park, MD 20740-3843. (301) 209-3100. 1978 to present. Annual. $45.00. ISSN: 0147-1821.

Research Centers Directory. Gale Research, 835 Penobscot Building, Detroit, MI 48226-4094. (313) 961-2242. (800) 877-4253. 20th edition. 1995. $485.00.

ENCYCLOPEDIAS AND DICTIONARIES

Concise Dictionary of Astronomy. Jacqueline Mitton. Oxford University Press, Inc., 200 Madison Avenue, New York, NY 10016. (212) 725-6000. (800) 334-4249. 1992. $24.95.# #*Encyclopedia of Astronomy and Astrophysics.* Stephen Maran, editor. Van Nostrand Reinhold, 115 Fifth Avenue, New York, NY 10003. (212) 254-3232. (800) 842-3636. 1992. $129.95.

McGraw-Hill Encyclopedia of Science and Technology. McGraw-Hill Book Company, Inc., 1221 Avenue of the Americas, New York, NY 10020. (212) 512-2000. (800) 262-4729. 7th edition. 20 volume set. 1992. $1900.00.

GENERAL WORKS

Astrophysics of the Sun. Harold Zinn. Cambridge University Press, 40 West 20th Street, New York, NY 10011-4211. (212) 924-3900. (800) 872-7423. 1988. $34.95 in paper.

Basic Plasma Processes On the Sun. E. R. Priest and V. Krishan, editors. Kluwer Academic Publishers, 101 Philip Drive, Assinippi Park, Norwell, MA 02061. (617) 871-6600. 1990. $144.00.

Cool Stars, Stellar Systems and the Sun. Jeffrey L. Linsky and R. E. Stencel, editors. Springer-Verlag New York, Inc., 175 Fifth Avenue, New York, NY 10010. (212) 460-1500. (800) 777-4643. 1988. $66.00.

Discovering the Secrets of the Sun. Rudolph Kippenhahn. John Wiley & Sons, Inc., 605 Third Avenue, New York, NY 10158-0012. (212) 850-6000. (800) 225-5945. 1994. $39.95in paper.

SOLAR WIND

Ency. of Physical Sciences and Engineering Info. Sources

Future of the Sun. Jean-Claude Pecker. McGraw-Hill Publishing Company, Inc., 1221 Avenue of the Americas, New York, NY 10020. (212) 512-2000. (800) 262-4729. 1992. $11.95.

Solar and Planetary Dynamics. M. R. Proctor. Cambridge University Press, 40 West 20th Street, New York, NY 10011- 4211. (212) 924-3900. (800) 872-7423. 1994. $9.95.

Solar Interior and Atmosphere. Arthur N. Cox et al, editors. University of Arizona Press, 1230 North Park Avenue, Number 102, Tucson, AZ 85719. (520) 621-1441. (800) 426-3797. 1992. $75.00.

Solar Magnetic Fields. Manfred Schussler and Wolfgang Schmidt, editors. Cambridge University Press, 40 West 20th Street, New York, NY 10011-4211. (212) 924-3900. (800) 872-7423. 1994. $59.95.

Solar Observations: Techniques and Interpretation. F. Sanches, et al, editors. Cambridge University Press, 40 West 20th Street, New York, NY 10011-4211. (212) 924-3900. (800) 872-7423. 1991. $79.95.

Solar System Plasma in Space and Time. J. Burch and J. H. Waite Jr., editors. American Geophysical Union, 2000 Florida Avenue, N.W., Washington, DC 20009. (202) 462-6903. (800) 966-2481. 1994. $57.00

The Solar Wind and the Earth. Syun-Ichi Akasofu and Y. Kamide, editors. Kluwer Academic Publishers, 101 Philip Drive, Assinippi Park, Norwell, MA 02061. (617) 871-6600. 1987. $163.50.

Solar Wind Sources of Magnetospheric Ultra-low Frequency Waves. M. Engebretson, et al editors. American Geophysical Union, 2000 Florida Avenue, N.W., Washington, DC 20009. (202) 462-6903. (800) 966-2481. 1994. $80.00.

Sun and Earth. H. Freidman. W. H. Freeman, 41 Madison Avenue, East 26th Avenue, 35th Floor, New York, NY 10010. (212) 576-0400. 1995.

Venus and Mars: Atmospheres, Ionospheres, and Solar Wind Interactions. J. G. Luhmann, et al, editors. American Geophysical, 2000 Florida Avenue, N.W., Washington, DC 20009. (202) 462-6903. (800) 966-2481. 1992. $59.00.

HANDBOOKS AND MANUALS

Astrophysical Data: Planets and Stars. Kenneth R. Lanb. Springer-Verlag New York, Inc., 175 Fifth Avenue, New York, NY 10010. (212) 460-1500. (800) 777-4643. 1993. $59.00.

Landolt-borenstein Numerical Data and Functional Relationships in Science and Technology: Astronomy, Astrophsyics and Space Research. GROUP VI. Springer-Verlag New York, Inc., 175 Fifth Avenue, New York, NY 10010. (212) 460-1500. (800) 777-4643. volumes priced individually.

Observing the Sun. Peter O. Taylor. Cambridge University Press, 40 West 20th Street, New York, NY 10011-4211. (212) 924-3900. (800) 872-7423. 1992. $29.95

ONLINE DATABASES AND CD-ROMS

Dissertation Abstracts. University Microfilms International, 300 North Zeeb Road, Ann Arbor, MI 48106. (800) 521-0600 or (313) 761-4700. Scope includes virtually all doctoral dissertations accepted at accredited American institutions from 1861 to present in 252 subject areas. Available on BRS, (800) 289-4277, DIALOG, (800) 334-2564, and OCLC EPIC, (800) 848-5878, online services. Also available on CD-ROM. Inquire as to cost and availability.

INSPEC. Institution of Electrical Engineers, Michael Faraday House, Six Hills Way, Stevenage, Herts. SG1 2AY, England. Telephone: 0438 313311 or FAX 0438 742840. Contains citations to the worldwide literature of physics, electronics and electrical engineering, computer technology, and related fields. Available on BRS, (800) 289-4277, DIALOG, (800) 334-2564, ORBIT, (800) 456-7248, and STN International, FIZ Karlsruhe, P.O. Box 2465, W-7500, Karlsruhe 1, Germany, online services. Inquire as to cost and availability.

NTIS Bibliographic Database. National Technical Information Service, 5285 Port Royal Road, Springfield, VA 22161. (703) 487-4929 or FAX (703) 321-8199. Broad coverage of government-sponsored science and technology research reports, 1964 to present. Available on BRS,(800) 289-4277, DIALOG, (800) 334-2564, ORBIT, (800) 456-7248, and STN International, FIZ Karlsruhe, P.O. Box 2465, W-7500, Karlsruhe 1, Germany, online services. Also available on CD-ROM. Inquire as to cost and availability.

SCISEARCH. Institute for Scientific Information, 3501 Market Street, Philadelphia, PA 19104. (800) 523-1850 or (215) 386-0100. Broad multidisciplinary title and author index to the international literature of science and technology, 1974 to present. Available on DIALOG, (800) 334-2564, and ORBIT, (800) 456-7248, online services. Also available on CD-ROM. Inquire as to cost and availability.

WILSONLINE. H.W. Wilson Company, 950 University Avenue, Bronx, NY 10452. (800) 367-6770 or (212) 588-8400. Makes available online versions of the printed H.W. Wilson indexes including Applied Science and Technology Index, Business Periodicals Index, General Science Index, and Readers' Guide to Periodical Literature. Period covered is generally 1983 to present. Available on BRS, (800) 289-4277, DIALOG, (800) 334-2564, and OCLC EPIC, (800) 848-5878, online services.Also available on CD-ROM. Inquire as to cost and availability.

PERIODICALS

Astronomical Journal. American Institute of Physics, One Physics Ellipse, College Park, MD 20740-3843. (301) 209-3000. Published for the American Astronomical Society. 1849 to present. Monthly. $280.00 per year. ISSN: 0004- 6256.

Astronomical Society of the Pacific Publications. Astronomical Society of the Pacific, 390 Ashton Avenue, San Francisco, CA 94112. (415) 337-1100. FAX (415) 337-5205. 1889 to present. Monthly. $175.00 per year. ISSN: 0004- 6280.

Astronomy. Kalmbach Publishing Company, Box 1612, Waukesha, WI 53187-1612. (414) 796-8776. FAX (414) 796-1142. 1973 to present. Monthly. $27.00 per year. ISSN: 0091-6358.

Astronomy and Astrophysics. Springer-Verlag New York, Inc., 44 Hartz Way, Secaucus, NJ 07096-2491. (201) 348-4033. FAX (201) 348-4505. 1969 to date. 36 issues per year in 12 volumes. $1733.00 per year. ISSN: 0004-6361.

Astronomy Reports. American Institute of Physics, One Physics Ellipse, College Park, MD 20740-3843. (301) 209- 3000. English translation of Astronomicheksii Zhurnal. Formerly Soviet Astronomy AJ. 1957 to present. Bimonthly. $1160.00 per year. ISSN: 1063-7729.

Astrophysical Journal; An International Review of Astronomy and Astronomical Physics. University of Chicago Press, Journals Division, 5720 South Woodlawn Avenue, Chicago, IL 60637. (312) 753-3347. FAX (312) 753-0811. 1895 to date. three issues per month. $740.00 per year. ISSN: 0004-637X.

Astrophysics and Space Science; An International Journal of Cosmic Physics. Kluwer Academic Publishers, Box 358, Accord Station, Hingham, MA 02018-0358. (617) 871-6600. FAX (617) 871-6528. 1968 to present. 24 issues per year. $2544.00 per year. ISSN: 0004-640X.

Icarus; International Journal of Solar System Studies. Academic Press, Inc., Journal Division, 525 B Street, Suite 1900, San Diego, CA 92101-4495. (619) 230-1840. FAX (619) 699-6800. 1962 to the present. Monthly. $1080.00. ISSN: 0019-1035.

Mercury. Astronomical Society of the Pacific. 390 Ashton Avenue, San Francisco, CA 94112. (415) 337-1100. FAX: (415) 337-5205. 1972 to present. Bimonthly. $175.00 per year. ISSN: 0047-6773.

Monthly Notices of the Royal Astronomical Society. Blackwell Scientific Publication LT., Osney Mead, Oxford OX2 OEL, England. TEL 0865-240201. FAX 0865-721205. 1827 to present. Fortnightly. $1899.00 per year. ISSN: 0035-8711.

Observatory. c/o Dr. D. J. Stickland Space and Astrophysics Division, Rutherford Appleton Laboratory, Chilton, Didcot, Oxon OX11 OQX England. FAX 0235-445848. 1877 to present. Bimonthly. $42.00 per year. ISSN: 0029-7704.

Planetary and Space Science. Elsevier Science Publishing Company, Inc., 660 White Plains Road, Terrytown, NY 10591- 5153. (914) 524-9200. FAX: (914) 333-2444. 1959 topresent. Monthly. $1355.00 per year. ISSN: 0032-0633.

Solar-geophysical Data, Part 1 (prompt Reports) and 2 (comprehensive Reports). National Geophysical Data Center, 325 Broadway, NOAA E/GC2, Boulder, CO 80303. (303) 497-6135. 1955 to present. Monthly. $60.00 for each part.

Solar Physics; A Journal For Solar and Solar-stellar Research and the Study of Solar Terrestrial Physics. Kluwer Academic Publishers, Box 358, Accord Station, Hingham, MA 02018-0358. (617) 871-6600. FAX (617) 871-6528. 1967 topresent. Fourteen issues per year. $1592.00 per year. ISSN: 0038-0938.

Space Science Reviews. Kluwer Academic Publishers, Box 358, Accord Station, Hingham, MA 02018-0358. (617) 871- 6600. FAX (617) 871-6528. 1962 to present. Sixteen issues per year. $940.00 per year. ISSN: 0038-6308.

Sky and Telescope. Sky Publishing Corporation, Box 9111, Belmont, MA 02178. (617) 864-7360. FAX (617) 864-6117. 1941 to present. Monthly. $27.00 per year. ISSN: 0037- 6604.

Vistas in Astronomy; An International Review Journal. Albert C. Beer, editor. Elsevier Science Publishing Company, Inc., 660 White Plains Road, Terrytown, NY 10591- 5153. (914) 524-9200. FAX: (914) 333-2444. 1958 to present. Quarterly. $415.00 per year. ISSN: 0083-6656.

RESEARCH CENTERS AND INSTITUTES

Astronomy Program. University of Maryland, 1207 Computer & Space Sciences Building, College Park, MD 20742. (301) 405- 3001. FAX (801) 314-9067.

Center For Space Plasma and Aeronomic Research. University of Alabama in Huntsville, Huntsville, AL 35899. (205) 895-6268. (205) 895-6382.

High Altitude Observatory. National Center for Atmospheric Research, P.O. Box 3000, Boulder, CO 80307. (303) 497-1500.

Lunar and Planetary Laboratory. University of Arizona, Tucson, AZ 85721. (602) 621-6963. (602) 621-4933.

National Solar Observatory, 950 North Cherry Avenue, Tucson, AZ 85719. (602) 325-9294. FAX (602) 325-9278.

Space Science Center. University of New Hampshire, Science and Engineering Research Building, Durham, NH 03824. (603) 962-2751.

Smithsonian Astrophysical Observatory, 60 Garden Street, Cambridge, MA 02138. (617) 495-7461. FAX (617) 495-7105.

SOLID STATE CHEMISTRY

See: PHYSICAL CHEMISTRY SOLID STATE PHYSICS

SOLID STATE PHYSICS

See also: MATERIALS SCIENCE, PHYSICS

ABSTRACT SERVICES AND INDEXES

Acoustics Abstracts. Parts A & B. Multi-Science Publishing Co., Ltd, Box 176, Avenel, NJ 07001. 1967 to present. Monthly. L265 per year. ISSN: 0001-4974.

Applied Science and Technology Index; A Cumulative Subject Index To English Language Periodicals in the Fields of Aeronautics and Space Science, Computer Technology, Chemistry, Construction Industry, Energy and Related Areas. H.W. Wilson Co., 950 University Avenue, Bronx, NY 10452. (212) 588-8400. (800) 367-6770. FAX (718) 590-1617. From 1958 to present. Monthly. Inquire about cost and availability. Also available on CD-ROM and online. ISSN: 0003-6986.

Chemical Abstracts. Chemical Abstracts Service, 2540 Olentangy River Road, Box 3012, Columbus, OH 43210-0012. (614) 447-3600. (800) 848-6538. FAX (614) 447-3713.From 1907 to present. Weekly. $16800.00 per year. Also available on CD-ROM and online. Inquire regarding cost and availability.

SOLID STATE PHYSICS

Ency. of Physical Sciences and Engineering Info. Sources

Current Contents: Physical, Chemical and Earth Sciences. Institute for Scientific Information, 3501 Market Street, Philadelphia, PA 19104. (215) 386-0100. FAX (215) 386-2291. From 1961 to present. Weekly. $442.00 per year. Also available on CD-ROM and online. Inquire regarding cost and availability. ISSN: 0163-2574.

Current Papers in Physics. Institute of Electrical Engineers, Michael Faraday House, Six Hill Way, Stevenage, Herts, SG1 2AY, England. Distributed by INSPEC/IEEE, Box 1331, 445 Hoes Lane, Piscataway, NJ 08855-1331. (908) 562-5549. 1966 to present. Fortnightly. $410.00 per year. ISSN: 0011-3786.

General Science Index. H.W. Wilson Company, 950 University Avenue, Bronx, NY 10452. (212) 588-8400. (800) 367-6770. FAX (718) 590-1617. From 1978 to present. Ten issues per year; quarterly and annual cumulations. Service basis. Available on CD-ROM and online. Inquire regarding cost and availability. ISSN: 0162-1963.

Government Reports Announcements and Index. U. S. National Technical Information Service (NTIS), 5285 Port Royal Road, Springfield, VA 22161. (703) 487-4650. FAX (703) 321-8547. From 1968 to present. Annual. $630.00 per year. Also available online as NTIS Bibliographic Database and on CD-ROM.

Journal of Current Laser Abstracts. PennWell Publishing Co., 10 Tara Boulevard, Nashua, NH 03062-2801. (603) 891-9177. FAX (603) 891-0539. From 1964 to present. Monthly. $450.00 per year. ISSN: 0022—0264.

Mathematical Reviews: A Review Journal Covering the World Literature of Mathematical Research. American Mathematical Society, P.O. Box 1571, Annex Station, Providence, RI 02901-9930. (800) 556-7774 or (401) 455-4000. 1940 to present. Monthly. $4594.00 per year. Also available via network (MATHSCINET), online and on CD-ROM. Inquire regardingcost and availability. ISSN: 0025-5629.

Meteorological and Geoastrophysical Abstracts. American Meteorological Society, c/o Inforonics, Inc., 550 Newtown Road, Littleton, MA 01460. (508) 486-8976. FAX (508) 486-0027. Covers literature in environmental sciences, meteorology, astrophysics, hydrology, glaciology, and physical oceanography. From 1950 to present. Monthly. $950.00 per year. Also available on CD-ROM and online. Inquire regarding cost and availability. ISSN: 0026-1130.

NTIS Alerts: Physics. U. S. National Technical Information Service. 5285 Port Royal Road, Springfield, VA 22161. (703) 487-4650. FAX (703) 321-8547. Weekly. $140.00 per year.

Pascal 10: Mechanique, Acoustique Et Transport DE Chaleur. Centre National de la Recherche Scientifique, Institue de l'Information Scientifique et Technique, 2 aliee du Parc de Brabois, 54514 Vandoeuvre-Les Nancy Cedex, France. TEC 83-50-46-00. FAX 83-50-46-50. 1984 to present. Ten issues per year. 1530 F per year. Also available on CE-ROM and online. ISSN: 1136-5107.

Physics Abstracts. INSPEC. Section A, Science Abstracts. Institution of Electrical Engineers (IEE). Available from INSPEC/IEEE - Institute of Electrical and Electronic Engineers,, Box 1331, 445 Hoes Lane, Piscaway NJ 08855-1331. (908) 562-5549. 1898 to present. 24 issues per year. $2835.00 per year. Also available on CD-ROM and online. ISSN: 0036-8091.

Physics Briefs (Physikalische Berichte). Information Center for Energy, Physics, Mathematics; German Physical Society. V C H Publishers, Inc., 220 East 23rd Street, New York, NY 10010-4606. (212) 683-8333. 1845 to present. 24 issues per year. $2390.00 per year. Also available online. ISSN: 0179-7434.

Science Citation Index. SCI. Institute for Scientific Information, 3501 Market Street, Philadelphia, PA 19104. (215) 386-0100. (800) 523-1850. FAX (215) 386-2991. 1961 to present. Six issues per year, plus annual cumulation. $11650.00 per year. Also available online and on CD-ROM. Inquire about price and availability. ISSN: 0036-827X.

Solid State and Superconductivity Abstracts: Covers theory, Production and Application of Solid State Materials. Cambridge Scientific Abstracts, 7200 Wisconsin Avenue, Bethesda, MD 20824. (301) 961-6750. FAX (301) 961-6720. 1957 to present. Bimonthly. $1320.00 per year. Also available online. ISSN: 0896-5900.

ANNUAL REVIEWS AND YEARBOOKS

Advances in Applied Mechanics. Academic Press, Inc., 6277 Sea Harbor Drive, Orlando, FL. (800) 321-5068. 1948 to present. Irregular. ISSN: 0065-2156.

Advances in Atomic, Molecular and Optical Physics. Academic Press, Inc., 6277 Sea Harbor Drive, Orlando, FL. (800) 321-5068. 1965 to present. Irregular. ISSN: 1049-250X. # #*Advances in Physics.* Taylor & Francis, Ltd., Rankine Road, Basingstoke, Hants RG24 8PR, England. 0256-840366. FAX 0256-47938. 1952 to present. Bimonthly. $511.00 per year. ISSN: 0001-8732.

Advances in Solid State Technology. Kluwer Academic Publishers, 101 Philip Drive, Assinippi Park, Norwell, MA 02061. (617) 871-6600. 1985 to present. Irregular. Price varies.

Annual Review of Nuclear and Particle Science. Annual Reviews, Inc., 4139 El Camino Way, Palo Alto, CA 94303-0139. (415) 493-4400. (800) 523-8635. Fax (415) 855-9815. 1952 to present. Annual. $62.00.

Critical Reviews in Solid State and Materials Sciences. CRC Press, Inc., 2000 Corporate Boulevard, NW, Boca Raton, FL 33431. (407) 994-0555. (800) 272-7737. FAX (407) 998-9784. 1970 to present. Bimonthly. $265.00 per year. ISSN: 1040-8436.

Progress in Crystal Growth and Characterization. Elsevier Science Inc., 660 White Plains Road, Tarrytown, NY 10591-5153. (914) 524-9200. FAX (914) 333-2444.

Progress in Quantum Electronics. Elsevier Science Inc., 660 White Plains Road, Tarrytown, NY 10591-5153. (914) 524-9200. FAX (914) 333-2444. 1969 to present. 5 issues per year. $395.00 per year. ISSN: 0079-6727.

Solid State Physics: Advances in Research and Applications. Academic Press, Inc., 6277 Sea Harbor Drive, Orlando, FL. (800) 321-5068. 1955 to present. Irregular. ISSN: 0081-1947.

Topics in Applied Physics. Springer-Verlag New York, Inc., 175 Fifth Avenue, New York, NY 10010. (212) 460-1500. (800) 777-4643. FAX (212) 473-6272. ISSN: 0303-4216.

Ency. of Physical Sciences and Engineering Info. Sources

SOLID STATE PHYSICS

ASSOCIATIONS AND PROFESSIONAL SOCIETIES

American Institute of Physics. One Physics Ellipse, College Park, MD 20740-3843. (301) 209-3100.

American Physical Society. One Physics Ellipse, College Park, MD 20740-3843. (301) 209-3200. FAX (301) 209-0865.

American Society of Mechanical Engineers. 345 East 47th Street, New York, NY 10017. (212) 705-7722.

IEEE (Institute of Electrical and Electronics Engineers) 345 East 47th Street, New York, NY 10017. (212) 705-7900. FAX (212) 705-4929.

Universities Research Association, 1111 19th Street Nw, Suite 400, Washington, DC 20036. (202) 293-1382.

BIBLIOGRAPHIES

Guide To the Archival Collectins in the Niels Bohr Libary At the American Institute of Physics. Niels Bohr Library Staff. American Institute of Physics, One Physics Ellipse, College Park, MD 20740-3843. (301) 209-3100. 1995. $140.00

Science Books and Films. American Association for the Advancement of Science, 1333 H Street NW, Washington, DC 20005. (202) 326-6454. Reviews of print, film and software materials in all sciences. 1965 to present. Nine issues per year. $40.00 per year. ISSN: 0098-342X

Scientific and Technical Books and Serials in Print, 1995. an index to literature in science and technology. R.R. Bowker Inc., 121 Chanlon Road, New Providence, NJ 07974. (908) 464-6800. (800) 521-8110. 4 volumes. Annual. 1972 to present. $299.95. Also available on compact disc and online. ISSN: 0000-054X

DIRECTORIES AND BIOGRAPHICAL SOURCES

American Association of Physics Teachers. Directory of members. American Association of Physics Teachers. 5112 Berwyn Road, College Park, MD 20704. (301) 345-4200. Triennal. Included in membership.

American Men and Women of Science: Physical and Biological Sciences. R. R. Bowker Inc., 121 Chanlon Road, New Providence, NJ 07974. (908) 464-6800. (800) 521-8110. 20th edition. 8 volumes. 1996. $850.00.

American Physical Society. Membership Directory Bulletin. American Physical Society, One Physics Ellipse, College Park, MD 20740-3843. (301) 209-3200. FAX (301) 209-0865. Biennial. $50.00.

The Biographical Dictionary of Scientists: Physicists. D. Abbott, editor. Peter Bedrick Books, Inc., 2112 Broadway, Room 318, New York, NY 10023. (212) 496-0751. 1984.

Directory of Physics and Astronomy Staff Members. American Institute of Physics, One Physics Ellipse, College Park, MD 20740-3843. (301) 209-3100. Annual. $45.00.

Graduate Programs in Physics, Astronomy and Related Fields. 1995-1996. American Institute of Physics, One Physics Ellipse,

College Park, MD 20740-3843. (301) 209-3100. Annual. $45.00. ISSN: 0147-1821.

Legends in their Own Time: A Century of American Physical Scientists. Anthony Serafini. Plenum Publishing Corp., 233 Spring Street, New York, NY. (212) 620-8000. (800) 2221-9369. FAX (212) 463-0742. 1993. $27.50. # #*The Nobel Prize Winners.* Frank N. Magill, editor. Salem Press, Inc., P.O. Box 1097, Englewood Cliffs, NJ 07632. (201) 871-3700. (800) 221-1592. 3 volumes. 1989. $210.00.

Research Centers Directory. Gale Research, 835 Penobscot Building, Detroit, MI 48226-4094. (313) 961-2242. (800) 877-4253. 20th edition. 1995. $485.00. ISSN: 0080- 1518.

Scientific and Technical Organizations and Agencies Directory. Gale Research, 835 Penobscot Building, Detroit, MI 48226-4094. (313) 961-2242. (800) 877-4253. 4th edition. 1996. $195.00.

World Guide To Scientific Associations and Learned Societies. K. G. Saur, 121 Chanlon Road, New Providence, NJ 07974. (908) 464-6800. (800) 521-8110. 5th edition. 1990. $225.00.

ENCYCLOPEDIAS AND DICTIONARIES

A Concise Dictionary of Physics. Oxford University Press, Inc., 200 Madison Avenue, New York, NY 10016. (212) 725-6000. (800) 334-4249. 1990. $10.95.

Concise Encyclopedia of Solid State Physics. Rita G. Lerner and George L. Trigg, editors. Addison-Wesley Publishing Co., Inc., 1 Jacob Way, Reading, MA 01867. (617) 944-3700. (800) 447-2226. 1983. $40.95 in paper.

Dictionary of Effects and Phenomena in Physics. Joachim Schubert. VCH Publications, Inc., 220 East 23rd Street, Suite 909, New York, NY 10010-4606. (212) 683-8333. (800) 422-8824. 1987. $35.00.

Dictionary of the Physical Sciences: Terms, Formulas, Data. Cesare Emiliani. Oxford University Press, Inc., 200 Madison Avenue, New York, NY 10016. (212) 725-6000. (800) 334-4249.1989. $19.95.

Encyclopedia of Applied Physics. George Trigg, editor. VCH Publications, Inc., 220 East 23rd Street, Suite 909, New York, NY 10010-4606. (212) 683-8333. (800) 422-8824. 20 volume set. 1991-. $5990.00.

Encyclopedia of Physics. Robert M. Besancon. Chapman & Hall, 1 Penn Plaza, New York, NY 10119. (212) 564-1060. 3rd edition. 1990. $54.95 in paper.

Facts On File Dictionary of Physics. John Daintith, editor. Facts-on-File, Inc., 460 Park Avenue South, New York, NY 10016-7382. (212) 683-2244. (800) 322-8755. Fax (800) 683-3633. Revised edition. 1990. $12.95.

Encyclopedia of Physical Science and Technology. Academic Press, Inc., 6277 Sea Harbor Drive, Orlando, FL. (800) 321-5068. 2nd edition. 18 volume set. 1992. $2625.00.

Landolt-boernstein Numerical Data and Functional Relationships in Science and Technology: Nuclear and Particle Physics. K. H. Hellwege editor. Springer Verlag New York, Inc., 175 Fifth Avenue, New York, NY 10010. (212) 460-1500. (800)

777-4643. FAX (212) 473-6272. 1993. Volumes priced separately.

McGraw-Hill Encyclopedia of Science and Technology. McGraw-Hill Book Company, Inc., 1221 Avenue of the Americas, New York, NY 10020. (212) 512-2000. (800) 262-4729. 7th edition. 20 volume set. 1992. $1900.00.

Penguin Dictionary of Physics. Valerie Illingworth, editor. Viking Penguin, 375 Hudson Street, New York, NY 10014-3657. (212) 366-2000. (800) 331-4624 2nd edition. 1991.

GENERAL WORKS

Electronic Structure of Materials. Adrian D. Sutton. Oxford University Press, Inc., 200 Madison Avenue, New York, NY 10016. (212) 725-6000. (800) 334-4249. 1993. $26.50 in paper.

Electronic Properties of Materials: An Introduction For Engineers. Rolf E. Hummel. Springer-Verlag New York, Inc., 175 Fifth Avenue, New York, NY 10010. (212) 460-1500. (800) 777-4643. FAX (212) 473-6272. 2nd edition. 1994. $49.00.

Elements of Solid State Physics. M.N. Rudden and J. Wilson. John Wiley & Sons, Inc., 605 Third Avenue, New York, NY 10158-0012. (212) 850-6000. (800) 225-5945. 2nd edition. 1993. $36.95 in paper.

Fundamental Particles and Forces. Richard A. Carrigan, Jr., and W. Peter Trower. W. H. Freeman & Co., 41 Madison Avenue, East 26th Avenue, 35th Floor, New York, NY 10010. (212) 576-0400. 1995. $13.95.

The History of Solid State Physics. Lillian Hoddeson. Oxford University Press, Inc., 200 Madison Avenue, New York, NY 10016. (212) 725-6000. (800) 334-4249. 1992. $75.00

Introduction to Solid State Physics. Charles Kittel. John Wiley & Sons, Inc., 605 Third Avenue, New York, NY 10158-0012. (212) 850-6000. (800) 225-5945. 6th edition. 1986.

Introduction To the Physics of Electrons in Solids. Brian K. Tanner. Cambridge University Press, 40 West 20th Street, New York, NY 10011-4211. (212) 924-3900. (800) 872-7423. 1995. $24.95 in paper.

Molecular Solid State Physics. G. G. Hall. Springer-Verlag New York, Inc., 175 Fifth Avenue, New York, NY 10010. (212) 460-1500. (800) 777-4643. FAX (212) 473-6272. 1991. $32.50 in paper.

Physics. David Haliday and Robert Resnick. John Wiley & Sons, Inc., 605 Third Avenue, New York, NY 10158-0012. (212) 850-6000. (800) 225-5945. 2 volumes. 4th edition. 1992.

Quantum theory of the Solid State. Joseph Callaway. Academic Press, Inc., 6277 Sea Harbor Drive, Orlando, FL. (800) 321-5068. 2nd edition. 1991. $69.95.

Solid State Physics. J. R. Hook and H. E. Hall. John Wiley & Sons, Inc., 605 Third Avenue, New York, NY 10158-0012. (212) 850-6000. (800) 225-5945. 2nd edition. 1993. $35.50 in paper.

Solid State Physics For Engineering and Materials Science. John P. McKelvey. Krieger Publishing Company, P.O. Box 9542, Melbourne, FL 32902-9542. (407) 724-9542. 1993. $49.50.

HANDBOOKS AND MANUALS

Chemical and Physical Data. Arthur M. James and M. P. Lord, Van Nostrand Reinhold, 115 Fifth Avenue, New York, NY 10003. (212) 254-3232. (800) 842-3636. 1992. $59.95.

Compendium of thermophysical Property Measurements Methods. Kosta D. Maglic, et al, editors. Plenum Publishing Corp., 233 Spring Street, New York, NY. (212) 620-8000. (800) 2221-9369. FAX (212) 463-0742. 1991. $165.00.

CRC Handbook of Chemistry and Physics. David R. Kide, editor. CRC Press, Inc., 2000 Corporate Boulevard, NW, Boca Raton, FL 33431. (407) 994-0555. (800) 272-7737. 77th edition. 1996. $99.95.

Handbook of Physical Quantities. Igor S. Grigoriev and Evgenil Z. Melikhov, editors. CRC Press, Inc., 2000 Corporate Boulevard, NW, Boca Raton, FL 33431. (407) 994- 0555. (800) 272-7737. 1995. $99.00.

Landolt-borenstein Numerical Data and Functional Relationships in Science and Technology: Solid State Physics. Group III. Springer-Verlag New York, Inc., 175 Fifth Avenue, New York, NY 10010. (212) 460-1500. (800) 777-4643. volumes priced individually.

McGraw-Hill Encyclopedia of Physics. Sybil P. Parker, editor. McGraw-Hill Publishing Company, Inc., 1221 Avenue of the Americas, New York, NY 10020. (212) 512-2000. (800) 262-4729. 2nd edition. 1993. $95.50

Newnes Engineering and Physical Science Pocketbook. J. O. Bird and P. J. Chivers. Butterworth-Heinmann, 313 Washington Avenue, Newton, MA 02158. (617) 928-2500. (800) 366-2665. 1993. $49.95. # #*A Physicist's Desk Reference.* Herbert L. anderson, editor. American Institute of Physics, One Physics Ellipse, College Park, MD 20740-3843. (301) 209-3100. 2nd edition. 1989. $70.00.

Tables of Physical and Chemical Constants and Some Mathematical Functions. G. W. Kaye and T. M. Laby. Halsted Press 605 Third Avenue, New York, 10158-0012. (212) 850- 6400. 15th edition. 1986. $74.95.

ONLINE DATABASES AND CD-ROMS

CA Search. Chemical Abstracts Service, P.O. Box 3012, Columbus, OH 43210-0012. (614) 447-3600. (800) 848-6533. FAX (614) 447-3709. Very comprehensive guide to worldwide chemical literature and related fields, 1972 to present. Available on BRS,(800) 289-4277, DIALOG, (800) 334-2564, ORBIT (800) 456-7248, and STN International, FIZ Karlsruhe, P.O. Box 2465, W-7500, Karlsruhe 1, Germany, online services. Inquire as to cost and availability.

Current Contents Search. Institute for Scientific Information, 3501 Market Street, Philadelphia, PA 19104. (215) 386-0100. FAX (215) 386-6362. Contains citations to articles listed in the table of contents of science and technology journals. Also articles in social sciences and life sciences journals. Available on

BRS, (800) 289-4277, DIALOG, (800) 334-2564, online services. Inquire as to cost and availability.

Dissertation Abstracts Online. University Microfilms International, 300 North Zeeb Road, Ann Arbor, MI 48106. (800) 521-0600 or (313) 761-4700. Scope includes virtually all doctoral dissertations accepted at accredited American institutions from 1861 to present in 252 subject areas. Available on BRS, (800) 289-4277, DIALOG, (800) 334-2564,and OCLC EPIC, (800) 848-5878, online services. Also available on CD-ROM. Inquire as to cost and availability.

INSPEC. Institution of Electrical Engineers, Michael Faraday House, Six Hills Way, Stevenage, Herts. SG1 2AY, England. Telephone: 0438 313311 or FAX 0438 742840. Contains citations to the worldwide literature of physics, electronics and electrical engineering, computer technology, and related fields. Available on BRS, (800) 289-4277, DIALOG, (800) 334-2564, ORBIT, (800) 456-7248, and STN International, FIZ Karlsruhe, P.O. Box 2465, W-7500, Karlsruhe 1, Germany, online services. Inquire as to cost and availability.

Mathsci and Mathscinet. American Mathematical Society, P.O. Box 6248, Providence, RI 02940. (800) 321-4667 or (401) 455-4000 or FAX (401) 331-3842. Scope includes pure and applied mathematics and related areas of physics, statistics, engineering, computer science, and operations research literature since 1959. Available on DIALOG,(800) 334-2564, online service. Also available on CD-ROM. Inquire as to cost and availability.

NTIS Bibliographic Database. National Technical Information Service, 5285 Port Royal Road, Springfield, VA 22161. (703) 487-4929 or FAX (703) 321-8199. Broad coverage of government-sponsored science and technology research reports, 1964 to present. Available on BRS,(800) 289-4277, DIALOG, (800) 334-2564, ORBIT, (800) 456-7248, and STN International, FIZ Karlsruhe, P.O. Box 2465, W-7500, Karlsruhe 1, Germany, online services. Also available on CD-ROM. Inquire as to cost and availability.

Physics Briefs. American Institute of Physics, 335 East 45th Street, New York, NY 10017. (212) 661-9260 or FAX (212) 661-2036. Contains citations with abstracts of the literature of physics and related fields, 1979 to present. Available on the STN International, FIZ Karlsruhe, P.O. Box 2465, W-7500, Karlsruhe 1, Germany, online service. Inquire as to cost and availability.

Scientific and Technical Books and Serials in Print. R.R. Bowker Inc., 121 Chanlon Road, New Providence, NJ 07974. (800) 521-8110 or (908) 464-6800. List of currently published books and serials in the physical and biological sciences, engineering and technology, with subject, author and titles indexes. Available on ORBIT, (800) 456-7248,online service. Inquire as to cost and availability.

SCISEARCH. Institute for Scientific Information, 3501 Market Street, Philadelphia, PA 19104. (800) 523-1850 or (215) 386-0100. Broad multidisciplinary title and author index to the international literature of science and technology, 1974 to present. Available on DIALOG, (800) 334-2564, and ORBIT, (800) 456-7248, online services. Also available on CD-ROM. Inquire as to cost and availability.

WILSONLINE. H.W. Wilson Company, 950 University Avenue, Bronx, NY 10452. (800) 367-6770 or (212) 588-8400. Makes available online versions of the printed H.W. Wilson indexes including Applied Science and Technology Index, Business Periodicals Index, General Science Index, and Readers' Guide to Periodical Literature. Period covered is generally 1983 to

present. Available on BRS, (800) 289-4277, DIALOG, (800) 334-2564, and OCLC EPIC, (800) 848-5878, online services.Also available on CD-ROM. Inquire as to cost and availability.

PERIODICALS

American Journal of Physics. American Association of Physics Teachers, 5112 Berywn Road, College Park, MD 20740. (301) 345-4200. 1933 to present. Monthly. $215.00 per year. ISSN: 0002-9505.

Applied Physics A: Solids and Surfaces. Springer-Verlag New York, 44 Harz Way, Secaucus, NJ 07096-2491. (201) 348-4033. FAX (201) 348-4505. 1973 to present. Monthly. $1155.00 per year. ISSN: 0721-7250. # #*Applied Physics Letters.* American Institute of Physics, One Physics Ellipse, College Park, MD 20740-3843. (301) 209- 3000. 1962 to present. Weekly. $1060.00 per year. ISSN: 0003-6951.

Canadian Journal of Physics. National Research Council of Canada. Research Journals, Ottawa K1A OR6, Canada. (613) 993-9084. FAX (613) 952-7656. 1929 to present. Monthly. $285.00 per year. ISSN: 0008-4204.

Chaos; An Interdisciplinary Journal of Nonlinear Science. American Institute of Physics, One Physics Ellipse, College Park, MD 20740-3843. (301) 209-3000. 1991 to present. Quarterly. $225.00 per year. ISSN: 1054-1500.

Contemporary Physics. Taylor & Francis, Ltd, Rankine Road, Basingstoke, Hants RG24 8PR, England. TEL 0256-840366. FAX 0256-479438. 1959 to present. Bimonthly. $318.00 per year. ISSN: 0010-7514.

European Journal of Physics. I O P Publishing. U. S. subscriptions to: American Institute of Physics, 500 Sunnyside Boulevard, Woodbury, NY 11797-2999. (516) 575- 2270. 1980 to present. Bimonthly. $339.00 per year. ISSN: 0143-0807.

International Journal of Modern Physics B. World Scientific Publishing Co., 1060 Main Street, Suite 18, River Edge, NJ 07661. (800) 227-7562. 1987 to present. 28 issues per year. $1200.00 per year. ISSN: 0217-9792..

Journal of Applied Physics. American Institute of Physics, One Physics Ellipse, College Park, MD 20740-3843. (301) 209-3000. 1931 to present. Semimonthly. $1655.00 per year. ISSN: 0021-8979.

Journal of Non-crystalline Solids. Elsevier Science Publishing Company, Inc., 655 Avenue of the Americas, New York, NY 10010. (212) 989-5800. FAX (212) 633-3990. 1969 to present. 45 issues per year in 15 volumes. $3535.00 per year. ISSN: 0022-3093.

Journal of Physics. A: Mathematical and General. I O P Publishing. U. S. orders to: American Institute of Physics, 500 Sunnyside Boulevard, Woodbury, NY 11797-2999. (516) 576-2200. 1968 to present. Twenty-four issues per year. $2487.00 per year. ISSN: 0305-4470.

Journal of Physics B: Atomic and Molecular Physics. I O P Publishing. U. S. orders to: American Institute of Physics, 500 Sunnyside Boulevard, Woodbury, NY 11797-2999. (516) 576-2200. 1968 to present. 24 issues per year. $2240.00 per year. ISSN: 0953-4075.

SOLID WASTE DISPOSAL

Ency. of Physical Sciences and Engineering Info. Sources

Journal of Physics: Condensed Matter. I O P Publishing. Orders to: American Institute of Physics, 500 Sunnyside Boulevard, Woodbury, NY 11797-2999. (516) 576-2200. 1968 to present. 50 issues per year. $4229.00 per year. ISSN: 0953-8984.

Landolt-boernstein, Zahlenwerte Und Funktionen Aus Naturwissenschaften Und Technik. Group 3: Crystal Physics. Springer-Verlag New York, Inc., 175 Fifth Avenue, New York, NY 10010. (212) 460-1500. FAX (212) 473-6272. 1965. Irregular. Price varies. ISSN: 0075-7896.

Journal of the Optical Society of America. Part B: Optical physics. Orders to: American Institute of Physics, 500 Sunnyside Boulevard, Woodbury, NY 11797-2999. (516) 576-2200.

Physical Review B (Condensed Matter). American Institute of Physics, One Physics Ellipse, College Park, MD 20740-3843. (301) 209-3000. 1970 to present. 48 issues per year. $3180.00 per year. ISSN: 0163-1829.

Physical Review Letters. American Institute of Physics, One Physics Ellipse, College Park, MD 20740-3843. (301) 209-3000. 1958 to present. Monthly. $1370.00. ISSN: 0031-9007.

Physics of the Solid State. American Institute of Physics, One Physics Ellipse, College Park, MD 20740-3843. (301) 209-3000. Formerly Soviet Physics - Solid State. 1959 to present. Monthly. $2205.00. ISSN: 1063-7834.

Physics Reports - A Review Section of Physics Letters. Elsevier Science Publishing Company, Inc., 655 Avenue of the Americas, New York, NY 10010. (212) 989-5800. FAX (212) 633-3990. 1971 to present. 72 issues per year in 12 volumes. $2342.00 per year. ISSN: 0370-1573.

Physics Today. American Institute of Physics, One Physics Ellipse, College Park, MD 20740-3843. (301) 209-3000. 1948 to present. Monthly. $140.00 per year. ISSN: 0031- 9228.

Reviews of Modern Physics. American Institute of Physics, One Physics Ellipse, College Park, MD 20740-3843. (301) 209-3000. 1929 to present. Quarterly. $300.00 per year. ISSN: 0034-6861.

Royal Society of London, Philosophical Transactions. Series A, Physical Sciences and Engineering. Royal Society of London, 6 Carlton House Terrace, London SW1Y 5AG, England. TEL 071-839-5561. FAX 071-976-1837. 1665 to present. Monthly. L485 per year. ISSN: 0962-8428.

Semiconductors. American Institute of Physics, One Physics Ellipse, College Park, MD 20740-3843. (301) 209-3000.

Formerly Soviet Physic - Semiconductors. 1967 to present. Monthly. $1990.00 per year. ISSN: 1063-7826.

Solid State Communications. Elsevier Science Publishing Company, Inc., 655 Avenue of the Americas, New York, NY 10010. (212) 989-5800. FAX (212) 633-3990. 1963 to present.48 issues per year in 4 volumes. $1790.00 per year. ISSN: 0038-1098.

RESEARCH CENTERS AND INSTITUTES

Center For Electronic Materials and Processing. Pennsylvania State University, 113 Electrical Engineering, West Building, University Park, PA 16802. (814) 865-0870.

Center For Integrated Systems. Stanford University, Stanford, CA 94305-4070. (415) 725-3620.

Clinton P. Anderson Meson Physics Facility. MS H850, Los Alamos National Laboratory, Los Alamos, NM 87545. (505) 667- 6000. FAX (505) 667-1712.

Engineering Experiment Station. Ohio State University, 2070 Neil Avenue, Columbus, OH 43210. (614) 422-2411.

Semiconductor Processing Facility. University of Arizona, College of Engineering and Mines, Tucson, AZ 85721. (602) 621-7496.

SOLID WASTE DISPOSAL

ABSTRACT SERVICES AND INDEXES

Applied Science and Technology Index; A Cumulative Subject Index To English Language Periodicals in the Fields of Aeronautics and Space Science, Computer Technology, Chemistry, Construction Industry, Energy and Related Areas. H.W. Wilson Co., 950 University Avenue, Bronx, NY 10452. (800) 367-6770 or (212) 588-8400. From 1958 to present. Monthly. Inquire about cost and availability. Also available on CD-ROM and online.

Current Contents: Engineering, Technology, and Applied Sciences. Institute for Scientific Information, 3501 Market Street, Philadelphia, PA 19104. (215) 386-0100. FAX (215) 386-2991. 1970 to present. Weekly. $442.00 per year.

Environmental Abstracts. R.R. Bowker Inc., 121 Chanlon Road, New Providence, NJ 07974. (800) 521-8110 or (908) 464-6800. Contains citations, most with abstracts, to literature relating to environmental issues and problems, from 1971 to present. Inquire as to cost and availability.

Environmental Engineering Abstracts, Cambridge Scientific Abstracts, 7200 Wisconsin Avenue, Bethesda, MD 20814-4823. (301) 961-6750. FAX (301) 961-6720. Monthly. Covers hazardous materials, environmental impact and protection, treatment of sewage and industrial wastes, hydroelectric power, tidal and wind power, artic and tropical engineering.

Science Citation Index. Institute for Scientific Information, 3501 Market Street, Philadelphia, PA 19104. (215) 386-0100. FAX (215) 386-6362. Inquire about availability and cost. Also available on CD-ROM.

ASSOCIATIONS AND PROFESSIONAL SOCIETIES

Air & Waste Management Association. 1 Gateway Circule, 3d floor, Pittsburgh, PA 15222. (800) 270-3444. FAX (412) 232-3450.

Association of State and Territorial Solid Waste Management officials. 444 N. Capitol Street NW, Suite 388, Washington, DC 20001. (202) 624-5828. FAX (202) 624-7875.

International Association of Environmental Managers. 243 W Main Street, Kutztown, PA 19530. (215) 683-5098. FAX (215) 683-3171.

Municipal Waste Management Association. 1620 I Street NW, 4th Floor, Washington, DC 20006. (202) 293-7330.

National Solid Wastes Management Association. Suite 1000, 1730 Rhode Island Avenue NW, Washington, DC 20036. (202) 659-4613. FAX (202) 659-0925.

Solid Waste Association of North America. 1100 Wayne Avenue, Suite 700, Box 7219, Silver Spring, MD 20910. (301) 585-2898. FAX (301) 585-7068.

BIBLIOGRAPHIES

Catalog of Hazardous and Solid Waste Publications. U.S. Environmental Protection Agency, office of Solid Waste and Emergency Response. 8th edition. Superintendent of Documents, U.S. Government Printing office, Box 371954, Pittsburgh, PA 15250-7954. (202) 783-3238. FAX (202) 512-2233. 1995. Inquire for cost and availability.

Refuse and Garbage Disposal: Index of New Information With Authors & Subjects. Bernice R. Schindler. ABBE Pubs Assn 1993. $49.50 (hardbound), $39.50 (paperbound).

DIRECTORIES AND BIOGRAPHICAL SOURCES

International Directory of Solid Waste Management: The Iswa Yearbook. Jeanne Moller, ed. International Books, Box 605, Herndon, VA 22070. (703) 435-7064. FAX (703) 689-0660. 1993 to present. Annual. $75.00.

ENCYCLOPEDIAS AND DICTIONARIES

Encyclopedia of Garbage. Steve Coffel. Facts-on-File, Inc., 460 Park Avenue South, New York, NY 10016-7382. (800) 322-8755. Fax (212) 213-4578. 1996. Inquire for price.

GENERAL WORKS

Administration of Public Solid Waste. F.T. Lancaster. WRP 1992. $37.00.# #*Composting: Solutions For Waste Management.* Intl City-Cnty Mgmt 1992. $28.00.

Municipal Solid Waste Management: Recycling Resource Recovery & Landfills: A Source Guide. Gordon Press Publishers, PO Box 459, Bowling Green Station, New York, NY 10004. (718) 624-8419. 1991. $75.00.

Municipal Waste Disposal. Bela G. Liptak. Chilton Book Co., One Chilton Way, Radnor PA 19089. (215) 964-4028 or (800) 695-1214. FAX (610) 964-4745. 1991. $69.95.

Solid Waste: State & Federal Efforts To Manage Non-hazardous Waste. Diane Publishing Company, 600 Upland Avenue, Upland, PA 19015. (610) 499-7415. FAX (610) 499-7429. 1995. $40.00.

Waste Treatment Plants. C.A. Sastry, et al., editors. John Wiley and Sons, Inc., 605 Third Avenue, New York, NY 10158. (800) 526-5368 or (212) 850-6000. 1996. $45.00.

Wastes Management International: Yearbook of Products and Services 1994. Jeremy Gambrill, ed. John Wiley and Sons, Inc., 605 Third Avenue, New York, NY 10158. (800) 526-5368 or (212) 850-6000. 1994. $125.00.

HANDBOOKS AND MANUALS

Handbook of Solid Waste Management. Frank Kreith. McGraw-Hill Publishing Company, 1221 Avenue of the Americas, New York, NY 10020. (800) 262-4729 or (212) 512-3825. 1994. $89.50.

ONLINE DATABASES AND CD-ROMS

Compendex Plus. Engineering Information, Inc., 345 East 47th Street, New York, NY 10017. (212) 705-7600 or (800) 221-1044. Contains citations with abstracts to worldwide literature in engineering and technology, from 1970 to present. Available on online BRS,(800) 289-4277, DIALOG, (800) 334-2564, ORBIT (800) 456-7248, and STN International, FIZ Karlsruhe, P.O. Box 2465, W-7500, Karlsruhe 1, Germany, online services. Also available on CD-ROM. Inquire as to cost and availability.

CSA Engineering. Cambridge Scientific Abstracts, 7200 Wisconsin Avenue, Suite 601, Bethesda, MD 20814. (301) 961-6750 or (800) 843-7751. Contains citations and abstracts of international periodicals and research literature covering all fields of engineering and science and technology,including computer and information science, electronics, mechanical engineering, solid state materials, 1981 to present. Available on BRS,(800) 289-4277, online service.Inquire as to cost and availability.

Current Contents Search. Institute for Scientific Information, 3501 Market Street, Philadelphia, PA 19104. (215) 386-0100. FAX (215) 386-6362. Contains citations to articles listed in the table of contents of science and technology journals. Also articles in social sciences and life sciences journals. Available on BRS,(800) 289-4277, DIALOG,(800) 334-2564, online services. Inquire as to cost and availability.

Dissertation Abstracts. University Microfilms International, 300 North Zeeb Road, Ann Arbor, MI 48106. (800) 521-0600 or (313) 761-4700. Scope includes virtually all doctoral dissertations accepted at accredited American institutions from 1861 to present in 252 subject areas. Available on BRS,(800) 289-4277, DIALOG,(800) 334-2564, and OCLC EPIC,(800) 848-5878, online services. Also available on CD-ROM. Inquire as to cost and availability.

NTIS Bibliographic Database. National Technical Information Service, 5285 Port Royal Road, Springfield, VA 22161. (703) 487-4929 or FAX (703) 321-8199. Broad coverage of government-sponsored science and technology research reports, 1964 to present. Available on BRS,(800) 289-4277, DIALOG, (800) 334-2564, ORBIT, (800) 456-7248, and STN International, FIZ Karlsruhe, P.O. Box 2465, W-7500, Karlsruhe 1, Germany, online services. Also available on CD-ROM. Inquire as to cost and availability.

Wasteinfo. United Kingdom Atomic Energy Authority, Building 7.12, Harwell Laboratory, Didcot, Oxon, OX11 ORA, England. Contains citations to worldwide literature on non-nuclear waste management, from 1973 to present. Available on ORBIT (800) 456-7248, online service. Also available on CD-ROM. Inquire as to cost and availability.

SOLID WASTE DISPOSAL

Ency. of Physical Sciences and Engineering Info. Sources

PERIODICALS

Biocycle. J G Press Inc., 419 State Avenue, Emmaus, PA 18049. (215) 967-4135. 1960 to present. Monthly. $63.00.

Environmental Waste Management. International Association of Environmental Managers, 243 W Main Street, Kutztown, PA 19530. (215) 683-5098. FAX (215) 683-3171. 1983 to present. Bi-monthly. $30.00 per year.

Journal of Solid Waste Technology and Management. National Center for Resource Recovery and Management, University of Pennsylvania, 220 S. 33rd Street, Towne Bldg., Room 229, Philadelphia, PA 19104-6315. (215) 898-2771. FAX (215) 573-2065. 1972 to present. Quarterly. $60.00 per year.

Municipal Solid Waste News. Solid Waste Association of North America, 1100 Wayne Avenue, Suite 700, Box 7219, Silver Spring, MD 20910. (301) 585-2898. FAX (301) 585-7068. 1977 to present. Monthly. $100.00 per year.

Solid Waste Report. Business Publishers Inc., 951 Pershing Drive, Silver Spring, MD 20910-4464. (301) 587-6300. FAX (301) 585-9075. 1970 to present. Weekly. $500.00 per year.

Solid Waste Technologies. HCI Publishers, 410 Archibald Street, Kansas City, MO 64111-3046. (816) 931-1311. FAX (816) 931-2015. 1987 to present. Bi-monthly. $49.00 per year.

Waste Age. National Solid Wastes Management Association, Suite 1000, 1730 Rhode Island Avenue NW, Washington, DC 20036. (202) 659-4613. FAX (202) 659-0925. 1970 to present. Monthly. $45.00 per year.

Waste Management. Elsevier Science [journals], 660 White Plains Rd., Tarrytown, NY 10159-5153. (919) 524-9200. FAX (919) 333-2444. 1980 to present. Eight times a year. $670.00 per year.

Waste Management and Research. Harcourt Brace & Co., Ltd., Foots Cray High Street, Sidcup, Kent DA14 5HP, England. Telephone 44-81-300-3322. FAX 44-81-309-0807. 1983 to present. Bi-monthly. Inquire for price.

Water and Waste Treatment. Faversham House Group Ltd., 111 St. Jame's Road, Croydon, Surrey CR9 2TH, England. 1950 to present. Monthly. Inquire for price.

World Wastes. Argus Inc., 6151 Powers Ferry Road NW, Atlanta, GA 30339-2941. (404) 955-2500. FAX (404) 955-0400. 1957 to present. Monthly. $48.00 per year.

RESEARCH CENTERS AND INSTITUTES

Center For Waste Reduction Technologies. American Institute of Chemical Engineers, 345 E 47th Street, New York, NY 10017. (212) 705-7407.

Cornell University, Center For Environmental Research. 470 Hollister Hall, Ithaca, NY 14853-3501. (607) 255-7535. FAX (607) 255-5945.

Environmental Research Foundation. PO Box 3541, Princeton, NJ 08543-3541. (609) 683-0707.

Miami University, Institute of Enviromental Science. Oxford, OH 45056. (513) 529-5811. FAX (513) 529-5814.

Renew America. 1400 16th Street NW, Suite 710, Washington, DC 20036. (202) 232-2252.

Tufts University, Center For Environmental Management. Curtis Hall, 474 Boston Avenue, Meford, MA 02155. (617) 381-3486. FAX (617) 381-3084.

SONAR

See also: ACOUSTICS, ANTENNAS, ELECTRICAL ENGINEERING, ELECTRONIC CIRCUITS and COMPONENTS, ELECTRONICS, ELECTRONICS ENGINEERING, MICROWAVES, RADAR, RADIO, SIGNAL PROCESSING

ABSTRACT SERVICES AND INDEXES

Applied Science and Technology Index; A Cumulative Subject Index To English Language Periodicals in the Fields of Aeronautics and Space Science, Computer Technology, Chemistry, Construction Industry, Energy and Related Areas. H.W. Wilson Co., 950 University Avenue, Bronx, NY 10452. (212) 588-8400. (800) 367-6770. FAX (718) 590-1617. From 1958 to present. Monthly. Inquire about cost and availability. Also available on CD-ROM and online. ISSN: 0003-6986.

Current Contents: Engineering, Technology and Applied Sciences. Institute for Scientific Information, 3501 Market Street, Philadelphia, PA 19104. (215) 386-0100. FAX (215) 386-6362. 1970 to present. Weekly. $442.00 per year. Also available on CD-ROM and online. Inquire regarding cost and availability. ISSN: 0095-7917.

Electrical and Electronics Abstracts. Institution of Electrical Engineers (IEE), London. Available from INSPEC/IEEE - Institute of Electrical and Electronic Engineers, Box 1331, Hoes Lane, Piscataway, NJ 08855-1331. (908) 562-5549. 1898 to present. Monthly. $2200.00 per year. Also available on CD-ROM and online as INSPEC. ISSN: 0036-8105.

Engineering Index Monthly; Indexes and Abstracts the World's Engineering and Technical Literature. Engineering Information, Inc., Castle Point on the Hudson, Hoboken, NJ 07030. (201) 216-8500. (800) 221-1044. FAX (201) 216-8532. Monthly. $2300.00 per year. Available online as COMPENDEX and also on CD-ROM. ISSN: 0742-1974.

General Science Index. H.W. Wilson Company, 950 University Avenue, Bronx, NY 10452. (212) 588-8400. (800) 367-6770. FAX (718) 590-1617. From 1978 to present. Ten issues per year; quarterly and annual cumulations. Service basis. Available on CD-ROM and online. Inquire regarding cost and availability. ISSN: 0162-1963.

Government Reports Announcements and Index. U. S. National Technical Information Service (NTIS), 5285 Port Royal Road, Springfield, VA 22161. (703) 487-4650. FAX (703) 321-8547. From 1968 to present. Annual. $630.00 per year. Also available online as NTIS Bibliographic Database and on CD-ROM.

Index To IEEE Publications. IEEE Service Center, 445 Hoes Lane, Piscataway, NJ 08855-1331. (908) 981-1393. (800) 678-

IEEE. FAX (908) 981-9667. 1973 to present. Annual. ISSN: 0099-1368.

Physics Abstracts. INSPEC. Section A, Science Abstracts. Institution of Electrical Engineers (IEE), London. Available from: INSPEC/IEEE - Institute of Electrical and Electronic Engineers, Box 1331, Hoes Lane, Piscataway, NJ 08855-1331. (908) 562-5549. 1898 to present. 24 issues per year. $2835.00 per year. Also available online and on CD-ROM. ISSN: 0036-8091.

Physics Briefs (Physikalische Berichte). Information Center for Energy, Physics, Mathematics; German Physical Society. V C H Publishers, Inc., 220 East 23rd Street, New York, NY 10010-4606. (212) 683-8333. 1845 to present. 24 issues per year. $2390.00 per year. Also available online. ISSN: 0179-7434.

Science Citation Index. SCI. Institute for Scientific Information, 3501 Market Street, Philadelphia, PA 19104. (215) 386-0100. (800) 523-1850. FAX (215) 386-2991. 1961 to present. Six issues per year, plus annual cumulation. $11650.00 per year. Also available online and on CD-ROM.Inquire about price and availability. ISSN: 0036-827X.

ANNUAL REVIEWS AND YEARBOOKS

Advances in Electronics and Electron Physics. Academic Press, Inc., 6277 Sea Harbor Drive, Orlando, FL 32887. (800) 321-5068. From 1948 to present. Irregular. Price varies, inquire. ISSN: 0065-2539.

ASSOCIATIONS AND PROFESSIONAL SOCIETIES

American Electronics Association. 5201 Great America Way, Suite 520, P.O. 52990, Santa Clara, CA 95056. (408) 987- 4200. FAX (408) 970-8565.

American Institute of Physics. One Physics Ellipse, College Park, MD 20740-3843. (301) 209-3100.

Electronics Industries Association. 2500 Wilson Boulevard, Arlington, VA 22201. (703) 907-7500. FAX (202) 457-4985.

IEEE Antennas and Propagation Society. c/o Institute of Electrical and Electronic Engineers, 345 East 47th Street. New York, NY 10017. (212) 705-7900. FAX (212) 705-4929.

IEEE (Institute of Electrical and Electronic Engineers). 345 East 47th Street. New York, NY 10017. (212) 705-7900. FAX (212) 705-4929,

National Association of Radio and Telecommunications Engineers. PO Box 678, Medway, MA 02053. (508) 533- 8333. FAX (508) 533-3815.

United States National Committee For the International Union of Radio Science. Board on Physics and Astronomy, National Research Council, 2101 Constitution Avenue NW, Washington, DC 20418. (202) 334-3520. FAX (202) 3343-2791.

DIRECTORIES AND BIOGRAPHICAL SOURCES

American Electronics Association Directory. American Electronics Association, 5201 Great America Way, Suite 520, P.O. 52990, Santa Clara, CA 95056. (408) 987-4200. FAX (408) 970-8565. Annual. $175.00.

American Men and Women of Science: Physical and Biological Sciences. R. R. Bowker Inc., 121 Chanlon Road, New Providence, NJ 07974. (908) 464-6800. (800) 521-8110. 20th edition. 8 volumes. 1996. $850.00.

E E M - Electronic Engineer's Master. Hearst Business Communications, Inc., 645 Stewart Avenue, Garden City NY 11530. (516) 227-1300. Annual. $90.00 ISSN: 0732-9016.

Engineering Research Centers: Incorporating Electronics Research Centers. Stockton Press, 345 Park Avenue, New York, NY 10010. (212) 689-9200. (800) 221-2123. 4th edition. 1995. $515.00.

IEEE Membership Directory. Institute of Electrical and Electronics Engineers, IEEE Service Center, 445 Hoes Lane, Piscataway, NY 08854. (908) 981-1393. (800) 678-IEEE. FAX (908) 981-9667. 2 volumes. Annual. $190.00.

International Directory of Abbreviations and Acroynms of Electronics, Electrical Engineering, Computer Technology and Information Processing. Peter Wennrich. K. G. Saur, 121 Chanlon Road, New Providence, NJ 07974. (908) 464-6800. (800) 521-8110 2 volumes. 1992. $230.00.

International Engineering Directory. American Consulting Engineers Council, 1015 15th Street, N.W. Suite 802, Washington, DC 20005-2670. (202) 347-7474. Annual. $10.00.

Research Centers Directory. Gale Research, 835 Penobscot Building, Detroit, MI 48226-4094. (313) 961-2242. (800) 877-4253. 20th edition. 1995. $485.00. ISSN: 0080- 1518.

Who's Who in Engineering. Gordon Davis, editor. American Association of Engineering Societies. 1111 19th Street, NY, Suite 608, Washington, DC 20036. (202) 296-2237. (800) 658-8897. 9th edition. 1995. $220.00.

Who's Who in Technology; Biographies and Index. Amy L. Unterburger, editor. Gale Research, 835 Penobscot Building, Detroit, MI 48226-4094. (313 961-2242. (800) 521-4253. 7th edition. 1995. $195.00. ISN 0-8103-7467-6.

ENCYCLOPEDIAS AND DICTIONARIES

The Audio Dictionary. Glenn D. White. University of Washington Press, P. O. Box 50096. Seattle, WA 98145- 5096. (206) 543-4050. (800) 441-4115. 2nd revised edition. 1991. $19.95 in paper.

Chambers Science and Technology Dictionary. Peter M.B. Walker, editor. Cambridge University Press, 40 West 20th Street, New York, NY 10011-4211. (212) 924-3900. 1988. $39.95.

IEEE Standard Dictionary of Electrical and Electronics Terms. Christopher J. Booth, editor. IEEE Service Center, 445 Hoes Lane, Piscataway, NJ 08855-1331. (908) 981-1393. (800) 678-IEEE. FAX (908) 981-9667. IEEE Standard 100- 1992. 5th edition. 1993. $150.00.

Illustrated Dictionary of Electronics. Stan Gibilsco. TAB Books, P.O. Box 40, Blue Summit, PA 17294-0850. (717) 794- 2191. (800) 233-1128. 7th edition. 1994. $34.95.

McGraw-Hill Electronics Dictionary. J. Markus. McGraw-HillBook Company, Inc., 1221 Avenue of the Americas, New York, NY 10020. (212) 997-3675. 5th edition. 1994. $49.95.

SONAR

Ency. of Physical Sciences and Engineering Info. Sources

McGraw-Hill Encyclopedia of Science and Technology. McGraw-Hill Book, Incorporated, 1221 Avenue of the Americas, New York, NY 10020. (212) 997-3675. (800) 262-4729. Seventh edition. Twenty volumes. 1992. $1900.00.

GENERAL WORKS

Antennas For Radar and Communications: A Polarimetric Approach. Harold Mott. John Wiley & Sons, Inc., 605 Third Avenue, New York, NY 10158-0012. (212) 850-6000. (800) 225-5945. 1992. $99.95.

Elements of Signal Detection and Estimation. Carl W. Helstrom. Prentice Hall General Reference & Travel, 15 Columbus Circle, New York, NY 10023. (212) 373-8500. (800) 223-2348. 1994. $63.00.

Introduction To the Theory and Design of Sonar Transducers. Oscar B. Wilson. Peninsula Publishing, PO Box 867, Los Altos, CA 94023-9912. (415) 948-2511. 1989. $40.95.

The Physics of Sound. Richard E. Berg and David G. Stork. Prentice-Hall, 113 Sylvan Avenue, Route 9W, Englewood Cliffs, NJ 07632. (201) 592-2000. (800) 922-0579. 2nd edition. 1994. $63.00.

Radar and Sonar. Richard E. Blahut, et al. Springer-Verlag New York, Inc., 175 Fifth Avenue, New York, NY 10010. (212) 460-1500. (800) 777-4643. FAX (212) 473-6272. Parts 1 & 2. 1991. $74.00 set.

Signals, Noise and Active Sensors, Radar, Sonar and Laser Radar. John R. Minkoff. John Wiley & Sons, Inc., 605 Third Avenue, New York, NY 10158-0012. (212) 850-6000. (800) 225-5945. 1992. $67.95.

Sonar Images. Harold E. Edgerton. Prentice Hall, 113 Sylvan Avenue, Route 9W, Englewood Cliffs, NJ 07632. (201) 592-2000. (800) 922-0579. 1986. $56.00.

Sonar Signal Processing. Richard O. Nielsen. Artech House, 685 Canton Street, Norwood, MA 02062. (617) 769-9750. (800) 225-9977. 1991. $95.00.

The Sonar of Dolphins. Willow W. Au. Springer-Verlag New York, Inc., 175 Fifth Avenue, New York, NY 10010. (212) 460-1500. (800) 777-4643. FAX (212) 473-6272. 1994. $79.00.

Sound Propagation in the Sea. R. J. Urick. Peninsula Publishing, PO Box 867, Los Altos, CA 94023-9912. (415) 948-2511. 1982. $36.95.

Sound Underwater Images; A Guide To the Generation and Interpretation of Side Scan Sonar Data. John P. Fish and H. Arnold Carr. Lower Cape Publishing, PO Box 901, Orleans MA 02653. (508) 255-2244. 1990. $125.00 in paper.

Theory and Application of Statistical Wave-period Processing. Albert A. Geriach. Gordon & Breach Science Publishers, Inc., P.O.Box 200029, Riverfront Plaza Station, Newark, NJ 07102-0301. (201) 643-7500. 3 volumes. 1970. $694.00 set.

HANDBOOKS AND MANUALS

Antenna Engineering Handbook. Richard C. Johnson. McGraw-Hill Book Company, 1221 Avenue of the Americas, New York,

NY 10020. (212) 512-2000. (800) 262-4729. 3rd edition, 1993. $119.50.

Electrical Engineer's Handbook. Donald Christiansen, editor. McGraw Book Company, 1221 Avenue of the Americas, New York, NY 10020. (212) 512-2000. (800) 262-4729. 4th edition. 1996. $110.00

Engineering Tables and Data. A.M. Howatson, P.G. Lund and J.D. Todd. Van Nostrand Reinhold, 115 Fifth Avenue, New York, NY 10003. (212) 254-3232. (800) 926-2665. Second Edition. 1991. $39.95.

Sonar Engineering Handbook. Harrison T. Loeser. Peninsula Publishing, PO Box 867, Los Altos, CA 94023-9912. (415) 948-2511. Reprint edition. 1993. $39.95.

Standard Handbook of Engineering Calculations. Tyler G. Hicks. McGraw-Hill Publishing Company, 1221 Avenue of the Americas, New York, NY 10020. (800) 262-4729 or (212) 512-3825. 3rd edition. 1994. $99.50. ISBN: 0-07-028812-7.

ONLINE DATABASES AND CD-ROMS

COMPENDEX PLUS. Engineering Information, Inc., 345 East 47th Street, New York, NY 10017. (212) 705-7600 or (800) 221-1044. Contains citations with abstracts to worldwide literature in engineering and technology, from 1970 to present. Available on online BRS,(800) 289-4277, DIALOG, (800) 334-2564, ORBIT (800) 456-7248, and STN International, FIZ Karlsruhe, P.O. Box 2465, W-7500, Karlsruhe 1, Germany, online services. Also available on CD-ROM. Inquire as to cost and availability.

Current Contents Search. Institute for Scientific Information, 3501 Market Street, Philadelphia, PA 19104. (215) 386-0100. FAX (215) 386-6362. Contains citations to articles listed in the table of contents of science and technology journals. Also articles in social sciences and life sciences journals. Available on BRS,(800) 289-4277, DIALOG,(800) 334-2564, online services. Inquire as to cost and availability.

Dissertation Abstracts. University Microfilms International, 300 North Zeeb Road, Ann Arbor, MI 48106. (800) 521-0600 or (313) 761-4700. Scope includes virtually all doctoral dissertations accepted at accredited American institutions from 1861 to present in 252 subject areas. Available on BRS,(800) 289-4277, DIALOG,(800) 334-2564, and OCLC EPIC,(800) 848-5878, online services. Also available on CD-ROM. Inquire as to cost and availability.

INSPEC. Institution of Electrical Engineers, Michael Faraday House, Six Hills Way, Stevenage, Herts. SG1 2AY, England. Telephone: 0438 313311 or FAX 0438 742840. Contains citations to the worldwide literature of physics, electronics and electrical engineering, computer technology, and related fields. Available on BRS,(800) 289-4277, DIALOG, (800) 334-2564, ORBIT, (800) 456-7248, and STN International, FIZ Karlsruhe, P.O. Box 2465, W-7500, Karlsruhe 1, Germany, online services. Inquire as to cost and availability. The National Aeronautics and Space Administration only.

NTIS Bibliographic Database. National Technical Information Service, 5285 Port Royal Road, Springfield, VA 22161. (703) 487-4929 or FAX (703) 321-8199. Broad coverage of government-sponsored science and technology research reports, 1964 to present. Available on BRS,(800) 289-4277, DIALOG, (800) 334-2564, ORBIT, (800) 456-7248, and STN International, FIZ Karlsruhe, P.O. Box 2465, W-7500, Karlsruhe 1, Germany,

online services. Also available on CD-ROM. Inquire as to cost and availability.

SCISEARCH. Institute for Scientific Information, 3501 Market Street, Philadelphia, PA 19104. (800) 523-1850 or (215) 386-0100. Broad multidisciplinary title and author index to the international literature of science and technology, 1974 to present. Available on DIALOG,(800) 334-2564, and ORBIT,(800) 456-7248, online services. Also available on CD-ROM. Inquire as to cost and availability.

WILSONLINE. H.W. Wilson Company, 950 University Avenue, Bronx, NY 10452. (800) 367-6770 or (212) 588-8400. Makes available online versions of the printed H.W. Wilson indexes including Applied Science and Technology Index, Business Periodicals Index, General Science Index, and Readers' Guide to Periodical Literature. Period covered is generally 1983 to present. Available on BRS,(800) 289-4277, DIALOG,(800) 334-2564, and OCLC EPIC,(800) 848-5878, online services. Also available on CD-ROM. Inquire as to cost and availability.

PERIODICALS

Acoustical Society of America Journal. American Institute of Physics, 500 Sunnyside Boulevard, Woodbury NY 11797-2999. (519) 576-2270. 1929 to present. Monthly. $760.00 per year. ISSN: 0001-4966.

Applied Acoustics. Elsevier Science, 660 White Plains Road, Tarrytown, NY 10591-5153. (914) 524-9200. FAX (914) 333-2444. 1968 to present. Monthly. $770.00 per year. ISSN: 0003-682X.

Audio Engineering Society. Journal. Audio Engineering Society, 60 East 42nd Street, New York, NY 10165. (212) 661-2355. 1953 to present. 10 issues per year. $125.00 per year. ISSN: 0004-7554.

Electronic Design. Penton Publishing, San Jose Gateway, Suite 354. 2025 Gateway Place, San Jose, CA 95110. (408) 441-0550. 1952 to present. Fortnightly. $95.00. ISSN: 0013-4872.

Electronics. Penton Publishing, San Jose Gateway, Suite 354. 2025 Gateway Place, San Jose, CA 95110. (408) 441-0550. 1930 to present. Semi-weekly. $98.00 per year. ISSN: 0883-4989.

Electronics World and Wireless World. Reed Business Publishing, Ltd., Quadrant House, the Quadrant, Sutton, Surrey SM2 5AS, England. 1911 to present. Monthly. L35. ISSN: 0266-3244.

IEEE - Signal Processing Magazine. Institute of Electrical and Electronics Engineers, Inc., Box 1331, 445 Hoes Lane, Piscataway, NJ 08855-1331. (908) 981-0060. 1984 to present. Quarterly. $70.00 per year. ISSN: 1053-5888.

IEEE Circuits and Devices Magazine. Institute of Electrical and Electronics Engineers, Inc., Box 1331, 445 Hoes Lane, Piscataway, NJ 08855-1331. (908) 981-0060. 1985 to present. Bi-monthly. $120.00 per year. ISSN: 8755-3996.

IEEE Transactions On Antennas and Propagation. Institute of Electrical and Electronics Engineers, Inc., Box 1331, 445 Hoes Lane, Piscataway, NJ 08855-1331. (908) 981-0060. 1952 to present. Monthly. $212.00 per year. ISSN: 0018-926X.

IEEE Transactions on Signal Processing. Institute of Electrical and Electronics Engineers, Inc., Box 1331, 445 Hoes Lane, Piscataway, NJ 08855-1331. (908) 981-0060.

1951 to present. Monthly. $400.00. ISSN: 1053-587X.

IEE Proceedings. Part H: Microwaves, Antennas and Propagation. Institution of Electrical Engineers. Available from: Institute of Electrical and Electronic Engineers (IEEE) Inc., Box 1331, 445 Hoes Lane, Piscataway, NJ 08855- 1331. (908) 981-0060. 1980 to present. Bi-monthly. ISSN:0950-107X.

International Journal of Microwave and Millimeter-wave Computer Aided Engineering. John Wiley and Sons, Inc., 605 Third Avenue, New York, NY 10158. (800) 526-5368 or (212) 850-6000. 1991 to present. Quarterly. $175.00 per year.

IEEE Transactions On Ultrasonics, Ferroelectrics and Frequency Control. Institute of Electrical and Electronics Engineers, Inc., Box 1331, 445 Hoes Lane, Piscataway, NJ 08855-1331. (908) 981-0060. 1954 to present. Bi-monthly. $230.00 per year. ISSN: 0885-3010.

Journal of Low Frequency Noise and Vibration. Multi-Science Publishing Co. Ltd., Box 176 Avenel, NJ 070001. 1982 to present. Quarterly. L102 per year. ISSN: 0263-0923.

Ultrasonics. Butterworth-Heinemann, 313 Washington Street, Newton, MA 02158. (618-928-2500. (800) 366-2665. 1963 to present. Bimonthly. L245 per year. ISSN: 0041-624X

RESEARCH CENTERS AND INSTITUTES

Applied Microelectronics Institute. 1127 Barrington Street, Halifax, NS, Canada B3H 2P8. (902) 421-1250. FAX (902) 429-9983.

Applied Research Laboratories. University of Texas at Austin, PO Box 8029, University Station, Austin, TX 78713-8029. (512) 835-3200. FAX (512) 835-3259.

Radar and Instrumentation Development Laboratory. Georgia Institute of Technology, Georgia Tech Research Institute, Atlanta, GA 30332. (404) 528-7741.

Radar Systems and Remote Sensing Laboratory. University of Kansas Center for Research, 2291 Irving Hill Road, Lawrence, KS 66045. (913) 864-4832.

SOUND RECORDING

See: AUDIO ENGINEERING

SPACECRAFT

Ency. of Physical Sciences and Engineering Info. Sources

SPACECRAFT

See also: AEROSPACE ENGINEERING, MANNED SPACE FLIGHT, ROCKETS, SPACE SHUTTLE

ABSTRACT SERVICES AND INDEXES

Applied Science and Technology Index; A Cumulative Subject Index To English Language Periodicals in the Fields of Aeronautics and Space Science, Computer Technology, Chemistry, Construction Industry, Energy and Related Areas. H.W. Wilson Co., 950 University Avenue, Bronx, NY 10452. (800) 367-6770 or (212) 588-8400. FAX (718) 590-1617. From 1958 to present. Monthly. Inquire about cost and availability. Also available on CD-ROM.

Astronomy and Astrophysics Abstracts. Springer-Verlag, 175 Fifth Avenue, New York, NY 10010. (212) 460-1500 or (800) 777-4643. FAX (212) 473-6272. Irregular. Price varies, inquire.

Current Contents: Engineering, Technology and Applied Sciences. Institute for Scientific Information, 3501 Market Street, Philadelphia, PA 19104. (215) 386-0100. FAX (215) 386-6362. Weekly. $360.00 per year.

Engineering Index Monthly. Engineering Information, Inc., Castle Point on the Hudson, Hoboken, NJ 07030. (800) 221-1044. FAX (212) 832-1857. Monthly. $2200.00 per year. Also available online as COMPENDEX, and also on CD-ROM.

Index To Scientific and Technical Proceedings. Institute for Scientific Information, 3501 Market St., Philadelphia, PA 19104. (215) 386-0100. FAX (215) 386-6362. Monthly. $500.00 per year.

Index To Scientific Reviews. Institute for Scientific Information, 3501 Market St., Philadelphia, PA 19104. (215) 386-0100. FAX (215) 386-6362. Semi-annual.

International Aerospace Abstracts. Technical Information Service, American Institute of Aeronautics and Astronautics, Inc., 555 West 57th St., New York, NY 10019. (212) 247-6500. FAX (212) 582-4861. Semi-monthly. $1295.00 per year.

NASA Patent Abstracts Bibliography. National Aeronautics and Space Administration, Center for Aerospace Information, Box 8757, BWI Airport, Baltimore, MD 21240. (301) 621-0153. Semi-annual. Inquire for price information.

Numerical/chronological/author Index. An index to articles and papers published by or for the American Astronautical Society, including: Journal of the Astronautical Sciences, Astronautical Sciences Review, Advances in the Astronautical Sciences, Science and Technology, AAS history series, and AAS microfiche series. Horace Jacobs and Robert H. Jacobs. Univelt, Inc. PO Box 28130, San Diego, CA 92198. (619) 746-4005. Inquire for price and availability.

Science Citation Index. Institute for Scientific Information, 3501 Market Street, Philadelphia, PA 19104. (215) 386-0100. FAX (215) 386-6362. Inquire about availability and cost. Also available on CD-ROM.

Scientific and Technical Aerospace Reports (star). National Aeronautics and Space Administration. NASA Center for Aerospace Information, Box 8757, BWI Airport, Baltimore, MD 21240. (301) 621-0153. Monthly. Inquire about availability and cost. Also available through the NASA online retrieval service (RECON), and through the Aerospace Database through DIALOG.

ASSOCIATIONS AND PROFESSIONAL SOCIETIES

Aerospace Industries Association of America. 1250 Eye Street NW, Washington, DC 20005. (202) 371-8400.

American Astronautical Society. 6352 Rolling Mill Place, Suite 102, Springfield, VA 22152.

American Institute of Aeronautics and Astronautics. 370 L'Enfant Promenade SW, Washington, DC 20024. (202) 646-7400.

British Interplanetary Society. 27/29 South Lambeth Road, London SW8 1SZ, England. (071) 735-3160. FAX (071) 820-1504.

Canadian Aeronautics and Space Institute. 222 Somerset Street West, Suite 601, Ottawa, Ontario, Canada K2P 0J1.

National Space Society. 922 Pennsylvania Avenue SE, Washington, DC 20003-2140. (202) 543-1900. FAX (202) 546-4189.

United States Space Foundation. 2860 S. Circle Drive, No. 2301. Colorado Springs, CO 80906. (719) 576-8000. FAX (719) 576-8801.

DIRECTORIES AND BIOGRAPHICAL SOURCES

American Astronautical Society Directory. American Astronautical Society, 6352 Rolling Mill Place, Suite 102, Springfield, VA 22152.

American Institute of Aeronautics and Astronautics Roster. American Institute of Aeronautics and Astronautics. 370 L'Enfant Promenade SW, Washington, DC 20024. (202) 646-7400.

Jane's Space Directory. A. Wilson. Jane's Information Group, 1340 Braddock Place, Suite 300, Alexandria, VA 22314-1651. (703) 683-3700. FAX (703) 836-0029. 1993. $245.00.

Research Centers Directory. Gale Research, 835 Penobscot Building, Detroit, MI 48226-4094. (313) 961-2242. (800) 877-4253. FAX (313) 961-6083. $485.00.

Scientific and Technical Organizations and Agencies Directory. Gale Research, 835 Penobscot Building, Detroit, MI 48226-4094. (313) 961-2242. (800) 877-4253. FAX (313) 961-6083. 4th ed. 1995. $195.00.# #*Who's Who in Engineering.* 7th ed. American Association of Engineering Societies, 1111 19th St. NW, Suite 608, Washington, DC 20036. (800)-658-8897. 1988. $200.00.

ENCYCLOPEDIAS AND DICTIONARIES

Concise Encyclopedia of Aeronautics and Space Systems. M. Pelegrin and W.M. Hollister. Pergamon Press Inc., Maxwell House, Fairview Park, Elmsford, NY 10523. (914) 592-7700. Fax (914) 592-3625. 1993. Inquire for cost and availability.

Ency. of Physical Sciences and Engineering Info. Sources

SPACECRAFT

Encyclopedia of Physical Science and Technology. Robert A. Meyers, ed. Academic Press, Inc., 6277 Sea Harbor Drive, Orlando, FL 32887. (800) 321-5068. 1992. $2100.00.

Jane's Aerospace Dictionary, B. Guston. 3rd ed. Jane's Information Group, 1340 Braddock Place, Suite 300, Alexandria, VA 22314-1651. (703) 683-3700. FAX (703) 836-0029. 1991. $45.00.

McGraw-Hill Encyclopedia of Science and Technology, Sybil P. Parker, ed. 7th ed. 20 vols. McGraw-Hill Publishing Company, 1221 Avenue of the Americas, New York, NY 10020. (800) 262-4729 or (212) 512-3825. $1900.00

SAE Dictionary of Aerospace Engineering. William H. Cubberly, ed. Society of Automotive Engineers, 400 Commonwealth Dr., Warrendale, PA 15096. (412) 776-4841. 1992. $85.00.

GENERAL WORKS

Astronautics and Aeronautics, 1979-1984: A Chronology. Bette R. Janson and Eleanor H. Ritchie. National Aeronautics and Space Administration, Scientific and Technical Information Division. 1990. Order from: Superintendent of Documents, U.S. Government Printing office, Washington, DC 20402-9325. (202) 783-3238. Inquire for cost and availability.

Europe in Space. Guy Collins. St. Martin's Press, Inc., 175 Fifth Avenue, New York, NY 10010. (212) 674-5151 or (800) 221-7945. 1991. $59.95.

Introduction To Space Flight. Francis J. Hale. Prentice Hall Inc., 113 Sylvan Avenue, Route 9W, Englewood Cliffs, NJ 07632. (201) 592-2000 or (800) 922-0579. 1994. $39.50 (hardback), $32.50 (paperback).

Prospects For Interstellar Travel. John H. Maudlin. Univelt, Inc. PO Box 28130, San Diego, CA 92198. (619) 746-4005. FAX (619) 746-3139. 1992. $27.00.

Rockets Into Space. Frank H. Winter. Harvard University Press, 79 Garden Street, Cambridge, MA 02138. (617) 495-2600. FAX (617) 495-8924. 1990. $12.95.

Space Almanac. Anthony R. Curtis, ed. 2d ed. Gulf Publishing Co., PO Box 2608, Houston, TX 77252. (713) 529-4301 or (800) 231-6275. FAX (713) 525-9647. 1992. $36.95 (hardback) and $24.95 (paperback).

Space Mission Analysis and Design. Wiley J. Larson and James R. Wertz, eds. 2d ed. Kluwer Academic Publishers, P.O. Box 358, Accord Station, Hingham, MA 02018-0358. (617) 871-6000. FAX (617) 871-6528. 1992. $99.00.

Spacecraft Systems Engineering. Peter W. Fortescue and John P.W. Stark. John Wiley and Sons, Inc., 605 Third Avenue, New York, NY 10158. (800) 526-5368 or (212) 850-6000. 1992. $64.95.

HANDBOOKS AND MANUALS

Space Exploration: A Reference Handbook. Mrinal Bali. ABC-CLIO, PO Box 1911, Santa Barbara, CA 93116-2546. (800) 422-2546. FAX (805) 685-9685. 1990. $39.50.

ONLINE DATABASES AND CD-ROMS

Aerospace Database. American Institute of Aeronautics and Astronautics, 370 L'Enfant Promenade, S.W., Washington, DC 20024. (202) 646-7400. Worldwide published literature on research and development in aerospace and related areas, with abstracts. Covers 1962 to present. Available on DIALOG, (800) 334-2564, online service. Also available on CD-ROM. Inquire as to cost and availability.

Compendex Plus. Engineering Information, Inc., 345 East 47th Street, New York, NY 10017. (212) 705-7600 or (800) 221-1044. Contains citations with abstracts to worldwide literature in engineering and technology, from 1970 to present. Available on online BRS,(800) 289-4277, DIALOG, (800) 334-2564, ORBIT (800) 456-7248, and STN International, FIZ Karlsruhe, P.O. Box 2465, W-7500, Karlsruhe 1, Germany, online services. Also available on CD-ROM. Inquire as to cost and availability.

CSA Engineering. Cambridge Scientific Abstracts, 7200 Wisconsin Avenue, Suite 601, Bethesda, MD 20814. (301) 961-6750 or (800) 843-7751. Contains citations and abstracts of international periodicals and research literature covering all fields of engineering and science and technology,including computer and information science, electronics, mechanical engineering, solid state materials, 1981 to present. Available on BRS,(800) 289-4277, online service.Inquire as to cost and availability.

Current Contents Search. Institute for Scientific Information, 3501 Market Street, Philadelphia, PA 19104. (215) 386-0100. FAX (215) 386-6362. Contains citations to articles listed in the table of contents of science and technology journals. Also articles in social sciences and life sciences journals. Available on BRS,(800) 289-4277, DIALOG,(800) 334-2564, online services. Inquire as to cost and availability.

Dissertation Abstracts. University Microfilms International, 300 North Zeeb Road, Ann Arbor, MI 48106. (800) 521-0600 or (313) 761-4700. Scope includes virtually all doctoral dissertations accepted at accredited American institutions from 1861 to present in 252 subject areas. Available on BRS,(800) 289-4277, DIALOG,(800) 334-2564, and OCLC EPIC,(800) 848-5878, online services. Also available on CD-ROM. Inquire as to cost and availability.

International Aerospace Abstracts. American Institute of Aeronautics and Astronautics, 370 L'Enfant Promenade, S.W., Washington, DC 20024. (202) 646-7400. Contains references and abstracts of journal and monograph literature relating to aerospace science and technology, from 1963 to present. Available through the NASA/RECON system of the National Aeronautics and Space Administration only.

NASA Database. American Institute of Aeronautics and Astronautics, 370 L'Enfant Promenade, S.W., Washington, DC 20024. (202) 646-7400. Citations and abstracts of aeronautics and astronautics literature, 1962 to present. Also contains citations from STAR, Scientific and Technical Aerospace Reports, and International Aerospace Abstracts. Available through NASA/RECON online service. Inquire as to cost and availability.

NTIS Bibliographic Database. National Technical Information Service, 5285 Port Royal Road, Springfield, VA 22161. (703) 487-4929 or FAX (703) 321-8199. Broad coverage of government-sponsored science and technology research reports, 1964 to present. Available on BRS,(800) 289-4277, DIALOG, (800) 334-2564, ORBIT, (800) 456-7248, and STN International, FIZ Karlsruhe, P.O. Box 2465, W-7500, Karlsruhe 1, Germany,

online services. Also available on CD-ROM. Inquire as to cost and availability.

SCISEARCH. Institute for Scientific Information, 3501 Market Street, Philadelphia, PA 19104. (800) 523-1850 or (215) 386-0100. Broad multidisciplinary title and author index to the international literature of science and technology, 1974 to present. Available on DIALOG,(800) 334-2564, and ORBIT,(800) 456-7248, online services. Also available on CD-ROM. Inquire as to cost and availability.

WILSONLINE. H.W. Wilson Company, 950 University Avenue, Bronx, NY 10452. (800) 367-6770 or (212) 588-8400. Makes available online versions of the printed H.W. Wilson indexes including Applied Science and Technology Index, Business Periodicals Index, General Science Index, and Readers' Guide to Periodical Literature. Period covered is generally 1983 to present. Available on BRS,(800) 289-4277, DIALOG,(800) 334-2564, and OCLC EPIC,(800) 848-5878, online services. Also available on CD-ROM. Inquire as to cost and availability.

PERIODICALS

Acta Astronautica. Pergamon Press, Pergamon Press, 660 White Plains Rd., Tarrytown, NY 10591-5153. 14 times a year. Inquire for price and availability.

Ad Astra. National Space Society, 922 Pennsylvania Avenue SE, Washington, DC 20003-2140. (202) 543-1900. FAX (202) 546-4189. 1989 to present. 6 times a year. $35.00 a year.

Advances in the Astronautical Sciences. American Astronautical Society, 6352 Rolling Mill Place, Suite 102, Springfield, VA 22152. (703) 866-0020. FAX (703) 866-3526. Quarterly. Write or call for price information.

Advances in Space Research. Pergamon Press, Pergamon Press, 660 White Plains Rd., Tarrytown, NY 10591-5153. 12 times a year. Inquire for price and availability.

Aeronautical Journal. Royal Aeronautical Society, PO Box 139, Tonbridge, Kent, TN9 1EW, England. 0732-770823. FAX 0723-361708. 1897 to present. Ten times a year. Write for price and availability.

Aerospace. Royal Aeronautical Society, PO Box 139, Tonbridge, Kent, TN9 1EW, England. 0732-770823. FAX 0723-361708. Monthly. Write for price and availability.

Aerospace America. American Institute of Aeronautics and Astronautics, 370 L'Enfant Promenade SW, Washington, DC 20024-2518. (202) 646-7471. 1932 to present. Monthly. $70 per year (non-members).

Aerospace Engineering. SAE, Inc., 400 Commonwealth Dr., Warrendale, PA 15096. (412) 776-4841. 1981 to present. Monthly. $48 a year.

Aviation & Aerospace. Baxter Publishing Company, 310 Dupont Street, Toronto, Ontario M5R 1V9 Canada. (416) 968-7252. FAX (416) 968-2377. Bi-monthly. $42.00 per year.

Aviation Week and Space Technology. 1221 Avenue of the Americas, New York, NY 10020. (212) 512-2999 or (800) 525-5003. 1916 to present. Weekly. $82.00 per year.

ESA Bulletin. European Space Agency, Publications Division, Keplerlaan 1, 22000 AG Nordwijk, Netherlands. Phone 31-1719-86555. FAX 31-1719-85433. Quarterly. Free.

ESA Journal. European Space Agency, Publications Division, Keplerlaan 1, 22000 AG Nordwijk, Netherlands. Phone 31-1719-86555. FAX 31-1719-85433. Quarterly. Free.

Journal of the British Interplanetary Society (JBIS). British Interplanetary Society, 27/29 South Lambeth Road, London SW8 1SZ, England. (071) 735-3160. FAX (071) 820-1504. Monthly. $370.00 per year.

Interavia/aerospace World. Aerospace Media Publishing, Swissair Centre, 31 Route de l'Aeroport, POB Box 437, 1215 Geneva 15, Switzerland. (022) 788 27 88. FAX (022) 788 27 26. Monthly. $128.00 per year (U.S.).

International Journal of Satellite Communications. John Wiley and Sons Ltd. Journals, Baffins Ln., Chichester, Sussex PO19 1UD, England. Phone 0243-77977. FAX 0243-775878. Bi-monthly. $525.00 per year.

Journal of the Astronautical Sciences. American Astronautical Society, 6352 Rolling Mill Place, Suite 102, Springfield, VA 22152. Quarterly. $35 per year.

Journal of Guidance, Control and Dynamics. American Institute of Aeronautics and Astronautics, 370 L'Enfant Promenade SW, Washington, DC 20024-2518. 1978 to present. Bimonthly. $42.00 (members) and $250.00 (non-members).

Journal of Propulsion and Power. American Institute of Aeronautics and Astronautics, 370 L'Enfant Promenade SW, Washington, DC 20024-2518. 1985 to present. Bimonthly. $38.00 (members) and $260.00 (non-members).

Journal of Spacecraft and Rockets. American Institute of Aeronautics and Astronautics, the Aerospace Center, 370 L'Enfant Promenade SW, Washington, DC 20024. (202) 646-7400. FAX (202) 646-7508. 1964 to present. Bi-monthly. $185.00 per year.

Planetary Report. Planetary Society, 65 N. Catalina, Pasadena, CA 91106-2301. (818) 793-5100. FAX (818) 793-5528. Bi-monthly. $25.00 per year.

Progress in Aerospace Sciences. Pergamon Press, 660 White Plains Rd., Tarrytown, NY 10591-5153. 1961 to present. Quarterly. $320.00 per year.

Propellants, Explosives, Pyrotechnics. VCH Publishers, Inc., 220 E. 23rd St., New York, NY 10010-4606. 6 times a year. $415.00 per year.

Space Science Reviews. Kluwer Academic Publishers, P.O. Box 358, Accord Station, Hingham, MA 02018-0358. (617) 871-6000. Monthly except March, June, September, December. $258.50 per volume.

Space Technology: Industrial and Commercial Applications. Pergamon Press, 660 White Plains Rd., Tarrytown, NY 10591-5153. Bimonthly. $535.00 per year.

Spaceflight. British Interplanetary Society, 27/29 South Lambeth Road, London SW8 1SZ, England. (071) 735-3160. FAX (071) 820-1504. Monthly. $69 annual membership includes subscription.

RESEARCH CENTERS AND INSTITUTES

Aerospace Research Applications Center, Indianapolis Center For Advanced Research, 611 N. Capitol Ave., Indianapolis, in 46204. (317) 262-5003. FAX (317) 262-5044.

Aerospace Research Center, 1250 Eye St., NW, Washington, DC 20005. (202) 371-8400. FAX (202) 371-8470.

Auburn University Space Power Institute, 231 Leach Center, Auburn, AL 36849. (205) 844-5894.

Canadian Space Agency, Space Science Program, C/o National Research Council of Canada, 100 Sussex Drive, Ottawa, Ontario, Canada K1a 0r6. (613) 990-0799. FAX (613) 952-0974.

Center For Aerospace Technology, Weber State College, Ogden, UT 84408-1805. (801) 626-7272.

Center For Space Research, Massachusetts Institute of Technology, 77 Massachusetts Ave., Cambridge, MA 02139. (617) 253-7501.

Center For Control and Systems Research, University of Texas At Austin, Aerospace Engineering and Engineering Mechanics, Wrw 408, Austin, TX 78712-1085. (512) 471-4908. FAX (512) 471-3788.

Charles Stark Draper Laboratory Inc., 555 Technology Square, Cambridge, MA 02139. (617) 258-1000.

Florida A&m University, Space Life Sciences Training Program, 106 Honor House, Tallahassee, FL 32307. (904) 599-3636.

Goddard Space Flight Center, Greenbelt, MD 20771. (301) 286-7351.

Houston Advanced Research Center, Space Technology and Research Center, 4802 Research Forest Dr., the Woodlands, TX 77381. (713) 363-7922. FAX (713) 363-7914.

Innovative Nuclear Space Power and Propulsion Institute, College of Engineering, University of Florida, Gainesville, FL 32611. (904) 392-1427. FAX (904) 392-8656.

Institute For Space and Security Studies, 5115 Highway A1a, Melbourne Beach, FL 32951. (407) 952-0600.

Instrumentation and Control Laboratory, Princeton University, Dept. of MAE, Engineering Quadrangle, Princeton, NJ 08544. (609) 452-5154.

Man-vehicle Laboratory, Massachusetts Institute of Technology, Rm. 37-219, Cambridge, MA 02139. (617) 253-7805.

George C. Marshall Space Flight Center, Huntsville, AL 35812. (205) 544-0024.

Ohio Aerospace Institute, 2001 Aerospace Parkway, Brook Park, OH 44142. (216) 891-2100. FAX (216) 891-2140.

Pennsylvania State University, Propulsion Engineering Research Center, 106 Research Bldg. E, Bigler Rd., University Park, PA 16802. (814) 863-6272. FAX (814) 865-3389.

University of Tennessee, Center For Advanced Space Propulsion, Po Box 1385, Tullahoma, TN 37388-8897. (615) 454-9294. FAX (615) 455-6167.

Space Science and Engineering Center, University of Wisconsin—madison, 1225 W. Dayton St., Madison, WI 53706. (608) 262-0544.

Thermal Sciences and Propulsion Center, Purdue University,Lafayette, in 47907. (317) 494-1503.

SPACE SHUTTLE

See also: AEROSPACE ENGINEERING, MANNED SPACE FLIGHT, SPACECRAFT, SPACE STATION, ROCKETS

ABSTRACT SERVICES AND INDEXES

Applied Science and Technology Index; A Cumulative Subject Index To English Language Periodicals in the Fields of Aeronautics and Space Science, Computer Technology, Chemistry, Construction Industry, Energy and Related Areas. H.W. Wilson Co., 950 University Avenue, Bronx, NY 10452. (800) 367-6770 or (212) 588-8400. FAX (718) 590-1617. From 1958 to present. Monthly. Inquire about cost and availability. Also available on CD-ROM.

Astronomy and Astrophysics Abstracts. Springer-Verlag, 175 Fifth Avenue, New York, NY 10010. (212) 460-1500. FAX (212) 473-6272. Irregular. Price varies, inquire.

Current Contents: Engineering, Technology and Applied Sciences. Institute for Scientific Information, 3501 Market Street, Philadelphia, PA 19104. (215) 386-0100. FAX (215) 386-6362. Weekly. $360.00 per year.

Engineering Index Monthly. Engineering Information, Inc., Castle Point on the Hudson, Hoboken, NJ 07030. (800) 221-1044. FAX (212) 832-1857. Monthly. $2200.00 per year. Also available online as COMPENDEX, and also on CD-ROM.

Index To Scientific and Technical Proceedings. Institute for Scientific Information, 3501 Market St., Philadelphia, PA 19104. (215) 386-0100. FAX (215) 386-6362. Monthly. $500.00 per year.# #*Index To Scientific Reviews.* Institute for Scientific Information, 3501 Market St., Philadelphia, PA 19104. (215) 386-0100. FAX (215) 386-6362. Semi-annual.

International Aerospace Abstracts. Technical Information Service, American Institute of Aeronautics and Astronautics, Inc., 555 West 57th St., New York, NY 10019. (212) 247-6500. FAX (212) 582-4861. Semi-monthly. $1295.00 per year.

NASA Patent Abstracts Bibliography. National Aeronautics and Space Administration, Center for Aerospace Information, Box 8757, BWI Airport, Baltimore, MD 21240. (301) 621-0153. Semi-annual. Inquire for price information.

Numerical/chronological/author Index. An index to articles and papers published by or for the American Astronautical Society, including: Journal of the Astronautical Sciences, Astronautical Sciences Review, Advances in the Astronautical Sciences, Science and Technology, AAS history series, and AAS microfiche series. Horace Jacobs and Robert H. Jacobs. Univelt, Inc. PO

Box 28130, San Diego, CA 92198. (619) 746-4005. FAX (619) 746-3139. Inquire for price and availability.

Science Citation Index. Institute for Scientific Information, 3501 Market Street, Philadelphia, PA 19104. (215) 386-0100. FAX (215) 386-6362. Inquire about availability and cost. Also available on CD-ROM.

Scientific and Technical Aerospace Reports (star). National Aeronautics and Space Administration. NASA Center for Aerospace Information, Box 8757, BWI Airport, Baltimore, MD 21240. (301) 621-0153. Monthly. Inquire about availability and cost. Also available through the NASA online retrieval service (RECON), and through the Aerospace Database through DIALOG.

ASSOCIATIONS AND PROFESSIONAL SOCIETIES

Aerospace Industries Association of America. 1250 Eye Street NW, Washington, DC 20005. (202) 371-8400.

American Astronautical Society. 6352 Rolling Mill Place, Suite 102, Springfield, VA 22152.

American Institute of Aeronautics and Astronautics. 370 L'Enfant Promenade SW, Washington, DC 20024. (202) 646-7400.

National Space Society, 922 Pennsylvania Avenue Se, Washington, DC 20003-2140. (202) 543-1900. FAX (202) 546-4189.

DIRECTORIES AND BIOGRAPHICAL SOURCES

American Astronautical Society Directory. American Astronautical Society, 6352 Rolling Mill Place, Suite 102, Springfield, VA 22152.

American Institute of Aeronautics and Astronautics Roster. American Institute of Aeronautics and Astronautics. 370 L'Enfant Promenade SW, Washington, DC 20024. (202) 646-7400.

Jane's Space Directory. Jane's Information Group, 1340 Braddock Place, Suite 300, Alexandria, VA 22314-1651. (703) 683-3700. FAX (703) 836-0029. Annual. $245.00. Also available on CD-ROM.

Research Centers Directory. Gale Research, 835 Penobscot Building, Detroit, MI 48226-4094. (313) 961-2242. (800) 877-4253. FAX (313) 961-6083. $485.00.

Scientific and Technical Organizations and Agencies Directory. Gale Research, 835 Penobscot Building, Detroit, MI 48226-4094. (313) 961-2242. (800) 877-4253. FAX (313) 961-6083. 4th ed. 1995. $195.00.

Who's Who in Engineering. 7th ed. American Association of Engineering Societies, 1111 19th St. NW, Suite 608, Washington, DC 20036. (800)-658-8897. 1988. $200.00.

ENCYCLOPEDIAS AND DICTIONARIES

Concise Encyclopedia of Aeronautics and Space Systems. M. Pelegrin and W.M. Hollister. Pergamon Press Inc., Maxwell

House, Fairview Park, Elmsford, NY 10523. (914) 592-7700. Fax (914) 592-3625. 1993. Inquire for cost.

Jane's Aerospace Dictionary, B Guston. 3rd ed. Jane's Information Group, 1340 Braddock Place, Suite 300, Alexandria, VA 22314-1651. (703) 683-3700. FAX (703) 836-0029. 1991. $45.00.

McGraw-Hill Encyclopedia of Science and Technology, Sybil P. Parker, ed. 7th ed. 20 vols. McGraw-Hill Publishing Company, 1221 Avenue of the Americas, New York, NY 10020. (800) 262-4729 or (212) 512-3825. $1900.00

SAE Dictionary of Aerospace Engineering. William H. Cubberly, ed. Society of Automotive Engineers, 400 Commonwealth Dr., Warrendale, PA 15096. (412) 776-4841. 1992. $85.00.

GENERAL WORKS

Astronautics and Aeronautics, 1979-1984: A Chronology. Bette R. Janson and Eleanor H. Ritchie. National Aeronautics and Space Administration, Scientific and Technical Information Division. 1990. Order from: Superintendent of Documents, U.S. Government Printing office, Box 371954, Pittsburgh, PA 15250-7954. (202) 783-3238. FAX (202) 512-2233. Inquire for price and availability.

Space Almanac. Anthony R. Curtis, ed. 2d ed. Gulf Publishing Co., PO Box 2608, Houston, TX 77252. (713) 529-4301 or (800) 231-6275. 1992. $36.95 (hardback) and $24.95 (paperback).

Space Mission Analysis and Design. Wiley J. Larson and James R. Wertz, eds. 2d ed. Kluwer Academic Publishers, P.O. Box 358, Accord Station, Hingham, MA 02018-0358. (617) 871-6000. FAX (617) 871-6528. 1992. $99.00.

Space Shuttle. Dennis R. Jenkins. Dennis R. Jenkins Publishing/ Motorbooks International, Box 2, Osceola, WI 54020. (800) 826-6600. 1992. $24.95.

Space Shuttle: Dawn of An Era. William F. Rector III & Paul A. Penzo, editors. 2 parts. Univelt, Inc. PO Box 28130, San Diego, CA 92198. (619) 746-4005. FAX (619) 746-3139. 1980. Part 1, $45.00 ($35.00 paperback); part 2, $55.00 ($40.00 paperback).

Space Shuttle Log: The First 25 Flights. Gene Gurney & Jeff Forte. TAB Books, P.O. Box 40, Blue Ridge Summit, PA 17294-0850. (717) 794-2191 or (800) 233-1128. FAX (717) 794-2080. 1987. Inquire for cost and availability.

The World's Fastest Aircraft: From Supermarine To the Space Shuttle. Martin W. Bowman. Patrick Stephens (G.B.), c/o Haynes Publications Inc., 861 Lawrence Dr., Newbury Park, CA 91320. (800) 442-9637. 1990. Inquire for cost and availability.

HANDBOOKS AND MANUALS

Space Exploration: A Reference Handbook. Mrinal Bali. ABC-CLIO, PO Box 1911, Santa Barbara, CA 93116-2546. (800) 422-2546. FAX (805) 685-9685. 1990. $39.50.

ONLINE DATABASES AND CD-ROMS

Aerospace Database. American Institute of Aeronautics and Astronautics, 370 L'Enfant Promenage, S.W., Washington, DC

Ency. of Physical Sciences and Engineering Info. Sources

SPACE SHUTTLE

20024. (202) 646-7400. Worldwide published literature on research and development in aerospace and related areas, with abstracts. Covers 1962 to present. Available on DIALOG, (800) 334-2564, online service. Also available on CD-ROM. Inquire as to cost and availability.

Compendex Plus. Engineering Information, Inc., 345 East 47th Street, New York, NY 10017. (212) 705-7600 or (800) 221-1044. Contains citations with abstracts to worldwide literature in engineering and technology, from 1970 to present. Available on online BRS,(800) 289-4277, DIALOG, (800) 334-2564, ORBIT (800) 456-7248, and STN International, FIZ Karlsruhe, P.O. Box 2465, W-7500, Karlsruhe 1, Germany, online services. Also available on CD-ROM. Inquire as to cost and availability.

CSA Engineering. Cambridge Scientific Abstracts, 7200 Wisconsin Avenue, Suite 601, Bethesda, MD 20814. (301) 961-6750 or (800) 843-7751. Contains citations and abstracts of international periodicals and research literature covering all fields of engineering and science and technology,including computer and information science, electronics, mechanical engineering, solid state materials, 1981 to present. Available on BRS,(800) 289-4277, online service.Inquire as to cost and availability. # #*Current Contents Search.* Institute for Scientific Information, 3501 Market Street, Philadelphia, PA 19104. (215) 386-0100. FAX (215) 386-6362. Contains citations to articles listed in the table of contents of science and technology journals. Also articles in social sciences and life sciences journals. Available on BRS,(800) 289-4277, DIALOG,(800) 334-2564, online services. Inquire as to cost and availability.

Dissertation Abstracts. University Microfilms International, 300 North Zeeb Road, Ann Arbor, MI 48106. (800) 521-0600 or (313) 761-4700. Scope includes virtually all doctoral dissertations accepted at accredited American institutions from 1861 to present in 252 subject areas. Available on BRS,(800) 289-4277, DIALOG,(800) 334-2564, and OCLC EPIC,(800) 848-5878, online services. Also available on CD-ROM. Inquire as to cost and availability.

International Aerospace Abstracts. American Institute of Aeronautics and Astronautics, 370 L'Enfant Promenade, S.W., Washington, DC 20024. (202) 646-7400. Contains references and abstracts of journal and monograph literature relating to aerospace science and technology, from 1963 to present. Available through the NASA/RECON system of the National Aeronautics and Space Administration only.

NASA Database. American Institute of Aeronautics and Astronautics, 370 L'Enfant Promenade, S.W., Washington, DC 20024. (202) 646-7400. Citations and abstracts of aeronautics and astronautics literature, 1962 to present. Also contains citations from Star, Scientific and Technical Aerospace Reports, and International Aerospace Abstracts. Available through NASA/RECON online service. Inquire as to cost and availability.

NTIS Bibliographic Database. National Technical Information Service, 5285 Port Royal Road, Springfield, VA 22161. (703) 487-4929 or FAX (703) 321-8199. Broad coverage of government-sponsored science and technology research reports, 1964 to present. Available on BRS,(800) 289-4277, DIALOG, (800) 334-2564, ORBIT, (800) 456-7248, and STN International, FIZ Karlsruhe, P.O. Box 2465, W-7500, Karlsruhe 1, Germany, online services. Also available on CD-ROM. Inquire as to cost and availability.

SCISEARCH. Institute for Scientific Information, 3501 Market Street, Philadelphia, PA 19104. (800) 523-1850 or (215) 386-0100. Broad multidisciplinary title and author index to the in-

ternational literature of science and technology, 1974 to present. Available on DIALOG,(800) 334-2564, and ORBIT,(800) 456-7248, online services. Also available on CD-ROM. Inquire as to cost and availability.

WILSONLINE. H.W. Wilson Company, 950 University Avenue, Bronx, NY 10452. (800) 367-6770 or (212) 588-8400. Makes available online versions of the printed H.W. Wilson indexes including Applied Science and Technology Index, Business Periodicals Index, General Science Index, and Readers' Guide to Periodical Literature. Period covered is generally 1983 to present. Available on BRS,(800) 289-4277, DIALOG,(800) 334-2564, and OCLC EPIC,(800) 848-5878, online services. Also available on CD-ROM. Inquire as to cost and availability.

PERIODICALS

Acta Astronautica. Pergamon Press, Pergamon Press, 660 White Plains Rd., Tarrytown, NY 10591-5153. 14 times a year. Inquire for price and availability.

Ad Astra. National Space Society, 922 Pennsylvania Avenue SE, Washington, DC 20003-2140. (202) 543-1900. FAX (202) 546-4189. 1989 to present. 6 times a year. $35.00 a year.

Advances in the Astronautical Sciences. American Astronautical Society, 6352 Rolling Mill Place, Suite 102, Springfield, VA 22152. (703) 866-0020. FAX (703) 866-3526. Quarterly. Write or call for price information.

Advances in Space Research. Pergamon Press, Pergamon Press, 660 White Plains Rd., Tarrytown, NY 10591-5153. 12 times a year. Inquire for price and availability.

Aeronautical Journal. Royal Aeronautical Society, PO Box 139, Tonbridge, Kent, TN9 1EW, England. 0732-770823. FAX 0723-361708. 1897 to present. Ten times a year. Write for price and availability.

Aerospace. Royal Aeronautical Society, PO Box 139, Tonbridge, Kent, TN9 1EW, England. 0732-770823. FAX 0723-361708. Monthly. Write for price and availability.

Aerospace America. American Institute of Aeronautics and Astronautics, 370 L'Enfant Promenade SW, Washington, DC 20024-2518. (202) 646-7471. 1932 to present. Monthly. $70 per year (non-members).

Aerospace Engineering. SAE, Inc., 400 Commonwealth Dr., Warrendale, PA 15096. (412) 776-4841. 1981 to present. Monthly. $48 a year.

Aviation & Aerospace. Baxter Publishing Company, 310 Dupont Street, Toronto, Ontario M5R 1V9 Canada. (416) 968-7252. FAX (416) 968-2377. Bi-monthly. $42.00 per year.

Aviation Week and Space Technology. 1221 Avenue of the Americas, New York, NY 10020. (212) 512-2999 or (800) 525-5003. 1916 to present. Weekly. $82.00 per year.

ESA Bulletin. European Space Agency, Publications Division, Keplerlaan 1, 22000 AG Nordwijk, Netherlands. Phone 31-1719-86555. FAX 31-1719-85433. Quarterly. Free. # #*ESA Journal.* European Space Agency, Publications Division, Keplerlaan 1, 22000 AG Nordwijk, Netherlands. Phone 31-1719-86555. FAX 31-1719-85433. Quarterly. Free.

SPACE SHUTTLE

Ency. of Physical Sciences and Engineering Info. Sources

Journal of the British Interplanetary Society (JBIS). British Interplanetary Society, 27/29 South Lambeth Road, London SW8 1SZ, England. (071) 735-3160. FAX (071) 820-1504. Monthly. $370.00 per year.

Interavia/aerospace World. Aerospace Media Publishing, Swissair Centre, 31 Route de l'Aeroport, POB Box 437, 1215 Geneva 15, Switzerland. (022) 788 27 88. FAX (022) 788 27 26. Monthly. $128.00 per year (U.S.).

International Journal of Satellite Communications. John Wiley and Sons Ltd. Journals, Baffins Ln., Chichester, Sussex PO19 1UD, England. Phone 0243-77977. FAX 0243-775878. Bimonthly. $525.00 per year.

Journal of the Astronautical Sciences. American Astronautical Society, 6352 Rolling Mill Place, Suite 102, Springfield, VA 22152. Quarterly. $35 per year.

Journal of Guidance, Control and Dynamics. American Institute of Aeronautics and Astronautics, 370 L'Enfant Promenade SW, Washington, DC 20024-2518. 1978 to present. Bimonthly. $42.00 (members) and $250.00 (non-members).

Journal of Propulsion and Power. American Institute of Aeronautics and Astronautics, 370 L'Enfant Promenade SW, Washington, DC 20024-2518. 1985 to present. Bimonthly. $38.00 (members) and $260.00 (non-members).

Journal of Spacecraft and Rockets. American Institute of Aeronautics and Astronautics, the Aerospace Center, 370 L'Enfant Promenade SW, Washington, DC 20024. (202) 646-7400. FAX (202) 646-7508. 1964 to present. Bi-monthly. $185.00 per year.

Planetary Report. Planetary Society, 65 N. Catalina, Pasadena, CA 91106-2301. (818) 793-5100. FAX (818) 793-5528. Bimonthly. $25.00 per year.

Progress in Aerospace Sciences. Pergamon Press, 660 White Plains Rd., Tarrytown, NY 10591-5153. 1961 to present. Quarterly. $320.00 per year.

Propellants, Explosives, Pyrotechnics. VCH Publishers, Inc., 220 E. 23rd St., New York, NY 10010-4606. 6 times a year. $415.00 per year.

Space Science Reviews. Kluwer Academic Publishers, P.O. Box 358, Accord Station, Hingham, MA 02018-0358. (617) 871-6000. Monthly except March, June, September, December. $258.50 per volume.# #*Space Technology: Industrial and Commercial Applications.* Pergamon Press, 660 White Plains Rd., Tarrytown, NY 10591-5153. Bimonthly. $535.00 per year.

Spaceflight. British Interplanetary Society, 27/29 South Lambeth Road, London SW8 1SZ, England. (071) 735-3160. FAX (071) 820-1504. Monthly. $69 annual membership includes subscription.

RESEARCH CENTERS AND INSTITUTES

Aerospace Research Applications Center, Indianapolis Center For Advanced Research, 611 N. Capitol Ave., Indianapolis, in 46204. (317) 262-5003. FAX (317) 262-5044.

Aerospace Research Center, 1250 Eye St., NW, Washington, DC 20005. (202) 371-8400. FAX (202) 371-8470.

Auburn University Space Power Institute, 231 Leach Center, Auburn, AL 36849. (205) 844-5894.

Canadian Space Agency, Space Science Program, C/o National Research Council of Canada, 100 Sussex Drive, Ottawa, Ontario, Canada K1a 0r6. (613) 990-0799. FAX (613) 952-0974.

Center For Aerospace Technology, Weber State College, Ogden, UT 84408-1805. (801) 626-7272.

Center For Space Research, Massachusetts Institute of Technology, 77 Massachusetts Ave., Cambridge, MA 02139. (617) 253-7501.

Center For Control and Systems Research, University of Texas At Austin, Aerospace Engineering and Engineering Mechanics, Wrw 408, Austin, TX 78712-1085. (512) 471-4908. FAX (512) 471-3788.

Charles Stark Draper Laboratory Inc., 555 Technology Square, Cambridge, MA 02139. (617) 258-1000.

Florida A&m University, Space Life Sciences Training Program, 106 Honor House, Tallahassee, FL 32307. (904) 599-3636.

Goddard Space Flight Center, Greenbelt, MD 20771. (301) 286-7351.

Houston Advanced Research Center, Space Technology and Research Center, 4802 Research Forest Dr., the Woodlands, TX 77381. (713) 363-7922. FAX (713) 363-7914.

Innovative Nuclear Space Power and Propulsion Institute, College of Engineering, University of Florida, Gainesville, FL 32611. (904) 392-1427. FAX (904) 392-8656.

Institute For Space and Security Studies, 5115 Highway A1a, Melbourne Beach, FL 32951. (407) 952-0600.

Instrumentation and Control Laboratory, Princeton University, Dept. of MAE, Engineering Quadrangle, Princeton, NJ 08544. (609) 452-5154.

Man-vehicle Laboratory, Massachusetts Institute of Technology, Rm. 37-219, Cambridge, MA 02139. (617) 253-7805.

George C. Marshall Space Flight Center, Huntsville, AL 35812. (205) 544-0024.

Ohio Aerospace Institute, 2001 Aerospace Parkway, Brook Park, OH 44142. (216) 891-2100. FAX (216) 891-2140.

Pennsylvania State University, Propulsion Engineering Research Center, 106 Research Bldg. E, Bigler Rd., University Park, PA 16802. (814) 863-6272. FAX (814) 865-3389.

University of Tennessee, Center For Advanced Space Propulsion, Po Box 1385, Tullahoma, TN 37388-8897. (615) 454-9294. FAX (615) 455-6167.

Space Science and Engineering Center, University of Wisconsin—madison, 1225 W. Dayton St., Madison, WI 53706. (608) 262-0544.

Thermal sciences and propulsion center, purdue university, Lafayette, in 47907. (317) 494-1503.

SPACE STATION
See also: AEROSPACE ENGINEERING, MANNED SPACE FLIGHT,
SPACE SHUTTLE

ABSTRACT SERVICES AND INDEXES

Applied Science and Technology Index; A Cumulative Subject Index To English Language Periodicals in the Fields of Aeronautics and Space Science, Computer Technology, Chemistry, Construction Industry, Energy and Related Areas. H.W. Wilson Co., 950 University Avenue, Bronx, NY 10452. (800) 367-6770 or (212) 588-8400. FAX (718) 590-1617. From 1958 to present. Monthly. Inquire about cost and availability. Also available on CD-ROM.

Current Contents: Engineering, Technology and Applied Sciences. Institute for Scientific Information, 3501 Market Street, Philadelphia, PA 19104. (215) 386-0100. FAX (215) 386-6362. Weekly. $360.00 per year.

Engineering Index Monthly. Engineering Information, Inc., Castle Point on the Hudson, Hoboken, NJ 07030. (800) 221-1044. FAX (212) 832-1857. Monthly. $2200.00 per year. Also available online as COMPENDEX, and also on CD-ROM.

Index To Scientific and Technical Proceedings. Institute for Scientific Information, 3501 Market St., Philadelphia, PA 19104. (215) 386-0100. FAX (215) 386-6362. Monthly. $500.00 per year.# #*Index To Scientific Reviews.* Institute for Scientific Information, 3501 Market St., Philadelphia, PA 19104. (215) 386-0100. FAX (215) 386-6362. Semi-annual.

International Aerospace Abstracts. Technical Information Service, American Institute of Aeronautics and Astronautics, Inc., 555 West 57th St., New York, NY 10019. (212) 247-6500. FAX (212) 582-4861. Semi-monthly. $1295.00 per year.

NASA Patent Abstracts Bibliography. National Aeronautics and Space Administration, Center for Aerospace Information, Box 8757, BWI Airport, Baltimore, MD 21240. (301) 621-0153. Semi-annual. Inquire for price information.

Numerical/chronological/author Index. An index to articles and papers published by or for the American Astronautical Society, including: Journal of the Astronautical Sciences, Astronautical Sciences Review, Advances in the Astronautical Sciences, Science and Technology, AAS history series, and AAS microfiche series. Horace Jacobs and Robert H. Jacobs. Univelt, Inc. PO Box 28130, San Diego, CA 92198. (619) 746-4005. Inquire for price and availability.

Science Citation Index. Institute for Scientific Information, 3501 Market Street, Philadelphia, PA 19104. (215) 386-0100. FAX (215) 386-6362. Inquire about availability and cost. Also available on CD-ROM.

Scientific and Technical Aerospace Reports (star). National Aeronautics and Space Administration. NASA Center for Aerospace Information, Box 8757, BWI Airport, Baltimore, MD 21240. (301) 621-0153. Monthly. Inquire about availability and cost. Also available through the NASA online retrieval service (RECON), and through the Aerospace Database through DIALOG.

ASSOCIATIONS AND PROFESSIONAL SOCIETIES

Aerospace Industries Association of America. 1250 Eye Street NW, Washington, DC 20005. (202) 371-8400.

American Astronautical Society. 6352 Rolling Mill Place, Suite 102, Springfield, VA 22152.

American Institute of Aeronautics and Astronautics. 370 L'Enfant Promenade SW, Washington, DC 20024. (202) 646-7400.

National Space Society, 922 Pennsylvania Avenue Se, Washington, DC 20003-2140. (202) 543-1900. FAX (202) 546-4189.

DIRECTORIES AND BIOGRAPHICAL SOURCES

American Astronautical Society Directory. American Astronautical Society, 6352 Rolling Mill Place, Suite 102, Springfield, VA 22152.# #*American Institute of Aeronautics and Astronautics Roster.* American Institute of Aeronautics and Astronautics. 370 L'Enfant Promenade SW, Washington, DC 20024. (202) 646-7400.

Jane's Space Directory. Jane's Information Group, 1340 Braddock Place, Suite 300, Alexandria, VA 22314-1651. (703) 683-3700. FAX (703) 836-0029. Annual. $245.00. Also available on CD-ROM.

Research Centers Directory. Gale Research, 835 Penobscot Building, Detroit, MI 48226-4094. (313) 961-2242. (800) 877-4253. FAX (313) 961-6083. $485.00.

Scientific and Technical Organizations and Agencies Directory. Gale Research, 835 Penobscot Building, Detroit, MI 48226-4094. (313) 961-2242. (800) 877-4253. FAX (313) 961-6083. 4th ed. 1995. $195.00.

Who's Who in Engineering. 7th ed. American Association of Engineering Societies, 1111 19th St. NW, Suite 608, Washington, DC 20036. (800)-658-8897. 1988. $200.00.

ENCYCLOPEDIAS AND DICTIONARIES

Concise Encyclopedia of Aeronautics and Space Systems. M. Pelegrin and W.M. Hollister. Pergamon Press Inc., Maxwell House, Fairview Park, Elmsford, NY 10523. (914) 592-7700. Fax (914) 592-3625. 1993. Inquire for cost.

Jane's Aerospace Dictionary, B. Guston. 3d ed. Jane's Information Group, 1340 Braddock Place, Suite 300, Alexandria, VA 22314-1651. (703) 683-3700. FAX (703) 836-0029. 1991. $45.00.

McGraw-Hill Encyclopedia of Science and Technology, Sybil P. Parker, ed. 7th ed. 20 vols. McGraw-Hill Publishing Company, 1221 Avenue of the Americas, New York, NY 10020. (800) 262-4729 or (212) 512-3825. $1900.00

SAE Dictionary of Aerospace Engineering. William H. Cubberly, ed. Society of Automotive Engineers, 400 Commonwealth Dr., Warrendale, PA 15096. (412) 776-4841. 1992. $85.00.

GENERAL WORKS

Astronautics and Aeronautics, 1979-1984: A Chronology. Bette R. Janson and Eleanor H. Ritchie. National Aeronautics and Space Administration, Scientific and Technical Information Division. 1990. Order from: Superintendent of Documents, U.S. Government Printing office, Box 371954, Pittsburgh, PA 15250-7954. (202) 783-3238. FAX (202) 512-2233. Inquire for cost and availability.

Space Age. William J. Walter. Random House, 201 E. 50th Street, 22nd floor, New York, NY 10022. (212) 751-2600 or (800) 733-3000. 1992. $29.50.# #*Space Almanac.* Anthony R. Curtis, ed. 2d ed. Gulf Publishing Co., PO Box 2608, Houston, TX 77252. (713) 529-4301 or (800) 231-6275. 1992. $36.95 (hardback) and $24.95 (paperback).]

Spacelab: An International Success Story. Douglas R. Lord. NASA/ Order from: Superintendent of Documents, U.S. Government Printing office, Box 371954, Pittsburgh, PA 15250-7954. (202) 783-3238. FAX (202) 512-2233. 1987. $33.00.

The Space Station Decision: Incremental Politics and Technological Choice. Howard E. McCurdy. Johns Hopkins University Press, 2715 North Charles St., Baltimore, MD 21218. (410) 516-6900 or (800) 537-5487. FAX (410) 516-6998. 1990. $42.50.

Space Stations & Platforms. G.R. Woodcock. Krieger Publishing Company, P.O. Box 9542, Melbourne, FL 32902-9542. (407) 724-9542. FAX (407) 951-3671. 1986. $52.50.

ONLINE DATABASES AND CD-ROMS

Aerospace Database. American Institute of Aeronautics and Astronautics, 370 L'Enfant Promenade, S.W., Washington, DC 20024. (202) 646-7400. Worldwide published literature on research and development in aerospace and related areas, with abstracts. Covers 1962 to present. Available on DIALOG, (800) 334-2564, online service. Also available on CD-ROM. Inquire as to cost and availability.

Compendex Plus. Engineering Information, Inc., 345 East 47th Street, New York, NY 10017. (212) 705-7600 or (800) 221-1044. Contains citations with abstracts to worldwide literature in engineering and technology, from 1970 to present. Available on online BRS,(800) 289-4277, DIALOG, (800) 334-2564, ORBIT (800) 456-7248, and STN International, FIZ Karlsruhe, P.O. Box 2465, W-7500, Karlsruhe 1, Germany, online services. Also available on CD-ROM. Inquire as to cost and availability.

CSA Engineering. Cambridge Scientific Abstracts, 7200 Wisconsin Avenue, Suite 601, Bethesda, MD 20814. (301) 961-6750 or (800) 843-7751. Contains citations and abstracts of international periodicals and research literature covering all fields of engineering and science and technology,including computer and information science, electronics, mechanical engineering, solid state materials, 1981 to present. Available on BRS,(800) 289-4277, online service.Inquire as to cost and availability.

Current Contents Search. Institute for Scientific Information, 3501 Market Street, Philadelphia, PA 19104. (215) 386-0100. FAX (215) 386-6362. Contains citations to articles listed in the table of contents of science and technology journals. Also articles in social sciences and life sciences journals. Available on BRS,(800) 289-4277, DIALOG,(800) 334-2564, online services. Inquire as to cost and availability.

Dissertation Abstracts. University Microfilms International, 300 North Zeeb Road, Ann Arbor, MI 48106. (800) 521-0600 or (313) 761-4700. Scope includes virtually all doctoral dissertations accepted at accredited American institutions from 1861 to present in 252 subject areas. Available on BRS,(800) 289-4277, DIALOG,(800) 334-2564, and OCLC EPIC,(800) 848-5878, online services. Also available on CD-ROM. Inquire as to cost and availability.

International Aerospace Abstracts. American Institute of Aeronautics and Astronautics, 370 L'Enfant Promenade, S.W., Washington, DC 20024. (202) 646-7400. Contains references and abstracts of journal and monograph literature relating to aerospace science and technology, from 1963 to present. Available through the NASA/RECON system of the National Aeronautics and Space Administration only.

NASA Database. American Institute of Aeronautics and Astronautics, 370 L'Enfant Promenade, S.W., Washington, DC 20024. (202) 646-7400. Citations and abstracts of aeronautics and astronautics literature, 1962 to present. Also contains citations from Star, Scientific and Technical Aerospace Reports, and International Aerospace Abstracts. Available through NASA/RECON online service. Inquire as to cost and availability.

NTIS Bibliographic Database. National Technical Information Service, 5285 Port Royal Road, Springfield, VA 22161. (703) 487-4929 or FAX (703) 321-8199. Broad coverage of government-sponsored science and technology research reports, 1964 to present. Available on BRS,(800) 289-4277, DIALOG, (800) 334-2564, ORBIT, (800) 456-7248, and STN International, FIZ Karlsruhe, P.O. Box 2465, W-7500, Karlsruhe 1, Germany, online services. Also available on CD-ROM. Inquire as to cost and availability.

SCISEARCH. Institute for Scientific Information, 3501 Market Street, Philadelphia, PA 19104. (800) 523-1850 or (215) 386-0100. Broad multidisciplinary title and author index to the international literature of science and technology, 1974 to present. Available on DIALOG,(800) 334-2564, and ORBIT,(800) 456-7248, online services. Also available on CD-ROM. Inquire as to cost and availability.

WILSONLINE. H.W. Wilson Company, 950 University Avenue, Bronx, NY 10452. (800) 367-6770 or (212) 588-8400. Makes available online versions of the printed H.W. Wilson indexes including Applied Science and Technology Index, Business Periodicals Index, General Science Index, and Readers' Guide to Periodical Literature. Period covered is generally 1983 to present. Available on BRS,(800) 289-4277, DIALOG,(800) 334-2564, and OCLC EPIC,(800) 848-5878, online services. Also available on CD-ROM. Inquire as to cost and availability.

PERIODICALS

Acta Astronautica. Pergamon Press, 660 White Plains Rd., Tarrytown, NY 10591-5153. 14 times a year. Inquire for price and availability.

Ad Astra. National Space Society, 922 Pennsylvania Avenue SE, Washington, DC 20003-2140. (202) 543-1900. FAX (202) 546-4189. 1989 to present. 6 times a year. $35.00 a year.

Advances in the Astronautical Sciences. American Astronautical Society, 6352 Rolling Mill Place, Suite 102, Springfield, VA 22152. (703) 866-0020. FAX (703) 866-3526. Quarterly. Write or call for price information.

Advances in Space Research. Pergamon Press, Pergamon Press, 660 White Plains Rd., Tarrytown, NY 10591-5153. 12 times a year. Inquire for price and availability.

Aeronautical Journal. Royal Aeronautical Society, PO Box 139, Tonbridge, Kent, TN9 1EW, England. 0732-770823. FAX 0723-361708. 1897 to present. Ten times a year. Write for price and availability.

Aerospace. Royal Aeronautical Society, PO Box 139, Tonbridge, Kent, TN9 1EW, England. 0732-770823. FAX 0723-361708. Monthly. Write for price and availability.

Aerospace America. American Institute of Aeronautics and Astronautics, 370 L'Enfant Promenade SW, Washington, DC 20024-2518. (202) 646-7471. 1932 to present. Monthly. $70 per year (non-members).

Aerospace Engineering. SAE Inc., 400 Commonwealth Dr., Warrendale, PA 15096. (412) 776-4841. 1981 to present. Monthly. $48 a year.

Aviation & Aerospace. Baxter Publishing Company, 310 Dupont Street, Toronto, Ontario M5R 1V9 Canada. (416) 968-7252. FAX (416) 968-2377. Bi-monthly. $42.00 per year.

Aviation Week and Space Technology. 1221 Avenue of the Americas, New York, NY 10020. (212) 512-2999 or (800) 525-5003. 1916 to present. Weekly. $82.00 per year.

Journal of the British Interplanetary Society (jbis). British Interplanetary Society, 27/29 South Lambeth Road, London SW8 1SZ, England. (071) 735-3160. FAX (071) 820-1504. Monthly. $370.00 per year.

Interavia/aerospace World. Aerospace Media Publishing, Swissair Centre, 31 Route de l'Aeroport, POB Box 437, 1215 Geneva 15, Switzerland. (022) 788 27 88. FAX (022) 788 27 26. Monthly. $128.00 per year (U.S.).

International Journal of Satellite Communications. John Wiley and Sons Ltd. Journals, Baffins Ln., Chichester, Sussex PO19 1UD, England. Phone 0243-77977. FAX 0243-775878. Bi-monthly. $525.00 per year.

Journal of the Astronautical Sciences. American Astronautical Society, 6352 Rolling Mill Place, Suite 102, Springfield, VA 22152. Quarterly. $35 per year.

Journal of Spacecraft and Rockets. American Institute of Aeronautics and Astronautics, the Aerospace Center, 370 L'Enfant Promenade SW, Washington, DC 20024. (202) 646-7400. FAX (202) 646-7508. 1964 to present. Bi-monthly. $185.00 per year.

Progress in Aerospace Sciences. Pergamon Press, 660 White Plains Rd., Tarrytown, NY 10591-5153. 1961 to present. Quarterly. $320.00 per year.

Space Science Reviews. Kluwer Academic Publishers, P.O. Box 358, Accord Station, Hingham, MA 02018-0358. (617) 871-6000. Monthly except March, June, September, December. $258.50 per volume.

Space Technology: Industrial and Commercial Applications. Pergamon Press, 660 White Plains Rd., Tarrytown, NY 10591-5153. Bimonthly. $535.00 per year.

Spaceflight. British Interplanetary Society, 27/29 South Lambeth Road, London SW8 1SZ, England. (071) 735-3160. FAX (071) 820-1504. Monthly. $69 annual membership includes subscription.

RESEARCH CENTERS AND INSTITUTES

Aerospace Research Applications Center, Indianapolis Center For Advanced Research, 611 N. Capitol Ave., Indianapolis, in 46204. (317) 262-5003. FAX (317) 262-5044.

Aerospace Research Center, 1250 Eye St., NW, Washington, DC 20005. (202) 371-8400. FAX (202) 371-8470.

Center For Aerospace Technology, Weber State College, Ogden, UT 84408-1805. (801) 626-7272.

Center For Space Research, Massachusetts Institute of Technology, 77 Massachusetts Ave., Cambridge, MA 02139. (617) 253-7501.

Charles Stark Draper Laboratory Inc., 555 Technology Square, Cambridge, MA 02139. (617) 258-1000.

Florida A&m University, Space Life Sciences Training Program, 106 Honor House, Tallahassee, FL 32307. (904) 599-3636.

Goddard Space Flight Center, Greenbelt, MD 20771. (301) 286-7351.

Houston Advanced Research Center, Space Technology and Research Center, 4802 Research Forest Dr., the Woodlands, TX 77381. (713) 363-7922. FAX (713) 363-7914.

Institute For Space and Security Studies, 5115 Highway A1a, Melbourne Beach, FL 32951. (407) 952-0600.

Man-vehicle Laboratory, Massachusetts Institute of Technology, Rm. 37-219, Cambridge, MA 02139. (617) 253-7805.

George C. Marshall Space Flight Center, Huntsville, AL 35812. (205) 544-0024.

Ohio Aerospace Institute, 2001 Aerospace Parkway, Brook Park, OH 44142. (216) 891-2100. FAX (216) 891-2140.

Space Science and Engineering Center, University of Wisconsin At Madison, 1225 W. Dayton St., Madison, WI 53706. (608) 262-0544.

SPACE STRUCTURES

See: SPACE STATION

SPECTROSCOPY

Ency. of Physical Sciences and Engineering Info. Sources

SPECTROSCOPY

See also: ANALYTICAL CHEMISTRY, ELECTRON SPECTROSCOPY, MASS SPECTROMETRY, NUCLEAR MAGNETIC RESONANCE

ABSTRACT SERVICES AND INDEXES

Applied Science and Technology Index; A Cumulative Subject Index To English Language Periodicals in the Fields ofaeronautics and Space Science, Computer Technology,chemistry, Construction Industry, Energy and Related Areas.

H.W. Wilson Co., 950 University Avenue, Bronx, NY 10452.(212) 588-8400. (800) 367-6770. FAX (718) 590-1617. From1958 to present. Monthly. Inquire about cost and availability. Also available on CD-ROM and online. ISSN: 0003-6986.

Chemical Abstracts. Chemical Abstracts Service. 2540 Olentangy River Road, Box 3012, Columbus, OH 43210-0012. (614) 447-3600. FAX (614) 447-3713. 1907 to present. Weekly. $16,800.00 per year. Available online and on CD-ROM. CA is also available in five section groupings. Inquire regarding cost and availability.

Current Contents: Physical, Chemical and Earth Sciences. Institute for Scientific Information, 3501 Market Street, Philadelphia, PA 19104. (215) 386-0100. FAX (215) 386-2291. From 1961 to present. Weekly. $442.00 per year. Also available on CD-ROM and online. Inquire regarding cost and availability. ISSN: 0163-2574.

Engineering Index Monthly; Indexes and Abstracts the World's Engineering and Technical Literature. Engineering Information, Inc., Castle Point on the Hudson, Hoboken, NJ 07030. (201) 216-8500. (800) 221-1044. FAX (201) 216-8532. Monthly. $2300.00 per year. Available online as COMPENDEX and also on CD-ROM. ISSN: 0742-1974.

General Science Index. H.W. Wilson Company, 950 University Avenue, Bronx, NY 10452. (212) 588-8400. (800) 367-6770. FAX (718) 590-1617. From 1978 to present. Ten issues per year; quarterly and annual cumulations. Service basis. Available on CD-ROM and online. Inquire regarding cost and availability. ISSN: 0162-1963.

PHYSICS ABSTRACTS. INSPEC. Section A, Science Abstracts. Institution of Electrical Engineers (IEE), London. Available from: INSPEC/IEEE - Institute of Electrical and Electronic Engineers, Box 1331, Hoes Lane, Piscataway, NJ 08855-1331. (908) 562-5549. 1898 to present. 24 issues per year. $2835.00 per year. Also available online and on CD-ROM. ISSN: 0036-8091.

Physics Briefs (Physikalische Berichte). Information Center for Energy, Physics, Mathematics; German Physical Society. V C H Publishers, Inc., 220 East 23rd Street, New York, NY 10010-4606. (212) 683-8333. 1845 to present. 24 issues per year. $2390.00 per year. Also available online. ISSN: 0179-7434.

Science Citation Index. SCI. Institute for Scientific Information, 3501 Market Street, Philadelphia, PA 19104. (215) 386-0100. (800) 523-1850. FAX (215) 386-2991. 1961 to present. Six issues per year, plus annual cumulation. $11650.00 per year. Also available online and on CD-ROM. Inquire about price and availability. ISSN: 0036-827X.

ANNUAL REVIEWS AND YEARBOOKS

Annual Reports On NMR Spectroscopy. Academic Press, Inc., 6277 Sea Harbor Drive, Orlando, FL 32887. (800) 321-5068. 1968 to present. Annual. ISSN: 0066-4103.

N M R: Basic Principles and Progress. Springer-Verlag New York, Inc., 175 Fifth Avenue, New York, NY 10010. (212) 460-1500. (800) 777-4643.1969 to present. Irregular. Price varies. ISSN: 0170-5989.

ASSOCIATIONS AND PROFESSIONAL SOCIETIES

American Institute of Physics. One Physics Ellipse, College Park, MD 20740-3843. (301) 209-3100. FAX (415) 855-2954.

American Society For Mass Spectroscopy. 1201 Don Diego, Santa Fe, NM 87505. (505) 989-4517.

Coblentz Society. c/o David W. Schiering, Perkin-Elmer Corporation, 761 Main Avenue, Norwalk, CT 06859-0240. (203) 761-2915. FAX (203) 761-2842.

Federation of Analytical Chemistry and Spectroscopoy Societies. c/o Jo Ann Brown, 201-B Broadway Street, Frederick, MD 21701. (301) 846-4797. FAX (301) 694-6860.

Society For Applied Spectroscopy. 201B Broadway Street, Frederick, MD 21701. (301) 694-8122. FAX (301) 694-6860.

DIRECTORIES AND BIOGRAPHICAL SOURCES

American Men and Women of Science: Physical and Biological Sciences. R. R. Bowker Inc., 121 Chanlon Road, New Providence, NJ 07974. (908) 464-6800. (800) 521-8110. 20th edition. 8 volumes. 1996. $850.00.

Directory of Engineering Societies and Related Organizations. American Association of Engineering Societies, 1111 19th Street, Suite 608, Washington, DC 20036-3603. (202) 296-2237. Semi-annual. $150.00.

Engineering Research Centers: Incorporating Electronics Research Centers. Stockton Press, 345 Park Avenue, New York, NY 10010. (212) 689-9200. (800) 221-2123. 4th edition. 1995. $515.00.

International Engineering Directory. American Consulting Engineers Council, 1015 15th Street, N.W. Suite 802, Washington, DC 20005-2670. (202) 347-7474. Annual. $10.00.

Research Centers Directory. Gale Research, 835 Penobscot Building, Detroit, MI 48226-4094. (313) 961-2242. (800) 877-4253. 20th edition. 1995. $485.00. ISSN: 0080-1518.

Scientific and Technical Organizations and Agencies Directory. Gale Research, 835 Penobscot Building, Detroit, MI 48226-4094. (313) 961-2242. (800) 877-4253. 4th edition. 1996. $195.00.

Society For Applied Spectroscopy Membership Directory. Society for Applied Spectroscopy, 201B Broadway Street, Frederick, MD 21701. (301) 694-8122. FAX (301) 694-6860. Irregular.

ENCYCLOPEDIAS AND DICTIONARIES

Chambers Nuclear Energy and Radiation Dictionary. Peter M.B. Walker, editor. Cambridge University Press, 40 West 20th Street, New York, NY 10011-4211. (212) 924-3900. 1992. $40.00.

Encyclopedia of Applied Physics. George Trigg, editor. VCH Publications, Inc., 220 East 23rd Street, Suite 909, New York, NY 10010-4606. (212) 683-8333. (800) 422-8824. 20 volume set. 1991 - . $5990.00.

McGraw-Hill Encyclopedia of Science and Technology. McGraw-Hill Book, Incorporated, 1221 Avenue of the Americas, New York, NY 10020. (212) 997-3675. (800) 262-4729. Seventh edition. Twenty volumes. 1992. $1900.00.

GENERAL WORKS

Algebraic theory of Molecules. F. Iachello and R. D. Levine. Oxford University Press, Inc., 200 Madison Avenue, New York, NY 10016. (212) 725-6000. (800) 334-4249. 1995. $59.95.

Application of Numerical Methods: Molecular Spectroscopy. Pelikan. CRC Press, Inc., 2000 Corporate Boulevard, NW, Boca Raton, FL 33431. (407) 994-0555. (800) 272-7737. 1993. $79.95.

Chemical, Biological and Industrial Applications of Infrared Spectroscopy. James R. Durig. John Wiley & Sons, Inc., 605 Third Avenue, New York, NY 10158-0012. (212) 850-6000. (800) 225-5945. 1985. $130.00.

Energy Levels in Atoms and Molecules. W. Graham Richards and Peter Scott. Oxford University Press, Inc., 200 Madison Avenue, New York, NY 10016. (212) 725-6000. (800) 334-4249. 1995. $9.95.

Engineering Applications of Correlation and Spectral Analysis. Julius Bendat and Allan G. Piersol. John Wiley & Sons, Inc., 605 Third Avenue, New York, NY 10158-0012. (212) 850-6000. (800) 225-5945.

Gamma and X-ray Spectrometry Techniques and Applications. K. Debertin and W. B. Mann. Elsevier Science Publishing Company, Inc., 655 Avenue of the Americas, New York, NY 10010. (212) 989-5800. FAX (914) 333-2444.

Internal Reflection Spectroscopy: Methods and Techniques. Mirabella. Marcel Dekker, Inc., 270 Madison Avenue, New York, NY 10016. (212) 696- 9000. (800) 228-1160. 1992. $160.00.

Introduction To Nmr Spectroscopy. R. J. Abraham, et al. John Wiley & Sons, Inc., 605 Third Avenue, New York, NY 10158-0012. (212) 850-6000. (800) 225-5945. 1992. $36.95.

Introductory Raman Spectroscopy. John R. Ferraro and Kazuo Nakamoto, editors. Academic Press, Inc., 6277 Sea Harbor Drive, Orlando, FL. (800) 321-5068. 1994. $69.95.

Modern NMR Spectroscopy: A Guide For Chemists. Jeremy K Sanders and Brian K. Hunter. Oxford University Press, Inc., 200 Madison Avenue, New York, NY 10016. (212) 725-6000. (800) 334-4249. 2nd edition. 1993. $80.00.

Quantum Optics and Spectroscopy. J. Fiutak, et al, editors. Nova Science Publishers, Inc., 6080 Jericho Turnpike, Suite 207, Commack, NY 11725-2808. (516) 499-3103. 1993. $83.00.

Solid State Phenomena As Seen By Muons, Protons and Excited Nuclei. Erik B. Karlsson. Oxford University Press, Inc., 200 Madison Avenue, New York, NY 10016. (212) 725-6000. (800) 334-4249.

HANDBOOKS AND MANUALS

CRC Handbook of Chemistry and Physics. David R. Kide, editor. CRC Press Inc., 2000 Corporate Blvd., NW, Boca Raton, FL 33431. (407) 994-0555. (800) 333-8300. 77th edition. 1996. $99.95.

Guide to Spectroscopic Identification of Organic Compounds. Karen Feinstein. CRC Press Inc., 2000 Corporate Blvd., NW, Boca Raton, FL 33431. (407) 994-0555. (800) 333-8300. 1994. $29.95.

Guide to the Complete Interpretation of Infrared Spectra of Organic Structures. Noel P. G. Roeges. John Wiley & Sons, Inc., 605 Third Avenue, New York, NY 10158-0012. (212) 850-6000. (800) 225-5945. 1994. $69.95.

Handbook of Inductively Coupled Plasma Atomic Emission Spectra. CRC Press Inc., 2000 Corporate Blvd., NW, Boca Raton, FL 33431. (407) 994-0555. (800) 333-8300. 1990. $207.95.

Handbook of Nuclear Spectrometry. Juhnai Kantele. Academic Press, Inc., 6277 Sea Harbor Drive, Orlando, FL. (800) 321-5068.1995. $79.95.

Handbook of X-ray Spectrometry: Methods and Techniques. Van Greiken, editor. Marcel Dekker, Inc., 270 Madison Avenue, New York, NY 10016. (212) 696-9000. (800) 228-1160. 1992. $240.00.

Organic Electronic Spectral Data. Henry Feuer, et al. John Wiley & Sons, Inc., 605 Third Avenue, New York, NY 10158-0012. (212) 850-6000. (800) 225-5945. 1992. $240.00.

Practical Handbook of Spectroscopy. James W. Robinson. CRC Press, Inc., 2000 Corporate Boulevard, NW, Boca Raton, FL 33431. (407) 994-0555. (800) 272-7737. 1991. $96.50.

Ramin-ir Atlas of Organic Compounds. B. Schrader. VCH Publications, Inc., 220 East 23rd Street, Suite 909, New York, NY 10010-4606. (212) 683-8333. (800) 422-8824. 1989. $595.00.

ONLINE DATABASES AND CD-ROMS

CA Search. Chemical Abstracts Service, P.O. Box 3012, Columbus, OH 43210-0012. (614) 447-3600. (800) 848-6533. FAX (614) 447-3709. Very comprehensive guide to worldwide chemical literature and related fields, 1972 to present. Available on BRS,(800) 289-4277, DIALOG, (800) 334-2564,ORBIT (800) 456-7248, and STN International, FIZ Karlsruhe, P.O. Box 2465, W-7500, Karlsruhe 1, Germany, online services. Inquire as to cost and availability.

Compendex Plus. Engineering Information, Inc., 345 East 47th Street, New York, NY 10017. (212) 705-7600 or (800) 221-1044. Contains citations with abstracts to worldwide literature in engineering and technology, from 1970 to present. Available on online BRS,(800) 289-4277, DIALOG, (800) 334-2564, ORBIT (800) 456-7248, and STN International, FIZ Karlsruhe, P.O. Box 2465, W-7500, Karlsruhe 1, Germany, online services. Also available on CD-ROM. Inquire as to cost and availability.

Current Contents Search. Institute for Scientific Information, 3501 Market Street, Philadelphia, PA 19104. (215) 386-0100. FAX (215) 386-6362. Contains citations to articles listed in the table of contents of science and technology journals. Also articles in social sciences and life sciences journals. Available on BRS,(800) 289-4277, DIALOG,(800) 334-2564, online services. Inquire as to cost and availability.

Dissertation Abstracts Online. University Microfilms International, 300 North Zeeb Road, Ann Arbor, MI 48106. (800) 521-0600 or (313) 761-4700. Scope includes virtually all doctoral dissertations accepted at accredited American institutions from 1861 to present in 252 subject areas. Available on BRS, (800) 289-4277, DIALOG, (800) 334-2564, and OCLC EPIC,(800) 848-5878, online services. Also available on CD-ROM. Inquire as to cost and availability.

INSPEC. Institution of Electrical Engineers, Michael Faraday House, Six Hills Way, Stevenage, Herts. SG1 2AY, England. Telephone: 0438 313311 or FAX 0438 742840. Contains citations to the worldwide literature of physics, electronics and electrical engineering, computer technology, and related fields. Available on BRS, (800) 289-4277, DIALOG, (800) 334-2564, ORBIT, (800) 456-7248, and STN International, FIZ Karlsruhe, P.O. Box 2465, W-7500, Karlsruhe 1, Germany, online services. Inquire as to cost and availability.

NTIS Bibliographic Database. National Technical Information Service, 5285 Port Royal Road, Springfield, VA 22161. (703) 487-4929 or FAX (703) 321-8199. Broad coverage of government-sponsored science and technology research reports, 1964 to present. Available on BRS,(800) 289-4277, DIALOG, (800) 334-2564, ORBIT, (800) 456-7248, and STN International, FIZ Karlsruhe, P.O. Box 2465, W-7500, Karlsruhe 1, Germany, online services. Also available on CD-ROM. Inquire as to cost and availability.

Physics Briefs. American Institute of Physics, 335 East 45th Street, New York, NY 10017. (212) 661-9260 or FAX (212) 661-2036. Contains citations with abstracts of the literature of physics and related fields, 1979 to present. Available on the STN International, FIZ Karlsruhe, P.O. Box 2465, W-7500, Karlsruhe 1, Germany, online service. Inquire as to cost and availability.

SCISEARCH. Institute for Scientific Information, 3501 Market Street, Philadelphia, PA 19104. (800) 523-1850 or (215) 386-0100. Broad multidisciplinary title and author index to the international literature of science and technology, 1974 to present. Available on DIALOG,(800) 334-2564, and ORBIT,(800) 456-7248, online services. Also available on CD-ROM. Inquire as to cost and availability.

WILSONLINE. H.W. Wilson Company, 950 University Avenue, Bronx, NY 10452. (800) 367-6770 or (212) 588-8400. Makes available online versions of the printed H.W. Wilson indexes including Applied Science and Technology Index, Business Periodicals Index, General Science Index, and Readers' Guide to Periodical Literature. Period covered is generally 1983 to present. Available on BRS,(800) 289-4277, DIALOG,(800) 334-2564, and OCLC EPIC,(800) 848-5878, online services. Also available on CD-ROM. Inquire as to cost and availability.

PERIODICALS

Applied Spectroscopy. Society for Applied Spectroscopy, Thomas Johnson Drive, Suite 2, Frederick, MD 21702. (301) 694-8122. 1946 to present. Monthly. $180.00 per year. ISSN: 0003-7028.

Applied Spectroscopy Reviews. Marcel Dekker Journals, 270 Madison Avenue, New York, NY 10016. (212) 696-9000. FAX (212) 685-4540. 1964 to present. Quarterly. $395.00 per year. ISSN: 0570-4928.

International Journal of Mass Spectrometry and Ion Processes. Elsevier Science Inc., Publishing Co., Inc., Box 882, Madison Square Station, New York, NY 10159. (212) 989-5800. (212) 633-3990. 1968 to present. 36 issues per year in 12 volumes. $2504.00 per year. ISSN: 0168-1176.

Journal of Electron Spectroscopy and Related Phenomena. Elsevier Science Inc., Publishing Co., Inc., Box 882, Madison Square Station, New York, NY 10159. (212) 989-5800. (212) 633-3990. 1972 to present. 20 issues per year. $1030.00 per year. ISSN: 0368-2048.

Journal of Molecular Spectroscopy. Academic Press, Inc., 525 B Street, Suite 1900, San Diego, CA 92101-4495. (619) 230-1840. 1957 to present. Monthly. $1216.00 per year. ISSN: 0022-2852.

Journal of Quantiative Spectroscopy and Radiative Transfer. Elsevier Science Inc., Publishing Co., Inc., Box 882, Madison Square Station, New York, NY 10159. (212) 989-5800. (212) 633-3990. 1961 to present. Monthly. $1815.00 per year. ISSN: 0022-4073.# #*Mass Spectrometry Reviews.* John Wiley and Sons, Inc., 605 Third Avenue, New York, NY 10158. (800) 526-5368 or (212) 850-6000. 1982 to present. Bi-monthly. $330.00 per year. ISSN: 0277-7037.

Spectrochemica Acta A: Molecular Spectroscopy. Elsevier Science Inc., Publishing Co., Inc., Box 882, Madison Square Station, New York, NY 10159. (212) 989-5800. (212) 633-3990. 1939 to present. 14 issues per year. $1270.00 per year. ISSN: 0584-8539.

Spectrochemica Acta B: Atomic Spectroscopy. Elsevier Science Inc., Publishing Co., Inc., Box 882, Madison Square Station, New York, NY 10159. (212) 989-5800. (212) 633-3990. 1939 to present. 14 issues per year. $1445.00 per year. ISSN: 0584-8547.

Spectroscopy Letters. Marcel Dekker Journals, 270 Madison Avenue, New York, NY 10016. (212) 696-9000. FAX (212) 685-4540. 1985 to present.9 issues per year. $59.00 per year. ISSN: 0887-6703.

RESEARCH CENTERS AND INSTITUTES

Atomic Physics and Chemical Physics Laboratory. University of Nevada - Reno, Department of Physics, Reno, NV 89507. (702) 784-4920.

George R. Harrison Spectroscopy Laboratory. Massachusetts Institute of Technology, 77 Massachusetts Avenue, Cambridge, MA 02139. (617) 253-7700.

Institute For Physical Science and Technology. University of Maryland, College Park, MD 20742. (301) 454-2636.

Institute of Optics. University of Rochester, Rochester, NY 14627. (716) 275-5428.

Nuclear Magnetic Resonance Facility. University of Missouri-Columbia, Chemical Building, Columbia, MO 65211. (314) 882-7725. FAX (314) 882-2754.

South Carolina Nuclear Magnetic Resonance Laboratory. University of South Carolina, Department of Chemistry, Columbia, SC 29208. (803) 777-7341.   * *SPECTRUM ANALYSIS
See: SPECTROSCOPY

SPELEOLOGY

See: CAVES and CAVING

SPIN GLASSES

See: SOLID STATE PHYSICS

STANDARDS

See: METEOROLOGY

STARS

See also: ASTRONOMY, ASTROPHYSICS, BINARY STARS, CELESTIAL MECHANICS, COSMOLOGY, GALAXIES, INTERSTELLAR MATTER, TELESCOPES

ABSTRACT SERVICES AND INDEXES

Applied Science and Technology Index; A Cumulative Subject Index To English Language Periodicals in the Fields of Aeronautics and Space Science, Computer Technology, Chemistry, Construction Industry, Energy and Related Areas. H.W. Wilson Co., 950 University Avenue, Bronx, NY 10452. (212) 588-8400. (800) 367-6770. FAX (718) 590-1617. From1958 to present. Monthly. Inquire about cost and availability. Also available on CD-ROM and online. ISSN: 0003-6986.

Astronomy and Astrophysics Abstracts. Springer-Verlag New York, 175 Fifth Avenue, New York, NY 10010. (212) 460-1500.FAX (212) 473-6272. Published for the Astronomisches Rechen-Institut. Comprehensive coverage of all aspects of astronomy, astrophysics and related fields. 1969 to present. Two parts per year. Annual. ISSN: 0067-0022.

General Science Index. H.W. Wilson Company, 950 University Avenue, Bronx, NY 10452. (212) 588-8400. (800) 367-6770. FAX (718) 590-1617. From 1978 to present. Ten issues per year; quarterly and annual cumulations. Service basis. Available on CD-ROM and online. Inquire regarding cost and availability. ISSN: 0162-1963.

Government Reports Announcements and Index. U. S. National Technical Information Service (NTIS), 5285 Port Royal Road, Springfield, VA 22161. (703) 487-4650. FAX (703) 321-8547.

From 1968 to present. Annual. $630.00 per year. Also available online as NTIS Bibliographic Database and on CD- ROM.

Meteorological and Geoastrophysical Abstracts. American Meteorological Society, c/o Inforonics, Inc., 550 Newtown Road, Littleton, MA 01460. (508) 486-8976. FAX (508) 486-0027. Covers literature in environmental sciences, meteorology, astrophysics, hydrology, glaciology, and physical oceanography. From 1950 to present. Monthly. $950.00 per year. Also available on CD-ROM and online. Inquire regarding cost and availability. ISSN: 0026-1130.

NTIS Alerts: Astronomy & Astrophysics. U. S. National Technical Information Service. 5285 Port Royal Road, Springfield, VA 22161. (703) 487-4650. FAX (703) 321-8547. Weekly. $140.00 per year. # #*Pascal 48: Environnement Cosmique Terrestre, Astronomie Et Geologie Extraterrestre.* Centre National de la Recherche Scientifique, Institut de Information Scientifique et Technique, 2 aliee du Parc de Brabois, 54514 Vandoeuvre Les Nancy Cedex, France. TEL 83- 50-46-00. FAX: 83-50-46-50. 1985 to present. Ten issues per year. 770 F. Also available on CD-ROM and online.

Physics Abstracts. INSPEC. Section A, Science Abstracts. Institute of Electrical Engineers, London, United Kingdom. Available from: INSPEC/IEEE - Institute of Electrical and Electronic Engineers, Box 1331, Hoes Lane, Piscataway, NJ 08855-1331. (908) 562-5549. 1898 to present. 24 issuesper year. $2835.00 per year. ISSN: 0036-8091. Also available online and on CD-ROM.

Science Citation Index. SCI. Institute for Scientific Information, 3501 Market Street, Philadelphia, PA 19104. (215) 386-0100. (800) 523-1850. FAX (215) 386-2991. 1961 to present. Six issues per year, plus annual cumulation. $11650.00 per year. Also available online and on CD-ROM. Inquire about price and availability. ISSN: 0036-827X.

STAR. Scientific and Technical Aerospace Reports. U.S. National Aeronautics and Space Administration. Distributed by U. S. Superintendent of Documents, Washington, DC 20402. 1963 to present. Semi-monthly, with semiannual and annual indexes. $114.00 per year. ISSN: 0036-8741. Also available online and on CD-ROM.

ANNUAL REVIEWS AND YEARBOOKS

The Astronomical Almanac. Superintendent of Documents, U. S. Government Printing office, Washington, DC 10402. (202) 783-3238. 1981 to present. Supercedes Astronomical Ephemeris and the American Ephermis and Nautical Almanac. Annual. $44.95. ISSN: 0737-6421.

Annual Review of Astronomy and Astrophysics; Original Reviews of Critical Literature and Current Developments in Astronomy and Astrophysics. Annual Reviews, Inc., 4139 El Camino Way, Palo Alto, CA 94306-0139. (415) 493-4400. (800) 523-8635. FAX (415) 855-9815. 1963 to date. Annual. $60.00. ISSN: 0066-4146.

Highlights of Astronomy. I. Appenzeller, editor. Kluwer Academic Publishers, Box 358, Accord Station, Hingham, MA 02018-0358. (617) 871-6600. International Astronomical Union Highlights series. ISBN: 07923-3553-8. Triennal.1995. Price varies.

ASSOCIATIONS AND PROFESSIONAL SOCIETIES

Amateur Astronomers Association. 1010 Park Avenue, New York, NY 10028. (212) 535-2922.

American Association of Variable Star Observers. 23 Birch Street, Cambridge, MA 02138. (617) 354-0484. FAX: (617)354-0665.

American Astronomical Society. 2000 Florida Avenue NW, Suite 400, Washington, DC 20009. (202) 328-2010. FAX:(202) 234-2560.

American Institute of Physics. One Physics Ellipse, College Park, MD 20740-3843. (301) 209-3100.

Association of University For Research in Astronomy. Suite 701, 1625 Massachusetts Avenue, NW, Washington, DC 20036. (202) 483-2101. FAX (202) 483-2106.

Astronomical League. 2112 Kingfisher Lane East, Rolling Meadows, IL 60008. (708) 398-0562.

Astronomical Society of the Pacific. 390 Ashton Avenue, San Francisco, CA 94112. (415) 337-1100. FAX: (415) 337-5205.

International Astronomical Union. 98 bis Boulevard Arago, F-75014 Paris, France. Tel 1-43-25-83-58. FAX 1-43-25-26-16.

Joint Institute for Laboratory Astrophyics. University of Colorado, Box 440, Boulder, CO 80309-0440. (303) 492-7780. FAX (303) 492-5235.

Royal Astronomical Society of Canada. 136 Dupont Street, Toronto. ON N5R 1V2 Canada. (416) 924-7973. FAX (416) 968-6687.

Webb Society. 1440 South Marmora Avenue, Tuscon, AZ 95713. (602) 628-1077.

BIBLIOGRAPHIES

Astronomy and Astrophysics: A Source Guide. Gordon Press, P.O. Box 459, Bowling Green Station, New York, NY 10004. (718) 624-8419. 1991. $250.00. .

A Bibliography of Astronomy, 1970-1979. R. A. Sea and S. S. Martin. Libraries Unlimited, Inc., Littleton, CO 80160. 1982.

Catalog of the Naval Observatory Library. Naval Observatory Staff. G. K. Hall & Company, 866 Third Avenue, New York, NY 10022. (212)702-6789. (800) 257-5755. 6 volumes. 1977. $655.00

Science Books and Films. American Association for the Advancement of Science, 1333 H Street NW, Washington, DC 20005. (202) 326-6454. 1965 to present. Nine issues per year. $40.00 per year. ISSN: 0098-342X

Scientific and Technical Books and Serials in Print; An Index To Literature in Science and Technology. R. R. Bowker. 121 Chanlon Road, New Providence, NJ 07974. (908) 464-6800. FAX (908) 665-3502. (800) 521-8110. 1972 to present. Annual. $299.95 per year. Also available on CD-ROM. ISSN: 0000-054X.

A Source Book On Astronomy and Astrophysics, 1900 - 1975. Kenneth Lang and Owen Gingerich, editors. Harvard University Press, 79 Garden Street, Cambridge MA. (617) 495-2600. 1980. $41.95.

DIRECTORIES AND BIOGRAPHICAL SOURCES

American Astronomical Society. Membership Directory. American Astronomical Society. 2000 Florida Avenue NW, Suite 400, Washington, DC 20009. (202) 328-2010. FAX (202) 234-2560 Annual. Included in membership dues. ISSN: 1061- 9038.

American Men and Women of Science: Physical and Biological Sciences. R. R. Bowker Inc., 121 Chanlon Road, New Providence, NJ 07974. (908) 464-6800. (800) 521-8110.20th edition. 8 volumes. 1996. $850.00.

The Astronomers. Donald Goldsmith. St. Martin's Press, Inc., 175 Fifth Avenue, New York, NY 10010. (212) 674-5151. (800) 221-7945. 1993. $14.95 in paper.

Astronomical Centers of the World. Kevin Krisciunas. Cambridge University Press, 40 West 20th Street, New York, NY 10011-4211. (212) 924-3900. (800) 872-7423. 1988. $34.95

The Biographical Dictionary of Scientists: Astronomers. D. Abbott, editor. Peter Bedrick Books, Inc., 2112 Broadway, Room 318, New York, NY 10023. (212) 496-0751. 1984.

Directory of Physics and Astronomy Staff Members. American Institute of Physics, One Physics Ellipse, College Park, MD 20740-3843. (301) 209-3100. 1975/76 to present. Annual. $60.00. ISSN: 0361-2228.

Earth and Astronomical Research Centers. Stockton Press, 345 Park Avenue South, New York, NY 10010. 4th edition. 1995. $515.00. ISBN: 1-56169-0967.

Graduate Programs in Physics, Astronomy and Related Fields. 1995-1996. American Institute of Physics, One Physics Ellipse, College Park, MD 20740-3843. (301) 209-3100. 1978 to present. Annual. $45.00. ISSN: 0147-1821.

Research Centers Directory. Gale Research, 835 Penobscot Building, Detroit, MI 48226-4094. (313) 961-2242. (800) 877-4253. 20th edition. 1995. $485.00.

ENCYCLOPEDIAS AND DICTIONARIES

Astronomy From A To Z: A Dictionary of Celestial Objects and Ideas. Charles A. Schweighauser. Illinois Issues, Sangamon State University, K-10, Springfield, IL 62794-9243. (217) 786-9243. 1991. $14.95 in paper.

Concise Dictionary of Astronomy. Jacqueline Mitton. Oxford University Press, Inc., 200 Madison Avenue, New York, NY 10016. (212) 725-6000. (800) 334-4249. 1992. $24.95.

Dictionary of Astronomical Names. Adrian Room. Routledge, Chapman and Hall, Inc., 29 West 35th Street, New York, NY 10001-3291. (212) 244-3336. 1988. $27.50.

Encyclopedia of Astronomy and Astrophysics. Stephen Maran, editor. Van Nostrand Reinhold, 115 Fifth Avenue, New York, NY 10003. (212) 254-3232. (800) 842-3636. 1992. $129.95.

Ency. of Physical Sciences and Engineering Info. Sources

STATISTICS

Facts On File Dictionary of Astronomy. Valerie Illingworth. Facts-on-File, Inc., 460 Park Avenue South, New York, NY 10016-7382. (212) 683-2244. (800) 322-8755. Fax (800) 683-3633. 1994. $27.95.

McGraw-Hill Encyclopedia of Science and Technology. McGraw-Hill Book Company, Inc., 1221 Avenue of the Americas, New York, NY 10020. (212) 512-2000. (800) 262-4729. 7th edition. 20 volume set. 1992. $1900.00.

Oxford Illustrated Encyclopedia. Volume 8, the Universe. Archie Roy, editor. Oxford University Press, Inc., 200 Madison Avenue, New York, NY 10016. (212) 725-6000. (800) 334-4249. 1993. $45.00.

OTHER SOURCES

Atlas of Deep-sky Spendors. Hans Vehrenberg. Sky Publishing Corp., 49 Bay State Road, Cambridge, MA 02138. (617) 864-7360. (800) 253-0245. 4th edition. 1983.$24.95.

Atlas of the Ultraviolet Sky. Richard C. Henry, Wayne B. Landsman, et al. Johns Hopkins University Press, 2715 North Charles Street, Baltimore, MD 21218-4319. (410) 516-6900. (800) 537-5487. 1988. $70.00.

Atlas of the Universe. Patrick Moore. Rand McNally & Company. 8255 North Central Park, Skokie, IL 60076-2970. (708) 673-9100. 1994. $29.95.

The Bright Star Catalog. Dorrit Hoffleit and Carlos Jaschek. Yale University Observatory, 260 Whitney Avenue, P.O. Box 6666, New Haven, CT 06511. (203) 432-3000. 1982. $35.00 in paper. Supplement. 1984. $12.00 in paper.

Cambridge Atlas of Astronomy. Jean Audouze, et al. Cambridge University Press, 40 West 20th Street, New York, NY 10011-4211. (212) 924-3900. (800) 872-7423. 3rd edition. 1994. $75.00.

Cambridge Guide To the Constellations. Michael E. Bakich. Cambridge University Press, 40 West 20th Street, New York, NY 10011-4211. (212) 924-3900. (800) 872-7423. 1995. $19.95 in paper.

Cambridge Star Atlas 2000.0. Wil Tirion. Cambridge University Press, 40 West 20th Street, New York, NY 10011- 4211. (212) 924-3900. (800) 872-7423. 2nd edition. 1996. $19.95.

Color Atlas of Galaxies. James Wray. Cambridge University Press, 40 West 20th Street, New York, NY 10011-4211. (212) 924-3900. (800) 872-7423. 1988. $94.95.

Exploring the Southern Sky: A Pictorial Atlas From the European Southern Observatory. S. Lausten, et al. Springer-Verlag, 175 Fifth Avenue, New York, NY 10010. (212) 460-1500. (800) 777-4643. 1987. $59.00.

The Hubble Atlas of Galaxies. Allan Sandage. Carnegie Institution of Washington, 1530 P Street NW, Washington, DC 20005. (202) 387-6411. Reprint edition. 1984. $35.00.

Observer's Sky Atlas. E. Karkoschka. Springer-Verlag New York, Inc., 175 Fifth Avenue, New York, NY 10010. (212) 460-1500. (800) 777-4643. 1995. $15.95 in paper.

NGC 2000.0: the Complete New General Catalogue and Index Catalogues of Nebulae and Star Clusters. J. L. E. Dreyer. editor. Cambridge University Press, 40 West 20th Street, New York, NY 10011-4211. (212) 924-3900. (800) 872-74231989. $21.95.

Norton's 2000.0 Star Atlas and Reference Handbook. Ian Ridpath, editor. Halsted Press, 605 Third Avenue, New York, 10158-0012. (212) 850-6400. 1989. $49.95.

Observing Handbook and Catalog of Deep-sky Objects. Christian B. Luginbuhl and Brian A. Skiff. Cambridge University Press, 40 West 20th Street, New York, NY 10011-4211. (212) 924-3900. (800) 872-7423. 1990. $49.95.

Starlist Two Thousand: A Quick Reference Star Catalog For Astronomers. Richard Dibon-Smith. John Wiley & Sons, Inc., 605 Third Avenue, New York, NY 10158-0012. (212) 850-6000. (800) 225-5945. 1992. $29.95.

The Sky: A User's Guide. David H. Levy. Cambridge University Press, 40 West 20th Street, New York, NY 10011- 4211. (212) 924-3900. (800) 872-7423. 1993. $14.95 in paper.

Sky Catalog 2000.0: Volume 1: Stars to Magnitute 8.0. Alan Hirshfeld, et al. Cambridge University Press, 40 West 20th Street, New York, NY 10011-4211. (212) 924-3900. (800) 872-7423. 2nd edition. 1991. $42.95 in paper.

STATICS

See: MECHANICS

STATISTICAL MECHANICS

See: MECHANICS

STATISTICAL MECHANICS

See: MECHANICS

STATISTICS

See also: FRACTALS, MARKOV PROCESSES, MAtheMATICS, PROBABILITY, STOCHASTIC PROCESSES

ABSTRACT SERVICES AND INDEXES

Applied Mechanics Reviews: An assessment of world literature in engineering sciences. American Society of Mechanical Engineers, 345 East 47th Street, New York, NY 10017. (212) 705-7703. 1948 to present. Monthly. $360.00 per year. ISSN: 0003-6900.

Applied Science and Technology Index; A Cumulative Subject Index To English Language Periodicals in the Fields of Aeronautics and Space Science, Computer Technology, Chemistry, Construction Industry, Energy and Related Areas. H. W. Wilson and Company, 950 University Avenue, Bronx, NY 10452.

(212) 588-8400. (800) 367-6770. FAX (718) 590-1617. 1958 to present. Monthly. Available on-line from BRS and WILSONLINE. Also available on CD-ROM and online. Inquire as to cost and availability. ISSN: 0003-6986.

Compactmath - Compact Mathematics Library. Cumulative CD-ROM edition of Zentralblatt fuer Mathematik - Mathematics Abstracts. Springer-Verlag New York, Inc. 44 Hartz Way. Secaucus. NJ. 07096-2491. (201) 348-4033. FAX (201) 348-4505. 1993 to present. Annual. Available only on CD- ROM. Inquire as to cost and availability. ISSN: 0938-3174.

American Statistical Association, 1429 Duke Street, Alexandria, VA 22314-3402. (703) 684-1221. FAX (703) 684- 2037. 1976 to present. Annual. $50.00 per year. Available on-line from BRS, Dialog, European Space Agency. ISSN: 0364-1228.

Current Mathematical Publications. American Mathematical Society. P.O. Box 1571, Annex Station, Providence, RI 02901-9930. (800) 556-7774 or (401) 455-4000. FAX (401) 331-3842. 1969 to present. Seventeen times per year. $377.00 per year. Available online from BRS, DIALOG, European Space Agency. Also available on CD-ROM. Inquire as to cost and availability. ISSN: 0361-4794.

General Science Index. H.W. Wilson Company, 950 University Avenue, Bronx, NY 10452. (212) 588-8400. (800) 367-6770. FAX (718) 590-1617. From 1978 to present. Ten issues per year; quarterly and annual cumulations. Service basis. Available on CD-ROM and online. Inquire regarding cost and availability. ISSN: 0162-1963.

Index To Mathematical Tables. Alan Fletcher. Addison- Wesley Publishing Co., Inc., 1 Jacob Way, Reading, MA 01867. (617) 944-3700. (800) 447-2226. 2nd edition. 1962. 2 volumes.

Mathematical Reviews: A Review Journal Covering the World Literature of Mathematical Research. American Mathematical Society, P.O. Box 1571, Annex Station, Providence, RI 02901-9930. (800) 556-7774 or (401) 455-4000. 1940 to present. Monthly. $4594.00 per year. Also available via network (MATHSCINET), online and on CD-ROM. Inquire regardingcost and availability. ISSN: 0025-5629.

N T I S Alerts: Mathematical Sciences. U.S. National Technical Information Service, 5285 Port Royal Road, Springfield, VA 22161. (703) 487-4650. FAX (703) 321-8547.Weekly. $140.00.

Science Citation Index. Institute for Scientific Information. 3501 Market Street, Philadelphia, PA 19104. (215) 386-0100. (800) 523-1850. FAX (215) 386-2991. 1961 to present. Six issues per year, plus annual cumulation. $11650.00 per year. Also available online and on CD-ROM. Inquire about price and availability. ISSN: 0036-827X.

Statistical theory and Method Abstracts. International Statistical Institute. Prinses Beatrixiaan 428, Postbus 950, 2270 AZ Voorburg, Netherlands. 31-70-3375737. FAX 31-70-3860025. 1959 to present. Quarterly. $191.82. ISSN: 0039-0518.

Zentralblatt Fuer Mathematik Und Ihre Grenzebiete/mathematics Abstracts. Heidelberger Akademie der Wissenschaften/ Springer-Verlag New York, Inc., 44 Hartz Way, Seacaucus, NJ 07096-2491. (201) 348-4033. FAX (201) 348-4505. 1931 to present. 30 issues per year. DM 8340. ISSN: 0044-4235. Also available online and on CD-ROM. Inquire regarding cost and availability.

ANNUAL REVIEWS AND YEARBOOKS

Advances in Applied Mathematics. Academic Press, Inc., 6277 Sea Harbor Drive, Orlando, FL 32821. (800) 321-5068. Irregular, Price varies.

C B M S -n S F Regional Conference Series in Applied Mathematics. Society for Industrial and Applied Mathematics, 3600 University City Science Center, Philadelphia, PA 19104-2688. (215) 382-9800. FAX (215) 386-7999. 1971 to present. Irregular. ISSN: 0163-9439.

Ergebnisse Der Mathematik Und Ihrer Grenzgebiete. Springer-Verlag New York, Inc. 44 Hartz Way. Secaucus. NJ. 07096-2491. (201) 348-4033. FAX (201) 348-4505. Irregular. Price varies.

Frontiers in Pure and Applied Probability. Books International, Inc., Box 605 Herndon, VA 22070. (703) 435- 7064. FAC (703) 689-0660. 1993 to present. Irregular. Price varies.

LECTURE NOTES in MAtheMATICS. Springer-Verlag New York, Inc. 44 Hartz Way. Secaucus. NJ. 07096-2491. (201) 348-4033. FAX (201) 348-4505. Irregular. Price varies.

Probability and Mathematical Statistics. Academic Press, Inc., 6277 Sea Harbor Drive, Orlando, FL (800) 321-5068.

1967 to present. Irregular. Price varies.

Probability: Pure and Applied. Marcel Dekker, Inc., 270 Madison Avenue, New York, NY 10016. (212) 696-9000. FAX (212) 685-4540. (800) 228-1160. 1984 to present. Irregular. Price varies.

Statistics: Textbooks and Monographs Series. Marcel Dekker, Inc., 270 Madison venue, New York, NY 10016. (212) 696-9000. FAX (212) 685-4540. 1972 to present. Irregular. Price varies.

ASSOCIATIONS AND PROFESSIONAL SOCIETIES

American Mathematical Society. P.O. Box 6248, Providence, RI 02940. (401) 455-4000. FAX (401) 331-3842.

American Statistical Association. 1429 Duke Street, Alexandria, VA 22314-3402. (703) 684-1221. FAX (703) 684- 2037.

Association For Women in Mathematics. 4114 Computer and Space Science Building, University of Maryland, College Park, MD 20742-2461. (301) 405-7892.

Biometric Society, Eastern North American Region. c/o Randy Price, 11250 Roger Bacon Drive, Suite 8, Reston, VA 22090-5202. (703) 525-1191. FAX (703) 276-8196.

Caucus For Women in Statistics. c/o Cyntha Struthers, St. Jerome's College, Waterloo, ON, Canada N2L 3G3. (519) 888-4801. FAX (519) 746-6530.

ECONOMETRIC SOCIETY. Northwestern University, Department of Economics, Evanston, IL 60208-2600. (708) 491-3615.

Institute of Mathematical Statistics. 3401 Investment Boulevard, Suite 7, Hayward CA 94545-3819. (510) 783-8141. FAX (510) 783-4131.

International Association For Statistical Computing. Prinses Beatrixiaan 428, Postbus 950, NL-2270 AZ Voorburg,Netherlands. PH: 70 3375737. FX: 70 3860025.

International Biometric Society. 1429 Duke Street, Suite 401, Alexandria, VA 22314. (703) 836-8311.

International Statistical Institute. Prinses Beatrixiaan 428, Postbus 950, NL-2270 AZ Voorburg, Netherlands, PH 70-3375737. FX 70 3860025.

Mathematical Association of America. 1529 18th Street, NW, Washington, DC 20036. (202) 387-5200. FAX (202) 265-2384.

NATIONAL INSTITUTE of STATISTICAL SCIENCES. PO Box 14162, Research Triangle Park, NC 27709. (919) 541-6255. FAX (919) 541-7102.

Operations Research Society of America. 1314 Guilford Avenue, Baltimore, MD. 21202. (410) 528-4146. FAX (410) 528-8556.

Royal Statistical Society. 12 Errol Street, London EC1Y 8LX, England. PH 171 7235882. FX 171 7061710.

Society For Industrial and Applied Mathematics. 3600 University City Science Center, Philadelphia, PA 19104-2688. (215) 382-9800. FAX (215) 386-7999.

BIBLIOGRAPHIES

Bibliography of Statistical Literature. Maurice Kendall and Alison Doig. Ayer Company Pubs., Inc., Lower Mill Road, North Stratford, NH 03590. (603) 922-5105. (800) 282- 5413. 1981. 3 volumes: Pre-1940, 1940 to 1949, 1950 to 1958. $82.95, set.

Chronological Annotated Bibliography of Order Statistics, Volume Vi. H. Leon Harter. American Sciences Press, 20 Cross Road, Syracuse, NY 13224-2144. 1991. $115.00.

Guide To Statistical Methods and To the Pertinent Literature. L. Sachs. Springer-Verlag New York, Inc., 175 Fifth Avenue, New York, NY 10010. (212) 460-1500. (800) 777-4643. 1986. $39.00 in paper.

How To Find Out About Statistics. Gillian Burrington. Franklin Book Co., Inc., Elkins Park, PA 19117. (215) 635- 5252. 1972. reprint edition. $74.00.

MAtheMATICAL JOURNALS: AN ANNOTATED GUIDE. Diana F. Liang. Scarecrow Press. Distributed by University Press of America, 4720 Boston Way, Lanham, MD 20706. (301) 459-3366. (800) 462-6420. $29.50. ISBN: 0-8108-2585-6.

State and Local <subhead2>statistics Sources. Sarojini Balachandran. Gale Research, Inc., 835 Penobscot Building, Detroit, MI 48226-4094. (313) 961- 2242. (800) 877-4253. 1993. 2nd edition. $145.00.

Stochastic Orders and Applications: A Classified Bibliography. Karl Mosler and Marco Scasini. Springer- Verlag New York, Inc., 175 Fifth Avenue, New York, NY 10010. (212) 460-1500. (800) 777-4643. 1993. wfi

DIRECTORIES AND BIOGRAPHICAL SOURCES

Assistantships and Graduate Fellowships in the Mathematical Sciences. American Mathematical Society, P.O. Box 6248, Providence, RI 02940. (401) 455-4000. (800) 321-4267. Annual. $19.00. ISSN: 1040-7650.

Combined Membership List. American Mathematical Society, P.O. Box 6248, Providence, RI 02940. (401) 455-4000. (800) 321-4267. Annual. $50.00. ISSN: 0569-6461.

Directory of Statisticians. American Statistical Association, 1429 Duke Street, Alexandria, VA 22314-3402. (703) 684-1221. FAX (703) 684-2037. Periodic.

Mathematical Science Professional Directory. American Mathematical Society, P.O. Box 6248, Providence, RI 02940. (401) 455-4000. (800) 321-4267. Annual. $45.00. ISBN: 0-8218-0173-2.

Research Centers Directory. Gale Research Company, 835 Penobscot Building, Detroit, MI 48226-4094. (313) 961- 2242. (800) 877-4253. 20th edition. 1995. Annual. $485.00. ISSN: 0080-1518.

World Directory of Mathematicians. G.D. Mostow, editor. International Mathematical Union, Helsinki. 1958 to present. 9th revised edition. 1990. $40.00. Available from: American Mathematical Society, P.O. Box 6248, Providence, RI 02940. (401) 455-4000. ISSN: 0512-2740.

ENCYCLOPEDIAS AND DICTIONARIES

Dictionary of Gaming, Modelling & Simulation. G. Ian Gibbs. Books on Demand, 300 North Zeeb Road, Ann Arbor, MI 48106-1346. (313) 761-4700. (800) 521-0600. $49.60 in paper.

Dictionary - Outline of Basic Statistics. John Freund and Frank J. Williams. Dover Publications, Inc., 180 Varick Street, New York, NY 10014. (212) 255-3755. (800) 223- 3130. Reprint edition. $5.95 in paper.

Dictionary of Statistical Terms. F. H. Marriott. Halsted Press, 605 Third Avenue, New York, 10158-0012. (212) 850- 6400. 5th edition. 1990. $76.95 in paper.

Dictionary of Statistics. Michael G. Mulhall. Gordon Press Publications, P.O. Box 459, Bowling Green Station, New York, NY 10004. (718) 624-8419 . 1972. $300.00.

Encyclopedia of Statistical Sciences. Samuel Kotz, Norman L. Johnson, and Campbell B. Read, editors. John Wiley & Sons, Inc., 605 Third Avenue, New York, NY 10158-0012. (212) 850-6000. (800) 225-5945. 1988. 9 volumes. ISBN: 0- 471-05544-1. $1440.00 set.

Encyclopedic Dictionary of Mathematics. Kiyosi Ito, editor. MIT Press, 55 Hayward Street, Cambridge, MA 02142. (617) 625-8569. 2nd edition, reprint. 1993. 2 volumes. ISBN: 0- 262-59020-4. $70.00.

Harpercollins Dictionary of Statistics. Roger Porkess. HarperCollins Publ. 10 East 53rd Street, New York, NY 10022-5299. (800) 331-3761. 1991. $14.00 in paper.

STATISTICS

Ency. of Physical Sciences and Engineering Info. Sources

International Encyclopedia of Statistics, 2. William H. Kruskal and Judith M. Tanur, editors. The Free Press, 1230 Avenue of the Americas, New York, NY 10020. (212) 698-7000. (800) 223-2348. 1978. $90.00.

Mathematics Dictionary. Robert C. James, editor. Van Nostrand Reinhold, 115 Fifth Avenue, New York, NY 10003. (212) 254-3232. (800)-842-3636. 5th edition. 1992. $42.95. ISBN: 0-442-00741-8.

Russian-english - Russian-english Dictionary On Probability, Statistics and Combinatorics. K. A. Borokov. Society for Industrial and Applied Mathematics, 3600 University City Science Center, Philadelphia, PA 19104-2688. (215) 382- 9800. FAX (215) 386-7999. (800) 447-7426. 1993. $47.50 in paper.

Topical Dictionary of Statistics: Advanced Industrial Terminology. Gary L. Teitjen. Chapman & Hall, 1 Penn Plaza, New York, NY 10119. (212) 564-1060. 1985. $24.50.

GENERAL WORKS

Applied Statistics and Probability For Engineers. Douglas C. Montgomery and George C. Runger. John Wiley & Sons, Inc., 605 Third Avenue, New York, NY 10158-0012. (212) 850-6000. (800) 225-5945. 1994. wfi.

Beating Murphy's Law: The Amazing Science of Risk. R. Berger. Dell Publishing Co., Inc., 1540 Broadway, New York, NY 10036-4094. (212) 354-6500. (800) 223-6934. 1994. $11.95.

The Collected Works of George E. P. BOX. George E. Tiao, editor. Chapman & Hall, 1 Penn Plaza, New York, NY 10119. (212) 564-1060. 1984. 2 volumes. $59.95 each.

Foundations of Statistics. Leonard J. Savage. Dover Publications, Inc., 180 Varick Street, New York, NY 10014. (212) 255-3755. (800) 223-3130. 1972. reprint edition. $8.95 in paper.

History of Statistics. Stephen M. Stigler. Belknap Press, 79 Garden Street, Cambridge, MA 02138. (617) 495-2600. 1986. $40.00.

An Introduction To Statistics. A. W. Bowman and D. R. Robinson. IOP Publishing, Public Ledger Building, Suite 1035, Independence Square, Philadelphia, PA 19106. (215) 627-0880. (800) 358-4677. 1987. $39.00 in paper.

Introduction To Statistics: A Calculus-based Approach. Howard B. Christensen. SCP: Third World Literature Publishing House, P.O. Box 482, Lithonia, GA 30058-0482. (404) 785-7725. 1992. $59.00.

Introduction To Probability and Statistics. Narayan C. Giri. Marcel Dekker, Inc., 270 Madison Avenue, New York, NY 10016. (212) 696-9000. (800) 228-1160. 1993. $65.00.

Linear Regression and Correlation Introduction. Allen L. Edwards. W. H. Freeman & Co., 41 Madison Avenue, East 26th Avenue, 35th Floor, New York, NY 10010. (212) 576-0400. 2nd edition. 1995. $wfi.

Mathematical Statistics and Data Analysis. John A. Rice. Wadsworth Publishing Co., 10 Davis Drive, Belmont, CA 94002. (415) 595-2350. (800) 354-9706. 1995. $63.95.

Probability and Statisics in Experimental Physics. Byron P. Roe. Springer-Verlag New York, Inc., 175 Fifth Avenue, New York, NY 10010. (212) 460-1500. (800) 777-4643. 1994. $39.00.

Probability, Statistics and Truth. Richard Von Mises. Dover Publications, Inc., 180 Varick Street, New York, NY 10014. (212) 255-3755. (800) 223-3130. 1981. reprint edition. $5.95 in paper.

Statistical Methods. George W. Snedecor and William G. Cochran. Iowa State University Press, 2121 South State Avenue, Ames, IA 50014-8300. (515) 292-0140. (800) 862-6657. 8th revised edition. 1989. $49.95.

Statistics Problem Solver. Research and Education Association, 61 Ethel Road, West Piscataway, NJ 08854. (908) 819-8880. 1994. $23.95 in paper.

HANDBOOKS AND MANUALS

Biometrika Tables For Statistics. E. S. Pearson and H. O. Hartley, editors. Lubrecht & Cramer, Ltd., 38 County Route 48, Forestburgh, NY 12777-6400. (914) 794-8539. volume 1, 3rd edition. volume 2, reprint edition. 1976. $75.00 each.

Computational Handbook of Statistics. James L. Bruning, and B. L. Kintz. HarpCollins Publ. 10 East 53rd Street, New York, NY 10022-5299. (800) 331-376. 3rd edition. 1990. $38.00 in paper.

Guide To Tables in Mathematical Statistics. Joseph A. Greenwood, and H. O. Hartley. Books on Demand, 300 North Zeeb Road, Ann Arbor, MI 48106-1346. (313) 761-4700. (800)521-0600 $180.00.

Handbook of Applicable Mathematics. Walter Ledermann, editor. John Wiley & Sons, Inc., 605 Third Avenue, New York, NY 10158-0012. (212) 850-6000. (800) 225-5945. 1985. $1496.00. 6 volumes. ISBN: 0-471-90804-5.

Handbook of Applied Mathematics; Selected Results and Methods. Carl E. Pearsen, editor. Chapman & Hall, 1 Penn Plaza, New York NY 10119. (212) 564-1060. 1990. $46.95. ISBN: 0-442-00521-0.

Handbook of Mathematical Functions With Formulas, Graphs, and Mathmetical Tables. Milton Abramowitz and Irene A. Stegun, editors. U.S. National Bureau of Standards. U.S. Government Printing office, Washington, D.C. 10th edition. 1972. $43.00. ISBN: 0-16-000202-8.

Handbook of Mathematical Sciences. William H. Beyer, editor. CRC Press, 2000 Corporate Boulevard, Boca Raton, FL 33431. (800) 272-7737. 6th edition. 1987. $91.95. ISBN: 0-8493-0656-6.

Handbook of Mathematics. I. N. Bronshtein and K. A. Semedyayev. Van Nostrand Reinhold, 115 Fifth Avenue, New York, NY 10003. (800)-842-3636. 20th edition. $59.95. ISBN: 0-442-21171-6.

Handbook of Tables For Probability and Statistics. William H. Beyer, editor. CRC Press, 2000 Corporate Boulevard, Boca Raton, FL 33431. (800) 272-7737. 2nd edition. 1968. $92.95 ISBN: 0-8493-0692-2.

Handbook of Writing For the Mathematical Sciences. Society for Industrial and Applied Mathematics, 3600 University City Science Center, Philadelphia, PA 19104-2688. (215) 382-9800. FAX (215) 386-7999. 1993. $21.50. ISBN: 0-89871-314-5.

Standard Probability and Statistics Tables and Formulas. William H. Beyer. CRC Press, Inc., 2000 Corporate Boulevard, NW, Coca Raton, FL 33431. (407) 994-0555. (800)272-7737. 1990. $34.95.

Standard Mathematical Tables and Formulas. William H. Beyer, editor. CRC Press, 2000 Corporate Boulevard, Boca Raton, FL 33431. (800) 272-7737. 29th edition. 1991. $41.95. ISBN: 084-930-6299.

Tables of Integrals, Series and Products: Corrected and Enlarged Edition. I. S. Gradshteyn and I. M. Ryzhik. Academic Press, Inc., 6277 Sea Harbor Drive, Orlando, FL 32821. (800) 321-5068. 5th edition. 1993. $61.00. ISBN: 0-12-294755-X.

ELECTRONIC RESOURCES

CA Search. Chemical Abstracts Service, P.O. Box 3012, Columbus, OH 43210-0012. (614) 447-3600. (800) 848-6533. FAX (614) 447-3709. Very comprehensive guide to worldwide chemical literature and related fields, 1972 to present. Available on BRS,(800) 289-4277, DIALOG, (800) 334-2564, ORBIT (800) 456-7248, and STN International, FIZ Karlsruhe, P.O. Box 2465, W-7500, Karlsruhe 1, Germany, online services. Inquire as to cost and availability.

Compendex Plus. Engineering Information, Inc., 345 East 47th Street, New York, NY 10017. (212) 705-7600 or (800) 221-1044. Contains citations with abstracts to worldwide literature in engineering and technology, from 1970 to present. Available on online BRS,(800) 289-4277, DIALOG, (800) 334-2564, ORBIT (800) 456-7248, and STN International, FIZ Karlsruhe, P.O. Box 2465, W-7500, Karlsruhe 1, Germany, online services. Also available on CD-ROM. Inquire as to cost and availability.

Current Contents Search. Institute for Scientific Information, 3501 Market Street, Philadelphia, PA 19104. (215) 386-0100. FAX (215) 386-6362. Contains citations to articles listed in the table of contents of science and technology journals. Also articles in social sciences and life sciences journals. Available on BRS,(800) 289-4277, DIALOG,(800) 334-2564, online services. Inquire as to cost and availability.

Dissertation Abstracts. University Microfilms International, 300 North Zeeb Road, Ann Arbor, MI 48106. (800) 521-0600 or (313) 761-4700. Scope includes virtually all doctoral dissertations accepted at accredited American institutions from 1861 to present in 252 subject areas. Available on BRS,(800) 289-4277, DIALOG,(800) 334-2564, and OCLC EPIC,(800) 848-5878, online services. Also available on CD-ROM. Inquire as to cost and availability.

INSPEC. Institution of Electrical Engineers, Michael Faraday House, Six Hills Way, Stevenage, Herts. SG1 2AY, England. Telephone: 0438 313311 or FAX 0438 742840. Contains citations to the worldwide literature of physics, electronics and electrical engineering, computer technology, and related fields. Available on BRS,(800) 289-4277, DIALOG, (800) 334-2564, ORBIT, (800) 456-7248, and STN International, FIZ Karlsruhe, P.O. Box 2465, W-7500, Karlsruhe 1, Germany, online services. Inquire as to cost and availability.

Mathsci and Mathscinet. American Mathematical Society, P.O. Box 6248, Providence, RI 02940. (800) 321-4667 or (401) 455-4000 or FAX (401) 331-3842. Scope includes pure and applied mathematics and related areas of physics, statistics, engineering, computer science, and operations research literature since 1959. Available on DIALOG,(800) 334-2564, online service. Also available on CD-ROM and MATHSCINET, on the Internet. Inquire as to cost and availability.

NTIS Bibliographic Database. National Technical Information Service, 5285 Port Royal Road, Springfield, VA 22161. (703) 487-4929 or FAX (703) 321-8199. Broad coverage of government-sponsored science and technology research reports, 1964 to present. Available on BRS,(800) 289-4277, DIALOG, (800) 334-2564, ORBIT, (800) 456-7248, and STN International, FIZ Karlsruhe, P.O. Box 2465, W-7500, Karlsruhe 1, Germany, online services. Also available on CD-ROM. Inquire as to cost and availability.

Scientific and Technical Books and Serials in Print. R.R. Bowker Inc., 121 Chanlon Road, New Providence, NJ 07974. (800) 521-8110 or (908) 464-6800. List of currently published books and serials in the physical and biological sciences, engineering and technology, with subject, author and titles indexes. Available on ORBIT,(800) 456-7248, online service.Inquire as to cost and availability.

SCISEARCH. Institute for Scientific Information, 3501 Market Street, Philadelphia, PA 19104. (800) 523-1850 or (215) 386-0100. Broad multidisciplinary title and author index to the international literature of science and technology, 1974 to present. Available on DIALOG,(800) 334-2564, and ORBIT,(800) 456-7248, online services. Also available on CD-ROM. Inquire as to cost and availability.

WILSONLINE. H.W. Wilson Company, 950 University Avenue, Bronx, NY 10452. (800) 367-6770 or (212) 588-8400. Makes available online versions of the printed H.W. Wilson indexes including Applied Science and Technology Index, Business Periodicals Index, General Science Index, and Readers' Guide to Periodical Literature. Period covered is generally 1983 to present. Available on BRS,(800) 289-4277, DIALOG,(800) 334-2564, and OCLC EPIC,(800) 848-5878, online services. Also available on CD-ROM. Inquire as to cost and availability.

PERIODICALS

Advances in Applied Probability. Applied Probability Trust, School of Mathematics, University of Sheffield, Sheffield S3 7RH, England. TEL 44-742-824269. FAX 44-742-729782. 1969 to present. Quarterly. $56.00 per year. ISSN: 0001-8678.

American Statistician. American Statistical Association. 1429 Duke Street, Alexandria, VA 22314-3402. (703) 684-1221. FAX (703) 684-2037. 1947 to present. Quarterly. $55.00 per year. ISSN: 0003-1305.

Annals of Statistics. Institute of Mathematical Statistics. 3401 Investment Boulevard, Suite 7, Hayward, CA 94545. (510) 783-8141. 1973 to present. Quarterly. $130.00 per year. ISSN: 0090-5364.

Biometrics. Biometric Society. 1429 Duke Street, Suite 401, Alexandria, VA 22314-3402. (703) 836-8311. 1945 to present. Quarterly. $90.00 per year. ISSN: 0006-341X.

Canadian Journal of Statistics/revue Canadienne DE Statistique. Statistical Society of Canada, Room 612 Dunton Tower, Carleton University, Ottawa, ON K1S 5B6, Canada. (613) 788-2167. 1973 to present. Quarterly. $98.00 per year. ISSN: 0319-5724.

Chance. Springer-Verlag, 175 Fifth Avenue, New York, NY 10010. (212) 460-1612. FAX (212) 473-6272. 1988 topresent. Quarterly. $69.00 per year. ISSN: 0933-2480.

Communications in Statistics, Part A: Theory and Methods. Marcel Dekker Journals, 270 Madison Avenue, New York, NY, 10016. (212) 696-9000. FAX (212) 685-4540. 1973 to present. Twelve issues per year. $1295.00 per year. ISSN: 0361-0926.

International Statistical Review. International Statistical Institute. Prinses Beatrixiaan 428 Postbus 950, 22270 Voorburg Netherlands. TEL 31-70-3375737. FAX 31-70- 3860025. 1933 to present. Three issues per year. $76.10 per year. ISSN: 0306-7734.

Institut Henri Poincare. Annales. *Probabilities et Statistics.* Gauthier-Villars, 15 rue Gossin, 92543 Montrouge Cedex, France. TEL 33-1-40-92-65-00. FAX 33-1- 40-92-65-97. 1930 to present. Quarterly. 1230F. ISSN: 0246-0203.

Jasa. Journal of the American Statistical Association. American Statistical Association, 1429 Duke Street, Alexandria, VA 22314-3402. (703) 684-1221. FAX (703) 684-2037. 1888 to present. Quarterly. $210.00 per year. ISSN: 0162-1259.

International Journal of Game theory. Physica-Verlag GmbH. Distributed by Springer-Verlag New York, Inc., 175 Fifth Avenue, New York, NY 10010. (212) 460-1500. 1971 topresent. Quarterly. DM 390. ISSN: 0020-7276.

Journal of Applied Statistics. Carfax Publishing Co. Box 2025, Dunnellon, FL 34430-2025. 1975 to present. Six issues per year. $398.00 per year. ISSN: 0266-4763.

Metrika; International Journal For theoretical and Applied Statistics. Physica-Verlag GmbH und Co. Distributed by Springer-Verlag New York, Inc., 175 Fifth Avenue, New York, NY 10010. (212) 460-1500. 1953 to present. Bimonthly. DM360. ISSN: 0026-1335.

Risk Analysis. Society for Risk Analysis. Plenum Publishing Corp. 233 Spring Street, New York, NY 10013- 1578. (212) 620-8000. FAX (212) 463-0742. 1980 to present. Bimonthly. $320.00 per year. ISSN: 0272-4332.

Royal Statistical Society. Journal. Series A - C. Basil Blackwell Ltd., 108 Cowley Road, Oxford OX4, 1JF, Eng. TEL 0865-791100. FAX 0865-791347. SeriesA: Statistics in Society; 1838 to present. Series B: Methodological; 1934 to present. Series C: Applied Statistics; 1952 to present. Each series, three issues per year. $73.00 per year, each. ISSN: 0035-9238, 9246, 9254.

Sankhya; Parts A and B. Indian Journal of Statistics. Statistical Publishing Society, 204-1 Barrackpore Trunk Road, Calcutta 700035, India. 1933 to present. Quarterly. ISSN: 0581-572X, 5738.

Siam Journal On Applied Mathematics. Society for Industrial and Applied Mathematics, 3600 University City Science Center, Philadelphia, PA 19104-2688. (215) 382- 9800. FAX (215) 386-7999. 1953 to present. Bimonthly. $242.00 per year. ISSN: 0036-1399.

Siam Journal On Numerical Analysis. Society for Industrial and Applied Mathematics, 3600 University City Science Center, Philadelphia, PA 19104-2688. (215) 382-9800. FAX (215) 386-7999. Bimonthly. $255.00 per year. ISSN: 0036-1429.

Siam Journal On Scientific Computing. Society for Industrial and Applied Mathematics, 3600 University City Science Center, Philadelphia, PA 19104-2688. (215) 382-9800. FAX (215) 386-7999. 1980 to present. Bimonthly. $260.00 per year. ISSN: 1064-8275.

Siam Review. Society for Industrial and Applied Mathematics, 3600 University City Science Center, Philadelphia, PA 19104-2688. (215) 382-9800. 1959 to present. Quarterly. $148.00 per year. ISSN: 0036-1445.

Statistical Science. Institute of Mathematical Statistics. 3401 Investment Boulevard, Suite 7, Hayward, CA 94545. (510) 783-8141. 1986 to present. Quarterly. $75.00 per year. ISSN: 0883-4237.

Statistics and Computing. Chapman & Hall, One Penn Plaza, 41st Floor, New York, NY 10019. (212) 564-1060/ FAX (212) 564-1505. Quarterly. $190.00 per year. ISSN: 0960-3174.

Technometrics. American Statistical Association, 1429 Duke Street, Alexandria, VA 22314-3402. (703) 684-1221. FAX(703) 684-2037. 1959 to present. Quarterly. $50.00. ISSN: 0040-1706.

RESEARCH CENTERS AND INSTITUTES

Institute for Mathematics and its Applications. University of Minnesota, 514 Vincent Hall, 206 Church Street SE, Minneapolis, MN 55455. (612) 642-6066. (FAX (612) 626- 7370.

Institute of Applied Mathematics. University of Missouri- Rolla, Rolla, MO 65401. (314) 341-4641.

Institute of Statistics. North Carolina State University, 110 Cox Hall, Box 8203, Raleigh, NC 27695. (919) 737-2528. FAX (919) 737-3787.

Laboratory For Research in Statistics and Probability. Carleton University, Dunton Tower, Room 611, Ottawa, ON, Canada K1S 5B6. (613) 788-2167.

Mathematical Sciences Institute. Cornell University, 201 Caldwell Hall, Ithaca, NY 14853-6201. (607) 255-8005.

Mathematical Statistics Program. University of Maryland, 1107 Mathematics Building, College Park, MD 20742. (301) 405 5061. FAX (301) 314-0827.

Operations Research Center. Massachusetts Institute of Technology, Cambridge, MA 02139. (617) 253-3601. FAX (617) 258-9214.

Statistical Laboratory. Iowa State University, 102 Snedecor Hall, Ames, IA 50011. (515) 294-3440. (FAX) 512) 294-4040.

Statistical Laboratory. Kansas State University. Dickens Hall, Manhattan, KS 66506-0802. (903) 532-6883. FAX (913) 532-7736.

Statistics Department. Rutgers University, Hill Center, New Brunswick, NJ 08903. (201) 932-2691. FAX (201) 932-3428.

STEAM ENGINEERING

See also: BOILERS, GEOtheRMAL ENERGY, HEAT EXCHANGERS, HEAT TRANSFER, MECHANICAL ENGINEERING, NUCLEAR ENGINEERING, PRESSURE VESSELS, TURBINES, THERMODYNAMICS

ABSTRACT SERVICES AND INDEXES

Applied Mechanics Reviews: An Assessment of World Literature in Engineering Sciences. 1948-present. American Society of Mechanical Engineers, 345 East 47th Street, New York, NY 10017. (212) 705-7703. Monthly. $360.00 per year.

Applied Science and Technology Index; A Cumulative Subject Index To English Language Periodicals in the Fields of Aeronautics and Space Science, Computer Technology, Chemistry, Construction Industry, Energy and Related Areas. H.W. Wilson Co., 950 University Avenue, Bronx, NY 10452. (800) 367-6770 or (212) 588-8400. FAX (718) 590-1617. From 1958 to present. Monthly. Inquire about cost and availability. Also available on CD-ROM and online.

Current Contents: Engineering, Technology, and Applied Sciences. Institute for Scientific Information, 3501 Market Street, Philadelphia, PA 19104. (215) 386-0100. FAX (215) 386-6362. 1970 to present. Weekly. $442.00 per year.

Engineering Index Monthly. Engineering Information, Inc., Castle Point on the Hudson, Hoboken, NJ 07030. (800) 221-1044. FAX (212) 832-1857. Monthly. $2300.00 per year. Also available online as COMPENDEX, and also on CD-ROM. Covers chemical engineering, computers, electrical engineering, civil engineering, metals and mining, industrial management, and mechanical engineering.

Mechanical Engineering Abstracts (formerly ISMEC), Cambridge Scientific Abstracts, 7200 Wisconsin Avenue, Bethesda, MD 20814-4823. (301) 961-6750. FAX (301) 961-6720. 1967 to present. Monthly. $895.00 per year. Summarizes world literature in mechanical engineering, production engineering, and engineering management. Also available online.

Physics Abstracts. Institute of Electrical Engineers, Michael Faraday House, Six Hill Way, Stevenage, Herts, SG1 2AY, England. Distributed by IEEE, 445 Hoes Lane, Piscataway, NJ 08854. (908) 562-5549. 1898 to present. Monthly. $2700.00 per year. Also available online as INSPEC.

Physics Briefs. American Institute of Physics, 335 East 45th Street, New York, NY 10017 (212) 661-9404 or FAX (516)349-9704. 1980 to present. Six times per year. $2600.00 per year.

Science Citation Index. Institute for Scientific Information, 3501 Market Street, Philadelphia, PA 19104. (215) 386-0100. FAX (215) 386-6362. Inquire about availability and cost. Also available on CD-ROM.

ASSOCIATIONS AND PROFESSIONAL SOCIETIES

American Boiler Manufacturers Association. 950 N. Glebe Road, Suite 160, Arlington, VA 22203. (703) 522-7350.

American Institute of Chemical Engineers. 345 East 47th Street, New York, NY 10017-2396. (212) 705-7663.

American Society of Heating, Refrigerating & Air-conditioning Engineers, 1791 Tullie Circle, N.E., Atlanta, GA 30329. (404) 636-8400. FAX (404) 321-5478.

American Society of Mechanical Engineers. 345 East 47th Street, New York, NY 10017-2398. (212) 705-7703.

International Association For the Properties of Water and Steam. EPRI, PO Box 10412, Palo Alto, CA 94303. (415) 955-2458.

National Association of Power Engineers. 5-7 Springfield Street, Chicopee, MA 01013. (413) 592-6273.

Society of Automotive Engineers. SAE International, 400 Commonwealth Drive, Warrendale, PA 15096-0001. (412) 776-4841.

Uniform Boiler & Pressure Vessel Laws Society. 308 N. Evergreen Road, Suite 240, Louisville, KY 40243-1010. (502) 244-6029. FAX (502) 244-6030.

DIRECTORIES AND BIOGRAPHICAL SOURCES

American Men & Women of Science, 1995-96. R.R. Bowker Staff, eds. 19th edition. 8 volumes. R.R. Bowker/Reed International Publishing Company, 121 Chanlon Road, New Providence, NJ 07974. (908) 464-6800 or (800) 521-8110. 1995. $850.00

Directory of Engineering Societies and Related Organizations. Gordon Davis. 13th edition. American Association of Engineering Societies, 1111 19th Street NW, Suite 608, Washington, DC 20036. (202) 296-2237 or (800) 658-8897. 1989. Inquire for price.

Plant Engineering Product Supplier Guide. Cahners Publishing Company (Des Plains), 1350 East Touhy Avenue, Des Plaines, IL 60017-5080. (800) 323-4958. FAX (708) 390-2779. Included with subscription to PLANT ENGINEERING (periodical—see below).

Research Centers Directory. Gale Research, Inc., 835 Penobscot Building, Detroit, MI 48226-4094. (313) 961-2242. (800) 877-4253. FAX (313) 961-6083. $485.00.

Scientific and Technical Organizations and Agencies Directory. Gale Research, Inc., 835 Penobscot Building, Detroit, MI 48226-4094. (313) 961-2242. (800) 877-4253. FAX (313) 961-6083. 4th edition. 1996. $195.00.

Who's Who in Engineering. American Association of Engineering Societies, 1111 19th Street NW, Suite 608, Washington, DC 20036. (202) 296-2237 or (800) 658-8897. 8th edition. 1991. Inquire for price.

ENCYCLOPEDIAS AND DICTIONARIES

Encyclopedia of Physical Science and Technology. Robert A. Meyers, ed. 18 volumes. Academic Press Inc., 6277 Sea Harbor Drive, Orlando, FL 32887. (800) 321-5068. 1992. $2100.00

McGraw-Hill Encyclopedia of Engineering. Sybil P. Parker, ed. 2nd edition. McGraw-Hill Publishing Company, 1221 Avenue of the Americas, New York, NY 10020. (800) 262-4729 or (212) 512-3825. 1993. $95.50.# #*McGraw-Hill Encyclopedia of Science and Technology.* Sybil P. Parker, ed. 7th edition. 20 volumes. McGraw-Hill Publishing Company, 1221 Avenue of the Americas, New York, NY 10020. (800) 262-4729 or (212) 512-3825. 1992. $1900.00

STEAM ENGINEERING

Ency. of Physical Sciences and Engineering Info. Sources

Thesaurus of Scientific, Technical, and Engineering Terms. Hemisphere Publishing Corporation, 1900 Frost Road, Suite 101, Bristol, PA 19007-1598. (215) 785-5800 or FAX (215) 785-5515. 1987. $173.00.

GENERAL WORKS

Managing Steam: An Engineering Guide To Commercial, Industrial, and Utility Systems. Jason Makansi, ed. Hemisphere Publishing Corporation, 1900 Frost Road, Suite 101, Bristol, PA 19007-1598. (215) 785-5800 or FAX (215) 785-5515. 1985. Inquire for cost and availability.

Marine Boilers. G.T.H. Flanagan. 3d ed. Butterworth-Heinemann, 313 Washington Street, Newton, MA 02158. (617) 928-2500 or (800) 366-2665. 1990. $27.95.

The Power of Steam: An Illustrated History of the World's Steam Age. Asa Briggs. University of Chicago Press, 5801 Ellis Avenue, 4th Floor, Chicago, IL 60637. (312) 702-7700. FAX (312) 702-9756. 1982. Inquire for cost and availability.

Steam Boiler Operation: Principles and Practice. J.J. Jackson. 2nd edition. Prentice Hall, 113 Sylvan Avenue, Route 9W, Englewood Cliffs, NJ 07632. (201) 592-2000 or (800) 922-0579. 1987. $47.00.

Steam: Its Generation and Use. S.C. Stultz & J.B. Kitto. 40th edition. Babcock & Wilcox, 20 S. Van Buren Avenue, PO Box 351, Barberton, OH 44203-0351. (216) 753-4511. 1992. $80.00.

Steam-plant Operation. E.B. Woodruff, et al. 6th ed. McGraw-Hill Publishing Company, 1221 Avenue of the Americas, New York, NY 10020. (800) 262-4729 or (212) 512-3825. 1992. Inquire for cost and availability.

HANDBOOKS AND MANUALS

Boiler and Pressure Vessel Code. American Society of Mechanical Engineers, 345 East 47th Street, New York, NY 10017-2398. (212) 705-7703. 1980. $55.00, $65.00 (looseleaf).

Handbook of Heat Transfer Applications. W.M. Rohsenow, et al., editors. McGraw-Hill Publishing Company, 1221 Avenue of the Americas, New York, NY 10020. (800) 262-4729 or (212) 512-3825. 1985. Inquire for cost and availability.

Heat Exchanger Sourcebook. J. W. Palen. Hemisphere Publishing Corporation, 1900 Frost Road, Suite 101, Bristol, PA 19007-1598. (215) 785-5800 or FAX (215) 785-5515. 1986. $81.00.

Stationary Engineering Handbook. K.L.Petrocelly. Fairmont Press, 700 Indian Trail, Lilburn, GA 30247. (404) 925-9388. 1989. $67.00.

Steam Plant Calculations Manual. V. Ganapathy. 2d ed. Marcel Dekker, Inc., 270 Madison Avenue, New York, NY 10016. (212) 696-9000. FAX (212) 685-4540. 1993. Inquire for cost and availability.

Steam Tables: Thermodynamic Properties of Water Including Vapor, Liquid & Solid Phases. Joseph H. Keenan, et al., editors. 2nd edition. Krieger Publishing Company, P.O. Box 9542, Melbourne, FL 32902-9542. (407) 724-9542. FAX (407) 951-3671. 1992 (reprint of 1978 edition). $64.50.

ONLINE DATABASES AND CD-ROMS

Compendex Plus. Engineering Information, Inc., 345 East 47th Street, New York, NY 10017. (212) 705-7600 or (800) 221-1044. Contains citations with abstracts to worldwide literature in engineering and technology, from 1970 to present. Available on online BRS,(800) 289-4277, DIALOG, (800) 334-2564, ORBIT (800) 456-7248, and STN International, FIZ Karlsruhe, P.O. Box 2465, W-7500, Karlsruhe 1, Germany, online services. Also available on CD-ROM. Inquire as to cost and availability.

CSA Engineering. Cambridge Scientific Abstracts, 7200 Wisconsin Avenue, Suite 601, Bethesda, MD 20814. (301) 961-6750 or (800) 843-7751. Contains citations and abstracts of international periodicals and research literature covering all fields of engineering and science and technology, including computer and information science, electronics, mechanical engineering, solid state materials, 1981 to present. Available on BRS,(800) 289-4277, online service. Inquire as to cost and availability.

Current Contents Search. Institute for Scientific Information, 3501 Market Street, Philadelphia, PA 19104. (215) 386-0100. FAX (215) 386-6362. Contains citations to articles listed in the table of contents of science and technology journals. Also articles in social sciences and life sciences journals. Available on BRS,(800) 289-4277, DIALOG,(800) 334-2564, online services. Inquire as to cost and availability.

Dissertation Abstracts. University Microfilms International, 300 North Zeeb Road, Ann Arbor, MI 48106. (800) 521-0600 or (313) 761-4700. Scope includes virtually all doctoral dissertations accepted at accredited American institutions from 1861 to present in 252 subject areas. Available on BRS,(800) 289-4277, DIALOG,(800) 334-2564, and OCLC EPIC,(800) 848-5878, online services. Also available on CD-ROM. Inquire as to cost and availability.

ISMEC: Mechanical Engineering Abstracts. Cambridge Scientific Abstracts, 7200 Wisconsin Avenue, Suite 601, Bethesda, MD 20814. (301) 961-6750 or (800) 843-7751. Contains citations to the literature in mechanical engineering, industrial and production engineering, energy, power, mechanics, devices and related areas, from 1973 to present. Available on the DIALOG,(800) 334-2564, online service. Inquire as to cost and availability.

NTIS Bibliographic Database. National Technical Information Service, 5285 Port Royal Road, Springfield, VA 22161. (703) 487-4929 or FAX (703) 321-8199. Broad coverage of government-sponsored science and technology research reports, 1964 to present. Available on BRS,(800) 289-4277, DIALOG, (800) 334-2564, ORBIT, (800) 456-7248, and STN International, FIZ Karlsruhe, P.O. Box 2465, W-7500, Karlsruhe 1, Germany, online services. Also available on CD-ROM. Inquire as to cost and availability.

SCISEARCH. Institute for Scientific Information, 3501 Market Street, Philadelphia, PA 19104. (800) 523-1850 or (215) 386-0100. Broad multidisciplinary title and author index to the international literature of science and technology, 1974 to present. Available on DIALOG,(800) 334-2564, and ORBIT,(800) 456-7248, online services. Also available on CD-ROM. Inquire as to cost and availability.

WILSONLINE. H.W. Wilson Company, 950 University Avenue, Bronx, NY 10452. (800) 367-6770 or (212) 588-8400. Makes available online versions of the printed H.W. Wilson indexes including Applied Science and Technology Index, Business Periodicals Index, General Science Index, and Readers' Guide

to Periodical Literature. Period covered is generally 1983 to present. Available on BRS,(800) 289-4277, DIALOG,(800) 334-2564, and OCLC EPIC,(800) 848-5878, online services. Also available on CD-ROM. Inquire as to cost and availability.

PERIODICALS

Ashrae Journal: Heating, Refrigerating, Air Conditioning, and Ventilating. American Society of Heating, Refrigeration and Air Conditioning Engineers, 1791 Tullie Circle NE, Atlanta, GA 30329. (404) 636-8400. FAX (404) 321-5478. 1914 to present. Monthly. $49.00 per year for non-members.

Ashrae Transactions. American Society of Heating, Refrigeration and Air Conditioning Engineers, 1791 Tullie Circle NE, Atlanta, GA 30329. (404) 636-8400. FAX (404) 321-5478. 1895 to present. Semi-annual. $170.00 per volume.

Experimental thermal and Fluid Science. Elsevier Science Publishing Company, Inc., 655 Avenue of the Americas, New York, NY 10010. (212) 989-5800. 1988 to present. 8 times a year. $484.00 per year.

Heat Transfer Engineering. Taylor and Francis, Inc., 79 Madison Avenue, Suite 1106, New York, NY 10016. (212) 725-1999 or (800) 821-8312. 1979 to present. Quarterly. $176.00 per year.

Heating, Piping, Air Conditioning. Penton Publishing, 110 Superior Ave., Cleveland, OH 44114-2543. 1929 to present. Monthly. $50.00 per year.

International Communications in Heat and Mass Transfer. Elsevier Science [journals], 660 White Plains Rd., Tarrytown, NY 10159-5153. (919) 524-9200. FAX (919) 333-2444. 1974 to present. 6 times a year. $430.00 per year.

International Journal of Pressure Vessels and Piping. Elsevier Science [journals], 660 White Plains Rd., Tarrytown, NY 10159-5153. (919) 524-9200. FAX (919) 333-2444. 1972 to present. 12 times a year. $1505.00 per year.

Journal of Heat Transfer. American Society of Mechanical Engineers, 345 E. 47th Street, New York, NY 10017. (212) 705-7722. 1970 to present. Quarterly. $155.00 per year.

Journal of Pressure Vessel Technology. American Society of Mechanical Engineers, 345 East 47th Street, New York, NY 10017-2398. (212) 705-7703. 1974 to present. Quarterly. $100.00 per year.

Numerical Heat Transfer A: Applications. Taylor and Francis, Inc., 79 Madison Avenue, Suite 1106, New York, NY 10016. (212) 725-1999 or (800) 821-8312. 1978 to present. 8 times a year. $530.00 per year.

Plant Engineering. Cahners Publishing Company (Des Plains), 1350 East Touhy Avenue, Des Plaines, IL 60017-5080. (800) 323-4958. FAX (708) 390-2779. 1947 to present. 19 times a year. $69.95 per year.

Power. McGraw-Hill Publishing Company, 1221 Avenue of the Americas, New York, NY 10020. (800) 262-4729 or (212) 512-3825. 1882 to present. Monthly. Inquire for cost.

Power Transmission Design. Penton Publishing, 110 Superior Ave., Cleveland, OH 44114-2543. 1959 to present. Monthly. $50.00 per year.

RESEARCH CENTERS AND INSTITUTES

Heat Transfer Laboratory. Massachusetts Institute of Technology, 77 Massachusetts Avenue, Cambridge, MA 02139. (716) 253-2248.

Heat Transfer Research. University of Wisconsin—Milwaukee, PO Box 784, Milwaukee, WI 53201. (414) 963-5001. FAX (414) 229-6958.

Institute of thermo-fluid Engineering and Science. Lehigh University, Bldg. A, Mountain Top Campus, Bethlehem, PA 18015. (215) 758-4091. FAX (215) 758-5057.

International Gas Turbine Institute, Asme. 5801 Peachtree Dunwoody Road, Suite 100, Atlanta, GA 30342-1503. (404) 847-0072.

Turbomachinery Laboratory. Texas A&M University, Mechanical Engineering Department, College Station, TX 77843. (409) 845-7417. FAX (409) 845-1835.

STEAM ENGINES

See: ENGINES, STEAM

STEAM TURBINES

See: STEAM, TURBINES

STEEL AND STEEL MAKING

See also: ALLOYS, CORROSION, FERROALLOYS, MACHINING, METALLURGICAL ENGINEERING, METALLURGY, METALS and METAL WORKING

ABSTRACT SERVICES AND INDEXES

Alloys Index. American Society for Metals, Metals Park, OH 44073. (216) 338-5151. FAX (216) 338-4634. 1974 to present. Monthly. $295.00.

Applied Mechanics Reviews: An Assessment of World Literature in Engineering Sciences. 1948-present. American Society of Mechanical Engineers, 345 East 47th Street, New York, NY 10017. (212) 705-7703. Monthly. $360.00 per year.

Applied Science and Technology Index; A Cumulative Subject Index To English Language Periodicals in the Fields of Aeronautics and Space Science, Computer Technology, Chemistry, Construction Industry, Energy and Related Areas. H.W. Wilson Co., 950 University Avenue, Bronx, NY 10452. (800) 367-6770 or (212) 588-8400. FAX (718) 590-1617. From 1958 to present. Monthly. Inquire about cost and availability. Also available on CD-ROM and online.

Corrosion Abstracts; Abstracts of the World's Literature On Corrosion and Corrosion Mitigation. National Association of Corrosion Engineers (NACE), Box 218340, Houston, TX 77218.

STEEL AND STEEL MAKING

Ency. of Physical Sciences and Engineering Info. Sources

(713) 492-0535 or FAX (713) 492-8254. 1962 to present. Bi-monthly. $195.00 per year. Also available on CD-ROM.

Current Contents: Engineering, Technology, and Applied Sciences. Institute for Scientific Information, 3501 Market Street, Philadelphia, PA 19104. (215) 386-0100. FAX (215) 386-6362. 1970 to present. Weekly. $442.00 per year.

Engineering Index Monthly. Engineering Information, Inc., Castle Point on the Hudson, Hoboken, NJ 07030. (800) 221-1044. FAX (212) 832-1857. Monthly. $2300.00 per year. Also available online as COMPENDEX, and also on CD-ROM. Covers chemical engineering, computers, electrical engineering, civil engineering, metals and mining, industrial management, and mechanical engineering.

I.M.M. ABSTRACTS and INDEX. Institution of Mining and Metallurgy, 44 Portland Pl., London W1N 4BR, England. 071-580-3802. FAX 071-436-5388. Bi-monthly. $364 for non-members.

Mechanical Engineering Abstracts (formerly ISMEC), Cambridge Scientific Abstracts, 7200 Wisconsin Avenue, Bethesda, MD 20814-4823. (301) 961-6750. FAX (301) 961-6720. 1967 to present. Monthly. $895.00 per year. Summarizes world literature in mechanical engineering, production engineering, and engineering management. Also available online.

Manufacturing and Process Engineering Abstracts, Cambridge Scientific Abstracts, 7200 Wisconsin Avenue, Bethesda, MD 20814-4823. (301) 961-6750. FAX (301) 961-6720. 1993 to present. Monthly. $385.00 per year. Covers concurrent engineering, quality control, automated manufacturing, petroleum engineering, oil field operations and equipment, energy management, metallurgy and metallography, foundry practice.

Metals Abstracts and Metals Abstracts Index. American Society for Metals, Metals Park, OH 44073. (216) 338-5151. 1968 to present. Monthly. Abstracts are $1500.00 per year and Index is $460.00 per year.

Physics Abstracts. Institute of Electrical Engineers, Michael Faraday House, Six Hill Way, Stevenage, Herts, SG1 2AY, England. Distributed by IEEE, 445 Hoes Lane, Piscataway, NJ 08854. (908) 562-5549. 1898 to present. Monthly. $2700.00 per year. Also available online as INSPEC.

SCIENCE CITATION INDEX. Institute for Scientific Information, 3501 Market Street, Philadelphia, PA 19104. (215) 386-0100. FAX (215) 386-6362. Inquire about availability and cost. Also available on CD-ROM.

ANNUAL REVIEWS AND YEARBOOKS

American Iron and Steel Institute Annual Statistical Report. American Iron and Steel Institute, 1101 17th St., NW, Washington, DC 20036-4700. (202) 452-7100. FAX (202) 463-6573. Annual. Price varies.

World Metal Statistics Yearbook. World Bureau of Metal Statistics, 272 High Street, Ware, Herts SG12 9BA, England. Telephone 0920-461274. FAX 0920-464258. 1984 to present. Annual. Inquire for cost and availability.

ASSOCIATIONS AND PROFESSIONAL SOCIETIES

American Institute of Steel Construction, Inc. One East Wacker Drive, Suite 3100, Chicago, IL 60601-2001. (312) 670-2400 or FAX (312) 670-5403.

American Iron and Steel Institute. 1101 17th St., NW, Washington, DC 20036-4700. (202) 452-7100. FAX (202) 463-6573.

American Powder Metallurgy Institute. 105 College Road E., Princeton, NJ 08540. (609) 452-7700.

American Society For Metals. Materials Park, OH 44073. (216) 338-5151 or FAX (216) 338-4634.

American Society For Testing and Materials. 1916 Race Street, Philadelphia, PA 19103. (215) 299-5585.

American Society of Mechanical Engineers. 345 East 47th Street, New York, NY 10017-2398. (212) 705-7703.

Association of Iron and Steel Engineers. Three Gateway Center, Suite 2350, Pittsburgh, PA 15222. (412) 281-6323.

International Iron and Steel Institute. Rue Colonel Bourg 120, B-1140 Brussels, Belgium. Telephone 32-2-7359075. FAX 32-2-7358012.

Iron and Steel Society. 410 Commonwealth Drive, Warrendale, PA 15086-7512. (412) 776-1535.

Metallurgical Society of the Aime (american Institute of Mining, Metallurgical, and Petroleum Engineers). 345 E.47th Street, 14th Floor, New York, NY 10017. (212) 705-7695.

National Association of Corrosion Engineers. Box 218340, Houston, TX 77218. (713) 492-0535. FAX (713) 492-8254.

Steel Founders Society of America. Cast Metals Federation Bldg., 455 State Street, Des Plaines, IL 60016. (708) 299-9160. FAX (708) 299-3105.

Steel Manufacturers Association. 1730 Rhode Island Avenue NW, Suite 801, Washington, DC 20036-3101. (202) 296-1515.

DIRECTORIES AND BIOGRAPHICAL SOURCES

Directory of Engineering Societies and Related Organizations. Gordon Davis. 13th edition. American Association of Engineering Societies, 1111 19th Street NW, Suite 608, Washington, DC 20036. (202) 296-2237 or (800) 658-8897. 1989. Inquire for price.

Directory of Steel Foundries. Steel Founders Society of America, Cast Metals Federation Bldg., 455 State Street, Des Plaines, IL 60016. (708) 299-9160. FAX (708) 299-3105. Bi-ennial. Inquire for cost and availability.

Dun's Industrial Guide: The Metalworking Directory, 1993-94. Dun & Bradstreet Information Services Staff. 3 volumes. Dun & Bradstreet Information Services, 899 Eaton Avenue, Bethlehem, PA 18025. (610) 882-7000 or (800) 526-0651. FAX (610) 882-7269. 1993. $795.00.

Iron and Steel Works of the World. Metal Bulletin Inc., 220 Fifth Avenue, New York, NY 10001-7781. (212) 213-6202. FAX (212) 213-1870. 1952 to present. Quadrennial. $52.00 for non-members.

Machinery Buyers Guide. Findlay Publishers Inc. Ltd., Franks Hall, Franks Lane, Horton Kirby, Kent DA4 9LL, England. 1926 to present. Annual. Inquire for cost and availability.

Metallurgical Society of Aime Membership List. American Institute of Mining, Metallurgical and Petroleum Engineers, 345 E.47th Street, 14th Floor, New York, NY 10017. (212) 705-7695. Inquire for availability.

Research Centers Directory. Gale Research, Inc., 835 Penobscot Building, Detroit, MI 48226-4094. (313) 961-2242. (800) 877-4253. FAX (313) 961-6083. $485.00.

Scientific and Technical Organizations and Agencies Directory. Gale Research, Inc., 835 Penobscot Building, Detroit, MI 48226-4094. (313) 961-2242. (800) 877-4253. FAX (313) 961-6083. 4th edition. 1996. $195.00.

Who's Who in Engineering. American Association of Engineering Societies, 1111 19th Street NW, Suite 608, Washington, DC 20036. (202) 296-2237 or (800) 658-8897. 8th edition. 1991. Inquire for price. ENCYLOPEDIAS and DICTIONARIES

McGraw-Hill Encyclopedia of Engineering. Sybil P. Parker, ed. 2nd edition. McGraw-Hill Publishing Company, 1221 Avenue of the Americas, New York, NY 10020. (800) 262-4729 or (212) 512-3825. 1993. $95.50.

McGraw-Hill Encyclopedia of Science and Technology. Sybil P. Parker, ed. 7th edition. 20 volumes. McGraw-Hill Publishing Company, 1221 Avenue of the Americas, New York, NY 10020. (800) 262-4729 or (212) 512-3825. 1992. $1900.00

Thesaurus of Scientific, Technical, and Engineering Terms. Hemisphere Publishing Corporation, 1900 Frost Road, Suite 101, Bristol, PA 19007-1598. (215) 785-5800 or FAX (215) 785-5515. 1987. $173.00.

GENERAL WORKS

Fundamentals of Steelmaking Metallurgy. Brahma Deo & Rob Boom. Prentice Hall, 113 Sylvan Avenue, Route 9W, Englewood Cliffs, NJ 07632. (201) 592-2000 or (800) 922-0579. 1992. $30.00.

A Nation of Steel. Thomas A. Misa. Johns Hopkins University Press, 2715 North Charles St., Baltimore, MD 21218. (410) 516-6900 or (800) 537-5487. FAX (410) 516-6998. 1995. Inquire for cost and availability.

Restructuring the U.S. STEEL INDUSTRY. Jose G. Lima. Westview Press, 5500 Central Avenue, Boulder, CO 80301-2847. (303) 444-3541 or (800) 456-1995. FAX (303) 449-3356. 1993. $52.50.

Stainless Steel. J.G. Parr, A. Hanson, and R.A. Lula. ASM International, Materials Information, Materials Park, OH 44073-0002. (216) 338-5151 or (800) 336-5152. FAX (216) 338-4634. 1985. $79.00.

Steels: Heat Treatment and Processing Principles. G. Krauss. ASM International, Materials Information, Materials Park, OH

44073-0002. (216) 338-5151 or (800) 336-5152. FAX (216) 338-4634. 1990. $118.00.

Steels: Metallurgy and Applications. D.T. Llewellyn. 2nd ed. Butterworth-Heinemann, 313 Washington Street, Newton, MA 02158. (617) 928-2500 or (800) 366-2665. FAX (617) 928-2620. 1994. Inquire for cost and availability.

Towards A Better Steelworks' Yield. International Iron and Steel Institute, Rue Colonel Bourg 120, B-1140 Brussels, Belgium. Telephone 32-2-7359075. FAX 32-2-7358012. 1992. Inquire for cost and availability.

HANDBOOKS AND MANUALS

Asm Handbook Volume 1: Properties and Selection: Irons, Steels, and High-performance Alloys. ASM International Handbook Committee. ASM International, Materials Information, Materials Park, OH 44073-0002. (216) 338-5151 or (800) 336-5152. FAX (216) 338-4634. 1990. $147.00.

ASM Handbook Volume 14: Forming and Forging. ASM International Handbook Committee. ASM International, Materials Information, Materials Park, OH 44073-0002. (216) 338-5151 or (800) 336-5152. FAX (216) 338-4634. 1988. $147.00.

ASM Handbook Volume 15: Casting. ASM International Handbook Committee. ASM International, Materials Information, Materials Park, OH 44073-0002. (216) 338-5151 or (800) 336-5152. FAX (216) 338-4634. 1988. $147.00.

ASM Specialty Handbook: Stainless Steels. J.R. Davis, editor. ASM International Handbook Committee. ASM International, Materials Information, Materials Park, OH 44073-0002. (216) 338-5151 or (800) 336-5152. FAX (216) 338-4634. 1994. $159.00.

Stahlschlussel (key To Steel). C.W. Wegst. 17th edition. ASM International, Materials Information, Materials Park, OH 44073-0002. (216) 338-5151 or (800) 336-5152. FAX (216) 338-4634. 1995. $140.00.

Metals Handbook. ASM International, Materials Information, Materials Park, OH 44073-0002. (216) 338-5151 or (800) 336-5152. FAX (216) 338-4634. $154.00.

Woldman's Engineering Alloys. J. Frick, editor. 8th edition. ASM International, Materials Information, Materials Park, OH 44073-0002. (216) 338-5151 or (800) 336-5152. FAX (216) 338-4634. 1994. $175.00.

ONLINE DATABASES AND CD-ROMS

Compendex Plus. Engineering Information, Inc., 345 East 47th Street, New York, NY 10017. (212) 705-7600 or (800) 221-1044. Contains citations with abstracts to worldwide literature in engineering and technology, from 1970 to present. Available on online BRS,(800) 289-4277, DIALOG, (800) 334-2564, ORBIT (800) 456-7248, and STN International, FIZ Karlsruhe, P.O. Box 2465, W-7500, Karlsruhe 1, Germany, online services. Also available on CD-ROM. Inquire as to cost and availability.

CSA Engineering. Cambridge Scientific Abstracts, 7200 Wisconsin Avenue, Suite 601, Bethesda, MD 20814. (301) 961-6750 or (800) 843-7751. Contains citations and abstracts of international periodicals and research literature covering all fields

STEEL AND STEEL MAKING

Ency. of Physical Sciences and Engineering Info. Sources

of engineering and science and technology, including computer and information science, electronics, mechanical engineering, solid state materials, 1981 to present. Available on BRS,(800) 289-4277, online service.Inquire as to cost and availability.

Current Contents Search. Institute for Scientific Information, 3501 Market Street, Philadelphia, PA 19104. (215) 386-0100. FAX (215) 386-6362. Contains citations to articles listed in the table of contents of science and technology journals. Also articles in social sciences and life sciences journals. Available on BRS,(800) 289-4277, DIALOG,(800) 334-2564, online services. Inquire as to cost and availability.

ISMEC: Mechanical Engineering Abstracts. Cambridge Scientific Abstracts, 7200 Wisconsin Avenue, Suite 601, Bethesda, MD 20814. (301) 961-6750 or (800) 843-7751. Contains citations to the literature in mechanical engineering, industrial and production engineering, energy, power, mechanics, devices and related areas, from 1973 to present. Available on the DIALOG,(800) 334-2564, online service. Inquire as to cost and availability.

Metadex. Jointly produced by ASM International and the Institute of Materials. Contains more than 925,000 records from the international literature on metals and alloys, concerning properties, processes, materials classes, applications, and metallurgical systems. Updated monthly. Available from ORBIT-QUESTEL (703) 442-0900.

NTIS Bibliographic Database. National Technical Information Service, 5285 Port Royal Road, Springfield, VA 22161. (703) 487-4929 or FAX (703) 321-8199. Broad coverage of government-sponsored science and technology research reports, 1964 to present. Available on BRS,(800) 289-4277, DIALOG, (800) 334-2564, ORBIT, (800) 456-7248, and STN International, FIZ Karlsruhe, P.O. Box 2465, W-7500, Karlsruhe 1, Germany, online services. Also available on CD-ROM. Inquire as to cost and availability.

SCISEARCH. Institute for Scientific Information, 3501 Market Street, Philadelphia, PA 19104. (800) 523-1850 or (215) 386-0100. Broad multidisciplinary title and author index to the international literature of science and technology, 1974 to present. Available on DIALOG,(800) 334-2564, and ORBIT,(800) 456-7248, online services. Also available on CD-ROM. Inquire as to cost and availability.

WILSONLINE. H.W. Wilson Company, 950 University Avenue, Bronx, NY 10452. (800) 367-6770 or (212) 588-8400. Makes available online versions of the printed H.W. Wilson indexes including Applied Science and Technology Index, Business Periodicals Index, General Science Index, and Readers' Guide to Periodical Literature. Period covered is generally 1983 to present. Available on BRS,(800) 289-4277, DIALOG,(800) 334-2564, and OCLC EPIC,(800) 848-5878, online services. Also available on CD-ROM. Inquire as to cost and availability.

PERIODICALS

Alloy Digest. Alloy Digest Inc., 27 Canfield Street, Orange, NJ 07050. (201) 677-9161. 1952 to present. Monthly. $140.00 per year.

Industry Week. Penton Publishing, 110 Superior Ave., Cleveland, OH 44114-2543. 1882 to present. 23 times a year. $60.00 per year.

Iron and Steel Engineer. Association of Iron and Steel Engineers, Three Gateway Center, Suite 2350, Pittsburgh, PA 15222. (412) 281-6323. 1924 to present. Monthly. $50.00 per year.

Iron & Steelmaker. Iron and Steel Society, 410 Commonwealth Drive, Warrendale, PA 15086. (412) 776-9460. FAX (412) 776-0430. 1974 to present. Monthly. $50.00 per year.

Ironmaking and Steelmaking. Institute of Materials, 1 Carlton House Terrace, London SW1Y 5DB, England. Telephone 071-839-4071. FAX 071-839-2078. 1974 to present. Bi-monthly. $420.00 per year for non-members.

Journal of Engineering For Industry. American Society of Mechanical Engineers, 345 E. 47th Street, New York, NY 10017. (212) 705-7722. 1970 to present. Quarterly. $100.00 per year.

J O M (JOURNAL of METALS). Minerals, Metals and Materials Society, 420 Commonwealth Drive, Warrendale, PA 15086. (412) 776-9080. 1949 to present. Monthly. $50.00 per year.

Metallurgical Transactions A: Physical Metallurgy and Materials Science. ASM International, Materials Park, OH 44073. (216) 338-5151 or FAX (216) 338-4634. 1970 to present. Monthly. $520.00 per year.

Metals Week. McGraw-Hill Publishing Company, 1221 Avenue of the Americas, New York, NY 10020. (800) 262-4729 or (212) 512-3825. 1930 to present. Weekly. $770.00 per year.

Metalworking Digest. Gordon Publications Inc., 301 Gibraltar Drive, Box 350, Morris Plains, NJ 07950-0650. (201) 292-5100. FAX (201) 898-9281. 1968 to present. 12 times a year. $48.00 per year.

Modern Steel Construction. American Institute of Steel Construction Inc., One East Wacker Drive, Suite 3100, Chicago, IL 60601-2001. (312) 670-2400 or FAX (312) 670-5403. 1944 to present. Monthly. $30.00 per year.

Steel Construction Today. Kingslea Press Ltd., 137 Newhall Street, Birmingham B3 1SF, England. Telephone 021-236-8112. FAX 021-200-1480. 1987 to present. Bi-monthly. Inquire for cost and availability.

Steel Digest. Intersteel Technology Inc., 8301 Univ. Exec. Park, Suite 130, Charlotte, NC 28262. (704) 549-4177. FAX (704) 542-5107. 1983 to present. Bi-monthly. Free.

Steel Research. Verlag Stahleisen mbH, Sohnstr. 65, 40237 Duesseldorf, Germany. Telephone 0211-6070-0. FAX 0211-6707517. 1927 to present. Monthly. Inquire.

Steel Technology International. Sterling Publications Ltd., 86-88 Edgeware Road, London W2 2YW. Telephone 01-258-0066. Annual. Inquire for cost and availability.

Steel Times. FMJ International Publications, Ltd., Queensway House, 2 Queensway, RedHill, Surrey RH1 1QS England. Telephone 0737-768611. FAX 0737-761685. 1866 to present. Monthly. $194.90.

World Metal Statistics. World Bureau of Metal Statistics, 272 High Street, Ware, Herts SG12 9BA, England. Telephone 0920-461274. FAX 0920-464258. 1948 to present. Monthly. Inquire for cost and availability.

RESEARCH CENTERS AND INSTITUTES

American Iron and Steel Institute. 1101 17th St., NW, Washington, DC 20036-4700. (202) 452-7100. FAX (202) 463-6573.

Carnegie Mellon University Center For Iron and Steel Making Research. MEMS Department, Pittsburgh, PA 15213. (412) 268-2677.

Colorado School of Mines, Advanced Steel Processing and Products Research Center. Golden, CO 80401. (303) 273-3774.

George Winter Structural Engineering Laboratory. Cornell University, Thurston Hall, Ithaca, NY 14853. (607) 255-4078.

University of Wisconsin—milwaukee Foundry & Solidification Processing Research Laboratory. PO Box 784, Mechanical Engineering Department, College of Engineering & Applied Sciences, Milwaukee, WI 53201. (414) 229-4987.

SPECIFICATIONS AND STANDARDS

Annual Book of Astm Standards, Volumes 01.01-01.05. American Society for Testing and Materials (ASTM), 1916 Race Street, Philadelphia, PA 19103. (215) 299-5585. Annual. Inquire for cost and availability.

STEREOCHEMISTRY

See: PHYSICAL CHEMISTRY

STOCHASTIC PROCESSES

See also: MARKOV PROCESSES, MATHEMATICS, PROBABILITY, STATISTICS

ABSTRACT SERVICES AND INDEXES

Applied Mechanics Reviews: An assessment of world literature in engineering sciences. American Society of Mechanical Engineers, 345 East 47th Street, New York, NY 10017. (212) 705-7703. 1948 to present. Monthly. $360.00 per year. ISSN: 0003-6900.

Applied Science and Technology Index; A Cumulative Subject Index To English Language Periodicals in the Fields of Aeronautics and Space Science, Computer Technology, Chemistry, Construction Industry, Energy and Related Areas. H. W. Wilson and Company, 950 University Avenue, Bronx, NY 10452. (212) 588-8400.(800) 367-6770. FAX (718) 590-1617. 1958 to present. Monthly. Available on-line from BRS and WILSONLINE. Also available on CD-ROM. Inquire as to cost and availability. ISSN: 0003-6986.

Compactmath - Compact Mathematics Library. Cumulative CD-ROM edition of Zentralblatt fuer Mathematik - Mathematics Abstracts. Springer-Verlag New York, Inc. 44 Hartz Way. Secaucus. NJ. 07096-2491. (201) 348-4033. FAX (201) 348-4505. 1993 to present. Annual. Available only on CD-ROM. Inquire as to cost and availability. ISSN: 0938-3174.

Current Index To Statistics; Applications-methods-theory. American Statistical Association, 1429 Duke Street, Alexandria, VA 22314-3402. (703) 684-1221. FAX (703) 684-2037. 1976 to present. Annual. $50.00 per year. Available on-line from BRS, Dialog, European Space Agency. ISSN: 0364-1228.

Current Mathematical Publications. American Mathematical Society. P.O. Box 1571, Annex Station, Providence, RI 02901-9930. (800) 556-7774 or (401) 455-4000. FAX (401) 331-3842. 1969 to present. Seventeen times per year. $377.00 per year. Available online from BRX, DIALOG, European Space Agency. Also available on CD-ROM. Inquire as to cost and availability. ISSN: 0361-4794.

Index To Mathematical Tables. Alan Fletcher. Addison-Wesley Publishing Co., Inc., 1 Jacob Way, Reading, MA 01867. (617) 944-3700. (800) 447-2226. 2nd edition. 1962. 2 volumes

Mathematical Reviews: A Review Journal Covering the World Literature of Mathematical Research. American Mathematical Society, P.O. Box 1571, Annex Station, Providence, RI 02901-9930. (800) 556-7774 or (401) 455-4000. 1940 to present. Monthly. $4594.00 per year. Also available via network (MATHSCINET), online and on CD-ROM. Inquire regarding cost and availability. ISSN: 0025-5629.

N T I S Alerts: Mathematical Sciences. U.S. National Technical Information Service, 5285 Port Royal Road, Springfield, VA 22161. (703) 487-4650. FAX (703) 321-8547. Weekly. $140.00.

Science Citation Index. SCI. Institute for Scientific Information. 3501 Market Street, Philadelphia, PA 19104. (215) 386-0100. (800) 523-1850. FAX (215) 386-2991. 1961 to present. Six issues per year, plus annual cumulation. $11650.00 per year. Also available online and on CD-ROM. Inquire about price and availability. ISSN: 0036-827X.

Statistical theory and Method Abstracts. International Statistical Institute. Prinses Beatrixiaan 428, Postbus 950, 2270 AZ Voorburg, Netherlands. 31-70-3375737. FAX 31-70-3860025. 1959 to present. Quarterly. $191.82. ISSN: 0039-0518.

ANNUAL REVIEWS AND YEARBOOKS

Advances in Probability and Related Topics. Marcel Dekker, Inc., 270 Madison Avenue, New York, NY 10016. (212) 696-9000. FAX (212) 685-4540. 1971 to present. 1984? Irregular.

Advances in Applied Mathematics. Academic Press, Inc., 6277 Sea Harbor Drive, Orlando, FL 32821. (800) 321-5068. Irregular, Price varies.

Ergebnisse Der Mathematik Und Ihrer Grenzgebiete. Springer-Verlag New York, Inc. 44 Hartz Way. Secaucus. NJ. 07096-2491. (201) 348-4033. FAX (201) 348-4505. Irregular.Price varies.

Frontiers in Pure and Applied Probability. Books International, Inc., Box 605 Herndon, VA 22070. (703) 435-7064. FAC (703) 689-0660. 1993 to present. Irregular. Price varies.

Lecture Notes in Mathematics. Springer-Verlag New York, Inc. 44 Hartz Way. Secaucus. NJ. 07096-2491. (201)348-4033. FAX (201) 348-4505. Irregular. Price varies.

STOCHASTIC PROCESSES

Ency. of Physical Sciences and Engineering Info. Sources

Probability and Mathematical Statistics. Academic Press, Inc., 6277 Sea Harbor Drive, Orlando, FL (800) 321-5068. 1967 to present. Irregular. Price varies.

Probability: Pure and Applied. Marcel Dekker, Inc., 270 Madison Avenue, New York, NY 10016. (212) 696-9000. FAX (212) 685-4540. (800) 228-1160. 1984 to present. Irregular. Price varies.

Statistics: Textbooks and Monographs Series. Marcel Dekker, Inc., 270 Madison venue, New York, NY 10016. (212) 696-9000. FAX (212) 685-4540. 1972 to present. Irregular. Price varies.

ASSOCIATIONS AND PROFESSIONAL SOCIETIES

American Mathematical Society. P.O. Box 6248, Providence, RI 02940. (401) 455-4000. FAX (401) 331-3842.

American Statistical Association. 1429 Duke Street, Alexandria, VA 22314-3402. (703) 684-1221. FAX (703) 684-2037.

Association For Women in Mathematics. 4114 Computer and Space Science Building, University of Maryland, College Park, MD 20742-2461. (301) 405-7892.

Biometric Society, Eastern North American Region. c/o Randy Price, 11250 Roger Bacon Drive, Suite 8, Reston, VA 22090-5202. (703) 525-1191. FAX (703) 276-8196.

Caucus For Women in Statistics. c/o Cyntha Struthers, St. Jerome's College, Waterloo, ON, Canada N2L 3G3. (519) 888-4801. FAX (519) 746-6530.

Econometric Society. Northwestern University, Department of Economics, Evanston, IL 60208-2600. (708) 491-3615.

Institute of Mathematical Statistics. 3401 Investment Boulevard, Suite 7, Hayward CA 94545-3819. (510) 783-8141. FAX (510) 783-4131.

International Association For Statistical Computing. Prinses Beatrixiaan 428, Postbus 950, NL-2270 AZ Voorburg, Netherlands. PH: 70 3375737. FX: 70 3860025.

International Biometric Society. 1429 Duke Street, Suite 401, Alexandria, VA 22314. (703) 836-8311.

International Statistical Institute. Prinses Beatrixiaan 428, Postbus 950, NL-2270 AZ Voorburg, Netherlands, PH 70-3375737. FX 70 3860025.

Mathematical Association of America. 1529 18th Street, NW, Washington, DC 20036. (202) 387-5200. FAX (202) 265-2384.

Operations Research Society of America. 1314 Guilford Avenue, Baltimore, MD. 21202. (410) 528-4146. FAX (410)528-8556.

Royal Statistical Society. 12 Errol Street, London EC1Y 8LX, England. PH 171 7235882. FX 171 7061710.

Society For Industrial and Applied Mathematics. 3600 University City Science Center, Philadelphia, PA 19104-2688. (215) 382-9800. FAX (215) 386-7999.

BIBLIOGRAPHIES

Bibliography of Statistical Literature. Maurice Kendall and Alison Doig. Ayer Company Pubs., Inc., Lower Mill Road, North Stratford, NH 03590. (603) 922-5105. (800) 282-5413. 1981. 3 volumes: Pre-1940, 1940 to 1949, 1950 to 1958. $82.95, set.

Chronological Annotated Bibliography of Order Statistics, Volume Vi. H. Leon Harter. American Sciences Press, 20 Cross Road, Syracuse, NY 13224-2144. 1991. $115.00.

Guide To Statistical Methods and To the Pertinent Literature. L. Sachs. Springer-Verlag New York, Inc., 175 Fifth Avenue, New York, NY 10010. (212) 460-1500. (800) 777-4643. 1986. $39.00 in paper.

How To Find Out About Statistics. Gillian Burrington. Franklin Book Co., Inc., Elkins Park, PA 19117. (215) 635-5252. 1972. reprint edition. $74.00.

Mathematical Journals: an Annotated Guide. Diana F. Liang. Scarecrow Press. Distributed by University Press of America, 4720 Boston Way, Lanham, MD 20706. (301) 459-3366. (800) 462-6420. $29.50. ISBN: 0-8108-2585-6.

Stochastic Orders and Applications: A Classified Bibliography. Karl Mosler and Marco Scasini. Springer-Verlag New York, Inc., 175 Fifth Avenue, New York, NY 10010. (212) 460-1500. (800) 777-4643. 1993. wfi

DIRECTORIES AND BIOGRAPHICAL SOURCES

Assistantships and Graduate Fellowships in the Mathematical Sciences. American Mathematical Society, P.O. Box 6248, Providence, RI 02940. Annual. (401) 455-4000. (800) 321-4267. $19.00. ISSN: 1040-7650.

Combined Membership List. American Mathematical Society, P.O. Box 6248, Providence, RI 02940. (401) 455-4000. (800) 321-4267. Annual. $50.00. ISSN: 0569-6461.

Directory of Statisticians. American Statistical Association, 1429 Duke Street, Alexandria, VA 22314-3402. (703) 684-1221. FAX (703) 684-2037. Periodic.

Mathematical Science Professional Directory. American Mathematical Society, P.O. Box 6248, Providence, RI 02940. (401) 455-4000. (800) 321-4267. Annual. $45.00. ISBN: 0-8218-0173-2.

Research Centers Directory. Gale Research Company, 835 Penobscot Building, Detroit, MI 48226-4094. (313) 961-2242. (800) 877-4253. 20th edition. 1995. Annual. $485.00. ISSN: 0080-1518.

World Directory of Mathematicians. G.D. Mostow, editor. International Mathematical Union, Helsinki. 1958 to present. 10th revised edition. 1995. $45.00. Available from: American Mathematical Society, P.O. Box 6248, Providence, RI 02940. (401) 455-4000. ISSN: 0512-2740.

ENCYCLOPEDIAS AND DICTIONARIES

Dictionary of Gaming, Modelling & Simulation. G. Ian Gibbs. Books on Demand, 300 North Zeeb Road, Ann Arbor, MI 48106-1346. (313) 761-4700. (800) 521-0600. $49.60 in paper.

Dictionary - Outline of Basic Statistics. John Freund and Frank J. Williams. Dover Publications, Inc., 180 Varick Street, New York, NY 10014. (212) 255-3755. (800) 223- 3130. Reprint edition. $5.95 in paper.

Dictionary of Statistical Terms. F. H. Marriott. Halsted Press, 605 Third Avenue, New York, 10158-0012. (212) 850- 6400. 5th edition. 1990. $76.95 in paper.

Dictionary of Statistics. Michael G. Mulhall. Gordon Press Publications, P.O. Box 459, Bowling Green Station, New York, NY 10004. (718) 624-8419 . 1972. $300.00.

Encyclopedia of Statistical Sciences. Samuel Kotz, Norman L. Johnson, and Campbell B. Read, editors. John Wiley & Sons, Inc., 605 Third Avenue, New York, NY 10158-0012. (212) 850-6000. (800) 225-5945. 1988. 9 volumes. ISBN: 0- 471-05544-1. $1440.00 set.

Encyclopedic Dictionary of Mathematics. Kiyosi Ito, editor. MIT Press, 55 Hayward Street, Cambridge, MA 02142. (617) 625-8569. 2nd edition, reprint. 1993. 2 volumes. ISBN: 0- 262-59020-4. $70.00.

Harpercollins Dictionary of Statistics. Roger Porkess. HarperCollins Publ. 10 East 53rd Street, New York, NY 10022-5299. (800) 331-3761. 1991. $14.00 in paper.

International Encyclopedia of Statistics, 2. William H. Kruskal and Judith M. Tanur, editors. The Free Press, 1230 Avenue of the Americas, New York, NY 10020. (212) 698-7000. (800) 223-2348. 1978. $90.00.

Mathematics Dictionary. Robert C. James, editor. Van Nostrand Reinhold, 115 Fifth Avenue, New York, NY 10003. (212) 254-3232. (800)-842-3636. 5th edition. 1992. $42.95. ISBN: 0-442-00741-8.

Russian-English - Russian-English Dictionary On Probability, Statistics and Combinatorics. K. A. Borokov. Society for Industrial and Applied Mathematics, 3600 University City Science Center, Philadelphia, PA 19104-2688. (215) 382- 9800. FAX (215) 386-7999. (800) 447-7426. 1993. $47.50 in paper.

Topical Dictionary of Statistics: Advanced Industrial Terminology. Gary L. Teitjen. Chapman & Hall, 1 Penn Plaza, New York, NY 10119. (212) 564-1060. 1985. $24.50.

GENERAL WORKS

Adventures in Stochastic Processes. S. Resnick. Springer-Verlag New York, Inc., 175 Fifth Avenue, New York, NY 10010. (212) 460-1500. (800) 777-4643. 1994. $64.50.

Applied Probability and Stochastic Processes in Engineering and the Physical Sciences. Michael K. Ochi. John Wiley & Sons, Inc., 605 Third Avenue, New York, NY 10158-0012. (212) 850-6000. (800) 225-5945. 1990. $114.00.

The Art of Probability For Scientists and Engineers. Richard W. Hamming. Addison-Wesley Publishing Co., Inc., 1 Jacob Way, Reading, MA 01867. (617) 944-3700. (800) 447-2226.1991. $45.95.

Chance and Chaos. David Ruelle. Princeton University Press, 41 William Street, Princeton, NJ 08540. (609) 258- 4900. (800) 777-4726. 1993. $35.00.

Chaotic and Stochastic Behavior in Automatic Production Lines. Max-Oliver Hongler. Springer-Verlag New York, Inc.,175 Fifth Avenue, New York, NY 10010. (212) 460-1500. (800) 777-4643.

Complex Stochastic Systems and Engineering. D. M. Titterington, editor. Oxford University Press, Inc., 200 Madison Avenue, New York, NY 10016. (212) 725-6000. (800) 334-4249. 1995. $80.00.

Contemporary Stochastic Analysis. G. O. Ekhagnere, editor. World Scientific Publishing Company, Inc., 1060 Main Street, River Edge, NJ 07661. (201) 487-9655. (800) 227-7562. 1991. $74.00.

Electrical Engineering Probability. Richard Williams. West Publishing Co., 610 Opperman Drive, P. O. Box 64526, Saint Paul, MN 55164-0526. (612) 687-8000. (800) 328-9424. 1991. $69.25.

The Empire of Chance: How Probability Changed Science and Everyday Life. Gerd Gigerenzer et al. Cambridge University Press, 40 West 20th Street, New York, NY 10011-4211. (212) 924-3900. 1990. $19.95 in paper.

Elementary Probability theory With Stochastic Processes. Kai L. Chung. Springer-Verlag New York, Inc., 175 Fifth Avenue, New York, NY 10010. (212) 460-1500. (800) 777- 4643. 1990. $39.95.

Essentials of Probability. Research and Education Association, 61 Ethel Road, West Piscataway, NJ 08854. (908) 819-8880. Revised edition. 1994. $5.95 in paper.

Fallacies in Dealing With Probabilities. E. C. Poulton. Cambridge University Press, 40 West 20th Street, New York,

NY 10011-4211. (212) 924-3900. 1994. $59.95.

Introduction To Stochastic Processes and Nonequilibrium Statistical Physics. Haracio S. Wio. World Scientific Publishing Company, Inc., 1060 Main Street, River Edge, NJ 07661. (201) 487-9655. (800) 227-7562. 1994. $48.00.

Mathematical Statistics and Data Analysis. John A. Rice. Wadsworth Publishing Co., 10 Davis Drive, Belmont, CA 94002. (415) 595-2350. (800) 354-9706. 1995. $63.95.

Nonlinear Dynamic and Stochastic Mehcanics. Wolfgang H. Kliemann. CRC Press Inc., 2000 Corporate Boulevard, NW, Coca Raton, FL 33431. (407) 994-0555. (800) 272-7737. 1994. $79.95.

Probability and Random Processes. Geoffery Grimmett and David Stirzaker. Oxford University Press, Inc., 200 Madison Avenue, New York, NY 10016. (212) 725-6000. (800) 334-4249.2nd edition. 1992. $45.95.

Random Vibration of Mechanical and Structural Systems. T. T. Soong and Mircea Grigoriu. Prentice-Hall, 15 Columbus Circle, New York, NY 10023. (212) 373-8500. (800) 223-2348.1992. $82.00

Regular and Chaotic Dynamics. A. J. Lichtenberg and M. A. Lichtenberg. Springer-Verlag New York, Inc., 175 Fifth Avenue, New York, NY 10010. (212) 460-1500. (800) 777-4643. 1992. $59.95.

STOCHASTIC PROCESSES

Ency. of Physical Sciences and Engineering Info. Sources

Stochastic Modeling: Analysis and Simulation. Barry L. Nelson. McGraw-Hill Publishing Company, 1221 Avenue of the Americas, New York, NY 10020. (212) 512-2000. (800) 262- 4729. 1995. $46.00.

Stochastic Modeling and Control. Mark H. Davis and Richard Vinter. Chapman & Hall, 1 Penn Plaza, New York, NY 10119. (212) 564-1060. 1985. $42.50.

HANDBOOKS AND MANUALS

Biometrika Tables For Statistics. E. S. Pearson and H. O. Hartley, editors. Lubrecht & Cramer, Ltd., 38 County Route 48, Forestburgh, NY 12777-6400. (914) 794-8539. volume 1, 3rd edition. volume 2, reprint edition. 1976. $75.00 each.

Computational Handbook of Statistics. James L. Bruning, and B. L. Kintz. HarpCollins Publ. 10 East 53rd Street, New York, NY 10022-5299. (800) 331-376. 3rd edition. 1990. $38.00 in paper.

Guide To Tables in Mathematical Statistics. Joseph A. Greenwood, and H. O. Hartley. Books on Demand, 300 North Zeeb Road, Ann Arbor, MI 48106-1346. (313) 761-4700. (800) 521-0600 $180.00.

Handbook of Applicable Mathematics. Walter Ledermann, editor. John Wiley & Sons, Inc., 605 Third Avenue, New York, NY 10158-0012. (212) 850-6000. (800) 225-5945. 1985. $1496.00 6 volumes. ISBN: 0-471-90804-5.

Handbook of Applied Mathematics; Selected Results and Methods. Carl E. Pearsen, editor. Chapman & Hall, 1 Penn Plaza, New York NY 10119. (212) 564-1060. 1990. $46.95. ISBN: 0-442-00521-0.

Handbook of Mathematical Functions With Formulas, Graphs, and Mathmetical Tables. Milton Abramowitz and Irene A. Stegun, editors. U.S. National Bureau of Standards. U.S. Government Printing office, Washington, D.C. 10th edition. 1972. $43.00. ISBN: 0-16-000202-8.

Handbook of Mathematical Sciences. William H. Beyer, editor. CRC Press, 2000 Corporate Boulevard, Boca Raton, FL 33431. (800) 272-7737. 6th edition. 1987. $91.95. ISBN: 0-8493-0656-6.

Handbook of Mathematics. I. N. Bronshtein and K. A. Semedyayev. Van Nostrand Reinhold, 115 Fifth Avenue, New York, NY 10003. (800)-842-3636. 20th edition. $59.95. ISBN: 0-442-21171-6.

Handbook of Stochastic Methods. C. W. Gardiner. Springer-Verlag New York, Inc., 175 Fifth Avenue, New York, NY 10010. (212) 460-1500. (800) 777-4643. 1994. $59.00 in paper.

Handbook of Tables For Probability and Statistics. William H. Beyer, editor. CRC Press, 2000 Corporate Boulevard, Boca Raton, FL 33431. (800) 272-7737. 2ndedition. 1968. $92.95. ISBN: 0-8493-0692-2.

Standard Probability and Statistics Tables and Formulas. William H. Beyer. CRC Press, Inc., 2000 Corporate Boulevard, NW, Coca Raton, FL 33431. (407) 994-0555. (800) 272-7737. 1990. $34.95.

Standard Mathematical Tables and Formulas. William H. Beyer, editor. CRC Press, 2000 Corporate Boulevard, Boca Raton, FL 33431. (800) 272-7737. 29th edition. 1991. $41.95. ISBN: 084-930- 6299.

Stochastic Models. D. P. Heyman and M. J. Sobel. Elsevier Science Publishing Company, Inc., 655 Avenue of the Americas, New York, NY 10010. (212) 989-5800. 1990. $115.00.

Tables of Integrals, Series and Products: Corrected and Enlarged Edition. I. S. Gradshteyn and I. M. Ryzhik. Academic Press, Inc., 6277 Sea Harbor Drive, Orlando, FL 32821. (800) 321-5068. 5th edition. 1993. $61.00. ISBN: 0-12-294755-X.

ONLINE DATABASES AND CD-ROMS

CA Search. Chemical Abstracts Service, P.O. Box 3012, Columbus, OH 43210-0012. (614) 447-3600. (800) 848-6533. FAX (614) 447-3709. Very comprehensive guide to worldwide chemical literature and related fields, 1972 to present. Available on BRS,(800) 289-4277, DIALOG, (800) 334-2564, ORBIT (800) 456-7248, and STN International, FIZ Karlsruhe, P.O. Box 2465, W-7500, Karlsruhe 1, Germany, online services. Inquire as to cost and availability.

Compendex Plus. Engineering Information, Inc., 345 East 47th Street, New York, NY 10017. (212) 705-7600 or (800) 221-1044. Contains citations with abstracts to worldwide literature in engineering and technology, from 1970 to present. Available on online BRS,(800) 289-4277, DIALOG, (800) 334-2564, ORBIT (800) 456-7248, and STN International, FIZ Karlsruhe, P.O. Box 2465, W-7500, Karlsruhe 1, Germany, online services. Also available on CD-ROM. Inquire as to cost and availability.

Current Contents Search. Institute for Scientific Information, 3501 Market Street, Philadelphia, PA 19104. (215) 386-0100. FAX (215) 386-6362. Contains citations to articles listed in the table of contents of science and technology journals. Also articles in social sciences and life sciences journals. Available on BRS,(800) 289-4277, DIALOG,(800) 334-2564, online services. Inquire as to cost and availability.

Dissertation Abstracts. University Microfilms International, 300 North Zeeb Road, Ann Arbor, MI 48106. (800) 521-0600 or (313) 761-4700. Scope includes virtually all doctoral dissertations accepted at accredited American institutions from 1861 to present in 252 subject areas. Available on BRS,(800) 289-4277, DIALOG,(800) 334-2564, and OCLC EPIC,(800) 848-5878, online services. Also available on CD-ROM. Inquire as to cost and availability.

INSPEC. Institution of Electrical Engineers, Michael Faraday House, Six Hills Way, Stevenage, Herts. SG1 2AY, England. Telephone: 0438 313311 or FAX 0438 742840. Contains citations to the worldwide literature of physics, electronics and electrical engineering, computer technology, and related fields. Available on BRS,(800) 289-4277, DIALOG, (800) 334-2564, ORBIT, (800) 456-7248, and STN International, FIZ Karlsruhe, P.O. Box 2465, W-7500, Karlsruhe 1, Germany, online services. Inquire as to cost and availability.

Mathsci and Mathscinet. American Mathematical Society, P.O. Box 6248, Providence, RI 02940. (800) 321-4667 or (401) 455-4000 or FAX (401) 331-3842. Scope includes pure and applied mathematics and related areas of physics, statistics, engineering, computer science, and operations research literature since 1959. Available on DIALOG,(800) 334-2564, online service. Also available on CD-ROM. Inquire as to cost and availability.

NTIS Bibliographic Database. National Technical Information Service, 5285 Port Royal Road, Springfield, VA 22161. (703) 487-4929 or FAX (703) 321-8199. Broad coverage of government-sponsored science and technology research reports, 1964 to present. Available on BRS,(800) 289-4277, DIALOG, (800) 334-2564, ORBIT, (800) 456-7248, and STN International, FIZ Karlsruhe, P.O. Box 2465, W-7500, Karlsruhe 1, Germany, online services. Also available on CD-ROM. Inquire as to cost and availability.

Scientific and Technical Books and Serials in Print. R.R. Bowker Inc., 121 Chanlon Road, New Providence, NJ 07974. (800) 521-8110 or (908) 464-6800. List of currently published books and serials in the physical and biological sciences, engineering and technology, with subject, author and titles indexes. Available on ORBIT,(800) 456-7248, online service.Inquire as to cost and availability.

SCISEARCH. Institute for Scientific Information, 3501 Market Street, Philadelphia, PA 19104. (800) 523-1850 or (215) 386-0100. Broad multidisciplinary title and author index to the international literature of science and technology, 1974 to present. Available on DIALOG,(800) 334-2564, and ORBIT,(800) 456-7248, online services. Also available on CD-ROM. Inquire as to cost and availability.

WILSONLINE. H.W. Wilson Company, 950 University Avenue, Bronx, NY 10452. (800) 367-6770 or (212) 588-8400. Makes available online versions of the printed H.W. Wilson indexes including Applied Science and Technology Index, Business Periodicals Index, General Science Index, and Readers' Guide to Periodical Literature. Period covered is generally 1983 to present. Available on BRS,(800) 289-4277, DIALOG,(800) 334-2564, and OCLC EPIC,(800) 848-5878, online services. Also available on CD-ROM. Inquire as to cost and availability.

PERIODICALS

Advances in Applied Probability. Applied Probability Trust, School of Mathematics, University of Sheffield, Sheffield S3 7RH, England. TEL 44-742-824269. FAX 44-742-729782. 1969 to present. Quarterly. $56.00 per year. ISSN: 0001-8678.

American Statistician. American Statistical Association. 1429 Duke Street, Alexandria, VA 22314-3402. (703) 684-1221. FAX (703) 684-2037. 1947 to present. Quarterly. $55.00 per year. ISSN: 0003-1305.

Annals of Probability. Institute of Mathematical Statistics. 3401 Investment Boulevard, Suite 7, Hayward, CA 94545. (510) 783-8141. 1973 to present. Quarterly. $130.00 per year. ISSN: 0090-5364.

Biometrics. Biometric Society. 1429 Duke Street, Suite 401, Alexandria, VA 22314-3402. (703) 836-8311. 1945 to present. Quarterly. $90.00 per year. ISSN: 0006-341X.

Canadian Journal of Statistics/Revue Canadienne de Statistique. Statistical Society of Canada, Room 612 Dunton Tower, Carleton University, Ottawa, ON K1S 5B6, Canada. (613) 788-2167. 1973 to present. Quarterly. $98.00 per year. ISSN: 0319-5724.

Chance. Springer-Verlag, 175 Fifth Avenue, New York, NY 10010. (212) 460-1612. FAX (212) 473-6272. 1988 topresent. Quarterly. $69.00 per year. ISSN: 0933-2480.

Communications in Statistics, Part A: Theory and Methods. Marcel Dekker Journals, 270 Madison Avenue, New York, NY, 10016. (212) 696-9000. FAX (212) 685-4540. 1973 to present. Twelve issues per year. $1295.00 per year. ISSN: 0361-0926.

Institut Henri Poincare. ANNALES. PROBABILITIES ET STATISTICS. Gauthier-Villars, 15 rue Gossin, 92543 Montrouge Cedex, France. TEL 33-1-40-92-65-00. FAX 33-1-40-92-65-97. 1930 to present. Quarterly. 1230F. ISSN: 0246-0203.

International Journal of Game theory. Physica-Verlag GmbH. Distributed by Springer-Verlag New York, Inc., 175 Fifth Avenue, New York, NY 10010. (212) 460-1500. 1971 to present. Quarterly. DM 390. ISSN: 0020-7276.

Jasa. Journal of the American Statistical Association. American Statistical Association, 1429 Duke Street, Alexandria, VA 22314-3402. (703) 684-1221. FAX (703) 684-2037. 1888 to present. Quarterly. $210.00 per year. ISSN: 0162-1259.

Journal of Applied Probability. Applied Probability Trust, University of Sheffield, Sheffield S3 7RH, England. TEL 44-742-824269. FAX 44-742-729782. 1964 to present. Quarterly. $168.00 per year. ISSN: 0021-9002.

Journal of Theoretical Probability. Plenum Publishing Corp., 233 Spring Street, New Yorik, NY 10013-1578. (212) 620-8000. FAX (212) 463-0742. 1988 to present. Quarterly. $225.00. ISSN: 0894-9846

Metrika; International Journal For theoretical and Applied Statistics. Physica-Verlag GmbH und Co. Distributed by Springer-Verlag New York, Inc., 175 Fifth Avenue, New York, NY 10010. (212) 460-1500. 1953 to present. Bimonthly. DM360. ISSN: 0026-1335.

Probability theory and Related Fields. Springer-Verlag New York, Inc., 175 Fifth Avenue, New York, NY 10010. (212) 460-1500. (800) 777-4643. 1962 to present. Monthly. $1553.00. ISSN: 0178-8051.

Random Structures and Algorithms. John Wiley & Sons, Inc., 605 Third Avenue, New York, NY 10158-0012. (212) 850-6000. FAX (212) 850-6088. 1990 to present. Quarterly. $185.00 per year. ISSN: 1042-9832.

Risk Analysis. Society for Risk Analysis. Plenum Publishing Corp. 233 Spring Street, New York, NY 10013- 1578. (212) 620-8000. FAX (212) 463-0742. 1980 to present. Bimonthly. $320.00 per year. ISSN: 0272-4332.

Royal Statistical Society. JOURNAL. SERIES A-C. Basil Blackwell Ltd., 108 Cowley Road, Oxford OX4, 1JF, Eng. TEL 0865-791100. FAX 0865-791347. SeriesA: Statistics in Society; 1838 to present. Series B: Methodological; 1934 to present. Series C: Applied Statistics. Each series, three issues per year; $73.00 per year, each. ISSN: 0035- 9238, 9246, 9254.

Siam Journal On Applied Mathematics. Society for Industrial and Applied Mathematics, 3600 University City Science Center, Philadelphia, PA 19104-2688. (215) 382- 9800. FAX (215) 386-7999. 1953 to present. Bimonthly. $242.00 per year. ISSN: 0036-1399.

Siam Journal On Control and Optimization. Society for Industrial and Applied Mathematics, 3600 University City Science Center, Philadelphia, PA 19104-2688. (215) 382- 9800. FAX

STOCHASTIC PROCESSES

Ency. of Physical Sciences and Engineering Info. Sources

(215) 386-7999. 1963 to present. Bimonthly. $287.00 per year. ISSN: 0363-0129.

Siam Journal On Scientific Computing. Society for Industrial and Applied Mathematics, 3600 University City Science Center, Philadelphia, PA 19104-2688. (215) 382-9800. FAX (215) 386-7999. 1980 to present. Bimonthly. $260.00 per year. ISSN: 1064-8275.

Siam Review. Society for Industrial and Applied Mathematics, 3600 University City Science Center, Philadelphia, PA 19104-2688. (215) 382-9800. 1959 to present. Quarterly. $148.00 per year. ISSN: 0036-1445.

Statistical Science. Institute of Mathematical Statistics. 3401 Investment Boulevard, Suite 7, Hayward, CA 94545. (510) 783-8141. 1986 to present. Quarterly. $75.00 per year. ISSN: 0883-4237.

Statistics and Computing. Chapman & Hall, One Penn Plaza, 41st Floor, New York, NY 10019. (212) 564-1060/ FAX (212) 564-1505. Quarterly. $190.00 per year. ISSN: 0960-3174.

Stochastic Analysis and Applications. Marcel Dekker Journals, 270 Madison Avenue, New York, NY, 10016. (212) 696-9000. FAX (212) 685-4540. 1983 to present. Five issues per year. $535.00 per year. ISSN: 0736-2992.

Stochastic Processes and their Applications. Elsevier Science, Inc., Box 882, Madison Square Station, New York, NY 10159. (212) 989-5800. FAX (212) 633-3990. 1973 to present. Monthly. $1171.00 per year. ISSN: 0304-4149.

Technometrics. American Statistical Association, 1429 Duke Street, Alexandria, VA 22314-3402. (703) 684-1221. FAX (703) 684-2037. 1959 to present. Quarterly. $50.00.

RESEARCH CENTERS AND INSTITUTES

Center for Applied Mathematics. University of Florida, 201 Walker Hall, Gainsville, FL 32611. (804) 392-0281.

Center For Mathematical Sciences. University of Wisconsin-Madison, 610 Walnut Street, Madison, WI 53705. (608) 263-2696. FAX (608) 263-2841.

Courant Institute of Mathematical Sciences. New York University, 251 Mercer Street, New York, NY 10012. (212) 998-3000.

Institute For Mathematics and Its Applications. University of Minnesota, 514 Vincent Hall, 206 Church Street, S.E., Minneapolis, MN 55455. (612) 624-6066. FAX (612) 626-7370.

Laboratory For Research in Statistics and Probability. Carleton University, Dunton Tower, Room 611, Ottawa, ON, Canada K1S 5B6. (613) 788-2167.

Lefschetz Center For Dynamical Systems. Brown University, 182 George Street, Providence, RI 02912. (401) 863-2358.

Mathematical Sciences Institute. Cornell University, 201 Caldwell Hall, Ithaca, NY 14853. (607) 255-8005.

Mathematical Statistics Program. University of Maryland, 1107 Mathematics Building, College Park, MD 20742. (301) 405 5061. FAX (301) 314-0827.

Operations Research Center. Massachusetts Institute of Technology. Cambridge, MA 02139. (617) 253-3601. FAX (617) 258-9214

Statistical Laboratory. Iowa State University, 102 Snedecor Hall, Ames, IA 50011. (515) 294-3440. FAX (515) 294-4040.

Statistical Laboratory. Kansas State University. Dickens Hall, Manhattan, KS 66506-0802. (913) 532-6882. (913) 532- 7736.

Statistics Center. Massachusetts Institute of Technology, 77 Massachussetts Avenue, E40-111, Cambridge, MA 02139. (617) 253-8722.

Statistics Department. Rutgers University, Hill Center, New Brunswick, NJ 08903. (201) 932-2691. FAX (201) 932-3428.

STOL-VTOL

See also: AERODYNAMICS, AEROSPACE ENGINEERING, AIRCRAFT

ABSTRACT SERVICES AND INDEXES

Aeronautical Engineering. Scientific and Technical Branch, National Aeronautics and Space Administration. National Technical Information Service (NTIS), 5285 Port Royal Road, Springfield, VA 22161. (703) 487-4650. FAX (703) 321-8547. Monthly. A selection of annotated references to unclassified reports and journal articles that were introduced into the NASA scientific and technical information system and announced in STAR and IAA.

Alloys Index. American Society for Metals, Metals Park, OH 44073. (216) 338-5151. FAX (216) 338-4634. 1974 to present. Monthly. $295.00.

Applied Mechanics Reviews: An Assessment of World Literature in Engineering Sciences. American Society of Mechanical Engineers, 345 East 47th Street, New York, NY 10017. (212) 705-7703. 1948 to present. Monthly. $360.00 per year.

Applied Science and Technology Index; A Cumulative Subject Index To English Language Periodicals in the Fields of Aeronautics and Space Science, Computer Technology, Chemistry, Construction Industry, Energy and Related Areas. H.W. Wilson Co., 950 University Avenue, Bronx, NY 10452. (800) 367-6770 or (212) 588-8400. FAX (718) 590-1617. From 1958 to present. Monthly. Inquire about cost and availability. Also available on CD-ROM.

Current Contents: Engineering, Technology, and Applied Sciences. Institute for Scientific Information, 3501 Market Street, Philadelphia, PA 19104. (215) 386-0100. FAX (215) 386-6362. Weekly. $360.00 per year.

Engineering Index Monthly. Engineering Information, Inc., Castle Point on the Hudson, Hoboken, NJ 07030. (800) 221-1044. FAX (212) 832-1857. Monthly. $2200.00 per year. Also available online as COMPENDEX, and also on CD-ROM.

International Aerospace Abstracts. Technical Information Service, American Institute of Aeronautics and Astronautics, Inc., 555 West 57th St., New York, NY 10019. (212) 247-6500. FAX (212) 582-4861. Semi-monthly. $1295.00 per year.

Metals Abstracts and Metals Abstracts Index. American Society for Metals, Metals Park, OH 44073. (216) 338-5151. 1968 to present. Monthly. Abstracts are $1500.00 per year and Index is $460.00 per year.

Science Citation Index. Institute for Scientific Information, 3501 Market Street, Philadelphia, PA 19104. (215) 386-0100. FAX (215) 386-6362. Inquire about availability and cost. Also available on CD-ROM.

Scientific and Technical Aerospace Reports (star). National Aeronautics and Space Administration. NASA Center for Aerospace Information, Box 8757, BWI Airport, Baltimore, MD 21240. (301) 621-0153. Monthly. Inquire about availability and cost. Also available through the NASA online retrieval service (RECON), and through the Aerospace Database through DIALOG.

World Aluminum Abstracts. Aluminum Association, Inc., 818 Connecticut Ave. NW, Washington, DC 20006. (202) 862-5156. Inquire about price and availability.

ASSOCIATIONS AND PROFESSIONAL SOCIETIES

American Institute of Aeronautics and Astronautics. 370 L'Enfant Promenade SW, Washington, DC 20024. (202) 646-7400.

Society of Automotive Engineers. SAE International, Inc., 400 Commonwealth Dr., Warrendale, PA 15096. (412) 776-4841.

BIBLIOGRAPHIES

Aeronautical Engineering: A Continuing Bibliography With Indexes. National Aeronautics and Space Administration, Center for Aerospace Information, Box 8757, BWI Airport, Baltimore, MD 21240. (301) 621-0153. Inquire for cost and availability.

DIRECTORIES AND BIOGRAPHICAL SOURCES

Research Centers Directory. Gale Research, Inc., 835 Penobscot Building, Detroit, MI 48226-4094. (313) 961-2242. (800) 877-4253. FAX (313) 961-6083. $485.00.

Who's Who in Engineering. American Association of Engineering Societies, 1111 19th St. NW, Suite 608, Washington, DC 20036. 7th ed. 1988. $200.00.

World Aviation Directory. McGraw-Hill, Aviation Week Group (NY), 1221 Avenue of the Americas, New York, NY 10020. (215) 237-4112. FAX (215) 586-3232. Semiannual. $250.00.

ENCYCLOPEDIAS AND DICTIONARIES

Dictionary of Aerospace Engineering. M.G. Kotik. Elsevier Science Publishing Company, Inc., 655 Avenue of the Americas, New York, NY 10010. (212) 989-5800. 1986. $225.75.

Encyclopedia of Physical Science and Technology. Robert A. Meyers, ed. 18 volumes. Academic Press, Inc., 6277 Sea Harbor Drive, Orlando, FL 32887. (800) 321-5068. 1992. $2100.00

Jane's Aerospace Dictionary, B. Guston. 3d ed. Jane's Information Group, 1340 Braddock Place, Suite 300, Alexandria, VA 22314-1651. (703) 683-3700. FAX (703) 836-0029. 1991. $45.00.

Jane's Encyclopedia of Aviation. Michael J. Taylor, comp. Outlet Book Co., 40 Engelhard Ave., Avenal, NJ 07001. (908) 827-2700 or (800) 223-6804. FAX (908) 827-2641. 1993. $34.99.

McGraw-Hill Encyclopedia of Engineering. Sybil P. Parker, ed. 2nd edition. McGraw-Hill Publishing Company, 1221 Avenue of the Americas, New York, NY 10020. (800) 262-4729 or (212) 512-3825. 1993. $95.50.

McGraw-Hill Encyclopedia of Science and Technology, Sybil P. Parker, ed. 7th ed. 20 vols. McGraw-Hill Publishing Company, 1221 Avenue of the Americas, New York, NY 10020. (800) 262-4729 or (212) 512-3825. $1900.00

SAE Dictionary of Aerospace Engineering. William H. Cubberly, ed. Society of Automotive Engineers, 400 Commonwealth Drive, Warrendale, PA 15096. (412) 776-4841. 1992. $85.00.

GENERAL WORKS

Harrier: The V/stol Warrior. John Dibbs and Tony Holmes. Osprey Aerospace, c/o Motorbooks International, Box 2, Osceola, WI 54020. (800) 826-6600. 1992. $17.95.

Introduction To V/stol Airplanes. David L. Kolman. Iowa State University Press, 2121 South State Avenue, Ames, IA 50010. (800) 862-6657. FAX (515) 292-3348. 1981. $22.00.

V/stol: An Update and Overview. SAE International, Inc., 400 Commonwealth Dr., Warrendale, PA 15096. (412) 776-4841. 1984. Inquire for cost and availability.

V/stol Design Concepts For the Custom Aircraft Builder. Philip Terpstra. Spirit Publishing, PO Box 23417, Tucscon, AZ 85734. (602) 822-2030. 1991. Inquire for cost and availability.

HANDBOOKS AND MANUALS

Jane's All the World's Aircraft. Jane's Information Group, 1340 Braddock Place, Suite 300, Alexandria, VA 22314-1651. (703) 683-3700. FAX (703) 836-0029. Annual. $245.00. Also available on CD-ROM.

Jane's Avionics. Jane's Information Group, 1340 Braddock Place, Suite 300, Alexandria, VA 22314-1651. (703) 683-3700. FAX (703) 836-0029. Annual. $245.00. Also available on CD-ROM.

ONLINE DATABASES AND CD-ROMS

Aerospace Database. American Institute of Aeronautics and Astronautics, 370 L'Enfant Promenade, S.W., Washington, DC 20024. (202) 646-7400. Worldwide published literature on research and development in aerospace and related areas, with abstracts. Covers 1962 to present. Available on DIALOG, (800) 334-2564, online service. Also available on CD-ROM. Inquire as to cost and availability.

Compendex Plus. Engineering Information, Inc., 345 East 47th Street, New York, NY 10017. (212) 705-7600 or (800) 221-

1044. Contains citations with abstracts to worldwide literature in engineering and technology, from 1970 to present. Available on online BRS,(800) 289-4277, DIALOG, (800) 334-2564, ORBIT (800) 456-7248, and STN International, FIZ Karlsruhe, P.O. Box 2465, W-7500, Karlsruhe 1, Germany, online services. Also available on CD-ROM. Inquire as to cost and availability.

CSA Engineering. Cambridge Scientific Abstracts, 7200 Wisconsin Avenue, Suite 601, Bethesda, MD 20814. (301) 961-6750 or (800) 843-7751. Contains citations and abstracts of international periodicals and research literature covering all fields of engineering and science and technology,including computer and information science, electronics, mechanical engineering, solid state materials, 1981 to present. Available on BRS,(800) 289-4277, online service.Inquire as to cost and availability.

Current Contents Search. Institute for Scientific Information, 3501 Market Street, Philadelphia, PA 19104. (215) 386-0100. FAX (215) 386-6362. Contains citations to articles listed in the table of contents of science and technology journals. Also articles in social sciences and life sciences journals. Available on BRS,(800) 289-4277, DIALOG,(800) 334-2564, online services. Inquire as to cost and availability.

Dissertation Abstracts. University Microfilms International, 300 North Zeeb Road, Ann Arbor, MI 48106. (800) 521-0600 or (313) 761-4700. Scope includes virtually all doctoral dissertations accepted at accredited American institutions from 1861 to present in 252 subject areas. Available on BRS,(800) 289-4277, DIALOG,(800) 334-2564, and OCLC EPIC,(800) 848-5878, online services. Also available on CD-ROM. Inquire as to cost and availability.

International Aerospace Abstracts. American Institute of Aeronautics and Astronautics, 370 L'Enfant Promenade, S.W., Washington, DC 20024. (202) 646-7400. Contains references and abstracts of journal and monograph literature relating to aerospace science and technology, from 1963 to present. Available through the NASA/RECON system of the National Aeronautics and Space Administration only.

*ISMEC: Mechanical Engineering Abstracts.*Cambridge Scientific Abstracts, 7200 Wisconsin Avenue, Suite 601, Bethesda, MD 20814. (301) 961-6750 or (800) 843-7751. Contains citations to the literature in mechanical engineering, industrial and production engineering, energy, power, mechanics, devices and related areas, from 1973 to present. Available on the DIALOG,(800) 334-2564, online service. Inquire as to cost and availability.

NASA Database. American Institute of Aeronautics and Astronautics, 370 L'Enfant Promenade, S.W., Washington, DC 20024. (202) 646-7400. Citations and abstracts of aeronautics and astronautics literature, 1962 to present. Also contains citations from *STAR, Scientific and Technical Aerospace Reports, and International Aerospace Abstracts.* Available through NASA/RECON online service. Inquire as to cost and availability.

NTIS Bibliographic Database. National Technical Information Service, 5285 Port Royal Road, Springfield, VA 22161. (703) 487-4929 or FAX (703) 321-8199. Broad coverage of government-sponsored science and technology research reports, 1964 to present. Available on BRS,(800) 289-4277, DIALOG, (800) 334-2564, ORBIT, (800) 456-7248, and STN International, FIZ Karlsruhe, P.O. Box 2465, W-7500, Karlsruhe 1, Germany, online services. Also available on CD-ROM. Inquire as to cost and availability.

SAE Global Mobility Database. Society of Automative Engineers (SAE), Electronic Publishing Division, 400 Commonwealth Drive, Warrendale, PA 15098. (412) 776-4841. Contains citations with abstracts to technical papers on automotive and aerospace technology and vehicular-related industries that have been presented at SAE conferences. Covers 1906 to present. Available on ORBIT,(800) 456-7248,online service. Inquire as to cost and availability.

Scientific and Technical Books and Serials in Print. R.R. Bowker Inc., 121 Chanlon Road, New Providence, NJ 07974. (800) 521-8110 or (908) 464-6800. List of currently published books and serials in the physical and biological sciences, engineering and technology, with subject, author and titles indexes. Available on ORBIT,(800) 456-7248, online service.Inquire as to cost and availability.

SCISEARCH. Institute for Scientific Information, 3501 Market Street, Philadelphia, PA 19104. (800) 523-1850 or (215) 386-0100. Broad multidisciplinary title and author index to the international literature of science and technology, 1974 to present. Available on DIALOG,(800) 334-2564, and ORBIT,(800) 456-7248, online services. Also available on CD-ROM. Inquire as to cost and availability.

Tris (transportation Research Information). National ACADemy of Science, 2101 Constitution Avenue, N.W., Washington, DC 20418. (202) 334-3313 or (800) 624-6242. Citations with abstracts of literature on transportation, including air, highway, rail, maritime and other modes from 1968 to present. Available on DIALOG,(800) 334-2564, online service. Inquire as to cost and availability.

WILSONLINE. H.W. Wilson Company, 950 University Avenue, Bronx, NY 10452. (800) 367-6770 or (212) 588-8400. Makes available online versions of the printed H.W. Wilson indexes including Applied Science and Technology Index, Business Periodicals Index, General Science Index, and Readers' Guide to Periodical Literature. Period covered is generally 1983 to present. Available on BRS,(800) 289-4277, DIALOG,(800) 334-2564, and OCLC EPIC,(800) 848-5878, online services. Also available on CD-ROM. Inquire as to cost and availability.

PERIODICALS

Aeronautical Journal. The Royal Aeronautical Society, 4 Hamilton Place, London W1V 0BQ, England. Monthly except June and August. 1897 to present. Inquire for price information in U.S.

Aerospace America. American Institute of Aeronautics and Astronautics, 370 L'Enfant Promenade SW, Washington, DC 20024-2518. (202) 646-7471. 1932 to present. Monthly. $70 annually (non-members).

Aerospace Daily. McGraw-Hill Inc., Aviation Week Group, 1200 G Street NW, Suite 200, Washington, DC 20005. (202) 383-2350. 1973 to present. Five times a week. $1340.00 per year.

Aerospace Engineering. Society of Automotive Engineers, 400 Commonwealth Dr., Warrendale, PA 15096. (412) 776-4841. 1981 to present. Monthly. $48.00 per year.

Aerospace Propulsion. McGraw-Hill Inc., Aviation Week Group, 1200 G Street NW, Suite 200, Washington, DC 20005. (202) 383-2350. 1963 to present. 25 times a year. $550.00 a year.

AIAA Journal. American Institute of Aeronautics and Astronautics, 370 L'Enfant Promenade SW, Washington, DC 20024-2518. (202) 646-7471. 1963 to present. Monthly. $52.00 per year (members) and $435.00 per year (non-members).

Aircraft Engineering and Aerospace Technology. BunHill Publications Ltd., 127 Stanstead Rd., London SE23 1JE, England. 1929 to present. Monthly. $102.00 per year (U.S.).

Airpower Journal. Airpower Research Institute, 401 Chennault Circle, Maxwell AFB, AL 36112-8930. Quarterly. Order from New Orders, Supt. of Documents, PO Box 371954, Pittsburg, PA 15250-7954. $13.00 per year.

Aviation Week and Space Technology. 1221 Avenue of the Americas, New York, NY 10020. (212) 512-2999 or (800) 525-5003. 1916 to present. Weekly. $82.00 per year.

Interavia/aerospace World. Aerospace Media Publishing, Swissair Centre, 31 Route de l'Aeroport, POB Box 437, 1215 Geneva 15, Switzerland. (022) 788 27 88. FAX (022) 788 27 26. 19987 to present. Monthly. $128.00 per year (U.S.).

Journal of Aircraft. American Institute of Aeronautics and Astronautics, 370 L'Enfant Promenade SW, Washington, DC 20024-2518. 1963 to present. Bimonthly. $42.00 (members) and $230.00 (non-members).

Journal of the American Helicopter Society. American Helicopter Society, Inc., 217 North Washington St., Alexandria, VA 22314. (703) 684-6777. 1956 to present. Quarterly. $35.00 per year.

Progress in Aerospace Sciences. Pergamon Press, 660 White Plains Rd., Tarrytown, NY 10591-5153. 1961 to present. Quarterly. $320.00 per year.

Rotor and Wing International. PHillips Publishing Inc., 7811 Montrose Rd., Potomac, MD 20854. (301) 340-2100. FAX (301) 309-3847. 1967 to present. Monthly. $49.00 a year.

Vertiflite. American Helicopter Society, 217 N. Washington St., Alexandria, VA 22314. (703) 684-6777. FAX (703) 739-9279. 1955 to present. Bimonthly. $25 per year (members) and $35 per year (non-members).

RESEARCH CENTERS AND INSTITUTES

Aerospace Research Applications Center, Indianapolis Center For Advanced Research, 611 N. Capitol Ave., Indianapolis, in 46204. (317) 262-5003. FAX (317) 262-5044.

Aerospace Research Center, 1250 Eye St., NW, Washington, DC 20005. (202) 371-8400. FAX (202) 371-8470.

Center For Aerospace Technology, Weber State College, Ogden, UT 84408-1805. (801) 626-7272.

Center For Control and Systems Research, University of Texas At Austin, Aerospace Engineering and Engineering Mechanics, Wrw 408, Austin, TX 78712-1085. (512) 471-4908. FAX (512) 471-3788.

Flight Research Laboratory, University of Kansas, Raymond Nichols Hall, Lawrence, KS 66045. (913) 864-3043. FAX (913) 864-7789.

Joint Institute For Advancement of Flight Sciences, George Washington University, Langley Research Center, Mail Stop 269, Hampton, VA 23665. (804) 864-1982. FAX (804) 864-5894.

Ohio Aerospace Institute, 2001 Aerospace Parkway, Brook Park, OH 44142. (216) 891-2100. FAX (216) 891-2140.

STRATIGRAPHY

See also: SEDIMENTATION

ABSTRACT SERVICES AND INDEXES

Bibliography and Index of Geology. American Geological Institute, 4220 King Street, Alexandria, VA 22302. (703) 379-2480. Fax (703) 379-7563. Monthly. $1295.00 per year. Also available online as GeoRef.

Current Contents: Physical, Chemical and Earth Sciences. Institute for Scientific Information, 3501 Market Street, Philadelphia, PA 19104. (215) 386-0100. FAX (215) 386-6362. Weekly. $360.00 per year.

Geodex Retrieval System For Geotechnical Abstracts. Geodex Retrieval Systems, 669 Broadway, Box 279, Sonoma, CA 95476. (707) 939-8476. FAX (707) 996-8734. 1970 to present. Three times a year. $645.00 per year includes Geotechnical Abstracts, as Geodex Systems-GRS.

Geological Abstracts. Elsevier - Geo Abstracts, Regency House, 34 Duke Street, Norwich NR3 3AP, England. Monthly. $760.00 per year. Also available online as GEOBASE.

Geological Society of America. ABSTRACTS WITH PROGRAMS. Geological Society of America. 3300 Penrose Place, P.O. Box 9140, Boulder, CO 80301-9140. (303) 447-2020. Abstracts and programs of the annual conference. Annual. $69.00.

Geotechnical Abstracts. Geodex Retrieval Systems, 669 Broadway, Box 279, Sonoma, CA 95476. (707) 939-8476. FAX (707) 996-8734. 1970 to present. Three times a year. $280.00 per year. Covers soil mechanics, foundation engineering, rock mechanics, and engineering geology.

ASSOCIATIONS AND PROFESSIONAL SOCIETIES

American Geological Institute. 4220 King Street, Alexandria, VA 22302. (703) 379-2480. Fax (703) 379-7563.

American Geophysical Union. 2000 Florida Avenue, N.W., Washington, DC 20009. (202) 462-6900. FAX (202) 328-0566.

American Institute of Professional Geologists. 7828 Vance Drive, Suite 103, Arvada, CO 80003. (303) 431-0831. FAX (303) 431-1332.

Association of Engineering Geologists. 323 Boston Post Road, Suite 2D, Sudbury, MA 01776. (508) 443-4639.

Geological Society of America. 3300 Penrose Place, PO Box 9140, Denver, CO 80301-9140. (303) 447-2020.

DIRECTORIES AND BIOGRAPHICAL SOURCES

American Men & Women of Science, 1995-96. R.R. Bowker Staff, eds. 19th edition. 8 volumes. R.R. Bowker/Reed International Publishing Company, 121 Chanlon Road, New Providence, NJ 07974. (908) 464-6800 or (800) 521-8110. 1995. $850.00

Directory of Engineering Societies and Related Organizations. Gordon Davis. 13th edition. American Association of Engineering Societies, 1111 19th Street NW, Suite 608, Washington, DC 20036. (202) 296-2237 or (800) 658-8897. 1989. Inquire for price.

International Research Centers Directory. Gale Research, Inc., 835 Penobscot Building, Detroit, MI 48226-4094. (313) 961-2242. (800) 877-4253. FAX (313) 961-6083. 8th edition. 1995. $410.00

Research Centers Directory. Gale Research, Inc., 835 Penobscot Building, Detroit, MI 48226-4094. (313) 961-2242. (800) 877-4253. FAX (313) 961-6083. $485.00.

Scientific and Technical Organizations and Agencies Directory. Gale Research, Inc., 835 Penobscot Building, Detroit, MI 48226-4094. (313) 961-2242. (800) 877-4253. FAX (313) 961-6083. 4th edition. 1996. $195.00.

Who's Who in Engineering. American Association of Engineering Societies, 1111 19th Street NW, Suite 608, Washington, DC 20036. (202) 296-2237 or (800) 658-8897. 8th edition. 1991. Inquire for price.

ENCYCLOPEDIAS AND DICTIONARIES

The Encyclopedia of the Solid Earth Sciences. Philip Kearey, editor-in-chief. Blackwell Scientific Publications, 238 Main Street, Cambridge, MA 02142. (617) 876-7000 or (800) 759-6102. FAX (617) 876-7022. 1993. $60.00.

Encyclopedia of Physical Science and Technology. Robert A. Meyers, ed. 18 volumes. Academic Press, Inc., 6277 Sea Harbor Drive, Orlando, FL 32887. (800) 321-5068. 1992. $2100.00

McGraw-Hill Dictionary of Scientific and Technical Terms. Sybil P. Parker, ed. 5th edition. McGraw-Hill Publishing Company, 1221 Avenue of the Americas, New York, NY 10020. (800) 262-4729 or (212) 512-3825. 1993. $110.50.

GENERAL WORKS

The Key To Earth History: An Introduction To Stratigraphy. Peter Doyle, et al. John Wiley and Sons, Inc., 605 Third Avenue, New York, NY 10158. (800) 526-5368 or (212) 850-6000. 1994. $24.95.

Practical Sequence Stratigraphy. D. Emery and K. Myers. Blackwell Scientific Publications, 238 Main Street, Cambridge, MA 02142. (617) 876-7000 or (800) 759-6102. FAX (617) 876-7022. 1994. Inquire for price.

Principles of Sedimentary Deposits: Stratigraphy and Sedimentation. Gerald M. Friedman, et al. Macmillan Publishing, 200 Old Tappan Road, Old Tappan, NJ 07675. (800) 223-2336. FAX (800) 445-6991. 1992. Inquire for cost and availability.

Principles of Sedimentology and Stratigraphy. Sam Boggs. 2nd edition. Macmillan Publishing, 200 Old Tappan Road, Old Tappan, NJ 07675. (800) 223-2336. FAX (800) 445-6991. 1994. Inquire for cost and availability.

Stratigraphy. Pierre Cotillon. Springer-Verlag, 175 Fifth Avenue, New York, NY 10010. (212) 460-1500. FAX (212) 473-6272. 1992. $40.00.

Stratigraphy and Sedimentation. William C. Krumbein and L.L. Sloss. 2nd edition. W.H. Freeman and Company, 41 Madison Avenue, East 26th, 35th Floor, New York, NY 10010. (212) 576-9400. FAX (212) 689-2383. 1995. Inquire for cost and availability.

ONLINE DATABASES AND CD-ROMS

Compendex Plus. Engineering Information, Inc., 345 East 47th Street, New York, NY 10017. (212) 705-7600 or (800) 221-1044. Contains citations with abstracts to worldwide literature in engineering and technology, from 1970 to present. Available on online BRS,(800) 289-4277, DIALOG, (800) 334-2564, ORBIT (800) 456-7248, and STN International, FIZ Karlsruhe, P.O. Box 2465, W-7500, Karlsruhe 1, Germany, online services. Also available on CD-ROM. Inquire as to cost and availability.

CSA Engineering. Cambridge Scientific Abstracts, 7200 Wisconsin Avenue, Suite 601, Bethesda, MD 20814. (301) 961-6750 or (800) 843-7751. Contains citations and abstracts of international periodicals and research literature covering all fields of engineering and science and technology,including computer and information science, electronics, mechanical engineering, solid state materials, 1981 to present. Available on BRS,(800) 289-4277, online service.Inquire as to cost and availability.

Current Contents Search. Institute for Scientific Information, 3501 Market Street, Philadelphia, PA 19104. (215) 386-0100. FAX (215) 386-6362. Contains citations to articles listed in the table of contents of science and technology journals. Also articles in social sciences and life sciences journals. Available on BRS,(800) 289-4277, DIALOG,(800) 334-2564, online services. Inquire as to cost and availability.

Dissertation Abstracts. University Microfilms International, 300 North Zeeb Road, Ann Arbor, MI 48106. (800) 521-0600 or (313) 761-4700. Scope includes virtually all doctoral dissertations accepted at accredited American institutions from 1861 to present in 252 subject areas. Available on BRS,(800) 289-4277, DIALOG,(800) 334-2564, and OCLC EPIC,(800) 848-5878, online services. Also available on CD-ROM. Inquire as to cost and availability.

Earth Sciences. U.S. Geological Survey, 12201 Sunrise Valley Drive, Reston, VA 22092-9998. (703) 648-4460. CD-ROM of earth science databases including the U.S. Geological Survey Library, Earth Science Data Directory, and GEOINDEX, citations to published geological maps. $350.00 per year with quarterly updates.

Geoarchive. Geosystems, P.O. Box 1024, Westminster, London, England, SW1 P 2JL. Citations to literature on geoscience, 1969 to present. Inquire as to online cost and availability.

Geobase. Elsevier - Geo Abstracts, Regency House, 34 Duke Street, Norwich NR3 3AP, England. Contains citations to the worldwide earth science literature from 1980 to date. Available on DIALOG, ORBIT online services. Inquire as to cost and availability.

Georef: Bibliography and Index of Geology. American Geological Institute, 4220 King Street, Alexandria, VA 22302. (703) 379-2480. Fax (703) 379-7563. Monthly. Inquire for price and availability.

SCISEARCH. Institute for Scientific Information, 3501 Market Street, Philadelphia, PA 19104. (800) 523-1850 or (215) 386-0100. Broad multidisciplinary title and author index to the international literature of science and technology, 1974 to present. Available on DIALOG,(800) 334-2564, and ORBIT,(800) 456-7248, online services. Also available on CD-ROM. Inquire as to cost and availability.

PERIODICALS

A A P G Bulletin. American Association of Petroleum Geologists, Box 979, Tulsa, OK 74101. (918) 584-2555. FAX (918) 584-0469. 1917 to present. Monthly. $135.00.

Earth Science Reviews. Elsevier Science Inc., Box 882, Madison Square Station, New York, NY 10159. (212) 989-5800. 1966 to present. Eight times a year. $439.00.

Geological Magazine. Cambridge University Press, Journals Department, 40 West 20th Street, New York, NY 10011-4211. (212) 924-3900. 1864 to present. Bi-monthly. $263.00 per year.

Geological Society Journal. Geological Society Publishing House, Unit 7, Brassmill Enterprise Centre, Brassmill Lane, Bath BA1 3JN, England. Telephone 0225-445046. FAX 0225-442836. 1845 to present. Six times a year. $524.00.

Geological Society of America Bulletin. 3300 Penrose Place, PO Box 9140, Denver, CO 80301-9140. (303) 447-2020. 1888 to present. Monthly. $205.00 per year.

Geology. Geological Society of America, 3300 Penrose Place, PO Box 9140, Denver, CO 80301-9140. (303) 447-2020. 1973 to present. Monthly. $170.00 per year for non-members.

Journal of Geology. University of Chicago Press, Journals Division, 5720 S. Woodlawn Avenue, Chicago, IL 60637. (312) 753-3347. FAX (312) 753-0811. 1893 to present. Bi-monthly. $45.00 per year.

Lethaia. Scandinavian University Press, 200 Meacham Avenue, Elmont, NY 11003. (516) 352-7300. 1968 to present. Quarterly. Inquire for price.

Sedimentary Geology. Elsevier Science Inc., Box 882, Madison Square Station, New York, NY 10159. (212) 989-5800. 1967 to present. 24 times a year. $1187.00.

Stratigraphy and Geological Correlations. Interperiodica, Box 1831, Birmingham, AL 35201-1831. (800) 633-4931. FAX (205) 995-1588.

Tectonics. American Geophysical Union, 2000 Florida Avenue, N.W., Washington, DC 20009. (202) 462-6900. FAX (202) 328-0566. 1982 to present. Bi-monthly. $330.00 for non-members.

RESEARCH CENTERS AND INSTITUTES

Institute of Sedimentary and Petroleum Geology. 3303 33rd Street NW, Calgary, AB, Canada T2L 2A7. (403) 292-7049. FAX (403) 292-5377.

Iowa Department of Natural Resources, Geological Survey Bureau. 123 N. Capitol Street, Iowa City, IA 52242. (319) 335-1575. FAX (319) 335-2754.

Iowa State University, Geotechnical/civil Engineering Materials Research Laboratories. Materials Lab/Sprangler Geotechnical Lab, Ames, IA 50011. (515) 294-9470.

University of Minnesota, Minnesota Geological Survey. 2642 University Avenue, St. Paul, MN 55114. (612) 627-4780. FAX (612) 627-4778.

University of South Carolina At Columbia, Earth Sciences and Resources Institute (esri). Byrnes International Center, 901 Sumter Street, Columbia, SC 29208. (803) 777- 6484. FAX (803) 777-6437.

University of Texas At Arlington, Cretaceous Stratigraphic Studies Center. Box 19049 UTA Station, Arlington, TX 76019. (817) 273-2987.

STRENGTH OF MATERIALS

See: MATERIALS SCIENCE

STRESS AND STRAIN

See also: AERONAUTICAL ENGINEERING, CIVIL ENGINEERING, DYNAMICS, FAILURE ANALYSIS, MATERIALS SCIENCE, MECHANICAL ENGINEERING, MECHANICS, STRUCTURAL ENGINEERING

ABSTRACT SERVICES AND INDEXES

Abstract Journal in Earthquake Engineering. University of California, Berkeley, Earthquake Engineering Research Center, 1301 South 46th Street, Richmond, CA 94804-4698. (510) 231-9413 or FAX (510) 231-9471. 1972 to present. Annual. $80.00 per year.

Applied Mechanics Reviews: An Assessment of World Literature in Engineering Sciences. 1948-present. American Society of Mechanical Engineers, 345 East 47th Street, New York, NY 10017. (212) 705-7703. Monthly. $360.00 per year.

Applied Science and Technology Index; A Cumulative Subject Index To English Language Periodicals in the Fields of Aeronautics and Space Science, Computer Technology, Chemistry, Construction Industry, Energy and Related Areas. H.W. Wilson Co., 950 University Avenue, Bronx, NY 10452. (800) 367-6770 or (212) 588-8400. FAX (718) 590-1617. From 1958 to present. Monthly. Inquire about cost and availability. Also available on CD-ROM and online.

ASCE Combined Annual Combined Index. American Society of Civil Engineers, 345 East 47th Street, New York, NY 10017-2398. (212) 705-7520 or (800) 548-2723. Annual. $48.00.

ASCE Publications Information. 1966-present. American Society of Civil Engineers, 345 East 47th Street, New York, NY 10017-2398. (212) 705-7520 or (800) 548-2723. Bi-monthly. $160.00 per year for non-members.

STRESS AND STRAIN

Ency. of Physical Sciences and Engineering Info. Sources

Civil and Structural Engineering Abstracts, Cambridge Scientific Abstracts, 7200 Wisconsin Avenue, Bethesda, MD 20814-4823. (301) 961-6750. FAX (301) 961-6720. 1993 to present. Monthly. $385.00 per year. Topics covered include structural design, construction equipment and methods, civil defense and military engineering, surveying, highway engineering, maritime and port structures, materials, land reclamation, and soil mechanics.

Current Contents: Engineering, Technology, and Applied Sciences. Institute for Scientific Information, 3501 Market Street, Philadelphia, PA 19104. (215) 386-0100. FAX (215) 386-6362. 1970 to present. Weekly. $442.00 per year.

Engineering Index Monthly. Engineering Information, Inc., Castle Point on the Hudson, Hoboken, NJ 07030. (800) 221-1044. FAX (212) 832-1857. Monthly. $2300.00 per year. Also available online as COMPENDEX, and also on CD-ROM. Covers chemical engineering, computers, electrical engineering, civil engineering, metals and mining, industrial management, and mechanical engineering.

International Civil Engineering Abstracts. CITIS Ltd., 2 Rosemount Terrace, Blackrock, Dublin, Ireland. Telephone 353-1-2886227. FAX 353-1-885-971. 1974 to present. Monthly. $660.00 per year. Also available on CD-ROM.

NTIS Alerts: Building Industry Technology. U.S. National Technical Information Service, 5825 Port Royal Road, Springfield, VA 22161. (703) 487-4630. FAX (703) 321-8547. Weekly. $135.00 per year.

Science Citation Index. Institute for Scientific Information, 3501 Market Street, Philadelphia, PA 19104. (215) 386-0100. FAX (215) 386-6362. Inquire about availability and cost. Also available on CD-ROM.

ASSOCIATIONS AND PROFESSIONAL SOCIETIES

American Society For Testing and Materials (astm). 1916 Race Street, Philadelphia, PA 19103. (215) 299-5585.

American Society of Civil Engineers. 345 East 47th Street, New York, NY 10017-2398. (212) 705-7520 or (800) 548-2723.

American Society of Mechanical Engineers. 345 East 47th Street, New York, NY 10017-2398. (212) 705-7703.

DIRECTORIES AND BIOGRAPHICAL SOURCES

Directory of Engineering Societies and Related Organizations. Gordon Davis. 13th edition. American Association of Engineering Societies, 1111 19th Street NW, Suite 608, Washington, DC 20036. (202) 296-2237 or (800) 658-8897. 1989. Inquire for price.

International Research Centers Directory. Gale Research, Inc., 835 Penobscot Building, Detroit, MI 48226-4094. (313) 961-2242. (800) 877-4253. 8th edition. 1995. $410.00

Research Centers Directory. Gale Research, Inc., 835 Penobscot Building, Detroit, MI 48226-4094. (313) 961-2242. (800) 877-4253. FAX (313) 961-6083. $485.00.

Scientific and Technical Organizations and Agencies Directory. Gale Research, Inc., 835 Penobscot Building, Detroit, MI 48226-

4094. (313) 961-2242. (800) 877-4253. FAX (313) 961-6083. 4th edition. 1996. $195.00.

Who's Who in Engineering. American Association of Engineering Societies, 1111 19th Street NW, Suite 608, Washington, DC 20036. (202) 296-2237 or (800) 658-8897. 8th edition. 1991. Inquire for price.

ENCYCLOPEDIAS AND DICTIONARIES

McGraw-Hill Dictionary of Scientific and Technical Terms. Sybil P. Parker, ed. 5th edition. McGraw-Hill Publishing Company, 1221 Avenue of the Americas, New York, NY 10020. (800) 262-4729 or (212) 512-3825. 1993. $110.50.

McGraw-Hill Encyclopedia of Engineering. Sybil P. Parker, ed. 2nd edition. McGraw-Hill Publishing Company, 1221 Avenue of the Americas, New York, NY 10020. (800) 262-4729 or (212) 512-3825. 1993. $95.50.

McGraw-Hill Encyclopedia of Science and Technology. Sybil P. Parker, ed. 7th edition. 20 volumes. McGraw-Hill Publishing Company, 1221 Avenue of the Americas, New York, NY 10020. (800) 262-4729 or (212) 512-3825. 1992. $1900.00

GENERAL WORKS

Experimental Stress Analysis. James W. Dally and William F. Riley. 3d ed. McGraw-Hill Publishing Company, 1221 Avenue of the Americas, New York, NY 10020. (800) 262-4729 or (212) 512-3825. 1991. $45.78.

Failure of Materials in Mechanical Design: Analysis, Prediction, Prevention. J.A. Collins. 2nd edition. John Wiley and Sons, Inc., 605 Third Avenue, New York, NY 10158. (800) 526-5368 or (212) 850-6000. 1993. $74.95.

Practical Stress Analysis in Engineering Design. A. Blake. 2nd revised and expanded edition. Marcel Dekker, Inc., 270 Madison Avenue, New York, NY 10016. (212) 696-9000. FAX (212) 685-4540. 1989. $69.75.

Solid Mechanics: An Introduction. J.P. Ward. Kluwer Academic Publishers, P.O. Box 358, Accord Station, Hingham, MA 02018-0358. (617) 871-6000. FAX (617) 871-6528. 1992. $89.00.

HANDBOOKS AND MANUALS

Civil Engineering Practice. Paul N. Cheremisinoff, et al. 5 volumes. Technomic Publishing Company, Inc., 851 New Holland Avenue, Box 3535, Lancaster, PA 17604. (717) 291-5609 or (800) 233-9936. FAX (717) 295-4538. 1987-88. $199.00.

Handbook of Mechanics, Materials, and Structures. Alexander Blake, editor. John Wiley and Sons, Inc., 605 Third Avenue, New York, NY 10158. (800) 526-5368 or (212) 850-6000. 1985. $120.00.

Marks' Standard Handbook For Mechanical Engineers. E.A. Avallone and T. Baumeister III. 9th edition. McGraw-Hill Publishing Company, 1221 Avenue of the Americas, New York, NY 10020. (800) 262-4729 or (212) 512-3825. 1987. $133.00.

Standard Handbook of Engineering Calculations. Tyler G. Hicks, ed. 2d ed. McGraw-Hill Publishing Company, 1221 Av-

Ency. of Physical Sciences and Engineering Info. Sources

STRESS AND STRAIN

enue of the Americas, New York, NY 10020. (800) 262-4729 or (212) 512-3825. 1985. $89.50.

Stress and Strain Data Handbook. Teng H. Hsu. Gulf Publishing Company, P.O. Box 2608, Houston, TX. (713) 529-4301 or (800) 231-6275. 1986. $49.00.

Structural Engineering Handbook. Edwin H. Gaylord Jr. and Charles N. Gaylord. 3d ed. McGraw-Hill Publishing Company, 1221 Avenue of the Americas, New York, NY 10020. (800) 262-4729 or (212) 512-3825. 1990. $89.50.

ONLINE DATABASES AND CD-ROMS

Compendex Plus. Engineering Information, Inc., 345 East 47th Street, New York, NY 10017. (212) 705-7600 or (800) 221-1044. Contains citations with abstracts to worldwide literature in engineering and technology, from 1970 to present. Available on online BRS,(800) 289-4277, DIALOG, (800) 334-2564, ORBIT (800) 456-7248, and STN International, FIZ Karlsruhe, P.O. Box 2465, W-7500, Karlsruhe 1, Germany, online services. Also available on CD-ROM. Inquire as to cost and availability.

CSA Engineering. Cambridge Scientific Abstracts, 7200 Wisconsin Avenue, Suite 601, Bethesda, MD 20814. (301) 961-6750 or (800) 843-7751. Contains citations and abstracts of international periodicals and research literature covering all fields of engineering and science and technology,including computer and information science, electronics, mechanical engineering, solid state materials, 1981 to present. Available on BRS,(800) 289-4277, online service.Inquire as to cost and availability.

Current Contents Search. Institute for Scientific Information, 3501 Market Street, Philadelphia, PA 19104. (215) 386-0100. FAX (215) 386-6362. Contains citations to articles listed in the table of contents of science and technology journals. Also articles in social sciences and life sciences journals. Available on BRS,(800) 289-4277, DIALOG,(800) 334-2564, online services. Inquire as to cost and availability.

Dissertation Abstracts. University Microfilms International, 300 North Zeeb Road, Ann Arbor, MI 48106. (800) 521-0600 or (313) 761-4700. Scope includes virtually all doctoral dissertations accepted at accredited American institutions from 1861 to present in 252 subject areas. Available on BRS,(800) 289-4277, DIALOG,(800) 334-2564, and OCLC EPIC,(800) 848-5878, online services. Also available on CD-ROM. Inquire as to cost and availability.

Georef: Bibliography and Index of Geology. American Geological Institute, 4220 King Street, Alexandria, VA 22302. (703) 379-2480. Fax (703) 379-7563. Monthly. Inquire for price and availability.# #*ISMEC: Mechanical Engineering Abstracts.* Cambridge Scientific Abstracts, 7200 Wisconsin Avenue, Suite 601, Bethesda, MD 20814. (301) 961-6750 or (800) 843-7751. Contains citations to the literature in mechanical engineering, industrial and production engineering, energy, power, mechanics, devices and related areas, from 1973 to present. Available on the DIALOG,(800) 334-2564, online service. Inquire as to cost and availability.

NTIS Bibliographic Database. National Technical Information Service, 5285 Port Royal Road, Springfield, VA 22161. (703) 487-4929 or FAX (703) 321-8199. Broad coverage of government-sponsored science and technology research reports, 1964 to present. Available on BRS,(800) 289-4277, DIALOG, (800) 334-2564, ORBIT, (800) 456-7248, and STN International, FIZ Karlsruhe, P.O. Box 2465, W-7500, Karlsruhe 1, Germany,

online services. Also available on CD-ROM. Inquire as to cost and availability.

SCISEARCH. Institute for Scientific Information, 3501 Market Street, Philadelphia, PA 19104. (800) 523-1850 or (215) 386-0100. Broad multidisciplinary title and author index to the international literature of science and technology, 1974 to present. Available on DIALOG,(800) 334-2564, and ORBIT,(800) 456-7248, online services. Also available on CD-ROM. Inquire as to cost and availability.

WILSONLINE. H.W. Wilson Company, 950 University Avenue, Bronx, NY 10452. (800) 367-6770 or (212) 588-8400. Makes available online versions of the printed H.W. Wilson indexes including Applied Science and Technology Index, Business Periodicals Index, General Science Index, and Readers' Guide to Periodical Literature. Period covered is generally 1983 to present. Available on BRS,(800) 289-4277, DIALOG,(800) 334-2564, and OCLC EPIC,(800) 848-5878, online services. Also available on CD-ROM. Inquire as to cost and availability.@ @Periodicals

Experimental Techniques. Society for Experimental Mechanics, 7 School Street, Bethel, CT 06801. (203) 790-6373. FAX (203) 790-4472. 1980 to present. Bi-monthly. $88.00 per year (includes subscription to EXPERIMENTAL MECHANICS).

Journal of Engineering Mechanics. American Society of Civil Engineers, Engineering Mechanics Division, 345 East 47th Street, New York, NY 10017-2398. (212) 705-7520 or (800) 548-2723. 1956 to present. Monthly. $284.00 per year for nonmembers.

Journal of Strain Analysis For Engineering. Mechanical Engineering Publications Ltd., Northgate Avenue, Bury St. Edmunds, Suffolk IP32 6BW, England. Telephone 0284-763277. FAX 0284-704006. 1965 to present. Quarterly. Inquire for cost and availability.# #*Journal of Structural Engineering.* American Society of Civil Engineers, Structural Engineering Division, 345 East 47th Street, New York, NY 10017-2398. (212) 705-7520 or (800) 548-2723. 1956 to present. Monthly. $300.00 per year for non-members.

Journal of Testing and Evaluation. American Society for Testing and Materials (ASTM), 1916 Race Street, Philadelphia, PA 19103. (215) 299-5585. FAX (215) 977-9679. 1966 to present. Bi-monthly. $103.00 per year for non-members.

Materials Engineering. Penton Publishing, 110 Superior Ave., Cleveland, OH 44114-2543. 1929 to present. Monthly. $50.00 per year.

SAMPE Journal. Society for the Advancement of Material and Process Engineering, Covina, CA 91724. (818) 331-0610. FAX (818) 332-8929. 1965 to present. Bi-monthly. $65.00 per year.

Structural Engineer. Structural Engineers Trading Organization Ltd., 11 Upper Belgrave Street, London SW1X 8BH, England. Telephone 071-235-4535. FAX 071-235-4294. 1922 to present. 24 times a year. Inquire for cost.

Structural Engineering International. International Association for Bridge and Structural Engineering, ETH-Hoenggerberg, CH-8093 Zurich, Switzerland. Telephone 01-3772647. FAX 01-3712131. Quarterly. Inquire for cost and availability.

STRESS AND STRAIN

Ency. of Physical Sciences and Engineering Info. Sources

RESEARCH CENTERS AND INSTITUTES

Structural Engineering Laboratory, University of Michigan. 2340 G. G. Brown Bldg., Ann Arbor, MI 48109. (313) 763-3046. FAX (313) 764-4292.

Structural Engineering Materials Laboratory, University of California At Berkeley. Davis Hall, Berkeley, CA 94720. (415) 642-3464.

Structural Stability Research Council. Fritz Engineering Laboratory, Lehigh University, Bethlehem, PA 18015. (215) 758-3522. FAX (215) 758-4522.

University of Wisconsin—madison, Center For Structural and Materials Testing. 1415 Johnson Street, Madison, WI 53706. (608) 262-3205.

STRIP MINING

See: MINING ENGINEERING

STRUCTURAL ENGINEERING

See also: BRIDGES, CIVIL ENGINEERING, CONCRETE, CONSTRUCTION ENGINEERING, FAILURE ANALYSIS, HIGHWAY ENGINEERING, STEEL and STEEL MAKING, STRUCTURES

ABSTRACT SERVICES AND INDEXES

Abstract Journal in Earthquake Engineering. University of California, Berkeley, Earthquake Engineering Research Center, 1301 South 46th Street, Richmond, CA 94804-4698. (510) 231-9413 or FAX (510) 231-9471. 1972 to present. Annual. $80.00 per year.

Applied Science and Technology Index; A Cumulative Subject Index To English Language Periodicals in the Fields of Aeronautics and Space Science, Computer Technology, Chemistry, Construction Industry, Energy and Related Areas. H.W. Wilson Co., 950 University Avenue, Bronx, NY 10452. (800) 367-6770 or (212) 588-8400. FAX (718) 590-1617. From 1958 to present. Monthly. Inquire about cost and availability. Also available on CD-ROM and online.

ASCE COMBINED ANNUAL COMBINED INDEX. American Society of Civil Engineers, 345 East 47th Street, New York, NY 10017-2398. (212) 705-7520 or (800) 548-2723. Annual. $48.00.

ASCE Publications Information. 1966-present. American Society of Civil Engineers, 345 East 47th Street, New York, NY 10017-2398. (212) 705-7520 or (800) 548-2723. Bi-monthly. $160.00 per year for non-members.

Civil and Structural Engineering Abstracts, Cambridge Scientific Abstracts, 7200 Wisconsin Avenue, Bethesda, MD 20814-4823. (301) 961-6750. FAX (301) 961-6720. 1993 to present. Monthly. $385.00 per year. Topics covered include structural design, construction equipment and methods, civil defense and military engineering, surveying, highway engineering, maritime and port structures, materials, land reclamation, and soil mechanics.

Current Contents: Engineering, Technology, and Applied Sciences. Institute for Scientific Information, 3501 Market Street, Philadelphia, PA 19104. (215) 386-0100. FAX (215) 386-6362. 1970 to present. Weekly. $442.00 per year.

Engineering Index Monthly. Engineering Information, Inc., Castle Point on the Hudson, Hoboken, NJ 07030. (800) 221-1044. FAX (212) 832-1857. Monthly. $2300.00 per year. Also available online as COMPENDEX, and also on CD-ROM. Covers chemical engineering, computers, electrical engineering, civil engineering, metals and mining, industrial management, and mechanical engineering.

Index To Scientific and Technical Proceedings. Institute for Scientific Information, 3501 Market St., Philadelphia, PA 19104. (215) 386-0100. FAX (215) 386-6362. Monthly. $500.00 per year.

International Civil Engineering Abstracts. CITIS Ltd., 2 Rosemount Terrace, Blackrock, Dublin, Ireland. Telephone 353-1-2886227. FAX 353-1-885-971. 1974 to present. Monthly. $660.00 per year. Also available on CD-ROM.

NTIS Alerts: Building Industry Technology. U.S. National Technical Information Service, 5825 Port Royal Road, Springfield, VA 22161. (703) 487-4630. FAX (703) 321-8547. Weekly. $135.00 per year.

Offshore Engineering Abstracts. STI Limited, 4 Kings Meadow, Ferry Hinksey Road, Oxford, OX2 ODU, England. Distributed in the United States by Air Science Company, Box 143, Corning NY 14830. (607) 962-5591. Contains citations on planning, design, and construction of offshore platforms and pipelines, from 1985 to present. Monthly. $400.00 per year.

Science Citation Index. Institue for Scientific Information, 3501 Market Street, Philadelphia, PA 19104. (215) 386-0100. FAX (215) 386-6362. Inquire about availability and cost. Also available on CD-ROM.

ASSOCIATIONS AND PROFESSIONAL SOCIETIES

American Society For Testing and Materials (astm). 1916 Race Street, Philadelphia, PA 19103. (215) 299-5585.

American Society of Civil Engineers. 345 East 47th Street, New York, NY 10017-2398. (212) 705-7520 or (800) 548-2723.

BIBLIOGRAPHIES

Scientific and Technical Books and Serials in Print. R.R. Bowker Inc., 121 Chanlon Road, New Providence, NJ 07974. (800) 521-8110 or (908) 464-6800. List of currently published books and serials in the physical and biological sciences, engineering and technology, with subject, author and titles indexes. Also available on ORBIT online service. Inquire as to cost and availability.

DIRECTORIES AND BIOGRAPHICAL SOURCES

American Society of Civil Engineers Directory: official Register. 345 East 47th Street, New York, NY 10017-2398. (212) 705-7520 or (800) 548-2723. Inquire for cost and availability.

Directory of Engineering Societies and Related Organizations. Gordon Davis. 13th edition. American Association of Engineering Societies, 1111 19th Street NW, Suite 608, Washington, DC 20036. (202) 296-2237 or (800) 658-8897. 1989. Inquire for price.

International Research Centers Directory. Gale Research, Inc., 835 Penobscot Building, Detroit, MI 48226-4094. (313) 961-2242. (800) 877-4253. FAX (313) 961-6083. 8th edition. 1995. $410.00

Research Centers Directory. Gale Research, Inc., 835 Penobscot Building, Detroit, MI 48226-4094. (313) 961-2242. (800) 877-4253. FAX (313) 961-6083. $485.00.

Scientific and Technical Organizations and Agencies Directory. Gale Research, Inc., 835 Penobscot Building, Detroit, MI 48226-4094. (313) 961-2242. (800) 877-4253. FAX (313) 961-6083. 4th edition. 1996. $195.00.

Who's Who in Engineering. American Association of Engineering Societies, 1111 19th Street NW, Suite 608, Washington, DC 20036. (202) 296-2237 or (800) 658-8897. 8th edition. 1991. Inquire for price.

ENCYCLOPEDIAS AND DICTIONARIES

Construction Glossary: An Encylopedic Reference and Manual. J. Stewart Stern. 2nd edition. John Wiley and Sons, Inc., 605 Third Avenue, New York, NY 10158. (800) 526-5368 or (212) 850-6000. 1993. $99.95.

Dictionary of Civil Engineering. V.N. Vazirani. South Asia Books, PO Box 502, Columbia, MO 65205. (314) 474-0116. FAX (314) 474-8124. 1992. $20.00.

Dictionary of Structural Engineering. Cyril M. Harris. McGraw-Hill Publishing Company, 1221 Avenue of the Americas, New York, NY 10020. (800) 262-4729 or (212) 512-3825. 1992. $14.95.

McGraw-Hill Dictionary of Scientific and Technical Terms. Sybil P. Parker, ed. 5th edition. McGraw-Hill Publishing Company, 1221 Avenue of the Americas, New York, NY 10020. (800) 262-4729 or (212) 512-3825. 1993. $110.50.

McGraw-Hill Encyclopedia of Engineering. Sybil P. Parker, ed. 2nd edition. McGraw-Hill Publishing Company, 1221 Avenue of the Americas, New York, NY 10020. (800) 262-4729 or (212) 512-3825. 1993. $95.50.

Thesaurus of Scientific, Technical, and Engineering Terms. Hemisphere Publishing Corporation, 1900 Frost Road, Suite 101, Bristol, PA 19007-1598. (215) 785-5800 or FAX (215) 785-5515. 1987. $173.00.

GENERAL WORKS

Basic Structural Behavior. Barry Hilson. Thomas Telford Services Ltd., Thomas Telford House, 1 Heron Quay, London E14

4JD, England. Telephone 071-987-6999. FAX 071-538-4101. 1993. Inquire for cost and availability.

Introduction To Structual Engineering Analysis and Design. John M. Biggs. Prentice Hall, 113 Sylvan Avenue, Route 9W, Englewood Cliffs, NJ 07632. (201) 592-2000 or (800) 922-0579. 1986. Inquire for cost and availability.

An Introduction To the History of Structural Mechanics. Edoardo Benvenuto. 2 vols. Springer-Verlag, 175 Fifth Avenue, New York, NY 10010. (212) 460-1500 or (800) 777-4643. FAX (212) 473-6272. 1991. $59.00 for each volume.

Progress in Structural Engineering. Donald E. Grierson, et al., editors. Kluwer Academic Publishers, P.O. Box 358, Accord Station, Hingham, MA 02018-0358. (617) 871-6000. FAX (617) 871-6528. 1991. $163.50.

Structural Engineering: The Nature of theory and Design. William Addis. E. Horwood Publishing, distributed by Prentice Hall, 113 Sylvan Avenue, Route 9W, Englewood Cliffs, NJ 07632. (201) 592-2000 or (800) 922-0579.1990. Inquire for cost and availability.

Structures: Fundamental theory and Behavior. Richard M. Gutkowski. 2nd edition. Van Nostrand Reinhold, 115 Fifth Avenue, New York, NY 10003. (212) 254-3232. FAX (212) 254-9499. 1990. $69.95.

HANDBOOKS AND MANUALS

Building Structural Design Handbook. Richard N. White & Charles G. Salmon, editors. Krieger Publishing Company, P.O. Box 9542, Melbourne, FL 32902-9542. (407) 724-9542. FAX (407) 951-3671. 1987. $95.00.

Civil Engineering Calculations Reference Guide. Tyler G. Hicks. McGraw-Hill Publishing Company, 1221 Avenue of the Americas, New York, NY 10020. (800) 262-4729 or (212) 512-3825. 1987. $46.00.

Civil Engineering Practice. Paul N. Cheremisinoff, et al. 5 volumes. Technomic Publishing Company, Inc., 851 New Holland Avenue, Box 3535, Lancaster, PA 17604. (717) 291-5609 or (800) 233-9936. FAX (717) 295-4538. 1987-88. $199.00.

Handbook of Concrete Engineering. Mark Fintel. 2nd edition. Routledge, Chapman and Hall, Inc., 29 West 35th Street, New York, NY 10001-2291. (212) 244-3336 or FAX (212) 563-2269. 1985. $115.00.

Handbook of Mechanics, Materials, and Structures. Alexander Blake, editor. John Wiley and Sons, Inc., 605 Third Avenue, New York, NY 10158. (800) 526-5368 or (212) 850-6000. 1985. $120.00.

Handbook On Structural Testing. Robert T. Reese and Wendell A. Kawahara. Fairmont Press, 700 Indian Trail, Lilburn, GA 30247. (404) 925-9388. 1993. $95.00.

Standard Handbook For Civil Engineers. Frederick S. Merritt, editor. 3rd edition. McGraw-Hill Publishing Company, 1221 Avenue of the Americas, New York, NY 10020. (800) 262-4729 or (212) 512-3825. 1983. $124.50.

Structural Engineering Handbook. Edwin H. Gaylord Jr. and Charles N. Gaylord. 3d ed. McGraw-Hill Publishing Company,

STRUCTURAL ENGINEERING

Ency. of Physical Sciences and Engineering Info. Sources

1221 Avenue of the Americas, New York, NY 10020. (800) 262-4729 or (212) 512-3825. 1990. $89.50.

Structural Masonry Designers' Manual. W.G. Curtin, et al. 2nd edition. Blackwell Scientific Publications, 238 Main Street, Cambridge, MA 02142. (617) 876-7000 or (800) 759-6102. FAX (617) 876-7022. 1995. Inquire for cost and availability.

ONLINE DATABASES AND CD-ROMS

Architectural periodicals Index. Royal Institute of British Architects, British Architectural Library, 66 Portland Place, London, W1N 4AD, England. Citations to worldwide literature on architecture, construction methods and standards, urban planning, and interior design, from 1978 to present. Available on DIALOG, (800) 334-2564, online service. Inquire as to cost and availability.

Compendex Plus. Engineering Information, Inc., 345 East 47th Street, New York, NY 10017. (212) 705-7600 or (800) 221-1044. Contains citations with abstracts to worldwide literature in engineering and technology, from 1970 to present. Available on online BRS, (800) 289-4277, DIALOG, (800) 334-2564, ORBIT (800) 456-7248, and STN International, FIZ Karlsruhe, P.O. Box 2465, W-7500, Karlsruhe 1, Germany, online services. Also available on CD-ROM. Inquire as to cost and availability.

CSA Engineering. Cambridge Scientific Abstracts, 7200 Wisconsin Avenue, Suite 601, Bethesda, MD 20814. (301) 961-6750 or (800) 843-7751. Contains citations and abstracts of international periodicals and research literature covering all fields of engineering and science and technology, including computer and information science, electronics, mechanical engineering, solid state materials, 1981 to present. Available on BRS, (800) 289-4277, online service. Inquire as to cost and availability.

Current Contents Search. Institute for Scientific Information, 3501 Market Street, Philadelphia, PA 19104. (215) 386-0100. FAX (215) 386-6362. Contains citations to articles listed in the table of contents of science and technology journals. Also articles in social sciences and life sciences journals. Available on BRS, (800) 289-4277, DIALOG, (800) 334-2564, online services. Inquire as to cost and availability.

Dissertation Abstracts. University Microfilms International, 300 North Zeeb Road, Ann Arbor, MI 48106. (800) 521-0600 or (313) 761-4700. Scope includes virtually all doctoral dissertations accepted at accredited American institutions from 1861 to present in 252 subject areas. Available on BRS, (800) 289-4277, DIALOG, (800) 334-2564, and OCLC EPIC, (800) 848-5878, online services. Also available on CD-ROM. Inquire as to cost and availability.

Georef: Bibliography and Index of Geology. American Geological Institute, 4220 King Street, Alexandria, VA 22302. (703) 379-2480. Fax (703) 379-7563. Monthly. Inquire for price and availability.

ICONDA (cib International Construction Database). Fraunhofer-Gesellschaft, Informationszentrum RAUM und BAU, Nobelstrasse, 12, D-7000, Stuttgart 80, Germany. Contains citations with abstracts to worldwide technical literature on construction and civil engineering, structural engineering, and engineering geology, from 1976 to present. Available on ORBIT, (800) 456-7248, and STN International, FIZ Karlsruhe, P.O. Box 2465, W-7500, Karlsruhe 1, Germany, online services. Inquire as to cost and availability.

NTIS Bibliographic Database. National Technical Information Service, 5285 Port Royal Road, Springfield, VA 22161. (703) 487-4929 or FAX (703) 321-8199. Broad coverage of government-sponsored science and technology research reports, 1964 to present. Available on BRS, (800) 289-4277, DIALOG, (800) 334-2564, ORBIT, (800) 456-7248, and STN International, FIZ Karlsruhe, P.O. Box 2465, W-7500, Karlsruhe 1, Germany, online services. Also available on CD-ROM. Inquire as to cost and availability.

SCISEARCH. Institute for Scientific Information, 3501 Market Street, Philadelphia, PA 19104. (800) 523-1850 or (215) 386-0100. Broad multidisciplinary title and author index to the international literature of science and technology, 1974 to present. Available on DIALOG, (800) 334-2564, and ORBIT, (800) 456-7248, online services. Also available on CD-ROM. Inquire as to cost and availability.

Tris (transportation Research Information). National ACADemy of Science, 2101 Constitution Avenue, N.W., Washington, DC 20418. (202) 334-3313 or (800) 624-6242. Citations with abstracts of literature on transportation, including air, highway, rail, maritime and other modes from 1968 to present. Available on DIALOG, (800) 334-2564, online service. Inquire as to cost and availability.

WILSONLINE. H.W. Wilson Company, 950 University Avenue, Bronx, NY 10452. (800) 367-6770 or (212) 588-8400. Makes available online versions of the printed H.W. Wilson indexes including Applied Science and Technology Index, Business Periodicals Index, General Science Index, and Readers' Guide to Periodical Literature. Period covered is generally 1983 to present. Available on BRS, (800) 289-4277, DIALOG, (800) 334-2564, and OCLC EPIC, (800) 848-5878, online services. Also available on CD-ROM. Inquire as to cost and availability.

PERIODICALS

American Concrete Institute Materials Journal. American Concrete Institute, Box 19150, Redford Station, Detroit, MI 48219. (313) 532-2600. FAX (313) 538-0655. 1929 to present. Bi-monthly. $101.00 per year for non-members.

American Concrete Institute Structural Journal. American Concrete Institute, Box 19150, Redford Station, Detroit, MI 48219. (313) 532-2600. FAX (313) 538-0655. 1929 to present. Bi-monthly. $101.00 per year for non-members.

Civil Engineering ASCE. American Society of Civil Engineers, 345 East 47th Street, New York, NY 10017-2398. (212) 705-7520 or (800) 548-2723. 1930 to present. Monthly. $85.00 per year.

Composite Structures. Elsevier Science [journals], 660 White Plains Rd., Tarrytown, NY 10159-5153. (919) 524-9200. FAX (919) 333-2444. 1983 to present. 12 times a year. $1315.00 per year.

Computers and Structures. Elsevier Science [journals], 660 White Plains Rd., Tarrytown, NY 10159-5153. (919) 524-9200. FAX (919) 333-2444. 1971 to present. 24 times a year. $2450.00 per year.

Concrete. The Concrete Society, Framewood Road, Wexham, Slough, Berks. SL3 6PJ England. Telephone 0753-662226. FAX 0753-662126. 1966 to present. Bi-monthly. $75.00 per year.

Earthquake Engineering and Structural Dynamics. J. Wiley & Sons Journals, Baffins Lane, Chichester, Sussex PO19 1UD, England. Telephone 0243-779773. FAX 0243-775878. 1972 to present. 12 times a year. $995.00 per year.

Engineering Journal. American Institute of Steel Construction, Inc., One East Wacker Drive, Suite 3100, Chicago, IL 60601-2001. (312) 670-2400 or FAX (312) 670-5403. 1964 to present. Quarterly. $15.00 per year.

Journal of Engineering Mechanics. American Society of Civil Engineers, Engineering Mechanics Division, 345 East 47th Street, New York, NY 10017-2398. (212) 705-7520 or (800) 548-2723. 1956 to present. Monthly. $284.00 per year for non-members.

Journal of Performance of Constructed Facilities. ASCE, Technical Council on Forensic Engineering, 345 East 47th Street, New York, NY 10017-2398. (212) 705-7520 or (800) 548-2723. 1987 to present. Quarterly. $80.00 per year.

Journal of Structural Engineering. American Society of Civil Engineers, Structural Engineering Division, 345 East 47th Street, New York, NY 10017-2398. (212) 705-7520 or (800) 548-2723. 1956 to present. Monthly. $300.00 per year for non-members.

Journal of Surveying Engineering. American Society of Civil Engineers, 345 East 47th Street, New York, NY 10017-2398. (212) 705-7520 or (800) 548-2723. 1983 to present.

Structural Engineer. Structural Engineers Trading Organization Ltd., 11 Upper Belgrave Street, London SW1X 8BH, England. Telephone 071-235-4535. FAX 071-235-4294. 1922 to present. 24 times a year. Inquire for cost.

Structural Engineering International. International Association for Bridge and Structural Engineering, ETH-Hoenggerberg, CH-8093 Zurich, Switzerland. Telephone 01-3772647. FAX 01-3712131. Quarterly. Inquire for cost and availability.

Structural Engineering Reviews. Elsevier Science [journals], 660 White Plains Rd., Tarrytown, NY 10159-5153. (919) 524-9200. FAX (919) 333-2444. 1988 to present. 4 times a year. $190.00 per year.

RESEARCH CENTERS AND INSTITUTES

George Winter Structural Engineering Laboratory. Cornell University, Thurston Hall, Ithaca, NY 14853. (607) 255-4078.

Phil M. Ferguson Structural Engineering Laboratory, University of Texas at Austin. Balcones Research Center, 10100 Burnet Road, Bldg. 24, Austin, TX 78758. (512) 471-3062.

Structural Engineering Laboratory, University of Michigan. 2340 G. G. Brown Bldg., Ann Arbor, MI 48109. (313) 763-3046. FAX (313) 764-4292.

Structural Engineering Materials Laboratory, University of California At Berkeley. Davis Hall, Berkeley, CA 94720. (415) 642-3464.

Structural Stability Research Council. Fritz Engineering Laboratory, Lehigh University, Bethlehem, PA 18015. (215) 758-3522. FAX (215) 758-4522.

University of Wisconsin—madison, Center For Structural and Materials Testing. 1415 Johnson Street, Madison, WI 53706. (608) 262-3205.

STRUCTURAL GEOLOGY

See: PHYSICAL GEOLOGY

STRUCTURAL MATERIALS

See: STRUCTURAL ENGINEERING

STRUCTURAL PETROLOGY

See: PETROLOGY

STRUCTURES

See also: BRIDGES, CIVIL ENGINEERING, CONCRETE, CONSTRUCTION ENGINEERING, FAILURE ANALYSIS, HIGHWAY ENGINEERING, STEEL AND STEEL MAKING, STRUCTURAL ENGINEERING

ABSTRACT SERVICES AND INDEXES

Abstract Journal in Earthquake Engineering. University of California, Berkeley, Earthquake Engineering Research Center, 1301 South 46th Street, Richmond, CA 94804-4698. (510) 231-9413 or FAX (510) 231-9471. 1972 to present. Annual. $80.00 per year.

Applied Science and Technology Index; A Cumulative Subject Index To English Language Periodicals in the Fields of Aeronautics and Space Science, Computer Technology, Chemistry, Construction Industry, Energy and Related Areas. H.W. Wilson Co., 950 University Avenue, Bronx, NY 10452. (800) 367-6770 or (212) 588-8400. FAX (718) 590-1617. From 1958 to present. Monthly. Inquire about cost and availability. Also available on CD-ROM and online.

ASCE Combined Annual Combined Index. American Society of Civil Engineers, 345 East 47th Street, New York, NY 10017-2398. (212) 705-7520 or (800) 548-2723. Annual. $48.00.

ASCE Publications Information. 1966-present. American Society of Civil Engineers, 345 East 47th Street, New York, NY 10017-2398. (212) 705-7520 or (800) 548-2723. Bi-monthly. $160.00 per year for non-members.

Civil and Structural Engineering Abstracts, Cambridge Scientific Abstracts, 7200 Wisconsin Avenue, Bethesda, MD 20814-4823. (301) 961-6750. FAX (301) 961-6720. 1993 to present. Monthly. $385.00 per year. Topics covered include structural design, construction equipment and methods, civil defense and military engineering, surveying, highway engineering, maritime and port structures, materials, land reclamation, and soil mechanics.

STRUCTURES

Ency. of Physical Sciences and Engineering Info. Sources

Current Contents: Engineering, Technology, and Applied Sciences. Institute for Scientific Information, 3501 Market Street, Philadelphia, PA 19104. (215) 386-0100. FAX (215) 386-6362. 1970 to present. Weekly. $442.00 per year.

Engineering Index Monthly. Engineering Information, Inc., Castle Point on the Hudson, Hoboken, NJ 07030. (800) 221-1044. FAX (212) 832-1857. Monthly. $2300.00 per year. Also available online as COMPENDEX, and also on CD-ROM. Covers chemical engineering, computers, electrical engineering, civil engineering, metals and mining, industrial management, and mechanical engineering.

Index To Scientific and Technical Proceedings. Institute for Scientific Information, 3501 Market St., Philadelphia, PA 19104. (215) 386-0100. FAX (215) 386-6362. Monthly. $500.00 per year.

International Civil Engineering Abstracts. CITIS Ltd., 2 Rosemount Terrace, Blackrock, Dublin, Ireland. Telephone 353-1-2886227. FAX 353-1-885-971. 1974 to present. Monthly. $660.00 per year. Also available on CD-ROM.

NTIS Alerts: Building Industry Technology. U.S. National Technical Information Service, 5825 Port Royal Road, Springfield, VA 22161. (703) 487-4630. FAX (703) 321-8547. Weekly. $135.00 per year.

Offshore Engineering Abstracts. STI Limited, 4 Kings Meadow, Ferry Hinksey Road, Oxford, OX2 ODU, England. Distributed in the United States by Air Science Company, Box 143, Corning NY 14830. (607) 962-5591. Contains citations on planning, design, and construction of offshore platforms and pipelines, from 1985 to present. Monthly. $400.00 per year.

Science Citation Index. Institue for Scientific Information, 3501 Market Street, Philadelphia, PA 19104. (215) 386-0100. FAX (215) 386-6362. Inquire about availability and cost. Also available on CD-ROM.

ASSOCIATIONS AND PROFESSIONAL SOCIETIES

American Society For Testing and Materials (astm). 1916 Race Street, Philadelphia, PA 19103. (215) 299-5585.

American Society of Civil Engineers. 345 East 47th Street, New York, NY 10017-2398. (212) 705-7520 or (800) 548-2723.

DIRECTORIES AND BIOGRAPHICAL SOURCES

American Society of Civil Engineers Directory: official Register. 345 East 47th Street, New York, NY 10017-2398. (212) 705-7520 or (800) 548-2723. Inquire for cost and availability.

Directory of Engineering Societies and Related Organizations. Gordon Davis. 13th edition. American Association of Engineering Societies, 1111 19th Street NW, Suite 608, Washington, DC 20036. (202) 296-2237 or (800) 658-8897. 1989. Inquire for price.

International Research Centers Directory. Gale Research, Inc., 835 Penobscot Building, Detroit, MI 48226-4094. (313) 961-2242. (800) 877-4253. FAX (313) 961-6083. 8th edition. 1995. $410.00.

Research Centers Directory. Gale Research, Inc., 835 Penobscot Building, Detroit, MI 48226-4094. (313) 961-2242. (800) 877-4253. FAX (313) 961-6083. $485.00.

Scientific and Technical Organizations and Agencies Directory. Gale Research, Inc., 835 Penobscot Building, Detroit, MI 48226-4094. (313) 961-2242. (800) 877-4253. FAX (313) 961-6083. 4th edition. 1996. $195.00.

Who's Who in Engineering. American Association of Engineering Societies, 1111 19th Street NW, Suite 608, Washington, DC 20036. (202) 296-2237 or (800) 658-8897. 8th edition. 1991. Inquire for price.

ENCYCLOPEDIAS AND DICTIONARIES

Construction Glossary: An Encylopedic Reference and Manual. J. Stewart Stern. 2nd edition. John Wiley and Sons, Inc., 605 Third Avenue, New York, NY 10158. (800) 526-5368 or (212) 850-6000. 1993. $99.95.

Dictionary of Civil Engineering. V.N. Vazirani. South Asia Books, PO Box 502, Columbia, MO 65205. (314) 474-0116. FAX (314) 474-8124. 1992. $20.00.

McGraw-Hill Dictionary of Scientific and Technical Terms. Sybil P. Parker, ed. 5th edition. McGraw-Hill Publishing Company, 1221 Avenue of the Americas, New York, NY 10020. (800) 262-4729 or (212) 512-3825. 1993. $110.50.

McGraw-Hill Encyclopedia of Engineering. Sybil P. Parker, ed. 2nd edition. McGraw-Hill Publishing Company, 1221 Avenue of the Americas, New York, NY 10020. (800) 262-4729 or (212) 512-3825. 1993. $95.50.

Thesaurus of Scientific, Technical, and Engineering Terms. Hemisphere Publishing Corporation, 1900 Frost Road, Suite 101, Bristol, PA 19007-1598. (215) 785-5800 or FAX (215) 785-5515. 1987. $173.00.

GENERAL WORKS

Basic Structural Behavior. Barry Hilson. Thomas Telford Services Ltd., Thomas Telford House, 1 Heron Quay, London E14 4JD, England. Telephone 071-987-6999. FAX 071-538-4101. 1993. Inquire for cost and availability.

Elementary theory of Structures. Yuan-Yu Hsieh. 3rd edition. Prentice Hall, 113 Sylvan Avenue, Route 9W, Englewood Cliffs, NJ 07632. (201) 592-2000 or (800) 922-0579. 1988. $58.00.

Structural Dynamics: Recent Advances. G.I. Schueller, editor. Springer-Verlag, 175 Fifth Avenue, New York, NY 10010. (212) 460-1500. FAX (212) 473-6272. 1991. $115.00.

Structures: Fundamental theory and Behavior. Richard M. Gutkowski. 2nd edition. Van Nostrand Reinhold, 115 Fifth Avenue, New York, NY 10003. (212) 254-3232. FAX (212) 254-9499. 1990. $69.95.

HANDBOOKS AND MANUALS

Building Structural Design Handbook. Richard N. White & Charles G. Salmon, editors. Krieger Publishing Company, P.O.

Box 9542, Melbourne, FL 32902-9542. (407) 724-9542. FAX (407) 951-3671. 1987. $95.00.

Civil Engineering Calculations Reference Guide. Tyler G. Hicks. McGraw-Hill Publishing Company, 1221 Avenue of the Americas, New York, NY 10020. (800) 262-4729 or (212) 512-3825. 1987. $46.00.

Civil Engineering Practice. Paul N. Cheremisinoff, et al. 5 volumes. Technomic Publishing Company, Inc., 851 New Holland Avenue, Box 3535, Lancaster, PA 17604. (717) 291-5609 or (800) 233-9936. FAX (717) 295-4538. 1987-88. $199.00.

Handbook of Concrete Engineering. Mark Fintel. 2nd edition. Routledge, Chapman and Hall, Inc., 29 West 35th Street, New York, NY 10001-2291. (212) 244-3336 or FAX (212) 563-2269. 1985. $115.00.

Handbook of Mechanics, Materials, and Structures. Alexander Blake, editor. John Wiley and Sons, Inc., 605 Third Avenue, New York, NY 10158. (800) 526-5368 or (212) 850-6000. 1985. $120.00.

Standard Handbook For Civil Engineers. Frederick S. Merritt, editor. 3rd edition. McGraw-Hill Publishing Company, 1221 Avenue of the Americas, New York, NY 10020. (800) 262-4729 or (212) 512-3825. 1983. $124.50.

Structural Engineering Handbook. Edwin H. Gaylord Jr. and Charles N. Gaylord. 3d ed. McGraw-Hill Publishing Company, 1221 Avenue of the Americas, New York, NY 10020. (800) 262-4729 or (212) 512-3825. 1990. $89.50.

ONLINE DATABASES AND CD-ROMS

Compendex Plus. Engineering Information, Inc., 345 East 47th Street, New York, NY 10017. (212) 705-7600 or (800) 221-1044. Contains citations with abstracts to worldwide literature in engineering and technology, from 1970 to present. Available on online BRS,(800) 289-4277, DIALOG, (800) 334-2564, ORBIT (800) 456-7248, and STN International, FIZ Karlsruhe, P.O. Box 2465, W-7500, Karlsruhe 1, Germany, online services. Also available on CD-ROM. Inquire as to cost and availability.

CSA Engineering. Cambridge Scientific Abstracts, 7200 Wisconsin Avenue, Suite 601, Bethesda, MD 20814. (301) 961-6750 or (800) 843-7751. Contains citations and abstracts of international periodicals and research literature covering all fields of engineering and science and technology,including computer and information science, electronics, mechanical engineering, solid state materials, 1981 to present. Available on BRS,(800) 289-4277, online service.Inquire as to cost and availability.

Current Contents Search. Institute for Scientific Information, 3501 Market Street, Philadelphia, PA 19104. (215) 386-0100. FAX (215) 386-6362. Contains citations to articles listed in the table of contents of science and technology journals. Also articles in social sciences and life sciences journals. Available on BRS,(800) 289-4277, DIALOG,(800) 334-2564, online services. Inquire as to cost and availability.

Dissertation Abstracts. University Microfilms International, 300 North Zeeb Road, Ann Arbor, MI 48106. (800) 521-0600 or (313) 761-4700. Scope includes virtually all doctoral dissertations accepted at accredited American institutions from 1861 to present in 252 subject areas. Available on BRS,(800) 289-4277, DIALOG,(800) 334-2564, and OCLC EPIC,(800) 848-5878,

online services. Also available on CD-ROM. Inquire as to cost and availability.

Georef: Bibliography and Index of Geology. American Geological Institute, 4220 King Street, Alexandria, VA 22302. (703) 379-2480. Fax (703) 379-7563. Monthly. Inquire for price and availability.

NTIS Bibliographic Database. National Technical Information Service, 5285 Port Royal Road, Springfield, VA 22161. (703) 487-4929 or FAX (703) 321-8199. Broad coverage of government-sponsored science and technology research reports, 1964 to present. Available on BRS,(800) 289-4277, DIALOG, (800) 334-2564, ORBIT, (800) 456-7248, and STN International, FIZ Karlsruhe, P.O. Box 2465, W-7500, Karlsruhe 1, Germany, online services. Also available on CD-ROM. Inquire as to cost and availability.

SCISEARCH. Institute for Scientific Information, 3501 Market Street, Philadelphia, PA 19104. (800) 523-1850 or (215) 386-0100. Broad multidisciplinary title and author index to the international literature of science and technology, 1974 to present. Available on DIALOG,(800) 334-2564, and ORBIT,(800) 456-7248, online services. Also available on CD-ROM. Inquire as to cost and availability.

Tris (transportation Research Information). National ACADemy of Science, 2101 Constitution Avenue, N.W., Washington, DC 20418. (202) 334-3313 or (800) 624-6242. Citations with abstracts of literature on transportation, including air, highway, rail, maritime and other modes from 1968 to present. Available on DIALOG,(800) 334-2564, online service. Inquire as to cost and availability.

WILSONLINE. H.W. Wilson Company, 950 University Avenue, Bronx, NY 10452. (800) 367-6770 or (212) 588-8400. Makes available online versions of the printed H.W. Wilson indexes including Applied Science and Technology Index, Business Periodicals Index, General Science Index, and Readers' Guide to Periodical Literature. Period covered is generally 1983 to present. Available on BRS,(800) 289-4277, DIALOG,(800) 334-2564, and OCLC EPIC,(800) 848-5878, online services. Also available on CD-ROM. Inquire as to cost and availability.

PERIODICALS

American Concrete Institute Materials Journal. American Concrete Institute, Box 19150, Redford Station, Detroit, MI 48219. (313) 532-2600. FAX (313) 538-0655. 1929 to present. Bi-monthly. $101.00 per year for non-members.

American Concrete Institute Structural Journal. American Concrete Institute, Box 19150, Redford Station, Detroit, MI 48219. (313) 532-2600. FAX (313) 538-0655. 1929 to present. Bi-monthly. $101.00 per year for non-members.

Civil Engineering ASCE. American Society of Civil Engineers, 345 East 47th Street, New York, NY 10017-2398. (212) 705-7520 or (800) 548-2723. 1930 to present. Monthly. $85.00 per year.

Concrete. The Concrete Society, Framewood Road, Wexham, Slough, Berks. SL3 6PJ England. Telephone 0753-662226. FAX 0753-662126. 1966 to present. Bi-monthly. $75.00 per year.

Engineering Journal. American Institute of Steel Construction, One E. Wacker Street, Suite 3100, Chicago, IL 60601-2001.

STRUCTURES

Ency. of Physical Sciences and Engineering Info. Sources

(312) 670-2400. FAX (312) 670-5403. 1964 to present. Quarterly. $15.00 per year.

Engineering Structures. Butterworth-Heinemann, Linacre House, Jordan Hill, Oxford OX2 8DP, England. Telephone 0865-310366. FAX 0865-310898. 1979 to present. Bi-monthly. Inquire for cost.

Journal of Engineering Mechanics. American Society of Civil Engineers, Engineering Mechanics Division, 345 East 47th Street, New York, NY 10017-2398. (212) 705-7520 or (800) 548-2723. 1956 to present. Monthly. $284.00 per year for non-members.

Journal of Structural Engineering. American Society of Civil Engineers, Structural Engineering Division, 345 East 47th Street, New York, NY 10017-2398. (212) 705-7520 or (800) 548-2723. 1956 to present. Monthly. $300.00 per year for non-members.

Structural Engineer. Structural Engineers Trading Organization Ltd., 11 Upper Belgrave Street, London SW1X 8BH, England. Telephone 071-235-4535. FAX 071-235-4294. 1922 to present. 24 times a year. Inquire for cost.

Structural Engineering International. International Association for Bridge and Structural Engineering, ETH-Hoenggerberg, CH-8093 Zurich, Switzerland. Telephone 01-3772647. FAX 01-3712131. Quarterly. Inquire for cost and availability.

Structural Engineering Reviews. Elsevier Science [journals], 660 White Plains Rd., Tarrytown, NY 10159-5153. (919) 524-9200. FAX (919) 333-2444. 1988 to present. 4 times a year. $190.00 per year.

RESEARCH CENTERS AND INSTITUTES

Phil M. Ferguson Structural Engineering Laboratory, University of Texas at Austin. Balcones Research Center, 10100 Burnet Road, Bldg. 24, Austin, TX 78758. (512) 471-3062.

Structural Engineering Laboratory, University of Michigan. 2340 G. G. Brown Bldg., Ann Arbor, MI 48109. (313) 763-3046. FAX (313) 764-4292.

Structural Engineering Materials Laboratory, University of California At Berkeley. Davis Hall, Berkeley, CA 94720. (415) 642-3464.

Structural Stability Research Council. Fritz Engineering Laboratory, Lehigh University, Bethlehem, PA 18015. (215) 758-3522. FAX (215) 758-4522.

University of Wisconsin—madison, Center For Structural and Materials Testing. 1415 Johnson Street, Madison, WI 53706. (608) 262-3205.

SUBMARINES

See also: NAVAL ARCHITECTURE

ABSTRACT SERVICES AND INDEXES

Applied Science and Technology Index; A Cumulative Subject Index To English Language Periodicals in the Fields of Aeronautics and Space Science, Computer Technology, Chemistry, Construction Industry, Energy and Related Areas. H.W. Wilson Co., 950 University Avenue, Bronx, NY 10452. (800) 367-6770 or (212) 588-8400. FAX (718) 590-1617.From 1958 to present. Monthly. Inquire about cost and availability. Also available on CD-ROM and online.

Civil and Structural Engineering Abstracts, Cambridge Scientific Abstracts, 7200 Wisconsin Avenue, Bethesda, MD 20814-4823. (301) 961-6750. FAX (301) 961-6720. 1993 to present. Monthly. $385.00 per year. Topics covered include structural design, construction equipment and methods, civil defense and military engineering, surveying, highway engineering, maritime and port structures, materials, land reclamation, and soil mechanics.

Current Contents: Engineering, Technology, and Applied Sciences. Institute for Scientific Information, 3501 Market Street, Philadelphia, PA 19104. (215) 386-0100. FAX (215) 386-6362. 1970 to present. Weekly. $442.00 per year.

Engineering Index Monthly. Engineering Information, Inc., Castle Point on the Hudson, Hoboken, NJ 07030. (800) 221-1044. FAX (212) 832-1857. Monthly. $2300.00 per year. Also available online as COMPENDEX, and also on CD-ROM. Covers chemical engineering, computers, electrical engineering, civil engineering, metals and mining, industrial management, and mechanical engineering.

Science Citation Index. Institute for Scientific Information, 3501 Market Street, Philadelphia, PA 19104. (215) 386-0100. FAX (215) 386-6362. Inquire about availability and cost. Also available on CD-ROM.

ASSOCIATIONS AND PROFESSIONAL SOCIETIES

Naval Submarine League. PO Box 1146, Annandale, VA 22003. (703) 256-0891. FAX (703) 642-5815.

Society of Naval Architects and Marine Engineers. 601 Pavonia Avenue, Suite 400, Jersey City, NJ 07306. (201) 798-4800. FAX (201) 798-4975.

DIRECTORIES AND BIOGRAPHICAL SOURCES

Directory of Engineering Societies and Related Organizations. Gordon Davis. 13th edition. American Association of Engineering Societies, 1111 19th Street NW, Suite 608, Washington, DC 20036. (202) 296-2237 or (800) 658-8897. 1989. Inquire for price.

International Research Centers Directory. Gale Research, Inc., 835 Penobscot Building, Detroit, MI 48226-4094. (313) 961-2242. (800) 877-4253. 8th edition. 1995. $410.00

Research Centers Directory. Gale Research, Inc., 835 Penobscot Building, Detroit, MI 48226-4094. (313) 961-2242. (800) 877-4253. FAX (313) 961-6083. $485.00.

Scientific and Technical Organizations and Agencies Directory. Gale Research, Inc., 835 Penobscot Building, Detroit, MI 48226-

4094. (313) 961-2242. (800) 877-4253. FAX (313) 961-6083. 4th edition. 1996. $195.00.

Who's Who in Engineering. American Association of Engineering Societies, 1111 19th Street NW, Suite 608, Washington, DC 20036. (202) 296-2237 or (800) 658-8897. 8th edition. 1991. Inquire for price.@ @Encyclopedias and Dictionaries

McGraw-Hill Dictionary of Scientific and Technical Terms. Sybil P. Parker, ed. 5th edition. McGraw-Hill Publishing Company, 1221 Avenue of the Americas, New York, NY 10020. (800) 262-4729 or (212) 512-3825. 1993. $110.50.

McGraw-Hill Encyclopedia of Engineering. Sybil P. Parker, ed. 2nd edition. McGraw-Hill Publishing Company, 1221 Avenue of the Americas, New York, NY 10020. (800) 262-4729 or (212) 512-3825. 1993. $95.50.

McGraw-Hill Encyclopedia of Science and Technology. Sybil P. Parker, ed. 7th edition. 20 volumes. McGraw-Hill Publishing Company, 1221 Avenue of the Americas, New York, NY 10020. (800) 262-4729 or (212) 512-3825. 1992. $1900.00

Thesaurus of Scientific, Technical, and Engineering Terms. Hemisphere Publishing Corporation, 1900 Frost Road, Suite 101, Bristol, PA 19007-1598. (215) 785-5800 or FAX (215) 785-5515. 1987. $173.00.

HANDBOOKS AND MANUALS

NSL Fact Book. Naval Submarine League, PO Box 1146, Annandale, VA 22003. (703) 256-0891. FAX (703) 642-5815. Annual. Inquire for availability information.

GENERAL WORKS

Building American Submarines, 1914-1940. Gary E. Weir. Naval Historical Center, Washington Navy Yard, Bldg. 57, Washington, DC 20374. (202) 433-3892. 1991. Inquire for cost and availability.

Concepts in Submarine Design. Roy Burcher and Louis Rydill. Cambridge University Press, 40 West 20th Street, New York, NY 10011-4211. (212) 924-3900 or (800) 872-7423. FAX (914) 937-4712. 1994. Inquire for cost and availability.

Guide To Manned Submersibles & Submarines, 1990-91. Deam Given & Patrick W. Smalls, eds. Windate Enterprises, PO Box 368, Spring Valley, CA 91976. (619) 660-0402. 1990. Inquire for cost and availability.

Navies in the Nuclear Age. Robert Gardiner, ed. Naval Institute Press, U.S. Naval Institute, Preble Hall, 118 Maryland Avenue, Annapolis, MD 21402-5035. (410) 268-6110 or (800) 233-8764. 1993. Inquire for cost and availability.

Sharks of Steel. Yogi Kaufman and Paul Stillwell. Naval Institute Press, U.S. Naval Institute, Preble Hall, 118 Maryland Avenue, Annapolis, MD 21402-5035. (410) 268-6110 or (800) 233-8764. 1993. $39.95.

Ships, Submarines, and the Sea. P.J. Gates and N.M. Lynn. Brassey's, 1313 Dolley Madison Blvd., Suite 101, McLean, VA 22102-3926. (703) 442-4535 or (800) 775-2518. FAX (703) 442-9848. 1990. $40.00 (hardback), $25.00 (paperback).

The Submarine Book. Chuck Lawliss. Thames and Hudson, 500 Fifth Avenue, New York, NY 10110. (212) 354-3763. FAX (212) 689-0856. 1991. Inquire for cost and availability.

Submarines of the World. David Miller. Orion Books, Crown Publishing Group, 201 E. 50th St., New York, NY 10022. (800) 726-0600 or (800) 733-3000. 1992. $30.00.

ONLINE DATABASES AND CD-ROMS

Compendex Plus. Engineering Information, Inc., 345 East 47th Street, New York, NY 10017. (212) 705-7600 or (800) 221-1044. Contains citations with abstracts to worldwide literature in engineering and technology, from 1970 to present. Available on online BRS,(800) 289-4277, DIALOG, (800) 334-2564, ORBIT (800) 456-7248, and STN International, FIZ Karlsruhe, P.O. Box 2465, W-7500, Karlsruhe 1, Germany, online services. Also available on CD-ROM. Inquire as to cost and availability.

CSA Engineering. Cambridge Scientific Abstracts, 7200 Wisconsin Avenue, Suite 601, Bethesda, MD 20814. (301) 961-6750 or (800) 843-7751. Contains citations and abstracts of international periodicals and research literature covering all fields of engineering and science and technology,including computer and information science, electronics, mechanical engineering, solid state materials, 1981 to present. Available on BRS,(800) 289-4277, online service.Inquire as to cost and availability.

Current Contents Search. Institute for Scientific Information, 3501 Market Street, Philadelphia, PA 19104. (215) 386-0100. FAX (215) 386-6362. Contains citations to articles listed in the table of contents of science and technology journals. Also articles in social sciences and life sciences journals. Available on BRS,(800) 289-4277, DIALOG,(800) 334-2564, online services. Inquire as to cost and availability.

Dissertation Abstracts. University Microfilms International, 300 North Zeeb Road, Ann Arbor, MI 48106. (800) 521-0600 or (313) 761-4700. Scope includes virtually all doctoral dissertations accepted at accredited American institutions from 1861 to present in 252 subject areas. Available on BRS,(800) 289-4277, DIALOG,(800) 334-2564, and OCLC EPIC,(800) 848-5878, online services. Also available on CD-ROM. Inquire as to cost and availability.

NTIS Bibliographic Database. National Technical Information Service, 5285 Port Royal Road, Springfield, VA 22161. (703) 487-4929 or FAX (703) 321-8199. Broad coverage of government-sponsored science and technology research reports, 1964 to present. Available on BRS,(800) 289-4277, DIALOG, (800) 334-2564, ORBIT, (800) 456-7248, and STN International, FIZ Karlsruhe, P.O. Box 2465, W-7500, Karlsruhe 1, Germany, online services. Also available on CD-ROM. Inquire as to cost and availability.

SAE Global Mobility Database. Society of Automative Engineers (SAE), Electronic Publishing Division, 400 Commonwealth Drive, Warrendale, PA 15098. (412) 776-4841. Contains citations with abstracts to technical papers on automotive and aerospace technology and vehicular-related industries that have been presented at SAE conferences. Covers 1906 to present. Available on ORBIT,(800) 456-7248,online service. Inquire as to cost and availability.

SCISEARCH. Institute for Scientific Information, 3501 Market Street, Philadelphia, PA 19104. (800) 523-1850 or (215) 386-0100. Broad multidisciplinary title and author index to the international literature of science and technology, 1974 to present.

SUBMARINES

Ency. of Physical Sciences and Engineering Info. Sources

Available on DIALOG,(800) 334-2564, and ORBIT,(800) 456-7248, online services. Also available on CD-ROM. Inquire as to cost and availability.

TRIS (Transportation Research Information). National ACADemy of Science, 2101 Constitution Avenue, N.W., Washington, DC 20418. (202) 334-3313 or (800) 624-6242. Citations with abstracts of literature on transportation, including air, highway, rail, maritime and other modes from 1968 to present. Available on DIALOG,(800) 334-2564, online service. Inquire as to cost and availability.

WILSONLINE. H.W. Wilson Company, 950 University Avenue, Bronx, NY 10452. (800) 367-6770 or (212) 588-8400. Makes available online versions of the printed H.W. Wilson indexes including Applied Science and Technology Index, Business Periodicals Index, General Science Index, and Readers' Guide to Periodical Literature. Period covered is generally 1983 to present. Available on BRS,(800) 289-4277, DIALOG,(800) 334-2564, and OCLC EPIC,(800) 848-5878, online services. Also available on CD-ROM. Inquire as to cost and availability.

PERIODICALS

Fathom. U.S. Navy Safety Center, Naval Air Station, Norfolk, VA 23511. (804) 444-6970. FAX (804) 444-7205. 1969 to present. Bi-monthly. $11.00 per year.

Submarine Review. Naval Submarine League, PO Box 1146, Annandale, VA 22003. (703) 256-0891. FAX (703) 642-5815. Quarterly. Inquire for cost and availability.

S U T News. Society for Underwater Technology, 76 Mark Lane, London EC3R 7JN, England. Telephone 071-481-0750. FAX 071-481-4001. Quarterly. Inquire for cost and availability.

SUN

See also: SOLAR SYSTEM, SOLAR WIND

ABSTRACT SERVICES AND INDEXES

Applied Science and Technology Index; A Cumulative Subject Index To English Language Periodicals in the Fields of Aeronautics and Space Science, Computer Technology, Chemistry, Construction Industry, Energy and Related Areas. H. W. Wilson Co., 950 University Avenue, Bronx, NY 10452- 9978. (212) 588-8400. (800) 367-6770. FAX (718) 590-1617.From 1958 to present. Monthly. Inquire about cost and availability. Also available online (BRS and WILSONLINE) and on CD-ROM. ISSN: 0003-6986.

Astronomy and Astrophysics Abstracts. Springer-Verlag New York, 175 Fifth Avenue, New York, NY 10010. (212) 460-1500. FAX (212) 473-6272. Published for the Astronomisches Rechen-Institut. Comprehensive coverage of all aspects of astronomy, astrophysics and related fields. 1969 to present. Two parts per year. Annual. ISSN: 0067-0022.

Current Contents: Physical, Chemical, and Earth Sciences. Institute for Scientific Information, 3501 Market Street, Philadelphia, PA 19104. (215) 386-0100. (800) 523-1850. FAX (215) 386-2291. 1961 to present. Weekly. $442.00 per year. Also available online (BRS, DIALOG) and on CD- ROM.

General Science Index. H.W. Wilson Company, 950 University Avenue, Bronx, NY 10452. (212) 588-8400. (800) 367-6770. FAX (718) 590-1617. From 1978 to present. quarterly and annual cumulations. Service basis. Available on CD-ROM and online. Inquire regarding cost and availability. ISSN: 0162-1963.

Physics Abstracts. INSPEC. Section A, Science Abstracts. Institute of Electrical Engineers, London, United Kingdom. Available from: INSPEC/IEEE - Institute of Electrical and Electronic Engineers, Box 1331, Hoes Lane, Piscataway, NJ 08855-1331. (908) 562-5549. 1898 to present. 24 issues per year. $2835.00 per year. ISSN: 0036-8091. Also available online and on CD-ROM.

Science Citation Index. SCI. Institute for Scientific Information, 3501 Market Street, Philadelphia, PA 19104. (215) 386-0100. (800) 523-1850. FAX (215) 386-2991. 1961 to present. Six issues per year, plus annual cumulation. $11650.00 per year. Also available online and on CD-ROM. Inquire about price and availability. ISSN: 0036-827X.

STAR. Scientific and Technical Aerospace Reports. U.S. National Aeronautics and Space Administration. Distributed by U. S. Superintendent of Documents, Washington, DC 20402. 1963 to present. Semi-monthly, with semiannual and annual indexes. $114.00 per year. ISSN: 0036-8741. Also available online and on CD-ROM.

ANNUAL REVIEWS AND YEARBOOKS

The Astronomical Almanac. Superintendent of Documents, U. S. Government Printing office, Washington, DC 10402. (202) 783-3238. 1981 to present. Superceeds Astronomical Ephemeris and the American Ephermis and Nautical Almanac. Annual. $44.95. ISSN: 0737-6421.

Annual Review of Astronomy and Astrophysics. Original reviews of critical literature and current developments in astronomy and astrophysics. Annual Reviews, Inc., 4139 El Camino Way, Palo Alto, CA 94303-0139. (415) 493-4400. (800)523-8635. Fax (415) 855-9815. 1963 to date. Annual. $60.00. ISSN: 0066-4146.

ASSOCIATIONS AND PROFESSIONAL SOCIETIES

Amateur Astronomers Association. 1010 Park Avenue, New York, NY 10028. (212) 535-2922.

American Astronomical Society. 2000 Florida Avenue NW, Suite 400, Washington, DC 20009. (202) 328-2010. FAX\;(202) 234-2560.

American Association of Variable Star Observers. 23 Birch Street, Cambridge, MA 02138. (617) 354-0484. FAX: (617)354-0665.

American Geophysical Union. 2000 Florida Avenue NW, Washington, DC 20009. (202) 462-6900. (800) 966-AGU1. FAX (202) 328-0566.

American Institute of Physics. 1 Physics Ellipse, College Park, MD 20740-3843. (301) 209-3100.

Association of Universities For Research in Astronomy, Inc. (AURA). Suite 701, 1625 Massachusetts Avenue, NW, Washington, DC 20036. (202) 483-2101. FAX (202) 483-2106.

Astronomical League. 2112 Kingfisher Lane East, Rolling Meadows, IL 60008. (708) 398-0562. # #Astronomical Society of the Pacific. 390 Ashton Avenue, San Francisco, CA 94112. (415) 337-1100. FAX: (415) 337-5205. @ @Bibliographies

Astronomy and Astrophysics: A Source Guide. Gordon Press, P.O. Box 459, Bowling Green Station, New York, NY 10004. (718) 624-8419. 1991. $250.00.

A Bibliography of Astronomy, 1970 - 1979. R. A. Sea and S. S. Martin. Libraries Unlimited, Inc., Littleton, CO 80160. 1982. $37.50.

Science Books & Films. American Association for the Advancement of Science, 1333 H Street NW, Washington, DC 20005. (202) 326-6434. Reviews of print, film and software materials in all sciences. 1965 to present. Nine issues per year. $40.00 per year. ISSN: 0098-342X.

Scientific and Technical Books and Serials in Print; An Index To Literature in Science and Technology. R. R. Bowker. 121 Chanlon Road, New Providence, NJ 07974. (908) 464-6800. FAX (908) 665-3502. (800) 521-8110. 1972 to present. Annual. $299.95 per year. Also available on CD-ROM. ISSN: 0000-054X.

DIRECTORIES AND BIOGRAPHICAL SOURCES

American Astronomical Society. Membership Directory. American Astronomical Society. 2000 Florida Avenue NW, Suite 400, Washington, DC 20009. (202) 328-2010. FAX (202) 234-2560 Annual. Included in membership dues. ISSN: 1061- 9038.

American Men and Women of Science: Physical and Biological Sciences. R. R. Bowker Inc., 121 Chanlon Road, New Providence, NJ 07974. (908) 464-6800. (800) 521-8110. 20th edition. 8 volumes. 1996. $850.00.

The Astronomers. Donald Goldsmith. St. Martin's Press, Inc., 175 Fifth Avenue, New York, NY 10010. (212) 674-5151. (800) 221-7945. 1993. $14.95 in paper.

Astronomical Centers of the World. Kevin Krisciunas. Cambridge University Press, 40 West 20th Street, New York,

NY 10011-4211. (212) 924-3900. (800) 872-7423. 1988. $34.95

Directory of Physics and Astronomy Staff. American Institute of Physics, One Physics Ellipse, College Park, MD 20740-3843. (301) 209-3100. 1975/76 to present. Annual. $60.00. ISSN: 0361-2228.

Earth and Astronomical Research Centers. Stockton Press, 345 Park Avenue South, New York, NY 10010. 4th edition. 1995. $515.00. ISBN: 1-56169-0967.

Graduate Programs in Physics, Astronomy and Related Fields. American Institute of Physics, One Physics Ellipse, College Park, MD 20740-3843. (301) 209-3100. 1978 to present. Annual. $45.00. ISSN: 0147-1821.

Research Centers Directory. Gale Research Inc., 835 Penobscot Building, Detroit, MI 48226-4094. (313) 961-2242. (800) 877-4253. 20th edition. 1995. $485.00.

ENCYCLOPEDIAS AND DICTIONARIES

Concise Dictionary of Astronomy. Jacqueline Mitton. Oxford University Press, Inc., 200 Madison Avenue, New York, NY 10016. (212) 725-6000. (800) 334-4249. 1992. $24.95.

Encyclopedia of Astronomy and Astrophysics. Stephen Maran, editor. Van Nostrand Reinhold, 115 Fifth Avenue, New York, NY 10003. (212) 254-3232. (800) 842-3636. 1992. $129.95.

McGraw-Hill Encyclopedia of Science and Technology. McGraw-Hill Book Company, Inc., 1221 Avenue of the Americas, New York, NY 10020. (212) 512-2000. (800) 262-4729. 7th edition. 20 volume set. 1992. $1900.00.

GENERAL WORKS

Astrophysics of the Sun. Harold Zinn. Cambridge University Press, 40 West 20th Street, New York, NY 10011-4211. (212) 924-3900. (800) 872-7423. 1988. $34.95 in paper.

Blinded By the Light. John Gribbin. Crown Publishing Group, 201 East 50th Street, New York, NY 10022. (212) 751- 2600. (800) 726-0600. 1991. $20.00.

Cool Stars, Stellar Systems and the Sun. Jeffrey L. Linsky and R. E. Stencel, editors. Springer-Verlag New York, Inc., 175 Fifth Avenue, New York, NY 10010. (212) 460-1500. (800) 777-4643. 1988. $66.00.

Discovering the Secrets of the Sun. Rudolph Kippenhahn.

John Wiley & Sons, Inc., 605 Third Avenue, New York, NY 10158-0012. (212) 850-6000. (800) 225-5945. 1994. $39.95in paper.

Exloring the Sun: Solar Science Since Galileo. Karl Hufbauer. John Hopkins University Press, 2715 North Charles Street, Baltimore, MD 21218-4319. (410) 516-6900. (800) 537-5487. 1991. $50.00.

Future of the Sun. Jean-Claude Pecker. McGraw-Hill Publishing Company, Inc., 1221 Avenue of the Americas, New York, NY 10020. (212) 512-2000. (800) 262-4729. 1992. $11.95.

Progress in Solar Physics. C. DeJager and Z. Svestka. Kluwer Academic Publishers, 101 Philip Drive, Assinippi Park, Norwell, MA 02061. (617) 871-6600. 1986. $181.50.

Solar and Planetary Dynamics. M. R. Proctor. Cambridge University Press, 40 West 20th Street, New York, NY 10011- 4211. (212) 924-3900. (800) 872-7423. 1994. $9.95.

Solar Interior and Atmosphere. Arthur N. Cox et al, editors. University of Arizona Press, 1230 North Park Avenue, Number 102, Tucson, AZ 85719. (520) 621-1441. (800) 426-3797. 1992. $75.00.

Solar Magnetic Fields. Manfred Schussler and Wolfgang Schmidt, editors. Cambridge University Press, 40 West 20th Street, New York, NY 10011-4211. (212) 924-3900. (800)872-7423. 1994. $59.95.

Solar Observations: Techniques and Interpretation. F. Sanches, et al, editors. Cambridge University Press, 40 West 20th Street,

SUN

Ency. of Physical Sciences and Engineering Info. Sources

New York, NY 10011-4211. (212) 924-3900. (800) 872-7423. 1991. $79.95.

The Sun. Time-Life Editors . Voyage Through the Universe series. Time-Life, Inc., 777 Duke Street, Alexandria, VA 22314. (703) 8388-7000. (800) 621-7026. 1990. $24.60.

Sun and Earth. H. Freidman. W. H. Freeman, 41 Madison Avenue, East 26th Avenue, 35th Floor, New York, NY 10010. (212) 576-0400. 1995.

The Sun, Our Star. Robert W. Noyes. Books on Demand, 300 North Zeeb Road, Ann Arbor, MI 48106-1346. (313) 761-4700.(800) 521-0600. Reprint edition. $77.00 in paper.

HANDBOOKS AND MANUALS

Astrophysical Data: Planets and Stars. Kenneth R. Lanb. Springer-Verlag New York, Inc., 175 Fifth Avenue, New York, NY 10010. (212) 460-1500. (800) 777-4643. 1993. $59.00.

Field Guide to the Stars and Planets. Jay M. Pasachoff and Donald H. Menzel. Houghton Mifflin Co., 222 Berkeley Street, Boston, MA 02116. (617) 351-5000. (800) 225-3362. Revised edition. 1992. $24.95.

Landolt-Borenstein Numerical Data and Functional Relationships in Science and Technology: Astronomy, Astrophsyics and Space Research. Group VI. Springer-Verlag New York, Inc., 175 Fifth Avenue, New York, NY 10010. (212) 460-1500. (800) 777-4643. volumes priced individually.

Observing the Sun. Peter O. Taylor. Cambridge University Press, 40 West 20th Street, New York, NY 10011-4211. (212) 924-3900. (800) 872-7423. 1992. $29.95

ONLINE DATABASES AND CD-ROMS

Dissertation Abstracts. University Microfilms International, 300 North Zeeb Road, Ann Arbor, MI 48106. (800) 521-0600 or (313) 761-4700. Scope includes virtually all doctoral dissertations accepted at accredited American institutions from 1861 to present in 252 subject areas. Available on BRS, (800) 289-4277, DIALOG, (800) 334-2564, and OCLC EPIC, (800) 848-5878, online services. Also available on CD-ROM. Inquire as to cost and availability.

INSPEC. Institution of Electrical Engineers, Michael Faraday House, Six Hills Way, Stevenage, Herts. SG1 2AY, England. Telephone: 0438 313311 or FAX 0438 742840. Contains citations to the worldwide literature of physics, electronics and electrical engineering, computer technology, and related fields. Available on BRS, (800) 289-4277, DIALOG, (800) 334-2564, ORBIT, (800) 456-7248, and STN International, FIZ Karlsruhe, P.O. Box 2465, W-7500, Karlsruhe 1, Germany, online services. Inquire as to cost and availability.

NTIS Bibliographic Database. National Technical Information Service, 5285 Port Royal Road, Springfield, VA 22161. (703) 487-4929 or FAX (703) 321-8199. Broad coverage of government-sponsored science and technology research reports, 1964 to present. Available on BRS,(800) 289-4277, DIALOG, (800) 334-2564, ORBIT, (800) 456-7248, and STN International, FIZ Karlsruhe, P.O. Box 2465, W-7500, Karlsruhe 1, Germany, online services. Also available on CD-ROM. Inquire as to cost and availability.

SCISEARCH. Institute for Scientific Information, 3501 Market Street, Philadelphia, PA 19104. (800) 523-1850 or (215) 386-0100. Broad multidisciplinary title and author index to the international literature of science and technology, 1974 to present. Available on DIALOG, (800) 334-2564, and ORBIT, (800) 456-7248, online services. Also available on CD-ROM. Inquire as to cost and availability.

WILSONLINE. H.W. Wilson Company, 950 University Avenue, Bronx, NY 10452. (800) 367-6770 or (212) 588-8400. Makes available online versions of the printed H.W. Wilson indexes including Applied Science and Technology Index, Business Periodicals Index, General Science Index, and Readers' Guide to Periodical Literature. Period covered is generally 1983 to present. Available on BRS, (800) 289-4277, DIALOG, (800) 334-2564, and OCLC EPIC, (800) 848-5878, online services.Also available on CD-ROM. Inquire as to cost and availability.

PERIODICALS

Astronomical Journal. American Institute of Physics, One Physics Ellipse, College Park, MD 20740-3843. (301) 209- 3000. Published for the American Astronomical Society. 1849 to present. Monthly. $280.00 per year. ISSN: 0004- 6256.

Astronomical Society of the Pacific. PUBLICATIONS. Astronomical Society of the Pacific, 390 Ashton Avenue, San Francisco, CA 94112. (415) 337-1100. FAX (415) 337-5205. 1889 to present. Monthly. $175.00 per year. ISSN: 0004- 6280.

Astronomy. Kalmbach Publishing Company, Box 1612, Waukesha, WI 53187-1612. (414) 796-8776. FAX (414) 796-1142. 1973 to present. Monthly. $27.00 per year. ISSN: 0091-6358.

Astronomy and Astrophysics. Springer-Verlag New York, Inc., 44 Hartz Way, Secaucus, NJ 07096-2491. (201) 348-4033. FAX (201) 348-4505. 1969 to date. 36 issues per year in 12 volumes. $1733.00 per year. ISSN: 0004-6361.

Astronomy Reports. American Institute of Physics, One Physics Ellipse, College Park, MD 20740-3843. (301) 209- 3000. English translation of Astronomicheksii Zhurnal. Formerly Soviet Astronomy AJ. 1957 to present. Bimonthly. $1160.00 per year. ISSN: 1063-7729.

Astrophysical Journal; An International Review of Astronomy and Astronomical Physics. University of Chicago Press, Journals Division, 5720 South Woodlawn Avenue, Chicago, IL 60637. (312) 753-3347. FAX (312) 753-0811. 1895 to date. three issues per month. $740.00 per year. ISSN: 0004-637X.

Astrophysics and Space Science; An International Journal of Cosmic Physics. Kluwer Academic Publishers, Box 358, Accord Station, Hingham, MA 02018-0358. (617) 871-6600. FAX (617) 871-6528. 1968 to present. 24 issues per year. $2544.00 per year. ISSN: 0004-640X.

Geochemica Et Cosmochimica Acta. Elsevier Science. 660 White Plains Road, Tarrytown, NY 10591-5153. (914) 524-9200. FAX (914) 333-2444. 1950 to the present. Biweekly. $895.00 per year. ISSN: 0016-7037.

Icarus; International Journal of Solar System Studies. Academic Press, Inc., Journal Division, 525 B Street, Suite 1900,

San Diego, CA 92101-4495. (619) 230-1840. FAX (619) 699-6800. 1962 to the present. Monthly. $1080.00. ISSN: 0019-1035.

Mercury. Astronomical Society of the Pacific. 390 Ashton Avenue, San Francisco, CA 94112. (415) 337-1100. FAX: (415) 337-5205. 1972 to present. Bimonthly. $175.00 per year. ISSN: 0047-6773.

Monthly Notices of the Royal Astronomical Society. Blackwell Scientific Publication LT., Osney Mead, Oxford OX2 OEL, England. TEL 0865-240201. FAX 0865-721205. 1827 to present. Fortnightly. $1899.00 per year. ISSN: 0035-8711.

Observatory. c/o Dr. D. J. Stickland Space and Astrophysics Division, Rutherford Appleton Laboratory, Chilton, Didcot, Oxon OX11 OQX England. FAX 0235-445848. 1877 to present. Bimonthly. $42.00 per year. ISSN: 0029-7704.

Planetary and Space Science. Elsevier Science Publishing Company, Inc., 660 White Plains Road, Terrytown, NY 10591- 5153. (914) 524-9200. FAX: (914) 333-2444. 1959 to present. Monthly. $1355.00 per year. ISSN: 0032-0633.

Solar-geophysical Data, Part 1 (prompt Reports) and 2 (comprehensive Reports). National Geophysical Data Center, 325 Broadway, NOAA E/GC2, Boulder, CO 80303. (303) 497-6135. 1955 to present. Monthly. $60.00 for each part.

Solar Physics; A Journal For Solar and Solar-stellar Research and the Study of Solar Terrestrial Physics. Kluwer Academic Publishers, Box 358, Accord Station, Hingham, MA 02018-0358. (617) 871-6600. FAX (617) 871-6528. 1967 to present. Fourteen issues per year. $1592.00 per year. ISSN: 0038-0938.

Space Science Reviews. Kluwer Academic Publishers, Box 358, Accord Station, Hingham, MA 02018-0358. (617) 871- 6600. FAX (617) 871-6528. 1962 to present. Sixteen issues per year. $940.00 per year. ISSN: 0038-6308.

Sky and Telescope. Sky Publishing Corporation, Box 9111, Belmont, MA 02178. (617) 864-7360. FAX (617) 864-6117. 1941 to present. Monthly. $27.00 per year. ISSN: 0037- 6604.

Vistas in Astronomy; An International Review Journal. Albert C. Beer, editor. Elsevier Science Publishing Company, Inc., 660 White Plains Road, Terrytown, NY 10591- 5153. (914) 524-9200. FAX: (914) 333-2444. 1958 to present. Quarterly. $415.00 per year. ISSN: 0083-6656.

RESEARCH CENTERS AND INSTITUTES

Astronomy Program. University of Maryland, 1207 Computer & Space Sciences Building, College Park, MD 20742. (301) 405- 3001. FAX (301) 314-9067.

Big Bear Solar Observatory. California Institute of Technology. 40396 North Shore Lane, Big Bear City, CA 92314. (818) 356-4014. FAX (818) 356-3814.

Harvard-smithsonian Center For Astrophysics, 60 Garden Street, Cambridge, MA 02138. (617) 495-7461. FAX (617)495-7326.

High Altitude Observatory. National Center for Atmospheric Research, P.O. Box 3000, Boulder, CO 80307. (303) 497-1500.

National Astronomy and Ionosphere Center. Cornell University. Space Sciences Building, Ithaca, NY 14853. (607) 255-3735. FAX (607) 255-8803.

National Solar Observatory, 950 North Cherry Avenue, Tucson, AZ 85719. (602) 325-9294. FAX (602) 325-9278.

San Fernando Oservatory. California State University, Northridge. Department of Physics and Astronomy, 18111 Nordhoff Street, Northridge, CA 91330. (818) 367-9333.

Santa Catalina Laboratory For Experimental Relativity By Astrometry. University of Arizona, Building 81, Department of Physics, Tucson, AZ 85721. (602) 621-6782.

SUPERCONDUCTING SUPERCOLLIDER

See: PARTICLE ACCELERATORS

SUPERCONDUCTIVITY

See also: CRYOGENICS, ELECTRICAL ENGINEERING, ELECTRICITY, PHYS CAL CHEMISTRY, PHYSICS, SOLID STATE PHYSICS

ABSTRACT SERVICES AND INDEXES

Applied Mechanics Reviews: An Assessment of World Literature in Engineering Sciences. American Society of Mechanical Engineers, 345 East 47th Street, New York, NY 10017. (212) 705-7703. 1948 to present. Monthly. $360.00 per year. ISSN: 0003-6900.

Applied Science and Technology Index; A Cumulative Subject Index To English Language Periodicals in the Fields of Aeronautics and Space Science, Computer Technology, Chemistry, Construction Industry, Energy and Related Areas. H.W. Wilson Co., 950 University Avenue, Bronx, NY 10452. (212) 588-8400. (800) 367-6770. FAX (718) 590-1617. From 1958 to present. Monthly. Inquire about cost and availability. Also available on CD-ROM and online. ISSN: 0003-6986.

Chemical Abstracts. Chemical Abstracts Service, 2540 Olentangy River Road, Box 3012, Columbus, OH 43210-0012. (614) 447-3600. (800) 848-6538. FAX (614) 447-3713. From 1907 to present. Weekly. $16800.00 per year. Also available on CD-ROM and online. Inquire regarding cost and availability. ISSN:

Current Contents: Engineering, Technology, and Applied Sciences. Institute for Scientific Information, 3501 Market Street, Philadelphia, PA 19104. (215) 386-0100. FAX (215) 386-2291. From 1961 to present. Weekly. $442.00 per year. Also available on CD-ROM and online. Inquire regarding cost and availability. ISSN: 0095-7917

.Current Papers in Physics. Institute of Electrical Engineers, Michael Faraday House, Six Hill Way, Stevenage, Herts, SG1 2AY, England. Distributed by INSPEC/IEEE, Box 1331, 445 Hoes Lane, Piscataway, NJ 08855-1331. (908) 562- 5549. 1966 to present. Fortnightly. $410.00 per year. ISSN: 0011-3786.

Engineering Index Monthly. Engineering Information, Inc., Castle Point on the Hudson, Hoboken, NJ 07030. (201) 216-

SUPERCONDUCTIVITY

Ency. of Physical Sciences and Engineering Info. Sources

8500. FAX (201) 216-8532. Monthly. $2200.00 per year. Also available online as COMPENDEX and on CD-ROM. ISSN: 0742-1974.

General Science Index. H.W. Wilson Company, 950 University Avenue, Bronx, NY 10452. (212) 588-8400. (800) 367-6770. FAX (718) 590-1617. From 1978 to present. Ten issues per year; quarterly and annual cumulations. Service basis. Available on CD-ROM and online. Inquire regarding cost and availability. ISSN: 0162-1963.

Government Reports Announcements and Index. U. S. National Technical Information Service (NTIS), 5285 Port Royal Road, Springfield, VA 22161. (703) 487-4650. FAX (703) 321-8547. From 1968 to present. Annual. $630.00 per year. Also available online as NTIS BIBLIOGRAPHIC DATABASE and on CD-ROM. ISSN:

*ISMEC: Mechanical Engineering Abstracts.*Cambridge Scientific Abstracts, 7200 Wisconsin Avenue, Suite 601, Bethesda, MD 20814. (301) 961-6750 or (800) 843-7751. Contains citations to the literature in mechanical engineering, industrial and production engineering, energy, power, mechanics, devices and related areas, from 1973 to present. Available on the DIALOG,(800) 334-2564, online service. Inquire as to cost and availability.

Physics Abstracts. INSPEC. Section A, Science Abstracts. Institution of Electrical Engineers (IEE). Available from INSPEC/ IEEE - Institute of Electrical and Electronic Engineers,, Box 1331, 445 Hoes Lane, Piscaway NJ 08855-1331. (908) 562-5549. 1898 to present. 24 issues per year. $2835.00 per year. Also available on CD-ROM and online. ISSN: 0036-8091.

Science Citation Index. SCI. Institute for Scientific Information, 3501 Market Street, Philadelphia, PA 19104. (215) 386-0100. (800) 523-1850. FAX (215) 386-2991. 1961 to present. Six issues per year, plus annual cumulation. $11650.00 per year. Also available online and on CD-ROM. Inquire about price and availability. ISSN: 0036-827X.

ASSOCIATIONS AND PROFESSIONAL SOCIETIES

American Chemical Society. 1155 16th Street, N.W., Washington, DC 20036. (202) 872-4600.

American Institute of Chemical Engineers. 345 East 47th Street, New York, NY 10017. (212) 705-7657. FAX (212) 752-3294.

American Institute of Physics. One Physics Ellipse, College Park, MD 20740-3843. (301) 209-3100.

American Society of Mechanical Engineers, 345 East 47th Street, New York, NY 10017. (212) 705-7722. FAX (212) 705- 7739.

Cryogenic Engineering Conference. c/o Dr. P. Kittel, MS 244-10, NASA/AMES, Moffett Field, CA 94035-1000. (415) 604-4297. FAX (415) 604-0673.

Cryogenic Society of America. c/o Huget Advertising, 1033 South Boulevard, Oak Park, IL 60302. (708) 383-6220. FAX (708) 383-9337.

Institute of Electrical and Electronics Engineers. 345 East 47th Street, New York, NY 10017. (212) 705-7900.

DIRECTORIES AND BIOGRAPHICAL SOURCES

American Men and Women of Science: Physical and Biological Sciences. R. R. Bowker Inc., 121 Chanlon Road, New Providence, NJ 07974. (908) 464-6800. (800) 521-8110. 20th edition. 8 volumes. 1996. $850.00.

International Directory of Engineering Societies and Related Organizations. American Society for Engineering Education, 1818 N Street NW, Suite 600, Washington, DC 20036. (202) 331-3526. 15th edition. 1996. $185.00.

International Engineering Directory. American Consulting Engineers Council, 1015 15th Street, N.W. Suite 802, Washington, DC 20005-2670. (202) 347-7474. Annual. $10.00

Research Centers Directory. Gale Research Inc., 835 Penobscot Building, Detroit, MI 48226-4094. (313) 961-2242. (800) 877-4253. 20th edition. 1995. $485.00. ISSN: 0080- 1518.

Scientific and Technical Organizations and Agencies Directory. Gale Research, 835 Penobscot Building, Detroit, MI 48226-4094. (313) 961-2242. (800) 877-4253. 4th edition. 1996. $195.00.

Who's Who in Engineering. Gordon Davis, editor. American Society for Engineering Education, 1111 19th Street, NW, Suite 608, Washington, DC 20036. 9th edition. 1995. $220.00.

Who's Who in Technology Gale Research, 835 Penobscot Building, Detroit, MI 48226-4094. (313 961-2242. (800) 521-4253. 7th edition. 1995. $195.00. ISBN 0-8103-7467-6.

ENCYCLOPEDIAS AND DICTIONARIES

Chambers Science and Technology Dictionary. Peter M.B. Walker, editor. Cambridge University Press, 40 West 20th Street, New York, NY 10011-4211. (212) 924-3900. 1988. $39.95.

A Concise Dictionary of Physics. Oxford University Press, Inc., 200 Madison Avenue, New York, NY 10016. (212) 725- 6000. (800) 334-4249. 1990. $10.95.

Concise Encyclopedia of Magentic and Superconducting Materials. J. E. Evetts, editor. Elsevier Science Publishing Company, Inc., 655 Avenue of the Americas, New York, NY 10010. (212) 989-5800. FAX (914) 333-2444. 1992. $290.00.

Encyclopedia of Applied Physics. George Trigg, editor. VCH Publications, Inc., 220 East 23rd Street, Suite 909, New York, NY 10010-4606. (212) 683-8333. (800) 422-8824. 20 volume set. 1991-. $5990.00.

Encyclopedia of Physical Science and Technology. Academic Press, Inc., 6277 Sea Harbor Drive, Orlando, FL. (800) 321-5068. 2nd edition. 18 volume set. 1992. $2625.00.

McGraw-Hill Encyclopedia of Science and Technology. McGraw-Hill Book Company, Inc., 1221 Avenue of the Americas, New York, NY 10020. (212) 512-2000. (800) 262-4729. 7th edition. 20 volume set. 1992. $1900.00.

GENERAL WORKS

APPLIED theRMODYANMICS FOR ENGINEERING TECH-NOLOGIES. Thomas D. Eastop and A. McConkey. Halsted Press, 605 Third Avenue, New York, 10158-0012. (212) 850-6400. 5th edition.1993. $57.95.

Fundamentals of Classical thermodyanamics. Gordon J. Van Wylen, et al. John Wiley & Sons, Inc., 605 Third Avenue, New York, NY 10158-0012. (212) 850-6000. (800) 225-5945.4th edition. 1993. $wfi.

Principles of Modern thermodynamics. B. M. Roy. IOP Publishing, Public Ledger Building, Suite 1035, Independence Square, Philadelphia, PA 19106. (215) 627-0880. (800) 358-4677. 1995. $100.00.

Superconducting Devices. Steven T. Ruggiero and David A. Rudman. Academic Press, Inc., 6277 Sea Harbor Drive, Orlando, FL. (800) 321-5068. 1990. $73.00.

Superconductivity: The Threshold of A New Technology. Jonathan L. Mayo. TAB Books, P.O. Box 40, Blue Summit, PA 17294-0850. (717) 794-2191. (800) 233-1128. 1988. $12.95 in paper.

Superconductivity and Cryoelectronics. W. Krech, et al, editors. World Scientific Publishing Company, Inc., 1060 Main Street, River Edge, NJ 07661. (201) 487-9655. (800) 227-7562. 1991. $61.00.

Superconductivity: The New Revolution? Gianfranco Vidali. Cambridge University Press, 40 West 20th Street, New York, NY 10011-4211. (212) 924-3900. (800) 872-7423. 1993. $17.95 in paper.

Superfluidity and Superconductivity. D. R. Tilley and J. Tilley. IOP Publishing, Public Ledger Building, Suite 1035, Independence Square, Philadelphia, PA 19106. (215) 627- 0880. (800) 358-4677. 3rd edition. 1990. $43.00 in paper.

Thermodynamics and the Design, Analysis and Improvement of Energy Systems. H. J. Richter, editor. ASME - American Society of Mechanical Engineers, 345 East 47th Street, New York, NY 10017. (212) 705-7722. (800) 843-2763. 1993. $80.00.

HANDBOOKS AND MANUALS

Gas Tables, Thermodynamics Properties of Air, Products of Combustion, and Component Gases. Joseph H. Keenan, et al. Krieger Publishing Company, P.O. Box 9542, Melbourne, FL 32902-9542. (407) 724-9542. 2nd edition; reprinted. 1992. 64.50.

Handbook of Applied thermodynamics. David A. Palmer, editor. CRC Press, Inc., 2000 Corporate Boulevard, NW, Boca Raton, FL 33431. (407) 994-0555. (800) 272-7737. 1987. $216.95.

Steam Tables: Thermodynamic Properties of Water Including Vapor, Liquid and Solid Phases. Joseph H. Keegan, et al. Krieger Publishing Company, P.O. Box 9542, Melbourne, FL 32902-9542. (407) 724-9542. 2nd edition, reprinted. 1992. $64.50.

ONLINE DATABASES AND CD-ROMS

CA Search. Chemical Abstracts Service, P.O. Box 3012, Columbus, OH 43210-0012. (614) 447-3600. (800) 848-6533. FAX (614) 447-3709. Very comprehensive guide to worldwide chemical literature and related fields, 1972 to present. Available on BRS,(800) 289-4277, DIALOG, (800) 334-2564, ORBIT (800) 456-7248, and STN International, FIZ Karlsruhe, P.O. Box 2465, W-7500, Karlsruhe 1, Germany, online services. Inquire as to cost and availability.

Compendex Plus. Engineering Information, Inc., 345 East 47th Street, New York, NY 10017. (212) 705-7600 or (800) 221-1044. Contains citations with abstracts to worldwide literature in engineering and technology, from 1970 to present. Available on online BRS,(800) 289-4277, DIALOG, (800) 334-2564, ORBIT (800) 456-7248, and STN International, FIZ Karlsruhe, P.O. Box 2465, W-7500, Karlsruhe 1, Germany, online services. Also available on CD-ROM. Inquire as to cost and availability.

Current Contents Search. Institute for Scientific Information, 3501 Market Street, Philadelphia, PA 19104. (215) 386-0100. FAX (215) 386-6362. Contains citations to articles listed in the table of contents of science and technology journals. Also articles in social sciences and life sciences journals. Available on BRS, (800) 289-4277, DIALOG, (800) 334-2564, online services. Inquire as to cost and availability.

Dissertation Abstracts Online. University Microfilms International, 300 North Zeeb Road, Ann Arbor, MI 48106. (800) 521-0600 or (313) 761-4700. Scope includes virtually all doctoral dissertations accepted at accredited American institutions from 1861 to present in 252 subject areas. Available on BRS, (800) 289-4277, DIALOG, (800) 334-2564, and OCLC EPIC, (800) 848-5878, online services. Also available on CD-ROM. Inquire as to cost and availability.

*ISMEC: Mechanical Engineering Abstracts.*Cambridge Scientific Abstracts, 7200 Wisconsin Avenue, Suite 601, Bethesda, MD 20814. (301) 961-6750 or (800) 843-7751. Contains citations to the literature in mechanical engineering, industrial and production engineering, energy, power, mechanics, devices and related areas, from 1973 to present. Available on the DIALOG,(800) 334-2564, online service. Inquire as to cost and availability.

INSPEC. Institution of Electrical Engineers, Michael Faraday House, Six Hills Way, Stevenage, Herts. SG1 2AY, England. Telephone: 0438 313311 or FAX 0438 742840. Contains citations to the worldwide literature of physics, electronics and electrical engineering, computer technology, and related fields. Available on BRS, (800) 289-4277, DIALOG, (800) 334-2564, ORBIT, (800) 456-7248, and STN International, FIZ Karlsruhe, P.O. Box 2465, W-7500, Karlsruhe 1, Germany, online services. Inquire as to cost and availability.

NTIS Bibliographic Database. National Technical Information Service, 5285 Port Royal Road, Springfield, VA 22161. (703) 487-4929 or FAX (703) 321-8199. Broad coverage of government-sponsored science and technology research reports, 1964 to present. Available on BRS,(800) 289-4277, DIALOG, (800) 334-2564, ORBIT, (800) 456-7248, and STN International, FIZ Karlsruhe, P.O. Box 2465, W-7500, Karlsruhe 1, Germany, online services. Also available on CD-ROM. Inquire as to cost and availability.

Physics Briefs. American Institute of Physics, 335 East 45th Street, New York, NY 10017. (212) 661-9260 or FAX (212) 661-2036. Contains citations with abstracts of the literature of

SUPERCONDUCTIVITY

Ency. of Physical Sciences and Engineering Info. Sources

physics and related fields, 1979 to present. Available on the STN International, FIZ Karlsruhe, P.O. Box 2465, W-7500, Karlsruhe 1, Germany, online service. Inquire as to cost and availability.

SCISEARCH. Institute for Scientific Information, 3501 Market Street, Philadelphia, PA 19104. (800) 523-1850 or (215) 386-0100. Broad multidisciplinary title and author index to the international literature of science and technology, 1974 to present. Available on DIALOG, (800) 334-2564, and ORBIT, (800) 456-7248, online services. Also available on CD-ROM. Inquire as to cost and availability.

WILSONLINE. H.W. Wilson Company, 950 University Avenue, Bronx, NY 10452. (800) 367-6770 or (212) 588-8400. Makes available online versions of the printed H.W. Wilson indexes including Applied Science and Technology Index, Business Periodicals Index, General Science Index, and Readers' Guide to Periodical Literature. Period covered is generally 1983 to present. Available on BRS, (800) 289-4277, DIALOG, (800) 334-2564, and OCLC EPIC, (800) 848-5878, online services.Also available on CD-ROM. Inquire as to cost and availability.

PERIODICALS

Cryogas International. J. R. Campbell & Associates, Inc., 5 Militia Drive, Lexington, MA 02173. (617) 863-9411. FAX (617)863-9411. 1963 to present. Monthly. $150.00. ISSN: 1052-0139.

Cryogenics: The International Journal of Low Temperature Engineering and Resarch. Butterworth-Heinemann; Turpin Transactions Ltd., Distribution Center, Blackhorse, Road, Letchworth, Herts SG6 1HN, England. Tel 0462-672555.

Experimental thermal and Fluid Science. Elsevier Science Inc., 655 Avenue of the Americas, New York, NY 10010. (212) 989-5800. Subscriptions to: Box 882, Madison Square Station, NY 10159-0882. 1988 to present. 8 issues per year. $484.00 per year. ISSN: 0894-1777.

Heat Transfer Engineering. Taylor and Francis, 1900 Frost Road, Suite 101, Bristol PA 190077-1598. (800) 821-8312. 1979 to present. Quarterly. $176 per year. ISSN: 0145-7632.

International Communications in Heat and Mass Transfer. Elsevier Science, 660 White Plains Road, Tarrytown NY 10591-5151. (914) 524-9200. FAX (914) 333-2444. 1974 to present. Bi-monthly. $430.00 per year. Combined subscription with International Journal of Heat and Mass Transfer; $2070.00. ISSN: 0735-1933.

International Journal of Heat and Fluid Flow. Butterworth-Heinemann, 313 Washington Street, Newton MA 02158. (617) 928- 2500. (800) 366-2665. 1979 to present. Bi-monthly. $450.00 per year. ISSN: 0142-727X. # #*Journal of Heat Transfer.* American Society of Mechanical Engineers, 345 East 47th Street, New York, NY 10017. (212) 705-7722. volume 92, 1970 to present. Quarterly. $155.00 per year. ISSN: 0022-1481.

Journal of Low Temperature Physics. Plenum Publishing Corp. 233 Spring Street, New York, NY 10013-1578. (212) 620-8000. FAX (212) 463-0742. 1969 to present. 24 issues per year in 3 volumes. $955.00 per year. ISSN: 0022-2291.

Journal of Superconductivity. Plenum Publishing Corp. 233 Spring Street, New York, NY 10013-1578. (212) 620-8000.FAX

(212) 463-0742. 1988 to present. Bi-monthly. $275.00per year. ISSN: 0896-1107.

Journal of thermophysics and Heat Transfer. American Institute of Aeronautics and Astronautics, 370 L'Enfant Prominade SW, Washington, DC 20024. (202) 646-7400. 1987 topresent. Quarterly. $165.00 per year. ISSN: 0887-8722.

Numerical Heat Transfer: An International Journal of Computation and Methodology. Taylor and Francis, 1900 Frost Road, Suite 101, Bristol, PA 19007-1598. (800) 821-8312. FAX (215) 785-5515. 1978 to present. 8 issues per year. Part A: Applications, $530.00 per year; ISSN: 1040-7782. Part B: Fundamentals, $330.00 per year; ISSN: 1040-7790.

RESEARCH CENTERS AND INSTITUTES

Applied Superconductivity Research Center. Univers of Wisconsin at Madison, Engineering Research Building, 1500 Johnson Drive, Madison, WI 53706. (608) 263-5026.

Center For thermodynamics. Brigham Young University, 226 ESC, Provo, UT 84602. (801) 378-3668.

Heat Transfer Laboratory. Massacchusetts Institute of Technology, 77 Massachusetts Avenue, Cambridge, MA 02139. (716) 253-2248.

Thermodynamics Research Center. Texas A&M University, Texas Engineering Experiment Station, College Station, TX 77843- 3111. (409) 845-4940.

SUPERNOVA

See: STARS

SUPERSONIC FLIGHT

See also: AERODYNAMICS, AEROSPACE ENGINEERING, AIR-CRAFT, AIRCRAFT ENGINES, JET PROPULSION, PROPELLANTS, PROPULSION SYSTEMS

ABSTRACT SERVICES AND INDEXES

Aeronautical Engineering. Scientific and Technical Branch, National Aeronautics and Space Administration. National Technical Information Service (NTIS), 5285 Port Royal Road, Springfield, VA 22161. (703) 487-4650. FAX (703) 321-8547. Monthly. A selection of annotated references to unclassified reports and journal articles that were introduced into the NASA scientific and technical information system and announced in STAR and IAA.

Alloys Index. American Society for Metals, Metals Park, OH 44073. (216) 338-5151. FAX (216) 338-4634. 1974 to present. Monthly. $295.00.

Applied Mechanics Reviews: An Assessment of World Literature in Engineering Sciences. American Society of Mechanical Engineers, 345 East 47th Street, New York, NY 10017. (212) 705-7703. 1948 to present. Monthly. $360.00 per year.

Applied Science and Technology Index; A Cumulative Subject Index To English Language Periodicals in the Fields of Aeronautics and Space Science, Computer Technology, Chemistry, Construction Industry, Energy and Related Areas. H.W. Wilson Co., 950 University Avenue, Bronx, NY 10452. (800) 367-6770 or (212) 588-8400. FAX (718) 590-1617.From 1958 to present. Monthly. Inquire about cost and availability. Also available on CD-ROM.

Current Contents: Engineering, Technology, and Applied Sciences. Institute for Scientific Information, 3501 Market Street, Philadelphia, PA 19104. (215) 386-0100. FAX (215) 386-6362. Weekly. $360.00 per year.

Engineering Index Monthly. Engineering Information, Inc., Castle Point on the Hudson, Hoboken, NJ 07030. (800) 221-1044. FAX (212) 832-1857. Monthly. $2200.00 per year. Also available online as COMPENDEX, and also on CD-ROM.

International Aerospace Abstracts. Technical Information Service, American Institute of Aeronautics and Astronautics, Inc., 555 West 57th St., New York, NY 10019. (212) 247-6500. FAX (212) 582-4861. Semi-monthly. $1295.00 per year.

Metals Abstracts and Metals Abstracts Index. American Society for Metals, Metals Park, OH 44073. (216) 338-5151. 1968 to present. Monthly. Abstracts are $1500.00 per year and Index is $460.00 per year.

Science Citation Index. Institute for Scientific Information, 3501 Market Street, Philadelphia, PA 19104. (215) 386-0100. FAX (215) 386-6362. Inquire about availability and cost. Also available on CD-ROM.

Scientific and Technical Aerospace Reports (star). National Aeronautics and Space Administration. NASA Center for Aerospace Information, Box 8757, BWI Airport, Baltimore, MD 21240. (301) 621-0153. Monthly. Inquire about availability and cost. Also available through the NASA online retrieval service (RECON), and through the Aerospace Database through DIALOG.

World Aluminum Abstracts. Aluminum Association, Inc., 818 Connecticut Ave. NW, Washington, DC 20006. (202) 862-5156.

ASSOCIATIONS AND PROFESSIONAL SOCIETIES

American Institute of Aeronautics and Astronautics, 370 L'enfant Promenade Sw, Washington, DC 20024-2518.

Canadian Aeronautics and Space Institute, 222 Somerset Street West, Suite 601, Ottawa, Ontario, Canada K2p 0j1.

Flight Safety Foundation, Inc., 2200 Wilson Blvd., Suite 500, Arlington, VA 22201-3306. (703) 522-8300. FAX (703) 525-6047.

Society of Automotive Engineers. SAE, Inc., 400 Commonwealth Dr., Warrendale, PA 15096. (412) 776-4841.

BIBLIOGRAPHIES

Aeronautical Engineering: A Continuing Bibliography With Indexes. National Aeronautics and Space Administration, Center for Aerospace Information, Box 8757, BWI Airport, Balti-

more, MD 21240. (301) 621-0153. Inquire for cost and availability.

Scientific and Technical Books and Serials in Print, R.R. Bowker Inc., 121 Chanlon Road, New Providence, NJ 07974. (800) 521-8110 or (908) 464-6800. 3 volumes, annually. $285.00. Also available on CD-ROM.

DIRECTORIES AND BIOGRAPHICAL SOURCES

Aas Directory. American Astronautical Society, 6352 Rolling Mill Place, Suite 102, Springfield, VA 22152-2354. (703) 866-0020. FAX (703) 866-3526. Inquire for cost and availability.

Research Centers Directory. Gale Research, Inc., 835 Penobscot Building, Detroit, MI 48226-4094. (313) 961-2242. (800) 877-4253. FAX (313) 961-6083. $485.00.

SAE Aerospace Sources and Suppliers Directory 1993. SAE International, Inc., 400 Commonwealth Dr., Warrendale, PA 15096. (412) 776-4841. 1993. Inquire for cost and availability.

Who's Who in Engineering. American Association of Engineering Societies, 1111 19th St. NW, Suite 608, Washington, DC 20036. 7th ed. 1988. $200.00.

World Aviation Directory. McGraw-Hill, Aviation Week Group (NY), 1221 Avenue of the Americas, New York, NY 10020. (215) 237-4112. FAX (215) 586-3232. Semiannual. $250.00.

ENCYCLOPEDIAS AND DICTIONARIES

Dictionary of Aeronautical Terms, Dale Crane, Comp. Aviation Supplies and ACADemics, Inc., 7005 132nd Place SE, Newcastle, WA 98059. (206) 235-1500 or (800) 426-8338. FAX (206) 235-0128. 1991. $15.95.

Dictionary of Aerospace Engineering. M.G. Kotik. Elsevier Science Publishing Company, Inc., 655 Avenue of the Americas, New York, NY 10010. (212) 989-5800. 1986. $225.75.

Encyclopedia of Physical Science and Technology. Robert A. Meyers, ed. Academic Press, Inc., 525 B Street, Suite 1900, San Diego, CA 92101-4495. (619) 231-0926. FAX (619) 699-6715. 1992. Inquire for price.

Jane's Aerospace Dictionary, B. Guston. 3d ed. Jane's Information Group, 1340 Braddock Place, Suite 300, Alexandria, VA 22314-1651. (703) 683-3700. FAX (703) 836-0029. 1991. $45.00.

Jane's Encyclopedia of Aviation. Michael J. Taylor, comp. Outlet Book Co., 40 Engelhard Ave., Avenal, NJ 07001. (908) 827-2700 or (800) 223-6804. 1993. $34.99.

McGraw-Hill Encyclopedia of Science and Technology, Sybil P. Parker, ed. 7th ed. 20 vols. McGraw-Hill Publishing Company, 1221 Avenue of the Americas, New York, NY 10020. (800) 262-4729 or (212) 512-3825. $1900.00

SAE Dictionary of Aerospace Engineering. William H. Cubberly, ed. Society of Automotive Engineers, 400 Commonwealth Drive, Warrendale, PA 15096. (412) 776-4841. 1992. $85.00.

SUPERSONIC FLIGHT

Ency. of Physical Sciences and Engineering Info. Sources

GENERAL WORKS

Faster Than Sound: The Story of Supersonic Flight. Bill Gunston. Patrick Stephens, c/o Haynes Publications, 861 Lawrence Drive, Newbury Park CA 91320. (800) 442-9637. 1992. Inquire for cost and availability.

Into the Unknown: The X-1 Story. Louis Rotondo. Smithsonian Institution Press, 470 L'Enfant Plaza, Suite 7100, Washington, DC 20560. (202) 287-3738. 1994. $29.95.

Mach 1 and Beyond: The Illustrated Guide To High-speed Flight. L.W. Reithmaier. TAB Books, P.O. Box 40, Blue Ridge Summit, PA 17294-0850. (717) 794-2191 or (800) 233-1128. FAX (717) 794-2080. 1995. Inquire for cost and availability.

Wings To Come: NASA's High-speed Research Program. National Aeronautics and Space Administration. Center for Aerospace Information, Box 8757, BWI Airport, Baltimore, MD 21240. (301) 621-0153. 1993. Inquire for cost and availability.

HANDBOOKS AND MANUALS

Jane's All the World's Aircraft. Jane's Information Group, 1340 Braddock Place, Suite 300, Alexandria, VA 22314-1651. (703) 683-3700. FAX (703) 836-0029. Annual. $245.00. Also available on CD-ROM.

ONLINE DATABASES AND CD-ROMS

Aerospace Database. American Institute of Aeronautics and Astronautics, 370 L'Enfant Promenade, S.W., Washington, DC 20024. (202) 646-7400. Worldwide published literature on research and development in aerospace and related areas, with abstracts. Covers 1962 to present. Available on DIALOG, (800) 334-2564, online service. Also available on CD-ROM. Inquire as to cost and availability.

Compendex Plus. Engineering Information, Inc., 345 East 47th Street, New York, NY 10017. (212) 705-7600 or (800) 221-1044. Contains citations with abstracts to worldwide literature in engineering and technology, from 1970 to present. Available on online BRS,(800) 289-4277, DIALOG, (800) 334-2564, ORBIT (800) 456-7248, and STN International, FIZ Karlsruhe, P.O. Box 2465, W-7500, Karlsruhe 1, Germany, online services. Also available on CD-ROM. Inquire as to cost and availability.

CSA Engineering. Cambridge Scientific Abstracts, 7200 Wisconsin Avenue, Suite 601, Bethesda, MD 20814. (301) 961-6750 or (800) 843-7751. Contains citations and abstracts of international periodicals and research literature covering all fields of engineering and science and technology,including computer and information science, electronics, mechanical engineering, solid state materials, 1981 to present. Available on BRS,(800) 289-4277, online service.Inquire as to cost and availability.

Current Contents Search. Institute for Scientific Information, 3501 Market Street, Philadelphia, PA 19104. (215) 386-0100. FAX (215) 386-6362. Contains citations to articles listed in the table of contents of science and technology journals. Also articles in social sciences and life sciences journals. Available on BRS,(800) 289-4277, DIALOG,(800) 334-2564, online services. Inquire as to cost and availability.

International Aerospace Abstracts. American Institute of Aeronautics and Astronautics, 370 L'Enfant Promenade, S.W., Wash-ington, DC 20024. (202) 646-7400. Contains references and abstracts of journal and monograph literature relating to aerospace science and technology, from 1963 to present. Available through the NASA/RECON system of the National Aeronautics and Space Administration only.

NASA Database. American Institute of Aeronautics and Astronautics, 370 L'Enfant Promenade, S.W., Washington, DC 20024. (202) 646-7400. Citations and abstracts of aeronautics and astronautics literature, 1962 to present. Also contains citations from STAR, SCIENTIFIC and TECHNICAL AEROSPACE REPORTS, and INTERNATIONAL AEROSPACE ABSTRACTS. Available through NASA/RECON online service. Inquire as to cost and availability.

NTIS Bibliographic Database. National Technical Information Service, 5285 Port Royal Road, Springfield, VA 22161. (703) 487-4929 or FAX (703) 321-8199. Broad coverage of government-sponsored science and technology research reports, 1964 to present. Available on BRS,(800) 289-4277, DIALOG, (800) 334-2564, ORBIT, (800) 456-7248, and STN International, FIZ Karlsruhe, P.O. Box 2465, W-7500, Karlsruhe 1, Germany, online services. Also available on CD-ROM. Inquire as to cost and availability.

SAE Global Mobility Database. Society of Automative Engineers (SAE), Electronic Publishing Division, 400 Commonwealth Drive, Warrendale, PA 15098. (412) 776-4841. Contains citations with abstracts to technical papers on automotive and aerospace technology and vehicular-related industries that have been presented at SAE conferences. Covers 1906 to present. Available on ORBIT,(800) 456-7248,online service. Inquire as to cost and availability.

SCISEARCH. Institute for Scientific Information, 3501 Market Street, Philadelphia, PA 19104. (800) 523-1850 or (215) 386-0100. Broad multidisciplinary title and author index to the international literature of science and technology, 1974 to present. Available on DIALOG,(800) 334-2564, and ORBIT,(800) 456-7248, online services. Also available on CD-ROM. Inquire as to cost and availability.

Tris (transportation Research Information). National ACADemy of Science, 2101 Constitution Avenue, N.W., Washington, DC 20418. (202) 334-3313 or (800) 624-6242. Citations with abstracts of literature on transportation, including air, highway, rail, maritime and other modes from 1968 to present. Available on DIALOG,(800) 334-2564, online service. Inquire as to cost and availability.

WILSONLINE. H.W. Wilson Company, 950 University Avenue, Bronx, NY 10452. (800) 367-6770 or (212) 588-8400. Makes available online versions of the printed H.W. Wilson indexes including Applied Science and Technology Index, Business Periodicals Index, General Science Index, and Readers' Guide to Periodical Literature. Period covered is generally 1983 to present. Available on BRS,(800) 289-4277, DIALOG,(800) 334-2564, and OCLC EPIC,(800) 848-5878, online services. Also available on CD-ROM. Inquire as to cost and availability.

PERIODICALS

Aeronautical Journal. The Royal Aeronautical Society, 4 Hamilton Place, London W1V 0BQ, England. 1897 to present. Monthly except June and August. Inquire for price information in U.S.# #*Aerospace America.* American Institute of Aeronautics and Astronautics, 370 L'Enfant Promenade SW, Washington, DC 20024-2518. (202) 646-7471. 1932 to present. Monthly. $70 annually (non-members).

Aerospace Daily. McGraw-Hill Inc., Aviation Week Group, 1200 G Street NW, Suite 200, Washington, DC 20005. (202) 383-2350. Five times a week. $1340.00 per year.

Aerospace Engineering. Society of Automotive Engineers, 400 Commonwealth Dr., Warrendale, PA 15096. (412) 776-4841. 1981 to present. Monthly. $48.00 per year.

Aerospace Propulsion. McGraw-Hill Inc., Aviation Week Group, 1200 G Street NW, Suite 200, Washington, DC 20005. (202) 383-2350. 1963 to present. 25 times a year. $550.00 a year.

AIAA Journal. American Institute of Aeronautics and Astronautics, 370 L'Enfant Promenade SW, Washington, DC 20024-2518. (202) 646-7471. 1962 to present. Monthly. $52.00 per year (members) and $435.00 per year (non-members).

Air Progress. Challenge Publications, Inc., 7950 Deering Ave., Canoga Park, CA 91304. (818) 887-0550. FAX (818) 884-1343. Monthly. $26.50 per year.

Aircraft Engineering and Aerospace Technology. BunHill Publications Ltd., 127 Stanstead Rd., London SE23 1JE, England. 1929 to present. Monthly. $102.00 per year (U.S.).

Aviation Week and Space Technology. 1221 Avenue of the Americas, New York, NY 10020. (212) 512-2999 or (800) 525-5003. Weekly. $82.00 per year.

Canadian Aeronautics and Space Journal. Canadian Aeronautics and Space Institute, Suite 818, the National Building, 130 Slater St., Ottawa, Canada K1P 6E2. 1955 to present. Monthly. $75.00 per year (U.S.).

Flying. Hachette Filipacchi Magazines, Inc., 1633 Broadway, New York, NY 10019. (203) 622-2700. FAX (203) 622-2725. Monthly. $24.00 per year.

IEEE Transactions On Aerospace and Electronic Systems. Institute of Electrical and Electronics Engineers, 445 Hoes Lane, PO Box 1331, Piscataway, NJ 08855-1331. (908) 981-0060. Quarterly. $10.00 per issue (members), $20.00 per issue (non-members).

Interavia/aerospace World. Aerospace Media Publishing, Swissair Centre, 31 Route de l'Aeroport, POB Box 437, 1215 Geneva 15, Switzerland. (022) 788 27 88. FAX (022) 788 27 26. Monthly. $128.00 per year (U.S.).

International Journal of Turbo and Jet Engines. Freund Publishing House, Ltd., PO Box 35010, Tel Aviv 63150, Israel. 972-2-5662925. FX 972-3-5605335. 1983 to present. Quarterly. $190.00 per year.

Journal of Aircraft. American Institute of Aeronautics and Astronautics, 370 L'Enfant Promenade SW, Washington, DC 20024-2518. 1963 to present. Bimonthly. $42.00 (members) and $230.00 (non-members).

Journal of Guidance, Control and Dynamics. American Institute of Aeronautics and Astronautics, 370 L'Enfant Promenade SW, Washington, DC 20024-2518. 1978 to present. Bimonthly. $42.00 (members) and $250.00 (non-members).

Journal of Propulsion and Power. American Institute of Aeronautics and Astronautics, 370 L'Enfant Promenade SW, Washington, DC 20024-2518. 1985 to present. Bimonthly. $38.00 (members) and $260.00 (non-members).

Progress in Aerospace Sciences. Pergamon Press, 660 White Plains Rd., Tarrytown, NY 10591-5153. 1961 to present. Quarterly. $320.00 per year.

RESEARCH CENTERS AND INSTITUTES

Aerospace Research Applications Center, Indianapolis Center For Advanced Research, 611 N. Capitol Ave., Indianapolis, in 46204. (317) 262-5003. FAX (317) 262-5044.

Aerospace Research Center, 1250 Eye St., NW, Washington, DC 20005. (202) 371-8400. FAX (202) 371-8470.

Aviation Safety Institute, 6797 N. High St., Suite 316, PO Box 304, Worthington, OH 43085. (614) 885-4242. FAX (614) 885-5891.

Center For Aerospace Technology, Weber State College, Ogden, UT 84408-1805. (801) 626-7272.

Center For Control and Systems Research, University of Texas At Austin, Aerospace Engineering and Engineering Mechanics, Wrw 408, Austin, TX 78712-1085. (512) 471-4908. FAX (512) 471-3788.

Flight Research Laboratory, University of Kansas, Raymond Nichols Hall, Lawrence, KS 66045. (913) 864-3043. FAX (913) 864-7789.

Joint Institute For Advancement of Flight Sciences, George Washington University, Langley Research Center, Mail Stop 269, Hampton, VA 23665. (804) 864-1982. FAX (804) 864-5894.

Machinery and Engine Technology Laboratory, Institute For Mechanical Engineering, Bldg. M-7, Montreal Rd., Ottawa, Ontario, Canada K1A 0R6. (613) 993-2425. (613) 957-3281.

Ohio Aerospace Institute, 2001 Aerospace Parkway, Brook Park, OH 44142. (216) 891-2100. FAX (216) 891-2140.

Pennsylvania State University, Propulsion Engineering Research Center, 106 Research Bldg. E, Bigler Rd., University Park, PA 16802. (814) 863-6272. FAX (814) 865-3389.

SURFACE CHEMISTRY

See also: CATALYSIS, CHEMICAL BONDS, CHEMISTRY, COLLOIDS, ELECTROCHEMISTRY, PHOTOCHEMISTRY, PHYSICAL CHEMISTRY, POLYMERS

ABSTRACT SERVICES AND INDEXES

Applied Science and Technology Index; A Cumulative Subject Index To English Language Periodicals in the Fields of Aeronautics and Space Science, Computer Technology, Chemistry, Construction Industry, Energy and Related Areas. H. W. Wilson Co., 950 University Avenue, Bronx, NY 10452- 9978. (212)

SURFACE CHEMISTRY

Ency. of Physical Sciences and Engineering Info. Sources

588-8400. (800) 367-6770. FAX (718) 590-1617. From 1958 to present. Monthly. Inquire about cost and availability. Also available online (BRS and WILSONLINE) and on CD-ROM. ISSN: 0003-6986.

Chemical Abstracts. Chemical Abstracts Service. 2540 Olentangy River Road, Box 3012, Columbus, OH 43210-0012. (614) 447-3600. FAX (614) 447-3713. 1907 to present. Weekly. $16,800.00 per year. Available online and on CD- ROM. CA is also available in five section groupings. Inquire regarding cost and availability.

Current Contents: Physical, Chemical and Earth Sciences. Institute for Scientific Information, 3501 Market Street, Philadelphia, PA 19104. (215) 386-0100. FAX (215) 386-2291. From 1961 to present. Weekly. $442.00 per year. Also available on CD-ROM and online. Inquire regarding cost and availability. ISSN: 0163-2574.

Engineering Index Monthly. Engineering Information, Inc., Castle Point on the Hudson, Hoboken, NJ 07030. (201) 216-8500. (800) 221-1044. FAX(201) 216-8532. Monthly. $2300.00 per year for monthly issues. $1980 per year for annual cumulation. Available on CD-ROM and online as COMPENDEX. ISSN: 0742-1974; 0360-8557.

General Science Index. H.W. Wilson Company, 950 University Avenue, Bronx, NY 10452. (212) 588-8400. (800) 367-6770. FAX (718) 590-1617. From 1978 to present. Ten issues per year; quarterly and annual cumulations. Service basis. Available on CD-ROM and online. Inquire regarding cost and availability. ISSN: 0162-1963.

Physics Abstracts. INSPEC. Section A, Science Abstracts. Institution of Electrical Engineers (IEE), London, United Kingdom. Available from: INSPEC/IEEE -Institute of Electrical and Electronic Engineers, Box 1331, 445 Hoes Lane, Piscataway, NJ 08855-1331. (908) 562-5549. 1898 to present. 24 issues per year. $2835.00 per year. ISSN: 0036-8091. Also available online and on CD-ROM.

Science Citation Index. SCI. Institute for Scientific Information, 3501 Market Street, Philadelphia, PA 19104. (215) 386-0100. (800) 523-1850. FAX (215) 386-2991. 1961 to present. Six issues per year, plus annual cumulation. $11650.00 per year. Also available online and on CD-ROM. Inquire about price and availability. ISSN: 0036-827X.

ANNUAL REVIEWS AND YEARBOOKS

Annual Review of Physical Chemistry. Annual Reviews, Inc., 4139 El Camino Way, Palo Alto, CA 94306-0897. (415) 493-4400. Fax (415) 855-9815. 1951 to present. Annual. $55.00 per year.

ASSOCIATIONS AND PROFESSIONAL SOCIETIES

American Chemical Society, 1155 16th Street, Nw, Washington, DC 20036. (202) 872-4600.

American Institute of Chemists. 501 Wythe Street, Alexandria, VA 22314-1917. (703) 836-2090. FAX (703) 836- 2091.

Association of Consulting Chemists and Chemical Engineers, 50 East 41st Street, Suite 92, New York, NY 10017. (212) 684-6255.

Division of Physical Chemistry (a Division of the American Chemical Society). Iowa State University, Department of Chemistry, Wilhelm Hall, Room 303, Ames, IA 50011-4111. (515) 294-4111.

Electrochemical Society. 10 South Main Street, Pennington, NJ 08534. (609) 737-1902.

BIBLIOGRAPHIES

Chemical Information. H. R. Collier, editor. Springer- Verlag New York, Inc., 175 Fifth Avenue, New York, NY 10010. (212) 460-1500. (800) 777-4643. 1990. $91.00.

Chemical Information Sources. Gary Wiggins. McGraw-Hill Publishing Company, 1221 Avenue of the Americas, New York, NY 10020. (800) 262-4729 or (212) 512-3825. 1991. $42.50.

Information Sources in Chemistry. R.T. Bottle and J.F.B Rowland, editors. R.R. Bowker Inc., 121 Chanlon Road, New Providence, NJ 07974. (800) 521-8110 or (908) 464-6800. Fourth edition. 1993. $75.00.

Scientific and Technical Books and Serials in Print; An Index To Literature in Science and Technology. R.R. Bowker Inc., 121 Chanlon Road, New Providence, NJ 07974. (908) 464-6800. (800) 521-8110. FAX (908) 665-3502. 1972 to present. Annual. 4 volumes. 1994. $299.95. Also available on compact disc and online. ISSN: 0000-054X.

DIRECTORIES AND BIOGRAPHICAL SOURCES

American Institute of Chemists - Professional Directory. American Institute of Chemists, 501 Wythe Street, Alexandria, VA 22314-1917. (703) 836-2090. FAX (703) 836-2091. Annual. $65.00.

American Men and Women of Science. Physical and Biological Sciences. R.R. Bowker Inc., 121 Chanlon Road, New Providence, NJ 07974. (908) 464-6800. (800) 521-8110. 20th edition. 8 volumes. 1996. $850.00

Chemical Sources International. Mike Desing and Kurt Gandenberger, editors. Chemical Sources International, PO Box 1824, Clemson, SC 29633. (803) 646-7840. 1994.$285.00.

Chemical Week — Buyers Guide Issue. Chemical Week Associates, 888 Seventh Avenue, New York, NY 10106. (212) 621-4900. FAX (212) 621-4949. Annual, October. $50.00.

Directory of Chemistry Software 1992. Wendy Warr, Peter Willett, Geoff Downs. American Chemical Society, 1155 16th Street, NW, Washington, DC 20036. (202) 872-4600. 1992. $35.95.

Research Centers Directory. Gale Research Company,Inc., 835 Penobscot Building, Detroit, MI 48226-4094. (313) 961- 2242. (800) 877-4253. 20th edition. 1995. $485.00. ISSN: 0080-1518.

ENCYCLOPEDIAS AND DICTIONARIES

Academic Press Dictionary of Science and Technology. Christopher Morris, editor. Academic Press, Inc., 1250 Sixth Avenue, San Diego, CA 92101. (619) 231-0926. FAX (619) 699-6715. 1991. $115.00.

Concise Encyclopedia Chemistry. Translated and revised by Mary Eagleson. Walter de Gruyter, Inc., 200 Saw Mill River Road, Hawthorne, New York, 10532. (914) 747-0110 or Fax (914) 747-1326. 1994. $69.95.

Dictionary of Chemical Names and Synomyms. Philip A. Howard and Michael W. Neil. Lewis Pub.,/ CRC Press, Inc., 2000 Corporate Boulevard, NW, Boca Raton, FL 33431. (407) 994- 0555. (800) 272-7737. 1991. $149.95.

Dictionary of Physical Chemistry. Satish Anand and Raj Kumar. South Asia Publications, PO Box 502 Collumbus, MO 65205. (314) 474-0116. 1990. $23.50.

Encyclopedia of Applied Physics. VCH Publishers, Inc., 303 Northwest 12th Avenue, Deerfield Beach, FL 33442. (800) 367-8249. 1991-. Twenty volumes. $6000.00.

Encyclopedia of Chemical Processing and Design. McKetta. Dekker, Inc., 270 Madison Avenue, New York, NY 10016. (212) 696-9000. (800) 228-1160. 1976 - . $175.00 per volume. ISSN: .

Grant & Hawks Chemical Dictionary. R. A. Grant. McGraw-Hill Publishing Company, Inc., 1221 Avenue of the Americas, New York, NY 10020. (212) 512-2000. (800) 262-4729. 5th edition. 1987. $64.50.

Hawley's Condensed Chemical Dictionary. Richard J. Lewis. Van Nostrand Reinhold, 115 Fifth Avenue, New York, NY 10003. (212) 254-3232. (800) 842-3636. 12th edition. 1993. $69.95.

Kirk-Othmer Encyclopedia of Chemical Technology. John Wiley and Sons, Inc., 605 Third Avenue, New York, NY 10158. (800) 526-5368 or (212) 850-6000. Fourth edition. 1991-. Twenty-seven volumes. $5400.00.

Language of Colloid and Interface Science: A Dictionary of Terms. Laurier L. Schramm. American Chemical Society, 1155 16th Street, NW, Washington, DC 20036. (202) 872-4600. . 1993. $39.95 in paper.

McGraw-Hill Encyclopedia of Chemistry. Sybil P.Parker, editor. McGraw-Hill Book, Incorporated, 1221 Avenue of the Americas, New York, NY 10020. (212) 997-3675. Second edition. 1993. $95.50.

McGraw-Hill Encyclopedia of Science and Technology. McGraw-Hill Book, Incorporated, 1221 Avenue of the Americas, New York, NY 10020. (212) 997-3675. Seventeenth edition. Twenty volumes. 1992. $1900.00.

GENERAL WORKS

Elementary Physicochemical Processes On Solid Surfaces. V. P. Zhdanov. Plenum Publishing Corp., 233 Spring Street, New York, NY. (212) 620-8000. (800) 221-9369. FAX (212) 463-0742. 1992. $95.00

Introduction To Colloid and Surface Chemistry. Duncan J. Shaw. Butterworth-Heinemann, 313 Washington Street, Newton, MA 02158. (618-928-2500). (800) 366-2665. 4th edition. 1992. $31.95 in paper.

Introduction To Surface Chemistry and Catalysis. Gabor J. Somorjai. John Wiley & Sons, Inc., 605 Third Avenue, New

York, NY 10158-0012. (212) 850-6000. (800) 225-5945.. 1994. $59.95.

Introduction To Surface Physics. M. Prutton. Oxford University Press, Inc., 200 Madison Avenue, New York, NY 10016. (212) 725-6000. (800) 334-4249. 1994. $24.95 in paper.

Modern Techniques of Surface Science. D. P. Woodruff and T. A. Delchar. Cambridge University Press, 40 West 20th Street, New York, NY 10011-4211. (212) 924-3900. (800) 872-7423. 2nd edition. 1994. $89.95.

Physical Chemistry of Surfaces. Thomas A. Adamson. John Wiley & Sons, Inc., 605 Third Avenue, New York, NY 10158-0012. (212) 850-6000. (800) 225-5945. 1990. $79.95.

Semiconductor Surface and Interfaces. Winfried Monch. Springer-Verlag New York, Inc., 175 Fifth Avenue, New York, NY 10010. (212) 460-1500. (800) 777-4643. FAX (212) 473-6272. 1993. $79.00.

Statistical thermodynamics of Surfaces, Interfaces and Membranes. Samuel A. Safran. Addison-Wesley Publishing Co., Inc., 1 Jacob Way, Reading, MA 01867. (617) 944-3700. (800) 447-2226. 1994. $55.95.

Structure of Electrified Interfaces. Jacek Lipowski and Philip N. Ross. VCH Pubications, Inc., 220 East 23rd Street, Suite 909, New York, NY 10010-4606. (212) 683-8333. (800) 422-8824. 1993. $145.00.

Surface and Colloid Chemistry in Natural Waters and Water Treatment. R. Beckett, editor. Plenum Publishing Corp., 233 Spring Street, New York, NY. (212) 620-8000. (800) 221-9369. FAX (212) 463-0742. 1990. $65.00.'

Surface Electron Transfer Processes. Dwayne Miller. VCH Publications, Inc., 220 East 23rd Street, Suite 909, New York, NY 10010-4606. (212) 683-8333. (800) 422-8824. 1995. wfi.

Surface Engineering: Processes and Applications. Ian Sare, et al, editors. Technomic, 851 New Holland Avenue, Box 3535, Lancaster, PA 17604. (717) 291-5609. (800) 233-9936.1994. $79.00 in paper.

Surface Science Techniques. Walls. Elsevier Science Publishing Company, Inc., 655 Avenue of the Americas, New York, NY 10010. (212) 989-5800. FAX (914) 333-2444. 1994. $62.00.

HANDBOOKS AND MANUALS

Chemical Formulary. H. Bennett, editor. Chemical Publishing Company, Inc., 80 Eighth Avenue, New York, NY 10011. (212) 255-1950. Volumes 1-30. $70.00 per volume.

Chemometrics: Chemical and Sensory Data. David R. Burgard and James T. Kuznicki. CRC Publishers, Inc., 2000 Corporate Blvd., N.W., Boca Raton, FL 33431. (407) 994-0555 or (800) 333-8300. 1990. $120.00.

Chemical Reference Handbook. Gordon Press, PO Box 459, Bowling Green Station, New York, NY 10004. (718) 624-8419. 1995. $260.00.

CRC Handbook of Chemistry and Physics. David R. Kide, editor. CRC Press Inc., 2000 Corporate Blvd., NW, Boca Raton,

FL 33431. (407) 994-0555. (800) 333-8300. 77th edition. 1996. $99.95.

Guide To Basic Chemical Compounds. D.R. Lide, Jr. CRC Publishers, Inc., 2000 Corporate Blvd., N.W., Boca Raton, FL 33431. (407) 994-0555. (800) 333-8300. 1993. $120.00.

Handbook of Surface Imaging and Visualization. Arthur T. Hubbard. CRC Press, Inc., 2000 Corporate Boulevard, NW, Boca Raton, FL 33431. (407) 994-0555. (800) 272-7737. . 1995. $159.95

Improving Safety in the Chemical Laboratory: A Practical Guide. Jay A. Young, editor. John Wiley and Sons, Inc., 605 Third Avenue, New York, NY 10158. (800) 526-5368 or (212) 850-6000. Second edition. 1991. $75.00.

Lange's Handbook of Chemistry. John A. Dean, editor. McGraw-Hill Publishing Company, 1221 Avenue of the Americas, New York, NY 10020. (800) 262-4729 or (212) 512-3825. Fourteenth edition. 1992. $79.50.

Riegel's Handbook of Industrial Chemistry. James A. Kent, editor. Van Nostrand Reinhold, 115 Fifth Avenue, New York, NY 10003. (212) 254-3232 or (800) 926-2665. Ninth edition. 1992. $114.95.

Surface Crystallographic Information Service: A Handbook of Surface Structures. J. M. Maclaren. Kluwer Academic Publishers, 101 Philip Drive, Assinippi Park, Norwell, MA 02061. (617) 871-6600. 1987. $80.50 in paper.

ONLINE DATABASES AND CD-ROMS

Analytical Abstracts Online. Royal Society of Chemistry, Information Services, Thomas Graham House, Science Park, Milton Road, Cambridge, CB4 4WF, England. Contains citations, mostly with abstracts, of the worldwide literature on analytical chemistry, from 1980 to present. Available on DIALOG, (800) 334-2564, and STN International, FIZ Karlsruhe, P.O. Box 2465, W-7500, Karlsruhe 1, Germany, online services. Inquire as to cost and availability.

CA Search. Chemical Abstracts Service, P.O. Box 3012, Columbus, OH 43210-0012. (614) 447-3600. (800) 848-6533. FAX (614) 447-3709. Very comprehensive guide to worldwide chemical literature and related fields, 1972 to present. Available on BRS,(800) 289-4277, DIALOG, (800) 334-2564, ORBIT (800) 456-7248, and STN International, FIZ Karlsruhe, P.O. Box 2465, W-7500, Karlsruhe 1, Germany, online services. Inquire as to cost and availability.

Chemical Journals of the American Chemical Society. American Chemical Society, 1155 16th Street, N.W., Washington, DC 20036. (202) 872-4381 or (800) 424-6747. Contains complete text of approximately 90,000 articles from 22 primary journals published by the American Chemical Society, from mostly 1982 to present. Available on STN International, FIZ Karlsruhe, P.O. Box 2465, W-7500, Karlsruhe 1, Germany, online service. Inquire as to cost and availability.

Compendex Plus. Engineering Information, Inc., 345 East 47th Street, New York, NY 10017. (212) 705-7600 or (800) 221-1044. Contains citations with abstracts to worldwide literature in engineering and technology, from 1970 to present. Available on online BRS,(800) 289-4277, DIALOG, (800) 334-2564, ORBIT (800) 456-7248, and STN International, FIZ Karlsruhe,

P.O. Box 2465, W-7500, Karlsruhe 1, Germany, online services. Also available on CD-ROM. Inquire as to cost and availability.

Current Contents Search. Institute for Scientific Information, 3501 Market Street, Philadelphia, PA 19104. (215) 386-0100. FAX (215) 386-6362. Contains citations to articles listed in the table of contents of science and technology journals. Also articles in social sciences and life sciences journals. Available on BRS,(800) 289-4277, DIALOG,(800) 334-2564, online services. Inquire as to cost and availability.

Dissertation Abstracts. University Microfilms International, 300 North Zeeb Road, Ann Arbor, MI 48106. (800) 521-0600 or (313) 761-4700. Scope includes virtually all doctoral dissertations accepted at accredited American institutions from 1861 to present in 252 subject areas. Available on BRS,(800) 289-4277, DIALOG,(800) 334-2564, and OCLC EPIC,(800) 848-5878, online services. Also available on CD-ROM. Inquire as to cost and availability.

Gmelin Database. Gmelin-Institut fur Anorganische Chemie und Grenzgebiete, Varrentrapperstrasse, 40-42, Carl-Bosch-Haus, D-6000, Frankfurt am Main 90, Germany. Contains structural and factual data relating to inorganic and organometallic chemistry. Provides data from the Gmelin Handbook of Inorganic and Organometallic Chemistry. Covers the period 1817 to 1975; 1988-89. Available on STN International, FIZ Karlsruhe, P.O. Box 2465, W-7500, Karlsruhe 1, Germany, online service. Inquire as to cost and availability.

Kirk-Othmer Encyclopedia of Chemical Technology. John Wiley and Sons, Inc., 605 Third Avenue, New York, NY 10158. (800) 526-5368 or (212) 850-6000. Contains the complete text of all chapters in the 27 volume fourth edition of the KIRK-Othmer ENCYCLOPEDIA of CHEMICAL TECHNOLOGY. 1991. Available on BRS,(800) 289-4277, DIALOG,(800) 334-2564, online services. Inquire as to cost and availability.

NTIS Bibliographic Database. National Technical Information Service, 5285 Port Royal Road, Springfield, VA 22161. (703) 487-4929 or FAX (703) 321-8199. Broad coverage of government-sponsored science and technology research reports, 1964 to present. Available on BRS,(800) 289-4277, DIALOG, (800) 334-2564, ORBIT, (800) 456-7248, and STN International, FIZ Karlsruhe, P.O. Box 2465, W-7500, Karlsruhe 1, Germany, online services. Also available on CD-ROM. Inquire as to cost and availability.

SCISEARCH. Institute for Scientific Information, 3501 Market Street, Philadelphia, PA 19104. (800) 523-1850 or (215) 386-0100. Broad multidisciplinary title and author index to the international literature of science and technology, 1974 to present. Available on DIALOG,(800) 334-2564, and ORBIT,(800) 456-7248, online services. Also available on CD-ROM. Inquire as to cost and availability.

WILSONLINE. H.W. Wilson Company, 950 University Avenue, Bronx, NY 10452. (800) 367-6770 or (212) 588-8400. Makes available online versions of the printed H.W. Wilson indexes including Applied Science and Technology Index, Business Periodicals Index, General Science Index, and Readers' Guide to Periodical Literature. Period covered is generally 1983 to present. Available on BRS,(800) 289-4277, DIALOG,(800) 334-2564, and OCLC EPIC,(800) 848-5878, online services. Also available on CD-ROM. Inquire as to cost and availability.

PERIODICALS

Chemical Physics. Elsevier Science Publishing Co., Inc., Box 882, Madison Square Station, New York, NY 10159. (212) 989-5800. FAX (212) 633-3990. 1973 to present. 33 issues per year in 11 volumes. $2759.00 per year. ISSN: 0301-0104.

Chemical Physics Letters. Elsevier Science Inc., Publishing Co., Inc., Box 882, Madison Square Station, New York, NY 10159. (212) 989-5800. (212) 633-3990. 1967 to present. 102 issues per year in 17 volumes. $5063.00 per year. ISSN: 0009-2614.

Chemical Reviews. American Chemical Society, Box 3337, Columbus, OH 43210. (614) 447-3776. 1924 to present. 8 issues per year. $346.00 per year. ISSN: 0009-2665.

Chemical Week; Includes Annual Buyers Guide. Chemical Week Associates, 888 Seventh Avenue, New York, NY 10106. (212) 621-4900. FAX (212) 621-4949. 1914 to present. Weekly. $99.00 per year. ISSN: 0009-272X.

Chemtech. American Chemical Society. Box 3337, Columbus, OH 43210. (614) 447-3776. 1970 to present. Monthly. $370.00 per year. ISSN: 0009-2703.

Colloids and Surfaces. Elsevier Science Inc., Publishing Co., Inc., Box 882, Madison Square Station, New York, NY 10159. (212) 989-5800. (212) 633-3990. 1980 to present. 36 issues per year in 12 volumes. $2364.00 per year. ISSN: 0927-7757.

Inorganic Chemistry. American Chemical Society, Box 3337, Columbus, OH 43210. (614) 447-3776. 1962 to present. Semimonthly. $500.00 per year. ISSN: 0020-1669.

International Journal of Chemical Kinetics. John Wiley and Sons, Inc., 605 Third Avenue, New York, NY 10158. (800) 526-5368 or (212) 850-6000. 1968 to present. Monthly. $730.00 per year. ISSN: 0538- 8066.

Journal of Catalysis. Academic Press, Inc., Box 620000, Orlando, FL 32891-8340. (800) 543-9534. 1962 to present. Monthly. $1296.00 per year. ISSN: 0021-9517.

Journal of Colloid and Interface Science. Academic Press, Inc., Box 620000, Orlando, FL 32891-8340. (800) 543-9534. 1946 to present. 14 issues per year. $1302.00 per year. ISSN: 0021-9797.

Journal of Physical Chemistry. American Chemical Society, Box 3337, Columbus, OH 43210. (614) 447-3776. (800) 333- 9511. FAX (614) 447-3671. 1896 to present. Weekly. $1140.00 per year to non-members. ISSN: 0894-3230.

Journal of Solid State Chemistry. Academic Press, Inc., Box 620000, Orlando, FL 32891-8340. 1969 to present. Monthly. $1248.00 per year. ISSN: 0022-4596. # #*Journal of the American Chemical Society.* American Chemical Society, Box 3337, Columbus, OH 43210. (614) 447-3776. 1879 topresent. Bi-weekly. $1055.00 per year. ISSN: 0002-7863.

Surface and Coatings Technology. Elsevier Science Inc., Publishing Co., Inc., Box 882, Madison Square Station, New York, NY 10159. (212) 989- 5800. (212) 633-3990. 1972 to present. 21 issues per year. $1810.00 per year. ISSN: 0257-8972.

RESEARCH CENTERS AND INSTITUTES

Institute For theoretical Chemistry. University of Texas at Austin, Department of Chemistry, WEL 3.237, Austin, TX 78712. (512) 471-3114.

Institute of Colloid and Surface Science. Clarkson University. Potsdam, NY 13699-5660. (315) 268-3820. FAX (315) 268-3841.

Quantum Institute. Rice University. PO Box 1892, Houston, TX 77251. (713) 527-6028.

SURVEYING

See also: AERIAL PHOTOGRAPHY, CARTOGRAPHY, GEODESY, PHOTOGRAMMETRY, REMOTE SENSING, SATELLITES— ARTIFICIAL, TOPOGRAPHIC MAPPING

ABSTRACT SERVICES AND INDEXES

Applied Science and Technology Index; A Cumulative Subject Index To English Language Periodicals in the Fields of Aeronautics and Space Science, Computer Technology, Chemistry, Construction Industry, Energy and Related Areas. H.W. Wilson Co., 950 University Avenue, Bronx, NY 10452. (800) 367-6770 or (212) 588-8400. FAX (718) 590-1617. From 1958 to present. Monthly. Inquire about cost and availability. Also available on CD-ROM and online.

ASCE Combined Annual Combined Index. American Society of Civil Engineers, 345 East 47th Street, New York, NY 10017-2398. (212) 705-7520 or (800) 548-2723. Annual. $48.00.

ASCE Publications Information. 1966-present. American Society of Civil Engineers, 345 East 47th Street, New York, NY 10017-2398. (212) 705-7520 or (800) 548-2723. Bi-monthly. $160.00 per year for non-members.

Bibliography and Index of Geology. American Geological Institute, 4220 King Street, Alexandria, VA 22302. (703) 379-2480. Fax (703) 379-7563. Monthly. $1295.00 per year. Also available online as GeoRef.

Civil and Structural Engineering Abstracts, Cambridge Scientific Abstracts, 7200 Wisconsin Avenue, Bethesda, MD 20814-4823. (301) 961-6750. FAX (301) 961-6720. 1993 to present. Monthly. $385.00 per year. Topics covered include structural design, construction equipment and methods, civil defense and military engineering, surveying, highway engineering, maritime and port structures, materials, land reclamation, and soil mechanics.

Current Contents: Engineering, Technology, and Applied Sciences. Institute for Scientific Information, 3501 Market Street, Philadelphia, PA 19104. (215) 386-0100. FAX (215) 386-6362. 1970 to present. Weekly. $442.00 per year.

Engineering Index Monthly. Engineering Information, Inc., Castle Point on the Hudson, Hoboken, NJ 07030. (800) 221-1044. FAX (212) 832-1857. Monthly. $2300.00 per year. Also available online as COMPENDEX, and also on CD-ROM. Covers chemical engineering, computers, electrical engineering, civil engineering, metals and mining, industrial management, and mechanical engineering.

Meteorological and Geoastrophysical Abstracts. American Meteorological Society, c/o Inforonics Inc., 550 Newtown Road, Littleton, MA 01460. (508) 486-8976. FAX (508) 486-0027. 1950 to present. Monthly. $950.00.

Remote Sensing of Earth Resources. University of New Mexico Earth Data Analysis Center, 2500 Yale Blvd., Suite 100, Albuquerque, NM 87131-6031. (505) 277-3622. FAX (505) 277-3614. 1974 to present (1991 to present under current title). Quarterly. $175.00 per year.

ASSOCIATIONS AND PROFESSIONAL SOCIETIES

American Association For Geodetic Surveying. 5410 Grosvenor Lane, Suite 100, Bethesda, MD 20814. (301) 493-0200.

American Cartographic Association. 5410 Grosvenor Lane, Suite 100, Bethesda, MD 20814. (301) 493-0200.

American Congress On Surveying and Mapping. 5410 Grosvenor Lane, Suite 100, Bethesda, MD 20814. (301) 493-0200.

American Society of Civil Engineers. 345 East 47th Street, New York, NY 10017-2398. (212) 705-7520 or (800) 548-2723.

American Society For Photogrammetry and Remote Sensing. 5410 Grosvenor Lane, Suite 210, Bethesda, MD 20814.

The International Society For Optical Engineering (spie), Po Box 10, 1000-20th Street, Bellingham, WA 98227-0010.

National Society of Professional Surveyors. 5410 Grosvenor Lane, Suite 100, Bethesda, MD 20814. (301) 493-0200.

North American Cartographic Information Society. PO Box 399, Milwaukee, WI 53201. (414) 229-6282.

Optical Society of America, 2010 Massachusetts Avenue Nw, Washington, DC 20036. (202) 223-8130.

Society of Photographic Scientists and Engineers. 7003 Kilworth Lane, Springfield, VA 22151. (703) 642-9090.

DIRECTORIES AND BIOGRAPHICAL SOURCES

Directory of Engineering Societies and Related Organizations. Gordon Davis. 13th edition. American Association of Engineering Societies, 1111 19th Street NW, Suite 608, Washington, DC 20036. (202) 296-2237 or (800) 658-8897. 1989. Inquire for price.

International Research Centers Directory. Gale Research, Inc., 835 Penobscot Building, Detroit, MI 48226-4094. (313) 961-2242. (800) 877-4253. FAX (313) 961-6083. 8th edition. 1995. $410.00

Research Centers Directory. Gale Research, Inc., 835 Penobscot Building, Detroit, MI 48226-4094. (313) 961-2242. (800) 877-4253. FAX (313) 961-6083. $485.00.

Scientific and Technical Organizations and Agencies Directory. Gale Research, Inc., 835 Penobscot Building, Detroit, MI 48226-4094. (313) 961-2242. (800) 877-4253. FAX (313) 961-6083. 4th edition. 1996. $195.00.

Who's Who in Engineering. American Association of Engineering Societies, 1111 19th Street NW, Suite 608, Washington, DC 20036. (202) 296-2237 or (800) 658-8897. 8th edition. 1991. Inquire for price.

GENERAL WORKS

Navigational & Surveying Instruments: Industry and Trade Summary. Sundar A. Shetty. Diane Publishing Company, 600 Upland Avenue, Upland, PA 19015. (215) 499-7415. FAX (610) 499-7429. 1994. $40.00.

Principles of Surveying. Charles A. Herubin. 4th ed. Prentice Hall, 113 Sylvan Avenue, Route 9W, Englewood Cliffs, NJ 07632. (201) 592-2000 or (800) 922-0579. 1991. $51.00.

Surveying Instruments and their Operational Principles. Lajos Fialovszky, ed. Elsevier Science Publishing Company, Inc., 655 Avenue of the Americas, New York, NY 10010. (212) 989-5800. 1991. $160.00.

Surveying: Principles and Applications. Barry F. Kavanagh and S.J. Glenn Bird. 3d ed. Prentice Hall, 113 Sylvan Avenue, Route 9W, Englewood Cliffs, NJ 07632. (201) 592-2000 or (800) 922-0579. 1992. $56.00.

The Surveyor's Pocketbook. John Clancy & John Dunn. Chapman & Hall, 115 Fifth Avenue, New York, NY 10211-0906. 1994. $20.95.

HANDBOOKS AND MANUALS

Manual of Photogrammetry. Chester C. Slama, ed. 4th edition. American Society for Photogrammetry and Remote Sensing, 5410 Grosvenor Lane, Suite 210, Bethesda, MD 20814. (301) 493-0290. FAX (301) 493-0208. 1980. Inquire for price and availability.

Manual of Remote Sensing. Robert N. Colwell, ed. 2nd edition. American Society for Photogrammetry and Remote Sensing, 5410 Grosvenor Lane, Suite 210, Bethesda, MD 20814. (301) 493-0290. FAX (301) 493-0208. 1983. Inquire for price and availability.

The Surveying Handbook. Russell C. Brinker & Roy Minnick. 2nd edition. Chapman & Hall, 115 Fifth Avenue, New York, NY 10211-0906. 1995. $99.95.

Surveying Ready Reference Manual. G.O. Stenstrom. McGraw-Hill Publishing Company, 1221 Avenue of the Americas, New York, NY 10020. (800) 262-4729 or (212) 512-3825. 1987. $38.50.

ONLINE DATABASES AND CD-ROMS

Compendex Plus. Engineering Information, Inc., 345 East 47th Street, New York, NY 10017. (212) 705-7600 or (800) 221-1044. Contains citations with abstracts to worldwide literature in engineering and technology, from 1970 to present. Available on online BRS,(800) 289-4277, DIALOG, (800) 334-2564, ORBIT (800) 456-7248, and STN International, FIZ Karlsruhe, P.O. Box 2465, W-7500, Karlsruhe 1, Germany, online services. Also available on CD-ROM. Inquire as to cost and availability.

Georef: Bibliography and Index of Geology. American Geological Institute, 4220 King Street, Alexandria, VA 22302. (703) 379-2480. Fax (703) 379-7563. Monthly. Inquire for price and availability.

NTIS Bibliographic Database. National Technical Information Service, 5285 Port Royal Road, Springfield, VA 22161. (703) 487-4929 or FAX (703) 321-8199. Broad coverage of government-sponsored science and technology research reports, 1964 to present. Available on BRS,(800) 289-4277, DIALOG, (800) 334-2564, ORBIT, (800) 456-7248, and STN International, FIZ Karlsruhe, P.O. Box 2465, W-7500, Karlsruhe 1, Germany, online services. Also available on CD-ROM. Inquire as to cost and availability.

SCISEARCH. Institute for Scientific Information, 3501 Market Street, Philadelphia, PA 19104. (800) 523-1850 or (215) 386-0100. Broad multidisciplinary title and author index to the international literature of science and technology, 1974 to present. Available on DIALOG,(800) 334-2564, and ORBIT,(800) 456-7248, online services. Also available on CD-ROM. Inquire as to cost and availability.

WILSONLINE. H.W. Wilson Company, 950 University Avenue, Bronx, NY 10452. (800) 367-6770 or (212) 588-8400. Makes available online versions of the printed H.W. Wilson indexes including Applied Science and Technology Index, Business Periodicals Index, General Science Index, and Readers' Guide to Periodical Literature. Period covered is generally 1983 to present. Available on BRS,(800) 289-4277, DIALOG,(800) 334-2564, and OCLC EPIC,(800) 848-5878, online services. Also available on CD-ROM. Inquire as to cost and availability.

PERIODICALS

Acsm Bulletin. American Congress on Surveying and Mapping, 5410 Grosvenor Lane, Suite 100, Bethesda, MD 20814. (301) 493-0200. 1950 to present. Bi-monthly. $75.00 per year.

American Congress On Surveying and Mapping Proceedings. American Congress on Surveying and Mapping, 5410 Grosvenor Lane, Suite 100, Bethesda, MD 20814. (301) 493-0200. 1942 to present. Semi-annual. Price varies.

IEEE Transactions On Geoscience and Remote Sensing. IEEE, 345 E. 47th Street, New York, NY 10017. (908) 981-0060. FAX (908) 981-9667. 1963 to present. Bi-monthly. $188.00 per year.

Isprs Journal of Photogrammetry and Remote Sensing. Elsevier Science [journals], 660 White Plains Rd., Tarrytown, NY 10159-5153. (919) 524-9200. FAX (919) 333-2444. 1989-present. Bi-monthly. $225.00 per year.

Journal of Imaging Science and Technology. Society for Imaging Science and Technology, 7003 Kilworth Lane, Springfield, VA 22151. (703) 642-9090. FAX (703) 642-9094. 1950 to present. Bi-monthly. $120.00 per year.

Journal of Surveying Engineering. American Society of Civil Engineers, 345 East 47th Street, New York, NY 10017-2398. (212) 705-7520 or (800) 548-2723. 1956 to present. Quarterly. $76.00 per year.

Marine Geodesy. Taylor & Francis, 1900 Frost Road, Suite 101, Bristol, PA 19007. (215) 785-5800. FAX (215) 785-5515. 1977 to present. Quarterly. $141.00 per year.

Photogrammetric Engineering and Remote Sensing. American Society for Photogrammetry and Remote Sensing, 5410 Grosvenor Lane, Suite 210, Bethesda, MD 20814. (301) 493-0290. FAX (301) 493-0208. 1934-present. Monthly. $160.00 per year.

Photogrammetric Record. Photogrammetric Society, Dept. of Photogrammetry & Surveying, University College London, Gower St., London WC1E 6BT England. Telephone 071-387-7050. FAX 071-380-0453. 1953-present. Semi-annual. $88.00 per year.

Remote Sensing of Environment. Elsevier Science Publishing Company, Inc., 655 Avenue of the Americas, New York, NY 10010. (212) 989-5800. 1969 to present. 12 times a year. $995.00 per year.

Remote Sensing Reviews. Harwood ACADemic, c/o Gordon & Breach Science Publishers, PO Box 786, Cooper Station, New York, NY 10276. (212) 206-8900. Quarterly. Inquire for cost.

Scientific and Applied Photography and Cinematography. Gordon and Breach Science Publishers, 820 Town Center Drive, Langhorne, PA 19047. (215) 750-2642. FAX (215) 750-6343. 6 times a year. Inquire for price.

Surveying and Land Information Systems. American Congress on Surveying and Mapping, 5410 Grosvenor Lane, Suite 100, Bethesda, MD 20814. (301) 493-0200. 1941 to present. Quarterly. $85.00.

RESEARCH CENTERS AND INSTITUTES

Boston University Center For Remote Sensing and Cartography. 725 Commonwealth Avenue, Boston, MA 02215. (617) 353-5081. FAX (617) 353-3200.

Cartographic Center, Ohio University. Porter Hall, Athens, OH 45701. (614) 593-1150.

Measurement & Control Engineering Center. University of Tennessee at Knoxville, College of Engineering, 101 Perkins Hall, Knoxville, TN 37996-2000. (615) 974-2375.

Michigan State University Center For Remote Sensing. 302 Berkey Hall, East Lansing, MI 48824. (517) 353-7195. FAX (517) 353-1821.

University of Delaware Center For Remote Sensing. College of Marine Studies, Newark, DE 19711. (302) 451-2336.

University of Idaho, Idaho Remote Sensing Research Unit. Moscow, ID 83843. (208) 885-7209.

University of Minnesota Remote Sensing Laboratory. 1530 N. Cleveland Avenue, St. Paul, MN 55108. (612) 624-7764. FAX (612) 625-5212.

University of North Dakota, Institute For Remote Sensing. Geography Department, Grand Forks, ND 58202-8274. (701) 777-4246.

University of Toronto Center For Surveying Science. Erindale Campus, Mississauga Road, Mississauga, ON, Canada L5L 1C6. (416) 828-5298. FAX (416) 828-5328.

SYNTHETIC FUELS

Ency. of Physical Sciences and Engineering Info. Sources

SWITCHING THEORY

See: ELECTRICAL ENGINEERING

SYNTHETIC FUELS

See also: CHEMICAL ENGINEERING, COAL, ENERGY, FUELS, NATURAL GAS, NUCLEAR ENERGY, PETROLEUM

ABSTRACT SERVICES AND INDEXES

Applied Science and Technology Index; A Cumulative Subject Index To English Language Periodicals in the Fields of Aeronautics and Space Science, Computer Technology, Chemistry, Construction Industry, Energy and Related Areas. H. W. Wilson Co., 950 University Avenue, Bronx, NY 10452- 9978. (212) 588-8400. (800) 367-6770. FAX (718) 590-1617.From 1958 to present. Monthly. Inquire about cost and availability. Also available online (BRS and WILSONLINE) and on CD-ROM. ISSN: 0003-6986.

Chemical Abstracts. Chemical Abstracts Service. 2540 Olentangy River Road, Box 3012, Columbus, OH 43210-0012. (614) 447-3600. FAX (614) 447-3713. 1907 to present. Weekly. $16,800.00 per year. Available online and on CD- ROM. CA is also available in five section groupings. Inquire regarding cost and availability.

Current Contents: Engineering, Technology and Applied Sciences. Institute for Scientific Information, 3501 Market Street, Philadelphia, PA 19104. (215) 386-0100. FAX (215) 386-2991. From 1970 to present. Weekly. $442.00 per year. Also available online. Inquire regarding cost and availability. ISSN: 0095-7917.

Energy Research Abstracts. U. S. Department of Energy, office of Scientific and Technical Information, Box 62, Oak Ridge, TN 37831. (615) 674-0733. Subscriptions to: U. S. Government Printing office, Box 371954, Pittsburgh, PA 15250-7954. (202) 783-3238. 1976 to present. Monthly. $164.00 per year. ISSN: 0160-3604.

Engineering Index Monthly. Engineering Information, Inc., Castle Point on the Hudson, Hoboken, NJ 07030. (201) 216-8500. (800) 221-1044. FAX (201) 216-8532. Monthly. $2300.00 per year for monthly issues. $1980 per year for annual cumulation. Available on CD-ROM and online as COMPENDEX. ISSN: 0742-1974; 0360-8557.

General Science Index. H.W. Wilson Company, 950 University Avenue, Bronx, NY 10452. (212) 588-8400. (800) 367-6770. FAX (718) 590-1617. From 1978 to present. Ten issues per year; quarterly and annual cumulations. Service basis. Available on CD-ROM and online. Inquire regarding cost and availability. ISSN: 0162-1963.

N T I S Alerts: Energy. U.S. National Technical Information Service, 5285 Port Royal Road, Springfield, VA 22161. (703) 487-4929. FAX (703) 321-8547. Weekly. $160.00 per year.

Physics Abstracts. INSPEC. Section A, Science Abstracts.Institution of Electrical Engineers (IEE), London, UnitedKingdom. Available from: INSPEC/IEEE - Institute of Electrical and Electronic Engineers, Box 1331, 445 Hoes Lane,

Piscataway, NJ 08855-1331. (908) 562-5549. 1898 to present. 24 issues per year. $2835.00 per year. ISSN: 0036-8091. Also available online and on CD-ROM.

Process and Chemical Engineering/chemical Engineering Abstracts. Royal Society of Chemistry, Information Services, Thomas Graham House, Science Park, Milton Road, Cambridge, CB4 4WF, England. Contains citations, mostly with abstracts, of the worldwide literature on chemical engineering; from 1982 to present. Monthly. $610.00 per year. Also available online. ISSN: 0960-5045.

Science Citation Index. SCI. Institute for Scientific Information, 3501 Market Street, Philadelphia, PA 19104. (215) 386-0100. (800) 523-1850. FAX (215) 386-2991. 1961 to present. Six issues per year, plus annual cumulation. $11650.00 per year. Also available online and on CD-ROM. Inquire about price and availability. ISSN: 0036-827X.

ANNUAL REVIEWS AND YEARBOOKS

Advances in Chemical Engineering. Academic Press, Inc., 6277 Sea Harbor Drive, Orlando, FL 32887. (800) 321-5068. 1956 to present. Irregular. ISSN: 0065-2377.

Advances in Energy Systems and Technology. Academic Press, Inc., 6277 Sea Harbor Drive, Orlando, FL 32887. (800) 321-5068. 1979 to present. Irregular. Inquire.

Annual Review of Energy and the Environment. Annual Reviews, Inc., 4139 El Camino Way, Palo Alto, CA 94303-0139. (415) 943-4400. (800) 523-8635. 1976 to present. Annual. $71.00. ISSN: 1056-3466.

ASSOCIATIONS AND PROFESSIONAL SOCIETIES

American Chemical Society, 1155 16th Street, Nw, Washington, DC 20036. (202) 872-4600.

American Institute of Chemical Engineers. 345 East 47th Street, New York, NY 10017. (212) 705-7338.

American Institute of Physics. One Physics Ellipse, College Park, MD 20740-3843. (301) 209-3100.

Association of Energy Engineers, 4025 Pleasantdale Road, Suite 340, Atlanta, GA 30340. (404) 447-5083.

Council On Alternate Fuels. 1110 North Glebe Road, Suite 610, Arlington, VA 22201. (703) 276-6655.

IEEE (Institute of Electrical and Electronics Engineers) 345 East 47th Street, New York, NY 10017. (212) 705-7900. FAX 212) 705-4929.

BIBLIOGRAPHIES

Information Sources in Chemistry. R.T. Bottle and J.F.B Rowland, editors. R.R. Bowker Inc., 121 Chanlon Road, New Providence, NJ 07974. (800) 521-8110 or (908) 464-6800. Fourth edition. 1993. $75.00.

Information Sources in Engineering. Peter Hicks, editor. Bowker-Saur, 121 Chanlon Road, New Providence, NJ 07974. (800) 521-8110 or (908) 464-6800. 3rd edition. 1996. $125.00.

Scientific and Technical Books and Serials in Print; An Index To Literature in Science and Technology. R.R. Bowker

Inc., 121 Chanlon Road, New Providence, NJ 07974. (908) 464-6800. (800) 521-8110. FAX (908) 665-3502. 1972 to present. Annual. 4 volumes. 1994. $299.95. Also available on compact disc and online. ISSN: 0000-054X.

DIRECTORIES AND BIOGRAPHICAL SOURCES

American Institute of Chemical Engineers. Directory. American Institute of Chemical Engineers. 345 East 47th Street, New York, NY 10017-2396. (212) 705-7663. Annual.

American Men and Women of Science. Physical and Biological Sciences. Fifteenth edition. R.R. Bowker Inc., 121 Chanlon Road, New Providence, NJ 07974. (908) 464-6800. (800) 521-8110. 20th edition. 8 volumes. 1996. $850.00

Energy and Nuclear Science International Who's Who. Gale Research, Inc., 835 Penobscot Building, Detroit, MI 48226-4094. (313) 961-2242. (800) 877-4253. Triennial. 1993. $325.00.

Energy Information Centers Directory. U. S. Council for Energy Awareness, 1776 I Street, NW, Suite 400, Washington, DC 20006. (202) 293-0770. Triennial. No charge.

Engineering Research Centers: Incorporating Electronics Research Centers. Stockton Press, 345 Park Avenue, New York, NY 10010. (212) 689-9200. (800) 221-2123. 4th edition. 1995. $515.00.

Research Centers Directory. Gale Research Company, Inc., 835 Penobscot Building, Detroit, MI 48226-4094. (313) 961-2242. (800) 877-4253. 20th edition. 1995. $485.00. ISSN: 0080-1518.

Synfuels Project Directory. Pasha Publications, Inc., 1401 Wildon Boulevard, Suite 1000, Arlington, VA 22209. (703) 528-1244. 1984. Inquire.

Who's Who in Engineering. Gordon Davis, editor. American Association of Engineering Societies. 1111 19th Street, NY, Suite 608, Washington, DC 20036. (202) 296-2237. (800) 658-8897. 9th edition. 1995. $220.00.

Who's Who in Technology; Biographies and Index. Amy L. Unterburger, editor. Gale Research Company, 835 Penobscot Building, Detroit, MI 48226-4094. (313) 961-2242. (800) 521-4253. 7th edition. 1995. $195.00. ISN 0-8103-7467-6.

World Energy and Nuclear Directory. Gale Research, 835 Penobscot Building, Detroit, MI 48226-4094. (313) 961-2242. (800) 877-4253. Biennial, even years. $450.00.

ENCYCLOPEDIAS AND DICTIONARIES

Dictionary of Named Processes in Chemical Technology. Alan E. Comyns. Oxford University Press, Inc., 200 Madison Avenue, New York, NY 10016. (212) 725-6000. (800) 334-4249. 1994. $75.00.

Encyclopedia of Applied Physics. VCH Publishers, Inc., 303 Northwest 12th Avenue, Decrfield Beach, FL 33442. (800) 367-8249. 1991-. Twenty volumes. $6000.00.

Encyclopedia of Chemical Processing and Design. McKetta Marcel Dekker, Inc., 270 Madison Avenue, New York, NY 10016. (212) 696-9000. (800) 228-1160. 1976 - . $175.00 per volume.

Kirk-Othmer Encyclopedia of Chemical Technology. John Wiley and Sons, Inc., 605 Third Avenue, New York, NY 10158. (800) 526-5368 or (212) 850-6000. Fourth edition. 1991 - . Twenty-seven volumes. $5400.00.

McGraw-Hill Encyclopedia of Science and Technology. McGraw-Hill Book, Incorporated, 1221 Avenue of the Americas, New York, NY 10020. (212) 997-3675. (800) 262-4729. Seventh edition. Twenty volumes. 1992. $1900.00.

Ullman's Encyclopedia of Industrial Chemistry. VCH Publications, Inc., 220 East 23rd Street, Suite 909, New York, NY 10010-4606. (212) 683-8333. (800) 422-8824. 5th edition. 1984 - . Price varies per volume.

GENERAL WORKS

Chemical Process Design. Robin Smith. McGraw-Hill Publishing Company, Inc., 1221 Avenue of the Americas, New York, NY 10020. (212) 512-2000. (800) 262-4729. 1995. $65.00.

Combustion of Synthetic Fuels. William Bartok, editor. American Chemical Society, 1155 16th Street, NW, Washington, DC 20036. (202) 872-4600. (800) 333-9511. FAX (614) 447- 3671. 1983. $38.95.

Numerical Modeling in Combustion. T. J. Chung, editor. Hemisphere Publishing Corp., 1900 Frost Road, Suite 101, Bristol PA 19007. (215) 785-5800. (800) 821-8312. 1993. $110.00.

Progress in Synthetic Fuels. G. Imarisio and J. M. Bemtgen. Kluwer Academic Publishers, 101 Philip Drive, Assinippi Park, Norwell, MA 02061. (617) 871-6600. 1989. $128.00.

Synthetic Fuel Technology Development in the United States: A Retrospective Assessment. Michael Crow, et al. Greenwood Press, 88 Post Road West, Box 5507, Westport, CT 06881. (203) 226-3571. (800) 225-5800. 1988. $55.00.

Synthetic Fuels. Ronald F. Probstein and R. Edwin Hicks. pH Press, MIT Branch, P.O. Box 397195 Cambridge, MA 02139. (617) 253-2240. Reprint edition. 1990. $39.00.

Synthetic Fuels From Coal: Status of the Technology. P. F. Paul, et al editors. Kluwer Academic Publishers, 101 Philip Drive, Assinippi Park, Norwell, MA 02061. (617) 871-6600. 1988. $125.00 in paper.

Transport and Global Climate Change. David L. Greene and Danilo J. Santini, editors. 1992. American Council for an Energy Efficient Economy, 1001 Connecticut Avenue, Suite 801, Washington, DC 20036. (202) 429-8873. 1993. $28.00 in paper.

The Unfulfilled Promise of Synthetic Fuels: Technological Failure, Policy Immobilism OR Commercial Illusion. Ernest J. Yanarella and William C. Green, editors. Greenwood Press, 88 Post Road West, Box 5507, Westport, CT 06881. (203) 226-3571. (800) 225-5800. 1987. $59.95.

Unit Operations in Chemical Engineering. Warren L. McCabe. McGraw-Hill Publishing Company, Inc., 1221 Avenue of the

Americas, New York, NY 10020. (212) 512-2000. (800) 262-4729. 5th edition. 1993.

HANDBOOKS AND MANUALS

Chemical Engineering Reference Manual. Randall N. Robinson. Professional Publications, Inc., 1250 Fifth Avenue, Belmont, CA 94002. (415) 593-0110. (800) 426-1178. 1988. $49.95.

CRC Handbook of Chemistry and Physics. David R. Kide, editor. CRC Press Inc., 2000 Corporate Blvd., NW, Boca Raton, FL 33431. (407) 994-0555. (800) 333-8300. 77th edition. 1996. $99.95.

Handbook of Chemical Engineering Calculations. Nicholas P. Chopey and Tyler G. Hicks, editors. McGraw-Hill Publishing Company, Inc., 1221 Avenue of the Americas, New York, NY 10020. (212) 512-2000. (800) 262-4729. 2nd edition. 1993. $69.50.

Perry's Chemical Engineers' Handbook. Robert H. Perry and Donald W. Green, editors. McGraw-Hill Publishing Company, Inc., 1221 Avenue of the Americas, New York, NY 10020. (212) 512-2000. (800) 262-4729. 6th edition. 1996. $129.50.

Synthetic Fuels: A Resource Guide. Gordon Press. Gordon Press Publications, P.O. Box 459, Bowling Green Station, New York, NY 10004. (718) 624-8419. 1991.$75.00.

Riegel's Handbook of Industrial Chemistry. James A. Kent, editor. Van Nostrand Reinhold, 115 Fifth Avenue, New York, NY 10003. (212) 254-3232 or (800) 926-2665. Ninth edition. 1992. $114.95.

Summary Description of Design Criteria, Codes, Standards, and Regulatory Provisions Typically Used For the Civil and Structural Design of Nuclear Fuel Cycle Facilities. American Society of Civil Engineers. 345 East 47th Street, New York, NY 10017. (212) 705-7496. (800) 548-2723. 1988. $17.00

ONLINE DATABASES AND CD-ROMS

CA Search. Chemical Abstracts Service, P.O. Box 3012, Columbus, OH 43210-0012. (614) 447-3600. (800) 848-6533. FAX (614) 447-3709. Very comprehensive guide to worldwide chemical literature and related fields. 1972 to present. Available on BRS,(800) 289-4277, DIALOG, (800) 334-2564, ORBIT (800) 456-7248, and STN International, FIZ Karlsruhe, P.O. Box 2465, W-7500, Karlsruhe 1, Germany, online services. Inquire as to cost and availability.

Compendex Plus. Engineering Information, Inc., 345 East 47th Street, New York, NY 10017. (212) 705-7600 or (800) 221-1044. Contains citations with abstracts to worldwide literature in engineering and technology, from 1970 to present. Available on online BRS,(800) 289-4277, DIALOG, (800) 334-2564, ORBIT (800) 456-7248, and STN International, FIZ Karlsruhe, P.O. Box 2465, W-7500, Karlsruhe 1, Germany, online services. Also available on CD-ROM. Inquire as to cost and availability.

Current Contents Search. Institute for Scientific Information, 3501 Market Street, Philadelphia, PA 19104. (215) 386-0100. FAX (215) 386-6362. Contains citations to articles listed in the table of contents of science and technology journals. Also articles in social sciences and life sciences journals. Available on

BRS,(800) 289-4277, DIALOG,(800) 334-2564, online services. Inquire as to cost and availability.

Dissertation Abstracts. University Microfilms International, 300 North Zeeb Road, Ann Arbor, MI 48106. (800) 521-0600 or (313) 761-4700. Scope includes virtually all doctoral dissertations accepted at accredited American institutions from 1861 to present in 252 subject areas. Available on BRS, (800) 289-4277, DIALOG, (800) 334-2564, and OCLC EPIC, (800) 848-5878, online services. Also available on CD-ROM. Inquire as to cost and availability.

INSPEC. Institution of Electrical Engineers, Michael Faraday House, Six Hills Way, Stevenage, Herts. SG1 2AY, England. Telephone: 0438 313311 or FAX 0438 742840. Contains citations to the worldwide literature of physics, electronics and electrical engineering, computer technology, and related fields. Available on BRS, (800) 289-4277, DIALOG, (800) 334-2564, ORBIT, (800) 456-7248, and STN International, FIZ Karlsruhe, P.O. Box 2465, W-7500, Karlsruhe 1, Germany, online services. Inquire as to cost and availability.

Kirk-Othmer Encyclopedia of Chemical Technology. John Wiley and Sons, Inc., 605 Third Avenue, New York, NY 10158. (800) 526-5368 or (212) 850-6000. Contains the complete text of all chapters in the 27 volume fourth edition of the KIRK-Othmer ENCYCLOPEDIA of CHEMICAL TECHNOLOGY. 1991. Available on BRS,(800) 289-4277, DIALOG,(800) 334-2564, online services. Inquire as to cost and availability.

NTIS Bibliographic Database. National Technical Information Service, 5285 Port Royal Road, Springfield, VA 22161. (703) 487-4929 or FAX (703) 321-8199. Broad coverage of government-sponsored science and technology research reports, 1964 to present. Available on BRS,(800) 289-4277, DIALOG, (800) 334-2564, ORBIT, (800) 456-7248, and STN International, FIZ Karlsruhe, P.O. Box 2465, W-7500, Karlsruhe 1, Germany, online services. Also available on CD-ROM. Inquire as to cost and availability.

SCISEARCH. Institute for Scientific Information, 3501 Market Street, Philadelphia, PA 19104. (800) 523-1850 or (215) 386-0100. Broad multidisciplinary title and author index to the international literature of science and technology, 1974 to present. Available on DIALOG, (800) 334-2564, and ORBIT, (800) 456-7248, online services. Also available on CD-ROM. Inquire as to cost and availability.

WILSONLINE. H.W. Wilson Company, 950 University Avenue, Bronx, NY 10452. (800) 367-6770 or (212) 588-8400. Makes available online versions of the printed H.W. Wilson indexes including Applied Science and Technology Index, Business Periodicals Index, General Science Index, and Readers' Guide to Periodical Literature. Period covered is generally 1983 to present. Available on BRS, (800) 289-4277, DIALOG, (800) 334-2564, and OCLC EPIC, (800) 848-5878, online services. Also available on CD-ROM. Inquire as to cost and availability.

Who's Who in Technology. Gale Research Inc., 835 Penobscot Building, Detroit, MI 48226-4094. (313) 961-2242. (800) 877-4253. Contains biographical information of contemporary American scieNTISts and engineers. Available on DIALOG,(800) 334-2564, online service. Inquire as to cost and availability.

PERIODICALS

AIChe Journal. American Institute of Chemical Engineers. 345 East 47th Street, New York, NY 10017. (212) 705-7338. 1955 to present. Monthly. $295.00 per year.

Coal and Synfuels Technology. Pasha Publications, Inc., 1616 North Fort Myer Drive, Arlington, VA 22209-3107. (703) 528-1244. 1979 to present. Weekly. $790.00 per year. ISSN: 0883-9735.

Energy and Fuels. American Chemical Society, Box 3337, Columbus, OH 43210. (614) 447-3774. 1987 to present. Bimonthly. $345.00 per year. ISSN: 0887-0624.

Energy Engineering. Association of Energy Engineers. Available from Fairmont Press, 700 Indian Trail, Lilburn, GA 30324. (404) 925-9388. 1904 to present. Bimonthly. $99.00 per year. ISSN: 0199-8595.

Energy Sources. Taylor and Francis, 1900 Forst Road, Stuie 101, Bristol, PA 19007-1598. (215) 785-5800. 1973 to present. Quarterly. $225.00. ISSN: 0090-8312.

Journal of Energy Resources Technology. American Society of Mechanical Engineers, 345 East 47th Street, New York, NY 10017. (212) 705-7703. 1979 to present. Quarterly. $100.00 per year. ISSN: 0195-0738.

Power Engineering (Tulsa). PenWell Publishing Co., 1421 Sheridan Road, Tulsa, OK 74101. (918) 8335-3161. 1896 to present. Monthly. $38.00 per year. ISSN: 0032-5961.

RESEARCH CENTERS AND INSTITUTES

Center for Applied Energy Research. University of Kentucky, 3572 Iron Works Pike, Lexington, KY (606) 257-0305. FAX (606) 257-0220.

Center For Energy Studies. University of Texas at Austin, 10100 Burnet Road, Building 133, Austin, TX 78758. (512) 471-4496. FAX (512) 471-0781.

Energy Research Laboratories. Canada Centre for Mineral and Energy Technology, 555 Booth Street, Ottawa, ON Canada KIA 0G1. (613) 996-8201. FAX (613) 995-9584.

Institute of Gas Technology. 3424 South State Street, Chicago, IL 60616. (312) 567-3650. FAX (312) 567-5209.

Oak Ridge National Laboratory. U. S. Department of Energy, P. O. Box 2008, Oak Ridge, TN 37831-6255. (615) 576-2900. FAX (615) 576-2912.

SYSTEMS ENGINEERING

See also: COMPUTER OPERATING SYSTEMS, SOFTWARE, SOFTWARE ENGINEERING

ABSTRACT SERVICES AND INDEXES

ACM Guide To Computing Literature. Association for Computing Machinery, 11 West 42nd Street, New York, NY 10036.
(212) 869-7440. Fax (212) 869-0481. 1964 to present. Annual. $175.00 per year.

Applied Science and Technology Index; A Cumulative Subject Index To English Language Periodicals in the Fields of Aeronautics and Space Science, Computer Technology, Chemistry, Construction Industry, Energy and Related Areas. H.W. Wilson Co., 950 University Avenue, Bronx, NY 10452. (800) 367-6770 or (212) 588-8400. FAX (718) 590-1617. From 1958 to present. Monthly. Inquire about cost and availability. Also available on CD-ROM and online.

Computer Abstracts, MCBUniversity Press Ltd., PO Box 10812, Birmingham, AL 35201-0812. (800) 633-4931. FAX (205) 995-1588. Monthly. Covers computer theory, data, hardware, systems, networks, human-computer interaction, artificial intelligence, as well as applications of computers in aerospace, business, CAD/CAM, cartography, civil engineering, electronics and electrical engineering, industrial engineering, mechanical engineering, medicine, structural engineering, etc.

Computer and Information Systems Abstracts Journal. Cambridge Scientific Abstracts, 7200 Wisconsin Avenue, Bethesda, MD 20814. (301) 961-6750. Fax (301) 961-6720. 1962 to present. Monthly. $1035 per year.

Computer Literature Index. Applied Computer Research Inc., Box 82266, Phoenix, AZ 85071-2266. (800) 234-2227. 1971 to present. Quarterly plus annual cumulation. $198.50 per year. Bibliography of books, articles, and reports.

Computing Journal Abstracts. Techgnosis Ltd., Blade House, Battersea Road, Stockport, Cheshire 3AE, England. Telephone 061-442-2639. FAX 061-443-1162. 1969 to present. Monthly. Inquire for price.

Computing Reviews, Association For Computing Machinery, 11 West 42nd Street, New York, NY 10036. (212) 869-7440. Fax (212) 869-0481. Monthly.

Engineering Index Monthly. Engineering Information, Inc., Castle Point on the Hudson, Hoboken, NJ 07030. (800) 221-1044. FAX (212) 832-1857. Monthly. $2300.00 per year. Also available online as COMPENDEX, and also on CD-ROM. Covers chemical engineering, computers, electrical engineering, civil engineering, metals and mining, industrial management, and mechanical engineering.

Index To Scientific and Technical Proceedings. Institute for Scientific Information, 3501 Market St., Philadelphia, PA 19104. (215) 386-0100. FAX (215) 386-6362. Monthly. $500.00 per year.

Index To Scientific Reviews. Institute for Scientific Information, 3501 Market St., Philadelphia, PA 19104. (215) 386-0100. FAX (215) 386-6362. Semi-annual.

Science Citation Index. Institute for Scientific Information, 3501 Market Street, Philadelphia, PA 19104. (215) 386-0100. FAX (215) 386-6362. Inquire about availability and cost. Also available on CD-ROM.

ASSOCIATIONS AND PROFESSIONAL SOCIETIES

Association For Computing Machinery. 11 West 42nd Street, New York, NY 10036. (212) 869-7440. Fax (212) 869-0481.

SYSTEMS ENGINEERING

Ency. of Physical Sciences and Engineering Info. Sources

IEEE (institute of Electrical and Electronic Engineers).445 Hoes Lane, Piscataway, NJ 08854. (908) 562-5545 or (800) 678-IEEE.

Society For Computer Simulation International. 4838 Ronson Court Lane, San Diego, CA 92111-1810. (619) 277-3888.

BIBLIOGRAPHIES

ACM Guide To Computing Literature. Association for Computing Machinery, 1515 Broadway, 17th Floor, New York, NY 10036-5701. (212) 869-7440. FAX (212) 944-1318. 1964 to present. Annual. $190.00 for non-members. Also available online via DIALOG.

DIRECTORIES AND BIOGRAPHICAL SOURCES

Computing Information Directory. Hildebrandt Inc., Box 285, Colville, WA 99114. (509) 684-2324. FAX (509) 684-2324. 1981 to present. Annual. $199.95 per year.

Directory of Engineering Societies and Related Organizations. Gordon Davis. 13th edition. American Association of Engineering Societies, 1111 19th Street NW, Suite 608, Washington, DC 20036. (202) 296-2237 or (800) 658-8897. 1989. Inquire for price.

International Research Centers Directory. Gale Research, Inc., 835 Penobscot Building, Detroit, MI 48226-4094. (313) 961-2242. (800) 877-4253. 8th edition. 1995. $410.00

Research Centers Directory. Gale Research, Inc., 835 Penobscot Building, Detroit, MI 48226-4094. (313) 961-2242. (800) 877-4253. FAX (313) 961-6083. $485.00.

Scientific and Technical Organizations and Agencies Directory. Gale Research, Inc., 835 Penobscot Building, Detroit, MI 48226-4094. (313) 961-2242. (800) 877-4253. FAX (313) 961-6083. 4th edition. 1996. $195.00.

Who's Who in Engineering. American Association of Engineering Societies, 1111 19th Street NW, Suite 608, Washington, DC 20036. (202) 296-2237 or (800) 658-8897. 8th edition. 1991. Inquire for price.

ENCYCLOPEDIAS AND DICTIONARIES

Computer Dictionary. Donald D. Spencer. 4th edition. Camelot Publishing Company, 709 SW 80th Blvd., Gainesville, FL 32607-1537. (904) 331-0952. 1993. $24.95.

Dictionary of Computing. S.M.H. Collin. 2nd edition. Peter Collin Publishing, 8 the Causeway, Teddington, TW11 0HE, England. FAX 0181-943-3386. 1994. Inquire for cost.

Encyclopedia of Computer Science and Engineering. Anthony Ralston & Edwin D. Reilly Jr., editors. 3rd revised edition. Van Nostrand Reinhold, 115 Fifth Avenue, New York, NY 10003. (212) 254-3232. FAX (212) 254-9499. 1993. $125.00.

Microsoft Press Computer Dictionary: The Comprehensive Standard For Business, School, Library, and Home. 2nd edition. Microsoft Press, One Microsoft Way, Redmond, WA 98052-6399. (206) 936-0055. FAX (206) 823-8101. 1994. $19.95.

GENERAL WORKS

Information Systems Engineering: An Introduction. A. Solvberg and D.C. Kung. Springer-Verlag, 175 Fifth Avenue, New York, NY 10010. (212) 460-1500. FAX (212) 473-6272. 1993. $69.00.

Introduction To Industrial and Systems Engineering. Wayne C. Turner, et al. 3rd edition. Prentice Hall, 113 Sylvan Avenue, Route 9W, Englewood Cliffs, NJ 07632. (201) 592-2000 or (800) 922-0579. 1992. $72.00.

Principles of Systems Programming. Robert M. Graham. Krieger Publishing Company, P.O. Box 9542, Melbourne, FL 32902-9542. (407) 724-9542. FAX (407) 951-3671. 1990. $56.50.

Systems Engineering: Principles and Practice of Computer-based Systems Engineering. Bernard Thome. John Wiley and Sons, Inc., 605 Third Avenue, New York, NY 10158. (800) 526-5368 or (212) 850-6000. 1992. $64.95.

HANDBOOKS AND MANUALS

McGraw-Hill Computing Essentials, 1993-1994. Timothy J. O'Leary, et al. McGraw-Hill Publishing Company, 1221 Avenue of the Americas, New York, NY 10020. (800) 262-4729 or (212) 512-3825. 1994. Inquire for price and availability.

ONLINE DATABASES AND CD-ROMS

Compendex Plus. Engineering Information, Inc., 345 East 47th Street, New York, NY 10017. (212) 705-7600 or (800) 221-1044. Contains citations with abstracts to worldwide literature in engineering and technology, from 1970 to present. Available on online BRS,(800) 289-4277, DIALOG, (800) 334-2564, ORBIT (800) 456-7248, and STN International, FIZ Karlsruhe, P.O. Box 2465, W-7500, Karlsruhe 1, Germany, online services. Also available on CD-ROM. Inquire as to cost and availability.

CSA Engineering. Cambridge Scientific Abstracts, 7200 Wisconsin Avenue, Suite 601, Bethesda, MD 20814. (301) 961-6750 or (800) 843-7751. Contains citations and abstracts of international periodicals and research literature covering all fields of engineering and science and technology,including computer and information science, electronics, mechanical engineering, solid state materials, 1981 to present. Available on BRS,(800) 289-4277, online service.Inquire as to cost and availability.

Current Contents Search. Institute for Scientific Information, 3501 Market Street, Philadelphia, PA 19104. (215) 386-0100. FAX (215) 386-6362. Contains citations to articles listed in the table of contents of science and technology journals. Also articles in social sciences and life sciences journals. Available on BRS,(800) 289-4277, DIALOG,(800) 334-2564, online services. Inquire as to cost and availability.

Dissertation Abstracts. University Microfilms International, 300 North Zeeb Road, Ann Arbor, MI 48106. (800) 521-0600 or (313) 761-4700. Scope includes virtually all doctoral dissertations accepted at accredited American institutions from 1861 to present in 252 subject areas. Available on BRS,(800) 289-4277, DIALOG,(800) 334-2564, and OCLC EPIC,(800) 848-5878, online services. Also available on CD-ROM. Inquire as to cost and availability.

NTIS Bibliographic Database. National Technical Information Service, 5285 Port Royal Road, Springfield, VA 22161. (703)

487-4929 or FAX (703) 321-8199. Broad coverage of government-sponsored science and technology research reports, 1964 to present. Available on BRS,(800) 289-4277, DIALOG, (800) 334-2564, ORBIT, (800) 456-7248, and STN International, FIZ Karlsruhe, P.O. Box 2465, W-7500, Karlsruhe 1, Germany, online services. Also available on CD-ROM. Inquire as to cost and availability.

SCISEARCH. Institute for Scientific Information, 3501 Market Street, Philadelphia, PA 19104. (800) 523-1850 or (215) 386-0100. Broad multidisciplinary title and author index to the international literature of science and technology, 1974 to present. Available on DIALOG,(800) 334-2564, and ORBIT,(800) 456-7248, online services. Also available on CD-ROM. Inquire as to cost and availability.

WILSONLINE. H.W. Wilson Company, 950 University Avenue, Bronx, NY 10452. (800) 367-6770 or (212) 588-8400. Makes available online versions of the printed H.W. Wilson indexes including Applied Science and Technology Index, Business Periodicals Index, General Science Index, and Readers' Guide to Periodical Literature. Period covered is generally 1983 to present. Available on BRS,(800) 289-4277, DIALOG,(800) 334-2564, and OCLC EPIC,(800) 848-5878, online services. Also available on CD-ROM. Inquire as to cost and availability.

PERIODICALS

RESEARCH CENTERS AND INSTITUTES

Arizona State University Center For Research in Engineering & Applied Sciences. Tempe, AZ 85287-5506. (602) 965-1725. FAX (602) 965-8296.

Rensselaer Polytechnic Institute, Information and Decision Systems Laboratory. Troy, NY 12180. (518) 266-6440.

Stanford University Center For Integrated Systems. Stanford, CA 94305-4070. (415) 725-3630.

University of California, Berkeley, Engineering Systems Research Center. 3115 Etcheverry Hall, Berkeley, CA 94720. (415) 642-4993. FAX (415) 643-8982.

University of Missouri—rolla, Intelligent Systems Center. Engineering Research Laboratory, Rolla, MO 65401. (314) 341-4350. FAX (314) 341-6512.

University of Texas At Austin, Bureau of Engineering Research. Cockrell Hall 10.340, Austin, TX 78712-1080. (512) 471-4325. FAX (512) 471-3955.

T

TECTONICS

See: PLATE TECTONICS

TEKTITES

See also: ASTROGEOLOGY, ASTEROIDS, METEORITES, PLANETARY SCIENCES, SOLAR SYSTEM

ABSTRACT SERVICES AND INDEXES

Applied Science and Technology Index; A Cumulative Subject Index To English Language Periodicals in the Fields of aeronautics and Space Science, Computer Technology, chemistry, Construction Industry, Energy and Related Areas. H.W. Wilson Co., 950 University Avenue, Bronx, NY 10452-9978. (212) 588-8400. (800) 367-6770. FAX (718) 590-1617. From 1958 to present. Monthly. Inquire about cost and availability. Also available online (BRS and WILSONLINE)and on CD-ROM. ISSN: 0003-6986.

Astronomy and Astrophysics Abstracts. Springer-Verlag New York, 175 Fifth Avenue, New York, NY 10010. (212) 460-1500.FAX (212) 473-6272. Published for the Astronomisches Rechen-Institut. Comprehensive coverage of all aspects of astronomy, astrophysics and related fields. 1969 to present. Two parts per year. Annual. ISSN: 0067-0022.

Bibliography and Index of Geology. American Geological Institute, 4220 King Street, Alexandria, VA 22302-1507.(703) 379-2480. FAX (703) 379-7563. 1969 to present. Monthly. $1295.00 per year. ISSN: 0098-2784. Also available as GEOREF online (CISTI, DIALOG, Orbit, STN) and on CD-ROM. Inquire about price and availability.

Meteorological and Geoastrophysical Abstracts. American Meteorological Society, c/o Inforonics, Inc., 550 Newtown Road, Box 458, Littleton, MA 01460. (508) 486-8976. FAX(508) 486-0027. Covers literature in environmental sciences, meteorology, astrophysics, hydrology, glaciology and physical oceanography. 1950 to present. Monthly. $950.00 per year. ISSN: 0026-1130. Also available online(DIALOG) and on CD-ROM.

Physics Abstracts. INSPEC. Section A, Science Abstracts. Institute of Electrical Engineers, London, United Kingdom. Available from: INSPEC/IEEE-Institute of Electrical and Electronic Engineers, Box 1331, Hoes Lane, Piscataway, NJ 08855-1331. (908) 562-5549. 1898 to present. 24 issues per year. $2835.00 per year. ISSN: 0036-8091. Also available online and on CD-ROM.

Science Citation Index. SCI. Institute for Scientific Information, 3501 Market Street, Philadelphia, PA 19104.(215) 386-0100. (800) 523-1850. FAX (215) 386-2991. 1961 to present. Six issues per year, plus annual cumulation. $11650.00 per year. Also available online and on CD-ROM. Inquire about price and availability. ISSN: 0036-827X.

STAR. Scientific and Technical Aerospace Reports. U.S. National Aeronautics and Space Administration. Distributed by U.S. Superintendent of Documents, Washington, DC 20402. 1963 to present. Semi-monthly, with semiannual and annual indexes. $114.00 per year. ISSN: 0036-8741. Also available online and on CD-ROM.

ANNUAL REVIEWS AND YEARBOOKS

The Astronomical Almanac. Superintendent of Documents, U.S. Government Printing Office, Washington, DC 10402. (202) 783-3238. 1981 to present. Supersedes *Astronomical Ephemeris and the American Ephermis and Nautical Almanac.* Annual. $44.95. ISSN: 0737-6421.

Annual Review of Astronomy and Astrophysics. Original reviews of critical literature and current developments in astronomy and astrophysics. Annual Reviews, Inc., 4139 El Camino Way, Palo Alto, CA 94303-0139. (415) 493-4400. (800) 523-8635. Fax (415) 855-9815. 1963 to date. Annual. $60.00. ISSN: 0066-4146.

Annual Review of Earth and Planetary Sciences. Annual Reviews, Inc., 4139 El Camino Way, Palo Alto, CA 94303-0139. (415) 493-4400. (800) 523-8635. Fax (415) 855-9815. 1973 to date. Annual. $62.00. ISSN: 0084-6597.

Proceedings of the Lunar and Planetary Science Conference. Lunar and Planetary Institute, 3600 Bay Area Boulevard, Houston, TX 77058. (713) 486-2143. 1970 to date. Annual. ISSN: 0270-9511.

ASSOCIATIONS AND PROFESSIONAL SOCIETIES

TECTITES

Ency. of Physical Sciences and Engineering Info. Sources

American Astronomical Society. 2000 Florida Avenue NW, Suite 400, Washington, DC 20009. (202) 328-2010. FAX (202) 234-2560.

American Geological Institute. 4220 King Street, Alexandria, VA 22302-1507. (703) 379-2480. FAX (703) 379-7563.

American Geophysical Union. 2000 Florida Avenue NW, Washington, DC 20009. (202) 462-6900. (800) 966-AGU1. FAX (202) 328-0566.

American Institute of Physics. 1 Physics Ellipse, College Park, MD 20740-3843. (301) 209-3100.

American Meteor Society. SUNY-Geneseo, One College Circle. Department of Physics and Astronomy, Geneseo, NY 14454. (716) 245-5282. FAX (716) 245-5288.

Association of Universities For Research in Astronomy, Inc. (AURA). Suite 701, 1625 Massachusetts Avenue, NW, Washington, DC 20036. (202) 483-2101. FAX (202) 483-2106.

Astronomical Society of the Pacific. 390 Ashton Avenue, San Francisco, CA 94112. (415) 337-1100. FAX: (415) 337-5205.

Geological Society of America. 3300 Penrose Place, PO Box 9140. Boulder, CO 80301-9140. (303) 447-2020. FAX (303) 447-1133.

Lunar and Planetary Institute. 3600 Bay Area Boulevard, Houston, TX 77058. (713) 486-2143.

Meteoritical Society. University of Massachusetts, 125 Marston Hall, Amherst, MA 01003. (413) 545-0300. (413) 545-0724.

Planetary Society. 65 North Catalina Avenue, Pasadena, CA 91106. (818) 793-5100. (800) WOW-MARS. FAX (818) 793-5528.

BIBLIOGRAPHIES

A Bibliography of Astronomy, 1970-1979. R.A. Sea and S.S. Martin. Libraries Unlimited, Inc., Littleton, CO 80160. 1982. $37.50.

Science Books & Films. American Association for the Advancement of Science, 1333 H Street NW, Washington, DC 20005. (202) 326-6434. Reviews of print, film and software materials in all sciences. 1965 to present. Nine issues per year. $40.00 per year. ISSN: 0098-342X.

Scientific and Technical Books and Serials in Print; An Index To Literature in Science and Technology. R. R. Bowker. 121 Chanlon Road, New Providence, NJ 07974. (908) 464-6800. FAX (908) 665-3502. (800) 521-8110. 1972 to present. Annual. $299.95 per year. Also available on CD-ROM. ISSN: 0000-054X.

DIRECTORIES AND BIOGRAPHICAL SOURCES

American Astronomical Society. Membership Directory. American Astronomical Society. 2000 Florida Avenue NW, Suite 400, Washington, DC 20009. (202) 328-2010. FAX (202) 234-2560 Annual. Included in membership dues. ISSN: 1061-9038.

American Men and Women of Science: Physical and Biological Sciences. R.R. Bowker Inc., 121 Chanlon Road, New Providence, NJ 07974. (908) 464-6800. (800) 521-8110. 20th edition. 8 volumes. 1996. $850.00.

The Astronomers. Donald Goldsmith. St. Martin's Press, Inc., 175 Fifth Avenue, New York, NY 10010. (212) 674-5151. (800) 221-7945. 1993. $14.95 in paper.

Astronomical Centers of the World. Kevin Krisciunas. Cambridge University Press, 40 West 20th Street, New York, NY 10011-4211. (212) 924-3900. (800) 872-7423. 1988. $34.95

Directory of Physics and Astronomy Staff. American Institute of Physics, One Physics Ellipse, College Park, MD 20740-3843. (301) 209-3100. 1975/76 to present. Annual. $60.00. ISSN: 0361-2228.

Graduate Programs in Physics, Astronomy and Related Fields. American Institute of Physics, One Physics Ellipse, College Park, MD 20740-3843. (301) 209-3100. 1978 to present. Annual. $45.00. ISSN: 0147-1821.

Earth and Astronomical Research Centers. Stockton Press, 345 Park Avenue South, New York, NY 10010. 4th edition. 1995. $515.00. ISBN: 1-56169-0967.

ENCYCLOPEDIAS AND DICTIONARIES

Concise Dictionary of Astronomy. Jacqueline Mitton. Oxford University Press, Inc., 200 Madison Avenue, New York, NY 10016. (212) 725-6000. (800) 334-4249. 1992. $24.95.

Encyclopedia of Astronomy and Astrophysics. Stephen Maran, editor. Van Nostrand Reinhold, 115 Fifth Avenue, New York, NY 10003. (212) 254-3232. (800) 842-3636. 1992. $129.95.

McGraw-Hill Encyclopedia of Science and Technology. McGraw-Hill Book Company, Inc., 1221 Avenue of the Americas, New York, NY 10020. (212) 512-2000. (800) 262-4729. 7th edition. 20 volume set. 1992. $1900.00.

New Guide To the Planets. Patrick Moore. Trans-Atlantic Publications, Inc., 311 Bainbridge Street, Philadelphia, PA 19147. (215) 925-5083. 1993. $37.50.

GENERAL WORKS

Catalogue of Meteorites. A. L. Graham, et al., editors. University of Arizona Press, 1230 North Park Avenue, Number 102, Tucson, AZ 85719. (520) 621-1441. (800) 426-3797. 4th edition. 1985. $75.00.

Chemistry and Physics of the Terrestrial Planets. Surendra K. Saxwna. Springer-Verlag New York, Inc., 175 Fifth Avenue, New York, NY 10010. (212) 460-1500. (800) 777-4643. 1986. $109.00.

Comets, Meteors and Asteroids: How They Affect the Earth. Stan Gibilisco. TAB Books, P.O. Box 40, Blue Summit, PA 17294-0850. (717) 794-2191. (800) 233-1128. 1985.

Evolution of the Earth and Planets. E. Takahashi, et al, editors. American Geophysical Union, 2000 Florida Avenue, N.W., Washington, DC 20009. (202) 462-6903. (800) 966-2481. 1993. $28.00.

The Geology of Multi-Ring Impact Basins: The Moon and Other Planets. P.D. Spudis. Cambridge University Press, 40 West 20th Street, New York, NY 10011-4211. (212) 924-3900. (800) 872-7423. Planetary Sciences Series. 1993. $59.95.

An Introduction To Cosmochemistry. Charles R. Cowley. Cambridge University Press, 40 West 20th Street, New York, NY 10011-4211. (212) 924-3900. (800) 872-7423. 1995. $29.95 in paper.

Meteorites: An Introduction. Fritz Heide. Springer-Verlag New York, Inc., 175 Fifth Avenue, New York, NY 10010. (212) 460-1500. (800) 777-4643. 1995. $24.00 in paper.

Meteorites and the Origin of Planets. John A. Wood. Books on Demand, 300 North Zeeb Road, Ann Arbor, MI 48106-1346. (313) 761-4700. (800) 521-0600. Reprint edition. $35.00 in paper.

Meteorites: Their Record On Early Solar System History. John T. Wasson. W. H. Freeman and Co., 41 Madison Avenue, East 26th Avenue, 35th Floor, New York, NY 10010. (212) 576-0400. 1985.

Moons and Planets. William K. Hartmann. Wadsworth Publishing Co., 10 Davis Drive, Belmont, CA 94002. (415) 595-2350. (800) 354-9706. 3rd edition. 1993. $49.95.

Resources of Near-Earth Space. John S. Lewis, et al. University of Arizona Press, 1230 North Park Avenue, Number 102, Tucson, AZ 85719. (520) 621-1441. (800) 426-3797. 1993. $75.00.

Rings: Discoveries From Galileo To Voyager. James C. Elliot and Richard Kerr. MIT Press, 55 Hayward Street, Cambridge, MA 02142. (617) 253-8569. Reprint edition. 1987. $9.95 in paper.

Tektites. Virgil Barnes and Mildred Barnes, editors. Van Nostrand Reinhold Co. Inc., 115 West 50th Street, New York, NY 10003. (212) 254-3232. (800) 842-3636. 1973.

Worlds Apart: A Textbook in Planetary Sciences. Guy Consolmagno. Prentice-Hall, 113 Sylvan Avenue, Route 9W, Englewood Cliffs, NJ 07632. (201) 592-2000. (800) 922-0579. 1994. $62.00.

HANDBOOKS AND MANUALS

Astrophysical Data: Planets and Stars. Kenneth R. Lamb. Springer-Verlag New York, Inc., 175 Fifth Avenue, New York, NY 10010. (212) 460-1500. (800) 777-4643. 1993. $59.00.

Field Guide to the Stars and Planets. Jay M. Pasachoff and Donald H. Menzel. Houghton Mifflin Co., 222 Berkeley Street, Boston, MA 02116. (617) 351-5000. (800) 225-3362. Revised edition. 1992. $24.95.

Handbook of Elemental Abundances in Meteorites: Reviews in Cosmochemistry and Allied Subjects. Brian Mason. Gordon and Breach Science Publishers, Inc., 50 West 23rd Street, New York, NY 10010. (212) 206-8900. 1971. $261.00 in paper.

Meteorite and Tektite Collectors Handbook. Phillip M. Bagnall. Willmann-Bell Inc., P.O. Box 35025, Richmond, VA 23235. (804) 320-7016. 1991. $24.95.

Planet Observer's Handbook. Fred W. Price. Cambridge University Press, 40 West 20th Street, New York, NY 10011-4211. (212) 924-3900. (800) 872-7423. 1994. $34.95.

ONLINE DATABASES AND CD-ROMS

CA Search. Chemical Abstracts Service, P.O. Box 3012, Columbus, OH 43210-0012. (614) 447-3600. (800) 848-6533. FAX (614) 447-3709. Very comprehensive guide to worldwide chemical literature and related fields, 1972 to present. Available on BRS, (800) 289-4277, DIALOG, (800) 334-2564, ORBIT (800) 456-7248, and STN International, FIZ Karlsruhe, P.O. Box 2465, W-7500, Karlsruhe 1, Germany, online services. Inquire as to cost and availability.

Georef. American Geological Institute, 4220 King Street, Alexandria, VA 22302. (800) 336-4764 or (703) 379-2480. Geology and geosciences literature, 1785 to present for North America. Available on DIALOG, (800) 334-2564, ORBIT (800) 456-7248, online services. Also available on CD-ROM. Inquire as to cost and availability.

NTIS Bibliographic Database. National Technical Information Service, 5285 Port Royal Road, Springfield, VA 22161. (703) 487-4929 or FAX (703) 321-8199. Broad coverage of government-sponsored science and technology research reports, 1964 to present. Available on BRS, (800) 289-4277, DIALOG, (800) 334-2564, ORBIT, (800) 456-7248, and STN International, FIZ Karlsruhe, P.O. Box 2465, W-7500, Karlsruhe 1, Germany, online services. Also available on CD-ROM. Inquire as to cost and availability.

SCISEARCH. Institute for Scientific Information, 3501 Market Street, Philadelphia, PA 19104. (800) 523-1850 or (215) 386-0100. Broad multidisciplinary title and author index to the international literature of science and technology, 1974 to present. Available on DIALOG, (800) 334-2564, and ORBIT, (800) 456-7248, online services. Also available on CD-ROM. Inquire as to cost and availability.

WILSONLINE. H.W. Wilson Company, 950 University Avenue, Bronx, NY 10452. (800) 367-6770 or (212) 588-8400. Makes available online versions of the printed H.W. Wilson indexes, including Applied Science and Technology Index, Business Periodicals Index, General Science Index, and Readers' Guide to Periodical Literature. Period covered is generally 1983 to present. Available on BRS, (800) 289-4277, DIALOG, (800) 334-2564, and OCLC EPIC, (800) 848-5878, online services. Also available on CD-ROM. Inquire as to cost and availability.

OTHER SOURCES

Atlas of the Planets. Vincent DeCallatay and Audouin Dollfus. Books on Demand, 300 North Zeeb Road, Ann Arbor, MI 48106-1346. (313) 761-4700. (800) 521-0600. $45.60.

Atlas of the Solar System. B. Yenne. Simon & Schuster, Inc. 1230 Avenue of the Americas, New York, NY 10020. (212) 698-7000. (800) 223-2348. 1987. $12.98.

Cambridge Atlas of Astronomy. Cambridge University Press, 40 West 20th Street, New York, NY 10011-4211. (212) 924-3900. (800) 872-7423. 3rd edition. 1994. $74.00.

The Sky: A User's Guide. David H. Levy. Cambridge University Press, 40 West 20th Street, New York, NY 10011-4211. (212) 924-3900. (800) 872-7423. 1993. $14.95.

PERIODICALS

Astronomical Journal. American Institute of Physics, One Physics Ellipse, College Park, MD 20740-3843. (301) 209-3000. Published for the American Astronomical Society. 1849 to present. Monthly. $280.00 per year. ISSN: 0004-6256.

Astronomy. Kalmbach Publishing Company, Box 1612, Waukesha, WI 53187-1612. (414) 796-8776. FAX (414) 796-1142. 1973 to present. Monthly. $27.00 per year. ISSN: 0091-6358.

Earth, Moon and Planets; An International Journal of Comparative Planetology. Kluwer Academic Publishers, Box 358, Accord Station, Hingham, MA 02018-0358. (617) 871-6600. FAX (617) 871-6528. 1969 to present. Monthly. $840.00 per year. ISSN: 0167-9295.

Geochemica Et Cosmochimica Acta. Elsevier Science. 660 White Plains Road, Tarrytown, NY 10591-5153. (914) 524-9200. FAX (914) 333-2444. 1950 to the present. Biweekly. $895.00 per year. ISSN: 0016-7037.

Icarus; International Journal of Solar System Studies. Academic Press, Inc., Journal Division, 525 B Street, Suite 1900, San Diego, CA 92101-4495. (619) 230-1840. FAX (619) 699-6800. 1962 to the present. Monthly. $1080.00. ISSN: 0019-1035.

International Comet Quarterly. Smithsonian Astrophysical Observatory. 60 Garden Street, Cambridge, MA 02138. (617) 495-7440. 1979 to date. Quarterly. $31.00. ISSN: 0736-6922.

JGR: Journal of Geophysical Research: Planets. American Geophysical Union, 2000 Florida Avenue, NW, Washington, CD 20009. (202) 462-6900. FAX (202) 328-0566. 1991 to present. Monthly. $597.00 per year. ISSN: 0148-0227.

Mercury. Astronomical Society of the Pacific. 390 Ashton Avenue, San Francisco, CA 94112. (415) 337-1100. FAX: (415) 337-5205. 1972 to present. Bimonthly. $175.00 per year. ISSN: 0047-6773.

Meteoritics. Meteoritical Society, Department of Chemistry, University of Arkansas, Fayetteville, AR 72701. (501) 575-7625. FAX (501) 575-7778. 1953 to present. Bimonthly. $210.00 per year. ISSN: 0026-1114.

Planetary and Space Science. Elsevier Science Publishing Company, Inc., 660 White Plains Road, Terrytown, NY 10591-5153. (914) 524-9200. FAX: (914) 333-2444. 1959 to present. Monthly. $1355.00 per year. ISSN: 0032-0633.

Special Publications. University of New Mexico. Institute of Meteoritics. Albuquerque, NM 87131. (505) 277-2747.FAX (505) 277-3577. 1970 to present. Irregular. $10.00 per issue. ISSN: 0085-3968.

Sky and Telescope. Sky Publishing Corporation, Box 9111, Belmont, MA 02178. (617) 864-7360. FAX (617) 864-6117. 1941 to present. Monthly. $27.00 per year. ISSN: 0037-6604.

RESEARCH CENTERS AND INSTITUTES

Center for Meteorite Studies. Arizona State University, Tempe, AZ 85287-2504. (602) 965-6511.

Center For Space Science and Astrophysics. Stanford University. 325 Durand Building, Stanford, CA 04305. (415) 723-3582.

Center For Space Sciences. University of Texas at Dallas, P. O. Box 830688, MS-F022, Richardson, TX 75083-0688. (214) 690-2851. FAX (214) 690-2848.

Institute For Astronomy. University of Hawaii at Manoa, 2680 Woodlawn Drive, Honolulu, HI 96822. (808) 956-8312. FAX (808) 988-2790.

Institute of Geophysics and Planetary Physics. University of California, Riverside, CA 92521. (714) 787-4503. FAX (714) 787-4529.

Laboratory For Planetary Geology. Arizona State University, Department of Geology, Tempe, AZ 85281. (601) 965-7029.

Laboratory For Planetary Studies. Cornell University, Space Sciences Building, Ithaca, NY 14853. (607) 255-4971.

Lunar and Planetary Laboratory. University of Arizona, Tucson, AZ 85721. (602) 621-6963. FAX (602) 621-4933.

McDonnell Center For the Space Sciences. Washington University, Box 1105, One Brookings Drive, St. Louis, MO 63130-4899. (314) 889-6255. FAX (314) 889-6219.

National Astronomy and Ionosphere Center. Cornell University. Space Sciences Building. Ithaca, NY 14853. (607) 255-3735. FAX (607) 255-8803.

Smithsonian Astrophysical Observatory. 60 Garden Street, Cambridge, MA 02138. (617) 495-7461. FAX (617) 495-7105.

TELECOMMUNICATIONS

See also: ANTENNAS, ARTIFICIAL SATELLITES, RADIO, REMOTE SENSING, TELEVISION

ABSTRACT SERVICES AND INDEXES

Applied Science and Technology Index; A Cumulative Subject Index to English Language Periodicals in the Fields of Aeronautics and Space Science, Computer Technology, Chemistry, Construction Industry, Energy and Related Areas. H.W. Wilson Co., 950 University Avenue, Bronx, NY 10452. (212) 588-8400. (800) 367-6770. FAX (718) 590-1617. From 1958 to present. Monthly. Inquire about cost and availability. Also available on CD-ROM and online. ISSN: 0003-6986.

Current Contents: Engineering, Technology and Applied Sciences. Institute for Scientific Information, 3501 Market Street, Philadelphia, PA 19104. (215) 386-0100. FAX (215) 386-6362. 1970 to present. Weekly. $442.00 per year. Also available on CD-ROM and online. Inquire regarding cost and availability. ISSN: 0095-7917.

Electrical and Electronics Abstracts. Institution of Electrical Engineers (IEE), London. Available from INSPEC/IEEE-Institute of Electrical and Electronic Engineers, Box 1331, Hoes Lane, Piscataway, NJ 08855-1331. (908) 562-5549. 1898 to present. Monthly. $2200.00 per year. Also available on CD-ROM and online as INSPEC.

Engineering Index Monthly; Indexes and Abstracts the World's Engineering and Technical Literature. Engineering Information, Inc., Castle Point on the Hudson, Hoboken, NJ 07030. (201) 216-8500. (800) 221-1044. FAX (201) 216-8532. Monthly. $2300.00 per year. Available online as COMPENDEX and also on CD-ROM. ISSN: 0742-1974.

General Science Index. H.W. Wilson Company, 950 University Avenue, Bronx, NY 10452. (212) 588-8400. (800) 367-6770. FAX (718) 590-1617. From 1978 to present. Ten issues per year; quarterly and annual cumulations. Service basis. Available on CD-ROM and online. Inquire regarding cost and availability. ISSN: 0162-1963.

Index To IEEE Publications. IEEE Service Center, 445 Hoes Lane, Piscataway, NJ 08855-1331. (908) 981-1393. (800) 678-IEEE. FAX (908) 981-9667. 1973 to present. Annual. ISSN: 0099-1368.

Physics Abstracts. INSPEC. Section A, Science Abstracts. Institution of Electrical Engineers (IEE), London. Available from: INSPEC/IEEE-Institute of Electrical and Electronic Engineers, Box 1331, Hoes Lane, Piscataway, NJ 08855-1331. (908) 562-5549. 1898 to present. 24 issues per year. $2835.00 per year. Also available online and on CD-ROM. ISSN: 0036-8091.

Physics Briefs (Physikalische Berichte). Information Center for Energy, Physics, Mathematics; German Physical Society. VCH Publishers, Inc., 220 East 23rd Street, New York, NY 10010-4606. (212) 683-8333. 1845 to present. 24 issues per year. $2390.00 per year. Also available online. ISSN: 0179-7434.

ASSOCIATIONS AND PROFESSIONAL SOCIETIES

American Electronics Association. 5201 Great America Parkway, Suite 520, PO Box 54990, Santa Clara, CA 95056. (408) 987-4200. FAX (408) 970-8565.

American Radio Relay League. 225 Main Street, Newington, CT 06111. (203) 666-1541. FAX (203) 665-7531.

Audio Engineering Society. 60 East 42nd Street, New York, NY 10017. (212) 661-8528.

Electronics Industries Association. 2500 Wilson Boulevard, Arlington, VA 22201. (703) 907-7500. FAX (202) 457-4985.

IEEE Broadcast Technology Society. c/o Institute of Electrical and Electronics Engineers, 445 Hoes Lane, Piscataway, NJ 08855-1331. (919) 752-7181. FAX (908) 981-1769.

IEEE Communications Society. 345 East 47th Street, New York, NY 10017. (212) 705-7018. FAX (212) 705-7865.

International Telecommunications Satellite Organization. 3400 International Drive, NW, Washington DC, 20008. (202) 944-6800. Fax (202) 944-7860.

National Amateur Radio Association. PO Box 201407, Arlington, TX 76006. (817) 860-0978. FAX (817) 860-0979.

National Association of Radio and Telecommunications Engineers. PO Box 678, Medway, MA 02053. (508) 533-8333. FAX (508) 533-3815.

Radio Amateur Satellite Corporation. 850 Sligo Avenue, Suite 600, Silver Spring, MD 20910. (301) 589-6062. FAX (301) 608-3410.

Society of Cable Telecommunications Engineers. 669 Exton Commons, Exton, PA 19341. (215) 363-6888. FAX (610) 363-5898.

Society of Motion Picture and Television Engineers. 595 West Hartsdale Avenue, White Plains, NY 10607. (914) 761-1100. FAX (914) 761-3115.

DIRECTORIES AND BIOGRAPHICAL SOURCES

Broadcast Engineering-Equipment Reference Manual. Intertec Publishing Corporation, Box 12901, Overland Park, KS 66212. (913) 888-4664. Annual. $20.00.

EEM-Electronic Engineer's Master. Hearst Business Communications, Inc., 645 Stewart Avenue, Garden City NY 11530. (516) 227-1300. Annual. $90.00. ISSN: 0732-9016.

IEEE Membership Directory. Institute of Electrical and Electronics Engineers, IEEE Service Center, 445 Hoes Lane, Piscataway, NY 08854. (908) 981-1393. (800) 678-IEEE. FAX (908) 981-9667. 2 volumes. Annual. $190.00. ISSN:

International Satellite Directory. Design Publishers, 800 Siesta Way, Sonoma, CA 95476. (707) 930-0306. Annual. $250.00.

Satellite Communications, Satellite Industry Directory Issue. Cardiff Publishing, 6300 South Syracuse Way, Suite 650, Englewood, CO 80111. (303) 220-0600. Annual, June. $20.00.

Who's Who in Engineering. Gordon Davis, editor. American Association of Engineering Societies. 1111 19th Street, NY, Suite 608, Washington, DC 20036. (202) 296-2237. (800) 658-8897. 9th edition. 1995. $220.00.

World Satellite Directory. Phillip Publishing, 7811 Montrose Road, Potomac, MD 20854. (800) 777-5006. (301) 340-2100. Annual. $227.00.

ENCYCLOPEDIAS AND DICTIONARIES

Acronyms and Abbreviations of Computer Technology and Telecommunications. David Tavaglione. Marcel Dekker, Inc., 270 Madison Avenue, New York, NY 10016. (212) 696-9000. (800) 228-1160. 1992. $49.75.

Amateur Radio Encyclopedia. Stan Gibilsco. McGraw-Hill Publishing Company, Inc., 1221 Avenue of the Americas, New York, NY 10020. (212) 512-2000. (800) 262-4729. 1993. $49.95.

Dictionary of Communications Technology: Terms, Definitions and Abbreviations. Gilbert Held. John Wiley & Sons, Inc., 605 Third Avenue, New York, NY 10158-0012. (212) 850-6000. (800) 225-5945. 1994. $49.95 in paper.

Facts On File Dictionary of Telecommunications. John Graham and Sue J. Lowe. Facts-On-File, Inc., 460 Park Avenue South, New York, NY 10016-7382. (212) 683-2244. (800) 322-8755. Fax (800) 683-3633. 1991. $24.95.

TELECOMMUNICATIONS

Ency. of Physical Sciences and Engineering Info. Sources

Encyclopedia of Telecommunications. Frohlich and Kent. Marcel Dekker, Inc., 270 Madison Avenue, New York, NY 10016. (212) 696-9000. (800) 228-1160. Volume 1-. 1991-. $185.00 per volume.

IEEE Standard Dictionary of Electrical and Electronics Terms. Christopher J. Booth, editor. IEEE Service Center, 445 Hoes Lane, Piscataway, NJ 08855-1331. (908) 981-1393. (800) 678-IEEE. FAX (908) 981-9667. IEEE Standard 100-1992. 5th edition. 1993. $150.00.

Illustrated Dictionary of Electronics. Stan Gibilsco. TAB Books, P.O. Box 40, Blue Summit, PA 17294-0850. (717) 794-2191. (800) 233-1128. 7th edition. 1994. $34.95.

GENERAL WORKS

Broadcast Transmission Engineering Practice. Wharton, et al. Butterworth-Heinemann, 313 Washington Street, Newton, MA 02158. (618) 928-2500. (800) 366-2665. 1992. $125.00.

Coherent Lightwave Communication Systems. Shiro Ryu. Artech House, 685 Canton Street, Norwood, MA 02062. (617) 769-9750. (800) 225-9977. 1994. $79.00.

Electronic Communications Systems: Fundamentals Through Advanced. Wayne Tomasi. Prentice-Hall, 113 Sylvan Avenue, Route 9W, Englewood Cliffs, NJ 07632. (201) 592-2000. (800) 922-0579. 2nd edition. 1993. $77.00.

Elements of Digital Communication. J.C. Bic, et al. John Wiley & Sons, Inc., 605 Third Avenue, New York, NY 10158-0012. (212) 850-6000. (800) 225-5945. 1991. $150.00.

Introduction To Fiber Optic Systems. John P. Powers. Irwin, 1333 Burr Ridge Parkway, Burr Ridge, IL 60521. (708) 789-4000. (800) 634-3961. 1993. $64.95.

Introduction To Radio Frequency and Microwave Engineering. Robert K. Feeney and David R. Hertling. TAB Books, P.O. Box 40, Blue Summit, PA 17294-0850. (717) 794-2191. (800) 233-1128. 1995. $44.50.

Low Power Communications. Richard H. Arland. Tiare Publications, P.O. Box 493, Lake Geneva, WI 53147. (800) 420-0579 volume 1: Basic QRP. 1992. $14.95.

Microwave Transmission For Telecommunications. Paul F. Combes. John Wiley & Sons, Inc., 605 Third Avenue, New York, NY 10158-0012. (212) 850-6000. (800) 225-5945. 1991. $174.00.

Practical Antenna Handbook. Joseph J. Carr. McGraw-Hill Publishing Company, Inc., 1221 Avenue of the Americas, New York, NY 10020. (212) 512-2000. (800) 262-4729. 1990. $32.95.

Telecommunication Wiring. Clyde N. Herrick and C. Lee McKim. Prentice-Hall, 113 Sylvan Avenue, Route 9W, Englewood Cliffs, NJ 07632. (201) 592-2000. (800) 922-0579. 1991. $45.00.

Telecommunications Circuit Design. Patrick D. Van der Puije. John Wiley & Sons, Inc., 605 Third Avenue, New York, NY 10158-0012. (212) 850-6000. (800) 225-5945. 1992. $79.95.

Telecommunications Primer: Signals, Building Blocks and Networks. E. Bryan Carne. Plenum Publishing Corp., 233 Spring Street, New York, NY. (212) 620-8000. (800) 2221-9369. FAX (212) 463-0742.

Telecommunications Switching, Traffic and Networks. J. E. Flood. Prentice-Hall, 113 Sylvan Avenue, Route 9W, Englewood Cliffs, NJ 07632. (201) 592-2000. (800) 922-0579. 1995. $55.00.

Telecommunications Transmission Engineering. Belcore and Bell Operating Companies. Technical Personnel Staff. Bellcord, 60 New England Avenue, Piscataway, NJ 08854-4196. (201) 699-5802. (800) 521-2673. 3 volumes. 3rd edition. 1990. $399.00 set.

HANDBOOKS AND MANUALS

AARL Handbook For Radio Amateurs. American Radio Relay League, 225 Main Street, Newington, CT 06111. (203) 666-1541. FAX (203) 665-7531. 78th edition. 1996. $30.00. ISSN: 0079-9440.

American Electricians Handbook. Terrell Croft and Wilford I. Summers. McGraw-Hill Book Company, 1221 Avenue of the Americas, New York, NY 10020. (212) 512-2000. (800) 262-4729. 12th edition. 1992. $77.50.

Antenna Engineering Handbook. Richard C. Johnson. McGraw-Hill Publishing Company, Inc., 1221 Avenue of the Americas, New York, NY 10020. (212) 512-2000. (800) 262-4729. 3rd edition. 1993. $119.50.

Electronics Handbook. Jerry C. Whitaker. CRC Press, Inc., 2000 Corporate Boulevard, NW, Boca Raton, FL 33431. (407) 994-0555. (800) 272-7737. 1996. $120.00.

Omnicom Index of Communications Standards. Harold C. Folts. McGraw-Hill Publishing Company, Inc., 1221 Avenue of the Americas, New York, NY 10020. (212) 512-2000. (800) 262-4729. 1989. $140.00.

Reference Manual For Telecommunications Engineering. Roger L. Freeman. John Wiley & Sons, Inc., 605 Third Avenue, New York, NY 10158-0012. (212) 850-6000. (800) 225-5945. 2nd edition. 1993. $195.00.

Standard Handbook For Electrical Engineers. Donald Fink. McGraw-Hill Publishing Company, 1221 Avenue of the Americas, New York, NY 10020. (212) 512-2000. (800) 262-4729. 13th edition. 1996. $110.00.

Telecommunications Handbook. Rector Press, Ltd., 130 Rattlesnake, Leverett, MA 01054-9726. (413) 548-9708.(800) 247-3473. 1995. $ 125.00

Traffic System Design Handbook: Timesaving Telecommunication Traffic Tables and Programs. James R. Boucher. Institution of Electrical Engineers/INSPEC, 445 Hoes Lane, Piscataway, NJ 08855-1331. (908) 562-5553. 1993. $39.95 in paper.

ONLINE DATABASES AND CD-ROMS

COMPENDEX PLUS. Engineering Information, Inc., 345 East 47th Street, New York, NY 10017. (212) 705-7600 or (800) 221-1044. Contains citations with abstracts to worldwide literature in engineering and technology, from 1970 to present. Available online BRS,(800) 289-4277, DIALOG, (800) 334-2564, OR-BIT (800) 456-7248, and STN International, FIZ Karlsruhe, P.O. Box 2465, W-7500, Karlsruhe 1, Germany, online services. Also available on CD-ROM. Inquire as to cost and availability.

INSPEC. Institution of Electrical Engineers, Michael Faraday House, Six Hills Way, Stevenage, Herts. SG1 2AY, England. Telephone: 0438 313311 or FAX 0438 742840. Contains citations to the worldwide literature of physics, electronics and electrical engineering, computer technology, and related fields. Available on BRS, (800) 289-4277, DIALOG, (800) 334-2564, ORBIT, (800) 456-7248, and STN International, FIZ Karlsruhe, P.O. Box 2465, W-7500, Karlsruhe 1, Germany, online services. Inquire as to cost and availability.

NTIS Bibliographic Database. National Technical Information Service, 5285 Port Royal Road, Springfield, VA 22161. (703) 487-4929 or FAX (703) 321-8199. Broad coverage of government-sponsored science and technology research reports, 1964 to present. Available on BRS,(800) 289-4277, DIALOG, (800) 334-2564, ORBIT, (800) 456-7248, and STN International, FIZ Karlsruhe, P.O. Box 2465, W-7500, Karlsruhe 1, Germany, online services. Also available on CD-ROM. Inquire as to cost and availability

WILSONLINE. H.W. Wilson Company, 950 University Avenue, Bronx, NY 10452. (800) 367-6770 or (212) 588-8400. Makes available online versions of the printed H.W. Wilson indexes, including Applied Science and Technology Index, Business Periodicals Index, General Science Index, and Readers' Guide to Periodical Literature. Period covered is generally 1983 to present. Available on BRS, (800) 289-4277, DIALOG, (800) 334-2564, and OCLC EPIC, (800) 848-5878, online services. Also available on CD-ROM. Inquire as to cost and availability.

PERIODICALS

Audio Engineering Society Journal. Audio Engineering Society, 60 East 42nd Street, New York, NY 10017. (212) 661-8528. 1953 to present. 10 issues per year. $125.00 per year. ISSN: 0004-7554.

Broadcasting and Cable. Cahners Publishing Company, Box 6399, Torrence, CA 90504. (800) 554-5729. 1931 to present. Weekly. $99.00 per year. ISSN: 1068-6827.

Db, the Sound Engineering Magazine. 203 Commack Road, Suite 1010, Commack, NY 1725. (516) 586-6530. 1967 to present. $15.00 per year. ISSN: 0011-7145.

Electronic Design. Penton Publishing, San Jose Gateway, Suite 354. 2025 Gateway Place, San Jose, CA 95110. (408) 441-0550. 1952 to present. Fortnightly. $95.00. ISSN: 0013-4872.

Electronics. Penton Publishing, San Jose Gateway, Suite 354. 2025 Gateway Place, San Jose, CA 95110. (408) 441-0550. 1930 to present. Semi-weekly. $98.00 per year. ISSN: 0883-4989.

Electronics Now. Gernsback Publications, Inc., 500-B Bi-County Boulevard, Farmingdale, NY 11735. (516) 293-3000. FAX (516) 293-3115. 1929 to present. Monthly. $19.95. ISSN: 0033-7862.

Electronics World and Wireless World. Reed Business Publishing, Ltd., Quadrant House, the Quadrant, Sutton, Surrey SM2 5AS, England. 1911 to present. Monthly. L35. ISSN: 0266-3244.

IEEE Antennas and Propagation Magazine. Institute of Electrical and Electronics Engineers, Inc., 445 Hoes Lane, Box 1331, Piscataway, NJ 08855-1331. (908) 981-0060. FAX (908) 981-9667. Bi-monthly. $85.00. ISSN: 1045-9243.

IEEE Communications Magazine. Institute of Electrical and Electronics Engineers, Inc., 445 Hoes Lane, Box 1331, Piscataway, NJ 08855-1331. (908) 981-0060. FAX (908) 981-9667. 1953 to present. Monthly. $23.00 per year. ISSN: 0163-6804.

IEEE Journal On Selected Areas in Communications. Institute of Electrical and Electronics Engineers, Inc., Box 1331, 445 Hoes Lane, Piscataway, NJ 08855-1331. (908) 981-0060. 1983 to present. 9 issues per year. $275.00 per year. ISSN: 0733-8716.

IEEE Transactions On Broadcasting. Institute of Electrical and Electronics Engineers, Inc., Box 1331, 445 Hoes Lane, Piscataway, NJ 08855-1331. (908) 981-0060. 1955 to present. Quarterly. $50.00 per year. ISSN: 0018-9316.

Microwaves and RF. Penton Publishing Co., Circulation Department, 1100 Superior Avenue, Cleveland, OH 44114. (201) 393-6060. FAX (201) 393-6297. From 1962 to present. Monthly. $60.00. ISSN: 0745-2992.

Satellite Communications. Argus Inc., Box 41528, Nashville, TN 37204. (615) 377-3322. 1977 to present. Monthly. $32.00 per year. ISSN: 0147-7439.

RESEARCH CENTERS AND INSTITUTES

Center For Advanced Technology in Telecommunications. Polytechnic University, 333 Jay Street, Brooklyn, NY 11202. (718) 643-5160.

Communications Research Laboratory. McMaster University, 1280 Main Street West, Hamilton, ON Canada L85 4K1. (416) 525-9140.

Communications Satellite Planning Center. Stanford University, ERL, Building, Room 202, Stanford, CA 94305-4053. (415) 723-3471. FAX (415) 723-7298.

TELEGRAPHY

See: TELECOMMUNICATIONS

TELEPHONES

See: TELECOMMUNICATIONS

TELESCOPES

See also: ASTRONOMY, ASTROPHYSICS, GALAXIES, OBSERVATORIES, OPTICS, RADIO TELESCOPES, SOLAR SYSTEM

TELESCOPES

Ency. of Physical Sciences and Engineering Info. Sources

ABSTRACT SERVICES AND INDEXES

Applied Science and Technology Index; A Cumulative Subject Index To English Language Periodicals in the Fields of Aeronautics and Space Science, Computer Technology, Chemistry, Construction Industry, Energy and Related Areas. H. W. Wilson Co., 950 University Avenue, Bronx, NY 10452-9978. (212) 588-8400. (800) 367-6770. FAX (718) 590-1617. From 1958 to present. Monthly. Inquire about cost and availability. Also available online (BRS and WILSONLINE) and on CD-ROM. ISSN: 0003-6986.

Astronomy and Astrophysics Abstracts. Springer-Verlag New York, 175 Fifth Avenue, New York, NY 10010. (212) 460-1500. FAX (212) 473-6272. Published for the Astronomisches Rechen-Institut. Comprehensive coverage of all aspects of astronomy, astrophysics and related fields. 1969 to present. Two parts per year. Annual. ISSN: 0067-0022.

Physics Abstracts. INSPEC. Section A, Science Abstracts. Institute of Electrical Engineers, London, United Kingdom. Available from: INSPEC/IEEE-Institute of Electrical and Electronic Engineers, Box 1331, Hoes Lane, Piscataway, NJ 08855-1331. (908) 562-5549. 1898 to present. 24 issues per year. $2835.00 per year. ISSN: 0036-8091. Also available online and on CD-ROM.

Physics Briefs (Physikalische Berichte). Information Center for Energy, Physics, Mathematics; German Physical Society. VCH Publishers, Inc., 220 East 23rd Street, New York, NY 10010-4606. (212) 683-8333. 1845 to present. 24 issues per year. $2390.00 per year. Also available online. ISSN: 0179-7434.

Science Citation Index. SCI. Institute for Scientific Information, 3501 Market Street, Philadelphia, PA 19104. (215) 386-0100. (800) 523-1850. FAX (215) 386-2991. 1961 to present. Six issues per year, plus annual cumulation. $11650.00 per year. Also available online and on CD-ROM. Inquire about price and availability. ISSN: 0036-827X.

STAR. Scientific and Technical Aerospace Reports. U.S. National Aeronautics and Space Administration. Distributed by U. S. Superintendent of Documents, Washington, DC 20402. 1963 to present. Semi-monthly, with semiannual and annual indexes. $114.00 per year. ISSN: 0036-8741. Also available online and on CD-ROM.

ANNUAL REVIEWS AND YEARBOOKS

The Astronomical Almanac. Superintendent of Documents, U. S. Government Printing Office, Washington, DC 10402. (202) 783-3238. 1981 to present. Supersedes Astronomical Ephemeris and the American Ephermis and Nautical Almanac. Annual. $44.95. ISSN: 0737-6421.

Annual Review of Astronomy and Astrophysics. Original reviews of critical literature and current developments in astronomy and astrophysics. Annual Reviews, Inc., 4139 El Camino Way, Palo Alto, CA 94303-0139. (415) 493-4400. (800) 523-8635. Fax (415) 855-9815. 1963 to date. Annual. $60.00. ISSN: 0066-4146.

ASSOCIATIONS AND PROFESSIONAL SOCIETIES

American Astronomical Society. 2000 Florida Avenue NW, Suite 400, Washington, DC 20009. (202) 328-2010.FAX(202) 234-2560.

American Institute of Physics. One Physics Ellipse, College Park, MD 20740-3843. (301) 209-3100.

American Association of Variable Star Observers. 23 Birch Street, Cambridge, MA 02138. (617) 354-0484. FAX: (617)354-0665.

Association of Universities For Research in Astronomy, Inc. (AURA). Suite 701, 1625 Massachusetts Avenue, NW, Washington, DC 20036. (202) 483-2101. FAX (202) 483-2106.

Astronomical League. 2112 Kingfisher Lane East, Rolling Meadows, IL 60008. (708) 398-0562.

Astronomical Society of the Pacific. 390 Ashton Avenue, San Francisco, CA 94112. (415) 337-1100. FAX: (415) 337-5205.

Institute of Electrical and Electronics Engineers (IEEE). 345 East 47th Street, New York, NY 10017. (212) 705-7900.

Optical Society of America. 1816 Jefferson Place, N.W., Washington, DC 20036. (202) 223-8130.

Society of Photo-Optical Instrumentation Engineers; The International Society of Optical Engineering. P.O. Box 10, 1022 19th Street, Bellingham, WA 98227. (206) 676-3290.

BIBLIOGRAPHIES

A Bibliography of Astronomy, 1970-1979. R. A. Sea and S. S. Martin. Libraries Unlimited, Inc., Littleton, CO 80160. 1982. $37.50.

Science Books & Films. American Association for the Advancement of Science, 1333 H Street NW, Washington, DC 20005. (202) 326-6434. Reviews of print, film and software materials in all sciences. 1965 to present. Nine issues per year. $40.00 per year. ISSN: 0098-342X.

Scientific and Technical Books and Serials in Print; An Index To Literature in Science and Technology. R. R. Bowker. 121 Chanlon Road, New Providence, NJ 07974. (908) 464-6800. FAX (908) 665-3502. (800) 521-8110. 1972 to present. Annual. $299.95 per year. Also available on CD-ROM. ISSN: 0000-054X

DIRECTORIES AND BIOGRAPHICAL SOURCES

American Astronomical Society Membership Directory. American Astronomical Society. 2000 Florida Avenue NW, Suite 400, Washington, DC 20009. (202) 328-2010. FAX (202) 234-2560 Annual. Included in membership dues. ISSN: 1061-9038.

American Men and Women of Science: Physical and Biological Sciences. R. R. Bowker Inc., 121 Chanlon Road, New Providence, NJ 07974. (908) 464-6800. (800) 521-8110. 20th edition. 8 volumes. 1996. $850.00.

The Astronomers. Donald Goldsmith. St. Martin's Press, Inc., 175 Fifth Avenue, New York, NY 10010. (212) 674-5151. (800) 221-7945. 1993. $14.95 in paper.

Astronomical Centers of the World. Kevin Krisciunas. Cambridge University Press, 40 West 20th Street, New York, NY 10011-4211. (212) 924-3900. (800) 872-7423. 1988. $34.95

Directory of Physics and Astronomy Staff. American Institute of Physics, One Physics Ellipse, College Park, MD 20740-3843. (301) 209-3100. 1975/76 to present. Annual. $60.00. ISSN: 0361-2228.

Earth and Astronomical Research Centers. Stockton Press, 345 Park Avenue South, New York, NY 10010. 4th edition. 1995. $515.00. ISBN: 1-56169-0967.

Graduate Programs in Physics, Astronomy and Related Fields. American Institute of Physics, One Physics Ellipse, College Park, MD 20740-3843. (301) 209-3100. 1978 to present. Annual. $45.00. ISSN: 0147-1821.

Research Centers Directory. Gale Research, 835 Penobscot Building, Detroit, MI 48226-4094. (313) 961-2242. (800) 347-4253. 20th edition. 1995. $485.00.

ENCYCLOPEDIAS AND DICTIONARIES

Concise Dictionary of Astronomy. Jacqueline Mitton. Oxford University Press, Inc., 200 Madison Avenue, New York, NY 10016. (212) 725-6000. (800) 334-4249. 1992. $24.95.

Encyclopedia of Astronomy and Astrophysics. Stephen Maran, editor. Van Nostrand Reinhold, 115 Fifth Avenue, New York, NY 10003. (212) 254-3232. (800) 842-3636. 1992. $129.95.

McGraw-Hill Encyclopedia of Science and Technology. McGraw-Hill Book Company, Inc., 1221 Avenue of the Americas, New York, NY 10020. (212) 512-2000. (800) 262-4729. 7th edition. 20 volume set. 1992. $1900.00.

GENERAL WORKS

Advanced Telescope Making Techniques. Alan Mackintosh, editor. Willmann-Bell, Inc., P.O. Box 35025, Richmond, VA 23235. (804)-320-7016. Volume 1: Optics; volume 2: Mechanical. 1986. $19.95 each volume in paper.

An Amateur Radio Telescope. George W. Swenson. Pachart Publishing House, 1130 San Lucas Circle, Tucson, AZ 85704. (520) 297-6760. 1980. $9.95 in paper.

The Backyard Astronomer's Guide. Terence Dickinson and Alan Dyer. Firefly Books Ltd., P.O. Box 1338, Ellicott Station, Buffalo, NY 14205. (800) 387-5085. 1991. $39.95.

Build Your Own Telescope. Richard Berry. Willmann-Bell, Inc., P.O. Box 35025, Richmond, VA 23235. (804) 320-7016. 1993.

Classics in Radio Astronomy. W. Sullivan. Kluwer Academic Publishers, 101 Philip Drive, Assinippi Park, Norwell, MA 02061. (617) 871-6600. 1982. $131.50.

The Cosmic Inquirers: Modern Telescopes and their Makers. Wallace Tucker and Karen Tucker. Harvard University Press, 79 Garden Street, Cambridge MA. (617) 495-2600. Reprint edition. 1989. $10.95 in paper.

Galactic and Extragalactic Radio Astronomy. G. K. Verschuur and K. Kellermann. Springer-Verlag New York, 175 Fifth Avenue, New York, NY 10010. (212) 460-1500. (800) 777-4643. 1991. $55.00 in paper.

Instrumentation and Techniques For Radio Astronomy. P. F. Goldsmith. Institute of Electrical and Electronic Engineers (IEEE), 345 East 47th Street, New York, NY 10017. (212) 705-7900. 1988. $59.95.

Interferometry in Radioastronomy and Radio Telescopes. R. Wohlleben, et al. Kluwer Academic Publishers, 101 Philip Drive, Assinippi Park, Norwell, MA 02061. (617) 871-6600. 1991. $92.00.

Extragalactic Radio Sources From Beams To Jets. Jacques Roland, et al, editors. Cambridge University Press, 40 West 20th Street, New York, NY 10011-4211. (212) 924-3900.(800) 872-7423.

History of Radio Astronomy and the National Radio Astronomy Observatory: Evolution Toward Big Science. Benjamin K. Malphrus. Kreiger Publishing Company, P.O. Box 9542, Melbourne, FL 32902-9542. (407) 724-9542. 1995. $wfi

The History of the Telescope. Henry C. King. Dover Publications, Inc., 180 Varick Street, New York, NY 10014. (212) 255-3755. (800) 223-3130. Reprint edition. 1979. $12.95 in paper.

National Radio Astronomy Observatory Very Large Array Sky Survey. J. J. Condon, et al. Rector Press, Ltd., 130 Rattlesnake, Leverett, MA 01054-9726. (413) 548-9708.(800) 247-3473. 1994. $45.00 in paper.

Optical Astronomy From the Earth and Moon. Diane M. Pyper and Ronald Angione, editors. Astronomical Society of the Pacific, 390 Ashton Avenue, San Francisco, CA 94112. (415) 337-1100. 1994. $40.00.

Pauper and Prince: Richey, Hale and Big American Telescopes. Donald E. Osterbrock. University of Arizona Press, 1230 North Park Avenue, Number 102, Tucson, AZ 85719. (520) 621-1441. (800) 426-3797. 1993. $45.00.

Planets and Perception: Telescopic Views and Interpretation, 1609-1909. University of Arizona Press, 1230 North Park Avenue, Number 102, Tucson, AZ 85719. (520) 621-1441. (800) 426-3797. 1988. $40.00.

Robotic Telescopes in the 1990's. A. Filipenko. Astronomical Society of the Pacific, 390 Ashton Avenue, San Francisco, CA 94112. (415) 337-1100. 1992. $40.00.

Seeing the Deep Sky: Telescopic Astronomy Projects Beyond the Solar System. Fred Schaaf. John Wiley & Sons, Inc., 605 Third Avenue, New York, NY 10158-0012. (212) 850-6000. (800) 225-5945. 1992. $14.95 in paper.

Space Technology and Planetary Astronomy. Joseph N. Tatarewicz. Indiana University Pres, 601 North Morton Street, Bloomington, in 47404-3797. (812) 855-6804. (800) 842-6796. 1990. $29.95.

Structural Mechanics of Optical Systems. Hatheway, editor. SPIE-International Society for Optical Engineering, 1000 20th Street, Bellingham, WA 98225. (360) 676-3290. 1987. $43.00.

Telescope Optics-Evaluation and Design. Rutten and Van Venrooij . Willman-Bell, Inc., P.O. Box 35025, Richmond, VA 23235. (804) 320-7016. 1993. $24.95.

TELESCOPES

Ency. of Physical Sciences and Engineering Info. Sources

Tools of Radio Astronomy. K. Rolhfs. Springer-Verlag, 175 Fifth Avenue, New York, NY 10010. (212) 460-1500. (800) 777-4643. 2nd edition. 1990. $45.00.

Unusual Telescopes. Peter L. Manly. Cambridge University Press, 40 West 20th Street, New York, NY 10011-4211. (212) 924-3900. (800) 872-7423. 1992. $39.95.

Visual Astronomy of the Deep Sky. Roger N. Clark. Cambridge University Press, 40 West 20th Street, New York, NY 10011-4211. (212) 924-3900. (800) 872-7423. 1991. $39.95.

HANDBOOKS AND MANUALS

Astrophysical Data: Planets and Stars. Kenneth R. Lanb. Springer-Verlag New York, Inc., 175 Fifth Avenue, New York, NY 10010. (212) 460-1500. (800) 777-4643. 1993. $59.00.

Planet Observer's Handbook. Fred W. Price. Cambridge University Press, 40 West 20th Street, New York, NY 10011-4211. (212) 924-3900. (800) 872-7423. 1994. $34.95.

ONLINE DATABASES AND CD-ROMS

Dissertation Abstracts. University Microfilms International, 300 North Zeeb Road, Ann Arbor, MI 48106. (800) 521-0600 or (313) 761-4700. Scope includes virtually all doctoral dissertations accepted at accredited American institutions from 1861 to present in 252 subject areas. Available on BRS, (800) 289-4277, DIALOG, (800) 334-2564, and OCLC EPIC, (800) 848-5878, online services. Also available on CD-ROM. Inquire as to cost and availability.

INSPEC. Institution of Electrical Engineers, Michael Faraday House, Six Hills Way, Stevenage, Herts. SG1 2AY, England. Telephone: 0438 313311 or FAX 0438 742840. Contains citations to the worldwide literature of physics, electronics and electrical engineering, computer technology, and related fields. Available on BRS, (800) 289-4277, DIALOG, (800) 334-2564, ORBIT, (800) 456-7248, and STN International, FIZ Karlsruhe, P.O. Box 2465, W-7500, Karlsruhe 1, Germany, online services. Inquire as to cost and availability.

NTIS Bibliographic Database. National Technical Information Service, 5285 Port Royal Road, Springfield, VA 22161. (703) 487-4929 or FAX (703) 321-8199. Broad coverage of government-sponsored science and technology research reports, 1964 to present. Available on BRS,(800) 289-4277, DIALOG, (800) 334-2564, ORBIT, (800) 456-7248, and STN International, FIZ Karlsruhe, P.O. Box 2465, W-7500, Karlsruhe 1, Germany, online services. Also available on CD-ROM. Inquire as to cost and availability.

SCISEARCH. Institute for Scientific Information, 3501 Market Street, Philadelphia, PA 19104. (800) 523-1850 or (215) 386-0100. Broad multidisciplinary title and author index to the international literature of science and technology, 1974 to present. Available on DIALOG, (800) 334-2564, and ORBIT, (800) 456-7248, online services. Also available on CD-ROM. Inquire as to cost and availability.

WILSONLINE. H.W. Wilson Company, 950 University Avenue, Bronx, NY 10452. (800) 367-6770 or (212) 588-8400. Makes available online versions of the printed H.W. Wilson indexes, including Applied Science and Technology Index, Business Periodicals Index, General Science Index, and Readers' Guide to Periodical Literature. Period covered is generally 1983 to

present. Available on BRS, (800) 289-4277, DIALOG, (800) 334-2564, and OCLC EPIC, (800) 848-5878, online services. Also available on CD-ROM. Inquire as to cost and availability.

OTHER SOURCES

Cambridge Atlas of Astronomy. Cambridge University Press, 40 West 20th Street, New York, NY 10011-4211. (212) 924-3900. (800) 872-7423. 3rd edition. 1994. $74.00.

The Sky: A User's Guide. David H. Levy. Cambridge University Press, 40 West 20th Street, New York, NY 10011-4211. (212) 924-3900. (800) 872-7423. 1993. $14.95.

PERIODICALS

Astronomical Journal. American Institute of Physics, One Physics Ellipse, College Park, MD 20740-3843. (301) 209-3000. Published for the American Astronomical Society. 1849 to present. Monthly. $280.00 per year. ISSN: 0004-6256.

Astronomical Society of the Pacific Publications. Astronomical Society of the Pacific, 390 Ashton Avenue, San Francisco, CA 94112. (415) 337-1100. FAX (415) 337-5205.1889 to present. Monthly. $175.00 per year. ISSN: 0004-6280.

Astronomy. Kalmbach Publishing Company, Box 1612, Waukesha, WI 53187-1612. (414) 796-8776. FAX (414) 796-1142. 1973 to present. Monthly. $27.00 per year. ISSN: 0091-6358.

Astronomy and Astrophysics. Springer-Verlag New York, Inc., 44 Hartz Way, Secaucus, NJ 07096-2491. (201) 348-4033. FAX(201) 348-4505. 1969 to date. 36 issues per year in 12 volumes. $1733.00 per year. ISSN: 0004-6361.

Astrophysical Journal. University of Chicago Press, 5720 South Woodlawn Avenue, Chicago, IL 60637. (312) 753-3347. FAX (312) 753-0811. 1895 to date. 3 issues per monthly. $740.00 per year. ISSN: 0004-637X.

Astrophysics and Space Science. Kluwer Academic Publishers, Box 358, Accord Station, Hingham MA 02018-0358. (617) 871-6600. FAX (617) 871-6528. 1968 to present. 24 issues per year. $2544.00 per year. ISSN: 0004-640X.

Celestial Mechanics; An International Journal of Space Dynamics. Kluwer Academic Publishers, Box 358, Accord Station, Hingham MA 02018-0358. (617) 871-6600. FAX (617) 871-6528. 1969 to date. 12 issues per year. $703.50 per year. ISSN: 0923-2958.

Icarus; International Journal of Solar System Studies. Academic Press, Inc., Journal Division, 525 B Street, Suite 1900, San Diego, CA 92101-4495. (619) 230-1840. FAX (619) 699-6800. 1962 to the present. Monthly. $1080.00. ISSN: 0019-1035.

Mercury. Astronomical Society of the Pacific. 390 Ashton Avenue, San Francisco, CA 94112. (415) 337-1100. FAX: (415) 337-5205. 1972 to present. Bimonthly. $175.00 per year. ISSN: 0047-6773.

Monthly Notices of the Royal Astronomical Society. Blackwell Scientific Publications, Inc., Osney Mead, Oxford OX2 QEL England. TEL 0865-240201. FAX 0865-721205. 1827 to present. Fortnightly. $1899.00 per year. ISSN: 0035-8711.

Ency. of Physical Sciences and Engineering Info. Sources

TELEVISION

Observatory. c/o Dr. D. J. Strickland, Space and Astrophysics Division, Rutherford Appleton Laboratory, Chilton, Didcot, Oxon, OX11 OQX, England. FAX 0235-445848. 1877 to present. Bimonthly. $42.00. ISSN: 0029-7704.

Planetary and Space Science. Elsevier Science Publishing Company, Inc., 660 White Plains Road, Terrytown, NY 10591-5153. (914) 524-9200. FAX: (914) 333-2444. 1959 to present. Monthly. $1355.00 per year. ISSN: 0032-0633.

Sky and Telescope. Sky Publishing Corporation, Box 9111, Belmont, MA 02178. (617) 864-7360. FAX (617) 864-6117.1941 to present. Monthly. $27.00 per year. ISSN: 0037-6604.

Space Science Reviews. Kluwer Academic Publishers, Box 358, Accord Station, Hingham MA 02018-0358. (617) 871-6600. FAX (617) 871-6528. 1962 to present. 16 issues per year. $940 issues per year. ISSN: 0038-6308.

Vistas in Astronomy. Elsevier Science Publishing Company, Inc., 660 White Plains Road, Terrytown, NY 10591-5153. (914) 524-9200. FAX: (914) 333-2444. 1958 to date. Quarterly. $415.00 per year. ISSN: 0083-6656.

RESEARCH CENTERS AND INSTITUTES

Arecibo Observatory. Cornell University, Box 995. Arecibo, PR 00613. (809) 878-2612. FAX (809) 878-1861.

Haystack Observatory. Massachusetts institute of Technology. Westford, MA 01886. (508) 692-4765. FAX (617) 981-0590.

Herzberg Institute of Astrophysics. National Research Council of Canada, 100 Sussex Drive, Ottawa, ON Canada K1A 0R6. (613) 990-0907. FAX (613) 952-6602.

National Astronomy and Ionosphere Center. Space Sciences Building, Ithaca, NY 14853. (607) 255-3735. FAX (607) 255-8803.

National Radio Astronomy Observatory. Edgemont Road, Charlottesville, VA 22903. (804) 296-0211.

Owens Valley Radio Observatory. California Institute of Technology. Pasadena, CA 91125. (818) 795-6811.

Radio Astronomy Laboratory. University of California, Berkeley, 601 Campbell Hall, Berkeley, CA 94720. (415) 642-5275. FAX (415) 642-3411.

Radio Astronomy Observatory. University of Michigan, 810 Dennison Building, Ann Arbor, MI 48109. (313) 426-8441.

Smithsonian Astrophysical Observatory. 60 Garden Street, Cambridge, MA 02138. (617) 495-7461. FAX (617) 495-7105.

TELEVISION

See also: ANTENNAS, ELECTRICAL ENGINEERING, ELECTRONIC CIRCUITS and COMPONENTS, ELECTRONICS ENGINEERING, RADAR, RADIO

ABSTRACT SERVICES AND INDEXES

Applied Science and Technology Index; A Cumulative Subject Index To English Language Periodicals in the Fields of Aeronautics and Space Science, Computer Technology, Chemistry, Construction Industry, Energy and Related Areas. H.W. Wilson Co., 950 University Avenue, Bronx, NY 10452. (212) 588-8400. (800) 367-6770. FAX (718) 590-1617. From 1958 to present. Monthly. Inquire about cost and availability. Also available on CD-ROM and online. ISSN: 0003-6986.

Current Contents: Engineering, Technology and Applied Sciences. Institute for Scientific Information, 3501 Market Street, Philadelphia, PA 19104. (215) 386-0100. FAX (215) 386-6362. 1970 to present. Weekly. $442.00 per year. Also available on CD-ROM and online. Inquire regarding cost and availability. ISSN: 0095-7917.

Electrical and Electronics Abstracts. Institution of Electrical Engineers (IEE), London. Available from INSPEC/IEEE-Institute of Electrical and Electronic Engineers, Box 1331, Hoes Lane, Piscataway, NJ 08855-1331. (908) 562-5549. 1898 to present. Monthly. $2200.00 per year. Also available on CD-ROM and online as INSPEC.

Engineering Index Monthly; Indexes and Abstracts the World's Engineering and Technical Literature. Engineering Information, Inc., Castle Point on the Hudson, Hoboken, NJ 07030. (201) 216-8500. (800) 221-1044. FAX (201) 216-8532. Monthly. $2300.00 per year. Available online as COMPENDEX and also on CD-ROM. ISSN: 0742-1974.

General Science Index. H.W. Wilson Company, 950 University Avenue, Bronx, NY 10452. (212) 588-8400. (800) 367-6770. FAX (718) 590-1617. From 1978 to present. Ten issues per year; quarterly and annual cumulations. Service basis. Available on CD-ROM and online. Inquire regarding cost and availability. ISSN: 0162-1963.

Index To IEEE Publications. IEEE Service Center, 445 Hoes Lane, Piscataway, NJ 08855-1331. (908) 981-1393. (800) 678-IEEE. FAX (908) 981-9667. 1973 to present. Annual. ISSN: 0099-1368.

Physics Abstracts. INSPEC. Section A, Science Abstracts. Institution of Electrical Engineers (IEE), London. Available from: INSPEC/IEEE-Institute of Electrical and Electronic Engineers, Box 1331, Hoes Lane, Piscataway, NJ 08855-1331. (908) 562-5549. 1898 to present. 24 issues per year. $2835.00 per year. Also available online and on CD-ROM. ISSN: 0036-8091.

Physics Briefs (Physikalische Berichte). Information Center for Energy, Physics, Mathematics; German Physical Society. V C H Publishers, Inc., 220 East 23rd Street, New York, NY 10010-4606. (212) 683-8333. 1845 to present. 24 issues per year. $2390.00 per year. Also available online. ISSN: 0179-7434.

ASSOCIATIONS AND PROFESSIONAL SOCIETIES

American Electronics Association. 5201 Great America Parkway, Suite 520, PO Box 54990, Santa Clara, CA 95056. (408) 987-4200. FAX (408) 970-8565.

Audio Engineering Society. 60 East 42nd Street, New York, NY 10017. (212) 661-8528.

TELEVISION

Ency. of Physical Sciences and Engineering Info. Sources

Electronics Industries Association. 2500 Wilson Boulevard, Arlington, VA 22201. (703) 907-7500. FAX (202) 457-4985.

IEEE (Institute of Electrical and Electronics Engineers). 345 East 47th Street, New York, NY 10017. (212) 705-7900. FAX (212 705-4929.

National Association of Radio and Telecommunications Engineers. PO Box 678, Medway, MA 02053. (508) 533-8333. FAX (508) 533-3815.

Society of Cable Telecommunications Engineers. 669 Exton Commons, Exton, PA 19341. (215) 363-6888. FAX (610) 363-5898.

Society of Motion Picture and Television Engineers. 595 West Hartsdale Avenue, White Plains, NY 10607. (914) 761-1100. FAX (914) 761-3115.

DIRECTORIES AND BIOGRAPHICAL SOURCES

Broadcast Engineering-Equipment Reference Manual. Intertec Publishing Corporation, Box 12901, Overland Park, KS 66212. (913) 888-4664. Annual. $20.00.

EEM-Electronic Engineer's Master. Hearst Business Communications, Inc., 645 Stewart Avenue, Garden City NY 11530. (516) 227-1300. Annual. $90.00. ISSN: 0732-9016.

IEEE Membership Directory. Institute of Electrical and Electronics Engineers, IEEE Service Center, 445 Hoes Lane, Piscataway, NY 08854. (908) 981-1393. (800) 678-IEEE. FAX (908) 981-9667. 2 volumes. Annual. $190.00.

International Satellite Directory. Design Publishers, 800 Siesta Way, Sonoma, CA 95476. (707) 930-0306. Annual. $250.00.

Satellite Communications, Satellite Industry Directory Issue. Cardiff Publishing, 6300 South Syracuse Way, Suite 650, Englewood, CO 80111. (303) 220-0600. Annual, June. $20.00.

Society of Motion Picture and Television Engineers - Membership Directory. SMPTE, 595 West Hartsdale Avenue, White Plains, NY 10607. (914) 761-1100. Members only.

Television and Cable Factbook. Warren Publishing, 2115 Ward Court NW, Washington, DC 20037. (202) 672-9200.

World Satellite Directory. PHILLIP Publishing, 7811 Montrose Road, Potomac, MD 20854. (800) 777-5006. (301) 340-2100. Annual. $227.00.

Who's Who in Engineering. Gordon Davis, editor. American Association of Engineering Societies. 1111 19th Street, NY, Suite 608, Washington, DC 20036. (202) 296-2237. (800) 658-8897. 9th edition. 1995. $220.00.

ENCYCLOPEDIAS AND DICTIONARIES

Facts On File Dictionary of Television, Cable and Video. Robert M. Reed. Facts-on-File, Inc., 460 Park Avenue South, New York, NY 10016-7382. (212) 683-2244. (800) 322-8755. Fax (800) 683-3633. 1994. $24.95.

IEEE Standard Dictionary of Electrical and Electronics Terms. Christopher J. Booth, editor. IEEE Service Center, 445 Hoes Lane, Piscataway, NJ 08855-1331. (908) 981-1393. (800) 678-IEEE. FAX (908) 981-9667. IEEE Standard 100-1992. 5th edition. 1993. $150.00.

Illustrated Dictionary of Electronics. Stan Gibilsco. TAB Books, P.O. Box 40, Blue Summit, PA 17294-0850. (717) 794-2191. (800) 233-1128. 7th edition. 1994. $34.95.

McGraw-Hill Electronics Dictionary. John Markus and Neil Sclater. McGraw-Hill-Hill Publishing Company, Inc., 1221 Avenue of the Americas, New York, NY 10020. (212) 512-2000. (800) 262-4729. 5th edition. 1994. $49.95. @ @General Works

Advanced Television and Electronic Imaging For Film and Video. Joyce R. Hurwitz. Society of Motion Picture and Television Engineers-SMPTE, 595 West Hartsdale Avenue, White Plains, NY 10607. (914) 761-1100. 1993. $25.00 in paper.

Broadcast Transmission Engineering Practice. Wharton, et al. Butterworth-Heinemann, 313 Washington Street, Newton, MA 02158. (618) 928-2500. (800) 366-2665. 1992. $125.00.

Color and Black and White Television Theory and Servicing. Alvin A. Liff and Sam Wilson. Prentice Hall, 113 Sylvan Avenue, Route 9W, Englewood Cliffs, NJ 07632. (201) 592-2000. (800) 922-0579. 3rd edition. 1993. $65.00.

Complete Guide To Digital Television Troubleshooting and Repair. John D. Lenk. Prentice Hall, 113 Sylvan Avenue, Route 9W, Englewood Cliffs, NJ 07632. (201) 592- 2000. (800) 922-0579.

Electric and Electronic Engineering For Scientists and Engineers. K. A. Krishnamurthy. Halsted Press, 605 Third Avenue, New York, 10158-0012. (212) 850-6400. 1944. $48.95.

Introduction To Radio Frequency and Microwave Engineering. Robert K. Feeney and David R. Hertling. TAB Books, P.O. Box 40, Blue Summit, PA 17294-0850. (717) 794-2191. (800) 233-1128. 1995. $44.50.

Low Power Communications. Richard H. Arland. Tiare Publications, P.O. Box 493, Lake Geneva, WI 53147. (800) 420-0579 volume 1: Basic QRP. 1992. $14.95.

Pocket Radio Operator's Manual. Glynn B. Rogers. CQ Communications, Inc., 76 North Broadway, Hicksville, NY 11801. (516) 681-2922. 1993. $15.95 in paper.

The Technique of Lighting For Television and Film. Gerald Millerson. Butterworth-Heinemann, 313 Washington Street, Newton, MA 02158. (618) 928-2500. (800) 366-2665. 3rd edition. 1991. $46.95.

Telecommunications Circuit Design. Patrick D. Van der Puije. John Wiley & Sons, Inc., 605 Third Avenue, New York, NY 10158-0012. (212) 850-6000. (800) 225-5945. 1992. $79.95.

Television and Video Systems. Charles Buscombe. Prentice- Hall, 113 Sylvan Avenue, Route 9W, Englewood Cliffs, NJ 07632. (201) 592-2000. (800) 922-0579. 1990. $70.00.

Troubleshooting and Repairing Solid-State TVS. Homer L. Davidson. TAB Books, P.O. Box 40, Blue Summit, PA 17294-0850. (717) 794-2191. (800) 233-1128.

Video Engineering: NTSC, EDTV and HDTV Systems. F. Inglis. McGraw-Hill Publishing Company, Inc., 1221 Avenue of the Americas, New York, NY 10020. (212) 512-2000. (800) 262-4729. 1992. $50.00.

HANDBOOKS AND MANUALS

AARL Handbook For Radio Amateurs. American Radio Relay League, 225 Main Street, Newington, CT 06111. (203) 666-1541. FAX (203) 665-7531. 78th edition. 1996. $30.00. ISSN: 0079-9440.

American Electricians Handbook. Terrell Croft and Wilford I. Summers. McGraw-Hill Book Company, 1221 Avenue of the Americas, New York, NY 10020. (212) 512-2000. (800) 262-4729. 12th edition. 1992. $77.50.

Antenna Engineering Handbook. Richard C. Johnson. McGraw-Hill Publishing Company, Inc., 1221 Avenue of the Americas, New York, NY 10020. (212) 512-2000. (800) 262-4729. 3rd edition. 1993. $119.50.

Complete Handbook of Amateur Radio. Clay Laster. McGraw-Hill Publishing Company, Inc., 1221 Avenue of the Americas, New York, NY 10020. (212) 512-2000. (800) 262-4729. 3rd edition. 1993. $21.95.

Electronics Handbook. Jerry C. Whitaker. CRC Press, Inc., 2000 Corporate Boulevard, NW, Boca Raton, FL 33431. (407) 994-0555. (800) 272-7737. 1996. $120.00.

Lenk's Television Handbook: Operation and Troubleshooting. John D. Lenk. McGraw-Hill Publishing Company, Inc., 1221 Avenue of the Americas, New York, NY 10020. (212) 512-2000. (800) 262-4729. 1993. $39.50.

Newnes Television and Video Engineer's Pocket Book. Eugene Trundle. Butterworth-Heinemann, 313 Washington Street, Newton, MA 02158. (618) 928-2500. (800) 366-2665. 1992.$32.95.

Practical Antenna Handbook. Joseph J. Carr. McGraw-Hill Publishing Company, Inc., 1221 Avenue of the Americas, New York, NY 10020. (212) 512-2000. (800) 262-4729. 1990. $32.95.

Standard Handbook For Electrical Engineers. Donald Fink. McGraw-Hill Publishing Company, 1221 Avenue of the Americas, New York, NY 10020. (212) 512-2000. (800) 262-4729. 13th edition. 1996. $110.00.

TAB Service Manual For CCTV and MATV. Robert L. Goodman. TAB Books, P.O. Box 40, Blue Summit, PA 17294-0850. (717) 794-2191. (800) 233-1128. 1991. $29.95.

Television and Video Engineer's Reference Book. K. Jackson and B. Townsend. Butterworth-Heinemann, 313 Washington Street, Newton, MA 02158. (618) 928-2500. (800) 366-2665. 1991. $79.95.

ONLINE DATABASES AND CD-ROMS

COMPENDEX PLUS. Engineering Information, Inc., 345 East 47th Street, New York, NY 10017. (212) 705-7600 or (800) 221-1044. Contains citations with abstracts to worldwide literature in engineering and technology, from 1970 to present. Available on online BRS,(800) 289-4277, DIALOG, (800) 334-2564, ORBIT (800) 456-7248, and STN International, FIZ Karlsruhe,

P.O. Box 2465, W-7500, Karlsruhe 1, Germany, online services. Also available on CD-ROM. Inquire as to cost and availability.

INSPEC. Institution of Electrical Engineers, Michael Faraday House, Six Hills Way, Stevenage, Herts. SG1 2AY, England. Telephone: 0438 313311 or FAX 0438 742840. Contains citations to the worldwide literature of physics, electronics and electrical engineering, computer technology, and related fields. Available on BRS, (800) 289-4277, DIALOG, (800) 334-2564, ORBIT, (800) 456-7248, and STN International, FIZ Karlsruhe, P.O. Box 2465, W-7500, Karlsruhe 1, Germany, online services. Inquire as to cost and availability.

NTIS Bibliographic Database. National Technical Information Service, 5285 Port Royal Road, Springfield, VA 22161. (703) 487-4929 or FAX (703) 321-8199. Broad coverage of government-sponsored science and technology research reports, 1964 to present. Available on BRS,(800) 289-4277, DIALOG, (800) 334-2564, ORBIT, (800) 456-7248, and STN International, FIZ Karlsruhe, P.O. Box 2465, W-7500, Karlsruhe 1, Germany, online services. Also available on CD-ROM. Inquire as to cost and availability.

WILSONLINE. H.W. Wilson Company, 950 University Avenue, Bronx, NY 10452. (800) 367-6770 or (212) 588-8400. Makes available online versions of the printed H.W. Wilson indexes, including Applied Science and Technology Index, Business Periodicals Index, General Science Index, and Readers' Guide to Periodical Literature. Period covered is generally 1983 to present. Available on BRS, (800) 289-4277, DIALOG, (800) 334-2564, and OCLC EPIC, (800) 848-5878, online services. Also available on CD-ROM. Inquire as to cost and availability.

PERIODICALS

Audio Engineering Society Journal. Audio Engineering Society, 60 East 42nd Street, New York, NY 10017. (212) 661-8528. 1953 to present. 10 issues per year. $125.00 per year. ISSN: 0004-7554.

Broadcast Engineering. Intertec Publishing Corporation, Box 12801, Overland Park, KS 66212. (913) 341-1300. 1959 to present. Monthly. $50.00 per year. ISSN: 0007-1994.

DB, the Sound Engineering Magazine. 203 Commack Road, Suite 1010, Commack, NY 1725. (516) 586-6530. 1967 to present. $15.00 per year. ISSN: 0011-7145.

Electronic Design. Penton Publishing, San Jose Gateway, Suite 354. 2025 Gateway Place, San Jose, CA 95110. (408) 441-0550. 1952 to present. Fortnightly. $95.00. ISSN: 0013-4872.

Electronics. Penton Publishing, San Jose Gateway, Suite 354. 2025 Gateway Place, San Jose, CA 95110. (408) 441-0550. 1930 to present. Semi-weekly. $98.00 per year. ISSN: 0883-4989.

Electronics Now. Gernsback Publications, Inc., 500-B Bi-County Boulevard, Farmingdale, NY 11735. (516) 293-3000. FAX (516) 293-3115. 1929 to present. Monthly. $19.95. ISSN: 0033-7862.

Electronics World and Wireless World. Reed Business Publishing, Ltd., Quadrant House, the Quadrant, Sutton, Surrey SM2 5AS, England. 1911 to present. Monthly. L35. ISSN: 0266-3244.

IEEE Antennas and Propagation Magazine. Institute of Electrical and Electronics Engineers, Inc., 445 Hoes Lane, Box 1331,

TELEVISION

Ency. of Physical Sciences and Engineering Info. Sources

Piscataway, NJ 08855-1331. (908) 981-0060. FAX (908) 981-9667. Bi-monthly. $85.00. ISSN: 1045-9243.

IEEE Circuits and Devices Magazine. Institute of Electrical and Electronics Engineers, Inc., Box 1331, 445 Hoes Lane, Piscataway, NJ 08855-1331. (908) 981-0060. 1985 to present. Bi-monthly. $120.00 per year. ISSN: 8755-3996.

IEEE Journal of Solid State Circuits. Institute of Electrical and Electronics Engineers, Inc., Box 1331, 445 Hoes Lane, Piscataway, NJ 08855-1331. (908) 981-0060. 1966 to present. Bi-monthly. $275.00 per year. ISSN: 0018-9200.

IEEE Transactions On Broadcasting. Institute of Electrical and Electronics Engineers, Inc., Box 1331, 445 Hoes Lane, Piscataway, NJ 08855-1331. (908) 981-0060. 1955 to present. Quarterly. $50.00 per year. ISSN: 0018-9316.

Microwaves and RF. Penton Publishing Co., Circulation Department, 1100 Superior Avenue, Cleveland, OH 44114. (201) 393-6060. FAX (201) 393-6297. From 1962 to present. Monthly. $60.00. ISSN: 0745-2992.

Television Broadcast. Globecom Publishing Ltd., 4451 West 107th, No. 210, Overland Park, KS 66207-4024. 1978 to present. Monthly. $38.00 per year. ISSN: 0898-767X.

RESEARCH CENTERS AND INSTITUTES

Center for Advanced Technology in Telecommunications. Polytechnic University, 333 Jay Street, Brooklyn, NY 11202. (718) 643-5160.

Communications Research Laboratory. McMaster University, 1280 Main Street West, Hamilton, ON Canada L85 4K1. (416) 525-9140.

Communications Satellite Planning Center. Stanford University, ERL, Building, Room 202, Stanford, CA 94305-4053. (415) 723-3471. FAX (415) 723-7298.

TEMPERATURE MEASUREMENT

See: THERMOMETERS

TERRAIN SENSING

See: REMOTE SENSING

TERRESTRIAL MAGNETISM

See: GEOPHYSICS

TEXTILES

See also: DYES AND DYEING, MATERIALS SCIENCE

ABSTRACT SERVICES AND INDEXES

Applied Science and Technology Index; A Cumulative Subject Index to English Language Periodicals in the Fields of Aeronautics and Space Science, Computer Technology, Chemistry, Construction Industry, Energy and Related Areas. H.W. Wilson Co., 950 University Avenue, Bronx, NY 10452. (212) 588-8400. (800) 367-6770. FAX (718) 590-1617. From 1958 to present. Monthly. Inquire about cost and availability. Also available on CD-ROM and online. ISSN: 0003-6986.

Chemical Abstracts. Chemical Abstracts Service, 2540 Olentangy River Road, Box 3012, Columbus, OH 43210-0012. (614) 447-3600. (800) 848-6538. FAX (614) 447-3713. From 1907 to present. Weekly. $16800.00 per year. Also available on CD-ROM and online. Inquire regarding cost and availability.

Engineering Index Monthly. Engineering Information, Inc., Castle Point on the Hudson, Hoboken, NJ 07030. (201) 216-8500. FAX (201) 216-8532. Monthly. $2200.00 per year. Also available online as COMPENDEX and on CD-ROM. ISSN: 0742-1974.

General Science Index. H.W. Wilson Company, 950 University Avenue, Bronx, NY 10452. (212) 588-8400. (800) 367-6770. FAX (718) 590-1617. From 1978 to present. Ten issues per year; quarterly and annual cumulations. Service basis. Available on CD-ROM and online. Inquire regarding cost and availability. ISSN: 0162-1963.

Science Citation Index. SCI. Institute for Scientific Information, 3501 Market Street, Philadelphia, PA 19104. (215) 386-0100. (800) 523-1850. FAX (215) 386-2991. 1961 to present. Six issues per year, plus annual cumulation. $11650.00 per year. Also available online and on CD-ROM. Inquire about price and availability. ISSN: 0036-827X.

Textile Technology Digest. Institute of Textile Technology, Charlottesville, VA 22902. (804) 296-5511. FAX (804) 977-5400. 1944 to present. Monthly. $475.00 per year. ISSN: 0040-5191.

World Textile Abstracts. Elsevier Science, 660 White Plains Road, Tarrytown, NY 10591-5151. (914) 524-9200. FAX (914) 333-2444. 1969 to present. Monthly. $645.00 per year. Also available online. Inquire about price and availability. ISSN: 0043-9118.

ASSOCIATIONS AND PROFESSIONAL SOCIETIES

American Association For Textile Technology. PO Box 99, Gastonia, NC 28053. (704) 824-3522. FAX (704) 824-0630.

American Association of Textile Chemists and Colorists. PO Box 12215. Research Triangle Park, NC 27709-2215. (919) 549-8141. FAX (919) 549-8933.

Industrial Fabrics Association International. 345 Cedar Street, Suite 800, St. Paul MN 55101. (612) 222-2508. FAX (612) 222-8215.

TRI/Princeton; (Textile Research Institute). PO Box 625, Princeton, NJ 08542. (609) 924-3150. (609) 683-7836.

DIRECTORIES AND BIOGRAPHICAL SOURCES

American Association of Textile Chemists and Colorists-Membership Directory. AATCC, PO Box 99, Gastonia, NC 28053. (704) 824-3522. FAX (704) 824-0630. Annual. $55.00.

America's Textiles International-Special Directory Issue. Billian Publishing, 2100 Powers Ferry Road, Suite 300, Atlanta, GA 30339. (404) 955-5656. Annual, July. $25.00.

Davison's Textile Blue Book. Davison Publishing Company, Inc., Box 477, Ridgewood, NJ 07451. (201) 445-3135. 1866 to present. Annual. $120.00 per year. ISSN: 0070-2951.

Research Centers Directory. Gale Research, 835 Penobscot Building, Detroit, MI 48226-4094. (313) 961-2242. (800) 347-4253. 20th edition. 1995. $485.00. ISSN: 0080-1518.

Who's Who in Technology: Biographies and Index. Gale Research, 835 Penobscot Building, Detroit, MI 48226-4094. (313) 961-2242. (800) 347-4253. 7th edition. 1995. $195.00.

ENCYCLOPEDIAS AND DICTIONARIES

Encyclopedia of Textiles. Judith Jerde. Facts-on-File, Inc., 460 Park Avenue South, New York, NY 10016-7382. (212) 683-2244. (800) 322-8755. Fax (800) 683-3633. 1992. $45.00.

Fairchild's Dictionary of Textiles. Phyllis B. Tortora, editor. Fairchild Books, 7 West 34th Street, New York, NY 10001. (212) 630-3880. (800) 247-6622.

Kirk-Othmer Encyclopedia of Chemical Technology. John Wiley and Sons, Inc., 605 Third Avenue, New York, NY 10158. (800) 526-5368 or (212) 850-6000. Fourth edition. 1991-. Twenty-seven volumes. $5400.00.

McGraw-Hill Encyclopedia of Science and Technology. McGraw-Hill Book Company, Inc., 1221 Avenue of the Americas, New York, NY 10020. (212) 512-2000. (800) 262-4729. 7th edition. 20 volume set. 1992. $1900.00.

Textile Terms and Definitions. Textile Institute. State Mutual Book Service, Ltd. 521 Fifth Avenue, New York, NY 10175. (212) 682-5844. 1991. $150.00.

GENERAL WORKS

Basics of Dyeing and Finishing. American Association of Textile Chemists and Colorists, PO Box 12215. Research Triangle Park, NC 27709-2215. (919) 549-8141. FAX (919) 549-8933. 1993. $65.00.

Comfort Properties of Textiles. K. Slater. State Mutual Book Service, Ltd. 521 Fifth Avenue, New York, NY 10175. (212) 682-5844. 1989. $90.00.

Geotextiles and Geomembranes: Definitions, Properties and Design. Jean-Pierre Giroud. Industrial Fabrics Association, Intl., 345 Cedar Street, Suite 800, St. Paul MN 55101. (612) 222-2508. FAX (612) 222-8215. 1984. $49.00.

High Performance Fibers, Composites and Engineering Textile Structure. John W. Hearle, editor. State Mutual Book Service,

Ltd. 521 Fifth Avenue, New York, NY 10175. (212) 682-5844. 1990. $125.00.

Illustrated History of Textiles. Neil Harvey. Random House, Inc., 201 East 50th Street, New York, NY 10022. (212) 751-2600. (800) 733-3000. 1991. $24.99.

Introductory Textile Science. Marjory L. Joseph. Harcourt Brace College Publications, 6277 Sea Harbor Drive, Orlando, FL 32887. (800) 782-4479. 6th edition. 1993.

Synthetic Fibre Materials. H. Brody, editor. Longman Publishing Group, 10 Bank Street, White Plains, NY 10606-1951. (914) 993-5000. (800) 266-8855. 1994. $175.00.

Understanding Textiles. Phyllis G. Tortora and Billie J. Collier. Macmillan Publishing Company, Inc,. 200 Old Tappan Road, Old Tappan, NJ 07675. (800) 233-2336. 5th edition. 1996. $60.00.

Woven Fabric Composites. Niranjan Naik. Technomic Publishing Co., 851 New Holland Avenue, Box 3535, Lancaster PA 17604. (800) 233-9936. 1993. $65.00 in paper.

HANDBOOKS AND MANUALS

AATCC Technical Manual. American Association of Textile Chemists and Colorists, PO Box 12215. Research Triangle Park, NC 27709-2215. (919) 549-8141. FAX (919) 549-8933. 1994. $95.00.

Thames & Hudson Manual of Dyes and Fabrics. Joyce Storey. Thames & Hudson, 500 Fifth Avenue, New York, NY 10110. (212) 354-3763. 1992. $14.95.

Textile Conservator's Manual. Sheila Landi. Butterworth-Heinemann, 313 Washington Street, Newton, MA 02158. (618) 928- 2500. (800) 366-2665. 2nd edition. 1992. $155.00

The Textile Industry: An Information Sourcebook. J. Thomas Vogel and Barbara Lowry. Ornyx Press, 4041 North Central Avenue, Phoenix, AZ 85012-3397. 1989. $52.00.

ONLINE DATABASES AND CD-ROMS

CA Search. Chemical Abstracts Service, P.O. Box 3012, Columbus, OH 43210-0012. (614) 447-3600. (800) 848-6533. FAX (614) 447-3709. Very comprehensive guide to worldwide chemical literature and related fields, 1972 to present. Available on BRS,(800) 289-4277, DIALOG, (800) 334-2564, ORBIT (800) 456-7248, and STN International, FIZ Karlsruhe, P.O. Box 2465, W-7500, Karlsruhe 1, Germany, online services. Inquire as to cost and availability.

COMPENDEX Plus. Engineering Information, Inc., 345 East 47th Street, New York, NY 10017. (212) 705-7600 or (800) 221-1044. Contains citations with abstracts to worldwide literature in engineering and technology, from 1970 to present. Available on online BRS,(800) 289-4277, DIALOG, (800) 334-2564, ORBIT (800) 456-7248, and STN International, FIZ Karlsruhe, P.O. Box 2465, W-7500, Karlsruhe 1, Germany, online services. Also available on CD-ROM. Inquire as to cost and availability.

Dissertation Abstracts Online. University Microfilms International, 300 North Zeeb Road, Ann Arbor, MI 48106. (800) 521-0600 or (313) 761-4700. Scope includes virtually all doctoral

TEXTILES

Ency. of Physical Sciences and Engineering Info. Sources

dissertations accepted at accredited American institutions from 1861 to present in 252 subject areas. Available on BRS, (800) 289-4277, DIALOG, (800) 334-2564, and OCLC EPIC,(800) 848-5878, online services. Also available on CD-ROM. Inquire as to cost and availability.

INSPEC. Institution of Electrical Engineers, Michael Faraday House, Six Hills Way, Stevenage, Herts. SG1 2AY, England. Telephone: 0438 313311 or FAX 0438 742840. Contains citations to the worldwide literature of physics, electronics and electrical engineering, computer technology, and related fields. Available on BRS, (800) 289-4277, DIALOG, (800) 334-2564, ORBIT, (800) 456-7248, and STN International, FIZ Karlsruhe, P.O. Box 2465, W-7500, Karlsruhe 1, Germany, online services. Inquire as to cost and availability.

NTIS Bibliographic Database. National Technical Information Service, 5285 Port Royal Road, Springfield, VA 22161. (703) 487-4929 or FAX (703) 321-8199. Broad coverage of government-sponsored science and technology research reports, 1964 to present. Available on BRS,(800) 289-4277, DIALOG, (800) 334-2564, ORBIT, (800) 456-7248, and STN International, FIZ Karlsruhe, P.O. Box 2465, W-7500, Karlsruhe 1, Germany, online services. Also available on CD-ROM. Inquire as to cost and availability.

SCISEARCH. Institute for Scientific Information, 3501 Market Street, Philadelphia, PA 19104. (800) 523-1850 or (215) 386-0100. Broad multidisciplinary title and author index to the international literature of science and technology, 1974 to present. Available on DIALOG, (800) 334-2564, and ORBIT, (800) 456-7248, online services. Also available on CD-ROM. Inquire as to cost and availability.

WILSONLINE. H.W. Wilson Company, 950 University Avenue, Bronx, NY 10452. (800) 367-6770 or (212) 588-8400. Makes available online versions of the printed H.W. Wilson indexes, including Applied Science and Technology Index, Business Periodicals Index, General Science Index, and Readers' Guide to Periodical Literature. Period covered is generally 1983 to present. Available on BRS, (800) 289-4277, DIALOG, (800) 334-2564, and OCLC EPIC, (800) 848-5878, online services. Also available on CD-ROM. Inquire as to cost and availability.

World Textiles. Geo Abstracts, Regency House, 34 Duke Street, Norwich NR3 3AP, England. Contains citations to the worldwide textile literature from 1970 to date. Available on DIALOG,(800) 334-2564, online service. Inquire as to cost and availability.

PERIODICALS

America's Textiles International. Billian Publishing, 2100 Powers Ferry Road, Suite 300, Atlanta, GA 30339. (404) 955-5656. 1887 to present. Monthly. $43.00 per year. ISSN: 0890-9970.

Disposables and Nonwovens. Chandler Publications zltd., 10 South Street, Totnes, Doven TP9 5DZ, England. Bi-monthly. $70.00 per year. ISSN: 0012-3811.

Fiber Organon. Fiber Economics Bureau, Inc., 101 Eisenhower Parkway, Roseland, NJ 07068. (201) 228-1107. 1930 to present. Monthly. $300.00 per year. ISSN: 0040-5132.

High Performance Textiles. Elsevier Science, 660 White Plains Road, Tarrytown, NY 10591-5151. (914) 524-9200. FAX (914) 333-2444. 1980 to present. Monthly. $395.00 per year. ISSN: 0144-5871.

Industrial Fabric Products Review. Industrial Fabrics Product Association International, 345 Cedar Street, St. Paul, MN 55101-1088. (612) 222-2508. 1915 to present. Monthly. $34.00 per year. ISSN: 0019-8307.

Journal of Coated Fabrics. Technomic Publishing Co., Inc., 851 New Holland Avenue, Box 3535, Lancaster, PA 17604. (717) 291-5609. 1971 to present. Quarterly. $205.00. ISSN: 0093-4658.

Textile Chemist and Colorist. American Association of Textile Chemists and Colorists. One Davis Drive, PO Box 12215. Research Triangle Park, NC 27709-2215. (919) 549-8141. FAX (919) 549-8933. 1969 to present. Monthly. $30.00 per year. ISSN: 0040-490X.

Textile Institute Journal. Textile Institute, 10 Blackfriars Street, Manchester M3 5DR, England. 1910 to present. Quarterly. L105 per year. ISSN: 0040-5000.

Textile Research Journal. Textile Research Institute, 601 Prospect Avenue, PO Box 625, Princeton, NJ 08542. (609) 924-3150. (609) 683-7836. 1930 to present. Monthly. $170.00 per year. ISSN: 00400-5175.

Textile World. MacLean Hunter Publishing Company, Textile Publications, 4170 Ashford-Dunwood Road, Atlanta, GA 30319. (404) 847-2770. 1868 to present. Monthly. $42.00 per year. ISSN: 0040-5213.

RESEARCH CENTERS AND INSTITUTES

American Association of Textile Chemists and Colorists. PO Box 12215. Research Triangle Park, NC 27709-2215. (919) 549-8141. FAX (919) 549-8933.

Apparel Manufacturing Technology Center. Georgia Tech Research Institute, O'Keefe Building, Room 215, Georgia Institute of Technology, Atlanta, GA 30332. (404) 894-3636. FAX (404) 853-9172.

Fibrous Materials Research Center. Drexel University, Department of Materials Engineering, 31st and Chestnut Streets, Philadelphia, PA 19104. (215) 895-1642. FAX (215) 895-6684.

Institute of Textile Technology. PO Box 391, Charlottesville, VA 22902. (804) 296-5511. (804) 296-2957.

International Center For Textile Research and Development. Texas Tech University, PO Box 5888, Lubbock, TX 79417. (806) 747-3790. (806) 747-3796.

International Fabricare Institute. 12251 Tech Road, Montgomery Industrial Park, Silver Spring, MD 20904. (301) 622-1900. FAX (301) 236-9320.

Museum of American Textile History. 800 Massachusetts Avenue, North Andover, MA 01845. (508) 686-0191.

Textile and Apparel Research. Philadelphia College of Textiles and Science, Schoolhouse Lane and Henry Avenue, Philadelphia, PA 19144. (215) 951-2750. (215) 951-2615.

Textile Research Institute. 601 Prospect Avenue, PO Box 625, Princeton, NJ 08542. (609) 924-3150. (609) 683-7836.

THEORETICAL PHYSICS

See: PHYSICS

THERMOCHEMISTRY

See also: CARBON, CHEMISTRY, CHEMICAL ENGINEERING, COMBUSTION

ABSTRACT SERVICES AND INDEXES

Analytical Abstracts. Royal Society of Chemistry, Information Services, Thomas Graham House, Science Park, Milton Road, Cambridge, CB4 4WF, England. Contains citations, mostly with abstracts, of the worldwide literature on analytical chemistry, from 1954 to present. Monthly. $636.00 per year. Also available online.

Applied Science and Technology Index; A Cumulative Subject Index To English Language Periodicals in the Fields of Aeronautics and Space Science, Computer Technology, Chemistry, Construction Industry, Energy and Related Areas. H.W. Wilson Co., 950 University Avenue, Bronx, NY 10452. (212) 588-8400. (800) 367-6770. FAX (718) 590-1617. From 1958 to present. Monthly. Inquire about cost and availability. Also available on CD-ROM and online. ISSN: 0003-6986.

Chemical Abstracts. Chemical Abstracts Service. 2540 Olentangy River Road, Box 3012, Columbus, OH 43210-0012. (614) 447-3600. FAX (614) 447-3713. 1907 to present. Weekly. $16,800.00 per year. Available online and on CD-ROM. CA is also available in five section groupings. Inquire regarding cost and availability.

Chemical Engineering Abstracts. Royal Society of Chemistry, Information Services, Thomas Graham House, Science Park, Milton Road, Cambridge, CB4 4WF, England. Contains citations, mostly with abstracts, of the worldwide literature on chemical engineering, from 1982 to present. Monthly. $450.00 per year. Also available online.

Current Contents: Physical, Chemical and Earth Sciences. Institute for Scientific Information, 3501 Market Street, Philadelphia, PA 19104. (215) 386-0100. FAX (215) 386-2291. From 1961 to present. Weekly. $442.00 per year. Also available on CD-ROM and online. Inquire regarding cost and availability. ISSN: 0163-2574.

Engineered Materials Abstracts. ASM International (American Society for Metals), 9639 Kinsman Road, Materials Park, OH 44073. (216) 338-5151 or FAX (216) 338-4634. Covers literature on technical developments in polymer, ceramic, and composite materials and engineering. 1986 to present. Monthly. $1175.00 per year. Also available on CD-ROM. ISSN: 0951-9998.

Engineering Index Monthly; Indexes and Abstracts the World's Engineering and Technical Literature. Engineering Information, Inc., Castle Point on the Hudson, Hoboken, NJ 07030. (201) 216-8500. (800) 221-1044. FAX (201) 216-8532. Monthly. $2300.00 per year. Available online as COMPENDEX and also on CD-ROM. ISSN: 0742-1974.

General Science Index. H.W. Wilson Company, 950 University Avenue, Bronx, NY 10452. (212) 588-8400. (800) 367-6770. FAX (718) 590-1617. From 1978 to present. Ten issues per year;

quarterly and annual cumulations. Service basis. Available on CD-ROM and online. Inquire regarding cost and availability. ISSN: 0162-1963.

Government Reports Announcements and Index. National Technical Information Service (NTIS), 5285 Port Royal Road, Springfield, VA 22161. (703) 487-4650. FAX (703) 321-8547. From 1968 to present. Annual. $630.00 per year. Also available online as NTIS Bibliographic Database and on CD-ROM.

Literature Abstracts: Catalysts and Catalysis. American Petroleum Institute, Central Abstracting and Information Services, 275 Seventh Avenue, New York, NY 10001. (212) 366-4040 or FAX (212) 366-4298. Contains citations with abstracts to literature relating to catalysts used in petroleum refining and the petrochemical industries, from 1985 to present. Weekly. $160.00 per year. Also available online.

Physics Abstracts. INSPEC. Section A, Science Abstracts. Institution of Electrical Engineers (IEE), London. Available from: INSPEC/IEEE-Institute of Electrical and Electronic Engineers, Box 1331, Hoes Lane, Piscataway, NJ 08855-1331. (908) 562-5549. 1898 to present. 24 issues per year. $2835.00 per year. Also available online and on CD-ROM. ISSN: 0036-8091.

Science Citation Index. SCI. Institute for Scientific Information, 3501 Market Street, Philadelphia, PA 19104. (215) 386-0100. (800) 523-1850. FAX (215) 386-2991. 1961 to present. Six issues per year, plus annual cumulation. $11650.00 per year. Also available online and on CD-ROM. Inquire about price and availability. ISSN: 0036-827X.

ANNUAL REVIEWS AND YEARBOOKS

Advances in Catalysis. Academic Press, Inc., 1250 Sixth Avenue, San Diego, CA 92101-4311. (619) 231-0926. FAX (619) 699-6715. 1948 to present. Irregular. Price varies.

Advances in Physical Organic Chemistry. Academic Press, Inc., 1250 Sixth Avenue, San Diego, CA 92101-4311. (619) 231-0926. FAX (619) 699-6715. 1963 to present. Irregular. Price varies

Annual Review of Physical Chemistry. Annual Reviews, Inc., 4139 El Camino Way, Palo Alto, CA 94306-0897. (415) 493-4400. Fax (415) 855-9815. 1951 to present. Annual. $55.00 per year.

ASSOCIATIONS AND PROFESSIONAL SOCIETIES

American Chemical Society, 1155 16th Street, NW, Washington, DC 20036. (202) 872-4600.

American Institute of Chemical Engineers. 345 East 47th Street, New York, NY 10017-2396. (212) 705-7663.

American Institute of Physics, One Physics Ellipse, College Park, MD 20740-3843. (301) 209-3100.

Association of Consulting Chemists and Chemical Engineers, 50 East 41st Street, Suite 92, New York, NY 10017. (212) 684-6255.

Association of official Analytical Chemists, 2200 Wilson Boulevard, Suite 400, Arlington, VA 22001. (703) 522-3032.

THERMOCHEMISTRY

Ency. of Physical Sciences and Engineering Info. Sources

Chemical Manufacturers Association. 2501 M Street, N.W., Washington, DC 20037. (202) 887-1182

Combustion Institute. 5001 Baum Boulevard, Pittsburgh, PA 15213-1851. (412) 687-1366.

BIBLIOGRAPHIES

Chemical Information Sources. Gary Wiggins. McGraw-Hill Publishing Company, 1221 Avenue of the Americas, New York, NY 10020. (800) 262-4729 or (212) 512-3825. 1991. $42.50.

Information Sources in Chemistry. R.T. Bottle and J.F.B Rowland, editors. R.R. Bowker Inc., 121 Chanlon Road, New Providence, NJ 07974. (800) 521-8110 or (908) 464-6800. Fourth edition. 1993. $75.00.

Handbooks and Tables in Science and Technology. Russell H. Powell, editor. Oryx Press, 4041 North Central, Suite 700, Phoenix, AZ 85012-3330. (602) 265-2651 or (800) 279-6799. Third edition. 1994. $65.00.

Scientific and Technical Books and Serials in Print; An Index To Literature in Science and Technology. R.R. Bowker Inc., 121 Chanlon Road, New Providence, NJ 07974. (908) 464-6800. (800) 521-8110. FAX (908) 665-3502. 1972 to present. Annual. 4 volumes. 1994. $299.95. Also available on compact disc and online. ISSN: 0000-054X.

DIRECTORIES AND BIOGRAPHICAL SOURCES

American Institute of Chemical Engineers Directory. American Institute of Chemical Engineers. 345 East 47th Street, New York, NY 10017-2396. (212) 705-7663. Annual.

American Men and Women of Science. Physical and Biological Sciences. R.R. Bowker Company, 121 Chanlon Road, New Providence, NJ 07974. (908) 464-6800. (800) 521-8110. 20th edition. 8 volumes. 1996. $850.00.

Chemical Engineering Faculties. American Institute of Chemical Engineers. 345 East 47th Street, New York, NY 10017-2396. (212) 705-7663. Annual. $75.00.

Consulting Services: Chemists and Chemical Engineers. Association of Consulting Chemists and Chemical Engineers, 50 East 41st Street, Suite 92, New York, NY 10017. (212) 684-6255. Biennial. $60.00.

Directory of Chemistry Software 1992. Wendy Warr, Peter Willett, Geoff Downs. American Chemical Society, 1155 16th Street, NW, Washington, DC 20036. (202) 872-4600. 1992. $35.95.

Engineering Research Centers: Incorporating Electronics Research Centers. Stockton Press, 345 Park Avenue, New York, NY 10010. (212) 689-9200. (800) 221-2123. 4th edition. 1995. $515.00.

Research Centers Directory. Gale Research, 835 Penobscot Building, Detroit, MI 48226-4094. (313) 961-2242. (800) 347-4253. 20th edition. 1995. $485.00. ISSN: 0080-1518.

Who's Who in Engineering. Gordon Davis, editor. American Association of Engineering Societies. 1111 19th Street, NY,

Suite 608, Washington, DC 20036. (202) 296-2237. (800) 658-8897. 9th edition. 1995. $220.00.

ENCYCLOPEDIAS AND DICTIONARIES

Academic Press Dictionary of Science and Technology. Christopher Morris, editor. Academic Press, Inc., 1250 Sixth Avenue, San Diego, CA 92101. (619) 231-0926. FAX (619) 699-6715. 1991. $115.00.

Dictionary of Named Processes in Chemical Technology. Alan E. Comyns. Oxford University Press, Inc., 200 Madison Avenue, New York, NY 10016. (212) 725-6000. (800) 334-4249. 1994. $75.00.

Encyclopedia of Applied Physics. VCH Publishers, Inc., 303 Northwest 12th Avenue, Deerfield Beach, FL 33442. (800) 367-8249. 1991-. Twenty volumes. $6000.00.

Encyclopedia of Chemical Processing and Design. McKetta Marcel Dekker, Inc., 270 Madison Avenue, New York, NY 10016. (212) 696-9000. (800) 228-1160. 1976-. $175.00 per volume.

Illustrated Chemistry Laboratory Terminology. Gerbert W. Ockerman. CRC Press, Inc., 2000 Corporate Boulevard, NW, Boca Raton, FL 33431. (407) 994-0555. (800) 272-7737. 1991. $32.95.

Encyclopedia of Applied Physics. VCH Publishers, Inc., 303 Northwest 12th Avenue, Deerfield Beach, FL 33442. (800) 367-8249. 1991-. Twenty volumes. $6000.00.

Industrial Chemical Thesaurus. Michael Ash and Irene Ash. VCH Publishers, Inc., 220 East 23rd Street, New York, NY 10010-4606. (800) 367-8249. 1992. $295.00.

Kirk-Othmer Encyclopedia of Chemical Technology. John Wiley and Sons, Inc., 605 Third Avenue, New York, NY 10158. (800) 526-5368 or (212) 850-6000. Fourth edition. 1991-. Twenty-seven volumes. $5400.00.

McGraw-Hill Encyclopedia of Chemistry. Sybil P. Parker, editor. McGraw-Hill Book, Incorporated, 1221 Avenue of the Americas, New York, NY 10020. (212) 997-3675. Second edition. 1993. $95.50.

McGraw-Hill Encyclopedia of Science and Technology. McGraw-Hill Book, Incorporated, 1221 Avenue of the Americas, New York, NY 10020. (212) 997-3675. (800) 262-4729. Seventh edition. Twenty volumes. 1992. $1900.00.

Ullman's Encyclopedia of Industrial Chemistry. VCH Publications, Inc., 220 East 23rd Street, Suite 909, New York, NY 10010-4606. (212) 683-8333. (800) 422-8824. 5th edition. 1984-. Price varies per volume.

GENERAL WORKS

Applied Combustion. Eugene L. Keating. Marcel Dekker, Inc., 270 Madison Avenue, New York, NY 10016. (212) 696-9000. (800) 228-1160. 1993. $165.00.

Chemical Thermodynamics. Peter A. Rock. University Science Books, 55D Gate Five Road, Sausalito, CA 94965. (415) 332-5390. 1983. $54.00.

Dynamics of Gaseous Combustion. Kuhl et al. American Institute of Aeronautics and Astronautics, 370 L'Enfant Promenade, SW, Washington, DC 20024. (202) 646-7400.(800) 682-2422. 1993. $99.95.

Fuel Combustion: A Source Book. M. Bartok. John Wiley & Sons, Inc., 605 Third Avenue, New York, NY 10158-0012. (212) 850-6000. (800) 225-5945. 1991. $145.00.

Flame and Combustion. J.N. Bradley. Chapman & Hall, 1 Penn Plaza, New York, NY 10119. (212) 564-1060. 2nd edition. 1985. $35.00 in paper.

Gases, Liquids and Solids: and Other States of Matter. D. Tabor. Cambridge University Press, 40 West 20th Street, New York, NY 10011-4211. (212) 924-3900. 1992. $80.00.

Introduction To Combustion Concepts and Applications. Stephen R. Turns. McGraw-Hill Publishing Company, Inc., 1221 Avenue of the Americas, New York, NY 10020. (212) 512-2000. (800) 262-4729. 1996.

Materials Chemistry At High Temperatures. John W. Hastie, editor. Humana Press, 999 Riverview Drive, Suite 209, Totowa, NJ 07512. (201) 256-1699. Volume 2: Thermochemistry & Models. 1990. $120.00.

Materials Thermochemistry. Ortrud, Kubachewski, et al. Elsevier Science Publishing Company, Inc., 655 Avenue of the Americas, New York, NY 10010. (212) 989-5800. FAX (914) 333-2444. 6th edition, revised. 1993. $47.00 in paper.

Thermochemical Data of Pure Substances. Ihsan Barin, et al. VCH Publications, Inc., 220 East 23rd Street, Suite 909, New York, NY 10010-4606. (212) 683-8333. (800) 422-8824. 2nd edition. 1992. $500.00.

Thermochemical Properties of Inorganic Substances. O. Kancke, et al, editors. Springer-Verlag New York, Inc., 175 Fifth Avenue, New York, NY 10010. (212) 460-1500. (800) 777-4643. FAX (212) 473-6272. 2 volumes. 1991. $381.00.

HANDBOOKS AND MANUALS

Chemometrics: Chemical and Sensory Data. David R. Burgard and James T. Kuznicki. CRC Publishers, Inc., 2000 Corporate Blvd., N.W., Boca Raton, FL 33431. (407) 994-0555 or (800) 333-8300. 1990. $120.00.

Catalyst Handbook. Martyn V. Twigg. CRC Press Inc., 2000 Corporate Blvd., NW, Boca Raton, FL 33431. (407) 994-0555. (800) 333-8300. 2nd edition. 1989. $104.00.

Chemical Engineering Reference Manual. Randall N. Robinson. Professional Publications, Inc., 1250 Fifth Avenue, Belmont, CA 94002. (415) 593-0110. (800) 426-1178. 1988. $49.95.

Chemical Formulary. H. Bennett, editor. Chemical Publishing, Co., Inc. 80 Eighth Avenue, New York, NY 10011. (212 255-1950. Volumes 1-30. $60.00 per volume.

CRC Handbook of Chemistry and Physics. David R. Kide, editor. CRC Press Inc., 2000 Corporate Blvd., NW, Boca Raton, FL 33431. (407) 994-0555. (800) 333-8300. 77th edition. 1996. $99.95.

Guide To Basic Chemical Compounds. D.R. Lide, Jr. CRC Press Inc., 2000 Corporate Blvd., N.W., Boca Raton, FL 33431. (407) 994-0555 or (800) 333-8300. 1993. $120.00.

Guide To IUPAC Nomenclature of Organic Compounds. J. C. Richer, et al, editors. CRC Press Inc., 2000 Corporate Blvd., NW, Boca Raton, FL 33431. (407) 994-0555. (800) 333-8300. 1993. $39.95.

Handbook of Chemical Engineering Calculations. Nicholas P. Chopey and Tyler G. Hicks, editors. McGraw-Hill Publishing Company, Inc., 1221 Avenue of the Americas, New York, NY 10020. (212) 512-2000. (800) 262-4729. 2nd edition. 1993. $69.50.

Handbook of Compositions At thermodynamic Equilibrium. Charles R. Noddings and Gary M. Mullett. Books on Demand, 300 North Zeeb Road, Ann Arbor, MI 48106-1346. (313) 761-4700. (800) 521-0600. $179.50.

Handbook of Homogeneous Catalysis. B. Cornils, editor. VCH Publications, Inc., 220 East 23rd Street, Suite 909, New York, NY 10010-4606. (212) 683-8333. (800) 422-8824. 1996. $345.00.

Handbook of Laboratory Safety. A. Keith Furr. CRC Press, Inc., 2000 Corporate Boulevard, NW, Boca Raton, FL 33431. (407) 994-0555. (800) 272-7737. 1995. 125.00.

Handbook of Industrial Chemical Additives. Michael Ash and Irene Ash. VCH Publications, Inc., 220 East 23rd Street, Suite 909, New York, NY 10010-4606. (212) 683-8333. (800) 422-8824. 1991. $195.00.

Handbook of Organic Chemistry. Hans Beyer and Wolfgang Walter. Prentice Hall General Reference & Travel, 15 Columbus Circle, New York, NY 10023. (212) 373-8500. (800) 223-2348. 1995. $96.00.

Handbook of Thermochemical Data for Compounds and Aqueous Species. Herbert E. Barner and Richard V. Scheuerman. Books on Demand. 300 North Zeeb Road, Ann Arbor, MI 48106-1346. (313) 761-4700. $47.50.

Handbook of Thermodynamics of Organic Compounds For Chemical Engineers. Elsevier Science Publishing Company, Inc., 655 Avenue of the Americas, New York, NY 10010. (212) 989-5800. FAX (914) 333-2444. 1987. $79.00.

Laboratory Handbook of Materials, Equipment, and Techniques. Gary S. Coyne. Prentice Hall, 15 Columbus Circle, New York, NY 10023. (212) 373-8500 or (800) 922-0579. 1992. $45.00.

Lange's Handbook of Chemistry. John A. Dean, editor. McGraw-Hill Publishing Company, 1221 Avenue of the Americas, New York, NY 10020. (212) 512-2000. (800) 262-4729. 14th edition. 1996. $99.95.

Perry's Chemical Engineers' Handbook. Robert H. Perry and Donald W. Green, editors. McGraw-Hill Publishing Company, Inc., 1221 Avenue of the Americas, New York, NY 10020. (212) 512-2000. (800) 262-4729. 6th edition. 1996. $129.50.

Riegel's Handbook of Industrial Chemistry. James A. Kent, editor. Van Nostrand Reinhold, 115 Fifth Avenue, New York, NY 10003. (212) 254-3232 or (800) 926-2665. Ninth edition. 1992. $114.95.

THERMOCHEMISTRY

Ency. of Physical Sciences and Engineering Info. Sources

Tables of Spectral Data For Structure Determination of Organic Compounds. E. Pretsch, et al. Springer-Verlag New York, Inc., 175 Fifth Avenue, New York, NY 10010. (212) 460-1500. (800) 777-4643. FAX (212) 473-6272. 1994. $49.95 in paper.

ONLINE DATABASES AND CD-ROMS

Analytical Abstracts Online. Royal Society of Chemistry, Information Services, Thomas Graham House, Science Park, Milton Road, Cambridge, CB4 4WF, England. Contains citations, mostly with abstracts, of the worldwide literature on analytical chemistry, from 1980 to present. Available on DIALOG, (800) 334-2564, and STN International, FIZ Karlsruhe, P.O. Box 2465, W-7500, Karlsruhe 1, Germany, online services. Inquire as to cost and availability.

Beilstein Online. Beilstein-Institut fur Literatur der Organischen Chemie, Varrentrapperstrasse, 40-42, D-6000, Frankfurt am Main 90, Germany. Contains data on carbon compounds from the Beilstein Handbook of Organic Chemistry. Structural and factual data for more than 3.4 million heterocyclic, isocyclic, and acyclic compounds included. Covers the period 1830 to present. Available on DIALOG, (800) 334-2564, ORBIT (800) 456-7248, and STN International, FIZ Karlsruhe, P.O. Box 2465, W-7500, Karlsruhe 1, Germany, online services. Inquire as to cost and availability.

CA Search. Chemical Abstracts Service, P.O. Box 3012, Columbus, OH 43210-0012. (614) 447-3600. (800) 848-6533. FAX (614) 447-3709. Very comprehensive guide to worldwide chemical literature and related fields, 1972 to present. Available on BRS,(800) 289-4277, DIALOG, (800) 334-2564, ORBIT (800) 456-7248, and STN International, FIZ Karlsruhe, P.O. Box 2465, W-7500, Karlsruhe 1, Germany, online services. Inquire as to cost and availability.

Chemical Journals of the American Chemical Society. American Chemical Society, 1155 16th Street, N.W., Washington, DC 20036. (202) 872-4381 or (800) 424-6747. Contains complete text of approximately 90,000 articles from 22 primary journals published by the American Chemical Society, from mostly 1982 to present. Available on STN International, FIZ Karlsruhe, P.O. Box 2465, W-7500, Karlsruhe 1, Germany, online service. Inquire as to cost and availability.

COMPENDEX Plus. Engineering Information, Inc., 345 East 47th Street, New York, NY 10017. (212) 705-7600 or (800) 221-1044. Contains citations with abstracts to worldwide literature in engineering and technology, from 1970 to present. Available on online BRS,(800) 289-4277, DIALOG, (800) 334-2564, ORBIT (800) 456-7248, and STN International, FIZ Karlsruhe, P.O. Box 2465, W-7500, Karlsruhe 1, Germany, online services. Also available on CD-ROM. Inquire as to cost and availability.

Current Contents Search. Institute for Scientific Information, 3501 Market Street, Philadelphia, PA 19104. (215) 386-0100. FAX (215) 386-6362. Contains citations to articles listed in the table of contents of science and technology journals. Also articles in social sciences and life sciences journals. Available on BRS,(800) 289-4277, DIALOG,(800) 334-2564, online services. Inquire as to cost and availability.

Dissertation Abstracts. University Microfilms International, 300 North Zeeb Road, Ann Arbor, MI 48106. (800) 521-0600 or (313) 761-4700. Scope includes virtually all doctoral dissertations accepted at accredited American institutions from 1861 to present in 252 subject areas. Available on BRS,(800) 289-4277,

DIALOG,(800) 334-2564, and OCLC EPIC,(800) 848-5878, online services. Also available on CD-ROM. Inquire as to cost and availability.

Engineered Materials Abstracts. Materials Information, ASM International, Materials Park, OH 44073. (216) 338-5151. Contains citations with abstracts of worldwide literature in the development, processing, and production of ceramic, composite, and polymeric materials for engineering uses. Available on DIALOG, (800) 334-2564, ORBIT (800) 456-7248, and STN International, FIZ Karlsruhe, P.O. Box 2465, W-7500, Karlsruhe 1, Germany, online services. Inquire as to cost and availability.

Gmelin Database. Gmelin-Institut fur Anorganische Chemie und Grenzgebiete, Varrentrapperstrasse, 40-42, Carl-Bosch-Haus, D-6000, Frankfurt am Main 90, Germany. Contains structural and factual data relating to inorganic and organometallic chemistry. Provides data from the Gmelin Handbook of Inorganic and Organometallic Chemistry. Covers the period 1817 to 1975; 1988-89. Available on STN International, FIZ Karlsruhe, P.O. Box 2465, W-7500, Karlsruhe 1, Germany, online service. Inquire as to cost and availability.

Kirk-Othmer Encyclopedia of Chemical Technology. John Wiley and Sons, Inc., 605 Third Avenue, New York, NY 10158. (800) 526-5368 or (212) 850-6000. Contains the complete text of all chapters in the 27-volume, fourth edition of the *KIRK-Othmer Encyclopedia of Chemical Technology.* 1991. Available on BRS,(800) 289-4277, DIALOG,(800) 334-2564, online services. Inquire as to cost and availability.

NTIS Bibliographic Database. National Technical Information Service, 5285 Port Royal Road, Springfield, VA 22161. (703) 487-4929 or FAX (703) 321-8199. Broad coverage of government-sponsored science and technology research reports, 1964 to present. Available on BRS,(800) 289-4277, DIALOG, (800) 334-2564, ORBIT, (800) 456-7248, and STN International, FIZ Karlsruhe, P.O. Box 2465, W-7500, Karlsruhe 1, Germany, online services. Also available on CD-ROM. Inquire as to cost and availability.

SCISEARCH. Institute for Scientific Information, 3501 Market Street, Philadelphia, PA 19104. (800) 523-1850 or (215) 386-0100. Broad multidisciplinary title and author index to the international literature of science and technology, 1974 to present. Available on DIALOG,(800) 334-2564, and ORBIT,(800) 456-7248, online services. Also available on CD-ROM. Inquire as to cost and availability.

WILSONLINE. H.W. Wilson Company, 950 University Avenue, Bronx, NY 10452. (800) 367-6770 or (212) 588-8400. Makes available online versions of the printed H.W. Wilson indexes, including Applied Science and Technology Index, Business Periodicals Index, General Science Index, and Readers' Guide to Periodical Literature. Period covered is generally 1983 to present. Available on BRS,(800) 289-4277, DIALOG,(800) 334-2564, and OCLC EPIC,(800) 848-5878, online services. Also available on CD-ROM. Inquire as to cost and availability.

PERIODICALS

AICHE Journal. American Institute of Chemical Engineers. 345 East 47th Street, New York, NY 10017. (212) 705-7338. FAX (212) 752-3294. 1955 to present. Monthly. $395.00 per year. ISSN: 0001-1541.

Applied Catalysis A: General. Elsevier Science Inc., Box 882, Madison Square Station, New York, NY 10159-0882. (212) 989-

5800. FAX (212) 633-3990. 1981 to present. 28 issues per year. $2921 per year. $3295 in combination with section B Environmental. ISSN: 0926-860X.

Catalysis Reviews: Science and Engineering. Marcel Dekker, Inc., 270 Madison Avenue, New York, NY 10016. (212) 696-9000. FAX (212) 685-4540. 1967 to present. Quarterly. $400.00 per year.

Chemical and Engineering News. American Chemical Society, 1155 16th Street, NW, Washington, DC 20036. (202) 872-4600. 1923 to present. Weekly. $115.00 per year. ISSN: 0009-2347.

Chemical Engineering; The Chemical and Process Industries Journal. Institution of Chemical Engineers/Taylor and Francis, Ltd., 1900 Frost Road, Suite 101, Bristol, PA 19007-1598. (215) 785-5800. 1923 to present. Fortnightly. $176.00 per year. ISSN: 0302-0797.

Chemical Engineering. McGraw-Hill Publishing Company, Inc., 1221 Avenue of the Americas, New York, NY 10020. (212) 512- 2000. (800) 262-4729. 1902 to present. Monthly. $35.50 per year. ISSN: 0009-2460.

Chemical Engineering Communications. Gordon and Breach Science Publishers, 820 Town Center Drive, Langhorne, PA 19047. (215) 750-2642. FAX (215) 750-6343. 1973 to present. 72 issues per year in 12 volumes. 142 ECU per volume per year. ISSN: 0098-6445.

Chemical Engineering Progress. American Institute of Chemical Engineers, 345 East 47th Street, New York, NY 10017. (212) 705-7338. FAX (212) 752-3294. 1947 to present. Monthly. $75.00 per year. ISSN: 0360-7275.

Chemical Processing. Putman Publishing Co., 301 East Erie Street, Chicago, IL 60611. (312) 644-2020. 1938 to present. Monthly. $45.00 per year. ISSN: 0009-2630.

Chemical Week; Includes Annual Buyers Guide. Chemical Week Associates, 888 Seventh Avenue, New York, NY 10106. (212) 621-4900. FAX (212) 621-4949. 1914 to present. Weekly. $99.00 per year. ISSN: 0009-272X.

Chemtech. American Chemical Society. Box 3337, Columbus, OH 43210. (614) 447-3776. 1970 to present. Monthly. $370.00 per year. ISSN: 0009-2703.

Chemie-Ingenieur-Technik; Zeitschrift fuer Technische Chemie, Verfahrenstechnik, Apparatewesen Und Biotechnologie. VCH Publishers, Inc., 220 East 23rd Street, Suite 909, New York, NY 10010-4606. (212) 683-8333. (800) 422-8824. 1928 to present. Monthly. $470.00 per year. ISSN: 0009-286X.

Industrial and Engineering Chemistry Research. American Chemical Society, Box 3337, Columbus, OH 43210. (614) 447-3776. (800) 333-9511. FAX (614) 447-3671. 1962 to present. Monthly. $567.00 per year. ISSN: 0888-5885.

RESEARCH CENTERS AND INSTITUTES

Center for Catalytic Science and Technology. University of Delaware. Departments of Chemical Engineering and Chemistry, Newark, DE 19716. (302) 451-8056. (302) 451-1048.

Chemical Laboratories. Harvard University. Oxford Street, Cambridge, MA 02138. (617) 495-4283. FAX (617) 496-5618.

Chemistry Laboratories. Rensselaer Polytechnic Institute. Cogswell Laboratory, Troy, NY 12180-3590. (518) 276-8981.

Engineering Research Program. Pennsylvania State University, 101 Hammond Building, University Park, PA 16802.

Institute For Systems Design and Optimization. Kansas State University. Department of Chemical Engineering, Manhattan, KS 66506. (913) 532-5584.

University/Industry Chemical Research Center. Mississippi State University, Department of Chemistry, P.O. Drawer CH, Mississippi State, MS 39762. (601) 325-3584.

PERIODICALS

AICHE Journal. American Institute of Chemical Engineers. 345 East 47th Street, New York, NY 10017. (212) 705-7338. FAX (212) 752-3294. 1955 to present. Monthly. $395.00 per year. ISSN: 0001-1541.

Analytical Chemistry. American Chemical Society, Drawer 1734, Atlanta, GA 30301-1734. (800) 227-5558. Monthly. $373.00 per year.

Analytical Methods and Instrumentation. John Wiley and Sons, Inc., 605 Third Avenue, New York, NY 10158. (800) 526-5368 or (212) 850-6000. Bimonthly. $250.00 per year.

Applied Catalysis A: General. Elsevier Science Inc., Box 882, Madison Square Station, New York, NY 10159-0882. (212) 989-5800. FAX (212) 633-3990. 1981 to present. 28 issues per year. $2921 per year. $3295 in combination with section B Environmental. ISSN: 0926-860X.

Catalysis Reviews: Science and Engineering. Marcel Dekker, Inc., 270 Madison Avenue, New York, NY 10016. (212) 696-9000. FAX (212) 685-4540. 1967 to present. Quarterly. $400.00 per year.

Chemical and Engineering News. American Chemical Society, 1155 16th Street, NW, Washington, DC 20036. (202) 872-4600. 1923 to present. Weekly. $115.00 per year. ISSN: 0009- 2347.

Chemical Engineering; The Chemical and Process Industries Journal. Institution of Chemical Engineers/Taylor and Francis, Ltd., 1900 Frost Road, Suite 101, Bristol, PA 19007-1598. (215) 785-5800. 1923 to present. Fortnightly. $176.00 per year. ISSN: 0302-0797.

Chemical Week. Chemical Week Associates, 810 7th Avenue, New York, NY 10019. Weekly. $30.00 per year.

Chemtech. American Chemical Society, Drawer 1734, Atlanta, GA 30301-1734. (800) 227-5558. Monthly. $300.00 per year.

Fuel Science and Technology International. Marcel Dekker, Inc., 270 Madison Avenue, New York, NY 10016. (212) 696-9000. FAX (212) 685-4540. 1982 to present. Nine times per year. $200.00 per year.

Journal of Catalysis. Academic Press, Inc., 1250 Sixth Avenue, San Diego, CA 92101-4311. (619) 231-0926. FAX (619) 699-6715. 1962 to present. Monthly. $1100.00 per year.

THERMOCHEMISTRY

Ency. of Physical Sciences and Engineering Info. Sources

Journal of Nondestructive Evaluation. Plenum Publishing Corporation, 233 Spring Street, New York, NY 10013. (212) 620-8000. FAX (212) 463-0742. 1980 to present. Quarterly. $100.00 per year.

Journal of Organic Chemistry. American Chemical Society, Drawer 1734, Atlanta, GA 30301-1734. (800) 227-5558. Bi-Weekly. $575.00 per year.

RESEARCH CENTERS AND INSTITUTES

Harvard University Chemical Laboratories. Oxford Street, Cambridge, MA 02138. (617) 495-4283. FAX (617) 496-5618.

Lawrence Berkeley Laboratory, Chemical Sciences Division. One Cyclotron Road, Building 66, Berkeley, CA 94720. (510) 486-6062. FAX (510) 486-4995.

Rensselaer Polytechnic Institute. Chemistry Laboratories. Cogswell Laboratory, Troy, NY 12180-3590. (518) 276-8981.

THERMOCOUPLES

See also: ELECTRICAL ENGINEERING, METROLOGY, THERM0DYNAMICS

ABSTRACT SERVICES AND INDEXES

Applied Science and Technology Index; A Cumulative Subject Index To English Language Periodicals in the Fields of Aeronautics and Space Science, Computer Technology, Chemistry, Construction Industry, Energy and Related Areas. H.W. Wilson Co., 950 University Avenue, Bronx, NY 10452. (212) 588-8400. (800) 367-6770. FAX (718) 590-1617. From 1958 to present. Monthly. Inquire about cost and availability. Also available on CD-ROM and online. ISSN: 0003-6986.

Current Contents: Engineering, Technology and Applied Sciences. Institute for Scientific Information, 3501 Market Street, Philadelphia, PA 19104. (215) 386-0100. FAX (215) 386-6362. 1970 to present. Weekly. $442.00 per year. Also available on CD-ROM and online. Inquire regarding cost and availability. ISSN: 0095-7917.

Current Papers in Electrical and Electronics Engineering. Institution of Electrical Engineers (IEE), London. Distributed by INSPEC/IEEE, Box 1331, 445 Hoes Lane, Piscataway, NJ 08855-1331. (908) 562-5549. 1969 to present. Monthly. $345.00 per year. ISSN: 0011-3778.

Electrical and Electronics Abstracts. Institution of Electrical Engineers (IEE), London. Available from INSPEC/IEEE-Institute of Electrical and Electronic Engineers, Box 1331, Hoes Lane, Piscataway, NJ 08855-1331. (908) 562-5549. 1898 to present. Monthly. $2200.00 per year. Also available on CD-ROM and online as INSPEC. ISSN: 0036-8105.

Engineering Index Monthly; Indexes and Abstracts the World's Engineering and Technical Literature. Engineering Information, Inc., Castle Point on the Hudson, Hoboken, NJ 07030. (201) 216-8500. (800) 221-1044. FAX (201) 216-8532. Monthly. $2300.00 per year. Available online as COMPENDEX and also on CD-ROM. ISSN: 0742-1974.

General Science Index. H.W. Wilson Company, 950 University Avenue, Bronx, NY 10452. (212) 588-8400. (800) 367-6770. FAX (718) 590-1617. From 1978 to present. Ten issues per year; quarterly and annual cumulations. Service basis. Available on CD-ROM and online. Inquire regarding cost and availability. ISSN: 0162-1963.

Index To IEEE Publications. IEEE Service Center, 445 Hoes Lane, Piscataway, NJ 08855-1331. (908) 981-1393. (800) 678-IEEE. FAX (908) 981-9667. 1973 to present. Annual. ISSN: 0099-1368.

Physics Abstracts. INSPEC. Section A, Science Abstracts. Institution of Electrical Engineers (IEE), London. Available from: INSPEC/IEEE-Institute of Electrical and Electronic Engineers, Box 1331, Hoes Lane, Piscataway, NJ 08855-1331. (908) 562-5549. 1898 to present. 24 issues per year. $2835.00 per year. Also available online and on CD-ROM. ISSN: 0036-8091.

Physics Briefs (Physikalische Berichte). Information Center for Energy, Physics, Mathematics; German Physical Society. VCH Publishers, Inc., 220 East 23rd Street, New York, NY 10010-4606. (212) 683-8333. 1845 to present. 24 issues per year. $2390.00 per year. Also available online. ISSN: 0179-7434.

Science Citation Index. SCI. Institute for Scientific Information, 3501 Market Street, Philadelphia, PA 19104. (215) 386-0100. (800) 523-1850. FAX (215) 386-2991. 1961 to present. Six issues per year, plus annual cumulation. $11650.00 per year. Also available online and on CD-ROM. Inquire about price and availability. ISSN: 0036-827X.

ASSOCIATIONS AND PROFESSIONAL SOCIETIES

American Electronics Association. 5201 Great America Way, Suite 520, P.O. 52990, Santa Clara, CA 95056. (408) 987-4200. FAX (408) 970-8565.

American Institute of Physics. One Physics Ellipse, College Park, MD 20740-3843. (301) 209-3100.

American Society For Testing and Materials. 1916 Race Street, Philadelphia, PA 19103-1180. (215) 299-5400.

Edison Electric Institute. 701 Pennsylvania Avenue NW, Washington, DC 20004-2696. (202) 508-5000, 5454. FAX (202) 508-5360.

Electric Power Research Institute. 3412 Hillview Avenue, Palo Alto, CA 94304. (415) 855-2000. FAX (415) 855-2954.

Electronics Industries Association. 2500 Wilson Boulevard, Arlington, VA 22201. (703) 907-7500. FAX (202) 457-4985.

IEEE (Institute of Electrical and Electronic Engineers). 345 East 47th Street. New York, NY 10017. (212) 705-7900. FAX (212) 705-4929.

Instrument Society of America. PO Box 12277, 67 Alexander Drive, Research Triangle Park, NC 27709. (919) 549-8411. FAX (919) 549-8288.

National Electrical Manufacturers Association. 1300 North 17th Street, Suite 1847, Rosslyn VA 22209. (703) 841-3200. FAX (703) 841-3300.

DIRECTORIES AND BIOGRAPHICAL SOURCES

American Men and Women of Science: Physical and Biological Sciences. R. R. Bowker Inc., 121 Chanlon Road, New Providence, NJ 07974. (908) 464-6800. (800) 521-8110. 20th edition. 8 volumes. 1996. $850.00.

Directory of Engineering Societies and Related Organizations. American Association of Engineering Societies, 1111 19th Street, Suite 608, Washington, DC 20036-3603. (202) 296-2237. Semi-annual. $150.00.

EEM-Electronic Engineer's Master. Hearst Business Communications, Inc., 645 Stewart Avenue, Garden City NY 11530. (516) 227-1300. ISSN: 0732-9016.

Engineering Research Centers: Incorporating Electronics Research Centers. Stockton Press, 345 Park Avenue, New York, NY 10010. (212) 689-9200. (800) 221-2123. 4th edition. 1995. $515.00.

IEEE Membership Directory. Institute of Electrical and Electronics Engineers, IEEE Service Center, 445 Hoes Lane, Piscataway, NY 08854. (908) 981-1393. (800) 678-IEEE. FAX(908) 981-9667. 2 volumes. Annual. $190.00.

International Directory of Abbreviations and Acronyms of Electronics, Electrical Engineering, Computer Technology and Information Processing. Peter Wennrich. K. G. Saur, 121 Chanlon Road, New Providence, NJ 07974. (908) 464-6800. (800) 521-8110 2 volumes. 1992. $230.00.

International Engineering Directory. American Consulting Engineers Council, 1015 15th Street, N.W. Suite 802, Washington, DC 20005-2670. (202) 347-7474. Annual. $10.00.

Research Centers Directory. Gale Research, 835 Penobscot Building, Detroit, MI 48226-4094. (313) 961-2242. (800) 347-4253. 20th edition. 1995. $485.00. ISSN: 0080-1518.

Who's Who in Engineering. Gordon Davis, editor. American Association of Engineering Societies. 1111 19th Street, NY, Suite 608, Washington, DC 20036. (202) 296-2237. (800) 658-8897. 9th edition. 1995. $220.00.

Who's Who in Technology; Biographies and Index. Gale Research Company, 835 Penobscot Building, Detroit, MI 48226-4094. (313 961-1242. (800) 521-4253. 7th edition. 1995. $195.00. ISBN 0-8103-7467-6.

ENCYCLOPEDIAS AND DICTIONARIES

Encyclopedia of Applied Physics. George Trigg, editor. VCH Publications, Inc., 220 East 23rd Street, Suite 909, New York, NY 10010-4606. (212) 683-8333. (800) 422-8824. 20 volume set. 1991-. $5990.00.

IEEE Standard Dictionary of Electrical and Electronics Terms. Christopher J. Booth, editor. IEEE Service Center, 445 Hoes Lane, Piscataway, NJ 08855-1331. (908) 981-1393. (800) 678-IEEE. FAX (908) 981-9667. IEEE Standard 100-1992. 5th edition. 1993. $150.00.

Illustrated Dictionary of Electronics. Stan Gibilsco. TAB Books, P.O. Box 40, Blue Summit, PA 17294-0850. (717) 794-2191. (800) 233-1128. 7th edition. 1994. $34.95.

GENERAL WORKS

Exploring Electronic Devices. Mark E. Hazen. SCP: Third World Literature Publishing House, P.O. Box 482, Lithonia, GA 30058-0482. (404) 785-7725. 1991. $50.75.

Modern Thermoelectrics. Prentice Hall , 113 Sylvan Avenue, Route 9W, Englewood Cliffs, NJ 07632. (201) 592-2000. (800) 922-0579. 1983. $48.00.

Schaum's Outline of Electrical Technology. Milton Kaufman and Peter Brooks. McGraw-Hill Publishing Company, Inc., 1221 Avenue of the Americas, New York, NY 10020. (212) 512-2000. (800) 262-4729. 2nd edition. 1996.

Thermal Power Cycles. George H. Cole. Routledge, Chapman & Hall, Inc., 29 West 35th Street, New York, NY 10001-3291. (212) 244-3336. 1992. $35.00 in paper.

Temperature Measurement. Bela G. Liptak. Chilton Book Company, 201 King of Prussia Road, Radnor PA 19089-0230. (610) 964-4000. (800) 695-1214. 1993. $29.95.

Thermal Storage System Design. Russell M. Keeler. McGraw-Hill Publishing Company, Inc., 1221 Avenue of the Americas, New York, NY 10020. (212) 512-2000. (800) 262-4729. 1994. $47.00.

Thermocouples: Theory and Practice. Daniel D. Pollock. CRC Press, Inc., 2000 Corporate Boulevard, NW, Boca Raton, FL 33431. (407) 994-0555. (800) 272-7737. 1991. $99.95.

Thermodynamics. Edward E. Anderson. PWS Publications, 20 Park Plaza, Boston, MA 02116. (617) 542-3377. (800) 354-9706. 1994. $67.95.

HANDBOOKS AND MANUALS

Active Electronic Component Handbook. Charles A. Harper and Harold C. Jones. McGraw-Hill Book Company, 1221 Avenue of the Americas, New York, NY 10020. (212) 512-2000. (800) 262-4729. 2nd edition. 1996. $79.50.

Compendium of Thermophysical Property Measurement Methods. Kosta D. Maglic, editor. Plenum Publishing Corp., 233 Spring Street, New York, NY. (212) 620-8000. (800) 2221-9369. FAX (212) 463-0742. 1993. $149.50.

Electrical Engineering Handbook. Richard C. Dorf, editor. CRC Press, Inc., 2000 Corporate Boulevard, NW, Boca Raton, FL 33431. (407) 994-0555. (800) 272-7737. 1993. $99.95.

Electrical Engineer's Handbook. Donald Christiansen, editor. McGraw Book Company, 1221 Avenue of the Americas, New York, NY 10020. (212) 512-2000. (800) 262-4729. 4th edition. 1996. $110.00.

Handbook of Electrical and Electronic Technology. Curtis Johnson. Prentice Hall , 113 Sylvan Avenue, Route 9W, Englewood Cliffs, NJ 07632. (201) 592-2000. (800) 922-0579. 1996. $88.00.

Handbook of thermodynamics of Organic Compounds For Chemical Engineers. Stanislaw Malanowski and R. Stephenson. Elsevier Science Publishing Company, Inc., 655 Avenue of the

THERMOCOUPLES

Ency. of Physical Sciences and Engineering Info. Sources

Americas, New York, NY 10010. (212) 989-5800. FAX (914) 333-2444. 1987. $79.00.

Industrial Electronics and Systems Handbook. Irwin, CRC Press, Inc., 2000 Corporate Boulevard, NW, Boca Raton, FL 33431. (407) 994-0555. (800) 272-7737. 1996. $129.95.

Manual On the Use of thermocouples in Temperature Measurement. Richard M. Parks, editor. American Society for Testing and Materials, 1916 Race Street, Philadelphia, PA 19103. (215) 299-5419. Manual series, number MNL 12. 1993. $49.00.

Standard Handbook For Electrical Engineers. Donald Fink. McGraw-Hill Publishing Company, 1221 Avenue of the Americas, New York, NY 10020. (212) 512-2000. (800) 262-4729. 13th edition. 1996. $110.00.

Temperature Measurement Thermocouples: An ANSI Approved Standard. MC96.1. Instrument Society of America Revised edition. 1982. $40.00.

ONLINE DATABASES AND CD-ROMS

CA Search. Chemical Abstracts Service, P.O. Box 3012, Columbus, OH 43210-0012. (614) 447-3600. (800) 848-6533. FAX (614) 447-3709. Very comprehensive guide to worldwide chemical literature and related fields, 1972 to present. Available on BRS,(800) 289-4277, DIALOG, (800) 334-2564, ORBIT (800) 456-7248, and STN International, FIZ Karlsruhe, P.O. Box 2465, W-7500, Karlsruhe 1, Germany, online services. Inquire as to cost and availability.

Compendex Plus. Engineering Information, Inc., 345 East 47th Street, New York, NY 10017. (212) 705-7600 or (800) 221-1044. Contains citations with abstracts to worldwide literature in engineering and technology, from 1970 to present. Available on online BRS,(800) 289-4277, DIALOG, (800) 334-2564, ORBIT (800) 456-7248, and STN International, FIZ Karlsruhe, P.O. Box 2465, W-7500, Karlsruhe 1, Germany, online services. Also available on CD-ROM. Inquire as to cost and availability.

Current Contents Search. Institute for Scientific Information, 3501 Market Street, Philadelphia, PA 19104. (215) 386-0100. FAX (215) 386-6362. Contains citations to articles listed in the table of contents of science and technology journals. Also articles in social sciences and life sciences journals. Available on BRS,(800) 289-4277, DIALOG,(800) 334-2564, online services. Inquire as to cost and availability.

Dissertation Abstracts Online. University Microfilms International, 300 North Zeeb Road, Ann Arbor, MI 48106. (800) 521-0600 or (313) 761-4700. Scope includes virtually all doctoral dissertations accepted at accredited American institutions from 1861 to present in 252 subject areas. Available on BRS, (800) 289-4277, DIALOG, (800) 334-2564, and OCLC EPIC,(800) 848-5878, online services. Also available on CD-ROM. Inquire as to cost and availability.

INSPEC. Institution of Electrical Engineers, Michael Faraday House, Six Hills Way, Stevenage, Herts. SG1 2AY, England. Telephone: 0438 313311 or FAX 0438 742840. Contains citations to the worldwide literature of physics, electronics and electrical engineering, computer technology, and related fields. Available on BRS, (800) 289-4277, DIALOG, (800) 334-2564, ORBIT, (800) 456-7248, and STN International, FIZ Karlsruhe, P.O. Box 2465, W-7500, Karlsruhe 1, Germany, online services. Inquire as to cost and availability.

NTIS Bibliographic Database. National Technical Information Service, 5285 Port Royal Road, Springfield, VA 22161. (703) 487-4929 or FAX (703) 321-8199. Broad coverage of government-sponsored science and technology research reports, 1964 to present. Available on BRS,(800) 289-4277, DIALOG, (800) 334-2564, ORBIT, (800) 456-7248, and STN International, FIZ Karlsruhe, P.O. Box 2465, W-7500, Karlsruhe 1, Germany, online services. Also available on CD-ROM. Inquire as to cost and availability.

Physics Briefs. American Institute of Physics, 335 East 45th Street, New York, NY 10017. (212) 661-9260 or FAX (212) 661-2036. Contains citations with abstracts of the literature of physics and related fields, 1979 to present. Available on the STN International, FIZ Karlsruhe, P.O. Box 2465, W-7500, Karlsruhe 1, Germany, online service. Inquire as to cost and availability.

SCISEARCH. Institute for Scientific Information, 3501 Market Street, Philadelphia, PA 19104. (800) 523-1850 or (215) 386-0100. Broad multidisciplinary title and author index to the international literature of science and technology, 1974 to present. Available on DIALOG,(800) 334-2564, and ORBIT,(800) 456-7248, online services. Also available on CD-ROM. Inquire as to cost and availability.

WILSONLINE. H.W. Wilson Company, 950 University Avenue, Bronx, NY 10452. (800) 367-6770 or (212) 588-8400. Makes available online versions of the printed H.W. Wilson indexes, including Applied Science and Technology Index, Business Periodicals Index, General Science Index, and Readers' Guide to Periodical Literature. Period covered is generally 1983 to present. Available on BRS,(800) 289-4277, DIALOG,(800) 334-2564, and OCLC EPIC,(800) 848-5878, online services. Also available on CD-ROM. Inquire as to cost and availability.

PERIODICALS

Electrical World. McGraw-Hill, Inc., Box 513 Hightstown, NJ 08520. (212) 512-3288. 1874 to present. Monthly. $55.00 per year. ISSN: 0013-4457.

Electronics. Penton Publishing, San Jose Gateway, Suite 354. 2025 Gateway Place, San Jose, CA 95110. (408) 441-0550. 1930 to present. Semi-weekly. $98.00 per year. ISSN: 0883-4989.

IEEE Transactions On Instrumentation and Measurement. Institute of Electrical and Electronics Engineers, Inc., Box 1331, 445 Hoes Lane, Piscataway, NJ 08855-1331. (908) 981- 0060. 1952 to present. Bi-monthly. $165.00 per year. ISSN: 0018-9456.

Review of Scientific Instruments. American Institute of Physics, 500 Sunnyside Boulevard, Woodbury, NY 11797-2999. (516) 576-2270. 1930 to present. Monthly. $800.00 per year. ISSN: 0034-6748.

RESEARCH CENTERS AND INSTITUTES

Edison Electric Institute. 701 Pennsylvania Avenue, NW, Washington, DC 20004. (202) 508-5000

Electrical Engineering Research Laboratories. Purdue University. Electrical Engineering Building, West Lafayette, in 47907. (317) 494-3536. FAX (317) 494-6440.

Electronics Design Center. Case Western Reserve University, Bingham Building, Cleveland, OH 44106. (216) 368-2934. (216) 368-8738.

Laboratory For Electromagnetic and Electronic Systems. Massachusetts Institute of Technology, 77 Massachusetts Avenue, Cambridge, MA 02139. (617) 253) 4631.

Weber Research Institute. Polytechnic University. Route 110, Farmingdale, NY 11735. (516) 755-4250, FAX (516) 755-4404.

THERMODYNAMICS

See also: CRYOGENICS, FLUID MECHANICS, HEAT TRANSFER, PHYSICS, THERMOCHEMISTRY

ABSTRACT SERVICES AND INDEXES

Applied Mechanics Reviews: An Assessment of World Literature in Engineering Sciences. American Society of Mechanical Engineers, 345 East 47th Street, New York, NY 10017. (212) 705-7703. 1948 to present. Monthly. $360.00 per year. ISSN: 0003-6900.

Applied Science and Technology Index; A Cumulative Subject Index To English Language Periodicals in the Fields of Aeronautics and Space Science, Computer Technology, Chemistry, Construction Industry, Energy and Related Areas. H.W. Wilson Co., 950 University Avenue, Bronx, NY 10452. (212) 588-8400. (800) 367-6770. FAX (718) 590-1617. From 1958 to present. Monthly. Inquire about cost and availability. Also available on CD-ROM and online. ISSN: 0003-6986.

Chemical Abstracts. Chemical Abstracts Service, 2540 Olentangy River Road, Box 3012, Columbus, OH 43210-0012. (614) 447-3600. (800) 848-6538. FAX (614) 447-3713. From 1907 to present. Weekly. $16800.00 per year. Also available on CD-ROM and online. Inquire regarding cost and availability.

Current Contents: Engineering, Technology, and Applied Sciences. Institute for Scientific Information, 3501 Market Street, Philadelphia, PA 19104. (215) 386-0100. FAX (215)386-2291. From 1961 to present. Weekly. $442.00 per year. Also available on CD-ROM and online. Inquire regarding cost and availability. ISSN: 0095-7917.

Current Papers in Physics. Institute of Electrical Engineers, Michael Faraday House, Six Hill Way, Stevenage, Herts, SG1 2AY, England. Distributed by INSPEC/IEEE, Box 1331, 445 Hoes Lane, Piscataway, NJ 08855-1331. (908) 562-5549. 1966 to present. Fortnightly. $410.00 per year. ISSN: 0011-3786.

Engineering Index Monthly. Engineering Information, Inc., Castle Point on the Hudson, Hoboken, NJ 07030. (201) 216-8500. FAX (201) 216-8532. Monthly. $2200.00 per year. Also available online as COMPENDEX and on CD-ROM. ISSN: 0742-1974.

General Science Index. H.W. Wilson Company, 950 University Avenue, Bronx, NY 10452. (212) 588-8400. (800) 367-6770. FAX (718) 590-1617. From 1978 to present. Ten issues per year; quarterly and annual cumulations. Service basis. Available on CD-ROM and online. Inquire regarding cost and availability. ISSN: 0162-1963.

Government Reports Announcements and Index. U.S. National Technical Information Service (NTIS), 5285 Port Royal Road, Springfield, VA 22161. (703) 487-4650. FAX (703) 321-8547.

From 1968 to present. Annual. $630.00 per year. Also available online as NTIS Bibliographic Database and on CD-ROM.

ISMEC: Mechanical Engineering Abstracts. Cambridge Scientific Abstracts, 7200 Wisconsin Avenue, Suite 601, Bethesda, MD 20814. (301) 961-6750 or (800) 843-7751. Contains citations to the literature in mechanical engineering, industrial and production engineering, energy, power, mechanics, devices and related areas, from 1973 to present. Available on the DIALOG, (800) 334-2564, online service. Inquire as to cost and availability.

Physics Abstracts. INSPEC. Section A, Science Abstracts. Institution of Electrical Engineers (IEE). Available from INSPEC/IEEE-Institute of Electrical and Electronic Engineers, Box 1331, 445 Hoes Lane, Piscataway, NJ 08855-1331. (908) 562-5549. 1898 to present. 24 issues per year. $2835.00 per year. Also available on CD-ROM and online. ISSN: 0036-8091.

Science Citation Index. SCI. Institute for Scientific Information, 3501 Market Street, Philadelphia, PA 19104. (215) 386-0100. (800) 523-1850. FAX (215) 386-2991. 1961 to present. Six issues per year, plus annual cumulation. $11650.00 per year. Also available online and on CD-ROM. Inquire about price and availability. ISSN: 0036-827X.

ASSOCIATIONS AND PROFESSIONAL SOCIETIES

American Chemical Society. 1155 16th Street, N.W., Washington, DC 20036. (202) 872-4600.

American Institute of Chemical Engineers. 345 East 47th Street, New York, NY 10017. (212) 705-7657. FAX (212) 752-3294.

American Institute of Physics. One Physics Ellipse, College Park, MD 20740-3843. (301) 209-3100.

American Society of Mechanical Engineers, 345 East 47th Street, New York, NY 10017. (212) 705-7722. FAX (212) 705-7739.

DIRECTORIES AND BIOGRAPHICAL SOURCES

American Men and Women of Science: Physical and Biological Sciences. R. R. Bowker Inc., 121 Chanlon Road, New Providence, NJ 07974. (908) 464-6800. (800) 521-8110. 20th edition. 8 volumes. 1996. $850.00.

American Physical Society Membership Directory Bulletin. American Physical Society, One Physics Ellipse, College Park, MD 20740-3843. (301) 209-3200. FAX (301) 209-0865. Biennial. $50.00.

Directory of Physics and Astronomy Staff Members. American Institute of Physics, One Physics Ellipse, College Park, MD 20740-3843. (301) 209-3100. Annual. $45.00.

International Directory of Engineering Societies and Related Organizations. American Society for Engineering Education, 1818 N Street NW, Suite 600, Washington, DC 20036. (202) 331-3526. 15th edition. 1996. $185.00.

International Engineering Directory. American Consulting Engineers Council, 1015 15th Street, N.W. Suite 802, Washington, DC 20005-2670. (202) 347-7474. Annual. $10.00

THERMODYNAMICS

Ency. of Physical Sciences and Engineering Info. Sources

Research Centers Directory. Gale Research, 835 Penobscot Building, Detroit, MI 48226-4094. (313) 961-2242. (800) 347-4253. 20th edition. 1995. $485.00. ISSN: 0080- 1518.

Scientific and Technical Organizations and Agencies Directory. Gale Research, 835 Penobscot Building, Detroit, MI 48226-4094. (313) 961-2242. (800) 347-4253. 4th edition. 1996. $195.00.

Who's Who in Engineering. Gordon Davis, editor. American Society for Engineering Education, 1111 19th Street, NW, Suite 608, Washington, DC 20036. 9th edition. 1995. $220.00.

Who's Who in Technology. Gale Research, 835 Penobscot Building, Detroit, MI 48226-4094. (313 961-2242. (800) 521-4253. 7th edition. 1995. $195.00. ISN 0-8103-7467-6.

ENCYCLOPEDIAS AND DICTIONARIES

Chamber's Science and Technology Dictionary. Peter M.B. Walker, editor. Cambridge University Press, 40 West 20th Street, New York, NY 10011-4211. (212) 924-3900. 1988. $39.95.

A Concise Dictionary of Physics. Oxford University Press, Inc., 200 Madison Avenue, New York, NY 10016. (212) 725-6000. (800) 334-4249. 1990. $10.95.

Encyclopedia of Applied Physics. George Trigg, editor. VCH Publications, Inc., 220 East 23rd Street, Suite 909, New York, NY 10010-4606. (212) 683-8333. (800) 422-8824. 20 volume set. 1991-. $5990.00.

Encyclopedia of Physical Science and Technology. Academic Press, Inc., 6277 Sea Harbor Drive, Orlando, FL. (800) 321-5068. 2nd edition. 18 volume set. 1992. $2625.00.

McGraw-Hill Encyclopedia of Science and Technology. McGraw-Hill Book Company, Inc., 1221 Avenue of the Americas, New York, NY 10020. (212) 512-2000. (800) 262-4729. 7th edition. 20 volume set. 1992. $1900.00.

GENERAL WORKS

Applied Chemical Engineering Thermodynamics. Dimitrios P. Tassios. Springer-Verlag New York, Inc., 175 Fifth Avenue, New York, NY 10010. (212) 460-1500. (800) 777-4643. FAX (212) 473-6272. 1993. $149.00.

Applied Thermodynamics For Engineering Technologies. Thomas D. Eastop and A. McConkey. Halsted Press, 605 Third Avenue, New York, 10158-0012. (212) 850-6400. 5th edition. 1993. $57.95.

Basic Engineering Thermodynamics. P. B. Whalley, Oxford University Press, Inc., 200 Madison Avenue, New York, NY 10016. (212) 725-6000. (800) 334-4249. 1993. $49.95.

Fundamentals of Classical Thermodynamics. Gordon J. Van Wylen, et al. John Wiley & Sons, Inc., 605 Third Avenue, New York, NY 10158-0012. (212) 850-6000. (800) 225-5945. 4th edition. 1993.

Principles of Modern Thermodynamics. B. M. Roy. IOP Publishing, Public Ledger Building, Suite 1035, Independence Square, Philadelphia, PA 19106. (215) 627-0880. (800) 358-4677. 1995. $100.00.

Survey of Thermodynamics. Martin Bailyn. American Institute of Physics, One Physics Ellipse, College Park, MD 20740-3843. (301) 209-3100. 1994. $65.00.

Thermodynamics and the Design, Analysis and Improvement of Energy Systems. H. J. Richter, editor. ASME-American Society of Mechanical Engineers, 345 East 47th Street, New York, NY 10017. (212) 705-7722. (800) 843-2763. 1993. $80.00.

Thermodynamics of Irreversible Processes. Bernard H. Lavenda. Dover Publications, Inc., 180 Varick Street, New York, NY 10014. (212) 255-3755. (800) 223-3130. Reprint edition. 1993. $7.95.

Understanding Energy: Energy, Entropy and Thermodynamics For Every Man. R. Stephen Berry. World Scientific Publishing Company, Inc., 1060 Main Street, River Edge, NJ 07661. (201) 487-9655. (800) 227-7562. 1991. $23.00.

HANDBOOKS AND MANUALS

Applied Fluid Dynamics Handbook. Robert D. Blevins. Krieger Publishing Company, P.O. Box 9542, Melbourne, FL 32902-9542. (407) 724-9542. 1992. $83.50.

Gas Tables: Thermodynamics Properties of Air, Products of Combustion, and Component Gases. Joseph H. Keenan, et al. Krieger Publishing Company, P.O. Box 9542, Melbourne, FL 32902-9542. (407) 724-9542. 2nd edition; reprinted. 1992. 64.50.

Handbook of Applied thermodynamics. David A. Palmer, editor. CRC Press Inc., 2000 Corporate Boulevard, NW, Boca Raton, FL 33431. (407) 994-0555. (800) 272-7737. 1987. $216.95.

Steam Tables: Thermodynamic Properties of Water, Including Vapor, Liquid and Solid Phases. Joseph H. Keegan, et al. Krieger Publishing Company, P.O. Box 9542, Melbourne, FL 32902-9542. (407) 724-9542. 2nd edition, reprinted. 1992. $64.50.

ONLINE DATABASES AND CD-ROMS

CA SEARCH. Chemical Abstracts Service, P.O. Box 3012, Columbus, OH 43210-0012. (614) 447-3600. (800) 848-6533. FAX (614) 447-3709. Very comprehensive guide to worldwide chemical literature and related fields, 1972 to present. Available on BRS,(800) 289-4277, DIALOG, (800) 334-2564, ORBIT (800) 456-7248, and STN International, FIZ Karlsruhe, P.O. Box 2465, W-7500, Karlsruhe 1, Germany, online services. Inquire as to cost and availability.

Compendex Plus. Engineering Information, Inc., 345 East 47th Street, New York, NY 10017. (212) 705-7600 or (800) 221-1044. Contains citations with abstracts to worldwide literature in engineering and technology, from 1970 to present. Available on online BRS,(800) 289-4277, DIALOG, (800) 334-2564, ORBIT (800) 456-7248, and STN International, FIZ Karlsruhe, P.O. Box 2465, W-7500, Karlsruhe 1, Germany, online services. Also available on CD-ROM. Inquire as to cost and availability.

Current Contents Search. Institute for Scientific Information, 3501 Market Street, Philadelphia, PA 19104. (215) 386-0100. FAX (215) 386-6362. Contains citations to articles listed in the table of contents of science and technology journals. Also articles in social sciences and life sciences journals. Available on BRS, (800) 289-4277, DIALOG, (800) 334-2564, online services. Inquire as to cost and availability.

Dissertation Abstracts Online. University Microfilms International, 300 North Zeeb Road, Ann Arbor, MI 48106. (800) 521-0600 or (313) 761-4700. Scope includes virtually all doctoral dissertations accepted at accredited American institutions from 1861 to present in 252 subject areas. Available on BRS, (800) 289-4277, DIALOG, (800) 334-2564, and OCLC EPIC, (800) 848-5878, online services. Also available on CD-ROM. Inquire as to cost and availability.

ISMEC: Mechanical Engineering Abstracts. Cambridge Scientific Abstracts, 7200 Wisconsin Avenue, Suite 601, Bethesda, MD 20814. (301) 961-6750 or (800) 843-7751. Contains citations to the literature in mechanical engineering, industrial and production engineering, energy, power, mechanics, devices and related areas, from 1973 to present. Available on the DIALOG,(800) 334-2564, online service. Inquire as to cost and availability.

INSPEC. Institution of Electrical Engineers, Michael Faraday House, Six Hills Way, Stevenage, Herts. SG1 2AY, England. Telephone: 0438 313311 or FAX 0438 742840. Contains citations to the worldwide literature of physics, electronics and electrical engineering, computer technology, and related fields. Available on BRS, (800) 289-4277, DIALOG, (800) 334-2564, ORBIT, (800) 456-7248, and STN International, FIZ Karlsruhe, P.O. Box 2465, W-7500, Karlsruhe 1, Germany, online services. Inquire as to cost and availability.

NTIS Bibliographic Database. National Technical Information Service, 5285 Port Royal Road, Springfield, VA 22161. (703) 487-4929 or FAX (703) 321-8199. Broad coverage of government-sponsored science and technology research reports, 1964 to present. Available on BRS,(800) 289-4277, DIALOG, (800) 334-2564, ORBIT, (800) 456-7248, and STN International, FIZ Karlsruhe, P.O. Box 2465, W-7500, Karlsruhe 1, Germany, online services. Also available on CD-ROM. Inquire as to cost and availability.

Physics Briefs. American Institute of Physics, 335 East 45th Street, New York, NY 10017. (212) 661-9260 or FAX (212) 661-2036. Contains citations with abstracts of the literature of physics and related fields, 1979 to present. Available on the STN International, FIZ Karlsruhe, P.O. Box 2465, W-7500, Karlsruhe 1, Germany, online service. Inquire as to cost and availability.

SCISEARCH. Institute for Scientific Information, 3501 Market Street, Philadelphia, PA 19104. (800) 523-1850 or (215) 386-0100. Broad multidisciplinary title and author index to the international literature of science and technology, 1974 to present. Available on DIALOG, (800) 334-2564, and ORBIT, (800) 456-7248, online services. Also available on CD-ROM. Inquire as to cost and availability.

WILSONLINE. H.W. Wilson Company, 950 University Avenue, Bronx, NY 10452. (800) 367-6770 or (212) 588-8400. Makes available online versions of the printed H.W. Wilson indexes, including Applied Science and Technology Index, Business Periodicals Index, General Science Index, and Readers' Guide to Periodical Literature. Period covered is generally 1983 to present. Available on BRS, (800) 289-4277, DIALOG, (800) 334-2564, and OCLC EPIC, (800) 848-5878, online services. Also available on CD-ROM. Inquire as to cost and availability.

PERIODICALS

Cryogas International. J. R. Campbell & Associates, Inc., 5 Militia Drive, Lexington, MA 02173. (617) 863-9411. FAX (617)863-9411. 1963 to present. Monthly. $150.00. ISSN: 1052-0139.

Cryogenics: The International Journal of Low Temperature Engineering and Research. Butterworth-Heinemann; Turpin Transactions Ltd., Distribution Center, Blackhorse, Road, Letchworth, Herts SG6 1HN, England. Tel 0462-672555.

Experimental Thermal and Fluid Science. Elsevier Science Inc., 655 Avenue of the Americas, New York, NY 10010. (212) 989-5800. Subscriptions to: Box 882, Madison Square Station, NY 10159-0882. 1988 to present. 8 issues per year. $484.00 per year. ISSN: 0894-1777.

Heat Transfer Engineering. Taylor and Francis, 1900 Frost Road, Suite 101, Bristol PA 190077-1598. (800) 821-8312. 1979 to present. Quarterly. $176 per year. ISSN: 0145-7632.

International Communications in Heat and Mass Transfer. Elsevier Science, 660 White Plains Road, Tarrytown NY 10591-5151. (914) 524-9200. FAX (914) 333-2444. 1974 to present. Bi-monthly. $430.00 per year. Combined subscription with *International Journal of Heat and Mass Transfer*; $2070.00. ISSN: 0735-1933.

International Journal of Heat and Fluid Flow. Butterworth-Heinemann, 313 Washington Street, Newton MA 02158. (617) 928-2500. (800) 366-2665. 1979 to present. Bi-monthly. $450.00 per year. ISSN: 0142-727X.

Journal of Heat Transfer. American Society of Mechanical Engineers, 345 East 47th Street, New York, NY 10017. (212) 705-7722. Volume 92, 1970 to present. Quarterly. $155.00 per year. ISSN: 0022-1481.

Journal of Low Temperature Physics. Plenum Publishing Corp. 233 Spring Street, New York, NY 10013-1578. (212) 620-8000. FAX (212) 463-0742. 1969 to present. 24 issues per year in 3 volumes. $955.00 per year. ISSN: 0022-2291.

Journal of Superconductivity. Plenum Publishing Corp. 233 Spring Street, New York, NY 10013-1578. (212) 620-8000. FAX (212) 463-0742. 1988 to present. Bi-monthly. $275.00 per year. ISSN: 0896-1107.

Journal of Thermophysics and Heat Transfer. American Institute of Aeronautics and Astronautics, 370 L'Enfant Prominade SW, Washington, DC 20024. (202) 646-7400. 1987 to present. Quarterly. $165.00 per year. ISSN: 0887-8722.

Numerical Heat Transfer: An International Journal of Computation and Methodology. Taylor and Francis, 1900 Frost Road, Suite 101, Bristol, PA 19007-1598. (800) 821-8312. FAX (215) 785-5515. 1978 to present. 8 issues per year. Part A: Applications, $530.00 per year; ISSN: 1040-7782. Part B: Fundamentals, $330.00 per year; ISSN: 1040-7790.

RESEARCH CENTERS AND INSTITUTES

Center for Thermodynamics. Brigham Young University, 226 ESC, Provo, UT 84602. (801) 378-3668.

Heat Transfer Laboratory. Massachusetts Institute of Technology, 77 Massachusetts Avenue, Cambridge, MA 02139. (716) 253-2248.

Thermodynamics Research Center. Texas A&M University, Texas Engineering Experiment Station, College Station, TX 77843-3111. (409) 845-4940.

THERMODYNAMICS

Ency. of Physical Sciences and Engineering Info. Sources

Aeronautical Laboratory. University of Washington, FS-10, Seattle, WA 98105. (206) 543-0439.

Computational Fluid Mechanics Laboratory. University of Arizona, Building #16, Room 312, Tucson, AZ 85721. (602) 621-4423.

Environmental Fluid Mechanics Laboratory. Stanford University. Department of Civil Engineering, Stanford, CA 94305. (415) 723-1825.

Fluid Mechanics Laboratory. Purdue University, School of Mechanical Engineering, West Lafayette, IN 47907. (317) 494-5633.

Fluid Properties Research, Inc. School of Chemical Engineering, Georgia Institute of Technology, Atlanta, GA 30332. (405) 894-3098.

Institute of Hydraulic Research. University of Iowa. Iowa City, IA 52242. (319) 353-4679.

Institute of Thermo-Fluid Engineering and Science. LeHigh University, Whitaker Laboratory, Bethleham, PA 18015. (215) 861-4091.

THERMOELECTRICITY

See: ELECTRICITY

THERMOMETERS

See also: INSTRUMENTATION

ABSTRACT SERVICES AND INDEXES

Applied Mechanics Reviews: An Assessment of World Literature in Engineering Sciences. 1948-present. American Society of Mechanical Engineers, 345 East 47th Street, New York, NY 10017. (212) 705-7703. Monthly. $360.00 per year.

Applied Science and Technology Index; A Cumulative Subject Index To English Language Periodicals in the Fields of Aeronautics and Space Science, Computer Technology, Chemistry, Construction Industry, Energy and Related Areas. H.W. Wilson Co., 950 University Avenue, Bronx, NY 10452. (800) 367-6770 or (212) 588-8400. FAX (718) 590-1617. From 1958 to present. Monthly. Inquire about cost and availability. Also available on CD-ROM and online.

Electrical and Electronics Abstracts. Institute of Electrical Engineers, Michael Faraday House, Six Hill Way, Stevenage, Herts, SG1 2AY, England. Distributed by IEEE, 445 Hoes Lane, Piscataway, NJ 08854. (908) 562-5549. 1898 to present. Monthly. $2200.00 per year. Also available on CD-ROM and online as INSPEC.

Engineering Index Monthly. Engineering Information, Inc., Castle Point on the Hudson, Hoboken, NJ 07030. (800) 221-1044. FAX (212) 832-1857. Monthly. $2300.00 per year. Also available online as COMPENDEX, and also on CD-ROM. Covers chemical engineering, computers, electrical engineering, civil engineering, metals and mining, industrial management, and mechanical engineering.

Index To Scientific and Technical Proceedings. Institute for Scientific Information, 3501 Market St., Philadelphia, PA 19104. (215) 386-0100. FAX (215) 386-6362. Monthly. $500.00 per year.

Index To Scientific Reviews. Institute for Scientific Information, 3501 Market St., Philadelphia, PA 19104. (215) 386-0100. FAX (215) 386-6362. Semi-annual.

Key Abstracts—Electronic Instrumentation. IEEE Service Center, 445 Hoes Lane, Piscataway, NJ 08854. (908) 562-5549. 1976 to present. Monthly. $160.00 per year.

Key Abstracts—Measurements in Physics. IEEE Service Center, 445 Hoes Lane, Piscataway, NJ 08854. (908) 562-5549. 1976 to present. Monthly. $160.00 per year.

Physics Abstracts. Institute of Electrical Engineers, Michael Faraday House, Six Hill Way, Stevenage, Herts, SG1 2AY, England. Distributed by IEEE, 445 Hoes Lane, Piscataway, NJ 08854. (908) 562-5549. 1898 to present. Monthly. $2700.00 per year. Also available online as INSPEC.

Physics Briefs. American Institute of Physics, 335 East 45th Street, New York, NY 10017 (212) 661-9404 or FAX (516)349-9704. 1980 to present. Six times per year. $2600.00 per year.

Science Citation Index. Institute for Scientific Information, 3501 Market Street, Philadelphia, PA 19104. (215) 386-0100. FAX (215) 386-6362. Inquire about availability and cost. Also available on CD-ROM.

Standard Actions. American National Standards Institute, 11 W. 42nd Street, 13th Floor, New York, NY 10036. (212) 642-4900. FAX (212) 302-1286. 1970 to present. Bi-weekly. $100.00 per year.

ASSOCIATIONS AND PROFESSIONAL SOCIETIES

American Institute of Physics. 335 East 45th Street, New York, NY 10017. (212) 661-9404 or (800) 445-6638.

American Society For Testing and Materials. 1916 Race Street, Philadelphia, PA 19103. (215) 299-5585.

Instrument Society of America. 67 Alexander Drive, Box 12277, Research Triangle Park, NC 27709. (919) 549-8411. FAX (919) 549-8288.

Standards Engineering Society. 1706 Darst Avenue, Dayton, OH 45403-3104. (513) 258-1955.

U.S. National Institute of Standards and Technology. U.S. Department of Commerce, Gaithersburg, MD 20899. (301) 975-3069.

DIRECTORIES AND BIOGRAPHICAL SOURCES

American Men & Women of Science, 1995-96. R.R. Bowker Staff, eds. 19th edition. 8 volumes. R.R. Bowker/Reed International Publishing Company, 121 Chanlon Road, New Providence, NJ 07974. (908) 464-6800 or (800) 521-8110. 1995. $850.00

Directory of Engineering Societies and Related Organizations. Gordon Davis. 13th edition. American Association of Engineering Societies, 1111 19th Street NW, Suite 608, Washington, DC 20036. (202) 296-2237 or (800) 658-8897. 1989. Inquire for price.

International Research Centers Directory. Gale Research, 835 Penobscot Building, Detroit, MI 48226-4094. (313) 961-2242. (800) 347-4253. FAX (313) 961-6083. 8th edition. 1995. $410.00

ISA Directory of Instrumentation. Instrument Society of America, 67 Alexander Drive, Box 12277, Research Triangle Park, NC 27709. (919) 549-8411. FAX (919) 549-8288. 1979 to present. Annual. $100.00.

Research Centers Directory. Gale Research, 835 Penobscot Building, Detroit, MI 48226-4094. (313) 961-2242. (800) 347-4253. FAX (313) 961-6083. $485.00.

Scientific and Technical Organizations and Agencies Directory. Gale Research, 835 Penobscot Building, Detroit, MI 48226-4094. (313) 961-2242. (800) 347-4253. FAX (313) 961-6083. 4th edition. 1996. $195.00.

Who's Who in Engineering. American Association of Engineering Societies, 1111 19th Street NW, Suite 608, Washington, DC 20036. (202) 296-2237 or (800) 658-8897. 8th edition. 1991. Inquire for price.

ENCYCLOPEDIAS AND DICTIONARIES

Concise Encyclopedia of Measurement and Instrumentation. L. Finkelstein and K.T.V. Grattan. Pergamon Press Inc., Maxwell House, Fairview Park, Elmsford, NY 10523. (914) 592-7700. Fax (914) 592-3625. 1993. Inquire for cost.

McGraw-Hill Dictionary of Scientific and Technical Terms. Sybil P. Parker, ed. 5th edition. McGraw-Hill Publishing Company, 1221 Avenue of the Americas, New York, NY 10020. (800) 262-4729 or (212) 512-3825. 1993. $110.50.

McGraw-Hill Encyclopedia of Engineering. Sybil P. Parker, ed. 2nd edition. McGraw-Hill Publishing Company, 1221 Avenue of the Americas, New York, NY 10020. (800) 262-4729 or (212) 512-3825. 1993. $95.50.

McGraw-Hill Encyclopedia of Science and Technology. Sybil P. Parker, ed. 7th edition. 20 volumes. McGraw-Hill Publishing Company, 1221 Avenue of the Americas, New York, NY 10020. (800) 262-4729 or (212) 512-3825. 1992. $1900.00

Thesaurus of Scientific, Technical, and Engineering Terms. Hemisphere Publishing Corporation, 1900 Frost Road, Suite 101, Bristol, PA 19007-1598. (215) 785-5800 or FAX (215) 785-5515. 1987. $173.00.

GENERAL WORKS

Annual Book of ASTM Standards, 1993, Vol. 14.03: Temperature Measurement. American Society for Testing and Materials (ASTM), 1916 Race Street, Philadelphia, PA 19103. (215) 299-5585. 1993. $54.00.

Principles and Methods of Temperature Measurement. Thomas D. McGee. John Wiley and Sons, Inc., 605 Third Avenue, New York, NY 10158. (800) 526-5368 or (212) 850-6000. 1988. $74.95.

Temperature Measurement. Bela G. Liptak. Chilton Book Co., One Chilton Way, Radnor PA 19089. (215) 964-4028 or (800) 695-1214. FAX (610) 964-4745. 1993. $29.95.

Temperature Measurement. L. Michalski, et al. John Wiley and Sons, Inc., 605 Third Avenue, New York, NY 10158. (800) 526-5368 or (212) 850-6000. 1991. $145.00.

HANDBOOKS AND MANUALS

Handbook of Measurement Science. P.H. Sydenham, editor. 3 volumes. John Wiley and Sons, Inc., 605 Third Avenue, New York, NY 10158. (800) 526-5368 or (212) 850-6000. 1982 (vol.1), 1983 (vol.2), 1992 (vol.3). $270.00 each for vols.1-2, $199.95 (vol.3).

ONLINE DATABASES AND CD-ROMS

Compendex Plus. Engineering Information, Inc., 345 East 47th Street, New York, NY 10017. (212) 705-7600 or (800) 221-1044. Contains citations with abstracts to worldwide literature in engineering and technology, from 1970 to present. Available on online BRS,(800) 289-4277, DIALOG, (800) 334-2564, ORBIT (800) 456-7248, and STN International, FIZ Karlsruhe, P.O. Box 2465, W-7500, Karlsruhe 1, Germany, online services. Also available on CD-ROM. Inquire as to cost and availability.

CSA Engineering. Cambridge Scientific Abstracts, 7200 Wisconsin Avenue, Suite 601, Bethesda, MD 20814. (301) 961-6750 or (800) 843-7751. Contains citations and abstracts of international periodicals and research literature covering all fields of engineering and science and technology, including computer and information science, electronics, mechanical engineering, solid state materials, 1981 to present. Available on BRS,(800) 289-4277, online service. Inquire as to cost and availability.

Current Contents Search. Institute for Scientific Information, 3501 Market Street, Philadelphia, PA 19104. (215) 386-0100. FAX (215) 386-6362. Contains citations to articles listed in the table of contents of science and technology journals. Also articles in social sciences and life sciences journals. Available on BRS,(800) 289-4277, DIALOG,(800) 334-2564, online services. Inquire as to cost and availability.

Dissertation Abstracts. University Microfilms International, 300 North Zeeb Road, Ann Arbor, MI 48106. (800) 521-0600 or (313) 761-4700. Scope includes virtually all doctoral dissertations accepted at accredited American institutions from 1861 to present in 252 subject areas. Available on BRS,(800) 289-4277, DIALOG,(800) 334-2564, and OCLC EPIC,(800) 848-5878, online services. Also available on CD-ROM. Inquire as to cost and availability.

ISMEC: Mechanical Engineering Abstracts. Cambridge Scientific Abstracts, 7200 Wisconsin Avenue, Suite 601, Bethesda, MD 20814. (301) 961-6750 or (800) 843-7751. Contains citations to the literature in mechanical engineering, industrial and production engineering, energy, power, mechanics, devices and related areas, from 1973 to present. Available on the DIALOG,(800) 334-2564, online service. Inquire as to cost and availability.

NTIS Bibliographic Database. National Technical Information Service, 5285 Port Royal Road, Springfield, VA 22161. (703)

THERMOMETERS

Ency. of Physical Sciences and Engineering Info. Sources

487-4929 or FAX (703) 321-8199. Broad coverage of government-sponsored science and technology research reports, 1964 to present. Available on BRS,(800) 289-4277, DIALOG, (800) 334-2564, ORBIT, (800) 456-7248, and STN International, FIZ Karlsruhe, P.O. Box 2465, W-7500, Karlsruhe 1, Germany, online services. Also available on CD-ROM. Inquire as to cost and availability.

SCISEARCH. Institute for Scientific Information, 3501 Market Street, Philadelphia, PA 19104. (800) 523-1850 or (215) 386-0100. Broad multidisciplinary title and author index to the international literature of science and technology, 1974 to present. Available on DIALOG,(800) 334-2564, and ORBIT,(800) 456-7248, online services. Also available on CD-ROM. Inquire as to cost and availability.

WILSONLINE. H.W. Wilson Company, 950 University Avenue, Bronx, NY 10452. (800) 367-6770 or (212) 588-8400. Makes available online versions of the printed H.W. Wilson indexes, including Applied Science and Technology Index, Business Periodicals Index, General Science Index, and Readers' Guide to Periodical Literature. Period covered is generally 1983 to present. Available on BRS,(800) 289-4277, DIALOG,(800) 334-2564, and OCLC EPIC,(800) 848-5878, online services. Also available on CD-ROM. Inquire as to cost and availability.

PERIODICALS

ASTM Standardization News. American Society for Testing and Materials (ASTM), 1916 Race Street, Philadelphia, PA 19103. (215) 299-5585. FAX (215) 977-9679. 1973 to present. Monthly. $18.00 per year.

Control Engineering. Cahners Publishing Company (Des Plaines), 1350 East Touhy Avenue, Des Plaines, IL 60017-5080. (708) 635-8800. FAX (708) 635-9950. 1954 to present. Monthly. $79.95 per year.

IEEE Transactions On Instrumentation and Measurement. IEEE (Institute of Electrical and Electronic Engineers), 345 E. 47th Street, New York, NY 10017-2394. 1952 to present. Bi-monthly. $165.00 per year for non-members.

Industrial and Scientific Instruments. Wilmington Publishing, Wilmington House, Church Hill, Dartford, Kent UA2 7EF, England. Telephone 0332-277788. FAX 0332-276476. 1960 to present. Monthly. $139.00 per year.

Instrumentation and Automation News. Chilton Co., One Chilton Way, Radnor PA 19089. (215) 964-4028. 1953 to present. Monthly. $35.00 per year.

Instrumentation and Control Systems. Chilton Co., One Chilton Way, Radnor PA 19089. (215) 964-4028. 1928 to present. Monthly. $60.00 per year.

Intech. Instrument Society of America, 67 Alexander Drive, Box 12277, Research Triangle Park, NC 27709. (919) 549-8411. FAX (919) 549-8288. 1954 to present. Monthly. $65.00 per year.

Measurement Science and Technology. IOP Publishing, Institute of Physics. Distributed by American Institute of Physics, Subscriber Services, 500 Sunnyside Boulevard, Woodbury, NY 11797-2999. 1968 to present. Monthly. $759.00 per year.

Measurement and Control. Institute of Measurement and Control, 87 Gower Street, London WC1E 6AA, England. Telephone 44-71-387-4949. FAX 44-71-388-8431. 1968 to present. Monthly. Inquire for cost.

Measurements and Control. Measurements and Data Corporation, 2994 W. Liberty Avenue, Pittsburgh, PA 15216. (412) 343-4666. 1967 to present. Bi-monthly. $22.00 per year.

Measurements and Control News. Measurements and Data Corporation, 2994 W. Liberty Avenue, Pittsburgh, PA 15216. (412) 343-4666. 6 times a year. Inquire for cost.

Metrologia. Bureau International des Poids et Mesures, Pavillon de Breteuil, 92312 Sevres Decex, France. FAX 45-34-20-21. 1965 to present. 6 times a year. $298.00 per year.

National Institute of Standards and Technology Journal of Research. U.S. National Institute of Standards and Technology, U.S. Department of Commerce, Gaithersburg, MD 20899. (301) 975-3069. 1928 to present. Bi-monthly. $27.00 per year.

Review of Scientific Instruments. American Institute of Physics, One Physics Ellipse, College Park, MD 20740. (301) 209-3000. 1930 to present. Monthly. $800.00 per year.

Standards Engineering. Standards Engineering Society, 1706 Darst Avenue, Dayton, OH 45403-3104. (513) 258-1955. Bi-monthly. $40.00 per year.

RESEARCH CENTERS AND INSTITUTES

Facilty For Advanced Instrumentation. University of California—Davis, Davis, CA 95616. (916) 752-0284.

Instrumentation and Control Laboratory. Princeton University, Department of MAE, Engineering Quadrangle, Princeton, NJ 08544. (609) 452-5154.

Major Analytical Instrumentation Center, University of Florida. College of Engineering, Rhines Hall, Gainesville, FL 32611. (904) 392-6985.

SPECIFICATIONS AND STANDARDS

Annual Book of ASTM Standards, Volume 14.03. American Society for Testing and Materials (ASTM), 1916 Race Street, Philadelphia, PA 19103. (215) 299-5585. Annual. Inquire for cost and availability.

THERMONUCLEAR FUSION

See: FUSION

THERMOPLASTICS

See also: GUMS and RESINS, MATERIALS SCIENCE, ORGANIC CHEMISTRY, PACKAGING, PLASTICS, POLYMERS,

ABSTRACT SERVICES AND INDEXES

Applied Science and Technology Index; A Cumulative Subject Index to English Language Periodicals in the Fields of Aeronautics and Space Science, Computer Technology, Chemistry, Construction Industry, Energy and Related areas. H.W. Wilson Co., 950 University Avenue, Bronx, NY 10452-9978. (212) 588-8400. (800) 367-6770. FAX (718) 590-1617. From 1958 to present. Monthly. Inquire about cost and availability. Also available online (BRS and WILSONLINE) and on CD-ROM. ISSN: 0003-6986.

Chemical Abstracts. Chemical Abstracts Service. 2540 Olentangy River Road, Box 3012, Columbus, OH 43210-0012. (614) 447-3600. FAX (614) 447-3713. 1907 to present. Weekly. $16,800.00 per year. Available online and on CD-ROM. CA is also available in five section groupings. Inquire regarding cost and availability.

Chemical Engineering Abstracts. Royal Society of Chemistry, Information Services, Thomas Graham House, Science Park, Milton Road, Cambridge, CB4 4WF, England. Contains citations, mostly with abstracts, of the worldwide literature on chemical engineering, from 1982 to present. Monthly. $450.00 per year. Also available online.

Current Contents: Physical, Chemical and Earth Sciences. Institute for Scientific Information, 3501 Market Street, Philadelphia, PA 19104. (215) 386-0100. (800) 523-1850. FAX (215) 386-2291. 1961 to present. Weekly. $442.00 per year. Also available online (BRS, DIALOG) and on CD-ROM. Inquire regarding cost and availability. ISSN: 0163-2574.

Engineered Materials Abstracts. ASM (American Society of Metals) International, Materials Information, Materials Park, OH 44073. (216) 338-5151 or FAX (216) 338-4634. Covers literature on technical developments in polymer, ceramic, and composite materials and engineering. 1986 to present. Monthly. $1175.00 per year. Also available on CD-ROM.

Engineering Index Monthly. Engineering Information, Inc., Castle Point on the Hudson, Hoboken, NJ 07030. (201) 216-8500. (800) 221-1044. FAX (201) 216-8532. Monthly. $2300.00 per year for monthly issues. $1980 per year for annual cumulation. Available on CD-ROM and online as COMPENDEX. ISSN: 0742-1974; 0360-8557.

General Science Index. H.W. Wilson Company, 950 University Avenue, Bronx, NY 10452. (212) 588-8400. (800) 367-6770. FAX (718) 590-1617. From 1978 to present. Ten issues per year; quarterly and annual cumulations. Service basis. Available on CD-ROM and online. Inquire regarding cost and availability. ISSN: 0162-1963.

Process and Chemical Engineering/chemical Engineering Abstracts. Royal Society of Chemistry, Information Services, Thomas Graham House, Science Park, Milton Road, Cambridge, CB4 4WF, England. Contains citations, mostly with abstracts, of the worldwide literature on chemical engineering; from 1982 to present. Monthly. $610.00 per year. Also available online. ISSN: 0960-5045.

Science Citation Index. SCI. Institute for Scientific Information, 3501 Market Street, Philadelphia, PA 19104. (215) 386-0100. (800) 523-1850. FAX (215) 386-2991. 1961 to present. Six issues per year, plus annual cumulation. $11650.00 per year. Also available online and on CD-ROM. Inquire about price and availability. ISSN: 0036-827X.

Theoretical Chemical Engineering. Royal Society of Chemistry, Information Services, Thomas Graham House, Science Park, Milton Road, Cambridge, CB4 4WF, England. Covers theoretical chemical engineering, including theory and laboratory experimentation. 1964 to present. Monthly. $235.00. ISSN: 0960-5053.

ANNUAL REVIEWS AND YEARBOOKS

Advances in Chemical Engineering. Academic Press, Inc., 6277 Sea Harbor Drive, Orlando, FL 32887. (800) 321-5068. 1956 to present. Irregular. ISSN: 0065-2377.

Progress in Colloid and Polymer Science Series. Springer-Verlag, 175 Fifth Avenue, New York, NY 10010. (212) 460-1500. Irregular, price varies.

ASSOCIATIONS AND PROFESSIONAL SOCIETIES

American Chemical Society, 1155 16th Street, NW, Washington, DC 20036. (202) 872-4600.

American Institute of Chemical Engineers. 345 East 47th Street, New York, NY 10017-2396. (212) 705-7663.

Association of Consulting Chemists and Chemical Engineers, 50 East 41st Street, Suite 92, New York, NY 10017. (212) 684-6255.

Chemical Manufacturers Association. 2501 M Street, N.W., Washington, DC 20037. (202) 887-1182.

Plastics Institute of America. 277 Fairfield Road, Suite 100, Fairfield, NJ 07004-1932. (201) 808-5950. FAX (201) 808-5953.

Society of Plastics Engineers, Inc., 14 Fairfield Drive, Brookfield, Center, CT 06805-0403. (203) 775-0471. FAX (203) 775-8490.

Society of the Plastics Industry. 1275 K Street NW, Suite 400, Washington, DC 20005. (202) 371-5200. FAX (202) 371-1022.

BIBLIOGRAPHIES

Information Sources in Polymers and Plastics. R.T. Adkins. Bowker-Saur, 121 Chanlon Road, New Providence, NJ 07974. (908) 464-6800. (800) 521-8110. 1989. $75.00.

Organic Reaction Mechanisms: An Annual Survey of Literature. A. C. Knipe and W. E. Watts, editors. John Wiley & Sons, Inc., 605 Third Avenue, New York, NY 10158-0012. (212) 850-6000. (800) 225-5945. vol. 92, 1992. price varies.

Scientific and Technical Books and Serials in Print; An Index To Literature in Science and Technology. R.R. Bowker Inc., 121 Chanlon Road, New Providence, NJ 07974. (908) 464-6800. (800) 521-8110. FAX (908) 665-3502. 1972 to present. Annual. 4 volumes. 1994. $299.95. Also available on compact disc and online. ISSN: 0000-054X.

THERMOPLASTICS

Ency. of Physical Sciences and Engineering Info. Sources

DIRECTORIES AND BIOGRAPHICAL SOURCES

American Institute of Chemical Engineers. Directory. American Institute of Chemical Engineers. 345 East 47th Street, New York, NY 10017-2396. (212) 705-7663. Annual.

American Men and Women of Science. Physical and Biological Sciences. R.R. Bowker Company, 121 Chanlon Road, New Providence, NJ 07974. (908) 464-6800. (800) 521-8110. 20th edition. 8 volumes. 1996. $850.00.

Chemical Engineering Faculties. American Institute of Chemical Engineers. 345 East 47th Street, New York, NY 10017-2396. (212) 705-7663. Annual. $75.00.

Consulting Services: Chemists and Chemical Engineers. Association of Consulting Chemists and Chemical Engineers, 40 East 45th Street, New York, NY 10017. (212) 983-3160. FAX (212) 983-3161. Biennial. $60.00.

Engineering Research Centers: Incorporating Electronics Research Centers. Stockton Press, 345 Park Avenue, New York, NY 10010. (212) 689-9200. (800) 221-2123. 4th edition. 1995. $515.00.

Modern Plastics Encyclopedia Directory of Trade Names. McGraw-Hill, 1221 Avenue of the Americas, New York, NY 10020. (212) 512-6266. Annual. $10.00.

Research Centers Directory. Gale Research Company, Inc., 835 Penobscot Building, Detroit, MI 48226-4094. (313) 961-2242. (800) 347-4253. 20th edition. 1995. $485.00. ISSN: 0080-1518.

Society of the Plastics Industry-Membership Directory and Buyers' Guide. 1275 K Street NW, Suite 400, Washington, DC 20005. (202) 371-5200. FAX (202) 371-1022. Annual. $180.00.

Who's Who in Engineering. Gordon Davis, editor. American Association of Engineering Societies. 1111 19th Street, NY, Suite 608, Washington, DC 20036. (202) 296-2237. (800) 658-8897. 9th edition. 1995. $220.00.

ENCYCLOPEDIAS AND DICTIONARIES

Atlas of Polymer and Plastic Analysis. Dieter O. Hummel. VCH Publications, Inc., 220 East 23rd Street, Suite 909, New York, NY 10010-4606. (212) 683-8333. (800) 422-8824. Volume 1, parts A and B. 3rd edition. 1991. $525.00.

Concise Encyclopedia of Polymer Science. Herman F. Mark. John Wiley & Sons, Inc., 605 Third Avenue, New York, NY 10158-0012. (212) 850-6000. (800) 225-5945. 1990. $199.00.

Dictionary of Plastics Technology. R. J. Heath and A. W. Birley. Chapman & Hall, 1 Penn Plaza, New York, NY 10119. (212) 564-1060. 1992. $59.95.

Encyclopedia of Chemical Processing and Design. McKetta Marcel Dekker, Inc., 270 Madison Avenue, New York, NY 10016. (212) 696-9000. (800) 228-1160. 1976-. $175.00 per volume.

Encyclopedia of Polymer Science and Engineering. Herman F. Mark et al, editors. John Wiley & Sons, Inc., 605 Third Avenue, New York, NY 10158-0012. (212) 850-6000. (800) 225-5945. 2nd edition. 1985-. 22 volumes. $4788.00.

Handbook of Polymer Science and Technology. Nicholas P. Cheremisinoff. Marcel Dekker, Inc., 270 Madison Avenue, New York, NY 10016. (212) 696-9000. (800) 228-1160. 4 volumes. 1987. $235.00 each.

Illustrated Chemistry Laboratory Terminology. Gerbert W. Ockerman. CRC Press, Inc., 2000 Corporate Boulevard, NW, Boca Raton, FL 33431. (407) 994-0555. (800) 272-7737. 1991. $32.95.

Kirk-Othmer Encyclopedia of Chemical Technology. John Wiley and Sons, Inc., 605 Third Avenue, New York, NY 10158. (800) 526-5368 or (212) 850-6000. Fourth edition. 1991-. Twenty-seven volumes. $5400.00.

Modern Plastics-Encyclopedia Issue. McGraw-Hill 1221 Avenue of the Americas, New York, NY 10020. (212) 512-6255. Annual; issued in October. $49.95.

Polymer Handbook: Polymerization And Depolymerization. J. Brandrup and E. H. Immergut. John Wiley & Sons, Inc., 605 Third Avenue, New York, NY 10158-0012. (212) 850-6000.(800) 225-5945. Volume 1. 3rd edition. 1989. $245.00.

Ullman's Encyclopedia of Industrial Chemistry. VCH Publications, Inc., 220 East 23rd Street, Suite 909, New York, NY 10010-4606. (212) 683-8333. (800) 422-8824. 5th edition. 1984-. Price varies per volume.

GENERAL WORKS

Advanced Thermoplastic Composites: Characteristics and Processing. Hans-Henning Kausch. Hanser-Gardner Publications, 6600 Clough Pike, Cincinnati, OH 45244-4090. (513)527-8977. (800) 950-8977. 1992. $94.50

Engineering Properties of Thermoplastics. R.M. Ogorkiewiez, editor. Books on Demand, 300 North Zeeb Road, Ann Arbor, MI 48106-1346. (313) 761-4700. (800) 521-0600. $94.10.

High Performance and Engineering Thermoplastic Composites. A. Brent Strong. Technomic Publishing Co., 851 New Holland Avenue, Box 3535, Lancaster PA 17604. (800) 233-9936. 1993. $55.00 in paper.

Introduction To Plastics. Hans-Georg Elias. VCH Publications, Inc., 220 East 23rd Street, Suite 909, New York, NY 10010-4606. (212) 683-8333. (800) 422-8824. 1993. $50.00.

Plastics Engineering. R. J. Crawford. Elsevier Science Publishing Company, Inc., 655 Avenue of the Americas, New York, NY 10010. (212) 989-5800. FAX (914) 333-2444. 2nd edition. 1987. $44.00 in paper.

Plastics Recycling: Products and Processes. Raymond J. Ehrig. Hanser-Gardner Publications, 6600 Clough Pike, Cincinnati, OH 45244-4090. (513)527-8977. (800) 950-8977. 1992. $62.50.

Polymer Processing: Principles and Design. Donald G. Baird. Butterworth-Heinemann, 313 Washington Street, Newton, MA 02158. (618) 928-2500. (800) 366-2665. 1995. $125.00.

Thermoplastics: Materials Engineering. L. Mascia. Elsevier Science Publishing Company, Inc., 655 Avenue of the Americas, New York, NY 10010. (212) 989-5800. FAX (914) 333-2444. 1989. $129.75.

HANDBOOKS AND MANUALS

Chemical Formulary. H. Bennett, editor. Chemical Publishing, Co., Inc. 80 Eighth Avenue, New York, NY 10011. (212) 255-1950. Volumes 1-30. $60.00 per volume.

CRC Handbook of Chemistry and Physics. David R. Kide, editor. CRC Press Inc., 2000 Corporate Blvd., NW, Boca Raton, FL 33431. (407) 994-0555. (800) 333-8300. 77th edition. 1996. $99.95.

Flammability Handbook For Plastics. Carlos J Hilado, editor. Technomic Publishing Co., 851 New Holland Avenue, Box 3535, Lancaster PA 17604. (800) 233-9936. 4th edition. 1990. $65.00.

Handbook of Plastics Materials and Technology. Irvin I. Rubin. John Wiley and Sons, Inc., 605 Third Avenue, New York, NY 10158. (800) 526-5368 or (212) 850-6000 1990. $185.00.

Handbook of Plastics, Elastomers and Composites. Charles A. Harper. McGraw-Hill Publishing Company, Inc., 1221 Avenue of the Americas, New York, NY 10020. (212) 512-2000. (800) 262-4729. 1992. $89.50.

Plastics Engineering Handbook of the Society of the Plastics Industry. Joel Frados, editor. Van Nostrand Reinhold, 115 Fifth Avenue, New York, NY 10003. (212) 254-3232. (800) 842-3636. 1985.

Plastics Handbook. Modern Plastics Magazine Staff. McGraw-Hill Publishing Company, Inc., 1221 Avenue of the Americas, New York, NY 10020. (212) 512-2000. (800) 262-4729. 1995. $55.50.

Plastics Technology Handbook. Manas Chanda and Salil K. Roy. 2nd edition. Marcel Dekker, Inc., 270 Madison Avenue, New York, NY 10016. (212) 696-9000. (800) 228-1160. 1992. $215.00.

Riegel's Handbook of Industrial Chemistry. James A. Kent, editor. Van Nostrand Reinhold, 115 Fifth Avenue, New York, NY 10003. (212) 254-3232 or (800) 926-2665. Ninth edition. 1992. $114.95.

ONLINE DATABASES AND CD-ROMS

Beilstein Online. Beilstein-Institut fur Literatur der Organischen Chemie, Varrentrapperstrasse, 40-42, D-6000, Frankfurt am Main 90, Germany. Contains data on carbon compounds from the Beilstein Handbook of Organic Chemistry. Structural and factual data for more than 3.4 million heterocyclic, isocyclic, and acyclic compounds included. Covers the period 1830 to present. Available on DIALOG, (800) 334-2564, ORBIT (800) 456-7248, and STN International, FIZ Karlsruhe, P.O. Box 2465, W-7500, Karlsruhe 1, Germany, online services. Inquire as to cost and availability.

CA Search. Chemical Abstracts Service, P.O. Box 3012, Columbus, OH 43210-0012. (614) 447-3600. (800) 848-6533. FAX (614) 447-3709. Very comprehensive guide to worldwide chemical literature and related fields, 1972 to present. Available on BRS,(800) 289-4277, DIALOG, (800) 334-2564, ORBIT (800) 456-7248, and STN International, FIZ Karlsruhe, P.O. Box 2465, W-7500, Karlsruhe 1, Germany, online services. Inquire as to cost and availability.

Chemical Journals of the American Chemical Society. American Chemical Society, 1155 16th Street, N.W., Washington, DC 20036. (202) 872-4381 or (800) 424-6747. Contains complete text of approximately 90,000 articles from 22 primary journals published by the American Chemical Society, from mostly 1982 to present. Available on STN International, FIZ Karlsruhe, P.O. Box 2465, W-7500, Karlsruhe 1, Germany, online service. Inquire as to cost and availability.

Compendex Plus. Engineering Information, Inc., 345 East 47th Street, New York, NY 10017. (212) 705-7600 or (800) 221-1044. Contains citations with abstracts to worldwide literature in engineering and technology, from 1970 to present. Available on online BRS,(800) 289-4277, DIALOG, (800) 334-2564, ORBIT (800) 456-7248, and STN International, FIZ Karlsruhe, P.O. Box 2465, W-7500, Karlsruhe 1, Germany, online services. Also available on CD-ROM. Inquire as to cost and availability.

Current Contents Search. Institute for Scientific Information, 3501 Market Street, Philadelphia, PA 19104. (215) 386-0100. FAX (215) 386-6362. Contains citations to articles listed in the table of contents of science and technology journals. Also articles in social sciences and life sciences journals. Available on BRS,(800) 289-4277, DIALOG,(800) 334-2564, online services. Inquire as to cost and availability.

Dissertation Abstracts. University Microfilms International, 300 North Zeeb Road, Ann Arbor, MI 48106. (800) 521-0600 or (313) 761-4700. Scope includes virtually all doctoral dissertations accepted at accredited American institutions from 1861 to present in 252 subject areas. Available on BRS,(800) 289-4277, DIALOG,(800) 334-2564, and OCLC EPIC,(800) 848-5878, online services. Also available on CD-ROM. Inquire as to cost and availability.

Engineered Materials Abstracts. Materials Information, ASM International, Materials Park, OH 44073. (216) 338-5151. Contains citations with abstracts of worldwide literature in the development, processing, and production of ceramic, composite, and polymeric materials for engineering uses. Available on DIALOG, (800) 334-2564, ORBIT (800) 456-7248, and STN International, FIZ Karlsruhe, P.O. Box 2465, W-7500, Karlsruhe 1, Germany, online services. Inquire as to cost and availability.

Kirk-Othmer Encyclopedia of Chemical Technology. John Wiley and Sons, Inc., 605 Third Avenue, New York, NY 10158. (800) 526-5368 or (212) 850-6000. Contains the complete text of all chapters in the 27 volume fourth edition of the *KIRK-Othmer Encyclopedia of Chemical Technology.* 1991. Available on BRS,(800) 289-4277, DIALOG,(800) 334-2564, online services. Inquire as to cost and availability.

NTIS Bibliographic Database. National Technical Information Service, 5285 Port Royal Road, Springfield, VA 22161. (703) 487-4929 or FAX (703) 321-8199. Broad coverage of government-sponsored science and technology research reports, 1964 to present. Available on BRS,(800) 289-4277, DIALOG, (800) 334-2564, ORBIT, (800) 456-7248, and STN International, FIZ Karlsruhe, P.O. Box 2465, W-7500, Karlsruhe 1, Germany, online services. Also available on CD-ROM. Inquire as to cost and availability.

Scientific and Technical Books and Serials in Print. R.R. Bowker Inc., 121 Chanlon Road, New Providence, NJ 07974. (800) 521-8110 or (908) 464-6800. List of currently published books and serials in the physical and biological sciences, engineering and technology, with subject, author and titles indexes. Available on ORBIT,(800) 456-7248, online service. Inquire as to cost and availability.

THERMOPLASTICS

Ency. of Physical Sciences and Engineering Info. Sources

SCISEARCH. Institute for Scientific Information, 3501 Market Street, Philadelphia, PA 19104. (800) 523-1850 or (215) 386-0100. Broad multidisciplinary title and author index to the international literature of science and technology, 1974 to present. Available on DIALOG,(800) 334-2564, and ORBIT,(800) 456-7248, online services. Also available on CD-ROM. Inquire as to cost and availability.

Who's Who in Technology. Gale Research, 835 Penobscot Building, Detroit, MI 48226-4094. (313) 961-2242. (800) 347-4253. Contains biographical information of contemporary American scientists and engineers. Available on DIALOG,(800) 334-2564, online service. Inquire as to cost and availability.

WILSONLINE. H.W. Wilson Company, 950 University Avenue, Bronx, NY 10452. (800) 367-6770 or (212) 588-8400. Makes available online versions of the printed H.W. Wilson indexes, including Applied Science and Technology Index, Business Periodicals Index, General Science Index, and Readers' Guide to Periodical Literature. Period covered is generally 1983 to present. Available on BRS,(800) 289-4277, DIALOG,(800) 334-2564, and OCLC EPIC,(800) 848-5878, online services. Also available on CD-ROM. Inquire as to cost and availability.

PERIODICALS

Advances in Polymer Technology. John Wiley and Sons, Inc., 605 Third Avenue, New York, NY 10158. (800) 526-5368 or (212) 850-6000. 1977 to present.

AICHE Journal. American Institute of Chemical Engineers, 345 47th Street, New York, NY 10017-2396. (212) 705-7663. 1955 to present. Monthly. $295.00 per year.

Canadian Plastics. Southam Magazine Group, 1450 Don Mills Road, Don Mills, ON M3B 2X7. Canada. (416) 445-6641. FAX (416) 442-2213. 1943 to present. 8 issues per year. $55.00 per year. ISSN: 0008-4778.

Chemical Week; Includes Annual Buyers Guide. Chemical Week Associates, 888 Seventh Avenue, New York, NY 10106. (212) 621-4900. FAX (212) 621-4949. 1914 to present. Weekly. $99.00 per year. ISSN: 0009-272X.

Chemtech. American Chemical Society. Box 3337, Columbus, OH 43210. (614) 447-3776. 1970 to present. Monthly. $370.00 per year. ISSN: 0009-2703.

High Performance Plastics. Elsevier Science Publishing Company, Inc., 655 Avenue of the Americas, New York, NY 10010. (212) 989-5800. 1983 to present. Monthly. $390.00 per year. ISSN: 0264-7753.

Industrial and Engineering Chemistry Research. American Chemical Society, Box 3337, Columbus, OH 43210. (614) 447-3776. (800) 333-9511. FAX (614) 447-3671. 1962 to present. Monthly. $567.00 per year. ISSN: 0888-5885.

Journal of Cellular Plastics. Technomic Publishing Company, Inc., 851 New Holland Avenue, Box 3535, Lancaster, PA 17604. (717) 291-5609. FAX (717) 295-4538. 1965 to present. Bi-monthly. $175.00 per year. ISSN: 0021-955X

Journal of Elastomers and Plastics. Technomic Publishing Company, Inc., 851 New Holland Avenue, Box 3535, Lancaster, PA 17604. (717) 291-5609. FAX (717) 295-4538. 1969 to present. Quarterly. $205.00 per year. ISSN: 0095-2443.

Journal of Materials Processing and Manufacturing. Technomic Publishing Company, Inc., 851 New Holland Avenue, Box 3535, Lancaster, PA 17604. (717) 291-5609. 1992 to present. Quarterly. $195.00 per year.

Journal of Plastic Film and Sheeting. Technomic Publishing Company, Inc., 851 New Holland Avenue, Box 3535, Lancaster, PA 17604. (717) 291-5609. FAX (717) 295-4538. 1985 to present. Quarterly. $265.00. per year. ISSN: 8756-0879.

Journal of Polymer Science. Part A: Polymer Chemistry. John Wiley and Sons, Inc., 605 Third Avenue, New York, NY 10158. (212) 850-6088. FAX (212) 850-6088. 1962 to present. Subscription includes part B: Polymer Physics, and Symposia Proceedings. 13 issues per year. $2540.00 per year. ISSN: 0887-624X; 0887-6266; 0360-8905.

Modern Plastics. McGraw-Hill, Inc., 1221 Avenue of the Americas, New York, NY 10020. (212) 512-6267. FAX (212) 512-6111. 1925 to present. Monthly. $41.75 per year. ISSN: 0026-8275.

Plastics Compounding: For Resin Producers, Formulators and Compounders. Advanstar Communications, Inc., 7500 Old Oak Boulevard, Cleveland, OH 44130. (216) 826-2839. FAX (216) 891-2726. 1977 to present. Bi-monthly. $40.00 per year. ISSN: 0148-9119.

Plastics Design Forum. Advanstar Communications, Inc., 7500 Old Oak Boulevard, Cleveland, OH 44130. (216) 826-2839. FAX (216) 891-2726. 1976 to present. 7 issues per year. $40.00 per year. ISSN: 0362-9376.

Plastics Engineering. Society of Plastics Engineers, Inc., 14 Fairfield Drive, Brookfield, Center, CT 06805-0403. (203) 775-0471. FAX (203) 775-8490. 1945 to present. Monthly. $50.00 per year. ISSN: 0091-9578.

Plastics Technology. Bill Communications Inc., 355 Park Avenue South, New York, NY 10011-1706. (212) 592-6570. 1955 to present. 13 issues per year. $65.00 per year. ISSN: 0032-1257.

Polymer Composites. Society of Plastics Engineers, Inc., 14 Fairfield Drive, Brookfield, Center, CT 06805-0403. (203) 775-0471. FAX (203) 775-8490. 1980 to present. Bi-monthly. $200.00 per year. ISSN: 0272-8397.

Processing of Advanced Materials. Routledge, Chapman and Hall, Inc., 29 West 35th Street, New York, NY 10001-2291. (212) 244-3336. 1991 to present. Quarterly. $145.00 per year.

RESEARCH CENTERS AND INSTITUTES

Center for Plastics Recycling Research. Rutgers University, Building 3529, Busch Campus, PO Box 1179, Piscataway, NJ 08855-1179. (809) 932-3683. FAX (98) 932-5636.

Institute of Polymer Engineering. University of Akron, Akron, OH 44325-0301. (216) 972-6865. (216) 258-2339.

Plastics Institute of America. 277 Fairfield Road, Suite 100, Fairfield, NJ 07004-1932. (201) 808-5950. FAX (201) 808-5953.

Polymer Processing Institute. Stevens Institute of Technology, Castle Point Station, Hoboken, NJ 07030.(201) 420-5819.

Ency. of Physical Sciences and Engineering Info. Sources

THUNDERSTORMS

Polymer Research Center. University of Cincinnati, Mail Location 172, Cincinnati, OH 45221. (513) 465-2453.

Polymer Research Laboratory. University of Michigan, 2046 Dow Building, Ann Arbor, MI 48109. (313) 763-2240. fax (313) 763-4788.

THIN FILMS

See: SEMICONDUCTORS

THUNDERSTORMS

See also: LIGHTNING, METEOROLOGY, RAIN, SNOW, TORNADOS, WIND

ABSTRACT SERVICES AND INDEXES

Applied Science and Technology Index; A Cumulative Subject Index To English Language Periodicals in the Fields of Aeronautics and Space Science, Computer Technology, Chemistry, Construction Industry, Energy and Related Areas. H.W. Wilson Co., 950 University Avenue, Bronx, NY 10452-9978. (212) 588-8400 (800) 367-6770. FAX (718) 590-1617. From 1958 to present. Monthly. Inquire about cost and availability. Also available online (BRS and WILSONLINE) and on CD-ROM. ISSN: 0003-6986.

Deep-Sea Research. Part A: Oceanographic Research Papers. Deep-Sea Research. Part B: Oceanographic Literature Review. Pergamon Press, Inc., Maxwell House, Fairview Park, Elmsford, NY 10523. (914) 592-7700. Fax (914) 592-3625. 1953 to present. Twelve times per year. $2000.00 per year for Parts A and B. Oceanographic Literature Review also available on CD-ROM. Inquire about price and availability.

General Science Index. H.W. Wilson Co., 950 University Avenue, Bronx, NY 10452. (212) 588-8400. (800) 367-6770. FAX (718) 590-1617. From 1978 to present. Ten issues per year; quarterly and annual cumulations. Service basis. Available on CD-ROM and online. Inquire regarding cost and availability. ISSN: 0162-1963.

Government Reports Announcements and Index. National Technical Information Service (NTIS), 5285 Port Royal Road, Springfield, VA 22161. (703) 487-4650. FAX (703) 321-8547. From 1968 to present. Annual. $630.00 per year. Also available online as NTIS Bibliographic Database and on CD-ROM.

Meteorological and Geoastrophysical Abstracts. American Meteorological Society, c/o Inforonics, Inc., 550 Newtown Road, Box 458, Littleton, MA 01460. (508) 486-8976. FAX (508) 486-0027. Covers literature in environmental sciences, meteorology, astrophysics, hydrology, glaciology, and physical oceanography. 1950 to present. Monthly. $950.00 per year. ISSN: 0026-1130. Also available online (DIALOG) and on CD-ROM.

Physics Abstracts. Institute of Electrical Engineers, London, United Kingdom. Available from: Institute of Electrical and Electronic Engineers (IEEE), 345 East 47th Street, New York, NY 10017. (212) 705-7900.

Science Citation Index. SCI. Institute for Scientific Information, 3501 Market Street, Philadelphia, PA 19104. (215) 386-0100. (800) 523-1850. FAX (215) 386-2991. 1961 to present. Six issues per year, plus annual cumulation. $11650.00 per year. Also available online and on CD-ROM. Inquire about price and availability. ISSN: 0036-827X.

Selected Water Resources Abstracts. U.S. Geological Survey. Distributed by National Technical Information Service (NTIS), 5285 Port Royal Road, Springfield, VA 22161. (703) 487-4650. Monthly. $115.00 per year.

ANNUAL REVIEWS AND YEARBOOKS

Developments in Atmospheric Science Series. Elsevier Science Publishing Company, Inc., 655 Avenue of the Americas, New York, NY 10010. (212) 989-5800. Irregular. Inquire about price and availability.

American Meteorological Society. Meteorological Monographs. 45 Beacon Street, Boston, MA 02108-3693. (617) 227-2425. FAX(617) 742-8718. Irregular. Price varies.

ASSOCIATIONS AND PROFESSIONAL SOCIETIES

American Association of State Climatologists. c/o Myron Moinau, University of Idaho, Agriculture Engineering Department, State Climate Service, Moscow, ID 834844-2040.

American Meteorological Society, 45 Beacon Street, Boston, MA 02108-3693. (617) 227-2425. FAX (617) 742-8718.

Association of American Weather Observers, 401 Whitney Boulevard, Box 455, Belvedere, IL 61008. (815) 544-5665. FAX (815) 544-6334.

Canadian Meteorological and Oceanographic Society, P.O. Box 334, Newmarket, Ontario, L3Y 4X7, Canada. (416) 898-1040. FAX (416) 898-7937.

International Association of Meteorology and Atmospheric Physics. c/o UCAR, P.O. Box 3000, Boulder CO 80307-3000. (303) 497-1673.

International Association of Severe Weather Specialists. c/o Warren Faidley, PO Box 31808, Tucson, AZ 85751. (520) 751-9964. FAX (520) 751-1185.

National Environmental Satellite Data, and Information Service. 2069 Federal Building 4 Room 2060, Washington, CD 20233. (301) 763-71990. FAX (301) 763-4011.

National Weather Association, 6704 Wolke Court, Montgomery, AL 36116-2134. (334) 213-0388. FAX (334) 213-0388.

University Corporation For Atmospheric Research. P.O. Box 3000, Boulder, CO 80307-3000. (303) 497-1650. FAX (303) 497-1654.

BIBLIOGRAPHIES

Information Sources in the Earth Sciences. David N. Wood, Joan E. Hardy, and Anthony P. Harvey, editors. Bowker-Saur/K.G. Saur. Distributed by R.R. Bowker, 121 Chanlon Road, New

Providence, NJ 07974. (800) 521-8110 or (908) 464-6800. Second edition. 1989. $85.00.

DIRECTORIES AND BIOGRAPHICAL SOURCES

American Men and Women of Science: Physical and Biological Sciences. Eighteenth edition. R.R. Bowker Company, 245 West 17th Street, New York, NY 10011. (800) 521-8810 or (212) 916-1600. $750.00.

American Meteorological Society. Professional Directory. American Meteorological Society. 45 Beacon Street, Boston, MA 02108. (617) 227-2425. Included in each issue of the Bulletin of the Society.

Meteorological Services of the World. American Meteorological Society, 45 Beacon Street, Boston, MA 02108-3693. (617) 227-2425. FAX (617) 742-8718. Annual. $70.00.

National Weather Service Offices and Stations. National Oceanic and Atmospheric Administration, Department of Commerce, Silver Spring, MD 20910. (301) 427-7698. Annual. Free.

Research Centers Directory. Gale Research, 835 Penobscot Building, Detroit, MI 48226-4094. (313) 961-2242. (800) 347-4253. 20th edition. 1995. $485.00. ISSN: 0080-1518.

ENCYCLOPEDIAS AND DICTIONARIES

Climates of the States. Bair. Gale Research, 835 Penobscot Building, Detroit, MI 48226-4094. (313) 961-2242. (800) 347-4253. 5th edition. 1995. $255.00.

Concise Oxford Dictionary of the Earth Sciences. Ailsa Allaby and Michael Allaby, editors. Oxford University Press, Inc., 200 Madison Avenue, New York, NY 10016. (800) 334-4249 or (212) 679-7300. 1990. $42.95.

Encyclopedia of Climatology. Rhodes W. Fairbridge, and John E. Oliver. Chapman & Hall, 1 Penn Plaza, New York, NY 10119. (212) 564-1060. 1986. $115.00.

Handbook of Applied Meteorology. David D. Houghton, editor. John Wiley & Sons, Inc. 605 Third Avenue, New York, NY 10158-0012. (212) 850-6000. (800) 225-5945. 1985.

McGraw-Hill Encyclopedia of Ocean and Atmospheric Sciences. Sybil P. Parker, editor. McGraw-Hill Publishing Company, 1221 Avenue of the Americas, New York, NY 10020. (800) 262-4729 or (212) 512-3825. 1979. $79.95.

McGraw-Hill Encyclopedia of Science and Technology. McGraw-Hill Publishing Company, 1221 Avenue of the Americas, New York, NY 10020. (800) 262-4729 or (212) 512-3825. Seventh edition. 1992. $1900.00. @ @General Works

Instruments and Techniques for Thunderstorm Observation and Analysis. Edwin Kessler. University of Oklahoma Press, 1005 Asp Avenue, Norman, OK 73019-0445. (405) 325-5111. (800) 627-7377. 1988. $85.00.

Instruments and Techniques For Thunderstorm Observations and Analysis. Edwin Kessler. University of Oklahoma Press, 1005 Asp Avenue, Norman, OK 73019-0445. (405) 325-5111 or (800) 627-7377. 1988. $80.00.

Mesoscale Meteorology and Forecasting. Peter S. Ray, editor. American Meteorological Society, 45 Beacon Street, Boston, MA 02108-3693. (617) 227-2425. FAX (617) 742-8718. 1986. $50.00.

Severe and Unusual Weather. Joe R. Eagleman. Trimedia Publishing Company, 12008 West 87th Street, Suite 117, Lenexa, KS 66215. (913) 599-0505. Second edition. 1990. $41.95.

The Thunderstorm in Human Affairs. Edwin Kessler, editor. University of Oklahoma Press, 1005 Asp Avenue, Norman, OK 73019-0445. (405) 325-5111. revised edition. 1988. $19.95.

Thunderstorm Morphology and Dynamics. Edwin Kessler. University of Oklahoma Press.1005 Asp Avenue, Norman, OK 73019-0445. (405) 325-5111. 2nd revised edition. 1992. $34.95.

HANDBOOKS AND MANUAL

Author's Guide To the Journals of the American Meteorological Society. American Meteorological Society, 45 Beacon Street, Boston, MA 02108-3693. (617) 227-2425. FAX (617) 742-8718. 1983. $15.00 in paper.

Handbook in Applied Meteorology. David D. Houghton, editor. John Wiley and Sons, Inc., 605 Third Avenue, New York, NY 10158. (800) 526-5368 or (212) 850-6000. 1985. $114.00.

ONLINE DATABASES AND CD-ROMS

Climate Assessment Database. National Weather Service, National Meteorological Center, 5200 Auth Road, Suite 101, Camp Springs, MD 20233. (301) 763-8016. Contains daily, weekly and monthly summaries of North American and world climatological data. Also provides five to ten weather forecasts, and 30 to 90 day weather outlook. Subscription required. Inquire as to cost and availability.

CA Search. Chemical Abstracts Service, P.O. Box 3012, Columbus, OH 43210-0012. (614) 447-3600. (800) 848-6533. FAX (614) 447-3709. Very comprehensive guide to worldwide chemical literature and related fields, 1972 to present. Available on BRS,(800) 289-4277, DIALOG, (800) 334-2564, ORBIT (800) 456-7248, and STN International, FIZ Karlsruhe, P.O. Box 2465, W-7500, Karlsruhe 1, Germany, online services. Inquire as to cost and availability.

Compendex Plus. Engineering Information, Inc., 345 East 47th Street, New York, NY 10017. (212) 705-7600 or (800) 221-1044. Contains citations with abstracts to worldwide literature in engineering and technology, from 1970 to present. Available on online BRS,(800) 289-4277, DIALOG, (800) 334-2564, ORBIT (800) 456-7248, and STN International, FIZ Karlsruhe, P.O. Box 2465, W-7500, Karlsruhe 1, Germany, online services. Also available on CD-ROM. Inquire as to cost and availability.

Current Contents Search. Institute for Scientific Information, 3501 Market Street, Philadelphia, PA 19104. (215) 386-0100. FAX (215) 386-6362. Contains citations to articles listed in the table of contents of science and technology journals. Also articles in social sciences and life sciences journals. Available on BRS,(800) 289-4277, DIALOG,(800) 334-2564, online services. Inquire as to cost and availability.

Dissertation Abstracts. University Microfilms International, 300 North Zeeb Road, Ann Arbor, MI 48106. (800) 521-0600 or (313) 761-4700. Scope includes virtually all doctoral disserta-

tions accepted at accredited American institutions from 1861 to present in 252 subject areas. Available on BRS,(800) 289-4277, DIALOG,(800) 334-2564, and OCLC EPIC,(800) 848-5878, online services. Also available on CD-ROM. Inquire as to cost and availability.

INSPEC. Institution of Electrical Engineers, Michael Faraday House, Six Hills Way, Stevenage, Herts. SG1 2AY, England. Telephone: 0438 313311 or FAX 0438 742840. Contains citations to the worldwide literature of physics, electronics and electrical engineering, computer technology, and related fields. Available on BRS,(800) 289-4277, DIALOG, (800) 334-2564, ORBIT, (800) 456-7248, and STN International, FIZ Karlsruhe, P.O. Box 2465, W-7500, Karlsruhe 1, Germany, online services. Inquire as to cost and availability.

International Aerospace Abstracts. American Institute of Aeronautics and Astronautics, 370 L'Enfant Promenade, SW, Washington, DC 20024. (202) 646-7400. Contains references and abstracts of journal and monograph literature relating to aerospace science and technology, from 1963 to present. Available through the NASA/RECON System of the National Aeronautics and Space Administration only.

Meteorological and Geoastrophysical Abstracts. American Meteorological Society, 45 Beacon Street, Boston, MA 02108-3693. (617) 227-2425. FAX (617) 742-8718. Contains citations and abstracts to the worldwide literature on significant research in meteorology and geoastrophysics. Related topics include physical oceanography, hydrology, environmental sciences and glaciology. Covers the period 1972 to present. Available on DIALOG,(800) 334-2564, online service. Inquire as to cost and availability.

NTIS Bibliographic Database. National Technical Information Service, 5285 Port Royal Road, Springfield, VA 22161. (703) 487-4929 or FAX (703) 321-8199. Broad coverage of government-sponsored science and technology research reports, 1964 to present. Available on BRS,(800) 289-4277, DIALOG, (800) 334-2564, ORBIT, (800) 456-7248, and STN International, FIZ Karlsruhe, P.O. Box 2465, W-7500, Karlsruhe 1, Germany, online services. Also available on CD-ROM. Inquire as to cost and availability.

Physics Briefs. American Institute of Physics, 335 East 45th Street, New York, NY 10017. (212) 661-9260 or FAX (212) 661-2036. Contains citations with abstracts of the literature of physics and related fields, 1979 to present. Available on the STN International, FIZ Karlsruhe, P.O. Box 2465, W-7500, Karlsruhe 1, Germany, online service. Inquire as to cost and availability.

SCISEARCH. Institute for Scientific Information, 3501 Market Street, Philadelphia, PA 19104. (800) 523-1850 or (215) 386-0100. Broad multidisciplinary title and author index to the international literature of science and technology, 1974 to present. Available on DIALOG and ORBIT online services. Also available on CD-ROM. Inquire as to cost and availability.

WILSONLINE. H.W. Wilson Company, 950 University Avenue, Bronx, NY 10452. (800) 367-6770 or (212) 588-8400. Makes available online versions of the printed H.W. Wilson indexes, including Applied Science and Technology Index, Business Periodicals Index, General Science Index, and Readers' Guide to Periodical Literature. Period covered is generally 1983 to present. Available on BRS,(800) 289-4277, DIALOG,(800) 334-2564, and OCLC EPIC,(800) 848-5878, online services. Also available on CD-ROM. Inquire as to cost and availability.

World Climate Disc. Chadwyck-Healey Inc., 1101 King Street, Alexandria, VA 22314. (703) 683-4890. FAX (703) 683-7589. Weather and climate data from approximately 5,000 weather stations worldwide, covering the years 1854 to 1990 on CD-ROM. First edition 1992. Approximately $1200.00 per year with annual updates.

World Weatherdisc. WeatherDisc Associates, Inc., 4584 N.E. 89th Street, Seattle, WA 98115. (206) 524-4314. FAX (206) 543-0308. Meteorological data on CD-ROM which describes the climate of the earth today and for the past few hundred years. First edition 1989. Approximately $295.00 per year with biannual updates.

PERIODICALS

American Meteorological Society. Bulletin. American Meteorological Society, 45 Beacon Street, Boston, MA 02108-3693. (617) 227-2425. FAX (617) 742-8718. Monthly. $60.00 per year.

American Meteorological Society. Meteorological Monographs. American Meteorological Society, 45 Beacon Street, Boston, MA 02108-3693. (617) 227-2425. FAX (617) 742-8718. Irregular. Price varies.

American Weather Observer. Association of American Weather Observers, 401 Whitney Boulevard, Box 455, Belvedere, IL 61008. (815) 544-5665. FAX (815) 544-6334. Monthly. $21.00 per year.

Atmosphere-Ocean. Canadian Meteorological and Oceanographic Society, P.O. Box 334, Newmarket, Ontario, L3Y 4X7, Canada. (416) 898-1040. FAX (416) 898-7937. Quarterly. $30.00 per year.

Dynamics of Atmospheres and Oceans. Elsevier Science Publishing Company, Inc., 655 Avenue of the Americas, New York, NY 10010. (212) 989-5800. Six times per year. $205.00 per year.

Journal of Applied Meteorology. American Meteorological Society, 45 Beacon Street, Boston, MA 02108-3693. (617) 227-2425. FAX (617) 742-8718. Monthly. $165.00 per year.

Journal of the Atmospheric Sciences. American Meteorological Society, 45 Beacon Street, Boston, MA 02108-3693. (617) 227-2425. FAX (617) 742-8718. Semi-monthly. $320.00 per year.

Monthly Weather Review. American Meteorological Society, 45 Beacon Street, Boston, MA 02108-3693. (617) 227-2425. FAX (617) 742-8718. Monthly. $205.00 per year.

National Weather Digest. National Weather Association, 4400 Stamp Road, Room 404, Temple Hills, MD 20748. (301) 899-3784.

Weather. Royal Meteorological Society, 104 Oxford Road, Reading, Berks RG1 7LJ, England. Monthly. $44.00 per year.

Weatherwise. Heldref Publications, 1319 Eighteenth Street, N.W., Washington, DC 20036-1802. (202) 296-6267. FAX (202) 296-5149. Bi-monthly. $28.00 per year.

THUNDERSTORMS

Ency. of Physical Sciences and Engineering Info. Sources

RESEARCH CENTERS AND INSTITUTES

Desert Research Institute--Air Motions Laboratory. University of Nevada at Reno, P.O. Box 60220, Reno, NV 89506. (702) 677-3201.

Institute of Atmospheric Physics. University of Arizona, Tucson, AZ 85721. (602) 621-6831.

Institute of Atmospheric Sciences. South Dakota School of Mines and Technology, 501 East St. Joseph Street, Rapid City, SD 57701-3995. (605) 394-2291.

Irving Langmiuir Laboratory For Atmospheric Research. New Mexico Institute of Mining and Technology. Socorro, NM 87801. (505) 835-5423.

National Center For Atmospheric Research. P.O. Box 3000, Boulder, CO 80307. (303) 496-1000.

National Severe Storms Forecast Center. 601 East 12th Street, Kansas City, MO 64106. (816) 374-5922.

TIDAL WAVES

See: TSUNAMIS

TIN

See also: ALLOYS, BRASS and BRONZE, COPPER, MATERIALS SCIENCE, ZINC

ABSTRACT SERVICES AND INDEXES

Alloys Index. American Society for Metals, Metals Park, OH 44073. (216) 338-5151. FAX (216) 338-4634. 1974 to present. Monthly. $380.00. Also available on CD-ROM and online via METADEX, STN International, and DIALOG.

Aluminum Industry Abstracts. Aluminum Association, Materials Park, OH 44073. (216) 338-5151. FAX (216) 338-4634. 1968 to present. Monthly. $525.00 per year.

Applied Mechanics Reviews: An Assessment of World Literature in Engineering Sciences. 1948-present. American Society of Mechanical Engineers, 345 East 47th Street, New York, NY 10017. (212) 705-7703. Monthly. $360.00 per year.

Applied Science and Technology Index; A Cumulative Subject Index To English Language Periodicals in the Fields of Aeronautics and Space Science, Computer Technology, Chemistry, Construction Industry, Energy and Related Areas. H.W. Wilson Co., 950 University Avenue, Bronx, NY 10452. (800) 367-6770 or (212) 588-8400. FAX (718) 590-1617. From 1958 to present. Monthly. Inquire about cost and availability. Also available on CD-ROM and online.

Current Contents: Engineering, Technology, and Applied Sciences. Institute for Scientific Information, 3501 Market Street, Philadelphia, PA 19104. (215) 386-0100. FAX (215) 386-6362. 1970 to present. Weekly. $442.00 per year.

Engineering Index Monthly. Engineering Information, Inc., Castle Point on the Hudson, Hoboken, NJ 07030. (800) 221-1044. FAX (215) 832-1857. Monthly. $2300.00 per year. Also available online as COMPENDEX, and also on CD-ROM. Covers chemical engineering, computers, electrical engineering, civil engineering, metals and mining, industrial management, and mechanical engineering.

I.M.M. Abstracts and Index. Institution of Mining and Metallurgy, 44 Portland Pl., London W1N 4BR, England. 071-580-3802. FAX 071-436-5388. Bi-monthly. $364 for non-members.

International Copper Information Bulletin. Copper Development Association, Orchard House, Mutton Lane, Potters Bar, Herts EN6 3AP, England. Telephone 0707-50711. FAX 0707-42769. 1976 to present. 3 times a year. Inquire for cost and availability.

Metals Abstracts and Metals Abstracts Index. American Society for Metals, Metals Park, OH 44073. (216) 338-5151. 1968 to present. Monthly. Abstracts are $1925.00 per year and Index is $460.00 per year.

ANNUAL REVIEWS AND YEARBOOKS

Annual Review of Materials Science. Annual Reviews Inc., 4139 El Camino Way, PO Box 10139, Palo Alto, CA 94303-0897. (800) 523-8635 or (415) 493-4400. FAX (415) 855-9815. Annual. $75.00 (1994 edition).

Non-Ferrous Metal Data. American Bureau of Metal Statistics, Inc., Box 1405 Plaza Station, 400 Plaza Drive, Secaucus, NJ 07094-0405. (201) 863-6900. FAX (201) 863-6050. 1920 to present. Annual. $350.00 per year.

Tin Statistics. International Tin Council, Haymarket House, 1 Oxendon Street, London SW1Y 4EQ, England. Telephone 01-930-0451. 1973 to present. Annual. Inquire for cost and availability.

ASSOCIATIONS AND PROFESSIONAL SOCIETIES

American Society For Metals. Materials Park, OH 44073. (216) 338-5151 or FAX (216) 338-4634.

American Tin Trade Association. PO Box 1347, New York, NY 10150. (908) 364-2280. FAX (908) 367-2304.

Brass and Bronze Ingot Manufacturers. 300 W. Washington, Room 1500, Chicago, IL 60606. (312) 236-2715.

Copper Development Association. 260 Madison Avenue, New York, NY 10016. (212) 251-7200.

Metallurgical Society of the AIME (American Institute of Mining, Metallurgical, and Petroleum Engineers). 345 E.47th Street, 14th Floor, New York, NY 10017. (212) 705-7695.

Minerals, Metals and Materials Society. 420 Commonwealth Drive, Warrendale, PA 15086. (412) 776-9080.

Tin Research Institute. 1353 Perry Street, Columbus, OH (614) 424-6200.

DIRECTORIES AND BIOGRAPHICAL SOURCES

Directory of Engineering Societies and Related Organizations. Gordon Davis. 13th edition. American Association of Engineering Societies, 1111 19th Street NW, Suite 608, Washington, DC 20036. (202) 296-2237 or (800) 658-8897. 1989. Inquire for price.

Dun's Industrial Guide: The Metalworking Directory, 1993-94. Dun & Bradstreet Information Services Staff. 3 volumes. Dun & Bradstreet Information Services, 899 Eaton Avenue, Bethlehem, PA 18025. (610) 882-7000 or (800) 526-0651. FAX (610) 882-7269 1993. $795.00.

International Research Centers Directory. Gale Research, 835 Penobscot Building, Detroit, MI 48226-4094. (313) 961-2242. (800) 347-4253. FAX (313) 961-6083. 8th edition. 1995. $410.00.

Research Centers Directory. Gale Research, 835 Penobscot Building, Detroit, MI 48226-4094. (313) 961-2242. (800) 347-4253. FAX (313) 961-6083. $485.00.

Scientific and Technical Organizations and Agencies Directory. Gale Research, 835 Penobscot Building, Detroit, MI 48226-4094. (313) 961-2242. (800) 347-4253. FAX (313) 961-6083. 4th edition. 1996. $195.00.

Who's Who in Engineering. American Association of Engineering Societies, 1111 19th Street NW, Suite 608, Washington, DC 20036. (202) 296-2237 or (800) 658-8897. 8th edition. 1991. Inquire for price.

ENCYCLOPEDIAS AND DICTIONARIES

Encyclopedia of Physical Science and Technology. Robert A. Meyers, ed. 18 volumes. Academic Press Inc., 6277 Sea Harbor Drive, Orlando, FL 32887. (800) 321-5068. 1992. $2100.00.

McGraw-Hill Dictionary of Scientific and Technical Terms. Sybil P. Parker, ed. 5th edition. McGraw-Hill Publishing Company, 1221 Avenue of the Americas, New York, NY 10020. (800) 262-4729 or (212) 512-3825. 1993. $110.50.

McGraw-Hill Encyclopedia of Science and Technology. Sybil P. Parker, ed. 7th edition. 20 volumes. McGraw-Hill Publishing Company, 1221 Avenue of the Americas, New York, NY 10020. (800) 262-4729 or (212) 512-3825. 1992. $1900.00

GENERAL WORKS

Tin (Metals & Minerals Series). Gordon Press Publishers, PO Box 459, Bowling Green Station, New York, NY 10004. (718) 624-8419. 1993. $250.95.

Tin and Tin Mining. R.L. Atkinson. State Mutual Book & Periodical Service Ltd., 521 Fifth Ave., 17th floor, New York, NY 10175. (718) 261-1704. FAX (516) 537-0412. 1989. $25.00.

Tin in Antiquity: Its Mining and Trade Throughout the Ancient World With Particular Reference To Cornwall. R.D. Penhallurick. Institute of Metals/ Ashgate Publishing Co., Old Post Rd., Brookfield, VT 05036. (802) 276-3162. FAX (802) 276-3837. 1986. $62.90.

Tin, Its Production and Marketing. William Robertson. Greenwood Publishing Group, 88 Post Road W., Box 5007, Westport, CT 06881. (203) 226-3571 or (800) 225-5800. FAX (203) 222-1502. 1982. $55.00.

Tinplate and Modern Canmaking Technology. E. Morgan. Pergamon Press Inc., Maxwell House, Fairview Park, Elmsford, NY 10523. (914) 592-7700. Fax (914) 592-3625. 1985. $27.00.

HANDBOOKS AND MANUALS

ASM Metals Reference Book. 3rd edition. ASM International, Materials Information, Materials Park, OH 44073-0002. (216) 338-5151 or (800) 336-5152. FAX (216) 338-4634. 1993. $111.00.

CRC Handbook of Chemistry and Physics. David R. Lide, editor. 75th edition. CRC Press, 2000 Corporate Blvd., N.W., Boca Raton, FL 33431. (407) 994-0555 or (800) 272-7737. FAX (407) 994-0949. $99.50.

Metals Handbook. ASM International, Materials Information, Materials Park, OH 44073-0002. (216) 338-5151 or (800) 336-5152. FAX (216) 338-4634. $154.00.

Smithell's Metals Reference Book. E.A. Brandes & G.B. Brook. 7th edition. Butterworth-Heinemann, 313 Washington Street, Newton, MA 02158. (617) 928-2500 or (800) 366-2665. FAX (617) 928-2620. 1992. $250.00.

ONLINE DATABASES AND CD-ROMS

Compendex Plus. Engineering Information, Inc., 345 East 47th Street, New York, NY 10017. (212) 705-7600 or (800) 221-1044. Contains citations with abstracts to worldwide literature in engineering and technology, from 1970 to present. Available on online BRS,(800) 289-4277, DIALOG, (800) 334-2564, ORBIT (800) 456-7248, and STN International, FIZ Karlsruhe, P.O. Box 2465, W-7500, Karlsruhe 1, Germany, online services. Also available on CD-ROM. Inquire as to cost and availability.

Current Contents Search. Institute for Scientific Information, 3501 Market Street, Philadelphia, PA 19104. (215) 386-0100. FAX (215) 386-6362. Contains citations to articles listed in the table of contents of science and technology journals. Also articles in social sciences and life sciences journals. Available on BRS,(800) 289-4277, DIALOG,(800) 334-2564, online services. Inquire as to cost and availability.

Metadex. Jointly produced by ASM International and the Institute of Materials. Contains more than 925,000 records from the international literature on metals and alloys, concerning properties, processes, materials classes, applications, and metallurgical systems. Updated monthly. Available from ORBIT-QUESTEL (703) 442-0900.

NASA Database. American Institute of Aeronautics and Astronautics, 370 L'Enfant Promenade, SW, Washington, DC 20024. (202) 646-7400. Citations and abstracts of aeronautics and astronautics literature, 1962 to present. Also contains citations from STAR, Scientific and Technical Aerospace Reports, and International Aerospace Abstracts. Available through NASA/RECON online service. Inquire as to cost and availability.

NTIS Bibliographic Database. National Technical Information Service, 5285 Port Royal Road, Springfield, VA 22161. (703)

487-4929 or FAX (703) 321-8199. Broad coverage of government-sponsored science and technology research reports, 1964 to present. Available on BRS,(800) 289-4277, DIALOG, (800) 334-2564, ORBIT, (800) 456-7248, and STN International, FIZ Karlsruhe, P.O. Box 2465, W-7500, Karlsruhe 1, Germany, online services. Also available on CD-ROM. Inquire as to cost and availability.

WILSONLINE. H.W. Wilson Company, 950 University Avenue, Bronx, NY 10452. (800) 367-6770 or (212) 588-8400. Makes available online versions of the printed H.W. Wilson indexes, including Applied Science and Technology Index, Business Periodicals Index, General Science Index, and Readers' Guide to Periodical Literature. Period covered is generally 1983 to present. Available on BRS,(800) 289-4277, DIALOG,(800) 334-2564, and OCLC EPIC,(800) 848-5878, online services. Also available on CD-ROM. Inquire as to cost and availability.

PERIODICALS

Alloy Digest. Alloy Digest Inc., 27 Canfield Street, Orange, NJ 07050. (201) 677-9161. 1952 to present. Monthly. $140.00 per year.

JOM (Journal of Metals). Minerals, Metals and Materials Society, 420 Commonwealth Drive, Warrendale, PA 15086. (412) 776-9080. 1949 to present. Monthly. $50.00 per year.

Metallurgical Transactions A: Physical Metallurgy and Materials Science. ASM (American Society for Metals) International, Materials Information, Materials Park, OH 44073. (216) 338-5151 or FAX (216) 338-4634. 1970 to present. Monthly. $520.00 per year for non-members.

Metallurgical Transactions B: Process Metallurgy. ASM (American Society for Metals) International, Materials Information, Materials Park, OH 44073. (216) 338-5151 or FAX (216) 338-4634. 1975 to present. Bi-monthly. $375.00 per year for non-members.

Metals Week. McGraw-Hill Publishing Company, 1221 Avenue of the Americas, New York, NY 10020. (800) 262-4729 or (212) 512-3825. 1930 to present. Weekly. $770.00 per year.

Mineral Industry Surveys, Tin Industry. U.S. Bureau of Mines, Production and Distribution, Cochrans Mill Rd., Box 18070, Pittsburgh, PA 15236. (412) 892-4411. Monthly. Inquire for price and availability.

Tin International. MIIDA Ltd., PO Box 21327, London NW10 6TN, England. Telephone 081-961-7407. FAX 081-961-7487. 1928 to present. Monthly. Inquire for cost and availability.

RESEARCH CENTERS AND INSTITUTES

Texas A&M University Mechanics and Materials Center. Civil Engineering Department, College Station, TX 77843. (409) 845-7512. (409) 845-6156.

Tin Research Institute. 1353 Perry Street, Columbus, OH (614) 424-6200.

University of Connecticut Institute of Materials Science. U-136, 97 N. Eagleville Road, Storrs, CT 06268. (203) 486-4623. (203) 486-4745.

University of Florida Department of Materials Science and Engineering. Gainesville, FL 32611. (904) 392-1454. (904) 392-6359.

TITANIUM

See also: ALLOYS, ALUMINUM, MATERIALS SCIENCE, METALLURGY

ABSTRACT SERVICES AND INDEXES

Alloys Index. American Society for Metals, Metals Park, OH 44073. (216) 338-5151. FAX (216) 338-4634. 1974 to present. Monthly. $380.00. Also available on CD-ROM and online via METADEX, STN International, and DIALOG.

Aluminum Industry Abstracts. Aluminum Association, Materials Park, OH 44073. (216) 338-5151. FAX (216) 338-4634. 1968 to present. Monthly. $525.00 per year.

Applied Mechanics Reviews: An Assessment of World Literature in Engineering Sciences. 1948-present. American Society of Mechanical Engineers, 345 East 47th Street, New York, NY 10017. (212) 705-7703. Monthly. $360.00 per year.

Applied Science and Technology Index; A Cumulative Subject Index To English Language Periodicals in the Fields of Aeronautics and Space Science, Computer Technology, Chemistry, Construction Industry, Energy and Related Areas. H.W. Wilson Co., 950 University Avenue, Bronx, NY 10452. (800) 367-6770 or (212) 588-8400. FAX (718) 590-1617. From 1958 to present. Monthly. Inquire about cost and availability. Also available on CD-ROM and online.

Current Contents: Engineering, Technology, and Applied Sciences. Institute for Scientific Information, 3501 Market Street, Philadelphia, PA 19104. (215) 386-0100. FAX (215) 386-6362.1970 to present. Weekly. $442.00 per year.

Engineering Index Monthly. Engineering Information, Inc., Castle Point on the Hudson, Hoboken, NJ 07030. (800) 221-1044. FAX (212) 832-1857. Monthly. $2300.00 per year. Also available online as COMPENDEX, and also on CD-ROM. Covers chemical engineering, computers, electrical engineering, civil engineering, metals and mining, industrial management, and mechanical engineering.

I.M.M. Abstracts and Index. Institution of Mining and Metallurgy, 44 Portland Pl., London W1N 4BR, England. 071-580-3802. FAX 071-436-5388. Bi-monthly. $364 for non-members.

Metals Abstracts and Metals Abstracts Index. American Society for Metals, Metals Park, OH 44073. (216) 338-5151. 1968 to present. Monthly. Abstracts are $1925.00 per year and Index is $460.00 per year.

ANNUAL REVIEWS AND YEARBOOKS

Annual Review of Materials Science. Annual Reviews Inc., 4139 El Camino Way, PO Box 10139, Palo Alto, CA 94303-0897. (800) 523-8635 or (415) 493-4400. FAX (415) 855-9815. Annual. $75.00 (1994 edition).

Titanium (Year): A Statistical Review. Titanium Development Association, 4141 Arapahoe Avenue, Suite 100, Boulder, CO 80303. (303) 443-7515. FAX (303) 443-4406. 1989 to present. Annual. Inquire for cost and availability.

ASSOCIATIONS AND PROFESSIONAL SOCIETIES

Aluminum Association. 900 19th Street NW, Suite 300, Washington, DC 20006. (202) 862-5100. FAX (202) 862-5164.

American Society For Metals. Materials Park, OH 44073. (216) 338-5151 or FAX (216) 338-4634.

Titanium Development Association. 4141 Arapahoe Avenue, Suite 100, Boulder, CO 80303. (303) 443-7515. FAX (303) 443-4406.

DIRECTORIES AND BIOGRAPHICAL SOURCES

Directory of Engineering Societies and Related Organizations. Gordon Davis. 13th edition. American Association of Engineering Societies, 1111 19th Street NW, Suite 608, Washington, DC 20036. (202) 296-2237 or (800) 658-8897. 1989. Inquire for price.

Dun's Industrial Guide: The Metalworking Directory, 1993-94. Dun & Bradstreet Information Services Staff. 3 volumes. Dun & Bradstreet Information Services, 899 Eaton Avenue, Bethlehem, PA 18025. (610) 882-7000 or (800) 526-0651. FAX (610) 882-7269. 1993. $795.00

Research Centers Directory. Gale Research, 835 Penobscot Building, Detroit, MI 48226-4094. (313) 961-2242. (800) 347-4253. FAX (313) 961-6083. $485.00.

Scientific and Technical Organizations and Agencies Directory. Gale Research, 835 Penobscot Building, Detroit, MI 48226-4094. (313) 961-2242. (800) 347-4253. FAX (313) 961-6083. 4th edition. 1996. $195.00.

Who's Who in Technology. Gale Research, 835 Penobscot Building, Detroit, MI 48226-4094. (313) 961-2242. (800) 347-4253. FAX (313) 961-6083. 1995. $195.00.

ENCYCLOPEDIAS AND DICTIONARIES

Encyclopedia of Physical Science and Technology. Robert A. Meyers, ed. 18 volumes. Academic Press Inc., 6277 Sea Harbor Drive, Orlando, FL 32887. (800) 321-5068. 1992. $2100.00

McGraw-Hill Dictionary of Scientific and Technical Terms. Sybil P. Parker, ed. 5th edition. McGraw-Hill Publishing Company, 1221 Avenue of the Americas, New York, NY 10020. (800) 262-4729 or (212) 512-3825. 1993. $110.50.

McGraw-Hill Encyclopedia of Science and Technology. Sybil P. Parker, ed. 7th edition. 20 volumes. McGraw-Hill Publishing Company, 1221 Avenue of the Americas, New York, NY 10020. (800) 262-4729 or (212) 512-3825. 1992. $1900.00

GENERAL WORKS

Geology of Titanium-Mineral Deposits. Eric R. Force. Geological Society of America Special Paper 259. Geological Society

of America, 3300 Penrose Place, Box 9140, Boulder, CO 80301. (303) 447-2020. FAX (303) 447-1133. 1991. $28.75.

The Physical Metallurgy of Titanium Alloys. E. W. Collings. ASM (American Society of Metals) International, Materials Information, Materials Park, OH 44073. (216) 338-5151 or FAX (216) 338-4634. 1984. Inquire for cost and availability.

Titanium (Metals & Minerals Series). Gordon Press Publishers, PO Box 459, Bowling Green Station, New York, NY 10004. (718) 624-8419. 1993. $250.95.

Titanium: A Technical Guide. Matthew J. Donachie Jr., ed. ASM International, Materials Information, Materials Park, OH 44073. (216) 338-5151 or FAX (216) 338-4634. 1988. $94.00.

Titanium Science and Technology. H.F. Froes & I. Caplan, editors. Minerals, Metals & Materials Society, 400 Commonwealth Dr., Warrendale, PA 15096. (412) 776-9000 or (800) 759-4867. 1993. $299.00.

HANDBOOKS AND MANUALS

ASM Metals Reference Book. 3rd edition. ASM International, Materials Information, Materials Park, OH 44073-0002. (216) 338-5151 or (800) 336-5152. FAX (216) 338-4634. 1993. $111.00.

CRC Handbook of Chemistry and Physics. David R. Lide, editor. 75th edition. CRC Press, 2000 Corporate Blvd., N.W., Boca Raton, FL 33431. (407) 994-0555 or (800) 272-7737. FAX (407) 994-0949. $99.50.

Metals Handbook. ASM International, Materials Information, Materials Park, OH 44073-0002. (216) 338-5151 or (800) 336-5152. FAX (216) 338-4634. $154.00.

Non-Ferrous Metal Data. American Bureau of Metal Statistics, Inc., Box 1405 Plaza Station, 400 Plaza Drive, Secaucus, NJ 07094-0405. (201) 863-6900. FAX (201) 863-6050. 1920 to present. Annual. $350.00 per year.

Smithell's Metals Reference Book. E.A. Brandes & G.B. Brook. 7th edition. Butterworth-Heinemann, 313 Washington Street, Newton, MA 02158. (617) 928-2500 or (800) 366-2665. FAX (617) 928-2620. 1992. $250.00.

ONLINE DATABASES AND CD-ROMS

Compendex Plus. Engineering Information, Inc., 345 East 47th Street, New York, NY 10017. (212) 705-7600 or (800) 221-1044. Contains citations with abstracts to worldwide literature in engineering and technology, from 1970 to present. Available on online BRS,(800) 289-4277, DIALOG, (800) 334-2564, ORBIT (800) 456-7248, and STN International, FIZ Karlsruhe, P.O. Box 2465, W-7500, Karlsruhe 1, Germany, online services. Also available on CD-ROM. Inquire as to cost and availability.

Current Contents Search. Institute for Scientific Information, 3501 Market Street, Philadelphia, PA 19104. (215) 386-0100. FAX (215) 386-6362. Contains citations to articles listed in the table of contents of science and technology journals. Also articles in social sciences and life sciences journals. Available on BRS,(800) 289-4277, DIALOG,(800) 334-2564, online services. Inquire as to cost and availability.

Metadex. Jointly produced by ASM International and the Institute of Materials. Contains more than 925,000 records from the international literature on metals and alloys, concerning properties, processes, materials classes, applications, and metallurgical systems. Updated monthly. Available from ORBIT-QUESTEL (703) 442-0900.

NASA Database. American Institute of Aeronautics and Astronautics, 370 L'Enfant Promenade, SW, Washington, DC 20024. (202) 646-7400. Citations and abstracts of aeronautics and astronautics literature, 1962 to present. Also contains citations from STAR, Scientific and Technical Aerospace Reports, and International Aerospace Abstracts. Available through NASA/RECON online service. Inquire as to cost and availability.

NTIS Bibliographic Database. National Technical Information Service, 5285 Port Royal Road, Springfield, VA 22161. (703) 487-4929 or FAX (703) 321-8199. Broad coverage of government-sponsored science and technology research reports, 1964 to present. Available on BRS,(800) 289-4277, DIALOG, (800) 334-2564, ORBIT, (800) 456-7248, and STN International, FIZ Karlsruhe, P.O. Box 2465, W-7500, Karlsruhe 1, Germany, online services. Also available on CD-ROM. Inquire as to cost and availability.

WILSONLINE. H.W. Wilson Company, 950 University Avenue, Bronx, NY 10452. (800) 367-6770 or (212) 588-8400. Makes available online versions of the printed H.W. Wilson indexes, including Applied Science and Technology Index, Business Periodicals Index, General Science Index, and Readers' Guide to Periodical Literature. Period covered is generally 1983 to present. Available on BRS,(800) 289-4277, DIALOG,(800) 334-2564, and OCLC EPIC,(800) 848-5878, online services. Also available on CD-ROM. Inquire as to cost and availability.

PERIODICALS

Alloy Digest. Alloy Digest Inc., 27 Canfield Street, Orange, NJ 07050. (201) 677-9161. 1952 to present. Monthly. $140.00 per year.

Engineering & Mining Journal. Maclean Hunter Publishing Company, 29 N. Wacker Dr., Chicago, IL 60606. (312) 726-2802. FAX (312) 726-2574. Monthly. $60.00 per year.

JOM (Journal of Metals). Minerals, Metals and Materials Society, 420 Commonwealth Drive, Warrendale, PA 15086. (412) 776-9080. 1949 to present. Monthly. $50.00 per year.

Light Metal Age. Fellom Publishing, 170 S. Spruce Avenue, Suite 120, San Francisco, CA 94080. (415) 588-8832. FAX (415) 588-0901. 1943 to present. Bi-monthly. $35.00 per year.

Mineral Industry Surveys, Titanium. U.S. Bureau of Mines, Production and Distribution, Cochrans Mill Rd., Box 18070, Pittsburgh, PA 15236. (412) 892-4411. Quarterly. Inquire for price and availability.

Titanium Digest. ASM International, Materials Information, Materials Park, OH 44073-0002. (216) 338-5151 or (800) 336-5152. FAX (216) 338-4634. 1976 to present. Monthly. $155.00 per year.

RESEARCH CENTERS AND INSTITUTES

Texas A&M University Mechanics and Materials Center. Civil Engineering Department, College Station, TX 77843. (409) 845-7512. (409) 845-6156.

TOPOGRAPHIC MAPPING

See also: AERIAL PHOTOGRAPHY, CARTOGRAPHY, GEODESY, PHOTOGRAMMETRY, REMOTE SENSING, SATELLITES—ARTIFICIAL, SURVEYING

ABSTRACT SERVICES AND INDEXES

Applied Science and Technology Index; A Cumulative Subject Index To English Language Periodicals in the Fields of Aeronautics and Space Science, Computer Technology, Chemistry, Construction Industry, Energy and Related Areas. H.W. Wilson Co., 950 University Avenue, Bronx, NY 10452. (800) 367-6770 or (212) 588-8400. FAX (718) 590-1617. From 1958 to present. Monthly. Inquire about cost and availability. Also available on CD-ROM.

Bibliography and Index of Geology. American Geological Institute, 4220 King Street, Alexandria, VA 22302. (703) 379-2480. Fax (703) 379-7563. Monthly. $1295.00 per year. Also available online as GeoRef.

Engineering Index Monthly. Engineering Information, Inc., Castle Point on the Hudson, Hoboken, NJ 07030. (800) 221-1044. FAX (212) 832-1857. Monthly. $2200.00 per year. Also available online as COMPENDEX, and also on CD-ROM.

Meteorological and Geoastrophysical Abstracts. American Meteorological Society, c/o Inforonics Inc., 550 Newtown Road, Littleton, MA 01460. (508) 486-8976. FAX (508) 486-0027. 1950 to present. Monthly. $950.00.

Remote Sensing of Earth Resources. University of New Mexico Earth Data Analysis Center, 2500 Yale Blvd., Suite 100, Albuquerque, NM 87131-6031. (505) 277-3622. FAX (505) 277-3614. 1974 to present (1991 to present under current title). Quarterly. $175.00 per year.

Science Citation Index. Institute for Scientific Information, 3501 Market Street, Philadelphia, PA 19104. (215) 386-0100. FAX (215) 386-6362. Inquire about availability and cost. Also available on CD-ROM.

ASSOCIATIONS AND PROFESSIONAL SOCIETIES

American Association For Geodetic Surveying. 5410 Grosvenor Lane, Suite 100, Bethesda, MD 20814. (301) 493-0200.

American Cartographic Association. 5410 Grosvenor Lane, Suite 100, Bethesda, MD 20814. (301) 493-0200.

American Congress On Surveying and Mapping. 5410 Grosvenor Lane, Suite 100, Bethesda, MD 20814. (301) 493-0200.

American Society For Photogrammetry and Remote Sensing, 5410 Grosvenor Lane, Suite 210, Bethesda, MD 20814. (301) 493-0290. FAX (301) 493-0208.

Ency. of Physical Sciences and Engineering Info. Sources

TOPOGRAPHIC MAPPING

Association of American Geographers. 1710 16th Street NW, Washington, DC 20009-3198. (202) 234-1450.

The International Society For Optical Engineering (SPIE), Po Box 10, 1000-20th Street, Bellingham, WA 98227-0010.

National Society of Professional Surveyors. 5410 Grosvenor Lane, Suite 100, Bethesda, MD 20814. (301) 493-0200.

Optical Society of America, 2010 Massachusetts Avenue NW, Washington, DC 20036. (202) 223-8130.

Society of Photographic Scientists and Engineers. 7003 Kilworth Lane, Springfield, VA 22151. (703) 642-9090.

BIBLIOGRAPHIES

Bibliography of Cartography. Library of Congress, Geography and Map Division. 5 volumes (1973), 2-volume supplement (1980). G.K. Hall, 866 Third Avenue, New York, NY 10022. (212) 702-6789 or (800) 257-5755. Inquire for cost and availability.

DIRECTORIES AND BIOGRAPHICAL SOURCES

Directory of Engineering Societies and Related Organizations. Gordon Davis. 13th edition. American Association of Engineering Societies, 1111 19th Street NW, Suite 608, Washington, DC 20036. (202) 296-2237 or (800) 658-8897. 1989. Inquire for price.

International Research Centers Directory. Gale Research, 835 Penobscot Building, Detroit, MI 48226-4094. (313) 961-2242. (800) 347-4253. FAX (313) 961-6083. 8th edition. 1995. $410.00

Research Centers Directory. Gale Research, 835 Penobscot Building, Detroit, MI 48226-4094. (313) 961-2242. (800) 347-4253. FAX (313) 961-6083. $485.00.

Scientific and Technical Organizations and Agencies Directory. Gale Research, 835 Penobscot Building, Detroit, MI 48226-4094. (313) 961-2242. (800) 347-4253. FAX (313) 961-6083. 4th edition. 1996. $195.00.

Who's Who in Engineering. American Association of Engineering Societies, 1111 19th Street NW, Suite 608, Washington, DC 20036. (202) 296-2237 or (800) 658-8897. 8th edition. 1991. Inquire for price.

GENERAL WORKS

Interpretation of Topographic Maps. V. Miller & M. Westerback. Macmillan Publishing, 200 Old Tappan Road, Old Tappan, NJ 07675. (800) 223-2336. FAX (800) 445-6991. 1989. Inquire for cost and availability.

Inventory of World Topographic Mapping, Volume One. Rolf Bohme, comp. Elsevier Science B.V., PO Box 945, Madison Square Station, New York, NY 10159-0945. (212) 633-3650. FAX (212) 633-3680. 1989. $185.00.

Surveying and Mapping For Field Scientists. William Ritchie, et al. John Wiley and Sons, Inc., 605 Third Avenue, New York, NY 10158. (800) 526-5368 or (212) 850-6000. 1988. $45.95.

Topographic Mapping. U.S. Department of the Interior, U.S. Geological Survey. Order from: Superintendent of Documents, U.S. Government Printing Office, Box 371954, Pittsburgh, PA 15250-7954. (202) 783-3238. FAX (202) 512-2233. 1992. Inquire for cost and availability.

HANDBOOKS AND MANUALS

Manual of Photogrammetry. Chester C. Slama, ed. 4th edition. American Society for Photogrammetry and Remote Sensing, 5410 Grosvenor Lane, Suite 210, Bethesda, MD 20814. (301) 493-0290. FAX (301) 493-0208. 1980. Inquire for price and availability.

Manual of Remote Sensing. Robert N. Colwell, ed. 2nd edition. American Society for Photogrammetry and Remote Sensing, 5410 Grosvenor Lane, Suite 210, Bethesda, MD 20814. (301) 493-0290. FAX (301) 493-0208. 1983. Inquire for price and availability.

ONLINE DATABASES AND CD-ROMS

Compendex Plus. Engineering Information, Inc., 345 East 47th Street, New York, NY 10017. (212) 705-7600 or (800) 221-1044. Contains citations with abstracts to worldwide literature in engineering and technology, from 1970 to present. Available on online BRS,(800) 289-4277, DIALOG, (800) 334-2564, ORBIT (800) 456-7248, and STN International, FIZ Karlsruhe, P.O. Box 2465, W-7500, Karlsruhe 1, Germany, online services. Also available on CD-ROM. Inquire as to cost and availability.

Georef: Bibliography and Index of Geology. American Geological Institute, 4220 King Street, Alexandria, VA 22302. (703) 379-2480. Fax (703) 379-7563. Monthly. Inquire for price and availability.

NTIS Bibliographic Database. National Technical Information Service, 5285 Port Royal Road, Springfield, VA 22161. (703) 487-4929 or FAX (703) 321-8199. Broad coverage of government-sponsored science and technology research reports, 1964 to present. Available on BRS,(800) 289-4277, DIALOG, (800) 334-2564, ORBIT, (800) 456-7248, and STN International, FIZ Karlsruhe, P.O. Box 2465, W-7500, Karlsruhe 1, Germany, online services. Also available on CD-ROM. Inquire as to cost and availability.

SCISEARCH. Institute for Scientific Information, 3501 Market Street, Philadelphia, PA 19104. (800) 523-1850 or (215) 386-0100. Broad multidisciplinary title and author index to the international literature of science and technology, 1974 to present. Available on DIALOG,(800) 334-2564, and ORBIT,(800) 456-7248, online services. Also available on CD-ROM. Inquire as to cost and availability.

WILSONLINE. H.W. Wilson Company, 950 University Avenue, Bronx, NY 10452. (800) 367-6770 or (212) 588-8400. Makes available online versions of the printed H.W. Wilson indexes, including Applied Science and Technology Index, Business Periodicals Index, General Science Index, and Readers' Guide to Periodical Literature. Period covered is generally 1983 to present. Available on BRS,(800) 289-4277, DIALOG,(800) 334-

2564, and OCLC EPIC,(800) 848-5878, online services. Also available on CD-ROM. Inquire as to cost and availability.

PERIODICALS

ACSM Bulletin. American Congress on Surveying and Mapping, 5410 Grosvenor Lane, Suite 100, Bethesda, MD 20814. (301) 493-0200. 1950 to present. Bi-monthly. $75.00 per year.

American Congress On Surveying and Mapping Proceedings. American Congress on Surveying and Mapping, 5410 Grosvenor Lane, Suite 100, Bethesda, MD 20814. (301) 493-0200. 1942 to present. Semi-annual. Price varies.

Association of American Geographers Annals. Basil Blackwell Inc., 238 Main Street, Cambridge, MA 02142. (617) 547-7110. FAX (617) 547-0789. 1911 to present. Quarterly. $70.00 per year.

Cartography and Geographic Information Systems. American Congress on Surveying and Mapping, 5410 Grosvenor Lane, Suite 100, Bethesda, MD 20814. (301) 493-0200. 1974 to present. 4 times a year. $85.00 per year.

IEEE Transactions On Geoscience and Remote Sensing. IEEE, 345 E. 47th Street, New York, NY 10017. (908) 981-0060. FAX (908) 981-9667. 1963 to present. Bi-monthly. $188.00 per year.

Journal of Imaging Science and Technology. Society for Imaging Science and Technology, 7003 Kilworth Lane, Springfield, VA 22151. (703) 642-9090. FAX (703) 642-9094. 1950 to present. Bi-monthly. $120.00 per year.

Marine Geodesy. Taylor & Francis, 1900 Frost Road, Suite 101, Bristol, PA 19007. (215) 785-5800. FAX (215) 785-5515. 1977 to present. Quarterly. $141.00 per year.

Photogrammetric Engineering and Remote Sensing. American Society for Photogrammetry and Remote Sensing, 5410 Grosvenor Lane, Suite 210, Bethesda, MD 20814. (301) 493-0290. FAX (301) 493-0208. 1934-present. Monthly. $160.00 per year.

Photogrammetric Record. Photogrammetric Society, Dept. of Photogrammetry & Surveying, University College London, Gower St., London WC1E 6BT England. Telephone 071-387-7050. FAX 071-380-0453. 1953-present. Semi-annual. $88.00 per year.

Remote Sensing of Environment. Elsevier Science Publishing Company, Inc., 655 Avenue of the Americas, New York, NY 10010. (212) 989-5800. 1969 to present. 12 times a year. $995.00 per year.

Remote Sensing Reviews. Harwood Academic, c/o Gordon & Breach Science Publishers, PO Box 786, Cooper Station, New York, NY 10276. (212) 206-8900. Quarterly. Inquire for cost.

Scientific and Applied Photography and Cinematography. Gordon and Breach Science Publishers, 820 Town Center Drive, Langhorne, PA 19047. (215) 750-2642. FAX (215) 750-6343. 6 times a year. Inquire for price.

RESEARCH CENTERS AND INSTITUTES

Boston University Center For Remote Sensing and Cartography. 725 Commonwealth Avenue, Boston, MA 02215. (617) 353-5081. FAX (617) 353-3200.

Cartographic Center, Ohio University. Porter Hall, Athens, OH 45701. (614) 593-1150.

Michigan State University Center For Remote Sensing. 302 Berkey Hall, East Lansing, MI 48824. (517) 353-7195. FAX (517) 353-1821.

University of Delaware Center For Remote Sensing. College of Marine Studies, Newark, DE 19711. (302) 451-2336.

University of Idaho, Idaho Remote Sensing Research Unit. Moscow, ID 83843. (208) 885-7209.

University of Minnesota Remote Sensing Laboratory. 1530 N. Cleveland Avenue, St. Paul, MN 55108. (612) 624-7764. FAX (612) 625-5212.

University of North Dakota, Institute For Remote Sensing. Geography Department, Grand Forks, ND 58202-8274. (701) 777-4246.

University of Wisconsin—Madison, University Cartographic Laboratory. Science Hall, Madison, WI 53706. (608) 262-1363.

U.S. Army Topographic Engineering Center, Geographic Information Laboratory. Ft. Belvoir, VA 22060-5546. (703) 355-2800. FAX (703) 355-3176.

U.S. Army Topographic Engineering Center, Simulization and Visualization Laboratory. 7701 Telegraph Road, Alexandria, VA 22310-3864. (703) 355-2700.

TOPOLOGY

See also: ALGEBRA, CALCULUS, GEOMETRY, STATISTICS

ABSTRACT SERVICES AND INDEXES

Applied Mechanics Reviews: An Assessment of World Literature in Engineering Sciences. American Society of Mechanical Engineers, 345 East 47th Street, New York, NY 10017. (212) 705-7703. 1948 to present. Monthly. $360.00 per year. ISSN: 0003-6900.

Applied Science and Technology Index; A Cumulative Subject Index To English Language Periodicals in the Fields of Aeronautics and Space Science, Computer Technology, Chemistry, Construction Industry, Energy and Related Areas. H. W. Wilson and Company, 950 University Avenue, Bronx, NY 10452. (212) 588-8400. (800) 367-6770. FAX (718) 590-1617. 1958 to present. Monthly. Available on-line from BRS and WILSONLINE. Also available on CD-ROM. Inquire as to cost and availability. ISSN: 0003-6986.

Compactmath-Compact Mathematics Library. Cumulative CD-ROM edition of Zentralblatt fuer Mathematik-Mathematics Abstracts. Springer-Verlag New York, Inc. 44 Hartz Way, Secaucus, NJ 07096-2491. (201) 348-4033. FAX (201) 348-4505. 1993 to

Ency. of Physical Sciences and Engineering Info. Sources

TOPOLOGY

present. Annual. Available only on CD-ROM. Inquire as to cost and availability. ISSN: 0938-3174.

Current Index To Statistics; Applications-Methods-Theory. American Statistical Association, 1429 Duke Street, Alexandria, VA 22314-3402. (703) 684-1221. FAX (703) 684- 2037. 1976 to present. Annual. $50.00 per year. Available on-line from BRS, Dialog, European Space Agency. ISSN: 0364-1228.

Current Mathematical Publications. American Mathematical Society. P.O. Box 1571, Annex Station, Providence, RI 02901-9930. (800) 556-7774 or (401) 455-4000. FAX (401) 331-3842. 1969 to present. Seventeen times per year. $377.00 per year. Available online from BRX, DIALOG, European Space Agency. Also available on CD-ROM. Inquire as to cost and availability. ISSN: 0361-4794.

Index To Mathematical Tables. Alan Fletcher. Addison- Wesley Publishing Co., Inc., 1 Jacob Way, Reading, MA 01867. (617) 944-3700. (800) 447-2226. 2nd edition. 1962. 2 volumes.

Mathematical Reviews: A Review Journal Covering the World Literature of Mathematical Research. American Mathematical Society, P.O. Box 1571, Annex Station, Providence, RI 02901-9930. (800) 556-7774 or (401) 455-4000. 1940 to present. Monthly. $4594.00 per year. Also available via network (MathSciNet), online and on CD-ROM. Inquire regarding cost and availability. ISSN: 0025-5629.

NTIS Alerts: Mathematical Sciences. U.S. National Technical Information Service, 5285 Port Royal Road, Springfield, VA 22161. (703) 487-4650. FAX (703) 321-8547. Weekly. $140.00.

Science Citation Index. Institute for Scientific Information. 3501 Market Street, Philadelphia, PA 19104. (215) 386-0100. (800) 523-1850. FAX (215) 386-2991. 1961 to present. Six issues per year, plus annual cumulation. $11650.00 per year. Also available online and on CD-ROM. Inquire regarding cost and availability. ISSN: 0036-827X.

Statistical Theory and Method Abstracts. International Statistical Institute. Prinses Beatrixiaan 428, Postbus 950, 2270 AZ Voorburg, Netherlands. 31-70-3375737. FAX 31-70-3860025. 1959 to present. Quarterly. $191.82. ISSN: 0039-0518.

Zentralblatt Fuer Mathematik Und Ihre Grenzebiete/Mathematics Abstracts. Heidelberger Akademie der Wissenschaften/Springer-Verlag New York, Inc., 44 Hartz Way, Secaucus, NJ 07096-2491. (201) 348-4033. FAX (201) 348-4505. 1931 to present. 30 issues per year. DM 8340. ISSN: 0044-4235. Also available online and on CD-ROM. Inquire regarding cost and availability.

ANNUAL REVIEWS AND YEARBOOKS

Advances in Applied Mathematics. Academic Press, Inc., 6277 Sea Harbor Drive, Orlando, FL 32821. (800) 321-5068. irregular, Price varies; inquire.

Ergebnisse Der Mathematik Und Ihrer Grenzgebiete. Springer-Verlag New York, Inc. 44 Hartz Way. Secaucus. NJ. 07096-2491. (201) 348-4033. FAX (201) 348-4505. Irregular. Price varies; inquire.

Lecture Notes in Mathematics. Springer-Verlag New York, Inc. 44 Hartz Way. Secaucus. NJ. 07096-2491. (201) 348-4033. FAX (201) 348-4505. Irregular. Price varies; inquire.

ASSOCIATIONS AND PROFESSIONAL SOCIETIES

American Mathematical Society. P.O. Box 6248, Providence, RI 02940. (401) 455-4000. FAX (401) 331-3842.

American Statistical Association. 1429 Duke Street, Alexandria, VA 22314-3402. (703) 684-1221. FAX (703) 684-2037.

Association For Symbolic Logic. University of Illinois, Department of Mathematics, 1409 West Green Street, Urbana IL 61801. (217) 244-7902. FAX (217) 333-9576.

Association For Women in Mathematics. 4114 Computer and Space Science Building, University of Maryland, College Park, MD 20742-2461. (301) 405-7892.

Mathematical Association of America. 1529 18th Street, NW, Washington, DC 20036. (202) 387-5200. FAX (202) 265-2384.

National Council of Teachers of Mathematics. 1906 Association Drive, Reston, VA 22091. (703) 620-9840. FAX (703) 475-2970.

Operations Research Society of America. 1314 Guilford Avenue, Baltimore, MD. 21202. (410) 528-4146. FAX (410) 528-8556.

Society For Industrial and Applied Mathematics. 3600 University City Science Center, Philadelphia, PA 19104-2688. (215) 382-9800. FAX (215) 386-7999.

Special Interest Group On Numerical Mathematics (SIGNUM). c/o ACM. 1515 Broadway, New York, NY 10036-5701. (212) 869-7440. FAX (212) 944-1318.

Special Interest Group For Symbolic and Algebraic Manipulation (SIGSAM). c/o ACM. 1515 Broadway, New York, NY 10036-5701. (212) 869-7440. FAX (212) 944-1318.

BIBLIOGRAPHIES

Books You Can Count On: Linking Mathematics & Literature. Rachel Griffiths and Margaret Clyne. Heinemann Publ. 361 Hanover Street, Portsmouth, NH 03801-3912. (603) 431-7894. (800) 541-2086. $15.00. ISBN: 0-435-08322-8.

Mathematical Book Review Index, 1800-1940. Louise S. Grinstein. Garland Publ. Fifth Avenue, Suite 2500, New York, NY 10022-8101. (212) 751-7447. $72.00. ISBN: 0-8240-4114-3.

Mathematical Journals: An Annotated Guide. Diana F. Liang. Scarecrow Press. Distributed by University Press of America, 4720 Boston Way, Lanham, MD 20706. (800) 462-6420. $29.50. ISBN: 0-8108-2585-6.

Reviews of Papers in Algebraic and Differential Topology, Topological Groups and Horological Algebra 1940-67. Norman E. Steenrod, editor. American Mathematical Society, P.O. Box 6248, Providence, RI 02940. (401) 455-4000. (800) 321- 4267. 1969. $95.00 in paper.

TOPOLOGY

Ency. of Physical Sciences and Engineering Info. Sources

DIRECTORIES AND BIOGRAPHICAL SOURCES

Assistantships and Graduate Fellowships in the Mathematical Sciences. American Mathematical Society, P.O. Box 6248, Providence, RI 02940. (401) 455-4000. (800) 321-4267. Annual. $19.00. ISSN: 1040-7650.

Combined Membership List. American Mathematical Society, P.O. Box 6248, Providence, RI 02940. (401) 455-4000. Annual. $50.00. ISSN: 0569-6461.

Mathematical Science Professional Directory. American Mathematical Society, P.O. Box 6248, Providence, RI 02940. Annual. $45.00. ISBN: 0-8218-0173-2.

Research Centers Directory. Gale Research Company, 835 Penobscot Building, Detroit, MI 48226-4094. (313) 961- 2242. (800) 347-4253. 20th edition. 1995. Annual. $485.00. ISSN: 0080-1518.

World Directory of Mathematicians. G.D. Mostow, editor. International Mathematical Union, Helsinki. 1958 to present. 9th revised edition. 1990. $40.00. Available from: American Mathematical Society, P.O. Box 6248, Providence, RI 02940. (401) 455-4000. ISSN: 0512-2740.

ENCYCLOPEDIAS AND DICTIONARIES

Beyond Numeracy: An Uncommon Dictionary of Mathematics. John A. Paulos. Alfred A. Knopf Inc., 201 East 50th Street, New York, NY 10022. (800) 733-3000. 1991. $21.50. ISBN: 0-394-58640-9.

Concise Oxford Dictionary of Mathematics. Christopher Clapham. Oxford University Press, 200 Madison Avenue, New York, NY 10016. (212) 725-6000. (800) 334-4249. 1990. $10.95. ISBN: 0-19-286103-4.

Encyclopedia of Mathematics and Its Applications. Addison-Wesley Publishing Co., Inc., 1 Jacob Way, Reading, MA 01867. (617) 944-3700. (800) 447-2226. 1976 to present. Irregular. Volumes are individually priced.

Encyclopedic Dictionary of Mathematics. Kiyosi Ito, editor. MIT Press, 55 Hayward Street, Cambridge, MA 02142. (617) 625-8569. 2nd edition, reprint. 1993. 2 volumes. ISBN: 0-262-59020-4.

HarperCollins Dictionary of Mathematics. E. J. Borowski and J. J. Borwein. HarperCollins Publ. 10 East 53rd Street, New York, NY 10022-5299. (800) 331-3761. 1990. $10.95. ISBN: 0-19-286103-4.

Mathematics Dictionary. Robert C. James, editor. Van Nostrand Reinhold, 115 Fifth Avenue, New York, NY 10003. (800) 842-3636. 5th edition. 1992. $42.95. ISBN: 0-442-00741-8.

Penguin Dictionary of Mathematics. J. Daintith and R. D. Nelson. Viking Penguin, 375 Hudson Street, New York, NY 10014-3657. (800) 331-4624. 1989. $12.00. ISBN: 0-14-051119-9.

The Words of Mathematics: An Etymological Dictionary of Mathematical Terms Used in English. Steven Schwartzman. Mathematical Association of America, 18th Street NW, Washington, DC 20036. (800) 331-1622. 1994. $29.50. ISBN: 0-88385-551.

GENERAL WORKS

Basic Topology. M. A. Armstrong. Springer-Verlag New York, Inc., 175 Fifth Avenue, New York, NY 10010. (212) 460-1500. (800) 777-4643. Revised edition. 1994. $34.00.

Classical Topology and Combinatorial Group Theory. John C. Stillwell. American Mathematical Society, P.O. Box 6248, Providence, RI 02940. (401) 455-4000. (800) 321-4267. 2nd edition. 1993. $49.90.

Classical Topology and Quantum States. A. P. Balachandran et al. World Scientific Publishing Company, Inc., 1060 Main Street, River Edge, NJ 07661. (201) 487-9655. (800) 227- 7562. 1991. $36.00 in paper.

Differential Geometry and Topology. R. Cadeo and F. Tricerri. World Scientific Publishing Company, Inc., 1060 Main Street, River Edge, NJ 07661. (201) 487-9655. (800) 227-7562. 1993. $105.00.

Elementary Topology. Michael C. Gemignani. Dover Publications, Inc., 180 Varick Street, New York, NY 10014. (212) 255-3755. (800) 223-3130. 1990. $6.95 in paper.

Essentials of Topology. Research and Education Association, 61 Ethel Road, West Piscataway, NJ 08854. (908) 819-8880. Revised edition. 1994. $5.95 in paper.

General Topology, V.1: Basic Concepts and Constructions, Dimensions Theory. A.V. Arkhangel'skii, et al, editors. Springer-Verlag New York, Inc., 175 Fifth Avenue, New York, NY 10010. (212) 460-1500. (800) 777-4643. 1990. $65.00.

Geometry and Topology. G. Stratopoulos and G. Rassias, editors. World Scientific Publishing Company, Inc., 1060 Main Street, River Edge, NJ 07661. (201) 487-9655. (800) 227-7562. 1989. $85.00

Geometry, Topology and Field theory. K. Fukaya. World Scientific Publishing Company, Inc., 1060 Main Street, River Edge, NJ 07661. (201) 487-9655. (800) 227-7562. 1994. $55.00.

General Topology. N. Bourbaki. Springer-Verlag New York, Inc., 175 Fifth Avenue, New York, NY 10010. (212) 460-1500. (800) 777-4643. 2 volumes (Chapters 1-4 and 5-10). 1989. $79.00 ea.

How Surfaces Interact in Space: An Introduction To Topology. J. S. Carter. World Scientific Publishing Company, Inc., 1060 Main Street, River Edge, NJ 07661. (201) 487-9655. (800) 227-7562. 1993. $23.00 in paper.

Introduction To General Topology. George L. Cain. Addison-Wesley Publishing Co., Inc., 1 Jacob Way, Reading, MA 01867.(617) 944-3700. (800) 447-2226. 1994. $58.25.

An Introduction To Topology and Homotopy. Sieradski. PWS Publications, 20 Park Plaza, Boston, MA 02116. (617) 542-3377. (800) 354-9706. 1992. $62.95.

Measure, Topology and Fractal Geometry. Gerald A. Edgar. Springer-Verlag New York, Inc., 175 Fifth Avenue, New York, NY 10010. (212) 460-1500. (800) 777-4643. 1992. $29.95.

Principles of Topology. Frederick Croom. SCP: Third World Literature Publishing House, P.O. Box 482, Lithonia, GA 30058-0482. (404) 785-7725. 1989. $56.00.

Space-Filling Curves. Hans Sagan. Springer-Verlag New York, Inc., 175 Fifth Avenue, New York, NY 10010. (212) 460-1500. (800) 777-4643. 1994. $29.95.

Topological Picturebook. G.K. A. Francis. Springer-Verlag New York, Inc., 175 Fifth Avenue, New York, NY 10010. (212) 460-1500. (800) 777-4643. 1988. $42.00

Topology of Surfaces. Christine Kinsey. Springer-Verlag New York, Inc., 175 Fifth Avenue, New York, NY 10010. (212) 460-1500. (800) 777-4643. 1993.

Topology Problem Solver. Research and Education Association, 61 Ethel Road, West Piscataway, NJ 08854. (908) 819-8880. 1994. $29.95 in paper.

Visual Geometry and Topology. Anatolij Fomenko. Springer-Verlag New York, Inc., 175 Fifth Avenue, New York, NY 10010.(212) 460-1500. (800) 777-4643. 1994. $89.00.

HANDBOOKS AND MANUALS

Handbook of Mathematical Sciences. William H. Beyer, editor. CRC Press, 2000 Corporate Boulevard, Boca Raton, FL 33431. (800) 272-7737. 6th edition. 1987. $91.95. ISBN: 0-8493-0656-6.

Handbook of Mathematics. I.N. Bronshtein and K.A. Semedyayev. Van Nostrand Reinhold, 115 Fifth Avenue, New York, NY 10003. (800)-842-3636. 20th edition. $59.95. ISBN: 0-442-21171-6.

Handbook of Set-Theoretic Topology. K. Kunen, and J. Vaughan, editors. Elsevier Science Publishing Company, Inc., 655 Avenue of the Americas, New York, NY 10010. (212) 989-5800. 1988. $72.00 in paper.

Handbook of Tables For Probability and Statistics. William H. Beyer, editor. CRC Press, 2000 Corporate Boulevard, Boca Raton, FL 33431. (800) 272-7737. 2nd edition. 1968. $92.95. ISBN: 0-8493-0692-2.

Handbook of Writing For the Mathematical Sciences. Society for Industrial and Applied Mathematics, 3600 University City Science Center, Philadelphia, PA 19104-2688. (215) 382-9800. FAX (215) 386-7999. 1993. $21.50.

Standard Mathematical Tables and Formulas. William H. Beyer, editor. CRC Press, 2000 Corporate Boulevard, Boca Raton, FL 33431. (800) 272-7737. 29th edition. 1991. $41.95. ISBN: 084-930-6299.

Tables of Integrals, Series and Products: Corrected and Enlarged Edition. I.S. Gradshteyn and I.M. Ryzhik. Academic Press, Inc., 6277 Sea Harbor Drive, Orlando, FL 32821. (800) 321-5068. 5th edition. 1993. $61.00. ISBN: 0-12-294755-X.

ONLINE DATABASES AND CD-ROMS

Compendex Plus. Engineering Information, Inc., 345 East 47th Street, New York, NY 10017. (212) 705-7600 or (800) 221-1044. Contains citations with abstracts to worldwide literature in engineering and technology, from 1970 to present. Available on online BRS,(800) 289-4277, DIALOG, (800) 334-2564, ORBIT (800) 456-7248, and STN International, FIZ Karlsruhe,

P.O. Box 2465, W-7500, Karlsruhe 1, Germany, online services. Also available on CD-ROM. Inquire as to cost and availability.

Dissertation Abstracts. University Microfilms International, 300 North Zeeb Road, Ann Arbor, MI 48106. (800) 521-0600 or (313) 761-4700. Scope includes virtually all doctoral dissertations accepted at accredited American institutions from 1861 to present in 252 subject areas. Available on BRS,(800) 289-4277, DIALOG,(800) 334-2564, and OCLC EPIC,(800) 848-5878, online services. Also available on CD-ROM. Inquire as to cost and availability.

INSPEC. Institution of Electrical Engineers, Michael Faraday House, Six Hills Way, Stevenage, Herts. SG1 2AY, England. Telephone: 0438 313311 or FAX 0438 742840. Contains citations to the worldwide literature of physics, electronics and electrical engineering, computer technology, and related fields. Available on BRS,(800) 289-4277, DIALOG, (800) 334-2564, ORBIT, (800) 456-7248, and STN International, FIZ Karlsruhe, P.O. Box 2465, W-7500, Karlsruhe 1, Germany, online services. Inquire as to cost and availability.

Mathsci and Mathscinet. American Mathematical Society, P.O. Box 6248, Providence, RI 02940. (800) 321-4667 or (401) 455-4000 or FAX (401) 331-3842. Scope includes pure and applied mathematics and related areas of physics, statistics, engineering, computer science, and operations research literature since 1959. Available on DIALOG,(800) 334-2564, online service. Also available on CD-ROM and Mathscinet, on the Internet. Inquire as to cost and availability.

NTIS Bibliographic Database. National Technical Information Service, 5285 Port Royal Road, Springfield, VA 22161. (703) 487-4929 or FAX (703) 321-8199. Broad coverage of government-sponsored science and technology research reports, 1964 to present. Available on BRS,(800) 289-4277, DIALOG, (800) 334-2564, ORBIT, (800) 456-7248, and STN International, FIZ Karlsruhe, P.O. Box 2465, W-7500, Karlsruhe 1, Germany, online services. Also available on CD-ROM. Inquire as to cost and availability.

SCISEARCH. Institute for Scientific Information, 3501 Market Street, Philadelphia, PA 19104. (800) 523-1850 or (215) 386-0100. Broad multidisciplinary title and author index to the international literature of science and technology, 1974 to present. Available on DIALOG,(800) 334-2564, and ORBIT,(800) 456-7248, online services. Also available on CD-ROM. Inquire as to cost and availability.

WILSONLINE. H.W. Wilson Company, 950 University Avenue, Bronx, NY 10452. (800) 367-6770 or (212) 588-8400. Makes available online versions of the printed H.W. Wilson indexes, including Applied Science and Technology Index, Business Periodicals Index, General Science Index, and Readers' Guide to Periodical Literature. Period covered is generally 1983 to present. Available on BRS,(800) 289-4277, DIALOG,(800) 334-2564, and OCLC EPIC,(800) 848-5878, online services. Also available on CD-ROM. Inquire as to cost and availability.

PERIODICALS

American Journal of Mathematics. Johns Hopkins University Press, 2715 North Charles Street, Baltimore, MD 21218. (410) 516-6987. FAX (410) 516-6968. 1878 to present. Bimonthly. $162.00 per year. ISSN: 0002-9327.

American Mathematical Monthly. Mathematical Association of America. 1529 Eighteenth Street, NW, Washington, DC 20036.

TOPOLOGY

Ency. of Physical Sciences and Engineering Info. Sources

(202) 387-5200. 1894 to present. Ten issues per year. $160.00. ISSN: 0002-9890.

Canadian Applied Mathematical Quarterly. Canadian Mathematical Society. Distributed by Rocky Mountain Mathematics Consortium, Department of Mathematics, Arizona State University, Tempe AZ 85287-1904. 1992 to present. $175.00 per year. ISSN: 1073-1849.

Communications On Pure and Applied Mathematics. Courant Institute of Mathematical Sciences. Distributed by John Wiley & Sons, Inc. 605 Third Avenue, New York, NY 10158-0012. (212) 850-6000. FAX (212) 850-6088. 1939 to present. $730.00 per year. ISSN: 0010-3640.

IMA Journal of Applied Mathematics. Oxford University Press, 2001 Evans Road, Cary, NC 27513. (919) 677-0977. 1981 to present. Bimonthly. $410.00 per year. ISSN: 0272-4960.

Industrial Mathematics. Industrial Mathematics Society, Box 159 Roseville, MI 48066. (313) 771-0403. 1950 to present. Semiannual. $15.00 per year.

Journal of Recreational Mathematics. Baywood Publishing Co. Inc., 26 Austin Avenue, Box 337, Amityville, NY 11701. (516) 691-1770. 1968 to present. Quarterly. $78.00. ISSN: 0022-412X.

Mathematical Intelligencer. Springer-Verlag, 175 Fifth Avenue, New York, NY 10010. (212) 460-1500. FAX (212) 473-6272. (800) 777-4643. 1978 to present. Quarterly. $39.00 per year.

Mathematics Magazine. Mathematical Association of America. 1529 Eighteenth Street, NW, Washington, DC 20036. (202) 387-5200. 1926 to present. Bimonthly. $80.00 per year. ISSN: 0025-570X.

Quarterly Journal of Mechanics and Applied Mathematics. Oxford University Press, 2001 Evans Road, Cary, NC 27513. (919) 677-0977. 1948 to present. Quarterly. $214.00 per year. ISSN: 0033-5614.

SIAM Journal On Applied Mathematics. Society for Industrial and Applied Mathematics, 3600 University City Science Center, Philadelphia, PA 19104-2688. (215) 382-9800. FAX (215) 386-7999. 1953 to present. Bimonthly. $242.00 per year. ISSN: 0036-1399.

SIAM Journal On Discrete Mathematics. Society for Industrial and Applied Mathematics, 3600 University City Science Center, Philadelphia, PA 19104-2688. (215) 382-9800. FAX (215) 386-7999. 1980 to present. Quarterly. $220.00 per year. ISSN: 0895-4801.

SIAM Journal On Mathematical Analysis. Society for Industrial and Applied Mathematics, 3600 University City Science Center, Philadelphia, PA 19104-2688. (215) 382-9800. FAX (215) 386-7999. 1970 to present. Bimonthly. $352.00 per year. ISSN: 0036-1410.

SIAM Journal On Numerical Analysis. Society for Industrial and Applied Mathematics, 3600 University City Science Center, Philadelphia, PA 19104-2688. (215) 382-9800. FAX (215) 386-7999. Bimonthly. $255.00 per year. ISSN: 0036-1429.

SIAM Review. Society for Industrial and Applied Mathematics, 3600 University City Science Center, Philadelphia, PA 19104-

2688. (215) 382-9800. 1959 to present. Quarterly. $148.00 per year. ISSN: 0036-1445.

RESEARCH CENTERS AND INSTITUTES

Center for Mathematical Science Research. Rutgers University. New Brunswick, NJ 08903. (201) 932-3117. FAX (201) 932-5530.

Center For Mathematical Sciences. University of Wisconsin-Madison. 610 Walnut Street, Madison, WI 53705. (608) 263-2696. FAX (608) 263-2841.

Center For Pure and Applied Mathematics. University of California, Berkeley. 977 Evans Hall, Berkeley, CA 94720. (415) 642-0116. FAX (415) 642-6726.

Courant Institute of Mathematical Sciences. New York University, 251 Mercer Street, New York, NY 10012. (212) 998-3000.

Institute of Applied Mathematics. University of Missouri-Rolla, Rolla, MO 65401. (314) 341-4641.

TORNADOES

See also: CYCLONES, JET STREAM, METEOROLOGY, RAIN, TORNADOS, WEATHER FORECASTING

ABSTRACT SERVICES AND INDEXES

Applied Science and Technology Index; A Cumulative Subject Index to English Language Periodicals in the Fields of Aeronautics and Space Science, Computer Technology, Chemistry, Construction Industry, Energy and Related Areas. H.W. Wilson Co., 950 University Avenue, Bronx, NY 10452-9978. (212) 588-8400. (800) 367-6770. FAX (718) 590-1617. From 1958 to present. Monthly. Inquire about cost and availability. Also available online (BRS and WILSONLINE) and on CD-ROM. ISSN: 0003-6986.

Deep-Sea Research. Part A: Oceanographic Research Papers. Deep-Sea Research. Part B: Oceanographic Literature Review. Pergamon Press, Inc., Maxwell House, Fairview Park, Elmsford, NY 10523. (914) 592-7700. Fax (914) 592-3625. 1953 to present. Twelve times per year. $2000.00 per year for Parts A and B. Oceanographic Literature Review also available on CD-ROM. Inquire about price and availability.

General Science Index. H.W. Wilson Co., 950 University Avenue, Bronx, NY 10452. (212) 588-8400. (800) 367-6770. FAX (718) 590-1617. From 1978 to present. Ten issues per year; quarterly and annual cumulations. Service basis. Available on CD-ROM and online. Inquire regarding cost and availability. ISSN: 0162-1963.

Government Reports Announcements and Index. National Technical Information Service (NTIS), 5285 Port Royal Road, Springfield, VA 22161. (703) 487-4650. FAX (703) 321-8547. From 1968 to present. Annual. $630.00 per year. Also available online as NTIS Bibliographic Database and on CD-ROM.

Meteorological and Geoastrophysical Abstracts. American Meteorological Society, c/o Inforonics, Inc., 550 Newtown Road, Box 458, Littleton, MA 01460. (508) 486-8976. FAX (508) 486-

0027. Covers literature in environmental sciences, meteorology, astrophysics, hydrology, glaciology, and physical oceanography. 1950 to present. Monthly. $950.00 per year. ISSN: 0026-1130. Also available online (DIALOG) and on CD-ROM.

Science Citation Index. SCI. Institute for Scientific Information, 3501 Market Street, Philadelphia, PA 19104. (215) 386-0100. (800) 523-1850. FAX (215) 386-2991. 1961 to present. Six issues per year, plus annual cumulation. $11650.00 per year. Also available online and on CD-ROM. Inquire about price and availability. ISSN: 0036-827X.

Selected Water Resources Abstracts. U.S. Geological Survey. Distributed by National Technical Information Service (NTIS), 5285 Port Royal Road, Springfield, VA 22161. (703) 487-4650. Monthly. $115.00 per year.

ANNUAL REVIEWS AND YEARBOOKS

American Meteorological Society. Meteorological Monographs. American Meteorological Society, 45 Beacon Street, Boston, MA 02108-3693. (617) 227-2425. FAX (617) 742-8718. Irregular. Price varies.

Annual Review of Earth and Planetary Sciences. Annual Reviews, Inc., 4139 El Camino Way, Palo Alto, CA 94306-0897. (415) 493-4400. (800) 523-8635. Fax (415) 855-9815. Annual. $55.00 per year.

Developments in Atmospheric Science Series. Elsevier Science Publishing Company, Inc., 655 Avenue of the Americas, New York, NY 10010. (212) 989-5800. Irregular. Inquire about price and availability.

ASSOCIATIONS AND PROFESSIONAL SOCIETIES

American Association of State Climatologists. c/o Myron Moinau, University of Idaho, Agriculture Engineering Department, State Climate Service, Moscow, ID 834844-2040.

American Meteorological Society, 45 Beacon Street, Boston, MA 02108-3693. (617) 227-2425. FAX (617) 742-8718.

Association of American Weather Observers, 401 Whitney Boulevard, Box 455, Belvedere, IL 61008. (815) 544-5665. FAX (815) 544-6334.

International Association of Meteorology and Atmospheric Physics. UCAR, P.O. Box 3000, Boulder CO 80307-3000. (303) 497-1673.

International Association of Severe Weather Specialists. c/o Warren Faidley, PO Box 31808, Tucson, AZ 85751. (520) 751-9964. FAX (520) 751-1185.

National Environmental Satellite Data, and Information Service. 2069 Federal Building 4 Room 2060, Washington, DC 20233. (301) 763-71990. FAX (301) 763-4011.

National Weather Association, 6704 Wolke Court, Montgomery, AL 36116-2134. (334) 213-0388. FAX (334) 213-0388.

University Corporation For Atmospheric Research. PO Box 3000, 1850 Table Mesa Drive, Boulder, CO 80307-3000. (303) 497-1673. FAX (303) 497-1654.

BIBLIOGRAPHIES

Information Sources in the Earth Sciences. David N. Wood, Joan E. Hardy, and Anthony P. Harvey, editors. Bowker-Saur/K.G. Saur. Distributed by R.R. Bowker, 121 Chanlon Road, New Providence, NJ 07974. (800) 521-8110 or (908) 464-6800. Second edition. 1989. $85.00.

Scientific and Technical Books and Serials in Print; An Index To Literature in Science and Technology. R.R. Bowker Co., 205 E. 42nd Street, New York, NY 10017. (800) 521-8110 or (212) 916-1600. 1992. $250.00.

DIRECTORIES AND BIOGRAPHICAL SOURCES

American Men and Women of Science: Physical and Biological Sciences. Eighteenth edition. R.R. Bowker Company, 245 West 17th Street, New York, NY 10011. (800) 521-8810 or (212) 916-1600. $750.00.

American Meteorological Society. Professional Directory. American Meteorological Society. 45 Beacon Street, Boston, MA 02108. (617) 227-2425. Included in each issue of the Bulletin of the Society.

Meteorological Services of the World. American Meteorological Society, 45 Beacon Street, Boston, MA 02108-3693. (617) 227-2425. FAX (617) 742-8718. Annual. $70.00.

National Weather Service Offices and Stations. National Oceanic and Atmospheric Administration, Department of Commerce, Silver Spring, MD 20910. (301) 427-7698. Annual. Free.

Research Centers Directory. Gale Research, 835 Penobscot Building, Detroit, MI 48226-4094. (313) 961-2242. (800) 347-4253. 20th edition. 1995. $485.00. ISSN: 0080-1518.

ENCYCLOPEDIAS AND DICTIONARIES

Climates of the States. Bair. Gale Research, 835 Penobscot Building, Detroit, MI 48226-4094. (313) 961-2242. (800) 347-4253. 5th edition. 1995. $255.00.

Concise Oxford Dictionary of the Earth Sciences. Ailsa Allaby and Michael Allaby, editors. Oxford University Press, Inc., 200 Madison Avenue, New York, NY 10016. (800) 334-4249 or (212) 679-7300. 1990. $42.95.

Encyclopedia of Climatology. Rhodes W. Fairbridge, and John E. Oliver. Chapman & Hall, 1 Penn Plaza, New York, NY 10119. (212) 564-1060. 1986. $115.00.

McGraw-Hill Encyclopedia of Science and Technology. McGraw-Hill Publishing Company, 1221 Avenue of the Americas, New York, NY 10020. (800) 262-4729 or (212) 512-3825. Seventh edition. 1992. $1900.00.

Smithsonian Meteorological Tables. Books on Demand, 300 North Zeeb Road, Ann Arbor, MI 48106-1346. (313) 761-4700. (800) 521-0600. Reprint edition. $153.90.

Weather Satellite Handbook. American Radio Relay League, 225 Main Street, Newington, CT 06111. (203) 666-1541 1994. $20.00.

TORNADOES

Ency. of Physical Sciences and Engineering Info. Sources

GENERAL WORKS

Early American Tornadoes, 1685-1870. David M. Ludlum. Am Meteorological Society, 45 Beacon Street, Boston, MA 02108-3693. (617) 227-2425. FAX (617) 742-8718. 1970. $20.00.

Hurricanes, Storms and Tornados: Geographic Characteristics and Geologic Activity. D. Nalivkin. Ashgate Publishing Co., Old Post Road, Brookfield, VT 05036. (802) 276-3162. 1983. $130.00.

1001 Questions Answered About Hurricanes, Tornadoes and Other Natural Air Disasters. Barbara Tuffy. Dover Publications, Inc., 180 Varick Street, New York, NY 10014. (212) 255-3755. (800) 223-3130. 1987. $7.95.

Significant Tornadoes, 1680-1991. Thomas P. Grazulis. Environmental Films, P.O. Box 302 Saint Johnsbury, VT 05819. (802) 748-2505. 1991. $12.95.

The Tornado. John E. Weems. Texas A&M University Press Drawer C, College Station, EX 77843-4354. (409) 845-1436. (800) 826-8911. 1991. $10.95 in paper.

Tornado Terror and Survival: The Andover Tornado. Howard Inglish. Butler County Counseling Center, 217 Ira Court, Andover, Kansas 67002. (316) 321-6036. 1991. $8.95.

Tornadoes, Dark Days, Anomalous Precipitation and Related Weather Phenomena. William Corless. Sourcebook Project, P.O. Box 107, Glen Arm, MD 21057. (410) 668 6047. 1983. $14.95.

ONLINE DATABASES AND CD-ROMS

Climate Assessment Database. National Weather Service, National Meteorological Center, 5200 Auth Road, Suite 101, Camp Springs, MD 20233. (301) 763-8016. Contains daily, weekly and monthly summaries of North American and world climatological data. Also provides five to ten weather forecasts, and 30 to 90 day weather outlook. Subscription required. Inquire as to cost and availability.

Current Contents Search. Institute for Scientific Information, 3501 Market Street, Philadelphia, PA 19104. (215) 386-0100. FAX (215) 386-6362. Contains citations to articles listed in the table of contents of science and technology journals. Also articles in social sciences and life sciences journals. Available on BRS,(800) 289-4277, DIALOG,(800) 334-2564, online services. Inquire as to cost and availability.

Dissertation Abstracts. University Microfilms International, 300 North Zeeb Road, Ann Arbor, MI 48106. (800) 521-0600 or (313) 761-4700. Scope includes virtually all doctoral dissertations accepted at accredited American institutions from 1861 to present in 252 subject areas. Available on BRS,(800) 289-4277, DIALOG,(800) 334-2564, and OCLC EPIC,(800) 848-5878, online services. Also available on CD-ROM. Inquire as to cost and availability.

Meteorological and Geoastrophysical Abstracts. American Meteorological Society, 45 Beacon Street, Boston, MA 02108-3693. (617) 227-2425. FAX (617) 742-8718. Contains citations and abstracts to the worldwide literature on significant research in meteorology and geoastrophysics. Related topics include physical oceanography, hydrology, environmental sciences and glaciology. Covers the period 1972 to present. Available on DIALOG,(800) 334-2564, online service. Inquire as to cost and availability.

NTIS Bibliographic Database. National Technical Information Service, 5285 Port Royal Road, Springfield, VA 22161. (703) 487-4929 or FAX (703) 321-8199. Broad coverage of government-sponsored science and technology research reports, 1964 to present. Available on BRS,(800) 289-4277, DIALOG, (800) 334-2564, ORBIT, (800) 456-7248, and STN International, FIZ Karlsruhe, P.O. Box 2465, W-7500, Karlsruhe 1, Germany, online services. Also available on CD-ROM. Inquire as to cost and availability.

SCISEARCH. Institute for Scientific Information, 3501 Market Street, Philadelphia, PA 19104. (800) 523-1850 or (215) 386-0100. Broad multidisciplinary title and author index to the international literature of science and technology, 1974 to present. Available on DIALOG and ORBIT online services. Also available on CD-ROM. Inquire as to cost and availability.

WILSONLINE. H.W. Wilson Company, 950 University Avenue, Bronx, NY 10452. (800) 367-6770 or (212) 588-8400. Makes available online versions of the printed H.W. Wilson indexes, including Applied Science and Technology Index, Business Periodicals Index, General Science Index, and Readers' Guide to Periodical Literature. Period covered is generally 1983 to present. Available on BRS,(800) 289-4277, DIALOG,(800) 334-2564, and OCLC EPIC,(800) 848-5878, online services. Also available on CD-ROM. Inquire as to cost and availability.

World Climate Disc. Chadwyck-Healey Inc., 1101 King Street, Alexandria, VA 22314. (703) 683-4890. FAX (703) 683-7589. Weather and climate data from approximately 5,000 weather stations worldwide, covering the years 1854 to 1990 on CD-ROM. First edition 1992. Approximately $1200.00 per year with annual updates.

World Weatherdisc. WeatherDisc Associates, Inc., 4584 N.E. 89th Street, Seattle, WA 98115. (206) 524-4314. FAX (206) 543-0308. Meteorological data on CD-ROM which describes the climate of the earth today and for the past few hundred years. First edition 1989. Approximately $295.00 per year with biannual updates.

PERIODICALS

Agricultural and Forest Meteorology. Elsevier Science Publishing Company, Inc., 655 Avenue of the Americas, New York, NY 10010. (212) 989-5800. Twenty times per year. $750.00 per year.

American Meteorological Society. Bulletin. American Meteorological Society, 45 Beacon Street, Boston, MA 02108-3693. (617) 227-2425. FAX (617) 742-8718. Monthly. $60.00 per year.

American Weather Observer. Association of American Weather Observers, 401 Whitney Boulevard, Box 455, Belvedere, IL 61008. (815) 544-5665. FAX (815) 544-6334. Monthly. $21.00 per year.

JGR: Journal of Geophysical Research: Atmosphere. American Geophysical Union, 2000 Florida Avenue, N.W., Washington, DC 20009. (202) 462-6903. Monthly. $90.00 per year to members.

Journal of Applied Meteorology. American Meteorological Society, 45 Beacon Street, Boston, MA 02108-3693. (617) 227-2425. FAX (617) 742-8718. Monthly. $165.00 per year.

Journal of the Atmospheric Sciences. American Meteorological Society, 45 Beacon Street, Boston, MA 02108-3693. (617) 227-2425. FAX (617) 742-8718. Semi-monthly. $320.00 per year.

Monthly Weather Review. American Meteorological Society, 45 Beacon Street, Boston, MA 02108-3693. (617) 227-2425. FAX (617) 742-8718. Monthly. $205.00 per year.

National Weather Digest. National Weather Association, 4400 Stamp Road, Room 404, Temple Hills, MD 20748. (301) 899-3784.

Weather. Royal Meteorological Society, 104 Oxford Road, Reading, Berks RG1 7LJ, England. Monthly. $44.00 per year.

Weatherwise. Heldref Publications, 1319 Eighteenth Street, N.W., Washington, DC 20036-1802. (202) 296-6267. FAX (202) 296-5149. Bi-monthly. $28.00 per year.

RESEARCH CENTERS AND INSTITUTES

Center for Analysis and Prediction of Storms. University of Oklahoma, 401 East Boyd, Norman, OK 73019. (405) 325-3041.

Cooperative Institute For Mesoscale Meteorological Studies. University of Oklahoma, 401 East Boyd, Norman, OK 73019. (405) 325-3041.

Institute For Disaster Research. Texas Tech University, P. O. Box 4089, Lubbock, TX 79409-1023. (806) 742-3476. (803) 742-3488.

National Center For Atmospheric Research. Box 3000, Boulder, CO 80307. (303) 497-1000. FAX (303) 497-1654.

National Severe Storms Laboratory. 1313 Halley Circle, Norman, OK 73069. (406) 360-3620. FAX (405) 360-0472.

Wind Research Laboratory. University of Chicago, 5711 Ellis Street, Chicago, IL 60637. (312) 702-8114.

TRACE ANALYSIS

See: ANALYTICAL CHEMISTRY

TRAFFIC ENGINEERING

See: HIGHWAY ENGINEERING

TRANSFORMERS

See: ELECTRIC POWER ENGINEERING

TRANSISTORS

See: SEMICONDUCTORS

TRANSMISSION LINES

See: ELECTRIC POWER ENGINEERING

TRANSMISSIONS

See: AUTOMOTIVE ENGINEERING

TRANSPORTATION

See also: AERONAUTICAL ENGINEERING, AUTOMOTIVE ENGINEERING, CIVIL ENGINEERING, HIGHWAY ENGINEERING, MARINE ENGINEERING, RAILROAD ENGINEERING

ABSTRACT SERVICES AND INDEXES

Applied Science and Technology Index; A Cumulative Subject Index To English Language Periodicals in the Fields of Aeronautics and Space Science, Computer Technology, Chemistry, Construction Industry, Energy and Related Areas. H.W. Wilson Co., 950 University Avenue, Bronx, NY 10452. (800) 367-6770 or (212) 588-8400. FAX (718) 590-1617. From 1958 to present. Monthly. Inquire about cost and availability. Also available on CD-ROM and online.

Automotive Literature Index. A. Wallace, 2307 Shoreland Avenue, Toledo, OH 43611. (419) 729-9065. 1981 to present. Quinquennial. $40.00.

Current Contents: Engineering, Technology, and Applied Sciences. Institute for Scientific Information, 3501 Market Street, Philadelphia, PA 19104. (215) 386-0100. FAX (215) 386-6362. 1970 to present. Weekly. $442.00 per year.

Energy Research Abstracts. 1976-present. U.S. Department of Energy, Office of Scientific and Technical Information, Box 62, Oak Ridge, TN 37831. (615) 574-0733. Subscriptions: Superintendent of Documents, U.S. Government Printing Office, Box 371954, Pittsburgh, PA 15250-7954. (202) 783-3238. FAX (202) 512-2233. Monthly. $164.00 per year. Abstracts all the scientific and technical reports, journal articles, conference proceedings, patents, theses, and monographs sponsored by the U.S. Energy Research and Development Administration.

Engineering Index Monthly. Engineering Information, Inc., Castle Point on the Hudson, Hoboken, NJ 07030. (800) 221-1044. FAX (212) 832-1857. Monthly. $2300.00 per year. Also available online as COMPENDEX, and also on CD-ROM. Covers chemical engineering, computers, electrical engineering, civil engineering, metals and mining, industrial management, and mechanical engineering.

Environmental Abstracts. R.R. Bowker Inc., 121 Chanlon Road, New Providence, NJ 07974. (800) 521-8110 or (908) 464-6800. Contains citations, most with abstracts, to literature relating to environmental issues and problems, from 1971 to present. Inquire as to cost and availability.

Fuel and Energy Abstracts. Butterworth-Heinemann, Linacre House, Jordan Hill, Oxford OX2 8DP, England. Telephone 0865-310366. FAX 0865-310898. Summary of world literature on all scientific, technical, commercial, and environmental aspects of fuel and energy. 1960 to present. Six times a year. Inquire for price in U.S.

TRANSPORTATION

Ency. of Physical Sciences and Engineering Info. Sources

Government Reports Announcements and Index. National Technical Information Service (NTIS), 5285 Port Royal Road, Springfield, VA 22161. (703) 487-4650. 1968 to present. Annual. $525.00 per year. Also available online as NTIS Bibliographic Database. Also available on CD-ROM.

International Aerospace Abstracts. Technical Information Service, American Institute of Aeronautics and Astronautics, Inc., 555 West 57th St., New York, NY 10019. (212) 247-6500. FAX (212) 582-4861. Semi-monthly. $1295.00 per year.

NTIS Alerts: Transportation. U.S. National Technical Information Service, 5825 Port Royal Road, Springfield, VA 22161. (703) 487-4630. FAX (703) 321-8547. Weekly. $135.00 per year.

Science Citation Index. Institute for Scientific Information, 3501 Market Street, Philadelphia, PA 19104. (215) 386-0100. FAX (215) 386-6362. Inquire about availability and cost. Also available on CD-ROM.

Scientific and Technical Aerospace Reports (STAR). National Aeronautics and Space Administration. NASA Center for Aerospace Information, Box 8757, BWI Airport, Baltimore, MD 21240. (301) 621-0153. Monthly. Inquire about availability and cost. Also available through the NASA online retrieval service (RECON), and through the Aerospace Database through DIALOG.

ANNUAL REVIEWS AND YEARBOOKS

Jane's Urban Transport Systems. Jane's Information Group, 1340 Braddock Place, Suite 300, Alexandria, VA 22314-1651. (703) 683-3700. FAX (703) 836-0029. 1982 to present. Annual. $285.00.

ASSOCIATIONS AND PROFESSIONAL SOCIETIES

Advanced Transit Association. 9019 Hamilton Drive, Fairfax, VA 22031. (703) 591-9328.

Air Transport Association of America. 1709 New York Avenue NW, Washington, DC 20006. (202) 626-4000.

American Association of State Highway and Transportation officials. 444 N. Capitol Street NW, Suite 249, Washington, DC 20001. (202) 624-5800. FAX (202) 624-5806.

American Society of Transportation and Logistics. 216 E. Church Street, Lock Haven, PA 17745. (717) 748-8515. FAX (717) 748-9118.

Institute of Transportation Engineers. 525 School Street SW, Suite 410, Washington, DC 20024-2797, (202) 554-8050. FAX (202) 863-5486.

Society of Automotive Engineers (SAE International). 400 Commonwealth Drive, Warrendale, PA 15096-0001. (412) 776-4841.

Transportation Research Board. 2101 Constitution Avenue NW, Washington, DC 20418. (202) 334-2934. FAX (202) 334-2003.

Transportation Research Forum. 1730 N. Lynn Street, Suite 502, Arlington, VA 22209. (703) 525-1191. FAX (703) 276-8196.

DIRECTORIES AND BIOGRAPHICAL SOURCES

American Men & Women of Science, 1995-96. R.R. Bowker Staff, eds. 19th edition. 8 volumes. R.R. Bowker/Reed International Publishing Company, 121 Chanlon Road, New Providence, NJ 07974. (908) 464-6800 or (800) 521-8110. 1995. $850.00

Directory of Engineering Societies and Related Organizations. Gordon Davis. 13th edition. American Association of Engineering Societies, 1111 19th Street NW, Suite 608, Washington, DC 20036. (202) 296-2237 or (800) 658-8897. 1989. Inquire for price.

Directory of Transpiration Professionals, 1995. National Association of Regulatory Utility Commissioners, 1102 ICC Bldg., PO Box 684, Washington, DC 20898-2200. 1995. $23.00.

International Research Centers Directory. Gale Research, 835 Penobscot Building, Detroit, MI 48226-4094. (313) 961-2242. (800) 347-4253. FAX (313) 961-6083. 8th edition. 1995. $410.00

Reference Book of Department Personnel. American Association of State Highway and Transportation officials, 444 N. Capitol Street NW, Suite 249, Washington, DC 20001. (202) 624-5800. FAX (202) 624-5806. 1993. $10.00.

Research Centers Directory. Gale Research, 835 Penobscot Building, Detroit, MI 48226-4094. (313) 961-2242. (800) 347-4253. FAX (313) 961-6083. $485.00.

Scientific and Technical Organizations and Agencies Directory. Gale Research, 835 Penobscot Building, Detroit, MI 48226-4094. (313) 961-2242. (800) 347-4253. FAX (313) 961-6083. 4th edition. 1996. $195.00.

Who's Who in Engineering. American Association of Engineering Societies, 1111 19th Street NW, Suite 608, Washington, DC 20036. (202) 296-2237 or (800) 658-8897. 8th edition. 1991. Inquire for price.

Who's Who in Technology. Gale Research, 835 Penobscot Building, Detroit, MI 48226-4094. (313) 961-2242. (800) 347-4253. FAX (313) 961-6083. 1995. $195.00.

ENCYCLOPEDIAS AND DICTIONARIES

Concise Encyclopedia of Traffic and Transportation Systems. Markos Papageorgiou, editor. Pergamon Press, 660 White Plains Road, Tarrytown, NY 10591-5153. 1991. $425.00.

Dictionary For Automotive Engineering. Jean de Coster. K.G. Saur, 121 Chanlon Road, New Providence, NJ 07974. (908) 464-6800 or (800) 521-8110. 1986. Inquire for cost and availability.

McGraw-Hill Encyclopedia of Engineering. Sybil P. Parker, ed. 2nd edition. McGraw-Hill Publishing Company, 1221 Avenue of the Americas, New York, NY 10020. (800) 262-4729 or (212) 512-3825. 1993. $95.50.

GENERAL WORKS

Contemporary Transportation. Donald F. Wood and James C. Johnson. Fourth edition. Macmillan Publishing, 200 Old Tappan

Road, Old Tappan, NJ 07675. (800) 223-2336. FAX (800) 445-6991. 1992. Inquire for price.

Highway and Transportation Engineering and Planning. Gavin Macpherson. John Wiley and Sons, Inc., 605 Third Avenue, New York, NY 10158. (800) 526-5368 or (212) 850-6000. 1993. $49.95.

Transportation Engineering and Planning. C.S. Papacostas and P.D. Prevedouros. Second edition. Prentice Hall, 113 Sylvan Avenue, Route 9W, Englewood Cliffs, NJ 07632. (201) 592-2000 or (800) 922-0579. 1992. $77.00.

Transportation Engineering Basics. A.S. Narasimha Nurthy and R. Henry Mohle. American Society of Civil Engineers, 345 East 47th Street, New York, NY 10017-2398. (212) 705-7520 or (800) 548-2723. 1993. $16.00.

Transportation Systems Theory and Application of Advanced Technology. Liu and Blosseville. 2 volumes. Elsevier Science Inc., Box 882, Madison Square Station, New York, NY 10159. (212) 989-5800. 1995. $180.00.

HANDBOOKS AND MANUALS

Traffic Engineering Handbook. Institute of Transportation Engineers staff. James L. Pline, editor. Fourth edition. Prentice Hall, 113 Sylvan Avenue, Route 9W, Englewood Cliffs, NJ 07632. (201) 592-2000 or (800) 922-0579. 1992. $75.00.

Transportation Planning Handbook. Institute of Transportation Engineers staff. John D. Edwards Jr., editor. Prentice Hall, 113 Sylvan Avenue, Route 9W, Englewood Cliffs, NJ 07632. (201) 592-2000 or (800) 922-0579. 1992. $76.00.

International Handbook of Transportation Policy. Tsuneo Akaha, editor. Greenwood Publishing Group, 88 Post Road W., Box 5007, Westport, CT 06881. (203) 226-3571 or (800) 225-5800. 1990. $89.50.

ONLINE DATABASES AND CD-ROMS

Compendex Plus. Engineering Information, Inc., 345 East 47th Street, New York, NY 10017. (212) 705-7600 or (800) 221-1044. Contains citations with abstracts to worldwide literature in engineering and technology, from 1970 to present. Available on online BRS,(800) 289-4277, DIALOG, (800) 334-2564, ORBIT (800) 456-7248, and STN International, FIZ Karlsruhe, P.O. Box 2465, W-7500, Karlsruhe 1, Germany, online services. Also available on CD-ROM. Inquire as to cost and availability.

CSA Engineering. Cambridge Scientific Abstracts, 7200 Wisconsin Avenue, Suite 601, Bethesda, MD 20814. (301) 961-6750 or (800) 843-7751. Contains citations and abstracts of international periodicals and research literature covering all fields of engineering and science and technology, including computer and information science, electronics, mechanical engineering, solid state materials, 1981 to present. Available on BRS,(800) 289-4277, online service. Inquire as to cost and availability.

Current Contents Search. Institute for Scientific Information, 3501 Market Street, Philadelphia, PA 19104. (215) 386-0100. FAX (215) 386-6362. Contains citations to articles listed in the table of contents of science and technology journals. Also articles in social sciences and life sciences journals. Available on BRS,(800) 289-4277, DIALOG,(800) 334-2564, online services. Inquire as to cost and availability.

Dissertation Abstracts. University Microfilms International, 300 North Zeeb Road, Ann Arbor, MI 48106. (800) 521-0600 or (313) 761-4700. Scope includes virtually all doctoral dissertations accepted at accredited American institutions from 1861 to present in 252 subject areas. Available on BRS,(800) 289-4277, DIALOG,(800) 334-2564, and OCLC EPIC,(800) 848-5878, online services. Also available on CD-ROM. Inquire as to cost and availability.

International Aerospace Abstracts. American Institute of Aeronautics and Astronautics, 370 L'Enfant Promenade, S.W., Washington, DC 20024. (202) 646-7400. Contains references and abstracts of journal and monograph literature relating to aerospace science and technology, from 1963 to present. Available through the NASA/RECON system of the National Aeronautics and Space Administration only.

NTIS Bibliographic Database. National Technical Information Service, 5285 Port Royal Road, Springfield, VA 22161. (703) 487-4929 or FAX (703) 321-8199. Broad coverage of government-sponsored science and technology research reports, 1964 to present. Available on BRS,(800) 289-4277, DIALOG, (800) 334-2564, ORBIT, (800) 456-7248, and STN International, FIZ Karlsruhe, P.O. Box 2465, W-7500, Karlsruhe 1, Germany, online services. Also available on CD-ROM. Inquire as to cost and availability.

SAE Global Mobility Database. Society of Automotive Engineers (SAE), Electronic Publishing Division, 400 Commonwealth Drive, Warrendale, PA 15098. (412) 776-4841. Contains citations with abstracts to technical papers on automotive and aerospace technology and vehicular-related industries that have been presented at SAE conferences. Covers 1906 to present. Available on ORBIT,(800) 456-7248, online service. Inquire as to cost and availability.

SCISEARCH. Institute for Scientific Information, 3501 Market Street, Philadelphia, PA 19104. (800) 523-1850 or (215) 386-0100. Broad multidisciplinary title and author index to the international literature of science and technology, 1974 to present. Available on DIALOG,(800) 334-2564, and ORBIT,(800) 456-7248, online services. Also available on CD-ROM. Inquire as to cost and availability.

TRIS (Transportation Research Information). National Academy of Science, 2101 Constitution Avenue, N.W., Washington, DC 20418. (202) 334-3313 or (800) 624-6242. Citations with abstracts of literature on transportation, including air, highway, rail, maritime and other modes from 1968 to present. Available on DIALOG,(800) 334-2564, online service. Inquire as to cost and availability.

PERIODICALS

Highways and Transportation. Institution of Highways and Transportation, 3 Lygon Place, Ebury Street, London SW1 0JS, England. Telephone 071-730-5245. FAX 071-730-1628. 1930 to present. Monthly. Inquire for cost.

ITE Journal. Institute of Transportation Engineers, 525 School Street SW, Suite 410, Washington, DC 20024. (202) 554-8050. 1930 to present. Monthly. $50.00 per year.

Journal of Advanced Transportation. Institute for Transportation, Inc., Duke University, Box 90304, Durham, NC 22706. (919) 660-5312. FAX (919) 660-8963. 1967 to present. Three times a year. $80.00.

TRANSPORTATION

Ency. of Physical Sciences and Engineering Info. Sources

Journal of Transportation Engineering. American Society of Civil Engineers, Air Transport, Highway, Pipeline, Urban Transportation Divisions, 345 East 47th Street, New York, NY 10017-2398. (212) 705-7520 or (800) 548-2723. 1969 to present. Bimonthly. $124.00 per year for non-members.

SAE Transactions. SAE International, 400 Commonwealth Drive, Warrendale, PA 15096-0001. (412) 776-4841. Annual. $850.00 per year.

Traffic Engineering & Control. Printerhall Ltd., 29 Newman Street, London W1P 3PE, England. Telephone 071-636-3956. FAX 071-436-7016. 1960 to present. Monthly. $106.00 per year.

Transportation. Kluwer Academic Publishers, P.O. Box 358, Accord Station, Hingham, MA 02018-0358. (617) 871-6000. 1972 to present. Four times a year. $187.00.

Transportation Builder. American Road and Transportation Builders Association, 1010 Massachusetts Avenue NW, Washington, DC 20001. (202) 488-2722. FAX (202) 488-3631. 1923 to present. 6 times a year. $50.00.

Transportation Journal. American Society of Transportation and Logistics, 216 E. Church Street, Lock Haven, PA 17745. (717) 748-8515. FAX (717) 748-9118. 1961 to present. Quarterly. $50.00.

Transportation Planning & Technology. Gordon and Breach Science Publishers, 820 Town Center Drive, Langhorne, PA 19047. (215) 750-2642. FAX (215) 750-6343. 1972 to present. Four times a year. Inquire for price.

Transportation Quarterly. Eno Transportation Foundation, 44211 Statestone Court, Leesburg, VA 22075. (703) 729-7200. FAX (703) 729-7219. 1947 to present. Quarterly. $40.00.

Transportation Research A: Policy and Practice. Elsevier Science [journals], 660 White Plains Rd., Tarrytown, NY 10159-5153. (919) 524-9200. FAX (919) 333-2444. 1967 to present. 6 times a year. $405.00 per year ($910.00 for A, B, and C).

Transportation Research B: Methodological. Elsevier Science [journals], 660 White Plains Rd., Tarrytown, NY 10159-5153. (919) 524-9200. FAX (919) 333-2444. 1967 to present. 6 times a year. $405.00 per year ($910.00 for A, B, and C).

Transportation Research C: Emerging Technologies. Elsevier Science [journals], 660 White Plains Rd., Tarrytown, NY 10159-5153. (919) 524-9200. FAX (919) 333-2444. 1993 to present. 4 times a year. $260.00 per year ($910.00 for A, B, and C).

Transportation Research Record. National Research Council, Transportation Research Board, 2101 Constitution Avenue NW, Washington, DC 20418. (202) 334-3218. FAX (202) 334-2519. 1963 to present. Irregular. $865.00 per year.

Transportation Science. Operations Research Society of America, 1314 Guilford Avenue, Baltimore, MD 21202. (410) 528-4146. FAX (301) 528-8556. 1967 to present. Quarterly. $37.00.

UMTRI Research Review. University of Michigan Transportation Research Institute, 2901 Baxter Road, Ann Arbor, MI 48109-2150. (313) 764-6504. FAX (313) 936-1081. 1970 to present. Bi-monthly. $35.00.

RESEARCH CENTERS AND INSTITUTES

Massachusetts Institute of Technology, Center For Transportation Studies. 77 Massachusetts Avenue (I-123), Cambridge, MA 02139. (617) 253-5320. FAX (617) 258-5942.

Northwestern University Transportation Center. 1936 Sheridan Road, Evanston, IL 60201. (708) 491-7287.

Texas A&M University, Texas Transportation Institute, System Division. College Station, TX 77843-3135. (409) 945-9871.

University of California At Berkeley, Institute of Transportation Studies. 109 McLaughlin Hall, Berkeley, CA 94720. (415) 642-3585. FAX (415) 642-1246.

University of Michigan Transportation Research Institute. 2901 Baxter Road, Ann Arbor, MI 48109-2150. (313) 764-6504. FAX (313) 936-1081.

TRIBOLOGY

See: LUBRICATION

TRIGONOMETRY

See: MATHEMATICS

TSUNAMIS

See also: EARTHQUAKE ENGINEERING, GEOLOGY, GEOPHYSICS, GEOTECHNICAL ENGINEERING, OCEANOGRAPHY, PLATE TECTONICS, SEISMOLOGY, VOLCANOLOGY

ABSTRACT SERVICES AND INDEXES

Abstract Journal in Earthquake Engineering. University of California, Berkeley, Earthquake Engineering Research Center, 1301 South 46th Street, Richmond, CA 94804-4698. (510) 231-9413 or FAX (510) 231-9471. 1972 to present. Annual. $80.00 per year.

Bibliography and Index of Geology. American Geological Institute, 4220 King Street, Alexandria, VA 22302. (703) 379-2480. Fax (703) 379-7563. Monthly. $1295.00 per year. Also available online as GeoRef.

Current Contents: Physical, Chemical and Earth Sciences. Institute for Scientific Information, 3501 Market Street, Philadelphia, PA 19104. (215) 386-0100. FAX (215) 386-6362. Weekly. $360.00 per year.

Geological Abstracts. Elsevier-Geo Abstracts, Regency House, 34 Duke Street, Norwich NR3 3AP, England. Monthly. $760.00 per year. Also available online as GEOBASE.

Geological Society of America. Abstracts with Programs. Geological Society of America. 3300 Penrose Place, P.O. Box 9140, Boulder, CO 80301-9140. (303) 447-2020. Abstracts and programs of the annual conference. Annual. $69.00.

Ency. of Physical Sciences and Engineering Info. Sources

TSUNAMIS

Index To Scientific and Technical Proceedings. Institute for Scientific Information, 3501 Market St., Philadelphia, PA 19104. (215) 386-0100. FAX (215) 386-6362. Monthly. $500.00 per year.

Index To Scientific Reviews. Institute for Scientific Information, 3501 Market St., Philadelphia, PA 19104. (215) 386-0100. FAX (215) 386-6362. Semi-annual.

Meteorological and Geoastrophysical Abstracts. American Meteorological Society, c/o Inforonics Inc., 550 Newtown Road, Littleton, MA 01460. (508) 486-8976. FAX (508) 486-0027. 1950 to present. Monthly. $950.00 per year. Current abstracts of books, reports, research papers, and miscellaneous literature on environmental sciences, meteorology, astrophysics, hydrology, glaciology, and physical oceanography.

Oceanic Abstracts. Cambridge Scientific Abstracts, 7200 Wisconsin Avenue, Bethesda, MD 20814. (301) 961-6750. Fax (301) 961-6720. Bimonthly. $995.00 per year.

Physics Abstracts. Institute of Electrical Engineers, Michael Faraday House, Six Hill Way, Stevenage, Herts, SG1 2AY, England. Distributed by IEEE, 445 Hoes Lane, Piscataway, NJ 08854. (908) 562-5549. 1898 to present. Monthly. $2700.00 per year. Also available online as INSPEC.

Physics Briefs. American Institute of Physics, 335 East 45th Street, New York, NY 10017 (212) 661-9404 or FAX (516)349-9704. 1980 to present. Six times per year. $2600.00 per year.

Science Citation Index. Institute for Scientific Information, 3501 Market Street, Philadelphia, PA 19104. (215) 386-0100. FAX (215) 386-6362. Inquire about availability and cost. Also available on CD-ROM.

ASSOCIATIONS AND PROFESSIONAL SOCIETIES

American Geophysical Union. 2000 Florida Avenue, N.W., Washington, DC 20009. (202) 462-6900. FAX (202) 328-0566.

Earthquake Engineering Research Institute. 499 14th Street, Suite 320, Oakland, CA 94612-1902. (510) 451-0905.

Geological Society of America. 3300 Penrose Place, PO Box 9140, Denver, CO 80301-9140. (303) 447-2020.

International Institute of Seismology and Earthquake Engineering. Building Research Institute—Ministry of Construction, 1 Tatehara, Tsukuba, Ibaraki 305, Japan.

International Seismological Centre. Pipers Lane, Thatcham, Newbury, Berks. RG13 4NS, England. Telephone 0635-61022. FAX 0635-72351.

Seismological Society of America. 201 Plaza Professional Bldg., El Cerrito, CA 94530. (510) 525-5474. FAX (510) 525-7204.

DIRECTORIES AND BIOGRAPHICAL SOURCES

American Men & Women of Science, 1995-96. R.R. Bowker Staff, eds. 19th edition. 8 volumes. R.R. Bowker/Reed International Publishing Company, 121 Chanlon Road, New Providence, NJ 07974. (908) 464-6800 or (800) 521-8110. 1995. $850.00

Directory of Engineering Societies and Related Organizations. Gordon Davis. 13th edition. American Association of Engineering Societies, 1111 19th Street NW, Suite 608, Washington, DC 20036. (202) 296-2237 or (800) 658-8897. 1989. Inquire for price.

International Research Centers Directory. Gale Research, 835 Penobscot Building, Detroit, MI 48226-4094. (313) 961-2242. (800) 347-4253. FAX (313) 961-6083. 8th edition. 1995. $410.00

Research Centers Directory. Gale Research, 835 Penobscot Building, Detroit, MI 48226-4094. (313) 961-2242. (800) 347-4253. FAX (313) 961-6083. $485.00.

Scientific and Technical Organizations and Agencies Directory. Gale Research, 835 Penobscot Building, Detroit, MI 48226-4094. (313) 961-2242. (800) 347-4253. FAX (313) 961-6083. 4th edition. 1996. $195.00.

ENCYCLOPEDIAS AND DICTIONARIES

Encyclopedia of Earthquakes and Volcanoes. David Ritchie. Facts-on-File, Inc., 460 Park Avenue South, New York, NY 10016-7382. (800) 322-8755. Fax (212) 213-4578. 1994. $40.00.

GENERAL WORKS

Tsunami Hazard: A Practical Guide For Tsunami Hazard Reduction. E.N. Bernard, editor. Kluwer Academic Publishers, P.O. Box 358, Accord Station, Hingham, MA 02018-0358. (617) 871-6000. FAX (617) 871-6528. 1991. $98.00.

Tsunamis: 1992-1994: Their Generation, Dynamics, and Hazards. Kenji Satake & Fumihiko Imamura, editors. Reprint edition. Birkhauser Boston, Division of Springer-Verlag, 675 Massachusetts Avenue, Cambridge, MA 02139. (617) 876-2333 or (800) 777-4643. FAX (212) 473-6272. 1995. $42.50.

Tsunamis: Their Science and Engineering. T. Iwasaki & K. Iida, editors. Kluwer Academic Publishers, P.O. Box 358, Accord Station, Hingham, MA 02018-0358. (617) 871-6000. FAX (617) 871-6528. 1983. $262.50.

ONLINE DATABASES AND CD-ROMS

Compendex Plus. Engineering Information, Inc., 345 East 47th Street, New York, NY 10017. (212) 705-7600 or (800) 221-1044. Contains citations with abstracts to worldwide literature in engineering and technology, from 1970 to present. Available on online BRS,(800) 289-4277, DIALOG, (800) 334-2564, ORBIT (800) 456-7248, and STN International, FIZ Karlsruhe, P.O. Box 2465, W-7500, Karlsruhe 1, Germany, online services. Also available on CD-ROM. Inquire as to cost and availability.

Earth Sciences. U.S. Geological Survey, 12201 Sunrise Valley Drive, Reston, VA 22092-9998. (703) 648-4460. CD-ROM of earth science databases, including the U.S. Geological Survey Library, Earth Science Data Directory, and GEOINDEX, citations to published geological maps. $350.00 per year with quarterly updates.

Geoarchive. Geosystems, P.O. Box 1024, Westminster, London, England, SW1 P 2JL. Citations to literature on geoscience, 1969 to present. Inquire as to online cost and availability.

TSUNAMIS

Ency. of Physical Sciences and Engineering Info. Sources

Geobase. Elsevier-Geo Abstracts, Regency House, 34 Duke Street, Norwich NR3 3AP, England. Contains citations to the worldwide earth science literature from 1980 to date. Available on DIALOG, ORBIT online services. Inquire as to cost and availability.

Georef: Bibliography and Index of Geology. American Geological Institute, 4220 King Street, Alexandria, VA 22302. (703) 379-2480. Fax (703) 379-7563. Monthly. Inquire for price and availability.

NTIS Bibliographic Database. National Technical Information Service, 5285 Port Royal Road, Springfield, VA 22161. (703) 487-4929 or FAX (703) 321-8199. Broad coverage of government-sponsored science and technology research reports, 1964 to present. Available on BRS,(800) 289-4277, DIALOG, (800) 334-2564, ORBIT, (800) 456-7248, and STN International, FIZ Karlsruhe, P.O. Box 2465, W-7500, Karlsruhe 1, Germany, online services. Also available on CD-ROM. Inquire as to cost and availability.

SCISEARCH. Institute for Scientific Information, 3501 Market Street, Philadelphia, PA 19104. (800) 523-1850 or (215) 386-0100. Broad multidisciplinary title and author index to the international literature of science and technology, 1974 to present. Available on DIALOG,(800) 334-2564, and ORBIT,(800) 456-7248, online services. Also available on CD-ROM. Inquire as to cost and availability.

PERIODICALS

Disasters. Basil Blackwell Inc., 238 Main Street, Cambridge, MA 02142. (617) 547-7110. FAX (617) 547-0789. 1977 to present. $89.50.

Earthquake History of the United States. National Oceanic and Atmospheric Administration, National Geophysical Data Center, 325 Broadway, Boulder, CO 80303-3328. (303) 497-6419. 1928 to present. Irregular (approximately every five years). Price varies. Also available on CD-ROM.

Earthquake Spectra. Earthquake Engineering Research Institute, 499 14th Street, Suite 320, Oakland, CA 94612-1902. (510) 451-0905. 1984 to present. Quarterly. $120.00.

Geophysical and Astrophysical Fluid Dynamics. Gordon and Breach Science Publishers, 820 Town Center Drive, Langhorne, PA 19047. (215) 750-2642. FAX (215) 750-6343. 1970 to present. 28 times a year. Inquire for price.

Geophysical Research Letters. American Geophysical Union, 2000 Florida Avenue, N.W., Washington, DC 20009. (202) 462-6900. 1974 to present. Semi-monthly. $590.00 per year for non-members.

International Seismological Centre Bulletin. Pipers Lane, Thatcham, Newbury, Berks. RG13 4NS, England. Telephone 0635-61022. FAX 0635-72351. 1964 to present. Monthly. Inquire for price in U.S.

International Seismological Centre Regional Catalogue of Earthquakes. Pipers Lane, Thatcham, Newbury, Berks. RG13 4NS, England. Telephone 0635-61022. FAX 0635-72351. 1964 to present. Semi-annual. Inquire for price in U.S.

Seismological Research Letters. Seismological Society of America, 201 Plaza Professional Bldg., El Cerrito, CA 94530.

(510) 525-5474. FAX (510) 525-7204. 1929 to present. Quarterly. $15.00.

Seismological Society of America Bulletin. 201 Plaza Professional Bldg., El Cerrito, CA 94530. (510) 525-5474. FAX (510) 525-7204. 1911 to present. Bi-monthly. $135.00.

Tsunami Newsletter. International Tsunami Information Center, Box 50027, Honolulu, HI 96850-4993. (808) 541-1658. FAX (808) 541-1678. 1968 to present. 2 times a year. Free.

Volcanology and Seismology. Gordon and Breach Science Publishers, 820 Town Center Drive, Langhorne, PA 19047. (215) 750-2642. FAX (215) 750-6343. Twelve times a year. Inquire for cost.

RESEARCH CENTERS AND INSTITUTES

Incorporated Research Institutions For Seismology. Suite 1440, 1616 N. Fort Myer Drive, Arlington, VA 22209. (702) 524-6222. (703) 527-7256.

International Institute of Seismology and Earthquake Engineering. Building Research Institute—Ministry of Construction, 1 Tatehara, Tsukuba, Ibaraki 305, Japan.

International Seismological Centre. Pipers Lane, Thatcham, Newbury, Berks. RG13 4NS, England. Telephone 0635-61022. FAX 0635-72351.

International Tsunami Information Center. Box 50027, Honolulu, HI 96850-4993. (808) 541-1658. FAX (808) 541-1678.

Joint Institute For Marine and Atmospheric Research. University of Hawaii at Manoa, 1000 Pope Road, MSB 312, Honolulu, HI 96822. (808) 956-8083. FAX (808) 956-4104.

University of Hawaii At Manoa, J.K.K. Look Laboratory of Oceanographic Engineering. 811 Olomehani Street, Honolulu, HI 96813. (808) 533-6412. FAX (808) 537-5607.

TURBINES

See also: AUTOMOTIVE ENGINEERING, GAS TURBINES, JET PROPULSION, PROPULSION SYSTEMS, STEAM ENGINEERING

ABSTRACT SERVICES AND INDEXES

Applied Mechanics Reviews: An Assessment of World Literature in Engineering Sciences. 1948-present. American Society of Mechanical Engineers, 345 East 47th Street, New York, NY 10017. (212) 705-7703. Monthly. $360.00 per year.

Applied Science and Technology Index; A Cumulative Subject Index To English Language Periodicals in the Fields of Aeronautics and Space Science, Computer Technology, Chemistry, Construction Industry, Energy and Related Areas. H.W. Wilson Co., 950 University Avenue, Bronx, NY 10452. (800) 367-6770 or (212) 588-8400. FAX (718) 590-1617. From 1958 to present. Monthly. Inquire about cost and availability. Also available on CD-ROM and online.

Engineering Index Monthly. Engineering Information, Inc., Castle Point on the Hudson, Hoboken, NJ 07030. (800) 221-1044. FAX

Ency. of Physical Sciences and Engineering Info. Sources

TURBINES

(212) 832-1857. Monthly. $2300.00 per year. Also available online as COMPENDEX, and also on CD-ROM. Covers chemical engineering, computers, electrical engineering, civil engineering, metals and mining, industrial management, and mechanical engineering.

International Aerospace Abstracts. Technical Information Service, American Institute of Aeronautics and Astronautics, Inc., 555 West 57th St., New York, NY 10019. (212) 247-6500. FAX (212) 582-4861. Semi-monthly. $1295.00 per year.

Mechanical Engineering Abstracts (formerly ISMEC), Cambridge Scientific Abstracts, 7200 Wisconsin Avenue, Bethesda, MD 20814-4823. (301) 961-6750. FAX (301) 961-6720. 1967 to present. Monthly. $895.00 per year. Summarizes world literature in mechanical engineering, production engineering, and engineering management. Also available online.

ASSOCIATIONS AND PROFESSIONAL SOCIETIES

American Institute of Aeronautics and Astronautics. 370 L'Enfant Promenade, S.W., Washington, DC 20024. (202) 646-7400.

American Society of Mechanical Engineers. 345 East 47th Street, New York, NY 10017-2398. (212) 705-7703.

International Gas Turbine Institute, ASME. 5801 Peachtree Dunwoody Road, Suite 100, Atlanta, GA 30342-1503. (404) 847-0072.

National Association of Power Engineers. 5-7 Springfield Street, Chicopee, MA 01013. (413) 592-6273.

Society of Automotive Engineers International. 400 Commonwealth Dr., Warrendale, PA 15096. (412) 776-4841.

DIRECTORIES AND BIOGRAPHICAL SOURCES

Diesel and Gas Turbine Worldwide Catalog. Diesel and Gas Turbine Publications, 13555 Bishop's Court, Brookfield, WI 53005-6286. (914) 784-9177. FAX (914) 784-9177. 1935 to present. Annual. $75.00 per year.

Directory of Engineering Societies and Related Organizations. Gordon Davis. 13th edition. American Association of Engineering Societies, 1111 19th Street NW, Suite 608, Washington, DC 20036. (202) 296-2237 or (800) 658-8897. 1989. Inquire for price.

International Research Centers Directory. Gale Research, 835 Penobscot Building, Detroit, MI 48226-4094. (313) 961-2242. (800) 347-4253. 8th edition. 1995. $410.00

Research Centers Directory. Gale Research, 835 Penobscot Building, Detroit, MI 48226-4094. (313) 961-2242. (800) 347-4253. FAX (313) 961-6083. $485.00.

Scientific and Technical Organizations and Agencies Directory. Gale Research, 835 Penobscot Building, Detroit, MI 48226-4094. (313) 961-2242. (800) 347-4253. FAX (313) 961-6083. 4th edition. 1996. $195.00.

Who's Who in Engineering. American Association of Engineering Societies, 1111 19th Street NW, Suite 608, Washington,

DC 20036. (202) 296-2237 or (800) 658-8897. 8th edition. 1991. Inquire for price.

ENCYCLOPEDIAS AND DICTIONARIES

McGraw-Hill Dictionary of Scientific and Technical Terms. Sybil P. Parker, ed. 5th edition. McGraw-Hill Publishing Company, 1221 Avenue of the Americas, New York, NY 10020. (800) 262-4729 or (212) 512-3825. 1993. $110.50.

McGraw-Hill Encyclopedia of Engineering. Sybil P. Parker, ed. 2nd edition. McGraw-Hill Publishing Company, 1221 Avenue of the Americas, New York, NY 10020. (800) 262-4729 or (212) 512-3825. 1993. $95.50.

McGraw-Hill Encyclopedia of Science and Technology. Sybil P. Parker, ed. 7th edition. 20 volumes. McGraw-Hill Publishing Company, 1221 Avenue of the Americas, New York, NY 10020. (800) 262-4729 or (212) 512-3825. 1992. $1900.00

Thesaurus of Scientific, Technical, and Engineering Terms. Hemisphere Publishing Corporation, 1900 Frost Road, Suite 101, Bristol, PA 19007-1598. (215) 785-5800 or (800) 821-8312. FAX (215) 785-5515. 1987. $173.00.

GENERAL WORKS

Aircraft Engines and Gas Turbines. Jack L. Kerrebrock. 2nd ed. MIT Press, 55 Hayward Street, Cambridge, MA 02142. (617) 625-8569 or (800) 356-0343. FAX (617) 253-1709. 1992. $45.00.

Elements of Gas Turbine Propulsion. Jack D. Mattingly. McGraw-Hill Publishing Company, 1221 Avenue of the Americas, New York, NY 10020. (800) 262-4729 or (212) 512-3825. 1996. Inquire for cost and availability.

Fundamentals of Gas Turbines. William W. Bathie. John Wiley and Sons, Inc., 605 Third Avenue, New York, NY 10158. (800) 526-5368 or (212) 850-6000. 1984. Inquire for cost and availability.

Steam and Gas Turbines. A. Stodola. 2 volumes. 6th edition. Peter Smith Publishing, 5 Lexington Avenue, Magnolia, MA 01930. (508) 525-3562. $48.00.

HANDBOOKS AND MANUALS

Gas Turbine Engineering Handbook. Meherwan P. Boyce. Gulf Publishing Company, P.O. Box 2608, Houston, TX. (713) 529-4301 or (800) 231-6275. 1982. $79.00.

Marks' Standard Handbook For Mechanical Engineers. E.A. Avallone and T. Baumeister III. 9th edition. McGraw-Hill Publishing Company, 1221 Avenue of the Americas, New York, NY 10020. (800) 262-4729 or (212) 512-3825. 1987. $133.00.

Machinery's Handbook. Erik Oberg, et al., editors. 24th edition. Industrial Press Inc., 200 Madison Avenue, New York, NY 10016. (212) 889-6330. FAX (212) 545-8327. 1992. $75.00.

Sawyer's Turbomachinery Maintenance Handbooks. John W. Sawyer & David Japikse, editors. 3rd edition. 3 volumes. Turbomachinery International Publications, Business Journals

TURBINES

Ency. of Physical Sciences and Engineering Info. Sources

Inc., Box 5550, Norwalk, CT 06856. (203) 853-6015. FAX (203) 853-8175. $248.50.

ONLINE DATABASES AND CD-ROMS

Compendex Plus. Engineering Information, Inc., 345 East 47th Street, New York, NY 10017. (212) 705-7600 or (800) 221-1044. Contains citations with abstracts to worldwide literature in engineering and technology, from 1970 to present. Available on online BRS,(800) 289-4277, DIALOG, (800) 334-2564, ORBIT (800) 456-7248, and STN International, FIZ Karlsruhe, P.O. Box 2465, W-7500, Karlsruhe 1, Germany, online services. Also available on CD-ROM. Inquire as to cost and availability.

CSA Engineering. Cambridge Scientific Abstracts, 7200 Wisconsin Avenue, Suite 601, Bethesda, MD 20814. (301) 961-6750 or (800) 843-7751. Contains citations and abstracts of international periodicals and research literature covering all fields of engineering and science and technology, including computer and information science, electronics, mechanical engineering, solid state materials, 1981 to present. Available on BRS,(800) 289-4277, online service. Inquire as to cost and availability.

Current Contents Search. Institute for Scientific Information, 3501 Market Street, Philadelphia, PA 19104. (215) 386-0100. FAX (215) 386-6362. Contains citations to articles listed in the table of contents of science and technology journals. Also articles in social sciences and life sciences journals. Available on BRS,(800) 289-4277, DIALOG,(800) 334-2564, online services. Inquire as to cost and availability.

International Aerospace Abstracts. American Institute of Aeronautics and Astronautics, 370 L'Enfant Promenade, S.W., Washington, DC 20024. (202) 646-7400. Contains references and abstracts of journal and monograph literature relating to aerospace science and technology, from 1963 to present. Available through the NASA/RECON system of the National Aeronautics and Space Administration only.

ISMEC: Mechanical Engineering Abstracts. Cambridge Scientific Abstracts, 7200 Wisconsin Avenue, Suite 601, Bethesda, MD 20814. (301) 961-6750 or (800) 843-7751. Contains citations to the literature in mechanical engineering, industrial and production engineering, energy, power, mechanics, devices and related areas, from 1973 to present. Available on the DIALOG,(800) 334-2564, online service. Inquire as to cost and availability.

NTIS Bibliographic Database. National Technical Information Service, 5285 Port Royal Road, Springfield, VA 22161. (703) 487-4929 or FAX (703) 321-8199. Broad coverage of government-sponsored science and technology research reports, 1964 to present. Available on BRS,(800) 289-4277, DIALOG, (800) 334-2564, ORBIT, (800) 456-7248, and STN International, FIZ Karlsruhe, P.O. Box 2465, W-7500, Karlsruhe 1, Germany, online services. Also available on CD-ROM. Inquire as to cost and availability.

SAE Global Mobility Database. Society of Automative Engineers (SAE), Electronic Publishing Division, 400 Commonwealth Drive, Warrendale, PA 15098. (412) 776-4841. Contains citations with abstracts to technical papers on automotive and aerospace technology and vehicular-related industries that have been presented at SAE conferences. Covers 1906 to present. Available on ORBIT,(800) 456-7248, online service. Inquire as to cost and availability.

WILSONLINE. H.W. Wilson Company, 950 University Avenue, Bronx, NY 10452. (800) 367-6770 or (212) 588-8400. Makes available online versions of the printed H.W. Wilson indexes, including Applied Science and Technology Index, Business Periodicals Index, General Science Index, and Readers' Guide to Periodical Literature. Period covered is generally 1983 to present. Available on BRS,(800) 289-4277, DIALOG,(800) 334-2564, and OCLC EPIC,(800) 848-5878, online services. Also available on CD-ROM. Inquire as to cost and availability.

PERIODICALS

Diesel and Gas Turbine Worldwide. Diesel and Gas Turbine Publications, 13555 Bishop's Court, Brookfield, WI 53005-6286. (914) 784-9177. FAX (914) 784-9177. 1969 to present. 10 times a year. $55.00 per year.

Institution of Diesel and Gas Turbine Engineers Transactions. Mechanical Engineering Publications Ltd., Northgate Avenue, Bury St. Edmunds, Suffolk IP32 6BW, England. Telephone 0284-763277. FAX 0284-704006. 6 times a year. Inquire for cost.

International Journal of Turbo and Jet Engines. Freund Publishing House, Ltd., PO Box 35010, Tel Aviv 63150, Israel. 972-2-5662925. FX 972-3-5605335. Quarterly. $190.00 per year.

Journal of Engineering For Gas Turbines and Power. American Society of Mechanical Engineers, 345 E. 47th Street, New York, NY 10017. (212) 705-7722. Quarterly. Inquire for cost.

Journal of Turbomachinery. American Society of Mechanical Engineers, 345 E. 47th Street, New York, NY 10017. (212) 705-7722. Quarterly. $100.00 per year for non-members.

Journal of Propulsion and Power. American Institute of Aeronautics and Astronautics, 370 L'Enfant Promenade SW, Washington, DC 20024-2518. Bimonthly. $38.00 (members) and $260.00 (non-members).

Turbomachinery International. Turbomachinery International Publications, Business Journals Inc., Box 5550, Norwalk, CT 06856. (203) 853-6015. FAX (203) 853-8175. 1959 to present. 7 times a year. $49.00.@ @Research Centers and Institutes

International Gas Turbine Institute, ASME. 5801 Peachtree Dunwoody Road, Suite 100, Atlanta, GA 30342-1503. (404) 847-0072.

Machinery and Engine Technology Laboratory. Institute for Mechanical Engineering, Bldg. M-7, Montreal Road, Ottawa, ON, Canada K1A 0R6. (613) 993-2425. FAX (613) 957-3281.

Sloan Automotive Laboratory, Massachusetts Institute of Technology. Room 3-340, Cambridge, MA 02139. (617) 253-2243.

Turbomachinery Laboratory. Texas A&M University, Mechanical Engineering Department, College Station, TX 77843. (409) 845-7417. FAX (409) 845-1835.

Virginia Polytechnic Institute Center For Turbomachinery and Propulsion Research. Department of Mechanical Engineering, Blacksburg, VA 24061-0238. (703) 231-6641. FAX (703) 231-7248.

U

ULTRALIGHT AIRCRAFT

See: AIRCRAFT

ULTASONICS

See: ACOUSTICS

ULTRAVIOLET ASTRONOMY

See: ASTRONOMY

ULTRAVIOLET SPECTROSCOPY

See: SPECTROSCOPY

UNCERTAINTY PRINCIPLE

See: QUANTUM MECHANICS

UNDERWATER PHOTOGRAPHY

See also: CAMERAS, PHOTOCHEMISTRY, PHOTOGRAPHIC FILM,
PHOTOGRAPHY

ABSTRACT SERVICES AND INDEXES

Applied Science and Technology Index; A Cumulative Subject
Index To English Language Periodicals in the Fields of Aero-
nautics and Space Science, Computer Technology, Chemistry,
Construction Industry, Energy and Related Areas. H.W. Wilson
Co., 950 University Avenue, Bronx, NY 10452. (800) 367-6770
or (212) 588-8400. FAX (718) 590-1617. From 1958 to present.
Monthly. Inquire about cost and availability. Also available on
CD-ROM and online.

*Chemical Abstracts - Applied Chemistry and Chemical Engi-
neering Sections.* Chemical Abstracts Service, 2540 Olentangy
River Road, Box 3012, Columbus, OH 43210. (800) 848-6538
or (614) 447-3600, FAX (614) 447-3713. Bi-weekly. $1410.00
per year.

Engineering Index Monthly. Engineering Information, Inc.,
Castle Point on the Hudson, Hoboken, NJ 07030. (800) 221-

1044. FAX (212) 832-1857. Monthly. $2300.00 per year. Also
available online as COMPENDEX, and also on CD-ROM. Covers chemical engineering, computers, electrical engineering, civil
engineering, metals and mining, industrial management, and
mechanical engineering.

*Imaging Abstracts (formerly Photographic Abstracts), Pergamon
Press Inc.*, Maxwell House, Fairview Park, Elmsford, NY 10523.
(914) 592-7700. Fax (914) 592-3625. Bimonthly. Inquire for
price.

ASSOCIATIONS AND PROFESSIONAL SOCIETIES

The International Society For Optical Engineering (spie), Po
Box 10, 1000-20th Street, Bellingham, WA 98227-0010.

Optical Society of America, 2010 Massachusetts Avenue Nw,
Washington, DC 20036. (202) 223-8130.

Photographic Society of America. 3000 United Founders Blvd.,
No.103, Oklahoma City, OK 73112-3940. (405) 843-1437. FAX
(405) 843-1438.

The Society For Imaging Science and Technology, 7003
Kilworth Lane, Springfield, VA 22151. (703) 642-9090.

Society of Motion Picture and Television Engineers, 595 W.
Hartsdale Avenue, White Plains, NY 10607.

Society of Photographic Scientists and Engineers. 7003 Kilworth
Lane, Springfield, VA 22151. (703) 642-9090.

DIRECTORIES AND BIOGRAPHICAL SOURCES

Directory of Engineering Societies and Related Organizations.
Gordon Davis. 13th edition. American Association of Engineering Societies, 1111 19th Street NW, Suite 608, Washington,

UNDERWATER PHOTOGRAPHY

Ency. of Physical Sciences and Engineering Info. Sources

DC 20036. (202) 296-2237 or (800) 658-8897. 1989. Inquire for price.

International Research Centers Directory. Gale Research, 835 Penobscot Building, Detroit, MI 48226-4094. (313) 961-2242. (800) 877-4253. FAX (313) 961-6083. 8th edition. 1995. $410.00.

PTN Master Buying Guide and Directory. PTN Publishing Corporation, 445 Broad Hollow Road, Suite 21, Melville, NY 11747-4722. (516) 845-2700. FAX (516) 845-7109. 1937 to present. Annual. Included with subscription to PTN (Photographic Trade News).

Research Centers Directory. Gale Research, 835 Penobscot Building, Detroit, MI 48226-4094. (313) 961-2242. (800) 877-4253. FAX (313) 961-6083. $485.00.

Scientific and Technical Organizations and Agencies Directory. Gale Research, 835 Penobscot Building, Detroit, MI 48226-4094. (313) 961-2242. (800) 877-4253. FAX (313) 961-6083. 4th edition. 1996. $195.00.

Who's Who in Engineering. American Association of Engineering Societies, 1111 19th Street NW, Suite 608, Washington, DC 20036. (202) 296-2237 or (800) 658-8897. 8th edition. 1991. Inquire for price.

ENCYCLOPEDIAS AND DICTIONARIES

The Focal Encyclopedia of Photography. Leslie Stroebel & Richard D. Zakia, editors. 3rd edition. Butterworth-Heinemann, 313 Washington Street, Newton, MA 02158. (617) 928-2500 or (800) 366-2665. FAX (617) 928-2620. 1993. $125.00.

Thesaurus of Photographic Science and Engineering. Society of Photographic Scientists and Engineers Staff. Books on Demand, 300 N. Zeeb Road, Ann Arbor, MI 48106-1346. (313) 761-4700 or (800) 521-0600. 1968. $34.30.

Thesaurus of Scientific, Technical, and Engineering Terms. Hemisphere Publishing Corporation, 1900 Frost Road, Suite 101, Bristol, PA 19007-1598. (215) 785-5800 or (800) 821-8312. FAX (215) 785-5515. 1987. $173.00.

GENERAL WORKS

Underwater Photography. Charles R. Seaborn. Watson-Guptill, 1515 Broadway, New York, NY 10036. (212) 536-5121. (800) 451-1741. FAX (212) 536-5359. 1988. $18.95.

Underwater Photography and Television For Scientists. J.D. George, G.I. Lythgoe, and J. N. Lythgoe, eds. Oxford University Press, Inc., 198 Madison Avenue, New York, NY 10016-4314. (212) 726-6000. FAX (212) 726-6446. 1985. $57.50.

Underwater Photography Camera Basic Equipment Care. Geri Murphy. PADI, Professional Association of Sports Educators, 1251 E. Dyer Road, No. 100, Santa Ana, CA 92705. (714) 540-7234 or (800) 729-7234. 1989. $9.95.

Underwater Photography: Scientific and Engineering Applications. Paul Ferris Smith, comp. Van Nostrand Reinhold, 115 Fifth Avenue, New York, NY 10003. (212) 254-3232. FAX (212) 254-9499. 1984. Inquire for cost and availability.

HANDBOOKS AND MANUALS

The Underwater Photographer's Handbook. Peter Rowlands. Van Nostrand Reinhold, 115 Fifth Avenue, New York, NY 10003. (212) 254-3232. FAX (212) 254-9499. 1983. Inquire for cost and availability.

Underwater Videographer's Handbook. Lynn Laymon. Amherst Media, 418 Homecrest Drive, Amherst, NY 14226. (716) 874-4450 or (800) 622-3278. FAX (716) 874-4508. 1992. $19.95.

ONLINE DATABASES AND CD-ROMS

Compendex Plus. Engineering Information, Inc., 345 East 47th Street, New York, NY 10017. (212) 705-7600 or (800) 221-1044. Contains citations with abstracts to worldwide literature in engineering and technology, from 1970 to present. Available on online BRS,(800) 289-4277, DIALOG, (800) 334-2564, ORBIT (800) 456-7248, and STN International, FIZ Karlsruhe, P.O. Box 2465, W-7500, Karlsruhe 1, Germany, online services. Also available on CD-ROM. Inquire as to cost and availability.

CSA Engineering. Cambridge Scientific Abstracts, 7200 Wisconsin Avenue, Suite 601, Bethesda, MD 20814. (301) 961-6750 or (800) 843-7751. Contains citations and abstracts of international periodicals and research literature covering all fields of engineering and science and technology,including computer and information science, electronics, mechanical engineering, solid state materials, 1981 to present. Available on BRS,(800) 289-4277, online service.Inquire as to cost and availability.

Current Contents Search. Institute for Scientific Information, 3501 Market Street, Philadelphia, PA 19104. (215) 386-0100. FAX (215) 386-6362. Contains citations to articles listed in the table of contents of science and technology journals. Also articles in social sciences and life sciences journals. Available on BRS,(800) 289-4277, DIALOG,(800) 334-2564, online services. Inquire as to cost and availability.

NTIS Bibliographic Database. National Technical Information Service, 5285 Port Royal Road, Springfield, VA 22161. (703) 487-4929 or FAX (703) 321-8199. Broad coverage of government-sponsored science and technology research reports, 1964 to present. Available on BRS,(800) 289-4277, DIALOG, (800) 334-2564, ORBIT, (800) 456-7248, and STN International, FIZ Karlsruhe, P.O. Box 2465, W-7500, Karlsruhe 1, Germany, online services. Also available on CD-ROM. Inquire as to cost and availability.

SCISEARCH. Institute for Scientific Information, 3501 Market Street, Philadelphia, PA 19104. (800) 523-1850 or (215) 386-0100. Broad multidisciplinary title and author index to the international literature of science and technology, 1974 to present. Available on DIALOG,(800) 334-2564, and ORBIT,(800) 456-7248, online services. Also available on CD-ROM. Inquire as to cost and availability.

WILSONLINE. H.W. Wilson Company, 950 University Avenue, Bronx, NY 10452. (800) 367-6770 or (212) 588-8400. Makes available online versions of the printed H.W. Wilson indexes including Applied Science and Technology Index, Business Periodicals Index, General Science Index, and Readers' Guide to Periodical Literature. Period covered is generally 1983 to present. Available on BRS,(800) 289-4277, DIALOG,(800) 334-2564, and OCLC EPIC,(800) 848-5878, online services. Also available on CD-ROM. Inquire as to cost and availability.

PERIODICALS

British Journal of Photography. Bouverie Publishing Company Ltd., 147-151 Temple Chambers, Temple Avenue, London EC47 0DT, England. Telephone 071-583-3030. FAX 071-583-4068. 1854 to present. Weekly. Inquire for price and availability.

Industrial Photography. PTN Publishing Corporation, 445 Broad Hollow Road, Melville, NY 11747-4722. (516) 845-2700. FAX (516) 845-7109. 1952 to present. Monthly. $60.00 per year.

Journal of Imaging Science and Technology. Society for Imaging Science and Technology, 7003 Kilworth Lane, Springfield, VA 22151. (703) 642-9090. FAX (703) 642-9094. 1950 to present. Bi-monthly. $120.00 per year.

Journal of Photographic Science. The Barn, Whitehall, Near Middle Marwood, Barnstaple, N. Devon EX31 4EQ, England. Telephone 0271-72482. FAX 0271-72482. 1952 to present. Bi-monthly. Inquire for price and availability.

Photogrammetric Engineering and Remote Sensing. American Society for Photogrammetry and Remote Sensing, 5410 Grosvenor Lane, Suite 210, Bethesda, MD 20814. (301) 493-0290. FAX (301) 493-0208. 1934-present. Monthly. $160.00 per year.

PSA Journal. Photographic Society of America, 3000 United Founders Blvd., No.103, Oklahoma City, OK 73112-3940. (405) 843-1437. FAX (405) 843-1438. 1934 to present. Monthly. Included with $35.00 annual membership.

PTN (Photographic Trade News). PTN Publishing Corporation, 445 Broad Hollow Road, Suite 21, Melville, NY 11747-4722. (516) 845-2700. FAX (516) 845-7109. 1937 to present. 24 times a year. $25.00 per year.

Scientific and Applied Photography and Cinematography. Gordon and Breach Science Publishers, 820 Town Center Drive, Langhorne, PA 19047. (215) 750-2642. FAX (215) 750-6343. 6 times a year. Inquire for price.

UNIX

See: COMPUTER OPERATING SYSTEMS

UPPER ATMOSPHERE

See: METEOROLOGY

URANUS

See also: NATURAL SATELLITES, PLANETARY SCIENCES, SOLAR SYSTEM

ABSTRACT SERVICES AND INDEXES

Applied Science and Technology Index; A Cumulative Subject Index To English Language Periodicals in the Fields of Aeronautics and Space Science, Computer Technology, Chemistry, Construction Industry, Energy and Related Areas. H. W. Wil-

son Co., 950 University Avenue, Bronx, NY 10452- 9978. (212) 588-8400. (800) 367-6770. FAX (718) 590-1617.From 1958 to present. Monthly. Inquire about cost and availability. Also available online (BRS and WILSONLINE) and on CD-ROM. ISSN: 0003-6986.

Astronomy and Astrophysics Abstracts. Springer-Verlag New York, 175 Fifth Avenue, New York, NY 10010. (212) 460-1500.FAX (212) 473-6272. Published for the Astronomisches Rechen-Institut. Comprehensive coverage of all aspects of astronomy, astrophysics and related fields. 1969 to present. Two parts per year. Annual. ISSN: 0067-0022.

General Science Index. H.W. Wilson Company, 950 University Avenue, Bronx, NY 10452. (212) 588-8400. (800) 367-6770. FAX (718) 590-1617. From 1978 to present. Ten issues per year; quarterly and annual cumulations. Service basis. Available on CD-ROM and online. Inquire regarding cost andavailability. ISSN: 0162-1963.

Meteorological and Geoastrophysical Abstracts. American Meteorological Society, c/o Inforonics, Inc., 550 NewtownRoad, Box 458, Littleton, MA 01460. (508) 486-8976. FAX (508) 486-0027. Covers literature in environmental sciences, meteorology, astrophsyics, hydrology, glaciology and physical oceanography. 1950 to present. Monthly. $950.00 per year. Also available online (DIALOG) and onCD-ROM. ISSN: 0026-1130.

NTIS Alerts: Astronomy & Astrophysics. U. S. National Technical Information Service. 5285 Port Royal Road,Springfield, VA 22161. (703) 487-4650. FAX (703) 321-8547.Weekly. $140.00 per year.

Physics Abstracts. INSPEC. Section A, Science Abstracts. Institute of Electrical Engineers, London, United Kingdom.Available from: INSPEC/IEEE - Institute of Electrical andElectronic Engineers, Box 1331, Hoes Lane, Piscataway, NJ08855-1331. (908) 562-5549. 1898 to present. 24 issuesper year. $2835.00 per year. ISSN: 0036-8091. Also available online and on CD-ROM.

Science Citation Index. SCI. Institute for ScientificInformation, 3501 Market Street, Philadelphia, PA 19104.(215) 386-0100. (800) 523-1850. FAX (215) 386-2991. 1961to present. Six issues per year, plus annual cumulation.$11650.00 per year. Also available online and on CD-ROM.Inquire about price and availability. ISSN: 0036-827X.

STAR. Scientific and Technical Aerospace Reports. U.S. National Aeronautics and Space Administration. Distributed by U. S. Superintendent of Documents, Washington, DC 20402. 1963 to present. Semi-monthly, with semiannual and annual indexes. $114.00 per year. ISSN: 0036-8741. Also available online and on CD-ROM.

ANNUAL REVIEWS AND YEARBOOKS

The Astronomical Almanac. Superintendent of Documents, U. S. Government Printing office, Washington, DC 10402. (202) 783-3238. 1981 to present. Superceeds Astronomical Ephemeris. Annual. $44.95. ISSN: 0737-6421.

Annual Review of Astronomy and Astrophysics. Original reviews of critical literature and current developments in astronomy and astrophysics. Annual Reviews, Inc., 4139 El Camino Way, Palo Alto, CA 94303-0139. (415) 493-4400. (800)523-8635. Fax (415) 855-9815. 1963 to date. Annual. $60.00. ISSN: 0066-4146.

URANUS

Ency. of Physical Sciences and Engineering Info. Sources

Annual Review of Earth and Planetary Sciences. Annual Reviews, Inc., 4139 El Camino Way, Palo Alto, CA 94303-0139. (415) 493-4400. (800) 523-8635. Fax (415) 855-9815. 1973 to date. Annual. $62.00. ISSN: 0084-6597.

ASSOCIATIONS AND PROFESSIONAL SOCIETIES

American Astronomical Society. 2000 Florida Avenue NW, Suite 400, Washington, DC 20009. (202) 328-2010. FAX (202) 234-2560.

American Geophysical Union. 2000 Florida Avenue NW, Washington, DC 20009. (202) 462-6900. (800) 966-AGU1. FAX (202) 328-0566.

American Institute of Physics. 1 Physics Ellipse, College Park, MD 20740-3843. (301) 209-3100.

Association of Universities For Research in Astronomy, Inc. (AURA). Suite 701, 1625 Massachusetts Avenue, NW, Washington, DC 20036. (202) 483-2101. FAX (202) 483-2106.

Astronomical Society of the Pacific. 390 Ashton Avenue, San Francisco, CA 94112. (415) 337-1100. FAX: (415) 337-5205.

Planetary Society. 65 North Catalina Avenue, Pasadena, CA 91106. (818) 793-5100. (800) WOW-MARS. FAX (818) 793-5528.

BIBLIOGRAPHIES

A Bibliography of Astronomy, 1970 - 1979. R. A. Sea and S. S. Martin. Libraries Unlimited, Inc., Littleton, CO 80160. 1982. $37.50.

DIRECTORIES AND BIOGRAPHICAL SOURCES

American Astronomical Society. Membership Directory. American Astronomical Society. 2000 Florida Avenue NW, Suite 400, Washington, DC 20009. (202) 328-2010. FAX (202)234-2560 Annual. Included in membership dues. ISSN: 1061- 9038.

American Men and Women of Science: Physical and Biological Sciences. R. R. Bowker Inc., 121 Chanlon Road, New Providence, NJ 07974. (908) 464-6800. (800) 521-8110. 20th edition. 8 volumes. 1996. $850.00.

The Astronomers. Donald Goldsmith. St. Martin's Press, Inc., 175 Fifth Avenue, New York, NY 10010. (212) 674-5151. (800) 221-7945. 1993. $14.95 in paper.

Astronomical Centers of the World. Kevin Krisciunas. Cambridge University Press, 40 West 20th Street, New York, NY 10011-4211. (212) 924-3900. (800) 872-7423. 1988. $34.95

Directory of Physics and Astronomy Staff Members. American Institute of Physics, One Physics Ellipse, College Park, MD 20740-3843. (301) 209-3100. 1975/76 to present. Annual. $60.00. ISSN: 0361-2228.

Earth and Astronomical Research Centers. Stockton Press, 345 Park Avenue South, New York, NY 10010. 4th edition. 1995. $515.00. ISBN: 1-56169-0967.

ENCYCLOPEDIAS AND DICTIONARIES

Concise Dictionary of Astronomy. Jacqueline Mitton. Oxford University Press, Inc., 200 Madison Avenue, New York, NY 10016. (212) 725-6000. (800) 334-4249. 1992. $24.95.

Encyclopedia of Astronomy and Astrophysics. Stephen Maran, editor. Van Nostrand Reinhold, 115 Fifth Avenue, New York, NY 10003. (212) 254-3232. (800) 842-3636. 1992. $129.95.

McGraw-Hill Encyclopedia of Science and Technology. McGraw-Hill Book Company, Inc., 1221 Avenue of the Americas, New York, NY 10020. (212) 512-2000. (800) 262-4729. 7th edition. 20 volume set. 1992. $1900.00.

Moons of the Solar System: An Illustrated Encyclopedia. John Stewart. McFarland & Company, Inc., Box 611, Jefferson, MC 28640. (910) 246-4460. (800) 253-2187. 1991. $49.95.

New Guide To the Planets. Patrick Moore. Trans-Atlantic Publications, Inc., 311 Bainbridge Street, Philadelphia, PA 19147. (215) 925-5083. 1993. $37.50.

Stars and Planets: The Sierra Club Guide To Sky Watching and Direction Finding. W. S. Kals. The Sierra Press, 4988 Gold Leaf Drive, Mariposa, CA 95338. (209) 966-5071. (800) 745-2631. $15.00.

GENERAL WORKS

A Distant Puzzle: The Planet Uranus. Isaac Asimov, et al. Gareth Stevens, Inc., River Center Building, 1555 North River Center Drive, Milwaukee, WI 53212. (414) 225-0333. (800) 341-3569. 1994. $18.60.

Evolution of the Earth and Planets. E. Takahashi, et al, editors. American Geophysical Union, 2000 Florida Avenue, N.W., Washington, DC 20009. (202) 462-6903. (800) 966-2481. 1993. $28.00.

Exploring the Planets. W. Kenneth Hamblin and Eric H. Christiansen. Macmillan Publishing Company, Inc,. 200 Old Tappan Road, Old Tappan, NJ 07675. (800) 233-2336. 2nd edition. 1995. $62.00.

Moons and Planets. William K. Hartmann. Wadsworth Publishing Co., 10 Davis Drive, Belmont, CA 94002. (415) 595-2350. (800) 354-9706. 3rd edition. 1993. $49.95.

Planetary Landscapes. Ronald Greeley. Chapman & Hall, 1 Penn Plaza, New York, NY 10119. (212) 564-1060. Div. of Routledge,.. 2nd edition. 1994. $49.95.

The Planetary System. Tobias Owen and David Morrison. Addison-Wesley Publishing Co., Inc., 1 Jacob Way, Reading, MA 01867. (617) 944-3700. (800) 447-2226. 1988. $49.50.

Space Technology and Planetary Astronomy. Joseph N. Tatarewicz. Indiana University Press, 601 North Morton Street, Bloomington, in 47404-3797. (812) 855-6804. (800) 842-6796. 1990. $29.95.

Uranus. Jay T. Bergstralh, et al, editors. University of Arizona Press, 1230 North Park Avenue, Number 102, Tucson, AZ 85719. (520) 621-1441. (800) 426-3797. 1991. $75.00.

Uranus and Neptune: the Distant Giants. Eric Burgess. Columbia University Press, 562 West 113th Street, New York, NY 10025. (212) 666-1000. (800) 944-8648. 1988. $40.00.

Worlds Apart: A Textbook in Planetary Sciences. Guy Consolmagno. Prentice-Hall, 113 Sylvan Avenue, Route 9W, Englewood Cliffs, NJ 07632. (201) 592-2000. (800) 922- 0579. 1994. $62.00.

HANDBOOKS AND MANUALS

Astrophysical Data: Planets and Stars. Kenneth R. Lanb. Springer-Verlag New York, Inc., 175 Fifth Avenue, New York, NY 10010. (212) 460-1500. (800) 777-4643. 1993. $59.00.

Field Guide to the Stars and Planets. Jay M. Pasachoff and Donald H. Menzel. Houghton Mifflin Co., 222 Berkeley Street, Boston, MA 02116. (617) 351-5000. (800) 225-3362. Revised edition. 1992. $24.95.

New Guide To the Planets. Patrick Moore. Trans-Atlantic Publications, Inc., 311 Bainbridge Street, Philadelphia, PA 19147. (215) 925-5083. 1993. $37.50.

Planet Observer's Handbook. Fred W. Price. Cambridge University Press, 40 West 20th Street, New York, NY 10011- 4211. (212) 924-3900. (800) 872-7423. 1994. $34.95.

ONLINE DATABASES AND CD-ROMS

Dissertation Abstracts. University Microfilms International, 300 North Zeeb Road, Ann Arbor, MI 48106. (800) 521-0600 or (313) 761-4700. Scope includes virtually all doctoral dissertations accepted at accredited American institutions from 1861 to present in 252 subject areas. Available on BRS, (800) 289-4277, DIALOG, (800) 334-2564,and OCLC EPIC, (800) 848-5878, online services. Also available on CD-ROM. Inquire as to cost and availability.

GEOREF. American Geological Institute, 4220 King Street, Alexandria, VA 22302. (800) 336-4764 or (703) 379-2480. Geology and geosciences literature, 1785 to present for North America. Available on DIALOG,(800) 334-2564, ORBIT (800) 456-7248, online services. Also available on CD-ROM. Inquire as to cost and availability.

INSPEC. Institution of Electrical Engineers, Michael Faraday House, Six Hills Way, Stevenage, Herts. SG1 2AY, England. Telephone: 0438 313311 or FAX 0438 742840. Contains citations to the worldwide literature of physics, electronics and electrical engineering, computer technology, and related fields. Available on BRS, (800) 289-4277, DIALOG, (800) 334-2564, ORBIT, (800) 456-7248, and STN International, FIZ Karlsruhe, P.O. Box 2465, W-7500, Karlsruhe 1, Germany, online services. Inquire as to cost and availability.

NTIS Bibliographic Database. National Technical Information Service, 5285 Port Royal Road, Springfield, VA 22161. (703) 487-4929 or FAX (703) 321-8199. Broad coverage of government-sponsored science and technology research reports, 1964 to present. Available on BRS,(800) 289-4277, DIALOG, (800) 334-2564, ORBIT, (800) 456-7248, and STN International, FIZ Karlsruhe, P.O. Box 2465, W-7500, Karlsruhe 1, Germany, online services. Also available on CD-ROM. Inquire as to cost and availability.

SCISEARCH. Institute for Scientific Information, 3501 Market Street, Philadelphia, PA 19104. (800) 523-1850 or (215) 386-0100. Broad multidisciplinary title and author index to the international literature of science and technology, 1974 to present. Available on DIALOG, (800) 334-2564, and ORBIT, (800) 456-7248, online services. Also available on CD-ROM. Inquire as to cost and availability.

WILSONLINE. H.W. Wilson Company, 950 University Avenue, Bronx, NY 10452. (800) 367-6770 or (212) 588-8400. Makes available online versions of the printed H.W. Wilson indexes including Applied Science and Technology Index, Business Periodicals Index, General Science Index, and Readers' Guide to Periodical Literature. Period covered is generally 1983 to present. Available on BRS, (800) 289-4277, DIALOG, (800) 334-2564, and OCLC EPIC, (800) 848-5878, online services.Also available on CD-ROM. Inquire as to cost and availability.

OTHER SOURCES

Atlas of the Solar System. B. Yenne. Simon & Schuster, Inc. 1230 Avenue of the Americas, New York, NY 10020. (212) 698-7000. (800) 223-2348. 1987. $12.98.

Atlas of Uranus. Garry E. Hunt and Patrick Moore. Cambridge University Press, 40 West 20th Street, New York, NY 10011-4211. (212) 924-3900. (800) 872-7423. 1989. $34.95.

The Sky: A User's Guide. David H. Levy. Cambridge University Press, 40 West 20th Street, New York, NY 10011- 4211. (212) 924-3900. (800) 872-7423. 1993. $14.95.

The View from Space: Photographic Exploration of the Planets. Merton Davies and Bruce C. Murray. Columbia University Press, 562 West 113th Street, New York, NY 10025. (212) 666-1000. (800) 944-8648. 1973. $17.50.

PERIODICALS

Association of Lunar and Planetary Observers. Journal. Association of Lunar and Planetary Observers, Box 16131, San Francisco, CA 94116. (415) 566-5786. FAX (415) 731-8242. 1947 to present. Quarterly. $14.00 per year. ISSN: 0039- 2502.

Astronomical Journal. American Institute of Physics, One Physics Ellipse, College Park, MD 20740-3843. (301) 209- 3000. Published for the American Astronomical Society. 1849 to present. Monthly. $280.00 per year. ISSN: 0004- 6256.

Astronomical Society of the Pacific. Publications. Astronomical Society of the Pacific, 390 Ashton Avenue, San Francisco, CA 94112. (415) 337-1100. FAX (415) 337-5205.1889 to present. Monthly. $175.00 per year. ISSN: 0004- 6280.

Astronomy. Kalmbach Publishing Company, Box 1612, Waukesha, WI 53187-1612. (414) 796-8776. FAX (414) 796-1142. 1973 to present. Monthly. $27.00 per year. ISSN: 0091-6358.

Astronomy and Astrophysics. Springer-Verlag New York, Inc., 44 Hartz Way, Secaucus, NJ 07096-2491. (201) 348-4033. FAX(201) 348-4505. 1969 to date. 36 issues per year in 12 volumes. $1733.00 per year. ISSN: 0004-6361.

Earth, Moon and Planets; An International Journal of Comparative Planetology. Kluwer Academic Publishers, Box 358,

Accord Station, Hingham, MA 02018-0358. (617) 871-6600. FAX (617) 871-6528. 1969 to present. Monthly. $840.00 per year. ISSN: 0167-9295.

Geochemica Et Cosmochimica Acta. Elsevier Science. 660 White Plains Road, Tarrytown, NY 10591-5153. (914) 524-9200. FAX (914) 333-2444. 1950 to the present. Biweekly. $895.00 per year. ISSN: 0016-7037.

Icarus; International Journal of Solar System Studies. Academic Press, Inc., Journal Division, 525 B Street, Suite 1900, San Diego, CA 92101-4495. (619) 230-1840. FAX (619) 699-6800. 1962 to the present. Monthly. $1080.00. ISSN: 0019-1035.

JGR: Journal of Geophysical Research: Planets. American Geophysical Union, 2000 Florida Avenue, NW, Washington, CD 20009. (202) 462-6900. FAX (202) 328-0566. 1991 to present. Monthly. $597.00 per year. ISSN: 0148-0227.

Lunar and Planetary Information Bulletin. Lunar and Planetary Institute. 3600 Bay Area Boulevard, Houston, TX 77058-1113. (713) 486-2175. FAX (713) 486-2125. 1970 to present. Quarterly. Free. Also available online.

Mercury. Astronomical Society of the Pacific. 390 Ashton Avenue, San Francisco, CA 94112. (415) 337-1100. FAX: (415) 337-5205. 1972 to present. Bimonthly. $175.00 per year. ISSN: 0047-6773.

Planetary and Space Science. Elsevier Science Publishing Company, Inc., 660 White Plains Road, Terrytown, NY 10591-5153. (914) 524-9200. FAX: (914) 333-2444. 1959 to present. Monthly. $1355.00 per year. ISSN: 0032-0633.

Planetary Report. Planetary Society. 65 North Catalina, Pasadena, CA 91106-2301. (818) 793-5100. FAX (818) 793-5528. 1980 to present. $25.00 per year. ISSN: 0736-3680.

Sky and Telescope. Sky Publishing Corporation, Box 9111, Belmont, MA 02178. (617) 864-7360. FAX (617) 864-6117. 1941 to present. Monthly. $27.00 per year. ISSN: 0037-6604.

RESEARCH CENTERS AND INSTITUTES.

Center For Space Plasma and Aeronomic Research. University of Alabama in Huntsville, Huntsville, AL 35899. (205) 895-6258. (205) 805-6382.

Center For Space Science and Astrophysics. Stanford University. 325 Durand Building, Stanford, CA 04305. (415) 723-3582.

Center For Space Sciences. University of Texas at Dallas, P. O. Box 830688, MS_F022, Richardson, TX 75083-0688. (214) 690-2851. FAX (214) 690-2848.

Institute For Astronomy. University of Hawaii at Manoa, 2680 Woodlawn Drive, Honolulu, HI 96822. (808) 956-8312. FAX (808) 988-2790.

Laboratory For Planetary Studies. Cornell University, Space Sciences Building, Ithaca, NY 14853. (607) 255-4971.

Lunar and Planetary Institute. 3303 NASA Road One, Houston, TX 77058-4399. (713) 486-2139. FAX 713-496-2162.

Lunar and Planetary Laboratory. University of Arizona, Tucson, AZ 85721. (602) 621-6963. FAX (602) 621-4933.

V

VACUUM METALLURGY

See: METALLURGY

VENTILATION

See: AIR CONDITIONING

VENUS

See also: ASTROGEOLOGY, PLANETARY SCIENCES, SATELLITES
(NATURAL), SOLAR SYSTEM

ABSTRACT SERVICES AND INDEXES

Applied Science and Technology Index; A Cumulative Subject
Index To English Language Periodicals in the Fields of aero-
nautics and Space Science, Computer Technology,chemistry,
Construction Industry, Energy and Related Areas.H.W. Wilson
Co., 950 University Avenue, Bronx, NY 10452. (212) 588-8400.
(800) 367-6770. FAX (718) 590-1617. From 1958 to present.
Monthly. Inquire about cost andavailability. Also available on
CD-ROM and online. ISSN:0003-6986.

Astronomy and Astrophysics Abstracts. Springer-Verlag New
York, 175 Fifth Avenue, New York, NY 10010. (212) 460-
1500.FAX (212) 473-6272. Published for the
AstronomischesRechen-Institut. Comprehensive coverage of all
aspects ofastronomy, astrophysics and related fields. 1969
topresent. Two parts per year. Annual. ISSN: 0067-0022.

Bibliography and Index of Geology. American Geological Institute, 4220 King Street, Alexandria, VA 22302-1507.(703)
379-2480. FAX (703) 379-7563. From 1969 to present.Monthly.
$1295.00 per year. ISSN: 0098-2784. Also available as GEOREF
online (CISTI, DIALOG, Orbit, STN) and on CD-ROM. Inquire about price and availability.

Chemical Abstracts. Chemical Abstracts Service, 2540Olentangy
River Road, Box 3012, Columbus, OH 43210-0012.(614) 447-
3600. (800) 848-6538. FAX (614) 447-3713.From 1907 to
present. Weekly. $16800.00. Available onlineand on CD-ROM.
CA is also available in five sectiongroupings. Inquire regarding
cost and availability.

General Science Index. H.W. Wilson Company, 950 University
Avenue, Bronx, NY 10452. (212) 588-8400. (800) 367-
6770.FAX (718) 590-1617. From 1978 to present. Ten issues
peryear; quarterly and annual cumulations. Service
basis.Available on CD-ROM and online. Inquire regarding cost
andavailability. ISSN: 0162-1963.

Government Reports Announcements and Index. U. S.
NationalTechnical Information Service (NTIS), 5285 Port Royal
Road, Springfield, VA 22161. (703) 487-4650. FAX (703) 321-
8547.From 1968 to present. Annual. $630.00 per year. Also
available online as NTIS Bibliographic Database and on CD-
ROM.

Meteorological and Geoastrophysical Abstracts. American
Meteorological Society, c/o Inforonics, Inc., 550 NewtownRoad,
Littleton, MA 01460. (508) 486-8976. FAX (508) 486-0027.
Covers literature in environmental sciences,meteorology, astro-
physics, hydrology, glaciology, andphysical oceanography. From
1950 to present. Monthly. $950.00 per year. Also available on
CD-ROM and online.Inquire regarding cost and availability.
ISSN: 0026-1130.

NTIS Alerts: Astronomy & Astrophysics. U. S. National Technical Information Service. 5285 Port Royal Road,Springfield, VA
22161. (703) 487-4650. FAX (703) 321-8547.Weekly. $140.00
per year. <bodylead>*Pascal 48: Environnement Cosmique
Terrestre, Astronomie Et Geologie Extraterrestre.* Centre National de la Recherche Scientifique, Institut de Information
Scientifique etTechnique, 2 aliee du Parc de Brabois, 54514
VandoeuvreLesNancy Cedex, France. TEL 83-50-46-00. FAX:
83-50-46-50. 1985 to present. Ten issuesper year. 770 F.
Alsoavailable on CD-ROM and online.

Physics Abstracts. INSPEC. Section A, Science Abstracts. Institute of Electrical Engineers, London, United Kingdom. Available from: INSPEC/IEEE - Institute of Electrical and Electronic
Engineers, Box 1331, Hoes Lane, Piscataway, NJ 08855-1331.
(908) 562-5549. 1898 to present. 24 issues per year. $2835.00
per year. ISSN: 0036-8091. Also available online and on CD-
ROM.

Physics Briefs (Physikalische Berichte). Information Centerfor
Energy, Physics, Mathematics; German Physical Society. V C
H Publishers, Inc., 220 East 23rd Street, New York, NY 10010-
4606. (212) 683-8333. 1845 to present. 24 issues per year.
$2390.00 per year. Also available online. ISSN: 0179-7434.

VENUS

Ency. of Physical Sciences and Engineering Info. Sources

Science Citation Index. SCI. Institute for ScientificInformation, 3501 Market Street, Philadelphia, PA 19104. (215) 386-0100. (800) 523-1850. FAX (215) 386-2991. 1961to present. Six issues per year, plus annual cumulation. $11650.00 per year. Also available online and on CD-ROM.Inquire about price and availability. ISSN: 0036-827X.

STAR. Scientific and Technical Aerospace Reports. U.S. National Aeronautics and Space Administration. Distributed by U. S. Superintendent of Documents, Washington, DC 20402. 1963 to present. Semi-monthly, with semiannual and annual indexes. $114.00 per year. ISSN: 0036-8741. Also available online and on CD-ROM.

ANNUAL REVIEWS AND YEARBOOKS

The Astronomical Almanac. Superintendent of Documents, U. S. Government Printing office, Washington, DC 10402. (202) 783-3238. 1981 to present. Superceeds Astronomical Ephemeris and the American Ephermis and Nautical Almanac. Annual. $44.95. ISSN: 0737-6421.

Annual Review of Astronomy and Astrophysics. Original reviews of critical literature and current developments in astronomy and astrophysics. Annual Reviews, Inc., 4139 El Camino Way, Palo Alto, CA 94303-0139. (415) 493-4400. (800) 523-8635. Fax (415) 855-9815. 1963 to date. Annual. $60.00. ISSN: 0066-4146.

Annual Review of Earth and Planetary Sciences. Annual Reviews, Inc., 4139 El Camino Way, Palo Alto, CA 94303-0139. (415) 493-4400. (800) 523-8635. Fax (415) 855-9815. 1973 to date. Annual. $62.00. ISSN: 0084-6597.

Proceedings of the Lunar and Planetary Science Conference. Lunar and Planetary Institute, 3600 Bay Area Boulevard, Houston, TX 77058. (713) 486-2143. 1970 to date. Annual. ISSN: 0270-9511.

ASSOCIATIONS AND PROFESSIONAL SOCIETIES

American Astronomical Society. 2000 Florida Avenue NW, Suite 400, Washington, DC 20009. (202) 328-2010. FAX\;(202) 234-2560.

American Geophysical Union. 2000 Florida Avenue NW, Washington, DC 20009. (202) 462-6900. (800) 966-AGU1. FAX (202) 328-0566.

American Institute of Physics. 1 Physics Ellipse, College Park, MD 20740-3843. (301) 209-3100.

Astronomical League. 2112 Kingfisher Lane East, Rolling Meadows, IL 60008. (708) 398-0562.

Association of Universities For Research in Astronomy, Inc. (AURA). Suite 701, 1625 Massachusetts Avenue, NW, Washington, DC 20036. (202) 483-2101. FAX (202) 483-2106.

Astronomical Society of the Pacific. 390 Ashton Avenue, San Francisco, CA 94112. (415) 337-1100. FAX: (415) 337-5205. Geological Society of America. 3300 Penrose Place, PO Box 9140. Boulder, CO 80301-9140. (303) 447-2020. FAX (303) 447-1133.

IEEE Geoscience and Remote Sensing Society. c/o Institute of Electrical and Electronics Engineers. 345 East 47th Street, New York, NY 10017. (212) 705-7900. FAX (212) 705-4929.

Meteoritical Society. University of Massachusetts, 125 Marston Hall, Amherst, MA 01003. (413) 545-0300. (413) 545-0724.

Planetary Society. 65 North Catalina Avenue, Pasadena, CA 91106. (818) 793-5100. (800) WOW-MARS. FAX (818) 793-5528.

BIBLIOGRAPHIES

Astronomy and Astrophysics: A Source Guide. Gordon Press, P.O. Box 459, Bowling Green Station, New York, NY 10004. (718) 624-8419. 1991. $250.00. .

A Bibliography of Astronomy, 1970-1979. R. A. Sea and S. S. Martin. Libraries Unlimited, Inc., Littleton, CO 80160. 1982.

Catalog of the Naval Observatory Library. Naval Observatory Staff. G. K. Hall & Company, 866 Third Avenue, New York, NY 10022. (212)702-6789. (800) 257-5755. 6 volumes. 1977. $655.00

Science Books and Films. American Association for the Advancement of Science, 1333 H Street NW, Washington, DC 20005. (202) 326-6454. 1965 to present. Nine issues per year. $40.00 per year. ISSN: 0098-342X

Scientific and Technical Books and Serials in Print, 1995. R.R. Bowker Inc., 121 Chanlon Road, New Providence, NJ 07974. (908) 464-6800. (800) 521-8110. 4 volumes. 1994. $299.95. Also available on compact disc and online.

A Source Book On Astronomy and Astrophysics, 1900 - 1975. Kenneth Lang and Owen Gingerich, editors. Harvard University Press, 79 Garden Street, Cambridge MA. (617) 495-2600. 1980. $41.95.

DIRECTORIES AND BIOGRAPHICAL SOURCES

American Astronomical Society. Membership Directory. American Astronomical Society. 2000 Florida Avenue NW, Suite 400, Washington, DC 20009. (202) 328-2010. FAX (202) 234-2560 Annual. Included in membership dues. ISSN: 1061- 9038.

American Men and Women of Science: Physical and Biological Sciences. R. R. Bowker Inc., 121 Chanlon Road, New Providence, NJ 07974. (908) 464-6800. (800) 521-8110. 20th edition. 8 volumes. 1996. $850.00.

The Astronomers. Donald Goldsmith. St. Martin's Press, Inc., 175 Fifth Avenue, New York, NY 10010. (212) 674-5151. (800) 221-7945. 1993. $14.95 in paper.

Astronomical Centers of the World. Kevin Krisciunas. Cambridge University Press, 40 West 20th Street, New York, NY 10011-4211. (212) 924-3900. (800) 872-7423. 1988. $34.95

The Biographical Dictionary of Scientists: Astronomers. D. Abbott, editor. Peter Bedrick Books, Inc., 2112 Broadway, Room 318, New York, NY 10023. (212) 496-0751. 1984.

Directory of Physics and Astronomy Staff Members. American Institute of Physics, One Physics Ellipse, College Park, MD 20740-3843. (301) 209-3100. Annual. $45.00.

Earth and Astronomical Research Centers. Stockton Press, 345 Park Avenue South, New York, NY 10010. 4th edition. 1995. $515.00. ISBN: 1-56169-0967.

Graduate Programs in Physics, Astronomy and Related Fields. 1995-1996. American Institute of Physics, One Physics Ellipse, College Park, MD 20740-3843. (301) 209-3100. 1978 to present. Annual. $45.00.

Research Centers Directory. Gale Research, 835 Penobscot Building, Detroit, MI 48226-4094. (313) 961-2242. (800) 877-4253. 20th edition. 1995. $485.00.

ENCYCLOPEDIAS AND DICTIONARIES

Concise Dictionary of Astronomy. Jacqueline Mitton. Oxford University Press, Inc., 200 Madison Avenue, New York, NY 10016. (212) 725-6000. (800) 334-4249. 1992. $24.95.

Encyclopedia of Astronomy and Astrophysics. Stephen Maran, editor. Van Nostrand Reinhold, 115 Fifth Avenue, New York, NY 10003. (212) 254-3232. (800) 842-3636. 1992. $129.95.

McGraw-Hill Encyclopedia of Science and Technology. McGraw-Hill Book Company, Inc., 1221 Avenue of the Americas, New York, NY 10020. (212) 512-2000. (800) 262-4729. 7th edition. 20 volume set. 1992. $1900.00.

Moons of the Solar System: An Illustrated Encyclopedia. John Stewart. McFarland & Company, Inc., Box 611, Jefferson, MC 28640. (910) 246-4460. (800) 253-2187. 1991. $49.95.

New Guide To the Planets. Patrick Moore. Trans-Atlantic Publications, Inc., 311 Bainbridge Street, Philadelphia, PA 19147. (215) 925-5083. 1993. $37.50.

Stars and Planets: The Sierra Club Guide To Sky Watching and Direction Finding. W. S. Kals. The Sierra Press, 4988 Gold Leaf Drive, Mariposa, CA 95338. (209) 966-5071. (800) 745-2631. $15.00.

GENERAL WORKS

The Evening Star: Venus Observed. Henry S. Cooper. Johns Hopkins University Press, 2715 North Charles Street, Baltimore, MD 21218-4319. (410) 516-6900. (800) 537-5487 1994. $22.00.

Evolution of the Earth and Planets. E. Takahashi, et al, editors. American Geophysical Union, 2000 Florida Avenue, N.W., Washington, DC 20009. (202) 462-6903. (800) 966- 2481. 1993. $28.00.

Exploration of Venus & Mars Atmospheres. G. M. Keating, editor. Pergamon Press, Inc., Maxwell House, Fairview Park, Elmsford, NY 10523. (914) 592-7700. Fax (914) 592-3625. 1995. 94.00.

Geology of the Terrestrial Planets. Michael H. Carr. U. S. Government Printing office, Superintendent of Documents, Washington, DC 20402-9325. (202) 783-3238. 1985. $16.00

Moons and Planets. William K. Hartmann. Wadsworth Publishing Co., 10 Davis Drive, Belmont, CA 94002. (415) 595-2350. (800) 354-9706. 3rd edition. 1993. $49.95.

The Near Planets. Time-Life Books Editors. Time-Life, Inc., 777 Duke Street, Alexandria, VA 22314. (703) 8388- 7000. (800) 621-7026. Voyage through the Universe Series. 1992.

Pioneer Venus: A Planet Unveiled. Richard Fimmel. NASA AMES Research Center, Moffett Field, CA 94035-1000. (415) 604-5658. $40.00.

Planetary Landscapes. Ronald Greeley. Chapman & Hall, 1 Penn Plaza, New York, NY 10119. (212) 564-1060. Div. of Routledge,2nd edition. 1994. $49.95.

Resources of Near-Earth Space. John S. Lewis, et al. University of Arizona Press, 1230 North Park Avenue, Number 102, Tucson, AZ 85719. (520) 621-1441. (800) 426-3797. 1993. $75.00.

Terraforming: Engineering Planetary Environments. Martyn J. Fogg. Society of Automotive Engineers (SAE), 400 Commonwealth Drive, Warrendale, PA 15096. (412) 776-4841. 1995. $49.00.

Venus. Donald M. Hunter, et al, editors. University of Arizona Press, 1230 North Park Avenue, Number 102, Tucson, AZ 85719. (520) 621-1441. (800) 426-3797. 1983. $70.00.

Venus: A New Geology. Peter Cattermole, Johns Hopkins University Press, 2715 North Charles Street, Baltimore, MD 21218-4319. (410) 516-6900. (800) 537-5487 1994. $49.95.

Venus Aeronomy. C. T. Russell. Kluwer Academic Publishers, 101 Philip Drive, Assinippi Park, Norwell, MA 02061. (617) 871-6600. 1991. $230.50.

Venus: An Errant Twin. Eric Burgess. Columbia University Press, 562 West 113th Street, New York, NY 10025. (212) 666-1000. (800) 944-8648. 1985. $41.00.

Venus and Mars: Atmospheres, Ionospheres, and Solar Wind Interactions. J. G. Luhman, et al, editors. American Geophysical Union, 2000 Florida Avenue N.W., Washington, DC 20009. (202) 462-6903. (800) 966-2481. 1992. $59.00.

Venus Geology, Geochemistry, and Geophysics: Research Results From the Soviet Union. V. L. Barsukov et al, editors. University of Arizona Press, 1230 North Park Avenue, Number 102, Tucson, AZ 85719. (520) 621-1441. (800) 426-3797. 1992. $75.00.

HANDBOOKS AND MANUALS

Astrophysical Data: Planets and Stars. Kenneth R. Lanb. Springer-Verlag New York, Inc., 175 Fifth Avenue, New York, NY 10010. (212) 460-1500. (800) 777-4643. 1993. $59.00.

Field Guide to the Stars and Planets. Jay M. Pasachoff and Donald H. Menzel. Houghton Mifflin Co., 222 Berkeley Street, Boston, MA 02116. (617) 351-5000. (800) 225-3362. Revised edition. 1992. $24.95.

New Guide To the Planets. Patrick Moore. Trans-Atlantic Publications, Inc., 311 Bainbridge Street, Philadelphia, PA 19147. (215) 925-5083. 1993. $37.50.

VENUS

Ency. of Physical Sciences and Engineering Info. Sources

Pioneer Venus: A Planet Unveiled. Richard Fimmel. NASA Ames Research Center, Moffett Field, CA 94035-1000. (415) 604-5658. 1995. $40.00.

Planet Observer's Handbook. Fred W. Price. Cambridge University Press, 40 West 20th Street, New York, NY 10011-4211. (212) 924-3900. (800) 872-7423. 1994. $34.95.

Planets and their Moons. Aububon Society Staff. Alfred A. Knopf, Inc. 201 E. 50th Street, New York, NY 10022. (800) 733-3000. 1995. $7.99.

The Venus Geologic Mapper's Handbook. K. L. Tanaka, et al. Diane Publishing Co., 600 Upland Avenue, Upland, PA 19015. (610) 499-7415. 1994. $40.00 in paper.

OTHER SOURCES

Atlas of the Planets. Vincent DeCallatay and Audouin Dollfus. Books on Demand, 300 North Zeeb Road, Ann Arbor, MI 48106-1346. (313) 761-4700. (800) 521-0600. $45.60.

Atlas of the Solar System. B. Yenne. Simon & Schuster, Inc. 1230 Avenue of the Americas, New York, NY 10020. (212) 698-7000. (800) 223-2348. 1987. $12.98.

Planetary, Lunar and Solar Positions Six Hundred-one B.C. to A.D. One at Five-day and Ten-day Intervals. Bryant Tuckerman. American Philosophical Society, 104 South 5th Street, Philadelphia, PA 19106-3387. (215) 440-3400. FAX (215) 440-3436. Memoirs Series, volume 56. 1979. $20.00.

Planetary, Lunar and Solar Positions, A.D.2 to A.D. 1649 at Five Day and Ten Day Intervals. Bryant Tuckerman. American Philosophical Society, 104 South 5th Street, Philadelphia, PA 19106-3387. (215) 440-3400. FAX (215)440-3436. Memoirs Series, volume 59. 1964. $30.00.

Planetary, Lunar and Solar Positions, 1650 - 1805. Owen Gingerich and Barbara L. Welther. American Philosophical Society, 104 South 5th Street, Philadelphia, PA 19106-3387. (215) 440-3400. FAX (215)440-3436. Memoirs Series, volume 59S. 1983. $20.00.

The Sky: A User's Guide. David H. Levy. Cambridge University Press, 40 West 20th Street, New York, NY 10011-4211. (212) 924-3900. (800) 872-7423. 1993. $14.95.

The View from Space: Photographic Exploration of the Planets. Merton Davies and Bruce C. Murray. Columbia University Press, 562 West 113th Street, New York, NY 10025. (212) 666-1000. (800) 944-8648. 1973. $17.50.

PERIODICALS

Association of Lunar and Planetary Observers Journal. Association of Lunar and Planetary Observers, Box 16131, San Francisco, CA 94116. (415) 566-5786. FAX (415) 731-8242. 1947 to present. Quarterly. $14.00 per year. ISSN: 0039-2502.

Astronomical Journal. American Institute of Physics, One Physics Ellipse, College Park, MD 20740-3843. (301) 209-3000. Published for the American Astronomical Society. 1849 to present. Monthly. $280.00 per year. ISSN: 0004-6256.

Astronomical Society of the Pacific. Publications. Astronomical Society of the Pacific, 390 Ashton Avenue, San Francisco, CA 94112. (415) 337-1100. FAX (415) 337-5205.1889 to present. Monthly. $175.00 per year. ISSN: 0004-6280.

Astronomy. Kalmbach Publishing Company, Box 1612, Waukesha, WI 53187-1612. (414) 796-8776. FAX (414) 796-1142. 1973 to present. Monthly. $27.00 per year. ISSN: 0091-6358.

Astronomy and Astrophysics. Springer-Verlag New York, Inc., 44 Hartz Way, Secaucus, NJ 07096-2491. (201) 348-4033. FAX(201) 348-4505. 1969 to date. 36 issues per year in 12 volumes. $1733.00 per year. ISSN: 0004-6361.

Earth, Moon and Planets; An International Journal of Comparative Planetology. Kluwer Academic Publishers, Box 358, Accord Station, Hingham, MA 02018-0358. (617) 871-6600. FAX (617) 871-6528. 1969 to present. Monthly. $840.00 per year. ISSN: 0167-9295.

Geochemica Et Cosmochimica Acta. Elsevier Science. 660 White Plains Road, Tarrytown, NY 10591-5153. (914) 524-9200. FAX (914) 333-2444. 1950 to the present. Biweekly. $895.00 per year. ISSN: 0016-7037.

Icarus; International Journal of Solar System Studies. Academic Press, Inc., Journal Division, 525 B Street, Suite 1900, San Diego, CA 92101-4495. (619) 230-1840. FAX (619) 699-6800. 1962 to the present. Monthly. $1080.00. ISSN: 0019-1035.

JGR: Journal of Geophysical Research: Planets. American Geophysical Union, 2000 Florida Avenue, NW, Washington, CD 20009. (202) 462-6900. FAX (202) 328-0566. 1991 to present. Monthly. $597.00 per year. ISSN: 0148-0227.

Lunar and Planetary Information Bulletin. Lunar and Planetary Institute. 3600 Bay Area Boulevard, Houston, TX 77058-1113. (713) 486-2175. FAX (713) 486-2125. 1970 to present. Quarterly. Free. Also available online.

Mercury. Astronomical Society of the Pacific. 390 Ashton Avenue, San Francisco, CA 94112. (415) 337-1100. FAX: (415) 337-5205. 1972 to present. Bimonthly. $175.00 per year. ISSN: 0047-6773.

Meteoritics. Meteoritical Society, Department of Chemistry, University of Arkansas, Fayetteville, AR 72701. (501) 575-7625. FAX (501) 575-7778. 1953 to present. Bimonthly. $210.00 per year. ISSN: 0026-1114.

Observatory. c/o Dr. D. J. Stickland Space and Astrophysics Division, Rutherford Appleton Laboratory, Chilton, Didcot, Oxon OX11 OQX England. FAX 0235-445848. 1877 to present. Bimonthly. $42.00 per year. ISSN: 0029-7704.

Planetary and Space Science. Elsevier Science Publishing Company, Inc., 660 White Plains Road, Terrytown, NY 10591-5153. (914) 524-9200. FAX: (914) 333-2444. 1959 to present. Monthly. $1355.00 per year. ISSN: 0032-0633.

Planetary Report. Planetary Society. 65 North Catalina, Pasadena, CA 91106-2301. (818) 793-5100. FAX (818) 793-5528. 1980 to present. $25.00 per year. ISSN: 0736-3680.

Sky and Telescope. Sky Publishing Corporation, Box 9111, Belmont, MA 02178. (617) 864-7360. FAX (617) 864-6117.1941 to present. Monthly. $27.00 per year. ISSN: 0037-6604.

RESEARCH CENTERS AND INSTITUTES.

Center For Space Science and Astrophysics. Stanford University. 325 Durand Building, Stanford, CA 04305. (415) 723-3582.

Earth and Planetary Remote Sensing Laboratory. Washington University, Department of Earth and Planetary Sciences. Campus Box 1169, 1 Brookings Drive, St. Louis, MO 63130. (314) 889-5679. FAX (314) 889-5799.

Institute For Astronomy. University of Hawaii at Manoa, 2680 Woodlawn Drive, Honolulu, HI 96822. (808) 956-8312. FAX (808) 988-2790.

Institute For Terrestrial and Planetary Atmospheres. State University of New York at Stony Brook, Stony Brook, NY 11794-2300. (516) 632-6170. FAX (516) 632-6251.

Institute of Geophysics and Planetary Physics. University of California, Riverside, CA 92521. (714) 787-4503. FAX (714) 787-4529.

Laboratory For Atmospheric and Space Physics. University of Colorado - Boulder, Boulder, CO 80309-0392. (303) 492-7677. (303) 492-6946.

Laboratory For Planetary Studies. Cornell University, 302 Space Sciences Building, Ithaca, NY 14853. (607) 255-4971.

Lunar and Planetary Institute. 3303 NASA Road One, Houston, TX 77058-4399. (713) 486-2139. FAX 713-496-2162.

VERTICAL TAKEOFF AND LANDING

See: STOL-VTOL

VERY LARGE SCALE INTEGRATION (VLSI)

See also: ELECTRICAL ENGINEERING,ELECTRICITY, ELECTRONIC CIRCUITS and COMONENTS, ELECTRONICS ENGINEERING, MICROELECTRICS, MICROPROCESSORS

ABSTRACT SERVICES AND INDEXES

Applied Science and Technology Index; A Cumulative Subject Index To English Language Periodicals in the Fields of Aeronautics and Space Science, Computer Technology, Chemistry, Construction Industry, Energy and Related Areas. H.W. Wilson Co., 950 University Avenue, Bronx, NY 10452. (212) 588-8400. (800) 367-6770. FAX (718) 590-1617. From 1958 to present. Monthly. Inquire about cost and availability. Also available on CD-ROM and online. ISSN: 0003-6986.

Current Contents: Engineering, Technology and Applied Sciences. Institute for Scientific Information, 3501 Market Street, Philadelphia, PA 19104. (215) 386-0100. FAX (215) 386-6362. 1970 to present. Weekly. $442.00 per year. Also available on CD-ROM and online. Inquire regarding cost and availability. ISSN: 0095-7917.

Current Papers in Electrical and Electronics Engineering. Institution of Electrical Engineers (IEE), London. Distributed by

INSPEC/IEEE, Box 1331, 445 Hoes Lane, Piscataway, NJ 08855-1331. (908) 562-5549. 1969 to present. Monthly. $345.00 per year. ISSN: 0011-3778.

Electrical and Electronics Abstracts. Institution of Electrical Engineers (IEE), London. Available from INSPEC/IEEE - Institute of Electrical and Electronic Engineers, Box 1331, Hoes Lane, Piscataway, NJ 08855-1331. (908) 562-5549. 1898 to present. Monthly. $2200.00 per year. Also available on CD-ROM and online as INSPEC.

Engineering Index Monthly; Indexes and Abstracts the World's Engineering and Technical Literature. Engineering Information, Inc., Castle Point on the Hudson, Hoboken, NJ 07030. (201) 216-8500. (800) 221-1044. FAX (201) 216-8532. Monthly. $2300.00 per year. Available online as COMPENDEX and also on CD-ROM. ISSN: 0742-1974.

General Science Index. H.W. Wilson Company, 950 University Avenue, Bronx, NY 10452. (212) 588-8400. (800) 367-6770. FAX (718) 590-1617. From 1978 to present. Ten issues per year; quarterly and annual cumulations. Service basis. Available on CD-ROM and online. Inquire regarding cost and availability. ISSN: 0162-1963.

Government Reports Announcements and Index. U. S. National Technical Information Service (NTIS), 5285 Port Royal Road, Springfield, VA 22161. (703) 487-4650. FAX (703) 321-8547. From 1968 to present. Annual. $630.00 per Year. Also Available online as NTIS Bibliographic Database and on CD-ROM. ISSN:

Index To IEEE Publications. IEEE Service Center, 445 Hoes Lane, Piscataway, NJ 08855-1331. (908) 981-1393. (800) 678-IEEE. FAX (908) 981-9667. 1973 to present. Annual. ISSN: 0099-1368.

Physics Abstracts. INSPEC. Section A, Science Abstracts. Institution of Electrical Engineers (IEE), London. Available from: INSPEC/IEEE - Institute of Electrical and Electronic Engineers, Box 1331, Hoes Lane, Piscataway, NJ 08855-1331. (908) 562-5549. 1898 to present. 24 issues per year. $2835.00 per year. Also available online and on CD-ROM. ISSN: 0036-8091.

Physics Briefs (Physikalische Berichte). Information Center for Energy, Physics, Mathematics; German Physical Society. V C H Publishers, Inc., 220 East 23rd Street, New York, NY 10010-4606. (212) 683-8333. 1845 to present. 24 issues per year. $2390.00 per year. Also available online. ISSN: 0179-7434.

Science Citation Index. SCI. Institute for Scientific Information, 3501 Market Street, Philadelphia, PA 19104. (215) 386-0100. (800) 523-1850. FAX (215) 386-2991. 1961 to present. Six issues per year, plus annual cumulation. $11650.00 per year. Also available online and on CD-ROM. Inquire about price and availability. ISSN: 0036-827X.

Solid State and Superconductivity Abstracts: Covers theory, Production and Application of Solid State Materials. Cambridge Scientific Abstracts, 7200 Wisconsin Avenue, Bethesda, MD 20824. (301) 961-6750. FAX (301) 961-6720. 1957 to present. Bimonthly. $1320.00 per year. Also available online. ISSN: 0896-5900.

ANNUAL REVIEWS AND YEARBOOKS

Advances in Electronics and Electron Physics. Academic Press, Inc., 6277 Sea Harbor Drive, Orlando, FL 32887. (800) 321-

VERY LARGE SCALE INTEGRATION (VSLI)

Ency. of Physical Sciences and Engineering Info. Sources

5068. From 1948 to present. Irregular. Price varies, inquire. ISSN: 0065-2539.

Critical Reviews in Solid State and Materials Sciences. CRC Press, Inc., 2000 Corporate Boulevard, NW, Boca Raton, FL

33431. (407) 994-0555. (800) 272-7737. FAX (407) 998- 9784. 1970 to present. Bimonthly. $265.00 per year. ISSN: 1040-8436.

ASSOCIATIONS AND PROFESSIONAL SOCIETIES

American Association of Engineering Societies, 1111 19th Street, Suite 608, Washington, DC 20036-3603. (202) 296-2237. FAX (202) 296-1151.

American Electronics Association. 5201 Great America Way, Suite 520, P.O. 52990, Santa Clara, CA 95056. (408) 987- 4200. FAX (408) 970-8565.

American Institute of Physics. One Physics Ellipse, College Park, MD 20740-3843. (301) 209-3100.

Edison Electric Institute. 701 Pennsylvania Avenue NW, Washington, DC 20004-2696. (202) 508-5000, 5454. FAX (202) 508-5360.

Electric Power Research Institute. 3412 Hillview Avenue, Palo Alto, CA 94304. (415) 855-2000. FAX (415) 855-2954.

Electronics Industries Association. 2500 Wilson Boulevard, Arlington, VA 22201. (703) 907-7500. FAX (202) 457-4985. <bodylead>IEEE Circuits and Systems Society. c/o Institute of Electrical and Electronic Engineers. 345 East 47th Street. New York, NY 10017. (212) 705-7900. FAX (212) 705-4929.

IEEE Solid-state Circuits Council. c/o Institute of Electrical and Electronic Engineers. 345 East 47th Street. New York, NY 10017. (212) 705-7900. FAX (212) 705-4929.

Institute For Interconnecting and Packaging Electronic Electronic Circuits. 2215 Sanders Road, Northbrook, IL 60062-6135. (708) 509-9700. (708) 509-9798.

Institute of Electrical and Electronics Engineers. 345 East 47th Street. New York, NY 10017. (212) 705-7900. FAX (212) 705-4929.

International Society for Hydrid Microelectronics. 1850 Centennial Park Drive, Suite 105, Reston, VA 22091. (703) 758-1060. FAX (703) 758-1066.

National Electrical Manufacturers Association. 1300 North 17th Street, Suite 1847, Rosslyn VA 22209. (703) 841-3200. FAX (703) 841-3300.

DIRECTORIES AND BIOGRAPHICAL SOURCES

American Electronics Association Directory. 5201 Great America Way, Suite 520, P.O. 52990, Santa Clara, CA 95056. (408) 987-4200. FAX (408) 970-8565. Annual. $175.00.

American Men and Women of Science: Physical and Biological Sciences. R. R. Bowker Inc., 121 Chanlon Road, New Providence, NJ 07974. (908) 464-6800. (800) 521-8110. 20th edition. 8 volumes. 1996. $850.00.

Directory of Engineering Societies and Related Organizations. American Association of Engineering Societies, 1111 19th Street, Suite 608, Washington, DC 20036-3603. (202) 296-2237. Semi-annual. $150.00.

E E M - Electronic Engineer's Master. Hearst Business Communications, Inc., 645 Stewart Avenue, Garden City NY 11530. (516) 227-1300. ISSN: 0732-9016.

Electrical and Electronics Trades Directory: The Blue Book. Institution of Electrical Engineers. c.o IEEE Service Center, 445 Hoes Lane, Piscataway, NY 08854. (908) 981-1393. (800) 678-IEEE. FAX (908) 981-9667. Annual. 1995. $140.00

Engineering Research Centers: Incorporating Electronics Research Centers. Stockton Press, 345 Park Avenue, New York, NY 10010. (212) 689-9200. (800) 221-2123. 4th edition. 1995. $515.00.

IC Master. Hearst Business Communications, Inc., 645 Stewart Avenue, Garden City, NY 11530. (516) 227-1300. 1975 to present. Annual. $170.00 per year. ISSN: 0894- 6809.

IEEE Membership Directory. Institute of Electrical and Electronics Engineers, IEEE Service Center, 445 Hoes Lane, Piscataway, NY 08854. (908) 981-1393. (800) 678-IEEE. FAX (908) 981-9667. 2 volumes. Annual. $190.00. ISSN:

International Directory of Abbreviations and Acroynms of Electronics, Electrical Engineering, Computer Technology and Information Processing. Peter Wennrich. K. G. Saur, 121 Chanlon Road, New Providence, NJ 07974. (908) 464-6800. (800) 521-8110 2 volumes. 1992. $230.00.

International Engineering Directory. American Consulting Engineers Council, 1015 15th Street, N.W. Suite 802, Washington, DC 20005-2670. (202) 347-7474. Annual. $10.00.

Research Centers Directory. Gale Research, 835 Penobscot Building, Detroit, MI 48226-4094. (313) 961-2242. (800) 877-4253. 20th edition. 1995. $485.00. ISSN: 0080- 1518.

Who's Who in Engineering. Gordon Davis, editor. American Association of Engineering Societies. 1111 19th Street, NY, Suite 608, Washington, DC 20036. (202) 296-2237. (800) 658-8897. 9th edition. 1995. $220.00.

Who's Who in Technology. Gale Research, 835 Penobscot Building, Detroit, MI 48226-4094. (313 961-2242. (800) 521-4253. 7th edition. 1995. $195.00. ISN 0-8103-7467-6.

ENCYCLOPEDIAS AND DICTIONARIES

Chambers Science and Technology Dictionary. Peter M.B. Walker, editor. Cambridge University Press, 40 West 20th Street, New York, NY 10011-4211. (212) 924-3900. 1988. $39.95.

Encyclopedia of Applied Physics. George Trigg, editor. VCH Publications, Inc., 220 East 23rd Street, Suite 909, New York, NY 10010-4606. (212) 683-8333. (800) 422-8824. 20 volume set. 1991 - . $5990.00.

Encyclopedia of Integrated Circuits. Walter A. Buchsbaum.

Prentice Hall, 113 Sylvan Avenue, Route 9W, Englewood Cliffs, NJ 07632. (201) 592-2000. (800) 922-0579. 1987. $39.95.

IEEE Standard Dictionary of Electrical and Electronics Terms. Christopher J. Booth, editor. IEEE Service Center, 445 Hoes Lane, Piscataway, NJ 08855-1331. (908) 981-1393. (800) 678-IEEE. FAX (908) 981-9667. IEEE Standard 100- 1992. 5th edition. 1993. $150.00.

Illustrated Dictionary of Electronics. Stan Gibilsco. TAB Books, P.O. Box 40, Blue Summit, PA 17294-0850. (717) 794- 2191. (800) 233-1128. 7th edition. 1994. $34.95.

Illustrated Encyclopedic Dictionary of Electronic Circuits. John Douglas-Young. Prentice Hall , 113 Sylvan Avenue, Route 9W, Englewood Cliffs, NJ 07632. (201) 592-2000. (800) 922-0579. 1983. $32.95.

International Encyclopedia of Integrated Circuits. Arthur A. Seidman. TAB Books, P.O. Box 40, Blue Summit, PA 17294-0850. (717) 794-2191. (800) 233-1128. 1991. $75.00.

McGraw-Hill Electronics Dictionary. J. Markus. McGraw-HillBook Company, Inc., 1221 Avenue of the Americas, New York, NY 10020. (212) 997-3675. 5th edition. 1994. $49.95.

McGraw-Hill Encyclopedia of Engineering. Sybil P. Parker, editor. McGraw-Hill Book Company, Inc. 1221 Avenue of the Americas, New York, NY 10020. (212) 997-3675. Second edition. 1993. $95.50. ISBN: 0-07-051392-9.

McGraw-Hill Encyclopedia of Science and Technology. McGraw-Hill Book, Incorporated, 1221 Avenue of the Americas, New York, NY 10020. (212) 997-3675. (800) 262-4729. Seventh edition. Twenty volumes. 1992. $1900.00.

GENERAL WORKS

Algorithms for VLSI Physcal Design. Naveed Sherwani. Kluwer Academic Publishers, 101 Philip Drive, Assinippi Park, Norwell, MA 02061. (617) 871-6600. 2nd edition. 1995. $120.00.

Basic VLSI Design: Systems and Circuits. Doug Pucknell and Kamram Eshraghian. Prentice-Hall, 113 Sylvan Avenue, Route 9W, Englewood Cliffs, NJ 07632. (201) 592-2000. (800) 922-0579. 3rd edition. 1994. $50.00.

Chipmakers. Time-Life Books Inc., 777 Duke Street, Alexandria, VA 22314. (703) 8388-7000. (800) 621-7026. Revised edition. 1990.

Computer Design: VLSI in Computers and Processors. IEEE Computer Society, 345 East 47th Street, New York, NY 10017. (212) 705-7900. 1991. $100.00.

Electric Circuits: Principles, Applications and Computer Analysis. David A. Bell. Prentice Hall , 113 Sylvan Avenue, Route 9W, Englewood Cliffs, NJ 07632. (201) 592- 2000. (800) 922-0579. 5th edition. 1995. $72.00.

Exploring Electronic Devices. Mark E. Hazen. SCP: Third World Literature Publishing House, P.O. Box 482, Lithonia, GA 30058-0482. (404) 785-7725. 1991. $50.75.

Introduction To Parallel Processing. Bruno Codenotti and Mauro Leoncini. Addison-Wesley Publishing Co., Inc., 1 Jacob Way, Reading, MA 01867. (617) 944-3700. (800) 447-2226. . 1993. $33.50 in paper.

Introduction To VLSI Techology. T. E. Price. Prentice Hall, 113 Sylvan Avenue, Route 9W, Englewood Cliffs, NJ 07632. (201) 592-2000. (800) 922-0579. 1994. $60.00.

Logic Synthesis. Devadas Srinivas, et al McGraw-Hill Publishing Company, Inc., 1221 Avenue of the Americas, New York, NY 10020. (212) 512-2000. (800) 262-4729. 1994. $50.00.

Multichip Module Technologies and Alternatives: The Basics. Daryl A. Doane. Van Nostrand Reinhold, 115 Fifth Avenue, New York, NY 10003. (212) 254-3232. (800) 842-3636. 1993. $89.95.

Physical Architecture of VLSI Systems. Rob Hannemann, et al, editors. John Wiley & Sons, Inc., 605 Third Avenue, New York, NY 10158-0012. (212) 850-6000. (800) 225-5945. 1994. $89.95.

VLSI Digital Signal Processors: An Introduction To Rapid Prototyping and Design Synthesis. Vijay Madisetti. Butterworth-Heinemann, 313 Washington Street, Newton, MA 02158. (618-928-2500. (800) 366-2665. 1994. $64.95.

VLSI Signal Processing Technology. M. A. Bayoumi. Kluwer Academic Publishers, 101 Philip Drive, Assinippi Park, Norwell, MA 02061. (617) 871-6600. 1994. $88.00.

HANDBOOKS AND MANUALS

Active Electronic Component Handbook. Charles A. Harper and Harold C. Jones. McGraw-Hill Book Company, 1221 Avenue of the Americas, New York, NY 10020. (212) 512-2000. (800) 262-4729. 2nd edition. 1996. $79.50.

EDN Designer's Companion. Ian Hickman and Bill Travis, editors. Butterworth-Heinemann, 313 Washington Street, Newton, MA 02158. (618-928-2500. (800) 366-2665. 1994.$29.95.

Electronics Circuits Handbook. M. Tooley. Butterworth-Heinemann, 313 Washington Street, Newton, MA 02158. (618-928-2500. (800) 366-2665. 1993. $49.95.

Electronics Handbook. Jerry C. Whitaker. CRC Press, Inc., 2000 Corporate Boulevard, NW, Boca Raton, FL 33431. (407) 994-0555. (800) 272-7737. 1996. $120.00.

Handbook of Electrical and Electronic Technology. Curtis Johnson. Prentice Hall , 113 Sylvan Avenue, Route 9W, Englewood Cliffs, NJ 07632. (201) 592-2000. (800) 922- 0579. 1996. $88.00.

Handbook of Quality Integrated Circuit Manufacturing. Robert Zurich. Academic Press, Inc., 6277 Sea Harbor Drive, Orlando, FL. (800) 321-5068. 1991. $59.95.

Handbook of VLSI Chip Design and Expert Systems. A. F. Schwartz. Academic Press, Inc., 6277 Sea Harbor Drive, Orlando, FL. (800) 321-5068.

Handbook of VLSI Microlithography: Principles, Technology and Applications. William B. Glendeining and John N. Helbert, editors. Noyes, 120 Mill Road, Park Ridge, NJ 07656. (201) 391-8484. 1992. $96.00.

Integrated Circuits Handbook. Rector Press, Ltd., 130 Rattlesnake, Leverett, MA 01054-9726. (413) 548-9708. (800) 247-3473. 3 volumes. 1995. $295.00 in paper.

VERY LARGE SCALE INTEGRATION (VLSI)

Ency. of Physical Sciences and Engineering Info. Sources

Semiconductor Materials and Process Technology Handbook: For Very Large Scale Integration (vlsi) and Ultra Large Sacle Integration (ulsi). Gary E. McGuire, editor. Noyes, 120 Mill Road, Park Ridge, NJ 07656. (201) 391-8484. 1988. $92.00.

Standard Handbook for Electrical Engineers. Donald Fink. McGraw-Hill Publishing Company, 1221 Avenue of the Americas, New York, NY 10020. (212) 512-2000. (800) 262-4729. 13th edition. 1996. $110.00.

ONLINE DATABASES AND CD-ROMS

COMPENDEX PLUS. Engineering Information, Inc., 345 East 47th Street, New York, NY 10017. (212) 705-7600 or (800) 221-1044. Contains citations with abstracts to worldwide literature in engineering and technology, from 1970 to present. Available on online BRS,(800) 289-4277, DIALOG, (800) 334-2564, ORBIT (800) 456-7248, and STN International, FIZ Karlsruhe, P.O. Box 2465, W-7500, Karlsruhe 1, Germany, online services. Also available on CD-ROM. Inquire as to cost and availability.

Current Contents Search. Institute for Scientific Information, 3501 Market Street, Philadelphia, PA 19104. (215) 386-0100. FAX (215) 386-6362. Contains citations to articles listed in the table of contents of science and technology journals. Also articles in social sciences and life sciences journals. Available on BRS,(800) 289-4277, DIALOG,(800) 334-2564, online services. Inquire as to cost and availability.

Dissertation Abstracts Online. University Microfilms International, 300 North Zeeb Road, Ann Arbor, MI 48106. (800) 521-0600 or (313) 761-4700. Scope includes virtually all doctoral dissertations accepted at accredited American institutions from 1861 to present in 252 subject areas. Available on BRS, (800) 289-4277, DIALOG, (800) 334-2564, and OCLC EPIC,(800) 848-5878, online services. Also available on CD-ROM. Inquire as to cost and availability.

INSPEC. Institution of Electrical Engineers, Michael Faraday House, Six Hills Way, Stevenage, Herts. SG1 2AY, England. Telephone: 0438 313311 or FAX 0438 742840. Contains citations to the worldwide literature of physics, electronics and electrical engineering, computer technology, and related fields. Available on BRS, (800) 289-4277, DIALOG, (800) 334-2564, ORBIT, (800) 456-7248, and STN International, FIZ Karlsruhe, P.O. Box 2465, W-7500, Karlsruhe 1, Germany, online services. Inquire as to cost and availability.

NTIS Bibliographic Database. National Technical Information Service, 5285 Port Royal Road, Springfield, VA 22161. (703) 487-4929 or FAX (703) 321-8199. Broad coverage of government-sponsored science and technology research reports, 1964 to present. Available on BRS,(800) 289-4277, DIALOG, (800) 334-2564, ORBIT, (800) 456-7248, and STN International, FIZ Karlsruhe, P.O. Box 2465, W-7500, Karlsruhe 1, Germany, online services. Also available on CD-ROM. Inquire as to cost and availability.

Physics Briefs. American Institute of Physics, 335 East 45th Street, New York, NY 10017. (212) 661-9260 or FAX (212) 661-2036. Contains citations with abstracts of the literature of physics and related fields, 1979 to present. Available on the STN International, FIZ Karlsruhe, P.O. Box 2465, W-7500, Karlsruhe 1, Germany, online service. Inquire as to cost and availability.

SCISEARCH. Institute for Scientific Information, 3501 Market Street, Philadelphia, PA 19104. (800) 523-1850 or (215) 386-0100. Broad multidisciplinary title and author index to the international literature of science and technology, 1974 to present. Available on DIALOG,(800) 334-2564, and ORBIT,(800) 456-7248, online services. Also available on CD-ROM. Inquire as to cost and availability.

WILSONLINE. H.W. Wilson Company, 950 University Avenue, Bronx, NY 10452. (800) 367-6770 or (212) 588-8400. Makes available online versions of the printed H.W. Wilson indexes including Applied Science and Technology Index, Business Periodicals Index, General Science Index, and Readers' Guide to Periodical Literature. Period covered is generally 1983 to present. Available on BRS,(800) 289-4277, DIALOG,(800) 334-2564, and OCLC EPIC,(800) 848-5878, online services. Also available on CD-ROM. Inquire as to cost and availability.

PERIODICALS

Circuits, Systems, and Signal Processing. Birkhauser, 675 Massachusetts Avenue, Cambridge, MA 02139-3309. FAX (201) 348-4505. 1982 to present. Quarterly. $184.00 per year. ISSN: 0278-081X.

CRC Critical Reviews in Solid State and Materials Science. CRC Publishers, Inc., 2000 Corporate Blvd., N.W., Boca Raton, FL 33431. (407) 994-0555 or (800) 333-8300. Quarterly. $295.00 per year.

Electrical World. McGraw-Hill, Inc., Box 513 Hightstown, NJ 08520. (212) 512-3288. 1874 to present. Monthly. $55.00 per year. ISSN: 0013-4457.

Electronic Design. Penton Publishing, San Jose Gateway, Suite 354. 2025 Gateway Place, San Jose, CA 95110. (408) 441-0550. 1952 to present. Fortnightly. $95.00. ISSN: 0013-4872.

Electronic Engineering Times. CMP Publications, Inc., 600 Community Drice, Manhasset, NY 11030. (516) 562-5000. FAX (516) 562-5325. 1972 to present. Weekly. $159.00 per year. ISSN: 0192-1541.

Electronics. Penton Publishing, San Jose Gateway, Suite 354. 2025 Gateway Place, San Jose, CA 95110. (408) 441-0550. 1930 to present. Semi-weekly. $98.00 per year. ISSN: 0883-4989.

Electronics World and Wireless World. Reed Business Publishing, Ltd., Quadrant House, the Quadrant, Sutton, Surrey SM2 5AS, England. 1911 to present. Monthly. L35. ISSN: 0266-3244.

IEEE Circuits and Devices Magazine. Institute of Electrical and Electronics Engineers, Inc., Box 1331, 445 Hoes Lane, Piscataway, NJ 08855-1331. (908) 981-0060. 1985 to present. Bi-monthly. $120.00 per year. ISSN: 8755-3996.

IEEE Journal of Solid State Circuits. Institute of Electrical and Electronics Engineers, Inc., Box 1331, 445 Hoes Lane, Piscataway, NJ 08855-1331. (908) 981-0060. 1966 to present. Bi-monthly. $275.00 per year. ISSN: 0018-9200.

IEEE Spectrum. Institute of Electrical and Electronics Engineers, Inc., Box 1331, 445 Hoes Lane, Piscataway, NJ 08855-1331. (908) 981-0060. 1964 to present. Monthly. $157.00 per year. ISSN: 0018-9235.

IEEE Transactions On Circuits and Systems. Part 1: Fundamental theory and Applications. Institute of

Electrical and Electronics Engineers, Inc., Box 1331, 445 Hoes Lane, Piscataway, NJ 08855-1331. (908) 981-0060. 1952 to present. Monthly. $241.00 per year. ISSN: 1057-7122.

IEEE Transactions On Electron Devices. Institute of Electrical and Electronics Engineers, Inc., Box 1331, 445 Hoes Lane, Piscataway, NJ 08855-1331. (908) 981-0060. 1952 to present. Monthly. $395.00 per year. ISSN: 0018-9393.

Institute of Electrical and Electronics Engineers. Proceedings. Institute of Electrical and Electronics Engineers, Inc., Box 1331, 445 Hoes Lane, Piscataway, NJ 08855-1331. (908) 981-0060. 1913 to present. Monthly. $275.00. ISSN: 0018-9219.

Microelectronic Engineering: An Interdisciplinary Journal of Semiconductor Manufacturing Technology. Elsevier Science, 660 White Plains Road, Tarrytown, NY 10591-5153. (914) 524-9200. FAX (914) 333-2444. 1983 to present. 16 issues per year. $845.00 per year. ISSN: 0167-9317.

Microelectronics Journal. Elsevier Science, 660 White Plains Road, Tarrytown, NY 10591-5153. (914) 524-9200. 1967 to present. 8 issues per year. $455.00 per year. ISSN: 0026-2692.

Semiconductor International. Cahners Publishing Co., 44 Cook Street, Denver, CO 80206. (708) 635-8000. FAX (708) 390-2770. (800) 662-7776. 1978 to present. 13 issues per year. $84.95 per year. ISSN: 0163-3767.

Solid State Electronics. Elsevier Science, 660 White Plains Road, Tarrytown, NY 10591-5153. (914) 524-9200. FAX (914) 333-2444. 1960 to present. Monthly. $1025.00 per year. ISSN: 0038-1101.

RESEARCH CENTERS AND INSTITUTES

Alabama Microelectronics Science and Technology Center. Auburn University, 200 Broun Hall, Auburn University, AL 36849. (205) 844-1871. (205) 844-1809.

Center For Integrated Sensors and Circuits. University of Michigan, 1246 EECS Building, Ann Arbor, MI 48109-2122. (313) 764-3346. FAX (313) 747-1781.

Center For Integrated Circuit Packaging. Lehigh University, Sherman Fairchild Laboratory, Building 161. Bethlehem, PA 18015. (215) 758-4409. FAX (215) 758-4561.

Center For Integrated Systems. Stanford University, School of Engineering, Stanford, CA 94305-4028. (415) 723-9038.

Electrical Engineering Research Laboratories. Purdue University. Electrical Engineering Building, West Lafayette, in 47907. (317) 494-3536. FAX (317) 494-6440.

Electronics Research Laboratory. University of California, Berkeley, 253 Cory Hall, Berkeley, CA 94720. (415) 642-2301.

Engineering Research Center. University of Maryland, Clark School of Engineering, Potomac Building, Room 2104. (301) 405-3906. FAX (301_ 403-4105.

Solid State Electronics Laboratory. North Carolina State University, 432 Daniels Hall, Raleigh, NC 27695. (919) 737-2336.

VIBRATION

See also: MACHINE DESIGN, MECHANICAL ENGINEERING

ABSTRACT SERVICES AND INDEXES

Applied Mechanics Reviews: An Assessment of World Literature in Engineering Sciences. 1948-present. American Society of Mechanical Engineers, 345 East 47th Street, New York, NY 10017. (212) 705-7703. Monthly. $360.00 per year.

Applied Science and Technology Index; A Cumulative Subject Index To English Language Periodicals in the Fields of Aeronautics and Space Science, Computer Technology, Chemistry, Construction Industry, Energy and Related Areas. H.W. Wilson Co., 950 University Avenue, Bronx, NY 10452. (800) 367-6770 or (212) 588-8400. FAX (718) 590-1617. From 1958 to present. Monthly. Inquire about cost and availability. Also available on CD-ROM and online.

Current Contents: Engineering, Technology, and Applied Sciences. Institute for Scientific Information, 3501 Market Street, Philadelphia, PA 19104. (215) 386-0100. FAX (215) 386-6362. 1970 to present. Weekly. $442.00 per year.

Engineering Index Monthly. Engineering Information, Inc., Castle Point on the Hudson, Hoboken, NJ 07030. (800) 221-1044. FAX (212) 832-1857. Monthly. $2300.00 per year. Also available online as COMPENDEX, and also on CD-ROM. Covers chemical engineering, computers, electrical engineering, civil engineering, metals and mining, industrial management, and mechanical engineering.

Mechanical Engineering Abstracts (formerly *ISMEC*), Cambridge Scientific Abstracts, 7200 Wisconsin Avenue, Bethesda, MD 20814-4823. (301) 961-6750. FAX (301) 961-6720. 1967 to present. Monthly. $895.00 per year. Summarizes world literature in mechanical engineering, production engineering, and engineering management. Also available online.

Physics Abstracts. Institute of Electrical Engineers, Michael Faraday House, Six Hill Way, Stevenage, Herts, SG1 2AY, England. Distributed by IEEE, 445 Hoes Lane, Piscataway, NJ 08854. (908) 562-5549. 1898 to present. Monthly. $2700.00 per year. Also available online as INSPEC.

Science Citation Index. Institute for Scientific Information, 3501 Market Street, Philadelphia, PA 19104. (215) 386-0100. FAX (215) 386-6362. Inquire about availability and cost. Also available on CD-ROM.

ASSOCIATIONS AND PROFESSIONAL SOCIETIES

American Society of Mechanical Engineers. 345 E. 47th Street, New York, NY 10017. (212) 705-7722.

Vibration Institute. 6262 S. Kingery Highway, Suite 212, Willowbrook, IL 60514. (708) 654-2254. FAX (708) 654-2271.

VIBRATION

Ency. of Physical Sciences and Engineering Info. Sources

DIRECTORIES AND BIOGRAPHICAL SOURCES

Directory of Engineering Societies and Related Organizations. Gordon Davis. 13th edition. American Association of Engineering Societies, 1111 19th Street NW, Suite 608, Washington, DC 20036. (202) 296-2237 or (800) 658-8897. 1989. Inquire for price.

Research Centers Directory. Gale Research, 835 Penobscot Building, Detroit, MI 48226-4094. (313) 961-2242. (800) 877-4253. FAX (313) 961-6083. $485.00.

Scientific and Technical Organizations and Agencies Directory. Gale Research, 835 Penobscot Building, Detroit, MI 48226-4094. (313) 961-2242. (800) 877-4253. FAX (313) 961-6083. 4th edition. 1996. $195.00.

Who's Who in Engineering. American Association of Engineering Societies, 1111 19th Street NW, Suite 608, Washington, DC 20036. (202) 296-2237 or (800) 658-8897. 8th edition. 1991. Inquire for price.

ENCYCLOPEDIAS AND DICTIONARIES

McGraw-Hill Encyclopedia of Engineering. Sybil P. Parker, ed. 2nd edition. McGraw-Hill Publishing Company, 1221 Avenue of the Americas, New York, NY 10020. (800) 262-4729 or (212) 512-3825. 1993. $95.50.

McGraw-Hill Encyclopedia of Science and Technology. Sybil P. Parker, ed. 7th edition. 20 volumes. McGraw-Hill Publishing Company, 1221 Avenue of the Americas, New York, NY 10020. (800) 262-4729 or (212) 512-3825. 1992. $1900.00

Thesaurus of Scientific, Technical, and Engineering Terms. Hemisphere Publishing Corporation, 1900 Frost Road, Suite 101, Bristol, PA 19007-1598. (215) 785-5800 or FAX (215) 785-5515. 1987. $173.00.

GENERAL WORKS

Advanced theory of Vibrations. J.S. Rao. Halsted Press (Division of John Wiley and Sons, Inc.), 605 Third Avenue, New York, NY 10158. (800) 526-5368 or (212) 850-6418. 1992. $44.95.

Noise and Vibration Control Engineering. Leo L. Beranek and Istvan L. Ver, eds. John Wiley and Sons, Inc., 605 Third Avenue, New York, NY 10158. (800) 526-5368 or (212) 850-6000. 1992. $89.95

Theory of Vibration. Ahmed A. Shabana. 2d ed. Springer-Verlag, 175 Fifth Avenue, New York, NY 10010. (212) 460-1500 or (800) 777-4643. FAX (212) 473-6272. 1991. $49.50.

Theory of Vibration With Applications. William T. Thomson. 4th ed. Prentice Hall, 113 Sylvan Avenue, Route 9W, Englewood Cliffs, NJ 07632. (201) 592-2000 or (800) 922-0579. 1993. $65.00.

HANDBOOKS AND MANUALS

Marks' Standard Handbook For Mechanical Engineers. E.A. Avallone and T. Baumeister III. 9th edition. McGraw-Hill Pub-

lishing Company, 1221 Avenue of the Americas, New York, NY 10020. (800) 262-4729 or (212) 512-3825. 1987. $133.00.

Machine Design Fundamentals: A Mechanical Designers' Workbook. Joseph E. Shigley & Charles R. Mischke, editors in chief. McGraw-Hill Publishing Company, 1221 Avenue of the Americas, New York, NY 10020. (800) 262-4729 or (212) 512-3825. 1989. $31.95.

Machine Design Data Handbook. K. Lingaiah. McGraw-Hill Publishing Company, 1221 Avenue of the Americas, New York, NY 10020. (800) 262-4729 or (212) 512-3825. 1994. $89.50.

Machinery's Handbook. Erik Oberg, et al., editors. 24th edition. Industrial Press Inc., 200 Madison Avenue, New York, NY 10016. (212) 889-6330. FAX (212) 545-8327. 1992. $75.00

Mechanical Design and Systems Handbook. Harold A. Rothbart, editor in chief. McGraw-Hill Publishing Company, 1221 Avenue of the Americas, New York, NY 10020. (800) 262-4729 or (212) 512-3825. 1985. Inquire for cost and availability.

Mechanical Engineer's Data Handbook. James Carvill. CRC Publishers, Inc., 2000 Corporate Blvd., N.W., Boca Raton, FL 33431. (407) 994-0555 or (800) 272-7737. FAX (407) 994-0949. 1993. $69.95.

Mechanical Engineer's Reference Book. Edward H. Smith, editor. 12th edition. SAE International, 400 Commonwealth Drive, Warrendale, PA 15096-0001. (412) 776-4970. FAX (412) 776-0790. 1994. $155.00.

Standard Handbook of Machine Design. Joseph E. Shigley & Charles R. Mischke, editors in chief. McGraw-Hill Publishing Company, 1221 Avenue of the Americas, New York, NY 10020. (800) 262-4729 or (212) 512-3825. 1986. $124.50.

ONLINE DATABASES AND CD-ROMS

Compendex Plus. Engineering Information, Inc., 345 East 47th Street, New York, NY 10017. (212) 705-7600 or (800) 221-1044. Contains citations with abstracts to worldwide literature in engineering and technology, from 1970 to present. Available on online BRS, (800) 289-4277, DIALOG, (800) 334-2564, ORBIT (800) 456-7248, and STN International, FIZ Karlsruhe, P.O. Box 2465, W-7500, Karlsruhe 1, Germany, online services. Also available on CD-ROM. Inquire as to cost and availability.

CSA Engineering. Cambridge Scientific Abstracts, 7200 Wisconsin Avenue, Suite 601, Bethesda, MD 20814. (301) 961-6750 or (800) 843-7751. Contains citations and abstracts of international periodicals and research literature covering all fields of engineering and science and technology, including computer and information science, electronics, mechanical engineering, solid state materials, 1981 to present. Available on BRS, (800) 289-4277, online service. Inquire as to cost and availability.

Current Contents Search. Institute for Scientific Information, 3501 Market Street, Philadelphia, PA 19104. (215) 386-0100. FAX (215) 386-6362. Contains citations to articles listed in the table of contents of science and technology journals. Also articles in social sciences and life sciences journals. Available on BRS, (800) 289-4277, DIALOG, (800) 334-2564, online services. Inquire as to cost and availability.

Dissertation Abstracts. University Microfilms International, 300 North Zeeb Road, Ann Arbor, MI 48106. (800) 521-0600 or (313) 761-4700. Scope includes virtually all doctoral disserta-

Ency. of Physical Sciences and Engineering Info. Sources

VIBRATION

tions accepted at accredited American institutions from 1861 to present in 252 subject areas. Available on BRS,(800) 289-4277, DIALOG,(800) 334-2564, and OCLC EPIC,(800) 848-5878, online services. Also available on CD-ROM. Inquire as to cost and availability.

ISMEC: Mechanical Engineering Abstracts. Cambridge Scientific Abstracts, 7200 Wisconsin Avenue, Suite 601, Bethesda, MD 20814. (301) 961-6750 or (800) 843-7751. Contains citations to the literature in mechanical engineering, industrial and production engineering, energy, power, mechanics, devices and related areas, from 1973 to present. Available on the DIALOG,(800) 334-2564, online service. Inquire as to cost and availability.

NTIS Bibliographic Database. National Technical Information Service, 5285 Port Royal Road, Springfield, VA 22161. (703) 487-4929 or FAX (703) 321-8199. Broad coverage of government-sponsored science and technology research reports, 1964 to present. Available on BRS,(800) 289-4277, DIALOG, (800) 334-2564, ORBIT, (800) 456-7248, and STN International, FIZ Karlsruhe, P.O. Box 2465, W-7500, Karlsruhe 1, Germany, online services. Also available on CD-ROM. Inquire as to cost and availability.

SCISEARCH. Institute for Scientific Information, 3501 Market Street, Philadelphia, PA 19104. (800) 523-1850 or (215) 386-0100. Broad multidisciplinary title and author index to the international literature of science and technology, 1974 to present. Available on DIALOG,(800) 334-2564, and ORBIT,(800) 456-7248, online services. Also available on CD-ROM. Inquire as to cost and availability.

WILSONLINE. H.W. Wilson Company, 950 University Avenue, Bronx, NY 10452. (800) 367-6770 or (212) 588-8400. Makes available online versions of the printed H.W. Wilson indexes including Applied Science and Technology Index, Business Periodicals Index, General Science Index, and Readers' Guide to Periodical Literature. Period covered is generally 1983 to present. Available on BRS,(800) 289-4277, DIALOG,(800) 334-2564, and OCLC EPIC,(800) 848-5878, online services. Also available on CD-ROM. Inquire as to cost and availability.

PERIODICALS

International Journal For Numerical Methods in Engineering. J. Wiley & Sons Ltd., Journals, Baffins Lane, Chichester, Sussex PO19 1UD, England. Telephone 0243-779777. FAX 0243-775878. 1964 to present. 24 times a year. $1795.00 per year.

Journal of Applied Mechanics. American Society of Mechanical Engineers, 345 E. 47th Street, New York, NY 10017. (212) 705-7722. 1935 to present. Quarterly. $120.00.

Journal of Engineering For Industry. American Society of Mechanical Engineers, 345 E. 47th Street, New York, NY 10017. (212) 705-7722. 1970 to present. Quarterly. $100.00 per year.

Journal of Heat Transfer. American Society of Mechanical Engineers, 345 E. 47th Street, New York, NY 10017. (212) 705-7722. 1970 to present. Quarterly. $155.00 per year.

Journal of Mechanical Design. American Society of Mechanical Engineers, 345 E. 47th Street, New York, NY 10017. (212) 705-7722. 1978 to present. Quarterly. $100.00 per year.

Journal of Sound and Vibration. Harcourt Brace & Company, Ltd., Foots Cray High Street, Sidcup, Kent DA14 5HP, England.

Telephone 44-81-300-3322. FAX 44-81-309-0807. 1964 to present. 50 times a year. Inquire for cost and availability.

Journal of Vibration and Acoustics. ASME, 22 Law Drive, Box 2300, Fairfield, NJ 07007-2300. (800) 321-2633. 1978 to present. Quarterly. $100.00 per year.

Machine Design. Penton Publishing, 110 Superior Ave., Cleveland, OH 44114-2543. 1929 to present. 6 times a year. $180.00 per year.

Mechanical Engineering. American Society of Mechanical Engineers, 345 E. 47th Street, New York, NY 10017. (212) 705-7722. 1906 to present. Monthly. $45.00 per year.

Sound and Vibration. Acoustical Publications Inc., Box 40416, Bay Village, OH 44140. (216) 835-0101. FAX (216) 835-9303. 1967 to present. Monthly. Inquire for cost and availability.

Sound and Vibration Digest. The Vibration Institute, Suite 212, 6262 S. Kingery Highway, Willowbrook, IL 60514. (708) 654-2254. FAX (708) 654-2271. 1969 to present. Bimonthly. $250.00 per year.

RESEARCH CENTERS AND INSTITUTES

Center For Acoustics and Vibration, Pennsylvania State University. 157 Hammond Bldg., University Park, PA 16802. (814) 865-2761. FAX (814) 863-7222.

Laboratory For Experimental Mechanics Research. SUNY, Stony Brook, NY 11794-2300. (516) 632-8311. FAX (516) 632-8720.

Mechanical Engineering Design Laboratory. University of Florida, 237 Mechanical Engineering Bldg., Gainesville, FL 32611. (904) 392-0827. FAX (904) 392-1071.

Mechanical Engineering Laboratories. Stevens Institute of Technology, Hoboken, NJ 07030. (201) 420-5591. FAX (201) 420-6978.

Mechanical Engineering Research Laboratory. Kansas State University, Durland Hall, Manhattan, KS 66506. (913) 532-5610.

Massachusetts Institute of Technology Acoustics & Vibration Laboratory. Department of Mechanical Engineering, Bldg. 3, Room 366, Cambridge, MA 02139. (617) 253-2214.

Vibration Institute. Suite 212, 6262 S. Kingery Highway, Willowbrook, IL 60514. (708) 654-2254. FAX (708) 654-2271.

VIDEO DISKS

See: VIDEO TECHNOLOGY

VIDEO TAPE

See: VIDEO TECHNOLOGY

VIDEO TECHNOLOGY

See also: ELECTRONIC CIRCUITS AND COMPONENTS, ELECTRONICS ENGINEERING, OPTICS, TELECOMMUNICATIOINS, TELEVISION

ABSTRACT SERVICES AND INDEXES

Applied Science and Technology Index; A Cumulative Subject Index To English Language Periodicals in the Fields of Aeronautics and Space Science, Computer Technology, Chemistry, Construction Industry, Energy and Related Areas. H.W. Wilson Co., 950 University Avenue, Bronx, NY 10452. (212) 588-8400. (800) 367-6770. FAX (718) 590-1617. From 1958 to present. Monthly. Inquire about cost and availability. Also available on CD-ROM and online. ISSN: 0003-6986.

Current Contents: Engineering, Technology and Applied Sciences. Institute for Scientific Information, 3501 Market Street, Philadelphia, PA 19104. (215) 386-0100. FAX (215) 386-6362. 1970 to present. Weekly. $442.00 per year. Also available on CD-ROM and online. Inquire regarding cost and availability. ISSN: 0095-7917.

Electrical and Electronics Abstracts. Institution of Electrical and Electronic Engineers (IEE), London. Available from INSPEC/IEEE - Institute of Electrical and Electronic Engineers, Box 1331, Hoes Lane, Piscataway, NJ 08855-1331. (908) 562-5549. 1898 to present. Monthly. $2200.00 per year. Also available on CD-ROM and online as INSPEC.

Engineering Index Monthly; Indexes and Abstracts the World's Engineering and Technical Literature. Engineering Information, Inc., Castle Point on the Hudson, Hoboken, NJ 07030. (201) 216-8500. (800) 221-1044. FAX (201) 216-8532. Monthly. $2300.00 per year. Available online as *Compendex* and also on CD-ROM. ISSN: 0742-1974.

General Science Index. H.W. Wilson Company, 950 University Avenue, Bronx, NY 10452. (212) 588-8400. (800) 367-6770.FAX (718) 590-1617. From 1978 to present. Ten issues per year; quarterly and annual cumulations. Service basis. Available on CD-ROM and online. Inquire regarding cost and availability. ISSN: 0162-1963.

Index To IEEE Publications. IEEE Service Center, 445 Hoes Lane, Piscataway, NJ 08855-1331. (908) 981-1393. (800) 678-IEEE. FAX (908) 981-9667. 1973 to present. Annual. ISSN: 0099-1368.

Physics Abstracts. INSPEC. Section A, Science Abstracts. Institution of Electrical Engineers (IEE), London. Available from: INSPEC/IEEE - Institute of Electrical and Electronic Engineers, Box 1331, Hoes Lane, Piscataway, NJ 08855-1331. (908) 562-5549. 1898 to present. 24 issues per year. $2835.00 per year. Also available online and on CD-ROM. ISSN: 0036-8091.

Physics Briefs (Physikalische Berichte). Information Center for Energy, Physics, Mathematics; German Physical Society. V C H Publishers, Inc., 220 East 23rd Street, New York, NY 10010-4606. (212) 683-8333. 1845 to present. 24 issues per year. $2390.00 per year. Also available online. ISSN: 0179-7434.

ASSOCIATIONS AND PROFESSIONAL SOCIETIES

American Electronics Association. 5201 Great America Parkway, Suite 520, PO Box 54990, Santa Clara, CA 95056. (408) 987-4200. FAX (408) 970-8565.

Audio Engineering Society. 60 East 42nd Street, New York, NY 10017. (212) 661-8528.

Electronics Industries Association. 2500 Wilson Boulevard, Arlington, VA 22201. (703) 907-7500. FAX (202) 457-4985.

IEEE (Institute of Electrical and Electronics Engineers). 345 East 47th Street, New York, NY 10017. (212) 705-7900. FAX (212 705-4929.

National Association of Radio and Telecommunications Engineers. PO Box 678, Medway, MA 02053. (508) 533-8333. FAX (508) 533-3815.

Society of Cable Telecommunications Engineers. 669 Exton Commons, Exton, PA 19341. (215) 363-6888. FAX (610) 363-5898.

Society of Motion Picture and Television Enginers. 595 West Hartsdale Avenue, White Plains, NY 10607. (914) 761-1100. FAX (914) 761-3115.

DIRECTORIES AND BIOGRAPHICAL SOURCES

Broadcast Engineering - Equipment Reference Manual. Intertec Publishing Corporation, Box 12901, Overland Park, KS 66212. (913) 888-4664. Annual. $20.00.

E E M - Electronic Engineer's Master. Hearst Business Communications, Inc., 645 Stewart Avenue, Garden City NY 11530. (516) 227-1300. Annual. $90.00. ISSN: 0732-9016.

IEEE Membership Directory. Institute of Electrical and Electronics Engineers, IEEE Service Center, 445 Hoes Lane, Piscataway, NY 08854. (908) 981-1393. (800) 678-IEEE. FAX (908) 981-9667. 2 volumes. Annual. $190.00. ISSN:

International Satellite Directory. Design Publishers, 800 Siesta Way, Sonoma, CA 95476. (707) 930-0306. Annual. $250.00.

Satellite Communications, Satellite Industry Directory Issue. Cardiff Publishing, 6300 South Syracuse Way, Suite 650, Englewood, CO 80111. (303) 220-0600. Annual, June.

$20.00.

Society of Motion Picture and Television Engineers - Membership Directory. SMPTE, 595 West Hartsdale Avenue, White Plains, NY 10607. (914) 761-1100. Members only.

Television and Cable Factbook. Warren Publishing, 2115 Ward Court NW, Washington, DC 20037. (202) 672-9200.

World Satellite Directory. PHillip Publishing, 7811 Montrose Road, Potomac, MD 20854. (800) 777-5006. (301) 340-2100. Annual. $227.00.

Who's Who in Engineering. Gordon Davis, editor. American Association of Engineering Societies. 1111 19th Street, NY, Suite 608, Washington, DC 20036. (202) 296-2237. (800) 658-8897. 9th edition. 1995. $220.00.

ENCYCLOPEDIAS AND DICTIONARIES

Facts on File Dictionary of Television, Cable and Video. Robert M. Reed. Facts-on-File, Inc., 460 Park Avenue South, New York, NY 10016-7382. (212) 683-2244. (800) 322-8755. Fax (800) 683-3633. 1994. $24.95.

IEEE Standard Dictionary of Electrical and Electronics Terms. Christopher J. Booth, editor. IEEE Service Center, 445 Hoes Lane, Piscataway, NJ 08855-1331. (908) 981-1393. (800) 678-IEEE. FAX (908) 981-9667. IEEE Standard 100- 1992. 5th edition. 1993. $150.00.

Illustrated Dictionary of Electronics. Stan Gibilsco. TAB Books, P.O. Box 40, Blue Summit, PA 17294-0850. (717) 794- 2191. (800) 233-1128. 7th edition. 1994. $34.95.

McGraw-Hill Electronics Dictionary. John Markus and Neil Sclater. McGraw-Hill-Hill Publishing Company, Inc., 1221 Avenue of the Americas, New York, NY 10020. (212) 512-2000. (800) 262-4729. 5th edition. 1994. $49.95.

GENERAL WORKS

Advanced Television and Electronic Imaging For Film and Video. Joyce R. Hurwitz. Society of Motion Picture and Television Engineers - SMPTE, 595 West Hartsdale Avenue, White Plains, NY 10607. (914) 761-1100. 1993. $25.00 in paper.

The Basics of Video Production. Lvyder and Swainson. Butterworth-Heinemann, 313 Washington Street, Newton, MA 02158. (618-928-2500. (800) 366-2665. 1995. $14.95.

Broadcast Transmission Engineering Practice. Wharton, et al. Butterworth-Heinemann, 313 Washington Street, Newton, MA 02158. (618-928-2500. (800) 366-2665. 1992. $125.00.

Color and Black and White Television theory and Servicing. Alvin A. Liff and Sam Wilson. Prentice Hall, 113 Sylvan Avenue, Route 9W, Englewood Cliffs, NJ 07632. (201) 592-2000. (800) 922-0579. 3rd edition. 1993. $65.00.

Complete Guide To Digital Television Troubleshooting and Repair. John D. Lenk. Prentice Hall, 113 Sylvan Avenue, Route 9W, Englewood Cliffs, NJ 07632. (201) 592- 2000. (800) 922-0579.

Electronic Displays: Technology, Design, and Applications. Jerry Whitaker. McGraw-Hill Publishing Company, Inc., 1221 Avenue of the Americas, New York, NY 10020. (212) 512-2000. (800) 262-4729. 1993. $55.00.

A Guide To Television and Video Technology. Eugene Trundie.Butterworth-Heinemann, 313 Washington Street, Newton, MA 02158. (618-928-2500. (800) 366-2665. 1995. $32.95.

The Technique of Lighting For Television and Film. Gerald Millerson. Butterworth-Heinemann, 313 Washington Street, Newton, MA 02158. (618-928-2500. (800) 366-2665. 3rd edition. 1991. $46.95.

Telecommunications Circuit Design. Patrick D. Van der Puije. John Wiley & Sons, Inc., 605 Third Avenue, New York, NY 10158-0012. (212) 850-6000. (800) 225-5945. 1992. $79.95.

Television and Video Systems. Charles Buscombe. Prentice- Hall, 113 Sylvan Avenue, Route 9W, Englewood Cliffs, NJ 07632. (201) 592-2000. (800) 922-0579. 1990. $70.00.

Using Digital Video. Arch C.Luther. Academic Press, Inc., 6277 Sea Harbor Drive, Orlando, FL. (800) 321-5068. 1994. $34.95.

Video Engineering: Ntsc, Edtv and Hdtv Systems. andrew F. Inglis. McGraw-Hill Publishing Company, Inc., 1221 Avenue of the Americas, New York, NY 10020. (212) 512-2000. (800) 262-4729. 1992. $50.00.

HANDBOOKS AND MANUALS

Lenk's Video Handbook: Operation and Troubleshooting. John D. Lenk. McGraw-Hill Publishing Company, Inc., 1221 Avenue of the Americas, New York, NY 10020. (212) 512-2000. (800) 262-4729. 2nd edition. 1996. $29.95.

Newnes Television and Video Engineer's Pocket Book. Eugene Trundle. Butterworth-Heinemann, 313 Washington Street, Newton, MA 02158. (618- 928-2500. (800) 366-2665. 1992. $32.95.

Standard Handbook For Electrical Engineers. Donald Fink. McGraw-Hill Publishing Company, 1221 Avenue of the Americas, New York, NY 10020. (212) 512-2000. (800) 262- 4729. 13th edition. 1996. $110.00.

Tab Service Manual For Cctv and Matv. Robert L. Goodman. TAB Books, P.O. Box 40, Blue Summit, PA 17294-0850. (717) 794-2191. (800) 233-1128. 1991. $29.95.

Television and Video Engineer's Reference Book. K. Jackson and B. Townsend. Butterworth-Heinemann, 313 Washington Street, Newton, MA 02158. (618-928-2500. (800) 366-2665. 1991. $79.95.

Videl Test and Measurement Handbook. Horn. CRC Press, Inc., 2000 Corporate Boulevard, NW, Boca Raton, FL 33431. (407) 994-0555. (800) 272-7737. 1995.

ONLINE DATABASES AND CD-ROMS

Compendex Plus. Engineering Information, Inc., 345 East 47th Street, New York, NY 10017. (212) 705-7600 or (800) 221-1044. Contains citations with abstracts to worldwide literature in engineering and technology, from 1970 to present. Available on online BRS,(800) 289-4277, DIALOG, (800) 334-2564, ORBIT (800) 456-7248, and STN International, FIZ Karlsruhe, P.O. Box 2465, W-7500, Karlsruhe 1, Germany, online services. Also available on CD-ROM. Inquire as to cost and availability.

INSPEC. Institution of Electrical Engineers, Michael Faraday House, Six Hills Way, Stevenage, Herts. SG1 2AY, England. Telephone: 0438 313311 or FAX 0438 742840. Contains citations to the worldwide literature of physics, electronics and electrical engineering, computer technology, and related fields. Available on BRS, (800) 289-4277, DIALOG, (800) 334-2564, ORBIT, (800) 456-7248, and STN International, FIZ Karlsruhe, P.O. Box 2465, W-7500, Karlsruhe 1, Germany, online services. Inquire as to cost and availability.

NTIS Bibliographic Database. National Technical Information Service, 5285 Port Royal Road, Springfield, VA 22161. (703) 487-4929 or FAX (703) 321-8199. Broad coverage of government-sponsored science and technology research reports, 1964 to present. Available on BRS,(800) 289-4277, DIALOG, (800)

VIDEO TECHNOLOGY

Ency. of Physical Sciences and Engineering Info. Sources

334-2564, ORBIT, (800) 456-7248, and STN International, FIZ Karlsruhe, P.O. Box 2465, W-7500, Karlsruhe 1, Germany, online services. Also available on CD-ROM. Inquire as to cost and availability

WILSONLINE. H.W. Wilson Company, 950 University Avenue, Bronx, NY 10452. (800) 367-6770 or (212) 588-8400. Makes available online versions of the printed H.W. Wilson indexes including Applied Science and Technology Index, Business Periodicals Index, General Science Index, and Readers' Guide to Periodical Literature. Period covered is generally 1983 to present. Available on BRS, (800) 289-4277, DIALOG, (800) 334-2564, and OCLC EPIC, (800) 848-5878, online services.Also available on CD-ROM. Inquire as to cost and availability.

PERIODICALS

Audio Engineering Society. JOURNAL. Audio Engineering Society, 60 East 42nd Street, New York, NY 10017. (212) 661-8528. 1953 to present. 10 issues per year. $125.00 per year. ISSN: 0004-7554.

Broadcast Engineering. Intertec Publishing Corporation, Box 12801, Overland Park, KS 66212. (913) 341-1300. 1959 to present. Monthly. $50.00 per year. ISSN: 0007-1994.

DB, the Sound Engineering Magazine. 203 Commack Road, Suite 1010, Commack, NY 1725. (516) 586-6530. 1967 to present. $15.00 per year. ISSN: 0011-7145. .

Document and Image Automation. Meckler Publishing Corporation, 11 Ferry Lane West, Westport, CT 06880-5808. (203) 226-6967. 1981 to present. Bi-monthly. $125.00 per year. ISSN: 1054-9692.

Electronic Design. Penton Publishing, San Jose Gateway, Suite 354. 2025 Gateway Place, San Jose, CA 95110. (408) 441-0550. 1952 to present. Fortnightly. $95.00. ISSN: 0013-4872.

Electronics. Penton Publishing, San Jose Gateway, Suite 354. 2025 Gateway Place, San Jose, CA 95110. (408) 441-0550. 1930 to present. Semi-weekly. $98.00 per year. ISSN: 0883-4989.

Electronics Now. Gernsback Publications, Inc., 500-B Bi-County Boulevard, Farmingdale, NY 11735. (516) 293-3000. FAX (516) 293-3115. 1929 to present. Monthly. $19.95. ISSN: 0033-7862.

Electronics World and Wireless World. Reed Business Publishing, Ltd., Quadrant House, the Quadrant, Sutton, Surrey SM2 5AS, England. 1911 to present. Monthly. L35. ISSN: 0266-3244.

IEEE Circuits and Devices Magazine. Institute of Electrical and Electronics Engineers, Inc., Box 1331, 445 Hoes Lane, Piscataway, NJ 08855-1331. (908) 981-0060. 1985 to present. Bi-monthly. $120.00 per year. ISSN: 8755-3996.

IEEE Journal of Solid State Circuits. Institute of Electrical and Electronics Engineers, Inc., Box 1331, 445 Hoes Lane, Piscataway, NJ 08855-1331. (908) 981-0060. 1966 to present. Bi-monthly. $275.00 per year. ISSN: 0018- 9200.

IEEE Transactions On Broadcasting. Institute of Electrical and Electronics Engineers, Inc., Box 1331, 445 Hoes Lane,

Piscataway, NJ 08855-1331. (908) 981-0060. 1955 to present. Quarterly. $50.00 per year. ISSN: 0018-9316.

Microwaves and RF. Penton Publishing Co., Circulation Department, 1100 Superior Avenue, Cleveland, OH 44114. (201) 393-6060. FAX (201) 393-6297. From 1962 to present. Monthly. $60.00. ISSN: 0745-2992.

Television Broadcast. Globecom Publishing Ltd., 4451 West 107th, No. 210, Overland Park, KS 66207-4024. 1978 to present. Monthly. $38.00 per year. ISSN: 0898-767X.

RESEARCH CENTERS AND INSTITUTES

Center for Advanced Technology in Telecommunications. Polytechnic University, 333 Jay Street, Brooklyn, NY 11202. (718) 643-5160.

Center For Imaging Science. Rochester Institute of Technology, One Lomb Memorial Drive, Rochester, NY 14623. (716) 475-2774.

Communications Research Laboratory. McMaster University, 1280 Main Street West, Hamilton, ON Canada L85 4K1. (416) 525-9140. <bodylead>Communications Satellite Planning Center. Stanford University, ERL, Building, Room 202, Stanford, CA 94305- 4053. (415) 723-3471. FAX (415) 723-7298.

Video Center. New York Institute of Technology. Northern Boulevard, Old Westbury, NY 11568. (516) 626-3570. FAX (516) 626-0716.

VIRTUAL REALITY

See also: ARTIFICIAL INTELLIGENCE, COMPUTER VISION, EXPERT SYSTEMS

ABSTRACT SERVICES AND INDEXES

ACM Guide To Computing Literature. Association for Computing Machinery, Association for Computing Machinery, 1515 Broadway, 17th Floor, New York, NY 10036-5701. (212) 869-7440. FAX (212) 944-1318. 1964 to present. Monthly. $190.00 per year for non-members.

Applied Science and Technology Index; A Cumulative Subject Index To English Language Periodicals in the Fields of Aeronautics and Space Science, Computer Technology, Chemistry, Construction Industry, Energy and Related Areas. H.W. Wilson Co., 950 University Avenue, Bronx, NY 10452. (800) 367-6770 or (212) 588-8400. FAX (718) 590-1617. From 1958 to present. Monthly. Inquire about cost and availability. Also available on CD-ROM and online.

Computer Abstracts, MCB University Press Ltd., PO Box 10812, Birmingham, AL 35201-0812. (800) 633-4931. FAX (205) 995-1588. Monthly. Covers computer theory, data, hardware, systems, networks, human-computer interaction, artificial intelligence, as well as applications of computers in aerospace, business, CAD/CAM, cartography, civil engineering, electronics and electrical engineering, industrial engineering, mechanical engineering, medicine, structural engineering, etc. Also available on CD-ROM.

Computer & Control Abstracts (INSPEC). INSPEC/IEEE, Box 1331, 445 Hoes Lane, Piscataway, NJ 08855-1331. (908) 562-5549. Abstracts organized by subjects of international technological information. Monthly. $1455.00 per year. Also available on CD-ROM and online via BRS Online Products, DIALOG Information Services, Orbit Search Service, and STN International.

Computer and Information Systems Abstracts Journal. Cambridge Scientific Abstracts, 7200 Wisconsin Avenue, Bethesda, MD 20814. (301) 961-6750. Fax (301) 961-6720. 1962 to present. Monthly. $1265.00 per year. Also available online via STN International.

Computer Literature Index. Applied Computer Research Inc., Box 82266, Phoenix, AZ 85071-2266. (800) 234-2227. 1971 to present. Quarterly plus annual cumulation. $198.50 per year. Bibliography of books, articles, and reports.

Computing Journal Abstracts. Techgnosis Ltd., Blade House, Battersea Road, Stockport, Cheshire 3AE, England. Telephone 061-442-2639. FAX 061-443-1162. 1969 to present. Monthly. Inquire for price.

Computing Reviews. Association for Computing Machinery, 1515 Broadway, 17th Floor, New York, NY 10036-5701. (212) 869-7440. FAX (212) 944-1318. 1960 to present. Monthly. $130.00 per year for non-members. Also available online via DIALOG Information Services.

Current Contents: Engineering, Technology, and Applied Sciences. Institute for Scientific Information, 3501 Market Street, Philadelphia, PA 19104. (215) 386-0100. FAX (215) 386-6362. 1970 to present. Weekly. $442.00 per year.

Engineering Index Monthly. Engineering Information, Inc., Castle Point on the Hudson, Hoboken, NJ 07030. (800) 221-1044. FAX (212) 832-1857. Monthly. $2300.00 per year. Also available online as *Compendex*, and also on CD-ROM. Covers chemical engineering, computers, electrical engineering, civil engineering, metals and mining, industrial management, and mechanical engineering.

Index To Scientific and Technical Proceedings. Institute for Scientific Information, 3501 Market St., Philadelphia, PA 19104. (215) 386-0100. FAX (215) 386-6362. Monthly. $500.00 per year.

Index To Scientific Reviews. Institute for Scientific Information, 3501 Market St., Philadelphia, PA 19104. (215) 386-0100. FAX (215) 386-6362. Semi-annual.

Science Citation Index. Institute for Scientific Information, 3501 Market Street, Philadelphia, PA 19104. (215) 386-0100. FAX (215) 386-6362. Inquire about availability and cost. Also available on CD-ROM.

ANNUAL REVIEWS AND YEARBOOKS

ACM Monograph Series. Association for Computing Machinery (ACM), 11 West 42nd Street, New York, NY 10036. (212) 869-7440. Fax (212) 869-0481. Irregular. Price varies, inquire.

Advances in Computers. Academic Press Inc., 6277 Sea Harbor Drive, Orlando, FL 32887. (800) 321-5068. Irregular. Price varies, inquire.

Advances in Computing Research. JAI Press, Inc., 55 Old Post Road, Number 2, Box 1678, Greenwich, CT 06830. (203) 661-7602. FAX (203) 661-0792. Annual. $68.50 per year.

ASSOCIATIONS AND PROFESSIONAL SOCIETIES

Association For Computing Machinery. 1515 Broadway, 17th Floor, New York, NY 10036-5701. (212) 869-7440. FAX (212) 944-1318.

IEEE Computer Society. 1730 Massachusetts Avenue NW, Washington, DC 20036-1992. (202) 371-0101. FAX (202) 728-9614.

Society For Computer Simulation International, 4838 Ronson Court Lane, San Diego, CA 92111-1810. (619) 277-3888.

BIBLIOGRAPHIES

ACM Guide To Computing Literature. Association for Computing Machinery, 1515 Broadway, 17th Floor, New York, NY 10036-5701. (212) 869-7440. FAX (212) 944-1318. 1964 to present. Annual. $190.00 for non-members. Also available online via DIALOG.

DIRECTORIES AND BIOGRAPHICAL SOURCES

American Men & Women of Science, 1995-96. R.R. Bowker Staff, eds. 19th edition. 8 volumes. R.R. Bowker/Reed International Publishing Company, 121 Chanlon Road, New Providence, NJ 07974. (908) 464-6800 or (800) 521-8110. 1995. $850.00

Computing Information Directory. Hildebrandt Inc., PO Box 576, Pullman, WA 99114. 1981 to present. Annual. $199.95 per year.

Directory of Engineering Societies and Related Organizations. Gordon Davis. 13th edition. American Association of Engineering Societies, 1111 19th Street NW, Suite 608, Washington, DC 20036. (202) 296-2237 or (800) 658-8897. 1989. Inquire for price.

International Research Centers Directory. Gale Research, 835 Penobscot Building, Detroit, MI 48226-4094. (313) 961-2242. (800) 877-4253. FAX (313) 961-6083. 8th edition. 1995. $410.00

Research Centers Directory. Gale Research, 835 Penobscot Building, Detroit, MI 48226-4094. (313) 961-2242. (800) 877-4253. FAX (313) 961-6083. $485.00.

Scientific and Technical Organizations and Agencies Directory. Gale Research, 835 Penobscot Building, Detroit, MI 48226-4094. (313) 961-2242. (800) 877-4253. FAX (313) 961-6083. 4th edition. 1996. $195.00.

Who's Who in Technology. Gale Research, 835 Penobscot Building, Detroit, MI 48226-4094. (313) 961-2242. (800) 877-4253. FAX (313) 961-6083. 1995. $195.00.

VIRTUAL REALITY

Ency. of Physical Sciences and Engineering Info. Sources

ENCYCLOPEDIAS AND DICTIONARIES

Computer Dictionary. Donald D. Spencer. 4th edition. Camelot Publishing Company, 709 SW 80th Blvd., Gainesville, FL 32607-1537. (904) 331-0952. 1993. $24.95.

Dictionary of Computer Graphics Technology & Applications. Roy Latham. 2nd edition. Springer-Verlag, 175 Fifth Avenue, New York, NY 10010. (212) 460-1500 or (800) 777-4643. FAX (212) 473-6272. $19.95.

Dictionary of Computing. S.M.H. Collin. 2nd edition. Peter Collin Publishing, 8 the Causeway, Teddington, TW11 0HE, England. FAX 0181-943-3386. 1994. Inquire for cost.

The Dictionary of Computer Graphics Technology and Applications. Roy Latham. 2nd edition. Springer-Verlag, 175 Fifth Avenue, New York, NY 10010. (212) 460-1500 or (800) 777-4643. FAX (212) 473-6272. 1995. $19.95.

Encyclopedia of Computer Science and Engineering. Anthony Ralston & Edwin D. Reilly Jr., editors. 3rd revised edition. Van Nostrand Reinhold, 115 Fifth Avenue, New York, NY 10003. (212) 254-3232 or FAX (212) 254-9499. 1993. $125.00.

Microsoft Press Computer Dictionary: The Comprehensive Standard For Business, School, Library, and Home. 2nd edition. Microsoft Press, One Microsoft Way, Redmond, WA 98052-6399. (206) 936-0055. FAX (206) 823-8101. 1994. $19.95.

GENERAL WORKS

Create Your Own Virtual Reality System. Joseph Levy. McGraw-Hill Publishing Company, 1221 Avenue of the Americas, New York, NY 10020. (800) 262-4729 or (212) 512-3825. 1995. $44.95 (book and diskette).

Designing 3-D Graphics: How To Create Real-Time 3-D Models For Games and Virtual Reality. Josh White. John Wiley and Sons, Inc., 605 Third Avenue, New York, NY 10158. (800) 526-5368 or (212) 850-6000. 1996. $39.95.

Experiments in Virtual Reality. David Harrison. Butterworth-Heinemann, 313 Washington Street, Newton, MA 02158. (617) 928-2500 or (800) 366-2665. FAX (617) 928-2620. 1996. Inquire for price.

Virtual Realities and their Discontents. Robert Markley, editor. Johns Hopkins University Press, 2715 North Charles St., Baltimore, MD 21218. (410) 516-6900 or (800) 537-5487. FAX (410) 516-6998. 1996. $38.50.

Virtual Reality Applications. R.A. Karnshaw, et al., editors. Academic Press Inc., 6277 Sea Harbor Drive, Orlando, FL 32887. (800) 321-5068. 1995. $44.95.

Virtual Reality: Applications and Explorations. Alan Wexelblat, editor. Academic Press Inc., 6277 Sea Harbor Drive, Orlando, FL 32887. (800) 321-5068. 1994. $29.95.

Virtual Reality: Markets and Opportunities. Business Communications Company, Inc., 25 Van Zant St., Norwalk, CT 06855. (203) 853-4266. FAX (203) 853-0348. 1996. Inquire for price.

Virtual Reality: Strategies For Intranet and World Wide Web Applications. Computer Technology Research Corporation, 6

N. Atlantic Wharf, Charleston, SC 29401-2150. (803) 853-6460. FAX (803) 853-7210. 1996. $315.00.

Virtual Reality Systems. John Vince. Addison-Wesley Publishing Company, 1 Jacob Way, Reading, MA 01867. (617) 944-3700 or (800) 447-2226. FAX 617-942-1117. 1995. $44.95.

HANDBOOKS AND MANUALS

The Virtual Reality Homebrewer's Handbook. Robin Hollands. John Wiley and Sons, Inc., 605 Third Avenue, New York, NY 10158. (800) 526-5368 or (212) 850-6000. 1996. $39.95.

ONLINE DATABASES AND CD-ROMS

Compendex Plus. Engineering Information, Inc., 345 East 47th Street, New York, NY 10017. (212) 705-7600 or (800) 221-1044. Contains citations with abstracts to worldwide literature in engineering and technology, from 1970 to present. Available on online BRS,(800) 289-4277, DIALOG, (800) 334-2564, ORBIT (800) 456-7248, and STN International, FIZ Karlsruhe, P.O. Box 2465, W-7500, Karlsruhe 1, Germany, online services. Also available on CD-ROM. Inquire as to cost and availability.

Compuscience. FIZ Karlsruhe, D-7514, Eggenstein-Leopoldshafen 2, Germany. Contains citations with abstracts to European and North American literature on computer science, 1972 to present. Available on STN International, FIZ Karlsruhe, P.O. Box 2465, W-7500, Karlsruhe 1, Germany, online service. Inquire as to cost and availability.

Computer and Information Systems Abstracts. Cambridge Scientific Abstracts, 7200 Wisconsin Avenue, Suite 601, Bethesda, MD 20814. (301) 961-6750 or (800) 843-7751. Contains citations to worldwide literature in theoretical and applied computer science and related areas, from 1981 to present.Inquire as to cost and availability.

Computer and Mathematics Search. Institute for Scientific Information, 3501 Market Street, Philadelphia, PA 19104. (215) 386-0100. FAX (215) 386-6362. Covers worldwide literature in computer science and mathematics, from 1980 to present. Available on BRS,(800) 289-4277, online service. Inquire as to cost and availability.

Computer Database. Information Access Company, 362 Lakeside Drive, Foster City, CA 94404. (415) 378-5000 or (800) 227-8431. Contains citations with abstracts to literature from trade journals covering the computer,telecommunications,and electronics industries. Available on the BRS, (800) 289-4277, Compuserve Information Service,(800) 848-8990, and DIALOG,(800) 334-2564, online services. Inquire as to cost and availability.

CSA Engineering. Cambridge Scientific Abstracts, 7200 Wisconsin Avenue, Suite 601, Bethesda, MD 20814. (301) 961-6750 or (800) 843-7751. Contains citations and abstracts of international periodicals and research literature covering all fields of engineering and science and technology,including computer and information science, electronics, mechanical engineering, solid state materials, 1981 to present. Available on BRS,(800) 289-4277, online service.Inquire as to cost and availability.

Current Contents Search. Institute for Scientific Information, 3501 Market Street, Philadelphia, PA 19104. (215) 386-0100. FAX (215) 386-6362. Contains citations to articles listed in the table of contents of science and technology journals. Also ar-

ticles in social sciences and life sciences journals. Available on BRS,(800) 289-4277, DIALOG,(800) 334-2564, online services. Inquire as to cost and availability.

Dissertation Abstracts. University Microfilms International, 300 North Zeeb Road, Ann Arbor, MI 48106. (800) 521-0600 or (313) 761-4700. Scope includes virtually all doctoral dissertations accepted at accredited American institutions from 1861 to present in 252 subject areas. Available on BRS,(800) 289-4277, DIALOG,(800) 334-2564, and OCLC EPIC,(800) 848-5878, online services. Also available on CD-ROM. Inquire as to cost and availability.

INSPEC. Institution of Electrical Engineers, Michael Faraday House, Six Hills Way, Stevenage, Herts. SG1 2AY, England. Telephone: 0438 313311 or FAX 0438 742840. Contains citations to the worldwide literature of physics, electronics and electrical engineering, computer technology, and related fields. Available on BRS,(800) 289-4277, DIALOG, (800) 334-2564, ORBIT, (800) 456-7248, and STN International, FIZ Karlsruhe, P.O. Box 2465, W-7500, Karlsruhe 1, Germany, online services. Inquire as to cost and availability.

Mathsci. American Mathematical Society, P.O. Box 6248, Providence, RI 02940. (800) 321-4667 or (401) 455-4000 or FAX (401) 331-3842. Scope includes pure and applied mathematics and related areas of physics, statistics, engineering, computer science, and operations research literature since 1959. Available on DIALOG,(800) 334-2564, online service. Also available on CD-ROM. Inquire as to cost and availability.

NTIS Bibliographic Database. National Technical Information Service, 5285 Port Royal Road, Springfield, VA 22161. (703) 487-4929 or FAX (703) 321-8199. Broad coverage of government-sponsored science and technology research reports, 1964 to present. Available on BRS,(800) 289-4277, DIALOG, (800) 334-2564, ORBIT, (800) 456-7248, and STN International, FIZ Karlsruhe, P.O. Box 2465, W-7500, Karlsruhe 1, Germany, online services. Also available on CD-ROM. Inquire as to cost and availability.

SCISEARCH. Institute for Scientific Information, 3501 Market Street, Philadelphia, PA 19104. (800) 523-1850 or (215) 386-0100. Broad multidisciplinary title and author index to the international literature of science and technology, 1974 to present. Available on DIALOG,(800) 334-2564, and ORBIT,(800) 456-7248, online services. Also available on CD-ROM. Inquire as to cost and availability.

Wilsonline. H.W. Wilson Company, 950 University Avenue, Bronx, NY 10452. (800) 367-6770 or (212) 588-8400. Makes available online versions of the printed H.W. Wilson indexes including Applied Science and Technology Index, Business Periodicals Index, General Science Index, and Readers' Guide to Periodical Literature. Period covered is generally 1983 to present. Available on BRS,(800) 289-4277, DIALOG,(800) 334-2564, and OCLC EPIC,(800) 848-5878, online services. Also available on CD-ROM. Inquire as to cost and availability.

PERIODICALS

ACM Transactions On Computer Systems. Association for Computing Machinery, 11 West 42nd Street, New York, NY 10036. (212) 869-7440. Fax (212) 869-0481. 1964 to present. Annual. $180.00 per year.

IEEE Computer Graphics and Applications. IEEE Computer Society, 10662 Vaqueros Circle, Box 3014, Los Alamitos, CA 90720. (714) 821-8380. Bimonthly. $120.00 per year.

Image and Vision Computing. Butterworth-Heinemann, Linacre House, Jordan Hill, Ocford OX2 8DP, England. Telephone 0865-310366. FAX 0865-310898. 1982 to present. Ten times a year. Inquire for price.

International Journal of Computer Vision. Kluwer Academic Publishers, P.O. Box 358, Accord Station, Hingham, MA 02018-0358. (617) 871-6000. 1987 to present. Six times a year. $314.00.

Virtual Reality Now. Media Dimension Inc., 1562 First Avenue, Suite 286, New York, NY 10028-4004. (212) 533-7481. FAX (212) 475-1209. 1992 to present. Quarterly. $65.00 per year.

Virtual Reality Report. Meckler Publishing Corporation, 11 Ferry Lane W., Westport, CT 06880-5808. (203) 226-6967. 1990 to present. 12 times a year. $327.00 per year.

Virtual Reality World. Meckler Publishing Corporation, 11 Ferry Lane W., Westport, CT 06880-5808. (203) 226-6967. 1993 to present. Bi-monthly. $33.00 per year.

VOLCANOLOGY

See also: EARTHQUAKE ENGINEERING,EARTHQUAKES, GEOLOGY, GEOPHYSICS, GEOTHERMAL ENERGY, IGNEOUS ROCKS, OCEANOGRAPHY, PLANETARY SCIENCE, PLATE TECTONICS

ABSTRACT SERVICES AND INDEXES

Bibliography and Index of Geology. American Geological Institute, 4220 King Street, Alexandria, VA 22302. (703) 379-2480. Fax (703) 379-7563. Monthly. $1295.00 per year. Also available online as GeoRef.

Current Contents: Physical, Chemical and Earth Sciences. Institute for Scientific Information, 3501 Market Street, Philadelphia, PA 19104. (215) 386-0100. FAX (215) 386-6362. Weekly. $360.00 per year.

Geological Abstracts. Elsevier - Geo Abstracts, Regency House, 34 Duke Street, Norwich NR3 3AP, England. Monthly. $760.00 per year. Also available online as GEOBASE.

Geological Society of America. ABSTRACTS WITH PROGRAMS. Geological Society of America. 3300 Penrose Place, P.O. Box 9140, Boulder, CO 80301-9140. (303) 447-2020. Abstracts and programs of the annual conference. Annual. $69.00.

Index To Scientific and Technical Proceedings. Institute for Scientific Information, 3501 Market St., Philadelphia, PA 19104. (215) 386-0100. FAX (215) 386-6362. Monthly. $500.00 per year.

Index To Scientific Reviews. Institute for Scientific Information, 3501 Market St., Philadelphia, PA 19104. (215) 386-0100. FAX (215) 386-6362. Semi-annual.

Meteorological and Geoastrophysical Abstracts. American Meteorological Society, c/o Inforonics Inc., 550 Newtown Road, Littleton, MA 01460. (508) 486-8976. FAX (508) 486-0027.

VOLCANOLOGY

Ency. of Physical Sciences and Engineering Info. Sources

1950 to present. Monthly. $950.00 per year. Current abstracts of books, reports, research papers, and miscellaneous literature on environmental sciences, meteorology, astrophysics, hydrology, glaciology, and physical oceanography.

Oceanic Abstracts. Cambridge Scientific Abstracts, 7200 Wisconsin Avenue, Bethesda, MD 20814. (301) 961-6750. Fax (301) 961-6720. Bimonthly. $995.00 per year.

Science Citation Index. Institute for Scientific Information, 3501 Market Street, Philadelphia, PA 19104. (215) 386-0100. FAX (215) 386-6362. Inquire about availability and cost. Also available on CD-ROM.

ASSOCIATIONS AND PROFESSIONAL SOCIETIES

American Geological Institute. 4220 King Street, Alexandria, VA 22302. (703) 379-2480. Fax (703) 379-7563.

American Geophysical Union. 2000 Florida Avenue, N.W., Washington, DC 20009. (202) 462-6900. FAX (202) 328-0566.

Earthquake Engineering Research Institute. 499 14th Street, Suite 320, Oakland, CA 94612-1902. (510) 451-0905.

Geological Society of America. 3300 Penrose Place, PO Box 9140, Denver, CO 80301-9140. (303) 447-2020.

International Association of Volcanology and Chemistry of the Earth's Interior. Australian Geological Survey Organization, GPO Box 378, Canberra, ACT 2601, Australia. Telephone 6-2499377. FAX 6-2499983.

International Institute of Seismology and Earthquake Engineering. Building Research Institute—Ministry of Construction, 1 Tatehara, Tsukuba-city, Ibaraki Prefecture 305, Japan.

International Seismological Centre. Pipers Lane, Thatcham, Newbury, Berks. RG13 4NS, England. Telephone 0635-61022. FAX 0635-72351.

Seismological Society of America. 201 Plaza Professional Bldg., El Cerrito, CA 94530. (510) 525-5474. FAX (510) 525-7204.

DIRECTORIES AND BIOGRAPHICAL SOURCES

American Men & Women of Science, 1995-96. R.R. Bowker Staff, eds. 19th edition. 8 volumes. R.R. Bowker/Reed International Publishing Company, 121 Chanlon Road, New Providence, NJ 07974. (908) 464-6800 or (800) 521-8110. 1995. $850.00

Directory of Engineering Societies and Related Organizations. Gordon Davis. 13th edition. American Association of Engineering Societies, 1111 19th Street NW, Suite 608, Washington, DC 20036. (202) 296-2237 or (800) 658-8897. 1989. Inquire for price.

International Research Centers Directory. Gale Research, 835 Penobscot Building, Detroit, MI 48226-4094. (313) 961-2242. (800) 877-4253. FAX (313) 961-6083. 8th edition. 1995. $410.00.

Research Centers Directory. Gale Research, 835 Penobscot Building, Detroit, MI 48226-4094. (313) 961-2242. (800) 877-4253. FAX (313) 961-6083. $485.00.

Scientific and Technical Organizations and Agencies Directory. Gale Research, 835 Penobscot Building, Detroit, MI 48226-4094. (313) 961-2242. (800) 877-4253. FAX (313) 961-6083. 4th edition. 1996. $195.00.

ENCYCLOPEDIAS AND DICTIONARIES

Encyclopedia of Earthquakes and Volcanoes. David Ritchie. Facts-on-File, Inc., 460 Park Avenue South, New York, NY 10016-7382. (800) 322-8755. Fax (212) 213-4578. 1994. $40.00.

The Encyclopedia of the Solid Earth Sciences. Philip Kearey, editor-in-chief. Blackwell Scientific Publications, 238 Main Street, Cambridge, MA 02142. (617) 876-7000 or (800) 759-6102. FAX (617) 876-7022. 1993. $60.00.

GENERAL WORKS

Earthquakes and Volcanoes: Readings From the Scientific American. W.H. Freeman and Company, 41 Madison Avenue, East 26th, 35th Floor, New York, NY 10010. (212) 576-9400. FAX (212) 689-2383. 1995. Inquire for price.

Quakes, Eruptions and Other Geologic Cataclysms. Jon Erickson. Facts-on-File, Inc., 460 Park Avenue South, New York, NY 10016-7382. (800) 322-8755. FAX (212) 213-4578. 1994. $24.95.

Volcano Monitoring Techniques. Rector Press Ltd., 130 Rattlesnake, Leverett, MA 01054-9726. (800) 247-3473. FAX (413) 367-2853. 1994. $295.00.

Volcanoes. Robert and Barbara Decker. Second edition. W.H. Freeman and Company, 41 Madison Avenue, East 26th, 35th Floor, New York, NY 10010. (212) 576-9400. FAX (212) 689-2383. 1995. Inquire for price.

Volcanoes: A Planetary Perspective. Peter Francis. Oxford University Press, Inc., 198 Madison Avenue, New York, NY 10016-4314. (212) 726-6000. FAX (212) 726-6446. 1993. $42.95.

Volcanoes of the World. Tom Simkin. Second edition. Geoscience Press, 12629 N. Tatum Blvd., Suite 201, Phoenix, AZ 85032. (602) 953-2330. FAX (602) 953-1926. 1994. $25.00.

Volcanology and Geothermic Energy. Kenneth Wohletz and Grant Heiken. University of California Press, 2120 Berkeley Way, Berkeley, CA 94720. (415) 642-4191 or FAX (415) 643-7127. 1992. $65.00.

ONLINE DATABASES AND CD-ROMS

Compendex Plus. Engineering Information, Inc., 345 East 47th Street, New York, NY 10017. (212) 705-7600 or (800) 221-1044. Contains citations with abstracts to worldwide literature in engineering and technology, from 1970 to present. Available on online BRS,(800) 289-4277, DIALOG, (800) 334-2564, ORBIT (800) 456-7248, and STN International, FIZ Karlsruhe, P.O. Box 2465, W-7500, Karlsruhe 1, Germany, online services. Also available on CD-ROM. Inquire as to cost and availability.

Earth Sciences. U.S. Geological Survey, 12201 Sunrise Valley Drive, Reston, VA 22092-9998. (703) 648-4460. CD-ROM of earth science databases including the U.S. Geological Survey Library, Earth Science Data Directory, and GEOINDEX, citations to published geological maps. $350.00 per year with quarterly updates. <bodylead>*Geoarchive.* Geosystems, P.O. Box 1024, Westminster, London, England, SW1 P 2JL. Citations to literature on geoscience, 1969 to present. Inquire as to online cost and availability.

Geobase. Elsevier - Geo Abstracts, Regency House, 34 Duke Street, Norwich NR3 3AP, England. Contains citations to the worldwide earth science literature from 1980 to date. Available on DIALOG, ORBIT online services. Inquire as to cost and availability. <bodylead>*Georef: Bibliography and Index of Geology.* American Geological Institute, 4220 King Street, Alexandria, VA 22302. (703) 379-2480. Fax (703) 379-7563. Monthly. Inquire for price and availability.

NTIS Bibliographic Database. National Technical Information Service, 5285 Port Royal Road, Springfield, VA 22161. (703) 487-4929 or FAX (703) 321-8199. Broad coverage of government-sponsored science and technology research reports, 1964 to present. Available on BRS,(800) 289-4277, DIALOG, (800) 334-2564, ORBIT, (800) 456-7248, and STN International, FIZ Karlsruhe, P.O. Box 2465, W-7500, Karlsruhe 1, Germany, online services. Also available on CD-ROM. Inquire as to cost and availability.

SCISEARCH. Institute for Scientific Information, 3501 Market Street, Philadelphia, PA 19104. (800) 523-1850 or (215) 386-0100. Broad multidisciplinary title and author index to the international literature of science and technology, 1974 to present. Available on DIALOG,(800) 334-2564, and ORBIT,(800) 456-7248, online services. Also available on CD-ROM. Inquire as to cost and availability.

PERIODICALS

Bulletin of Volcanology. Springer-Verlag, 44 Hartz Way, Secaucus, NJ 07096-2491. 1924 to present. Eight times a year. $443.00.

Geophysical and Astrophysical Fluid Dynamics. Gordon and Breach Science Publishers, 820 Town Center Drive, Langhorne, PA 19047. (215) 750-2642. FAX (215) 750-6343. 1970 to present. 28 times a year. Inquire for price.

Geophysical Research Letters. American Geophysical Union, 2000 Florida Avenue, N.W., Washington, DC 20009. (202) 462-6900. 1974 to present. Semi-monthly. $590.00 per year for non-members.

Geotimes. American Geological Institute, 4220 King Street, Alexandria, VA 22302. (703) 379-2480. Fax (703) 379-7563. 1956 to present. Monthly. $24.95.

International Seismological Centre Bulletin. Pipers Lane, Thatcham, Newbury, Berks. RG13 4NS, England. Telephone 0635-61022. FAX 0635-72351. 1964 to present. Monthly. Inquire for price in U.S.

International Seismological Centre Regional Catalogue of Earthquakes. Pipers Lane, Thatcham, Newbury, Berks. RG13 4NS, England. Telephone 0635-61022. FAX 0635-72351. 1964 to present. Semi-annual. Inquire for price in U.S.

Journal of Volcanology and Geothermal Research. Elsevier Science Inc., Box 882, Madison Square Station, New York, NY 10159. (212) 989-5800. FAX (212) 633-3990. 1976 to present. 24 times a year. $1155.00.

Physics of the Earth and Planetary Interiors. Elsevier Science Inc., Box 882, Madison Square Station, New York, NY 10159. (212) 989-5800. 1967 to present. 24 times a year. $1317.00.

Seismological Research Letters. Seismological Society of America, 201 Plaza Professional Bldg., El Cerrito, CA 94530. (510) 525-5474. FAX (510) 525-7204. 1929 to present. Quarterly. $15.00.

Seismological Society of America Bulletin. 201 Plaza Professional Bldg., El Cerrito, CA 94530. (510) 525-5474. FAX (510) 525-7204. 1911 to present. Bi-monthly. $135.00.

Tectonophysics. Elsevier Science Inc., Box 882, Madison Square Station, New York, NY 10159. (212) 989-5800. FAX (212) 633-3990. 1964 to present. 56 times a year. $2429.00.

Volcanology and Seismology. Gordon and Breach Science Publishers, 820 Town Center Drive, Langhorne, PA 19047. (215) 750-2642. FAX (215) 750-6343. Twelve times a year. Inquire for cost.

RESEARCH CENTERS AND INSTITUTES

Geothermal Resources Council. PO Box 1350, Davis, CA 95617-1350. (916) 758-2360. FAX (916) 758-2839.

Incorporated Research Institutions For Seismology. Suite 1440, 1616 N. Fort Myer Drive, Arlington, VA 22209. (702) 524-6222. (703) 527-7256.

International Seismological Centre. Pipers Lane, Thatcham, Newbury, Berks. RG13 4NS, England. Telephone 0635-61022. FAX 0635-72351.

University of California, Institute of Geophysics and Planetary Physics. UC-San Diego, A-025, LA Jolla, CA 92093. (619) 534-2870.

University of California—Santa Cruz, Charles F. Richter Seismological Laboratory. Institute of Tectonics, Santa Cruz, CA 95064. (408) 459-4137.

W

WANKEL ENGINE

See: ENGINES

WATER

See also: GROUND WATER, GROUND WATER POLLUTION, HYDROLOGY, WATER POLLUTION, WATER RESOURCES, WATER TREATMENT

ABSTRACT SERVICES AND INDEXES

Aqualine Abstracts. Water Research Centre, PO Box 85, Frankland Rd., Blagrove, Swindon, Wilts. SN5 8YF, England. 1927 to present. 26 times a year. Inquire for price in U.S. Also available online.

Bibliography and Index of Geology. American Geological Institute, 4220 King Street, Alexandria, VA 22302. (703) 379-2480. Fax (703) 379-7563. Monthly. $1295.00 per year. Also available online as GeoRef.

Current Contents: Physical, Chemical and Earth Sciences. Institute for Scientific Information, 3501 Market Street, Philadelphia, PA 19104. (215) 386-0100. FAX (215) 386-6362. Weekly. $360.00 per year.

Fluid Abstracts: Civil Engineering. Elsevier Science [journals], 660 White Plains Rd., Tarrytown, NY 10159-5153. (919) 524-9200. FAX (919) 333-2444. 1991 to present. Monthly plus annual cumulation. $645.00 per year. Also available online as FLUIDEX. Covers civil engineering applications of fluid mechanics, hydraulics, flow metering and measuring, offshore engineering, environmental hydraulics, and related aspects of wind energy, the atmosphere, and aerodynamics.

Geological Abstracts. Elsevier - Geo Abstracts, Regency House, 34 Duke Street, Norwich NR3 3AP, England. Monthly. $760.00 per year. Also available online as GEOBASE.

Geological Society of America. Abstracts with programs. Geological Society of America. 3300 Penrose Place, P.O. Box 9140, Boulder, CO 80301-9140. (303) 447-2020. Abstracts and programs of the annual conference. Annual. $69.00.

Hydro-Abstracts (formerly Water Resources Abstracts). HydroScience Press, 2527 Jackson Street, N.E., Minneapolis, MN 55418. (612) 781-9081. Monthly. $150.00 per year.

Meteorological and Geoastrophysical Abstracts. American Meteorological Society, c/o Inforonics Inc., 550 Newtown Road, Littleton, MA 01460. (508) 486-8976. FAX (508) 486-0027. 1950 to present. Monthly. $950.00 per year. Current abstracts of books, reports, research papers, and miscellaneous literature on environmental sciences, meteorology, astrophysics, hydrology, glaciology, and physical oceanography.

Science Citation Index. Institute for Scientific Information, 3501 Market Street, Philadelphia, PA 19104. (215) 386-0100. FAX (215) 386-6362. Inquire about availability and cost. Also available on CD-ROM.

Selected Water Resources Abstracts. U.S. Geological Survey, Water Resources Scientific Information Center, 425 National Center, Reston, VA 22092. (703) 648-6820. Monthly. $115.00 peryear. Also available on CD-ROM.

ANNUAL REVIEWS AND YEARBOOKS

Advances in Hydrosciences. Academic Press Inc., 6277 Sea Harbor Drive, Orlando, FL 32887. (800) 321-5068. Inquire for cost and availability.

ASSOCIATIONS AND PROFESSIONAL SOCIETIES

American Institute of Hydrology. 3416 University Avenue SE, Minneapolis, MN 55414-3328. (612) 379-1030. FAX (612) 379-0169.

American Society of Civil Engineers. 345 East 47th Street, New York, NY 10017-2398. (212) 705-7520 or (800) 548-2723.

American Society of Limnology and Oceanography. School of Oceanography, WB-10, University of Washington, Seattle, WA 98195.

American Water Resources Association. 5410 Grosvenor Lane, Suite 220, Bethesda, MD 20814-2192. (301) 493-8600. FAX (301) 493-5844.

WATER

Ency. of Physical Sciences and Engineering Info. Sources

American Water Works Association. 6666 W. Quincy Avenue, Denver, CO 80235. (303) 794-7711. FAX (303) 794-7310.

Association of Ground Water Scientists and Engineers. A division of the National Water Well Association, 6375 Riverside Drive, Dublin, OH 43017. (614) 761-1711. FAX (614) 761-3446.

Ground Water Protection Council. 827 NW 63rd Street, Suite 103, Oklahoma City, OK 73116-7639. (405) 848-0690. FAX (405) 848-0722.

National Ground Water Association, 6375 Riverside Drive, Dublin, OH 43017. (614) 761-1711. FAX (614) 761-3446.

National Water Resources Association. 3800 N. Fairfax Drive, Suite 4, Arlington, VA 22203-1703. (703) 524-1544. FAX (703) 524-1548.

Universities Council On Water Resources. 4543 Faner Hall, Department of Geography, Southern Illinois University at Carbondale, Carbondale, IL 62901-4526. (618) 536-7571. FAX (618) 453-2671.

Water Environment Federation. 601 Wythe Street, Alexandria, VA 22314-1994. (703) 684-2400. FAX (703) 684-2492. Formerly the Water Pollution Control Federation.

Water Quality Association. 4151 Naperville Road, Lisle, IL 60532. (708) 505-0160. FAX (708) 505-9637.

BIBLIOGRAPHIES

Reading About the Environment: An Introductory Guide. P.E. Jansma. Libraries Unlimited, PO Box 6633, Englewood, CO 80155-6633. (800) 237-6124. FAX (303) 220-8843. 1993. $27.50.

Geraghty & Miller's Groundwater Bibliography. Fifth edition. Lewis Publishers, 121 S. Main St., PO Box 519, Chelsea, MI 48118. (313) 475-8619 or (800) 272-7737. 1991. $79.95.

DIRECTORIES AND BIOGRAPHICAL SOURCES

Directory of Engineering Societies and Related Organizations. Gordon Davis. 13th edition. American Association of Engineering Societies, 1111 19th Street NW, Suite 608, Washington, DC 20036. (202) 296-2237 or (800) 658-8897. 1989. Inquire for price.

International Research Centers Directory. Gale Research, 835 Penobscot Building, Detroit, MI 48226-4094. (313) 961-2242. (800) 877-4253. FAX (313) 961-6083. 8th edition. 1995. $410.00

Research Centers Directory. Gale Research, 835 Penobscot Building, Detroit, MI 48226-4094. (313) 961-2242. (800) 877-4253. FAX (313) 961-6083. $485.00.

Scientific and Technical Organizations and Agencies Directory. Gale Research, 835 Penobscot Building, Detroit, MI 48226-4094. (313) 961-2242. (800) 877-4253. FAX (313) 961-6083. 4th edition. 1996. $195.00.

ENCYCLOPEDIAS AND DICTIONARIES

The Encyclopedia of the Solid Earth Sciences. Philip Kearey, editor-in-chief. Blackwell Scientific Publications, 238 Main Street, Cambridge, MA 02142. (617) 876-7000 or (800) 759-6102. FAX (617) 876-7022. 1993. $60.00.

Encyclopedia of Physical Science and Technology. Robert A. Meyers, ed. 18 volumes. Academic Press Inc., 6277 Sea Harbor Drive, Orlando, FL 32887. (800) 321-5068. 1992. $2100.00.

McGraw-Hill Dictionary of Scientific and Technical Terms. Sybil P. Parker, ed. 5th edition. McGraw-Hill Publishing Company, 1221 Avenue of the Americas, New York, NY 10020. (800) 262-4729 or (212) 512-3825. 1993. $110.50.

GENERAL WORKS

Chemistry of Water and Water Pollution. Jan Dojildo and Gerald Best. Prentice Hall, 113 Sylvan Avenue, Route 9W, Englewood Cliffs, NJ 07632. (201) 592-2000 or (800) 922-0579. 1994. $73.00.

Hydrology, Hydraulics, and Water Quality. Susan Brown, editor. National ACADemy Press, Division of the National ACADemy of Science, 2101 Constitution Avenue, N.W., Lockbox 285, Washington, DC 20055. (800) 624-6242. FAX (202) 334-2793. 2994. $25.00.

Water: A Comprhensive Treatise. Felix Franks, editor. 6 volumes. Plenum Publishing Corporation, 233 Spring Street, New York, NY 10013-1578. (212) 620-8000. FAX (212) 463-0742. Inquire for cost and availability.

Water Analysis. W. Fresenius, et al., editors. Springer-Verlag, 175 Fifth Avenue, New York, NY 10010. (212) 460-1500 or (800) 777-4643. FAX (212) 473-6272. 1991. $115.00.

Water: Its Global Nature. Michael Allaby. Facts-on-File, Inc., 460 Park Avenue South, New York, NY 10016-7382. (800) 322-8755. Fax (212) 213-4578. 1992. $35.00.

HANDBOOKS AND MANUALS

Groundwater: Managing the Unseen Resources: A Handbook For States. Edwin H. Clark and Philip J. Cherry. World Wildlife Fund & the Conservation Foundation, 1250 24th Street NW, Washington, DC 20037. (202) 293-4800. FAX (410) 516-6998. 1992. $8.50.

Handbook of Hydrology. David R. Maidment, editor in chief. McGraw-Hill Publishing Company, 1221 Avenue of the Americas, New York, NY 10020. (800) 262-4729 or (212) 512-3825. 1993. $115.00.

Practical Groundwater Hydrology. Michael E. Renz. Lewis Publishers, 121 S. Main St., PO Box 519, Chelsea, MI 48118. (313) 475-8619 or (800) 272-7737. 1995. Inquire for cost and availability.

ONLINE DATABASES AND CD-ROMS

Compendex Plus. Engineering Information, Inc., 345 East 47th Street, New York, NY 10017. (212) 705-7600 or (800) 221-1044. Contains citations with abstracts to worldwide literature in engineering and technology, from 1970 to present. Available on online BRS,(800) 289-4277, DIALOG, (800) 334-2564, ORBIT (800) 456-7248, and STN International, FIZ Karlsruhe, P.O. Box 2465, W-7500, Karlsruhe 1, Germany, online services. Also available on CD-ROM. Inquire as to cost and availability.

Current Contents Search. Institute for Scientific Information, 3501 Market Street, Philadelphia, PA 19104. (215) 386-0100. FAX (215) 386-6362. Contains citations to articles listed in the table of contents of science and technology journals. Also articles in social sciences and life sciences journals. Available on BRS,(800) 289-4277, DIALOG,(800) 334-2564, online services. Inquire as to cost and availability.

Dissertation Abstracts. University Microfilms International, 300 North Zeeb Road, Ann Arbor, MI 48106. (800) 521-0600 or (313) 761-4700. Scope includes virtually all doctoral dissertations accepted at accredited American institutions from 1861 to present in 252 subject areas. Available on BRS,(800) 289-4277, DIALOG,(800) 334-2564, and OCLC EPIC,(800) 848-5878, online services. Also available on CD-ROM. Inquire as to cost and availability.

Earth Sciences. U.S. Geological Survey, 12201 Sunrise Valley Drive, Reston, VA 22092-9998. (703) 648-4460. CD-ROM of earth science databases including the U.S. Geological Survey Library, Earth Science Data Directory, and GEOINDEX, citations to published geological maps. $350.00 per year with quarterly updates.

Geoarchive. Geosystems, P.O. Box 1024, Westminster, London, England, SW1 P 2JL. Citations to literature on geoscience, 1969 to present. Inquire as to online cost and availability.

Geobase. Elsevier - Geo Abstracts, Regency House, 34 Duke Street, Norwich NR3 3AP, England. Contains citations to the worldwide earth science literature from 1980 to date. Available on DIALOG, ORBIT online services. Inquire as to cost and availability.

Georef: Bibliography and Index of Geology. American Geological Institute, 4220 King Street, Alexandria, VA 22302. (703) 379-2480. Fax (703) 379-7563. Monthly. Inquire for price and availability.

NTIS Bibliographic Database. National Technical Information Service, 5285 Port Royal Road, Springfield, VA 22161. (703) 487-4929 or FAX (703) 321-8199. Broad coverage of government-sponsored science and technology research reports, 1964 to present. Available on BRS,(800) 289-4277, DIALOG, (800) 334-2564, ORBIT, (800) 456-7248, and STN International, FIZKarlsruhe, P.O. Box 2465, W-7500, Karlsruhe 1, Germany, online services. Also available on CD-ROM. Inquire as to cost and availability.

Water Resources Abstracts (WRA). U.S. Geological Survey, Water Resources Scientific Information Center, 12201 Sunrise Valley Drive, Reston, VA 22092-9998. (703) 648-4460. Contains citations with abstracts to scientific and technical literature on the water-sources-related aspects of the physical, social and life sciences, from 1968 to present. Available on DIALOG,(800) 334-2564, and ORBIT,(800) 456-7248, online services. Inquire as to cost and availability.

Waternet. American Water Works Association, Technical Library. Available on DIALOG online services. Citations to literature on water quality, water utility management, water analysis, water pollution, and related areas, 1971 to present. Avail-

able on DIALOG,(800) 334-2564, online service. Inquire as to cost and availability.

PERIODICALS

American Water Works Association Journal. 6666 W. Quincy Avenue, Denver, CO 80235. (303) 794-7711. FAX (303) 794-7310. 1914 to present. Monthly. $85.00 to libraries and government agencies only.

Ground Water. Ground Water Publishing Company, National Well Water Association, 6375 Riverside Drive, Dublin, OH 43017. (614) 761-1711. FAX (614) 761-3446. 1963 to present. Bi-monthly. $90.00 per year.

Ground Water Monitoring and Remediation. Ground Water Publishing Company, National Well Water Association, 6375 Riverside Drive, Dublin, OH 43017. (614) 761-1711. FAX (614) 761-3446. 1981 to present. Quarterly. $30.00 per year.

Hydrological Science and Technology. American Institute of Hydrology, 3416 University Avenue SE, Minneapolis, MN 55414-3328. (612) 379-1030. FAX (612) 379-0169. 1985 to present. Quarterly. $75.00 per year.

Journal of Hydrology. Elsevier Science Inc., Box 882, Madison Square Station, New York, NY 10159. (212) 989-5800. FAX (212) 633-3990. 1963 to present. 48 times a year. $2309.00.

Journal of Water Resources Planning and Management. American Society of Civil Engineers, Water Resources Planning and Management Division, 345 East 47th Street, New York, NY 10017-2398. (212) 705-7520 or (800) 548-2723. Bi-monthly. $112.00 per year.

Limnology and Oceanography. American Society of Limnology and Oceanography, School of Oceanography, WB-10, University of Washington, Seattle, WA 98195. 1956 to present. Eight times a year. $160.00 per year.

Water Resources Bulletin. American Water Resources Association, 5410 Grosvenor Lane, Suite 220, Bethesda, MD 20814-2192. (301) 493-8600. FAX (301) 493-5844. 1965 to present. Bi-monthly. $15.00.

Water Resources Research. American Geophysical Union,2000 Florida Avenue, N.W., Washington, DC 20009. (202) 462-6900. FAX (202) 328-0566. 1965 to present. Monthly. $660.00 for non-members.

RESEARCH CENTERS AND INSTITUTES

University of California Berkeley Hydraulic Laboratories. 412 O'Brien Hall, Berkeley, CA 94720. (415) 642-6777.

University of Florida, Water Resources Research Center. 424 Black Hall, Gainesville, FL 32611. (904) 392-0840. FAX (904) 392-3076.

University of Hawaii At Manoa, Water Resources Research Center. 2540 Dole Street, Honolulu, HI 96822. (808) 948-7847. FAX (808) 956-5044.

University of Massachusetts, Massachusetts Water Resources Research Center. Blaisdell House, Amherst, MA 01003. (413) 545-2842. FAX (413) 545-2304.

WATER POLLUTION

Ency. of Physical Sciences and Engineering Info. Sources

WATER ANALYSIS

See: ENVIRONMENTAL ENGINEERING

WATER POLLUTION

See also: GROUND WATER, GROUND WATER POLLUTION, HYDROGEOLOGY, HYDROLOGY, POLLUTION, WATER RESOURCES

ASBTRACT SERVICES AND INDEXES

Aqualine Abstracts (formerly *Water Pollution Abstracts*) Water Research Centre, Po Box 85, Frankland Rd., Blagrove, Swindon, Wilts. SN5 8YF, England. 1927 to present. 26 times a year. Inquire for price in U.S. Also available online.

CA Selects: Environmental Pollution. Chemical Abstracts Service, 2540 Olentangy River Road, Box 3012, Columbus, OH 43210. (800) 848-6538 or (614) 447-3600, FAX (614) 447-3713. Semi-weekly. $210.00 per year.

CA Selects: Pollution Monitoring. Chemical Abstracts Service, 2540 Olentangy River Road, Box 3012, Columbus, OH 43210. (800) 848-6538 or (614) 447-3600, FAX (614) 447-3713. Semi-weekly. $210.00 per year.

Current Contents: Engineering, Technology, and Applied Sciences. Institute for Scientific Information, 3501 Market Street, Philadelphia, PA 19104. (215) 386-0100. FAX (215) 386-2991. 1970 to present. Weekly. $442.00 per year.

Current Contents: Physical, Chemical and Earth Sciences. Institute for Scientific Information, 3501 Market Street, Philadelphia, PA 19104. (215) 386-0100. FAX (215) 386-6362. Weekly. $360.00 per year.

Engineering Index Monthly. Engineering Information, Inc., Castle Point on the Hudson, Hoboken, NJ 07030. (800) 221-1044. FAX (201) 216-8532. Monthly. $2300.00 per year. Also available online as COMPENDEX, and also on CD-ROM. Covers chemical engineering, computers, electrical engineering, civil engineering, metals and mining, industrial management, and mechanical engineering.

Environmental Abstracts. R.R. Bowker Inc., 121 Chanlon Road, New Providence, NJ 07974. (800) 521-8110 or (908) 464-6800. Contains citations, most with abstracts, to literature relating to environmental issues and problems, from 1971 to present. Inquire asto cost and availability.

Environmental Engineering Abstracts, Cambridge Scientific Abstracts, 7200 Wisconsin Avenue, Bethesda, MD 20814-4823. (301) 961-6750. FAX (301) 961-6720. Monthly. Covers hazardous materials, environmental impact and protection, treatment of sewage and industrial wastes, hydroelectric power, tidal and wind power, arctic and tropical engineering.

Hydro-Abstracts (formerly Water Resources Abstracts). HydroScience Press, 2527 Jackson Street, N.E., Minneapolis, MN 55418. (612) 781-9081. Monthly. $150.00 per year.

NTIS Alerts: Environmental Pollution and Control. National Techical Information Service (NTIS), 5285 Port Royal Road,
Springfield, VA 22161. (703) 487-4650. Weekly. $150.00 per year.

Pollution Abstracts, Cambridge Scientific Abstracts, 7200 Wisconsin Avenue, Bethesda, MD 20814-4823. (301) 961-6750. FAX (301) 961-6720. Inquire for cost and availability.

Science Citation Index. Institute for Scientific Information, 3501 Market Street, Philadelphia, PA 19104. (215) 386-0100. FAX (215) 386-6362. Inquire about availability and cost. Also available on CD-ROM.

Selected Water Resources Abstracts. U.S. Geological Survey, Water Resources Scientific Information Center, 425 National Center, Reston, VA 22092. (703) 648-6820. Monthly. $115.00 per year. Also available on CD-ROM.

ASSOCIATIONS AND PROFESSIONAL SOCIETIES

American Academy of Environmental Engineering. 130 Holiday Ct., Suite 100, Annapolis, MD 21401. (410) 266-3311. FAX (410) 266-7653.

American Public Works Association. 106 West 11th Street, Suite 1800, Kansas City, MO 64015-1806. (816) 472-6100. FAX (816) 472-1610.

American Society of Civil Engineers. 345 East 47th Street, New York, NY 10017-2398. (212) 705-7520 or (800) 548-2723.

Association of Ground Water Scientists and Engineers. A division of the National Water Well Association, 6375 Riverside Drive, Dublin, OH 43017. (614) 761-1711. FAX (614) 761-3446.

Association of State and Interstate Water Pollution Control Adminstrators. 750 First Street SE, Suite 910, Washington, DC 20002. (202) 898-0905.

Clean Water Action. 1320 18th Street NW, Suite 300, Washington, DC 20036. (202) 457-1286. FAX (202) 457-0287.

Environmental Defense Fund. 257 Park Avenue South, New York, NY 10010. (212) 505-2100. FAX (212) 505-2375.

Federal Water Quality Association. PPO Box 44163, Washington, DC 20026. (202) 720-4925.

International Center For the Solution of Environmental Problems. 535 Lovett Blvd., Houston, TX 77006. (713) 527-8711. FAX (713) 527-8025.

National Association of Environmental Professionals. 5165 MacArthur Blvd. NW, PO Box 9400, Washington, DC 20016-3315. (202)966-1500. FAX (202) 966-1977.

National Water Resources Association. 3800 N. Fairfax Drive, Suite 4, Arlington, VA 22203-1703. (703) 524-1544. FAX (703) 524-1548.

National Water Well Association. 6375 Riverside Drive, Dublin, OH 43017. (614) 761-1711. FAX (614) 761-3446.

Water Environment Federation. 601 Wythe Street, Alexandria, VA 22314-1994. (703) 684-2400. FAX (708) 684-2492.

Water Quality Association. 4151 Naperville Road, Lisle, IL 60532. (708) 505-0160. FAX (708) 505-9637.

BIBLIOGRAPHIES

Environmental Periodicals Bibliography. International Academy at Santa Barbara, Environmental Studies Institute, 800 Garden Street, Suite D, Santa Barbara, CA 93101-1552. (805) 965-5010. 1972 to present. Bi-monthly. Price varies.

Reading About the Environment: An Introductory Guide. P.E. Jansma. Libraries Unlimited, PO Box 6633, Englewood, CO 80155-6633. (800) 237-6124. FAX (303) 220-8843. 1993. $27.50.

GENERAL WORKS

Chemistry of Water & Water Pollution. Jan Dojilido & Gerald Best. Prentice Hall, 113 Sylvan Avenue, Route 9W, Englewood Cliffs, NJ 07632. (201) 592-2000 or (800) 922-0579. 1994. $73.00.

Drinking Water Quality: Problems & Solutions. N.F. Gray. John Wiley and Sons, Inc., 605 Third Avenue, New York, NY 10158. (800) 526-5368 or (212) 850-6000. 1994. Inquire for cost.

Soil and Water Contamination and Remediation. Van Nostrand Reinhold, 115 Fifth Avenue, New York, NY 10003. (212) 254-3232. FAX (212) 254-9499.1994. 1994. Inquire for price.

Water: The Vital Resource. Norma H. Jones, et al. Information Plus, 2812 Exchange Street, Wylie, TX 75098. (800) 463-6757. FAX (214) 442-1189.

HANDBOOKS AND MANUALS

Environmental Science and Technology Handbook. Porter C. Knowles, et al. Government Institutes Inc., 4 Research Place, Suite 200, Rockville, MD 20850. (301) 921-2355. 1994. $75.00.

Water Pollution (Handbook of Environmental Chemistry Series, Volume 5 Part A). O. Hutzinger, editor. Springer-Verlag, 175 Fifth Avenue, New York, NY 10010. (212) 460-1500 or (800) 777-4643. FAX (212) 473-6272. 1991. $128.00.

ONLINE DATABASES AND CD-ROMS

CA Search. Chemical Abstracts Service, P.O. Box 3012, Columbus, OH 43210-0012. (614) 447-3600. (800) 848-6533. FAX (614) 447-3709. Very comprehensive guide to worldwide chemical literature and related fields, 1972 to present. Available on BRS,(800) 289-4277, DIALOG, (800) 334-2564, ORBIT (800) 456-7248, and STN International, FIZ Karlsruhe, P.O. Box 2465, W-7500, Karlsruhe 1, Germany, online services. Inquire as to cost and availability.

NTIS Bibliographic Database. National Technical Information Service, 5285 Port Royal Road, Springfield, VA 22161. (703) 487-4929 or FAX (703) 321-8199. Broad coverage ofgovernment-sponsored science and technology research reports, 1964 to present. Available on BRS,(800) 289-4277, DIALOG, (800) 334-2564, ORBIT, (800) 456-7248, and STN International, FIZ Karlsruhe, P.O. Box 2465, W-7500, Karlsruhe

1, Germany, online services. Also available on CD-ROM. Inquire as to cost and availability.

SCISEARCH. Institute for Scientific Information, 3501 Market Street, Philadelphia, PA 19104. (800) 523-1850 or (215) 386-0100. Broad multidisciplinary title and author index to the international literature of science and technology, 1974 to present. Available on DIALOG,(800) 334-2564, and ORBIT,(800) 456-7248, online services. Also available on CD-ROM. Inquire as to cost and availability.

Waternet. American Water Works Association, Technical Library. Available on DIALOG online services. Citations to literature on water quality, water utility management, water analysis, water pollution, and related areas, 1971 to present. Available on DIALOG,(800) 334-2564, online service. Inquire as to cost and availability.

PERIODICALS

Air & Water Pollution Control. Bureau of National Affairs, 1231 25th Street NW, Washington, DC 20037. (202) 452-4200. FAX (202) 822-8092. 1986 to present. Bi-weekly. $301.00 per year.

American Water Works Association Journal. 6666 W. Quincy Avenue, Denver, CO 80235. (303) 794-7711. FAX (303) 794-7310. 1914 to present. Monthly. $85.00 to libraries and government agencies only.

Bulletin of Environmental Contamination and Toxicology. Springer-Verlag, 175 Fifth Avenue, New York, NY 10010. (212) 460-1500 or (800) 777-4643. FAX (212) 473-6272. 1966 to present. Monthly. $399.00 per year.

Environmental Engineering. Mechanical Engineering Publications Ltd., Northgate Avenue, Bury St. Edmunds, Suffolk IP32 6BW, England. Telephone 0284-763277. FAX 0284-704006. 1972 to present. Quarterly. Inquire for price.

Environmental Monitoring and Assessment. Kluwer Academic Publishers, P.O. Box 358, Accord Station, Hingham, MA 02018-0358. (617) 871-6000. FAX (617) 871-6528. 1981 to present. 15 times a year. $765.00 per year.

Environmental Pollution. Elsevier Science [journals], 660 White Plains Rd., Tarrytown, NY 10159-5153. (919) 524-9200. FAX (919) 333-2444. 1967 to present. 1970 to present. 12 times a year. $1250.00 per year.

Environmental Science and Technology. American Chemical Society, 1155 16th Street, N.W., Washington, DC 20036. (202) 872-4381 or (800) 333-9511. FAX (614) 447-3671. 1967 to present. Monthly. $89.00 per year for non-members.

Environmental Toxicology and Water Quality. John Wiley and Sons, Inc., 605 Third Avenue, New York, NY 10158. (800) 526-5368 or (212) 850-6000. 1986 to present. Four times a year.$250.00 per year (to institutions).

EPA Journal. Superintendent of Documents, U.S. Government Printing office, Box 371954, Pittsburgh, PA 15250-7954. (202) 783-3238. FAX (202) 512-2233. 1975 to present. Bi-monthly. $10.00 per year.

Ground Water. Ground Water Publishing Company, National Well Water Association, 6375 Riverside Drive, Dublin, OH

WATER POLLUTION

Ency. of Physical Sciences and Engineering Info. Sources

43017. (614) 761-1711. FAX (614) 761-3446. 1963 to present. Bi-monthly. $90.00 per year.

Journal of Contaminant Hydrology. Elsevier Science B.V., PO Box 945, Madison Square Station, New York, NY 10160-0757. (212) 633-3650. FAX (212) 633-3680. 1986 to present. 12 times a year. $553.00.

Journal of Environmental Engineering. American Society of Civil Engineers, Environmental Engineering Division, 345 East 47th Street, New York, NY 10017-2398. (212) 705-7520 or (800) 548-2723. FAX (212) 705-7712. 1956 to present. Bi-monthly. $136.00 per year for non-members.

Journal of Hydrology. Elsevier Science Inc., Box 882, Madison Square Station, New York, NY 10159. (212) 989-5800. FAX (212) 633-3990. 1963 to present. 48 times a year. $2309.00.

Journal of Irrigation and Drainage. American Society of Civil Engineers, Irrigation and Drainage Division, 345 East 47th Street, New York, NY 10017-2398. (212) 705-7520 or (800) 548-2723. 1956 to present. Bi-monthly. $136.00 per year.

Journal of Water Resources Planning and Management. American Society of Civil Engineers, Water Resources Planning and Management Division, 345 East 47th Street, New York, NY 10017-2398. (212) 705-7520 or (800) 548-2723. Bi-monthly. $112.00 per year.

Limnology and Oceanography. American Society of Limnology and Oceanography, School of Oceanography, WB-10, University of Washington, Seattle, WA 98195. 1956 to present. Eight times a year. $160.00 per year.

Marine Pollution Bulletin. Elsevier Science [journals], 660 White Plains Rd., Tarrytown, NY 10159-5153. (919) 524-9200. FAX (919) 333-2444. 1970 to present. 24 times a year. $430.00 per year.

Pollution Engineering. Cahners Publishing Company (Des Plains), 1350 East Touhy Avenue, Des Plaines, IL 60017-5080. (800) 323-4958. FAX (708) 390-2779. 1969 to prsent. 13 times a year. $70.00 per year.

Techlink Water Pollution. National Technical Information Service (NTIS), 5285 Port Royal Road, Springfield, VA 22161. (703) 487-4650. FAX (703) 321-8547. 1993 to present. Monthly. $300.00 per year.

Water, Air and Soil Pollution. Kluwer Academic Publishers, P.O. Box 358, Accord Station, Hingham, MA 02018-0358. (617) 871-6000. FAX (617) 871-6528. 1971 to present. 28 times a year. $1519.00.

Water Environment Research. Water Environment Federation, 601 Wythe Street, Alexandria, VA 22314-1994. 1928 to present.Monthly. $144.00 per year.

Water Pollution Control. Bureau of National Affairs, 1231 25th Street NW, Washington, DC 20037. (202) 452-4200. FAX (202) 822-8092. 1979 to present. Bi-weekly. $612.00 per year.

Water Quality International. Elsevier Science, 660 White Plains Rd., Tarrytown, NY 10159-5153. (919) 524-9200. FAX (919) 333-2444. 1987 to present. 4 times a year. $125.00 per year.

Water Resources Bulletin. American Water Resources Association, 5410 Grosvenor Lane, Suite 220, Bethesda, MD 20814-2192. (301) 493-8600. FAX (301) 493-5844. 1965 to present. Bi-monthly. $15.00.

Water Resources Research. American Geophysical Union, 2000 Florida Avenue, N.W., Washington, DC 20009. (202) 462-6900. FAX (202) 328-0566. 1965 to present. Monthly. $660.00 for non-members.

RESEARCH CENTERS AND INSTITUTES

Academy of Natural Sciences of Philadelphia, Division of Environmental Research. 19th Street and the Parkway, Philadelphia, PA 19103. (215) 299-1081.

California Institute of Technology, W.M. Keck Laboratory of Hydraulics and Water Resources. 138-78, Pasadena, CA 91125. (818) 356-4404. FAX (818) 356-2940.

University of Illinois, Institute For Environmental Studies. 1101 W Peabody Drive, Urbana, IL 61801. (217) 333-4178. FAX (217) 333-8046.

Lenox Institute of Water Technology. 101 Yokun Avenue, PO Box 1639, Lenox, MA 01240. (413) 637-3025. (413) 637-0768.

National Center For Ground Water Research. University of Oklahoma, 200 Telgar Street, Room 127, Norman, OK 73019-0470. (405) 325-5202.

Pennsylvania State University, Environmental Resources Research Institute. 100 Land and Water Resource Bldg., University Park, PA 16802. (814) 863-0291. (814) 865-3378.

Resources For the Future, Inc., Quality of the Environment Division. 1616 P Street NW, Washington, DC 20036. (202) 328-5000. FAX (202) 265-8069.

University of Alaska—fairbanks, Water Research Center. Fairbanks, AK 99775-1760. (907) 474-7350. FAX (907) 474-6087.

WATER RESOURCES

See also: GROUND WATER, GROUND WATER POLLUTION,HYDROGEOLOGY, HYDROLOGY

ABSTRACT SERVICES AND INDEXES

Aqualine Abstracts. Water Research Centre, PO Box 85, Frankland Rd., Blagrove, Swindon, Wilts. SN5 8YF, England. 1927 to present. 26 times a year. Inquire for price in U.S. Also available online.

ASCE Combined Annual Combined Index. American Society of Civil Engineers, 345 East 47th Street, New York, NY 10017-2398. (212) 705-7520 or (800) 548-2723. Annual. $48.00.

ASCE Publications Information. 1966-present. AmericanSociety of Civil Engineers, 345 East 47th Street, New York, NY 10017-2398. (212) 705-7520 or (800) 548-2723. Bi-monthly. $160.00 per year for non-members.

Bibliography and Index of Geology. American Geological Institute, 4220 King Street, Alexandria, VA 22302. (703) 379-2480. Fax (703) 379-7563. Monthly. $1295.00 per year. Also available online as GeoRef.

Biological and Agricultural Index. H.W. Wilson Company, 950 University Avenue, Bronx, NY 10452. (800) 367-6770. FAX (718) 590-1617. 1964 to present. Monthly. Inquire for price. Also available online and on CD-ROM.

Current Contents: Physical, Chemical and Earth Sciences. Institute for Scientific Information, 3501 Market Street, Philadelphia, PA 19104. (215) 386-0100. FAX (215) 386-6362. Weekly. $360.00 per year.

Engineering Index Monthly. Engineering Information, Inc., Castle Point on the Hudson, Hoboken, NJ 07030. (800) 221-1044. FAX (212) 832-1857. Monthly. $2300.00 per year. Also available online as COMPENDEX, and also on CD-ROM. Covers chemical engineering, computers, electrical engineering, civil engineering, metals and mining, industrial management, and mechanical engineering.

Environmental Abstracts. R.R. Bowker Inc., 121 Chanlon Road, New Providence, NJ 07974. (800) 521-8110 or (908) 464-6800. Contains citations, most with abstracts, to literature relating to environmental issues and problems, from 1971 to present. Inquire as to cost and availability.

Environmental Engineering Abstracts, Cambridge Scientific Abstracts, 7200 Wisconsin Avenue, Bethesda, MD 20814-4823. (301) 961-6750. FAX (301) 961-6720. Monthly. Covers hazardous materials, environmental impact and protection, treatment of sewage and industrial wastes, hydroelectric power, tidal and wind power, artic and tropical engineering.

Geological Abstracts. Elsevier - Geo Abstracts, Regency House, 34 Duke Street, Norwich NR3 3AP, England. Monthly. $760.00 per year. Also available online as GEOBASE.

Geological Society of America. Abstracts with programs. Geological Society of America. 3300 Penrose Place, P.O. Box 9140, Boulder, CO 80301-9140. (303) 447-2020. Abstracts and programs of the annual conference. Annual. $69.00.

Hydro-Abstracts (formerly Water Resources Abstracts). HydroScience Press, 2527 Jackson Street, N.E., Minneapolis, MN 55418. (612) 781-9081. Monthly. $150.00 per year.

Meteorological and Geoastrophysical Abstracts. American Meteorological Society, c/o Inforonics Inc., 550 Newtown Road, Littleton, MA 01460. (508) 486-8976. FAX (508) 486-0027. 1950 to present. Monthly. $950.00 per year. Current abstracts of books, reports, research papers, and miscellaneous literature on environmental sciences, meteorology, astrophysics, hydrology, glaciology, and physical oceanography.

Science Citation Index. Institute for ScientificInformation, 3501 Market Street, Philadelphia, PA 19104. (215) 386-0100. FAX (215) 386-6362. Inquire about availability and cost. Also available on CD-ROM.

Selected Water Resources Abstracts. U.S. Geological Survey, Water Resources Scientific Information Center, 425 National Center, Reston, VA 22092. (703) 648-6820. Monthly. $115.00 per year. Also available on CD-ROM.

ANNUAL REVIEWS AND YEARBOOKS

Advances in Hydrosciences. Academic Press Inc., 6277 Sea Harbor Drive, Orlando, FL 32887. (800) 321-5068. Inquire for cost and availability.

I W S A Yearbook. International Water Supply Association, c/o L.R. Bays, Seoy. General, 1 Queen Anne's Gate, London SW1H 9BT, England. Telephone 071-957-4567. FAX 071-222-7243. 1979 to present. Annual. Inquire for price in U.S.

ASSOCIATIONS AND PROFESSIONAL SOCIETIES

American Institute of Hydrology. 3416 University Avenue S.E., Minneapolis, MN 55414-3329. (612) 379-1030. FAX (379-0169.

American Society of Civil Engineers. 345 East 47th Street, New York, NY 10017-2398. (212) 705-7520 or (800) 548-2723.

American Water Foundation. 1616 17th Street, Suite 376, Denver, CO 80215. (303) 628-5516. FAX (303) 236-5151.

American Water Resources Association. 5410 Grosvenor Lane, Suite 220, Bethesda, MD 20814-2192. (301) 493-8600. FAX (301) 493-5844.

American Water Works Association. 6666 W. Quincy Avenue, Denver, CO 80235. (303) 794-7711.

Association of Ground Water Scientists and Engineers. A division of the National Water Well Association, 6375 Riverside Drive, Dublin, OH 43017. (614) 761-1711. FAX (614) 761-3446.

International Water Resources Association. University of Illinois, 205 N. Mathews Avenue, Urbana, IL 61801. (217) 333-0536. FAX (217) 244-6633.

International Water Supply Association. c/o L.R. Bays, Seoy. General, 1 Queen Anne's Gate, London SW1H 9BT, England. Telephone 071-957-4567. FAX 071-222-7243.

National Ground Water Association, 6375 Riverside Drive, Dublin, OH 43017. (614) 761-1711. FAX (614) 761-3446.

National Water Resources Association. 3800 N. Fairfax Drive, Suite 4, Arlington, VA 22203-1703. (703) 524-1544. FAX (703) 524-1548.

National Water Well Association. 6375 Riverside Drive, Dublin, OH 43017. (614) 761-1711. FAX (614) 761-3446.

Universities Council On Water Resources. 4543 Faner Hall, Department of Geography, Southern Illinois University at Carbondale, Carbondale, IL 62901-4526. (618) 536-7571. FAX (618) 453-2671.

Water Environment Federation. 601 Wythe Street, Alexandria, VA 22314-1994. (703) 684-2400. FAX (703) 684-2492.

Water Quality Association. 4151 Naperville Road, Lisle,IL 60532. (708) 505-0160. FAX (708) 505-9637.

BIBLIOGRAPHIES

Geraghty & Miller's Groundwater Bibliography. Fifth edition. Lewis Publishers, 121 S. Main St., PO Box 519, Chelsea, MI 48118. (313) 475-8619 or (800) 272-7737. 1991. $79.95.

DIRECTORIES AND BIOGRAPHICAL SOURCES

American Men & Women of Science, 1995-96. R.R. Bowker Staff, eds. 19th edition. 8 volumes. R.R. Bowker/Reed International Publishing Company, 121 Chanlon Road, New Providence, NJ 07974. (908) 464-6800 or (800) 521-8110. 1995. $850.00

Directory of Engineering Societies and Related Organizations. Gordon Davis. 13th edition. American Association of Engineering Societies, 1111 19th Street NW, Suite 608, Washington, DC 20036. (202) 296-2237 or (800) 658-8897. 1989. Inquire for price.

International Research Centers Directory. Gale Research, 835 Penobscot Building, Detroit, MI 48226-4094. (313) 961-2242. (800) 877-4253. FAX (313) 961-6083. 8th edition. 1995. $410.00

Research Centers Directory. Gale Research, 835 Penobscot Building, Detroit, MI 48226-4094. (313) 961-2242. (800) 877-4253. FAX (313) 961-6083. $485.00.

Scientific and Technical Organizations and Agencies Directory. Gale Research, 835 Penobscot Building, Detroit, MI 48226-4094. (313) 961-2242. (800) 877-4253. FAX (313) 961-6083. 4th edition. 1996. $195.00.

ENCYCLOPEDIAS AND DICTIONARIES

Dictionary of Civil Engineering. V.N. Vazirani. South Asia Books, Box 502, Columbia, MO 65205. (314) 474-0116. FAX (314) 474-8124. 1992. $20.00.

The Encyclopedia of the Solid Earth Sciences. Philip Kearey, editor-in-chief. Blackwell Scientific Publications, 238 Main Street, Cambridge, MA 02142. (617) 876-7000 or (800) 759-6102. FAX (617) 876-7022. 1993. $60.00.

Encyclopedia of Physical Science and Technology. Robert A. Meyers, ed. 18 volumes. Academic Press Inc., 6277 Sea Harbor Drive, Orlando, FL 32887. (800) 321-5068. 1992. $2100.00

International Glossary of Hydrology. UNESCO staff. UNIPUB, 4611-F Assembly Drive, Lanham, MD 20706-4391. (800) 274-4888. FAX (301) 459-0056. 1993. $45.00.

GENERAL WORKS

Statistical Analysis in Water Resources Engineering. Mamdouth Shahin, editor. Ashgate Publishing Co., Old Post Rd., Brookfield, VT 05036. (802) 276-3162. FAX (802) 276-3837. 1993. $60.00 (hardbound) and $35.00 (paperbound).

Water Resources Engineering. Ray K. Linsley. Fourth edition. McGraw-Hill Publishing Company, 1221 Avenue of the Americas, New York, NY 10020. (800) 262-4729 or (212) 512-3825. 1992. Inquire for price.

Water Resources—issues and Strategies. Adrian McDonald and David Kay. John Wiley and Sons, Inc., 605 Third Avenue, New York, NY 10158. (800) 526-5368 or (212) 850-6000. 1989. $53.95.

Water Resources Planning. andrew A. Dzurik. Rowman & Littlefield, 4720 Boston Way, Suite A, Lanham, MD 20706. (800) 462-6420. FAX (301) 459-2118.

Water Resources Planning & Management & Urban Water Resources. Jerry L. anderson, editor. American Society of Civil Engineers, 345 East 47th Street, New York, NY 10017-2398. (212) 705-7520 or (800) 548-2723. FAX (212) 705-7712. 1991. $90.00 (paperback).

Water Supply. A.C. Twort, et al. 4th edition. Chapman & Hall, 115 Fifth Avenue, New York, NY 10211-0906. 1995. $62.95.

Water Treatment Principles and Design. J.M. Montgomery. John Wiley and Sons, Inc., 605 Third Avenue, New York, NY 10158. (800) 526-5368 or (212) 850-6000. 1985. $95.00.

HANDBOOKS AND MANUALS

Groundwater: Managing the Unseen Resources: A Handbook For States. Edwin H. Clark and Philip J. Cherry. World Wildlife Fund & the Conservation Foundation, 1250 24th Street NW, Washington, DC 20037. (202) 293-4800. FAX (410) 516-6998. 1992. $8.50.

Handbook of Hydrology. David R. Maidment, editor in chief. McGraw-Hill Publishing Company, 1221 Avenue of the Americas, New York, NY 10020. (800) 262-4729 or (212) 512-3825. 1993. $115.00.

ONLINE DATABASES AND CD-ROMS

Compendex Plus. Engineering Information, Inc., 345 East 47th Street, New York, NY 10017. (212) 705-7600 or (800) 221-1044. Contains citations with abstracts to worldwide literature in engineering and technology, from 1970 to present. Available on online BRS,(800) 289-4277, DIALOG, (800) 334-2564, ORBIT (800) 456-7248, and STN International, FIZ Karlsruhe, P.O. Box 2465, W-7500, Karlsruhe 1, Germany, online services. Also available on CD-ROM. Inquire as to cost and availability.

CSA Engineering. Cambridge Scientific Abstracts, 7200 Wisconsin Avenue, Suite 601, Bethesda, MD 20814. (301) 961-6750 or (800) 843-7751. Contains citations and abstracts of international periodicals and research literature covering all fields of engineering and science and technology, including computer and information science, electronics, mechanical engineering, solid state materials, 1981 to present. Available on BRS,(800) 289-4277, online service. Inquire as to cost and availability.

Current Contents Search. Institute for Scientific Information, 3501 Market Street, Philadelphia, PA 19104. (215) 386-0100. FAX (215) 386-6362. Contains citations to articles listed in the table of contents of science and technology journals. Also articles in social sciences and life sciences journals. Available on BRS,(800) 289-4277, DIALOG,(800) 334-2564, online services. Inquire as to cost and availability.

Dissertation Abstracts. University Microfilms International, 300 North Zeeb Road, Ann Arbor, MI 48106. (800)521-0600 or (313) 761-4700. Scope includes virtually all doctoral dissertations accepted at accredited American institutions from 1861 to present in 252 subject areas. Available on BRS,(800) 289-4277, DIALOG,(800) 334-2564, and OCLC EPIC,(800) 848-5878, online services. Also available on CD-ROM. Inquire as to cost and availability.

Earth Sciences. U.S. Geological Survey, 12201 Sunrise Valley Drive, Reston, VA 22092-9998. (703) 648-4460. CD-ROM of earth science databases including the U.S. Geological Survey Library, Earth Science Data Directory, and GEOINDEX, citations to published geological maps. $350.00 per year with quarterly updates.

Geoarchive. Geosystems, P.O. Box 1024, Westminster, London, England, SW1 P 2JL. Citations to literature on geoscience, 1969 to present. Inquire as to online cost and availability.

Geobase. Elsevier - Geo Abstracts, Regency House, 34 Duke Street, Norwich NR3 3AP, England. Contains citations to the worldwide earth science literature from 1980 to date. Available on DIALOG, ORBIT online services. Inquire as to cost and availability.

Georef: Bibliography and Index of Geology. American Geological Institute, 4220 King Street, Alexandria, VA 22302. (703) 379-2480. Fax (703) 379-7563. Monthly. Inquire for price and availability.

NTIS Bibliographic Database. National Technical Information Service, 5285 Port Royal Road, Springfield, VA 22161. (703) 487-4929 or FAX (703) 321-8199. Broad coverage of government-sponsored science and technology research reports, 1964 to present. Available on BRS,(800) 289-4277, DIALOG, (800) 334-2564, ORBIT, (800) 456-7248, and STN International, FIZ Karlsruhe, P.O. Box 2465, W-7500, Karlsruhe 1, Germany, online services. Also available on CD-ROM. Inquire as to cost and availability.

SCISEARCH. Institute for Scientific Information, 3501 Market Street, Philadelphia, PA 19104. (800) 523-1850 or (215) 386-0100. Broad multidisciplinary title and author index to the international literature of science and technology, 1974 to present. Available on DIALOG,(800) 334-2564, and ORBIT,(800) 456-7248, online services. Also available on CD-ROM. Inquire as to cost and availability.

Waternet. American Water Works Association, Technical Library. Available on DIALOG online services. Citations to literature on water quality, water utility management, water analysis, water pollution, and related areas, 1971 to present. Available on DIALOG,(800) 334-2564, online service. Inquire as to cost and availability.

Water Resources Abstracts (WRA). U.S. Geological Survey, Water Resources Scientific Information Center, 12201 Sunrise Valley Drive, Reston, VA 22092-9998. (703) 648-4460. Contains citations with abstracts to scientific and technical literature on the water-sources-related aspects of the physical, social and life sciences, from 1968 to present. Available on DIALOG,(800) 334-2564,and ORBIT,(800) 456-7248, online services. Inquire as to cost and availability.

PERIODICALS

Advances in Water Resources. Elsevier Science [journals], 660 White Plains Rd., Tarrytown, NY 10159-5153. (919) 524-9200. FAX (919) 333-2444. 1978 to present. Six times a year. $345.00.

American Water Works Association Journal. American Water Works Association, 6666 W. Quincy Avenue, Denver, CO 80235. (303) 794-7711. 1914 to present. Monthly. $85.00 to libraries and governmental agencies only.

AWWA Mainstream. American Water Works Association, 6666 W. Quincy Avenue, Denver, CO 80235. (303) 794-7711. 1955 to present. Monthly. $13.00 to non-members.

Ground Water. Ground Water Publishing Company, National Well Water Association, 6375 Riverside Drive, Dublin, OH 43017. (614) 761-1711. FAX (614) 761-3446. 1963 to present. Bi-monthly. $90.00 per year.

Hydrological Science and Technology. American Institute of Hydrology, 3416 University Avenue SE, Minneapolis, MN 55414-3328. (612) 379-1030. FAX (612) 379-0169. 1985 to present. Quarterly. $75.00.

Journal of Hydrology. Elsevier Science Inc., Box 882, Madison Square Station, New York, NY 10159. (212) 989-5800. FAX (212) 633-3990. 1963 to present. 48 times a year. $2309.00.

Journal of Irrigation and Drainage. American Society of Civil Engineers, Irrigation and Drainage Division, 345 East 47th Street, New York, NY 10017-2398. (212) 705-7520 or (800) 548-2723. 1956 to present. Bi-monthly. $136.00 per year.

Journal of Water Chemistry and Technology. Allerton Press Incorporated, 150 Fifth Avenue, New York, NY 10011. (212) 924-3950. FAX (212) 463-9684. 1981 to present. Monthly. $915.00.

Journal of Water Resources Planning and Management. American Society of Civil Engineers, Water Resources Planning and Management Division, 345 East 47th Street, New York, NY 10017-2398. (212) 705-7520 or (800) 548-2723. Bi-monthly. $112.00 per year.

Limnology and Oceanography. American Society of Limnology and Oceanography, School of Oceanography, WB-10, University of Washington, Seattle, WA 98195. 1956 to present. Eight times a year. $160.00 per year.

Water Engineering and Management. Scranton Gillette Communications, Inc., 380 East Northwest Highway, Des Plaines, IL 60016. (708) 298-6622. 1882 to present. Monthly. $25.00.

Water International. International Water Resources Association, University of Illinois, 205 N. Mathews Avenue, Urbana, IL 61801. (217) 333-0536. FAX (217) 244-6633. 1975 to present. Quarterly. $75.00.

Water Research. Elsevier Science [journals], 660 White Plains Rd., Tarrytown, NY 10159-5153. (919) 524-9200. FAX (919) 333-2444. 1967 to present. 12 times a year. $1615.00.

Water Resources Bulletin. American Water ResourcesAssociation, 5410 Grosvenor Lane, Suite 220, Bethesda, MD 20814-2192. (301) 493-8600. FAX (301) 493-5844. 1965 to present. Bi-monthly. $115.00.

Water Resources Management. Kluwer Academic Publishers, P.O. Box 358, Accord Station, Hingham, MA 02018-0358. (617) 871-6000. 1987 to present. Four times a year. $174.00.

WATER RESOURCES

Ency. of Physical Sciences and Engineering Info. Sources

Water Resources Research. American Geophysical Union, 2000 Florida Avenue, N.W., Washington, DC 20009. (202) 462-6900. 1965 to present. Monthly. $660.00 for non-members.

Water Science and Technology. Elsevier Science [journals], 660 White Plains Rd., Tarrytown, NY 10159-5153. (919) 524-9200. FAX (919) 333-2444. 1972. 24 times a year. $2310.00.

World Water and Environmental Engineer. Thomas Telford Services Ltd., Thomas Telford House, 1 Heron Quay, London E14 4JD, England. Telephone 071-987-6999. FAX 071-538-4101. 1978 to present. Monthly. Inquire for price in U.S.

RESEARCH CENTERS AND INSTITUTES

Indiana Water Resources Research Center. Purdue University, 1284 School of Civil Engineering, W. Lafayette, in 47907-1284. (317) 494-8041. FAX (317) 494-0395.

Water Resources Center. University of Illinois at Urbana-Champaign, 2535 Hydrosystems Lab, 205 Mathews Avenue, Urbana, IL 61801. (217) 333-0536. FAX (217) 244-6633.

Water Resources Research Institute. University of North Carolina, Box 7912, Raleigh, NC 27695-7912. (919) 515-2815. FAX (919) 515-7802.

Water Resources Research Institute. Auburn University, 202 Hargis Hall, Auburn, AL 36849. (205) 826-5075.

WATER TREATMENT

See also: GROUND WATER, GROUND WATER POLLUTION, HYDROGEOLOGY, HYDROLOGY, WATER POLLUTION, WATER RESOURCES

ABSTRACT SERVICES AND INDEXES

Current Contents: Physical, Chemical and Earth Sciences. Institute for Scientific Information, 3501 Market Street, Philadelphia, PA 19104. (215) 386-0100. FAX (215) 386-6362. Weekly. $360.00 per year.

Engineering Index Monthly. Engineering Information, Inc., Castle Point on the Hudson, Hoboken, NJ 07030. (800) 221-1044. FAX (201) 216-8532. Monthly. $2300.00 per year. Also available online as COMPENDEX, and also on CD-ROM. Covers chemical engineering, computers, electrical engineering, civil engineering, metals and mining, industrial management, and mechanical engineering.

Hydro-Abstracts (formerly Water Resources Abstracts). HydroScience Press, 2527 Jackson Street, N.E., Minneapolis, MN 55418. (612) 781-9081. Monthly. $150.00 per year.

Pollution Abstracts, Cambridge Scientific Abstracts, 7200 Wisconsin Avenue, Bethesda, MD 20814-4823. (301) 961-6750. FAX (301) 961-6720. Inquire for cost and availability.

Science Citation Index. Institute for Scientific Information, 3501 Market Street, Philadelphia, PA 19104. (215)386-0100. FAX (215) 386-6362. Inquire about availability and cost. Also available on CD-ROM.

Selected Water Resources Abstracts. U.S. Geological Survey, Water Resources Scientific Information Center, 425 National Center, Reston, VA 22092. (703) 648-6820. Monthly. $115.00 per year. Also available on CD-ROM.

ASSOCIATIONS AND PROFESSIONAL SOCIETIES

American Public Works Association. 106 West 11th Street, Suite 1800, Kansas City, MO 64015-1806. (816) 472-6100. FAX (816) 472-1610.

American Society of Civil Engineers. 345 East 47th Street, New York, NY 10017-2398. (212) 705-7520 or (800) 548-2723.

American Water Resources Association. 5410 Grosvenor Lane, Suite 220, Bethesda, MD 20814-2192. (301) 493-8600. FAX (301) 493-5844.

American Water Works Association. 6666 W. Quincy Avenue, Denver, CO 80235. (303) 794-7711. FAX (303) 794-7310.

Association of Ground Water Scientists and Engineers. A division of the National Water Well Association, 6375 Riverside Drive, Dublin, OH 43017. (614) 761-1711. FAX (614) 761-3446.

Association of State Drinking Water Administrators. 1120 Connecticut Avenue NW, Suite 1060, Washington, DC 20036. (202) 293-7655. FAX (202) 293-7656.

National Water Resources Association. 3800 N. Fairfax Drive, Suite 4, Arlington, VA 22203-1703. (703) 524-1544. FAX (703) 524-1548.

National Water Well Association. 6375 Riverside Drive, Dublin, OH 43017. (614) 761-1711. FAX (614) 761-3446.

Water Quality Association. 4151 Naperville Road, Lisle, IL 60532. (708) 505-0160. FAX (708) 505-9637.

BIBLIOGRAPHIES

Geraghty & Miller's Groundwater Bibliography. Fifth edition. Lewis Publishers, 121 S. Main St., PO Box 519, Chelsea, MI 48118. (313) 475-8619 or (800) 272-7737. 1991. $79.95.

DIRECTORIES AND BIOGRAPHICAL SOURCES

Directory of Engineering Societies and Related Organizations. Gordon Davis. 13th edition. American Association of Engineering Societies, 1111 19th Street NW, Suite 608, Washington, DC 20036. (202) 296-2237 or (800) 658-8897. 1989. Inquire for price.

International Research Centers Directory. Gale Research, 835 Penobscot Building, Detroit, MI 48226-4094. (313) 961-2242. (800) 877-4253. FAX (313) 961-6083. 8th edition. 1995. $410.00

Research Centers Directory. Gale Research, 835 Penobscot Building, Detroit, MI 48226-4094. (313) 961-2242. (800) 877-4253. FAX (313) 961-6083. $485.00.

Scientific and Technical Organizations and Agencies Directory. Gale Research, 835 Penobscot Building, Detroit, MI 48226-4094. (313) 961-2242. (800) 877-4253. FAX (313) 961-6083. 4th edition. 1996. $195.00.

ENCYCLOPEDIAS AND DICTIONARIES

Encyclopedia of Physical Science and Technology. Robert A. Meyers, ed. 18 volumes. Academic Press Inc., 6277 Sea Harbor Drive, Orlando, FL 32887. (800) 321-5068. 1992. $2100.00.

McGraw-Hill Dictionary of Scientific and Technical Terms. Sybil P. Parker, ed. 5th edition. McGraw-Hill Publishing Company, 1221 Avenue of the Americas, New York, NY 10020. (800) 262-4729 or (212) 512-3825. 1993. $110.50.

GENERAL WORKS

An Engineer's Guide To Water Treatment. George S. Solt and Chris B. Shirley. Ashgate Publishing Co., Old Post Rd., Brookfield, VT 05036. (802) 276-3162. FAX (802) 276-3837. 1991. $92.95.

Water Quality & Treatment. 4th edition. McGraw-Hill Publishing Company, 1221 Avenue of the Americas, New York, NY 10020. (800) 262-4729 or (212) 512-3825. 1990. $99.50.

Water Treatment: Industrial, Commercial & Municipal. D.C. Brandvold. 3d edition. DCB Enterprises, 8286-C2 Western Way Circle, Jacksonville, FL 32256-8389. (904) 636-0758. 1994. $15.00.

Water Treatment and Water Recovery: Advanced Technology and Applications. Paul N. Cheremisinoff and Nicholas P. Cheremisinoff. Prentice Hall, 113 Sylvan Avenue, Route 9W, Englewood Cliffs, NJ 07632. (201) 592-2000 or (800) 922-0579. 1993. $69.00.

Water Treatment Processes: Simple Options. S. Vigneswaran and C. Visanathan. CRC Press, 2000 Corporate Blvd., N.W., Boca Raton, FL 33431. (407) 994-0555 or (800) 272-7737. FAX (407) 994-0949. 1995. $69.95.

HANDBOOKS AND MANUALS

Environmental Science and Technology Handbook. Porter C. Knowles, et al. Government Institutes Inc., 4 Research Place, Suite 200, Rockville, MD 20850. (301) 921-2355. 1994. $75.00.

ONLINE DATABASES AND CD-ROMS

CA Search. Chemical Abstracts Service, P.O. Box 3012, Columbus, OH 43210-0012. (614) 447-3600. (800) 848-6533. FAX (614) 447-3709. Very comprehensive guide to worldwide chemical literature and related fields, 1972 to present. Available on BRS,(800) 289-4277, DIALOG, (800) 334-2564, ORBIT (800) 456-7248, and STN International, FIZ Karlsruhe, P.O. Box 2465, W-7500, Karlsruhe 1, Germany, online services. Inquire as to cost and availability.

NTIS Bibliographic Database. National Technical Information Service, 5285 Port Royal Road, Springfield, VA 22161. (703) 487-4929 or FAX (703) 321-8199. Broad coverage of government-sponsored science and technology research reports, 1964 to present. Available on BRS,(800) 289-4277, DIALOG, (800) 334-2564, ORBIT, (800) 456-7248, and STN International, FIZ Karlsruhe, P.O. Box 2465, W-7500, Karlsruhe 1, Germany, online services. Also available on CD-ROM. Inquire as to cost and availability.

SCISEARCH. Institute for Scientific Information, 3501Market Street, Philadelphia, PA 19104. (800) 523-1850 or (215) 386-0100. Broad multidisciplinary title and author index to the international literature of science and technology, 1974 to present. Available on DIALOG,(800) 334-2564, and ORBIT,(800) 456-7248, online services. Also available on CD-ROM. Inquire as to cost and availability.

PERIODICALS

American Water Works Association Journal. 6666 W. Quincy Avenue, Denver, CO 80235. (303) 794-7711. FAX (303) 794-7310. 1914 to present. Monthly. $85.00 to libraries and government agencies only.

Ground Water. Ground Water Publishing Company, National Well Water Association, 6375 Riverside Drive, Dublin, OH 43017. (614) 761-1711. FAX (614) 761-3446. 1963 to present. Bi-monthly. $90.00 per year.

Journal of Environmental Engineering. ASCE, Enviromental Engineering Division, 345 East 47th Street, New York, NY 10017-2398. (212) 705-7520 or (800) 548-2723. FAX (212) 705-7712. 1956 to present. Bi-monthly. $136.00 per year for non-members.

Journal of Water Resources Planning and Management. American Society of Civil Engineers, Water Resources Planning and Management Division, 345 East 47th Street, New York, NY 10017-2398. (212) 705-7520 or (800) 548-2723. Bi-monthly. $112.00 per year.

Water, Air and Soil Pollution. Kluwer Academic Publishers, P.O. Box 358, Accord Station, Hingham, MA 02018-0358. (617) 871-6000. FAX (617) 871-6528. 1971 to present. 28 times a year. $1519.00 per year.

Water Engineering and Management. Scranton Gillette Communications, Inc., 380 East Northwest Highway, Des Plaines, IL 60016. (708) 298-6622. FAX (708) 390-0408. 1882 to present. Monthly. $25.00 per year.

Water Environment Research. Water Environment Federation, 601 Wythe Street, Alexandria, VA 22314-1994. (703) 684-2400. FAX (708) 684-2492. 1928 to present. Monthly. $144.00 per year.

Water Pollution Control. Bureau of National Affairs, 1231 25th Street NW, Washington, DC 20037. (202) 452-4200. FAX (202) 822-8092. 1979 to present. Bi-weekly. $612.00 per year.

Water Quality International. Elsevier Science [journals], 660 White Plains Rd., Tarrytown, NY 10159-5153. (919) 524-9200. FAX (919) 333-2444. 1987 to present. 4 times a year. $125.00 per year.

Water Resources Bulletin. American Water Resources Association, 5410 Grosvenor Lane, Suite 220, Bethesda, MD 20814-2192. (301) 493-8600. FAX (301) 493-5844. 1965 to present. Bi-monthly. $15.00.

WATER TREATMENT

Ency. of Physical Sciences and Engineering Info. Sources

Water Resources Research. American Geophysical Union, 2000 Florida Avenue, N.W., Washington, DC 20009. (202) 462-6900. FAX (202) 328-0566. 1965 to present. Monthly. $660.00 for non-members.

RESEARCH CENTERS AND INSTITUTES

California Institute of Technology,W.M. Keck Laboratoryof Hydraulics and Water Resources. 138-78, Pasadena, CA 91125. (818) 356-4404. FAX (818) 356-2940.

University of Illinois, Institute For Environmental Studies. 1101 W Peabody Drive, Urbana, IL 61801. (217) 333-4178. FAX (217) 333-8046.

Lenox Institute of Water Technology. 101 Yokun Avenue, PO Box 1639, Lenox, MA 01240. (413) 637-3025. (413) 637-0768.

Pennsylvania State University Environmental Engineering Laboratory. 212 Sackett Bldg., University Park, PA 16802. (814) 863-4385. FAX (814) 863-7304.

WAVE OPTICS

See: OPTICS

WAVES

See: OCEANOGRAPHY, TSUANAMIS

WEATHER

See also: CLIMATE, METEOROLOGY

ABSTRACT SERVICES AND INDEXES

Applied Science and Technology Index; A Cumulative Subject Index To English Language Periodicals in the Fields of Aeronautics and Space Science, Computer Technology, Chemistry, Construction Industry, Energy and Related Areas. H.W. Wilson Co., 950 University Avenue, Bronx, NY 10452-9978. (212) 588-8400. (800) 367-6770. FAX (718) 590-1617. From 1958 to present. Monthly. Inquire about cost and availability. Also available online (BRS and WILSONLINE) and on CD-ROM. ISSN: 0003-6986.

Deep-Sea Research. Part A: Oceanographic Research Papers. Deep-sea Research. Part B: Oceanographic Literature Review. Pergamon Press, Inc., Maxwell House, Fairview Park, Elmsford, NY 10523. (914) 592-7700. Fax (914) 592-3625. Twelve times per year. $1370.00 per year for Parts A and B. Oceanographic Literature Review also available on CD-ROM. Inquire about price and availability.

General Science Index. H.W. Wilson Co., 950 University Avenue, Bronx, NY 10452. (800) 367-6770 or (212) 588-8400. Inquire about price and availability. (212) 588-8400. (800) 367-6770. FAX (718) 590-1617. From 1978 to present. Ten issues per year; quarterly and annual cumulations. Service basis. Avail-

able on CD-ROM and online. Inquire regarding cost and availability. ISSN: 0162-1963.

Government Reports Announcements and Index. National Technical Information Service (NTIS), 5285 Port Royal Road, Springfield, VA 22161. (703) 487-4650. 1968 to present. Annual. $525.00 per year. Also available online as *NTIS Bibliographic Database* and on CD-ROM. FAX (703) 321-8547. From 1968 to present. Annual. $630.00 per year. Also available online as NTIS BIBLIOGRAPHIC DATABASE and on CD-ROM. ISSN:

Meteorological and Geoastrophysical Abstracts. American Meteorological Society, c/o Inforonics, Inc., 550 Newtown Road, Box 458, Littleton, MA 01460. (508) 486-8976. FAX (508) 486-0027. Covers literature in environmental sciences, meteorology, astrophysics, hydrology, glaciology, and physical oceanography. 1950 to present. Monthly. $950.00 per year. ISSN: 0026-1130. Also available online (DIALOG) and on CD-ROM.

Oceanic Abstracts. Cambridge Scientific Abstracts, 7200 Wisconsin Avenue, Bethesda, MD 20814. (301) 961-6750. Fax (301) 961-6720. Bimonthly. $995.00 per year.

Physics Abstracts. Institute of Electrical Engineers, London, United Kingdom. Available from: Institute of Electrical and Electronic Engineers (IEEE), 345 East 47th Street, New York, NY 10017. (212) 705-7900.

Science Citation Index. SCI. Institute for Scientific Information, 3501 Market Street, Philadelphia, PA 19104. (215) 386-0100. (800) 523-1850. FAX (215) 386-2991. 1961 to present. Six issues per year, plus annual cumulation. $11650.00 per year. Also available online and on CD-ROM. Inquire about price and availability. ISSN: 0036-827X.

Selected Water Resources Abstracts. U.S. Geological Survey. 7Distributed by National Technical Information Service (NTIS), 5285 Port Royal Road, Springfield, VA 22161. (703) 487-4650. Monthly. $115.00 per year.

ANNUAL REVIEWS AND YEARBOOKS

Annual Review and Earth and Planetary Sciences. Annual Reviews, Inc., 4139 El Camino Way, Palo Alto, CA 94306-0897. (415) 493-4400. Fax (415) 855-9815. Annual. $55.00 per year.

Developments in Atmospheric Science Series. Elsevier Science Publishing Company, Inc., 655 Avenue of the Americas, New York, NY 10010. (212) 989-5800. Irregular. Inquire about price and availability.

ASSOCIATIONS AND PROFESSIONAL SOCIETIES

American Meteorological Society, 45 Beacon Street, Boston, MA 02108-3693. (617) 227-2425. FAX (617) 742-8718. Irregular. Price varies.

Association of American Weather Observers, 401 Whitney Boulevard, Box 455, Belvedere, IL 61008. (815) 544-5665. FAX (815) 544-6334.

Canadian Meteorological and Oceanographic Society, P.O. Box 334, Newmarket, Ontario, L3Y 4X7, Canada. (416) 898-1040. FAX (416) 898-7937.

National Weather Association, 6704 Wolke Court, Montgomery, AL 36116-2134. (334) 213-0388. FAX (334) 213-0388.

BIBLIOGRAPHIES

Information Sources in the Earth Sciences. David N. Wood, Joan E. Hardy, and Anthony P. Harvey, editors. Bowker-Saur/K.G. Saur. Distributed by R.R. Bowker, 121 Chanlon Road, New Providence, NJ 07974. (800) 521-8110 or (908) 464-6800. Second edition. 1989. $85.00.

Scientific and Technical Books and Serials in Print; An index to Literature in Science and Technology. R.R. Bowker Co., 205 E. 42nd Street, New York, NY 10017. (800) 521-8110 or (212) 916-1600. 1992. $250.00.

DIRECTORIES AND BIOGRAPHICAL SOURCES

American Men and Women of Science: Physical and Biological Sciences. Eighteenth edition. R.R. Bowker Company, 245 West 17th Street, New York, NY 10011. (800) 521-8810 or (212) 916-1600. $750.00.

American Meteorological Society Professional Directory. American Meteorological Society. 45 Beacon Street, Boston, MA 02108. (617) 227-2425. Included in each issue of the Bulletin of the Society.

Meteorological Services of the World. American Meteorological Society, 45 Beacon Street, Boston, MA 02108- 3693. (617) 227-2425. FAX (617) 742-8718. Annual. $70.00.

National Weather Service Offices and Stations. National Oceanic and Atmospheric Administration, Deparatment of Commerce, Silver Spring, MD 20910. (301) 427-7698. Annual. Free

Research Centers Directory. Gale Research, 835 Penobscot Building, Detroit, MI 48226-4094. (313) 961-2242. (800) 877-4253. 20th edition. 1995. $485.00.

ENCYCLOPEDIAS AND DICTIONARIES

Climates of the States. Bair. Gale Research, 835 Penobscot Building, Detroit, MI 48226-4094. (313) 961-2242. (800) 877-4253. 5th edition. 1995. $255.00.

Concise Oxford Dictionary of the Earth Sciences. Ailsa Allaby and Michael Allaby, editors. Oxford University Press, Inc., 200 Madison Avenue, New York, NY 10016. (800) 334-4249 or (212) 679-7300. 1990. $42.95.

Encyclopedia of Climatology. Rhodes W. Fairbridge, and John E. Oliver. Chapman & Hall, 1 Penn Plaza, New York, NY 10119. (212) 564-1060. 1986. $115.00.

Handbook of Applied Meteorology. David D. Houghton, editor. John Wiley & Sons, Inc., 605 Third Avenue, New York, NY 10158- 0012. (212) 850-6000. (800) 225-5945. 1985.

McGraw-Hill Encyclopedia of Ocean and Atmospheric Sciences. Sybil P. Parker, editor. McGraw-Hill Publishing Company, 1221 Avenue of the Americas, New York, NY 10020. (800) 262-4729 or (212) 512-3825. 1979. $79.95.

McGraw-Hill Encyclopedia of Science and Technology. McGraw-Hill Publishing Company, 1221 Avenue of the Americas, New York, NY 10020. (800) 262-4729 or (212) 512-3825. Seventh edition. 1992. $1900.00.

Magill's Survey of Science: Earth Science Series. Salem Press, Inc., P.O. Box 1097, Englewood Cliffs, NJ 07632. (800) 221-1592 or (201) 871-3700. Five volumes. 1990. $400.00 for the set.

One Thousand One Questions Answered About the Weather. Frank H. Forrester. Dover Publications, Inc., 180 Varick Street, New York, NY 10014. (212) 255-3755. (800) 223- 3130. 1982. Reprint edition. $6.95 in paper.

Smithsonian Meteorological Tables. Books on Demand, 300 North Zeeb Road, Ann Arbor, MI 48106-1346. (313) 761-4700. (800)521-0600. reprint edition. $153.90.

Weather Concepts and Terminology. Joe R. Engleman. Trimedia Publishing Co., 12008 West 87th, Suite 117, Lenexa, KS 66215. (913) 599-0505. 1989. $13.95.

Weather Almanac. Gale Research, 835 Penobscot Building, Detroit, MI 48226-4094. (313) 961-2242. (800) 877- 4253. 7th edition. 1996. $130.00.

GENERAL WORKS

THE AMERICAN WEATHER BOOK. DAVID LUDLUM. AMERICAN METEOROLOGICAL SOCIETY. 45 BEACON STREET, BOSTON, MA 02108- 3693. (617) 227-2425. FAX (617) 742-8718. REPRINT EDITION. 1989. $20.00.

The Atmosphere: An Introduction To Meteorology. Frederick K. Lutgens and Edward J. Tarbuck. Prentice Hall (Division of Simon and Schuster), 15 Columbus Circle, New York, NY 10023. (212) 373-8500 or (800) 922-0579. Fifth edition. 1991. $49.00.

Atmosphere, Weather and Climate. Roger G. Barry and Richard J. Chorley. Routledge, Chapman & Hall, Inc., 29 West 35th Street, New York. 1993. $99.95.

Cloud Dynamics. Robert A. Houze. Academic Press, Inc., 1250 Sixth Avenue, San Diego, CA 92101-4311. (619) 231-0926. FAX (619) 699-6715. 1993. $149.00.

Clouds, Rain, and Rainmaking. Basil J. Mason. Books on Demand, Division of University Microfilms International, 300 North Zeeb Road, Ann Arbor, MI 48106-1346. (313) 761-4700 or (800) 521-0600. 1974. $49.50.

Doppler Radar and Weather Observations. Richard J. Doviak and Dusan S. Zrnic. Academic Press, Inc., 1250 Sixth Avenue, San Diego, CA 92101-4311. (619) 231-0926. FAX (619) 699-6715. Second edition. 1993. $95.00.

Early American Winters, 1821 - 1879. David M. Ludlum. American Meteorological Society. 45 Beacon Street, Boston, MA 02108-3693. (617) 227-2425. FAX (617) 742-8718.1968. $20.00.

Meteorology and Environmental Sciences. R. Guzzi and others. World Scientific Publishing Company, Inc., 1060 Main Street, Unit B, River Edge, NJ 07661. (800) 227-7562 or (201) 487-9655. 1990. $120.00.

Severe and Unusual Weather. Joe R. Eagleman. Trimedia Publishing Company, 12008 West 87th Street, Suite 117, Lenexa, KS 66215. (913) 599-0505. Second edition. 1990. $41.95.

Weather Companion: An Album of Meteorological History, Science, Legend, and Folklore. Gary Lockhart. John Wiley and Sons, Inc., 605 Third Avenue, New York, NY 10158. (800) 526-5368 or (212) 850-6000. 1988. $12.95.

Weather Cycles: Real or Imaginary? William J. Burroughs. Cambridge University Press, 40 West 20th Street, New York, NY 10011-4211. (212) 924-3900. (800) 872-7423. 1995. $19.95.

WEAtheR SOURCEBOOK. Ronald L. Wagner and Bill Adler Jr. Globe Pequot Press, P.O. Box 833 Old Saybrook, CT 06475. (203) 395-0440. (800) 234-0495. 1994. $12.95.

HANDBOOKS AND MANUAL

Author's Guide To the Journals of the American Meteorological Society. American Meteorological Society, 45 Beacon Street, Boston, MA 02108-3693. (617) 227-2425. FAX (617) 742-8718. 1983. $15.00 in paper.

Basic Meteorology Lab Manual. Thomas A. Leavy. Allegheny Press, P.O. Box 220, Elgin, PA 19413. (814) 664-8504. Second edition. 1969. $7.95.

Handbook in Applied Meteorology. David D. Houghton, editor. John Wiley and Sons, Inc., 605 Third Avenue, New York, NY 10158. (800) 526-5368 or (212) 850-6000. 1985. $168.00.

Weather Handbook. H. McKinley Conway and Linda L. Liston. Conway Data, Inc., 40 Technology Park, Atlanta. Number 200, Norcross, GA 30092. (404) 446-6996. (800) 554-5686. . revised edition. 1990. $39.95

ONLINE DATABASES AND CD-ROMS

Climate Assessment Database. National Weather Service, National Meteorological Center, 5200 Auth Road, Suite 101, Camp Springs, MD 20233. (301) 763-8016. Contains daily, weekly and monthly summaries of North American and world climatological data. Also provides five to ten weather forecasts, and 30 to 90 day weather outlook. Subscription required. Inquire as to cost and availability.

CA Search. Chemical Abstracts Service, P.O. Box 3012, Columbus, OH 43210-0012. (614) 447-3600. (800) 848-6533. FAX (614) 447-3709. Very comprehensive guide to worldwide chemical literature and related fields, 1972 to present. Available on BRS,(800) 289-4277, DIALOG, (800) 334-2564, ORBIT (800) 456-7248, and STN International, FIZ Karlsruhe, P.O. Box 2465, W-7500, Karlsruhe 1, Germany, online services. Inquire as to cost and availability.

Compendex Plus. Engineering Information, Inc., 345 East 47th Street, New York, NY 10017. (212) 705-7600 or (800) 221-1044. Contains citations with abstracts to worldwide literature in engineering and technology, from 1970 to present. Available on online BRS,(800) 289-4277, DIALOG, (800) 334-2564,

ORBIT (800) 456-7248, and STN International, FIZ Karlsruhe, P.O. Box 2465, W-7500, Karlsruhe 1, Germany, online services. Also available on CD-ROM. Inquire as to cost and availability.

Current Contents Search. Institute for Scientific Information, 3501 Market Street, Philadelphia, PA 19104. (215) 386-0100. FAX (215) 386-6362. Contains citations to articles listed in the table of contents of science and technology journals. Also articles in social sciences and life sciences journals. Available on BRS,(800) 289-4277, DIALOG,(800) 334-2564, online services. Inquire as to cost and availability.

Dissertation Abstracts. University Microfilms International, 300 North Zeeb Road, Ann Arbor, MI 48106. (800) 521-0600 or (313) 761-4700. Scope includes virtually all doctoral dissertations accepted at accredited American institutions from 1861 to present in 252 subject areas. Available on BRS,(800) 289-4277, DIALOG,(800) 334-2564, and OCLC EPIC,(800) 848-5878,online services. Also available on CD-ROM. Inquire as to cost and availability.

INSPEC. Institution of Electrical Engineers, Michael Faraday House, Six Hills Way, Stevenage, Herts. SG1 2AY, England. Telephone: 0438 313311 or FAX 0438 742840. Contains citations to the worldwide literature of physics, electronics and electrical engineering, computer technology, and related fields. Available on BRS,(800) 289-4277, DIALOG, (800) 334-2564, ORBIT, (800) 456-7248, and STN International, FIZ Karlsruhe, P.O. Box 2465, W-7500, Karlsruhe 1, Germany, online services. Inquire as to cost and availability.

International Aerospace Abstracts. American Institute of Aeronautics and Astronautics, 370 L'Enfant Promenade, S.W., Washington, DC 20024. (202) 646-7400. Contains references and abstracts of journal and monograph literature relating to aerospace science and technology, from 1963 to present. Available through the NASA/RECON system of the National Aeronautics and Space Administration only.

Meteorological and Geoastrophysical Abstracts. American Meteorological Society, 45 Beacon Street, Boston, MA 02108-3693. (617) 227-2425. FAX (617) 742-8718. Contains citations and abstracts to the worldwide literature on significant research in meteorology and geoastrophysics. Related topics include physical oceanography, hydrology, environmental sciences and glaciology. Covers the period 1972 to present. Available on DIALOG,(800) 334-2564, online service. Inquire as to cost and availability.

NTIS Bibliographic Database. National Technical Information Service, 5285 Port Royal Road, Springfield, VA 22161. (703) 487-4929 or FAX (703) 321-8199. Broad coverage of government-sponsored science and technology research reports, 1964 to present. Available on BRS,(800) 289-4277, DIALOG, (800) 334-2564, ORBIT, (800) 456-7248, and STN International, FIZ Karlsruhe, P.O. Box 2465, W-7500, Karlsruhe 1, Germany, online services. Also available on CD-ROM. Inquire as to cost and availability.

Physics Briefs. American Institute of Physics, 335 East 45th Street, New York, NY 10017. (212) 661-9260 or FAX (212) 661-2036. Contains citations with abstracts of the literature of physics and related fields, 1979 to present. Available on the STN International, FIZ Karlsruhe, P.O. Box 2465, W-7500, Karlsruhe 1, Germany, online service. Inquire as to cost and availability.

SCISEARCH. Institute for Scientific Information, 3501 Market Street, Philadelphia, PA 19104. (800) 523-1850 or (215) 386-0100. Broad multidisciplinary title and author index to the in-

ternational literature of science and technology, 1974 to present. Available on DIALOG and ORBIT online services. Also available on CD-ROM. Inquire as to cost and availability.

WILSONLINE. H.W. Wilson Company, 950 University Avenue, Bronx, NY 10452. (800) 367-6770 or (212) 588-8400. Makes available online versions of the printed H.W. Wilson indexes includingApplied Science and Technology Index, Business Periodicals Index, General Science Index, and Readers' Guide to Periodical Literature. Period covered is generally 1983 to present. Available on BRS,(800) 289-4277, DIALOG,(800) 334-2564, and OCLC EPIC,(800) 848-5878, online services. Also available on CD-ROM. Inquire as to cost and availability.

World Climate Disc. Chadwyck-Healey Inc., 1101 King Street, Alexandria, VA 22314. (703) 683-4890. FAX (703) 683-7589. Weather and climate data from approximately 5,000 weather stations worldwide, covering the years 1854 to 1990 on CD-ROM. First edition 1992. Approximately $1200.00 per year with annual updates.

World Weatherdisc. WeatherDisc Associates, Inc., 4584 N.E. 89th Street, Seattle, WA 98115. (206) 524-4314. FAX (206) 543-0308. Meteorological data on CD-ROM which describes the climate of the earth today and for the past few hundred years. First edition 1989. Approximately $295.00 per year with biannual updates.

PERIODICALS

Agricultural and Forest Meteorology. Elsevier Science Publishing Company, Inc., 655 Avenue of the Americas, New York, NY 10010. (212) 989-5800. Twenty times per year. $750.00 per year.

American Meteorological Society. BULLETIN. American Meteorological Society, 45 Beacon Street, Boston, MA 02108-3693. (617) 227-2425. FAX (617) 742-8718. Monthly. $60.00 per year.

American Meteorological Society. METEOROLOGICAL MONOGRAPHS. American Meteorological Society, 45 Beacon Street, Boston, MA 02108-3693. (617) 227-2425. FAX (617) 742-8718. Irregular. Price varies.

American Weather Observer. Association of American Weather Observers, 401 Whitney Boulevard, Box 455, Belvedere, IL 61008. (815) 544-5665. FAX (815) 544-6334. Monthly. $21.00 per year.

Atmosphere - Ocean. Canadian Meteorological and Oceanographic Society, P.O. Box 334, Newmarket, Ontario, L3Y 4X7, Canada. (416) 898-1040. FAX (416) 898-7937. Quarterly. $30.00 per year.

Boundary-layer Meteorology: An International Journal of Physical and Biological Processes in the Atmospheric Boundary Layer. Kluwer Academic Publishers, P.O. Box 358, Accord Station, Hingham, MA 02018-0358. (617) 871-6000. Sixteen per year. $785.00 per year.

Climate Change. Kluwer Academic Publishers, P.O. Box 358, Accord Station, Hingham, MA 02018-0358. (617) 871-6000. Six times per year. $327.00 per year.

Dynamics of Atmospheres and Oceans. Elsevier Science Publishing Company, Inc., 655 Avenue of the Americas, New York,

NY 10010. (212) 989-5800. Six times per year. $205.00 per year.

International Journal of Climatology. Royal Meteorological Society, distributed by John Wiley and Sons, Inc., 605 Third Avenue, New York, NY 10158. (800) 526-5368 or (212) 850-6000. Eight times per year. $425.00 per year.

JGR: Journal of Geophysical Research: Atmosphere. American Geophysical Union, 2000 Florida Avenue, N.W., Washington, DC 20009. (202) 462-6903. Monthly. $90.00 per year to members.

JGR: Journal of Geophysical Research: Oceans. American Geophysical Union, 2000 Florida Avenue, N.W., Washington, DC 20009. (202) 462-6903. Monthly. $1545.00 per year.

Journal of Applied Meteorology. American Meteorological Society, 45 Beacon Street, Boston, MA 02108-3693. (617) 227-2425. FAX (617) 742-8718. Monthly. $165.00 per year.

Journal of Climate. American Meteorological Society, 45 Beacon Street, Boston, MA 02108-3693. (617) 227-2425. FAX (617) 742-8718. Monthly. $175.00 per year.

Journal of Atmospheric Sciences. American Meteorological Society, 45 Beacon Street, Boston, MA 02108-3693. (617) 227-2425. FAX (617) 742-8718. Semi-monthly. $320.00 per year.

Monthly Weather Review. American Meteorological Society, 45 Beacon Street, Boston, MA 02108-3693. (617) 227-2425. FAX (617) 742-8718. Monthly. $205.00 per year.

National Weather Digest. National Weather Association, 4400 Stamp Road, Room 404, Temple Hills, MD 20748. (301) 899-3784.

Royal Meteorological Society Quarterly Journal. Royal Meteorological Society, 104 Oxford Road, Reading, Berks RG1 7LJ, England. Six times per year. $250.00 per year.

Weather. Royal Meteorological Society, 104 Oxford Road, Reading, Berks RG1 7LJ, England. Monthly. $44.00 per year.

Weatherwise. Heldref Publications, 1319 Eighteenth Street, N.W., Washington, DC 20036-1802. (202) 296-6267. FAX (202) 296-5149. Bi-monthly. $28.00 per year.

RESEARCH CENTERS AND INSTITUTES

Atlantic Oceanographic and Meteorological Laboratory. 4301 Rickenbacker Causeway, Miami, FL 33149. (305) 361-4300.

Climate Analysis Center. National Weather Service, 5200 Auth Road, Camp Springs, MD 20746. (301) 763-8167.

Cooperative Institute For Mesoscale Meteorological Studies. University of Oklahoma, 401 East Boyd, Norman, OK 73019. (405) 325-3041.

Geophysical Fluid Dynamics Laboratory. Princeton University, P.O. Box 308, Princeton, NJ 08542. (609) 452-6500.

Goddard Space Flight Center. Laboratory for Atmospheres, Mail Code 610, Greenbelt, MD 20771. (301) 286-5002.

WEATHER

Ency. of Physical Sciences and Engineering Info. Sources

Hurricane Research Division. Atlantic Oceanographic and Meteorological Laboratory. 4301 Rickenbacker Causeway, Miami, FL 33149. (305) 361-4400.

Joint Institute For Marine and Atmospheric Research. University of Hawaii, 1000 Pope Road, Honolulu, HI 96822. (808) 541-2876.

National Center For Atmospheric Research. P.O. Box 3000, Boulder, CO 80307. (303) 496-1000.

National Hurricane Center. 1320 South Dixie Highway, Coral Gables, FL 33146. (305) 666-4612.

National Meteorological Center. World Weather Building, Room 101, 5200 Auth Road, Camp Springs, MD 20746. (301) 763-8016.

National Severe Storms Forecast Center. 601 East 12th Street, Kansas City, MO 64106. (816) 374-5922.

National Weather Service. 1325 East-West Highway, Silver Spring, MD 20910. (301) 427-7689.

WEATHER FORECASTING

See also: CLIMATE, HURRICANES, METEOROLOGY, RAIN, SNOW, THUNDERSTORMS, TORNADOS, WEATHER

ABSTRACT SERVICES AND INDEXES

Applied Science and Technology Index; A Cumulative Subject Index To English Language Periodicals in the Fields of Aeronautics and Space Science, Computer Technology, Chemistry, Construction Industry, Energy and Related Areas. H.W. Wilson Co., 950 University Avenue, Bronx, NY 10452. (800) 367-6770 or (212) 588-8400. From 1958 to present. Monthly. Inquire about cost and availability. Also available on CD-ROM.

Deep-sea Research. Part A: Oceanographic Research Papers. Deep-sea Research. Part B: Oceanographic Literature Review. Pergamon Press, Inc., Maxwell House, Fairview Park, Elmsford, NY 10523. (914) 592-7700. Fax (914) 592-3625. Twelve times per year. $1370.00 per year for Parts A and B.

General Science Index. H.W. Wilson Co., 950 University Avenue, Bronx, NY 10452. (800) 367-6770 or (212) 588-8400. Inquire about price and availability.

Government Reports Announcements and Index. National Technical Information Service (NTIS), 5285 Port Royal Road, Springfield, VA 22161. (703) 487-4650. 1968 to present. Annual. $525.00 per year. Also available online as NTIS BIBLIOGRAPHIC DATABASE and on CD-ROM.

Meteorological and Geoastrophysical Abstracts. American Meteorological Society, c/o Inforonics, Inc., 550 Newtown Road, Box 458, Littleton, MA 01460. (508) 486-8976. Monthly. $750.00 per year.

Oceanic Abstracts. Cambridge Scientific Abstracts, 7200 Wisconsin Avenue, Bethesda, MD 20814. (301) 961-6750. Fax (301) 961-6720. Bimonthly. $995.00 per year.

Physics Abstracts. Institute of Electrical Engineers, London, United Kingdom. Available from: Institute of Electrical and Electronic Engineers (IEEE), 345 East 47th Street, New York, NY 10017. (212) 705-7900.

Science Citation Index. Institute for Scientific Information, 3501 Market Street, Philadelphia, PA 19104. (800) 523-1850 or (215) 386-0100. Inquire about price and availability.

Selected Water Resources Abstracts. U.S. Geological Survey. Distributed by National Technical Information Service (NTIS), 5285 Port Royal Road, Springfield, VA 22161. (703) 487-4650. Monthly. $115.00 per year.

ANNUAL REVIEWS AND YEARBOOKS

Annual Review and Earth and Planetary Sciences. AnnualReviews, Inc., 4139 El Camino Way, Palo Alto, CA 94306-0897. (415) 493-4400. Fax (415) 855-9815. Annual. $55.00 per year.

Developments in Atmospheric Science Series. Elsevier Science Publishing Company, Inc., 655 Avenue of the Americas, New York, NY 10010. (212) 989-5800. Irregular. Inquire about price and availability.

Ocean Yearbook. Elisabeth M. Borgese, et al, editors. University of Chicago Press, 5801 Ellis Avenue, Chicago, IL 60637. (312) 702-7700. 1995. $77.00.

ASSOCIATIONS AND PROFESSIONAL SOCIETIES

American Meteorological Society, 45 Beacon Street, Boston, MA 02108-3693. (617) 227-2425. FAX (617) 742-8718. Irregular. Price varies.

Association of American Weather Observers, 401 Whitney Boulevard, Box 455, Belvedere, IL 61008. (815) 544-5665. FAX (815) 544-6334.

Canadian Meteorological and Oceanographic Society, P.O. Box 334, Newmarket, Ontario, L3Y 4X7, Canada. (416) 898-1040. FAX (416) 898-7937.

International Association of Meteorology and Atmospheric Physics. c/o UCAR, P.O. Box 3000, Boulder CO 80307-3000. (303) 497-1673.

International Association of Severe Weather Specialists. c/o Warren Faidley, PO Box 31808, Tuscon, AZ 85751. (520) 751-9964. FAX (520) 751-1185.

National Environmental Satellite Data, and Information Service. 2069 Federal Building 4 Room 2060, Washington, DC 20233. (301) 763-71990. FAX (301) 763-4011.

National Weather Association, 6704 Wolke Court, Montgomery, AL 36116-2134. (334) 213-0388. FAX (334) 213-0388.

University Corporation For Atmospheric Research. PO Box 3000, 1850 Table Mesa Drive, Boulder, CO 80307-3000. (303) 497-1673. FAX (303) 497-1654.

Weather Modification Association. PO Box 26926, Fresno, CA 93729-6926. (209) 434-3486.

BIBLIOGRAPHIES

Information Sources in the Earth Sciences. David N. Wood, Joan E. Hardy, and Anthony P. Harvey, editors. Bowker-Saur/K.G. Saur. Distributed by R.R. Bowker, 121 Chanlon Road, New Providence, NJ 07974. (800) 521-8110 or (908) 464-6800. Second edition. 1989. $85.00.

Scientific and Technical Books and Serials in Print; An index to Literature in Science and Technology. R.R. Bowker Co., 205 E. 42nd Street, New York, NY 10017. (800) 521-8110 or (212) 916-1600. 1992. $250.00.

DIRECTORIES AND BIOGRAPHICAL SOURCES

American Men and Women of Science: Physical and Biological Sciences. Eighteenth edition. R.R. Bowker Company, 245 West 17th Street, New York, NY 10011. (800) 521-8810 or (212) 916-1600. $750.00.

Research Centers Directory. Gale Research, 835 Penobscot Building, Detroit, MI 48226-4094. (800) 877-4253 or (313) 961-2242. 17th edition, 1992. $400.00.

ENCYCLOPEDIAS AND DICTIONARIES

Climates of the States. Bair. Gale Research Inc. 835 Penobscot Building, Detroit, MI 48226-4094. (313) 961-2242. (800) 877-4253. 5th edition. 1995. $255.00.

Concise Oxford Dictionary of the Earth Sciences. Ailsa Allaby and Michael Allaby, editors. Oxford University Press, Inc., 200 Madison Avenue, New York, NY 10016. (800) 334-4249 or (212) 679-7300. 1990. $42.95.

McGraw-Hill Encyclopedia of Ocean and Atmospheric Sciences. Sybil P. Parker, editor. McGraw-Hill Publishing Company, 1221 Avenue of the Americas, New York, NY 10020. (800) 262-4729 or (212) 512-3825. 1979. $79.95.

McGraw-Hill Encyclopedia of Science and Technology. McGraw-Hill Publishing Company, 1221 Avenue of the Americas, New York, NY 10020. (800) 262-4729 or (212) 512-3825. Seventh edition. 1992. $1900.00.

Magill's Survey of Science: Earth Science Series. Salem Press, Inc., P.O. Box 1097, Englewood Cliffs, NJ 07632. (800) 221-1592 or (201) 871-3700. Five volumes. 1990. $400.00 for the set.

GENERAL WORKS

The Basic Essentials of Weather Forecasting. Michael Hodgson. ICS Books, P.O. Box 10767, Merrillville, in 46411-0767. 9210) 769-0585. (800) 541-7323. . 1992. $5.99.

Clouds, Rain, and Rainmaking. Basil J. Mason. Books on Demand, 300 North Zeeb Road, Ann Arbor, MI 48106-1346. (313) 761-4700 or (800) 521-0600. 2nd edition. $56.20.

Doppler Radar and Weather Observations. Richard J. Doviak and D. S. Zrinc. Academic Press, Inc., 6277 Sea Harbor Drive, Orlando, FL. (800) 321-5068. 2nd edition. 1993. $69.95.

Images in Weather Forecasting: A Practical Guide For Interpreting Satellite and Radar Imagery. M. J. Bader. Cambridge University Press, 40 West 20th Street, New York, NY 10011-4211. (212) 924-3900. (800) 872-7423. 1994.

Mesoscale Meteorology and Forecasting. Peter S. Ray. American Meteorological Society, 45 Beacon Street, Boston, MA 02108-3693. (617) 227-2425. FAX (617) 742-8718. reprint edition. $66.25.

Principles and Methods of Extended Period Forecasting. Robert P. Harnack. National Weather Service, National Meteorological Center, 5200 Auth Road, Suite 101, Camp Springs, MD 20233. (301) 763-8016. 1986. $12.00 in paper.

Severe and Unusual Weather. Joe R. Eagleman. Trimedia Publishing Company, 12008 West 87th Street, Suite 117, Lenexa, KS 66215. (913) 599-0505. Second edition. 1990. $41.95.

Weather: Understanding the Forces of Nature. Louise Quayle. Random House Value Publishing, Inc., 40 Englehard Avenue,Avenal NJ 07001. (908) 827-2700. (800) 223-6804. 1991. $19.50.

Weather Radar and Flood Forecasting. C. Kirby and V. K. Cullings. John Wiley & Sons, Inc., 605 Third Avenue, New York, NY 10158-0012. (212) 850-6000. (800) 225-5945.

1987. $225.00.

The Weather Revolution: Innovations and Imminent Breakthroughs in Accurate Forecasting. J. Fishman, and R. Kalish. Plenum Publishing Corp., 233 Spring Street, New York, NY. (212) 620-8000. (800) 2221-9369. FAX (212) 463-0742. 1994. $27.95.

HANDBOOKS AND MANUALS

Author's Guide To the Journals of the American Meteorological Society. American Meteorological Society, 45 Beacon Street, Boston, MA 02108-3693. (617) 227-2425. FAX (617) 742-8718. 1983. $15.00 in paper.

Basic Meteorology Lab Manual. Thomas A. Leavy. Allegheny Press, P.O. Box 220, Elgin, PA 19413. (814) 664-8504. Second edition. 1969. $7.95.

Handbook in Applied Meteorology. David D. Houghton, editor. John Wiley and Sons, Inc., 605 Third Avenue, New York, NY 10158. (800) 526-5368 or (212) 850-6000. 1985. $114.00.

Weather Almanac. Gale Research, 835 Penobscot Building, Detroit, MI 48226-4094. (313) 961-2242. (800) 877- 4253. 7th edition. 1996. $130.00.

Weather Satellite Handbook. American Radio Relay League, 225 Main Street, Newington CT -6111. (203) 666-1541. 1994. $20.00.

ONLINE DATABASES AND CD-ROMS

Climate Assessment Database. National Weather Service, National Meteorological Center, 5200 Auth Road, Suite 101, Camp Springs, MD 20233. (301) 763-8016. Contains daily, weekly and monthly summaries of North American and world climatological data. Also provides five to ten weather forecasts, and 30

WEATHER FORECASTING

Ency. of Physical Sciences and Engineering Info. Sources

to 90 day weather outlook. Subscription required. Inquire as to cost and availability.

CA Search. Chemical Abstracts Service, P.O. Box 3012, Columbus, OH 43210-0012. (614) 447-3600. (800) 848-6533. FAX (614) 447-3709. Very comprehensive guide to worldwide chemical literature and related fields, 1972 to present. Available on BRS,(800) 289-4277, DIALOG, (800) 334-2564, ORBIT (800) 456-7248, and STN International, FIZ Karlsruhe, P.O. Box 2465, W-7500, Karlsruhe 1, Germany, online services. Inquire as to cost and availability.

Compendex Plus. Engineering Information, Inc., 345 East 47th Street, New York, NY 10017. (212) 705-7600 or (800) 221-1044. Contains citations with abstracts to worldwide literature in engineering and technology, from 1970 to present. Available on online BRS,(800) 289-4277, DIALOG, (800) 334-2564, ORBIT (800) 456-7248, and STN International, FIZ Karlsruhe, P.O. Box 2465, W-7500, Karlsruhe 1, Germany, online services. Also available on CD-ROM. Inquire as to cost and availability.

Current Contents Search. Institute for Scientific Information, 3501 Market Street, Philadelphia, PA 19104. (215)386-0100. FAX (215) 386-6362. Contains citations to articles listed in the table of contents of science and technology journals. Also articles in social sciences and life sciences journals. Available on BRS,(800) 289-4277, DIALOG,(800) 334-2564, online services. Inquire as to cost and availability.

Dissertation Abstracts. University Microfilms International, 300 North Zeeb Road, Ann Arbor, MI 48106. (800) 521-0600 or (313) 761-4700. Scope includes virtually all doctoral dissertations accepted at accredited American institutions from 1861 to present in 252 subject areas. Available on BRS,(800) 289-4277, DIALOG,(800) 334-2564, and OCLC EPIC,(800) 848-5878, online services. Also available on CD-ROM. Inquire as to cost and availability.

INSPEC. Institution of Electrical Engineers, Michael Faraday House, Six Hills Way, Stevenage, Herts. SG1 2AY, England. Telephone: 0438 313311 or FAX 0438 742840. Contains citations to the worldwide literature of physics, electronics and electrical engineering, computer technology, and related fields. Available on BRS,(800) 289-4277, DIALOG, (800) 334-2564, ORBIT, (800) 456-7248, and STN International, FIZ Karlsruhe, P.O. Box 2465, W-7500, Karlsruhe 1, Germany, online services. Inquire as to cost and availability.

International Aerospace Abstracts. American Institute of Aeronautics and Astronautics, 370 L'Enfant Promenade, S.W., Washington, DC 20024. (202) 646-7400. Contains references and abstracts of journal and monograph literature relating to aerospace science and technology, from 1963 to present. Available through the NASA/RECON system of the National Aeronautics and Space Administration only.

Meteorological and Geoastrophysical Abstracts. American Meteorological Society, 45 Beacon Street, Boston, MA 02108-3693. (617) 227-2425. FAX (617) 742-8718. Contains citations and abstracts to the worldwide literature on significant research in meteorology and geoastrophysics. Related topics include physical oceanography, hydrology, environmental sciences and glaciology. Covers the period 1972 to present. Available on DIALOG,(800) 334-2564, online service. Inquire as to cost and availability.

NTIS Bibliographic Database. National Technical Information Service, 5285 Port Royal Road, Springfield, VA 22161. (703) 487-4929 or FAX (703) 321-8199. Broad coverage of government-sponsored science and technology research reports, 1964 to present. Available on BRS,(800) 289-4277, DIALOG, (800) 334-2564, ORBIT, (800) 456-7248, and STN International, FIZ Karlsruhe, P.O. Box 2465, W-7500, Karlsruhe 1, Germany, online services. Also available on CD-ROM. Inquire as to cost and availability.

Physics Briefs. American Institute of Physics, 335 East 45th Street, New York, NY 10017. (212) 661-9260 or FAX (212) 661-2036. Contains citations with abstracts of the literature of physics and related fields, 1979 to present. Available on the STNInternational, FIZ Karlsruhe, P.O. Box 2465, W-7500, Karlsruhe 1, Germany, online service. Inquire as to cost and availability.

SCISEARCH. Institute for Scientific Information, 3501 Market Street, Philadelphia, PA 19104. (800) 523-1850 or (215) 386-0100. Broad multidisciplinary title and author index to the international literature of science and technology, 1974 to present. Available on DIALOG and ORBIT online services. Also available on CD-ROM. Inquire as to cost and availability.

WILSONLINE. H.W. Wilson Company, 950 University Avenue, Bronx, NY 10452. (800) 367-6770 or (212) 588-8400. Makes available online versions of the printed H.W. Wilson indexes including Applied Science and Technology Index, Business Periodicals Index, General Science Index, and Readers' Guide to Periodical Literature. Period covered is generally 1983 to present. Available on BRS,(800) 289-4277, DIALOG,(800) 334-2564, and OCLC EPIC,(800) 848-5878, online services. Also available on CD-ROM. Inquire as to cost and availability.

World Climate Disc. Chadwyck-Healey Inc., 1101 King Street, Alexandria, VA 22314. (703) 683-4890. FAX (703) 683-7589. Weather and climate data from approximately 5,000 weather stations worldwide, covering the years 1854 to 1990 on CD-ROM. First edition 1992. Approximately $1200.00 per year with annual updates.

World Weatherdisc. WeatherDisc Associates, Inc., 4584 N.E. 89th Street, Seattle, WA 98115. (206) 524-4314. FAX (206) 543-0308. Meteorological data on CD-ROM which describes the climate of the earth today and for the past few hundred years. First edition 1989. Approximately $295.00 per year with biannual updates.

PERIODICALS

Agricultural and Forest Meteorology. Elsevier Science Publishing Company, Inc., 655 Avenue of the Americas, New York, NY 10010. (212) 989-5800. Twenty times per year. $750.00 per year.

American Meteorological Society. BULLETIN. American Meteorological Society, 45 Beacon Street, Boston, MA 02108-3693. (617) 227-2425. FAX (617) 742-8718. Monthly. $60.00 per year.

American Meteorological Society. METEOROLOGICAL MONOGRAPHS. American Meteorological Society, 45 Beacon Street, Boston, MA 02108-3693. (617) 227-2425. FAX (617) 742-8718. Irregular. Price varies.

American Weather Observer. Association of American Weather Observers, 401 Whitney Boulevard, Box 455, Belvedere, IL 61008. (815) 544-5665. FAX (815) 544-6334. Monthly. $21.00 per year.

Atmosphere - Ocean. Canadian Meteorological and Oceanographic Society, P.O. Box 334, Newmarket, Ontario, L3Y 4X7, Canada. (416) 898-1040. FAX (416) 898-7937. Quarterly. $30.00 per year.

Boundary-layer Meteorology: An International Journal of Physical and Biological Processes in the Atmospheric Boundary Layer. Kluwer Academic Publishers, P.O. Box 358, Accord Station, Hingham, MA 02018-0358. (617) 871-6000. Sixteen per year. $785.00per year.

Climate Change. Kluwer Academic Publishers, P.O. Box 358, Accord Station, Hingham, MA 02018-0358. (617) 871-6000. Six times per year. $327.00 per year.

Dynamics of Atmospheres and Oceans. Elsevier Science Publishing Company, Inc., 655 Avenue of the Americas, New York, NY 10010. (212) 989-5800. Six times per year. $205.00 per year.

International Journal of Climatology. Royal Meteorological Society, distributed by John Wiley and Sons, Inc., 605 Third Avenue, New York, NY 10158. (800) 526-5368 or (212) 850-6000. Eight times per year. $425.00 per year.

JGR: Journal of Geophysical Research: Atmosphere. American Geophysical Union, 2000 Florida Avenue, N.W., Washington, DC 20009. (202) 462-6903. Monthly. $90.00 per year to members.

JGR: Journal of Geophysical Research: Oceans. American Geophysical Union, 2000 Florida Avenue, N.W., Washington, DC 20009. (202) 462-6903. Monthly. $1545.00 per year.

Journal of Applied Meteorology. American Meteorological Society, 45 Beacon Street, Boston, MA 02108-3693. (617) 227-2425. FAX (617) 742-8718. Monthly. $165.00 per year.

Journal of Climate. American Meteorological Society, 45 Beacon Street, Boston, MA 02108-3693. (617) 227-2425. FAX (617) 742-8718. Monthly. $175.00 per year.

Journal of Atmospheric Sciences. American Meteorological Society, 45 Beacon Street, Boston, MA 02108-3693. (617) 227-2425. FAX (617) 742-8718. Semi-monthly. $320.00 per year.

Monthly Weather Review. American Meteorological Society, 45 Beacon Street, Boston, MA 02108-3693. (617) 227-2425. FAX (617) 742-8718. Monthly. $205.00 per year.

National Weather Digest. National Weather Association, 4400 Stamp Road, Room 404, Temple Hills, MD 20748. (301) 899-3784.

Royal Meteorological Society. QUARTERLY JOURNAL. Royal Meteorological Society, 104 Oxford Road, Reading, Berks RG1 7LJ, England. Six times per year. $250.00 per year.

Weather. Royal Meteorological Society, 104 Oxford Road, Reading, Berks RG1 7LJ, England. Monthly. $44.00 per year.

Weatherwise. Heldref Publications, 1319 Eighteenth Street, N.W., Washington, DC 20036-1802. (202) 296-6267. FAX (202) 296-5149. Bi-monthly. $28.00 per year.

RESEARCH CENTERS AND INSTITUTES

Cooperative Institute for Mesoscale Meteorological Studies. University of Oklahoma, 401 East Boyd, Norman, OK 73019. (405) 325-3041.

Hurricane Research Division. Atlantic Oceanographic and Meteorological Laboratory. 4301 Rickenbacker Causeway, Miami, FL 33149. (305) 361-4400.

Joint Institute For Marine and Atmospheric Research. University of Hawaii, 1000 Pope Road, Honolulu, HI 96822. (808) 541-2876.

National Center For Atmospheric Research. P.O. Box 3000,Boulder, CO 80307. (303) 496-1000.

National Hurricane Center. 1320 South Dixie Highway, Coral Gables, FL 33146. (305) 666-4612.

National Meteorological Center. World Weather Building, Room 101, 5200 Auth Road, Camp Springs, MD 20746. (301) 763-8016.

National Severe Storms Forecast Center. 601 East 12th Street, Kansas City, MO 64106. (816) 374-5922.

National Weather Service. 1325 East-West Highway, Silver Spring, MD 20910. (301) 427-7689.

WEATHER MODIFICATION

See: WEATHER

WELDING

See also: MACHINING, METALLURGICAL ENGINEERING, METALLURGY, METALS and METAL WORKING

ABSTRACT SERVICES AND INDEXES

Alloys Index. American Society for Metals, Metals Park, OH 44073. (216) 338-5151. FAX (216) 338-4634. 1974 to present. Monthly. $380.00. Also available on CD-ROM and online via METADEX, STN International, and DIALOG.

Aluminum Industry Abstracts. Aluminum Association, Materials Park, OH 44073. (216) 338-5151. FAX (216) 338-4634. 1968 to present. Monthly. $525.00 per year.

Applied Mechanics Reviews: An Assessment of World Literature in Engineering Sciences. 1948-present. American Society of Mechanical Engineers, 345 East 47th Street, New York, NY 10017. (212) 705-7703. Monthly. $360.00 per year.

Applied Science and Technology Index; A Cumulative Subject Index To English Language Periodicals in the Fields of Aeronautics and Space Science, Computer Technology, Chemistry, Construction Industry, Energy and Related Areas. H.W. Wilson Co., 950 University Avenue, Bronx, NY 10452. (800) 367-6770 or (212) 588-8400. FAX (718) 590-1617. From 1958 to

WELDING

Ency. of Physical Sciences and Engineering Info. Sources

present. Monthly. Inquire about cost and availability. Also available on CD-ROM and online.

Engineering Index Monthly. Engineering Information, Inc., Castle Point on the Hudson, Hoboken, NJ 07030. (800) 221-1044. FAX (212) 832-1857. Monthly. $2300.00 per year. Also available online as COMPENDEX, and also on CD-ROM. Covers chemical engineering, computers, electrical engineering, civil engineering, metals and mining, industrial management, and mechanical engineering.

I.M.M. Abstracts and Index. Institution of Mining and Metallurgy, 44 Portland Pl., London W1N 4BR, England. 071-580-3802. FAX 071-436-5388. Bi-monthly. $364 for non-members.

Leadscan (formerly Lead Abstracts), Lead Development Association, 42 Weymouth Street, London W1X 3lQ, England. 1958 to present. Quarterly. Inquire for cost.

Metals Abstracts and Metals Abstracts Index. American Society for Metals, Metals Park, OH 44073. (216) 338-5151. 1968 to present. Monthly. Abstracts are $1925.00 per year and Index is $460.00 per year.

Physics Abstracts. Institute of Electrical Engineers, Michael Faraday House, Six Hill Way, Stevenage, Herts, SG1 2AY, England. Distributed by IEEE, 445 Hoes Lane, Piscataway, NJ 08854. (908) 562-5549. 1898 to present. Monthly. $2700.00 per year. Also available online as INSPEC.

Science Citation Index. Institute for Scientific Information, 3501 Market Street, Philadelphia, PA 19104. (215) 386-0100. FAX (215) 386-6362. Inquire about availability and cost. Also available on CD-ROM.

ANNUAL REVIEWS AND YEARBOOKS

Welding Research Council Yearbook. Welding Research Council, United Engineering Center, 345 E. 47th Street, New York, NY 10017. (212) 705-7956. FAX (212) 371-9622. 1936 to present. Annual. Included in membership dues.

ASSOCIATIONS AND PROFESSIONAL SOCIETIES

Aluminum Association. 900 19th Street NW, Suite 300, Washington, DC 20006. (202) 862-5100. FAX (202) 862-5164.

American Society For Metals. Materials Park, OH 44073. (216) 338-5151. FAX (216) 338-4634.

American Society For Testing and Materials (astm). 1916 Race Street, Philadelphia, PA 19103. (215) 299-5585.

American Society of Mechanical Engineers. 345 East 47th Street, New York, NY 10017. (212) 705-7703.

American Welding Institute. 10628 Dutchtown Road, Knoxville, TN 37932. (615) 675-2150. FAX (615) 675-6081.

American Welding Society. 550 N.W. Le Jeune Road, Box 351040, Miami, FL 33126. (305) 443-9353. FAX (305) 443-7559.

Metallurgical Society of the Aime (american Institute of Mining, Metallurgical, and Petroleum Engineers). 345 East 47th Street, New York, NY 10017. (212) 705-7695.

Resistance Welder Manufacturers Association. 1900 Arch Street, Philadelphia, PA 19103-1498. (215) 564-3484. FAX (215) 963-9785.

DIRECTORIES AND BIOGRAPHICAL SOURCES

Directory of Engineering Societies and Related Organizations. Gordon Davis. 13th edition. American Association of Engineering Societies, 1111 19th Street NW, Suite 608, Washington, DC 20036. (202) 296-2237 or (800) 658-8897. 1989. Inquire for price.

Dun's Industrial Guide: The Metalworking Directory, 1993-94. Dun & Bradstreet Information Services Staff. 3 volumes. Dun & Bradstreet Information Services, 899 Eaton Avenue, Bethlehem, PA 18025. (610) 882-7000 or (800) 526-0651. FAX (610) 882-7269 1993. $795.00.

International Research Centers Directory. Gale Research, 835 Penobscot Building, Detroit, MI 48226-4094. (313) 961-2242. (800) 877-4253. FAX (313) 961-6083. 8th edition. 1995. $410.00.

Metallurgical Society of AIME—Membership List. 345 East 47th Street, New York, NY 10017. (212) 705-7695. Inquire for cost and availability.

Research Centers Directory. Gale Research, 835 Penobscot Building, Detroit, MI 48226-4094. (313) 961-2242. (800) 877-4253. FAX (313) 961-6083. $485.00.

Scientific and Technical Organizations and Agencies Directory. Gale Research, 835 Penobscot Building, Detroit, MI 48226-4094. (313) 961-2242. (800) 877-4253. FAX (313) 961-6083. 4th edition. 1996. $195.00.

Welding & Fabricating Data Book. Penton Publishing, 110 Superior Ave., Cleveland, OH 44114-2543. (800) 321-7003. 1958 to present. Biennial. $30.00.

Who's Who in Engineering. American Association of Engineering Societies, 1111 19th Street NW, Suite 608, Washington, DC 20036. (202) 296-2237 or (800) 658-8897. 8th edition. 1991. Inquire for price.

GENERAL WORKS

Advanced Welding Processes. John Norrish. A. Hilger Publishers. Distributed by the American Institute of Physics, 335 East 45th Street, New York, NY 10017. (212) 661-9404 or (800) 445-6638. 1992. Inquire for cost and availability.

Automating the Welding Process: Successful Implementation of Automated Welding Systems. James M. Berge. Industrial Press, Inc., 200 Madison Avenue, New York, NY 10016. (212) 889-6330. FAX (212) 545-8327. 1994. Inquire for cost and availability.

Introduction To the Physical Metallurgy of Welding. Kenneth Easterling. 2nd edition. Butterworth-Heinemann, 313 Washington Street, Newton, MA 02158. (617) 928-2500 or (800) 366-2665. FAX (617) 928-2620. 1992. $42.95.

Metallurgy of Welding. J.F. Lancaster. 5th edition. Chapman & Hall, 115 Fifth Avenue, New York, NY 10211-0906. 1993. $55.95.

The Science and Practice of Welding. A.C. Davies. 2 volumes. 10th edition. Cambridge University Press, 40 West 20th Street, New York, NY 10011-4211. (212) 924-3900 or (800) 872-7423. FAX (914) 937-4712. 1992. $69.95 (volume 1), $79.95 (volume 2), $29.95 (volume 2, paperback).

Welding: Theory and Practice. David L. Olson, et al., editors. North-Holland/ Elsevier Science B.V., PO Box 945, Madison Square Station, New York, NY 10159-0945. (212) 633-3650. FAX (212) 633-3680. 1990. $141.00.

HANDBOOKS AND MANUALS

Handbook of Structural Welding. John Lancaster. McGraw-Hill Publishing Company, 1221 Avenue of the Americas, New York, NY 10020. (800) 262-4729 or (212) 512-3825. 1993. Inquire for cost and availability.

Metals Handbook. ASM International, Materials Information, Materials Park, OH 44073-0002. (216) 338-5151 or (800) 336-5152. FAX (216) 338-4634. $154.00.

Smithell's Metals Reference Book. E.A. Brandes & G.B. Brook. 7th edition. Butterworth-Heinemann, 313 Washington Street, Newton, MA 02158. (617) 928-2500 or (800) 366-2665. FAX (617)928-2620. 1992. $250.00.

Welding Handbook. Leonard P. O'Connor, editor. 8th edition. American Welding Society, 550 N.W. Le Jeune Road, Box 351040, Miami, FL 33126. (305) 443-9353. FAX (305) 443-7559. 1987

ONLINE DATABASES AND CD-ROMS

Compendex Plus. Engineering Information, Inc., 345 East 47th Street, New York, NY 10017. (212) 705-7600 or (800) 221-1044. Contains citations with abstracts to worldwide literature in engineering and technology, from 1970 to present. Available on online BRS,(800) 289-4277, DIALOG, (800) 334-2564, ORBIT (800) 456-7248, and STN International, FIZ Karlsruhe, P.O. Box 2465, W-7500, Karlsruhe 1, Germany, online services. Also available on CD-ROM. Inquire as to cost and availability.

CSA Engineering. Cambridge Scientific Abstracts, 7200 Wisconsin Avenue, Suite 601, Bethesda, MD 20814. (301) 961-6750 or (800) 843-7751. Contains citations and abstracts of international periodicals and research literature covering all fields of engineering and science and technology,including computer and information science, electronics, mechanical engineering, solid state materials, 1981 to present. Available on BRS,(800) 289-4277, online service.Inquire as to cost and availability.

Current Contents Search. Institute for Scientific Information, 3501 Market Street, Philadelphia, PA 19104. (215) 386-0100. FAX (215) 386-6362. Contains citations to articles listed in the table of contents of science and technology journals. Also articles in social sciences and life sciences journals. Available on BRS,(800) 289-4277, DIALOG,(800) 334-2564, online services. Inquire as to cost and availability.

ISMEC: Mechanical Engineering Abstracts. Cambridge Scientific Abstracts, 7200 Wisconsin Avenue, Suite 601, Bethesda,

MD 20814. (301) 961-6750 or (800) 843-7751. Contains citations to the literature in mechanical engineering, industrial and production engineering, energy, power, mechanics, devices and related areas, from 1973 to present. Available on the DIALOG,(800) 334-2564, online service. Inquire as to cost and availability.

Metadex. Jointly produced by ASM International and the Institute of Materials. Contains more than 925,000 records from the international literature on metals and alloys, concerning properties, processes, materials classes, applications, and metallurgical systems. Updated monthly. Available from ORBIT-QUESTEL (703) 442-0900.

NTIS Bibliographic Database. National Technical Information Service, 5285 Port Royal Road, Springfield, VA 22161. (703) 487-4929 or FAX (703) 321-8199. Broad coverage of government-sponsored science and technology research reports, 1964 to present. Available on BRS,(800) 289-4277, DIALOG, (800) 334-2564, ORBIT, (800) 456-7248, and STN International, FIZ Karlsruhe, P.O. Box 2465, W-7500, Karlsruhe 1, Germany, online services. Also available on CD-ROM. Inquire as to cost and availability.

WILSONLINE. H.W. Wilson Company, 950 University Avenue, Bronx, NY 10452. (800) 367-6770 or (212) 588-8400. Makes available online versions of the printed H.W. Wilson indexes including Applied Science and Technology Index, Business Periodicals Index, General Science Index, and Readers' Guide to Periodical Literature. Period covered is generally 1983 to present. Available on BRS,(800) 289-4277, DIALOG,(800) 334-2564, and OCLC EPIC,(800) 848-5878, online services. Also available on CD-ROM. Inquire as to cost and availability.

PERIODICALS

J O M (Journal of Metals). Minerals, Metals and Materials Society, 420 Commonwealth Drive, Warrendale, PA 15086. (412) 776-9080. 1949 to present. Monthly. $50.00 per year.

Metals Week. McGraw-Hill Publishing Company, 1221 Avenue of the Americas, New York, NY 10020. (800) 262-4729 or (212) 512-3825. 1930 to present. Weekly. $770.00 per year.

Metalworking Digest. Gordon Publications Inc., 301 Gibraltar Drive, Box 350, Morris Plains, NJ 07950-0650. (201) 292-5100. FAX (201) 898-9281. 1968 to present. 12 times a year. $48.00 per year.

TWI Journal. The Welding Institute, Abington Hall, Abington, Cambridge CB1 6AH, England. Telephone 0223-891358. FAX 0223-893694. 1992 to present. Inquire for cost and availability.

Welding and Metal Fabrication. International Business & Technical Magazines Inc., Queensway House, 2 Queensway, RedHill, Surrey RH1 1QS England. Telephone 0737-768611. FAX 0737-761685. 1933 to present. 10 times a year. Inquire for cost and availability.

Welding-Brazing-Soldering Digest. ASM International, Materials Information, Materials Park, OH 44073-0002. (216) 338-5151 or (800) 336-5152. FAX (216) 338-4634. Monthly. $155.00 per year.

Welding Design and Fabrication. Penton Publishing, 110 Superior Ave., Cleveland, OH 44114-2543. 1930 to present. Monthly. $50.00 per year.

WELDING

Ency. of Physical Sciences and Engineering Info. Sources

Welding International. Woodhead Publishing Ltd., Abington Hall, Abington, Cambridge CB1 6AH, England. Telephone 0223-891358. FAX 0223-893694. 1987 to present. Monthly. Inquire for cost and availability.

Welding Journal. American Welding Society, 550 N.W. Le Jeune Road, Box 351040, Miami, FL 33126. (305) 443-9353. FAX (305) 443-7559. 1922 to present. Monthly. $30.00 per year.

WRC Progress Reports. Welding Research Council, United Engineering Center, 345 E. 47th Street, New York, NY 10017. (212) 705-7956. FAX (212) 371-9622. 6 times a year. $1100.00.

RESEARCH CENTERS AND INSTITUTES

Center For Welding & Joining Research. Colorado School of Mines, Golden, CO 80401. (303) 273-3025. FAX (303) 273-3795.

Twi, the Welding Institute. Abington Hall, Abington, Cambridge CB1 6AH, England. Telephone 0223-891358. FAX 0223-893694.

Welding Research Council. United Engineering Center, 345 E. 47th Street, New York, NY 10017. (212) 705-7956. FAX (212) 371-9622.

WELLS

See: HYDROLOGY, PETROLEUM ENGINEERING

WIND POWER

ABSTRACTS AND INDEXES

Energy Research Abstracts. 1976-present. U.S. Department of Energy, office of Scientific and Technical Information, Box 62, Oak Ridge, TN 37831. (615) 574-0733. Subscriptions: Superintendent of Documents, U.S. Government Printing office, Box 371954, Pittsburgh, PA 15250-7954. (202) 783-3238. FAX (202) 512-2233. Monthly. $164.00 per year. Abstracts all the scientific and technical reports, journal articles, conference proceedings, patents, theses, and monographs sponsored by the U.S. Energy Research and Development Administration.

Wind Energy Abstracts. 1983-present. WindBooks Inc., Box 4008, St. Johnsbury, VT 05819-4008. (802) 748-5148. FAX (802) 748-3286. Bi-monthly. $195.00 per year.

Wind Energy Technology. National Technical Information Service (NTIS), 5285 Port Royal Road, Springfield, VA 22161. (703) 487-4650. Bi-monthly. $135.00 per year.

Wind Engineering Abstracts. Multi-Science Publishing Co. Ltd., 107 High St., Brentwood, Essex CM14 4RX, England. Telephone 0277-224632. FAX 0277-224632. (U.S. subscription to: Box 176, Avenel, NJ 07001). 4 times a year. Inquire for price.

ASSOCIATIONS AND PROFESSIONAL SOCIETIES

American Wind Energy Association. 122 C St., NW, Suite 400, Washington, DC 20001. (202) 408-8988. FAX (202) 408-8536.

European Wind Energy Association. Via Bormida 2, I-00198, Rome, Italy. Telephone 6-8552329. FAX 6-8411933.

DIRECTORIES AND BIOGRAPHICAL SOURCES

American Wind Energy Association Membership Directory. American Wind Energy Association, 122 C St., NW, Suite 400, Washington, DC 20001. (202) 408-8988. FAX (202) 408-8536. Annual. Inquire for cost and availability.

Directory of Engineering Societies and Related Organizations. Gordon Davis. 13th edition. American Association of Engineering Societies, 1111 19th Street NW, Suite 608, Washington, DC 20036. (202) 296-2237 or (800) 658-8897. 1989. Inquire for price.

European Directory of Renewable Energy Suppliers and Services. Bruce Cross, editor. 3rd edition. Books International Inc., PO Box 605, Herndon, VA 22070-0605. (800) 359-7340. 1993. $75.00.

The International Directory of New and Renewable Energy Information Sources and Research Centres. 3rd edition. Books International Inc., PO Box 605, Herndon, VA 22070-0605. (800) 359-7340. 1993. $135.00.

International Research Centers Directory. Gale Research, 835 Penobscot Building, Detroit, MI 48226-4094. (313) 961-2242. (800) 877-4253. FAX (313) 961-6083. 8th edition. 1995. $410.00.

Renewable Energy Manufacturer's Lists: Wind and Hydro Power. Synerjy, Box 1854 Cathedral Station, New York, NY 10025. Annual. $5.00.

Research Centers Directory. Gale Research, 835 Penobscot Building, Detroit, MI 48226-4094. (313) 961-2242. (800) 877-4253. FAX (313) 961-6083. $485.00.

Scientific and Technical Organizations and Agencies Directory. Gale Research, 835 Penobscot Building, Detroit, MI 48226-4094. (313) 961-2242. (800) 877-4253. FAX (313) 961-6083. 4th edition. 1996. $195.00.

Synerjy. 1974-present. Synerjy, Box 1854 Cathedral Station, New York, NY 10025. Semi-annual. $50.00.

Who's Who in Engineering. American Association of Engineering Societies, 1111 19th Street NW, Suite 608, Washington, DC 20036. (202) 296-2237 or (800) 658-8897. 8th edition. 1991. Inquire for price.

Wind Energy in the U.S.: A State by State Survey. American Wind Energy Association, 122 C St., NW, Suite 400, Washington, DC 20001. (202) 408-8988. FAX (202) 408-8536. 1994. Inquire for cost and availability.

Wind Power: A Source Guide. Gordon Press Publishers, PO Box 459, Bowling Green Station, New York, NY 10004. (718) 624-8419. 1991. $76.00.

ENCYCLOPEDIAS AND DICTIONARIES

McGraw-Hill Dictionary of Scientific and Technical Terms. Sybil P. Parker, ed. 5th edition. McGraw-Hill Publishing Company, 1221 Avenue of the Americas, New York, NY 10020. (800) 262-4729 or (212) 512-3825. 1993. $110.50.

McGraw-Hill Encyclopedia of Engineering. Sybil P. Parker, ed. 2nd edition. McGraw-Hill Publishing Company, 1221 Avenue of the Americas, New York, NY 10020. (800) 262-4729 or (212) 512-3825. 1993. $95.50.

McGraw-Hill Encyclopedia of Science and Technology. Sybil P. Parker, ed. 7th edition. 20 volumes. McGraw-Hill Publishing Company, 1221 Avenue of the Americas, New York, NY 10020. (800) 262-4729 or (212) 512-3825. 1992. $1900.00

Reference Book On Wind Energy. European Wind Energy Association, Via Bormida 2, I-00198, Rome, Italy. Telephone 6-8552329, FAX 6-8411933. 1987. Inquire for cost and availability.

GENERAL WORKS

Earth, Wind, Fire, and Water: Energy Resources Utilization. Ben W. Ebenhack. PennWell Books, 1421 S. Sheridan Road, Box 1260, Tulsa, OK 74101. (918) 835-3161 or (800) 752-9764. FAX (918) 831-9555. 1995. Inquire for cost and availability.

Introduction To Wind Energy. Vaughn Nelson. 4th edition. Alternative Energy Institute, PO Box 248, Canyon, TX 79016. (806) 656-2295. 1994. Inquire for cost and availability.

Power From Wind: A History of Windmill Technology. Richard Leshe Hills. Cambridge University Press, 40 West 20th Street, New York, NY 10011-4211. (212) 924-3900 or (800) 872-7423. FAX (914) 937-4712. 1994. Inquire for cost and availability.

Understanding Your Wind Resource. American Wind Energy Association, 122 C St., NW, Suite 400, Washington, DC 20001. (202) 408-8988. FAX (202) 408-8536. 1994. Inquire for cost and availability.

Wind-Diesel Systems: A Guide To the Technology and Its Implementation. Ray Hunter & George Elliott, editors. Cambridge University Press, 40 West 20th Street, New York, NY 10011-4211. (212) 924-3900 or (800) 872-7423. FAX (914) 937-4712. 1994. $59.95.

Wind Energy Comes of Age. Paul Gipe. John Wiley and Sons, Inc., 605 Third Avenue, New York, NY 10158. (800) 526-5368 or (212) 850-6000. 1995. Inquire for cost and availability.

Wind Power Equipment. D.F. Warne. 2nd edition. Chapman & Hall, 115 Fifth Avenue, New York, NY 10211-0906. 1996. $120.95.

Wind Power For Home and Business. Paul Gipe. Chelsea Green Publishing Company, PO Box 428, Gates-Briggs Bldg., White River Junction, VT 05001. (802) 295-6300. FAX (802) 265-6444. 1993. $35.00.

Windmill Construction and Generating Power. F.E. Powell. Regal Publications, PO Box 1071, Provo, UT 84603. (801) 377-5367. 1992 [originally published as *Windmills and Wind Motors*, 1910] $20.00 (hardback), $10.00 (paperback).

HANDBOOKS AND MANUALS

Alternative Energy Handbook. Paul Rosenberg. Prentice Hall, 113 Sylvan Avenue, Route 9W, Englewood Cliffs, NJ 07632. (201) 592-2000 or (800) 922-0579. 1993. $75.00.

ONLINE DATABASES AND CD-ROMS

Compendex Plus. Engineering Information, Inc., 345 East 47th Street, New York, NY 10017. (212) 705-7600 or (800) 221-1044. Contains citations with abstracts to worldwide literature in engineering and technology, from 1970 to present. Available on online BRS,(800) 289-4277, DIALOG, (800) 334-2564, ORBIT (800) 456-7248, and STN International, FIZ Karlsruhe, P.O. Box 2465, W-7500, Karlsruhe 1, Germany, online services. Also available on CD-ROM. Inquire as to cost and availability.

CSA Engineering. Cambridge Scientific Abstracts, 7200 Wisconsin Avenue, Suite 601, Bethesda, MD 20814. (301) 961-6750 or (800) 843-7751. Contains citations and abstracts of international periodicals and research literature covering all fields of engineering and science and technology,including computer and information science, electronics, mechanical engineering, solid state materials, 1981 to present. Available on BRS,(800) 289-4277, online service.Inquire as to cost and availability.

Current Contents Search. Institute for Scientific Information, 3501 Market Street, Philadelphia, PA 19104. (215) 386-0100. FAX (215) 386-6362. Contains citations to articles listed in the table of contents of science and technology journals. Also articles in social sciences and life sciences journals. Availableon BRS,(800) 289-4277, DIALOG,(800) 334-2science and technology,including computer and information science, electronics, mechanical engineering, solid state materials, 1981 to present. Available on BRS,(800) 289-4277, online service.Inquire as to cost and availability.

Current Contents Search. Institute for Scientific Information, 3501 Market Street, Philadelphia, PA 19104. (215) 386-0100. FAX (215) 386-6362. Contains citations to articles listed in the table of contents of science and technology journals. Also articles in social sciences and life sciences journals. Available on BRS,(800) 289-4277, DIALOG,(800) 334-2564, online services. Inquire as to cost and availability.

Dissertation Abstracts. University Microfilms International, 300 North Zeeb Road, Ann Arbor, MI 48106. (800) 521-0600 or (313) 761-4700. Scope includes virtually all doctoral dissertations accepted at accredited American institutions from 1861 to present in 252 subject areas. Available on BRS,(800) 289-4277, DIALOG,(800) 334-2564, and OCLC EPIC,(800) 848-5878, online services. Also available on CD-ROM. Inquire as to cost and availability.

ISMEC: Mechanical Engineering Abstracts. Cambridge Scientific Abstracts, 7200 Wisconsin Avenue, Suite 601, Bethesda, MD 20814. (301) 961-6750 or (800) 843-7751. Contains citations to the literature in mechanical engineering, industrial and production engineering, energy, power, mechanics, devices and related areas, from 1973 to present. Available on the DIALOG,(800) 334-2564, online service. Inquire as to cost and availability.

NTIS Bibliographic Database. National Technical Information Service, 5285 Port Royal Road, Springfield, VA 22161. (703) 487-4929 or FAX (703) 321-8199. Broad coverage of government-sponsored science and technology research reports, 1964 to present. Available on BRS,(800) 289-4277, DIALOG, (800)

334-2564, ORBIT, (800) 456-7248, and STN International, FIZ Karlsruhe, P.O. Box 2465, W-7500, Karlsruhe 1, Germany, online services. Also available on CD-ROM. Inquire as to cost and availability.

SCISEARCH. Institute for Scientific Information, 3501 Market Street, Philadelphia, PA 19104. (800) 523-1850 or (215) 386-0100. Broad multidisciplinary title and author index to the international literature of science and technology, 1974 to present. Available on DIALOG, (800) 334-2564, and ORBIT, (800) 456-7248, online services. Also available on CD-ROM. Inquire as to cost and availability.

WILSONLINE. H.W. Wilson Company, 950 University Avenue, Bronx, NY 10452. (800) 367-6770 or (212) 588-8400. Makes available online versions of the printed H.W. Wilson indexes including Applied Science and Technology Index, Business Periodicals Index, General Science Index, and Readers' Guide to Periodical Literature. Period covered is generally 1983 to present. Available on BRS, (800) 289-4277, DIALOG, (800) 334-2564, and OCLC EPIC, (800) 848-5878, online services. Also available on CD-ROM. Inquire as to cost Suite 217, Redding, CA 96099-6007. Monthly. $65.00 per year.

PERIODICALS

Awea Wind Energy Weekly. American Wind Energy Association, 122 C St., NW, Suite 400, Washington, DC 20001. (202) 408-8988. FAX (202) 408-8536. 1982 to present. 50 times a year. $450.00 per year.

Wind Energy News. WindBooks Inc., Box 4008, St.Johnsburg, VT 05819-4008. (802) 748-5148. FAX (802) 748-3286. 1987 to present. Monthly. $96.00 per year.

Wind Engineering. Multiscience Publishing Company Ltd., 107 High Street, Brentwood, Essex CM14 4RX, England. Telephone 0277-224632. FAX 0277-224632. 1977 to present. 6 times a year. Inquire for cost and availability.

Windmiller's Gazette. T. Lindsay Baker, Box 507, Rio Vista, TX 76093. 1982 to present. Quarterly. $12.00 per year.

Windpower Monthly. Torgny Moeller, Vrinners Hoved, DK-8420 Knebel, Denmark. Telephone 86-36-59-00. FAX 86-36-56-26. 1985 to present. Monthly. $65.00 per year.

RESEARCH CENTERS AND INSTITUTES

National Wind Technology Center. 1617 Cole Blvd., Golden, CO 80401-3393.

U.S. Department of Energy, Office of Renewable Energy Technologies, Wind/hydro/ocean Technologies Division. Mail Stop CE-121, 1000 Independence Avenue SW, Room 511047, Washington, DC 20585. (202) 586-5630. FAX (202) 586-5124.

West Texas State University, Alternative Energy Institute. PO Box 248, Canyon, TX 79016. (806) 656-2295.

WIND TUNNELS

See: AERONAUTICAL ENGINEERING

WINDS

See also: CYCLONE, HURRICANES, JET STREAM, RAIN, THUNDER-STORMS, TORNADOS,

ABSTRACT SERVICES AND INDEXES

Applied Science and Technology Index; A Cumulative Subject Index To English Language Periodicals in the Fields of Aeronautics and Space Science, Computer Technology, Chemistry, Construction Industry, Energy and Related Areas. H.W. Wilson Co., 950 University Avenue, Bronx, NY 10452-9978. (212) 588-8400. (800) 367-6770. FAX (718) 590-1617. From 1958 to present. Monthly. Inquire about cost and availability. Also available online (BRS and WILSONLINE) and on CD-ROM. ISSN: 0003-6986.

Deep-Sea Research. Part A: Oceanographic Research Papers. Deep-Sea Research. Part B: Oceanographic Literature Review. Pergamon Press, Inc., Maxwell House, Fairview Park, Elmsford, NY 10523. (914) 592-7700. Fax (914) 592-3625. for Parts A and B. Twelve times per year. $2000.00 per year for Parts A and B.Oceanographic Literature Review also available on CD-ROM. Inquire about price and availability

General Science Index. H.W. Wilson Co., 950 UniversityAvenue, Bronx, NY 10452. (212) 588-8400. (800) 367-6770. FAX (718) 590-1617. From 1978 to present. Ten issues per year; quarterly and annual cumulations. Service basis. Available on CD-ROM and online. Inquire regarding cost and availability. ISSN: 0162-1963.

Government Reports Announcements and Index. National Technical Information Service (NTIS), 5285 Port Royal Road, Springfield, VA 22161. (703) 487-4650. FAX (703) 321-8547. From 1968 to present. Annual. $630.00 per year. Also available online as NTIS BIBLIOGRAPHIC DATABASE and on CD-ROM. ISSN:

Meteorological and Geoastrophysical Abstracts. American Meteorological Society, c/o Inforonics, Inc., 550 Newtown Road, Box 458, Littleton, MA 01460. (508) 486-8976. FAX (508) 486-0027. Covers literature in environmental sciences, meteorology, astrophysics, hydrology, glaciology, and physical oceanography. 1950 to present. Monthly. $950.00 per year. ISSN: 0026-1130. Also available online (DIALOG) and on CD-ROM.

Oceanic Abstracts. Cambridge Scientific Abstracts, 7200 Wisconsin Avenue, Bethesda, MD 20814. (301) 961-6750. Fax (301) 961-6720. Bimonthly. $995.00 per year.

Science Citation Index. SCI. Institute for Scientific Information, 3501 Market Street, Philadelphia, PA 19104. (215) 386-0100. (800) 523-1850. FAX (215) 386-2991. 1961 to present. Six issues per year, plus annual cumulation. $11650.00 per year. Also available online and on CD-ROM. Inquire about price and availability. ISSN: 0036-827X.

Selected Water Resources Abstracts. U.S. Geological Survey. Distributed by National Technical Information Service (NTIS), 5285 Port Royal Road, Springfield, VA 22161. (703) 487-4650. Monthly. $115.00 per year.

Ency. of Physical Sciences and Engineering Info. Sources

WINDS

ANNUAL REVIEWS AND YEARBOOKS

American Meteorological Society. Meteorological Monographs. American Meteorological Society, 45 Beacon Street, Boston, MA 02108-3693. (617) 227-2425. FAX (617) 742-8718.

Annual Review of Earth and Planetary Sciences. Annual Reviews, Inc., 4139 El Camino Way, Palo Alto, CA 94306-0897. (415) 493-4400. (800) 523-8635. FAX (415) 855-9815. Annual. $55.00 per year.

Developments in Atmospheric Science Series. Elsevier Science Publishing Company, Inc., 655 Avenue of the Americas, New York, NY 10010. (212) 989-5800. Irregular. Inquire about price and availability.

Ocean Yearbook. Elisabeth M. Borgese, et al, editors. University of Chicago Press, 5801 Ellis Avenue, Chicago, IL 60637. (312) 702-7700. 1995. $77.00.

ASSOCIATIONS AND PROFESSIONAL SOCIETIES

American Association of State Climatologists. c/o Myron Moinau, University of Idaho, Agriculture Engineering Department, State Climate Service, Moscow, ID 834844-2040.

American Meteorological Society, 45 Beacon Street, Boston, MA 02108-3693. (617) 227-2425. FAX (617) 742-8718.Irregular. Price varies.

Association of American Weather Observers, 401 Whitney Boulevard, Box 455, Belvedere, IL 61008. (815) 544-5665. FAX (815) 544-6334.

Canadian Meteorological and Oceanographic Society, P.O. Box 334, Newmarket, Ontario, L3Y 4X7, Canada. (416) 898-1040. FAX (416) 898-7937.

International Association of Meteorology and Atmospheric Physics. c/o UCAR, P.O. Box 3000, Boulder CO 80307-3000. (303) 497-1673.

National Environmental Satellite Data, and Information Service. 2069 Federal Building 4 Room 2060, Washington, CD 20233. (301) 763-71990. FAX (301) 763-4011.

National Weather Association, 4400 Stamp Road, Room 404, Temple Hills, MD 20748. (301) 899-3784.

University Corporation For Atmospheric Research. PO Box 3000, 1850 Table Mesa Drive, Boulder, CO 80307-3000. (303) 497-1673. FAX (303) 497-1654.

BIBLIOGRAPHIES

Information Sources in the Earth Sciences. David N. Wood, Joan E. Hardy, and Anthony P. Harvey, editors. Bowker-Saur/K.G. Saur. Distributed by R.R. Bowker, 121 Chanlon Road, New Providence, NJ 07974. (800) 521-8110 or (908) 464-6800. Second edition. 1989. $85.00.

International Bibliography of Meteorology: from the Beginning of Printing to 1889. Oliver L. Fassig. Diane Publishing Co., 600 Upland Avenue, Upland, PA 19015. (610) 499-7415. Reprint edition. 1994. $195.00

Scientific and Technical Books and Serials in Print; An Index To Literature in Science and Technology. R.R. Bowker Co., 205 E. 42nd Street, New York, NY 10017. (800) 521-8110 or (212) 916-1600. 1992. $250.00.

DIRECTORIES AND BIOGRAPHICAL SOURCES

American Men and Women of Science: Physical and Biological Sciences. Eighteenth edition. R.R. Bowker Company, 245 West 17th Street, New York, NY 10011. (800) 521-8810 or (212) 916-1600. $750.00.

American Meteorological Society. Professional Directory. American Meteorological Society. 45 Beacon Street, Boston, MA 02108. (617) 227-2425. Included in each issue of the Bulletin of the Society.

Meteorological Services of the World. American Meteorological Society, 45 Beacon Street, Boston, MA 02108- 3693. (617) 227-2425. FAX (617) 742-8718. Annual. $70.00.

National Weather Service offices and Stations. National Oceanic and Atmospheric Administration, Deparatment of Commerce, Silver Spring, MD 20910. (301) 427-7698. Annual. Free.

Research Centers Directory. Gale Research, 835 Penobscot Building, Detroit, MI 48226-4094. (313) 961-2242. (800) 877-4253. 20th edition. 1995. $485.00. ISSN: 0080

ENCYCLOPEDIAS AND DICTIONARIES

Climates of the States. Bair. Gale Research Inc. 835 Penobscot Building, Detroit, MI 48226-4094. (313) 961-2242. (800) 877-4253. 5th edition. 1995. $255.00.

Concise Oxford Dictionary of the Earth Sciences. Ailsa Allaby and Michael Allaby, editors. Oxford University Press, Inc., 200 Madison Avenue, New York, NY 10016. (800) 334-4249 or (212) 679-7300. 1990. $42.95.

Encyclopedia of Climatology. Rhodes W. Fairbridge, and John E. Oliver. Chapman & Hall, 1 Penn Plaza, New York, NY 10119. (212) 564-1060. 1986. $115.00.

Handbook of Applied Meteorology. David D. Houghton, editor. John Wiley & Sons, Inc. 605 Third Avenue, New York, NY 10158-0012. (212) 850-6000. (800) 225-5945. 1985.

McGraw-Hill Encyclopedia of Science and Technology. McGraw-Hill Publishing Company, 1221 Avenue of the Americas, New York, NY 10020. (800) 262-4729 or (212) 512-3825. Seventh edition. 1992. $1900.00.

Smithsonian Meteorological Tables. Books on Demand, 300 North Zeeb Road, Ann Arbor, MI 48106-1346. (313) 761-4700. (800) 521-0600. reprint edition. $153.90.

Weather Satellite Handbook. American Radio Relay League, 225 Main Street, Newington CT -6111. (203) 666-1541. 1994. $20.00.

GENERAL WORKS

Atlas of the Oceans' Wind and Wave Climate. I. R. Young and G. J. Holland, compilers. Elsevier Science, Inc., 655 Avenue

WINDS

Ency. of Physical Sciences and Engineering Info. Sources

of the Americas, New York, NY 10010. (212) 989- 5800. 1995. $240.00.

Atlas of Three Hundred Millibar Wind Characteristics For the Northern Hemisphere. James F. Lahey, et al. University of Wisconson Press, 114 North Nurray Street, Madison, WI 53715-1199. 1960. $50.00.

Physical Properties in Atmospheric Models. D. R. Sikka and S.S. Singh. Halsted Press, 605 Third Avenue, New York, NY 10158-0012. (212) 850-6400. 1992. $59.95.

Sea Breeeze and Local Winds. John E. Simpson. Cambridge University Press, 40 West 20th Street, New York, NY 10011. (212) 924-3900. (800) 872-7423. 1994. $42.95.

Severe and Unusual Weather. Joe R. Eagleman. Trimedia Publishing Company, 12008 West 87th Street, Suite 117, Lenexa, KS 66215. (913) 599-0505. Second edition. 1990. $41.95.

Wind and the Built Environment: U.s. Needs in Wind Engineering and Hazard Mitigation. National Academy Press, 2101 Constitution Avenue NW, Washington, D.C. 20055. (202) 334-3313. (800) 624-6242.

Wind Effects On Structures; An Introduction To Wind Engineering. John Wiley & Sons, Inc., 605 Third Avenue, New York, NY 10158-0012. (212) 850-6000. (800) 225-

5945. 2nd edition. 1986. $89.95.

Wind Energy. Wolfgang Palz and W. Schnell. Kluwer Academic Publishers, 101 Philip Drive, Assinippi Park, Norwell, MA 02061. (617) 871-6600. 1993. $70.00.

Wind Engineering. N. I. Cook, editor. American Society of Civil Engineers, 345 East 47th Street, New York, 10017- 2723. (212) 705-7496. (800) 548-2723. 1993. $115.00.

HANDBOOKS AND MANUALS

Author's Guide To the Journals of the American Meteorological Society. American Meteorological Society, 45 Beacon Street, Boston, MA 02108-3693. (617) 227-2425. FAX (617) 742-8718. 1983. $15.00 in paper.

Handbook in Applied Meteorology. David D. Houghton, editor. John Wiley and Sons, Inc., 605 Third Avenue, New York, NY 10158. (800) 526-5368 or (212) 850-6000. 1985. $168.00.

The Hurricane Handbook A Practical Guide For Residents of the Hurricane Belt. Sharon M. Carpenter and Toni G. Carpenter. Tailored Tours Publications, Inc., Box 22861, Lake Buena Vista, FL 32830. (407) 345-912. 1993. $9.95.

ONLINE DATABASES AND CD-ROMS

Climate Assessment Database. National Weather Service, National Meteorological Center, 5200 Auth Road, Suite 101, Camp Springs, MD 20233. (301) 763-8016. Contains daily, weekly and monthly summaries of North American and world climatological data. Also provides five to ten weather forecasts, and 30 to 90 day weather outlook. Subscription required. Inquire as to cost and availability.

Current Contents Search. Institute for Scientific Information, 3501 Market Street, Philadelphia, PA 19104. (215) 386-0100. FAX (215) 386-6362. Contains citations to articles listed in the table of contents of science and technology journals. Also articles in social sciences and life sciences journals. Available on BRS,(800) 289-4277, DIALOG,(800) 334-2564, online services. Inquire as to cost and availability.

Dissertation Abstracts. University Microfilms International, 300 North Zeeb Road, Ann Arbor, MI 48106. (800) 521-0600 or (313) 761-4700. Scope includes virtually all doctoral dissertations accepted at accredited American institutions from 1861 to present in 252 subject areas. Available on BRS,(800) 289-4277, DIALOG,(800) 334-2564, and OCLC EPIC,(800) 848-5878, online services. Also available on CD-ROM. Inquire as to cost and availability.

Meteorological and Geoastrophysical Abstracts. American Meteorological Society, 45 Beacon Street, Boston, MA 02108-3693. (617) 227-2425. FAX (617) 742-8718. Contains citations and abstracts to the worldwide literature on significant research in meteorology and geoastrophysics. Related topics include physical oceanography, hydrology, environmental sciences and glaciology. Covers the period 1972 to present. Available on DIALOG,(800) 334-2564, online service. Inquire as to cost and availability.

NTIS Bibliographic Database. National Technical Information Service, 5285 Port Royal Road, Springfield, VA 22161.(703) 487-4929 or FAX (703) 321-8199. Broad coverage of government-sponsored science and technology research reports, 1964 to present. Available on BRS,(800) 289-4277, DIALOG, (800) 334-2564, ORBIT, (800) 456-7248, and STN International, FIZ Karlsruhe, P.O. Box 2465, W-7500, Karlsruhe 1, Germany, online services. Also available on CD-ROM. Inquire as to cost and availability.

SCISEARCH. Institute for Scientific Information, 3501 Market Street, Philadelphia, PA 19104. (800) 523-1850 or (215) 386-0100. Broad multidisciplinary title and author index to the international literature of science and technology, 1974 to present. Available on DIALOG and ORBIT online services. Also available on CD-ROM. Inquire as to cost and availability.

WILSONLINE. H.W. Wilson Company, 950 University Avenue, Bronx, NY 10452. (800) 367-6770 or (212) 588-8400. Makes available online versions of the printed H.W. Wilson indexes including Applied Science and Technology Index, Business Periodicals Index, General Science Index, and Readers' Guide to Periodical Literature. Period covered is generally 1983 to present. Available on BRS,(800) 289-4277, DIALOG,(800) 334-2564, and OCLC EPIC,(800) 848-5878, online services. Also available on CD-ROM. Inquire as to cost and availability.

World Climate Disc. Chadwyck-Healey Inc., 1101 King Street, Alexandria, VA 22314. (703) 683-4890. FAX (703) 683-7589. Weather and climate data from approximately 5,000 weather stations worldwide, covering the years 1854 to 1990 on CD-ROM. First edition 1992. Approximately $1200.00 per year with annual updates.

World Weatherdisc. WeatherDisc Associates, Inc., 4584 N.E. 89th Street, Seattle, WA 98115. (206) 524-4314. FAX (206) 543-0308. Meteorological data on CD-ROM which describes the climate of the earth today and for the past few hundred years. First edition 1989. Approximately $295.00 per year with biannual updates.

PERIODICALS

Agricultural and Forest Meteorology. Elsevier Science Publishing Company, Inc., 655 Avenue of the Americas, New York, NY 10010. (212) 989-5800. Twenty times per year. $750.00 per year.

American Meteorological Society. BULLETIN. American Meteorological Society, 45 Beacon Street, Boston, MA 02108-3693. (617) 227-2425. FAX (617) 742-8718. Monthly. $60.00 per year.

American Weather Observer. Association of American Weather Observers, 401 Whitney Boulevard, Box 455, Belvedere, IL 61008. (815) 544-5665. FAX (815) 544-6334. Monthly. $21.00 per year.

Atmosphere - Ocean. Canadian Meteorological and Oceanographic Society, P.O. Box 334, Newmarket, Ontario, L3Y 4X7, Canada. (416) 898-1040. FAX (416) 898-7937. Quarterly. $30.00 per year.

Climate Change. Kluwer Academic Publishers, P.O. Box 358, Accord Station, Hingham, MA 02018-0358. (617) 871-6000. Six times per year. $327.00 per year.

Dynamics of Atmospheres and Oceans. Elsevier Science Publishing Company, Inc., 655 Avenue of the Americas, New York, NY 10010. (212) 989-5800. Six times per year. $205.00 per year.

JGR: Journal of Geophysical Research: Atmosphere. American Geophysical Union, 2000 Florida Avenue, N.W., Washington, DC 20009. (202) 462-6903. Monthly. $90.00 per year to members.

Journal of Applied Meteorology. American Meteorological Society, 45 Beacon Street, Boston, MA 02108-3693. (617) 227-2425. FAX (617) 742-8718. Monthly. $165.00 per year.

Journal of Atmospheric Sciences. American Meteorological Society, 45 Beacon Street, Boston, MA 02108-3693. (617) 227-2425. FAX (617) 742-8718. Semi-monthly. $320.00 per year.

Monthly Weather Review. American Meteorological Society, 45 Beacon Street, Boston, MA 02108-3693. (617) 227-2425. FAX (617) 742-8718. Monthly. $205.00 per year.

National Weather Digest. National Weather Association, 4400 Stamp Road, Room 404, Temple Hills, MD 20748. (301) 899-3784.

Weather. Royal Meteorological Society, 104 Oxford Road, Reading, Berks RG1 7LJ, England. Monthly. $44.00 per year.

Weatherwise. Heldref Publications, 1319 Eighteenth Street, N.W., Washington, DC 20036-1802. (202) 296-6267. FAX (202) 296-5149. Bi-monthly. $28.00 per year.

RESEARCH CENTERS AND INSTITUTES

Atmospheric Science Flight Facility. University of Wyoming, P.O. Box 3038, Laramie, WY 82071. (307) 766-3245.

Center for Analysis and Prediction of Storms. University of Oklahoma. 401 East Boyd, Norman OK 73019. (405) 325-3041.

Coastal and Oceanographic Engineering Laboratory. University of Florida, 336 Weil Hall, Gainsville, FL 32607. (904) 392-1436.

Cooperative Institute For Mesoscale Meteorological Studies. University of Oklahoma, 401 East Boyd, Norman, OK 73019. (405) 325-3041.

National Center For Atmospheric Research. Box 3000, Boulder, CO 80307. (303) 497-1000. FAX (303) 497-1654.

National Severe Storms Laboratory. 1313 Halley Circle, Norman, OK 73069. (406) 360-3620. FAX (405) 360-0472.

WIRE AND CABLE

See also: WIRING--ELECTRICAL

ABSTRACT SERVICES AND INDEXES

Applied Science and Technology Index; A Cumulative Subject Index To English Language Periodicals in the Fields of Aeronautics and Space Science, Computer Technology, Chemistry, Construction Industry, Energy and Related Areas. H.W. Wilson Co., 950 University Avenue, Bronx, NY 10452. (800) 367-6770 or (212) 588-8400. FAX (718) 590-1617. From 1958 to present. Monthly. Inquire about cost and availability. Also available on CD-ROM and online.

Current Contents: Engineering, Technology, and Applied Sciences. Institute for Scientific Information, 3501 Market Street, Philadelphia, PA 19104. (215) 386-0100. FAX (215) 386-6362. 1970 to present. Weekly. $442.00 per year.

Engineering Index Monthly. Engineering Information, Inc., Castle Point on the Hudson, Hoboken, NJ 07030. (800) 221-1044. FAX (212) 832-1857. Monthly. $2300.00 per year. Also available online as COMPENDEX, and also on CD-ROM. Covers chemical engineering, computers, electrical engineering, civil engineering, metals and mining, industrial management, and mechanical engineering.

International Copper Information Bulletin. Copper Development Association, Orchard House, Mutton Lane, Potters Bar, Herts EN6 3AP, England. Telephone 0707-50711. FAX 0707-42769. 1976 to present. 3 times a year. Inquire for cost and availability.

Materials Science and Engineering Abstracts, Cambridge Scientific Abstracts, 7200 Wisconsin Avenue, Bethesda, MD 20814-4823. (301) 961-6750. FAX (301) 961-6720. 1993 to present. Monthly. $385.00 per year. Focuses on mechanical and physical properties of materials and commercial or industrial applications for materials, methods for strength testing, effects of vibration and other stresses, corrosion and protective coatings, storage and handling, ceramics, composites, metals, wood, plastics, and polymers.

Metals Abstracts and Metals Abstracts Index. American Society for Metals, Metals Park, OH 44073. (216) 338-5151. 1968 to present. Monthly. Abstracts are $1925.00 per year and Index is $460.00 per year.

WIRE AND CABLE

Ency. of Physical Sciences and Engineering Info. Sources

ASSOCIATIONS AND PROFESSIONAL SOCIETIES

American Society For Metals. Materials Park, OH 44073. (216) 338-5151 or FAX (216) 338-4634.

American Wire Producers Association. 1101 Connecticut Avenue NW, Suite 700, Washington, DC 20036-4303. (202) 857-1155. FAX (202) 429-5154.

Associated Wire Rope Fabricators. PO Box 20126, Lehigh Valley, PA 18002-0126. (215) 974-9974. FAX (215) 691-6833.

Insulated Cable Engineers Association. PO Box 440, South Yarmouth, MA 02664. (508) 394-4424.

International Wire & Machinery Association. 46 Holly Walk, Leamington Spa, Warwickshire CV32 4HY, England. Telephone 926-334137. FAX 926-314755.

Wire Association International. 1570 Boston Post Road, PO Box H, Guilford, CT 06437. (203) 453-2777.

Wire Fabricators Association. 710 E. Ogden Avenue, Suite 113, Napierville, IL 60503. (708) 369-2406.

Wire Industry Suppliers Association. 7297 Lee Highway, Suite N, Falls Church, VA 22042. (703) 533-9530.

DIRECTORIES AND BIOGRAPHICAL SOURCES

Dun's Industrial Guide: The Metalworking Directory, 1993-94. Dun & Bradstreet Information Services Staff. 3 volumes. Dun & Bradstreet Information Services, 899 Eaton Avenue, Bethlehem, PA 18025. (610) 882-7000 or (800) 526-0651. FAX (610) 882-7269. 1993. $795.00

Research Centers Directory. Gale Research, 835 Penobscot Building, Detroit, MI 48226-4094. (313) 961-2242. (800) 877-4253. FAX (313) 961-6083. $485.00.

Who's Who in Engineering. American Association of Engineering Societies, 1111 19th Street NW, Suite 608, Washington, DC 20036. (202) 296-2237 or (800) 658-8897. 8th edition. 1991. Inquire for price.

Who's Who in Technology. Gale Research, 835 Penobscot Building, Detroit, MI 48226-4094. (313) 961-2242. (800) 877-4253. FAX (313) 961-6083. 1995. $195.00.

Wire Industry Yearbook. Publex International Ltd., 110 Station Road E., Oxted, Surrey RH8 0QA, England. Telephone 0883-717755. FAX 0883-714554. 1951 to present. Annual. Inquire for cost.

ENCYCLOPEDIAS AND DICTIONARIES

McGraw-Hill Dictionary of Scientific and Technical Terms. Sybil P. Parker, ed. 5th edition. McGraw-Hill Publishing Company, 1221 Avenue of the Americas, New York, NY 10020. (800) 262-4729 or (212) 512-3825. 1993. $110.50.

McGraw-Hill Encyclopedia of Engineering. Sybil P. Parker, ed. 2nd edition. McGraw-Hill Publishing Company, 1221 Avenue of the Americas, New York, NY 10020. (800) 262-4729 or (212) 512-3825. 1993. $95.50.

McGraw-Hill Encyclopedia of Science and Technology. Sybil P. Parker, ed. 7th edition. 20 volumes. McGraw-Hill Publishing Company, 1221 Avenue of the Americas, New York, NY 10020. (800) 262-4729 or (212) 512-3825. 1992. $1900.00

GENERAL WORKS

Cables and Wiring. AVO Multi-Amp Institute Staff. Delmar Publishers, Division of Thomson Educational Publishing, Inc., 3 Columbia Circle, Box 15015, Albany, NY 12205. (518) 464-3500 or (800) 347-7707. FAX (518) 464-0358. 1993. $25.95.

The Electrical & Electronic Wire & Cable Industry. Leading Edge Reports, 2171 Jericho Turnpike, No. 200, Commack, NY 11725-2900. (800) 866-4648. FAX (216) 791-0333. 1991. $1950.00.

The U.S. Insulated Wire and Cable Industry: Past Performance, Current Trends, and Strategies for the Future. Business Trend Analysts, 2171 Jericho Turnpike, Commack, NY 11725. 1991. $1195.00.

U.S. Insulated Wire and Cable Markets. Leading Edge Reports, 2171 Jericho Turnpike, No. 200, Commack, NY 11725-2900. (800) 866-4648. FAX (216) 791-0333. 1992. $1750.00.

HANDBOOKS AND MANUALS

American Electricians Handbook. C. Terrell & W.I. Summers. 12th edition. McGraw-Hill Publishing Company, 1221 Avenue of the Americas, New York, NY 10020. (800) 262-4729 or (212) 512-3825. 1992. $74.50.

Electric Cables Handbook. D. McAllister. 2nd edition. Sheridan House Inc., 145 Palisade Street, Dobbs Ferry, NY 10522. (914) 693-2410. FAX (914) 693-0776. 1990. $150.00.

Electronic Materials and Processes Handbook. Charles A. Harper and Ronald N. Sampson. 2d ed. McGraw-Hill Publishing Company, 1221 Avenue of the Americas, New York, NY 10020. (800) 262-4729 or (212) 512-3825. 1994. $79.50.

Wire and Cable For Electronics: A User's Handbook. Neil Sclater. TAB Books, P.O. Box 40, Blue Ridge Summit, PA 17294-0850. (717)794-2191 or (800) 233-1128. FAX (717) 794-2080. 1991. $29.95.

ONLINE DATABASES AND CD-ROMS

Compendex Plus. Engineering Information, Inc., 345 East 47th Street, New York, NY 10017. (212) 705-7600 or (800) 221-1044. Contains citations with abstracts to worldwide literature in engineering and technology, from 1970 to present. Available on online BRS,(800) 289-4277, DIALOG, (800) 334-2564, ORBIT (800) 456-7248, and STN International, FIZ Karlsruhe, P.O. Box 2465, W-7500, Karlsruhe 1, Germany, online services. Also available on CD-ROM. Inquire as to cost and availability.

CSA Engineering. Cambridge Scientific Abstracts, 7200 Wisconsin Avenue, Suite 601, Bethesda, MD 20814. (301) 961-6750 or (800) 843-7751. Contains citations and abstracts of international periodicals and research literature covering all fields of engineering and science and technology,including computer and information science, electronics, mechanical engineering,

solid state materials, 1981 to present. Available on BRS,(800) 289-4277, online service.Inquire as to cost and availability.

Current Contents Search. Institute for Scientific Information, 3501 Market Street, Philadelphia, PA 19104. (215) 386-0100. FAX (215) 386-6362. Contains citations to articles listed in the table of contents of science and technology journals. Also articles in social sciences and life sciences journals. Available on BRS,(800) 289-4277, DIALOG,(800) 334-2564, online services. Inquire as to cost and availability.

ISMEC: Mechanical Engineering Abstracts. Cambridge Scientific Abstracts, 7200 Wisconsin Avenue, Suite 601, Bethesda, MD 20814. (301) 961-6750 or (800) 843-7751. Contains citations to the literature in mechanical engineering, industrial and production engineering, energy, power, mechanics, devices and related areas, from 1973 to present. Available on the DIALOG,(800) 334-2564, online service. Inquire as to cost and availability.

Metadex. Jointly produced by ASM International and the Institute of Materials. Contains more than 925,000 records from the international literature on metals and alloys, concerning properties, processes, materials classes, applications, and metallurgical systems. Updated monthly. Available from ORBIT-QUESTEL (703) 442-0900.

NASA Database. American Institute of Aeronautics and Astronautics, 370 L'Enfant Promenade, S.W., Washington, DC 20024. (202) 646-7400. Citations and abstracts of aeronautics and astronautics literature, 1962 to present. Also contains citations from STAR, SCIENTIFIC and TECHNICAL AEROSPACE REPORTS, and INTERNATIONAL AEROSPACE ABSTRACTS. Available through NASA/RECONonline service. Inquire as to cost and availability.

NTIS Bibliographic Database. National Technical Information Service, 5285 Port Royal Road, Springfield, VA 22161. (703) 487-4929 or FAX (703) 321-8199. Broad coverage of government-sponsored science and technology research reports, 1964 to present. Available on BRS,(800) 289-4277, DIALOG, (800) 334-2564, ORBIT, (800) 456-7248, and STN International, FIZ Karlsruhe, P.O. Box 2465, W-7500, Karlsruhe 1, Germany, online services. Also available on CD-ROM. Inquire as to cost and availability.

WILSONLINE. H.W. Wilson Company, 950 University Avenue, Bronx, NY 10452. (800) 367-6770 or (212) 588-8400. Makes available online versions of the printed H.W. Wilson indexes including Applied Science and Technology Index, Business Periodicals Index, General Science Index, and Readers' Guide to Periodical Literature. Period covered is generally 1983 to present. Available on BRS,(800) 289-4277, DIALOG,(800) 334-2564, and OCLC EPIC,(800) 848-5878, online services. Also available on CD-ROM. Inquire as to cost and availability.

PERIODICALS

J O M (Journal of Metals). Minerals, Metals and Materials Society, 420 Commonwealth Drive, Warrendale, PA 15086. (412) 776-9080. 1949 to present. Monthly. $50.00 per year.

Metals Week. McGraw-Hill Publishing Company, 1221 Avenue of the Americas, New York, NY 10020. (800) 262-4729 or (212) 512-3825. 1930 to present. Weekly. $770.00 per year.

Wire. Postfach 2069, 96011 Bamberg, Germany. Telephone 0951-861-135. FAX 0951-861-158. 1951 to present. Bi-monthly. Inquire for cost.

Wire Industry. Publex International Ltd., 110 Station Road E., Oxted, Surrey RH8 0QA, England. Telephone 0883-717755. FAX 0883-714554. 1934 to present. Monthly. Inquire for cost.

Wire Industry News. Business Information Services, 7 Hampden Road, Stafford Springs, CT 06076-31404. (203) 684-5877. (203) 684-9158. 1973 to present. Bi-weekly. $250.00 per year.

Wire Journal International. Wire Journal Inc., 1570 Boston Post Road, Box H, Guilford, CT 06437. (203) 453-2777. FAX (203) 453-8384. 1968 to present. Monthly. $60.00 per year.

Wire Technology International. Initial Publications Inc., 3869 Darrow Road, Suite 101, Stow, OH 44224. (216) 686-9544. (216) 686-9563. 1972 to present. Bi-monthly. $30.00 per year.

Wireworld. Vogel Europublishing Inc., 20092 Gibbs Drive, Sonora, CA 95370. (209) 533-3555. FAX (209) 533-9555. 1959 to present. 6 times a year. Inquire for cost.

RESEARCH CENTERS AND INSTITUTES

International Copper Association Ltd. 708 Third Avenue, New York, NY 10017. (212) 697-9355. FAX (212) 697-5417.

SPECIFICATIONS AND STANDARDS

Annual Book of Astm Standards, Volume 01.03. American Society for Testing and Materials (ASTM), 1916 Race Street,Philadelphia, PA 19103. (215) 299-5585. Annual. Inquire for cost and availability.

WIRING (ELECTRICAL)

See also: WIRE AND CABLE

ABSTRACT SERVICES AND INDEXES

Applied Science and Technology Index; A Cumulative Subject Index To English Language Periodicals in the Fields of Aeronautics and Space Science, Computer Technology, Chemistry, Construction Industry, Energy and Related Areas. H.W. Wilson Co., 950 University Avenue, Bronx, NY 10452. (800) 367-6770 or (212) 588-8400. FAX (718) 590-1617. From 1958 to present. Monthly. Inquire about cost and availability. Also available on CD-ROM and online.

Current Contents: Engineering, Technology, and Applied Sciences. Institute for Scientific Information, 3501 Market Street, Philadelphia, PA 19104. (215) 386-0100. FAX (215)- 6362. 1970 to present. Weekly. $442.00 per year.

Engineering Index Monthly. Engineering Information, Inc., Castle Point on the Hudson, Hoboken, NJ 07030. (800) 221-1044. FAX (212) 832-1857. Monthly. $2300.00 per year. Also available online as COMPENDEX, and also on CD-ROM. Covers chemical engineering, computers, electrical engineering, civil engineering, metals and mining, industrial management, and mechanical engineering.

WIRING (ELECTRICAL)

Ency. of Physical Sciences and Engineering Info. Sources

Metals Abstracts and Metals Abstracts Index. American Society for Metals, Metals Park, OH 44073. (216) 338-5151. 1968 to present. Monthly. Abstracts are $1925.00 per year and Index is $460.00 per year.

ANNUAL REVIEWS AND YEARBOOKS

E C and M's Electrical Products Yearbook. Intertec Publishing Corporation, 9800 Metcalf, Overland Park, KS 66212-2215. (913) 341-1300. FAX (913) 967-1898. Annual. $10.00.

ASSOCIATIONS AND PROFESSIONAL SOCIETIES

American Society For Metals. Materials Park, OH 44073. (216) 338-5151 or FAX (216) 338-4634.

American Wire Producers Association. 1101 Connecticut Avenue NW, Suite 700, Washington, DC 20036-4303. (202) 857-1155.

Insulated Cable Engineers Association. PO Box 440, South Yarmouth, MA 02664. (508) 394-4424.

Wire Association International. 1570 Boston Post Road, PO Box H, Guilford, CT 06437. (203) 453-2777.

Wire Industry Suppliers Association. 7297 Lee Highway, Suite N, Falls Church, VA 22042. (703) 533-9530.

DIRECTORIES AND BIOGRAPHICAL SOURCES

Dun's Industrial Guide: The Metalworking Directory, 1993-94. Dun & Bradstreet Information Services Staff. 3 volumes. Dun & Bradstreet Information Services, 899 Eaton Avenue, Bethlehem, PA 18025. (610) 882-7000 or (800) 526-0651. FAX (610) 882-7269. 1993. $795.00

Research Centers Directory. Gale Research, 835 Penobscot Building, Detroit, MI 48226-4094. (313) 961-2242. (800)877-4253. FAX (313) 961-6083. $485.00.

Who's Who in Engineering. American Association of Engineering Societies, 1111 19th Street NW, Suite 608, Washington, DC 20036. (202) 296-2237 or (800) 658-8897. 8th edition. 1991. Inquire for price.

Who's Who in Technology. Gale Research, 835 Penobscot Building, Detroit, MI 48226-4094. (313) 961-2242. (800) 877-4253. FAX (313) 961-6083. 1995. $195.00.

Wire Industry Yearbook. Publex International Ltd., 110 Station Road E., Oxted, Surrey RH8 0QA, England. Telephone 0883-717755. FAX 0883-714554. 1951 to present. Annual. Inquire for cost.

ENCYCLOPEDIAS AND DICTIONARIES

McGraw-Hill Dictionary of Scientific and Technical Terms. Sybil P. Parker, ed. 5th edition. McGraw-Hill Publishing Company, 1221 Avenue of the Americas, New York, NY 10020. (800) 262-4729 or (212) 512-3825. 1993. $110.50.

McGraw-Hill Encyclopedia of Engineering. Sybil P. Parker, ed. 2nd edition. McGraw-Hill Publishing Company, 1221 Avenue of the Americas, New York, NY 10020. (800) 262-4729 or (212) 512-3825. 1993. $95.50.

McGraw-Hill Encyclopedia of Science and Technology. Sybil P. Parker, ed. 7th edition. 20 volumes. McGraw-Hill Publishing Company, 1221 Avenue of the Americas, New York, NY 10020. (800) 262-4729 or (212) 512-3825. 1992. $1900.00

GENERAL WORKS

Electrical Installation Practice. Henry A. Miller. 5th ed. Blackwell Scientific Publications, 238 Main Street, Cambridge, MA 02142. (617) 876-7000 or (800) 759-6102. FAX (617) 876-7022. 1993. Inquire for cost and availability.

Electrical Wiring, Commercial. Ray C. Mullin. 8th edition. Delmar Publishers, Division of Thomson Educational Publishing, Inc., 3 Columbia Circle, Box 15015, Albany, NY 12205. (518) 464-3500 or (800) 347-7707. FAX (518) 464-0358. 1992. $29.95.

Electrical Wiring, Residential. Ray C. Mullin. 11th edition. Delmar Publishers, Division of Thomson Educational Publishing, Inc., 3 Columbia Circle, Box 15015, Albany, NY 12205. (518) 464-3500 or (800) 347-7707. FAX (518) 464-0358. 1992. $29.95.

Practical Electrical Wiring: Residential, Farm, and Industrial. H.P. Richter and W.C. Schwan. 16th ed. McGraw-Hill Publishing Company, 1221 Avenue of the Americas, New York, NY 10020. (800) 262-4729 or (212) 512-3825. 1993. $35.00.

Standard For Electrical Wires, Cables, and Flexible Cords. 2d rev ed. Underwriters' Laboratories, 333 Pfingsten Rd., Northbrook, IL 60062-2096. (708) 272-8800. FAX (312) 272-8129. 1994. Inquire for cost and availability.

HANDBOOKS AND MANUALS

American Electricians Handbook. C. Terrell & W.I. Summers. 12th edition. McGraw-Hill Publishing Company, 1221 Avenue of the Americas, New York, NY 10020. (800) 262-4729 or (212)512-3825. 1992. $74.50.

Electric Cables Handbook. D. McAllister. 2nd edition. Sheridan House Inc., 145 Palisade Street, Dobbs Ferry, NY 10522. (914) 693-2410. FAX (914) 693-0776. 1990. $150.00.

Electric Wire Handbook. K. Gillett & M. Suba. Wire Association International, 1570 Boston Post Road, PO Box H, Guilford, CT 06437. (203) 453-2777. 1983. $50.00.

Guide To the 1993 National Electrical Code. Roland E. Palmquist. Macmillan Publishing, 200 Old Tappan Road, Old Tappan, NJ 07675. (800) 223-2336. FAX (800) 445-6991. 1993. Inquire for cost and availability.

Lineman's and Cableman's Handbook. E.B. Kurtz. 8th edition. McGraw-Hill Publishing Company, 1221 Avenue of the Americas, New York, NY 10020. (800) 262-4729 or (212) 512-3825. 1992. $69.50.

ONLINE DATABASES AND CD-ROMS

Compendex Plus. Engineering Information, Inc., 345 East 47th Street, New York, NY 10017. (212) 705-7600 or (800) 221-1044. Contains citations with abstracts to worldwide literature in engineering and technology, from 1970 to present. Available on online BRS,(800) 289-4277, DIALOG, (800) 334-2564, ORBIT (800) 456-7248, and STN International, FIZ Karlsruhe, P.O. Box 2465, W-7500, Karlsruhe 1, Germany, online services. Also available on CD-ROM. Inquire as to cost and availability.

ISMEC: Mechanical Engineering Abstracts. Cambridge Scientific Abstracts, 7200 Wisconsin Avenue, Suite 601, Bethesda, MD 20814. (301) 961-6750 or (800) 843-7751. Contains citations to the literature in mechanical engineering, industrial and production engineering, energy, power, mechanics, devices and related areas, from 1973 to present. Available on the DIALOG,(800) 334-2564, online service. Inquire as to cost and availability.

Metadex. Jointly produced by ASM International and the Institute of Materials. Contains more than 925,000 records from the international literature on metals and alloys, concerning properties, processes, materials classes, applications, and metallurgical systems. Updated monthly. Available from ORBIT-QUESTEL (703) 442-0900.

NTIS Bibliographic Database. National Technical Information Service, 5285 Port Royal Road, Springfield, VA 22161. (703) 487-4929 or FAX (703) 321-8199. Broad coverage of government-sponsored science and technology research reports, 1964 to present. Available on BRS,(800) 289-4277, DIALOG, (800) 334-2564, ORBIT, (800) 456-7248, and STN International, FIZ Karlsruhe, P.O. Box 2465, W-7500, Karlsruhe 1, Germany, online services. Also available on CD-ROM. Inquire as to cost and availability.

SCISEARCH. Institute for Scientific Information, 3501 Market Street, Philadelphia, PA 19104. (800) 523-1850 or (215) 386-0100. Broad multidisciplinary title and author index to the international literature of science and technology, 1974 topresent. Available on DIALOG,(800) 334-2564, and ORBIT,(800) 456-7248, online services. Also available on CD-ROM. Inquire as to cost and availability.

WILSONLINE. H.W. Wilson Company, 950 University Avenue, Bronx, NY 10452. (800) 367-6770 or (212) 588-8400. Makes available online versions of the printed H.W. Wilson indexes including Applied Science and Technology Index, Business Periodicals Index, General Science Index, and Readers' Guide to Periodical Literature. Period covered is generally 1983 to present. Available on BRS,(800) 289-4277, DIALOG,(800) 334-2564, and OCLC EPIC,(800) 848-5878, online services. Also available on CD-ROM. Inquire as to cost and availability.

PERIODICALS

Electrical Construction and Maintenance. Intertec Publishing Corporation, 9800 Metcalf, Overland Park, KS 66212-2215. (913) 341-1300. FAX (913) 967-1898. 1901 to present. Monthly. $30.00 per year.

J O M (Journal of Metals). Minerals, Metals and Materials Society, 420 Commonwealth Drive, Warrendale, PA 15086. (412) 776-9080. 1949 to present. Monthly. $50.00 per year.

Metals Week. McGraw-Hill Publishing Company, 1221 Avenue of the Americas, New York, NY 10020. (800) 262-4729 or (212) 512-3825. 1930 to present. Weekly. $770.00 per year.

Wire. Postfach 2069, 96011 Bamberg, Germany. Telephone 0951-861-135. FAX 0951-861-158. 1951 to present. Bi-monthly. Inquire for cost.

Wire Industry. Publex International Ltd., 110 Station Road E., Oxted, Surrey RH8 0QA, England. Telephone 0883-717755. FAX 0883-714554. 1934 to present. Monthly. Inquire for cost.

Wire Industry News. Business Information Services, 7 Hampden Road, Stafford Springs, CT 06076-31404. (203) 684-5877. (203) 684-9158. 1973 to present. Bi-weekly. $250.00 per year.

Wire Journal International. Wire Journal Inc., 1570 Boston Post Road, Box H, Guilford, CT 06437. (203) 453-2777. FAX (203) 453-8384. 1968 to present. Monthly. $60.00 per year.

Wire Technology International. Initial Publications Inc., 3869 Darrow Road, Suite 101, Stow, OH 44224. (216) 686-9544. (216) 686-9563. 1972 to present. Bi-monthly. $30.00 per year.

Wireworld. Vogel Europublishing Inc., 20092 Gibbs Drive, Sonora, CA 95370. (209) 533-3555. FAX (209) 533-9555. 1959 to present. 6 times a year. Inquire for cost.

RESEARCH CENTERS AND INSTITUTES

International Copper Association Ltd. 708 Third Avenue, New York, NY 10017. (212) 697-9355. FAX (212) 697-5417.

XYZ

X-RAY ASTRONOMY

See: ASTRONOMY

X RAYS

See also: ASTRONOMY, CRYSTALLOGRAHY, MATERIALS
SCIENCE, NONDESTRUCTIVE TESTING, PARTICLE PHYSICS

ABSTRACT SERVICES AND INDEXES

*Applied Science and Technology Index; A Cumulative Subject
Index To English Language Periodicals in the Fields of Aero-
nautics and Space Science, Computer Technology, Chemistry,
Construction Industry, Energy and Related Areas.* H.W. Wilson
Co., 950 University Avenue, Bronx, NY 10452. (212) 588-8400.
(800) 367-6770. FAX (718) 590-1617. From 1958 to present.
Monthly. Inquire about cost and availability. Also available on
CD-ROM and online. ISSN: 0003-6986.

Astronomy and Astrophysics Abstracts. Springer-Verlag New
York, 175 Fifth Avenue, New York, NY 10010. (212) 460-1500.
FAX (212) 473-6272. Published for the Astronomisches Rechen-
Institut. Comprehensive coverage of all aspects of astronomy,
astrophysics and related fields. 1969 to present. Two parts per
year. Annual. ISSN: 0067-0022.

Chemical Abstracts. Chemical abstracts Service. 2540
Oleantangy River Road, Box 3012, Columbus, OS 43210-0012.
(614) 447-3600. (800) 848-6538. FAX (614) 447-3713. 1907 to
present. Weekly. $16,800.00 per year. Available online and on
CD-ROM. CA is also available in five section groupings. Inquire regarding cost and availability. ISSN:

Engineering Index Monthly. Engineering Information, Inc.,
Castle Point on the Hudson, Hoboken, NJ 07030. (201) 216-
8500. (800) 221-1044. FAX (201) 216-8532. Monthly. $2300.00
per year for monthly issues. $1980 per year for annual cumula-
tion. Available on CD-ROM and online as COMPENDEX. ISSN:
0742-1974; 0360-8557.

General Science Index. H.W. Wilson Company, 950 University
Avenue, Bronx, NY 10452. (212) 588-8400. (800) 367-6770.
FAX (718) 590-1617. From 1978 to present. Ten issues per year;
quarterly and annual cumulations. Service basis. Available on
CD-ROM and online. Inquire regarding cost and availability.
ISSN: 0162-1963.

NTIS Alerts: Astronomy & Astrophysics. U. S. National Technical Information Service. 5285 Port Royal Road, Springfield,
VA 22161. (703) 487-4650. FAX (703) 321-8547. Weekly.
$140.00 per year.

Physics Abstracts. INSPEC. Section A, Science Abstracts. Institute of Electrical Engineers, London, United Kingdom. Available from: INSPEC/IEEE - Institute of Electrical and Electronic
Engineers, Box 1331, Hoes Lane, Piscataway, NJ 08855-1331.
(908) 562-5549. 1898 to present. 24 issues per year. $2835.00
per year. ISSN: 0036-8091. Alsoavailable online and on CD-
ROM.

Physics Briefs (Physikalische Berichte). Information Center for
Energy, Physics, Mathematics; German Physical Society.VCH
Publishers, Inc., 220 East 23rd Street, New York, NY 10010-
4606. (212) 683-8333. 1845 to present. 24 issues per year.
$2390.00 per year. Also available online. ISSN: 0179-7434.

Science Citation Index. SCI. Institute for Scientific Information, 3501 Market Street, Philadelphia, PA 19104. (215) 386-
0100. (800) 523-1850. FAX (215) 386-2991. 1961 to present.
Six issues per year, plus annual cumulation. $11650.00 per year.
Also available online and on CD-ROM.Inquire about price and
availability. ISSN: 0036-827X.

Star. Scientific and Technical Aerospace Reports. U.S. National
Aeronautics and Space Administration. Distributed by U. S.
Superintendent of Documents, Washington, DC 20402. 1963
to present. Semi-monthly, with semiannual and annual indexes.
$114.00 per year. ISSN: 0036-8741. Also available online and
on CD-ROM.

X-ray Diffraction Abstracts. PRM Science and Technology
Agency, LTd., 787 High Road, North Finchley, London, N12
BJT, England. 1973 to present. Quarterly. L45 per year. ISSN:
0309-5212.

ANNUAL REVIEWS AND YEARBOOKS

*Annual Review of Astronomy and Astrophysics; Original Re-
views of Critical Literature and Current Developments in Astronomy and Astrophysics.* Annual Reviews, Inc., 4139 El
Camino Way, Palo Alto, CA 94306-0139. (415) 493-4400. (800)

X RAYS

Ency. of Physical Sciences and Engineering Info. Sources

523-8635. FAX (415) 855-9815. 1963 to date. Annual. $60.00. ISSN: 0066-4146.

ASSOCIATIONS AND PROFESSIONAL SOCIETIES

American Astronomical Society. 2000 Florida Avenue NW, Suite 400, Washington, DC 20009. (202) 328-2010. FAX: (202) 234-2560.

American Chemical Society, 1155 16th Street, Nw, Washington, DC 20036. (202) 872-4600.

American Institute of Physics. One Physics Ellipse, College Park, MD 20740-3843. (301) 209-3100.

American Society For Nondestructive Testing. 1711 Arlingate Lane, Columbus, OH 43228-0518. (614) 274-6003. FAX (614) 274-6899.

American Society For Testing and Materials. 1916 Race Street, Philadelphia, PA 19103. (215) 299-5400. FAX (215) 977-9679. <bodylead>Association of Universities for Research in Astronomy. Suite 701, 1625 Massachusetts Avenue, NW, Washington, DC 20036. (202) 483-2101. FAX (202) 483-2106.

National Council On Radiation Protection and Measurements. 7910 Woodmont Avenue, Suite 1016, Bethesda, MD 10814. (301) 657-2652.

Radiation Research Society. 2021 Spring Road, Suite 600, Oak Brook Road, IL 60521. (708) 571-2881. (708) 571-7837.

ANNUAL REVIEWS AND YEARBOOKS

Advances in X-ray Analysis. Plenum Publishing Corp., 233 Spring Street, New York, NY 10013-1578. (212) 620-8000. 1960 to present. Annual. ISSN: 0096-8490.

BIBLIOGRAPHIES

Science Books and Films. American Association for theAdvancement of Science, 1333 H Street NW, Washington, DC 20005. (202) 326-6454. 1965 to present. Nine issues per year. $40.00 per year. ISSN: 0098-342X

Scientific and Technical Books and Serials in Print, 1995. R.R. Bowker Inc., 121 Chanlon Road, New Providence, NJ 07974. (908) 464-6800. (800) 521-8110. 4 volumes. 1994. $299.95. Also available on compact disc and online.

DIRECTORIES AND BIOGRAPHICAL SOURCES

American Men and Women of Science: Physical and Biological Sciences. R. R. Bowker Inc., 121 Chanlon Road, New Providence, NJ 07974. (908) 464-6800. (800) 521-8110. 20th edition. 8 volumes. 1996. $850.00.

Directory of Physics and Astronomy Staff Members. American Institute of Physics, One Physics Ellipse, College Park, MD 20740-3843. (301) 209-3100. Annual. $45.00.

Earth and Astronomical Research Centers. Stockton Press, 345 Park Avenue South, New York, NY 10010. 4th edition. 1995. $515.00. ISBN: 1-56169-0967.

Graduate Programs in Physics, Astronomy and Related Fields. 1995-1996. American Institute of Physics, One Physics Ellipse, College Park, MD 20740-3843. (301) 209-3100. 1978 to present. Annual. $45.00.

Research Centers Directory. Gale Research, 835 Penobscot Building, Detroit, MI 48226-4094. (313) 961-2242. (800) 877-4253. 20th edition. 1995. $485.00.

ENCYCLOPEDIAS AND DICTIONARIES

McGraw-Hill Encyclopedia of Science and Technology. McGraw-Hill Book Company, Inc., 1221 Avenue of the Americas, New York, NY 10020. (212) 512-2000. (800) 262-4729. 7th edition. 20 volume set. 1992. $1900.00.

GENERAL WORKS

Dynamic Scattering of X-rays in Crystals. Z. G. Pinsker. Springer-Verlag New York, Inc., 175 Fifth Avenue, New York, NY 10010. (212) 460-1500. (800) 777-4643. FAX (212) 473-6272. 1978. $58.00.

Funadmentals of X-ray and Radium Physics. Joseph Selman. C. C. Thomas Publ. 2600 South First Street, Springfield, IL 62794-9265. (217) 789-8980. (800) 258-8980. 8th edition.1994. $46.95.

Gamma and X-ray Spectrometry with Semiconductor Detectors. K. Debertin and R. G. Helmer. Elsevier Science Publishing Company, Inc., 655 Avenue of the Americas, New York, NY 10010. (212) 989-5800. FAX (914) 333-2444. 1988. $92.50.

Gravitation, Heat and X-rays. Bernard I. Cohen, editor. Ayer Company Pubs., Inc., Lower Mill Road, North Stratford, NH 03590. (603) 922-5105. (800) 282-5413. 1981. $38.95.

UV and X-ray Spectroscopy of Laboratory and Astrophysical Plasmas. Eric H. Silver and Steven M. Kahn, editors. Cambridge University Press, 40 West 20th Street, New York, NY 10011-4211. (212) 924-3900. (800) 872-7423. 1993. $79.95

X-ray Binaries and Recycled Pulsars. E. P. Van den Heuvel and S. A. Rappaport, editors. Kluwer Academic Publishers, 101 Philip Drive, Assinippi Park, Norwell, MA 02061. (617) 871-6600. 1992. $222.00.

X-ray Diffraction in Crystals, Imperfect Crystals and Amorphous Bodies. A. Guinier. Dover Publications, Inc., 180 Varick Street, New York, NY 10014. (212) 255-3755. (800) 223-3130. Reprint edition. 1994. $11.95

X-ray Science and Technology. A. G. Michette and C. Buckley. IOP Publishing, Public Ledger Building, Suite 1035, Independence Square, Philadelphia, PA 19106. (215) 627-0880. (800) 358-4677. 1993. $58.00.

X-rays in Atomic and Nuclear Physics. N. Dyson. Cambridge University Press, 40 West 20th Street, New York, NY 10011-4211. (212) 924-3900. (800) 872-7423. 2nd editon. 1990. $115.00.

HANDBOOKS AND MANUALS

Crystallography in Modern Chemistry: a Resource Book of Crystal Structures. Thomas C. Mak and Zhou Gongdu. John Wiley

& Sons, Inc., 605 Third Avenue, New York, NY 10158-0012. (212) 850-6000. (800) 225-5945. 1992. $195.00.

Handbook of Partial Attenuation Coefficients of Characteristic X-ray Radiation. O. S. Marenkov, editor. Nova Science Publishers, Inc., 6080 Jericho Turnpike, Suite 207, Commack, NY 11725-2808. (516) 499-3103. 1994. $115.00.

Handbook of X-ray Spectrometry: Methods and Techniques. Van Grieken and Markowica, editors. Marcel Dekker, Inc., 270 Madison Avenue, New York, NY 10016. (212) 696-9000. (800) 228-1160. 1992. $215.00.

International Tables For X-ray Crystallography. John Kasper and Kathaleen Lonsdale, editors. Kluwer Academic Publishers, 101 Philip Drive, Assinippi Park, Norwell, MA 02061. (617) 871-6600. 2nd edition. 1985. $85.00.

ONLINE DATABASES AND CD-ROMS

CA Search. Chemical Abstracts Service, P.O. Box 3012, Columbus, OH 43210-0012. (614) 447-3600. (800) 848-6533. FAX (614) 447-3709. Very comprehensive guide to worldwide chemical literature and related fields, 1972 to present. Available on BRS,(800) 289-4277, DIALOG, (800) 334-2564, ORBIT (800) 456-7248, and STN International, FIZ Karlsruhe, P.O. Box 2465, W-7500, Karlsruhe 1, Germany, online services. Inquire as to cost and availability.

Dissertation Abstracts Online. University Microfilms International, 300 North Zeeb Road, Ann Arbor, MI 48106. (800) 521-0600 or (313) 761-4700. Scope includes virtually all doctoral dissertations accepted at accredited American institutions from 1861 to present in 252 subject areas. Available on BRS, (800) 289-4277, DIALOG, (800) 334-2564, and OCLC EPIC, (800) 848-5878, online services. Also available on CD-ROM. Inquire as to cost and availability.

Georef. American Geological Institute, 4220 King Street, Alexandria, VA 22302. (800) 336-4764 or (703) 379-2480. Geology andgeosciences literature, 1785 to present for North America. Available on DIALOG,(800) 334-2564, ORBIT (800) 456- 7248, online services. Also available on CD-ROM. Inquire as to cost and availability.

INSPEC. Institution of Electrical Engineers, Michael Faraday House, Six Hills Way, Stevenage, Herts. SG1 2AY, England. Telephone: 0438 313311 or FAX 0438 742840. Contains citations to the worldwide literature of physics, electronics and electrical engineering, computer technology, and related fields. Available on BRS, (800) 289-4277, DIALOG, (800) 334-2564, ORBIT, (800) 456-7248, and STN International, FIZ Karlsruhe, P.O. Box 2465, W-7500, Karlsruhe 1, Germany, online services. Inquire as to cost and availability.

NTIS Bibliographic Database. National Technical Information Service, 5285 Port Royal Road, Springfield, VA 22161. (703) 487-4929 or FAX (703) 321-8199. Broad coverage of government-sponsored science and technology research reports, 1964 to present. Available on BRS,(800) 289-4277, DIALOG, (800) 334-2564, ORBIT, (800) 456-7248, and STN International, FIZ Karlsruhe, P.O. Box 2465, W-7500, Karlsruhe 1, Germany, online services. Also available on CD-ROM. Inquire as to cost and availability.

Physics Briefs. American Institute of Physics, 335 East 45th Street, New York, NY 10017. (212) 661-9260 or FAX (212) 661-2036. Contains citations with abstracts of the literature of physics and related fields, 1979 to present. Available on the STN International, FIZ Karlsruhe, P.O. Box 2465, W-7500, Karlsruhe 1, Germany, online service. Inquire as to cost and availability.

SCISEARCH. Institute for Scientific Information, 3501 Market Street, Philadelphia, PA 19104. (800) 523-1850 or (215) 386-0100. Broad multidisciplinary title and author index to the international literature of science and technology, 1974 to present. Available on DIALOG, (800) 334-2564, and ORBIT, (800) 456-7248, online services. Also available on CD-ROM. Inquire as to cost and availability.

WILSONLINE. H.W. Wilson Company, 950 University Avenue, Bronx, NY 10452. (800) 367-6770 or (212) 588-8400. Makes available online versions of the printed H.W. Wilson indexes including Applied Science and Technology Index, Business Periodicals Index, General Science Index, and Readers' Guide to Periodical Literature. Period covered is generally 1983 to present. Available on BRS, (800) 289-4277, DIALOG, (800) 334-2564, and OCLC EPIC, (800) 848-5878, online services.Also available on CD-ROM. Inquire as to cost and availability.

PERIODICALS

Applied Optics. Optical Society of America. 2010 Massachusetts Avenue NW, Washington, DC 20036-1023. (202) 223-8130. 1962 to present. 36 issues per year. $1090.00 per year. ISSN: 0003-6935.

Applied Spectroscopy. Society for Applied Spectroscopy,198 Thomas Johnson Drive, Suite 2, Frederick, MD 21702. (301) 694-8122. FAX (301) 694-6860. 1946 to present. Monthly. $180.00 per year. ISSN: 0003-7028.

Journal of the Optical Society of America. Part A: Optics and Image Science. Optical Society of America, Inc., 2010 Massachusetts Avenue, NW, Washington, DC 20036-1023. (202) 223-8130. 1917 to present. Monthly. $610.00 per year. ISSN: 0740-3232.

Journal of the Optical Society of America. Part B: Optical Physics. Optical Society of America, Inc., 2010 Massachusetts Avenue, NW, Washington, DC 20036-1023. (202) 223-8130. 1917 to present. Monthly. $610.00 per year. ISSN: 0740-3224.

Optical Engineering. SPIE - International Society for Optical Engineering, Box 10, 1000 20th Street, Bellingham, WA 98227-0010. (206) 676-3290. FAX (206) 647-1445. 1962 topresent. Monthly. $170.00 per year. ISSN: 0091-3286.

Optics Communications. Elsevier Science Inc., Box 882, Madison Square Station, New York, NY 10159. (212) 989-5800. FAX (212) 633-4990. 1969 to present. 54 issues per year. $2121.00. per year. ISSN: 0030-4018.

Optics Letters. Optical Society of America, Inc., 2010 Massachusetts Avenue, NW, Washington, DC 20036-1023. (202) 223-8130. 1977 to present. Semi-Monthly. $625.00 per year. ISSN: 0146-9592.

XRS X-ray Spectrometry. John Wiley & Sons, Baffins Lane, Chichester, Sussex PO19 1UD, England. TEL 0243-779777. 1972 to present. Bi-monthly. $745.00 per year. ISSN: 0049-8246.

X RAYS

Ency. of Physical Sciences and Engineering Info. Sources

RESEARCH CENTERS AND INSTITUTES

Center For Astrophysics and Space Physics. University of California, San Diego, C-0111, LA Jolla, CA 92093-0111. (619) 534-3460. FAX (619) 534-2294.

Center For Engineering Applications of Radioisotopes. North Carolina State University, Box 7909, Department of Nuclear Engineering, Raleigh, NC 27695-7909. (919) 737-3378. FAX(919) 737-3928.

Center For X-ray Optics. Lawrence Berkeley Laboratory, 1 Cyclotron Road, Berkeley, CA 94720. (415) 486-4985.

Fred Lawrence Whipple Observatory. P. O. Box 97, Amado, AZ 85645. (602) 670-6741. FAX (602) 670-6779.

Major Analytical Instrumentation Center. University of Florida, College of Engineering, Rhines Hall, Gainsville, FL 32611. (904) 392-6985.

Nuclear Physics Laboratory, University of Washington, Seattle, WA 98195. (206) 543-4080.

ZINC

See also: ALLOYS, BRASS AND BRONZE, MATERIALS SCIENCE, TIN

ABSTRACT SERVICES AND INDEXES

Alloys Index. American Society for Metals, Metals Park, OH 44073. (216) 338-5151. FAX (216) 338-4634. 1974 to present.Monthly. $380.00. Also available on CD-ROM and online via METADEX, STN International, and DIALOG.

Applied Mechanics Reviews: An Assessment of World Literature in Engineering Sciences. 1948-present. American Society of Mechanical Engineers, 345 East 47th Street, New York, NY 10017. (212) 705-7703. Monthly. $360.00 per year.

Applied Science and Technology Index; A Cumulative Subject Index To English Language Periodicals in the Fields of Aeronautics and Space Science, Computer Technology, Chemistry, Construction Industry, Energy and Related Areas. H.W. Wilson Co., 950 University Avenue, Bronx, NY 10452. (800) 367-6770 or (212) 588-8400. FAX (718) 590-1617. From 1958 to present. Monthly. Inquire about cost and availability. Also available on CD-ROM and online.

Current Contents: Engineering, Technology, and Applied Sciences. Institute for Scientific Information, 3501 Market Street, Philadelphia, PA 19104. (215) 386-0100. FAX (215) 386-6362. 1970 to present. Weekly. $442.00 per year.

Engineering Index Monthly. Engineering Information, Inc., Castle Point on the Hudson, Hoboken, NJ 07030. (800) 221-1044. FAX (212) 832-1857. Monthly. $2300.00 per year. Also available online as COMPENDEX, and also on CD-ROM. Covers chemical engineering, computers, electrical engineering, civil engineering, metals and mining, industrial management, and mechanical engineering.

International Copper Information Bulletin. Copper Development Association, Orchard House, Mutton Lane, Potters Bar, Herts EN6 3AP, England. Telephone 0707-50711. FAX 0707-42769. 1976 to present. 3 times a year. Inquire for cost and availability. Recent reports, publications and abstracts on copper, its alloys and compounds.

Metals Abstracts and Metals Abstracts Index. American Society for Metals, Metals Park, OH 44073. (216) 338-5151. 1968 to present. Monthly. Abstracts are $1925.00 per year and Index is $460.00 per year.

Zincscan. C&C Associates, 4 Newmans Row, Lincolns Road, High Wycombe, Bucks HP12 3RE, England. 1943 to present. Quarterly. Inquire for cost. Formerly ZINC ABSTRACTS Review of recent technical literature on the uses of zinc and its products.

ASSOCIATIONS AND PROFESSIONAL SOCIETIES

American Society For Metals. Materials Park, OH 44073. (216) 338-5151 or FAX (216) 338-4634.

American Zinc Association. 1112 16th Street NW, Suite 240, Washington, DC 20036. (202) 835-0164. FAX (202) 835-0155.

Copper Development Association. Orchard House, Mutton Lane, Potters Bar, Herts EN6 3AP, England. Telephone 0707-50711. FAX 0707-42769.

Independent Zinc Alloyers Association. 1000 16th Street NW, Suite 400, Washington, DC 20036. (202) 785-0558. FAX (202) 785-0210.

International Lead Zinc Research Organization Inc., 2525Meridian Parkway, PO Box 12036, Research Triangle Park, North Carolina 27709. (919) 361-4647.

International Lead and Zinc Study Group. 58 St.Jame's Street, London SW1A 1LD, England. Telephone 071-499-9373.

Metallurgical Society of the Aime (american Institute of Mining, Metallurgical, and Petroleum Engineers). 345 E.47th Street, 14th Floor, New York, NY 10017. (212) 705-7695.

DIRECTORIES AND BIOGRAPHICAL SOURCES

Dun's Industrial Guide: The Metalworking Directory, 1993-94. Dun & Bradstreet Information Services Staff. 3 volumes. Dun & Bradstreet Information Services, 899 Eaton Avenue, Bethlehem, PA 18025. (610) 882-7000 or (800) 526-0651. FAX (610) 882-7269. 1993. $795.00

International Research Centers Directory. Gale Research, 835 Penobscot Building, Detroit, MI 48226-4094. (313) 961-2242. (800) 877-4253. FAX (313) 961-6083. 8th edition. 1995. $410.00

Research Centers Directory. Gale Research, 835 Penobscot Building, Detroit, MI 48226-4094. (313) 961-2242. (800) 877-4253. FAX (313) 961-6083. $485.00.

Scientific and Technical Organizations and Agencies Directory. Gale Research, 835 Penobscot Building, Detroit, MI 48226-4094. (313) 961-2242. (800) 877-4253. FAX (313) 961-6083. 4th edition. 1996. $195.00.

Who's Who in Engineering. American Association of Engineering Societies, 1111 19th Street NW, Suite 608, Washington, DC 20036. (202) 296-2237 or (800) 658-8897. 8th edition. 1991. Inquire for price.

ENCYCLOPEDIAS AND DICTIONARIES

Encyclopedia of Physical Science and Technology. Robert A. Meyers, ed. 18 volumes. Academic Press Inc., 6277 Sea Harbor Drive, Orlando, FL 32887. (800) 321-5068. 1992. $2100.00

McGraw-Hill Dictionary of Scientific and Technical Terms. Sybil P. Parker, ed. 5th edition. McGraw-Hill Publishing Company, 1221 Avenue of the Americas, New York, NY 10020. (800) 262-4729 or (212) 512-3825. 1993. $110.50.

McGraw-Hill Encyclopedia of Science and Technology. Sybil P. Parker, ed. 7th edition. 20 volumes. McGraw-Hill Publishing Company, 1221 Avenue of the Americas, New York, NY 10020. (800) 262-4729 or (212) 512-3825. 1992. $1900.00

GENERAL WORKS

Zinc (Metals & Minerals Series). Gordon Press Publishers, PO Box 459, Bowling Green Station, New York, NY 10004. (718) 624-8419. 1993. $250.95.

Zinc and Its Alloys and Compounds. S.W.K. Morgan. E. Horwood, distributed by Halsted Press (Division of John Wiley and Sons, Inc.), 605 Third Avenue, New York, NY 10158. (800) 526-5368 or (212) 850-6418. 1985. Inquire for cost and availability.

Zinc-Based Steel Coatings. G. Krauss & D.K. Matlock, editors. Minerals, Metals & Materials Society, 400 Commonwealth Dr., Warrendale, PA 15096. (412) 776-9000 or (800) 759-4867. 1990.$104.00.

Zinc in Human Biology. Colin F. Mills, ed. Springer-Verlag, 175 Fifth Avenue, New York, NY 10010. (212) 460-1500 or (800) 777-4643. FAX (212) 473-6272. 1988. Inquire for cost and availability.

Zinc Plating. H. Geduld. State Mutual Book and Periodical Service, 521 Fifth Avenue, 17th Floor, New York, NY 10175. (718) 261-1704. FAX (516) 537-0412. 1991. $295.00.

Zinc, the Science and Technology of the Metal, Its Alloys, and Compounds. C.H. Mathewson. Books on Demand, Division of University Microfilms International, 300 North Zeeb Road, Ann Arbor, MI 48106-1346. (313) 761-4700 or (800) 521-0600. $160.00.

HANDBOOKS AND MANUALS

CRC Handbook of Chemistry and Physics. David R. Lide, editor. 75th edition. CRC Press, 2000 Corporate Blvd., N.W., Boca Raton, FL 33431. (407) 994-0555 or (800) 272-7737. FAX (407) 994-0949. $99.50.

Metals Handbook. ASM International, Materials Information, Materials Park, OH 44073-0002. (216) 338-5151 or (800) 336-5152. FAX (216) 338-4634. $154.00.

Smithell's Metals Reference Book. E.A. Brandes & G.B. Brook. 7th edition. Butterworth-Heinemann, 313 Washington Street, Newton, MA 02158. (617) 928-2500 or (800) 366-2665. FAX (617) 928-2620. 1992. $250.00.

Zinc Handbook Properties, Processing, and Use in Design. Frank C. Porter. Marcel Dekker, Inc., 270 Madison Avenue, New York, NY 10016. (212) 696-9000. FAX (212) 685-4540. 1991. $175.00.

ONLINE DATABASES AND CD-ROMS

Compendex Plus. Engineering Information, Inc., 345 East 47th Street, New York, NY 10017. (212) 705-7600 or (800) 221-1044. Contains citations with abstracts to worldwide literature in engineering and technology, from 1970 to present. Available on online BRS,(800) 289-4277, DIALOG, (800) 334-2564, ORBIT (800) 456-7248, and STN International, FIZ Karlsruhe, P.O. Box 2465, W-7500, Karlsruhe 1, Germany, online services. Also available on CD-ROM. Inquire as to cost and availability.

Current Contents Search. Institute for Scientific Information, 3501 Market Street, Philadelphia, PA 19104. (215) 386-0100. FAX (215) 386-6362. Contains citations to articles listed in the table of contents of science and technology journals. Also articles in social sciences and life sciences journals. Available on BRS,(800) 289-4277, DIALOG,(800) 334-2564, online services. Inquire as to cost and availability.

Metadex. Jointly produced by ASM International and the Institute of Materials. Contains more than 925,000 records from the international literature on metals and alloys, concerning properties, processes, materials classes, applications, and metallurgical systems. Updated monthly. Available from ORBIT-QUESTEL (703) 442-0900.

NASA Database. American Institute of Aeronautics andAstronautics, 370 L'Enfant Promenade, S.W., Washington, DC 20024. (202) 646-7400. Citations and abstracts of aeronautics and astronautics literature, 1962 to present. Also contains citations from *STAR, Scientific and Technical Aerospace Reports, and International Aerospace Abstracts.* Available through NASA/RECON online service. Inquire as to cost and availability.

NTIS Bibliographic Database. National Technical Information Service, 5285 Port Royal Road, Springfield, VA 22161. (703) 487-4929 or FAX (703) 321-8199. Broad coverage of government-sponsored science and technology research reports, 1964 to present. Available on BRS,(800) 289-4277, DIALOG, (800) 334-2564, ORBIT, (800) 456-7248, and STN International, FIZ Karlsruhe, P.O. Box 2465, W-7500, Karlsruhe 1, Germany, online services. Also available on CD-ROM. Inquire as to cost and availability.

WILSONLINE. H.W. Wilson Company, 950 University Avenue, Bronx, NY 10452. (800) 367-6770 or (212) 588-8400. Makes available online versions of the printed H.W. Wilson indexes including Applied Science and Technology Index, Business Periodicals Index, General Science Index, and Readers' Guide to Periodical Literature. Period covered is generally 1983 to present. Available on BRS,(800) 289-4277, DIALOG,(800) 334-2564, and OCLC EPIC,(800) 848-5878, online services. Also available on CD-ROM. Inquire as to cost and availability.

X RAYS

Ency. of Physical Sciences and Engineering Info. Sources

PERIODICALS

Alloy Digest. Alloy Digest Inc., 27 Canfield Street, Orange, NJ 07050. (201) 677-9161. 1952 to present. Monthly. $140.00 per year.

J O M (Journal of Metals). Minerals, Metals and Materials Society, 420 Commonwealth Drive, Warrendale, PA 15086. (412) 776-9080. 1949 to present. Monthly. $50.00 per year.

Lead and Zinc Statistics. International Lead and Zinc Study Group, 58 St.Jame's Street, London SW1A 1LD, England. Telephone 071-499-9373. 1961 to present. Monthly. $75.00 per year.

Metallurgical Transactions A: Physical Metallurgy and Materials Science. ASM (American Society for Metals) International, Materials Information, Materials Park, OH 44073. (216) 338-5151 or FAX (216) 338-4634. 1970 to present. Monthly. $520.00 per year for non-members.

Metallurgical Transactions B: Process Metallurgy. ASM (American Society for Metals) International, Materials Information, Materials Park, OH 44073. (216) 338-5151 or FAX (216) 338-4634. 1975 to present. Bi-monthly. $375.00 per year for non-members.

Metals Week. McGraw-Hill Publishing Company, 1221 Avenue of the Americas, New York, NY 10020. (800) 262-4729 or (212) 512-3825. 1930 to present. Weekly. $770.00 per year.

Mineral Industry Surveys, Zinc Industry. U.S. Bureau of Mines, Production and Distribution, Cochrans Mill Rd., Box 18070, Pittsburgh, PA 15236. (412) 892-4411. Monthly. Inquire for price and availability.

R&D Focus. International Lead Zinc Research Organization Inc., 2525 Meridian Parkway, PO Box 12036, Research Triangle Park, North Carolina 27709. (919) 361-4647. 1986 to present. 3 times a year. Free.

RESEARCH CENTERS AND INSTITUTES

Copper Development Association. 260 Madison Avenue, New York, NY 10016. (212) 251-7200.

International Lead Zinc Research Organization Inc., 2525 Meridian Parkway, PO Box 12036, Research Triangle Park, North Carolina 27709. (919) 361-4647.

University of Connecticut Institute of Materials Science. U-136, 97 N. Eagleville Road, Storrs, CT 06268. (203) 486-4623. FAX (203) 486-4745.

University of Florida Department of Materials Science and Engineering. Gainesville, FL 32611. (904) 392-1454. FAX (904) 392-6359.

SPECIFICATIONS AND STANDARDS

*Annual Book of ASTM Standards, Volume 02.*04. American Society for Testing and Materials (ASTM), 1916 Race Street, Philadelphia, PA 19103. (215) 299-5585. Annual. Inquire for cost and availability.

ZIRCONIUM

See also: ELEMENTS (CHEMICAL)

ABSTRACT SERVICES AND INDEXES

Applied Science and Technology Index; A Cumulative Subject Index To English Language Periodicals in the Fields of Aeronautics and Space Science, Computer Technology, Chemistry, Construction Industry, Energy and Related Areas. H.W. Wilson Co., 950 University Avenue, Bronx, NY 10452. (800) 367-6770 or (212) 588-8400. FAX (718) 590-1617. From 1958 to present. Monthly. Inquire about cost and availability. Also available on CD-ROM and online.

Bibliography and Index of Geology. American Geological Institute, 4220 King Street, Alexandria, VA 22302. (703) 379-2480. Fax (703) 379-7563. Monthly. $1295.00 per year. Also available online as GeoRef.

Chemical Abstracts - Physical, Inorganic and Analytical Chemistry. Chemical Abstracts Service, 2540 Olentangy River Road, Box 3012, Columbus, OH 43210. (800) 848-6538 or (614) 447-3600, FAX (614) 447-3713. Bi-weekly. $1410.00 per year.

Current Contents: Engineering, Technology, and Applied Sciences. Institute for Scientific Information, 3501 Market Street, Philadelphia, PA 19104. (215) 386-0100. FAX (215) 386-6362. 1970 to present. Weekly. $442.00 per year.

Current Contents: Physical, Chemical and Earth Sciences. Institute for Scientific Information, 3501 Market Street, Philadelphia, PA 19104. (215) 386-0100. FAX (215) 386-6362. Weekly. $360.00 per year.

Index To Scientific and Technical Proceedings. Institute for Scientific Information, 3501 Market St., Philadelphia, PA19104. (215) 386-0100. FAX (215) 386-6362. Monthly. $500.00 per year.

Index To Scientific Reviews. Institute for Scientific Information, 3501 Market St., Philadelphia, PA 19104. (215) 386-0100. FAX (215) 386-6362. Semi-annual.

Science Citation Index. Institute for Scientific Information, 3501 Market Street, Philadelphia, PA 19104. (215) 386-0100. FAX (215) 386-6362. Inquire about availability and cost. Also available on CD-ROM.

ASSOCIATIONS AND PROFESSIONAL SOCIETIES

American Chemical Society. 1155 16th Street NW, Washington, DC 20036. (202) 872-4414.

American Institute of Chemical Engineers. 345 E. 47th Street, New York, NY 10017. (212) 705-7338.

American Institute of Physics. 335 East 45th Street, New York, NY 10017. (212) 661-9404 or (800) 445-6638.

BIBLIOGRAPHIES

Scientific and Technical Books and Serials in Print. R.R. Bowker Inc., 121 Chanlon Road, New Providence, NJ 07974. (800) 521-8110 or (908) 464-6800. List of currently published books and serials in the physical and biological sciences, engineering and technology, with subject, author and titles indexes. Also available on ORBIT online service,(800) 456-7248. Inquire as to cost and availability.

DIRECTORIES AND BIOGRAPHICAL SOURCES

Directory of Engineering Societies and Related Organizations. Gordon Davis. 13th edition. American Association of Engineering Societies, 1111 19th Street NW, Suite 608, Washington, DC 20036. (202) 296-2237 or (800) 658-8897. 1989. Inquire for price.

International Research Centers Directory. Gale Research, 835 Penobscot Building, Detroit, MI 48226-4094. (313) 961-2242. (800) 877-4253. FAX (313) 961-6083. 8th edition. 1995. $410.00

Research Centers Directory. Gale Research, 835 Penobscot Building, Detroit, MI 48226-4094. (313) 961-2242. (800) 877-4253. FAX (313) 961-6083. $485.00.

Scientific and Technical Organizations and Agencies Directory. Gale Research, 835 Penobscot Building, Detroit, MI 48226-4094. (313) 961-2242. (800) 877-4253. FAX (313) 961-6083. 4th edition. 1996. $195.00.

Who's Who in Engineering. American Association of Engineering Societies, 1111 19th Street NW, Suite 608, Washington, DC 20036. (202) 296-2237 or (800) 658-8897. 8th edition. 1991. Inquire for price.

ENCYCLOPEDIAS AND DICTIONARIES

Encyclopedia of Physical Science and Technology. Robert A. Meyers, ed. 18 volumes. Academic Press Inc., 6277 Sea Harbor Drive, Orlando, FL 32887. (800) 321-5068. 1992. $2100.00.

McGraw-Hill Dictionary of Scientific and Technical Terms. Sybil P. Parker, ed. 5th edition. McGraw-Hill Publishing Company,1221 Avenue of the Americas, New York, NY 10020. (800) 262-4729 or (212) 512-3825. 1993. $110.50.

McGraw-Hill Encyclopedia of Engineering. Sybil P. Parker, ed. 2nd edition. McGraw-Hill Publishing Company, 1221 Avenue of the Americas, New York, NY 10020. (800) 262-4729 or (212) 512-3825. 1993. $95.50.

McGraw-Hill Encyclopedia of Science and Technology. Sybil P. Parker, ed. 7th edition. 20 volumes. McGraw-Hill Publishing Company, 1221 Avenue of the Americas, New York, NY 10020. (800) 262-4729 or (212) 512-3825. 1992. $1900.00

GENERAL WORKS

ASTM Manual On Zirconium and Hafnium. J.H. Schemel. American Society for Testing and Materials (ASTM), 1916 Race Street, Philadelphia, PA 19103. (215) 299-5585. 1977. $9.50.

Industrial Applications of Zirconium. C.S. Young & J.C. Durham. American Society for Testing and Materials (ASTM), 1916 Race Street, Philadelphia, PA 19103. (215) 299-5585. 1986. $50.00.

Science & Technology of Zirconia. A.H. Heuer & L.W. Hobbs, eds. American Ceramic Society, Inc., 757 Brooksedge Plaza Drive, Westerville, OH 43081-2821. (614) 890-6136. 1981. $65.00.

Zirconium & Hafnium (Metals & Minerals Series). Gordon Press Publishers, PO Box 459, Bowling Green Station, New York, NY 10004. (718) 624-8419. 1993. $250.95.

Zirconium: International Strategic Minerals Inventory Summary Report. Diane Publishing Company, 600 Upland Avenue, Upland, PA 19015. (215) 499-7415. FAX (610) 499-7429. 1994. $40.00.

Zirconium in the Nuclear Industry: Tenth International Symposium. Anand M. Garde & E. Ross Bradley, editors. American Society for Testing and Materials (ASTM), 1916 Race Street, Philadelphia, PA 19103. (215) 299-5585. 1993. Inquire for cost.

Zirconium Production, Processing, Harding & Storage. National Fire Protection Association, Batterymarch Park, Quincy, MA 02269. (617) 770-3000 or (800) 344-3555. 1987. $15.50.

HANDBOOKS AND MANUALS

CRC Handbook of Chemistry and Physics. David R. Lide, editor. 75th edition. CRC Press, 2000 Corporate Blvd., N.W., Boca Raton, FL 33431. (407) 994-0555 or (800) 272-7737. FAX (407) 994-0949. $99.50.

ONLINE DATABASES AND CD-ROMS

Compendex Plus. Engineering Information, Inc., 345 East 47th Street, New York, NY 10017. (212) 705-7600 or (800) 221-1044. Contains citations with abstracts to worldwide literature in engineering and technology, from 1970 to present. Available on online BRS,(800) 289-4277, DIALOG, (800) 334-2564, ORBIT (800) 456-7248, and STN International, FIZ Karlsruhe, P.O. Box 2465, W-7500, Karlsruhe 1, Germany, online services. Also available on CD-ROM. Inquire as to cost and availability.

CSA Engineering. Cambridge Scientific Abstracts, 7200 Wisconsin Avenue, Suite 601, Bethesda, MD 20814. (301) 961-6750 or (800) 843-7751. Contains citations and abstracts of internationalperiodicals and research literature covering all fields of engineering and science and technology,including computer and information science, electronics, mechanical engineering, solid state materials, 1981 to present. Available on BRS,(800) 289-4277, online service.Inquire as to cost and availability.

Current Contents Search. Institute for Scientific Information, 3501 Market Street, Philadelphia, PA 19104. (215) 386-0100. FAX (215) 386-6362. Contains citations to articles listed in the table of contents of science and technology journals. Also articles in social sciences and life sciences journals. Available on BRS,(800) 289-4277, DIALOG,(800) 334-2564, online services. Inquire as to cost and availability.

Dissertation Abstracts. University Microfilms International, 300 North Zeeb Road, Ann Arbor, MI 48106. (800) 521-0600 or (313) 761-4700. Scope includes virtually all doctoral dissertations accepted at accredited American institutions from 1861 to present in 252 subject areas. Available on BRS,(800) 289-4277, DIALOG,(800) 334-2564, and OCLC EPIC,(800) 848-5878, online services. Also available on CD-ROM. Inquire as to cost and availability.

ZIRCONIUM

Ency. of Physical Sciences and Engineering Info. Sources

Georef: Bibliography and Index of Geology. American Geological Institute, 4220 King Street, Alexandria, VA 22302. (703) 379-2480. Fax (703) 379-7563. Monthly. Inquire for price and availability.

NTIS Bibliographic Database. National Technical Information Service, 5285 Port Royal Road, Springfield, VA 22161. (703) 487-4929 or FAX (703) 321-8199. Broad coverage of government-sponsored science and technology research reports, 1964 to present. Available on BRS,(800) 289-4277, DIALOG, (800) 334-2564, ORBIT, (800) 456-7248, and STN International, FIZ Karlsruhe, P.O. Box 2465, W-7500, Karlsruhe 1, Germany, online services. Also available on CD-ROM. Inquire as to cost and availability.

SCISEARCH. Institute for Scientific Information, 3501 Market Street, Philadelphia, PA 19104. (800) 523-1850 or (215) 386-0100. Broad multidisciplinary title and author index to the international literature of science and technology, 1974 to present. Available on DIALOG,(800) 334-2564, and ORBIT,(800) 456-7248, online services. Also available on CD-ROM. Inquire as to cost and availability.

WILSONLINE. H.W. Wilson Company, 950 University Avenue, Bronx, NY 10452. (800) 367-6770 or (212) 588-8400. Makes available online versions of the printed H.W. Wilson indexes including Applied Science and Technology Index, Business Periodicals Index, General Science Index, and Readers' Guide to Periodical Literature. Period covered is generally 1983 to present. Available on BRS,(800) 289-4277, DIALOG,(800) 334-2564, and OCLC EPIC,(800) 848-5878, online services. Also available on CD-ROM. Inquire as to cost and availability.